Wild Mammals of North America

George A. Feldhamer is professor of zoology at Southern Illinois University at Carbondale (SIUC), where he also serves as director of the undergraduate Environmental Studies Program. A mammalogist, he is the senior author of the textbook *Mammalogy: Adaptation, Diversity, and Ecology* now in the 2nd edition. He is past president of the Illinois chapter of the Wildlife Society, a former associate editor for the *Wildlife Society Bulletin,* and currently on the editorial board for the Forest Biology and Ecology section of the *Journal of Forest Research*. Dr. Feldhamer teaches courses in game mammal management as well as mammalogy and is a past recipient of the Outstanding Teacher Award from the College of Science at SIUC.

Bruce C. Thompson became director of the New Mexico Department of Game and Fish in April 2003. During compilation of this book, he was a research wildlife biologist with the U.S. Geological Survey (Biological Resources) and leader of the New Mexico Cooperative Fish and Wildlife Research Unit at New Mexico State University, Las Cruces. He has previously held natural resource management, research, and administrative positions with the U.S. Marine Corps, Washington Department of Game, and Texas Parks and Wildlife Department. Dr. Thompson has worked with a variety of wildlife during the past 25 years, including extensive work with migratory birds, colonial breeding waterbirds, furbearing animals, alligators, and many nongame species. He is a former editor-in-chief of the *Wildlife Society Bulletin.*

Joseph A. Chapman is president of North Dakota State University, where he is also professor of biology. A wildlife ecologist who has worked extensively on rabbits and hares, he is a fellow of the Institute of Biology (London), a fellow of the Explorers Club (New York), a member of the Cosmos Club (Washington, D.C.), and an honorary member of the Species Survival Commission, IUCN, headquartered in Switzerland. Dr. Chapman has produced numerous books and edited volumes, monographs, book chapters, symposia proceedings, journal publications, reports, and popular articles.

WILD MAMMALS OF NORTH AMERICA

Biology, Management, and Conservation

SECOND EDITION

EDITED BY GEORGE A. FELDHAMER, BRUCE C. THOMPSON, AND JOSEPH A. CHAPMAN

THE JOHNS HOPKINS UNIVERSITY PRESS
Baltimore and London

Printed in the United States of America on acid-free paper
9 8 7 6 5 4 3 2 1

The Johns Hopkins University Press
2715 North Charles Street
Baltimore, Maryland 21218-4363
www.press.jhu.edu

The first edition of this work appeared in 1982, under the title *Wild Mammals of North America: Biology, Management, and Economics,* and was edited by Joseph A. Chapman and George A. Feldhamer.

Library of Congress Cataloging-in-Publication Data

Wild mammals of North America : biology, management, and conservation / edited by George A. Feldhamer, Bruce C. Thompson, and Joseph A. Chapman.— 2nd ed.
p. cm.
Includes bibliographical references and index.
ISBN 0-8018-7416-5 (hardcover : alk. paper)
1. Mammals—North America. 2. Wildlife management—North America. 3. Wildlife conservation—North America. I. Feldhamer, George A. II. Thompson, Bruce Carlyle, 1949– III. Chapman, Joseph A.
QL715.W56 2003
639.97′9′097—dc21

2003007662

A catalog record for this book is available from the British Library.

Contents

SECTION VII. HOOFED MAMMALS

SECTION VIII. INTRODUCED MAMMALS

Preface and Acknowledgments

This revised edition of *Wild Mammals of North America* probably is long overdue. The 55 chapters in this volume include pertinent life history and management information for mammalian species or species groups of management significance in North America. This revision reviews and reflects changes in the knowledge base of the scientific discipline of mammalian wildlife ecology as well as changes in approaches to conservation and management of wildlife resources since the original volume was published in 1982. We were encouraged to proceed with this revision based on the overwhelming number of positive comments on the usefulness of the original volume from resource professionals, university faculty, students, and other users. The accounts for some species may have changed only moderately since the first edition. For most species or species groups, however, it is apparent that significant changes have occurred in the amount of research completed and the information available. With this revised edition, we have attempted to provide a thoughtful and useful distillation of the salient literature, management approaches and controversies, conservation initiatives, and research needs.

As with the original volume, a basic question central to this revision was what to include and what to exclude. Coverage, format, and subject matter naturally represent a compromise between an ideal and what was feasible given constraints of volume length. Two chapters on foxes in the original volume and two chapters on skunks have been consolidated into single chapters on each of these groups in the revision. We have added chapters for two other species groups that are of management and conservation significance—prairie dogs and woodrats—and have deleted the chapter on Old World rats and mice. We have retained the general format from the 1982 volume for each chapter: distribution, description, physiology, reproduction, ecology, feeding habits, habitat, behavior, mortality factors, management and conservation status, and research and management needs. Although we have attempted to maintain consistency of format between chapters, the differences between accounts on a single species and those on species groups, as well as preferences of different authors, sometimes dictated minor deviations in the structure of chapters.

Common and scientific names used in this revision generally follow D. E. Wilson and F. R. Cole *Common Names of Mammals of the World* (Smithsonian Institution Press, Washington, DC, 2000). However, if inconsistencies arose, the preferences of chapter authors were honored.

We hope this revised edition of *Wild Mammals of North America* will maintain the tradition of the original volume in contributing to the field of wildlife ecology as a valuable reference for students, scientists, and resource management professionals for many years to come. It is a tribute to the 102 contributors whose efforts are reflected in this work. Given the size of this volume and the wide array of scientific and other technical literature drawn upon, we worked diligently with each contributor to maintain accuracy and guard against overgeneralizations and inaccurate inferences or conclusions. Naturally, however, as editors we accept full responsibility for any inconsistencies or errors of fact or omission.

Lisa Russell, of the Environmental Studies Program at Southern Illinois University at Carbondale, provided invaluable secretarial assistance throughout the entire process of preparing this book, including scanning chapters, maintaining databases, and checking literature citations for many of the chapters. Connie Alsworth, Joanna Steele, and Julie Cummings, of New Mexico State University, also reviewed many lines of text and literature citations of draft manuscripts. We thank them all for their diligence, patience, and dedication to this project. Several of the skull drawings by Wilma Martin and line drawings by Frances Younger from the original volume appear in this edition, along with new artwork by J/B Woolsey and Associates. Justin T. Sipiorski, of the Zoology Department at Southern Illinois University at Carbondale, prepared the drawings of the baleen whales, Caribbean monk seal, and larga seal.

The publication of this book was made possible through the generous support of the Rocky Mountain Elk Foundation as well as support from Southern Illinois University at Carbondale, New Mexico State University, and North Dakota State University.

Introduction: A Twenty-Year Retrospective

Bruce C. Thompson
George A. Feldhamer

In this introductory chapter, we provide a brief overview of the technological and institutional changes that have occurred in wildlife management and conservation biology as they relate to North American mammals. In addition, we explore some of the management and conservation challenges that accrue with an increased knowledge base and societal constraints. A broadened knowledge base, numerous technological advances, and evolving management approaches since the initial 1982 edition of *Wild Mammals of North America* are reflected throughout this second edition.

TECHNOLOGICAL ADVANCES

Since 1982, wildlife management and conservation biology have become much more complex and sophisticated disciplines. More powerful computers, geographic information systems (GIS), and telemetry technologies, including global positioning systems (GPS), have been in the forefront of a shift in emphasis toward more quantitative approaches. The advent of GPS for widespread public use in the late 1980s and 1990s rapidly established a process for highly accurate spatial referencing of biological data. This innovation, along with dramatic, rapid development of computing capabilities, has revolutionized analysis and modeling of biological systems. These refinements also have produced rapid evolution of GIS for metapopulation and conservation analyses, radiotelemetry advances including GPS and satellite data transmission, and aerial videography with associated georeferencing capability (Rossi et al. 1992; Bobbe et al. 1993; Villard and Maurer 1996). Other technologies have been adapted to provide important advances in quantitative assessment of animal populations. These include passive integrated transponder (PIT) tags, refinement of tooth sectioning and cementum analysis for more reliable age estimation, portable acoustic bat detectors to augment traditional mist-netting techniques, and portable sonography tools for physiology and body condition research on mammalian species that are large and difficult to handle.

Many species can be handled more humanely and safely because of advances in immobilization drugs (Kreeger 1996). Advances in telemetry allow more individuals to be monitored for longer time periods. The larger data sets that are generated can be analyzed with more powerful and sophisticated computers and software programs. Analyses of DNA, often with non- or minimally invasive sampling, including microsatellites and mitochondrial DNA, have allowed for revolutionary advances in assessment of parentage, systematics, metapopulation analysis, estimation of reproductive success, description of social structure, and delineation of dietary features of many species (Avise 1994; Goldstein and Pollock 1997; Karp et al. 1998).

Computer enhancements and analytical software have expanded analysis of home range, population and survivorship models, and dispersal models using more sophisticated and powerful statistics programs. Especially important has been development of databases and spatial information layers that support wildlife–habitat relationship modeling. Such modeling must be approached carefully (Morrison et al. 1997; Morrison 2001) and is the underpinning of major conservation assessment initiatives such as gap analysis (Scott et al. 1996). Data collection and transfer are facilitated by Internet and personal data assistant applications. All of these advances are coupled with a burgeoning printed and electronic literature base that ever increases the difficulty of "keeping up with the literature." Often, the best we can hope for is to fall behind gracefully!

As professionals, researchers are justifiably proud of integrating and adapting technological advances into wildlife resource research and management endeavors. Nonetheless, this shift in emphasis often has been at the expense of more traditional natural history areas, including such basic skills as identification of plants and animals (B. D. Leopold 2000). During preparation of this book, Frank Miller—author of the chapter on caribou (*Rangifer tarandus*) and a long-time Canadian wildlife biologist commented:

> "I think we caused the demise of naturalists and general behaviorists before their time! We made the person on the hillside with field book and binoculars a want-to-be quantitative biologist before we obtained a satisfactory foundation in basic qualitative and limited quantitative behavior and activities. Now, we are forced to fill the gaps with subjective reasoning. Unfortunately, it is unlikely that agency mandates or biologists' ambitions will encourage anyone to carry out the kind of studies that are needed to strengthen what should be the bastion of our profession—knowing an animal in its natural world."

Many academic institutions find it increasingly difficult to fit the technological requirements and the needed natural history background into most undergraduate and graduate programs (Matter and Steidl 2000; Porter and Baldassarre 2000). Retaining and nurturing a healthy curiosity in natural history while also addressing the necessary expertise in quantitative skills will be a significant challenge in the immediate future for wildlife students as well as working professionals.

As statistics and quantitative methods become more central to wildlife research, it becomes increasingly necessary to consider proper experimental design, analytical procedures, and adequate sample sizes. A great deal more attention should be paid to designing projects so that they meet essential sampling requirements (Czaplewski et al. 1983). Wildlife biologists must be aware that assumptions that are axiomatic in all statistical analyses render many hypothesis tests superfluous (Nester 1996) and that statistical hypothesis tests frequently confuse interpretation of data (Johnson 1999; Anderson et al. 2001). Hypothesis testing has often been misunderstood. Therefore, out of necessity, comparative evaluations have been made using fragmentary data sets that are limited spatially and temporally. As noted by Steidl et al. (2000:518), wildlife research efforts sometimes have been limited by poor understanding of "the differences between research hypotheses and statistical hypotheses, and between biological and statistical significance." The need to do better has been recognized and the tools are available—implementation remains.

INSTITUTIONAL AND SOCIETAL CHANGES

Concurrent with technological advances, we have seen a change from "game management" to "ecosystem management," with resultant larger scales, longer time frames, and increased complexity of variables (Agee and Johnson 1988; Stanford and Poole 1996; Yaffee et al. 1996; Slocombe 1998; Thomas and Pletscher 2000). A more holistic perspective on all wildlife has necessitated expertise not only in biology, but also in economic, social, legal, and human dimensions aspects of conservation (Decker et al. 2001). This perspective is especially apparent to those responsible for designing and conducting research that meets societal requirements imposed through institutional animal care and use committees to promote attention to animal welfare in the quest for scientific understanding. As a result, multidisciplinary and interdisciplinary teams of individuals now often replace the lone field biologist of yesterday.

Concerns for escalating loss of biological diversity voiced in the 1960s continue unabated into the twenty-first century. The first issue of *Conservation Biology* was published 5 years after the original edition of *Wild Mammals of North America*. This was followed by several comprehensive texts regarding principles of conservation biology (e.g., Primack 1993; Meffe and Carroll 1994). Since then, wildlife biology and conservation biology as separate disciplines have merged in many ways. This merger was concurrent with a shift away from "species-specific" research and management—programs often focused on game species—toward more comprehensive landscape-level approaches, applied ecology, and "ecosystem management." This disciplinary evolution occurred in tandem with the 1980s emphasis on nongame programs (Thompson 1987), development of large-scale biotic databases through state natural heritage programs, and a broader consciousness for recognizing the importance of subtle as well as charismatic biotic elements and conserving them (Noss and Cooperrider 1994). As noted by Thomas and Pletscher (2000:547), this change resulted from the "political necessity to deal with broad-scale assessments and management plans with special attention to threatened or endangered species, retention of biodiversity, and political direction to take an ecosystem management approach." This societal view has endured with stronger application of the federal Endangered Species Act, legislation recognized as being directed at the most desperate conditions. More recently, broad-based public support has arisen for agricultural programs such as the U.S. Farm Bill and various amendments and other legislation (e.g., the Conservation and Reinvestment Act) that provide more diverse conservation benefits to people and a variety of animal species.

Much of the "necessity" for ecosystem management on the part of state and federal agencies has been mandated by court decisions (Thomas and Pletscher 2000) and a public much more interested and involved in decisions affecting wildlife populations than previous generations have been. Unfortunately, political expediency often prevails at the expense of what is biologically sound, as in conflicts with native users and resource managers concerning caribou or whales, for example. Native hunters are not convinced that they ever misused resources or at least will not admit to it. Instead, various anthropogenic factors and biologists' activities are blamed for declining numbers and changes in the distribution of traditionally harvested mammals. Current sociopolitical stands by native groups over aboriginal rights and land claims and the strong desire for self-rule require firm adherence to this conviction. The problem of harvest control is unlikely to be resolved in the near future and will never be solved unless native users have full participation in the process. However, effective comanagement of mammals subject to subsistence taking is a problematic and uncertain step in the process.

MANAGEMENT AND CONSERVATION CHALLENGES

The past 20 years have seen increased competition for land and decreased wildlife habitat caused by increasing human populations. There are fewer hunters, reduced open land, and increased numbers of appreciative user groups. The adjective *appreciative* has replaced *nonconsumptive*, with recognition that essentially all use of wildlife directly or indirectly includes a consumptive component. For example, a traditional "game" species that receives significant attention from wildlife biologists and the general public is white-tailed deer (*Odocoileus virginianus*). Sociopolitical issues surrounding whitetails involve recreational hunting, vehicular accidents, crop damage, and forest regeneration (Waller and Alverson 1997; Feldhamer 2002). Whitetail populations in many areas have increased beyond "cultural carrying capacity," with increased interface between wildlife and urban areas [see the special issue of *Wildlife Society Bulletin* on deer overabundance (Warren 1997)]. Whereas hunting is the control method favored by resource managers, nonlethal methods are favored by the public (Curtis et al. 1995; Stout et al. 1997), and much effort has been invested in immunocontraceptive techniques (DeNicola et al. 1997). As another example, there has also been a significant decline in the past 20 years in the number of people who participate in recreational fur trapping, and trapping does not enjoy strong public support (U.S. Fish and Wildlife Service 1997; Armstrong and Rossi 2000). Many furbearers, such as foxes, were formerly considered as pests or vermin to be eradicated. The public now considers these "mesocarnivores" as critical components in the proper functioning of ecosystems. Population densities of many species of foxes, as well as raccoons (*Procyon lotor*) and other furbearers, have increased as the number of trappers, and fur prices, have declined. Each of these changes illustrates the increasing need to consider human dimensions associated with wildlife conservation (Decker et al. 2001). Especially critical for promoting any successful management program are good verbal and written communication skills to facilitate effective interaction with a concerned and increasingly vocal public (Jacobson 1999).

Many significant and interesting conservation challenges lie ahead. One of the most critical is the proliferation of exotic wildlife in North America and throughout the world, including "game ranching." Potential for disease transmission and for genetic introgression with native species are among numerous concerns associated with the introduction of exotics (Hawley 1993). Other challenges include continued exploration of oceans and the undersea environment, which can negatively affect marine mammals; environmental statutory compliance; public interest in wildlife contraceptives versus lethal control; subsistence hunting; and trends toward increased litigation in conservation issues. In the past as well as today, land areas within states, provinces, and territories often are divided into wildlife management regions, zones, areas, or units based on geopolitical boundaries with little or no ecological consideration given to the divisions. Although such an approach may be administratively expedient, wildlife resources may not always be best served by these jurisdictional boundaries.

Interestingly, despite the technological advances and societal changes noted above, many of the same questions that are asked and the answers that are needed for effective management of mammalian wildlife are those voiced by A. Leopold (1933) in *Game Management*. However, we must move forward with more systematic and inclusive conservation planning in mind (Margules and Pressey 2000). The various chapters in this book describe the many nuances of enhanced technology, societal change, and conservation challenges related to the mammalian species or species groups reviewed. Fortunately, whether through management efforts or increased public awareness and shifts in attitudes, most mammalian populations, economically important or not, may be more secure now than they were throughout most of the twentieth century.

LITERATURE CITED

Agee, J. K., and D. R. Johnson. 1988. Ecosystem management for parks and wilderness. University of Washington Press, Seattle.

Anderson, D. R., W. A. Link, D. H. Johnson, and K. P. Burnham. 2001. Suggestions for presenting the results of data analyses. Journal of Wildlife Management 65:373–78.

Armstrong, J. B., and A. N. Rossi. 2000. Status of avocational trapping based on the perspectives of state furbearer biologists. Wildlife Society Bulletin 28:825–32.

Avise, J. C. 1994. Molecular markers, natural history and evolution. Chapman and Hall, New York.

Bobbe, T., D. Reed, and J. Schramek. 1993. Georeferencing airborne imagery. Journal of Forestry 91(8):34–37.

Curtis, P. D., R. J. Stout, and L. A. Myers. 1995. Citizen task force strategies for suburban deer management: The Rochester experience. Pages 143–49 *in* J. B. McAninch, ed. Urban deer: A manageable resource? Symposium Proceedings 55th Midwest Fish and Wildlife Conference.

Czaplewski, R. L., D. M. Crowe, and L. L. McDonald. 1983. Sample sizes and confidence intervals for wildlife population ratios. Wildlife Society Bulletin 11:121–28.
Decker, D. J., T. L. Brown, and W. F. Siemer, eds. 2001. Human dimensions of wildlife management in North America. The Wildlife Society, Bethesda, MD.
DeNicola, A. J., D. J. Kesler, and R. K. Swihart. 1997. Remotely delivered prostaglandin $F_{2\alpha}$ implants terminate pregnancy in white-tailed deer. Wildlife Society Bulletin 25:527–31.
Feldhamer, G. A. 2002. Acorns and white-tailed deer (*Odocoileus virginianus*): Interrelationships in forest ecosystems. Pages 215–23 *in* W. J. McShea and W. M. Healy, eds. The ecology and management of oaks for wildlife. Johns Hopkins University Press, Baltimore.
Goldstein, D. B., and D. D. Pollock. 1997. Launching microsatellites: A review of mutation processes and methods of phylogenetic inference. Journal of Heredity 88:335–42.
Hawley, A. W. L., ed. 1993. Commercialization and wildlife management: Dancing with the devil. Krieger, Malabar, FL.
Jacobson, S. K. 1999. Communication skills for conservation professionals. Island Press, Washington, D.C.
Johnson, D. H. 1999. The insignificance of statistical significance testing. Journal of Wildlife Management 63:763–72.
Karp, A., P. G. Isaac, and D. S. Ingram. 1998. Molecular tools for screening biodiversity: Plants and animals. Chapman and Hall, New York.
Kreeger, T. J. 1996. Handbook of wildlife chemical immobilization. Wildlife Pharmaceuticals, Fort Collins, CO.
Leopold, A. 1933. Game management. Scribner's, New York.
Leopold, B. D. 2000. They are our future. Wildlife Society Bulletin 28:489.
Margules, C., and B. Pressey. 2000. Systematic conservation planning. Nature 405:243–53.
Matter, W. J., and R. J Steidl. 2000. University undergraduate curricula in wildlife: Beyond 2000. Wildlife Society Bulletin 28:503–7.
Meffe, G. K., and C. R. Carroll. 1994. Principles of conservation biology. Sinauer Associates, Sunderland, MA.
Morrison, M. L. 2001. A proposed research emphasis to overcome the limits of wildlife–habitat relationship studies. Journal of Wildlife Management 65:613–23.
Morrison, M. L., B. G. Marcot, and R. W. Mannan. 1997. Wildlife–habitat relationships: Concepts and applications, 2nd ed. University of Wisconsin Press, Madison.
Nester, M. R. 1996. An applied statistician's creed. Journal of Royal Statistical Society (Applied Statistics) 45:401–10.
Noss, R. F., and A. Y. Cooperrider. 1994. Saving nature's legacy: Protecting and restoring biodiversity. Island Press, Washington, D.C.
Porter, W. F., and G. A. Baldassarre. 2000. Future directions for the graduate curriculum in wildlife biology: Building on our strengths. Wildlife Society Bulletin 28:508–13.
Primack, R. B. 1993. Essentials of conservation biology. Sinauer Associates, Sunderland, MA.
Rossi, R. E., D. J. Mulla, A. G. Journel, and E. H. Franz. 1992. Geostatistical tools for modeling and interpreting ecological spatial dependence. Ecological Monographs 62:277–314.
Scott, J. M., T. H. Tear, and F. W. Davis, eds. 1996. Gap analysis: A landscape approach to biodiversity planning. American Society for Photogrammetry and Remote Sensing, Bethesda, MD.
Slocombe, D. S. 1998. Defining goals and criteria for ecosystem-based management. Environmental Management 22:483–93.
Stanford, J. A., and G. C. Poole. 1996. A protocol for ecosystem management. Ecological Applications 6:741–44.
Steidl, R. J., S. DeStefano, and W. J. Matter. 2000. On increasing the quality, reliability, and rigor of wildlife science. Wildlife Society Bulletin 28:518–521.
Stout, R. J., B. A. Knuth, and P. D. Curtis. 1997. Preferences of suburban landowners for deer management techniques: A step towards better communication. Wildlife Society Bulletin 25:348–59.
Thomas, J. W., and D. H. Pletscher. 2000. The convergence of ecology, conservation biology, and wildlife biology: Necessary or redundant? Wildlife Society Bulletin 28:546–49.
Thompson, B. C. 1987. Attributes and implementation of nongame and endangered species programs in the United States. Wildlife Society Bulletin 15:210–16.
U.S. Fish and Wildlife Service. 1997. 1996 national survey of hunting, fishing, and wildlife associated recreation. U.S. Government Printing Office, Washington, D.C.
Villard, M., and B. A. Maurer. 1996. Geostatistics as a tool for examining hypothesized declines in migratory birds. Ecology 77:59–68.
Waller, D. M., and W. S. Alverson. 1997. The white-tailed deer: A keystone herbivore. Wildlife Society Bulletin 25:217–26.
Warren, R. J., ed. 1997. Special issue: Deer overabundance. Wildlife Society Bulletin 25(2).
Yaffee, S. L., A. F. Phillips, I. C. Frentz, P. W. Hardy, S. M. Maleki, and B. E. Thorpe. 1996. Ecosystem management in the United States: An assessment of current experience. Island Press, Washington, D.C.

I

Opossum, Moles, Bats, and Armadillo

1

Opossum

Didelphis virginiana

Alfred L. Gardner
Melvin E. Sunquist

NOMENCLATURE

Common Names. Virginia opossum, opossum, possum
Scientific Name. *Didelphis virginiana*
Subspecies. *D. v. virginiana, D. v. pigra, D. v. californica,* and *D. v. yucatanensis* (see Gardner 1973, 1993)

Commonly, but erroneously, considered a "living fossil" (Reynolds 1953) inhabiting North America since Cretaceous times, *Didelphis virginiana* is the only native member of the family Didelphidae found north of Mexico. The earliest known remains likely representing the Virginia opossum date from the Sangamon Interglacial Stage of the Pleistocene (Hibbard et al. 1965). Post-Wisconsin remains are widespread in the United States and Mexico. The earliest fossil record for *Didelphis* is from Pliocene deposits in South America.

Marsupials were a conspicuous part of the Cretaceous and early Tertiary fauna of North America and persisted there until the Miocene (for summary, see Clemens 1977). Clemens (1968) did not regard *Didelphis* as an archetypal remnant of the Tertiary fauna. Instead, on the basis of the derived nature of several morphological characters, particularly dental, he considered *Didelphis* a relatively late evolutionary product of a South American marsupial radiation. Chromosomal data support this point of view (Reig et al. 1977).

Gardner (1973) hypothesized that the Virginia opossum evolved from the common opossum (*Didelphis marsupialis*) relatively late in the Pleistocene. However, information from DNA/DNA hybridization (Kirsch et al. 1993) and mitochondrial DNA sequences (Patton et al. 1996) suggests that the Virginia opossum represents an earlier radiation more closely related to the complex of white-eared opossums (*Didelphis albiventris*) of South America. These data support the conclusion that the common opossum is a relative newcomer to Central America and Mexico. Remains from archaeological sites (Guilday 1958) and evidence in numerous other reports (e.g., Grinnell et al. 1937; Hamilton 1958) demonstrate that the Virginia opossum has moved northward in historic times and, aided by human activities, continues to expand its range.

The taxonomy of the Virginia opossum has had an intricately complicated history. Traditionally, the name *virginiana* has been associated with the populations in the United States and Canada. Much of the recent literature has followed Hershkovitz (1951), who treated *D. virginiana* as a subspecies of *D. marsupialis*. As Gardner (1973) demonstrated, the Virginia opossum is distinct from its widespread, sympatric neotropical congener; the two occur together from northeastern Mexico to northwestern Costa Rica. Nomenclature and taxonomy of the genus in North and Middle America presented here follow the review by Gardner (1973).

DISTRIBUTION

The Virginia opossum occurs from southern Ontario and British Columbia, Canada, through much of the United States and Mexico to northwestern Costa Rica (Fig. 1.1). When Europeans began to settle North America, this species ranged as far north as northern Ohio and northern West Virginia (Guilday 1958). Since then, the species has gradually moved northward (Fig. 1.2). Much of this advance has been encouraged by human activities, which include introductions. All U.S. and Canadian populations west of the Great Plains and the Rocky Mountains are the result of transplants from the eastern United States. The Virginia opossum has probably moved southward in Central America during historic time, but documentary evidence is lacking. Introduced populations in California have spread to northern Baja California, Mexico (Gardner and Cortés-Calvo 1999).

Reports of distributional records are numerous: Arizona (Hock 1952; Hoffmeister and Goodpaster 1954), Massachusetts and New Hampshire (Kennard 1925), Minnesota (Hazard 1963), New Mexico (Sands 1960), New York (Coleman 1929; Stoner 1939; Severinghaus 1975; Manuel 1977), Ontario (J. H. Smith 1935; Peterson and Downing 1956), Oregon (Jewett and Dobyns 1929), South Dakota (Findley 1956), Texas (Bowers and Judd 1969), Vermont (Kirk 1921; Osgood 1938; Davis 1938), Washington (Scheffer 1943), Wisconsin (Hollister 1908; Long and Copes 1968), and Wyoming (Brown 1965; C. A. Long 1965; Crowe 1986). Summaries of the natural and human-caused spread of the species are found in state, regional, and faunal reports (Miller 1899; Grinnell et al. 1937; Hamilton 1958; Jones 1964; Packard and Judd 1968; Armstrong 1972; Godin 1977). Virginia opossums are indicated for 42 states in the United States and for Ontario and British Columbia on the basis of survey questionnaires returned by colleges, universities, and state and provincial wildlife management agencies (Deems and Pursley 1978). New Mexico did not list the species, but Arizona and Maine did. Arizona populations may have perished. Burmudez et al. (1995) summarized recent records for New Mexico. Most New Mexican localities are associated with the Rio Grande and its tributaries; some represent the Canadian River watershed. We have been unable to find any record of specimens from Maine, although Godin (1977) included the southwestern corner of the state within the range of New England populations.

The Virginia opossum occurs in almost any habitat from sea level to elevations over 3000 m. Northern and elevational limits appear to be controlled by climate and availability of den sites and winter food.

DESCRIPTION

Size and Weight. Newborn young are about 14 mm in snout–rump length and weigh 0.13 g. By weaning age, the average head and body (snout–rump) length has increased to near 200 mm and weight to near 160 g. Weights and measurements of weaned opossums vary greatly. Hartman (1928) listed an extremely small (650 g) but sexually mature female (had pouch young) that was 345 mm in head and body length. He considered 400 mm and 1300 g the approximate size of an average reproductively mature female.

Adult males are larger than adult females (Table 1.1), a size distinction that begins at sexual maturity (Gardner 1973). Sexual dimorphism in certain dimensions (such as length of canine) has a primary genetic basis, but most size differences between sexes probably result secondarily from reproductive activity. Opossums grow throughout life (Washburn 1946; Lowrance 1949); however, energy demands

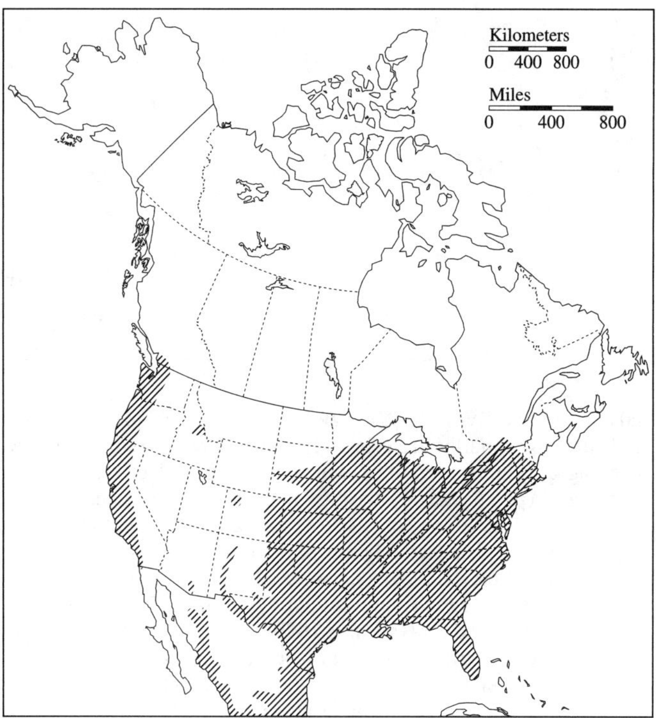

FIGURE 1.1. Distribution of the Virginia opossum (*Didelphis virginiana*).

on females rearing young result in smaller size and lower weight (Table 1.2).

Reported external measurements do not reveal any trend in size among U.S. opossums, except for tail length (shorter in more northern populations). On the average, Latin American populations have longer tails (see Table 1.1), but weigh less than animals in the United States. Cranial measurements of Latin American *D. virginiana* were reported by Gardner (1973).

Few reports include weights, and few of these distinguish sexes and ages. Peterson (1966) said Virginia opossums weigh up to 6.35 kg, and Jackson (1961) gave 6.12 kg as attainable by excessively fat, old individuals. Because the average female weighs less (see Table 1.2), these weights probably represent males. Audubon and Bachman's (1851) weight of 12 lb (5.4 kg) for a female seems excessive. Females at little more than half that weight are unusually large. The heaviest wild-caught male reported (from Illinois) weighed 5.9 kg (Pippitt 1976). Brocke's (1970) mention of a 7.9-kg specimen in Michigan did not include information on sex or whether it was wild caught at that weight or was an obese captive. Comparisons of weights in Table 1.2 show a general trend toward lighter animals in southern populations.

Weight is not as closely correlated with size in Virginia opossums as it is in other New World marsupials (Eisenberg and Wilson 1981). This opossum is the only didelphid capable of putting on large stores of body fat. Body weights are influenced by season; animals are heaviest in fall and early winter. The total lipid fraction of body weight can be as high as 31% in animals taken in winter (Brocke 1970) and should be even higher in the fall. In northern latitudes, overwintering opossums can survive a 40–45% loss of body weight (Fitch and Sandidge 1953; Brocke 1970); 30–40% of this loss is fat, and the remaining 60–70% is due to a combination of carbohydrate and protein catabolism.

External Characteristics. "*An Opassom hath a head like a Swine,* & a taile like a *Rat*, and is of the Bignes of a Cat. Under her belly shee hath a bagge, wherein shee lodgeth, carrieth, and sucketh her young" (Smith 1612:14). One of the earliest published accounts, Captain John Smith's description of animals from Virginia needs but few details to be complete.

The whitish to pale gray conical head tapers to a pointed snout. The cheek invariably is white, bordered above by a grayish to dusky black eye stripe and eye ring, and behind by darker colored sides of the head and neck. The darker color of the dorsum often extends forward over the crown in a narrow V-shaped wedge to between the eyes. The naked, leathery ears are black with white or flesh-colored tips. The lower legs are black, and the feet are black with white toes; the first toe of the hind foot is large, opposable, and lacks a claw or nail. The long, scaly, and scantily haired tail is black at the base but otherwise flesh colored to dirty white. The body fur is long and dense, the hairs white

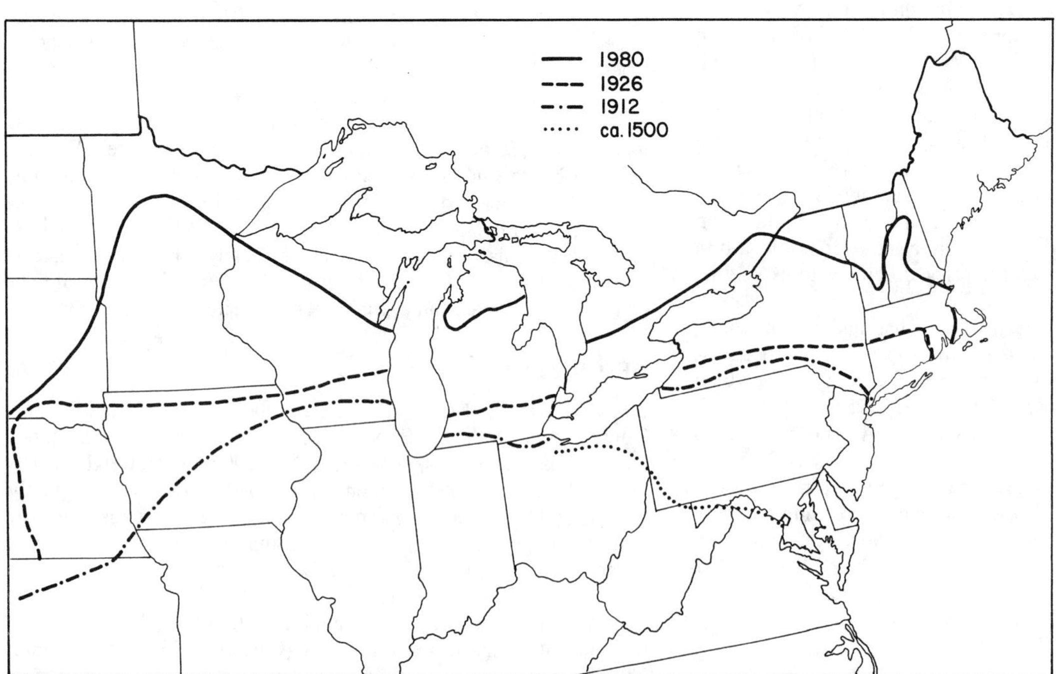

FIGURE 1.2. Changes in the northern distributional limits of the Virginia opossum (*Didelphis virginiana*) in eastern North America. SOURCE: Data on distribution ca. 1500 from Guilday (1958); data on 1912 from Cory (1912); data on 1926 from Ashbrook and Arnold (1927); data on 1980 from references in text.

TABLE 1.1. External measurements (mm) of *Didelphis virginiana*

Location	Sex	N	Total Length X̄	(Range)	Tail X̄	(Range)	Hind Foot X̄	(Range)	Ear X̄	(Range)	Reference
Pennsylvania	♂	10	779	(698–883)	291	(221–356)	69	(63–75)	54	(44–60)	Blumenthal and Kirkland 1976
	♀	18	753	(666–828)	289	(216–319)	64	(60–68)	55	(49–60)	
Ohio	♀	5	729	(600–817)	298	(255–355)	63	(55–70)			Preble 1942
Arkansas	♂	11	694		280		58		49		Sealander 1979
	♀	7	661		274		55		49		
Georgia	♂	6	560	(513–692)	254	(224–313)	53	(48–63)			Golley 1962[a]
	♀	5	583	(352–808)	261	(222–321)	57	(49–71)			
Texas	♂♂		782		324		66				W. B. Davis 1974
	♀♀		710		320		63				
Louisiana	♂	22	821	(751–940)	310	(223–380)	62	(50–85)	51	(40–60)	Lowery 1974
	♀	22	732	(613–900)	302	(260–329)	60	(38–75)	50	(42–58)	
Sinaloa–Sonora	♂	8	805	(740–870)	376	(346–425)	65	(57–70)	53	(48–60)	Gardner 1982[b]
	♀	8	765	(680–860)	389	(325–470)	58	(50–63)	50	(46–53)	
Oaxaca	♂	22	827	(675–947)	374	(305–446)	65	(50–80)			Gardner 1982[b]
	♀	18	778	(687–890)	358	(295–412)	62	(52–70)			
Nicaragua	♂	16	824	(730–910)	377	(310–445)	63	(48–73)	52	(40–60)	Gardner 1982[b]
	♀	20	760	(678–838)	352	(255–409)	60	(50–74)	48	(34–59)	

SOURCE: Measurements of samples from the United States are from the literature. Mexican (Sonora, Sinaloa, and Oaxaca) and Central American (Nicaragua) samples are part of the Latin American series reported on by Gardner (1973).

[a]Either not all adults as stated or errors in these measurements.

[b]Adults (age classes 5 and 6).

basally, dark brown to black terminally, and interspersed with long white guard hair (gray phase) or with broadly black-tipped guard hair (black phase).

The foregoing description best fits *D. v. virginiana,* which in northern parts of its range may have an entirely white face, white forefeet, and little or no black pigmentation at the base of the tail. Northern opossums that have overwintered often have lost the tips of the ears and end of the tail from frostbite.

Virginia opossums from Gulf Coast states (*D. v. pigra*) are darker, often with all-black ears and hind feet. Opossums from southern Texas through Nicaragua (*D. v. californica* and *D. v. yucatanensis*) are even darker and have all black ears and feet. In these populations, the white cheek is conspicuous against the darker face. They also have proportionally longer tails (see Table 1.1), of which half or more of the naked portion is black.

Females have an external, fur-lined abdominal pouch (marsupium) enclosing the teats, which are most commonly arranged in a 12-teat horseshoe arc with an additional teat in the center. Teat number varies from 9 to 17 (Hamilton 1958). Males have permanently descended testes in a pendulous scrotum anterior to a bifurcated penis. A urinogenital sinus adjoins the rectum in both sexes; a common sphincter muscle controls both openings. The chest area of males is often stained yellow to orange from skin glands over the manubrium (anterior sternum). Sweat glands are restricted to tail and plantar surfaces of feet (Fortney 1973). Friction ridges are present on plantar surfaces of feet; the tail is prehensile.

Cranial and Skeletal Characteristics. The braincase is small, the postorbital width of the skull is less than interorbital width, and the sagittal crest is well developed. The nasals are long, narrow anteriorly,

TABLE 1.2. Weights (kg) of male and female *Didelphis virginiana*

Location[a]	Males N	Males X̄	Males (Range)	Females N	Females X̄	Females (Range)	Reference
Michigan	40	3.6	(2.8–4.6)	12	2.4	(2.0–3.2)	Brocke 1970[b]
Michigan	66	2.2	(0.8–3.3)	36	1.6	(0.7–3.1)	Brocke 1970[c]
New York	83	2.8		60	1.9		Hamilton 1958[b]
New York	10	4.1	(3.4–5.0)	9	2.7	(2.5–3.2)	Hamilton 1958[d]
Pennsylvania	10	3.4	(2.2–4.0)	18	2.4	(1.8–3.0)	Blumenthal and Kirkland 1976
Iowa	5	3.1		10	1.8		Wiseman and Hendrickson 1950
Illinois	9	3.8	(1.7–5.9)	7	2.4	(1.8–3.1)	Pippitt 1976
Indiana	1	5.4					Lindsay 1960[d]
Kansas				18[e]	1.4	(0.8–2.0)	Fitch and Sandidge 1953
Kansas	3	2.9	(2.6–3.4)	4	2.6	(2.2–3.2)	Pippitt 1976
Georgia				5	1.1	(0.9–3.7)	Golley 1962
Louisiana	105	2.0	(0.9–3.0)	74	1.8	(0.3–1.6)	Edmunds et al. 1978
Louisiana	22		(2.7–4.7)	22		(1.2–2.2)	Lowery 1974
Nicaragua	20	1.6	(0.9–2.3)	21	1.2	(0.5–2.0)	USNM[f]

[a]Locations arranged north to south.

[b]Adults.

[c]Juveniles.

[d]Heaviest of each sex.

[e]Females with pouch young taken in March.

[f]From specimens in National Museum of Natural History.

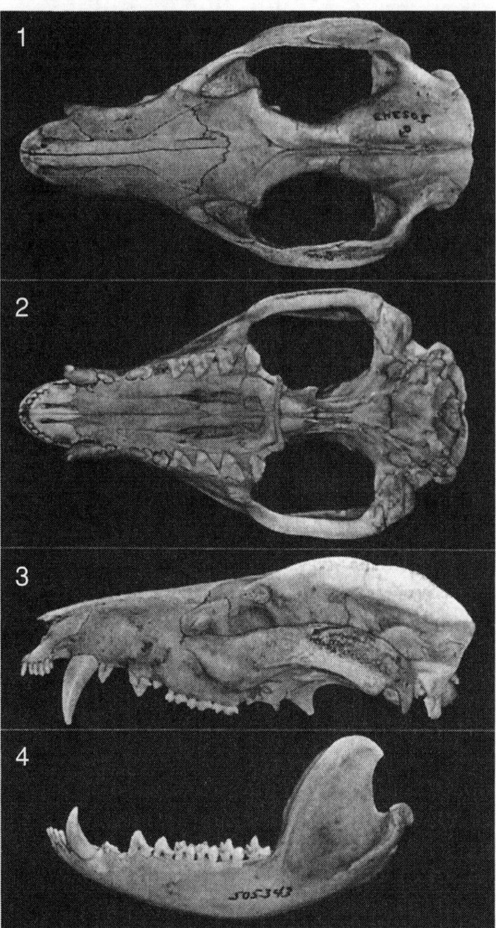

FIGURE 1.3. Skull of the Virginia opossum (*Didelphis virginiana*). From top to bottom: dorsal view of cranium, ventral view of cranium, lateral view of cranium, lateral view of mandible.

wide posteriorly (Fig. 1.3), and often in contact with the lacrimal, which recedes from the outer margin of the jugal before terminating. The orbital extension of the palatine is usually broad, and there are two pairs of palatine vacuities. The angular process of the dentary is inflected medially. Adults have 50 teeth, for which the dental formula is I 5/4, C 1/1, P 3/3, M 4/4; only the third (last) premolar has a deciduous (milk) precursor. Epipubic bones are present in both sexes; the male lacks a baculum.

Color Phases. Gray is the most common color pattern in the Virginia opossum. The black phase is uncommon to rare north of Georgia and the Gulf Coast states (including Texas), but is common in the southeastern United States and southward through Mexico to Costa Rica. Judging from museum specimens, black predominates in some Mexican populations (Gardner 1973). The main difference between black and gray phases is the color of the guard hair, which is all white in the latter and terminally one-third or more black in the former. The iris is black in both. Black-phase opossums often have a few all-white guard hairs scattered through the pelage.

Cinnamon-colored animals have been reported from the United States, but not from Latin American populations. This rare variant differs from gray and black animals in that it is cinnamon brown wherever the latter two are normally black (including ears, irides, and base of tail), and has cinnamon-colored guard hair. Overall, the fur is softer, perhaps because of the shorter guard hair. Hartman (1922) suggested that two kinds of cinnamon mutant occur, one with guard hair and the other without.

Two white mutants are known. True albinos with pink (pigmentless) ears, lips, eyelids, irides, feet, and tail are known throughout the range of the species. The other mutant, called albinotic by Hartman (1952:42), has white fur, but normally pigmented skin (i.e., black ears, eyelids, and lips). Albinotic animals, although extremely rare, have been reported from several populations.

The genetic basis for these color variants and mutants is incompletely known. The allele for black guard hair may be recessive and expressed only in the homozygous condition. This would explain finding black- and gray-phase animals in the same litter when one or both parents are gray. The breeding experiments necessary to verify this hypothesis have not been done.

GENETICS

Schneider (1977) summarized the knowledge of the karyology and cytogenetics of *D. virginiana*. His sections on cell cycles and DNA, RNA, and protein synthesis are excellent reviews, but the section on chromosomal evolution omitted the studies by Gardner (1973) and Reig et al. (1977).

Although the chromosomes of the Virginia opossum have been studied since Jordan (1911) reported an erroneous diploid number of 17, Shaver (1962) was the first to describe the karyotype correctly. The diploid number is 22, the fundamental number (number of autosomal arms) is 32. Autosomes consist of three pairs of large subtelocentric, three pairs of medium-sized subtelocentric, and four pairs of medium-sized acrocentric chromosomes. The X chromosome is a smaller, medium-sized submetacentric, the Y a small acrocentric.

The *D. virginiana* karyotype differs markedly from the 22-acrocentric-chromosome karyotype of its congeners. The chromosomal differences are complex and unique among American marsupials because they are not explainable by whole-arm translocations. Gardner (1973) hypothesized the attainment of reproductive isolation from a *D. marsupialis*-like ancestor through a series of rapidly fixed chromosomal rearrangements acquired by an isolate of the parental stock.

ANATOMY

The most distinctive features of marsupials are in the anatomy of the reproductive tract. Females have two uteri, each connected to the median vagina by lateral vaginae, which receive sperm. A temporary median birth canal forms at parturition to permit direct passage of neonates through the median vagina to the urogenital sinus. The female reproductive tract was first described and illustrated by Tyson (1698). The male also was described by Tyson (1704). These were again described by Hill and Fraser (1925) for females and Chase (1939) for males.

Studies of the skeletal system include those by Washburn (1946) and Nesslinger (1956) on the development of ossification centers, by Cheng (1955) on the pectoral girdle, by Elftman (1929) on the pelvic girdle, and by Lowrance (1949, 1957) on skeletal variability and growth. There are 7 cervical, 13 thoracic, 6 lumbar, and 2 sacral vertebrae. Caudal vertebrae vary from 26 to 29. The coracoid becomes fused to the scapula by the time an individual attains reproductive maturity. Certain sutures of the cranial and appendicular skeleton never fuse; therefore, some parts of the opossum skeleton grow throughout life.

Extensive summaries on the anatomy of marsupials (including *D. virginiana*) are found in chapters by Barbour, Parker, and Crompton et al. in Stonehouse and Gilmore (1977). Other works include Ellsworth (1975a, 1975b) on musculature and Johnson (1977) on the central nervous system. Bryant (1977) presented a synthesis of the literature on the lymphatic system, the immunohematopoietic complex, and the development of immune mechanisms. Little is known of the structure and function of the endocrine system aside from Kingsbury's (1940) work on the thymus, McDonald's (1977) review of adrenocortical function, and Hearn's (1977) review of pituitary function. Structure and function of the testes and associated reproductive glands were summarized by Setchell (1977). Morphology of the penis and spermatozoa was covered by Biggers (1966). Selenka (1887) was the first to report paired (copulatory) spermatozoa, which he found in the lateral vaginae of a recently bred female. This peculiar configuration (didelphid type) results from pairing of spermatozoa in the epididymides (Biggers and Creed 1962; Biggers and DeLamater 1965). Paired spermatozoa move in a straight line, but if separated, each of the two swims in a circle.

PHYSIOLOGY

Morrison and Petajan (1962) and Petajan and Morrison (1962) showed that pouch young begin to develop thermoregulatory ability by 55 days of age. By 94 days of age, young opossums are able to maintain deep body temperatures at air temperatures as low as 5°C for up to 2 hr. Dills (1972) and Dills and Manganiello (1973), using surgically implanted transmitters, demonstrated average circadian temperature fluctuations of about 3°C correlated with the activity state of the animal. Reported body temperatures are 35.0–35.5°C (Higgenbotham and Koon 1955; Morrison and Petajan 1962; McManus 1969). The general conclusion from these studies and those by Nardone et al. (1955), Brocke (1970), and Pippitt (1976) is that body temperatures fluctuate daily and seasonally. Telemetered body temperatures of free-ranging Virginia opossums in northern Illinois ranged from 32.2°C to 37.9°C in winter and from 35.0°C to 37.0°C in summer (Pippitt 1976). The range of body temperature recorded under different circumstances varied according to activity, body size, nutritional state, ambient temperature, and manipulative methods used by the investigators. Pippitt (1976) found that body temperatures of laboratory-held opossums averaged 35.75°C (32.2–38.0°C). Large (>2.5 kg) winter-acclimated animals maintained body temperatures at ambient temperatures as low as −5°C. One large Virginia opossum gradually raised its body temperature from about 35.7°C to near 36.3°C when exposed to an ambient temperature of 5°C over a 120-min period. Under the same conditions, the body temperature of a small (<2.3 kg) opossum decreased from about 35.9°C to near 34.7°C. Body temperatures of small (<2.5 kg), winter-acclimated animals were higher than those of larger opossums over the same range of ambient temperatures, except at −5°C, when body temperatures were significantly lower (Pippitt 1976). Small opossums had difficulty regulating body temperatures when ambient temperatures were below 0°C, whereas large animals could tolerate air temperatures as low as −20°C for short periods. Vasoconstriction, piloerection, shivering, and behavioral avoidance of low temperatures are the most important thermoregulatory responses to cold.

McNab (1978) found thermal conductance in *D. virginiana* was much lower than in Panamanian *D. marsupialis,* although basal metabolic rates were about the same. Thermal conductance levels for McNab's opossums from Florida were close to those reported by Pippitt (1976) for winter-acclimated opossums of comparable weights from northern Illinois. Pippitt also showed that opossums can lower conductance, probably by reducing blood flow to peripheral parts of the body, as the ambient temperature declines. He found, however, that conductance increases at air temperatures ≤0°C. McNab (1978) reported basal metabolic rates about double those reported by Brocke (1970) and Lustick and Lustick (1972). Pippitt (1976), whose values also exceeded those of Brocke, pointed out that some opossums have lower basal rates in winter than in summer. Brocke's measurements, as low as 0.15 ml O_2/g/hr in a 4.7 kg male, were made on sleeping animals. Pippitt's free-ranging summer animals had a mean oxygen consumption rate of 0.73 ml O_2/g/hr. Mean rates for animals in winter varied from 0.51 to 0.70 ml O_2/g/hr, depending on body weight.

Opossums experience heat stress at ambient temperatures exceeding the thermal neutral zone (30°C). Typical responses are panting and spreading saliva over the body. Eccrine sweat glands are located only on plantar surfaces and the skin of the tail (Fortney 1973). Because Virginia opossums are nocturnal, they are unlikely to encounter stressfully high ambient temperatures. Water balance was studied by Plakke (1970) and Plakke and Pfeiffer (1965).

Pippitt (1976) subdermally implanted two-channel radiotransmitters to monitor heart rate and body temperature in laboratory-held and free-ranging opossums. Heart rate in free-ranging opossums averaged 154.5 (109–220) beats/min in small animals during summer and from 120.4 (75–310) to 149.8 (89–214) beats/min, depending on weight class, in animals during winter. Francq (1970) measured heart rate in alert and death-feigning "young adult" opossums. QRS- and T-wave deflection were the same in both activity states. Heart rate averaged 2.76 (2.2–3.6) beats/sec in alert and 2.97 (1.7–4.6) beats/sec in death-feigning animals. Wilber (1955) recorded 200 beats/min for an opossum under light anesthesia. The higher average rates in laboratory tests (Wilber 1955; Francq 1970) probably resulted from the agitated state of the subjects.

Studies on gas-transporting capacity, tolerance to high and low oxygen levels, and wound healing were summarized by McManus (1974). Black (1935) discussed wound healing as evidenced by the high frequency of healed broken bones.

Tamar (1961) found that opossums have low taste sensitivity. Although they are able to perceive color (Friedman 1967) and can discriminate black and white objects under test conditions (James 1960), opossums respond more quickly to audio than to visual cues when pursuing small moving prey (Langley 1979).

REPRODUCTION

Breeding Season. The breeding season, defined as the period beginning with earliest conception and ending with weaning the latest litter, generally begins in January and extends to November. Most published accounts document the onset of reproductive activity, but few offer adequate information on later and terminal phases. Abundant evidence confirms two distinct breeding periods: a short, early period of high breeding activity extending 6–7 weeks, followed by a longer, less intense period beginning 2–4 weeks later and extending at least 2 months.

The earliest record for probable conception is mid-December (Edmunds et al. 1978) in northern Louisiana. Hartman (1928) also found evidence of ovulation and conception in December, but recorded an ovulation peak during the third and fourth weeks of the year (after 1 January) in the vicinity of Austin, Texas. Jurgelski and Porter (1974) found that about 30% of North Carolinian females showed signs of estrus from 20 December to 8 January. About half of these cycles were abortive. Most females entered estrus during the 3 weeks between 8 and 31 January. Early breeding dates were summarized by Grote and Dalby (1973). Accurate fixing of latest conception is difficult from the literature. Hamilton's (1958) data indicate successful breeding as late as August (Fig. 1.4).

The shift from December–January onset of breeding in the southern United States to January–February at northern latitudes is shown in Figure 1.4. The breeding season is unknown for Latin American populations, but may be inferred from Figure 1.5.

Estrous Cycle. Hartman (1921a, 1923b) and Reynolds (1952) determined the onset of estrus from vaginal smears. Palpation of the mammary area was suggested for determining estrus (Hartman 192la), but Reynolds (1952) and Jurgelski and Porter (1974) found it unreliable. The vaginal-smear method is lucidly described and illustrated by Jurgelski and Porter (1974).

Reynolds (1952) concluded that estrus lasts about 36 hr and fertile matings occur only during the first 12 hr. Spontaneous ovulation occurs during estrus. Ova are fertilized in the fallopian tubes before the addition of a coat of albumen and a thin shell membrane on their 24-hr trip to the uterus.

In California, Reynolds (1952) found variability of 22–38 days (average-29.5) in the length of estrous cycles in Virginia opossums. Jurgelski and Porter (1974) reported variation of 17–38 days (average-25.5). Reynolds (1952) noted that the time of estrus of sister littermates did not vary more than 5 days, suggesting genetic variation in length of estrus. The Virginia opossum is polyestrous, and five or six (Reynolds 1952) or seven (Jurgelski and Porter 1974) cycles are possible during a breeding season in the absence of pregnancy. Reynolds (1952) found that older (second-year) opossums have fewer (three to four) cycles, but that the length of the estrous cycle is not influenced by age. Females

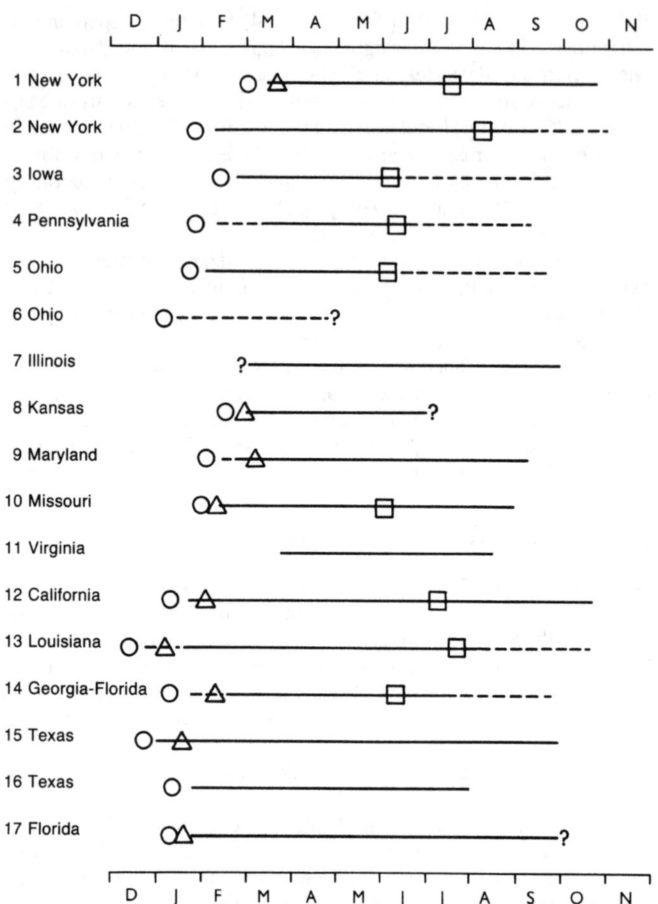

FIGURE 1.4. Breeding season of the Virginia opossum (*Didelphis virginiana*). Circles indicate the time of earliest conception; triangles, the time of birth peaks during the first breeding period; squares, the time of latest recorded birth; a solid line, the time when females had pouch young; and a broken line, the time females probably had pouch young. Numbers correspond to the following references, arranged by state and latitude from north to south: 1, VanDruff 1971 (43°N); 2, Hamilton 1958 (42° 30′ N); 3, Wiseman and Hendrickson 1950 (40° 50′ N); 4, Blumenthal and Kirkland 1976 (40° N); 5, Petrides 1949 (40° N); 6, Grote and Dalby 1973 (40° N); 7, Sanderson 1961 (40° N); 8, Fitch and Sandidge 1953 (30° 08′ N); 9, Llewellyn and Dale 1964 (39° 06′ N); 10, Reynolds 1945 (38° 58′ N); 11, Stout and Sonenshine 1974 (37° 28′ N); 12, Reynolds 1952 (ca. 36° N); 13, Edmunds et al. 1978 (32° 30′ N); McKeever 1958 (30° 40′ N); 15, Hartman 1928 (30° 15′ N); 16, Lay 1942 (30° N); 17, Burns and Burns 1957 (29° 40′ N).

reenter estrus within 2–8 days following loss of young. Estrous females breed only once in each cycle. A temporary vaginal plug, which is shed within 36 hr, forms after copulation.

Opossums examined by Reynolds (1952) and Jurgelski and Porter (1974) were all in anestrum by the third week of July. Some opossums in New York (Hamilton 1958) and Nicaragua (Biggers 1966), however, were still fertile in August. Males have spermatozoa in either the testes, epididymides, or both throughout the year, but testis weight is usually lower from September to December (in Pennsylvania), coincident with the anestrous period of females (Biggers 1966).

Gestation. Gestation is about 12 days, 18 hr long (Reynolds 1952; Burns and Burns 1957). Most embryonic development occurs while enclosed in the shell membrane, which begins to disintegrate permitting implantation on day 10 (Harder et al. 1993). Young are born in embryologic stage 34 (McGrady 1938) during the first half of day 13, or in stage 35 during the second half of day 13. Stage 34 embryos are recognizable by their blocklike, elongate heads and fully developed oral shields. The head is rounded or ovoid, and the oral shield has disappeared by stage 35, only 12 hr later. McGrady (1938) did the definitive study on the embryology of the Virginia opossum and reviewed the earlier literature on intrauterine development.

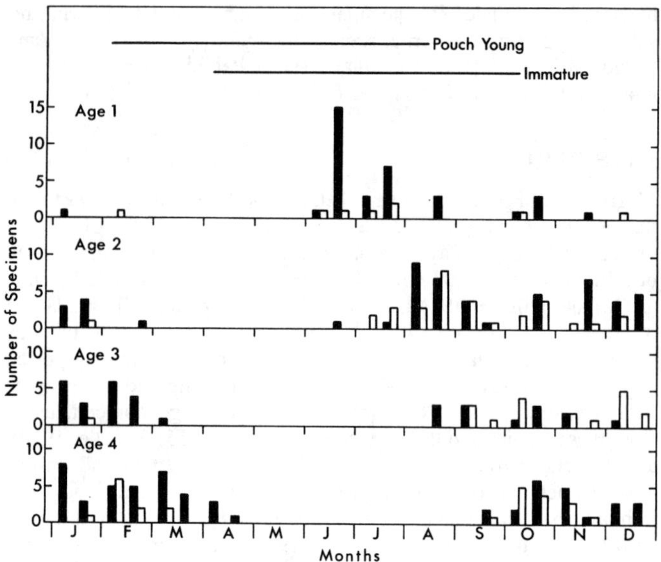

FIGURE 1.5. Seasonal distribution of age classes of the Virginia opossum (*Didelphis virginiana*) in Latin America. Mexican specimens are represented by a solid bar, Central American specimens by an open bar.

Birth. A few minutes before giving birth, the female usually shows preparturient behavior: restlessness, ears folded against head, and tail drawn forward between the legs. She then assumes a humped sitting position with the tail forward, and licks the vulva as the young are expelled. All young are born within 12 min (Reynolds 1952). Most neonates emerge already free of fetal membranes and swinging their forelimbs in a swimming motion. They grasp the hairs of the belly with their forefeet and climb upward to the pouch. The distance to be traveled is only 40–50 mm because of the posture of the mother. Upward orientation is maintained because the newborn can use only the forelimbs to crawl; therefore, the weight of the body controls direction. Once the pouch is reached, each neonate attaches to a teat or dies. Hartman (1952) mentioned finding 21 young in the pouch. Twelve of these were attached to teats; the remaining 9 perished.

Litter Size. The size of a litter is determined by the number of young reaching the pouch and the number of functional teats. Litter sizes of 15 and 16 (Barton 1823; Hamilton 1958) and 17 (Bailey 1923) are known. Because 13 is the normal number of teats, pouch young can exceed this number only when more teats are available. Reynolds (1952) believed that only 7–10 of the normal number of 13 teats were functional in most females from central California. He observed births of 4, 13, 15, and 25 young (total of 57), but only 34 (60%) reached the pouch, where some failed to find teats.

There is a general trend toward larger litters in northern populations (Table 1.3). Burns and Burns (1957) found the smallest mean litter size (6.3) in Florida. Forty-four percent of the litters they examined contained fewer than 7, the modal number. We found no clear trend toward a higher frequency of smaller litters indicated in other reports. Most investigators did not report modal numbers of pouch young.

Although of general interest, litters of single young (Table 1.3) are of little biological importance because these are not usually carried through to weaning age. The stimulus of at least two suckling young is normally required to maintain lactation (Reynolds 1952). Two (3 litters) was the lowest number of pouch young among the 346 litters Hamilton (1958) examined from New York. Litter sizes of Virginia opossums from Mexico, based on label information from museum specimens, ranged from 1 to 14 (mean = 7.6, mode = 6, $N = 15$; Gardner 1982).

Llewellyn and Dale (1964) found a mean of 8.5 in 26 litters of young ≤4.0 mm in size (up to 25 days old) and a mean of 6.6 in

TABLE 1.3. Litter sizes of *Didelphis virginiana* in the United States

Location	*N*	Litter Size: Mean	Mode	Range	Reference
New York	156	9.0		1–15	VanDruff 1971
New York	346	8.7	9 (22%)	2–15	Hamilton 1958
Pennsylvania	20	9.2		1–14	Blumenthal and Kirkland 1976
Iowa	7	9.0		6–12	Wiseman and Hendrickson 1950
Missouri	42	8.9	8 (33%)	5–13	Reynolds 1945
Nebraska	23	8.6			Reynolds 1945
Ohio	5	9.6			Petrides 1949
Illinois	85	7.9			A. C. V. Holmes and Sanderson 1965
Maryland	57	7.7[a]			Llewellyn and Dale 1964
Kansas	28	7.5	7 (25%)	1–12	Fitch and Sandidge 1953
Virginia	3	7.3		6–9	Stout and Sonenshine 1974
California	44	7.2	7 (25%)	4–11	Reynolds 1952
California	6	10.0		8–12	Grinnell et al. 1937
Georgia–Florida	143	7.1		1–11	McKeever 1958
Louisiana	68	6.8		1–10	Edmunds et al. 1978
Texas	65	6.8			Lay 1942
Florida	50	6.3	7 (44%)	3–9	Burns and Burns 1957

NOTE: Mode followed by percentage of litters (in parentheses). Sample size (*N*) is number of litters. Locations are arranged from north to south.

[a]Mean of 8.5 in 26 litters of young ≤40 mm; 6.6 in 23 litters of young ≥60 mm.

23 litters of young ≥60 mm in size (≥40 days old). These data suggest progressive mortality of pouch young as litters mature. Fitch and Sandidge (1953) found progressively fewer pouch young being carried by some females when recaptured with the same litter, and reported losses of 23%. Sanderson (1961) calculated pouch young mortality as 12%. Hamilton (1958) found no evidence of mortality among pouch young and believed that if such losses occurred, they were trivial. Enders (1966) said that the critical period when losses occur begins at the time hair appears. This is about the time when pouch young begin to release their hold on the nipple; therefore, losses may increase because of their newly acquired relative freedom.

Burns and Burns (1957) found no correlation between the weight of the mother and the number of young in the pouch. Austad and Sunquist (1986) also found no correlation between body weight of female *D. marsupialis* and litter size. However, Hossler et al. (1994) reported that litter size in a New York population was correlated with hind leg fat index, a measure of nutritional condition. Reynolds (1952) found evidence that older females had decreased fertility, were less able to rear young, and showed a tendency to return to anestrum earlier than young opossums. These older females were probably in their second breeding season. That females in their second reproductive season have smaller litters or fail to reproduce is known for opossums in Virginia (Seidensticker et al. 1987), South Carolina (Austad 1993), and Florida (Sunquist and Eisenberg 1993; Wright et al. 1995). Austad (1993), however, found no reproductive decline in the second year for females on Sapelo Island, a predator-free island off the Georgia coast. Of the few females surviving into a third reproductive year, none had pouch young, suggesting they were postreproductive.

Number of Litters per Season. Audubon and Bachman (1851), Hartman (1928), and McManus (1974) suggested that three litters might occasionally be produced during a single year by some females. Although rare, this has been confirmed by Sunquist and Eisenberg (1993). Tyndale-Biscoe and Mackenzie (1976) found two litters/year in Colombian *D. marsupialis,* except in Valle del Cauca, where there were birth peaks in January, May, and August. They interpreted the third birth peak as evidence of the production of three litters/year. Biggers (1966), who did not distinguish between *D. marsupialis* and *D. virginiana* in his sample from Nicaragua, also noted three peaks of reproductive activity (January, May, and August). Three peaks in reproduction are also reported for *D. virginiana* in Florida and *D. marsupialis* in Venezuela, but only 7% of females known to have reared two litters attempted to raise a third in the same season. At each location, only one female successfully reared three litters in a single calendar year (Sunquist and Eisenberg 1993).

At the onset of the first breeding season for Floridian *D. virginiana,* Sunquist and Eisenberg (1993) found 96.8% (179 of 185) of adult females had young in the pouch. In May, 91.0% (142 of 156) of adult females had young in the pouch.

To determine whether Latin American Virginia opossums produced >2 litters/yr, we examined the data on Mexican and Middle American *D. virginiana* gathered during Gardner's (1973) systematic review of the genus. All records for animals that had not yet acquired a full set of teeth, including 60 specimens from Nicaragua, were segregated by age class, and the date of capture (from specimen labels) was plotted by half-month intervals (see Fig. 1.5). Younger age classes are poorly represented in the sample, possibly reflecting lower trap susceptibility of younger animals and bias by the preparator. Figure 1.5 shows evidence of births in August or early September, but shows no third peak that would indicate intensive breeding activity in August.

Biggers's (1966) data for Nicaraguan opossums may reflect breeding season differences between sympatric *D. virginiana* and *D. marsupialis.* The western region of Nicaragua has a seasonal climate with a well-marked dry season, whereas the eastern zone is much more humid and seasonality of rainfall is less extreme. These ecological differences favor *D. virginiana* in the drier Pacific region and undoubtedly influence reproductive success in both species.

The simplest explanation for the three breeding peaks found by Biggers (1966) and Tyndale-Biscoe and Mackenzie (1976), as well as the evidence in Figure 1.5 for births in August or early September, is as follows. The first peak (January) represents the start of normal breeding activity synchronized by the October–December anestrous period. January and early February are when most females give birth to their first litter. The second peak (May) represents return to estrus at the time of weaning the first litter and includes some contribution by females whose earlier breeding efforts were unsuccessful. Females giving birth to first or second litters in May or June are not likely to wean them in time to breed again in August. The third peak (August) is caused by sexually precocial (6- to 7-month-old) females born in January and early February. Females who were previously unable to rear young successfully and females weaning litters born in March or April could also contribute to the August peak. An examination of the timing of breeding peaks (Biggers 1966; Tyndale-Biscoe and Mackenzie 1976) shows a 4- to 4.5-month interval between the first and second, but a shorter 2.5- to 3-month span between the second and third, which argues against any female successfully rearing three litters a year. An analysis of the ages and the conditions of the marsupium in the cohort breeding in August is necessary to test this hypothesis. Sunquist and Eisenberg (1993) found no evidence for sexually precocial females in Florida; however, they did find a small percentage of first-litter females of *D. marsupialis* in Venezuela breeding in the year of their birth. These females gave birth in August and September.

Although females can easily produce two litters/year, Hamilton (1958) showed that not all do. Based on the presence of pouch young, he found barren females in each month of the breeding season. The frequency of barren females varied from a low of 20% in April to a high of 46% in August. Jackson (1961) said that one litter/year was typical in Wisconsin and that two was the exception. Data usually are insufficient to estimate the percentage of females that rear two litters/year. In central New York, VanDruff (1971) found that 21 of 40 (52%) recaptured females had produced a second litter. Lay (1942) believed that two litters/year was normal in eastern Texas; however, he had positive evidence from only three females.

Numbers of litters born per female each season is unknown. Although two litters can be raised to weaning age, a female may actually give birth to three or more during a season. Females that have single young and others who happen to lose their litters, either at birth or after the young are established in the pouch, return to estrus and breed again. Reynolds (1952) fed an inadequate diet to three females carrying pouch young and discovered that each subsequently lost its litter and was back in estrus 3–15 days after the young had died. Reynolds's experimental result may mimic natural losses of litters while a female is unable to find adequate food when temporarily incapacitated by injury or during severe drought or harsh spring weather.

In their comparison of the reproductive strategies of *D. marsupialis* and *D. virginiana,* Sunquist and Eisenberg (1993) found that 16 of 37 (43%) Venezuelan female *D. marsupialis* reared one litter in their lifetime, 17 reared two litters (46%), and four (11%) had three litters. In Florida, 12 of 39 (31%) *D. virginiana* reared one litter in their lifetime, 17 (43%) had two, 6 (15%) had three, 3 (8%) had four, and 1 (3%) had five litters. This last female reared three litters her first year and successfully reared two litters in her second reproductive year, an extraordinary feat.

It was once believed that a lag of several weeks was required after the young were weaned before estrus recurred because lactation was presumed to inhibit recurrence of estrus. Also, time was presumedly required to allow the enlarged nipples to regress to a size a newborn opossum could take into its mouth (Hartman 1923a, 1923b; Reynolds 1945). Reynolds (1952) later showed that females may breed while still producing milk and that nipples rapidly regress within a few days, but instead of receding to their original size, each develops a small papilla at the tip to which newborn young attach. These papillae were illustrated by Hamilton (1958:37).

Sexual Maturity. A 6-month-old (186 days) female that Reynolds (1952) reared in captivity gave birth to and successfully reared six young. She and a littermate, which was in estrus at the same time, had been weaned only 90 days earlier. Reynolds, however, claimed that females born in the earliest litter recorded in central California (29 January) would not have sexually matured by the end of the mating period (early July) of the same season. A female that weighed 660 g and was 345 mm in snout–rump length (S–R) had embryos that were 10.5 mm in crown–rump length (C–R) on 11 February (Hartman 1928). She could not have been younger than 6.5–7 months of age if the latest breeding in Texas is in July. The earliest age of first reproduction of females studied by Sunquist and Eisenberg (1993) in Florida was 245 days. The youngest reproductively mature female cited by Reynolds (1952) bred at 173 days of age when she was only 366 mm (S–R), and represents the minimum known breeding age (5.5 months) for the Virginia opossum. Sunquist and Eisenberg (1993) reported a Venezuelan *D. marsupialis* 172 days of age that had young in the pouch.

Males studied by Reynolds (1952) reached sexual maturity (had sperm in the epididymides) by 8.5 months of age. One inseminated a female when he was 247 days old. Biggers (1966), using Reynolds's data, stated that sexual maturity in the male is reached at approximately 240 days (8 months) of age.

Sex Ratios. Sex ratios of opossums in U.S. populations are summarized in Table 1.4. With the exception of the ratio given by Petrides (1949), sex ratios among pouch young are usually at parity. Though the population-wide sex ratio at birth may be unbiased, there is evidence that the reproductive allocation of individual parents may be sexually biased. Wright et al. (1995) found that female *D. virginiana* produced significantly more males in their first litters. Although more females were produced in their second litters, the trend was not significant. Labeled the "first-cohort advantage hypothesis," the explanation is that females should bias the sex ratio of their first litter toward males because first-cohort males will achieve more matings than second-cohort males. During their first reproductive year, first-cohort males are larger than second-cohort males, and larger male opossums have higher breeding success than do smaller second-cohort males (Ryser 1992; Wright et al. 1995). Given the higher mortality rate of males, their first year may be their only reproductive year.

Wright et al. (1995) found no significant correlation between maternal age and condition and the sex ratios in their litters. Their data did not support Trivers and Willard's (1973) maternal condition hypothesis. Austad and Sunquist (1986) had earlier field tested the maternal condition hypothesis on *D. marsupialis* in Venezuela by supplementing the diets of females. They predicted that mothers receiving additional food would produce litters with male-biased sex ratios because among polygynous mammals, when males have higher variance in reproductive success than do females, sex ratios of litters

TABLE 1.4. Sex ratios of *Didelphis virginiana* in the United States

	Percentage Males												
	Pouch Young		Immature			Adult			Immature and Adult				
Location	*N*	%	S	*N*	%	S	*N*	%	S	*N*	%	Reference
Wisconsin									R, T, H	161	53	Knudsen and Hale 1970
Michigan			T	153	61	T	48	50	T	201	59	Stuewer 1943
New York						T, R, C	434	27				VanDruff 1971
New York	811	52	T	149	56	T	146	20	T	295	38	VanDruff 1971
New York	911	52							N	846	53	Hamilton 1958
Pennsylvania	183	51							T	62	34[a]	Blumenthal and Kirkland 1976
Missouri									P	2185	57	Bennitt and Nagel 1937
Missouri									T, D	116	55	Reynolds 1945
Missouri									P	1076	58	Reynolds 1945
Ohio									P	330	54	Petrides 1949
Ohio	48	56							T	48	42	Petrides 1949
Illinois	363	49	T	337	56	T	180	42	T	519	51	A. C. V. Holmes and Sanderson 1965[b]
Maryland			T	114	62	T	106	56	T	220	54	Llewellyn and Dale 1964
Kansas									P	426	56	Sandidge 1953
Kansas									T	62	44	Sandidge 1953
Kansas	16	50							T	56	50	Sandidge 1953
Virginia			T	42	52	T	29	52	T	71	52	Stout and Sonenshine 1974
Louisiana			T	30	60	T	179	59	T	209	59	Edmunds et al. 1978
Texas									T	117	56	Lay 1942

NOTE: Percentages are of males in sample (*N*). Specimens (S) examined were pelts (P) or whole animals acquired by trapping (T), shooting (H), or predator control (C), caught with aid of dogs (D), found dead on road (R), or not segregated on basis of source (N). Locations are arranged approximately north to south.

[a]Percentage of males (38.9%) reported was an error; correct value was 33.9% (G. L. Kirkland, Jr., personal communication).

[b]Includes data from Sanderson 1961.

produced by high-quality females should favor males. This is predicated on knowing that some males breed more than one female, whereas others have no mates. Consequently, high-quality males will have greater reproductive success than any single female, and females producing high-quality offspring have the potential to contribute more to the population if they produce males. Conversely, if the capacity to invest is low, females should produce females because essentially all females find mates.

In the experimental group, females were located every other day at their sleeping dens and provisioned with about 125 g of sardines. Provisioning was initiated before mating began and continued until the end of the breeding season. Females in the control group were not fed. Austad and Sunquist (1986) found that the overall offspring sex ratio of provisioned females was significantly more male biased than was the sex ratio of litters from control females. Although their results supported the maternal condition hypothesis, they were criticized by Wright et al. (1995) because their analysis was based on overall sex ratios rather than on the litter itself as the sampling unit. Hardy (1997) independently reanalyzed the data used by Austad and Sunquist (1986) and confirmed that the sex ratios of litters of provisioned females were more male biased than were litters of control females.

A similar study of the response to dietary supplementation on *D. virginiana* in Florida (Sunquist and Eisenberg 1993) resulted in male-biased offspring sex ratios and significantly more male-biased litters in a year with normal rainfall, but not during a drought year. Even provisioned females were not in good condition during the drought. Although their findings tended to support the "first-cohort advantage hypothesis" (Wright et al. 1995), this hypothesis predicts an advantage, regardless of female age and condition, for producing male-biased litters at the start of the breeding season. Although first-cohort males are larger than second-cohort males at the onset of the first reproductive season and larger males have greater mating success, without a female in good condition, the sex-ratio data on litters may be misleading. The nutritional status of females is likely dependent on environmental conditions. Sunquist and Eisenberg (unpublished data) found no significant male bias in 59 first litters (29 male biased, 24 female biased, and 6 unbiased) examined in Florida between 1993 and 1999. Over the same period, 33 second litters were somewhat more female biased (11 male biased, 14 female biased, and 8 unbiased). Furthermore, ovarian senescence that occurs in *Didelphis* means that absolute age is important because litter size declines with age (Eisenberg 1988) and there is a trend for older females to produce more female-biased litters. Thus, condition and age of females may be operating simultaneously to influence sex ratios.

Deviations from a 1:1 sex ratio are common in immature and adult opossums. The percentage of males varies from 20% (VanDruff 1971) and 34% (Blumenthal and Kirkland 1976) to 62% (LIewellyn and Dale 1964). Free-ranging they are more susceptible than females to capture because of their greater cruising range and the higher probability that they will be active during the winter because their larger bulk imparts greater tolerance to colder temperatures.

Under certain circumstances, these traits may contribute to disproportionately high mortality among males. When local configurations of roads and barriers to dispersal, for example, expose the wider ranging males to mortality from highway traffic, there is a lower proportion of males to females in the population. Highway mortality may have caused the lower proportion of males reported by VanDruff (1971) and Blumenthal and Kirkland (1976), particularly if that mortality factor was biased. That more males than females are active during the winter was shown by Blumenthal and Kirkland (1976), who reported greater numbers of males in their winter sample, yet the percentage of males for all seasons combined was 34% (see Table 1.4). Petrides (1949) had a higher winter (December and January) proportion of males (54%) based largely on pelts, but had a lower ratio (42%) based on year- round livetrapping. Whatever the cause of higher mortality among males in central New York (VanDruff 1971), the decline occurred among adults because the relative numbers of immature males compares well with results from other studies (see Table 1.4).

DEVELOPMENT

Timing of developmental events is reasonably fixed and consistent in most mammals. Opossums are an exception. They exhibit a frustrating degree of variation in almost every phase of growth and development. Several reports have contributed to the understanding of the sequence of these events in the Virginia opossum. Among the more important are Hartman's (1928) review of the breeding season and rate of intrauterine and postnatal development; McGrady's (1938) treatise on embryology, which includes information on postnatal development; Petrides's (1949) report on sex and age determination; and Reynolds's (1952) study on reproduction. Although some of the information in these four reports is contradictory, we have extensively used them and a few others summarized by McManus (1974) to construct the outline presented here.

Growth of Young. The tiny, naked, and altricial newborn opossum is about 14 mm in C–R length and 0.13 g in weight. Each neonate attaches to a nipple, where it remains for approximately 60 days. The end of the nipple swells in the mouth, and from the third day through the first few weeks of life the young must be removed carefully to avoid tearing the lips. Young so removed after the third day are difficult to reattach. Young become easier to reattach to the nipple shortly before the age when jaws open. The nipple gradually lengthens (from about 1 mm to approximately 35 mm) and acts as a tether permitting the attached young freedom of movement in the pouch. Marked variation in size, weight, and developmental stage occurs among littermates. Petrides (1949) constructed a growth table based on lengths and weights from Hartman (1928), Moore and Bodian (1940), and Reynolds (1942). Reynolds (1952) also constructed a growth table for young opossums 1–95 days of age. His table does not include weight. The data of Petrides (1949) and Reynolds (1952) at 5-day intervals are presented in Table 1.5.

Age estimation based on size alone becomes more difficult as the young grow older. Reynolds (1952) included a growth curve to show the magnitude of individual variation among pouch young of known age; Hunsaker (1977) presented a similar growth curve. Based on averages of the S–R lengths of littermates, the accuracy limits of age estimates range from 2–3 days for pouch young <20 days of age to approximately 2 weeks for animals at weaning age (95–100 days old).

TABLE 1.5. Growth of pouch young of *Didelphis virginiana* from the United States

	Petrides (1949)		Reynolds (1952)	
Age (days)	S–R[a] (mm)	Weight (g)	$\overline{X}$ S–R (mm)	Range (mm)
1	13	0.13	13.8	13.0–15.4
5	17	0.4	18.8	17.7–19.6
10	24	0.9	24.5	22.5–27.0
15	30	1.3	30.4	27.2–32.0
20	33	1.7	37.1	33.0–40.0
25	40	2.4	41.7	37.0–44.6
30	45	3.9	48.4	42.6–52.7
35	53	5.4	52.8	50.0–56.0
40	60	7.0	60.7	54.4–65.2
45			68.4	58.4–72.1
50	75	13.0	71.9	64.0–80.8
55			79.6	70.3–93.8
60	100	25.0	91.6	81.2–100.8
65			106.0	89.0–117.3
70	125	45.0	115.8	101.8–123.0
75			128.0	110.0–140.9
80	150	80.0	148.0	134.1–165.0
85			161.1	147.0–183.0
90			185.1	165.0–200.0
95			195.2	171.6–217.7
100	180	125.0		

[a]The snout–rump (S–R) measurement given by Petrides is distance between snout and anus.

The S–R length measurement (essentially the same as snout–anus length and length of head and body) is difficult to apply to attached young. The measurement can always be expected to become more variable as size increases because it is almost impossible to align the spinal column in the same way every time the animal is measured. Tyndale-Biscoe and Mackenzie (1976) used the length of the head as a measure of size. This technique has not been applied to the Virginia opossum; it could prove a more accurate estimator of age than S–R length.

The weight (mass) of pouch young can be reliably predicted from tail length of opossums, as Sunquist and Eisenberg (1993) demonstrated in *Didelphis* in Florida and Venezuela. These data can then be associated with the appearance of certain morphological characteristics of the young to estimate age.

Development of Young. Although a newborn opossum is embryonic in appearance, it has functional but relatively undeveloped circulatory, respiratory, and digestive systems; a functional mesonephros; a partly functional muscular system; and a wholly cartilagenous skeleton. The ear and eye are nonfunctional and the central nervous system is incomplete. Forelimbs are developed and the fingers, which bear deciduous claws, are capable of gripping hairs and climbing, but the hind limbs and tail are rudimentary. The circular mouth grasps a nipple, which in 3 days enlarges within the mouth, anchoring the young to the mother while it suckles for another 55–60 days. The sequence of developmental stages follows:

Days 1–3. If removed, young can reattach themselves to nipples.
Days 11–15. Follicles for rostral vibrissae appear.
Day 16. Vibrissae and juvenile body hair appear on nose.
Day 17. Sex organs are visible on all young. Sex of some young may be determined as early as day 11 (McGrady 1938).
Day 20. Cheek vibrissae begin to appear; hind limbs can be used to shift position, but hind toes cannot be flexed.
Days 20–25. Pinnae become free.
Days 28–34. Pigmentation appears on lateral margins of scrotum. Females still show no pigmentation.
Days 33–41. Pigmentation appears on base of tail in both sexes.
Days 34–35. Downy body hair is visible using a hand lens; vibrissae begin to lengthen.
Day 37. Young cry when handled; mouth is still sealed.
Day 40. Pigmentation appears on upper neck and shoulders.
Day 43. Young able to flex toes on hind feet.
Day 48. Pigmentation is spread over entire dorsum, sparse dark hair emerges on neck and shoulders. This is the earliest age when young begin releasing the nipple; mouth still sealed.
Day 50. Hair short and sparse. This is the youngest age when the eyes and mouth begin to open. Young now are approximately the size of house mice.
Days 55–68. Mouth opens.
Days 58–72. Eyes open.
Day 60. Body completely covered with short dark hair; the first tooth erupts (deciduous third premolar); nipples (now about 25 mm long) permit young to lie outside of the pouch while suckling.
Day 70. Young conspicuously furry; juvenile hair (sometimes called juvenile guard hair) is about 15 mm long and beginning to shed; dark underfur about 10 mm long. Young commence leaving the mother and crawling short distances before returning to pouch. Gray and black phases cannot be distinguished until the juvenile guard hair is shed.
Days 70–85. Young usually leave den in mother's pouch and if left behind, begin crying. The female normally makes a clicking sound when the young cry and the young respond by going to her. If the female does not make the sound, some or all of the young may remain in the den.
Day 75. Second premolar is present; young capable of feigning death.
Days 75–85. Last four upper incisors, the canines, and the first premolar erupt.
Day 77. Descent of testes is completed.
Days 81–82. Young are commonly left in the den while female forages. All young cry if separated from the mother before day 83.
Days 83–91. Young do not cry when left in den.
Day 87. Weaning begins; young show strong interest in solid food, but are still suckling. Petrides (1949) and Hartman (1928) said that young are weaned by day 80. This remark may have been based on finding females without their litter (see days 81 and 82), which does not mean that the young have been weaned. Some or all young are still left in the den when the mother forages. If young leave the den with the female, they travel on her back or run beside her.
Days 87–92. Young produce clicking noise.
Day 95. First incisor, first lower molar, and sometimes the first upper molar are present.
Days 92–107. All young are weaned. All young show excited interest in solid food after day 96, at which time they are nursed infrequently. The young are largely independent of the mother after day 96, but will run to her in response to clicks as late as day 108. They now may leave the den alone and may sleep elsewhere. By weaning age, the young have developed their full range of hearing (about eight octaves). The most acute hearing is at the upper end of the scale, and the young respond more readily to sounds made by rustling leaves and grass than to the human voice (McGrady 1938).

Sunquist and Eisenberg (1993) found the duration of lactation for first litters of provisioned and control females of Floridian *D. virginiana* and Venezuelan *D. marsupialis* was approximately the same (92–94 days). However, duration of lactation for second litters of provisioned and control females at both locations was 1–2 weeks longer, and young were fully weaned at between 99 and 107 days. Provisioning did not shorten duration of lactation with either litter.

SEX DETERMINATION

Pouch Young. Sex can be determined from the age of 11 (McGrady 1938) to 15 (Reynolds 1952) days, when the rudimentary pouch and scrotum first become visible. These features were illustrated by McGrady (1938:200) and Petrides (1949:367).

Free-Living Opossums and Carcasses. The presence of a prominent scrotum in males and a pouch in females is clear evidence of the sex. Sex of carcasses can be determined by noting the presence or absence of a penis, which is often withdrawn into the urogenital sinus.

Skin. Most opossum skins encountered will be cased pelts collected during the trapping season (late fall and winter). Evidence of a scrotum or a pouch will indicate the sex. Sex can be difficult to determine on some skins because the identifying features have been torn away or discarded during preparation. Bare patches of black-pigmented skin near the ventral edge of the pelt indicate a male even if the scrotum has been removed. The abdominal skin is usually thinner in females and the outline of a pouch can often be seen from the flesh side. If the abdominal skin has been torn away, nipples can sometimes be found along the margin of the tear, verifying the skin as female. Subadult females and those that did not rear litters during the previous breeding season have small, easily overlooked pouches lined with unstained fur. Conspicuous teats within a patch of sparse, tan to orange-brown hair indicate an adult female that has reared young. Winter skins of adult females collected after January have white-haired pouches because the stained pelage has been replaced by new hair. Pouches of adults are conspicuously larger than those of subadults. Petrides (1949:368) illustrated skin and pouch characteristics.

Skull. The sex of adult (age classes 4 and 5; Table 1.6) Virginia opossums can be determined by the size of the canines, which are consistently longer and heavier in the male. To verify this, we measured the anterior–posterior length of the crown at the level of the palate and the height of the crown (often incorrectly called the length of the tooth)

TABLE 1.6. Age classes and approximate ages in months for *Didelphis* spp. based on eruption sequence of molars and replacement of deciduous (d) third (last) premolar

	Molars				Age Class			Age in Months			
Premolar (third)	1	2	3	4	Tyndale-Biscoe and MacKenzie[a] (1976)	Gardner[b] (1973)	Lowrance[c] (1949)	Gilmore[a] (1943)	Petrides[c] (1949)	VanDruff[c] (1971)	Gardner (1982)[d]
d3/d3	0/(1)	0/0	0/0	0/0	1[e]	Immature			80 days+		<4
d3/d3	(1)/1	0/0	0/0	0/0							
d3/d3	1/1	0/(2)	0/0	0/0	2	1			4		4
d3/d3	1/1	(2)/2	0/3	0/0				Juvenile 6–8			
d3/(3)	1/1	2/2	(3)/3	0/(4)	3	2	1		5–8.5	4–6 5–7	5 5–6
(3)/3	1/1	2/2	3/3	0/4	4	3	2	Subadult 8–10	7–11	7–8	6–7
3/3	1/1	2/2	3/3	(4)/4		4	3			9–10	7–9
3/3	1/1	2/2	3/3	4/4	5	5	4	Adult 10+	10+	10+	10+
Wear on M^{1-2}					6						
Wear on all molars					7	6					

NOTE: Parentheses around tooth numbers indicate erupting teeth.
[a] *D. marsupialis* and *D. albiventris.*
[b] *D. marsupialis* and *D. virginiana.*
[c] *D. virginiana.*
[d] Approximate ages based in part on data in Figure 1.5.
[e] See Tyndale-Briscoe and Mackenzie (1976:252).

from the level of the palate on 64 *D. virginiana* selected at random from specimens in the collections of the National Museum of Natural History (USNM) representing populations from several localities in the United States. Although many of the canines are well worn and a few have broken tips, there is little overlap in the measurements. The sex of each specimen always was identified correctly when both dimensions of the canine were used for individuals in the same age class. The mean and range of the measurements (in mm) for the length and height of the crown, respectively, are as follows: age class 4, males 6.6 (6.3–6.7) and 13.5 (10.7–14.8), and females 5.1 (4.6–5.6) and 10.3 (9.4–11.3); age class 5, males 7.7 (6.4–9.8) and 16.3 (13.7–20.5), and females 5.5 (4.5–6.4) and 11.2 (9.4–12.8).

AGE ESTIMATION

Pouch young, immature or juvenile, and adult are the age categories used in most reports. The term *pouch young* is self-explanatory. Present subadult categories are ambiguous, and thus far in this chapter we have simply used the term *immature* when referring to information on animals out of the pouch that have been called immature, juvenile, or subadult in the technical literature (see Table 1.4). We prefer to use the term *immature* for animals that still lack an erupted first upper molar (M^1). This age class consists of weaning-age opossums. All other age classes (1–6) are based on the molar eruption sequence and the replacement of the deciduous last premolar. Some age class 1 opossums may be still of weaning age. These age classes and their corresponding probable ages in months are summarized in Table 1.6.

Reasonably accurate criteria for estimating the ages of pouch young have been published (see Table 1.5), but those for determining the ages of older opossums are less exact. Petrides (1949) used a growth curve based on weight and total length of known-age animals to arrive at age estimates for tooth-eruption stages. Similar categories were developed by Lowrance (1949), VanDruff (1971), Gardner (1973, 1982), and Tyndale-Biscoe and Mackenzie (1976).

The dental criteria for the age classes in Table 1.6 and illustrated in Figure 1.6 are the tooth eruption and replacement sequence in maxillary and mandibular toothrows. The topmost toothrows in Figure 1.6 (immature age class) show the upper molariform deciduous premolar. The remaining toothrows (age classes 1–6) are aligned on the third premolar. The example for age class 3 lacks the crown of the upper deciduous premolar on the side shown here (probably lost during the skull-cleaning process), but retains it in the left toothrow (not shown). Age class 5 opossums are >10 months old. The wear facets on the molars of the age class 6 example are typical for older adult opossums. The crowns of the first and second upper premolars in this specimen have been worn away, leaving the separate peglike, double roots of each tooth. The minimum age when the last upper molar shows wear has not been determined.

The series of photographs used by Tyndale-Biscoe and Mackenzie (1976) to illustrate the sequence of tooth eruption, replacement, and wear is misleading if their figure was the basis for their system of age classes. Their "dental class 1" toothrow shows no upper molars; therefore, it corresponds to our "immature" age class. Their dental classes 2, 3, and 4 correspond to our age classes 1, 2, and 3, respectively. Other inconsistencies are mainly differences in interpretation.

Petrides (1949) examined the usefulness of the sagittal crest, radial epiphyses, and epipubic bones as guides for estimating age in opossums. He believed that the length of epipubic bones had potential for age determination because these bones were shorter in specimens still lacking the upper last molar (M^4) than in animals with complete

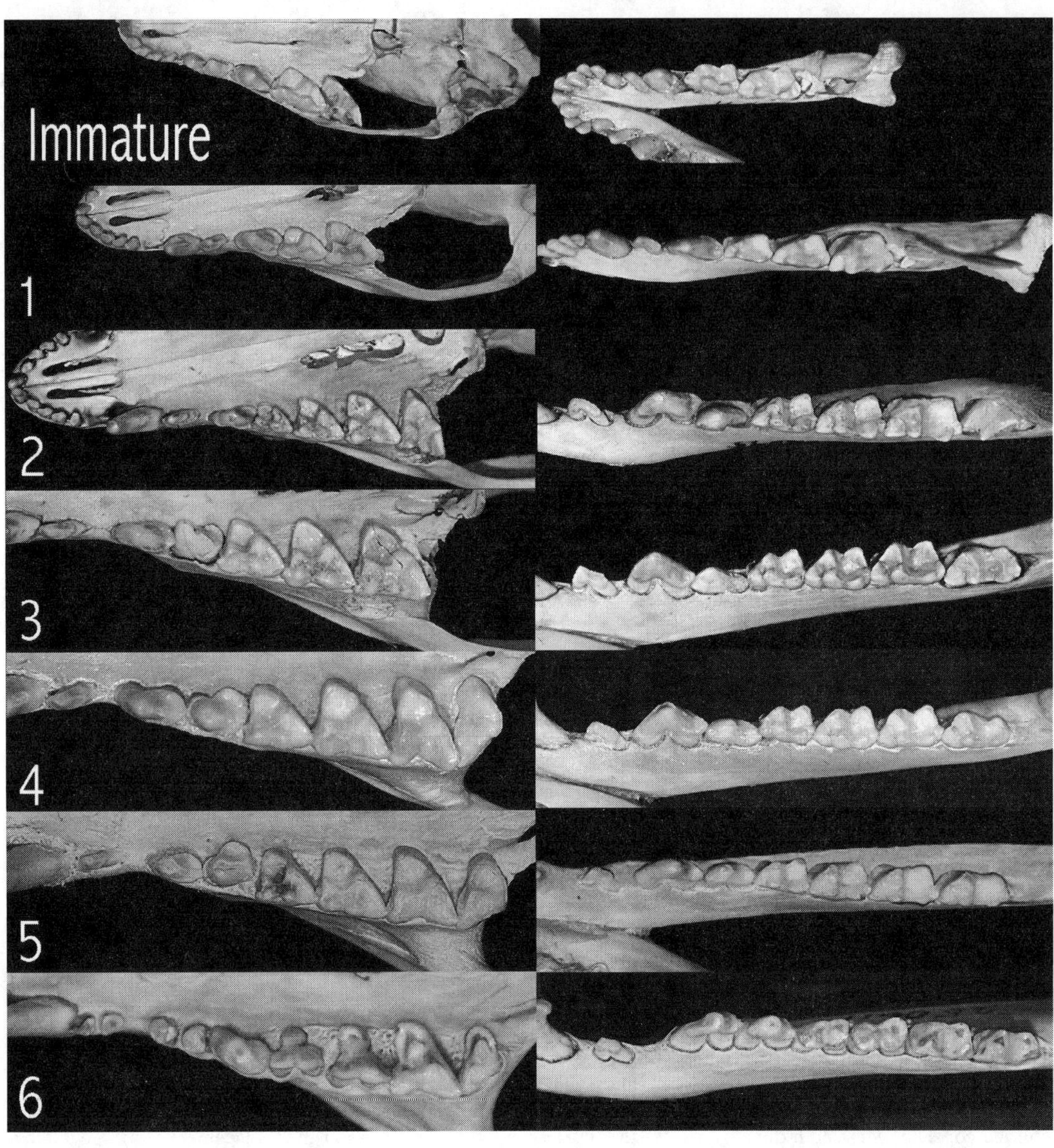

FIGURE 1.6. Right upper and lower toothrows of the different age classes of the Virginia opossum (*Didelphis virginiana*). Age classes are arranged in sequence from immature at the top to age class 6 at the bottom.

dentition. Lowrance (1949) considered the length of the epipubic bones to be the most variable linear dimension she examined.

The ages when the coracoid fuses with the scapula (fused by 7–8 months of age) and the ilium fuses with the pubis and ischium (fused by 10 months of age) are variable (Lowrance 1949) and of limited value when estimating age of skeletons.

Some age class 3 and all age class 4 and older females are sexually mature. Males younger than age class 4 (average age is 8 months) are not reproductively mature. Age categories based on tooth eruption and replacement sequence need to be calibrated against dental stages of known-age, free-ranging opossums. Samples from throughout the range of the species should be contrasted to determine what effect sex, climate, birth period, and breeding history may have had on dental-stage ages.

ECOLOGY

Habitat Preference. Opossums prefer deciduous woodlands in association with streams, but all habitats within their range of ecological tolerance are used (Lay 1942; Reynolds 1945; Llewellyn and Dale 1964; Stout and Sonenshine 1974). The broad ecological tolerances of the Virginia opossum contrast markedly with those of its congener, the common opossum (*D. marsupialis*) of Latin America. In the northern extremes of its range, the common opossum is restricted to the humid tropical and subtropical habitats of eastern and southern Mexico, generally below 1000 m. The Virginia opossum, however, ranges from sea level to over 3000 m in elevation throughout most of Mexico in a variety of humid to arid habitats.

In the United States, opossums occur in marshlands (VanDruff 1971) and in an array of forested, grassland, agricultural, and suburban habitats (Yeager 1936; Sandidge 1953; Fitch and Sandidge 1953; Verts 1963; Wood and Odum 1964). Lowest numbers are in residential, agricultural, and grassland habitats, in that order (Verts 1963; Llewellyn and Dale 1964; Blumenthal and Kirkland 1976). Opossums in Cumberland Valley, Pennsylvania, showed marked preference for agricultural edges (Blumenthal and Kirkland 1976), but the mosaic of land-use patterns in this valley suggests that the animals were responding to habitat diversity and edge effect. Yeager (1936) believed that small farms improved habitat conditions. Packard and Judd (1968) and Bowers and Judd (1969) noted that distribution in the Llano Estacado of Texas was associated with wooded canyons and escarpments that extend from the essentially treeless high plains eastward to the mesquite plains, where opossums are common. Distribution within these habitats is influenced by availability of dens and water and by locations of seasonally abundant foods, especially fruits and berries.

Daytime dens include holes, cracks, and crevices in and under trees, stumps, hollow logs, haystacks, vine tangles, rock outcrops, road culverts, attics and foundations of buildings, and piles of rock, brush, and debris (Yeager 1936; Hamilton 1958; Holmes and Sanderson 1965; Fitch and Shirer 1970; Shirer and Fitch 1970). Use of crow and squirrel tree nests (Yeager 1936; Reynolds 1945) suggests that opossums may also modify the arboreal nests of the eastern woodrat (*Neotoma floridana*). Opossums use holes dug by turtles (Rand and Host 1942) and commonly use burrows made by skunks, armadillos (*Dasypus novemcinctus*), and woodchucks (*Marmota monax*). Enlargement of these dens by the red fox (*Vulpes vulpes*) probably benefits larger opossums in the northern parts of their range.

Radio-tagged opossums in southeastern New York relied primarily on abandoned woodchuck burrows for dens (Hossler et al. 1994). Weaning dens (where young are left while mother forages) were all in burrows and commonly located in dense vegetative cover. In contrast, 87% of pre- and postweaning dens were in burrows, about half of which were in shrub–brush associations. At the microhabitat scale, weaning dens were strongly associated with tall horizontal cover, canopy cover, and proximity to a stone wall. Among nonweaning dens, only canopy cover appeared to be an important habitat variable (Hossler et al. 1994).

Winter dens in the central and northern United States are ground burrows, which afford maximum protection in cold weather. Opossums on poorly drained soils must have an adequate supply of dens at or above ground level. Lay (1942) found that ground dens were often flooded during the rainy season in east Texas.

Llewellyn and Dale (1964) found higher concentrations of opossums in woodlands near water than in dry upland areas. Lay (1942) found that animals temporarily abandoned certain areas and moved several hundred yards to seek water during droughts. The greatest distance from water recorded by Reynolds (1945) was 228.6 m. Actual distances may have been greater because Reynolds was using dogs to capture opossums, which ran toward the nearest pond or stream when pursued. Sandidge (1953) said the greatest distance between any den and a source of drinking water was approximately 366 m. Captives require considerable amounts of water to avoid dessication, and accessibility of surface water may be critical to suitable opossum habitat.

Home Range. Opossums are not territorial and do not maintain separate home ranges. Because they are generally solitary wanderers and rarely remain in any one area for long periods of time, home ranges (activity ranges) are difficult to define.

Lay (1942) based his home-range estimates on 29 (of 117 marked) opossums that were recaptured three or more times. One of these 29 was taken 48 times at 11 trap stations, another was taken 30 times at 13 stations, and a third opossum was taken 20 times at 7 stations. The average of the greatest distances traveled was 445 m. Based on a mean travel radius of 223 m, Lay calculated a circular home range of 15.5 ha. Lay concluded that the home range of an average opossum was between 4.7 and 15.5 ha.

Wiseman and Hendrickson (1950) said that recaptures of three opossums marked the year before indicated a yearly mobility of 402 m. Recaptures of four others were all within 804 m of the tagging sites.

Fitch and Sandidge (1953) estimated the usual home range as near 20.2 ha or a little less. They found no discernible difference in distances traveled by males and females or by adults and young of the year. Thirteen of their marked opossums were captured ≥ 5 times, at an average distance between trapping stations of 596 m, indicating a home range of nearly 28.4 ha. This, they believed, was too large and reflected the greater time intervals between captures during which the opossums had shifted their activity into new areas.

Verts (1963) caught 13 opossums ≥ 2 times in a study in northwestern Illinois where about 98% of the area was under cultivation. The mean distance between captures was 512.4 m, and the mean of the longest distances traveled by each opossum was 830.8 m. He used these data to calculate home ranges of between 54.3 and 82.2 ha and surmised that opossums living in intensively cultivated regions had to travel farther as they foraged than those living in wooded areas elsewhere. Verts calculated the home ranges as circular, but stated that home ranges determined on the basis of elliptical shape might be more accurate. Therefore, he used the method of Stumpf and Mohr (1962) on his own data and on those of Fitch and Sandidge (1953) from Kansas, and found that home ranges in Illinois were about 2.9 times longer than wide and those in Kansas were about 2.7 times longer than wide. This method, applied to Lay's (1942) data on Texas opossums, yielded home ranges about 2.7 times longer than wide (Stumpf and Mohr 1962). Using elliptical home ranges as a model, Verts derived opossum home ranges of about 12.5 ha in Texas, about 13.4 ha in Kansas, and about 38.9 ha in intensively cultivated Carroll County, Illinois.

Llewellyn and Dale (1964) believed that ranges tended to be long and narrow rather than circular. They reported a male that had been caught nine times in 8 months within an area about 3.2 km long and 0.6 km wide.

Fitch and Shirer (1970), using radiotelemetry to study movements in opossums in Kansas, noted that most of the home ranges tended toward a circular shape. Opossums distributed activity in different directions from whichever den was in use at the time instead of moving toward a particular foraging area. When intensity of activity relative to distance from the den was examined, found that 51.1% of the activity was concentrated within the central 10% (12 ha) of the hypothetical circular home range. Fitch and Shirer (1970:178) also commented on what they called the "compound nature of the home range," and suggested that the home range may consist of many overlapping segments, each having a den as its focal point. They also pointed out that the total area increased with time and suggested that a home range be considered as either a lifetime range, a range from one den, or the area used during some arbitrary time period (such as 1 month). They calculated an average home range of approximately 120 ha on the basis of a circular configuration. Opossums changed dens often; the number of days a den was in use averaged 2.2 (range = 1–26 days), based on 220 den shifts. The den-to-den distance averaged 301 m and was not significantly different for the two sexes.

Shirer and Fitch (1970) reported 223 den shifts averaging 302 m (range = 9–903 m) between successive dens. Seventy-five percent (168) of the 265 den-use periods were for only 1 day. Distances traveled from the den on each night's foray varied from 9.1 to 617.6 m. A radius of 617.6 m from a single den yielded a potential home range of 119.9 ha. For opossums trailed to several dens, the potential home range, again based on a circle, increased to an average >243 ha.

VanDruff (1971) found that some of the opossums he studied on the Montezuma National Wildlife Refuge in central New York were more sedentary than others. One radio-tracked female yearling remained within the same 4.9-ha area for >4 months. Activity ranges varied in size from approximately 4 ha for a juvenile female to >40.5 ha for a large adult male. No opossum had a circular home range, although the ranges of juveniles and subadults were more circular than those of adults. As the activity range increased in size, the area became elliptical. The activity ranges of six radio-tracked adults averaged about 2.7 times longer than wide, similar to the dimensions determined by Verts (1963). The Montezuma Refuge opossums showed a preference for ecotonal or edge habitats. Fitch and Sandidge (1953) noted the same preference in Kansas. Opossums most often foraged along the edges of fields and along creeks and gullies in nonforested areas and along streams and rock ledges in woodlands.

Radio-tracked opossums in northern Illinois selected den sites near areas of ripening fruit during late summer and fall (Pippitt 1976). Pippitt

(1976) found seasonal differences in the amount of time spent in a den and the rate at which den shifts were made. During winter, opossums <4 kg usually remained in the same den if the air temperature was below freezing. Larger opossums shifted dens even when the air temperature was −11°C to −20°C.

Opossums show little inclination to follow trails and tend to take circuitous and erratic routes as they forage. Wiseman and Hendrickson (1950) trailed five opossums in light snow over the entire course of a night's travel. The total distance traveled each night per opossum ranged from approximately 0.8 to 3.2 km. Brocke (1970) followed an opossum that had traveled 2.5 km in 15 cm of snow during a single night. The straight-line distance from any point on this trail to the den never exceeded 400 m.

Ryser (1995) collected data on the temporal and spatial activities of opossums from January 1986 to February 1988 in north-central Florida by radio-tracking them at night during the breeding season. He located individuals at hourly intervals throughout the night and routinely located denned opossums during the day at all seasons (Ryser 1992). Over the course of the study he obtained information from 113 radio-tagged *D. virginiana* (70 males and 43 females).

These opossums were strictly nocturnal, with activity levels highest between 2000 hr and 0200 hr. Nightly activity was greatly reduced when temperatures fell below 8°C. Average nightly activity differed little from August to December or between males and females. Opossums lived in well-defined and seasonally stable home ranges with moderate to extensive overlap. Females exhibited more exclusive space use and greater stability of home range than did males. Home range size for males and females averaged 142 and 64 ha, respectively, except during the breeding season, when males greatly expanded their home ranges to an average of 318 ha. Males often shifted their home ranges following the mating season. Average distances traveled per hour were 234 m for males and 178 m for females. Nightly movements of males and females averaged 1835 and 1465 m, respectively, with respective maxima of 4665 and 3975 m. In neighboring Georgia, Allen et al. (1985) reported average home range sizes of radio-tagged males and females as 78 and 22 ha, respectively. Based on locating tagged individuals at 2-hr intervals, their data showed average nightly movements of 1278 m for males and 1026 m for females.

In northern Virginia, Seidensticker et al. (1987) found nightly foraging distances of radio-tagged opossums varied from 1137 to 1555 m, depending on sex and season. Their tracking data also revealed that den sites were located adjacent to foraging areas, that individuals did not forage in distinct, predictable spatial patterns, and that activities were distributed over the entire nightly foraging range. As food availability changed spatially, opossums shifted their foraging areas and den sites, a behavior Seidensticker et al. (1987) interpreted as avoiding the costs and risks associated with long movements between dens and foraging areas. They termed this a "pulse-matching" spatial-use pattern.

During April–November, Gillette (1980) recorded average home range sizes of radio-tagged opossums in Wisconsin as 108 ha for males and 51 ha for females. Nightly foraging ranges (based on farthest point from den) during nonwinter months averaged 946 and 413 m for males and females, respectively. With the onset of cold weather in December, he found that home range size decreased to 30 ha for males and 24 ha for females, and no opossum moved beyond its previously established home range between mid-November and April. As ambient temperatures decreased below freezing, foraging distance decreased dramatically. Between temperatures of 0°C and −10°C, some opossums made short (up to 100 m) forays from the den, but no opossum went >15 m from the den entrance when temperatures were below −10°C at sunset.

Density. Lay (1942) estimated an average of 1 opossum/1.6 ha in poorly drained coastal pine–hardwood forest in eastern Texas. The less productive sandy coastal–prairie habitat had an estimated density of 1/5.9 ha based on the excavation of all dens found on 1.6-ha sample plots.

Wiseman and Hendrickson (1950) determined opossum density was about 1/43.2 ha in mixed pasture, woodland, and agricultural habitat in Iowa. Fitch and Sandidge (1953) roughly estimated the density of the fall population in Kansas as 1/8.1 ha and said the population was reduced to about half that by late spring. Verts (1963) estimated density as 1/25.9 ha in a region of intensively cultivated farmland in northwestern Illinois. Seidensticker et al. (1987) recorded mean densities of 3.9/km^2 (range = 0.5–7.8/km^2) in deciduous forest habitat in northern Virginia.

Sanderson (1961) and Holmes and Sanderson (1965) estimated population densities in an isolated 1.1-ha patch of mature oak–hickory forest and in old-field habitats surrounded by cultivated farmland in Illinois. Densities were calculated by two methods: (1) the number of young marked while in the pouch and later trapped as independent animals and (2) the ratio of marked to unmarked opossums retrapped the year following initial capture. The first method assumed no mortality of juveniles. Both methods used 1 July as an arbitrary date and were based on the average litter size determined for each year and the assumptions that sex ratios were even and that movements into and out of the area were negligible. Using the first method, they estimated 1 opossum/ha in 1958, 1/0.6 ha in 1959, 1/1.9 ha in 1960, and 1/ha in 1961. Using the second method, they estimated 1 opossum/0.4 ha in 1957, 1/1.1 ha in 1958, and 1/0.8 ha in 1959. The numbers actually live trapped per 259.2 ha (1 mi^2) varied from 11.8 to 16.8% of the population sizes estimated for 1 year.

VanDruff (1971) found the average density of adult opossums was 1/7 ha in waterfowl nesting habitat in the Montezuma National Wildlife Refuge. Fall populations, which included juveniles, may have been as high as 1/0.9–1.0 ha.

Stout and Sonenshine (1974) reported an average density of 1 opossum/20.2 ha in an area of heterogeneous habitat near Richmond, Virginia. Minimum numbers known to be alive were lowest in the first quarter of the year and highest in the second or third quarter. Estimates ranged from as low as zero in spring 1965 to a high of approximately 1 opossum/5.6 ha in autumn 1967. Their data, although covering only 6 years, showed peak densities in 1964 and 1967, an interval of 4 years. Apgar (1934) noted periodicity in the numbers of live Virginia opossums presented to the Philadelphia Zoological Garden over a period of 26 years. On the average, noticeable increases in accessions occurred about every 6 years, indicating high levels in the wild population at those times. Hunsaker (1977) reviewed and compared the results presented in most of the pre-1975 reports outlined above.

Longevity and Population Composition. Maximum longevity in captivity is in excess of 4 years, 5 months (Crandall 1964). There is little information on longevity in the wild. For individuals whose birth and death dates are known (Sunquist and Eisenberg 1993), longevity of female opossums in Florida averaged 559 days (SD = 187 days). Females provided with supplemental food at their dens did not show increased longevity. Wright et al. (1995) reported that 37% of female and 25% of male opossums survived to their second reproductive year in north-central Florida. Sunquist and Eisenberg (1993) found 26% of females survived into the second breeding year in Florida. Seidensticker et al. (1987) reported that only 8% of females lived into their second year of reproduction in northern Virginia.

An interesting contrast in life history traits of *D. virginiana* was revealed by Austad's (1993) study of a population on Sapelo Island off the Georgia coast, where females survived significantly longer than those on the adjacent mainland. Of 34 breeding females studied on the island, 50% survived into their second reproductive year and 9% survived into a third year. Longevity of island females averaged 24.6 months (SD = 7 months) with a maximum of 45 months. For mainland females, 27% survived into their second reproductive year; longevity averaged 20 months (SD = 5 months) with a maximum of 31 months. Litter sizes were significantly smaller for all ages on the island (mean = 5.7 island vs. 7.6 mainland) and pouch young of island females grew significantly more slowly than those of mainland females. Island opossums also aged more slowly than mainland females, as measured by chemical changes in tendon fibers in the tail. Austad (1993) suggested that these traits in the island population (greater survivorship, smaller litters, slower growth of young, and slower aging rate of tendon collagen in the tail) are not explained by environmental factors, but are consistent with the "evolutionary senescence theory," which predicts retarded senescence in populations evolving in "safer" environments.

The mean life span determined by Lay (1942) from trapping results was little more than 15 months. The adult male mentioned by Reynolds (1945:375) was at least 2 years old when killed, assuming that it was about 1 year old when originally tagged. Petrides (1949) estimated the average longevity to be 1.3 years and the probable turnover period as 4.8 years. This estimate was considerably less than the 7+ years estimated by Hartman (1923a) based on captives. Fitch and Sandidge (1953) listed a female marked as a pouch young on 6 May 1950 that was recaptured on 28 February 1952, indicating an age of 22+ months. They reported that 70.8% of the opossums trapped during a 3-year period were yearlings. The breeding population of females consisted mainly of individuals born the preceding spring, and they estimated that 95% of the 1949/50 breeding population was replaced by the following breeding season. Sanderson (1961) trapped no opossum older than 15 months after their first capture, and concluded that few live beyond the summer following their birth. Reynolds (1952) and Jurgelski and Porter (1974) believed that females in their second reproductive season were less fecund. Only 3% of the opossums reported on by Llewellyn and Dale (1964) were recorded for longer than 1 year. The longest record was for a male estimated to be about 3 years old when last trapped. One female was estimated as 27 months old; another was about 24 months old when last caught. Their results indicated that Petrides's (1949) 1.3 years as the average life expectancy and 4.8 years as the turnover period were too high. VanDruff (1971) calculated average ecological longevity as 1.1 years and population turnover as 3.5 years. Young of the year made up 87.7% of the late-fall population; the average annual winter carryover rate for marked opossums was 9.2%.

The structure of the population of opossums in northern Virginia studied by Seidensticker et al. (1987) followed a fairly predictable annual cycle. Numbers were lowest during winter and early spring, when only adults were present. These included young from the previous year (6–12 months of age) and a small number of individuals entering their second reproductive year. The first independent young appear in early summer, and total numbers peak in autumn, when second-cohort young become independent. During summer and fall, young females constituted 56% and 64% of the female population, respectively. In the same two time periods, young males made up 74% and 85% of the male population, respectively. The number of resident adult females in the area was more constant throughout the year than was the number of adult resident males.

In general terms, the fewest opossums are found from January through April (the first period of the breeding season), when the female population consists entirely of reproductively adult animals (not counting pouch young). Assuming that births occurred from January through July in the preceding year, we find that the cohort of females reproducing for the first time would be from 6 to 12 months old in January. Females entering their second breeding season would be from 20 to 26 months old in January. The reproductively adult segment of the male population in January would consist of 8- to 12-month-old animals breeding for the first time and males >20 months of age that may or may not have bred the previous year. Relative ages of the breeding animals and the onset of breeding activity shift 1 month or more depending on climate and latitude (see Fig. 1.4).

Independent young of the year should first appear in the population from late April through May and peak population levels including all age classes can be expected in the fall after the latest litters of the season are weaned. Latest-born young in northern extremes of the range may be at a decided disadvantage because of the lower winter-survival potential of smaller animals. Brocke (1970) found that all animals weighing <1 kg were absent from his Michigan population after mid-January.

In a detailed study of mortality and dispersal in Florida opossums, Wright (1989) radio-collared 68 juveniles at weaning and another 67 postweaning opossums. Mortality was highest (60–75%) within the first 4 weeks after weaning, regardless of sex. Most known deaths were due to predation by owls. Young opossums spent more time in tree cavities and nests, presumably to reduce their vulnerability to predators. Wright (1989) estimated that 10–11% of opossums in north Florida survived to their first breeding season. Hossler et al. (1994) estimated survival of opossums to their first breeding season to be <10% in New York.

Studies relying on recapturing marked animals are based on the premises that all animals are equally susceptible to traps and are lifetime residents on the study area. Ample evidence proves the opposite as far as the Virginia opossum is concerned. Some animals become habituated to traps, but the majority is caught only once. Wiseman and Hendrickson (1950) noted that some opossums passed about 1 m from freshly baited traps without showing apparent interest. Opossums often shift dens and their home range (activity area) every few days. Therefore, long periods between recaptures may mean only that the animal was absent from the study area during that interval.

The difficulty in accurately determining the age of postpouch opossums is the major reason the age class composition of populations is so poorly known. Perhaps another reason is that investigators have mistakenly attributed a long life span to the Virginia opossum. Persuasive evidence indicates that opossums are not long-lived animals. Although adults in wretched physical condition (lacked parts of ears, tail, and limbs; had worn, broken, and carious teeth) studied by VanDruff (1971) appeared to be several years old, some had been tagged as juveniles the year before. Jurgelski and Porter (1974) considered the effective reproductive life as 1 year for females. They distinguished females in their second reproductive year by the worn and broken condition of the canines and by the wider skin flaps on the pouch (Petrides 1949). According to Hunsaker (1977), the longevity and reproductive activity of opossums are short compared with similar-sized mammals. He indicated that they were mature at 9 months and reproductively fit for at least 18 months, but probably not beyond 3 years. The effective breeding life of males is unknown. Most of the available life history information simply does not permit the determination of the age composition of adults.

Dispersal Patterns. Reynolds (1945) surmised that most opossums are nomadic on the basis of recovering only 5 of 68 marked animals, the record of an adult male that moved 11.2 km in 9 months, and the rapid repopulation of an area cleared of opossums by hunters.

Fitch and Sandidge (1953) found that young males, after becoming independent, usually wander more widely than young females and tend to settle in areas removed from the mother's home range. All recaptured females marked as pouch young had moved, on average, <365 m.

The longest single move recorded by Holmes and Sanderson (1965) was 1738.5 m for an adult male. Ten captures for 8 juvenile females and 33 captures for 20 juvenile males on successive nights yielded average distances of 88 and 162 m, respectively. The 29 captures of 9 adult females on successive nights averaged 279 m apart; 5 captures of 4 adult males averaged 653 m apart. Some juvenile females stayed in the general area where they were born; three generations of females were caught in the same area during a period of 25 months.

Of juveniles known to be alive, Wright (1989) found that 96% of males dispersed, whereas 86% of females were philopatric. A small number of first-cohort males dispersed in September at about the time second-litter young became independent. The majority of males left their natal ranges in mid- to late January, coincident with onset of the breeding season. Dispersal movements were rapid and fairly direct, and most were completed in one or two nights; only 1 male died during dispersal. Of 2 radio-tagged females that dispersed, both moved long distances (5.5 and 5.75 km, respectively). Dispersal by 21 males ranged from 1.25 to 6.5 km; 3 moved 7 km, the longest dispersal distance Wright (1989) recorded.

Fitch and Shirer (1970) noted that immatures wandered less widely than adults, but occasionally dispersed into new areas. Based on radiotelemetric observations, they found that young opossums progressively ranged farther as they matured and that adult males moved more widely than adult females.

In Wisconsin, Gillette (1980) reported that first- and second-litter young spent their first 3 postweaning months increasing the size of their foraging areas as they gradually moved away from the area where weaned. Most first-cohort young dispersed between late August and mid-October. Four juveniles traveled long distances, moving an average of 1.7 km on the first night; no juveniles dispersed during winter. Gillette (1980) also noted dispersal by several adults, including females with pouch young. On the first night, eight adult females moved an average of 1689 m (range 648–2438 m), distances similar to those recorded for dispersing juveniles.

The longest straight-line distance recorded by VanDruff (1971) was 8050 m during a 15-month period by a male originally marked as a juvenile. An adult female moved 4930 m in 10 days in June, a second was carrying 9 pouch young when she moved 3560 m in 7 days, and a third carrying 10 pouch young traveled 3460 m in 4 days. The longest single move recorded was that of an adult male who moved 3.2 km from one den to another in a single night. Most opossums did not travel as extensively. Juveniles moved less than adults and some remained in an area for a few weeks before moving on. Males moved farther than females. The longest and most rapid moves occurred in late spring and early summer. Understanding the dispersal pattern of newly independent young was difficult, partly because the location where they last left the mother could not be determined.

Interspecific Interactions. Because opossums do not dig burrows, they must rely on ground dens dug by other animals. Opossums have been found sharing dens with an armadillo (Lay 1942), a woodchuck (Sandidge 1953), and a raccoon (*Procyon lotor*) (Stuewer 1943). In addition to the last two species, Pippitt (1976) found opossums sharing dens with cottontail rabbits (*Sylvilagus floridanus*). A complex burrow system in Michigan was simultaneously occupied by an opossum, a woodchuck, a raccoon, and a striped skunk (*Mephitis mephitis*) (Brocke 1970). Shrews, mice, weasels, snakes, and other small vertebrates as well as a host of insects and other invertebrates may cohabit parts of the den used by opossums. Some dens are used communally. During the year, the same burrow may be used at different times by red fox, raccoon, woodchuck, opossum, striped skunk, cottontail rabbit, and gopher turtle (*Gopherus polyphemus*).

Shirer and Fitch (1970) found no apparent competition for dens by raccoons, striped skunks, and opossums. Striped skunks dug the ground dens used by the others. Brocke (1970) and Pippitt (1976) suggested that clearing a forest area resulted in higher numbers of woodchucks and striped skunks, thereby increasing the number of underground dens that seem to be critical for the winter survival of opossums. Kissell and Kennedy (1992) found no evidence for mutual exclusion between co-occurring raccoons and Virginia opossums in Tennessee based on spatial attributes or habitat affinity. They concluded that the two species occur independently of each other. However, also in Tennessee, Ladine (1997) reported that where the two species co-occur, they temporally partition the habitat, and opossums shift their foraging times to avoid raccoons.

Opossums prey on small vertebrates and invertebrates. They kill cottontail rabbits and chickens, and take eggs and nestlings of ground-nesting birds including waterfowl. This marsupial is among the few mammals that regularly preys on shrews, moles, and reptiles, including poisonous snakes. Studies of feeding habits, however, demonstrate that most of the other mammals and birds consumed are scavenged as carrion.

Distributional Limits. Minimally adequate habitat for opossums must include accessible water and a sufficient number of dens. Food usually is not limiting where these requirements are met, except in regions where winter climate hampers foraging activity.

Tyndale-Biscoe (1973) used the −7°C January isotherm to predict the eventual northern distributional limit. Brocke (1970) suggested that the southern edge of the pine–hemlock (*Pinus–Tsuga*) ecotone approximates the northern distributional limit for the Virginia opossum in eastern North America. Brocke believed that the lack of foraging success imposed by a combination of the number of days with subfreezing temperatures and those with accumulated snow depths exceeding 28 cm was critical. He found that the northern distributional limit in Michigan coincided with a winter severity level of 70 days of enforced inactivity. Brocke's hypothesis explains the periodic reductions of populations along the northern periphery of the range following severe winters and suggests that the general warming trend over the past century has had as much to do with the northern expansion of the distribution of the Virginia opossum, as have human activities in providing den sites and winter food.

FEEDING HABITS

Dietary studies confirm omnivory in the Virginia opossum. They are well known to scavenge carrion and garbage, and are prone to use the most abundant foods available. Opossums also eat grass and other green vegetation, seem to prefer maggots gleaned from rotting flesh over the putrid flesh itself, dig and consume mushrooms from beneath snow, kill and eat poisonous snakes, cannibalize their less fortunate brethren (thereby aiding survival during harsh northern winters), and eat great quantities of earthworms, insects, and other invertebrates. However, studies of feeding habits show that opossums do not live up to their reputation as rapacious raiders of chicken coops or as serious predators of rabbits, raccoons, pheasant (*Phasianus colchicus*), waterfowl, and other game.

Several studies have been based on stomach analysis alone. Others dealt with scats or various combinations of scat, stomach, and intestinal contents. The most extensive analysis was by Hamilton (1958), who examined 461 stomachs. His study demonstrated the need to segregate results on a monthly basis to understand feeding habits, food availability, and dietary requirements. Hamilton (1951, 1958) and others have pointed out the omissions possible when only feces are analyzed in feeding habits studies. Determinations of foods easily identified in the stomach may not be possible from gut contents or from feces, where green vegetation and many soft-bodied invertebrates may go unrecognized. Food items should be reported by frequency and volume to avoid the biases inherent in each method.

Some reports are contradictory. For example, Worth (1975:517) retrieved "opossum fecal droppings consisting entirely of persimmon seeds held together by . . . undigested pulp" and, after finding that the seeds germinated, commented on the persimmon (*Diospyros virginiana*)-disseminating ability of the Virginia opossum. Reynolds (1945:373), however, said that 19 field-collected scats with remains of persimmons did not contain seeds. Examination of 63 scats recovered from 11 captive opossums fed exclusively on persimmons for a week revealed only 1 with a seed.

Several accounts of feeding habits mentioned paper, cellophane, leaves, and other trash and detritus found in the feces or intestinal tracts. Some accounts also noted egg shells, isolated feathers, and insect exoskeletons, much of the latter eaten during winter (Hamilton 1958). When gleaning for food, opossums ingest a variety of minute items. Although a few of these items may be consumed accidentally, perhaps most are leaves that held clusters of insect eggs, or are blood- or grease-soiled dirt, leaves, or trash. Except when fighting a trap, opossums probably do not ingest seemingly extraneous items unless they contain something desired as food.

Thus far, studies of feeding habits have not explored the possible effect of sex, age, or reproductive status on the diet of the Virginia opossum. The diets of Latin American populations are unknown.

Audubon and Bachman (1851) identified the foods of opossums as corn, chestnuts, acorns (*Quercus* sp.) and other nuts, small tubers, young briar (*Smilax* sp.) shoots, blackberries (*Rubus* sp.), wild cherries (*Prunus* sp.), persimmons and other fruits, mice, cotton rats (*Sigmodon* sp.), bird eggs, nestling birds, and broods of young rabbits. Sperry (1933) reported an opossum in North Carolina that had eaten a red bat (*Lasiurus borealis*). Additional information on the foods consumed by opossums has been reported for specific states.

Michigan. Dearborn (1932) gave volumetric percentages of foods consumed by 40 opossums as 30.4% insects, 24.7% birds, and 23.2% mammals. Two thirds of the mammalian remains consisted of opossum. Other animal food items of lesser importance were crayfish (*Cambarus* sp.), snakes, bird eggs, and frogs.

Stuewer (1943) analyzed 15 stomachs and nine scats and found that opossums had eaten rabbits, short-tailed shrews (*Blarina brevicauda*), mice, a red squirrel (*Tamiasciurus hudsonicus*), chicken, pheasant, frogs, caterpillars, and small slugs. Plant food items included corn, buckwheat, berries, wild grape (*Vitis* sp.), and several kinds of seed.

Taube (1947) examined the stomach contents of 6 opossums collected in September and 49 collected in November and December 1941 in his evaluation of the predatory status of Michigan populations. He added data on 78 stomachs collected in the fall of 1933 and 8 collected in the spring of 1934. Eastern cottontail (*Sylvilagus floridanus*) made up the greatest volume (31%) and occurred in 19% of the stomachs. The autumn 1933 sample contained rabbit in 40% (12% by volume) of the stomachs. Other mammals in order of their presence in the diet were opossum (9% by volume, in 4 stomachs), vole (*Microtus* sp., in 6% of the stomachs), eastern mole (*Scalopus aquaticus*), shrew, fox squirrel (*Sciurus niger*), and striped skunk. Other animal foods were pheasant (1 stomach), chicken (4 stomachs), miscellaneous birds, snakes, amphibians (frogs, toads, and salamanders), insects, snails and slugs, and earthworms. Of the autumn 1941 opossums, insects were in about 90% of the stomachs. The volume of grasshoppers was 32% in September, but only 4% in November and December, although they were present in 88% of the samples. Earthworms occurred in 27% of the stomachs and made up 8% of the food consumed. Plant materials included fruit (about 11% of volume, in over 50% of the samples) and grass (5% of contents, but in 69% of the stomachs). Rabbits and pheasants were eaten in the fall, probably as carrion.

Brocke (1970) examined 20 stomachs collected in January, February, and March. He found remains of striped skunk, muskrat (*Ondatra zibethicus*), cottontails, white-tailed deer (*Odocoileus virginianus*), opossums, shrews, mice, a garter snake (*Thamnophis* sp.), a leopard frog (*Rana* sp.), insects, earthworms, grass and plant fibers, and unidentified mammals. Remains of mammals, much of which were carrion, occurred in 85% of the stomachs, and plant material in 35%.

Texas. Lay (1942) estimated the volumetric percentages of the following food items recovered from 16 opossums taken in September in eastern Texas: insects and worms, 45%; fruit, 11.8%, green leaves, 11.0%; leaf and log litter, 10.6%; mammals, 7.0%; acorns, 4.7%; birds, 4.3%; crayfish, 3.3%; snails, 0.8%; and trace amounts of cellophane and grass seeds.

Wood (1954) analyzed the contents of 23 scats and 25 digestive tracts of opossums collected in oak woodlands of east Texas and recovered 39 food items. Thirty-six of these were in digestive tracts, 26 more than were recorded from scats, indicating that digestive tracts (particularly stomachs) contain more identifiable food items than scats. The volumetric percentages were as follows: plants (fruit and green vegetation), 44.8%; insects, 25.0%; mammals, 14.9%; amphibians and reptiles, 7.4%; birds, 3.8%; unidentified remains, 3.2%; and a trace of noninsect invertebrates. Cottontails were the most common mammal identified (in 12% of the digestive tracts and 4.3% of the scats). Other mammals identified were opossums, hispid cotton rats (*Sigmodon hispidus*), and white-footed mice (*Peromyscus leucopus*). Insects, second in prevalence as food, occurred in 88.0% of the digestive tracts and 69.5% of the scats. Copperheads (*Agkistrodon contortix*) were the most common reptile in the diet (5.7% by volume). Acorns (31.4% by volume) were the most important plant food. Other identified plant items were persimmon, pear, blackberry (*Rubus* sp.), French mulberry (*Morus* sp.), grass, watermelon, grape, and hackberry (*Celtis occidentalis*). Although poorly represented in the diet, birds included the yellow-bellied sapsucker (*Sphyrapicus varius*), mourning dove (*Zenaida macroura*), spotted towhee (*Pipilo maculatus*), and chicken.

Missouri. Reynolds (1945) identified 69 food items in 259 scats of opossums from central Missouri collected between September and May. Frequencies of the food categories were as follows: insects, 87.6%; fruits, 50.6%; noninsect invertebrates, 32.4%; mammals, 28.2%; reptiles, 18.9%; grains and seeds, 12.7%; and birds and bird eggs, 8.9%. Reynolds also recorded 52 food items in the stomach contents of 68 opossums collected during the 6 months from December to May. Of these, he tabulated the following categories by percentage volume: insects, 34.2%; mammals, 32.3%; reptiles, 10.0%; grain and seeds, 7.3%; fruits, 6.8%; birds and bird eggs, 4.9%; and noninsect invertebrates, 4.5%. With the exception of crickets (Gryllidae), squash bugs (Coreidae), and stink bugs (Pentatomidae), more fall and spring scats contained insects than did winter scats. Fruits generally were more commonly eaten during the autumn. Frequencies of land snails were 39% and 26% for fall and spring scats, respectively. Cottontail rabbits, moles, and fox squirrels were more common in winter scats than in fall or spring scats. Reptiles were most numerous in the spring, and grains and seeds were most frequently encountered in the fall scats, except corn, which was present in 29% of the winter sample. Chickens or eggs, although present in only 12 of 259 scats, were more common in the winter. Cottontail rabbits were the most important food by volume (12.9%) in the stomachs. Next in order of importance by volume were carabid beetles, 8.7%; scarabaeid beetles, 7.5%; corn, 7.3%; eastern moles, 6.4%; stink bugs, 5.3%; opossums, 4.9%; squash bugs, 4.8%; and short-horned grasshoppers (Locustidae), 4.4%.

Pennsylvania. Gifford and Whitebread (1951) found insects, earthworms, mice, an anuran, rabbit hair, snake scales, feathers, and pokeweed (*Phytolacca americana*) seeds in 4 opossum stomachs from south-central Pennsylvania. Grimm and Roberts (1950) noted the percentage occurrence of the following foods in 18 fall and winter stomachs provided by trappers from southeastern Pennsylvania: insects and spiders, 83.3%; fruits, 44.4%; cottontail rabbits, 27.7%; birds other than poultry, 22.2%; mice, 22.2%; carrion, 22.2%; shrews and moles, 11.1%; poultry, 11.1%; woodchuck, 11.1%; frogs and toads, 5.1%; and earthworms, 5.1%. They believed that most if not at all of the rabbit, poultry, and woodchuck was carrion. Some of these items were used by trappers for bait.

Blumenthal and Kirkland (1976) examined the stomach contents of 62 opossums from the Cumberland Valley. Percentages of the frequency and volume, respectively, of the 15 food categories they listed were as follows: mammals, 98 and 26; grasses and grains, 90 and 13; insects, 85 and 9; leaves, 78 and 8; fruits and seeds, 76 and 12; sand and stones, 72 and 7; stems, 44 and 4; earthworms, 43 and 8; mollusks, 39 and 4; birds, 30 and 4; egg shells, 8 and <1; arachnids, 8 and <1; fish, 6 and <1; trash, 6 and <1; and amphibians, 1 and <1. Mammals, grasses and grains, and insects were found in 75% of the stomachs. Mammals, grasses and grains, and fruits and seeds made up 51% of the total food volume. Seasonal shifts in some foods probably reflected availability (e.g., earthworms in the spring and birds and insects in the summer). Weed seeds were most commonly found in the winter samples and fruits and berries were most common in the summer and fall. Mammals were the most important food at all seasons and included the Virginia opossum, short-tailed shrew, white-footed mouse, house mouse (*Mus musculus*), and meadow vole (*Microtus pennsylvanicus*).

Iowa. Wiseman and Hendrickson (1950) examined 81 samples of fecal material recovered at traps and six scats found in the field in southeastern Iowa during winter (January–March) and spring (April–June). Frequencies of the food categories were as follows: insects, 87%; grain and plant items, 66%; fruits, 31%; crayfish, 24%; mammals, 22%; millipedes, 13%; birds and bird eggs, 8%; and reptiles, 1%. In order of frequency, the most common winter foods were corn, ground beetles (Carabidae), grasshoppers, miscellaneous seeds, ground cherries (*Physalis* sp.), unidentified insects, and mammals (rabbit). The most frequent spring foods in order of frequency were ground beetles, crayfish, mulberry (*Morus* sp.), corn, mammals (voles), millipedes, and miscellaneous seeds.

New York. Hamilton (1951, 1958) presented the most detailed analysis of food habits published to date. The large number of stomachs (180) whose contents were reported in 1951 was increased to 461 by the time of his 1958 report, which included more than 118 identified food items. Percentages of the frequency and volume, respectively, of the 16 major food categories he listed are as follows: insects, 42.9 and 7.9; mammals, 42.1 and 22.6; green vegetation, 38.6 and 8.1; fruits, 33.4 and 14.1; earthworms, 23.6 and 10.3; amphibians, 23.2 and 9.3; birds, 18.9 and 7.2; mollusks, 15.6 and 3.0; reptiles, 10.8 and 5.6; carrion, 7.8 and 6.0; grain and mast, 3.5 and 1.9; millipedes, 2.2 and trace; centipedes, 1.3 and trace; crustaceans, 0.3 and 0.9; fungi, 0.3 and trace; and undetermined, 4.6 and 3.0. Mammals, fruit, earthworms, amphibians, green vegetation, and insects made up 72.3% of the volume consumed. The frequencies of these six major food categories varied from 23.0% to 42.7%. The frequency of mammals varied seasonally from a high of 77.8% in December to a low of 25.0% in August; of insects, from 77.1% in August to 10.3% in March; of fruits, from 42.3% in October to 5.6% in April; of amphibians, from 50.0% in August to zero in February; of earthworms, from 51.7% in March to 9.1% in February; and of green vegetation, from 58.6% in March to 31.3% in August.

Principal insects included grasshoppers, beetles, and lepidopteran larvae. The most frequently encountered mammal was the meadow vole (in 17.6% of the stomachs); next was the short-tailed shrew (8.0%). Green grasses, clover, and other ground vegetation were a natural and important part of the diet and were found in 178 animals. Grapes were the main fruit eaten; much of the fruit was dried, and fallen berries and drupes were recovered long after their time of ripening. Earthworms were present in 20% or more of the stomachs in January, March, April, October, November, and December. Toads (*Bufo* sp.) were found in 10% of the stomachs. Most of the birds consumed were probably carrion, except for nestlings, which were probably taken from their nests. Slugs and snails were eaten. The snapping turtle (*Chelydra serpentina*) and painted turtle (*Chrysemys* sp.) that had been eaten were recently hatched. Carrion was probably underestimated, and many of the animal foods may have been gleaned from roadsides or were unrecovered hunter kills. Grains and acorns, found in 3.7% of the stomachs, were not as important as other studies have indicated. Hamilton did not report the frequency of trap debris, dirt, stones, dead leaves, and trash because he believed these were probably not ingested purposely.

Opossums were responsible for 15 (24%) of 63 duck nests destroyed during a 3-year study on predation on waterfowl nests in the Montezuma Wildlife Refuge in central New York (VanDruff 1971). However, 437 (48%) of 916 animals killed during 9 years (1961–1969) of the toxic-egg predator reduction program at the Montezuma Refuge were opossums.

Maryland. Llewellyn and Uhler (1952) examined 37 stomachs and 66 scats of opossums from the Patuxent Wildlife Research Center in Maryland. A variety of insects, vertebrates, and plant items as well as snails and millipedes were recovered. Plant foods included persimmons, wild grapes, apples, pokeberries (*Phytolacca americana*), briar, corn, beechnuts (*Fagus grandifolia*), nightshade (*Solanum nigrum*), cherries, dewberries (*Rubus* sp.), and blackberries. The last two made up about 20% of the foods consumed in June and July, whereas the first three were among the most important fall foods. Animal food matter (about one third insects) made up 86% of the estimated volume.

Kansas. Sandidge (1953) based his analysis of opossum feeding habits on field observations and the examination of 62 digestive tracts collected from September to March. The percentages of volume, weight (dried), and frequency, respectively, of the major food categories were as follows: insects, 42.2, 42.6, and 86.7; mammals, 41.4, 39.4, and 33.3; fruits, 8.6, 10.3, and 13.3; and birds, 3.1, 2.8, and 21.7.

The most common insects were carabid beetles and short-horned grasshoppers, the only insects found in winter samples. Other insects present were metallic wood borers (Buprestidae), lady bird beetles (Coccinellidae), horned passalus (Passalidae), lamellicorn beetles (Scarabaeidae), carrion beetles (Silphidae), stink bugs, assassin bugs (Reduviidae), and crickets. Cottontail rabbits were the most important mammalian food. Other mammals identified were the white-footed mouse, muskrat, prairie vole (*Microtus ochrogaster*), eastern mole, and Virginia opossum. Flesh-eating beetles (Silphidae) accompanied rabbit, muskrat, and opossum in 12 of 19 occurrences, thereby indicating carrion.

Fruits occurred in 8 digestive tracts. Pears were greatest in volume and weight in late fall and winter. Fruits and seeds included winter wheat, goosefoot (*Chenopodium* sp.), wild grape, apple, and pear.

Chicken (feathers, bone, and egg shells) was in 9 of the 13 digestive tracts containing birds, but was likely gleaned from garbage or eaten as carrion. Other birds (one each recorded) were the yellow-shafted flicker (*Colaptes auratus*), cardinal (*Cardinalis cardinalis*), meadowlark (*Sturnella* sp.), and starling (*Sturnus vulgaris*).

The five-lined skink (*Eumeces fasciatus*), DeKay's snake (*Storeria dekayi*), worm snake (*Carphophis amoenus*), and leopard frog were the only herpetozoa identified. Centipedes made up 1.6% of the total weight and 1.6% of the volume, and occurred in 38.3% of the samples. Land snails (*Triodopsis albolabris*) and crayfish (*Orconetes* sp.) were identified in 1 and 2 stomachs, respectively.

Fitch and Sandidge (1953) examined scats collected on the University of Kansas Natural History Reservation from August to November 1951 and in January, September, October, and November 1952. Wild fruits (principally wild grape and hackberry) made up the bulk of the diet. Blackberry was the important summer food. Crayfish was the most common animal item found. Other foods were wild plum (*Prunus americanus*), wild crabapple (*Pyrus ioensis*), cherry, corn, insects (beetles, cicadas, grasshoppers, yellow jackets—*Vespula* sp.), blue jay (*Cyanocitta cristata,* one feather), fox squirrel, eastern cottontail, copperheads, and a snail. A large male opossum was observed as it killed and had begun to eat a 150-g young rabbit.

Illinois. Stieglitz and Klimstra (1962) reported on the contents of 131 opossum digestive tracts collected from 1958 through 1960 during the 6-month period from August to February in southern Illinois. Of the 75 animal and 66 plant items recorded, only 24 animal and 11 plant food items contributed $\geq$0.5% by volume to the diet. Animal food items made up 76.2% of the total volume, plant foods 23.8%. Mammals made up 48.7% of the total volume. Opossum filled 4 stomachs and contributed 16.3% of the volume. Cottontail rabbit was second, with 14.7% of the volume, and occurred in 15.3% of the digestive tracts. Other mammals included the prairie vole, deer mouse (*Peromyscus maniculatus*), gray fox (*Urocyon cinereoargenteus*), striped skunk, short-tailed shrew, eastern mole, and woodchuck. Other vertebrates included chicken (7.1% of volume), grackle (*Quiscalus quiscala*), meadowlark, domestic pigeon, towhee, junco (*Junco* sp.), cardinal, Carolina wren (*Thryotherus ludovicianus*), blue racer (*Coluber constrictor*), turtle, frog, and toad. Insects appeared in 93.1% of the tracts, but contributed only 6.3% of the volume. The most important insects were scarabaeid beetles (larvae), short-horned grasshoppers, and lepidopteran larvae. Fly larvae appeared in several tracts, but not always in association with carrion. Earthworms appeared in only 5 digestive tracts.

Plant food items were dominated by persimmon (8.1% of total volume), which ranked third among all foods. Other plants were corn, grasses, pokeberry, wild grape, and other fruits.

The 10 top-ranking foods by volume were opossum, cottontail rabbit, persimmon, chicken, prairie vole, pokeberry, grackle, gray fox, frog, and scarabaeid larvae. The ranking by percentage frequency was as follows: grasses, 82.4%; short-horned grasshopper, 54.2%; opossum, 52.7%; ground beetle, 38.9%; snail, 31.3%; pokeberry, 25.2%; nightshade, 25.2%; stinkbug, 22.9%; persimmon, 21.4%; and cottontail rabbit, 15.3%. Most of the seasonal variation was seen in the decrease in plant materials from fall through winter and the increase in mammalian food consumed during the same period.

Wisconsin. Knudsen and Hale (1970) examined the contents of 151 opossum stomachs represented by at least 5 from each month of the year. They found 65 different identifiable food items, which they grouped in the following major categories (values are percentages of occurrence and total volume, respectively): mammals, 25 and 41; birds, 12 and 24; invertebrates, 19 and 10; reptiles, amphibians, and fish, 10 and 12; plants, 7 and 6; and garbage, trash, and litter, 27 and 6.

Cottontail rabbits and several species of mice were the important mammalian foods. Other mammals were shrews, moles, gray squirrels (*Sciurus carolinensis*), fox squirrels, muskrats, rats (*Rattus* sp.), dogs, house cats, and swine. Most of the animals were taken as carrion, as confirmed by the decayed condition of the flesh and the presence of maggots in the stomachs. Next in importance were birds, of which chicken was most abundant. Other birds represented were ducks, bobwhite quail (*Colinus virginianus*), pheasant, domestic pigeon (*Columba livia*), screech owl (*Otus asio*), crow (*Corvus brachyrhynchos*), and several species of songbirds. Game birds were in 10 stomachs, but in small amounts.

Invertebrates included earthworms, crayfish, snails, and several kinds of insect. Earthworms, although absent from the winter samples, were the most frequent food in the other three seasons and in the annual total, where they made up 8% of the volume.

The reptiles, amphibians, and fish included small frogs, toads, garter snakes, and a number of fish believed to have been discarded by fishermen. Plant items were fruits, grains, vegetables, and mast. Apples were the important food in this group.

Mammals and insects were the important winter foods. Most of the garbage was noted during the winter, indicating the raiding of refuse and garbage by opossums when foods are scarce. Earthworms, insects, songbirds, frogs, snakes, and chickens were the most commonly eaten foods in the summer. There was no indication that opossums were a serious threat to game bird populations.

Indiana. Whitaker et al. (1977) identified 71 categories of food from the contents of 83 stomachs. Volumes of the following major food categories were as follows: mammals, 22.2%; birds, 21.3%; other vertebrates, 9.0%; insects, 11.7%; other invertebrates, 13.6%; vegetation, 19.0%; and garbage, 3.1%. Most of the vertebrate remains, which included shrews, voles, opossums, eastern chipmunks (*Tamias striatus*), cottontail rabbits, squirrels, deer mice, house mice, bobwhite quail, robins (*Turdus migratorius*), chicken, flickers, toads, turtle, and salamander, were probably acquired as carrion. Carabid beetles, crickets, grasshoppers, and lepidopteran larvae were the most important insects in the diet. Earthworms, at 10.9% of the volume and in 34.9% of the stomachs, were exceeded in importance only by unidentified birds and unidentified plants. Plant food items included grass, wheat, corn and other grain and seeds, and apples and other fruit.

BEHAVIOR

Activity. Opossums are nocturnal except in winter, when they are occasionally active in the daytime during warm periods in otherwise bitterly cold weather (Brocke 1970; Fitch and Shirer 1970; Pippitt 1976). McManus (1971) found seasonal shifts in general activity; activity was greatest in spring and summer. He also noted seasonal variation in the relative frequency of certain activities. Greatest locomotor activity was in the summer and least in the fall. Feeding activity was greatest in the fall; nest-building activity was greater in the fall and winter than at other times of the year. Hunsaker and Shupe (1977) discussed other aspects of opossum activity patterns.

Feigning Death. Feigning death, although not unique to opossums, is so clearly associated with the animal that the term *playing possum* has become idiomatic. Opossums assume a highly stereotyped catatonic state when faced with inescapable threat situations, as when attacked by dogs or when approached if caught in a steel trap. Tactile stimuli are usually required to produce the response. The opossum falls on its side and lies still with the body slightly flexed. The corners of the mouth are drawn back from the slightly opened jaws, from which drools copious saliva. The animal often defecates and discharges a greenish, foul-smelling substance from its anal glands. The eyes remain open and the ears may twitch at sharp sounds, otherwise there is little or no response to movements or prodding. Recovery may be immediate, but appears to be slow as the animal assesses the situation. Added stimulation during this time causes the animal to return to its death-feigning state. Habituated captives often do not feign death.

Behavioral and electrocardiographic features of death feigning were examined by Francq (1969, 1970). He found no significant differences between alert and death-feigning states in heart rate or height of T-wave deflection, QRS-wave configuration, and QT interval.

Locomotion. The Virginia opossum has terrestrial, arboreal, and aquatic locomotor patterns. Limb posture and the characteristic quadrupedal type of terrestrial locomotion were studied cineradiographically by Jenkins (1971) to track limb bone positions as the animal walked. Although it is not an agile or rapid climber, the opossum's arboreal locomotion is aided by the opposable hallux, prehensile tail, and palmar and plantar friction ridges. Opossums climb trees to forage, escape danger, and rest during the evening and to seek tree holes and arboreal leaf nests as daytime retreats. Opossums are slow, strong swimmers and can swim underwater (Moore 1955; McManus 1970). Wilber and Weidenbacher (1961) found that a female swam 430 min in a 22°C water bath, indicating the ability to swim across moderately large expanses of water. Locomotor speed and behavior were reviewed by Hunsaker and Shupe (1977).

Grooming. The hind foot is extensively used in grooming the back of the head, ears, and sides and upper parts of the body. The hair is combed by the hind toes and claws, which are licked clean after each bout. Opossums lick the forefeet and then rub them over the face and snout in a catlike manner. The abdominal and genital areas are groomed by licking. Females extensively lick the pouch, especially if young are present. Males lick the penis after sexual activity. Additional grooming behavior was covered extensively by Hunsaker and Shupe (1977).

Nest Building. Nest-building materials are passed under the body and packed into the coiled tail, in which they are carried to the nest (Pray 1921; Smith 1941). Hopkins (1977) noted this behavior in 88- to 97-day-old captives.

Opossums enlarge abandoned crow and squirrel nests by adding grass, leaves, and corn husks (Reynolds 1945). Reynolds (1945) also used the presence or absence of nest-building materials as a criterion for identifying ground dens used by opossums. Pippitt (1976) observed that fall and winter dens were plugged with leaves. The plug in fall dens was loose, but the leaves and other materials were so tightly packed in the entrances of winter dens that he was unable to probe the nest cavity to sample CO_2 levels.

Vocalizations. The Virginia opossum makes four distinct vocal sounds (clicking, hissing, growling, and a screech) and all may be used in aggressive interactions. Clicking sounds, uttered by young and adults, are used by males in mating behavior, during aggressive encounters between adults, and by females in the presence of young (McManus 1970). These clicks, described by Hartman (1923a:354) as having a "metallic ring," were discussed by Hunsaker and Shupe (1977), who believed that the sound was made by the upper lips. Although some clicklike sounds may be produced by the lips or other parts of the mouth, we believe that the rapid, metallic-sounding clicks reported most often are made by clashing the canines together. Adults, especially males, have flattened areas worn on the anterior faces of the upper canines. These are evident in the lateral view of the rostrum in Figure 1.7, even though this skull is of a young adult (age class 4). Hunsaker and Shupe (1977) described the other opossum vocalizations and the contexts in which they are used.

Social Behavior. Opossums are basically solitary and most interactions with other animals are avoided except when mating, caring for young, and personal defense are involved.

Copulation and Sexual Behavior. Copulation has been described by several observers including McManus (1967) and Hunsaker and Shupe (1977). The male is tolerated only briefly by the female. While in pursuit of a female, the male clicks continuously. When she is receptive,

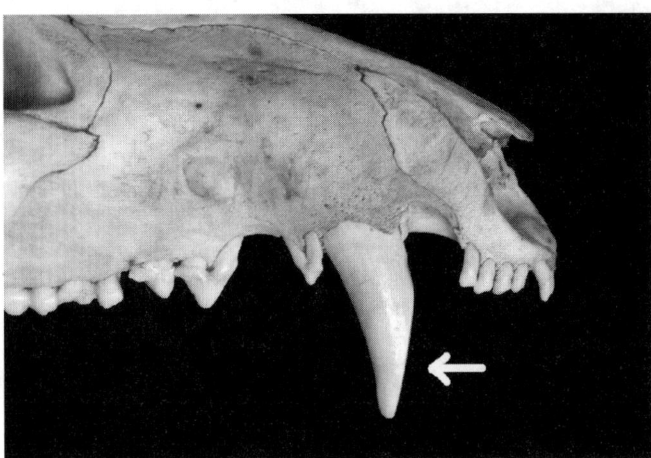

FIGURE 1.7. Right side of the rostrum of a male Virginia opossum (*Didelphis virginiana*), showing the worn, flattened anterior face of the upper canine.

the male mounts by placing his forefeet on her shoulders, grasping her neck in his teeth and shifting his whole weight onto her, and clasping each of her hind legs with one of his feet. The curious aspect of this behavior is the almost universal observation that the mating pair then tumble over onto their right sides. In the instances when the pair remains upright or falls to their left, subsequent examination failed to find sperm in the female's genital tract (Reynolds 1952). After copulation, which lasts approximately 20 min, all additional attempts to mount the female are rebuffed. King (1960) noted an example of Davian behavior (Dickerman 1960) when he observed a male attempting to copulate with a dead female lying in the middle of a highway.

Pippitt (1976) suggested that males may produce a sex-attractant pheromone, which assists an estrous female in locating a potential mate. He reached this conclusion after noting that three radio-tagged females moved relatively long distances during unfavorable weather to dens occupied by males where mating evidently took place.

Based on his work with radio-tagged opossums, Ryser (1992) found that males began visiting females at least 10 days before a female was in estrus. Except in one case where a male stayed with a female continuously for 4 days before mating, males did not guard females. On the night of estrus, females were followed by one to five males, who competed for access. Dominance and mating success was related to body mass, and successful males ($n = 20$) were larger, averaging 3.18 kg, in contrast to unsuccessful males ($n = 30$), whose average weight was 2.65 kg.

Maternal Behavior. Although the female supplies milk and provides warmth and protection, her only active contribution to the welfare of her offspring from birth until the time they begin to leave the pouch is cleaning the pouch and contained young. The young travel to the pouch, attach to nipples, and suckle milk without assistance. Each young remains at the nipple during most of this time, anchored by the swollen nipple tip held in the mouth with the aid of a series of small projections on the lips, tongue, and palate. After the young are able to release the nipple and move by themselves, the female responds with the clicking vocalization, which serves to orient the young to her. Weaning may be largely a passive activity by the female. The young probably wean themselves as they acquire more and more interest in solid food and less interest in suckling (Reynolds 1952). The normal lactation period of about 100 days was lengthened by 1.5 months when Reynolds (1952) substituted 60-day-old young for the 94-day-old young of a captive female. Another female, after having been penned for 1 week without her young and with a drying pouch, adopted and successfully nursed another's litter (Wiseman and Hendrickson 1950).

Agonistic Behavior. The Virginia opossum either avoids or shows aggressive displays toward other individuals. Males usually are aggressive toward other males, but rarely toward females. Females are tolerant of each other unless one is in estrus, which elicits aggression by nonestrous females (Reynolds 1952). Females aggressively repulse males unless sexually receptive, and females have killed males much larger than themselves when penned together in small enclosures. Reynolds (1952) described a fighting dance or dominance display by aggressive males and females. The shuffle-walk dance is silent, but is preceded and followed by clicking.

Opened-mouthed threat behavior and aggressive growling and screeching are common in male encounters. Males stand face to face, the head and shoulders of each weaving from side to side as they hold the mouth open and flatten the ears and vibrissae against the head. Each proceeds to feint and lunge at the other. This bluffing routine either is followed by the turning away and retreat of the submissive male or accelerates into savage biting, snapping, and slashing until one gives ground or is killed. Fighting may result in death feigning by the weaker individual (McManus 1974). At least among captive opossums, death feigning may not be an effective strategy to avoid being killed because of the opossum's propensity for cannibalistic behavior, even when other food is available. Active fighting and cannibalism occur among weaned littermates if they are housed together; young have been known to kill their mother, and vice versa (Raven 1929).

When wild opossums are caught by hand, they growl, defecate, and emit a foul-smelling greenish substance from the pair of anal glands. Although the Virginia opossum is more docile and easier to handle than the common opossum (*D. marsupialis*) of Latin America, it bites when handled.

Scent Marking. The male opossum alternately licks and rubs the sides of his head against an object. Although Reynolds (1952) noted this behavior throughout the year, marking reached its height at the time of breeding and was elicited by the odor of other males and by the odor of estrous females. Other males respond to marked objects by licking and rubbing them and by performing the fighting dance. Females also show interest in the marks and may engage in a similar, but less vigorous dance. The marked sites probably serve to advertise the presence of a male and may attract estrous females.

Cohabitation. Opossums usually den alone; however, females (Reynolds 1945) or occasionally a male and female may share the same den (Lay 1942; Fitch and Shirer 1970; Pippitt 1976). Recently weaned littermates may share occupancy for a few days or weeks. Holmes (1987) found communal denning was common in captive Virginia opossums housed outdoors in seminatural conditions.

MORTALITY

Several authors have commented on the poor or weakened appearance of many opossums. Opossums in regions subjected to severe winter weather are often missing half or more of the ears and tail due to frostbite and subsequent sloughing of the frozen tissue (Smiley 1938; Stuewer 1943; Wiseman and Hendrickson 1950; Hamilton 1958). Opossums often have excessive numbers of cuts, scratches, ripped ears, lost toes, broken teeth, and broken bones as well as internal parasites and ectoparasites (Black 1935; Lay 1942; Fitch and Sandidge 1953). Fitch and Sandidge (1953) commented that the toe-clip wounds of marked animals were slow to heal, but did not indicate that marking contributed to mortality. The process of studying populations sometimes causes death, usually through exposure to cold (Fitch and Sandidge 1953) or heat (Llewellyn and Dale 1964) while the opossum is in a trap.

Opossums are preyed on by dogs, coyotes (*Canis latrans*), foxes, raccoons, bobcats (*Lynx rufus*), raptors, and large snakes. Of these, dogs and horned owls are probably the most serious predators. Rand and Host (1942) mentioned an opossum skull found in an eagle's nest in Florida. Some opossums, especially juveniles, may be preyed on by larger adults. Nevertheless, it is likely that losses through natural predation are of little consequence.

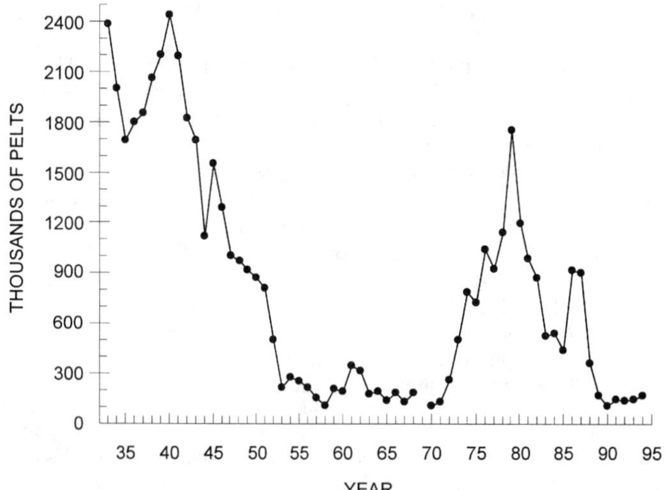

FIGURE 1.8. Reported annual fur harvest of the Virginia opossum (*Didelphis virginiana*) in the United States from the 1933/34 season to the 1994/95 season. SOURCE: Data from Deems and Pursley (1978), Linscombe (1997), and records of the annual fur catch in the United States from the files of the Biological Survey Unit, USGS Patuxent Wildlife Research Center, National Museum of Natural History, Washington, D.C.

Hunting and trapping activities take a high toll (Fig. 1.8), and there is a tradition among farmers to kill all opossums encountered because of their reputation for killing poultry. Although human activities have generally stimulated the expansion of opossum populations, rapid urbanization and concomitant construction of roads and highways have both reduced available habitat and, without doubt, encouraged the most serious cause of mortality, motor vehicle traffic.

Interest in the Virginia opossum as an experimental animal and as a reservoir of zoonoses infecting humans and domestic animals has resulted in a large number of reports on opossum diseases and parasites. Most of these reports have been summarized by Barr (1963) and Potkay (1970, 1977). Those reports not reviewed by Potkay (1977) are mainly taxonomic (e.g., Pence and Little 1972; Pence 1973; Joseph 1974; Prestwood 1976), descriptive (Long et al. 1975), or experimental (Stone et al. 1972).

Diseases. Viral diseases found in the Virginia opossum include rabies and a number of arboviruses. Experimental infection with rabies virus is difficult, and young opossums are more susceptible than adults. The natural incidence of rabies is low and opossums are not considered to be important reservoirs of the disease. Twelve other viral diseases found or tested in opossums (Potkay 1977) include yellow fever, herpes virus (pseudorabies), and a series of encephalitis-causing viruses. Although refractory to most of these pathogens, pseudorabies, vesicular stomatitis, and an unspecified group B arbovirus prove fatal to opossums.

Mycobacterial (tuberculosis), *Borrelia recurrentis* (relapsing fever), *Francisella tularensis* (tularemia), and *Salmonella* spp. (enteritis, typhoid fever) infections are known in opossums. Captives with bacterial endocarditis harbored streptococcal bacteria. Other diseases in opossums include *Bordetella, Stapholococcus, Proteus, Aerobactor, Pseudomonas,* and leptospiral infections. Opossums are considered an important natural reservoir of leptospirosis in wildlife and humans.

Opossums are natural hosts for spotted fever and murine endemic typhus rickettsiae and often show no clinical symptoms of these diseases. Virginia opossums are known to harbor *Toxoplasma, Sarcocystis, Besnoitia, Coccidia, Trichomonas,* and *Trypanosoma.* Animals carrying *Trypanosoma cruzi* (Chagas disease) have been found in Texas, Georgia, Florida, and Alabama. Ringworm and histoplasmosis infect opossums.

Captive opossums often suffer from cage paralysis (usually rickets or osteomalacia) if they receive an inadequate diet. A number of neoplasms have been reported in the Virginia opossum. The urogenital, respiratory, and reticuloendothelial systems are those most commonly afflicted by tumors. Fibrous osteodystrophy of the jaws was described by Long et al. (1975). Lesions have been noted in all parts of the digestive, urogenital, respiratory, cardiovascular, musculoskeletal, hematopoietic, central nervous, and endocrine systems. Many of these lesions are presumed to be caused by parasites. The occurrence of these conditions was summarized in tabular form by Potkay (1977).

Ectoparasites. Opossums are host to a number of ticks, mites, bugs (Hemiptera), lice, fleas, and pentastomids. Many of these ectoparasites are vectors of diseases such as spotted fever, typhus, and Chagas disease.

Although Hamilton (1958) said that opossums were relatively free of ectoparasites, Hock (1952) considered them to be among the most heavily parasitized mammals in North America. Morlan (1952) examined 349 opossums from Georgia and recorded two species of lice, nine of fleas, nine of mites, five of ticks, and two of chiggers. The cat flea (*Ctenocephalides felis*) and dog tick (*Dermacentor variabilis*) infested the greatest number of individuals. The dog tick was also found on opossums in Texas (Lay 1942), Kansas (Fitch and Sandidge 1953), Missouri (Reynolds 1945), and Indiana (Whitaker et al. 1977). Other records were summarized by Potkay (1977). Blumenthal and Kirkland (1976) noted seasonal differences in ectoparasite loads. Highest infestations were on winter animals, a finding that they interpreted as the result of cold-season denning and cohabitation with other mammals. Whitaker et al. (1977), noting that some of the ectoparasites found are considered specific on other hosts, suggested that opossums acquired these parasites because of their carrion-eating habit. Experimental transfer of sarcoptic mange mites (*Sarcoptes scabiei*) to opossums from red foxes failed to produce infestations (Stone et al. 1972).

Endoparasites. An unknown number of protozoan parasites and at least 30 nematodes, 26 trematodes, 7 cestodes, and 7 acanthocephalans inhabit tissues and organs of opossums (Pence and Little 1972; Potkay 1977). Roundworms (*Physaloptera turgida*), common nematode parasites of the digestive tract, were more abundant in females carrying pouch young than in females of the same age class without pouch young. Pinworms (*Cruzia americana*), on the other hand, showed no differences in infection loads, even when hosts were compared by habitat, age class, sex, and reproductive condition (Blumenthal and Kirkland 1976). Virginia opossums are definitive hosts of a number of endoparasites that cause serious disease in other animals including the protozoan *Sarcocystis neurona* (Dubey et al. 2001).

Response to Snake Venoms. The remarkable resistance of the Virginia opossum to envenomation by poisonous snakes was reported on by Kilmon (1976) and Werner and Vick (1977). Kilmon stressed 15 anesthetized opossums by subjecting them to bites by the eastern diamondback rattlesnake (*Crotalus adamanteus*), timber rattlesnake (*Crotalus horridus*), cottonmouth (*Agkistrodon piscivorus*), Russell's viper (*Vipera russelli*), and common Asiatic cobra (*Naja naja*). None of the opossums showed any tissue reaction other than the fang punctures. There was a decrease in arterial blood pressure of 5 mm Hg (from a norm of 140/105 to 135/100 mm Hg) in another anesthetized, 4.1-kg opossum immediately following the bite of a cottonmouth. The heart rate increased from 160 to 180 beats/min, but the respiration rate was unchanged (12/min). Blood pressure recovered after 10 min, and by 30 min the heart rate had returned to normal limits. A smaller (3.6-kg) opossum received a dose of cottonmouth venom in the caudal vein equivalent to 5 times the lethal dosage for a 15-kg dog. A decrease in blood pressure and a rise in heart rate were the immediate responses; recovery was complete within 30 min. The opossum was then given almost double the first dosage, with similar results. No organ damage was found 24 hr later, when this animal was necropsied.

Werner and Vick (1977) stressed opossums by actual snake bite or by intravenous and intramuscular injection of 4–60 times the dosage normally lethal to susceptible mammals. The venoms used were from pit vipers (Crotalidae), cobras and coral snakes (Elapidae), puff adder

(Viperidae), and sea snake (Hydrophidae). All opossums, with the exception of those given crotalid venoms, died within 24 hr. Those given crotalid venoms in high doses responded with small changes in blood pressure and heart rate, but all recovered in 15–30 min. None died or showed signs of local hemorrhaging, swelling, or tissue damage. Apparently, only adults were used in these experiments. Although young opossums may not be as immune to snake venom as the adults, the general conclusion is that opossums are unaffected by the bites of pit vipers, but are sensitive to venoms of other poisonous snakes. Therefore, opossums can safely prey on copperheads, rattlesnakes, and cottonmouths. In a series of reports on the energy metabolism of Virginia opossum red blood cells, Bethlenfalvay et al. (1984, 1988, 1989, 1990) revealed physiological characteristics that may protect erythrocytes from lysis in the presence of pit viper venom. McKellar and Pérez (2002) exposed Virginia opossums to western diamondback rattlesnake (*Crotalus atrox*) venom to determine if they were capable of producing antibodies. They found no evidence of antibody production and concluded that the opossums' natural proteinase inhibitors rendered the venom harmless before an antibody response could occur.

ECONOMIC STATUS, MANAGEMENT, AND CONSERVATION

Fur. Hamilton (1958) pointed out the numerically high rank opossums have held in the fur trade in the United States. The number of opossum pelts reported between 1933 and 1995 is graphed in Figure 1.8. The high number of opossum pelts taken in the 1933/34, 1940/41, and 1979/80 seasons (2.40, 2.42, and 1.75 million, respectively) represent only part of the annual harvest during those years. Ignoring the unknown numbers killed for food, sport, and because they are considered predators, only 15 of the 31 states then known to have harvestable populations of opossums reported the number of pelts taken in the 1933/34 season. The number of opossum pelts reported for the 1940/41 season was based on information from only 21 states. Some states, such as Texas and New York, indicated that their actual harvest may have been underestimated by 50% or more. The numbers for the 1979/80 season probably would have exceeded 2 million if Michigan had reported its take. Of the 6 states not reporting opossums for 1979/80, only Michigan normally reported high numbers and ranked second or third among states in five of the nine seasons it reported opossums in the 1980–1995 period (Linscombe 1997). The steady decline from 1940 through 1953 and the continued low number reported through 1971 and again from 1989 to the present reflect the low prices paid and the general unpopularity of long-haired furs during that time. The average price paid for opossum pelts between 1934 and 1971 varied from approximately 30¢ to 58¢ per pelt. The average price per skin increased from 85¢ for furs taken in the 1970/71 season to $2.50 the next season and peaked at $3.27 in the 1979/80 season (Deems and Pursley 1978; Linscombe 1997). The price per pelt from 1982 to 1995 averaged $1.00 (Linscombe 1997). The latest values range from $2.00 to $3.00 for better skins (extra large silver), with smaller sizes and lower grades not wanted (e.g., Schroeder 2002).

For comparison, the number of opossum pelts and the average price paid per pelt in Louisiana (Lowery 1974) between 1913 and 1972 range from a high of 518,295 at $1.10 apiece in the 1928/29 season to a low of 3009 at 30¢ apiece in the 1967/68 trapping season. The lowest average price paid was 10¢ in the 1938/39, 1940/41, and 1958/59 seasons.

According to the North American fur harvest statistics for the 1979/80 season, 34 of the 40 states assumed to have harvestable populations of opossums reported a combined harvest of 1,750,338 pelts with a total estimated value of $5,715,928 (Linscombe 1997). The average price per pelt ($3.27) was the lowest paid for any North American furbearer, except for weasels ($1.62). Nevertheless, opossums ranked first in numbers of pelts reported by 2 states (Alabama and Oklahoma), second in 9 states (Arkansas, Florida, Georgia, Kansas, Mississippi, Missouri, South Carolina, Tennessee, and Texas), third in 11 (Delaware, Illinois, Indiana, Kentucky, Maryland, New Jersey, North Carolina, Ohio, Rhode Island, Virginia, and West Virginia), and fourth in 3 (Louisiana, Pennsylvania, and New York).

In terms of the numbers of pelts reported nationwide in the United States (except Alaska) from the 1970/71 season through the 1994/95 season (the latest for which complete data are available), the 10 most important furbearers were muskrat, raccoon, nutria (*Myocastor coypus*), opossum, red fox, mink (*Mustela vison*), beaver (*Castor canadensis*), coyote, gray fox, and striped skunk. During these 25 trapping seasons, opossums ranked third in 4 states (behind muskrat and raccoon) and fourth in 13 (behind muskrat, raccoon, and nutria). Trappers often consider opossums as bycatch. Consequently, during years of low fur prices, most are either released, used as bait, or discarded, except by recreational trappers for whom profit is a secondary motive for trapping. Opossum fur is dyed and plucked to simulate more expensive furs or is used in its natural state, most commonly as trim.

Food. Opossums are eaten in many regions of the United States, and historically were available in markets in some states and the District of Columbia (Bailey 1923). Although eating opossum long has been a tradition in the South, where "possum and taters" is an esteemed dish (Lay 1942; Hartman 1952; Lowery 1974), the meat lacks appeal in other areas, particularly in the Northeast and the Great Lake states. Hamilton (1958), for example, found the meat greasy and lacking flavor.

Many recipes for the preparation of opossum recommend scalding in hot lye water and scraping off the hair. The obvious purpose is to retain as much fat as possible. The Virginia opossum is the only member of the genus *Didelphis* that accumulates heavy layers of fat, and also is the only didelphid commonly eaten by humans. Many accounts of travels in South America have mentioned the unpalatability of the common opossum, *D. marsupialis* (Hartman 1952:150). Yucatecan Indians reported that dark-colored specimens taste bad, but that paler ones are delectable (Hatt 1938). It is likely that the paler opossums were *D. virginiana* (see description of cheek color) and the others were *D. marsupialis.*

People who have spent several weeks or more in field camps, where meat is usually lean and hard to come by, often appreciate the flavor of fat meat. Once out of the field and enjoying a richer diet, fat meat loses its appeal and may taste greasy instead of delicious. Now that larger segments of a primarily urban society enjoy a richer and more varied diet, opossum meat is not as popular as it once was. This may be particularly true where opossums are hunted for sport instead of for meat and where peoples tastes are offended at the thought of eating any furbearer, especially one as odd-appearing as an opossum.

Sport. Night hunting for opossums, with or without dogs, is a common activity in some parts of the country. However, as pointed out by Allen (1940), Stuewer (1943), Taube (1947), and Hamilton (1958), many sportsmen prefer to consider opossums as an impediment to good raccoon hunting and a detriment to more highly valued wildlife than as an object worthy of pursuit.

Predation. Opossums have enjoyed a long, if not entirely justified, reputation as rapacious destroyers of poultry (Hartman 1952:57). It is true that they kill chickens and prey on young cottontail rabbits as well as on the eggs and broods of ground-nesting birds. Although these depredations on poultry and game may alarm farmers and sportsmen, the opossum's impact is negligible. Analyses of feeding habits show the Virginia opossum is an omnivorous gleaner and scavenger of a wide variety of foods. Most of the mammals and birds eaten (including poultry) are encountered as carrion. Opossums also eat large numbers of voles and mice as well as poisonous snakes.

Those farmers, sportsmen, and game biologists who irrationally consider the animal a serious pest or predator may find it difficult to accept the fact that opossums have comparatively little impact on economically important wildlife. Anthropomorphic generalizing such as "I don't like him either. He is a sluggish, smelly, disreputable critter without a semblance of character or self respect" (Allen 1940:4) can hardly be construed as an objective evaluation of the attributes of the species.

RESEARCH NEEDS

Opossums have long fascinated biologists interested in marsupials. Much of the initial stimulus to do research on the Virginia opossum was the belief that the animal was a "living fossil" that retained all of its Cretaceous qualities and characteristics. The major obstacle to using opossums as research animals was the lack of efficient and effective methods for their husbandry. The reports by Jurgelski (1974), Jurgelski et al. (1974), and Jurgelski and Porter (1974) indicate that most of the problems have been solved. Although opossums may be expected to be important in laboratory research, recent animal welfare concerns and regulations governing animal maintenance facilities and research on laboratory animals mitigate against widespread use of these marsupials for that purpose.

Despite great interest, as indicated by the many published reports on the biology and ecology of opossums, relatively little is known about their behavior, dispersal patterns, and population structure. Austad and Sunquist (1986), Sunquist and Eisenberg (1993), and Austad (1993) explored the effect of age, nutritional status, and freedom from predation on longevity, reproductive output, and sex ratios of pouch young. The following questions suggest promising areas of research:

1. How do diet, climate, and population density influence the breeding season and reproductive potential?
2. What is the reproductive life of a free-ranging opossum?
3. What is the age structure of populations and how does it vary seasonally?
4. What are the exact causes of mortality?
5. How do mortality factors affect sex and age-class composition of populations?
6. Do the feeding habits and feeding behaviors of newly weaned and juvenile age classes differ from those of subadults and adults?
7. What is the role played by opossums in suburban and urban wildlife populations?

One obstacle to the field study of opossums has been the lack of uniformly applicable methods for determining the age of individuals. Methods and the developmental criteria used for determining age are outlined and discussed in the sections on Development and Age Determination. Other, more effective methods might be developed with additional study and a little imagination. Measuring the length of the head (Tyndale-Biscoe and Mackenzie 1976) instead of snout–rump length on pouch young to monitor growth rate has not been done on North American populations.

Other problems concern monitoring activity and movements. Live-trapping has not been nearly as effective as radiotelemetry in determining the activity ranges of opossums. Radiotelemetry was used by A. C. V. Holmes and Sanderson (1965), Fitch and Shirer (1970), Shirer and Fitch (1970), and VanDruff (1971). Pippitt (1976) used two-channel implanted radios to monitor physiological functions in free-living animals.

Of the several marking techniques available, those most commonly used have been toe clipping and the affixing of different kinds of ear tags (Lay 1942; Fitch and Sandidge 1953; Sanderson 1961). Another method using a tag encircling the Achilles tendon was described by Cook (1943). Radioisotope labels (Wolff and Holleman 1978) have never been tried on opossums.

Several studies of feeding habits have included descriptions of the methods used (Hamilton 1958). A more detailed description of processing procedures for studies of feeding habits was given by Korschgen (1971, 1980). Techniques for evaluating feeding habits and the nutritional value of foods consumed can be found in references in Haufler and Servello (1996) and Litvaitis et al. (1996).

The Virginia opossum has limited food and sport value. Its value as a furbearer results from the large numbers of pelts taken and the apparent ability of populations to maintain themselves despite high mortality from vehicular traffic and heavy hunting and trapping intensity. Much of the concern on the part of game management agencies is due to prejudice against opossums by farmers and sportsmen. All serious studies, however, indicate that opossums have little impact on poultry and wildlife. Instead, opossums probably serve a positive role by reducing some noxious species of wildlife and suppressing some wildlife diseases by removing dead animals.

In areas where predation on ground-nesting upland game and waterfowl is a potential problem, opossums are rarely the primary predator. Opossums rely on ground dens. Therefore, reducing the availability of den sites may be the most effective long-term control method and also would reduce depredations caused by other predators that rely on ground dens. Other control methods are the same as those used to reduce raccoon, skunk, and fox populations.

MISCONCEPTIONS

The Virginia opossum has played a prominent role in the folklore of the southern United States, and is the subject of numerous songs, rhymes, fables, and misconceptions (Hartman 1921b, 1952; Lowery 1974). Some of the following more common misconceptions still find their way into the literature:

1. *False:* The Virginia opossum is a living fossil, a holdover in North America from the age of dinosaurs.
   True: All marsupials originally inhabiting North America became extinct by the mid-Tertiary. The first record for the Virginia opossum is late Pleistocene.
2. *False:* The opossum is slow, dull-witted, and stupid.
   True: Opossums are inhibited animals, especially under lighted conditions. Although somewhat misleading, results from some learning and discrimination tests rank opossums above dogs and more or less on a par with pigs in intelligence (Hunsaker and Shupe 1977).
3. *False*: Male opossums copulate with the female's nose.
   True: The forked penis permits deposition of spermatozoa into the lateral vaginae of the female, not in the nostrils.
4. *False:* Young opossums are not born as in other mammals, but are formed at the ends of the nipples like buds of a plant.
   True: Young are born as in other mammals, although in a relatively undeveloped condition.
5. *False:* Young are blown through the mother's nostrils into the pouch.
   True: Same as 4 above.
6. *False:* The mother pumps milk into the young.
   True: The young suckle without aid from the mother (Enders 1966).
7. *False:* The lips and tongue fuse to the nipple.
   True: The pouch young become firmly attached to the nipple, but no fusion occurs. The false observation is based on the bloody, sometimes torn mouth and nipple that results when young are not carefully removed. To avoid damage to the young, researchers have suggested gently twisting to remove them from the nipple.
8. *False:* A female normally rears three litters a year.
   True: There are very few examples where the Virginia opossum successfully rears more than two litters a year. Females may give birth to three or more litters during a season if earlier litters are lost.

LITERATURE CITED

Allen, C. H., R. L. Marchinton and W. Mac Lentz 1985. Movement, habitat use and denning of opossums in the Georgia Piedmont. American Midland Naturalist 113:408–12.

Allen, D. L. 1940. Nobody loves the 'possum. Michigan Conservation 1940 (March):4, 10.

Apgar, C. S., Jr. 1934. Analysis of life records of *Didelphis virginiana*. Pages 51–55 *in* H. Fox. Report of the Laboratory and Museum of Comparative Pathology of the Zoological Society of Philadelphia. Philadelphia.

Armstrong, D. M. 1972. Distribution of mammals in Colorado (Monograph 3). University of Kansas, Museum of Natural History, Lawrence, KS.

Ashbrook, F. G., and B. M. Arnold. 1927. Fur-bearing animals of the United States: The opossum. Fur Journal 1:28–29.

Audubon, J. J., and J. Bachman. 1851. The quadrupeds of North America, Vol. 2. V. G. Audubon, New York.

Austad, S. N. 1993. Retarded senescence in an insular population of Virginia opossums (*Didelphis virginiana*). Journal of Zoology (London) 229:695–708.

Austad, S. N., and M. E. Sunquist. 1986. Sex-ratio manipulation in the common opossum. Nature 324:58–60.

Bailey, V. 1923. Mammals of the District of Columbia. Proceedings of the Biological Society of Washington 36:103–38.

Barr, T. R. B. 1963. Infectious diseases in the opossum, a review. Journal of Wildlife Management 27:53–71.

Barton, B. S. 1823. Facts, observations, and conjectures relative to the generation of the opossum of North America. Annals of Philosophy, New Series 6:349–54.

Bennitt, R., and W. O. Nagel. 1937. A survey of the resident game and furbearers of Missouri. University of Missouri Studies 12, 215 pp.

Bethlenfalvay, N. C., J. E. Lima, and T. Waldrup. 1984. Studies on the energy metabolism of opossum (*Didelphis virginiana*) erythrocytes. I. Utilization of carbohydrates and purine nucleosides. Journal of Cellular Physiology 120:69–74.

Bethlenfalvay, N. C., J. E. Lima, E. Chadwick, and I. Stewart. 1988. Studies on the energy metabolism of opossum *Didelphis virginiana* erythrocytes—III. Metabolic depletion with 2-deoxyglucose markedly accelerates methemoglobin reduction in opossum but not in human erythrocytes. Comparative Biochemistry and Physiology 89A(2):119–24.

Bethlenfalvay, N. C., E. Chadwick, and J. E. Lima. 1989. Studies on the energy metabolism of opossum *Didelphis virginiana* erythrocytes—IV. Red cells have low adenosine deaminase activity and high levels of deoxyadenosine nucleotides. Life Sciences 44:963–70.

Bethlenfalvay, N. C., J. C. White, E. Chadwick, and J. E. Lima. 1990. Studies on the energy metabolism of opossum (*Didelphis virginiana*) erythrocytes—VI. *De novo* purine nucleotide biosynthesis is limited to the final steps of the pathway *in vitro*. Comparative Biochemistry and Physiology 97B(1):193–96.

Biggers, J. D. 1966. Reproduction in male marsupials. Pages 251–80 *in* I. W. Rowlands, ed. Comparative biology of reproduction in mammals (Symposium 15). Zoological Society of London, London.

Biggers, J. D., and R. F. S. Creed. 1962. Conjugate spermatozoa of the North American opossum. Nature 196:1112–13.

Biggers, J. D., and E. D. DeLamater. 1965. Marsupial spermatozoa: Pairing in the epididymis of the American forms. Nature 208:402–4.

Black, J. D. 1935. Vitality of the Virginia opossum as exhibited in the skeleton. Journal of Mammalogy 16:223.

Blumenthal, E. M., and G. L. Kirkland, Jr. 1976. The biology of the opossum, *Didelphis virginiana,* in southcentral Pennsylvania. Proceedings of the Pennsylvania Academy of Sciences 50:81–85.

Bowers, J. H., and F. W. Judd. 1969. Notes on the distribution of *Didelphis marsupialis* and *Citellus spilosoma* in western Texas. Texas Journal of Science 20:277.

Brocke, R. H. 1970. The winter ecology and bioenergetics of the opossum, *Didelphis marsupialis,* as distributional factors in Michigan. Ph.D. Dissertation, Michigan State University, East Lansing.

Brown, L. N. 1965. Status of opossum, *Didelphis marsupialis,* in Wyoming. Southwestern Naturalist 10:142–43.

Bryant, B. J. 1977. The development of the lymphatic and immunohematopoietic systems. Pages 349–86 *in* D. Hunsaker II, ed. The biology of marsupials. Academic Press, New York.

Burmudez, F. C., J. N. Stuart, J. K. Frey, and R. Valdez. 1995. Distribution and status of the Virginia opossum (*Didelphis virginiana*) in New Mexico. Southwestern Naturalist 40:336–40.

Burns, R. K., and L. M. Burns. 1957. Observations on the breeding of the American opossum in Florida. Revue Suisse de Zoologie 64:595–605.

Chase, E. B. 1939. The reproductive system of the male opossum *Didelphis virginiana* Kerr and its experimental modification. Journal of Morphology 65:215–39.

Cheng, C. 1955. The development of the shoulder region of the opossum *Didelphis virginiana* Kerr with special reference to the musculature. Journal of Morphology 97:415–72.

Clemens, W. A. 1968. Origin and early evolution of marsupials. Evolution 22:1–18.

Clemens, W. A. 1977. Phylogeny of the marsupials. Pages 51–68 *in* B. Stonehouse and D. Gilmore, eds. The biology of marsupials. University Park Press, Baltimore.

Coleman, R. H. 1929. Opossum in the lower Hudson Valley, New York. Journal of Mammalogy 10:250.

Cook, A. H. 1943. A technique for marking mammals. Journal of Mammalogy 24:45–47.

Cory, C. B. 1912. The mammals of Illinois and Wisconsin. Field Museum of Natural History, Zoological Series 11:1–502.

Crandall, L. S. 1964. The management of wild mammals in captivity. University of Chicago Press, Chicago.

Crowe, D. M. 1986. Furbearers of Wyoming. Wyoming Game and Fish Department, Cheyenne.

Davis, G. W. 1938. Virginia opossum in Vermont. Journal of Mammalogy 19:499.

Davis, W. B. 1974. The mammals of Texas. Texas Parks and Wildlife Department, Austin, Bulletin 41:1–294.

Dearborn, N. 1932. Foods of some predatory fur-bearing animals in Michigan. University of Michigan, School of Forestry and Conservation, Bulletin 1:1–52.

Deems, E. F., Jr., and D. Pursley. 1978. North American furbearers. International Association of Fish and Wildlife Agencies, Washington, DC.

Dickerman, R. W. 1960. "Davian behavior complex" in ground squirrels. Journal of Mammalogy 41:403.

Dills, G. 1972. Telemetered thermal responses of a specimen of the Virginia opossum. Journal of the Alabama Academy of Science 43:55–62.

Dills, G., and T. Manganiello. 1973. Diel temperature fluctuations of the Virginia opossum (*Didelphis virginiana virginiana*). Journal of Mammalogy 54:763–65.

Dubey, J. P., D. S. Lindsay, W. J. A. Saville, S. M. Reed, D. E. Granstrom, and C. A. Speer. 2001. A review of *Sarcocystis neurona* and equine protozoal myeloencephalitis (EPM). Veterinary Parasitology 95:89–131.

Edmunds, R. M., J. W. Goertz, and G. Linscombe. 1978. Age ratios, weights, and reproduction of the Virginia opossum in north Louisiana. Journal of Mammalogy 59:884–85.

Eisenberg, J. F. 1988. Reproduction in polyprotodont marsupials and similar-sized eutherians with a speculation concerning the evolution of litter size in mammals. Pages 291–310 *in* M. A. Boyce, ed. Evolution of life histories of mammals. Yale University Press, New Haven, CT.

Eisenberg, J. F., and D. E. Wilson. 1981. Relative brain size and demographic strategies in didelphid marsupials. American Naturalist 118:110–26.

Elftman, H. O. 1929. Functional adaptations of the pelvis in marsupials. Bulletin of the American Museum of Natural History 58:189–232.

Ellsworth, A. F. 1975a. Atlas of the North American opossum, *Didelphis.* R. Krieger, Huntington, NY.

Ellsworth, A. F. 1975b. Reassessment of muscle homologues and nomenclature in conservative amniotes. R. Krieger, Huntington, NY.

Enders, R. K. 1966. Attachment, nursing and survival of young in some didelphids. Pages 195–204 *in* I. W. Rowlands, ed. Comparative biology of reproduction in mammals (Symposium 15). Zoological Society of London, London.

Findley, J. S. 1956. Mammals of Clay County, South Dakota. University of South Dakota Publications, Biology 1:1–45.

Fitch, H. S., and L. L. Sandidge. 1953. Ecology of the opossum on a natural area in northeastern Kansas. University of Kansas Publications, Museum of Natural History 7:305 38.

Fitch, H. S., and H. W. Shirer. 1970. A radiotelemetric study of spatial relationships in the opossum. American Midland Naturalist 48:170–86.

Fortney, J. A. 1973. Cytology of eccrine sweat glands in the opossum. American Journal of Anatomy 136:205–19.

Francq, E. N. 1969. Behavioral aspects of feigned death in the opossum *Didelphis marsupialis.* American Midland Naturalist 81:556–68.

Francq, E. N. 1970. Electrocardiograms of the opossum, *Didelphis marsupialis,* during feigned death. Journal of Mammalogy 51:395.

Friedman, H. 1967. Colour vision in the Virginia opossum. Nature 213:835–36.

Gardner, A. L. 1973. The systematics of the genus *Didelphis* (Marsupialia: Didelphidae) in North and Middle America. Special Publications, the Museum, Texas Tech University 4:1–81.

Gardner, A. L. 1982. Virginia opossum. Pages 3–36 *in* J. A. Chapman and G. A. Feldhamer, eds. Wild mammals of North America. Johns Hopkins University Press, Baltimore.

Gardner, A. L. 1993. Order Didelphimorphia. Pages 15–24 *in* D. E. Wilson and D. M. Reeder, eds. Mammal species of the world: A taxonomic and geographic reference. Smithsonian Institution Press, Washington, DC.

Gardner, A. L., and P. Cortés-Calvo. 1999. Didelphidae. Pages 29–38 *in* S. T. Alvarez-Castañeda and J. L. Patton, eds. Mamíferos del noroeste de México. Centro de Investigaciones Biológicas del Noroeste, S.C., La Paz, Baja California Sur, Mexico.

Gifford, C. L., and R. Whitebread. 1951. Mammal survey of south central Pennsylvania. Pennsylvania Game Commission, Harrisburg.

Gillette, L. N. 1980. Movement patterns of radio-tagged opossums in Wisconsin. American Midland Naturalist 104:1–12.

Gilmore, R. M. 1943. Mammalogy in an epidemiological study of jungle yellow fever in Brazil. Journal of Mammalogy 24:144–162.

Godin, A. J. 1977. Wild mammals of New England. Johns Hopkins University Press, Baltimore.

Golley, F. B. 1962. Mammals of Georgia. University of Georgia Press, Athens.

Grimm, W. C., and H. A. Roberts. 1950. Mammal survey of southwestern Pennsylvania. Pennsylvania Game Commission, Harrisburg.

Grinnell, J. J., S. Dixon, and J. M. Linsdale. 1937. Fur-bearing mammals of California, Vol. 1. University of California Press, Berkeley.

Grote, J. C., and P. L. Dalby. 1973. An early litter for the opossum (*Didelphis marsupialis*) in Ohio. Ohio Journal of Science 73:240–41.

Guilday, J. E. 1958. The prehistoric distribution of the opossum. Journal of Mammalogy 39:39–43.

Hamilton, W. J., Jr. 1951. The food of the opossum in New York State. Journal of Wildlife Management 15:258–64.

Hamilton, W. J., Jr. 1958. Life history and economic relations of the opossum (*Didelphis marsupialis virginiana*) in New York State. Cornell University, Agricultural Experiment Station, Memoir 354:1–48.

Harder, J. D., M. J. Stonerook, and J. Pondy. 1993. Gestation and placentation in two New World opossums: *Didelphis virginiana* and *Monodelphis domestica*. Journal of Experimental Zoology 266:463–79.

Hardy, I. C. W. 1997. Opossum sex ratios revisited: Significant or nonsignificant. American Naturalist 150:420–24.

Hartman, C. G. 1921a. Dioestrous changes in the mammary gland of the opossum and the diagnosis of pregnancy. American Journal of Physiology 55:308–9.

Hartman, C. G. 1921b. Traditional belief concerning the generation of the opossum (*Didelphis virginiana* L.). Journal of American Folklore 34:321–23.

Hartman, C. G. 1922. A brown mutation in the opossum (*Didelphis virginiana*) with remarks upon the gray and the black phases in this species. Journal of Mammalogy 3:146–49.

Hartman, C. G. 1923a. Breeding habits, development, and birth of the opossum. Smithsonian Report [for 1921], pp. 347–63.

Hartman, C. G. 1923b. The oestrous cycle in the opossum. American Journal of Anatomy 32:353–421.

Hartman, C. G. 1928. The breeding season of the opossum (*Didelphis virginiana*) and the rate of intrauterine and postnatal development. Journal of Morphology 46:143–215.

Hartman, C. G. 1952. Possums. University of Texas Press, Austin.

Hatt, R. T. 1938. Notes concerning mammals collected in Yucatan. Journal of Mammalogy 19:333–37.

Haufler, J. B., and F. A. Servello. 1996. Techniques for wildlife nutritional analyses. Pages 307–23 *in* T. A. Bookhout, ed. Research and management techniques for wildlife and habitats. Wildlife Society, Bethesda, MD.

Hazard, E. B. 1963. Records of the opossum in northern Minnesota. Journal of Mammalogy 44:118.

Hearn, J. P. 1977. Pituitary function in marsupial reproduction. Pages 337–44 *in* B. Stonehouse and D. Gilmore, eds. The biology of marsupials. University Park Press, Baltimore.

Hershkovitz, P. 1951. Mammals from British Honduras, Mexico, Jamaica and Haiti. Fieldiana Zoology 31:547–69.

Hibbard, C. W., C. E. Ray, D. E. Savage, and D. W. Taylor, and J. E. Guilday. 1965. Quaternary mammals of North America. Pages 509–25 *in* H. E. Wright, Jr. and D. O. Frey, eds. The Quaternary of the United States. Princeton University Press, Princeton, NJ.

Higgenbotham, A. C., and W. E. Koon. 1955. Temperature regulation in the Virginia opossum. American Journal of Physiology 181:69–71.

Hill, J. P., and E. A. Fraser. 1925. Some observations on the female urogenital organs of Didelphyidae. Proceedings of the Zoological Society of London 1925:189–219.

Hock, R. J. 1952. The opossum in Arizona. Journal of Mammalogy 33:464–70.

Hoffmeister, D. F., and W. W. Goodpaster. 1954. The mammals of the Huachuca Mountains, southeastern Arizona. Illinois Biological Monographs 24:1–152.

Hollister, N. 1908. Notes on Wisconsin mammals. Bulletin of the Wisconsin Natural History Society 4:137–42.

Holmes, A. C. V., and G. C. Sanderson 1965. Populations and movements of opossums in east-central Illinois. Journal of Wildlife Management 29:287–95.

Holmes, D. 1987. Social complexity and potential for chemocommunication in captive Virginia opossums, *Didelphis virginiana* Kerr. Ph.D. Dissertation, Bowling Green State University, Bowling Green, OH.

Hopkins, D. 1977. Nest-building behavior in the immature Virginia opossum (*Didelphis virginiana*). Mammalia 41:361–62.

Hossler, R. J., J. B. McAninch, and J. D. Harder. 1994. Maternal denning behavior and survival of juveniles in opossums in southeastern New York. Journal of Mammalogy 75:60–70.

Hunsaker II, D. 1977. Ecology of New World marsupials. Pages 95–156 *in* D. Hunsaker II, ed. The biology of marsupials. Academic Press, New York.

Hunsaker II, D., and Shupe, D. 1977. Behavior of New World marsupials. Pages 279–348 *in* D. Hunsaker II, ed. The biology of marsupials. Academic Press, New York.

Jackson, H. H. T. 1961. Mammals of Wisconsin. University of Wisconsin Press, Madison.

James, W. T. 1960. A study of visual discrimination in the opossum. Journal of Genetic Psychology 97:127–30.

Jenkins, F. A., Jr. 1971. Limb posture and locomotion in the Virginia opossum (*Didelphis marsupialis*) and in other non-cursorial mammals. Journal of Zoology (London) 165:303–15.

Jewett, S. G., and Dobyns, H. W. 1929. The Virginia opossum in Oregon. Journal of Mammalogy 10:351.

Johnson, J. I., Jr. 1977. Central nervous system of marsupials. Pages 157–278 *in* D.Hunsaker II, ed. The biology of marsupials. Academic Press, New York.

Jones, J. K., Jr. 1964. Distribution and taxonomy of mammals of Nebraska. University of Kansas Publications, Museum of Natural History 16:1–356.

Jordan, H. E. 1911. The spermatogenesis of the opossum (*Didelphis virginiana*) with special reference to the accessory chromosome and chondriosomes. Archiv für Zellforschung 7:41–86.

Joseph, T. 1974. *Eimeria indianensis* sp. n. and an *Isospora* sp. from the opossum *Didelphis virginiana* (Kerr). Journal of Protozoology 21:12–15.

Jurgelski,W., Jr. 1974. The opossum (*Didelphis virginiana* Kerr) as a biomedical model. I. Research perspective, husbandry, and laboratory techniques. Laboratory Animal Science 24:376–403.

Jurgelski, W., Jr., and M. E. Porter. 1974. The opossum (*Didelphis virginiana* Kerr) as a biomedical model. III. Breeding the opossum in captivity: Methods. Laboratory Animal Science 24:412–25.

Jurgelski, W., Jr., W. Forsythe, D. Dahl, L. D. Thomas, J. A. Moore, P. Kotin, H. L. Falk, and F. S.Vogel. 1974. The opossum (*Didelphis virginiana* Kerr) as a biomedical model. II. Breeding the opossum in captivity: Facility design. Laboratory Animal Science 24:404–11.

Kennard, F. H. 1925. The Virginia opossum in Massachusetts and New Hampshire. Journal of Mammalogy 6:196.

Kilmon, J. A., Sr. 1976. High tolerance to snake venom by the Virginia opossum, *Didelphis virginiana*. Toxicon 14:337–40.

King, O. M. 1960. A note on opossum behavior. Journal of Mammalogy 42:397.

Kingsbury, B. F. 1940. The development of the pharyngeal derivatives of the opossum (*Didelphis virginiana*) with special reference to the thymus. American Journal of Anatomy 67:393–435.

Kirk, G. L. 1921. Opossum in Vermont. Journal of Mammalogy 2:109.

Kirsch, J. A. W., R. E. Bleiweiss, A. W. Dickerman, and O. A. Reig. 1993. DNA/DNA hybridization studies of carnivorous marsupials. III. Relationships among species of *Didelphis* (Didelphidae). Journal of Mammalian Evolution 1:75–97.

Kissell, R. E., Jr., and M. L. Kennedy. 1992. Ecologic relationships of co-occurring populations of opossums (*Didelphis virginiana*) and raccoons (*Procyon lotor*) in Tennessee. Journal of Mammalogy 73:808–13.

Knudsen, G. J., and J. B. Hale. 1970. Food habits of opossums in southern Wisconsin. Wisconsin Department of Natural Resources, Report 61:1–11.

Korschgen, L. J. 1971. Procedures for food-habits analyses. Pages 233–50 *in* R. H. Giles, Jr., ed. Wildlife management techniques, 3rd ed. Wildlife Society, Washington, DC.

Korschgen, L. J. 1980. Procedures for food-habits analyses. Pages 113–27 *in* S. D. Schemnitz, ed. Wildlife management techniques manual, 4th ed. Wildlife Society, Bethesda, MD.

Ladine, T. A. 1997. Activity patterns of co-occurring populations of Virginia opossums (*Didelphis virginiana*) and raccoons (*Procyon lotor*). Mammalia 61:345–54.

Langley, W. M. 1979. Preference of the striped skunk and opossum for auditory over visual prey stimuli. Carnivore 2:31–38.

Lay, D. W. 1942. Ecology of the opossum in eastern Texas. Journal of Mammalogy 23:147–59.

Lindsay, D. M. 1960. Mammals of Ripley and Jefferson counties, Indiana. Journal of Mammalogy 41:253–262.

Linscombe, G. 1997. U.S. fur harvest (1970–1995) and fur value (1974–1995) statistics by state and region. International Association of Fish and Wildlife Agencies, Washington, DC.

Litvaitis, J. A., K.Titus, and E. M. Anderson. 1996. Measuring vertebrate use of terrestrial habitats and foods. Pages 254–74 *in* T. A. Bookhout, ed. Research and management techniques for wildlife and habitats, 5th ed. Wildlife Society, Bethesda, MD.

Llewellyn, L. M., and F. H. Dale. 1964. Notes on the ecology of the opossum in Maryland. Journal of Mammalogy 45:113–22.

Llewellyn, L. M., and F. M. Uhler. 1952. The food of fur animals of the Patuxent Research Refuge, Maryland. American Midland Naturalist 48:193–203.

Long, C. A. 1965. The mammals of Wyoming. University of Kansas Publications, Museum of Natural History 14:493–758.

Long, C. A., and F. A. Copes. 1968. Note on the rate of dispersion of the opossum in Wisconsin. American Midland Naturalist 80:283–84.

Long, G. G., J. L. Stookey, T. G. Terrell, and G. D. Whitney. 1975. Fibrous osteodystrophy in an opossum. Journal of Wildlife Diseases 11:221–23.

Lowery, G. H., Jr. 1974. The mammals of Louisiana and its adjacent waters. Louisiana State University Press, Baton Rouge.

Lowrance, E. W. 1949. Variability and growth of the opossum skeleton. Journal of Morphology 85:569–93.

Lowrance, E. W. 1957. Correlations of certain ponderal and linear skeletal measurements with skull weight and skull length in the opossum. Anatomical Record 128:69–76.

Lustick, S., and D. D. Lustick. 1972. Energetics in the opossum, *Didelphis marsupialis virginiana*. Comparative Biochemistry and Physiology 43:643–47.

Manuel, B. J. 1977. Occurrence of the opossum on the Tug Hill Plateau. New York Fish and Game Journal 24:98.

McDonald, I. R. 1977. Adrenocortical function in marsupials. Pages 345–78 *in* B. Stonehouse and D. Gilmore, eds. The biology of marsupials. University Park Press, Baltimore.

McGrady, E., Jr. 1938. The embryology of the opossum. American Anatomical Memoir 16:1–233.

McKeever, S. 1958. Reproduction in the opossum in southwestern Georgia and northwestern Florida. Journal of Wildlife Management 22:303.

McKeller, M. R., and J. C. Pérez. 2002. The effects of western diamondback rattlesnake (*Crotalus atrox*) venom on the production of antihemorrhagins and/or antibodies in the Virginia opossum (*Didelphis virginiana*). Toxicon 40:427–439.

McManus, J. J. 1967. Observations on sexual behavior of the opossum, *Didelphis marsupialis*. Journal of Mammalogy 48:486–87.

McManus, J. J. 1969. Temperature regulation in the opossum, *Didelphis marsupialis virginiana*. Journal of Mammalogy 50:550–58.

McManus, J. J. 1970. Behavior of captive opossums, *Didelphis marsupialis virginiana*. American Midland Naturalist 84:144–69.

McManus, J. J. 1971. Activity of captive *Didelphis marsupialis*. Journal of Mammalogy 52:846–48.

McManus, J. J. 1974. *Didelphis virginiana*. Mammalian Species 40:1–6.

McNab, B. K. 1978. The comparative energetics of Neotropical marsupials. Journal of Comparative Physiology, B 125:115–28.

Miller, G. S. 1899. Preliminary list of New York mammals. Bulletin of the New York State Museum 6:271–390.

Moore, C. R., and D. Bodian. 1940. Opossum pouch young as experimental material. Anatomical Record 76:319–27.

Moore, J. C. 1955. Opossum takes refuge under water. Journal of Mammalogy 36:559–61.

Morlan, H. B. 1952. Host relationships and seasonal abundance of some southwest Georgia ectoparasites. American Midland Naturalist 48:74–93.

Morrison, P. R., and J. H. Petajan. 1962. The development of temperature regulation in the opossum, *Didelphis marsupialis virginiana*. Physiological Zoology 35:52–65.

Nardone, R. M., C. G. Wilber, and X. J. Musacchia. 1955. Electrocardiogram of the opossum during exposure to cold. American Journal of Physiology 181:352–56.

Nesslinger, C. L. 1956. Ossification centers and skeletal development in the postnatal Virginia opossum. Journal of Mammalogy 37:382–94.

Osgood, F. L. 1938. The mammals of Vermont. Journal of Mammalogy 19:435–41.

Packard, R. L., and F. W. Judd. 1968. Comments on some mammals from western Texas. Journal of Mammalogy 49:535–38.

Patton, J. L., S. F. dos Reis, and M. N. F. da Silva. 1996. Relationships among didelphid marsupials based on sequence variation in the mitochondrial cytochrome *b* gene. Journal of Mammalian Evolution 3:3–29.

Pence, D. B. 1973. Notes on two species of hypopial nymphs of the genus *Marsupialichus* (Arcarina: Glycyphagidae) from mammals in Louisiana. Journal of Medical Entomology (Honolulu) 10:329–32.

Pence, D. B., and M. D. Little. 1972. *Anatrichosoma buccalis* sp. n. (Nematoda: Trichosomoididae) from the buccal mucosa of the common opossum, *Didelphis marsupialis* L. Journal of Parasitology 58:767–73.

Petajan, J. H., and P. R. Morrison. 1962. Physical and physiological factors modifying the development of temperature regulation in the opossum. Journal of Experimental Zoology 149:45–57.

Peterson, R. L. 1966. The mammals of eastern Canada. Oxford University Press, Toronto.

Peterson, R. L., and S. C. Downing. 1956. Distribution records of the opossum in Ontario. Journal of Mammalogy 37:431–35.

Petrides, G. A. 1949. Sex and age determination in the opossum. Journal of Mammalogy 30:364–78.

Pippitt, D. D. 1976. A radiotelemetric study of the winter energetics of the opossum *Didelphis virginiana* Kerr. Ph.D. Dissertation, University of Kansas, Lawrence.

Plakke, R. K. 1970. Urea, electrolyte and total solution excretion following water deprivation in the opossum (*Didelphis marsupialis virginiana*). Comparative Biochemistry and Physiology 34:325–32.

Plakke, R. K., and E. W. Pfeiffer. 1965. Influence of plasma urea on urine concentration in the opossum (*Didelphis marsupialis virginiana*). Nature 207:866–67.

Potkay, S. 1970. Diseases of the opossum (*Didelphis marsupialis*): A review. Laboratory Animal Care 20:502–11.

Potkay, S. 1977. Diseases of marsupials. Pages 415–506 *in* D. Hunsaker, II, ed. The biology of marsupials. Academic Press, New York.

Pray, L. 1921. Opossum carries leaves with its tail. Journal of Mammalogy 2:109–10.

Preble, N. A. 1942. Notes on the mammals of Morrow County, Ohio. Journal of Mammalogy 23:82–86.

Prestwood, A. 1976. *Didelphostrongylus hayesi* gen. et sp. n. (Melastrongyloidea: Filaroididae) from the opossum, *Didelphis marsupialis*. Journal of Parasitology 62:272–75.

Rand, A. L., and P. Host. 1942. Results of the Archbold Expeditions No. 45. Mammal notes from Highland County, Florida. Bulletin of the American Museum of Natural History 80:1–21.

Raven, H. C. 1929. A case of matricide in the opossum. Journal of Mammalogy 10:168.

Reig, O. A., A. L. Gardner, N. O. Bianchi, and J. L. Patton. 1977. The chromosomes of the Didelphidae (Marsupialia) and their evolutionary significance. Biological Journal, Linnaean Society of London 4:191–216.

Reynolds, H. C. 1942. A contribution to the life history and ecology of the opossum *Didelphis virginiana* (Kerr), in central Missouri. M.S. Thesis, University of Missouri, Columbia. [Not seen, cited in Petrides, 1949.]

Reynolds, H. C. 1945. Some aspects of the life history and ecology of the opossum in central Missouri. Journal of Mammalogy 26:361–79.

Reynolds, H. C. 1952. Studies on reproduction in the opossum (*Didelphis virginiana*). University of California Publications in Zoology 52:223–84.

Reynolds, H. C. 1953. The opossum. Scientific American 188:88–94.

Ryser, J. 1992. The mating system and male mating success of the Virginia opossum (*Didelphis virginiana*) in Florida. Journal of Zoology (London) 228:127–39.

Ryser, J. 1995. Activity, movement and home range of Virginia opossums (*Didelphis virginiana*) in Florida. Bulletin of the Florida Museum of Natural History, Biological Sciences 38:177–94.

Sanderson, G. C. 1961. Estimating opossum populations by marking young. Journal of Wildlife Management 25:20–27.

Sandidge, L. L. 1953. Food and dens of the opossum (*Didelphis virginiana*) in northeastern Kansas. Transactions of the Kansas Academy of Science 56:97–106.

Sands, J. L. 1960. The opossum in New Mexico. Journal of Mammalogy 41:393.

Scheffer, J. B. 1943. The opossum settles in Washington state. Murrelet 24:27–28.

Schneider, L. K. 1977. Marsupial chromosomes, cell cycles, and cytogenetics. Pages 51–94 *in* D. Hunsaker II, ed. The biology of marsupials. Academic Press, New York.

Schroeder, G. 2002. Fur market report. Fur-Fish-Game 99(2):49–53.

Sealander, J. A. 1979. A guide to Arkansas mammals. River Road Press, Conway, AR.

Seidensticker, J., M. A. O'Connell, and A. J. T. Johnsingh. 1987. Virginia opossum. Pages 246–63 *in* M. Novak, J. A. Baker, M. E. Obbard, and B. Malloch, eds. Wild furbearer management and conservation in North America. Ontario Trappers Association, Toronto.

Selenka, E. 1887. Studien uber Entwicklungsgeschichte. Vol. 4, Das Opossum (*Didelphys virginiana*). C. W. Kreidel, Wiesbaden, Germany.

Setchell, P. P. 1977. Reproduction in male marsupials. Pages 411–58 *in* B. Stonehouse and D. Gilmore, eds. *The biology of marsupials.* University Park Press, Baltimore.

Severinghaus, C. W. 1975. Occurrence of the opossum in the central Adirondacks. New York Fish and Game Journal 22:80.

Shaver, E. L. 1962. The chromosomes of the opossum, *Didelphis virginiana*. Canadian Journal of Genetics and Cytology 4:62–68.

Shirer, H. W., and H. S. Fitch. 1970. Comparison from radiotracking of movements and denning habits of the raccoon, striped skunk, and opossum in northeastern Kansas. Journal of Mammalogy 51:491–503.

Smiley, D., Jr. 1938. An opossum in New York State feels the effects of winter. Journal of Mammalogy 19:499.

Smith, J. 1612. A map of Virginia. J. Barnes, Oxford, England.

Smith, J. H. 1935. The opossum in Kent County, Ontario. Canadian Field-Naturalist 49:109.

Smith, L. 1941. An observation on the nest-building behavior of the opossum. Journal of Mammalogy 22:201–2.

Sperry, C. C. 1933. Opossum and skunk eat bats. Journal of Mammalogy 14:152–53.

Stieglitz, W. O., and W. D. Klimstra. 1962. Dietary pattern of the Virginia opossum, *Didelphis marsupialis virginianus* Kerr, late summer–winter, southern Illinois. Transactions of the Illinois Academy of Science 55:198–208.

Stone, W. B., Jr., E. Parks, B. L.Weber, and F. V. Parks. 1972. Experimental transfer of sarcoptic mange from red foxes and wild canids to captive wildlife and domestic animals. New York Fish and Game Journal 19:1–11.

Stonehouse, B., and D. Gilmore, eds. 1977. The biology of marsupials. University Park Press, Baltimore.

Stoner, D. 1939. Remarks on abundance and range of the opossum. Journal of Mammalogy 20:250–51.

Stout, J., and D. E. Sonenshine. 1974. Ecology of an opossum population in Virginia, 1963–69. Acta Theriologica 19:235–45.

Stuewer, F. W. 1943. Raccoons: Their habits and management in Michigan. Ecological Monographs 13:203–58.

Stumpf, W. A., and C. O. Mohr. 1962. Linearity of home ranges of California mice and other animals. Journal of Wildlife Management 26:149–54.

Sunquist, M. E., and J. F. Eisenberg. 1993. Reproductive strategies of female *Didelphis.* Bulletin of the Florida Museum of Natural History, Biological Sciences 36:109–40.

Tamar, H. 1961. Taste reception in the opossum and the bat. Physiological Zoology 34:86–91.

Taube, C. M. 1947. Food habits of Michigan opossums. Journal of Wildlife Management 11:97–103.

Trivers, R. L., and D. E. Willard. 1973. Natural selection of parental ability to vary the sex ratio of offspring. Science 179:90–92.

Tyndale-Biscoe, and C.H. 1973. Life of marsupials. American Elsevier, New York.

Tyndale-Biscoe, C. H., and R. B. Mackenzie. 1976. Reproduction in *Didelphis marsupialis* and *D. albiventris* in Colombia. Journal of Mammalogy 57:249–65.

Tyson, E. 1698. Cariqueya *seu* marsupiale Americanum. Or, The anatomy *of an* opossum. Philosophical Transactions of the Royal Society of London 20:105–64.

Tyson, E. 1704. Carigueya *seu* marsupiale Americanum masculum. Or, The anatomy of a male opossum. Philosophical Transactions of the Royal Society of London 24:1565–75.

VanDruff, L. W. 1971. The ecology of the raccoon and opossum, with emphasis on their role as waterfowl nest predators. Ph.D. Dissertation, Cornell University, Ithaca, NY.

Verts, B. J. 1963. Movements and populations of opossums in a cultivated area. Journal of Wildlife Management 27:127–29.

Washburn, S. L. 1946. The sequence of epiphysial union in the opossum. Anatomical Record 95:353–63.

Werner, R. M., and J. A. Vick. 1977. Resistance of the opossum (*Didelphis virginiana*) to envenomation by snakes of the family Crotalidae. Toxicon 15:29–33.

Whitaker, J. O., Jr., G. S. Jones, and R. J. Goff. 1977. Ectoparasites and food habits of the opossum, *Didelphis virginiana,* in Indiana. Proceedings of the Indiana Academy of Science 86:501–7.

Wilber, C. G. 1955. Electrocardiographic studies on the opossum. Journal of Mammalogy 36:284–86.

Wilber, C. G., and G. H. Weidenbacher. 1961. Swimming capacity of some wild mammals. Journal of Mammalogy 42:428–29.

Wiseman, G. L., and G. O. Hendrickson. 1950. Notes on the life history of the opossum in southeast Iowa. Journal of Mammalogy 31:331–37.

Wolff, J. O., and D. F. Holleman. 1978. Use of radioisotope labels to establish genetic relationships in free-ranging small mammals. Journal of Mammalogy 59:859–60.

Wood, J. E. 1954. Food habits of furbearers of the upland post oak region in Texas. Journal of Mammalogy 35:406–15.

Wood, J. E., and Odum, E. P. 1964. A nine-year history of furbearer populations on the AEC Savannah River Plant area. Journal of Mammalogy 45:540–51.

Worth, C. B. 1975. Virginia opossums (*Didelphis virginiana*) as disseminators of the commom persimmon (*Diospyros virginiana*). Journal of Mammalogy 56:517.

Wright, D. D. 1989. Mortality and dispersal of juvenile opossums, *Didelphis virginiana.* M.S. Thesis, University of Florida, Gainesville.

Wright, D. D., J. T. Ryser, and R. A. Kiltie. 1995. First-cohort advantage hypothesis: A new twist on facultative sex ratio adjustment. American Naturalist 145:133–45.

Yeager, L. H. 1936. Winter daytime dens of opossums. Journal of Mammalogy 17:410–11.

Alfred L. Gardner, U.S. Geological Survey, Patuxent Wildlife Research Center, National Museum of Natural History, Smithsonian Institution, Washington, D.C. 20560-0111. Email: gardner.alfred@nmnh.si.edu.

Melvin E. Sunquist, Department of Wildlife Ecology and Conservation, University of Florida, Gainesville, Florida 32611-0430. Email: sunquist@mail.ifas.ufl.edu.

2

Moles

Talpidae

Gregory D. Hartman
Terry L. Yates

NOMENCLATURE

ORDER. Insectivora
FAMILY. Talpidae

Seven species of moles are recognized in North America: *Condylura cristata,* the star-nosed mole; *Neürotrichus gibbsii,* the shrew mole; *Parascalops breweri,* the hairy-tailed mole; *Scalopus aquaticus,* the eastern mole; *Scapanus latimanus,* the broad-footed mole; *Scapanus orarius,* the coast mole; and *Scapanus townsendii,* Townsend's mole.

Moles of the family Talpidae are diverse and widespread in the United States, ranging from the tiny shrew mole of the Pacific Northwest to the bizarre, semiaquatic, star-nosed mole of New England. The meandering surface runways made by these animals are familiar sights to many Americans, yet few people have more than vague ideas concerning the nature of the creatures that inhabit them. Relatively speaking, little is known scientifically about these animals. Moles likely are the least studied and understood major component of the North American mammalian fauna.

DISTRIBUTION

Modern talpids are distributed over much of the northern hemisphere's temperate region and also are found in parts of Southeast Asia (Yates 1978; Gorman and Stone 1990). Of the five genera represented in North America, *Condylura, Parascalops,* and *Scalopus* occur east of the Rocky Mountains and *Neürotrichus* and *Scapanus* are restricted to areas west of the Rocky Mountains (Hall 1981).

The eastern mole, *Scalopus aquaticus,* has the widest range of any North American mole, occurring throughout much of the eastern United States where soils are favorable (Fig. 2.1). *S. aquaticus* ranges from northern Tamaulipus, Mexico, northward to southeastern South Dakota, Minnesota, and Michigan, eastward to Massachusetts and much of southern New England, and south to the southernmost tip of Florida (Yates and Schmidly 1978; Hall 1981). Lowery (1974) reported that the eastern mole occurs throughout the upland portions of Louisiana, but is not common in coastal situations. Two relict populations have been reported from northern Coahuila (Baker 1951) and Presidio County, Texas (Allen 1891), yet neither has been verified since originally reported.

The star-nosed mole, *Condylura cristata,* and the hairy-tailed mole, *Parascalops breweri,* occur in the northeastern United States and southeastern Canada, where they are sympatric in many areas (Figs. 2.2 and 2.3). The range of *C. cristata* is more extensive than that of *P. breweri*. In the northern part of its range, the star-nosed mole is found from extreme eastern Manitoba and Minnesota to as far northeast as Labrador and Nova Scotia; this species ranges southeastward through much of Wisconsin, northern Indiana and Ohio, along the Atlantic coast as far south as southeastern Georgia, and in the Appalachian Mountains to eastern Tennessee and western South Carolina (Petersen and Yates 1980; Hall 1981; Beane 1995; Laerm et al. 1997). The hairy-tailed mole ranges from southern Ontario across southern Quebec, possibly into New Brunswick (Peterson 1966), then south to central Ohio and Connecticut and along the Appalachian Mountains to western North Carolina (Hallett 1978; Hall 1981).

The remaining four species of North American moles are restricted to areas west of the Rocky Mountains. The broad-footed mole, *Scapanus latimanus,* is found from southern Oregon south along the Pacific coastal regions to the San Pedro Martir Mountains in Baja California (see Fig. 2.3); its range also extends throughout much of northern California east into western Nevada, then south in California east of the Central Valley (Yates 1978; Hall 1981; Verts and Carraway 2001). The other two species of *Scapanus* have more northerly distributions and occur sympatrically throughout much of their ranges. The coast mole (*S. orarius*) occurs from southwestern British Columbia through the western portions of Washington and Oregon to coastal northwestern California (see Fig. 2.2). Coast moles also occur in parts of eastern Washington and Oregon and in one area of extreme west-central Idaho (Hall 1981; Hartman and Yates 1985). Townsend's mole (*S. townsendii*) is more restricted in distribution (see Fig. 2.3), and is found from extreme northwestern California along the coastal regions of Oregon and Washington to extreme southwestern British Columbia (Yates 1978; Hall 1981; Carraway et al. 1993). Individuals typically are lowland animals, with the exception of one population that occurs in subalpine meadows in the Olympic Mountains of Washington State (Johnson and Yates 1980). The shrew-mole, *Neürotrichus gibbsii,* is restricted to western regions of North America from Santa Cruz County, California, north through western Oregon and Washington to southern British Columbia (see Fig. 2.1); a small population of shrew moles also is known from Destruction Island, Jefferson County, Washington (Dalquest 1948; Hall 1981; Carraway and Verts 1991).

DESCRIPTION

Many of the unique morphological features of North American moles can be attributed to their being highly fossorial (adapted for digging) and leading a largely subterranean existence (Jackson 1915; Slonaker 1920; Campbell 1939; Dalquest 1948; Gorman and Stone 1990). Although heterogeneous as a family, all talpids in North America possess numerous characteristics that distinguish them from their closest North American relatives, the shrews (family Soricidae). Many of the "moles" caught by lay persons or family pets and then brought to mammalogists for identification prove to be shrews.

Externally, moles present a streamlined body well structured for life underground. Unlike shrews, there are no ear pinnae, and the external appendages are short and remain close to the body (Fig. 2.4), all features that reduce drag when burrowing or moving about in the confines of a tunnel. The eyes are minute and probably useless except for light detection. In *Neürotrichus gibbsii* and *Scalopus aquaticus,* the eyes are completely covered with skin. In most species, the forepaws are broader than they are long and the terminal phalanges of the forefeet are bifurcate, providing additional support for stout claws, which are used in digging. The zygomata and auditory bullae, which are absent in shrews, always are present in moles. In addition, the enamel covering the teeth of moles is white and lacks the reddish-brown or mahogany

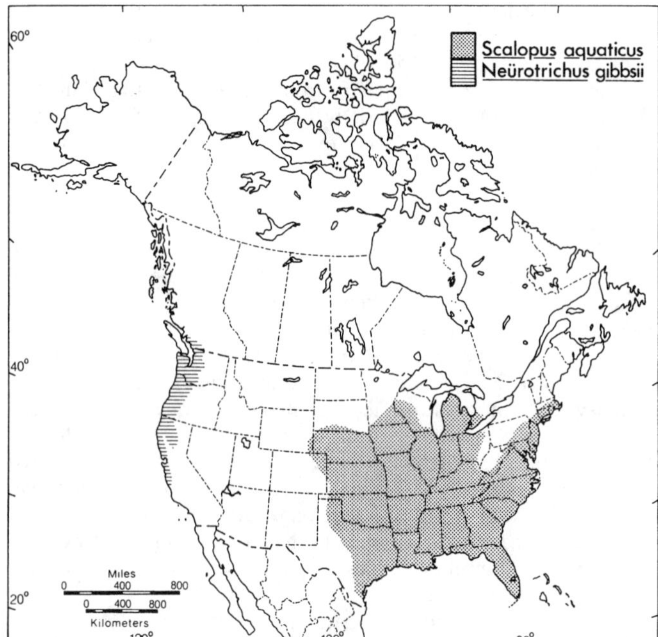

FIGURE 2.1. Distribution of the eastern mole (*Scalopus aquaticus*) and shrew mole (*Neürotrichus gibbsii*).

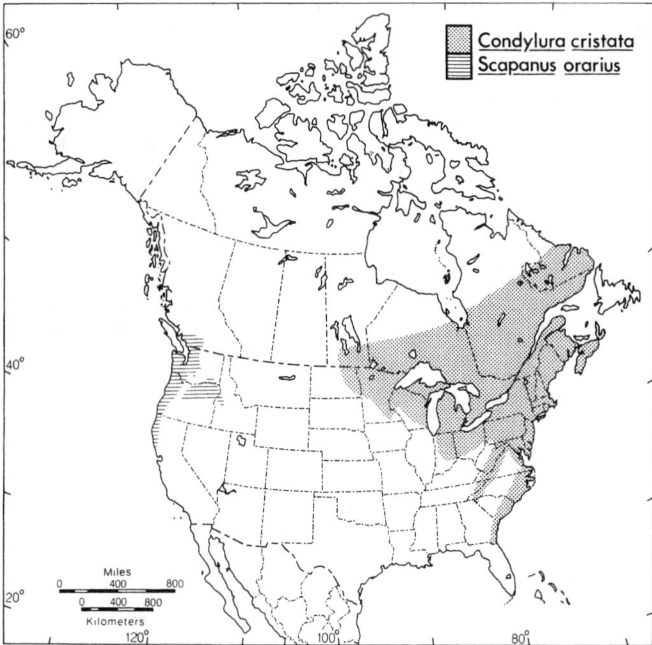

FIGURE 2.2. Distribution of the star-nosed mole (*Condylura cristata*) and coast mole (*Scapanus orarius*).

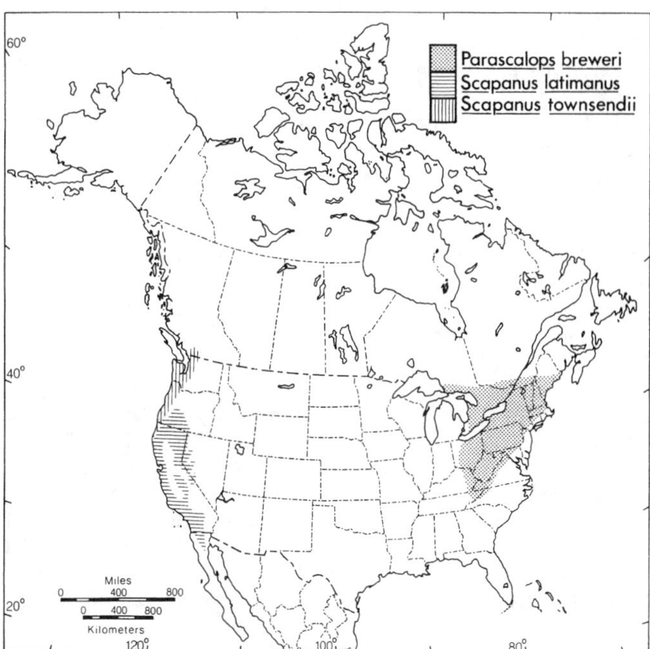

FIGURE 2.3. Distribution of the hairy-tailed mole (*Parascalops breweri*), broad-footed mole (*Scapanus latimanus*), and Townsend's mole (*Scapanus townsendii*).

FIGURE 2.4. Eastern mole (*Scalopus aquaticus*), showing typical body form.

staining that is characteristic of all North American shrews. A penis bone (*os baculum*) is present in *Parascalops breweri* (Eadie 1947), *Scapanus orarius, S. townsendii* (Maser and Brown 1972), and some but not all males of *Scalopus aquaticus* (G. D. Hartman, pers. obs.), but is lacking in *Condylura cristata* (Eadie 1948b) and *Neürotrichus gibbsii* (Eadie 1951). The form of the penis bone of moles is highly variable, even within a species.

North American moles often are confused with another group of fossorial subterranean mammals, the pocket gophers (Rodentia: Geomyidae). Although moles and gophers have numerous similarities as a result of their subterranean habits, the two groups exhibit striking differences in their skeletomuscular adaptations for burrowing. Pocket gophers and most other fossorial mammals dig with the forepaws held beneath the body. Talpids, on the other hand, have evolved a system of lateral-stroke burrowing that is unique among Recent mammals; this has been accompanied by substantial modification of the pectoral girdle and bones of the forelimbs. The pelvic girdle is relatively small, narrow, and unmodified. A most unusual feature of the pectoral girdle in moles is the presence of a humeroclavicular joint. The clavicle is a small cubical bone that articulates directly with the humerus instead of the scapula. The humeri are massive rectangular bones unlike those of other mammalian groups. The humerus provides a large surface area for the attachment of the well-developed musculature used in digging. The scapulae of talpids are extremely elongated bones: Their lengths are as much as 4 times the greatest width of the blade (Gaughran 1954). The unusual pectoral girdle and associated muscles obscure the short neck region and cause the palmar surfaces of forepaws to be rotated laterally (see Fig. 2.4). Many talpids lack the ability to bring the forefeet under the body as is seen in a more standard mammalian stance. The forepaws are broad and covered with a thick layer of skin. Many species have the surface area of the forepaws increased by the presence of a semicircular sesamoid bone (*os falciforme*), which supports an additional flap of skin. It has been suggested that the unusual anatomy of the forelimbs originally may have arisen as an adaptation for swimming and then secondarily for tunneling (Campbell 1939; Whidden 1999); however, other authors have felt that the semiaquatic lifestyle of star-nosed moles and some European talpids was acquired secondarily (Grand et al. 1998; Campbell et al. 1999).

Although all North American species possess the same general adaptations described above, *Neürotrichus gibbsii* and *Condylura cristata* have undergone various specializations related to lifestyles other than purely subterranean. The star-nosed mole is a semiaquatic species characterized by a ring that normally consists of 22 fleshy appendages around the nose. Perhaps 5% or more of individuals of *C. cristata* have stars consisting of other than the modal number of 22 rays (Catania et al. 1999). The tail of a star-nosed mole is relatively longer than that of the more purely subterranean species and its average length is approximately the same as that of the body; although they are not webbed, the hind feet are larger than in other North American moles, a modification useful in swimming.

The shrew mole is a semisubterranean species that often moves about above ground (Dalquest and Orcutt 1942). The pectoral girdle and forefeet are specialized for digging, but not as much as in other, more purely subterranean talpids. Unlike other North American moles, the reproductive tract of male shrew moles includes paired ampullary glands, which typically occur in shrews (Eadie 1951). The forefeet are longer than they are wide, a characteristic that causes shrew moles to be mistaken for shrews on many occasions.

The overall size of moles is highly variable. Adult weights range from 10 g for *Neürotrichus gibbsii* (Dalquest and Orcutt 1942) to 171 g for *Scapanus townsendii* (Pedersen 1963). *Scapanus latimanus* and *Scalopus aquaticus* exhibit the most extensive amount of geographic variation in size (Palmer 1937; Yates 1978). Total length in *S. latimanus* females ranges from 135 mm in specimens from Baja California to 193 mm in specimens from northern California (Yates 1978); males average slightly larger. A similar latitudinal variation occurs in *S. aquaticus*.

Many species of talpid in North America exhibit significant sexual size dimorphism; on average, males are larger than females. The typical secondary sexual variation found in moles is illustrated in Table 2.1 for two samples of *Scalopus aquaticus* from Texas. Interestingly, the two species of mole that are not strictly subterranean, *Condylura cristata* and *Neürotrichus gibbsii,* do not exhibit significant size differences between the sexes.

Pelage. A striking characteristic of the pelage of moles is its soft, velvetlike appearance and texture. The velvet texture is more pronounced in more purely subterranean species, where the hairs are nearly equal in length and no distinct underfur is present (Jackson 1915). In *Neürotrichus gibbsii* and *Condylura cristata,* some of the hairs are thicker, longer, and coarser than others, producing a less silky appearance. Jackson (1915) reported that the basal pelage consisted of a series of transverse vermiculations (most pronounced in *Scalopus aquaticus* and species of *Scapanus,* least in *Neürotrichus gibbsii*), which were due to structural as well as chromatic differences. Each hair consisted of alternating normally pigmented cylindrical sections 1–2 mm long with finer flat sections 0.2–0.5 mm long that were unpigmented. The flat sections act as hinges, allowing the hair to be bent forward or backward. This condition allows moles to move forward or backward in their tunnels with relative ease.

Molt lines, appearing as a sharp line of demarcation between old and new pelage, often are seen in all species of North American moles. The time of molt appears to be related to climatic factors and is highly variable among genera (Jackson 1915). In most species there is a molt twice a year roughly corresponding to spring and autumn.

Brown, yellow, orange, and olivaceous tints sometimes occur on the snout, chin, wrist, venter, and other parts of the bodies of moles. This led may authors to characterize, partially or completely, various subspecific and specific forms on the basis of these chromatic variations (Yates and Schmidly 1977). Among the North American genera, pelage spotting is most common in *Condylura, Parascalops, Scalopus,* and *Scapanus* and usually is more pronounced in males, especially during the breeding season. With the exception of white spots and lines, which Jackson (1915) referred to as "partial albinism," Eadie (1954) regarded these chromatic variations as temporary stains produced by sudoriferous and perineal glands, not genetic variation in pigmentation. However, there is convincing evidence that pelage spotting observed on individuals of *Scapanus townsendii* and possibly other species may not be due solely to staining by sudoriferous glands (Carraway and Verts 1991).

TABLE 2.1. Analysis of variance of anatomical characteristics between males and females for two samples of *Scalopus aquaticus*

	Conroe, Texas					Rockport, Texas				
	Males		Females			Males		Females		
Variate	*N*	Mean	*N*	Mean	*F*	*N*	Mean	*N*	Mean	*F*
Total length	19	151.8	12	140.2	30.7*	8	140.0	10	131.0	11.8*
Tail length	19	24.4	12	23.2	1.1	8	24.2	10	21.6	12.5*
Hind foot length	19	19.5	12	18.5	7.1*	10	17.6	10	16.3	7.8*
Width of forepaw	19	16.3	12	14.9	9.8*	10	14.8	12	13.4	13.3*
Length of forepaw	19	21.1	12	19.6	25.8*	10	18.4	12	17.9	1.2
Greatest length of skull	19	32.8	12	31.7	20.2*	10	31.0	12	30.3	6.3**
Basilar length	19	27.3	12	26.3	16.3*	10	25.9	12	24.8	17.3*
Mastoidal breadth	19	17.3	12	16.8	10.3*	10	16.6	12	16.2	10.7*
Interobital breadth	19	7.1	12	7.0	0.8	10	6.8	12	6.6	2.0
Length of maxillary toothrow	19	10.1	12	9.7	16.5*	10	10.2	12	9.8	9.0*
Length of palate	19	14.2	12	13.7	11.7*	10	14.1	12	13.5	6.2**
Width across M2–M2	19	8.9	12	8.6	11.3*	10	9.6	12	8.7	3.9***
Width across canines	19	3.8	12	3.7	3.6***	10	3.8	12	3.6	3.9***
Depth of skull	19	9.8	12	9.4	22.4*	10	9.2	12	9.2	0.1

SOURCE: Adapted from Yates and Schmidly 1977.
NOTE: mean values in millimeters.
*$p < 0.01$. **$p < 0.05$. ***$p < 0.10$.

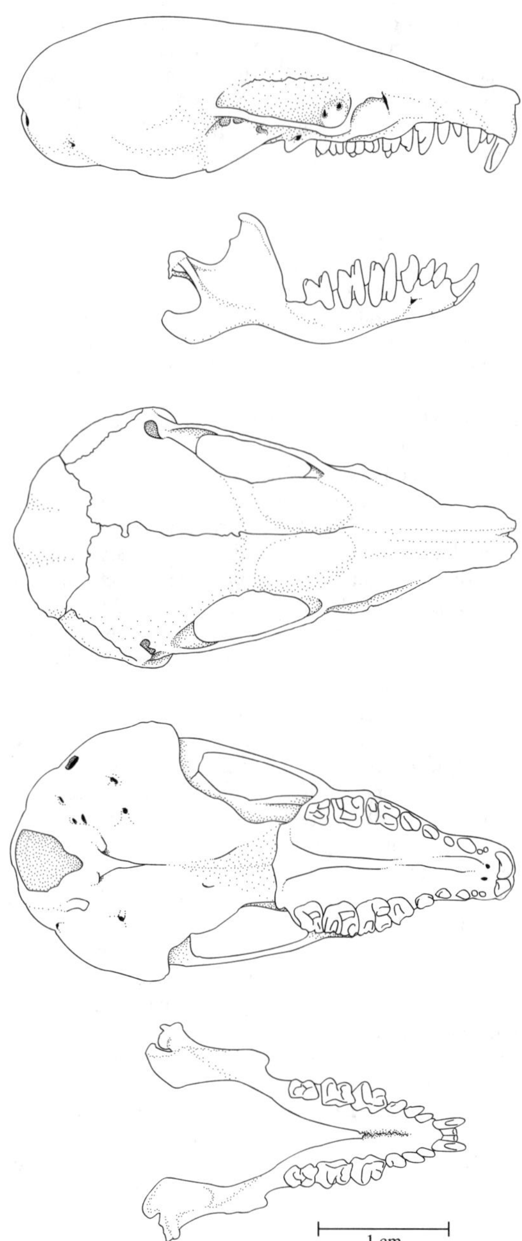

FIGURE 2.5. Skull of the eastern mole (*Scalopus aquaticus*). From top to bottom: lateral view of cranium, lateral view of mandible, dorsal view of cranium, ventral view of cranium, dorsal view of mandible.

Skull and Dentition. The skulls and dentition of all talpids, although differing in detail, exhibit the same basic plan (Fig. 2.5). The cranium is dorsoventrally compressed (less so in *Condylura*) and the rostrum is long and narrow in most species. Sutures ossify early in most species, making age estimation using suture closure unreliable. Ossification of the maxillary bones occurs only in adults, and the roots of the upper molars usually are exposed in immature specimens; however, the rate at which this ossification occurs varies among individuals. The zygomatic arch is complete in all species, but lacks jugal bones. The auditory bullae, although present, may be complete (*Scalopus, Scapanus*) or incomplete (*Condylura, Neürotrichus, Parascalops*).

The first upper incisor is flattened in all species and lacks the elongated crown that is characteristic of shrew incisors. The number of teeth is 36 in *Neürotrichus gibbsii* and *Scalopus aquaticus* and 44 in all other North American species. Some individuals of *S. aquaticus* have 1 or 2 rudimentary mandibular teeth in addition to the more typical

TABLE 2.2. Summary of numbers and forms of chromosomes for selected moles

Taxon	Males	Females	Biarmed	Acrocentric	2*N*	FN[a]
Condylura cristata (Meylan 1968)					34	64
Neürotrichus gibbsii (R. M. Brown and Waterbury 1971)	0	1	38	0	38	72
Parascalops breweri (Gropp 1969)	2	0	30	2	34	62
Scalopus aquaticus (Yates and Schmidly 1975)	22	1	32	0	34	64
Scapanus latimanus (Lynch 1971)	1	1	32	0	34	64
Talpa europaea (Gropp 1969)	10	9	32	0	34	64
Urotrichus talpoides (Tsuchiya 1979)					34	

[a]FN, Fundamental number.

compliment of 36 total teeth. These rudimentary teeth correspond to C and/or P_1 (Conaway and Landry 1958), generally do not break through the gums, but can be seen easily in carefully cleaned skulls. The dental formula for North American talpids is I3/2–3, C1/0–1, P2–4/2–4, M3/3. All talpids have premolars that are sharply differentiated from the molars, and the upper molars always are dilambdodont.

GENETICS

The secretive subterranean existence of many moles has made it particularly difficult to acquire sufficient numbers of specimens for detailed genetic studies. Although several authors (Meylan 1968; Gropp 1969; Brown and Waterbury 1971; Lynch 1971; Yates and Schmidly 1975; Tsuchiya 1979) have reported on the karyology of various species of talpid, knowledge of the karyotypes of moles is limited for the most part to fundamental and diploid numbers of a few specimens with only implied homology of elements (Table 2.2). The only published C- and G-banded karyotypes of North American moles were by Yates et al. (1976).

Cytological investigations of many fossorial subterranean rodents (such as *Geomys, Thomomys, Spalax,* and *Ctenomys*) have revealed substantial chromosomal variation and polymorphism within and among genera and species. Moles, however, appear to be karyotypically conservative (Yates and Schmidly 1975). All North American species have a diploid number (2*N*) of 34, except *Neürotrichus gibbsii,* which has a diploid number of 38. It is noteworthy that many mole species from Europe and Japan also have 34 chromosomes. The fundamental number (FN) in North American species ranges from 62 in *Parascalops breweri* to 72 in *N. gibbsii* (see Table 2.2). No intraspecific karyotypic difference has been published for any North American species. We have, however, observed the occurrence of an acrocentric pair of chromosomes in specimens of *Scalopus aquaticus* from Gainesville, Florida (G. D. Hartman and T. L. Yates, unpublished data).

The sex chromosomes in all species examined are the expected XX in females and XY in males. The Y chromosome is minute and appears to be similar in all talpid species. Yates et al. (1976) reported a secondary constriction corresponding to the nucleolar organizing region on a pair of autosomes in *Scalopus aquaticus*. Similar constrictions have been reported for *Parascalops breweri* and *Talpa europaea* (Gropp 1969). The presence of a heterochromatic polymorphism has been found in *S. aquaticus* (Yates et al. 1976), along with large amounts of heterochromatin (Fig. 2.6).

Reasons for the uniformity of mole karyotypes are difficult to explain. It may be that genetic factors affect the symmetry of the karyotype. From an evolutionary standpoint, talpids are a much older group than are subterranean rodents such as geomyids or ctenomyids.

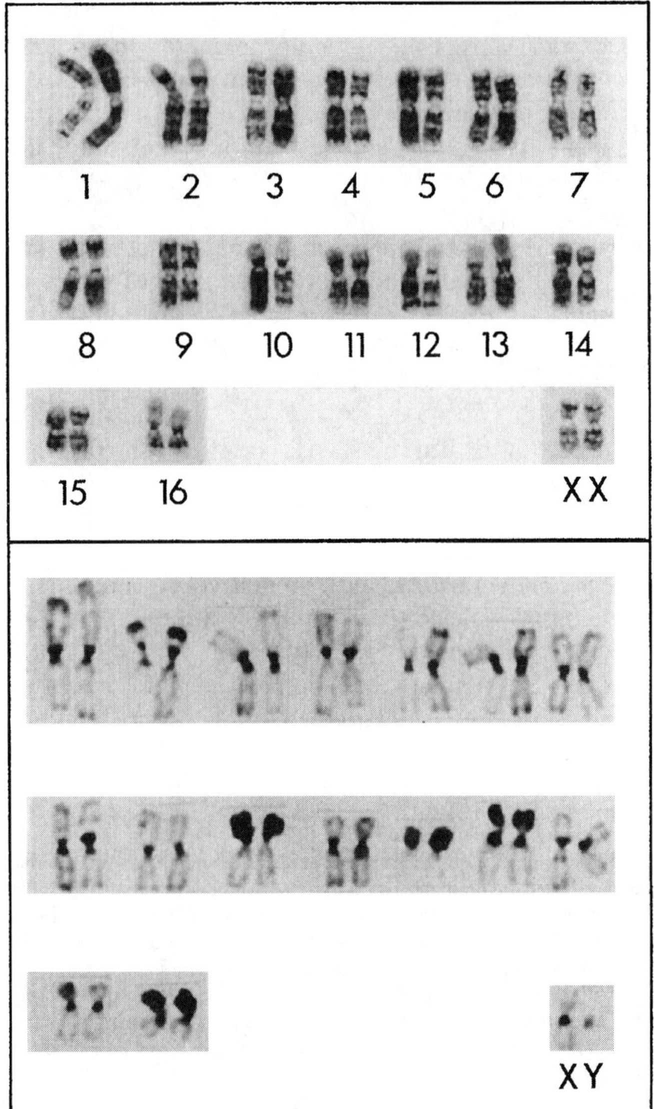

FIGURE 2.6. (Top) G-banded and (bottom) C-banded karyotype of the Eastern mole (*Scalopus aquaticus*).

Bickham and Baker (1979) proposed that because Talpidae is such an old group, the karyotype may have become canalized.

Estimates of genic variability based on protein electrophoresis have been published for all species of North American moles (Yates and Greenbaum 1982; Tolliver et al. 1985; Moore 1986; Yates and Moore 1990; Hartman 1996) (Table 2.3). The only study where population-level variation has been investigated in a North American mole was Hartman's (1996) study of *Scalopus aquaticus* in South Carolina. Even over short distances, he found a great deal of spatial heterogeneity in mean direct-count heterozygosity, the mean number of alleles per locus, and the percentage of polymorphic loci (Table 2.4). Within *S. aquaticus,* genetic variation is greater within the southern portion of the range; populations in the northern portion of the range appear to have low levels of intra- and interpopulation variation, even over great distances (Hartman 1996).

Yates and Moore (1990) presented data suggesting that among talpids, the more purely subterranean species may have lower levels of genetic variation. Some authors have argued that subterranean mammals as a group have levels of genetic variation lower than those of aboveground mammals (Nevo 1979; Nevo et al. 1984, 1990). In pocket gophers, low levels of genetic variation usually are linked to small, patchily distributed populations within which genetic drift and fixation of alleles occur (Patton and Smith 1990). Subterranean animals commonly have been assumed to perceive their environments as stable and fine-grained. Stability of subterranean environments frequently has been invoked to explain low levels of genetic variation reported for moles and other subterranean mammals, but there is little empirical basis for this commonly held belief (Hartman 1996). Data on food habits of *Scalopus aquaticus* throughout the year (Hartman et al. 2000) and the patterns of genetic variation that have been observed for this species (Hartman 1996) suggest that some subterranean environments are heterogeneous both temporally and spatially. In the different studies where allozyme variation in moles has been investigated, different electrophoretic buffer systems were used and different loci, different numbers of loci, and different numbers of individuals were examined. This lack of uniformity among research protocols and a general lack of information on population structure in moles has made it difficult to ensure meaningful comparisons regarding intra- and interspecific or population genetic variation.

Richards et al. (1996) reported that kallikrein (see section on Physiology) from *Scalopus aquaticus* appears to be a member of a single-gene or small-multigene family. Data on immunoreactivity and kininogenase activity both suggest that mole kallikrein more closely resembles that of humans than that of "rat" (presumably a laboratory strain of *Rattus norvegicus* or *R. rattus*). For these reasons, it has been proposed that

TABLE 2.3. Estimates of genetic variation in North American moles

Species	Locality	*N*	N_L	*P*	*H*	Source
Condylura cristata	South Carolina	3	18	0.0	0.000	Tolliver et al. 1985
Condylura cristata	Ohio	1	18	0.0	0.000	Yates and Greenbaum 1982
Condylura cristata	Maine, Massachusetts, Ohio, Vermont	31	23	4.4	0.011	Moore 1986
Neürotrichus gibbsii	Oregon, Washington	10	18	16.7	0.050	Yates and Greenbaum 1982
Neürotrichus gibbsii	Oregon, Washington	18	23	8.7	0.012	Moore 1986
Parascalops breweri	Massachusetts, Ohio, Pennsylvania	24	18	5.6	0.003	Yates and Greenbaum 1982
Parascalops breweri	Massachusetts, Ohio	24	23	0.0	0.000	Moore 1986
Scalopus aquaticus	Kentucky, Massachusetts, Michigan, Tennessee, Texas	15	18	11.1	0.015	Yates and Greenbaum 1982
Scalopus aquaticus	Florida, Texas	15	23	21.7	0.009	Moore 1986
Scalopus aquaticus	Missouri	30	18	33.3	0.050	Tolliver et al. 1985
Scalopus aquaticus	South Carolina	32	18	22.2	0.050	Tolliver et al. 1985
Scalopus aquaticus	South Carolina	382	26	30.8	0.072	Hartman 1996
Scapanus latimanus	California (Fresno)	4	18	5.6	0.014	Yates and Greenbaum 1982
Scapanus latimanus	California (Hilton)	8	18	5.6	0.007	Yates and Greenbaum 1982
Scapanus latimanus	California, Mexico	19	23	13.0	0.007	Moore 1986
Scapanus orarius	California, Oregon	6	18	0.0	0.000	Yates and Greenbaum 1982
Scapanus orarius	California, Oregon	7	23	4.4	0.005	Moore 1986
Scapanus townsendii	Oregon, Washington	12	18	0.0	0.000	Yates and Greenbaum 1982
Scapanus townsendii	Oregon, Washington	17	23	17.4	0.013	Moore 1986

NOTE. *N*, Number of individuals examined; N_L, number of presumptive loci examined; *P*, percentage of polymorphic loci; *H*, mean direct-count heterozygosity.

TABLE 2.4. Measures of genetic variability for *Scalopus aquaticus* collected from 16 sampling grids within a 23-km^2 area in South Carolina

	Grid (*n*)															
	A (21)	B (23)	C (22)	D (54)	E (20)	F (20)	G (21)	H (22)	I (47)	J (20)	K (22)	L (22)	M (20)	O (5)	R (21)	Z (22)
P	26.92	15.38	19.23	23.07	23.07	23.07	19.23	23.07	19.23	19.23	26.92	23.07	15.38	15.38	15.38	23.07
A	1.38	1.23	1.23	1.35	1.27	1.27	1.19	1.27	1.27	1.19	1.35	1.27	1.23	1.15	1.15	1.23
H	0.088	0.069	0.080	0.064	0.087	0.071	0.059	0.072	0.067	0.067	0.096	0.075	0.067	0.062	0.044	0.093

SOURCE: Hartman 1996.
NOTE: *P*, Percentage of polymorphic loci; *A*, mean number of alleles observed per locus; *H*, mean Direct-count heterozygosity.

S. aquaticus may provide advantages over rodent models for biomedical research.

PHYSIOLOGY

Moles generally have metabolic rates equivalent to or slightly higher than those of many other mammals of comparable mass; at the same time, the body temperature T_b of moles is slightly lower than average (Table 2.5). The average rate of secretion of L-thyroxine in eastern moles at 25.6°C has been estimated as 1.96 μg/100 g body weight/day (range = 1.5–2.0) (Leach et al. 1962). In the same study, no statistically significant differences were detected among males and females or adults and juveniles. Mean thyroxine rate increased by almost 68% in moles that had been maintained at 4.4°C for 22 days. A great deal of heat energy is liberated when the powerful forelimb muscles are actively being used for digging. The relatively low body temperatures and metabolic rates of many moles likely reflect the need to dissipate heat within the rather confining quarters of a tunnel system (McNab 1979; Campbell et al. 1999).

Star-nosed moles and the aptly named shrew moles have metabolic rates that are higher than those of other North American moles, both relatively and absolutely (Campbell et al. 1999; Campbell and Hochachka 2000; McIntyre 2000); their body temperatures are higher as well. Metabolic rates in individuals of *Neürotrichus gibbsii* are comparable to those of north-temperate shrews. In star-nosed moles, elevated metabolic rate and body temperature likely aid in thermoregulation when the animals are immersed in water. Most talpids appear to precisely regulate body temperature; an exception to this pattern is seen in *Neürotrichus gibbsii,* where T_b is labile and varies linearly with ambient temperature both within and below the thermoneutral zone (Campbell and Hochachka 2000). Furthermore, there is evidence that shrew moles may exhibit torpor, an energy-saving tactic not known to occur in any other talpid (Campbell and Hochachka 2000).

The zone of thermoneutrality in individuals of *Scalopus aquaticus* extends from about 26°C to 33°C (McNab 1979); for *Condylura cristata,* from 24°C to 33°C (Campbell et al. 1999; McIntyre 2000); and for *Neürotrichus gibbsii,* from 24.9°C to 32°C (Campbell and Hochachka 2000). Brain temperature in individuals of *S. aquaticus* ranges from 33°C to 36°C and typically is highest during sleep and lowest when the animals are digging. When in contact with cooler soil, air, or water, a mole's snout likely allows a great deal of heat loss (Allison and Van Twyver 1970a; Campbell et al. 1999) and may aid in cooling the brain's arterial blood supply (Allison and Van Twyver 1970a).

Pedersen (1963) reported no unusual physiological abnormality in urine analyses performed on Townsend's moles from Tillamook County, Oregon. The results of these analyses are presented in Table 2.6. Albumin was present in the urine in amounts of 3+ according to the qualitative Clinitest.

When underground, moles are subjected to conditions of hypoxia, hypercapnia (high CO_2 levels) (Schaefer and Sadlier 1979), and high levels of humidity (Nevo 1979). As an aid to coping with such environmental conditions, the lungs of moles tend to have large volumes and sometimes can constitute more than 20% of the body weight, the hematocrit tends to be somewhat elevated, and the blood has high O_2-binding affinity and capacity (Bartels et al. 1969; Quilliam et al. 1971; Jelkmann et al. 1981; McIntyre 2000; K. Campbell, University of Manitoba, pers. commun., 2000) (Table 2.7). Large red blood cell (RBC) volume and densely packed hemoglobin within these cells contribute to the increased O_2-carrying capacity of the blood, and in some moles, the blood's high O_2 affinity is linked partly to a reduced interaction of hemoglobin with 2,3-diphosphoglycerate (2,3-DPG) (Jelkmann

TABLE 2.5. Metabolic rates and body temperatures of North American moles

Species	Metabolic Rate (cm^3 O_2/g per hour)	T_b (°C)	Source
Condylura cristata	4.2–4.5		Pearson 1947
	1.84–2.37[a]		Wiegert 1961
	2.25 ± 0.05[b]	37.7 ± 0.05[b]	Campbell et al. 1999
	2.35 ± 0.05[b]	37.7 ± 0.05[b]	McIntyre 2000
Neürotrichus gibbsii	3.94 ± 0.04[b]	38.4 ± 0.02[b]	Campbell and Hochachka 2000
Parascalops breweri	2.70 ± 0.51[c,d]		Jensen 1983
	2.27 ± 0.51[c,e]		Jensen 1983
Scalopus aquaticus	1.6 (1.4–1.8)		Leach et al. 1962
	1.41 ± 0.053[b]	36.0	McNab 1979
Scapanus latimanus	1.25	37.1	Contreras and McNab 1990
Scapanus orarius	1.02		Kenagy and Vleck 1982
	1.33 ± 0.03[b]		McIntyre 2000
Scapanus townsendii	0.82		Kenagy and Vleck 1982

[a]Juvenile animals.
[b]±*SE*.
[c]In soil.
[d]±95% confidence interval.
[e]In artificial tunnels.

TABLE 2.6. Urinalysis of *Scapanus townsendii* from Oregon

Specimen Number	pH	Sugar[a]	Albumin[b]	Miscellaneous
22	8.0	Negative	3	
40	8.0	Negative	4	
42	6.5	Negative	3	
56	6.5	Negative	4	
62	8.0	Negative	4	
97	6.5	Negative	3	Triple phosphate and uric acid crystals
137	7.0	1+	3	
138	6.5	Trace	3	
140	7.0	Trace	3	
199	7.0	2+	4	
200	6.0	Trace	3	

[a]Robert's test.
[b]Clinitest (qualitative).

TABLE 2.7. Characteristics of blood of *Condylura cristata, Scapanus orarius,* and *Scapanus townsendii*

Taxon	Hemoglobin (g/100 cm^3 blood)	Hematocrit[a]	Blood O_2 Capacity (vol %)	WBC	RBC	White Blood Cells PMN	SL	Mono
Scapanus townsendii ($n = 7$)	16.92 ± 1.16	46.43 ± 2.97		5721.43 [2750–8150]	6.12[b] [5.48–7.03]	48[b] [25–70]	49.6[b] [30–74]	2.2[b] [0–7]
Scapanus orarius ($n = 11$)	17.42 ± 0.84	46.75 ± 1.98	23.35 ± 1.12					
Condylura cristata								
Juveniles ($n = 2$)	15.50 ± 0.35	49.92 ± 4.58	20.77 ± 0.47					
Adults ($n = 7$)	17.17 ± 1.24	50.51 ± 2.57[c]	23.01 ± 1.66					

SOURCE: Data from Pedersen 1963 and McIntyre 2000.

NOTE: Values are means $\pm SE$. Values in brackets are ranges. WBC, white blood cells; RBC, red blood cells in millions/mm^3 of blood; PMN, polymorphonuclear neutrophils/100 cells; SL, small lymphs/100 cells; Mono, monocytes/100 cells.

[a] Packed cell volume in percent.

[b] $n = 5$.

[c] $n = 8$.

et al. 1981; K. Campbell, University of Manitoba, pers. commun., 2000).

Whole-body stores of O_2 are relatively higher in moles than in other terrestrial mammals. These large O_2 stores in part are due to high concentrations of myoglobin in skeletal muscle. McIntyre (2000) determined myoglobin concentrations in forelimb, hind limb, and cardiac muscles in individuals of *Condylura cristata, Neürotrichus gibbsii,* and *Scapanus orarius*. The highest concentration of myoglobin was in forelimb muscles, followed by hind limb muscles and then cardiac muscle. Not surprisingly, the concentrations of myoglobin in skeletal muscle of star-nosed moles and coast moles were greater than those in nonaquatic and less fossorial *N. gibbsii*. The concentration of glycogen was greater in hind limb muscles than forelimb muscles both in *C. cristata* and *S. orarius,* with values higher in the latter species. However, skeletal muscles of moles have neither particularly large glycogen stores nor great buffering capacity, which suggests that the cells comprising these muscles do not routinely respire anaerobically. Normoxic (atmosphere approximately 21% O_2 by volume) respiratory frequency is lower than expected and heart rate tends to be arrhythmic and lower than expected (Armsby et al. 1966; Allison and Van Twyver 1970a).

Star-nosed moles are accomplished divers, and rival some other endothermic diving vertebrates that are substantially greater in size. The average dive duration reported by McIntyre (2000) was 9.2 sec, with a maximum of 47 sec. However, both the frequency with which star-nosed moles dive and the average duration of dives decrease with decreasing water temperature. In addition, these moles are not able to maintain body temperature when immersed in water colder than 30°C and the rate of heat loss is even more substantial at temperatures less than 20°C.

Diving poses metabolic and thermoregulatory challenges for mammals, particularly those as small as star-nosed moles. When breathing is not possible, there either must be sufficient O_2 stores for aerobic cellular respiration to be sustained or the buffering capacity of skeletal and cardiac muscles must be sufficient to resist changes in pH that accompany the accumulation of lactic acid produced during anaerobic respiration. Body O_2 stores typically are elevated in diving vertebrates, and, as already noted, the same is true of moles. Star-nosed moles have total muscle O_2-storage capacities that are about 13.5% higher than those of coast moles, even though individuals of *Condylura cristata* on average are about 21% smaller than coast moles and have slightly less of their total mass attributable to muscle. The large total body O_2 stores of diving rodents and pinnipeds typically are associated with the blood; however, in star-nosed moles such stores primarily are possible because of large lungs and muscle myoglobin. Lungs of adult individuals of *C. cristata* account for about 46% of total O_2 stores as compared to only about 25% in coast moles. Average myoglobin concentrations in star-nosed moles are about 16% higher than those of coast moles. Duration of dives by star-nosed moles rarely exceeds the time that O_2 stores allow for aerobic respiration in muscles (McIntyre 2000). Because surface area-to-volume ratios of moles are relatively large, thermal conductance is expected to be high when the animals are in water and especially if the water is cold. The somewhat elevated metabolic rate of star-nosed moles may be linked to the metabolic demands associated with their aquatic activities. The rate of body heat loss when in water is attenuated by the pelage, but this is highly dependent on the maintenance of dead-air space between the skin and the water; if this air space is eliminated by wetting, the rate of body heat loss increases by more than 40% (McIntyre 2000).

Richards et al. (1996) isolated and characterized a kallikrein enzyme from the salivary glands of *Scalopus aquaticus*. This enzyme has a molecular mass of 30 kDa, optimal activity at pH 9.0, and an isoelectric point of 5.3. Tissue kallikreins are trypsine-like serine proteinases and serve as processing enzymes for activation of growth factors and hormones in mammals. Mole kallikrein releases kinin from kininogen, a plasma globulin. Kinins are small peptides that play important roles in blood pressure homeostasis, functioning to relax vascular smooth muscle and enhance vascular permeability. Salivary gland kallikrein is released from acinar cells as the result of parasympathetic activity and plays a role in the dilation and increased permeability of arterioles and capillaries serving the salivary glands.

There have been no formal studies on water relations in moles. The high humidity levels in tunnel systems may create problems of water loading (E. Toolson, University of New Mexico, pers. commun., 1983). It is uncertain whether drinking is important, but water does not appear to be available in most natural situations (Mellanby 1971) and dependence on the intake of free water likely is a function of the type of food consumed and atmospheric conditions (Godfrey and Crowcroft 1960). Captive eastern moles will drink water regularly from a bottle or dish when fed a diet of meat or mealworms, but drink rarely, if at all, when earthworms are provided as the only source of food (G. D. Hartman, pers. obs.).

Moles exhibit many stages of sleep corresponding to those observed in humans, including rapid eye movement (REM = paradoxical) sleep, and there is a rest–activity cycle of about 4 hr (Allison and Van Twyver 1970a). The deep stages of sleep exhibited by moles may be the result of the safety afforded them by their subterranean habits. The depth of sleep exhibited by animals is significantly correlated with predation pressure and the relative amount of danger while asleep; mammalian species that are subject to more danger when asleep (e.g., ungulates, lagomorphs) tend to be light sleepers, whereas those subject to less danger (carnivores, moles) are deep sleepers (Allison and Cicchetti 1976).

Senses. As a group, it appears that talpids have a somatosensory cortex that is anteroposteriorly oriented, a condition that is unusual among mammals. The sensory cortex is dominated by representation of the facial region and the forelimbs, and relatively little cortex is devoted to the hind limbs and trunk of the body. The organization and amount of cortex devoted to different structures in different species in part is related to differences in sensory structures associated with the facial region (Catania and Kaas 1997a).

Moles have eyes that are greatly reduced in size and appear as small black spheres. The eyes do not occupy bony sockets, but lie far forward on the maxillary regions of the skull. Extrinsic ocular muscles are present (at least in *Neürotrichus gibbsii* and *Scalopus aquaticus*), but do not attach to any bones. There are no oculomotor nerves; however, the optic nerve is present (Gaughran 1954; Lewis 1983). The optic nerve cells sometimes form a layer several cells thick, or they may be crowded into a mass (Arlton 1936). Depending on the species, the eyes may be totally useless or at best allow the animals to distinguish light from dark (Slonaker 1902; Arlton 1936; Godfrey and Crowcroft 1960). Dalquest and Orcutt (1942) found that shrew moles did not respond to bright light flashed 2.5 cm from their eyes, which is not too surprising because their eyes lack a vitreous body, rods, or cones, and have a pigmented extension of the retina that covers the anterior surface of the lens (Lewis 1983). In *S. aquaticus,* the eye has a vitreous body (Slonaker 1902), but lacks rods and cones in the retina (Grim 1990). Although the retina does appear to have functional sensory cells (based on electron microscopic studies, Grim 1990), Allison and Van Twyver (1970b) and Catania and Kaas (1997a) concluded that visual cortex is lacking in *S. aquaticus*. In *Condylura cristata,* where the eyes are exposed and not covered by skin, the presumptive visual cortex is small (Catania and Kaas 1995). European moles (*Talpa europaea*), which also have eyes that are not covered by skin, can be trained in a maze based on light discrimination (Johannesan-Gross 1988).

Hearing appears to be reasonably well developed in most moles (Slonaker 1920; Hamilton 1931; Arlton 1936; Gaughran 1954). With the exception of loss of the pinna, no atrophy of the ear has occurred. The structures of the middle and inner ear are normal and relatively large (Arlton 1936; Stroganov 1945); a stapedial muscle is present but *Musculus tensor tympani* is absent (Whidden 2000). The degree of acuteness among the various species is not known. Dalquest and Orcutt (1942) found that *Neürotrichus gibbsii* did not respond to the sound of the ordinary human voice, but high-pitched sounds in the range of 8000–30,000 vibrations per second (8–30 kHz) produced an immediate response. Pedersen (1963) found that Townsend's moles responded to tapping on the side of their cage, evidence of sensing vibration, but not necessarily hearing. Arlton (1936) noted that captive eastern moles responded to household noises (e.g., turning of a door knob, closing of a door, whispering, walking across a floor, or turning the pages of a notebook). Allison and Van Twyver (1970b) reported that they were able to record auditory-evoked potentials in the cortex of eastern moles, whereas Catania and Kaas (1995, 1997a) reported that cortex responsive to auditory stimuli was present in star-nosed moles, but not in eastern moles. McVean (1999) suggested that relatively enlarged semicircular canals in moles and other subterranean mammals may confer increased sensitivity to the animals, possibly in response to increased demand presented by the need to navigate horizontal mazes.

Reports on the sense of smell in North American moles have been anecdotal. Olfaction may play some role in recognition of food, but it does not appear to be of much use in locating it. Various authors have reported no response until food was placed within a few centimeters of the nose, and in most cases, actual contact was required (Hamilton 1931; Arlton 1936; Dalquest and Orcutt 1942; Catania 1995a; Catania and Kaas 1997a). However, such reports demonstrate only that smell may be of little use in locating food, not that the sense of smell is poorly developed or that it is of little importance to a mole. There is a large body of evidence that in European moles, olfaction plays an important and key role in territory recognition and the social structuring of populations (Gorman and Stone 1990) (see section on Home Range and Social Structure).

Tactile ability is well developed in all species. All North American moles possess numerous tactile hairs on the snout, on the dorsal surface of the head, along the borders of the forepaws, and possibly along the tail (Hamilton 1931; Eadie 1939; Dalquest and Orcutt 1942). In addition, highly sensitive tactile organs called Eimer's organs (Eimer 1871) occur on the snout of all species of North American mole except *Scalopus aquaticus* (Giacometti and Machida 1965; Van Vleck 1965; Catania 1995b, 2000). Eimer's organs function as mechanoreceptors similar to Pacinian corpuscles (Van Vleck 1965) and are very sensitive to light touch (Catania and Kaas 1995; Catania 1996). Eimer's organs are known to occur only in talpids (Catania 2000). A highly developed tactile sense assuredly must provide a substantial adaptive advantage to a blind subterranean mammal by providing a wide array of information regarding orientation to the surroundings, the presence of food items, and potential danger.

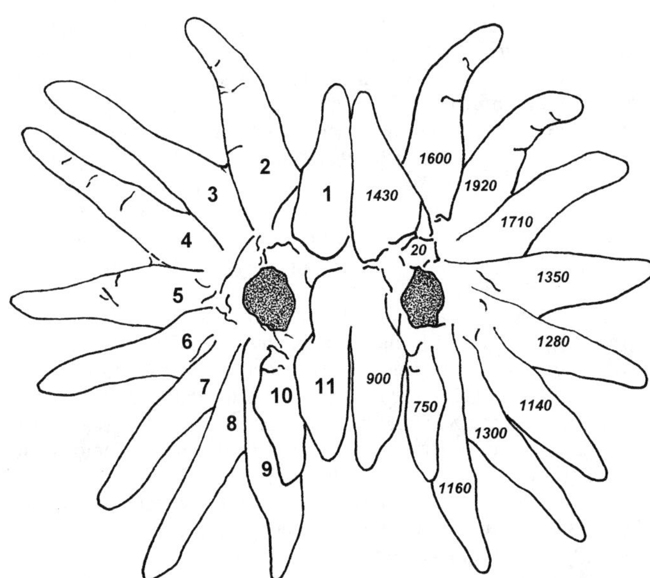

FIGURE 2.7. Drawing of the star of the star-nosed mole (*Condylura cristata*), showing (left) standardized numbering of individual rays and (right) estimates of numbers of Eimer's organs associated with particular parts of the star. SOURCE: After Grand et al. 1998 and Catania 1995a.

Whereas most moles may have a few thousand Eimer's organs present around their nostrils, star-nosed moles have as many as 25,000–30,000 (Catania 1995a). Interestingly, this species makes great efforts to keep the nose clean by frequent washings in available water. Before entering water, a star-nosed mole always will dip its ventral rays (pairs 8–10) (Fig. 2.7) onto the surface of the water (Gould et al. 1993). When the animal is feeding in shallow water or soil, its head and snout move side to side while the rays of the star are swept forward and backward. The rays can be moved fore and aft about 90 deg, from a fanned-out position to a fully closed position wherein the rays project forward and the proboscis can be elevated to 90 deg, a "snorkeling" position that the animals use when swimming submerged. Movement of the rays of the star is accomplished using a series of tendons attached to three nasolabial and zygomatic muscles on each side of the face (Grand et al. 1998).

Rays are not used to capture prey, manipulate objects, or position food that is being eaten (Catania 1995a; Catania and Kaas 1995); rather, they are used to explore items by touch. Star-nosed moles will neither capture nor bite prey until it first has come into contact with the star (Catania 2000). Once contact by the rays has been made, the nose is shifted immediately so that ray number 11 (see Fig. 2.7) repeatedly comes in contact with the item; if it is a food item, ray 11 repeatedly contacts it before and as it is being eaten (Catania and Kaas 1996). Ray 11 frequently shows signs of abrasion and is a behavioral focus of the nose, which is used preferentially for discriminative examination of an object (Catania and Kaas 1997b; Catania 1999). Its use is similar to that of the *fovea centralis* of an eye: important stimuli frequently are first detected via peripheral vision and then examined more closely with the fovea. The rays of the star have corresponding morphological specializations in somatosensory cortex, represented as a cortical pinwheel of 11 stripes similar to the "cortical barrels" that have been reported to be associated with mystacial vibrissae in some rodents (Catania and Kaas

1995). Ray 11 is relatively small, but has a disproportionately large cortical representation in the somatosensory area (Catania 1995c; Catania and Kaas 1995); in fact, it has the largest cortical representation, often 25% that of the entire nose (Catania 1995c). Although ray 11 has fewer Eimer's organs than almost any other ray (Catania 1995a; see Fig. 2.7), it has the greatest area of cortex representation per sensory organ, the highest innervation density per sensory organ, and the highest average area of cortex per primary afferent neuron. The relative sizes of cortical representation of rays are not proportional to their innervation densities (Catania and Kaas 1997b). Behavioral observations in one study suggested that the star of *Condylura cristata* also may allow the animals to sense electrical stimuli (Gould et al. 1993); if true, this would be a capability that is unique among eutherian mammals. Whether star-nosed moles actually are capable of electroreception has been a point of controversy (see Grand et al. [1998] and Catania [2000] for opposing arguments). One camp has suggested that appropriate neurophysiological studies have yet to be performed that would conclusively demonstrate whether the star has electroreceptive properties (Grand et al. 1998); however, achieving this goal is difficult because, by their very nature, all neuronal fibers respond to electrical currents (K. Catania, Vanderbilt University, pers. commun., 2000).

The manner of development of the rays of the star of star-nosed moles differs fundamentally from that of any other type of animal appendage. Whereas animal appendages typically develop as outgrowths of the body wall, in the development of the star there is no outgrowth stage. The rays develop as 22 separate anteroposteriorly oriented epidermal cylinders along the side of the face; the posterior portion of each cylinder then erupts from the face and then rotates anteriorly. The anterior portion of a cylinder becomes the base of the ray in the adult structure. Adults of all species of *Scapanus* exhibit a pattern of raised epidermal subdivisions on their snouts that is reminiscent of the embryonic star in individuals of *Condylura cristata* (Catania et al. 1999; Catania 2000).

REPRODUCTION

Breeding. Breeding in moles occurs only once a year, usually during late winter or early spring. The peak of the breeding season varies geographically and may differ by several months in a species with an extensive geographic range. Conaway (1959) reported the peak of the breeding season for eastern moles in Wisconsin as the last week in March and the first week in April. Hairy-tailed moles from New Hampshire have a similar breeding season (Eadie 1939). Davis (1942) and Lowery (1974) reported that the breeding season of the eastern mole in Texas and Louisiana began in early February; Yates and Schmidly (1977) believed it began as early as January in those regions. Moore (1939) stated that Townsend's mole bred between early February and early March, whereas male star-nosed moles are reported to be reproductively active from January through June (Hamilton 1931; Eadie and Hamilton 1956). The breeding season of shrew moles apparently is more extensive, ranging from late February until August (Dalquest and Orcutt 1942), and breeding may occur throughout the year (Dalquest 1948).

Both sexes of *Condylura cristata* and *Parascalops breweri* have been reported to breed when <1 year old (Eadie 1939; Eadie and Hamilton 1956). Conaway (1959) found no evidence that females of *Scalopus aquaticus* breed during the year they are born; however, he found no females ≥1 year old to have passed through a breeding season without breeding.

In star-nosed moles, an annual swelling of the tail occurs during the winter and spring months (Hamilton 1931). The function of this swelling is not known, but it appears to be caused by deposition of fat in the tail. Eadie and Hamilton (1956) noted that the great majority of males had swollen tails before and during the breeding season, but most did not after the breeding season. This led these authors to postulate that the function of this fatty tissue in males may be to act as a temporary reservoir of energy during the breeding season. However, females also exhibit periodic enlargement of the tail during winter and spring (Hamilton 1931).

Gestation, Litter Size, and Neonates. The exact gestation period is not known for any North American species of mole; it generally is assumed to be from 4 to 6 weeks. Within most species, litter size typically ranges from two to five (Eadie 1939; Dalquest 1948; Scheffer 1949; Conaway 1959; Kuhn et al. 1966; Leftwich 1972; Hartman 1992), although in *Condylura cristata,* some individuals may have up to seven young (D. E. Davis and Peek 1970); one *Parascalops breweri* was taken with eight embryos (Richmond and Rosland 1949). Kuhn et al. (1966) estimated the average weight of neonate Townsend's moles from Oregon to be about 5 g. Hartman and Gottschang (1983) reported the weights of two male neonates of *Scalopus aquaticus* from Ohio to be 5.35 and 5.36 g, respectively. Newborns were naked except for numerous mystacial and mandibular vibrissae, had eyes covered by skin, and auditory meatuses that had not yet opened.

Reproductive Cycle. The reproductive cycles of American moles, though differing in some details, appear to be quite similar; however, few species have been researched in depth. In all species, the vagina remains sealed until follicles with antra appear in the ovaries. This condition, coupled with the lack of a scrotum in males, makes accurate sex determination of nonbreeding moles difficult without dissection. Checking for elongate teats sometimes can be of help in identifying sex of nonbreeding adult females that have previously bred.

The distribution of follicles within the ovaries differs among species. As was reported for *Talpa europaea* (Matthews 1935) and *T. occidentalis* (Jiménez et al. 1990, 1996), the follicles of *Condylura cristata* are confined to a specific region of an ovary and the rest of the organ is comprised of a steroid (androgen)-secreting interstitial gland, which enlarges during autumn when the size of the ovarian component decreases (Whitworth et al. 1999). This *ovotestis* condition is not seen in *Scalopus aquaticus,* where the ovaries appear normal (Bedford et al. 1999); some preliminary studies have indicated that *Scapanus latimanus* and *S. orarius* also have normal ovaries (N. Osypka-Rubenstein, University of California-Berkeley, pers. commun., 2000). Conaway (1959) suggested that ovulation in *S. aquaticus* likely is induced.

In males, the testes and associated glands become greatly enlarged before mating. There is a well-defined annual testicular cycle, which involves changes in weight (Conaway 1959; Glendenning 1959; Hartman 1995a) (Fig. 2.8) and cytology (Conaway 1959). The male cycle in *Scalopus aquaticus* also is manifest in the anatomy of the penis. During the breeding season, the glans becomes covered with short and proximally recurved epidermal spines (Bedford et al. 1999) (Fig. 2.9); however, outside the breeding season, the surface of the phallus is glabrous.

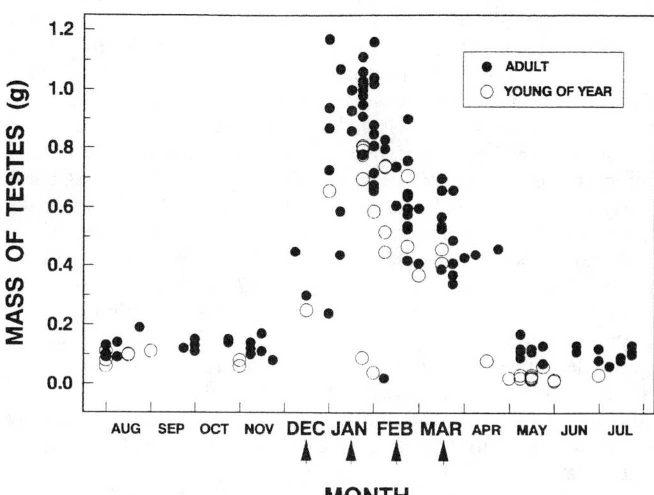

FIGURE 2.8. Combined weight of testes of 136 individuals of the eastern mole (*Scalopus aquaticus*) collected throughout the year in South Carolina. An arrow below the month denotes the capture of significantly greater numbers of males.

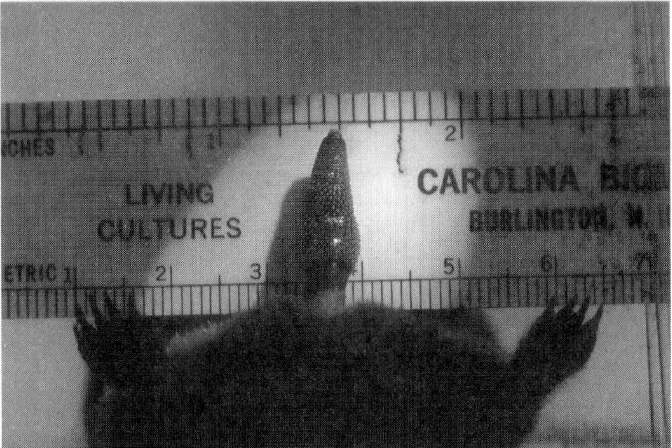

FIGURE 2.9. The extruded and erect phallus of an eastern mole (*Scalopus aquaticus*) from South Carolina during the breeding season. From top to bottom: anterior, posterior, and lateral views. In photos showing a scale, the smallest increments on the bottom of the rule are millimeters.

Spines begin to form with the onset of increased mass of the testes during fall, are present throughout the breeding season, and then are sloughed off at about the same time the testes return to weights corresponding to the nonbreeding condition (G. D. Hartman unpublished data). Spines also occur on the phallus of individuals of *Condylura cristata* and *Neürotrichus gibbsii* (Eadie 1951; Bedford et al. 1999), but apparently are lacking in *Parascalops breweri* (Eadie 1947). Whether these spines play a role in inducing ovulation is unknown.

In describing the male reproductive anatomy in *Condylura cristata* and *Parascalops breweri,* Eadie (1948a, 1948b) reported that the reproductive tract and associated glands may constitute as much as 14% of the total body weight in a breeding male. Eadie (1948a) described a system to help ensure male mating success in *C. cristata* and *P. breweri*. Corpora amylacea are secreted by the paired prostate glands and a substance secreted from the Cowper's gland causes it to coagulate. This causes the formation of a copulatory plug in the female, which retains the seminal mass and might act as a deterrent to mating attempts by other males. Copulatory plugs also have been reported to occur in *Scalopus aquaticus* (Hartman 1992) and *Scapanus townsendii* (Moore 1939).

The factors that trigger the complex reproductive cycle in these typically blind, subterranean mammals are not known. It generally has been assumed that neither temperature nor relative day length is involved. Conaway (1959) suggested that internal rhythms may be of primary importance in regulating the cycles. Pévet et al. (1976) suggested that, at least in the European mole, the pineal gland may play an important role in regulating the reproductive cycle.

ECOLOGY

Habitat. Moles occur in a variety of habitats, from subalpine meadows in the case of some populations of *Scapanus townsendii* (Johnson and Yates 1980) to lowland swamps in the case of *Condylura cristata*. Individuals of *Neürotrichus gibbsii* generally prefer moist, forested areas and soils with high humus content (Dalquest and Orcutt 1942; Terry 1981). The highly subterranean moles tend to prefer drier soils than do star-nosed or shrew moles, but all prefer soils where burrowing is easy, and usually are absent from heavy clay and stony or gravelly soils (Jackson 1915). The degree of rockiness appears to be unimportant to hairy-tailed moles, but soils with high clay or moisture content are avoided (Eadie 1939). In addition to soil type, the soil condition, moisture, and availability of food items appear to be the most important factors affecting the presence of moles (Arlton 1936). For example, it has been shown that coast moles prefer light, wet soils with a large number of earthworms (Schaefer and Sadlier 1981). Moles tend to be absent altogether from arid lands (Silver and Moore 1941). Soil types otherwise suitable for habitation that are too wet or too dry often are avoided by the highly subterranean moles such as *Scalopus aquaticus* (Davis 1942; Glass 1943).

Although *Scapanus orarius* and *S. townsendii* occur sympatrically throughout much of their ranges, individuals of these two species appear to prefer different microenvironments. Individuals of *S. orarius* seem to prefer deeper burrows and better drained soils and are less colonial than Townsend's mole (Dalquest 1948). Coast moles also are less common in agricultural lands and often are found in dense thickets and deciduous woods, which the larger Townsend's mole appears to avoid.

Burrows, Tunnels, and Nests. All species construct two basic types of tunnel: deep, more permanent tunnels and shallower surface runways. Differences in the extent and nature of these tunnels occur within and among species. *Neürotrichus gibbsii* most commonly constructs shallow burrows beneath decaying leaf litter; these tunnels form complex interconnecting networks (Dalquest and Orcutt 1942). Although these burrows are little more than shallow troughs, shrew moles do construct some tunnels similar to those of other mole species; these tunnels seldom are found more than 30 cm below the surface, are less extensive than those of other species, and have open entrances (Dalquest and Orcutt 1942). Ventilation of tunnels largely is dependent on air flow above the surface of the ground, and, at least in European moles, air enters the tunnel system predominantly through newly formed mounds or "molehills" (Olszewski and Skoczeń 1965).

The tunnels of star-nosed moles are similar to those of the more purely subterranean species, but typically are constructed near marshy areas or streams (Hamilton 1931; Rust 1966). Molehills are formed during the construction of deep tunnels. Tunnels sometimes open near or even under the surface of water; the openings associated with these tunnels usually are not plugged.

Parascalops breweri and *Scalopus aquaticus* construct deep and surface tunnels similar to those of *Condylura cristata,* although loamy, well-drained soils are preferred. Deep tunnels range from 25 to 45 cm below the surface in *P. breweri* and from 15 to 60 cm deep in *S. aquaticus* (Hisaw 1923b). Some of the surface tunnels constructed during foraging are used only once, whereas others are used frequently and for many years. Wright (1945) reported that some tunnels of *P. breweri* may be in use for up to 8 years. We have observed tunnels used by individuals of *S. aquaticus* for 10 and more years. Breaks in tunnels usually are repaired when they first are encountered. Individuals of *S. aquaticus* will not tolerate openings in the burrow system and repair all breaks in their tunnels that they encounter. Eadie (1939) reported that artificial breaks are repaired by hairy-tailed moles, but some openings in the tunnel, possibly used by shrews and mice, were not repaired even though the tunnels were used by moles.

The types of tunnel constructed by the three species of *Scapanus* are similar to those of *Scalopus aquaticus*. Surface and deep tunnels are constructed, although *S. townsendii* constructs two types of deep tunnel. The most extensive permanent system is an intricate interconnected system 15–20 cm deep. Less commonly, tunnels are formed under fence rows, building foundations, and roadbeds. These tunnels may be 1–3 m below the surface, frequently are used by more than one individual, and serve as major travel routes between areas. Deep runs are used much more frequently by Townsend's moles than are surface runs. The result is a more complex network of deep tunnels than is common in most species and a large number of molehills in areas where the tunnels occur. Pederson (1963) reported mounds of this species numbering up to 805/ha. Silver and Moore (1941) reported that a Townsend's mole constructed 302 mounds in 77 days. Glendenning (1959) reported that from October to March, a single coast mole may construct between 200 and 400 mounds. Where *S. orarius* and *S. townsendii* are sympatric, tunnel diameter and mound height and width all average significantly larger for Townsend's mole. These measurements are useful for discriminating which species is responsible for the workings (Sheehan and Galindo-Leal 1997). Coast moles construct more mounds during autumn and winter than during summer months (Schaefer 1978; Schaefer and Sadlier 1981).

The methods used to construct burrows appear to be the same in all species. Hisaw (1923b) provided a detailed account of this process for the eastern mole. He found that moles do not "swim" through the soil or use the head and/or neck as was once believed commonly; instead, burrows primarily are dug with the powerful forelimbs. In constructing the familiar surface runs, a lateral-stroke type of digging is employed with the anterior part of the body rotated approximately 45 deg alternatively to the right and left. Each time after the body has been rotated, the forepaws are brought together several times to position the claws. If the body was rotated to the right, the left forepaw is thrust upward rapidly; at the same time the right foot, which is braced against the burrow wall, is extended to create more force. The soil thus is forced upward, forming a surface ridge; the body then is rotated to the left and the process is repeated. The nose is not used in loosening soil, but serves a tactile function in directing the forepaws. The procedure for forming deeper runways basically is the same except that the forefeet are used to loosen soil, which then either is brought to the surface to form mounds or deposited in abandoned surface tunnels. Mounds are formed by pushing soil upward through a packed column of soil; as soil is added and pushed up from below, an equal amount is added to the mound on the soil surface.

Mounds are formed less frequently by individuals of *Scalopus aquaticus* within the southern portions of the species range as compared with those in the north. In the south, surface ridges frequently can be found with no open tunnel beneath them and it appears that the moles are using abandoned surface tunnels for depositing the soil from the excavation of deep tunnels. Hickman (1984b) excavated a single tunnel system of an individual of *S. aquaticus* in Florida and reported the occurrence of several mounds, stating that mound formation by other moles in the study area was prominent during July and November. However, Brown (1972) excavated 25 tunnel systems in Florida over a period of about 2 years and reported finding only two mounds. At least some of this apparent discrepancy may be explained by individual variation in behavior. Some individuals of *S. aquaticus* have a predilection for mound building (Leftwich 1972, G. D. Hartman, pers. obs.). However, both Brown (1972) and Hickman (1984b) found that deep tunnels constituted only a minor portion of the entire tunnel system. The lower frequency of mound building by southern *S. aquaticus* simply may be a reflection of the lower frequency of deep tunnel formation. Bradshaw and Goldberg (1989) reported that mounds of *S. aquaticus* in Michigan usually had significantly lower moisture, nutrient, and organic matter contents than undisturbed adjacent soil, but were subject to higher levels of incident light.

Although most people can tell the difference among moles, gophers, voles, and mice "in hand," their telltale sign frequently is misinterpreted. Molehills easily can be distinguished from gopher mounds by the characteristic shape and position of the burrow exit. Molehills are volcano-like in shape. The soil is pushed from the burrow exit to form a symmetric cone with no readily apparent sign of an exit point. Only careful excavation of the molehill reveals an exit point located in the center of the cone at ground level. A gopher expels soil from its tunnel system by pushing the soil from the burrow and depositing a mound in a fan-shaped or crescentic configuration, and the burrow entrance is not located at the center of the mound. The gopher burrow entrance has a visible plug of soil, whereas the mole burrow is unmarked by a visible soil plug. Meadow mice do not deposit soil mounds; their presence can be identified by small burrow openings, 1–3 cm in diameter, interconnected by a series of surface runways. Careful examination of the runways will reveal accumulations of fecal droppings at several points in the interconnected surface runway system.

Most talpids construct nest and rest chambers. The form and structure vary with species, but the functional aspects of these chambers—rearing young and resting—remain rather uniform. *Scapanus townsendii* appears to be the most intricate nest builder. Nest mounds of this species are located easily during March by the aboveground appearance of a single mound 51–76 cm in diameter or larger or an aggregation of small mounds (Kuhn et al. 1966). A single nest mound was found in Tillamook County, Oregon, 142 × 71 × 36 cm in size (Pedersen 1963). A nest chamber 15–20 cm below ground level is most common, but chambers have been located 76–127 cm deep. The nest chamber averages 30–36 cm in diameter, with as many as 11 exits. Frequently, 1 exit descends vertically immediately below the chamber (Kuhn et al. 1966). The nest is composed of an inner layer of dry grass and an outer layer of green grass, which periodically is replaced during the time young occupy the nest (Pedersen 1963).

Harvey (1976) located nests of *Scalopus aquaticus* 15–25 cm beneath the soil surface composed of coarse grass and/or leaves. The nests were 18–22 cm long and 10–12 cm wide, and were enlargements of deep tunnels. He found that some moles used two to seven nests. Only one to three nests have been found associated with tunnel systems of *S. aquaticus* in Florida and these do not always contain nesting material (Brown 1972; Hickman 1984b).

Nests of *Condylura cristata* and *Parascalops breweri* are constructed in a similar fashion in shallow and deep tunnels and are composed of well-packed leaves and grass (Hallett 1978). Nests for the young are larger than those used for resting, but are similar in construction. Nest sites of *C. cristata* are by necessity carefully chosen to be above any high-water flooding (Hamilton 1931; Davis and Peek 1970).

Neürotrichus gibbsii apparently is the only North American mole that may nest above ground. Dalquest and Orcutt (1942) found that shrew moles easily climbed shrubby vegetation. One nest composed of damp willow leaves was found 0.6 m above ground in an old alder (*Alnus*) stump. Racey (1929) noted that shrew moles frequently enlarged a portion of the tunnel system to form a small sleeping chamber with a vent hole; no vegetation was found in these structures.

Digging a system of tunnels requires a great deal of energy. However, once this energy has been expended, a tunnel system affords a mole an efficient and economical means of foraging over a wide array of temperatures and prey densities (Jensen 1986b). Soil around mole tunnels may have prey biomass that is up to 80% less than that of undisturbed soil. This reduction in biomass apparently is not solely the result of predation by moles; in and around tunnels from which moles have been excluded, prey densities continue to be lower than in undisturbed soil even after 1 year has elapsed (Jensen 1986a). The lower prey densities in and around tunnels likely reflect changes in edaphic factors. In spite of lower prey density, it generally is more efficient energetically for a mole to forage in an existing tunnel than to dig a new one (Jensen 1986b). This fact at least in part explains the tendency for a mole to use the abandoned tunnels of another mole (Schaefer 1978; Hartman and Gottschang 1983; Elshoff 1989; Gorman and Stone 1990).

Mole tunnels provide subterranean pathways for a variety of invertebrate and vertebrate creatures other than moles, including camel crickets and snakes (Hartman, pers. obs.), salamanders (J. D. Krenz, Minnesota State University, Mankato, pers. commun., 2001), various species of *Microtus* and other rodents (Couch 1924; Eadie 1939; Yates et al. 1979; Hartman and Wike 1998), and shrews (Eadie 1939; Hartman and Wike 1998; Hartman et al. 2001). Southern short-tailed shrews (*Blarina carolinensis*) are captured more frequently in traps placed in mole tunnels than in traps placed on the soil surface (Hartman et al. 2001).

The fruiting bodies of certain fungi in the genus *Hebeloma* are reliable indicators of nest and latrine sites of European moles in England and of various mole species in Japan (Sagara 1989, 1999; Sagara et al. 1989). The fruiting bodies of *H. radicosum* are specifically associated with mole and mouse latrines (Sagara 1999). Fungal and floral associations with tunnel systems of North American moles have not been investigated. We have noticed that in mixed pine–scrub oak habitats in South Carolina, fruiting bodies of coral fungi (Clavariaceae) almost always occur in association with surface tunnels of *Scalopus aquaticus*. Furthermore, in pine–savannah habitats, the surface ridges associated with old, well-traveled tunnels often will have lichens or mosses growing on them, but not on the surrounding soil.

Home Range and Social Structure. The only comprehensive study of home range in a North American species was for the eastern mole (Harvey 1976) and only 12 specimens were involved in that study. Male moles have considerably larger home ranges (Table 2.8) than do females (Harvey 1976), and moles may range over larger areas than other fossorial and subterranean mammals. The average home range of the eastern moles in Harvey's (1976) study exceeded that of many rodents. The home range of a male eastern mole averaged almost 23 times that of a male Plains pocket gopher (*Geomys bursarius*), 42 times as large as that of a male Botta's pocket gopher (*Thomomys bottae*), and five times as large as that of a male Ord's kangaroo rat (*Dipodomys ordii;* Table 2.8). Home range in North American moles likely varies as a function of season, but this has not been studied.

TABLE 2.8. Estimates of mean home range size of *Scalopus aquaticus, Talpa europaea,* and five species of rodent

Species	Home Range (m^2)	
	Male	Female
Scalopus aquaticus (Harvey 1976)	10,640	2,749
Talpa europaea[a] (Gorman and Stone 1990)	7,343	1,314
Talpa europaea[b] (Gorman and Stone 1990)	2,679	1,655
Talpa europaea[c] (Gorman and Stone 1990)	7,718	2,086
Talpa europaea[d] (Gorman and Stone 1990)	3,364	1,945
Talpa europaea (Macdonald et al. 1997)	2,324	2,324
Dipodomys ordii (Garner 1973)	1,952	2,231
Dipodomys elator (Roberts and Packard 1973)	791	791
Reithrodontomys fulvescens (Packard 1968)	1,859	2,334
Geomys bursarius (Wilks 1963)	468	145
Thomomys bottae (Howard and Childs 1959)	251	129

[a]Breeding individuals in pasture.
[b]Nonbreeding individuals in pasture.
[c]Breeding individuals in woodland.
[d]Nonbreeding individuals in woodland.

Hamilton (1931) suggested that star-nosed moles are gregarious and perhaps colonial, although these local aggregations of animals may relate more to food supply than to social structuring. The same reasoning may explain the local concentrations of shrew moles, which Dalquest and Orcutt (1942) described as traveling in "loose bands." Their conclusion was based on the fact that multiple individuals routinely are trapped within the same runway (also reported by Campbell and Hochachka 2000).

Individuals of *Parascalops breweri, Scalopus aquaticus,* and all species of *Scapanus* appear to be solitary except during the breeding season. Giger (1973) found that movements of *S. townsendii* were confined to isolated tunnel systems or portions of larger systems, and Eadie (1939) reported similar findings for *P. breweri*. Harvey (1976) reported that the ranges of individual *S. aquaticus* overlapped, but multiple captures from a single tunnel system were rare. Eadie (1939) captured numerous males from female burrow systems during the breeding season, a pattern that also occurs for livetrapped eastern moles (G. D. Hartman unpublished data). In late summer, after the young have left the nests, both sexes appear to associate freely for a short period of time.

Although it generally is assumed that moles are solitary and territorial, this has yet to be conclusively demonstrated for any North American talpid. Little or nothing is known about the social structuring of populations in North American genera. This is not the case for European moles. In an elegant series of field experiments, R. David Stone (reported in Gorman and Stone 1990) showed that the home ranges of individual moles largely are exclusive. For the most part, home ranges of males do not overlap with those of other males, but male home ranges may overlap slightly with those of several females. Home ranges of females may overlap those of other females as well. Areas where home ranges overlap tend to be used by the different individuals at different times. If an animal is removed from its tunnel system, neighboring moles begin to enter the vacated tunnels fairly rapidly. The secretions of "preputial" glands (actually anal glands; Brown et al. 1994), which are present in males and females, play a major role in olfactory communication of territorial boundaries. Composition of anal gland secretions remains essentially the same in males throughout the year. In females, secretions are similar to those of males outside the breeding season, but change with the onset of the breeding season. These changes in female secretions may represent either a cue for males to enter a female's territory or the lack of signal to stay out (Khazanehdari et al. 1996). Gorman and Stone (1990) reported that the locations where moles urinate in their tunnel systems appears to be nonrandom, always occurring at the junctures of tunnels. The precise role that urination may play in social structuring is unknown.

Dispersal and Homing. Above-ground dispersal is rare in the most subterranean genera and usually involves juveniles exclusively (Giger 1965; Leftwich 1972). Flooding also may account for some apparent dispersal. In *Scapanus townsendii* and *Scalopus aquaticus,* dispersal occurs during the month after young are weaned. Individuals of *Condylura cristata* and *Neürotrichus gibbsii* may disperse above ground more commonly, but data on home range are not available for these species.

The one study in which dispersal in a North American talpid was quantified suggests that moles are philopatric. Giger (1965) reported that 61% of juvenile Townsend's moles had dispersed <152 m and 87% of them had dispersed <305 m from their birth sites. The greatest dispersal distance recorded was approximately 722 m, and three individuals of *Scapanus townsendii* dispersed across paved roads. No significant difference between dispersal distances of males and females was observed.

It is doubtful that small bodies of water such as rivers and streams present significant barriers to dispersal because most moles are good swimmers (Hamilton 1931; Schmidt 1931; Arlton 1936; Foote 1941; Dalquest and Orcutt 1942; Reed and Riney 1943; Giger 1973; Hickman 1984a). Giger (1973) found that Townsend's moles that had been displaced from their home ranges were able to traverse barriers including small canals, paved elevated highways, and, in one instance, a river. Yates and Schmidly (1975, 1978) felt that the heavy clay soils associated with certain river systems, instead of the rivers themselves, form the real barriers to dispersal. The only species known to take to water freely is the star-nosed mole. Individuals of *Condylura cristata* have been reported swimming under the ice in winter and frequently are caught in muskrat (*Ondatra zibethicus*) and minnow traps set in streams and lakes (Hamilton 1931).

Moles exhibit site fidelity and homing behavior. Giger (1973) artificially displaced individuals of *Scapanus townsendii* and reported that the greatest distance moved by an individual (a female) in returning to its home range was 455 m. Homing behavior also has been observed for individuals of *Scalopus aquaticus* (Leftwich 1972).

Sex Ratio. Population samples of moles frequently are comprised of more males than females, although the reverse sometimes is reported. Estimates of population sex ratio based on male:female captures may accurately reflect sex ratio or may reflect sampling bias. For example, the larger ranges of males and their higher activity levels increase the likelihood that males will encounter traps (Townsend 1935).

In *Parascalops breweri, Scalopus aquaticus,* and *Talpa europaea,* significantly more males than females are captured during the first half of the calendar year, whereas during the second half, numbers of males and females captured do not differ statistically. The time period when significantly more males are captured generally corresponds to the breeding season and, more specifically, to the months during which the testes have increased in mass (Hartman 1995a; see Fig. 2.8). The marked tendency to capture more males during the breeding season may be related to behavioral changes mediated by changes in hormone levels. During the breeding season, males exhibit increased activity and expanded ranges. However, changes in female behavior during the breeding season also likely play a role in more males being captured during this time. Females become more wary during the breeding season (Eadie 1939; Skoczeń 1966; Katanova 1972); for example, live traps placed within the known home ranges of female *S. aquaticus* frequently show signs of moles having visited the traps without having entered them; however, as the breeding season comes to an end, females begin to be captured in these same traps (Hartman 1995a).

A few studies on moles reported possible variation in population sex ratio among different habitats (Godfrey and Crowcroft 1960; Katanova 1972), but these studies were not specifically designed to assess any relationship between sex ratio and habitat. Differential mortality among the sexes also may result in an observed bias in the capture or sex ratio (Davis and Choate 1993). To ensure accuracy when assessing sex ratio in a mole population, the best course of action clearly would be to enumerate the population. However, because accounting for every individual in a population of moles often is at best a difficult task, trapping at least should be conducted outside the breeding season. Even when trapping outside the breeding season, habitat, demography, and trapping effort all are factors that should be considered when interpreting capture data as they may relate to population sex ratio.

Population Density. As a result of relatively large home range size, largely carnivorous habits, and apparent territoriality, population densities of moles generally do not approach those attained by many rodents. Population density estimates for *Scalopus aquaticus* from Missouri ranged from 0.59 to 1.55 animals/ha (Leftwich 1972). Hartman and Krenz (1993) estimated population densities of eastern moles in pine plantations in South Carolina using a trapping grid and assessment lines and reported densities ranging from 1.71 to 3.02 moles/ha. Estimated population density reported for *Parascalops breweri* averages about 3/ha, but sometimes may be as high as 30/ha (Eadie 1939; Hamilton 1939). The highest population densities tend to be those estimated for shrew moles. Dalquest and Orcutt (1942) estimated about 12–15 animals/ha in suitable habitats and in one instance, a rather remarkable 247 individuals/ha were estimated after all other small mammals had been removed.

FEEDING HABITS

In many of the early studies of North American moles, feeding habits frequently received more attention than other aspects of their biology. In general, the stomach contents of moles appear to reflect the availability and abundance of food items within the soil (Opperman 1968; Funmilayo 1979). The diet is highly variable both within and among species, but earthworms, insects, and other arthropods and invertebrates usually are consumed in the greatest quantities (Table 2.9). Vegetation constitutes a portion of the diet in most species and there are sufficient data (Scheffer 1917; Wight 1928; Moore 1933; Pedersen 1963; Whitaker and Schmeltz 1974; Whitaker et al. 1979; Hartman et al. 2000) to suggest that it is not accidentally ingested. Vegetation consumed by moles frequently includes mycorrhizal fungi (Endogonaceae).

TABLE 2.9. Consumption of selected foods within five species of North American moles

Species	Mean Percentage Volume of Diet			
	Earthworms	Vegetation	Insects	Seeds
Neürotrichus gibbsii				
Dalquest and Orcutt 1942	42.0		12.0	
Whitaker et al. 1979	48.5	0.6	43.3	
Parascalops breweri				
Eadie 1939	34.0	2.0	47.0	
Scalopus aquaticus				
West 1910	31.0	13.0	52.0	
Whitaker and Schmeltz 1974	26.8	12.0	47.8	4.5
Hartman et al. 2000	2.9	3.9	32.6	2.3
Scapanus orarius				
Moore 1933	69.9	1.2	28.7	
Glendenning 1959	93.0		3.0	
Whitaker et al. 1979	56.2	2.3	30.0	2.0
Scapanus townsendii				
Moore 1933	76.1	15.9	7.4	
Pedersen 1963	72.0	28.0		
Whitaker et al. 1979	54.9	9.4	13.9	6.8

Earthworms appear to represent the most important food item in the diet of most species of mole. Whitaker et al. (1979) found this true for *Neürotrichus gibbsii, Scapanus orarius,* and *S. townsendii* in Oregon. Rust (1966) reported that the stomach contents of eight star-nosed moles from Wisconsin contained almost 84% earthworms, whereas Hamilton (1931) found star-nosed moles feeding almost exclusively on aquatic annelids and insects. The difference between the two analyses may be related to where the animals were caught. Hamilton captured moles near large bodies of water, whereas Rust's specimens came from areas with only small ponds as sources of aquatic food.

Eadie (1939) examined the stomach contents of 100 individuals of *Parascalops breweri* and found that 47% of the diet consisted of insects and 34% of earthworms. Brooks (1923) reported that hairy-tailed moles in West Virginia often destroyed the nests of ground-dwelling wasps and fed heavily on the larvae and pupae. Brooks (1908) maintained captive individuals of *P. breweri* on bird eggs and meat, but found that they starved when fed only vegetable matter.

Townsend's moles appear to be limited to a narrow range of food items, primarily earthworms and insects (Table 2.9). Pedersen (1963) examined the stomachs of 106 males and 76 females from Tillamook County, Oregon, and found that vegetation in the form of fleshy-fibrous grass roots was a year-round component; 81% of the stomachs contained 200–1800 mg of vegetation. A "root ball" was found in 18% of the stomachs; this appeared as a dark brown mass of compacted roots, averaging about 10 × 5 mm and shaped like a kidney bean. Earthworms were found in 79% of all stomachs and ranged in amount from 200 to 4000 mg. Insects were found in only trace amounts in 6 stomachs, and 1 stomach contained 100% mollusks.

The diet of *Neürotrichus gibbsii* in Oregon was reported as 48.5% earthworms, 13.6% unidentified insects, 11.4% Coleoptera, 9.9% Diptera, 4.3% centipedes, 4.1% snails and slugs, and 0.6% unidentified vegetation (Whitaker et al. 1979). In the Cascade Mountains (Washington), the diet of shrew moles was 75–78% invertebrates during

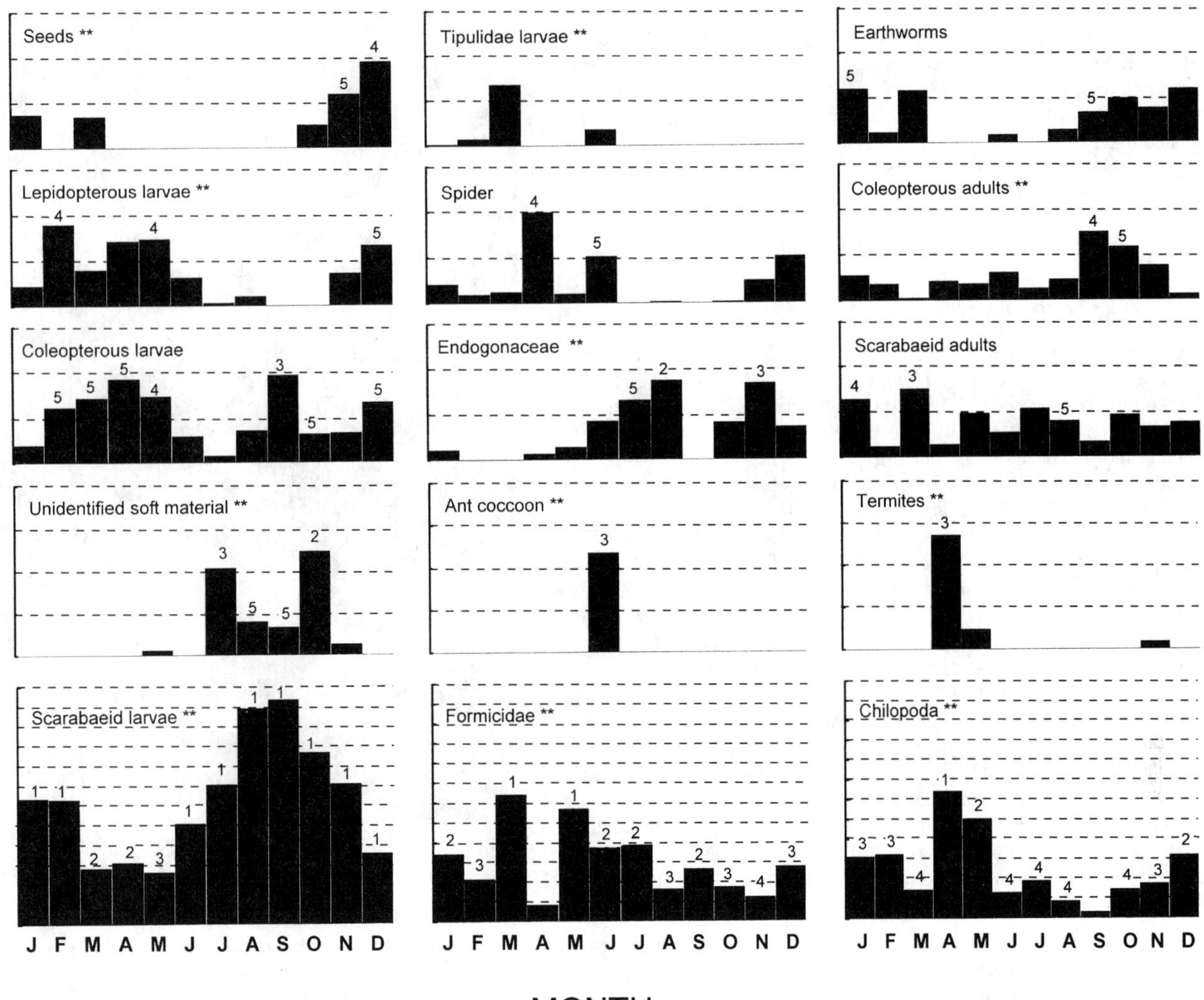

FIGURE 2.10. Monthly variation in selected food items in the diet of eastern moles (*Scalopus aquaticus*) from South Carolina. Food items marked with a double asterisk had mean values that varied significantly across months. Numbers above bars represent the rank of a food item among the top five foods for that month. SOURCE: Hartman et al. 2000.

September, but in July, 36% of the diet consisted of conifer seeds and 32% of lichens (Gunther et al. 1983).

Variation in the diet of all species of mole likely occurs as a function of locality, time of year, and prevalence of locally occurring food items. This is best illustrated by the diet of eastern moles, for which food habits have been studied most extensively (Dyche 1903; Scheffer 1910; West 1910; Whitaker and Schmeltz 1974; Hartman et al. 2000). Composition of the diet can vary a great deal from month to month and if sampling is restricted to only 1 or a few months, an incomplete or biased account of feeding habits may result (Fig. 2.10).

The relative size of the stomach in moles and other insectivores is inversely proportional to body weight (Myrcha 1967), and the intestinal tracts of talpids are relatively short, digestion is fairly rapid, and digesta pass through the gut relatively quickly. The general activity pattern for many moles appears to consist of regular bouts of foraging for 3–4 hr followed by periods of similar duration spent in the nest resting and digesting. Star-nosed moles may use the supplementary heat that is liberated during the mechanical and biochemical processing of a meal (heat increment of feeding) to warm their nest chamber. The heat increment of feeding is maximal in star-nosed moles about 120 min after a bout of foraging (Campbell et al. 2000).

On average, the stomach contents of male eastern moles weigh significantly more than those of females, which is not terribly surprising given the sexual size dimorphism in this species (Hartman et al. 2000). In that same study, weight of the stomach contents of females averaged greater than that of males during months roughly corresponding to the time when females are expected to be lactating. Very similar seasonal patterns of food consumption also have been reported for European moles (Godfrey and Crowcroft 1960; Gorman and Stone 1990).

Hisaw (1923a) found that captive eastern moles preferred earthworms and white grubs, insect larvae, adult insects, and vegetable matter, in that order of preference. Whitaker and Schmeltz (1974) reported similar findings (see Table 2.9). In captivity, individuals of *Scalopus aquaticus* eat anything from ground beef (Hisaw 1923a) to mice and small birds (Christian 1950). Yates and Schmidly (1977) found that this species did well on Alpo dog food. Captive eastern moles have voracious appetites and consume large percentages of their body weights in food each day. The average daily consumption rate in captivity ranges from 32.1% (Hisaw 1923a) to 55.0% (Christian 1950) of body weight. Hisaw (1923a) showed that this rate increased to 66.6% when an eastern mole had not been fed for 24 hr. Fay (1954) found that captive hairy-tailed moles weighing 41, 49, and 45 g ate an average of 158.9, 132.4, and 116.0 g of earthworms per day, respectively. Shrew moles are capable of eating 1.4 times their body weight in 12 hr (Dalquest and Orcutt 1942). Glendenning (1959) reported that captive coast moles consume almost twice their body mass in earthworms daily. At least some of these reports of food consumption in captivity likely represent overestimates of food consumption in the wild. Newly captured moles frequently are frantic, which increases their energy requirements and thus their food consumption. In addition, in natural settings prey usually are mobile and relatively scarce; for this reason, it has been suggested that moles may be "hardwired" to snatch up virtually every prey item available to them (Jensen 1983). Reports of food consumption of captive moles frequently have been used as the basis for arguing that these animals have exceptionally high metabolic rates; however, such assertions largely are unfounded (Jensen 1983).

LIFE SPAN AND MORTALITY

Age Estimation. Most studies on moles, particularly those in North America, have been hampered by the lack of reliable data on age. Various methods have been employed to attempt to place moles into age classes. Glendenning (1959) used body mass and skin texture to estimate age of coast moles, but did not state his methods. Other researchers have used criteria based on tooth wear and/or other cranial characteristics (Godfrey and Crowcroft 1960; Hoslett and Imaizumi 1966; Skoczeń 1966; Leftwich 1972; Funmilayo 1976; Schaefer 1978; Yates 1978; Davis and Choate 1993). Ages of European moles have been estimated using the number of cementum annuli surrounding the roots of canine (Lodal and Grue 1985; Pankakoski et al. 1993; Yokohata 1999) or molar teeth (Katanova 1972). Using dental characteristics to estimate age of moles is appealing intuitively because successive cohorts always differ by 10–12 months of age. However, in none of the above studies did the authors have any known-age individuals to assess whether age had been determined accurately. Using a small series of known-age individuals, Hartman (1995b) found that ages of individuals of *Scalopus aquaticus* could be accurately assessed using wear on the upper cheek teeth (Fig. 2.11 and Table 2.10). The cementum annulus method has not been attempted with any species of North American mole; reliability of this method and the use of tooth wear criteria needs to be compared.

Age Structure. Population samples comprised of large numbers of moles collected from relatively large areas generally exhibit a pattern of successively fewer individuals in older age classes, which gives the impression of approximating a stable-age type of distribution. However, on a more local scale, mole populations are not in demographic equilibrium (i.e., do not approximate a stable age distribution) and the proportions of young individuals and adults vary among years and among habitats (Skoczeń 1966; Funmilayo 1976; Gorman and Stone 1990; Hartman 1995b). Age structures observed in local breeding populations of eastern moles (Hartman 1995b) are remarkably similar to those reported for Botta's pocket gopher (*Thomomys bottae*) (Patton and Smith 1990), where some populations consist of large numbers of relatively young individuals and very few old individuals, or the reverse. Populations of subterranean mammals tend to be limited by the number of territories available. Local dynamics of mole populations appear to involve the establishment of the members of one or a few cohorts in an area, with recruitment into the population then restricted until the current residents begin to die.

Survivorship. As a result of the lack of reliable estimates of age for most moles, there are relatively few data on survivorship and/or mortality in North American talpids. In *Scalopus aquaticus,* about 50% of individuals in a cohort are estimated to die within 6 months of birth (Hartman 1995b), adult survivorship is relatively high (Table 2.11), mortality is age dependent, and there is differential mortality between the sexes (Davis and Choate 1993; Hartman 1995b). Low survivorship in the youngest age class of moles likely is due at least in part to risks associated with leaving the maternal home range to find and establish a territory (Godfrey and Crowcroft 1960; Skoczeń 1962; Pedersen 1963; Giger 1965). Juvenile eastern moles disperse from the maternal home range by 6 months after birth (Leftwich 1972) and increased survivorship after this time may reflect the fact that individuals have established territories. Higher estimates of mortality in the youngest age class are

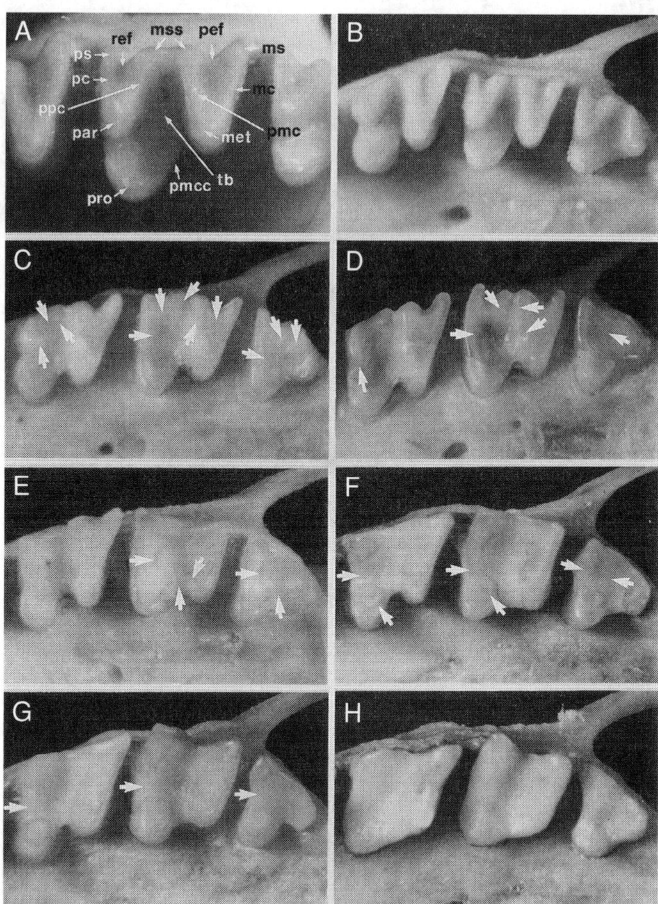

FIGURE 2.11. Representative year-interval tooth wear morphs of an eastern mole (*Scalopus aquaticus*) from South Carolina. Tooth wear morphs shown are representative of individuals captured during late winter and early spring. In each frame, left is anterior, right is posterior, top is buccal, and bottom is lingual. (A) Upper left second molar of juvenile, showing cusps and cristae; (B) juvenile; (C) 1-year-old; (D) 2-year-old; (E) 3-year-old; (F) 4-year-old; (G) 5-year-old; (H) 6-year-old. Nomenclature: mc, metacrista; met, metacone; ms, metastyle, mss, anterior and posterior mesostyles; par, paracone; pc, paracrista; pef, postectoflexus; pmc, premetacrista; pmcc, postmetaconule crista; ppc, postparacrista; pro, protocone; ps, parastyle; ref, pre-ectoflexus; tb, trigon basin. Arrows in panels c–g point to key characteristcs used for estimating age; see Table 2.10. SOURCE: Hartman 1995b:112. Reprinted with permission of Cambridge University Press.

reported for mole populations more centrally located on large continents, where winters tend to be more severe.

Longevity. There have been only a few studies on North American moles providing reliable estimates of longevity. In a sample of 23 coast moles from British Columbia, Schaefer (1978) reported that the mean age of males was greater than that of females (estimated mean ages not stated); the oldest individuals were estimated to be between 3 and 4 years old. The maximum longevity of eastern moles in Kansas is estimated to be >3.5 years (Davis and Choate 1993) and the oldest individual of this species that Hartman (1995b) observed in his study in South Carolina had entered its seventh year of life. On average, female eastern moles live longer than males in Kansas and in South Carolina. Some female *S. aquaticus* in South Carolina live 6.2 years or more and some males, 5.9 years or more. Interestingly, male European moles live longer on average than do females (Katanova 1972; Funmilayo 1976; Lodal and Grue 1985). Several of the characteristics of eastern and European moles suggest that the breeding system in both these species is polygyny (Gorman and Stone 1990; Hartman 1992; Davis and Choate 1993). Males in polygynous mating systems are predicted to have shorter life spans than females and thus male European moles

TABLE 2.10. Tooth wear criteria used to place individuals of *Scalopus aquaticus* from South Carolina into year-interval age classes

Juvenile
All cusps and cristae distinct and covered with enamel; pre- and postectoflexi sometimes with very slight exposure of dentine; if dI^1 present, long, slender and lying along anterior medial border of pI^1

One year old
M^3: Paracone obliterated; postparacrista with some enamel usually still present; if not, then still discernible as a ridge of dentin; premetacrista barely discernible or obliterated; trigon basin with dentin exposed, occlusal surface beginning to appear shallow, bowl-like
M^2: Paracone obliterated or very nearly so (if present, only base persists); pre-ectoflexus confluent with trigon basin, dentin exposed; postparacrista obliterated or nearly so; postectoflexus with dentin exposed; premetacrista partially obliterated, but to lesser degree than postparacrista
M^1: Paracone worn, but with lingual face of enamel still present at base; pre-ectoflexus partially to completely confluent with trigon basin; postparacrista extremely worn or obliterated

Two years old
M^3: Occlusal surface worn smooth, bowl-shaped
M^2: Paracone and postparacrista obliterated; postectoflexus confluent with trigon basin (but see description of premetacrista); premetacrista obliterated, or sometimes with small portion persisting near mesostyle
M^1: Paracone essentially obliterated, no lingual face of enamel present, or face extending posteriorly about 1 mm or less

Three years old
M^3: Trigon basin bowl-like with distinctive ring differentiating dentin layers; postmetaconule crista worn away leaving furrow of dentin between protocone and metacone
M^2: Trigon basin bowl-like with distinctive ring differentiating dentin layers; postmetaconule crista worn away leaving furrow of dentin between protocone and metacone; premetacrista obliterated except for lingual-most portion adjoining metacone

Four years old
M^3: Trigon basin appearing as furrow of exposed dentin extending from lingual to buccal faces of tooth; paracrista persisting as ridge of dentin, dentin of ridge differentiated from exposed dentin in trigon basin
M^2: Buccal face of protocone a series of concentric rings; paracrista obliterated or very nearly so, leaving furrow of dentin between protocone and parastyle
M^1: Buccal face of protocone a series of concentric rings; paracrista obliterated, leaving furrow of dentin between protocone and parastyle

Five years old
M^3, M^2, and M^1: Paracrista obliterated, leaving furrow of dentin between protocone and parastyle and not differentiated from dentin of trigon basin

Six years old or older
M^3, M^2, and M^1: Enamel on occlusal surface, if present, restricted to protocone, parastyle, and/or metastyle; all teeth showing extreme wear. First incisors and canines blunt-tipped, sometimes with pulp cavities exposed or very nearly so

SOURCE: Adapted from Hartman 1995b.

NOTE: Descriptions are based on specimens collected during late winter and spring. M, M2, and M3, First, second, and third upper molars, respectively; dI^1, deciduous first upper incisor; pI^1, permanent first upper incisor. See Figure 2.11 for identification of various cusps, cristae, and representative tooth wear morphs.

would be predicted to have shorter life spans than females. Why male European moles live longer than females is not clear. The possibility that reduced life span of female European moles might be related to their production of androgens has not been investigated. Longevity of mammals scales allometrically with body mass (Calder 1984), and eastern moles from Kansas and European moles are almost twice as heavy as *Scalopus aquaticus* from South Carolina. On this basis, eastern moles from South Carolina are expected to have shorter life spans than moles either in Kansas or Europe, which they do not.

Predation. Because of their subterranean habits, moles have few major predators except humans. They are, however, taken on occasion by a variety of organisms, especially when forced from their burrow systems or during dispersal periods. Hawks and owls are the most significant predators of North American moles (Hamilton 1931; Silver 1933; Arlton 1936; von Bloeker 1937; Cowan 1942; Dalquest and Orcutt 1942; Selleck and Glading 1943; Roest 1952; Parmalee 1954; Giger 1965; Maser and Brodie 1966; Craighead and Craighead 1969; Forsman and Maser 1970; Choate 1971; Giger 1973; Hafner 1974; Hallett 1978). Various mammalian carnivores known to prey on moles are red (*Vulpes vulpes*) and gray foxes (*Urocyon cinereoargenteus*) (Hamilton et al. 1937; Eadie 1939; Scott 1943; Davis 1951), coyotes (*Canis latrans*) (Toweill and Anthony 1988), weasels (*Mustela* sp.) (Silver 1933; Hall 1951), pine martens (*Martes americana*) (Zielinski et al. 1983), fishers (*Martes pennanti*) (Giuliano et al. 1989), raccoons (*Procyon lotor*) (Dalquest and Orcutt 1942), and skunks (*Mephitis* spp.) (Hamilton 1931). Dogs and cats kill moles fairly frequently, but rarely if ever eat them (Hamilton 1939; Dalquest and Orcutt 1942; Glendenning 1959; Gordon and Bailey 1963; Maser et al. 1981). There also have been reports of various venomous and nonvenomous snakes consuming moles (Saylor 1938; Dalquest and Orcutt 1942; Hamilton 1943; Davis 1951; Maser et al. 1981). Krull (1969) reported observing an eastern chipmunk (*Tamias striatus*) feeding on a star-nosed mole. Bullfrogs (*Rana catesbeiana*) (Heller 1927) and opossums (*Didelphis virginiana*) (Sperry 1933) have been reported feeding on hairy-tailed moles. In addition, Eadie (1939) speculated that short-tailed shrews (*Blarina*) may be predators of nestling hairy-tailed moles.

Parasites. Little comprehensive work on the parasites of moles is available. Ectoparasites have received the most attention; a summary is provided in Table 2.12. Yates et al. (1979) provided the only account concerning the ectoparasites of all North American species of mole. They found that host specificity was not pronounced among the species of ectoparasite examined and suggested that most are acquired secondarily by moles from other rodents and insectivores that inhabit mole tunnels.

Fleas and mites are the most common ectoparasites found on moles. Twenty-three species of flea and at least 76 species of mite have been identified from North American moles (see Table 2.12). The mites *Androlaelaps fahrenholzi* and *Haemogamasus liponyssoides* have been reported from all species of mole in North America. Other ectoparasites reported from moles include four species of beetle in the genus *Leptinus,* at least two species of louse, and three species of tick. Densities on most hosts have been reported as light, although Yates et al. (1979) found 477 specimens of the chigger *Neotrombicula brennani* on an individual *Scapanus townsendii*. Eadie (1939) reported up to 24 fleas and 20 mites per host on specimens of *Parascalops breweri*.

Far less is known regarding the endoparasites of moles. Twenty species of coccidian protozoans have been reported (Duszynski 1985, 1989; Ford and Duszynski 1988, 1989). Among helminth parasites, at least five species of nematode, two species of trematode, three species

TABLE 2.11. Static life tables for moles captured during autumn

Species, Locality	Age Class (years)	n_x	l_x	q_x	e_x
Scalopus aquaticus, South Carolina ($N = 101$)	0–1	37	1.00	0.32	2.23
	1–2	25	0.68	0.20	2.06
	2–3	20	0.54	0.40	1.45
	3–4	12	0.32	0.50	1.08
	4–5	6	0.16	0.83	0.66
	5–6	1	0.02	1.00	0.50
Scalopus aquaticus, South Carolina ($N = 101 + 102$)[a]	0.0	102	1.00	0.64	1.49
	0.5	37	0.36	0.32	2.23
	1.5	25	0.24	0.20	2.06
	2.5	20	0.20	0.40	1.45
	3.5	12	0.12	0.50	1.08
	4.5	6	0.06	0.83	0.66
	5.5	1	0.01	1.00	0.50
Scalopus aquaticus, Kansas[c] ($N = 235$)	0–1	116	1.00	0.36	1.53
	1–2	74	0.64	0.51	1.11
	2–3	36	0.31	0.75	0.75
	3–4	9	0.08	1.00	0.50
Talpa europaea, The Netherlands ($N = 109$)	0–1	70	1.00	0.64	1.17
	1–2	20	0.36	0.51	1.38
	2–3	11	0.18	0.58	1.30
	3–4	4	0.07	0.50	1.39
	4–5	1	0.04	0.24	1.27
	5–6	3	0.03	1.00	0.50
Talpa europaea, Russia ($N = 71$)	0–1	47	1.00	0.83	1.01
	1–2	8	0.17	0.13	2.50
	2–3	7	0.15	0.43	1.79
	3–4	4	0.09	0.25	1.75
	4–5	3	0.06	0.33	1.12
	5–6	2	0.04	1.00	0.50

SOURCE: South Carolina, Hartman 1995b; Kansas, F. W. Davis and Choate 1993; The Netherlands, Lodal and Grue 1985; Russia, Katanova 1972.

NOTE: Column designations: n_x, number of survivors at start of age interval x; l_x, proportion of animals surviving to start of age interval x; q_x, rate of mortality during age interval x to $x + 1$; e_x, mean expectation of life for animals alive at start of age interval x.

[a]Includes an estimate of the size of the cohort born the previous spring.

of cestode, and two species of acanthocephalan are known to occur in moles (see Table 2.12). Elshoff (1989) reported finding a specimen of *Scalopus aquaticus* with parasitic hepatitis; she hypothesized that the malady most likely was caused by a "*Schistosoma*-type" parasite.

ECONOMIC STATUS AND MANAGEMENT

The economic status of moles has been debated for years with little consensus as to whether they are harmful or beneficial. Because of the diverse lifestyles and wide range of habitats occupied by these creatures, any discussion of economic status of North American moles must, by necessity, center on individual species. *Condylura cristata, Neürotrichus gibbsii,* and *Parascalops breweri* must be considered economically neutral, if not beneficial. All three tend to be rather localized in distribution and usually occur in areas not suitable for cultivation. The occasional damage by star-nosed and hairy-tailed moles to lawns, flower beds, and golf courses is more than offset by their destruction of harmful insects and tilling and aeration of soils.

When and where they do not come in contact with humans, eastern moles and the various species of *Scapanus* should be viewed as being beneficial for the same reasons discussed above. However, these moles have been associated with some economic loss by farmers, commercial bulb growers, produce growers, and home gardeners. Major damage by Townsend's moles is incurred as a result of one to four molehills discharged per day as a mole extends it tunnel system in search of food. Economic loss to dairy farmers in Tillamook County, Oregon, was estimated to be $100,000/year as a result of forage loss, increased equipment breakdown, and poor silage quality (Wick 1962). Forage production in pastures can be reduced by 10–50% because of being covered by molehills. During harvest operations, soil and rocks in the molehills cause equipment breakdown and stoppage, and the resultant mix of forage and dirt affects silage quality. Commercial bulb growers, produce growers, and home gardeners frequently experience losses as a direct result of damage to bulbs and roots and the indirect effect of underground portions of plants being desiccated. Moles frequently extend tunnel systems under row crops in search of earthworms. This problem is especially pronounced in areas where *Scalopus aquaticus* is common. During the process of tunneling, some bulbs and roots may be ingested as food, but more frequently damage to the plant results from the removal of soil around the bulb or root, leading to desiccation. Lawns and golf courses also are subject to mole damage. In addition, molehills provide an excellent medium for "weed" seed germination. Mole damage often is confused with damage done by gophers (*Thomomys* sp.), voles (*Microtus* sp.), and other small mammals. As noted earlier, the use of mole tunnels by other mammals is common and frequent.

Inspection of damaged plants usually will help in determining which type of animal is responsible. Damage to plant material by moles is confined to underground parts: damaged bulbs will appear to have been shredded without visible incisor tooth marks. Gophers, because of their incisiform tooth structure and habits of "barking" or "clipping" plants, often cause considerable damage. Incisiform tooth marks and angular stem clippings are characteristic of gopher damage. Voles occasionally will girdle small woody plants and sometimes the roots of fruit trees; however, their incisiform tooth marks are different, and appear as very narrow striations.

Control Methods. We are contacted routinely by persons frustrated to the point of being "at the end of their rope" as the result of trying to rid their property of moles. It is primarily for this reason, and not necessarily because of widespread and large-scale economic impact, that general treatises on moles so often emphasize methods for trapping, repelling, or removing them. Methods for controlling subterranean mammals vary with the species causing the damage, making it important to identify the target species.

There are and have been a myriad of control methods applied to moles. We knew of a woman in Ohio some years ago who would wait to see a mole digging a new tunnel in her yard; she then would strike the soil with a squarely placed axe blade an inch or two behind the area where she saw the soil being raised, both killing the mole and eliminating the need to handle it. An Oregon native waited in a rocking chair with a shotgun with similar results. Although reasonably effective, such control methods can be time-consuming and can have their own particular hazards associated with them.

The most common control method employed in noncommercial situations is trapping. Trapping moles is practical for home owners and on relatively small properties, but is time-consuming and uneconomical for large commercial acreage. Trapping requires locating a portion of the tunnel system either visually or by probing, excavating a trap hole (for certain types of traps), and placing the traps. Traps commonly employed are "cinch," "squeeze," or "scissors" traps, spear or harpoon traps, and choker-loop traps. All require aligning the trap with the burrow system and properly placing the trigger on a collapsed or blocked portion of tunnel. The mole is killed when it attempts to clear the runway system, discharging the trap.

Members of North American genera of moles other than *Scalopus* and *Scapanus* often can be livetrapped successfully by sinking a 1-gallon container below the mole tunnel and covering the top of the excavated tunnel with a board. Individuals of all of the more purely subterranean species, including star-nosed moles, can be taken in specially constructed livetraps (Moore 1940; Yates and Schmidly 1975; Jensen 1982). Star-nosed moles also can be captured using metal box-type rodent livetraps placed in tunnels. Livetrapping moles likely will be considered too time-consuming by the person who simply wishes to rid their property of the animals. Most types of traps have the disadvantage

TABLE 2.12. Ectoparasites, endoparasites, and phoronts reported for North American moles

Parasite	Host						
	Condylura cristata	*Neürotrichus gibbsii*	*Parascalops breweri*	*Scalopus aquaticus*	*Scapanus latimanus*	*Scapanus orarius*	*Scapanus townsendii*
Protozoans							
Cyclospora ashtabulensis			20				
C. megacephali				19			
C. parascalopi			20				
Eimeria aethiospora			20				
E. aquatici				19			
E. condylurae	10						
E. heterocapita		9					
E. motleiensis				19			
E. neurotrichi		9					
E. parastiedica		9					
E. scalopi				19			
E. titthus			20				
Isospora aquatici				19			
I. ashtabulensis			20				
I. condylurae	10						
I. cristatae	10						
I. lamoillensis	10						
I. motleiensis				19			
I. neurotrichi		9					
I. parascalopi			20				
Nematoda							
Undetermined species	23		11				
Capillaria maseri		49					
Physaloptera limbata				46			
Porrocaecum americanum			67	67			
Rictularia scalopis				67			
Tricholinstowia maseri						8	
Trematoda							
Brachylaema condylura	45						
Ityogonimus scalopi				56			
Cestoda	23					21	
Hymenolepis scalopi				50			
Hymenolepis sp.		5		46			
Rodentolepis olsoni							44
Staphylocystis bacillaris				66			
Acanthocephala							
Macracanthorhynchus ingens			68				
Moniliformis clarki			1	1, 46			
Moniliformis sp.			11				
Mites							
Androlaelaps casalis						64	64
A. fahrenholzi	43, 55, 57, 59, 69	69	69	57, 62, 63, 69	54, 69	64	54, 61, 64, 69
Anoetidae	57						
Atricholaelaps glasgowi				46			
Bakerdania sp.		64	60				64
B. jonesi		41					
B. plurisetosa			60	41, 57			
Cyrtolaelaps sp.	57		60				
Dermacarus hypudaei	15						
Eadiea brevihamata		30, 64					
E. condylurae	43, 57, 59						
E. neurotrichus		30, 39, 64					
E. scapanus							16, 61, 64
Echinonyssus blarinae			60	62, 69		69	69
E. isabellinus	59						
E. obsoletus						64	24, 61, 64
E. talpae	43, 55						
Eulaelaps stabularis	43, 55, 57, 59			57, 62, 63		64	61, 64
Euryparasitus sp.	57	64	60				64
Eushoengastia trigenuala				57			
Eutalpacarus peltatus		30, 61, 64					
Glycyphagus hypudaei	59						60, 61, 64
Haemogamasus ambulans	43, 55, 57, 59, 63			57, 62		64	34, 35, 64

(*Continues*)

Table 2.12—*Continued*

Parasite	Host						
	Condylura cristata	*Neürotrichus gibbsii*	*Parascalops breweri*	*Scalopus aquaticus*	*Scapanus latimanus*	*Scapanus orarius*	*Scapanus townsendii*
H. harperi				35, 57, 62, 63, 65			
H. keegani		33, 34, 35, 64			69		33, 61, 65
H. liponyssoides	35, 57, 59	48	35, 60	35, 57, 62, 63, 69	48, 63, 69	69	35, 69
H. occidentalis		35, 64, 65				35, 53, 64, 65	35, 61, 64, 65
H. reidi			69			64	61, 64
Hirstionyssus sp.			31				
H. blarinae			24	24, 57		69	
H. obsoletus		32				64	24
H. talpae	57						
H. utahensis		32, 34, 64					
Histiostomatidae			60				
Hypoaspis miles			60				
Labidophorus nearcticus[a]			18, 60				
Laelaps alaskensis			60				
L. kochi		61, 64					
Listrophorus mexicanus							61, 64
Machrocheles sp.	57			57			
M. ontariensis				36			
M. polypunctatus							36
Neotrombicula brennani							69
N. cavicola							2
N. microti			60				
Ornithonyssus bacoti				57, 62			
O. sylvarium				57			
Orycteroxenus canadensis	43, 59, 63						
O. soricis	59						
Proctolaelaps sp.			60				17
Protomyobia brevisetosa							61, 64
Pseudoparasitus sp.		61, 64					
Pygmephorus sp.			60				64
P. brevicaudae			60				
P. brevipes						4	
P. designatus			60	7, 51, 57		4	
P. erlangensis		51	60				
P. faini	51						
P. forcipatus		4					
P. hamiltoni			60				
P. hastatus	7		60	40, 41, 51, 57			
P. horridus	41, 51, 58, 59	4, 51	60	40, 51, 57		4, 64	
P. johnstoni		61, 64	60				
P. lutterloughae	58, 59		60				51, 61
P. moreohorridus	41, 51, 57, 58, 59						
P. muhunkai				51, 57		4	51, 61
P. plurispinosus		4					
P. proctorae			52, 60				
P. rackae			60	7		4	
P. scalopi				7, 40, 41, 51, 57			
P. spinosus	40, 41, 51		60	6		4	
P. stammeri			60				
P. tamiasi			60				
P. whartoni			60	51, 57			
P. whitakeri	51	4	40, 51, 60	7, 40, 41, 51, 57		4, 51	
P. wrenschae			60				
Radfordia arborimus							61
Scalopacarus obesus				57, 62, 63			
S. scapanus							17
Xenoryctes latiporus				57, 62, 63			
X. nudus	59						
Ticks							
Ixodes angustus		69			69	69	13, 69
I. pacificus						22	
I. soricis		13					
Fleas							
Catallagia charlottensis							37
C. decipiens						69	

TABLE 2.12—*Continued*

Parasite	Host						
	Condylura cristata	*Neürotrichus gibbsii*	*Parascalops breweri*	*Scalopus aquaticus*	*Scapanus latimanus*	*Scapanus orarius*	*Scapanus townsendii*
C. sculleni		12, 26, 27, 38, 69					37
Cediopsylla simplex				57			
Ceratophyllus wickhami	23						
Corypsylla jordani		12, 26, 27, 38, 69			69		
C. ornata		69			69	21, 69	12, 27, 29, 39, 69
Ctenopthalmus pseudagyrtes	23, 57		3, 11, 25, 31, 60	46, 57			
C. wenmanni	23						
Doratopsylla blarinae			3, 60				
Epitedia jordani		28					
E. scapani		12, 38, 69				21	28, 37
Foxella ignota							37
Histrichopsylla tahavuana			3, 25, 60				
Megabothris abantis							37
M. acerbus			25				
M. asio			60				
Nearctopsylla genalis			60	57			37
N. hygini			11				
N. jordani						69	27, 29, 37, 69
N. martyoungi							37
Peromyscopsylla hesperomys			3				
Stenoponia americana				57			
Lice							
Euhaematopinus abnormis			11				
Haematopinoides squamosus			60	57			
Beetles							
Leptinus americanus			31	57			
L. occidentamericanus		47					
L. orientamericanus	14						
L. testaceus		42		46			

SOURCE: 1, Benton 1954; 2, Brennan and Wharton 1950; 3, Conner 1960; 4, Cudmore et al. 1987; 5, Dalquest and Orcutt 1942; 6, Dastych et al. 1991; 7, Dastych et al. 1992; 8, Durrete-Desset and Vaucher 1974; 9, Duszynski 1985; 10, Duszynski 1989; 11, Eadie 1939; 12, Easton 1983; 13, Easton and Goulding 1974; 14, Eckerlin and Painter 1993; 15, Fain and Whitaker 1973; 16, Fain and Whitaker 1975; 17, Fain and Whitaker 1981; 18, Fain and Whitaker 1985; 19, Ford and Duszynski 1988; 20, Ford and Duszynski 1989; 21, Glendenning 1959; 22, Gregson 1956; 23, Hamilton 1931; 24, Herrin 1970; 25, Holland and Benton 1968; 26, Hubbard 1940a; 27, Hubbard 1940b; 28, Hubbard 1940c; 29, Hubbard 1949; 30, Jameson 1949; 31, Jameson 1950a; 32, Jameson 1950b; 33, Jameson 1952; 34, Jameson and Brennan 1957; 35, Keegan 1951; 36, Krantz and Whitaker 1988; 37, R. E. Lewis et al. 1988; 38, R. E. Lewis and Maser 1981; 39, Lukoschus et al. 1980; 40, Mahunka 1973; 41, Mahunka 1975; 42, Maser and Hooven 1971; 43, Mumford and Whitaker 1982; 44, Neiland and Senger 1952; 45, Odlaug 1952; 46, Olive 1950; 47, Peck 1982; 48, Radovsky 1960; 49, Rausch and Rausch 1973; 50, Schultz 1939; 51, Smiley and Whitaker 1979; 52, Smiley and Whitaker 1984; 53, Spencer 1941; 54, Strandtmann 1949; 55, Timm 1975; 56, Turner and McKeever 1980; 57, Whitaker 1982; 58, Whitaker et al. 1982; 59, Whitaker and French 1982; 60, Whitaker and French 1988; 61, Whitaker and Maser 1985; 62, Whitaker and Schmeltz 1974; 63, Whitaker and Wilson 1974; 64, Whitaker et al. 1979; 65, Williams et al. 1978; 66, Yamaguti 1959; 67, Yamaguti 1961; 68, Yamaguti 1963; 69, Yates et al. 1979.

[a]Reported as *L. talpae* by Fain and Whitaker 1973.

of being conspicuous, which can make them subject to theft or tampering; in addition, injury to children or pets can result if appropriate precautions are not observed.

Moles also can be captured alive by spading or by hand. These techniques work well only when dealing with animals whose workings include surface ridges. The animal is captured when found digging a new surface tunnel or repairing a tunnel that had been pressed down. A spade is placed just behind the digging animal, blocking the mole's entry back into the newly repaired tunnel; the mole then quickly is brought up to the surface with the spade. Alternatively, after blocking the newly repaired tunnel behind the creature, the turf can be quickly torn open to expose the mole, which then can be picked up by its fur. Like other live capture techniques, spading and capture by hand both require some skill, probably will be considered by the average individual to be too time-consuming, and leave the trapper with a live mole to be disposed of somehow.

Chemical control agents, including sulfate, strychnine, sodium monofluoroacetate (1080), and phosphorus compounds, formerly were the chemicals most commonly used commercially to control moles. Recent federal and state restrictions on bait manufacturers and commercial applicators have diminished the use and manufacture of these chemicals. Thallium sulfate, 1080, and phosphorus compounds no longer are available to the general public or commercial bait manufactures. Strychnine baits are available in some states for both commercial and public use. However, application of these agents often is restricted by a rigid licensing system and lacks specificity.

Baits used to control moles usually are treated cereal grains or composite mixtures of cereal grains, binders, and other ingredients particular to the individual bait producer. A common control agent found in commercial baits involves an anticoagulant chemical (e.g., chlorophacinone) that inhibits normal platelet function in the blood and results in internal hemorrhaging and death. Application of chemical control agents usually involves depositing the bait material into the runway system or excavating the tunnel and placing the bait. The most efficient and productive method is bait placement in the tunnel system using a probe. The degree of success in mole control usually is dependent on proper bait placement, intensity of effort, soil conditions, and bait acceptance (Elshoff 1989). Baits have the advantage of being relatively inconspicuous, but they are potentially toxic to children, pets, or other nontarget species, and often kill other occupants of the mole tunnels in greater numbers than the moles.

Other approaches to mole control include attempting to repel the animals or to remove their food source. Renardine is a bone-oil-based compound commonly used in the United Kingdom for repelling problem mammals. Atkinson and Macdonald (1994) reported that moles are effectively repelled by trenching around a prescribed area, placing Renardine in the trench, and then refilling it; unfortunately, to be effective this entire process must repeated about every 30 days. Such a method clearly can become expensive in terms of time and/or money, with the added problem that the product gives off an unpleasant odor. Applying insecticides or other chemicals to the soil to kill and eliminate prey for moles usually provides limited success. This type of practice potentially is hazardous, is ecologically unsound, and does not necessarily ensure that moles will cease to enter an area in search of food or potential mates during the breeding season.

Other methods of mole control that frequently are encountered are derived from folklore, home remedies, or other unproven methods. One of the most common home remedies used involves a noisemaker device placed in the yard to scare moles from the premises. One of these methods involves placing empty soft drink bottles at an angle with the bottom in the mole tunnel, necks sticking out. Supposedly, the wind blowing in the bottles make a piping sound that causes the mole to desert the runway. Other home remedies involve placing in the tunnel system an offensive odoriferous material or an injurious substance or mechanical device that causes physical impairment to the mole. These include engine exhaust fumes, moth balls, chewing gum, broken glass, razor blades, and thorns. Unfortunately, these remedies sometimes prove more hazardous to the applicator than to the moles. Results usually are nonevident.

Whether conflicts with humans are counterbalanced by the beneficial aspects of moles probably varies with each situation. In the vast majority of the cases, moles probably will be judged to be more beneficial than harmful. In cases where the reverse is true, control is difficult at best.

Although moleskins are of little value today, trade in them once was a viable business in Europe and the United States. From the seventeenth through nineteenth centuries, moleskins were used for caps, purses, tobacco pouches, and trimmings for garments. The demand for moleskins was such in nineteenth-century Germany that applications were made for state protection of the mole to prevent its extinction (Godfrey and Crowcroft 1960). Today, moleskins no longer are used in the U.S. clothing industry, although millions once were imported from Europe. Godfrey and Crowcroft (1960) reported that during 1959 in Britain, approximately 1 million moles were trapped for their skins. It is possible that a commercial outlet for the large Townsend's mole would enhance and complement any other control efforts.

CONSERVATION AND RESEARCH NEEDS

Conservation efforts currently do not appear to be warranted for any of the North American moles, largely because of the catholic diets and relatively wide range of habitats occupied by these creatures. The same cannot be said for desmans (*Desmana moschata* and *Galemys pyrenaicus*); these talpids, also known as water moles, are highly aquatic and once were common across much of Europe before the Pleistocene. Currently, *D. moschata* is restricted to a few rivers and lakes of European Russia and *G. pyrenaicus* occurs only in fast-flowing mountain streams in the northern Iberian peninsula and the French side of the Pyrenees Mountains. Pollution and habitat destruction have contributed to the continued reduction of population numbers of both species of desman.

Research is needed in virtually all aspects of the biology of North American moles. Moles compose a major component of the North American mammalian fauna and often are of considerable economic importance, yet published information on most species of this important group remains, at best, fragmented and incomplete. The lack of information concerning talpid biology may be a major reason that control and management have been essentially ineffective.

Detailed studies are needed on the demographic, social, and genetic structuring of mole populations. The validity of many such studies in large part will rely on the development and documentation of reliable criteria for determining the age of individuals. Data on life history, social interactions, juvenile dispersal, emigration and immigration, population size and density, and the degree to which any of these may vary geographically are not available for most North American talpids. After more than 200 years of research on mammals in North America, a seemingly simple characteristic like length of the gestation period remains unknown. Similarly, many of the basic physiological adaptations to the subterranean niche remain poorly understood. How well developed are the nontactile senses within and among species? How do moles navigate through their tunnel systems? How are moles able to exhibit homing, sometimes even across canals and rivers? Why do females of some species possess ovotestes and some not, and what is the functional significance of these organs? Within a given geographic area, individuals in mole populations exhibit amazing synchrony in their reproductive cycles. The environmental and/or endogenous cues that underlie these cycles remain unknown.

Another area for research that largely has been ignored, particularly in North America, is the role that the presence or absence of moles and their tunnels plays in the structure of local plant and animal communities. Because moles frequently consume mycorrhizal fungi, they may act as important dispersing agents. Virtually nothing is known about how the presence or absence of mole tunnels relates to the distribution and abundance of the many nontalpid animals that use them.

Because of their trophic position, moles appear good candidates for tracking various environmental pollutants, such as metals; the few studies that have been done along these lines have borne this out (Hegstrom and West 1989; Pankakoski et al. 1993). Even "simple" feeding habit studies have proved to be important in the documentation of biological diversity, indicating the presence of species not previously known to occur in an area (Hartman et al. 2000).

Subterranean mammals are similar in many features of their biology, but in some ways moles seem to differ from other mammals that live underground, such as pocket gophers. There is a clear need for studies to document the extent to which all the various aspects of the biology of moles compare with those of other fossorial and subterranean mammals. These kinds of data are critical in delineating those factors that, from an evolutionary standpoint, ultimately result in the patterns of variation observed in subterranean species. Finally, there is a need to extend this knowledge to the ecosystem level and to delineate the role of these species in the normal functioning of the below- and aboveground environments. It will be this understanding of the biological complexity of systems that will enable researchers to make predictions necessary for sound management decisions.

LITERATURE CITED

Allen, J. A. 1891. Allen on mammals from Texas and Mexico. Bulletin of the American Museum of Natural History 3:221.

Allison, T., and D. V. Cicchetti. 1976. Sleep in mammals: Ecological and constitutional correlates. Science 194:732–34.

Allison, T., and H. Van Twyver. 1970a. Sleep in the moles, *Scalopus aquaticus* and *Condylura cristata*. Experimental Neurology 27:564–78.

Allison, T., and H. Van Twyver. 1970b. Sensory representation in the neocortex of the mole, *Scalopus aquaticus*. Experimental Neurology 27:554–63.

Arlton, A. V. 1936. An ecological study of the mole. Journal of Mammalogy 7:349–71.

Armsby, A.,T. A. Quilliam, and H. Soehnle. 1966. Some observations on the ecology of the mole. Journal of Zoology (London) 149:110–12.

Atkinson, R. P. D., and D. W. Macdonald. 1994. Can repellents function as a nonlethal means of controlling moles (*Talpa europaea*)? Journal of Applied Ecology 31:731–36.

Baker, R. H. 1951. Two new moles (genus *Scalopus*) from Mexico and Texas. University of Kansas Publications, Museum of Natural History 5:17–24.

Bartels, H., R. Schmelzle, and S. Ulrich. 1969. Comparative studies of the respiratory function of mammalian blood. V. Insectivora: Shrew, mole and nonhibernating and hibernating hedgehog. Respiration Physiology 7:278–86.

Beane, J. C. 1995. New distributional records for the star-nosed mole, *Condylura cristata* (Insectivora: Talpidae), in North Carolina, with comments on its occurrence in the Piedmont region. Brimleyana 22:77–86.

Bedford, J. M., O. B. Mock, S. K. Nagdas, V. P. Winfrey, and G. E. Olson. 1999. Reproductive features of the eastern mole (*Scalopus aquaticus*) and the star-nose mole (*Condylura cristata*). Journal of Reproduction and Fertility 117:345–53.

Benton, A. H. 1954. Notes on *Moniliformis clarki* (Ward) in eastern New York (Moniliformidae: Acanthocephala). Journal of Parasitology 40:102–3.

Bickham, J. W., and R. J. Baker. 1979. Canalization model of chromosomal evolution. Carnegie Museum of Natural History , Bulletin 13:70–84.

Bradshaw, L., and D. E. Goldberg. 1989. Resource levels in undisturbed vegetation and mole mounds in old fields. American Midland Naturalist 121:176–83.

Brennan, J. M., and G. W. Wharton. 1950. Studies on North American chiggers, No. 3. The subgenus *Neotrombicula*. American Midland Naturalist 44:153–97.

Brooks, F. E. 1908. Notes on the habits of mice, moles and shrews. Bulletin of the West Virginia University Agricultural and Forestry Experiment Station 113:87–133.

Brooks, F. E. 1923. Moles destroy wasps' nests. Journal of Mammalogy 4:183.

Brown, J. C., A. J. Buglass, J. R. Flowerdew, C. Khazanehdari, and J. S. Waterhouse. 1994. Identity of the enlarged inguinal glands of the mole (*Talpa europaea*)—anal or preputial glands? Journal of Zoology (London) 234:674–77.

Brown, L. N. 1972. Unique features of tunnel systems of the eastern mole in Florida. Journal of Mammalogy 53:394–95.

Brown, R. M., and A. M. Waterbury. 1971. Karyotype of a female shrew-mole *Neürotrichus gibbsii gibbsii*. Mammalian Chromosomes Newsletter 12:45.

Calder, W. A., III. 1984. Size, function, and life history. Harvard University Press, Cambridge, MA.

Campbell, B. 1939. The shoulder anatomy of the moles. A study of phylogeny and adaptation. American Anatomist 64:1–39.

Campbell, K. L., and P. W. Hochachka. 2000. Thermal biology and metabolism of the American shrew-mole , *Neürotrichus gibbsii*. Journal of Mammalogy 81:578–85.

Campbell, K. L., I. W. McIntyre, and R. A. MacArthur. 1999. Fasting metabolism and thermoregulatory competence of the star-nosed mole, *Condylura cristata* (Talpidae: Condylurinae). Comparative Biochemistry and Physiology A 123:293–98.

Campbell, K. L., I. W. McIntyre, and R. A. MacArthur. 2000. Postprandial heat increment does not substitute for active thermogenesis in cold-challenged star-nose moles (*Condylura cristata*). Journal of Experimental Biology 203:301–10.

Carraway, L. N., and B. J. Verts. 1991. Pattern and color aberrations in pelages of *Scapanus townsendii*. Northwest Science 65:16–21.

Carraway, L. N., L. F. Alexander, and B. J. Verts. 1993. *Scapanus townsendii*. Mammalian Species 434:1–7.

Catania, K. C. 1995a. Structure and innervation of the sensory organs on the snout of the star-nose mole. Journal of Comparative Neurology 351:536–48.

Catania, K. C. 1995b. A comparison of the Eimer's organs of the three North American moles: The hairy-tailed mole (*Parascalops breweri*), the star-nosed mole (*Condylura cristata*), and the eastern mole (*Scalopus aquaticus*). Journal of Comparative Neurology 354:150–60.

Catania, K. C. 1995c. Magnified cortex in star-nosed moles. Nature 375:453–54.

Catania, K. C. 1996. Ultrastructure of the Eimer's organ of the star-nosed mole. Journal of Comparative Neurology 365:343–54.

Catania, K. C. 1999. A nose that looks like a hand and acts like an eye: The unusual mechanosensory system of the star-nosed mole. Journal of Comparative Physiology A 185:367–72.

Catania, K. C. 2000. Epidermal sensory organs of moles, shrew-moles, and desmans: A study of the family Talpidae with comments on the function and evolution of Eimer's organ. Brain Behavior and Evolution 56:146–74.

Catania, K. C., and J. H. Kaas. 1995. Organization of the somatosensory cortex of the star-nosed mole. Journal of Comparative Neurology 351:549–67.

Catania, K. C., and J. H. Kaas. 1996. The unusual nose and brain of the star-nosed mole. Bioscience 46:578–86.

Catania, K. C., and J. H. Kaas. 1997a. Organization of somatosensory cortex and distribution of corticospinal neurons in the eastern mole (*Scalopus aquaticus*). Journal of Comparative Neurology 378:337–53.

Catania, K. C., and J. H. Kaas. 1997b. Somatosensory fovea in the star-nose mole: Behavioral use of the star in relation to innervation patterns and cortical representation. Journal of Comparative Neurology 387:215–33.

Catania, K. C., R. G. Northcutt, and J. H. Kaas. 1999. The development of a biological novelty: A different way to make appendages as revealed in the snout of the star-nosed mole *Condylura cristata*. Journal of Experimental Biology 202:2719–26.

Choate, J. R. 1971. Notes on geographic distribution and habitats of mammals eaten by owls in southern New England. Transactions of the Kansas Academy of Science 74:212–16.

Christian, J. J. 1950. Behavior of the mole (*Scalopus*) and the shrew (*Blarina*). Journal of Mammalogy 31:281–87.

Conaway, C. H. 1959. The reproductive cycle of the eastern mole. Journal of Mammalogy 40:180–94.

Conaway, C. H., and S. O. Landry, Jr. 1958. Rudimentary mandibular teeth of *Scalopus aquaticus*. Journal of Mammalogy 39:58–64.

Conner, P. F. 1960. The small mammals of Otsego and Schoharie Counties, New York. New York State Museum and Science Service Bulletin 382:1–84.

Contreras, L. C., and B. K. McNab. 1990. Thermoregulation and energetics in subterranean mammals. Pages 231–50 *in* E. Nevo and O. A. Reig, eds. Progress in clinical and biological research. Vol. 335, Evolution of subterranean mammals at the organismal and molecular levels. Wiley-Liss, New York.

Couch, L. K. 1924. Mice and moles. Journal of Mammalogy 5:264.

Cowan, I. M. 1942. Food habits of the barn owl in British Columbia. Murrelet 23:48–53.

Craighead, J. J., and F. C. Craighead, Jr. 1969. Hawks, owls and wildlife. Dover, New York.

Cudmore, W. W., J. O. Whitaker, Jr., and R. L. Smiley. 1987. Mites of the genus *Pygmephorus* in Oregon. Acaralogia 28:331–32.

Dalquest, W. W. 1948. Mammals of Washington. University of Kansas Publications, Museum of Natural History 2:1–444.

Dalquest, W. W., and D. R. Orcutt. 1942. The biology of the least shrew-mole, *Neürotrichus gibbsii minor.* American Midland Naturalist 27:387–401.

Dastych, H., G. Rack, and N. Wilson. 1991. Notes on mites of the genus *Pygmephorus* (Acari: Heterostigmata) associated with North American mammals. Zoologisches Institut und Zoologisches Museum Hamburg Entomologische Mitteilungen 88:161–74.

Dastych, H., G. Rack, and N. Wilson. 1992. Notes on mites of the genus *Pygmephorus* (Acari, Heterostigmata) associated with North American mammals (Part II). Zoologisches Institut und Zoologisches Museum Hamburg Entomologische Mitteilungen 89:141–56.

Davis, D. E., and F. Peek. 1970. Litter size of the star-nosed mole (*Condylura cristata*). Journal of Mammalogy 51:156.

Davis, W. B. 1942. The moles (genus *Scalopus*) of Texas. American Midland Naturalist 27:380–86.

Davis, W. B. 1951. Eastern mole eaten by cottonmouth and gray fox. Journal of Mammalogy 32:114–15.

Davis, F. W., and J. R. Choate. 1993. Morphologic variation and age structure in a population of the eastern mole, *Scalopus aquaticus*. Journal of Mammalogy 74:1014–25.

Durette-Desset, M. C., and C. Vaucher. 1974. Nématodes héligmasomes parasites d'insectivores talpidés de la région holarctique. Annales de Parasitologie Humaine et Comparee 49:191–200. (In French, with English summary.)

Duszynski, D. W. 1985. Coccidian parasites (Apicomplexa: Eimeriidae) from Insectivores: New species from shrew moles (Talpidae) in the United States. Journal of Protozoology 32:577–80.

Duszynski, D. W. 1989. Coccidian parasites (Apicomplexa: Eimeriidae) from insectivores. VIII. Four new species from the star-nosed mole, *Condylura cristata*. Journal of Parasitology 75:514–18.

Dyche, L. L. 1903. Food habits of the common garden mole. Transactions of the Kansas Academy of Science 18:183–86.

Eadie, W. R. 1939. A contribution to the biology of *Parascalops breweri*. Journal of Mammalogy 20:150–73.

Eadie, W. R. 1947. The accessory reproductive glands of *Parascalops* with notes on homologies. Anatomical Record 97:239–52.

Eadie, W. R. 1948a. Corpora amylacea in the prostatic secretion and experiments on the formation of a copulatory plug in some insectivores. Anatomical Record 102:259–72.

Eadie, W. R. 1948b. The male accessory reproductive glands of *Condylura* with notes on a unique prostatic secretion. Anatomical Record 101:59–79.

Eadie, W. R. 1951. A comparative study of the male accessory genital glands of *Neürotrichus*. Journal of Mammalogy 32:36–43.

Eadie, W. R. 1954. Skin gland activity and pelage descriptions in moles. Journal of Mammalogy 35:186–96.

Eadie, W. R., and W. J. Hamilton, Jr. 1956. Notes on reproduction in the star-nosed mole. Journal of Mammalogy 37:223–31.

Easton, E. R. 1983. Ectoparasites in two diverse habitats in western Oregon. III. Interrelationships of fleas (Siphonaptera) and their hosts. Journal of Medical Entomology 20:216–19.

Easton, E. R., and R. L. Goulding. 1974. Ectoparasites in two diverse habitats in western Oregon. I. *Ixodes* (Acarina: Ixodidae). Journal of Medical Entomology 11:413–18.

Eckerlin, R. P., and H. F. Painter. 1993. The star-nosed mole, *Condylura cristata,* a new host for *Leptinus orientamericanus* (Coleoptera, Leptinidae) in Virginia. Proceedings of the Entomological Society of Washington 95:639.

Eimer, T. 1871. Die Schnauze des Maulwurfes als Tastwerkzeug. Archiv für Mikroskopische Anatomie 7:181–91.

Elshoff, D. K. 1989. The effectiveness of Orco mole bait in controlling mole damage. M.S. Thesis, Michigan State University, East Lansing.

Fain, A., and J. O. Whitaker, Jr. 1973. Phoretic hypopi of North American mammals (Acarina: Sarcoptiformes, Glycyphagidae). Acaralogia 15:144–70.

Fain, A., and J. O. Whitaker, Jr. 1975. Two new species of Myobiidae from North American mammals (Acarina). Société Royal Belge d'Entomologie, Bulletin et Annales 111:57–65.

Fain, A., and J. O. Whitaker, Jr. 1981. *Scalopacarus scapanus* sp. n. (Acari: Glycyphagidae), a new fur mite from *Scapanus townsendii* in the USA. Journal of Parasitology 67:111–12.

Fain, A., and J. O. Whitaker, Jr. 1985. *Labidophorus nearcticus* n. sp. (Astigmata: Labidophorinae) a new glycyphagid mite from *Parascalops breweri* in the United States. Journal of Parasitology 71:327–30.

Fay, F. H. 1954. Quantitative experiments on food consumption of *Parascalops breweri.* Journal of Mammalogy 35:107–9.

Foote, L. E. 1941. A swimming hairy-tailed mole. Journal of Mammalogy 22:452.

Ford, P. L., and D. W. Duszynski. 1988. Coccidian parasites (Apicomplexa: Eimeriidae) from Insectivores. VI. Six new species from the eastern mole, *Scalopus aquaticus.* Journal of Protozoology 35:223–26.

Ford, P. L., and D. W. Duszynski. 1989. Coccidian parasites (Apicomplexa: Eimeriidae) from insectivores. VII. Six new species from the hairy-tailed mole, *Parascalops breweri.* Journal of Parasitology 75:508–13.

Forsman, E., and C. Maser. 1970. Saw-whet owl preys on red tree mice. Murrelet 51:10.

Funmilayo, O. 1976. Age determination, age distribution, and sex ratio in mole [*sic*] population. Acta Theriologica 21:207–15.

Funmilayo, O. 1979. Food consumption, preferences, and storage in the mole. Acta Theriologica 25:379–89.

Garner, H. W. 1973. Population dynamics, reproduction, and activities of the kangaroo rat, *Dipodomys ordii,* in west Texas. Graduate Studies, Texas Tech University 7:1–28.

Gaughran, G. R. L. 1954. A comparative study of the osteology and myology of the cranial and cervical regions of the shrew, *Blarina brevicauda,* and the mole, *Scalopus aquaticus.* Miscellaneous Publications, Museum of Zoology, University of Michigan 80:1–82.

Giacometti, I., and H. Machida. 1965. The skin of the mole (*Scapanus townsendii*). Anatomical Record 153:31–39.

Giger, R. D. 1965. Surface activity of moles as indicated by remains in barn owl pellets. Murrelet 46:32–36.

Giger, R. D. 1973. Movements and homing in Townsend's mole near Tillamook, Oregon. Journal of Mammalogy 54:648–59.

Giuliano, W. M., J. A. Litvaitus, and C. L. Stevens. 1989. Prey selection in relation to sexual dimorphism of fishers (*Martes pennanti*) in New Hampshire. Journal of Mammalogy 70:639–41.

Glass, B. P. 1943. Factors governing the distribution of *Scalopus aquaticus.* Manuscript located at W. B. Davis private library, Department of Wildlife and Fisheries Science, Texas A&M University, College Station.

Glendenning, R. 1959. Biology and control of the coast mole, *Scapanus orarius orarius* True, in British Columbia. Canadian Journal of Animal Science 39:34–44.

Godfrey, G., and P. Crowcroft. 1960. The life of the mole (*Talpa europaea* Linnaeus). Latimer, Trend, and Co., London.

Gordon, R. E., and J. R. Bailey. 1963. The occurrence of *Parascalops breweri* on the Highlands (North Carolina) Plateau. Journal of Mammalogy 44:580–81.

Gorman M. L., and R. D. Stone. 1990. The natural history of moles. Comstock, Ithaca, NY.

Gould, E.,W. McShea, and T. Grand. 1993. Function of the star in the star-nosed mole, *Condylura cristata.* Journal of Mammalogy 74:108–16.

Grand, T., E. Gould, and R. Montali. 1998. Structure of the proboscis and rays of the star-nosed mole, *Condylura cristata.* Journal of Mammalogy 79:492–501.

Gregson, J. D. 1956. The Ixodoidea of Canada. Canada Department of Agriculture, Science Service, Entomology Division 930:1–92.

Grim, J. N. 1990. Whorl-like outer segments in the retina of the mole (*Scalopus aquaticus*). Acta Anatomica 138:261–64.

Gropp, A. M. 1969. Cytologic mechanisms of karyotype evolution in insectivores. Pages 247–66 *in* K. Bernirscke, ed. Comparative mammalian cytogenetics. Springer-Verlag, New York.

Gunther, P. M., B. S. Horn, and G. D. Babb. 1983. Small mammal populations and food selection in relation to timber harvest practices in the western Cascade Mountains. Northwest Science 57:32–44.

Hafner, J. C. 1974. Seasonal predation on moles by the red-tailed hawk. Condor 76:225.

Hall, E. R. 1951. American weasels. University of Kansas Publications, Museum of Natural History 4:1–466.

Hall, E. R. 1981. The mammals of North America, Vol. 1, 2nd ed. John Wiley, New York.

Hallett, J. G. 1978. *Parascalops breweri.* Mammalian Species 98:1–4.

Hamilton, W. J., Jr. 1931. Habits of the star-nosed mole, *Condylura cristata.* Journal of Mammalogy 12:345–55.

Hamilton, W. J., Jr. 1939. Activity of Brewer's mole (*Parascalops breweri*). Journal of Mammalogy 20:307–10.

Hamilton, W. J., Jr. 1943. The mammals of eastern United States. Comstock, Ithaca, New York.

Hamilton, W. J., Jr., N. W. Howley, and A. E. MacGregor. 1937. Late summer and early fall flood foods of the red fox in central Massachusetts. Journal of Mammalogy 18:366–67.

Hartman, G. D. 1992. Demographic and population genetic structure in the eastern mole, *Scalopus aquaticus howelli* (Insectivora: Talpidae). Ph.D. Dissertation, University of New Mexico, Albuquerque.

Hartman, G. D. 1995a. Seasonal effects on sex ratios in moles collected by trapping. American Midland Naturalist 133:293–303.

Hartman, G. D. 1995b. Age determination, age structure, and longevity in the mole, *Scalopus aquaticus* (Mammalia: Insectivora). Journal of Zoology (London) 237:107–22.

Hartman, G. D. 1996. Genetic variation in a subterranean mammal, *Scalopus aquaticus* (Insectivora: Talpidae). Biological Journal of the Linnean Society 59:115–25.

Hartman, G. D., and J. L. Gottschang. 1983. Notes on sex determination, neonates, and behavior of the eastern mole, *Scalopus aquaticus*. Journal of Mammalogy 64:539–40.

Hartman, G. D., and J. D. Krenz. 1993. Estimating population density of moles *Scalopus aquaticus* using assessment lines. Acta Theriologica 38:305–14.

Hartman, G. D., and L. D. Wike. 1998. An assessment of the vertebrate fauna associated with the tunnels of eastern moles (*Scalopus aquaticus*) on the Savannah River Site (WSRC-RP-98–00714). Westinghouse Savannah River Co., Aiken, SC.

Hartman, G. D., and T. L. Yates. 1985. *Scapanus orarius.* Mammalian Species 253:1–5.

Hartman, G. D., J. O. Whitaker, Jr., and J. R. Munsee. 2000. Diet of the mole *Scalopus aquaticus* from the Coastal Plain region of South Carolina. American Midland Naturalist 144:342–51.

Hartman, G. D., A. M. White, and L. D. Wike. 2001. Seasonal differences in the use of mole tunnels by short-tailed shrews *Blarina carolinensis*. American Midland Naturalist 145:358–66.

Harvey, M. J. 1976. Home range, movements, and diel activity of the eastern mole, *Scalopus aquaticus*. American Midland Naturalist 95:436–45.

Hegstrom, L. J., and S. D. West. 1989. Heavy metal accumulation in small mammals following sewage sludge application to forests. Journal of Environmental Quality 18:345–49.

Heller, J. A. 1927. Brewer's mole as food of the bullfrog. Copeia 165:116.

Herrin, C. S. 1970. A systematic revision of the genus *Hirstionyssus* (Acari: Mesostigmata) of the Nearctic region. Journal of Medical Entomology 7:391–437.

Hickman, G. C. 1984a. Swimming ability of talpid moles, with particular reference to the semi-aquatic *Condylura cristata.* Mammalia 48:505–504.

Hickman, G. C. 1984b. An excavated burrow of *Scalopus aquaticus* from Florida, with comments on Nearctic talpid/geomyid burrow structure. Säugetierkundliche Mitteilungen 31:243–49.

Hisaw, F. L. 1923a. Feeding habits of moles. Journal of Mammalogy 4:9–20.

Hisaw, F. L. 1923b. Observations on the burrowing habits of moles. Journal of Mammalogy 4:79–88.

Holland, G. P., and A. H. Benton. 1968. Siphonaptera from Pennsylvania mammals. American Midland Naturalist 80:252–61.

Hoslett, S. A., and Y. H. Imaizumi. 1966. Age structure of a Japanese mole population. Journal of the Mammalogical Society of Japan 2:151–56.

Howard, W. E., and H. E. Childs, Jr. 1959. Ecology of pocket gophers with emphasis on *Thomomys bottae mewa*. Hilgardia 29:277–358.

Hubbard, C. A. 1940a. A check list of fleas of the Pacific Northwest (Washington, Oregon, northern California and northwestern Nevada) with notes from southern California. Pacific University Bulletin 37(4):1–4.

Hubbard, C. A. 1940b. West Coast crested fleas *Corypsylla* and *Nearctopsylla*. Pacific University Bulletin 37(1):1–10.

Hubbard, C. A. 1940c. American mole and shrew fleas. Pacific University Bulletin 37(2):1–12.

Hubbard, C. A. 1949. New fleas and records from the western states. Bulletin of the Southern California Academy of Sciences 48:47–54.

Jackson, H. H. T. 1915. A review of the American moles. North American Fauna 38:1–100.

Jameson, E. W., Jr. 1949. Myobiid mites (Acarina: Myobiidae) from *Condylura cristata* (Linnaeus) and *Neürotrichus gibbsii* (Baird) (Mammalia: Talpidae). Journal of Parasitology 35:423–30.

Jameson, E. W., Jr. 1950a. The external parasites of the short-tailed shrew, *Blarina brevicauda* (Say). Journal of Mammalogy 31:138–45.

Jameson, E. W., Jr. 1950b. *Hirstionyssus obsoletus,* a new mesostigmatic mite from small mammals of the western United states (Acarina). Proceedings of the Biological Society of Washington 63:31–34.

Jameson, E. W., Jr. 1952. *Euheamogamasus keegani,* new species, a parasitic mite from western North America (Acarina: Laelaptidae, Heamogamasinae). Annals of the Entomological Society of America 45:600–604.

Jameson, E. W., Jr., and J. M. Brennan. 1957. An environmental analysis of some ectoparasites of small forest mammals in the Sierra Nevada, California. Ecological Monographs 27:45–54.

Jelkmann, W., W. Oberthür, T. Kleinschmidt, and G. Braunitzer. 1981. Adaptation of hemoglobin function to subterranean life in the mole, *Talpa europaea*. Respiration Physiology 46:7–16.

Jensen, I. M. 1982. A new live trap for moles. Journal of Wildlife Management 46:249–52.

Jensen, I. M. 1983. Metabolic rates of the hairy-tailed mole, *Parascalops breweri* (Bachman, 1842). Journal of Mammalogy 64:453–62.

Jensen, I. M. 1986a. Foraging strategies of the mole (*Parascalops breweri* Bachman, 1842). I. The distribution of prey. Canadian Journal of Zoology 64:1727–33.

Jensen, I. M. 1986b. Foraging strategies of the mole (*Parascalops breweri* Bachman, 1842). II. The economics of finding prey. Canadian Journal of Zoology 64:1734–38.

Jiménez, R., M. Burgos, A. Sánchez, and R. Díaz de la Guardia. 1990. The reproductive cycle of *Talpa occidentalis* in the southeastern Iberian Peninsula. Acta Theriologica 35:165–69.

Jiménez, R., F. J. Alarcón, A. Sánchez, M. Burgos, and R. Díaz de la Guardia. 1996. Ovotestis variability in young and adult females of the mole *Talpa occidentalis* (Insectivora, Mammalia). Journal of Experimental Zoology 274:130–37.

Johannesan-Gross, C. von 1988. Lernversuche in einer Zweifachwahlapparatur zum Hell-Dunkel-Sehen des Maulwurfs (*Talpa europaea* L.). Zeitschrift für Säugetierkunde 53:193–201.

Johnson, M. L., and T. L. Yates. 1980. A new Townsend's mole (Insectivora: Talpidae) from the state of Washington. Occasional Papers, The Museum, Texas Tech University 63:1–6.

Katanova, L. N. 1972. Some peculiarities of sex and age structure of a population of *Talpa europaea*. Zoologicheskii Zhurnal 51:1214–18. (In Russian, with English summary.)

Keegan, H. L. 1951. The mites of the subfamily Haemogamasinae (Acari: Laelaptidae). Proceedings of the U.S. National Museum 101:203–68.

Kenagy, G. J., and D. Vleck. 1982. Daily temporal organization of metabolism in small mammals: Adaptation and diversity. Pages 322–38 *in* J. Aschoff, S. Daan, and G. Groos, eds. Vertebrate circadian systems. Springer-Verlag, Berlin.

Khazanehdari, C., A. J. Buglass, and J. S. Waterhouse. 1996. Anal gland secretion of European mole: Volatile constituents and significance in territorial maintenance. Journal of Chemical Ecology 22:383–92.

Krantz, G. W., and J. O. Whitaker, Jr. 1988. Mites of the genus *Machrocheles* (Acari: Machrochelidae) associated with small mammals in North America. Acarologia 29:225–59.

Krull, J. N. 1969. Observation of *Tamias striatus* feeding upon *Condylura cristata*. Illinois State Academy of Science, Transactions 62:221.

Kuhn, L. W., W. Q. Wick, and R. J. Pedersen. 1966. Breeding nests of Townsend's mole in Oregon. Journal of Mammalogy 47:239–49.

Laerm, J., G. Livingston, C. Spencer, and B. Stuart. 1997. *Condylura cristata* (Insectivora: Talpidae) in the Blue Ridge Province of western South Carolina. Brimleyana 24:46–49.

Leach, B. J., T. R. Bauman, and C. W. Turner. 1962. Thyroxine secretion rate of Missouri Valley mole, *Scalopus aquaticus*. Proceedings of the Society for Experimental Biology and Medicine 110:681–82.

Leftwich, B. H. 1972. Population dynamics and behavior of the eastern mole, *Scalopus aquaticus machrinoides*. Ph.D. Dissertation, University of Missouri, Columbia.

Lewis, R. E., and C. Maser. 1981. Invertebrates of the H. J. Andrews Experimental Forest, Western Cascades, Oregon. I. An annotated checklist of fleas. U.S. Department of Agriculture Forest Service Research Note PNW-378:1–10.

Lewis, R. E., J. H. Lewis, and C. Maser. 1988. The fleas of the Pacific Northwest. Oregon State University Press, Corvallis.

Lewis, T. H. 1983. The anatomy and histology of the rudimentary eye of *Neürotrichus*. Northwest Science 57:8–15.

Lodal, J., and H. Grue. 1985. Age determination and age distribution in populations of moles (*Talpa europaea*) in Denmark. Acta Zoologica Fennica 173:279–81.

Lowery, G. H. 1974. The mammals of Louisiana and its adjacent waters. Louisiana State University Press, Baton Rouge.

Lukoschus, F. S., J. S. H. Klompen, and J. O. Whitaker, Jr. 1980. *Eadiea neurotrichus,* n. sp. (Prostigmata: Myobiidae) from *Neürotrichus gibbsii* (Insectivora: Talpidae). Journal of Medical Entomology 17:498–501.

Lynch, J. R. 1971. The chromosomes of the California mole (*Scapanus latimanus*). Mammalian Chromosomes Newsletter 12:83–84.

Macdonald, D. W., R. P. D. Atkinson, and G. Blanchard. 1997. Spatial and temporal patterns in the activity of European moles. Oecologia 109:88–97.

Mahunka, S. 1973. *Pygmephorus* species (Acari, Tarsonemida) from North American small mammals. Parasitologia Hungarica 6:247–59.

Mahunka, S. 1975. Further data to the knowledge of Tarsonemida (Acari) living on small mammals in North America. Parasitologia Hungarica 8:85–94.

Maser, C., and E. D. Brodie, Jr. 1966. A study of owl pellet contents from Linn, Benton, and Polk Counties, Oregon. Murrelet 47:9–14.

Maser, C., and C. Brown. 1972. Bacula of two western moles. Northwest Science 46:319–21.

Maser, C., and E. F. Hooven. 1971. New host and locality records for *Leptinus testaceus* Müller in western Oregon (Coleoptera: Leptinidae). Coleopterists Bulletin 25:119–20.

Maser, C., B. R. Mate, J. F. Franklin, and C. T. Dyrness. 1981. Natural history of Oregon coast mammals. U.S. Department of Agriculture Forest Service Report PNW-133:1–496.

Matthews, L. H. 1935. The œstrous cycle and intersexuality in the female mole (*Talpa europaea* Linn.). Proceedings of the Zoological Society of London 1935:347–83.

McIntyre, I. W. 2000. Diving energetics and temperature regulation of the star-nosed mole (*Condylura cristata*) with comparisons to non-aquatic talpids and the water shrew (*Sorex palustris*). M.Sc. Thesis, University of Manitoba, Winnipeg, Canada.

McNab, B. K. 1979. The influence of body size on the energetics and distribution of fossorial and burrowing mammals. Ecology 60:1010–21.

McVean, A. 1999. Are the semicircular canals of the European mole, *Talpa europaea,* adapted to a subterranean habitat? Comparative Biochemistry and Physiology A 123:173–78.

Mellanby, K. 1971. The mole. Collins, Glasgow, Scotland.

Meylan, A. 1968. Formules chromosomiques de quelques petits mammifères nord-américains. Revue Suisse de Zoologie 75:691–96.

Moore, A. W. 1933. Food habits of Townsend and coast moles. Journal of Mammalogy 14:36–40.

Moore, A. W. 1939. Notes on the Townsend mole. Journal of Mammalogy 20:499–501.

Moore, A. W. 1940. A live mole trap. Journal of Mammalogy 21:223–25.

Moore, D. W. 1986. Systematics and biogeographic relationships among the Talpinae (Insectivora: Talpidae). Ph.D. Dissertation, University of New Mexico, Albuquerque.

Mumford, R. E., and J. O. Whitaker, Jr. 1982. Mammals of Indiana. Indiana University Press, Bloomington.

Myrcha, A. 1967. Comparative studies on the morphology of the stomach in the Insectivora. Acta Theriologica 12:223–44.

Neiland, K. A., and C. M. Senger. 1952. Helminths of northwestern mammals. Part I. Two new species of *Hymenolepis*. Journal of Parasitology 38:409–14.

Nevo, E. 1979. Adaptive convergence and divergence of subterranean mammals. Annual Review of Ecology and Systematics 10:269–308.

Nevo, E., A. Beiles, and R. Ben-Shlomo. 1984. The evolutionary significance of genetic diversity: Ecological, demographic and life history correlates. Lecture Notes in Biomathematics 53:13–213.

Nevo, E., M. G. Filippucci, and A. Beiles. 1990. Genetic diversity and its ecological correlates in nature: Comparisons between subterranean, fossorial, and aboveground small mammals. Pages 347–66 *in* E. Nevo and O. A. Reig, eds. Progress in clinical and biological research. Vol. 335, Evolution of subterranean mammals at the organismal and molecular levels. Wiley-Liss, New York.

Odlaug, T. O. 1952. *Brachylaima condylura* n. sp. from the star-nosed mole, *Condylura cristata*. Transactions of the American Microscopical Society 71:344–46.

Olive, J. R. 1950. Some parasites of the prairie mole, *Scalopus aquaticus machrinus*. Ohio Journal of Science 50:263–66.

Olszewski, J. L., and S. Skoczeń 1965. The airing of burrows of the mole, *Talpa europaea* Linnaeus, 1758. Acta Theriologica 10:181–93.

Opperman, J. 1968. Die Nahrung des Maulwurfs (*Talpa europaea* L., 1758). Pedobiologia 8:59–74.

Packard, R. L. 1968. An ecological study of the fulvous harvest mouse in eastern Texas. American Midland Naturalist 79:68–88.

Palmer, F. G. 1937. Geographic variation in the mole *Scapanus latimanus*. Journal of Mammalogy 18:280–314.

Pankakoski, E., H. Hyvärinen, M. Jalkanen, and I. Koivisto. 1993. Accumulation of heavy metals in the mole in Finland. Environmental Pollution 80:9–16.

Parmalee, P. W. 1954. Food of the great horned owl and barn owl in east Texas. Auk 71:469–70.

Patton, J. L., and M. F. Smith. 1990. The evolutionary dynamics of the pocket gopher *Thomomys bottae,* with emphasis on California populations. University of California Publications, Zoology 123:1–161.

Pearson, O. P. 1947. The rate of metabolism of some small mammals. Ecology 28:127–45.

Peck, S. B. 1982. A review of the ectoparasite *Leptinus* beetles of North America (Coleoptera: Leptinidae). Canadian Journal of Zoology 60:1517–27.

Pedersen, R. J. 1963. The life history and ecology of Townsend's mole *Scapanus townsendii* (Bachman) in Tillamook County, Oregon. M.S. Thesis, Oregon State University, Corvallis.

Petersen, K. E., and T. L. Yates. 1980. *Condylura cristata*. Mammalian Species 129:1–4.

Peterson, R. L. 1966. The mammals of eastern Canada. Oxford University Press, Toronto.

Pévet, P., M. T. Juillard, A. R. Smith, and J. Kappeers. 1976. The pineal gland of the mole (*Talpa europaea* L.). Part 3: A fluorescence histochemical study. Cell and Tissue Research 165:297–306.

Quilliam, T. A., J. A. Clarke, and A. J. Salsbury. 1971. The ecological significance of certain new haematological findings in the mole and hedgehog. Comparative Biochemistry and Physiology Part A 40:89–102.

Racey, K. 1929. Observation on *Neürotrichus gibbsii gibbsii*. Murrelet 10:61–62.

Radovsky, F. J. 1960. *Haemogamasus liponyssoides hesperus,* n. ssp., with a discussion of the *H. liponyssoides* complex (Acarina: Haemogamasidae). Journal of Parasitology 46:401–9.

Rausch, R. L. and R. V. Rausch. 1973. *Capillaria maseri* sp. n. (Nematoda) from insectivores (Soricidae and Talpidae) in Oregon. Proceedings of the Helminthological Society of Washington 40:107–12.

Reed, C. A., and T. Riney. 1943. Swimming, feeding and locomotion of a captive mole. American Midland Naturalist 30:790–91.

Richards, G. P., C. Zintz, J. Chao, and L. Chao. 1996. Purification and characterization of salivary kallikrein from an insectivore (*Scalopus aquaticus*): Substrate specificities immunoreactivity, and kinetic analyses. Archives of Biochemistry and Biophysics 329:104–12.

Richmond, N. D., and H. R. Rosland. 1949. Mammal survey of northwestern Pennsylvania. Pennsylvania Game Commission, Harrisburg.

Roberts, J. D., and R. L. Packard. 1973. Comments on movements, home range, and ecology of the Texas kangaroo rat, *Dipodomys elater* Merriam. Journal of Mammalogy 54:957–62.

Roest, A. I. 1952. Red-tailed hawk as a mole predator. Journal of Mammalogy 33:110.

Rust, C. C. 1966. Notes on the star-nosed mole (*Condylura cristata*). Journal of Mammalogy 47:538.

Sagara, N. 1989. European record of the presence of a mole's nest indicated by a particular fungus. Mammalia 53:301–5.

Sagara, N. 1999. Mycological approach to the natural history of talpid moles—A review with new data and proposal of habitat-cleaning symbiosis. Pages 33–55 *in* Y. Yokohata and S. Nakamura, eds. Recent advances in the biology of Japanese Insectivora. Hiba Society of Natural History, Hiroshima, Japan.

Sagara, N., S. Kobayashi, H. Ota, T. Itsubo, and H. Okabe. 1989. Finding *Euroscaptor mizura* (Mammalia: Insectivora) and its nest from under *Hebeloma radicosum* (Fungi: Agaricales) in Ashiu, Kyoto, with data of possible continuous occurrences of three talpine species in this region. Contributions of the Biology Laboratory of Kyoto University 27:261–72.

Saylor, L. W. 1938. Hairy-tailed mole in Virginia. Journal of Mammalogy 19:247.

Schaefer, V. H. 1978. Aspects of habitat selection in the coast mole (*Scapanus orarius* True), in British Columbia. Ph.D. Dissertation, Simon Fraser University, Burnaby, British Columbia.

Schaefer, V. H., and R. M. F. S. Sadlier. 1979. Concentrations of carbon dioxide and oxygen in mole tunnels. Acta Theriologica 24:267–71.

Schaefer, V. H., and R. M. F. S. Sadlier. 1981. Factors influencing molehill construction by the coast mole (*Scapanus orarius* True). Mammalia 45:31–38.

Scheffer, T. H. 1910. The common mole. Kansas State Agricultural College Experimental Station Bulletin 168:1–36.

Scheffer, T. H. 1917. The common mole of eastern United States. U.S. Department of Agriculture Farmers' Bulletin 583:1–12.

Scheffer, T. H. 1949. Ecological comparisons of three genera of moles. Transactions of the Kansas Academy of Science 52:30–37.

Schmidt, F. J. W. 1931. Mammals of western Clark County. Journal of Mammalogy 12:99–117.

Schultz, R. L. 1939. *Hymenolepis scalopi* n. sp. American Midland Naturalist 21:641–44.

Scott, T. G. 1943. Some food coactions of the Northern Plains red fox. Ecological Monographs 13:427–79.

Selleck, D. M., and B. Glading. 1943. Food habits of nesting barn owls and marsh hawks at Dune Lakes, California, as determined by the "cage nest" method. California Fish and Game 29:122–31.

Sheehan, S. T., and C. Galindo-Leal. 1997. Identifying coast moles, *Scapanus orarius,* and Townsend's moles, *Scapanus townsendii,* from tunnel and mound size. Canadian Field-Naturalist 111:463–65.

Silver, J. 1933. Mole control. U.S. Department of Agriculture Farmer's Bulletin 1716:1–17.

Silver, J., and A. W. Moore. 1941. Mole control (U.S. Department of Interior Conservation Bulletin 16). U.S. Government Printing Office, Washington, DC.

Skoczeń, S. 1962. Age structure of skulls of the mole, *Talpa europaea* Linnaeus 1758, from the food of the buzzard (*Buteo buteo*). Acta Theriologica 6:1–9.

Skoczeń, S. 1966. Age determination, age structure and sex ratio in mole, *Talpa europaea* Linnaeus, 1758 populations. Acta Theriologica 11:523–36.

Slonaker, J. R. 1902. The eye of the common mole, *Scalopus aquaticus machrinus*. Journal of Comparative Neurology 12:335–66.

Slonaker, J. R. 1920. Some morphological changes for adaptation in the mole. Journal of Morphology 34:335–63.

Smiley, R. L., and J. O. Whitaker, Jr. 1979. Mites of the genus *Pygmephorus* (Acari: Pygmephoridae) on small mammals in North America. Acta Zoologica Academiae Scientiarum Hungarica 25:383–408.

Smiley, R. L., and J. O. Whitaker, Jr. 1984. Key to New and Old World *Pygmephorus* species and descriptions of six new species (Acari: Pygmephoridae). International Journal of Acarology 10:59–73.

Spencer, G. J. 1941. Ectoparasites of birds and mammals in British Columbia. VI. A preliminary list of parasitic mites. Journal of the Entomological Society of British Columbia 37:14–18.

Sperry, C. C. 1933. Opossum and skunk eat bats. Journal of Mammalogy 14:152–53.

Strandtmann, R. W. 1949. The blood-sucking mites of the genus *Haemolaelaps* (Acarina: Laelaptidae) in the United States. Journal of Parasitology 35:325–52.

Stroganov, S. U. 1945. Morphological characters of the auditory ossicles of Recent Talpidae. Journal of Mammalogy 26:412–20.

Terry, C. J. 1981. Habitat differentiation among three species of *Sorex* and *Neürotrichus gibbsii* in Washington. American Midland Naturalist 106:119–25.

Timm, R. M. 1975. Distribution, natural history and parasites of mammals of Cook County, Minnesota (Occasional Papers No. 14). University of Minnesota, Bell Museum of Natural History, Minneapolis, MN.

Tolliver, D. K., M. H. Smith, and R. H. Leftwich. 1985. Genetic variability in Insectivora. Journal of Mammalogy 66:405–10.

Toweill, D. E., and R. G. Anthony. 1988. Coyote foods in a coniferous forest in Oregon. Journal of Wildlife Management 52:507–12.

Townsend, M. T. 1935. Studies on some of the small mammals of central New York. Roosevelt Wildlife Bulletin 4:1–120.

Turner, H. M., and S. McKeever. 1980. *Ityogonimus scalopi* sp. n. (Trematoda: Brachylaemidae) from the eastern mole, *Scalopus aquaticus* (Linnaeus 1758). Journal of Parasitology 66:823–24.

Tsuchiya, K. 1979. A contribution to the chromosome study in Japanese mammals. Japan Academy, Proceedings, Series B 55:191–95.

Van Vleck, D. B. 1965. The anatomy of the nasal rays on *Condylura cristata*. Journal of Mammalogy 46:248–53.

Verts, B. J., and L. N. Carraway. 2001. *Scapanus latimanus*. Mammalian Species 666:1–7.

von Bloeker, J. C., Jr. 1937. Mammal remains from detritus of raptorial birds in California. Journal of Mammalogy 18:360–61.

West, J. A. 1910. A study of the food of moles in Illinois. Bulletin of the Illinois State Laboratory of Natural History 9:14–22.

Whidden, H. P. 1999. The evolution of locomotor specializations in moles. American Zoologist 39(5):135A [Abstract].

Whidden, H. P. 2000. Comparative myology of moles and the phylogeny of the Talpidae (Mammalia, Lipotyphla). American Museum Novitates 3294:1–53.

Whitaker, J. O., Jr. 1982. Ectoparasites of mammals of Indiana (Monograph 4). Indiana Academy of Science, Indianapolis, IN.

Whitaker, J. O., Jr., and T. W. French. 1982. Ectoparasites and other associates of some insectivores and rodents from New Brunswick. Canadian Journal of Zoology 60:2787–97.

Whitaker, J. O., Jr., and T. W. French. 1988. Ectoparasites and other arthropod associates of the hairy-tailed mole, *Parascalops breweri*. Great Lakes Entomologist 21:39–41.

Whitaker, J. O., Jr., and C. Maser. 1985. Mites (excluding chiggers) of mammals of Oregon. Great Basin Naturalist 45:67–76.

Whitaker, J. O., Jr., and L. L. Schmeltz. 1974. Food and external parasites of the eastern mole, *Scalopus aquaticus,* from Indiana. Proceedings of the Indiana Academy of Science 83:478–81.

Whitaker, J. O., Jr., and N. Wilson. 1974. Host and distribution lists of mites (Acari), parasitic and phoretic, in the hair of wild mammals of North America, north of Mexico. American Midland Naturalist 91:1–67.

Whitaker, J. O., Jr., C. Maser, and R. J. Pedersen. 1979. Food and ectoparasitic mites of Oregon moles. Northwest Science 53:268–73.

Whitaker, J. O., Jr., T. W. French, and R. L. Smiley. 1982. Notes on host relationships and host specificity of mites of the genus *Pygmephorus* (Acari: Pygmephoridae) on insectivores and rodents from Mount Carleton Provincial Park, New Brunswick. International Journal of Acarology 8:233–35.

Whitworth, D. J., P. Licht, P. A. Racey, and S. E. Glickman. 1999. Testis-like steroidogenesis in the ovotestis of the European mole, *Talpa europaea*. Biology of Reproduction 60: 413–18.

Wick, W. Q. 1962. Control moles for $1 an acre. Oregon's Agricultural Progress 9:10–11.

Wiegert, R. G. 1961. Nest construction and oxygen consumption of *Condylura*. Journal of Mammalogy 42:528–29.

Wight, H. M. 1928. Food habits of Townsend's mole *Scapanus townsendii* (Bachman). Journal of Mammalogy 9:19–23.

Wilks, B. J. 1963. Some aspects of the ecology and population dynamics of the pocket gopher (*Geomys bursarius*) in southern Texas. Texas Journal of Science 15:241–83.

Williams, G. L., R. L. Smiley, and B. C. Redington. 1978. A taxonomic study of the genus *Haemogamasus* in North America, with descriptions of two new species (Acari: Mesostigmata, Laelaptidae). International Journal of Acarology 4:235–73.

Wright, P. L. 1945. *Parascalops* tunnel in use after eight years. Journal of Mammalogy 26:438–39.

Yamaguti, S. 1959. Systema Helminthum. Vol. II, The cestodes of vertebrates. Interscience, New York.

Yamaguti, S. 1961. Systema Helminthum. Vol. III, Part I, The nematodes of vertebrates. Interscience, New York.

Yamaguti, S. 1963. Systema Helminthum. Vol. V, Acanthocephala. Interscience, New York.

Yates, T. L. 1978. Systematics and evolution of North American moles (Insectivora: Talpidae). Ph.D. Dissertation, Texas Tech University, Lubbock.

Yates, T. L., and I. F. Greenbaum. 1982. Biochemical systematics of North American moles (Insectivora: Talpidae). Journal of Mammalogy 63:368–74.

Yates, T. L., and D. W. Moore. 1990. Speciation and evolution in the family Talpidae (Mammalia: Insectivora). Pages 1–22 *in* E. Nevo and O. A. Reig, eds. Progress in clinical and biological research. Vol. 335, Evolution of subterranean mammals at the organismal and molecular levels. Wiley-Liss, New York.

Yates, T. L., and D. J. Schmidly. 1975. Karyotype of the eastern mole (*Scalopus aquaticus*), with comments on the karyology of the family Talpidae. Journal of Mammalogy 56:902–5.

Yates, T. L., and D. J. Schmidly. 1977. Systematics of *Scalopus aquaticus* (Linnaeus) in Texas and adjacent states. Occasional Papers, the Museum, Texas Tech University 45:1–36.

Yates, T. L., and D. J. Schmidly. 1978. *Scalopus aquaticus*. Mammalian Species 105:1–4.

Yates, T. L., A. D. Stock, and D. J. Schmidly. 1976. Chromosome banding patterns and the nucleolar organizer region of the eastern mole (*Scalopus aquaticus*). Experientia 32:1276–77.

Yates, T. L., D. B. Pence, and G. K. Lauchbaugh. 1979. Ectoparasites from seven species of North American moles (Insectivora: Talpidae). Journal of Medical Entomology 16:166–68.

Yokohata, Y. 1999. What the [*sic*] Yukawa collection of talpids inform us? Biology of the shrew-mole and moles in Hiwa, Hiroshima Prefecture, Japan. Pages 1–13 *in* Y. Yokohata and S. Nakamura, eds. Recent advances in the biology of Japanese Insectivora. Hiba Society of Natural History, Hiroshima, Japan.

Zielinski, W. J., W. D. Spencer, and R. H. Barrett. 1983. Relationship between food habits and activity patterns of pine martens. Journal of Mammalogy 64:387–96.

Gregory D. Hartman, W. A. K. Seale Vertebrate Museum, Department of Biological and Environmental Sciences, McNeese State University, Lake Charles, Louisiana 70609. Email: ghartman@mail.mcneese.edu.

Terry L. Yates, Museum of Southwestern Biology, Department of Biology, University of New Mexico, Albuquerque, New Mexico 87131. Email: tyates@unm.edu.

3

Bats

Vespertilionidae, Molossidae, Phyllostomidae

William L. Gannon

NOMENCLATURE

Vespertilionidae
COMMON NAME. Pallid bat
SCIENTIFIC NAME. *Antrozous pallidus* (Le Conte)
SUBSPECIES. *A. p. bunkeri, A. p. koopmani, A. p. minor, A. p. pacificus, A. p. packardi,* and *A. p. pallidus;* of these six subspecies, three occur north of Mexico

COMMON NAME. Big brown bat
SCIENTIFIC NAME. *Eptesicus fuscus* (Palisot de Beauvois)
SUBSPECIES. *E. f. bahamensis, E. f. bernardinus, E. f. dutertreus, E. f. fuscus, E. f. miradorensis, E. f. osceola, E. f. pallidus,* and *E. f. peninsulae;* of the 10 recognized subspecies, 8 occur in North America north of Mexico

COMMON NAME. Hoary bat
SCIENTIFIC NAME. *Lasiurus cinereus* (Palisot de Beauvois)
SUBSPECIES. *L. c. cinereus, L. c. villosissimus,* and *L. c. semotus. L. c. cinereus* is North American, *L. c. villosissimus* was described from Paraguay, and *L. c. semotus* is endemic to Hawaii

COMMON NAME. Spotted bat
SCIENTIFIC NAME. *Euderma maculatum* (J. A. Allen)
SUBSPECIES. None; this is a monotypic species

COMMON NAME. Gray bat
SCIENTIFIC NAME. *Myotis grisescens* A. H. Howell
SUBSPECIES. None; this is a monotypic species

COMMON NAME. Indiana bat
SCIENTIFIC NAME. *Myotis sodalis* Miller and Allen
SUBSPECIES. None; this is a monotypic species

Molossidae
COMMON NAMES. Free-tailed bat, Brazilian free-tailed bat, Mexican free-tailed bat, guano bat
SCIENTIFIC NAME. *Tadarida brasiliensis* (I. Geof. St.-Hilaire)
SUBSPECIES. *T. b. bahamensis, T. b. cynocephala,* and *T. b. mexicana;* all three subspecies occur, at least partly, in North America north of Mexico; nine subspecies were listed by Wilkins (1989)

Phyllostomidae
COMMON NAME. California leaf-nosed bat, big-eared bat
SCIENTIFIC NAME. *Macrotus californicus* Baird
SUBSPECIES. None

These eight species were selected for this chapter from approximately 50 North American bat species because they broadly represent the bats of North America. In the previous edition (1982), Humphrey chose bat species about which a graduate student or wildlife manager might commonly seek information. Similar criteria were adopted for the content of this chapter, except that a broader representation of species present in North America was added. These species encompass three chiropteran families: Vespertilionidae, Molossidae, and Phyllostomidae.

The vespertilionid (mouse-eared, plain-nosed, or evening bats) bats are the largest family of bats in terms of number of species (35 genera, 320 species). The genus *Myotis* is the largest, with about 85 species. Most North American bats belong to this family. All are insectivorous (animalivorous) and have a well-developed sense of echolocation. They have plain noses, a tragus, and a tail that extends slightly beyond the hind edge of the uropatagial membrane. The front of the skull is flattened and there is a space between the two front incisors. A few species make very long migrations to their wintering grounds, and most make at least a short migration. Many exhibit delayed implantation, mating usually in the fall, but sometimes in the spring and winter. Delayed fertilization may allow energy savings during hibernation. At parturition, neonates drop into the uropatagium and then move to the breast. Many species hibernate and readily enter torpor on cool days during the nonhibernation season as a means of energy conservation. Life history events include winter hibernation, spring awakening or migration with consummate foraging, summer birthing, fall migration or swarming before entering winter hibernation, and mating. Males are often solitary or form small bachelor groups. Females form maternity groups in roost structures; these aggregations are sometimes small (10 individuals) or very large (100,000 or more).

Only a few species of molossids occur in North America. As their vernacular name suggests, the tail of free-tailed bats is partially contained in and partially free from the uropatagium. These bats have long, narrow wings and are fast, high fliers. Their ears are thick and leathery and meet on top of the head; in *Tadarida brasiliensis,* the ears meet at the forehead. Members of this family have long, stiff tactile hairs on the toes. This family includes about 80 species in 12 genera. Some species make extensive seasonal migrations.

The American leaf-nosed bats, phyllostomids, have 52 genera and 154 species, mostly found in tropical and subtropical regions of the Western Hemisphere. There are only four species in North America within three genera. This group of bats is under intensive analysis and debate with respect to phylogeny and taxonomy (Nowak 1994). As the common name implies, a nose-leaf is usually present and probably functions morphologically in the echolocation system when feeding (Arita 1990). Most phyllostomids feed on insects, fruit, nectar, and pollen. Notorious members of the family, the vampire bats (*Desmodus rotundus* and two other species), feed on fresh blood.

DISTRIBUTION

Distribution maps are only an estimation of where species occur; this may be especially true of bats. Maps herein reflect museum collection records or previously published records. Shading is used rather than points to depict where individuals of a species could occur irrespective of habitat. Habitat affinity is perhaps the most difficult aspect of bat biology to determine because bats often move through various habitats but may not be using those habitats for any biological function.

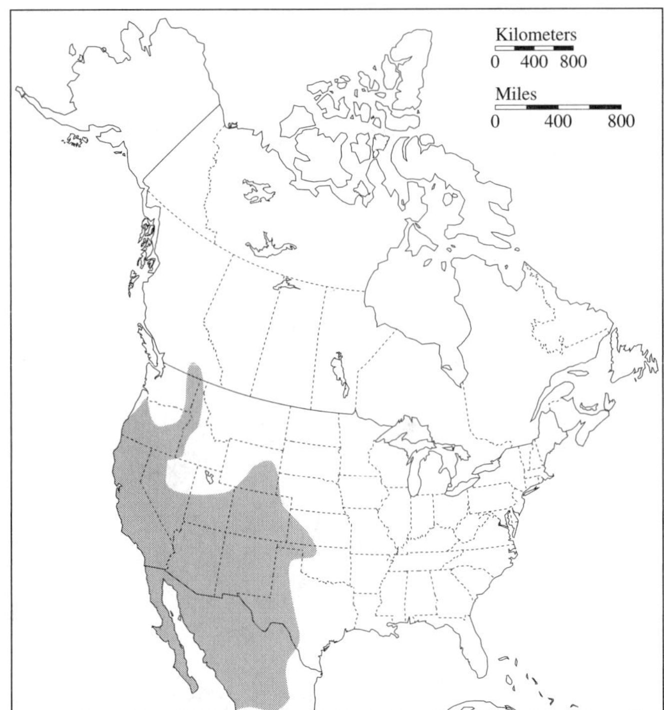

FIGURE 3.1. Distribution of the pallid bat (*Antrozous pallidus*).

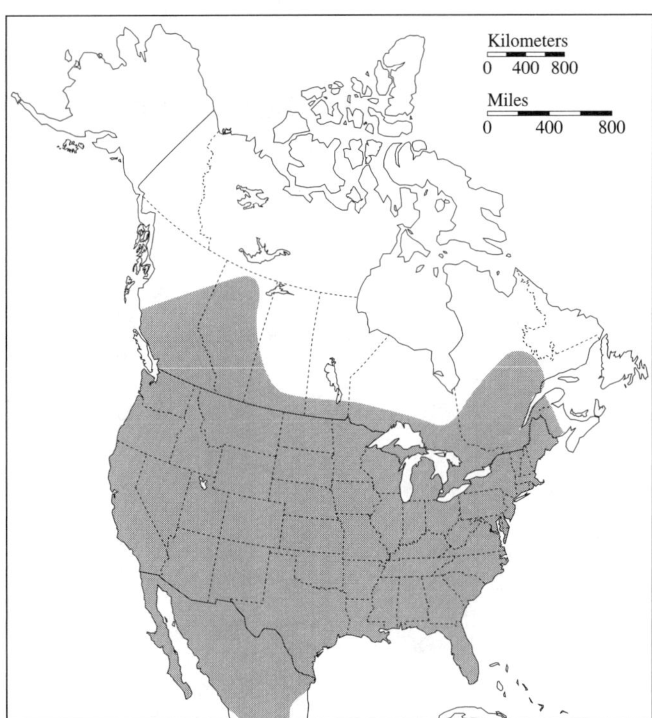

FIGURE 3.2. Distribution of the big brown bat (*Eptesicus fuscus*).

Antrozous pallidus is distributed from southern British Columbia and Montana to central Mexico and east to Texas, Oklahoma, and Kansas (Fig. 3.1). An isolated population also occurs on Cuba (Orr and Taboada 1960). The pallid bat occurs in habitats ranging from rocky arid deserts to grasslands into higher elevation coniferous forests. They are most abundant in the arid Sonoran life zones below 1800 m elevation, but occur up to 3000 m in the Sierra Nevada of California. The big brown bat (*Eptesicus fuscus*) is found in almost every habitat in North America and the northern part of South America and islands of the West Indies. It is extremely common in urban areas (Tuttle 1988), and human-made structures are often used for roosts. Nevertheless, there are few detailed studies on the extent of their distribution. Generally, this species occurs from the Pacific to the Atlantic coasts, north into Canada (including northern British Columbia and Alberta), and south through Mexico into Central America (Fig. 3.2). Similar to *Eptesicus fuscus,* the hoary bat (*Lasiurus cinereus*) occurs from coast to coast throughout the United States except for portions of the southeast, north into Canada (Fig. 3.3), and south into Mexico and Central America. Unlike *E. fuscus* and many other bat species, the hoary bat roosts mostly in trees of coniferous and some deciduous forests. Furthermore, it migrates seasonally into Central America. On its northward return to North America, female *L. cinereus* populate the eastern half and males populate the western half of North America. Areas in the central portion of the continent have both sexes (Findley and Jones 1964). This is the only bat species with a population on Hawaii. The spotted bat, *Euderma maculatum,* ranges from sea level in California to 3230 m in New Mexico, south into southern Mexico, and north to Idaho, and edges into Canada (Fig. 3.4). This species is widespread, but uncommon in most areas. Its rareness, measured by traditional mist-net studies, has resulted in its listed status in many western states (Bogan 2001) (listed as threatened by the New Mexico Department of Game and Fish and the Texas Department of Parks and Wildlife; Jones and Schmitt 1996). Spotted bats are considered a Species of Concern by the U.S. Fish and Wildlife Service. The gray bat, *Myotis grisescens,* is found throughout the limestone region of the southern midwest and southeast states (Fig. 3.5). Because of roosting requirements and human interactions, this species also is federally listed as endangered. The Indiana bat, *Myotis sodalis,* occurs throughout much of the eastern United States from the central Mississippi Valley, eastern Alabama, to northern Florida and north to New England, avoiding the Atlantic Coast. To the west, it occurs in Missouri, Arkansas, and Oklahoma (Fig. 3.6). Because roosts are limited to only a few large colonies that are very susceptible to disturbance by humans, this species is federally listed as endangered. The Brazilian free-tailed bat, *Tadarida brasiliensis,* is one of the most widely distributed mammalian species in the Western Hemisphere. Northern limits extend

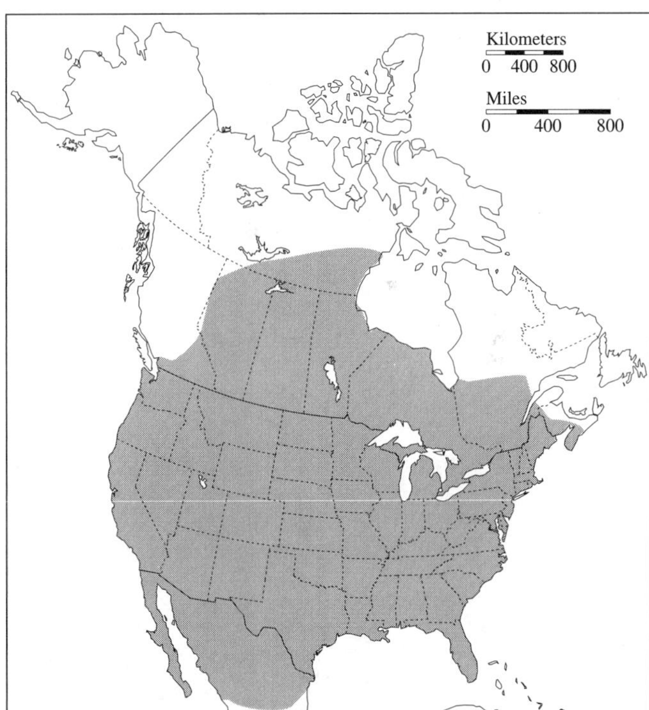

FIGURE 3.3. Distribution of the hoary bat (*Lasiurus cinereus*).

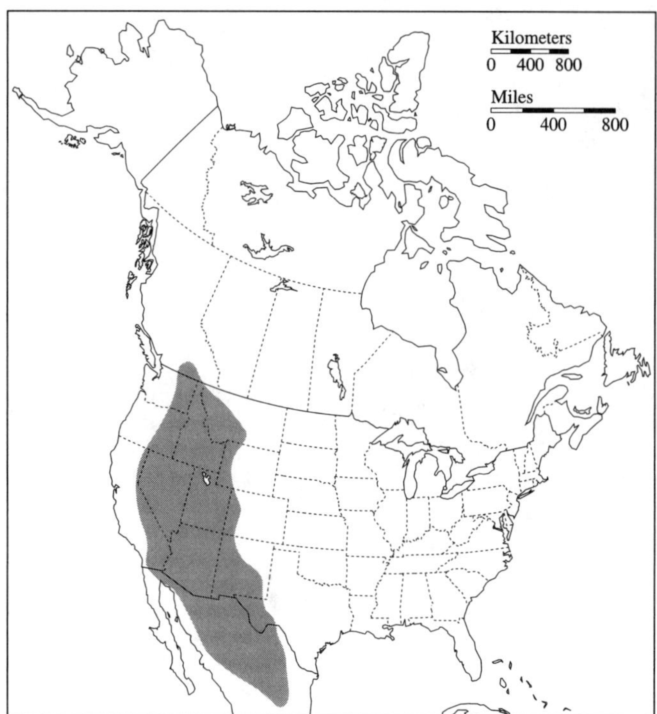

FIGURE 3.4. Distribution of the spotted bat (*Euderma maculatum*).

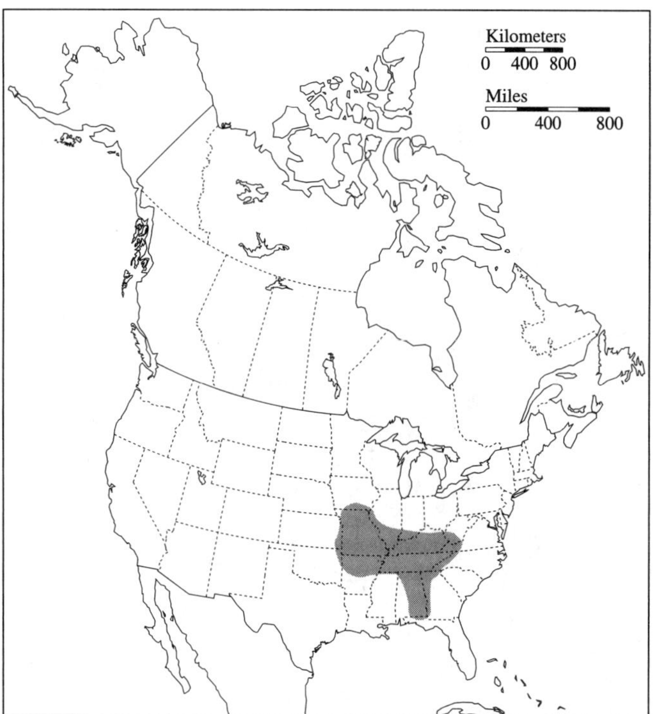

FIGURE 3.5. Distribution of the gray bat (*Myotis grisescens*).

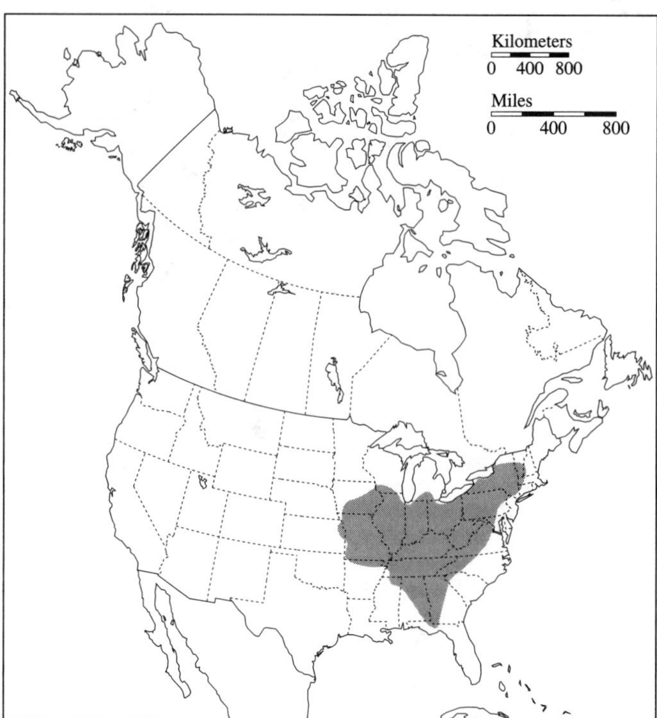

FIGURE 3.6. Distribution of the Indiana bat (*Myotis sodalis*).

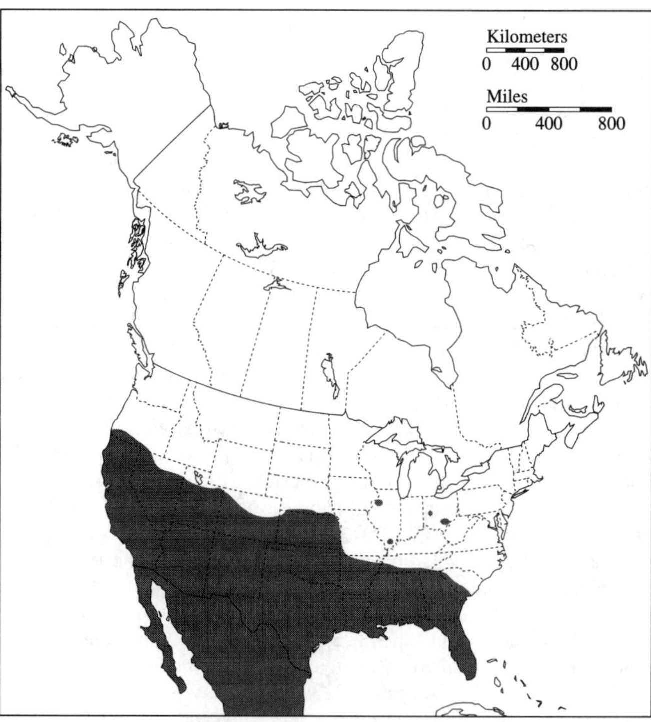

FIGURE 3.7. Distribution of the free-tailed bat (*Tadarida brasiliensis*).

from central Oregon to Nevada, to Utah, across to Nebraska and Alabama, and to Georgia and North Carolina (Fig. 3.7). *T. brasiliensis* ranges south through Central America and into South America. The South American distribution is not well known except that this species is found through the western countries of the continent to at least central Argentina. Some records occur in southern Brazil; no occurrence is known from Amazonia. The California leaf-nosed bat, *Macrotus californicus,* occurs in the southwestern United States and northwestern Mexico including Baja California (Fig. 3.8). They usually occupy arid lowlands. Caves are frequent roosting sites, but this species also is found in mines and tunnels. They may be more abundant now in the southwestern United States because of mining (Brown and Berry 1991). Fossil material from Terlingua, Texas, suggests that as recent as 10,000 years before the present, *Macrotus californicus* ranged as far east as southern Texas and northeastern Mexico (Ray and Wilson 1979). This bat is the only species of phyllostomid to remain active in the United States year-round, although many members migrate south into warmer regions during winter (Hoffmeister 1986).

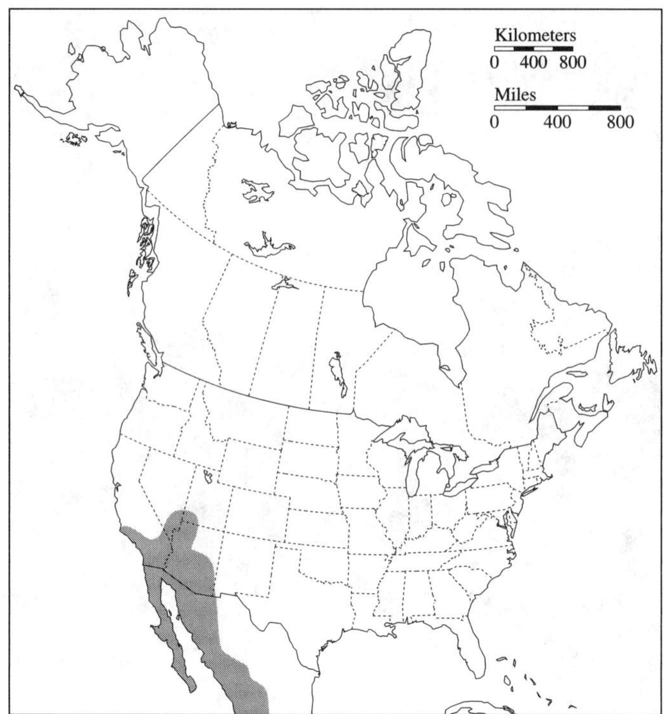

FIGURE 3.8. Distribution of the California leaf-nosed bat (*Macrotus californicus*).

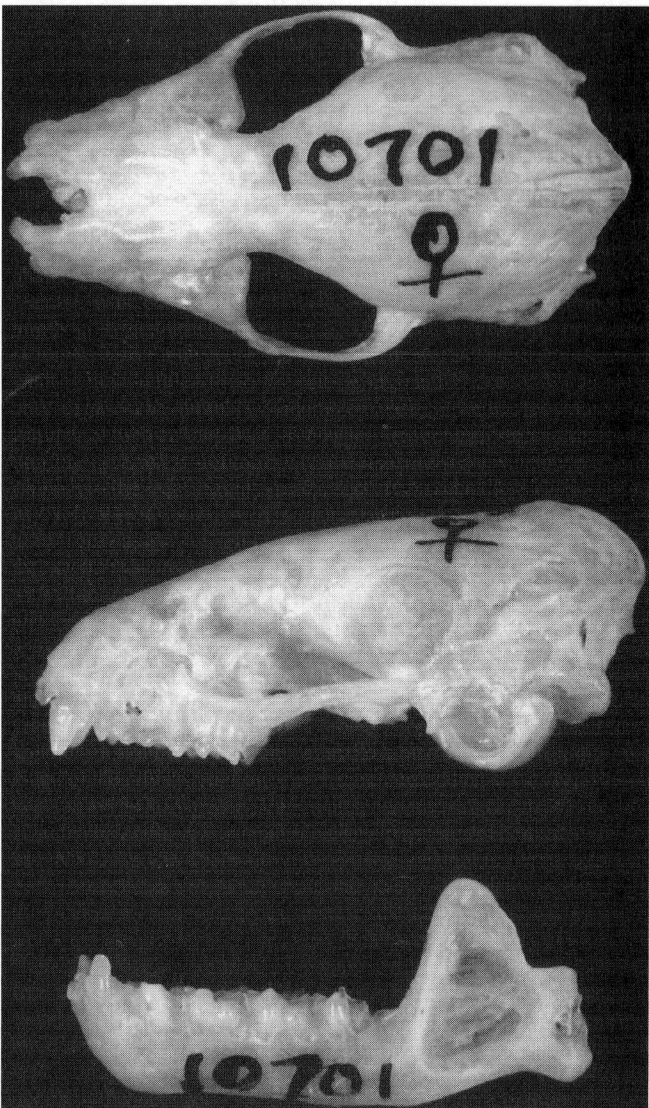

FIGURE 3.9. Skull of the pallid bat (*Antrozous pallidus*). From top to bottom: dorsal view of cranium, lateral view of cranium, lateral view of mandible. Greatest length of skull = 21.32 mm, cranial height = 9.13 mm, width of braincase at distal zygomata = 9.38 mm.

DESCRIPTION

Bats of North America in general comprise several categories: free-tailed bats (Molossidae), where the tail extends far beyond the uropatagium; long-eared bats, which include bats of the genera *Euderma, Idionycteris, Corynorhinus* (Vespertilionidae), and *Macrotus* (Phyllostomidae); and the more common species, which include the genera *Myotis, Eptesicus, Lasiurus,* and *Pipistrellus* (Vespertilionidae). Although identifying a bat in hand or on the wing to the level of species may be difficult, several published keys and photographs are available, including Harvey et al. (1999). For exact identifications, taxonomic experts and curators at natural history museum collections should be consulted.

The pallid bat is a large (forearm length = 45–60 mm), pale cream-colored bat with large ears, a blunt snout with a ridge across the top, and a distinctive skunklike musk odor. Measurements (in mm) are as follows: total length, 95–135; length of tail, 35–53; length of hind foot, 11–16; length of ear, 21–37; and length of forearm 45–60; mass (in g) is 13.6–24.1 for males and 13.9–28.9 for females (Fig. 3.9; Hermanson and O'Shea 1983).

The big brown bat is large for North America, especially compared to similar-looking brown bats in the genus *Myotis*. The wingspan of *Eptesicus fuscus* is 335–350 mm, among the largest of North American bats. Its long fur has deep brown outer hairs and a black base. The wings have no hair. The ears are short, black, furred at the base, and with a broad and rounded tragus. The calcar is keeled. Measurements (in mm) are as follows: total length, 90–138; length of tail, 35–46; length of hind foot, 11–13; length of ear, 14–18; and length of forearm, 39–51; mass is 12–25 g (Fig. 3.10). Although this species could be confused with *Myotis,* only a few *Myotis* have keeled calcars, including the long-legged myotis (*M. volans*), the California myotis (*M. californicus*), the western small-footed myotis (*M. ciliolabrum*), the Indiana bat (*M. sodalis*), and perhaps a few others, but the tragus of *Eptesicus* is blunt, whereas the tragus is pointed in species of *Myotis*.

The hoary bat is also a large bat by North American standards. It has long and pointed wings with a wingspan of 380–410 mm. This species is quite striking, with a yellowish ventral surface about the neck, a brown chest, and a white belly. The rest of the body is covered in long fur with the dorsal hairs black at the base, then yellow, then brown with a white tip, giving a frosted or hoary appearance. The uropatagium and other wing surfaces are furred ventrally. Measurements (in mm) are as follows: total length 120–145; length of tail, 49–60; length of hind foot, 9–12; length of ear, 12–18; and length of forearm, 46–56; mass is 18–32 g (Fig. 3.11).

The spotted bat is beautiful and unmistakable. It is a medium-sized bat with enormous (45–50 mm) pink ears, a patch of white hairs at their base, a conspicuous black dorsum with three large (15 mm) white spots, and a venter with a frosted appearance. Ranges of measurements (in mm) are as follows: total length, 105–119; length of tail, 44–50; length of hind foot, 10–12; length of ear, 37–50; and length of forearm, 48–52; mass is 13–24 g (Fig. 3.12).

The gray bat is distinguished from all other *Myotis* by the uniformly colored gray fur without a dark base. The wing membrane inserts at the tarsus rather than the foot. Measurements average (in mm) as follows: total length, 88.6 (80.2–96); tail length, 38 (32.8–44.2); length of hind foot, 9.9 (8.4–11.2); forearm length, 43.1 (40.6–45.8); and ear length, 16; mass is 7.6–10 g (Fig. 3.13; Whitaker and Hamilton 1998).

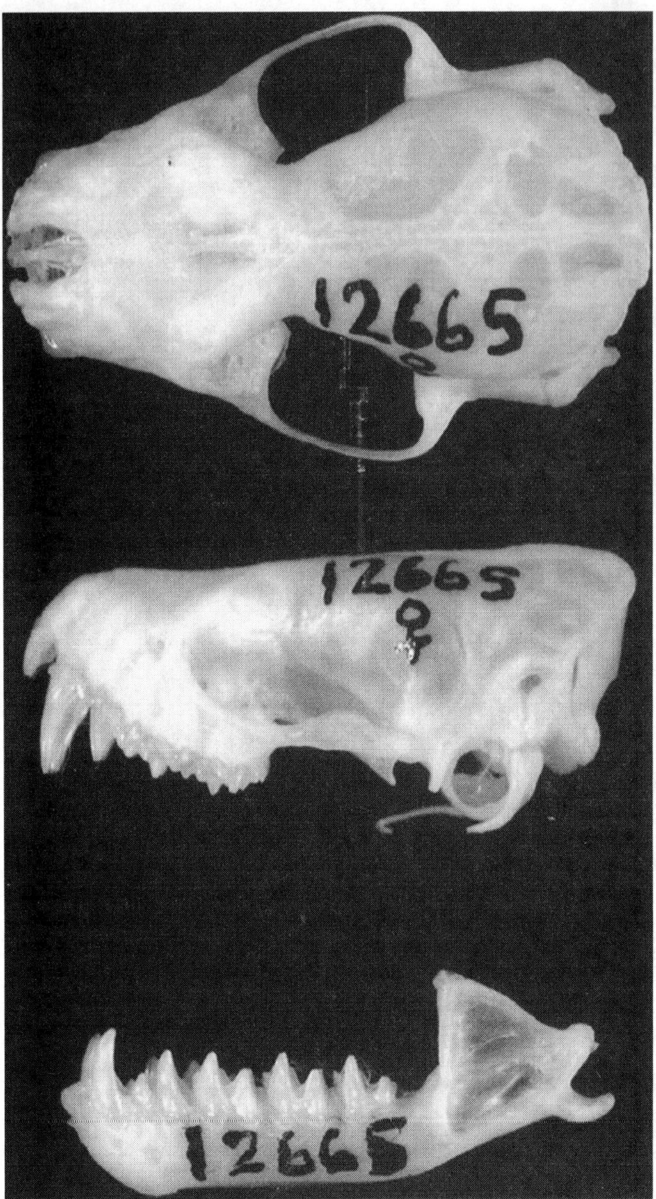

FIGURE 3.10. Skull of the big brown bat (*Eptesicus fuscus*). From top to bottom: dorsal view of cranium, lateral view of cranium, lateral view of mandible. Greatest length of skull = 20.31 mm, cranial height = 7.50 mm, width of braincase at distal zygomata = 9.52 mm.

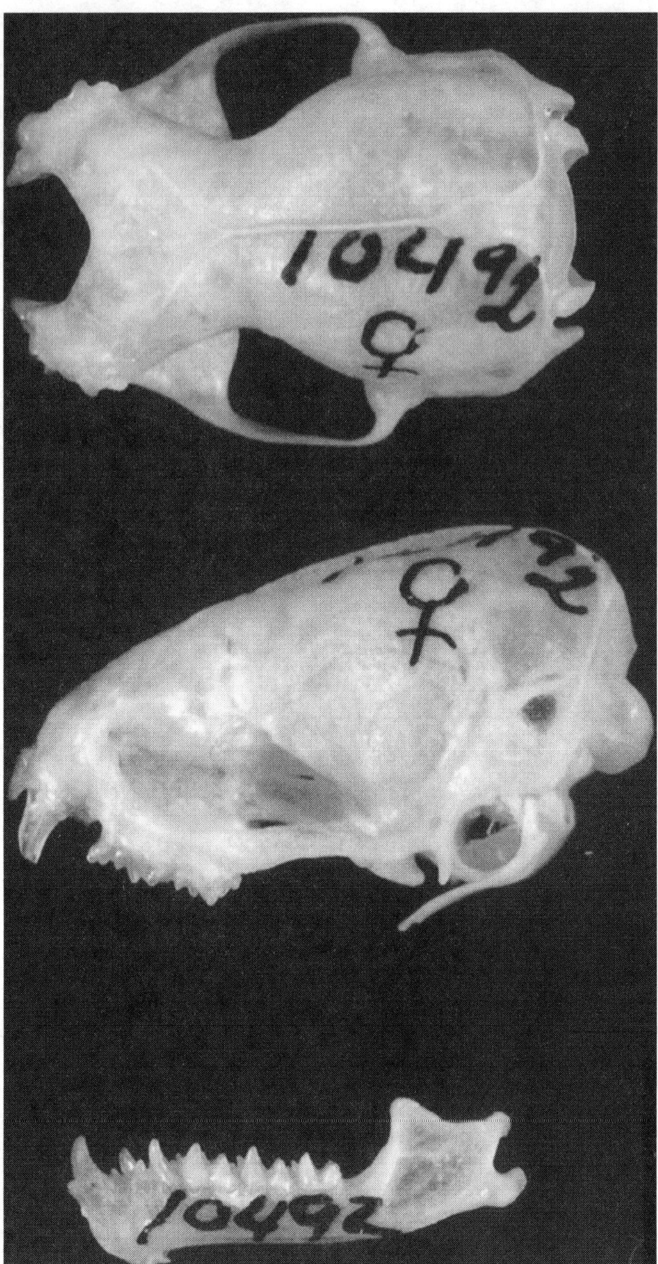

FIGURE 3.11. Skull of the hoary bat (*Lasiurus cinereus*). From top to bottom: dorsal view of cranium, lateral view of cranium, lateral view of mandible. Greatest length of skull = 18.08 mm, cranial height = 9.82 mm, width of braincase at distal zygomata = 10.05 mm.

Because the Indiana bat bears such a strong resemblance to the little brown bat (*Myotis lucifugus*), it was not described as a species until 1928. It is distinguished by short, inconspicuous toe hairs, a small foot (9–10 mm), keeled calcar, and a long forearm (between 35 and 41 mm). The feet are small and contrast with those of *M. lucifugus* by lacking hairs that extend beyond the claws. The Indiana bat has been described as having banded fur with a superficial chestnut color, except at the shoulder, where the hairs lack chestnut tips and appear dark brown to a uniform, distinct brown-pinkish (Whitaker and Hamilton 1998). Measurements average (in mm) as follows: total length, 81.7 (70.8–90.6); tail length, 36.4 (27–43.8); length of hind foot, 7.9 (7.2–8.6); forearm length, 38.8 (36–40.6); and ear length 12.9 (10–15); mass is 5–11 g (Fig. 3.14; Whitaker and Hamilton 1998).

The Brazilian or Mexican free-tailed bat has short, uniformly brown or gray fur. The forearm is 36–46 mm long and the wings are very narrow and long. Half of the tail extends beyond the interfemoral membrane (uropatagium). The ears attach to one another at the forehead between the eyes. Stiff hairs as long as the feet extend from the toes. Prominent vertical wrinkles or grooves are present on the upper lip. The calcar is not keeled. Measurements from Oregon to Alabama average (in mm) as follows: total length, 91.9 (81–109); tail length, 33.2 (27–44); length of hind foot, 10 (8.7–14); ear length, 13–20; and forearm length, 43.1 (36–45.5); mass is 8–14 g (Fig. 3.15) (Wilkins 1989; Verts and Carraway 1998).

The California leaf-nosed bat has brownish or grayish dorsal pelage and brown or buff ventral pelage generally with a whitish or silvery wash. Measurements for the California leaf-nosed bat average (in mm) as follows: total length, 85–110; tail length, 35–41; length of hind foot 13–15; ear length, 33–36, and forearm length, 45–58; mass is 12–20 g. It has a prominent nose-leaf, large ears, and broad wings (Fig. 3.16).

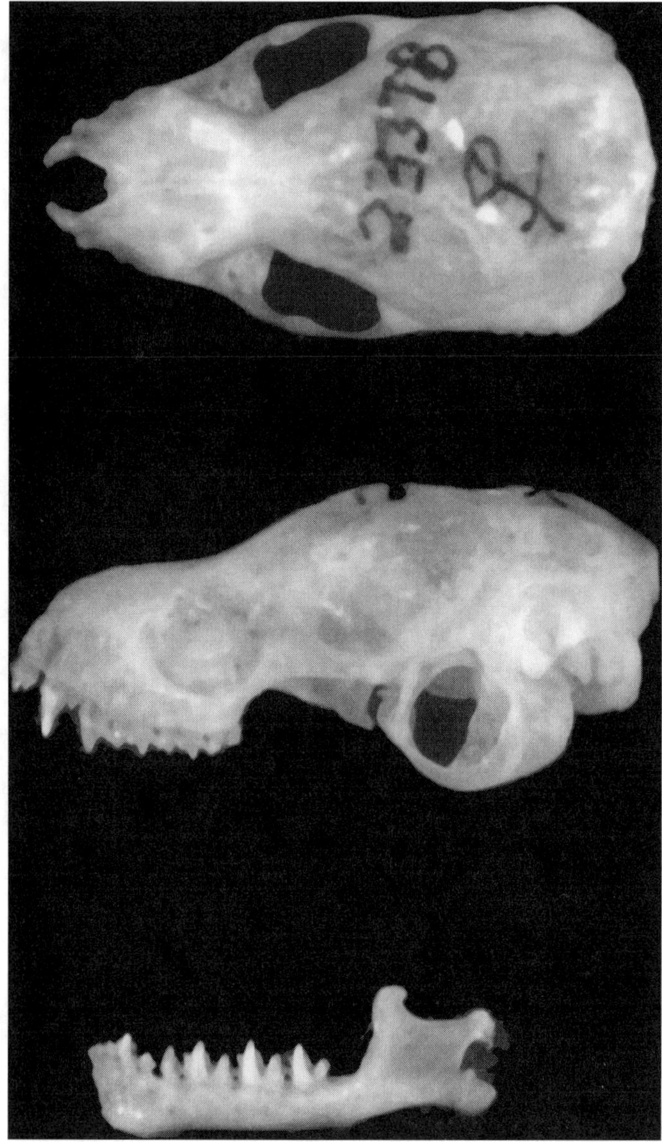

FIGURE 3.12. Skull of the spotted bat (*Euderma maculatum*). From top to bottom: dorsal view of cranium, lateral view of cranium, lateral view of mandible. Greatest length of skull = 19.53 mm, cranial height = 7.29 mm, width of braincase at distal zygomata = 9.70 mm.

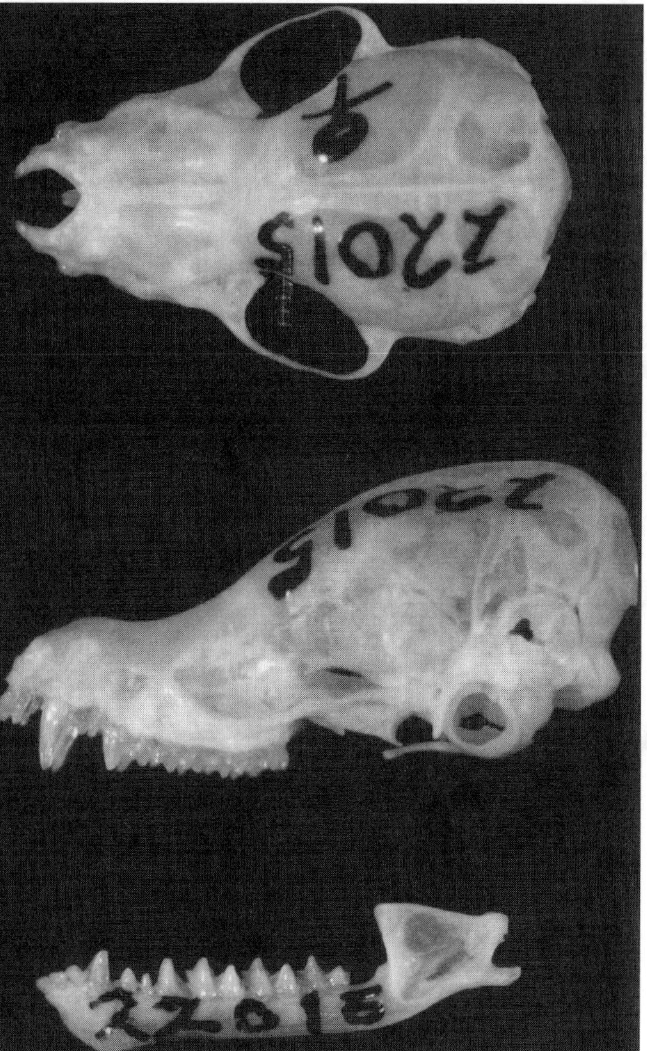

FIGURE 3.13. Skull of the gray bat (*Myotis grisescens*). From top to bottom: dorsal view of cranium, lateral view of cranium, lateral view of mandible. Greatest length of skull = 16.11 mm, cranial height = 7.18 mm, width of braincase at distal zygomata = 7.74 mm.

PHYSIOLOGY

Thermoregulation. McNab (1983) reviewed the physiological ecology of bats and showed that for the most part, food supply, not temperature, limits the northward dispersal of tropical bats. Because bats are chiefly tropical, it may be difficult to understand how they are successful in temperate regions, where there is a lower average temperature. Many female temperate-zone vespertilionids become homeothermic during the season of parturition and lactation. Rainy and cloudy weather, which inhibits insect activity, also inhibits foraging by bats. Male *Myotis* remain heterothermic in summer, but are able to become hypothermic when feeding is not possible. Females are unable to forage enough in cold places to maintain homeothermy and are excluded from places that are constantly cloudy and rainy (Thomas and Bell 1986). For instance, female little brown bats stop regulating their body temperatures from the normal 37°C for the last 18 days of pregnancy and during lactation; infants are also poor thermoregulators (Studier and O'Farrell 1972). Cessation of thermoregulation late in gestation is associated with a rapid increase in energy use during embryo growth. The energy demand is much greater for lactation than for late pregnancy. Consequently, ambient roost temperature has a direct effect on the growth rate of embryos and young (Studier et al. 1973; Tuttle 1975). North American bats can tolerate ambient temperatures several degrees higher than most other small mammals. Hot environments are tolerated by accepting a heat load of 1–2.4°C above ambient temperature and undertaking evaporative cooling by using metabolic water to pant, salivate, and lick exposed skin (Licht and Leitner 1967). However, a temperature of 43.3°C is lethal within 1 or 2 hr. Geluso (1978) showed that pallid bats are capable of concentrating urine as a way of reducing their need for free water. Pallid bat kidneys are modified with enlarged medulla and renal papillae for increased urine concentration abilities. Also, they lose 0.44 mg H_2O/ml of O_2 consumed, which is a low value especially when compared with desert rodents (Chew and White 1960).

Water is an important feature in the lives of most mammals; however bats are able to prosper in some of the driest climates on Earth. In situations where surface water is limited, large numbers of bats may appear suddenly over water holes (Findley 1993). These appearances are usually in the early and late evening and are likely correlated to pre- and postforaging activity. Although water holes are an attractant for bats, in most parts of the world, water does not appear to limit bats or to affect bat community structure (Findley 1993). As a way of radiating

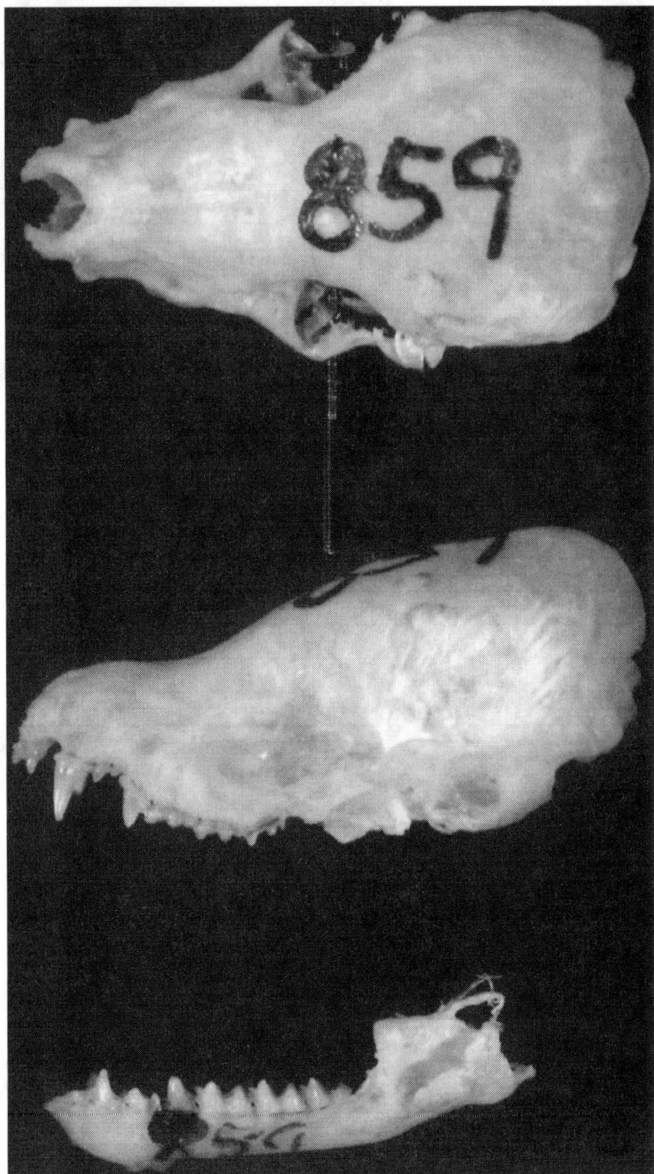

FIGURE 3.14. Skull of the Indiana bat (*Myotis sodalis*). From top to bottom: dorsal view of cranium, lateral view of cranium, lateral view of mandible. Greatest length of skull = 14.31 mm, cranial height = 5.95 mm, width of braincase at distal zygomata = 7.14 mm.

body heat loads, the ears of long-eared bats may assist in eliminating heat. When cooled, the large ears of spotted bats roll up into a ram's head position and conserve heat (Constantine 1961).

Hibernation. The prevailing thought is that many bat species accumulate a large supply of body fat, enter a roost with a cool, stable microclimate, and go into torpor. The body temperature declines to conform to the temperature of the major heat sink (such as a cave wall; McNab 1974), and the heart rate slows from about 600 to 10–80 beats/min (Lyman 1970). Metabolism slows during hibernation (0.004 ml O_2/g per hour recorded in a laboratory; the lowest known for a mammal; Humphrey 1982). Deep hibernation of bats can last, without arousal, for 2–3 months, but as long as 8 months at high elevations (Folk 1940). By the end of winter, almost no fat remains and protein may be catabolized as an energy source (Dodgen and Blood 1956; Humphrey 1982). If ambient temperatures approach freezing, bats can respond either by arousing and moving to a warmer site or remaining in hibernation but increasing their metabolic rate. Several species can withstand brief

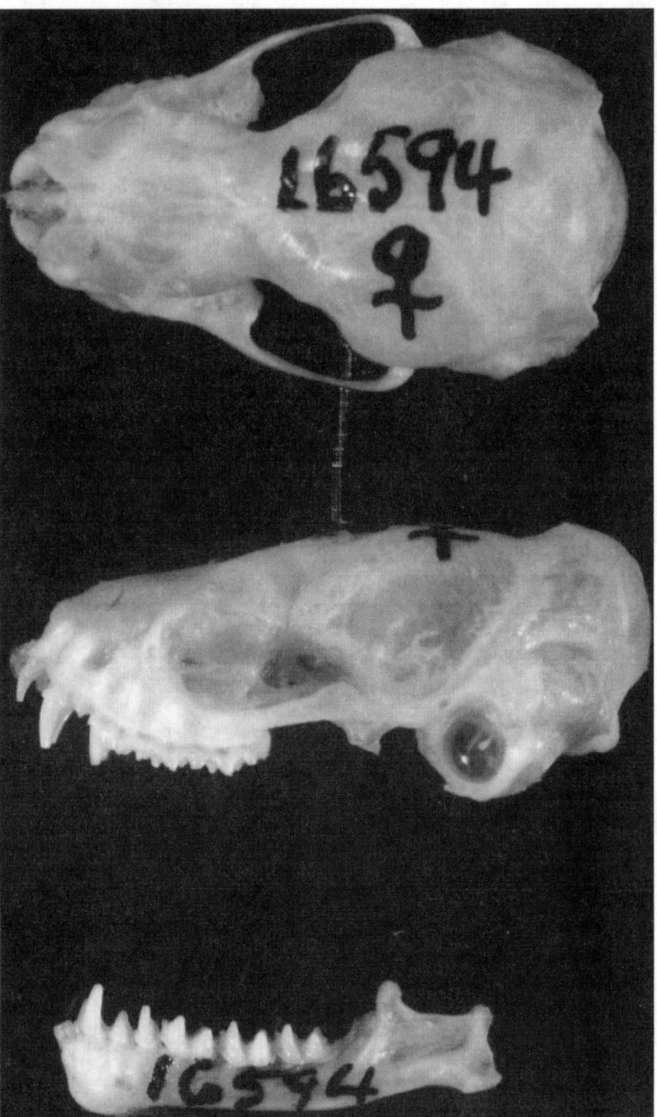

FIGURE 3.15. Skull of the free-tailed bat (*Tadarida brasiliensis*). From top to bottom: dorsal view of cranium, lateral view of cranium, lateral view of mandible. Greatest length of skull = 16.92 mm, cranial height = 7.25 mm, width of braincase at distal zygomata = 8.53 mm.

periods of below-freezing body temperatures without ill effects. Lasiurine bats, such as the hoary bat and red bat (*Lasiurus borealis*), raise their rate of respiration, become more rounded in shape, and stretch the furred tail membrane over the ventral body surface when exposed to subfreezing temperatures. During warm winter days, these bats may not arouse until ambient temperatures rise to about 20°C. Arousal from hibernation to a metabolic rate that can sustain flight can occur in as little as 7 min. Heat is generated in several ways during arousal as the heart rate increases to 700–800 beats/min, shivering of the skeletal muscles ensues, and brown fat metabolic activity occurs (Lyman 1970; Humphrey 1982). Bodies of brown fat provide a source of energy that accelerates arousal from hibernation (Smalley and Dryer 1963). Big brown bats have a 24-hr biological clock that triggers arousals during hibernation (Twente and Twente 1987).

REPRODUCTION

Anatomy. In bats, the uterus of females is bicornuate. Males in reproductive condition have externally discernible descended testes and enlarged epididymides. In *Myotis,* although both ovaries are functional,

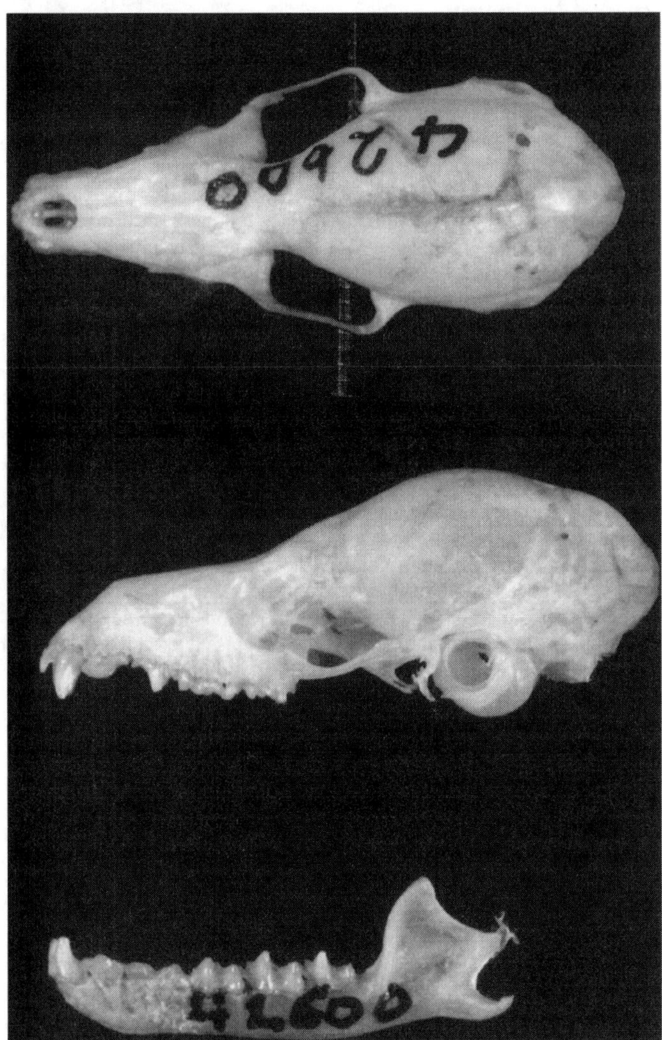

FIGURE 3.16. Skull of the California leaf-nosed bat (*Macrotus californicus*). From top to bottom: dorsal view of cranium, lateral view of cranium, lateral view of mandible. Greatest length of skull = 23.75 mm, cranial height = 9.44 mm, width of braincase at distal zygomata = 8.96 mm.

the developing embryo uses the right uterine horn because the left horn is greatly reduced. Usually one young is born per year. In *Tadarida,* only the right ovary and the right uterine horn are functional. They also typically have one young per year. Jones and Findley (1965) noted that in a spotted bat the right uterine horn was large and flaccid and contained one placental scar. Other bats from this species showed fairly uniform ovary size. In eastern North America, female big brown bats typically bear twins, whereas a single young is typical in other parts of the range. Lasiurines have multiple young per year; *L. cinereus* average 2, and *L. borealis* average 3.5 (Hoffmeister 1986). Lactating females are easily recognized with one pair of enlarged pectoral mammary glands (Bogan 1972).

Physiology. Reproduction is seasonal and tied to the specializations of hibernation. Often, sufficient information about age and reproductive status of bats can be recorded readily in hand. Spermatogenesis takes place in summer and fall. Sexually active males, especially of temperate species, may have epididymides so swollen with semen that they extend obviously into the interfemoral membrane. Similarly, sexually inactive males display no obvious epididymis and the testes are miniscule. Epididymides fill with sperm in early fall. They remain full through winter, and empty by spring. Male accessory organs reach maximum development in fall, and continue to be active during mating and hibernation, but become reduced shortly after hibernation (Humphrey 1982). Females remain in breeding condition from fall through hibernation. Large mature follicles survive through the winter, which are structurally specialized and associated with larger amounts of glycogen. Ovulation is delayed until bats emerge from hibernation in spring. Spermatozoa are stored in the uterus and remain fertile through winter. Spermatozoa are metabolically active and have access to the glycogen in the epithelium and to a variety of sugars. As in other mammals, luteinizing hormone regulates ovulation. Large amounts of this hormone are produced only near the end of the hibernation period (Humphrey 1982). In nonhibernating *Tadarida,* spermatogenesis occurs from September through winter, with spermatozoa available from 20 January through 18 April (Sherman 1937). Thus, the period of sperm storage in the epididymides is much briefer than in hibernating species. Furthermore, female *Tadarida* lack the specialized follicles of hibernating bats. Testis size in *Euderma maculatum* ranged from 2 by 4 mm to 3 by 7 mm (Jones and Findley 1965; Poché and Bailey 1974). Female big brown bats may release five to seven eggs per ovulation and fertilization, but only one or two develop to term (Kurta and Baker 1990). Lactating female bats have obviously enlarged nipples that will express milk with little prodding. Females in advanced pregnancy are conspicuous with a greatly distended posterior ventrum. In the California leaf-nosed bat, the estrous cycle commences with ovulation in late September and early October (Carter and Bleier 1988). Ovulation occurs exclusively in the right ovary, but the left ovary is functional if the right ceases to function (Crichton and Krutzsch 2000). Multiple ovulations have been observed when fertilization is not successful (Carter and Bleier 1988). Spermatogenesis occurs in summer and fall; insemination and fertilization coincide with ovulation (Bleier 1975). Gestation is from 7 to 9 months and embryonic development is delayed; limb buds are not recognized on the embryo until March and birth may occur in June (Bleier 1975). Twinning has been observed, but single births are more common (Bradshaw 1962).

Breeding Season. For hibernating species, specializations including sperm storage in male epididymides, long survival of spermatozoa in uteri of females, and delayed fertilization allow for a prolonged breeding season. Because of this, breeding may occur in fall, winter, and early spring. Most mating occurs during fall in hibernacula with additional efforts during winter. Pregnant female pallid bats gather in summer maternity colonies within warm rock crevices, abandoned mines, caves, hollow trees, and attics of buildings. Copulation takes place between October and February, with parturition generally occurring between May and July depending on local climatic variables. Females can give birth to a single pup, commonly twins, rarely triplets. Young are generally weaned in mid- to late August. Maternity colonies disband between August and October. The bats are relatively inactive during the winter, but occasional winter activity has been reported in southern portions of the range. They have been caught in nets or recorded acoustically throughout the year (Orr 1954; Hermanson and O'Shea 1983; W. L. Gannon, unpublished data). Mating in big brown bats occurs in fall and winter as indicated by spermatozoa stored in the female's reproductive tract in winter. In males, testes are scrotal in August. Pregnant females form nursery colonies during summer; males form bachelor colonies separate from females and young. Newborn bats are naked and nearly immobile; eyes and ears open within hours of birth. Nursery colonies are from 12 to several hundred individuals. Considerable shifting of colonies occurs from hollow trees, to buildings and other human-made structures, to caves and mines (Kunz 1974). Females produce milk for 32–40 days and young begin to first fly at 18–35 days of age (Kurta and Baker 1990). Mating in hoary bats is thought to occur during winter when they are in the southern wintering area, although males with descended testes have been captured in North America in June or July, as have pregnant females (Findley and Jones 1964; Shump and Shump 1982). Parturition in spotted bats occurs in late May to mid-June. A netted pregnant female gave birth to a single male weighing 4 g on 11 June at Big Bend National Park, Texas. The newborn was born naked with no indication of the striking adult color pattern (Schmidly 1991). Mexican free-tailed bats breed in early spring similar to other nonhibernating species. Although males have active sperm throughout the winter, they experience a short breeding season in the spring (mid-February to April), the only time spermatozoa are in the epididymis. Ovulation in females occurs in late March (Hill and Smith 1984). The California leaf-nosed bat forms gregarious, interacting colonies of dozens or hundreds of individuals. The sexes are present together in the early spring (March and April), but segregate later in the season. Ovulation, insemination, and fertilization in *Macrotus californicus* occur in September and October. Young females are able to mate in their first season, but later (late October) than the adults. Males do not mate until their second season (Carter and Bleier 1988).

Gestation. Length of gestation was estimated at 50–60 days for the gray bat (Kurta and Whitaker 1998). Gestation time for the hoary bat is 80 days (Shump and Shump 1982). Gestation in the Mexican free-tailed bat takes between 77 and 100 days (Fenton 2001). Parturition in big brown bats occurs 60 days after fertilization. In the California leaf-nosed bat, the embryo undergoes delayed development. It grows slowly through winter until March, and then grows rapidly until a single young is born in early May. Births in a colony may continue through July after a total pregnancy of about 8 months (Anderson 1969).

Reproductive Rate. Many species of bats form maternity roosts where the reproductive rate is nearly 100%. In a study of *Myotis sodalis,* 23 of 25 (92%) females reproduced (Humphrey and Cope 1977). Pregnancy rates of 92% are reported for *Eptesicus fuscus,* but only 25–50% of yearling females are parous (Schowalter and Gunson 1979). Davis et al. (1962) found that 96.4% of 1710 *Tadarida brasiliensis* were pregnant. Pups may reach densities of 5000 individuals/m^2 (Loughry and McCracken 1991).

Litter Size. Litter size is typically one in most species of bats and more than that is rare. Big brown bats produce one young west of the Great Plains and two to the east (Cockrum 1955; Christian 1956). Two young are typical of hoary bats, but are rare for vespertilionids; red bats typically have three or four young/year (Bogan 1974; LaVal and LaVal 1979). Litter size in spotted bats, free-tailed bats, and gray bats is one. Twinning has been observed in California leaf-nosed bats, although rare (Bradshaw 1962). Neonates weigh approximately 6 g, and pups are weaned after 1 month when permanent dentition is present and the young are able to catch their own food (Bleier 1975; Bradshaw 1962). A single pup is usual for *T. brasiliensis,* with a full-term crown-to-rump length of neonates of 25 mm. While giving birth, *T. brasiliensis* remain suspended head down. Unlike vespertilionids, the young is not received into the uropatagium at birth. Rather, the mother scratches the amnion until the young drops and hangs at the end of the umbilical cord. Because the wings are as yet useless, the young hangs there until it is able to climb up to the breast. The placenta remains attached to the young for up to 2 days, when it dries up and falls off (Sherman 1937).

Breeding Synchrony. All North American bat species produce young in late spring or early summer. Parturition tends to occur later in northern latitudes. Local variations occur in this pattern due to differences in roost temperatures, which may affect embryonic growth rates. For *A. pallidus,* time of parturition appears to be determined by local climate. The length of period of embryonic development is therefore variable, but is likely around 9 weeks (Orr 1954). Lactating *E. fuscus* have been captured from June through September, but most occur in July (Kurta and Baker 1990). Lactating *E. maculatum* have been captured from June until mid-August in various parts of its range (Poché and Bailey 1984).

Breeding Age. Most females of North American bat species, including all species in this chapter, produce their first young at the age of 1 year. Specifically, yearling *Tadarida brasiliensis* can produce young (Sherman 1937) and *Eptesicus fuscus* produce young at the age of 1 year (Kunz 1974). *Macrotus californicus* could breed the first fall after their birth (at an age of 7 months), but males of this species are not sexually mature until the following spring (as yearlings) (Anderson 1969).

ECOLOGY

Habitat. It is difficult to assign a particular habitat to a particular species of bat because they are volant; however, several general statements are possible. For instance, pallid bats are characteristic of desert areas and seem to be most abundant in Sonoran life zones (Orr 1954). Big brown bats forage for insects continuously, consuming prey on the wing. They search for prey over woodlands, fields, urban parklands, or water during foraging times that may last as long at 120 min (Kurta and Baker 1990). Spotted bats forage continuously mostly alone in open habitat, flying 5–10 m above the ground, usually after midnight. Acoustic surveys, which detect their audible call, led some to believe that spotted bats fly all night; this has been confirmed with results from radio-tracking studies. They leave their roosts at dusk and return to the same day roost just before dawn (Fenton 2001). These bats have been captured in montane Ponderosa pine (*Pinus ponderosa*) forests, pinyon–juniper (*Pinus edulis–Juniperus monosperma*) woodlands, and open semidesert scrublands. Rocky cliffs are thought to be necessary to provide cracks for suitable roosting. Access to water also may be an important feature for many bats of arid regions. Gray bats are true cave obligates as they hibernate and give birth in caves. The Indiana bat gathers during summer in small colonies in semiwooded areas of upland or bottomland forests under loose tree bark (Kurta et al. 1993). The Brazilian free-tailed bat occurs at lower elevations in the southwestern United States in pinyon–juniper woodlands, grasslands, and semidesert scrublands. Roosts are in caves, mines, buildings, bridges, and, in the southeast United States, hollow black mangrove trees (*Avicennia germinans*). They have a characteristic musky odor, which can identify roosts from a distance. The California leaf-nosed bat occurs in arid lowlands roosting in caves and mines.

Roosting Habitat. The roosting ecology of bats has been well reviewed by Kunz (1982). Many bats have roosting requirements that are determined by physiological demands of the adults or the young, predation pressures, social structure, or morphology. For some species, roosting requirements are so specific that the absence of suitable roosts precludes their occurrence. Pallid bats roost in rock crevices, tree hollows, mines, caves, and a variety of anthropogenic structures, including vacant and occupied buildings. Tree roosting has been documented in large conifer snags, including ponderosa pine, inside basal hollows of redwoods and giant sequoias (*Sequoia sempervirens* and *S. gigantea,* respectively), and in bole cavities in oaks (*Quercus* spp). They also have been reported roosting in stacks of burlap sacks and piles of stones (Orr 1954). The hoary bat is a solitary, wide-ranging species, which uses a variety of trees as roosts. It appears to favor deciduous trees in the east, but conifers in the western United States. Roosts are located 4–5 m above ground and are protected by good leaf or branch cover, but allow for unobstructed flight in and out of the roost. Roosts are frequently associated with margins or edge habitat. Relative abundance is usually low, but can be clumped (catches in mist nets are single individuals or groups up to 20; Gannon, unpublished data), as small groups can be encountered during migration. Indiana bats congregate in winter to hibernate in huge numbers in cold sinkhole caves. These caves number <20 in Tennessee, Arkansas, and Missouri, with only 3 of them acting as points of congregation for most Indiana bats (Tuttle 1976). The summer ecology of *Myotis sodalis* is not well understood. Migrating bats leave their summer range and move toward 1 of these 3 caves, making stopovers on their way. When moving northward, Indiana bats in Florida make a broad western sweep. The large winter colonies disperse, forming smaller maternity groups using tree roosts (Callahan et al. 1997).

Often migrating in large numbers, gray bats reach winter hibernation caves between August and October, with females arriving first. Females enter hibernation first after copulation. Males remain active until hibernation in November. Females emerge first from hibernacula in late March through early April, followed by subadults, then males. Trees used for roosts include slippery elm (*Ulmus rubra*), red oak (*Quercus rubra*), hickory (*Carya glabra*), cottonwood (*Populus deltoides*), green ash (*Sorbus americana*), sycamore (*Platanus occidentalis*), red maple

(*Acer rubrum*), and white oak (*Quercus alba*), and under exfoliating bark of mostly dead elm (*Ulmus americana;* Kurta et al. 1993). Spotted bats may roost during the day under loose rocks or boulders, but are more likely found in crevices of high cliffs. More commonly, they roost in high cliffs and drop to forage in pine forest (*Pinus ponderosa*) or over open meadows adjacent to coniferous forest (Poché and Bailey 1974; Fenton 2001). Spotted bats may also show seasonal preference for ponderosa pine woodlands in the reproductive season and higher elevations at other times of the year. *Tadarida brasiliensis* typically roost in caves, mines, rock fissures, buildings, and bridges. In the southeastern United States, these bats roost mostly in buildings. The species is gregarious and forms colonies of up to 20 million individuals, although most roosts in buildings or bridges may number only in the tens of thousands. There are spectacular cave exit-flights from places such as Carlsbad Caverns, New Mexico, or Bracken Cave, Texas. Although some colonies show a trend of reducing number, some sites, such as the Orient Mine, Saguache County, Colorado, show considerable increase in density. *Macrotus californicus* primarily roost in caves and hollows, but also frequent abandoned mines and tunnels. Colonies may number in the hundreds, but smaller numbers of individuals are more common. This species may be more abundant now in the southwestern United States because of mining (Brown and Berry 1991). They roost within 10–30 m of the opening of a cave in areas that are fairly illuminated. *Macrotus californicus* may roost with other species such as *Antrozous pallidus, Corynorhinus townsendii,* and several species of *Myotis* (Vaughan 1959). They have also been found in geothermally heated roosts, which have allowed bats to roost in cooler regions of their range in winter. Although Burt (1934) found California leaf-nosed bats in the far recesses of a mine at a depth of 560 m, relatively constant and warm temperatures were cited as important roost requirements. Most California leaf-nosed bats migrate south into warmer regions; some individuals stay and overwinter in small caves in the southwestern United States (Bell et al. 1986; Hoffmeister 1986). Although semidormant, they are not considered true hibernators.

FEEDING HABITS

Insectivorous bats adjust their reproductive cycles so that young are born during periods of food abundance (Findley 1993). Although foraging areas themselves may not be limiting, foraging habitat specialization may result from competitive interactions among species. Pallid bats are primarily insectivorous, feeding on large prey that are taken on the ground, or sometimes in flight. Prey items also include flightless arthropods such as scorpions (Scorpionida), ground crickets (Gryllidae: Nemobiinae), and cicadas (Cicadidae). Foraging is concentrated in 2 periods at the beginning and end of the nocturnal cycle of activity during most of the active season. Carbon stable-isotope techniques were used to find that *Antrozous* obtains substantial amounts of carbon from plants of the Cactaceae and Agavaceae that use the crassulacean acid metabolism photosynthetic pathway. It is likely that carbon is ingested as pollen or nectar from these plants while the bats are foraging for insects around them (Herrera et al. 1993). Gray bats feed on mayflies (Ephemerida), beetles (Carabidae, Chrysomelidae, and Scarabaeidae), Trichoptera, Diptera, and Lepidoptera (Lacki et al. 1995). The big brown bat usually emerges at dusk, and flies directly to foraging areas. They are thought to visit specific foraging areas regularly and patrol at about 10 m above the ground. They are efficient foragers and complete foraging on mostly beetles in about 1 hr. *E. fuscus* uses night roosts to rest between foraging bouts, which may last as long at 120 min (Kurta and Baker 1990; Whitaker and Hamilton 1998). Hoary bats emerge well after dark and begin foraging for mostly large moths, but will also take beetles and wasps (Orr 1950; Black 1974). The diet of *E. maculatum* consists of moths, grasshoppers, beetles (including Scarabaeidae), katydids, and other insects. Among moths, spotted bats prefer larger noctuid moths, and pluck off the head and wings and eat the large, meaty abdomen (Poché and Bailey 1974). Evening foraging of *T. brasiliensis* starts shortly after sunset. Free-tailed bats may forage as much as 40–50 km from the roost. Flight speed can exceed 40 km/hr; speeds to 95 km/hr have been documented (Davis et al. 1962). Swarms of bats can be tracked using Doppler radar, and it was found that they follow high-flying swarms of insects $\geq$3500 m in altitude. They almost exclusively forage on small moths taken in flight. Whitaker et al. (1996) found that *Tadarida* in Texas fed largely on coleopterans and lygaeid bugs during evening feeding, and mostly on moths during morning feeding bouts. Flight is erratic. Average time foraging is 4 hr and bats may return to day roosts without having used any night roosts. Although bats emerge in impressive numbers over a short time to begin foraging, they return singly to the roost throughout the night, with the last stragglers returning as the sun rises. *Macrotus californicus* emerges later in the evening than most bats, at 90–120 min after sunset (Anderson 1969). Feeding is done mostly in flight because the California leaf-nosed bat is not good ambulating on the ground. The diet consists of insects such as crickets, moths, beetles, and a variety of other arthropods. They also consume fruits, including those of cacti (Burt 1938). These bats take their insect prey to their roost to eat, dropping items such as legs and wings to the ground. *Euderma maculatum* has been observed foraging on the ground and seeking cover under loose rocks or boulders. Although commonly reported active after midnight, they begin to forage shortly after dark with consistent activity levels throughout the evening; they maintain distinct foraging areas (Navo et al. 1992).

BEHAVIOR

Site Attachment and Homing. Pallid bats are gregarious, and often roost in colonies of between 20 and several hundred individuals. They are extremely versatile at walking on the ground and use a variety of strides and gaits (Bell 1982). They are less maneuverable than other vespertilionids in flight, but also may hover or glide or swoop in figure eights. Distances flown during foraging are largely unknown for pallid bats, although movements of 30 km between night roosts have been recorded. Nightly movements of these bats were estimated at $\leq$3 km (Bell 1982). Homing experiments with pallid bats indicated a maximum return distance of 174 km (Davis 1968). A study of homing ability in *Myotis sodalis* found that they were less able to return to their roost with increasing release distance (Hassell 1963).

Migration. Gray bats show strong ties to particular cave systems during summer and winter, although they may migrate up to 525 km between these cave systems in winter and summer. Pallid bats are not known to migrate and are believed to hibernate as solitary individuals or in small groups. Although *E. fuscus* spend winter underground in mines and caves, and occasionally hibernate in buildings, numbers of individuals found active in the winter are much less than during summer. They are either extremely elusive or move to warmer latitudes to overwinter. Hoary bats migrate seasonally in waves, with females migrating earlier than males. Northward movements occur in May and June. Although little is known about their winter range, southward movement occurs in late August to September (Shump and Shump 1982; Cryan 2003). Seasonal migrations of most populations of *Tadarida brasiliensis* typically occur from their summer range in North America to their winter range in Central and South America. Some populations do not migrate, and overwinter in the western United States. Populations in the southeast United States are sedentary and also may not migrate. Migrations of 1200–1700 km are typical in free-tailed bats from Oklahoma and New Mexico that winter in Mexico or further south. These migrations take about 2 months to complete. Most free-tailed bats that are going to migrate do so by the end of September or early October. In winter, spotted bats disappear from where they may have been observed during summer; it is unclear whether individuals hibernate locally or migrate to moderate climates (Fenton 2001). However, at least on one occasion, spotted bats were active in their summer range In winter (Gannon and Sherwin in press). Many *Macrotus californicus* migrate south into warmer regions, although some individuals stay and remain active in caves or mines in the southwestern United States (Hoffmeister 1986).

Mother–Young Relationships. Mother and young *Antrozous* may emerge to forage in unison. In *E. fuscus,* young are born naked and helpless, but are capable of flight at about 1 month of age (Kunz 1974). Young *T. brasiliensis* are left in large nursery groups while the mothers fly off to forage. Returning mothers find their young among thousands of neonates by a combination of visual cues, olfaction, and their infant's call (Gustin and McCracken 1987; Balcombe 1990). How this is accomplished is unclear.

Calls and Acoustics. Bats emit high-frequency sounds (echolocation) for communication, orientation, and prey capture. Most species produce the majority of calls in the ultrasonic frequency range above human perception. Presumably, the bat perceives an acoustic scene in some sense as humans perceive a visual scene. Thus, in this scene the bat would have information about insect targets, foliage, other bats, and general ground clutter. The bat is able to select and focus on a single insect from amid all these distractors and track it through clutter.

Echolocation calls are usually ultrasonic and range in frequency from 20 to 200 kHz, whereas the maximum for human hearing normally is 20 kHz. Even so, humans can hear echolocation clicks from some bats, such as the spotted bat and portions of the calls of hoary bats and Mexican free-tailed bats (Poché and Bailey 1974). In general, echolocation calls are characterized by their frequency, their intensity in decibels (dB), and their duration in milliseconds (msec). In terms of pitch, bats produce echolocation calls with both constant frequencies (CF calls) and varying frequencies that are frequently modulated (FM calls). Most bats produce a complicated sequence of calls, combining CF and FM components. Although low-frequency sound travels further than high-frequency sound, calls at higher frequencies give bats more detailed information, such as size, range, position, speed, and direction of a prey's flight. Thus, these sounds are used more often. In terms of loudness, bats emit calls as low as 50 dB and as high as 120 dB, which is louder than a smoke detector at 10 cm. A call that loud could be damaging to human hearing. The Indiana bat can emit such an intense sound. Fortunately, because the call is of ultrasonic frequencies, humans are unable to hear it. The ears and brain cells in bats are especially tuned to the frequencies of the sounds they emit and the echoes that result. A concentration of receptor cells in their inner ear makes bats extremely sensitive to frequency changes. Some horseshoe bats (Rhinolophidae) can detect differences as slight as 0.0001 kHz (Fenton 2001). For bats to listen to the echoes of their original emissions and not be temporarily deafened by the intensity of their own calls, the middle ear muscle (stapedius) contracts to separate the bones, malleus, incus, and stapes and reduce the hearing sensitivity. This contraction occurs about 6 msec before the larynx muscles (crycothyroid) begin to contract. The middle ear muscle relaxes 2–8 msec later. At this point, the ear is ready to receive the echo of an insect 1 m away, which takes only 6 msec (Fenton 2001). Many of the structural features on the heads of bats—nose leafs, tragus, wrinkling around the mouth or ears—are related to echolocation. The external structure of the ears of bats also plays an important role in receiving echoes. The large variation in sizes, shapes, folds, and wrinkles are thought to aid in the reception and funneling of echoes and sounds emitted from prey. Gannon et al. (2001) described a significant relationship between the height of ear pinnae and echolocation call characters for *Eptesicus fuscus, Myotis californicus,* and *M. ciliolabrum*.

A call produced by a big brown bat with a frequency of 30 kHz is well suited for prey detection. The wavelength (distance from one peak amplitude to the next in a sound wave) of this call is about 11 mm, roughly equivalent to the total length of a small moth (Feldhamer et al. 1999). Bats also deal with the effects of environmental variables such as temperature, wind, or plant community structure, which act to impede or attenuate sound propagation. When a bat is pursuing an insect, the calls that it emits change with the bat's behavior. Typically a bat will produce a commuting type of call, which is generally the flattest (longest in duration, sweeping through the fewest frequencies) call that it will produce (e.g., Fig. 3.17B). As the bat detects a target, whether a potential prey or an obstacle, the animal gets more information about the area in front by emitting more calls of shorter duration, perhaps adding harmonics, and covering a broader range of frequencies by frequency modulation (e.g., Fig. 3.17C). Although for the sake of classification, calls are placed in differing categories corresponding to bat behavior (search, pursuit or approach, and terminal or buzz phases), these calls are normally a continuum and all can be heard throughout the night from a single bat. It may seem that there is too much variation and that all these factors change with the bat's behavior. However, each species likely has a suite of characters that is species-specific. This is true in North America, where members of all species have been recorded extensively (O'Farrell and Gannon 1999). In fact, there have been several cases where new species have been discovered and later described after first noting their unique phonotypes (Jones and van Parus 1993).

In the last few years, portable acoustic detectors have allowed easy detection of calls in the field. Echolocation calls, especially those not used for pursuit and capture of prey, such as the search or commuting calls, can be used to identify species. Species can be placed into acoustic guilds depending on their foraging and acoustic behavior. High-, fast-flying species such as *T. brasiliensis* are aerial insectivores with calls with minimum frequency in the range of 20–25 kHz and low slope. Conversely, many *Myotis* species, such as *M. sodalis* and *M. grisescens,* are agile insectivores in cluttered habitats such as forest. They have high minimum-frequency calls of near 40 kHz or greater and very steep slope. Intermediate mixed strategists such as *E. fuscus* have midrange minimum-frequency calls from 35 to 25 kHz and a duration of 4–10 msec, which begin with a steep slope, but then, after a distinct elbow, flatten at midfrequency range (Fig. 3.17), (Sherwin et al. 2000a; Gannon et al. 2001). Variation in calls may depend on the relative amount of clutter, where an aerial hawking species may produce calls with a minimum frequency of 23 kHz in open meadow and a steeper, sloped call of minimum frequency of 28 kHz in moderately cluttered forest (Fig. 3.17B). Using their echolocation, big brown bats can detect a 19-mm-diameter sphere at a distance of 5 m and perceive the texture of that sphere (Kurta and Baker 1990). After an initial foraging period, pallid bats with full stomachs gather at night roosts and join in clusters locating each other by vocalizations (O'Shea and Vaughan 1977). Calls of this species are mostly linear (Fig. 3.17A), but they can also produce sweeping calls into the human–audible range, which are probably communication calls to conspecifics (Fig. 3.18) (W. L. Gannon, unpublished data).

Hoary bats demonstrate calls that fluctuate in minimum frequency within a sequence, a pattern common to other species within the genus (Fig. 3.17C). For the dominant harmonic, the maximum frequency is about 50 kHz, with minimum frequency alternating between 25 and 20 kHz from one call to the next. Calls form a distinct reverse-J shape when the bat is foraging in moderate clutter (typical coniferous forest with some understory). A study of geographic variation in the calls of this species showed that many variables could affect outcomes of acoustic analysis. Comparisons of calls recorded under differing conditions suggest that there is little geographic variation for calls of hoary bats (O'Farrell et al. 2000; W. L. Gannon unpublished data).

O'Farrell (1999) was able to discern *Myotis sodalis* from *Myotis lucifugus* acoustically in a blind test. Search calls in an uncluttered environment for *Myotis sodalis* had a minimum frequency of 40 kHz (tending to drift toward 50 kHz; Fig. 3.19A Page 69), whereas calls of *Myotis lucifugus* may start at a minimum frequency of 40 kHz, but then drift to 35 kHz. Calls of *Myotis sodalis* were also curvilinear, with a short, flat terminal segment (or toe). These calls were also fairly steep (120 octaves/sec) compared with those of *Myotis lucifugus* (60 octaves/sec). California leaf-nosed bats use echolocation to find targets, even ground-based prey. Calls are of low intensity and broadband (Fenton 2001). Combined with vision, prey-generated sounds are also used by these predators to locate and capture prey. These bats can see adequately in light as low as 0.0002 mLux, approximately equal to available light on a clear, moonless night (Bell and Fenton 1986).

Aside from the large white spots on a black background of its pelage, calls of *Euderma maculatum* are perhaps the most distinctive identifying character of this species. The call sounds like a soft, high-pitched, metallic squeak similar to knocking two coins against each

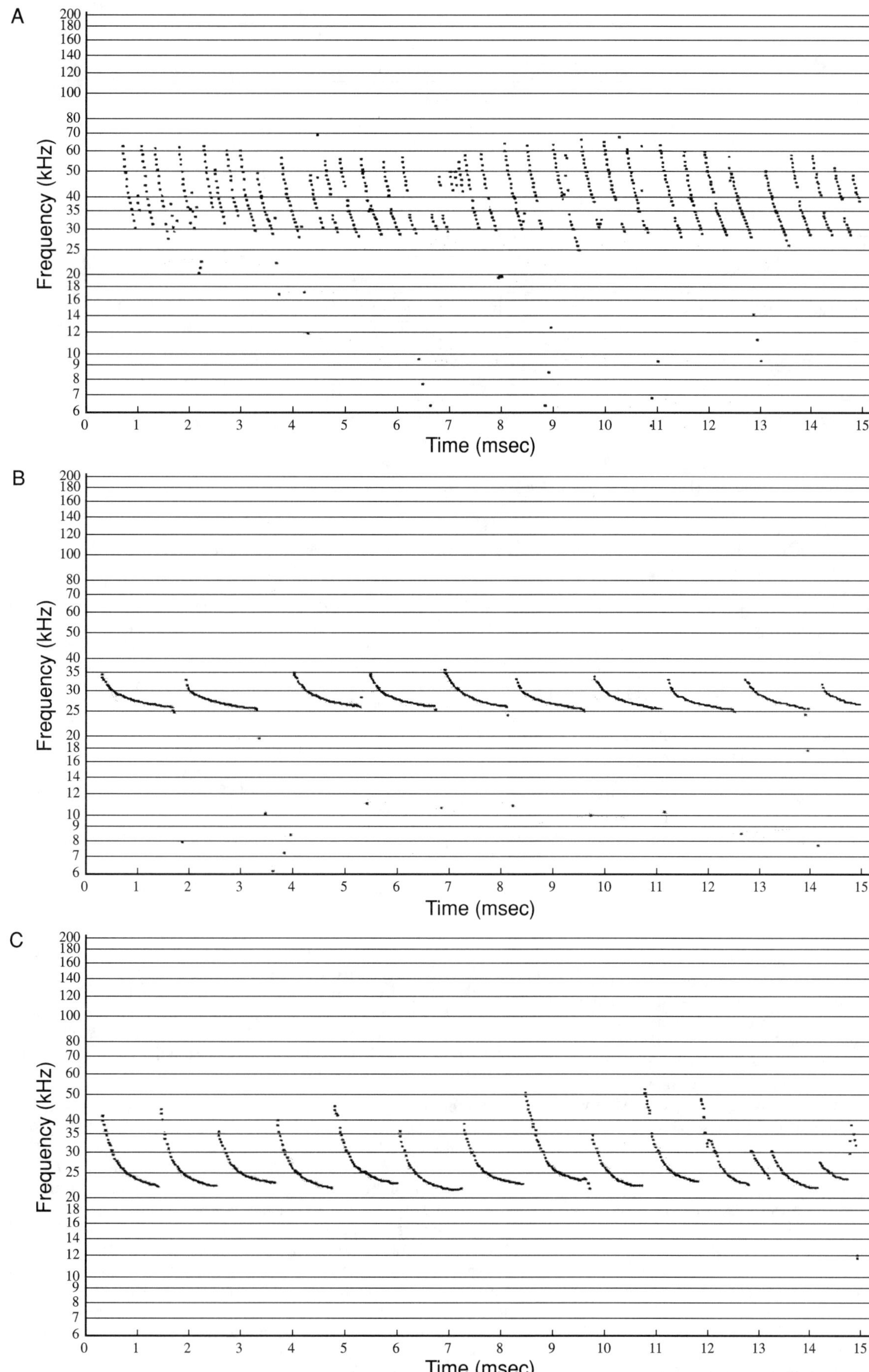

FIGURE 3.17. Calls of (A) the pallid bat (*Antrozous pallidus*), (B) the big brown bat (*Eptesicus fuscus*), and (C) the hoary bat (*Lasiurus cinereus*).

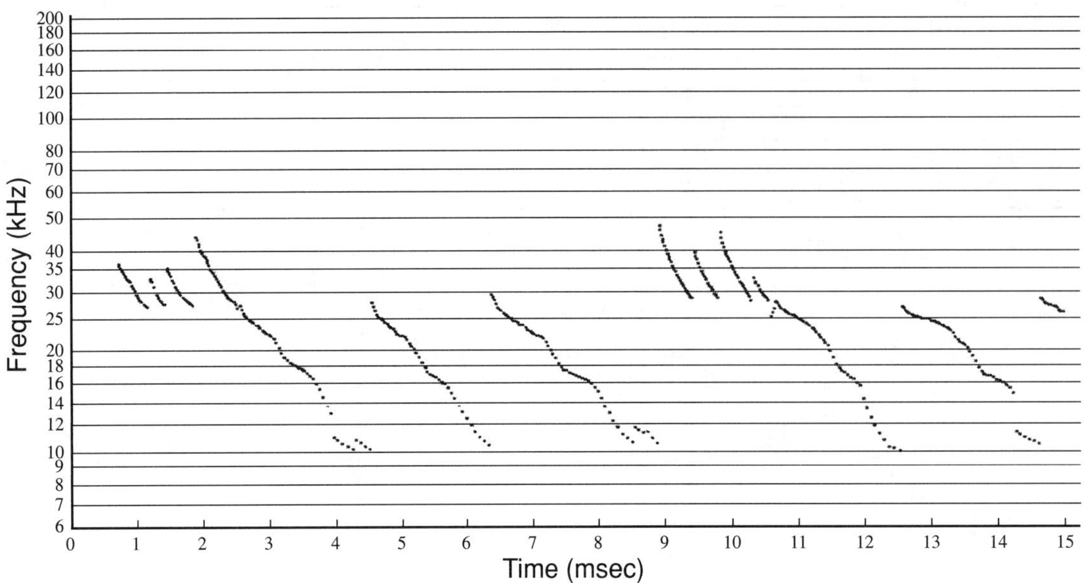

FIGURE 3.18. Human-audible calls recorded from the pallid bat (*Antrozous pallidus*), which likely serve as communication calls to conspecifics.

other at a rate of about 2 calls/sec (Fig. 3.19B). Entirely audible to humans, a call of a spotted bat has a duration that averages 5 msec and sweeps from 15 to 9 kHz; minimum frequency may range from 8 to 12 kHz. Leonard and Fenton (1984) stated that these audible calls allow individuals to space themselves. Navo et al. (1992) believed that the low-frequency calls characteristic of spotted bats suggest that structural features of the environment are related to density of clutter and may be more predictive of habitat suitability and foraging space than other indicators such as diet.

A significant characteristic of *T. brasiliensis* is that a sequence of its calls appears varied (Fig. 3.19C). In a sequence, calls may begin very flat and at low frequency (20 kHz). As the sequence proceeds, the slope of each call increases so that the minimum frequency may reach 28 kHz, the maximum frequency reaches 40 kHz, and the duration between calls becomes 0.1 msec. Other molossids also vary the form of their calls predictably, but at difference suites of frequency than the Mexican free-tailed bat. The call of the big free-tailed bat (*Nyctinomops macrotis*), for instance, is shifted lower to about a minimum frequency of 16 kHz. Calls of *Macrotus californicus* are also species-specific (Fig. 3.20A). Depending on the complexity of the habitat, a call of a bat may differ strictly due to environmental effects on the call.

The big brown bat may be primarily an aerial hawking species foraging on prey in open areas, but it also forages freely in areas of increasing clutter. Calls recorded in more open habitat appear flatter, with a minimum frequency of about 25 kHz. Calls from the same bat recorded in more structural clutter are steeper and with a minimum frequency of about 30 kHz (Fig. 3.20B, 3.20C) (Gannon et al. 2003). Although these calls may appear to be different, the context in which they are recorded may be as important as the species identity.

MORTALITY

Juvenile and Adult Mortality Rates. Bats generally have few enemies. The little that is known of specific mortality is broken into life history categories. Humphrey (1982) gave an excellent review of prenatal mortality rates and mortality rates from birth to weaning for bats. Once bats reach adulthood, after their first winter, they have a 50–80% chance of surviving through the next year (Hill and Smith 1984). Fifteen temperate-zone vespertilionid species for which there are data (summarized by Gaisler 1979) had a mean survivorship of 71%, ranging from 57% to 86%. Survival rates are much lower (5–45%) in species of tropical rodents of comparable size. Some banded big brown bats have survived 19 years in the wild; more commonly, individuals live 10 years or more (Kurta and Baker 1990).

Mortality Factors. In *Antrozous,* a number of wing defects has been found, likely because of injuries sustained while group foraging (Davis 1968). Despite some severe abnormalities, these individuals continue to fly and forage successfully. Undoubtedly, predators catch pallid bats while the bats forage on the ground. Significant mortality of juvenile gray bats occurs during migration, from unknown causes (Tuttle and Stevenson 1977). Easterla (1973) observed a kestrel (*Falco sparverius*), peregrine falcon (*F. peregrinus*), and red-tailed hawk (*Buteo jamaicensis*) diving after released *E. maculatum* (also see Black 1976). The Mexican free-tailed bat is preyed on by a variety of raptors. They have also been taken by cave-dwelling mammals, which include raccoons (*Procyon lotor*), skunks (*Mephitus mephitus*), opossums (*Didelphis virginiana*), and several species of snakes. Great horned owls (*Bubo virginianus*) and barn owls (*Tyto alba*) are also considered predators. *Macrotus californicus* have been found dead pierced by spines of cacti (Carter and Bleier 1988).

The commonest factor affecting survival of bats is believed to be human disturbance. Disturbance can occur in the form of individual people walking into and destroying key roosts or structures. Disturbance can also result from commercial exploitation of mines or caves for mineral and guano content, or be in the form of pollution or other environmental contaminants. *T. brasiliensis* populations were severely affected by the widespread use of pesticides in the United States and Mexico (Geluso et al. 1965).

Parasites and Disease. The behavior of ground-foraging bats such as *A. pallidus* may relate to the large ectoparasite load, including streblids, bat flies, ticks, and fleas, that they carry (Steinlein et al., 2001). Species within genera of fleas (*Myodopsylla*) and mites (*Macronyssus, Spiturnix,* and *Olabidocarpus*) are among the main ectoparasites of gray bats. Internal parasites include cestodes, trematodes, and nematodes (Whitaker and Hamilton 1998). Whitaker and Easterla (1974) and Poché and Keirans (1975) reported ectoparasites including *Crytonyssus* sp., *Basalia rondanii,* and *Ornithodorus* sp. on spotted bats from west Texas, and *Basalia forcipata* from a bat from New Mexico. Parasites are abundant on *T. brasiliensis,* the mite *Chiroptonyssus robustus* (Macronyssidae) usually being the most common. Mites of the genus *Radfordia* and a flea, *Sternopsylla distincta,* were collected from free-tailed bats in Florida. Histoplasmosis is commonly associated with cavern-dwelling species in the southwestern United States.

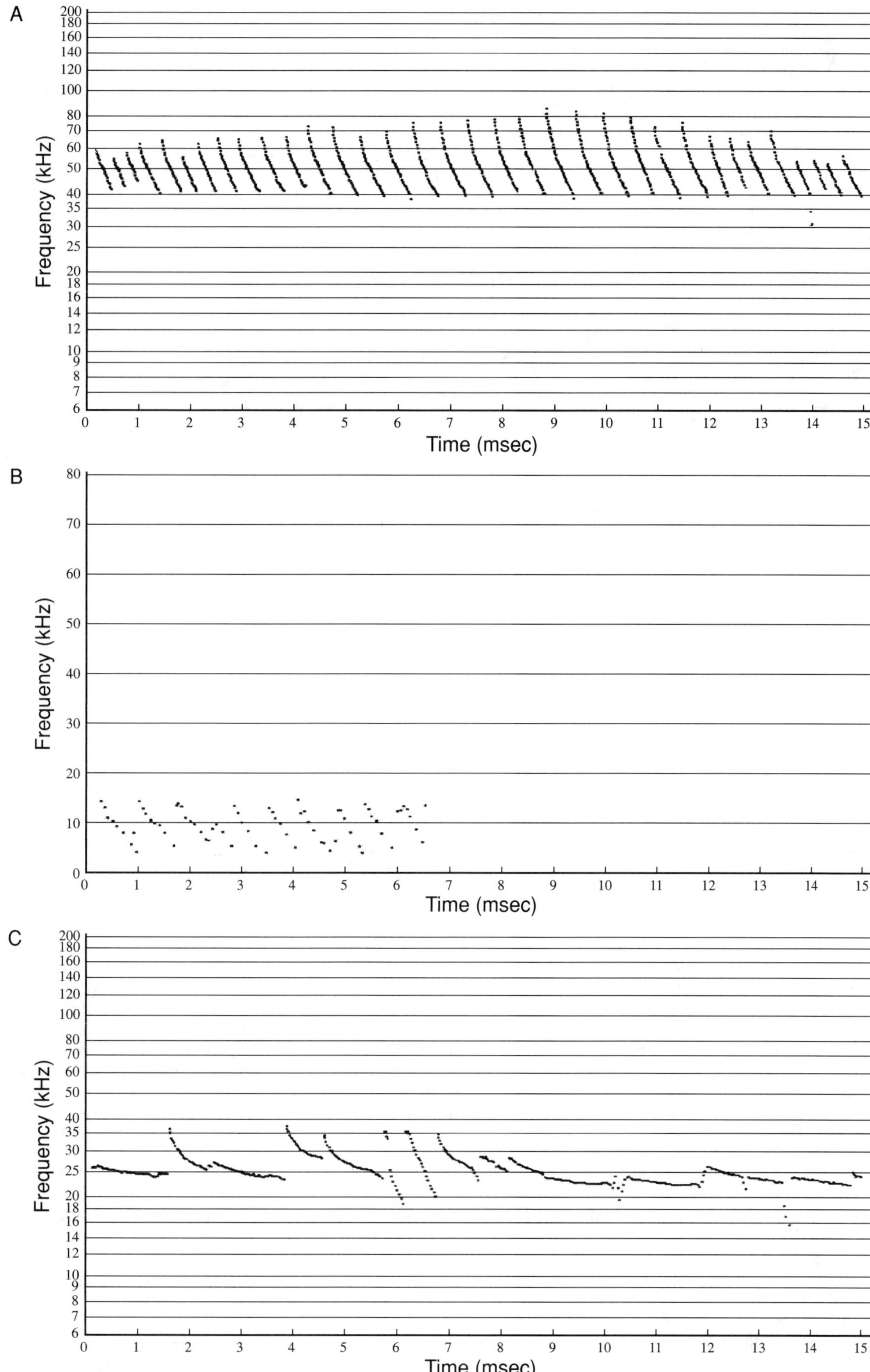

FIGURE 3.19. Calls of (A) the gray bat (*Myotis grisescens*), (B) the spotted bat (*Euderma maculatum*), and (C) the free-tailed bat (*Tadarida brasiliensis*).

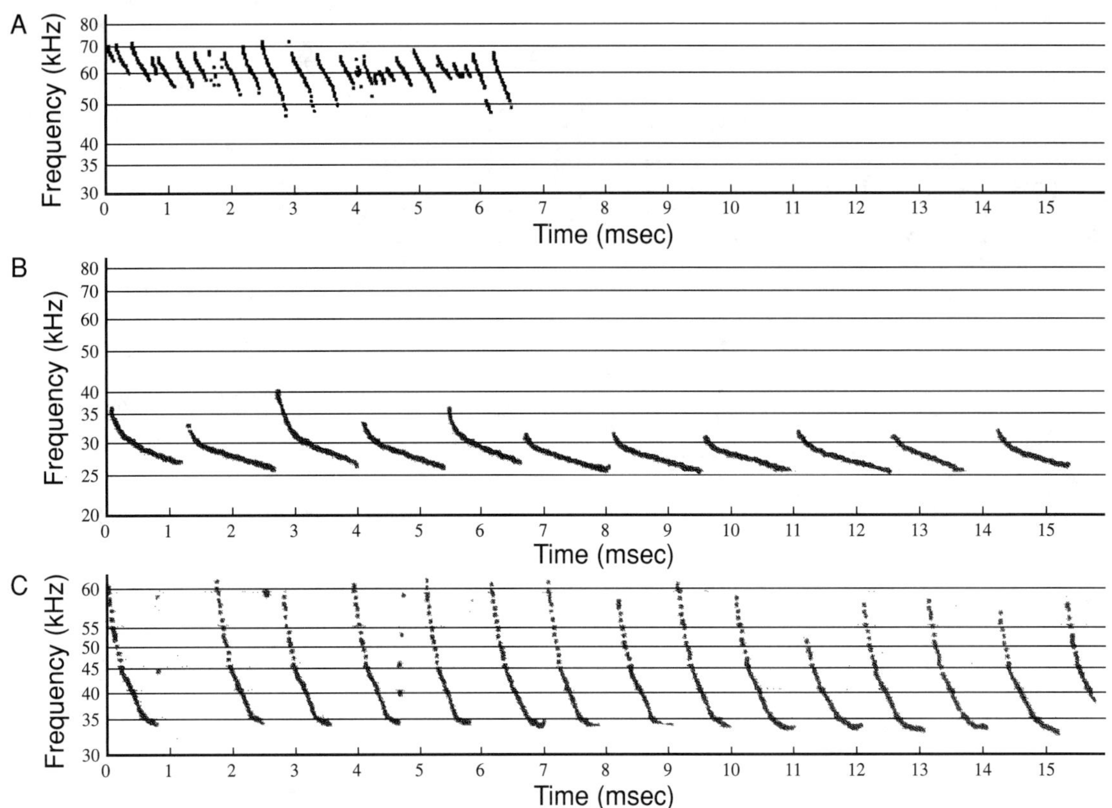

FIGURE 3.20. Calls of (A) the California leaf-nosed bat (*Macrotus californicus*) and (B, C) the big brown bat (*Eptesicus fuscus*), showing variation when calls are recorded in open (B, meadow) and cluttered habitats (C, forest). Differences in the call are due to the bats' response to influences of the structure of different habitats.

Rabies in Bats. Rabies is a disease of the nervous system caused by rhabdovirus, a family of virus containing the rod-shaped genus *Lyssavirus;* rabies is one of seven variants within this genus (Centers for Disease Control and Prevention [CDC] 2001). It is fatal to all mammals that contract it. When an animal contracts the virus, it becomes progressively ill, usually dying in 3 days. The virus is transported in peripheral nerve axons to the central nervous system, where it causes brain inflammation, delirium, and death. It is rare in humans, although it has been estimated to affect 1 in every 1 million bats (Fenton 2001). Animals may become paralyzed distally to the medial trunk, become "crazed" and act unpredictably violent, or both. Rabies causes a constriction of the muscles of the anterior neck and throat that promote swallowing. Rabid animals cannot swallow and appear hydrophobic because they are unable to swallow water. This often gives the foamy mouth and drooling behavior attributed to rabid animals. The crazed bat may be less dangerous than the rabid animal that is immobilized from the virus. A person may use less care when interacting with an animal that is not outwardly aggressive, so may not seek medical care if the animal happens to infect the person with its saliva. The virus survives in, and is transmitted through, the saliva. The latest ideas on how rabies is transmitted include a bite by an infected animal, ingestion of infected tissue, transmission by bites from ectoparasites infected with blood of an infected host, transmission by aerosolized urine of an infected animal, or inhalation of aerosolized virus in humid caves. Bats are often thought of as carriers of rabies to the exclusion of many other species. Fenton (2001) reported that 5 of 236 (2.6%) cases of rabies in humans documented in the United States between 1946 and 1965 were caused by bats, compared with 57% of these cases caused by dogs. From 1980 to 1997, 36 reported cases of rabies in humans were due to 12 dogs and 21 bats. It appears that human cases overall have declined in recent years, but those cases that did occur were contracted via a greater number of bats. Why there is such an increase in the percentage of cases attributed to bats is not clear. In a large sample of bats tested for rabies in New York, 4.6% (312 of 6810) were positive; of that number, 2 were silver-haired bats (Hunt and Bhatnagar 1997). This is the bat species predominantly reported as responsible for human rabies cases. It is perplexing why this small species with a narrow gape would be so often found at state epidemiological test labs. In New Mexico, although no humans were infected, 6, 9, and 21 samples were positive in 1998, 1999, and 2000, respectively (17 bats, 5 dogs; CDC 2001). Aside from silver-haired bats, big brown bats are often brought to state health labs for rabies testing because they are one of the more common house bats. It could be that any time that a bat is found near human activity it is suspected of rabies and collected for testing whether exhibiting symptoms of rabies or not. Much more information is needed before any general management action can occur. Moreover, public perception of bats and education about the role bats might play in the transmission of rabies must be very carefully articulated. Although rabies has been found at one time or another in all species discussed in this chapter, it is relatively uncommon.

MANAGEMENT AND CONSERVATION

Excluding Bats. Since the early 1980s, an increased awareness by the public has occurred relative to bats. In urban areas, this usually means pest control of bats, as they invade attics and other structures where humans reside. In less populated areas, mines that have been abandoned are an attractive resource for many cavernicolous bat species, but are also a liability and health hazard when curious humans happen into them. For the problem of abandoned mines, exclusionary structures or "bat gates" have been constructed, which allow bats to traverse the mine opening, but exclude humans. In the United States, it is estimated that 367,538 mine openings exist (Meier and Garcia 2001). Of the top seven states with large mining districts, approximately 80,000, 48,948,

18,000, 20,071, 165,000, and 20,043 abandoned mines exist in Arizona, California, Colorado, New Mexico, Nevada, and Utah, respectively (R. E. Sherwin, University of New Mexico, pers. commun., 2001). Of these mines, approximately (in the same order) 83 (68 gated), an unknown number (198 gated), 5254 (321 gated), 1252 (127 gated), 5615 (28 gated), and 4500 (300 gated) are closed. The success of gating these openings has only recently been tested. Thus far, researchers have done a poor job assessing the effectiveness of gating and the thoroughness of mine closure protocols (Sherwin et al. 2000b, 2000c; Sherwin and Altenbach 2001).

To exclude bats from urban structures such as houses and barns, several guides are available (Greenhall 1982). In general, the best method of preventing bats from roosting in houses or other buildings is simply to close the openings through which they enter after they have exited. Bats are likely to be gone from a structure between October and February in North America. Hanging garden netting over exit sites with the lower portion of it unattached will allow most bats to exit, but not re-enter a roost site. Bat proofing a home also results in energy savings and conservation.

Many urban areas are beginning to encourage the presence of bat populations. Because they often roost in homes and barns, *T. brasiliensis* often come into contact with humans. Although its construction as a bat-friendly structure was accidental, the Congress Avenue Bridge in the center of town in Austin, Texas, harbors 1.5 million bats. This has become a major tourist attraction and symbol of the symbiosis that human and insect-eating bats can attain.

Effect of Human Disturbance. Humans can easily disturb gray bats at summer colonies; these caves should not be entered during this time. Cave users should be exceedingly cautious about disturbing either summer or winter colonies of any species of bat and should avoid entering caves at all when maternity roosts are present. In most caverniculus bat species, human disturbance can cause neonates to drop from the walls or to lose their grip from their mothers when flying. When young, if nonvolant bats hit a cave floor, they do not usually survive because the roost floor is often awash in scavenger arthropods, which consume the fallen young quickly.

The Mexican free-tailed bat is especially affected by the widespread use of pesticides by humans. The large population of *T. brasiliensis* at Carlsbad Cavern, New Mexico, declined from an estimated population of 8–9 million bats to about 300,000 over a 50-year period. Bracken Cave in Texas is estimated to have 20 million bats. Large-scale application of pesticides and other chemicals for human agriculture can have dramatic effects on populations of bats. Dieldrin (insecticide residue) has been directly linked to mortality of young *Myotis grisescens* (Clark et al. 1978). Gray bats spend summer in large colonies in several states, especially in Illinois, Missouri, Tennessee, Kentucky, Alabama, and portions of Florida (Tuttle 1976). They are extremely vulnerable to human disturbance since many winter and summer caves used by gray bats have been commercialized, destroyed, or overused. The largest of the concentrations of this species are in several cave complexes. Although most counts of gray bats show dramatic decreases in number, one site, the Jesse James Cave in Kentucky, had an increase from 230,000 bats in 1983 to 300,000 bats in 1997 (Whitaker and Hamilton 1998). Species such as this are especially vulnerable when many individuals congregate in one spot. By putting all of their "eggs in one basket," a species can be fatally affected by a major disturbance at one of these highly concentrated sites.

Conservation. Monitoring bat populations and assessing their habitat needs are a challenge. To address these conservation problems, one must understand the limitations of techniques used and be aware of peculiarities of bat behavior and natural history. Two important factors make bat populations especially vulnerable. One is the tendency of individuals of a species to concentrate, at least seasonally, into a few large colonies. This makes them especially susceptible to vandals and natural events such as fire or predation. One such occasion occurred in October near Socorro, New Mexico, in the early 1990's, where several people set fire to timbers of an abandoned mine. This mine was a roost to perhaps thousands of Townsend's big-eared bats (*Corynorhinus townsendii*). After the fire, many bats were found dead, singed and burned in the charred remains of the workings (J. S. Altenbach, University of New Mexico, pers. commun., 2001). Although today parts of the mine contain bats from a number of species (such as *Eptesicus fuscus, Myotis eiliolabrum*) animals that concentrate are at risk of catastrophic actions from humans). Bats disturbed during hibernation may use limited energy reserves for arousal, which may hasten adult death. Low reproductive rate inhibits chances of species to recover from sudden increased mortality. In recent times, bats have become increasingly dependent on human structures and habitats. Regulating human access to natural caves, preclosure surveys for mines, and inventory of highway bridges are all necessary exercises that should be routine. Mitigation should follow after these surveys (Sherwin et al. 2000b). Surveys using acoustic techniques can assist in locating concentrations of bats (O'Farrell and Gannon 1999).

Installation of boxes designed to attract colonies of bats provides additional safe roosting opportunities for some species. Bat Conservation International monitors bat house use. In 1998, of 778 houses, approximately 100,000 bats of 11 species occupied 445 houses in 31 U.S. states, Puerto Rico, three Canadian provinces, and the Cayman Islands. The most frequent users were big brown bats, little brown bats, and Mexican free-tailed bats. One person placed 100 houses in Florida, Georgia, and Alabama and had an occupation rate of 30% in 1997 to 65% in 1999 (http://batcon.org/). Some larger houses have been constructed for commercial use to hold 1600–1800 bats, with the intention of reducing pest insects on pecans such as the hickory shuckworm (*Laspeyresia caryana*). Finally, the University of Florida in Gainesville has an extra-large bat house, which houses mostly Mexican free-tailed bats, evening bats (*Nycticeius humeralis*), and southeastern myotis (*Myotis austroriparius*). The numbers have steadily grown since the house was constructed in 1993, such that by 1999 the population was estimated at 75,000 bats. Over the summer season, thousands of people gather nearby to witness the nightly emergence of these animals.

Use of mines by pallid bats puts them in jeopardy due to mine closure projects. Additional threats include human vandalism within roosts, roost destruction, extermination of bats in buildings, and increasing use of pesticides. Loss of tree roosts could occur through commercial timber harvest and loss of oak woodlands to suburban development and expansion. Gray bats still occur in large numbers, but have declined dramatically from previous populations densities partly because of their habit of congregating in large numbers in a few caves (especially in winter). Gray bats were placed on the federal endangered species list as of 28 April 1976. Many of the caves used by this species in winter have vertical openings that are difficult for humans to enter. Some people consider difficult access to caves a challenging sport. Resulting disturbance could have devastating effects on this species. One cave in Kentucky housed 500,000 bats; now about 61,000 are found (Rabinowitz and Tuttle 1980). In Alabama and Tennessee, 22 summer colonies declined 76% from 1,199,000 before 1968 to 293,600 in 1983, with a slight rise 15 years later to 300,000 in 1997. However, some populations are increasing dramatically in other states, and absolute movement of all bats is not known.

Many species have broad ranges (Kunz and Fenton 2003). Because bats cover a large amount of territory, damage to local populations may not be considered important. In fact, there are a number of factors that are extremely important and have profound conservation implications. These factors are known taxonomy, full knowledge of the range of species including historical ranges, range extension and contraction, and seasonal distributions (see Cryan 2003). Knowing what species are being considered is not as easy as it may seem. Taxonomists are always refining systematic relationships to the point that most researchers need to contact natural history collection managers for current names. Names are important. Recently, Morales and Bickham (1995) revised the genus *Lasiurus* based on restriction maps of the mitochondrial ribosomal genes (also see Baker et al. 1988). Their conclusions distinguished the western red bat (*L. blossevillii*) in the western United States from *L. borealis* in the eastern United States. Similar treatments have been applied to some members of the genus *Myotis* (for instance, the Arizona myotis [*M. occultus*] from *M. lucifigus;* Valdez et al. 1999, Piaggio et al. 2002). Aside from revision, some species are difficult to distinguish in hand. For instance, *Myotis californicus* and the western small-footed myotis (*M. ciliolabrum*) can be differentiated in some parts of their range by the extension of the tail beyond the uropatagium, or by the minimum frequency of their echolocation call (Constantine 1998, Gannon et al. 2001). Among the species in this treatment, it is unlikely to have identification problems for *Antrozous pallidus, Eptesicus fuscus, Lasiurus cinereus, Euderma maculatum, Tadarida brasiliensis,* and *Macrotus californicus*. However, *Myotis sodalis* bears such a strong resemblance to *Myotis lucifugus* that it was not described as a species until 1928. Indiana bats are best distinguished by short, inconspicuous toe hairs, a small foot (9–10 mm), a keeled calcar, and a forearm length between 35 and 41 mm.

For some time, distributions of mammalian species in North America have been defined by the extent of their marginal records as found by Hall (1981). Marginal records based on museum–voucher specimens are an excellent way to outline the extent of terrestrial mammals. However, caution in inferring the range of a species internal to those marginal records calls for additional information, such as habitat requirements and energetic needs. For bats, marginal records are equally important as with other mammals except that different criteria must be applied in assuming how bat species fill in their distributions. Also, distributional information is not well known for bats compared with other mammals. Often, range maps show the entire distribution of a bat species, which includes their seasonal ranges. This may be misleading. For *Antrozous pallidus* and *Eptesicus fuscus,* broad-ranging depictions of these species' ranges may be fairly certain. In the case of *Lasiurus cinereus;* however, they spend winter throughout Central America and parts of South America (see Figs. 3.1–3.3). In spring, individuals move northward and males populate the western portions of North American and females the eastern portion (Findley and Jones 1964; Cryan 2003). Most maps showing the range of this species, including the one here, include the entire area where this bat is found (see Fig. 3.3). Species that extend beyond political boundaries require international cooperation for putting conservation plans in place and for their enforcement. Other examples of this include *Tadarida brasiliensis* and *Macrotus californicus;* both species have extensive populations (at least seasonally) in both the United States and Mexico.

Where are bats in winter in North America? Not much attention has been paid to this aspect of bat biology. Gannon et al. (Submitted) found significant bat activity during winter throughout New Mexico. Pallid bats were active on several occasions in January and February. Big brown bats and silver-haired bats were located during every survey period (n = 37) over 3 years during winter. Most bats were detected over water or foraging along roadways. Insect activity was low, but there was some activity. A spotted bat was found active and roosting in a building in Albuquerque each year for 3 years in February (Gannon and Sherwin, In press). Although there are records of bats during winter, obviously they do not approach the levels of activity for bats during summer because of thermal constraints and behavioral modifications during hibernation.

A number of conservation initiatives have resulted in the study and protection of bats internationally. Maps showing the distribution of *Myotis grisescens* as blanketing much of southeastern North America are misleading (see Fig. 3.5). This species is a cavern-obligate found only in nine caves in karst formations within its range (Rabinowitz and Tuttle 1980). Most (87%) of all known *Myotis sodalis* have been documented at only seven hibernation sites (see Fig. 3.6), (Humphrey and Cope 1977). *Euderma maculatum,* with a published range covering much of western North America (see Fig. 3.5), is known from <300 specimens since it was first described in 1891, and only about 12 sites are known where they can commonly be captured (Pierson 1998). Perhaps because researchers have collected so few spotted bats, they were listed as "Rare" in the *IUCN Red Data Book* (1985). It was once estimated that California leaf-nosed bats were distributed across southern California and were especially common at the Salton Sea (see Fig. 3.8) (Howell 1920). They may not be common there; however, it has been estimated that overall numbers of bats have increased because of the prevalence of open mines. Although bats may be more abundant than before mining, this species is now more susceptible to disturbance because of mine expansion projects and their tolerance for warm mines that are frequented by humans (Bell et al. 1986).

RESEARCH AND MANAGEMENT NEEDS

Data are lacking regarding the seasonal movements of pallid bats. Additional information is required regarding winter activity patterns including use of roost sites. More information is needed on feeding by gray bats as well as the status of cave roosts. More information is needed on *T. brasiliensis,* especially from the eastern United States and South America. What are the winter habits? At the same latitude, why do some populations migrate and others remain over winter? It seems that some summer roosts are declining in numbers of individuals, whereas other areas show an increase; what is roost fidelity in this species? Densities of *Macrotus californicus* are declining because of their preference for warm mines and caves; do all of these sites have a high possibility of human disturbance? All these species have the common issue of deleterious human impact. It is critical that researchers and managers assess human disturbance and propose mitigation measures to reduce the severity of these impacts.

ACKNOWLEDGMENTS

I am grateful to the many who have worked so hard at understanding the biology of bats. Bats' complexity and distinction among mammals make them one of the most difficult groups to explore scientifically. Two anonymous reviewers as well as A. Brown, R. Sherwin, G. Racz, and D. Tinnin made constructive comments on earlier versions on this chapter. L. T. Arciniega reviewed literary content and readability for nontechnical readers. N. M. Gannon managed to keep her 2-year-old fingers off the disks containing the manuscript files. R. Sherwin, C. Corben, M. O'Farrell, and K. Livingood supplied additional calls and/or data. The Division of Mammals, Museum of Southwestern Biology, provided partial support of this work.

LITERATURE CITED

Anderson, S. 1969. *Macrotus waterhousii.* Mammalian Species 1:1–4.

Arita, H. T. 1990. Noseleaf morphology and ecological correlates in phyllostomid bats. Journal of Mammalogy 71:36–47.

Baker, R. J., J. C. Patton, H. H. Genoways, and J. W. Bickham. 1988. Genic studies of *Lasiurus* (Chiroptera: Vespertilionidae). Occasional Papers; the Museum, Texas Tech University 117:1–15.

Balcombe, J. P. 1990. Vocal recognition of pups by mother Mexican free-tailed bats *Tadarida brasiliensis mexicana.* Animal Behaviour 39:960–66.

Bell, G. P. 1982. Behavioural and ecological aspects of gleaning by a desert insectivorous bat, *Antrozous pallidus* (Chiroptera: Vespertilionidae). Behavioural Ecology and Sociobiology 10:217–23.

Bell, G. P., and M. B. Fenton. 1986. Visual acuity, sensitivity, and binocularity in a gleaning insectivorous bat, *Macrotus californicus* (Chiroptera: Phyllostomidae). Animal Behaviour 34:409–14.

Bell, G. P., G. A. Bartholomew, and K. A. Nagy. 1986. The roles of energetics, water economy, foraging behavior, and geothermal refugia in the distribution of the bat, *Macrotus californicus.* Journal of Comparative Physiology B 156:441–50.

Black, H. L. 1974. A north temperate bat community: Structure and prey populations. Journal of Mammalogy 55:138–57.

Black, H. L. 1976. American kestrel predation on the bats *Eptesicus fuscus, Euderma maculatum* and *Tadarida brasiliensis*. Southwestern Naturalist 21:250–51.

Bleier, W. J. 1975. Early embryology and implantation in the California leaf-nosed bat *Macrotus californicus*. Anatomical Record 182:237–53.

Bogan, M. A. 1972. Observations on parturition and development in the hoary bat, *Lasiurus cinereus*. Journal of Mammalogy 53:611–14.

Bogan, M. A. 1974. Identification of *Myotis californicus* and *M. leibii* in southwestern North America. Proceedings of the Biological Society of Washington 87:49–56.

Bogan, M. A. 2001. Western bats and mining. Pages 86–98 *in* K. C. Vories and D. Throgmorton, eds. Proceedings of Bat Conservation and Mining: A technical interactive forum. U.S. Department of Interior, Office of Surface Mining, Alton, IL.

Bradshaw, G. V. R. 1962. Reproductive cycle of the California leaf-nosed bat, *Macrotus californicus*. Science 136:645–46.

Brown, P. E., and R. D. Berry. 1991. Bats, habitats, impacts, and migration. Pages 26–30 *in* Thorne Ecological Institute proceedings V: Issues and technology in the management of impacted wildlife, Thorne Ecological Institute, Boulder, CO.

Burt, W. H. 1934. The mammals of southern Nevada. Transactions of the San Diego Society of Natural History 7:375–427.

Burt, W. H. 1938. Faunal relationships and geographic distribution of mammals in Sonora, Mexico. Miscellaneous Publications of the Museum of Zoology, University of Michigan 39:1–77.

Callahan, E. V., R. D. Drobney, and R. L. Clawson. 1997. Selection of summer roosting sites by Indiana bats (*Myotis sodalis*) in Missouri. Journal of Mammalogy 78:818–25.

Carter, F. D., and W. J. Bleier. 1988. Sequential multiple ovulations in *Macrotus californicus*. Journal of Mammalogy 69:386–88.

Centers for Disease Control and Prevention. 2001. Available at http://wonder.cdc.gov/mmwr—resp.asp.

Chew, R. M., and H. E. White. 1960. Evaporative water losses of the Pallid bat. Journal of Mammalogy 41:452–58.

Christian, J. J. 1956. The natural history of a summer aggregation of the big brown bat *Eptesicus fuscus fuscus*. American Midland Naturalist 55:66–95.

Clark, D. R., Jr., R. K. LaVal, and D. M. Swineford. 1978. Dieldrin induced mortality in an endangered species, the gray bat (*Myotis grisescens*). Science 199:1357–59.

Cockrum, E. L. 1955. Reproduction in North American bats. Transactions of the Kansas Academy of Sciences 58:487–511.

Constantine, D. G. 1961. Spotted bat and big free-tailed bat in northern New Mexico. Southwestern Naturalist 6:92–97.

Cryan, P. M. 2003. Seasonal distribution of migratory tree bats (*Lasiurus* and *Lasionycteris*) in North America. Journal of Mammalogy 84:579–593.

Crichton, E. G., and P. H. Krutzsch, eds. 2000. Reproductive biology of bats. Academic Press, New York.

Davis, R. 1968. Wing defects in a population of pallid bats. American Midland Naturalist 79:388–95.

Davis, R. B., C. F. Herreid II, and H. L. Short. 1962. Mexican free-tailed bats in Texas. Ecological Monographs 32:311–46.

Dodgen, C. L., and F. R. Blood. 1956. Energy sources in the bat. American Journal of Physiology 187:151–54.

Easterla, D. A. 1973. Ecology of the 18 species of Chiroptera at Big Bend National Park, Texas. Northwest Missouri State University Studies 34:1–165.

Feldhamer, G. A., L. C. Drickamer, S. H. Vessey, and J. F. Merritt. 1999. Mammalogy; adaptation, diversity, and ecology. McGraw-Hill, Boston.

Fenton, M. B. 2001. Bats, rev. ed. Checkmark, New York.

Findley, J. S. 1993. Bats: A community perspective. Cambridge University Press, Cambridge.

Findley, J. S., and C. Jones. 1964. Seasonal distribution of the hoary bat. Journal of Mammalogy 45:461–70.

Folk, G. E., Jr. 1940. Shift of population among hibernating bats. Journal of Mammalogy 59:299–304.

Gaisler, J. 1979. Ecology of bats. Pages 281–342 *in* D. M. Stoddart, ed. Ecology of small mammals. Chapman and Hall, London.

Gannon, W. L., R. E. Sherwin, T. N. deCarvalho, and M. J. O'Farrell. 2001. Pinnae and echolocation differences between *Myotis californicus* and *M. ciliolabrum* (Chiroptera: Vespertilionidae). Acta Chiropterologica 3:77–91.

Gannon, W. L., and R. E. Sherwin. In press. Spotted bat (*Euderma maculatum*) winter activity in urban setting. Southwestern Naturalist.

Gannon, W. L., R. E. Sherwin, and S. Haymond. 2003. On the importance of articulating assumptions while conducting acoustic studies of habitat use by bats. Wildlife Society Bulletin, 31:45–61.

Gannon, W. L., R. E. Sherwin, and M. S. Burt. Submitted. Winter activity of bats in New Mexico.

Geluso, K. N. 1978. Urine concentrating ability and renal structure of insectivorous bats. Journal of Mammalogy 59:312–323.

Geluso, K. N., J. S. Altenbach, and D. E. Wilson. 1965. Bat mortality: Pesticide poisoning. Science 194:184–86.

Greenhall, A. M. 1982. House bat management. U.S. Fish and Wildlife Service, Resource Publication 143:1–97.

Gustin, M. K., and G. F. McCracken 1987. Scent recognition between females and pups in the bat *Tadarida brasiliensis mexicana*. Animal Behaviour 35:13–19.

Hall, E. R. 1981. The mammals of North America. John Wiley, New York.

Harvey, M. J., J. S. Altenbach, and T. L. Best. 1999. Bats of the United States. Arkansas Game and Fish Commission, Little Rock, AR.

Hassell, M. D. 1963. A study of homing in the Indiana bat *Myotis sodalis*. Transactions of the Kentucky Academy of Science 24:1–4.

Hermanson, J. W., and T. J.O'Shea. 1983. *Antrozous pallidus*. Mammalian Species 213:1–8.

Herrera, L. G., T. H. Fleming, and J. S. Findley. 1993. Geographic variation in carbon composition of the pallid bat, *Antrozous pallidus,* and its dietary implications. Journal of Mammalogy 74:601–6.

Hill, J. E., and J. D. Smith. 1984. Bats: A natural history. British Museum (Natural History), London.

Hoffmeister, D. F. 1986. Mammals of Arizona. University of Arizona Press, Tucson.

Howell, A. B. 1920. Some Californian experiences with bat roosts. Journal of Mammalogy 1:169–77.

Humphrey, S. R. 1982. Bats, Vespertilionidae and Molossidae. Pages 52–70 *in* J. A. Chapman and G. A. Feldhamer, eds. Wild mammals of North America: Biology, management, and economics. Johns Hopkins University Press, Baltimore.

Humphrey, S. R., and J. B. Cope. 1977. Survival rates of the endangered Indiana bat, *Myotis sodalis*. Journal of Mammalogy 58:32–36.

Hunt, L. A., and K. P. Bhatnagar. 1997. Human rabies and silver-haired bats in the United States. Bat Research News 38:85–89.

IUCN Red Data Book, 3rd ed. 1985. IUCN, Cambridge.

Jones, C., and J. S. Findley. 1965. Ecological distribution and activity patterns of bats of the Mogollon Mountains area of New Mexico and adjacent Arizona. Tulane Studies in Zoology 12:93–100.

Jones, C. J., and C. G. Schmitt. 1996. Mammal species of concern in New Mexico. Pages 179–205 *in* T. L. Yates, W. L. Gannon, and D. E. Wilson, eds. Life among the muses: Papers in honor of J. S. Findley (Special Publication No. 3). Museum of Southwestern Biology, Albuquerque, NM.

Jones, G., and S. M. van Parus. 1993. Bimodal echolocation in pipistrelle bats: Are cryptic species present? Proceedings of the Royal Society of London B 251:119–25.

Kunz, T. H. 1974. Reproduction, growth, and mortality of the vespertilionid bat, *Eptesicus fuscus,* in Kansas. Journal of Mammalogy 55:1–13.

Kunz, T. H. 1982. Roosting ecology. Pages 1–56 *in* T. H. Kunz, ed. Ecology of bats. Plenum Press, New York.

Kunz, T. H., and M. B. Fenton (editors). 2003. Bat ecology. University of Chicago Press, Chicago, Illinois.

Kurta, A. J., and R. H. Baker. 1990. *Eptesicus fuscus*. Mammalian Species 356:1–10.

Kurta, A., and J. O. Whitaker, Jr. 1998. Diet of the endangered Indiana bat (*Myotis sodalis*) on the northern edge of its range. American Midland Naturalist 140:280–86.

Kurta, A. J., J. Kath, E. L. Smith, R. Foster, M. W. Orick, and R. Ross. 1993. A maternity roost of the endangered Indiana bat (*Myotis sodalis*) in an unshaded, hollow sycamore tree (*Platanus occidentalis*). American Midland Naturalist 130:405–7.

Lacki, M. J., L. S. Burford, and J. O. Whitaker, Jr. 1995. Food habits of gray bats in Kentucky. Journal of Mammalogy 76:1256–59.

LaVal, R. K., and M. L. LaVal. 1979. Notes on reproduction, behavior, and abundance of the red bat, *Lasiurus borealis*. Journal of Mammalogy 60:209–12.

Leonard, M. L., and M. B. Fenton. 1984. Echolocation calls of *Euderma maculatum* (Chiroptera: Vespertilionidae): Use in orientation and communication. Journal of Mammalogy 65:122–26.

Licht, P., and P. Leitner. 1967. Physiological responses to high environmental temperatures in three species of microchiropteran bats. Comparative Biochemical Physiology 22:371–87.

Loughry, W. J., and G. F. McCracken. 1991. Factors influencing female-pup scent recognition in Mexican free-tailed bats. Journal of Mammalogy 72:624–6.

Lyman, C. P. 1970. Thermoregulation and metabolism in bats. Pages 301–30 *in* W. A. Wimsatt, ed. Biology of bats, Vol. 1. Academic Press, New York.

McNab, B. K. 1974. The behavior of temperate cave bats in a subtropical environment. Ecology 55:844–46.

McNab, B. K. 1983. Evolutionary alternatives in the physiological ecology of bats. Pages 151–200 *in* T. H. Kunz, ed. Ecology of bats. Plenum Press, New York.

Meier, L., and J. Garcia. 2001. Importance of mines for bat conservation. Pages 26–52 *in* K. C. Vories and D. Throgmorton, eds. Proceedings of bat conservation and mining: A technical interactive forum. U.S. Department of Interior, Office of Surface Mining, Alton, IL.

Morales, J. C., and J. W. Bickham. 1995. Molecular systematics of the genus *Lasiurus* (Chiroptera: Vespertilionidae) based on restriction-site maps of the mitochondrial ribosomal genes. Journal of Mammalogy 76:730–49.

Navo, K. W., J. A. Gore, and G. T. Skiba. 1992. Observations on the spotted bat, *Euderma maculatum,* in northwestern Colorado. Journal of Mammalogy 73:547–51.

Nowak, R. M. 1994. Walker's bats of the world. Johns Hopkins University Press, Baltimore.

O'Farrell, M. J. 1999. Blind test for ability to discriminate vocal signatures of the little brown bat *M. lucifugus* and the Indiana bat *Myotis sodalis*. Bat Research News 40:44–48.

O'Farrell, M. J., and W. L. Gannon. 1999. A comparison of acoustic versus capture techniques for the inventory of bats. Journal of Mammalogy 80:24–30.

O'Farrell, M. J., C. Corben, and W. L. Gannon. 2000. Geographic variation in the echolocation calls of the hoary bat (*Lasiurus cinereus*). Acta Chiropterologica 2:75–83.

Orr, R. T. 1950. Unusual behavior and occurrence of a hoary bat. Journal of Mammalogy 31:456–57.

Orr, R. T. 1954. Natural history of the pallid bat, *Antrozous pallidus*. Proceedings of the California Academy of Sciences 28:165–264.

Orr, R. T. and G. S. Taboada. 1960. A new species of bat from the genus *Antrozous* from Cuba. Proceedings of the Biological Society of Washington 73:83–86.

O'Shea, T. J., and T. A. Vaughan. 1977. Nocturnal and seasonal activities of the pallid bat, *Antrozous pallidus*. Journal of Mammalogy 58:268–84.

Piaggio, A. J., E. W. Valdez, M. A. Bogan, and G. S. Spicer. 2002. Systematics of *Myotis occultus* (Chiroptera: Vespertilionidae) inferred from sequences of two mitochondrial genes. Journal of Mammalogy, 83:386–395.

Pierson, E. D. 1998. Tall trees, deep holes, and scarred landscapes: Conservation biology of North American bats. Pages 309–25 *in* T. H. Kunz and P. A. Racey, eds. Bat biology and conservation. Smithsonian Insitution Press, Washington, DC.

Poché, R. M., and G. L. Bailey. 1974. Notes on the spotted bat (*Euderma maculatum*) from southwest Utah. Great Basin Naturalist 34:254–56.

Poché, R. M., and J. E. Keirans. 1975. Report of a tick, *Ornithodorus rossi* (Acarina: Argasidae) from the spotted bat, *Euderma maculatum* (Chiroptera: Vespertilionidae). Journal of Medical Entomology 12:503.

Rabinowitz, A., and M. D. Tuttle. 1980. Status of summer colonies of the endangered gray bat in Kentucky. Journal of Wildlife Management 44:955–60.

Ray, C. E., and D. E. Wilson. 1979. Evidence for *Macrotus californicus* from Terlingua, Texas. Occasional Papers, the Museum, Texas Tech University 57:1–10.

Schowalter, D. B., and J. R. Gunson. 1979. Reproductive biology of the big brown bat (*Eptesicus fuscus*) in Alberta. Canadian Field-Naturalist 93:48–54.

Schmidly, D. J. 1991. The bats of Texas. Texas A&M University Press, College Station.

Sherman, H. B. 1937. Breeding habits of the free-tailed bat. Journal of Mammalogy 18:176–87.

Sherwin, R. E., and J. S. Altenbach. 2001. Success of bat gates. Pages 165–73 *in* K. C. Vories and D. Throgmorton, eds. Proceedings of bat conservation and mining: A technical interactive forum. U.S. Department of Interior, Office of Surface Mining, Alton, IL.

Sherwin, R. E.,W. L. Gannon, and S. Haymond. 2000a. The efficacy of acoustic techniques to infer differential use of habitat by bats. *Acta Chiropterologica* 2:145–53.

Sherwin, R. E., W. L. Gannon, J. S. Altenbach, and D. Stricklan. 2000b. Roost fidelity of Townsend's big-eared bat in Utah and Nevada. Transactions of the Western Section of the Wildlife Society 36:15–20.

Sherwin, R. E., D. Stricklan, and D. S. Rogers. 2000c. Roosting affinities of Townsend's big-eared bat (*Corynorhinus townsendii*) in northeast Utah. Journal of Mammalogy 81:939–47.

Shump, K. A., Jr., and A. U. Shump. 1982. *Lasiurus cinereus*. Mammalian Species 185:1–5.

Smalley, R. L., and R. L. Dryer. 1963. Brown fat: Thermogenic effect during arousal from hibernation in the bat. Science 140:1333–34.

Steinlein, D. B., L. A. Durden, and W. L. Gannon. 2001. Tick (*Acari*) infestations of bats in new Mexico. Journal of Medical Entomology, 38:609–611.

Studier, E. H., and M. J. O'Farrell. 1972. Biology of *Myotis thysanodes* and *M. lucifugus* (Chiroptera: Vespertilionidae). Part 1: Thermoregulation. Comparative Biochemical Physiology 44A:567–95.

Studier, E. H., V. L. Lysengen, and M. J. O'Farrell. 1973. Biology of *Myotis thysanodes* and *M. lucifugus* (Chiroptera: Vespertilionidae). Part 2: Bioenergetics of pregnancy and lactation. Comparative Biochemical Physiology 44A:467–71.

Thomas, D. W., and G. P. Bell. 1986. Thermoregulatory strategies and distributions of bats along climatic gradients. Bat Research News 27:39.

Tuttle, M. D. 1975. Population ecology of the gray bat (*Myotis grisescens*): Philopatry, timing and patterns of movement, weight loss during migration, and seasonal adaptive strategies. University of Kansas, Museum of Natural History, Occasional Paper 54:1–38.

Tuttle, M. D. 1976. Population ecology of the gray bat (*Myotis grisescens*): Factors influencing growth and survival of newly volant young. Ecology 57:587–95.

Tuttle, M. D. 1988. America's neighborhood bats. University of Texas Press, Austin.

Tuttle, M. D., and D. E. Stevenson. 1977. An analysis of migration as a mortality factor in the gray bat based on public recoveries of banded bats. American Midland Naturalist 97:235–40.

Twente, J. W., and J. Twente. 1987. Biological alarm clock arouses hibernating big brown bats, *Eptesicus fuscus*. Canadian Journal of Zoology 65:1668–74.

Valdez, E. W., J. R. Choate, M. A. Bogan, and T. L. Yates. 1999. Taxonomic status of *Myotis occultus*. Journal of Mammalogy 80:545–52.

Vaughan, T. A. 1959. Functional morphology of three bats: *Eumops, Myotis,* and *Macrotus*. University of Kansas, Publications of the Museum of Natural History 12:1–153.

Verts, B. J., and L. N. Carraway. 1998. Land mammals of Oregon. University of California Press, Berkeley.

Whitaker, J. O., Jr., and D. A. Easterla. 1974. Batflies (Streblidae and Nycteribiidae) in the eastern United States, and a nycteribiid record from Saskatchewan. Entomological News 85:221–23.

Whitaker, J. O., Jr., and W. J. Hamilton, Jr. 1998. Mammals of the Eastern United States. Cornell University Press, Ithaca, NY.

Whitaker, J. O., Jr., C. Neefus, and T. H. Kunz. 1996. Dietary variation in the Mexican free-tailed bat (*Tadarida brasiliensis mexicana*). Journal of Mammalogy 77:716–24.

Wilkins, K. T. 1989. *Tadarida brasiliensis*. Mammalian Species 331:1–10.

William L. Gannon, Museum of Southwesteern Biology, University of New Mexico, Albuquerque, New Mexico 87131. Email: wgannon@unm.edu.

4

Armadillo

Dasypus novemcinctus

James N. Layne

NOMENCLATURE

COMMON NAMES. Nine-banded armadillo, common long-nosed armadillo
SCIENTIFIC NAME. *Dasypus novemcinctus*
SUBSPECIES. *D. n. mexicanus*

Dasypus novemcinctus is one of six currently recognized species in the genus (Gardner 1993) and most recently has been assigned to the subgenus *Hyperoambon* (Wetzel 1985). The type locality of the species was restricted to Pernambuco, Brazil (Cabrera 1958), and that of the subspecies to Matamoros, Tamaulipas, Mexico (Hollister 1925). Six subspecies of *D. novemcinctus* are recognized (McBee and Baker 1982), although Galbreath (1982) stated the need for a reanalysis of the subspecific taxonomy. Storrs and Burchfield (1990) noted differences in morphological characteristics, protein composition, and susceptibility to leprosy bacilli (*Mycobacterium leprae*) between samples from Florida and Venezuela and suggested that the two populations might warrant separate specific rank. *D. novemcinctus* and *D. hybridus* appear to be closely related based on external morphology, occurrence of polyembryony, karyotype, and electrophoretic mobilities of proteins and liver enzymatic activities (Jorge et al. 1985; Ryder and Davis 1985). As in Galbreath (1982), all references to the "armadillo" in this account refer to the subspecies *D. n. mexicanus* in the United States unless otherwise noted. Based on osteological similarity, except for larger size, *Dasypus bellus* appears to be the Pleistocene counterpart of *D. novemcinctus* (Slaughter 1961).

The common name armadillo is from the Spanish "little armored one," in reference to the armorlike covering of the body. The generic name given by Linnaeus is believed to have been derived from the Greek word for hare or rabbit, representing Linnaeus's attempt to translate into Latin the Aztec name for the species, *azotochtli,* meaning tortoise or turtle-rabbit (Smith and Doughty 1984). An alternative interpretation of the meaning of the generic name, *Dasypus,* is that it is derived from two Greek words meaning hairy-footed, the applicability of which is questionable unless it is assumed to imply a thick-footed or rough-footed condition (Lowery 1974). The specific name refers to the nine (*novem*) movable bands (*cinctus*) of the carapace.

DISTRIBUTION

The nine-banded armadillo is the only species in the order Xenarthra (edentates) to occur in North America in the Recent geological period. It also has the widest geographic distribution of any species in the order. It occurs from the southern United States southward through Mexico and Central America to northwestern Peru on the west side of the Andes and all of South America to northern Argentina east of the Andes, including the islands of Grenada (Lesser Antilles), Margarita, Trinidad, and Tobago (Wetzel 1985).

The range of the subspecies *D. n. mexicanus* extends from the southern United States through Mexico to Honduras (Hall 1981) and has undergone a remarkable northward and eastward expansion in the United States since the late 1800s. As a result of the armadillo's distinctive appearance, conspicuousness when active, ease of being captured or shot, and vulnerability to being run over on highways, the history of its range expansion has been especially well documented. The extensive literature dealing with its spread has been summarized in detail by a number of authors, including Fitch et al. (1952), Talmage and Buchanan (1954), Buchanan (1958), Cleveland (1970), Humphrey (1974), McBee and Baker (1982), Smith and Doughty (1984), and Taulman and Robbins (1996). Humphrey (1974) and Taulman and Robbins (1996) incorporated results of mail surveys as well as published data in their analyses. The earliest account of the armadillo in the United States is that of Audubon and Bachman (1849), who stated that it was found in areas of the lower Rio Grande Valley of southern Texas. The species subsequently steadily extended its range to the north and east. It had expanded throughout the southern half of Texas by 1914; advanced northward to southern Oklahoma and eastward to eastern Louisiana by 1954; and extended north to mid-Oklahoma and Arkansas and east to western Mississippi, southwestern Alabama, and extreme west Florida by 1972 (Humphrey 1974). A separate population center was established in peninsular Florida through intentional or accidental introductions. The first documented records from Florida were a male captured in 1922 and a female with four young killed in 1924, both in the Miami area in the vicinity in which a captive pair from Texas was reportedly released at the end of World War I (Bailey 1924). Additional occurrences were reported through the mid-1940s. By 1954, the armadillo was established throughout much of the eastern half of the peninsula. By 1972, it had spread throughout the state east of the Aucilla River and into southeast Georgia, leaving a gap of approximately 200 km between the ranges of the introduced Florida population and the largely naturally expanding population from the west, with only scattered records in the intervening area. Two scenarios have been advanced for the subsequent closure of the gap. Humphrey (1974) assumed that the population in west Florida marked the advancing front of the natural range expansion and that closure of the gap in the Tallahassee region of Florida involved merging of this population with the introduced population expanding from the east. According to the alternative scenario, based on evidence that armadillos in western Florida were derived from an introduction near Foley, Alabama (Wolfe 1968; Stevenson and Crawford 1974), closure of the gap involved the merger of three rather than two populations and the zone of contact between introduced and naturally expanding populations was in Alabama or Mississippi rather than Florida (Layne 1997). As the species appears to be still expanding its range and it is not always easy to determine whether locality records at the periphery of the range represent established populations or pioneering or released individuals, it is difficult to precisely define the range limits at any point in time. The map of the current distribution (Fig. 4.1) is based on that of Taulman and Robbins (1996) with emendations noted below. Established populations now occur in the southern United States from western Texas northward and eastward through most of Oklahoma except the extreme western panhandle area, southeastern Nebraska, southern Missouri, extreme southwestern Tennessee, southern parts of Alabama and

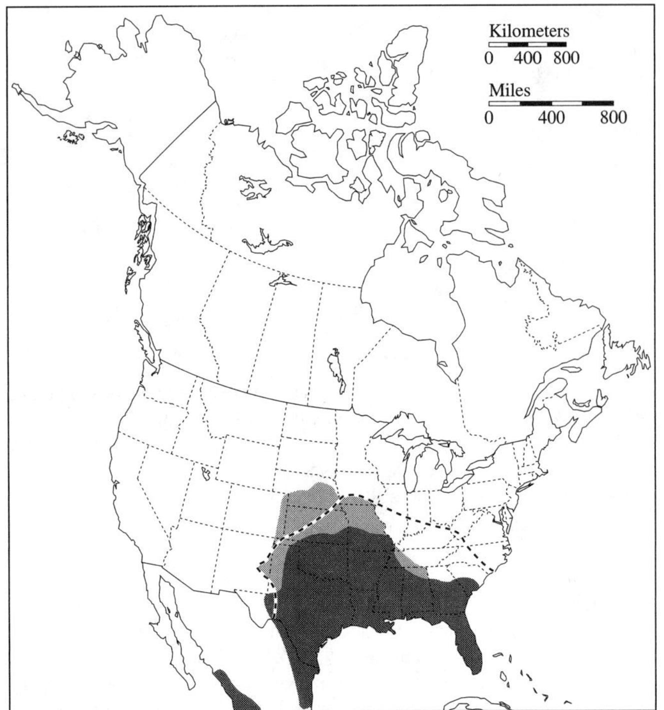

Dasypus novemcinctus (established populations)

Dasypus novemcinctus (scattered individuals)

---- *Dasypus bellus*

FIGURE 4.1. Distribution of the armadillo (*Dasypus novemcinctus*) and the approximate northern limits of the range of the extinct *Dasypus bellus* in the late Pleistocene. SOURCE: Adapted from Taulman and Robbins (1996) and modified to include all of peninsular Florida based on Layne (1984) and peripheral records from extreme west Texas and New Mexico. Northern range limits of the extinct *Dasypus bellus* in the late Pleistocene are from Klippel and Parmalee (1984).

Georgia, and the southwest corner of South Carolina. Records considered to represent pioneering or introduced individuals peripheral to the range limits of assumed established populations occur from extreme west Texas, eastern Colorado, southern Nebraska, northern Missouri, western Tennessee, and northeastern Alabama. Although Taulman and Robbins (1996) indicated armadillos were absent in the Everglades–Big Cypress region of Florida, as have other recent armadillo range maps, the species has occurred in that region since at least 1959 (Layne 1984). Also included in the present distribution map are locality records from New Mexico (Bailey 1931; Hendricks 1963). Additional records from southeastern Nebraska were cited by Freeman and Genoways (1998). It should also be noted that, either through introduction, crossing bridges, or actually swimming, armadillos have become established on some of the barrier islands along the coast of the Gulf of Mexico and the Atlantic coast of Georgia and Florida. It is also conceivable that in some instances armadillos reached barrier islands by being caught in coastal rivers during floods and carried out to sea.

The rate of range expansion by the armadillo up to 1972 was estimated as 4–10 km/year, which is comparable to that of such species as the cotton rat (*Sigmodon hispidus*) and masked shrew (*Sorex cinereus*) with less potential mobility (Humphrey 1974). Taulman and Robbins (1996) compared estimated rates of range expansion in different directions for 1972–1995 with those for 1880–1995. Expansion to the northwest has been much more limited than to the north and northeast as a result of an aridity barrier. Range extension to the northwest was greatest from 1880 to 1905, with no consistent trend of further increase since 1905. Although the range expanded at a modest rate of 1.7 km/year between 1972 and 1995, the gain in area was still less than that occupied from 1905 to 1954, indicating diminished expansion likely related to increasing aridity (Humphrey 1974). The rate of advance to the north was 11 km/year in 1971–1995 compared to the overall rate of 7.8 km/year. The rate of range expansion during the historic period has been highest to the northeast, with the rate for 1972–1995, 11.2 km/year, being comparable to the overall rate of 10.3 km/year. The rate of closure of the hiatus between the western and eastern segments of the range was estimated to be about 4.3 km/year for each population, but this estimate depends on which scenario of the range closure, as noted above, is accepted.

Based on climatic factors believed to determine the potential range of the armadillo, Taulman and Robbins (1996) concluded that the boundaries of the range in west Texas and the Great Plains now approach the precipitation- and temperature-defined boundaries, respectively. They predicted further expansion up the East Coast to about 41°N latitude and northward in the midwestern states to about 39°N latitude.

Expansion of the armadillo's range in the Recent geological period apparently parallels that of the extinct *Dasypus bellus,* whose range (Fig. 4.1) during equable climatic intervals in late Pleistocene extended as far north as Iowa, Indiana, West Virginia, Virginia, and North Carolina (Klippel and Parmalee 1984). This closely approximates the predicted northern range limit of the armadillo (Taulman and Robbins 1996).

A number of factors have probably played a role in facilitating the range expansion of the armadillo in the United States. Climate, specifically winter temperature and precipitation, probably determines the ultimate limits of the range. Humphrey (1974) noted contraction of the range in west Texas coincident with decreasing precipitation and northward expansion of the range with warmer winters. Taulman and Robbins (1996) predicted that annual precipitation of $\geq$38 cm and winter temperatures of >-2°C or $<$24 annual freeze days will determine the limits of the future distribution. Fitch et al. (1952) concluded that habitat changes associated with human settlement, especially reduction or local elimination of large carnivores, was a major factor. As an example of a possible effect of habitat change, expansion of the range of the armadillo and other species in Texas has been more or less synchronous with the extensive invasion of former coastal prairie by the northward and eastward expansion of brush land since 1870 (Price 1942). Riparian corridors in semiarid regions probably also have served as dispersal routes in some cases (Humphrey 1974). Freeman and Genoways (1998) suggested that extensive ditch and center-pivot irrigation may have facilitated movement of armadillos into eastern Nebraska. A labile diel activity pattern and well-developed nest-building behavior, which allow it to tolerate a broader range of environmental conditions (Layne and Glover 1985), and more generalized feeding habits than suggested by its morphological specializations for feeding on ants and termites (Smith and Redford 1990) may also have contributed to its success in expanding its geographic range. Based on data from Panama (Wislocki 1933) and Texas (Johansen 1961), Moore (1968) speculated that population differences in thermoregulatory mechanisms may have played a role in the northward expansion of the armadillo's range. Delayed implantation, which shifts the birth of young from midwinter to spring, when greater food supply provides better conditions for nursing and growth of the young, would be advantageous in northern areas with more severe winters (Talmage and Buchanan 1954). Superdelayed parturition, considered by Storrs et al. (1988, 1989) to be a facultative survival mechanism induced by stress, may also contribute to the success of pregnant females in producing young in recently occupied areas by reducing energy expenditure for embryonic development until they have adjusted to new conditions. Similarly, delayed fertilization (Storrs et al. 1988) may also facilitate establishment of populations in newly invaded areas. Loughry et al. (1998a) noted that the low reproductive output in a Florida armadillo population, assuming it accurately reflected lifetime reproductive success, was difficult to reconcile with the fact of rapid range expansion. They suggested that reproductive productivity of a population might decline with time since colonization as a result of habitat saturation.

In addition to climatic, biotic, and physiological factors that may have contributed to the northward expansion of the armadillo, accidental and deliberate relocations of animals to unoccupied areas by humans probably also has played a major role in its spread (Fitch et al. 1952). A record of an armadillo killed with a bow and arrow in a field in central New York during the mid-1960s may be an extreme example of transport by humans (J. N. Layne, pers. obs.). Factors aiding the establishment of new populations as a result of relocation by people may be the relative stability of home ranges, weak homing tendency, low dispersal rate (Layne and Glover 1977), and such reproductive features as superdelayed parturition and delayed fertilization (Storrs et al. 1988, 1989). In addition to creation of favorable habitats through forest clearing, agriculture, and other land use practices, humans have further facilitated the spread of armadillos through construction of levees, ditch berms, elevated roadbeds, bridges, and other features that allow occupation of, or movement through, otherwise unsuitable environments. For example, Suttkus and Jones (1999) postulated that the armadillo gained access to the lower Mississippi River delta via the Mississippi River levees, and its penetration of the extensive wetland habitats of the Everglades–Big Cypress region of Florida was undoubtedly facilitated by elevated roadbeds, levees, and spoil banks of canals.

DESCRIPTION

External Morphology. The armadillo cannot be confused with any other North American mammal (Fig. 4.2). The body is covered by a carapace (shell) made up of small scales or scutes consisting of a bony plate attached to a tough epidermal layer by connective tissue. The carapace has three main divisions: an anterior scapular shield covering the shoulder area, a posterior pelvic shield covering the hip region, and a middle section comprising a series of bands connected by soft, infolded skin between the bands. The bands overlap slightly, with the posterior margin of a band overlapping the anterior edge of the one behind, which together with the scaleless skin connecting the bands provides some telescopic movement. The number of bands varies from 7 to 10, with 8 being the usual number in the northern and southern part of the range and 9 in the middle region (Wetzel 1985). Variation also exists in the number of complete, partial, and fused bands. The carapace, even in adults, is somewhat flexible in living animals. The top of the head is covered with thick scales closely adherent to the skull, and the exposed parts of the legs are protected by scales. The tail is covered by a series of 12–15 rings, which decrease progressively in size from base to tip, which is covered with irregular scales. The skin between the scutes of the head and the scapular shield and between the pelvic shield and first tail ring is soft and infolded. The lateral edges of the carapace are also connected to the body with soft skin, giving some capacity for expansion and contraction of the carapace. The scutes of the carapace are closely attached to the skeleton at two points in the pelvic region and the tail rings are attached to the vertebral processes. The general coloration is grayish brown with many yellowish tan scales on the sides of the carapace. A possible albino specimen captured near San Angelo, Texas, and escaped near Boerne, Texas, was reported by Strecker (1927). The ears, underparts, and parts of the head and limbs not covered with scales are protected by tough skin. The skin of the ears is blackish and has a pebble-grain texture, and that of the underparts is

FIGURE 4.2. Armadillo (*Dasypus novemcinctus*). SOURCE: Photo by J. N. Layne.

sparsely clothed with long, coarse hairs arising in groups from scalelike nuclei. The hair tends to be most dense on the insides of the limbs and the sides of the abdomen protected by the overlapping edges of the carapace. According to Kalmbach (1943), hair is usually more profuse in males than females. The carapace has sparse hairs arising largely from the posterior edges of individual scales.

The head is relatively small with an elongate snout terminating in a soft, flesh-colored, piglike nose. The eyes are small, and the ears are large (40–50% of head length) and closely set. The short and robust limbs have four toes on the forefeet and five on the hind feet equipped with large, slightly curved claws, which are best developed on the second and third toes of the forefoot and the third digit of the hind foot. The tail is long, making up about 70% or more of the combined length of head and body. Females have four teats, with one pair located in the pectoral and one pair in the inguinal region. Both sexes possess prominent anal glands, which are protruded when the animal is excited and produce a strong and distinctive odor (Kalmbach 1943).

McBee and Baker (1982) summarized data on measurements and mass as follows: total length, 615–800 mm; length of tail, 245–370 mm; length of hind foot, 75–107 mm; length of ear from notch, 40 mm; mass, 5.5–7.7 kg for males and 3.6–6.0 kg for females. As the mass data indicate, males tend to be larger than females (Kalmbach 1943; Hall 1955). Data from Oklahoma provide an example of the magnitude of the sex difference in a given population (Zimmerman 1982). Mean measurements and mass of males and females, respectively, were as follows: total length, 778.0 and 752.0 mm; length of tail, 348.8 and 340.9 mm; length of body, 452.4 and 432.5 mm; length of hind foot, 68.4 and 65.3 mm; and mass, 4.66 and 4.32 kg. Differences in body length, hind foot length, and mass were statistically significant.

Skull and Dentition. The skull is somewhat tubular in shape, with smooth contours (Fig. 4.3). The rostrum is elongated, accounting for 55–63% of the length of the skull (Wetzel and Mondolfi 1979), and the cranial portion is relatively small, as reflected in a low encephalization quotient of 0.371, compared with 1.00 for a hypothetical "average" mammal (Appendix 6 in Eisenberg 1981). The palate is long, 66–71% of the length of the skull, with rounded rather than keeled posterolateral borders. The auditory region is represented by a tympanic ring, with no ossified external auditory meatus or tympanic bulla (Wetzel and Mondolfi 1979; Wetzel 1985). The lower mandible is long and slender, without a distinct maxillary fossa and with an obliquely ascending ramus. The teeth are simple, peglike, single-rooted structures numbering 7 or 8 in each jaw, with a total number of 28, 30, or 32. The 7 anteriormost teeth are deciduous and are regarded as premolars, whereas the 8th tooth is not preceded by a deciduous tooth and thus is considered to be a molar (Stangl et al. 1995). The teeth of adults lack enamel, but a functional enamel organ is present in the fetus and the teeth are covered with enamel at birth (Talmage and Buchanan 1954). Deviation from the normal 8/8 formula, with the upper molars most frequently missing, is assumed to represent tooth loss due to age or other factors, although failure to erupt is another possibility. Aberrant dental variations attributable to developmental or genetic factors have also been documented (Stangl et al. 1995).

Skull measurements were given by Stangl et al. (1995) for specimens from Texas and Oklahoma categorized as subadult, adult, and old adult age classes. Means of selected measurements for the adult category were as follows (in mm): maximum length of skull, 101.01; maximum zygomatic width, 40.91; rostral width, 18.63; occipital breadth, 23.92; maximum width of nasals, 10.15; length of palatine, 18.29; length of maxillary toothrow, 25.64; and length of mandibular toothrow, 26.88. Although sex differences exist in body measurements and mass, there were no significant differences in measurements of adult or old adult skulls related to sex.

Other Characteristics. As in other members of the order, armadillos are characterized by having additional articular surfaces (zygapophyses) on trunk vertebrae. Gaudin and Biewener (1992) attributed the greater resistance of the armadillo backbone to dorsal and lateral, but not ventral, bending compared with a generalized mammal such as the opossum (*Didelphis virginiana*) to the xenarthrous vertebrae, which presumably reflect an adaptation for digging. At least the second and third and sometimes additional cervical vertebrae are fused into a mesocervical bone, and anterior caudal vertebrae are incorporated into the strong sacrum. The olfactory mucosa of *Dasypus* (*D. hybridus*) described by Ferrari et al. (2000) generally is similar to that of other mammals. This contrasts with the large hairy armadillo (*Chaetophractus villosus*), which exhibits more complex ultrastructural features, a difference perhaps associated with the more specialized digging behavior of *Chaetophractus* compared with *Dasypus*. The palatine mucosa of the armadillo was studied by light and scanning electron microscopy by Martinez et al. (1998). Differences found in the morphology of the hard and soft palate were presumed to reflect their respective roles in mastication and swallowing. The testes are permanently abdominal, and the placenta is hemochorial, with three layers of tissue between the maternal and the fetal circulation (Asdell 1965). McBee and Baker (1982) presented a detailed review of the literature pertaining to the morphology of the species.

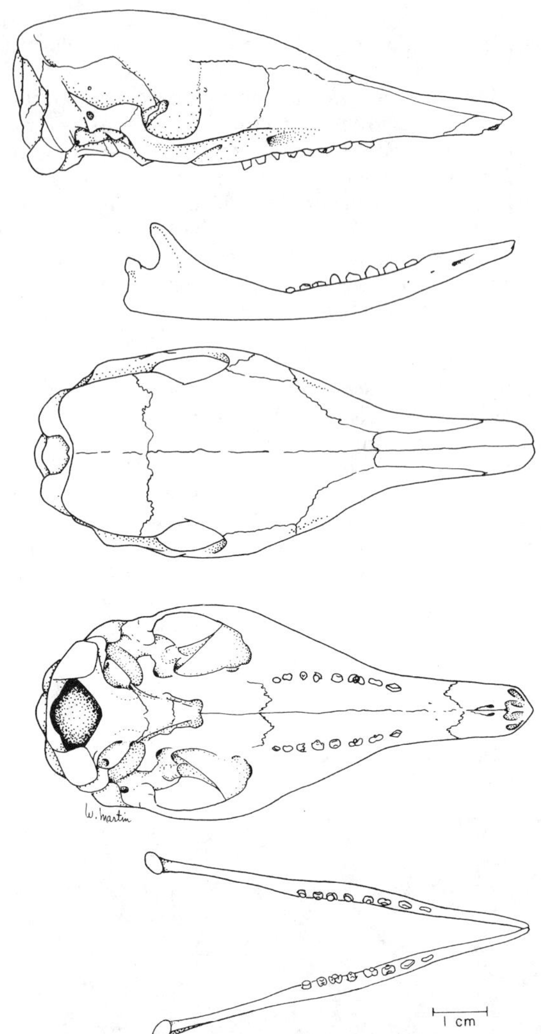

FIGURE 4.3. Skull of the armadillo (*Dasypus novemcinctus*). From top to bottom: lateral view of cranium, lateral view of mandible, dorsal view of cranium, ventral view of cranium, dorsal view of mandible.

GENETICS

The diploid number of chromosomes is 64, with 46 acrocentric elements (Benirschke et al. 1969). There is variation in specimens from Texas, Colombia, and Brazil in morphology of particular chromosome pairs (Jorge et al. 1985). Ramsey and Grigsby (1985) investigated genic variation at 28 loci encoding for 19 proteins involving samples from 13 localities in the United States. They found a very low level of heterozygosity. They suggested that the source population of armadillos originally expanding into the United States may have had a larger pool of genetic variation, and that this pool became reduced as the result of genetic bottlenecks associated with establishment of new populations from relatively few individuals combined with inbreeding along the advancing range-front.

PHYSIOLOGY

Body Temperature and Metabolism. Like xenarthrans in general, the armadillo has a low basal metabolic rate, relatively low body temperature, and high thermal conductance. McNab (1980) recorded mean body temperature T_b of 34.5°C, mean basal metabolic rate of 0.243 $cm^3 O_2/g/hr$, and mean minimal thermal conductance of 0.039 cm^3 $O_2/g/hr$. The body temperature is well below the 36–38°C range typical of other mammals. Relative to expected values based on other mammals of comparable body mass, basal metabolic rate and minimal thermal conductance are 54% below and 225% above, respectively. Low basal rate and high conductance have been interpreted as an adaptation to reduce the tendency to overheat in burrows (McNab 1979, 1980), although it should be noted that well-furred, burrow-dwelling mammals of similar body size lack such adaptations. An additional adaptive value of the armadillo's low metabolic rate suggested by McNab (1980) may be the reduction of energy requirements during periods of reduced food availability due to low temperatures or dry conditions. On the other hand, the high thermal conductance resulting from the sparsely haired venter and carapace is reflected in an increase in the lower limit of thermoneutrality. This requires an increase in metabolism even under relatively mild ambient temperatures, which in turn presumably is a major factor limiting the northward expansion of the range (Talmage and Buchanan 1954; Humphrey 1974; McNab 1980, 1985).

Respiration. Mean respiration rate of males and females in the laboratory reported in several studies (cited in Burns and Waldrip 1971) ranged from 18 to 32 breaths/min. The armadillo can incur a large oxygen debt, which may be related to vigorous digging activity with the nose buried in soil for protracted periods with their breathing interrupted (Scholander et al. 1943). Burns et al. (1975) also found that certain aspects of lactate dehydrogenase (LDH) activity in armadillo muscle and heart tissue were related to burrowing activity, which may lead to oxygen debt. Maximum LDH activity in muscle and heart tissue was associated with relatively high pyruvate concentrations, a characteristic shared with other vertebrates subjected to periods of sustained hypoxia. Comparison of LDH activity at pH 7.2 and 7.6 showed that muscle was more metabolically active than heart tissue at the lower pH, indicating the ability of muscle tissue to remain functional under anaerobic conditions with resultant buildup of lactic acid.

Cardiac Function. Wilber (1964) reported heart rates of 95 and 170 beats/min for two individuals in the laboratory. Burns and Waldrip (1971) found a significant difference in mean heart rate of males (126/min) and females (84/min). Electrocardiogram measurements in the two studies were comparable, with marked differences between males and females in the P–R and Q–T time intervals corresponding to the different heart rates. The overall electrocardiagram of the armadillo is similar to that of other mammals.

Blood Chemistry. Ramsey et al. (1981) measured a number of chemical components of the blood of captive and wild armadillos. Cholesterol and alkaline phosphatase concentrations in the wild population exhibited higher levels in spring (May) than in fall (November) or winter (February). Wild and laboratory individuals differed significantly in five of the nine other parameters measured: calcium, phosphorus, glucose, urea nitrogen, uric acid, total protein, albumin, glutamic oxalacetic transaminase, and lactate dehydrogenase. Plasma albumin level in wild individuals was significantly correlated with gross energy, the protein fraction of stomach contents, and gross fecal energy, whereas uric acid concentration was correlated with fecal energy and protein levels. Herbst et al. (1989) reported packed cell volume and plasma and erythrocyte cholinesterase values for armadillos from a wild population in Florida and in laboratory animals. Packed cell volume ranged from 24% to 50%, plasma cholinesterase activity from 105 to 549 U/liter, and erythrocyte cholinesterase from 2915 to 15,126 U/liter. Packed cell volumes did not differ significantly between sexes, seasons, or wild and laboratory animals, whereas significant differences occurred between wild and captive individuals for plasma cholinesterase activity and between seasons for erythrocyte cholinesterase activity.

Excretory System. Unlike the case of the desert-inhabiting little hairy armadillo (*Chaetophractus vellerosus*), the renal papilla of the nine-banded armadillo kidney does not protrude into the renal pelvis. The average relative medullary thickness of the kidney is 4.9, which falls well below the expected value for a xeric-adapted mammal and somewhat above that for a mesic-adapted species of similar body mass. Armadillos in the laboratory that were fed a diet of partially dehydrated horse meat (40.6% water content) lost an average of 15.28% of original body mass within an 11-day period. Even when dehydrated, they continued to urinate copiously and drank their urine. The relatively poor concentrating ability of the armadillo kidney probably is an important factor limiting its ability to exist in xeric environments (Greegor 1975, 1985).

REPRODUCTION

Reproductive Tract Morphology. The male lacks a scrotum, the testes descending to the pelvis at the entrance to the inguinal canal. Mean maximum mass and length of the adult testis are about 2.4 g and 31 mm, respectively (see Figs. 2 and 3 in McCusker 1985). The penis is relatively large and lacks a baculum (os penis). It is somewhat thickened at the base and terminates in a pointed tip that projects between two lateral lobes (Wetzel and Mondolfi 1979). Based on artificial inflation of the corpus cavernosum, the penis increases an average of 29.0% in length, 16% in dorsoventral diameter, and 17% in lateral diameter during erection (Kelly 1999). As in other mammals, the collagen fibers in the tunica albuginea are arranged in two major layers. Fibers are arranged parallel to the long axis of the penis in the outer layer and circumferentially around the corpus cavernosum in the inner layer. This gives the penis a reproducible shape, maximum size, and resistence to compression, bending, and tensile forces during erection (Kelly 1997). Networks of collagenous trabeculae extending through the vascular space of the corpus cavernosum and connecting the dorsal and ventral sides of the corpus also aid in maintaining the noncircular, cross-sectional shape of the corpus cavernosum during erection so as to prevent restriction of the urethra and consequently reduced semen transport.

Anatomy of the female reproductive tract exhibits a mix of primitive and specialized characteristics. The kidney-shaped ovaries are of approximately equal size in immature individuals and adults during anestrus. The uterus is simplex, grossly resembling that of primates, and apparently evolved through a bicornuate stage. A primitive feature of the female tract is the absence of a true vagina. The uterus opens into a urogenital sinus, which functions as both a vagina and urethra and opens to the outside through a long vestibular cleft (Talmage and Buchanan 1954).

Age at Maturity. Talmage and Buchanan (1954) stated that males mature at about 6 months of age, and McCusker (1985) found developed spermatids in males estimated to be 7–12 months old. There is less agreement among authors on the age at first breeding of females, estimates varying from 1 year (Talmage and Buchanan 1954) to 2 years (Galbreath 1980; Gause 1980) and up to 29 months after birth for ovulation in the majority of nulliparous females (Enders 1966). As in the case of other reproductive parameters, age at puberty in armadillo populations may be influenced by environmental factors such as prolonged drought or unusually low winter temperatures.

Breeding Season. Females are monestrus and typically produce a single ovum (Enders 1966). The time of ovulation may vary spatially and temporally due to environmental conditions, particularly drought. This variability is exemplified by data from Texas. Hamlett (1932) cited 15 July as the average date of mating, but Talmage and Buchanan (1954) believed the period of ovulation could only be narrowed to the June–August period. Estrus occurred from June to July in the generalized reproductive cycle presented by Enders (1966), who also noted that ovulation in nulliparous individuals occurred later than in adults. Most multiparous females ovulated by September, compared with November for nulliparous individuals. McCusker (1985) documented an annual cycle in testis mass and length in armadillos from north-central Texas, with larger and heavier testes occurring from April to September. This trend, however, was not clearly associated with changes in spermatogenic activity, as spermatids were present throughout the year. He concluded that continuous spermatogenesis, in contrast to the well-marked seasonality of ovulation, may be a secondary result of maintenance of a high gonadotrophic level associated with aggressive behavior or territoriality. Seasonal increase in testis size and mass may be an energy storage mechanism at times of increased food availability, although increased testicular lipid storage does not appear to be involved. In males from southern Georgia and north and south Florida, testis volume, seminiferous tubule area, and Leydig cell volume were greatest in summer and lowest in winter. Changes in these parameters and in abundance of spermatozoa in testes and epididymides were most pronounced in the more northern populations (Gause 1980). Galbreath (1982) stated that mature males in south-central Florida typically produced sperm throughout the year. Storrs et al. (1977) observed seasonal variation in sperm motility in a laboratory colony and wild population.

Incidence of paired males and females in a study by McDonough (2000) indicated a well-defined breeding season extending from June through November, with 85% of pairings occurring from June through August. Such variability in the breeding season has also been reported from other parts of the range. Gause (1980) stated that in Florida and southern Georgia ovulation occurred from late May to August, with a peak in mid-July. Breeding behavior usually began in late February and actual copulation occurred from June to August. In both sexes, variation in reproductive parameters in different years was correlated with environmental factors. In Mississippi, Jacobs (1979) frequently observed pairs from April through October, with the highest number in June and July. She assumed this was evidence of an extended breeding season, although whether copulation was observed throughout this period was not stated.

Gestation. The normal gestation period is approximately 8–9 months. This includes an embryonic diapause (delayed implantation) at the blastocyst stage of approximately 14 weeks as determined by Hamlett (1932) and generally accepted as 3–4 months (from August to November) in the normal reproductive cycle in Texas, as summarized by Enders (1966). However, Storrs et al. (1988, 1989) documented superdelayed parturition in the armadillo, a phenomenon not previously known among mammals. Some wild-caught females brought into the laboratory and isolated from males had litters 13–24 months after capture, indicating a greatly prolonged period of embryonic diapause. Storrs et al. (1988) also documented two instances in which females isolated from males since capture became pregnant in two successive years, indicating either that (1) delayed fertilization had occurred, an ovum being produced each year and fertilized by viable sperm stored in the reproductive tract, or (2) that two ova (diovulation) were produced the first year, a rare phenomenon in the armadillo, and one fertilized ovum was stored in the uterus for 15 months while the other developed. The latter explanation appears to fit a case described by Buchanan (1967) in which a female had both an implantation site with normal embryos as well as a large unimplanted blastocyst, but only a single corpus luteum.

Season of Birth. The time of parturition extends over at least a 6-month period, February–July. Talmage and Buchanan (1954) observed recently born litters from February to May in Texas. Gause (1980) stated that parturition occurred from early March to May in the southern Georgia–northern Florida region, with a probable peak in late March. In another Florida population, the mean estimated birth date of litters (calculated by backdating to an assumed neonatal mass of 100 g and a postnatal growth rate of 10.6 g/day) was 9 March for females and 18 March for males, with no significant difference between sexes (Loughry et al. 1998c). Jacobs (1979) reported estimated months of births in Mississippi extended from February to July, with 81% occurring during April–June and a sharp peak in May. The Mississippi population also exhibited a significant difference between male and female litters in the timing of births, with male litters tending to occur earlier.

Litter Size. The armadillo exhibits monozygotic polyembryony in which a single fertilized egg normally gives rise to four separate embryos at the blastula stage following division of the inner cell mass. The genus *Dasypus* is the only vertebrate taxon in which this reproductive phenomenon is known to occur. Each embryo develops its own amniotic sac and independent placental attachment, although the combined placentas form a single large, lobed disc with no circulatory interconnections (Storrs 1978). The embryological evidence for the clonal relationship of sibling armadillos has been confirmed by molecular genetic analysis (Prodöhl et al. 1996). Storrs et al. (1988, 1989) noted that an adaptive advantage of polyembryony in a species producing only a single ovum per year and with minimal maternal care is the fourfold increase in birth rate while insuring preservation of individual genomes. Armadillos as a group also appear to be unique among the xenarthrans in having greater fecundity (litter size >1) that is unrelated to metabolic rate (McNab 1985).

Galbreath (1980) postulated that the origin of polyembryony first involved specialization of the uterus to maximize successful implantation under conditions of delayed implantation by ensuring placement of a single blastocyst in an area of the uterus that provided optimal conditions for development as described by Buchanan (1967). This stage was subsequently followed by selective pressure to increase litter size, which could be achieved only through polyembryony because of the prior adaptation of the uterus to accommodate a single blastocyst. Although the typical number of embryos is four, infrequently there are three, representing the loss of an embryo, or five, and, in one known case, six plus an additional amniotic sac (Newman 1913; Buchanan 1957). Galbreath (1980) recorded a litter in which all four fetuses had severe skeletal deformities and two were necrotic.

DEVELOPMENT

Embryonic development was described in detail by Newman and Patterson (1910). Suttkus and Jones (1999) described fetal growth and details of development, including body shape, scalation, jaw structure, pigmentation, eyes, feet and digits, and external genitalia, in a series of litters forming a progressive series of body size, although actual gestational ages were unknown. Relative growth of the tail and trunk of fetuses exceeded that of head, hind foot, and ear pinnae.

Armadillos give birth to precocial young. The neonates have a mass of approximately 85–113.5 g (Storrs 1967; Storrs in Jacobs 1979; McDonough et al. 1998). The eyes are open at birth, and the young become active soon after birth. The carapace is much thinner and more pliable than that of adults. The ages in days at which selected developmental events occurred in a male and a female litter born in captivity were as follows, for males and females, respectively: 20 and 22, first day out of the nest; 21 and 25, begin to drink water; 35 and 42, begin to consume solid food; 71and 74, begin to include insects in the diet; 82–140 and 89–162, period of weaning (McDonough et al. 1998). Although these results suggest that males are somewhat more precocious in development than females, they are inconclusive because of small sample size. The young begin to accompany the mother outside the burrow at about 2–3 months of age and become self-sufficient within 3–4 months (Storrs 1978). An observation by Moseley (1971) indicates that young are capable of feeding in the typical fashion at an early age. When she placed recently weaned young on a fire ant mound, they immediately began rooting and feeding in the characteristic manner.

Jacobs (1979) estimated the mean growth rate of young as 10.6 g/day based on wild juveniles recaptured at various intervals from June to November. Growth in mass of one male and one female litter in the laboratory was linear from birth through about 400 days of age, at which time they had reached adult mass of 3.5–4.0 kg (McDonough et al. 1998). There was a suggestion of further, but slower growth in the second year beyond 700 days of age to about 4.5 kg, and the capture of even heavier individuals in the wild suggests continued growth beyond 2 years. In the absence of growth data for linear measurements such as body length, there is a question as to the extent to which increased mass after 2 years of age represents actual increase in size or accumulation of fat deposits. Seasonal variation in body mass also complicates its use as a measure of growth in wild individuals because it may vary as much as 25% between summer and winter in adults (Jacobs 1979). Storrs (1978) and Galbreath (1980) stated that full size was attained at 2 years and 3–4 years, respectively.

ECOLOGY

Habitat. The armadillo occurs in a wide range of natural and human-modified terrestrial habitats. Its habitat versatility is undoubtedly one of the major factors contributing to its success in expanding its range. It generally avoids or is scarce in very wet and very dry habitat types, although it has been assisted in invading wetlands by construction of levees, spoil banks, and elevated road beds. Given its feeding and burrowing habits, habitat suitability may depend more on the characteristics of the substrate than on vegetation type. Armadillos appear to prefer mesic habitats and may be particularly abundant in riparian woodlands, which in some cases have probably served as corridors for their spread. Natural habitat types occupied by armadillos in different parts of their geographic range include pine (*Pinus*)–oak (*Quercus*) woodlands, oak–elm (*Ulmus*) woodlands, pine forests, mixed pine–hardwood forests, bottomland forests, riparian woodlands, mesic hardwood forests, scrub, chaparral–mixed grass, inland and coastal prairies, salt marsh, coastal dunes, and coastal strand. Human-dominated habitats include pastures, parkland, cemeteries, golf courses, citrus groves, pine plantations, plant nurseries, cut-over pineland, and various croplands (Taber 1945; Clark 1951; Fitch et al. 1952; Manaro 1961; Ehrhart 1976; Layne 1976; Jacobs 1979; Zimmerman 1982; Breece and Dusi 1985; Holler et al. 1989; Sikes et al. 1990; Suttkus and Jones 1999; McDonough 2000).

Home Range and Dispersal. Armadillos have relatively small home ranges compared with carnivores of similar body size such as the striped skunk (*Mephitis mephitis*), presumably because of the greater biomass per unit area of their primary foods—insects and other small invertebrates—compared with the larger, more dispersed prey of the carnivores. Mean home range size of adults estimated by the minimum polygon method in 11 studies at localities in Alabama, Florida, Georgia, Louisiana, Mississippi, Missouri, Oklahoma, and Texas and in various habitat types ranged from 0.6 to 9.2 ha, with an overall mean of 4.2 ha (mean sample size = 8.2, range = 1–21) (Clark 1951; Layne and Glover 1977; Jacobs 1979; Galbreath 1980; Thomas 1980; Zimmerman 1982; Breece and Dusi 1985; Schnell 1994; Suttkus and Jones 1999; McDonough 2000; Bond et al. 2001). Minimum home range estimates of 3–9 ha for adult females in north-central Florida (Herbst and Redford 1991) fall within this range. A home range estimate of 19.6 ha for one individual in Louisiana (Fitch et al. 1952) is not directly comparable to the preceding values, as it was based on the assumption that the home range was circular with diameter equal to the maximum observed range length (503 m). At least some of the variation in reported home range sizes probably reflects habitat differences and associated food resources. However, small sample sizes and high variance of estimates in most studies preclude critical analysis of the relation of home range size to habitat or other environmental factors. In the study by Layne and Glover (1977), home ranges tended to be smaller in moister areas, where feeding also was more concentrated.

Home ranges appear to be relatively stable over time. In the case of six individuals present in both years of Jacob's (1979) study, the home range boundaries of a given individual in the second year overlapped those of the first year by an average of 45%. Layne and Glover (1977) did not observe any correlation in estimated home range size and the length of time, up to 778 days, during which individuals were monitored, indicating that range boundaries were stable. Individuals were recaptured after 6 and 24 months within 30 and 442 m of the original capture site, well within the mean minimum home range length of adults. Home range stability was also indicated by the study of Loughry and McDonough (1998a), in which both within- and between-year distances moved by individuals were generally <200 m. There was no difference due to sex in either juveniles or adults in distances moved within or among years, although the mean within-year distance between sightings was greater for adults than juveniles.

There appears to be no consistent difference in home range size of males and females, although sample sizes in most cases are too small to permit definitive statistical testing for between-sex differences. Young individuals have smaller home ranges than adults (Clark 1951; Layne and Glover 1977; Breece and Dusi 1985; McDonough 2000). Home ranges also vary seasonally. In a Mississippi population, home ranges of males and females decreased between April–June and July–September (Jacobs 1979). McDonough (2000) found that home ranges of adults and 1- to 2-year-old individuals were larger during the breeding season than nonbreeding season, and breeding males had larger ranges than nonbreeding males. Movements are reduced in response to cold weather in winter (Burns and Waldrip 1971; Bond et al. 2001).

Home range boundaries of armadillos generally overlap those of other individuals. In a Texas study, 73.3% of the home ranges of adult females overlapped those of other individuals to varying degree (other females, 29.8%; breeding males, 44.8%, nonbreeding males, 25.5%), whereas breeding males and nonbreeding males exhibited less range overlap with other individuals (30.8% and 49.6%, respectively). Among breeding males, greatest home range overlap was with females (32.5%) and the least (11.8%) with other breeding males (McDonough 2000). Jacobs (1979) also found that the amount of range overlap was greater between females than males in 1 year (63% vs. 31%). The situation was reversed the following year, however, with 44% overlap among males versus 25% among females. Females in a Florida study shared an average of about 25% of their home ranges with other females (Herbst and Redford 1991). Layne and Glover (1977) found ranges of most adults and subadults overlapped. Of 23 individuals in a Louisiana population, 22 shared portions of their ranges; only 1 male occupied an exclusive area (Suttkus and Jones 1999). Although home range overlap appears to be the general rule in armadillo populations, adult females may occupy mutually exclusive ranges under some conditions (Layne and Glover 1977; Galbreath 1980; Zimmerman 1982), perhaps reflecting increased aggressiveness associated with pregnancy or presence of a litter (McDonough 1994). An obvious qualification in assessing the degree of overlap or exclusivity of home ranges is the accuracy of home range boundaries, which involves both the number of observations as well as the method used to delineate the boundaries.

Despite their seemingly aimless travels within the home range, armadillos appear to remain well oriented as to their position at any time. For example, if alarmed while foraging, an animal usually takes a straight course to the burrow. Observations suggest that at least under some conditions they use olfactory cues to determine their position relative to the burrow or other feature within the home range. When Clark (1951) released an adult male at the point of capture after holding it in captivity for a 24-hr period, it briefly moved around slowly on an apparently random course with its snout pressed to the ground, then suddenly began to run on a direct course to the only hole under a fence enclosing the area. Jacobs (1979) observed similar behavior in which individuals released after handling would walk very slowly from the release point with their noses to the ground for 0.5–2 min, stop abruptly, then make a straight dash to the burrow. Haas (1972) observed that armadillos tended to use the same paths when foraging, even at some distance from the burrow, so that they know the best escape route if disturbed. As an example, a juvenile released at the point of capture, after being taken to the laboratory for marking and examination, ran off without hesitation on exactly the same path it had used before capture.

Jacobs (1979) believed that considerable immigration, particularly dispersal of juveniles from the natal area, occurred in late summer and fall based on the appearance of unmarked individuals in her Mississippi study area. The presence of a few marked animals remaining in the study area the following spring also suggested considerable emigration, but it was not possible to distinguish between losses, other than car-related mortality, due to overwinter mortality and movement off the study area. Little is known of the homing ability of armadillos. An individual in Texas allegedly returned three times to the original capture site after being relocated at distances of 1.2, 11.3, and 37.0 km, respectively. (Chamberlain 1980). A Florida individual that escaped from an outdoor cage returned to the original capture site about 930 m away within a 3-week period. However, two other escaped individuals settled in the immediate vicinity instead of returning to their original capture sites 300 and 1986 m away, respectively. (Layne and Glover 1977).

Homesites Armadillos typically inhabit burrows of their own construction. The average width of the burrow entrance is about 20–22 cm, but varies depending on size of the occupant, age of the burrow and frequency of use, soil type, degree of slope, and extent of reinforcement by tree roots or other features (Clark 1951; Zimmerman 1982). The width of the tunnel beyond the entrance is about 18–19 cm with little variation (Kalmbach 1943; Clark 1951; Zimmerman 1982). Burrows tend to be slightly oval in cross section with a greater vertical than horizontal axis, reflecting the cross-sectional shape of the armadillo's body. For example, mean width and height of the entrance and deeper portion of burrows measured by Zimmerman (1982) were 19.8 and 21.9 cm and 17.5 and 20.5 cm, respectively. Burrows range in length from 0.3 to 7.3 m, with mean length of 1.3 and 1.4 m in two Texas studies (Lehmann 1934; Taber 1945; Clark 1951). The burrow slopes down gradually to a depth of as much as 1.2 m and may change direction for no apparent reason or to avoid roots or other obstructions. Burrows may have two or three branches and up to four entrances, although a single entrance is the rule and if more than a single entrance is present, one appears to be used more than the others (Kalmbach 1943; Taber 1945; Clark 1951).

Burrows used as dens have an enlarged nest chamber with an average diameter of about 34 cm and tend to be longer (1.9 vs. 0.9 m) and more complicated than those dug for other purposes (Kalmbach 1943; Clark 1951). The nest is a bulky mass of dried plant debris crammed into the nest chamber and lacks any obvious structure, the animal simply pushing its way into the interior (Kalmbach 1943; Taber 1945; Clark 1951). Masses of decomposing nest material are sometimes found at the entrance of well-used burrows, suggesting that the animals periodically remove old material and replace it with new, dry material (Clark 1951).

Burrow entrances are often associated with clumps of shrubs or other perennial vegetation and may be located under roots of trees or logs, in bases of hollow trees, or under buildings. There seems to be no significant tendency for the entrances to be oriented in any particular direction (Taber 1945; Clark 1951; Chamberlain 1980; Galbreath 1980; Zimmerman 1982; Breece and Dusi 1985). The entrance to a burrow under the roots of a cypress tree (*Taxodium distichum*) in Florida went through a large accumulation of sticks and other dried vegetation, which appeared to have been gathered there by the armadillo, perhaps to restrict the size of the space within the roots (J. N. Layne, pers. obs.). Burrows are frequently located on sloping ground, sandy knolls, sides of spoil banks and levees, road shoulders, and cut banks of streams, which facilitate surface drainage of rain water away from the entrance and reduce chances of flooding in poorly drained areas (Kalmbach 1943; Taber 1945; Fitch et al. 1952; Zimmerman 1982; Suttkus and Jones 1999).

Unlike some burrowing mammals, individual armadillos tend to have a relatively high number of burrows. In Texas, Taber (1945) recorded an average of 4.5 and 8.5 burrows per individual in two habitats, respectively. An individual studied by Thomas (1980) used 8 burrows from one to five times each during a 55-day period. Adults in Mississippi used an average of 2.1 burrows (range = 1–4+) during periods of 2–12 days (Jacobs 1979). Females in Florida used 17–32 different burrows during a 12-month period of monitoring (Herbst and Redford 1991). Individuals on Cumberland Island, Georgia, used an average of 10.9 burrows during an 11-month period (Bond et al. 2001). In addition to the obvious function of burrows in providing protection from the elements and predators, Taber (1945) suggested that burrows might also serve as a food trap, which would be of particular benefit during periods of drought or cold. Clark (1951) observed burrows in Texas that contained masses of thousands of camel-crickets (*Ceuthophilus*), which were almost certainly fed on by the armadillos. Armadillo nests in Texas were inhabited by at least 31 species of arthropods, including larval and adult insects, mites, spiders, and centipedes (Clark 1951). Burrows and nests are used by a variety of vertebrates, including the opossum (*Didelphis virginiana*), eastern cottontail rabbit (*Sylvilagus floridanus*), cotton rat, striped skunk, mink (*Mustela vison*), and burrowing owl (*Athene cunicularia*) in Texas (Taber 1945) and the eastern indigo snake (*Drymarchon corais couperi*) and Florida mouse (*Podomys floridanus*) in Florida (Layne 1990; Layne and Steiner 1996). Still other species may take temporary refuge in the burrows. Burrows or nests may be used concurrently by armadillos and other species. Taber (1945) reported a case of a cotton rat and a cottontail rabbit sharing a burrow and nest with an armadillo. In another case, an armadillo and an opossum used the same nest.

In digging their burrows, armadillos use both the snout and forefeet to loosen the soil and push it beneath the abdomen. After a slight pause, the animal balances on the forefeet and tail with the body strongly arched and the hind feet over the pile, then propels the soil a good distance to the rear with a sudden straightening of the body and a powerful backward thrust of the hind feet (Taber 1945). They dig with surprising rapidity. An individual observed by Taber (1945) buried itself so completely in 2 min that a pick was required to dig it out of the hard-packed soil.

Armadillos living in uplands in the Texas Hill Country use fissures and caves in the shallow limestone soils for den sites that involve little digging (Taber 1945), and in Florida they often use burrows of the gopher tortoise (*Gopherus polyphemus*). Usual signs of occupancy of a tortoise burrow by an armadillo are its tracks on the sand mound at the entrance. Also, the burrow entrance is usually altered from a more or less half-moon shape with greater width than height to a somewhat oval shape with a greater vertical than horizontal dimension, reflecting the cross-sectional shapes of the tortoise and the armadillo, respectively.

Armadillos in captivity build nests of straw or dry grasses on the floor of indoor pens or on the ground surface in an outdoor enclosure (Johansen 1961; Anderson and Benirschke 1966; Moore 1968). Similar nests also have been recorded in the wild in Texas (Clark 1951), in Florida (Layne and Waggener 1984), and on Cumberland Island, Georgia (Bond et al. 2001). Ground-surface nests in Florida (Fig. 4.4) usually were located in areas with poorly drained soils, which become saturated or flooded during periods of high rainfall. Nests resembled a miniature haystack when built in open grassland habitats. The nests typically consisted of a shallow depression or form the size of the animal's body, roofed over with a mass of dry plant material. There was no well-defined nest chamber, although the interior was loosely filled with finer material than that making up the outer layers. Most nests had a single entrance, and in the few cases where there were two or three entrances, one appeared to be used most. Nests in open grassland habitats were situated directly on the ground surface adjacent to tussocks of broom sedge (*Andropogon* sp.), and had a mean maximum height of 49 cm and mean maximum and minimum diameters of 101 and 71 cm, respectively (Layne and Waggener 1984). Some nests were located in slightly elevated clumps of saw palmetto (*Serenoa repens*) in wooded areas. These had a greater variety of dried plant materials and smaller average size than those in open grassland sites. Most of the occupied nests were found in winter and spring. Except for one nest with three

FIGURE 4.4. Surface nest of an armadillo (*Dasypus novemcinctus*) in open grassland habitat in south-central Florida. SOURCE: Photo by J. N. Layne.

adults and another with an adult female and three young, active nests had a single occupant. In some cases, typical burrows were located near the surface nest. The single surface nest recorded in Texas was located beneath an accumulation of water-borne debris on a creek bank (Clark 1951). No details were given for those reported by Bond et al. (2001) from Georgia.

Nest-building behavior of armadillos in captivity involved raking up nesting material with the forefeet, holding it against the hind legs, and transporting it to the nest box by hopping backward (Taber 1945; Eisenberg 1961). Wild armadillos also have been observed gathering and transporting nest material in a similar manner in Texas (Clark 1951), Arkansas (Taulman 1994), and Florida (C. E. Winegarner and P. J. Cone, pers. commun., 2001). The individuals observed in Arkansas and Florida made multiple forays to collect nest material at distances of 12–14 m.

Population Dynamics The prenatal sex ratio of litters examined by Galbreath (1980) in Florida was 1:0.78 (male litters:female litters), with a marked difference in samples from two different years (1:0.25 vs. 1:1.07). Adult sex ratios (males:females) reported in different parts of the range include 1:1 in Texas (Newman 1913), 1:0.58 in Oklahoma (Zimmerman 1982), 1:1.13 in Mississippi (Jacobs 1979), and 1:1.38 in Florida (Loughry and McDonough 1998a). In the Oklahoma study, factors suggested as contributing to the marked predominance of males were higher mortality of young or pregnant females during prolonged periods of cold weather and a sampling bias resulting from a greater probability of males being killed on roads as a result of their larger home ranges. Jacobs (1979) in Mississippi and Loughry and McDonough (1998a) in Florida found a higher proportion of females among juveniles than adults (1:1.42 vs. 1:1.13 and 1:1.50 vs. 1.38, respectively), which may reflect higher mortality of juvenile females or a greater probability of capturing adult males.

Zimmerman (1982) reported the distribution of 95 males and females among seven age classes. Age classes (from 0 to 6) did not reflect actual chronological age, although class 0 included animals from birth to about 5 months of age based on degree of fusion of skull sutures. Age class 1 predominated, classes 2–5 were much smaller and about equal, class 0 was smaller than 2–5, and class 6 was the smallest of all. These data were based on road-killed individuals and may not reflect the actual population age class distribution. Loughry and McDonough (1996) in Florida found a significant difference in the age composition of road-killed armadillos and live-captured individuals from an adjacent population, with almost no juveniles in the road-killed sample. Juveniles constituted 28.6% and adults 71.4% in a Mississippi population (Jacobs 1979).

Loughry et al. (1998b) found low reproductive success in a north Florida study area. The majority of the adults in the population failed to produce any surviving offspring; those that did produced young in only 1 of the 4 years of the study. They found no evidence of differential reproductive success of males based on any morphological parameters considered. For females, however, larger body mass was associated with a higher number of surviving young.

Reported estimates of armadillo population densities (single values or means of multiple samples for a given study site or region) are highly variable and give no indication of any overall geographic trend. Twelve estimates from various localities and habitats in Texas ranged from 0.8 to 145/100 ha with a mean of 26.9/100 ha (Kalmbach 1943; Taber 1945; McDonough 2000). Chamberlain (1980) reported densities for 12 Texas counties with major human population centers ranging from 0.2–0.4 to 6.4–8.0 armadillos/100 ha. Densities reported elsewhere in the geographic range include 0.1 and 10.0/100 ha in Florida and 5.7 and 281.4/100 ha in Louisiana (Fitch et al. 1952; Bushnell 1952; Layne and Glover 1977; Suttkus and Jones 1999). Although habitat is undoubtedly a major determinant of abundance, populations are affected by so many other variables that there is no clear overall relation between density and habitat in the data. Some suggestion, however, of the effect of habitat on density was provided by Taber (1945) in Texas, where density estimates were 13.3/100 ha in pine woodlands with sandy soils; 0.8, 4.0, and 80.0/100 ha in post oak (*Quercus stellata*) association with heavy clay soils (the maximum value reflecting the abundance of streams in the area); 8.8/100 ha in coastal prairie with scattered mounds of sandier soil than the surrounding area; and extremely low (value not given) in clay blackland soils. Reported burrow densities, which are undoubtedly higher than actual densities of individuals because of use of multiple dens by at least some individuals, include 219–370/100 ha (mean = 304) in Texas, 317/100 ha in Oklahoma, and 540/ha in Louisiana (Taylor 1946; Fitch et al. 1952; Zimmerman 1982). In lieu of more labor-intensive efforts to estimate the actual number of individuals in a given area, counts of active burrows may provide an acceptable index of abundance for some purposes.

Although numbers in a particular region may fluctuate yearly as the result of drought, severe winters, or other factors, there is evidence of a general decline in populations in at least some parts of the range in recent years. In 1999, R. H. Baker (Michigan State University, pers. commun., 2001) conducted a mail survey and interviews involving more than 100 knowledgeable observers in Texas regarding the status of the armadillo; 80–90% of the respondents indicated numbers had declined. In Florida, mean track counts/day in July on a standard census route in the south-central region of the state were 48.5 for 1974–1986, 28.9 during 1992–1996, and 7.8 in 2000–2001 (J. N. Layne, unpublished data). The cause of the apparent decline is unknown and does not appear to be associated with long-term weather conditions or habitat changes.

Activity. Armadillo populations in widely separated parts of the range typically exhibit seasonal variation in the diel activity pattern, with increased diurnal activity in winter as a response to low temperatures and largely nocturnal activity in summer (Taber 1945; Fitch et al. 1952; Moore 1968; Burns and Waldrip 1971; Layne and Glover 1977, 1985; Galbreath 1980; McCusker 1985). Zimmerman (1982) noted that in Oklahoma the shift to daylight activity in winter was less pronounced than in other parts of the range. In Alabama, feeding activity in summer usually began after sunset on days when peak temperature exceeded 32°C. Activity decreased markedly when ambient temperature fell below 10°C (Breece and Dusi 1985). In south-central Florida, Layne and Glover (1985) found that peak activity occurred in the 20–25°C temperature range throughout the year, with activity compressed into a shorter time span in winter. Track counts indicated that there was not only a shift to more diurnal activity in winter, but a reduction in the overall level of activity as well. A similar seasonal trend was documented in an experimental outdoor enclosure study in Texas (Moore 1968). In Georgia, Bond et al. (2001) found that armadillos were located more frequently at dens and used fewer dens in winter than in summer. A reduction in highway mortality in Florida in winter (Inbar and Mayer 1999; J. N. Layne, unpublished data) also may indicate an overall reduction in activity level in cooler weather, although restriction of movements or shifts in habitat use also may be involved. In contrast, peak incidence of road mortality in Mississippi occurred in November and December (Jacobs 1979).

The major adaptive advantage of shifting to diurnal activity in cold weather is presumably energy conservation, but increased food availability also may be a factor. Based on ambient temperatures and estimated metabolic rates of active armadillos in January and July in south-central Florida, Layne and Glover (1985) concluded that an animal shifting to diurnal activity in winter would expend 50–60% less energy for thermoregulation than if it adhered to the summer nocturnal activity pattern. Higher daytime soil temperatures in winter also may result in greater abundance of invertebrates and small vertebrates near the surface than at night. In an experimental study in an outdoor enclosure, Moore (1968) determined that the difference in the total cost of foraging in late February–early March and in June was relatively small (15.2 vs. 9.1 Cal/individual/hr) as a result of the timing of activity in the two seasons to avoid temperature extremes. Gause (1980) and Layne and Glover (1985) found that adult females and immature animals were more diurnal than adult males in cold weather, possibly reflecting lower fat reserves (Gause 1980) coupled with higher metabolic demands of young animals and pregnant females (McNab 1980). Lower assimilation of energy in females, as suggested by the greater fecal energy in fall and winter in females than males (Ramsey et al. 1981), may be an additional factor leading to the tendency for females to exhibit more diurnal activity than males, particularly in cooler weather. Armadillos may bask during the winter to conserve energy (McCusker 1985).

Factors other than temperature also may influence activity patterns. Increased activity during the spring–early summer period (Moore 1968; Layne and Glover 1985), when temperature is presumably not a factor, was attributed to breeding activity. Increased numbers due to recruitment of young might also have been involved (Layne and Glover 1985). Lehmann (1934) in Texas noted that among adult females in summer, individuals with young commenced activity earlier. Breece and Dusi (1985) and McDonough and Loughry (1997a) also found that juveniles commenced activity about 2 hr earlier each day than adults. In Texas, Hamilton (1946) observed armadillos actively foraging for black persimmon (*Diospyros texana*) fruits during the hottest part of the day in August, suggesting that in times of food shortage due to dry soil conditions or other factors armadillos may be forced to extend foraging activity into the daylight hours even in summer. Armadillos also may shift activity to more mesic habitats during dry periods (Taber 1945) or drastically reduce activity during prolonged droughts (Clark 1951; Zimmerman 1982).

Daily weather conditions within any season also may affect activity. McDonough and Loughry (1997a) found that activity was positively correlated with cloud cover during the day and drier, warmer conditions at night. Armadillos may continue foraging during rain, even ignoring thunder and lightning (Taber 1945; Moore 1968; Zimmerman 1982; Breece and Dusi 1985), but reduce activity and appear more wary under windy conditions (Breece and Dusi 1985). In Oklahoma, armadillos ceased activity for a period of days following snow (Zimmerman 1982).

Seasonal variation in physiological parameters such as body temperature and heart rate (Burns and Waldrip 1971), hematopoeisis in bone marrow (Weiss and Wislocki 1956), and body fat reserves (Gause 1980) may be associated with changes in activity patterns. Armadillos are largely nocturnal in captivity (Johansen 1961; Anderson and Benirschke 1966; Yaksh 1967; Prudom and Klemm 1973).

FEEDING HABITS

Morphological Specializations. Redford (1985) recognized four groups among the nine genera of armadillos based on feeding habits, with the genus *Dasypus* being the sole member of the terrestrial generalist insectivore group. Correlating with its highly insectivorous diet, the armadillo's stomach is specialized for dealing with an excessive amount of chitin in the food (Owen 1830). It has an extensible saclike anterior portion capable of receiving a large quantity of food, a heavily muscled pyloric region for pulverizing food, and a specialized valvelike structure that prevents movement of ingesta into the small intestine until it has been sufficiently reduced. Baker (1943) and Kalmbach (1943) gave the mean volume of stomach contents as 64.2 and 82.3 cm^3, respectively, and Kalmbach (1943) and Fitch et al. (1952) reported mean values of the mass of stomach contents as 62.8 and approximately 50 g, respectively. Also correlated with the high chitin content of the armadillo's diet is the production of the enzyme chitinase in the gastric tissues, which aids in the digestion of chitinous exoskeletons, and the possible presence in the gut of microorganisms that degrade chitin (Smith et al. 1998). Features of the cranial and dental morphology of *Dasypus,* including reduction of jaw musculature, teeth, and facial bones, reflect specialization for a myrmecophagous diet (Wetzel and Mondolfi 1979; Smith and Redford

1990), although the actual diet is much more generalized, especially in North America. Smith and Redford (1990) suggested several evolutionary scenarios, not necessarily exclusive, to account for the discrepancy between the armadillo's morphological specialization for feeding on ants and termites with the much broader variety of foods actually eaten. These include the following possibilities: (1) primitive members of the genus were specialists on ants and termites and developed appropriate morphological specializations, but may be reevolving ecological generality, (2) *D. novemcinctus* has broadened its diet in association with its expansion into North America in response to reduction in ants and termites and increased availability of other potential foods, and (3) the species evolved morphological specialization for a myrmecophagous diet as a response to competition from other litter feeders in Central and South America, but was able to broaden its feeding niche in North America as a result of reduced competition.

Dietary Variation. Major studies of armadillo food habits in various habitats in different parts of its North American range include those of Breece and Dusi (1985) in Alabama; Sikes et al. (1990) in Arkansas; Bushnell (1952), Nesbitt et al. (1979), and Wirtz et al. (1985) in Florida; Fitch et al. (1952) in Louisiana; Zimmerman (1982) in Oklahoma; and Baker (1943) and Kalmbach (1943) in Texas. Complicating detailed comparison of the results of different studies is the use of different methods of analysis, whether frequency of occurrence or percentage of volume. The relative abundance of various taxa in the diet may reflect differences in geographic region, habitat, and season as well as among years in a given locality. An example of an effect of habitat on the composition of the diet is the occurrence of many food items in Florida armadillo stomachs normally associated with aquatic or wetland habitats, including crayfish, leeches, snails, adults or larvae of at least 10 families of insects, and such vertebrates as frogs, water snakes, and fish (Wirtz et al. 1985). Although armadillos have been seen foraging in very shallow water in Florida (J. N. Layne, pers. obs.), aquatic organisms in the diet probably largely reflect foraging in recently flooded areas or in wet margins of aquatic habitats. Differences between studies in the relative importance of a particular taxon in the diet also may reflect some special local condition or rare event, such as emergence of a brood of 17-year cicadas, leading to unusually high incidence of a particular species in the diet (Zimmerman 1982). Seasonal variation in the composition of the diet in different studies ranges from slight to pronounced and may reflect seasonal variation in abundance of different life cycle stages or total populations of different animal groups, fruiting of plant species, or soil moisture as well as the effect of temperature on activity of potential prey species such as cold-blooded vertebrates, which may influence their susceptibility to capture. Wirtz et al. (1985) found seasonal differences in species richness and numbers of prey items in Florida armadillo stomachs. Greatest variety of food items (74) occurred in October–December and highest mean number of prey items per stomach (1535) was in April–June. Seasonal variation in the nutritive properties of the food of armadillos in the wild was demonstrated by Ramsey et al. (1981). They found that the energy content of the food was greatest in summer and fall, the highest percentage of protein and the lowest percentage of ash occurred in spring, and the percentages of fat and fiber were lowest in winter. As an example of the effect of seasonal variation in foods eaten on dietary composition, low fiber in winter was associated with reduction of plant material in stomachs and low ash in spring with a decrease in heavily sclerotinized arthropods. Trends in fecal composition generally followed those for the diet, with males having overall lower fecal energy than females, the difference being most pronounced in fall. Digestive efficiencies estimated for wild and laboratory animals were 84% and 86%, respectively.

Taxa Consumed. The armadillo is an opportunistic feeder. Roughly 90% of the food items consist of invertebrates, primarily insects, occurring in soil, leaf litter, and decaying wood. Small vertebrates and plant material make up the remainder of the diet. The feeding habits study by Wirtz et al. (1985) in Florida, involving 167,039 food items in 186 stomachs, study provides a good example of the generalized nature of the armadillo's diet. Animal foods included 11 orders, 75 families, and 90 genera of insects; 4 phyla, 6 classes, and 10 orders of other invertebrates; and 2 classes, 2 orders, 9 families, and 11 genera of vertebrates plus unidentified fish. An example of the high number of individual prey items that may occur in armadillo stomachs is the 40,000 ants of several species contained in a stomach examined by Kalmbach (1943).

Insects constitute the major component of the diet in all armadillo feeding-habits studies. Insects occurred in 100% of the stomachs analyzed by Wirtz et al. (1985) and Sikes et al. (1990) and made up from 54.0% to 82.9% (mean = 71.3%) of the diet on a volumetric basis in seven other studies (Baker 1943; Kalmbach 1943; Bushnell 1952; Fitch et al. 1952; Nesbitt et al. 1979; Zimmerman 1982; Breece and Dusi 1985). Adult and larval beetles (Coleoptera) made up the largest proportion of the insects consumed in all studies, followed by ants (Hymenoptera) in most cases. Kalmbach (1943) suggested that where ants are abundant they are a preferred food. Redford (1985) speculated that ants and termites may constitute a greater proportion of the armadillo's diet in South and Central America as a result of greater abundance compared with North America. The remaining nine insect orders recorded in armadillo stomachs include Orthoptera (grasshoppers, crickets, and roaches), Lepidoptera (moths and butterflies), Diptera (flies), Isoptera (termites), Hemiptera (true bugs), Homoptera (cicadas and coccids), Dermaptera (earwigs), and Odonata (dragonflies and damselflies). All insect orders in the diet are represented by adult and immature stages (larvae or pupae), with larval stages predominating among Lepidoptera and Diptera. Although there is considerable variation among studies in the ranking of these orders in the diet, Lepidoptera, Diptera, and Orthoptera are consumed most frequently.

Seasonal variation in the taxonomic composition of insects in the diet has been observed in most studies. In Texas, Kalmbach (1943) reported a tendency toward overall reduction of insects in the diet in the colder months. A similar trend was evident in Oklahoma, with beetles and ants showing the greatest seasonal difference in percentage volume (highest in summer, lowest in winter) (Zimmerman 1982). The most marked seasonal shift in insect foods in Arkansas was dominance of dipteran larvae in winter and of adult and larval beetles in the other seasons (Sikes et al. 1990). Although Fitch et al. (1952) did not find striking seasonal trends in insect foods in Louisiana, beetles were relatively more important in summer and fall than in winter and spring, and Lepidoptera and Diptera were most prevalent in spring. Among insect orders occurring in the stomachs of Florida armadillos, Coleoptera as a whole exhibited relatively little seasonal variation, although individual families showed considerable differences in seasonal occurrence. Other orders prominently represented in the diet (Lepidoptera, Hymenoptera, Orthoptera, Diptera) showed highest use in January–June and October–December and lowest use in July–September (Wirtz et al. 1985).

Major invertebrate taxa other than insects ranged from 3.2% to 71.0% (mean = 22.3%) in frequency of occurrence in armadillos from Florida (Wirtz et al. 1985). Invertebrates exclusive of insects had a combined frequency of occurrence of 37.1% in Arkansas (Sikes et al. 1990), and averaged 17.5% (range = 7.9–19.1%) of the volume of stomach contents in other studies (Baker 1943; Kalmbach 1943; Bushnell 1952; Fitch et al. 1952; Nesbitt et al. 1979; Zimmerman 1982; Breece and Dusi 1985). A variety of arthropods other than insects are consumed, including spiders, scorpions, and harvestmen (Arachnida); sowbugs (Isopoda); crayfish (Decapoda); centipedes (Chilopoda); and millipedes (Myriopoda). Other invertebrates besides arthropods recorded in armadillo stomachs include earthworms and leeches (Annelida), slugs and snails (Gastropoda), and roundworms (Nematoda), the latter possibly as endoparasites.

Incidence of annelids in stomachs is probably correlated with soil moisture, and was greatest in January in Texas, in winter in Arkansas, and in spring in Oklahoma and Louisiana (Kalmbach 1943; Fitch et al. 1952; Zimmerman 1982; Sikes et al. 1990). Spiders were significant components of the diet of Florida armadillos in all seasons, whereas earthworms, centipedes, and millipedes were most important in July–September or October–December (Wirtz et al. 1985).

Occurrence of Vertebrates. Although a minor component of the diet, vertebrates occur in all studies of armadillo feeding habits. Mean overall frequency of occurrence of vertebrates was 22.4% in stomachs collected in southern Georgia in winter (Gause 1980), and mean frequency of occurrence of different vertebrate taxa in Florida stomachs was 4.7% (Wirtz et al. 1985). Vertebrates made up 0.6–5.1% of the volume of stomach contents in other studies (Baker 1943; Kalmbach 1943; Bushnell 1952; Fitch et al. 1952; Nesbitt et al. 1979; Zimmerman 1982; Breece and Dusi 1985; Sikes et al. 1990). Small amphibians and reptiles predominate among the vertebrates consumed by armadillos, with secretive or burrowing species tending to be most frequently represented. Genera identified in stomachs include five taxa of salamanders, seven taxa of frogs and toads, eight of lizards, and six of snakes (Table 4.1). The diversity of amphibians and reptiles recorded in the diet of armadillos in Florida exceeds that of any other geographic region. Reptile eggs in stomachs examined by Fitch et al. (1952) were believed to be mostly those of the green anole, but also may have included those of skinks, fence lizards, and racerunners, as well as one presumably from a snake. Lizard eggs, probably those of the green anole, also were found in armadillo stomachs in Alabama (Breece and Dusi 1985). Wirtz et al. (1985) reported unidentified reptile eggs in 7.5% of stomachs from Florida. Kalmbach (1943) found a higher frequency of eggs than individual lizards in stomachs of Texas armadillos, with one case of a stomach containing 12 lizard eggs.

Incidence of amphibians and reptiles in the diet tends to increase in colder weather, suggesting that these poikilotherms are more susceptible to capture when slowed down or inactive at low temperature. Kalmbach (1943) recorded maximum abundance of reptiles in stomachs of armadillos collected in January in Texas. He suggested that more complete sampling in winter might have confirmed a higher proportion of vertebrates taken in that season as a result of increased vulnerability to capture because of cool weather. The highest volumes of

Table 4.1. Amphibians and reptiles recorded from stomachs of the armadillo in different parts of its geographic range

Taxon	Location	Source
Amphibians		
Amphiuma sp. (amphiuma)	Louisiana	Fitch et al. 1952
Ambystoma sp. (ambystomatid salamander)	Texas	Kalmbach 1943
Notophthalmus sp. (newt)	Texas	Kalmbach 1943
Plethodon glutinosus (slimy salamander)	Alabama	Breece and Dusi 1985
Eurycea quadridigitata (dwarf salamander)	Louisiana	Fitch et al. 1952
Unidentified salamanders	Texas	Baker 1943
Scaphiopus sp. (spadefoot toad)	Texas	Kalmbach 1943
Bufo sp. (toad)	Florida	Wirtz et al. 1985
	Texas	Kalmbach 1943
Unidentified toads	Florida	Nesbitt 1978
Acris gryllus (cricket frog)	Florida	Wirtz et al. 1985
Hyla crucifer (spring peeper)	Florida	Gause 1980
Hyla squirella (squirrel treefrog)	Florida	Gause 1980
Hyla sp. (treefrog)	Alabama	Breece and Dusi 1985
	Florida	Wirtz et al. 1985
Pseudacris sp. (chorus frog)	Texas	Kalmbach 1943
Unidentified tree frogs (Hylidae)	Florida	Nesbitt 1978
Gastrophryne carolinensis (eastern narrowmouth toad)	Florida	Wirtz et al. 1985
Gastrophryne sp. (narrowmouth toad)	Texas	Kalmbach 1943
Rana catesbeiana (bullfrog)	Florida	Bushnell 1952
Rana utricularia (southern leopard frog)	Florida	Bushnell 1952
Rana sp. (ranid frog)	Florida	Wirtz et al. 1985
	Texas	Kalmbach 1943
Reptiles		
Rhineura floridana (Florida worm lizard)	Florida	Layne 1976, Gause 1980, Wirtz et al. 1985
Ophisaurus attenuatus (slender glass lizard)	Florida	Wirtz et al. 1985
Anolis carolinensis (green anole)	Alabama	Breece and Dusi 1985
	Florida	Gause 1980, Wirtz et al. 1985
	Louisiana	Fitch et al. 1952
	Texas	Kalmbach 1943
Sceloporus woodi (Florida scrub lizard)	Florida	Layne 1976
Eumeces fasciatus (five-lined skink)	Louisiana	Fitch et al. 1952
Neoseps reynoldsi (sand skink)	Florida	Layne 1976
Scincella lateralis (ground skink)	Florida	Gause 1980, Wirtz et al. 1985
	Louisiana	Fitch et al. 1952
	Texas	Kalmbach 1943
Unidentified skink	Arkansas	Sikes et al. 1990
Unidentified skinks	Florida	Bushnell 1978
Cnemidophorus sexlineatus (six-lined racerunner)	Arkansas	Sikes et al. 1990
	Florida	Gause 1980
	Louisiana	Fitch et al. 1952
Unidentified lizards	Texas	Baker 1943
Leptotyphlops dulcis (Texas blind snake)	Texas	Kalmbach 1943
Virginia striatula (rough earth snake)	Alabama	Breece and Dusi 1985
	Louisiana	Fitch et al. 1952
Diadophis punctatus (ringneck snake)	Florida	Gause 1980
Nerodia sp. (water snake)	Florida	Wirtz et al. 1985
Tantilla relicta (peninsula crowned snake)	Florida	Layne 1976
Tantilla gracilis (flathead snake)	Texas	Kalmbach 1943
Thamnophis sp. (garter or ribbon snake)	Florida	Wirtz et al. 1985

amphibians and reptiles in armadillo stomachs from Alabama (Breece and Dusi 1985) and southern Georgia (Gause 1980) occurred in winter. Fitch et al. (1952) also recorded a substantially higher proportion of amphibians and reptiles in stomachs in winter than at other seasons, although occurrence of reptile eggs was confined to spring and summer. In Florida, greatest diversity of anurans in stomachs occurred from January to March, with the highest number of individuals from April to June. Hatchling to adult reptiles had the highest frequency of occurrence in October–December, whereas 18 of 24 reptile eggs occurred in April–June (Wirtz et al. 1985).

Among other vertebrates, the only record of fish in the diet is that of a single, small, unidentified specimen in stomachs from Florida (Wirtz et al. 1985). Other than young cotton rats (*Sigmodon hispidus*) and "neonatal mice" found in armadillo stomachs in Florida and Arkansas, respectively (Nesbitt et al. 1979; Sikes et al. 1990), mammalian remains in armadillo stomachs have been considered to represent carrion. The incidence of bird remains other than eggs is limited to traces of feather fragments, which could have been ingested accidentally, in two stomachs examined by Kalmbach (1943). Based on stomach contents, incidence of feeding on bird eggs is low. Kalmbach (1943) recorded shell fragments in 3 of 281 stomachs analyzed in Texas, and Wirtz et al. (1985) found shell fragments of unidentified species in only 3 of 186 stomachs from Florida. Also in Florida, Bushnell (1952) recorded remains of bobwhite quail (*Colinus virginianus*) eggs in 6 of 139 stomachs examined. Gause (1980) recorded a chicken embryo in the stomach of an armadillo collected in a farm yard with free-ranging chickens. Although armadillos swallow both shell and contents of quail eggs (Bushnell 1952), they may break open eggs and only eat the contents. Thus, the frequency of shells in stomachs may underestimate the rate of predation on ground-nesting birds (Breece and Causey 1973).

In addition to data from stomach analyses, armadillos have been observed feeding on vertebrates. In Texas, they have been recorded digging up and feeding on reptile eggs, including those of snakes and a turtle, destroying nests of bobwhite quail and wild turkeys (*Meleagris galapavo*) (Kalmbach 1943), and preying on young quail (Lehmann 1934) and small chickens (Kalmbach 1943). Newman and Baker (1942) recorded predation on nestling eastern cottontail rabbits, and captives consumed "young mice" (*Mus musculus*?) (Redford 1985). H. H. Newman (1913) reported apparent cannibalism. In Florida, Gause (1980) found evidence of armadillos excavating and eating eggs in unidentified turtle nests. Bushnell (1952) observed feeding on softshell turtle (*Apalone ferox*) eggs, cited reports of predation on loggerhead turtle (*Caretta caretta*) eggs, and recorded destruction of bobwhite quail and wild turkey nests. Aycrigg et al. (1996) observed an armadillo preying on a black swamp snake (*Seminatrix pygaea*). Armadillos were recorded eating adults and eggs of tree snails (*Liguus*) in southern Florida (Duever et al. 1979). Consumption of carrion in the wild and in captivity was reported by Kalmbach (1943), Clark (1951), and Fitch et al. (1952). An armadillo in Florida was observed feeding on the carcass of an eastern cottontail rabbit (J. N. Layne, pers. obs.).

Vegetation Consumed. Considering plant material in armadillo stomachs, Kalmbach (1943) noted that it is not always possible to distinguish between items that are intentionally consumed and those accidentally ingested. Plant material assumed to have been consumed for its nutritive value occurred in 14.0% of stomachs studied by Wirtz et al. (1985). Vegetation constituted an average of 3.5% (range = 0.5–10.2%) of volume in six other studies (Baker 1943; Kalmbach 1943; Bushnell 1952; Fitch et al. 1952; Nesbitt et al. 1979; Breece and Dusi 1985). Louisiana armadillo stomachs contained appreciably higher amounts of plant material in all seasons compared with studies elsewhere (Fitch et al. 1952). Kalmbach (1943) identified mushrooms and other fungi, particularly puffballs, and fruits of blackberry (*Rubus*), mulberry (*Morus*), French mulberry (*Callicarpa*), plum (*Prunus*), holly (*Ilex*), and juniper (*Juniperus*), the latter possibly accidentally ingested, in armadillo stomachs in Texas. He cited observations of armadillos feeding on persimmons and reports of depredation on cantaloupes (*Cucumis melo*). Hamilton (1946) recorded pulp and seeds of black persimmon in 80% of armadillo stomachs collected in August near San Antonio, Texas. In contrast, Sikes et al. (1990) found persimmon (*D. virginiana*) remains in only one stomach in Arkansas, although specimens were collected in areas where the fruits were abundant. Fitch et al. (1952) identified mushrooms; fruits of dewberry (*Rubus trivialis*), muscadine grape (*Vitis munsoniana*), and white bay (*Magnolia virginiana*); and seeds of longleaf pine (*Pinus palustris*) among plant material in armadillo stomachs in Louisiana. Wild grapes were the only plant foods recorded in Alabama (Breece and Dusi 1985). Plants identified in stomachs in Florida include huckleberries (*Gaylussacia*), fragment of orange skin (*Citrus sinensis*), blackberries (*Rubus* sp.), and saw palmetto berries (*Serenoa repens*) (Bushnell 1952; Nesbitt 1978). B. Crawford (pers. comm.) watched an armadillo feeding on palmetto fruits. It would stand up on its hind legs sniffing at the stalk with fruits, select one, take it in its mouth, apparently chew on it and swallow the seed. In addition, an armadillo was observed feeding on the pulp of rotting oranges lying on the ground (J. N. Layne, pers. obs.). When broken open, watermelons and cantaloupes were eaten by captive armadillos (Taber 1945).

Higher amounts of plant material in the diet were recorded in May, October, and December than in other months in Texas (Kalmbach 1943); in May–June and August–September in Oklahoma (Zimmerman 1982); and in fall in Louisiana (Fitch et al. 1952); these presumably reflect periods of increased availability of fruits. Hamilton (1946) found high use of fruits in late summer in a Texas locality, and Breece and Dusi (1985) recorded a 61.1% frequency of occurrence of wild grapes in armadillo stomachs in Alabama in summer.

As a result of their method of feeding, armadillos often ingest considerable amounts of sand or soil; inedible plant materials such as pine needles, grass, decayed wood fragments, and twigs; and droppings and other material from termite galleries (Kalmbach 1943; Galbreath 1982). Recorded amounts of such debris ranged from 0.3% to 8.8% (mean = 4.5%) of the volume of stomach contents (Baker 1943; Kalmbach 1943; Bushnell 1952; Nesbitt et al. 1979).

Armadillos presumably obtain much of their moisture requirements from the diet, although Taber (1945) believed they required water. He recorded visits to small ponds for drinking and a tendency of individuals to congregate around water holes during periods of drought. In Florida, an armadillo was observed drinking from a dripping water spigot (J. N. Layne, pers. obs.). They drink by lapping the water with their long, protrusible tongues (Taber 1945).

BEHAVIOR

Foraging Behavior. A major portion of the time armadillos are active outside the burrow is devoted to feeding. They tend to begin foraging as they emerge from the burrow and move at a slow pace following an often erratic course. Estimated rates of travel of four individuals while foraging were 0.22, 0.26, 0.54, and 1.05 km/hr, respectively (Lehmann 1934; Layne and Glover 1977). Lehmann (1934) estimated normal travel speed without foraging as slightly less than 1.6 km/hr. Haas (1972) observed two foraging individuals that remained within areas of 225 and 2700 m^2, respectively, during a 1-hr period. An intensively foraging armadillo often can be closely approached and even touched before it reacts to the observer. Prey is apparently detected by smell, although observations by Bushnell (1952) indicated that sound may also play a role. Armadillos capture prey by rooting in the substratum with the snout, assisted, as needed, with vigorous digging with the well-developed claws on the forefeet. Typical foraging activity involves quickly probing here and there with the snout while appearing to sniff almost continually and pausing at intervals to briefly dig for prey. The pits dug by foraging individuals range from about 2.5 to 16.5 cm deep, depending on soil type, and are a characteristic sign of armadillo presence. Foraging pits are often most concentrated in areas with moist soil, presumably reflecting prey availability and/or abundance. In a study of armadillos in an outdoor enclosure, the distribution of foraging pits was significantly correlated with radiolocation foraging records (Moore 1968). Armadillos also probe through leaf litter, sometimes becoming almost completely covered and leaving furrows that mark their path. They tear apart rotting logs and stumps with their strong claws to reach prey. Small invertebrates are captured with the extensible tongue, which has tiny, recurved papillae to which they adhere, and are usually swallowed whole. Larger, hard-bodied insects, small vertebrates such as frogs or lizards, and plant foods such as berries or mushrooms are masticated to varying degree (Kalmbach 1943). The feet and claws may be employed to hold down and tear apart larger food items, such as carrion (Taber 1945). Armadillos often utter soft piglike grunts when engaged in intensive digging or rooting. At intervals, a foraging individual will stop and assume an upright posture supported by the hind legs and tail and sniff the air (Kalmbach 1943; Clark 1951). The bipedal stance may also be used to reach food items such as berries located above ground level (Kalmbach 1943). Sometimes a posture intermediate between quadrupedal and bipedal is assumed; the animal shifts its weight to the hind limbs, with the forefeet only lightly resting on the ground. An armadillo foraging on a steeply sloping spoil bank appeared to use its tail as a prop to keep from sliding backward (J. N. Layne, pers. obs.). Armadillos in areas where they are regularly harassed may be more wary than those in areas where they are disturbed less frequently (Lehmann 1934).

Sensory Acuity. Armadillos appear to rely more heavily on their sense of smell than hearing or vision for information about their environment. As noted above under Home Range and Movements, observations by several observers have indicated a strong reliance by armadillos on olfactory cues for navigating around their home range (Clark 1951; Haas 1972; Jacobs 1979). An experimental study by Loughry and McDonough (1994) demonstrated the ability of armadillos to discriminate between their own odor and that of other individuals, suggesting that this capability may be used for spatial orientation in the wild. Marking behavior involving dragging the anal region on the ground or defecating (presumably depositing scent from the well-developed anal glands) has been observed in single individuals while foraging, in members of a pair, in aggressive interactions, and in breeding males marking the boundaries of their home range (McDonough 1997, 2000). Hearing appears to be relatively acute, as the large ears would suggest. Armadillos sometimes ignore sounds that should seemingly alert them. This is often the case when they are actively foraging and probably reflects their intense concentration on the activity in which they are engaged rather than lack of auditory acuity (Talmage and Buchanan 1954). Haas (1972) noted a tendency for animals to respond more readily to some sounds than others. For example, slight rustling of palmettos by an observer would cause an armadillo to become alert or run off, whereas the sound of a nearby passing train would elicit no response. He also noted that members of a pair paid no attention to the rustling noises made by each other as they foraged close together in heavy palmettos, yet when Haas made a rustling noise from nearby, both armadillos instantly became alert. Individuals once alerted by a sound tend to remain wary for some time afterward. Vision is obviously less important than olfactory or auditory acuity, as reflected in the very small eyes in proportion to the size of the body and the fact that the eyes are often covered when the animal is probing for food. Although armadillos usually react to the movement of an observer if they are not heavily engrossed in rooting, they appear to be unable to detect a motionless observer and may even come up and sniff his or her shoes or blunder into the observer while foraging.

Response to Danger. When alerted to potential danger, an armadillo may take one of several actions, depending on the strength of the stimulus. It may remain motionless for a few moments, then resume foraging if not further disturbed, or it may try to hide under brush or leaves, in a shallow depression, or beneath a log, where it may only be partially concealed (Taber 1945; Clark 1951). A disturbed individual may move to nearby cover with a bouncing gait. If sufficiently alarmed, it may run to its burrow with a surprising burst of speed (Lehmann 1934). When highly alarmed, armadillos leap vertically into the air like a bucking horse before charging off to the burrow. This sudden upward leap may startle a predator enough to give the armadillo a head start in escaping. When held by the tail with feet on the ground, armadillos give powerful bucks in an effort to escape. If held off the ground, they give surprisingly strong jerks by alternately flexing and straightening the body. These actions, coupled with the smooth, tough carapace, may help an armadillo escape if seized by a predator. Although armadillos do not attempt to bite or display any other deliberate defensive behavior when seized by a human or, presumably, a predator, they may scratch with their well-developed claws in their vigorous struggle to escape. When an armadillo is excited, such as when struggling to escape when captured, the anal glands are often protruded and give off a strong odor. Although this might be considered to have a defensive function, there appears to be no evidence of a deterrent effect on potential predators. The shell may afford some defense against smaller predators, although certainly not larger ones such as mountain lions (*Puma concolor*), black bears (*Ursus americanus*), and alligators (*Alligator mississipiensis*), but its primary function is probably to protect the animal in the manner of a cowboy's chaps when it plunges through or forages in dense cover (Talmage and Buchanan 1954). Unlike some other armadillo species, the nine-banded armadillo cannot curl into a compact ball to protect its vulnerable ventral surface.

Climbing and Swimming. Despite their strong specialization for terrestrial and fossorial life, armadillos have been observed climbing up wire netting in captivity (Crandall 1964). On several occasions, wild individuals in Florida unhesitatingly climbed about 0.5 m up livestock fences in order to get through the larger meshes at that height (J. N. Layne, pers. obs.). Armadillos use 2 modes of locomotion in water. They may traverse shallow bodies of water by walking on the bottom completely submerged or with only the back exposed and infrequently with the snout held above water. More often, however, they swim using the typical mammalian dog-paddle, although as a result of their high specific gravity, the body is almost entirely submerged (Kalmbach 1943; Taber 1945). Buoyancy may be increased by ingestion of air to inflate the stomach and intestines, which is accompanied by a distinct change in the shape of the ventral surface of the body from concave to barrel-like (Kalmbach 1943). In one individual the stomach and intestines were inflated to twice normal size and the girth increased about 20% following several swimming tests (Aldous, cited in Kalmbach 1943). Bushnell (1952) watched an armadillo in Florida swim across a river >137 m wide after considerable gulping of air. Lehmann (1934) stated that typical swimming speed was about 0.8 km/hr. Taber (1945) noted that the occurrence of armadillos on Bloodworth Island, Texas, suggested that they occasionally swim in salt water.

In Texas, Taber (1945) noted that tracks in mud around small ponds indicated that armadillos took mud baths, and Taulman (1994) observed an individual apparently bathing in a puddle in a one-lane truck trail in Arkansas. Moseley (1971) observed recently weaned young wallow in a puddle under a dripping faucet, then roll in grass to wipe off the sand. Captives sometimes took baths in their drinking trough on hot days (Taber 1945).

Vocalizations. Christensen and Waring (1980) summarized vocalizations previously reported for armadillos and described two additional types. Sounds include "low wheezy grunts" associated with digging and rooting, a "wheezy grunt" uttered by recently captured individuals, an "audible buzzing noise" resembling an active swarm of bees given when highly alarmed and fleeing, and a "piglike squeal" given by frightened individuals (Kalmbach 1943; Lowery 1974; Clark 1951). Christensen and Waring (1980) recorded a "chuck" sound—a low-frequency (mean peak frequency = 1.85 kHz), low-amplitude (mostly <1.0 kHz) sound of 0.02- to 0.14-sec duration and given as bouts of 5–8 notes. This vocalization was given by the male and female of pairs foraging in close proximity in August, suggesting that it might function as a contact call between paired individuals during the breeding season. McDonough (1997) recorded this vocalization in only 18% (5 of 28) pairs she observed. However, as the vocalization was audible only within 5 m of a pair, the actual frequency with which it is given might have been higher. Another sound produced by armadillos is a "weak purring" heard from a young of the year attempting to nurse from an unrelated female

(Christensen and Waring 1980). Whether this sound was produced vocally or mechanically could not be determined.

Learning. In laboratory tests of learning ability, subjects were unable to discriminate between vertical and horizontal bar patterns in a Y-maze, probably because of visual inability to resolve the bar shapes rather than a lack of discrimination learning ability. Tests involving extended reversal learning and operant conditioning demonstrated that the species was capable of learning simple discrimination and of forming a learning set (Yaksh 1967).

Defecation Sites. The feces (scats) are typically spherical, about the size of a marble, and have the consistency of mud. Fragments of insects and other invertebrates eaten are held together in a matrix of soil ingested during foraging. Feces of animals that have been feeding in moist, muddy areas such as pond margins may contain much mud and have the consistency of a thick liquid. Taber (1945) found no evidence of the use of particular sites for defecation; scats were observed within burrows and scattered along runways leading from a burrow, with no more than two or three together. Jacobs (1979) recorded captive juveniles burying their feces and urinating on the site before covering it up. Taber (1945) observed a partially tame individual dig a hole in an ant bed, defecate in it, and cover the feces. Three free-ranging, recently weaned juveniles observed by Moseley (1971) consistently buried their feces at some distance from where they slept. However, whether wild armadillos routinely bury their feces is not known.

Social Behavior. Armadillos are primarily solitary animals with social behavior mainly restricted to the mother–young group and breeding activity. There is little detailed information on behavioral interactions within family groups. Following weaning, the young forage together for at least part of their first summer and appear to be more or less on their own (Jacobs 1979; Loughry et al. 1998b). Three recently weaned orphaned young observed outdoors and unconfined for 10 days dug their own burrow. Whenever a human observer appeared, the young immediately lined up in single file and closely followed the person, suggesting they may have been imprinted on the mother before her death (Moseley 1971). Loughry and McDonough (1998a) recorded association between littermates in 26.1% of 69 litters. An observation by Neck (1976) suggested that in some instances while the young are still associated with the mother she may leave them to find a good foraging site and then return and lead them to the area. Denson (1979) also observed what appeared to be play behavior among littermates. The young rolled down a slope for a distance of about 1.5 m, then climbed back up and repeated the performance. All four young engaged in this activity, occasionally jostling each other as they rolled down the slope together. Littermates in captivity huddled together when sleeping (Jacobs 1979). In scent discrimination tests, infants exhibited the ability to discriminate between their siblings and an unrelated juvenile (Loughry and McDonough 1994), although in another study involving staged encounters between juveniles there was no difference in behavioral interactions between sibling–sibling and sibling–nonsibling combinations (Loughry et al. 1998a). The fact that odors of siblings can be differentiated is of interest from the standpoint of their identical genotypes, but the adaptive signficance of olfactory kin recognition is not clear.

Lehman (1934) excavated a burrow that contained 12 young, and Jacobs (1979) observed young of two litters <2 months old go into the same burrow, indicating that under some circumstances adult females with young may associate. Tolerance of mothers to the young of other females suggested by these observations is supported by Jacobs's (1979) observation of foster brooding in captive armadillos. Adult females, as well as young of both sexes, when introduced to an unweaned litter, allowed the young to attempt to nurse. Prodöhl et al. (1998), through assessment of maternity and paternity by means of microsatellite DNA markers, found that average geographic distances between inferred parents and offspring were significantly less than between random individuals. There was no difference in distances between assumed mothers and fathers, although other evidence suggested offspring tended to occur closer to their presumed mothers than fathers. Loughry and McDonough (1998a) also observed that juveniles of the same sex were closer to one another than those of opposite sex, suggesting a tendency for littermates to remain in proximity. These data, however, do not necessarily reflect any direct social interactions between related adults and offspring or between littermates.

Adult Pairing. A close association of a male and female as a pair is one of the major types of social interactions among adult armadillos. In a Mississippi population studied by Jacobs (1979), pairing (individuals within 2 m of each other) was observed in all months except March and April. First sightings of different pairs occurred from January to July, and the number of pairs increased progressively from a low in January to a peak in June. Individual males were associated with 1–3 females (means = 1.7 and 1.4 in 2 years) and females with 1 or 2 males (mean = 1.3 in both years). The mean duration of association of pairs was 3.0 months, with a range of 1–4.5 months. Short-term pair associations, in which a given male and female were observed only once, or during a short period, or at widely separated intervals, were most frequent later in the season. The most frequently recorded activity of pairs was foraging, the individuals usually remaining in close proximity and sometimes even in actual contact. Behavior interpreted as play between members of a pair was observed infrequently. It involved the animals jumping at each other, springing into the air, and tumbling and wrestling together on the ground. Haas (1972) observed a pair of armadillos in Florida using the same burrow during the summer.

In a Texas study, McDonough (2000) observed the same pairs over time periods ranging from 2 to 64 days, although in most cases the pair was not together continuously. Most males paired with more than one female during the breeding season, most females paired with a single male, and a few older males monopolized females, indicating a polygynous mating system. Difference in body size of males appeared to be involved in adult pair formation. Paired males were larger than unpaired males, suggesting that pairing may have a mate-guarding function and that larger males are better able to attain access to females (Loughry et al. 1998c). However, genetic evidence and the lack of size differences between fathers and nonfathers of litters indicated that, although smaller males might be less successful in pairing with females, they are not excluded from breeding. Thus, reproductive success in males is essentially random with respect to body size. Genetic evidence for a Florida population did not support a polygynous mating system, suggesting that social organization may differ among armadillo populations. Types of behavior associated with pairing recorded by McDonough (1997) included dorsal touches, sniffing, tail lifting, tail wagging, approaches, retreats, vigilance, and contact. Occurrences of some of these behaviors were significantly associated; for example, dorsal touches on a female by a male elicited tail wagging by the female, which probably relayed olfactory information about her receptivity. Males also were more likely to exhibit contact and sniffing behavior after tail wagging by the female. Males took the initiative in maintaining the pair association, resulting in a cost in lost foraging time.

Another type of male–female interaction recognized by Jacobs (1979) was "running pairs"—males chasing females—which she interpreted as agonistic in nature. If the male overtook the female during such a chase, he might scramble on top of her and she would respond by jumping into the air. Sometimes the male would also jump up and the two individuals would twist and scratch at each other while off the ground.

Aggressive Interactions. Interactions between unpaired neighbors active within 50 m of each other were also recorded by Jacobs (1979). When foraging in the same field, these individuals maintained a uniform distance, indicating they were cognizant of each other. Neighbor associations were most frequent in May and June, with fewer female–female than male–female or male–male associations from July to October. Several observed interactions between female neighbors involved chases and were interpreted as agonistic behavior. In one case, the same individual approached the other twice within a 5-min interval, jumped on its back, and then chased it. At the second encounter, the two females faced off and jumped repeatedly, clawing at each other's venter, before the chase began. Denson (1979) also observed aggressive behavior between adults that involved one individual chasing another away from the area in which it was foraging. Distances of the chases ranged from 40 to 70 m and involved the attacking animal delivering solid kicks with the hind feet to the other's carapace. Although the sexes of the individuals were not known, Denson (1979) suggested that the behavior might have represented aggression or territorial defense by females near parturition or with young.

Males were the aggressor in all agonistic encounters observed by Herbst and Redford (1991) in Florida. The greater overlap of female than male home ranges, including shared use of burrows by females in the overlap area, also reflected lower aggression between females than between males.

McDonough (1994) examined the effects of sex and age on aggression in a wild population in Texas. Adult males and females were equally likely to be the initiators and recipients of aggression, and adults initiated significantly more aggressive interactions than juveniles. Adult females were more aggressive than expected toward juveniles and other adult females during the period of late pregnancy and lactation, whereas adult males exhibited a higher level of aggression toward 1- and 2-year-old individuals. Breeding males were significantly more likely to take the initiative in aggressive interactions than nonbreeders. Younger individuals (juveniles and 1- to 2-year-olds) initiated proportionately more intensive aggressive behavior (fights vs. chases) than adults, and heavier individuals won significantly more aggressive encounters in cases where the two individuals were of different size. These observations suggested that male aggression functions to ensure access to receptive females, whereas female aggression is related to defense of the current litter and promoting dispersal of young from the previous year.

Aggressive behavior observed in the laboratory by Yaksh (1967) typically involved two individuals jockeying for position and ending up side by side. The slightest movement of either animal elicited an attack consisting of throwing the body sideways and kicking with both hind legs. A well-placed kick could move an animal up to 1.8 m. Other possibly aggressive behaviors involved "ear licking" and "chewing" when animals were in close proximity while feeding, or one individual scratching the scapular area of the carapace of another. The last behavior also is observed in the context of mating. In groups of males and females housed together, aggressive behavior was largely restricted to males, with the rare instances of male attacks on females appearing to represent displacement behavior.

Mating Behavior. Mating behavior involves the male persistently following the female and attempting to probe her hindquarters or mount her. If unreceptive, the female kicks backward with her hind legs or simply walks away. Between attempts to mount, the pair may forage quietly side by side. When receptive, the female moves backward toward the male while wagging the tail and hind quarters from side to side, and the male responds by nuzzling her hind quarters (Gause 1980). Knowledge of actual copulation behavior of armadillos in the wild is limited to Newman's (1913) report that copulation occurred with the female on her back because of the carapace and position of the vulva. Yaksh (1967) provided additional details of copulatory behavior in the laboratory. The typical sequence of events involved: (1) a male approaching and sniffing at the anal region of a female; (2) the female assuming a flattened position on the floor and frequently closing her eyes, usually concomitant with short bouts of shivering; (3) increased sniffing and licking of the female's anal region by the male; (4) the male mounting the female's dorsum, generally as he scratched on her scapular shield with his claws, resulting in the female assuming a lordotic position and exposing the anal region to more direct licking; (5) the male scratching on the side of the female's carapace to cause her to turn partially on her side; (6) intromission with the male straddling the female's tail or moving over it to her ventral surface. Whether intromission occurred or not, the male dismounted occasionally to lick and scratch the female's anal region, which eventually caused the female to turn completely on her back with legs in the air. The total duration of copulatory activity varied from about 3 to 15 min, during which time the male either remained mounted or dismounted to lick and scratch at the female's anal region. Copulation involved a series of short, rapid pelvic thrusts usually <5 sec in duration, but rarely continuing for 10–15 sec. Laboratory observations by Haas (1972) of copulatory behavior of a pair of Florida animals closely agree with those of Yaksh (1967). One copulation bout involved three thrusts within 5 sec. The female exhibited more energetic post-copulatory behavior than the male, indulging in much vigorous sniffing of the male and climbing on him, which may have stimulated the male to remount. Once while the male was mounted and the female was on her side, her anal glands were seen to be everted and pulsating. Both observers saw males occasionally attempting to mount females from the front during mating bouts.

The armadillo has also attracted attention as a model for testing such broad ecological and behavioral hypotheses as the Trivers–Willard hypothesis relating sex ratio of young to the condition of females and kin-selected altruistic behavior. Galbreath (1980) found no support for the Trivers–Willard hypothesis in his Florida study of the armadillo. Although on theoretical grounds the genetically identical young of armadillos would appear to be good candidates for selection for cooperative behavior, field studies of the pre- and postweaning behavior and spatial dispersion of litter mates have produced no evidence for kin selection in the species (Jacobs 1979; Prodöhl et al. 1996; Loughry et al. 1998b).

MORTALITY

Prenatal mortality involving the loss of one of the normal four embryos is well documented, and Galbreath (1980) recorded presumed mortality of all four embryos. There are, however, no quantitative data on the incidence of prenatal mortality in wild populations in relation to environmental factors. Knowledge of postnatal mortality and longevity is also sparse. A marked female in Florida originally captured as an adult in February 1973 was last captured in March 1980, giving a minimum age of between 7 and 8 years (Layne, unpublished data). Jacobs (1979) recorded a yearly loss of 93.5% of juveniles and 20.0% of adults in Mississippi. In Alabama, 16 of 26 adults (61.5%) and 11 of 25 juveniles (44.0%) marked in summer–fall in one year were present on the study area the following spring and summer (Breece and Dusi 1985). Based on surveys of carapaces of dead armadillos in Texas, McDonough and Loughry (1997b) found that mortality of juveniles (young of the year) was higher than that of adults. Juveniles made up about 33% of the population, but 66% of the carapaces collected. Crandall (1964) noted that the nine-banded armadillo appeared to be less viable in captivity than South American species and gave maximum survival times of adults in zoos as about 6 1/2 years. Smith and Doughty (1984) noted that some captives have survived, and in rare cases even reproduced, for up to 10 years. Storrs (1978) estimated a minimum life span of 12–15 years based on the long gestation period and age at sexual maturity and noted that several wild-caught individuals in her laboratory colony were assumed to be approximately 11 years old. McDonough (1994) mentioned a captive 22 years of age, which was still alive at the time of her report.

Road Mortality. In areas with roads and highways inhabited by armadillos, road kills may constitute a major source of mortality. Armadillos were the most frequently killed vertebrates (1.18 individuals/100 km) on highways in north-central Florida (Cristoffer 1991). Inbar and Mayer (1999) reported an average of 2.19 road kills/100 km during summer compared with 1.15/100 km in winter in central Florida. Armadillos were the most frequently killed (3.14/100 km) of 20 mammalian species recorded in a 1974–1981 survey of vertebrate road mortality along 347 km of highways in south-central Florida (J. N. Layne, unpublished data). They had a much higher road-kill rate in summer (2.67/100 km) than in winter (0.47/100 km). Bushnell (1952) recorded minimum mortality of 212 individuals from February to September on a 29-km stretch of road in Florida. Jacobs (1979), in a 2-year period, documented road mortality of 4.3% of juveniles and actual and probable mortality of 20.0% and 30.4%, respectively, of adults in Mississippi, with highest frequency of road kills in November and December. Loughry and McDonough (1996) also observed a low incidence of road-killed juveniles compared with adults, probably because of smaller home ranges of juveniles and their tendency to remain in the natal area. The tendency to leap vertically into the air when startled increases the vulnerability of armadillos to being killed by vehicles. A high proportion of road-killed specimens have damage to the back received when they jumped up as a vehicle passed over them.

Predation. Data on the actual extent of predation on armadillos by large carnivores are few, but suggest that predation generally does not play a major role in population regulation in the species. Reports of the Texas Parks and Wildlife Department have listed coyotes (*Canis latrans*), dogs, black bears (*Ursus americanus*), bobcats (*Lynx rufus*), foxes, and raccoons (*Procyon lotor*) as predators on armadillos in Texas, although without actual observations of coyote predation (Smith and Doughty 1984). A red-tailed hawk (*Buteo jamaicensis*) nest in Texas contained remains of a juvenile armadillo (McDonough and Loughry 1997b), which may indicate predation, although red-tail hawks also feed on road-killed mammals. A mountain lion preyed on an armadillo in Texas (Kalmbach 1943). In Oklahoma, armadillo remains occurred in <1% of coyote stomachs examined by Best et al. (1981). The higher frequency of puncture wounds in the carapace of juveniles compared with adults suggested that the higher mortality in young in the Texas study of McDonough and Loughry (1997b) was attributable to predation, although the identity of the predators was not determined. Mortality of juveniles also increased in late summer, when they were becoming more independent of their siblings and perhaps at greater risk of predation. Rollin Baker (Michigan State University, pers. commun., 2001) noted a decline during the 1990s in recently weaned armadillos, which coincided with an increase in feral hogs (*Sus scrofa*) at a site in Texas, suggesting the hogs preyed on young armadillos. Known predators on armadillos in Georgia and Florida include panther, black bear, bobcat, alligator, and dogs (Bushnell 1952; Layne 1976; Carr 1982; Maehr and Brady 1984; Maehr and DeFazio 1985; Shoop and Ruckdeschel 1990; Dalrymple and Bass 1996; Maehr 1997; Roof 1997; Stratman and Pelton 1999). Maehr and DeFazio (1985) found that armadillos were the most common vertebrates in Florida black bear stomachs and scats collected throughout the state. They suggested that where bears occur, they may be important predators on armadillos. Wassmer et al. (1988) found no evidence of armadillos in analyses of scats of bobcats in south-central Florida, where armadillos were abundant. They reported a bobcat stalking an armadillo by the sounds of its foraging, then ignoring it after seeing what it was, suggesting that armadillos are not favored prey of bobcats. Remains of armadillos are frequently found in and around crested caracara (*Caracara plancus*) nests in Florida (J. N. Layne, pers. obs.). Adult remains most likely represent dead individuals scavenged from roads. However, at least some juvenile remains probably represent actual captures because their smaller size makes them easier prey and they are seldom killed on roads.

Climatic Factors. Low winter temperatures cause mortality in armadillo populations in different parts of the range. Fitch et al. (1952) estimated that the Louisiana population was reduced by about 80% by a severe freeze in 1948. Breece and Dusi (1985) recorded the disappearance over winter of 61.5% of 26 marked adults and 44.0% of 25 juveniles on an Alabama study area, although the extent to which the losses represented mortality or movements off the study area was not determined. Although no large-scale mortality from cold weather has been reported in Florida, Layne and Glover (1985) observed emaciated and weakened individuals after prolonged cold periods in south-central Florida. A reported reduction of evidence of armadillos in west Florida following the severe winter of 1976/77 suggested high mortality from cold weather. Bushnell (1952) observed dead armadillos in burrows that were apparently killed by cold during a particularly severe winter in Florida. Jacobs (1979) recorded a mean weight loss of about 10% in adults over an unusually cold winter in Mississippi as well as reduced reproduction the following year. Analysis of the relationships between body mass, fat content, and ambient temperature to survival time for tolerance of starvation indicated that immatures and lean adults have the lowest survival times at low ambient temperatures (McNab 1980). Large fat deposits in pregnant females compensate for increased energetic demands of pregnancy.

Drought conditions resulting in reduction of soil invertebrates or drying of the soil to the point where probing is difficult also may be a cause of increased mortality in armadillo populations, particularly in the more arid regions of the range. Taber (1945) and Clark (1951) cited reduction of populations in areas of Texas as a result of prolonged, severe drought. Also, in a closely monitored Texas population, McDonough and Loughry (1997b) recorded a loss through mortality or emigration of 88% of the adults during the 4 months of the most severe period of a drought. In addition to their direct impact on mortality rate, unusually cold or dry conditions also may have deleterious effects on reproduction.

Environmental Contaminants. Residues from spraying of mirex have been detected in armadillo tissues in Texas and Florida (Wheeler et al. 1975, 1977). Effect of the pesticide on individuals or populations is unknown (Forrester 1992).

Injuries. Evidence of injuries observed by Zimmerman (1982) in a sample of 95 road-killed and 13 captured individuals in Oklahoma included 6 individuals with scars on the carapace and 66 with healed broken tails. These data indicate a relatively high incidence of injuries in wild populations from various causes and suggest that mortality from injuries may be relatively high. Frostbite and predator attacks have been suggested as causes of tail injuries (Galbreath 1980, Zimmerman 1982).

Parasites and Diseases. Known parasites of the armadillo in North American include fungi, viruses, protozoans, helminths, and arthropods. Additional parasite species, particularly among arthropods, have been recorded from armadillos in South America. In contrast to other mammals of comparable size such as opossums, raccoons, and skunks, armadillos have a low incidence of parasites (Taber 1945; Chandler 1954; Forrester 1992). The carapace and sparsely haired venter are factors limiting the ectoparasite fauna (Chandler 1954; Storrs 1971). The armadillo is host to approximately 50 species of parasitic organisms throughout its range (Chandler 1954). The following summary is limited to parasites reported in armadillos in the United States. Renal lesions, largely associated with the cortex, that could not be associated with any parasite were present in 68% of 50 armadillos examined in Louisiana (Stuart et al. 1977).

Wenker et al. (1998) described an infection of the dimorphic fungus *Sporothrix schenckii* in an adult armadillo, which represented the third case known for the species, all of which involved specimens originating from Florida. The fungus exists as a mold on vegetation or in soils and as a yeastlike form in infected animals. The infected specimen exhibited ulcerative skin lesions on the carapace, multiple hemorrhages in the lungs, and an enlarged spleen. Sporotrichosis, a chronic disease principally involving skin and subcutaneous tissues, caused by *S. schenckii* has been reported from a wide range of mammals, including humans. Texas armadillos have a high incidence of granulomatous lesions in the lungs, most likely caused by *Emmonsia crescens* (Anderson and Benirschke 1966).

The first case of naturally acquired rabies in the armadillo in the United States was reported from Texas in 1987 (Leffingwell and Neill 1989), and antibodies of Everglades virus and St. Louis encephalitis virus have been detected in peninsular Florida populations (Forrester 1992). Armadillos may be involved in St. Louis encephalitis virus amplification and transmission cycles in Florida (Day et al. 1995).

The first report of a naturally acquired leprosy (*Mycobacterium leprae*) infection in the armadillo was in 1975 (Walsh et al. 1975). Areas in which the disease is known to occur in armadillos include (prevalence in parentheses) Louisiana (6.8, 10%), Texas (4.7%), Mississippi (0.5%), and Mexico (1.0%), whereas surveys for leprosy in Alabama, Georgia, and Florida yielded no positive cases (Howerth et al. 1990; Forrester 1992). Hypotheses offered to account for the apparent geographic differences in the incidence of leprosy in wild armadillo populations include the following: (1) sample sizes from different parts of the range are inadequate and/or biased, (2) the higher incidence in Louisiana reflects the fact that the disease has long been endemic in humans in that state, and/or that the habitat may be more favorable for the survival of the bacteria in the soil or conducive to transmission, and (3) armadillos in Louisiana may have greater genetic susceptibility to leprosy infections (Storrs 1978). A genetic basis for the apparent difference in susceptibility to leprosy infections in different parts of the range also was suggested by Ramsey and Grigsby (1985). They noted that the southwestern population and the Florida population, presumably derived from introductions, differed in the alleles for isocitrate dehydrogenase. They suggested that the appearance of leprosy in the presently leprosy-free Florida population following contact between the two populations, as reflected in the appearance of the southwestern isocitrate dehydrogenase allele in the original Florida range, would provide circumstantial evidence for the inheritance-of-susceptibility hypothesis. Low body temperature presumably related to an inefficient immune system may be a significant factor in increasing the susceptibility of the armadillo to the leprosy bacillus and other pathogens (Purtilo et al. 1974). Leprosy in armadillos is usually severe and fatal, and the symptoms tend to develop more rapidly than in humans (Storrs 1982). There is no evidence of transmission of leprosy in armadillos to humans (Filice et al. 1977).

Five other types of bacterial infection have been reported from armadillos in Florida (Forrester 1992). These include 10 serovars, 3 of which are also known from Louisiana, of *Leptospira interrogans; Francisella tularensis,* the etiologic agent of tularemia; *Edwardsiella tarda,* possibly associated with hemorrhagic enteritis in some species; five serovars of *Salmonella* spp.; and *Notocardia* sp., a fungus-like bacterium known to cause respiratory disease in some mammals and most likely acquired by armadillos from contaminated soil through digging activity.

Protozoan parasites known from armadillos in North America include *Toxoplasma gondii, Trypanosoma cruzi,* and the etiologic agent of equine protozoal myeloencephalitis. Nineteen percent of armadillos in a Florida sample were positive for *T. gondii,* the cause of toxoplasmosis. Serologic evidence of *T. cruzi* infection, the cause of Chagas disease, was found in 1.7% of armadillos sampled in Florida and in 25% of a sample from the New Orleans area in Louisiana (Forrester 1992). According to Anderson and Benirschke (1966), armadillos in Texas usually have infections of sarcosporidia in the tongue, diaphragm, and probably other muscle tissues.

The fluke *Brachylaemus virginianus,* in which the infective stage occurs in land snails, was recorded from Texas (Chandler 1946). Records of tapeworms in armadillos from Texas include larvae of *Oochoristica* sp. encysted in mesenteries, *Mathevotaenia surinamensis,* and a questionable record of "*Taenia* sp." (Chandler 1954; Buchanan 1956). Three, and possibly four, species of acanthocephalans occur in Texas armadillos, including *Hamanniella* (?) *tortuosa; Oncicola canis,* typically found in dogs and coyotes; *Echinorhynchus* sp.; and possibly *Echinopardalis* sp. (Chandler 1946, 1954). All but the first species occurred as encysted larvae in mesenteries. Nematodes known from the armadillo in North America include the oxyurid *Aspidodera fasciata* and species of the spiruroid genera *Physaloptera, Physocephalus,* and *Ascarops* in Texas (Chandler 1946) and *Aspidodera* sp. in Florida (Herbst and Greiner in Forrester 1992). In addition, a case of gnathostomiasis (meningocephalitis caused by the aberrant migration of the larval form of the spiruroid *Gnathostoma* sp.) was documented in a wild-caught armadillo from Louisiana (Cockman-Thomas et al. 1993). Fitch et al. (1952) speculated that armadillos in Louisiana might be a vector of trichinosis as a result of becoming infected with *Trichinella spiralis* through feeding on dead hogs.

Chandler (1954) and Storrs (1971) noted the low incidence of ectoparasite infestations in armadillos. Although 13 ectoparasitic arthropods have been recorded from armadillos throughout the range, only 3 species are apparently known from the United States. Hightower et al. (1953) collected a specimen of the flea *Echidnophaga gallinacea* from a single armadillo in Texas. In Florida, Baskerville and Francis (1981) observed a mangelike condition, presumably contracted in the wild, associated with mites (*Echimyopus dasypus*), and recorded a second mite (*Ornithonyssus* sp.) not associated with skin lesions. Herbst (in Forrester 1992) also observed a mangelike condition in 2 of 130 armadillos examined in north-central Florida, which also may have been caused by *E. dasypus,* although no specimens were recovered from scrapings. Infestations of nymphs of the hypopial mite, *Marsupialichus johnstoni,* in and around the ears of armadillos were reported from Louisiana (Pence 1973). The ticks *Amblyomma concolor* and *A. pseudoconcolor,* which appear to be true parasites of the armadillo (Chandler 1954), apparently have not been reported in the United States.

ECONOMIC STATUS, MANAGEMENT, AND CONSERVATION

Field Study Methods. Armadillos are not readily captured in baited mammal live traps, but unbaited live traps set to block the burrow entrance so that the animal is forced to enter the trap when exiting are reasonably effective. Open-ended live traps with planks extending at a 20- to 45-deg angle from the entrance and placed where armadillos travel or forage also have been used (Marsh and Howard 1978; Chamberlain 1980). A mesh net placed in burrow entrances (Zimmerman 1982) and a trap of South American design (Carter et al. 1981) have also been used in armadillo studies in the United States. Capturing armadillos by hand is generally more effective than trapping. Individuals foraging during the day or night usually can be closely approached and caught by hand or with a long-handled net. If they are alarmed and running for cover, they can sometimes be overtaken and captured, but once they have entered the burrow, they are difficult to extract even with a firm grip on the tail. Trained dogs also have been used to capture armadillos (Anderson and Benirschke 1966). Armadillos subjected to repeated capture attempts may become more easily alarmed and less readily observed. Although armadillos do not attempt to bite when captured or handled, they can inflict severe scratches with their claws in their struggles to escape. An effective way to handle an armadillo is to hold it upside down with a tight hold on the tail near the base with one hand and a firm grip on the carapace with the other. Three intramuscular anesthetic combinations (tiletamine/zolazepam, ketamine/xylazine, ketamine/medetomidine) tested by Fournier-Chambrillon et al. (2000) proved safe and effective for immobilizing armadillos, with ketamine/medetomidine being preferable under field conditions. Collecting methods for studies requiring data from dead specimens include shooting, hand capture and sacrifice, and salvage of road kills.

Armadillos have been marked for permanent identification with numbered metal ear tags, passive integrated transponder tags, tatoos in the ears, and numbered plastic tags or metal bands attached to the edge of the carapace. For individual identification in the field, disks of different colored plastic attached to the ears, patches of paint on different parts of the carapace, numbers painted on the carapace, different color combinations of plastic tape wrapped around the base of the tail, reflective tape, and natural identifying marks such as scars, damaged ears, or stub tails have been used (Taber 1945; Clark 1951; Fitch et al. 1952; Layne and Glover 1977; Jacobs 1979; Zimmerman 1982; Suttkus and Jones 1999; McDonough 2000). Radio transmitters have been implanted in the body cavity or attached to the back of the pelvic shield or tail at or near its base (Moore 1968; Jacobs 1979; Thomas 1980; Zimmerman 1982; Herbst 1991; Herbst and Redford 1991).

Care in Captivity. Captive armadillos require housing either outdoors or indoors under conditions that do not exceed their thermal tolerance. Pens or cages containing a nest box and a floor covered by sand, soil, or straw to absorb excreta and to avoid injury to the feet and claws provide satisfactory housing conditions (Anderson and Benirschke 1966). Storrs et al. (1989) housed armadillos singly in stacked plastic dog kennels. Various diets have been provided for captives including milk, eggs, and finely chopped raw meat; a combination of canned dog food supplemented with raw chopped meat and cod-liver oil; commercial cat chow with a vitamin–mineral–protein supplement; canned fish, milk, and dry high-protein dog food supplemented with insects; and canned dog food (Crandall 1964; Anderson and Benirschke 1966; Moore 1968; Ramsey et al. 1981). Drinking water is required for armadillos in captivity (Taber 1945; Crandall 1964; Anderson and Benirschke 1966). Armadillos have proved difficult to breed in captivity, and apparently the first clearly documented cases of successful captive breeding were reported by Storrs and Burchfield (1987). Armadillos born in zoos have a low survival rate (Crandall 1964).

Deleterious Impacts. In Texas, armadillos damage cantaloupes, presumably penetrating the rind with their claws (Kalmbach 1943). Damage to such crops as corn or peanuts through their digging for insects may be cases where armadillos are more beneficial through destruction of insect pests than harmful to the plants (Kalmbach 1943). They have been implicated as predators on eggs and/or young of bobwhite quail, turkey, and other ground-nesting birds, although evidence indicates that they do not have a significant impact on populations of these species (Lehmann 1934; Kalmbach 1943; Bushnell 1952; Breece and Causey 1973; Baker 1999). Other deleterious impacts of armadillos in various parts of the range include damage to lawns, golf courses, vegetable gardens, and ornamental vegetation from their probing for prey; weakening and promoting erosion of ditch banks, dikes, and levees through burrow construction; and burrowing under houses and farm buildings (Fitch et al. 1952; Chamberlain 1980; Layne 1997). A major source of complaints is the damage to lawns, shrubs, and other plants in urban or suburban settings, although the overall extent of the problem is comparatively modest. For example, over a 5-year period in 12 metropolitan areas in Texas, 362 damage complaints were received with a total estimated loss of $20,316.50 (Chamberlain 1980). Burrows may be a danger to livestock when exposed by land clearing (Chamberlain 1980), and some Florida ranchers consider armadillo burrows a hazard to horses and riders (Layne 1997). In addition, their foraging activity is believed to disrupt the structure and productivity of the leaf-mold layer of forest soils, and they serve as known or potential vectors or hosts of diseases and parasites of humans and domestic mammals (Chamberlain 1980; Carr 1982; Forrester 1992; Layne 1997). In Florida, armadillos are known to prey on the eggs or adults of endemic or rare and endangered species such as the scrub lizard (*Sceloporus woodi*), sand skink (*Neoseps reynoldsi*), peninsula crowned snake (*Tantilla relicta*), loggerhead turtle, and tree snail (Bushnell 1952; Duever et al. 1979; Layne 1997). In some areas in Florida, armadillos serve as a major mammalian blood source for the mosquito *Culex nigripalpus*, a vector of human encephalitis (Edman and Taylor 1968). Galbreath (1982) speculated that armadillos sometimes may actively overturn gopher tortoises or evict them and take over their burrows. However, the only documented cause of gopher tortoises being overturned in the wild is as a consequence of fighting between males (Hailman et al. 1991). Disruption of archaeological sites also can be added to the list of deleterious impacts of armadillos. A survey in the Big Cypress National Preserve in Florida revealed damage to 90 of 402 midden sites or burial mounds from armadillo burrowing, with the amount of destruction ranging from 10% to 100% (J. E. Ehrenhard, letter to J. Stevenson, 1981). Problems with armadillos and control measures were reviewed by Chamberlain (1980).

Positive Impacts. Armadillos serve as prey for native predators such as the endangered Florida panther (*P. c. coryi*). In Florida, road-killed armadillos are an increasingly important food resource for scavengers such as the crested caracara, black vulture (*Coragyps atratus*), and turkey vulture (*Cathartes aura*). Other sources of carrion have been reduced because of improved ranching practices, screwworm control, and changing land-use patterns (Layne 1997). Many of the arthropods eaten by armadillos are insect pests (Kalmbach 1943), including, in the southeast, love bug larvae (*Plecia nearctica*) and fire ants (*Solenopsis invicta*) (Moseley 1971; Storrs 1982; Breece and Dusi 1985; Layne 1997). Burrows are used by various native invertebrate and vertebrate species, including the eastern indigo snake, a federally listed Threatened Species, and the Florida mouse, a state-listed Species of Special Concern (Taber 1945; Layne 1990; Layne and Steiner 1996). Armadillo burrows also may serve as an important temporary refuge from wildland fires for non-burrow-dwelling snakes, lizards, and small mammals as well as various invertebrates. Taber (1945) noted that decomposing nest material in abandoned burrows or cleaned out of burrows following flooding added humus to the soil. The armadillo's digging activity may contribute to the breakdown and addition of dead wood to the soil and create soil disturbance favorable to the establishment of plants beneficial to wildlife (Fitch et al. 1952).

Baskets, lampshades, and other novelty items are made from the dried "shell." Newman (1913) noted that one dealer shipped 40,000 baskets in a 6-year period. Commercial use of such products has declined in recent years. The meat, which resembles pork in taste, is now used to a limited extent in outdoor cookery, although during the depression years of the 1930s armadillos were a staple food in rural Texas, being referred to as "poor man's hog" or "Hoover hogs" (Kalmbach 1943; Smith and Doughty 1984). The meat has been canned for human consumption and used as poultry food (Kalmbach 1943). Baker (1971) provided a recipe for armadillo sausage.

Armadillos have been captured and raised commercially for their products and for sale to zoos, research scientists, and as pets (Smith and Doughty 1984). The armadillo enjoys considerable popularity in Texas. Its image appears on T-shirts, plates, and coffee mugs; it serves as an unofficial mascot of the University of Texas at Austin; it appears on television as a symbol for a brand of beer; and armadillo races are popular events in various communities (Chamberlain 1980).

The discovery that the dermal ossicles accumulate strontium-90, presumably through the food chain, raises the possibility of using the armadillo as a biological indicator of the level of radioactive fallout (Jackson et al. 1972). The species also may be useful for monitoring pesiticide residues in the southern United States by virtue of its position at the top of the soil invertebrate food chain (Wheeler et al. 1975, 1977). Since precipitation and temperature are major factors determining the limits of its range, the armadillo is a prime candidate for monitoring effects of climatic warming.

The armadillo has been an important subject for biomedical research. Leprosy-free armadillos from Florida have played an important role in research on leprosy; and Lepromin, a reagent used for prognosis of leprosy in humans, is produced from tissues of infected armadillos (Storrs 1982). Anderson and Benirschke (1966) summarized a number of studies attesting to its considerable potential as a model in experimental biology. Examples of research on the armadillo that have broad implications include the studies of its penile structure (Kelly 1997, 1999), which are applicable to mammals in general. These studies have increased understanding of the hydrostatic systems found in other vertebrates and invertebrates. Nakakura and Lasley (1985) suggested that the armadillo might serve as a model for the study of the fetal pituitary tropic factors that likely control the fetal adrenal and gonadal steroid production during the middle–late stages of gestation.

RESEARCH AND MANAGEMENT NEEDS

As noted by Galbreath (1982), a great deal remains to be learned about many aspects of the basic life history and ecology of the armadillo. More detailed data on the interactions between armadillos and indigenous species are necessary to fully evaluate its impacts, both positive and negative, on native biota and environments. Comparative studies, such as that of Loughry and McDonough (1998b), involving Florida and Brazilian populations, of armadillos at the periphery and in the core areas of the range in the United States would be particularly useful in providing further insight into the factors involved in range expansion and the possible evolutionary implications of differences between populations.

Additional research is needed to determine whether the introduction-derived armadillo population of Florida was free from leprosy infection as a result of natural immunity or the absence of a source of the bacteria in soil. Continued monitoring of the Florida population to detect the appearance of leprosy infections is particularly important now that the gap between the introduced and expanding western populations has closed. The degree to which genetic relatedness is reflected in behavioral interactions and population biology is a question of considerable scientific importance. Although previous studies have not produced clear evidence of kin selection in armadillos, further investigation of the behavioral and ecological implications of clonal reproduction in the armadillo is warranted. Ideally, such studies should involve long-term monitoring of related and unrelated individuals for comparative data on behavioral interactions, reproductive success, longevity, home range size, dispersal, and other relevant factors. The potential of the armadillo as a bioindicator of global warming warrants more intensive and systematic surveys to more precisely document future changes in its range boundaries.

The armadillo's low body temperature and metabolic rate, unique reproductive characteristics, physiological adaptations for digging, and other aspects of its biology, including a potentially long life span, make it a valuable subject for biomedical research. Its usefulness would be further increased if husbandry techniques could be developed to ensure reliable breeding in the laboratory.

LITERATURE CITED

Anderson, J. M., and K. Benirschke. 1966. The armadillo, *Dasypus novemcinctus,* in experimental biology. Laboratory Animal Care 16:202–16.

Audubon, J. J., and J. Bachman. 1849. Quadrupeds of North America, Vol. 3. George R. Lockwood, New York.

Asdell, S. A. 1965. Reproduction and development. Pages 1–41 *in* W. V. Mayer and R. G. Van Gelder, eds. Physiological mammalogy, Vol. 2. Academic Press, New York.

Aycrigg, A. D., T. M. Farrell, and P. G. May. 1996. *Seminatrix pygaea pygaea* (black swamp snake) predation. Herpetological Review 27:84.

Bailey, H. H. 1924. The armadillo in Florida and how it reached there. Journal of Mammalogy 5:264–65.

Bailey, V. 1931. Mammals of New Mexico. North American Fauna 53:1–412.

Baker, R. H. 1943. May food habits of armadillos in eastern Texas. American Midland Naturalist 29:379–80.

Baker, R. H. 1971. Armadillo sausage. Pages 8–9 *in* L. A. Douglas, ed. The explorers cookbook. Caxton Printers, Caldwell, ID.

Baker, R. H. 1999. Texans clamored for quail in the 1930s. East Texas Historical Association 37:42–46.

Baskerville, A., and L. Francis. 1981. Mange in newly-imported armadillos (*Dasypus novemcinctus*). Laboratory Animals 15:305–7.

Benirschke, K., R. J. Low, and V. H. Ferm. 1969. Cytogenetic studies of some armadillos. Pages 330–44 *in* K. Benirschke, ed. Comparative mammalian cytogenetics. Springer-Verlag, New York.

Best, T. L., B. Hoditschek, and H. H. Thomas. 1981. Foods of coyotes (*Canis latrans*) in Oklahoma. Southwestern Naturalist 26:67–69.

Bond, B. T., M. I. Nelson, and R. J. Warren. 2001. Home range dynamics and den use by armadillos on Cumberland Island, Georgia. Abs. No. 301. Abstracts of the 81st Annual Meeting of the American Society of Mammalogists. Missoula, Montana.

Breece, G. A., and M. K. Causey. 1973. Armadillo depredation on "dummy" bobwhite quail nests in southwest Alabama. Proceedings of the Annual Conference of Southeastern Game and Fish Commissioners 27:18–22.

Breece, G. A., and J. L. Dusi. 1985. Food habits and home ranges of the common long-nosed armadillo *Dasypus novemcinctus* in Alabama. Pages 419–27 *in* G. G. Montgomery, ed. The evolution and ecology of armadillos, sloths, and vermilinguas. Smithsonian Institution Press, Washington, DC.

Buchanan, G. D. 1956. Occurrence of the cestode *Mathevotaenia surinamensis* (Cohn, 1902) Spasskii, 1951 in a North American armadillo. Journal of Parasitology 42:34–38.

Buchanan, G. D. 1957. Variation in litter size of nine-banded armadillos. Journal of Mammalogy 38:529.

Buchanan, G. D. 1958. The current range of the armadillo *Dasypus novencinctus mexicanus* in the United States. Texas Journal of Science 10:349–51.

Buchanan, G. D. 1967. The presence of two conceptuses in the uterus of a nine-banded armadillo. Journal of Reproductive Fertility 13:329–331.

Burns, T. A., and E. B. Waldrip. 1971. Body temperature and electrocardiographic data for the nine-banded armadillo (*Dasypus novemcinctus*). Journal of Mammalogy 52:472–73.

Burns, T. A., G. Chen, and J. C. Lin. 1975. Electrophoretic patterns and kinetics of heart and muscle lactate dehydrogenase of the nine-banded armadillo, *Dasypus novemcinctus*. Comparative Biochemistry and Physiology 51A:385–87.

Bushnell, R. III. 1952. The place of the armadillo in Florida wildlife communities. M.S. Thesis, Stetson University, DeLand, FL.

Cabrera, A. 1958. Catalogo de los mamiferos de America del Sur. Revista Museo de Naturales *"Bernardino Rivadavia"* e Instituto Nacional de Investigation de las Ciencias Naturales Zoologia 4:1–107.

Carr, A. F. 1982. Armadillo dilemma. Animal Kingdom 85:40–43.

Carter, T. S., J. H. Shaw, and B. P. Glass. 1981. Capture, handling, and marking of armadillos and anteaters. Abs. No. 166, Abstracts of the Annual Meeting of the American Society of Mammalogists. Miami University, Oxford, OH.

Chamberlain, P. A. 1980. Armadillos: Problems and control. Pages 163–69 *in* J. P. Clark, ed. Proceedings of the 9th Vertebrate Pest Conference. University of California, Davis.

Chandler, A. C. 1946. Helminths of armadillos, *Dasypus novemcinctus,* in eastern Texas. Journal of Parasitology 32:237–41.

Chandler, A. C. 1954. Parasites of armadillos. Pages 43–66 *in* R. V. Talmage and G. D. Buchanan. The armadillo (*Dasypus novemcinctus*): A review of its natural history, ecology, anatomy, and reproductive physiology (Rice Institute Pamphlet, Monograph in Biology, Vol. 41, No. 2). Rice University, Houston, TX.

Christensen, C. G., and G. H. Waring. 1980. The "chuck" sound of the nine-banded armadillo (*Dasypus novemcinctus*). Journal of Mammalogy 61:737–38.

Clark, W. K. 1951. Ecological life history of the armadillo in the Edwards Plateau region. American Midland Naturalist 46:337–58.

Cleveland, A. G. 1970. The current geographic distribution of the armadillo in the United States. Texas Journal of Science 22:90–92.

Cockman-Thomas, R. A., C. A.Colleton, C. H.Gardiner, and W. M.Meyers. 1993. Gnathostomiasis in a wild-caught nine-banded armadillo (*Dasypus novemcinctus*). Laboratory Animal Science 43:630–32

Crandall, L. S. 1964. The management of wild mammals in captivity. University of Chicago Press, Chicago.

Cristoffer, C. 1991. Road mortalities of northern Florida vertebrates. Florida Scientist 54:65–68.

Dalrymple, G. H., and O. L.Bass, Jr. 1996. The diet of the Florida panther in Everglades National Park, Florida. Bulletin of the Florida State Museum of Natural History 39:173–93.

Day, J. F., E. E. Storrs, L. M. Stark, A. L. Lewis, and S. Williams. 1995. Antibodies to St. Louis encephalitis virus in armadillos from southern Florida. Journal of Wildlife Diseases 31: 10–14.

Denson, R. D. 1979. Aggression and tumbling among armadillos. Southwestern Naturalist 24:697–98.

Duever, M. J., J. E. Carlson, J. F. Meeder, L. C. Duever, L. H. Gunderson, L. A. Riopelle, T. R. Alexander, R. L. Myers, and D. P. Spangler. 1979. The Big Cypress National Preserve. (Research Report 8). National Audubon Socicty, Ncw York.

Edman, J. D., and D. J. Taylor. 1968. *Culex nigripalpus:* Seasonal shift in the bird–mammal feeding ratio in a mosquito vector of human encephalitis. Science 161:67–68.

Ehrhart, L. M. 1976. Mammal studies (Final report to the National Aeronautics and Space Administration Kennedy Space Center). Office of Graduate Studies and Research, Florida Technological University, Orlando.

Eisenberg, J. F. 1961. Observations on the nest-building behavior of armadillos. Proceedings of the Zoological Society of London 137:322–24.

Eisenberg, J. F. 1981. The mammalian radiations. University of Chicago Press, Chicago.

Enders, A. C. 1966. The reproductive cycle of the nine-banded armadillo (*Dasypus novemcinctus*). Pages 295–310 *in* I. W. Rowlands, ed. Comparative biology of reproduction in mammals. Academic Press, London.

Ferrari,C. C., P. D.Carmanchahi, H. J. Aldana Marcos, and J. M. Affani. 2000. Ultrastructural characterisation of the olfactory mucosa of the armadillo *Dasypus hybridus* (Dasypodidae, Xenarthra). Journal of Anatomy 196:269–78.

Filice, G. A., R. N. Greenberg, and D. W. Fraser. 1977. Lack of observed association between armadillo contact and leprosy in humans. American Journal of Tropical Medicine and Hygiene 26:137–39.

Fitch, H. S.,P. Goodrum, and C. Newman. 1952. The armadillo in the southeastern United States. Journal of Mammalogy 33:21–37.

Forrester, D. J. 1992. Parasites and diseases of wild mammals of Florida. University Press of Florida, Gainesville.

Fournier-Chambrillon, C. I. Vogel, P. Fournier, B. de Thoisy, and J.-C. Vié. 2000. Immobilization of free-ranging nine-banded and great long-nosed armadillos with three anesthetic combinations. Journal of Wildlife Diseases 36:131–40.

Freeman, P. W., and H. H. Genoways. 1998. Recent northern records of the nine-banded armadillo (*Dasypodidae*) in Nebraska. Southwestern Naturalist 43:491–504.

Galbreath, G. J. 1980. Aspects of natural selection in *Dasypus novemcinctus*. Ph.D. Dissertation, University of Chicago, Chicago.

Galbreath, G. J. 1982. Armadillo. Pages 71–79 *in* J. A. Chapman and G. A. Feldhamer, eds. Wild mammals of North America. Johns Hopkins University Press, Baltimore.

Gardner, A. L. 1993. Order Xenarthra. Pages 63–68 *in* D. E. Wilson and D. M. Reeder, eds. Mammal species of the world, 2nd ed. Smithsonian Institution Press, Washington, DC.

Gaudin, T. J., and A. A. Biewener. 1992. The functional morphology of xenarthrous vertebrae in the armadillo *Dasypus novemcinctus* (Mammalia, Xenarthra). Journal of Morphology 214:63–81.

Gause, G. E. 1980. Physiological and morphometric responses of the nine-banded armadillo (*Dasypus novemcinctus*) to environmental factors. Ph.D. Dissertation, University of Florida, Gainesville.

Greegor, D. H., Jr. 1975. Renal capabilities of an Argentine desert armadillo. Journal of Mammalogy 56:626–32.

Greegor, D. H., Jr. 1985. Ecology of the little hairy armadillo *Chaetophractus vellerosus*. Pages 397–405 *in* G. G. Montgomery, ed. The evolution and ecology of armadillos, sloths, and vermilinguas. Smithsonian Institution Press, Washington, DC.

Haas, W. E. 1972. Observations on the nine-banded armadillo, *Dasypus novemcinctus,* in south central Florida. Unpublished report, Archbold Biological Station, Lake Placid, FL.

Hailman, J. P., J. N. Layne, and R. Knapp. 1991. Notes on aggressive behavior of the gopher tortoise. Herpetological Review 22:87–88.

Hall, E. R. 1955. Handbook of mammals of Kansas (Miscellaneous Publication 7). University of Kansas, Museum of Natural History, Lawrence.

Hall, E. R. 1981. The mammals of North America, Vol. 1, 2nd ed. John Wiley, New York.

Hamilton, W. J., Jr. 1946. The black persimmon as a summer food of the Texas armadillo. Journal of Mammalogy 27:175.

Hamlett, G. W. D. 1932. The reproductive cycle in the armadillo. Zeitschrift für Wissenschaftliche Zoologie 141:143–57.

Hendricks, L. J. 1963. Observation of armadillo in east-central New Mexico. Journal of Mammalogy 44:581.

Herbst, L. H. 1991. Pathological and reproductive effects of intraperitoneal telemetry devices on female armadillos. Journal of Wildlife Management 55:628–31.

Herbst, L., and K. Redford. 1991. Home range among female armadillos. Research and Exploration 7:236–37.

Herbst, L. H., A. I. Webb, R. M. Clemmons, M. R. Dorsey-Lee, and E. E. Storrs. 1989. Plasma and erythrocyte cholinesterase values for the common long-nosed armadillo, *Dasypus novemcinctus*. Journal of Wildlife Diseases 25:364–69.

Hightower, B. G., V. W. Lehman, and R. B. Eads. 1953. Ectoparasites from mammals and birds on a quail preserve. Journal of Mammalogy 34:268.

Holler, N. R., R. M. Dawson, T. Simons, and M. C. Wooten. 1989. Reestablishment of the Perdido Key beach mouse (*Peromyscus polionotus trissyllepsis*) on Gulf Island National Seashore. Conservation Biology 3:397–404.

Hollister, N. 1925. The systematic name of the Texas armadillo. Journal of Mammalogy 6:60.

Howerth, E. W.,D. E. Stallknecht, W. R. Davidson, and E. J. Wentworth. 1990. Survey for leprosy in nine-banded armadillos (*Dasypterus novemcinctus*) from the southeastern United States. Journal of Wildlife Diseases 26:112–15.

Humphrey, S. R. 1974. Zoogeography of the nine-banded armadillo (*Dasypus novencinctus*) in the United States. BioScience 24:457–62.

Inbar, M., and R. T. Mayer. 1999. Spatio-temporal trends in armadillo diurnal activity and road-kills in central Florida. Wildlife Society Bulletin 27:865–72.

Jackson, C. G., Jr., C. M. Holcomb, and M. M. Jackson. 1972. Strontium-90 in the exoskeletal ossicles of *Dasypus novemcinctus*. Journal of Mammalogy 53:921–22.

Jacobs, J. F. 1979. Behavior and space usage patterns of the nine-banded armadillo (*Dasypus novemcinctus*) in southwestern Mississippi. M.S. Thesis, Cornell University, Ithaca, NY.

Johansen, K. 1961. Temperature regulation in the nine-banded armadillo (*Dasypus novemcinctus*). Physiological Zoology 34:126–44.

Jorge, W., A. T. Orsi-Souza, and R. Best. 1985. The somatic chromosomes of Xenarthra. Pages 121–29 *in* G. G. Montgomery, ed. The evolution and ecology of armadillos, sloths, and vermilinguas. Smithsonian Institution Press, Washington, DC.

Kalmbach, E. R. 1943. The armadillo: Its relationship to agriculture and game. Game, Fish, and Oyster Commission, Austin, TX.

Kelly, D. A. 1997. Axial orthogonal fiber reinforcement in the penis of the nine-banded armadillo (*Dasypus novemcinctus*). Journal of Morphology 233:249–55.

Kelly, D. A. 1999. Expansion of the tunica albuginea during penile inflation in the nine-banded armadillo (*Dasypus novemcinctus*). Journal of Experimental Biology 292:253–65.

Klippel, W. E., and P. W. Parmalee 1984. Armadillos in North American late Pleistocene contexts. Pages 149–60 *in* H. H. Genoways and M. R. Dawson, eds. Contributions in Quaternary vertebrate paleontology: A volume in memorial to John E. Guilday (Special Publication 8). Carnegie Museum of Natural History, Pittsburgh, PA.

Layne, J. N. 1976. The armadillo: One of Florida's oddest animals. Florida Naturalist 42:8–12.

Layne, J. N. 1984. The land mammals of south Florida. Pages 269–96 *in* P. J. Gleason, ed. Environments of south Florida present and past, 2nd ed. Miami Geological Society, Coral Gables, FL.

Layne, J. N. 1990. The Florida mouse. Pages 1–21 *in* C. K. Dodd, Jr., R. E. Ashton, Jr., R. Franz, and E. Wester, eds. Burrow associates of the gopher tortoise. Proceedings of the 8th Annual Meeting of the Gopher Tortoise Council. Gainesville, FL.

Layne, J. N. 1997. Nonindigenous mammals. Pages 157–86 *in* D. Simberloff, D. C. Schmitz, and T. C. Brown, eds. Strangers in paradise: Impact and management of nonindigenous species in Florida. Island Press, Washington, DC.

Layne, J. N., and D. Glover. 1977. Home ranges of the armadillo in Florida. Journal of Mammalogy 58:411–13.

Layne, J. N., and D. Glover. 1985. Activity patterns of the common long-nosed armadillo *Dasypus novemcinctus* in south-central Florida. Pages 407–17 *in* G. G. Montgomery, ed. The evolution and ecology of armadillos, sloths, and vermilinguas. Smithsonian Institution Press, Washington, DC.

Layne, J. N., and T. M. Steiner. 1996. Eastern indigo snake (*Drymarchon corais couperi*): Summary of research conducted on Archbold Biological Station. U.S. Fish and Wildlife Service, Jackson, MI.

Layne, J. N., and A. M. Waggener, Jr. 1984. Above-ground nests of the nine-banded armadillo in Florida. Florida Field Naturalist 12:58–61.

Leffingwell, L. M., and S. L. Neill. 1989. Naturally acquired rabies in an armadillo (*Dasypus novemcinctus*) in Texas. Journal of Clinical Microbiology 27:174–75.

Lehmann, V. 1934. Armadillo investigations June 1–June 30, 1934. Denver Food Habits Research Laboratory, U.S. Fish and Wildlife Service, Denver, CO.

Loughry, W. J., and C. M. McDonough. 1994. Scent discrimination by infant nine-banded armadillos. Journal of Mammalogy 75:1033–39.

Loughry, W. J., and C. M. McDonough. 1996. Are road kills valid indicators of armadillo population structure? American Midland Naturalist 135:53–59.

Loughry, W. J., and C. M. McDonough. 1998a. Spatial patterns in a population of nine-banded armadillos (*Dasypus novemcinctus*). American Midland Naturalist 140:161–69.

Loughry, W. J., and C. M. McDonough. 1998b. Comparisons between nine-banded armadillo (*Dasypus novemcinctus*) populations in Brazil and the United States. Revista de Biologia Tropical 46:1173–83.

Loughry, W. J.,G. M. Dwyer, and C. M. McDonough. 1998a. Behavioral interactions between juvenile nine-banded armadillos (*Dasypus novemcinctus*) in staged encounters. American Midland Naturalist 139:125–32.

Loughry, W. J., A. Prodöhl, C. M. McDonough, and J. C. Avise. 1998b. Polyembryony in armadillos. American Scientist 86:274–80.

Loughry, W. J., A. Prodöhl, C. M. McDonough, W. S. Nelson, and J. C. Avise. 1998c. Correlates of reproductive success in a population of nine-banded armadillos. Canadian Journal of Zoology 76:1815–21.

Lowery, G. H., Jr. 1974. The mammals of Louisiana and its adjacent waters. Louisiana State University Press, Baton Rouge.

Maehr, D. S. 1997. The Florida panther: Life and death of a vanishing carnivore. Island Press, Washington, DC.

Maehr, D. S., and J. R. Brady. 1984. Food habits of Florida black bears. Journal of Wildlife Management 48:230–35.

Maehr, D. S., and J. T. DeFazio, Jr. 1985. Foods of black bears in Florida. Florida Field Naturalist 13:8–12.

Manaro, A. J. 1961. Observations on the behavior of the spotted skunk in Florida. Quarterly Journal of the Florida Academy of Sciences 24:59–63.

Marsh, R. E., and W. E. Howard. 1978. Armadillos: Pest control. Vertebrate Pest Control Manual 46:22–23.

Martinez, M., F. E. Martinez, and Il-Sei Watanabe. 1998. Light and scanning electron microscopic study of the palatine mucosa of nine-banded armadillo (*Dasypus novemcinctus*). European Journal of Morphology 36:97–104.

McBee, K., and R. J. Baker. 1982. *Dasypus novemcinctus*. Mammalian Species 162:1–9.

McCusker, J. S. 1985. Testicular cycles of the common long-nosed armadillo *Dasypus novemcinctus* in north central Texas. Pages 255–61 *in* G. G. Montgomery, ed. The evolution and ecology of armadillos, sloths, and vermilinguas. Smithsonian Institution Press, Washington, DC.

McDonough, C. M. 1994. Determinants of aggression in nine-banded armadillos. Journal of Mammalogy 75: 189–94.

McDonough, C. M. 1997. Pairing behavior of the nine-banded armadillo (*Dasypus novemcinctus*). American Midland Naturalist 138:290–98.

McDonough, C. M. 2000. Social organization of nine-banded armadillos (*Dasypus novemcinctus*) in a riparian habitat. American Midland Naturalist 144:139–51.

McDonough, C. M., and W. J. Loughry. 1997a. Influences on activity patterns in a population of nine-banded armadillos. Journal of Mammalogy 78: 932–41.

McDonough, C. M., and W. J. Loughry. 1997b. Patterns of mortality in a population of nine-banded armadillos, *Dasypus novemcinctus*. American Midland Naturalist 138:299–305.

McDonough, C. M.,S. A. McPhee, and W. J. Loughry. 1998. Growth rates of juvenile nine-banded armadillos. Southwestern Naturalist 43:462–68.

McNab, B. K. 1979. The influences of body size on the energetics and distribution of fossorial and burrowing mammals. Ecology 60:1010–21.

McNab, B. K. 1980. Energetics and the limits to a temperate distribution in armadillos. Journal of Mammalogy 61:606–27.

McNab, B. K. 1985. Energetics, population biology, and distribution of xenarthrans, living and extinct. Pages 219–32 *in* G. G. Montgomery, ed. The evolution and ecology of armadillos, sloths, and vermilinguas. Smithsonian Institution Press, Washington, DC.

Moore, A. M. 1968. A radiolocation study of armadillo foraging with respect to environmental variables. Ph.D. Dissertation, University of Texas, Austin.

Moseley, K. 1971. The armored dullard. Florida Wildlife 24:4–6.

Nakakura, K. C., and B. L. Lasley. 1985. Fetal endocrinology of the common long-nosed armadillo *Dasypus novemcinctus*. Pages 247–53 *in* G. G. Montgomery, ed. The evolution and ecology of armadillos, sloths, and vermilinguas. Smithsonian Institution Press, Washington, DC.

Neck, R. W. 1976. Possible adaptive significance of certain aspects of armadillo foraging behavior. Southwestern Naturalist 21:242–43.

Nesbitt, S. 1978. Armadillo: Extending its range in Florida since 1920. Florida Wildlife 31:14–16.

Nesbitt, S. A., W. M. Hetrick, L. E. Williams, Jr., and D. H. Austin. 1979. Foods of the nine-banded armadillo in Florida. Proceedings of the Annual Conference of the Southeastern Association of Fish and Wildlife Agencies 31:57–61.

Newman, C. C., and R. H. Baker. 1942. Armadillo eats young rabbits. Journal of Mammalogy 23:450.

Newman, H. H. 1913. The natural history of the nine-banded armadillo in Texas. American Naturalist 47:513–39.

Newman, H. H., and J. T. Patterson. 1910. The development of the nine-banded armadillo from primitive streak to birth, with special references to the question on polyembryony. Journal of Morphology 21:359–423.

Owen, R. 1830. Notes on the anatomy of the nine-banded armadillo (*Dasypus peba* Desin). Proceedings of the Zoological Society of London 1: 141–44.

Pence, D. B. 1973. Notes on two species of hypopial nymphs of the genus *Marsupialichus* (Acarina: Glycyphagidae) from mammals in Louisiana. Journal of Medical Entomology 10:329–32.

Price, W. A. 1942. Certain recent geological and biological changes in south Texas, with consideration of probable causes. Proceedings of the Texas Academy of Science 26:138–56.

Prodöhl, P. A., W. J. Loughry, C. M. McDonough, W. S. Nelson, and J. C. Avise. 1996. Molecular documentation of polyembryony and micro-spatial dispersion of clonal sibships in the nine-banded armadillo, *Dasypus novemcinctus*. Proceedings of the Royal Society of London B 263:1643–49.

Prodöhl, P. A., W. J. Loughry, C. M. McDonough, W. S. Nelson, E. A. Thompson, and J. C. Avise. 1998. Genetic maternity and paternity in a local population of armadillos assessed by microsatellite DNA markers and field data. American Naturalist 151:7–19.

Prudom, A. E., and W. R. Klemm. 1973. Electrographic correlates of sleep behavior in a primitive mammal, the armadillo *Dasypus novemcinctus*. Physiology and Behavior 10:275–82.

Purtilo, D. T., G. P. Walsh, E. E. Storrs, and I. S. Banks. 1974. Impact of cool temperature and transformation of human and armadillo lymphocytes (*Dasypus novemcinctus* Linn.) as related to leprosy. Nature 248: 450–52.

Ramsey, P. R., and B. A. Grigsby. 1985. Protein variation in populations of *Dasypus novemcinctus* and comparisons to *D. hybridus, D. sabanicola* and *Chaetophractus villosus*. Pages 131–41 *in* G. G. Montgomery, ed. The evolution and ecology of armadillos, sloths, and vermilinguas. Smithsonian Institution Press, Washington, DC.

Ramsey, P. R., D. F. Tyler, Jr., J. R. Waddil, and E. E. Storrs. 1981. Blood chemistry and nutritional balance of wild and captive armadillos (*Dasypus novemcinctus* L.). Comparative Biochemistry and Physiology 69A: 517–21.

Redford, K. H. 1985. Food habits of armadillos (Xenarthra: Dasypodidae). Pages 429–37 *in* G. G. Montgomery, ed. The evolution and ecology of armadillos, sloths, and vermilinguas. Smithsonian Institution Press, Washington, DC.

Roof, J. C. 1997. Black bear food habits in the lower Wekiva River basin of central Florida. Florida Field Naturalist 25:92–97.

Ryder, O. A., and R. L. Davis. 1985. Genetic comparison of *Dasypus novemcinctus* and *D. hybridus* based on electrophoretic studies. Pages 143–46 *in* G. G. Montgomery, ed. The evolution and ecology of armadillos, sloths, and vermilinguas. Smithsonian Institution Press, Washington, DC.

Schnell, P. T. 1994. Home range, activity period, burrow use, and body temperatures of the nine-banded armadillo (*Dasypus novemcinctus*) on the northern edges of its range. M.S. Thesis, Southwest Missouri State University, Springfield [cited from McDonough 2000].

Scholander, P. F.,L. Irving, and S. W. Grinnell. 1943. Respiration of the armadillo, with possible implications as to its burrowing. Journal of Cellular and Comparative Physiology 21:53–63.

Shoop, C. R., and C. A. Ruckdeschel. 1990. Alligators as predators on terrestrial mammals. American Midland Naturalist 124:407–12.

Sikes, R. S., G. A. Heidt, and D. A. Elrod. 1990. Seasonal diets of the nine-banded armadillo (*Dasypus novemcinctus*) in a northern part of its range. American Midland Naturalist 123:383–89.

Slaughter, B. H. 1961. The significance of *Dasypus bellus* (Simpson) in Pleistocene local faunas. Texas Journal of Science 13:311–15.

Smith, K. K., and K. H. Redford. 1990. The anatomy and function of the feeding apparatus in two armadillos (Dasypoda): Anatomy is not destiny. Journal of the Zoological Society of London 222:27–47.

Smith, L. L., and R. W. Doughty. 1984. The amazing armadillo. University of Texas Press, Austin.

Smith, S. A., L. W. Robbins, and J. G. Steiert. 1998. Isolation and characterization of a chitinase from the nine-banded armadillo, *Dasypus novemcinctus*. Journal of Mammalogy 79:486–91.

Stangl, F. B., S. L. Beauchamp, and N. G. Konermann. 1995. Cranial and dental variation in the nine-banded armadillo, *Dasypus novemcinctus*, from Texas and Oklahoma. Texas Journal of Science 47:89–100.

Stevenson, H. M., and R. L. Crawford. 1974. Spread of the armadillo into the Tallahassee–Thomasville area. Florida Field Naturalist 2:8–10.

Storrs, E. E. 1967. Individuality in monozygotic quadruplets of the armadillo *Dasypus novemcinctus*. Ph.D. Dissertation, University of Texas, Austin.

Storrs, E. E. 1971. The nine-banded armadillo: A model for leprosy and other biomedical research. International Journal of Leprosy 39:703–14.

Storrs, E. E. 1978. The life and habitat of the *Dasypus novemcinctus*. Pages 3–11 *in* The armadillo as an experimental model in biomedical research (Scientific Publication 366). Pan American Health Organization, Washington, DC.

Storrs, E. E. 1982. The astonishing armadillo. National Geographic 161: 820–30.

Storrs, E. E., and H. P. Burchfield. 1987. Successful breeding of the common long-nosed armadillo (*Dasypus novemcinctus*) in captivity. Abs. No. 87, Abstracts of the 67th Annual Meeting of the American Society of Mammalogists. Albuquerque, NM.

Storrs, E. E., and H. P. Burchfield. 1990. New armadillo species? Abs. No. 24, Abstracts of the 70th Annual Meeting of the American Society of Mammalogists. Frostburg, MD.

Storrs, E. E., G. H. D'Addamio, and J. D. Rousssel. 1977. Seasonal variation in semen quality in feral and colony adapted nine-banded armadillos. Abs. No. 163, Abstracts of the 57th Annual Meeting of the American Society of Mammalogists. East Lansing, MI.

Storrs, E. E., H. P. Burchfield, and J. W. Rees. 1988. Superdelayed parturition in armadillos: A new mammalian survival strategy. Leprosy Review 59:11–15.

Storrs, E. E., H. P. Burchfield, and J. W. Rees. 1989. Reproduction delay in the common long-nosed armadillo, *Dasypus novemcinctus* L. Advances in Neotropical Mammalogy 1989:535–48.

Stratman, M. R., and M. R. Pelton. 1999. Feeding ecology of black bears in northwest Florida. Florida Field Naturalist 27:95–102.

Strecker, J. K. 1927. A possible albino armadillo. Journal of Mammalogy 8:60.

Stuart, B. P., W. A. Crowell, W. V. Adams, and J. C. Carlisle. 1977. Spontaneous renal disease in Louisiana armadillos (*Dasypus novemcinctus*). Journal of Wildlife Diseases 13:240–44.

Suttkus, R. D., and C. Jones. 1999. Observations on the nine-banded armadillo, *Dasypus novemcinctus,* in southern Louisiana. Tulane Studies in Zoology and Botany 31:1–22.

Taber, F. W. 1945. Contribution to the life history and ecology of the nine-banded armadillo. Journal of Mammalogy 26:211–26.

Talmage, R. V., and G. D. Buchanan. 1954. The armadillo (*Dasypus novemcinctus*): A review of its natural history, ecology, anatomy, and reproductive physiology (Monograph in Biology Vol. 41, No. 2). Rice University, Houston, TX.

Taulman, J. F. 1994. Observations of nest construction and bathing behaviors in the nine-banded armadillo, *Dasypus novemcinctus.* Southwestern Naturalist 39:378–80.

Taulman, J. F., and L. W. Robbins. 1996. Recent range expansion and distributional limits of the nine-banded armadillo (*Dasypus novemcinctus*) in the United States. Journal of Biogeography 23:635–48.

Taylor, W. P. 1946. Armadillos abundant in Kerr County, Texas. Journal of Mammalogy 27:273.

Thomas, W. D. 1980. Summer movement of a male armadillo in central Texas. Texas Journal of Science 32:363–65.

Walsh, G. P., E. E. Storrs, H. P. Burchfield, E. H. Cottrell, M. F. Vidrine, and C. H. Binford. 1975. Leprosy-like disease occurring naturally in armadillos. Journal of the Reticuloendothelial Society 18:347–51.

Wassmer, D. A., D. D. Guenther, and J. N. Layne. 1988. Ecology of the bobcat in south-central Florida. Bulletin of the Florida State Museum, Biological Sciences 33:159–228.

Weiss, L. P., and G. B. Wislocki. 1956. Seasonal variations in hematopoiesis in the dermal bones of the nine-banded armadillo. Anatomical Record 126:143–62.

Wenker, C. J., L. Kaufman, L. N. Bacciarini, and N. Robert. 1998. Sporotrichosis in a nine-banded armadillo (*Dasypus novemcinctus*). Journal of Zoo and Wildlife Medicine 29:474–78.

Wetzel, R. M. 1985. The identification and distribution of Recent Xenarthra (= Edentata). Pages 5–21 *in* G. G. Montgomery, ed. The evolution and ecology of armadillos, sloths, and vermilinguas. Smithsonian Institution Press, Washington, DC.

Wetzel, R. M., and E. Mondolfi. 1979. The subgenera and species of long-nosed armadillos, genus *Dasypus.* Pages 43–63 *in* J. F. Eisenberg, ed. Vertebrate ecology in the northern neotropics. Smithsonian Institution Press, Washington, DC.

Wheeler, R. L., D. F. Friday, H. P. Burchfield, and E. E. Storrs. 1975. Use of the nine-banded armadillo for sampling soil insects for pesticide residue analysis. Environmental Quality and Safety 3(Suppl.):129–34.

Wheeler, W. B., D. P. Jouvenaz, D. P. Wojcik, W. A. Banks, C. H. VanMiddelen, C. S. Lofgren, S. Nesbitt, L. Williams, and R. Brown. 1977. Mirex residues in nontarget organisms after application of 10–5 bait for fire ant control, Northeast Florida—1972–74. Pesticide Monitoring Journal 11:146–56.

Wilber, C. G. 1964. Electrocardiogram of the armadillo. Journal of Mammalogy 45:642.

Wirtz, W. O., D. H. Austin, and G. W. Dekle. 1985. Food habits of the common long-nosed armadillo *Dasypus novemcinctus* in Florida, 1960–61. Pages 439–51 *in* G. G. Montgomery, editor. The evolution and ecology of armadillos, sloths, and vermilinguas. Smithsonian Institution Press, Washington, DC.

Wislocki, G. B. 1933. Location of the testes and body temperatue in mammals. Quarterly Review of Biology 8:385–96.

Wolfe, J. L. 1968. Armadillo distribution in Alabama and northwest Florida. Quarterly Journal of the Florida Academy of Sciences 31:209–12.

Yaksh, T. L. 1967. Observations of behavior and learning in the armadillo (*Dasypus novemcinctus mexicanus*). M.S. Thesis. Georgia Institute of Technology, Athens, Georgia.

Zimmerman, J. W. 1982. The common long-nosed armadillo (*Dasypus novemcinctus*) in northcentral Oklahoma. M.S. Thesis, Oklahoma State University, Stillwater.

James N. Layne, Archbold Biological Station, Lake Placid, Florida 33852. Email: jlayne@strato.net.

II

Rabbits and Hares

5

Eastern Cottontail

Sylvilagus floridanus and Allies

Joseph A. Chapman
John A. Litvaitis

NOMENCLATURE

Common Names. Eastern cottontail, Florida cottontail
Scientific Name. *Sylvilagus floridanus*
Subspecies North of Mexico. *S. f. similis, S. f. mearnsii, S. f. llanensis, S. f. alacer, S. f. mallurus, S. f. hitchensi. S. f. floridanus, S. f. ammophilus, S. f cognatus, S. f. chapmani, S. f. holzneri, S. f. hesperius,* and *S. f. paulsoni*

Common Names. Desert cottontail, Audubon's cottontail
Scientific Name. *Sylvilagus audubonii*
Subspecies North of Mexico. *S. a. audubonii, S. a. vallicola, S. a. sanctidiegi, S. a. arizonae, S. a. baileyi, S. a. cedrophilus, S. a. neomexicanus, S. a. minor,* and *S. a. parvulus*

Common Name. Brush rabbit
Scientific Name. *Sylvilagus bachmani*
Subspecies North of Mexico. *S. b. ubericolor, S. b. tehamae, S. b. macrorhinus, S. b. riparius, S. b. mariposae; S. b. bachmani, S. b. virgulti,* and *S. b. cinerascens*

Common Names. Nuttall's cottontail, mountain cottontail
Scientific Name. *Sylvilagus nuttallii*
Subspecies. *S. n. nuttallii, S. n. grangeri,* and *S. n. pinetis*

Common Names. Swamp rabbit, canecutter
Scientific Name. *Sylvilagus aquaticus*
Subspecies. *S. a. aquaticus* and *S. a. littoralis*

Common Name. Marsh rabbit
Scientific Name. *Sylvilagus palustris*
Subspecies. *S. p. paludicola, S. p. palustris,* and *S. p. hefneri*

Common Name. New England cottontail
Scientific Name. *Sylvilagus transitionalis*
Subspecies. There are currently no recognized subspecies of *S. transitionalis*

Common Name. Appalachian cottontail
Scientific Name. *Sylvilagus obscurus*
Subspecies. There are currently no recognized subspecies of *S. obscurus*

Common Name. Davis Mountain cottontail
Scientific Name. *Sylvilagus robustus*
Subspecies. There are currently no recognized subspecies of *S. robustus*

Common Name. Pygmy rabbit
Scientific Name. *Brachylagus idahoensis*
Subspecies. There are currently no recognized subspecies of *B. idahoensis*

Subspecific designations are based on Hall (1981).

DISTRIBUTION

Cottontails are widely distributed throughout the United States and extreme southern Canada. The most widely distributed of the cottontails is *S. floridanus*. It inhabits diverse areas over broad geographic provinces from southern Canada through the United States into Mexico and further south. This rabbit occurs sympatrically with eight species of *Sylvilagus,* six species of *Lepus,* and *Brachylagus*. The range of no other species of rabbit overlaps those of so many other leporids (Chapman et al. 1980; Chapman and Flux 1990; Wilson and Ruff 1999). The eastern cottontail has been widely transplanted, and for this reason subspecific designations are somewhat meaningless, particularly in eastern North America (Chapman and Morgan 1973). This species has also expanded its range northward, particularly in New England. Populations of eastern cottontails have been introduced and established in Washington (Dalquest 1941) and Oregon (Graf 1955) (Fig. 5.1).

Audubon's cottontail is found throughout the arid southwest United States and the high deserts into northern Montana (Fig. 5.2). The mountain cottontail occurs in the intermountain region, and its range broadly overlaps that of Audubon's cottontail in Montana, Wyoming, Utah, and Colorado (Fig. 5.3). The ranges of both these species are stable, although the eastern cottontail seems to be displacing *S. nuttallii* in southwestern North Dakota (Genoways and Jones 1972). The Davis Mountain cottontail is found in the Davis Mountains, Jeff Davis County, Texas (Fig. 5.4). Only recently was *S. robustus* described as a distinct species (Ruedas 1998).

The brush rabbit is confined to the Pacific coast of North America south of the Columbia River (Fig. 5.4). The range of the species has not changed in recent years. The pygmy rabbit is found mainly in the Great Basin region and is associated with big sagebrush (*Artemisia tridentata*). There is an isolated population of pygmy rabbits in southeastern Washington, the status of which is poorly known (Dobler and Dixon 1990).

Swamp and marsh rabbits are confined to the southeastern United States (Fig. 5.5). The range of the marsh rabbit has apparently changed little from historic times. However, the swamp rabbit's range has begun to diminish southward, apparently due to drainage of wetlands and other habitat alterations. Potential swamp rabbit habitat in Missouri decreased from 850,000 ha in 1870 to <40,000 ha by 1973, primarily due to the conversion of lowland hardwood forests to row crops (Korte and Fredrickson 1977).

New England cottontails are restricted to disjunct populations west of the Hudson River to southern Maine (Litvaitis and Litvaitis 1996). Apparently, the ancestral leporid of *S. transitionalis* and *S. obscurus* retreated southward during the cooling climate and vegetative changes associated with the Pleistocene glaciation (Chapman et al. 1992). When the warming trend began and vegetative changes reversed, the ancestral leporid in its extended range may have been vulnerable to competition from the more ubiquitous *S. floridanus. S. obscurus* was gradually restricted to mountain balds and higher elevations that simulated more northern environments (Fig. 5.6). *S. transitionalis* was confined to the northern climes of New England (Chapman et al. 1992). Since 1960, the range of the New England cottontails has been reduced by

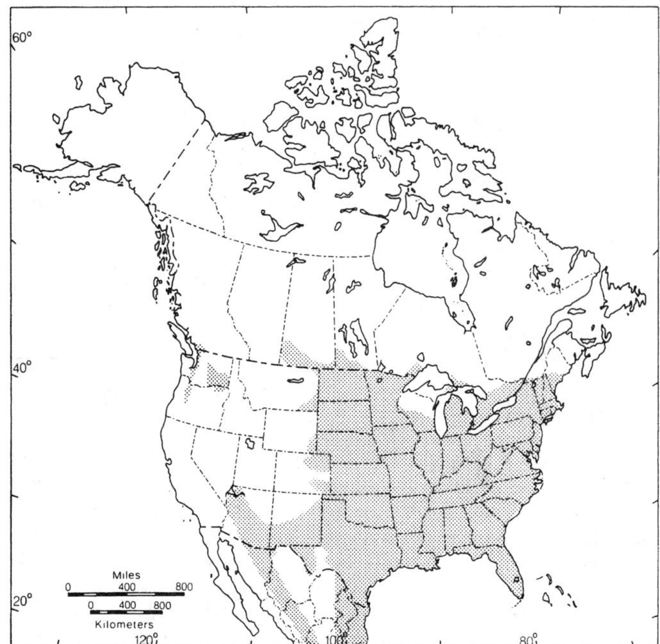

FIGURE 5.1. Distribution of the eastern cottontail (*Sylvilagus floridanus*).

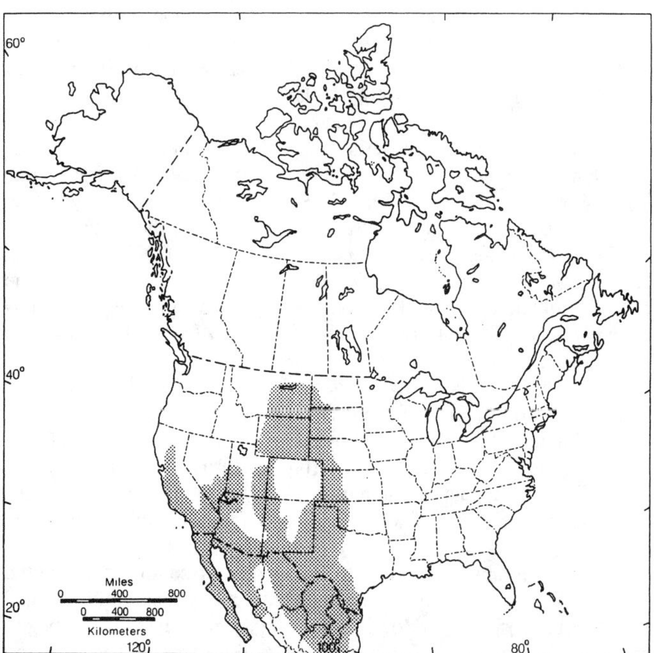

FIGURE 5.2. Distribution of the desert cottontail (*Sylvilagus audubonii*).

approximately 75% (Litvaitis and Litvaitis 1996), prompting discussions by federal and state agencies to consider listing it as threatened.

DESCRIPTION

Cottontails are true rabbits; the young are altricial and born naked in a nest. They are members of the family Leporidae, which includes four North American genera: *Sylvilagus,* the cottontails; *Lepus,* the hares and jackrabbits; *Brachylagus,* the monotypic pygmy rabbit of the Great Basin; and *Romerolagus,* the monotypic volcano rabbit (*R. diazi*) found only on volcanic slopes in central Mexico. The genera *Sylvilagus, Brachylagus,* and *Romerolagus* are all restricted to the New World.

Cottontails vary in size from the small *B. idahoensis* to the largest member of the genus, *S. aquaticus*. Ranges of body measurements for the genus are as follows: total length 250–538 mm, tail length

FIGURE 5.3. Distribution of Nuttall's cottontail (*Sylvilagus nuttallii*).

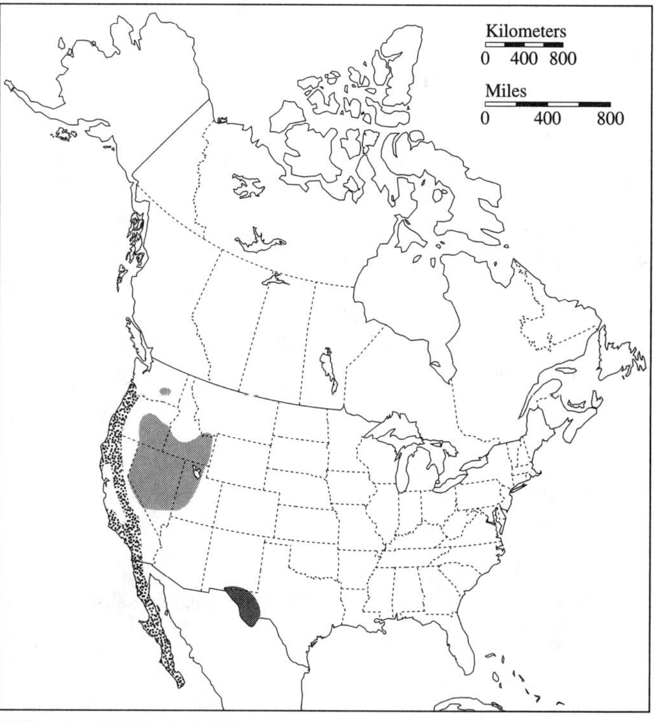

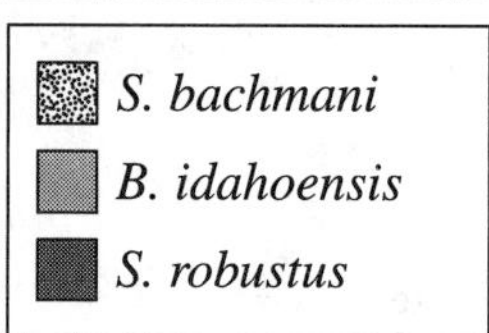

FIGURE 5.4. Distribution of the brush rabbit (*Sylvilagus bachmani*), Davis Mountain cottontail (*S. robustus*), and pygmy rabbit (*B. idahoensis*).

18–73 mm, hind foot length 65–110 mm, and ear length from notch (dry) 36–74 mm. All cottontails have relatively large ears and feet. The skulls of *Sylvilagus* and *Brachylagus* are typically rabbitlike, with a highly fenestrated maxillary bone. The animals possess a supraorbital process of the frontal bone, a straight cutting edge on the upper incisors,

FIGURE 5.5. Distribution of the swamp rabbit (*Sylvilagus aquaticus*) and marsh rabbit (*S. palustris*).

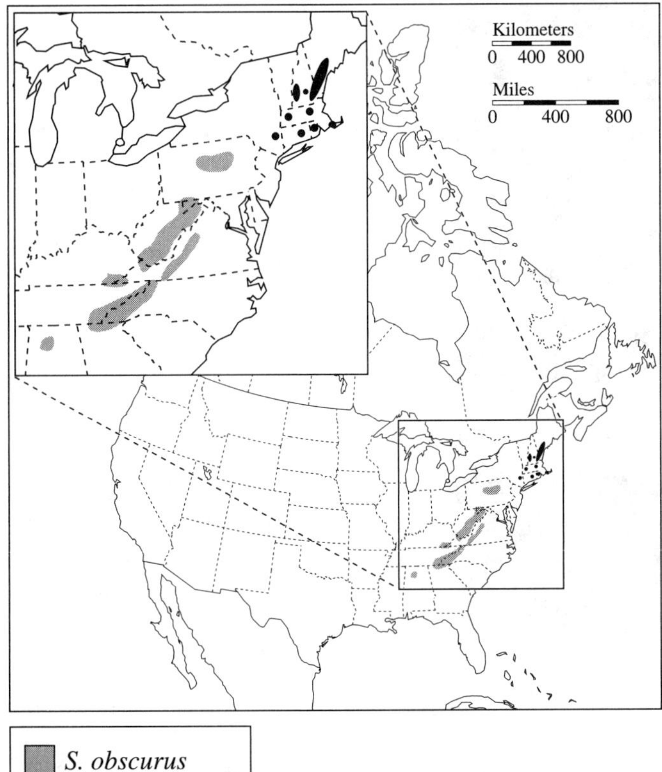

FIGURE 5.6. Distribution of the New England cottontail (*Sylvilagus transitionalis*) and Appalachian cottontail (*S. obscurus*).

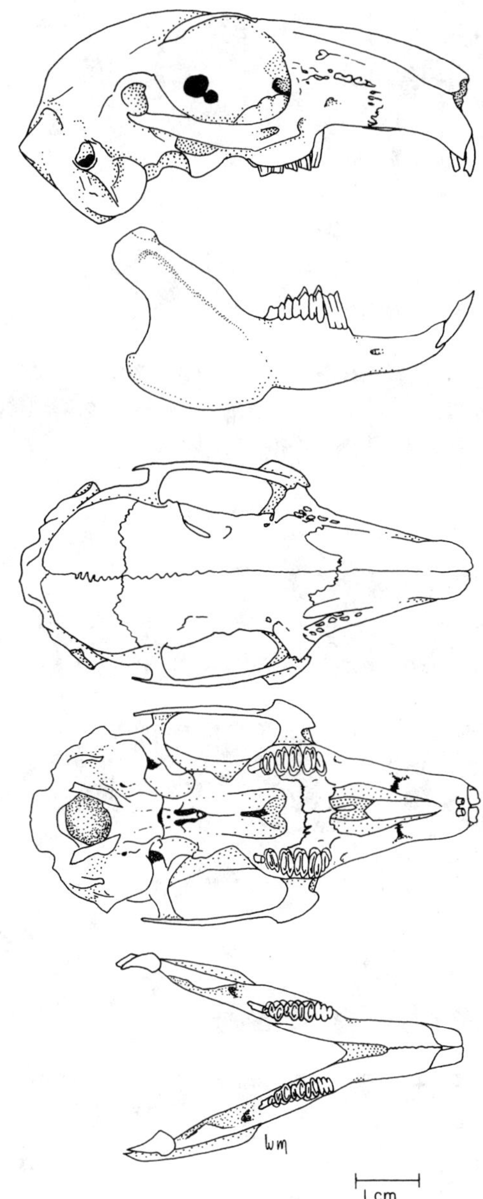

FIGURE 5.7. Skull of the eastern cottontail (*Sylvilagus floridanus*). From top to bottom: lateral view of cranium, lateral view of mandible, dorsal view of cranium, ventral view of cranium, dorsal view of mandible.

and a second set of "pegged" teeth directly behind the upper incisors. The dental formula is I 2/1, C 0/0, P 3/2, M 3/3. The presence of an interparietal bone distinguishes the genera *Sylvilagus* and *Brachylagus* from the genus *Lepus* (Figs. 5.7 and 5.8). Sexual dimorphism occurs in most species, with females about 1–10% larger than males (Orr 1940; Chapman 1971a; Chapman and Morgan 1973). Below is a brief description of each of the 10 species covered in this chapter.

Sylvilagus floridanus. The eastern cottontail is a large rabbit. The pelage is dense and long, and is brown to gray, with a white underside of the body and tail (Fig. 5.9A). The posterior extension of the supraorbital process of the frontal is transversely thickened. This cottontail has the widest distribution of any of the *Sylvilagus*. Diagnostic characteristics vary according to locality, but it is generally easy to distinguish this cottontail from other, sympatric species. On the basis of pelage, it may be difficult to distinguish *S. floridanus* from *S. transitionalis* in New England and from *S. obscurus* south along the Appalachians. The eastern cottontail is lighter in color and may possess a white spot on the forehead. The New England cottontail and the Appalachian cottontail rarely possesses a white spot on the forehead and often have a black spot between the ears. However, the eastern cottontail is readily distinguished from these two species by cranial characteristics (see Figs. 5.8A and 5.8B). Detailed discussions of body measurements and cranial characteristics are given in Nelson (1909), Hall (1951), Chapman et al. (1980), and Chapman (1999a).

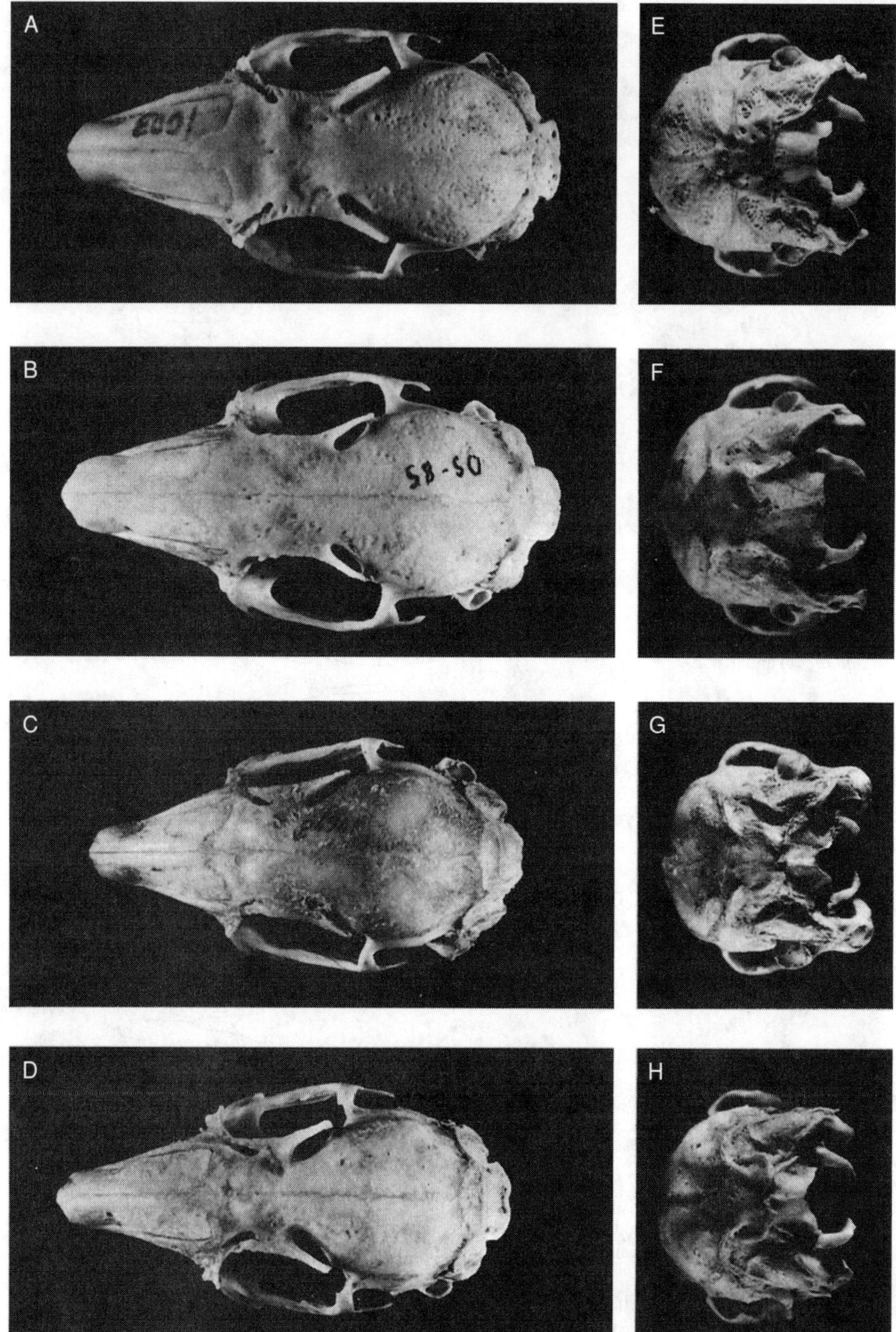

FIGURE 5.8. Dorsal and occipital views of skulls of some North American *Sylvilagus:* (A) *S. floridanus,* (B) *S. obscurus;* (C) *S. audubonii,* (D) *S. bachmani,* (E) *S. nuttallii,* (F) *S. aquaticus,* (G) *S. palustris,* (H) *B. idahoensis*.

Sylvilagus audubonii. Audubon's cottontail is large for the genus *Sylvilagus* (total length 370–400 mm). The ears are quite long (55–67 mm) and sparsely haired on the inner surfaces, and the feet are also sparsely haired. The upper body and large tail are gray and the ventral surfaces are white (see Fig. 5.9B). The rostrum is long (about 31 mm) and the skull has substantially upturned supraorbital processes. An identifying characteristic is the broad postorbital extension of the supraorbital process (see Fig. 5.8C). Detailed descriptions are given in Orr (1940), Hall (1951), Chapman and Willner (1978), and Chapman (1999b).

Sylvilagus bachmani. The brush rabbit is small to medium in size (total length 303–369 mm). The color varies from dark brown to gray brown dorsally to white ventrally. The hind feet are small (71–86 mm) and the legs are short. The feet are not covered with long, dense hair. The ears (45–63 mm) and tail (10–30 mm) are small. The rostrum is short and the supraorbital processes are small and well separated from the cranium; the preorbital notch is prominent (Fig. 5.8D). Detailed descriptions are given in Orr (1940), Hall (1951), and Chapman (1971a, 1974, 1999c).

Sylvilagus nuttallii. The mountain cottontail is medium to large in size (total length 338–390 mm). The feet are covered with long, dense hair and the legs are long. The ears are short and rounded at the tip. This rabbit is grayish dorsally and white ventrally, with a large (30–54 mm) grizzled tail. The rostrum is long. Supraorbital processes are small and

FIGURE 5.9. Four North American species of *Sylvilagus:* (A) *S. floridanus,* (B) *S. audubonii,* (C) *S. palustris,* (D) *S. transitionalis.* SOURCE: (A) Photo by G. L. Twiest, Mammal Images Library, American Society of Mammalogists; (B, C, D) Photos by Leonard Lee Rue III.

have very pointed anterior projections, with long postorbital processes (see Fig. 5.8E). Descriptions are given in Orr (1940), Hall (1951), and Chapman (1975a, 1999d).

Sylvilagus aquaticus. The swamp rabbit is the largest member of the genus (total length 452–552 mm). The dorsal surface is black to rusty brown and the ventral surface is white. Width of the rostrum over the anterior maxillary teeth is greater than the interorbital width. The anterior and posterior projections of the supraorbital process are lightly fused to the skull or lacking (see Fig. 5.8F). Descriptions are given in Hall (1951), Golley (1962), Lowery (1974), Chapman and Feldhamer (1981), and Chapman (1999e).

Sylvilagus palustris. The marsh rabbit is a large cottontail (total length 425–440 mm; see Fig. 5.9C). The dorsal surface is blackish brown to reddish brown, whereas the belly is brownish or gray, but never white. The ears, tail, and feet are all very small for the size of the rabbit. The dark color of the ventral surface of the tail distinguishes this species from all other sympatric cottontails. The anterior and posterior extensions of the supraorbital processes are attached to the skull along the full length (Hall 1951) (see Fig. 5.8G). Descriptions are given in Hall (1951), Chapman and Willner (1981), Lazell (1984), and Chapman (1999f).

Sylvilagus transitionalis. The New England cottontail (Fig. 5.9D) is medium to large in size (total length 388–439 mm). The dorsal surface is dark brown to buff overlain with a blackwash, which gives a penciled effect. The anterior edges of the ears are covered with black hair. There is a distinct black spot between the ears. To the untrained observer, these pelage characteristics may be difficult to distinguish from those of *S. floridanus*.

The supraorbital process is short or missing and the postorbital process is long and slender, rarely touching the skull. The suture between the frontal and the nasals are irregular or jagged in outline. The skull of *S. transitionalis* is very similar to that of *S. obscurus* (see Fig. 5.8B). Descriptions are given in Hall (1951, 1981), Chapman and Paradiso (1972), Chapman and Morgan (1973), Chapman (1975b, 1999g), Chapman and Stauffer (1981), and Hall (1981).

Sylvilagus obscurus. The Appalachian cottontail is medium to large in size (total length 386–430 mm) and similar in appearance to the New England cottontail. Positive identification of these two species is based on cranial morphometrics, chromosome analysis, or geographic distribution. *S. obscurus* is difficult to distinguish from *S. transitionalis* on the basis of pelage or gross examination of the skull alone. Descriptions are given in Chapman et al. (1992) and Chapman (1999h).

Sylvilagus robustus. The Davis Mountain cottontail is a large cottontail (average total length of five adults is 458 mm). The species is buffy gray, with an iron gray rump patch and large, white, thickly furred feet. The legs are rusty cinnamon in color and the ears are large and gray. This species is best distinguished from *S. floridanus* using cranial morphometrics and geographic distribution (Ruedas 1998). Descriptions are given in V. Bailey (1905) and Nelson (1909).

Brachylagus idahoensis. The pygmy rabbit is the smallest cottontail (total length 250–290 mm). The hind legs are short and the hind feet are broad and heavily haired. The ears are short, rounded, and densely

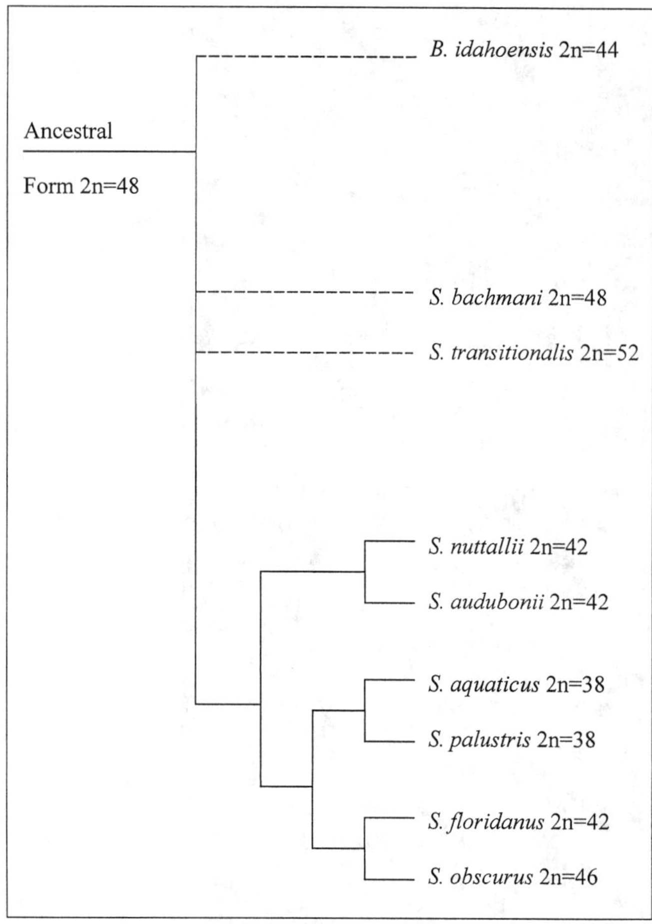

FIGURE 5.10. Composite dendrogram of the systematic relationships among some of the North American cottontails (*Sylvilagus* and *Brachylagus*). SOURCE: The basic dendrogram is taken from Robinson et al. (1984) and Chapman and Cabellos (1990) and adapted using material from Holden and Eabry (1970), Robinson et al. (1983), Rueda et al. (1989), Chapman et al. (1992), and Halaych and Robinson (1997).

haired. The dorsal surface is gray, the tail is small, and the ventral surface is buff rather than white. The rostrum is short and pointed and the anterior and posterior supraorbital processes are long and pointed. The skull appears immature (see Fig. 5.8H). Descriptions are given in Hall (1951), Green and Flinders (1980), and Flinders (1999).

GENETICS

The diploid chromosome numbers for nine North American species of *Sylvilagus* vary from 38 to 52, with most species apparently having $2n = 42$ (Fig. 5.10). There are several studies of the phylogenetic relationships among the New World rabbits (e.g., Robinson et al. 1983, 1984; Ruedas et al. 1989; Chapman et al. 1992; Halaych and Robinson 1997; Litvaitis et al. 1997).

Sequence data for cottontails obtained from mitochondrial DNA (12S region) have been used to investigate phylogenetic relationships among several species (Halaych and Robinson 1997). The monotypic genus *Brachylagus* is warranted. Cottontails with overlapping ranges tend to be more closely related. This is apparent for *S. nuttallii* and *S. audubonii* ($2n = 42$), and *S. aquaticus* and *S. palustris* ($2n = 38$) (Halaych and Robinson 1997) (Figs. 5.2, 5.3, 5.5, and 5.10).

There have been electrophoretic studies of the serum proteins of several species of *Sylvilagus* (M. L. Johnson and Wicks 1964; M. L. Johnson 1968; Scribner and Chesser 1993). An extensive study of the serum proteins of *S. floridanus, S. audubonii,* and *S. obscurus* was conducted by Chapman and Morgan (1973) and R. P. Morgan and Chapman (1981). Twenty serum proteins were resolved for *S. audubonii,* of which transferrin was polymorphic. This was expected because many desert species exhibit varying degrees of polymorphism. On the other extreme, *S. obscurus* had no polymorphism for 18 proteins resolved. The patterns of the serum proteins were very consistent among individuals for each species.

Four populations of *S. floridanus* were examined by Chapman and Morgan (1973). Three were considered native: *S. f. alacer* (Kansas), *S. f. mearnsii* (Missouri), and *S. f. chapmani* (Texas). The other population was considered to be of intergraded origin, including *S. f. mallurus* and many other subspecies of *S. floridanus*. In *S. f. alacer,* 23 serum proteins and a polymorphic transferrin were observed. A pretransferrin polymorphism was also reported. Twenty-two serum proteins were observed in *S. f. mearnsii,* with a pretransferrin polymorphism that was presumably the same as observed in *S. f. alacer*. There were 20 serum proteins in *S. f. chapmani,* but the pattern was distinct from those of *S. f. alacer* and *S. f. mearnsii*.

The serum proteins of the intergrade *S. floridanus* from Maryland are quite variable (Chapman and Morgan 1973). This increased variation is attributed to the massive introductions of *S. floridanus* and other *Sylvilagus* species into Maryland from 1920 to 1950. In the Maryland *S. floridanus,* 18 serum proteins were usually found. Ten polymorphic regions were observed, including two pretransferrins, the transferrin, and a posttransferrin. In addition, eight other pretransferrin and two posttransferrin patterns were observed. Three "hybrid" patterns were observed, suggesting that some Maryland *S. floridanus* may have crossbred with another species that had been introduced into the area. R. P. Morgan and Chapman (1981) examined the serum proteins of eastern cottontails introduced from western Maryland to St. Clements Island. They found that the island population exhibited previously unobserved serum protein patterns and a remarkable degree of polymorphism.

Changes in spatial associations of genotypes of *S. floridanus* were found by Scribner and Chesser (1993). They believed that these changes were related to "episodic exchange of individuals from semi-isolated populations that differ in allele frequency, resulting in a paucity of heterozygotes in adult, premating populations; subsequent restoration of expected genotypic proportions in offspring through random mating, which results in age differences in genotypic frequencies; a patchy distribution of habitat types within playa basins; and non-random placement of offspring (nest sites) in specific habitat types." Spatial distribution of genotypes is related to one or more of these factors and may be common processes contributing to seasonal population changes (Scribner and Chesser 1993:1142).

PHYSIOLOGY

Physiological Cycles. Four morphological indicators have been used to assess the physiological responses of cottontails to environmental factors. These include the adrenal index, the spleen index, the body fat index, and the condition index (J. A. Bailey 1968; Chapman et al. 1977; Bittner and Chapman 1981).

Adrenal and Splenic Indices. Adrenal index values of *S. floridanus* are high in the spring during the peak of the breeding season (Chapman et al. 1977; Bittner and Chapman 1981). Male adrenal indices were consistently higher than those of females. Adrenal hypertrophy has been related to a variety of factors, including increasing and high population densities, breeding activity, varying environmental factors, disease, and increased competition for food (Christian 1963).

Chapman et al. (1977) found that adrenal values in a population of *S. floridanus* from three counties in western Maryland were high in the winter, during the period of the most inclement weather. However, the index values for *S. obscurus* (see Chapman et al. [1992] for description of *S. obscurus* originally reported to be *S. transitionalis*) from the same region were highest from April to June, during the breeding season. They concluded that *S. obscurus* was better adapted to the cold boreal environment than *S. floridanus*.

On St. Clements Island, Maryland, Bittner and Chapman (1981) found that mean adrenal index values were high in the spring, which was the height of the breeding season. Male adrenal index values were also consistently higher than those of females, as in western Maryland (Chapman et al. 1977). It has been shown that social hierarchy is stronger in males than in females (Marsden and Holler 1964; Chapman et al. 1977), with the "pressure to breed" probably the underlying cause of higher values among males.

However, adrenal hypertrophy has been related to factors other than breeding pressure. Selye (1973) found that adrenal hypertrophy resulted when organisms were subjected to a variety of stressors. Specific factors causing stress in animals included increasing and high population densities, varying environmental factors, disease, and increased competition for food (Christian 1963). High density and decreased food availability were evident on St. Clements Island, yet Bittner and Chapman (1981) reported that high stress as indicated by adrenal hypertrophy was not apparent under such conditions.

Physiological responses have been used to determine whether stress factors were acting on animal populations. Most of this work has centered around the adrenal gland and its associated hormones (Selye 1956; Christian 1963; Myers 1967; Chapman et al. 1977). The spleen has also been used to describe responses to stress, but results vary among species (Selye 1956; Conaway and Wight 1962; Christian 1963; Willner et al. 1979). Bittner and Chapman (1981) contradicted some of the studies on the adrenal gland because stress, as evidenced by adrenal hypertrophy, was not apparent. In fact, adrenal index values for St. Clements Island were significantly lower than those reported by Chapman et al. (1977) for eastern cottontail populations in western Maryland subjected to much lower densities. This suggests that adrenal weight alone is not sufficient evidence of stress when comparing lagomorph populations from two different locations.

Body Fat Indices. Body fat indices have been used to assess the general condition of *S. floridanus* (Lord 1963; Chapman et al. 1977; Bittner and Chapman 1981) and *S. obscurus* (Chapman et al. 1977). In western Maryland, *S. floridanus* and *S. obscurus* experienced little variation in the body fat index except for two instances: Eastern cottontails sharply increased their body fat index between summer and fall, and Appalachian cottontails decreased their body fat index between winter and spring (Chapman et al. 1977). Both species had a relatively high body fat index in the winter. Lord (1963) found that *S. floridanus* in Illinois had more body fat in the winter, as did Bittner and Chapman (1981) for the St. Clements Island population.

Condition Indices. J. A. Bailey (1968) used length–weight ratios to assess the condition of *S. floridanus*. He believed that winter weight loss was normal for eastern cottontails in northern latitudes. Loss of weight (condition) in eastern cottontails has been reported by several investigators (Allen 1939; Elder and Sowls 1942; Haugen 1942; Chapman et al. 1977). Even with supplemental feeding, Lord and Casteel (1960) were unable to prevent winter weight loss in cottontails.

Chapman et al. (1977) found that *S. obscurus,* in contrast to *S. floridanus,* was in its best condition in winter. Differences in condition index values of these two species indicated that *S. obscurus* was better adapted to the colder temperatures characteristic of northern climates. *S. obscurus* responded in a totally different way than any of the northern-latitude *S. floridanus* populations studied.

Index Interrelationships. Male and female eastern cottontails showed some correlations between condition indices on St. Clements Island (Bittner and Chapman 1981). Adrenal and splenic index values peaked in the spring, when the condition index and the body fat index were relatively low. As expected, the body fat index was highest when the condition index was highest.

Chapman et al. (1977) found that *S. floridanus* exhibited a consecutive peaking of physiological indices: peak body fat in fall, peak adrenal index in winter, and peak condition in spring. The seasonal pattern for attaining these peaks was the same for all indices, but occurred one season out of phase. Therefore, it seemed that the eastern cottontail physiologically anticipated the stressful winter period, increasing its body fat to allow it to better survive cold and heavy snow and to enter the breeding season in late winter or early spring in good condition. However, *S. obscurus* tended to be more varied but immediate in its responses to stress, the primary stress apparently being reproduction.

This tendency toward immediate reaction to stress (as seen by the correlation between the body fat index and the inverse of the adrenal index) indicated that *S. obscurus* was slightly unstable with respect to change. It responded well to small stresses by rapidly metabolizing fat, but not favorably to larger stress factors, presumably because it had metabolized large amounts of body fat, reducing its condition index.

Cottontails on St. Clements Island bred while in their best physical condition (Bittner and Chapman 1981). Eastern cottontails in western Maryland also exhibited their best condition in the breeding season, which occurred during the most moderate weather (Chapman et al. 1977).

Condition Indices and Environmental Factors. In general, in western Maryland and nearby West Virginia, environmental conditions appeared favorable during spring and summer for *S. floridanus* and during fall and winter for *S. obscurus* (Chapman et al. 1977). This was manifested in a peaking of the condition index in spring for *S. floridanus* and a peaking of the body fat index in the fall for *S. obscurus*. This was supported by the low adrenal index of *S. obscurus* in the fall and winter. Myers (1967) reported that the European rabbit (*Oryctolagus cuniculus*) had its highest adrenal weight in winter and a decline in summer, similar to *S. floridanus*. Therefore, the physiological responses of *S. obscurus* to seasonal change were different from those of either *S. floridanus* or *O. cuniculus*.

According to Chapman et al. (1977), measurement of physiological indices indicated the following general conclusions: (1) *S. obscurus* was better adapted to colder weather, (2) *S. floridanus* was more eurythermal than *S. obscurus,* and (3) reproduction was inherently stressful, as both species were in the best condition just before the breeding season.

Flux (1964:485), working on introduced hares, stated that "fat deposits are probably for maintaining breeding condition rather than acting merely as a winter food store." The observations of Chapman et al. (1977) on *S. floridanus* and *S. obscurus* support that theory. However, Lord (1963) believed that *S. floridanus* in Illinois increased fat reserves as a physiological response to advancing winter.

Tomich (1962) found that adrenal glands of male California ground squirrels (*Spermophilus beecheyi*) increased during and after the breeding season. He felt this indicated postbreeding recovery from fighting and crowding. For male ground squirrels, summer was the time of greatest stress, but not so for *S. floridanus*. Chapman et al. (1977) felt that crowding, if it caused an increase in adrenal size (Christian and Davis 1955), would most likely occur in July–September. However, this did not occur in either *S. floridanus* or *S. obscurus*. Thus, male/male interaction as described by Marsden and Holler (1964) and cold temperatures seemed to be probable causes for increased male adrenal weight in winter in western Maryland and nearby West Virginia. Tomich (1962) found a postbreeding stressor in the California ground squirrel, and a similar phenomenon may occur for *S. obscurus*. Herrick (1965) found that pregnancy was the greatest stress factor identified in female black-tailed jackrabbits (*Lepus californicus*). Tomich (1962) found that pregnancy was a severe stressor for female ground squirrels.

The onset of reproduction of male *S. floridanus* and *S. obscurus* was correlated with the end of adverse weather (Chapman et al. 1977). Males of both species were in breeding condition well before the first warm days of spring. Increased day length induces reproductive behavior, complete libido, and spermatogenesis in *S. transitionalis* (Bissonnette and Csech 1939). In western Maryland and nearby West Virginia, male breeding activity was observed well in advance of the breeding season of female *S. obscurus* (Chapman et al. 1977). Early male reproductive activity preceding the actual onset of breeding allows cottontail populations to take advantage of yearly fluctuations in the onset of breeding. There is minimum delay in breeding after good weather finally arrives. This is particularly critical in late springs and toward the northern limits of occupied range.

Thermoregulation. As has been noted, *S. floridanus* is more eurythermal than *S. obscurus* (Chapman et al. 1977). However, there have been few studies on cottontail thermoregulation.

Hinds (1973) studied thermoregulation in *S. audubonii*. He reported that the body temperature was 38.3°C at ambient temperatures below 30°C, and was the same regardless of season. Body temperature equaled the ambient temperature at 41.9°C. Furthermore, *S. audubonii* has a relatively high lethal temperature and evaporative cooling capacity. These adaptations allow this species to survive the hot conditions of the open desert for short periods of time. According to Hinds (1973:708), "*Sylvilagus audubonii* survives in the desert by taking advantage of every possibility to minimize the heat load and water expenditure. A relatively high evaporative cooling capacity and high lethal body temperature of 44.8°C provides a safety factor for desert cottontails if avoidance is not possible."

Katzner et al. (1997) found that pygmy rabbits are thermally stressed during harsh winters. However, this species exhibited metabolic and insulative characteristics similar to those of other leporids. *B. idahoensis* body temperature (38.47°C) was similar to that of *S. audubonii* (38.3°C) (Hinds 1973). However, both species had lower body temperatures than species of *Lepus* that have been studied. Pygmy rabbits showed no circadian rhythms in body temperature regulation associated with activity or ambient temperature (Katzner et al. 1997).

Metabolism. Younger *S. floridanus* have a tendency to choose foods that contain more digestible energy and protein. This assists juvenile and subadult rabbits in satisfying their energy needs. Rose (1974:476) found that "total assimilation and respiration are greater for larger rabbits but that the rates of assimilation and respiration per gram of body weight decrease with increasing body size." Even though *S. floridanus* consumes a relatively small percentage of the vegetation available, digestible energy may be limiting in the winter and early spring (Rose 1973).

S. audubonii showed a definite seasonal change in basal metabolic rate (Hinds 1973). However, *S. audubonii* did not show a reduction in metabolic rate relative to other mammals. The basal metabolic rate was 0.651 ml O_2/g per hour during the summer, and was 18% lower than the basal metabolic rate of 0.790 ml O_2/g per hour during winter. High rates of metabolism may be associated with demands for thermoregulation in winter and reproduction in spring and early summer. High basal metabolic rates would be a disadvantage in bad weather.

The pygmy rabbit is a relatively small mammal with a high metabolic rate living in a thermally stressful environment. In the winter, these rabbits compensate by eating only sagebrush, which has a high-energy content. This suggests that they are vulnerable to habitat alteration, especially on the periphery of their distribution (Katzner et al. 1997).

Organ Weights. In *S. floridanus*, the left adrenal gland is heavier than the right (J. A. Bailey and Schroeder 1967), and a similar relationship has been reported for *S. bachmani* (Chapman 1971a). The left kidney of female *S. bachmani* was significantly heavier than that of males. The right kidney of male *S. bachmani* was heavier than the left. Chapman (1971a) also weighed the liver, heart, and spleen, but found no significant difference in the weight of those organs between the sexes.

Hormonal Studies. Male *S. floridanus* pituitary glands contain more follicle-stimulating hormone (FSH) than those of females. Pituitaries from cottontails <1 year old contain significantly less FSH than those of older rabbits. Because the pituitary secretes very little gonadotropin until several weeks of age, variations with age are expected. Cottontail pituitaries collected during the breeding season (March–August) contain higher quantities of FSH than those collected at other months of the year (Stevens 1962).

During the breeding season, luteinizing hormone (LH) levels are apparently higher than at other times. Thyroid gland activity also seems to become progressively higher from January to September. Thyroid gland activity seems to be correlated with seasonal temperatures (Stevens 1962). Numerous studies on a variety of mammals have emphasized interactions that may occur among hormones from the pituitary, thyroid, adrenal cortex, and ovary to modify the responses of the female reproductive system. However, ovarian responses are not just the result of the independent action of the gonadotropins (Stevens 1962).

Sodium deficiency caused a marked response of the adrenal glands of *S. floridanus*. In sodium-poor environments, the availability of salt may be a valid management concern (McCreedy and Weeks 1992).

REPRODUCTION AND DEVELOPMENT

Anatomy. The reproductive system of male *Sylvilagus* is similar to that of the domestic rabbit (Wells 1964). Males possess a scrotal sac that is entirely covered with hair and is visible only during the breeding season. The paired testes of adults of some species are large (>10 g) and are scrotal during the breeding season. The penis is cylindrical and normally withdrawn into a sheath. The seminal vesicle, vesicular gland, prostate, paraprostates, and bulbourethal glands are similar in gross morphology and histology to those of other lagomorphs. In *S. obscurus* and *S. floridanus*, histological comparisons of the testes revealed no difference between the two species, with the tunica albuginea 37–62.5 μm thick in both species (Chapman et al. 1977).

The female reproductive system of *Sylvilagus* also is comparable to that of the domestic rabbit (Wells 1964). *Sylvilagus* possesses corpora lutea of varying sizes, sometimes reaching 2.48 mm in diameter, compared to about 5 mm in the domestic rabbit. Chapman et al. (1977) gave detailed descriptions of the gross morphology and the histology of the ovaries of *S. obscurus* and *S. floridanus*. The uterus is duplex, each side having its own distinct cervical canal. The size and condition of the uterus are dependent on the reproductive condition of the animal. Nonpregnant, nonparous females tend to have a smoother, unstriated, less convoluted, and less muscular uterus than pregnant or parous females. After parturition, the uterus shrinks rapidly in size. The vagina is a single tube leading from the cervices. The clitoris is near the external opening. The color of the vulva is variable, ranging from pinkish white to dark purple. Females possess four to five pairs of mammae; pectoral, thoracic, and abdominal mammae are present. These are supplied by two strips of mammary tissue lying on each side of the midline of the abdomen.

Physiology. Ecke (1955) gave an excellent account of the physiology of the reproductive cycle of female *S. floridanus*, and the following discussion is drawn largely from his work.

Female *S. floridanus* are in anestrus (quiescent period) during winter. At this time, follicular growth in the ovaries is greatly suppressed. The follicles may develop slightly, but maturation does not occur. As the breeding season approaches, an external stimulus (probably increasing day length in combination with temperature) stimulates the pituitary gland to begin secreting FSH into the blood stream, which in turn reaches the ovaries. FSH acts as a somatic nutrient and stimulates the growth of follicles and the development of ova. The ova develop to a submature stage, at which time the rabbit is in heat. Heat is maintained until copulation occurs (Ecke 1955).

The stimulation of copulation results in the pituitary's secreting LH, which results in rapid growth of the follicle and ovulation. Ovulation occurs 10 hr after copulation. Ova are fertilized in the fallopian tube. On about the fourth day, the fertilized ova enter the uterus. On or about the seventh day, they are implanted. The blastocysts are at the 1- to 5-mm stage when implanted (Ecke 1955).

The cottontail may remain in pre-estrus for long periods of time if copulation does not occur. While in pre-estrus, submature follicles are present and old follicles are either replaced or remain (Ecke 1955).

Cottontails do not experience a true estrous cycle, because they are induced ovulators. However, occasionally ovulation is induced by one doe attempting to copulate with another or by copulation with an infertile male. This may result in pseudopregnancy. At this time, follicular growth is retarded and corpora lutea are formed (Ecke 1955).

Once ovulation has occurred, corpora lutea form where the follicles ruptured and are present through the entire pregnancy. The corpora lutea secrete progesterone, which prevents the formation of mature ova. Toward the end of the pregnancy, the placenta begins to secrete progesterone and the corpora lutea decrease in size. Reduction in corpora lutea signifies reduced progesterone secretion. With reduction in ovarian progesterone, follicles begin to mature. At parturition, fully mature follicles are present (except lost litters). Breeding occurs immediately after parturition. Sex pheromones are secreted before parturition, attracting males. These pheromones are probably by-products of ovarian estrogen from maturing follicles. Thus, cottontails usually copulate again immediately following parturition (Ecke 1955). It is this phenomena that results in breeding synchrony in cottontail populations.

The hormones that stimulate follicular and fetal development are also involved in mammary function. Domestic rabbits that become pregnant immediately following parturition may not be able to supply nutrients to suckle young and support fetal growth. This can result in total litter resorption between the 8th and 15th days (Ecke 1955).

Total litter resorption is rare in *Sylvilagus*. Total litter resorption occurred in only 1.7% of the litters of *S. floridanus* in western Maryland (Chapman et al. 1977), and only two complete resorptions occurred in a large sample of *S. palustris* from southern Florida (Holler and Conaway 1979). No total litter resorptions were reported for *S. bachmani* in Oregon (Chapman and Harman 1972) or *S. obscurus* in West Virginia (Chapman et al. 1977). Ecke (1955) believed total litter resorption was rare because wild cottontails have relatively small litters and their nursing period is shorter than that of the domestic rabbit.

Verts et al. (1997) noted that fetal and placental mass was significantly greater for males in eastern cottontails. This was not related to intrauterine position. They believed that this represents a differential parental investment in male offspring.

Breeding Season. In *S. floridanus,* initial reproductive activity occurs later in higher elevations and in higher latitudes (Conaway et al. 1974; Chapman et al. 1977). Onset of breeding begins as early as the first week of January in Alabama (Barkalow 1962) to as late as the last week in March in southern Wisconsin (Rongstad 1966). The breeding season lasts from mid-March to mid-September in Connecticut (Dalke 1942), from February through September in New York (Schierbaum 1967), and from late February through August in western Maryland (Chapman et al. 1977). In the southern states, breeding seasons are of longer duration, as in Georgia, which has a 9-month season (Pelton and Provost 1972), and in southern Texas, with a year-round breeding season (Bothma and Teer 1977). An introduced population of *S. floridanus* in western Oregon began to breed in late January and ceased breeding by early September (Trethewey and Verts 1971).

Onset of breeding varies among different populations and within the same population from year to year (Conaway and Wight 1962; Hill 1966; Chapman et al. 1977). Hill (1966) suggested that temperature, rather than diet, is the primary factor controlling onset (date) of breeding each year. Many researchers have correlated severe weather with delays in onset of the breeding season (Hamilton 1940; Wight and Conaway 1961; Conaway and Wight 1962).

Ecke (1955) concluded that the limits of the breeding season are closely related to the availability of succulent vegetation. To a degree, onset of breeding anticipates the availability of succulent green foods 28 days later. Changes in photoperiod are an important factor regulating cottontail breeding seasons (Bissonnette and Csech 1939), perhaps more so in more northern latitudes. The major environmental factors controlling breeding activity in southern Texas are temperature and rainfall. Rainfall affects the amount of succulent vegetation available (Bothma and Teer 1977). Onset of male reproductive activity for *S. floridanus* and *S. obscurus* was closely correlated with temperature (Chapman et al. 1977).

In *S. audubonii,* the breeding season is also variable. In California, it begins in December and ends in June (Orr 1940), and in Arizona, it begins in January and ends in September (Sowls 1957; Stout 1970). Ingles (1941) reported on a California population that bred year around. The breeding season in Texas did not begin until late February or early March (Chapman and Morgan 1974). In *S. nuttallii* in northeastern California, the breeding season began about April and ended in July (Orr 1940). In Oregon, the breeding season lasted from February to July (Powers and Verts 1971).

The length of the breeding season of *S. bachmani* seems to be about the same in northern and southern parts of the range. The breeding season in Oregon lasted from February through August (Chapman and Harman 1972), and in California from December through about June (Mossman 1955). The breeding season of *S. obscurus* lasted from March to September in Maryland and West Virginia (Chapman et al. 1977).

Apparently in some regions of Texas, the swamp rabbit breeds year around. The breeding season is longest in the south-central United States and becomes shorter with increasing latitude (Hunt 1959). In Louisiana, breeding has been reported in all months except October (Svihla 1929). In Missouri, the breeding season lasts from February through June (Sorensen et al. 1968). In southern Florida the marsh rabbit breeds year around (Holler and Conaway 1979), whereas in northern Florida an anestrous period occurs from October to March (Blair 1936).

The breeding season of *B. idahoensis* is very short. In Idaho, it lasts from March through May (Wilde et al. 1976), and in Utah, from February through March (Janson 1946). Orr (1940) believed that the breeding season of *B. idahoensis* was limited to the spring.

Gestation. The mean gestation period of the eastern cottontail is about 28 days (range-25–35 days) (Dice 1929; Dalke 1942; Bruna 1952; Evans 1962; Marsden and Conaway 1963). The gestation period of *S. bachmani* is 27 ± 3 days (Chapman and Harman 1972), of *S. nuttallii* is 29 ± 1 days (Cowan and Guiquet 1956), of *S. transitionalis* is 28 days (Dalke 1942), of *S. obscurus* is 28 days (Tefft and Chapman 1987), of *S. aquaticus* is 37 ± 3 days (Hunt 1959; Holler et al. 1963; Sorensen et al. 1968), and of *B. idahoensis* is 26–28 days (Dobler and Dixon 1990).

Chapman (1984) reported on the relationship between latitude and gestation period in New World rabbits. The correlation is highest for northern limits of distribution. Factors believed to affect gestation period are weather related. Chapman (1984) believed it would be adaptive in multiparous rabbits with northern distributions to have shorter gestation periods so the maximum number of young could be produced during periods of suitable weather and vegetation growth.

Breeding Synchrony. Like other lagomorphs, *Sylvilagus* is an induced ovulator, ovulation occurring only after copulation or other suitable stimulus has occurred. If pregnancy does not follow ovulation, the cottontail may exhibit pseudopregnancy, which may last for about one half of the normal gestation period (Conaway and Wight 1962). Total loss of a litter through abortion or resorption also may produce effects similar to pseudopregnancy and may be accompanied by lactation.

Cottontails exhibit a relatively well synchronized breeding season, and conception usually follows almost immediately after parturition of the previous litter (Casteel 1967; Tefft and Chapman 1987). The breeding season can be divided into conception periods on the basis of this breeding synchrony (Evans 1962; Conaway et al. 1963; Chapman and Harman 1972; Pelton and Provost 1972; Chapman et al. 1977).

Synchrony apparently begins to break down after late June or early July (Trethewey and Verts 1971; Pelton and Provost 1972; L. W. Johnson 1973). Breakdown in synchrony is associated with reduced attention of preparturient females and the failure of females to breed for several days or not at all postpartum (L. W. Johnson 1973). This suggests the failure of ovarian development during the final week of pregnancy, thus probable reduced pituitary function. Timing with respect to the summer solstice suggests that photoperiod may be involved.

Litter Size. As with many other mammals, there is an inverse relationship between litter size and latitude in some *Sylvilagus,* which is compensated for by an increase in the length of the breeding season at lower latitudes (Lord 1960). Conaway et al. (1974) gave a review of latitude and litter size in *S. floridanus*. The size of first litters varies from 2.95 to 5.10 over the range of this species (Table 5.1) and the mean litter size varies from 3.06 to 5.60 (Table 5.2). There is considerable variation in the mean litter size among the species of *Sylvilagus* (Tables 5.1–5.3). The most fecund species is clearly *S. floridanus;* however, the genus as a whole produces relatively large litters.

In most species, the number of young per litter varies with the time of the year. Usually the first and last litters of the year are smaller

TABLE 5.1. Regional comparisons of the mean size of first litters of *Sylvilagus floridanus*

Location	Mean Size of Litter	Sample Size	Year(s)	Source
United States (40–45°N)	5.10	36	1964–65	Conaway et al. 1974
United States (35–40°N)	4.20	158	1964–65	Conaway et al. 1974
United States (30–35°N)	3.40	50	1964–65	Conaway et al. 1974
Western Oregon	3.87	16	1969	Trethewey and Verts 1971
North Dakota	5.00	55	1964–65	Conaway et al. 1974
Western Maryland	4.50	20	1971–72	Chapman et al. 1977
St. Clement's Island, Maryland	3.00	1	1976–77	Bittner and Chapman 1981
Iowa	4.80	63	1958–61	Kline 1962
Northern Iowa	4.9	43	1958–61	Kline 1962
Southern Iowa	4.6	17	1958–61	Kline 1962
Missouri	3.30	17	1938–40	Schwartz 1942
Northern Missouri	4.10	55	1962	Evans et al. 1965
Southern Missouri	3.60	38	1962	Evans et al. 1965
Tennessee Valley, Alabama	3.72	103	1959–67	Hill 1972
Piedmont plateau, Alabama	3.00	16	1959–67	Hill 1972
Central Alabama	3.65	40	1959–67	Hill 1972
Upper coastal plains, Alabama	3.13	30	1959–67	Hill 1972
Lower coastal plains, Alabama	2.95	80	1959–67	Hill 1972
Southwestern Texas	4.00	3	1973	Chapman and Morgan 1974

(Tables 5.1 and 5.2). Litters gradually increase in size and peak in midseason and then decline toward the end of the breeding season (Chapman et al. 1977).

Variation in litter size has also been associated with age of the rabbit. In *S. floridanus,* adults seem to produce larger litters (Chapman et al. 1977); conversely, in *S. aquaticus,* younger females produce larger litters (Sorensen et al. 1968). Big does tend to have more young than small does of the same genotype.

The mean litter sizes from areas of high-fertility soils are significantly larger than those from areas of low-fertility soils. Exceedingly large litter sizes have been reported by Rongstad (1966) and Barkalow (1961) (a litter of 9 young), Lemke (1957) (a litter of 10), and Kirkpatrick (1960) (12 young in a litter). *Brachylagus idahoensis* produces three litters per year. Litter size varies from 5 to 8 and averages 6 (Dobler and Dixon 1990).

Ovulations and Resorptions. The mean ovulation rate for *S. floridanus* in western Maryland was 5.8, whereas the mean litter size was 5.01. Thus, 12.9% of the eggs ovulated either failed to implant or were resorbed (Chapman et al. 1977). For *S. palustris,* a preimplantation loss of 8% was reported by Holler and Conaway (1979), for *S. aquaticus,* 2% (Hill 1967).

Conaway and Wight (1962) found that luteinized follicles occurred in only about 1% of the ovaries of Missouri cottontails. They found no polyovular follicles past primary stages. Chapman et al. (1977) reported that luteinized follicles occurred in <1% of *S. floridanus,* and none were found in the *S. obscurus* they examined. The polyovular condition was found in one *S. floridanus* and four *S. obscurus.* Six percent of the visible fetuses of *S. floridanus* examined in western Maryland were being resorbed and there was considerable variation in the resorption rate by month. In fact, the resorption rate was inversely related to litter size. Total litter resorption was not common. Trethewey and Verts (1971) reported that 28.3% of the introduced female *S. floridanus* in Oregon contained resorbing fetuses.

In *S. obscurus,* 8% of the visible fetuses were resorbed (Chapman et al. 1977). Sowls (1957) reported finding only 1 case of embryo resorption in 56 female *S. audubonii* he examined in Arizona. Similarly, Holler and Conaway (1979) reported that resorption of visible fetuses of *S. palustris* accounted for only 3% of the ovulated ova. They also found two cases of complete litter resorption. Hill (1967) found that 2% of the visible fetuses were resorbed in *S. aquaticus* from Alabama. Conversely, Stout (1970) reported embryos being resorbed in 40% of the female *S. audubonii* he examined in Arizona.

TABLE 5.2. Regional comparisons of mean annual litter sizes of *Sylvilagus floridanus*

Location	Mean Size of Litter	Sample Size	Year(s)	Source
Western Oregon	5.10	106	1969	Trethewey and Verts 1971
New York	4.50	28	—	Hamilton 1940
Pennsylvania	5.42	26	1939	Beule 1940
Michigan	5.10	11	1935–37	Allen 1939
Wisconsin[a]	4.95	20	1961–63	Rongstad 1966
Western Maryland	5.02	65	1971–72	Chapman et al. 1977
St. Clements Island, Maryland	3.57	21	1976–77	Bittner and Chapman 1981
Maryland	4.80	35	1955	Sheffer 1957
Illinois	5.60	31	1947–48	Ecke 1955
Illinois	5.31	469	1957–59	Lord 1961
Missouri	4.40	42	1968	Schwartz 1942
Virginia	4.70	21	1939–41	Lleyellyn and Handley 1945
Georgia (mountain)	3.06	16	1966–67	Pelton and Jenkins 1971
Georgia (piedmont)	3.11	85	1966–67	Pelton and Jenkins 1971
Georgia (coastal plain)	3.18	108	1966–67	Pelton and Jenkins 1971
Alabama	3.47	611	—	Hill 1972
South Texas	3.30	279	1965–68	Bothma and Teer 1977

[a]Studies conducted in pens.

TABLE 5.3. Mean size of litters of *Sylvilagus* and *Brachylagus* as reported in technical literature

Species	Location	Mean Size of Litter[a]	Sample Size	Year(s)	Source
S. obscurus	West Virginia	3.22 (F)	9	1971–75	Chapman et al. 1977
S. obscurus	West Virginia	4.00 (S)	7	1971–75	Chapman et al. 1977
S. obscurus	West Virginia	3.56 (C)	16	1971–75	Chapman et al. 1977
S. transitionalis	Connecticut	5.20 (C)	19	1936	Dalke 1937
S. audubonii	Texas	2.60 (F)	10	1973	Chapman and Morgan 1974
S. audubonii	Arizona	2.90 (C)	56	1951–55	Sowls 1957
S. audubonii	Arizona	3.30 (C)	10	1967–68	Stout 1970
S. audubonii	California	3.33 (C)	119	1966–67	Graves and Asserson 1967
S. audubonii	California	3.60 (C)	19	1939–41	Fitch 1947
S. audubonii	California	3.30 (C)	19	—	Orr 1940
S. bachmani	Oregon	2.86 (C)	15	1968–69	Chapman and Harman 1972
S. bachmani	California	4.00 (C)	14	1950–51	Mossman 1955
S. bachmani	California	3.50 (C)	11	—	Orr 1940
S. aquaticus	Alabama	2.89 (F)	95	1960–67	Hill 1967
S. aquaticus	Alabama	3.17 (S)	17	1960–67	Hill 1967
S. aquaticus	Missouri	2.80 (C)	14	1956–57	Toll et al. 1960
S. aquaticus	Texas	2.83 (C)	28	—	Hunt 1959
S. nuttallii	Oregon	4.00 (F)	5	1969	Powers and Verts 1971
S. nuttallii	Oregon	4.30 (C)	31	1969	Powers and Verts 1971
S. palustris	Florida	4.00 (C)	3	—	Blair 1936
S. palustris	Southern Florida	2.82 (C)	121	1968–69	Holler and Conaway 1979
B. idahoensis	Utah	5.90 (C)	14	—	Janson 1946

[a]F, First litter; S, subsequent litters; C, combined.

Breeding Age. Juvenile breeding is well documented in *S. floridanus*. Juvenile females accounted for 3.9% of the pregnancies in western Maryland (Chapman et al. 1977). Conversely, 52.4% of the juvenile females were sexually active in western Oregon (Trethewey and Verts 1971). In both Maryland and Oregon, juvenile females had smaller mean litter size than older females. Casteel and Edwards (1964) reported two instances of multiparous juvenile female cottontails.

About 18% of juvenile female *S. obscurus* bred in West Virginia (Chapman et al. 1977), whereas only 6.7% of juvenile female *S. nuttallii* bred in eastern Oregon (Powers and Verts 1971). In South Florida, 22% of juvenile female marsh rabbits 6–9 months of age were parus (Holler and Conaway 1979). Sowls (1957) believed that juvenile female breeding in *S. audubonii* was an important factor in offsetting the low litter size he found for the species in Arizona.

Juvenile male reproductive activity is reported for *Sylvilagus*, but is considered insignificant in terms of populations because of the polygamous nature of the genus. Chapman et al. (1977) reported that a few juvenile male *S. floridanus* were sexually active from July to September in western Maryland. They also reported that all *S. obscurus* males were potentially reproductively active during the breeding season following the one in which they were born, but not before.

L. W. Johnson (1973) observed apparent reduced libido in adult male cottontails in July and speculated on the basis of their behavior that early-born males may participate in late-season breeding.

Reproductive Rate. The number of litters per year in *Sylvilagus* varies among species as well as on a latitudinal gradient within the same species. However, the number of litters produced per year can be misleading because many species have small litter sizes. In terms of the number of young produced, *S. floridanus* is clearly the most fecund cottontail. Trethewey and Verts (1971) reported that introduced *S. floridanus* in western Oregon produced 39 young in eight litters. Conaway et al. (1963) in Missouri reported 35 young in seven litters.

For other cottontails, total productivity varies considerably. *S. nuttallii* produced 22 young in five litters (Powers and Verts 1971). *S. obscurus* produced about 23 young in about six litters (Chapman et al. 1977). *S. bachmanii* produced about 15 young in about six litters (Chapman and Harman 1972). *S. palustris* produced about 14–19 young in a year-round breeding season (Holler and Conaway 1979), and *B. idahoensis* produced about 18 young in three litters (Dobler and Dixon 1990).

Nests and Newborns. Female cottontails build elaborate nests in which they give birth to their altricial young. Nests of *S. floridanus* are slanting holes in the ground with average length of 18.03 cm, width of 12.57 cm, and depth of 11.94 cm (Friley 1955). In Texas, five nests averaged 12.5 cm long, 10.4 cm wide, and 9.1 cm deep (Bothma and Teer 1977). Casteel (1966) found the average measurements from 21 nests were 14.6 cm long, 12.1 cm wide, and 10.2 cm deep. Bothma and Teer (1977) reported no relationship between the size of the nest and the size of the litter. Friley (1955) found that nest holes in southern Michigan farming areas contained an outer lining of grass or herbaceous stems covering all sides and a heavy inner layer of belly or side fur from the female. Illinois nests were lined first with leaves, then with an inner lining of fur plucked from the female (Ecke 1955). Casteel (1966) determined that residual vegetative cover was the preferred nest material, especially grass stems when available. Females pulled fur from almost every part of the body except the abdomen.

Nests of *S. aquaticus* are built on the ground surface and constructed of stalks of dead weeds around an inner lining of fur (Goodpaster and Hoffmeister 1952). Nests have a side entrance and are 4–7 cm deep, 15 cm wide, and 18 cm long (Lowe 1958; Holler et al. 1963). Nests of *S. audubonii* are pear-shaped excavations in the ground about 15–25 cm deep. They are lined with grass and rabbit fur (Ingles 1941). *B. idahoensis* is the only cottontail that digs its own burrows, but no one has yet verified the use of burrows as nesting sites or for rearing young (Weiss and Verts 1984).

Ecke (1955) gave the following description and average measurements for neonates of *S. floridanus:* weight 35–45 g, total length 90–110 mm, and hind foot length 21–23 mm. Young at birth are covered with fine hair, eyes are tightly closed, and legs are developed enough for them to crawl into their nests. Eyes opened on the 4th or 5th day and young were able to leave the nest between 14 and 16 days after birth. Kentucky nestlings opened their eyes at 7–8 days of age and were able to move out of the nest at 14 days (Bruna 1952). Bothma and Teer (1977) found that nestling rabbits in southeastern Texas opened their eyes at 6 or 7 days of age and moved away from the nest at 12 days. Cottontails begin to eat green plants on about the 8th day and are weaned by the 15th day. Tefft and Chapman (1983) conducted studies on confined Appalachian cottontails and developed a key to determine age of nestlings to 16 days (Table 5.4). In general, the newborn young of all *Sylvilagus* are similar in appearance. Hendrickson (1943) concluded there may be occasional common usage of nests by two or more females.

TABLE 5.4. Mean body measurements (±*SD*) and descriptions correlated with age for Appalachian cottontail nestlings born 9 May 1980

Age (days)	Length (mm ± *SD*) Hind Foot	Tail	Ear	Crown–Rump	Comments
Birth	10.5[a]	13.5[a]	9.6[a]	73.8[a]	Eyes tightly closed, premature movement
1	22.3±1.37	13.5	10.9	77.3±7.12	Eyes tightly closed, premature movement
4	27.8±0.74	14.8±1.70	14.8±0.74	81.4±4.00	Eyes tightly closed, squirming movement only
7	34.3±1.0	16.8±1.50	18.2±0.74	94.2±3.80	Eyes begin to open half, some coordination evident, movement weak
10	39.3±1.20	21.5±1.04	21.5±1.04	105.8±5.80	Eyes fully open, ability to sit rabbit-like, some hopping, first gender determination possible
16	49.7±1.20	22.7±2.30	30.7±0.63	133.2±5.00	Kittens entirely independent of nest, excellent coordination, ability to run and hop quickly
Mean growth (mm/day)	1.85	0.400	1.31	3.52	

SOURCE: Tefft and Chapman (1983), Table 1. See Chapman et al. (1992) for clarification on taxonomy of *S. obscurus*.
[a]Calculated values based on growth rate.

ECOLOGY

Sex Ratios. Adult and embryonic sex ratios of some species of *Sylvilagus* are given in Table 5.5. Proportions of females among embryonic young were not different for the seven studies cited. However, the proportion of females was greater than males for adult cottontails ($\chi^2 = 23.2, p < .05$) in the 14 studies cited. Chapman et al. (1977) suggested that there was selective mortality of males in a Maryland population of *S. floridanus*. However, a phenomenon of selective mortality does not seem typical for the genus. Edwards (1962a) found no difference in the rates of hunter harvests of male and female cottontails in mid-November or among cottontails trapped in October and early November. However, small patches of habitat ($\leq$2.5 ha) were occupied by male New England cottontails (*S. transitionalis*) more often than by female cottontails in human-dominated landscapes (Barbour and Litvaitis 1993). This difference may have been a result of differential dispersal capabilities. Alternatively, females may not have considered small patches to be suitable habitat.

Age Structure. Age ratios of cottontail populations may vary spatially and temporally (Edwards 1962a, 1962b). In general, cottontail populations having large juvenile:adult ratios reflect proportionally more young from late litters (Chapman et al. 1982). Edwards (1962a) reported that mid-November juvenile:adult ratios of *S. floridanus* in Ohio ranged from 1.4:1 to 10.9:1, and estimated that net annual productivity was 3–20 young per adult female. Large juvenile:adult ratios generally indicate high productivity (Chapman et al. 1982). However, they also may indicate low adult-survival rates (Caughley 1974), but adult-biased mortality among cottontails seems unlikely. Edwards (1962a, 1962b) examined age structure of >8000 cottontails killed by hunters in Ohio over 3 years. The proportion of juvenile rabbits in these samples was consistently 0.83 each year.

Abundance, Density, and Periodicity. In general, the highest densities of cottontails occur on islands or enclosures where emigration is restricted and predation often is reduced. Populations of cottontails (especially *S. floridanus*) may occasionally reach densities of 20/ha, but usually are considerably less. For example, the autumn density of a confined population of *S. floridanus* on the 210-ha Urban Wildlife Area in Ohio was 14.9/ha (Leite 1965), and peak density of *S. floridanus* on St. Clements Island, Maryland, reached 10.2/ha (Bittner and Chapman 1981). In comparison, the average density in autumn was 3.1 rabbits/ha

TABLE 5.5. Adult and embryonic sex ratios for *Sylvilagus*

Species	Location	Sex Ratio (Males/Females)	Sample Size	Year(s)	Source
Adult					
S. floridanus	Wisconsin	108/100	398	—	Elder and Sowls 1942
S. floridanus	Western Maryland	83/100	444	1971–73	Chapman et al. 1977
S. floridanus	St. Clements Island, Maryland	90/100	207	1976–77	Bittner and Chapman 1981
S. floridanus	Western Oregon	97/100	486	1968–69	Trethewey and Verts 1971
S. floridanus	Michigan	103/100	383	1935–37	Allen 1939
S. audubonii	Arizona	110/100	407	1951–55	Sowls 1957
S. audubonii	Texas	80/100	18	1973	Chapman and Morgan 1974
S. obscurus	Western Maryland/West Virginia	58/100	88	1971–74	Chapman et al. 1977
S. nuttallii	Central Oregon	84/100	266	1968–69	Powers and Verts 1971
S. aquaticus	Alabama	87/100	438	1960–67	Hill 1967
S. aquaticus	Texas	74/100	152	1954–55	Hunt 1959
S. aquaticus	Missouri	127/100	191	1957–59	Holten and Toll 1960
S. aquaticus	Mississippi	78/100	142	1981–88	Palmer et al. 1991
S. bachmani	Western Oregon	82/100	102	1967–69	Chapman and Harman 1972
S. bachmani	California	128/100	82	1950–51	Mossman 1955
Embryonic					
S. floridanus	Western Maryland	133/100	205	1971–73	Chapman et al. 1977
S. floridanus	St. Clements Island, Maryland	65/100	28	1976–77	Bittner and Chapman 1981
S. floridanus	Western Oregon	116/100	173	1968–69	Trethewey and Verts 1971
S. obscurus	Western Maryland/West Virginia	50/100	6	1971–74	Chapman et al. 1977
S. nuttallii	Central Oregon	84/100	76	1968–69	Powers and Verts 1971
S. bachmani	Western Oregon	85/100	26	1967–69	Chapman and Harman 1972
S. bachmani	California	92/100	27	1950–51	Mossman 1955

for a free-ranging population of eastern cottontails in Ohio (Chapman et al. 1982).

In the Florida Keys, densities of *S. palustris* ranged from 0.1 to 7.4/ha on remnant patches of habitat (Forys and Humphrey 1999a). Barbour and Litvaitis (1993) found that densities of local populations of *S. transitionalis* were affected by the size of the habitat patch that rabbits occupied. Rabbit density averaged 2.2/ha on small patches (<2.5 ha), whereas on large patches (>5 ha), density averaged 1 rabbit/ha. Differences in density among species of cottontails can be partially attributed to body size and geographic differences in habitat productivity. In the canebrakes of southwestern Indiana, autumn densities of *S. aquaticus* reached 0.4/ha (Terrel 1972). Flinders and Hansen (1973) estimated densities of *S. audubonii* in winter at 0.02/ha in the shortgrass prairie region of northeastern Colorado. McKay and Verts (1978) reported wide fluctuations of 0.06–2.5 rabbits/ha from monthly estimates of *S. nuttallii* in central Oregon. In British Columbia, densities of the same species varied from 0.2 to 0.4/ha (Sullivan et al. 1989). In the mountains of Tennessee, densities of *S. obscurus* reached 0.8/ha at one site, but cottontails were too sparse to estimate density in several other areas (Laseter 1999).

Numerous workers have reported cyclic tendencies in North American leporids (Leopold 1931; Grange 1949; Wight 1959; Keith 1963). J. A. Bailey (1968) suggested an 8- to 9-year pattern in regional populations of cottontails. Sadler (1981) also suggested population highs occurred at 10-year intervals. Unfortunately, the long-term data sets needed to rigorously test any temporal pattern in abundance are not available. Additionally, any influence of periodicity in cottontail populations may be much less than population changes (mostly declines) associated with the ramifications of human populations on land-use patterns (see section, Responses to Land-Use Changes).

Weather. The relationships of cottontail populations and weather have been considered by many investigators (Allen 1939; Grinnell 1939; A. M. Johnson and Hendrickson 1958; Wight and Conaway 1961). However, these relationships are typically nebulous and difficult to substantiate. Havera (1973) observed that the weather parameter most consistently related to cottontail harvests in Illinois was snowfall during the previous winter, especially in February and March. Snowfall and rain may affect cottontail populations by influencing production and survival of the first litter in early spring, and subsequently the potential for early-born young that contribute later as breeders in autumn populations (Havera 1973; Applegate and Trout 1976).

Among New England cottontails, Brown and Litvaitis (1995) observed a strong correlation between days with snow coverage and predation rates on cottontails. The consequences of snow coverage can have substantial implications for small, disjunct populations of this species, when correlation in mortality rates among populations of rabbits may result in regional extirpation (*sensu* Harrison and Quinn 1989). Along this line of thought, Keith and Bloomer (1993) showed that persistent snow cover and low temperatures, combined with the brown pelage and the large foot-loading of eastern cottontails (resulting in a conspicuous rabbit with limited mobility in deep snow), apparently made them vulnerable to intense predation in winter. The combination of severe winter weather and limited adaptations to snow likely determines the northern range limit of most cottontails.

Activity, Movements, and Home Ranges. Cottontails are active throughout the day, but activity and movement are substantially less between sunrise and sunset (Mech et al. 1966; Holler and Marsden 1970) (Fig. 5.11). Weather extremes also affect activity. Among eastern cottontails in Illinois, activity was decreased during periods with high ambient temperature (maximum daily temperature >29°C) (Lepitzki 1990). In winter, activity also was noticeably reduced among New England cottontails when minimum daily temperatures remained less than 10°C for several days (J. Litvaitis, pers. obs.).

In general, cottontails do not maintain territories. Home ranges of different age and sex classes of cottontails overlap during much of the year, particularly in late fall and winter, when rabbits tend to congregate

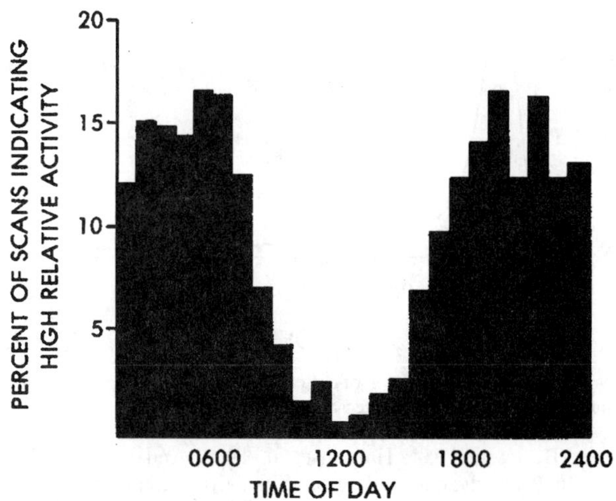

FIGURE 5.11. Daily relative activity of eastern cottontail (*Sylvilagus floridanus*) in southern Illinois (July–November). Histograms are a composite of 9660 four-minute scans of nine rabbits. SOURCE: Lepitzki 1990. Reproduced with permission from the author and *Canadian Field-Naturalist*.

in areas containing a combination of food and escape cover (Chapman et al. 1982). However, Trent and Rongstad (1974) found little or no overlap in the ranges of adult females during the breeding season.

Estimates of home range size are plagued by a variety of problems and considerations, including age and sex of animals monitored, method of estimation, number of locations, sampling interval, and study period (White and Garrott 1990; Larkin and Halkin 1994; Powell 2000). As a result, the utility of home range size as a useful parameter in understanding the ecology of cottontails may be limited to site-specific applications where home range estimates are used to address specific questions. To understand patterns of resource use by introduced *S. floridanus* and endemic *S. bachmani,* Dixon et al. (1981) compared home range size and shape using activity isopleths (Dixon and Chapman 1980). Differences in home range size seemed to be a consequence of differences in habitat use by the two species. Home ranges of *S. bachmani* were characterized by core areas of activity that were associated with clumps of *Rubus* and limited activity outside of the clumps (Chapman 1971b; Dixon and Chapman 1980). Conversely, *S. floridanus* occupied larger home ranges without well-defined core areas. This pattern supported observations in the same study area based on mark–recapture techniques (Chapman and Trethewey 1972) indicating that *S. floridanus* used more diverse habitats than *S. bachmani.* In Florida, Forys and Humphrey (1996) observed that marsh rabbits moved among patches of suitable habitat as juveniles. As adults, marsh rabbits usually became more sedentary and remained in one patch until their deaths. In other regions, however, home ranges of adults may not be static. Home ranges of swamp rabbits were dynamic apparently in response to seasonal flooding of riparian habitats (Conaway et al. 1960; Zollner et al. 2000).

HABITAT

Cottontails are widely distributed across North America. As a result, no single vegetative community provides habitat for this group. Although the requirements of each species are different, there are some basic similarities. Cottontails are found in a wide variety of disturbed, early successional, or shrub-dominated habitats, which include an abundance of forage and dense understory cover.

Species-Specific Habitat Associations. Based on their association with specific plants and structural elements of habitat, populations of cottontails respond spatially and temporally to variation in habitat suitability. In forests, eastern cottontails are often associated with early-successional habitats, and local populations decline as forests mature (Fig. 5.12). Abandoned agricultural fields, shrublands, and shrub–woodland are considered among the most productive habitats for this

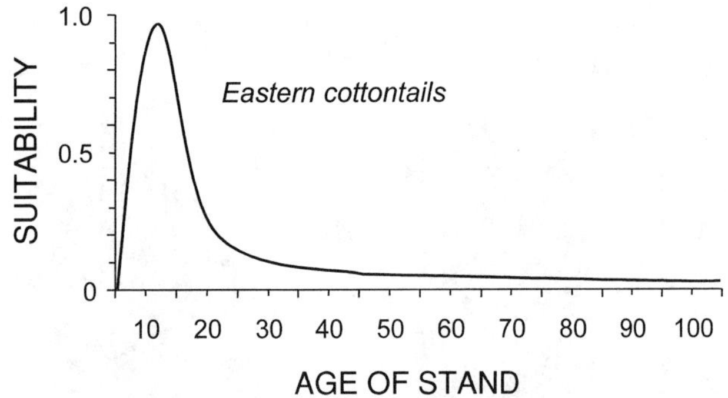

FIGURE 5.12. Suitability function of eastern cottontail (*Sylvilagus floridanus*) in forest habitats. SOURCE: Adapted from Giles 1978.

species (Beckwith 1954; Hanson et al. 1969; Anderson and Pelton 1976). In Pennsylvania, underground dens (including those dug by woodchucks, *Marmota monax*) often provide escape cover and shelter, especially among female eastern cottontails (Althoff et al. 1997). Use of underground dens was most frequent when air temperature was less than 5°C (Althoff et al. 1997). In the same region, aboveground bedding sites (forms) of eastern cottontails were characterized by dense, woody vegetation within 0.5 m of ground level. As a result, the microclimate of forms included lower temperatures, less net radiation, and lower wind speeds (Althoff et al. 1997). K. A. Morgan and Gates (1983) reported that hedgerows of multiflora rose (*Rosa multiflora*) provided good cover in agricultural landscapes of Maryland. Althoff et al. (1997) also indicated that some shrubs, especially Tartarian honeysuckle (*Lonicera tatarica*), were important because of their tendency for early leaf emergence in comparison to other shrubs. The selection of dense vegetation as resting sites has prompted wildlife managers to advocate construction of brushpiles as a method to augment cover for cottontails (but see section, Management) or to plant dense hedgerows. However, multiflora rose and Tartarian honeysuckle are exotic species. Therefore, they may be inappropriate for planting because of their potential to reduce abundance of native shrub species and disrupt community dynamics (Schmidt and Whelan 1999).

New England cottontails seem more restricted in habitat selection than eastern cottontails, especially with regard to their affinity for understory cover (D. F. Smith and Litvaitis 2000). In New Hampshire, Barbour and Litvaitis (1993) investigated selection of microhabitats in winter among *S. transitionalis*. Because coniferous stems provided approximately three times the visual obstruction of deciduous stems in winter (Litvaitis et al. 1985), these investigators converted understory stems (<7.5-cm-diameter breast height) to stem-cover units to generate a metric with a standard value where stem-cover units/ha = deciduous stems plus (3 × coniferous stems/ha). Selection indices (percentage of use minus percentage of availability) showed an obvious use of sites with >50,000 stem-cover units/ha (Fig. 5.13). Furthermore, New England cottontails were reluctant to venture >5 m from cover (Barbour and Litvaitis 1993).

Sommer (1997) also indicated that Appalachian cottontails in western Maryland (reported as information on New England cottontails) avoided sites with less than ~60,000 stems/ha and sites >5 m from cover. In this region, habitats were characterized by mature mixed-oak (*Quercus* spp.) forests with dense ericaceous vegetation (especially *Kalmia latifolia* and *Vaccinium* spp.) or clearcuts with dense raspberry canes. Laseter (1999) also indicated that dense understory vegetation was a critical component to the habitat of Appalachian cottontails in Tennessee.

Desert, Nuttall's, and pygmy cottontails all occupy arid habitats in western North America. Orr (1940) indicated that desert cottontails selected stream bottoms with abundant willows (*Salix* spp.). Nuttall's cottontails occupy rocky ravines and sagebrush-covered hills (Dice 1926). In British Columbia, Sullivan et al. (1989) reported that sagebrush habitats with at least 30% vegetative cover represented good habitat for Nuttall's cottontails. Sagebrush also is a common

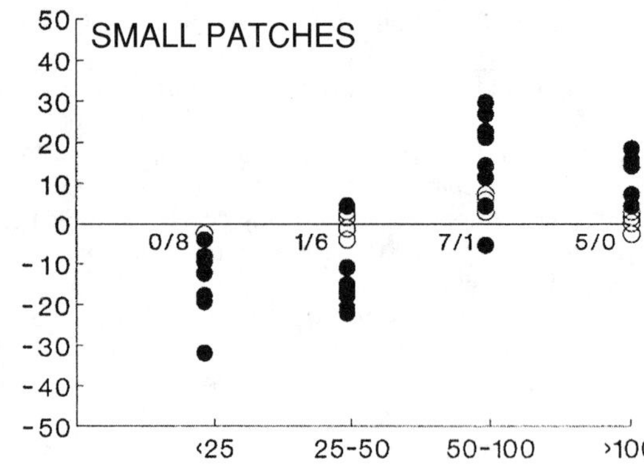

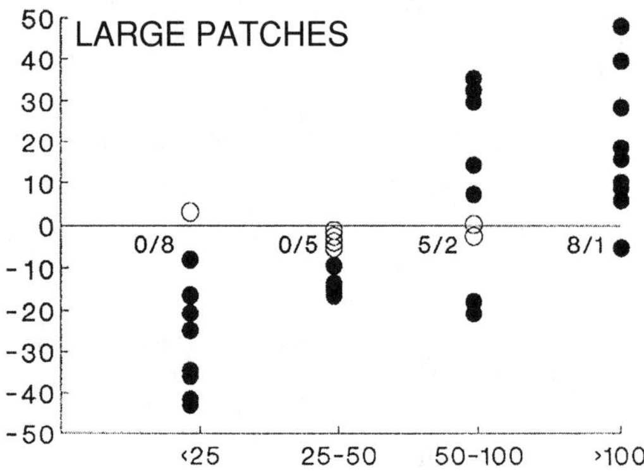

FIGURE 5.13. Habitat selection by New England cottontail (*Sylvilagus transitionalis*) on 12 small (<2.5 ha) and 9 large (>5 ha) patches of habitat in southern New Hampshire. Selection was based on fecal pellet distributions in relation to understory density [stem-cover units = deciduous stems + (3 × coniferous stems)]. Filled circles indicate use that differed from expected, and numerals associated with a column are the number of patches where use was significantly greater (+) or less (−) than expected. SOURCE: Barbour and Litvaitis 1993. Reproduced with permission from *Oecologia,* ©Springer-Verlag.

component of habitats occupied by pygmy rabbits (Severaid 1950; Green and Flinders 1980; Weiss and Verts 1984; Gabler et al. 2000). In Wyoming, Katzner and Parker (1997) found that tall, wide sagebrush plants were more abundant within home ranges of pygmy rabbits than in adjacent nonused areas. The abundance of sagebrush affected the size of home range core areas (Katzner and Parker 1997) and may have determined local density of pygmy rabbits.

In contrast to western species, in the southeastern United States, swamp and marsh rabbits are never far from water. Terrel (1972) found that swamp rabbits limit their home ranges to within 2 km of a major body of water; this species is usually associated with forested wetlands (Korte 1975; Kjolhaug and Woolf 1988). Porath (1997) found that sites occupied by swamp rabbits had lower tree densities and more ground cover than unoccupied sites. Others also have reported that swamp rabbits frequently occupy canopy gaps within bottomland forests (Terrel 1972; Korte 1975; Fredrickson 1980). Zollner et al. (2001) observed that daytime resting sites were distinguished from random sites by having more ground covered by shrubs, down treetops, and herbaceous vegetation. These observations have prompted some to recommend silvicultural practices, such as selective cuts and small clearcuts, that increase the abundance of food and cover used by swamp rabbits (Garner 1969; Mullin 1982; Hurst and Smith 1986). Latrines also seem to be an important habitat component of swamp rabbits (Lowe 1958; Terrel 1972;

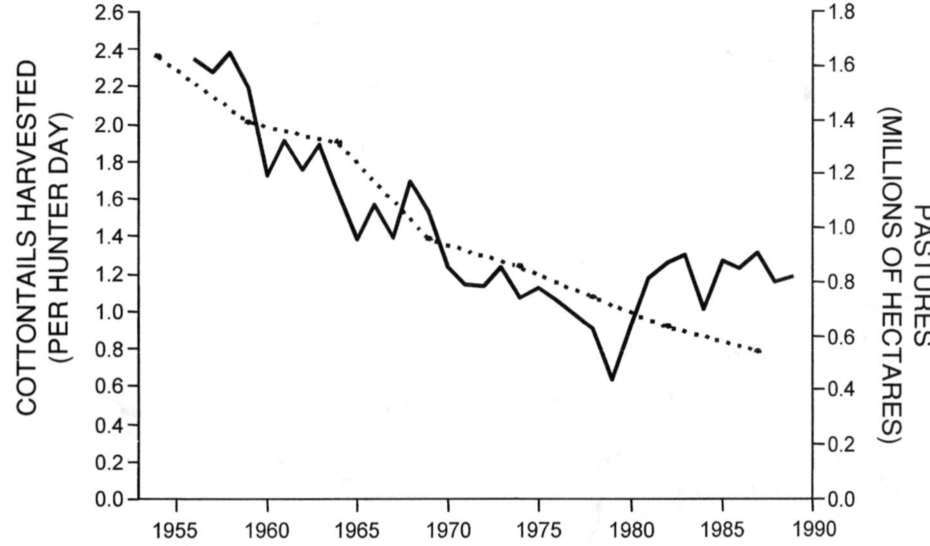

FIGURE 5.14. Relationship between eastern cottontail (*Sylvilagus floridanus*) (indexed by hunter harvests; solid line) and loss of agricultural pastures (dashed line) to other land uses, largely conversion to corn and soybean fields SOURCE: Mankin and Warner 1999a. Figure redrawn by P. Mankin.

Heuer and Perry 1976). Because the frequency of fecal pellet deposition on logs by swamp rabbits is correlated with rabbit density (Martinson et al. 1961), some investigators have used inventories of pellets to survey swamp rabbit populations (e.g., Porath 1997). Whitaker and Abrell (1986) speculated that pellets were deposited by rabbits on logs because they spend considerable time on logs as observation posts. However, latrines may be used to delineate territories (Zollner et al. 1996), similar to the marking behavior of European rabbits (Sneddon 1991). However, conclusive evidence supporting either hypothesis on the role of latrines is lacking (Zollner et al. 1996).

Habitat requirements of marsh rabbits have not been extensively studied. In Florida, Blair (1936) indicated that marsh rabbits occur in hammocks (small stands of woody vegetation in extensive marshes or swamps) and among cattail (*Typha* spp.) stands along pond edges. Latrines occur on logs, stumps, or along active runways (Blair 1936). In the Florida Keys, Forys and Humphrey (1999a) considered marsh rabbit habitat to be a transition zone along the edges of marshes that were dominated by thick grasses and shrubs. A consistent habitat component of marsh rabbits is pooled water (Chapman and Willner 1981). Like swamp rabbits, marsh rabbits frequently swim while moving between feeding and resting sites or avoiding predators.

Responses to Land-Use Changes. As Europeans colonized North America, clearing of eastern forests for agriculture substantially altered wildlife habitats. The resulting interspersion of agricultural fields, brushy edges, and woodlots provided cottontails with an abundance of food and cover, and populations increased (Chapman and Morgan 1973). However, subsequent changes in agriculture have reversed this trend.

In the midwestern United States, farming practices have shifted to large fields dedicated to single crops and more intensive clearing of idle areas and hedgerows (Vance 1976; Mankin and Warner 1999a). These changes have resulted in dramatic declines among populations of eastern cottontails (Fig. 5.14). In Illinois, populations of eastern cottontails declined >70% from 1956 to 1978 (Edwards et al. 1981) as diversified, rotation farming shifted to intensive monocultures dominated by row crops (Edwards et al. 1981; Mankin and Warner 1999a). Although cottontails are able to persist in such habitats, they are substantially restricted to small portions of the landscape where suitable cover is available (Swihart and Yahner 1982; Mankin and Warner 1999b). During the nongrowing season, shelterbelts and other woody vegetation associated with farmsteads provide the remaining cover that is critical to cottontail survival (Mankin and Warner 1999b).

Similar to eastern cottontails, populations of New England cottontails have responded to changes in land use. Throughout the range of this species, forests were cleared for agriculture by European settlers. But unlike the midwestern United States, forests have returned to the Northeast. In New Hampshire, for example, land that was forested was reduced from an estimated 95% at the time of European colonization to <50% by 1880 (Harper 1918). However, by the mid-1800s, farmers in New Hampshire (and adjacent states) were unable to compete with more productive farms in the Midwest (Irland 1982). As a result, these lands were abandoned during the late 1800s and the first half of the twentieth century (Black 1950). Much of the abandoned farmland was subsequently colonized by second-growth forests, and within 100 years, >85% of New Hampshire was forested (Litvaitis 1993). New England cottontails responded to regeneration of forests and relatively short-term availability of early-successional habitats. During the years of peak abundance of young forests (approximately 1910–1960; Litvaitis 1993), populations of New England cottontails spanned approximately 60% of New Hampshire (ca. 1950) (C. L. Stevens, University of New Hampshire, unpublished data; illustrated in Jackson 1973). However, as these forests matured and understory cover thinned, populations of cottontails declined rapidly (Litvaitis 1993; Litvaitis et al. 1999). Currently, small, disjunct populations span <20% of the state (Litvaitis 1993). This pattern of population expansion followed by contraction occurred throughout the range of New England cottontails (Litvaitis and Litvaitis 1996).

Remaining populations of New England cottontails are disjunct and restricted to patches of habitat in what are considered induced metapopulations (Fig. 5.15). Litvaitis and Villafuerte (1996) indicated that New England cottontails are structured in "mainland–island" systems. In such metapopulations, large patches (mainlands) are responsible for net recruitment into the population, whereas small patches (islands) support few cottontails with limited resources, which prevents successful recruitment, and residents have low survival. As a result, small patches function as sink habitat (Lidicker 1975; Van Horne 1982; Pulliam 1988) and are dependent on dispersing individuals from larger patches for recruits. Such spatially structured populations should not be confused with "patchy populations," where individuals move freely across the landscape and simply congregate in resource-rich patches of habitat (Harrison 1994; Hanski et al. 1995). Field evidence supports this description of spatial structure among New England cottontails. In southern New Hampshire, rabbits on small patches (<3 ha) encountered winter food shortages and were killed by predators at approximately twice the rate as cottontails on large patches (Barbour

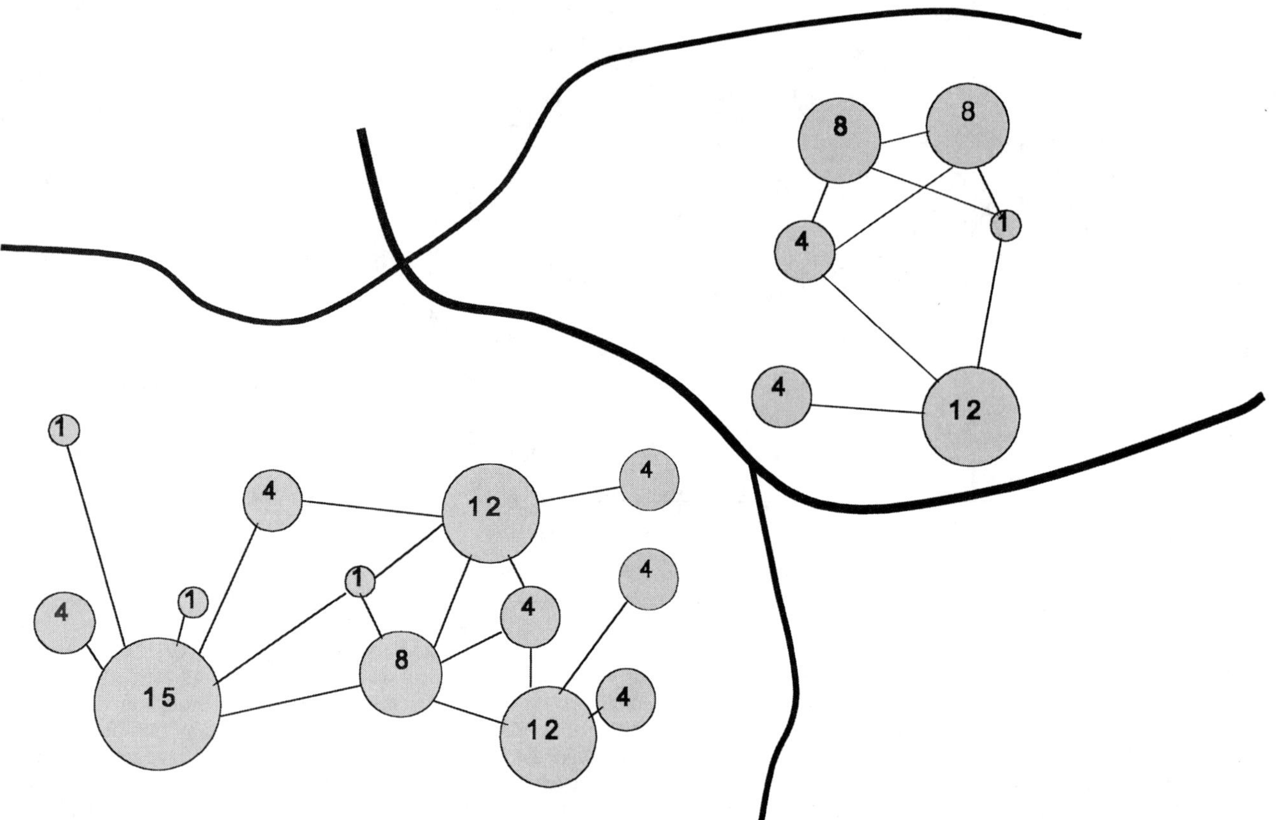

FIGURE 5.15. Schematic representation of an induced metapopulation of New England cottontail (*Sylvilagus transitionalis*). Lines that connect subpopulations represent possible dispersal routes, and numbers are carrying capacities for each patch. Bold lines represent a network of roads that effectively prevents exchanges between the two metapopulations, making the smaller population more vulnerable to extinction. SOURCE: Adapted from Litvaitis and Villafuerte 1996.

and Litvaitis 1993; Brown and Litvaitis 1995; Villafuerte et al. 1997). Concurrent with the loss of early-successional habitats, populations of generalist predators (especially coyotes, *Canis latrans;* and red foxes, *Vulpes vulpes*) have increased in response to the conversion of forests to other land uses (Oehler and Litvaitis 1996). In these landscapes, presence of large patches (>10 ha) seems critical to the survival of local metapopulations of cottontails (Litvaitis and Villafuerte 1996). However, current trends in land use and land-ownership patterns indicate that such habitats will become increasingly rare (Litvaitis et al. 1999; Litvaitis 2001). As a result, concern over long-term viability of New England cottontails has prompted several conservation organizations to petition the U.S. Fish and Wildlife Service to list the species as threatened or endangered.

Populations of at least three other species of cottontails also are in decline. Sagebrush habitats occupied by pygmy rabbits have been converted to cropland or used for livestock grazing (Gabler et al. 2000). Grazing is the dominant land use of shrubland habitats (West 1983), and overgrazing can reduce sagebrush cover and reduce native grasses. Such changes can make an area unsuitable for pygmy rabbits (Gabler et al. 2000). Degradation of sagebrush habitats has been substantial. In Washington, approximately 40% of shrub–steppe habitats have been converted to other cover types (Dobler 1992, cited in Gabler et al. 2000).

Bottomland forests occupied by swamp rabbits also are being eliminated at a rapid rate. Within the Mississippi River floodplain, agriculture and development have cleared approximately 80% of the riparian forest (Creasman et al. 1992, cited in Zollner et al. 2001). A restricted geographic range, combined with intensive development, has eliminated the lower Keys marsh rabbit from 9 of the 12 Florida Keys that it once occupied (Forys and Humphrey 1999b). As a result, the U.S. Fish and Wildlife Service has listed this subspecies of marsh rabbit as endangered (*Federal Register* 1990). Stabilizing and reversing the declines in habitats important to cottontails and other species will become a major conservation challenge.

FEEDING HABITS

The diets of cottontails have been studied extensively and vary greatly, depending on the species of rabbit and the locality and availability of palatable plants. Feeding habits of cottontails are cosmopolitan. In northern regions, herbaceous plants are consumed during the growing season and woody species are eaten during the dormant season.

Based on the large geographic range and diverse habitats occupied by eastern cottontails, it is not surprising this species consumes a wide variety of plants. According to R. H. Smith (1950), the most frequently consumed herbaceous species by cottontails in New York were Kentucky bluegrass (*Poa pratensis*), Canada bluegrass (*P. compressa*), timothy (*Phleum pratense*), quack grass (*Agropyron repens*), orchard grass (*Dactylis glomerata*), red clover (*Trifolium pratense*), and wild carrot (*Daucus carota*). In the same study, the most frequently consumed woody plants were apple (*Malus pumila*), staghorn sumac (*Rhus typhina*), red maple (*Acer rubrum*), blackberry (*Rubus allegheniensis*), and red raspberry (*R. strigosus*) (R. H. Smith 1950).

Dalke and Sime (1941) conducted extensive research on cottontail feeding habits in Connecticut, but made no distinction between *S. floridanus* and *S. transitionalis*. Spring and summer diets of these species consisted of herbaceous plants, mainly clover, timothy, and alfalfa. November and December was a transition period, when cottontails switched to woody plants. Winter diets consisted mainly of small trees, including gray birch (*Betula populifolia*), red maple, apple, aspen (*Populus tremuloides*), choke cherry (*Prunus virginiana*), and black cherry (*P. serotina*), and shrubs or vines, especially blackberry, dewberry (*Rubus villosus*), willow (*Salix* spp.), black alder (*Ilex beticillata*), maleberry (*Lyonia ligustrina*), and highbush blueberry (*Vaccinium corymbosum*).

Barbour and Litvaitis (1993) found that preferences for woody plants in winter by *S. transitionalis* varied with plant availability. On small patches of habitat (<2.5 ha), winter forage was less abundant per rabbit and individuals consumed a greater variety of available plants than cottontails on large patches (>5 ha). Diet quality in this study was indexed by twig diameter at point of browsing (dpb) because protein concentration declines with twig diameter (Wolff 1980). On small patches, stems with a dpb 3 mm represented 31% of the clipped twigs encountered, compared to 20% of browsed twigs on large patches. The incidence of bark use by rabbits also varied; 13% of plots sampled on small patches containing evidence of bark use, but only 2% of sample plots on large patches had evidence of bark use (Barbour and Litvaitis 1993). Because rabbits on small patches had lower body weight, consumption of bark may have been a reaction to food limitation.

Diets of sympatric Appalachian and eastern cottontails overlapped substantially (Spencer and Chapman 1986). However, eastern cottontails consumed more grasses than did Appalachian cottontails.

The seasonal availability of edible plants seems to be the most important influence on the diet of *S. audubonii* (Fitch 1947). This conclusion is supported by the work of Orr (1940), who found that *S. audubonii* seasonally fed on grasses, sedges, rushes, willows, oaks, blackberries, wild roses, and California mugwort (*Artemisia vulgaris*). The xeric habitats occupied by this cottontail may have caused rabbits to consume moisture-rich cactus and forbs as a source of water (Turkowski 1975), and may be the factor that determines forage selection during summer (Scribner and Krysl 1982).

Pygmy rabbits may be the most dietary specialized cottontails. They consume sagebrush (*Artemisia* spp.) throughout the year. In Idaho, sagebrush was eaten almost exclusively during winter, and was supplemented with grasses and forbs during other seasons (Green and Flinders 1980). Grasses were the most important food items of brush rabbits throughout their range (Orr 1940). However, when available, clover (*Trifolium involucratum*) was the preferred food. Brush rabbits also fed on the stems and berries of woody plants, such as blackberry (Orr 1940). In contrast, *S. aquaticus* fed on plants in proportion to their abundance (Toll et al. 1960). Sedges and grasses seemed to be a common food throughout the range of this species (Terrel 1972). Marsh rabbits, on the other hand, fed extensively on herbaceous and woody plants (Blair 1936). In the Florida keys, two grasses (*Sporobolus virginicus* and *Spartina spartinae*), a succulent shrub (*Borrichia frutescens*), and one tree species (*Laguncularia racemosa*) made up >70% of the diet of marsh rabbits. In general, *S. palustris* fed on plants that were most abundant in their habitat, and diet did not vary substantially among seasons (Forys 1999).

Coprophagy (reingestion of fecal pellets) has been reported for most species of *Sylvilagus*. Two types of pellets are excreted: hard, brown fecal pellets and soft, green food pellets. About 60% of the total fecal excretion of hard pellets is nutrients (Hamilton 1955; J. A. Bailey 1969). Soft pellets are produced in the cecum and provide vitamin B supplementation. Soft pellets are eaten directly from the anus before they touch the ground, usually two or three pellets at a time (Kirkpatrick 1956).

BEHAVIOR

Numerous observational studies have been conducted on cottontails. These include investigations on *S. floridanus* (Lord 1964; Marsden and Holler 1964; Casteel 1966), *S. audubonii* (Orr 1940; Ingles 1941), *S. nuttallii* (Orr 1940), *S. bachmani* (Chapman and Verts 1969; Zoloth 1969), *S. transitionalis* (Dalke 1942; Olmstead 1970), *S. obscurus* (Tefft and Chapman 1987), *S. idahoensis* (Orr 1940), *S. aquaticus* (Marsden and Holler 1964), and *S. palustris* (Blair 1936). Among the most comprehensive studies was that conducted by Marsden and Holler (1964) on *S. aquaticus* and *S. floridanus*. In confined populations, both species had a dominance hierarchy, especially among males. In these populations, aggressive interactions were usually initiated by dominant males. Among *S. aquaticus,* males maintained territories around the immediate area occupied by females. The frequency of male–female interactions among swamp rabbits was correlated with the status of the male, and the majority of observed copulations were by the dominant male. These traits were not observed in *S. floridanus*. In contrast, dominant male *S. floridanus* were more far-ranging than subordinate males (Marsden and Holler 1964), perhaps as a way to maximize encounters with females.

During the early 1900s, natural resource agencies and hunting clubs translocated thousands of eastern cottontails to the northeast (Johnston 1972; Chapman and Morgan 1973) and northwest United States (Dalquest 1941; Graf 1955). Chapman and Morgan (1973) indicated that cottontails transplanted in the Northeast included individuals from several subspecies, and that subsequent interbreeding by these rabbits resulted in offspring that were genetically diverse. The authors speculated that increased genetic variability had substantial ramifications on behavioral interactions among endemic and transplanted cottontails. Specifically, Chapman and Morgan (1973) speculated that offspring of transplanted eastern cottontails were able to expand their realized niche by moving into habitats that this species did not occupy before translocations. These habitats included those that were formerly occupied exclusively by *S. transitionalis* (and populations now considered to be *S. obscurus*). Sympatric populations of *S. floridanus* and *S. transitionalis* use similar foods (Dalke and Sime 1941) and cover (Eabry 1968; Linkkila 1971; Johnston 1972). Because *S. floridanus* is approximately 20% heavier than *S. transitionalis* (Litvaitis et al. 1991), interference competition (Case and Gilpin 1974) has been proposed as an explanation for the regional decline of *S. transitionalis* (Fay and Chandler 1955, Reynolds 1975). This conjecture was supported by roughly simultaneous expansion of populations of *S. floridanus* and decline among populations of *S. transitionalis* in the northeastern United States (Chapman and Morgan 1973; Probert and Litvaitis 1996), implying that expanding populations of eastern cottontails had a detrimental effect on populations of New England cottontails.

The ability of *S. floridanus* to dominate and exclude *S. transitionalis* from sites that *S. transitionalis* occupied exclusively was evaluated by Probert and Litvaitis (1996). These investigators examined interactions between individuals in small enclosures and also monitored use of microhabitats in a large enclosure where abundance of cover and food varied. Among behavioral dyads (dyad = an individual of each species released into a small enclosure) where one individual dominated another, *S. floridanus* was the dominant rabbit in 58% of the trials (Probert and Litvaitis 1996). This was not significantly different from random expectation. In addition, no consistent differences were detected in microhabitat selection in the large enclosure. Probert and Litvaitis (1996) concluded that interference competition did not explain the change in distribution and abundance of cottontails in the northeast United States. Eastern cottontails may benefit by being able to occupy habitats with less cover than required by New England cottontails. As a result, eastern cottontails may colonize disturbance patches sooner and maintain access to these habitats simply on a system of "prior rights" (Probert and Litvaitis 1996).

This speculation was supported by a subsequent comparison of foraging strategies used by eastern and New England cottontails (D. F. Smith and Litvaitis 1999). The response of both species was monitored in a large enclosure where cover was available in one portion of the enclosure and proximity of food (quality and quantity) in relation to cover varied among trials. In trials with low-quality food in cover and high-quality food in open areas, *S. transitionalis* sacrificed food quality for safety by remaining in or close to cover. *S. floridanus,* on the other hand, avoided low-quality food and maintained physical condition by foraging at sites away from cover that contained high-quality food. When all food was removed from cover, *S. transitionalis* was reluctant to forage in the open, lost a greater proportion of body weight, and succumbed to higher rates of predation than did *S. floridanus*. The behavioral differences between these species of cottontails may be partly attributed to morphological differences. Specifically, D. F. Smith and Litvaitis (1999) reported that *S. floridanus* had a larger exposed surface area of the eyes than *S. transitionalis*. This difference apparently enabled *S. floridanus* to detect a simulated predator (moving model of an owl) at greater distances ($\bar{X} = 21.2$ m) than *S. transitionalis* ($\bar{X} = 9.4$ m, D. F. Smith and Litvaitis 1999:58). When results of the foraging trials were applied to habitat-use patterns of free-ranging cottontails, D. F. Smith and Litvaitis (2000) estimated that *S. transitionalis* could exploit only 32% of the available habitat without experiencing intense predation, whereas *S. floridanus* could exploit 99% of the habitats. These results suggest that the consequences of widespread forest maturation and fragmentation in the northeastern United States that have been detrimental to *S. transitionalis* apparently have been less deleterious to *S. floridanus*. These differences in behavior and morphology probably explain the shift in composition of *Sylvilagus* populations in the northeastern United States.

MORTALITY

Edwards et al. (1981) advanced the thesis that for the eastern cottontail in the midwestern United States, population regulation depends primarily on survival. This view stems in part from the lack of evidence for significant density-dependent regulation of reproduction. To support that contention, they noted the following features pertaining to cottontails:

1. Synchronous breeding (Conaway and Wight 1962; Stevens 1962; Casteel 1966; Kirkpatrick and Baldwin 1974; Chapman et al. 1977)
2. Onset of breeding dependent on weather (Wight and Conaway 1961; Chapman et al. 1977)
3. Postpartum breeding (Conaway and Wight 1962; Casteel 1966)
4. Breeding by most of the sexually mature individuals (Wight and Conaway 1962b; Stevens 1962; Chapman et al. 1977; McKay and Verts 1978)
5. Consistency of litter size (Lord 1960; Wight and Conaway 1962b; Stevens 1962; Chapman et al. 1977)
6. Subadult breeding (Sowls 1957; Lord 1961; Stevens 1962; Casteel and Edwards 1964; Kibbe and Kirkpatrick 1971; Chapman et al. 1977; Bothma and Teer 1977)
7. Limited curtailment of breeding in response to density (Sheffer 1957; Kirkpatrick and Baldwin 1974; Bittner and Chapman 1981)
8. Minimal energy invested in rearing young (28 days gestation, rapid development, free living at 10–14 days, nursing only one or two times daily) (Chapman et al. 1982)

Understanding all the parameters of population regulation for cottontails is difficult (Chitty 1957, 1971; Watson and Moss 1970; Krebs 1971; Krebs and Myers 1974). However, as Gibb (1977) pointed out, no single factor is common to all situations. In a broad sense, regulation of cottontail abundance can be seen as a hierarchical system of interrelated multiple controls, the principal components of which include the following:

1. Limits of tolerance to extreme fluctuations in environmental factors
2. Innate dispersal that is independent of density
3. Denial of access to available essential resources by inter- and intraspecific competition, thus facilitating density-dependent dispersal
4. Innate density-dependent regulation of natality (this mechanism is either not present or not effectively operative in the cottontail)
5. Vulnerability to predation
6. Susceptibility to disease and infection
7. Consequences of exhaustion of available resources, principally nutritional inadequacies

The level of population control becomes increasingly severe as additional components of the system become more effective and as density increases. Starvation and stress-related symptoms are seen as physiological consequences, not innate regulatory mechanisms (Chitty 1960; Christian and Davis 1964). Disease also plays a role in regulating cottontail populations (Woolf et al. 1993). Population regulation in the cottontail has a density-dependent survival component, with dispersion contributing an element of self-regulation to the control process. Conaway and Wight (1962) and Bittner and Chapman (1981) believed that high density was a factor in reducing the reproductive rate of *S. floridanus*.

Predation is an important factor affecting abundance of cottontails (Pearson 1971; Maclean et al. 1974; Gibb 1977). Keith and Windberg (1978) felt that winter food shortage was the factor responsible for initiating the decline in the hare cycle, with predation prolonging that decline. Keith and Bloomer (1993) believed that eastern cottontail vulnerability to predators explained the absence of this species from the forests of northern Wisconsin. Based on extensive theoretical modeling, Holling (1966) concluded that predation is discontinuous, with search time important in determining when predation will effectively cease. Predation is probably the major factor in the death of most cottontails, and is the primary direct cause of cottontail population regulation. Many canids, felids, mustelids, and raptors prey on cottontails, as do certain snakes (see Table 5.8 in Chapman et al. 1982).

Parasites. Cottontails are hosts of a variety of parasites. Those ectoparasites that have received the most attention are ticks because they are a vector for Rocky Mountain spotted fever (*Rickettsia rickettsi*). Many of the tick-borne diseases are readily transmissible to humans and domestic pets, and thus are of concern from the public health standpoint. Among the more common ectoparasites of cottontails are ticks of the family Ixodidae, the fleas Pulicidae and Leptopsyllidae, and warbles, Cuterebridae (Chapman et al. 1982; Lepitzki et al. 1992). A summary of the most common ectoparasites of cottontails is given in Table 5.9 of Chapman et al. (1982).

Cottontails also host a variety of endoparasites (Chapman et al. 1982). Common endoparasites include nematodes of genera *Obeliscoides, Trichostrongylus, Longistriata,* and *Trichuris;* the cestodes, *Mosgovoyia* and *Taenia;* and coccidian parasites (Chapman et al. 1982; Lepitzki et al. 1992).

Viral, Bacterial, and Rickettsial Diseases. Cottontails are known reservoirs of tularemia, although the causative agent, *Francisella tularensis,* has been documented in >100 mammalian species (Hassell 1982). In Illinois, *S. floridanus* is the source of >90% of the human cases of tularemia (Yeatter and Thompson 1952). The main vector of tularemia is the rabbit tick (*Haemaphysalis leporis-palustris*), but other ticks and fleas may also carry the disease. Humans contract tularemia by coming in direct contact with the flesh and blood of infected rabbits or by eating infected rabbits that are not properly cooked. Tularemia occurs widely throughout North America, and the number of cases in various regions fluctuates yearly and seasonally (Yeatter and Thompson 1952). However, tularemia is generally most prevalent in the spring and fall. The disease is always fatal to the rabbit. The fatality rate for untreated cases in humans is 7%.

When infected cottontails are encountered in the field, they may appear somewhat sluggish. Internally, the disease can be recognized by a peppering of tiny white spots on the liver and spleen. If a rabbit with this condition is found, it should be burned and one's hands washed and disinfected. Cuts or abrasions should be treated with iodine. Cottontails may not exhibit any signs of tularemia in early stages (Yeatter and Thompson 1952).

Ectoparasites of cottontails also carry Rocky Mountain spotted fever and many other rickettsial diseases of humans (Philip 1946; Parker et al. 1952, Shirai et al. 1961; Burgdorfer et al. 1966; Sonenshine et al. 1966; Burgdorfer 1969; Cooney and Burgdorfer 1974). Rocky Mountain spotted fever is generally thought of as a disease of western North America; however, the disease is widely distributed. The primary natural means of infection in humans is through the bite of a tick, in the case of the cottontail, mainly *H. leporis-palustris, Ixodes dentatus* (Parker et al. 1952), and *Amblyomma americanum* (Philip 1946). Other diseases in which the cottontails have been implicated include *Staphylococcus aureus* (McCoy and Steenbergen 1969), eastern encephalitis (Hayes et al. 1964), and papilloma virus (Han et al. 1999, 2000).

AGE ESTIMATION

A number of techniques have been used to estimate age of cottontails. Those used most often include (1) judging prenatal growth, (2) epiphyseal cartilage closure, (3) body measurements, (4) bone growth, and (5) eye lens weight.

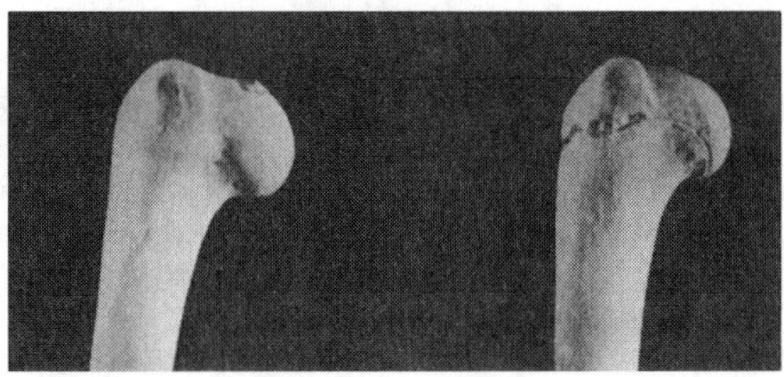

FIGURE 5.16. The proximal humerus of the eastern cottontail (*Sylvilagus floridanus*). (Left) Adult and (right) juvenile. Note the epiphyseal groove present on the humerus of the juvenile and the smooth, fully ossified humerus of the adult. SOURCE: Photo by Michael J. Reber.

Prenatal Development. The age of prenatal *S. floridanus* can be estimated (±1 day) from 5 days to term based on nose-to-tail length, hind foot length, and weight. Rongstad (1969) developed a scale based on known-age embryos.

Epiphyseal Cartilage Closure. Thomsen and Mortensen (1946) were the first to study the epiphyseal groove as an age criterion. They found that young of the year could be differentiated from adults by the presence of the epiphyseal cartilage. In newborn cottontails, a thick plate of epiphyseal cartilage occurs between the diaphysis, or shaft, and the epiphysis, or head, of long bones (Fig. 5.16). The cartilage area is a growth center where proliferation of cartilage and its replacement by bony tissue results in elongation of bone. As a rabbit grows older, the cartilage plate gradually decreases in thickness due to a regression of cell division and eventually is replaced entirely by bone during a rabbit's first winter. Bones of adult cottontails have no epiphyseal cartilage (Hale 1949).

Bones may be examined by radiographs or directly from carcasses. Epiphyseal cartilage disappears in the humerus of cottontails at approximately 9 months of age. Bothma et al. (1972) concluded that the epiphyseal groove was a useful tool for judging ages up to 9 months.

Body Measurements. Hill (1971) studied several body measurements that could be used for age determination in *S. floridanus*. Based on 151 known-age rabbits, he concluded that tarsus length provided the best estimate of age in young cottontails. Length of the hind foot also can be used as a criterion of age (Beule and Studholme 1942). However, Petrides (1951) believed that length of the hind foot was not useful beyond the age of 3.5 months. Body weight is not useful in determining the age of cottontails (Bothma et al. 1972).

Bone Growth. Cottontails may be placed into five age categories based on growth of the skull (Hoffmeister and Zimmerman 1967). In rabbits <21 days old, the teeth are not fully erupted; in rabbits 21–92 days old, there is no fusion anywhere along the exoccipital–supraoccipital suture; in rabbits 93–105 days old, the suture is visible, but slightly fused; in rabbits 106–170 days old, the suture is fused, but visible; and in rabbits older than 170 days, the suture is not visible.

Periosteal Zonations. Sullins et al. (1976) found that annuli in the periosteal zone of mandibles from *S. floridanus* and *S. nuttallii* are useful for separating age classes among adults. Although the periosteal zonation technique has not been thoroughly tested, it seems that in combination with eye lens weight this technique may provide a means by which age structures of cottontail populations can be established more precisely.

Eye Lens Weight. Weight of the dried eye lens is clearly the most useful measurement for age determination in cottontails. Using eye lens weight as an indicator of age was first suggested by Lord (1959). The method enables estimation of age past the period of epiphyseal closure. An algebraic equation (Edwards 1967) derived from the lens growth model of Dudzinski and Mykytowycz (1961) can be used to estimate ages of cottontails in days from dry lens weights. Edwards (1967) gave the details of the procedure for removing and fixing the lenses and a lens weight–age table for *S. floridanus* scaled to 371 days. Epiphyseal growth was compared to lens weight by Wight and Conaway (1962a) and Pelton (1969).

In practice, the lens technique is generally satisfactory for cottontails up to 9–10 months of age. For populations sampled through November and early December, it is possible to distinguish young of the year from "adults." Older year classes are not differentiated satisfactorily.

It is best to take weights of both lenses and use the heavier weight. Lenses that seem atypical in color or shape or those with evident loss of tissue should be discarded. Lens weights are grouped in 5-mg classes.

It is very important to standardize the period for collecting lenses if differences among years or areas are to be evaluated. Eberhardt et al. (1963) and Edwards (1962a) demonstrated that juveniles are more vulnerable to hunting and presumably other causes of mortality than adults. If hunter-killed rabbits provide a source of lenses, the opening 1–3 days of the hunting season are often a desirable sampling period, provided that the hunting season opens on about the same date each year at each area.

ECONOMIC STATUS

The eastern cottontail is the principal game animal in the United States. Extremely high reproductive rates enable the species to withstand heavy hunting pressures. The pelt is used for clothing and the meat is considered a wild food delicacy. Cottontails also are in demand by beagle clubs across the United States, as they provide excellent sport and challenge for trailing hounds and the people who train and breed them.

Occasionally, cottontails may become depredators of succulent garden crops, household landscaping, and nursery and orchard seedlings. Mature orchards also are subject to debarking in the winter. Overall, the recreational and economic benefits of cottontails from sport hunters greatly outweigh the minor damage done to crops, nurseries, and orchards.

MANAGEMENT AND CONSERVATION

For decades, cottontails were considered the most important game animal based on hunter participation and harvests (Chapman et al. 1982). Because of that popularity, substantial efforts have been directed toward maintaining or bolstering local populations of cottontails. As indicated previously, massive stocking programs by state natural resource agencies, private hunting clubs, and concerned individuals were conducted in an attempt to offset habitat deficiencies in the early 1900s (Johnston 1972; Chapman and Morgan 1973). In Pennsylvania alone, 50,000 cottontails were imported annually during the 1930s. Gerstell (1937a, 1937b) indicated that such efforts were largely unsuccessful in enhancing hunter harvests. Additional concerns were raised about the introduction of new diseases and parasites. As a result, stocking programs were substantially reduced by the 1950s and are essentially nonexistent as a method to increase hunting opportunities.

Habitat management remains a viable approach to increasing cottontails. A description of generic habitats of most cottontails includes escape cover interspersed with grassy openings that contain a diversity of forbs. In eastern North America, setting back forest succession is the major approach taken to improve habitats of cottontails. A variety of methods (used individually or in combination) have been applied to arrest or reverse succession. These include prescribed burning, mowing, livestock grazing, and selective application of herbicides. For eastern cottontails, treatments often are spaced at 3- to 5-year intervals (Chapman et al. 1982). Lochmiller et al. (1995) evaluated the effectiveness of habitat-induced changes on the nutritional ecology of *S. floridanus* in Oklahoma. These investigators postulated that the quality of forage (based on protein content) used by cottontails was responsive to disturbance schedules, and their results supported this speculation. Vegetation in habitats 4–5 years after disturbance (herbicides applied followed by burning) contained higher concentrations of essential amino acids and protein than did vegetation at sites that were 7–9 years after disturbance. Deficiencies in one or more of these amino acids can result in reduced rates of growth, reproductive success, and survival in domestic rabbits (Schultze et al. 1988). A reduction in breeding success in response to quality of diet has been observed among

S. floridanus (Trethewey and Verts 1971; Bothma and Teer 1977; Chapman et al. 1977). Lochmiller et al. (1995) suggested that the decline in essential amino acids may explain the preference for recently disturbed habitats.

Probably the most common recommendation to landowners interested in enhancing habitat quality for cottontails is the creation of brushpiles to increase cover (Madson 1959). Although cottontails do use such cover, the addition of brushpiles alone may not increase overwinter survival of cottontails (Cox et al. 1997). Increasing understory stem density may be a more effective, but more labor-intensive approach to improving escape cover.

Increasingly, wildlife managers will be confronted with declining populations of all species of cottontails. Urbanization and maturing forests are reducing the amount of cottontail habitat in the eastern United States, whereas intensive agriculture has reduced habitat in the Midwest. Remaining populations often are small and disjunct (e.g., Litvaitis and Litvaitis 1996). Probably the greatest challenge in dealing with the decline of early-successional habitats is convincing the public that action is needed and that such efforts are not contradictory to maintaining diverse ecosystems. It is important to note that lagomorphs are significant components of many biotic communities (e.g., Wagner 1981). In addition, populations of a number of insects, birds, mammals, and reptiles that are obligates of early-successional and thicket habitats also are declining in response to the loss of habitat (Litvaitis et al. 1999). Efforts to sustain or increase populations of cottontails will need to take a landscape perspective to assure demographic exchanges between populations and long-term survival (Litvaitis and Villafuerte 1996).

RESEARCH NEEDS

Cottontails are excellent subjects for a variety of ecological, physiological, behavioral, and medical studies. They are widely distributed and easy to obtain, and many species are abundant. They include important game species, prey species, threatened species, disease vectors, medical research subjects, crop depredators, and watchable wildlife. Despite their abundance and availability, they remain fertile ground for new discoveries. For example, in the last 10 years, two new species have been described in the United States, *S. obscurus* (Chapman et al. 1992) and *S. robustus* (Ruedas 1998). In addition, cottontails have been used extensively in medical research involving several important diseases in humans (e.g., Han et al. 1999; Christensen et al. 2000). In 1931, Leopold stated, "After surveying eight important rabbit states, I am convinced that the characteristics of rabbit populations, and the factors determining their abundance and scarcity, are more difficult to decipher, and are receiving less thought and study from sportsmen and naturalists, than is the case in any other species of small game" (p. 89). Strange as it may seem, even after more than 70 years, relatively little progress has been made on research directed to an understanding of natural regulation. Emigration and dispersal are important aspects of cottontail ecology that warrant serious attention.

The basis of cottontail management is habitat management. New approaches to wildlife habitat analysis are developing rapidly, based on computer modeling. The key element is vegetative structure. Discriminant and logistic analyses are being used to identify important characteristics. These techniques can and should be directed to an understanding of the complex habitat relations of cottontails. This will not be a brief, simple, or inexpensive form of research.

The cottontail uses primarily successional vegetation and occurs through a range of early and midsuccessional series. In managing cottontails, it is necessary to devise ways to periodically disrupt or inhibit natural succession, as progression to woodland is often rapid, particularly east of the Mississippi River. There is much to learn about the various potential types of disturbance as they relate to successional processes in different ecological situations.

Soil fertility is a principal factor in determining plant species composition and rates of change in secondary succession (Edwards 1975). Fertility management has considerable significance to cottontail management and warrants research attention, particularly on areas where soils tend to be infertile or eroded. One of the keys to good cottontail research is continuity of involvement. This does not mean that the same study should continue forever (although certain basic inventories should be conducted annually). What is important is that individuals with an interest in cottontails and an understanding of community ecology be given sustained support to maintain active programs of research related to the biology and ecology of cottontails in relation to their environments.

New biochemical and morphometric analysis techniques also afford the opportunity to better understand the systematic status of cottontails. Given the importance and complexity of the management of these species, further efforts are needed to help answer important phylogenetic questions.

LITERATURE CITED

Allen, D. L. 1939. Michigan cottontails in winter. Journal of Wildlife Management 3:307–22

Althoff, D. P., G. L. Storm, and D. R. DeWalle. 1997. Daytime habitat selection by cottontails in central Pennsylvania. Journal of Wildlife Management 61:450–59.

Anderson, B. F., and M. R. Pelton. 1976. Movements, home range, and cover use: Factors affecting the susceptibility of cottontails to hunting. Proceedings of the Southeastern Association of Game and Fish Commissioners 30:525–35.

Applegate, J. E., and J. R. Trout. 1976. Weather and the harvest of cottontails in New Jersey. Journal of Wildlife Management 40:658–62.

Bailey, J. A. 1968. A weight–length relationship for evaluating physical condition of cottontails. Journal of Wildlife Management 32:835–41.

Bailey, J. A. 1969. Quantity of soft pellets produced by caged cottontails. Journal of Wildlife Management 33:421.

Bailey, J. A., and R. E. Schroeder. 1967. Weights of left and right adrenal glands in cottontails. Journal of Mammalogy 48:475.

Bailey, V. 1905. Biological survey of Texas: Life zones, with characteristic species of mammals, birds, reptiles and plants. North American Fauna 25:1–222.

Barbour, M. S., and J. A. Litvaitis. 1993. Niche dimensions of New England cottontails in relation to habitat patch size. Oecologia 95:321–27.

Barkalow, F. S., Jr. 1961. A large cottontail litter. Journal of Mammalogy 42:254.

Barkalow, F. S., Jr. 1962. Latitude related to reproduction in the cottontail rabbit. Journal of Wildlife Management 26:32–37.

Beckwith, S. L. 1954. Ecological succession on abandoned farmlands and its relationship to wildlife management. Ecological Monographs 24:349–76.

Beule, J. D. 1940. Cottontail nesting-study in Pennsylvania. Transactions of the North American Wildlife Conference 5:320–28.

Beule, J. D, and A. T. Studholme. 1942. Cottontail rabbit nests and nestlings. Journal of Wildlife Management 6:133–40.

Bissonnette, T. H., and A. G. Csech. 1939. Modified sexual periodicity in cottontail rabbits. Biological Bulletin 17:364–67.

Bittner, S. L., and J. A. Chapman. 1981. Reproductive and physiological cycles in an island population of *Sylvilagus floridanus*. Pages 182–203 *in* K. Myers and C. D. MacInnes, eds. Proceedings of the world lagomorph conference. University of Guelph, Ontario, Canada.

Black, J. J. 1950. The rural economy of New England. Harvard University Press, Cambridge, MA.

Blair, W. F. 1936. The Florida marsh rabbit. Journal of Mammalogy 17:197–207.

Bothma, J. Du P., and J. G. Teer. 1977. Reproduction and productivity in South Texas cottontail rabbits. Mammalia 41:253–81.

Bothma, J. Du P., J. G. Teer, and C. E. Gates. 1972. Growth and age determination of the cottontail in South Texas. Journal of Wildlife Management 36:1209–10.

Brown, A. L., and J. A. Litvaitis. 1995. Habitat features associated with predation of New England cottontails: What scale is appropriate? Canadian Journal of Zoology 73:1005–11.

Bruna, J. F. 1952. Kentucky rabbit investigations (Federal Aid Project 26-R). Frankfurt, KY.

Burgdorfer, W. 1969. Ecology of tick vectors of American spotted fever. Bulletin of the World Health Organization 40:375–81.

Burgdorfer, W., K. T. Friedhoff, and J. L. Lancaster. 1966. Natural history of tick-borne spotted fever in the USA. Bulletin of the World Health Organization 35:149–53.

Case, T. J., and M. R. Gilpin. 1974. Interference competition and niche theory. Proceedings of the National Academy of Science of the USA 71:3073–77.

Casteel, D. A. 1966. Nest building, parturition, and copulation in the cottontail rabbit. American Midland Naturalist 75:160–67.

Casteel, D. A. 1967. Timing of ovulation and implantation in the cottontail rabbit. Journal of Wildlife Management 31:194–97.

Casteel, D. A., and W. R. Edwards. 1964. Two instances of multiparous juvenile cottontails. Journal of Wildlife Management 28:858–59.

Caughley, G. 1974. Interpretation of age ratios. Journal of Wildlife Management 38:557–362.

Chapman, J. A. 1971a. Organ weights and sexual dimorphism of the brush rabbit. Journal of Mammalogy 52:453–55.

Chapman, J. A. 1971b. Orientation and homing of the brush rabbit (*Sylvilagus bachmani*). Journal of Mammalogy 52:686–99.

Chapman, J. A. 1974. *Sylvilagus bachmani.* Mammalian Species 34:1–4.

Chapman, J. A. 1975a. *Sylvilagus nuttallii.* Mammalian Species 56:1–3.

Chapman, J. A. 1975b. *Sylvilagus transitionalis.* Mammalian Species 55:1–4.

Chapman, J. A. 1984. Latitude and gestation period in new world rabbits (Leporidae: *Sylvilagus* and *Romerolagus*). American Naturalist 124:442–45.

Chapman, J. A. 1999a. Eastern cottontail *Sylvilagus floridanus*. Pages 687–88 *in* D. E. Wilson and S. Ruff, eds. The Smithsonian book of North American mammals. Smithsonian Institution Press, Washington, DC.

Chapman, J. A. 1999b. Desert cottontail *Sylvilagus audubonii.* Pages 684–85 *in* D. E. Wilson and S. Ruff, eds. The Smithsonian book of North American mammals. Smithsonian Institution Press, Washington, DC.

Chapman, J. A. 1999c. Brush rabbit *Sylvilagus bachmani.* Pages 685–86 *in* D. E. Wilson and S. Ruff, eds. The Smithsonian book of North American mammals. Smithsonian Institution Press, Washington, DC.

Chapman, J. A. 1999d. Mountain cottontail *Sylvilagus nuttallii.* Pages 689–90 *in* D. E. Wilson and S. Ruff, eds. The Smithsonian book of North American mammals. Smithsonian Institution Press, Washington, DC.

Chapman, J. A. 1999e. Swamp rabbit *Sylvilagus aquaticus.* Pages 683–84 *in* D. E. Wilson and S. Ruff, eds. The Smithsonian book of North American mammals. Smithsonian Institution Press, Washington, DC.

Chapman, J. A. 1999f. Marsh rabbit *Sylvilagus palustris.* Pages 691–92 *in* D. E. Wilson and S. Ruff, eds. The Smithsonian book of North American mammals. Smithsonian Institution Press, Washington, DC.

Chapman, J. A. 1999g. New England cottontail *Sylvilagus transitionalis.* Pages 692–93 *in* D. E. Wilson and S. Ruff, eds. The Smithsonian book of North American mammals. Smithsonian Institution Press, Washington, DC.

Chapman, J. A. 1999h. Appalachian cottontail *Sylvilagus obscurus.* Pages 690–91 *in* D. E. Wilson and S. Ruff, eds. The Smithsonian book of North American mammals. Smithsonian Institution Press, Washington, DC.

Chapman, J. A., and G. Ceballos. 1990. The cottontails. Pages 95–110 *in* J. A. Chapman and J .E. C. Flux, eds. Rabbits, hares and pikas: Status survey and conservation action plan. IUCN, Gland, Switzerland.

Chapman, J. A., K. L. Cramer, N. J. Dippenaar, and T. J. Robinson. 1992. Systematics and biogeography of the New England cottontail, Sylvilagus transitionalis (Bangs, 1895), with the description of a new species from the Appalachian Mountains. Proceedings of the Biological Society of Washington 105:841–66.

Chapman, J. A., and G. A. Feldhamer. 1981. *Sylvilagus aquaticus.* Mammalian Species 151:1–4.

Chapman, J. A. and J. E. C. Flux, eds. 1990. Rabbits, hares and pikas: Status survey and conservation action plan. IUCN, Gland, Switzerland.

Chapman, J. A., and A. L. Harman. 1972. The breeding biology of a brush rabbit population. Journal of Wildlife Management 36:816–23.

Chapman, J. A., A. L. Harman, and D. E. Samuel. 1977. Reproductive and physiological cycles in the cottontail complex in western Maryland and nearby West Virginia. Wildlife Monographs 56:1–73.

Chapman, J. A., J. G. Hockman, and M. M. Ojeda. 1980. Sylvilagus floridanus. Mammalian Species 136:1–8.

Chapman, J. A., J. G. Hockman, and W. R. Edwards. 1982. Cottontails (Sylvilagus floridanus and allies). Pages 83–123 *in* J. A. Chapman and G. A. Feldhamer, eds. Wild mammals of North America. Johns Hopkins University Press, Baltimore.

Chapman, J. A., and R. P. Morgan. 1973. Systematic status of the cottontail complex in western Maryland and nearby West Virginia. Wildlife Monographs 36:1–54.

Chapman, J. A., and R. P. Morgan. 1974. Onset of the breeding season and size of first litters in two species of cottontails from southwestern Texas. Southwestern Naturalist 19:277–80.

Chapman, J. A., and J. L. Paradiso. 1972. First records of the New England cottontail (*Sylvilagus transitionalis*) from Maryland. Chesapeake Science 13:17–18.

Chapman, J. A., and J. R. Stauffer. 1981. The status and distribution of the New England cottontail. Pages 973–83 *in* K.Myers and C. D. MacInnes, eds. Proceedings of the world lagomorph conference. University of Quelph, Ontario, Canada.

Chapman, J. A., and D. E. C. Trethewey. 1972. Movements within a population of introduced eastern cottontail rabbits. Journal of Wildlife Management 36:155–58.

Chapman, J. A., and B. J. Verts. 1969. Interspecific aggressive behavior in rabbits. Murrelet 50:17–18.

Chapman, J. A., and G. R. Willner. 1978. *Sylvilagus audubonii.* Mammalian Species 106:1–4.

Chapman, J. A., and G. R. Willner. 1981. *Sylvilagus palustris.* Mammalian Species 153:1–3.

Chitty, D. 1957. Self-regulation of numbers through changes in viability. Cold Spring Harbor Symposium Quantitative Biology 22:277–80.

Chitty, D. 1960. Population processes in the vole and their relevance to general theory. Canadian Journal of Zoology 38:99–113.

Chitty, D. 1971. The natural selection of self-regulatory behavior in animal populations. Pages 136–70 *in* I. A. McLaren, ed. Natural regulation of animal populations. Atherton Press, New York.

Christensen, N. D., N. M. Cladel, C. A. Reed, and R. Han. 2000. Rabbit oral papillomavirus complete genome sequence and immunity following initial infection. Virology 269:451–61.

Christian J. J. 1963. Endocrine adaptive mechanisms and the physiologic regulation of population growth. Pages 189–353 *in* M. U. Mayer and R. G. VanGelder, eds. Physiological mammalogy, Vol. 1. Academic Press, New York.

Christian, J. J. 1964. Endocrines, behavior, and population. Science 146:1550–60.

Christian J. J., and D. E. Davis. 1955. Reduction of adrenal weight in rodents by reducing population size. Transactions of the North American Wildlife Conference 20:177–89.

Cockrum, E. L. 1949. Range extension of the swamp rabbit in Illinois. Journal of Mammalogy 30:427–29.

Conaway, C. H., T. S. Baskett, and J. E. Toll. 1960. Embryo resorption in the swamp rabbit. Journal of Wildlife Management 24:197–202.

Conaway, C. H., and H. M. Wight. 1962. Onset of reproductive season and first pregnancy of the season in cottontails. Journal of Wildlife Management 26:278–90.

Conaway, C. H., H. M. Wight, and K. C. Sadler. 1963. Annual production by a cottontail population. Journal of Wildlife Management 27:171–75.

Conaway, C. H., K. C. Sadler, and D. H. Hazelwood. 1974. Geographic variation in litter size and onset of breeding in cottontails. Journal of Wildlife Management 38:473–81.

Cooney, J. C., and W. Burgdorfer. 1974. Zoonotic potential (Rocky Mountain spotted fever and tularemia) in the Tennessee Valley region. Part 1: Ecological studies of ticks infesting mammals in the Land-between-the-Lakes. American Journal of Tropical Medicine and Hygiene 23:99–108.

Cowan, I. M., and C. J. Guiquet. 1956. The mammals of British Columbia. British Columbia Provincial Museum, Victoria, Canada.

Cox, E. W., R. A. Garrott, and J. R. Cary. 1997. Effect of supplemental cover on survival of snowshoe hares and cottontail rabbits in patchy habitat. Canadian Journal of Zoology 75:1357–63.

Creasman, L. N.,J. Craig, and M. Swan. 1992. The forested wetlands of the Mississippi River: An ecosystem in crisis. Louisiana Nature Conservancy, Baton Rouge.

Cushing, J. E. 1939. The relation of some observations upon predation to theories of protective coloration. Condor 41:100–111.

Dalke, P. D. 1937. A preliminary report of the New England cottontail studies. Transactions of the North American Wildlife Conference 2:542–48.

Dalke, P. D. 1942. The cottontail rabbits in Connecticut. Bulletin of the Connecticut Geological and Natural History Survey 65:1–97.

Dalke, P. D., and P. R. Sime. 1941. Food habits of the eastern and New England cottontails. Journal of Wildlife Management 5:216–28.

Dalquest, W. W. 1941. Distribution of cottontail rabbits in Washington State. Journal of Wildlife Management 5:408–11.

Dice, L. R. 1926. Notes on Pacific coast rabbits and pikas. Occasional Papers of the Museum of Zoology, University of Michigan 166:1–28.

Dice, L. R. 1929. An attempt to breed cottontail rabbits in captivity. Journal of Mammalogy 10:225–29.

Dixon, K. R., and J. A. Chapman. 1980. Harmonic mean measure of animal activity areas. Ecology 61:1040–44.

Dixon, K. R., J. A. Chapman, O. J. Rongstad, and K. M. Orhelein. 1981. A comparison of home range size in *Sylvilagus floridanus* and *S. bachmani.* Pages 541–48 *in* K. Myers and C. D. MacInnes, eds. Proceedings of the world lagomorph conference. University of Guelph, Ontario, Canada.

Dobler, F. C. 1992. The shrub–steppe ecosystem of Washington: A brief summary of knowledge and nongame wildlife conservation needs. Washington Department of Wildlife, Olympia.

Dobler, F. C., and K. R. Dixon. 1990. The pygmy rabbit *Brachylagus idahoensis.* Pages 111–15 *in* J. A. Chapman and J. E. C. Flux, eds. Rabbits, hares and pikas: Status survey and conservation action plan. IUCN, Gland, Switzerland.

Dudzinski, M. L., and R. Mykytowycz. 1961. The eye lens as an indicator of the age of the wild rabbit in Australia. CSIRO Wildlife Research 6:156–59.

Eabry, H. S. 1968. An ecological study of *Sylvilagus transitionalis* and *S. floridanus* of northeastern Connecticut. M.S. Thesis, University of Connecticut, Storrs.

Eberhardt, L. L., T. J. Peterle, and R. Schofield. 1963. Problems in a rabbit population study. Wildlife Monographs 10:1–51.

Ecke, D. H. 1955. The reproductive cycle of the Mearns cottontail in Illinois. American Midland Naturalist 53:294–311.

Edwards, W. R. 1962a. Age structure of Ohio cottontail populations from weights of lenses. Journal of Wildlife Management 26:125–32.

Edwards, W. R. 1962b. Farm game hunter questionnaire survey, 1959. Game Research in Ohio 1:13–15.

Edwards, W. R. 1963. Fifteen cottontails in a nest. Journal of Mammalogy 44:416–17.

Edwards, W. R. 1967. Tables for estimating ages and birth dates of cottontail rabbits. Illinois Natural History Survey Biological Notes 59:1–4.

Edwards, W. R. 1975. Soil fertility and competition in first-year secondary succession after cropping. Ph.D. Dissertation. University of Illinois, Urbana-Champaign.

Edwards, W. R., S. P. Havera, R. F. Labisky, J. A. Ellis, and R. E. Warner. 1981. The abundance of cottontails in relation to agricultural land use in Illinois (U.S.A.), 1956–1978, with comments on mechanisms of regulation. Pages 761–98 *in* K. Myers and C. D. MacInnes, eds. Proceedings of the world lagomorph conference. University of Guelph, Ontario, Canada.

Elder, W. H., and L. K. Sowls. 1942. Body weight and sex ratio of cottontail rabbits. Journal of Wildlife Management 6:203–7.

Evans, R. D. 1962. Breeding characteristics of southeastern Missouri cottontails. Proceedings of the Annual Conference of the Southeastern Association of Game Fish Commissioners 16:140–42.

Evans, R. D., K. C. Sadler, C. H. Conaway, and T. S. Baskett. 1965. Regional comparisons of cottontail reproduction in Missouri. American Midland Naturalist 74:176–84.

Fay, F. H., and E. H. Chandler. 1955. The geographical and ecological distribution of cottontail rabbits in MA. Journal of Mammalogy 36:415–24.

Federal Register. 1990. Endangered and threatened wildlife and plants; endangered status for the Lower Keys rabbit and threatened status for the squirrel chimney cave shrimp. Federal Register 55:25588–91.

Fitch, H. S. 1947. Ecology of a cottontail rabbit (*Sylvilagus audubonii*) population in central California. California Fish and Game Journal 33:159–84.

Flinders, J. T. 1999. Pygmy rabbit *Brachylagus idahoensis.* Pages 681–83 *in* D. E. Wilson and S. Ruff, eds. The Smithsonian book of North American mammals. Smithsonian Institution Press, Washington, DC.

Flinders, J. T., and R. M. Hansen. 1973. Abundance and dispersion of leporids within a shortgrass ecosystem. Journal of Mammalogy 54:287–91.

Flux, J. E. C. 1964. Hare reproduction in New Zealand. New Zealand Journal of Agriculture 109:483–86.

Forys, E. A. 1999. Food habits of the lower Florida keys marsh rabbit (*Sylvilagus palustris hefneri*). Florida Academy of Sciences 62:106–10.

Forys, E. A., and S. R. Humphrey. 1996. Home range and movements of the lower keys marsh rabbit in a highly fragmented habitat. Journal of Mammalogy 77:1042–48.

Forys, E. A., and S. R. Humphrey. 1999a. The importance of patch attributes and context to the management and recovery of an endangered lagomorph. Landscape Ecology 14:177–85.

Forys, E. A., and S. R. Humphrey. 1999b. Use of population viability analysis to evaluate management options for the endangered lower keys marsh rabbit. Journal of Wildlife Management 63:251–60.

Fredrickson, L. H. 1980. Management of lowland hardwood wetlands for wildlife: Problems and potential. Transactions of the North American Wildlife and Natural Resources Conference 45:376–86.

Friley, C. E. 1955. A study of cottontail habitat preferences on a southern Michigan farming area (Federal Aid to Wildlife Restoration Project W-48-R).

Gabler, K. I.,J. W. Laundre, and L. T. Heady. 2000. Predicting the suitability of habitat in southeast Idaho for pygmy rabbits. Journal of Wildlife Management 64:759–64.

Garner, G. W. 1969. Short term succession of vegetation following habitat manipulation of bottomland hardwoods for swamp rabbits in Louisiana. M.S. Thesis, Louisiana State University, Baton Rouge.

Genoways, H. H., and J. K. Jones, Jr. 1972. Mammals from southwestern North Dakota. Occasional Papers, the Museum, Texas Tech University 6:1–36.

Gerstell, R. 1937a. The management of the cottontail rabbit in Pennsylvania. Pennsylvania Game News 7:6, 7, 27.

Gerstell, R. 1937b. The management of the cottontail rabbit in Pennsylvania. Pennsylvania Game News 8:8–20.

Gibb, J. A. 1977. Factors affecting population density in the wild rabbit, *Oryctolagus cuniculus* (L.), and their relevance to small mammals. Pages 33–46 *in* B. Stonehouse and C. Perrins, eds. Evolutionary ecology. University Park Press, Baltimore.

Giles, R. H., Jr. 1978. Wildlife management. Freeman, San Francisco.

Golley, F. B. 1962. Mammals of Georgia. University of Georgia Press, Athens.

Goodpaster, W. W., and D. F. Hoffmeister. 1952. Notes on the mammals of western Tennessee. Journal of Mammalogy 33:362–71.

Graf, W. 1955. Cottontail rabbit introductions and distribution in western Oregon. Journal of Wildlife Management 19:184–88.

Grange, W. B. 1949. The way to game abundance with explanation of game cycles. Charles Scribner's Sons, New York.

Graves, W. C., and W. C. Asserson III. 1967. Cottontail rabbit investigations. (Job Completion Report, P-R Project W-47-R-15). California Department of Fish Game, Sacramento, CA.

Green, J. S., and J. T. Flinders. 1980. Habitat and dietary relationships of the pygmy rabbit. Journal of Range Management 33:136–42.

Grinnel, J. 1939. Effects of a wet year on mammalian populations. Journal of Mammalogy 20:62–64.

Halaych, K. M., and T. J. Robinson. 1997. Phylogenetic relationships of cottontails (*Sylvilagus,* Lagomorpha): Congruence of 12SrDNA and cytogenetic data. Molecular Phylogenetics and Evolution 7:294–302.

Hale, J. B. 1949. Aging cottontail rabbits by bone growth. Journal of Wildlife Management 13:216–25.

Hall, E. R. 1951. A synopsis of the North American lagomorpha. University Kansas Publication, Museum of Natural History 5:119–202.

Hall, E. R. 1981. The mammals of North America, Vol. I. John Wiley, New York.

Hamilton, W. J. 1940. Breeding habits of the cottontail rabbit in New York State. Journal of Mammalogy 21:8–11.

Hamilton, W. J. 1955. Coprophagy in the swamp rabbit. Journal of Mammalogy 36:303–4.

Han, R., N. M. Cladel, C. A. Reed, X. Peng, and N. D. Christensen. 1999. Protection of rabbits from viral challenge by gene gun-based intracutaneous vaccination with a combination of cottontail rabbit papillomavirus E1, E2, E6, and E7 genes. Journal of Virology 73:7039–43.

Han, R., N. M. Cladel, C. A. Reed, X. Peng, L. R. Budgeon, M. Pickel, and N. D. Christensen. 2000. DNA vaccination prevents and/or delays carcinoma development of papillomavirus-induced skin papillomas on rabbits. Journal of Virology 74:9712–16.

Hanski, I., T. Pakkala, M. Kuussaari, and G. Lei. 1995. Metapopulation persistence of an endangered butterfly in a fragmented landscape. Oikos 72:21–28.

Hanson, J. C., J. A. Bailey, and R. J. Siglin. 1969. Activity and use of habitat by radio-tagged cottontails during winter. Transactions of Illinois State Academy of Science 62:294–302.

Harper, R. M. 1918. Changes in the forest area of New England in three centuries. Journal of Forestry 16:442–52.

Harrison, S. 1994. Metapopulations and conservation. Pages 111–28 *in* P. J. Edwards, R. May, and N. R. Webb, eds. Large-scale ecology and conservation biology. Blackwell, Boston.

Harrison, S., and J. F. Quinn. 1989. Correlated environments and persistence of metapopulations. Oikos 56:293–98.

Hassell, M. P. 1982. Impact of infectious diseases on host populations: Group report. Pages 15–35 *in* R. M. Anderson and R. M. May, eds. Population biology of infectious diseases. Springer-Verlag, New York.

Haugen, A. O. 1942. Life history studies of the cottontail rabbit in southwestern Michigan. American Midland Naturalist 28:204–44.

Havera, S. P. 1973. The relationship of Illinois weather and agriculture to the eastern cottontail rabbit (Technical Report 4). Illinois State Water Survey, Champaign, IL.

Hayes, R. O., J. B. Daniels, H. T. Mixfield, and R. E. Wheeler. 1964. Field and laboratory studies of eastern encephalitis in warm- and cold-blooded vertebrates. American Journal of Tropical Medicine and Hygiene 13:595–606.

Hendrickson, G. O. 1943. Mearns cottontail investigations in Iowa. Ames Forester 21:59–73.

Herrick, E. H. 1965. Endocrine studies. Pages 7375 *in* The black-tailed jackrabbit in Kansas (Technical Bulletin 140). Kansas State University, Agricultural Experiment Station, Manhattan, KS.

Heuer, E. T., Jr., and H. R. Perry, Jr., 1976. Squirrel and rabbit abundances in the Atchafalaya Basin, Louisiana. Proceeding of the Southeastern Association of Fish and Wildlife Agencies 30:552–59.

Hill, E. P. 1966. Some effects of weather on cottontail reproduction in Alabama. Proceedings of the Annual Conference of Southeastern Game Fish Commissioners 19:48–57.

Hill, E. P. 1967. Notes on the life history of the swamp rabbit in Alabama. Proceedings of the Annual Conference of Southeastern Game and Fish Commissioners 21:117–23.

Hill, E. P. 1971. An evaluation of several body measurements for determining age in live juvenile cottontails. Proceedings of the Annual Conference of Southeastern Association Game and Fish Commissioners 25:269–81.

Hill, E. P. 1972. The cottontail rabbit in Alabama (Bulletin 440). Auburn University, Alabama Agricultural Experiment Station, Auburn, AL.

Hinds, D. S. 1973. Acclimation of thermoregulation in the desert cottontail, *Sylvilagus audubonii.* Journal of Mammalogy 54:708–28.

Hoffmeister, D. F., and E. G. Zimmerman. 1967. Growth of the skull in the cottontail (*Sylvilagus floridanus*) and its application to age-determination. American Midland Naturalist 78:198–206.

Holden, H. E., and H. S. Eabry. 1970. Chromosomes of *Sylvilagus floridanus* and *S. transitionalis.* Journal of Mammalogy 51:166–68.

Holler, N. R., and C. H. Conaway. 1979. Reproduction of the marsh rabbit (*Sylvilagus palustris*) in south Florida. Journal of Mammalogy 60:769–77.

Holler, N. R., and H. M. Marsden. 1970. Onset of evening activity of swamp rabbits and cottontails in relation to sunset. Journal of Wildlife Management 34:349–53.

Holler, N. R.,T. S. Baskett, and J. P. Rogers. 1963. Reproduction in confined swamp rabbits. Journal of Wildlife Management 27:179–83.

Holling, C. S. 1966. The functional responses of invertebrate predators to prey density. Memoirs of the Entomological Society of Canada 48:1–86.

Holten, J. W., and J. E. Toll. 1960. Winter weights of juvenile and adult swamp rabbits in southeastern Missouri. Journal of Wildlife Management 24:229–30.

Hunt, T. P. 1959. Breeding habits of the swamp rabbit with notes on its life history. Journal of Mammalogy 40:82–96.

Hurst, G. A., and M. W. Smith. 1986. Effect of silvicultural practices on swamp rabbit habitat. Pages 334–40 *in* Proceedings of the fourth biennial southern silvicultural research conference (General Technical Report SE-2). USDA Forest Service, Washington, DC.

Ingles, L. G. 1941. Natural history observations on the Audubon cottontail. Journal of Mammalogy 22:227–50.

Irland, L. C. 1982. Wildlands and woodlots: A story of New England's forests. University Press of New England, Hanover, NH.

Jackson, S. N. 1973. Distribution of cottontail rabbits (*Sylvilagus* spp.) in northern New England. M.S. Thesis, University of Connecticut, Storrs.

Janson, R. G. 1946. A survey of the native rabbits of Utah with reference to their classification, distribution, life histories and ecology. M.S. Thesis, Utah State University, Logan.

Johnson, A. M., and G. O. Hendrickson. 1958. Effects of weather conditions on the winter activity of Mearns cottontail. Proceedings of the Iowa Academy of Science 65:554–58.

Johnson, L. W. 1973. A model for the synchronous breeding of the cottontail. M.S. Thesis, University of Illinois, Urbana.

Johnson, M. L. 1968. Application of blood protein electrophoretic studies to problems in mammalian taxonomy. Systematic Zoology 17:23–30.

Johnson, M. L.,. and M. J. Wicks 1964. Serum-protein electrophoresis in mammals: significance in the higher taxonomic categories. Pages 681–94 *in* C. A. Leone, ed. Taxonomic biochemistry and serology. Ronald Press, New York.

Johnston, J. E. 1972. Identification and distribution of cottontail rabbits in southern New England. M.S. Thesis, University of Connecticut, Storrs.

Katzner, T. E., and K. L. Parker. 1997. Vegetative characteristics and size of home ranges used by pygmy rabbits (*Brachylagus idahoensis*) during winter. Journal of Mammalogy 78:1063–72.

Katzner, T. E., K. L. Parker, and H. H. Harlow. 1997. Metabolism and thermal response in winter-acclimatized pygmy rabbits (*Brachylagus idahoensis*). Journal of Mammalogy 78:1053–62.

Keith, L. B. 1963. Wildlife's ten-year cycle. University of Wisconsin Press, Madison.

Keith, L. B., and S. E. M. Bloomer. 1993. Differential mortality of sympatric snowshoe hares and cottontail rabbits in central Wisconsin. Canadian Journal of Zoology 71:1694–97.

Keith, L. B., and L. A. Windberg. 1978. A demographic analysis of the snowshoe hare cycle. Wildlife Monographs 58:1–70.

Kibbe, D. P., and D. L. Kirkpatrick. 1971. Systematic evaluation of late summer breeding in juvenile cottontails, *Sylvilagus floridanus.* Journal of Mammalogy 52:465–67.

Kirkpatrick, C. M. 1956. Coprophagy in the cottontail. Journal of Mammalogy 37:300.

Kirkpatrick, C. M. 1960. Unusual cottontail litter. Journal of Mammalogy 41:119–20.

Kirkpatrick, R. L., and D. M. Baldwin. 1974. Population density and reproduction in penned cottontail rabbits. Journal of Wildlife Management 38:482–87.

Kjolhaug, M. S., and A. Woolf. 1988. Home range of the swamp rabbit in southern Illinois. Journal of Mammalogy 39:194–97.

Kline, P. D. 1962. Vernal breeding of cottontails in Iowa. Journal of the Iowa Academy of Science 69:244–52.

Korte, P. A. 1975. Distribution and habitat requirements of the swamp rabbit in Missouri. M.S. Thesis, University of Missouri, Columbia.

Korte, P. A., and L. H. Fredrickson. 1977. Swamp rabbit distribution in Missouri. Transactions of the Missouri Academy of Sciences 10/11:72–77.

Krebs, C. J. 1971. Genetic and behavioral studies on fluctuating vole populations. Pages 243–56 *in* P. J. Den Boer and G. R. Gradwell, eds. Proceedings of the Advanced Studies Institute. Dynamics of numbers in populations. Pudoc, Wageningen, the Netherlands.

Krebs, C. J., and J. H. Myers. 1974. Population cycles in small mammals. Advances in Ecological Research 8:267–399.

Larkin, R. P., and D. Halkin. 1994. Wildlife software: A review of software packages for estimating animal home ranges. Wildlife Society Bulletin 22:274–87.

Laseter, B. R. 1999. Estimates of population density for the Appalachian cottontail (*Sylvilagus obscurus*) in eastern Tennessee. M.S. Thesis, University of Memphis, Tennessee.

Lazell, J. D. 1984. A new marsh rabbit (*Sylvilagus palustris*) from Florida's lower keys. Journal of Mammalogy 65:26–33.

Leite, E. A. 1965. Relation of habitat structure to cottontail rabbit production, survival and harvest rates, (Job Progress Report, Project W-103-R-8, Job 12). Ohio Division of Natural Resources, Columbus.

Lemke, G. W. 1957. An unusually late pregnancy in a Wisconsin cottontail. Journal of Mammalogy 38:275.

Leopold, A. 1931. Report on a game survey of the north central states. Sporting Arms and Ammunition Manufacturers Institute, Madison, WI.

Lepitzki, D. A. W. 1990. Summer and fall activity patterns of cottontail rabbits, *Sylvilagus floridanus,* in southern Illinois. Canadian Field-Naturalist 104:552–56.

Lepitzki, D. A. W., A. Woolf, and B. M. Bunn. 1992. Parasites of cottontail rabbits of southern Illinois. Journal of Parasitology 78:1080–83.

Lidicker, W. J., Jr. 1975. The role of dispersal in the demography of small mammals. Pages 103–28 *in* F. B. Golley, K. Petrusewicz, and L. Ryszkowski, eds. Small mammals: Their productivity and population dynamics. Cambridge University Press, New York.

Linkkila, T. E. 1971. Influence of habitat upon changes within the interspecific Connecticut cottontail populations. M.S. Thesis, University of Connecticut, Storrs.

Litvaitis, J. A. 1993. Response of early successional vertebrates to historic changes in land use. Conservation Biology 7:866–73.

Litvaitis, J. A. 2001. Importance of early successional habitats to mammals in eastern forests. Wildlife Society Bulletin 29:466–73.

Litvaitis, M. K., and J. A. Litvaitis. 1996. Using mitochondrial DNA to inventory the distribution of remnant populations of New England cottontails. Wildlife Society Bulletin 24:725–30.

Litvaitis, J. A., and R. Villafuerte. 1996. Factors affecting the persistence of New England cottontail metapopulations: The role of habitat management. Wildlife Society Bulletin 24:686–93.

Litvaitis, J. A., J. A. Sherburne, and J. A. Bissonette. 1985. Influence of understory characteristics on snowshoe hare habitat use and density. Journal of Wildlife Management 49:866–73.

Litvaitis, J. A., D. L. Verbyla, and M. K. Litvaitis. 1991. A field method to differentiate New England and eastern cottontails. Transactions of the Northeast Section of the Wildlife Society 48:11–14.

Litvaitis, M. K., J. A. Litvaitis, W. Lee, and T. D. Kocher. 1997. Variation in the mitochondrial DNA of the Sylvilagus complex occupying the northeastern United States. Canadian Journal of Zoology 75:595–605.

Litvaitis, J. A., D. L. Wagner, J. L. Confer, M. D. Tarr, and E. J. Snyder. 1999. Early successional forests and shrub-dominated habitats: Land-use artifact or critical community in the northeastern United States? Northeast Wildlife 54:101–18.

Llewellyn, L. M., and C. O. Handley. 1945. The cottontail rabbits of Virginia. Journal of Mammalogy 26:379–90.

Lochmiller, R. L., D. G. Peitz, D. M. Leslie, Jr., and D. M. Engle. 1995. Habitat-induced changes in essential amino-acid nutrition in populations of eastern cottontails. Journal of Mammalogy 76:1164–77.

Lord, R. D. 1959. The lens as an indication of age in cottontail rabbits. Journal of Wildlife Management 23:358–60.

Lord, R. D. 1960. Litter size and latitude in North American mammals. American Midland Naturalist 64:488–99.

Lord, R. D. 1961. Magnitudes of reproduction in cottontail rabbits. Journal of Wildlife Management 25:28–33.

Lord, R. D. 1963. The cottontail rabbit in Illinois. Technical Bulletin Illinois Department of Conservation 3:1–94.

Lord, R. D. 1964. Seasonal changes in the activity of penned cottontail rabbits. Animal Behavior 12:38–41.

Lord, R. D., and D. A. Casteel. 1960. Importance of food to cottontail winter mortality. Transactions of the North American Wildlife and Natural Resources Conference 25:267–74.

Lowe, C. E. 1958. Ecology of the swamp rabbit in Georgia. Journal of Mammalogy 39:116–27.

Lowery, G. H., Jr. 1974. The mammals of Louisiana and its adjacent waters. Louisiana State University Press, Baton Rouge.

Maclean, S. F., Jr., B. M. Fitzgerald, and F. A. Pitelka. 1974. Population cycles in Arctic lemmings: Winter reproduction and predation by weasels. Arctic and Alpine Research 6:1–12.

Madson, J. 1959. The cottontail rabbit. Olin Mathieson Chemical Corporation, East Alton, IL.

Mankin, P. C., and R. E. Warner. 1999a. A regional model of the eastern cottontail and land use changes in Illinois. Journal of Wildlife Management 63:956–63.

Mankin, P. C., and R. E. Warner. 1999b. Responses of eastern cottontails to intensive row-crop farming. Journal of Mammalogy 80:940–49.

Marsden, H. M., and C. H. Conaway. 1963. Behavior and the reproductive cycle in the cottontail. Journal of Wildlife Management 27:161–70.

Marsden, M. M., and N. R. Holler. 1964. Social behavior in confined populations of the cottontail and swamp rabbit. Wildlife Monographs 13:1–39.

Martinson, R. K., J. W. Holten, and G. K. Braklege. 1961. Age criteria and population dynamics of the swamp rabbit in Missouri. Journal of Wildlife Management 25:271–80.

McCoy, R. H., and F. Steenbergen. 1969. *Staphylococcus* epizootic in western Oregon cottontails. Bulletin of the Wildlife Disease Association 5:11.

McCreedy, C. D., and H. P. Weeks. 1992. Sodium provision and wild cottontail rabbits: Morphological change in adrenal glands. Journal of Wildlife Management 56:669–76.

McKay, D. O., and B. J. Verts. 1978. Estimates of some attributes of a population of Nuttall's cottontails. Journal of Wildlife Management 42:159–68.

Mech, L. D., K. L. Heezen, and D. B. Siniff. 1966. Onset and cessation of activity in cottontail rabbits and snowshoe hares in relation to sunset and sunrise. Animal Behaviour 14:410–13.

Morgan, K. A., and J. E. Gates. 1983. Use of forest edge and strip vegetation by eastern cottontails. Journal of Wildlife Management 47:259–64.

Morgan II, R. P., and J. A. Chapman. 1981. The serum proteins of the *Sylvilagus* complex. Pages 64–72 *in* K. Myers and C. D. MacInnes, eds. Proceedings of the world lagomorph conference, University of Guelph, Ontario, Canada.

Mossman, A. S. 1955. Reproduction of the brush rabbit in California. Journal of Wildlife Management 19:177–84.

Mullin, K. D. 1982. Aspects of the ecology of the swamp rabbit in disturbed bottomland hardwoods and associated pinewoods in west central Louisiana. M.S. Thesis, Northwestern Louisiana State University, Natchitoches.

Myers, K. 1967. Morphological changes in the adrenal glands of wild rabbits. Nature 213:147–50.

Nelson, E. W. 1909. The rabbits of North America. North American Fauna 29:1–314.

Oehler, J. D., and J. A. Litvaitis. 1996. The role of spatial scale in understanding responses by medium-sized carnivores to forest fragmentation. Canadian Journal of Zoology 74:2070–79.

Olmstead, D. L. 1970. Behavioral comparisons of two species of cottontails (*Sylvilagus floridanus* and *Sylvilagus transitionalis*). Transactions of the Northeastern Section of the Wildlife Society 27:115–26.

Orr, R. T. 1940. The rabbits of California. Occasional Papers of the California Academy of Science 19:1–227.

Palmer, W. E., G. A. Hurst, B. D. Leopold, and D. C. Cotton. 1991. Body weights and sex and age ratios for the swamp rabbit in Mississippi. Journal of Mammalogy 72:620–22.

Parker, E. R., J. F. Bell, W. S. Chalgren, F. B. Thrailkill, and M. T. McKee. 1952. The recovery of strains of Rocky Mountain spotted fever and tularemia from ticks of the eastern United States. Journal of Infectious Diseases 91:231–37.

Pearson, O. P. 1971. Additional measurements of the impact of carnivores on California voles (*Microtus californicus*). Journal of Mammalogy 52:41–49.

Pelton, M. R. 1969. The relationship between epiphyseal groove closure and age of the cottontail rabbit (*Sylvilagus floridanus*). Journal of Mammalogy 50:624–25.

Pelton, M. R., and J. H. Jenkins. 1971. Productivity of Georgia cottontails. Proceedings of the Annual Conference of the Southeastern Association of Game and Fish Commissioners 25:261–68.

Pelton, M. R., and E. E. Provost. 1972. Onset of breeding and breeding synchrony by Georgia cottontails. Journal of Wildlife Management 36:544–49.

Petrides, G. A. 1951. The determination of sex and age ratios in the cottontail rabbit. American Midland Naturalist 46:312–36.

Philip, C. B. 1946. Rickettsial diseases in man. Pages 97–112 *in* Symposium on Medical Science, American Association for the Advancement of Science, Boston.

Porath, J. W. 1997. Swamp rabbit (*Sylvilagus aquaticus*) status, distribution, habitat characteristics in southern Illinois. M.S. Thesis, Southern Illinois University, Carbondale.

Powell, R. A. 2000. Animal home ranges and territories and home range estimators. Pages 65–110 *in* L. Boitani and T. K. Fuller, eds. Research techniques in animal ecology: Controversies and consequences. Columbia University Press, New York.

Powers, R. A., and B. J. Verts. 1971. Reproduction in the mountain cottontail rabbit in Oregon. Journal of Wildlife Management 35:605–13.

Probert, B. L., and J. A. Litvaitis. 1996. Behavioral interactions between invading and endemic lagomorphs: Implications for conserving a declining species. Biological Conservation 76:289–96.

Pulliam, R. 1988. Sources, sinks, population regulation. American Naturalist 132:652–61.

Reynolds, D. 1975.

Robinson, T. J., F. F. B. Elder, and J. A. Chapman. 1983. Evolution of chromosomal variation in cottontails, genus *Sylvilagus* (Mammalia: Lagomorpha): *S. aquaticus, S. floridanus,* and *S. transitionalis.* Cytogenetics and Cell Genetics 35:216–22.

Robinson, T. J., F. F. B. Elder, and J. A. Chapman. 1984. Evolution of chromosomal variation in cottontails, genus *Sylvilagus* (Mammalia: Lagomorpha): *Sylvilagus audubonii, B. idahoensis, S. nuttallii,* and *S. palustris.* Cytogenetics and Cell Genetics 38:282–89.

Rongstad, O. J. 1966. Biology of penned cottontail rabbits. Journal of Wildlife Management 30:312–19.

Rongstad, O. J. 1969. Gross prenatal development of cottontail rabbits. Journal of Wildlife Management 33:164–68.

Rose, G. B. 1973. Energy metabolism of adult cottontail rabbits, Sylvilagus floridanus, in simulated field conditions. American Midland Naturalist 89:473–78.

Rose, G. B. 1974. Energy dynamics of immature cottontail rabbits. American Midland Naturalist 91:473–77.

Ruedas, L. A. 1998. Systematics of *Sylvilagus* Gray, 1867 (Lagomorpha:Leporidae) from southwestern North America. Journal of Mammalogy 79:1355–78.

Ruedas, L. A., R. C. Dowler, and E. Aita. 1989. Chromosomal variation in the New England cottontail, *Sylvilagus transitionalis.* Journal of Mammalogy 70:860–64.

Sadler, K. C. 1981. Thirty-one years of rabbit population data in Missouri. Pages 892–98 *in* K. Myers and C. D. MacInnes, eds. Proceedings of the world lagomorph conference. University of Guelph, Ontario, Canada.

Schierbaum, D. 1967. Job completion report, evaluation of cottontail rabbit productivity (Pittman-Robertson Project W-84-R-12). Albany, NY.

Schmidt, K. A., and C. J. Whelan. 1999. Effects of exotic *Lonicera* and *Rhamnus* on songbird nest predation. Conservation Biology 13:1502–6.

Schultze, W. H., W. C. Smith, and P. J. Moughan. 1988. Amino acid requirements of the growing meat rabbit: II. Comparative growth performance on practical diets of equal lysine concentration but decreasing levels of other amino acids. Animal Production 47:303–10.

Schwartz, C. W. 1942. Breeding season of the cottontail in central Missouri. Journal of Mammalogy 23:1–16.

Scribner, K. T., and R. K. Chesser. 1993. Environmental and demographic correlates of spatial and seasonal genetic structure in the eastern cottontail (*Sylvilagus floridanus*). Journal of Mammalogy 74:1164–71.

Scribner, K. T., and L. J. Krysl. 1982. Summer foods of the Audubon's cottontail (*Sylvilagus auduboni*: Leporidae) on Texas Panhandle playa basins. Southwestern Naturalist 27:460–63.

Selye, H. 1956. The stress of life. McGraw-Hill, New York.

Selye, H. 1973. The evolution of the stress concept. American Scientist 61:692–99.

Severaid, J. H. 1950. The pygmy rabbit (*Sylvilagus idahoensis*) in Mono County, California. Journal of Mammalogy 31:1–4.

Sheffer, D. E. 1957. Cottontail rabbit propagation in small breeding pens. Journal of Wildlife Management 21:90.

Shirai, A., F. M. Boxeman, S. Perri, J. W. Humphries, and H. S. Fuller. 1961. Ecology of Rocky Mountain spotted fever. Part 1: *Rickettsia rickettsi* recovered from a cottontail rabbit in Virginia. Experimental Biology and Medicine 107:211–14.

Smith, D. F., and J. A. Litvaitis. 1999. Differences in eye size and predator–detection distances of New England and eastern cottontails. Northeast Wildlife 54:55–60.

Smith, D. F., and J. A. Litvaitis. 2000. Foraging strategies of sympatric lagomorphs: Implications for differential success in fragmented landscapes. Canadian Journal of Zoology 78:2134–41.

Smith, R. H. 1950. Cottontail rabbit investigations (Final Report, Pittman-Robertson Project 1-R). New York Department of Environmental Conservation, Albany, NY.

Sneddon, I. A. 1991. Latrine use by the European rabbit (*Oryctolagus cuniculus*). Journal of Mammalogy 72:769–75.

Sommer, M. A. 1997. Distribution, habitat, and home range of the New England cottontail (*Sylvilagus transitionalis*) in western Maryland. M.S. Thesis, Frostburg State University, Frostburg, MD.

Sonenshine, D. E., E. L. Atwood, and J. T. Lamb. 1966. The ecology of ticks transmitting Rocky Mountain spotted fever in a study area in Virginia. Annals of the Entomological Society of America 59:1234–62.

Sorensen, M. F., J. P. Rogers, and T. S. Baskett. 1968. Reproduction and development in confined swamp rabbits. Journal of Wildlife Management 32:520–31.

Sowls, L. K. 1957. Reproduction in the Audubon cottontail in Arizona. Journal of Mammalogy 38:234–43.

Spencer, R. K., and J. A. Chapman. 1986. Seasonal feeding habits of New England and eastern cottontails. Proceedings of the Pennsylvania Academy of Science 60:157–60.

Stevens, V. C. 1962. Regional variation in productivity and reproduction physiology of the cottontail rabbit in Ohio. Transactions of the North American Wildlife and Natural Resources Conference 27:243–53.

Stout, G. G. 1970. The breeding biology of the desert cottontail in the Phoenix region, Arizona. Journal Wildlife Management 34:47–51.

Sullins, G. L., D. O. McKay, and B. J. Verts. 1976. Estimating ages of cottontails by periosteal zonations. Northwest Science 50:17–22.

Sullivan, T. P., B. Jones, and D. S. Sullivan. 1989. Population ecology and conservation of the mountain cottontail, *Sylvilagus nuttallii,* in southern British Columbia. Canadian Field-Naturalist 103:335–40.

Svihla, R. D. 1929. Habits of *Sylvilagus aquaticus* littoralis. Journal of Mammalogy 10:315–19.

Swihart, R. K., and R. H. Yahner. 1982. Habitat features influencing use of farmstead shelterbelts by the eastern cottontail (*Sylvilagus floridanus*). American Midland Naturalist 107:411–14.

Tefft, B. C., and J. A. Chapman. 1983. Growth and development of nestling New England cottontails, *Sylvilagus transitionalis.* Acta Theriologica 28:317–21.

Tefft, B. C., and J. A. Chapman. 1987. Social behavior of the New England cottontail, *Sylvilagus transitionalis* (Bangs), with a review of social behavior in New World rabbits (Mammalia: Leporidae). Revue d'Ecologie: La Terre et la Vie 42:235–76.

Terrel, T. L. 1972. The swamp rabbit (*Sylvilagus aquaticus*) in Indiana. American Midland Naturalist 87:283–95.

Thomsen, J. P., and O. A. Mortensen. 1946. Bone growth as an age criterion in the cottontail rabbit. Journal of Wildlife Management 10:171–74.

Toll, J. E., T. S. Baskett, and C. H. Conaway. 1960. Home range, reproduction, and foods of the swamp rabbit in Missouri. American Midland Naturalist 63:398–412.

Tomich, P. Q. 1962. The annual cycle of the California ground squirrel *Citellus beecheyi.* University of California Publications in Zoology 65:213–82.

Trent, T. T., and O. J. Rongstad. 1974. Home range and survival of cottontail rabbits in southwestern Wisconsin. Journal of Wildlife Management 38:459–72.

Trethewey, D. E. C., and B. J. Verts. 1971. Reproduction in eastern cottontails in western Oregon. American Midland Naturalist 86:463–76.

Turkowski, F. J. 1975. Dietary adaptability of the desert cottontail. Journal of Wildlife Management 39:748–56.

Van Horne, B. 1982. Niches of adult and juvenile deer mice in several stages of coniferous forest. Ecology 63:992–1003.

Vance, D. R. 1976. Changes in land use and wildlife populations in southeastern Illinois. Wildlife Society Bulletin 4:11–15.

Verts, B. J., L. N. Carraway, and R. L. Greene. 1997. Sex-bias in prenatal parental investment in the eastern cottontail (*Sylvilagus floridanus*). Journal of Mammalogy 78:1164–71.

Villafuerte R., J. A. Litvaitis, and D. F. Smith. 1997. Physiological responses by lagomorphs to resource limitations imposed by habitat fragmentation: Implications to condition-sensitive predation. Canadian Journal of Zoology 75:148–51.

Wagner, F. H. 1981. The role of lagomorphs in ecosystems. Pages 668–94 *in* K. Myers and C. D. MacInnes, eds. Proceedings of the world lagomorph conference. University of Guelph, Ontario, Canada.

Watson, A., and R. Moss. 1970. Dominance, spacing behavior and aggression in relation to population limitation in vertebrates. Pages 167–218 *in* A. Watson, ed. Animal populations in relation to their food resources. Blackwell, Oxford.

Weiss, N. T., and B. J. Verts. 1984. Habitat and distribution of pygmy rabbits (*Sylvilagus idahoensis*) in Oregon. Great Basin Naturalist 44:563–71.

Wells, T. A. G. 1964. The rabbit. Dover, New York.

West, N. E. 1983. Western intermountain sagebrush steppe. Pages 351–74 *in* N. E. West, ed. Temperate deserts and semi-deserts. Elsevier, New York.

Whitaker, J. O., Jr., and B. Abrell. 1986. The swamp rabbit, *Sylvilagus aquaticus,* in Indiana, 1984–1985. Indiana Academy of Science 95:563–70.

White, G. C., and R. A. Garrott. 1990. Analysis of wildlife radio-tracking data. Academic Press, New York.

Wight, H. M. 1959. Eleven years of rabbit population data in Missouri. Journal of Wildlife Management 23:34–39.

Wight, H. M., and C. H. Conaway. 1961. Weather influences on the onset of breeding in Missouri cottontails. Journal of Wildlife Management 25:87–89.

Wight, H. M., and C. H. Conaway. 1962a. A comparison of methods for determining age of cottontails. Journal of Wildlife Management 26:160–63.

Wight, H. M., and C. H. Conaway. 1962b. Determination of pregnancy rates of cottontail rabbits. Journal of Wildlife Management 26:93–95.

Wilde, D. B., J. S. Fisher, and B. L. Keller. 1976. A demographic analysis of the pygmy rabbit, Sylvilagus idahoensis. Pages 88–105 *in* O. D. Markam, ed. 1975 Progress Report, Idaho National Engineering Laboratory Site Radioecology-Ecology Programs. U.S. Energy Research and Development Administration, Idaho Falls, ID.

Willner, G. R., J. A. Chapman, and D. Pursley. 1979. Reproduction, physiological responses, food habits, and abundance of nutria on Maryland marshes. Wildlife Monographs 65:1–43.

Wilson, D. E., and S. Ruff, eds. 1999. The Smithsonian book of North American mammals. Smithsonian Institution Press, Washington, DC.

Wolff, J. O. 1980. The role of habitat patchiness in the population dynamics of snowshoe hares. Ecological Monographs 50:111–30.

Woolf, A., D. A. Shoemaker, and M. Cooper. 1993. Evidence of tularemia regulating a semi-isolated cottontail rabbit population. Journal of Wildlife Management 57:144–57.

Yeatter, R. E., and D. H. Thompson. 1952. Tularemia, weather, and rabbit populations. Illinois Natural History Survey Bulletin 25:351–82.

Zollner, P. A., W. P. Smith, and L. A. Brennan. 1996. Characteristics and adaptive significance of latrines of swamp rabbits (*Sylvilagus aquaticus*). Journal of Mammalogy 77:1049–58.

Zollner, P. A., W. P. Smith, and L. A. Brennan. 2000. Home range use by swamp rabbits (*Sylvilagus aquaticus*) in a frequently inundated bottomland forest. American Midland Naturalist 143:64–69.

Zollner, P. A., W. P. Smith, and L. A. Brennan. 2001. Microhabitat characteristics of sites used by swamp rabbits (*Sylvilagus aquaticus*). Wildlife Society Bulletin 29:1003–11.

Zoloth, S. R. 1969. Observations of the population of brush rabbits on Ano Nuevo Island, California. Wassman Journal of Biology 27:149–61.

JOSEPH A. CHAPMAN, Office of the President, North Dakota State University, Fargo, North Dakota 58105-5167. Email: joseph@ndsu.nodak.edu.

JOHN A. LITVAITIS, Department of Natural Resources, University of New Hampshire, Durham, New Hampshire 03824. Email: john@christa.unh.edu.

6

Black-tailed Jackrabbit

Lepus californicus and Allies

Jerran T. Flinders
Joseph A. Chapman

NOMENCLATURE

Common Names. Black-tailed jackrabbit, California jackrabbit
Scientific Name. *Lepus californicus*
Subspecies North of Mexico. *L. c. bennettii, L. c. californicus, L. c. deserticola, L. c. eremicus, L. c. melanotis, L. c. merriami, L. c. richardsonii, L. c. texianus,* and *L. c. wallawalla*

Common Names. White-tailed jackrabbit, prairie hare
Scientific Name. *Lepus townsendii*
Subspecies. *L. t. campanius* and *L. t. townsendii*

DISTRIBUTION

The black-tailed jackrabbit is the most common jackrabbit in the western United States. There may be as many as 17 subspecies, with 9 of those in the United States (Hall 1981), but a conservative numerical taxonomic approach would synonymize all western subspecies as *L. californicus californicus* and the eastern populations as *L. californicus texianus* (Dixon et al. 1983). Range of the black-tailed jackrabbit extends from the Pacific coast to western Missouri and Arkansas, and from the prairie and grassland regions of South Dakota southward to Texas. In the West, the black-tail ranges from Washington and Idaho in the north into at least central Mexico (Cervantes et al. 1997). In the western United States, they are important taxa within steppe grasslands and shrublands as well as the cold and hot desert regions (Hall 1981; Zeveloff 1988) (Fig. 6.1). Black-tails occur at elevations as high as 3750 m in mountain shrub zones and also below sea level in the Mojave Desert within Death Valley, California (Hall 1946; Best 1996).

The black-tailed jackrabbit was successfully introduced into a number of eastern states. DeVos et al. (1956) reported *L. californicus* was introduced to Nantucket and Martha's Vineyard, Massachusetts. The black-tailed jackrabbit is established in southern Florida, where it uses pasture and sand prairie habitats (Layne 1965). Clapp et al. (1976) and Chapman and Sandt (1977) reported black-tails were successfully introduced on the eastern shore of Virginia and Maryland and also New Jersey.

White-tailed jackrabbits are traditional inhabitants of the plains and open mountainous regions. Two subspecies are recognized; *L. t. campanius* is the eastern subspecies, and is roughly separated from the western subspecies, *L. t. townsendii,* by the Continental Divide and the eastern slopes of the Rocky Mountains (Hall 1981). There are few if any geographic barriers between these subspecies, and R. M. Hansen and Bear (1963) reported intergrades along the Continental Divide. There may be few significant differences between the subspecies (Long 1965; Armstrong 1972; Rogowitz and Wolfe 1991). Numerical taxonomic studies by Dixon et al. (1983) did not identify taxonomic differences between these two subspecies. The range of the species extends from the prairies of the midwestern states and southern Canada westward through sagebrush–grass habitat to the high mountain slopes of the Rockies (Braun and Streeter 1968), Cascades, and Sierra Nevada (Fig. 6.2.). Historically, white-tails may have ranged south to the northern borders

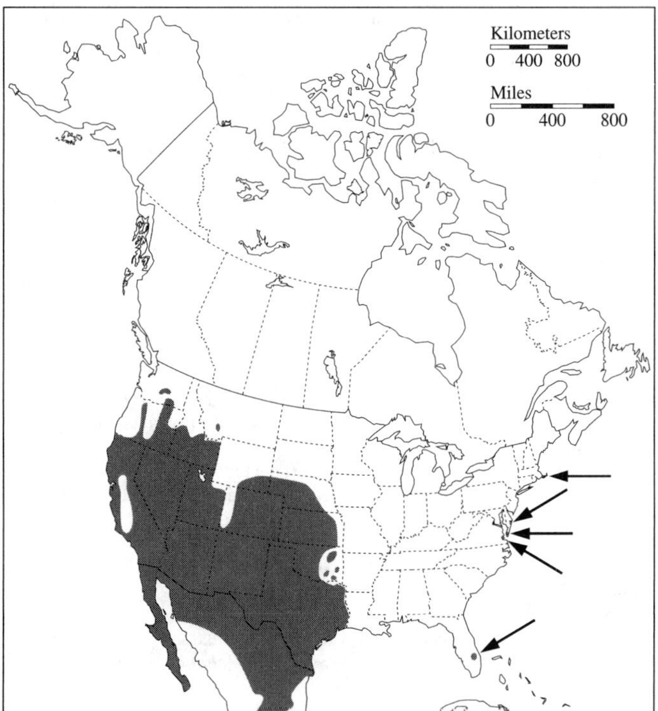

Figure 6.1. Distribution of the black-tailed jackrabbit (*Lepus californicus*).

of New Mexico and Arizona (Hoffmeister 1986). Lim (1987) noted white-tailed jackrabbits have the greatest elevational range of all native hares, being found from 4319 m on the summit of Mount Bross, Colorado, to just 30 m on the Columbia River plains.

During the early part of the 1900s, *L. townsendii* was reported to be extending its range as far east as Wisconsin, Iowa, and Missouri (Leopold 1945). Apparently the conversion of forests to farmland and parkland created acceptable habitats favoring the northeasterly range expansion of white-tailed jackrabbits in Minnesota and Manitoba (DeVos 1964). However, in recent years most populations have been declining due to many interactive, causative factors: habitat destruction, habitat fragmentation, competition with black-tailed jackrabbits, conversion of grasslands to shrublands, and unregulated hunting. In the basin and range region of southern Idaho, Nevada and western Utah, white-tailed jackrabbits are now mostly isolated in the upper regions of the mountain ranges with black-tailed jackrabbits found in the basin and foothill, shrub dominated, portions of the region. The white-tailed jackrabbit is considered extirpated from Kansas (Hall 1955) and southern Nebraska (Jones 1964), and though rare in Missouri in the 1970s (Watkins and Novak 1973), recent maps consider them not in Missouri (Hall 1981; Fitzgerald et al. 1994). Brown (1947a) believed the cultivation of land that was formerly open prairie may have caused *L. townsendii* to disappear in Kansas.

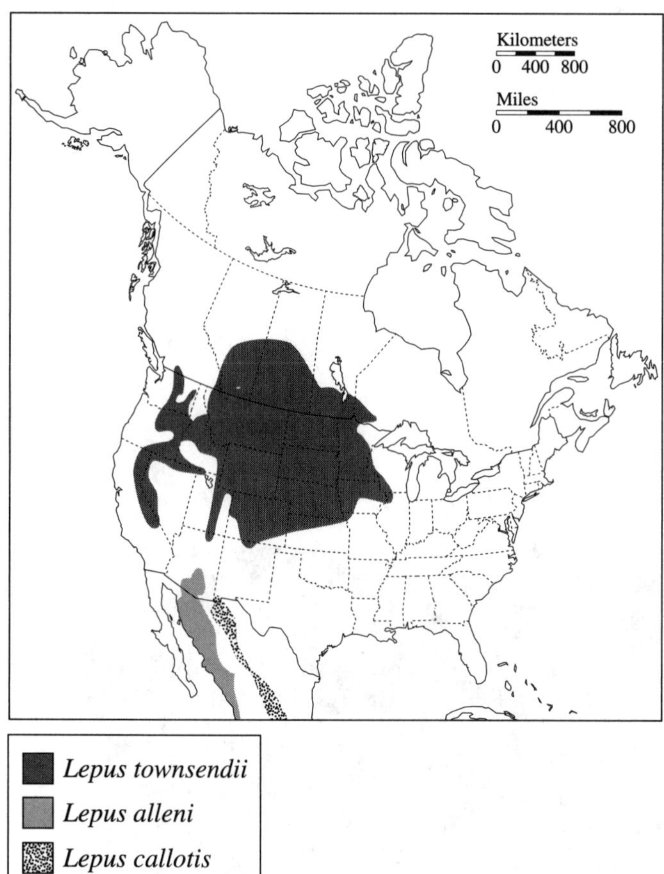

FIGURE 6.2. Distribution of the antelope jackrabbit (*Lepus alleni*), white-sided jackrabbit (*L. callotis*), and white-tailed jackrabbit (*L. townsendii*).

DESCRIPTION

Jackrabbits are hares; the young are born fully haired, whereas rabbits are born blind, naked, and remain in a nest until their eyes open. Jackrabbits, along with other hares and rabbits, belong to the family Leporidae, within the order Lagomorpha. The large ears and feet, the presence of a supraorbital process of the frontal bone, and a straight cutting edge on the upper incisor distinguish leporids. Absence of the interparietal bone differentiates the genus *Lepus* from the genus *Sylvilagus* (Glass 1973). The dental formula for all North American jackrabbits is I 2/1, C 0/0, P 3/2, M 3/3. The second pair of upper incisors ("peg teeth") are characteristic of all lagomorphs.

The black-tailed jackrabbit (*L. californicus*) can be distinguished from other hares by its tail, which has a broad, black, mid-dorsal stripe extending onto the back (Fig. 6.3, left). The upper body parts are gray to blackish. The black-tailed jackrabbit can be distinguished from the white-tailed jackrabbit by the black-edged ears and the less pronounced area of white on the sides of the body (Hall and Kelson 1959). The hind foot of *L. californicus* averages <135 mm in length, whereas that of *L. townsendii* averages 148 mm (Orr 1940).

These two species of jackrabbit also can be distinguished on the basis of cranial characters. *L. californicus* has a relatively long and slender skull, with the supraorbital process being narrower than in *L. townsendii* and extending only slightly above the frontal plane (Fig. 6.4, found on page 129). *L. townsendii* possesses a shorter, more arched skull, with the supraorbital process broader than in *L. californicus* and noticeably elevated above the frontal plane (Orr 1940). The pattern of infolding on the first upper incisor of *L. townsendii* can be used to distinguish it from *L. californicus* east of the Rocky Mountains. *L. californicus* has bifurcated tooth enamel, whereas *L. townsendii* has only a simple groove on the anterior surface of the tooth (Hall and Kelson 1959).

In Colorado, the mean adult body weight of black-tailed jackrabbits was 2.54 kg (Flinders and Hansen 1972). In Arizona, Vorhies and Taylor (1933) found the mean weight of 23 adult males was 2.3 kg. Female body weight was 14% greater and total length was 2.7% greater than those of males (Tiemeier 1965). Haskell and Reynolds (1947) reported the black-tailed jackrabbit attains its maximum body weight at 32 weeks of age. Body measurements for *L. californicus* were as follows: total length, 465–630 mm; tail, 50–112 mm; hind foot length, 112–145 mm; and ear length from notch, 99–131 mm (Hall and Kelson 1959).

The white-tailed jackrabbit is the only jackrabbit that exhibits two annual molts (Fig. 6.3, right), although other leporids have a second annual molt. The summer pelage is grayish brown on the dorsum, with the tail all white or with a very narrow, dark mid-dorsal stripe. The winter pelage is paler than that in the summer, and hares in the northern part of the range turn completely white. Winter guard hairs are completely white (Orr 1940; Hall and Kelson 1959).

The white winter coat is usually acquired in the part of the geographic range where snow persists throughout the winter. After the spring molt, the white coat is replaced with a darker summer pelage. Where snow cover is nonexistent or sparse, summer and winter coats are similar. Hansen and Bear (1963) classified winter overfur of white-tails into five categories from (1) white to (5) dark and concluded that this population represented intergradation between a mostly white population where there was heavy snow cover and a lower elevational population where winter snow cover was lacking or not consistent. They suggested winter pelage color of white-tails is controlled by two or more alleles.

Similarity between the snowshoe hare (*L. americanus*) in its winter pelage and the white-tailed jackrabbit can lead to misidentification in the field, although these hares occupy distinctly different habitats. Body weight of the white-tail is more than twice that of the snowshoe hare. Pelage of white-tails appears more coarse (Banfield 1974), and when the guard hair is completely white, the underfur may vary in color from iron gray to buff to reddish-brown (Hansen and Bear 1963). Banfield (1974) described snowshoe hare pelage as having three layers: first, the short, dense, silky, slate-gray underfur; then a thin coat of longer, buff-tipped hairs; and finally, the long, black-tipped (in summer) coarser guard hair.

Body measurements of adult white-tailed jackrabbits are as follows: total length, 565–655 mm; tail, 66–112 mm; hind foot length, 145–172 mm; and ear length from notch, 96–113 mm (Hall and Kelson 1959). The white-tailed jackrabbit is somewhat larger than the black-tailed jackrabbit. Mean weights of white-tailed jackrabbits range from 2.5 to 3.4 kg in California (Orr 1940). Sexual dimorphism occurs in *L. townsendii,* with females larger in total length and weight (Orr 1940; Bear and Hansen 1966). Rogowitz and Wolfe (1991) found females were 1–2% larger in body mass and appendage length, but thought correspondence with appendage size was genetically linked with body mass. An apparent east–west cline of reducing body mass of adult male and female white-tailed jackrabbits was considered an epigenetic response to available nutrients (Rogowitz and Wolfe 1991). Indications of a cline were suggested by 3.4- and 3.6-kg mean weights of males and females, respectively, in Iowa (Kline 1963) and 2.8 and 2.9 kg, respectively, in Wyoming (Rogowitz and Wolfe 1991).

GENETICS

Worthington and Sutton (1966) found the diploid (2*N*) chromosome number of *L. californicus* was 48. This species has 2 pairs of metacentric autosomes, 14 pairs of submetacentric autosomes, and 7 pairs of acrocentric autosomes. The X sex chromosome is a large submetacentric, and the Y is a "dot" chromosome.

In a study conducted in North Dakota, Jalal et al. (1967) reported 2*N* for the white-tailed jackrabbit was 48, with 23 pairs of autosomes and 1 pair of sex chromosomes. Chromosome length was 1.5–6.5 μm in the male. Hansen and Bear (1963) indicated the winter pelage of *L. townsendii* may be genetically determined. They postulated that

FIGURE 6.3. (Left) A black-tailed jackrabbit (*Lepus californicus*) and (right) a white-tailed jackrabbit (*L. townsendii*). The black tail of *L. californicus* readily distinguishes it from *L. townsendii*. Whereas *L. townsendii* may turn white during winter, *L. californicus* does not. SOURCE: Left, Photo by T. L. Best, Mammal Images Library, American Society of Mammalogists; right, Photo by C. D. Grondahl, Mammal Images Library, American Society of Mammalogists

incomplete dominance of two or more alleles may be responsible for the polytypic winter pelage of this species. Early work showed that the chromosomes of hares are very similar (Stock 1976; Robinson et al. 1983). Surprisingly, this similarity is evident in the karyotypes of the European hare (*L. europaeus*) and *L. californicus deserticola* (Schroder et al. 1978) as well as *L. alleni* and *L. townsendii* (Rogowitz and Wolfe 1991). There is evidence for hybridization between at least some *Lepus* species (Flux and Angermann 1990; Thulin et al. 1997). Flinders (in Flux 1983) reported hybrids between *L. californicus* and *L. townsendii* in traditional habitat of white-tails recently colonized by black-tailed jackrabbits. This leads to the supposition that historic isolation mechanisms are mainly geographic, behavioral, or ecological (Flux 1983; Halanych et al. 1999). Apparently, mechanisms of speciation for *Lepus* do not involve the chromosomal models often used for other mammals (Reig 1989; Dannelid 1991; Halanych et al. 1999). Work with mitochondrial cytochrome *b* DNA indicates North American hares are derived from two ancestral stocks and are therefore not monophyletic (Halanych et al. 1999). *L. townsendii* is more closely related to hares in an arctic clade, whereas a western American clade includes *L. californicus, L. alleni,* and *L. callotis* as well as other Mexican species (Halanych et al. 1999).

PHYSIOLOGY

Metabolism and Thermoregulation. The basal metabolic rate of adult black-tailed jackrabbits was estimated to be 54.11 kcal/kg body weight/24 hr, and a similar value of 53.14 kcal/kg body weight/24 hr was reported for the white-tailed jackrabbit (Flinders and Hansen 1972). In the Mojave Desert, where extremes in temperature occur, Shoemaker et al. (1976) estimated an annual energy expenditure of 55,200 kcal/kg for *L. californicus*. Nagy et al. (1976) suggested that jackrabbits were capable of regulating water losses from evaporation through physiological mechanisms. *L. californicus* exhibit a diurnal shift in body temperature. An adult jackrabbit at rest has a body temperature of 37–38°C (Shoemaker et al. 1976). To maintain such body temperatures, jackrabbits would normally require substantial amounts of water for evaporative cooling (Costa et al. 1976). However, they avoid this water demand by allowing body temperature to rise to 41°C during the heat of the day, thereby storing heat that otherwise would have to be dissipated (Shoemaker et al. 1976). Jackrabbits also possess some ability to concentrate urine and thereby reduce additional water loss. This may be accomplished through the precipitation of calcium salts in the urine. Jackrabbits can also reduce water loss by excreting dry feces. Fecal water loss was estimated at $\leq$15% of the total water loss (Nagy et al. 1976). Another physiological adaption for heat loss is the increased flow of blood to the ears, which results in greater convective and radiative heat loss (Schmidt-Nielsen et al. 1965). Behavioral mechanisms may aid in the dissipation of heat. Jackrabbits actively seek shade and may adjust their posture to maximize the surface-to-volume ratio for increasing convective heat loss (Schmidt-Nielsen et al. 1965).

Reese and Haines (1978) examined the effects of dehydration on black-tailed jackrabbits. They found that severe water restriction, which may occur seasonally, resulted in a decreased metabolic rate and a reduction in evaporative water loss. When deprived of water, black-tails

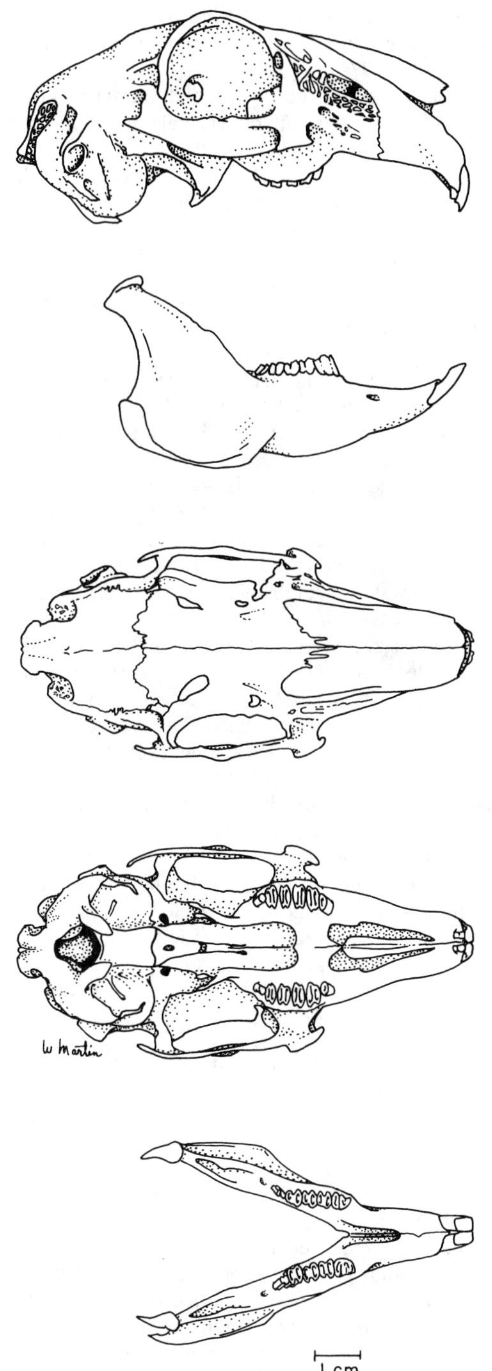

FIGURE 6.4. Skull of the black-tailed jackrabbit (*Lepus californicus*). From top to bottom: Lateral view of cranium, lateral view of mandible, dorsal view of cranium, ventral view of cranium, dorsal view of mandible.

reduced total evaporative water loss by 70% (Nagy 1994). In comparison with some other desert mammals, black-tails in the Mojave Desert do not seem well adapted physiologically or behaviorally to regular drought conditions, since large numbers die each year (Nagy 1994). Survival occurs at the population level via the high reproductive rate of those few individuals having year-round access to sufficient green vegetation (Nagy 1994). To meet water needs for survival, Westoby (1980) estimated that black-tails in July and August within the Great Basin of Utah needed foods with a water content about 500% that of dry weight. He did not identify foods providing the amount of water suggested, but found that the rabbits fed heavily upon succulent, salt-pumping plants with high oxalate concentrations.

With an experimentally induced food reduction of 25%, black-tails in Texas responded with lower body mass and kidney fat indices, higher fat in femur marrow, increased serum bilirubin and cortisol levels, increased width of the adrenal cortex, depressed immune response, reduced packed cell volume of red blood cells, altered white cell fractions, and irregularly shaped red blood cells (Henke and Demarais 1990). In this study, no single index of body condition could be used to adequately indicate nutritional stress.

Acclimatization to winter conditions is accomplished by mechanisms that reduce heat loss. Winter pelage is longer and thicker, which increases the total insulation of the body. There is also a slight increase in metabolism during winter (Hinds 1977; Harris et al. 1985).

Although true comparative studies are lacking, we would expect adaptive physiological differences between black-tailed and white-tailed jackrabbits because the latter species is aligned with an arctic clade of hares (Halanych et al. 1999). A rapid homeothermic response would seem requisite for white-tails. Thus, Rogowitz (1992a) monitored metabolic responses along with rectal temperatures in hares 4–81 days of age in relationship to induced variations in ambient temperature. He found that 4- to 5-day-old leverets responded to low ambient temperature by increasing metabolism, but in spite of this, body temperatures declined at lower ambient temperatures. By 80–81 days of age, rectal temperatures were stable over a wide range of ambient temperatures. Studies with adult white-tails showed high wind speed and low ambient temperature interact to increase metabolic rate. This effect is greater in summer than winter because, as noted, pelage of white-tailed jackrabbits is thicker in winter and the coefficient of heat transfer is less (Rogowitz and Gessaman 1990). Apparently, they expend minimal energy for thermoregulation at ambient temperatures in their natural habitats. It was experimentally shown that when ambient temperatures were set at 15°C below the lower critical temperature for white-tails, metabolic energy increased only to 1.50 times the standard metabolic rate in summer and 1.35 times in winter (Rogowitz 1990).

Adrenal Weights. Adrenal weights and morphology have been used to determine the endocrine response of the black-tailed jackrabbit (Herrick 1965). Pregnant females exhibited heavier adrenal weights than non-pregnant females, suggesting that pregnancy was a major stress factor. Comparison of biotic and environmental factors with adrenal weight indicated that temperature, age, and body weight also were important factors affecting adrenal weight (Herrick 1965).

REPRODUCTION AND DEVELOPMENT

Anatomy. The reproductive system of male black-tailed and white-tailed jackrabbits is very similar to that of the domestic rabbit. Jackrabbits have a scrotal sac that is entirely covered with hair and is visible only during the breeding season. Scrotums of older individuals usually have a dark blue color, which may serve as an indicator of age. The paired testes of adults are large (>8 cm^3), and are scrotal during the breeding season. The penis is cylindrical and is normally withdrawn into the sheath. The seminal vesicle, vesicular gland, prostate, paraprostates, and bulbourethral glands are similar in gross morphology and histology to those of other lagomorphs (Lechleitner 1959a).

The female reproductive system is comparable to that of the domestic rabbit. Jackrabbits have large corpora lutea, sometimes reaching 12 mm in diameter, compared to about 5 mm in domestic rabbits. The uterus is duplex, each side having its own distinct cervix. The size and condition of the uterus are dependent on the reproductive condition of the animal. Nonpregnant, nonparous females tend to have a smoother, unstriated, less convoluted, and less muscular uterus than that of pregnant or parous females. After parturition, the uterus shrinks rapidly. The vagina is a single tube leading from each cervix. The clitoris is near the external opening. The color of the vulva is variable, ranging from almost white to a dark purple. Females usually have six mammae, two thoracic and four abdominal. These are supplied by two strips of mammary tissue extending down the midline of the abdomen (Lechleitner 1959a; James and Seabloom 1969a).

Breeding Season. Length of the breeding season in black-tailed jackrabbits is variable, depending on latitude and various environmental factors. In the northern part of its range in Idaho, French et al. (1965) reported a 128-day breeding season, approximately from February through May. In California, the breeding season occurs from late January through August, with occasional breeding possible during any month of the year (Orr 1940; Haskell and Reynolds 1947; Lechleitner 1959a). In Utah, Gross et al. (1974) reported the breeding season was January–July. Over 75% of adult females were pregnant during February–April. In Kansas, Bronson and Tiemeier (1958a) reported a breeding season of 220 days, extending from January through August. Sperm production in males began in December, and all adult males contained sperm from December until late August. Sixty-five to 90% of the adult females were pregnant during March–August (Tiemeier 1965).

In the southwestern United States, breeding begins as early as January and ends by late August or September (Griffing and Davis 1976). In Arizona, Vorhies and Taylor (1933) reported jackrabbits breeding 10 months of the year. Two peaks of breeding, which corresponded to the annual cycle of rainfall and the greening of vegetation, were reported in California (Lechleitner 1959a), Arizona (Vorhies and Taylor 1933), and New Mexico (Davis et al. 1975). Comparison of the lengths of the breeding seasons from California, Kansas, Utah, and Arizona indicates a shorter breeding season for those areas located at higher latitudes with more severe winter climates (French et al. 1965).

Length of the breeding season of *L. townsendii* appears to be relatively constant throughout its range. In Iowa, North Dakota, and Wyoming, it averaged about 148 days in length and extended from late February through mid-July (Kline 1963; James and Seabloom 1969a; Rogowitz 1992b). In southern Colorado, ovarian size and weight along with uterine width reached a maximum during June (Bear and Hansen 1966).

Testes of white-tailed jackrabbits from North Dakota and Colorado were scrotal from late December until late July. They then ascended into the abdominal cavity during the remainder of the year. Sperm production peaked between March and June and then rapidly decreased during July (Bear and Hansen 1966; James and Seabloom 1969a).

Gestation. Haskell and Reynolds (1947) reported that gestation ranged from 41 to 47 days, with mean gestation of 43 days for black-tailed jackrabbits in California. This differed slightly from the mean gestation of 40 days reported by Gross et al. (1974) for a black-tailed jackrabbit population in Utah. Kline (1963) estimated that gestation was 42 days for white-tailed jackrabbits in Iowa.

Reproductive Rate. The number of litters conceived by black-tailed jackrabbits range from two litters/year in Idaho, where there is a 6-month breeding season (Feldhamer 1979), to seven litters/year in Arizona, where a 10-month breeding season occurs (Vorhies and Taylor 1933). The average annual production of young per female ranges from 8 in Arizona (Vorhies and Taylor 1933) to 16.7 for a population in Utah (Gross et al. 1974). The average annual production throughout the range of the black-tailed jackrabbit is about 14 young per female.

White-tailed jackrabbits show some variation regarding total number of litters produced. In North Dakota, the species averaged 3.29 litters/year (James and Seabloom 1969a). Kline (1963) reported females in Iowa may produce 3 or 4 litters/year. Bear and Hansen (1966) reported *L. townsendii* in southern Colorado had only 1 litter/year. James and Seabloom (1969a) estimated total annual production of the white-tail in North Dakota to be 14.7 young per female.

Breeding Synchrony. The jackrabbit, like other lagomorphs, is an induced ovulator; ovulation occurs only after copulation or other suitable stimulus has occurred. If pregnancy does not follow ovulation, a female may exhibit pseudopregnancy, which may last for about one half of the normal gestation period. The total loss of a litter through abortion or resorption also may produce effects similar to pseudopregnancy and may be accompanied by lactation (Lechleitner 1959a).

The black-tailed jackrabbit exhibits a relatively well synchronized breeding season, with conception possible immediately following parturition of the previous litter. The breeding season can be divided into four or five conception periods on the basis of this breeding synchrony (Gross et al. 1974).

Synchronous breeding occurs for both subspecies of the white-tailed jackrabbit. James and Seabloom (1969a) reported a synchronous breeding pattern in North Dakota, with four well-defined breeding peaks. In Wyoming, postpartum breeding served to synchronize the number of litters with weather variables. Thus, breeding could begin in late February or mid-March and end in July, with earlier breeding dates noted in more open, mild winters (Rogowitz 1992b).

Litter Size. Black-tailed jackrabbits show a direct relationship between litter size and latitude, with an increase in the length of the breeding season at lower latitudes (French et al. 1965). In the northern part of the range, litters were 4.9 in Idaho (Feldhamer 1979) and 3.8 in Utah (Gross et al. 1974). Tiemeier (1965) reported a litter size of 2.8 in southwest Kansas. In the more southerly portions of the range, average litter sizes were 2.3 in California (Lechleitner 1959a), 2.2 in Arizona (Vorhies and Taylor 1933), and 1.9 in New Mexico (Griffing and Davis 1976).

There is some change in the number of young per litter associated with the yearly breeding cycle. Litter sizes are small at the onset of breeding. They gradually increase and peak in midseason and then decline toward the end of the breeding season (Lechleitner 1959a; Tiemeier 1965; Gross et al. 1974).

Average litter size throughout the range of the white-tailed jackrabbit is not consistent. Lord (1960) found no significant correlation between litter size and latitude. In Colorado, the species averages 5 young per litter (Bear and Hansen 1966), whereas an average litter from North Dakota was 4.6 young (James and Seabloom 1969a). Kline (1963) found that *L. townsendii* in Iowa averaged 3.6 young per litter.

Variation of litter sizes of white-tails from successive litter periods was found in North Dakota. The greatest number of young per litter occurred during the second litter period. This was due to the variation in ova production and failure of the ova to implant during the first, third, and fourth litter periods (James and Seabloom 1969a). In Wyoming, three synchronized litters with means of four, six, and four viable fetuses, respectively, were observed. In this study, litter size was mediated by prenatal mortality, which in turn was a function of available nutrients as well as available water (Rogowitz 1992b).

Breeding Age. Young male black-tailed jackrabbits are first capable of producing spermatozoa between 5 and 7 months of age, when they weigh approximately 1900 g (Bronson and Tiemeier 1958a; Lechleitner 1959a). However, it is doubtful that males can breed at this age, because their testes do not contain a sufficiently high level of sperm. High sperm levels are not generally attained until the animal reaches a weight close to 2100 g, at between 7 and 8 months of age (Lechleitner 1959a).

Breeding by juvenile females is negligible in most populations (Haskell and Reynolds 1947; Bronson and Tiemeier 1958a). However, breeding by juvenile females was reported in a jackrabbit population in Utah. Juvenile females born during the first conception period in the spring were capable of breeding in their first year. In this study, twenty-seven percent of these young-of-the-year females had bred (Gross et al. 1974).

White-tailed jackrabbits normally breed in the spring following the year of their birth (Kline 1963; Bear and Hansen 1966; James and Seabloom 1969a). Some juvenile males may reach sexual maturity by late summer, but it is doubtful that they breed until the following year. There is no evidence of breeding in juvenile females (James and Seabloom 1969a; Rogowitz 1992b).

Nests and Newborn Hares. Female jackrabbits build "forms" in which they give birth to their precocial young (Tiemeier 1965). These forms are merely shallow excavations or depressions in the soil no more than a few centimeters deep, usually under some sort of protective cover. Tiemeier (1965) observed that litters may be dropped in depressions with little or no form preparation.

Newborn jackrabbits are very precocial. They are born fully furred and with their eyes open and are able to move about readily at birth.

The young leave the nest shortly after birth, usually within 24 hr of parturition. Tiemeier's (1965) observations suggested that litters may stay together for a week or more after leaving the nest.

Neonatal black-tailed jackrabbits have the characteristic black-tipped ears and black tail. Their pelage is darker than that of adults and is gradually replaced by a lighter coat. Often a small white patch is present on the forehead, but it disappears by 3 months of age. Newborn leverets (young hares) of white-tailed jackrabbits also have pelage that resembles the adult, except with a prominent white spot over the nasal–frontal junction. This spot may still be evident in November when older leverets are in advanced molt to white pelage. Adult pelage in both species is attained between 6 and 9 months of age (Haskell and Reynolds 1947; Bear and Hansen 1966).

Mean measurements of newborn black-tailed jackrabbits are total length, 168 mm; hind foot length, 43 mm; and weight, 110 g. The ear and hind foot reach maximum size at 15 weeks, total length at 28 weeks, and full weight at 32 weeks of age (Haskell and Reynolds 1947).

It appears that neonatal *L. townsendii* and *L. californicus* grow at the same rate. Average measurements from 1-day-old white-tailed jackrabbits in Colorado were hind foot length, 58.3 mm; ear length from notch, 40.7 mm; and weight, 105.3 g (Bear and Hansen 1966). Growth rates for leveret *L. townsendii* between birth and two-thirds asymptotic body mass indicated averages of 20–23 g/day, with the fastest rate being 30 g/day (Bear and Hansen 1966; Rogowitz and Wolfe 1991).

Tiemeier (1965) indicated a black-tailed jackrabbit dam suckled her leverets for 17–20 days. Sparks (1968) reported milk in the stomachs of black-tailed and white-tailed jackrabbits estimated to be 12 and 13 weeks of age, respectively. During the Septembers, of two years in Colorado, J. T. Flinders (unpublished data) found that 27.8% of white-tailed jackrabbits <18 weeks old had milk in their stomachs. A female 16–18 weeks old had milk constituting 13.5% of the weight of stomach contents. Milk was found in older white-tail leverets from August through October. In this same study, milk was not often found in stomachs of black-tailed jackrabbits with one exception. In May a male more than 24 weeks old had just less than 5% milk in his stomach contents.

ECOLOGY

Sex Ratio. The sex ratio in the black-tailed jackrabbit is constant among regions. A 1:1 sex ratio for adults and juveniles was reported for California (Lechleitner 1959a), New Mexico (Griffing and Davis 1976), Kansas (Tiemeier 1965), Colorado (Flinders and Hansen 1972), and Utah (Gross et al. 1974). Fetal sex ratios are also 1:1 (Lechleitner 1959a; Tiemeier 1965; Gross et al. 1974; Feldhamer 1979). A 1:1 sex ratio was noted for adult and young-of-the-year white-tailed jackrabbits (*L. t. campanius*) in southwestern Wyoming (Rogowitz and Wolfe 1991) and also for adult and juvenile *L. t. townsendii* in northeastern Colorado (Flinders and Hansen 1972).

Home Range. Home range size and shape of all the jackrabbits is affected by available nutrients, cover, water, predators, competition, and climatic variables. Lechleitner (1958b) reported the black-tailed jackrabbit in California had a well-defined home range, usually <20.2 ha.

French et al. (1965) reported a mean home range of <16.2 ha in Idaho while Tiemeier (1965) determined the mean home range of adults in Kansas was 16.2 ha. Intensive studies of radio-equipped black-tailed jackrabbits in Curlew Valley, Utah (Smith 1990), showed: first, that males had slightly larger home ranges than females; second, sizes of areas used did not differ between adults and juveniles; third, jackrabbit activity changed daily and seasonally; and fourth, in fall, some jackrabbits left summer home ranges for wintering areas. Smith (1990) used the 80% contour area of the harmonic mean method to calculate home range size and adult male home ranges varied from 73 to 183 ha, whereas adult female ranges were from 52 to 105 ha. These detailed studies indicated seasonal movements of black-tails were ≥35 km and 63% of these movements were ≤10 km (Smith et al. 2002).

There is a paucity of data concerning home range sizes of the white-tailed jackrabbit. Donoho (1972) found an average of 89.4 ha; males had larger home ranges. White-tailed jackrabbits seemed more nocturnal in their activities where they occurred with black-tailed jackrabbits on the shortgrass prairie of Colorado. Borell and Ellis (1934) also observed this for sympatric white-tailed and black-tailed jackrabbits in Nevada.

Fluctuations in Population Density. Gross et al. (1974) reported on a pattern of population fluctuation for the black-tailed jackrabbit in the Curlew Valley of northern Utah. Three 9- to 10-year population cycles were observed during the period 1951–1970, and jackrabbit densities ranged from 0.1/ha to as high as 1.0/ha. Although there was some variation in synchrony, density estimates were constant over a broad region of northern Utah and southern Idaho (Gross et al. 1974). Evans et al. (1970) commented on population fluctuations in which peak densities occurred every 6–10 years, and Clark (1975) reported peaks varied from 5 to 10 years in California.

Changes in jackrabbit population densities have been reported by many investigators. Bronson and Tiemeier (1959) observed that black-tailed jackrabbits were not cyclic in Kansas, but periods of drought concentrated jackrabbits in cultivated fields. Changes in the amount of food available were believed to be responsible for locally heavy concentrations where densities reached 34.6 hares/ha. During normal precipitation, densities were 0.4–0.7 jackrabbits/ha (Bronson and Tiemeier 1959). Similarly, changes in density in Arizona were related to grazing by cattle. Overgrazing, in combination with drought, created more favorable habitat by creating more open, weedy areas, which replaced the natural climax grasses (Taylor et al. 1935). Hayden (1966b) found maximum densities of jackrabbits in the Mojave Desert occurred around sources of water during winter. These seasonal densities decreased by 85% from spring to summer.

Density estimates of black-tailed jackrabbits in the arid southwest were from 0.2/ha in Nevada (Hayden 1966b), to 0.9/ha in the Great Salt Lake Desert of Utah (Woodbury 1955), to 1.2/ha in Arizona (Vorhies and Taylor 1933). A decline in range condition may favor these hares, for Daniel et al. (1993) found 0.23–1.00 black-tail/ha on good-condition rangeland and 0.46–1.09/ha on fair-condition rangeland in the Chihuahuan Desert. In this desert, black-tailed jackrabbits were more numerous in habitats supporting midseral rather than late seral stages of vegetation (Nelson et al. 1997). In the more temperate regions, densities ranged from 3.0/ha for California (Lechleitner 1958b) to as high as 34.6/ha for the agricultural areas of Kansas (Bronson and Tiemeier 1959). In Idaho, populations of black-tails go through regular fluctuations in abundance and peak at intervals of 7–12 years; peaks were seen in 1970–1971, 1980–1982, and 1990–1992 (Johnson and Peek 1984; Knick and Dyer 1997).

Black-tailed jackrabbit abundance may increase in response to extensive control programs for coyotes (*Canis latrans*) (Wagner and Stoddart 1972; Hoffmeister 1986; Henke and Bryant 1999). Stoddart (1983) found in northern Utah that mortality of juvenile black-tails was the most important factor affecting population abundance and that this mortality was positively correlated with coyote population density.

Several basic methods have been used to estimate population density or relative abundance of jackrabbits. These include line transects along roads and counts (number/area) conducted either during the day, when the hares are inactive (Wywialowski and Stoddart 1988), or when they are most active, at night (Flinders and Hansen 1973; Smith and Nydegger 1985). Relative abundance is inferred by counting leporid fecal pellets accumulated in plots during a given time period (Arnold and Reynolds 1943; Flinders and Crawford 1977). Johnson and Anderson (1984) combined fecal pellet census, diet analyses of fecal pellets, and habitat sampling in southern Idaho. They concluded that black-tailed jackrabbits selected open, grass-dominated habitats for feeding at night and then retreated to shrub cover during the day.

Where black-tailed and white-tailed jackrabbits occurred together on the short-grass prairie of Colorado, December population densities were 0.07 black-tail/ha and just more than 0.01 white-tail/ha. (Flinders and Hansen 1973). In this ecotype, habitat preferences were shown. Black-tailed jackrabbits were more abundant (>0.05/ha) in pastures, under moderate and light summer grazing by cattle and preferred the lowland pastures, where shrub cover was more prevalent. Responses of white-tailed jackrabbits to grazing treatments were not significant ($p = .08$), but they occurred more commonly in the upland portions of pastures with low-profile, sparse shrubs along with typical short-grass prairie grasses and forbs (Flinders and Hansen 1975).

A population density of 0.07 white-tailed jackrabbit/ha was not significantly different over the years from 1985 to 1987 in western Wyoming (Rogowitz and Wolfe 1991). Population density of white-tailed jackrabbits may vary greatly relative to factors not usually identified (e.g., succulent vegetation). In Iowa, average population densities ranged from 0.03 to 0.09/ha, but were reported as high as 0.7/ha (Kline 1963). White-tailed jackrabbit density in Minnesota averaged 0.06–0.12/ha, but peaks of 0.43/ha were noted (Mohr and Mohr 1936).

Although many mammals have a potential life span about five times their age at first reproduction, jackrabbits generally do not. Averages life spans of 1.4 and 1.8 years were reported for black-tailed jackrabbits in California (Lechleitner 1959a) and Kansas (Tiemeier 1965), respectively. A life span of >7 years would be unusual for a black-tailed jackrabbit (Haskell and Reynolds 1947). The life span of white-tailed jackrabbits was estimated at <1 year in Wyoming; just <1% of this sample was >2 years old (Rogowitz and Wolfe 1991). Jackson (1961) found an 8-year-old white-tailed jackrabbit; this must be near maximum longevity.

HABITAT

The black-tailed jackrabbit occupies many diverse habitats. Tiemeier (1965) found black-tails in the Midwest associated with short-grass or mixed-grass regions. He believed these hares increased in abundance and distribution in relationship to overgrazing by domestic livestock and with nonclimax stages of vegetational succession, including agriculture. In the Great Basin of the western states, black-tails are especially associated with plant communities dominated by greasewood (*Sarcobatus vermiculatus*), sagebrush, and shadscale (Linsdale 1938; Hall 1946; Currie and Goodwin 1966). In Arizona, black-tails were very evident in habitats dominated by mesquite (*Prosopis gladulosa*) and various species of cacti (Taylor et al. 1935).

In the Mojave Desert of California, jackrabbits used the sagebrush–creosote regions (Orr 1940). They also occurred in many cultivated agricultural areas of California as well as rangelands. In southeastern New Mexico, black-tails occupied typical desert shrub communities dominated by mesquite, snakeweed, and soap-tree yucca (*Yucca elata*) (Davis et al. 1975). Black-tails were important herbivores in the northern desert shrub region of northwestern Utah, often in association with juniper (*Juniperus osteosperma*) and sagebrush (Gross et al. 1974). The black-tailed jackrabbit has adapted well to many agricultural crops and waste areas in the drier western states. During periods of population abundance, they seriously depredate crops such as alfalfa and cereal grains (Johnson and Peek 1984).

Concern for prey of golden eagles (*Aquila chrysaetos*) and effects of recurring fire on native shrubs prompted examination of black-tailed jackrabbit habitat on a landscape scale in southwestern Idaho (Knick and Dyer 1997; Knick and Rotenberry 1998). The global positioning system locations of black-tails along dirt roads were plotted in association with the Snake River Birds of Prey National Conservation Area. The resultant geographic information system (GIS) map represented the similarity of GIS variables to the set of habitat characteristics associated with previous sightings of black-tailed jackrabbits (Knick and Dyer 1997). Verification counts and locations of black-tails in summer 1994 and winters 1993/1994 and 1994/1995 indicated that 50% of jackrabbits were located in the best 7% of habitats (Knick and Dyer 1997).

White-tailed jackrabbits are associated with prairie grassland habitats and the grassland–shrubland steppe habitats of the western high plains and mountains. In Bailey's (1936) survey of Oregon, he noted that white-tails occurred on the Columbia River plains in the upper Sonoran life zone and flourished in the more open transitional ecological zones to above timberline in the arctic–alpine zone. He noted that they strictly avoided timbered zones. In Wisconsin and Iowa, where little native prairie remains, white-tails were associated with the larger expanses of cropland and pasture with scattered brushy fencerows (Dubke 1973; Schwartz 1973). In Colorado, *L. townsendii* preferred grassland habitats, using both upland and lowland areas (Bear and Hansen 1966). In this study, foraging by these hares was at least four times greater on grasslands than in irrigated meadows, rabbitbrush, or sagebrush habitats. White-tailed jackrabbits often forage in open habitats adjacent to dense escape cover; the hares have freedom of movement and greater visibility in the open while feeding at night, but occupy more dense cover during their less active period of the day (Fautin 1946; Lechleitner 1958a). In Colorado, Braun and Streeter (1968) found white-tailed jackrabbits in the alpine zone above timberline, where they used alpine tundra vegetation.

Immigration of black-tailed jackrabbits into areas formerly occupied by white-tailed jackrabbits has been well documented (Couch 1927; Carter 1939; Brown 1947a). Habitat change brought about by cultivation of the prairie has been suggested as a major factor in this expansion (Carter 1939, Brown 1947a). Flinders and Hansen (1975) reported that when both species were in sympatry on the short-grass prairie, *L. townsendii* tended to select more sparsely vegetated upland habitats. The black-tailed jackrabbit was judged more efficient than the white-tailed jackrabbit in using a feeding site. Thus, the black-tailed jackrabbit may be able to displace the white-tailed jackrabbit due to its greater adaptability and efficiency in using different habitats (Hansen and Flinders 1969).

Through their foraging behavior, which includes ingestion of ripe seeds of grasses, forbs, and woody vegetation, jackrabbits have the potential for modifying their habitat in ways favorable to future foraging. Studies by Brown (1947b) showed that black-tailed jackrabbits distributed 14.3 kg/ha of seed of sand dropseed (*Sporobolus cryptandrus*) on prairie and prairie farmland in Kansas. Numerous mature seeds of pricklypear cactus have been found in feces or stomach contents of black-tailed jackrabbits and this may help explain the rapid spread of this cactus on prairie rangelands (Riegel 1941; Timmons 1942; Flinders and Hansen 1972). White-tailed jackrabbits also ingest pricklypear seed, but also ingest and thus perhaps disperse seeds of the pin cushion cactus (*Mammillaria vivipara*) on the shortgrass prairie of Colorado (Flinders and Hansen 1972).

FEEDING HABITS

Feeding habits of the black-tailed jackrabbit have been studied extensively by many biologists and are highly variable depending on the locality as well as availability of palatable plants.

Black-tailed jackrabbits prefer green succulent vegetation when available. Depending on the habitat, their diet is mainly grasses and forbs during summer. Diet changes to shrubs during winter, when snow or frost makes succulents unavailable (Hansen and Flinders 1969). Grasses and sedges are just as important as forbs and shrubs in the yearly diet of *L. californicus*. According to Flinders and Hansen (1972), grasses and sedges made up 49% of the overall diet of the species in northeastern Colorado.

In the sagebrush (*Artemisia tridentata*) and bitterbrush (*Purshia tridentata*) communities of Washington, forbs provided 75% of the blacktail's diet (Uresk 1978). In this study, yarrow (*Achillea millefolium*) was the most common food item, making up over 25% of the diet. Needle-and-thread grass (*Stipa comata*) was the preferred plant in the sagebrush habitat, whereas yarrow was preferred in the bitterbrush habitat.

On the salt desert ranges of Utah, Currie and Goodwin (1966) found black-tailed jackrabbits ate grasses, forbs, and shrubs in early spring. Grasses were preferred throughout spring and summer, and shrubs during late fall and winter. In that study, the more common grass species used were Indian ricegrass (*Oryzopsis hymenoides*) and squirreltail (*Sitanion hystrix*). The more prominent shrub species were winterfat (*Eurotia lanata*), shadscale (*Atriplex confertifolia*), saltsage (*A. nuttallii*), and sagebrush.

Black-tails in southeastern Idaho fed selectively on winterfat and perennial grasses as staple foods; these constituted about 80% of diets. These hares seemed to be selective foragers, since species composition of diets was more similar than was the vegetative species composition in the various study areas (Johnson and Anderson 1984). In southeastern Idaho, cluster analyses of black-tail diets identified just two dietary periods, spring–summer and fall–winter. Spring–summer diets were 43% grasses, 10% forbs, and 46% shrubs, whereas fall–winter diets were 9% grasses, 6% forbs, and 85% shrubs. In this latter period, winterfat and big sagebrush were especially important forages (MacCracken and Hansen 1982a, 1982b).

In northeastern Colorado, Flinders and Hansen (1972) observed that seven plant species accounted for 64% of the dry weight of plants eaten by black-tailed jack rabbits. In order of importance, those species were western wheatgrass (*Agropyron smithii*), alfalfa (*Medicago sativa*), summer cypress (*Kochia scoparia*), winter wheat (*Triticum aestivum*), crested wheatgrass (*Agropyron cristatum*), rubber rabbitbrush (*Chrysothamnus nauseosus*), and sun sedge (*Carex heliophila*). Sparks (1968) also noted the importance of western wheatgrass in the diet of blacktails located on sandhill rangelands of Colorado. When they are available, blacktails feed heavily on cultivated crops (winter wheat, alfalfa, and crested wheatgrass) during winter months (Flinders and Hansen 1972).

Black-tailed jackrabbits in the Gray Lodge Waterfowl Management Area of California fed on nearly all species of grasses and forbs, but herbaceous weed species were not preferred items in diets (Lechleitner 1958a). Their preferred foods were various cereal crops, especially young barley (*Hordeum* sp.). In October, Orr (1940) observed jackrabbits in the Great Basin region of California feeding mainly on sagebrush.

In Arizona, Vorhies and Taylor (1933) determined the feeding habits of black-tailed jackrabbits from analyses of stomach contents. Mesquite (*Prosopis juliflora*) made up 54% of the annual diet, grasses 24%, and cacti 3.3%. The amount of food available was related to the cycle of rainy and dry seasons in Arizona.

Examination of stomach contents of *L. californicus* from New Mexico also indicated a preference for mesquite leaves and seedpods. In addition, soaptree yucca (*Yucca glauca*), snakeweed (*Gutierrezia sarothrae*), croton (*Croton* sp.), and spurge (*Euphorbia* sp.) were present in the stomachs analyzed (Griffing and Davis 1976). Hayden (1966a) also determined the feeding habits of black-tailed jackrabbits in Nevada from analyses of stomach contents. Nineteen species of perennial shrub, six species of annual, and three species of grass were identified in the diet. The annual diet of these jackrabbits could be divided into two major periods. In late summer through winter, the diet was dominated by creosotebush (*Larrea* sp.) and winterfat, and in spring through early summer, the diet was dominated by Indian ricegrass, brome (*Bromus* sp.), and needle-and-thread grass (Hayden 1966a).

White-tailed jackrabbits fed mainly on plants that were in the prereproductive or early reproductive stages of development (Flinders and Hansen 1972). These plants usually have the greatest nutritive value and contain relatively high proportions of water and crude protein.

For white-tails in northeastern Colorado, seven plant species constituted >67% of the diet. Those species, in order of importance, were western wheatgrass, winter wheat, summer cypress, legumes (*Oxytropis* sp. and *Astragalus* sp.), sun sedge, rubber rabbitbrush, and crested wheatgrass (Flinders and Hansen 1972).

In southern Colorado, the diet of the white-tailed jackrabbit during spring was 87% shrubs, with Parry's rabbitbrush (*C. parryi*) constituting 70% of the diet and fringed sage (*A. frigida*) 15%. Grasses and forbs made up the remaining material (Bear and Hansen 1966). In the short-grass prairie region of northeastern Colorado, where farms are often adjacent to native grasslands, western wheatgrass and winter wheat constituted most of the diet during early spring. In this season, winter wheat is green and succulent and highly palatable to jackrabbits (Flinders and Hansen 1972).

Summer diet of white-tails was 70% forbs, 19% grasses, and 7% shrubs (Bear and Hansen 1966). The four most common species, in order of importance, were clover (*Trifolium* sp.), dandelion (*Taraxacum officinale*), dryland sedge (*C. obtusala*), and Indian paintbrush (*Castilleja integra*). In the short-grass prairie region, western wheatgrass was the most important food item during the summer, constituting 21% of the diet (Flinders and Hansen 1972).

The autumn diet was mostly grasses and forbs. Bear and Hansen (1966) reported that grasses made up 43% of the fall diet, forbs 34%, and shrubs 14%. Important plant species included dryland sedge, goosefoot (*Chenopodium* sp.), fringed sage, and winterfat. Western wheatgrass was an important fall food item, constituting as much as 46% of the diet (Flinders and Hansen 1972).

The winter diet of white-tails in mountains of Colorado was 76% shrubs, with rabbitbrush making up 72% and clover 12% (Bear and Hansen 1966). At lower elevations, white-tails fed on winter wheat and crested wheatgrass during winter months (Flinders and Hansen 1972; Pittman 1977). They also may feed on ears of corn left in fields (Findley 1956).

Black-tailed jackrabbits were believed to use mineral licks in Arizona (Vorhies and Taylor 1933). They dug or bit portions of earth, which were then ingested. Hares probably have mineral needs similar to those of rabbits (Cheeke 1987). In southeastern Idaho, geophagic behavior of black-tails was investigated to determine rates of uptake and importance of essential, toxic, and radioactive elements. For these hares, the mean (and 95% confidence interval) soil intake rate was 9.7 (9.0–10.6) g/day, with seasonal peaks in March–May and August–October. Geophagy was obligatory when foraging on some low-growing plants, but also was thought to be functional in correcting elemental imbalances in forage, especially sodium (Arthur and Gates 1988).

Studies of dietary habits of jackrabbits in relationship to forage availability show they are selective for particular species, tissues, and nutritional fractions while still being able to effectively use very different floristic communities in a wide range of habitats. They also commonly forage on plant species and tissues known to be poisonous to domestic livestock on the same rangelands (Flinders and Hansen 1972; Wansi et al. 1992; Alipayou et al. 1993; Ernest 1994). Diets of black-tailed jackrabbits have been well studied. These hares seasonally switch foraging emphasis from grasses, forbs, and shrubs, including succulents such as cacti, to maximize intake of nutrients, minerals, and water or to minimize intake of poisonous, antiherbivory, plant secondary compounds (Uresk 1978; Hoffman et al. 1993; Ernest 1994). Commonly, when feeding on shrubs considered poisonous, such as creosotebush and snakeweed (*Gutierrezia sarothrae*), black-tails cut a branch, then feed on the new growth portions while allowing the leaves and lignified wood to fall to the ground (Steinberger and Whitford 1983). In winter, when feeding on big sagebrush and rubber rabbitbrush, blacktails cut stems, then feed on the older tissues and drop the current year's growth. This is a common feeding strategy on many shrubs. The discarded stems and leaves become an important source of organic litter. Apparently, jackrabbits, especially black-tails, have the digestive physiology requisite for successfully handling many poisonous compounds, but this would entail costly allocations of other nutrients and water, and thus there is selective avoidance. Whether jackrabbits make pharmacological uses of ingested plant secondary compounds to combat internal and even external parasites is unknown, but their selected use of poisonous plants deserves investigation.

Feeding Activity. Feeding activities of jackrabbits are affected by various environmental factors such as weather, time of the year, and light (Orr 1940; Lechleitner 1958a). Feeding usually occurs during the early morning and evening hours and throughout the night. Black-tailed jackrabbits also may feed during daylight hours of cloudy days; feeding activity may be delayed several hours during a full moon. Calm, dry evenings are the most favorable for jackrabbit activity. Wind, falling temperatures, and precipitation were factors that significantly altered this feeding behavior (Lechleitner 1958a; Tiemeier 1965).

The amount and type of food present are the most important factors governing selection of a feeding site (Orr 1940). Black-tailed jackrabbits prefer to feed in areas where they are inconspicuous, yet where they can detect danger from a moderate distance. During the time of the year when woody plants make up a larger portion of the diet, black-tails use margins of brush tracts adjacent to open areas (Orr 1940). Longland (1991) found black-tails preferred to take alfalfa pellets from bait stations under the canopy of cold desert shrubs rather than from open sites located 5 and 10 m from the nearest shrub. His conclusion was that these hares were constrained by the risk of predation from foraging in the open sites.

Cecaphagy. Jackrabbits, and probably all leporids, produce two kinds of fecal pellets: hard and soft. The soft ones are reingested. Cecaphagy may be a term more appropriate than coprophagy when describing this natural, daily behavior of jackrabbits wherein a hare ingests material directly from the anus by placing the head between the hind legs while in a squatting position (Lechleitner 1957). Thacker and Brandt (1955) suggested hard fecal pellets are formed by material that has passed by the base of the cecum without being mixed with the main contents of the cecum. They described soft feces being formed by the emptying of the major portion of the cecum in a cyclic manner by a strong contraction of the spiral muscle. Composition of soft feces is comparable to that of cecal contents in regard to crude protein, crude fiber, and other proximate nutrients (Eden 1940; Thacker and Brandt 1955). Herndon and Hove (1955) reported 41.8% and 14.8% crude protein in soft and hard pellets, respectively and cecoctomized rabbits did no excrete typical soft feces. Microorganisms in the cecum digest, and thus break down, cellulose into usable carbohydrates, volatile fatty acids, and vitamins and also liberate minerals (Hansen and Flinders 1969). Many of these nutrients are first incorporated in the cells of bacteria. Griffiths and Davies (1963) concluded that 81% of the crude protein in soft feces of rabbits was in the form of bacterial cells. Reingestion subjects these nutritious cells and fractions to enzymatic digestion for the benefit of the leporid. Detailed studies of black-tailed jackrabbits in March by Steigers et al. (1982) showed that the transition from hard to soft pellets by both sexes began at 0414 hr to approximate completion by 0715 hr. Males began this transition before females. The transition from soft to hard pellets was completed for many by 1330 hr and all had made this transition by 1700 hr. Although specific data for white-tailed jackrabbits are lacking, the same kind of reingestion behavior has been noted (Bear and Hansen 1966). Soft feces and milk were found in stomachs of young white-tailed and black-tailed jackrabbits (J. T. Flinders, unpublished data). Young, precocial leverets probably obtain essential bacterial inoculum for their own digestive systems by ingesting the soft feces produced by the dam (Hansen and Flinders 1969).

BEHAVIOR

Social Behavior. Lim (1987) observed that white-tailed jackrabbits were one of the least sociable of hares. Feeding aggregations of 50–110 were noted in cultivated fields in winter, but these grouping were considered unusual (Lahrman 1980; Brunton 1981). There are few data on this subject, but in February 1979, J. T. Flinders (unpublished data) observed an aggregation of more than 150 *L. townsendii* on native rangelands in the Red Desert region of south-central Wyoming. It was a severe winter and these hares were congregated on a vegetated sand dune area dominated by serviceberry (*Amelanchier alnifolia*) and rubber rabbitbrush (*Chrysothamnus nauseosus*) with scattered clumps of wildrye grass (*Elymus cinereus*). During the first approach to the area, two golden eagles departed from large snow drifts on the edge of the vegetated dune. The area was crescent shaped and by crowding the hares into one point of the crescent, they could be roughly counted as they fled across the opening to the other arm of the crescent. True allelomimetic behavior was observed for these hares, since there was some degree of mutual stimulation and coordination of movement. The white-tails moved as a herd, but no leadership was evident. This could be considered a survival group. Hares took advantage of the lower risk of predation by being a member of a large group (Hamilton 1971).

Mating Behavior. Black-tailed and white-tailed jackrabbits have complex mating behaviors involving long chases, jumping, and frequent fighting between males and females. Lechleitner (1958a) observed that sexually excited males traveled along trails with their nose to the ground. They covered large areas usually corresponding to the size of the animal's home range. When a female was encountered, two types of reaction occurred. If the male approached in a hesitant manner, the female jumped at him and struck with the forefeet until he retreated. If the male approached in a persistent manner, either the female or the male jumped in the air, while the other ran underneath. During such jumping activity, urine emission by the male was frequently observed.

White-tailed jackrabbits usually exhibit more pronounced jumping behavior than black-tailed jackrabbits (Lechleitner 1958a; Blackburn 1968; Pontrelli 1968). Jumping behavior often was followed by the sexual chase, during which the male pursued the female while she ran in a rapid zigzag fashion. During the chase, the male often tried to mount the female. Usually this response was met by aggressive action on the part of the female. The purpose of sexual chase probably is to stimulate the female, after which she is receptive to mounting attempts by the male. Usually only two jackrabbits are involved in the chase; however, three or more have been observed in the chase (Lechleitner 1958a; Pontrelli 1968).

Copulation occurs immediately following the chase (Lechleitner 1958a). The female lowers her head and ears and elevates her hindquarters as the male mounts. Copulation lasts only a few seconds and is accompanied by rapid vibrations of the male's hind quarters. After ejaculation, the pair separate and any further attempts to mount are rebuffed by the female.

Aggressive Behavior. It is difficult to separate the sexual-aggressive behavior patterns of the jackrabbit from those of an intraspecific aggressive nature. Most agonistic responses take the form of head butting, biting, jumping, running in circles around another animal, or avoidance reactions (Tiemeier 1965).

Lechleitner (1958a) observed that black-tail males engaged in the sexual chase sometimes rebuffed other males in the immediate area. Such encounters were of short duration, consisting of a charge by one animal and a rapid retreat by the other.

Females also exhibit a high degree of antagonism during the breeding season. They charged other jackrabbits that came within 5–10 m of their position. This charging behavior resulted in the dispersion of females during the breeding season (Lechleitner 1958a).

Often, aggressive behavior ended very suddenly. If an individual was rebuffed by a male or female during the sexual chase, it often initiated displacement behavior. Feeding behavior was the most common displacement activity. Another was digging in soil with the forefeet and then "dusting" by rolling in the loose material (Lechleitner 1958a).

Escape Behavior. Jackrabbits apparently depend on hearing at least as much as sight to detect danger. At the first sign of danger, the ears are raised to a vertical position, with shifting and turning until the source of the disturbance is detected. Two types of escape response were noted in *L. californicus*. If the source of danger was distant and the animal was in dense vegetation, it tried to sneak away from the intruder. The jackrabbit ran with the body close to the ground and the ears lowered (Lechleitner 1958a). Jackrabbits also responded by "freezing" if surprised at close distances or in their forms. The head may be pressed to the ground, ears lowered, and hind quarters slightly elevated in this "freeze" position. If approached still closer, black-tails respond by rapidly running from the intruder. With its ears laid down close to the back, such running consists of long leaps with an occasional leap higher and longer than the rest. Jackrabbits can reach speeds up to 56 km/hr and cover 2–3 m with each bound (Orr 1940; Lechleitner 1958a). Black-tails and white-tails also may "freeze" in the more upright sitting position. The lack of facial and ear "comfort movements" helps identify this behavior.

When closely pursued or cornered, both species of jackrabbit may enter water and readily swim. Only the front feet seem to be used during swimming (Orr 1940; Lechleitner 1958a). Black-tailed jackrabbits were observed feeding on emergent vegetation in water up to 5 cm deep and swimming across ditches even when not pursued (Lechleitner 1958a).

Use of Forms and Trails. Use of forms has been reported for most if not all species of jackrabbits (West et al. 1961; Flux and Angermann 1990; Best 1996, 1999a, 1999b). Jackrabbits frequently enter forms or shallow depressions during resting periods. However, in the Mojave Desert, Costa et al. (1976) reported black-tailed jackrabbits entered short burrows that were dug either by the jackrabbit itself or by tortoises (*Gopherus agassizi*) and then enlarged by the jackrabbit. This rare behavior was noted only on hot summer days, when temperatures sometimes exceeded 42°C. Black-tailed jackrabbits were least active from 0800 to 1700 hr during spring, summer, and fall (Smith 1990). This includes the time when they are reingesting fecal matter derived from the cecum (Steigers et al. 1982). Black-tails occupy natural or modified openings (forms) in shrubs, forbs, or grasses during the inactive periods of the day. These forms provide shade, block effects of wind, and obscure the outline of the hare (Flinders and Elliott 1979). Although black-tailed and white-tailed jackrabbits may run for considerable distances when flushed from daytime forms, they readily move longer distances at night, often along well-established trails. Peak activity periods for both species are at night (Flinders and Hansen 1973, 1975).

During daylight hours, white-tailed jackrabbits are usually in forms that may be merely shallow depressions at the base of shrubs, posts, or rocks. The size of the form has been described as 46–61 cm long, 20–30 cm wide, and up to 20 cm deep (Lim 1987). White-tailed jackrabbits may use snow tunnels as resting sites (Orr 1940; Bear and Hansen 1966). When tracking these hares in snow, tracks will occasionally lead into abandoned badger (*Taxidea taxus*) holes, where hares might find refuge from predators and extreme weather conditions. They may even construct up to 1-m tunnels in snow, which serve as forms (Lim 1987).

When population density is high, jackrabbits leave visible, well-defined trails of compacted soil or trampled vegetation as well as paths in snow (Orr 1940; Crouch 1973). Trails are established and used when hares move about the habitat, flee from danger, and travel to and from distant food sources. Flinders and Hansen (1972) noted that black-tailed and white-tailed jackrabbits moved along converging systems of trails as they went from prairie rangelands to alfalfa and winter wheat fields.

Miscellaneous Behavior. Using motion-sensitive radio-transmitting collars on white-tailed jackrabbits, Rogowitz (1997) documented locomotor and sedentary activity and how various external events affected established daily patterns. Individual hares were highly synchronous in onset and cessation of locomotion. Movement began about 1 hr after sunset and ceased before sunrise throughout the year. Thus, duration of activity varied inversely with photoperiod. Distance, duration, and frequency of locomotion increased during the early breeding season, March and April. Low ambient temperature, cloud cover, and precipitation had no significant effect on locomotion or foraging activity; however, white-tails moved less when snow cover was present. In response to threat of predation, white-tails reduced movement during highly moonlit nights. They increased speed of movement in response to coyote vocalizations.

A high-pitched scream is produced when jackrabbits are handled or captured as prey. This seems to be a leporid, and not species-specific, behavior, although jackrabbits produce more of the raspy, guttural sounds. A low growl or grunt was heard from females acting aggressively toward males (Lechleitner 1958a).

When fleeing from a perceived threat, a black-tailed jackrabbit will stop periodically to view its back trail. If in thick cover when stopped, it may suddenly jump up in the air with the body still more or less parallel with the ground. They seem to jump just high enough to view their surroundings and back trail over the proximal shrubs. One aspect of white-tailed jackrabbit behavior is similar to that of the arctic hare (*L. arcticus*) (Flux and Angermann 1990). When fleeing from a perceived threat such as a terrestrial predator, white-tails will often run up slight-to-steeper slopes in the available terrain. As the hare ascends the upper portion of the slope, it may stand upright while in full stride and continue up the slope with several well-coordinated bipedal hops. This behavior also seems to allow the hare to inspect its back trail and surroundings and thus assess the proximity of a perceived threat. It then continues on in its characteristic, very swift, quadrupedal gait.

Jackrabbits spend part of their resting time in body maintenance activities. Black-tailed jackrabbits groom themselves by licking the body with the tongue. Those areas inaccessible to the tongue are washed by licking the feet and then using the feet to moisten the remaining body surfaces (Lechleitner 1958a). Tiemeier (1965) often observed jackrabbits dusting in shallow depressions dug in the soil. Such behavior probably brings relief from ticks and other ectoparasites and keeps the fur from becoming oily.

MORTALITY

Prenatal Mortality Rates. Prenatal mortality is that which occurs either before or after implantation of ova. Lechleitner (1959a) reported a preimplantation mortality rate of 6.7% for all ova shed and a postimplantation mortality rate of 6.2% for black-tailed jackrabbits in California. In Kansas, Tiemeier (1965) found 9.4% preimplantation mortality and a resorption rate of 5.1% for implanted embryos, for a total intrauterine mortality rate of 14.5%. Gross et al. (1974) estimated preimplantation and postimplantation mortality rates to be 8.0% and 3.0%, respectively. Feldhamer (1979) calculated an intrauterine mortality rate of 46.3% (16.3% preimplantation, 30.0% postimplantation loss) for the black-tailed jackrabbit in Idaho.

In white-tailed jackrabbits, James and Seabloom (1969a) found preimplantation losses affected 6.7–28.7% of all ova shed, with a 2-year mean of 16.7%. Resorption of embryos affected 4.6% of all ova shed and 19% of all litters. Mean prenatal loss from implantation failure and resorption was estimated at 21% of all ova shed. Total prenatal mortality from all causes was 28%.

Juvenile Mortality Rates. Estimates of juvenile mortality from birth to 12 months of age show little variation in the black-tailed jackrabbit. In Kansas, Tiemeier (1965) reported that juvenile losses ranged from 35% to 67%, with a mean of 63%, for a 6-year period. The highest juvenile mortality occurred during October, November, and December. Gross et al. (1974) reported that juvenile mortality rates in Utah ranged from 24% to 71%, with a mean of 59%. By the fall, more than half of the juveniles had been removed from the population. A first-year mortality rate of 91% was estimated by Feldhamer (1979) for black-tailed jackrabbits in Idaho.

Adult Mortality Rates. Adult mortality rate of the black-tailed jackrabbit was 57% from March to October in Utah (Gross et al. 1974). Mortality estimates ranging from 9% to 87% were obtained for an 8-year period. The October–March mean mortality rate was estimated to be 56%, with values ranging from 34% to 68%. There was no significant difference in mortality rates between the two periods.

Mortality Factors. Jackrabbits are subjected to predation from a variety of mammalian, reptilian, and avian predators. In many ecosystems, they function as essential prey of predators, some of which are now threatened with extinction. Many avian predators have been observed feeding on jackrabbits, and their remains have been present in regurgitated casts, feces, and within or near nests. These species include the golden eagle, bald eagle (*Haliaeetus leucocephalus*), rough-legged hawk (*Buteo lagopus*), Swainson's hawk (*B. swainsoni*), ferruginous hawk (*B. regalis*), red-tailed hawk (*B. jamaicensis*), northern harrier (*Circus cyaneus*), and great horned owl (*Bubo virginianus*) (Orr 1940; Tiemeier 1965; Wagner and Stoddart 1972; Zimmerman et al. 1996). An insightful evaluation by Woffinden and Murphy (1989) of the 20-year decline of ferruginous hawk populations linked this decline to reduced numbers of black-tailed jackrabbits and to the long-term ecological effects of introduced cheatgrass (*Bromus tectorum*), which has displaced many native plant species essential to this leporid. Similarly, the future of the golden eagle is solidly linked to perpetuation of robust populations of black-tailed jackrabbits (Steenhof and Kochert 1988; Marzluff et al. 1997; Steenhof et al. 1997).

Of the mammalian predators of jackrabbits, the coyote is clearly the most important. Coyote predation is the major mortality factor operating on black-tail populations in northern Utah (Stoddart 1970). F. W. Clark (1972) estimated that, in terms of volume, black-tails made up three fourths of the diet of coyotes in northern Utah. Stoddart (1970), using radiotelemetry to determine mortality rates, reported that 64% of transmitter-equipped black-tails fell victim to coyote predation. In their optimal foraging model for coyotes based on three prey species, MacCracken and Hansen (1987) reported a negative correlation between diet diversity of coyotes and abundance of *L. californicus*. At peak hare density, coyotes demonstrated functional and numerical responses. With a more diverse prey base in southern Texas, Windberg and Mitchell (1990) still found coyote dietary preference for leporids (*L. californicus* and *Sylvilagus audubonii*) linked to their abundance. However, abundance of cotton rats (*Sigmodon hispidus*) had greater influence on consumption of leporids than leporid abundance. The red fox (*Vulpes vulpes*) and gray fox (*Urocyon cinereoargenteus*) are known to take jackrabbits (Errington 1935; W. J. Hamilton 1935; Scott 1955).

The bobcat (*Lynx rufus*) is another predator, although of lesser importance than the coyote (Wagner and Stoddart 1972). Gashwiler et al. (1960) determined from stomach analysis of bobcats that leporids made up 45.2% of their diet in Utah and Nevada. Although 74.8% of leporid remains identified in feces of bobcats in California were *L. californicus*, Delibes and Hiraldo (1987) found *S. audubonii* was preferred. In southwestern Idaho, where the prey base is limited in diversity, bobcats specialize in taking leporids and may cease reproduction, disperse, or starve when leporid populations are low (Knick 1990; Knick et al. 1993). House cats (*Felis catus*) are known to prey on young black-tailed jackrabbits. In a study in the Sacramento Valley of California, dogs (*Canis familaris*) were the only predator that killed substantial numbers of jackrabbits (Lechleitner 1958b). Badgers preyed on jackrabbits in Kansas (Tiemeier 1965). Studies of feeding habits of the cougar (*Puma concolor*) in Utah and Nevada showed jackrabbits ranked fourth among their prey items (Robinette et al. 1959). Raccoons (*Procyon lotor*) and striped skunks (*Mephitis mephitis*) probably are capable of capturing newborn or very young leverets. Reptilian predators include garter snakes (*Thamnophis sirtalis*) and gopher snakes (*Pituophis catenifer*) as well as rattlesnakes (*Crotalus* sp.) (Vorhies and Taylor 1933; Lechleitner 1958b). In spite of all the predators, Evans et al. (1970) believed predation played a very limited role in regulating jackrabbit numbers to tolerable levels in agricultural situations.

Humans are an important mortality factor relative to jackrabbits. Where jackrabbits cause crop damage, people often take remedial action such as poisoning, shooting, or trapping, thereby locally contributing substantially to mortality. The jackrabbit is recognized as a game or huntable nongame species in many states, and jackrabbit hunting is a popular sport in many regions. Hunters may take substantial numbers annually. Tiemeier (1965) reported as many as 1800 black-tailed jackrabbits being removed by hunters during the winter from a 777 ha farm in Kansas.

Parasites. Many species of parasite occur on and within jackrabbits. Most of the same parasites also infect cottontails that are sympatric with jackrabbits (Pfaffenberger and Valencia 1988b). Though not well studied, all species of jackrabbits share, or may share, the same ecto- and endoparasites. Many ectoparasites are vectors of pathological diseases affecting leporids, livestock, and/or humans. Some of the more common ectoparasites of the black-tailed jackrabbit are rabbit fleas (*Euhoplopsyllus glacialis* and *E. foxi*) and others, the common rabbit tick (*Haemaphysalis leporispalustris*), the Rocky Mountain tick (*Dermacentor parumapertus*) and others, one species of louse (*Haemodipus setoni*), and larvae of the robust bot fly (*Cuterebra lepusculi, C. ruficrus,* and *C. mirabilis*). All are abundant parasites of the black-tailed jackrabbit (Phillip et al. 1955; Lechleitner 1959b; Hansen et al. 1965; Lipson and Krausman 1988; Pfaffenberger and Valencia 1988a). Up to 50% of black-tails studied in Arizona (Vorhies and Taylor 1933) and 27.5% of those in southern Texas (Crenshaw and Henke 1997) were infected with bot fly larvae. Multiple infestations on leporids are common. Bot flies (*Cuterebra* sp.) are myiasis-producing parasites, which may be debilitating at higher levels of infestation. In southern Texas, Crenshaw and Henke (1997) found significantly more *Cuterebra* larvae in female than in male black-tails. They also found more larvae in gravid than in nongravid black-tails.

Ectoparasites of the white-tailed jackrabbit include the fleas *Cediopsylla inaequalis, C. simplex, Euhoplopsyllus glacialis affinis,* and *Pulex irritans; Opisocrostis bruneri;* and the ticks *Dermacentor andersoni* and *Haemaphysalis leporispalustris* (Voth and James 1965; Kietzmann and Hugghins 1984).

Jackrabbits are host to a wide variety of endoparasites. Some cestode tapeworms include *Raillietina loeweni, R. retractulus, Taenia pisiformis, Mosqovoyia pectinata,* and *Multiceps* sp. (Bartel and Hansen 1964). A trematode fluke, *Fasciola hepatica,* common to cattle also is found in black-tails as a reservoir host. The roundworm nematodes in jackrabbits include *Passalurus nonanulatus, Micipsella brevicauda, Nematodirus leporis,* and *Dermatoxys veligera,* among others. Voth and James (1965) identified parasites of the white-tailed jackrabbit in North Dakota.

Viral, Bacterial, and Rickettsial Diseases. Black-tailed jackrabbits have been implicated as reservoirs for such diseases as tularemia, equine encephalitis, Lyme borreliosis, brucellosis, Q fever, and Rocky Mountain spotted fever. Jackrabbits may occur in large numbers and their proximity to humans and domestic animals makes them an important health issue (McMahon 1965; Burgess and Windberg 1989; Henke et al. 1990; Reisen et al. 1990; Monsen et al. 1991). Occurrence of diseases in the white-tailed jackrabbit is poorly known, but these are perhaps the same or similar to those reported for other leporids.

Occurrences of tularemia in jackrabbits is known to be sporadic, with ticks serving as vectors (Lechleitner 1959b). Phillip et al. (1955) were able to isolate the tularemia bacterium (*Francisella tularensis*) from ticks taken from jackrabbits. Jackrabbits injected with cultures of *F. tularensis* died within 6 days. Jackrabbits acquiring the disease in the wild die quickly, explaining the lack of antibodies observed for *F. tularensis* (Lechleitner 1959b).

Rocky Mountain spotted fever (*Rickettsia rickettsii*) infects *L. californicus* (Phillip et al. 1955; Lechleitner 1959b). The animals acquire the disease from infected ticks (McMahon 1965).

Presence of antibodies against Q fever (*Coxiella burnetti*) has been reported in jackrabbits in California. Jackrabbits also gave positive titers for western equine encephalitis. Brucellosis (*Brucella abortus*) has been reported in the jackrabbit in Nevada. However, it is not known whether the jackrabbit can serve as a reservoir for brucellosis that could affect cattle (Phillip et al. 1955; Lechleitner 1959b).

Studies by Zaugg and Newman (1985) found that black-tails play no role in the epizootiology of Florida *Anaplasma marginale* infections in cattle. As shown, the wide range of distribution of the black-tailed jackrabbit allows for monitoring the locational occurrence and movements of parasites and disease agents. In California, Clemons et al. (2000) conducted a sweeping evaluation of ecto- and endoparasites of black-tails. They found no differences in the prevalences or intensities of infection rates between males and females. Adult black-tails had greater parasite loads of the nematodes *Biogastranema leporis* and *Rauschia triangularis* than juveniles. Many first occurrences were reported (Clemmons et al. 2000). For black-tail epizootiology, this was the first known occurrence of *Trichostrongylus calcaratus, R. triangularis, Tichuris sylvilagi,* and *Dermacentor variabilis*. This was also the first report for *T. calcaratus* and *T. sylvilagi* in the western United States. Furthermore, *Ixodes spinipalpis,* a vector for Lyme disease, was another first for black-tails in California.

AGE ESTIMATION

Several methods have been used to estimate age in the jackrabbit, including the closure of the epiphysis of the proximal end of the humerus. Lechleitner (1959a) was able to distinguish three age classes for black-tailed jackrabbits: class 1 (2–9 months old) had a distinct epiphyseal groove, class 2 (10–12 months old) had an epiphyseal groove that was almost closed, and class 3 (>1 year old) had no evidence of an epiphyseal groove present. Similarly, James and Seabloom (1969b) found the white-tailed jackrabbit could be separated into three age classes very similar to those of *L. californicus*. The change from age class 1 to age class 2 occurred at 6–7 months of age. The change from age class 2 to age class 3 occurred when the animal was 13–14 months of age.

Weight of the dried eye lens has been used to determine age in black-tailed and white-tailed jack-rabbits (Tiemeier and Plenert 1964; Connolly et al. 1969; James and Seabloom 1969b; Gross et al. 1974). The lens weight technique is superior to other methods because it estimates the month in which the animal was born. However, lens weights may vary throughout the range of the species, possibly because of nutritional differences.

Juvenile males can be separated from adult males by the size and morphology of the penis. In black-tailed jackrabbits <9 months of age, the penis cannot be everted, whereas in older males it can. Adult penis length is from 3.5 to 5.0 cm. Males ≥1 year old have warty protuberances on the penis, which readily identify them as older adults (Lechleitner 1959a). In both species of jackrabbit, individuals >10 months old tend to have a dark blue scrotal sac, which is characteristic of sexually mature adults (Lechleitner 1959a; James and Seabloom 1969b).

Young-of-the-year females can be classified as juveniles by the condition of the reproductive tract. The presence of longitudinal striations in the uterus can be used to separate parous adults from juvenile, nonparous females. In addition, the ovaries of a juvenile *L. californicus* female consists mostly of interstitial tissue, whereas the ovaries of adults contain large follicles and corpora lutea (Lechleitner 1959a).

ECONOMIC STATUS

Economically, jackrabbits are both desirable and undesirable, depending on the situation and the relationship of humans to the animals. On a positive side, they provide sport hunting for a great many people. If one were to compute the monetary worth of this activity using even the conservative "travel cost method" (Feltus and Langenau 1984), the amount would be impressive; this has yet to be done. Hunters usually make no use of killed jackrabbits, but leave them in the field. Although not considered as desirable to eat as cottontails in North America, hares are favored as food by Europeans. Black-tails shipped to Italy were highly favored as food by wealthy hunters. Classified as a game or huntable nongame species, black-tailed and white-tailed jackrabbits are popular among sport hunters. Year-round hunting seasons with no bag limits exist in many states, and in others bag limits and seasons are quite liberal. Jackrabbits also provide falconers with a sporting challenge.

In North America, market hunting of jackrabbits for human consumption does not exist. In the 1800s and early 1900s, it occurred on a relatively large scale, with jackrabbits being sent to the larger cities throughout the United States. Residents of San Francisco were thought to consume the most rabbits, with 100–150 dozen rabbits sold each day in the winter of 1894/95 at 50¢ to $1 per dozen (Palmer 1897). In 1971 and 1972, black-tails in southern Idaho were at a their cyclic high point of population density. Commercial buyers from Italy were paying $1.85 each for these hares. They were shipped to Italy for hunting by members of private clubs. Members paid up to $50 for the privilege of shooting a black-tailed jackrabbit and then taking it home for dinner. Black-tails not killed were held on special farms, where handlers expected about eight leverets per female. Flux (1981) used market demand and results from existing facilities to project high profitability for well-managed "hare farms" in Europe and New Zealand. Black-tailed jackrabbits may even be preferred for these farms.

Black-tailed jackrabbit pelts, although not highly valued as fur, were sold for felt making in the early days of market hunting. However, the superior fur of other leporids, especially those from colder climates, which had dense winter pelts, was economically more important. Today, jackrabbits have a minor role in the fur industry. Bear and Hansen (1966) reported that in Colorado, when dense populations of white-tailed jackrabbits existed, hunters marketed unskinned rabbits. Pelts were used to make felt and carcasses were used for mink food. In some years as many as 65,000 jackrabbits were bought at one Colorado collection point.

On the negative side, black-tailed and white-tailed jackrabbits are considered important pests in some regions and situations, especially at high densities. In the winter of 1981/82, black-tails in eastern Idaho caused an estimated $10 million damage to agriculture (Evans et al. 1982). Hegdal (1966) documented jackrabbit damage in 17 western states. On pastures and rangeland, they may compete for forage with grazing livestock. Vorhies and Taylor (1933) estimated that 148 black-tails ate as much forage as one cow. Similarly, Currie and Goodwin (1966) reported that 6 black-tails could destroy as much forage as one sheep could eat in a day. Bear and Hansen (1966) estimated that 15 white-tailed jackrabbits ate as much forage as one sheep. In Colorado, dietary overlap between jackrabbits and cattle seemed to be greatest in early spring (Hansen and Flinders 1969). They can cause damage to a variety of orchard trees and other crops (Bronson and Tiemeier 1958b).

The most important economic conflict with jackrabbits concerns their periodic role in crop depredations. Jackrabbits forage on many crops. They may damage crops including trees, grain, alfalfa, cotton, and a variety of vegetables. In Kansas, Bronson and Tiemeier (1958b) reported heavy damage to wheat, sorghum, and alfalfa from black-tails. Crop damage is often most severe in fields adjacent to land that hares use for resting (Fagerstone et al. 1980; Johnson and Peek 1984). In the Southwest, Vorhies and Taylor (1933) reported that black-tails fed on alfalfa, orchard trees, cotton, and other crops. Droughts often contributed to excessive crop damage by drawing rabbits from rangeland or uncultivated fields to cultivated crops, especially those under irrigation (Bickler and Shoemaker 1975). Roundy et al. (1985) investigated the effects of a very high population density (1.4–2.9/ha) of black-tailed jackrabbits on three areas seeded to crested wheatgrass in Nevada. They found that high densities of jackrabbits and associated high forage use did not reduce seedling survival on mesic or xeric sites with below-average spring precipitation. Depending on the population density of black-tails and frequency of clipping, these hares may have a detrimental or beneficial effect on crested wheatgrass (McAdoo et al. 1987). Fagerstone et al. (1980) reported the successful use of relatively unpalatable potato plants as a protective buffer crop between rangelands and crops such as barley.

Jackrabbits frequent some open-forest habitats, but have little direct adverse effect on forest trees except during forest regeneration efforts that coincide with high points in hare population cycles. Crouch (1973) reported jackrabbit damage to ponderosa pine (*Pinus ponderosa*) seedlings in south-central Oregon. Hares clipped the main stems on 43% of 1080 tree seedlings planted following wildfire in a pine–bitterbrush–needlegrass plant community. Read (1971) reported similar damage to ponderosa pine seedlings on a tree plantation in Nebraska. However, damage from jackrabbits in reforestation efforts is not widespread.

The significance of rabbit diseases transmissible to people is of some concern (Hull 1963; McMahon 1965). Hunters always need to take into consideration the possibility of contracting tularemia (a debilitating but rarely fatal disease) when handling carcasses of jackrabbits and other leporids. In North America, 70% of tularemia cases in humans are attributed to contact with hares and rabbits. Eastern cottontails (*Sylvilagus floridanus*) are the direct source of over 55% of all cases of the disease in humans (McDowell et al. 1964). Although ticks and deer flies may serve as vectors for tularemia, this route of transmission is less important than direct contact with an infected animal while skinning or dressing (McDowell et al. 1964). Rubber gloves should be used to avoid infection through the skin, especially through abrasions or broken skin on hands. Well-cooked rabbit meat presents no tularemia hazard. With the use of a few commonsense precautions, the danger of contracting tularemia from hares or rabbits is remote.

Cases of bubonic plague in humans, though more often associated with ground squirrels (*Spermophilus* spp.) or other rodents, have been linked to leporids. Some species, including the black-tailed jackrabbit, have been infected with plague, although human cases have been associated primarily with the hunting and cleaning of cottontails (Graves et al.

1978). Fleas almost always serve as vectors in transmitting plague from animals to people, particularly when ground squirrels are involved.

Transmission of Rocky Mountain spotted fever from jackrabbits to humans is only a remote possibility because rabbit ticks do not ordinarily feed on humans (Lechleitner 1959b). Jackrabbits potentially may transmit certain livestock diseases such as Q fever and brucellosis, but little is known of the actual threat to livestock production (McMahon 1965).

MANAGEMENT

Management of jackrabbits in the United States has been directed primarily toward decreasing their populations because of their crop depredation activities, in contrast to the common objective of increasing the numbers of more desirable species such as eastern cottontails. Traditional methods of habitat management have not come into play with jackrabbits because these hares, when in suitable habitats, are usually in adequate numbers to satisfy hunters. Private, state, and federal lands are important habitats for hares and for hunters. Also important are certain military reservations, preserves, wildlife refuges, parks, and other recreational areas where, depending on management policies, hares may be hunted or are protected from harvest by hunters. Jackrabbits, if not overhunted, are adaptable enough to coexist very well with people and diverse land management practices. As would be expected, in states where jackrabbits are viewed as a serious and widely distributed pest to agriculture, little emphasis is placed on their management as desirable wildlife. This attitude may change as they are appreciated as essential prey for raptors.

Construction of numerous airports has unintentionally and substantially benefitted the black-tailed jackrabbit. The managed and mowed grasses and forbs between runways at many airports in the western United States support relatively high numbers of jackrabbits and these populations are often controlled (Rohe et al. 1963). Airport populations of jackrabbits attract both avian and mammalian predators, thus creating other hazards for aircraft. Although jackrabbits are occasionally run over on runways, this creates little hazard, even for light aircraft. However, dogs or other large mammalian predators chasing rabbits on runways do constitute a hazard, particularly for light aircraft. Avian predators may be sucked into jet engines, causing malfunctions, or may penetrate windshields or damage other parts of planes. In western airports, black-tailed jackrabbits are commonly controlled by fencing, shooting, poisoning, or other means.

Early western farmers quickly became aware of the native jackrabbits. Although they often provided fresh meat, jackrabbits were usually considered more of a liability as a pest to crops than an asset as food. Before farmers could grow crops successfully in many areas, it was essential to have some economical way to reduce crop losses from hares. Jackrabbits were controlled in the late 1800s usually with large "rabbit drives" (Palmer 1897). This period in American history was a time of agricultural expansion in many areas of the Midwest and West, and black-tailed jackrabbit populations over wide areas grew to devastating numbers. Introduced agriculture, along with irrigation, within prime jackrabbit habitat apparently triggered an enormous upward surge in hare populations (Fagerstone et al. 1980; Johnson and Peek 1984). In some regions, these populations reached levels where hares ate entire crops (Palmer 1897).

In a typical rabbit drive, the majority of the farm population of a district turned out and surrounded a tract of land, often several kilometers in expanse, and drove jackrabbits toward a central corral bordered by wide wing-type fences. After they were concentrated in such an enclosure, jackrabbits were clubbed to death by the hundreds or thousands. The largest rabbit drive on record was held about 24 km southwest of Fresno, California, on 12 March 1892. Approximately 8000 people were present, and an estimated 20,000–30,000 jackrabbits were killed (Palmer 1897). Hares killed were used as food by some people, fed to hogs, scalped for bounty, and/or discarded.

These drives undoubtedly saved many crops, but they were not always highly effective. Use of large rabbit drives as the principal control method reached its peak roughly between 1887 and 1892 (Palmer 1897) and then gradually gave way to hunting and the use of rabbit-proof fencing, protective tree trunk guards, chemical repellents, and poison baits. Because black-tailed jackrabbits exhibit high levels of population density every 7–10 years and populations were seldom at very high levels for more than 1 year, the actual mortality represented by these drives was probably just compensatory (D. R. Johnson and Peek 1984).

Organized group hunts, also popular, outlived rabbit drives. This method of hunting requires an organized group of hunters to systematically walk parallel to each other and flush jackrabbits (and other wildlife) from cover in front of them. Efficient hunters kill many jackrabbits using this method of hunting. On public lands, groups of hunters may cover the same transects several times in a weekend, thus heavily affecting the jackrabbit and cottontail populations. In most western states, except California, there are no season restrictions or bag limits on hunting jackrabbits. Following the late fall closure of most hunts for big game and upland game, sportsmen concentrate on hunting jackrabbits, especially on public land. Thus, many jackrabbit populations receive constant hunting well into the reproductive season. We have no data regarding the effect of this uncontrolled hunting on populations of hares, but this mortality, as well as loss of native habitat, may have interrupted the population cycles (periodicity) so well documented in the past.

Attempted control of jackrabbits by rabbit drives has essentially been abandoned. As noted, damage is now reduced largely with the use of fences, tree trunk protectors, chemical repellents, and other means of population reduction (Evans et al. 1982; J. P. Clark 1994; Knight 1994; Salmon and Gorenzel 1997). When population reduction is required to prevent damage, it most often involves either poisoning or shooting and only rarely trapping. Sport hunting, if well managed, may offer recreation in addition to providing some relief from crop depredations (Evans et al. 1982).

Jackrabbits, as well as other species, may be excluded from fields and gardens with the use of rabbit-proof fences. Woven-wire fence or poultry netting is used. Mesh size usually is not larger than 3.8 cm, fencing is 91–121 cm high, and the bottom is turned 15 cm outward and buried at least 15 cm in the ground. Jackrabbits and cottontails ordinarily will not jump a 60-cm fence, although jackrabbits can easily clear such a height when pursued by humans or dogs. Tight-fitting gates with sills or other means of preventing rabbits from digging beneath are essential, and gates must be kept closed except when actually in use (Evans et al. 1982; Clark 1994; Knight 1994; Salmon and Gorenzel 1997).

Hare and rabbit fences are expensive to build and ordinarily are impractical and uneconomical unless the crop is of high value and extensive damage is likely. For example, in a study of the protection of large fields of alfalfa in California, fencing was economical only when rabbit damage was considered severe by the landowner (Wetherbee 1967; Bickler and Shoemaker 1975). These fences are most likely to be constructed around specialty plantings such as experimental plots, nurseries, vegetable and flower seed crops, and other high-value crops. Fences also are used around noncommercial plantings such as home vegetable and flower gardens (Salmon and Gorenzel 1997).

Over the years, farmers have used protectors to guard trunks of individual young orchard trees against hares (Garlough et al. 1942; Salmon and Gorenzel 1997). Homemade and commercial wrap-around trunk protectors have been made from a variety of materials including plastic, cardboard, paper, and aluminum. These are valuable aids in preventing hare damage as well as sun scalding and freezing damage to sensitive trees (Evans et al. 1982; Salmon and Gorenzel 1997).

Cylinders of poultry netting make excellent protectors for trunks of young orchard trees. Cylinders formed from 30- to 45-cm sections of 60- to 91-cm-wide poultry netting (2.5-cm mesh) are used commonly to protect trees. Cylinders should be staked away from trees so that hares cannot press them against trees and gnaw between the wire mesh (Storer 1958). Wrapping the base of a haystack with 91-cm-high poultry netting also provides good protection from jackrabbits (Evans et al. 1982).

Various repellents have been used to discourage leporids from feeding on plants and gnawing on bark and twigs of trees and vines. Chemical repellents may provide relief from leporid damage and thus are valuable when other approaches are not practical (Hildreth and Brown 1955). Applications must often be repeated, however, to protect new growth or renew repellency lost through rain, snow, or sprinkler irrigation. Leporid repellents are somewhat unpredictable in effectiveness. They may offer some protection as long as leporids have acceptable alternative untreated food, but when food is scarce, repellents may have little value.

The most effective repellents for jackrabbits include products made with one of the following chemicals: tetramethyl thiuram disulfide, trinitrobenzene-anilene, and zinc dimethyldithiocarbamate cyclohexylamine (Evans et al. 1970). Repellents containing other active ingredients are also available. Control methods used for black-tailed jackrabbits will also be applicable for white-tailed jackrabbits where they are problems.

Where jackrabbits cause serious depredation on large fields of crops, poison baits (where safe to use) are the most effective and practical means of reducing the damage. Strychnine-poisoned bait was used extensively (Johnson 1964; Evans et al. 1970). However, following the development of the safer anticoagulant rodenticides, these more expensive poisons have been used for jackrabbit control (Clark 1975; Johnson 1978; Johnson and Peek 1984). Zinc phosphide-poisoned bait has been successfully used, although it also kills nontarget species (Evans et al. 1970). Conover (2001) lists chlorophacinone and diphacinone as toxicants approved for use with jackrabbits by the U.S. Environmental Protection Agency. In many localities, it is not safe to use poisons of any type. In other situations, properly conducted poisoning of jackrabbits may present minimal hazard to nontarget species (Johnson and Peek 1984).

"Catch crops" or buffer crops planted around borders of crops to reduce rabbit damage have been tried with limited success (Lewis 1946; Evans et al. 1970; Fagerstone et al. 1980). The buffer crop method of damage control is seldom practical or economically sound. Controlling rabbits biologically with disease or reproductive chemosterilants has been considered but ruled out because of potentially undesirable biological consequences (Evans et al. 1970; D. R. Johnson and Peek 1984).

Jackrabbits living on rangelands prefer more open areas lacking dense, tall vegetation. Thus, livestock grazing often helps provide favorable jackrabbit habitat (Vorhies and Taylor 1933; Arnold 1942; Bronson and Tiemeier 1958b). It is not clear, however, whether rabbits prefer one type of site over another because of openness, the type and amount of food available, or a combination of factors (Vorhies and Taylor 1933). A variety of factors such as shallow or poor soil and low rainfall may also influence the plant community and contribute to sparse and low vegetation. Evans et al. (1970) pointed out the inconsistencies in the influence of grazing intensity on jackrabbits. On the shortgrass prairie of Colorado, under grazing management for cattle, black-tailed jackrabbits were more abundant on pastures with light-summer and moderate-summer grazing treatment (Flinders and Hansen 1975). In the same study, white-tailed jackrabbits showed no strict preference for any grazing treatment, but preferred the upland habitats in pastures.

There is no evidence that lowering livestock stocking rates on rangeland in hopes of increasing vegetation cover will reduce jackrabbit numbers enough to compensate for the loss in livestock production. However, where ranges are fully stocked and jackrabbit numbers become high, consideration should be given to the possibility that livestock, in combination with a dense population of rabbits, will cause overuse of some forage plants (Taylor et al. 1935). In such situations, if the rabbit population does not crash from natural causes within a couple of years, then good range management practices would dictate that either the jackrabbit population be reduced artificially or the stocking rate for livestock be lowered to prevent permanent range deterioration. To increase success, rangeland seeding projects should be timed to low population levels of jackrabbits. If this is not done, then barrier fencing or population reduction practices may be necessary to prevent overuse by jackrabbits. Multiple uses and benefits of public rangelands must be considered in these instances.

As noted, *L. californicus* was introduced for hunting and other sporting purposes into a number of states, including Massachusetts, Maryland, Virginia, and Florida, as well as other countries, such as Italy. The history of these jackrabbit releases suggests that such introductions can have lasting undesirable ecological consequences. Jackrabbits may not only compete with existing fauna in an area, but have a profound and unfavorable influence on native vegetation. Equally important, introduced species may become pests, competing with efforts to grow crops and establish forests (Howard 1964, 1965). Proposed introductions of any rabbit or hare into new regions should be viewed very critically. Some states have laws or regulations prohibiting such introductions. The air shipment of black-tailed jackrabbits from Idaho to Italy for release for sport hunting is an example of the introduction of nonnative wildlife into a new region without assessment of short- or long-term consequences to native biota.

RESEARCH AND CONSERVATION NEEDS

Jackrabbits are ecologically important species, affecting various habitats as well as serving as essential prey for many raptor and mammalian predators. Because of their importance, funding is needed for long-term research to better understand their population dynamics, population genetics, and ecological relationships. Modeling of hare populations in relationship to their predators and landscape-level ecological perturbations could help predict problems within food webs as well as biotic communities. As we continue to lose essential habitat to human development, agriculture, and exotic plant-dominated landscapes, the "rabbit problem" of the future may well be a cascading loss of historical jackrabbit populations and their dependent predators.

We need more data on the link between essential water for metabolic needs of jackrabbits and the sources of dietary water in critical seasons in relationship to available habitats as well as how these and factors such as plant secondary compounds interact to effect hare survival. Even with the excellent genetic-based studies of hare systematics so far completed, we need a more interactive effort that incorporates genetics and behavior as well as traditional and numerical taxonomy. The wide range of opinion regarding the number of subspecies in each species of jackrabbit needs to be resolved to appreciate how jackrabbit control programs, unregulated sport hunting, and such things as extensive wildfire are modeled into conservation planning for each taxa.

When we consider the central ecological position of these hares, we must hope they continue to thrive in the future as in the past. Historically, jackrabbits at very high population densities have caused serious damage to agricultural crops. If we anticipate these occurrences, efforts may be made to forecast when, where, and how many jackrabbits will be a problem. Long-term monitoring of jackrabbit populations should facilitate these kinds of useful predictions. When practical, barriers around crops should be the preferred method to control depredations. There are limited data regarding the effectiveness of behavior-modifying devices or practices to prevent jackrabbit depredations on crops, as there are for birds. Lethal means of control such as poisons must be specific, short-lived in the environment, cause no secondary poisoning, and minimize exposure to nontarget species. Prescriptive hunting of jackrabbits to control depredations may well be the most effective method in certain circumstances. Nevertheless, jackrabbits are vital to many wintering populations of raptors, including golden eagle, bald eagle, ferruginous hawk, red-tailed hawk, rough-legged hawk, and others (Arnold 1989; Woffinden and Murphy 1989; Marzluff et al. 1997; Steenhof et al. 1997). Thus, regular (not control) hunting by people should be limited to an autumn season with reasonable bag limits.

We need to know more about the critical factors that operate when black-tailed and white-tailed jackrabbits become sympatric. Similar dynamic situations arise when the black-tailed jackrabbit becomes sympatric with *L. alleni* and *L callotis*. Maps showing the general distribution for species cannot depict the various-sized gaps in distribution where each species once occurred but now is extirpated, scarce, or engaged in a specific competitive or ecological struggle. Compilation of these data is past due and even conservation management plans for raptors need this vital information.

OTHER JACKRABBITS

Common Names. Antelope jackrabbit, white-sided jackrabbit, Allen's jackrabbit.
Scientific Name. *Lepus alleni*
Subspecies North of Mexico. *L. a. alleni*

Antelope Jackrabbit. The antelope jackrabbit, *Lepus alleni,* is also referred to as the Allen's, blanket, or saddle jackrabbit. Three subspecies occur in North America: *L. a. alleni, L. a. palitans,* and *L. a. tiburonensis*. Its range (Fig. 6.2) extends from the arid portions of southern Arizona southward along the coastal plain of the Gulf of California to the Mexican state of Nayarit (Best and Henry 1993b; Best 1999a).

The antelope jackrabbit is the largest of the North American jackrabbits. Mean adult male weight is 3.69 kg and that of nonpregnant females is 3.60 kg. In Arizona, average measurements of adult males and females, respectively, were as follows: total length, 619 and 625 mm; tail length, 57 and 59 mm; hind foot length, 140 and 141 mm; and length of ear from notch, 163 and 161 mm (Vorhies and Taylor 1933; Best and Henry 1993b). This species can easily be identified by its large, white-edged ears and the white rump patch (see Fig. 6.5). It also lacks the black-tipped ears characteristic of *L. californicus*. The top and sides of the head are buff colored, and the tail is white below, with a black mid-dorsal stripe extending onto the rump. The shoulders and flanks are a uniform pale gray (Hall and Kelson 1959).

This jackrabbit occurs exclusively in lower warm desert habitats, where it occupies elevations ranging from sea level to 1200 m. In Arizona, the preferred habitat consists of grassy slopes with moderate elevations. It also inhabits the cactus belt, where species of mesquite, grass, and catclaw (*Acacia greggi*) predominate. Smaller populations also may inhabit the creosote bush desert and valley bottoms (Vorhies and Taylor 1933). In Pinal County Arizona, early to late morning habitat selection by sympatric antelope jackrabbits and black-tailed jackrabbits showed *L. alleni* chose shade and cover for forms in woody vegetation composed primarily of mesquite (*Prosopis* sp.) and creosote bush (*Larrea tridentata*) (Brown and Krausman 2003). Unlike black-tailed jackrabbits, antelope jackrabbits selected for vegetative form cover disproportionately to availability (Brown and Krausman 2003).

There appears to be little difference between the breeding season of the antelope jackrabbit and that of the black-tailed jackrabbit where they are sympatric. Vorhies and Taylor (1933) reported a 10-month breeding season for both species in Arizona. Mating occurred every month of the year except November and December. The antelope jackrabbit averages 1.93 young per litter, with seven litters possible

Figure 6.5. Antelope jackrabbit (*Lepus alleni*). Note the pale color, in contrast with the darker dorsal surface, and the long ears and hind legs. Source: Leonard Lee Rue III.

per year. Vorhies and Taylor (1933) suggested that *L. alleni,* like *L. californicus,* usually place neonates in excavated, shallow, globe-shaped nests lined with hair plucked from the dam. The dam apparently suckles the young only at night. When not with the neonates, the dam closes the top of the nest with plant debris and soil. Vorhies and Taylor (1933) described three nests for *L. alleni,* one within the hollowed-out shell of a barrel cactus (*Echinocactus* sp.). After this neonatal period, leverets in litters are dispersed to forms in the same vicinity, where they remain alone and secretive. The precocial young possess the white-edged ears characteristic of the species, but lack the white rump patch associated with adults. Four individual neonates from the same litter averaged 121.4 g in weight. Average measurements of 5-day-old *L. alleni* were total length, 150 mm; tail length, 8.2 mm; hind foot length, 42.5 mm; and ear length from notch, 38.5 mm (Vorhies and Taylor 1933). Fetal sex ratios have not been determined, but those of adults did not differ significantly from a 1:1 ratio.

L. alleni does not exhibit marked fluctuations in population density. Densities ranged from 0.2 hare/ha in the semidesert habitat to 0.5 hare/ha in the mesa habitat (Vorhies and Taylor 1933).

The antelope jackrabbit is slightly gregarious during all times of the year, but especially during the breeding season, when small groups of males attend a female and compete for breeding rights. Groups of >100 individuals were observed together when they congregated in low areas to escape cold winter winds (Vorhies and Taylor 1933). Daily movements of 1–3 km usually involve moving to and from feeding areas during the night. Adult home ranges are probably several kilometers in diameter (Vorhies and Taylor 1933).

When alarmed and fleeing from potential danger, *L. alleni* may reach speeds of 72 km/hr and thus may be the fastest hare (Best 1999a). When in taller grass, *L. alleni* begins running with its body nearly on a horizontal plane. It then stands up vertically and takes four or five long, bipedal hops on the hind legs before dropping down to continue its quadrupedal gait. Besides having a zigzag pattern difficult for a predator to follow, these hares easily jump over shrubs 1 m in height. They also can clear 1.5-m fences in stride, and one broad jump of 7 m was recorded (Best and Henry 1993b).

The escape behavior exhibited by *L. alleni* may be the most spectacular of the jackrabbits. When this hare runs from a potential predator, it conspicuously flashes its whitish rump patch toward the predator while running in a zigzag fashion. These white "ruptive" marks on the rump are flashed toward the predator by a set of muscles under the skin. The skin of the hindquarters is pulled up over the back while the hairs are everted to expose the large white patch. Vorhies and Taylor (1933) concluded that the purpose of the rump flash was to confuse a predator, and obliterate the outline of the jackrabbit. This display seems similar to stotting behavior in gazelle, antelope, and deer, as well as the hare like rodent, the Patagonian mara (*Dolichotis patagonum*). The "message" may be at least twofold. First, advertisement of fitness in the face of a potential predator, and second, flashing of a danger signal to other potentially related conspecifics.

Feeding habits of the antelope jackrabbit are related to the alternating dry and rainy seasons and the apparent need for this species to obtain all water from ingested vegetation. Following the winter rains, grasses produce new leaves, which are succulent and highly palatable to jackrabbits, and thus may make up 80% of diets. As the grasses dry out, these hares turn to foraging on shrubs such as mesquite. When feeding on mesquite, they use their considerable height while standing on their hind feet to reach the shrub canopy. They eat tufts of green leaves in axils of mesquite spines. To obtain water and nutrients, *L. alleni* feed on the cladode pads and seed pods (tunas) of cacti, especially *Opuntia engelmanni*. These and other cacti may have water content from 78% to 94% (Best and Henry 1993b). Analyses of the stomach contents of *L. alleni* have shown 36% of the diet composed of mesquite, 45% grass, and 7.8% cactus, with various forbs and shrubs constituting the remainder of the diet (Vorhies and Taylor 1933). Species of grass consumed by this jackrabbit include three awn (*Aristida* sp.), grama (*Bouteloua* sp.), love grass (*Eragrostis* sp.), spike grass (*Trichloris crinita*), panic grass (*Panicum* sp.), sandbur (*Cenchrus* sp.), drop seed (*Sporobolus* sp.), redtop (*Agrostis* sp.), barnyard grass (*Echinochloa* sp.), and finger grass (*Chloris* sp.) (Vorhies and Taylor 1933).

The antelope jackrabbit is undoubtedly subjected to the same predators and many of the same parasites and diseases as those reported for the black-tailed jackrabbit. Tularemia is the only disease of any significance reported to affect antelope jackrabbits in Arizona (Vorhies and Taylor 1933).

This species has caused damage to crops and may compete with livestock for grasses on rangelands. This is particularly true during periods of drought, when food sources are limited (Vorhies and Taylor 1933). The same damage prevention or control measures that are applied to the black-tailed jackrabbit can be applied to the antelope jackrabbit.

In Arizona, the antelope jackrabbit is considered a nongame mammal with no closed season or bag limit. As such, it may be hunted throughout the year. It is not considered abundant or rare, and is frequently hunted along with black-tailed jackrabbits.

More comprehensive research is needed on the antelope jackrabbit. The fine work by Vorhies and Taylor (1933) should serve as a template for studies of existing populations of this jackrabbit. Long-term studies are needed to monitor the population dynamics and behavioral ecology of these hares in relationship to human-caused and natural changes in their habitats. We need a better understanding of the ecological niche of this hare and how the species functions as an agent in plant propagation. We also need to understand how antelope jackrabbits affect levels of nesting success and survival of native raptors. Genetic studies could assess population heterogeneity while we still have fairly robust populations with some degree of connectedness with metapopulations in Mexico.

COMMON NAMES. White-sided jackrabbit, Gaillard's jackrabbit, snow sides.
SCIENTIFIC NAME. *Lepus callotis*
SUBSPECIES NORTH OF MEXICO. *L. c. gaillardi*

White-Sided Jackrabbit. The white-sided jackrabbit was first described as *Lepus gaillardi* by Mearns (1896) from a specimen collected near Whitewater, Chihuahua, Mexico (Best and Henry 1993a). It has since been reclassified as a subspecies of *L. callotis* by Anderson and Gaunt (1962). Two subspecies of the white-sided jackrabbit are now recognized, *L. c. callotis* and *L. c. gaillardi*. Only the subspecies *L. c. gaillardi* occurs in the United States, as described below. The karyotype of *L. callotis* is $2N = 48$ as in other jackrabbits. The X chromosome is a large metacentric, and differs from that of *L. alleni,* the Tehuantepec jackrabbit (*L. flavigularis*), and *L. californicus*. The species *L. callotis, L. alleni,* and *L. californicus* all have a small, telocentric Y chromosome (Gonzalez and Cervantes 1996).

The range of *L. callotis* extends northward from the Mexican state of Oaxaca along the Sierra Madre to Chihuahua and eastern Sonora (see Fig. 6.2). In the United States, *L. callotis* is considered rare and is restricted to an area of about 120 km^2 in southern Hidalgo County, New Mexico (Anderson and Gaunt 1962; Bogan and Jones 1975; Findley et al. 1975; Bednarz 1977; Bednarz and Cook 1984). Populations of *L. callotis* and *L. alleni* are completely allopatric, although the range of *L. californicus* does overlap the northern distribution of both species. Anderson and Gaunt (1962) suggested that *L. callotis* probably evolved from an isolated population of *L. californicus* in Mexico. Part of the newly formed species *L. callotis* was then isolated on the western coastal plain, where it evolved into *L. alleni*.

The distinctive characters of *L. callotis* are its pale buffy to blackish hue, brown to blackish nape, white sides and underparts, and gray rump with black on the upper part of the tail (Anderson and Gaunt 1962). Where it occurs with *L. californicus,* the two species can be separated by examining overall skull characteristics. The skull of *L. callotis* has a higher nasal aperature, a more inclined parietal, smaller auditory bullae, more prominent supraorbital processes, smaller auditory meatuses, and a less constricted basioccipital (Anderson and Gaunt 1962). Average measurements from four adults in New Mexico were total length, 543.2 mm; tail length, 81.0 mm; hind foot length,125.8 mm; and ear length, 137.7 mm. The average weight was 2.7 kg (Bednarz 1977).

Information on reproduction in the white-sided jackrabbit is limited. Black-tailed jackrabbits in southeastern New Mexico have an 8-month breeding season extending from January to late August (Griffing and Davis 1976). A similar breeding season probably occurs for *L. callotis* at about the same latitude. Bednarz (1977) estimated from a limited number of pregnant females that the absolute minimum breeding season would be 18 weeks, extending from mid-April to mid-August. The average number of young per litter, from a sample of 10 females, was 2.2 (Anderson 1972; Bogan and Jones 1975; Bednarz 1977).

In New Mexico, *L. callotis* is an obligatory species native to the largely flat, desert grasslands of the region. They may be the best mammalian indicator species of this grassland type. They are important prey for raptors and other carnivores as well as serving as seed dispersers for native grasses, thus helping to perpetuate desert grasslands. Prominent grass species in preferred habitats include blue grama (*Bouteloua gracilis*), black grama (*B. eriopoda*), buffalo grass (*Buchloe dactyloides*), wolftail (*Lycurus phleoides*), and tobosa grass (*Hilaria mutica*). Bednarz (1977) found a positive correlation between the density of *L. callotis* and percent grass composition of the habitats. This jackrabbit preferred habitats composed of $\geq$65% grasses, $\leq$25% forbs, and <1% shrubs. Densities were estimated at 1 white-sided jackrabbit/31.6 ha, with the maximum population in the United States being 340 hares in 1976 (Bednarz 1977). Surveys in 1976 and 1981 along a 63-km census route in the Animas Valley of New Mexico found *L. callotis* declined from 15 ($\pm$7.1) to 7.6 ($\pm$4.1), respectively, whereas populations of *L. californicus* increased from 14.0 ($\pm$3.9) to 34.1 ($\pm$13.6), respectively, along with increases in *Sylvilagus auduboni* (Bednarz and Cook 1984). No doubt, the long-term survival of *L. callotis* is in jeopardy within the United States (Best 1999b). Competition is perhaps minimal between *L. callotis* and *L. californicus*. Only *L. callotis* is found on desert grasslands in good to excellent condition. *L. californicus* invades these habitats as the result of overgrazing by livestock and the subsequent invasion of shrubs and forbs. A change to unsuitable habitat, not necessarily competition with *L. californicus,* leads to a decrease in population density and distribution of *L. callotis* (Bednarz 1977; Bednarz and Cook 1984).

The white-sided jackrabbit has several unique behavioral features when compared to other species of jackrabbit. This hare is more nocturnal than *L californicus, L. townsendii,* and *L. alleni*. Bednarz (1977) reported that most activity for *L. callotis* occurred between 2200 and 0500 hr.

The escape behavior of the white-sided jackrabbit is similar to that of *L. alleni*. When flushed from its form, *L. callotis* alternately flashes its white side patches while running away from the intruder. Another escape behavior observed was that of leaping straight up into the air while extending the hind legs and flashing the white sides. This behavior occurs when individuals are startled or alarmed by a predator (Bednarz 1977).

The most conspicuous trait of the white-sided jackrabbit is its tendency to occur in male–female pairs (Anderson 1972; Bogan and Jones 1975; Conway 1976; Bednarz 1977). These strong pairings occur throughout the breeding season, but not beyond this period (Bednarz 1977). With the exception of the male–female pair bond, the breeding behavior of this species is similar to that of the black-tailed jackrabbit (Lechleitner 1958a; Pontrelli 1968). Once the pair bond is established, the male takes on the aggressive duties of defending the pair from intruding males. The function of pair bonding may be to keep the sexes together during the breeding season, because normal densities are quite low for this hare (Bednarz 1977).

The white-sided jackrabbit constructs forms and uses them during the day. These are slightly larger than forms reported for *L. californicus,* averaging 37 cm in length, 18.3 cm in width, and 6.3 cm in depth. A tangled canopy of tobosa grass usually surrounds forms. *L. callotis* forced to leave their form during the day fled into stands of tobosa grass (Bogan and Jones 1975; Bednarz 1977). Bednarz (1977) reported flushing *L. callotis* from an abandoned kit fox (*Vulpes macrotis*) den, but this was considered rare.

Three types of vocalization were found in *L. callotis*. One, an alarm or fear reaction, was characterized by a high-pitched scream. The second occurred when an intruding male approached a pair of jackrabbits. The male of the pair produced harsh grunts until the intruder left or was chased away. The last vocalization occurred during the sexual chase and consisted of a trilling grunt. It was not determined which member of the pair produced this sound (Bednarz 1977).

The feeding habits of *L. callotis* were described by Bednarz (1977). The diet consisted of >99% grass. The only nongrass fed on was the sedge nutgrass (*Cyperus rotundus*). Plant species observed being consumed were buffalo grass, tobosa grass, fiddleneck (*Amsinckia* sp.), wolftail, blue grama, vine mesquite (*Panicum obtusum*), ring muhly (*Muhlenbergia torreyi*), woolly Indian wheat (*Plantago purshi*), and Wright buckwheat (*Eriogonum wrightii*). Bogan and Jones (1975) also reported that a grazed thistle flower (*Cirsium* sp.) occurred in a form of *L. callotis*. As noted, *L. californicus* and *L. alleni* forage on specific plants to obtain water essential for life. Because *L. callotis* feeds primarily on grasses, how this hare satisfies its need for water in late summer and fall when grasses dry out is unknown. Because the plant species complement changes in native grasslands, lack of forage with essential amounts of water could be a serious limiting factor for *L. callotis*.

No information is available on mortality rates for *L. callotis* or on instances of predation. Many predators of the black-tailed jackrabbit occur throughout the range of *L. callotis* and probably feed on this species to some degree. Five species of pathogenic bacterium have been isolated from *L. callotis*. Those include four respiratory pathogens: *Pneumococcus* sp., *Pseudomonas pseudomallei, Klebsiella ozanae,* and a *Moraxella*-like organism. A potential pathogen of the mesentaries and lymph nodes, *Yersinia pseudotuberculosis,* also was isolated. An unidentified coccidiosis infection was reported (Bednarz 1977). The tick *Dermacentor paramapertus* and the flea *Pulex simulans* were ectoparasites found in New Mexico. No serious disease or epidemic has been reported in *L. callotis* populations (Bednarz 1977).

The economic effects of *L. callotis* populations are probably beneficial. Any dietary overlap between cattle and *L. callotis* cannot be viewed as competition because *L. callotis* likely distributes grass seed in feces and thus serves to perpetuate the grasslands favored by cattle. Because normal population densities were probably never high, their occurrence should be viewed as a positive indicator of good to excellent range conditions. In contrast to this, Bednarz (1977) indicated that when *L. californicus* takes over former habitat of *L. callotis,* they may reach population densities 20 times greater than *L. callotis*. Practices of rangeland management that encourage maintenance of native grasslands with a very high percent of perennial grasses and a low percent of shrubs while maintaining native forbs would best serve the needs of *L. callotis* (Bednarz 1977). Habitat preservation is crucial to the survival of this species. As noted, Bednarz (1977) estimated that the U.S. population was 340 hares in 1976. Native grasslands in Mexico critical to these hares have experienced the same kind of human-induced changes as in New Mexico. In New Mexico, *L. callotis* is a state-threatened species, not subject to sport hunting.

All aspects of the biology and ecology of the white-sided jackrabbit need more study. Organizing research to identify all limiting factors operating on the extant population seems critical. Removing identified limiting factors should be the goal of a recovery effort. All forms of mortality should be identified and classsified as additive or compensatory. Because *L. callotis* is ecologically tied to native desert grasslands, cooperative agreements, purchase of easements, or other arrangements with private land owners must be extended to preserve essential habitat.

LITERATURE CITED

Alipayou, D., J. L. Holechek, R. Valdez, L. S. Tembo, M. Rusco, and M. Cardenas. 1993. Range condition influences on Chihuahuan Desert cattle and jackrabbit diets. Journal of Range Management 46:296–301.

Anderson, S. 1972. Mammals of Chihuahua: Taxonomy and distribution. American Museum of Natural History Bulletin 148:149–410.

Anderson, S., and A. S. Gaunt. 1962. A classification of the white-sided jackrabbits of Mexico. American Museum Novitates 2088:1–16.

Armstrong, D. M. 1972. Distribution of mammals in Colorado. University of Kansas Museum of Natural History Monograph 3:1–415.

Arnold, J. F. 1942. Forage consumption and preferences of experimentally fed Arizona and antelope jack rabbits. (Technical Bulletin 98). University of Arizona, Agriculture Experiment Station, Tucson.

Arnold, J. F., and H. G. Reynolds. 1943. Droppings of Arizona and antelope jack rabbits and the "pellet census." Journal of Wildlife Management 7:322–27.

Arnold, T. W. 1989. Sex ratios of fledgling golden eagles and jackrabbit densities. Auk 106:521–22.

Arthur III, W. J., and R. J. Gates. 1988. Trace element intake via soil ingestion in pronghorns and in black-tailed jackrabbits. Journal of Range Management 41:162–66.

Bailey, V. 1936. The mammals and life zones of Oregon. North American Fauna 55:1–416.

Banfield, A. W. F. 1974. The mammals of Canada. University of Toronto Press, Toronto.

Bartel, M. H., and M. F. Hansen. 1964. *Raillietina loeweni* sp. from the hare in Kansas, with notes on *Raillietina* of North American mammals. Journal of Parasitology 50:448–53.

Bear, G. D., and R. M. Hansen. 1966. Food habits, growth, and reproduction of white-tailed jackrabbits in southern Colorado (Technical Bulletin 90). Colorado State University of Agriculture Experiment Station, Fort Collins.

Bednarz, J. 1977. The white-sided jackrabbit in New Mexico: Distribution, numbers, and biology in the grasslands of Hidalgo County (Research report). New Mexico Department of Game and Fish, Endangered Species Program, Santa Fe.

Bednarz, J., and J. A. Cook. 1984. Distribution and numbers of white-tailed jackrabbits in New Mexico. Southwestern Naturalist 29:358–60.

Best, T. L. 1996. *Lepus californicus.* Mammalian Species 530:1–10.

Best, T. L. 1999a. Antelope jackrabbit, *Lepus alleni.* Pages 693–95 *in* D. E. Wilson and S. Rue, eds. The Smithsonian book of North American mammals. Smithsonian Institution Press, Washington, DC.

Best, T. L. 1999b. White-sided jackrabbit, *Lepus callotis.* Pages 701–2 *in* D. E. Wilson and S. Rue, eds. The Smithsonian book of North American mammals. Smithsonian Institution Press, Washington, DC.

Best, T. L., and T. H. Henry. 1993a. *Lepus callotis.* Mammalian Species 442:1–6.

Best, T. L., and T. H. Henry. 1993b. *Lepus alleni.* Mammalian Species 424:1–8.

Bickler, P. E., and Shoemaker, V. H. 1975. Alfalfa damage by jackrabbits in southern California deserts. California Agriculture 29(7):10–12.

Blackburn, D. F. 1968. Courtship behavior among white-tailed and black-tailed jackrabbits. Great Basin Naturalist 33:203.

Bogan, M. A., and C. Jones. 1975. Observations on *Lepus callotis* in New Mexico. Proceedings of the Biological Society (Washington, DC) 88:45–50.

Borell, A. E., and R. Ellis. 1934. Mammals of the Ruby Mountains region of northeastern Nevada. Journal of Mammalogy 15:12–44.

Braun, C. E., and R. G. Streeter. 1968. Observations on the occurrence of the white-tailed jackrabbit in the alpine zone. Journal of Mammalogy 49:160–61.

Bronson, F. H., and O. W. Tiemeier. 1958a. Reproduction and age distribution of black-tailed jackrabbits in Kansas. Journal of Wildlife Management 22:409–14.

Bronson, F. H., and O. W. Tiemeier. 1958b. Notes on crop damage by jackrabbits. Transactions of the Kansas Academy of Science 49:455–56.

Bronson, F. H., and O. W. Tiemeier. 1959. The relationship of precipitation and black-tailed jackrabbit populations in Kansas. Ecology 40:194–98.

Brown, C. F., and P. R. Krausman. 2003. Habitat characteristics of 3 leporid species in southeastern Arizona. Journal of Wildlife Management. 67:83–89.

Brown, H. L. 1947a. Why has the white-tailed jackrabbit (*Lepus townsendii campanius*) become scarce in Kansas? Transactions of the Kansas Academy of Science 49:455–56.

Brown, H. L. 1947b. Coaction of jack rabbit, cottontail, and vegetation in a mixed prairie. Transactions of the Kansas Academy of Science 50:28–44.

Brunton, D. F. 1981. Nocturnal aggregations of white-tailed jackrabbits at Rimbey, Alberta. Blue Jay 39:120–21.

Burgess, E. C., and L. A. Windberg. 1989. *Borrelia* sp. in coyotes, black-tailed jackrabbits and desert cottontails in southern Texas. Journal of Wildlife Diseases 25:47–51.

Carter, F. L. 1939. A study in jackrabbit shifts in range in western Kansas. Transactions of the Kansas Academy of Science 42:431–35.

Cervantes, F. A., C. Lorenzo, and M. D. Engstrom. 1997. New records of the eastern cottontail (*Sylvilagus floridanus*) and black-tailed jackrabbit (*Lepus californicus*) in Mexico. Texas Journal of Science 49:75–77.

Chapman, J. A., and J. L. Sandt. 1977. The black-tailed jackrabbit, *Lepus californicus,* in Maryland. Chesapeake Science 18:319.

Cheeke, P. R. 1987. Rabbit feeding and nutrition. Academic Press, New York.

Clapp, R. B., J. S. Weske, and T. C. Clapp. 1976. Establishment of the black-tailed jackrabbit on the Virginia eastern shore. Journal of Mammalogy 57:180–81.

Clark, D. O. 1975. Vertebrate pest control handbook. California Department of Food and Agriculture, Sacramento.

Clark, F. W. 1972. Influence of jackrabbit density on coyote population change. Journal of Wildlife Management 36:343–56.

Clark, J. P. 1994. Vertebrate pest control handbook, 4th ed. Sacramento: Division of Plant industry, California Department of Food and Agriculture.

Clemons, C., L. G. Rickard, J. E. Keirans, and R. G. Botzler. 2000. Evaluation of host preferences by helminths and ectoparasites among black-tailed jackrabbits in northern California. Journal of Wildlife Diseases 36:555–58.

Connolly, G. E., M. L. Dudzinski, and W. M. Longhurst. 1969. The eye lens as an indicator of age in the black-tailed jackrabbit. Journal of Wildlife Management 33:159–64.

Conover, M. 2001. Resolving human–wildlife conflicts: The science of wildlife damage management. CRC Press, Boca Raton, FL.

Conway, M. C. 1976. A rare hare. New Mexico Wildlife 21:21–23.

Costa, R., K. A. Nagy, and V. H. Shoemaker. 1976. Observations of behavior on black-tailed jackrabbits in the Mojave Desert. Journal of Mammalogy 57:399–402.

Couch, L. K. 1927. Migrations of the Washington black-tailed jackrabbit. Journal of Mammalogy 8:313–14.

Crenshaw, D. K., and S. E. Henke. 1997. Prevalence of cuterebrid (Diptera: Cuterebridae) parasitism among black-tailed jackrabbits in southern Texas. Texas Journal of Science 49:335–38.

Crouch, G. L. 1973. Jackrabbits injure ponderosa pine seedlings. Tree Planters' Notes 24(3):15–17.

Currie, P. O., and D. L. Goodwin. 1966. Consumption of forage by black-tailed jackrabbits on salt desert ranges of Utah. Journal of Wildlife Management 30:304–11.

Daniel, A., J. Holechek, P. Valdez, A. Tembo, L. Saiwana, M. Fusco, and M. Cardenas. 1993. Jackrabbit densities on fair and good condition Chihuahuan desert range. Journal of Range Management 46:524–28.

Dannelid, E. 1991. The genus *Sorex* (Mammalia, Soricidae)—Distribution and evolutionary aspects of Eurasian shrews. Mammal Review 21:1–20.

Davis, C. A., J. A. Medlin, and J. P. Griffing. 1975. Abundance of black-tailed jackrabbits, desert cottontail rabbits, and coyotes in southeastern New Mexico (Research Report 293). New Mexico State University Agriculture Experiment Station, Las Cruces.

Delibes, M., and F. Hiraldo. 1987. Food habits of the bobcat in two habitats of the southern Chihuahuan Desert. Southwestern Naturalist 32:457–61.

DeVos, A. 1964. Range changes of mammals in the Great Lakes region. American Midland Naturalist 71:210–31.

DeVos, A., R. H. Manville, and R. G. Van Gelder. 1956. Introduced mammals and their influence on native biota. Zoologica 41:172.

Dixon, K. R., J. A. Chapman, G. R. Willner, D. E. Wilson, and W. Lopez-Forment. 1983. The New World jackrabbits and hares (genus *Lepus*)—2. Numerical taxonomic analysis. Acta Zoologica Fennica 174:53–56.

Donoho, H. S. 1972. Dispersion and dispersal of white-tailed and black-tailed jackrabbits, Pawnee National Grasslands. M.S. Thesis, Colorado State University, Fort Collins.

Dubke, R. T. 1973. The white-tailed jackrabbit in Wisconsin (Final Report, Pittman/Robertson Study 106, Project W-141-R-8). Wisconsin Department of Natural Resources, Madison.

Eden, A. 1940. Origin of night feces. Nature 145:628–29.

Ernest, K. A. 1994. Resistance of creosotebush to mammalian herbivory: Temporal consistency and browsing-induced changes. Ecology 75:1684–92.

Errington, P. L. 1935. Food habits of mid-west fox. Journal of Mammalogy 16:192–200.

Evans, J., P. L. Hegdal, and R. E. Griffith, Jr. 1970. Methods of controlling jackrabbits. Pages 109–16 *in* R. H. Dana, ed. Proceedings of 4th vertebrate pest conference. West Sacramento, CA.

Evans, J.,P. L. Hegdal, and R. E. Griffith, Jr. 1982. Wire fencing for controlling jackrabbit damage (Bulletin 618). University of Idaho, Moscow.

Fagerstone, K. A., G. K. Lavoie, and R. E. Griffith, Jr. 1980. Black-tailed jackrabbit diet and density on rangeland near agricultural crops. Journal of Range Management 33:229–33.

Fautin, R. W. 1946. Biotic communities of the Northern Desert Shrub biome in western Utah. Ecological Monographs 16:252–310.

Feldhamer, G. A. 1979. Age, sex ratios and reproductive potential in black-tailed jackrabbits. Mammalia 43:473–78.

Feltus, D. G., and E. E. Langenau, Jr. 1984. Optimization of firearm deer hunting and timber values in nothern lower Michigan. Wildlife Society Bulletin 12:6–12.

Findley, J. S. 1956. Mammals of Clay County, South Dakota. University of South Dakota Publications in Biology 1:45.

Findley, J. S., A. H. Harris, D. E. Wilson, and C. Jones. 1975. Mammals of New Mexico. University of New Mexico Press, Albuquerque.

Fitzgerald, J. P., C. A. Meaney, and D. M. Armstrong. 1994. Mammals of Colorado. Colorado Associated University Press, Niwot.

Flinders, J. T., and J. A. Crawford. 1977. Composition and degradation of jackrabbit and cottontail fecal pellets, Texas High Plains. Journal of Range Management 30:217–20.

Flinders, J. T., and C. L. Elliot. 1979. Abiotic characteristics of black-tailed jackrabbit forms and a hypothesis concerning form function. Encyclia 56:34–38.

Flinders, J. T., and R. M. Hansen. 1972. Diets and habitats of jackrabbits in northeastern Colorado (Science Series 12). Range Science Department, Colorado State University, Fort Collins.

Flinders, J. T., and R. M. Hansen. 1973. Abundance and dispersion of leporids within a shortgrass ecosystem. Journal of Mammalogy 54:287–91.

Flinders, J. T., and R. M. Hansen. 1975. Spring population responses of cottontails and jackrabbits to cattle grazing shortgrass prairie. Journal of Range Management 28:290–93.

Flux, J. E. C. 1981. Prospects for hare farming in New Zealand. New Zealand Agricultural Science 15:24–29.

Flux, J. E. C. 1983. Introduction to the taxonomic problems in hares. Acta Zoologica Fennica 174:7–10.

Flux, J. E. C., and R. Angermann. 1990. The hares and jackrabbits. Pages 61–97 *in* J. A. Chapman and J. E. C. Flux, eds. Rabbits, hares and pikas: Status survey and conservation action plan. IUCN/SSC, Gland, Switzerland.

French, N. R., R. McBride, and J. Detmer. 1965. Fertility and population density of the black-tailed jackrabbit. Journal of Wildlife Management 29:14–26.

Garlough, F. E., J. F. Welch, and H. J. Spencer. 1942. Rabbits in relation to crops (Conservation Bulletin 11). U.S. Fish and Wildlife Service, Washington, DC.

Gashwiler, J. S., W. L. Robinette, and O. W. Morris. 1960. Foods of bobcats in Utah and eastern Nevada. Journal of Wildlife Management 24:226–28.

Glass, B. P. 1973. A key to the skulls of North American mammals. Oklahoma State University Press, Stillwater.

Gonzalez, F. X., and F. A. Cervantes. 1996. Karyotype of the white-sided jackrabbit (*Lepus callotis*). Southwestern Naturalist 41:93–95.

Graves, G. N., W. C. Bennett, J. R. Wheelers, B. E. Miller, and D. L. Forcum. 1978. Sylvatic plague studies in southeast New Mexico. Part 2, Relationship of the desert cottontail and its fleas. Journal of Medical Entomology 14:511–22.

Griffing, J. P., and C. A. Davis. 1976. Black-tailed jackrabbits in southeastern New Mexico: Population structure, reproduction, feeding and use of forms (Research report). New Mexico State University Experiment Station, Las Cruces.

Griffiths, M., and D. Davies. 1963. The role of soft pellets in the production of lactic acid in the rabbit stomach. Journal of Nutrition 80:171–80.

Gross, J. E., L. C. Stoddart, and F. H. Wagner. 1974. Demographic analysis of a northern Utah jackrabbit population. Wildlife Monographs 40:1–68.

Halanych, K. M., J. R. Demboski, B. J. van Vuuren, D. R. Klein, and J. A. Cook. 1999. Cytochrome *b* phylogeny of North American hares and jackrabbits (*Lepus*, Lagomorpha) and the effects of saturation of outgroup taxa. Molecular Phylogenetics and Evolution 11:213–21.

Hall, E. R. 1946. Mammals of Nevada. University of California Press, Berkeley.

Hall, E. R. 1955. Handbook of mammals of Kansas (Miscellaneous Publication 7). University of Kansas Museum of Natural History, Lawrence.

Hall, E. R. 1981. The mammals of North America, Vol. I, 2nd ed. John Wiley, New York

Hall, E. R., and K. R. Kelson. 1959. The mammals of North America, Vol. 1. Ronald Press New York.

Hamilton, W. D. 1971. Geometry for the selfish herd. Journal of Theoretical Biology 31:295–311.

Hamilton, W. J., Jr. 1935. Notes on food of red foxes in New York and New England. Journal of Mammalogy 16:16–21.

Hansen, M. F., M. H. Bartel, E. T. Lyon, and B. M. El-Rawi. 1965. Helminth and arthropod parasites. Pages 41–61 *in* The black-tailed jackrabbit in

Kansas (Kansas State University Agriculture and Applied Science, Agriculture Experiment Station Technical Bulletin 140). Kansas State University, Manhattan.

Hansen, R. M., and G. D. Bear. 1963. Winter coats of white-tailed jackrabbits in southwestern Colorado. Journal of Mammalogy 44:420–21.

Hansen, R. M., and J. T. Flinders. 1969. Food habits of North American hares (Range Science Department, Science Series 1), Colorado State University, Fort Collins.

Harris, G. D., H. D. Huppi, and J. A. Gessman. 1985. The thermal conductance of winter and summer pelage of *Lepus californicus*. Journal of Thermal Biology 10:79–81.

Haskell, H. S., and H. G. Reynolds. 1947. Growth, developmental food requirements and breeding of the California jackrabbit. Journal of Mammalogy 28:129–36.

Hayden, P. 1966a. Food habits of the black-tailed jackrabbit in southern Nevada. Journal of Mammalogy 47:42–46.

Hayden, P. 1966b. Seasonal occurrence of jackrabbits on Jackass Flat, Nevada. Journal of Wildlife Management 30:835–38.

Hegdal, P. L. 1966. Jackrabbit damage in the western United States (Supplemental Report F-42.2). Denver Wildlife Research Center, Jackrabbit Research Station, Idaho.

Henke, S. E., and F. C. Bryant. 1999. Effects of coyote removal on the faunal community in western Texas. Journal of Wildlife Management 63:1066–81.

Henke, S. E., and S. Demarais. 1990. Effect of diet on condition indices in black-tailed jackrabbits. Journal of Wildlife Diseases 26:28–33.

Henke, S. E., D. B. Pence, S. Demarais, and J. R. Johnson. 1990. Serologic survey of selected zoonotic disease agents in black-tailed jackrabbits from western Texas. Journal of Wildlife Diseases 26:107–11.

Herndon, J. E., and E. L. Hove. 1955. Surgical removal of the cecum and its effect on digestion and growth in rabbits. Journal of Nutrition 57:261–70.

Herrick, E. H. 1965. Endocrine studies. Pages 73–75 *in* The black-tailed jackrabbit in Kansas (Kansas State University Agriculture and Applied Science, Agricultural Experiment Station, Technical Bulletin 140). Kansas State University, Manhattan.

Hildreth, A. C., and G. B. Brown. 1955. Repellents to protect trees and shrubs from damage by rabbits (Technical Bulletin 1134). U.S. Department of Agriculture, Washington, DC.

Hinds, D. S. 1977. Acclimatization of thermoregulation in desert inhabiting jackrabbits (*Lepus alleni* and *Lepus californicus*). Ecology 58:246–64.

Hoffman, M. T., C. D. James, G. I. H. Kerley, and W. G. Whitford. 1993. Rabbit herbivory and its effect on cladode, flower and fruit production of *Opuntia violacea* var. *macrocentra* (*Cactaceae*) in the northern Chihuahuan Desert, New Mexico. Southwestern Naturalist 38:309–15.

Hoffmeister, D. F. 1986. Mammals of Arizona. University of Arizona Press and Arizona Game and Fish Department, Tucson.

Howard, W. E. 1964. Introduced browsing mammals and habitat stability in New Zealand. Journal of Wildlife Management 28:421–29.

Howard, W. E. 1965. Interaction of behavior, ecology, and genetics of introduced mammals. Pages 461–84 *in* H. G. Baker and G. L. Stebbins, eds. The genetics of colonizing species. Academic Press, New York.

Hull, T. G. 1963. The role of different animals and birds in diseases transmitted to man. Pages 876–924 *in* T. G. Hull, ed. Diseases transmitted from animals to man, 5th ed. Charles C. Thomas, Springfield, IL.

Jackson, H. H. T. 1961. Mammals of Wisconsin. University of Wisconsin Press, Madison.

Jalal, S. M.,T. R. James, and R. W. Seabloom. 1967. Karyotype of the white-tailed jackrabbit. Proceedings of the North Dakota Academy of Science 21:92–98.

James, T. R., and R. W. Seabloom. 1969a. Reproductive biology of the white-tailed jackrabbit in North Dakota. Journal of Wildlife Management 33:558–68.

James, T. R., and R. W. Seabloom. 1969b. Aspects of growth in the white-tailed jackrabbit. Proceedings of the North Dakota Academy of Science 22:7–14.

Johnson, D. R., and J. M. Peek. 1984. The black-tailed jackrabbit in Idaho: Life history, population dynamics and control (Cooperative Extension Service Bulletin No. 637). University of Idaho College of Agriculture, Moscow.

Johnson, J. C. 1978. Anticoagulant baiting for jackrabbit control. Pages 152–53 *in* W. E. Howard, ed. Proceedings of 8th vertebrate pest conference. Sacramento, CA.

Johnson, R. D., and J. E. Anderson. 1984. Diets of black-tailed jackrabbits in relation to population density and vegetation. Journal of Range Management 37:79–83.

Johnson, W. V. 1964. Rabbit control. Pages 90–96 *in* Proceedings of 2nd vertebrate pest control conference.

Jones, J. K. 1964. Distribution and taxonomy of mammals of Nebraska. University of Kansas Museum of Natural History Publication 16:1–356.

Kietzmann, G. E., Jr., and E. J. Hugghins. 1984. Ectoparasites of white-tailed jackrabbits and eastern cottontail rabbits in South Dakota. Proceedings of the South Dakota Academy of Science 63:42–47.

Kline, P. D. 1963. Notes on the biology of the jackrabbit in Iowa. Proceedings of the Iowa Academy of Science 70:196–204.

Knick, S. T. 1990. Ecology of bobcats relative to exploitation and a prey decline in southeastern Idaho. Wildlife Monographs 108:1–42.

Knick, S. T., and D. L. Dyer. 1997. Distribution of black-tailed jackrabbit habitat determined by GIS in southwestern Idaho. Journal of Wildlife Management 61:75–85.

Knick, S. T., and J. T. Rotenberry. 1998. Limitations to mapping habitat use areas in changing landscapes using the Mahalanobis distance statistic. Journal of Agriculture, Biology, and Environmental Statistics 3:311–22.

Knick, S. T., E. C. Hellgren, and U. S. Seal. 1993. Hematologic, biochemical, and endocrine characteristics of bobcats during a prey decline in southeastern Idaho. Canadian Journal of Zoology 71:1448–53.

Knight, J. E. 1994. Jackrabbits and other hares. Pages D81–D86 *in* E. Hygnstrom, R. M. Timm, and G. E. Larson, eds. Prevention and control of wildlife damage, Vol. 2. University of Nebraska Cooperative Extension, Lincoln.

Lahrman, F. W. 1980. A concentration of white-tailed jackrabbits. Blue Jay 38:130.

Layne, J. N. 1965. Occurrence of black-tailed jackrabbits in Florida. Journal of Mammalogy 46:502.

Lechleitner, R. R. 1957. Reingestion in the black-tailed jackrabbit. Journal of Mammalogy 38:481–85.

Lechleitner, R. R. 1958a. Certain aspects of behavior of the black-tailed jackrabbit. American Midland Naturalist 60:145–55.

Lechleitner, R. R. 1958b. Movements, density, and mortality in a black-tailed jackrabbit population. Journal of Wildlife Management 22:371–84.

Lechleitner, R. R. 1959a. Sex ratio, age classes, and reproduction of the black-tailed jackrabbit. Journal of Mammalogy 40:63–81.

Lechleitner, R. R. 1959b. Some parasites and infectious diseases in a black-tailed jackrabbit population in the Sacramento Valley of California. California Fish and Game 45:83–91.

Leopold, A. 1945. The distribution of Wisconsin hares. Transactions of the Wisconsin Academy of Science 37:1–14.

Lewis, J. H. 1946. Planting practice to reduce crop damage by jackrabbits. Journal of Wildlife Management 10:277.

Lim, B. K. 1987. *Lepus townsendii.* Mammalian Species 288:1–6.

Lindsdale, J. M. 1938. Environmental responses of vertebrates in the Great Basin. American Midland Naturalist 19:1–206.

Lipson, M. P., and P. R. Krausman. 1988. Parasites of desert leporids in the Picacho Mountains, Arizona. Southwestern Naturalist 33:487–88.

Long, C. A. 1965. The mammals of Wyoming. University of Kansas Publications, Museum of Natural History 14:493–58.

Longland, W. 1991. Risk of predation and food consumption by black-tailed jackrabbits. Journal of Range Management 44:447–50.

Lord, R. D. 1960. Litter size and latitude in North American mammals. American Midland Naturalist 64:488–99.

MacCraken, J. G., and R. M. Hansen. 1982a. Herbaceous vegetation of habitat used by blacktail jackrabbits and Nuttall cottontails in southeastern Idaho. American Midland Naturalist 107:180–84.

MacCraken, J. G., and R. M. Hansen. 1982b. Seasonal foods of blacktail jackrabbits and Nuttall cottontails in southeastern Idaho. Journal of Range Management 37:256–59.

MacCraken, J. G., and R. M. Hansen. 1987. Coyote feeding strategies in southeastern Idaho: Optimal foraging by an opportunistic predator? Journal of Wildlife Management 51:278–85.

Marzluff, J. M., S. T. Knick, M. S. Vekasy, L. S. Schueck, and T. J. Zarriello. 1997. Spatial use and habitat selection of golden eagles in southerwestern Idaho. *Auk* 114:673–87.

McAdoo, J. K., W. S. Longland, G. J. Cluff, and D. A. Klebenow. 1987. Use of new rangeland seedings by black-tailed jackrabbits. Journal of Range Management 40:520–24.

McDowell, J. W., H. G. Scott, C. J. Stojanovich, and H. B. Weinburgh. 1964. *Tularemia.* U.S. Deptartment of Health Education and Welfare, Public Health Service, Atlanta, GA.

McMahon, K. J. 1965. Bacterial and rickettsial diseases. Pages 65–71 *in* The black-tailed jackrabbit in Kansas (Kansas State University Agriculture and Applied Science, Agriculture Experiment Station, Technical Bulletin 140). Kansas State University, Manhattan.

Mearns, E. A. 1896. Preliminary description of a new subgenus and six new species and subspecies of hares, from the Mexican border of the United States. Proceedings of the U.S. National Museum 18:551–65.

Mohr, W. P., and C. O. Mohr. 1936. Recent jackrabbit populations at Rapidan, Minnesota. Journal of Mammalogy 17:112–14.

Monsen, S. E., W. E. Hazeltine, S. D. Thomas, and N. R. Scott. 1991. Preliminary testing of the black-tailed jackrabbit as a possible sentinel for western equine encephalomyelitis and Lyme borreliosis. Proceeding, Annual Conference of California Mosquito and Vector Control Association 58:138–43.

Nagy, K. A. 1994. Seasonal water, energy and food use by free-living arid-habitat mammals. Australian Journal of Zoology 42:55–63.

Nagy, K. A., V. H. Shoemaker, and W. R. Costa. 1976. Water, electrolyte, and nitrogen budgets in jackrabbits (*Lepus californicus*) in the Mojave Desert. Physiological Zoology 49:351–62.

Nelson, T., J. L. Holechek, R. Valdez, and M. Cardenas. 1997. Wildlife numbers on late and mid seral Chihuahuan Desert rangelands. Journal of Range Management 50:593–99.

Orr, R. T. 1940. The rabbits of California (Occasional Papers of the California Academy of Sciences 19). San Francisco, CA.

Palmer, T. S. 1897. The jackrabbits of the United States. USDA, Division of Biological Services, Bulletin 8:1–87.

Pfaffenberger, G. S., and V. B. Valencia. 1988a. Ectoparasites of sympatric cottontails (*Sylvilagus auduboni* Nelson) and jackrabbit (*Lepus californicus* Mearns) from the high plains of eastern New Mexico. Journal of Parasitology 74:842–46.

Pfaffenberger, G. S., and V. B. Valencia. 1988b. Helminths of sympatric black-tailed jackrabbits (*Lepus californicus*) and desert cottontails (*Sylvilagus auduboni*) from the high plains of eastern New Mexico. Journal of Wildlife Diseases 24:375–77.

Phillip, C. B., J. F. Bell, and C. L. Larson. 1955. Evidence of infectious diseases and parasites in a peak population of black-tailed jackrabbits in Nevada. Journal of Wildlife Management 19:225–33.

Pittman, V. J. 1977. Winter wheat seedling preference by grazing rodents. Canada Journal of Plant Science 57:1009–12.

Pontrelli, M. J. 1968. Mating behavior of the black-tailed jackrabbit. Journal of Mammalogy 49:785–86.

Read, R. A. 1971. Browsing preference by jackrabbits in a ponderosa pine provenance plantation (USDA Forest Service Reserve Note RM-186). Rocky Mountain Forest and Range Experiment Station, Fort Collins, Co.

Reese, J. B., and H. Haines. 1978. Effects of dehydration on metabolic rate and fluid distribution in the jackrabbit, *Lepus californicus*. Physiology Zoologica 51:155–65.

Reig, O. A. 1989. Karyotypic repatterning as one triggering factor in cases of explosive speciation. Pages 246–49 *in* A. Fontdevlia, ed. Evolutionary biology of transient unstable populations. Springer-Verlag, Berlin.

Reisen, W. K., J. L. Hardy, W. C. Reeves, S. B. Presser, M. M. Milby, and R. P. Meyer. 1990. Persistence of mosquito-borne viruses in Kern County, California, 1983–1988. American Journal of Tropical Medicine and Hygiene 43:419–37.

Riegel, A. 1941. Some coactions of rabbits and rodents with cactus. Transactions of the Kansas Academy of Science 44:96–101.

Robinette, W. L., J. S. Gashwiler, and O. W. Morris. 1959. Food habits of the cougar in Utah and Nevada. Journal of Wildlife Management 23:261–72.

Robinson, T. J., F. F. B. Elder, and J. A. Chapman. 1983. Karyotypic conservatism in the genus *Lepus* (order Lagomorpha). Canadian Journal of Genetics and Cytology 25:540–44.

Rogowitz, G. L. 1990. Seasonal energetics of the white-tailed jackrabbit (*Lepus townsendii*). Journal of Mammalogy 71:277–85.

Rogowitz, G. L. 1992a. Postnatal variation in metabolism and body temperature in a precocial lagomorph. Functional Ecology 6:666–71.

Rogowitz, G. L. 1992b. Reproduction of white-tailed jackrabbits on semi-arid range. Journal of Mammalogy 56:676–84.

Rogowitz, G. L. 1997. Locomotor and foraging activities of the white-tailed jackrabbit (*Lepus townsendii*). Journal of Mammalogy 78:1172–81.

Rogowitz, G. L., and J. A. Gessaman. 1990. Influence of air temperature, wind and irradiance on metabolism of white-tailed jackrabbits. Journal of Thermal Biology 15:125–31.

Rogowitz, G. L., and M. L. Wolfe. 1991. Intraspecific variation on life-history traits of the white-tailed jackrabbit (*Lepus townsendii*). Journal of Mammalogy 72:796–806.

Rohe, D. L., N. Dallas, and L. S. Estes. 1963. Control of jack rabbits at a California municipal airport. California Vector Views 10:73–75.

Roundy, B. A., G. J. Cluff, J. K. McAdoo, and R. A. Evans. 1985. Effects of jackrabbit grazing, clipping, and drought on crested wheatgrass seedlings. Journal of Range Management 38:551–55.

Salmon, T. P., and W. P. Gorenzel. 1997. Pest notes: Rabbits. *in* B. Ohlendorf, ed. UC pest management guidelines (Publication 7447). University of California, Division of Agriculture and Natural Resources.

Schmidt-Nielsen, K., T. J. Dawson, H. T. Hammel, D. Hinds, and D. C. Jackson. 1965. The jackrabbit: A study in its desert survival. Hvalradets Skrifter 48:125–42.

Schroder, J., J. Antoni, and J. van der Loo. 1978. Comparison of the karyotype in the jack rabbit (*Lepus californicus deserticola*) and the European hare (*Lepus europaeus*). Hereditas 89:134–35.

Scott, T. G. 1955. Dietary patterns of red and gray foxes. Ecology 36:366–67.

Shoemaker, V. H., K. A. Nagy, and W. R. Costa. 1976. Energy utilization and temperature regulation by jack-rabbits (*Lepus californicus*) in the Mojave Desert. Physiological Zoology 49:364–75.

Smith, G. W. 1990. Home range and activity patterns of black-tailed jackrabbits. Great Basin Naturalist 50:249–56.

Smith, G. W., and N. C. Nydegger. 1985. A spotlight, line-transect method for surveying jack rabbits. Journal of Wildlife Management 49:699–702.

Smith, D. F., L. C. Stoddart, and F. F. Knowlton. 2002. Long-distance movements of black-tailed jackrabbits. Journal of Wildlife Management. 66:463–469.

Smith, H. N. 1948. Rabbits spread cedar menace. Cattleman 35(3):97–98.

Sparks, D. R. 1968. Diets of black-tailed jackrabbits on the Sandhill Rangeland of Colorado. Journal of Range Management 21:203–8.

Steenhof, K., and M. N. Kochert. 1988. Dietary responses of three raptor species to changing prey densities in a natural environment. Journal of Animal Ecology 57:37–48.

Steenhof, K., M. N. Kochert, and T. L. McDonald. 1997. Interactive effects of prey and weather on golden eagle reproduction. Journal of Animal Ecology 66:350–62.

Steigers, W. D., Jr., J. T. Flinders, and S. M. White. 1982. Rhythm of fecal production and protein content for black-tailed jackrabbits. Great Basin Naturalist 42:567–71.

Steinberger, Y., and W. G. Whitford. 1983. The contribution of shrub pruning by jackrabbits to litter input in a Chihuahuan Desert ecosystem. Journal of Arid Environment 6:183–87.

Stock, A. D. 1976. Chromosome banding pattern relationships of hares, rabbits, and pikas (order Lagomorpha). Cytogenetics and Cell Genetics 17:78–88.

Stoddart, L. C. 1970. A telemetric method for detecting jackrabbit mortality. Journal of Wildlife Management 34:501–7.

Stoddart, L. C. 1983. Relative abundance of coyotes, lagomorphs, and rodents on the Idaho National Engineering Laboratory. Pages 268–77 *in* O. D. Markham, ed. 1983 Progress Report Idaho National Engineering Laboratory Radioecology and Ecology Programs, Department of Energy, Idaho Falls, ID.

Storer, T. I. 1958. Controlling field rodents in California (Circular 434, Rev.). University of California, California Agricultural Experiment Station, Extension Service.

Taylor, W. P., C. T. Vorhies, and P. B. Lister. 1935. The relation of jackrabbits to grazing in southern Arizona. Journal of Forestry 33:490–98.

Thacker, E. J., and C. S. Brandt. 1955. Coprophagy in the rabbit. Journal of Nutrition 55:375–85.

Thulin, C. G., M. Jaarola, and H. Tegelstrom. 1997. The occurrence of mountain hare mitochondrial DNA in wild brown hares. Molecular Ecology 6:463–67.

Tiemeier, O. W. 1965. Bionomics. Pages 5–37 *in* The black-tailed jackrabbit in Kansas (Kansas State University of Agriculture and Applied Science, Agricultural Experiment Station, Technical Bulletin 140). Kansas State University, Manhattan.

Tiemeier, O. W., and M. L. Plenert. 1964. A comparison of three methods for determining the age of black-tailed jackrabbits. Journal of Mammalogy 45:409–16.

Timmons, F. L. 1942. The dissemination of prickly pear seed by jack rabbits. Journal of the American Society of Agronomy 34:513–20.

Uresk, D. W. 1978. Diets of the black-tailed hare in steppe vegetation. Journal of Range Management 31:439–42.

Vorhies, C. I., and W. P. Taylor. 1933. The life histories and ecology of the jackrabbits *Lepus alleni* and *Lepus californicus* in relation to grazing in Arizona. University of Arizona Agriculture Experiment Station Technical Bulletin 49:1–117.

Voth, D. R., and T. R. James. 1965. Parasites of the white-tailed jackrabbit in southwestern North Dakota. Proceedings of the North Dakota Academy of Science 19:15–18.

Wagner, F. H., and L. C. Stoddart. 1972. Influence of coyote predation on black-tailed jackrabbit populations in Utah. Journal of Wildlife Management 36:329–42.

Wansi, T., R. D. Pieper, R. F. Beck, and L. W. Murray. 1992. Botanical content of black-tailed jackrabbit diets on semidesert rangeland. Great Basin Naturalist 52:300–308.

Watkins, L. C., and R. M. Novak. 1973. The white-tailed jackrabbit in Missouri. Southwestern Naturalist 18:341–57.

West, R. R., M. H. Bartel, and M. L. Plenert. 1961. Use of forms by black-tailed jackrabbits in southwestern Kansas. Transactions of the Kansas Academy of Science 64:344–48.

Westoby, M. 1980. Black-tailed jackrabbit diets in Curlew Valley, northern Utah. Journal of Wildlife Management 44:1980.

Wetherbee, F. A. 1967. A method of controlling jack rabbits on a range rehabilitation project in California. Pages 111–17 *in* Proceedings of the 3rd vertebrate pest control conference.

Windberg, L. A., and C. D. Mitchell. 1990. Winter diets of coyotes in relation to prey abundance in southern Texas. Journal of Mammalogy 71:439–47.

Woffinden, N. D., and J. R. Murphy. 1989. Decline of a ferruginous hawk population: A 20-year summary. Journal of Wildlife Management 53:1127–32.

Woodbury, A. M. 1955. Ecology of the Great Salt Lake Desert. Ecology 36:353–56.

Worthington, D. H., and D. A. Sutton. 1966. Chromosome numbers and analysis in three species of Leporidae. Mammalian Chromosome Newsletter 22:194–95.

Wywialowski, A. P., and L. C. Stoddart. 1988. Estimation of jackrabbit density: Methodology makes a difference. Journal of Wildlife Management 52:57–59.

Zaugg, J. L., and B. A. Newman. 1985. Evaluation of jackrabbits as nonruminant hosts for *Anaplasma marginale*. American Journal of Veterinarian Research 46:669–70.

Zeveloff, S. I. 1988. Mammals of the intermountain west. University of Utah Press, Salt Lake City.

Zimmerman, G., P. Stapp, and B. Van Horne. 1996. Seasonal variation in the diet of great horned owls (*Bubo virginianus*) of shortgrass prairie. American Midland Naturalist 136:149–56.

Jerran T. Flinders, Department of Integrative Biology, Brigham Young University, Provo, Utah 84602. Email: jerran_flinders@byu.edu

Joseph A. Chapman, Office of the President, North Dakota State University, Fargo, North Dakota 58105-5167. Email: joseph@ndsu.nodak.edu

7

Snowshoe Hare and Other Hares

Lepus americanus and Allies

Dennis L. Murray

NOMENCLATURE

Common Names. Snowshoe hare, varying hare, snowshoe rabbit
Scientific Name. *Lepus americanus*
Subspecies. *L. a. americanus, L. a. bairdii, L. a. cascadensis, L. a. columbiensis, L. a. dalli, L. a. klamathensis, L. a. oregonus, L. a. pallidus, L. a. phaeonotus, L. a. pineus, L. a. seclusus, L. a. struthopus, L. a. tahoensis, L. a. virginianus,* and *L. a. washingtonii* (Hall 1981)

The snowshoe hare originally was described by Erxleben in 1777 from specimens taken from Hudson's Bay, Canada. *L. a. dalli* occurs in northwestern Canada and Alaska and was formerly recognized as *L. a. macfarlani* (Merriam 1900, cited in Bittner and Rongstad 1982). Dalquest (1942) reclassified *L. washingtonii tahoensis* and *L. bairdii oregonus* as *L. americanus tahoensis* and *L. americanus oregonus,* respectively (Hall 1981); lesser changes to the *L. americanus* classification also were described by Hall (1981). Morphometric analyses and pelage color descriptions previously were used to classify snowshoe hares into subspecies. Validity of the existing classification scheme is questionable in light of recent findings showing similarity among cranial features of several recognized subspecies (Nagorsen 1985).

Other Hares. The snowshoe hare is a member of the genus *Lepus,* which includes several species of hares and jackrabbits. Other native North American hares in this genus include arctic (*L. arcticus*) and Alaskan or tundra (*L. othus*) hares. The arctic hare is widely distributed across the tundra of northern Canada and Greenland. Although nine subspecies are recognized (Hall 1981; Best and Henry 1994a), a revision is warranted because of morphometric similarities between subspecies (Baker et al. 1978). Two subspecies of Alaskan hare are found along coastal Alaska (Howell 1936), but the status of Alaskan hare as a distinct species remains in doubt; *L. arcticus, L. othus,* and their European congener, *L. timidus,* may be conspecific (Baker et al. 1983; Chapman et al. 1983; Dixon et al. 1983; Flux 1983; Halanych et al. 1999). European hares (*L. europaeus, L. capensis*) originally were introduced to New England and the Great Lakes region (Dean and de Vos 1965). Although some populations initially thrived, most appear to have remained stationary or become extirpated. Accordingly, the current distribution of European hares in North America is poorly understood, but probably is patchy and highly restricted (Dean and de Vos 1965; Bittner and Rongstad 1982).

DISTRIBUTION

The snowshoe hare is common throughout the boreal and aspen parkland forests of Canada and Alaska (Fig. 7.1). The range extends north to tree line and south into the coniferous forests of the Pacific coast, through the Cascade Mountains of Washington and Oregon, and into northern California and western Nevada (Orr 1933, 1934). The southwestern distribution also includes high-elevation areas in Utah, Colorado, and New Mexico. The midwestern distribution includes northern hardwood–coniferous forests in the Great Lakes region of Minnesota, Wisconsin, and Michigan. Eastern populations occur throughout the Maritime Provinces and New England, and into the mixed forests of the mountains of North Carolina, Virginia, West Virginia, and Tennessee. The central and northern distribution is largely continuous, whereas the discontinuity of forested habitat in the southern range results in disjunct distribution at lower latitudes. Snowshoe hares allegedly were introduced to Newfoundland during 1860–1870 and occurred on most of the island by the end of the nineteenth century (Dodds 1960). Some southeastern hare populations have declined or were extirpated during the last 50–100 years, and attempts at recovery have been largely unsuccessful (Dalby 1985; Fies 1991, 1993). The extent of the recent change in distribution along the southern range boundary is not well known.

DESCRIPTION

The snowshoe hare is a member of the order Lagomorpha and the family Leporidae. Like most other leporids, it is characterized by elongated ears and hind feet (Fig. 7.2). The snowshoe hare's hind foot length (11–14 cm) is particularly large relative to body mass, and this adaptation facilitates travel in snowy environments through reduced foot-loading (Keith 1990; Murray and Boutin 1991; Keith and Bloomer 1993). Although hares resemble rabbits, they differ in terms of skeletal and cranial morphology and the fact that hares give birth to precocial rather than altricial young. Unlike rabbits, juvenile hares are not born in underground burrows, but rather in sheltered areas on the ground surface called "forms." Snowshoe hares typically are 36–52 cm in total length, with mean tail length 2.5–5.5 cm (Banfield 1974; Hall 1981). Body mass of adult snowshoe hares averages 1.3 kg with a range between 0.9 and 2.3 kg, and fluctuates seasonally and annually (Grange 1932a; Rowan and Keith 1959). Female snowshoe hares weigh about 10–25% more than males. During late spring and summer, females become visibly gravid with near-term embryos.

Pelage. During summer, snowshoe hare pelage is grayish-brown or reddish-brown; the nose, forehead, and ear tips are dark brown to black. The chin and nostrils are white, whereas the stomach, tail and limbs are grayish white (Grange 1932b). During winter, the pelage is almost pure white, except for black-tipped ears and grayish feet (Grange 1932b; Doutt et al. 1967). The seasonal change in snowshoe hare pelage color likely is a result of natural selection favoring cryptic coloration, where the timing of the molt is correlated with the occurrence of snowcover (Nagorsen 1983). Two subspecies, *L. a. oregonus* and *L. a. washingtonii,* occur at low elevations in the southwestern range, where snowfall is absent or negligible during most winters. These subspecies retain their brown coloration throughout the year (Dalquest 1942), and may cohabit the same sites as other snowshoe hare subspecies (Nagorsen 1983). It has been speculated that seasonal changes in coloration are detrimental to hares residing along the southeastern range because of the frequent absence of continuous snowcover during winter, which results in their higher visibility to predators (Brooks 1955; Fies 1993). However, the selection for and retention of one or both color morphs at low latitudes remains a fascinating research question in need of closer attention. Rare cases of melanistic snowshoe hares have been reported in New York (Gordon 1954).

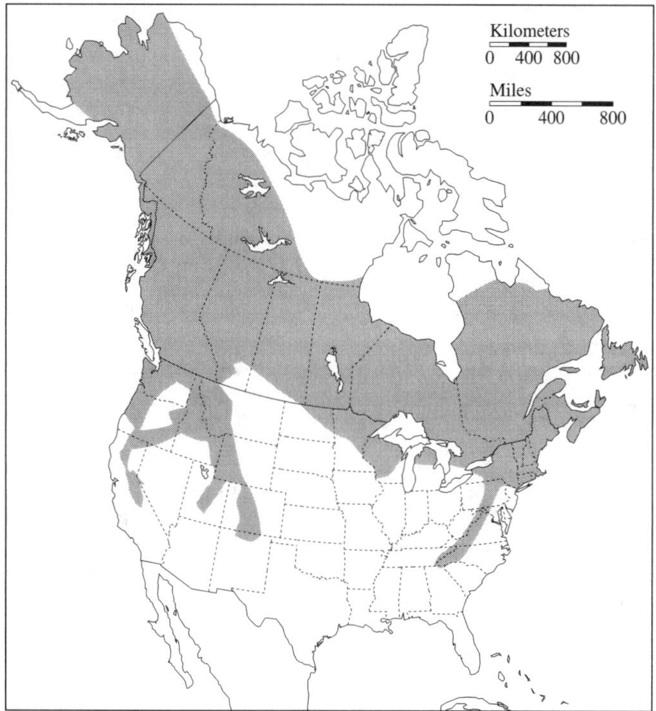

FIGURE 7.1. Distribution of the snowshoe hare (*Lepus americanus*).

FIGURE 7.2. Snowshoe hare (*Lepus americanus*). SOURCE: Photo by J. O. Wolff, Mammal Images Library, American Society of Mammalogists.

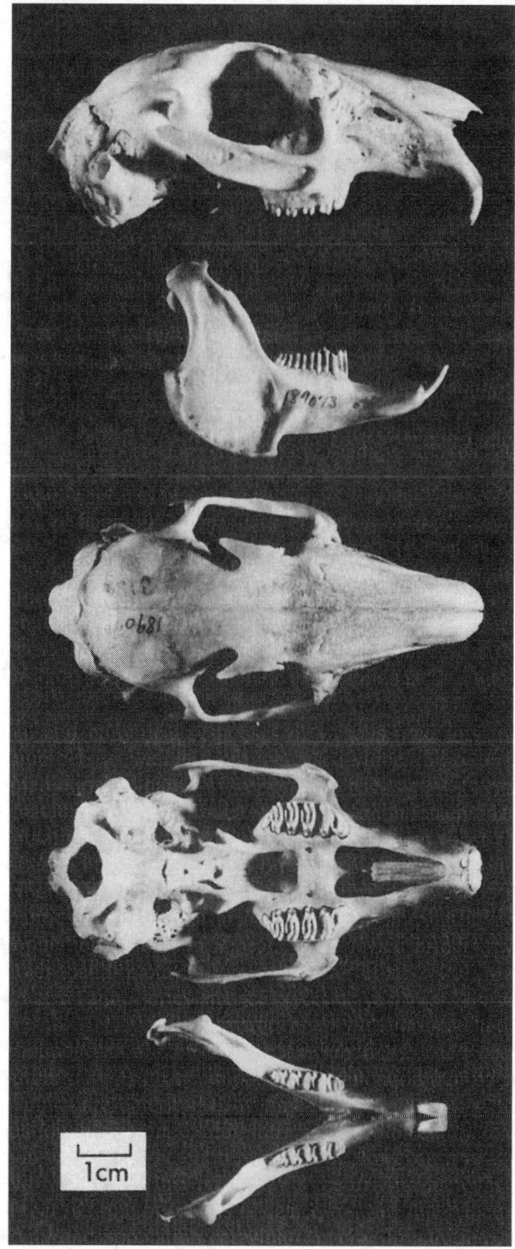

FIGURE 7.3. Skull of the snowshoe hare (*Lepus americanus*). From top to bottom: lateral view of cranium, lateral view of mandible, dorsal view of cranium, ventral view of cranium, dorsal view of mandible.

Dentition. Snowshoe hares have a dental formula that is characteristic of leporids: I 2/1, C 0/0, P 3/2, M 3/3. A second pair of small incisors is found behind the first set. These "peg teeth" occur in all lagomorphs. The cutting edge of the first upper incisors is straight (Fig. 7.3).

Morphology. Twenty-six species of hares are recognized globally; the snowshoe hare is the smallest (Keith 1990). Analysis of morphological variation in snowshoe hares revealed that skull size is largest in northwestern and eastern subspecies and smallest in populations from the southwestern United States (Nagorsen 1985). Size appears to be related to the average primary productivity in an area as well as its inter- and intra-annual variability in productivity. Larger hares occur where winter temperatures are lower, strong seasonality occurs in temperature and evapotranspiration, annual evapotranspiration is high, and July rainfall also is high. Presumably, larger body size in such environments is genetically determined and facilitates coping with increased variability in food resources (Nagorsen 1985), although it is important to note that the above relationships are based exclusively upon post-hoc correlations between hare body size and environmental variables.

PHYSIOLOGY

Molts. Snowshoe hares undergo molts during spring, late summer, and midautumn (Lyman 1943), although major molts occur during spring and fall. The late summer molt provides hares with a more luxuriant pelage. The midautumn molt provides hares with longer, thicker, denser pelage for winter. The vernal molt begins in March–April, and the autumnal molt commences in late September. Molting usually is complete after 70–90 days, and females tend to molt more rapidly than males (Grange 1932b; Severaid 1945a). Photoperiod and corresponding changes in hormonal profiles are closely associated with onset of molting. Snowshoe hare pelage consists of guard hair, pile (underhair), and fur. Molting is controlled by altered levels of pituitary gonadotropins, which are influenced by seasonal changes in photoperiod. Increased levels of gonadotropins result in brown pelage, whereas reduced gonadotropin induces white coloration (Lyman 1943). The autumnal molt follows the following sequence: ears, wrists, and feet; lower rump and tail; side of legs and upper rump; side of the face, chest, and lower back; upper back, shoulder, and crown (Grange 1932b). The chronological sequence of the vernal molt is the same as that of the autumnal molt. Hares may become more secretive before or after molting if pelage color does not correspond with snow/ground cover (Litvaitis 1991).

Metabolism. Snowshoe hare fur provides effective insulation from inclement weather, yet its insulative properties vary seasonally. Fur is approximately 30% more insulative during winter than summer, whereas heat conductance is 35% greater during summer than winter (Hart et al. 1965). Metabolic activities also undergo seasonal changes, with increased metabolism and altered epinephrine and norepinephrine production during winter (Feist and Rosenmann 1975; but see Hart et al. 1965). These changes may be important in regulating shivering thermogenesis because hares appear to be more susceptible to hypothermia during summer. It was estimated that lower critical temperatures are reached at 10°C during summer, compared to −5°C during winter (Hart et al. 1965). Hares may reduce radiant heat loss in winter through use of well-sheltered forms and snow burrows (O'Farrell 1965; Pietz and Tester 1983).

The basal metabolic rate of snowshoe hares was estimated at 0.68 ml O_2/g of body mass per hour, which is equivalent to 105 kcal/day (Irving et al. 1957). In captivity, snowshoe hares allegedly experience reduced oxygen uptake during winter, which apparently coincides with a 16% reduction in caloric intake (Hart et al. 1965). This finding may suggest that in the wild snowshoe hare activity and/or metabolic rates decrease in response to cold temperatures.

Body Mass. Body mass of adult hares undergoes pronounced variation. Adult mass increases during late summer–fall and reaches a peak around November, declining thereafter until spring (Rowan and Keith 1959; Newson and de Vos 1964; Keith and Windberg 1978; Smith et al. 1980). Body mass differential between sexes becomes most pronounced during April–August, when females are pregnant (Keith and Windberg 1978).

Average body mass of adult hares on a site can vary as much as 17% between years (Keith and Windberg 1978). Variations are associated mainly with changes in population density and food availability per capita, with lowest and highest mass occurring during years of population peaks and lows, respectively. The importance of food availability to hare body mass has been further demonstrated through food supplementation experiments using natural browse or commercial rabbit food pellets. Results from such studies indicate that during peaks in hare abundance, body mass is lighter among individuals not receiving supplemental food (Windberg and Keith 1976a; Vaughan and Keith 1981; Boutin 1984a; Krebs et al. 1986b; Smith et al. 1988). In contrast, during periods of low hare densities and presumably adequate natural food availability, or when food was not supplemented in large quantities so as not to create food superabundance, supplemental food elicited negligible changes in body mass (Krebs et al. 1986a; O'Donoghue and Krebs 1992; Murray et al. 1998).

Other factors also may contribute to changes in mean body mass. An experiment designed to evaluate the effects of predation risk on hares (Hik 1995) determined that during a natural population decline caused mainly by predation, average mass of females declined despite an apparent abundance of winter forage. Concurrently, hares found in study areas where terrestrial predators were excluded maintained body mass. The mechanism for such differences seems to be the impact that perceived predation risk has on hare foraging behavior (see Foraging Behavior). Hare body mass also may be related to levels of parasitism (Keith et al. 1985, 1986; Keith and Cary 1990; Murray et al. 1998) (see Parasites and Diseases).

Mass of juvenile hares is similar between sexes until late winter or spring of the year following birth, at which point females become heavier through the influence of pregnancy. Juvenile mass increases rapidly from summer until about November, after which growth rates subside and juvenile mass gradually approaches that of adults (Keith and Windberg 1978; Murray et al. 1998). Juveniles tend to be indistinguishable from adults by March–April of the year following birth, or when they reach about 10 months of age. Juvenile mass may be increased by supplemental feeding when natural food availability is limited (Windberg and Keith 1976a; Vaughan and Keith 1981; Smith et al. 1988), but not when natural food is abundant (O'Donoghue and Krebs 1992; Murray et al. 1998). Juvenile growth rates are influenced by maternal condition, with offspring from adequately fed mothers growing more rapidly than those from underfed females (Vaughan and Keith 1981).

Body Condition. Like most small mammals, snowshoe hares have relatively low body fat content. Thus, reported changes in mass largely reflect changes in protein rather than fat. Fat reserves in snowshoe hares have been estimated at 1–4% of body mass (Litvaitis and Mautz 1976, 1980; Davidson et al. 1978; Powers et al. 1989; Wirsing et al. 2002a), which, when combined with protein reserves, may be sufficient to sustain hares for up to 4 days during winter and 6 days during summer (Whittaker and Thomas 1983; Thomas 1987). Such changes may correspond to 15–30% changes in body mass (Pease et al. 1979). These findings imply that hares are highly susceptible to starvation either if food availability is limited or environmental factors preclude daily feeding (see Mortality Factors). Bone marrow fat, the last fat store to be mobilized before starvation, fluctuates seasonally in hares (Murray et al. 1997). Keith et al. (1984) proposed that during a cyclic decline in numbers, a high proportion of living hares had bone marrow fat and liver glycogen levels indicative of severe malnutrition (see Population Fluctuations). Hares not faced with dramatic numerical declines (and apparently not limited by food) also may experience low bone marrow fat (Sievert and Keith 1985; Rohner and Krebs 1996; Murray et al. 1997).

Body condition of live hares can be determined by scaling body mass to structural size [an index of structural size may be obtained by measurement of hind foot length (Keith et al. 1968; see also O'Donoghue and Krebs 1992)]. Using this index, researchers often deem juvenile hares to be in poor, and pregnant females in good, condition (Hodges et al. 1999b; Murray 2002). Like the annual trends observed in hare body mass, body condition also may deteriorate concurrently with a population decline (Boonstra et al. 1998a; Hodges et al. 1999b), show substantive seasonal variation (Murray 2002), and respond positively to addition of supplemental food (O'Donoghue and Krebs 1992; Murray 2002). As noted, predators may elicit reductions in hare body mass and condition through the effects of perceived predation risk on foraging behavior and stress levels, but such effects appear to be temporally inconsistent (Hik 1995; Boonstra et al. 1998a; Hodges et al. 1999a, 1999b). Alternatively, even small declines in body condition may increase hare vulnerability to predation, likely through compensatory foraging and state-dependent predation risk augmentation (Murray 2002, but see Wirsing et al. 2002b).

REPRODUCTION AND DEVELOPMENT

Anatomy. As with other leporids, the reproductive tract of female snowshoe hares is duplex, and consists of a single vagina, two cervixes, and two uteri. Newson (1964) described the uteri as being ribbonlike (4–6 mm wide) and weighing 0.4–3.3 g. Uteri of parous females are wider, striated, and more flaccid than those of nulliparous individuals. Female hares have eight mammae. Males have paired testes and epididymes. When males become scrotal during late winter, their testes lie anterior to the penis (DeBlase and Martin 1980).

Breeding Season. The snowshoe hare's breeding season may extend up to 7 months (March–September), and the annual reproductive cycle is strongly controlled by photoperiod (Severaid 1945a; Davis and Meyer 1972; 1973a; 1973b). Increasing day length after the winter solstice stimulates production of pituitary gonadotropins (follicle-stimulating hormone, and luteinizing hormone), which initiate breeding condition.

During January or February, male testes descend and activities associated with reproduction are initiated shortly thereafter (Adams 1959; Bookhout 1965a). In Alaska and Idaho, testes recrudescence occurs as late as March (O'Farrell 1965; Wirsing and Murray 2002). After the summer solstice, gonadotropin levels decline and gonadal regression begins. In a Yukon hare population, testes regression was discernible among many individuals by July, and complete for the entire population by late August (O'Donoghue and Krebs 1992). Manipulation of day length in the laboratory results in altered duration of gonadal development and recrudescence (Davis and Meyer 1972), which further emphasizes the linkage between photoperiod and reproductive condition.

Levels of follicle-stimulating hormone correlate with ovulation rates in female hares (Davis and Meyer 1972). Typically, two to four litters are produced per year. Conception tends to be synchronized within populations and immediate postpartum mating is normal. Accordingly, young hares form distinct litter groups in the wild (Keith 1981, 1990), likely as a result of selection for reproductive synchrony and predator swamping (O'Donoghue and Boutin 1995). Onset of breeding in female hares, as indicated by conception dates of first litters from necropsied carcasses, usually occurs in March or April (Grange 1932a; Aldous 1937; Adams 1959; Rowan and Keith 1956) (Table 7.1).

A number of environmental and ecological factors appears to affect the onset of female breeding. Keith et al. (1966) and Murray (2000) surmised that onset of the breeding season was similar across the snowshoe hares' geographic range, or perhaps slightly delayed at higher latitudes or elevations. Levels of illumination, as measured by the degree of cloud cover during midwinter, also may influence timing of reproduction, with overcast weather (and presumably less radiant heat) causing delays in onset of conception (Meslow and Keith 1971; Cary and Keith 1979). Overwinter body mass loss in female hares also is correlated with delays in conception dates of first litters (Vaughan and Keith 1981), implying that maternal nutritional status affects breeding condition. Indeed, availability of overwinter browse may influence onset of reproductive activities and duration of the breeding season. For example, females from food-supplemented populations conceive up to 2 weeks earlier in spring than nonsupplemented controls (Vaughan and Keith 1981; Krebs et al. 1986b). Accordingly, the breeding season varies according to food abundance, which is negatively related to population densities.

For males, breeding condition usually begins in January–February, and may continue until August in some populations. Spermatogenesis was discernible in about 33% of necropsied carcasses collected during February in Michigan, followed by 70% and 100% of those collected during March and April, respectively (Bookhout 1965a). Maximum mass of testes peaked during May in Ontario (Newson 1964) and Michigan (Bookhout 1965a) and June in Newfoundland (Dodds 1965). In a cyclic hare population (Cary and Keith 1979), testes mass showed cyclic changes, which preceded the cycle in hare abundance by about 3 years. Later attainment of male breeding condition, lower peak testes mass, and earlier cessation of breeding in summer all are possible reproductive outcomes of food limitation (Vaughan and Keith 1981; Boutin 1984a; Murray et al. 1998). However, food supplementation experiments do not always elicit male reproductive responses (O'Donoghue and Krebs 1992), which leaves unclear the specific conditions under which food plays a role in determining breeding condition of male hares.

Gestation. Gestation in the snowshoe hare ranges from 34 to 40 days (Severaid 1945b; Meslow and Keith 1971). Median gestation appears to be 37–38 days, as shown in studies from Alberta (Keith et al. 1966), Ontario (Newson 1964), New York (Dell and Schierbaum 1974), and Utah and Colorado (Dolbeer and Clark 1975). Average gestation may vary by several days between years on a given site (Severaid 1945b; Bittner and Rongstad 1982), perhaps in response to overwinter changes in food abundance or other ecological factors.

Number of Litters. Snowshoe hares are seasonally polyestrus and exhibit immediate postpartum breeding (Keith et al. 1966; Bittner and

TABLE 7.1. Reproductive attributes for snowshoe hare populations across their geographic range

Location	Years	Mean Date of First Conception	Litters		Female Productivity		Reference
			First	Later	Litters	Young	
Alaska	2	—	3.72	4.33	1.9	7.6	Tovey unpublished (in Keith 1990)
Alaska	4	April 25	4.19	4.97	2.1	9.7	Ernest 1974
Alaska	3	—	—	4.63	1.7	—	Trapp 1962
Alaska	1	—	—	4.91	—	—	Philip 1939
Alberta	16	April 7	2.74	4.82	3.0	12.7	Cary and Keith 1979
Alberta	6	—	2.82	4.21	3.2	12.1	Keith unpublished (in Keith 1990)
Colorado	3	May 10	3.00	4.75	2.1	8.2	Dolbeer and Clark 1975
Idaho	2	—	2.20	2.40	3.2	5.6	Wirsing and Murray 2002
Maine	4	March 20	2.50	3.40	—	8.2	Severaid 1942
Manitoba	13	—	2.52	4.21	—	—	Criddle 1938
Manitoba	1	—	—	4.08	—	—	MacLulich 1937
Manitoba	2	March 27	—	4.50	—	—	Murray et al. 1998
Maritimes	4	April 17	2.32	3.08	2.5	7.2	Wood and Munroe 1977
Michigan	6	April 12	—	2.48	2.7	7.5	Bookhout 1965c
Minnesota	7	—	2.21	3.28	2.6	7.5	Green and Evans 1940a
Minnesota	5	—	2.04	2.79	—	—	Aldous 1937
Montana	2	March 17	—	—	2.9	—	Adams 1959
Newfoundland	7	April 9	2.40	4.18	3.2	11.6	Dodds 1965
Ontario	3	April 10	2.16	3.17	2.9	8.1	Newson 1964
Ontario	2	—	2.25	3.39	2.8	8.4	MacLulich 1937
Québec	2	April 10	2.20	3.05	2.8	7.6	Alain 1967
Utah	2	April 20	3.56	5.91	2.3	11.5	Dolbeer and Clark 1975
Wisconsin	2	March 13	1.92	3.06	3.8	10.8	Kuvlesky and Keith 1983
Wisconsin	2	—	—	—	3.3	7.0	Keith et al. 1993
Wisconsin	3	April 9	2.25	2.46	2.8	6.4	Kuvlesky and Keith 1983
Yukon	2	April 18	3.70	5.30	—	12.1	O'Donoghue and Krebs 1992
Yukon	8	April 24	3.54	5.55	2.8	13.4	Hodges et al. 2001

SOURCE: Adapted in part from Keith (1990) and Murray (2000).
NOTE: Timing of first conception and reproductive output were calculated as per Green and Evans (1940a) and Cary and Keith (1979).

Rongstad 1982) and these two factors enable hares to achieve high potential natality rates. The reported average number of litters produced per female ranges from 1.9 in Alaska to 3.8 in Wisconsin (Table 7.1). The average number of litters produced in Alberta varied from roughly 3 near a population peak to about 4 during a numerical low (Meslow and Keith 1968; Cary and Keith 1979), suggesting that nutritional status affects litter production. Similar trends were reported for a hare population cycle in the Yukon (Hodges et al. 2001; Stefan and Krebs 2001). Keith et al. (1966) first proposed that number of litters was largest in the central part of the range and declined in peripheral populations. In a later analysis, Murray (2000) determined that the average number of litters per female declined along a westward gradient, but the reason for this trend is unclear.

Pregnancy Rate. Snowshoe hare pregnancy rates are variable during the onset of the breeding season, with ranges of 25–81% of the female population being pregnant during March–April in Minnesota (Green and Evans 1940c) and 11–91% in Michigan (Bookhout 1965a). Pregnancy rates tend to decline with successive litters. Average May–July pregnancy rate for a hare population in British Columbia was 64% (Sullivan and Sullivan 1988). In a cyclic population from Alberta, pregnancy rates for the first two of four possible litter groups ranged from 78% to 100% (Keith and Windberg 1978; Cary and Keith 1979). Pregnancy rates for the latter two litters were similarly high during the population low and increase, but declined to <20% in the case of the fourth-litter group during the numerical peak and decline. In Alberta, fourth-litter pregnancy rates varied cyclically and preceded hare population changes by 3 years (Cary and Keith 1979), whereas in the Yukon, pregnancy rate declines preceded hare numerical changes by 2 years and remained low well into the low phase of the cycle (O'Donoghue and Krebs 1992; Stefan and Krebs 2001; Hodges et al. 2001). Experimental manipulation of food augments pregnancy rates in some (Windberg and Keith 1976a; O'Donoghue and Krebs 1992) but not all (Hodges et al. 1999b) hare populations.

Litter Size. Litter size is notably variable, with the first litter produced each year consistently being smaller than later litters (Keith 1981). The third litter may be slightly larger (Cary and Keith 1979) or slightly smaller (O'Donoghue and Krebs 1992) than the second, but a modest decrease in litter size seems common for the fourth litter (Cary and Keith 1979; Hodges et al. 2001; Stefan and Krebs 2001). Litter size varies across the species range, with mean size of the first litter ranging from 1.9 in Wisconsin to 4.2 in Alaska (Table 7.1). Average size of later litters ranges from 2.2 in Idaho to 5.6 in the Yukon (Table 7.1). Average number of young produced per female during a breeding season also is geographically variable, ranging from 5.6 young per female in Idaho to 13.4 in the Yukon (Table 7.1). Keith and Windberg (1978) found that litter sizes tended to be smaller during years when adults experienced overwinter weight loss, which was related to food availability per capita. Total productivity, a product of pregnancy rate, litter size, and number of litters produced, may change across a snowshoe hare population fluctuation (Meslow and Keith 1968; Cary and Keith 1979) and may be closely associated with female condition during the previous winter. Accordingly, there could exist maternal effects that, when passed onto offspring, result in protracted low levels of production despite environmental conditions that should be conducive to population growth (Krebs et al. 2001a).

Early analyses of the geographic variability in hare litter sizes suggested that they were smaller along the southern periphery (Bookhout 1965a; Keith et al. 1966). Meslow and Keith (1971) reported that litter size was positively correlated with snow depth during the previous winter, likely due to increased access to high-quality food when snow accumulation allowed hares to access elevated browse. Evidence that natural browse availability and/or female nutritional status limits hare litter sizes includes the fact that litter size variation corresponds to cyclic phase, and may in fact precede the cycle (Cary and Keith 1979; Hodges et al. 2001; Stefan and Krebs 2001), and also that litter sizes increase when natural browse is supplemented experimentally (Vaughan and Keith 1981), but not when natural foods are seemingly abundant (O'Donoghue and Krebs 1992). Across the geographic range of the species, litter size may increase along a westward gradient, but the varying methodologies and unknown baseline population trends among research studies renders such analyses tenuous (Murray 2000). However, average number of young produced per female is not strongly related to geographic location (Keith et al. 1966; Murray 2000) (Table 7.1). Although observed differences in productivity among populations may be related principally to local food quality and quantity, genetic differences may contribute to the geographic disparity in litter sizes (Keith et al. 1966).

Prenatal Growth. Prenatal growth of snowshoe hares was studied extensively by early researchers (e.g., Bookhout 1964; Newson 1964) and summarized by Bittner and Rongstad (1982). Briefly, digits are formed 15 days into gestation, nails by day 19, and fat by day 21. During days 12–18, body length and mass increase at 2.7 mm and 0.24 g/day, respectively. Between days 30 and 34, increases in body length and mass average 4.8 mm and 7.0 g/day, respectively (Bookhout 1964; Newson 1964). Embryos are visible in necropsied carcasses after 6–8 days (Green and Evans 1940a; Dodds 1965), and are readily palpable in live animals after about 13 days (Keith et al. 1968).

Neonates. Snowshoe hares are born precocial, fully furred and with eyes open, in a nest consisting merely of a shallow depression on the ground (Aldous 1937). On average, newborn hares (called leverets) weigh 40 g in Montana (Adams 1959), 67 g in Maine (Severaid 1942), and 59–62 g in Yukon (Graf and Sinclair 1987; O'Donoghue and Krebs 1992). In New York, mean neonatal weight is 82 g and total length is 105 mm (range = 97–113) (Dell and Schierbaum 1974). Behavioral development in young hares is relatively rapid compared to that of most other small mammals (Nice et al. 1956). Mean body mass of newborn hares may be negatively related to litter size (O'Donoghue and Krebs 1992; but see Severaid 1945b). Juvenile growth rates average 16–17 g/day for the first 60 days following birth (Severaid 1942; Graf and Sinclair 1987). Juvenile growth rates are strongly influenced by the phase of the hare population cycle, with juveniles from declining populations experiencing lower growth rates than those from increasing populations (Keith and Windberg 1978). Under optimal conditions, hares may attain adult mass as early as 3 months following birth (Aldous 1937; O'Farrell 1965), but typically adult mass is reached after 9–11 months (Keith and Windberg 1978). Maternal condition may influence growth rates in juvenile hares since juvenile mass gain is negatively related to the magnitude of winter mass loss of adult females (Keith and Windberg 1978). Because the negative relationship is manifest through juvenile mass during summer (i.e., alleged period of food superabundance), this relationship is not related to natural food availability per capita, but rather to maternal condition during the previous winter. This conclusion is supported by the observation that food supplementation during summer fails to increase juvenile hare growth rates (O'Donoghue and Krebs 1992; Boonstra et al. 1998b).

Neonatal hares are very secretive, with littermates separating within a few days after birth to occupy hiding cover until fully weaned. This behavior likely reduces leveret predation risk. Mothers and littermates gather once per day at 1–2 hr after sunset to suckle for 2–10 min. After suckling, neonates and mothers separate, with neonates returning to hiding cover (Rongstad and Tester 1971; Graf and Sinclair 1987). Hare milk consists of high concentrations of protein and fat (Baker et al. 1970). Weaning occurs after 25–28 days except in later litters, where nursing may last up to 2 months (Severaid 1942; Rongstad and Tester 1971; Graf and Sinclair 1987).

Breeding Age. Male and female snowshoe hares reach reproductive maturity in the spring following their birth (9–11 months old). Juvenile breeding before the first fall may occur but is uncommon, with only nine reported cases during the course of an intensive 16-year study in Rochester, Alberta (Keith and Meslow 1967; Vaughan and Keith 1980). It is notable that cases of juvenile breeding occurred exclusively in first-litter females and during years with superabundant food following cyclic lows. Onset of reproductive condition in juvenile males may follow the same pattern as for females, with a low proportion of first-litter juveniles possibly showing testes recrudescence in late summer (O'Donoghue and Krebs 1992).

Sex Ratio. In general, the sex ratio in snowshoe hare populations is equal to 1 (Webb 1937; Newson and de Vos 1964; Dolbeer and Clark 1975; Kuvlesky and Keith 1983; Hodges et al. 2001) or else slightly male biased (Aldous 1937; Dodds 1965). An Alberta hare population had a strong female-biased sex ratio during 1 of 3 years (Rowan and Keith 1956). Sex ratios of juvenile hares may be slightly male (Adams 1959; Newson and de Vos 1964; Dodds 1965; Meslow and Keith 1968) or female (Rowan and Keith 1956) biased. Fetal sex ratio appears to be at parity (Newson and de Vos 1964; Dodds 1965).

ECOLOGY

Home Range. Many studies of snowshoe hare home range size have been hampered by small sample sizes or have used livecaptures or unnatural observational information (i.e., flushing hares from cover) to determine location. Also, home range sizes have been measured using a variety of estimators, which could further bias interstudy comparisons. Overall, it appears that home range size of hares is related to several ecological factors, including food and cover availability, conspecific density, age, and perhaps sex. Keith (1990) summarized data on hare home ranges and concluded that average year-round home range sizes were 5.2 and 6.7 ha for adult females and males, respectively; he also noted marked variability in home range size across studies. In Alaska and Utah, home range sizes tended to be similar between sexes (O'Farrell 1965; Dolbeer and Clark 1975), whereas in Montana (Adams 1959), Ontario (Austin 1960), Québec (Bider 1961; Ferron and Ouellet 1992), and Minnesota (Rongstad and Tester 1971), male home ranges were larger than those of females. Home range size increases as juvenile hares become older during late summer–early fall. Limited evidence suggests that home range size of adult females increases during pregnancy (Rongstad and Tester 1971). Home range size also seems inversely related to population density and thus per capita food resources. However, the role of food availability on home range size remains unclear, with conflicting results between food addition studies on the same site at different points in the hare population cycle (Boutin 1984b; Hodges 1999). Extensive overlap in home ranges between individuals is normal, yet home range size may not increase following removal of adjacent conspecifics even though animals may concentrate activities in areas in proximity of the removals (Boutin 1980). These findings suggest that aggressive behavior between conspecifics probably contributes little to home range size determination in snowshoe hares.

Although home ranges may be largely fixed for an individual's life span, home range shifts in excess of 400 m have been reported in several study areas (Adams 1959; O'Farrell 1965; Keith 1966). Hare populations can move seasonally from winter to summer home ranges based on snow accumulation and attendant habitat change (Adams 1959; Bider 1961; Bookhout 1965b, 1965c; Wolff 1980a; Wolfe et al. 1982). Surprisingly, Keith et al. (1993) failed to detect a relationship between hare home range size and habitat quality in northern Wisconsin.

Dispersal. Snowshoe hare dispersal distances may range from several hundred meters (which probably does not constitute true dispersal) up to 20 km (Aldous 1937; O'Farrell 1965; Windberg and Keith 1976b; Gillis and Krebs 1999). Although adults may disperse, juveniles are particularly prone to emigration (Windberg and Keith 1976b; Keith et al. 1984; Boutin et al. 1985). For juveniles, dispersal may occur as early as 1 month of age (O'Donoghue and Bergman 1992), and dispersal pulses for juveniles and adults seem to occur in fall and spring. Both sexes may disperse at similar rates (Windberg and Keith 1976b; Boutin et al. 1985; Burton et al. 2002), but during hare population peaks, male-biased dispersal may be most prevalent (O'Donoghue and Krebs 1992). During a cyclic increase in hare numbers in the Yukon, Gillis and Krebs (1999) found that 50% of juveniles dispersed by February of the year following their birth. In contrast, other studies found overall rates of dispersal (age classes combined) as low as 28% (Boutin et al. 1985) or 22% (Keith et al. 1993) per year. Keith et al. (1993) documented a higher rate of dispersal from small rather than large habitat patches. Juveniles born from later litters may disperse at a later age than do early-litter leverets (Gillis and Krebs 1999).

Dispersing hares tend to weigh less and have more scars and lighter adrenal mass than nondispersers (Windberg and Keith 1976b; Boutin et al. 1985). Thus, either the dispersing cohort consists of substandard individuals or dispersal promotes intraspecific strife with residents (see Aggressive Behavior). Some early reports suggested that population-level dispersal or emigration may occur in response to widespread food limitation (Cox 1936; Criddle 1938; Henshaw 1966), but food supplementation in Yukon failed to mitigate extensive dispersal (Boutin 1984a; Gillis and Krebs 1999; see also Murray 1999). Although in some cases naturally dispersing hares have similar survival rates as nondispersers (Boutin 1984a; Gillis and Krebs 2000), experimental transplant of hares to new habitat to simulate dispersal has caused notable reductions in survival (Sievert and Keith 1985; but see Wirsing et al. 2002c). Such results imply that loss of site familiarity and consequent increases in vulnerability to predation are important detrimental results of dispersal. In the southern portion of their range, hares may disperse at a high rate from areas having poor cover in search of sites with increased cover (Sievert and Keith 1985; Wirsing et al. 2002c). Notably, hares transplanted from their home ranges exhibit homing and often return to their native home range (Keith and Waring 1956).

Dispersal is important to recolonization of habitat where local hare populations have been reduced (Keith and Surrendi 1971; Windberg and Keith 1976b), and thus may contribute to the numerical synchrony observed in cyclic populations. Hodges (2000a), in a literature review, concluded that data do not strongly support the contention that hare dispersal rates vary through a population cycle.

Population Densities. Hare population densities have been estimated using a variety of procedures (see Population Estimation). Densities may vary seasonally, annually, with the hare population cycle, and geographically. In the southern part of the range, high-density populations may exceed 2 hares/ha in spring, whereas low-density populations may have fewer than 0.05 hares/ha (Table 7.2). Spring density estimates are almost always lower than fall estimates for the same study area (Table 7.2), which illustrates the potential demographic significance of juvenile recruitment on hare population size. Minimum densities are considerably lower in declining or stationary populations than in those that are increasing. Fall densities also seem to decline along northward and westward gradients (Murray 2000). Hare densities tend to be higher among food-supplemented populations (Krebs et al. 1986b; 1995; Hodges et al. 1999a), but such experiments do not always result in positive numerical responses (Boutin 1984a; Krebs et al. 1986a; Murray 1999).

Population Fluctuations. The well-documented 8- to 11-year cyclic fluctuations in abundance of snowshoe hares and their predators have served as the focus of extensive debate and research effort during the last half-century (Elton and Nicholson 1942; Cole 1951; Moran 1953a; 1953b; Rowan 1954; Keith 1963; Royama 1992). This debate has centered around the cyclic nature of snowshoe hare and lynx (*Lynx canadensis*) harvest statistics, the spatial influence on cycle synchrony and amplitude, and mechanistic factors and modeling exercises explaining the cycle.

Cyclic hare population trends are broadly synchronous over a vast geographic area from Alaska to central Canada, with numerical peaks tending to spread in wavelike manner over 2–3 years across most of the boreal forest (Keith 1963; Smith 1983; Keith and Rusch 1988; Ranta et al. 1997). The epicenter of such cycles appears to be in west-central Canada, and may have shifted southeast over the last 100 years (Smith and Davis 1981). During the twentieth century, synchrony in the hare population cycle across North America appears to have declined, but the causes of such decoupling and the topic of population synchrony in general, are currently the focus of considerable research efforts. (Stenseth et al. 1997; Blasius et al. 1999; Haydon and Greenwood 2000; Bjørnstad and Grenfell 2001; King and Schaffer 2001).

TABLE 7.2. Snowshoe hare densities (numbers of individuals/ha) from populations across North America

Location	Years	Population Trend[a]	Fall Density	Spring Density	Reference
Alaska	1	D/S	0.20	—	Staples 1995
Alaska	1	D/S	0.17	—	Wolff 1980a
Alberta	1	D/S	0.70	—	Nellis et al. 1972
Alberta	1	D/S	0.34	—	Brand et al. 1976
Colorado	3	I	0.73	—	Dolbeer and Clark 1975
Maine	2	D/S	1.45	1.50	Litvaitis et al. 1985a
Maine	2	D/S	0.55	0.90	Litvaitis et al. 1985a
Maine	2	D/S	0.50	0.80	Litvaitis et al. 1985a
Maine	2	D/S	0.15	0.25	Litvaitis et al. 1985a
Manitoba	2	D/S	0.49	0.74	Murray 1999
Montana	1	D/S	1.61	—	Adams 1959
New Brunswick	4	D/S	0.47	0.89	Wood and Munroe 1977
New York	1	I	—	4.45	Crissey and Darrow 1949
New York	1	D	0.14	0.03	Brocke 1982
New York	3	D/S	0.03	0.02	Richmond and Chien 1976
Nova Scotia	3	D/S	—	0.97	Parker et al. 1983
Oregon	2	D/S	1.45	2.95	Black 1965
Utah	1	D/S	0.46	—	Dolbeer and Clark 1975
Utah	2	D/S	0.36	1.40	Andersen et al. 1980
Wisconsin	2	D/S	1.70	—	Keith et al. 1993
Wisconsin	2	D/S	0.30	0.46	Kuvlesky and Keith 1983
Wisconsin	2	D/S	1.63	1.94	Kuvlesky and Keith 1983
Yukon	1	D/S	0.10	—	O'Donoghue et al. 1997
Yukon	1	D/S	0.20	—	Ward and Krebs 1985

SOURCE: Adapted in part from Keith (1990) and Murray (2000).

NOTE: Data from cyclic populations were obtained during numerical lows. Spring density refers to mark–recapture estimates including only adults, whereas fall estimates also include juveniles of the year.

[a]D/S, Declining/stationary; I, increasing.

Reported amplitude of the hare cycle in the Yukon during the early 1990s was 26- to 44-fold (Boutin et al. 1995), whereas the cyclic amplitude on the same area during the early 1980s ranged from 108- to 182-fold (Krebs et al. 1986b). In Alberta, cyclic amplitude ranged from 17- to 50-fold (Keith and Windberg 1978). More southerly populations in the Great Lakes area also may be cyclic, but with amplitude <15-fold (Green and Evans 1940a; Marshall 1954; Brooks 1955). Recent attenuation in amplitude of snowshoe hare harvest statistics may be indicative of cycle stabilization in some portions of the south-central range (Keith and Rusch 1988; Keith 1990; but see Hodges 2000a).

Hare populations in southwestern Canada and northwestern United States may not cycle or else cycle, but with reduced amplitude. A fundamental problem associated with determining the specific nature of hare population trends in the southern range is related to the paucity of long-term data sets of hare abundance (Keith 1981, 1990; Murray 2000). Howell (1923) examined fragmentary records and concluded that western U.S. populations were largely stationary. Later, the efforts of Helen and Dennis Chitty and colleagues (Chitty and Elton 1937; Chitty 1938, 1939, 1940; 1943a, 1943b, 1946, 1948, 1950; Chitty and Chitty 1942; Chitty and Nicholson 1943) focused mainly on trapper questionnaires and suggested that some hare populations in the southern range cycled, but with reduced amplitude. A statistical reanalysis of these data (Smith 1983) concluded that hare populations in southern British Columbia and parts of the Maritime Provinces exhibited numerical fluctuations that failed to follow a repeatable pattern. In contrast, a recent index of hare abundance in Montana is suggestive of a cyclic fluctuation in numbers with a modest amplitude and roughly 7-year period (Malloy 2000). In an attempt to summarize the current state of knowledge regarding southern hare population trends, Hodges (2000b) conducted a qualitative review of existing time series and concluded that populations at lower latitudes were weakly cyclic. However, from a statistical perspective it remains difficult to demonstrate cyclicity when numeric variability is low and time series data sets are short in length. Clearly, the enigma of whether or not southern hare populations undergo truly cyclic fluctuations in numbers remains to be fully addressed.

Causes of Population Cyclicity. Numerous hypotheses have been proposed to explain the causes of the snowshoe hare cycle. These may be classified into exogenous versus endogenous causes of hare population regulation (Keith 1963; Royama 1992) (Table 7.3). Early hypotheses tended to favor exogenous factors, probably because early researchers considered that biological systems were so complex as to rule out the chance of an intrinsically produced cycle (Keith 1963). Accordingly, snowshoe hare population cycles were attributed to periodic fluctuations in sunspots, ultraviolet rays, or weather patterns, through the effects of such factors on climate, plant production, and food availability (Table 7.3). Ultimately, each of these hypotheses was rejected because there was limited periodicity in the exogenous factors in question and poor correlation between the factors and hare numbers. However, there remains the possibility that exogenous factors synchronize, entrain, or further destabilize hare cycles which are driven by endogenous forces (Moran 1953b; Cole 1954; Sinclair et al. 1993; Sinclair and Gosline 1997). However, the linkage between such factors and hare demographic change remains to be clearly demonstrated (Krebs et al. 2001a).

Central to the current theory of animal population cycles is the idea that endogenous (i.e. density-dependent) factors act in a time-delayed manner to suppress survival and/or reproduction and thereby cause stable numerical oscillations (May 1975; Hassell et al. 1976). Accordingly, recent research efforts on snowshoe hares have focused on identifying factors that may cause such delayed density-dependent regulation (Table 7.3). Despite some early reports of hare population declines being caused by infectious or noninfectious diseases, later studies have failed to identify disease as a prevalent cause of either direct mortality (Keith et al. 1984; Boutin et al. 1986; Krebs et al. 1995) or lower production (Cary and Keith 1979; Sovell and Holmes 1996; Murray et al. 1998). However, in theory, sublethal parasitism may destabilize hare populations and thus contribute to numerical cyclicity through increased predation risk (Gilpin 1973; Ives and Murray 1997), although empirical support for this hypothesis is scant (see Parasites and Diseases).

Christian's (1950) hypothesis regarding fluctuations in physiological stress effects has not received widespread support for snowshoe

TABLE 7.3. Hypotheses explaining the snowshoe hare cycle

Hypothesis	Explanatory Factor	Reference
Exogenous		
Sunspots	Cycles in plant productivity associated with sunspots	Elton 1924, Sinclair et al. 1993
Weather	Changes in habitat quality due to weather	Grange 1949, Butler 1953
Ultraviolet rays	Changes in plant succession/composition related to ultraviolet radiation	Rowan 1950
Forest fires	Changes in plant succession/composition related to fire	Grange 1965, Fox 1978
Plant nutrients	Naturally cyclic variation in plant composition	Lauckhart 1957
Endogenous		
Infectious disease	Epidemics in high-density populations	MacLulich 1937
Noninfectious (shock)	Epidemics in high-density populations	Green and Larson 1938, Green and Evans 1940a, 1940b, 1940c
Physiological stress	Stress in high-density populations reduces fitness	Christian 1950
Polymorphic behavior	Spacing behavior in high-density populations reduces fitness	Chitty 1967, Krebs et al. 1992
Food abundance	Overbrowsing in high-density populations	Lack 1954
Induced plant defenses	Herbivory induces plant defense and causes food limitation	Fox and Bryant 1984, Bryant et al. 1985
Predation	Elevated predation in high-density populations	Moran 1953a, Trostel et al. 1987, Royama 1992
Food–predation	Interactive effects of winter food shortage and predation	Keith 1974, 1981, Krebs et al. 1995; Sinclair et al. 2000

SOURCE: Adapted in part from Keith (1963) and Royama (1992).

hares, but Boonstra and Singleton (1993) and Boonstra et al. (1998a, 1998b) expanded on this theme to posit that predator-induced stress during population declines/lows could further depress hare fitness and thus contribute to population cyclicity. Because free-ranging hares are not known to demonstrate spacing behavior or intensify agonistic interactions at high population densities, genetically determined self-regulating mechanisms consistent with the Chitty hypothesis are unlikely to drive their numerical cycles (Krebs 1986, 1996). Despite this, agonistic interactions between captive hares may intensify at high population densities and thereby cause food limitation. This finding implies that in the wild, food limitation may be elicited through social interactions (Quenette et al. 1997). However, it also is notable that hare populations undergo substantive genetic change during the course of a cycle, likely the result of selective forces favoring particular demographic traits (Lidicker et al. 2000). Accordingly, hare populations doubtless are subject to varying but substantive selective pressures, which, may contribute to numerical fluctuations.

Declining hare populations may experience food limitation and/or deterioration in food quality (Pease et al. 1979; Keith et al. 1984; Bryant et al. 1985). Such changes are not likely to explain hare population cyclicity alone, however, because food limitation during declines is not consistent geographically and declining plant quality may only be manifest after, rather than immediately before, a population decline (Sinclair et al. 1988a; Smith et al. 1988 but see Bascompte et al. 1997). Nonetheless, poor food quality may contribute to hare population dynamics by extending the low phase of the cycle (Fox and Bryant 1984; Krebs et al. 2001a). Observations of Gilpin (1973) and Weinstein (1977) on the premature decline in predator numbers while hare densities remain high have been rejected as evidence for the absence of predator regulation (and by inference indicative of possible food regulation) because hare and lynx data sets often came from different jurisdictions and thus their synchrony could not be compared (Finerty 1980).

Trostel et al. (1987) and Royama (1992) concluded that the time-delayed numerical responses of facultative and obligate hare predators alone were sufficient to drive hare population cycles, even in the absence of food shortage. However, observational (Keith et al. 1984) and experimental (Krebs et al. 1995; Hodges et al. 1999a; Krebs et al. 2001a, 2001d) work strongly suggests that food limitation and predation act in concert to destabilize hare populations and cause numerical oscillations. The details of this apparent interaction are subject to varying interpretations: Keith's (1974, 1990) conceptual model of the hare cycle invoked the deleterious effect of food limitation on hare reproduction at peak densities, combined with increased starvation rates during the early decline phase. It follows that predation rates also become aggravated during the decline and early low phase due to the delayed functional and numerical responses of various predators to changes in hare numbers as well as from increased vulnerability to predation among malnourished hares. More recently, Krebs et al. (1995, 2001a) argued that the relationship between food limitation, predation, and hare population demography may occur throughout most of the numerical cycle, but that predation probably supercedes food in terms of contributing to cyclic dynamics. Indeed, the snowshoe hare system dynamics are such that at both the plant and herbivore trophic levels population regulation appears to be levied principally by top-down rather than bottom-up forces (Sinclair et al. 2000). Mathematical models and time series analyses (Akçakaya 1992; Stenseth 1995; Stenseth et al. 1997; 1998; Blasius et al. 1999; Bjørnstad and Grenfell 2001; Choquenot et al. 2001; King and Schaffer 2001; Ruesink et al. 2002; Turchin 2003) further support the contention that hare population dynamics are determined by time-delayed regulation from food and predation, but again, the full intricacies of this complex tritrophic interaction have yet to be entirely revealed. However, it is apparent to most field researchers that predation is the driving factor in hare cycles and that the role of food is relegated to affecting principally hare productivity and population carrying capacity. In addition, predation may act to synchronize hare population cycles over broad geographic regions through the long-distance migration of predators to new areas following local hare population crashes (Krebs et al. 2001a). However, a similar synchronizing role of food also has been proposed (Bascompte et al. 1997).

Causes of Population Stabilization. The predominant hypothesis for the lower densities and increased stability observed among southern hare populations is related to the effects of strong and consistent predation by a suite of facultative and generalist predators, combined with the greater patchiness of suitable hare habitat (Wolff 1980a, 1981). Under this model, hares are heavily and continuously preyed on by several predator species which are less abundant (or more numerically variable) at higher latitudes (see Mortality Factors). Southern predators can rely extensively on alternate prey and thus do not respond numerically to hare population declines as they do at higher latitudes. Thus, predators should exert a more consistently heavy rate of predation on southern hare populations. It follows that greater discontinuity of suitable hare habitat in the southern range causes dispersing hares to experience high predation, thereby further limiting population increase and expansion.

Empirical support for this hypothesis comes from short-term studies showing restricted movements and high mortality among hares residing in small patches of suitable habitat along the southern range boundary (Dolbeer and Clark 1975; Buehler and Keith 1982; Sievert and Keith 1985; Keith et al. 1993). However, two studies (Keith et al. 1993; Wirsing et al. 2002c) have failed to detect higher predation rates among naturally dispersing individuals, leading to speculation that high on-site predation in suitable hare habitat, instead of high predation rate among dispersers, may be the principal population-stabilizing mechanism. Alternatively, the possibility that overwinter food abundance or food quality limits hare populations at lower latitudes is not supported, given the similar reproductive outputs observed between populations in suitable and marginal habitat (Kuvlesky and Keith 1983) and across a latitudinal gradient (Murray 2000). However, the notably low reproductive rates reported for some southern hare populations (Wirsing and Murray 2002) begs for more extensive assessment of the demographic implications of hare productivity at low latitudes. In fact, because northern and southern hare populations do not exhibit clear disparity in survival and/or reproductive rates, it is unclear which demographic factor is responsible for the alleged difference in population dynamics across the geographic range (Murray 2000).

HABITAT

Snowshoe hares occupy a variety of habitats across their geographic range, including boreal forest, aspen parkland, mixed deciduous forest, coniferous forest, commercial tree plantations, orchards, and shrub areas (Wolff 1980a; Pietz and Tester 1983; Hodges 2000b). The mechanism apparently driving/underlying hare habitat choice is the preference for areas with substantial vegetative structure (Buehler and Keith 1982; Orr and Dodds 1982; Wolfe et al. 1982; Carreker 1985; MacCracken et al. 1988). Vegetative structure may consist of shrubs and/or immature trees, but must provide thermal shelter and particularly hiding cover from predators. Understory is the primary source of cover that is relevant to hares (Bookhout 1965c; Wolff 1980a; Carreker 1985), but to a lesser extent overstory cover also may influence hare habitat use (Wolfe et al. 1982). Such cover may be visually obtrusive primarily in either horizontal (Brocke 1975; Wolfe et al. 1982; Carreker 1985) or vertical (Wolff 1980a; Pietz and Tester 1983) dimensions, perhaps depending on whether risk of predation at a particular site is mostly from terrestrial or avian predators. The importance of dense vegetative cover for concealment from predators is highlighted by the high predation rate and/or high dispersal rate observed among animals transplanted to marginal habitat lacking cover (Sievert and Keith 1985; Barta et al. 1989; Wirsing et al. 2002c), or the preponderance of hare mortalities occurring in open rather than closed cover (Rohner and Krebs 1996).

Snowshoe hare cover requirements are best met among forests that are in intermediate successional stages, typically those that are 20–25 years of age (Cook and Robeson 1945; Dodds 1960; Parker 1984; 1986; Litvaitis et al. 1985a; Koehler 1990a; Bendell and Bendell-Young 1993). Estimated stand density required to provide hares with adequate cover is reported to range from 11,590 to 33,210 stems/ha (Brocke 1975; Wolff 1980a; Parker 1984; 1986; Litvaitis et al. 1985a; Monthey 1986; Koehler 1990a), although such estimates are based largely on qualitative analyses of habitat preference. It has been well documented that hares generally avoid open areas lacking vegetative structure (Gashweiler 1959; Pietz and Tester 1983; Murray et al. 1994), and hares occupying treeless subarctic ecosystems are restricted to riparian areas with abundant shrub cover (St.-Georges et al. 1995). Hares may use small clearcuts for travel or feeding if intact stands are nearby (Conroy et al. 1979; Monthey 1986; Scott and Yahner 1989; Thomas et al. 1997). Snowshoe hares also occur in ecotones between various habitat types (Ferron and Ouellette 1992), and may return to previously burned habitat shortly after the emergence of new growth (Gashweiler 1970; Keith and Surrendi 1971).

Selection by hares for stands with tall understory plants also may be related to local snow accumulation during winter. In New Brunswick, hares select habitats with understory cover height ranging from 1 to 3 m, whereas areas dominated by understory height <1 m are not selected because such heights typically are exceeded by snow (Parker 1986).

Studies across the hare's range have determined that they prefer habitats dominated by coniferous vegetation (Severaid 1942; de Vos 1964; Bookhout 1965b, 1965c; Orr and Dodds 1982; Buehler and Keith 1982; Krenz 1988), whereas others suggest that deciduous cover is selected (Bailey 1946; Brooks 1955; Tompkins and Woehr 1979; Wolfe et al. 1982). Deciduous cover may be preferred during summer because of the role of leaves in reducing hare visibility to predators. It also appears that hares select habitats where food resources are clumped and cover is uniformly distributed within an individual's home range (Krenz 1988).

In most cases, forage quantity is correlated with hiding cover availability. Exceptions may occur if mature trees prevent growth of smaller food species, and consequently forage becomes most abundant in relatively open habitat (Adams 1959; Richmond and Chien 1976; Orr and Dodds 1982; Hik 1995). However, numerous studies have shown that selection for dense cover occurs regardless of plant species composition (Wolff 1980a; O'Donoghue 1983; Litvaitis et al. 1985a; Parker 1986; Rogowitz 1988; Ferron and Ouellette 1992; Thomas et al. 1997). Thus, the dominant force in habitat choice by hares surely is cover rather than food abundance, which implies that hares will forsake access to browse for the safety provided by vegetation. In Pennsylvania, hares selected areas with substantive amounts of coarse woody debris because of the cover it provided (Scott and Yahner 1989).

Habitat use patterns by hares may shift in association with numerical changes in their abundance. Keith (1966) observed that hares remaining after a population decline were restricted to dense spruce forest stands to the exclusion of habitats providing less food. Similarly, Wolff (1980a), Fuller and Heisey (1986), and Hik (1995) observed that hares favored the thickest forest habitats when numbers were low, but expanded their distribution to include marginal habitat with sparser cover as densities increased. It may be that males use habitats with denser understory cover, whereas females occupy sites with greater forage availability (Litvaitis 1990). Juveniles may use open habitat more extensively than adults (Dolbeer and Clark 1975), which could partly explain the heavier predation mortality typically incurred by the former cohort (see Juvenile Mortality Rates).

FEEDING HABITS

Hodges (2000b) reviewed in detail snowshoe hare feeding habits across the geographic range of the species. Hares are herbivorous, and rely mainly on leafy vegetation during summer and woody browse during winter. Hares begin feeding on leafy foods shortly after spring greenup and remain on this diet until after the first frost, at which time woody vegetation becomes the predominant food type. In both summer and winter, hares are selective generalists and thus use a variety of plants while showing distinct preferences for species of substantive nutritional value (Tahvanainen et al. 1991).

Summer Diet. During summer, hare diets include such foods as grasses (Graminae), sedges (Cyperaceae), ferns (Polypodiaceae), and forbs (Aldous 1936; Dodds 1960; Bider 1961; Mozejko 1971). Where available, clover (*Trifolium* spp.), lupine (*Lupinus latifolus*), dandelion (*Taraxacum* sp.), jewelweed (*Impatiens biflora*), marsh marigold (*Caltha palustris*), and the energy-rich flowers of spotted cat's ear (*Hypochoeris radicata*) are among the species that are consumed during summer (Grange 1932a; Dodds 1960; Bider 1961; Radwan and Campbell 1968; Mozejko 1971). In Alaska, summer diet consists mainly of sedges and other herbaceous plants as well as leaves from birch (*Betula* spp.) or willow (*Salix* spp.) trees (Klein 1977, Wolff 1978, Suominen et al. 1999). There is no evidence suggesting that hare populations are subject to food limitation during the summer, and in the Yukon, summer foods appear to be largely resource rather than herbivore regulated (John and Turkington 1997). However, when hare numbers are at peak densities, hare herbivory may substantially affect the growth and survival of establishing grasses and sedges (Dlott and Turkington 2000).

Winter Diet. Winter diet varies considerably according to local plant species composition. Also, the more limited array of foods available to hares during winter necessarily results in narrowed dietary breadth.

In Nova Scotia, winter diets included principally blueberry (*Vaccinium* spp.), maple (*Acer* spp.), and balsam fir (*Abies balsamea*) in one study area (Dodds 1987), and spruce (*Picea* spp.), birch, and maple in another (Telfer 1972). In Newfoundland, winter diet consists of birch, fir, and spruce (Dodds 1960). In Pennsylvania, the winter diet consists of raspberry (*Rubus* spp.), maple (*Acer pennsylvanicum*), and birch (*B. alleghaniensis*) (Brown 1984; Scott and Yahner 1989). Appalachian hare populations appear to rely extensively on cranberry (*Vaccinium erythrocarpum*) (Brooks 1955), whereas those in Oregon select disproportionately red huckleberry (*V. parvifolium*) (Mazejko 1971). In Minnesota, birch, willow, and aspen formed the bulk of winter–spring snowshoe hare diets (Aldous 1936), whereas in Ontario, maple, aspen, prune (*Prunus* spp.), shepherdia (*Shepherdia canadensis*), and hazel (*Corylus cornuta*) were heavily used, and larch (*Larix laricina*) and spruce were only lightly browsed (de Vos 1964). In Idaho, hare winter diet included mostly pines (*Pinus contorta, P. ponderosa*) and firs (*Abies lasiocarpa, A. grandis, Pseudotsuga menziesii*) (Wirsing and Murray 2002). In Alaska, hare populations consumed mainly twigs and needles from spruce, fir, and willow, with modest consumption of forbs and grasses (Wolff 1978, 1980a). In the Yukon, hares during a peak in abundance consumed several species of browse, including alder, birch, willow, and spruce (Smith et al. 1988). Some hare populations from Québec had a markedly diverse diet, which included several ornamental species found in plantations (Bergeron and Tardif 1988).

Overall, hares tend to prefer browse from deciduous rather than coniferous species (Bookhout 1965c; Klein 1977; Bryant and Kuropat 1980; Bryant 1981b). However, in areas where deciduous browse is limited, young pine (*P. resinosa, P. banksiana, P. strobus*) or cedar (*Thuja occidentalis*) are favorite foods (Aldous and Aldous 1944; Cook and Robeson 1945; Hough 1949; Krefting and Stoeckler 1953; Krefting 1975; Wirsing and Murray 2002), whereas young spruce are avoided due to their unpalatability (Sinclair and Smith 1984a; Rodgers and Sinclair 1997).

Hares browse twigs <4 mm in diameter, although twigs up to 1 cm may be browsed and saplings and mature trees may be girdled if food availability is limited (de Vos 1964; Vowles 1972; Wolff 1980a; Pietz and Tester 1983). Hares experience lower digestibility and thus fail to maintain weight on coarser browse. Among captive hares, daily energy requirement is about 300 g of browse per day, which for some species may correspond to a standing crop biomass of about 3000 g of plant material (Mautz et al. 1976; Bookhout 1965b; Pease et al. 1979). Dietary requirements of hares are about 6% of their body mass per day when fed ad lib amounts of a high-protein diet, which represents minimum daily needs of 110 kcal/$W^{0.75}$ of energy or 2.6 g/$W^{0.75}$ of digestible protein, where W = weight in kilograms (Holter et al. 1974).

Browse Depletion. Hare food abundance on a particular site remains difficult to measure because of the variability in hare feeding patterns and the disparity in browse quality. Intensity of winter browse consumption tends to be closely associated with hare density, with low-density populations consuming only modest (2–4%) amounts of the total available browse biomass in Michigan (Bookhout 1965b, 1965c), Minnesota (Krefting 1975), Pennsylvania (Scott and Yahner 1989), Alberta (Pease et al. 1979), and the Yukon (Smith et al. 1988). In contrast, high-density populations can remove considerable food biomass overwinter. In Newfoundland, 21–81% of stems <2 m above ground were consumed during winter (Dodds 1960). Browsing intensity of favored plant species reached 80–90% in the Yukon (Smith et al. 1988) and 100% in Alaska (Wolff 1980a) during years of population peaks. Similar results occurred for Alberta hare populations during cyclic peaks and early declines (Windberg and Keith 1978b; Pease et al. 1979; Keith et al. 1984). Surprisingly, during a cyclic decline in the Yukon, exclosure experiments revealed that the overall effects of hare herbivory on browse biomass were modest (John and Turkington 1995; Turkington et al. 2002), and in the Great Lakes area, browse depletion during a cyclic peak was only severe for a small proportion of the available plant species (de Vos 1964; Bookhout 1965b). Nonetheless, experiments simulating intense hare browsing on willow (Bryant 1987; McAvinchey 1991), which is typically observed in Alaska during cyclic highs, have shown that such herbivory can alter substantially habitat structure, composition, and quality. Furthermore, belowground root production also may be curtailed by actual or simulated hare herbivory (Ruess et al. 1998).

Food Competition with Other Species. Species composition and size and height classes of browse consumed by snowshoe hares may cause dietary overlap and potential competition with moose (*Alces alces*) (Dodds 1960; Wolff 1980b) or deer (Hough 1949; Telfer 1972). However, hares and ungulates do not always compete for food despite considerable dietary overlap, either because of food superabundance or differential use of browse height classes (Bookhout 1965b, 1965c; Oldemeyer 1983). In some areas, hares may benefit from porcupine (*Erethizon dorsatum*) feeding activities by consuming unbrowsed twigs that have been clipped by the latter species and have fallen to the ground (Ferguson and Merriam 1978).

Dietary Preferences. Hare dietary preferences may be related to a variety of factors, including plant composition, palatability, and digestibility. Typically, winter browse has <10% protein content (Walski and Mautz 1977; Pease et al. 1979; Pehrson 1984; Rogowitz 1988; Wirsing and Murray 2002), which apparently is inadequate to maintain hare body mass (Sinclair et al. 1982; Sinclair and Smith 1984b). Intermediate levels of crude protein in the diet (16–20%) are associated with maximum digestibility in hares (Holter et al. 1974), and free-ranging animals may forage selectively on specific plant parts or plant species so as to maximize protein intake. Highest protein content tends to occur in the buds and finer twigs of deciduous woody browse (Grigal and Moody 1980).

Winter browse also may have low energy (<6.0 kcal/dry g) and high fiber (27–32%) content, both of which could contribute to food being a limiting resource (Walski and Mautz 1977; Pease et al. 1979; Pehrson 1984; Rogowitz 1988). In the Yukon, hares prefer browse with higher digestibility of dry matter and energy (Rodgers and Sinclair 1997). The energy, ash, and phosphorous contents of hare browse tend to decline, whereas fiber content tends to increase, with increasing stem diameter (Grigal and Moody 1980). Thus, hares can consume less small material to obtain the same quantity of energy and nutrients, which is why they select small-diameter browse.

A suite of plant secondary compounds may act as feeding deterrents or be toxic to snowshoe hares. The most widely studied groups have been carbon-based phenolics, which are the most prevalent defensive substances produced by the deciduous plants on which boreal hare populations subsist, and terpenes, which are contained primarily in the defensive secretions of evergreens (Bryant et al. 1983a; 1983b; Sinclair et al. 1988b; Jogia et al. 1989; Reichardt et al. 1990a; 1990b; Bryant et al. 1991a; 1991b; Hjalen and Palo 1992; Lewinsohn et al. 1993; Sinclair et al. 1996). Feeding experiments have shown that the presence of these compounds may influence hare feeding preferences (Bryant 1981a; 1981b; Rodgers and Sinclair 1997) and, moreover, that hares will voluntarily reduce their food intake rates well below maintenance levels when fed browse heavily defended by these chemicals (Reichardt et al. 1984; Sinclair et al. 1988b). It follows that in the field, hares may be forced to avoid defended species unless alternatives are rare (Sinclair and Smith 1984a; Reichardt et al. 1990a; Schmitz et al. 1992).

Captive hares fed highly defended browse species have been shown to experience a nitrogen deficit due to the metabolic costs associated with detoxifying, as well as the antimicrobial activity of, certain secondary compounds (Bryant and Kuropat 1980; Reichardt et al. 1984; Sinclair et al. 1988b). The resulting losses of sodium and nitrogen in urine (Reichardt et al. 1984) are indicative of disrupted kidney function and energetic stress. Other diets may elicit loss of sodium and potassium (Pehrson 1983a, 1983b); because hares tend to occupy sodium-poor habitats, these losses potentially may be demographically significant (Smith et al. 1978). Thus, under conditions of natural browse limitation or low browse diversity, free-ranging hares may incur notable fitness costs because they necessarily must select a diet made up largely of heavily defended foods. The predicament of free-ranging hares under these conditions may be exacerbated by the fact that heavy browsing can induce increased production of defensive compounds, a phenomenon that may limit herbivory on either new growth or previously browsed mature stems that have subsequently reverted to their juvenile form (Klein 1977; Bryant 1981a; Bryant et al. 1985). However, certain defensive compounds actually may stimulate herbivory, perhaps because they are used by hares in selecting individual plants with limited toxicity or adequate nutrient content (Tahvanainen et al. 1991). Nitrogen fertilization of plants may stimulate hare feeding (Sullivan and Sullivan 1982b; Nams et al. 1996).

Clearly, woody plant resistance to herbivory is heritable (Dimock et al. 1976; Silen et al. 1986; Rousi et al. 1991). It follows that the role of inducible plant defenses in limiting snowshoe hare feeding patterns should be accentuated in the boreal forest rather than along the southern range of the species's distribution because of the lesser intensity of hare herbivory, and thus reduced selection for chemical defense, at lower latitudes. This supposition has been supported empirically; browse from lower latitudes is more palatable to hares than that from northern areas (Bryant et al. 1989, 1994; Swihart et al. 1994), and inducible defenses are less important in affecting hare browse choice at lower latitudes (Wirsing and Murray 2002). It also appears that at higher latitudes hares have coevolved at the regional scale with local plant populations to more effectively detoxify heavily defended browse (Bryant et al. 1989).

Many plant species appear to be largely unpalatable to hares, apparently due to the overwhelming toxicity of their chemical defenses (Bookhout 1965c; Keith et al. 1984; Carreker 1985). One such species, black spruce (*P. mariana*), may be consumed extensively by hares after fire has charred its bark and denatured its resins and waxes (Stephenson 1985). Constitutive plant defenses, such as cellulose, lignin, pectin, and silica, further limit hare herbivory in some browse species, although they may also promote gut motility and prevention of enterotoxemia (Cheeke 1983, 1987). Hares digest dietary fiber through microbial fermentation, which takes place in the enlarged cecum.

Hares are cecotrophic (also called coprophageous), and produce soft, mucus-coated fecal pellets (cecotropes), which are picked directly from the anus and reingested (Bookhout 1959). Cecotropes are the result of bacterial fermentation in the cecum, and contain both forage and bacterial protein rich in essential amino acids and certain B vitamins. Cecotropes usually are produced during daylight hours (Hansen and Flinders 1969), and up to 50% of snowshoe hare stomach contents may be of fecal origin (Taylor 1939; Spencer 1955). Among mountain hares (*L. timidus*), ingested cecotropes may constitute up to 40% of the total nitrogen intake (Pehrson 1983c), which implies that cecotrophy enables leporids to maintain nitrogen balance while feeding on low-protein food (Cheeke 1987; Fraga 1998). Plant parts that are resistant to digestion and of low nutritive value move rapidly through the gut and are excreted as hard fecal pellets. Protein intake also may be increased through feeding on animal carrion (Brooks 1955; Keith and Meslow 1966).

Hares meet their water requirements mainly by consuming leafy vegetation in summer, although in captivity, hares fed a diet of herbaceous material may drink up to 250 ml of water per day (Severaid 1945b; Dodds 1960; Hansen and Flinders 1969). In winter, water requirements are met by consuming snow. In mountain hares, increased water intake accompanies increases in food consumption (Pehrson 1983a, 1983b).

BEHAVIOR

Mating Behavior. Because hares are difficult to observe in the wild, our understanding of the species's mating behavior is derived largely from captive animals. Forcum (1966) described hare courtship as repeated approaches to prospective female mates by males, followed by several jumps in the air by individuals of either sex. Urination by one individual on the other seems to be a consistent characteristic of the hare mating ritual, as is chasing and running activity (Bookhout 1965a; Forcum 1966). Females do not build nests, nor do males partake in any aspect of parental care (Grange 1932a; Severaid 1942; Graf and Sinclair 1987).

Aggressive Behavior. Although hares tend to be solitary, aggressive behavior can occur particularly during the breeding season or when densities are high such as at feeding sites (Grange 1932a; Quenette et al. 1997). Aggressive displays and interactions typically involve chasing, jumping, and body collisions, which may culminate in scratching and biting. Males are most dominant over conspecifics during winter, whereas females are dominant during summer (Graf 1985; Fitzgerald and Keith 1990; Ferron 1993). Nonresident hares introduced to a captive setting often are involved in agonistic interactions with residents, usually over access to food, water, or resting sites, or during encounters along runways; nonresidents typically lose about 90% of such interactions (Graf and Sinclair 1987). Antagonistic interactions between captives are more prevalent when food is limited (Ferron 1993), leading to reductions in body mass among the nonresident cohort (Sinclair 1986).

Outside of captive situations, hares rarely have been observed in aggressive interactions. Limited observations have shown that age is a significant factor in determining hare dominance, with older individuals being more dominant (Graf and Sinclair 1987). Existence of intraspecific strife in free-ranging hares may be inferred by the degree of scarring observed in captured individuals. During numerical peaks, up to 20% of the population may be scarred on the ears, face, or hindquarters (Windberg and Keith 1976a). Such injuries are most common during the breeding season, which implies that they are the result of mating-related interactions either with prospective mates or, more likely, with competing individuals of the same sex. Scarring also is more prevalent among dispersing hares, likely because of the lower social rank of dispersers (Windberg and Keith 1976a).

Escape Behavior. Snowshoe hares remain motionless to avoid initial detection by predators. Once detected, however, hares typically flee by running at speeds up to 17 km/hr on familiar runways (Grange 1932a; Banfield 1974). They may even reach 50 km/hr when being chased through open areas (Terres 1941; Garland 1983). Typically, they use agile turns to flee in the direction of the nearest vegetative cover. During winter, terrestrial predators chase hares for distances of about 10 m through snow before either a kill is made or the hare escapes (Parker 1981; Murray et al. 1994; 1995). However, predator chases may become longer during cyclic lows and increases, perhaps because of increased predator motivation when hare abundance is limited (O'Donoghue et al. 1998). Adult hares may be more likely to flee from predators than are juveniles, probably because adults have greater wariness and running speed (Adams 1959); this could further contribute to differences observed in mortality rates between age classes. On occasion, hares also may swim or enter burrows to escape predators (Johnson 1925; Hunt 1950; Adams 1959).

Vocalization. Early researchers reported a number of vocalizations common to hares, including clicking sounds, whines, grunts, and squeals (Grange 1932a; Trapp and Trapp 1965). Most of these sounds appear to be associated with fear and displeasure among captured animals. High-pitched squeals in particular are characteristic of distress in hares, and likely startle predators and thus increase chances of escape. In theory, distress calls also may serve either to increase inclusive fitness by warning kin of predator presence (Sherman 1977) or, by attracting competing predators, to facilitate the prey's escape (Chivers et al. 1996).

Foraging Behavior. Snowshoe hares are predominantly active at dawn and dusk (Keith 1964; Mech et al. 1966; Foresman and Pearson 1999), and usually venture only at night to feeding areas that may lack substantive cover. Hare activity patterns also may vary seasonally, with animals typically spending more time foraging when temperature and wind speed are low or barometric pressure is high (Theau and Ferron 2000, 2001). Hares dig craters during winter to gain access to plants found beneath the snow (Gilbert 1990), and such activities are less common during full moon, presumably because of higher predation risk during such times (Gilbert and Boutin 1991). This pattern suggests that hares forage in a manner that is highly sensitive to predation risk

(Hik 1995; Rohner and Krebs 1996), although direct evidence in support of this assertion tends to be scant and conflicting. Indeed, K. Hodges (unpublished data) found similar diets and movement patterns between hares on predator-removal versus control study areas during a cyclic low in numbers, and Hodges (2001) did not detect disproportionately high mortality among predator-naive hares when first exposed to predation risk. Murray (2002) failed to find a clear relationship between hare activity levels and vulnerability to predation. Furthermore, foraging versus predation risk tradeoffs may differ across the species's distribution, depending on whether hare food is present in or near escape cover (e.g., aspen parkland of Alberta) or away from cover (e.g., spruce forest in Yukon). Thus, the specific role of predation risk on hare foraging behavior remains to be fully elucidated. Belovsky (1984) used simulations to predict that hares should forage as energy maximizers, through balancing the conflicting demands of feeding time, digestive capacity, and energy and protein requirements, rather than in specific response to predation risk. Under laboratory conditions, hare food intake rate seems to be determined principally by their ability to efficiently crop and chew woody browse (Shipley and Spalinger 1992).

MORTALITY

Prenatal Mortality Rates. Prenatal mortality can be inferred from differences in the number of corpora lutea and embryos counted among necropsied hare carcasses. Sources of prenatal mortality include the loss of zygotes before implantation, which may occur in 6–8% of the total number of zygotes produced (Newson 1964; Cary and Keith 1979). Prenatal mortality also may occur among embryos either shortly after implantation (through resorption or death) or during the more advanced stages of gestation. Overall, prenatal mortality rates average 9–11% (Dolbeer and Clark 1975; Cary and Keith 1979).

Juvenile Mortality Rates. Preweaning mortality among juvenile hares has only recently been studied, following the development of miniature radiotransmitters which can be glued to the back of small leverets (O'Donoghue 1994). In the Yukon, the mortality rate of young hares was notably high, with the majority of leverets dying during the first month, and >70% of such mortality occurring during the first 5 days after birth (O'Donoghue 1994; Hodges et al. 2001). Individual mortality rates were negatively related to body mass at birth and litter size. In contrast, in Idaho no mortality was observed in a small sample of leverets radio-monitored between the ages of 2 and 6 weeks (Wirsing et al. 2002c). In the Yukon, preweaning survival rates varied >10-fold and were highest during the increase phase of the hare cycle and lowest during the end of the decline (Stefan 1998; Hodges et al. 2001; Krebs et al. 2001a).

Until recently, postweaning mortality rates in hares have been inferred from disappearance of tagged individuals, a procedure which fails to differentiate between losses due to death versus those from dispersal or trapping bias. In general, juvenile mortality rates estimated using such means tend to be higher than those for adult hares, ranging from 64% to 99% per year (Table 7.4). This finding is consistent with results from most radiotelemetry survival studies (Keith 1990; Hodges et al. 2001), although in some cases juvenile and adult survival rates are similar (Keith et al. 1993). Juvenile mortality is correlated with population trend, and declining or stationary populations have about twice the juvenile mortality as increasing populations (Table 7.4). Juvenile mortality also varies geographically, with higher mortality occurring in populations from the southern and western regions (Table 7.4) (Murray 2000).

Postweaning juvenile mortality rates play a critical role in determining hare population dynamics (Meslow and Keith 1968; Keith 1981, 1990; Haydon et al. 1999), with declining populations having particularly low fall/winter juvenile survival (Keith and Windberg 1978; Keith et al. 1984; Hodges et al. 2001). High juvenile survival likely is the principal determinant of hare population growth rates during all phases of the hare cycle, except the decline (Haydon et al. 1999; Hodges et al. 2001). Not surprisingly, food supplementation experiments have demonstrated the importance of natural food shortage on mortality rates of juvenile hares (Windberg and Keith 1976a; Vaughan and Keith 1981). However, when natural food availability is not limited, juvenile survival is not improved by adding food (Gillis 1998). Juvenile hares from earlier litters tend to experience lower fall–winter mortality than those from later litters (Keith and Windberg 1978; Gillis 1998).

Adult Mortality Rates. Mortality of adult snowshoe hares, inferred from the disappearance of tagged animals, varies considerably both temporally and geographically (Meslow and Keith 1968; Keith 1990).

TABLE 7.4. Mortality rates in snowshoe hare populations in North America

Location	Years	Population Trend[a]	Mortality Rate		Reference
			Adult	Juvenile	
Alaska	1	D/S	0.83	0.93	Wolff 1980a
Alaska	1	I	—	0.64	Ernest 1974
Alaska	1	D/S	—	0.92	Ernest 1974
Alaska	1	D/S	—	0.64	Trapp 1962
Alberta	8	I	0.65	0.83	Keith and Windberg 1978
Alberta	7	D/S	0.83	0.97	Keith and Windberg 1978
Alberta	1	I	0.40	0.93	Windberg and Keith 1976a
Alberta	3	D/S	0.68	0.96	Windberg and Keith 1976a
British Columbia	3	D/S	0.88	0.99	Sullivan 1996
Colorado	2	I	0.55	0.81	Dolbeer and Clark 1975
Idaho	2	D/S	0.81	0.88	Wirsing et al. 2002c
Minnesota	1	I	0.59	0.56	Green and Evans 1940a, 1940c
Minnesota	5	D/S	0.74	0.81	Green and Evans 1940a, 1940c
Montana	1	D/S	0.78	0.81	Adams 1959
New Brunswick	3	D/S	0.73	0.89	Wood and Munroe 1977
Oregon	1	D/S	0.72	0.86	Black 1965
Wisconsin	1	D/S	0.69	0.93	Kuvlesky and Keith 1983
Yukon	3	I	0.78	0.90	Krebs et al. 1986b
Yukon	3	D/S	0.98	0.98	Krebs et al. 1986b

SOURCE: Adapted in part from Keith (1990) and Murray (2000).

NOTE: Estimates for adults refer to annual mortality, whereas for postweaning juveniles estimates represent summer-to-spring values. All mortality estimates were obtained via mark–recapture survival estimation and thus fail to account for dispersal.

[a]D/S, Declining/stationary; I, increasing.

Such estimates, however, overestimate mortality relative to telemetry-based survival monitoring (Boutin and Krebs 1986). Overall, annual adult mortality rate averages about 81% (Table 7.4). Rates appear to be substantially higher in declining/stationary than in increasing populations; these results are confirmed by long-term studies estimating survival via radiotelemetry (Haydon et al. 1999; Hodges et al. 2001). In the Yukon, increase in hare mortality rates preceded the population peak by 2 years (Hodges 2000a; Krebs et al. 2001a), which suggests that the survival response of hares is not strictly density dependent. Haydon et al. (1999) concluded that low survival of adult hares is the principal demographic attribute characterizing cyclic hare population declines. An annual survival rate exceeding 28% (mortality rate <72% per year) seems to be required for hare population growth in the Yukon (Hodges et al. 2001), where reproductive rates can be high relative to most other hare populations. This important summary statistic implies that, assuming similar productivity across the geographic range, some trapping estimates of mortality reported for increasing hare populations (Table 7.4) must be biased high. Mortality rates also vary seasonally, with more adults dying during winter or early spring (Green and Evans 1940b; Keith and Windberg 1978; Keith et al. 1984; Boutin et al. 1986; Krebs et al. 1986b). Some studies have identified habitat as a factor influencing hare mortality rates (Dolbeer and Clark 1975; Sievert and Keith 1985; Keith et al. 1993), whereas others have not (Keith and Bloomer 1993; Cox et al. 1997). Food supplementation experiments have provided a range of effects on adult hare survival (Hodges et al. 1999a) and have implied that the role of food on mortality probably is temporally and/or spatially inconsistent.

Mortality Factors. Predation is the overwhelming proximate cause of death in snowshoe hares (Morse 1939; Brand et al. 1975; Keith et al. 1984, 1993; Boutin et al. 1986; Murray et al. 1997). Mammalian and avian predators commonly kill hares, with total losses from predation >75% among all studies where a sample of animals was monitored intensively via telemetry for cause of death (Table 7.5). In the Yukon, the most important predators contributing to neonate mortality were red squirrels (*Tamiasciurus hudsonicus*) and arctic ground squirrels (*Spermophilus parryii*), with fewer leverets preyed on by great-horned owls (*Bubo virginianus*), red-tailed hawks (*Buteo jaimaicensis*), and goshawks (*Accipiter gentilis*) (O'Donoghue 1994). Predation can be the proximate cause of death for >80% of leverets (O'Donoghue 1994; Hodges et al. 2001).

The most common mammalian predators of postweaning juvenile and adult snowshoe hares include lynx, coyote (*Canis latrans*), and fisher (*Martes pennanti*) (Keith et al. 1977, 1984; Adamcik et al. 1978; Raine 1987; Kuehn 1989; Wirsing et al. 2002b). Less common predators include bobcat (*Lynx rufus*), wolf (*Canis lupus*), red fox (*Vulpes vulpes*), black bear (*Ursus americanus*), wolverine (*Gulo gulo*), marten (*Martes americana*), weasel (*Mustela erminea, M. frenata*), mink (*Mustela vison*), barred owl (*Strix varia*), and raven (*Corvus corax*) (Morse 1939; Adams 1959; Litvaitis et al. 1986; Raine 1987; Poole and Graf 1996; Otto 1998). On occasion, domestic dogs (*Canis familiaris*), domestic cats (*Felis catus*), and even northern short-tailed shrews (*Blarina brevicada*) killed hares (O'Farrell 1965; Rongstad 1965; Doucet 1973). Predators may vary their use of hares either seasonally (Thompson and Colgan 1987) or annually (Adamcik et al. 1978; Kuehn 1989; Thompson and Colgan 1990; Poole and Graf 1996), in response to temporal changes in either hare abundance and/or their relative vulnerability to predation.

Starvation may be an important source of mortality in hare populations when subject to severe food limitation over winter. Such deaths tend to be concentrated during winter and spring, and may constitute a significant proportion of mortalities occurring during cyclic population declines (Keith et al. 1984; Boutin et al. 1986). Relative to predation, starvation may be a more common cause of hare death during the increase and peak phases of the hare cycle (Hodges et al. 2001). However, in most instances starvation is a negligible source of overwinter mortality for hares, given their high rates of predation. However, malnutrition itself ultimately may be responsible for a number of hare deaths by increasing hare susceptibility to predation either through inhibited escape behavior or compensatory foraging and reduced predation risk sensitivity (Keith et al. 1984; Murray 2002). Overnutrition also may be associated with increased predation due to reduced risk sensitivity among energy-maximizing individuals (Rohner and Krebs 1996). Low temperatures (less than −30°C) appear to be deleterious to snowshoe hare health (Keith et al. 1984), and hare population declines often coincide with the coldest winter periods (Keith 1963; Höhn and Stelfox 1977). The combination of a high incidence of malnutrition in food-limited populations and severely cold temperatures, which aggravate thermoregulatory requirements, likely exerts substantive demands on the limited energy reserves of hares (Irving et al. 1957; Hart et al. 1965). It follows that an interaction between cold temperatures and malnutrition can either cause direct hare

TABLE 7.5. Predation rates and percentage mortality attributable to predation in showshoe hares in North America

Location	Population Trend[a]	Duration (years)	Annual Predation Rate	Percent Mortality due to Predation	Reference
Alberta	D/S	2.5	0.79	81 (26)	Brand et al. 1975
Alberta	D/S	0.4	0.89	92 (61)	Keith et al. 1984
Idaho	D/S	2	0.82	93 (70)	Wirsing et al. 2002c
Manitoba	D/S	2	0.68	95 (318)	Murray et al. 1997
Quebec	D/S	0.92	0.31	80 (10)	Ferron et al. 1998
Virginia[b]	?	0.6	1.00	100 (20)	Fies 1993
Wisconsin[b]	D/S	1.5	0.73	87 (67)	Sievert and Keith 1985
Wisconsin	D/S	3	0.82	96 (122)	Keith et al. 1993
Wisconsin	D/S	0.5	0.74	88 (8)	Cox et al. 1997
Yukon	D/S	1.5	0.68	77 (216)	Boutin et al. 1986
Yukon[c]	I	0.75	0.42	86 (43)	Gillis 1998
Yukon	D/S	0.30	0.53[d]	81 (170)	O'Donoghue 1994
Yukon	I	5.0	—	93 (461)[e]	Hodges et al. 2001
Yukon	D/S	3.0	—	99 (405)[e]	Hodges et al. 2001

NOTE: Total number of natural mortalities (predation and nonpredation) is in parentheses. Survival was monitored via radiotelemetry, and cause of death was determined at the kill site. Where hare populations were manipulated experimentally, estimates were calculated for all populations. Unless otherwise noted, predation rates are converted to annual rates even if studies did not last the entire year.

[a]D/S, Declining/stationary; I, increasing.

[b]Includes hares released as part of a translocation program.

[c]Juveniles only.

[d]Preweaned juveniles only, corresponds to 30-day predation rate.

[e]Hares residing in study areas where predators were excluded experimentally are omitted.

starvation or increase hare vulnerability to predation through compensatory foraging or impeded escape abilities. Also, cold and wet summer temperatures during the first 45 days after birth may cause neonates to die from exposure (Meslow and Keith 1971; O'Donoghue 1994).

Because of the overwhelming importance of predation, and to a lesser extent starvation, in hare mortality rates, other causes of natural death necessarily play a minor role. Hares may occasionally succumb to chronic infections or ulcers (Brand et al. 1975; Murray et al. 1997). Hares dying during a population crash in Minnesota were suspected of succumbing largely to shock disease (Green and Larson 1938; Green et al. 1938), although prevalence of this mortality source has not been reported in more recent studies and probably is not widespread (see Parasites and Diseases). Several microbes have been shown to induce hare mortality directly (see Parasites and Diseases), but they probably are only demographically significant to natural populations where predation rates are depressed.

Snowshoe hares are hunted by humans. Historically, large numbers of hares were harvested annually in the more populated regions of the species's distribution (Leopold 1937; Dolbeer 1972). Snowshoe hare harvest rates have corresponded with population abundance, and thus harvest rates often have revealed cyclic patterns and regional synchrony (Hodges 2000a; Murray 2000). Studies of snowshoe hare survival have reported modest rates of hunting-related mortality (Sievert and Keith 1985; Gillis 1998). However, the importance of hares as a game animal appears to be in decline, and human-caused mortality probably is not a significant cause of death across most of the geographic range (Hodges 2000a).

PARASITES AND DISEASES

Ectoparasites. Several species of ectoparasite have been associated with the snowshoe hare. During spring–summer, adult rabbit ticks (*Haemophysalis leporispalustris*) and Rocky Mountain wood ticks (*Dermacentor andersoni*) commonly are found near the facial region, whereas larvae and nymphs may occur anywhere on the body (Phillip 1938; Dodds and Mackiewicz 1961). After engorging, ticks drop off their host. Adult female ticks may lay up to 3000 eggs and then die; juveniles molt and await a new host (Campbell et al. 1980). Most ticks drop off during daylight while hares are resting in forms or near dusting sites (Campbell and Glines 1979), which facilitates tick transmission to other hares (Green et al. 1943; Keith and Cary 1990). Reduced snowshoe hare immunity to tick infestation compared to that of other leporids (McGowan et al. 1982; McGowan 1985) may explain the common occurrence of heavy infestations among certain individuals. Tick densities may fluctuate with the hare population cycle, and in Alberta, heavy tick infestation among female hares apparently was associated with litter size reductions (Keith and Cary 1990). Other ectoparasites recovered from snowshoe hares include fleas (*Hoplopsyllus* spp., *Ceratophyllus* spp., and *Spelopsyllus* spp.) and black flies (*Simulium* spp.) (Green et al. 1939; Burgdorfer et al. 1961; Dodds and Mackiewicz 1961). Although the importance of ectoparasites to hare populations is unclear, their primary significance may be as vectors for secondary infections (Iversen et al. 1970; Hoff et al. 1971b).

Macroparasites. Helminths infecting snowshoe hares include several species of nematodes, cestodes, and trematodes, most of which are found in the gastrointestinal tract or lungs. A stomachworm (*Obeliscoides cuniculi*) with a direct life cycle is common among hares throughout much of the species's range (Manweiler 1938; Erickson 1944; Dodds and Mackiewicz 1961; Bookhout 1971; Bloomer et al. 1995), but appears to be absent in populations from Idaho (D. Murray, unpublished data) and the Yukon (Sovell and Holmes 1996). *O. cuniculi* has high (>85%) prevalence among hares during spring–summer, but is rarely found in winter-killed animals because of the arrested development of larvae (Gibbs et al. 1977; Murray et al. 1996). Similar prevalence of *O. cuniculi* infection between adult and juvenile hares suggests that acquired immunity does not limit infection rates (Keith et al. 1986). Synchrony between the cycle in hare abundance and prevalence of *O. cuniculi* implies that transmission is density dependent (Keith et al. 1985). Several early authors (MacLulich 1937; Erickson 1944; Dodds and Mackiewicz 1961; Bookhout 1971) speculated that the pathology caused by the stomachworm could cause hare mortality, but more recent work shows that natural infection levels of *O. cuniculi* are unlikely to be lethal (Murray et al. 1997). Despite this, *O. cuniculi* infections appear to increase hare vulnerability to predation (Keith et al. 1986; Murray et al. 1997; but see Bloomer et al. 1995).

Other common nematodes found in the intestinal tract include pinworms (*Passalurus nonanulatus, P. ambiguus*), whipworms (*Trichuris leporis*), and several species of *Nematodirus* (*N. triangularis, N. leporis*) and *Trichostrongylus* (*T. affinis, T. calcaratus*) (Green et al. 1939; Erickson 1944; Bookhout 1971; van Nostrand 1971 cited in Dodds 1987; Maltais and Ouellette 1983). It is notable that intensities of *N. triangularis* and *T. leporis* were correlated with loss of hare body mass (Keith et al. 1986). Less common intestinal nematodes infecting the snowshoe hare intestinal tract were tabulated by Bittner and Rongstad (1982).

Nematodes also occur in other organs or tissues. The lungworm, *Protostrongylus boughtoni,* is common across the snowshoe hare's range (Goble and Dougherty 1943; Goble and Cheatum 1944; Olsen 1954) and is transmitted via an intermediate gastropod host that is inadvertently ingested by hares while feeding (Kralka and Samuel 1984, 1990). *P. boughtoni* undergoes seasonal fluctuations in abundance (Keith et al. 1985; Murray et al. 1996) and causes hares to lose body mass (Keith et al. 1986). The hookworm (*Dirofilaria scapiceps*) infects the connective tissue surrounding tendons of the ankle region and is transmitted through a mosquito vector (Bartlett 1984a, 1984b). It occurs throughout the range of the snowshoe hare and other leporids (Highby 1943; Penner 1954; Bartlett 1983). Its prevalence fluctuates with a 1-year delay with the snowshoe hare cycle (Keith et al. 1985).

Hares also are infected with a variety of other nematodes, cestodes, and trematodes (Dodds and Mackiewicz 1961; Bittner and Rongstad 1982). The most common cestodes infecting hares include *Taenia* spp., *Multiceps* spp., and *Cittotaenia* spp. (Green et al. 1939; Erickson 1944; Bursey and Burt 1970). Prevalence of *Taenia* spp. commonly is low (<15%) among hare populations (Boughton 1932; Philip 1938; Bookhout 1971; Keith et al. 1985; Bloomer et al. 1995). Although *P. boughtoni, D. scapiceps,* and *Taenia* spp. were considered by early researchers as potentially deleterious to hare survival (Erickson 1944; Dodds and Mackiewicz 1961; Bookhout 1971), more recent empirical work has not supported this assertion (Keith et al. 1986).

Microparasites. Snowshoe hares are infected by at least nine species of protozoa (*Eimeria* spp.) (Boughton 1932; Samoil and Samuel 1977a, 1977b), but none are known to reduce hare survival either directly or through predation (Keith et al. 1986). Several viruses also have been reported (silverwater, California encephalitis, encephalomyocarditis, buttonwillow, Powassan, eastern equine encephalomyelitis, and western equine encephalomyelitis), but their effect on hares is largely unknown (Burgdorfer et al. 1961; Yuill and Hanson 1964; Hoff et al. 1969, 1971a). The most prevalent viral infections are silverwater virus and California encephalitis virus. Silverwater virus has been studied most extensively, but is not associated with any clinical signs of disease (Hoff and Hanson 1973; Hoff et al. 1970, 1971c). However, Yuill et al. (1969), based on a rapid decline in virus antibody prevalence in a free-ranging hare population from Alberta, speculated that some viruses may predispose hares to other causes of mortality. Snowshoe hare virus is one of a large group of arboviruses from the California serogroup. It may cause illness in humans, but is not associated with clinical signs in hares (Grimstad 1988, 1994; McLean 1983). Human disease symptoms can include meningitis and encephalitis, but most infections probably go unnoticed (Artsob 1983). Transmission probably occurs via mosquito bites, and serological surveys have revealed levels of exposure up to 32% in some rural human populations (Artsob 1983).

Snowshoe hares are infected by at least 17 species of bacteria, several of which may be implicated in hare disease (van Nostrand 1971 cited in Dodds 1987). The bacterium causing tularemia, *Pasteurella tularensis,* has been recovered from several hare populations; although it appears to occur infrequently it may be the cause of some disease epidemics (Green and Bell 1939; Jellison et al. 1961; Hoff et al. 1970; van Nostrand 1971 cited in Dodds 1987). *Staphylococcus aureus* also is alleged to contribute to hare population die-offs, particularly after exaggerated population peaks (MacLulich 1937). Several species of the bacterium *Chlamydia* were recovered from hare carcasses found showing signs of severe illness during a population crash in Saskatchewan (Spalatin et al. 1966). Subsequent experimental exposure of captive hares to recovered chlamydial organisms resulted in fever, morbidity, opisthotonosis, convulsions, hypoglycemia, and, ultimately, death. The disease appears to be epizootic in hares and may be contracted through transmission from muskrats (*Ondatra zibethicus*), via ectoparasite bites, or through carcass scavenging (Iversen et al. 1970). Another bacterium found in hares (*Listeria monocytogenes),* appears to contribute to juvenile mortality (Dodds and Mackiewicz 1961). Although *Pasteurella pseudotuberculosis* and *Salmonella* spp. also have been recovered, their importance to hares is unclear (Green et al. 1939). Hares also can be infected by fungi (*Trichophyton mentagrophytes*) (Adams et al. 1955), and captive animals may die of coccidiosis, caused by the fungus *Coccidia* spp., when reared in cages where fecal material can accumulate (Severaid 1942, 1945b). Ricketsial organisms also have been recovered from hares, including *Rickettsia rickettsii* and *Coxiella burnettii,* the causative agents of Rocky Mountain spotted fever and Q fever, respectively (Dodds 1987). Parasitism and infectious disease, although not often noted as the proximate cause of hare deaths, need to be further evaluated as factors predisposing hares to predation and other causes of death.

Noninfectious Diseases. Two noninfectious diseases have been reported for snowshoe hares, both of which appear to result from stress. Opisthotonic shock disease, first described from a declining hare population in Minnesota (Green and Larson 1937, 1938; Green et al. 1938), consists of the rapid onset of convulsions, followed by a comatose state and death. The disease appears to be hypoglycemic. Low body mass, depressed levels of blood sugar and liver glycogen, darkened and hypotrophic liver and spleen, inflamed adrenal glands, and fluid accumulation in body cavities are diagnostic (Green et al. 1939; Christian and Ratcliffe 1952). Hares dying of starvation tend to be hypoglycemic (Keith et al. 1968, Iversen et al. 1972), and during a population decline substantial numbers of live hares exhibit low liver glycogen (Keith et al. 1984). This suggests a relationship between shock disease and chronic malnutrition. However, Chitty (1959) rejected the idea of shock disease as a prevalent source of mortality in natural hare populations because of the paucity of reports of symptoms among animals not subject to handling stress. Myers et al. (1981) reexamined shock disease in rabbits (*Oryctolagus cuniculus*) and concluded that it was indeed a natural phenomenon and could arise from the conflicting energetic demands of thermoregulation and osmoregulation. During hare population declines, increased corticosteroid activity and cortisol production may be common (Feist 1979, 1980; Dietrich and Feist 1980; Boonstra and Singleton 1993; Boonstra et al. 1998a), which indicates that shock disease ultimately may accompany deteriorating body condition and/or stress responses. However, the relationships among shock disease, stress, predation risk, and hare population demography have yet to be reconciled due to several conflicting reports (Windberg and Keith 1976a; Fevold and Drummond 1976; Höhn and Stelfox 1977; Vaughan and Keith 1981).

Trap sickness is similar to shock disease except that animals dying from the former have gastric ulcers and lack liver atrophy and body fluid accumulation (Iversen et al. 1972). Free-ranging hares confined to livetraps for an extended period of time exhibit this condition. Because subcutaneous administration of glucose (and even heating at low oven temperature) sometimes can reverse the disease, low blood sugar may be proximally involved (Keith et al. 1968). Thus, trap sickness and shock disease probably share similar etiology.

AGE ESTIMATION

Age of snowshoe hares can be estimated several ways. Juveniles may be distinguished qualitatively from adults during summer–fall by their darker hind feet (Keith et al. 1968). Until about March of the year following their birth, juvenile hares also are generally smaller, have lighter mass, and possess smaller hind feet than adults (Keith et al. 1968). Size and mass of juveniles become comparable to those of adults by 8–11 months (Keith and Windberg 1978). The epiphysis of the humerus may facilitate differentiation of juvenile and adult hare carcasses until closure occurs in juveniles at about 7 months of age (Newson and de Vos 1964; Dodds 1965). Eye lens mass increases with age and may be used to differentiate older age classes (Keith et al. 1968; Keith and Cary 1979).

Reproductive organs of juvenile hares differ from those of adults. Juvenile males possess a blunt penis, which is barely eversible, whereas that of adults is pointed and easily everted (Keith et al. 1968). For juvenile females, the uteri are notably narrower and more threadlike than they are for adults, and the length of teats is consistently less than 3 mm (Newson 1964; Keith et al. 1968).

ECONOMIC STATUS

Bittner and Rongstad (1982) reviewed in detail the economic importance of snowshoe hares. Most of the early economic value attributed to this species centered around tree damage caused by its winter feeding activities (Baker et al. 1921; Corson and Cheyney 1928; Cox 1938). Such damage may be severe in timber stands and ornamental tree plantations, particularly during winters of peak hare densities when natural browse availability is limited. However, in some stands, hare browsing actually may play a beneficial role by thinning small trees, which otherwise would have to be removed manually (Roe and Stoeckeler 1950; Hansen and Flinders 1969).

Hares were once an important game species in eastern United States and Canada (Leopold 1937; Dolbeer 1972). In Michigan alone, an average of over 300,000 hares were killed by hunters per year over a 61-year period (Murray 2000). Hare harvest estimates exceeded 500,000 per year for a 5-year period in Nova Scotia (Dodds and Thurber 1965; Dodds 1987). However, harvest numbers in other areas were substantially lower (Keith 1963; Hodges 2000b; Murray 2000), and hares in the mountainous western United States never achieved popularity as a game animal (Dolbeer 1972). It is unclear to what extent hare harvest statistics reflect hunting efforts directed specifically at hares versus incidental kills, and hunter effort and other potential biases may further affect these statistics. Thus, the accuracy of harvest data for indexing hare population trends probably is questionable outside of detecting dramatic numerical fluctuations.

Although Bittner and Rongstad (1982) predicted an increased importance of snowshoe hares as a game species (see also Dolbeer 1972), this has not occurred and seems unlikely. Most jurisdictions report a recent decline in the abundance of hares killed annually (D. Murray, unpublished data), probably due to reduced harvest effort (Royama 1992). This decline is consistent with the general declining trend for hunting among other game species (e.g., Enck et al. 1993). However, hares remain an important game species in some jurisdictions such as Newfoundland (Hodges 2000a).

The importance of snowshoe hares as prey to predators is another notable economic value of this species. Over much of their range, hare population fluctuations cause dramatic numerical variations in lynx, coyote, red fox, fisher, and marten as well as lesser changes in several other carnivore species (Keith 1963; Bulmer 1974, 1975; Finerty 1980). Because most of these carnivores are trapped for their fur, snowshoe hares have a strong, albeit indirect, importance to furbearer harvest rates, particularly in northern regions. The apparent decline in carnivore trapping across most of the North American continent (Novak 1987) may reduce this particular economic value of snowshoe hares in the future. The fur from snowshoe hares themselves is not resistant to wear and is of no commercial value.

MANAGEMENT AND CONSERVATION

Damage Control. Snowshoe hare feeding damage can be severe in some commercial timber plantations or managed timber stands, particularly when hare densities are high and bark damage is substantive (Cook and Robeson 1945; Lloyd-Smith and Piene 1981; Sullivan and Sullivan 1982a, 1986). Accordingly, considerable research effort has been expended to develop procedures for mitigating stand damage (e.g., Aldous and Aldous 1944; Krefting and Stoeckeler 1953; Keith 1972; Radvanyi 1987). The propensity for high immigration rates where hares have been removed may eliminate the efficacy of direct control at large scales, though removal may be effective in small plantations or orchards (Aldous and Aldous 1944; Keith 1972). Timing tree planting during periods when hare numbers are predictably low or refraining from planting slow-growing species also may effectively reduce hare feeding damage (Keith 1972; Roy et al. 1991). Because natural resistance to feeding damage by hares and other herbivores has genetic origins (Dimock et al. 1976; Silen et al. 1986; Rousi et al. 1991), herbivore-resistant strains of some of the most vulnerable species could be further developed for limited use in plantations, although such strains probably should be prevented from achieving more widespread distribution. Habitat management, in the form of cover removal, buffer strips, and stand alteration via scarification, may successfully reduce hare densities to a limited extent (Keith 1972; Sullivan and Moses 1986). In commercial plantations, herbicides may be used to reduce vegetative cover and availability of summer foods to hares, but the efficacy of such treatments on hare populations and habitat use is uncertain (Borrecco 1976; Sullivan 1994, 1996). Several types of repellents may deter hare feeding on trees (Aldous and Aldous 1944; Keith 1972; Szukeil 1973; Rangen et al. 1993), but the efficacy of most repellents is short-lived because of rapid breakdown and/or hare habituation. Application of repellents derived from predator odors may hold most promise for deterring snowshoe hare herbivory, particularly if encapsulated for time-controlled release (Sullivan 1986; Sullivan and Crump 1984; Sullivan et al. 1985). Repellents formulated from plant antifeedants (pinosylvin) also may be effective (Zimmerling and Zimmerling 1996).

Habitat Requirements. Timber stand management may cause immediate impacts to hare movement behavior, feeding activity, body condition, habitat use, and forage preference (Conroy et al. 1979; Sullivan and Sullivan 1982b, 1988; Ferron et al. 1998; de Bellefeuille et al. 2001a, 2001b), but rigorous experimental studies have not been conducted to evaluate timber harvest strategies that will sustain viable hare populations. Koehler and Brittell (1990) and Keith et al. (1993) suggest that forest stands should be a minimum of 8–10 ha to effectively sustain hare populations, whereas Thomas et al. (1997) suggested that clearcuts should not exceed 16 ha. Hares travel through clearcuts if they are small, and if they are never more than 200–400 m from canopy cover (Brocke 1975; Conroy et al. 1979). It is unclear whether a landscape matrix made up of alternating small clearcuts and buffer strips of variable shapes and sizes can sustain hare populations across a broader spatial scale (Ferron et al. 1998). Over the short term (1–6 years), precommercial thinning of timber stands may incur negative effects on hares (Skinner 1995), although this assertion clearly needs to be tested more extensively across habitat types and hare density gradients. Requirements for cover are particularly important along the southern range boundary, where fragmentation and high predation risk probably act in concert to limit hare populations. If hare population maintenance is an important factor in habitat management programs, it may be desirable to promote a high density of shrubs and immature trees, or to limit extensive timber harvest, fire suppression, or brush removal (Richmond and Chien 1976; Conroy et al. 1979; Buehler and Keith 1982).

Stocking and Artificial Propagation. Because of the former importance of snowshoe hare as a game species, stocking was used extensively to supplement naturally low or declining populations along the southeastern periphery of the species's range (Grange 1949; Boyle 1955; Fitzpatrick 1957). In Massachusetts alone, over 150,000 hares were stocked during 1919–1969, at a cost ranging from $3.25 to $4.50 per individual animal (McDonough 1969). Similar programs were initiated in Maine (Severaid 1942, 1945b), Pennsylvania (Glazer 1959), Virginia (August 1974), and Connecticut (Behrend 1960), but the success of such efforts in hare population restoration is questionable. More recent stocking programs involving the release of wild hares also failed to reestablish viable populations, largely because of heavy mortality incurred from predation (Schultz 1980; Fies 1991, 1993). Wild hares transplanted to new areas along the southern range boundary typically exhibit low survival rates and extensive movements (Sievert and Keith 1985; Wirsing et al. 2002c), which implies that transplant programs will remain questionable when habitat quality is marginal (also see Dalby 1985). Because transplanted populations initially may be subject to a "predator pit," in theory it may be possible to successfully restore hare populations to poor habitat if predator numbers are temporarily reduced.

Artificial propagation of hares for future release was investigated in Maine and found to be reasonably successful (Severaid 1942, 1945b), but captive-born animals released to foreign habitat where predators abound probably have very low survival rates. Accordingly, restoration of natural hare populations via captive breeding programs is unlikely.

Predator Control. Because predation is the overwhelming source of mortality in the majority of snowshoe hare populations, predator control should enhance population restoration, particularly where predators are abundant and suitable cover is limited. However, lethal predator control is challenging because of the federal protection afforded to avian predators and the difficulties associated with controlling elusive carnivores such as coyotes and bobcats. Although nonlethal control measures, including predator exclosures, may help reduce hare predation by mammalian and perhaps avian predators in experimental situations (e.g., Krebs et al. 1995), such measures are difficult to implement on a large scale and thus may not be practical (Boutin et al. 2001). The lethal and nonlethal control of predators for hare population restoration should only be considered as a small-scale interim solution until permanent habitat improvement can be achieved (Fies 1991). However, increased cover probably is best achieved through partial clearing of mature stands and subsequent understory regeneration, since artificial brushpile creation may not necessarily improve hare survival (Cox et al. 1997).

Prey for Carnivores. The close relationship between snowshoe hare numbers and the diet and abundance of their avian and mammalian predators has involved hares in the conservation of forest carnivores (Koehler and Brittell 1990; Ruggiero et al. 2000). Understanding hare population ecology in this context is particularly relevant in areas where hare or predator numbers are low, yet predator diet is made up largely of hares (e.g., Koehler 1990b). The species recovery plan for lynx, which recently were listed as threatened under the Endangered Species Act (*Federal Register* 2000), necessarily will involve habitat preservation and possibly habitat enhancement measures that benefit hare populations in areas where lynx may reside. Similar measures ultimately may be implemented as other predator species are protected.

Laws and Regulations. Across most of their range, snowshoe hares are considered as small game and thus can be hunted with limited restrictions on bag limits and hunting seasons (Dolbeer 1972). Because hare abundance is dictated principally by food and predation and hare populations have a propensity to grow rapidly, it is unlikely that hunting alone can be sufficiently density dependent so as to substantially depress hare numbers. However, during hare population lows and along the periphery of the species's range, hunting may play an important role in reducing hare numbers. In Virginia and probably elsewhere in the eastern United States, hare hunting seasons have been closed at least until depressed populations show signs of recovery.

Population Estimation. Snowshoe hare population size may be estimated or indexed using a variety of procedures. Winter track counts have been used to index hare habitat use and relative abundance, particularly in the southern range, where densities are low and habitat requirements are poorly understood (Hartman 1960; Litvaitis et al. 1985b; Byrne 1998). This method may correlate well with other techniques of measuring relative abundance (Litvaitis et al. 1985b; Thompson et al. 1989). However, winter track counts are restricted to periods of snowcover and thus may suffer from the effects of weather or other factors that influence track observability (Godbout 1999). The potential also exists for mistaking tracks of other leporids with those of snowshoe hares where they are sympatric. In addition, this method assumes that hare travel distances, and thus track density, are consistent across population density, which may not be the case if social interactions, predator densities, or food abundance affect movement patterns. Thus, whereas track transects may be useful for determining leporid presence in a given area, this method likely will remain limited for most purposes. Counts along line transects also have been used to index hare populations (MacLulich 1937, Webb 1942), but the low rate of detection of hares in the field limits the use of this method. Also, although Litvaitis et al. (1986) found that vegetative stem density was closely correlated with hare density in Maine, it is unlikely that vegetation density alone can be used to predict hare density across habitat types and geographic areas.

The most common method for indexing or estimating hare numbers across larger scales is based on determining the abundance of fecal pellets accumulated on plots (e.g., Hartman 1960; Wolff 1982; Swayze 1994; Eaton 1995; Thomas et al. 1997; Bartmann and Byrne 2001). This method may provide actual density estimates (e.g., Koehler 1990b; Poole 1994; Slough and Mowat 1996) and correlates well with known hare densities (Krebs et al. 1987, 2001b; Murray et al. 2002). Advantages of this estimation procedure are its low cost and effort level compared to other techniques and the fact that it can be implemented on a large geographic scale (Godbout 1999). However, this method may be biased if defecation patterns are influenced by hare diet or time of day, and therefore it may be best suited for evaluating broad-scale densities rather than microhabitat selection patterns. Also, potential observer bias in pellet counts may occur depending on plot size and shape (McKelvey et al. 2002; Murray et al. 2002). Because the rate of fecal pellet degradation may differ across sites and habitat types, hare density estimation using this procedure should involve preclearing of old pellets on transects 1 year before counts (Krebs et al. 1987). However, in some areas pellets degrade in less than a year, thereby requiring more frequent plot checks or count adjustment (D. Murray, unpublished data). To date, the pellet plot method for snowshoe hares has not received sufficient validation to warrant large-scale application across forest types or geographic regions.

Hares can be captured using a variety of means, including livetraps snares, and drive nets (Aldous 1946; Keith 1965; Keith et al. 1968; Brocke 1972; Cushwa and Burnham 1974), although the former method is clearly the most widely used technique. By livecapturing hares and releasing them after marking, it is possible to estimate numbers based on minimum number alive or standard mark–recapture procedures. Populations have been indexed using mimimum number alive estimation when densities are elevated (Sullivan and Sullivan 1983), but such estimates have been found to be conservative (Hilborn et al. 1976; Efford 1992) or high (Boutin 1984c), and therefore should be used with caution. However, when hare densities are low, this procedure may closely approximate estimates derived from other methods (Krebs et al. 1986b), particularly if combined with information from radio-monitored animals (Murray et al. 2002; Wirsing et al. 2002c).

Earlier studies determined that hare livetrapping success is largely random, providing unbiased estimates of population size (Hartman 1960; Keith and Meslow 1968). Accordingly, hare population estimation on small study areas has centered around mark–recapture estimation (Keith et al. 1968; Gendron and Bergeron 1975; Boulanger and Krebs 1994, 1996) mainly through the use of Jolly–Seber estimation and the suite of estimators available from PROGRAM CAPTURE (Rextad and Burnham 1991). Many of these estimators provide realistic and unbiased estimates of hare populations. Estimates can be improved by using telemetry-based estimators incorporating information on movements and survival to calculate the size of the previously marked cohort (Boutin 1984c; Murray 1999). Mark–recapture population estimation for hares is costly and labor intensive, and thus may only be feasible in intensive research situations. Also, more work is needed on hare space use and individual responses to trapping, to improve the conversion of hare population size estimates derived from mark–recapture to hare density estimates for a given area.

RESEARCH, CONSERVATION, AND MANAGEMENT NEEDS

Bittner and Rongstad (1982) cited the need to test hypotheses potentially explaining the causes of snowshoe hare population cycles as a critical research priority. In the two decades since then, much work has been done to address this need, largely through the completion of long-term field studies directed by Lloyd Keith in Alberta (e.g., Keith and Windberg 1978; Keith et al. 1984; Keith 1990) and Charles Krebs and colleagues in Yukon (Krebs et al. 1995, 2001a, 2001c). Although food and predation have emerged as the principal factors in the hare cycle in both study areas, it is unclear to what extent the two factors act sequentially versus synergistically, whether such interactions cooccur through stress or habitat use in response to perceived predation risk, exactly how plant quality and quantity may contribute to food-related effects, and how the food–predation interaction may differ in nature and magnitude across the geographic range where cycles exist (Keith 1990; Boonstra et al. 1998a; Dambacher et al. 1999; Krebs et al. 2001d). Recent models (Choquenot et al. 2001; King and Shaffer 2001) provide explicit predictions regarding hare population cyclicity and the nature of food–predation interactions, and these should be evaluated empirically in several geographic areas. Similarly, the importance of predator movements, sunspots, and regional weather patterns on the spatial synchrony of hare population cycles requires further research attention (Blasius et al. 1999; Krebs et al. 2001a).

Uncertainty surrounding the nature and cause of snowshoe hare numerical trends at lower latitudes has surfaced as a major impediment to the proper management and conservation of both hares and associated carnivore populations in the contiguous United States (Kloor 1999; Ruggiero et al. 2000). Clearly, the duration of previous studies at low latitudes was insufficient to assess either the weakly cyclic versus stationary nature of southern hare population dynamics or the reasons for the alleged differential numerical trends relative to northern populations. Experimental manipulations of hare predation pressure and food abundance, similar to those undertaken in the Yukon study (Krebs et al. 1995, 2001c), would be valuable for examining factors involved in hare population regulation at low latitudes. Ideally, such studies should span roughly a 10-year period to correspond to a possible numerical cycle. Hodges (2000b) provided an explicit framework for testing hypotheses related to facultative predation and habitat refugia as causes of greater stability in southern populations. Additional studies examining other factors limiting hare distribution along the southern range periphery (e.g., Buehler and Keith 1982; Sievert and Keith 1985; Keith et al. 1993) and the genetic and demographic linkage between isolated populations of hares (see Burton et al. 2002) should be prioritized. This latter field of study is particularly relevant in the context of hare population conservation and restoration along the southern distribution (see Stocking and Artificial Propagation), and can be achieved in part via experimental transplant and monitoring studies.

Long-term monitoring of hare population trends via fecal pellet counts would provide a low-cost means of evaluating numerical trends and habitat choice in southern populations. Such monitoring programs should be established on national forest and other public and private lands across the historic range in the contiguous United States. To facilitate geographic comparisons, standard protocols related to plot shape, size, and density should be adopted. Also, the relationship between pellet accumulation and degradation on monitoring plots should be further tested to ensure that correlations derived from northern (Krebs et al. 1987, 2001b) or southern (Murray et al. 2002) areas are suitable for widespread use.

Snowshoe hare habitat selection varies according to local availability of food and cover. Accordingly, where conservation of either hare populations themselves or the community of plants and animals that are shaped by hares is important, it is necessary to understand the forest management practices required for maintaining desired population densities. Obviously, clearcuts are not beneficial to hares in the short term, while the long-term effects of fire suppression also should be equally unfavorable. However, more specifically, research efforts should evaluate the impact of specific forest management protocols on hare population demography and dynamics. This line of investigation will require sound experimental design, including appropriate and sufficient replication (Krebs et al. 2001d). Although the specific stand attributes that are important to hares will vary locally, models transcending regional habitat variables should be developed and used to guide stand management in sensitive areas (Carreker 1985). Such work should be particularly important where forests are subject to notable harvest pressure or other anthropogenic effects and where conservation of threatened carnivores has become a priority.

OTHER HARES

Arctic Hare. Arctic hares (*Lepus arcticus*) are widely distributed in the tundra habitat of northern Canada, including the plateaus and mountains of western Newfoundland (Fig. 7.4). Most arctic hares spend summers north of treeline, but may move to forested habitat during winter (Bittner and Rongstad 1982; Flux and Angermann 1990; Best and Henry 1994a). Arctic hares may sometimes be observed on sea ice up to 5 km from land (Armstrong 1857 cited in Best and Henry 1994a). They are well adapted to harsh northern environments with their thick fur and reduced surface-to-area body ratio relative to other hare species (Wang et al. 1973). Such adaptations enable arctic hares to maintain normal body temperature and low metabolic rate in an environment where energy conservation is a premium. Arctic hares may respond behaviorally to inclement weather by altering their posture or orientation to reduce heat loss, or by digging snow dens (Gray 1993). Arctic hares also have feet that are heavily padded with fur, and their claws are well designed for digging through snow (Best and Henry 1994a). Their incisors are designed for feeding on snow-covered arctic plants. In addition, arctic hares have a locomotive efficiency that is appropriate to arctic landscapes, given their long legs relative to body size (Parker 1977; Klein and Bay 1994).

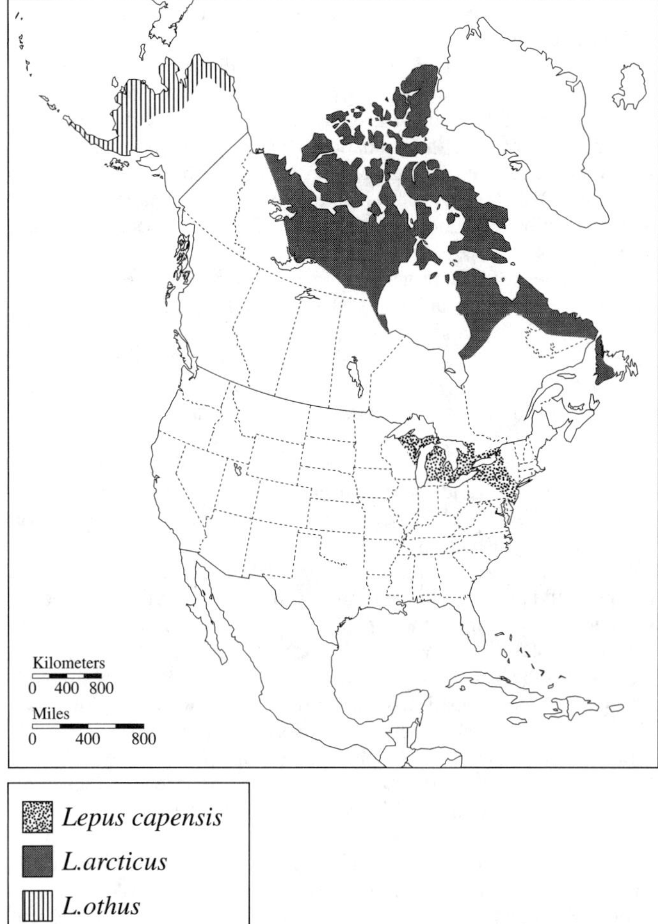

FIGURE 7.4. Distribution of European hares (*Lepus europaeus, L. capensis*), arctic hare (*L. arcticus*), and Alaskan hare (*L. othus*).

Arctic hares do not undergo the dramatic seasonal change in pelage coloration common to other hares, and in winter are white except for the black tips of the ears. The summer molt replaces the long winter coat with shorter fur that is grayish-white in southern subspecies and white in the others (Best and Henry 1994a). Females molt earlier than males (Banfield 1974). The size of arctic hares increases with increasing latitude, which likely is a response to selection initiated by local climatic conditions (Handley 1952; Baker et al. 1978). Body mass ranges from 3.9 to 4.5 kg and is lower among males, and overwinter declines in mass are the norm (Parker 1977). Like other hares, *L. arcticus* has a low body fat content (Thomas 1987) and relatively high daily energy requirements (Wang et al. 1973).

Mating occurs in April or May and gestation is roughly 53 days (Parker 1977; Best and Henry 1994a). Male testes may not regress until September, but late summer breeding probably is infrequent, given the rarity of pregnant females after July (Banfield 1974; Parker 1977). Females give birth in a nest lined with vegetation, which consists either of a simple depression in the ground or a sheltered spot under rocks (Aniskowicz et al. 1990). A single litter is born in June–July and averages five to six young, but may range from two to eight individuals (Parker 1977, Small et al. 1992). Litter sizes are smaller where day length at conception is shorter (Hearn et al. 1987). Arctic hare leverets remain with their mother for 2–3 weeks and are fully weaned by 8–9 weeks (Aniskowicz et al. 1990; Best and Henry 1994a). In years of limited food availability, recruitment of juveniles to the population appears not to occur (Mech 2000). Mass of the eye lens has been used to estimate age in arctic hares (Hearn and Mercer 1988).

The diet of arctic hares is varied, considering the limited plant species richness found in tundra habitats. On Banks Island, Northwest Territories, the winter diet consists mainly of willow (*S. arctica*) twigs with lesser amounts of legumes (*Astragalus alpinus, Oxytropis maydelliana*) (Larter 1999). During summer, the diet is dominated by legumes, but also includes mountain aven (*Dryas integrifolia*), sedges (*Carex aquatilis, Eriophorum scheuchzeri*), and tundra grass (*Dupontia fisheri*). In a population on Axel Heiberg Island, Northwest Territories, year-round use of willow twigs and leaves was observed with little use of legumes or sedges/grasses (Parker 1977). Sedges are underrepresented in hare diet relative to their availability in arctic habitats, likely because of their poor digestibility (Klein and Bay 1994, 1995). Arctic hares occasionally eat meat (Johnsen 1953). Limited evidence suggests that during winter, arctic hares may compete for food with moose, muskox (*Ovibos moschatus*), or caribou (*Rangifer tarandus*), but dietary overlap between these species usually is minimal (Wood 1974). Competition with snowshoe hares is unlikely given the apparent dominance of arctic hares (Fitzgerald and Keith 1990) and the tendency for differential habitat preference between the species (Barta et al. 1989).

Arctic hare population trends in North America generally are poorly understood (Keith 1983), but several reports point to large annual variations in density that may be food related (Howell 1936; Banfield 1974; Bonnyman 1975; Parker 1982; Mech 2000). Densities on Newfoundland average 1 hare/km^2 (Mercer et al. 1981). Although usually solitary, arctic hares can be more gregarious than most hare species and may form foraging groups of up to 300 individuals (Parker 1977; Klein 1999). Gregarious behavior is most pronounced in winter. Social interactions and group size may be further influenced by food abundance, predation risk, and snow conditions (Small et al. 1991). However, arctic hares appear to form breeding-season pairs and are allegedly monogamous (Howell 1936; Hearn et al. 1987). Home range size is variable (9–290 ha) and depends on habitat quality (Hearn et al. 1987; Barta et al. 1989; Small and Keith 1992). Arctic hares are parasitized by several species of *Eimeria,* nematodes, lice, and fleas (Best and Henry 1994a and references therein).

Strong selection for open habitat is related to the arctic hare's predator avoidance and escape behavior. Hares may sit upright to survey the tundra for approaching predators, and can escape a predatory attack by running at speeds exceeding 60 km/hr (Garland 1983). Groups of arctic hares may confuse predators by running erratically when chased, thereby limiting the predator's ability to focus on a single individual. In Newfoundland, arctic hare distribution and numbers declined dramatically following the increased abundance of lynx on the island, with losses occurring presumably in forested habitat where lynx occur (Bergerud 1967). However, other authors have contended that the distribution of arctic hare in Newfoundland has changed little (Mercer et al. 1981; Barta et al. 1989) and that fox predation may be more important in limiting their populations (Hearn et al. 1987; Small and Keith 1992). In the absence of elevated predator densities, arctic hares may occupy forested habitat with limited risk of predation, but when forced to use such habitats exclusively, they soon experience declining body mass and ultimately starve to death (Barta et al. 1989). In Eurasia, *L. timidus* has expanded its range southward into the boreal forest, but the absence of similar movements in *L. arcticus* may be due to the presence of *L. americanus* in the boreal forest and the latter species' effect on the number and diversity of predators occurring therein. Clearly, many aspects of arctic hare population biology remain to be fully elucidated.

Alaskan Hare. The taxonomic status of the Alaskan hare (*Lepus othus*) is unclear, with some authorities suggesting, based on cranial or other morphometric measurements, that they are conspecific with *L. arcticus*, and/or *L. timidus* (Hall 1951; Baker et al. 1983; Chapman et al. 1983; Dixon et al. 1983; Flux 1983). *L. othus* and *L. arcticus* also are known to share similar behavioral and ecological characteristics (Klein 1999). However, *L. othus* is geographically isolated and possesses different skull and incisor morphology from *L. arcticus*, and thus may warrant distinct taxonomic status (Rausch 1953; Jones et al. 1986; Best and Henry 1994b; Hoffmann 1993; Ramos 1998). The chromosome karyotype of *L. othus* is similar to that of other hares, including even *L. americanus* (Rausch and Ritter 1974; Stock 1976; Robinson et al. 1983). More recent phylogenetic analyses involving mitochondrial DNA sequence analysis (Halanych et al. 1999) further advocate the consolidation of *L. othus*, *L. arcticus*, and *L. timidus* into a single species.

The Alaskan hare has not been studied extensively, probably because of its restricted and isolated distribution along the Arctic tundra region of western Alaska (Fig. 7.4). The distribution is almost entirely north of treeline, including the North Slope of Alaska (Howell 1936), although verifiable records of this range extension are lacking (Anderson 1978). Alaskan hares have strongly recurved upper incisors, and their claws are effective for digging through snow to search for food (Best and Henry 1994b). During winter, Alaskan hares are white, except for black ear tips. During summer, the upper parts of the face and body turn brown (Walkinshaw 1947).

Alaskan hares are mainly solitary, but during the breeding season may form groups of 20 or more. Alaskan hares conceive between April and June, and gestation is about 46 days. The single litter averages 6.3 young (Anderson and Lent 1977). Preimplantation and postimplantation losses can be 35% and 8%, respectively, and are related to environmental conditions, food availability, and female age (Anderson 1974). The larger litter size and lower number of litters per year of *L. othus* compared to congeneric species likely are related to selection pressures exerted by a short growing season (Flux 1981). Not surprisingly, leverets grow rapidly and become fully weaned by about 5–9 weeks of age (Walkinshaw 1947; Anderson and Lent 1977). Adults and juveniles may be differentiated by body mass as well as through epiphyseal closure of the femur and the developmental stage of the anterior supraorbital process of the skull (Anderson 1974; Anderson and Lent 1977). Alaskan hares are among the largest species of *Lepus*. Body mass of adults usually ranges from 3.9 to 4.8 kg (Anderson and Lent 1977) and may sometimes exceed 7.2 kg (Howell 1936). A positive relationship exists between Alaskan hare body mass and increasing latitude (Anderson 1974).

Alaskan hares live in dense willow or alder (*Alnus* spp.) thickets and feed primarily on willow buds and leaves and crowberries (*Empetrum nigrum*) (Best and Henry 1994b). The two subspecies occupy disjunct populations; *L. othus othus* occupies tundra or alluvial plains and *L. othus poadromus* occurs primarily in coastal lowlands along the Alaskan island chain (Bittner and Rongstad 1982). A 42-year record of Alaskan hare pelt shipments suggests that their populations experience long-term fluctuations of high amplitude that may correspond to a 10-year cycle in abundance (Buckley 1954).

European Hare. European hares (*Lepus europaeus, L. capensis*) were introduced to North America as a game species, initially to New Jersey from England in 1888, and then to New York from Hungary in 1893 (Silver 1924, Stuttard 1981). Arrival of the European hare in southern Ontario in the early 1900s likely was the result of further introductions rather than range expansion (Reynolds 1952; Dean and de Vos 1965). The distribution of European hares is restricted to the Great Lakes region of Canada and the United States, east to the seaboard of New Jersey (Fig. 7.4). Because of its propensity to use open habitat and its speed and agility when being chased, the European hare became a highly prized game species in the early 1900s, with thousands being shot annually in Ontario alone (Howitt 1925; Stuttard 1981). European hares are no longer the focus of extensive sporting efforts, but are considered pests in forest plantations and fruit orchards throughout their range (Bittner and Rongstad 1982). Food habits of North American populations of the European hare are poorly known (Hansen and Flinders 1969).

Pelage of European hares is primarily rusty brown to gray. They molt particularly during spring and fall, but in some parts of their range they may molt year-round (Flux 1967). Adult body mass ranges between 3.0 and 4.0 kg, and in Europe, breeding takes place from January to September. Gestation averages 41 days (Raczynski 1964; Martinet et al. 1970; Flux 1981). European hares from Ontario averaged 1.6 fetuses in the early litters (January–March) and 3.8 in later litters (April–June); females may produce three to five litters per year (Reynolds and Stinson 1959; Flux 1981). Juveniles rarely breed during their first year. The home range reportedly is quite large, although few actual estimates of home range size are available.

Introduced European hare populations once achieved wide geographic distribution in the Great Lakes area of North America, but more recently their numbers have remained low and populations are isolated to small patches of suitable habitat (Dean and de Vos 1965; Hansen and Flinders 1969). Canadian populations may be higher than those found in the United States (Dean and de Vos 1965). Some populations in Ontario reportedly reach 0.08 hare/ha, and during the early 1900s may have undergone substantial numerical fluctuations (Reynolds 1955; Eabry 1970). In reviewing introductions of European hares, Jezierski (1968) concluded that competition with other leporids and lack of habitat suitability likely were responsible for the lack of increase in North American populations, but this assertion remains to be tested empirically.

ACKNOWLEDGMENTS

Support during preparation of this chapter was provided by the University of Idaho and the Idaho Department of Fish and Game. I am very grateful to D. Speten for assistance with library work, as well as to A. Wirsing and G. Feldhamer for constructive criticism of an earlier draft.

LITERATURE CITED

Adamcik, R. S., A. W. Todd, and L. B. Keith. 1978. Demographic and dietary responses of great horned owls during a snowshoe hare cycle. Canadian Field-Naturalist 92:156–66.

Adams, L. 1959. An analysis of a population of snowshoe hares in southwestern Montana. Ecological Monographs 29:141–70.

Adams, L., S. B. Salvin, and W. J. Hadlow. 1955. Ringworm in a population of snowshoe hare. Journal of Mammalogy 37:94–99.

Akçakaya, H. R. 1992. Population cycles of mammals: Evidence for a ratio-dependent predation hypothesis. Ecological Monographs 62:119–42.

Alain, G. 1967. Ecologie du lièvre d'Amérique, (*Lepus americanus struthopus*, Bangs) dans la region de Québec. B.S. Thesis, Laval University, Québec City, QC.

Aldous, C. M. 1936. Food habits of *Lepus americanus phaeonotus*. Journal of Mammalogy 17:175–76.

Aldous, C. M. 1937. Notes on the life history of the snowshoe hare. Journal of Mammalogy 18:46–57.

Aldous, C. M. 1946. Box trap for snowshoe hares and small rodents. Journal of Wildlife Management 10:71–72.

Aldous, C. M., and S. E. Aldous. 1944. The snowshoe hare, a serious enemy of forest plantations. Journal of Forestry 42:88–94.

Andersen, D. C., J. A. MacMahon, and M. L. Wolfe. 1980. Herbivorous mammals along a montane sere: Community structure and energetics. Journal of Mammalogy 61:500–519.

Anderson, H. L. 1974. Natural history and systematics of the tundra hare (*Lepus othus* Merriam) in western Alaska. M.S. Thesis, University of Alaska, Fairbanks, AK.

Anderson, H. L. 1978. Range of the tundra hare. Murrelet 59:72–74.

Anderson, H. L., and P. C. Lent. 1977. Reproduction and growth of the tundra hare (*Lepus othus*). Journal of Mammalogy 58:53–57.

Aniskowicz, B. T., H. Hamilton, D. R. Gray, and C. Downes. 1990. Nursing behaviour of arctic hares (*Lepus arcticus*). Pages 643–64 *in* C. R. Harrington, ed. Canada's missing dimension: Science and history in the Canadian Arctic islands. Canadian Museum of Man and Nature, Ottawa, ON.

Artsob, H. 1983. Distribution of California serogroup viruses and virus infections in Canada. Archives of Virology (Supplement) 1:249–58.

August, J. B. 1974. A study of liberation of snowshoe hare on ancestral range in Virginia. M.S. Thesis, Virginia Polytechnic Institute, Blacksburg, VA.

Austin, D. 1960. Some environmental relationships of the snowshoe hare, *Lepus americanus* Erxl., in Wellington County, Ontario. M.S. Thesis, Ontario Agricultural College, University of Toronto, Toronto, ON.

Bailey, R. A. 1946. Reading rabbit population cycles from pines. Wisconsin Conservation Bulletin 11:14–17.

Baker, A. J., R. L. Peterson, J. L. Eger, and T. H. Manning. 1978. Statistical analysis of geographic variation in the skull of the arctic hare (*Lepus arcticus*). Canadian Journal of Zoology 56:2067–82.

Baker, A. J., J. L. Eger, R. L. Peterson, and T. H. Manning. 1983. Geographic variation and taxonomy of arctic hares. Acta Zoologica Fennica 174:45–48.

Baker, B. E., H. W. Cook, J. R. Bider, and A. M. Pearson. 1970. Snowshoe hare (*Lepus americanus*) milk. I. Gross composition, fatty acid, and mineral constitution. Canadian Journal of Zoology 48:1349–52.

Baker, F. S., C. F. Korstian, and N. J. Feterolf. 1921. Snowshoe rabbits and conifers in the Wasatch Mountains of Utah. Ecology 2:304–10.

Banfield, A. W. F. 1974. The mammals of Canada. University of Toronto Press, Toronto, ON.

Barta, R. M., L. B. Keith, and S. M. Fitzgerald. 1989. Demography of sympatric arctic and snowshoe hare populations: An experimental assessment of interspecific competition. Canadian Journal of Zoology 67:2762–75.

Bartlett, C. M. 1983. Zoogeography and taxonomy of *Dirofilaria scapiceps* (Leidy 1886) and *D. uniformis*, Price 1957 (Nematoda: Filarioidea) of lagomorphs in North America. Canadian Journal of Zoology 61:1011–22.

Bartlett, C. M. 1984a. Pathology and epizootiology of *Dirofilaria scapiceps* (Leidy, 1886) (Nematoda: Filarioidea) in *Sylvilagus floridanus* (J. A. Allen) and *Lepus americanus* Erxleben. Journal of Wildlife Diseases 20:197–206.

Bartlett, C. M. 1984b. Development of *Dirofilaria scapiceps* (Leidy, 1886) (Nematoda: Filarioidea) in lagomorphs. Canadian Journal of Zoology 62:965–79.

Bartmann, R. M. and G. Byrne. 2001. Analysis and critique of the 1998 snowshoe hare pellet survey. Colorado Division of Wildlife Report No. 20, Denver, CO.

Bascompte, J., R. V. Solé, and N. Martínez. 1997. Population cycles and spatial patterns in snowshoe hares: Individual-oriented simulation. Journal of Theoretical Biology 187:213–222.

Behrend, D. F. 1960. A contribution to the ecology and management of the snowshoe hare in Connecticut. M.S. Thesis, University of Connecticut, Storrs, CT.

Belovsky, G. E. 1984. Snowshoe hare optimal foraging and its implications for population dynamics. Theoretical Population Biology 25:235–64.

Bendell, J. F., and L. I. Bendell-Young. 1993. Populations and habitats of snowshoe hares, ruffed, and spruce grouse in the southern boreal pine forest of Ontario. Pages 1–11 *in* D. Jenkins, ed. Proceedings of the international grouse symposium. World Pheasant Association, Reading, UK.

Bergeron, J.-M., and J. Tardif. 1988. Winter browsing preferences of snowshoe hares for coniferous seedlings and its implication in large-scale reforestation programs. Canadian Journal of Forest Research 18:280–82.

Bergerud, A. T. 1967. The distribution and abundance of arctic hares in Newfoundland. Canadian Field-Naturalist 81:242–48.

Best, T. L., and T. H. Henry. 1994a. *Lepus arcticus*. Mammalian Species 457:1–9.

Best, T. L., and T. H. Henry. 1994b. *Lepus othus*. Mammalian Species 458:1–5.

Bider, J. R. 1961. An ecological study of the hare *Lepus americanus*. Canadian Journal of Zoology 39:81–103.

Bittner, S. L., and O. J. Rongstad. 1982. Snowshoe hares and allies. Pages 146–63 *in* J. A. Chapman and G. A. Feldhamer, eds. Wild mammals of North America. Johns Hopkins University Press, Baltimore, MD.

Bjørnstad, O. N., and B. T. Grenfell. 2001. Noisy clockwork: Time series analysis of population fluctuations in animals. Science 293:638–43.

Black, H. C. 1965. An analysis of a population of snowshoe hares, *Lepus americanus washingtonii* Baird, in western Oregon. Ph.D. Dissertation, Oregon State University, Corvallis, OR.

Blasius, B., A. Huppert, and L. Stone. 1999. Complex dynamics and phase synchronization in spatially extended ecological systems. Nature 399:354–58.

Bloomer, S. E. M., T. Willebrand, I. M. Keith, and L. B. Keith. 1995. Impact of helminth parasitism on a snowshoe hare population in central Wisconsin: A field experiment. Canadian Journal of Zoology 73:1891–98.

Bonnyman, S. G. 1975. Behavioral ecology of *Lepus arcticus*. M.S. Thesis, Carleton University, Ottawa, ON.

Bookhout, T. A. 1959. Reingestion by the snowshoe rabbit. Journal of Mammalogy 40:250.

Bookhout, T. A. 1964. Prenatal development of snowshoe hares. Journal of Wildlife Management 28:338–45.

Bookhout, T. A. 1965a. Breeding biology of snowshoe hares in Michigan's Upper Peninsula. Journal of Wildlife Management 29:296–303.

Bookhout, T. A. 1965b. Feeding coactions between snowshoe hares and white-tailed deer in northern Michigan. Transactions of the North American Wildlife and Natural Resources Conference 30:321–35.

Bookhout, T. A. 1965c. The snowshoe hare in upper Michigan: Its biology and feeding coactions with white-tailed deer (Research and Development Report 438). Michigan Department of Conservation.

Bookhout, T. A. 1971. Helminth parasites in snowshoe hares from northern Michigan. Journal of Wildlife Diseases 7:246–48.

Boonstra, R., and G. R Singleton. 1993. Population declines in the snowshoe hare and the role of stress. General and Comparative Endocrinology 91:126–43.

Boonstra, R., D. Hik, G. R. Singleton, and A. Tinnikov. 1998a. The impact of predator-induced stress on the snowshoe hare cycle. Ecological Monographs 79:371–94.

Boonstra, R., C. J. Krebs, and N. C. Stenseth. 1998b. Population cycles in small mammals: The problem of explaining the low phase. Ecology 79:1479–88.

Borreco, J. E. 1976. Controlling damage by forest rodents and lagomorphs through habitat manipulation. Pages 203–10 *in* C. S. Siebe, ed. Proceedings of the 7th Vertebrate Pest Conference. Monterey, CA.

Boughton, R. V. 1932. The influence of helminth parasitism on the abundance of the snowshoe rabbit in western Canada. Canadian Journal of Research 7:524–47.

Boulanger, J. G., and C. J. Krebs. 1994. Comparison of capture–recapture estimators of snowshoe hare populations. Canadian Journal of Zoology 72:1800–1807.

Boulanger, J. G., and C. J. Krebs. 1996. Robustness of capture–recapture estimators to sample biases in a cyclic snowshoe hare population. Journal of Applied Ecology 33:530–42.

Boutin, S. 1980. Effect of spring removal experiments on the spacing behavior of female snowshoe hares. Canadian Journal of Zoology 58:2167–74.

Boutin, S. 1984a. Effect of late winter food addition on numbers and movements of snowshoe hares. Oecologia 62:393–400.

Boutin, S. 1984b. Effect of conspecifics on juvenile survival and recruitment of snowshoe hares. Journal of Animal Ecology 53:623–37.

Boutin, S. 1984c. Home range size and methods of estimating snowshoe hare densities. Acta Zoologica Fennica 171:275–78.

Boutin, S., and C. J. Krebs. 1986. Estimating survival rates of snowshoe hares. Journal of Wildlife Management 50:592–94.

Boutin, S., B. S. Gilbert, C. J. Krebs, A. R. E. Sinclair, and J. N. M. Smith. 1985. The role of dispersal in the population dynamics of snowshoe hares. Canadian Journal of Zoology 63:106–15.

Boutin, S., C. J. Krebs, A. R. E. Sinclair, and J. N. M. Smith. 1986. Proximate causes of losses in a snowshoe hare population. Canadian Journal of Zoology 64:606–10.

Boutin, S., C. J. Krebs, R. Boonstra, M. R. T. Dale, S. J. Hannon, K. Martin,

A. R. E. Sinclair, J. N. M. Smith, R. Turkington, M. Blower, A. Byrom, F. I. Doyle, C. Doyle, D. Hik, L. Hofer, A. Hubbs, T. Karels, D. L. Murray, M. O'Donoghue, C. Rohner, and S. Schweiger. 1995. Population changes of the vertebrate community during a snowshoe hare cycle in Canada's boreal forest. Oikos 74:69–80.

Boutin, S., C. J. Krebs, V. O. Nams, A. R. E. Sinclair, R. Boonstra, M. O'Donoghue, and C. Doyle. 2001. Experimental design and practical problems of implementation. Pages 50–66 *in* C. J. Krebs, S. Boutin, and R. Boonstra, eds. Ecosystem dynamics in the boreal forest: The Kluane project. Oxford University Press, New York, NY.

Boyle, J. D. 1955. An evaluation of stocking New Brunswick hare, *Lepus americanus,* in Massachusetts mixed forest habitat. M.S. Thesis, University of Massachusetts, Amherst, MA.

Brand, C. J., R. H. Vowles, and L. B. Keith. 1975. Snowshoe hare mortality monitored by telemetry. Journal of Wildlife Management 39:741–47.

Brand, C. J., L. B. Keith., and C. A. Fischer. 1976. Lynx responses to changing snowshoe hare densities in central Alberta. Journal of Wildlife Management 40:416–28.

Brocke, R. H. 1972. A live snare for trap-shy snowshoe hares. Journal of Wildlife Management 36:988–91.

Brocke, R. H. 1975. Preliminary guidelines for snowshoe hare habitat management in the Adirondacks. Transactions of the Northeast Fish and Wildlife Conference 32:46–66.

Brocke, R. H. 1982. Restoration of the lynx (*Lynx canadensis*) in Adirondack Park: A problem analysis and recommendations (New York State Department of Environmental Conservation Federal Aid Project E-1-3 and W-105-R). Albany, NY.

Brooks, M. 1955. An isolated population of the Virginia varying hare. Journal of Wildlife Management 19:54–61.

Brown, D. F. 1984. Snowshoe hare populations, habitat, and management in northern hardwood forest regeneration areas. M.S. Thesis, Pennsylvania State University, University Park, PA.

Bryant, J. P. 1981a. Phytochemical deterrence of snowshoe hare browsing by adventitious shoots of four Alaskan trees. Science 213:889–90.

Bryant, J. P. 1981b. The regulation of snowshoe hare feeding behavior during winter by plant antiherbivore chemistry. Pages 720–31 *in* K. Meyers and C. D. MacInnes, eds. Proceedings of the world lagomorph conference. University of Guelph, Guelph, ON.

Bryant, J. P. 1987. Feltleaf willow-snowshoe hare interactions: Plant carbon/nutrient balance and floodplain succession. Ecology 68:1319–27.

Bryant, J. P., and P. J. Kuropat. 1980. Selection of winter forage by subarctic browsing vertebrates: The role of plant chemistry. Annual Review of Ecology and Systematics 11:261–85.

Bryant, J. P., F. S. Chapin III, and D. R. Klein. 1983a. Carbon/nutrient balance of boreal plants in relation to vertebrate herbivory. Oikos 40:357–68.

Bryant, J. P., G. D. Wieland, P. B. Reichardt, V. E. Lewis, and M. C. McCarthy. 1983b. Pinosylvin methyl ether deters snowshoe hare feeding on green alder. Science 222:1023–25.

Bryant, J. P., G. D. Wieland, T. Clausen, and P. Kuropat. 1985. Interactions of snowshoe hare and feltleaf willow in Alaska. Ecology 66:1564–73.

Bryant, J. P., J. Tahvainen, M. Sulkinoia, R. Julken-Titto, P. Reichardt, and T. Green. 1989. Biogeographic evidence for the evolution of chemical defense by boreal birch and willow against mammalian browsing. American Naturalist 134:20–34.

Bryant, J. P., F. D. Provenza, J. Pastor, P. B. Reichardt, T. P. Clausen, and J. T. du Toit. 1991a. Interactions between woody plants and browsing mammals mediated by secondary metabolites. Annual Review of Ecology and Systematics 22:431–46.

Bryant, J. P., F. D. Provenza, P. B. Reichardt, and T. P. Clausen. 1991b. The effects of mammal browsing upon the chemistry of deciduous woody plants. Pages 135–54 *in* D. W. Pallamy and M. J. Raup, eds. Phytochemical induction by herbivores. John Wiley, New York, NY.

Bryant, J. P., R. K. Swihart, P. B. Reichardt, and L. Newton. 1994. Biogeography of woody plant chemical defense against snowshoe hare browsing: Comparison of Alaska and eastern North America. Oikos 70:385–95.

Buckley, J. L. 1954. Animal population fluctuations in Alaska: A history. Transactions of the North American Wildlife and Natural Resources Conference 19:338–57.

Buehler, D. A., and L. B. Keith. 1982. Snowshoe hare distribution and habitat use in Wisconsin. Canadian Field-Naturalist 96:19–29.

Bulmer, M. G. 1974. A statistical analysis of the 10-year cycle in Canada. Journal of Animal Ecology 43:701–18.

Bulmer, M. G. 1975. Phase relations in the ten-year cycle. Journal of Animal Ecology 44:609–21.

Burgdorfer, W., V. F. Newhouse, and L. A. Thomas. 1961. Isolation of California encephalitis virus from the blood of the snowshoe hare (*Lepus americanus*) in western Montana. American Journal of Hygiene 73:344–49.

Bursey, C. C., and M. D. B. Burt. 1970. *Taenia macrocystis* (Diesing, 1850), its occurrence in Eastern Canada and Maine, U.S.A., and its life cycle in wild felines (*Lynx rufus* and *L. canadensis*) and hares (*Lepus americanus*). Canadian Journal of Zoology 48:1287–93.

Burton, C., C. J. Krebs, and E. B. Taylor. 2002. Population genetic structure of the cyclic snowshoe hare (*Lepus americanus*) in southwestern Yukon, Canada. Molecular Ecology 11:1689–1702.

Butler, L. 1953. The nature of cycles in populations of Canadian mammals. Canadian Journal of Zoology 31:242–62.

Byrne, G. 1998. A Colorado winter track survey for snowshoe hares and other species. Colorado Division of Wildlife, Glenwood Springs.

Campbell, A., and M. V. Glines. 1979. Development, survival, and oviposition of the rabbit tick, *Haemaphysalis leporispalustris* (Packard) (Acari: Ixodidae), at constant temperatures. Journal of Parasitology 65:777–82.

Campbell, A., R. M. Ward, and M. B. Garvie. 1980. Seasonal activity and frequency distributions of ticks (Acari: Ixodidae) infesting snowshoe hares in Nova Scotia, Canada. Journal of Medical Entomology 17:22–29.

Carreker, R. G. 1985. Habitat suitability index models: Snowshoe hare (Biological Report 82 [10.101]). U.S. Fish and Wildlife Service.

Cary, J. R., and L. B. Keith. 1979. Reproductive change in the 10-year cycle of snowshoe hares. Canadian Journal of Zoology 57:375–90.

Chapman, J. A., K. R. Dixon, W. Lopez-Forment, and D. E. Wilson. 1983. The New World jackrabbits and hares (genus *Lepus*)., 1. Taxonomic history and population status. Acta Zoologica Fennica 174:49–51.

Cheeke, P. R. 1983. The significance of fiber in rabbit nutrition. Journal of Applied Rabbit Research 6:103–6.

Cheeke, P. R. 1987. Rabbit feeding and nutrition. Academic Press, New York, NY.

Chitty, D. 1938. The snowshoe rabbit enquiry, 1936–37. Canadian Field-Naturalist 52:63–72.

Chitty, D. 1939. The snowshoe rabbit enquiry, 1937–38. Canadian Field-Naturalist 53:63–71.

Chitty, D. 1940. The snowshoe rabbit enquiry, 1938–39. Canadian Field-Naturalist 54:117–24.

Chitty, D. 1959. A note on shock disease. Ecology 40:728–31.

Chitty, D. 1967. The natural selection of self-regulatory behaviour in animal populations. Proceedings of the Ecological Society of Australia 2:51–78.

Chitty, D. and H. Chitty., 1942. The snowshoe rabbit enquiry, 1939–40. Canadian Field-Naturalist 56:17–21.

Chitty, D. H., and C. Elton. 1937. The snowshoe rabbit enquiry, 1935–36. Canadian Field-Naturalist 51:63–73.

Chitty, D., and M. Nicholson. 1943. The snowshoe hare rabbit enquiry, 1940–41. Canadian Field-Naturalist 57:64–68.

Chitty, H. 1943a. The snowshoe rabbit enquiry, 1941–42. Canadian Field-Naturalist 57:136–41.

Chitty, H. 1943b. The snowshoe rabbit enquiry, 1943–46. Journal of Animal Ecology 17:39–44.

Chitty, H. 1946. The snowshoe rabbit enquiry, 1942–43. Canadian Field-Naturalist 60:67–70.

Chitty, H. 1948. The snowshoe rabbit enquiry, 1943–46. Journal of Animal Ecology 17:39–44.

Chitty, H. 1950. The snowshoe rabbit enquiry, 1946–48. Journal of Animal Ecology 19:15–20.

Chivers, D. P., G. E. Brown, and R. J. F. Smith. 1996. The evolution of chemical alarm signals: Attracting predators benefits alarm signal senders. American Naturalist 148:649–59.

Choquenot, D., C. J. Krebs, A. R. E. Sinclair, R. Boonstra, and S. Boutin. 2001. Vertebrate community structure in the boreal forest. Pages 437–62 *in* C. J. Krebs, S. Boutin, and R. Boonstra, eds. Ecosystem dynamics in the boreal forest: The Kluane project. Oxford University Press, New York, NY.

Christian, J. J. 1950. The adreno-pituitary system and population cycle in mammals. Journal of Mammalogy 31:247–59.

Christian, J. J., and H. L. Ratcliffe. 1952. "Shock disease" in captive wild animals. American Journal of Pathology 28:725–37.

Cole, L. C. 1951. Population cycles and random oscillations. Journal of Wildlife Management 15:233–52.

Cole, L. C. 1954. Some features of random population cycles. Journal of Wildlife Management 18:2–24.

Conroy, M. J., L. W. Gysel, and G. R. Duderar. 1979. Habitat components of clear-cut areas for snowshoe hares in Michigan. Journal of Wildlife Management 43:680–90.

Cook, D. B., and S. B. Robeson. 1945. Varying hare and forest succession. Ecology 26:405–10.

Corson, C. W., and E. G. Cheyney. 1928. Injury by rabbits to coniferous reproduction. Journal of Forestry 26:539–43.

Cox, E. W., R. A. Garrott, and J. R. Cary. 1997. Effect of supplemental cover on survival of snowshoe hares and cottontail rabbits in patchy habitat. Canadian Journal of Zoology 75:1357–63.

Cox, W. T. 1936. Snowshoe rabbit migration, tick infestation, and weather cycles. Journal of Mammalogy 17:216–21.

Cox, W. T. 1938. Snowshoe hare useful in thinning forest stands. Journal of Forestry 36:1107–9.

Criddle, S. 1938. A study of the snowshoe rabbit. Canadian Field-Naturalist 52:31–40.

Crissey, W. F., and R. W. Darrow. 1949. A study of predator control of Valcour Island. New York State Conservation Department, Division of Fish and Game, Research Service 1, Albany, NY.

Cushwa, C. T., and K. P. Burnham. 1974. An inexpensive live-trap for snowshoe hares. Journal of Wildlife Management 38:939–41.

Dalby, P. L. 1985. Snowshoe hare. Pages 367–70 *in* H. H. Genoways and F. J. Brenner, eds. Species of special concern in Pennsylvania. Carnegie Museum of Natural History, Pittsburgh, PA.

Dalquest, W. W. 1942. Geographic variation in northwestern snowshoe hares. Journal of Mammalogy 23:166–83.

Dambacher, J. M., H. W. Li., J. O. Wolff, and P. A. Rossignol. 1999. Parsimonious interpretation of the impact of vegetation, food, and predation on snowshoe hare. Oikos 84:530–32.

Davidson, R. P., W. W. Mautz, H. H. Hayes, and J. B. Holter. 1978. The efficiency of food utilization and energy requirements of captive female fishers. Journal of Wildlife Management 42:811–21.

Davis, G. J., and R. K. Meyer. 1972. The effect of daylength on pituitary FSH and LH and gonadal development of snowshoe hares. Biological Reproduction 6:264–69.

Davis, G. J., and R. K. Meyer. 1973a. FSH and LH in the snowshoe hare during the increasing phase of the 10-year cycle. General and Comparative Endocrinology 20:53–60.

Davis, G. J., and R. K. Meyer. 1973b. Seasonal variation in LH and FSH of bilaterally castrated snowshoe hares. General and Comparative Endocrinology 20:61–68.

Dean, P. B., and A. de Vos. 1965. The spread and present status of the European hare, *Lepus europaeus hybridus* (Desmarest), in North America. Canadian Field-Naturalist 79:38–48.

de Bellefeuille, S., L. Bélanger, and J. Huot. 2001a. Clear-cutting and regeneration practices in Québec boreal balsam fir forest: Effects on snowshoe hare. Canadian Journal of Forest Research 31:41–51.

de Bellefeuille, S., N. Gagné, L. Bélanger, J. Huot, A. Cimon, S. Déry, and J.-P. Jette. 2001b. Effects de trois scenarios de régéneration de la sapinière boreale sur les passereaux nicheurs, les petits mammifères, et le lièvre d'Amérique. Canadian Journal of Forest Research 31:1312–25.

DeBlase, A. F., and R. E. Martin. 1980. A manual of mammalogy, 2nd ed. William C. Brown, Dubuque, IA.

Dell, J., and D. L. Schierbaum. 1974. Prenatal development of snowshoe hares. New York Fish and Game Journal 21:89–104.

de Vos, A. 1964. Food utilization of snowshoe hares on Manitoulin Island, Ontario. Journal of Forestry 62:238–44.

Dietrich, R. A., and D. D. Feist. 1980. Hematology of Alaskan snowshoe hares (*Lepus americanus macfarlandi*) during years of population decline. Comparative Biochemistry and Physiology 66A:545–47.

Dimock II, E. J., R. R. Silen, and V. E. Allen. 1976. Genetic resistance in Douglas-fir to damage by snowshoe hare and black-tailed deer. Forest Science 22:106–21.

Dixon, K. R., J. A. Chapman, G. R. Willner, D. E. Wilson, and W. Lopez-Forment. 1983. The New World jackrabbits and hares (genus *Lepus*)., 2. Numerical taxonomic analysis. Acta Zoologica Fennica 174:53–56.

Dlott, F., and R. Turkington. 2000. Regulation of boreal forest understory vegetation: The roles of resources and herbivores. Plant Ecology 151:239–51.

Dodds, D. G. 1960. Food competition and range relationships of moose and snowshoe hare in Newfoundland. Journal of Wildlife Management 24:52–60.

Dodds, D. G. 1965. Reproduction and productivity of snowshoe hares in Newfoundland. Journal of Wildlife Management 29:303–15.

Dodds, D. G. 1987. Nova Scotia's snowshoe hare: Life history and management. Department of Lands and Forests, Halifax, Nova Scotia, Canada.

Dodds, D. G., and J. Mackiewicz. 1961. Some parasites and diseases of snowshoe hares in Newfoundland. Journal of Wildlife Management 25:409–14.

Dodds, D. G., and H. G. Thurber. 1965. Snowshoe hare (*Lepus americanus struthopus*) harvests on Long Island, Nova Scotia. Canadian Field-Naturalist 79:130–33.

Dolbeer, R. A. 1972. The snowshoe hare in western United States: Its status and potential as a game animal. Pages 437–51 *in* Proceedings of the 52nd annual conference of the Western Association of State Game and Fish Commissions. Portland, OR.

Dolbeer, R. A., and W. R. Clark. 1975. Population ecology of snowshoe hares in the central Rocky Mountains. Journal of Wildlife Management 39:535–49.

Doucet, G. J. 1973. House cat as predator of snowshoe hare. Journal of Wildlife Management 37:591.

Doutt, J. K., C. A. Heppenstall., and J. E. Guilday. 1967. Mammals of Pennsylvania. Pennsylvania Game Commission, Harrisburg, PA.

Eabry, H. S. 1970. A feasibility study to investigate and evaluate the possible future directions of European hare management in New York (Federal Aid Pitman-Robertson Project, W-84-R-17).

Eaton, B. R. 1995. Estimates of snowshoe hare abundance from fecal pellet plot counts: A critical evaluation. M.S. Thesis, Acadia University, Wolfville, NS.

Efford, M. 1992. Comment—Revised estimates of the bias in the "minimum number alive" estimator. Canadian Journal of Zoology 70:628–31.

Elton, C. 1924. Fluctuations in the numbers of animals: Their causes and effects. British Journal of Experimental Biology 2:119–63.

Elton, C., and M. Nicholson. 1942. The ten-year cycle in numbers of the lynx in Canada. Journal of Animal Ecology 11:215–44.

Enck, J. W., B. L. Swift, and D. J. Decker. 1993. Reasons for decline in duck hunting: Insights from New York. Wildlife Society Bulletin 21:10–21.

Erickson, A. B. 1944. Helminth infections in relation to population fluctuations in snowshoe hares. Journal of Wildlife Management 8:134–53.

Ernest, J. 1974. Snowshoe hare studies (Final Report, Federal Aid in Wildlife Restoration Project, W-17-4, W-17-6, Jobs 10.7R and 10.8R). Alaska Department of Fish and Game, Fairbanks, AK.

Federal Register. 2000. Determination of threatened status for the contiguous U.S. distinct population segment of the Canada lynx and related rule. Federal Register 65:16052–86.

Feist, D. D. 1979. Adrenal catecholamines in Alaskan snowshoe hares during years of decline in population density. Comparative Biochemistry and Physiology 64A:441–43.

Feist, D. D. 1980. Corticosteroid release by adrenal tissue of Alaskan snowshoe hares in a year of population decline. Journal of Mammalogy 61:134–36.

Feist, D. D., and M. Rosenmann. 1975. Seasonal sympathoadrenal and metabolic responses to cold in the Alaskan snowshoe hare (*Lepus americanus macfarlani*). Comparative Biochemistry and Physiology 51A:449–55.

Ferguson, M. A. D., and H. G. Merriam. 1978. A winter feeding relationship between snowshoe hares and porcupines. Journal of Mammalogy 59:878–80.

Ferron, J. 1993. How do population density and food supply influence social behaviour in the snowshoe hare (*Lepus americanus*)? Canadian Journal of Zoology 71:1084–89.

Ferron, J., and J.-P. Ouellette. 1992. Daily partitioning of summer habitat and use of space by the snowshoe hare in southern boreal forest. Canadian Journal of Zoology 70:2178–83.

Ferron, J. F., F. Potvin, and C. Dussault. 1998. Short-term effects of logging on snowshoe hares in the boreal forest. Canadian Journal of Forest Research 28:1335–43.

Fevold, H. R., and H. B. Drummond. 1976. Steroid biosynthesis by adrenal tissue of snowshoe hares (*Lepus americanus*) collected in a year of peak population density. General and Comparative Endocrinology 28:113–17.

Fies, M. L. 1991. Snowshoe hare. Pages 576–78 *in* K. Terwilliger, ed. Virginia's endangered species. McDonald and Woodward, Blacksburg, VA.

Fies, M. L. 1993. Survival and movements of relocated snowshoe hares in western Virginia. Virginia Department of Game and Inland Fisheries.

Finerty, J. P. 1980. The population ecology of cycles in small mammals. Yale University Press, New Haven, CT.

Fitzgerald, S. M., and L. B. Keith. 1990. Intra- and interspecific dominance relationships among arctic and snowshoe hares. Canadian Journal of Zoology 68:457–64.

Fitzpatrick, W. A. 1957. The survival and movements of live-trapped and introduced hares (*Lepus americanus* Erxleben) in Massachusetts. M.S. Thesis, University of Massachusetts, Amherst, MA.

Flux, J. E. C. 1967. Reproduction and body weights of the hare *Lepus europaeus* Pallas, in New Zealand. New Zealand Journal of Science 10:357–401.

Flux, J. E. C. 1981. Reproductive strategies in the genus *Lepus*. Pages 155–74 *in* K. Meyers and C. D. MacInnes, eds. Proceedings of the world lagomorph conference. University of Guelph, Guelph, ON.

Flux, J. E. C. 1983. Introduction to taxonomic problems in hares. Acta Zoologica Fennica 174:7–10.

Flux, J. E. C., and R. Angermann. 1990. The hares and jackrabbits. Pages 61–94 *in* J. A. Chapman and J. E. C. Flux, eds. Rabbits, hares, and pikas: Status survey and action plan. IUCN, Gland, Switzerland.

Forcum, D. L. 1966. Postpartum behavior and vocalizations of snowshoe hares. Journal of Mammalogy 47:543.

Foresman, K. R., and D. E. Pearson. 1999. Activity patterns of American martens, *Martes americana,* snowshoe hares, *Lepus americanus,* and red squirrels, *Tamiasciurus hudsonicus,* in westcentral Montana. Canadian Field-Naturalist 113:386–89.

Fox, J. F. 1978. Forest fires and the snowshoe hare–Canada lynx cycle. Oecologia 31:349–74.

Fox, J. F., and J. P. Bryant. 1984. Instability of the snowshoe hare and woody plant interaction. Oecologia 63:128–35.

Fraga, M. J. 1998. Protein digestion. Pages 39–53 *in* C. de Blas and J. Wiseman, eds. The nutrition of the rabbit. CABI, New York, NY.

Fuller, T. K., and D. M. Heisey. 1986. Density-related changes in winter distribution of snowshoe hares in northcentral Minnesota. Journal of Wildlife Management 50:261–64.

Garland, T., Jr. 1983. The relation between maximal running speed and body mass in terrestrial mammals. Journal of Zoology 199:157–70.

Gashwiler, J. S. 1959. Small mammal study in west-central Oregon. Journal of Mammalogy 40:128–39.

Gashwiler, J. S. 1970. Plant and mammal changes on a clearcut in west-central Oregon. Ecology 51:1018–26.

Gendron, J.-C., and J.-M. Bergeron. 1975. Comparaison de quatre méthodes statistiques pour estimer les niveaux de population du lièvre d'Amérique, *Lepus americanus.* Canadian Journal of Zoology 53:657–60.

Gibbs, H. C., W. J. Crenshaw, and M. Mowatt. 1977. Seasonal changes in stomach worms (*Obeliscoides cuniculi*) in snowshoe hares in Maine. Journal of Wildlife Diseases 13:327–32.

Gilbert, B. S. 1990. Use of winter feeding craters by snowshoe hares. Canadian Journal of Zoology 68:1600–1602.

Gilbert, B. S., and S. Boutin. 1991. Effect of moonlight on winter activity of snowshoe hares. Arctic and Alpine Research 23:61–65.

Gillis, E. A. 1998. Survival of juvenile hares during a cyclic population increase. Canadian Journal of Zoology 76:1949–56.

Gillis, E. A., and C. J. Krebs. 1999. Natal dispersal of snowshoe hares during a cyclic population increase. Journal of Mammalogy 80:933–39.

Gillis, E. A., and C. J. Krebs. 2000. Survival of dispersing versus philopatric juvenile snowshoe hares: Do dispersers die? Oikos 90:343–46.

Gilpin, M. E. 1973. Do hares eat lynx? American Naturalist 107:727–30.

Glazer, R. B. 1959. An evaluation of a snowshoe hare restocking program in Centre County, Pennsylvania. M.S. Thesis, Pennsylvania State University, University Park, PA.

Goble, F. C., and E. L Cheatum. 1944. Notes on the lungworms of North American Leporidae. Journal of Parasitology 30:119–20.

Goble, F. C., and E. C. Dougherty. 1943. Notes on the lungworms (genus *Protostrongylus*) of varying hares (*Lepus americanus*) in eastern North America. Journal of Parasitology 29:397–404.

Godbout, G. 1999. Détermination de la présence d'un cycle de population du lièvre d'amérique (*Lepus americanus*) au Québec et des méthodes de suivi applicables à cette espèce. Faune et Parcs, Québec City, QC.

Gordon, D. C. 1954. Melanism in the varying hare, *Lepus americanus virginianus.* Journal of Mammalogy 35:122.

Graf, R. P. 1985. Social organization of snowshoe hares. Canadian Journal of Zoology 63:468–74.

Graf, R. P., and A. R. E. Sinclair. 1987. Parental care and adult aggression toward juvenile snowshoe hares. Arctic 40:175–78.

Grange, W. B. 1932a. Observations of the snowshoe hare, *Lepus americanus phaeonotus,* Allen. Journal of Mammalogy 13:1–19.

Grange, W. B. 1932b. The pelages and color changes of the snowshoe hare, *Lepus americanus phaeonotus,* Allen. Journal of Mammalogy 12:99–116.

Grange, W. B. 1949. The way to game abundance. Scribner's, New York, NY.

Grange, W. B. 1965. Fire and tree growth relationships to snowshoe rabbits. Proceedings of Annual Tall Timbers Fire Ecology Conference 4:110–25.

Gray, D. R. 1993. Behavioural adaptations to arctic winter: Shelter seeking by arctic hare (*Lepus arcticus*). Arctic 46:340–53.

Green, R. G., and J. F. Bell. 1939. Nonfatal infection with *Pasteurella tularensis* in the snowshoe hare. Journal of Bacteriology 38:114.

Green, R. G., and C. A. Evans. 1940a. Studies on a population cycle of snowshoe hares on the Lake Alexander area. Part 3, Effects of reproduction and mortality of young hares on the cycle. Journal of Wildlife Management 4:247–58.

Green, R. G., and C. A. Evans. 1940b. Studies on a population cycle of snowshoe hares on the Lake Alexander area. Part 1, Gross annual census. Journal of Wildlife Management 4:220–38.

Green, R. G., and C. A. Evans. 1940c. Studies on a population cycle of snowshoe hares on the Lake Alexander area. Part 2, Mortality according to age groups and seasons. Journal of Wildlife Management 4:267–78.

Green, R. G., and C. L. Larson. 1937. Shock disease of wild snowshoe rabbits. American Journal of Physiology 119:319–20.

Green, R. G., and C. L. Larson. 1938. Shock disease and the snowshoe hare cycle. Science 87:298–99.

Green, R. G., C. L. Larson, and D. W. Mather. 1938. The natural occurrence of shock disease in hares. Transactions of the North American Wildlife and Natural Resources Conference 3:877–81.

Green, R. G., C. L. Larson, and J. F. Bell. 1939. Shock diseases as the cause of the periodic decimation of the snowshoe hare. American Journal of Hygiene 30:83–102.

Green, R. G., C. A. Evans, and C. L. Larson. 1943. A ten-year population study of the rabbit tick *Haemaphysalis leporispalustris.* American Journal of Hygiene 38:260–81.

Grigal, D. F., and N. R. Moody. 1980. Estimation of browse by size classes for snowshoe hare. Journal of Wildlife Management 44:34–40.

Grimstad, P. R. 1988. California serogroup viruses. Pages 99–136 *in* T. P. Monath, ed. The arboviruses: Epidemiology and ecology. CRC Press, Boca Raton, FL.

Grimstad, P. R. 1994. California serogroup viral infections. Pages 71–79 *in* G. W. Beran, ed. Handbook of zoonoses: Section B Viral. CRC Press, Boca Raton, FL.

Halanych, K. M., J. R. Demboski, B. J. van Vuuren, D. R. Klein, T. J. Robinson, and J. A. Cook. 1999. Cytochrome *b* phylogeny of North American hares and jackrabbits (*Lepus,* Lagomorpha) and the effects of saturation in outgroup taxa. Molecular Phylogenetics and Evolution 11:213–21.

Hall, E. R. 1951. A synopsis of the North American Lagomorpha. University of Kansas Museum of Natural History Publications 5:119–202.

Hall, E. R. 1981. The mammals of North America. John Wiley, New York, NY.

Handley, C. O. 1952. A new hare (*Lepus arcticus*) from northern Canada. Proceedings of the Biological Society of Washington 65:199–200.

Hansen, R. M., and J. T. Flinders. 1969. Food habits of North American hares (Science Series No. 1). Range Science Department, Colorado State University, Fort Collins, CO.

Hart, J. S., H. Pohl, and J. S. Tener. 1965. Seasonal acclimatization in varying hare (*Lepus americanus*). Canadian Journal of Zoology 43:731–44.

Hartman, F. H. 1960. Census techniques for snowshoe hares. M.S. Thesis, Michigan State University, East Lansing, MI.

Hassell, M. P., J. H. Lawton, and R. M. May. 1976. Patterns of dynamical behaviour in single-species populations. Journal of Animal Ecology 45:471–86.

Haydon, D. T., and P. E. Greenwood. 2000. Spatial coupling in cyclic population dynamics: Models and data. Theoretical Population Biology 58:239–54.

Haydon, D. T., E. A. Gillis, C. I. Stefan, and C. J. Krebs. 1999. Biases in the estimation of the demographic parameters of a snowshoe hare population. Journal of Animal Ecology 68:501–12.

Hearn, B. J., and W. E. Mercer. 1988. Eye-lens weight as an indicator of age in Newfoundland arctic hares. Wildlife Society Bulletin 16:426–29.

Hearn, B. J., L. B. Keith, and O. J. Rongstad. 1987. Demography and ecology of the arctic hare (*Lepus arcticus*) in southwestern Newfoundland. Canadian Journal of Zoology 65:852–61.

Henshaw, J. 1966. Mass movements by snowshoe rabbits, *Lepus americanus.* Canadian Field-Naturalist 80:181.

Highby, P. R. 1943. Vectors, transmission, development, and incidence of *Dirofilaria scapiceps* (Leidy, 1886) (Nematoda) from the snowshoe hare in Minnesota. Journal of Parasitology 29:253–59.

Hik, D. S. 1995. Does risk of predation influence population dynamics? Evidence from the cyclic decline of snowshoe hares. Wildlife Research 22:115–29.

Hilborn, R., J. A. Redfield, and C. J. Krebs. 1976. On the reliability of enumeration for mark and recapture census of voles. Canadian Journal of Zoology 54:1019–24.

Hjalen, J., and T. Palo. 1992. Selection of deciduous trees by free ranging voles and hares in relation to plant chemistry. Oikos 63:447–84.

Hodges, K. E. 1999. Proximate factors affecting snowshoe hare movements during a cyclic population low phase. Ecoscience 6:487–96.

Hodges, K. E. 2000a. The ecology of snowshoe hares in northern boreal forests. Pages 117–61 *in* L. F. Ruggiero, K. B. Aubry, S. W. Buskirk, G. M. Koehler, C. J. Krebs, K. S. McKelvey, and J. R. Squires, eds. Ecology and conservation of lynx in the United States. University Press of Colorado and U.S. Department of Agriculture, Rocky Mountain Research Station, Denver, CO.

Hodges, K. E. 2000b. Ecology of snowshoe hares in southern boreal and montane forests. Pages 163–206 *in* L. F. Ruggiero, K. B. Aubry, S. W. Buskirk, G. M. Koehler, C. J. Krebs, K. S. McKelvey, and J. R. Squires, eds. Ecology and conservation of lynx in the United States. University Press of Colorado and U.S. Department of Agriculture, Rocky Mountain Research Station, Denver, CO.

Hodges, K. E. 2001. Differential predation by coyotes on snowshoe hares. Canadian Journal of Zoology 79:1878–84.

Hodges, K. E., C. J. Krebs, and A. R. E. Sinclair. 1999a. Snowshoe hare demography during a cyclic population low. Journal of Animal Ecology 68:581–94.

Hodges, K. E., C. I. Stefan, and E. A. Gillis. 1999b. Does body condition affect fecundity in a cyclic population of snowshoe hares? Canadian Journal of Zoology 77:1–6.

Hodges, K. E., C. J. Krebs, D. Hik, C. I. Stefan, E. A. Gillis, and C. E. Doyle. 2001. Snowshoe hare demography. Pages 141–78 *in* C. J. Krebs, S. Boutin, and R. Boonstra, eds. Ecosystem dynamics in the boreal forest: The Kluane project. Oxford University Press, New York, NY.

Hoff, G. L., and R. P. Hanson. 1973. Selected physiological parameters of free-ranging snowshoe hares. Journal of Mammalogy 54:1007–8.

Hoff, G. L., T. M. Yuill, J. O. Iversen, and R. P. Hanson. 1969. Snowshoe hares and the California encephalitis virus group in Alberta, 1961–1968. Bulletin of the Wildlife Disease Association 5:254–60.

Hoff, G. L., T. M. Yuill, J. O. Iversen, and R. P. Hanson. 1970. Selected microbial agents in snowshoe hares and other vertebrates in Alberta. Journal of Wildlife Diseases 6:472–78.

Hoff, G. L., A. J. Spalatin, and R. P. Hanson. 1971a. Isolation of Montana snowshoe hare serotype of California encephalitis virus group from a snowshoe hare and *Aedes* mosquitoes. Journal of Wildlife Diseases 7:28–34.

Hoff, G. L., J. O. Iversen, T. M. Yuill, R. O. Anslow, J. O. Jackson, and R. P. Hanson. 1971b. Isolations of silverwater virus from naturally infected snowshoe hares and *Haemaphysalis* ticks from Alberta and Wisconsin. American Journal of Tropical Medicine and Hygiene 20:320–25.

Hoff, G. L., T. M. Yuill, J. O. Iversen, and R. P. Hanson. 1971c. Silverwater virus serology in snowshoe hares and other vertebrates. American Journal of Tropical Medicine and Hygiene 20:326–30.

Hoffmann, R. S. 1993. Order Lagomorpha. Pages 807–82 *in* D. E. Wilson and D. M. Reeder, eds. Mammal species of the world: A taxonomic and geographic reference. Smithsonian Institution Press, Washington, DC.

Höhn, E. O., and J. G. Stelfox. 1977. Snowshoe hare adrenal weights in relation to population density. Canadian Journal of Zoology 55:634–37.

Holter, J. B., G. Tyler, and T. Walski. 1974. Nutrition of the snowshoe hare (*Lepus americanus*). Canadian Journal of Zoology 52:1553–58.

Hough, A. F. 1949. Deer and rabbit browsing and available winter forage in Allegheny hardwood forests. Journal of Wildlife Management 13:135–41.

Howell, A. B. 1923. Periodic fluctuations in the numbers of small mammals. Journal of Mammalogy 4:149–55.

Howell, A. B. 1936. A revision of the American arctic hares. Journal of Mammalogy 17:315–37.

Howitt, H. 1925. Another invasion of Canada. Canadian Field-Naturalist 39:153–60.

Hunt, G. S. 1950. Aquatic activity of a snowshoe hare. Journal of Mammalogy 31:193–94.

Irving, L., J. Krog, H. Krog, and M. Monson. 1957. Metabolism of varying hare in winter. Journal of Mammalogy 38:527–29.

Iversen, J. O., J. Spalatin, C. E. O. Fraser, R. P. Hanson, and D. T. Berman. 1970. The susceptibility of muskrats and snowshoe hares to experimental infection with a chlamydial agent. Canadian Journal of Comparative Medicine 34:80–89.

Iversen, J. O., G. L. Hoff, T. M. Yuill, and R. P. Hanson. 1972. Gastric lesions in the snowshoe hare. Journal of Wildlife Diseases 8:7–9.

Ives, A. R., and D. L. Murray. 1997. Can sublethal parasitism destabilize predator-prey population dynamics? A model of snowshoe hares, predators and parasites. Journal of Animal Ecology 66:265–78.

Jellison, W. L., C. R. Owen, J. F. Bell, and G. M. Kohls. 1961. Tularemia and animal populations: Ecology and epizootiology. Wildlife Diseases 17:1–22.

Jezierski, W. 1968. Some ecological aspects of introduction of the European hare. Acta Theriologica 13:1–30.

Jogia, M. K., A. R. E. Sinclair, and R. J. Anderson. 1989. An antifeedant in balsam poplar inhibits browsing by snowshoe hares. Oecologia 79:189–92.

John, E., and R. Turkington. 1995. Herbaceous vegetation in the understorey of the boreal forest: Does nutrient supply or snowshoe hare herbivory regulate species composition and abundance? Journal of Ecology 83:581–90.

John, E., and R. Turkington. 1997. A five-year study of the effects of nutrient availability and herbivory on two boreal forest herbs. Journal of Ecology 85:419–30.

Johnsen, P. 1953. Mammals observed on Amdrup's journeys to East Greenland. 1899–1900. Meddelelser om Grønland 29:1–62.

Johnson, C. E. 1925. The jack and snowshoe rabbits as swimmers. Journal of Mammalogy 6:245–49.

Jones, J. K., D. C. Carter, H. H. Genoways, R. S. Hoffmann, D. W. Rice, and C. Jones. 1986. Revised checklist of North American mammals north of Mexico, 1986. Occasional Papers, the Museum, Texas Tech University 107:1–22.

Keith, I. M.,L. B. Keith, and J. R. Cary. 1986. Parasitism in a declining population of snowshoe hares. Journal of Wildlife Diseases 22:349–63.

Keith, L. B. 1963. Wildlife's ten year cycle. University of Wisconsin Press, Madison, WI.

Keith, L. B. 1964. Daily activity pattern of snowshoe hares. Journal of Mammalogy 45:626–27.

Keith, L. B. 1965. A live snare and a tagging snare for rabbits. Journal of Wildlife Management 29:877–80.

Keith, L. B. 1966. Habitat vacancy during a snowshoe hare decline. Journal of Wildlife Management 30:828–32.

Keith, L. B. 1972. Snowshoe hare populations and forest regeneration in northern Alberta. Alberta Forest Service, Edmonton, AB.

Keith, L. B. 1974. Some features of population dynamics in mammals. Proceedings of the International Congress of Game Biologists 11:17–58.

Keith, L. B. 1981. Population dynamics of hares. Pages 395–440 *in* K. Meyers and C. D. MacInnes, eds. Proceedings of the world lagomorph conference. University of Guelph, Guelph, ON.

Keith, L. B. 1983. Role of food in hare population cycles. Oikos 40:385–95.

Keith, L. B. 1990. Dynamics of snowshoe hare populations. Pages 119–95 *in* H. H. Genoways, ed. Current mammalogy. Plenum Press, New York, NY.

Keith, L. B., and S. E. M. Bloomer. 1993. Differential mortality of sympatric snowshoe hares and cottontail rabbits in central Wisconsin. Canadian Journal of Zoology 71:1694–97.

Keith, L. B., and J. R. Cary. 1979. Eye lens weights from free-living adult snowshoe hares of known age. Journal of Wildlife Management 43:965–69.

Keith, L. B., and J. R. Cary. 1990. Interaction of the tick (*Haemaphysalis leporispalustris*) with a cyclic snowshoe hare (*Lepus americanus*) population. Journal of Wildlife Diseases 26:427–34.

Keith, L. B., and E. C. Meslow. 1966. Animals using runways in common with snowshoe hares. Journal of Mammalogy 47:541.

Keith, L. B., and E. C. Meslow. 1967. Juvenile breeding in the snowshoe hare. Journal of Mammalogy 48:327.

Keith, L. B., and E. C. Meslow. 1968. Trap response by snowshoe hares. Journal of Wildlife Management 32:795–801.

Keith, L. B., and D. H. Rusch. 1988. Predation's role in the cyclic fluctuations of ruffed grouse. Pages 699–732 *in* Acta XIX Congress of International Ornithologists.

Keith, L. B., and D. C. Surrendi. 1971. Effect of fire on a snowshoe hare population. Journal of Wildlife Management 35:16–26.

Keith, L. B., and J. D. Waring. 1956. Evidence of orientation and homing in snowshoe hares. Canadian Journal of Zoology 34:579–82.

Keith, L. B., and L. A. Windberg. 1978. A demographic analysis of the snowshoe hare cycle. Wildlife Monographs 58:1–70.

Keith, L. B., O. J. Rongstad, and E. C. Meslow. 1966. Regional differences in reproductive traits of the snowshoe hare. Canadian Journal of Zoology 44:953–61.

Keith, L. B., E. C. Meslow, and O. J. Rongstad. 1968. Techniques for snowshoe hare population studies. Journal of Wildlife Management 32:801–12.

Keith, L. B., A. W. Todd, C. J. Brand, R. S. Adamcik, and D. H. Rusch. 1977. An analysis of predation during a cyclic fluctuation of snowshoe hares. Proceedings of the International Congress of Game Biologists 13:151–75.

Keith, L. B., J. R. Cary, O. J. Rongstad, and M. C. Brittingham. 1984. Demography and ecology of a declining snowshoe hare population. Wildlife Monographs 90:1–43.

Keith, L. B., J. R. Cary, T. M. Yuill, and I. M. Keith. 1985. Prevalence of helminths in a cyclic snowshoe hare population. Journal of Wildlife Diseases 21:233–53.

Keith, L. B., S. E. M. Bloomer, and T. Willebrand. 1993. Dynamics of a snowshoe hare population in fragmented habitat. Canadian Journal of Zoology 71:1385–92.

King, A. A., and W. M. Schaffer. 2001. The geometry of a population cycle: A mechanistic model of snowshoe hare demography. Ecology 82:814–30.

Klein, D. R. 1977. Winter food preferences of snowshoe hares (*Lepus americanus*) in interior Alaska. Proceedings of the International Congress of Game Biologists 13:266–75.

Klein, D. R. 1999. Comparative social learning among arctic herbivores: The caribou, muskox, and arctic hare. Pages 126–40 *in* H. O. Box and K. R. Gibson, eds. Mammalian social learning (Zoological Society of London, Symposium 73). Cambridge University Press, Cambridge, UK.

Klein, D. R., and C. Bay. 1994. Resource partitioning by mammalian herbivores in the high Arctic. Oecologia 97:439–50.

Klein, D. R., and C. Bay. 1995. Digestibility of forage types by arctic hares. Ecoscience 2:100–102.

Kloor, K. 1999. Lynx and biologists try to recover after disastrous start. Science 285:320–21.

Koehler, G. M. 1990a. Snowshoe hare, *Lepus americanus,* use of forest successional stages and population changes during 1985–1989 in north-central Washington. Canadian Field-Naturalist 105:291–93.

Koehler, G. M. 1990b. Population and habitat characteristics of lynx and snowshoe hares in north central Washington. Canadian Journal of Zoology 68:845–51.

Koehler, G. M., and J. D. Brittell. 1990. Managing spruce–fir habitat for lynx and snowshoe hares. Journal of Forestry 88:10–14.

Kralka, R. A., and W. M. Samuel. 1984. Experimental life cycle of *Protostrongylus boughtoni* (Nematoda: Metastrongyloidea), a lungworm of snowshoe hares, *Lepus americanus.* Canadian Journal of Zoology 62:473–79.

Kralka, R. A., and W. M. Samuel. 1990. The lungworm *Protostrongylus boughtoni* (Nematoda, Metastrongyloidea) in gastropod intermediate hosts and the snowshoe hare, *Lepus americanus.* Canadian Journal of Zoology 68:2567–75.

Krebs, C. J. 1986. Are lagomorphs similar to other small mammals in their population ecology? Mammal Review 16:187–94.

Krebs, C. J. 1996. Population cycles revisited. Journal of Mammalogy 77:8–24.

Krebs, C. J., S. Boutin, and B. S. Gilbert. 1986a. A natural feeding experiment on a declining snowshoe hare population. Oecologia 70:194–97.

Krebs, C. J., B. S. Gilbert, S. Boutin, A. R. E. Sinclair, and J. N. M. Smith. 1986b. Population biology of snowshoe hares. I. Demography of supplemented populations in the southern Yukon. Journal of Animal Ecology 55:963–82.

Krebs, C. J., B. S. Gilbert, S. Boutin, and R. Boonstra. 1987. Estimation of snowshoe hare population density from turd transects. Canadian Journal of Zoology 65:565–67.

Krebs, C. J., R. Boonstra, S. Boutin, M. Dale, S. Hannon, K. Martin, A. R. E. Sinclair, J. N. M. Smith, and R. Turkington. 1992. What drives the snowshoe hare cycle in Canada's Yukon? Pages 886–96 *in* D. M. McCullough and R. Barrett, eds. Wildlife 2001: populations. Elsevier, London, UK.

Krebs, C. J., S. Boutin, R. Boonstra, A. R. E. Sinclair, J. N. M. Smith, M. R. T. Dale, K. Martin, and R. Turkington. 1995. Impact of food and predation on the snowshoe hare cycle. Science 269:1012–15.

Krebs, C. J., R. Boonstra, S. Boutin, and A. R. E. Sinclair. 2001a. What drives the 10-year cycle of snowshoe hares? Bioscience 51:25–35.

Krebs, C. J., R. Boonstra, V. Nams, M. O'Donoghue, K. E. Hodges, and S. Boutin. 2001b. Estimating snowshoe hare population density from pellet counts: A further evaluation. Canadian Journal of Zoology 79:1–4.

Krebs, C. J., S. Boutin, and R. Boonstra, eds. 2001c. Ecosystem dynamics of the boreal forest: The Kluane project. Oxford University Press, New York, NY.

Krebs, C. J., R. Boonstra, S. Boutin, and A. R. E. Sinclair. 2001d. Conclusions and future directions. Pages 491–501 *in* C. J. Krebs, S. Boutin, and R. Boonstra, eds. Ecosystem dynamics in the boreal forest: The Kluane project. Oxford University Press, New York, NY.

Krefting, L. W. 1975. The effect of white-tailed deer and snowshoe hare browsing on trees and shrubs in northern Minnesota. (Technical Bulletin 302–1975, Forestry Series 18). Agricultural Experiment Station, University of Minnesota, Minneapolis, MN.

Krefting, L. W., and J. H. Stoeckeler. 1953. Effect of simulated snowshoe hare and deer damage on planted conifers in the lake states. Journal of Wildlife Management 17:487–94.

Krenz, J. D. 1988. Effect of vegetation dispersion on the density of wintering snowshoe hares (*Lepus americanus*) in northern Minnesota. M.S. Thesis, University of Minnesota, Minneapolis, MN.

Kuehn, D. W. 1989. Winter foods of fishers during a snowshoe hare decline. Journal of Wildlife Management 53:688–92.

Kuvlesky, W. P., and L. B. Keith. 1983. Demography of snowshoe hare populations in Wisconsin. Journal of Mammalogy 64:233–44.

Lack, D. 1954. Cyclic mortality. Journal of Wildlife Management 18:25–37.

Larter, N. C. 1999. Seasonal changes in arctic hare, *Lepus arcticus,* diet composition and differential digestibility. Canadian Field-Naturalist 113:481–86.

Lauckhart, J. B. 1957. Animal cycles and food. Journal of Wildlife Management 21:230–34.

Leopold, A. 1937. Game management. Scribner's, New York, NY.

Lewinsohn, E., T. J. Savage, M. Gijzen, and R. Croteau. 1993. Simultaneous analysis of monoterpenes and diterpenes of conifer oleoresin. Phytochemical Analysis 4:220–25.

Lidicker, W. J., J. O. Wolff, and R. A. Mowrey. 2000. Genetic change in a cyclic population of snowshoe hares. Ecoscience 7:247–55.

Litvaitis, J. A. 1990. Differential habitat use by sexes of snowshoe hares (*Lepus americanus*). Journal of Mammalogy 71:520–23.

Litvaitis, J. A. 1991. Habitat use by snowshoe hares, *Lepus americanus,* in relation to pelage color. Canadian Field-Naturalist 105:275–77.

Litvaitis, J. A., and W. W. Mautz. 1976. Energy utilization of 3 diets fed to a captive red fox. Journal of Wildlife Management 40:365–68.

Litvaitis, J. A., and W. W. Mautz. 1980. Food and energy use by captive coyotes. Journal of Wildlife Management 44:56–61.

Litvaitis, J. A., J. A. Sherburne, and J. A. Bissonette. 1985a. Influence of understory characteristics on snowshoe hare habitat use and density. Journal of Wildlife Management 49:866–73.

Litvaitis, J. A., J. A. Sherburne, and J. A. Bissonette. 1985b. A comparison of methods used to examine snowshoe hare habitat use. Journal of Wildlife Management 49:693–95.

Litvaitis, J. A., J. A. Sherburne, and J. A. Bissonette. 1986. Bobcat habitat use and home range size in relation to prey density. Journal of Wildlife Management 50:110–17.

Lloyd-Smith, J., and H. Piene. 1981. Snowshoe hare girdling of balsam fir on the Cape Breton Highlands (Information Report M-X-124). Canadian Forest Service, Maritimes Forest Research Centre.

Lyman, C. P. 1943. Control of coat color in the varying hare *Lepus americanus.* Bulletin of the Museum of Comparative Zoology (Harvard University) 93:391–461.

MacCracken, J. G.,W. D. Steigers, Jr., and P. V. Mayer. 1988. Winter and early spring habitat use by snowshoe hares, *Lepus americanus,* in south-central Alaska. Canadian Field-Naturalist 102:25–30.

MacLulich, D. A. 1937. Fluctuations in the numbers of the varying hare (*Lepus americanus*) (University of Toronto Studies, Series No. 43). University of Toronto Press, Toronto, ON.

Malloy, J. C. 2000. Snowshoe hare, *Lepus americanus,* fecal pellet fluctuations in western Montana. Canadian Field-Naturalist 114:409–12.

Maltais, P. M., and E. A. Ouellette. 1983. Helminth parasites of the snowshoe hare (*Lepus americanus*) in New Brunswick. Naturaliste Canadien 110:103–5.

Manweiler, J. 1938. Parasites of the snowshoe hare. Journal of Mammalogy 19:379.

Marshall, W. H. 1954. Ruffed grouse and snowshoe hare populations on the Cloquet Experimental Forest, Minnesota. Journal of Wildlife Management 18:109–12.

Martinet, L., J.-J. Legouis, and B. Moret. 1970. Quelques observations sur la reproduction du lièvre européen (*Lepus europaeus* Pallas) en captivité; influence du photopériodisme. Transactions of the International Congress of Game Biologists 10:553–61.

Mautz, W. W., T. W. Walski, and W. E. Urban. 1976. Digestibility of fresh frozen pelleted browse by snowshoe hares. Journal of Wildlife Management 40:496–99.

May, R. M. 1975. Stability and complexity in model ecosystems. Princeton University Press, Princeton, NJ.

McAvinchey, R. J. P. 1991. Winter herbivory by snowshoe hares and moose as a process affecting primary succession on an Alaskan floodplain. M.S. Thesis, University of Alaska, Fairbanks, AK.

McDonough, J. J. 1969. The snowshoe hare stocking program in Massachusetts. Transactions of the Northeast Section of the Wildlife Society Fish and Wildlife Conference 28:57–68.

McGowan, J. M. 1985. Relationship between skin-sensitizing antibody production in the snowshoe hare, *Lepus americanus,* and infestations by the rabbit tick, *Haemaphysalis leporispalustris* (Acari: Ixodidae). Journal of Parasitology 71:513–15.

McGowan, J. M., R. W. McNew, J. T. Homer, and J. H. Camins. 1982. Relationship between skin-sensitizing antibody production in the eastern cottontail, *Sylvilagus floridanus,* and infestations by the rabbit tick, *Haemaphysalis leporispalustris,* and the American dog tick, *Dermacentor variabilis* (Acari: Ixodidae). Journal of Medical Entomology 19:198–203.

McKelvey, K. S., G. W. McDaniel, L. S. Mills, and P. C. Griffin 2002. Effects of plot size and shape on pellet density estimates for snowshoe hares. Wildlife Society Bulletin 30:751–755.

McLean, D. M. 1983. Yukon isolates of snowshoe hare virus, 1972–1982. Pages 247–56 *in* C. H. Calisher and W. H. Thompson, eds. California serogroup viruses. Alan R. Liss, New York, NY.

Mech, L. D. 2000. Lack of reproduction in muskoxen and arctic hares caused by early winter? Arctic 53:69–71.

Mech, L. D., K. L. Heezen, and D. B. Siniff. 1966. Onset and cessation of activity of cottontail rabbits and snowshoe hares in relation to sunset and sunrise. Animal Behavior 14:410–13.

Mercer, W. E., B. J. Hearn, and C. Finlay. 1981. Arctic hare populations in insular Newfoundland. Pages 450–68 *in* K. Meyers and C. D. MacInnes, eds. Proceedings of the world lagomorph conference. University of Guelph, Guelph, ON.

Meslow, E. C., and L. B. Keith. 1968. Demographic parameters of a snowshoe hare population. Journal of Wildlife Management 32:812–34.

Meslow, E. C., and L. B. Keith. 1971. A correlation analysis of weather versus snowshoe hare population parameters. Journal of Wildlife Management 35:1–15.

Monthey, R. W. 1986. Responses of snowshoe hares, *Lepus americanus,* to timber harvesting in northern Maine. Canadian Field-Naturalist 100:568–70.

Moran, P. A. P. 1953a. The statistical analysis of the Canada lynx cycle. I. Structure and prediction. Australian Journal of Zoology 1:163–73.

Moran, P. A. P. 1953b. The statistical analysis of the Canada lynx cycle. II. Synchronization and meteorology. Australian Journal of Zoology 1:291–98.

Morse, M. 1939. A local study of predation upon hares and grouse during the cyclic decimation. Journal of Wildlife Management 3:203–11.

Mozejko, A. T. 1971. A food habits study of the snowshoe hare, *Lepus americanus washingtonii* Baird, in western Oregon. Ph.D. Dissertation, Oregon State University, Corvallis, OR.

Murray, D. L. 1999. The role of overwinter food limitation on a snowshoe hare population at a cyclic low. Oecologia 120:50–58.

Murray, D. L. 2000. A geographic analysis of snowshoe hare population demography. Canadian Journal of Zoology 78:1207–17.

Murray, D. L. 2002. Differential body condition and vulnerability to predation in snowshoe hares. Journal of Animal Ecology 71:614–625.

Murray, D. L., and S. Boutin. 1991. The influence of snow on lynx and coyote movements: Does morphology affect behavior? Oecologia 88:463–69.

Murray, D. L., S. Boutin, and M. O'Donoghue. 1994. Winter habitat selection by lynx and coyotes in relation to snowshoe hare abundance. Canadian Journal of Zoology 72:1444–51.

Murray, D. L., S. Boutin, M. O'Donoghue, and V. O. Nams. 1995. Hunting behaviour of a sympatric felid and canid in relation to vegetative cover. Animal Behaviour 50:1203–10.

Murray, D. L., L. B. Keith, and J. R. Cary. 1996. The efficacy of anthelmintic treatment on the parasite abundance of free-ranging snowshoe hares. Canadian Journal of Zoology 74:1604–11.

Murray, D. L., J. R. Cary, and L. B. Keith. 1997. Interactive effects of sublethal nematodes and nutritional status on snowshoe hare vulnerability to predation. Journal of Animal Ecology 66:250–64.

Murray, D. L., L. B. Keith, and J. R. Cary. 1998. Do parasitism and nutritional status interact to affect production in snowshoe hares? Ecology 79:1209 22.

Murray, D. L., J. D. Roth, E. Ellsworth, A. J. Wirsing, and T. D. Steury. 2002. Estimating low density snowshoe hare populations using fecal pellet counts. Canadian Journal of Zoology 80:771–781.

Myers, K., H. G. Bults, and N. Gilbert. 1981. Stress in the rabbit. Pages 103–36 *in* K. Meyers and C. D. MacInnes, eds. Proceedings of the world lagomorph conference. University of Guelph, Guelph, ON.

Nagorsen, D. W. 1983. Winter pelage in snowshoe hares (*Lepus americanus*) from the Pacific Northwest. Canadian Journal of Zoology 61:2313–18.

Nagorsen, D. W. 1985. A morphometric study of geographic variation in the snowshoe hare (*Lepus americanus*). Canadian Journal of Zoology 63:567–79.

Nams, V. O., N. F. G. Folkard, and J. N. M. Smith. 1996. Nitrogen fertilization stimulates herbivory by snowshoe hares in the boreal forest. Canadian Journal of Zoology 74:196–99.

Nellis, C. H., S. P. Wetmore, and L. B. Keith. 1972. Lynx-prey interactions in central Alberta. Journal of Wildlife Management 36:320–29.

Newson, J. 1964. Reproduction and prenatal mortality of snowshoe hares on Manitoulin Island, Ontario. Canadian Journal of Zoology 42:987–1004.

Newson, R., and A. de Vos. 1964. Population structure and body weights of snowshoe hares on Manitoulin Island, Ontario. Canadian Journal of Zoology 42:975–86.

Nice, M. M., C. Nice, and D. Ewers. 1956. Comparison of behavior development in snowshoe hares and red squirrels. Journal of Mammalogy 37:64–74.

Novak, M. 1987. The future of trapping. Pages 89–97 *in* M. Novak, J. A. Baker, M. E. Obbard, and B. Malloch, eds. Wild furbearrer management and conservation in North America. Ontario Ministry of Natural Resources, Toronto, ON.

O'Donoghue, M. 1983. Seasonal habitat selection by snowshoe hare in eastern Maine. Transactions of the Northeast Fish and Wildlife Conference 40:100–107.

O'Donoghue, M. 1994. Early survival of juvenile snowshoe hares. Ecology 75:1582–92.

O'Donoghue, M. and C. M. Bergman., 1992. Early movements and dispersal of juvenile snowshoe hares. Canadian Journal of Zoology 70:1787–91.

O'Donoghue, M., and S. Boutin. 1995. Does reproductive synchrony affect juvenile survival rates of northern mammals? Oikos 74:115–21.

O'Donoghue, M., and C. J. Krebs. 1992. Effects of supplemental food on snowshoe hare reproduction and juvenile growth at a cyclic population peak. Journal of Animal Ecology 61:631–41.

O'Donoghue, M., S. Boutin, C. J. Krebs, and E. J. Hofer. 1997. Numerical responses of coyotes and lynx to the snowshoe hare cycle. Oikos 80:150–62.

O'Donoghue, M., S. Boutin, C. J. Krebs, G. Zutela, D. L. Murray, and E. J. Hofer. 1998. Functional responses of coyotes and lynx to the snowshoe hare cycle. Ecology 79:1193–1208.

O'Farrell, T. P. 1965. Home range and ecology of snowshoe hares in interior Alaska. Journal of Mammalogy 46:406–18.

Oldemeyer, J. L. 1983. Browse production and its use by moose and snowshoe hares at the Kenai Moose Research Center, Alaska. Journal of Wildlife Management 47:486–96.

Olsen, O. W. 1954. Occurrence of the lungworm *Protostrongylus boughtoni* Goble and Dougherty, 1943 in snowshoe hares (*Lepus americanus bairdii*) in Colorado. Proceedings of the Helminthological Society of Washington 21:52.

Orr, R. T. 1933. A new race of snowshoe rabbit from California. Journal of Mammalogy 14:54–56.

Orr, R. T. 1934. Description of a new snowshoe rabbit from eastern Oregon, with notes on its life history. Journal of Mammalogy 15:152–54.

Orr, C. D., and D. G. Dodds. 1982. Snowshoe hare habitat preferences in Nova Scotia spruce–fir forests. Wildlife Society Bulletin 10:147–50.

Otto, R. D. 1998. Attempted predation on a snowshoe hare, *Lepus americanus,* by an American marten, *Martes americana,* and a northern raven, *Corvus corax.* Canadian Field-Naturalist 112:333–34.

Parker, G. R. 1977. Morphology, reproduction, diet, and behavior of the Arctic hare *(Lepus arcticus monstrabilis)* on Axel Heiberg Island, Northwest Territories. Canadian Field-Naturalist 91:8–18.

Parker, G. R. 1981. Winter habitat use and hunting activities of lynx (*Lynx canadensis*) on Cape Breton Island, Nova Scotia. Pages 221–48 *in* J. A. Chapman and D. Pursley, eds. Proceedings of the worldwide furbearer conference, Frostburg, MD.

Parker, G. R. 1982. Hordes of hopping hares: An arctic enigma. Canadian Geographic 102:30–33.

Parker, G. R. 1984. Use of spruce plantations by snowshoe hares in New Brunswick. Forestry Chronicle 60:162–66.

Parker, G. R. 1986. The importance of cover on use of conifer plantations by snowshoe hares in northern New Brunswick. Forestry Chronicle 62:159–63.

Parker, G. R., J. W. Maxwell, L. D. Morton, and G. E. J. Smith. 1983. The ecology of the lynx (*Lynx canadensis*) on Cape Breton Island. Canadian Journal of Zoology 61:770–86.

Pease, J. L., R. H. Vowles, and L. B. Keith. 1979. Interaction of snowshoe hares and woody vegetation. Journal of Wildlife Management 43:43–60.

Pehrson, Å. 1983a. Digestibility and retension of food components in caged mountain hares *Lepus timidus* during the winter. Holarctic Ecology 6:395–403.

Pehrson, Å. 1983b. Maximal winter browse intake in captive mountain hares. Finnish Game Research 41:45–55.

Pehrson, Å. 1983c. Caecotrophy in caged mountain hares (*Lepus timidus*). Journal of Zoology 199:563–74.

Pehrson, Å. 1984. Faecal nitrogen as an index of hare browse quality. Canadian Journal of Zoology 62:510–13.

Penner, L. R. 1954. *Dirofilaria scapiceps* from snowshoe hare in Connecticut. Journal of Mammalogy 35:458–59.

Philip, C. B. 1938. A parasitological reconnaissance in Alaska with particular reference to varying hares. II. Parasitological data. Journal of Parasitology. 24:483–88.

Philip, C. B. 1939. A parasitological reconnaissance in Alaska with particular reference to varying hares. I. Some biological considerations. Journal of Mammalogy 20:82–86.

Pietz, P. J., and J. R. Tester. 1983. Habitat selection by snowshoe hares in north central Minnesota. Journal of Wildlife Management 47:686–96.

Poole, K. G. 1994. Characteristics of an unharvested lynx population during a snowshoe hare decline. Journal of Wildlife Management 58:608–18.

Poole, K. G., and R. P. Graf. 1996. Winter diet of marten during a snowshoe hare decline. Canadian Journal of Zoology 74:456–66.

Powers, J. G., W. M. Mautz, and P. J. Pekins. 1989. Nutrient and energy assimilation of prey by bobcats. Journal of Wildlife Management 53:1004–8.

Quenette, P. Y., J. Ferron, and L. Sirois. 1997. Group foraging in snowshoe hares (*Lepus americanus*): Aggregation or social group? Behavioural Processes 41:29–37.

Raczynski, J. 1964. Studies on the European hare. V. Reproduction. Acta Theriologica 9:305–52.

Radvanyi, A. 1987. Snowshoe hares and forest plantations: A literature review and problem analysis. Canadian Forestry Service, Northern Forestry Centre, Edmonton, AB.

Radwan, M. A., and D. L. Campbell. 1968. Snowshoe hare preference for spotted catsear flowers in western Washington. Journal of Wildlife Management 32:104–8.

Raine, R. M. 1987. Winter food habits and foraging behavior of fishers (*Martes pennanti*) and martens (*Martes americana*) in southeastern Manitoba. Canadian Journal of Zoology 65:745–47.

Ramos, C. N. 1998. Evolution and biogeography of North American Leporidae. Ph.D. Dissertation, University of Colorado, Boulder, CO.

Rangen, S. A., A. W. Hawley, and R. J. Hudson. 1993. Response of captive snowshoe hares to thiram-treated conifers. Journal of Wildlife Management 57:648–51.

Ranta, E., J. Lindström, V. Kaitala, H. Kokko, H. Lindén, and E. Helle. 1997. Solar activity and hare dynamics: A cross continental comparison. American Naturalist 149:765–75.

Rausch, R. L. 1953. On the status of some Arctic mammals. Arctic 6:91–143.

Rausch, V. R., and D. G. Ritter. 1974. Karyotype of the arctic hare, *Lepus othus* Merriam (Lagomorpha: Leporidae). Mammalian Chromosome Newsletter 15:7–9.

Reichardt, P. B., J. P. Bryant, T. P. Clausen, and G. D. Wieland. 1984. Defense of winter-dormant Alaska paper birch against snowshoe hares. Oecologia 65:58–69.

Reichardt, P. B., J. P. Bryant, B. J. Anderson, D. Phillips, and T. P. Clausen. 1990a. Germacrone defends Labrador tea from browsing by snowshoe hares. Journal of Chemical Ecology 16:1961–70.

Reichardt, P., J. P. Bryant, B. R. Mattes, T. P. Clausen, and F. S. Chapin III. 1990b. The winter chemical defense of balsam poplar against snowshoe hares. Journal of Chemical Ecology 16:1941–60.

Rextad, E., and K. P. Burnham. 1991. User's guide for interactive program CAPTURE. Colorado Cooperative Fish and Wildlife Research Unit, Fort Collins, CO.

Reynolds, J. K. 1952. The biology of the European hare (*Lepus europaeus* Pallas) in southwestern Ontario. Ph.D. Dissertation, University of Western Ontario, London, ON.

Reynolds, J. K. 1955. Distribution and populations of the European hare in southern Ontario. Canadian Field-Naturalist 69:14–20.

Reynolds, J. K., and R. H. Stinson. 1959. Reproduction in the European hare in southern Ontario. Canadian Journal of Zoology 37:627–31.

Richmond, M. E., and C.-Y. Chien. 1976. Status of the snowshoe hare on the Connecticut Hill wildlife management area. New York Fish and Game Journal 23:1–12.

Robinson, T. J., F. F. B. Elder, and J. A. Chapman. 1983. Karyotypic conservatism in the genus *Lepus* (order Lagomorpha). Canadian Journal of Genetics and Cytology 25:540–44.

Rodgers, A. R., and A. R. E. Sinclair. 1997. Diet choice and nutrition of captive snowshoe hares (*Lepus americanus*): Interactions of energy, protein, and plant secondary compounds. Ecoscience 4:163–69.

Roe, E. I., and J. H. Stoeckeler. 1950. Thinning over-dense jack pine seedlings in the lake states. Journal of Forestry 48:861–65.

Rogowitz, G. L. 1988. Forage quality and use of reforested habitats by snowshoe hares. Canadian Journal of Zoology 66:2080–83.

Rohner, C., and C. J. Krebs. 1996. Owl predation on snowshoe hares: Consequences of antipredator behaviour. Oecologia 108:303–10.

Rongstad, O. J. 1965. Short-tailed shrew attacks young snowshoe hare. Journal of Mammalogy 46:221–26.

Rongstad, O. J. and J. R. Tester., 1971. Behavior and maternal relations of young snowshoe hares. Journal of Wildlife Management 35:338–46.

Rousi, M., J. Tahvanainen, and I. Uotila. 1991. Mechanism of resistance to hare browsing in winter-dormant European white birch *Betula pendula*. American Naturalist 137:64–82.

Rowan, W. 1950. Canada's premier problem of animal conservation. New Biology 9:38–57.

Rowan, W. 1954. Reflections on the biology of animal cycles. Journal of Wildlife Management 18:52–60.

Rowan, W., and L. B. Keith. 1956. Reproductive potential and sex ratios of snowshoe hares in northern Alberta. Canadian Journal of Zoology 34:273–81.

Rowan, W., and L. B. Keith. 1959. Monthly weights of snowshoe hares from north-central Alberta. Journal of Mammalogy 40:221–26.

Roy, L. D., M. W. Barrett, and J. W. Nolan. 1991. Evaluation of four methods to reduce depredation to spruce seedlings by snowshoe hares. Wildlife Biology Branch, Alberta Environmental Centre, Vegreville, AB.

Royama, T. 1992. Analytical population dynamics. Chapman and Hall, New York, NY.

Ruesink, J. L., K. E. Hodges, and C. J. Krebs. 2002. Mass-balance analyses of boreal forest population cycles: Merging demographic and ecosystem approaches. Ecosystems 5:138–158.

Ruess, R. W., R. L. Hendrick, and J. P. Bryant. 1998. Regulation of fine root dynamics by mammalian browsers in early successional Alaskan taiga forests. Ecology 79:2706–20.

Ruggiero, L. F., K. B. Aubry, S. W. Buskirk, G. M. Koehler, C. J. Krebs, K. S. McKelvey, and J. R. Squires. 2000. The scientific basis for lynx conservation: Qualified insights. Pages 443–54 *in* L. F. Ruggiero, K. B. Aubry, S. W. Buskirk, G. M. Koehler, C. J. Krebs, K. S. McKelvey, and J. R. Squires, eds. Ecology and conservation of lynx in the United States. University Press of Colorado and U.S. Department of Agriculture, Rocky Mountain Research Station, Denver, CO.

Samoil, H. P., and W. M. Samuel. 1977a. Description of nine species of *Eimeria* (Protozoa, Eimeriidae) in the snowshoe hare, *Lepus americanus*, of central Alberta. Canadian Journal of Zoology 55:1671–83.

Samoil, H. P., and W. M. Samuel. 1977b. Experimental study of *Eimeria robertsoni* (Protozoa: Eimeriidae) in the snowshoe hare, *Lepus americanus*. Journal of Parasitology 63:203–5.

Schmitz, O. J., D. S. Hik, and A. R. E. Sinclair. 1992. Plant chemical defense and twig selection by snowshoe hares. Oikos 65:295–300.

Schultz, W. C. 1980. Extent and causes of mortality in stocked snowshoe hares. Journal of Wildlife Management 44:716–19.

Scott, D. P., and R. H. Yahner. 1989. Winter habitat and browse use by snowshoe hares, *Lepus americanus*, in a marginal habitat in Pennsylvania. Canadian Field-Naturalist 103:560–63.

Severaid, J. H. 1942. The snowshoe hare its life history and artificial propagation. Maine Department of Inland Fisheries and Game, Augusta, ME.

Severaid, J. H. 1945a. Pelage changes in the snowshoe hare (*Lepus americanus struthopus* Bangs). Journal of Mammalogy 26:41–63.

Severaid, J. H. 1945b. Breeding potential and artificial propagation of the snowshoe hare. Journal of Wildlife Management 9:290–95.

Sherman, P. W. 1977. Nepotism and the evolution of alarm calls. Science 197:1246–53.

Shipley, L. A., and D. E. Spalinger. 1992. Mechanics of browsing in dense food patches: Effects of plan and animal morphology on intake rate. Canadian Journal of Zoology 70:1743–52.

Sievert, P. R., and L. B. Keith. 1985. Survival of snowshoe hares at a geographic range boundary. Journal of Wildlife Management 49:854–66.

Silen, P. R., W. K. Randall, and N. L. Mandell. 1986. Estimates of genetic parameters for deer browsing of Douglas fir. Forestry Science 32:178–84.

Silver, J. 1924. The European hare (*Lepus europaeus* Pallas) in North America. Journal of Agricultural Research 28:1133–37.

Sinclair, A. R. E. 1986. Testing multi-factor causes of population limitation: An illustration using snowshoe hares. Oikos 47:360–64.

Sinclair, A. R. E., and J. M. Gosline. 1997. Solar activity and mammal cycles in the Northern Hemisphere. American Naturalist 149:776–84.

Sinclair, A. R. E., and J. N. M. Smith. 1984a. Do plant secondary compounds determine feeding preferences of snowshoe hares? Oecologia 61:403–10.

Sinclair, A. R. E., and J. N. M. Smith. 1984b. Protein digestion in snowshoe hares. Canadian Journal of Zoology 62:520–21.

Sinclair, A. R. E., C. J. Krebs, and J. N. M. Smith. 1982. Diet quality and food limitation in herbivores: The case of the snowshoe hare. Canadian Journal of Zoology 60:889–97.

Sinclair, A. R. E., C. J. Krebs, J. N. M. Smith, and S. Boutin. 1988a. Population biology of snowshoe hares. III. Nutrition, plant secondary compounds and food limitation. Journal of Animal Ecology 57:787–806.

Sinclair, A. R. E., M. K. Jogia, and R. J. Andersen. 1988b. Camphor from juvenile white spruce as an antifeedant for snowshoe hares. Journal of Chemical Ecology 14:1505–14.

Sinclair, A. R. E., J. M. Gosline, G. Holdsworth, C. J. Krebs, S. Boutin, J. N. M. Smith, R. Boonstra, and M. Dale. 1993. Can the solar cycle and climate synchronize the snowshoe hare cycle in Canada? Evidence from tree rings and ice cores. American Naturalist 141:173–98.

Sinclair, A. R. E., D. E. Williams, R. J. Andersen, and J. Pain. 1996. Feeding avoidance of balsam poplar by snowshoe hares is related to abundance of buds. Ecoscience 3:223–25.

Sinclair, A. R. E., C. J. Krebs, J. M. Fryxell, R. Turkington, S. Boutin, R. Boonstra, P. Seccombe Hett, P. Lundberg, and L. Oksanen. 2000. Testing hypotheses of trophic level interactions: A boreal forest ecosystem. Oikos 89:313–28.

Skinner, W. R. 1995. The use of pre-commercially thinned stands by snowshoe hare in the Western Newfoundland Model Forest. Newfoundland and Labrador Wildlife Division, St. Johns, NF.

Slough, B. G., and G. Mowat. 1996. Lynx population dynamics in an untrapped refugium. Journal of Wildlife Management 60:946–61.

Small, R. J., and L. B. Keith. 1992. An experimental study of red fox predation on arctic and snowshoe hares. Canadian Journal of Zoology 70:1614–21.

Small, R. J., L. B. Keith, and R. M. Barta. 1991. Dispersion of introduced arctic hares (*Lepus arcticus*) on islands off Newfoundland's south coast. Canadian Journal of Zoology 69:2618–23.

Small, R. J., L. B. Keith, and R. M. Barta. 1992. Demographic responses of arctic hares *Lepus arcticus* placed on two predominantly forested islands in Newfoundland. Ecography 15:161–65.

Smith, C. H. 1983. Spatial trends in Canadian snowshoe hare, *Lepus americanus*, population cycles. Canadian Field-Naturalist 97:151–60.

Smith, C. H., and J. M. Davis. 1981. A spatial analysis of wildlife's ten-year cycle. Journal of Biogeography 8:27–35.

Smith, J. N. M., C. J. Krebs, A. R. E. Sinclair, and R. Boonstra. 1988. Population biology of snowshoe hares. II. Interactions with winter food plants. Journal of Animal Ecology 57:269–86.

Smith, M. C., J. F. Leatherland, and K. Myers. 1978. Effects of seasonal availability of sodium and potassium on the adrenal cortical function of a wild population of snowshoe hares, *Lepus americanus*. Canadian Journal of Zoology 56:1869–76.

Smith, R. L., D. J. Hubbartt, and R. L. Shoemaker. 1980. Seasonal changes in weight, cecal length, and pancreatic function of snowshoe hares. Journal of Wildlife Management 44:719–24.

Sovell, J. R., and J. C. Holmes. 1996. Efficacy of ivermectin against nematodes infecting field populations of snowshoe hares (*Lepus americanus*) in Yukon, Canada. Journal of Wildlife Diseases 32:23–30.

Spalatin, J., C. E. O. Fraser, R. Connell, R. P. Hanson, and D. T. Berman. 1966. Agents of *Psittacosis lymphogranuloma venerum* group isolated from muskrats and snowshoe hares in Saskatchewan. Canadian Journal of Comparative Medicine 30:260–64.

Spencer, J. L. 1955. Reingestion in three species of Lagomorphs. Lloydia 18:197–99.

Staples, W. 1995. Lynx and coyote diet and habitat relationships during a low hare population on the Kenai Peninsula, Alaska. M.S. Thesis, University of Alaska, Fairbanks, AK.

Stefan, C. I., and C. J. Krebs. 2001. Reproductive changes in a cyclic population of snowshoe hares. Canadian Journal of Zoology 79:2101–2108.

Stenseth, N. C. 1995. Snowshoe hare populations: Squeezed from below and above. Science 269:1061–62.

Stenseth, N. C., W. Falck, O. N. Bjørnstad, and C. J. Krebs. 1997. Population regulation in snowshoe hare and Canadian lynx: Asymmetric food web configurations between hare and lynx. Proceedings of the National Academy of Sciences of the USA 94:5147–52.

Stenseth, N. C., W. Falck, K. Chan, O. N. Bjørnstad, M. O'Donoghue, H. Tong, R. Boonstra, S. Boutin, C. J. Krebs, and N. G. Yoccoz 1998. From patterns to processes: Phase and density dependencies in the Canadian lynx cycle. Proceedings of the National Academy of Sciences 26:15430–15435.

Stephenson, D. E. 1985. The use of charred black spruce bark by snowshoe hare. Journal of Wildlife Management 49:296–300.

St.-Georges, M., S. Nadeau, D. Lambert, and R. Décarie. 1995. Winter habitat use by ptarmigan, snowshoe hares, red foxes, and river otters in the boreal forest–tundra transition zone of western Québec. Canadian Journal of Zoology 73:755–64.

Stock, A. D. 1976. Chromosome banding pattern relationships of hares, rabbits, and pikas (order Lagomorpha). Cytogenetics and Cell Genetics 17:78–88.

Stuttard, R. M. 1981. The hare as an object of sport. Pages 907–16 *in* K. Meyers and C. D. MacInnes, eds. Proceedings of the World Lagomorph Conference. University of Guelph, Guelph, ON.

Sullivan, T. P. 1986. Influence of wolverine (*Gulo gulo*) odor on feeding behavior of snowshoe hares (*Lepus americanus*). Journal of Mammalogy 67:385–88.

Sullivan, T. P. 1994. Influence of herbicide-induced habitat alteration on vegetation and snowshoe hare populations in sub-boreal spruce forest. Journal of Applied Ecology 31:717–30.

Sullivan, T. P. 1996. Influence of forest herbicide on snowshoe hare population dynamics: Reproduction, growth, and survival. Canadian Journal of Forest Resources 26:112–19.

Sullivan, T. P., and D. R. Crump. 1984. Influence of mustelid scent gland compounds on suppression of feeding densities of snowshoe hares. Journal of Chemical Ecology 10:1809–21.

Sullivan, T. P., and R. A. Moses. 1986. Demographic and feeding responses of a snowshoe hare population to habitat alteration. Journal of Applied Ecology 23:53–63.

Sullivan, T. P., and D. S. Sullivan. 1982a. Barking damage by snowshoe hares and red squirrels in lodgepole pine stands in central British Columbia. Canadian Journal of Forest Research 12:443–48.

Sullivan, T. P., and D. S. Sullivan. 1982b. Influence of fertilization on feeding attacks to lodgepole pine by snowshoe hares and red squirrels. Forestry Chronicle 58:263–66.

Sullivan, T. P., and D. S. Sullivan. 1983. Use of index lines and damage assessments to estimate population densities of snowshoe hares. Canadian Journal of Zoology 61:163–67.

Sullivan, T. P., and D. S. Sullivan. 1986. Impact of feeding damage by snowshoe hares on growth rates of juvenile lodgepole pine in central British Columbia. Canadian Journal of Forest Research 16:1145–49.

Sullivan, T. P., and D. S. Sullivan. 1988. Influence of stand thinning on snowshoe hare population dynamics and feeding damage in lodgepole pine forest. Journal of Applied Ecology 25:791–805.

Sullivan,T. P., L. O. Nordstrom, and D. S. Sullivan. 1985. The use of predator odors as repellents to reduce feeding damage by herbivores. I. Snowshoe hares (*Lepus americanus*). Journal of Chemical Ecology 11:903–20.

Suominen, O., K. Danell, and J. P. Bryant. 1999. Indirect effects of mammalian browsers on vegetation and ground–dwelling insects in an Alaskan floodplain. Ecoscience 6:505–10.

Swayze, L. A. 1994. Snowshoe hare use patterns in selected lodgepole pine stands in north central Washington. Okanogan National Forest, Twisp, WA.

Swihart, R. K., J. P. Bryant, and L. Newton. 1994. Latitudinal patterns in consumption of woody plants by snowshoe hares in the eastern United States. Oikos 70:427–34.

Szukiel, E. 1973. The effect of repellents on the food preferences of hares. Acta Theriologica 18:481–88.

Tahvanainen, J., P. Niemelä, and H. Henttonen. 1991. Chemical aspects of herbivory in boreal forest—Feeding by small rodents, hares, and cervids. Pages 115–31 *in* R. T. Palo and C. T. Robbins, eds. Plant defenses against mammalian herbivory. CRC Press, Boca Raton, FL.

Taylor, E. L. 1939. Does the rabbit chew the cud? Nature 143:981–83.

Telfer, E. S. 1972. Browse selection by deer and hare. Journal of Wildlife Management 36:1344–49.

Terres, J. K. 1941. Speed of the varying hare. Journal of Mammalogy 22:453–54.

Theau, J., and J. Ferron. 2000. Influence des conditions climatiques sur le comportement du lièvre d'Amérique (*Lepus americanus*) en semi-liberté. Canadian Journal of Zoology 78:1126–36.

Theau, J., and J. Ferron. 2001. Influence de paramètres climatiques sur les patrons d'activité saisonniers et journaliers du lièvre d'Amerique, *Lepus americanus*, en semi-liberté. Canadian Field-Naturalist 115:43–51.

Thomas, J. A., J. G. Hallett, and M. A. O'Connell. 1997. Habitat use by snowshoe hares in managed landscapes of northeastern Washington. Unpublished report, Washington State University, Pullman, WA.

Thomas, V. G. 1987. Similar winter energy strategies of grouse, hares and rabbits in northern biomes. Oikos 50:206–12.

Thompson, I. D., and P. W. Colgan. 1987. Numerical responses of martens to a food shortage in northwestern Ontario. Journal of Wildlife Management 51:824–35.

Thompson, I. D., and P. W. Colgan. 1990. Prey choice by marten during a decline in prey abundance. Oecologia 83:443–51.

Thompson, I. D., I. J. Davidson, S. O'Donnell, and F. Brazeau. 1989. Use of track transects to measure relative occurrence of some small boreal mammals in uncut forest regeneration stands. Canadian Journal of Zoology 67:1816–23.

Tompkins, D. B., and J. R. Woehr. 1979. Influence of habitat on movements and densities of snowshoe hares. Transactions of the Northeast Fish and Wildlife Conference 36:169–75.

Trapp, G. R. 1962. Snowshoe hares in Alaska. II. Home range and ecology during an early population increase. M.S. Thesis, University of Alaska, Fairbanks, AK.

Trapp, G. R., and C. Trapp. 1965. Another vocal sound made by snowshoe hares. Journal of Mammalogy 46:705.

Trostel, K., A. R. E. Sinclair, C. J. Walters, and C. J. Krebs. 1987. Can predation cause the 10-year hare cycle? Oecologia 74:185–92.

Turchin, P. 2003. Complex population dynamics: A theoretical/empirical synthesis. Princeton University Press, Princeton, NJ.

Turkington, R., E. John, S. Watson, and P. Seccombe-Hett. 2002. The effects of fertilization and herbivory on the herbaceous vegetation of the boreal forest in North-Western Canada: A 10-Year Story. Journal of Ecology 90:335–337.

Vaughan, M. R., and L. B. Keith. 1980. Breeding by juvenile snowshoe hares. Journal of Wildlife Management 45:354–80.

Vaughan, M. R., and L. B. Keith. 1981. Demographic response of experimental snowshoe hare populations to overwinter food shortage. Journal of Wildlife Management 45:354–80.

Vowles, R. H. 1972. Snowshoe hare–vegetation interactions at Rochester, Alberta. M.S. Thesis, University of Wisconsin, Madison, WI.

Walkinshaw, L. H. 1947. Notes on the Arctic hare. Journal of Mammalogy 28:353–75.

Walski, T. W., and W. W. Mautz. 1977. Nutritional evaluation of three winter browse species of snowshoe hares. Journal of Wildlife Management 41:144–47.

Wang, L. C. H., D. L. Jones, R. A. MacArthur, and W. A Fuller. 1973. Adaptation to cold: Energy metabolism in an atypical lagomorph, the arctic hare (*Lepus arcticus*). Canadian Journal of Zoology 51:841–46.

Ward, R. M. P., and C. J. Krebs. 1985. Behavioural responses of lynx to declining snowshoe hare abundance. Canadian Journal of Zoology 63:2817–24.

Webb, W. L. 1937. Notes on the sex ratio of the snowshoe rabbit. Journal of Mammalogy 18:343–47.

Webb, W. L. 1942. Notes on a method for censusing snowshoe hare populations. Journal of Wildlife Management 12:153–61.

Weinstein, M. S. 1977. Hares, lynx, and trappers. American Naturalist 111:806–8.

Whittaker, M. E., and V. G. Thomas. 1983. Seasonal levels of fat and protein reserves of snowshoe hares in Ontario. Canadian Journal of Zoology 61:1339–45.

Windberg, L. A., and L. B. Keith. 1976a. Snowshoe hare population response to artificial high densities. Journal of Mammalogy 57:523–53.

Windberg, L. A., and L. B. Keith. 1976b. Experimental analyses of dispersal in snowshoe hare populations. Canadian Journal of Zoology 54:2061–81.

Wirsing, A. J., and D. L. Murray. 2002. Patterns in consumption of woody plants by snowshoe hares in the northwestern United States. Ecoscience 9: 440–449.

Wirsing, A. J., T. D. Steury, and D. L. Murray. 2002a. Non-invasive estimation of body composition in small mammals: A comparison of conductive and morphometric techniques. Physiological and Biochemical Zoology 75:489–497.

Wirsing, A. J., T. D. Steury, and D. L. Murray. 2002b. The relationship between body condition and vulnerability to predation in red squirrels and snowshoe hares. Journal of Mammalogy 83:707–715.

Wirsing, A. J., T. D. Steury, and D. L. Murray. 2002c. A demographic analysis of a southern snowshoe hare population in a fragmented habitat: Evaluating the refugium model. Canadian Journal of Zoology 80:169–77.

Wolfe, M. L., N. V. Debyle, C. S. Winchell, and T. R. McCabe. 1982. Snowshoe hare cover relationships in northern Utah. Journal of Wildlife Management 46:662–70.

Wolff, J. O. 1978. Food habits of snowshoe hares in interior Alaska. Journal of Wildlife Management 42:148–53.

Wolff, J. O. 1980a. The role of habitat patchiness in the population dynamics of snowshoe hares. Ecological Monographs 50:111–29.

Wolff, J. O. 1980b. Moose–snowshoe hare competition during peak hare densities. Proceedings of the North American Moose Conference 16:238–54.

Wolff, J. O. 1981. Refugia, dispersal, predation, and geographic variation in snowshoe hare cycles. Pages 441–48 *in* K. Meyers and C. D. MacInnes, eds. Proceedings of the world lagomorph conference. University of Guelph, Guelph, ON.

Wolff, J. O. 1982. Snowshoe hare. Pages 140–41 *in* CRC handbook of census methods for terrestrial vertebrates. CRC Press, Boca Raton, FL.

Wood, T. J. 1974. Competition between arctic hares and moose in Gros Morne National Park, Newfoundland. Alces 10:215–37.

Wood, T. J., and S. A. Munroe. 1977. Dynamics of snowshoe hare populations in the maritime Provinces (Occasional Paper 30). Canadian Wildlife Service.

Yuill, T. M., and R. P. Hanson. 1964. Serological evidence of California encephalitis virus and Western equine encephalitis virus in snowshoe hares. Zoonoses Research 3:153–64.

Yuill, T. M., J. O. Iversen, and R. P. Hanson. 1969. Evidence of arbovirus infections in a population of snowshoe hares: A possible mortality factor. Bulletin of the Wildlife Disease Association 5:248–53.

Zimmerling, T. N., and L. M. Zimmerling. 1996. A comparison of the effectiveness of predator odor and plant antifeedant in deterring small mammal feeding damage on lodgepole pine seedlings. Journal of Chemical Ecology 22:2123–2132.

Dennis, L. Murray, Departments of Biology and Environmental Resource Studies, Trent University, Petersborough, Ontario, Canada K9L 7B8. Email: dennismurray@trentu.ca.

III

Rodents

8

Mountain Beaver

Aplodontia rufa

George A. Feldhamer
James A. Rochelle
Clifford D. Rushton

NOMENCLATURE

Common Names. Mountain beaver, boomer, sewellel, whistler, chehalis, mountain rat
Scientific Name. *Aplodontia rufa;* the genus name means "simple tooth," referring to cheek teeth; the species name is based on the reddish skins of mountain beavers obtained by Lewis and Clark from Indians in 1806 (Verts and Carraway 1998)
Subspecies. Seven subspecies were recognized by Hall (1981): *A. r. californica* (Sierra Nevada Mountains of central California, western Nevada), *A. r. humboldtiana* (northern coastal California), *A. r. nigra* (restricted to Point Arena, Mendocino County, California), *A. r. pacifica* (Coast Range of western Oregon and northwest California), *A. r. phaea* (restricted to Point Reyes, Marin County, California), *A. r. rainieri* (Cascade Range of Washington, Oregon, and northern California), and *A. r. rufa* (western Washington and southwest British Columbia, Canada)

The common name "mountain beaver" is a misnomer; the species generally prefers lower elevation areas and phylogenetically is not closely related to true beavers (family Castoridae). The mountain beaver is the only extant species of the family Aplodontidae. Its zygomasseteric structure is unique among extant rodents. The origin of the masseter muscle is completely on the zygomatic arch with no part anterior to the rostrum; the species generally is regarded as the most primitive member of the order Rodentia (Feldhamer et al. 1999). The fossil record in North America extends from the late Eocene or early Oligocene to Recent (McLaughlin 1984). Except for increasing hypsodonty, mountain beavers have changed little morphologically since the Miocene (Shotwell 1958; Rensberger 1982).

DISTRIBUTION

The mountain beaver is endemic to western North America, and extends from extreme southern British Columbia south to central California and east to the Cascade Mountains and the Sierra Nevada Mountains (Fig. 8.1). An apparent southeast range extension for *A. r. californica* was noted by Bleich and Racine (1991). Although it may range to elevations of 2200 m, mountain beavers more commonly occur in humid, densely vegetated areas at lower elevations (Nowak 1991). Limits to distribution are associated with rainfall and edaphic conditions, which promote relatively high humidity within burrows, adequate soil drainage, succulent forage, and an acceptable thermal regime (Voth 1968; Beier 1989). In British Columbia, the Fraser River prevents further northward distribution, and climate is too dry north and east of the Cascade Mountains (Gyug 2000).

DESCRIPTION

Size and Weight. The general body conformation of mountain beavers is compact, thickset, and cylindrical. Total length of mature individuals ranges from about 300 to 500 mm, which includes a stumpy, fully furred tail 10–25 mm long. Forelimbs and hind limbs are pentadactyl and of approximately equal length. The pollex is opposable. Mean adult body weight for both sexes is about 0.8 kg. Maximum adult body weight reported by Maser et al. (1981) was 1.4 kg. Adult body weights vary seasonally, with a high in July, when forage is most abundant, and lows in March or April, during or immediately after the reproductive season (Lovejoy and Black 1974).

Pelage. In adults, the dark reddish to blackish brown fur color is similar in both sexes. It is coarse, dense, and short with numerous interspersed guard hairs. The guard hairs on the dorsum and limbs are dark, whereas those on the ventral and lateral surfaces are lighter. Juveniles have fine, gray fur with numerous white-tipped hairs (Taylor 1918). Yearlings may retain some of their typical juvenile pelage, but generally have attained adult pelage (Lovejoy and Black 1974). All mountain beavers have a small white spot at the base of each ear. Albino and melanistic animals, although apparently rare, have been reported (Godin 1964). During all seasons, females have a nearly circular area of brownish black hair about 15 mm in diameter, which surrounds each of the three pairs of mammae. These patches of hair contrast with the generally lighter coloration of the ventral surface and do not occur in males.

Mountain beavers undergo a single annual molt, which generally begins in July or August and continues for 2–3 months. Pelage replacement begins on the anterior dorsal surface and posterior lateral surfaces. The ventral area is the last to complete the molt (Taylor 1918; Stangl and Grimes 1987).

Skull and Dentition. The skull of the adult mountain beaver is unusually broad and flattened, and there is no postorbital process. Characteristic features include flask-shaped auditory bullae, a palate which extends posteriorly to the third upper molar, and a mandible with a greatly inflected angular process and a relatively high coronoid process (Fig. 8.2). Condylobasal length is about 65 mm, with *A. r. pacifica* and *phaea* generally smaller. In his revision of the genus *Aplodontia,* Taylor (1918) noted "considerable" variation in the size of adult skulls from the same locality.

The cheekteeth are modified hypsodont, rootless (ever growing), and prismatic. The anteriormost upper premolar (P^3) is small and peglike. The remaining cheekteeth are set obliquely and decrease in size from the second premolar (P^4) to the last molar. With the exception of P^3, each cheektooth has a unique, spinelike projection. These projections occur on the labial side of the upper cheekteeth and the lingual side of the lower cheekteeth (Fig. 8.2). The dental formula is I 1/1, C 0/0, P 2/1, M 3/3.

Druzinsky (1995) provided details of the jaw and muscle activities that occur as mountain beavers use their incisors for gnawing. Most gnawing activity does not involve contact of the upper and lower incisors. This contrasts with sharpening of the lower incisors, in which the lingual surfaces are dragged across the tips of the upper incisors to produce a wear facet. Druzinsky (1995) reported that sharpening cycles were extremely rapid in *A. rufa,* with a mean of 11 cycles/sec.

GENETICS

McMillan and Sutton (1972) examined metaphase chromosome preparations of five *A. r. californica* and four *A. r. phaea.* Karyotypes were

FIGURE 8.1. Distribution of the mountain beaver (*Aplondontia rufa*).

identical for both subspecies. The diploid number (2*N*) was 46 chromosomes; 6 pairs were metacentric and 16 pairs were submetacentric. The Y chromosome of males was also submetacentric. Although the species is primitive in relation to other living rodents, McMillan and Sutton (1972:308) suggested that because their 2*N* number was not large, mountain beavers are "advanced within (their) own lineage."

The α and β subunits of the enzyme lactate dehydrogenase were investigated electrophoretically for 34 species of rodents, including mountain beavers, by Baur and Pattie (1968). Mountain beavers possessed both subunits, as did 11 species of hystricomorph rodents tested. Conversely, many myomorph rodents genetically suppressed expression of the β subunit.

PHYSIOLOGY

Body Temperature. Johnson (1971) found that the mean body temperature of free-ranging animals was 38°C, and reported hyperthermia at ambient temperatures above 29°C. A lethal body temperature of 42°C resulted from 2 hr of exposure to ambient temperatures between 32°C and 35°C. Limited ability of mountain beavers to physiologically thermoregulate to minimize heat stress may be another factor contributing to their restricted geographic distribution (Johnson 1971). Although mountain beavers may remain underground during severe winter weather, they do not hibernate.

Renal. Because it is the most primitive living rodent, the renal physiology of the mountain beaver has been thoroughly investigated. Structure of the kidney is very primitive. Pfeiffer (1968) noted that no extension of the renal pelvis penetrated between the outer zone of the medulla and the cortex. Seventy percent of the nephrons lay entirely in the cortex. Long-looped nephrons apparently are minimal (Pfeiffer 1968; Schmidt-Nielsen and Pfeiffer 1970), and those present do not have a thin segment (Pfeiffer et al. 1960). Thus, the ability to produce hypertonic urine is limited, and the mountain beaver is one of the least efficient mammals in its ability to conserve water (Dicker and Eggleton 1964; Nungesser and Pfeiffer 1965). Daily urine production in mountain beavers is so extensive that it amounts to one fourth to one third of the body mass (Nungesser et al. 1960). As a result, mountain beavers must consume considerable amounts of free water or succulent vegetation. As noted, this is a factor that limits their geographic distribution to regions of abundant rainfall and vegetation.

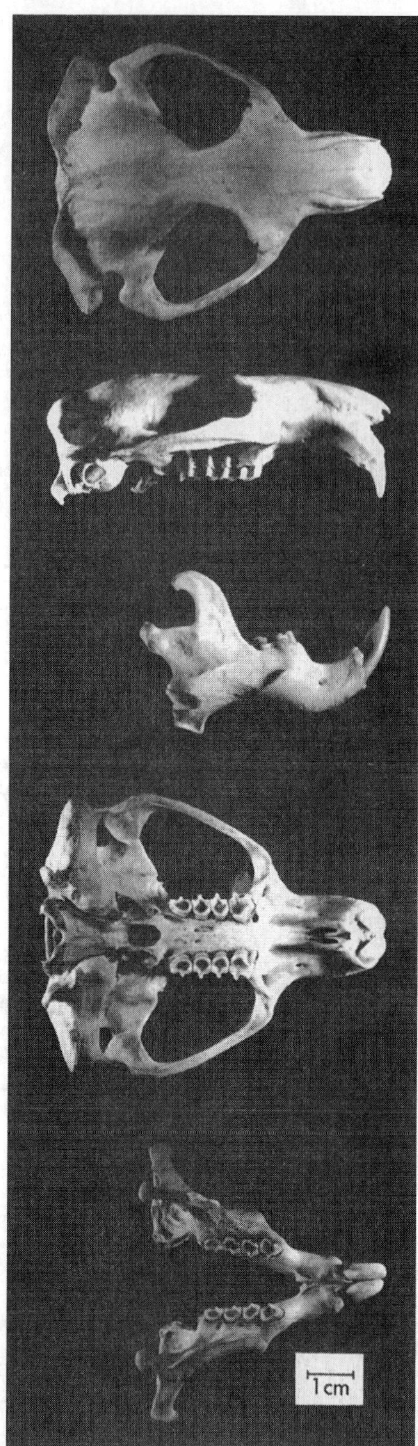

FIGURE 8.2. Skull of the mountain beaver (*Aplondontia rufa*). From top to bottom: dorsal view of cranium, lateral view of cranium, lateral view of mandible, ventral view of cranium, dorsal view of mandible.

Neural. Visual and auditory acuity of mountain beavers is poor. Their tactile and olfactory senses are highly attuned, however. As a result, they rarely move far from their burrow system. In an examination of brainstem auditory nuclei of 106 mammalian species, Merzenich et al. (1973) found that mountain beavers had a large, unique cochlear nuclear complex four to seven times larger than those of 17 other species of rodents investigated. They suggested that this system is specialized to detect slow changes in air pressure, which they believed would be of value to a tunnel-dwelling species. No similar specialization of this complex was found in other fossorial species, however, including pocket gophers (Geomyidae) and moles (Talpidae). Although all members of these families inhabit tunnels, gophers and moles plug all surface openings, whereas mountain beavers do not. Furthermore, only mountain beavers are active aboveground for extended periods. These factors may be associated with their particular neural specialization.

REPRODUCTION AND DEVELOPMENT

Anatomy. Ovaries of mountain beavers are elliptical; no follicular development is seen macroscopically in anestrous females. Ovaries of females in estrus, however, have follicles up to 3 mm in diameter, and the corpora lutea are reddish (Pfeiffer 1958). Externally, parous females may be distinguished from nulliparous individuals by their long, pendantlike nipples. Uteri of parous animals are characterized by their compressed, ribbon-like, more vascularized appearance and are much larger than the threadlike uteri of nulliparous females. After parturition, placental scars 1 cm in diameter are visible. These gradually regress until early fall, when they "become obscured by the gradual hypertrophy and hyperplasia of the uterus as estrus approaches" (Pfeiffer 1958:234).

The male reproductive tract was described by Pfeiffer (1956) and closely resembles that of certain sciurids. During the breeding season, the testes may be semiscrotal. They are abdominal during the remainder of the year. Testes and accessory sex organs vary in size throughout the year and attain their maximum size during the breeding season. For the remainder of the year, seminiferous tubules are much reduced in size and relatively few spermatogonia and primary spermatocytes are present.

Breeding. In Oregon, testes of mountain beavers descend in late December or early January. Females enter estrus from mid-February to early May (Lovejoy 1972; Lovejoy and Black 1974), although within populations, breeding is fairly synchronous (Pfeiffer 1958). They apparently remain in breeding condition for a relatively short duration. Only animals 2 years old form corpora lutea of pregnancy. Although estrus may occur in yearling females, apparently they do not conceive. Reproductive status likely is related to nutritional status. D. L. Nolte (USDA, Animal and Plant Health Inspection Service, pers. commun., 2001) observed year-old pups raised in captivity that were larger than some adults captured in the field.

Parturition. Parturition generally occurs from late March to early April, after gestation of 28–30 days. Cramblet and Ridenhour (1956) observed parturition in a captive mountain beaver. For 1.5 hr before birth, the animal periodically gnashed its teeth while pressing its forefeet against the genital region. Immediately before expulsion, it sat on its haunches and lowered its head between its legs. The first fetus was a caudal presentation, which the female assisted by pressing its forefoot around the vulva and licking the area. Two fetuses followed, one a cephalic presentation, and the actual birth process was completed in about 33 min. Neonates began to nurse within 20 min.

Litter Size and Neonatal Development. One litter per year is produced. Litter size ranges from two to six; two or three is most common. Neonates are altricial; vibrissae and pinnae are visible, although the latter are not extended, and the tail is apparent. Claws develop by 5 days of age. Fine, downy hairs develop by 1 week of age, and the complete pelage about 1 week later. Incisors erupt by about 1 month of age (Hooven 1977; Lovejoy et al. 1978). Eyes do not open until young are 7–8 weeks old. The average body length of three neonates described by Cramblet and Ridenhour (1956) was 87.7 mm and body weight was 27.0 g. The weights of two day-old young born in captivity in Oregon were 18 and 22 g, respectively. Mean weight was 110 g at 1 month of age and 347 g at 2 months. Young mountain beavers grow rapidly. They attain approximately 40% of the mean adult body weight during April and May, when adult females are lactating. The young are weaned at 2 months of age, although they may begin eating vegetation when they are 5–6 weeks old. They emerge from the burrow when about 10 weeks of age. By 4 months of age, they reach close to 70% of the mean adult body weight. Growth rate eventually declines, and yearling animals attain about 88% of the mean adult weight (Lovejoy 1972; Lovejoy and Black 1974).

ECOLOGY

Habitat. Mountain beavers inhabit densely vegetated areas with high annual precipitation. They commonly occur in the initial seral stages after clearcutting of forest areas. In western Oregon, Hacker and Coblentz (1993) observed that mountain beavers recolonizing clearcuts selected areas with large amounts of small- (<25 cm) and large-diameter (>25 cm) woody debris, forage plants, and uprooted stumps. They were likely to recolonize areas that had soft soils in drainage areas. Todd (1992) quantified habitat characteristics of areas with and without mountain beavers in the drier climate of Yosemite National Park, California. He found that areas occupied by mountain beavers had significantly greater shrub and herbaceous cover, narrower and shallower streams with steeper gradients, greater soil depth, and less susceptibility to flooding than unoccupied areas. Beier (1989:652) found similar physical and vegetation characteristics were associated with mountain beavers in the Sierra Nevada Mountains and suggested that habitat use by mountain beavers "involves strict requirements for an appropriate thermal regime and adequate soil drainage, and somewhat more flexible requirements for food."

Population Densities and Sex Ratios. Mountain beavers are not gregarious, and concentrations are a function of suitable habitat rather than sociability of the species. Lovejoy (1972) estimated that population density varied from 6.7 to 8.7/ha on a logged area in the Coast Range of Oregon, much the same as the 7.3–8.3/ha estimated by Motubu (1978). Voth (1968) estimated 41 animals/ha in a similar area. He felt this probably was an overestimation, however, even though the habitat was excellent. Hooven (1977) felt that population densities on areas with new growth may increase from <1/ha to 15–20/ha.

During a 2-year period, Lovejoy and Black (1974, 1979) trapped 109 adults, with a ratio of 1.6:1.0 males to females. This may have been the result of trapping bias, however, as the sex ratio of 72 juveniles from three successive breeding seasons was 1:1. Nonetheless, Lovejoy and Black (1979:88) considered the preponderance of males a "true ratio and not an artifact of trapping." Extensive trapping during a 5-year period in the Pacific Northwest resulted in an adult male:female sex ratio of 60:40 (Borrecco and Anderson 1980).

Daily and Seasonal Activity. Mountain beavers primarily are nocturnal, but may be active any time of the day. Ingles (1959) reported six or seven activity periods every 24 hr during the summer. These varied in duration to a maximum of 2.8 hr. Also, 50–60% more activity occurred at night, at least during the summer. Increased activity may be stimulated by a combination of environmental factors, such as changes in light intensity, temperature, relative humidity, or air movement. Increased nocturnal activity probably did not result from an attempt to minimize potential predation, as the two most common predators in the area, coyotes (*Canis latrans*) and great horned owls (*Bubo virginianus*), were themselves most active at night (Ingles 1959). Burrowing activity is most prevalent during the summer. Aboveground activity essentially ceases during late fall and winter, although, as noted, mountain beavers do not hibernate.

Home Range and Movements. Martin (1971) found that the home ranges of adult mountain beavers varied in size from 0.03 to 0.2 ha, with no apparent difference in mean range (= 0.12 ha) of males and females. Over 90% of the telemetry-determined positions he recorded for 10 animals occurred within 24 m of their nest sites. Other home range values reported are comparably restricted and within the expected range of variation caused by differences in population density, habitat, and other factors. Lovejoy and Black (1979) found that adult males had an average home range of 0.3 ha (range = 0.09–0.7), adult females about 0.2 ha (0.05–0.4), and juveniles about 0.1 ha (0.04–0.2). Again, individuals were caught most often near their nest sites. Although individual home ranges may overlap extensively, mountain beavers defend their nest sites.

Dispersing subadults can make extensive linear movements of >500 m. In an effort to establish permanent home ranges, they follow existing burrows and travel aboveground. Martin (1971) reported that after subadults establish nest sites, their movements are comparable to those of adults. Mountain beavers apparently are faithful to their site selection. Two males remained in the vicinity of their nests for at least 31 and 44 months, respectively.

Burrow System. Mountain beavers burrow in moderately firm soil with adequate drainage. They dig by scooping with the forefeet and pushing the soil underneath the body. The incisors also may be used to loosen packed soil and stones. They push the soil toward an opening with the head and shoulders and ultimately expel it from the burrow by scraping with the hind feet (Voth 1968). The diameter of a tunnel may range between 10 and 20 cm (Camp 1918), depending on the size of the animal and the texture of the soil. Tunnels are seldom >120 cm below the surface. Deep tunnels lead to or connect chambers; shallow tunnels, <25 cm from the surface, lead to burrow openings. An individual network of tunnel systems may be 100 m in diameter, with an opening every 6 or 7 m.

The microclimate within burrow systems is cool and stable. The maximum annual fluctuation in burrow temperature is only about 45% of the fluctuation in ambient temperature, and Johnson (1971) found that weekly temperature variation never exceeded 4°C within the burrows.

Within the burrow system, Voth (1968) described five types of chambers: nest, feeding, refuse, fecal pellet, and earth ball. The nest chamber, often 50–60 cm in diameter and 36 cm high, usually is located at sites where drainage is good, and, together with the feeding chamber, forms a hub from which other tunnels and chambers radiate (Martin 1971). Nests are roughly circular. Those of adults contain as much as a bushel of vegetative material, whereas those of subadults are usually smaller. Where available, salal (*Gaultheria shallon*) often makes up the soft, dry central portion of the nest. Sword fern (*Polystichum munitum*), bracken fern (*Pteridium aquilinum*), and other readily available vegetative materials are common components of the coarser, sometimes moist outer shell (Johnson and Martin 1969; Martin 1971). D. L. Nolte (USDA, Animal and Plant Health Inspection Service, pers. commun., 2001), in assessing effects of damage to the burrow system on mountain beaver behavior, noted that reoccupation took longer following destruction of the nest than other parts of the system.

Feeding chambers are adjacent to the nest chamber and often are as large or larger. Besides feeding, mountain beavers use these areas to store large caches of wilted and recently cut vegetation. Decayed plant material from the feeding areas is placed in the refuse chambers. These are sometimes only blind tunnels, occasionally no larger than 12 cm in diameter, which open from feeding chambers.

Fecal pellet chambers also may open from feeding chambers, or pellets may be deposited in a portion of a feeding chamber. Voth (1968) reported that the mean diameter of pellet chambers was 18 cm. He found as many as six pellet chambers associated with a single burrow system, each containing fecal pellets in a different state of decomposition.

The earth ball chambers are storage areas for stones or compacted dirt retained by the animals. Voth (1968) felt that these "balls," which averaged 8 cm in diameter and 200 g in weight, were used to plug entrances to nest and feeding chambers. Because the diet of mountain beavers consists mostly of succulent vegetation, the earth balls also may serve as abrasives on which they sharpen their incisors.

Besides potential predators, numerous other rodents, lagomorphs, and insectivores also have been caught in mountain beaver burrows, presumably as they use them for easy travel lanes. As noted by Pfeiffer (1953), these include bushy-tailed woodrats (*Neotoma cinerea*), dusky-footed woodrats (*N. fuscipes*), Douglas's squirrels (*Tamiasciurus douglasii*), jumping mice (*Zapus princeps*), deer mice (*Peromyscus maniculatus*), California vole (*Microtus californicus*), water vole (*M. richardsoni*), Botta's pocket gopher (*Thomomys bottae*), and brush rabbits (*Sylvilagus bachmani*). Whitaker et al. (1979), reported several of these species in mountain beaver tunnels, as well as shrew-moles (*Neurotrichus gibbsii*), coast moles (*Scapanus orarius*), snowshoe hares (*Lepus americanus*), Townsend's chipmunks (*Tamias townsendii*), California red-backed voles (*Clethrionomys californicus*), Oregon voles (*M. oregoni*), and Pacific jumping mice (*Z. trinotatus*).

FEEDING HABITS

The mountain beaver has a functional cecum, which, together with the stomach, provides ample room for large volumes of ingested vegetation. The small intestine averages 1.3 m in length and the large intestine, 0.9 m. In free-ranging individuals, the entire digestive tract may account for 25–50% of the body weight (Voth 1968). Nonetheless, because of the relatively low energy content of its food, at least 75% of the activity time of this species is spent gathering and ingesting food (Ingles 1959). Thus, mountain beavers harvest substantial quantities of vegetation, much of which decays before it is eaten. Voth (1968) felt that 2.5 times as much vegetation was cut by this species as was actually eaten. However, a portion of this may be used as nesting material. Vegetation is stored during winter for nesting and consumption, but actual caching is minimal.

The mountain beaver is strictly herbivorous. Sword fern was the main component of the diet of males during winter and spring in the population studied by Voth (1968). Lactating females incorporated conifers and high-protein grasses in the spring. During the summer and fall, bracken fern was the preferred food item for both sexes. The diet of juveniles after weaning is similar to that of adults. The general categories of vegetation in the diets of adult males, females, and juveniles are given in Table 8.1.

Although pteridophytes make up the preferred food species throughout the year, during winter in areas of heavy snowfall, bark and twigs are readily taken. Shrubby vegetation such as salmonberry (*Rubus spectabilis*) and blackberry (*Rubus ursinus*) commonly are eaten during the summer. Crouch (1968) found heavy clipping by mountain beavers on vine maple (*Acer circinatum*), red huckleberry (*Vaccinium parvifolium*), and red alder (*Alnus rubra*). Five other species of hardwood received moderate to light use. These included big-leaf maple (*Acer macrophyllum*), cascara (*Rhamnus purshiana*), hazel (*Corylus cornuta*), ocean spray (*Holodiscus discolor*), and willow (*Salix* spp.).The remaining species of shrub that were common on the study area showed no evidence of use. These included rose (*Rosa gymnocarpa*), cherry (*Prunus emarginata*), currant (*Ribes sanguineum*), and red-stem ceanothus (*Ceanothus sanguineus*). In Yosemite National Park, Todd (1992) identified 28 species of herbs, 14 species of shrubs, and 6 species of trees harvested by mountain beavers within 30 m of their burrow systems. Red fir (*Abies magnifica*), dogwood (*Cornus* spp.), willow (*Salix* spp.), and corn lily (*Veratrum californicum*) were taken most frequently. Harvesting of coniferous tree branches by mountain beavers was studied

TABLE 8.1. General categories of food items in the diet of mountain beavers as determined from counts of epidermal fragments from fecal pellets

	Age and Sex Group[a]		
		Females	
Vegetation Category	Males and Nonpregnant Females ($N = 12$)	Lactating ($N = 3$)	Juveniles ($N = 4$)
Pteridophytes (ferns)	84.0	37.7	90.7
Conifers	3.4	33.9	0.0
Grasses	2.5	18.4	4.6
Forbs	1.9	4.8	2.6
Hardwoods	5.4	1.3	1.3
Mosses	1.0	3.5	0.9
Shrubs	1.1	0.0	0.0

SOURCE: Adapted from Voth (1968).
[a]Numbers given are percentages of total for each age sex group.

by O'Brien (1988) at two sites above 1600 m elevation in the Sierra Nevada, California. They removed limbs from trees between September and May. Few limbs were removed between June and August. Conifer branches accounted for none of the mass of food caches in June, July, August, and September; 31% in October; 63% in November; and 100% in December, January, and February. Conifer branches were cut only in winter when other food plants were unavailable.

Interestingly, mountain beavers consume species that are unpalatable or even toxic to other herbivores, including bracken fern, rhododendron (*Rhododendron* spp.), devil's club (*Oplopanax horridus*), and stinging nettle (*Urtica dioica*). Conversely, they rarely harvest toxic digitalis plants (*Digitalis purpurea*) or even preferred foods treated with water extracts of digitalis (Nolte et al. 1995a, 1995b). Fitts (1996) discussed two exotic plant species important to Point Arena mountain beavers, *A. r. nigra,* which are listed in the United States as endangered. One introduced species, ice plant (*Carpobrotus edulis*), was considered a succulent food source, which provided water throughout the year. Free-ranging mountain beavers drink water when it is available. In fact, wet or humid habitats and succulent vegetation are considered essential because of the poor urine-concentrating ability of mountain beavers. Fisler (1965) reported captive mountain beavers were easily maintained on a diet of succulent vegetation in the absence of free-flowing water. Captive animals appear to "wash" food. If they are given dry food they will "soak" it and consume it later (D. L. Nolte, USDA, Animal and Plant Health Inspection Service, pers. commun., 2001). Carraway and Verts (1993) summarized the feeding ecology of mountain beavers.

BEHAVIOR

Foraging and Feeding. The burrows of mountain beavers often open directly into suitable vegetation, and the animals move only short distances while foraging. Although they may consume vegetation while aboveground, forage often is carried or dragged into the burrow, where it is eaten or stored (Martin 1971). The animals may climb small trees and shrubs to a height of 7 m while foraging, clipping branches as they ascend. Descent is made headfirst (Ingles 1960) or by simply releasing the grip and falling to the ground. While feeding, mountain beavers may sit on their haunches with the hind legs extended in front. Fisler (1965) observed this posture in a captive mountain beaver and noted that food was manipulated mainly with the forefeet, only rarely with the hind feet.

"Haymaking" is a phenomenon noted by numerous early observers (Godin 1964:17). Mountain beavers stack fresh vegetation near burrow entrances in piles that may be 60 cm high. Voth (1968) suggested that haymaking may be related to improved succulence or nutritional quality of the vegetation. Also, it may reduce the number of times the nest and feeding chambers are opened to bring in vegetation.

Mountain beavers reingest certain fecal pellets. As it defecates, a mountain beaver takes each pellet in its mouth and tosses it onto a fecal pile. Occasionally, discarded pellets are retrieved and chewed or eaten (Ingles 1961). Such reingestion (coprophagy) has been documented in several other species of rodents as well as lagomorphs and shrews (Feldhamer et al. 1999) as a method of extracting the maximum amount of nutrients from food material.

Other Behaviors. Although the hind feet are not webbed, mountain beavers enter pools or streams without hesitation and are excellent swimmers. Nowak (1991) stated that the animals will swim through flooded tunnels, although, as noted, a significant factor in habitat use is good soil drainage.

Grooming apparently is minimal. Fisler (1965) reported that a captive mountain beaver groomed infrequently, using its teeth and front claws, for periods of 2 min or less. He felt that the species was hampered in this behavior by the inflexibility of the neck and forelegs.

Although generally silent, several types of vocalization have been attributed to mountain beavers. These include growls, whistles, whines, "coughs," and squeals.

Nolte et al. (1993) used artificial burrow systems to study interactions between mountain beavers and the role of chemical communication. Although mountain beavers exhibited scent-marking behaviors, the marks did not prevent intruders from entering and exploring burrow systems. Encounters were agonistic and resident animals aggressively evicted male or female intruders.

MORTALITY

Little is known concerning mortality rates of mountain beavers, although numerous predators have been reported. Lovejoy and Black (1979) estimated that longevity in the wild of 5–6 years was common. Habitat management activities can be a source of mortality. Approximately 51% of the mountain beavers in a 7-ha area were killed by a controlled broadcast burn that left little residual slash. On a similar 9-ha area that had 1.8 ha of slash and brush left unburned, only 20% of the resident population was killed (Motubu 1978).

Predators. Predators cited by previous authors include most of the carnivores, although documentation is often lacking and there has been no determination of the effect of predation on population densities of mountain beavers. Taylor (1918) believed that striped skunks (*Mephitis mephitis*), spotted skunks (*Spilogale putorius*), mink (*Mustela vison*), gray foxes (*Urocyon cinereoargenteus*), raccoons (*Procyon lotor*), badgers (*Taxidea taxus*), bobcats (*Lynx rufus*), and fishers (*Martes pennanti*) preyed on mountain beavers. The extent of bobcat predation was quantified by Witmer and DeCalesta (1986) and Toweill and Anthony (1988a). Toweill and Maser (1985) documented cougars (*Puma concolor*) preying on mountain beavers, and long-tailed weasels (*Mustela frenata*) may prey on young (Godin 1964). As noted previously, coyotes are significant predators on mountain beavers (Toweill and Anthony 1988b). Over 70% of the coyote scats collected by Witmer and DeCalesta (1986) contained mountain beaver remains. Thrailkill et al. (2000) reported that mountain beavers constituted 3% of the prey taken by two northern goshawks (*Accipiter gentilis*) in western Oregon. They are also taken by other raptors, including golden eagles (*Aquila chrysaetos*) and great horned owls.

Parasites. Twenty-nine species of ectoparasite and 2 species of cestode were noted by Canaris and Bowers (1992) in their review of metazoan parasites from mountain beavers. Included were 12 species of fleas (order Siphonaptera), 14 species of mites (order Acari), and 2 species of hard ticks (order Ixodida). Two species of large mites, *Alphalaelaps aplodontiae* and *Patrinyssus hubbardi,* were common on 48 specimens examined by Whitaker et al. (1979) and were considered to be host-specific to mountain beavers. Three species of tiny mites recorded by Whitaker et al. (1979), *Aplodontochirus borealis, Microlabidopus americanus,* and *Aplodontopus latus,* are also host-specific to mountain beavers. Fain and Baker (1983) described a new genus and species of mite, taken on a flea, from *Aplodontia rufa*. Because of the phylogeny and ecology of mountain beavers, a host-specific mite community might be expected. The only endoparasites reported by Canaris and Bowers (1992) were cysticerci of the tapeworms *Taenia tenuicollis* and *T. pisiformis*. As they noted, it is unusual that no nematodes have been reported from mountain beavers given their humid, fossorial habitat and generally mild climates.

AGE ESTIMATION

Pelage characteristics may be used to differentiate juveniles and adults (see Description). Pfeiffer (1958) estimated the age of immature mountain beavers on the basis of body weight, using growth curves of known-age juveniles as a standard. However, Lovejoy (1972) found this method impractical for mature animals because of the extreme variation within age classes.

In males, the length of the baculum may be used to determine age class. Pfeiffer (1956) found that the mean length of the baculum in juveniles was about 11 mm, whereas it was 21 mm in yearlings and about 30 mm in adults. He did not indicate the amount of variation within age classes, however. For males and females, Pfeiffer (1958) considered the degree of closure of the epiphyseal femoral suture a "fairly reliable" method of age determination in mountain beavers. He also noted that the third upper molar is rooted, and used the degree of wear on M^3 as a supplemental determinant of age.

ECONOMIC STATUS, MANAGEMENT, AND CONSERVATION

The mountain beaver is considered neither a game animal nor a furbearer. Its meat is not sought after and even prime pelts historically sold for only about 20 cents each (Ingles 1965). The economic importance of the species involves the extensive damage it does to conifer seedlings and saplings and the resulting losses incurred by the forest products industry.

Interference with forest management, and particularly with reforestation efforts, by mountain beavers has been documented for some time (Couch 1925; Scheffer 1929; Staebler et al. 1954). Following a survey of forest landowners and managers in Washington, Oregon, and northern California, Borrecco et al. (1979) reported that damage by mountain beavers was occurring on about 111,000 ha of forest land. About 75% of this damage occurred in Douglas-fir (*Pseudotsuga menziesii*) forest types; western hemlock (*Tsuga heterophylla*) and several other conifer species were also damaged.

Mountain beaver damage to conifers takes several forms, depending on tree size. Burrowing activities may occasionally uproot or bury seedlings (Voth 1968), but feeding activities are the primary cause of injury (Hooven 1977; Black et al. 1979). Three primary categories of mountain beaver-caused injury to conifers were described by Lawrence et al. (1961): (1) stem clippings or cutting of seedlings, (2) branch cutting, and (3) basal girdling (removal of bark). Mortality, suppression of growth, and deformity of trees result from injury. Although damage may occur to trees from <1 to >20 years of age, the most common (Borrecco et al. 1979) and most serious form of damage is clipping of small seedlings. This damage, which occurs from immediately after planting and up to 4 years after planting, often results in seedling mortality and subsequent understocking or failure of plantations. Clipping or basal girdling of larger seedlings does not often result in mortality, but causes deformities and growth losses. Root girdling and undermining of trees by burrowing may contribute to tree fall under snow loads (Cafferata 1992). Borrecco and Anderson (1980) displayed the importance of seedling size to severity and extent of damage. From a 6000-tree sample taken in 24 randomly selected plantations in western Washington, they observed that 2 years after planting, of those trees sustaining damage from mountain beavers, 2-0 seedlings had sustained 53% mortality and 2-1 seedlings had sustained 36% mortality. (Seedlings designated 2-0 were grown in nursery seed beds for 2 years at high densities. Seedlings designated 2-1 were also in seed beds at high densities for 2 years and then transplanted and grown for another year at wider spacing before outplanting.) Mortality to 2-1 seedlings occurred only in the first year after planting, whereas 2-0 seedlings were killed over the 2-year period. The impact on height growth of surviving damaged seedlings was greater on 2-0 than on 2-1 seedlings (Fig. 8.3) and when damage occurred in the first year rather than the second year after planting (Fig. 8.4). Mountain beaver caused relatively greater damage on height growth of Douglas fir seedlings compared to clipping by snowshoe hare and browsing by deer (*Odocoileus* spp.) (Fig. 8.3).

In studies of levels of seedling damage relative to abundance of alternative foods and mountain beaver population densities, D. L. Nolte (USDA, Animal and Plant Health Inspection Service, pers. commun., 2001) found that damage intensity was directly related to population density and inversely related to food resources if populations were held constant. However, when given the freedom to move, animals moved to areas with greater food resources, and these areas sustained more or equivalent levels of damage than areas with fewer food resources.

The economic impacts of the mountain beaver and other types of wildlife damage to forest crops have been difficult to determine primarily because of the lack of information on long-term growth effects of

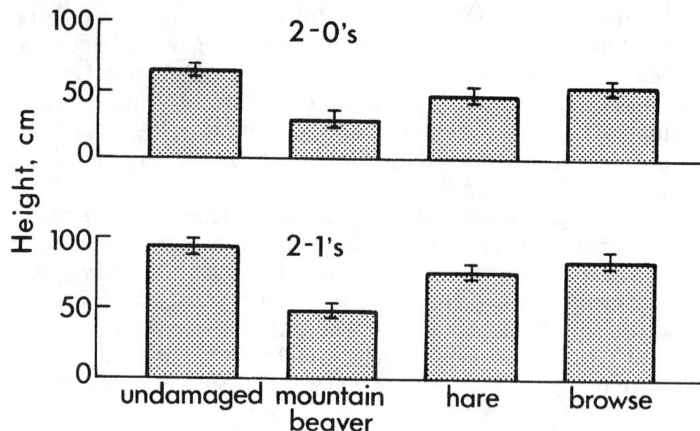

FIGURE 8.3. Mean heights and standard errors of damaged and undamaged 2-0 and 2-1 seedlings after 2 years in the field.

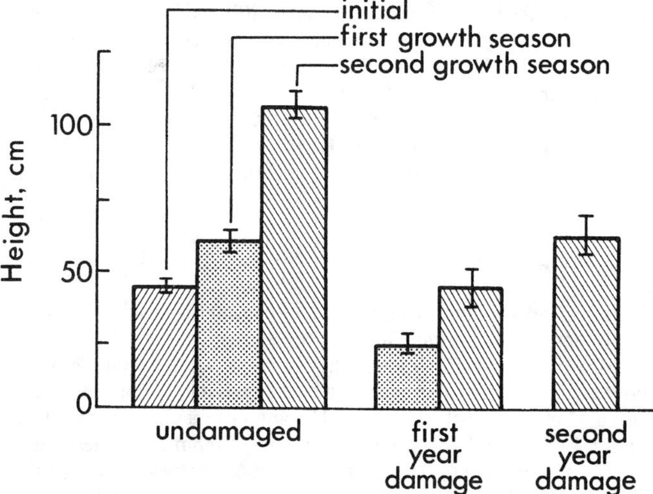

FIGURE 8.4. Mean heights and standard errors of undamaged Douglas-fir 2-1 nursery stock compared to stock damaged in the first and second years after planting, respectively.

damage occurring to trees at juvenile growth stages. Lawrence (1959) estimated wildlife-caused losses on Weyerhaeuser Company lands in the Pacific Northwest at about $900,000 per year; a substantial portion of this total was due to mountain beaver. In a later assessment, Weyerhaeuser Company estimated annual mountain beaver-caused losses at over $1 million on company lands. Black et al. (1979) presented information on long-term growth effects of wildlife damage to conifer plantations, and Brodie et al. (1979) estimated the economic impacts of these damage levels, but did not separate out the effect of mountain beavers in particular. Although there have been no recent estimates of economic losses, damage by mountain beavers continues to be a significant obstacle to forest regeneration in the Pacific Northwest (P. Heide, Washington Forest Protection Association, Olympia, pers. commun., 2001).

Control. As a result of the significant levels of economic loss caused by mountain beavers in managed forests, many approaches have been taken to control damage. Direct control methods involve the use of toxicants or trapping to reduce populations in problem areas and the use of physical or chemical barriers to feeding. Indirect damage control methods involve the use of silvicultural techniques to reduce the suitability of the habitat and discourage the expansion of populations.

Initial direct control efforts have focused on toxicants, and through time, several bait formulations have been developed for application in the burrow system. A number of challenges, including developing palatable baits and maintaining their acceptability to mountain beavers under

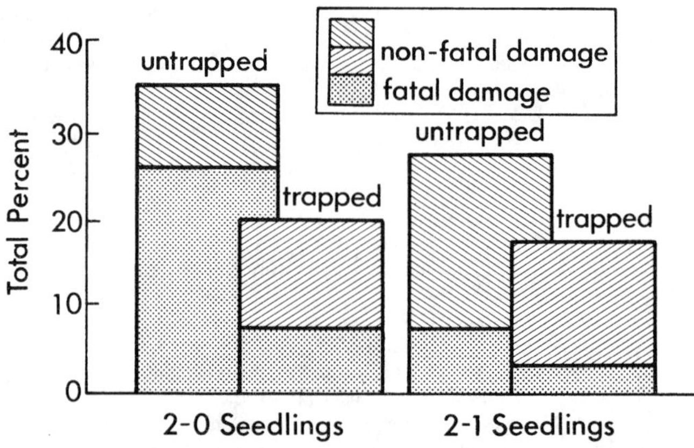

FIGURE 8.5. Trapping with Conibear traps is the most effective method of controlling populations of mountain beaver (*Aplondontia rufa*). Shown is the mortality (stippling) and nonfatal damage (cross hatching) in trapped and untrapped areas of Douglas-fir 2-0 and 2-1 nursery stock 2 years after planting.

the high-moisture microclimate of the burrow system, have limited the success of toxicants. Although some toxicant formulations have shown promise (Campbell and Evans 1988), the extensive and costly data requirements associated with development, testing, and Environmental Protection Agency registration, coupled with the relatively small potential market for these materials, have limited the development of new baits. No toxicants are registered for mountain beaver control.

Trapping using Conibear #110 traps has been the principal and most effective method of direct control of mountain beaver in recent years (Fig. 8.5). Trained crews place traps in main runways of burrow systems at rates of ≥20 traps/ha, depending on mountain beaver density. Trapping costs average $20–100/ha under good conditions, that is, moderate mountain beaver densities, moderate topography, and relatively slash-free sites (Borrecco and Anderson 1980). The high labor requirement for trapping, which competes for labor with other forestry activities, and the associated high costs are the major disadvantages of trapping. As the result of a year 2000 ballot initiative in Washington State, the use of body-gripping traps, including the Conibear traps used for mountain beaver control, has been banned and this method is no longer available. Trapping remains an available method of mountain beaver control in Oregon.

Chemical repellents applied to tree seedlings either in the nursery or in the field have been used with some success for several wildlife species causing damage. A number of materials have been tested for repellency to mountain beaver with some indications of efficacy, but practical formulations suitable for field use have yet to be developed. Deer-Away Big-Game Repellent (IntAgra, Inc., Minneapolis, MN) in powder form significantly reduced mountain beaver damage to planted Douglas-fir seedlings following the placement of treated cull seedlings in the burrow system (Campbell and Evans 1988). Several natural products, derived from nonpalatable or toxic plants or based on predator odors, have shown efficacy in controlled experiments. Epple et al. (1992) observed reduced retrieval of food by mountain beaver from feeding bowls scented with mustelid gland secretions or components thereof and/or predator urine. The more complex natural predator scents had greater repellency than their major volatile components. Observing that mountain beaver avoided consuming digitalis, a common but toxic plant in their range, Nolte et al. (1995a, 1995b) evaluated the mechanisms for this avoidance, and determined that it was not the toxic cardiac glycosides, but other chemosensory cues that were responsible.

Physical barriers continue to be among the few effective tools available for reducing mountain beaver damage to conifer seedlings. Campbell and Evans (1975) and Campbell et al. (1988) demonstrated that tubes of polypropylene plastic netting placed around individual seedlings significantly reduced damage caused by several wildlife species, including mountain beaver. For mountain beaver specifically, Borrecco and Anderson (1980) observed damage levels of >44% for trees without tubes compared to 3% for trees with protective tubes. Cost of the tubing and its application and the need for inspection and maintenance of "tubed" seedlings in the field are the major factors limiting widespread use of this technique. Recent experience in western Oregon (S. Cafferata, Weyerhaeuser Company, Qualdel Company, Coos Bay, Oregon, pers. commun., 2001) indicates total costs of material, transportation, and installation to be about 30–40 cents/tree, or ≥$375/ha, depending on plantation stocking levels.

A key strategy for controlling mountain beaver damage is to manage stands to maintain populations at low levels. Silvicultural approaches such as intensive site preparation, rapid postharvest regeneration with large planting stock, and timing of thinning to minimize development of understory vegetation can maintain stands and sites in a condition that does not encourage population expansion. However, this will not always be sufficient. Direct control methods, including tree protection and direct population control, may also be needed when populations are high and damage levels unacceptable. In most instances, a combination of approaches will be most effective.

RESEARCH AND MANAGEMENT NEEDS

Improved methods of controlling mountain beaver damage to conifer plantations are needed. Approaches that emphasize indirect control rather than population reduction will be better accepted by the general public and have greater potential for authorization by regulatory agencies. Although chemical repellents have been examined with little success, availability of a nursery-applied material to provide first-year protection in the field would be of significant value. The observations of Nolte et al. (1995a, 1995b) suggest potential sources of repellent in native plants (*Digitalis purpurea),* and those of Epple et al. (1995) suggest predator urines could be useful. Approaches involving the use of pheromones to modify behavior might ultimately provide effective damage control techniques, but require major basic research commitments. Unique physiological characteristics of mountain beavers, such as their inability to concentrate urine—thus the requirement for free water and highly succulent forage—might be exploited for chemical interference with controlling mechanisms. Knowledge of the limited thermoregulatory capability of mountain beavers led Campbell and Evans (1988) to test the potential of reserpine, a tranquilizer used to treat hypertension in humans, for producing hypothermia. High levels of mountain beaver mortality occurred, with low secondary hazards to nontarget wildlife, suggesting opportunities for development of reserpine as a control agent. An improved understanding of the digestive processes and nutritional requirements of mountain beavers also might provide insights into new control approaches. Although reproductive inhibitors have proven relatively unsuccessful in most rodent control applications, the single litter and relatively small size of mountain beavers suggest a possible control opportunity. An improved understanding of population dynamics, and particularly natural mortality factors, is needed.

Ecological approaches involving modification of habitat to reduce its capability to support mountain beaver populations are a potentially fruitful area of research. Voth (1968) suggested modification of plant communities to exclude preferred plant species through the introduction of insects or the use of herbicides. Cafferata (1992) noted instances where sheep grazing has had an indirect impact on mountain beaver by collapsing runways, and observed that both mammalian and avian predators focused their activities in these grazed areas. Neal and Borrecco (1981) determined that levels of damage in sapling stands were positively correlated with the size of openings. They concluded that stands should be managed to reduce the number and size of openings to reduce mountain beaver damage.

Cafferata (1992) suggested some modifications to standard silvicultural practices that could reduce damage levels; these warrant research attention. They include adjusting the timing of precommercial and commercial thinning and other partial cuts to minimize understory vegetation and associated high mountain beaver densities and reduce damage to the postharvest plantation. Size of slash piles created in site preparation following logging also seems to affect damage levels, with less damage occurring when piles are small. The ability to predict the responses of mountain beaver populations to these and other silvicultural activities, a better understanding of their association with other habitat elements, and the responses of populations and damage levels to alternative control approaches would provide valuable direction to management strategies to minimize damage. The application of integrated approaches, including vegetation manipulation through site preparation, selection of appropriately sized planting stock, attention to timing of trapping in relation to plantation establishment, individual tree protection, and careful and regular surveys of plantation areas before and after planting should continue to provide acceptable levels of damage control.

LITERATURE CITED

Bauer, E. W., and D. L. Pattie. 1968. Lactate dehydrogenase genes in rodents. Nature 218:341–43.

Beier, P. 1989. Use of habitat by mountain beaver in the Sierra Nevada. Journal of Wildlife Management 53:649–54.

Black, H. C., E. J. Dimock, J. A. Evans, and J. A. Rochelle. 1979. Animal damage to coniferous plantations in Oregon and Washington. Part 1: A survey, 1963–1975. Oregon State University, Forest Research Laboratory, Research Bulletin 25:1–44.

Bleich, V. C., and D. Racine. 1991. Mountain beaver (*Aplodontia rufa*) from Inyo County, California. California Fish and Game 77:153–55.

Borrecco, J. E., and R. J. Anderson. 1980. Mountain beaver problems in the forests of California, Oregon, and Washington. Proceedings of the Vertebrate Pest Conference 9:135–42.

Borrecco, J. E., H. W. Anderson, H. C. Black, J. Evans, K. S. Guenther, G. D. Lindsey, R. P. Matthews, and T. K. Moore. 1979. Survey of mountain beaver damage to forests in the Pacific Northwest, 1977 (DNR Note 26). Department of Natural Resources, Olympia, WA.

Brodie, D., H. C. Black, E. J. Dimock, J. Evans, C. Kao, and J. A. Rochelle. 1979. Animal damage to coniferous plantations in Oregon and Washington. Part 2: An economic evaluation, 1979. Oregon State University, Forest Research Laboratory, Research Bulletin 26:1–22.

Cafferata, S. L. 1992. Mountain beaver. Pages 231–51 *in* Silvicultural approaches to animal damage management in Pacific Northwest forests (General Technical Report PNW-GTR-287). USDA Forest Service, Pacific Northwest Research Station, Portland, OR.

Camp, D. L. 1918. Excavations of burrows of the rodent *Aplodontia,* with observations on habitat of the animal. University of California Publications in Zoology 17:517–36.

Campbell, D. L., and J. Evans. 1975. "Vexar" seedling protectors to reduce wildlife damage to Douglas-fir (Wildlife Leaflet 508). U.S. Department of the Interior, Washington, DC.

Campbell, D. L., and J. Evans. 1988. Recent approaches to controlling mountain beavers (*Aplodontia rufa*) in Pacific Northwest forests. Proceedings of the Vertebrate Pest Conference 13:183–87.

Campbell, D. L., J. Evans, and G. B. Hartman. 1988. Evaluation of seedling protection materials in western Oregon. (Technical Note OR-5). U.S. Department of the Interior, Bureau of Land Management, Washington, DC.

Canaris, A. G., and D. Bowers. 1992. Metazoan parasites of the mountain beaver, *Aplodontia rufa* (Rodentia: Aplodontidae), from Washington and Oregon, with a checklist of parasites. Journal of Parasitology 78:904–6.

Carraway, L. N., and B. J. Verts. 1993. *Aplodontia rufa*. Mammalian Species, 431:1–10.

Couch, L. K. 1925. Rodent damage to young forests. Murrelet 6:39.

Cramblet, H. M., and R. L. Ridenhour. 1956. Parturition in *Aplodontia*. Journal of Mammalogy 37:87–90.

Crouch, G. L. 1968. Clipping of woody plants by mountain beaver. Journal of Mammalogy 49:151–52.

Dicker, S. E., and M. G. Eggleton. 1964. Renal function in the primitive mammal, *Aplodontia rufa,* with some observations on squirrels. Journal of Physiology 170:186–94.

Druzinsky, R. E. 1995. Incisal biting in the mountain beaver (*Aplodontia rufa*) and woodchuck (*Marmota monax*). Journal of Morphology 226:79–101.

Epple, G., R. Mason, D. Nolte, and D. Campbell. 1992. Biologically based repellents reduced food consumption in a forest pest, the mountain beaver. Chemical Senses 17:619.

Epple, G., J. R. Mason, E. Aronov, D. L. Nolte, R. A. Hartz, R. Kaloostian, D. Campbell, and A. B. Smith III. 1995. Feeding responses to predator-based repellents in the mountain beaver (*Aplodontia rufa*). Ecological Appplications 5:1163–70.

Fain, A., and G. T. Baker. 1983. *Trichopsyllopus oregonensis* g.n., sp. n. (Acari, Acaridae), a hypopus phoretic on a flea *Trichopsylloides oregonensis,* parasite on the rodent *Aplodontia rufa* in the United States. Canadian Journal of Zoology 61:928–29.

Feldhamer, G. A., L. C. Drickamer, S. H. Vessey, and J. F. Merritt. 1999. Mammalogy: Adaptation, diversity, and ecology. McGraw-Hill, New York.

Fisler, G. F. 1965. A captive mountain beaver. Journal of Mammalogy 46:707–9.

Fitts, K. M. 1996. Observations on the use of two non-native plants by the Point Arena mountain beaver. California Fish and Game 82:59–60.

Godin, A. J. 1964. A review of the literature on the mountain beaver. United States Fish and Wildlife Service Special Scientific Report 78:1–52.

Gyug, L. W. 2000. Status, distribution, and biology of the mountain beaver, *Aplodontia rufa,* in Canada. Canadian Field-Naturalist 114:476–90.

Hacker, A. L., and B. E. Coblentz. 1993. Habitat selection by mountain beavers recolonizing Oregon Coast Range clearcuts. Journal of Wildlife Management 57:847–53.

Hall, E. R. 1981. The mammals of North America, 2nd ed. John Wiley, New York.

Hooven, E. F. 1977. The mountain beaver in Oregon: Its life history and control. Oregon State University, Forest Research Laboratory, Research Paper 30:1–20.

Ingles, L. G. 1959. A quantitative study of mountain beaver activity. American Midland Naturalist 61:419–23.

Ingles, L. G. 1960. Tree climbing beavers. Journal of Mammalogy 41:120–21.

Ingles, L. G. 1961. Reingestion in the mountain beaver. Journal of Mammalogy 42:411–12.

Ingles, L. G. 1965. Mammals of the Pacific states. Stanford University Press, Stanford, CA.

Johnson, N. E., and P. Martin. 1969. *Amydria effrentella* from nests of mountain beaver, *Aplodontia rufa*. Annals of the Entomological Society of America 62:396–99.

Johnson, S. R. 1971. The thermal regulation, microclimate and distribution of the mountain beaver, *Aplodontia rufa pacifica* Merriam. Ph.D. Dissertation, Oregon State University, Corvallis.

Lawrence, W. H. 1959. Wildlife-damage control problems on Pacific Northwest tree farms. Transactions of the North American Wildlife Conference 23:146–52.

Lawrence, W. H., N. B. Kverno, and H. D. Hartwell. 1961. Guide to wildlife feeding injuries on conifers in the Pacific Northwest. Western Forest Conservation Association, Portland, OR.

Lovejoy, B. P. 1972. A capture–recapture analysis of a mountain beaver population in western Oregon. Ph.D. Dissertation, Oregon State University, Corvallis.

Lovejoy, B. P., and H. C. Black. 1974. Growth and weight of the mountain beaver, *Aplodontia rufa pacifica*. Journal of Mammalogy 55:364–69.

Lovejoy, B. P., and H. C. Black. 1979. Movements and home range of the Pacific mountain beaver, *Aplodontia rufa pacifica*. American Midland Naturalist 101:393–402.

Lovejoy, B. P., H. C. Black, and E. F. Hooven. 1978. Reproduction, growth and development of the mountain beaver (*Aplodontia rufa pacifica*). Northwest Science 52:323–28.

Martin, P. 1971. Movements and activities of the mountain beaver (*Aplodontia rufa*). Journal of Mammalogy 52:717–23.

Maser, C., B. R. Mate, J. F. Franklin, and C. T. Dyrness. 1981. Natural history of Oregon coast mammals. U.S. Department of Agriculture, Forest Service, General Technical Report, PNW-133:1–496.

McLaughlin, C. A. 1984. Protrogomorph, Sciuromorph, Castorimorph, Myomorph (Geomyoid, Anomaluroid, Pedetoid, and Ctenodactyloid) rodents. Pages 267–88 *in* S. Anderson and J. K. Jones, Jr., eds. Orders and families of recent mammals of the world. John Wiley, New York.

McMillin, J. H., and D. A. Sutton. 1972. Additional information on chromosomes of *Aplodontia rufa* (Sciuridae). Southwestern Naturalist 17:307–8.

Merzenich, M. M., L. Kitzes, and L. Aitkin. 1973. Anatomical and physiological evidence for auditory specialization in the mountain beaver (*Aplodontia rufa*). Brain Research 58:331–44.

Motubu, D. A. 1978. Effects of controlled slash burning on the mountain beaver (*Aplodontia rufa rufa*). Northwest Science 52:92–99.

Neal, F. D., and J. E. Borrecco. 1981. Distribution and relationship of mountain beaver to openings in sapling stands. Northwest Science 55:79–86.

Nolte, D. L., G. Epple, D. L. Campbell, and J. R. Mason. 1993. Response of mountain beaver (*Aplodontia rufa*) to conspecifics in their burrow system. Northwest Science 67:251–55.

Nolte, D. L., K. L. Kelly, B. A. Kimball, and J. J. Johnston. 1995a. Herbivore avoidance of digitalis extracts is not mediated by cardiac glycosides. Journal of Chemical Ecology 21:1447–55.

Nolte, D. L., B. A. Kimball, K. L. Kelly, Z. Zhang, and D. L. Campbell. 1995b. Herbivore avoidance of a simple digitalis extract. Journal of Agricultural and Food Chemistry 43:830–32.

Nowak, R. M. 1991. Walker's Mammals of the world. 5th ed. Johns Hopkins University Press, Baltimore.

Nungesser, W. C., and E. W. Pfeiffer. 1965. Water balance and maximum concentrating capacity in the primitive rodent, *Aplodontia rufa*. Comparative Physiology and Biochemistry 14:289–97.

Nungesser, W. C., E. W. Pfeiffer, D. A. Iverson, and J. F. Wallerius. 1960. Evaluation of renal countercurrent hypothesis in *Aplodontia*. Federation Proceedings 19:362.

O'Brien, J. P. 1988. Seasonal selection of coniferous trees by the sewellel, *Aplodontia rufa*. Mammalia 52:325–30.

Pfeiffer, E. W. 1953. Animals trapped in mountain beaver (*Aplodontia rufa*) runways, and the mountain beaver in captivity. Journal of Mammalogy 34:396.

Pfeiffer, E. W. 1956. The male reproductive tract of a primitive rodent, *Aplodontia rufa*. Anatomical Record 124:629–35.

Pfeiffer, E. W. 1958. The reproductive cycle of the female mountain beaver. Journal of Mammalogy 39:223–35.

Pfeiffer, E. W. 1968. Comparative anatomical observations of the mammalian renal pelvis and medulla. Journal of Anatomy 102:321–31.

Pfeiffer, E. W., N. C. Nungesser, D. A. Iverson, and J. R. Wallerius. 1960. The renal anatomy of the primitive rodent, *Aplodontia rufa,* and a consideration of its functional significance. Anatomical Record 137:227–35.

Rensberger, J. M. 1982. Patterns of dental change in two locally persistent successions of fossil aplodontid rodents. Pages 323–49 *in* B. Kurten, ed. Teeth: Form, function, and evolution. Columbia University Press, New York.

Scheffer, T. H. 1929. Mountain beavers in the Pacific Northwest: Their habits, economic status, and control (Farmer's Bulletin 1958). U.S. Department of Agriculture, Washington, DC.

Schmidt-Nielson, B., and E. W. Pfeiffer. 1970. Urea and urinary concentrating ability in the mountain beaver *Aplodontia rufa*. American Journal of Physiology 218:1370–75.

Shotwell, J. A. 1958. Evolution and biogeography of the aplodontid and mylagaulid rodents. Evolution 12:451–84.

Staebler, G. R., P. G. Lauterbach, and A. W. Moore. 1954. Effect of animal damage on young coniferous plantations in southwest Washington. Journal of Forestry 52:730–33.

Stangl, F. B., Jr., and J. V. Grimes. 1987. Phylogenetic implications of comparative pelage morphology in Aplodontidae and the Nearctic Sciuridae, with observations on seasonal pelage variation. Occasional Papers, The Museum, Texas Tech University 112:1–21.

Taylor, W. P. 1918. Revision of the rodent genus *Aplodontia*. University of California Publications in Zoology 17:435–504.

Thrailkill, J. A., L. S. Andrews, and R. M. Claremont. 2000. Diet of breeding northern goshawks in the Coast Range of Oregon. Journal of Raptor Research 34:339–40.

Todd, P. A. 1992. Mountain beaver habitat use and management implications in Yosemite National Park. Natural Areas Journal 12:26–31.

Toweill, D. E., and R. G. Anthony. 1988a. Annual diet of bobcats in Oregon's Cascade Range. Northwest Science 62:99–103.

Toweill, D. E., and R. G. Anthony. 1988b. Coyote foods in a coniferous forest in Oregon. Journal of Wildlife Management 52:507–12.

Toweill, D. E., and C. Maser. 1985. Food of cougars in the Cascade Range of Oregon. Great Basin Naturalist 45:77–80.

Verts, B. J., and L. N. Carraway. 1998. Land mammals of Oregon. University of California Press, Berkeley.

Voth, E. H. 1968. Food habits of the Pacific mountain beaver, *Aplodontia rufa pacifica* Merriam. Ph.D. Dissertation, Oregon State University, Corvallis.

Whitaker, J. O., Jr., C. Maser, and W. M. Wallace. 1979. Parasitic mites of the mountain beaver (*Aplodontia rufa*) from Oregon. Northwest Science 53:264–67.

Witmer, G. W., and D. S. DeCalesta. 1986. Resource use by unexploited sympatric bobcats and coyotes in Oregon. Canadian Journal of Zoology 64:2333–38.

George A. Feldhamer, Department of Zoology, Southern Illinois University, Carbondale, Illinois 62901-6501. Email: feldhamer@zoology.siu.edu.

James A. Rochelle, Rochelle Environmental Forestry Consulting, Olympia, Washington 98516-1400. Email: rochellej@olywa.net.

Clifford D. Rushton, Washington Department of Ecology, P.O. Box 47600, Olympia, Washington 98504-7600. Email: drus461@ecy.wa.gov.

9

Marmots

Marmota monax and Allies

Kenneth B. Armitage

NOMENCLATURE

Scientific and Common Names. Six species of marmot occur in North America: yellow-bellied marmot (*M. flaviventris*), woodchuck (*M. monax*), hoary marmot (*M. caligata*), Olympic marmot (*M. olympus*), arctic or Alaskan marmot (*M. broweri*), and Vancouver Island marmot (*M. vancouverensis*); scientific names follow Hoffmann et al. (1993); nearly all marmots are called groundhogs, whistle pigs, or whistlers; those living in the mountains commonly are called rockchucks

Subspecies. No subspecies have been described for *M. broweri, M. olympus,* and *M. vancouverensis;* there are 11 recognized subspecies of *M. flaviventris* (Howell 1915; Frase and Hoffmann 1980): *M. f. avara, M. f. dacota, M. f. engelhardti, M. f. flaviventris, M. f. fortirostris, M. f. luteola, M. f. nosophora, M. f. notioros, M. f. obscura, M. f. parvula,* and *M. f. sierrae*. Nine subspecies of *M. monax* are recognized (Hall 1981): *M. m. bunkeri, M. m. canadensis, M. m. ignava, M. m. johnsoni, M. m. monax, M. m. ochracea, M. m. petrensis, M. m. preblorum,* and *M. m. rufescens*. Eight subspecies of *M. caligata* are recognized (Hall 1981): *M. c. caligata, M. c. cascadensis, M. c. nivaria, M. c. okanagana, M. c. oxytona, M. c. raceyi, M. c. sheldoni,* and *M. c. vigilis*

Other Marmots. Fourteen species of marmot are recognized worldwide (Barash 1989; Hoffmann et al. 1993); eight occur in Eurasia. Two species occur in Europe: the alpine marmot, *M. marmota,* in the Alps, Carpathians, Pyrenees, and Apennines; and the steppe marmot, *M. bobac,* in the steppe regions of Russia, Kazakhstan, and Ukraine. The other six species occur in the mountainous regions of Asia: the Himalayan marmot, *M. himalayana;* the tarbagan, *M. sibirica;* the black-capped marmot, *M. camtschatica;* Menzbier's marmot, *M. menzbieri;* the gray marmot, *M. baibacina;* and the red or long-tailed marmot, *M. caudata* (see Armitage [2000] for distribution of Eurasian species). Much of the Russian literature was summarized by Bibikow (1996). Information on marmots is available in the proceedings of three international conferences (Bassano et al. 1992; Le Berre et al. 1996; Armitage and Rumiantsev 2002) and in a comprehensive bibliography of 4034 entries complied by Ramousse (1997). Some of the recognized Eurasian subspecies may deserve specific status, such as *M. caudata caudata* and *M. c. aurea.*

PHYLOGENY

The first squirrel, *Protosciurus,* appeared in the Oligocene in North America and the first of the true ground squirrels, *Miospermophilus,* appeared in the late Oligocene (Black 1972). During the Miocene, well-developed genera of sciurids derived from *Miospermophilus* included *Marmota, Cynomys,* and *Spermophilus* (Hafner 1984). These modern genera are grouped in the Tribe Marmotini (Rodentia, Sciuridae). Within this tribe, the marmots form a monophyletic group (Kruckenhauser et al. 1998; Steppan et al. 1999) in subtribe Marmotina along with *Paenemarmota* from the middle or late Pliocene (Hafner 1984; Mein 1992). Cladistic analysis with allozyme data indicates that marmots were derived from *Spermophilus* (Hafner 1984). The first species, *M. vetus,* occurred in North America about 9.5 million years ago (Steppan et al. 1999). Marmots reached Eurasia at the beginning of the Quaternary via the Bering land bridge. Rapid divergence in the Pleistocene led to the present array of species.

Steppan et al. (1999) proposed that the 14 species of marmot be placed in two subgenera: (1) subgenus *Marmota,* which includes all the Eurasian species plus *M. monax* and *M. broweri,* and (2) subgenus *Petromarmota,* which includes *M. caligata, M. vancouverensis, M. flaviventris,* and *M. olympus*. This analysis, based on a complete cytochrome *b* sequence, rejects the two major hypotheses concerning the phylogenetic position of *M. broweri;* it is not related to *M. caligata* nor is it a sister species to *M. camtschatica* of eastern Siberia. However, this analysis does not explain how *M. broweri* came to host a flea, *Oropsylla silantiewi,* which otherwise occurs only on Palearctic marmots (Rausch and Rausch 1971). The presence of *O. silantiewi* on *M. broweri* supports the hypothesis that the arctic marmot reinvaded North America from Eurasia, but does not reject the hypothesis that *M. broweri* is a relict species, which was a member of a paraphyletic North American grade that was possibly sister to the Palearctic group (Steppan et al. 1999). *M. monax* probably represents a surviving member of the subgenus *Marmota,* which diverged from the *Petromarmota* in North America before crossing into and radiating in Eurasia.

DISTRIBUTION

The woodchuck, the widest ranging North American marmot, occurs at low elevations from eastern Alaska through southern Canada to southern Labrador and south across the eastern United States to northern Georgia and Alabama and west to eastern Kansas and Nebraska (Kwiecinski 1998). It is absent from the prairie regions of western Canada and the United States, but extends south from Alaska into northern Idaho (Fig. 9.1).

Yellow-bellied marmots are widely distributed in the western United States (Frase and Hoffmann 1980) and extend northward into south-central British Columbia and southern Alberta, where the species occurs at low to middle elevations in relatively warm, arid habitats. It also occurs in arid habitats in eastern Washington and Oregon and portions of northwestern Idaho and Nevada (Couch 1930). Its lower elevation limit becomes progressively higher to the south, where it reaches its range limits in the Sangre de Cristo Mountains in New Mexico, in the southern Sierra Nevada and White Mountains of California, and the Toquina and Pine Valley Mountains in the Great Basin of Nevada (Fig. 9.2). Because yellow-bellied marmots may occur only above 2000 m elevation, southern populations are likely isolated from each other, as intervening valleys act as dispersal barriers. These "island" populations may have been genetically isolated for thousands of years.

The hoary marmot occurs from central Alaska (south of the Brooks Range) and the Yukon and Northwest Territories of Canada southward to western and northeastern Washington, central Idaho, and western Montana. The Alaskan marmot occurs in the Brooks Range of northern Alaska and possibly also in the northern Yukon of Canada. Olympic marmots are restricted to the Olympic Mountains of western

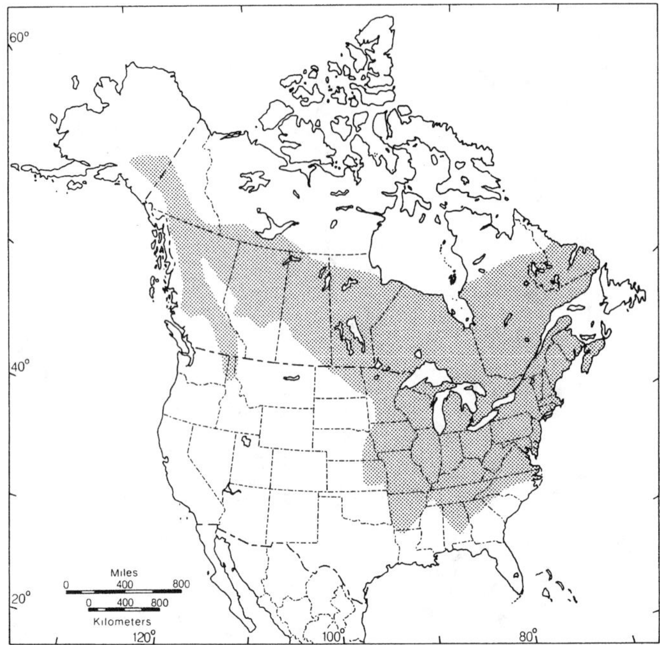

FIGURE 9.1. Distribution of the woodchuck (*Marmota monax*). SOURCE: Adapted from Hall and Kelson (1959).

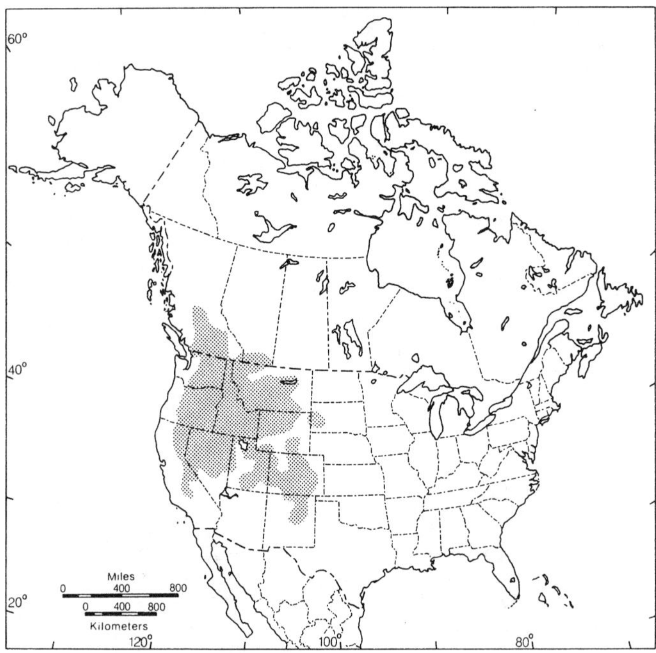

FIGURE 9.2. Distribution of the yellow-bellied marmot (*Marmota flaviventris*). SOURCE: Adapted from Hall and Kelson (1959).

Washington, whereas *M. vancouverensis* is endemic to Vancouver Island (Fig. 9.3).

DESCRIPTION

Marmots are the largest deep hibernator (Lyman et al. 1982). There are differences of 1.2 times in average body length and two times emergence mass between the smallest and the largest North American marmots. Typically, body mass is minimal at emergence from hibernation and maximal at the time of immergence (Table 9.1). Body length is significantly correlated with emergence mass ($p = .038$, $R^2 = .34$) and immergence mass ($p = .008$, $R^2 = .49$) (Armitage 1999). Males are larger than females (Armitage 1981). Yellow-bellied marmots from arid regions are smaller than those from more mesic montane habitats (Frase and Hoffmann 1980; Armitage et al. 1990).

FIGURE 9.3. Distribution of the hoary marmot (*Marmota caligata*), Alaska marmot (*M. broweri*), Vancouver marmot (*M. vancouverensis*), and Olympic marmot (*M. olympus*).

The feet have five claw-bearing digits, although the thumb is rudimentary. The claws are thick and slightly curved; those in the front feet are heavier. The palms are naked with five pads, and the soles are naked with six pads. The head is short and broad; all legs are short and thick set, and the densely haired, slightly flattened tail is one-fifth to one-third the total length. The ears are short, broad, rounded, and well haired. The eyes are circular and small. *Marmota monax* has four pairs of mammae, whereas all other species have five pairs. A cheek pouch is present but is rudimentary and lacks retractor muscles. (Lee and Funderburg 1982:177–78).

Pelage and Molt. The color of marmot species varies, but shades of brown and gray, often with splotches or streaks of yellow, reddish-brown, white, or black, predominate on the dorsal surfaces (Barash 1989). Ventral areas are usually paler and are yellow in *M. flaviventris*. In some *M. vancouverensis*, white hairs form a streak on the center of the chest (Nagorsen 1987). White facial markings typically occur in *M. flaviventris* and *M. vancouverensis*. Although the rich, brown color of *M. vancouverensis* was interpreted as melanism (Hoffmann et al. 1979), the color is not black as it is in melanistic *M. flaviventris* (Fryxell 1928). In the Teton Range, Wyoming, melanistic individuals constituted 23% of the population (Armitage 1961). Melanistic *M. monax* also were reported (Anderson 1934). Fur texture ranges from coarse to soft; all North American marmots except *M. broweri* have a coarse pelage (Frase and Hoffmann 1980). Hair length, depth, density, and diameter probably also vary among species; these hair properties strongly affect heat loss (Melcher 1987) and energy use during hibernation in *M. flaviventris* (Armitage et al. 2003). A single, annual molt occurs during the summer (Bibikow 1996).

Molt patterns are highly variable and molts may only be partial. In woodchucks, new fur appears on the hind quarters and proceeds anteriorly (Hamilton 1934; Davis 1966). An animal in partial molt has bright, new fur caudally and old fur anteriorly, giving it a two-tone appearance. Partial molt or bleaching may account for the color pattern in *M. olympus*, although Barash (1973a) suggested that the Olympic marmot molted twice annually. Two annual molts seem unlikely because of the

TABLE 9.1. Standard measurements and mean body masses of North American marmots

Species	Length (mm)			Body Mass (g)	
	Total	Tail	Hind Foot	Immergence	Emergence
M. flaviventris[1]	470–700	130–220	70–90	3431	2422
M. monax[2,3]	510–700	100–180	68–88	4718	3356
M. broweri[4]	553–600	150–162	86–88	3500[a]	3414
M. caligata[3]	630–820	170–252	90–113	6187	3283
M. vancouverensis[5]	650–708	187–236	93–99	5328	3899
M. olympus[2,3]	670–785	180–252	91–112	7100	3350

SOURCES: For length, (1) Frase and Hoffmann 1980, (2) E. R. Hall 1981, (3) Howell 1915, (4) Bee and Hall 1956, and (5) Nagorsen 1987. Body mass values are modified from Armitage and Blumstein (2002).

NOTE: Species listed in order of increasing total length.

[a]This value undoubtedly is too low, but only one value, from a captive adult female, was available. Three adult males averaged 4260 g in early spring (Rausch and Bridgens 1989) and must have averaged at least 5400 g at immergence.

high energy costs of molting (Bibikow 1996). Nonreproductive marmots initiate molt in late spring or early summer, whereas reproductive females initiate molt at about the time of weaning. Marmots lose hair and increase conductance after emergence from hibernation. Reduced insulation apparently is essential for coping with the thermal load of summer (Melcher et al. 1989, 1990). After the molt, conductance and oxygen consumption decrease (Armitage and Salsbury 1993). Pelage of adult and juvenile *M. broweri* and *M. caligata* are described by Hoffmann et al. (1979).

Skull and Dentition. Descriptions and drawings are included in species accounts for *M. flaviventris* (Frase and Hoffmann 1980), *M. monax* (Kwiecinski 1998), and *M. vancouverensis* (Nagorsen 1987).

Photographs and cranial measurements of a series of North American species and subspecies are available in Howell (1915). Condylobasal length is normally 75–100 mm. Rostrum and cranium are subequal, with the interorbital region being much wider than the postorbital region (Fig. 9.4). On all marmots, the second upper premolar (p4) is as large as or larger than the first molar (m1). The cheek teeth are highly crowned and metaloph is complete on each upper molar. The dental formula is I 1/1, C 0/0, P 2/1, M 3/3. Hoffmann et al. (1979) provided comparative cranial dimensions for 24 characters for *M. camtschatica, M. broweri, M. caligata, M. olympus,* and *M. vancouverensis* and discussed cranial sexual dimorphism and geographic and interspecific variation. (Lee and Funderburg 1982:178)

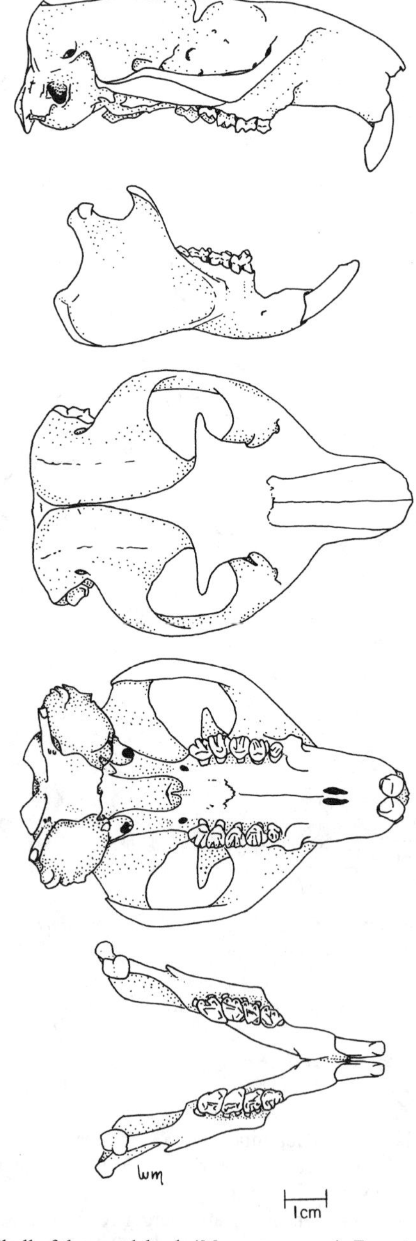

FIGURE 9.4. Skull of the woodchuck (*Marmota monax*). From top to bottom: lateral view of cranium, lateral view of mandible, dorsal view of cranium, ventral view of cranium, dorsal view of mandible.

GENETICS

The karyotypes of Eurasian marmots are similar, with a diploid number $2N = 38$, except for *M. camtschatica,* $2N = 40$. Karyotypes of North American marmots are diverse: *M. broweri,* $2N = 36$; *M. monax,* $2N = 38$; *M. olympus,* $2N = 40$; *M. flaviventris, M. caligata,* and *M. vancouverensis,* $2N = 42$. Apparently the fundamental number FN (the number of autosmal arms) is 66 in all species of marmots (Rausch and Rausch 1971).

Allozyme variation was explored in yellow-bellied marmot populations in Gunnison County, Colorado. Blood samples were collected from 88 marmots in North Pole Basin (3400 m elevation) and 203 marmots in the East River Valley. Tissue samples were obtained from 30 marmots each at elevations of 2680, 3170, and 3660 m (Schwartz and Armitage 1981). Allozyme variation occurred at 8 of 20 loci, whereas in *M. vancouverensis,* 4 of 22 loci were polymorphic (Bryant 1990). Neither gene frequencies nor heterozygosity were associated with age, altitude, habitat, litter size, sex, survivorship, and a suite of behavioral variables. There was significant heterogeneity in genetic variation among colonies (Schwartz and Armitage 1980). Heterogeneity among the social groups was promoted by the restriction of mating to those in the social group, the low rate of exchange between social groups, and the preferential recruitment of juvenile females from their natal colony.

Gene flow between colonies occurred primarily by male immigration, which prevents fixation of gene frequencies. By contrast, overall differentiation among colonies was much less in *M. vancouverensis* (Bryant 1990).

Genetic variation was examined in woodchuck populations with high and low prevalence rates of woodchuck hepatitis virus, which is endemic in areas of the mid-Atlantic states, but apparently is absent from populations in New York and much of New England. Eighteen monomorphic and six polymorphic loci were identified from blood samples from 40 woodchucks from New York and Delaware (Wright et al. 1987). Average heterozygosity of 0.066 in New York and 0.039 in Delaware was similar to the 0.073 for *M. vancouverensis* (Bryant 1990) and 0.075 for *M. flaviventris* (Schwartz and Armitage 1981). Significant heterogeneity between Delaware and New York samples was observed at two loci. It is not known whether the genetic differences were in response to the virus.

In general, studies of population variation in marmots are few. There is no explanation for the nature and pattern of subspecific variation, although Howell (1914) remarked that the subspecies of *M. flaviventris* are connected "by almost perfect series of intergrades." Physiological differences in montane and lowland yellow-bellied marmots were noted; because these differences persisted in animals maintained in the laboratory for ≥ 1 year, they may be genetic. The annual cycle of marmots in hot, dry areas may be shifted by several months; yellow-bellied marmots in Capitol Reef National Park in Utah emerge in March and hibernate in July (P. Hopkinson, pers. commun., 1999), whereas in eastern Washington, they emerge in late February and immerge in June (Couch 1930).

PHYSIOLOGY

The Circannual Cycle. The annual cycle consists of the heterothermal period when marmots hibernate and the homeothermal period in which mating, gestation, lactation, growth, and fattening occur. The sequence of events in the annual cycle is driven by an endogenous circannual clock (Davis 1976). Lack of activity in one phase of the cycle does not prevent activity in the next phase. For example, the cycles of testis size and testosterone concentrations are not altered in woodchucks by the absence of hibernation (Baldwin et al. 1985). Food consumption by wild woodchucks varies seasonally from low in March to a maximum in May and declines throughout the summer (Fall 1971). This annual rhythm of appetite is not altered by increasing the time in torpor or by early arousal; woodchucks adjust food intake and body mass to reassume the normal schedule (Davis 1970). Woodchucks maintained on restricted diet failed to gain mass until fed ad lib in August. Body mass did not reach the July peak of fed animals, but reached the mass of the fed animals in September (Young 1984). The circannual rhythm of mass gain and loss (Ward and Armitage 1981a), of reproductive cycles (Concannon et al. 1992), and of food and water consumption and urine production (Zatzman et al. 1984) persist under a constant photoperiod of 12 hr light:12 hr dark. Photoperiod is the likely environmental factor that entrains the rhythm. *M. monax* sent to Australia from Pennsylvania shifted the annual phase of maximum and of minimum body mass by 6 months (Davis and Finnie 1975). Reversal of daily changes in photoperiod can alter reproductive and metabolic cycles within 3–5 months in juveniles and within 6–8 months in adults and can reverse these cycles in 20–24 months (Concannon et al. 1993).

As might be expected, the annual cycle of mass gain and loss and of reproduction is associated with circannual rhythms in other physiological activities. For woodchucks, maximal values of serum blood fractions of total protein, prealbumin, and alpha and beta fractions occurred just before and into the hibernating period. Minimal values occurred just before and for a few weeks following emergence (Wenberg and Holland 1972). Most endocrine glands involute in preparation for hibernation. The 17-keto steroids have an annual cycle with low levels during hibernation; 17-hydroxy steroids have high values in spring and summer. There were no significant differences between males and females. Maximal values of urinary estrogen occurred from April to June, and epinephrine and norepinephrine were low during late summer, lowest in December and January, then increased in February and March to a peak in June (Wenberg and Holland 1973a). Epinephrine and norephinephrine concentrations did not differ between sexes nor between adults and young. In the spring, the thyroid gland of *M. monax* is extremely heterogeneous in appearance; some follicular and parafollicular cells appear quite active (Frink et al. 1977; Krupp et al. 1977). Iodine-131 uptake declines during the spring (Wenberg and Holland 1973b) and serum T_4 and T_3 and T_4-binding protein (TBG) are low (Young et al. 1986) and remain low during the summer, when the follicles are uniformly small. Plasma concentrations of T_4 and T_3 and T_4-binding protein levels peak in late winter and early postemergence (Wenberg and Holland 1973b; Young et al. 1986). Uptake levels of I^{131} are greatest during the late-winter months of hibernation and just before arousal. The increased serum T_4 and T_3 concentrations and increased T_4-binding protein in the winter may result, in part, from fasting (Young et al. 1986). High thyroid activity occurs at the same time of year as high metabolic activity, and nonhibernating, fed woodchucks have lowest metabolic rates during the normal hibernating season (Bailey 1965a). Fed woodchucks have lowest T_3 concentrations during the time of rapid mass gain (Young 1984). The thyroid gland seems to have a major role in the body mass cycle, but the significance of the pattern of thyroid activity and the thyroid-binding proteins is unclear. The circannual cycle may be driven by metabolic rate. In *M. flaviventris,* the maximum and minimum values of metabolic rate precede the maximum and minimum values of food consumption by at least 1 month and values of body mass by about 2 months (Ward and Armitage 1981a).

Hibernation. All species of marmots hibernate (Barash 1989; Bibikow 1996). Immergence occurs in autumn: *M. flaviventris* in late August to mid-September; *M. broweri, M. olympus,* and *M. caligata,* early to mid-September; *M. vancouverensis,* early October; and *M. monax,* early to late October. North American marmots hibernate for about 7–8 months, except for *M. monax.* Woodchucks hibernate for about 5.5 months in Canada (de Vos and Gillespie 1960; Ferron 1996), for about 5 months in upstate New York (Hamilton 1934), for about 4.0–4.5 months in Pennsylvania (Snyder et al. 1961) and Missouri (Twichell 1939), and for 3.5–4.0 months in Maryland (Grizzell 1955). Among free-ranging woodchucks with surgically implanted temperature sensitive radio transmitters, males had shorter hibernation periods (mean = 104.8 days) than females (mean = 121.8 days) (Zervanos and Salsbury 2003). When weather conditions are mild, woodchucks may be active in any month (Anthony 1962). Hibernation is an adaptation for conserving energy in response to a seasonal lack of food (Lyman et al. 1982; French 1986; Weiner 1989; Armitage et al. 2003). Thus, food-deprived woodchucks, but not fed woodchucks, become torpid (Davis 1967a). Yellow-bellied marmots in Colorado and Olympic marmots enter torpor when vegetation senesces (Wood 1973; Armitage 1996a). No food is stored; marmots rely on body fat as the source of energy during hibernation. All age classes of marmots gain mass during the homeothermal period. Mass gain may be delayed by cold and/or stormy weather in the spring; these conditions may cause mass loss after hibernation is terminated, as in *M. flaviventris* (Armitage 1996a), *M. monax* (Snyder et al. 1961), and *M. caligata* and *M. olympus* (Barash 1989). All age–sex groups gain mass linearly and it varies among years, among litters, between sexes, and at different elevations (Andersen et al. 1976; Armitage et al. 1976; Armitage 1996a; Lenihan and Van Vuren 1996). Mass gain for *M. monax* and *M. flaviventris* averages 11.5 to 13.0 g/day (Davis 1967b; Armitage et al. 1976). Mass gain in *M. flaviventris* does not occur throughout the active season and occurs for periods from about 43 days in postlactation females to about 93 days in yearling females. The period of stasis following mass gain may be as long as 5 weeks in adults. Yearling and 2-year-old females have longer growth periods than adults, probably because they increase size as well as accumulate fat (Armitage 1996a). Reproductive females fail to gain mass and may lose mass during lactation. When only the period of mass gain is considered, yellow-bellied marmots gain mass at the mean rate of 16–26 g/day. Summer drought reduces mass gain, which may result in increased mortality during the subsequent hibernation (Hamilton 1934; Armitage 1994). Mass loss during hibernation varies among species; *M. flaviventris, M. monax,* and *M. vancouverensis* lose 26–30% of their immergence mass, whereas *M. caligata* and *M. olympus* lose 46–53%. Mass loss per day ranges from 3.9 g in *M. flaviventris* to 16.7 g in *M. olympus*. Mass specific loss (mg/g immergence mass per day) varies among species: 1.14 in *M. flaviventris,* 1.37 in *M. vancouverensis,* 2.14 in *M. monax,* 2.23 in *M. caligata,* and 2.35 in *M. olympus* (Armitage and Blumstein 2002). In Ontario, *M. monax* lose more mass during hibernation than woodchucks in Pennsylvania, but at about the same rate (Ferron 1996). Rates of mass loss are much greater in euthermic, food-deprived woodchucks (20.7 g/day) than in hibernating woodchucks (3.0 g/day). Most of the mass loss of hibernators occurs during arousal, 13.7–17.9 g/day (Bailey and Davis 1965). Energetic efficiency varies among species; during a torpor bout, metabolism is significantly less in *M. flaviventris* than in *M. monax* (Armitage et al. 2000). *M. flaviventris* spends significantly more time in torpor than does *M. monax;* as a consequence, energy savings for the entire hibernation period are 43.2% for *M. monax* and 83.2% for *M. flaviventris*. Both *M. monax* and *M. flaviventris* hibernate singly, although in *M. flaviventris,* littermates (Lenihan and Van Vuren 1996) and high-elevation populations (Johns and Armitage 1979) may hibernate in groups. All other North American species hibernate in groups (Blumstein and Armitage 1999) and may reduce energetic expenditures by social thermoregulation (Arnold 1993).

Energy savings are possible because of the marked decrease in metabolism during torpor. In *M. monax,* heart rate decreases from 80–95 beats/min in awake, fasting animals to 3/min in deep hibernation. Oxygen consumption decreases from 0.32 to 0.98 ml/g/hour to 0.01–0.05ml/g/hour (Lyman 1958). Body temperature decreases to 1–3°C above ambient temperature (Benedict and Lee 1938). Similar results were reported for *M. flaviventris* (Hock 1969). Frequency and duration of arousals progressively decrease in the autumn, remain low through most of dormancy, and increase toward the end of dormancy (Albert and Panuska 1981; French 1990). Male woodchucks maintained a higher body temperature and shorter torpor bouts than females; as a consequence, the cost of hibernation was greater for males (Zervanos and Salsbury 2003). Two polyunsaturated fatty acids, linoleic and linolenic, are essential for successful hibernation (Florant 1998). Yellow-bellied marmots on diets deficient in essential fatty acids have bout lengths of 6.3 days compared to 11.0 days for controls, and thus essential fatty acid-deficient marmots spend more time in the energetically costly euthermic condition (Florant et al. 1993). Furthermore, metabolic rate during entrance into torpor, during deep torpor, and during arousal is higher in essential fatty acid-deficient animals (Thorp et al. 1994).

Entrance into torpor in *M. monax* and *M. flaviventris* is initiated by a rapid decline in oxygen consumption followed by a slow decline in body temperature (Lyman 1958; Armitage et al. 2000; Woods et al. 2002). This pattern of decline suggests that metabolism is actively suppressed and heat is lost passively. Hibernation terminates spontaneously in adult marmots, but juveniles may continue to hibernate until they become emaciated from starvation or are fed (French 1990). Frequent arousals in the spring allow marmots to assess variable weather conditions (French 1986). Emergence of *M. monax* (Davis 1977) and *M. flaviventris* (Nee 1969) is correlated with periods of warm weather, and *M. broweri* emerge when the hibernation plug is thawed sufficiently to allow the animals to dig out (Rausch and Bridgens 1989). Yellow-bellied marmot emergence in Colorado has become significantly earlier from 1976 to 1999; the regressed values for first emergence have changed 23 days. Earlier emergence is significantly correlated with warmer temperatures in April (Inouye et al. 2000).

The shift from euthermic, fat-storing animals to torpid, fat-catabolizing animals is associated with numerous other physiological changes. The half-saturation pressure of oxygen for woodchuck blood decreased from an average of 23.4 to 22.3 mm Hg and 2,3-diphosphoglycerate decreased from 5.2 to 3.4 mmol/ml erythrocytes during hibernation. The oxygen equilibrium curve shifts further to the left and hemoglobin and hematocrit increase in early hibernation. Normal values are restored about 1 month after arousal (Hock 1967; Wenberg et al. 1973; Harkness et al. 1974). Carotid systolic blood pressure and cardiac temperature at cardiac arrest are lower and resistance to cooling is greater in hibernating than in nonhibernating woodchucks. Woodchucks selectively lower the temperature of certain tissues and thus retard cooling of essential tissues (Albert and Panuska 1978). During hibernation in *M. flaviventris,* the decline of mean arterial pressure and the heart rate and the almost fourfold increase in total peripheral resistance retard flow to peripheral areas and should greatly retard heat loss and conserve energy (Zatzman and Thornhill 1986). Arterial pH of *M. monax* and *M. flaviventris* is 7.52–7.75 during hibernation and 7.38–7.50 when they are euthermic. Standard bicarbonate values are twice as high in hibernators as in euthermic individuals (Goodrich 1973). The hypothalamic regulation of body temperature in *M. flaviventris* is active throughout hibernation; the threshold hypothalamic temperature probably rises gradually several hours to a day before actual arousal (Florant and Heller 1977). In euthermic woodchucks, the auditory brainstem response consists of four waves. In hibernating woodchucks, waves I and II were traced to the lowest body temperature. Activity remains in the VIII cranial nerve, and the functional integrity of the brainstem auditory pathway is restored during arousal. Woodchucks arouse quickly to auditory stimulation (Katbamna et al. 1992).

Reduction in renal blood flow during hibernation is not accompanied by a decrease in active transport; renal function is maintained at about 10% of its euthermic level (Zatzman and South 1972). Urine osmolalities of hibernating *M. flaviventris* are about half those of euthermic animals, but are hyperosmotic to plasma. Urine flow is about 0.05–0.1 ml/day. Excretion rates increased by several orders of magnitude when body temperature reached 32°C during arousal (Zatzman and South 1975). Insulin affects food intake during periods of mass gain and loss (Florant and Bauman 1985; Florant et al. 1991; Tokuyama et al. 1991), and plasma glucose appears to be regulated during hibernation (Florant and Greenwood 1986; Florant et al. 1986). Plasma levels of cortisol (Florant and Weitzman 1980) and catecholamines (Florant 1981) are related to arousal.

Marmots plug their burrows at the start of the hibernation period (Bibikow 1996); thus one can expect hypoxia to occur. A family (three adults and five young) of *M. broweri* that hibernated in artificial outdoor dens sealed their nest box and were exposed to carbon dioxide levels as high as 13.5% (CO_2 levels were usually below 4%) and oxygen levels as low as 4% (but usually above 15%). Levels of CO_2 and O_2 and temperature cycled, with peak changes occurring every 5–8 days (Williams and Rausch 1973). Temperature was high when concentration of CO_2 was high and that of O_2 was low. This pattern probably represented torpor bouts, with the peak temperature and CO_2 concentration occurring during arousal. The response of euthermic *M. monax* to induced hypoxia is similar to that during hibernation: decreased cardiac output and heart rate, increased total peripheral resistance, decreased blood flow in all tissues, and increased ventilation rate, which decreased alkalosis (Burlington et al. 1971). Hypoxic woodchucks have a large Bohr coefficient and low bicarbonate levels, they do not have an elevated buffering capacity, and they maintain a high arterial partial pressure of CO_2 associated with a low arterial pH (Boggs and Birchard 1989). Woodchucks have a reduced ventilatory responsiveness to CO_2. Because this response is ubiquitous among burrowing mammals (Boggs et al. 1984), all species of marmots likely have a similar response. The reduced ventilatory response could represent an energy-conserving tactic because increased ventilation is likely of little utility in a high-CO_2 environment (Boggs et al. 1984).

Blood characteristics of *M. monax* and *M. flaviventris* differ. Hemoglobin concentrations are higher in *M. flaviventris* than in *M. monax* (Hall 1965; Bullard et al. 1966; Armitage 1983). The hematocrit of woodchucks living at low altitudes (38.9%, Hall 1965; 39.4%, Boggs and Birchard 1989) is much lower than that of high-altitude yellow-bellied marmots (49.4%, Bullard et al. 1966; 50.8%, Armitage 1983). Yellow-bellied marmots acclimatized to low altitude have lower hematocrit values (sea level, 42.4%; high altitude, 47%; Winders et al. 1974). Hematocrit values of yellow-bellied marmots transported from

Colorado to Kansas were about 4.3% lower than the high-altitude values and did not differ significantly from values for marmots captured in Washington and maintained in the laboratory (Armitage 1983). Hematocrit levels are inversely related to body mass for 11 species of semifossorial sciurids. Hematocrit values of *M. monax* are about as predicted from body mass, but those of *M. flaviventris* are much higher than predicted. This difference, attributed to chronic exposure to hypoxia by *M. flaviventris,* persists when yellow-bellied marmots are acclimatized to sea level, which suggests that at least part of the difference is genetic (Armitage 1983). Hematocrit and hemoglobin values are lower in juvenile *M. monax* (Wenberg et al. 1973) and *M. flaviventris* (Armitage 1983), but are similar to adult values at the time of hibernation. In general, oxygen equilibrium curves of marmots are shifted to the left in comparison to nonhibernators, such as the laboratory rat (Hall 1965; Bullard et al. 1966). Electrophoresis of hemoglobin produces single and double bands in *M. caligata* (Blumberg et al. 1960) and single bands in *M. flaviventris* and *M. monax* (Harkness et al. 1974).

During hibernation, the fasting marmots are on a lipid metabolism. The major storage fatty acids are palmitic, oleic, and linoleic in yellow-bellied marmots (Florant et al. 1990) and palmitic, oleic, and linolenic in woodchucks (Davis and McCarthy 1965). Polyunsaturated essential fatty acids are metabolized more slowly than other fatty acids (Florant et al. 1990; Xia et al. 1993). Adipose mass in *M. flaviventris* may be controlled by the seasonal regulation of genes involved in lipid storage or use (Wilson et al. 1992).

Metabolism and Water Balance. Oxygen consumption of montane–mesic and lowland–xeric yellow-bellied marmots is significantly affected by ambient temperature. The shape of the metabolism–temperature curve differed from that of a typical mammal. A thermoneutral zone occurs between 15°C and 20°C. Decrease in metabolism at 25°C may represent thermal stress. Metabolism decreases at higher ambient temperatures except for montane–mesic marmots on ad lib water, for which metabolism increases at 30°C, then decreases at 34°C. Oxygen consumption increases linearly with decreasing ambient temperature below 15°C (Armitage et al. 1990). By contrast, minimal oxygen consumption of *M. monax* occurs at 28°C and increases linearly as temperatures decrease (Benedict and Lee 1938). However, the metabolism–temperature curve of posthibernation woodchucks was similar to that reported for *M. flaviventris* (Armitage et al. 2000). Lowland–xeric yellow-bellied marmot populations had a lower metabolic rate at high ambient temperatures and a higher metabolic rate at lower temperatures than montane–mesic animals. Oxygen consumption was lower on a restricted-water regimen. Body temperature was significantly affected by ambient temperature (it was high at low and high environmental temperatures), water regimen (higher on ad lib water), and habitat (higher in lowland–xeric marmots). Conductance generally was lower in the montane–mesic than in the lowland–xeric marmots. Both populations increased conductance at high ambient temperatures. Metabolic water production equaled or exceeded evaporative water loss at 5–20°C, which indicates that metabolic water could satisfy water requirements during hibernation (Armitage et al. 2000). Metabolic rates of wild-caught yellow-bellied marmots were lower for all ages and sexes than predicted from body-mass equations (Armitage and Salsbury 1992). Resting metabolic rates of adult males were higher than those of adult females in the first month following emergence and then declined throughout the summer.

Resting metabolic rates of lactating females were about 1.3 times greater than those of nonreproductive adult females at the same time of year. Mass-specific metabolic rates of young were much greater than those of adults. The decrease in resting metabolism in all age and sex groups following the annual molt reduced maintenance costs and permitted more energy to be allocated to fat production (Armitage and Salsbury 1993). Although resting metabolic rate of males decreased in early summer, field metabolic rates significantly increased through the season. High field metabolic rates coincided with the peak of intrusions by adult males into male territories (Salsbury and Armitage 1994a). Energy expenditure varied among adult males and increased with the number and dispersion of defended females (Salsbury and Armitage 1995). Daily energy expenditure, excluding production, of yellow-bellied marmots varied from 539 kJ/day for a yearling female in June to 1017 kJ/day for a lactating female in July. Foraging accounted for 11–51% of daily energy expenditure; sitting, for 1–28%; time in the burrow, for 41–60%; and thermoregulation, for 1–6% (Melcher et al. 1989). Yellow-bellied marmots conserve energy (Armitage 1998); energy conservation is reflected in the high tissue growth efficiency of 16.8%, about five times greater than that of typical homeotherms (Kilgore and Armitage 1978). Wood (1973) calculated a tissue growth efficiency of 4.3% for all age classes of *M. olympus*.

The major source of water for marmots under natural conditions is vegetation, but *M. olympus* will eat snow or drink water (Barash 1973a) and *M. monax* will occasionally drink water (Kwiecinski 1998). Water influx in *M. flaviventris* varied with age (body size), season, reproductive status, and site (Melcher et al. 1989). High mass-specific water influx occurred in lactating females. In general, mass-specific water influx was up to five times greater than the influx of laboratory-housed animals on *ad lib* water. The major source of water for laboratory populations of montane–mesic and lowland–xeric yellow-bellied marmots was drinking (about 74% of total water); metabolic water contributed 23%, and water from food, 3% (Ward and Armitage 1981b). Water output was primarily urine loss; urine concentration was higher and urine volume lower in the lowland–xeric population. When provided restricted amounts of water, both populations reduced urinary and fecal losses of water. The lowland–xeric marmots were adapted to dry conditions; they required significantly less total water and were more capable of mobilizing evaporative water when heat stressed than were montane–mesic marmots.

Endocrine Response. Christian (1962) measured the mass of adrenal glands of 872 woodchucks. Adrenal mass, considered a measure of stress, was linearly related to body length. Mean adrenal mass increased by 60% from early March (immediate posthibernation) to early July, then declined sharply. It increased again by 48% at the end of August, and finally declined by October (time of hibernation immergence) to 15% below the March level. The initial rise in spring coincided with a high level of aggressiveness, and the subsequent decline was associated with the end of the breeding season, when the level of aggressiveness was low. The August peak was associated with movement of young woodchucks, when the potential for conflict increased. Adrenal mass seemed most responsive to aggressiveness in the population; the physiological process of reproduction had no detectable influence on adrenal mass. Adrenals of females were about 8% heavier than those of males. When trapping and removal of woodchucks increased the percentage of young in the population, adrenal weight and cortical mass generally decreased. These decreases were attributed to decreased levels of social pressure (Lloyd et al. 1964).

Total corticosteroid concentrations of yellow-bellied marmots were highest in young, intermediate in yearlings, and lowest in adults. Neither sex nor reproductive status affected their concentrations. Population density was not consistently related to corticosteroid concentrations; when significantly related, only 22–34% of the variation was explained. Social behavior and social status were the major factors affecting corticosteroid concentrations (Armitage 1991a).

REPRODUCTION AND DEVELOPMENT

Marmots breed shortly after emerging from hibernation (Nee 1969). An exception is *M. broweri,* which mates in the burrow before coming aboveground when the stomach and intestine are empty and much contracted (Rausch and Rausch 1971). Young are born about 2 weeks after emergence (Rausch and Bridgens 1989). Males of *M. monax* (Kwiecinski 1998) and *M. flaviventris* (Armitage 1991b) emerge before females. Males of *M. olympus, M. caligata,* and *M. vaucouverensis* probably emerge at the same time as the females because all group members of these species hibernate together (Blumstein and Armitage 1999). Because spermatogenesis in ground-dwelling squirrels occurs only during euthermy (Barnes 1996), euthermy probably is a requirement for spermatogenesis in marmots. Probably all male marmots terminate heterothermy earlier than females and may spend part or all of this early euthermic period quiescent in their burrows. Early emergence aboveground in *M. monax* and *M. flaviventris* may be related to their requirement to locate females, which is unnecessary in the group hibernators.

Anatomy and Physiology. The male reproductive tract of *M. monax* was superficially described by Mossman et al. (1932). The apparently typical sciurid tract consists of a pair of coiled, relatively little branched seminal vesicles; a simple, compact, highly compound tubular prostate gland opening by a single pair of primary ducts; and a pair of very large compound tubular Cowper's glands. These glands have wide spiral ducts, which constrict as they enter the corpus spongiosum of the bulb, but then expand and unite to form the bulbar gland. This gland drains through a single glandular penile duct, which runs distalward through the corpus spongiosum ventral to the urethra to the ventral flexure of the penis, where it joins the urethra from the ventral side. The bulbar gland and penile duct are the most characteristic features of all sciurid genera except *Tamiasciurus*. In *M. monax,* the genital system is considerably smaller relative to body size than in *Spermophilus,* but the glands have the same relative size. The testes are abdominal most of the year. Three anal glands are present, but are not part of the reproductive system. Males can be distinguished from females at all ages by the distance between the anal and genital openings. The normal length of the perineum is ≥2 times longer in males than in females (Grizzell 1955); in yellow-bellied marmots, the perineum of males typically is 25–45 mm and that of females, 5–10 mm. Nipples of pregnant yellow-bellied marmots enlarge and reach maximal size during lactation (Armitage and Wynne-Edwards 2002). Nipples develop in females that initiate breeding but fail to wean a litter. Nipple size remains smaller than that of successful females, regresses sooner, and decreases to the size of nonreproductive females, whereas nipples of successful females remain large through the first 5 weeks of postlactation. Nipples of reproductive *M. monax* are swollen during lactation and diminish slowly thereafter (Snyder and Christian 1960).

Mean mass of testes in known-aged adult, wild *M. monax* is greatest at emergence from hibernation in February and declines thereafter to a low in September. Mean mass of testes in yearlings increases in the month following emergence to about one half the size of the adult testes, then declines slowly through August, when testes masses of adults and yearlings coincide (Christian et al. 1972). The size of the seminiferous tubules and spermatogenic activity is greatest at emergence and least during autumn. Decline in testes mass and maturation depletion is evident before reproductive activity in March. Mean testes mass declines less rapidly than tubule size and spermatogenic activity. In October, the size of the seminiferous tubules increases, the germinal epithelium is active, and primary spermatocytes at the pachytene stage are present. Mature spermatozoa are present in the testes of some yearlings at emergence. The testes of young have spermatogonia and primary spermatocytes appear by September. No spermatozoa are found at the time of hibernation, which implies that spermatogenesis is completed at the termination of the heterothermal period.

The estrous cycle of captive woodchucks was examined by taking daily vaginal smears for 2 consecutive months. Seven of 10 adult females exhibited readily identified estrus, in which 83% of the cells were cornified. The animals were monoestrous and remained in a prolonged estrous period, that ranged from 12 to 27 days, with a mean of 18.1 days (Sinha Hikim et al. 1991), then became anestrous. Levels of estradiol were elevated before and during estrus and declined significantly 1 week after termination of estrus. Elevated levels of estradiol preceded and coincided with the maximal degree of vaginal cornification. Females with predominantly cornified cells were always receptive to males regardless of the time interval from onset of estrus and mated within 24 hr of pairing (Sinha Hikim et al. 1992). Ovulation occurred 20–30 hr after copulation. Induction of ovulation was associated with a fourfold increase in progesterone. Although ovulation is induced, there is evidence that spontaneous ovulation also occurs (Sinha Hikim et al. 1991). Progesterone levels of wild and laboratory woodchucks are highest during midpregnancy and decrease about 25% in the last third of pregnancy. Postpartum progesterone levels are elevated above basal values in lactating and nonlactating animals for 2–3 months postpartum (Concannon et al. 1983, 1984). The corpora lutea are either maintained or rejuvenated during the postpartum period, and nonreproductive females spontaneously form multiple corpora lutea, which secrete progesterone when parturient females are lactating (Concannon et al. 1984). In wild yellow-bellied marmots, progesterone levels of nonreproductive females are low during the season of early gestation and slowly decline to basal levels at about the fourth week of the postlactation period. Progesterone levels of reproductive females and those that initiated reproduction but failed to wean a litter are high during gestation, decrease during late gestation, increase to a peak in midlactation, and then decline steadily until reaching basal levels about 4 weeks postlactation (Armitage and Wynne-Edwards 2002). Progesterone levels are correlated with nipple enlargement. High levels of progesterone occur in yearling females during the lactation season and in juveniles in the late postlactation season. The elevated progesterone levels in nonbreeding females may function to (1) prevent ovulation and/or mating at an inappropriate time of year (Concannon et al. 1990) and (2) integrate the reproductive cycle into the overall annual cycle or be involved in critical priming events that occur in the year before reproduction (Armitage and Wynne-Edwards 2002).

Age of First Reproduction. Only *M. monax* breeds as a yearling (that is, in its second summer of life). About 20% of yearling females reproduce, but the percentage increases to 42% when the proportion of adults in the population is decreased (Snyder 1962). The yellow-bellied marmot first breeds at 2 years of age, only 22.6% of 2-year-old females reproduce, and the median age of first reproduction is 3 years (Schwartz et al. 1998). No known 2-year-old female *M. vancouverensis* bred; 3-year-old females can reproduce, but the mean age of first reproduction is 4.3 years (Bryant 1996). Both *M. caligata* (Barash 1974a) and *M. olympus* (Barash 1973a) first breed at age 3 years. The age of first reproduction of *M. broweri* is unknown, but is likely 3 years. The age of first reproduction occurs in the year after the maturity index reaches 0.65. This index is calculated as the emergence mass of yearlings or 2-year-olds/emergence mass of adults (Blumstein and Armitage 1999). The maturity index indicates that all species of marmots could breed at 2 years of age; delay beyond age 2 years may represent reproductive suppression in social marmots (Armitage 1999; Blumstein and Armitage 1999).

Mating. Mating occurs early in the homeothermal period (Synder and Christian 1960; Armitage 1965; Nee 1969; Barash 1973a; Holmes 1984a; Rausch and Bridgens 1989; Bryant 1996). In yellow-bellied and Olympic marmots, a male usually approached a female by sniffing her face or rump; sniffing was often accompanied with pawing, especially near the rump. The "approach" usually was followed by a grasp in which the male seized the female with his forelegs placed anterior to her hind legs. Attempted copulation occasionally followed a grasp, and began when the male lunged quickly and seized with his jaw the skin in the middle of the female's back while simultaneously grasping her. Woodchuck behavior was similar (Hoyt and Hoyt 1950). The female's tail was pushed aside and the male extended his hindquarters up under the female and began thrusting. During attempted copulation the female crouched quietly (Armitage 1965). Copulation attempts lasted from 1 to 23 min in yellow-bellied marmots and 7 to 14 min in *M. monax* (Sinha Hikim et al. 1992). Two yearling woodchucks copulated three times for 8, 5, and 3 min, respectively (Hoyt and Hoyt 1950). Mounts never occurred in females that did not breed in that year (Barash 1973a). In *M. flaviventris* and *M. olympus,* the male sniff could be followed by agonistic behavior by the female, or she might move away and the male might follow. Often the female initiated contact with the male. Sometimes the male chased the female. Fights occurred between male and female Olympic marmots, but were not observed in yellow-bellied marmots. In both species, sexual behavior declined rapidly after the first or second week of reproductive activity.

Gestation. The gestation period of *M. monax* was reported as about 28 days (Hamilton 1934), 31 days in a yearling female (Hoyt and Hoyt 1950), and 30–32 days in a laboratory population (Sinha Hikim et al. 1992). Captive woodchucks had gestation periods of 31–32 days when artificially inseminated and 32–33 days when mated (Grizzell 1955). Gestation period was estimated at 30 days for yellow-bellied marmots (Nee 1969; Armitage and Downhower 1974). Probably the gestation period of other species of marmots is about 30–32 days.

Parturition and Lactation. Parturition occurs in the burrow. Newborn *M. flaviventris* are 111.0 mm long and weigh 33.8 g (n = 3) (Frase and Hoffmann 1980). Newborn *M. monax* weigh 23.7 g (n = 22) in Ontario (Ferron and Ouellet 1991), 26.5 g (n = 20) in central New York (Hamilton 1934), 32.3 g (n = 3) in central Pennsylvania (Snyder et al. 1961), and 27.2 g (n = 11) in Maryland (Grizzell 1955). Young *M. monax* are 105 mm long (Hamilton 1934). There is no sexual dimorphism in young woodchucks before weaning (Ferron and Ouellet 1991). Weaning in woodchucks occurs at 42 (Ferron and Ouellet 1991) to 44 days (Snyder and Christian 1960), in yellow-bellied marmots, at 25 (Armitage 1981) to 35 days (Nee 1969), in the hoary marmots, at 25–30 days (Holmes 1984a), and in the Olympic marmot, at about 30 days (Armitage 1981). Generally, weaning is essentially completed when young first appear aboveground, but some lactation may continue. Thus, total lactation time calculated by Nee (1969) is greater than the time of emergence. The lactation period is shorter in marmots that have a short active season compared to the woodchuck, which has a long active season.

Litter size at birth is known only for laboratory-held animals. In *M. monax,* litter size averaged 4.8 (n = 5) in Ontario (Ferron and Ouellet 1991), 4.5 (n = 4) in Illinois (Sinha Hikim et al. 1992), and 4.3 (n = 7) in New York (Concannon et al. 1984). Litter size in a captive colony of *M. broweri* varied from 3 to 7 and averaged 5 (Rausch and Bridgens 1989). Snyder and Christian (1960) calculated mean litter size at birth by counting embryos, fetuses, and placental scars of 382 necropsied *M. monax* and subtracting the average loss of embryos or fetuses. They estimated a mean value of 3.42 in 1957 and 3.97 in 1958. Litter size typically is determined at weaning. Mean litter sizes are as follows: *M. flaviventris,* 4.1 (n = 265) (Schwartz et al. 1998); *M. olympus,* 4.6 (n = 14) (Barash 1973a); *M. caligata,* 3.3 (n = 6) (Barash 1975); *M. vancouverensis,* 3.36 (n = 36) (Bryant 1996); *M. broweri,* about 4.0 (Rausch and Bridgens 1989); and *M. monax,* 4.6 (n = 29) in Maryland (Grizzell 1955) and 3.75 (n = 18) in Ontario (de Vos and Gillespie 1960). Litter size is smaller in yearling *M. monax,* which averaged 3.13 implantations, whereas adults averaged 4.54 (Synder and Christian 1960). Two-year-old *M. flaviventris* had a smaller mean litter size than older age classes (O. A. Schwartz et al. 1998).

Sex ratio at weaning does not differ from 1:1 in *M. monax* (Synder and Christian 1960; Ferron and Ouellet 1991), *M. vancouverensis* (Bryant 1996), and *M. flaviventris* (Schwartz et al. 1998). Sex ratio in *M. flaviventris* does not vary with density of adult females, litter size, stress, or social environments. It varies with age; young females wean significantly more females than males (Armitage 1987). Age and social structure interact; young females living with at least one additional female in the only matriline on a habitat patch produce about twice as many daughters as sons. These females recruit more daughters per litter, and more daughters per female young weaned than expected. The female-biased litters are produced by females with a high probability of successful recruitment of daughters into the local population.

Growth and Development. Postnatal preweaning physical and behavioral development were described in detail only for *M. monax* (Hamilton 1934; Ferron and Ouellet 1991). Young are born blind and naked with the ear pinnae up. Vibrissae (2.5 mm long) and short hair on the muzzle, chin, and head are present. The external auditory meatus is closed. Backward and forward linear crawling, circular crawling, righting motion, rooting, and teat seeking occur.

Physical development does not differ between the sexes. Increase in tail, body, and hind foot length are relatively constant. Instantaneous growth rates diminish continuously. Calculated from birth to weaning, growth rates were 0.072 in Pennsylvania, 0.058 in New York, and 0.056 in Ontario. Hair appears gradually from head to tail and more rapidly on the back than on the belly. Physical development precedes the appearance of social behavioral patterns, which appear in the last third of development (Table 9.2).

TABLE 9.2. Physical and behavioral development in *M. monax*

Age (days)	Development
12	Grasping by forepaws
20	Lower incisors erupt
21	Eyes partly open
25	Regular walking begins
26	Upper incisors erupt
26	Alert postures appear
26	Olfactory exploration begins
28	Body grooming activities begin
28	Whistling
31	Hearing occurs
32	Feeding on solid food
33	Emergence from the nest
33	Tooth chattering
35	Digging
37	Biting and nose-to-nose smelling
39	Jumping begins
39	Social grooming
41	Anal gland full evagination
42	Weaning occurs
42	Play-fighting

SOURCE: Data from Ferron and Ouellet (1991).

Mother–infant contact peaks when young woodchucks are about 1 week old and declines thereafter to no contact at weaning (Barash 1974b). Aggressive responses by the mother toward the young increase over the same time period. Adult woodchucks are agonistic to the young (Anthony 1962). Aggressiveness by the adult female is associated with early dispersal in this species (Barash 1989). By contrast, only amicable social interactions were observed between mother and young *M. flaviventris* (Webb 1981; Rayor and Armitage 1991), and social interactions of young *M. olympus* with adults were primarily amicable (Barash 1973a). Young of both species hibernate in their natal social groups.

ECOLOGY

The woodchuck is a forest-edge species. It occupies meadow hedgerows and lives in old field associations, pastures, or cultivated crops interspersed with small woodlots. Burrows may be in fields or woodland in well-drained soil (Hamilton 1934; de Vos and Gillespie 1960; Ferron and Ouellet 1989). Sixty-three percent of 132 burrows occurred on slopes of 30 deg or greater and most were in rocky soil (Twichell 1939). The yellow-bellied marmot occupies grass–forb meadows with a mean slope of 33 deg and with rock outcrops or talus. This species avoids meadows where rocks are too tightly packed for digging or where large subsurface rock volume occurs. Grassy meadows and flats are inhabited in lowland inland valleys (Couch 1930; Svendsen 1974; Carey 1985a). Hoary marmots may occupy rock ledges and talus slopes with adjacent subalpine meadows (Barash 1974a) or relatively flat meadows with short mesophytic grassland vegetation above timberline (Holmes 1984a). The Olympic marmot lives in subalpine to alpine meadows and talus slopes just above and below timberline at 1500–1750 m elevation. Most colonies are oriented between southeast and southwest (Barash 1973a). About 81% of Vancouver Island marmots live at elevations between 1000 and 1400 m in clearcuts or grass–forb alpine meadows. Most (74%) are on south- to west-facing slopes; natural habitat patches are small, scattered, and typically in avalanche bowls (Bryant and Janz 1996). The Alaskan marmot occupies precipitous sides of canyons and valleys and lives at the base of active talus with extensive meadows

(Bee and Hall 1956). Hibernacula are on exposed ridges that become snow-free relatively early in the spring (Rausch and Rausch 1971). In general, habitats of mountain-dwelling marmots have the following characteristics: (1) meadow or grassland for foraging, (2) steep to moderate slope with good drainage, (3) a southern to eastern exposure where snow melts early, (4) a soil structure that both permits burrowing and supports burrows, and (5) elevation above or near timberline or lower elevation forest openings (Armitage 2000).

Activity Cycles and Time Budgets. Marmots are typically diurnal, but some activity may occur at night (Hamilton 1934; Grizzell 1955; Hayes 1976). For *M. olympus,* early season activity is unimodal, with some tendency for a late afternoon peak (Barash 1973a). Early and late in the active season, activity cycles of *M. monax* and *M. flaviventris* are unimodal, with peaks occurring in early or late afternoon. During most of the season, the activity of all three species peaks in the morning and late afternoon and declines at midday (Anthony 1962; Armitage 1962, 1965; Pattie 1967; Hayes 1976). The afternoon peak may be much higher in *M. monax* (Bronson 1962) and some populations of *M. flaviventris,* but the peak is higher in the morning in other populations (Armitage et al. 1996). The reduced midday activity during the summer is a consequence of high ambient temperatures and heat stress (Webb 1980). Melcher et al. (1990) analyzed the thermal environments of microhabitats of yellow-bellied marmots in the Elk Mountains of Colorado using the standard operative temperature (T_{es}) method. Higher wind speeds over rocks produced T_{es} up to 10°C less than in meadows. Marmots responded to high, stressful T_{es} by reducing aboveground activity, reducing the length of foraging bouts, sitting on rocks, and tolerating transient increases in body temperature. When body temperature reached about 40°C, marmots ceased foraging. Because young were more affected by convection than adults, they could be more active at midday, but were less active early and late, when T_{es} was low. In general, marmots avoided foraging during stressful T_{es}, but sometimes did so when necessary to meet energy demands.

An extensive time budget is available only for yearling and adult *M. flaviventris* (Armitage et al. 1996), based on 17 behaviors for three colonies in the Upper East River Valley of Colorado. Marmots allocated 40–60% (110–265 min daily) of their aboveground activity to sitting/lying and 12–23% (37– 94 min daily) to foraging. Vigilance/alert varied from 1.1% to 14.5% (12.0–71.7 min) daily. Social status affected the time budget, and time allocated to vigilance/alert increased in subordinate yearlings. Locomotion, social behavior, play, grooming, chirping, digging, investigation, and gathering grass each used <5% of the time. The adult-male cohort spent significantly more time aboveground and reproductive females spent significantly more time foraging. Aboveground activity was lowest during gestation, increased during lactation, remained high during early postlactation, and declined during late summer. On average, 28.5% of each day was spent aboveground during the 4-month active period. Since marmots hibernate underground for 8 months, only about 9.5% of the year is spent aboveground. Significant relationships common to the three colonies suggest species characteristics or common environmental influences. More time is allocated to foraging in the afternoon, and more time to foraging/alert, alert, and locomotion during gestation and lactation than during postlactation. Feeding is minimal early after emergence (Snyder and Christian 1960; Rausch and Bridgens 1989), peaks during gestation, and then decreases progressively through postlactation and before immergence (Anthony 1962; Fall 1971; Barash 1976, 1989; Arsenault and Romig 1985; Concannon et al. 1990; Armitage et al. 1996). Because marmots allocate so much time to sitting/lying and being below ground in the burrow, energy budgets may be constrained by time required to process food and not by foraging time. Time spent vigilant/alert did not seem to constrain energy intake; marmots often were alert while chewing. Social behavior was not limited by time. Reproductive hoary marmot females spend more time foraging, forage more on bad-weather days, and feed significantly farther from the nearest burrow during gestation and lactation periods than do nonreproductive females during the same time periods (Barash 1980).

Home Range and Movements. Mean home range of adult female *M. flaviventris* varied from 0.13 to 1.02 ha (Armitage 1975). Within a locality, mean home range area varied at least 300% among years. Mean home range size was not related to the mean number of adult females, but the percentage overlap of home ranges was significantly and positively correlated with the number of adult females. The degree of home range overlap ($\bar{X} = 42.3\%$) was highest among females whose degree of kinship was 0.5; at all *r* values below 0.5, the degree of overlap was less than that predicted from the assumption that overlap is directly related to the degree of kinship (Armitage 1996b). For colonial male yellow-bellied marmots, home range size varied from 0.2 to 1.98 ha (Armitage 1974). Range was smaller when two or more males occupied a habitat patch and was larger when one male was present. There was virtually no overlap in adjacent home ranges. When noncolonial males are included, home range size varied from 0.06 to 47.5 ha (Salsbury and Armitage 1994b). Home range size of males was not significantly correlated with the number of adult females (range = 0–7), but was significantly correlated with the maximum distance between females within a male's home range. Female home ranges expand subsequent to mating to include newly available foraging areas or to avoid agonistic behavior with other females (Armitage 1965; Andersen et al. 1976).

Home ranges of adult female *M. monax* did not overlap in early spring, but overlapped minimally (<10%) in late spring and summer (Meier 1992). Home ranges of resident males did not overlap those of other males, but did overlap those of one to three adult females. Mean size of female home ranges was 0.25 ha and for resident males was 1.6 ha. In Connecticut, overlap of home ranges was 36% for reproductive females and males, 13.5% for reproductive females, and 31% among males (Swihart 1992). Home range area changed seasonally in Ontario. Before dispersal, mean home range, based on burrows occupied ≥2 times, was 11.6 ha for males, 0.1 ha for females with young, and 2.7 ha for females without young. Following natal dispersal, home ranges of males and females without young decreased to 1.9 and 0.6 ha, respectively. Mean home range of females with young increased to 0.7 ha (Ferron and Ouellet 1989). Home ranges of males (2.8 ha) were larger than those of females (0.83 ha) in Iowa (Trump 1950) and 1.8 times larger in Connecticut (Swihart 1992).

Although home ranges of male *M. flaviventris* and *M. monax* are larger than those of females, there is no difference between sexes in home range for species living in restricted family groups (Armitage 2000). For *M. caligata* (Holmes 1984a), *M. olympus* (Barash 1973a), and *M. vancouverensis* (Nagorsen 1987), home range is a family home range. The adults defend the family range (= territory); there is a tendency in *M. caligata* for resident females to respond to female intrusions and the resident male to respond to male intrusions (Holmes 1984a). Intruder males are repulsed by the resident male in *M. olympus* (Barash 1973a).

Size of home range is affected by food resources. For *M. monax,* adult feeding area was 0.92 ha in woodland, 0.09 ha in a clover field, and 0.3 ha in clover-brush (Anthony 1962). Mean movement of *M. monax* on marginal habitat was 337 m and was 246 m on good habitat (de Vos and Gillespie 1960). There apparently is an inverse relationship between home range area and vegetation biomass. For example, in habitat with a vegetation biomass of 383 g/m^2, *M. flaviventris* had a home range of ≤1.0 ha, whereas with a vegetation biomass of 117 g/m^2, *M. caligata* had a home range of 13.8 ha (Armitage 2000).

Woodchucks frequently move between burrows. Adults used an average of eight burrow systems in a Connecticut population. Mean duration at a burrow was 4.5 days for males and 5.8 days for females. Simultaneous use of a burrow usually involved an adult male and a reproductive female (Swihart 1992). In Ontario, before natal dispersal, the mean number of burrows used by adult males (20.5) was significantly greater than the 5.0 used by reproductive females (Ferron and Ouellet 1989). During natal dispersal, the mean number of burrows used did not differ significantly between males (8.7) and females (11). The mean number of burrows used daily ranged from 1.7–2.8. Distances traveled between burrows by males averaged 319 m before natal dispersal and 145 m during natal dispersal; for females, mean distances were 45 m before natal dispersal and 118 m during natal dispersal. The increased movement by females during natal dispersal may be related to breaking the maternal bond, as distance is greater during the first 10 days of dispersal (Ferron and Ouellet 1989). Comparable movements are not recorded for other North American marmots. Female yellow-bellied marmots may move young to different burrows within their home range, but such movements are typically less than 30 m (K. B. Armitage unpublished data). Male yellow-bellied marmots (Salsbury and Armitage 1994b) and male hoary marmots (Barash 1981) may make short excursions (<1000 m) from their home ranges. Male excursions by yellow-bellied marmots occurred after the breeding season. Males did not seek extra copulations, but may have tried to locate nearby undefended females for an expanded harem. Excursions by hoary marmots occurred primarily during the mating season and presumably served to obtain extra copulations. Intruders were repulsed by residents, but successfully copulated when the colony male was removed (Barash 1981). Yellow-bellied marmots also made long excursions (>1000 m); the longest was about 4270 m. They occurred in mid- to late summer. Half the time, these males returned to their home range either before hibernation or at the start of the next active season; fates of those that did not return are unknown. The tendency to make long excursions was not related to the number of females in the male's home range. Since these males deserted established territories with reproductive females, the movements are an enigma.

Merriam (1963) used radiotelemetry to follow woodchuck movements. Most animals were juveniles, which almost always moved along permanent trails that interconnected most of the burrow systems. Some juveniles remained in a clump of burrows, whereas others wandered more widely. Dispersing young may occupy empty burrows (de Vos and Gillespie 1960; Anthony 1962). Repeated movements from the den to foraging areas or to other burrows result in the formation of trails (woodchucks, Hamilton 1934; Anthony 1962; yellow-bellied marmots, Armitage 1962; Olympic marmots, Barash 1989).

Marmots may move long distances from their hibernation sites to their summer dens. As the snow melted, a colony of *M. caligata* moved a distance of 500 m and an elevation of 200 m (Barash 1974a). When an outcropping of rock along a ridge became available because of the removal of the intervening forest, *M. flaviventris* adopted a pattern of long-distance movement (0.73–2.61 km) between the ridge and their summer food source (Thompson 1979b). *M. monax* regularly moved between summer and winter dens (Grizzell 1955). Marmots may disperse from their home ranges in time of drought (Armitage 2000).

Home range use is strongly influenced by foraging behavior and risk. *M. flaviventris* avoided areas where dense growth of avens (*Geum rossi*), a nonfood plant, was located (Andersen et al. 1976). Patch use by *M. flaviventris* in California was positively related to high food biomass and negatively related to soil moisture and vegetation height and volume (Carey 1985a) and associated with concentrations of clover (Stallman and Holmes 2002). Hoary marmots living on a talus slope associated with a meadow of herbs, grasses, and sedges overselected snow-free old-growth areas for foraging (Tyser and Moermond 1983). When foraging areas were fertilized, patch use by *M. caligata* increased by 62.5% (Holmes 1984b). When frequency of plants was chosen as the forage factor and number of burrows per patch and distance to talus served as risk factors, risk factor was more important (Holmes 1984b). Patch use by *M. flaviventris* shifted throughout the summer in response to plant phenology (Frase and Armitage 1984). In addition, use of space is influenced by kinship (only closely related animals share space) and individual behavioral characteristics. For example, subordinate individuals may be forced to a different burrow system or to a different foraging area (Frase and Armitage 1984).

Burrows. Burrows are a critical resource for marmots. Yellow-bellied marmots on large habitat patches (= colonies) use an average of 14 burrows, whereas those on small habitat patches (= satellites) use an average of 2.3 burrows (Svendsen 1974). The mean resident marmot population (3.16 at colonies, 1.39 at satellites) closely matches the number of resident burrows (3.0 at colonies, 1.14 at satellites). Burrows typically are located in rock outcrops, under boulders, in talus, and under tree or shrub roots. On a talus slope, nearly every suitable rock is used. For example, one colony had 78 burrows in various stages of use in an area of about 0.85 ha (Svendsen 1974), although only 6–8 burrows were used in each year (K. B. Armitage unpublished data). Burrows may be shared. Adult burrowmates consist of mother:daughter:sister kin groups. Burrows are of three types: home burrow or den, where a marmot spends the night, rears young, and retreats to when danger threatens; auxiliary, flight, or escape burrow, which serves as a refuge when the marmot cannot safely return to its home burrow (Armitage 1962); and hibernaculum, the burrow in which the marmot hibernates. The hibernaculum may be used as the home burrow or may be separate and located a considerable distance from the summer burrows (Thompson 1979b). The hibernaculum is shared by the entire social group in *M. olympus* (Barash 1973a), *M. caligata* (Holmes 1984a), *M. broweri* (Rausch and Bridgens 1989), *M. vancouverensis* (Bryant 1990; Bryant and Blood 1999) and populations of *M. flaviventris* at high elevation (Andersen et al. 1976; Johns and Armitage 1979). *M. monax* (Hamilton 1934; Ferron 1996) and *M. flaviventris* hibernate singly, although littermate young yellow-bellied marmots (Lenihan and Van Vuren 1996) and closely related members of a matriline (K. B. Armitage unpublished data) may hibernate together. Only certain burrows are suitable as hibernacula for *M. monax*. In Maryland, hibernacula are in woody or brushy habitats on well-drained sites on a southern exposure and, when possible, under the roots of a tree or stump (Grizzell 1955). In Ohio, hibernacula are usually located in areas of rock outcrops. Some young do not disperse in their first summer and probably share their mother's hibernaculum (Meier 1992). Some burrows are used repeatedly as hibernacula (de Vos and Gillespie 1960; Meier 1992).

Woodchuck burrows are located along woodland edges and brushy fence rows two to three times more often than expected based on availability (Swihart 1992). The spatial distribution of 112 woodchuck burrows was clustered. Part of the clustering resulted from a preference for steep slopes and well-drained soils; a greater area of imperfectly drained soils was available (Merriam 1971). Six hundred interburrow movements were recorded. Movements between burrows in a cluster are more frequent than movement between clusters. High levels of movement occur at midday and continue later in the evening than all other activity (Merriam 1966).

Burrow structure was described by Hamilton (1934), Grizzell (1955), Svendsen (1976), and Barash (1989). Burrows of *M. flaviventris, M. monax,* and *M. olympus* typically have dirt mounds formed by the excavated soil. Home burrows or dens usually have multiple entrances, up to 11 in *M. monax* (Merriam 1971). Only *M. olympus* and *M. monax* excavate burrows in meadows. However, *M. flaviventris* digs flight burrows in meadows when foraging areas are expanded (Armitage 1962). Digging is done with the forelegs; the dirt is thrown between the hind legs. Small rocks are carried to the surface in the mouth and dropped on the dirt mound. When dirt and rocks pile up at the entrance, the marmot lies at the edge of the pile and pushes the dirt away with its chest and forelegs (Armitage 1962). All species build a nest of grass and other vegetation and may also have latrine chambers (Grizzell 1955; Barash 1989). Yellow-bellied marmots gather nest material by tearing dry grass loose with the mouth and position it in their mouth with their forelegs. Gathering grass occurs frequently near the time of hibernation (Armitage 1962) and by reproductive females primarily during lactation (Armitage et al. 1996).

Yellow-bellied marmots rarely dig new burrows. In 40 years of marmot research, I observed that the same burrows are used year after year. As a result, burrows are renovated; soil, rocks, grass, and even marmot skulls are cast out on the surface. Merriam (1971), in a study of 99 burrow systems of woodchucks, reported only four new systems were dug in 4 years. New burrow systems may be dug primarily by juveniles or yearlings (de Vos and Gillespie 1960). Burrow design varies considerably. Flight burrows are short and shallow (Andersen and Johns 1977), whereas nest or home burrows are complex. There may be multiple nest chambers and blind pockets. The hibernation chamber may be so concealed that it and the hibernator may be missed by excavators

(Grizzell 1955). Woodchucks typically prepare wide entrance holes that narrow abruptly shortly below the surface. A second type of entrance, the plunge hole, descends vertically or near vertically and may terminate in less than 1 m in a turn-around chamber or connect to the shaft. The plunge hole is considerably smaller than the main entrance (Grizzell 1955). The main shaft may descend steeply or slope gently, and length varies, but is about 4 m for *M. flaviventris* (Svendsen 1976) and up to 15 m for woodchucks (Grizzell 1955). The main shaft may have multiple branches. Burrows of both species are shallow, usually about 1 m in *M. flaviventris* and 1–2 m in *M. monax*. The shaft usually is deeper than the nest chamber, which may function to drain water away from the nesting marmot. Temperature change is slow; in yellow-bellied marmot burrows, temperature increased from an average of 9.0°C in June to 11.3°C in late August, then declined to 9.1°C in early October (Kilgore and Armitage 1978). Probably all marmot species plug the hibernaculum chamber with a mixture of soil and feces (Bibikow 1996).

A population of yellow-bellied marmots used 11 standing and 1 fallen, horizontal cottonwood trees (*Populus* sp.) as burrow sites in Montana. Most of the marmots's vertical movements were made inside the trees. Most had entrances near the ground, and piles of dead grass and feces were found in several hollows (Garrott and Jenni 1978).

FEEDING HABITS

Marmots are herbivores but may eat some animal matter. June bugs (*Phyllophaga* sp.) occurred in the scat of *M. monax* (Gianini 1925); a *M. flaviventris* preyed on a pika, *Ochotona princeps* (Petterson 1992); and twice *M. olympus* was observed carrying a dead chipmunk, *Eutamias townsendi* (Barash 1973a). Marmots eat a variety of plants and are generalist herbivores (Frase and Armitage 1989; Swihart and Picone 1991a). *M. monax* ate 37 species of plants, of which 3 were grasses (Hamilton 1934). Woodchucks in Maryland consumed 24 different items; red clover (*Trifolium pratense*), white clover (*Trifolium repens*), grass, chickweed (*Stellaria media*), and alfalfa (*Medicago sativa*) were eaten most often and in large amounts (Grizzell 1955). Vancouver Island marmots ate 26 (grasses grouped as one species) of 88 available species. Four species accounted for 87.2% of the grazing: *Phlox diffusa* (42.2%), grasses (30.3%), carices (10.6%), and *Lupinus latifolius* (4.1%). Diet selection was not based on abundance (Milko 1984). The diet was similar among colonies; *Lupinus* spp. and *Eriophyllum lanatum* were major foods later in the summer (Martell and Milko 1986). For *M. caligata,* 28 species were identified, but vetches (*Oxytropis, Astragulus*), sedges, fleabanes (*Erigeron*), and fescues (*Festuca*) formed more than 80% of the diet (Hansen 1975). These hoary marmots also did not forage in proportion to relative abundance. Olympic marmots preferred new plant growth and the upper 6–10 cm of plants when feeding and selected inflorescences of most species when available (Wood 1973). Although grasses are important in marmot diets, forbs may be essential for a normal diet. Forbs were preferred over graminoids, especially in mid- and early summer, by *M. flaviventris* in California (Carey 1985b; Stallman and Holmes 2002) and in Colorado (Frase and Armitage 1989) and by *M. vancouverensis* in late summer (Martell and Milko 1986).

Woodchucks in Pennsylvania consumed 46 species of plants and selected dicots, especially clover, much more frequently than monocots (Arsenault and Romig 1985). Selectivity experiments support a preference for forbs. Woodchucks in Missouri selected wild lettuce (*Lactuca*), white clover, red clover, and grasses in that order (Twichell 1939); dandelion (*Taraxacum officinale*) and common plantain (*Plantago major*) were most commonly selected in cafeteria-style feeding trials (Swihart 1990). Woodchucks climb trees (Gianini 1925) and feed on leaves (Weeks and Kirpatrick 1978) of hackberry (*Celtis occidentalis*), red mulberry (*Morus rubra*), Norway maple (*Acer platanoides*), and peach (*Prunus persica*) (Swihart and Picone 1991a). I observed *M. caligata* feeding on leaves of cottonwood (*Populus trichocarpa*). Woodchucks housed in outdoor pens were presented five common components of orchard ground cover and leaves of four tree species. The percentage of foliage eaten by woodchucks varied from 100% for dandelion to 1.2% for orchard grass (*Dactylis glomerata*). Multiple comparisons indicated that dandelions, red clover, and red mulberry were most palatable; Norway maple and orchard grass were least palatable; and common strawberry (*Fragaria virginiana*), hackberry, birdvetch (*Vicia cracca*), and peach were moderately palatable (Swihart and Picone 1991a).

Food choice is based on additional considerations. In food choice experiments, yellow-bellied marmots reject plants or parts of plants containing defensive compounds (Armitage 1979a). Food plants may be chosen to fill mineral needs. Yellow-bellied marmots were observed licking the mud surface at salt licks (Armitage 2000). Woodchucks use small mineral licks and lick road surfaces for residues of salt applied during winter. When they were provided with salt-impregnated wooden pegs, pegs containing sodium compounds were more highly gnawed than pegs with calcium, potassium, or magnesium. Water-soaked pegs were not gnawed (Weeks and Kirkpatrick 1978). A woodchuck feeding on aquatic plants may have been seeking salt (Fraser 1979). Two principal forbs eaten by *M. flaviventris* had a calcium content two to three times greater than any other plant species. Forbs are significantly higher in phosphorus, calcium, and sodium, and significantly lower in cellulose, than graminoids (Carey 1985b). Food choice may be based on protein content (Frase and Armitage 1989) or essential fatty acid content (Florant 1998; Frank et al. 1998). High concentrations of essential fatty acids occur in the leaves of dandelion, cinquefoil (*Potentilla gracilis*), cow parsnip (*Heracleum lanatum*), and grasses (Hill and Florant 1999), all of which are common elements in the diet of yellow-bellied marmots (Frase and Armitage 1989).

Marmots have a quantitative and qualitative impact on vegetation. Yellow-bellied marmots used only 2–6.4% of the available net primary production (Kilgore and Armitage 1978); however, a colony of *M. olympus* used 30.4% of the net production. Seasonal standing crop of vegetation was greater in enclosed versus open plots (Wood 1973). In an old field inhabited by *M. monax,* total plant cover increased with distance from burrows. Species richness was low near and distant from burrows and relatively high at intermediate distances (English and Bowers 1994). Horse nettle (*Solanum carolinense*), Kentucky bluegrass (*Poa pratensis*), and fescue (*Festuca elatior*) increased and orchard grass decreased with distance from burrows. The strongest effects were limited to a 4-m radius around the burrows. Woodchucks function as important agents in creating a vegetational mosaic. Species close to burrows tend to be early-successional, mostly unpalatable, and early-colonizing annual and biennial species (English and Bowers 1994). Around *M. flaviventris* burrows, unpalatable fireweed (*Epilobium angustifolium*), Rocky Mountain pentstemon (*Pentstemon stricta*), nettle (*Urtica dioica*), and composites are conspicuous (Armitage 2000). In hayfields, forb biomass increased and grass biomass decreased as a function of distance from *M. monax* burrows. In a grass hayfield, orchard grass increased about 2.6% because of woodchuck feeding (Swihart 1991a). In a hayfield with alfalfa, the biomass of alfalfa decreased by an average of 2.5% and orchard grass increased 7.4% because of woodchuck activity. In feeding trials, alfalfa was selected on average 10 times more often than orchard grass. Orchard grass was more lush around burrows and contained significantly greater levels of crude protein than orchard grass 15 m distant (Swihart 1991a). Alfalfa stem totals were about three times greater 6 m or more from a woodchuck burrow than they were within 1.5 m of a burrow. Because woodchucks prefer legumes, differences in alfalfa versus grass stems can be attributed to selective foraging (Merriam and Merriam 1965). A lush green zone occurred next to the burrow; soil nitrogen was 1.7 times more concentrated near the burrows than in the field. Feces, and probably urine, deposited in the burrow area are the likely source of nitrogen, producing lush growth.

In meadows used by *M. olympus,* primarily unpalatable plant species occur on burrow mounds. Overall species richness was greater in the meadow than on mounds (Del Moral 1984). Marmots fed selectively, enhanced plot diversity, and reduced the dominance of common species.

DEMOGRAPHY AND MORTALITY

Population densities of yellow-bellied marmots are relatively stable (Armitage 1991b). Annual fluctuations occur primarily through reproduction and dispersal of yearlings. Populations increase when yearling daughters are recruited and decrease because of mortality during hibernation and predation (Armitage 1996c). Life-table statistics for a cohort of yellow-bellied marmots based on 1532 known-age individuals revealed that females had significantly better survivorship than males beyond the first-year age class. Thus, adult sex ratio became progressively female biased (Schwartz et al. 1998). Rates of survivorship were constant and the probability of mortality did not increase with age. Female generation length was 4.49 years. Reproductive values were approximately equal across reproductive age classes and the net reproductive rate was 0.85. There was significant yearly variation in births and deaths such that time-specific and cohort life tables were nonequivalent (Schwartz and Armitage 1998). Litter sizes were not equal among years, but were equal among age classes. Virtually all adult males were immigrants. The percentage of adult females that were immigrants varied from 16.7% at age 2 years to 40.6% at age 7 years (Schwartz et al. 1998). Immigrants varied from 22.3% to 96.9% among populations (Armitage 1988). Yellow-bellied marmot social systems are closed. Immigration occurs primarily when space is vacant because of overwinter mortality and no recruits are present. In only 5 of 141 colony-years did immigration and recruitment occur in the same colony. No introduction of adult female marmots into an occupied habitat patch was successful (Armitage 1991b).

Male survivorship appears to be lower than that of females for *M. olympus* (Barash 1973a) and *M. vancouverensis* (Bryant 1998). Survivorship is significantly greater at all ages beyond age 1 year for the Olympic and Vancouver Island marmots compared to yellow-bellied marmots (Armitage 1998; Blumstein et al. 2002). The difference probably results from dispersal at an earlier age in yellow-bellied marmots and the higher mortality of dispersers than of residents (Van Vuren and Armitage 1994). There may be a tradeoff between survival and reproduction, as yellow-bellied marmots reproduce at an earlier age than the other two species.

A woodchuck population on 4000 ha declined 80% between 1955, when the area was primarily grain and hay agriculture, and 1970, when the area became primarily old field. A low birth rate of 1.3 in (young per female in the population) 1955 was attributed to the small (20%) proportion of yearlings breeding. Birth rate increased to 1.7 by 1958, when 53% of the yearling females reproduced (Snyder and Christian 1960; Davis and Ludwig 1981). Rapid decline occurred from 1963 to 1966, when the survival of young declined. Drought occurred during this period, the vegetation dried prematurely, and young failed to gain the critical mass needed to survive hibernation. The population stabilized after 1967 (Davis and Ludwig 1981). Woodchucks were removed from two areas with mainly old-field vegetation to determine the effects of exploitation (Davis et al. 1964). The percentages of young and yearlings in the population increased greatly. High immigration of primarily young into one area, and the tendency of young animals born there to remain, resulted in an increase in the population despite removal of 1040 animals. Natality initially increased on this area, but then decreased because the proportion of adults declined and yearlings have a lower birth rate (Davis et al. 1964).

Weather affects survival and reproduction. For *M. flaviventris,* colony size, survival, and litter size were most affected by the length of winter, length of the growing season, and precipitation (Schwartz and Armitage 2002). Colony size was the larger, the earlier snowmelt occurred and the longer the growing season lasted. Extensive reproduction occurred in the year after a rainy year. Drought decreased survivorship of reproductive females and young in the following hibernation, and reproduction decreased the following summer (Armitage 1994). The effect of drought operates through reduced growth during the active season (Hamilton 1934; Davis and Ludwig 1981; Armitage 1994; Lenihan and Van Vuren 1996). The later snowmelt occurs, the fewer litters per female and fewer young per litter are produced (Van Vuren and Armitage 1991).

Dispersal. Typical age of dispersal is as follows: *M. monax,* young; *M. flaviventris,* yearling, *M. caligata* and *M. olympus,* age 2 years (Armitage 1999; Blumstein and Armitage 1999). Although dispersal reduces the density of the disperser's natal population, little is known about the fate of dispersers or the causes of dispersal. In yellow-bellied marmots, essentially all juvenile males and about one half of the females disperse (Armitage 1991b). Dispersal of yearling males was delayed when they were underweight, when the rates of amicable behavior between them and adults was high, and when there were many yearling males in the colony. Dispersal of yearling females was independent of the number of adult females present. It was delayed when rates of adult aggression were low and when adults behaved amicably toward them (Downhower and Armitage 1981). Dispersal probably is not genetically determined; when adults were removed from two populations, yearling females did not disperse (Brody and Armitage 1985). Dispersal is socially mediated by adults interacting with same-sex yearlings (Downhower and Armitage 1981). Dispersal distance is highly skewed; median dispersal distance for females was 400 m and for males was 1500 m (Van Vuren 1990). Marmots display three dispersal patterns: (1) 40% abandon their home range in a single, abrupt move to a new locality; (2) 33% gradually extend their home ranges until a new one is established; and (3) 27% disperse in two stages by first establishing new home ranges nearby and then, on average 41 days later, moving again. Females use the second pattern more than males and males move more often by the third pattern. Survival of dispersers was 0.73, not much less than that (0.87) of philopatric marmots; there was essentially no difference in survivorship (0.87 vs. 0.91) in the subsequent hibernation (Van Vuren and Armitage 1994).

Predation. The major predators of marmots are canids, mustelids, bears, and raptors. Predation rarely is observed, but coyotes (*Canis latrans*) and badgers (*Taxidea taxus*) were observed preying on yellow-bellied marmots (Verbeek 1965; Andersen and Johns 1977; Thompson 1979b; Armitage 1982). A coyote and a cougar (*Puma concolor*) were observed to prey on Olympic marmots (Barash 1973a). Predation by golden eagles (*Aquila chrysaetos*) was documented by examining prey remains at nests and in pellets. Yellow-bellied marmots contributed 71.4% (Knight and Erickson 1978) and 73.3% (Marr and Knight 1983) of the prey biomass found at golden eagle nests along the Columbia River, and 2.4% in southwestern Idaho (Collopy 1983). In these studies, biomass of marmots was estimated from mean weights of museum specimens. These estimates may be too high because yearlings were not distinguished from adults. All skulls and other remains of marmots found at an eagle nest in Colorado were yearlings (G. E. Svendsen, pers. commun., 1971) and a golden eagle was observed feeding on a yearling. Of six marmots with radio transmitters killed by eagles, three were yearlings and three were adults (Van Vuren 1990). Thus, the smaller yearlings seem to form a substantial proportion of marmots predated by eagles.

Van Vuren (1990) reported that 59% of the predator kills of yellow-bellied marmots with radio transmitters in Colorado were by coyotes; badgers were responsible for 13%. Other predators were eagles (11%), black bears (*Ursus americanus,* 7%), and martens (*Martes americana,* 5%). Yellow-bellied marmot remains occurred in 17% of 395 coyote scats collected during the summer in Colorado, thus verifying that marmots are frequent prey (Van Vuren 1991).

Golden eagles, wolves (*Canis lupus*), and cougars prey on the Vancouver Island marmot (Bryant 1998), and wolves, grizzly bears (*Ursus arctos*), red foxes (*Vulpes vulpes*), and golden eagles prey on the hoary marmot (Murie 1944). Hoary marmots were minor items in the diet of bears and foxes, and occurred in 6% of eagle pellets and 8% of wolf scats. Marmots were more abundant in wolf scats (37%) when caribou (*Rangifer tarandus*) were scarce.

Marmots respond to some predators by chasing them. Yellow-bellied marmots chased long-tailed weasels (*Mustela frenata*) and marten (Travis and Armitage 1972). A marmot observed by Waring (1965) made one or two "barking whistle" sounds per second while chasing less than 1 m behind a marten. An adult woodchuck chased a fox (de Vos and Gillespie 1960). After coyote predation on a female yearling, yellow-bellied marmots modified their home ranges and spent less time in the area where predation occurred (Armitage 1982). Marmots typically respond to an approaching predator by emitting an alarm call, a high-pitched tone produced by the vocal cords (Barash 1989). Marmots also may run to a burrow without calling. Predators and nonpredators may evoke calls. For example, yellow-bellied marmots responded to deer (*Odocoileus hemionus*), moose (*Alces alces*), black bears, horses (*Equus cabellus*), coyotes, sandhill cranes (*Grus canadensis*), ravens (*Corvus corax*), great blue herons (*Ardea herodias*), swallows (*Tachyaneta bicolor*), and northern harriers (*Circus cyaneus*) (Armitage 1962). Woodchucks when frightened aboveground fled to their burrow and had marked tachycardia; once in the burrow, the heart rate declined. When frightened in the burrow, there was an onset of bradycardia, which persisted as long as the stimulus lasted (Smith and Woodruff 1980).

Hoary marmots responded to a model golden eagle by giving an alarm call while simultaneously running to a burrow when >5 m away or called and visually tracked the model if <1 m from the burrow. Other marmots responded to the call by running to a burrow; 28% of these emitted a call after reaching a refuge. When at the refuge, all marmots tracked the model and 9 of 32 emitted calls (Noyes and Holmes 1979).

Marmots may flee from a predator at a distance that allows them to reach a refuge ahead of the predator with some margin of safety. An observer approached juvenile woodchucks from the direction opposite to the burrow. Flight distance increased directly with the distance to the burrow, but was not affected by the speed of approach. The rate of increase in flight initiation distance was higher when the burrow was between the observer and the woodchuck than when the woodchuck was between the observer and the burrow. Escape velocity was slower for woodchucks closer to the burrow. When distant, a woodchuck stopped before entering; when close, it usually entered directly (Bonenfant and Kramer 1996; Kramer and Bonenfant 1997).

Parasites and Diseases. Major external parasites of marmots are fleas, mites, and ticks; major internal parasites are tapeworms and roundworms (Bibikov 1992; Bibikow 1996; Bassano 1996). Ticks were almost always present on woodchucks in Missouri (Twichell 1939). Eight species of Protozoa were taken from the cecum of *M. flaviventris;* six were described from *M. monax* (Gabel 1961). Bassano (1996) summarized much of the information on the distribution of parasites among marmot species; additional information for the amphiberingian marmots is summarized by Rausch and Rausch (1971). Marmots may be widely infected with large cestodes and nematodes during autumn, but these are eliminated at hibernation when the stomach and intestines contract (Rausch and Rausch 1971). Incidence of parasitism may vary widely; one population of woodchucks was reported to be free of internal parasites (McTaggart-Cowan 1933). Although there is little evidence that internal parasites affect survival and reproduction, infected juvenile yellow-bellied marmots gain little or no mass and do not survive hibernation (Armitage 2000). Ectoparasites also may reduce fitness. Two seriously underweight woodchucks were heavily infected with mites, probably *Androlaelaps* sp. (Grizzell 1955). Yearling *M. flaviventris* with greater flea infestations grew more slowly, animals that died during hibernation had more fleas than survivors, and adult females that failed to reproduce had more fleas than those that reproduced (Van Vuren 1996).

Viral and bacterial diseases of North American marmots were described primarily for *M. monax* (Bassano 1996). Woodchuck hepatitis virus (WHV) was discovered in laboratory animals and subsequently found in wild animals in southeastern Pennsylvania, central New Jersey, and north-central Maryland (Tyler et al. 1981). Evidence of infection was not found in woodchucks <3 months old. No cases were reported from woodchucks captured in Vermont or Massachusetts (Young and Sims 1979). Because the virus belongs to the same class of viruses as hepatitis B, woodchucks serve as a laboratory model for the study of hepatitis. Other diseases reported from woodchucks are rabies, Rocky Mountain spotted fever, tularemia, leptospirosis, and human encephalitis (Young and Sims 1979).

Plague (*Yersinia pestis*) was reported in North American marmots (Bibikow 1996). The fleas *Thrassis acamantis* and *T. stanfordi* are primary vectors of plague. Both occur on *M. flaviventris,* but little is known about marmot susceptibility to plague (Stark 1970), which does not appear to have any significant demographic effects. In Eurasia, plague in *M. bobac, M. baibacina,* and *M. sibirica* typically occurs in localized areas (foci). In plague foci, only about 2% of the animals are infected. Rabies increased in woodchuck populations during 1980–1984 in association with an epizootic of rabies in raccoons (*Procyon lotor*). Although rabies in woodchucks may be a public health concern, there is no information on its effects on woodchucks (Fishbein et al. 1986).

Reproductive Success. Reproductive success should be measured as the number of reproductive descendants produced by an individual during its lifetime. However, lifetime reproductive success is usually unknown and this discussion will focus on annual reproductive success. Reproductive success of a male yellow-bellied marmot depends on the number of reproductive females and the length of time he is associated with them (Armitage 1991b); the numbers of both young and yearlings produced increase linearly with the number of females in the harem (Armitage 1986a). A male is resident for an average of 2.24 years (Armitage 1974), but residency may reach 4–6 years (Armitage 1984). An average resident male produces 11.1 young during his lifetime (Armitage 1991b).

Female reproductive success is affected by environmental harshness (Armitage 2000; Armitage and Blumstein 2002) and reproductive suppression by conspecifics (Wasser and Barash 1983; Blumstein and Armitage 1999). As a consequence of these two factors, the estimated percentage of adult females that breed is 72% for *M. monax* (Snyder and Christian 1960), 53% for *M. vancouverensis* (Bryant 1996), 52% for *M. flaviventris* (Armitage 1991b), 43% for *M. caligata* (Barash 1989), and 41% for *M. olympus* (Barash 1973a). No female yellow-bellied marmot living at a high elevation (3400 m) bred in successive years (Johns and Armitage 1979), whereas females at a lower elevation (2900 m) frequently breed in successive years (Armitage 1984). Hoary and Olympic marmots breed biennially (Barash 1973a, 1974a); in an Alaskan population, a mean of 3.3 years occurred between breedings (Holmes 1984a). The average time between breedings by female Vancouver Island marmots is 2 years (Bryant 1998).

Reproductive suppression occurs when mature females fail to breed in the presence of older, dominant individuals. Thus, in a captive population of *M. broweri* maintained in the laboratory for 6 years, only the dominant pair bred despite the presence of other individuals of both sexes up to 5 years of age (Rausch and Bridgens 1989). Occasionally a second nonparous female is present in a family of Olympic marmots (Barash 1973a), which suggests possible reproductive inhibition. In bigamous families of hoary marmots, females may skip an additional year when both would be expected to breed. When both females breed, the subordinate individual produces about one half as many young and one fifth as many yearlings as the dominant female (Wasser and Barash 1983). The frequency of reproduction by 2-year-old *M. flaviventris* is significantly lower when adult females, including their mothers, are present (Armitage 1998). Weaning success is lower in a single female living in proximity to a matriline of two or more females than when she lives solitarily or adjacent to a matriline of one female (Armitage 1986a). Net reproductive rate and survivorship increase with the increase in matriline size, then decrease in the largest groups (Armitage and Schwartz 2000). Females increase their reproductive success by leaving large matrilines.

Reproductive success of female *M. flaviventris* varies with behavioral phenotype. Over a 2-year period, submissive females produced few young, and social females produced about one third of the young (Svendsen 1974). Social females recruited more yearling daughters than did asocial females (Armitage 1984). Sociable females had higher lifetime values for number of female yearlings, number of recruits, and number of 2-year-old daughters than females categorized as "avoidance" or "approach" (Armitage 1986b).

BEHAVIOR

Basic behavioral patterns are common to all species of marmots, but the social context and frequency of expression vary among species (Barash 1989; Bibikow 1996). Social interactions may be broadly classified as amicable, which includes greeting and allogrooming, and agonistic, which includes chase, flight, fight, and avoidance (Armitage 1973; Johns and Armitage 1979). Sexual behavior includes genital sniffing, grasping, grappling, and mounting (Armitage 1965). Rates of amicable and agonistic behavior are related to population density, the age–sex structure of the population, length of shared residency, and individual behavioral phenotypes. Yearly changes in agonistic and amicable behavior among yellow-bellied marmots are not correlated within or among colonies (Armitage 1977). Barash (1973b) suggested that high-elevation colonies of *M. flaviventris* may be more social than medium elevation colonies as an adaptation to a shorter growing season. However, the age–sex structure and population density of these colonies varied and could account for the observed behavioral differences. Typically, social interactions are high in the spring and decline thereafter (Armitage 1962; Bronson 1964; Barash 1973a, 1974a; Heard 1977; Meier 1992). Social behaviors occurred more frequently in the morning and evening in Olympic and hoary marmots. In yellow-bellied marmots, social interactions occurred more frequently at midday in one colony and in the morning in another (Armitage et al. 1996). Social behaviors tended to be more frequent in the early postlactation period when young emerged and decreased by 3 weeks of postlactation.

Amicable Behaviors. A greeting occurs when two marmots approach head-on with tails arched followed by nose-to-nose, nose-to-cheek, or nose-to-mouth contact or when they "sniff" at each other without contact (Armitage 1962; Barash 1989). A greeting often is followed by allogrooming in which one marmot chews at the neck, back, or flanks of the other. Mutual allogrooming occurs when the "groomee" becomes the "groomer." Social interactions among woodchucks are rare (Bronson 1964; Ferron and Ouellet 1989). Amicable behavior occurs between all combinations of age and sex except among adult females or adult males (Barash 1989; Meier 1992). Adult males are more socially active in hoary (Barash 1974a; Holmes 1984a) and Olympic (Barash 1973a) marmots, but less active in yellow-bellied marmots (Armitage and Johns 1982). The rate of greetings is much greater in the restricted family groups of Vancouver Island, Olympic, and hoary marmots than in the matrilineal yellow-bellied marmot groups (Barash 1973a, 1974a; Armitage 1977; Heard 1977). Amicable behavior among yellow-bellied marmots varies with age–sex groups and kinship (Armitage and Johns 1982; Armitage 1986a, 1989). Amicable behavior predominates among females related by 0.5; and occurs much less than expected when relatedness is 0.25 or less. Female yearlings interact amicably with both parents, and male yearlings may behave amicably with their mothers (Armitage 1974; Armitage and Johns 1982).

Agonistic Behavior. Agonistic behavior characterizes social interactions among woodchucks (Bronson 1964). Subordinate penned woodchucks gain less body mass than the dominant animals (Bailey 1965b). Woodchucks may be organized into a complex of dominance–subordinate relationships that is maintained regardless of the location of the interaction. Yearlings are subordinate to adults. When a woodchuck threatens another animal, the tail is raised with hairs erect and often flipped up and down, the back is arched, and the head directed toward the opponent with mouth open and incisors showing (Bronson 1964). Home range size and social rank were not significantly related in woodchucks. Seasonal changes in aggressiveness were more significant than changes in population density (Bronson 1963). Agonistic behavior among Olympic marmots occurs primarily when the resident male chases intruders or when satellite males avoid the residents (Barash 1974a). Chases occurred among several age classes of Vancouver Island marmots, but were relatively rare (Heard 1977). Agonistic interactions characterize adult male–adult male and adult male–yearling male behavior in yellow-bellied marmots (Armitage 1974). Otherwise, agonistic behavior is strongly related to kinship. It occurs much less frequently than expected among kin related by 0.5 and more frequently than expected among more distant kin and unrelated individuals. However, adult females may be highly agonistic to yearlings that are nonlittermate full sibs and to nieces; some females are agonistic toward their daughters (Armitage and Johns 1982; Armitage 1986a). Agonistic behavior, especially among adult males, predominated in a captive population of *M. broweri*. Three adult females died and a fourth sustained serious injury; all had extensive wounds (Loibl 1983). This pattern of agonism is atypical for species with an extended family social system (Armitage 1996c). Perhaps confinement resulted in atypical behavior.

Infanticide can be considered an extreme expression of agonistic behavior and has been reported only for yellow-bellied marmots. Typically the young was either unrelated or distantly related to the adult perpetrator (Armitage 1979b; Brody and Melcher 1985).

Vigilance and Defense. Vigilance is characterized by a marmot sitting upright on its haunches, forelegs held in front of the chest, and surveying the surroundings (Armitage et al. 1996). All species of marmots express this behavior. Marmot defense also includes alert behavior in which the marmot has head up and looks around while sitting, lying, or foraging. Alert and vigilance combined occupied 1.1–14.5% and 12.0–71.7 min/day among three populations of *M. flaviventris* in Colorado (Armitage et al. 1996) and 10.2% of foraging time in a California population (Carey and Moore 1986). Reproductive females and adult males have high levels of vigilance (Barash 1973a; Armitage et al. 1996). Vigilance may be higher in marmots where human disturbance is frequent (Armitage et al. 1996) or when feeding near the periphery of a colony (Armitage 1962). It may be reduced when most individuals are members of a closely related kin group (Armitage et al. 1996). Woodchucks become vigilant and scan when in feeding areas (Anthony 1962). Adult and 2-year-old *M. caligata* were alert less often than yearlings. Marmots were alert less often when foraging near other individuals and when foraging near talus (Holmes 1984b). In yellow-bellied marmots, foraging group size, alarm call occurrence, and age significantly affected the time spent alert (Carey and Moore 1986). Young and yearlings were alert more while foraging than adults. Yellow-bellied marmots reduced the potential cost of time spent alert by chewing while alert or vigilant (Armitage et al. 1996).

Communication. Marmots use visual, auditory, and olfactory senses for communication. An individual running to the burrow frequently results in other marmots running to their burrows. Tail flicking, which frequently occurs when an animal is moving or initiating movement, may serve as an intention movement (Armitage 1962) or to coordinate group activities. Adults tail flick more frequently than young or yearlings (Barash 1973a). Adult male yellow-bellied marmots wave their tails conspicuously when patrolling their territories or when approaching another animal; tail flagging may signify dominance or serve to communicate presence (Armitage 1974).

All marmots have sudoriferous facial glands (Walro et al. 1983; Rausch and Bridgens 1989), which are used for scent marking by rubbing the cheeks and oral angles against rocks, tree roots, or other objects (Barash 1973a; Armitage 1976; Hebert and Prescott 1983; Taulman 1990; Brady and Armitage 1999). Scent marking rates do not differ between sexes for *M. monax* (Ouellet and Ferron 1988), but *M. flaviventris* males mark more frequently than females (Brady and Armitage 1999). Scent marking decreases significantly during the summer (Barash 1973a; Hebert and Prescott 1983; Brady and Armitage 1999). Both *M. flaviventris* and *M. monax* spend more time marking or investigating foreign than familiar scents (Hebert and

Barrette 1989; Meier 1991; Brady and Armitage 1999). Scent marking has multiple functions: territorial marking, dominance, burrow occupancy, individual identity, familiarity with the home range, and a possible self-reassurance role.

All marmots emit a variety of vocalizations including alarm calls, shrieks, tooth chattering, growls, and whines (Waring 1966; Lloyd 1972; Barash 1973a; Taulman 1977; Hoffmann et al. 1979; Blumstein and Armitage 1997a). Shrieks are heard during play of young or from a submissive animal during an encounter (Armitage 1962; Waring 1966). Tooth chattering occurs during agonistic encounters (Waring 1966; Taulman 1977). Growl and whine are directed at other marmots and represent threat and submissiveness, respectively (Armitage 1962; Taulman 1977). Alarm calling is the most intensively studied vocalization. Calls generally center on 2–4 kHz (Waring 1966; Hoffmann et al. 1979, Blumstein 1999). Yellow-bellied marmots produce three acoustically distinct alarm vocalizations; whistle, trill, and chuck (Blumstein and Armitage 1997a). The whistle is elicited by both aerial and terrestrial predators in addition to suddenly appearing birds. Submissive yellow-bellied marmots and woodchucks often call when chased (Anthony 1962; Armitage 1962). Trills are preceded by whistles. Marmots trill as they enter a burrow after being pursued by a predator, when suddenly surprised, or when fleeing a conspecific. Marmots chuck infrequently; chucks are produced when minimally alarmed and often follow whistles (Blumstein and Armitage 1997a). *M. monax* has a single whistle (Hamilton 1934; Lloyd 1972). *M. olympus* and *M. caligata* have four alarm vocalizations: ascending, flat, descending, and trill; *M. vancouverensis* produce those four sounds plus a fifth, the kee-aw (Blumstein 1999). Alarm calls in these species are elicited by a variety of birds and mammals and by conspecific aggression (Barash 1973a; Heard 1977; Blumstein 1999). About 54% of the variation in the repertoire size of marmot alarm calls was explained by a measure of social complexity, and significant variation was explained when controlling for phylogenetic effects (Blumstein and Armitage 1997b). Aerial and terrestrial stimuli do not elicit unique alarm calls in *M. flaviventris, M. olympus, M. caligata,* and *M. vancouverensis*. Rather, calls vary according to risk. In *M. flaviventris,* repetition rate increases with risk and this pattern does not vary geographically (Blumstein and Armitage 1997a). The other three species with similar vocalizations communicate relative predation risk differently. *M. olympus* varies the number and rate of alarm calling; *M. caligata* produces quickly paced calls in response to aerial stimuli and longer calls uttered at a slow rate, to terrestrial stimuli; *M. vancouverensis* varies call duration and bout ($\geq$1) composition (Blumstein 1999).

After young are weaned, adult female *M. flaviventris* with young call more than other age–sex classes; before weaning, calling rate does not differ among age–sex groups. Over the entire study period, about 42% of the variation in rate of calling was attributable to females that weaned young and 13% to the presence of older offspring. Degree of relatedness is not a significant factor (Blumstein et al. 1997). Indirect fitness apparently played a minor role and direct fitness a major role in the evolution of alarm calling in *M. flaviventris*. Although calling may function to alert offspring and other conspecifics, alarm calls may be directed toward the predator (Blumstein and Armitage 1997a, 1997b).

Yellow-bellied marmots may respond to the alarm calls of golden-mantled ground squirrels (*Spermophilus lateralis*) (Shriner 1998) and rock squirrels (*S. variegatus*) (Blumstein and Armitage 1997a), but not to bird calls used as controls. *M. olympus* may call in response to hearing conspecific alarm calls (Blumstein 1999). Some individuals are more highly excited than others and continue calling for >30 min, long after other animals are alerted (Armitage 1962; Waring 1966). Finally, when detecting a predator, a marmot may flee silently to its burrow and not alarm call (Waring 1966; Blumstein 1999).

Grooming. Marmots may scratch an anterior part of their body with a hind foot or chew their fur, usually while sitting on their haunches. Among yellow-bellied marmots, grooming, including allogrooming, accounts for 0.8–3.4% of the time aboveground (Armitage et al. 1996). Grooming generally occurs throughout the day. Reproductive females and adult males tend to groom more than other age–sex groups. A greeting typically precedes grooming in yellow-bellied and Olympic marmots (Barash 1973a). Woodchucks also groom (Anthony 1962); probably all marmot species do.

Play. This behavior includes a variety of motor patterns, including hide/seek, grapple, chase/flee, and mouth-spar. Jamieson and Armitage (1987) described and illustrated play motor patterns, transition motor patterns such as escaping, and play-associated motor patterns, which include allogroom, approach, greeting, autogroom, social investigation, and withdraw. Play is often called play-fighting and may escalate into aggressiveness, especially in those species in which adults play. Thus, 39% of the upright play fights between adult *M. caligata* were followed by a chase (Barash 1974a) and adult play fights between *M. olympus* adults often were more characteristic of fights (Barash 1973a).

All species of marmots play. It typically occurs only among young *M. monax* for a few days after emergence (Barash 1989). However, four instances of play were observed between an adult female and yearling female and eight instances between an adult male and yearling female (Meier 1992). These play bouts may have been conflict. I have frequently observed conflict between two *M. flaviventris* that has some of the same behavioral motor patterns as play. Play in yellow-bellied marmots occurs among young and among yearlings; male young and yearlings play more than female young and yearlings (Armitage 1974; Nowicki and Armitage 1979; Jamieson and Armitage 1987). Both young and yearlings play with the same sex at rates higher than expected and males initiate play with females more than the reverse. Role reversal (or flips) occurs whenever animals exchange top and bottom positions in wrestling. Role reversal occurs more often when two males are involved and much less often when a male–female pair or two females are playing. Young males engage in mouth-spar, grapple, wrestle, and allogroom significantly more than females (Nowicki and Armitage 1979). Yearling females are more likely to be in the bottom or flee position, and males in the top or chase position. Females are more likely to terminate play. Pouncing occurs as a transition from one bout to another; males are responsible for about 70% of the pounces (Jamieson and Armitage 1987). Play rarely occurs between young and yearlings, does not occur among adult *M. flaviventris* (Nowicki and Armitage 1979), and rarely between mothers and their offspring (Armitage et al. 1996). Play occurs more frequently in the morning and early in the season (Armitage et al. 1996). Yearling play rarely occurs beyond mid-July. Play contains motor patterns similar to those observed in adults during agonistic behavior. These patterns, plus the differences in play activity between males and females, suggest that play functions in social integration by facilitating social dominance and the coordination of agonistic behavior.

Adult male *M. vancouverensis* play with adults, 2-year-olds, and yearlings; adult females play with 2-year-old females and yearlings, but not with adult females or infants. Infants play with other infants, and 2-year-old females play with yearlings (Heard 1977). Play is common among yearling and infant *M. olympus* and occurs in 2-year-olds and adults (Barash 1973a). Young, yearling, and 2-year-old *M. caligata* play most often; adults play at low rates (Barash 1974a). None of these studies analyzed play behavioral patterns.

Individuality. Yellow-bellied marmots can be broadly classified as social or asocial; the asocial animals are either submissive avoiders or aggressive. Behavioral phenotypes appear to be relatively stable among adults (Svendsen and Armitage 1973; Svendsen 1974), but may change during ontogeny (Armitage 1986c). Adult female hoary marmots transplanted to new colonies tend to maintain their pretransplant levels of aggressiveness (Barash 1989).

Social Organization. Marmots can be grouped into four social systems: solitary, female kin groups, restricted family, and extended family (Armitage 2000). Only *M. monax* is solitary. Adult and yearling females may share the same burrow in early spring, but not after adult females give birth (Snyder 1962; Meier 1992). Spatial structure of woodchuck populations indicates a low degree of sociality: The frequency of burrow sharing is low, the degree of overlap between neighboring reproductive females is low, the presence of nonreproductive females at natal burrows is scarce, and juveniles disperse (Swihart 1992). The uniform spacing of burrows is attributed to agonistic behavior, primarily among juveniles during rapid invasion of new habitat (Hendersen and Gilbert 1978). Young may chase or fight with other young, and adults may chase young (Anthony 1962). The mating system is polygynous (Meier 1992).

Female yellow-bellied marmots are organized into mother: daughter:sister kin groups, which persist through time as matrilines (Armitage 1984, 1998). All male yearlings disperse. Adult males associate with ≥1 female matrilines to form a harem polygynous mating system. About 50% of the female yearlings are recruited into their natal matriline; the others disperse.

Olympic, hoary, and Vancouver Island marmots live in restricted family groups into which the male is integrated such that the social and mating systems are congruent (Armitage 2000). Family groups typically consist of an adult male, one or three females, yearlings, and one litter of young (Barash 1973a, 1974a; Heard 1977). Some populations of *M. caligata* are monogamous (Holmes 1984a). Dispersal is typically at age 2 years. An individual adult female typically is associated with either young or yearlings, but not both.

All remaining species of marmots form extended family groups; *M. broweri* is the only representative of this social system in North America. The typical family consists of an adult territorial pair that does all the breeding, subordinate nonbreeding adults ≥2 years old, yearlings, and young. Polyandry may occur. Dispersal occurs at age 3 years or older (i.e., beyond the age of first reproduction). Because all members of the family hibernate in the same burrow (Rausch and Rausch 1971), alloparental care may occur (Arnold 1993; Blumstein and Armitage 1999).

Barash (1974c) first pointed out the relationship between marmot sociality and the length of the growing season. These ideas were extended to all ground-dwelling sciurids (Armitage 1981). More recently, the following scenario was proposed for the evolution of marmot sociality (Armitage 1999, 2000). Marmots evolved in harsh landscapes and developed hibernation to cope with periods of food shortage or absence. The evolution of large body size as a means of maximizing the efficiency of storage and use of fat, coupled with a short active season, resulted in young requiring ≥1 years of additional growth to reach maturity. Young were retained in their natal area; retention of young within the parental home range formed social groups in all species except *M. monax*. Retention of young led to delayed dispersal, which provided the potential for cooperative breeding with its attendant alloparental care (Blumstein and Armitage 1999) and communal nesting (Armitage and Gurri-Glass 1994). Delayed dispersal probably evolved because habitats were saturated; marmots disperse at an older age, when the chances of success are greater (Armitage 1996c). Some consequences of increased sociality are a delay in the age of first reproduction, smaller litter size, and a smaller proportion of breeding females (Blumstein and Armitage 1998). At least some of the reproductive costs of sociality result from reproductive suppression.

AGE ESTIMATION

The most useful criterion for estimating age of marmots is body mass 30–60 days after emergence from hibernation, when body fat is minimal. Yearling, 2-year-old, and 3-year-old or older *M. flaviventris* (Armitage et al. 1976) and *M. olympus* (Barash 1973a) and yearling and adult *M. vancouverensis* (Bryant 1998) and *M. monax* (Davis 1964) can be distinguished by differences in body mass. Young of all species are easily identified because of their small size. In *M. monax,* the size difference between young and adults becomes less by late summer. Characteristics of pelage, shape and color of incisors, and shape of head are important for assessing chronological age and distinguishing yearlings from adults in the spring (Davis 1964). Older animals cannot be distinguished by length or body mass for any marmot species. Eye lens weight provides a good estimate of age of *M. monax,* but requires killing the animals. In *M. flaviventris,* premolar wear increases linearly with age through age 4 years, but is unsuitable for estimation beyond 3 years of age (Van Vuren and Salsbury 1992).

ECONOMIC STATUS, MANAGEMENT, AND CONSERVATION

Economic Status. Eurasian marmots are hunted for fur, fat, and flesh; during famines and World War II, marmots saved thousands of people from starvation (Bibikow 1996). Marmot fat is rendered to obtain oil, which is believed to be medicinal and is used to treat ailments such as burns and arthritis. In the former Soviet Union in 1991, a marmot was worth about 200 rubles, several times the monthly salary of a worker. No comparable economic use of North American marmots is known. Marmots are hunted for food (Bollengier 1994), and woodchuck fur and hide formerly were used for cheap coats, patching leatherwork, and straps (Schwartz and Schwartz 1981). Minor economic activity centers on the groundhog in Pennsylvania as a prognosticator of winter duration ("Punxsutawney Phil"). Marmots are used as symbols for commercial enterprises (Marmot Mountain Ltd.) and as characters in children's books. The hoary marmot was prized by northwest native Americans for clothing. Areas with marmots were privately owned, huts or cabins were built at the sites, and the marmots were hunted in the autumn after the molt (Drucker 1950). The hides were a chief article used in potlatches, and wealth among Tlingit and Gitksan was directly measured in hoary marmot skins. Okanagan Indians derived food and clothing from *M. c. okanagana* (Anderson 1934). Cut marks on bones and artifacts recovered from subalpine caves indicate that the Vancouver Island marmot was used as food by aboriginal peoples (Nagorsen et al. 1996). Native Americans in the Great Basin use marmots for food (C. Floyd, pers. commun., 2000). Marmots are a tourist attraction, especially in the western states, but this economic potential has not been developed to the extent that it has in Europe (Ramousse and Giboulet 2002). Woodchuck burrows are used by other small mammals of economic importance, such as skunks (*Mephitis mephitis*), cottontail rabbits (*Sylvilagus floridanus*), raccoons, and red fox (Eadie 1954; Grizzell 1955; Schmeltz and Whitaker 1977).

Only *M. flaviventris* and *M. monax* occur at elevations low enough to potentially negatively affect agriculture. Yellow-bellied marmots prefer alfalfa and may forage in irrigated meadows (Thompson 1979b) and consume much of the crop (Couch 1930). Marmots may also burrow into and cause breaks in levees and dikes (Eadie 1954). Damage by yellow-bellied marmots appears to be minimal, localized, and generally is not discussed in books on wildlife damage.

Woodchucks gnaw woody stems, primarily in the spring. There is greater stem mortality near burrows. Sugar maple was the species most severely gnawed, possibly because woodchucks may feed on the sap (Swihart and Picone 1991b).

Woodchucks may damage a variety of crops such as grains, clover, alfalfa, hay grasses, beans, peas, corn, and apple trees (Eadie 1954). In Connecticut, woodchucks damage apple trees near their burrows; 17% of trees within 1.5 m of a burrow were dead or recently replaced compared to none at distances greater than 12 m (Swihart and Picone 1994). Woodchucks gnawed 96% of 48 trees near their burrows; none were gnawed distant from burrows. Gnawed trees were smaller and produced 43% fewer apples, and individual apples were smaller. Up to 1.5% of the trees were affected by woodchucks, whose activity reduced the yield of apples by a minimum of 0.6%.

Both species may climb into automobiles and gnaw on electrical wiring or rubber hoses. They also cause electrical outages by gnawing on underground power cables (Bollengier 1994). Yellow-bellied marmots may excavate burrows under buildings, gnaw through a floor, and cause considerable disarray in a building. Usually, burrowing under a building is more of a nuisance because of odors from feces and rotting vegetation in the burrow than an economic liability.

Although marmots carry diseases, transmission of disease from marmots to humans or their domesticated animals is of little or no importance. For example, leptospirosis in woodchucks in New York was not associated with bovine leptospirosis (Fleming 1979).

Management. Control of marmots falls into three categories: exclusion, removal, and killing. A fence about 1 m high that extends 23–30 cm in the ground and includes modifications that prevent marmots from climbing it is useful for home gardens and obviates the necessity of ongoing removal or killing of marmots (Bollengier 1994). Repellents were tried on several common garden crops (Swihart and Conover 1991). Consumption of acorn and zucchini squash treated with Hot Sauce Animal Repellent was slightly reduced, but more than two thirds of the foliage was consumed. Over an 11-day period, woodchucks gradually developed an aversion to tomatoes treated with emetine dihydrochloride, but the woodchucks sampled the vegetation daily and lost the aversion. Consumption of romaine lettuce was not affected by Hinder, Cygon, or Sevin. In general, repellents have minimal success and are most effective on plant species of lower palatability. Gnawing of fruit trees occurs primarily during the spring in Connecticut and is associated with scent-marking behavior. Providing hardwood stakes as alternative scent-marking substrates delayed the onset of damage, but did not reduce it (Swihart 1991b). Spraying bobcat urine weekly on the lower 75 cm of the fruit trees near woodchuck burrows greatly reduced the number of trees damaged and the extent of the damage. A single application of thiram was ineffective in reducing damage.

Marmots may be removed by livetrapping. Traps should be placed in a runway or at the opening of the burrow and baited with apple slices, lettuce, peanut butter, peanuts, or soybeans (Thompson 1979a). Traps can be set at any time of day and left open overnight (Davis 1982). During the hot part of the day, traps should be closed or checked hourly to prevent heat death. Captured animals may be released into suitable habitat where damage is not a problem.

Marmots may be killed with poisons, shooting, or trapping consistent with local regulations on such actions. Cartridges of carbon monoxide may be placed in a burrow and the entrances sealed. Bollengier (1994) provided detailed directions for this procedure. Other procedures, such as use of aluminum phosphide tablets, require a certified pesticide applicator. A cartridge with poisonous gas is highly effective. When 165 burrows were treated on a 40-ha area, all woodchucks were killed (de Vos and Merrill 1957). However, within 1 month, the area was repopulated. Ploughing the area was more effective; after 2 months, none of the burrows that were ploughed were opened up. Woodchucks rapidly recolonize a depauperated area (Davis 1962). General broad-scale removal may be ineffective and control should focus on individual woodchucks that are doing damage. Shooting may keep a population at an acceptably low level, but should not be used where safety is a concern. In general, control measures should not be initiated without consulting with the proper local authorities. Body-gripping kill traps kill the animal quickly, but should not be used where domestic or other wild animals might be captured. Use of body-gripping kill traps may be limited by state or local laws; local authorities should be contacted (Bollengier 1994).

Although management traditionally focuses on eliminating or reducing marmot populations, management in some areas is concerned with increasing populations or recolonizing areas where they have died out (Bibikow 1996). Since 1970, the Vancouver Island marmot has declined in numbers. Natural subalpine meadows are small and scattered. Recently, *M. vancouverensis* colonized logged habitats within five adjacent watersheds on south-central Vancouver Island (Bryant and Janz 1996). The initial colonization of logged habitats and ski runs increased marmot populations (Bryant 1998), but the population decreased to about 40 animals by 2000 (A. A. Bryant, Marmot Recovery Foundation, pers. commun., 2000). Persistence of marmots at natural sites is higher than at logged sites (65% vs. 48%). Females live longer on natural sites; maximum age was 9 years versus 5 years on logged sites. No adult female inhabiting a clearcut weaned more than one litter; 5 of 14 females in natural habitats produced at least two litters during the 9 years of the study (Bryant 1996). The logged sites may act as sinks, which capture dispersing marmots and prevent them from recolonizing natural habitats. Reasons for the rapid decline and extinctions on logged and natural habitats are unclear, but disease and predation, both of which may have been enhanced by human activities, are implicated (Bryant 1997). A recovery plan was developed by the Vancouver Island Marmot Recovery Team in 1994 and updated in 2000. An international workshop made recommendations for conserving the species (Elner 2000) and a captive breeding program was established in zoos in Toronto and Calgary (Canadian Wildlife Service 1999–2000).

RESEARCH NEEDS

The study of marmot biology has progressed considerably since 1980. However, little is known about the population and social biology of *M. broweri*. Other species require long-term research to determine the role of kinship in social organization, the relationship between life-table statistics and social organization, the effect of weather and climate change on survival and reproduction, and the factors that contribute to lifetime reproductive success in males and females. Other areas deserving investigation include the basis and importance of geographic variation, the role of nutrition in food choice and habitat selection, the extent and mechanism of reproductive suppression, the function of alarm calling (warn kin, talking to predators ?) and the reason marmots frequently do not call, the significance of species-specific alarm calls, the importance of parasites and disease and the genetic and physiological mechanisms marmots use to combat them, and the annual profiles of hormones and their role in dominance relationships, reproductive and territorial behavior, and environmental stress.

Marmot body size is diverse. Little is known of the role of body size in the allocation of energy to maintenance during torpor or to euthermy in the burrow before emergence, or about the relationship between social organization and energetic efficiency and social thermoregulation during hibernation. Because marmots have low reproductive rates, it would be useful to know the energetic basis of reproductive skipping and how that might be managed to increase reproductive output in declining marmot populations.

Most economic problems appear to be local and are best managed at that level with the help of local authorities. Development of effective repellents could simplify control by eliminating the need for trapping and/or killing.

ACKNOWLEDGMENTS

The author thanks Dirk Van Vuren for providing materials on control and Jana Ross and Berry Clemens for typing the manuscript.

LITERATURE CITED

Albert, T. F., and J. A. Panuska. 1978. Regional heterothermy and cardiovascular responses during induced hypothermia in non-hibernated and hibernated woodchucks, *Marmota monax*. Comparative Biochemistry and Physiology 60A:1–6.

Albert, T. F., and J. A. Panuska. 1981. Comparison of complete and incomplete arousals from hibernation in a woodchuck, *Marmota monax*. Pages 316–24 *in* J. A. Chapman and D. Purseley, eds. Proceedings of the worldwide furbearer conference, Vol. 1. Frostburg, Md.

Andersen, D. C., and D. W. Johns. 1977. Predation by badger on yellow-bellied marmot in Colorado. Southwestern Naturalist 22:283–84.

Andersen, D. C., K. B. Armitage, and R. S. Hoffmann. 1976. Socioecology of marmots: Female reproductive strategies. Ecology 57:552–60.

Anderson, R. M. 1934. Notes on the distribution of the hoary marmots. Canadian Field-Naturalist 48:61–63.

Anthony, M. 1962. Activity and behavior of the woodchuck in southern Illinois. Occasional Papers of the C. C. Adams Center for Ecological Studies 6:1–25.

Armitage, K. B. 1961. Frequency of melanism in the golden-mantled marmot. Journal of Mammalogy 42:100–101.

Armitage, K. B. 1962. Social behaviour of a colony of the yellow-bellied marmot (*Marmota flaviventris*). Animal Behaviour 10:318–31.

Armitage, K. B. 1965. Vernal behaviour of the yellow-bellied marmot (*Marmota flaviventris*). Animal Behaviour 13:59–68.

Armitage, K. B. 1973. Population changes and social behavior following colonization by the yellow-bellied marmot. Journal of Mammalogy 54:842–54.

Armitage, K. B. 1974. Male behaviour and territoriality in the yellow-bellied marmot. Journal of Zoology (London) 172:233–65.
Armitage, K. B. 1975. Social behavior and population dynamics of marmots. Oikos 26:341–54.
Armitage, K. B. 1976. Scent marking by yellow-bellied marmots. Journal of Mammalogy 57:583–84.
Armitage, K. B. 1977. Social variety in the yellow-bellied marmot: A population–behavioural system. Animal Behaviour 25:585–93.
Armitage, K. B. 1979a. Food selectivity by yellow-bellied marmots. Journal of Mammalogy 60:628–29.
Armitage, K. B. 1979b. Cannibalism among yellow-bellied marmots. Journal of Mammalogy 60:205–7.
Armitage, K. B. 1981. Sociality as a life history tactic of ground squirrels. Oecologia 48:36–49.
Armitage, K. B. 1982. Marmots and coyotes: Behavior of prey and predator. Journal of Mammalogy 63:503–5.
Armitage, K. B. 1983. Hematological values for free-ranging yellow-bellied marmots. Comparative Biochemistry and Physiology 74A:89–93.
Armitage, K. B. 1984. Recruitment in yellow-bellied marmot populations: Kinship, philopatry, and individual variability. Pages 377–403 *in* J. O. Murie and G. R. Michener, eds. Biology of ground-dwelling squirrels. University of Nebraska Press, Lincoln.
Armitage, K. B. 1986a. Marmot polygyny revisited: Determinants of male and female reproductive strategies. Pages 303–31 *in* D. S. Rubenstein and R. W. Wrangham, eds. Ecological aspects of social evolution. Princeton University Press, Princeton, NJ.
Armitage, K. B. 1986b. Individuality, social behavior, and reproductive success in yellow-bellied marmots. Ecology 67:1186–93.
Armitage, K. B. 1986c. Individual differences in the behavior of juvenile yellow-bellied marmots. Behavioral Ecology and Sociobiology 18:419–24.
Armitage, K. B. 1987. Do female yellow-bellied marmots adjust the sex ratios of their offspring? American Naturalist 129:501–19.
Armitage, K. B. 1988. Resources and social organization of ground-dwelling squirrels. Pages 131–55 *in* C. N. Slobodchikoff, ed. The ecology of social behavior. Academic Press, New York.
Armitage, K. B. 1989. The function of kin discrimination. Ethology Ecology and Evolution 1:111–21.
Armitage, K. B. 1991a. Factors affecting corticosteroid concentrations in yellow-bellied marmots. Comparative Biochemistry and Physiology 98A:47–54.
Armitage, K. B. 1991b. Social and population dynamics of yellow-bellied marmots: Results from long-term research. Annual Review of Ecology and Systematics 22:379–407.
Armitage, K. B. 1994. Unusual mortality in a yellow-bellied marmot population. Pages 5–13 *in* V. Rumiantsev, ed. Actual problems of marmots investigation. ABF, Moscow.
Armitage, K. B. 1996a. Seasonal mass gain in yellow-bellied marmots. Pages 223–26 *in* M. LeBerre, R. Ramousse, and L. LeGuelte, eds. Biodiversity in marmots. International Marmot Network, Lyon.
Armitage, K. B. 1996b. Resource sharing and kinship in yellow-bellied marmots. Pages 129–34 *in* M. LeBerre, R. Ramousse, and L. LeGuelte, eds. Biodiversity in marmots. International Marmot Network, Lyon.
Armitage, K. B. 1996c. Social dynamics, kinship, and population dynamics of marmots. Pages 113–28 *in* M. LeBerre, R. Ramousse, and L. LeGuelte, eds. Biodiversity in marmots. International Marmot Network, Lyon.
Armitage, K. B. 1998. Reproductive strategies of yellow-bellied marmots: Energy conservation and differences between the sexes. Journal of Mammalogy 79:385–93.
Armitage, K. B. 1999. Evolution of sociality in marmots. Journal of Mammalogy 80:1–10.
Armitage, K. B. 2000. The evolution, ecology, and systematics of marmots. Oecologia Montana 9:1–18.
Armitage, K. B., and D. T. Blumstein. 2002. Body-mass diversity in marmots. Pages 22–40 *in* K. B. Armitage and V. Yu. Rumiantsev, eds. Holarctic marmots as a factor of biodiversity. ABF, Moscow.
Armitage, K. B., D. T. Blumstein, and B. C. Woods. 2003. Energetics of hibernating yellow-bellied marmots (*Marmota flaviventris*). Comparative Biochemistry and Physiology 103A:729–737.
Armitage, K. B., and J. F. Downhower. 1974. Demography of yellow-bellied marmot populations. Ecology 55:1233–45.
Armitage, K. B., J. F. Downhower, and G. E. Svendsen. 1976. Seasonal changes in weights of marmots. American Midland Naturalist 96:36–51.
Armitage, K. B., and G. E. Gurri-Glass. 1994. Communal nesting in yellow-bellied marmots. Pages 14–26 *in* V. Rumiantsev, ed. Actual problems of marmots investigation. ABF, Moscow.
Armitage, K. B., and D. W. Johns. 1982. Kinship, reproductive strategies and social dynamics of yellow-bellied marmots. Behavioral Ecology and Sociobiology 11:55–63.
Armitage, K. B., J. C. Melcher, and J. M. Ward, Jr. 1990. Oxygen consumption and body temperature in yellow-bellied marmot populations from montane–mesic and lowland–xeric environments. Journal of Comparative Physiology B 160:491–502.
Armitage, K. B., and V. Yu. Rumiantsev, eds. 2002. Holarctic marmots as a factor of biodiversity. ABF, Moscow.
Armitage, K. B., and C. M. Salsbury. 1992. Factors affecting oxygen consumption in wild-caught yellow-bellied marmots (*Marmota flaviventris*). Comparative Biochemistry and Physiology 103A:729–37.
Armitage, K. B., and C. M. Salsbury. 1993. The effect of molt on oxygen consumption of yellow-bellied marmots (*Marmota flaviventris*). Comparative Biochemistry and Physiology 106A:667–70.
Armitage, K. B., C. M. Salsbury, E. L. Barthelmess, R. C. Gray, and A. Kovach. 1996. Population time budget for the yellow-bellied marmot. Ethology Ecology and Evolution 8:67–95.
Armitage, K. B., and O. A. Schwartz. 2000. Social enhancement of fitness in yellow-bellied marmots. Proceedings of the National Academy of Sciences of the USA 97:12149–52.
Armitage, K. B., B. C. Woods, and C. M. Salsbury. 2000. Energetics of hibernation in woodchucks (*Marmota monax*). Pages 73–80 *in* G. Heldmaier and M. Klingenspor, eds. Life in the cold. Springer-Verlag, Berlin.
Armitage, K. B., and K. E. Wynne-Edwards. 2002. Progesterone concentrations in wild-caught yellow-bellied marmots. Pages 41–47 *in* K. B. Armitage and V. Yu. Rumiantsev, eds. Holarctic marmots as a factor of biodiversity. ABF, Moscow.
Arnold, W. 1993. Energetics of social hibernation. Pages 65–80 *in* C. Carey, G. L. Florant, B. A. Wunder, and B. Horwitz, eds. Life in the cold. Westview Press, Boulder, CO.
Arsenault, J. R., and R. F. Romig. 1985. Plants eaten by woodchucks in three northeast Pennsylvania counties. Proceedings of the Pennsylvania Academy of Sciences 59:131–34.
Bailey, E. D. 1965a. Seasonal changes in metabolic activity of nonhibernating woodchucks. Canadian Journal of Zoology 43:905–9.
Bailey, E. D. 1965b. The influence of social interaction and season on weight change in woodchucks. Journal of Mammalogy 46:438–45.
Bailey, E. D., and D. E. Davis. 1965. The utilization of body fat during hibernation in woodchucks. Canadian Journal of Zoology 43:701–7.
Baldwin, B. H., B. C. Tennant, T. J. Reimers, R. G. Cowan, and P. W. Concannon. 1985. Circannual changes in serum testosterone concentrations of adult and yearling woodchucks (*Marmota monax*). Biology of Reproduction 32:804–12.
Barash, D. P. 1973a. The social biology of the Olympic marmot. Animal Behaviour Monographs 6:173–245.
Barash, D. P. 1973b. Social variety in the yellow-bellied marmot (*Marmota flaviventris*). Animal Behaviour 21:579–84.
Barash, D. P. 1974a. The social behaviour of the hoary marmot (*Marmota caligata*). Animal Behaviour 22:256–61.
Barash, D. P. 1974b. Mother–infant relations in captive woodchucks (*Marmota monax*). Animal Behaviour 22:446–48.
Barash, D. P. 1974c. The evolution of marmot societies: A general theory. Science 185:415–20.
Barash, D. P. 1975. Ecology of paternal behavior in the hoary marmot (*Marmota caligata*): An evolutionary interpretation. Journal of Mammalogy 56:613–18.
Barash, D. P. 1976. Pre-hibernation behavior of free-living hoary marmots, *Marmota caligata*. Journal of Mammalogy 57:182–85.
Barash, D. P. 1980. The influence of reproductive status on foraging by hoary marmots (*Marmota caligata*). Behavioral Ecology and Sociobiology 7:201–5.
Barash, D. P. 1981. Mate guarding and gallivanting by male hoary marmots (*Marmota caligata*). Behavioral Ecology and Sociobiology 9:187–93.
Barash, D. P. 1989. Marmots: Social behavior and ecology. Stanford University Press, Stanford, CA.
Barnes, B. M. 1996. Relationships between hibernation and reproduction in male ground squirrels. Pages 71–80 *in* F. Geiser, A. J. Hulbert, and S. C. Nicol, eds. Adaptations to the cold. University of New England Press, Armidale, Australia.
Bassano, B. 1996. Sanitary problems related to marmots–other animals cohabitation in mountain areas. Pages 75–88 *in* M. LeBerre, R. Ramousse, and L. LeGuelte, eds. Biodiversity in marmots. International Marmot Network, Lyon.

Bassano, B., P. Durio, V. Gallo Orsi, and E. Macchi, eds. 1992. First international symposium on alpine marmot and on genus *Marmota*. Torino, Italy.

Bee, J. W., and E. R. Hall. 1956. Mammals of northern Alaska on the Arctic Slope. University of Kansas Museum of Natural History Miscellaneous Publications 8:1–309.

Benedict, F. G., and R. C. Lee. 1938. Hibernation and marmot physiology. Carnegie Institution of Washington, Washington, DC.

Bibikov, D. I. 1992. Marmots are zoonosis provoking carriers. Pages 25–29 *in* B. Bassano, P. Durio, U. Gallo Orsi, and E. Macchi, eds. First international symposium on alpine marmot and on genus *Marmota*. Torino, Italy.

Bibikow, D. I. 1996. Die Murmeltiere der Welt. Westarp Wissenschaften, Magdeburg, Germany.

Black, C. C. 1972. Holarctic evolution and dispersal of squirrels (Rodentia: Sciuridae). Evolutionary Biology 6:305–22.

Blumberg, B. S., A. C. Allison, and B. Garry. 1960. The haptoglobins, hemoglobins and serum proteins of the Alaskan fur seal, ground squirrel and marmot. Journal of Cellular and Comparative Physiology 55:61–71.

Blumstein, D. T. 1999. Alarm calling in three species of marmots. Behaviour 136:731–57.

Blumstein, D. T., and K. B. Armitage. 1997a. Alarm calling in yellow-bellied marmots: I. The meaning of situationally variable alarm calls. Animal Behaviour 53:143–71.

Blumstein, D. T., and K. B. Armitage. 1997b. Does sociality drive the evolution of communicative complexity? A comparative test with ground-dwelling sciurid alarm calls. American Naturalist 150:179–200.

Blumstein, D. T., and K. B. Armitage. 1998. Life history consequences of social complexity: A comparative study of ground-dwelling sciurids. Behavioral Ecology 9:8–19.

Blumstein, D. T., and K. B. Armitage. 1999. Cooperative breeding in marmots. Oikos 84:369–82.

Blumstein, D. T., J. Steinmetz, K. B. Armitage, and J. C. Daniel. 1997. Alarm calling in yellow-bellied marmots: II. The importance of direct fitness. Animal Behaviour 53:173–84.

Blumstein, D. T., J. C. Daniel, and W. Arnold. 2002. Survivorship of golden marmot (*Marmota caudata aurea*) in Pakistan. Pages 82–85 *in* K. B. Armitage and V. Yu. Rumiantsev, eds. Holarctic marmots as a factor of biodiversity. ABF, Moscow.

Boggs, D. F., and G. F. Birchard. 1989. Cardiorespiratory responses of the woodchuck and porcupine to CO_2 and hypoxia. Journal of Comparative Physiology B 159:641–48.

Boggs, D. F., D. L. Kilgore, Jr., and G. F. Birchard. 1984. Respiratory physiology of burrowing mammals and birds. Comparative Biochemistry and Physiology 77A:1–7.

Bollengier, R. M., Jr., 1994. Woodchucks. Pages B183–87 *in* S. E. Hygnstrom, R. M. Timm, and G. E. Larson, eds. Prevention and control of wildlife damage. Great Plains Agricultural Council Wildlife Committee, Lincoln, NB.

Bonenfant, M., and D. L. Kramer. 1996. The influence of distance to burrow on flight initiation distance in the woodchuck, *Marmota monax*. Behavioral Ecology 7:299–303.

Brady, K. M., and K. B. Armitage. 1999. Scent-marking in the yellow-bellied marmot (*Marmota flaviventris*). Ethology Ecology and Evolution 11:35–47.

Brody, A. K., and K. B. Armitage. 1985. The effects of adult removal on dispersal of yearling yellow-bellied marmots. Canadian Journal of Zoology 63:2560–64.

Brody, A. K., and J. C. Melcher. 1985. Infanticide in yellow-bellied marmots. Animal Behaviour 33:673–74.

Bronson, F. H. 1962. Daily and seasonal activity patterns in woodchucks. Journal of Mammalogy 43:425–27.

Bronson, F. H. 1963. Some correlates of interaction rate in natural populations of woodchucks. Ecology 44:637–43.

Bronson, F. H. 1964. Agonistic behaviour in woodchucks. Animal Behaviour 12:470–78.

Bryant, A. A. 1990. Genetic variability and minimum viable populations in the Vancouver Island marmot (*Marmota vancouverensis*). M. S. Thesis, University of Calgary, Alberta, Canada.

Bryant, A. A. 1996. Reproduction and persistence of Vancouver Island marmots (*Marmota vancouverensis*) in natural and logged habitats. Canadian Journal of Zoology 74:678–87.

Bryant, A. A. 1997. Vancouver Island marmot *Marmota vancouverensis* (Status report on species at risk in Canada). Committee on the Status of Endangered Wildlife in Canada.

Bryant, A. A. 1998. Metapopulation ecology of Vancouver Island marmots (*Marmota vancouverensis*). Ph.D. Dissertation, University of Victoria, Victoria, Canada.

Bryant, A. A., and D. A. Blood. 1999. Vancouver Island marmot (Wildlife at risk in British Columbia). Ministry of Environment, Lands and Parks, Victoria, Canada.

Bryant, A. A., and D. W. Janz. 1996. Distribution and abundance of Vancouver Island marmots (*Marmota vancouverensis*). Canadian Journal of Zoology 74:667–77.

Bullard, R. W., C. Broumand, and F. R. Meyer. 1966. Blood characteristics and volume in two rodents native to high altitude. Journal of Applied Physiology 21:994–98.

Burlington, R. F., J. A.Vogel, T. M. Burton, and I. A. Salkovitz. 1971. Cardiac output and regional blood flow in hypoxic woodchucks. American Journal of Physiology 220:1565–68.

Canadian Wildlife Service. 1999–2000. Recovery of nationally endangered wildlife. Environment Canada, Canadian Wildlife Service.

Carey, H. V. 1985a. The use of foraging areas by yellow-bellied marmots. Oikos 44:273–79.

Carey, H. V. 1985b. Nutritional ecology of yellow-bellied marmots in the White Mountains of California. Holarctic Ecology 8:259–64.

Carey, H. V., and P. Moore. 1986. Foraging and predation risk in yellow-bellied marmots. American Midland Naturalist 116:267–75.

Christian, J. J. 1962. Seasonal changes in the adrenal glands of woodchucks. Endocrinology 71:431–47.

Christian, J. J., E. Steinberger, and T. D. McKinney. 1972. Annual cycle of spermatogenesis and testis morphology in woodchucks. Journal of Mammalogy 53:708–16.

Collopy, M. W. 1983. A comparison of direct observations and collections of prey remains in determining the diet of golden eagles. Journal of Wildlife Management 47:360–68.

Concannon, P., B. Baldwin, J. Lawless, W. Hornbuckle, and B. Tennant. 1983. Corpora lutea of pregnancy and elevated serum progesterone during pregnancy and postpartum anestrus in woodchucks (*Marmota monax*). Biology of Reproduction 29:1128–34.

Concannon, P., B. Baldwin, and B. Tennant. 1984. Serum progesterone profiles and corpora lutea of pregnant, postpartum, barren, and isolated females in a laboratory colony of woodchucks (*Marmota monax*). Biology of Reproduction 30:945–51.

Concannon, P., B. Baldwin, P. Roberts, and B. Tennant. 1990. Endocrine correlates of hibernation-independent gonadal recrudescence and the limited late-winter breeding season in woodchucks, *Marmota monax*. Journal of Experimental Zoology Supplement 4:203–6.

Concannon, P., J. E. Parks, P. J. Roberts, and B. C. Tennant. 1992. Persistent free-running circannual reproductive cycles during prolonged exposure to a 12L:12D photoperiod in laboratory woodchucks (*Marmota monax*). Laboratory Animal Science 42:382–91.

Concannon, P., P. Roberts, B. Baldwin, H. Erb, and B. Tennant. 1993. Alteration of growth, advancement of puberty, and season-appropriate circannual breeding during 28 months of photoperiod reversal in woodchucks (*Marmota monax*). Biology of Reproduction 48:1057–70.

Couch, L. K. 1930. Notes on the pallid yellow-bellied marmot. Murrelet 11:3–7.

Davis, D. E. 1962. The potential harvest of woodchucks. Journal of Wildlife Management 26:144–49.

Davis, D. E. 1964. Evaluation of characters for determining age of woodchucks. Journal of Wildlife Management 28:9–15.

Davis, D. E. 1966. The moult of woodchucks (*Marmota monax*). Mammalia 30:640–44.

Davis, D. E. 1967a. The role of environmental factors in hibernation of woodchucks (*Marmota monax*). Ecology 48:683–89.

Davis, D. E. 1967b. The annual rhythm of fat deposition in woodchucks (*Marmota monax*). Physiological Zoology 40:391–402.

Davis, D. E. 1970. Failure of schedule of torpor to alter annual rhythm of appetite of woodchucks (*Marmota monax*). Mammalia 34:542–44.

Davis, D. E. 1976. Hibernation and circannual rhythms of food consumption in marmots and ground squirrels. Quarterly Review of Biology 51:477–514.

Davis, D. E. 1977. Role of ambient temperature in emergence of woodchucks (*Marmota monax*) from hibernation. American Midland Naturalist 97:224–29.

Davis, D. E. 1982. Woodchucks. Page 147 *in* D. E. Davis, ed. CRC handbook of census methods for terrestrial vertebrates. CRC Press, Boca Raton, FL.

Davis, D. E., and E. P. Finnie. 1975. Entrainment of circannual rhythm in weight of woodchucks. Journal of Mammalogy 56:199–203.

Davis, D. E., and J. Ludwig. 1981. Mechanism for decline in a woodchuck population. Journal of Wildlife Management 45:658–68.

Davis, D. E., and R. D. McCarthy. 1965. Major fatty acids in blood serum and adipose tissue of woodchucks (*Marmota monax*). BioScience 15:749–50.

Davis, D. E., J. J. Christian, and F. Bronson. 1964. Effect of exploitation on birth, mortality, and movement rates in a woodchuck population. Journal of Wildlife Management 28:1–9.

Del Moral, R. 1984. The impact of the Olympic marmot on subalpine vegetation. American Journal of Botany 71:1228–36.

de Vos, A., and D. I. Gillespie. 1960. A study of woodchucks on an Ontario farm. Canadian Field-Naturalist 74:130–45.

de Vos, A., and H. A. Merrill. 1957. Results of a woodchuck control experiment. Journal of Wildlife Management 21:454–56.

Downhower, J. F., and K. B. Armitage. 1981. Dispersal of yearling yellow-bellied marmots (*Marmota flaviventris*). Animal Behaviour 29:1064–69.

Drucker, P. 1950. Northwest Coast. Anthropological Record 9:157–294.

Eadie, W. R. 1954. Animal control in field, farm, and forest. Macmillan, New York.

Elner, R. W. 2000. Proceedings: International workshop for the conservation of Vancouver Island marmot (Technical Report Series No. 346). Canadian Wildlife Service, Pacific and Yukon Region, British Columbia.

English, E. I., and M. A. Bowers. 1994. Vegetational gradients and proximity to woodchuck (*Marmota monax*) burrows in an old field. Journal of Mammalogy 75:775–80.

Fall, M. W. 1971. Seasonal variation in the food consumption of woodchucks (*Marmota monax*). Journal of Mammalogy 52:370–75.

Ferron, J. 1996. How do woodchucks (*Marmota monax*) cope with harsh winter conditions? Journal of Mammalogy 77:412–16.

Ferron, J., and J.-P. Ouellet. 1989. Temporal and intersexual variations in the use of space with regard to social organization in the woodchuck (*Marmota monax*). Canadian Journal of Zoology 67:1642–49.

Ferron, J., and J.-P. Ouellet. 1991. Physical and behavioral postnatal development of woodchucks (*Marmota monax*). Canadian Journal of Zoology 69:1040–47.

Fishbein, D. B., A. J. Belotto, R. E. Pacer, J. S. Smith, W. G. Winkler, S. R. Jenkins, and K. M. Porter. 1986. Rabies in rodents and lagomorphs in the United States, 1971–1984: Increased cases in the woodchuck (*Marmota monax*) in mid-Atlantic states. Journal of Wildlife Diseases 22:151–55.

Fleming, W. J. 1979. Serologic evidence of leptospirosis in woodchucks (*Marmota monax*) in central New York state. Journal of Wildlife Diseases 15:245–51.

Florant, G. L. 1981. Plasma catecholamine and cortisol concentrations in euthermic and hibernating woodchucks (*M. monax*). Acta Universitatis Carolinae Biologica 1979:249–52.

Florant, G. L. 1998. Lipid metabolism in hibernators: The importance of essential fatty acids. American Zoologist 38:331–40.

Florant, G. L., and W. A. Bauman. 1985. Seasonal variations in carbohydrate metabolism in mammalian hibernators: Insulin and body weight changes. Advances in Obesity Research 4:57–64.

Florant, G. L., and M. R. C. Greenwood. 1986. Seasonal variations in pancreatic function in marmots: The role of pancreatic hormones and lipoprotein lipase in fat deposition. Pages 273–80 *in* H. C. Heller, X. J. Musacchia, and L. C. Wang, eds. Living in the cold: Physiological and biochemical adaptations. Elsevier, New York.

Florant, G. L., and H. C. Heller. 1977. CNS regulation of body temperature in euthermic and hibernating marmots (*Marmota flaviventris*). American Journal of Physiology 232:R203–8.

Florant, G. L., and E. D. Weitzman. 1980. Diurnal and episodic pattern of plasma cortisol during fall and spring in young and old woodchucks (*Marmota monax*). Comparative Biochemistry and Physiology 66A:575–81.

Florant, G. L., R. Hoo-Paris, C. Castex, W. A. Bauman, and B. C. Sutter. 1986. Pancreatic A and B cell stimulation in euthermic and hibernating marmots (*Marmota flaviventris*): Effects of glucose and arginine administration. Journal of Comparative Physiology B 156:309–14.

Florant, G. L., L. C. Nuttle, D. E. Mullinex, and D. A. Rintoul. 1990. Plasma and white adipose tissue lipid composition in marmots. American Journal of Physiology 258:R1123–31.

Florant, G. L., R. D. Richardson, S. Mahan, L. Singer, and S. C. Woods. 1991. Seasonal changes in CSF insulin levels in marmots: Insulin may not be a satiety signal for fasting in winter. American Journal of Physiology 260:R712–16.

Florant, G. L., L. Hester, S. Ameenuddin, and D. A. Rintoul. 1993. The effect of a low essential fatty acid diet on hibernation in marmots. American Journal of Physiology 264:R747–53.

Frank, C. L., E. S. Dierenfeld, and K. B. Storey. 1998. The relationship between lipid peroxidation, hibernation, and food selection in mammals. American Zoologist 38:341–49.

Frase, B. A., and K. B. Armitage. 1984. Foraging patterns of yellow-bellied marmots: Role of kinship and individual variability. Behavioral Ecology and Sociobiology 16:1–10.

Frase, B. A., and K. B. Armitage. 1989. Yellow-bellied marmots are generalist herbivores. Ethology Ecology and Evolution 1:353–66.

Frase, B. A., and R. S. Hoffmann. 1980. *Marmota flaviventris*. Mammalian Species 135:1–8.

Fraser, D. 1979. Aquatic feeding by a woodchuck. Canadian Field-Naturalist 93:309–10.

French, A. R. 1986. Patterns of thermoregulation during hibernation. Pages 393–402 *in* H. C. Heller, X. J. Musacchia, and L.C.H. Wang, eds. Living in the cold: Physiological and biochemical adaptations. Elsevier, New York.

French, A. R. 1990. Age-class differences in the pattern of hibernation in yellow-bellied marmots, *Marmota flaviventris*. Oecologia 82:93–96.

Frink, R., P. P. Krupp, and R. A. Young. 1977. Seasonal variation in the morphology of thyroid parafollicular (C) cells in the woodchuck, *Marmota monax*: A light and electron microscopic study. Anatomical Record 189:397–412.

Fryxell, F. M. 1928. Melanism among the marmots of the Teton Range, Wyoming. Journal of Mammalogy 9:336–37.

Gabel, J. R. 1961. Protozoa of the mountain marmot, *Marmota flaviventris* Audubon & Bachman, 1841. Transactions of the American Microscopical Society 80:43–53.

Garrott, R. A., and D. A. Jenni. 1978. Arboreal behavior of yellow-bellied marmots. Journal of Mammalogy 59:433–34.

Gianini, C. A. 1925. Tree-climbing and insect-eating woodchucks. Journal of Mammalogy 6:281–82.

Goodrich, C. A. 1973. Acid–base balance in euthermic and hibernating marmots. American Journal of Physiology 224:1185–89.

Grizzell, R. A. 1955. A study of the southern woodchuck, *Marmota monax monax*. American Midland Naturalist 53:257–93.

Hafner, D. J. 1984. Evolutionary relationships of the Nearctic Sciuridae. Pages 3–23 *in* J. O. Murie and G. R. Michener, eds. The biology of ground-dwelling squirrels. University of Nebraska Press, Lincoln.

Hall, E. R. 1981. The mammals of North America, 2nd ed. Wiley Interscience, New York.

Hall, E. R., and Kelson, K. R. 1959. The mammals of North America, Ronald Press, New York.

Hall, F. G. 1965. Hemoglobin and oxygen affinities in seven species of Sciuridae. Science 148:1350–51.

Hamilton, W. J., Jr. 1934. The life history of the rufescent woodchuck, *Marmota monax rufescens* Howell. Annals of Carnegie Museum 23:85–178.

Hansen, R. M. 1975. Foods of the hoary marmot on Kenai Peninsula, Alaska. American Midland Naturalist 94:348–53.

Harkness, D. R., S. Roth, and P. Goldman. 1974. Studies on the red blood cell oxygen affinity and 2,3-diphosphoglyceric acid in the hibernating woodchuck (*Marmota monax*). Comparative Biochemistry and Physiology 48A:591–99.

Hayes, S. R. 1976. Daily activity and body temperature of the southern woodchuck, *Marmota monax monax*, in northwestern Arkansas. Journal of Mammalogy 57:291–99.

Heard, D. C. 1977. The behaviour of Vancouver Island marmots, *Marmota vancouverensis*. M.S. Thesis, University of British Columbia, Vancouver, Canada.

Hebert, P., and C. Barrette. 1989. Experimental demonstration that scent marking can predict dominance in the woodchuck, *Marmota monax*. Canadian Journal of Zoology 67:575–78.

Hebert, P., and J. Prescott. 1983. Etude de marquage olfactif chez la marmotte commune (*Marmota monax*) en captivité. Canadian Journal of Zoology 61:1720–25.

Henderson, J. A., and F. F. Gilbert. 1978. Distribution and density of woodchuck burrow systems in relation to land-use. Canadian Field-Naturalist 92:128–36.

Hill, V. L., and G. L. Florant. 1999. Patterns of fatty acid composition in free-ranging yellow-bellied marmots (*Marmota flaviventris*) and their diet. Canadian Journal of Zoology 77:1494–1503.

Hock, R. J. 1967. Seasonal hematologic changes in high altitude hibernators. Federation Proceedings 26:719.

Hock, R. J. 1969. Thermoregulatory variations of high-altitude hibernators in relation to ambient temperature, season, and hibernation. Federation Proceedings 28:1047–52.

Hoffmann, R. S., J. W. Koeppl, and C. F. Nadler. 1979. The relationships of the Amphiberingian marmots (Mammalia: Sciuridae). Occasional Papers of the Museum of Natural History, University of Kansas 83:1–56.

Hoffmann, R. S., C. G. Anderson, R. W. Thorington, and L. Heaney. 1993. Family Sciuridae. Pages 419–65 *in* D. E. Wilson and D. M. Reeder, eds. Mammal species of the world: A taxonomic and geographic reference, 2nd ed. Smithsonian Institution Press, Washington, DC.

Holmes, W. G. 1984a. The ecological basis of monogamy in Alaskan hoary marmots. Pages 250–74 *in* J. O. Murie and G. R. Michener, eds. The biology of ground-dwelling squirrels. University of Nebraska Press, Lincoln.

Holmes, W. G. 1984b. Predation risk and foraging behavior of the hoary marmot in Alaska. Behavioral Ecology and Sociobiology 15:293–301.

Howell, A. H. 1914. Ten new marmots from North America. Proceedings of the Biological Society of Washington 27:13–18.

Howell, A. H. 1915. Revision of the American marmots. North American Fauna 37:1–80.

Hoyt, S. Y., and S. F. Hoyt. 1950. Gestation period of the woodchuck, *Marmota monax*. Journal of Mammalogy 31:454.

Inouye, D. W., B. Barr, K. B. Armitage, and B. D. Inouye. 2000. Climate change is affecting altitudinal migrants and hibernating species. Proceedings of the National Academy of Sciences of the USA 97:1630–33.

Jamieson, S. H., and K. B. Armitage. 1987. Sex differences in the play behavior of yearling yellow-bellied marmots. Ethology 74:237–53.

Johns, D. W., and K. B. Armitage. 1979. Behavioral ecology of alpine yellow-bellied marmots. Behavioral Ecology and Sociobiology 5:133–57.

Katbamna, B., C. Thodi, J. B. Senturia, and D. A. Metz. 1992. Auditory-evoked brainstem responses in the hibernating woodchuck (*Marmota monax*). Comparative Biochemistry and Physiology 102A:513–17.

Kilgore, D. L., Jr. and K. B. Armitage. 1978. Energetics of yellow-bellied marmot populations. Ecology 59:78–88.

Knight, R. L. and A. W. Erikson. 1978. Marmots as a food source of Golden Eagles along the Columbia River. Murrelet 59:28–30.

Kramer, D. L., and M. Bonenfant. 1997. Direction of predator approach and the decision to flee to a refuge. Animal Behaviour 54:289–95.

Kruckenhauser, L., W. Pinsker, E. Haring, and W. Arnold. 1998. Marmot phylogeny revisited: Molecular evidence for a diphyletic origin of sociality. Journal of Zoological Systematics and Evolutionary Research 37:49–56.

Krupp, P. P., R. A. Young, and R. Frink. 1977. The thyroid gland of the woodchuck, *Marmota monax*: A morphological study of seasonal variation in follicular cells. Anatomical Record 187:495–514.

Kwiecinski, G. G. 1998. *Marmota monax*. Mammalian Species 591:1–8.

Le Berre, M., R. Ramousse, and L. LeGuelte, eds. 1996. Biodiversity in marmots. International Marmot Network, Lyon.

Lee, D. S., and J. B. Funderburg. 1982. Marmots *Marmota monax* and allies. Pages 176–91 *in* J. A. Chapman and G. A. Feldhamer, eds. Wild mammals of North America. Johns Hopkins University Press, Baltimore.

Lenihan, C., and D. Van Vuren. 1996. Growth and survival of juvenile yellow-bellied marmots (*Marmota flaviventris*). Canadian Journal of Zoology 74:297–302.

Lloyd, J. A., J. J. Christian, D. E. Davis, and F. H. Bronson. 1964. Effects of altered social structure on adrenal weights and morphology in populations of woodchucks (*Marmota monax*). General and Comparative Endocrinology 4:271–76.

Lloyd, J. E. 1972. Vocalization in *Marmota monax*. Journal of Mammalogy 53:214–16.

Loibl, M. F. 1983. Social and non-social behaviors of arctic marmots, *Marmota broweri,* housed outdoors. M.S. Thesis, University of Maryland, College Park.

Lyman, C. P. 1958. Oxygen consumption, body temperature and heart rate of woodchucks entering hibernation. American Journal of Physiology 194:83–91.

Lyman, C. P., J. S. Willis, A. Malan, and L. C. H. Wang. 1982. Hibernation and torpor in mammals and birds. Academic Press, New York.

Marr, N. V., and R. L. Knight. 1983. Food habits of golden eagles in eastern Washington. Murrelet 64:73–77.

Martell, A. M., and R. J. Milko. 1986. Seasonal diets of Vancouver Island marmots, *Marmota vancouverensis*. Canadian Field-Naturalist 100:241–45.

McTaggart-Cowan, I. 1933. The British Columbian woodchuck *Marmota monax petrensis* Howell. Canadian Field-Naturalist 47:57.

Meier, P. T. 1991. Response of adult woodchucks (*Marmota monax*) to oral-gland scents. Journal of Mammalogy 72:622–24.

Meier, P. T. 1992. Social organization of woodchucks (*Marmota monax*). Behavioral Ecology and Sociobiology 31:393–400.

Mein, P. 1992. Taxonomy. Pages 6–12 *in* B. Bassano, P. Durio, V. Gallo Orsi, and E. Macchi, eds. First international symposium on alpine marmot and on genus *Marmota*. Torino, Italy.

Melcher, J. C. 1987. The influence of thermal energy exchange on the activity and energetics of yellow-bellied marmots. Ph.D. Dissertation, University of Kansas, Lawrence.

Melcher, J. C., K. B. Armitage, and W. P. Porter. 1989. Energy allocation by yellow-bellied marmots. Physiological Zoology 62:429–48.

Melcher, J. C., K. B. Armitage, and W. P. Porter. 1990. Thermal influences on the activity and energetics of yellow-bellied marmots (*Marmota flaviventris*). Physiological Zoology 63:803–20.

Merriam, H. G. 1963. Low frequency telemetric monitoring of woodchuck movements. Pages 155–71 *in* L. E. Slater, ed. Bio-Telemetry. Pergamon Press, New York.

Merriam, H. G. 1966. Temporal distribution of woodchuck interburrow movements. Journal of Mammalogy 47:103–10.

Merriam, H. G. 1971. Woodchuck burrow distribution and related movement patterns. Journal of Mammalogy 52:732–46.

Merriam, H. G., and A. Merriam. 1965. Vegetation zones around woodchuck burrows. Canadian Field-Naturalist 79:177–80.

Milko, R. J. 1984. Vegetation and foraging ecology of the Vancouver Island marmot (*Marmota vancouverensis*). M.S. Thesis, University of Victoria, Victoria, Canada.

Mossman, H. W., J. W. Lawlah, and J. A. Bradley. 1932. The male reproductive tract of the Sciuridae. American Journal of Anatomy 51:89–155.

Murie, A. 1944. The wolves of Mt. McKinley. U.S. Government Printing Office, Washington, DC.

Nagorsen, D. W. 1987. *Marmota vancouverensis*. Mammalian Species 270:1–5.

Nagorsen, D. W., G. Keddie, and T. Luszcz. 1996. Vancouver Island marmot bones from subalpine caves: Archaeological and biological significance (Occasional Paper Number 4). British Columbia Ministry of Environment, Lands and Parks, Victoria, Canada.

Nee, J. A. 1969. Reproduction in a population of yellow-bellied marmots (*Marmota flaviventris*). Journal of Mammalogy 50:756–65.

Nowicki, S., and K. B. Armitage. 1979. Behavior of juvenile yellow-bellied marmots: Play and social integration. Zeitschrift für Tierpsychologie 51:85–105.

Noyes, D. H., and W. G. Holmes. 1979. Behavioral responses of free-living hoary marmots to a model golden eagle. Journal of Mammalogy 60:408–11.

Ouellet, J.-P., and J. Ferron. 1988. Scent-marking behavior by woodchucks (*Marmota monax*). Journal of Mammalogy 69:365–68.

Pattie, D. L. 1967. Observations on an alpine population of yellow-bellied marmots (*Marmota flaviventris*). Northwest Science 41:96–102.

Petterson, J. R. 1992. Yellow-bellied marmot, *Marmota flaviventris*, predation on pikas, *Ochotona princeps*. Canadian Field-Naturalist 106:130–31.

Ramousse, R. 1997. Bibliographia marmotarum. International Marmot Network, Lyon.

Ramousse, R., and O. Giboulet. 2002. In the name of marmot and chamois: Value of trade use of their names. Pages 323–327 *in* K. B. Armitage and V. Yu. Rumiantsev, eds. Holarctic marmots as a factor of biodiversity. ABF, Moscow.

Rausch, R. L., and J. G. Bridgens. 1989. Structure and function of sudoriferous facial glands in Nearctic marmots, *Marmota* spp. (Rodentia: Sciuridae). Zoologischer Anzeiger 223:265–82.

Rausch, R. L., and V. R. Rausch. 1971. The somatic chromosomes of some North American marmots (Sciuridae), with remarks on the relationships of *Marmota broweri* Hall and Gilmore. Mammalia 35:85–101.

Rayor, L. S., and K. B. Armitage. 1991. Social behavior and space-use of young of ground-dwelling squirrel species with different levels of sociality. Ethology Ecology and Evolution 3:185–205.

Salsbury, C. M., and K. B. Armitage. 1994a. Resting and field metabolic rates of adult male yellow-bellied marmots, *Marmota flaviventris*. Comparative Biochemistry and Physiology 108A:579–88.

Salsbury, C. M., and K. B. Armitage. 1994b. Home-range size and exploratory excursions of adult, male yellow-bellied marmots. Journal of Mammalogy 75:648–56.

Salsbury, C. M., and K. B. Armitage. 1995. Reproductive energetics of adult male yellow-bellied marmots (*Marmota flaviventris*). Canadian Journal of Zoology 73:1791–97.

Schmeltz, L. L., and J. O. Whitaker, Jr. 1977. Use of woodchuck burrows by woodchucks and other mammals. Transactions of the Kentucky Academy of Sciences 38:79–82.

Schwartz, C. W., and E. R. Schwartz. 1981. The wild mammals of Missouri, rev. ed. University of Missouri Press, Columbia.

Schwartz, O. A., and K. B. Armitage. 1980. Genetic variation in social mammals: The marmot model. Science 207:665–67.

Schwartz, O. A., and K. B. Armitage., 1981. Social substructure and dispersion of genetic variation in the yellow-bellied marmot (*Marmota flaviventris*). Pages 139–59 *in* M. H. Smith and J. Joule, eds. Mammalian population genetics. University of Georgia Press, Athens.

Schwartz, O. A., and K. B. Armitage. 1998. Empirical considerations on the stable age distribution. Oecologia Montana 7:1–6.

Schwartz, O. A. and K. B. Armitage. 2002. Correlations between weather factors and life-history traits of yellow-bellied marmots. Pages 345–351 *in* K. B. Armitage and V. Yu. Rumiantsev, eds. Holarctic marmots as a factor of biodiversity. ABF, Moscow.

Schwartz, O. A., K. B. Armitage, and D. Van Vuren. 1998. A 32-year demography of yellow-bellied marmots (*Marmota flaviventris*). Journal of Zoology (London) 246:337–46.

Shriner, W. M. 1998. Yellow-bellied marmot and golden-mantled ground squirrel responses to heterospecific alarm calls. Animal Behaviour 55:529–36.

Sinha Hikin, A. P., A. Woolf, A. Bartke, and A. G. Amador. 1991. The estrous cycle of captive woodchucks (*Marmota monax*). Biology of Reproduction 44:733–38.

Sinha Hikin, A. P., A. Woolf, A. Bartke, and A. G. Amador. 1992. Further observations on estrus and ovulation in woodchucks (*Marmota monax*) in captivity. Biology of Reproduction 46:10–16.

Smith, E. N., and R. A. Woodruff. 1980. Fear bradycardia in free-ranging woodchucks, *Marmota monax*. Journal of Mammalogy 61:750–53.

Snyder, R. L. 1962. Reproductive performance of a population of woodchucks after a change in sex ratio. Ecology 43:506–15.

Snyder, R. L., and J. J. Christian. 1960. Reproductive cycle and litter size of the woodchuck. Ecology 41:647–56.

Snyder, R. L., D. E. Davis, and J. J. Christian. 1961. Seasonal changes in weights of woodchucks. Journal of Mammalogy 42:297–312.

Stallman, E. L., and W. G. Holmes. 2002. Selective foraging and food distribution of high-elevation yellow-bellied marmots (*Marmota flaviventris*). Journal of Mammalogy 83:576–84.

Stark, H. E. 1970. A revision of the flea genus *Thrassis* Jordon 1933 (Siphonaptera: Ceratophyllidae). University of California Publications in Entomology 53:1–184.

Steppan, S. J., M. R. Akhverdyan, E. A. Lyapunova, D. G. Fraser, N. N. Vorontsov, R. S. Hoffmann, and M. J. Braun. 1999. Molecular phylogeny of the marmots (Rodentia: Sciuridae): Tests of evolutionary and biogeographic hypotheses. Systematic Biology 48:715–34.

Svendsen, G. E. 1974. Behavioral and environmental factors in the spatial distribution and population dynamics of a yellow-bellied marmot population. Ecology 55:760–71.

Svendsen, G. E. 1976. Structure and location of burrows of yellow-bellied marmot. Southwest Naturalist 20:487–94.

Svendsen, G. E., and K. B. Armitage. 1973. An application of mirror-image stimulation to field behavioral studies. Ecology 54:623–27.

Swihart, R. K. 1990. Common components of orchard ground cover selected as food by captive woodchucks. Journal of Wildlife Management 54:412–17.

Swihart, R. K. 1991a. Influence of *Marmota monax* on vegetation in hayfields. Journal of Mammalogy 72:791–95.

Swihart, R. K. 1991b. Modifying scent-marking behavior to reduce woodchuck damage to fruit trees. Ecological Applications 1:98–103.

Swihart, R. K. 1992. Home-range attributes and spatial structure of woodchuck populations. Journal of Mammalogy 73:604–18.

Swihart, R. K., and M. R. Conover. 1991. Responses of woodchucks to potential garden crop repellents. Journal of Wildlife Management 55:177–81.

Swihart, R. K., and P. M. Picone. 1991a. Arboreal foraging and palatability of tree leaves to woodchucks. American Midland Naturalist 125:372–74.

Swihart, R. K., and P. M. Picone. 1991b. Effects of woodchuck activity on woody plants near burrows. Journal of Mammalogy 72:607–11.

Swihart, R. K., and P. M. Picone. 1994. Damage to apple trees associated with woodchuck burrows in orchards. Journal of Wildlife Management 58:357–60.

Taulman, J. F. 1977. Vocalizations of the hoary marmot, *Marmota caligata*. Journal of Mammalogy 58:681–83.

Taulman, J. F. 1990. Observations on scent marking in hoary marmots, *Marmota caligata*. Canadian Field-Naturalist 104:479–82.

Thompson, M. P. 1979a. Most preferred woodchuck grains subject of study. Pest Control 47:22–26.

Thompson, S. E. 1979b. Socioecology of the yellow-bellied marmot (*Marmota flaviventris*) in central Oregon. Ph.D. Dissertation, University of California, Berkeley.

Thorp, C. R., P. K. Ran, and G. L. Florant. 1994. Diet alters metabolic rate in the yellow-bellied marmot (*Marmota flaviventris*) during hibernation. Physiological Zoology 67:1213–229.

Tokuyama, K., H. L. Galantino, R. Green, and G. L. Florant. 1991. Seasonal glucose uptake in marmots (*Marmota flaviventris*): The role of pancreatic hormones. Comparative Biochemistry and Physiology 100A:925–30.

Travis, S. E., and K. B. Armitage. 1972. Some quantitative aspects of the behavior of marmots. Transactions of the Kansas Academy of Science 75:308–21.

Trump, R. F. 1950. Home range of the southern woodchuck. Iowa Academy of Science 57:537–40.

Twichell, A. R. 1939. Notes on the southern woodchuck in Missouri. Journal of Mammalogy 20:71–74.

Tyler, G. V., J. W. Summers, and R. L. Snyder. 1981. Woodchuck hepatitis virus in natural woodchuck populations. Journal of Wildlife Diseases 17:297–301.

Tyser, R. W., and T. C. Moermond. 1983. Foraging behavior in two species of different-sized sciurids. American Midland Naturalist 109:240–45.

Van Vuren, D. 1990. Dispersal of yellow-bellied marmots. Ph.D. Dissertation, University of Kansas, Lawrence.

Van Vuren, D. 1991. Yellow-bellied marmots as prey of coyotes. American Midland Naturalist 125:135–39.

Van Vuren, D. 1996. Ectoparasites, fitness, and social behaviour of yellow-bellied marmots. Ethology 102:686–94.

Van Vuren, D., and K. B. Armitage. 1991. Duration of snow cover and its influence on life history variation in yellow-bellied marmots. Canadian Journal of Zoology 69:1755–58.

Van Vuren, D., and K. B. Armitage. 1994. Survival of dispersing and philopatric yellow-bellied marmots: What is the cost of dispersal? Oikos 69:179–81.

Van Vuren, D., and C. M. Salsbury. 1992. The relation between premolar wear and age in yellow-bellied marmots, *Marmota flaviventris*. Canadian Field-Naturalist 106:134–36.

Verbeek, N. A. M. 1965. Predation by badger on yellow-bellied marmot in Wyoming. Journal of Mammalogy 46:506.

Walro, J. M., P. T. Meier, and G. E. Svendsen. 1983. Anatomy and histology of the scent glands associated with the oral angle in woodchucks. Journal of Mammalogy 64:701–3.

Ward, J. M., Jr. and K. B. Armitage. 1981a. Circannual rhythms of food consumption, body mass, and metabolism in yellow-bellied marmots. Comparative Biochemistry and Physiology 69A:621–26.

Ward, J. M., Jr. and K. B. Armitage. 1981b. Water budgets of montane–mesic and lowland–xeric populations of yellow-bellied marmots. Comparative Biochemistry and Physiology 69A:627–30.

Waring, G. H. 1965. Behavior of a marmot toward a marten. Journal of Mammalogy 46:631.

Waring, G. H. 1966. Sounds and communications of the yellow-bellied marmot (*Marmota flaviventris*). Animal Behaviour 14:177–83.

Wasser, S. K., and D. P. Barash. 1983. Reproductive suppression among female mammals: Implications for biomedicine and sexual selection theory. Quarterly Review of Biology 58:513–38.

Webb, D. R. 1980. Environmental harshness, heat stress, and *Marmota flaviventris*. Oecologia 44:390–95.

Webb, D. R. 1981. Macro-habitat patch structure, environmental harshness, and *Marmota flaviventris*. Behavioral Ecology and Sociobiology 8:175–82.

Weeks, H. P., Jr. and G. M. Kirkpatrick. 1978. Salt preferences and sodium drive phenology in fox squirrels and woodchucks. Journal of Mammalogy 59:531–42.

Weiner, J. 1989. Metabolic constraints to mammalian energy budgets. Acta Theriologica 34:3–35.

Wenberg, G. M., and J. C. Holland. 1972. The circannual variations on the serum protein fractions of the woodchuck (*Marmota monax*). Comparative Biochemistry and Physiology 42A:989–97.

Wenberg, G. M. and J. C. Holland., 1973a. The circannual variations of some of the hormones of the woodchuck (*Marmota monax*). Comparative Biochemistry and Physiology 46A:523–35.

Wenberg, G. M., and J. C. Holland. 1973b. The circannual variations of thyroid activity in the woodchuck (*Marmota monax*). Comparative Biochemistry and Physiology 44A:775–80.

Wenberg, G. M., J. C. Holland, and J. Sewell. 1973. Some aspects of the hematology and immunology of the hibernating and nonhibernating woodchuck (*Marmota monax*). Comparative Biochemistry and Physiology 46A:513–21.

Williams, D. D., and R. L. Rausch. 1973. Seasonal carbon dioxide and oxygen concentrations in the dens of hibernating mammals (Sciuridae). Comparative Biochemistry and Physiology 44A:1227–35.

Wilson, B. E., S. Deeb, and G. L. Florant. 1992. Seasonal changes in hormone-sensitive and lipoprotein lipase mRNA concentrations in marmot white adipose tissue. American Journal of Physiology 262:R177–81.

Winders, R. L., M. O. Farber, K. F. Atkinson, and F. Manfredi. 1974. Parameters of oxygen delivery in the species *Marmota flaviventris* at sea level and 12,000 feet. Comparative Biochemistry and Physiology 49A:287–90.

Wood, W. A. 1973. Habitat selection and energetics of the Olympic marmot. M.S. Thesis, Western Washington State College, Bellingham.

Woods, B. C., K. B. Armitage, and D. T. Blumstein. 2002. Yellow-bellied marmots depress metabolism to enter torpor. Pages 400–404 *in* K. B. Armitage and V. Yu. Rumiantsev, eds. Holarctic marmots as a factor of biodiversity. ABF, Moscow.

Wright, J., B. C. Tennant, and B. May. 1987. Genetic variation between woodchuck populations with high and low prevalence rates of woodchuck hepatitis virus infection. Journal of Wildlife Diseases 23:186–91.

Xia, T., N. Mostafa, B. G. Bhat, G. L. Florant, and R. A. Coleman. 1993. Selective retention of essential fatty acids: The role of hepatic monoacylglycerol acyltransferase. American Journal of Physiology 265:R414–19.

Young, R. A. 1984. Interrelationships between body weight, food consumption and plasma thyroid hormone concentration cycles in the woodchuck, *Marmota monax*. Comparative Biochemistry and Physiology 77A:533–36.

Young, R. A., and E. A. H. Sims. 1979. The woodchuck, *Marmota monax*, as a laboratory animal. Laboratory Animal Science 29:770–80.

Young, R. A., R. Rajatanavin, L. E. Braverman, and B. C. Tennant. 1986. Seasonal changes in serum thyroid hormone binding proteins in the woodchuck (*Marmota monax*). Endocrinology 119:967–71.

Zatzman, M. L., and F. E. South. 1972. Renal function of the awake and hibernating marmot *Marmota flaviventris*. American Journal of Physiology 222:1035–39.

Zatzman, M. L., and F. E. South. 1975. Concentration of urine by the hibernating marmot. American Journal of Physiology 228:1326–40.

Zatzman, M. L., and G. V. Thornhill. 1986. Seasonal changes in blood pressure and cardiac output in marmots. Pages 453–59 *in* H. C. Heller, X. J. Musacchia, and L. C. Wang, eds. Living in the cold: Physiological and biochemical adaptations. Elsevier, New York.

Zatzman, M. L., G. V. Thornhill, W. J. Ray, and M. R. Ellersiek. 1984. Seasonal changes of food and water consumption and urine production of the marmot, *Marmota flaviventris*. Comparative Biochemistry and Physiology 77A:735–43.

Zervanos, S. M., and C. M. Salsbury. 2003. Seasonal body temperature fluctuations and energetic strategies in free-ranging eastern woodchucks (*Marmota monax*). Journal of Mammalogy 84:299–310.

Kenneth B. Armitage, Department of Ecology and Evolutionary Biology, University of Kansas, Lawrence, Kansas 66045-7534. Email: marmots@ukans.edu.

10

Ground Squirrels

Spermophilus and *Ammospermophilus* species

Eric Yensen
Paul W. Sherman

NOMENCLATURE

Scientific and Common Names. The North American ground-dwelling squirrels are classified into four closely related genera: *Spermophilus* (true ground squirrels), *Ammospermophilus* (antelope ground squirrels), *Marmota* (marmots; see Chapter 9), and *Cynomys* (prairie dogs; see Chapter 11). In the nineteenth century, *Spermophilus* Cuvier, 1825, which means "seed lover," was used as the generic name of the true ground squirrels. However, Allen (1902) discovered that an earlier name, *Citellus* Oken, 1816, had priority. Consequently, *Citellus* was used during the first half of the twentieth century. However, Hershkovitz (1949) pointed out that Oken's names were inconsistently binomial and hence invalid. The International Commission on Zoological Nomenclature (1956, Opinion 417) concurred, and validated the next oldest appropriate name, *Spermophilus*.

Of the 31 North American species of *Ammospermophilus* and *Spermophilus,* 17 are monotypic and the other 14 are divided into 87 subspecies, which differ in their degrees of distinctiveness (Table 10.1) (Hall 1981; Wilson and Ruff 1999). There are 5 species of antelope ground squirrels ("*juancito*" in Mexico) and 26 species of true ground squirrels (locally called picket pins, whistle pigs, and gophers).

GENETICS

Karyotypes are known for nearly all species (Nadler 1962, 1966a, 1966b; Nadler and Sutton 1963; Birney and Genoways 1973; Uribe-Alcocer et al. 1979; Nadler et al. 1982, 1984; Hafner and Yates 1983). All *Ammospermophilus* species have diploid chromosome numbers ($2N$) of 32. Most *Spermophilus* subgenera also have stable chromosome numbers: in *Otospermophilus,* $2N = 38$; in *Notocitellus,* $2N = 32$; in *Poliocitellus,* $2N = 42$; in *Callospermophilus,* $2N = 42$; in *Xerospermophilus,* $2N = 36$ or 38; and in *Ictidomys,* $2N = 32$ or 34. Chromosome numbers are most variable in the subgenus *Spermophilus,* where $2N = 30-46$. Various hypotheses of karyotype evolution have been proposed (Nadler 1968; Lyapunova and Vorontsov 1970; Nadler et al. 1982), but these are in poor agreement with each other and with a recent mitochondrial DNA phylogeny for the genus *Spermophilus* (R. G. Harrison, S. M. Bogdanowicz, R. S. Hoffmann, E. Yensen, and P. W. Sherman, in litt.).

Karyotype morphology and Giemsa bands, in conjunction with allozymes and morphological data, have helped clarify species-level taxonomy in *Spermophilus*. For example, *S. richardsonii* consisted of four subspecies, one with $2N = 36$ (now the species *S. richardsonii*) and three with $2N = 34$ (now all *S. elegans*) (Nadler et al. 1971). The former *S. townsendii* was discovered to consist of three species, with $2N = 36$ (*S. townsendii*), 38 (*S. mollis*), and 46 (*S. canus*) (Nadler 1968; Nadler et al. 1984; Hoffmann et al. 1993). The arctic ground squirrel (*S. undulatus*) is now known to be two species, *S. undulatus* ($2N = 36$), which is restricted to Asia, and *S. parryii* ($2N = 34$), which occurs in Alaska, Canada, and Siberia (Nadler et al. 1975).

Detailed studies of morphology (cranial characteristics, bacula) also have helped clarify species relationships. For example, *S. mexicanus* and *S. tridecemlineatus* are genetically very similar (i.e., genetic distance = 0.92, based on allozymes) but morphologically differentiated. For comparison, the genetic similarity of *S. mexicanus* to its next closest relative, *S. spilosoma,* is 0.77 (Cothran et al. 1977; Cothran 1983; Cothran and Honeycutt 1984). Although *S. mexicanus* and *S. tridecemlineatus* can hybridize, backcross hybrids are rare and found only in association with the former species. Two subspecies of the Idaho ground squirrel, *S. brunneus brunneus* and *S. b. endemicus,* also are quite similar genetically (0.89) (Gill and Yensen 1992; E. Yensen et al., unpublished data), but they are sufficiently different morphologically and phenologically to qualify as species.

Populations of most ground squirrels have patchy distributions due to microgeographic variations in habitat suitability and food abundance. The amount of gene flow among adjacent subpopulations (i.e., the degree to which local demes are metapopulations) also is highly variable. At one extreme, male *S. columbianus* disperse up to 8.5 km (although the average distance is <2 km), resulting in high gene flow ($N_{em} = 13.5$ effective migrants per generation) and low population differentiation ($F_{ST} = 0.03$) (Dobson 1994). At the other extreme is *S. b. brunneus,* in which males rarely disperse >1 km (<200 m on average), resulting in low gene flow among the 31 remaining, widely separated populations and considerable differentiation ($F_{ST} = 0.17$) (Gavin et al. 1999). *S. mollis* represents the intermediate situation because males sometimes disperse several kilometers (515 m on average), the mean deme size is 62.2 ha, and $F_{ST} = 0.07$ (Antolin et al. 2001).

PHYLOGENY

There have been numerous attempts to reconstruct the phylogenetic relationships and evolutionary history of the ground squirrels (Durrant and Hansen 1954; Nadler 1966b; Stangl and Grimes 1987; Mercer and Roth 2003). Bryant (1945) used anatomical features and fossils to propose a Miocene separation of *Marmota* from the *Spermophilus–Cynomys* lineage. Under his scenario, *Ammospermophilus* separated in late Miocene, followed by subgenera *Otospermophilus* and *Callospermophilus*. He considered subgenus *Spermophilus* and genus *Cynomys* as the most derived members of the clade, thus making subgenus *Spermophilus* paraphyletic. Hafner (1984) also proposed a late Miocene radiation of Marmotini, with *Ammospermophilus* and *Marmota* separating from the *Spermophilus–Cynomys* lineage, and the *Spermophilus–Cynomys* line then radiating into four lineages—*Otospermophilus–Callospermophilus, Xerospermophilus–Ictidomys–Poliocitellus, Spermophilus,* and *Cynomy*—late in the late Pliocene.

Nadler et al. (1982, 1984) used morphological, biochemical, and chromosomal data to establish species groups and propose a phylogeny for the subgenus *Spermophilus*. MacNeil and Strobeck (1987) used mtDNA sequences to hypothesize relationships among *S. parryii, S. columbianus,* and *S. richardsonii*; their arrangement differed substantially from that of Nadler et al. (1982, 1984). Molecular evidence suggests that both *Cynomys* and *Marmota* may have originated within the *Spermophilus* radiation rather than being early offshoots (Thomas and Martin 1993; Giboulet et al. 1997).

Recently, a new phylogeny was proposed based on sequence similarities of the cytochrome *b* gene of all species of *Cynomys* and

TABLE 10.1. Summary of distribution and measurements of 31 species of North American ground squirrels

Scientific Name	Common Name	Ssp.	Range[a]	Standard Lengths (mm) Total	Tail	HFt	GL Skull
Ammospermophilus							
harrisii	Harris's antelope squirrel	2	AZ, NM, Mexico	220–267	70–94	38–43	38.2–41.9
insularis	Espiritu Santo antelope squirrel	0	Baja California	210–240	71–83	36–40	40.3–42.4
leucurus	White-tailed antelope squirrel	9	Great Basin to Baja California	194–239	54–87	35–43	37.0–41.8
interpres	Texas antelope squirrel	0	NM, TX, Mexico	220–235	68–84	36–40	37.7–40.5
nelsoni	San Joaquin antelope squirrel	0	CA	230–267	66–78	37–44	41.5–43.1
Spermophilus							
Subgenus *Notocitellus*							
adocetus	Lesser tropical ground squirrel	2	Mexico	244–366	112–163	38–50	40.0–48.0
annulatus	Ring-tailed ground squirrel	2	Mexico	383–470	186–222	50–61	51.9–55.4
Subgenus *Otospermophilus*							
atricapillus	Baja California rock squirrel	0	Baja California	387–486	156–217	50–60	50.2–58.2
beecheyi	California ground squirrel	8	WA, OR, CA, Baja California	357–500	145–200	50–64	51.6–62.4
variegatus	Rock squirrel	9	ID, CO south to Mexico	430–525	172–252	53–65	56.0–67.7
Subgenus *Callospermophilus*							
lateralis	Golden-mantled ground squirrel	13	Canada to CO, CA, AZ	230–308	63–118	35–46	39.6–45.6
madrenis	Sierra Madre mantled squirrel	0	Mexico	215–244	52–66	35–40	41.1–44.1
saturatus	Cascades golden-mantled squirrel	0	WA, BC	286–315	92–118	43–49	44.0–48.3
Subgenus *Spermophilus*							
armatus	Uinta ground squirrel	0	ID, WY, MT, UT	280–303	63–81	42–46	46.3–48.5
beldingi	Belding ground squirrel	3	CA, OR, NV, ID, UT	254–300	55–76	40–47	41.3–46.3
brunneus	Idaho ground squirrel	2	Idaho	209–258	39–65	32–39	36.1–42.5
canus	Merriam's ground squirrel	2	OR, ID	188–257	35–52	29–37	34.3–40.6
columbianus	Columbian ground squirrel	2	WA, BC, AL, MT, ID, OR	327–410	80–116	48–58	49.5–57.0
elegans	Wyoming ground squirrel	3	OR, ID, NV, MT, WY, CO	188–315	66–100	39–48	42.0–48.6
mollis	Piute [Great Basin] ground squirrel	3	ID, OR, NV, UT, CA	167–271	32–72	29–38	32.4–43.3
parryii	Arctic ground squirrel	10[b]	AK, Yukon, NWT	332–495	77–153	50–68	50.7–65.8
richardsonii	Richardson's ground squirrel	0	MT, ND, SD, sc Canada, MN	277–306	65–83	43–47	45.1–48.4
townsendii	Townsend's ground squirrel	2	Washington	200–232	39–54	32–37	37.1–38.7
washingtoni	Washington ground squirrel	0	WA, OR	185–245	32–65	30–38	30.5–41.4
Subgenus *Xerospermophilus*							
mohavensis	Mohave ground squirrel	0	CA	210–230	57–72	32–38	38.1–40.0
tereticaudus	Round-tailed ground squirrel	4	NV, AZ, CA, Sonora, Baja California	204–266	60–107	32–40	34.9–39.3
Subgenus *Poliocitellus*							
franklinii	Franklin's ground squirrel	0	Upper Midwest, sc Canada	381–397	136–153	53–57	52.1–55.1
Subgenus *Ictidomys*							
mexicanus	Mexican ground squirrel	2	TX, NM, Mexico	280–380	110–166	38–51	41.0–52.5
perotensis	Perote ground squirrel	0	Mexico	243–261	57–78	38–40	42.2–44.5
spilosoma	Spotted ground squirrel	13	SD to AZ, TX, Mexico	185–253	55–92	28–38	34.1–42.7
tridecemlineatus	Thirteen-lined ground squirrel	10	Great Plains to OH, TX, UT	170–297	60–132	27–41	34.0–45.8

SOURCE: Data from Howell (1938), Hall (1981), Michener and Koeppl (1985), Best et al. (1990a, 1990b), Best (1995c), Yensen and Sherman (1997), Verts and Carraway (1998), Yensen and Valdés-Alarcón (1999).

NOTE: Ssp., Number of subspecies; HFt, hind foot; GL Skull, greatest length of the skull.

[a]BC, British Columbia; NWT, Northwest Territories, sc, south-central.

[b]Two additional subspecies in Siberia.

Spermophilus worldwide as well as representatives of *Ammospermophilus* and *Marmota* (R. G. Harrison, S. M. Bogdanowicz, R. S. Hoffmann, E. Yensen, and P. W. Sherman, in litt.). Results indicate that *Ammospermophilus* separated first, followed by a clade with two tropical species, *S. adocetus* and *S. annulatus* (i.e., *Notocitellus*), (Howell 1938). Then *Otospermophilus* and *Callospermophilus* split off, forming a clade that is a sister group to *Marmota* and the remaining ground squirrels. The Eurasian and North American members of current subgenus *Spermophilus* represent different clades. The sister clade to the North American subgenus *Spermophilus* includes the present genus *Cynomys* and subgenera *Ictidomys, Poliocitellus,* and *Xerospermophilus*. Although a close relationship between *Spermophilus* and *Cynomys* was suspected by Bryant (1945), it is interesting that *Cynomys* and *Marmota* probably were imbedded within the *Spermophilus* radiation, a result recently confirmed by Mercer and Roth (2003).

Although the phylogenetic hypothesis based on cytochrome *b* makes biogeographic sense, it differs substantially from phylogenies currently in use. If the mitochondrial DNA phylogeny is accepted, a number of nomenclatorial changes will have to be addressed. For example, (1) *Cynomys* has priority for the clade containing the three aforementioned ground squirrel subgenera, (2) the name *Spermophilus* belongs to the Eurasian clade (not the North American group) because the type species of *Spermophilus* is from Eurasia, and (3) several current subgenera (e.g., *Otospermophilus, Ictidomys*) may deserve recognition as genera.

DISTRIBUTION

Worldwide, the genus *Spermophilus* comprises 48 named species. It is Holarctic in distribution and ranges from Austria across Eurasia to Siberia, western North America from Alaska south to west central Mexico, and as far east as Ohio (Hoffmann et al. 1993). *Ammospermophilus* is endemic to the southwestern Nearctic, and occurs from Oregon south to Texas, northern Mexico, and Baja California.

In North America, *Spermophilus* subgenera occupy a diversity of habitats including tropical deciduous forests (*Notocitellus,* tropical ground squirrels), rocky areas (*Otospermophilus,* rock squirrels and California ground squirrel), evergreen forests and small meadows (*Callospermophilus,* mantled ground squirrels), intermontane grasslands and large meadows (*Spermophilus,* western ground squirrels), tall grass prairie (*Poliocitellus,* Franklin's ground squirrel), short grass prairie (*Ictidomys,* prairie-dwelling ground squirrels), and deserts (*Xerospermophilus,* round-tailed and Mohave ground squirrels) (Hafner 1984; Hoffmann et al. 1993; Yensen and Valdés-Alarcón 1999).

Range maps of individual North American species based on historical records are presented as Figures 10.1–10.6. However, as will be discussed (see Population Status), these maps give a misleading picture of where the animals occur today. Because of habitat fragmentation, many species exist as small, isolated populations (merely points on the maps) rather than being spread evenly across their ancestral range.

Figures 10.1–10.6 suggest that similar-sized species have parapatric distributions, but that in many areas a larger and a smaller species are sympatric. When this occurs, there actually is microgeographic separation suggestive of competitive exclusion: The smaller species usually is confined to the poorer, more xeric microhabitats. In further support of interspecific competition, the large-bodied, aggressive *S. columbianus* is rapidly replacing several smaller congeners, including *S. brunneus* in Idaho (Yensen and Sherman 1997), *S. beldingi* in northwestern Oregon (Turner 1972), and *S. richardsonii* in Montana (Howell 1938; Michener 1977).

FIGURE 10.1. Distribution of the white-tailed antelope squirrel (*Ammospermophilus leucurus*), Nelson's antelope squirrel (*A. nelsoni*), Harris's antelope squirrel (*A. harrisii*), and Texas antelope squirrel (*A. interpres*).

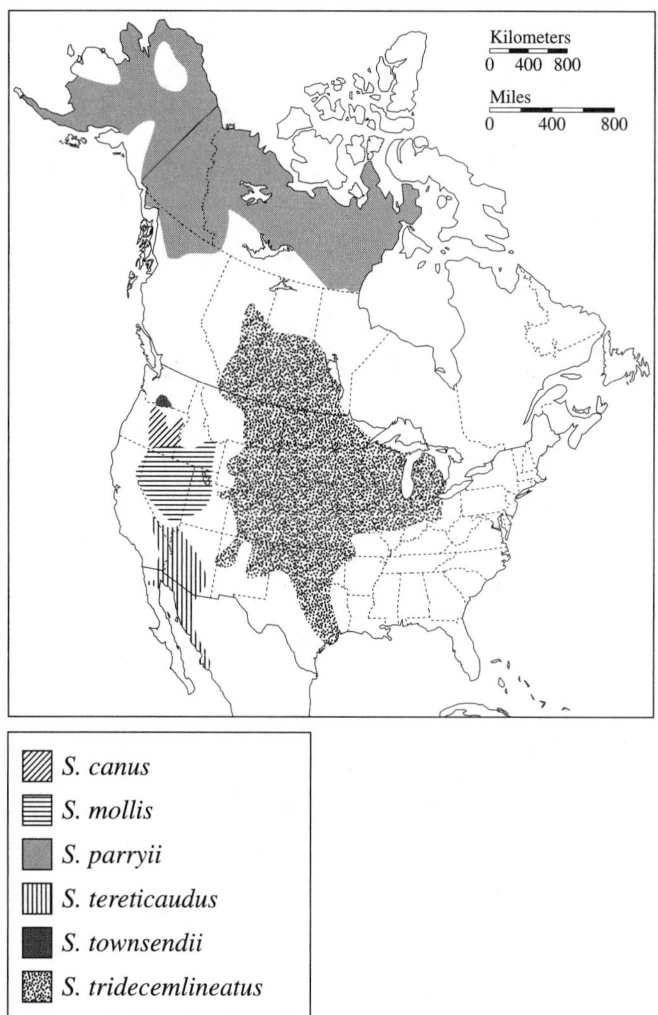

FIGURE 10.2. Distribution of the arctic ground squirrel (*Spermophilus parryii*), 13-lined ground squirrel (*S. tridecemlineatus*), Townsend's ground squirrel (*S. townsendii,* including *nancyae*) Merriam's ground squirrel (*S. canus*), Piute [=Great Basin] ground squirrel (*S. mollis*), and round-tailed ground squirrel (*S. tereticaudus*).

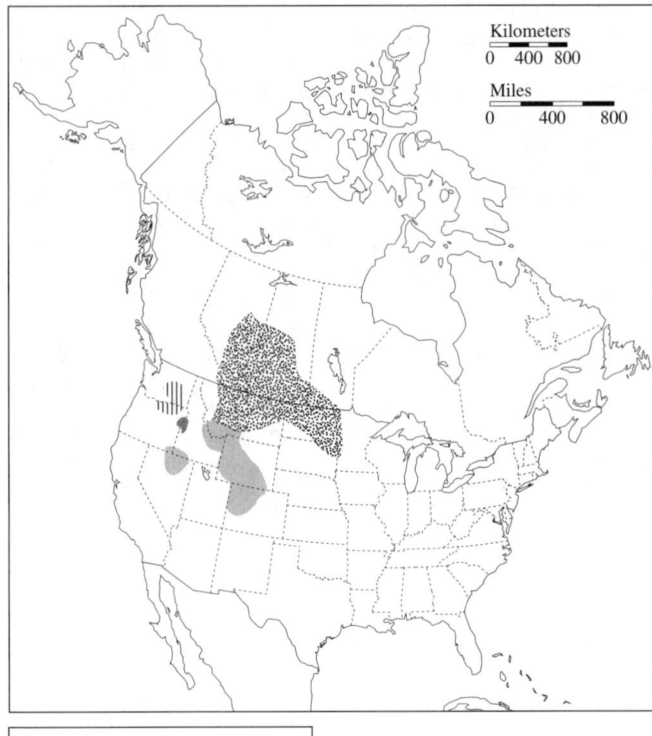

FIGURE 10.3. Distribution of the Richardson's ground squirrel (*Spermophilus richardsonii*), Wyoming ground squirrel (*S. elegans*), Washington ground squirrel (*S. washingtoni*), and Idaho ground squirrel (*S. brunneus*, including *endemicus*).

FIGURE 10.4. Distribution of the Columbian ground squirrel (*Spermophilus columbianus*), Franklin's ground squirrel (*S. franklinii*), California ground squirrel (*S. beecheyi*), and rock squirrel (*S. variegatus*).

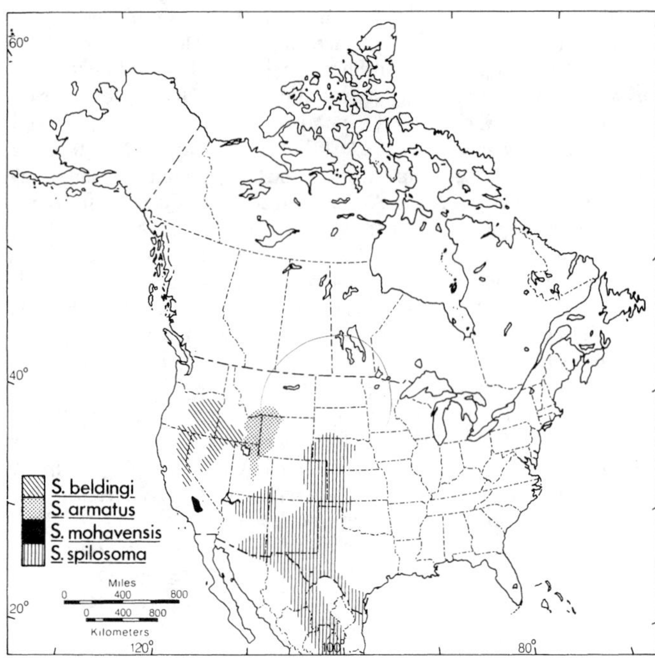

FIGURE 10.5. Distribution of the Belding's ground squirrel (*Spermophilus beldingi*), Uinta ground squirrel (*S. armatus*), Mohave ground squirrel (*S. mohavensis*), and spotted ground squirrel (*S. spilosoma*).

FIGURE 10.6. Distribution of the golden-mantled ground squirrel (*Spermophilus lateralis*), Cascade mantled ground squirrel (*S. saturatus*), and Mexican ground squirrel (*S. mexicanus*).

DESCRIPTION

Ground squirrels are diurnal and highly conspicuous, especially when they stand on their hind feet to scan for predators or give alarm calls. Although most species can climb (Shaw 1925b), they live primarily on and in the ground. Ground squirrels have cylindrical bodies, large eyes, chisel-shaped incisors, and long whiskers. Their short, powerful forelimbs are equipped with sharp claws for burrowing; their longer hind limbs propel rapid movements. They have four toes on their forefeet and five on their hind feet. Soft, convoluted pads on the soles of their feet enable them to grip the substrate and food items. Their tails are furry, varying from long and bushy (*Notocitellus, Otospermophilus*) to short and thin (*Xerospermophilus*). Their ear pinnae vary from erect and triangular to small and rounded. In some species (e.g., *S. tereticaudus, S. townsendii*) the pinnae are reduced to a rim of tissue around the external auditory meatus (Grinnell and Dixon 1918; Bryant 1945). In most species of *Spermophilus,* males are larger than females (Armitage 1981; Morton 1975; Rickart 1986, Yensen 1991). (See also Table 10.1)

Most ground squirrels have grayish or brownish pelage, darker dorsally than ventrally, and a light-colored eye ring. Some species also have dorsal spots distributed evenly (*S. spilosoma, S. mexicanus, S. washingtoni, S. brunneus*) or in rows (*S. tridecemlineatus*). Others have variegated pelage (*S. beecheyi, S. variegatus*), sometimes with flecks of lighter color (*S. townsendii, S. mollis*). Antelope ground squirrels are distinctively colored, with a white stripe on their sides. These desert dwellers have tails with white undersides, which they carry over their back when running, presumably to reflect the sun and provide shade. The desert-dwelling *S. mohavensis* also carries its tail, white side up, over its back. The mantled ground squirrels (*S. lateralis, S. saturatus, S. madrensis*) have a light lateral stripe with dark borders; they are forest dwellers and do not carry their tail over their back. The rock squirrels (*S. atricapillus, S. beecheyi, S. variegatus*) are large bodied, with longer legs and bushy tails, resembling tree squirrels.

There are three molt patterns in ground squirrels (Hansen 1954): (1) diffuse hair replacement, which occurs once per year, (2) complete molt once per year with a distinct molt line, and (3) complete molt twice per year with distinct molt lines each time.

Skulls of ground-dwelling squirrels (Fig. 10.7) have zygomatic arches that are widest at the posterior and twist toward horizontal anteriorly; this differentiates their skulls from those of tree squirrels. The thin postorbital process projects posteroventrally in the curvature of the orbit and tapers to a fine point. *Spermophilus* skulls have a tubercle for attachment of the masseter muscle ventral and lateral to the oval or triangular infraorbital foramen, whereas the tubercle is directly below a narrow foramen in *Ammospermophilus*. The cranium of *Ammospermophilus* is slightly more rectangular when viewed dorsally, and they lack the clavobrachialis muscle. The dental formula is I 1/1, C 0/0, P 1–2/1, M 3/3 in both genera (Hall 1981). *Spermophilus* and *Ammospermophilus* have zygomatic breadths of <48 mm (marmots, >48 mm), their maxillary toothrows do not converge posteriorly (prairie dogs have convergent toothrows), and they possess an antorbital canal (lacking in chipmunks, *Tamias*). Like other rodents, they have a single pair of incisors in each jaw, a large gap (diastema) in front of the premolars, and no canine teeth. Their incisors grow continuously and are worn back by use. Their lower incisors can operate independently. Their cheek teeth are rooted and have abrasive chewing (occlusal) surfaces. Membranous internal cheek pouches are known in *S. beecheyi, S. beldingi, S. elegans, S. lateralis, S. parryii, S. tereticaudis, S. tridecemlineatus,* and *S. variegatus* (Grinnell and Dixon 1918; Howell 1938), and may occur in other species.

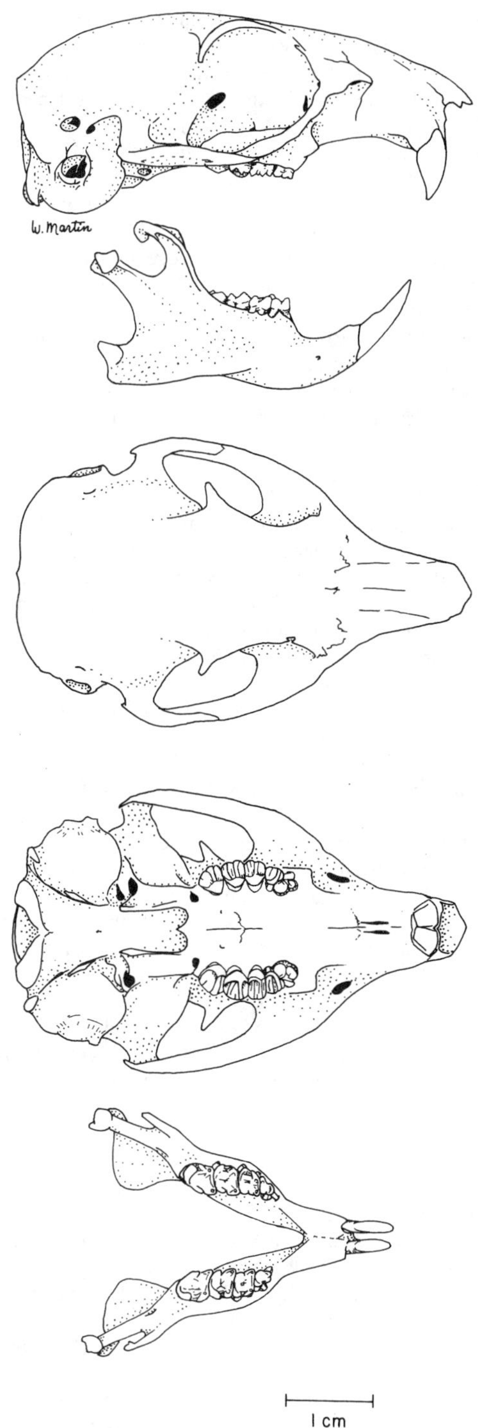

FIGURE 10.7. Skull of the California ground squirrel (*Spermophilus beecheyi*). From top to bottom: lateral view of cranium, lateral view of mandible, dorsal view of cranium, ventral view of cranium, dorsal view of mandible.

ANNUAL CYCLE

Spermophilus and *Ammospermophilus* species differ in their annual cycles and physiology. The latter are active all year, and there is no annual cycle of fat deposition. These desert dwellers are remarkable for their ability to remain active in the summer heat and the winter cold. In summer, they cool down by using their burrow walls as a heat sink; to save energy on winter nights, they become hypothermic, allowing their body temperature to decrease to 32–33°C. If it becomes too cold, they can enter torpor to further save energy (Chappell and Bartholomew 1981). However, antelope ground squirrels cannot arouse from torpor at nest temperatures of <10°C (Kramm 1972).

Some *Spermophilus* species that inhabit low elevations and latitudes also are active all year, for example, *S. atricapillus* in Baja California, especially near oases (Alvarez-Castañeda et al. 1996; Yensen and Valdés-Alarcón 1999), and *S. annulatus* and *S. adocetus* in Mexico, although they may estivate during the hottest months (Best 1995a, 1995b). Moving northward, some low-elevation populations of *S. beecheyi* in central California were active all year, and pregnant females were found from January to May, but in other areas the entire population estivated from May to November (Grinnell and Dixon 1918; Tomich 1962; Boyer and Barnes 1999).

All the other North American *Spermophilus* species are obligate hibernators, but the timing of their annual cycles differs within and among populations. For example, in *S. beldingi, S. lateralis, S. columbianus,* and *S. richardsonii,* dates of emergence from hibernation at the same site vary annually from several days to 6 weeks

(Sherman 1976; Bronson 1980; Murie and Harris 1982; Rickart 1982; Michener 1984, 1998). These variations correlate with winter weather (Michener 1984), spring snow cover (Phillips 1984), latitude, and elevation. At high elevations and latitudes, active seasons generally correspond with the summer months (May–September) (Kenagy and Barnes 1988). Populations that are near each other but that differ considerably in elevation can have annual cycles that are similar in length but staggered in time; at lower elevations, they emerge and immerge earlier (Bronson 1980, Dobson and Murie 1987; Holekamp 1983, 1986). In xeric habitats, where availability of green vegetation depends on spring rains and summers are hot and dry, ground squirrels also emerge from hibernation earlier (January–March) and enter seasonal torpor in early to midsummer (May–July) (Bailey 1936; Rickart 1982; Smith and Johnson 1985; Rickart and Yensen 1991; Yensen 1991; Yensen and Sherman 1997). However, regardless of when populations emerge and immerge, in all species individual females rear only one litter of young per year.

HIBERNATION PHYSIOLOGY

North American ground squirrels spend the majority of their lives out of sight, hibernating in solitarily burrows. They become torpid to conserve energy during lengthy, predictable food shortages and inclement weather, usually in winter (hibernation), but also during the summer in xeric habitats (estivation). Most *Spermophilus* species depend on fat reserves for survival during hibernation. They start hibernating when they have accumulated sufficient fat (Morton et al. 1974), and overwinter survival is correlated with accumulated fat reserves in many species including *S. armatus* (Slade and Balph 1974; Rieger 1996), *S. richardsonii* (Michener 1978), *S. mollis* (Van Horne et al. 1998), and *S. columbianus* (Murie and Boag 1984). In a few species, such as *S. columbianus* (Shaw 1925c), *S. richardsonii* (Michener 1998), and *S. parryii* (Buck and Barnes 1999), males store seeds in their hibernacula. The seeds are consumed at spring arousal, which helps to improve the male's condition and enhance competitive abilities.

White and brown adipose tissues constitute the main energy stores used by hibernating ground squirrels. High levels of plasma insulin and corticosterone (Nunes et al. 2002) and steady glucagon levels cause lipid storage in white adipose cells by stimulating synthesis and inhibiting breakdown of glycogen, triacylglycerides, and protein (Boyer and Barnes 1999). Insulin also enhances synthesis and activity of lipoprotein lipase, which increases lipid storage. During hibernation, some species also catabolize protein. For example, *S. parryii* may lose up to 43% of their fat-free mass (Buck and Barnes 1999). Body weight and lipid composition follow parallel trajectories, with both being highest at entry into hibernation and lowest at spring emergence (Morton 1975; Davis 1976; Michener 1984). Loss of body mass during hibernation can be considerable. *S. beldingi* and *S. mollis,* respectively, lose 33–43% and 45–51% of body mass over winter (Morton 1975; Rickart 1982). Interestingly, adult female *S. parryii* lose 33% of their body mass over winter, but males emerge just 3% lighter than when they entered hibernation, presumably because of their food caches (Buck and Barnes 1999).

Dawe and Spurrier (1969) reported inducing hibernation in *S. tridecemlineatus* by injecting them with blood from hibernating conspecifics. However, attempts to replicate this experiment have yielded inconsistent results, and it is unclear whether hibernation-induction molecules exist (Boyer and Barnes 1999). In general, ground squirrels become lethargic before hibernation, and concentrations of leukocytes decrease about 90%. Normal shivering and metabolic thermogenic responses are deactivated as the animal cools and settles into torpor at a new, lower hypothalamic set point (Boyer and Barnes 1999). Once in torpor, a ground squirrel's body temperature remains only 0.2–1.0°C above ambient, its heart rate and metabolism stabilize at about 1% of normal, and its breathing becomes irregular (several shallow breaths followed by up to 1 min of apnea). This reduces energy requirements by about 87% compared to what is needed for normal activity.

Hibernating ground squirrels periodically arouse spontaneously, and their body temperatures return to normal for a few hours or days. *S. parryii* may arouse as often as 10–12 times during a 7- to 8-month hibernation period (Boyer and Barnes 1999). In *S. beldingi* and *S. richardsonii,* bouts of torpor last 3–4 days early in the hibernation period, gradually lengthen to 18–19 days in midwinter, then shorten again to 5–6 days before emergence (Wang 1979; French 1988; Michener 1998). Arousal uses brown adipose tissue and it therefore is costly in terms of stored energy. Indeed, of the energy expended during hibernation, most (84%) was used during brief arousals. It is not clear why ground squirrels spontaneously arouse (Boyer and Barnes 1999), because aroused individuals do not venture from their burrow. Indeed, they spend most of their time sleeping (Knight et al. 2000) and they also void waste products. A clue to the function of arousal is that DNA translation decreases dramatically during hibernation in a regulated manner, not just due to the decline in metabolic rate. It is possible that arousal functions to enable synthesis and replacement of necessary proteins (Knight et al. 2000).

Most *Spermophilus* species also arouse when ambient temperatures approach 0°C, presumably to avoid freezing. However one species, *S. parryii,* has long been known to survive subfreezing temperatures while hibernating (Svihla 1958). Investigations by Buck and Barnes (1999) have shown that these animals actively thermoregulate while hibernating. Although most brain activity ceases, they maintain a constant body temperature of –2°C, most likely through nonshivering thermogenesis using brown adipose tissue, even when the temperature of the surrounding soil is much lower. Indeed, burrow temperatures can decrease to –18°C and the body temperature of the animals can fall to as low as –2.9°C without killing them. Remarkably, they avoid tissue damage during freezing by supercooling (Boyer and Barnes 1999).

As spring approaches, female and juvenile ground squirrels arouse more frequently. What actually triggers their emergence is not well understood, but soil temperature probably is important. Eventually they burrow to the surface and assess weather conditions and food availability. However, they do not completely terminate hibernation for the year until they have fed (French 1988). If late-season snowstorms cover the vegetation (Morton and Sherman 1978), females and juveniles can reenter hibernation, where they remain until their fat reserves run out. In contrast, adult males terminate yearly hibernation before their fat stores are depleted, regardless of the weather or whether they can feed on the surface (French 1988). For example, male *S. beldingi* dig out through deep snow in the spring (Sherman 1976; Morton and Sherman 1978). Early, spontaneous termination of hibernation enables adult males to prepare physiologically and behaviorally for the emergence of receptive females. However, it also renders them vulnerable to starvation if late-season storms cover their food supplies (Morton and Sherman 1978).

REPRODUCTION

Reproductive Activity and Mating. Toward the end of hibernation, the testes descend into the scrotum, which becomes darkly pigmented (Buck and Barnes 1999). Penises of mature males often are darkly pigmented, whereas penises of immature males are light colored. Males spend a few days in their burrow before emergence, during which time sperm maturation occurs. Once males emerge, they begin searching for females and, in some species, establishing mating territories (Michener 1983, 1998). Females typically begin to appear a few days to 2 weeks after males, and they soon become sexually receptive. In some cases, adult females emerge nearly simultaneously with adult males (Michener 1984). This occurs in populations of *S. b. brunneus* (Sherman 1989) and *S. tridecemlineatus* (Schwagmeyer 1995) where succulent vegetation is reliably available early in the spring.

When there is no further possibility of mating in a season, males' testes decrease in size and are withdrawn into the body cavity (Shaw 1925a). Then follows a period of intense foraging and fat deposition preparatory to hibernation. Typically, adult males cease aboveground activity first, followed by adult females, yearlings, and finally juveniles. As a result of "staggered" spring emergence and summer/fall immergence, ground squirrel populations as a whole (considering all age and sex classes) always are active for longer than any individual animal. For example, in some high-altitude and high-latitude populations of *S. beldingi* (Sherman 1976) and *S. richardsonii* (Michener 1983b,1998), adult males sometimes hibernate so early that their active seasons barely overlap those of late-weaning juveniles of that year. By contrast, in *S. beecheyi* (Tomich 1962), *S. armatus* (Knopf and Balph 1977), and *S. parryii* (Buck and Barnes 1999), adult males remain active later than adult females and nearly as long as juveniles; these males may be maintaining their territorial boundaries against late-season male intruders.

In both sexes, age of sexual maturity varies depending on when body mass at emergence from hibernation is sufficient for successful reproduction (i.e., after the first or the second winter). In areas with long growing seasons, individuals mature at younger ages. Both sexes of the desert-dwelling *Ammospermophilus* become reproductively active when they are 1 year old (Belk and Smith 1991), but some high-latitude and high-elevation *Spermophilus* populations do not mature until they are 2 or 3 years old (Morton and Sherman 1978; Festa-Bianchet 1981). In many *Spermophilus* species, maturation is sexually dimorphic. Because males are larger than females, it may take them 1 year longer than females to achieve appropriate mass, as in *S. beldingi* (Bushberg and Holmes 1985). In high-elevation populations of *S. columbianus*, males (but not females) mature at older ages than conspecifics in low-elevation populations (Dobson and Murie 1987).

Female ground squirrels are ready to mate soon after they emerge from hibernation. Indeed, in *S. b. brunneus*, mating occurs on the afternoon of the day females emerge (Sherman 1989). In other species, such as *S. beldingi* (Morton and Sherman 1978), *S. richardsonii* (Michener 1985), and *S. armatus* (Rieger 1996), females mate 3–6 days after spring emergence. Inclement weather can delay mating for several weeks, for example, in *S. tridecemlineatus* (Wade 1927) and *S. beldingi* (Morton and Sherman 1978).

Females are in estrus when the vagina is fully open, the labia are swollen and crenulated, and the percentage of cornified epithelial cells exceeds the percentage of leukocytes in vaginal levages. Female receptive period generally is short: <12 hr in *S. beldingi* (Hanken and Sherman 1981), *S. tridecemlineatus* (Schwagmeyer and Parker 1987), *S. armatus* (Rieger 1996), and *S. parryii* (Lacey et al. 1997), and 2–3 days in *S. columbianus* (Shaw 1925a) and *S. richardsonii* (Michener 1980). Females that do not become pregnant after their first mating sometimes become sexually receptive again, after 5 days in *S. richardsonii* (Michener 1980) and after 14–15 days in *S. columbianus* (Shaw 1925a).

Most ground squirrels copulate underground, presumably to avoid harassment by predators. However, a few species, such as *S. beldingi, S. columbianus,* and *S. tridecemlineatus,* regularly copulate on the surface (Sherman 1976, 1989; Hanken and Sherman 1981; Davis 1982, Schwagmeyer and Parker 1987; Dobson et al. 1999). Presence of a copulatory plug indicates that mating has occurred recently in many species including *S. beecheyi* (Tomich 1962), *S. columbianus, S. parryii* (Murie and McLean 1980), *S. richardsonii* (Michener 1984), *S. beldingi* (Sherman 1976), and *S. brunneus* (Sherman 1989). These plugs, which result from coagulation of the ejaculate, fill the vaginal lumen and adhere to the vaginal walls, and remain in place 15–17 hr. Their function, presumably, is to hinder mating with subsequent males (Michener 1984). However, plugs do not form for several hours after mating and females often remove them (Sherman 1976).

Ground squirrel mating systems vary from female-defense polygyny (territoriality) as in *S. armatus* (Rieger 1996), *S. b. parryii* (Lacey et al. 1997), *S. columbianus* (Murie and Harris 1978), and *S. beecheyi* (Dobson 1983; Boellstorff et al. 1994), to scramble-competition polygyny in *S. tridecemlineatus* (Schwagmeyer 1988), to mate guarding in *S. b. brunneus* (Sherman 1989), to resource-based leks in *S. beldingi* (Sherman 1976; Sherman and Morton 1984). Typically, only a few males have most of the reproductive success. For example, in *S. beldingi,* one fourth of the males had two thirds of the copulations (Sherman 1980). In *S. beldingi, S. columbianus* (Murie and Harris 1978), and *S. b. brunneus* (Sherman 1989), the most successful males were older and heavier and superior in male–male competition. Females may use the outcome of male–male competition to choose mates that are strong, vigorous, and successful in surviving hibernation in good condition. In every *Spermophilus* species that has been studied, females mate with multiple males and multiple paternity occurs (Hanken and Sherman 1981; Boellstorff et al. 1994; Murie 1995; Lacey et al. 1997). In *S. beldingi, S. parryii,* and *S. tridecemlineatus,* a female's first mate typically sires the most young (Lacey et al. 1997), and in *S. brunneus,* the male that guarded the female the longest and mated most frequently has the advantage (Sherman 1989).

Litter Sizes. Litter sizes in *Spermophilus* species average 4–9 pups (Armitage 1981). In general, small species that live at lower latitudes and elevations have larger litters than large-bodied, high-elevation, high-latitude congeners (Barash 1989; Yensen 1991). Mean litter sizes are 4.4 pups (range = 1–11) in *S. beldingi* (Sherman and Morton 1984), 4.0 pups (range = 2–6) in *S. parryii* (Karels et al. 2000), 4.0 pups (range = 2–7 pups) in *S. saturatus* (Kenagy et al. 1990), 5.2 pups (range = 1–9) in *S. brunneus* (Sherman 1989), 7.5 pups (range = 1–14) in *S. richardsonii* (Michener 1985), 7.1 pups (range = 4–9) in *S. beecheyi* (Boellstorff et al. 1994), and 8.6 pups (range = 6–12) in *S. mollis* (Rickart 1986).

Within species, litter sizes vary systematically among individuals, depending on the mother's age and condition, and among years and sites, depending on elevation and early-season weather. For example, in *S. columbianus,* females in any given population bear 1–9 pups (Murie et al. 1980), but mean litter sizes vary from 2.0 to 5.4 pups among populations. Litter sizes and proportions of adult reproductive females are greater at low elevations, presumably because of the longer growing season and the greater abundance of food (Dobson and Murie 1987). In *S. beldingi,* supplementing food supplies in areas inhabited by pregnant and lactating females increased litter sizes and juvenile weights (Trombulak 1991). Similarly, for *S. parryii,* adding food and excluding predators from an area of boreal forest resulted in significant increases in female body condition, proportions of females weaning litters, litter sizes, and population densities (Karels et al. 2000).

Gestation and Lactation. Gestation and lactation together generally take 50–60 days in *Spermophilus* species, (Armitage 1981; Michener 1985; Sherman 1989). Gestation usually lasts 3–4 weeks; for example, 21–24 days in *S. richardsonii* (Michener 1985), 23–24 days in *S. columbianus* (Shaw 1925a), 25–26 days in *S. beldingi* (Sherman 1976), and 27–28 days in *S. tridecemlineatus* (Wade 1927) and *S. lateralis* (Cameron 1967). Lactation typically lasts 3–4 weeks, and pups first emerge from their nest burrow (roughly coincident with weaning) at 24–26 days of age in *S. beldingi* (Sherman 1976), 29–31 days in *S. richardsonii* (Michener 1985), and 26–28 days in *S. columbianus* (Dobson et al. 1999). Among species, the duration of gestation increases with body size (Armitage 1981) and the duration of lactation increases with the length of the active season (Armitage 1981; Blumstein and Armitage 1998). Within a species, the duration of lactation varies directly with litter size. This is because lactation is energetically expensive, so it takes longer to rear a larger litter to weaning size (Trombulak 1991). For example, in *S. saturatus,* the mean energy demand of lactation was 336 kJ/day. Females nursing large litters had even higher energy demands, resulting in mass losses of 2.3 g/day during late lactation (Kenagy et al. 1990). In *S. columbianus,* females that weaned litters had higher mortality than those that did not reproduce (Neuhaus and Pelletier 2001).

DEVELOPMENT

Among species, masses of individual pups at birth are directly correlated with adult body sizes, but within species, masses of individual pups vary inversely with litter sizes. For example, *S. columbianus* pups weigh 7.6–12 g at birth (Shaw 1925a), whereas in the smaller *S. mollis,* newborns weigh 3.9 g (Rickart 1982). Neonatal *Spermophilus* pups are altricial, naked, blind, toothless, and with the external auditory meatus closed (Shaw 1925a). The instantaneous growth rate is rapid for the first 30–40 days, then plateaus in most species, including *S. armatus,*

S. columbianus, S. elegans, and *S. richardsonii* (Koeppl and Hoffmann 1981). No sexual dimorphism is apparent until 8–10 weeks, but thereafter males are heavier (Rickart 1986). In *S. elegans* and *S. beldingi,* hair first appears when pups are 10 days old, and by day 25, hair color and pattern are adult-like. In both species, vocalizations are first heard at day 14, eyes open at 21–22 days, and weaning occurs at 28–35 days of age. In the former, hind feet reach adult size at age 42 days, tail length at 56 days, and total length at 63 days (Clark 1970b). In *S. tridecemlineatus* (Bridgewater 1966) and *S. beldingi* (Morton and Tung 1971), growth of the hind feet, tail, and body is complete at 10 weeks, but weight gain continues until hibernation. Growth is more rapid in species that hibernate than in those that do not (Morton and Tung 1971), and, among hibernators, growth is more rapid in species with shorter active seasons (Kiell and Millar 1978). Thus, high-latitude *S. parryii* has the shortest active season and the fastest pup growth rate. Rapid growth rates (6–10 fold body size increase in 5–6 weeks) enable pups to reach a size adequate to survive hibernation even during a short active season.

ECOLOGY AND BEHAVIOR

Population Biology. Ground squirrels live at a wide range of densities, depending on availability of suitable habitat and food. Typical population densities of adults are <20 individuals/ha until juveniles are weaned, at which point densities temporarily increase to >50 animals/ha (Struebel and Fitzgerald 1978; Jenkins and Eschelman 1984; Zegers 1984; Michener 1985; Best 1995c; Lacey et al. 1997). Densities of ground squirrel populations have often been related to local food supplies, as in *S. tereticaudus* (Reynolds and Turkowski 1972), *S. columbianus* (Dobson and Kjelgaard 1985a, 1985b; Dobson 1995), and *S. parryii* (Hubbs and Boonstra 1997). Near rich food resources, such as pastures and alfalfa fields, population densities of adults can become very high, for example, up to 331/ha in *S. mollis* (Rickart 1988). When food supplies are artificially supplemented, as in *S. columbianus,* substantial increases in population density often are observed (Dobson and Oli 2001).

Individuals' home range sizes are usually <1 ha, and range from 0.1 to 7.9 ha (Clark 1970a; Sherman 1976; Recht 1977; Struebel and Fitzgerald 1978; Heaney 1984; Smith and Johnson 1985; Ortega 1990). Home range sizes differ among species and, within species, with the season, sex, and local food abundance. In many species, including *S. columbianus, S. richardsonii, S. b. brunneus, S. variegatus,* and *S. tridecemlineatus,* males enlarge their home ranges during the mating season as they search for receptive females (Murie and Harris 1978; Michener 1985; Sherman 1989; Ortega 1990; Schwagmeyer 1995). Home ranges of males shrink after mating is over, and become tiny just before hibernation. In contrast, home ranges of females are smallest just before emergence of their litters, and usually increase only after pups have dispersed. An exception to this general pattern is a population of *S. beecheyi* in which females had larger home ranges than males, and home ranges did not change significantly after the breeding season (Boellstorff et al. 1994; Boellstorff and Owings 1995). Density of food resources is an important determinant of home range size for both sexes. In *S. parryii,* home range sizes were two to seven times smaller on grids where food was artificially supplemented, regardless of the presence of predators (Hubbs and Boonstra 1997), and *S. variegatus* had smaller home ranges near corn fields than in oak (*Quercus*) savanna habitats (Ortega 1990).

Ground squirrel populations often fluctuate in size from year to year. For example, Boag and Murie (1981) reported interannual variations of 1.18–2.67 times in population sizes of six species. These fluctuations may relate to winter weather and availablility of hibernacula (Carl 1971), infectious diseases such as plague, predator abundance (below), or annual variations in food supplies, for example, in *S. parryii* (Byrom et al. 2000) and possibly *S. columbianus* (Murie 1992). Karels and Boonstra (2000) demonstrated both density-dependent (food supply) and density-independent (winter snow depth) regulation of *S. parryii* populations. These factors may underlie the 10- to 11-year population cycles that occur in some *S. parryii* (Byrom et al. 2000) and *S. franklinii* populations (Erlien and Tester 1984) in the northern portions their ranges. These cycles may also be linked to the well-known 11-year cycles of snowshoe hare (*Lepus americanus*) populations, for example, if predators switch to ground squirrels when hares are rare.

Burrows. Ground squirrels inhabit subterranean burrows. They usually excavate these themselves, although they occasionally use burrows constructed by other species. Typically, only a single adult or a mother with nursing young occupy a burrow system. Burrows usually are excavated in areas of soft soils with good drainage, often on gentle slopes. There are three general types: nest, auxiliary, and hibernation burrows. Howell (1938) described the burrow architecture of various species.

Ground squirrels use nest burrows for refuge during the active season and at night as well as for rearing young. In general, there are multiple open entrance holes, several enlarged chambers, and one or two blind exits that end just beneath the soil surface (to enable escape from predators). In deep, soft soils, nest burrow systems are longer and more elaborate than in shallow, hard soils. In general, nest burrows of males are simpler and shorter than those constructed by pregnant and lactating females (Shaw 1925c; Michener 1993b).

Ground squirrels prefer to excavate nest burrows in sites with good visibility (Owings and Borchert 1975; Ortega 1987) rather than visually occluded sites (Schooley et al. 1996). In *S. parryii,* survival is higher at sites with good visibility due to better predator detection (Karels and Boonstra 1999). Main burrow entrances are placed under trees, rocks, logs, shrubs, or other obstructions when they are available (Owings and Borchert 1975; Yensen et al. 1991); in a few species, burrow entrances are surrounded by low mounds of excavated soil (Shaw 1925c; Howell 1938; Yensen et al. 1991).

Nest burrow entrances often are the most inconspicuous. In *S. columbianus,* they are difficult to find and often located at the periphery of major burrow systems. Their entrances are plugged with soil for several days at a time during gestation and females are secretive around them. Possibly the plug serves to make these "nursery" burrows inconspicuous to predators, including marauding conspecific males (McLean 1978) and females (Sherman 1981b). Other species, such as *S. mexicanus* and *S. tridecemlineatus,* plug their nest burrows daily, not just when nursing pups are present (McLean 1978). In some species, such as *S. beecheyi,* nursery burrows may be 30–60 m long, with multiple nests and interconnecting tunnels (Grinnell and Dixon 1918). Nursery burrows of the smaller bodied *S. b. brunneus* and *S. mollis* are much shorter (up to 2.6 and 8.6 m long, respectively); they extend to depths of 1.2–1.4 m and have total volumes of 23–30 L (Reynolds and Wakkinen 1987; Reynolds and Laundré 1988; Yensen et al. 1991). Burrow diameters of *S. mollis* and *S. elegans* are 1.4–1.5 times wider than they are tall (Laundré 1989).

Ground squirrels use auxiliary burrows as refugia at various distances from their nest burrow. There are four kinds: short burrows under rocks and logs, runways under shrubs, tunnels constructed by other species such as pocket gophers and modified by ground squirrels, and burrows dug in open areas away from obstructions. The latter may consist of a short, blind-ended tunnel that slopes steeply downward with a small chamber at the end; these are most useful for escaping from raptors (Yensen et al. 1991).

Ground squirrels use hibernation burrows only during prolonged dormancy. They are >2 m long and gradually slope downward. There is one entrance and usually two chambers or branches, for the nest and bodily waste. Length and depth of the tunnels and chambers increase with the age and size of the owner (Young 1990). To minimize flooding, hibernation burrows on flat ground often are J-shaped with a drain below the nest chamber, whereas those on hillsides may lack drains, but have a nest chamber constructed above the access burrow (Shaw 1926; Edge 1934; Young 1990; Michener 1993b). Hibernation burrows are difficult to find because the animals are surreptitious while constructing them, the opening is never surrounded by an earthen mound, and the tunnel is plugged from inside as soon as the owner enters to commence dormancy.

Territoriality. Territoriality is more prominent in *Spermophilus* than *Ammospermophilus*. For example, *A. leucurus* is nonterritorial and has linear dominance hierarchies, which are maintained using a variety of visual and tactile signals (Fisler 1976). However, in most *Spermophilus* species, lactating females defend territories surrounding their nursery burrows, as in *S. armatus* (Slade and Balph 1974), *S. beldingi* (Sherman 1980), and *S. columbianus* (Festa-Bianchet and Boag 1982). In many species, males are territorial before and during the mating period, as in *S. beecheyi* (Owings and Coss, 1977), *S. columbianus* (Murie and Harris 1978), *S. richardsonii* (Yeaton 1972), and *S. parryii* (Carl 1971). Lactating females defend territories to sequester food resources near their burrows and to protect pups against infanticide by conspecifics (Sherman 1981b; Ebensberger 1998), whereas males defend territories to sequester sexually receptive females and exclude competitors for matings, food resources, and preferred hibernacula. In species in which females do not defend territories, such as *S. richardsonii* (Michener 1979) and *S. b. brunneus* (Sherman 1989), females are widely spaced and infanticide has not been reported.

Dispersal. Dispersal is "complete and permanent emigration from an individual's home range" (Holekamp and Sherman 1989:232). There are two main types: natal dispersal, which involves emigration of immatures, and breeding dispersal, which involves emigration of adults after mating. Both types are energetically expensive and risky (Michener 1983). Dispersal distances vary among species and sites, being shorter in continuous habitat and longer in fragmented habitats. In one survey, the greatest dispersal distance was 9.6 km (in *S. richardsonii*), but no species had a mean dispersal distance >0.55 km (Holekamp 1984). Although short-distance dispersers in *S. mollis* suffered no greater mortality than nondispersers (Olson and Van Horne 1998), the mortality rate of juvenile *S. parryii* increased over the longer distances they dispersed; most deaths were due to predation (Byrom and Krebs 1999). The dangers notwithstanding, the ubiquity of natal dispersal in ground squirrels indicates that there must be associated advantages.

A clue to what they are is that in most *Spermophilus* species, as in other terrestrial mammals, males are more dispersive than females (Sherman 1977, 1980; Holekamp 1984). Even in the few species in which both sexes disperse (e.g., *S. lateralis* and *S. franklinii*), males go farther than females. Sexual asymmetries in dispersal probably relate to minimizing nuclear family incest and local mate competition between brothers. By dispersing, males also avoid competing for females (sisters, mother) who might refuse to mate with them in any case (Dobson 1982; Hoogland 1982). But why is it that females remain near home? Probably because the location and quality of the nest burrow is so important for reproduction (Holekamp 1984; Holekamp and Sherman 1989; Byrom and Krebs 1999). Females require a secure nest burrow to rear young, but males do not. The site where a female was raised obviously is suitable and safe, so females remain near home. As adults, the sedentary females should prefer to mate with unfamiliar males to avoid deleterious consequences of inbreeding. Thus, young and sometimes adult males disperse to search for females who will be more receptive to them.

In most ground squirrel species, males disperse as juveniles, as in *S. armatus, S. beecheyi, S. elegans, S. richardsonii, S. tereticaudus,* and *S. tridecemlineatus* (reviewed by Pfeifer 1982). For example, 85–90% of juvenile male *S. beldingi* disperse before their first winter (Sherman 1977, 1980), and 100% disperse by the end of their yearling summer (Nunes et al. 1997). However, in *S. columbianus*, males disperse as juveniles in more southern locations, where the active season is longer, and as yearlings where the active season is shorter (Wiggett and Boag 1989a; Festa-Bianchet and King 1984). Organizational effects of male steroid hormones underlie male dispersal (Holekamp et al. 1984), the precise timing of which is dependent on attaining a sufficient body mass to enable individuals to survive the rigors of dispersal and subsequent hibernation (Holekamp and Sherman 1989). Indeed, body mass and time of year are the best predictors of when young male *S. armatus, S. tereticaudus,* and *S. beldingi* will disperse (Nunes and Holekamp 1996, Nunes et al. 1998). In *S. beldingi,* juveniles disperse when they reach a weight of about 125 g, which usually occurs at 7–9 weeks of age. However, in 1995; when the active season was delayed by heavy spring snow, juvenile males accumulated significantly less fat and 90% postponed dispersal until the following year (Nunes et al. 1998). Seasonal fattening preparatory to hibernation took precedence over natal dispersal, for obvious reasons. Conversely, when supplemental food was provided, juvenile male *S. beldingi* emigrated at younger ages. This also indicates that young males were not dispersing in search of food. Females, by contrast, were more likely to disperse when supplemental food was provided in *S. beldingi* (Nunes et al. 1998) and *S. parryii* (Byrom and Krebs 1999), probably because of increased competition for space due to frequent immigration by hungry, unrelated females.

Wiggett and Boag (1989a, 1989b) radio-collared individual *S. columbianus* and followed their dispersal movements. Dispersers departed from their natal burrow systems early in the day, and they were extremely wary and surreptitious. They sought shelter under logs, in shrubs, and in dense vegetation, made quick movements between temporary refuges, and scanned for long periods before moving on. They followed pathways such as valley floors, stream banks, game trails, and roads. Sometimes dispersers retraced their route back to the natal colony at night, then followed the same route out again the next day, gradually extending their excursions until they established a new home range (Wiggett and Boag 1989a, 1989b; see Holekamp 1984 for similar observations on *S. beldingi*).

Dispersers settle when they encounter suitable unused habitat or burrow systems (Hackett 1987) or when they are stopped by insurmountable topographic barriers. Although ground squirrels swim well and readily enter water (Fredrickson 1972), swift rivers apparently form barriers to gene flow (Gavin et al. 1999). For example, in southern Idaho, the Payette River separates *S. mollis* from *S. b. endemicus* and the Snake River separates *S. mollis* from *S. canus*. The Snake River also forms a barrier to *S. beldingi* (Davis 1939). However, *S. beecheyi* succeeded in crossing the Columbia River near The Dalles, Oregon, in about 1915, and expanded its range into Washington at a rate of about 2.4 km/year (Broadbrooks 1961).

Predator Defense. Active ground squirrels are vulnerable to numerous predators, and their antipredator behaviors are well developed. They are extremely vigilant, often standing or sitting up to listen and scan for danger (Sharpe and Van Horne 1998). They are especially vulnerable toward the end of the active season because, due to prehibernation fat deposition, running speed declines. For example, in *S. beldingi* with a relatively lean body mass of 225 g, running speed was 5 m/sec, but when mass increased to 425 g, running speed decreased to 3.5 m/sec (Trombulak 1989). To partly compensate, the animals spend less time on the surface, stay closer to their burrows, and become more wary, disappearing underground at the slightest provocation.

Some species (*S. beldingi, S. beecheyi*) discriminate predators from harmless species, fast-moving predators (usually hawks) from slow-moving predators (usually mammals), and may even recognize whether predators are actively hunting (Sherman 1977; Robinson 1980; Leger et al. 1984; Owings et al. 2001). Most species give loud calls when predators approach, and they have different calls for different types of predators (Blumstein and Armitage 1997). For example, *S. beldingi* responds to fast-moving (aerial) predators by giving single-note whistles, but responds to slow-moving (terrestrial) predators with multiple-note trills (Sherman 1977, 1985; Leger et al. 1984). These calls apparently have different functions. In the presence of terrestrial predators, females call more frequently than males. Calling is dangerous, and predators kill more calling than noncalling ground squirrels. The evolutionary reason for the apparent altruism of callers is protecting relatives: Due to female philopatry, the individuals that are warned by trills are often close female kin. By contrast, in the presence of aerial predators, females and males whistled equally often (Sherman 1985). The closer the animal was to a hawk and the further from a burrow, the more apt it was to whistle. Nonetheless, individuals that whistled were less likely to be captured than individuals that did not whistle. Apparently, noncallers were unaware of the raptor before it was too late. The advantage of whistling probably derives from the predator-confusing effects associated with numerous calling ground squirrels scurrying for cover simultaneously. This causes pandemonium and creates a group in which the whistler can hide.

In a fascinating series of papers, Owings, Coss, and their colleagues reported on the antisnake defenses of *S. beecheyi*. These animals discriminate gopher snakes from western diamondback rattlesnakes (*Crotalus viridis*), a major predator on juveniles (Coss and Biardi 1997). They also assess the snake's size and warmth (attributes of risk) using acoustical signals in the rattle. Lactating females invest more energy in assessing the size of the snake and in monitoring it than do males, and snakes closer to burrows elicit stronger defensive reactions from females (Swaisgood et al. 1999). Harassment behaviors include tail flagging to attract the snake's attention, throwing dirt in the face of the snake with the forepaws, remaining outside of striking range, and pouncing on the snake and biting it (Owings and Coss 1977; Coss and Owings 1978). Even populations that do not encounter rattlesnakes today nonetheless retain many elements of the antisnake behaviors, suggesting a strong genetic component. In addition, the blood sera of various *S. beecheyi* populations inhibit venom proteases of the *Crotalis* species that is historically relevant to each population (Biardi et al. 2000). Recently, Owings et al. (2001) reported that *S. variegatus* from wilderness areas also discriminate gopher snakes from rattlesnakes, but *S. variegatus* from urban areas do not. Therefore, unlike *S. beecheyi*, these ground squirrels apparently learn antisnake behaviors through experience.

Social Behavior. Species are "adaptively" social when the benefits of aggregation (e.g., reduced predation, social facilitation of foraging) outweigh the costs (e.g., parasite and disease transmission, competition for space, food, or mates). Species are "nonadaptively" social when suitable habitats are limited in size or patchy in distribution, forcing individuals to live near each other. Ground squirrels vary considerably in their degrees of sociality, which reflect variations in the costs and benefits of social living due to differences in habitats, food and space requirements, predators, and foraging strategies (Blumstein and Armitage 1998).

Armitage (1981) and Michener (1983a:534) distinguished five grades in a continuum of increasing sociality among ground-dwelling sciurids. Grade 1, characterized as "nonsociality," is exemplified by *Ammospermophilus* spp. and *S. franklinii*. In these species, males and females do not share territories or scent mark, juveniles disperse shortly after weaning, juvenile home ranges are distinct from those of their mother and littermates, and social interactions are rare and usually agonistic, even among kin. Individuals derive no benefits from proximity to conspecifics, but groups occasionally form due to resource patchiness.

Mothers and daughters are the fundamental units of social grades 2–5. Daughters remain within the mother's home range, and mothers, daughters, and sisters behave cooperatively. Males do not give parental care. Grade 2, which Michener (1983a) characterized as "single family female kin clusters," describes most species of *Spermophilus*, including *S. armatus, S. beldingi, S. elegans, S. richardsonii,* and *S. tereticaudus*. Males and females are polygamous and they occupy distinct home ranges after the breeding season. Whereas juvenile males disperse, females share the mother's home range and nearby areas with close female kin. In *S. beldingi* female kin cooperate to defend territories and juveniles against distantly related and unrelated females (Sherman 1980, 1981a). Females recognize each other using odors, that vary precisely with relatedness (Mateo 2002) and the memory of how kin smell is retained through hibernation (Sherman and Holmes 1985; Mateo and Johnston 2000). Female *Spermophilus* have three types of scent glands: oral glands on the face just posterior to the angle of the jaw, dorsal glands in various patterns on the back, and anal glands associated with three extrusible papillae in the anus. Chemicals from these glands constitute the "signatures" used in territory marking (Kivett et al. 1976), discrimination of familiar and unfamiliar individuals (Harris and Murie 1982; Hare 1998), and kin recognition by phenotype matching (Holmes 1986a; Mateo 2002), which results in nepotism (Sherman 1980, 1981a).

Grade 3, characterized as "female kin clusters with male territoriality," is exemplified by *S. columbianus, S. parryii,* and *S. variegatus*. In these species, juvenile males disperse and females are philopatric. Males and females are polygamous, and adult males defend territories during and after the breeding season. Territories of males encompass overlapping home ranges of several clusters, of female kin. Within kin clusters interactions are amicable and cooperative, especially among the closest relatives. Michener's (1983a) social grades 4 and 5, which consist of isolated and colonial harem-polygyny, respectively, occur in marmots and prairie dogs, but not among ground squirrels.

Body size is an important correlate of sociality. As Armitage (1981:36) pointed out, "species are more social when large body size combined with a relatively short growing season is associated with delayed dispersal and occurs in those species typically breeding for the first time at age two or older." Other life history consequences of increasing social complexity are increased survival of offspring through the first year, greater reproductive skew (not all females breed), reduced reproductive effort per year, and decreased average litter size (Barash 1989).

FEEDING HABITS

Ground squirrels are primarily, but not exclusively, herbivorous. *Ammospermophilus* species eat mainly green vegetation as well as seeds, berries, fruits, insects, and occasionally small vertebrates (Hoffmeister 1986; Best et al. 1990a, 1990b; Belk and Smith 1991). Animal matter constitutes >25% of the diet of *A. leucurus* in some seasons (Belk and Smith 1991).

Spermophilus species eat a variety of grasses and forbs early in the active season, but consume flower and grass seeds when those become available. Many species have cheek pouches for storing seeds, and pouch sizes vary with the degree of granivory (Grinnell and Dixon 1918). Juveniles learn which plant species to eat by observing adult conspecifics, especially their mother (Peacock and Jenkins 1988). *Spermophilus* species also consume berries, insects, bird eggs, baby mice, carrion, and sometimes conspecific young (Howell 1938; Sherman 1981b; Rickart 1982; Yensen and Quinney 1992; Frank 1994; Dyni and Yensen 1996; Van Horne et al. 1998).

There are variations on this basic diet. *S. lateralis* (subgenus *Callospermophilus*) is more of a forest dweller than most other temperate-zone ground squirrels, and it also is more omnivorous, consuming forbs, pine nuts, certain shrubs, insects, eggs, small vertebrates, and carrion (Bartels and Thompson 1993). The closely related, forest-dwelling *S. saturatus* is primarily fungivorous. This species consumes a high proportion of sporocarps of hypogeous fungi, which are abundant and readily detectable via strong odors, but offer a relatively low digestibility of about 60% (Cork and Kenagy 1989). They also consume grasses, forbs, bark, seeds, and berries (Trombulak 1988).

Species in subgenera *Ictidomys* and *Xerospermophilus* are more insectivorous and carnivorous. For example, insects such as lepidoptera larvae and grasshoppers constitute 39–70% of the diet of *S. tridecemlineatus* (Struebel and Fitzgerald 1978). *S. mexicanus* seasonally consumes many insects and may prey on small vertebrates up to the size of *Sylvilagus* (Packard 1958), although its main dietary items are leaves and mesquite beans (*Prosopis* spp., Young and Jones 1982). *S. tereticaudus* consumes mainly plant material and seeds, but it also preys on insects and small vertebrates up to house sparrow (*Passer domesticus*) size (Ernest and Mares 1987). *S. mohavensis* specializes on forbs (50–85% of the diet) and insects (5–8%), but switches to mycorrhizal fungi, and fruits, flowers, and seeds of yucca (*Yucca* spp.) when available (Best 1995b).

As hibernation approaches, ground squirrels seek foods that are high in polyunsaturated fatty acids, such as seeds. Linoleic acid and alpha-linoleic acid in particular promote earlier entry into hibernation, permit lower body temperatures during hibernation, and enable hibernating individuals to endure longer bouts of torpor. White adipose tissue of *S. beldingi* contains six different fatty acids, among which alpha-linoleic acid is especially abundant (Frank et al. 1998). Ground squirrels do not synthesize fatty acids, but acquire them, in the appropriate amounts, in their food (Frank 1991). However, oxidation of polyunsaturated fatty acids produces toxic lipid peroxides, which can damage tissues and even cause death. Thus, individuals must select dietary items with intermediate levels of polyunsaturated fatty acids to achieve a compromise between maximizing the hibernation-related benefits of polyunsaturated fatty acids and minimizing the damage caused by lipid peroxides (Frank et al. 1998).

For example, in *S. lateralis,* the optimum diet contained 33 mg/g of polyunsaturated fatty acids (Frank et al. 1998). To maintain this diet, *S. lateralis* balanced pine seeds, which are high in polyunsaturated fatty acids, with herbaceous plant parts, which are low in polyunsaturated fatty acids. When linoleic acid intake increases above the optimum, the animals must increase vitamin E intake to offset it (Frank et al. 1998). *S. lateralis* that were maintained on low-linoleic acid diets (18 mg/g) were less likely to enter hibernation and had higher mortality during hibernation than those maintained on a more optimal intake of 23–25 mg/g (Frank 1994).

MORTALITY

Ground squirrels live <6 years on average and <12 years at the maximum (Sherman and Morton 1984; Sherman and Runge 2002). For example, in a long-term study of individually marked *S. beldingi* (Sherman and Morton 1984), males lived 2–3 years and females lived 3–4 years on average and 7 years (males) and 12 years (females) at the maximum. In *S. b. brunneus* males lived 1–2 years and females lived 2–3 years on average and 5 years (males) or 7 years (females) at the maximum (Sherman and Runge 2002). Major mortality sources for *Spermophilus* include predators, diseases (e.g., plague; hemorrhagic gastritis in *S. mollis*; Wilber et al. 1996), and starvation or freezing (Morton and Sherman 1978; Neuhaus et al. 1999). Young of both sexes are particularly vulnerable during dormancy (Michener 1979). Among 10 *Spermophilus* species, mortality during the first year (weaning to emergence from hibernation the following spring) ranged from 97% to 54%.

In all *Spermophilus* species that have been studied, longevity is sexually dimorphic: females live longer than males on average and at the maximum (Sherman and Morton 1984; Sherman and Runge 2002). This difference often is manifested in population sex ratios. Whereas sex ratios are usually close to 1:1 among juveniles, they rapidly shift in favor of females among older age classes. For example, in *S. beldingi* 3- to 4-year-old females outnumbered males by ratios of 2.2:1 to 6.0:1 in different populations (Sherman and Morton 1984; Verts and Costain 1988). Similar skewed sex ratios were observed in *S. tridecemlineatus* (2.1 and 2.8:1; Struebel and Fitzgerald 1978) and *S. mollis* (1.4:1; Rickart 1982), and even more female-biased sex ratios have been observed in *S. richardsonii* (5:1 to 10:1; Michener and Michener 1977). Skewed sex ratios are apparently due to differential male mortality (Michener and Michener 1977; Sherman and Morton 1984; Neuhaus and Pelletier 2001). This occurs because of the dangers associated with male dispersal, injuries in fights with other males over sexual access to females (Sherman 1976; Michener 1983b), predation during the mating season due to male mobility and visibility; higher overwinter mortality because dispersing males may not find as satisfactory hibernacula as females, who remain near their birth site (Carl 1971); and senescence, which results from all the aforementioned extrinsic mortality sources (Williams 1957).

The short active season of many species makes the timing of reproduction crucial because, before entering hibernation, juveniles must accumulate sufficient body mass to carry them through the long winter (Morton 1975; Michener 1984; Murie and Boag 1984). In *S. columbianus* (Dobson et al. 1999) and *S. armatus* (Rieger 1996), females that were heavier at spring emergence (older females) bore larger litters of heavier pups than lighter (younger) females. The heavier females weaned litters earlier in the season, giving the young a head start on gaining weight preparatory to hibernation. Date of weaning and weight at weaning were correlated with prehibernation weight, which in turn was correlated with overwinter survival.

Predators. Major predators of various species of ground squirrels include badgers (*Taxidea taxus*), long-tailed weasels (*Mustela frenata*), coyotes (*Canis latrans*), red foxes (*Vulpes vulpes*), gray foxes (*Urocyon cinereoargenteus*), prairie falcons (*Falco mexicanus*), Cooper's hawks (*Accipiter cooperi*), goshawks (*A. gentilis*), red-tailed hawks (*Buteo jamaicensis*), Swainson's hawks (*B. swainsoni*), northern harriers (*Circus cyaneus*), golden eagles (*Aquila chrysaetos*), ravens (*Corvus corax*), gopher snakes (*Pituophis catenifer*), and western diamondback rattlesnakes (*Crotalis viridis*) (Michener and Michener 1977; Sherman 1977; Armitage 1981; Sherman and Morton 1984; Jenkins and Eschelman 1984; Zegers 1984; Trombulak 1988; Elliott and Flinders 1991; Yensen et al. 1992; Yensen and Sherman 1997). Among these predators, badgers are specialists on ground squirrels (Murie 1992). Badgers can dig very rapidly and they use a tactic of sealing the prey in by plugging secondary entrances to a burrow system before excavating the main entrance (Knopf and Balph 1969).

Conspecifics are important predators in some *Spermophilus* species, such as *S. beecheyi, S. beldingi, S. columbianus, S. parryii,* and *S. tridecemlineatus* (Sherman 1981b; Stevens 1998). In *S. beldingi,* an estimated 8% of pups were victims of infanticide (Sherman 1981b), and the main perpetrators were nonlactating females and yearling males. In *S. columbianus,* the chief aggressors were lactating females (Stevens 1998). In both species, victims were nearly or recently weaned pups, and attacks usually occurred close to their natal burrow. In *S. beldingi,* females that killed pups did not eat them, but did settle near their victims' burrows. This implies that infanticide functioned to remove future competitors for preferred burrow sites. By contrast, males that committed infanticide ate their victims, which suggests that they killed to obtain food. In *S. columbianus,* infanticide was concentrated in certain maternal lineages, and may serve to create space for occupancy by daughters of the infanticidal females.

Parasites. Ground squirrels serve as hosts for a variety of ecto- and endoparasites. For example, a survey of *Ammospermophilus leucurus* and five species of *Spermophilus* in Utah found 109 species of parasites including 7 protozoans, 1 trematode, 7 cestodes, 13 nematodes, 1 worm, 42 fleas, 5 lice, 8 mites, 5 ticks, and 20 fly larvae (Jenkins and Grundmann 1973). Another survey found 21 species of "true" ground squirrel fleas and 14 more that were considered "accidental" on *S. richardsonii, S. franklinii,* and *S. tridecemlineatus* (Galloway and Christie 1990). In small, fragmented ground squirrel populations, the proportion of individuals harboring fleas, ticks, and lice and the ectoparasite load per individual were lower than in larger, more continuously distributed populations, presumably because of more frequent transmission in the latter (Yensen et al. 1996).

Much of the interest in parasites of ground squirrels has been fueled by concern over their role as reservoirs of sylvatic plague (*Yersinia pestis*). Plague entered North America from Asia in 1899, and was noticed in wild rodents near Berkeley, California, in 1908; from there, the exotic bacterium spread rapidly. Ground squirrels and prairie dogs are often hosts of plague-carrying fleas (Link 1955; Barnes 1982), and the disease may, but does not always, devastate their populations (Svihla 1939; Hansen 1956; Messick et al. 1983; Hoogland 1995). The spread of plague can be monitored by observing ground squirrel die-offs and by checking antibody titers of their major predators, such as badgers and coyotes (Messick et al. 1983; Dyer and Huffman 1999). Plague is spread when ground squirrels disperse, and there is some evidence that fleas target dispersers. In *S. beecheyi,* juvenile males have more fleas (*Oropsylla montana* and *Hoplopsyllus anomalus*) than juvenile females, and juveniles of both sexes have more fleas than adults. The fleas were thus infesting individuals that were most apt to disperse and to disperse the farthest (Bursten et al. 1997).

Infestations of some fly larvae can also be lethal to ground squirrels. *Neobellieria citellivora* (Diptera: Sarcophagidae) have killed *S. richardsonii* and *S. columbianus* in Alberta (Michener 1993a). Other flies parasitic on Richardson's ground squirrels include *Liopygia argyrostoma* (Sarcophagidae) and *Calliphora coloradensis* and *Phormia regina* (Calliphoridae). The latter were apparently attracted to the wounds made by *Neobellieria citellivora.*

Other ground squirrel parasites are more benign: nematode eye worms (*Rhabditis orbitalis*) have been found in *S. b. brunneus* (Yensen et al. 1996), and a species of *Demodex* (Acari, Demodecidae) occurs in the ear canals of *S. beecheyi*. Although *Demodex canis* can cause mange in dogs, the squirrels were not symptomatic (Waggie and Marion 1997). The benign protozoan parasite *Sarcocystus campestris* (Sarcocystidae) occurs in *S. richardsonii* and *S. tridecemlineatus* (Lindsay et al. 2000). Coccidians (*Eimeria*) also infect several species of ground squirrels (Todd and Hammond 1968; Stanton et al. 1992; Wilber et al. 1994). *S. elegans* had stable levels of infection by six species of coccidians, three commensal and three parasitic; the latter can adversely affect nutrient assimilation (Seville et al. 1992, 1996).

MANAGEMENT

Field Techniques. Native Americans caught ground squirrels for food by flooding occupied burrows and grabbing escapees (Alcorn 1940). This works best when an animal enters a short, blind-ended auxiliary burrow. Native Americans also caught ground squirrels by "fishing" (i.e., dangling a hook baited with meat down a nest burrow).

Today, researchers employ various methods to enumerate populations, including focal animal livetrapping, grid live- and snaptrapping, counts of burrows and alarm calls, track stations, and infrared imaging. Among all methods, total mark–recapture is the most direct and accurate. This involves livetrapping, tagging, and dye-marking individuals, observing the population regularly, and trapping unmarked animals until, eventually, every individual has been marked (Sherman 1976, 1981a, 1989; Hoogland 1995). Of course, the disadvantage of this method is the time and effort necessary to completely mark the population—and to re-mark it when the animals molt in midseason. Grid live- or snaptrapping requires less intensive effort, but is less accurate because certain individuals consistently avoid traps and it inevitably results in greater mortality.

Among the indirect census methods, abundance of tracks at baited powder stations (Drennan et al. 1998; Hubbs et al. 2000) and infrared images of burrow entrances have been correlated with numbers of ground squirrels trapped in those areas. Infrared imaging was the more accurate method, but it required low ambient temperatures and assumed burrows were not plugged (Hubbs et al. 2000). Responses to recorded alarm calls also have been used to estimate population sizes. Lishak (1977) detected 83% of known residents in an *S. tridecemlineatus* population using playbacks. Accuracy of vocal censuses varies with the stage in the annual breeding cycle (lactating females are most responsive) as well as the time of day and weather. Finally, burrow counts have been used as indices of ground squirrel abundance (Nydegger and Smith 1986). However, the ratios of auxiliary burrows to nest burrows, numbers of ground squirrels to total burrows, and population densities to total burrows vary among species, habitats, and different stages in the annual cycle. Burrow counts thus provide only relative indices of population size. Whereas an area with hundreds of fresh burrows certainly contains more animals than an area with only a few dozen, hundreds of burrows do not necessarily mean hundreds of ground squirrels.

To directly enumerate or study populations, researchers can capture ground squirrels alive in box and walk-in traps or kill them with snap traps (Horn and Fitch 1946). Most researchers catch them in single-door wire mesh box traps such as Tomahawk livetraps (Sherman 1976, 1981a, 1989; Hoogland 1995). Placing traps near burrow entrances is much more effective than grid or broadcast trapping (Melchior and Iwen 1965). However, success is greatest using focal animal trapping (Sherman and Morton 1984; Sherman and Runge 2002). This involves walking slowly toward a particular individual until it submerges, totally blocking that burrow entrance with four box traps (two open and two shut) or inserting a tubular trap with a hinged door into the burrow entrance (Prychodko 1952; Wobeser and Leighton 1979), temporarily blocking all other nearby surface burrows with sticks or rocks, and waiting at a distance for the animal to re-emerge and enter a trap. To avoid killing captives, researchers must never leave traps unattended for >30 min, especially on warm, sunny days.

Some *Spermophilus* species are attracted to baited traps. Sullins and Verts (1978) compared the effectiveness of crimped oats, oat groats, whole wheat, rolled wheat, rye, barley, chopped apples, and chopped lettuce for attracting *S. beldingi*. Chopped apples were preferred in all seasons, but attractiveness of other baits varied seasonally, with succulent baits preferred early and grains in midseason. Other researchers have had success in baiting traps with peanut butter (e.g., Sherman 1976, 1980, 1981a, 1981b; Lacey et al. 1997; Trombulak 1991), rolled oats, oat groats (Verts and Costain 1988), and carrots (Buck and Barnes 1999). However, bait is unnecessary for some species, especially when using focal animal trapping [e.g., *S. armatus* (Balph 1968) and *S. brunneus* (Gavin et al. 1999)].

Ground squirrels can be removed from livetraps using gloves or a cloth bag. Anesthesia is not necessary for routine operations like tagging, weighing, marking, and obtaining hair follicles or buccal scrapings for genetic analyses (May et al. 1997). Gentle treatment and covering the animal's eyes minimizes mortality and subsequent trap aversion. Ground squirrels can be restrained for weighing by rolling them in a cloth holding bag fastened with a Velcro strip or drawstring. Blood samples can be obtained from the suborbital sinus (Hanken and Sherman 1981) or the tip of the tail. Cardiac puncture has been used, but it is considerably riskier.

Ground squirrels can be tagged for long-term identification with numbered eartags made of noncorrosive metal (e.g., monel fish-fingerling tags) or with passive integrated transponder (PIT) tags (Sherman and Morton 1984; Schooley et al. 1993; Sherman and Runge 2002). The former are inexpensive and easy to attach, but they sometimes rip through the ear flesh and are lost. Use of duplicate tags, one for each ear, generally solves this problem. PIT tags, which can be inserted under the skin of the back, are not easily lost, but they are expensive and involve some uncertainty. That is, if the PIT-tag reader shows no response, it is uncertain whether the animal was never tagged or the tag was not found in the animals' body.

Anesthesia may be necessary for fitting radiocollars or other procedures. Numerous anesthetics have been tried with varying success (Olson and McCabe 1986). Injectable ketamine hydrochloride produces deep anesthesia, but lasts up to 2 hr and may affect stress physiology. Among inhalants, methoxyflurane is the most expensive and has the longest recovery times, but a simple bell jar or in-circuit vaporizer may be used to control concentration for induction and maintenance of anesthesia, and overdosing is seldom a problem. Halothane or isoflurane require an out-of-circuit vaporizer to monitor concentration levels, but they are cheaper than methoxyflurane and allow rapid recovery. Methoxyflurane is probably the anesthesia of choice for field studies because of its relative safety, but isoflurane may be used if a battery-operated vaporizor is available (McColl and Boonstra 1999). If euthanasia becomes necessary, nembutol is effective and humane, but its use is restricted. Potassium chloride kills quickly and painlessly when injected directly into the heart.

Age structure of a ground squirrel population can be established unequivocally only by tagging juveniles or yearlings and subsequently recapturing them yearly until they all disappear (Sherman and Morton 1984; Hoogland 1995; Sherman and Runge 2002). Ages at time of death of museum specimens can be estimated from tooth wear classes (Yensen 1991) or by counting annuli in tooth cementum (Montgomery et al. 1971) or annual adhesion lines in periosteal bones and mandibles (Klevezal and Kleinenberg 1967; Costain and Verts 1982; Rickart 1982). Ages of live individuals have been estimated using the strength of tail collagen fibers (Sherman et al. 1985). This technique classified the ages of 30% of *S. beldingi* correctly, 62% within ±1 year of true age, and 86% within ±2 years of known age.

Body fat was traditionally measured by euthanizing individuals, homogenizing their tissues, extracting fat using chloroform or petroleum ether and Soxhlet funnels, and drying and weighing the residues (Rickart 1982). More recently, lipid content has been measured using portable total body electric conductance machines (Nunes et al. 1998; Buck and Barnes 1999). The advantage of this technique is that it is nondestructive, but the machine's readings must initially be calibrated for each species using the Soxhlet funnel technique.

Maintenance in Captivity. Ground squirrels can be kept in captivity during the active season by providing a variety of vegetables and seeds for food and avoiding extreme heat. Several species of ground squirrels, including *S. beldingi* (Holmes and Sherman 1982; Mateo and Johnston 2000) and *S. lateralis* (Pengelley and Fisher 1961), also have been kept through hibernation. The *S. beldingi* were housed individually in cages with plenty of nesting material, and fed green vegetables and fat-laden foods (e.g., seeds and lab chow) *ad libitum*. In September, animals were placed in temperature-controlled rooms (5°C) with filtered air, where they hibernated for 6–7 months; overwinter mortality was minimal. Ground squirrel species that have been bred in captivity include *S. columbianus* (Shaw 1925a), *S. beldingi* (Holmes 1986b), *S. beecheyi* (Howard 1959; Marsh and Howard 1968), *S. tridecemlineatus* (Wade 1927; Bridgewater 1966; Barr and Musacchia 1968), and *S. lateralis* (Cameron 1967). Shaw (1925a) used large, natural outdoor enclosures, whereas Marsh and Howard (1968), Barr and Musacchia (1968), and Holmes (1986b) used ca. 1 × 2 × 2 m outdoor pens or cages; Bridgewater (1966) used a large indoor cage for breeding. In all cases, one to three females were caged with a single male until they mated. Then the females were isolated and fed a mixture of fresh vegetation, seeds, and commercial lab chow.

Based on these successes, captive breeding might be an option for helping conserve threatened and endangered ground squirrels. However, there are some formidable obstacles, including overwinter mortality in cold rooms, the need to synchronize male and female reproductive readiness in the spring, the short period of female receptivity, and the low annual reproductive rate with a single reproductive period per year and a small litter size.

Control Methods. Ground squirrels have been long considered nuisances to agriculture (Howell 1938). The conversion of native grassland habitats to agricultural uses and the introduction of palatable plant species to rangelands provided ground squirrels with multiple new food sources, and their populations increased sharply in agricultural lands (Marsh 1998). This initiated a continuing battle with farmers and ranchers, which is especially intense during drought years, when plant productivity is poor (Marsh 1998). Burrows may cause inability to control water spread in irrigated pasturelands and result in reduced yields, and can also result in loss of water and ditchbank erosion. Ground squirrels can consume or spoil certain crops (e.g., grains, alfalfa) and they also may compete directly with livestock for forage (Howard et al. 1959). Two hundred *S. beecheyi* have been estimated to consume as much as one 450-kg steer (Grinnell and Dixon 1918), but this needs documentation. In a recent year, crop damage caused by *S. beldingi* and *S. beecheyi* in California and Oregon was estimated at $12–16 million, and economic losses of all types to ground squirrels were estimated at $20–28 million (Marsh 1998).

Other putative economic impacts of ground squirrels are harder to document. These include damaged farm machinery from running over burrow mounds, increased erosion and gully formation on overgrazed lands (Longhurst 1957), and injured livestock from tripping in burrow entrances (Marsh 1998). Although broken legs of livestock are often mentioned by ranchers, its occurrence has undoubtedly been overrated (Hoogland 1995).

From the 1900s until the 1950s, ground squirrels were routinely controlled with strychnine (Marsh 1998). However, strychnine is a dangerous, nonselective poison, and there is considerable risk of killing other wildlife species that do not avoid bitter tastes (El Hani et al. 1998). Compound 1080 (monofluoroacetic acid or sodium monofluoracetate) was used for ground squirrel control in the 1950s through 1970s, but it, too, was nonspecific and resulted in mortality of nontarget mammals, birds, and insects (Hegdal et al. 1986; Whisson et al. 1999). As a result, 1080 has been deregistered for use on ground squirrels.

Currently, agriculturalists eliminate ground squirrels with anticoagulant baits (e.g., zinc phosphide, chlorophacinone, diphacinone), fumigation (aluminum phosphide, acrolein), trapping, cultural methods, habitat modification, and shooting (Marsh 1994). Each technique has its limitations. Shooting is time-consuming and ineffective when animals remain underground. Trapping is labor-intensive. Zinc phosphide treated bait (e.g., wheat) kills 85–90% of ground squirrels that eat it, but survivors avoid treated baits (Matschke and Fagerstone 1982; Matschke et al. 1983) and some populations seem to have an aversion to baits. Fumigating many burrows in a large population is labor-intensive and may not be practical economically. Aluminum phosphide is an inexpensive fumigant, which suffocates the animals. It is toxic to all burrow-dwelling animals and very effective. However, labor costs ($500–750/ha) make it too expensive for many situations (Askham 1994; L. Bangerter, pers. commun., 2001). Acrolein is costly (up to $208/ha for dense populations). Placing baits on elevated platforms (Whisson 1999) or special bait stations (Marsh 1994) may reduce poisoning of smaller nontarget species.

Each state registers pesticides it deems necessary and justifiable for use on specific species. All pesticides mentioned here are restricted in their use, and those that are lawful in one state may not be in another. For example, chlorophacinone and diphacinone are registered in only some states. Each state department of agriculture maintains a list of legal applications for specific species, and the applicator must be state licensed.

A new population control technique, use of the chemosterilant mestranol, was tried on *S. richardsonii* in Alberta (Michener 1996). This steroid hormone inhibited female reproduction, but reduced fertility of the target population was offset by immigration from nearby populations. As a result, the local authorities continued to recommend traditional methods (Michener 1996).

There are cultural methods to discourage ground squirrels that do not involve chemical applications. First, most species prefer short vegetation (only *S. franklinii* inhabits tall-grass prairies), so planting taller growing species such as grasses or corn discourages them. Second, because wet soils are unsuitable for burrowing, periodic flooding or irrigation will make fields unacceptable (Marsh 1994). Third, ground squirrels do not persist in fields that are tilled, nor survive long on land converted to row crops (Marsh 1998). Of course, the problem with all three methods is that ground squirrels may persist at the margins of agricultural fields and enter them only to forage. Construction of physical barriers (fences) to discourage immigration usually is not effective because the ground squirrels tunnel underneath (Marsh 1994; Askham 1994). However, planting trees or shrubs (windbreaks) in peripheral areas will reduce their suitability for habitation, and occasional use of baited kill-traps can reduce remnant populations and immigration.

Although ground squirrels certainly can cause economic losses, the amount has been overestimated. Because ground squirrels are large, diurnal, and conspicuous, they have been blamed, inappropriately, for all animal-caused damage, when the true culprits may be smaller nocturnal or subterranean animals like voles (*Microtus* spp.), gophers (*Thomomys* spp.), mice (*Peromyscus* spp.), raccoons (*Procyon lotor*), and skunks (*Mephitis mephitis*). Moreover, there are many fewer ground squirrels now than there were 100 years ago. Because of the need for ground squirrel conservation (see Conservation), control measures should be reserved for situations where they are truly warranted.

Translocation. Ground squirrels can be trapped and translocated successfully. Translocation has been used as a management tool to remove the animals and as a conservation technique to re-establish extirpated populations (Van Vuren et al. 1997). The most successful translocations involve moving multiple animals to suitable habitat in the middle of the species's range, especially to places where there already are conspecifics or where they lived until recently (Griffith et al. 1989). Soft releases, in which animals are placed in a restraining cage for protection and allowed to acclimate to the new site before release, yield better results (lower emigration and mortality) than hard releases, in which animals are simply trapped, moved, and released (Salmon and Marsh 1981; Van Vuren et al. 1997). Translocated ground squirrels may attempt to return home; their homing success is directly related to the size of their home ranges and inversely related to the distance between the sites of capture and release (Van Vuren et al. 1997).

There is no agreed-upon best time for translocations and releases. Salmon and Marsh (1981) captured *S. columbianus* in the middle of the active season and maintained them until just before hibernation. They then placed the animals in release cages that were partially filled with soft soil, where the ground squirrels dug burrows and hibernated. The following spring, the cages were opened and the ground squirrels established residency. Ground squirrels also have been translocated during the active season, with acclimation times of a few days or weeks, as in *S. richardsonii* (Michener 1996) and *S. columbianus* (Wiggett and Boag 1986). Success may be improved by giving translocated individuals access to uninhabited burrows. Also, capturing pregnant females, allowing them to give birth in captivity, and translocating females and litters together may improve success because it reduces chances that translocated females will emigrate and gives pups opportunities to learn about their natal environment before hibernation.

CONSERVATION

Ecological Value. Ground squirrels, marmots, and prairie dogs are keystone species in their ecosystems (Yensen et al. 1992; Vander Haegen et al. 2001). Their loss may therefore have serious consequences. Burrowing rodents are important food sources for many mammalian, avian, and reptilian predators. Far more importantly, however, ground squirrels loosen, aerate, move, and mix soils. For example, an *S. columbianus* population in the Canadian Rockies brought 1.1–1.4 metric tons/ha/yr to the surface (D. J. Smith and Gardner 1985). A population of *S. parryii* in Yukon brought up 6.9 metric tons/ha/yr (Price 1971). In arid and semiarid regions, burrowing by mammals is the primary mechanism for bringing nutrients (especially insoluble nutrients) from deep soil layers to the surface (Abaturov 1972). Therefore, burrowing strongly influences soil fertility, plant species composition, and primary productivity. Indeed, burrowing mammals are essential to many western ecosystems where earthworms are rare (Grinnell 1923).

Burrowing also encourages water infiltration, alters soil ion exchange capacity, water holding capacity, organic matter content, and inorganic nutrient levels (Laundré 1998). In southeastern Idaho, which is semiarid, infiltration of water into the soil was 21% higher in the vicinity of burrows of *S. mollis* and *S. elegans* than in nearby control areas without burrows. Water also penetrated deeper into the soil in the vicinity of burrows. The additional water enhances decomposition of organic matter, plant survival, and seed germination rates (Laundré 1993). As a result, productivity of western wheatgrass (*Agropyron smithii*) and big sagebrush (*Artemisia tridentata*) increased 20% in areas adjacent to *S. mollis* burrows (Laundré 1998).

Burrowing brings to the surface seeds that have been buried in the seed bank, encouraging their germination and increasing local plant species diversity (Whitford and Kay 1999). Burrowing also increases microhabitat heterogeneity in several ways. Burrows trap seeds and plant litter, forming nutrient-rich pockets for germination and loose soils for root growth. Ground squirrels make runways and dig small holes while excavating bulbs and roots. These small-scale sources of soil disturbance and local heterogeneity trap seeds and litter (Whitford and Kay 1999). Furthermore, ground squirrels attract badgers, which are even more prodigious diggers. Badger digs greatly increase the amounts of soil moved, aeration, nutrient turnover, seed germination, and water infiltration and also provide nest sites for rabbits (*Sylvilagus* spp.) and burrowing owls (*Athene cunicularia*).

Eighty years ago, Grinnell stated (1923:148–49), "We grant that the farmer must combat the gopher and ground squirrel in his fields and gardens; we sympathize with him for yearning for the total eradication of the rodents there; and we will help him in every conceivable way to control them. But we do not agree with the policy of wholesale extermination advocated by some persons for all areas alike. ... On wild land the burrowing rodent is one of the necessary factors in the system of well being."

Grinnell's wisdom has largely been ignored. It is time to reverse this situation. Although ground squirrels can sometimes cause problems in agricultural situations, nuisance populations can be dealt with through focused control measures. Immediate negative responses to ground squirrels and attempts to destroy them, particularly on rangelands, are no longer justifiable economically or ecologically. The presence of ground squirrels brings ecological benefits that can far outweigh short-term economic losses, especially on arid lands. Therefore, it is ironic that farmers and ranchers in the American West still expend so much time and money attempting to eradicate animals that are crucial for the health and survival of the very ecosystems on which their own livelihoods depend.

Population Status. Populations of many ground squirrels, especially western shrub–steppe species, have declined throughout their historical ranges. The main reasons are (1) conversion of grassland habitats to cultivation or urban use; (2) deterioration of rangeland and forest meadows due to invasion of exotic annual plants, often resulting from overgrazing; (3) shrinking of meadows due to plant succession or conifer encroachment, the result of 100 years of fire suppression; and (4) decades of persecution through poisoning, trapping, and shooting (Yensen et al. 1992; Hafner et al. 1998; Gavin et al. 1999). In North America, 15 *Spermophilus* species or subspecies are of conservation concern (Hafner et al. 1998; E. Yensen and P. W. Sherman, unpublished data) (Table 10.2), and 2 more species are of concern in Mexico

TABLE 10.2. Conservation status of 15 taxa of North American ground squirrels

Species	Status[a]	Source
Ammospermophilus nelsoni	State Threatened, IUCN Endangered	1
Spermophilus brunneus brunneus	ESA Threatened; IUCN Critically Endangered	2
S. brunneus endemicus	ESA Candidate; IUCN Vulnerable	3
S. canus	Conservation concern	3
S. elegans nevadensis	IUCN Data Deficient	1
S. franklinii	State Endangered	4
S. lateralis wortmanni	IUCN Data Deficient	1
S. mohavensis	IUCN Vulnerable	1
S. mollis artemisiae	Conservation concern	3
S. parryii kodiacensis	IUCN Data Deficient	1
S. parryii lyratus	IUCN Data Deficient	1
S. parryii nebulicola	IUCN Data Deficient	1
S. townsendii	IUCN Data Deficient	1, 3
S. tridecemlineatus alleni	IUCN Data Deficient	1
S. washingtoni	ESA Candidate; IUCN Vulnerable	1, 3

SOURCE: 1, Hafner et al. (1998); 2, *Federal Register* (2000); 3, E. Yensen and P. W. Sherman, unpublished data; 4, Pergams and Nyberg (2001); Ostroff and Finck (2003).

[a]ESA, U.S. Endangered Species Act; IUCN, International Union for Conservation of Nature (Hafner et al. 1998).

(Valdéz and Ceballos 1997; Yensen and Valdés-Alarcón 1999). Five notable examples are discussed briefly.

The northern Idaho ground squirrel (*S. b. brunneus*) is the rarest spermophile in North America (Fig. 10.3). In 1999, it was listed as "threatened" under the Endangered Species Act (Table 10.2). Within the species's range, years of fire suppression have resulted in proliferation of coniferous trees, leading to shrinking and isolation of the their once contiguous meadow habitats (Truksa and Yensen 1990). Even in extant meadows, succession to lignified and woody (indigestible) plant species have led to population declines (Yensen and Sherman 1997; Gavin et al. 1999; Sherman and Runge 2002). To restore *S. b. brunneus* habitats, in 1997 the U.S. Forest Service began cutting conifers at edges of inhabitable meadows, conducting controlled burns to eliminate saplings and woody plants, and reseeding burned areas with native grasses and forbs. In cooperation with the U.S. Fish and Wildlife Service, several translocations and soft releases into previously inhabited meadows were conducted in 1998–2000. Although this program is in its initial stages, results are promising; several of the populations have increased sharply and others that were in decline have stabilized.

The Washington ground squirrel (*S. washingtoni*) also has experienced drastic population declines, primarily because much of the area in its historical range (Fig. 10.3), especially the most suitable microhabitats, has been converted to agriculture. Remaining rangeland also has been degraded (Vander Haegen et al. 2001). A detailed survey conducted in 1987–1989 revealed that *S. washingtoni* had disappeared from 74% of sites where it was known to have occurred historically in Washington, and 77% of such sites in Oregon (Betts 1990). Only 10 years later, in 1998, there had been a further decrease of inhabited sites, by 69% in Oregon and 27% in Washington (Betts 1999).

Populations of southern Idaho (*S. b. endemicus*), Merriam's (*S. canus*), and Townsend's ground squirrels (*S. townsendii*) also have declined and demes have become increasingly isolated due to agricultural conversion and degradation of rangeland habitats. Overgrazing in the second half of the nineteenth century enabled shrubs such as big sagebrush to increase and exotic grasses and forbs to invade native shrub–steppe habitats. In turn, this has provided fuel that increased the fire frequency (Yensen et al. 1992). Loss of shrubs to fires reduces protective cover, changes the microclimate, and may result in loss of food items. As a result, ground squirrel density declines when shrub cover decreases (Johnson et al. 1996; Higgins and Stapp 1997; Sharpe and Van Horne 1999). The exotic grasses that invade after fires are predominantly annuals, which depend on precipitation for survival and growth. Thus, their productivity is more variable than that of native perennial grasses and forbs, and their leaves and especially seeds do not provide sufficient nutrition to support stable populations of ground squirrels (Van Horne et al. 1998). Differences in the variability of food supplies may explain why *S. mollis* populations were relatively stable in native vegetation, but nearby populations in exotic annual-dominated vegetation fluctuated 12.5-fold (Yensen et al. 1992).

The result of anthropogenic habitat disturbances is that most of the small-bodied *Spermophilus* species no longer occur across the broad geographic "range" depicted in Figures 10.2 and 10.3. These range maps, which were made by connecting the peripheral-most sites where a species has ever been seen or captured, no longer paint a realistic picture. Instead, because suitable habitats are so discontinuous, the sites where the animals are known to occur today probably represent most of their population sites. In other words, the dots on maps that denote extant populations actually represent the species's range, not its boundaries (E. Yensen and P. W. Sherman, unpublished data).

As of 2003, *Spermophilus washingtoni, S. b. brunneus, S. b. endemicus, S. townsendii,* and *S. canus,* and probably others, exist as a series of small, widely separated populations, most of which are separated by unsuitable habitat. These populations are isolated demographically and genetically. If one population is eliminated by stochastic natural processes such as bad weather, badger predation, and drought, or anthropogenic disturbances such as poisoning, it probably will not be reestablished by immigration because there is no nearby source population to resupply it. The old adage "where there's one ground squirrel, there's bound to be lots more," must be replaced with "where there's one ground squirrel, there's a place we ought to protect because probably there aren't many more places with squirrels." This rather critical situation is just beginning to be recognized by conservation organizations and management agencies. Resource management professionals must move away from "controlling" ground squirrels to conserving them, and ranchers and farmers should see that ground squirrels can be allies, not always enemies.

RESEARCH NEEDS

The literature on ground squirrels obviously is extensive, and we have not covered it all. Nonetheless, there still is much to learn. For example, we are just beginning to realize the extent of intraspecific variations in annual cycles, social behavior, mating systems, genetics, demography, food preferences, and digestive and hibernation physiology. We do not yet understand the adaptive significance of these variations. We also do not understand the biotic and abiotic factors that affect population dynamics, nor do we know which anthropocentric disturbances are beneficial and which are deleterious to ground squirrel populations. For management, determining distribution, population trends, and conservation status of species and local populations is a high priority. Ground squirrel habitats have been so subdivided by urbanization and agricultural conversion that sites where populations now occur may represent the majority of places where a given species exists. Surveys are needed to document this. We also need to assess whether food supplementation is effective for maintaining endangered populations in degraded habitats and, if not, attempt to perfect methods for successfully translocating the animals. Finally, competition with domestic livestock merits further investigation because much of our information is out of date and of poor quality.

LITERATURE CITED

Abaturov, B. D. 1972. The role of burrowing animals in the transport of mineral substances in the soil. Pedobiologia 12:261–66.

Alcorn, J. R. 1940. Life history notes on the Piute ground squirrel. Journal of Mammalogy 21:160–70.

Allen, J. A. 1902. Mammal names proposed by Oken in his 'Lehrbuch der Zoologie.' Bulletin of the American Museum of Natural History 16:373–79.

Alvarez-Castañeda, S. T., G. Arnaud, and E. Yensen. 1996. *Spermophilus atricapillus*. Mammalian Species 531:1–3.

Antolin, M. F., B. Van Horne, M. D. Berger, Jr. A. K. Holloway, J. L. Roach, and R. D. Weeks, Jr. 2001. Effective population size and genetic structure of a Piute ground squirrel (*Spermophilus mollis*) population. Canadian Journal of Zoology 79:26–34.

Armitage, K. B. 1981. Sociality as a life history tactic of ground squirrels. Oecologia 48:36–49.

Askham, L. R. 1994. Franklin, Richardson, Columbian, Washington, and Townsend ground squirrels: Damage prevention and control methods. Pages B159–64 *in* S. E. Hygnstrom, R. M. Timm, and G. E. Larson, eds. Prevention and control of wildlife damage, 1994. Cooperative Extension Division, University of Nebraska, Lincoln.

Bailey, V. 1936. The mammals and life zones of Oregon. North American Fauna 55:1–416.

Balph, D. F. 1968. Behavioral responses of unconfined Uinta ground squirrels to trapping. Journal of Wildlife Management 32:778–94.

Barash, D. P. 1989. Marmots: Social behavior and ecology. Stanford University Press, Stanford, CA.

Barnes, A. M. 1982. Surveillance and control of bubonic plague in the United States. Symposium of the Zoological Society of London 50:237–70.

Barr, R. E., and X. J. Musacchia. 1968. Breeding among captive *Citellus tridecemlineatus*. Journal of Mammalogy 49:343–44.

Bartels, M. A., and D. P. Thompson. 1993. *Spermophilus lateralis*. Mammalian Species 440:1–8.

Belk, M. C., and H. D. Smith. 1991. *Ammospermophilus leucurus*. Mammalian Species 368:1–8.

Best, T. L. 1995a. *Spermophilus adocetus*. Mammalian Species 504:1–4.

Best, T. L. 1995b. *Spermophilus annulatus*. Mammalian Species 508:1–4.

Best, T. L. 1995c. *Spermophilus mohavensis*. Mammalian Species 509:1–7.

Best, T. L., C. L. Lewis, K. Caesar, and A. S. Titus. 1990a. *Ammospermophilus interpres*. Mammalian Species 365:1–6.
Best, T. L., A. S. Titus, C. L. Lewis, and K. Caesar. 1990b. *Ammospermophilus nelsoni*. Mammalian Species 367:1–7.
Betts, B. J. 1990. Geographic distribution and habitat preferences of Washington ground squirrels (*Spermophilus washingtoni*). Northwestern Naturalist 71:27–37.
Betts, B. J. 1999. Current status of Washington ground squirrels in Oregon and Washington. Northwestern Naturalist 80:35–38.
Biardi, J. E., R. G. Coss, and D. G. Smith. 2000. California ground squirrels (*Spermophilus beecheyi*) blood sera inhibits crotalid venom proteolytic activity. Toxicon 38:713–21.
Birney, E. C., and H. H. Genoways. 1973. Chromosomes of *Spermophilus adocetus* (Mammalia: Sciuridae), with comments on the subgeneric affinities of the species. Experientia 29:228–29.
Blumstein, D. T., and K. B. Armitage. 1997. Does sociality drive the evolution of communicative complexity? A comparative test with ground dwelling sciurid alarm calls. American Naturalist 150:179–200.
Blumstein, D. T., and K. B. Armitage. 1998. Life history consequences of social complexity: A comparative study of ground-dwelling sciurids. Behavioral Ecology 9:8–19.
Boag, D. A., and J. O. Murie. 1981. Population ecology of Columbian ground squirrels in southwestern Alberta. Canadian Journal of Zoology 59:2230–40.
Boellstorff, D. E., and D. H. Owings. 1995. Home range, population structure, and spatial organization of California ground squirrels. Journal of Mammalogy 76:551–61.
Boellstorff, D. E., D. H. Owings, M. C. T. Penedo, and M. J. Hersek. 1994. Reproductive behaviour and multiple paternity of California ground squirrels. Animal Behaviour 47:1057–64.
Boyer, B. B., and B. M. Barnes. 1999. Molecular and metabolic aspects of mammalian hibernation. BioScience 49:713–24.
Bridgewater, D. D. 1966. Laboratory breeding, early growth, development and behavior of *Citellus tridecemlineatus* (Rodentia). Southwestern Naturalist 11:325–37.
Broadbooks, H. E. 1961. California ground squirrel invades Washington. Journal of Mammalogy 42:257–58.
Bronson, M. T. 1980. Altitudinal variation in emergence time of golden-mantled ground squirrels (*Spermophilus lateralis*). Journal of Mammalogy 61:124–26.
Bryant, M. D. 1945. Phylogeny of Nearctic Sciuridae. American Midland Naturalist 33:257–390.
Buck, C. L., and B. M. Barnes. 1999. Annual cycle of body composition and hibernation in free-living arctic ground squirrels. Journal of Mammalogy 80:430–42.
Bursten, S. N., R. B. Kimsey, and D. H. Owings. 1997. Ranging of male *Oropsylla montana* fleas via male California ground squirrel (*Spermophilus beecheyi*) juveniles. Journal of Parasitology 83:804–9.
Bushberg, D. M., and W. G. Holmes. 1985. Sexual maturity in male Belding's ground squirrels: Influence of body weight. Biology of Reproduction 33:302–8.
Byrom, A. E., and C. J. Krebs. 1999. Natal dispersal of juvenile arctic ground squirrels in the boreal forest. Canadian Journal of Zoology 77:1048–59.
Byrom, A. E., T. J. Karels, C. J. Krebs, and R. Boonstra. 2000. Experimental manipulation of predation and food supply of arctic ground squirrels in the boreal forest. Canadian Journal of Zoology 78:1309–19.
Cameron, D. M., Jr. 1967. Gestation period of the golden-mantled ground squirrel (*Citellus lateralis*). Journal of Mammalogy 48:492–93.
Carl, E. A. 1971. Population control in arctic ground squirrels. Ecology 52:395–413.
Chappell, M. A., and G. A. Bartholomew. 1981. Activity and thermoregulation of the antelope ground squirrel *Ammospermophilus leucurus* in winter and summer. Physiological Ecology 54:215–23.
Clark, T. W. 1970a. Richardson's ground squirrel (*Spermophilus richardsonii*) in the Laramie Basin, Wyoming. Great Basin Naturalist 30:55–70.
Clark, T. W. 1970b. Early growth, development, and behavior of the Richardson ground squirrel (*Spermophilus richardsonii elegans*). American Midland Naturalist 83:197–205.
Cork, S. J., and G. J. Kenagy. 1989. Nutritional value of hypogeous fungus for a forest-dwelling ground squirrel. Ecology 70:577–86.
Coss, R. G., and J. E. Biardi. 1997. Individual variation in the antisnake behavior of California ground squirrels (*Spermophilus beecheyi*). Journal of Mammalogy 73:294–310.
Coss, R. G., and D. H. Owings. 1978. Snake-directed behavior by snake naive and experienced California ground squirrels in a simulated burrow. Zeitschrift für Tierpsychologie 48:421–35.
Costain, D. B., and B. J. Verts. 1982. Age determination and age-specific reproduction in Belding's ground squirrels. Northwest Science 56:230–35.
Cothran, E. G. 1983. Morphologic relationships of hybridizing ground squirrels *Spermophilus mexicanus* and *S. tridecemlineatus*. Journal of Mammalogy 64:591–602.
Cothran, E. G., and R. L. Honeycutt. 1984. Chromosomal differentiation of hybridizing ground squirrels (*Spermophilus mexicanus* and *S. tridecemlineatus*). Journal of Mammalogy 65:118–22.
Cothran, E. G., E. G. Zimmerman, and C. F. Nadler. 1977. Genic differentiation and evolution in the ground squirrel subgenus *Ictidomys* (genus *Spermophilus*). Journal of Mammalogy 58:610–22.
Davis, D. E. 1976. Hibernation and circannual rhythms of food consumption in ground squirrels and marmots. Quarterly Review of Biology 51:477–514.
Davis, L. S. 1982. Copulatory behavior of Richardson's ground squirrels (*Spermophilus richardsonii*) in the wild. Canadian Journal of Zoology 60:2953–55.
Davis, W. B. 1939. The Recent mammals of Idaho. Caxton, Caldwell, ID.
Dawe, A. R., and W. A. Spurrier. 1969. Hibernation induced in ground squirrels by blood transfusion. Science 163:298–300.
Dobson, F. S. 1982. Competition for mates and predominant juvenile male dispersal in mammals. Animal Behaviour 30:1183–92.
Dobson, F. S. 1983. Agonism and territoriality in the California ground squirrel. Journal of Mammalogy 64:218–25.
Dobson, F. S. 1994. Measures of gene flow in the Columbian ground squirrel. Oecologia 100:190–95.
Dobson, F. S. 1995. Regulation of population size: Evidence from Columbian ground squirrels. Oecologia 102:44–51.
Dobson, F. S., and J. D. Kjelgaard. 1985a. The influence of food resources on population dynamics in Columbian ground squirrels. Canadian Journal of Zoology 63:2095–2104.
Dobson, F. S., and J. D. Kjelgaard. 1985b. The influence of food resources on life history in Columbian ground squirrels. Canadian Journal of Zoology 63:2105–9.
Dobson, F. S., and J. O. Murie. 1987. Interpretation of intraspecific life history patterns: Evidence from Columbian ground squirrels. American Naturalist 129:382–97.
Dobson, F. S., and M. K. Oli. 2001. The demographic basis of population regulation in Columbian ground squirrels. American Naturalist 158:236–47.
Dobson, F. S., T. S. Risch, and J. O. Murie. 1999. Increasing returns in the life history of Columbian ground squirrels. Journal of Animal Ecology 68:73–86.
Drennan, J. E., P. Beier, and N. L. Dodd. 1998. Use of track stations to index abundance of sciurids. Journal of Mammalogy 79:352–59.
Durrant, S. D., and R. M. Hansen. 1954. Distribution patterns and phylogeny of some western ground squirrels. Systematic Zoology 3:82–85.
Dyer, N. W., and L. E. Huffman. 1999. Plague in free–ranging mammals in western North Dakota. Journal of Wildlife Diseases 35:600–602.
Dyni, E. J., and E. Yensen. 1996. Dietary similarity in sympatric Idaho and Columbian ground squirrels (*Spermophilus brunneus* and *S. columbianus*). Northwest Science 70:99–108.
Ebensberger, L. A. 1998. Strategies and counterstrategies to infanticide in mammals. Biological Reviews 73:321–46.
Edge, E. R. 1931. Seasonal activity and growth in the Douglas ground squirrel. Journal of Mammalogy 12:194–200.
Edge, E. R. 1934. Burrows and burrowing habits of the Douglas ground squirrel. Journal of Mammalogy 15:189–193.
El Hani, A., J. R. Mason, D. L. Nolte, and R. H. Schmidt. 1998. Flavor avoidance learning and its implications in reducing strychnine baiting hazards to nontarget animals. Physiology and Behavior 64:585–89.
Elliott, C. L., and J. T. Flinders. 1991. *Spermophilus columbianus*. Mammalian Species 372:1–9.
Erlien, D. A., and J. R. Tester. 1984. Population ecology of sciurids in northwestern Minnesota. Canadian Field-Naturalist 98:1–6.
Ernest, K. A., and M. A. Mares. 1987. *Spermophilus tereticaudus*. Mammalian Species 274:1–9.
Federal Register. 2002. Federal Register 65:17779–86.
Festa-Bianchet, M. 1981. Reproduction in yearling female Columbian ground squirrels (*Spermophilus columbianus*). Canadian Journal of Zoology 59:1032–35.
Festa-Bianchet, M., and D. A. Boag. 1982. Territoriality in adult female Columbian ground squirrels. Canadian Journal of Zoology 60:1060–66.
Festa-Bianchet, M., and W. J. King. 1984. Behavior and dispersal of yearling Columbian ground squirrels. Canadian Journal of Zoology 62:161–167.

Fisler, G. F. 1976. Agonistic signals and hierarchy changes of antelope squirrels. Journal of Mammalogy 57:94–102.

Frank, C. L. 1991. Adaptations for hibernation in the depot fats of a ground squirrel (*Spermophilus beldingi*). Canadian Journal of Zoology 69:2707–11.

Frank, C. L. 1994. Polyunsaturate content and diet selection by ground squirrels (*Spermophilus lateralis*). Ecology 75:458–63.

Frank, C. L., E. S. Dierenfeld, and K. B. Storey. 1998. The relationship between lipid peroxidation, hibernation, and food selection in mammals. American Zoologist 38:341–49.

Fredrickson, L. F. 1972. Swimming ability of *Spermophilus richardsonii*. Journal of Mammalogy 53:190–91.

French, A. R. 1988. The patterns of mammalian hibernation. American Scientist 76:568–75.

Galloway, T. D., and J. E. Christie. 1990. Fleas (Siphonaptera) associated with ground squirrels (*Spermophilus* spp.) in Manitoba, Canada. Canadian Entomologist 122:449–58.

Gavin, T. A., P. W. Sherman, E. Yensen, and B. May. 1999. Population genetic structure of the northern Idaho ground squirrel (*Spermophilus brunneus brunneus*). Journal of Mammalogy 80:156–68.

Giboulet, O., P. Cheret, R. Ramousse, and F. Catzeflis. 1997. DNA–DNA hybridization evidence for the recent origin of marmots and ground squirrels (Rodentia: Sciuridae). Journal of Mammalian Evolution 4:271–84.

Gill, A. E., and E. Yensen. 1992. Biochemical differentiation in the Idaho ground squirrel, *Spermophilus brunneus* (Rodentia: Sciuridae). Great Basin Naturalist 52:155–59.

Griffith, B., J. M. Scott, J. W. Carpenter, and C. Reed. 1989. Translocation as a species conservation tool: Status and strategy. Science 245:477–80.

Grinnell, J. 1923. The burrowing rodents of California as agents in soil formation. Journal of Mammalogy 4:137–49.

Grinnell, J., and J. Dixon. 1918. Natural history of the ground squirrels of California. Monthly Bulletin, California State Commission on Horticulture 7:597–708.

Hackett, D. 1987. Dispersal of yearling Columbian ground squirrels. Ph.D. Dissertation, University of Alberta, Edmonton, Canada.

Hafner, D. J. 1984. Evolutionary relationships of the Nearctic Sciuridae. Pages 13–23 *in* J. O. Murie and G. R. Michener, eds. The biology of ground-dwelling squirrels. University of Nebraska Press, Lincoln.

Hafner, D. J., and T. L. Yates. 1983. Systematic status of the Mojave ground squirrel, *Spermophilus mohavensis* (subgenus *Xerospermophilus*). Journal of Mammalogy 64:397–404.

Hafner, D. J., E. Yensen, and G. L. Kirkland, Jr. eds. 1998. North American rodents: Status survey and conservation action plan. IUCN, Gland, Switzerland.

Hall, E. R. 1981. The mammals of North America, Vol. 1. John Wiley, New York.

Hanken, J., and P. W. Sherman. 1981. Multiple paternity in Belding's ground squirrel litters. Science 212:351–53.

Hansen, R. M. 1954. Molt patterns in ground squirrels. Proceedings of the Utah Academy of Science, Arts, and Letters 31:57–60.

Hansen, R. M. 1956. Decline in Townsend ground squirrels in Utah. Journal of Mammalogy 37:123–24.

Hare, J. F. 1998. Juvenile Richardson's ground squirrels (*Spermophilus richardsonii*) manifest both littermate and neighbor/stranger discrimination. Ethology 104:991–1002.

Harris, M. A., and J. O. Murie. 1982. Responses to oral gland scents from different males in Columbian ground squirrels. Animal Behaviour 30:140–48.

Heaney, L. R. 1984. Climatic influences on life-history tactics and behavior of North American tree squirrels. Pages 43–78 *in* J. O. Murie and G. R. Michener, eds. The biology of ground-dwelling squirrels. University of Nebraska Press, Lincoln.

Hegdal, P. L., K. A. Fagerstone, T. A. Gatz, J. F. Glahn, and G. H. Matschke. 1986. Hazards to wildlife associated with 1080 baiting for California ground squirrels. Wildlife Society Bulletin 14:11–21.

Hershkovitz, P. 1949. Status of names credited to Oken, 1816. Journal of Mammalogy 30:289–301.

Higgins, L. C., and P. Stapp. 1997. Abundance of thirteen-lined ground squirrels in shortgrass prairie. Prairie Naturalist 29:25–37.

Hoffmann, R. S., C. G. Anderson, R. W. Thorington, Jr., and L. R. Heany. 1993. Family Sciuridae. Pages 419–65 *in* D. E. Wilson and D. M. Reeder, eds. Mammal species of the world: A taxonomic and geographic reference. Smithsonian Institution Press, Washington, DC.

Hoffmeister, D. F. 1986. Mammals of Arizona. University of Arizona Press, Tucson.

Holekamp, K. E. 1983. Proximal mechanisms of natal dispersal in Belding's ground squirrels (*Spermophilus beldingi beldingi*). Ph.D Dissertation, University of California, Berkeley.

Holekamp, K. E. 1984. Dispersal in ground-dwelling sciurids. Pages 297–320 *in* J. O. Murie and G. R. Michener, eds. The biology of ground-dwelling squirrels. University of Nebraska Press, Lincoln.

Holekamp, K. E. 1986. Proximal causes of natal dispersal in Belding's ground squirrels (*Spermophilus beldingi*). Ecological Monographs 56:365–91.

Holekamp, K. E., and P. W. Sherman. 1989. Why male ground squirrels disperse. American Scientist 77:232–39.

Holekamp, K. E., L. Smale, H. B. Simpson, and N. A. Holekamp. 1984. Hormonal influences on natal dispersal in free-living Belding's ground squirrels (*Spermophilus beldingi*). Hormones and Behavior 18:465–83.

Holmes, W. G. 1986a. Kin recognition by phenotype matching in female Belding's ground squirrels. Animal Behaviour 34:38–47.

Holmes, W. G. 1986b. Identification of paternal half siblings by captive Belding's ground squirrels. Animal Behaviour 34:321–27.

Holmes, W. G., and P. W. Sherman. 1982. The ontogeny of kin recognition in two species of ground squirrels. American Zoologist 22:491–517.

Hoogland, J. L. 1982. Prairie dogs avoid extreme inbreeding. Science 215:1639–41.

Hoogland, J. L. 1995. The black-tailed prairie dog. University of Chicago Press, Chicago.

Horn, E. E., and H. S. Fitch. 1946. Trapping the California ground squirrel. Journal of Mammalogy 27:220–24.

Howard, W. E. 1959. California ground squirrels breeding in captivity. Journal of Mammalogy 40:445–46.

Howard, W. E., K. A. Wagnon, and J. R. Bentley. 1959. Competition between ground squirrels and cattle for forage. Journal of Range Management 12:110–15.

Howell, A. H. 1938. Revision of the North American ground squirrels, with a classification of the North American Sciuridae. North American Fauna 56:1–256.

Hubbs, A. H., and R. Boonstra. 1997. Effects of food and predators on the home-range sizes of arctic ground squirrels (*Spermophilus parryii*). Canadian Journal of Zoology 76:592–96.

Hubbs, A. H., T. Karels, and R. Boonstra. 2000. Indices of population size for burrowing animals. Journal of Wildlife Management 64:296–301.

International Commission on Zoological Nomenclature., 1956. Rejection for nomenclatorial purposes of volume 3 (*Zoologie*) of the work by Lorenz Oken entitled *Okens Lehrbuch der Naturgeschichte* published in 1815–1816. Opinions and Declarations Rendered by the International Commission on Zoological Nomenclature 14:1–42.

Jenkins, E. R., and A. W. Grundmann. 1973. The parasitology of the ground squirrels of western Utah. Proceedings of the Helminthological Society of Washington 40:76–86.

Jenkins, S. H., and B. D. Eschelman. 1984. *Spermophilus beldingi*. Mammalian Species 221:1–8.

Johnson, K. H., R. A. Olson, and T. D. Whitson. 1996. Composition and diversity of plant and small mammal communities in tebuthiuron-treated big sagebrush (*Artemisia tridentata*). Weed Technology 10:404–16.

Karels, T. J., and R. Boonstra. 1999. The impact of predation on burrow use by arctic ground squirrels in the boreal forest. Proceedings Royal Society of London B 266:2117–23.

Karels, T. J., and R. Boonstra. 2000. Concurrent density dependence and independence in populations of arctic ground squirrels. Nature 408:460–63.

Karels, T. J., A. E. Byrom, R. Boonstra, and C. J. Krebs. 2000. The interactive effects of food and predators on reproduction and overwinter survival of arctic ground squirrels. Journal of Animal Ecology 69:235–47.

Kenagy, G. J., and B. M. Barnes. 1988. Seasonal reproductive patterns in four coexisting rodent species from the Cascade Mountains, Washington. Journal of Mammalogy 69:274–92.

Kenagy, G. J., D. Masman, S. M. Sharbaugh, and K. A. Nagy. 1990. Energy expenditure during lactation in relation to litter size in free-living golden-mantled ground squirrels. Journal of Animal Ecology 59:73–88.

Kiell, D. J., and J. S. Millar. 1978. Growth of juvenile arctic ground squirrels (*Spermophilus parryii*) at McConnell River, N. W. T. Canadian Journal of Zoology 56:1475–78.

Kivett, V. K., J. O. Murie, and A. L. Steiner. 1976. A comparative study of scent-gland location and related behavior in some northwestern Nearctic ground squirrel species (Sciuridae): An evolutionary approach. Canadian Journal of Zoology 54:1294–1306.

Klevezal, G. A., and S. E. Kleinenberg. 1967. Age determination in mammals from annual layers in teeth and bones. Israel Program for Scientific Translations, Jerusalem.

Knight, J. E., E. N. Narus, S. L. Martin, A. Jacobson, B. M. Barnes, and

B. B. Boyer. 2000. mRNA stability and polysome loss in hibernating arctic ground squirrels (*Spermophilus parryii*). Molecular and Cellular Biology 20:6374–79.

Knopf, F. L., and D. F. Balph. 1969. Badgers plug burrows to confine prey. Journal of Mammalogy 50:635–36.

Koeppl, J. W., and R. S. Hoffmann. 1981. Comparative postnatal growth of four ground squirrel species. Journal of Mammalogy 62:41–57.

Kramm, K. R. 1972. Body temperature regulation and torpor in the antelope ground squirrel, *Ammospermophilus leucurus*. Journal of Mammalogy 53:609–11.

Lacey, E. A., J. R. Wieczorek, and P. K. Tucker. 1997. Male mating behaviour and patterns of sperm precedence in arctic ground squirrels. Animal Behaviour 53:767–79.

Laundré, J. W. 1989. Horizontal and vertical diameter of burrows of five small mammal species in southeastern Idaho. Great Basin Naturalist 49:646–49.

Laundré, J. W. 1993. Effects of small mammal burrows on water infiltration in a cool desert environment. Oecologia 94:43–48.

Laundré, J. W. 1998. Effect of ground squirrel burrows on plant productivity in a cool desert environment. Journal of Range Management 51:638–43.

Leger, D. W., S. D. Berney-Key, and P. W. Sherman. 1984. Vocalizations of Belding's ground squirrels (*Spermophilus beldingi*). Animal Behaviour 32:753–64.

Lindsay, D. S., R. D. McKeown, and J. P. Dubey. 2000. *Sarcocystis campestris* from naturally infected 13-lined ground squirrels, *Spermophilus tridecemlineatus tridecemlineatus*, from Nebraska. Journal of Parasitology 86:1159–61.

Link, V. B. 1955. A history of plague in the United States. Public Health Monograph 26:1–120.

Lishak, R. S. 1977. Censusing 13-lined ground squirrels with adult and young alarm calls. Journal of Wildlife Management 41:755–59.

Longhurst, W. M. 1957. A history of squirrel burrow gully formation in relation to grazing. Journal of Range Management 10:182–84.

Lyapunova, E. A., and N. N. Vorontsov. 1970. Chromosomes and some issues of the evolution of the ground squirrel genus *Citellus* (Rodentia: Sciuridae). Experientia 26:1033–38.

MacNeil, D., and C. Strobeck. 1987. Evolutionary relationships among colonies of Columbian ground squirrels as shown by mitochondrial DNA. Evolution 41:873–81.

Marsh, R. E. 1994. Belding's, California, and rock ground squirrels: Damage prevention and control methods. Pages B151–58 *in* S. E. Hygnstrom, R. M. Timm, and G. E. Larson, eds. Prevention and control of wildlife damage. Cooperative Extension Division, University of Nebraska, Lincoln.

Marsh, R. E. 1998. Historical review of ground squirrel crop damage in California. International Biodeterioration and Biodegradation 42:93–99.

Marsh, R. E., and W. E. Howard. 1968. Breeding ground squirrels, *Spermophilus beecheyi*, in captivity. Journal of Mammalogy 49:781–83.

Mateo, J. M. 2002. Kin recognition abilities and nepotism as a function of sociality. Proceeding of the Royal Society of London B 269:721–27.

Mateo, J. M., and R. E. Johnston. 2000. Retention of social recognition after hibernation in Belding's ground squirrels. Animal Behaviour 59:491–99.

Matschke, G. H., and K. A. Fagerstone. 1982. Population reduction of Richardson's ground squirrels with zinc phosphide. Journal of Wildlife Management 46:671–77.

Matschke, G. H., M. P. Marsh, and D. L. Otis. 1983. Efficacy of zinc phosphide broadcast baiting for controlling Richardson's ground squirrels on rangeland. Journal of Range Management 36:504–6.

May, B., T. A. Gavin, P. W. Sherman, and T. M. Korves. 1997. Characterization of microsatellite loci in the northern Idaho ground squirrel *Spermophilus brunneus brunneus*. Molecular Ecology 6:399–400.

McColl, C. J., and R. Boonstra. 1999. Physiological effects of three inhalant anesthetics on arctic ground squirrels. Wildlife Society Bulletin 27:946–51.

McLean, I. G. 1978. Plugging of nest burrows by female *Spermophilus columbianus*. Journal of Mammalogy 59:437–39.

Melchior, H. R., and F. A. Iwen. 1965. Trapping, restraining, and marking arctic ground squirrels for behavioral observations. Journal of Wildlife Management 29:671–78.

Mercer, J. M., and V. L. Roth. 2003. The effects of Cenozoic global change on squirrel phylogeny. Science 299:1568–1572.

Messick, J. P., G. W. Smith, and A. M. Barnes. 1983. Serologic testing of badgers to monitor plague in southwestern Idaho. Journal of Wildlife Diseases 19:1–6.

Michener, G. R. 1977. Effect of climatic conditions on the annual activity and hibernation cycle of Richardson's ground squirrels and Columbian ground squirrels. Canadian Journal of Zoology 55:693–703.

Michener, G. R. 1978. Effect of age and parity on weight gain and entry into hibernation in Richardson's ground squirrels. Canadian Journal of Zoology 56:2573–77.

Michener, G. R. 1979. Spatial relationships and social organization of Richardson's ground squirrels in southern Alberta. Journal of Mammalogy 57:125–39.

Michener, G. R. 1980. Estrous and gestation periods in Richardson's ground squirrels. Journal of Mammalogy 61:531–34.

Michener, G. R. 1983a. Kin identification, matriarchies, and the evolution of sociality in ground-dwelling sciurids. Pages 528–572 *in* J. F. Eisenberg and D. G. Kleiman, eds. Recent advances in the study of mammalian behavior. American Society of Mammalogists.

Michener, G. R. 1983b. Spring emergence schedules and vernal behavior of Richardson's ground squirrels: Why do males emerge from hibernation before females? Behavioral Ecology and Sociobiology 14:29–38.

Michener, G. R. 1984. Age, sex, and species differences in the annual cycles of ground-dwelling sciurids: Implications for sociality. Pages 81–107 *in* J. O. Murie and G. R. Michener, eds. The biology of ground-dwelling squirrels. University of Nebraska Press, Lincoln.

Michener, G. R. 1985. Chronology of reproductive events for female Richardson's ground squirrels. Journal of Mammalogy 66:280–88.

Michener, G. R. 1993a. Lethal myiasis of Richardson's ground squirrels by the sarcophagid fly *Neobellieria citellivora*. Journal of Mammalogy 74:148–55.

Michener, G. R. 1993b. Sexual differences in hibernaculum contents of Richardson's ground squirrels: males store food. Pages 109–18 *in* C. Carey, G. L. Florant, B. A. Wunder, and B. Horowitz, eds. Life in the cold: Ecological, physiological, and molecular mechanisms. Westview Press, Boulder, CO.

Michener, G. R. 1996. Establishment of a colony of Richardson's ground squirrels in southern Alberta. Pages 303–8 *in* W. D. Williams and J. F. Dormaar, eds. Proceedings of the fourth prairie conservation and endangered species workshop (Natural History Occasional Paper No. 23). Provincial Museum of Alberta, Alberta, Canada.

Michener, G. R. 1998. Sexual differences in reproductive effort of Richardson's ground squirrels. Journal of Mammalogy 79:1–19.

Michener, G. R., and J. W. Koeppl. 1985. *Spermophilus richardsonii*. Mammalian Species 243:1–8.

Michener, G. R., and D. R. Michener. 1977. Population structure and dispersal in Richardson's ground squirrels. Ecology 58:359–68.

Montgomery, S. J., D. F. Balph, and D. M. Balph. 1971. Age determination of Uinta ground squirrels by teeth annuli. Journal of Wildlife Management 35:836–39.

Morton, M. L. 1975. Seasonal cycles of body weights and lipids in Belding's ground squirrels. Bulletin of the Southern California Academy of Sciences 74:128–43.

Morton, M. L., and P. W. Sherman. 1978. Effects of a spring snowstorm on behavior, reproduction, and survival of Belding's ground squirrels. Canadian Journal of Zoology 56:2578–90.

Morton, M. L., and H. L. Tung. 1971. Growth and development in the Belding ground squirrel (*Spermophilus beldingi beldingi*). Journal of Mammalogy 52:611–16.

Morton, M. L., C. S. Maxwell, and C. E. Wade. 1974. Body size, body composition, and behavior of juvenile Belding's ground squirrels. Great Basin Naturalist 34:121–134.

Murie, J. O. 1992. Predation by badgers on Columbian ground squirrels. Journal of Mammalogy 73:385–94.

Murie, J. O. 1995. Mating behavior of Columbian ground squirrels. 1. Multiple mating by females and multiple paternity. Canadian Journal of Zoology 73:1819–1826.

Murie, J. O., and D. A. Boag. 1984. The relationship of body weight to overwinter survival in Columbian ground squirrels. Journal of Mammalogy 65:688–90.

Murie, J. O., and M. A. Harris. 1978. Territoriality and dominance in male Columbian ground squirrels (*Spermophilus columbianus*). Canadian Journal of Zoology 56:2402–12.

Murie, J. O., and M. A. Harris. 1982. Annual variation of spring emergence and breeding in Columbian ground squirrels (*Spermophilus columbianus*). Journal of Mammalogy 63:431–39.

Murie, J. O., and I. G. McLean. 1980. Copulatory plugs in ground squirrels. Journal of Mammalogy 61:355–56.

Murie, J. O., D. A. Boag, and V. K. Kivett. 1980. Litter size in Columbian ground squirrels (*Spermophilus columbianus*). Journal of Mammalogy 61:237–44.

Nadler, C. F. 1962. Chromosome studies in certain subgenera of *Spermophilus*. Proceedings of the Society of Experimental Biology and Medicine 10:785–88.

Nadler, C. F. 1966a. Chromosomes of *Spermophilus franklini* and taxonomy of the ground squirrel genus *Spermophilus*. Systematic Zoology 15:199–206.

Nadler, C. F. 1966b. Chromosomes and systematics of American ground squirrels of the subgenus *Spermophilus*. Journal of Mammalogy 47:579–96.

Nadler, C. F. 1968. The chromosomes of *Spermophilus townsendii* (Rodentia: Sciuridae) and report of a new subspecies. Cytogenetics 7:144–57.

Nadler, C. F., and D. A. Sutton. 1962. Mitotic chromosomes of some North American Sciuridae. Proceedings of the Society of Experimental Biology and Medicine 10:36–38.

Nadler, C. F., R. S. Hoffmann, and K. Greer. 1971. Chromosomal divergence during evolution of ground squirrel populations (Rodentia: *Spermophilus*). Systematic Zoology 20:298–305.

Nadler, C. F., E. A. Lyapunova, R. S. Hoffmann, N. N. Vorontsov, and N. A. Malygina. 1975. Chromosomal evolution in Holarctic ground squirrels (*Spermophilus*). 1. Giemsa band homologies in *Spermophilus columbianus* and *S. undulatus*. Zeitschrift für Säugetierkunde 40:1–7.

Nadler, C. F., R. S. Hoffmann, N. N. Vorontsov, J. W. Koeppl, L. Deutsch, and R. I. Sukernik. 1982. Evolution in ground squirrels. II. Biochemical comparisons in Holarctic populations of *Spermophilus*. Zeitschrift für Säugetierkunde 47:198–215.

Nadler, C. F., E. I. Lyapunova, R. S. Hoffmann, N. N. Vorontsov, L. L. Shaitorova, and Y. M. Borisov. 1984. Chromosomal evolution in Holarctic ground squirrels. II. Giemsa band homologies of chromosomes, and the tempo of evolution. Zeitschrift für Säugetierkunde 49:78–90.

Neuhaus, P., and N. Pelletier. 2001. Mortality in relation to season, age, sex, and reproduction in Columbian ground squirrels (*Spermophilus columbianus*). Canadian Journal of Zoology 79:465–70.

Neuhaus, P., R. Bennett, and A. Hubbs. 1999. Effects of a late snowstorm and rain on survival and reproductive success in Columbian ground squirrels (*Spermophilus columbianus*). Canadian Journal of Zoology 77:879–84.

Nunes, S., and K. E. Holekamp. 1996. Mass and fat influence the timing of natal dispersal in Belding's ground squirrels. Journal of Mammalogy 77:807–17.

Nunes, S., E. M. Muecke, and K. E. Holekamp. 2002. Seasonal effects of food provisioning on body fat, insulin, and corticosterone in free-living juvenile Belding's ground squirrels (*Spermophilus beldingi*). Canadian Journal of Zoology 80:366–371.

Nunes, S., P. A. Zugger, A. L. Engh, K. O. Reinhart, and K. E. Holekamp. 1997. Why do female Belding's ground squirrels disperse away from food sources? Behavioral Ecology and Sociobiology 40:199–207.

Nunes, S., C. D. T. Ha, P. J. Garrett, E. M. Meuke, L. Smale, and K. E. Holekamp. 1998. Body fat and time of year interact to mediate dispersal behaviour in ground squirrels. Animal Behaviour 55:605–14.

Nydegger, N. C., and G. W. Smith. 1986. Prey populations in relation to *Artemisia* vegetation types in southwestern Idaho. Pages 152–56 *in* E. D. MacArthur and B. L. Welch, eds. Proceedings—Symposium on the biology of *Artemisia* and *Chrysothamnus*. U.S. Forest Service, Intermountain Research Station, Ogden, UT.

Olson, G. S., and B. Van Horne. 1998. Dispersal patterns of juvenile Townsend's ground squirrels in southwestern Idaho. Canadian Journal of Zoology 76:2084–89.

Olson, M. E., and K. McCabe. 1986. Anesthesia in the Richardson ground squirrel: Comparison of ketamine, ketamine and xylazine, droperidol, and fentanyl, and sodium phenobarbital. Journal of the American Veterinary Medical Association 189:1035–37.

Ortega, J. C. 1987. Den site selection by the rock squirrel (*Spermophilus variegatus*) in southeastern Arizona. Journal of Mammalogy 68:792–98.

Ortega, J. C. 1990. Home range size of adult rock squirrels (*Spermophilus variegatus*) in southeastern Arizona. Journal of Mammalogy 71:171–76.

Ostroff, A. C., and E. F. Finck. 2003. *Spermophilus franklinii*. Mammalian Species. 724:1–5.

Owings, D. H., and M. Borchert. 1975. Correlates of burrow location in Beechey ground squirrels. Great Basin Naturalist 35:402–4.

Owings, D. H., and R. G. Coss. 1977. Snake mobbing by California ground squirrels—Adaptive variation and ontogeny. Behaviour 62:50–69.

Owings, D. H., R. G. Coss, D. McKernon, M. P. Rowe, and P. C. Arrowood. 2001. Snake-directed anti-predator behavior of rock squirrels (*Spermophilus variegatus*): Population differences and snake-species discrimination. Behaviour 138:575–95.

Packard, R. L. 1958. Carnivorous behavior in the Mexican ground squirrel. Journal of Mammalogy 39:154.

Peacock, M. M., and S. H. Jenkins. 1988. Development of food preferences: Social learning by Belding's ground squirrels *Spermophilus beldingi*. Behavioral Ecology and Sociobiology 22:393–99.

Pengelley, E. T., and K. C. Fisher. 1961. Rhythmical arousal from hibernation in the golden-mantled ground squirrel, *Citellus lateralis tescorum*. Canadian Journal of Zoology 39:105–20.

Pergams, O. R. W., and D. Nyberg. 2001. Museum collections of mammals corroborate the exceptional decline of prairie habitat in the Chicago region. Journal of Mammalogy 82:984–92.

Pfeifer, S. 1982. Disappearance and dispersal of *Spermophilus elegans* juveniles in relation to behavior. Behavioral Ecology and Sociobiology 10:237–43.

Phillips, J. A. 1984. Environmental influences on reproduction in the golden-mantled ground squirrel. Pages 108–24 *in* J. O. Murie and G. R. Michener, eds. The biology of ground-dwelling squirrels. University of Nebraska Press, Lincoln.

Price, L. W. 1971. Geomorphic effect of the arctic ground squirrel in an alpine environment. Geografiska Annaler 53A:100–106.

Prychodko, W. 1952. A live trap for ground squirrels. Journal of Mammalogy 33:497.

Recht, M. A. 1977. The biology of the Mohave ground squirrel, *Spermophilus mohavensis*; home range, daily activity, foraging and weight gain and thermoregulatory behavior. Ph.D. Dissertation, University of California, Los Angeles.

Reynolds, H. G., and F. Turkowski. 1972. Reproductive variations in the round-tailed ground squirrel as related to winter rainfall. Journal of Mammalogy 53:893–98.

Reynolds, T. D., and J. W. Laundré. 1988. Vertical distribution of soil removed by four species of burrowing rodents in disturbed and undisturbed soils. Health Physics 54:445–50.

Reynolds, T. D., and W. L. Wakkinen. 1987. Characteristics of the burrows of four species of rodents in undisturbed soils in southeastern Idaho. American Midland Naturalist 118:245–50.

Rickart, E. A. 1982. Annual cycles of activity and body composition in *Spermophilus townsendii mollis*. Canadian Journal of Zoology 60:3298–3306.

Rickart, E. A. 1986. Postnatal growth of the Piute ground squirrel (*Spermophilus mollis*). Journal of Mammalogy 67:412–16.

Rickart, E. A. 1988. Population structure of the Piute ground squirrel (*Spermophilus mollis*). Southwestern Naturalist 33:91–96.

Rickart, E. A., and E. Yensen. 1991. *Spermophilus washingtoni*. Mammalian Species 371:1–5.

Rieger, J. F. 1996. Body size, litter size, timing of reproduction, and juvenile survival in the Uinta ground squirrel, *Spermophilus armatus*. Oecologia 107:463–68.

Robinson, S. R. 1980. Antipredator behaviour and predator recognition in Belding's ground squirrels. Animal Behaviour 28:840–52.

Salmon, T. P., and R. E. Marsh. 1981. Artificial establishment of a ground squirrel colony. Journal of Wildlife Management 45:1016–18.

Schooley, R. L., B. Van Horne, and K. P. Burnham. 1993. Passive integrated transponders for marking free-ranging ground squirrels. Journal of Mammalogy 74:480–84.

Schooley, R. L., P. B. Sharpe, and B. Van Horne. 1996. Can shrub cover increase predation risk for a desert rodent? Canadian Journal of Zoology 74:157–63.

Schwagmeyer, P. L. 1988. Scramble-competition polygyny in an asocial mammal: Male mobility and mating success. American Naturalist 131:885–92.

Schwagmeyer, P. L. 1995. Searching today for tomorrow's mates. Animal Behaviour 50:759–67.

Schwagmeyer, P. L., and G. A. Parker. 1987. Queuing for mates in thirteen-lined ground squirrels. Animal Behaviour 35:1015–25.

Seville, R. S., J. H. Harlow, N. L. Stanton, and M. L. Wagner. 1992. Effects of eimerian (Apicomplexa: Eimeriidae) infections on nutrient assimilation in the Wyoming ground squirrel. Journal of Parasitology 78:881–85.

Seville, R. S., N. L. Stanton, and K. Gerrow. 1996. Stable parasite guilds: Coccidia in spermophiline rodents. Oikos 75:365–72.

Sharpe, P. B., and B. Van Horne. 1998. Influence of habitat on behavior of Townsend's ground squirrels (*Spermophilus townsendii*). Journal of Mammalogy 79:906–18.

Sharpe, P. B., and B. Van Horne. 1999. Relationships between the thermal environment and activity of Piute ground squirrels (*Spermophilus mollis*). Journal of Thermal Biology 24:265–78.

Shaw, W. T. 1925a. Breeding and development of the Columbian ground squirrel. Journal of Mammalogy 6:106–13.

Shaw, W. T. 1925b. A life history problem and a means for its solution. Journal of Mammalogy 6:157–62.

Shaw, W. T. 1925c. The hibernation of the Columbian ground squirrel. Canadian Field-Naturalist 39:56–61, 79–82.

Shaw, W. T. 1926. Age of the animal and slope of the ground surface, factors modifying the structure of hibernation dens of ground squirrels. Journal of Mammalogy 7:91–96.

Sherman, P. W. 1976. Natural selection among some group-living organisms. Ph.D. Dissertation, University of Michigan, Ann Arbor.

Sherman, P. W. 1977. Nepotism and the evolution of alarm calls. Science 197:1246–53.

Sherman, P. W. 1980. The limits of ground squirrel nepotism. Pages 505–44 *in* G. W. Barlow and J. Silverberg, eds. Sociobiology: Beyond nature/nurture? Westview Press, Boulder, CO.

Sherman, P. W. 1981a. Kinship, demography, and Belding's ground squirrel nepotism. Behavioral Ecology and Sociobiology 8:251–59.

Sherman, P. W. 1981b. Reproductive competition and infanticide in Belding's ground squirrels and other organisms. Pages 311–31 *in* R. D. Alexander and D. W. Tinkle, eds. Natural selection and social behavior. Chiron Press, New York.

Sherman, P. W. 1985. Alarm calls of Belding's ground squirrels to aerial predators: nepotism or self-preservation? Behavioral Ecology and Sociobiology 17:313–323.

Sherman, P. W. 1989. Mate guarding as paternity insurance in Idaho ground squirrels. Nature 338:418–20.

Sherman, P. W., and W. G. Holmes. 1985. Kin recognition: Issues and evidence. Fortschritte de Zoologie 31:437–60.

Sherman, P. W., and M. L. Morton. 1984. Demography of Belding's ground squirrels. Ecology 65:1617–28.

Sherman, P. W., and M. C. Runge. 2002. Demography of a population collapse: The northern Idaho ground squirrel (*Spermophilus brunneus brunneus*). Ecology 83:2816–2831.

Sherman, P. W., M. L. Morton, L. M. Hoops, J. Bochantin, and J. M. Watt. 1985. The use of tail collagen strength to estimate age in Belding's ground squirrels. Journal of Wildlife Management 49:874–79.

Slade, N. A., and D. F. Balph. 1974. Population ecology of Uinta ground squirrels. Ecology 55:989–1003.

Smith, D. J., and J. S. Gardner. 1985. Geomorphic effects of ground squirrels in the Mount Rae area, Canadian Rocky Mountains. Arctic and Alpine Research 17:205–10.

Smith, G. W., and D. R. Johnson. 1985. Demography of a Townsend ground squirrel population in southwestern Idaho. Ecology 66:171–78.

Stangl, F. B., Jr., and J. V. Grimes. 1987. Phylogenetic implications of comparative pelage morphology in Aplodontidae and the Nearctic Sciuridae, with observations on seasonal pelage variation. Occasional Papers, The Museum, Texas Tech University 112:1–21.

Stanton, N. L., L. M. Shults, M. Parker, and R. S. Seville. 1992. Coccidian assemblages in the Wyoming ground squirrel, *Spermophilus elegans*. Journal of Parasitology 78:323–28.

Stevens, S. D. 1998. High incidence of infanticide by lactating females in a population of Columbian ground squirrels (*Spermophilus columbianus*). Canadian Journal of Zoology 76:1183–87.

Struebel, D. P., and J. P. Fitzgerald. 1978. *Spermophilus tridecemlineatus*. Mammalian Species 103:1–5.

Sullins, G. L., and B. J. Verts. 1978. Baits and baiting techniques for control of Belding's ground squirrels. Journal of Wildlife Management 42:890–96.

Svihla, A. 1939. Breeding habits of Townsend's ground squirrel. Murrelet 20:6–10.

Svihla, A. 1958. Subfreezing body temperatures in dormant ground squirrels. Journal of Mammalogy 39:296–98.

Swaisgood, R. R., M. P. Rowe, and D. H. Owings. 1999. Assessment of rattlesnake dangerousness by California ground squirrels: Exploitation of cues from rattling sounds. Animal Behaviour 57:1301–10.

Thomas, W. K., and S. L. Martin. 1993. A recent origin of marmots. Molecular Phylogenetics and Evolution 2:330–36.

Todd, K. S., Jr., and D. M. Hammond. 1968. Life cycle and host specificity of *Eimeria callospermophili* Henry, 1932 from the Uinta ground squirrel, *Spermophilus armatus*. Journal of Protozoology 15:1–8.

Tomich, P. Q. 1962. The annual cycle of the California ground squirrel *Citellus beecheyi*. University of California Publications in Zoology 65:213–82.

Trombulak, S. C. 1988. *Spermophilus saturatus*. Mammalian Species 322:1–4.

Trombulak, S. C. 1989. Running speed and body mass in Belding's ground squirrels. Journal of Mammalogy 70:194–97.

Trombulak, S. C. 1991. Maternal influence on juvenile growth rates in Belding's ground squirrel (*Spermophilus beldingi*). Canadian Journal of Zoology 69:2140–45.

Truksa, A. S., and E. Yensen. 1990. Photographic evidence of vegetation changes in Adams County, Idaho. Journal of the Idaho Academy of Sciences 26:18–40.

Turner, L. W. 1972. Habitat differences between *Spermophilus beldingi* and *Spermophilus columbianus* in Oregon. Journal of Mammalogy 53:914–917.

Uribe-Alcocer, M., A. Ahumada-Medina, A. Laguarda-Figueras, and F. Rodgiguez-Romero. 1979. The karyotype of *Spermophilus perotensis*. Mammalian Chromosome Newsletter 20:139–41.

Valdéz [=Valdés], M., and G. Ceballos. 1997. Conservation of endemic mammals of Mexico: The Perote ground squirrel (*Spermophilus perotensis*). Journal of Mammalogy 78:74–82.

Vander Haegen, W. M., S. M. McCorquodale, C. R. Peterson, G. A. Green, and E. Yensen. 2001. Wildlife communities of eastside shrubland and grassland habitats. Pages 349–77 *in* D. H. Johnson and T. A. O'Neil, eds. Wildlife–habitat relationships in Oregon and Washington. Oregon State University Press, Corvallis.

Van Horne, B., R. L. Schooley, and P. B. Sharpe. 1998. Influence of habitat, sex, age, and drought on the diet of Townsend's ground squirrels. Journal of Mammalogy 79:521–37.

Van Vuren, D., A. J. Kuenzi, I. Loredo, A. L. Leider, and M. L. Morrison. 1997. Translocation as a nonlethal alternative for managing California ground squirrels. Journal of Wildlife Management 61:351–59.

Verts, B. J., and L. N. Carraway. 1998. Land mammals of Oregon. University of California Press, Berkeley.

Verts, B. J., and D. B. Costain. 1988. Changes in sex ratios of *Spermophilus beldingi* in Oregon. Journal of Mammalogy 69:186–90.

Wade, O. 1927. Breeding habits and early life of the thirteen-striped ground squirrel, *Citellus tridecemlineatus* (Mitchill). Journal of Mammalogy 8:269–76.

Waggie, K. S., and P. L. Marion. 1997. *Demodex* sp. in California ground squirrels. Journal of Wildlife Diseases 33:368–70.

Wang, L. C. H. 1979. Time patterns and metabolic rates of natural torpor in the Richardson's ground squirrel. Canadian Journal of Zoology 57:149–55.

Whisson, D. A. 1999. Modified bait stations for California ground squirrel control in endangered kangaroo rat habitat. Wildlife Society Bulletin 27:172–77.

Whisson, D. A., S. B. Orloff, and D. L. Lancaster. 1999. Alfalfa yield loss from Belding's ground squirrels in northeastern California. Wildlife Society Bulletin 27:178–83.

Whitford, W. G., and F. R. Kay. 1999. Bioperturbation by mammals in deserts: A review. Journal of Arid Environments 41:203–30.

Wiggett, D. R., and D. A. Boag. 1986. Establishing colonies of ground squirrels during their active season. Wildlife Society Bulletin 14:288–91.

Wiggett, D. R., and D. A. Boag. 1989a. Intercolony natal dispersal in the Columbian ground squirrel. Canadian Journal of Zoology 67:42–50.

Wiggett, D. R., and D. A. Boag. 1989b. Movements of intercolony natal dispersers in the Columbian ground squirrel. Canadian Journal of Zoology 67:1447–52.

Wilber, P. G., B. Hanelt, B. Van Horne, and D. W. Duszynski. 1994. Two new species and temporal changes in the prevalence of eimerians in a free-living population of Townsend's ground squirrels (*Spermophilus townsendii*) in Idaho. Journal of Parasitology 80:251–59.

Wilber, P. G., D. W. Duszynski, and B. Van Horne. 1996. Hemorrhagic gastritis in free-living rodents in Idaho. Journal of Wildlife Diseases 32:665–69.

Williams, G. C. 1957. Pleiotropy, natural selection, and the evolution of senescence. Evolution 11:398–411.

Wilson, D. E., and and S. Ruff, 1999. The Smithsonian book of North American mammals. Smithsonian Institution Press, Washington, DC.

Wobeser, G. A., and F. A. Leighton. 1979. A simple burrow entrance trap for ground squirrels. Journal of Wildlife Management 43:571–72.

Yeaton, R. I. 1972. Social behavior and social organization in Richardson's ground squirrel (*Spermophilus richardsonii*) in Saskatchewan. Journal of Mammalogy 53:139–47.

Yensen, E. 1991. Taxonomy and distribution of the Idaho ground squirrel, *Spermophilus brunneus*. Journal of Mammalogy 72:583–600.

Yensen, E., and D. L. Quinney. 1992. Can Townsend's ground squirrels survive on a diet of exotic annuals? Great Basin Naturalist 52:269–77.

Yensen, E., and P. W. Sherman. 1997. *Spermophilus brunneus*. Mammalian Species 560:1–5.

Yensen, E., and M. Valdés-Alarcón. 1999. Sciuridae. Pages 239–320 *in* S. T. Alvarez-Castañeda and J. L. Patton, eds. Mamíferos del noroeste de México. Centro de Investigaciones Biológicas del Noroeste, S. C., La Paz, Baja California Sur, Mexico.

Yensen, E., M. P. Luscher, and S. Boyden. 1991. Structure of burrows used by the Idaho ground squirrel, *Spermophilus brunneus*. Northwest Science 65:93–100.

Yensen, E., D. L. Quinney, K. Johnson, K. Timmerman, and K. Steenhof. 1992. Fire, vegetation changes, and population fluctuations of Townsend's ground squirrels. American Midland Naturalist 128:299–312.

Yensen, E., C. R. Baird, and P. W. Sherman. 1996. Larger ectoparasites of the Idaho ground squirrel (*Spermophilus brunneus*). Great Basin Naturalist 56:237–46.

Young, C. J., and J. K. Jones, Jr. 1982. *Spermophilus mexicanus*. Mammalian Species 164:1–4.

Young, P. J. 1990. Structure, location and availability of hibernacula of Columbian ground squirrels (*Spermophilus columbianus*). American Midland Naturalist 123:357–64.

Zegers, D. A. 1984. *Spermophilus elegans*. Mammalian Species 214:1–7.

ERIC YENSEN, Department of Biology, Albertson College, Caldwell, Idaho 83605. Email: eyensen@albertson.edu.

PAUL W. SHERMAN, Department of Neurobiology and Behavior, Cornell University, Ithaca, New York 14853. Email: pws6@cornell.edu.

11

Black-tailed Prairie Dog

Cynomys ludovicianus and Allies

John L. Hoogland

NOMENCLATURE

Common and Scientific Names. The five species are black-tailed prairie dogs (*Cynomys ludovicianus*), Gunnison's prairie dogs (*C. gunnisoni*), Mexican prairie dogs (*C. mexicanus*), Utah prairie dogs (*C. parvidans*), and white-tailed prairie dogs (*C. leucurus*) (Hollister 1916; Pizzimenti 1975). Throughout most of their range (except the northeast section), Gunnison's prairie dogs are sometimes called Zuni prairie dogs (Taylor and Loftfield 1924; Burnett and McCampbell 1926; Scheffer 1947). Other common names for the black-tailed prairie dog include wishtonwish, barking ground squirrel, yaprat, tousa, and sod poodle (Hollister 1916; Hoogland 1995).

Subspecies. Hollister (1916) recognized two subspecies of black-tailed prairie dogs (*C. l. ludovicianus* and *C. l. arizonensis*) and two subspecies of Gunnison's prairie dogs (*C. g. gunnisoni* and *C. g. zuniensis*). Pizzimenti (1975:64) argued, however, that "there is no reason to support subspecific designation, and *C. ludovicianus* should be considered monotypic." Similarly, Pizzimenti (1975:64) asserted that "color alone is insufficient for . . . recognition [of subspecies]" and suggested that "*C. gunnisoni* [should] be regarded as a monotypic species."

Prairie dogs belong to the order Rodentia, suborder Protrogomorpha, family Sciuridae, subfamily Sciuirnae, tribe Marmotini, subtribe Spermophilina, genus *Cynomys* (Hafner 1984). In morphology and appearance, the five species of prairie dogs are similar. Because they have long (71–115 mm), black-tipped tails, Hollister (1916) grouped black-tailed prairie dogs with Mexican prairie dogs into the subgenus *Cynomys*. The other three species, all with shorter (30–70 mm), white- or gray-tipped tails, belong to the subgenus *Leucocrossuromys* (Hollister 1916; Clark et al. 1971; Pizzimenti 1975). Salient differences between the two subgenera include the following: *Leucocrossuromys* hibernate (for about 4 months each year), but *Cynomys* do not; *Leucocrossuromys* live at elevations of 1500–3000 m above sea level, but *Cynomys* live at lower elevations of 700–2200 m; shrubs and herbs within colonies of *Leucocrossuromys* are commonly $\geq$50 cm tall, but vegetation within colonies of *Cynomys* is rarely >20 cm tall (Scheffer 1947; King 1955; Tileston and Lechleitner 1966; Hoogland 1995). Furthermore, *Leucocrossuromys* have smaller molar teeth and thinner jugal bones than do *Cynomys* (Hollister 1916; Clark et al. 1971; Pizzimenti 1975).

Similarities between black-tailed and Mexican prairie dogs are striking, as are similarities between Utah and white-tailed prairie dogs. Indeed, classification of Mexican and Utah prairie dogs as separate species, rather than as relict populations of black-tailed and white-tailed prairie dogs, respectively, is somewhat arbitrary (Hollister 1916; Kelson 1949; Pizzimenti 1975; McCullough et al. 1987). The five species thus fall into three distinct groups: Gunnison's prairie dogs, black-tailed and Mexican prairie dogs, and white-tailed and Utah prairie dogs. The best characters for discriminating among prairie dog specimens include length and color of tail, presence or absence of a black or dark brown line above each eye, karyotype, serum proteins, and skeletal measurements (Hollister 1916; Lechleitner 1969; Clark 1973; Pizzimenti and Hoffmann 1973; Pizzimenti and Collier 1975; Pizzimenti 1976a, 1976b). In the wild, nonoverlapping geographic ranges promote easy identification (Fig. 11.1). Furthermore, territorial and antipredator calls easily distinguish the three groups (Waring 1970; Pizzimenti and McClenaghan 1974; Clark 1977; Wright-Smith 1978; Rayor 1988; Slobodchikoff et al. 1991; Hoogland 1995, 1996a; see below).

The earliest prairie dog fossils are from the Pleistocene (Black 1963; Clark et al. 1971; Pizzimenti 1975; Goodwin 1993, 1995). Prairie dogs differ anatomically from ground squirrels (*Spermophilus* spp. and *Ammospermophilus* spp.), their closest taxonomic relatives, by having larger body size, larger teeth with higher crowns, and broader skulls (Hollister 1916; Drearden 1953; Clark 1973; Hall 1981; Hafner 1984).

So why do we call them prairie dogs ? The first part of the common name refers to their grassland habitat (Hollister 1916; Clark 1977). The second part refers to the antipredator call of the black-tailed prairie dog, which reminded early settlers of the bark of a domestic dog (*Canis familiaris*) (Smith et al. 1976, 1977; Clark 1979).

DISTRIBUTION

The geographic distributions of the five species of prairie dogs do not overlap (Fig. 11.1). Colonies of black-tailed prairie dogs occur north to southern Saskatchewan in Canada, south to northern Chihuahua in Mexico, east to eastern Nebraska, and west to western Montana and eastern Arizona. *C. ludovicianus* thus has the largest geographic range and is the most common prairie dog. When scientists or nonscientists use the term "prairie dog," they almost invariably are referring to *C. ludovicianus*. Mexican prairie dogs inhabit the Mexican states of Coahuila, Nuevo Leon, Zacatecas, and San Luis Potosi. Gunnison's prairie dogs inhabit the Four Corners regions of northeastern Arizona, northwestern New Mexico, southeastern Utah, and southwestern Colorado. White-tailed prairie dogs occur in south-central Montana, western Wyoming, northeastern Utah, and northwestern Colorado. Utah prairie dogs occur only in southwestern Utah.

Figure 11.1 shows the geographic ranges of about 150 years ago for the different species of prairie dogs. Shooting, poisoning, destruction of habitat, and bubonic plague (*Yersinia [Pasteurella] pestis*) have led to precipitous declines. All five species are now rare, and persist in small colonies scattered throughout the former ranges (see Conservation and Management).

DESCRIPTION

Prairie dogs (Fig. 11.2) are diurnal, colonial, burrowing squirrels. Taxonomic relatives of the family Sciuridae include ground squirrels, marmots (*Marmota* spp.), tree squirrels (*Sciurus* spp. and *Tamiasciurus* spp.), flying squirrels (*Glaucomys* spp.), and chipmunks (*Tamias* spp. and *Eutamias* spp.) (Hafner 1984).

Total body length of adult prairie dogs ($\geq$20 months old) and yearlings (8–19 months old) (Hoogland 1995) ranges from 305 to 430 mm (Table 11.1). Standing adults and yearlings (which commonly breed; see below) are about 30 cm tall and distinctly pear-shaped. Body mass

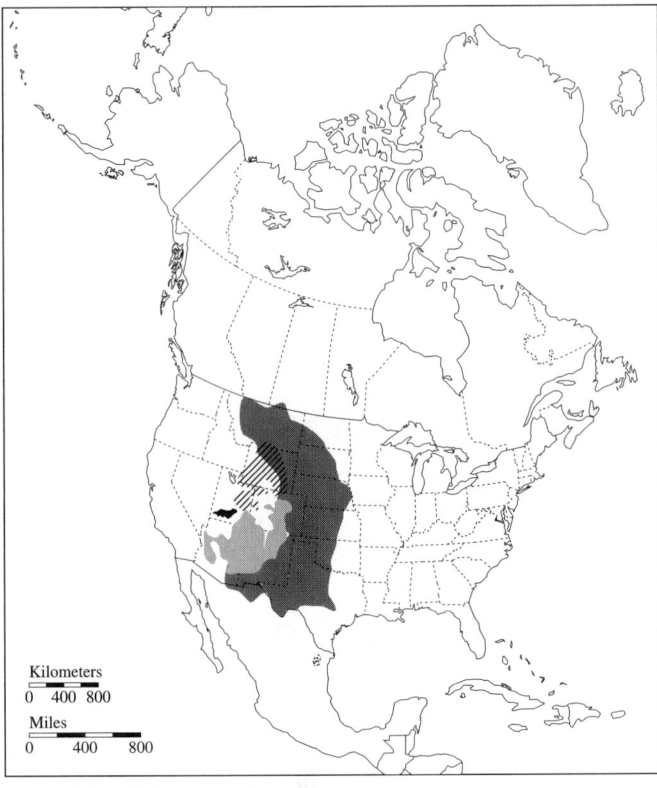

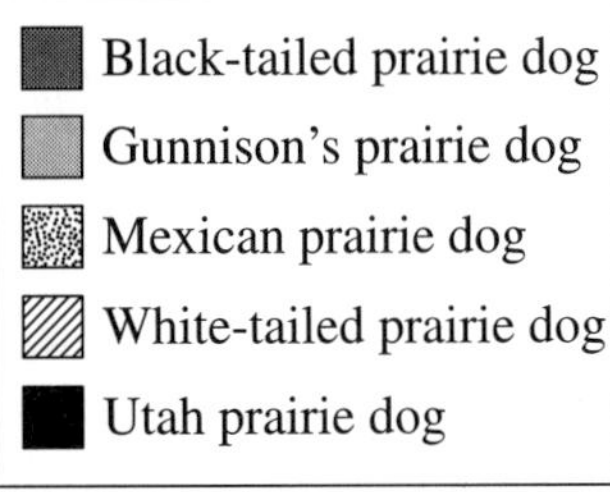

FIGURE 11.1. Distribution of the five species of prairie dog. This figure shows the distribution of prairie dogs about 150 years ago, before massive declines for all species resulting from shooting, poisoning, destruction of habitat, and bubonic plague. SOURCE: Data from Clark et al. (1971), Pizzimenti and Hoffmann (1973), Pizzimenti and Collier (1975), Hall (1981), Ceballos et al. (1993), and Hoogland (1996a).

varies seasonally for all five species (Fig. 11.3), and ranges from 247 to 1532 g for adults and yearlings (Table 11.1). During the breeding season, adult and yearling males are about 5% heavier than females for black-tailed prairie dogs, and about 25%–35% heavier than females for Gunnison's, Utah, and white-tailed prairie dogs (Fig. 11.3) (Hoogland 1995, 2003).

TABLE 11.1. Range of standard body measurements for adult and yearling prairie dogs

Species	Tail Length (mm)	Total Length (mm)	Body Mass (g)
Black-tailed	71–115	350–415	253–1390
Gunnison's	39–70	309–373	247–1075
Mexican	83–115	380–430	746–1303
Utah	30–60	305–360	328–1532
White-tailed	40–65	329–388	440–1440

SOURCE: Data from Hollister (1916); Clark et al. (1971); Collier and Spillett (1972, 1973, 1975); Pizzimenti and Hoffmann (1973); Pizzimenti (1975); Pizzimenti and Collier (1975); Ceballos and Wilson (1985); Trevino-Villarreal (1990); and Hoogland (1995, 1996a; 2003 also J. L. Hoogland, pers. obs.).

FIGURE 11.2. White-tailed prairie dog (*Cynomys leucurus*). Note the short, white tail and the black line over the eye. Utah prairie dogs look the same. Gunnison's prairie dogs are similar, but without the black line over the eye. Black-tailed and Mexican prairie dogs also do not have the black line over the eye, but have longer, black-tipped tails. SOURCE: Photo by Judy G. Hoogland.

Prairie dogs have five digits with thick, black, slightly curved claws on each foot. The head is short and broad, and the large, dark brown eyes are at the top of the head. Black whiskers are numerous, and reach 3 cm in length. The pinnae are short and leathery. Utah and white-tailed prairie dogs have a black or dark brown line above each eye, but individuals of the other species do not. Females of the subgenus *Cynomys* have 8 teats, but females of the subgenus *Leucocrossuromys* have 10 teats (Hollister 1916).

Pelage. Color of prairie dog fur ranges from yellowish to reddish to dark brown (Coues and Allen 1877; Hollister 1916). Assignment of individuals to different species from color alone is thus unreliable. Because dirt works its way into the pelage, the color of prairie dogs sometimes resembles the color of the local soil. Albinism occurs rarely among black-tailed and Gunnison's prairie dogs (Hollister 1916; Tate 1947; Costello 1970).

Adults and yearlings of all species molt the entire pelage twice each year (Hollister 1916; King 1955; Smith 1967; Hoogland 1995). In the change from long, thick winter fur to shorter, sparser summer fur, molting starts on the underside and moves to the dorsal side near the eyes and progresses posteriorly. The transition from summer fur to winter fur is reversed, progressing from tail to head to underside. From start to finish, molting and replacement of fur require about 2 weeks.

Fast-growing juveniles (<8 months old) of all species molt the entire pelage two or more times in their first summer before acquiring the winter fur in late August or September (Hollister 1916).

Skull. Prairie dog skulls (Fig. 11.4) are noteworthy for their wide zygomatic arches and conspicuous processes (Hollister 1916; Hafner 1984). Skulls for all five species are depicted in Hollister (1916) and Hall (1981) and also in the relevant *Mammalian Species* accounts (Clark et al. 1971; Pizzimenti and Hoffmann 1973; Pizzimenti and Collier 1975; Ceballos and Wilson 1985; Hoogland 1995).

Dentition. The dental formula for all prairie dog species is I 1/1, C 0/0, P 2/1, M 3/3 (Stockrahm and Seabloom 1990). Incisors of adults and yearlings are white or pale yellow.

Sequence of eruption for the permanent cheek teeth of *C. ludovicianus* is "M1 and m1, M2 and m2, M3 and m3, P3, p4, followed by P4 lagging somewhat behind the other premolars" (Stockrahm and Seabloom 1990:107). Within 2–3 months after first emergence from the natal burrow, juveniles of all species have acquired complete permanent dentition.

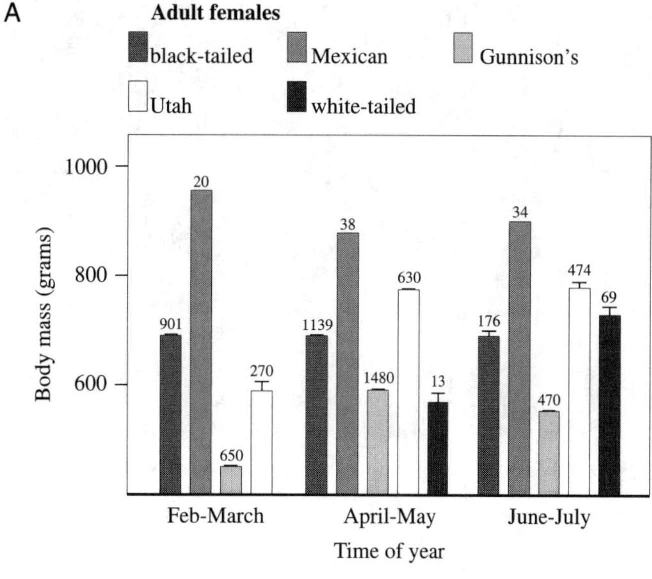

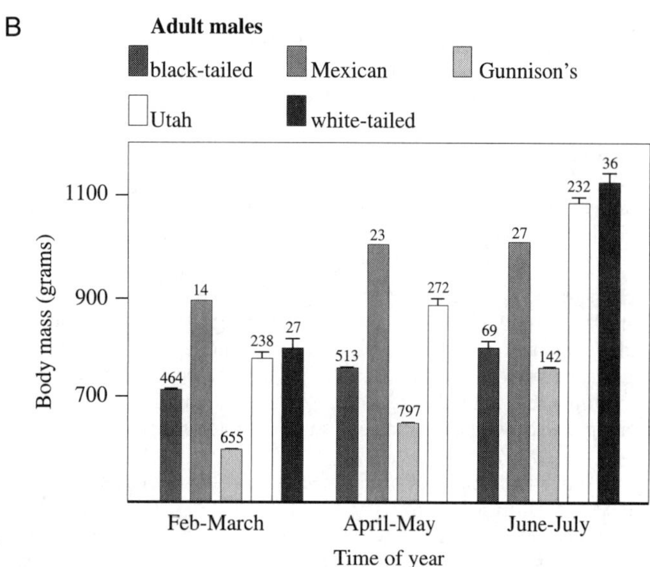

FIGURE 11.3. Seasonal variation in body mass for adult prairie dogs: (A) female and (B) male. Shown are means ±1 standard error; the number above each standard error line indicates the number of individuals weighed. SOURCE: Data from Trevino-Villarreal (1990) and Hoogland (1995, 2003).

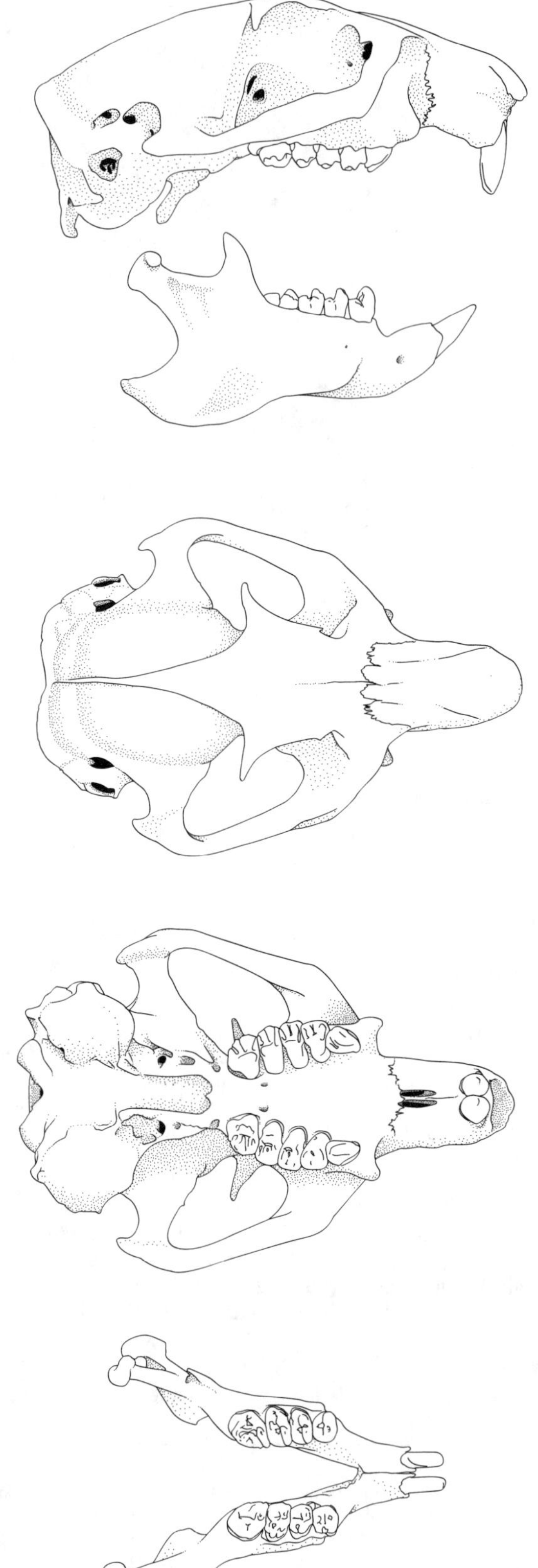

FIGURE 11.4. Skull of the black-tailed prairie dog (*Cynomys ludovicianus*). From top to bottom: lateral view of cranium, lateral view of mandible, dorsal view of cranium, ventral view of cranium, dorsal view of mandible. Greatest length of cranium (this specimen) is 61 mm. SOURCE: Data from Hoogland (1996a).

GENETICS

Gunnison's prairie dogs have a diploid number ($2N$) of 40 chromosomes. Black-tailed, Mexican, and Utah prairie dogs have a diploid number of 50 chromosomes. White-tailed prairie dogs usually have $2N = 50$ chromosomes, but occasionally have 48, 49, or 51 chromosomes (Nadler et al. 1971; Pizzimenti 1975).

Using blood samples from *C. ludovicianus* sampled at Wind Cave National Park, South Dakota, researchers have used starch-gel electrophoresis to examine over 60 loci for genetic polymorphism. Seven of these loci are polymorphic (Foltz and Hoogland 1983; Foltz et al. 1988; Daley 1992; see also Chesser 1983). Mexican prairie dogs show comparable genetic polymorphism (McCullough and Chesser 1987), but electrophoretic variation is rarer for Gunnison's, Utah, and white-tailed prairie dogs (Pizzimenti 1975, 1976b; Nichols and Nash 1980; Chesser 1984; Travis et al. 1996). Travis et al. (1996) and Haynie (2001) documented variation at microsatellite loci for Gunnison's and Utah prairie dogs.

PHYSIOLOGY

Hibernation. Gunnison's, Utah, and white-tailed prairie dogs all hibernate. For the latter two species, different age and sex groups enter hibernation at different times, which overlap. Adult and yearling males usually enter hibernation first, sometimes as early as July. Nonlactating females enter hibernation somewhat later, followed by lactating females. Juveniles usually do not enter hibernation until September or October (Stockard 1929, 1930; Egoscue 1975; Clark 1977; J. L. Hoogland, pers. obs.). For Gunnison's prairie dogs, entry of individuals into hibernation is more synchronous. Specifically, adults and yearlings of both sexes often remain active with juveniles into October and early November (Rayor et al. 1987; C. N. Slobodchikoff, Northern Arizona University, pers. commun., 2000).

For all three hibernating species, males start to emerge from hibernation in late February or early March. For white-tailed and Utah prairie dogs, the first females arouse about 2–3 weeks later (Stockard 1930; Clark 1977). In 1976, for example, I captured 22 white-tailed prairie dog males at the Arapaho National Wildlife Refuge in Walden, Colorado, from 6 to 23 March before I captured the first female on 24 March. Similarly, in 1998 at Bryce Canyon National Park, Utah, I captured 12 Utah prairie dog males in early March before I captured the first female on 12 March. The sexual asymmetry in time of arousal is less pronounced for Gunnison's prairie dogs, with the first females appearing aboveground about 1 week after the first males (Rayor et al. 1987; Rayor 1988; Hoogland 1999). For all three hibernating species, arousals of early females overlap with arousals of late males (Clark 1977; Rayor 1988; Hoogland 1999; J. L. Hoogland, pers. obs.).

After appearing to enter hibernation in July or early August, adult and yearling white-tailed and Utah prairie dogs sometimes reappear in late August for a few days (Clark 1977; J. L. Hoogland, pers. obs.). Following final submergences of juveniles for hibernation by late October or early November, however, all white-tailed and Utah prairie dogs usually remain underground until males start emerging about 4 months later (Bakko and Brown 1967; Clark 1977; J. L. Hoogland, pers. obs.; but see Fitzgerald et al. 1994). In contrast, 10–15 marked Gunnison's prairie dogs per study colony sometimes appear aboveground on warm days in December and January near Flagstaff, Arizona (C. N. Slobodchikoff, pers. commun., 2000; see also Rayor et al. 1987).

Partly in response to the metabolic changes associated with hibernation, body mass shows substantial seasonal variation for Gunnison's, Utah, and white-tailed prairie dogs (Fig. 11.3) (Clark 1977; Wright-Smith 1978; Rayor 1988). For adult and yearling males of these three species, body mass is greatest in July or August, just before the onset of hibernation. Male body mass declines to a minimum in late March or early April, at the end of the breeding season. For females, body mass also is greatest just before the onset of hibernation, declines to a minimum around the time of copulation, then peaks again just before parturition (Hoogland 1995, 2003).

Even though they do not hibernate (Pizzimenti and McClenaghan 1974; Harlow and Menkens 1986; Bakko et al. 1988; Trevino-Villarreal 1990), black-tailed and Mexican prairie dogs depend on fat reserves for survival during the winter months. Consequently, seasonal fluctuation in body mass for *C. ludovicianus* and *C. mexicanus* is similar to that for the three hibernating species (Fig. 11.3) (Harlow 1997). Individual black-tailed prairie dogs sometimes remain underground for several consecutive days during inclement weather in late autumn and winter (Koford 1958; Thomas and Riedesel 1975; Hamilton and Pfeiffer 1977; Harlow and Menkens 1986; Harlow 1995). Rarely, an individual or small group of individuals will remain underground for 1 month or more during severe winter weather (Hoogland 1995).

Hibernation is usually a mechanism by which individuals avoid cold weather, when food is scarce (Hudson 1978). Black-tailed prairie dogs, however, remain active throughout the winter in colonies as far north as Canada. In contrast, Gunnison's prairie dogs invariably hibernate in colonies as far south as Arizona and New Mexico (Hoogland 1995, 1999). Perhaps shortage of water has been more important than cold weather in the evolution of hibernation among prairie dogs (Bakko 1977; Bintz 1984; Stephens and Fevold 1993; Harlow 1995).

Gallstones. Black-tailed prairie dogs frequently develop gallstones under laboratory conditions (Chapman et al. 1998). This finding has led to a better understanding of gallstones and gall bladder disease (Brenneman et al. 1972; Holzbach et al. 1976; Gurll and DenBesten 1978; Broughton et al. 1991, Roslyn et al. 1991; Lamorte et al. 1993).

REPRODUCTION

Sex Determination. The external genitalia of male and female prairie dogs are surprisingly similar. Distance between the external genitalia and the anus is sexually distinctive, however. Specifically, the vulva and the anus are contiguous for all species. The penis and the anus are 1–1.5 cm apart in juveniles. For adult and yearling males, the penis and anus are 2–3 cm apart for black-tailed and Mexican prairie dogs and 3–4 cm apart for Gunnison's, Utah, and white-tailed prairie dogs (Hoogland 1995; J. L. Hoogland, pers. obs.).

During the breeding season, the scrotum of reproductive males of all species is black or dark gray (Hoogland 1995, 1999). The scrotum of nonreproductive males is pink or light gray. The testes of reproductive males are sometimes descended, but often recede into the abdominal cavity, especially during handling by humans (Hoogland 1995). A pigmented scrotum is thus a more reliable indicator of breeding condition than is the presence of descended testes. A female's vulva is usually swollen and closed before copulation, but is always open for several days after copulation (Hoogland 1995, 1998a). Teats are long and swollen from late pregnancy through lactation for all species. When a female prematurely loses her litter to predation or infanticide, her teats remain long for about 1 month but lose all turgidity within 3–4 days (Hoogland 1995). No quick, reliable method exists for sex determination of unmarked prairie dogs from a distance (King 1955; Chace 1976).

Age of Sexual Maturity. For black-tailed prairie dogs at Wind Cave National Park, the most common age of sexual maturity for both sexes, as measured by the age of first copulation, is 2 years (Hoogland 1995; see also King 1955; Koford 1958; Tileston and Lechleitner 1966; Halpin 1987; Knowles 1987; Stockrahm and Seabloom 1988). More precisely, individuals usually first copulate in their second February or March, approximately 23 months after birth and 22 months after first emergence from the natal burrow. Some black-tailed prairie dogs first copulate as yearlings, with the probability being higher for females than for males (35% vs. 6%). The probability of producing emergent juveniles as a yearling also is higher for females (9% vs. 2%) (Hoogland 1995). Conversely, some black-tailed prairie dogs delay sexual maturation until the third year. Again, a sexual asymmetry prevails, with males being more likely than females to delay (24% vs. 5%) (Hoogland 1995).

For Mexican, Gunnison's, Utah, and white-tailed prairie dogs, females usually copulate as yearlings (Tileston and Lechleitner 1966; Fitzgerald and Lechleitner 1974; Clark 1977; Rayor 1988; Trevino-Villarreal 1990). For Gunnison's prairie dogs at Petrified Forest National Park, Arizona, for example, 100% of 358 females attained sexual maturity and copulated as yearlings (Hoogland 1999). Males of Mexican, Gunnison's, Utah, and white-tailed prairie dogs sometimes copulate as yearlings, but commonly defer copulation until they are 2-year-olds (Rayor 1988; Trevino-Villarreal 1990; Hoogland 1995, 1999, 2001). For black-tailed and Gunnison's prairie dogs, males are more likely to copulate as yearlings in colonies with copious vegetation and minimal competition (Garrett et al. 1982; Rayor 1985b).

Copulation. Black-tailed, Gunnison's, Utah, and white-tailed prairie dog females come into estrus and are sexually receptive on only 1 day of the breeding season each year (Hoogland 1995, 1998a, 1998b, 2001). Females occasionally come into estrus a second time in the same year if they do not conceive in the first estrus (Hoogland 1995; J. L. Hoogland, pers. obs.).

Ninety-eight percent of black-tailed prairie dog females, 92% of Gunnison's prairie dog females, about 80% of Utah prairie dog females, and about 75% of white-tailed prairie dog females copulate underground (Hoogland 1995, 1998a; J. L. Hoogland, pers. obs.). Four diagnostic aboveground behaviors before or after an underground consortship are common to all four species : frequent sniffing of the estrous female's vulva by breeding males, a unique mating call by the males that copulate, self-licking of the genitals by both sexes, and late final submergence at the end of the day by the estrous female (Hoogland 1995, 1998a, 2001). Courting black-tailed prairie dog males sometimes take mouthfuls of dry grass into a burrow before copulating there, and both sexes of Gunnison's prairie dogs commonly roll around in the dirt (i.e., take a "dust bath") after an underground consortship. Copulations can occur as early as 0800 hr or as late as 1800 hr, but most occur in midafternoon (Hoogland 1995, 1998a).

Sixty-seven percent of estrous black-tailed prairie dog females copulate with only one male (Hoogland 1995). In contrast, ≥65% of estrous Gunnison's and Utah prairie dog females copulate with two or more males (Travis et al. 1996; Hoogland 1998a; 1998b; J. L. Hoogland, pers. obs.). The frequency of multiple paternity within litters (same mother, different fathers) is about 10% for black-tailed prairie dogs (Hoogland 1995), but is >25% for Gunnison's and Utah prairie dogs (Travis et al. 1996; Haynie 2001; Haynie et al. 2003).

Black-tailed, Gunnison's, and white-tailed prairie dog females avoid copulations with close kin such as fathers, sons, and full-brothers, but commonly copulate with more distant kin such as first and second cousins (Hoogland 1995, 1999). Copulation with close kin is more common in colonies of Utah prairie dogs (J. L. Hoogland, pers. obs.). Females of all four species are more likely to copulate with a male of the home harem than with outside males, but cuckoldry is nonetheless common (Hoogland 1995; Travis et al. 1996; Haynie 2001).

For black-tailed prairie dogs at Wind Cave National Park, copulation can occur as early as 16 February or as late as 13 April, with a mean ± *SD* of 7 March ± 8 days ($n = 534$) (Hoogland 1995; see also King 1955; Pfeiffer 1972). For Gunnison's prairie dogs at Petrified Forest National Park, copulations occur from 12 March through 13 April, with a mean ± *SD* of 28 March ± 5 days ($n = 308$) (Hoogland 1998a). For Utah prairie dogs at Bryce Canyon National Park and white-tailed prairie dogs at the Arapaho National Wildlife Refuge, most copulations also occur in late March and early April (J. L. Hoogland, pers. obs.). For Mexican prairie dogs, copulations occur as early as December or as late as April (Pizzimenti and McClenaghan 1974; Trevino-Villarreal 1990; Mellink and Madrigal 1993).

Utah and white-tailed prairie dog females usually copulate 2–5 days after arousal from hibernation, but Gunnison's prairie dog females usually wait 5–10 days after arousal before copulating (Rayor 1988; J. L. Hoogland pers. obs.).

Gestation. Gestation for black-tailed prairie dogs ranges from 33 to 38 days, with a mean ± *SD* of 34.6 ± 0.73 days; 88% of gestations are either 34 or 35 days ($n = 225$) (Hoogland 1995). Gestation for Gunnison's prairie dogs ranges from 28 to 31 days, with a mean ± *SD* of 29.3 ± 0.53 days; 97% of gestations are either 29 or 30 days ($n = 124$) (Hoogland 1997). Gestation for Utah prairie dogs is most commonly 28, 29, or 30 days (J. L. Hoogland, pers. obs.).

Parturition. Parturition occurs underground for all species, and is identifiable in three ways for black-tailed, Gunnison's, and Utah prairie dogs (Hoogland 1995, 1997). First, throughout pregnancy a female's vagina is open and white. She then abruptly shows a distinctly pink or red open vagina, usually coupled with bloodstains on the surrounding fur. Second, on the first day that the vagina is pink, body mass suddenly drops precipitously, sometimes by >100 g. Third, a dramatic behavioral change accompanies the first appearance of the pink vagina and the large decrease in body mass. Pregnant females usually emerge early in the morning, spend the entire day foraging aboveground, and are among the last to submerge for the night. On the first day of the pink vagina, however, the female is usually one of the last to appear aboveground in the morning, sometimes emerging as much as 4–5 hr after all other individuals. Furthermore, the female usually makes one or more long (0.5–2.0 hr) daytime visits to the home nursery burrow on the first day of the pink vagina and she is one of the first to submerge for the night. These striking changes in behavior are presumably indicative of parturition and maternal care (especially nursing) of neonates. The altered behaviors usually continue for about 1–2 weeks, after which mothers resume the routine typical of pregnant females. Most births occur early in the morning or late in the afternoon (Hoogland 1995, 1997). Parturition in late afternoon coincides with an unusually early final submergence by the mother for the night, as much as 3–4 hr before other colony residents.

Not every prairie dog female that copulates gives birth 4–5 weeks later. Failure to give birth can result from either failure to conceive or premature death of all embryos/fetuses (with or without resorption) (Anthony and Foreman 1951; Knowles 1987). The probability of giving birth after copulation is higher for adults than for yearlings for black-tailed, Gunnison's, and Utah prairie dogs (Rayor 1988; Hoogland 1995, 1999, 2001). Pregnant black-tailed and white-tailed prairie dog females sometimes resorb some implanted embryos or fetuses, but later give birth to the others (Anthony and Foreman 1951; Foreman 1962, 1981; Tileston and Lechleitner 1966; Bakko and Brown 1967; Knowles 1987).

Length of gestation varies inversely with litter size at first juvenile emergence for Gunnison's prairie dogs, but not for black-tailed prairie dogs (Hoogland 1995, 1997).

Lactation. Before they first emerge from the natal burrow, prairie dog juveniles depend primarily on nursing for nourishment, but sometimes eat plants brought underground by the mother. Conversely, emergent juveniles depend primarily on their own foraging, but sometimes continue to nurse as well (Longhurst 1944; King 1955; Hoogland et al. 1989; Hoogland 1999). To complicate matters, nursing almost always occurs underground. Furthermore, black-tailed, Gunnison's, and Utah prairie dog mothers sometimes nurse not only their own emergent offspring, but also those of other mothers via communal nursing (Hoogland 1995; J. L. Hoogland, pers. obs.). Weaning and duration of lactation thus are difficult to specify for prairie dogs. One approximate estimate for the length of lactation is the duration between parturition and the first emergence of juveniles from the natal burrow. For black-tailed prairie dogs, this duration ranges from 37 to 51 days, with a mean ± *SD* of 41.3 ± 2.5 days ($n = 149$) (Hoogland 1995). The range from parturition to first emergence for Gunnison's prairie dogs is 34–45 days, with a mean ± *SD* of 38.6 ± 2.1 days ($n = 112$) (Hoogland 1997). Utah and white-tailed prairic dogs havc ranges and means similar to those for *C. gunnisoni* (J. L. Hoogland, pers. obs.). For black-tailed and Gunnison's prairie dogs, the number of days between parturition and first emergence of juveniles varies inversely with litter size (Hoogland 1995, 1997).

For about 1–2 weeks after parturition, as noted above, mothers usually visit their offspring once or twice during daylight hours. These visits presumably involve nursing. For the last 2–3 weeks of lactation, however, mothers routinely forage aboveground continuously from first emergence at dawn until final submergence at dusk. All nursing of older preemergent offspring thus must occur after sunset and before dawn.

Litter Size. Because parturition occurs underground for all prairie dog species, estimates of litter size at birth come from laboratory research and from necropsies of pregnant and lactating females. Estimates of maximum litter size at birth are 8 for black-tailed prairie dogs (Wade 1928; Foreman 1962; Tileston and Lechleitner 1966; Knowles 1987), 6 for Mexican prairie dogs (Pizzimenti and McClenaghan 1974), 8 for Gunnison's prairie dogs (Aldous 1935; Longhurst 1944), and 10 for white-tailed prairie dogs (Stockard 1930; Bakko and Brown 1967).

Litter size when juveniles first appear aboveground is easier to determine than litter size at birth, and roughly coincides with litter size at weaning. Mean ± *SD* litter size at first juvenile emergence is 3.08 ± 1.06 for black-tailed prairie dogs (range = 1–6, $n = 361$), 3.77 ± 1.21 for Gunnison's prairie dogs (range = 1–7, $n = 340$), and 3.88 ± 1.26 for Utah prairie dogs (range = 1–7, $n = 161$) (Hoogland 1995, 2001). Neonates of all species are pink, blind, hairless, and weigh about 15–20 g (Stockard 1930; Foreman 1962, Pizzimenti and McClenaghan

1974; Hoogland 1995). For black-tailed prairie dogs, fur appears about 3 weeks after birth, and the eyes open about 2 weeks later (Johnson 1927).

Some litters contain only males and others contain only females, but mothers collectively rear approximately equal numbers of males and females to first juvenile emergence for black-tailed, Gunnison's, Utah, and white-tailed prairie dogs. Males of these four species are slightly heavier than females at first emergence (Hoogland 1995, 2003).

Factors That Affect Reproductive Success. Reproductive success varies curvilinearly with age for male and female black-tailed prairie dogs (Hoogland 1995). Age and body mass also vary curvilinearly for black-tailed prairie dogs (Hoogland 1995). Other factors that affect annual and lifetime reproductive success for black-tailed, Gunnison's, and Utah prairie dogs include precipitation, reproductive synchrony, and harem size (Knowles 1987; Hoogland 1995; J. L. Hoogland pers. obs.). More than any other factor, longevity enhances lifetime reproductive success for all species (Hoogland 1995, 1999).

ECOLOGY

Colonies. Prairie dogs live in aggregations called colonies, towns, or villages (Coues and Allen 1877; King 1955; Costello 1970). When unsuitable habitat such as a hill, tall vegetation, or a stream divides a colony, the resulting subcolonies are called wards (King 1955). Residents of one ward usually can see and hear residents of an adjacent ward, but movements and communications between wards are uncommon. Size and density of colonies vary for all species, but are usually higher for black-tailed and Mexican prairie dogs than for Gunnison's, Utah, and white-tailed prairie dogs (Scheffer 1947; Tileston and Lechleitner 1966; Wright-Smith 1978; Hoogland 1981a, 1995; Rayor 1985b; Trevino-Villarreal 1990; Travis et al. 1995).

Vegetation differentiates black-tailed prairie dog colonies from surrounding areas in two ways. First, the height of vegetation is markedly shorter within colonies (King 1955; Koford 1958; Tileston and Lechleitner 1966). Short vegetation results not only from normal foraging, but also because black-tailed prairie dogs prefer to colonize areas where the vegetation is already low (Koford 1958; Clark 1979; Snell 1985; Knowles 1986b). In addition, black-tailed prairie dogs use their teeth to clip tall plants (greater than about 20 cm) at the base without consuming them (King 1955; Hoogland 1995). Such clipping facilitates the detection of predators. Second, composition of the plant community is radically different within colonies (Koford 1958; Klatt and Hein 1978; Agnew et al. 1986; Whicker and Detling 1988; Foster and Hygnstrom 1990). Certain plants, such as scarlet globemallow (*Sphaeralcea coccinea*), black nightshade (*Solanum nigrum*), pigweed (*Amaranthus retroflexus*), and prairie dog weed (*Dyssodia papposa*), almost never occur outside colonies of *C. ludovicianus* (King 1955). Mainly because of their short vegetation, colonies of black-tailed and Mexican prairie dogs are conspicuous from afar. Colonies of Gunnison's, Utah, and white-tailed prairie dogs, by contrast, are less obvious.

Aboveground Activity. Prairie dogs appear aboveground only during daylight hours. Consistent with the abundance of cones and rarity of rods in their retinas (Walls 1941, 1942; Jacobs and Yolton 1972; Green and Dowling 1975; West and Dowling 1975), individuals released from livetraps after sunset sometimes seem disoriented (King 1955). In warm weather, individuals first emerge from burrows at about sunrise and remain aboveground until about sunset (King 1955; Fitzgerald and Lechleitner 1974; Clark 1977). When midday temperatures exceed about 27°C, individuals return to their burrows, presumably to cool off (Bakko et al. 1988). Black-tailed prairie dogs submerge repeatedly for 5–30 min at a time in hot weather (King 1955; Hoogland 1995). Gunnison's, Mexican, Utah, and white–tailed prairie dogs, however, sometimes remain underground for several consecutive hours on a hot day before reappearing aboveground in the coolness of late afternoon (Fitzgerald and Lechleitner 1974; Clark 1977; Rayor 1988; Trevino-Villarreal 1990; J. L. Hoogland, pers. obs.). In cold weather (e.g., less than about 4°C), individuals usually remain underground until 1–3 hr after sunrise, and submerge for the night 1–3 hr before sunset. In addition, individuals sometimes submerge for short periods, presumably to warm up, in early afternoon on cold days (J. L. Hoogland, pers. obs.). For all species, first emergences in the morning for different individuals commonly span 1–2 hr, as do final submergences for the night (Hoogland 1995; J. L. Hoogland pers. obs.).

Even though they spend so much time aboveground during daylight hours, prairie dogs submerge for certain critical behaviors. Black-tailed, Gunnison's, Utah, and white-tailed prairie dog females, for example, go underground to copulate and to nurse their own and other's offspring (Hoogland 1995, 1997). Infanticide and cannibalism within colonies of black-tailed prairie dogs also occur underground (Hoogland 1995).

Burrows. Prairie dogs depend on elaborate underground tunnels for protection from weather, predators, and darkness. Burrows typically are 10–30 cm in diameter at the entrances and narrower underground (Merriam 1902; King 1955; Clark 1971; Sheets et al. 1971; Stromberg 1978; Cooke and Swiecki 1992). Usually they are 5–10 m long and 2–3 m deep, but some burrows are as long as 33 m and as deep as 5 m (Longhurst 1944; King 1955; Sheets et al. 1971; Clark 1977; Burns et al. 1989; Hoogland 1995). This variation probably explains why burrows differ in suitability for spending the night, rearing offspring, and serving as refuge from weather and predators (Hoogland 1995). Number and positioning of burrow entrances change little through time (King 1955, 1984; Clark 1977; Hoogland 1995; J. L. Hoogland, pers. obs.). Prairie dogs sometimes create aboveground trails of worn-down vegetation between burrow entrances (King 1955).

Burrows used for rearing juveniles (nursery burrows) and burrows used for final submergence at sunset (sleeping burrows) contain ≥1 elliptical nest chamber(s) packed with dry grass. For black-tailed prairie dogs, each chamber is approximately 30 cm high and 50 cm wide (Sheets et al. 1971; Gunderson 1978; Hoogland 1995). To build nests, prairie dogs take mouthfuls of dry grass underground. Yearlings and adults of both sexes collect dry grass, but pregnant and lactating females of all species are the most conspicuous nest builders. Female black-tailed prairie dogs typically start to collect dry grass for the nursery burrow on the first day after copulation, but females of other species usually do not initiate nest building until they are farther along in pregnancy (Hoogland 1995, 1999; J. L. Hoogland, pers. obs.).

Besides nest chambers, burrows of black-tailed and white-tailed prairie dogs frequently contain smaller chambers approximately 1 m below the surface (Merriam 1902; Sheets et al. 1971; Clark 1977; Hoogland 1995). Individuals use these chambers to turn themselves around (Scheffer 1947). While allowing individuals to hear aboveground vocalizations, the chambers also provide safety from enemies that cannot easily enter burrows, such as coyotes (*Canis latrans*), bobcats (*Lynx* [*Felis*] *rufus*), and avian predators. Some black-tailed, Gunnison's, and white-tailed prairie dog burrows harbor hundreds of fecal pellets, but others contain almost none (Merriam 1902; Whitehead 1927; Longhurst 1944; King 1955; Sheets et al. 1971; Clark 1977; Hoogland 1995).

Prairie dog burrows have temperatures of 5–10°C in winter and 15–20°C in summer (Wilcomb 1954; Costello 1970; Clark 1977; Gunderson 1978). Burrows thus allow individuals to warm up in winter and cool down in summer. Relative humidity is always higher within burrows than at the surface, and is usually >80% (Wilcomb 1954; Costello 1970; Gunderson 1978).

For black-tailed and Gunnison's prairie dogs, most burrows have only one or two entrances, but some have three, and a few have as many as five or six (Scheffer 1937; Wilcomb 1954; Stromberg 1978; Hoogland 1995; J. L. Hoogland, pers. obs.). Burrows of Utah and white-tailed prairie dogs are more complex, and sometimes have ≥15 entrances (J. L. Hoogland, pers. obs.).

For some prairie dog burrows, two or more entrances connect continuously for several consecutive years. For other burrows, two or more entrances connect for a while, then do not connect, then connect again (King 1955; Hoogland 1995). Perhaps to defend against certain predators or to reduce heat loss in cold weather, prairie dogs use dirt, feces, and dry grass to plug burrows at one or more points beneath the surface (Longhurst 1944; Clark 1977; Martin et al. 1984).

Prairie dogs sometimes use dirt to close an entrance to a burrow. Black-tailed and Utah prairie dogs, for example, sometimes plug one or more entrances to burrows that contain long-tailed weasels (*Mustela frenata*), and black-tailed prairie dogs also plug entrances to burrows containing snakes and black-footed ferrets (Halpin 1983; Clark et al. 1984; Loughry 1987b; Hoogland 1995; J. L. Hoogland, pers. obs.). Gunnison's prairie dogs, by contrast, evidently do not plug entrances to burrows that contain snakes (Longhurst 1944; J. L. Hoogland, pers. obs.). After chasing a conspecific into a burrow entrance, black-tailed, Gunnison's, and Utah prairie dogs sometimes plug that entrance (Hoogland 1995; J. L. Hoogland, pers. obs.). Finally, black-tailed and Utah prairie dog mothers sometimes plug one or more of the entrances to their nursery burrows to reduce the probability of infanticide (Hoogland 1995; J. L. Hoogland, pers. obs.).

Especially near the colony's periphery, entrances to prairie dog burrows sometimes do not have a conspicuous mound of dirt. Other entrances are surrounded by wide, unstructured mounds of dirt called dome craters (King 1955; Sheets et al. 1971; Vogel et al. 1973; Flath 1979). Dome craters are 1–3 m in diameter and 0.2–1.0 m high. In colonies of black-tailed, Mexican, and Gunnison's prairie dogs, most dome craters have only one burrow entrance (King 1955; Tileston and Lechleitner 1974; Trevino-Villarreal 1990). In colonies of Utah and white-tailed prairie dogs, by contrast, dome craters commonly have two or three burrow entrances and sometimes as many as seven or eight (Hoogland 1981a, 1995; J. L. Hoogland, pers. obs.). In colonies of black-tailed and Mexican prairie dogs, some burrow entrances are surrounded by a high mound of molded dirt that resembles a miniature volcano (King 1955; Tileston and Lechleitner 1966; Trevino-Villarreal 1990). These mounds are called rim craters. Rim craters usually are 0.5–1.5 m in diameter and 0.2–1.0 m high. Especially when the ground is wet after rain, individuals reshape the mounds of rim craters by digging, scraping, and pushing the surrounding soil with their front and rear legs. With arched backs, they pound the soil into place with their noses (King 1955; Cincotta 1989; Hoogland 1995). Individuals usually work alone on rim craters, but sometimes toil together in groups of two to four.

Dome and rim craters function in at least three ways. First, they protect against flooding of burrows during severe rainstorms (Foster 1924; Whitehead 1927; King 1955; Costello 1970). Second, prairie dogs frequently use craters as vantage points to scan for predators and trespassing conspecifics (King 1955; 1959; Hoogland 1979b). Third, if a burrow has a low mound at one entrance and a higher mound at another entrance, then the slightest aboveground breeze will create a partial underground vacuum via Bernoulli's Principle (Vogel et al. 1973; Vogel 1989). Such improved ventilation might be important when burrows are especially long and deep, or when $\geq$14 prairie dogs spend the night in the same burrow (Hoogland 1995).

Perhaps in response to the ease or difficulty of excavating in different types of soil, the density of burrow entrances within colonies of prairie dogs shows great variation. The density ranges from 10 to 250 burrow entrances/ha for black-tailed prairie dogs ($N = 24$ colonies), for example, and from 3 to 223 burrow entrances/ha for white-tailed prairie dogs ($N = 191$ colonies) (Martin and Schroeder 1978, 1980; Campbell and Clark 1981; Hoogland 1981a, 1995). Density of burrow entrances does not accurately predict density of prairie dogs (King 1955; K. L. Powell et al. 1994; Hoogland 1995; Severson and Plumb 1998).

FEEDING HABITS

Prairie dogs are herbivorous, as shown by observations of foraging individuals and by analyses of stomach contents and feces (Stockard 1930; Kelso 1939; Bonham and Lerwick 1976; Fagerstone 1982; Hasenyager 1983; Uresk 1984; Shalaway and Slobodchikoff 1988). Exceptions to herbivory include an occasional meal of insects such as cicadas (Cicadidae), cutworms (Noctuidae), ground beetles (Carabidae), and short-horned grasshoppers (Acrididae) (Smith 1915; Burnett and McCampbell 1926; Whitehead 1927; Kelso 1939; Longhurst 1944; Costello 1970; Crocker-Bedford and Spillett 1981; O'Meilia et al. 1982).

Prairie dogs eat plants within or just outside colony boundaries, but not all plants are equally suitable. Black-tailed prairie dogs, for example, are selectively herbivorous, and their preferences show seasonal variation (King 1955; Koford 1958; Costello 1970; Fagerstone et al. 1981; Rogers-Wydeven and Dahlgren 1982). Favorite foods in the summer include wheatgrass (*Agropyron* spp.), grama (*Bouteloua* spp.), buffalo grass (*Buchloe dactyloides*), scarlet globemallow, and rabbitbrush (*Chrysothamnus* spp.) (Summers and Linder 1978). Favorites in the winter include prickly pear cactus (*Opuntia* spp.) and thistle (*Cirsium* spp.). Eating underground roots is also more common in winter (King 1955; Costello 1970; Summers and Linder 1978). Common plants within colony boundaries that black-tailed prairie dogs usually avoid include sagebrush (*Artemisia* spp.), threeawn (*Aristida* spp.), prairie dog weed, and horseweed (*Conyza ramosissima*) (King 1955; Costello 1970; Summers and Linder 1978).

Gunnison's, Utah, and white-tailed prairie dogs also are selectively herbivorous with seasonal variation. Preferred plants for Gunnison's prairie dogs include grama, umbrella plant (*Eriogonum* spp.), aster (*Aster* spp.), wheatgrass, sagebrush, koeleria grass (*Koeleria* spp.), muhly grass (*Muhlenbergia* spp.), fescue grass (*Festuca* spp.), and rabbitbrush (Taylor and Loftfield 1924; Longhurst 1944, Fitzgerald and Lechleitner 1974; Shalaway and Slobodchikoff 1988). Favorite plants for Utah prairie dogs include wheatgrass, meadow grass (*Poa* spp.), feathergrass (*Stipa* spp.), sedge (*Carex* spp.), umbrella plant, ragwort (*Senecio* spp.), and aster (Crocker Bedford and Spillett 1981; Jacquart 1986). Preferred plants for white-tailed prairie dogs include muhly grass, wheatgrass, sedge, goosefoot (*Chenopodium* spp.), saltbush (*Atriplex* spp.), rabbitbrush, and sagebrush (Tileston and Lechleitner 1966; Fitzgerald et al. 1994). On emergence from hibernation, white-tailed prairie dogs sometimes eat roots and tubers (Clark 1977).

Male and female black-tailed prairie dogs and male Utah prairie dogs show one outstanding deviation from herbivory: cannibalism. They frequently kill and cannibalize juvenile conspecifics (Hoogland 1985, 1995; J. L. Hoogland, pers. obs.). In addition, black-tailed and Utah prairie dogs cannibalize adults and yearlings that die aboveground. Infanticide and cannibalism are rare among Gunnison's prairie dogs (Fitzgerald and Lechleitner 1974; Hoogland 1999).

DEMOGRAPHY AND MORTALITY

After birth, prairie dog juveniles remain underground in the nursery burrow for about 5–6 weeks before they first appear aboveground. One important cause of mortality for preemergent juvenile black-tailed and Utah prairie dogs is infanticide. Such infanticide partially or totally eliminates 39% of black-tailed prairie dog litters and about 20% of Utah prairie dog litters (Hoogland 1995, 2001).

Predation of preemergent prairie dog juveniles also occurs, but is difficult to document. At a colony of Utah prairie dogs in 1998, long-tailed weasels killed >40 preemergent juveniles (J. L. Hoogland, pers. obs.). American badgers (*Taxidea taxus*) and black-footed ferrets also prey on preemergent prairie dogs (Tileston and Lechleitner 1966; R. E. Smith 1967; Hillman 1968; Henderson et al. 1969; Fitzgerald and Lechleitner 1974). Snakes, especially bull snakes (*Pituophis melanoleucus*) and several species of rattlesnakes (*Crotalus* spp.), prey on preemergent juveniles as well (Scheffer 1945; Owings and Owings 1979; Halpin 1983; Owings and Loughry 1985; Loughry 1987a, 1987b). Other mortality factors for preemergent juveniles include disease, genetic defects, and abandonment by the mother (Hoogland 1995).

Once juveniles emerge from the nursery burrow, researchers can eartag and mark them, and then track their survivorship through time (Hoogland 1995). For black-tailed, Gunnison's, Utah, and white-tailed prairie dogs, survivorship of emergent juveniles is approximately 50% in the first year and slightly higher in later years (Hoogland 1995, 2001). For these four species, juvenile body mass at first emergence and survivorship in the first year vary inversely with litter size (Hoogland 1995, 1999; J. L. Hoogland, pers. obs.).

For black-tailed prairie dogs, survivorship varies curvilinearly with age (Hoogland 1995). Male black-tailed and Gunnison's prairie dogs do not live longer than 5 years under natural conditions, but male Utah prairie dogs sometimes live ≥6 years. Female Gunnison's and Utah prairie dogs sometimes live ≥6 years, and female black-tailed prairie dogs sometimes live 8 years (Young 1944; Hoogland 1995, 1999, 2001).

Estimating ages of recently killed black-tailed prairie dogs is possible from masses of eye lenses, with the assumption that lens mass varies directly with age (Stockrahm et al. 1996). Because high molar cusps indicate young individuals and low molar cusps indicate older individuals, approximate aging of live black-tailed, Gunnison's, and Utah prairie dogs is possible from an ocular examination of molar attrition (Hoogland and Hutter 1987; Garrett and Franklin 1988; Cox and Franklin 1990; J. L. Hoogland, pers. obs.). Precise determination of prairie dog age requires tracking of individuals ear-tagged as juveniles.

Predation. As noted above, several enemies prey on preemergent prairie dog juveniles. Predators of adults, yearlings, and emergent juveniles include American badgers, black-footed ferrets, coyotes, bobcats, red foxes (*Vulpes vulpes*), swift foxes (*Vulpes velox*), common gray foxes (*Urocyon cinereoargenteus*), grizzly bears (*Ursus arctos*), and mountain lions (*Puma concolor*) (Sperry 1934; Longhurst 1944; Scheffer 1945; King 1955; Halloran 1972; Uresk and Sharps 1986; Goodrich and Buskirk 1998). Black-footed ferrets are specialists and usually prey exclusively on prairie dogs (Hillman 1968; Sheets and Linder 1969; Fortenberry 1972; Clark 1976, 1989; Clark et al. 1985; Campbell et al. 1987). Avian predators include golden eagles (*Aquila chrysaetos*), northern harriers (*Circus cyaneus*), peregrine falcons (*Falco peregrinus*), prairie falcons (*F. mexicanus*), northern goshawks (*Accipiter gentilis*), Cooper's hawks (*A. cooperi*), red-tailed hawks (*Buteo jamaicensis*), ferruginous hawks (*B. regalis*), and several other species of accipiter and buteo hawks (R. E. Smith 1967; Stromberg 1974; Olendorff 1976; Clark 1977; Trevino-Villarreal 1990). In response to so many enemies, prairie dogs frequently stand up and scan for predators (Fitzgerald and Lechleitner 1974; Clark 1977; Wright-Smith 1978; Hoogland 1979b; Trevino-Villarreal 1990; Loughry 1992, 1993a, 1993b; Kildaw 1995; Beam 2001). Such vigilance commonly consumes ≥25% of a prairie dog's time aboveground each day.

Burrowing owls (*Athene cunicularia*) are small, diurnal owls, which frequently live in burrows originally excavated by prairie dogs (Scheffer 1937, 1945; Desmond et al. 2000). Contrary to previous reports (e.g., Swenk 1915; Hollister 1916; Allen 1967; Costello 1970), burrowing owls usually do not attack prairie dog adults or young.

The ultimate benefit of coloniality for all prairie dog species is probably reduced predation on adults, yearlings, and juveniles (Hoogland 1981a, 1995). This benefit is substantial, so that researchers rarely observe predation on prairie dogs living under natural conditions (King 1955; Fitzgerald and Lechleithner 1974; Clark 1977; Wright-Smith 1978; Hoogland 1981a, 1995; Trevino-Villarreal 1990).

Powell (1982) argued that black-footed ferrets occur only in habitats of *C. ludovicianus*. If so, then the presence of ferrets ultimately might explain why black-tailed prairie dogs form larger, more densely populated colonies than white-tailed prairie dogs. Powell's argument stems from the black-footed ferret's range map depicted in Hall (1981). Hall's map, however, does not include findings of black-footed ferrets in white-tailed prairie dog colonies located several hundred kilometers from the nearest black-tailed prairie dog colonies (Martin and Schroeder 1978, 1980; Clark and Campbell 1981; Anderson and Inkley 1985, Biggins and Schroeder 1988; Seal et al. 1989). Protective cover is more likely than the presence of the black-footed ferret to explain interspecific differences in coloniality (Hoogland 1982b, 1995). Protective cover is minimal in habitats of black-tailed (and Mexican) prairie dogs, and individuals form large, densely populated colonies to maximize quick detection of predators. By contrast, tall vegetation is more common in habitats of white-tailed (and Gunnison's and Utah) prairie dogs, and individuals form smaller, more sparsely populated colonies to compromise between the benefits of hiding and of quick detection of predators (Hoogland 1981a, 1995).

Overwinter Mortality. Because of predation or insufficient body mass for survival during the winter months, Gunnison's, Utah, and white-tailed prairie dogs commonly die during hibernation (Clark 1977; Wright-Smith 1978; Rayor 1988; Hoogland 1999; J. L. Hoogland, pers. obs.). Even though they do not hibernate, black-tailed and Mexican prairie dogs in poor condition also die during the winter months (Trevino-Villarreal 1990; Hoogland 1995). For black-tailed, Gunnison's, Utah, and white-tailed prairie dogs, overwinter survivorship varies directly with body mass for adults, yearlings, and juveniles (Hoogland 1995, 1999; J. L. Hoogland, pers. obs.).

Dispersal. Natal dispersal is the permanent emigration of a young individual from the area of birth, and breeding dispersal is the emigration of a sexually mature individual from the area where it copulated (Greenwood 1980; Holekamp 1984). Natal dispersal is male biased for all species of prairie dogs (King 1955; Clark 1977; Rayor 1988; Trevino-Villarreal 1990; Hoogland 1995, 1999; J. L. Hoogland, pers. obs.). Specifically, females usually remain in the natal territory for life, but males disperse before first copulation. Female philopatry (matrilocality) is most pronounced in black-tailed prairie dogs, less pronounced in Gunnison's prairie dogs, and least pronounced in Utah and white-tailed prairie dogs (Hoogland 1995, 1999; J. L. Hoogland, pers. obs.).

Natal dispersal is most commonly to nearby territories within the home colony, but young males sometimes emigrate to different colonies (Garrett and Franklin 1988; Hoogland 1995, 1999). Breeding dispersal is also male biased for all species of prairie dogs, and once again most dispersers move to nearby territories (Trevino-Villarreal 1990; Hoogland 1995, 1999; J. L. Hoogland, pers. obs.). Intercolonial dispersal temporarily displaces individuals from burrows and vigilant conspecifics, and thereby increases susceptibility to predation. Lucky survivors sometimes travel as far as 5 km (Knowles 1985; Garrett and Franklin 1988). For black-tailed prairie dogs, long-distance dispersers commonly traverse roads and ungulate trails as they search for new colonies (Knowles 1985, 1986b; Garrett and Franklin 1988).

Age of dispersers and timing of dispersal differ for the various species of prairie dogs. For black-tailed and Mexican prairie dogs, males do not disperse as juveniles. Most dispersals by adult and yearling black-tailed prairie dog males occur in late spring and early summer, just before and after first emergences of juveniles from their natal burrows (Knowles 1985; Garrett and Franklin 1988; Hoogland 1995). Male Gunnison's, Utah, and white-tailed prairie dogs, by contrast, frequently disperse as juveniles in their first summer. Yearling and adult males of the latter three species also commonly disperse after emerging from hibernation, just before the breeding season (Clark 1977; Rayor 1988; Hoogland 1999; J. L. Hoogland, pers. obs.).

Diseases. White-tailed prairie dogs sometimes contract tularemia (Davis 1935). Information about other diseases in colonies of prairie dogs is almost nonexistent, except for bubonic plague (also called sylvatic plague or wild rodent plague). Plague first arrived in North America about 100 years ago, probably via fleas (Siphonaptera) on animals unloaded from European ships (Pollitzer 1951; Olsen 1981; Barnes 1982, 1993; Cully 1993; Fitzgerald 1993). Prairie dogs have evolved little or no immunity to plague, so the disease usually eliminates most or all colony residents within weeks after introduction (Eskey and Haas 1940; Lechleitner et al. 1962, 1968; Clark 1977; Rayor 1985a; Hasenyager et al. 1988; Ubico et al. 1988; Cully et al. 1997; Turner 2000). Fleas are the major vectors, but lice (Anoplura) and ticks (Acarina) also might transmit plague (Pollitzer 1952; Hirst 1953; Kartman et al. 1962; Barnes 1982, 1993; Anderson and Williams 1997). Killing fleas by treating prairie dog burrows with insecticide reduces the probability of an outbreak of plague (Barnes et al. 1972; J. L. Hoogland, pers. obs.).

Although bubonic plague devastates other sciurid populations as well, prairie dogs seem especially susceptible (Eskey and Haas 1940;

Pollitzer and Meyer 1961; Olsen 1981). The most likely explanation for this difference is that prairie dogs are more densely colonial than other squirrels (Lechleitner 1969; Michener 1983; Hoogland 1995).

Ectoparasites. Prairie dogs of all species frequently harbor ectoparasites such as fleas, lice, and ticks (Pizzimenti 1975; Hoogland 1979a). These ectoparasites depress individual fitness in three ways (Hoogland 1995). First, as noted above, ectoparasites transmit debilitating or fatal diseases such as bubonic plague. Second, ectoparasites frequently remove blood. Third, their bites, especially those of ticks, sometimes damage the integument and thereby promote infections.

Fleas are the most common ectoparasites of prairie dogs, and occur on the animals themselves, at burrow entrances, and in underground nests. The most common flea species are *Hoplosyllus anomalus, Leptopsylla segnis, Oropsylla* [*Opisocrostis*] *hirsuta, O. labis, O. tuberculata, Pulex irritans, P. simulans, Thrassis fotus, T. francisi,* and *T. stanfordi* (Jellison 1939, 1945; Ecke and Johnson 1952; Smit 1958; Costello 1970; Pizzimenti 1975; Trevino-Villarreal et al. 1998). Fleas are especially common just before the breeding season, when individual prairie dogs sometimes harbor as many as 300 (Hoogland 1979a; J. L. Hoogland, pers. obs.).

The most common species of ticks that parasitize prairie dogs are *Atricholaelaps glasgowi, Dermacentor andersoni, Ixodes kingi, I. sculptus,* and *Ornithodoros* spp. (King 1955; Pizzimenti 1975; Tyler and Buscher 1975). Warble flies (*Cuterebra* spp.) sometimes parasitize Gunnison's prairie dogs (Burnett and McCampbell 1926; Longhurst 1944). Evidently nobody has identified the lice that parasitize black-tailed and white-tailed prairie dogs (Hoogland 1979a, 1995).

Probably in response to ectoparasites, prairie dogs of all species frequently scratch, bite, or lick different parts of their own bodies (autogrooming) (Hoogland 1979a). Black-tailed and Mexican prairie dogs commonly scratch and lick conspecifics as well (King 1955; Trevino-Villarreal 1990), but such allogrooming is rare among Gunnison's, Utah, and white-tailed prairie dogs (Tileston and Lechleitner 1966; Fitzgerald and Lechleitner 1974; Hoogland 1979a; J. L. Hoogland, pers. obs.).

In addition to ectoparasites, prairie dogs also harbor numerous endoparasites, including protozoans (Protozoa), tapeworms (Cestoda), roundworms (Nematoda), and spiny-headed worms (Acanthocephala) (Vetterling 1964; Buscher and Tyler 1975; D. M. Thomas and Stanton 1994; Seville 1997). The effect of these endoparasites on survivorship and reproduction of prairie dogs is unknown.

BEHAVIOR

Social Organization Within Colonies. Within colonies, black-tailed prairie dogs live in harem-polygynous, territorial family groups called coteries (King 1955; Hoogland 1981b, 1983a, 1995). The number of adults and yearlings in a coterie ranges from 1 (the rare isolated individual) to 26, with a mean $\pm$ *SD* of 6.13 $\pm$ 3.53 ($n = 273$) (Hoogland 1995). The typical coterie contains one breeding male, two or three adult females, and one or two yearlings of each sex. Large coteries sometimes contain two or more breeding males. The males of such multimale coteries are commonly close kin, such as father-son or full-brothers (Hoogland 1995). Conversely, a single breeding male sometimes dominates two small, adjacent groups of females.

Coterie members have a well-defined territory, which they defend from prairie dogs of other coteries (King 1955; Tileston and Lechleitner 1966; Halpin 1987; Hoogland 1995). Boundaries between coterie territories sometimes coincide with unsuitable habitat containing tall vegetation, rocks, or poor drainage. More commonly, though, coterie territories are contiguous, and their boundaries are undetectable to human observers from physical features alone (King 1955). The area and configuration of the coterie territory usually remain constant across generations (King 1955). Several coterie territories at Wind Cave National Park, for example, remained the same size and shape for 14 consecutive years (Hoogland 1995).

Coterie territories range in size from 0.05 to 1.01 ha, with a mean $\pm$ *SD* of 0.31 $\pm$ 0.17 ha ($n = 273$) (Hoogland 1995; see also King 1955; Tileston and Lechleitner 1966). Larger territories support larger coteries. The number of burrow entrances per coterie territory ranges from 5 (for an isolated prairie dog) to 214, with a mean $\pm$ *SD* of 69.0 $\pm$ 37.6 ($n = 273$) (Hoogland 1995). Burrow entrances within the same coterie territory often connect by a common tunnel, but entrances from different territories do not connect. Except when pregnant or lactating females are defending nursery burrows, coterie members usually have access to all burrow entrances within the home territory (King 1955; Hoogland 1995). To spend the night or to escape precipitation and predation, coterie members commonly group together in the same burrow.

The resident breeding male is primarily responsible for the defense of the coterie territory against intruders, but adult females and yearlings of both sexes also help (King 1955). Trespassing is rare, so that black-tailed prairie dogs obtain >99% of their food and other resources from the home coterie territory (Hoogland 1995).

Like black-tailed prairie dogs, Mexican prairie dogs live in coteries (Trevino-Villarreal 1990). Gunnison's, Utah, and white-tailed prairie dogs live in harem-polygynous, territorial family groups called clans (Tileston and Lechleitner 1966; Clark 1977; Wright-Smith 1978; Rayor 1988; Hoogland 1999). Superficially, clans resemble coteries. A major difference is that members of different clans frequently forage together in common feeding areas (Fitzgerald and Lechleitner 1974; Clark 1977; Wright-Smith 1978; Travis and Slobodchikoff 1993; Hoogland 1999). Another important difference is that clan territories frequently change in area and configuration from year to year (Hoogland 1999; J. L. Hoogland, pers. obs.). For Gunnison's prairie dogs, clan size ranges from 1 to 19 individuals, with a mean $\pm$ *SD* of 5.30 $\pm$ 3.38 ($n = 159$) (Hoogland 1999). Clan territories for Gunnison's prairie dogs range from 0.16 to 1.82 ha, with a mean $\pm$ *SD* of 0.67 $\pm$ 0.34 ha ($n = 121$) (Hoogland 1999). Clans and clan territories are larger for Utah prairie dogs (J. L. Hoogland, pers. obs.).

Vocalizations. Black-tailed, Gunnison's, and white-tailed prairie dogs have numerous vocalizations (Waring 1970). The calls of Mexican and Utah prairie dogs are almost indistinguishable from those of black-tailed and white-tailed prairie dogs, respectively (Pizzimenti and Nadler 1972; Pizzimenti and McClenaghan 1974; Trevino-Villarreal 1990; W. J. Loughry, Valdosta State University, pers. commun., 2000).

Territorial calls are common to all prairie dog species and function in territorial disputes, as an "all-clear" signal following the disappearance of a predator, and in other contexts as well (Waring 1970). The territorial call of black-tailed and Mexican prairie dogs—also called the jump–yip display (Smith et al. 1976)—occurs as the caller jumps vertically into the air (King 1955; Waring 1970; Pizzimenti and McClenaghan 1974; Trevino-Villarreal 1990; Hoogland 1995). Territorial calls among Gunnison's and white-tailed prairie dogs are called raspy chatters and laughing barks, respectively (Waring 1970; Fitzgerald and Lechleitner 1974). For all species, a territorial call by one individual often initiates a chain reaction of territorial calls by nearby conspecifics (King 1955; Waring 1970).

Prairie dogs of all species give alarm calls—also called antipredator calls or repetitious barks—in response to predators. Males and females commonly call in colonies of black-tailed and Gunnison's prairie dogs, but only females routinely call in colonies of Utah prairie dogs (King 1955; Waring 1970; Hoogland 1995, 1996a; J. L. Hoogland, pers. obs.). Gunnison's and Utah prairie dog females call most often to warn emergent juvenile offspring (Hoogland 1996b; J. L. Hoogland, pers. obs.), but black-tailed prairie dogs also call to warn nondescendant kin (Hoogland 1983b, 1995).

Prairie dogs do not have distinct alarm calls for different predators, except for black-tailed prairie dogs in response to snakes (see below). Subtle variation in alarm calls might indicate different types of predators or different levels of urgency, however (Koford 1958; Costello 1970; Slobodchikoff et al. 1991; Ackers and Slobodchikoff 1999). Black-tailed, Gunnison's, Utah, and white-tailed prairie dogs seem to call at a faster rate (i.e., more barks per minute) when danger is especially imminent, for example (King 1955; Hoogland 1995, 1996b; J. L. Hoogland, pers. obs.).

In response to both poisonous and nonpoisonous snakes, black-tailed prairie dogs give antisnake calls, which superficially resemble territorial calls (King 1955; Halpin 1983; Owings and Loughry 1985; Loughry 1987a, 1987b, 1988). Black-tailed prairie dogs sometimes show repeated foot thumping in response to snakes as well (Owings and Owings 1979), as do Gunnison's prairie dogs (J. L. Hoogland, pers. obs.).

Just before or after copulating with an estrous female, black-tailed, Gunnison's, Utah, and white-tailed prairie dog males commonly give a unique vocalization called the mating call. For these four species, the mating call is superficially similar to the alarm call (Grady and Hoogland 1986; Hoogland 1995, 1997; J. L. Hoogland, pers. obs.).

By clicking the tips of the lower incisors against the upper incisors, prairie dogs of all species sometimes emit "tooth chatters" during aggressive interactions with conspecifics (Waring 1970). Tooth chatters from black-tailed prairie dogs frequently are audible >100 m away (King 1955), but tooth chatters from Gunnison's, Utah, and white-tailed prairie dogs are barely audible 5 m away (Waring 1970; J. L. Hoogland, pers. obs.).

When a larger black-tailed prairie dog from another coterie attempts to invade the home territory, an adult female or yearling of either sex commonly screams a distinctive defense bark (King 1955; Smith et al. 1977). The defense bark resembles the alarm call, but is softer (less intense) with a slower rate. The defense bark usually attracts the resident breeding male, who first chases the caller and then confronts the potential interloper (King 1955; Hoogland 1995).

Prairie dogs of all species also have several other calls such as snarls, growls, and screams. The functions of these calls are unclear (Waring 1970).

Behavioral Interactions. Frequent behavioral interactions occur within prairie dog colonies. Amicable interactions include kissing (mouth-to-mouth contact), anal sniffing (nose-to-anus contact), play, and, for black-tailed and Mexican prairie dogs, allogrooming. Play most commonly involves two juveniles, less commonly a yearling or adult with a juvenile, and least commonly one yearling or adult with another yearling or adult (King 1955; Tileston and Lechleitner 1966; Fitzgerald and Lechleitner 1974; J. L. Hoogland, pers. obs.). Hostile interactions commonly begin with a kiss or anal sniff, which turns into a fight, chase, or territorial dispute. Territorial disputes occur at territorial boundaries and include staring, bluff charges, flaring of tails, chattering of teeth, and, for black-tailed prairie dogs, reciprocal sniffing of everted, three-lobed perianal scent glands (King 1955; Smith 1967; Fitzgerald and Lechleitner 1974; Clark 1977; Jones and Plakke 1981). Male–male aggression in all species results primarily from competition for estrous females. Female–female aggression results mainly from competition for nursery burrows (Hoogland 1995, 1999). For both sexes of all species, heavy individuals are more dominant in aggressive interactions than are lighter individuals (Hoogland 1995; J. L. Hoogland, pers. obs.).

Black-tailed, Gunnison's, Utah, and white-tailed prairie dogs interact more amicably with kin than with nonkin. When interacting with adults and yearlings, however, individuals do not seem to discriminate between close kin and more distant kin (Hoogland 1995, 1999; J. L. Hoogland, pers. obs.).

MANAGEMENT AND CONSERVATION

Prairie Dogs as Keystone Species. A keystone species has a large overall effect on an ecosystem (Paine 1969, 1980; Mills et al. 1993; Power et al. 1996). Prairie dogs of all species, and especially black-tailed prairie dogs, qualify as keystone species for several reasons (Miller et al. 1990, 1994, 2000; Sharps and Uresk 1990; Kotliar et al. 1999; but see Stapp 1998). Their foraging, for example, decreases the height of vegetation, changes the floral composition, and increases landscape heterogeneity (Coppock et al. 1983a, 1983b; Whicker and Detling 1988; Cincotta et al. 1989; Cid et al. 1991; Weltzin et al. 1997). In addition, prairie dog excavations increase the mixing of topsoil and subsoil and promote uptake of nitrogen by plants (King 1955; Archer et al. 1987; Holland and Detling 1990). Further, because they attract myriad invertebrates and vertebrates (Miller et al. 1994, 2000; Hoogland 1995; Barko et al. 1999; Kotliar et al. 1999), colonies frequently increase biological diversity and species richness (Hansen and Gold 1977; Clark et al. 1982; O'Meilia et al. 1982; Agnew et al. 1986; Reading et al. 1989; Mellink and Madrigal 1993; Ceballos et al. 1999). Burrowing owls, mountain plovers (*Charadrius montana*), and black-footed ferrets, for example, are among vertebrates that are especially dependent on prairie dog colonies for survival (Clark et al. 1982; Knowles et al. 1982; Plumpton and Lutz, 1993; Desmond and Savidge 1996; Kotliar et al. 1999).

Prairie Dogs Pelts and Consumption by Humans. Even though the winter pelage of prairie dogs is long and thick, demand for their pelts is negligible.

When omnipresent long ago, prairie dogs were important food items for certain Native American tribes, and, more recently, for White explorers and settlers (Scheffer 1945; Wedel 1961; Costello 1970; Gorman 1974). Because they do not hibernate, black-tailed and Mexican prairie dogs probably were especially valuable as food to Native Americans during winter. Perhaps because they are now so rare, human consumption of prairie dogs currently is trivial.

Prairie Dogs as Pets. If captured as they first emerge from the nursery burrow or before, black-tailed prairie dogs make excellent, engaging pets (Cates 1927; Dale 1947; Ferrara 1985). Pets seem to "imprint" to their human owners, and current interest in black-tailed prairie dogs as pets is enormous (Bouasserie 1999; J. W. Vanderpool, pers. commun., 2000). Juveniles of the other prairie dog species probably would make good pets as well (e.g., Fitzgerald and Lechleitner 1974). Transmission of bubonic plague via pets is unlikely, and injuries from scratches or bites are rare.

Prairie dogs do not readily breed in captivity (Stockard 1929, 1930; Anthony and Foreman 1951; Foreman 1962, Pizzimenti and McClenaghan 1974). Most pets therefore must come from wild colonies. On one hand, the pet industry thus poses yet another obstacle to the long-term survival of prairie dogs. To reduce the detrimental effects, many collectors only capture potential pets at colonies that are doomed anyway because of plans for poisoning or destruction of habitat (J. W. Vanderpool, pers. commun., 2000). On the other hand, pets increase the public awareness of the charm and importance of prairie dogs. Pet owners are thus more likely to support efforts to save prairie dogs from extinction. The overall net effect of the pet industry on the conservation of prairie dogs is unclear.

Management and Current Status. Mainly because ranchers view them as range and agricultural pests, prairie dogs have been targets of intensive eradication programs, which involve shooting, poisoning, drowning, and destruction of habitat (Marsh 1984; Uresk et al. 1986; Cincotta et al. 1987; Clark 1989; Roemer and Forrest 1996; Vosburgh and Irby 1998). As many as 125,000 people once worked to eliminate prairie dogs (Merriam 1902; Swenk 1915; Randall 1976a, 1976b; Clark 1979; Garrett and Franklin 1983). More recently, bubonic plague has killed millions of prairie dogs as well. All five species are now rare. The Mexican prairie dog is currently on the list of endangered species (Trevino-Villarreal and Grant 1998). The Utah prairie dog is listed as threatened, but is under consideration for the list of endangered species (Roberts et al. 2000). The black-tailed prairie dog currently occupies <2% of its former range, and also is under consideration for the list of threatened species (Biodiversity Legal Foundation et al. 1998; National Wildlife Federation 1998; Miller et al. 2000).

Species dependent on prairie dog colonies for food or shelter also have declined (Clark et al. 1982; Knowles et al. 1982; Desmond and Savidge 1996; Kotliar et al. 1999). The black-footed ferret, for example, is on the brink of extinction. No known natural populations exist, and survival of the species will depend on introductions of laboratory-reared ferrets into the wild (Chadwick 1993; Biggins et al. 1998; Vargas et al. 1998; Dobson and Lyles 2000).

Ranchers disdain prairie dogs for two reasons (Costello 1970; Jameson 1973; Chace 1976; Petzal 1993; Zinn and Andelt 1999). First, they worry that domestic horses (*Equus caballus*) and domestic cows (*Bos taurus*) will step into prairie dog burrows and break a leg. Second, they believe that prairie dogs compete with livestock for food. In addition, ranchers fret that prairie dogs are prolific breeders from which they are not safe until every single pest is gone. Are these concerns legitimate ? The answer here is yes and no. Livestock do sometimes incur leg fractures after stepping into prairie dog burrows, but such fractures are exceedingly rare (Hoogland 1995; Aschwanden 2001). Prairie dogs do eat some of the same plants as horses and cattle, but competition is minimal. Specifically, prairie dogs avoid numerous plants that livestock prefer and prefer numerous plants that livestock avoid (O'Meilia et al. 1982; Coppock et al. 1983a, 1983b; Uresk 1984). In addition, the presence of prairie dogs improves the quality of certain plants, so that American bison (*Bison bison*), wapiti (*Cervus elaphus*), pronghorn antelope (*Antilocapra americana),* and livestock commonly prefer to forage at colony sites (Koford 1958; O'Meilia et al. 1982; Coppock et al. 1983a, 1983b; Knowles 1986b). Regarding reproduction, prairie dogs are not the prolific breeders that ranchers allege. Males of all species and females of one species commonly defer breeding until their second year, for example. In many years, <50% of the adult and yearling females rear emergent juveniles, and litter size at first emergence is usually only three or four (Rayor 1985b; Hoogland 1995, 1999, 2001). Further, because of widespread infanticide in at least two species, prairie dogs are often their own worst enemies (Hoogland 1985, 1995; J. L. Hoogland, pers. obs.).

Prairie dogs are especially likely to colonize areas that livestock have already overgrazed (Bond 1945; Osborn and Allan 1949; Koford 1958; Clark 1968, 1977, 1979; Costello 1970; Snell 1985; Knowles 1986b). Thus, prairie dogs are frequently the effect, rather than the cause, of overgrazing. Further, prairie dogs survive and reproduce better, and colonies therefore quickly repopulate, after reductions caused by shooting, poisoning, or plague (Crosby and Graham 1986; Knowles 1986a; Menkens and Anderson 1991; Radcliffe 1992; Cully 1997; Turner 2000). Finally, consider the scene 200 years ago when millions of American bison lived sympatrically with billions of prairie dogs. Because domestic cattle and American bison are so similar, large numbers of cattle and prairie dogs should be able to coexist in the same areas. For all these reasons, the financial costs of trying to eliminate prairie dogs usually exceed the benefits (O'Meilia et al. 1982; Collins et al. 1984; Uresk 1985; Knowles 1986a; Hygnstrom and VerCauteren 2000; Aschwanden 2001). The inescapable conclusion is that attempts to eradicate prairie dogs have been misguided and inappropriate (McNulty 1971; Miller et al. 1994, 2000).

Future Research and Conservation. Prairie dogs are fascinating social animals, and have contributed to our understanding of several key issues in behavioral ecology, including coloniality (King 1955; Hoogland 1979a, 1979b, 1981a), demography (Rayor 1985b; Halpin 1987), alarm calling (Hoogland 1983b, 1996b; Slobodchikoff et al. 1991; Ackers and Slobodchikoff 1999), communal nursing (Grossman 1987; Hoogland et al. 1989), kin recognition (Hoogland 1986), and sex ratio selection (Hoogland 1995). Additional comparative research with prairie dogs is likely to increase our understanding of other important issues such as multiple mating by females (Hoogland 1998a), levels of inbreeding (Hoogland 1982a, 1992, 1995), and infanticide (Hoogland 1985, 1995).

Recent interest in prairie dogs among wildlife managers and conservation biologists has skyrocketed. A major reason is that ecologists have recognized that prairie dogs are keystone species of the grassland ecosystems of North America (Miller et al. 1990, 1994, 2000; Kotliar et al. 1999). Another reason is that prairie dogs of all species have disappeared, and continue to disappear, at an alarming rate (Hoogland 1995; Wuerthner 1997; Miller et al. 2000). More research on ecology and social behavior will be necessary to save prairie dogs and prairie ecosystems from extinction. Also imperative will be public education on issues such as keystone species, competition between prairie dogs and livestock, drastic declines of prairie dog populations, and rate of prairie dog reproduction (Reading and Kellert 1993; Miller et al. 1994, 2000).

LITERATURE CITED

Ackers, S. H., and C. N. Slobodchikoff. 1999. Communication of stimulus size and shape in alarm calls of Gunnison's prairie dogs, *Cynomys gunnisoni.* Ethology 105:149–62.

Agnew, W., D. W. Uresk, and R. M. Hansen. 1986. Flora and fauna associated with prairie dog colonies and adjacent ungrazed mixed-grass prairie in western South Dakota. Journal of Range Management 39:135–39.

Aldous, S. E. 1935. Some breeding notes on rodents. Journal of Mammalogy 16:128–31.

Allen, D. 1967. The life of prairies and plains. McGraw-Hill, New York.

Anderson, S., and D. Inkley, eds. 1985. Black-footed ferret workshop proceedings. Wyoming Game and Fish Publications, Cheyenne.

Anderson, S. H., and E. S. Williams. 1997. Plague in a complex of white-tailed prairie dogs and associated small mammals in Wyoming. Journal of Wildlife Diseases 33:720–32.

Anthony, A., and D. Foreman. 1951. Observations on the reproductive cycle of the black-tailed prairie dog (*Cynomys ludovicianus*). Physiological Zoology 24:242–48.

Archer, S., M. G. Garrett, and J. K. Detling. 1987. Rates of vegetation change associated with prairie dog (*Cynomys ludovicianus*) grazing in North American mixed-grass prairie. Vegetatio 72:159–66.

Aschwanden, C. 2001. Learning to live with prairie dogs. National Wildlife Magazine. 39(3):20–29.

Bakko, E. B. 1977. Field water balance performance in prairie dogs (*Cynomys leucurus* and *C. ludovicianus*). Comparative Biochemical Physiology 56:443–51.

Bakko, E. B., and L. N. Brown. 1967. Breeding biology of the white-tailed prairie dog, *Cynomys leucurus,* in Wyoming. Journal of Mammalogy 48:100–12.

Bakko, E. B., W. P. Porter, and B. A. Wunder. 1988. Body temperature patterns in black-tailed prairie dogs in the field. Canadian Journal of Zoology 66:1783–89.

Barko, V. A., J. H. Shaw, and D. M. Leslie. 1999. Birds associated with black-tailed prairie dog colonies in southern shortgrass prairie. Southwestern Naturalist 44:484–89.

Barnes, A. M. 1982. Surveillance and control of bubonic plague in the United States. Symposia of the Zoological Society of London 50:237–70.

Barnes, A. M. 1993. A review of plague and its relevance to prairie dog populations and the black-footed ferret. Pages 28–37 *in* J. L. Oldemeyer, D. E. Biggins, B. J. Miller, and R. Crete, eds. Proceedings of the symposium on the management of prairie dog complexes for the reintroduction of black-footed ferrets (Biological Report No. 93). U.S. Fish and Wildlife Service, Washington, DC.

Barnes, A. M., L. J. Ogden, and E. G. Campos. 1972. Control of the plague vector, *Opisocrostis hirsutis,* by treatment of prairie dog (*Cynomys ludovicianus*) burrows with 2% carbaryl dust. Journal of Medical Entomology 9:330–33.

Beam, M. E. 2001. Vigilance in Utah prairie dogs. Master's thesis, Frostburg State University, Frostburg, MD.

Biggins, D., and M. H. Schroeder. 1988. Historical and present status of the black-footed ferret. Pages 93–97 *in* D. W. Uresk, G. L. Schenbeck, and R. Cefkin, eds. tech. coordi. Eighth Great Plains wildlife damage control workshop proceedings (General Technical Report RM-154). Fort Collins, CO.

Biggins, D. E., J. L. Godbey, L. R. Hanebury, B. Luce, P. E. Marinari, M. R. Matchett, and A. Vargas. 1998. The effect of rearing methods on survival of reintroduced black-footed ferrets. Journal of Wildlife Management 62:643–53.

Bintz, G. L. 1984. Water balance, water stress, and the evolution of season torpor in ground-dwelling squirrels. Pages 142–65 *in* J. O. Murie and G. R. Michener, eds. The biology of ground-dwelling squirrels. University of Nebraska Press, Lincoln.

Biodiversity Legal Foundation, J. C. Sharps, and Predator Project. 1998. Black-tailed prairie dog (*Cynomys ludovicianus*). Unpublished petition to U.S. Fish and Wildlife Service, Region 6, Denver, CO.

Black, C. C. 1963. A review of the North American Tertiary Sciuridae. Museum of Comparative Zoology, Harvard University, Bulletin 130:109–248.

Bond, R. M. 1945. Range rodents and plant succession. Transactions of the North American Wildlife Conference 10:229–34.

Bonham, C. D., and A. Lerwick. 1976. Vegetation changes induced by prairie dogs on shortgrass range. Journal of Range Management 29:221–25.

Boussarie, D. 1999. The prairie dog in consultation. Pratique Medicale et Chirurgicale de l'Animal de Compagnie 34:43–54.
Brenneman, D. E., W. E. Connor, E. L. Forker, and L. DenBesten. 1972. The formation of abnormal bile and cholesterol gallstones from dietary cholesterol in the prairie dog. Journal of Clinical Investigation 51:1495–1503.
Broughton, G., A. Tseng, R. Fitzgibbons, S. Tyndall, G. Stanislav, and E. Rongone. 1991. The prevention of cholelithiasis with infused chenodeoxycholate in the prairie dog (*Cynomys ludovicianus*). Comparative Biochemical Physiology 99A:609–13.
Burnett, W. L., and S. C. McCampbell. 1926. The Zuni prairie dog in Montezuma County, Colorado. Colorado Agricultural College 49:1–15.
Burns, J. A., D. L. Flath, and T. W. Clark. 1989. On the structure and function of white-tailed prairie dog burrows. Great Basin Naturalist 49:517–24.
Buscher, H. N., and J. D. Tyler. 1975. Parasites of vertebrates inhabiting prairie dog towns in Oklahoma. II. Helminths. Proceedings of the Oklahoma Academy of Science 55:108–11.
Campbell, T. M., and T. W. Clark. 1981. Colony characteristics and vertebrate associates of white-tailed and black-tailed prairie dogs in Wyoming. American Midland Naturalist 105:269–76.
Campbell, T. M., T. W. Clark, L. Richardson, S. C. Forrest, and B. R. Houston. 1987. Food habits of Wyoming black-footed ferrets. American Midland Naturalist 117:208–10.
Cates, E. C. 1927. Notes concerning a captive prairie-dog. Journal of Mammalogy 8:33–37.
Ceballos, G., and D. E. Wilson. 1985. *Cynomys mexicanus*. Mammalian Species 248:1–4.
Ceballos, G., E. Mellink, and L. Hanebury. 1993. Distribution and conservation status of prairie dogs (*Cynomys mexicanus* and *Cynomys ludovicianus*) in Mexico. Biological Conservation 63:105–12.
Ceballos, G., J. Pacheco, and R. List. 1999. Influence of prairie dogs (*Cynomys ludovicianus*) on habitat heterogeneity and mammalian diversity in Mexico. Journal of Arid Environments 41:161–72.
Chace, G. E. 1976. Wonders of prairie dogs. Dodd, Mead, New York.
Chadwick, D. H. 1993. The American prairie. National Geographic 184:90–119.
Chapman, W. C., J. Fisk, D. Schot, J. P. Debelak, M. K. Washington, R. F. Bluth, D. Pierce, and L. F. Williams. 1998. Establishment and characterization of primary gallbladder epithelial cell cultures in the prairie dog. Journal of Surgical Research 80:35–43.
Chesser, R. K. 1983. Genetic variability within and among populations of the black-tailed prairie dog. Evolution 37:320–31.
Chesser, R. K. 1984. Study of genetic variation in the Utah prairie dog (Report prepared for the U.S. Fish and Wildlife Service, Contract No. 14-16-006-83-049). Texas Tech University, Lubbock.
Cid, M. S., J. K. Detling, A. D. Whicker, and M. A. Brizuela. 1991. Vegetational responses of a mixed-grass prairie site following exclusion of prairie dogs and bison. Journal of Range Management 44:100–105.
Cincotta, R. P. 1989. Note on mound architecture of the black-tailed prairie dog. Great Basin Naturalist 49:621–23.
Cincotta, R. P., D. W. Uresk, and R. M. Hansen. 1987. Demography of black-tailed prairie dog populations reoccupying sites treated with rodenticide. Great Basin Naturalist 47:339–43.
Cincotta, R. P., D. W. Uresk, and R. M. Hansen. 1989. Plant compositional changes in a colony of black-tailed prairie dogs in South Dakota. Pages 171–77 *in* A. J. Bjugstad, D. W. Uresk, and R. H. Hamre, eds. Ninth Great Plains wildlife damage control workshop proceedings (General Technical Report RM-171). U.S. Fish and Wildlife Service, Fort Collins, CO.
Clark, T. W. 1968. Ecological roles of prairie dogs (No. 261). Wyoming Range Management, Laramie.
Clark, T. W. 1971. Notes on white-tailed prairie dog (*Cynomys leucurus*) burrows. Great Basin Naturalist 31:115–24.
Clark, T. W. 1973. A field study of the ecology and ethology of the white-tailed prairie dog (*Cynomys leucurus*): With a model of *Cynomys* evolution. Ph.D. dissertation, University of Wisconsin, Madison.
Clark, T. W. 1976. The black-footed ferret. Oryx 13:275–80.
Clark, T. W. 1977. Ecology and ethology of the white-tailed prairie dog (*Cynomys leucurus*). (Publications in Biology and Geology, No. 3), Milwaukee Public Museum, Milwaukee, WI.
Clark, T. W. 1979. The hard life of the prairie dog. National Geographic 156:270–81.
Clark, T. W. 1989. Conservation biology of the black-footed ferret, *Mustela nigripes* (Special Scientific Report, No. 3). Wildlife Preservation Trust International, Philadelphia.
Clark, T. W., and T. M. Campbell. 1981. Additional black-footed ferret (*Mustela nigripes*) reports in Wyoming. Great Basin Naturalist 41:360–61.
Clark, T. W., R. S. Hoffmann, and C. F. Nadler. 1971. *Cynomys leucurus*. Mammalian Species 7:1–4.
Clark, T. W., T. M. Campbell, D. G. Socha, and D. E. Casey. 1982. Prairie dog colony attributes and associated vertebrate species. Great Basin Naturalist 42:572–82.
Clark, T. W., L. Richardson, D. Casey, T. M. Campbell, and S. C. Forrest. 1984. Seasonality of black-footed ferret diggings and prairie dog burrow plugging. Journal of Wildlife Management 48:1441–44.
Clark, T. W., S. C. Forrest, T. M. I. Campbell, D. E. Casey, and K. A. Fagerstone. 1985. Black-footed ferret prey base. Pages 7.1–7.14 *in* S. H. Anderson and D. B. Inkley, eds. Black-footed ferret workshop proceedings. Wyoming Game and Fish Department, Laramie.
Collier, G. D., and J. J. Spillett. 1972. Status of the Utah prairie dog. Proceedings of the Utah Academy of Sciences, Arts, and Letters 49:27–39.
Collier, G. D., and J. J. Spillett. 1973. The Utah prairie dog: Decline of a legend. Utah Science 34:83–87.
Collier, G. D., and J. J. Spillett. 1975. Factors influencing the distribution of the Utah prairie dog, *Cynomys parvidens* (Sciuridae). Southwestern Naturalist 20:151–58.
Collins, A. R., J. P. Workman, and D. W. Uresk. 1984. An economic analysis of black-tailed prairie dog [*Cynomys ludovicianus*] control. Journal of Range Management 37:358–61.
Cooke, L. A., and S. R. Swiecki. 1992. Structure of a white-tailed prairie dog burrow. Great Basin Naturalist 52:288–89.
Coppock, D. L., J. K. Detling, J. E. Ellis, and M. I. Dyer. 1983a. Plant–herbivore interactions in a North American mixed-grass prairie. I. Effects of black-tailed prairie dogs on intraseasonal aboveground plant biomass and nutrient dynamics and plant species diversity. Oecologia 56:1–9.
Coppock, D. L., J. E. Ellis, J. K. Detling, and M. I. Dyer. 1983b. Plant–herbivore interactions in a North American mixed-grass prairie. II. Responses of bison to modification of vegetation by prairie dogs. Oecologia 56:10–15.
Costello, D. F. 1970. The world of the prairie dog. Lippincott, Philadelphia.
Coues, E., and J. A. Allen. 1877. Monographs of North American Rodentia (U. S. Survey of the Territories, Vol., 11). U.S. Government Printing Office, Washington, DC.
Cox, M. K., and W. L. Franklin. 1990. Premolar gap technique for aging live black-tailed prairie dogs. Journal of Wildlife Management 54:143–46.
Crocker-Bedford, D. C., and J. J. Spillett. 1981. Habitat relationships of the Utah prairie dog (Report 14-16-008-1117). U.S. Fish and Wildlife Service, Logan, UT.
Crosby, L. A., and R. Graham. 1986. Population dynamics and expansion rates of black-tailed prairie dogs. Pages 112–15 *in* T. P. Salmon, ed. Proceedings of the 12th vertebrate pest conference. University of California, Davis.
Cully, J. F. 1993. Plague, prairie dogs, and black-footed ferrets: Implications for management. Pages 38–49 *in* L. Oldemeyer, D. E. Biggins, B. J. Miller, and R. Crete, eds. Biological proceedings of the symposium on the management of prairie dog complexes for the reintroduction of black-footed ferrets (Biological Report 93). U.S. Fish and Wildlife Service.
Cully, J. F. 1997. Gunnison's prairie dog growth and life-history change after a plague epizootic. Journal of Mammalogy 78:146–57.
Cully, J. F., A. M. Barnes, T. J. Quan, and G. Maupin. 1997. Dynamics of plague in a Gunnison's prairie dog complex from New Mexico. Journal of Wildlife Diseases 33:706–19.
Dale, H. F. 1947. Prairie dogs as pets. Outdoor Nebraska 24:22.
Daley, J. G. 1992. Population reductions and genetic variability in black-tailed prairie dogs. Journal of Wildlife Management 56:212–20.
Davis, G. E. 1935. Tularemia. Susceptibility of the white-tailed prairie dog, *Cynomys leucurus* Merriam. U.S. Public Health Report 50:731–32.
Desmond, M. J., and J. A. Savidge. 1996. Factors influencing burrowing owl (*Speotyto cunicularia*) nest densities and numbers in western Nebraska. American Midland Naturalist 136:143–48.
Desmond, M. J., J. A. Savidge, and K. M. Eskridge. 2000. Correlations between burrowing owl and black-tailed prairie dog declines: A 7-year analysis. Journal of Wildlife Management 64:1067–75.
Dobson, A., and A. Lyles. 2000. Black-footed ferret recovery. Science 288:985–88.
Drearden, L. C. 1953. The gross anatomy of the viscera of the prairie dog. Journal of Mammalogy 34:15–26.
Ecke, D. H., and C. W. Johnson. 1952. Plague in Colorado. U.S. Public Health Monograph 6:1–37.
Egoscue, H. J. 1975. Abnormal juvenile pelages and estivation in the Utah prairie dog, *Cynomys parvidens*. Southwestern Naturalist 20:139–42.
Eskey, C. R., and V. H. Haas. 1940. Plague in the western part of the United States. U.S. Public Health Bulletin 254:1–83.

Fagerstone, K. A. 1982. A review of prairie dog diet and its variability among animals and colonies. Pages 178–84 *in* R. M. Timm and R. J. Johnson, eds. Proceedings of the fifth Great Plains wildlife damage control workshop. University of Nebraska, Institute of Agriculture and Natural Resources, Lincoln.

Fagerstone, K. A., H. P. Tietjen, and O. Williams. 1981. Seasonal variation in the diet of black-tailed prairie dogs. Journal of Mammalogy 62:820–24.

Ferrara, J. 1985. Prairie home companions. National Wildlife 23:48–53.

Fitzgerald, J. P. 1993. The ecology of plague in Gunnison's prairie dogs and suggestions for the recovery of black-footed ferrets. Pages 50–59 *in* J. L. Oldemeyer, D. E. Biggins, B. J. Miller, and R. Crete, eds. Proceedings of the symposium on the management of prairie dog complexes for the reintroduction of black-footed ferrets (Biological Report No. 93). U.S. Fish and Wildlife Service, Washington, DC.

Fitzgerald, J. P. and R. R. Lechleitner. 1974. Observations on the biology of Gunnison's prairie dog in central Colorado. American Midland Naturalist 92:146–63.

Fitzgerald, J. P., C. A. Meaney, and D. M. Armstrong. 1994. Mammals of Colorado. Denver Museum of Natural History and University Press of Colorado, Denver.

Flath, D. L. 1979. Mound characteristics of white-tailed prairie dog maternity burrows. American Midland Naturalist 102:395–98.

Foltz, D. W., and J. L. Hoogland. 1983. Genetic evidence of outbreeding in the black-tailed prairie dog (*Cynomys ludovicianus*). Evolution 37:273–81.

Foltz, D. W., J. L. Hoogland, and G. M. Koscielny. 1988. Effects of sex, litter size, and heterozygosity on juvenile weight in black-tailed prairie dogs (*Cynomys ludovicianus*). Journal of Mammalogy 69:611–14.

Foreman, D. 1962. The normal reproductive cycle of the female prairie dog and the effects of light. Anatomical Record 142:391–405.

Foreman, D. 1981. Follicular dynamics in a monestrous annually breeding mammal: Prairie dog (*Cynomys ludovicianus*). Pages 245–51 *in* N. B. Schwartz and M. Hunzicker-Dunn, eds. Dynamics of ovarian function. Raven Press, New York.

Fortenberry, D. K. 1972. Characteristics of the black-footed ferret (Bureau of Sport Fisheries and Wildlife Resource Publication No. 109). U.S. Department of the Interior, Fish and Wildlife Service, Rapid City, SD.

Foster, B. E. 1924. Provision of prairie-dog to escape drowning when town is submerged. Journal of Mammalogy 5:266–68.

Foster, N. S., and S. E. Hygnstrom. 1990. Prairie dogs and their ecosystem. Department of Forestry, Fisheries, and Wildlife, University of Nebraska, Lincoln.

Garrett, M. G., and W. L. Franklin. 1983. Diethylstilbestrol as a temporary chemosterilant to control black-tailed prairie dog populations. Journal of Range Management 36:753–56.

Garrett, M. G., and W. L. Franklin. 1988. Behavioral ecology of dispersal in the black-tailed prairie dog. Journal of Mammalogy 69:236–50.

Garrett, M. G., J. L. Hoogland, and W. L. Franklin. 1982. Demographic differences between an old and a new colony of black-tailed prairie dogs (*Cynomys ludovicianus*). American Midland Naturalist 108:51–59.

Goodrich, J. M., and S. W. Buskirk. 1998. Spacing and ecology of North American badgers (*Taxidea taxus*) in a prairie dog (*Cynomys leucurus*) complex. Journal of Mammalogy 79:171–79.

Goodwin, H. T. 1993. Subgeneric identification and biostratigraphic utility of late Pleistocene prairie dogs (*Cynomys,* Sciuridae) from the Great Plains. Southwestern Naturalist 38:105–10.

Goodwin, H. T. 1995. Systematic revision of fossil prairie dogs with descriptions of two new species. Miscellaneous Publications, University of Kansas Natural History Museum 86:1–38.

Gorman, R. C. 1974. Baked prairie dog. Pages 102–3 *in* C. Counter and K. Tani, eds. Palette in the kitchen. Sunstone Press, Santa Fe, NM.

Grady, R. M., and J. L. Hoogland. 1986. Why do male black-tailed prairie dogs (*Cynomys ludovicianus*) give a mating call? Animal Behaviour 34:108–12.

Green, D. G., and J. E. Dowling. 1975. Electrophysiological evidence for rod-like receptors in the gray squirrel, ground squirrel, and prairie dog retinas. Journal of Comparative Neurology 159:461–71.

Greenwood, P. J. 1980. Mating systems, philopatry and dispersal in birds and mammals. Animal Behaviour 28:1140–62.

Grossmann, J. 1987. A prairie dog companion. Audubon 89:52–67.

Gunderson, H. L. 1978. Under and around a prairie dog town. Natural History 87:57–66.

Gurll, N., and L. DenBesten. 1978. Animal models of human cholesterol gallstone disease: A review. Laboratory and Animal Science 28:428–32.

Hafner, D. J. 1984. Evolutionary relationships of the Nearctic Sciuridae. Pages 3–23 *in* J. O. Murie and G. R. Michener, eds. The biology of ground-dwelling squirrels. University of Nebraska Press, Lincoln.

Hall, E. R. 1981. The mammals of North America. John Wiley, New York.

Halloran, A. F. 1972. The black-tailed prairie dog: Yesterday and today. Great Plains Journal 11:138–44.

Halpin, Z. T. 1983. Naturally occurring encounters between black-tailed prairie dogs (*Cynomys ludovicianus*) and snakes. American Midland Naturalist 109:50–54.

Halpin, Z. T. 1987. Natal dispersal and the formation of new social groups in a newly established town of black-tailed prairie dogs (*Cynomys ludovicianus*). Pages 104–18 *in* B. D. Chepko-Sade and Z. T. Halpin, eds. Mammalian dispersal patterns: The effects of social structure on population genetics. University of Chicago Press, Chicago.

Hamilton, J. D., and E. W. Pfeiffer. 1977. Effects of cold exposure and dehydration on renal function in black-tailed prairie dogs. Journal of Applied Physiological, Respiratory, and Environmental Exercise Physiology 42:295–99.

Hansen, R. M., and I. K. Gold. 1977. Blacktail prairie dogs, desert cottontails, and cattle trophic relations on shortgrass range. Journal of Range Management 30:210–14.

Harlow, H. J. 1995. Fasting biochemistry of representative spontaneous and facultative hibernators: the white-tailed prairie dog and the black-tailed prairie dog. Physiological Zoology 68:915–34.

Harlow, H. J. 1997. Winter body fat, food consumption and nonshivering thermogenesis of representative spontaneous and facultative hibernators: The white-tailed and black-tailed prairie dog. Journal of Thermal Biology 22:21–30.

Harlow, H. J., and G. E. Menkens. 1986. A comparison of hibernation in the black-tailed prairie dog, white-tailed prairie dog, and Wyoming ground squirrel. Canadian Journal of Zoology 64:793–96.

Hasenyager, R. N. 1983. Diet selection of the Utah prairie dog (*Cynomys parvidens*) as determined by histological fecal analysis. Utah Division of Wildlife Research, Logan.

Hasenyager, R. N., T. Ball, and T. W. Clark. 1988. Utah prairie dog recovery plan. U.S. Fish and Wildlife Service, Denver, CO.

Haynic, M. L., R. A. van den Bussche, J. L. Hoogland, and D. A. Gilbert. 2003. Parentage, multiple paternity, and reproductive success in Gunnison's and Utah prairie dogs. Journal of Mammalogy, 84:(in press).

Haynie, M. L. 2001. Using microsatellites to resolve parentage in populations of Gunnison's and Utah prairie dogs. Master's thesis, Oklahoma State University, Stillwater.

Henderson, F. R., P. F. Springer, and R. Adrian. 1969. The black-footed ferret in South Dakota. (Technical Bulletin 4). South Dakota Department of Game, Fish, and Parks, Pierre.

Hillman, C. N. 1968. Field observations of black-footed ferrets in South Dakota. Transactions of the North American Wildlife and Natural Resources Conference 33:433–43.

Hirst, L. F. 1953. The conquest of plague. Clarendon Press, Oxford.

Holckamp, K. E. 1984. Dispersal in ground-dwelling sciurids. Pages 297–320 *in* J. O. Murie and G. R. Michener, eds. The biology of ground-dwelling squirrels. University of Nebraska Press, Lincoln.

Holland, E. A., and J. K. Detling. 1990. Plant response to herbivory and below ground nitrogen cycling. Ecology 71:1040–49.

Hollister, N. 1916. A systematic account of the prairie dogs. North American Fauna 40:1–37.

Holzbach, R. T., C. Corbusier, M. Marsh, and K. Naito. 1976. The process of cholesterol cholelithiasis induced by diet in the prairie dog: A physicochemical characterization. Journal of Laboratory and Clinical Medicine 87:987–98.

Hoogland, J. L. 1979a. Aggression, ectoparasitism, and other possible costs of prairie dog (Sciuridae: *Cynomys* spp.) coloniality. Behaviour 69:1–35.

Hoogland, J. L. 1979b. The effect of colony size on individual alertness of prairie dogs (Sciuridae: *Cynomys* spp.) Animal Behaviour 27:394–407.

Hoogland, J. L. 1981a. The evolution of coloniality in white-tailed and black-tailed prairie dogs (*Cynomys leucurus* and *C. ludovicianus*). Ecology 62:252–72.

Hoogland, J. L. 1981b. Nepotism and cooperative breeding in the black-tailed prairie dog (Sciuridae: *Cynomys ludovicianus*). Pages 283–310 *in* R. D. Alexander and D. W. Tinkle, eds. Natural selection and social behavior. Chiron Press, New York.

Hoogland, J. L. 1982a. Prairie dogs avoid extreme inbreeding. Science 215:1639–41.

Hoogland, J. L. 1982b. Reply to a comment by Powell. Ecology 63:1968–69.

Hoogland, J. L. 1983a. Black-tailed prairie dog coteries are cooperatively breeding units. American Naturalist 121:275–80.

Hoogland, J. L. 1983b. Nepotism and alarm calling in the black-tailed prairie dog (*Cynomys ludovicianus*). Animal Behaviour 31:472–79.

Hoogland, J. L. 1985. Infanticide in prairie dogs: Lactating females kill offspring of close kin. Science 230:1037–40.
Hoogland, J. L. 1986. Nepotism in prairie dogs (*Cynomys ludovicianus*) varies with competition but not with kinship. Animal Behaviour 34:263–70.
Hoogland, J. L. 1992. Levels of inbreeding among prairie dogs. American Naturalist 139:591–602.
Hoogland, J. L. 1995. The black-tailed prairie dog: Social life of a burrowing mammal. University of Chicago Press, Chicago.
Hoogland, J. L. 1996a. *Cynomys ludovicianus*. Mammalian Species 535:1–10.
Hoogland, J. L. 1996b. Why do Gunnison's prairie dogs give anti-predator calls? Animal Behaviour 51:871–80.
Hoogland, J. L. 1997. Duration of gestation and lactation for Gunnison's prairie dogs. Journal of Mammalogy 78:173–80.
Hoogland, J. L. 1998a. Estrus and copulation for Gunnison's prairie dogs. Journal of Mammalogy 79:887–97.
Hoogland, J. L. 1998b. Why do Gunnison's prairie dog females copulate with more than one male? Animal Behaviour 55:351–59.
Hoogland, J. L. 1999. Philopatry, dispersal, and social organization of Gunnison's prairie dogs. Journal of Mammalogy 80:243–51.
Hoogland, J. L. 2001. Black-tailed, Gunnison's, and Utah prairie dogs all reproduce slowly. Journal of Mammalogy. 82:917–27.
Hoogland, J. L. 2003. Sexual dimorphism in five species of prairie dogs. Journal of Mammalogy 84:(in press).
Hoogland, J. L., and J. M. Hutter. 1987. Aging live prairie dogs from molar attrition. Journal of Wildlife Management 51:393–94.
Hoogland, J. L., R. H. Tamarin, and C. K. Levy. 1989. Communal nursing in prairie dogs. Behavioral Ecology and Sociobiology 24:91–95.
Hudson, J. W. 1978. Shallow daily torpor: A thermoregulatory adaptation. Pages 67–108 *in* L. C. H. Wang and J. W. Hudson, eds. Strategies in cold: Natural torpidity and thermogenesis. Academic Press, New York.
Hygnstrom, S. E., and K. C. VerCauteren. 2000. Cost-effectiveness of five burrow fumigants for managing black-tailed prairie dogs. International Biodeterioration and Biogradation 45:159–68.
Jacobs, G. H., and R. L. Yolton. 1972. Some characteristics of the eye and the electroretinogram of the prairie dog. Experimental Neurology 37:538–49.
Jacquart, H. C. 1986. Prescriptive transplanting and monitoring of Utah prairie dog (*Cynomys parvidens*) populations. M.S. thesis, Brigham Young University, Provo, UT.
Jameson, W. C. 1973. On the eradication of prairie dogs: A point of view. Bios 44:129–35.
Jellison, W. L. 1939. Notes on the fleas of ie dogs, with description of a new subspecies. U.S. Public Health Reports 54:840–44.
Jellison, W. L. 1945. Siphonaptera: The genus *Oropsylla* in North America (*Citellus, Callospermophilus, Cynomys,* and *Marmota* parasitized). Journal of Parasitology 31:83–97.
Johnson, G. E. 1927. Observations on young prairie dogs (*Cynomys ludovicianus*) born in the laboratory. Journal of Mammalogy 8:110–15.
Jones, T. R., and R. K. Plakke. 1981. The histology and histochemistry of the perianal scent gland of the reproductively quiescent black-tailed prairie dog (*Cynomys ludovicianus*). Journal of Mammalogy 62:362–68.
Kartman, L., S. F. Quan, and R. R. Lechleitner. 1962. Die-off of a Gunnison's prairie dog colony in central Colorado. II. Retrospective determination of plague infection in flea vectors, rodents, and man. Zoonoses Research l:201–24.
Kelso, L. H. 1939. Food habits of prairie dogs (Circular No. 529). U.S. Department of Agriculture, Washington, DC.
Kelson, K. R. 1949. Speciation of rodents in the Colorado River drainage of eastern Utah. Ph.D. dissertation, University of Utah, Salt Lake City.
Kildaw, S. D. 1995. The effect of group size manipulations on the foraging behavior of black-tailed prairie dogs. Behavioral Ecology 6:353–58.
King, J. A. 1955. Social behavior, social organization, and population dynamics in a black-tailed prairiedog town in the Black Hills of South Dakota (Contribution No. 67). Laboratory of Vertebrate Biology, University of Michigan, Ann Arbor.
King, J. A. 1959. The social behavior of prairie dogs. Scientific American 201:128–40.
King, J. A. 1984. Historical ventilations on a prairie dog town. Pages 447–56 *in* J. O. Murie and G. R. Michener, eds. The biology of ground-dwelling squirrels. University of Nebraska Press, Lincoln.
Klatt, L. E., and D. Hein. 1978. Vegetative differences among active and abandoned towns of black-tailed prairie dogs (*Cynomys ludovicianus*). Journal of Range Management 31:315–17.
Knowles, C. J. 1985. Observations on prairie dog dispersal in Montana. Prairie Naturalist 17:33–40.
Knowles, C. J. 1986a. Population recovery of black-tailed prairie dogs following control with zinc phosphide. Journal of Range Management 39:249–51.
Knowles, C. J. 1986b. Some relationships of black-tailed prairie dogs to livestock grazing. Great Basin Naturalist 46:198–203.
Knowles, C. J. 1987. Reproductive ecology of black-tailed prairie dogs in Montana. Great Basin Naturalist 47:202–6.
Knowles, C. J., C. J. Stoner, and S. P. Gieb. 1982. Selective use of black-tailed prairie dog towns by mountain plovers. Condor 84:71–74.
Koford, C. B. 1958. Prairie dogs, whitefaces, and blue grama. Wildlife Monographs 3:1–78.
Kotliar, N. B., B. W. Baker, A. D. Whicker, and G. Plumb. 1999. A critical review of assumptions about the prairie dog as a keystone species. Environmental Management 24:177–92.
Lamorte, W. W., D. P. Oleary, M. L. Booker, and T. E. Scott. 1993. Increased dietary-fat content accelerates cholesterol gallstone formation in the cholesterol-fed prairie dog. Hepatology 18:1498–1503.
Lechleitner, R. R. 1969. Wild mammals of Colorado. Pruett, Boulder, CO.
Lechleitner, R. R., J. V. Tileston, and L. Kartman. 1962. Die-off of a Gunnison's prairie dog colony in central Colorado. I. Ecological observations and description of the epizootic. Zoonoses Research 1:185–99.
Lechleitner, R. R., L. Kartman, M. I. Goldberg, and B. W. Hudson. 1968. An epizootic of plague in Gunnison's prairie dogs (*Cynomys gunnisoni*) in south-central Colorado. Ecology 49:734–43.
Longhurst, W. 1944. Observations on the ecology of the Gunnison prairie dog in Colorado. Journal of Mammalogy 25:24–36.
Loughry, W. J. 1987a. Differences in experimental and natural encounters of black-tailed prairie dogs with snakes. Animal Behaviour 35:1568–70.
Loughry, W. J. 1987b. The dynamics of snake harassment by black-tailed prairie dogs. Behaviour 103:27–48.
Loughry, W. J. 1988. Population differences in how black-tailed prairie dogs deal with snakes. Behavioral Ecology and Sociobiology 22:61–67.
Loughry, W. J. 1992. Ontogeny of time allocation in black-tailed prairie dogs. Ethology 90:206–24.
Loughry, W. J. 1993a. Determinants of time allocation by adult and yearling black-tailed prairie dogs. Behaviour 124:23–43.
Loughry, W. J. 1993b. Mechanisms of change in the ontogeny of black-tailed prairie dog time budgets. Ethology 95:54–64.
Marsh, R. E. 1984. Ground squirrels, prairie dogs, and marmots as pests on rangelands. Pages 195–208 *in* Proceedings of the conference for organization and practice of vertebrate pest control, 1982 (Hampshire, United Kingdom). Plant Protection Division, Fernherst, UK.
Martin, S. J., and M. H. Schroeder. 1978. Black-footed ferret surveys on seven coal occurrence areas in southwestern and southcentral Wyoming, June 8 to September 25, 1978: Final report. U.S. Fish and Wildlife Service, Denver, CO.
Martin, S. J., and M. H. Schroeder. 1980. Black-footed ferret surveys on seven coal occurrence areas in Wyoming, February–September, 1979: Final report. Wyoming State Office, U.S. Bureau of Land Management, Cheyenne.
Martin, S. J., M. H. Schroeder, and H. Tietjen. 1984. Burrow plugging by prairie dogs in response to Siberian polecats. Great Basin Naturalist 44:447–49.
McCullough, D. A., and R. A. Chesser. 1987. Genetic variation among populations of the Mexican prairie dog. Journal of Mammalogy 68:555–60.
McCullough, D. A., R. A. Chesser, and R. D. Owen. 1987. Immunological systematics of prairié dogs. Journal of Mammalogy 68:561–68.
McNulty, F. 1971. Must they die? The strange case of the prairie dog and the black-footed ferret. Doubleday, Garden City, NY.
Mellink, E., and H. Madrigal. 1993. Ecology of Mexican prairie dogs, *Cynomys mexicanus,* in El Manantial, northeastern Mexico. Journal of Mammalogy 74:631–35.
Menkens, G. E., and S. H. Anderson. 1991. Population dynamics of white-tailed prairie dogs during an epizootic of sylvatic plague. Journal of Mammalogy 72:328–31.
Merriam, C. H. 1902. The prairie dog of the Great Plains. Pages 257–70 *in* Yearbook of United States Department of Agriculture (1901). U.S. Department of Agriculture, Washington, DC.
Michener, G. R. 1983. Kin identification, matriarchies, and the evolution of sociality in ground-dwelling sciurids. Pages 528–72 *in* J. F. Eisenberg and D. G. Kleiman, eds. Recent advances in the study of mammalian behavior (Special Publication No. 7). American Society of Mammalogists.
Miller, B., C. Wemmer, D. Biggins, and R. Reading. 1990. A proposal to conserve black-footed ferrets and the prairie dog ecosystem. Environmental Management 14:763–69.

Miller, B., G. Ceballos, and R. P. Reading. 1994. The prairie dog and biotic diversity. Conservation Biology 8:677–81.

Miller, B., R. Reading, J. Hoogland, T. Clark, G. Ceballos, R. List, S. Forrest, L. Hanebury, P. Manzano, J. Pacheco, and D. Uresk. 2000. The role of prairie dogs as keystone species: A response to Stapp. Conservation Biology 14:318–21.

Mills, L. S., M. E. Soule, and D. F. Doak. 1993. The history and current status of the keystone species concept. Bioscience 43:219–24.

Nadler, C. F., R. S. Hoffmann, and J. J. Pizzimenti. 1971. Chromosomes and serum proteins of prairie dogs and a model of *Cynomys* evolution. Journal of Mammalogy 52:545–55.

National Wildlife Federation., 1998. Petition for rule listing the black-tailed prairie dog (*Cynomys ludovicianus*) as threatened throughout its range. Submitted to U.S. Fish and Wildlife Service, Denver, CO.

Nichols, J. B., and D. J. Nash. 1980. Biochemical variations in three species of prairie dogs (*Cynomys*). Comparative Biochemical Physiology 65A:155–58.

Olendorff, R. R. 1976. The food habits of North American golden eagles. American Midland Naturalist 95:231–36.

Olsen, P. F. 1981. Sylvatic plague. Pages 232–43 *in* J. W. Davis, L. H. Karstadt, and D. O. Trainer, eds. Infectious diseases of wild animals. Iowa State University Press, Ames.

O'Meilia, M. E., F. L. Knopf, and J. C. Lewis. 1982. Some consequences of competition between prairie dogs (*Cynomys ludovicianus*) and beef cattle. Journal of Range Management 35:580–85.

Osborn, B., and P. F. Allan. 1949. Vegetation of an abandoned prairie dog town in tallgrass prairie. Ecology 30:322–32.

Owings, D. H., and W. J. Loughry. 1985. Variation in snake-elicited jump-yipping by black-tailed prairie dogs: Ontogeny and snake specificity. Zeitschrift für Tierpsychologie 70:177–200.

Owings, D. H., and S. C. Owings. 1979. Snake-directed behavior by black-tailed prairie dogs (*Cynomys ludovicianus*). Zeitschrift für Tierpsychologie 49:35–54.

Paine, R. T. 1969. A note on trophic complexity and community stability. American Naturalist 103:91–93.

Paine, R. T. 1980. Food webs: Linkage, interaction strength and community infrastructure. Journal of Animal Ecology 49:667–85.

Petzal, D. E. 1993. Doggin' it. Field and Stream 97:82–90.

Pfeiffer, D. G. 1972. Effects of diethylstilbestrol on reproduction in the black-tailed prairie dog. Master's thesis, South Dakota State University, Brookings.

Pizzimenti, J. J. 1975. Evolution of the prairie dog genus *Cynomys*. Occasional Papers of the Museum of Natural History, University of Kansas 39:1–73.

Pizzimenti, J. J. 1976a. Genetic divergence and morphological convergence in prairie dogs, *Cynomys gunnisoni* and *Cynomys leucurus*. I. Morphological and ecological analyses. Evolution 30:345–66.

Pizzimenti, J. J. 1976b. Genetic divergence and morphological convergence in prairie dogs, *Cynomys gunnisoni* and *Cynomys leucurus*. II. Genetic analyses. Evolution 30:367–79.

Pizzimenti, J. J., and G. D. Collier. 1975. *Cynomys parvidens*. Mammalian Species 52:1–3.

Pizzimenti, J. J., and R. S. Hoffmann. 1973. *Cynomys gunnisoni*. Mammalian Species 25:1–4.

Pizzimenti, J. J., and L. R. McClenaghan. 1974. Reproduction, growth and development, and behavior in the Mexican prairie dog, *Cynomys mexicanus* (Merriam). American Midland Naturalist 92:130–45.

Pizzimenti, J. J., and C. F. Nadler. 1972. Chromosomes and serum proteins of the Utah prairie dog, *Cynomys parvidens* (Sciuridae). Southwestern Naturalist 17:279–86.

Plumpton, D. L., and R. S. Lutz. 1993. Nesting habitat use by burrowing owls in Colorado. Journal of Raptor Research 27:175–79.

Pollitzer, R. 1951. Plague studies. l. A summary of the history and a survey of the present distribution of the disease. World Health Organization Bulletin 4:475–533.

Pollitzer, R. 1952. Plague studies. 7. Insect vectors. World Health Organization Bulletin 7:231–342.

Pollitzer, R., and K. F. Meyer. 1961. The ecology of plague. Pages 433–501 *in* J. M. May, ed. Studies of disease ecology. Hafner, New York.

Powell, K. L., R. J. Robel, K. E. Kemp, and M. D. Nellis. 1994. Aboveground counts of black-tailed prairie dogs: Temporal nature and relationship to burrow entrance density. Journal of Wildlife Management 58:361–66.

Powell, R. A. 1982. Prairie dog coloniality and black-footed ferrets. Ecology 64:1967–69.

Power, M. E., D. Tilman, J. A. Estes, B. A. Menge, W. J. Bond, L. S. Mills, G. Daily, J. C. Castilla, J. Lubchenco, and R. T. Paine. 1996. Challenges in the quest for keystones. BioScience 466:9–20.

Radcliffe, M. C. 1992. Repopulation of black-tailed prairie dog (*Cynomys ludovicianus*) colonies after artificial reduction. Master's thesis, Frostburg State University, Frostburg, MD.

Randall, D. 1976a. Poison the damn prairie dogs. Defenders 51:381–83.

Randall, D. 1976b. Shoot the damn prairie dogs. Defenders 51:378–81.

Rayor, L. S. 1985a. Dynamics of a plague outbreak in Gunnison's prairie dog. Journal of Mammalogy 66:194–96.

Rayor, L. S. 1985b. Effects of habitat quality on growth, age of first reproduction, and dispersal in Gunnison's prairie dogs (*Cynomys gunnisoni*). Canadian Journal of Zoology 63:2835–40.

Rayor, L. S. 1988. Social organization and space-use in Gunnison's prairie dog. Behavioral Ecology and Sociobiology 22:69–78.

Rayor, L. S., A. K. Brody, and C. Gilbert. 1987. Hibernation in the Gunnison's prairie dog. Journal of Mammalogy 68:147–50.

Reading, R. P., and S. R. Kellert. 1993. Attitudes toward a proposed reintroduction of black-footed ferrets (*Mustela nigripes*). Conservation Biology 7:569–80.

Reading, R. P., S. R. Beissinger, J. J. Grensten, and T. W. Clark. 1989. Attributes of black-tailed prairie dog colonies in northcentral Montana, with management recommendations for the conservation of biodiversity. Pages 13–27 *in* T. W. Clark, D. Hinckley, and T. Rich, eds. The prairie dog ecosystem: Managing for biological diversity (Wildlife Technical Bulletin No. 2). Montana Bureau of Land Management.

Roberts, W. M., J. P. Rodriguez, T. C. Good, and A. P. Dobson. 2000. Population viability analysis of the Utah prairie dog (Environmental Defense Report). Department of Ecology and Evolutionary Biology, Princeton University, Princeton, NJ.

Roemer, D. M., and S. C. Forrest. 1996. Prairie dog poisoning in northern Great Plains: An analysis of programs and policies. Environmental Management 20:349–59.

Rogers-Wydeven, P., and R. B. Dahlgren. 1982. A comparison of prairie dog stomach contents and feces using a microhistological technique. Journal of Wildlife Management 46:1104–8.

Roslyn, J. J., M. Z. Abedin, K. D. Saunders, J. A. Cates, S. D. Strichartz, M. Alperin, M. Fromm, and C. E. Palant. 1991. Uncoupled basal sodium absorption and chloride secretion in prairie dog (*Cynomys ludovicianus*) gallbladder. Comparative Biochemical Physiology 100A:335–41.

Scheffer, T. H. 1937. Study of a small prairie-dog town. Transactions of the Kansas Academy of Science 40:391–95.

Scheffer, T. H. 1945. Historical encounter and accounts of the plains prairie dog. Kansas History Quarterly 13:527–37.

Scheffer, T. H. 1947. Ecological comparisons of the plains prairie dog and the Zuni species. Transactions of the Kansas Academy of Science 49:401–6.

Seal, U. S., E. T. Thorne, M. A. Bogan, and S. H. Anderson, eds. 1989. Conservation biology and the black-footed ferret. Yale University Press, New Haven, CT.

Severson, K. E., and G. E. Plumb. 1998. Comparison of methods to estimate population densities of black-tailed prairie dogs. Wildlife Society Bulletin 26:859–66.

Seville, R. S. 1997. Eimeria spp. (Apicomplexa: Eimeriidae) from black- and white-tailed prairie dogs (*Cynomys ludovicianus* and *Cynomys leucurus*) in central and southeast Wyoming. Journal of Parasitology 83:166–68.

Shalaway, S., and C. N. Slobodchikoff. 1988. Seasonal changes in the diet of Gunnison's prairie dog. Journal of Mammalogy 69:835–41.

Sharps, J. C., and D. W. Uresk. 1990. Ecological review of black-tailed prairie dogs and associated species in western South Dakota. Great Basin Naturalist 50:339–45.

Sheets, R. G., and R. L. Linder. 1969. Food habits of the black-footed ferret (*Mustela nigripes*) in South Dakota. Proceedings of the South Dakota Academy of Science 48:58–61.

Sheets, R. G., R. L. Linder, and R. B. Dahlgren. 1971. Burrow systems of prairie dogs in South Dakota. Journal of Mammalogy 52:451–53.

Slobodchikoff, C. N., J. Kiriazis, C. Fischer, and E. Creef. 1991. Semantic information distinguishing individual predators in the alarm calls of Gunnison's prairie dogs. Animal Behaviour 42:713–19.

Smit, F. G. A. M. 1958. A preliminary note on the occurrence of *Pulex irritans* and *Pulex simulans* in North America. Journal of Parasitology 44:523–26.

Smith, H. E. 1915. The grasshopper outbreak in New Mexico during the summer of 1913 (Bulletin 293). U.S. Department of Agriculture, Washington, DC.

Smith, R. E. 1967. Natural history of the prairie dog in Kansas. Miscellaneous Publications of the Museum of Natural History, University of Kansas 49:1–39.

Smith, W. J., S. L. Smith, J. G. deVilla, and E. L. Oppenheimer. 1976. The jump-yip display of the black-tailed prairie dog, *Cynomys ludovicianus*. Animal Behaviour 24:609–21.

Smith, W. J., S. L. Smith, E. C. Oppenheimer, and J. G. deVilla. 1977. Vocalizations of the black-tailed prairie dog, *Cynomys ludovicianus*. Animal Behavior 25:152–64.

Snell, G. P. 1985. Results of control of prairie dogs. Rangelands 7:30.

Sperry, C. C. 1934. Winter food habits of coyotes: A report of progress, 1933. Journal of Mammalogy 15:286–90.

Stapp, P. 1998. A reevaluation of the role of prairie dogs in the Great Plains grasslands. Conservation Biology 12:1253–59.

Stephens, J. R., and H. R. Fevold. 1993. Biochemical adaptations in the black-tailed prairie dog (*Cynomys ludovicianus*) during long-term food and water deprivation. Comparative Biochemistry and Physiology A Physiology 104:613–18.

Stockard, A. H. 1929. Observations of reproduction in the white-tailed prairie dog (*Cynomys leucurus*). Journal of Mammalogy 10:209–12.

Stockard, A. H. 1930. Observations on the seasonal activities of the white-tailed prairie dog, *Cynomys leucurus*. Papers of the Michigan Academy of Science, Arts and Letters 11:471–79.

Stockrahm, D. M. B., and R. W. Seabloom. 1988. Comparative reproductive performance of black-tailed prairie dog populations in North Dakota. Journal of Mammalogy 69:160–64.

Stockrahm, D. M. B., and R. W. Seabloom. 1990. Tooth eruption in black-tailed prairie dogs from North Dakota. Journal of Mammalogy 71:105–8.

Stockrahm, D. M. B., B. J. Dickerson, S. L. Adolf, and R. W. Seabloom. 1996. Aging black-tailed prairie dogs by weight of eye lenses. Journal of Mammalogy 77:874–81.

Stromberg, M. R. 1974. Group response in black-tailed prairie dogs to an avian predator. Journal of Mammalogy 55:850–51.

Stromberg, M. R. 1978. Subsurface burrow connections and entrance spatial pattern of prairie dogs. Southwestern Naturalist 23:173–80.

Summers, C. A., and R. L. Linder. 1978. Food habits of the black-tailed prairie dog in western South Dakota. Journal of Range Management 31:134–36.

Swenk, M. H. 1915. The prairie dog and its control. Nebraska Agricultural Experiment Station Bulletin 154:3–38.

Tate, G. H. H. 1947. Albino prairie dog. Journal of Mammalogy 28:62.

Taylor, W. P., and J. V. G. Loftfield. 1924. Damage to range grasses by the Zuni prairie dog (Bulletin 1227). U.S. Department of Agriculture, Washington, DC.

Thomas, D. M., and N. L. Stanton. 1994. Eimerian species (Apicomplexa, Eimeriidae) in Gunnisons's prairie dogs (*Cynomys gunnisoni zuniensis*) and rock squirrels (*Spermophilus variegates grammurus*) from southeastern Utah. Journal of the Helminthological Society of Washington 61:17–21.

Thomas, T. H., and M. L. Riedesel. 1975. Evidence of hibernation in the black-tailed prairie dog *Cynomys ludovicianus*. Cryobiology 12:559.

Tileston, J. V., and R. R. Lechleitner. 1966. Some comparisons of the black-tailed and white-tailed prairie dogs in north-central Colorado. American Midland Naturalist 75:292–316.

Travis, S. E., and C. N. Slobodchikoff. 1993. Effects of food resource distribution on the social system of Gunnison's prairie dogs. Canadian Journal of Zoology 71:1186–92.

Travis, S. E., C. N. Slobodchikoff, and P. Keim. 1995. Ecological and demographic effects on intraspecific variation in the social system of prairie dogs. Ecology 76:1794–1803.

Travis, S. E., C. N. Slobodchikoff, and P. Keim. 1996. Social assemblages and mating relationships in prairie dogs: A DNA fingerprint analysis. Behavioral Ecology 7:95–100.

Trevino-Villarreal, J. 1990. The annual cycle of the Mexican prairie dog (*Cynomys mexicanus*). Occasional Papers of the Museum of Natural History, University of Kansas 139:1–27.

Trevino-Villarreal, J., and W. E. Grant. 1998. Geographic range of the endangered Mexican prairie dog (*Cynomys mexicanus*). Journal of Mammalogy 79:1273–87.

Trevino-Villarreal, J., I. M. Berk, A. Aguirre, and W. E. Grant. 1998. Survey for sylvatic plague in the Mexican prairie dog (*Cynomys mexicanus*). Southwestern Naturalist 43:147–54.

Turner, G. 2000. Recovery of prairie dog colonies following epizootics of sylvatic plague. Master's thesis, Frostburg State University, Frostburg, MD.

Tyler, J. D., and H. N. Buscher. 1975. Parasites of vertebrates inhabiting prairie dog towns in Oklahoma. I. Ectoparasites. Proceedings of the Oklahoma Academy of Science 55:166–68.

Ubico, S. R., G. O. Maupin, K. A. Fagerstone, and R. G. McLean. 1988. A plague epizootic in the white-tailed prairie dogs (*Cynomys leucurus*) of Meeteetse, Wyoming. Journal of Wildlife Diseases 24:399–406.

Uresk, D. W. 1984. Black-tailed prairie dog food habits and forage relationships in western South Dakota. Journal of Range Management 37:325–29.

Uresk, D. W. 1985. Effects of controlling black-tailed prairie dogs on plant production. Journal of Range Management 38:466–68.

Uresk, D. W., and J. C. Sharps. 1986. Denning habitat and diet of the swift fox in western South Dakota. Great Basin Naturalist 46:249–53.

Uresk, D. W., R. M. King, A. D. Apa, and R. L. Linder. 1986. Efficacy of zinc phosphide and strychnine for black-tailed prairie dog control. Journal of Range Management 39:298–99.

Vargas, A., M. Lockart, P. Marinari, and P. Gober. 1998. Preparing captive-raised black-footed ferrets *Mustela nigripes* for survival after release. Dodo 34:76–83.

Vetterling, J. M. 1964. Coccidea (*Eimeria*) from the prairie dog, *Cynomys ludovicianus,* in northern Colorado. Journal of Protozoology 11:89–91.

Vogel, S. 1989. Life's devices: The physical world of animals and plants. Princeton University Press, Princeton, NJ.

Vogel, S., C. P. Ellington, and D. L. Kilgore. 1973. Wind-induced ventilation of the burrow of the prairie-dog, *Cynomys ludovicianus*. Journal of Comparative Physiology 85:1–15.

Vosburgh, T. C., and L. R. Irby. 1998. Effects of recreational shooting on prairie dogs. Journal of Wildlife Management 62:363–72.

Wade, O. 1928. Notes on the time of breeding and the number of young of *Cynomys ludovicianus*. Journal of Mammalogy 9:149.

Walls, G. L. 1941. The vertebrate visual system. Cranbrook Institute of Science, Bloomfield Hills, MI.

Walls, G. L. 1942. The vertebrate eye and its adaptive radiation. Cranbrook Institute of Science, Bloomfield Hills, MI.

Waring, G. H. 1970. Sound communications of black-tailed, white-tailed, and Gunnison's prairie dogs. American Midland Naturalist 83:167–85.

Wedel, W. R. 1961. Prehistoric man on the great plains. University of Oklahoma Press, Norman.

West, R. W., and J. E. Dowling. 1975. Anatomical evidence for cone and rod-like receptors in the gray squirrel, ground squirrel, and prairie dog retinas. Journal of Comparative Neurology 159:439–59.

Weltzlin, J. F., S. L. Downhower, and R. K. Heitschmidt. 1997. Prairie dog effects on plant community structure in southern mixed-grass prairie. Southwestern Naturalist 42:251–58.

Whicker, A. D., and J. K. Detling. 1988. Ecological consequences of prairie dog disturbances. Bioscience 38:778–85.

Whitehead, L. C. 1927. Notes on prairie-dogs. Journal of Mammalogy 8:58.

Wilcomb, M. J. 1954. A study of prairie dog burrow systems and the ecology of their arthropod inhabitants in central Oklahoma. Ph.D. dissertation, University of Oklahoma, Norman.

Wright-Smith, M. A. 1978. The ecology and social organization of *Cynomys parvidens* (Utah prairie dog) in south central Utah. Master's thesis, Indiana University, Bloomington.

Wuerthner, G. 1997. The black-tailed prairie dog: Headed for extinction? Journal of Range Management 50:459–66.

Young, S. D. 1944. Longevity and other data in a male and female prairie dog kept as pets. Journal of Mammalogy 25:317–19.

Zinn, H. C., and W. F. Andelt. 1999. Attitudes of Fort Collins, Colorado, residents toward prairie dogs. Wildlife Society Bulletin 27:1098–1106.

JOHN L. HOOGLAND, Appalachian Laboratory, University of Maryland, Frostburg, Maryland 21532. Email: hoogland@al.umces.edu.

12

Fox and Gray Squirrels

Sciurus niger and *S. carolinensis*

John Edwards
Mark Ford
David Guynn

NOMENCLATURE

Common Names. Fox squirrel, cat squirrel, red squirrel, stump-eared squirrel
Scientific Name. *Sciurus niger*
The fox squirrel was first described by Linnaeus (1758:64); ten subspecies of *S. niger* are recognized (Hall 1981)
Subspecies. *S. n. avicennia, S. n. bachmani, S. n. cinereus, S. n. limitis, S. n. ludovicianus, S. n. niger, S. n. rufiventer, S. n. shermani, S. n. subauratus*, and *S. n. vulpinus*

Common Names. Gray squirrel, cat squirrel, migratory squirrel
Scientific Name. *Sciurus carolinensis*
The gray squirrel was first described by Gmelin (1788:148); five subspecies of *S. carolinensis* are recognized (Hall 1981)
Subspecies. *S. c. carolinensis, S. c. extimus, S. c. fuliginosus, S. c. hypophaeus*, and *S. c. pennsylvanicus*

DISTRIBUTION

Fox and gray squirrels occur sympatrically throughout most of eastern North America (Hall 1981). Fox squirrels range slightly farther west than gray squirrels; however, the range of the gray squirrel extends more northward (Figs. 12.1 and 12.2). Fox squirrels have been successfully introduced in several western cities in California, Oregon, and Washington (Fig. 12.1) (Flyger and Gates 1982). Gray squirrel introductions have occurred in California, Oregon, Washington, and Montana in the United States, and in British Columbia, Manitoba, and Ontario in Canada (Fig. 12.2) (Flyger and Gates 1982).

DESCRIPTION

Fox and gray squirrels are medium-sized tree squirrels in the order Rodentia and family Sciuridae. Neither species exhibits sexual dimorphism in color or size. Ranges for external measurements of adult fox squirrels are as follows: total length, 454–698 mm; tail length, 200–330 mm, and hind foot length, 51–82 mm (Hall and Kelson 1959); adult body mass ranges from 507 to 1361 g (Flyger and Gates 1982). Fox squirrels east of the Appalachian Mountains show a size cline in reverse of Bergmann's rule; body size is largest in northern Florida (Weigl et al. 1989, 1998). Body size in fox squirrels west of the Appalachians follows Bergmann's rule (Weigl et al. 1989). An east-to-west size cline also exists, with smaller fox squirrels in the western portion of the range (Weigl et al. 1998). Ranges for external measurements of adult gray squirrels are as follows: total length, 383–525 mm; tail length, 150–243 mm; and hind foot length, 53–76 mm (Flyger and Gates 1982); adult body mass ranges from 338 to 750 g (Barkalow and Shorten 1973). In contrast to fox squirrels, gray squirrel body mass follows Bergmann's rule throughout its range (Barnett 1977).

Pelage. Fox squirrel pelage is highly variable, locally and regionally. The unusual polymorphism largely accounts for it subdivision into numerous subspecies (Hall 1981). Fox squirrel subspecies can be divided into two distinctive but intergrading coloration groups (Weigl et al. 1998). *S. n. vulpinus* (easternmost), *S. n. cinereus, S. n. niger, S. n. shermani, S. n. avicennia*, and *S. n. bachmani* are characterized by silver, gray, agouti, and melanistic forms with tan, gold, or reddish venter. Animals usually possess black head markings, white or gray noses, and white ears and feet. The other group consists of *S. n. vulpinus* (western), *S. n. subauratus, S. n. ludovicianus*, and *S. n. limitis* and is characterized by a distinctly reddish, orange, or tan agouti coloration, a grizzled or black nose, and no white markings on the head or feet. Melanism is common in the southern portions of the range (Lowery 1974; Kiltie 1989). However, Turner and Laerm (1993) concluded that pelage characteristics are too varied and subjective to permit consistent determination of subspecies in the southeastern portions of the range.

Gray squirrel coloration is less variable than in fox squirrels. The dorsum is grizzled dark to light gray, usually with buffy underfur (Hall and Kelson 1959). Brown or cinnamon patches may be present on the flanks behind the forelegs and on the cheeks, and sometimes on the dorsal surface of the feet (Flyger and Gates 1982). Ears are buff to gray to white (behind); chin, throat, and venter are white (Flyger and Gates 1982). Melanistic gray squirrels are common in the northern portion of the range. Albinism is rare, although white squirrels are found locally in Olney, Illinois, and Marionville, Missouri (Koprowski 1994b).

Skull and Dentition. Fox and gray squirrel skulls (Fig. 12.3) are short, with broad, expanded zygomata (Koprowski 1994a, 1994b). The braincase is strongly depressed posteriorly and lacking prominent ridges; palate is broad and square posteriorly, and terminates immediately behind toothrows (Hall 1981). The rostrum is laterally compressed, the frontal area is flattened, and auditory bullae are moderately inflated (Koprowski 1994a, 1994b). Dental formulas of the fox and gray squirrel are I 1/1, C 0/0, P 1/1, M 3/3 and I 1/1, C 0/0, P 2/1, M 3/3, respectively (Flyger and Gates 1982). The anterior premolar (P3) in gray squirrels is small and peglike, and may be absent (Hall 1981). Incisor growth is continuous; molariform cheek teeth are brachyodont and bunodont (Flyger and Gates 1982; Koprowski 1994a, 1994b).

PHYSIOLOGY

Molts. Fox and gray squirrels molt twice annually, once in the spring and again in the fall. In March, the spring molt begins on the head and proceeds posteriorly; the fall molt begins in September on the flanks and progresses anteriorly (Flyger and Gates 1982). The annual tail molt occurs in July and August (Flyger and Gates 1982). It begins at the base and continues toward the tip (Gurnell 1987). Lactating females molt after young are weaned (Baumgartner 1943a). Juveniles molt into adult pelage at approximately 3 months of age (Gurnell 1987).

Metabolism. Fox and gray squirrels are nonhibernating homeotherms, which requires them to maintain a relatively stable core body temperature under various environmental conditions. Body temperature of tree squirrels normally varies from 37°C to 40°C (Gurnell 1987); rectal temperatures of adult gray squirrels range from 37.5°C to 39.9°C (Bolls and Perfect 1972). Nonshivering thermogenesis accounts for

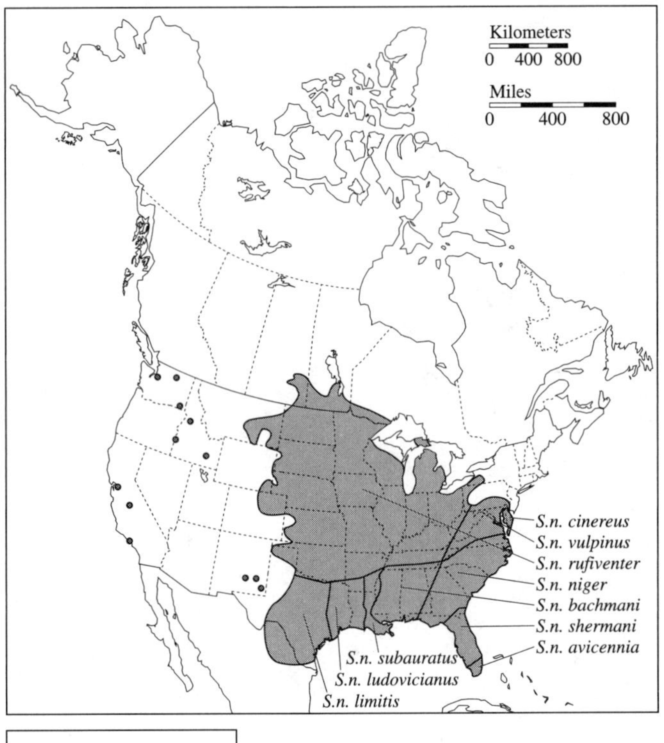

FIGURE 12.1. Distribution of the fox squirrel (*Sciurus niger*): *S. n. avicennia, S. n. bachmani, S. n. cinereus, S. n. limitis, S. n. ludovicianus, S. n. niger, S. n. rufiventer, S. n. shermani, S. n. subauratus,* and *S. n. vulpinus*. SOURCE: Adapted from Hall (1981), Flyger and Gates (1982), and Koprowski (1994a).

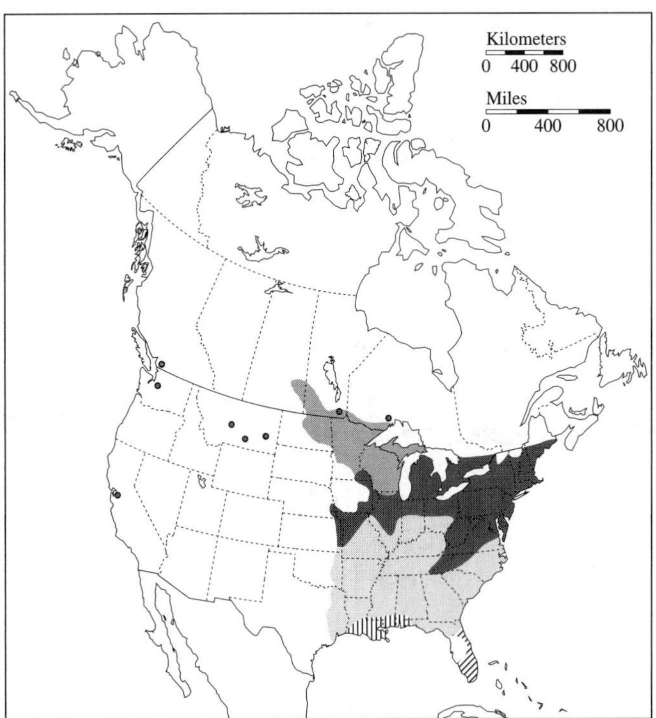

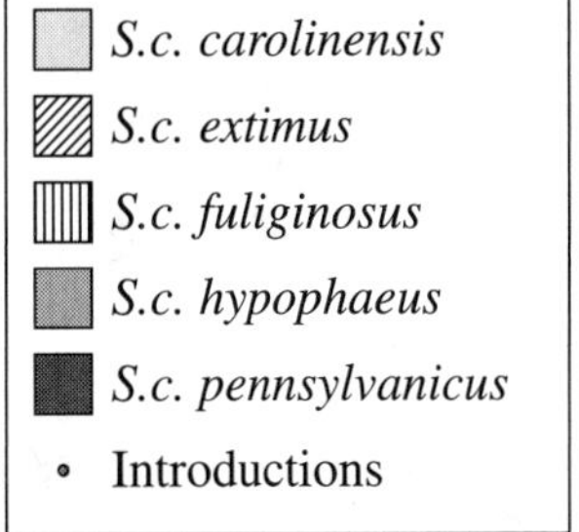

FIGURE 12.2. Distribution of the gray squirrel (*Sciurus carolinensis*): *S. c. carolinensis, S. c. extimus, S. c. fuliginosus, S. c. hypophaeus,* and *S. c. pennsylvanicus*. SOURCE: Adapted from Hall (1981), Flyger and Gates (1982), Koprowski (1994b).

~25% of the cold-induced heat production in winter gray squirrels, with the remainder (~75%) attributed to shivering. Shivering gray squirrels can generate heat at a rate 10.3 times greater than their predicted standard metabolic rate (Ducharme et al. 1989). Their ability to generate heat via shivering is among the highest reported for homeotherms. Positioning of the tail above the dorsum can increase the insulative value of the pelage (17.8% in fox squirrels) or act as a heat shield and reduce the uptake of heat from the environment (Muchlinski and Shump 1979). Countercurrent heat exchange near the base of the tail may also serve a thermoregulatory function (Muchlinski and Shump 1979). During winter, melanistic gray squirrels have a lower basal metabolic rate and lower energy cost of thermoregulation. Black morphs at temperatures of less than −10°C had 18% lower heat loss, 20% lower basal metabolic rates, and a nonshivering thermogenesis capacity 11% higher than gray morphs (Innes and Lavigne 1979; Ducharme et al. 1989). No differences in basal metabolic rate, energy cost of thermoregulation, or thermal conductance are exhibited between color morphs in summer pelage (Innes and Lavigne 1979). Pelage color does not appear to be the causal factor in improved thermoregulation, only a correlate to significant temperature-regulating adaptations or other physiological functions (Ducharme et al. 1989). Climatic correlates of melanism in fox squirrels are opposite those of the gray squirrel. The thermoenergetic advantage suggested as favoring melanistic gray squirrels in cold climates does not seem relevant to melanistic fox squirrels, which are more common in the relatively mild climate of the southeastern United States (Kiltie 1989).

Energy equilibrium (i.e., no net mass gain or loss) in adult fox squirrels in Michigan was maintained with a daily dietary level of 162 kcal of metabolizable energy/$W^{0.75}$, where W is weight in kg (Husband 1976). Adult gray squirrels in Virginia required 167 kcal of metabolizable energy/$W^{0.75}$ per day to maintain energy equilibrium (Ludwick et al. 1969). Food consumption in fox and gray squirrels peaks in late spring and autumn, with the highest levels occurring in September and October (Short and Duke 1971; Montgomery et al. 1975). Adult gray squirrels are able to endure several days of food deprivation under near-freezing conditions (Merson et al. 1978). Fox and gray squirrel body mass peaks in late fall following lipogenesis (Short and Duke 1971; Montgomery et al. 1975; Nixon et al. 1991). In Illinois, percentage body fat of fox squirrels was highest in February and lowest in August; carcass fat averaged 6.9% (range = 1.3–21.6%) for wild-caught animals that appeared to be in good condition (Havera 1977).

Vision and Hearing. Fox and gray squirrels have retinae with two anatomically distinct photoreceptors, which permit photopic and scotopic vision (Gouras 1964; Yolton 1975). Dichromatic vision in fox and gray squirrels allows for red–green discrimination (Jacobs 1974; MacDonald 1992; Yokoyama and Radlwimmer 1998). Fox squirrels hear tones ranging from 113 H to 49 kHz at 60 dB sound pressure level or less, with their best sensitivity of 1 dB occurring at 8 kHz (Jackson et al. 1997).

REPRODUCTION AND DEVELOPMENT

Anatomy. Female fox and gray squirrels have a duplex uterus characterized by two uteri, each with a cervix opening into the vagina. Nulliparous females have threadlike uteri 1.5 mm in diameter; in parous females, the uteri become significantly larger (Nixon and McClain

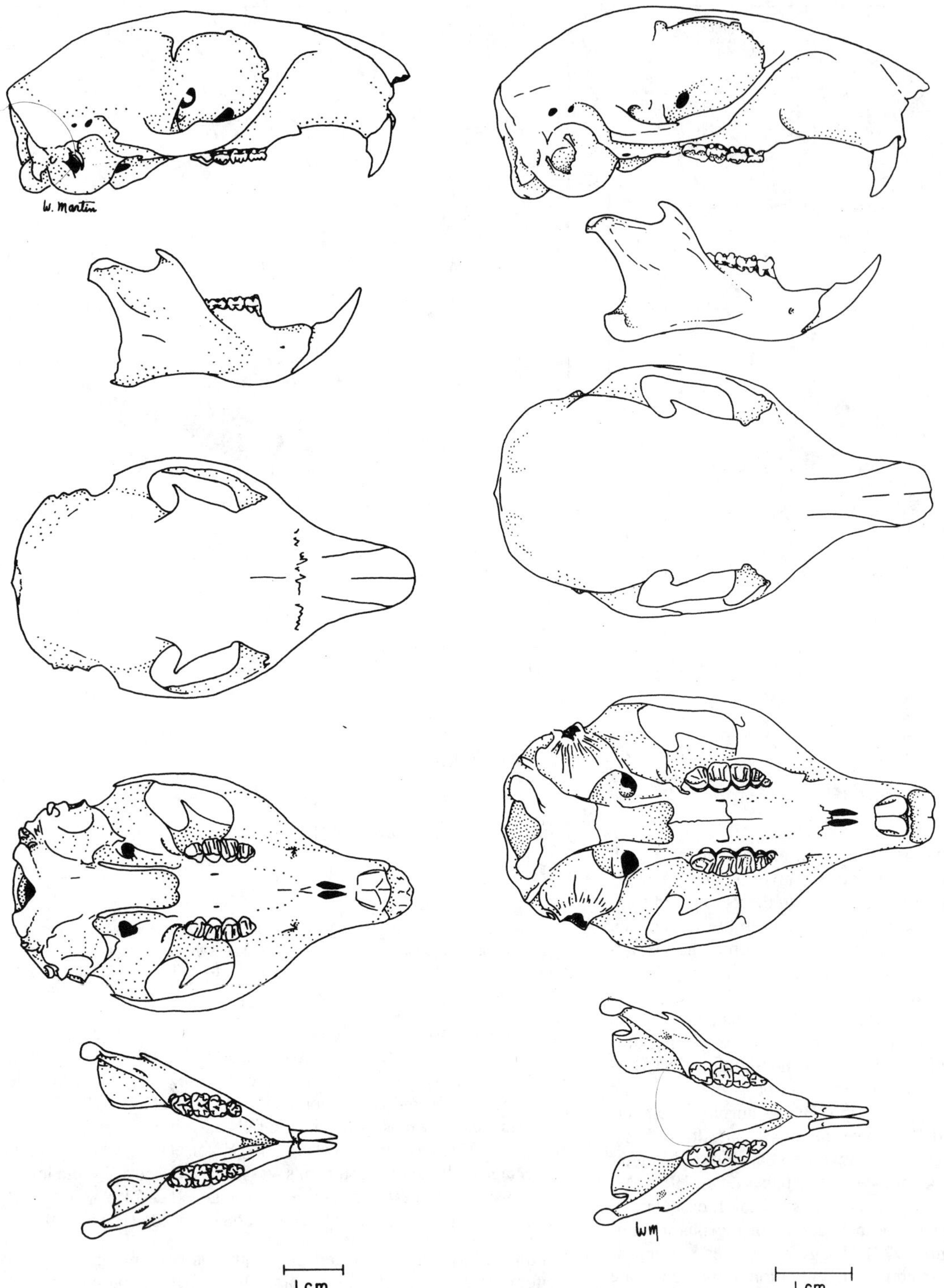

FIGURE 12.3. Skulls of (left) the fox squirrel (*Sciurus niger*) and (right) the gray squirrel (*S. carolinensis*), From top to bottom: lateral view of cranium, lateral view of mandible, dorsal view of cranium, ventral view of cranium, dorsal view of mandible.

1975; Flyger and Gates 1982). Females have eight mammae, one inquinal pair, one abdominal pair, and two pectoral pairs. In prepubertal females, the vulva is imperforate and the teats are small and pink; parous females have perforate vulvae and pigmented teats (Nixon and McClain 1975).

Reproductive organs of male fox and gray squirrels are similar in morphology (Mossman et al. 1955). Adult males undergo seasonal cycles in testicular development and in the activity of associated accessary glands, which include functional, quiescent, and redevelopment stages (Kirkpatrick and Hoffman 1960). Functional stages in glands correlate with winter and summer breeding peaks. Degeneration of glands and testes occurs in late summer and early autumn; regeneration occurs only in autumn (Mossman et al. 1955). Histological changes in the testes and in Cowper's and prostate glands from one stage to another have been documented by Kirkpatrick (1955) and Mossman et al. (1955). Zielinski (1996) found no significant correlation between testosterone level and reproductive condition in male gray squirrels. External genitalia only grossly reflect current physiological conditions, and therefore are a poor indicator of breeding condition (Kirkpatrick and Hoffman 1960). Baculum length ranges from 9.2 to 12.3 mm in gray squirrels (Burt 1960) and from 12.4 to 13.3 mm in fox squirrels (Wade and Gilbert 1942; Long and Frank 1968).

Breeding Age. Fox and gray squirrels rarely breed before 10 months of age (Allen 1942; Brown and Yeager 1945; Kirkpatrick 1955; Kirkpatrick and Hoffman 1960). Subadults born in spring usually enter estrus the following spring at 10–14 months of age. Subadults born in summer may breed for the first time the following summer (Allen 1942; Brown and Yeager 1945). Early maturation (6–8 months of age) and precocious breeding, however, have been reported in female fox and gray squirrels (Smith and Barkalow 1967; McCloskey and Vohs 1971; Nixon and McClain 1975). Reproductive longevity in both species may exceed 12 years in females and 6 years in males (Barkalow and Soots 1975; Koprowski et al. 1988).

Breeding Season. Fox and gray squirrels are dioestrus throughout their ranges, with variation in peaks in breeding activity influenced by latitude, environmental conditions, age, nutrition, and population density (Hoffman and Kirkpatrick 1959). Winter breeding in midwestern populations of fox and gray squirrels occurs in January–February, whereas summer breeding extends from May to July (Brown and Yeager 1945; Kirkpatrick and Hoffman 1960; Nixon and McClain 1975). Nupp (1992) reported peaks in gray squirrel breeding in Alabama during January and June. Fox squirrel breeding activity peaks in December–February and May–August in Florida (Moore 1957; Wooding 1997). In North Carolina, peaks are in December–February and July–August (Weigl et al. 1989). Although seasonal peaks occur, Nixon and McClain (1975) reported pregnant gray squirrels in all months except October, November, and December in Ohio. Layne (1998) reported gray squirrel births in southern Florida in all months except May, with peaks in January–February and July. In Indiana, Kirkpatrick and Hoffman (1960) found male gray squirrels in reproductive condition during all months except September; however, the occurrence was minimal in August and October.

Pregnancy Rate and Litter Size. Reproductive success in fox and gray squirrel populations is highly variable and dependent on demographic and environmental factors (Nixon and McClain 1969, 1975). Yearling females (10–14 months) usually produce only one litter in their first breeding year. The percentage of yearling female fox squirrels that breed ranged from 4% in Oklahoma (Chesemore 1975) to 11% in Illinois (Harnishfeger et al. 1978). The percentage of yearling gray squirrels breeding ranged from 56% in Ohio (Nixon and McClain 1975) to 64% in North Carolina (Smith 1967). The percentage of adult female fox squirrels that reproduce annually varied from 30% in Illinois (Harnishfeger et al. 1978) to 56% in Oklahoma (Chesemore 1975). In contrast, >90% of adult female gray squirrels reproduced annually in Ohio and North Carolina (Smith 1967; Nixon and McClain 1975). Although spring breeding was more prevalent in fox squirrel populations in Illinois, Iowa, Maryland, and North Carolina (McCloskey and Vohs 1971; Harnishfeger et al. 1978; Weigl et al. 1989; Larson 1990), gray squirrel reproduction was similar between spring and summer breeding periods in Ohio (Nixon and McClain 1975). Adult females (>14 months) may breed twice annually, once in the spring and again in summer. The number of adult females conceiving two litters per year is highly dependent on food availability. In Illinois and North Carolina, <3% of fox squirrels produced two litters per year (Harnishfeger et al. 1978; Weigl et al. 1989), although Allen (1942) suggested a higher occurrence in Michigan. The percentage of gray squirrels producing two litters per year averaged 27% during a 3-year period in Ohio (Nixon and McClain 1975), and was as high as 40% during years with abundant mast production in West Virginia (Uhlig 1955). Severe food shortage (e.g., mast failure) may inhibit normal reproduction in all age classes (Nixon and McClain 1969).

Fox squirrel litter size in midwestern populations averaged 2.1–3.0 young (McCloskey and Vohs 1971; Harnishfeger et al. 1978). In the southeast United States, litter size averaged 1.7–2.6 young (Larson 1990; Wooding 1997). Gray squirrel litter size averaged 2.3 young in Ohio (Nixon and McClain 1975) to 2.9 young in Alabama and North Carolina (Barkalow and Shorten 1973; Nupp 1992). Harnishfeger et al. (1978) and Nupp (1992) found no significant difference between spring and summer litter size in fox and gray squirrels.

Development of Young. Following a 44- to 45-day gestation period (Brown and Yeager 1945; Webley and Johnson 1983), fox and gray squirrel neonates are naked at birth except for vibrissae, and weigh 13–18 g (Brown and Yeager 1945; Koprowski 1994b). Hair is present on the dorsum of body and tail, and lower incisors erupt by 3 weeks of age; eyes open, upper incisors erupt, and hair is visible on the ventral surface of tail and body by week 5 (Brown and Yeager 1945; Shorten 1951; Uhlig 1955). Weaning occurs at 8–10 weeks of age in gray squirrels and 8–12 weeks in fox squirrels (Allen 1942; Flyger and Gates 1982). Gray squirrels reach adult body mass at 8–9 months of age (Shorten 1951; Horwich 1972). After 1 year, fox squirrels have reached approximately 92% of maximum body mass (Nixon et al. 1991). Sex ratios in fox and gray squirrel litters approximate 1:1 (Brown and Yeager 1945; Moore 1957; Smith 1967; Weigl et al. 1989).

ECOLOGY

Home Range. Home range estimates of both species vary considerably depending on geographic region, sex, age, population density, and method of determination. Fox squirrel home ranges are largest in the southeastern United States (Table 12.1). Adult male home ranges are generally larger than those of adult females (Don 1983) (Table 12.1). Juvenile gray squirrel (≤6 months) home ranges are smaller than those of adult males, but may be equal to or larger than those of adult females (Don 1983). Thompson (1978a) reported larger home ranges for adult gray squirrels in spring than in summer. Sheperd and Swihart (1995) found significantly smaller home ranges in the nongrowing season (i.e., plant growing season; 1 October–30 April) than in the growing season (1 May–30 September). Although most studies report seasonal changes in home ranges, it is generally not possible to remove the confounding effects of population age structure and breeding behavior (Don 1983). Home range area is also thought to be negatively correlated with population density in gray squirrels (Don 1983; Kenward 1985). However, this relation may not be independent of resource availability, which may be affected by squirrel density (Gurnell 1987). Sheperd and Swihart (1995) reported no significant relation between home range area and density in fox squirrels.

Activity and Movements. Fox and gray squirrels demonstrate a crepuscular, bimodal activity pattern in spring, summer, and fall and a unimodal activity peak of morning through midday in winter (Bakken 1959; Thompson 1977a). Gray squirrels generally are more active under lower light conditions than are fox squirrels (Brown and Yeager 1945; Packard 1956). Derge and Yahner (2000) used number of squirrel sightings to measure differences between morning and midday activity in Pennsylvania. They found no significant differences in fox squirrel activity between morning and midday surveys for all seasons; gray squirrel activity was significantly greater in morning surveys. Normal activity patterns of squirrels may be altered or suppressed by weather conditions, including high winds, temperature extremes, and heavy precipitation (Brown and Yeager 1945; Bakken 1959; Doebel and McGinnes 1974). The unimodal activity in winter may reflect an avoidance of the colder temperatures associated with early morning and late afternoon. Similarly, bimodal activity in summer allows for decreased activity during the warmest portion of the day.

Fox and gray squirrel activity levels also vary seasonally. Peak activity (on the basis of number of squirrels observed per unit time) occurred in September in Indiana (Allen 1954) and in October in Wisconsin (Bakken 1952). Increased levels of activity during early fall may correlate with availability of mast and foraging and caching activities. Nonetheless, Thompson (1977a) and Gurnell (1987) cautioned

TABLE 12.1. Home range estimates for fox squirrels (*Sciurus niger*) and gray squirrels (*S. carolinensis*) in the United States

State	Home Range Estimate (ha)	Sex (n)	Method of Determination[a]	Reference
Fox squirrel				
Nebraska	3.6	Female (17)	Bivariate ellipse	Adams 1976
	7.6	Male (20)		
Indiana	3.7	Female (19)	MCP	Sheperd and Swihart 1995
	3.0	Male (22)		
Maryland	9.4	Female (3)	Adaptive kernel	Paglione 1996
	22.1	Male (4)		
Virginia	2.5	Female (5)	MMA	Larson 1990
	6.1	Male (4)		
North Carolina	16.2	Female (10)	MCP	Weigl et al. 1989
	22.8	Male (5)		
South Carolina	19.3	Female (4)	MCP	Edwards 1986
	31.6	Male (9)		
Alabama	11.6	Female (9)	MCP	Powers 1993
	38.0	Male (14)		
Florida	33.0	Female (12)	MCP	Wooding 1997
	79.5	Male (17)		
Gray squirrel				
Maryland	0.5	Female (18)	MCP	Flyger 1960
	0.8	Male (20)		
Virginia	0.4	Female	Center of activity	Doebel and McGinnes 1974
	0.5	Male		
North Carolina	1.3	Female	MCP	Cordes and Barkalow 1972
	1.7	Male		
Alabama	1.3	Female	MCP	Fischer 1989
	5.4	Male		

[a]MCP, Minimum convex polygon; MMA, modified minimum-area method.

that increased observations of squirrels in fall may result from changes in population size and not from changes in individual activity levels. Hougart and Flyger (1981) measured time outside of nest or den of radiocollared gray squirrels in Maryland and found activity was greatest in July and least in February. Thompson (1977a) reported gray squirrel activity was significantly lower in winter (22 November–21 March) than in other seasons; females were significantly more active than males during spring and summer. A period of reduced activity during summer has been observed in gray and fox squirrels (Baker 1946; Weigl et al. 1989; Bendel and Therres 1994). Possible factors contributing to lower activity include increased temperatures, absence of breeding activity, reduced availability of mast, and lower energetic demands (Gurnell 1987; Weigl et al. 1989).

Dispersal in fox and gray squirrels occurs from spring through fall (Baumgartner 1943b; Brown and Yeager 1945; Allen 1954; Jordan 1971a; Wooding 1997). Subadults (8–11 months of age) constitute the majority of dispersers; males have a greater likelihood of long-distance movements (Mosby 1969; Cordes and Barkalow 1972). Although dispersal distance rarely exceeds 3.2 km in gray squirrels (Cordes and Barkalow 1972), fox squirrels have dispersed up to 64 km (Allen 1943). Koprowski (1996) found gray squirrels more likely to remain in their natal area (females 37%; males 6%), whereas no fox squirrels showed natal philopatry. Thompson (1978b) suggested that dispersal is the primary factor regulating gray squirrel populations, and is mediated by food availability.

Extensive emigrations of gray squirrels were reported in the 1700s and 1800s in North America (Seton 1920). These mass movements involved millions of squirrels and usually occurred in the fall. Squirrels reportedly swam large rivers, crossed open, treeless areas, and devastated crops. Mass movements of fox and gray squirrels were occasionally recorded in the 1900s, but on a much reduced scale (Schorger 1949; Flyger 1969).

Density and Population Fluctuations. Fox and gray squirrel densities vary considerably rangewide (Table 12.2). Fox squirrel densities throughout the southeast United States are substantially lower than those in the midwest. Gray squirrel populations, although highly variable, do not exhibit similar regional differences in density. Gray squirrel densities are generally higher than corresponding fox squirrel densities throughout the range.

Substantial regional differences in fox and gray squirrel densities are attributed to habitat quality. Brown and Yeager (1945) reported fox squirrel densities of 210–510/km^2 in high-quality habitats and 5/km^2 in low-quality habitats in Illinois. Fischer (1989) found gray squirrel densities in Alabama ranged from 18 to 169/km^2 in low- to high-quality habitats, respectively. Wooding (1997) suggested that differences in densities between southeastern and midwestern fox squirrel populations result from habitat quality, specifically the increased availability of hard mast. Differences in densities can also be attributed, in part, to variability in estimation procedures (e.g., time–area counts versus mark–recapture).

TABLE 12.2. Density estimates for fox squirrels (*Sciurus niger*) and gray squirrels (*S. carolinensis*) in the United States

State	Density estimate (per km^2)	Reference
Fox squirrel		
Iowa	12–37	Hicks 1949
Illinois	130–510	Brown and Yeager 1945
	210–220	Hansen et al. 1986
Maryland	39	Larson 1990
North Carolina	0–35	Weigl et al. 1989
South Carolina	75	Lee 1999
Georgia	20	Hilliard 1979
	15–18	Tappe 1991
	12–19	Conner et al. 1999
Florida	38	Moore 1957
	12	Kantola 1986
	7–12	Wooding 1997
Gray squirrel		
Illinois	368	Brown and Yeager 1945
West Virginia	52–348	Uhlig 1956
Maryland	297–964	Flyger 1959
Virginia	1110–1408	Mosby 1969
North Carolina	141	Barkalow et al. 1970
Alabama	18–169	Fischer 1989

Densities of both species fluctuate irregularly. In a 6-year study of gray squirrels, Uhlig (1956) found density ranges of 82–366, 52–111, 64–208, and 106–250/km^2 on four respective study areas in West Virginia. Fluctuations in densities were attributed to changes in food supply, weather conditions, predation, and disease. Fox squirrel densities in North Carolina fluctuated between 0 and 35/km^2 over an 8-year period as a result of variations in litter size, frequency, and possibly litter survival (Weigl et al. 1989). Reproductive parameters such as litter size, percentage breeding, age of sexual maturity, and recruitment of juveniles into the population are closely linked to food conditions (Smith 1967; Nixon and McClain 1969; Weigl et al. 1989). A poor mast crop in the fall normally results in lower productivity the succeeding year, whereas good mast years result in an increase in the number of young produced (Smith and Barkalow 1967; Barkalow et al. 1970; Nixon et al. 1975). In Ohio, Nixon and McClain (1969) reported an 83% decrease in fox and gray squirrel density following a severe spring frost. Available food stocks were substantially reduced and the population exhibited lowered body weights and reproduction and increased dispersal/mortality. Mosby (1969), however, found that mast conditions had no influence on gray squirrel densities during a 6-year period in Virginia.

Habitat Selection. Fox squirrels and gray squirrels are syntopic over a large extent of their range (Flyger and Gates 1982). Although both species possess similar ecological requirements, slight differences exist in habitat preference (Flyger and Gates 1982; Gurnell 1987). Fox squirrels select mature, upland forest with sparse understory (Nixon and Hansen 1987). In contrast, gray squirrels inhabit extensive, mature hardwood forests, often with dense undergrowth (Flyger and Gates 1982). Both species, however, use a variety of habitat types throughout their range.

Midwestern fox and gray squirrels occur primarily in habitats dominated by oaks (*Quercus* spp.). The presence and abundance of either species vary with the stand density, stand size, and past land-use practices (Bakken 1952). In Ohio, fox squirrels inhabit small (2–121 ha) farm woodlots, larger blocks of timber along the forest edge, and open ridgetops in the forested southeastern region of the state (Baumgartner 1943b). Oak–hickory (*Carya* spp.) forest types are preferred, but the American beech (*Fagus grandifolia*)–maple (*Acer* spp.) type also is occupied. Fox squirrels predominate if a high percentage of oak is present; gray squirrels are more abundant where oak is a lesser component (Baumgartner 1943b). Gray squirrels in Illinois select mature forested stands with closed canopies and well-developed understory structure above and below 1.5 m from ground level (Nixon et al. 1978). Activities that reduce overstory and understory densities and increase the number of early successional species, such as extensive timber harvesting, heavy grazing, and repeated burning, favor fox squirrels over gray squirrels (Allen 1954; Nixon et al. 1978). Brown and Batzli (1984), however, tested the hypothesis that gray squirrels prefer areas with denser undergrowth than do fox squirrels (L. G. Brown and Yeager 1945; Nixon et al. 1978). They found that the density of understory trees was not important in determining the distributions of fox and gray squirrels in Illinois. They suggested that the size of the forest patch may be an important factor in gray squirrel distributions. Understory cover may simply be correlated with patch size and the likelihood that larger patches are seldom grazed. Also in Illinois, Nixon and Hansen (1987) reported gray squirrels were most common in large (>40 ha) forested tracts along rivers and streams where 20% of area is forested, and where a diverse woody understory was present. Fox squirrels preferred upland forests with sparse, woody understories that offered abundant "edge" between forest and field. Fox and gray squirrel density was correlated with the abundance of winter-storable food. Production and availability of mast within habitats also was the primary factor determining seasonal and annual densities of gray squirrels in Wisconsin; gray squirrel densities were highest in areas of mature, productive northern red oak (*Q. rubra*) (Riege 1991).

In the northeastern portion of the range, extensive overlap and habitat partitioning occurs. In Pennsylvania, Derge and Yahner (2000) found habitat use by sympatric fox squirrels and gray squirrels differed primarily by proximity to forest–farmland edge habitats. Fox squirrels were observed most frequently in open habitat with few shrubs along forest edges; gray squirrels were more common farther from edges in areas with fewer understory trees (mean distance to edge: gray = 99.4 m; fox = 65.3 m). Fox squirrels were commonly seen in fields or fence rows, whereas gray squirrels were not observed beyond the forest edge. Gray squirrels in West Virginia were observed more often than expected in mixed hardwoods and white oak (*Q. alba*)–red oak–hickory forest types (Society of American Foresters 1967) and less often in white pine (*Pinus strobus*), oak–pine (*Pinus* spp.), and yellow pine types (Sanderson et al. 1976).

The Delmarva fox squirrel (*S. n. cinereus*) is listed as endangered by the U.S. Fish and Wildlife Service and occurs in four counties in eastern Maryland, representing <10% of its historical range, which formerly included Pennsylvania, Delaware, Virginia, and possibly New Jersey (U.S. Fish and Wildlife Service 1993). Delmarva fox squirrels prefer small stands of large timber and open understory, largely in agricultural areas (Taylor 1973). Dueser et al. (1988) found Delmarva fox squirrels occurred at sites with a higher percentage of trees >30 cm diameter at breast height, a lower percentage of shrub ground cover, and a slightly lower understory density than where fox squirrels were absent. In a reintroduced population in the Chincoteague National Wildlife Refuge in Virginia, Delmarva fox squirrels were most abundant in areas with larger overstory trees, higher densities of soft mast-producing hardwoods, and lower densities of pines (Larson 1990).

In the southeast United States, fox and gray squirrels also occupy a wide range of habitats and syntopy is common. Mature and open upland–pine and pine–hardwood associations are considered preferred habitats of fox squirrels (Edwards et al. 1989; Loeb and Lennartz 1989; Weigl et al. 1989). Moreover, southeastern fox squirrels differ substantially from midwestern subspecies in their ecological requirements; they select more pine-dominated habitats compared to deciduous habitats selected by fox squirrels in the Midwest (Weigl et al. 1989). In North Carolina, fox squirrels prefer open, mature, pine-oak habitats, especially longleaf pine (*P. palustris*)–turkey oak (*Q. laevis*), and the ecotones between pine and other vegetation types. Fox squirrels in the coastal plain of South Carolina selected hardwood, mixed pine–hardwood, pine with a hardwood midstory, and ecotone habitats (Edwards et al. 1989). Lee (1999) found fox squirrels on a coastal island in South Carolina preferred a mixture of hardwoods and pines, an open understory, and an open or moderately open canopy. Stands where >80% of overstory species composition was pine were used less than expected. In Georgia, Edwards (1995) reported greater stand use at higher slope positions by fox squirrels, with less understory, and greater herbaceous ground cover. Sympatric gray squirrel stand use was higher where greater numbers of overstory and midstory hardwood stems existed. Conner et al. (1999) found fox squirrel capture sites in Georgia had denser ground cover and higher pine and lower hardwood basal areas than gray squirrel capture sites. Moreover, herbaceous ground cover dominated at fox squirrel capture sites, whereas woody ground cover was dominant at gray squirrel sites. In Florida, Moore (1957) reported Sherman fox squirrels (*S. n. shermani*) most common in longleaf pine–turkey oak habitat with sparse herbaceous vegetation. Kantola and Humphrey (1990) suggested that the highest quality fox squirrel habitat was along the edge of longleaf pine savanna and live oak (*Q. virginiana*) forest associations. Big Cypress fox squirrels (*S. n. avicennia*) in southern Florida are most common in bald cypress (*Taxodium distichum*)–slash pine (*P. elliottii*) savanna or at the interface of these habitats (Williams and Humphrey 1979). Fox and gray squirrels are most abundant in bottomland habitats (bottomland hardwood, stream-side management zone) in Louisiana and Mississippi (Heuer and Perry 1976; Warren and Hurst 1980). In contrast, Fischer and Holler (1991) found gray squirrel abundance in Alabama did not differ between hardwood and mixed pine–hardwood habitats, but that squirrel abundance was lower in even-aged pine habitats. Gray squirrels, however, also selectively used hardwood "stringers" within mixed pine–hardwood and pine habitats in Alabama. In Mississippi, Ross (1996) found gray squirrel abundance positively correlated with stand age, hard mast abundance, presence of woody plants, and debris in the understory. Fox squirrel abundance was negatively correlated with percentage grass present in the understory and positively correlated with horizontal understory visibility, percentage hardwood, and basal area. Fox squirrels were most common in stands approaching 100 years old, having a hardwood species composition near 60% and a mean basal area of approximately 11.2 m^2.

Nest Selection. Fox and gray squirrels construct leaf nests and use tree cavities for protection from predators and inclement weather, escape, and rearing young. Leaf nests occur in a variety of tree species including bald cypress, pine, eastern hemlock (*Tsuga canadensis*), ash (*Fraxinus* spp.), aspen (*Populus* spp.), basswood (*Tilia americana*), American beech, elm (*Ulmus* spp.), gum (*Nyssa* spp.), hickory, maple, oak, sassafras (*Sassafras albidum*), sycamore (*Platanus occidentalis*), and willow (*Salix* spp.). In Michigan, fox squirrels constructed leaf nests in black oak (*Q. velutina*) and white pine more often than expected based on tree species availability (Allen 1942). Gray and fox squirrels in Indiana built leaf nests most often in white oak, American beech, scarlet oak (*Q. coccinea*), and sassafras (Allen 1954). In West Virginia, Sanderson et al. (1976) found gray squirrel leaf nests in 26 of 35 tree species recorded on the study area; 54% of nests were in the 3 most abundant species of oak. Gray squirrels in Alabama were located via radiotelemetry 84% of the time in leaf nests in hardwood trees: laurel oak (*Q. laurifolia*), water oak (*Q. nigra*), southern red oak (*Q. falcata*), and black gum (*N. sylvatica*) (Nupp 1992). In South Carolina, Edwards et al. (1989) reported fox squirrel leaf nests in oaks (43%), black gum (35%), pine (15%), and cypress (7%). Fox squirrels selected oaks and black gum more often than expected based on tree species availability. In Florida, Kantola (1986) found fox squirrel leaf nests in turkey oak (69%), longleaf pine (18%), live oak (5%), post oak (*Q. stellata*) (4%), laurel oak (3%), and slash pine (1%). Edwards and Guynn (1995) reported that fox squirrels in Georgia occupied leaf nests in pines more often than expected and in dogwood (*Cornus alternifolia*), hickory, winged elm (*U. alata*), and sweetgum (*Liquidambar styraciflua*) less than expected. Gray squirrels used leaf nests in oaks more often than expected and in dogwood less than expected according to availability. Other species of tree were used in proportion to their occurrence.

Nixon and Hansen (1987) suggested that leaf nests are usually built in or near favored food trees in late summer. Although this may be important during certain seasons, structural characteristics of the nest tree and its surrounding site conditions may be equally important. Fox squirrels selected larger diameter trees than expected in which to construct leaf nests in Georgia ($\overline{X} = 39.1$ cm diameter at breast height [dbh]; Hilliard 1979) and South Carolina ($\overline{X} = 40.4$ cm; Edwards et al. 1989). Gray and fox squirrels in Illinois also selected larger dbh (30–51+ cm) trees for leaf nests (Nixon and Hansen 1987). In Georgia, fox squirrel leaf nests were located higher, and in trees with larger dbh, than were gray squirrel leaf nests (Edwards and Guynn 1995). Vines such as grape (*Vitis* spp.) are an important component in nest site selection in midwestern fox and gray squirrels (Allen 1954; Sanderson et al. 1980). Leaf nests supported by vines are more resistant to damage (Baumgartner 1943b; Nixon and Donohoe 1979); vines may also serve as an early-warning system against potential intruders (J. Edwards, pers. obs.). In Georgia, however, Edwards and Guynn (1995) found fox squirrels rarely incorporated vines in their leaf nests (8% of nests), whereas 59% of gray squirrel nests contained vines. Hilliard (1979) reported 22% of fox squirrel leaf nests in Georgia contained vines.

Natural cavities are important to fox and gray squirrels over much of their range. Cavities preferred by both species have entrance holes 5–20 cm in diameter, are 30–51 cm deep, are situated in the trunk of the tree or in a large limb, and offer protection from wet and cold weather (Nixon and Hansen 1987). Fox and gray squirrels prefer to use cavities in live trees (Sanderson et al. 1976; Robb et al. 1996), but will readily use cavities in snags. Natural cavities used by gray squirrels in West Virginia averaged 7.9 m above ground level. They were most often located in the bole (80%) and less often in the base (10%) or in limbs (10%) (Sanderson et al. 1976). Fox squirrels in South Carolina selected cavities with an average height of 10.3 m and 45.6 cm dbh (Edwards et al. 1989). Edwards and Guynn (1995) found no differences in height and dbh of natural and nest box cavities used by sympatric fox and gray squirrels in Georgia. Use of natural or nest box cavities by gray squirrels in Louisiana was negatively correlated with age of the tree, dbh, height, and angle of cavity entrance from the horizontal. Use was positively correlated with distance to the nearest cavity tree and number of trees per 0.02 ha surrounding the cavity tree. Use of cavities by fox squirrels was positively correlated to age of cavity tree and cavity height, and negatively correlated with number of cavities per tree, size of cavity entrance, and distance to the nearest cavity tree (McComb and Noble 1981).

Both species use cavities in a variety of tree species throughout the range, including ash, aspen, basswood, beech, cypress, elm, hickory, maple, oak, pine, and sycamore (Allen 1942, 1954; Moore 1957; Hilliard 1979; Kantola 1986; Nixon and Hansen 1987). Tree species such as American beech, ash, basswood, black gum, elm, laurel oak, sugar maple, sweetgum, sycamore, and water oak are considered cavity prone, whereas hickory and southern yellow pines such as loblolly (*P. taeda*), longleaf, shortleaf (*P. echinata*), and slash are unlikely to contain cavities (Goodrum 1937; Allen 1954; McComb et al. 1986; Nixon and Hansen 1987).

Fox and gray squirrel use of leaf nests and tree cavities varies seasonally. Use of natural and artificial cavities is greatest for both species during winter and spring (Fowler and Dimmick 1983; Christisen 1985; Ivey and Frampton 1987; Nixon and Hansen 1987; Nupp 1992; Edwards and Guynn 1995; Layne 1998). Increased use of cavities by squirrels during winter has been attributed to increased thermal protection from wind, rain, and extreme temperatures (Barkalow and Soots 1965; Burger 1969). In Illinois, the mean temperature in winter was 25.9°C warmer inside an occupied nest box than the ambient temperature (Havera 1979). Female gray and fox squirrel use of cavities during the winter/spring reproductive period also is important for the survival of nestlings (Barkalow and Soots 1965; Burger 1969; Weigl et al. 1989). Cavity use at this time provides protection from inclement weather and predators, while confining young within the nest. Fitzwater and Frank (1944) proposed that heat and nest parasites may lead to increased use of leaf nests for raising litters during summer.

A lack of natural cavities can be a limiting resource in fox and gray squirrel populations. Barkalow and Soots (1965) in North Carolina, Burger (1969) in Maryland, and Nixon et al. (1984) in Illinois demonstrated that the addition of artificial cavities increased carrying capacity through increased survival of young and adults. Hilliard (1979) and Weigl et al. (1989) also suggested that the absence of suitable cavity trees may be a critical factor in litter survival and subsequent recruitment of fox squirrels in Georgia and North Carolina. However, others have found no evidence that an absence of cavities was limiting fox squirrels in Maryland, Florida, or Georgia (Lustig and Flyger 1975; Kantola 1986; Edwards and Guynn 1995). Moreover, only minimal use of cavities has been reported for fox squirrels in the southeast United States (Moore 1957; Kantola 1986; Edwards et al. 1989; Edwards and Guynn 1995).

FEEDING HABITS

Much of the information on fox and gray squirrel feeding habits has been gained through observations (i.e., qualitative data). Few studies have determined squirrel diets quantitatively via analyses of stomach contents (Baumgartner 1939; Nixon et al. 1968; Korschgen 1981). The difficulty in collecting adequate samples during closed hunting seasons and in identifying finely masticated and fragmented plant material makes this approach arduous and time-consuming (Korschgen 1981).

Extensive information on feeding habits is only available from populations of fox and gray squirrels in the Midwest. In Ohio, Nixon et al. (1968) reported the principal foods of fox and gray squirrels (Table 12.3). Hickory was consumed most often during August–October and least in late spring and early summer. Beechnuts were an

TABLE 12.3. Principal food items found in the diets of fox and gray squirrels in Ohio and Illinois

Ohio (Annual)	Illinois			
	Spring	Summer	Fall	Winter
Hickory nuts	Oak buds, flowers, acorns	Hickory nuts	Acorns	Acorns
Beechnuts	Black walnut flowers, nuts	Acorns	Hickory nuts	Black walnuts
Acorns	Elm buds, flowers, seeds	Black walnuts	Black walnuts	Osage orange seeds
Fungi	Hickory flowers, nuts	Mulberry (*Morus rubra*)	Osage orange seeds (*Maclura pomifera*)	Corn grain
Black walnuts (*Juglans nigra*)	Maple buds, flowers, seeds	Grape	Grape	Hickory nuts
Green vegetation	Fungi	Fungi	Corn grain	Elm buds
Yellow buckeye nuts (*Aesculus octandra*)		Maple seeds	Fungi	Soybeans
Yellow-poplar samaras (*Liriodendron tulipifera*)		Corn grain	Maple seed	
Flowering dogwood drupes (*C. florida*)		Hackberry seeds (*Celtis occidentalis*)	Dogwood drupes	
Ironwood nuts (*Carpinus caroliniana*)				
Hop hornbeam nuts (*Ostrya virginiana*)				

SOURCE: Ohio, data from Nixon et al. (1968). Illinois, data from Nixon and Hansen (1987).

important food in fall through late winter. Acorns were eaten throughout the year, with the greatest amounts from August to February. Hickory nuts, beechnuts, and acorns occurred in 76% of stomach samples and made up 67% of the total volume of foods eaten (winter, 62%; spring, 26%; summer, 53%; fall, 92% of total observations). Other principal foods were consumed in lesser amounts and on a more seasonal basis. Many additional seasonal food items were found, which constituted <1.0% of the annual diet. Baumgartner (1939) also reported hickory nuts, beechnuts, acorns, black walnuts, corn, and possibly buckeye nuts as important foods of fox squirrels throughout the year in Ohio. Nixon and Hansen (1987) reported the seasonally important foods of fox and gray squirrels in Illinois (Table 12.3). Brown and Yeager (1945) also reported hickory, oak, walnut, elm, mulberry, and corn as important foods of gray and fox squirrels in Illinois on the basis of feeding observations.

In Missouri, Korschgen (1981) found annual diets of fox squirrels included 109 plant parts, 18 of which were considered principal items and accounted for 82% of all foods eaten (Table 12.4). Gray squirrels consumed 97 plant and 18 animal foods, 18 of which were principal and made up 87% of their diet (Table 12.4). Korschgen (1981) considered fox and gray squirrel feeding habits similar, with differences reflecting habitat use and foraging behavior. Fox squirrel use of open forests, forest edge, woodlots, and fencerows allowed them to supplement hickory and oak (52% of diet) with corn, osage orange, white elm, and wheat. Gray squirrel use of dense and contiguous forests was reflected in their reliance on hickory and oak (73% of diet).

Additional observations of feeding in western populations include fox squirrels in Kansas eating cedar (*Juniper virginiana*) berries, honey locust (*Gleditsia triacanthos*) pods, black walnuts, wild gourd (*Cucurbita foetidissima*) seeds, hackberry seeds, American elm buds and seeds, and Russian olive (*Elaeagnus angustifolia*) fruits (Bugbee and Riegel 1945). Gray squirrels fed on thorn apple (*Crataegus mollis*) fruit in Ohio (Dambach 1942). Fox squirrels in Illinois ate pith from terminal twigs of Ohio buckeye (*A. glabra*) trees (Havera et al. 1976), and gray squirrels were predators on a nestling house wren (*Troglodytes aedon*) in Kansas (Pitts et al. 1996). In Michigan, Reichard (1976) found fox squirrels during late winter and spring consumed red maple (*A. rubrum*) buds and seeds, silver maple buds and seeds, sugar maple (*A. saccharum*) flowers and buds, bur oak (*Q. macrocarpo*) flowers, willow catkins, cottonwood (*Populus deltoides*) catkins, hackberry flowers, and beech flowers.

No quantitative feeding habits studies have been conducted on eastern populations of fox squirrels or gray squirrels. Feeding observations in both species, however, suggest diverse diets and seasonally important food items. In Kentucky, Barber (1954) reported both species consumed 54 different plant species throughout the year. Hickory nuts, acorns, beechnuts, and black walnuts, however, supplied 74% of the

TABLE 12.4. Seasonal diets of fox and gray squirrels in Missouri

Spring	Summer	Fall	Winter
Fox squirrel			
Corn grain	Red mulberry fruit	Shagbark hickory nuts	Corn grain
Black walnut flowers, buds	Black walnuts	White oak acorns	Osage orange seeds
White elm buds, flowers, seeds (*U. americana*)	Shagbark hickory nuts	Black walnuts	White oak acorns
Post oak acorns	Wheat grain	Black oak acorns	Black walnuts
Shagbark hickory flowers, nuts (*C. ovata*)	Apple fruit	Osage orange seeds	Black oak acorns
	Pecan nuts (*C. illinoensis*)	Pin oak acorns (*Q. palustris*)	Red oak acorns
			Post oak acorns
Gray squirrel			
Black oak acorns	Red mulberry fruit	Shagbark hickory nuts	White oak acorns
Black walnuts	Shagbark hickory nuts	Black oak acorns	Black oak acorns
White oak flowers, acorns	Black walnuts	White oak acorns	Post oak acorns
Shagbark hickory flowers, nuts	Apple fruit	Pin oak acorns	Sorghum grain
Silver maple seeds (*A. saccharinum*)	Shellbark hickory nuts (*C. laciniosa*)	Black walnuts	Black walnuts
	Bitternut hickory nuts (*C. cordiformis*)		

SOURCE: Data from Korschgen (1981).

annual diet. Other seasonally important food items included buds and other vegetative parts of maple, oak, yellow-poplar, and hickory during February–May; yellow-poplar seeds in August and September; mulberry fruit in June; and dogwood berries in January. No differences in food preferences were observed between gray and fox squirrels.

Diets of gray squirrels in Virginia predominantly (85%) comprised oak buds, flowers, and acorns; hickory buds, flowers, and nuts; elm buds and flowers; sugar maple buds and flowers; apple buds and fruit; May apple (*Podophyllum peltatum*) fruits and seeds; dogwood drupes; fungi; grasses; and various forbs (Dudderar 1967). Other foods consumed by gray squirrels include cherry (*Prunus* spp.) and apple fruit in Massachusetts (Woods 1941); honey locust seeds, horse chestnuts (*A. hippocastanum*), and sycamore buds in New York (Nichols 1958); southern magnolia (*Magnolia grandiflora*) fruits in Florida (Blair 1935); various insects in New York and Florida (Layne and Woolfenden 1958; Shealer et al. 1999); an adult house finch (*Carpodacus mexicanus*) in Kentucky (Eason 1998); and an eastern chipmunk (*Tamias striatus*) in Massachusetts (Faccio 1996).

Fox squirrels in the southeast use the following foods in North Carolina and South Carolina: mast from turkey oak, southern red oak, blackjack oak (*Q. marylandica*), bluejack oak (*Q. incana*), post oak, live oak, pecan, pignut hickory (*C. glabra*), Allegheny chinkapin (*Castanea pumila*), sweetgum, gum, pine, cypress, and maples; pine buds and staminate cones; berries from holly (*Ilex* spp.), redbay (*Persea borbonia*), grape, and persimmon (*Diospyros virginiana*); various fruits (e.g., *Rubus* spp.); hypogeous and epigeous fungi; and insects (Weigl et al. 1989; Lee 1999). Longleaf pine seeds and turkey oak acorns are considered particularly important in the North Carolina and Florida coastal plain and sandhills regions (Moore 1957; Ha 1983; Kantola 1986; Weigl et al. 1989). Big Cypress fox squirrels inhabiting golf courses in Florida most frequently consumed cabbage palm (*Sabal palmetto*) fruits, seeds from cypress cones, figs (*Ficus* spp.), silk oak (*Grevillea robusta*) flowers, bottlebrush (*Callistemon* spp.), slash pine seed, and queen palm (*Cocos plumosa*) fruit (Jodice and Humphrey 1992). In Virginia, Delmarva fox squirrels shifted from a diet of primarily pine and oak mast in fall and early winter to one of hardwood buds and seeds in late winter and spring (Larson 1990). Important spring hardwood species included oak, red maple, sassafras, and black cherry (*Prunus serotina*). Other spring foods were wax myrtle (*Myrica* spp.) fruits, acorns, and mushrooms.

BEHAVIOR

Communication. Fox and gray squirrel intraspecific communication occurs through a variety of visual, tactile, auditory, and olfactory interactions (Gurnell 1987). Visual communication includes tail movements, changes in body posture, and piloerection. Social grooming in squirrels is an important tactile behavior, which maintains close relationships between individuals, especially siblings (Horwich 1972).

Squirrel vocalizations can be classified into four main categories of calls: (1) nestling, (2) mating chase and mating, (3) agonistic, and (4) alarm or warning (Bakken 1959; Zelley 1971; Lishak 1982a, 1982b; 1984) (also see Gurnell [1987] for a detailed discussion of individual call structure and function). Individual calls may be difficult to distinguish and often are a graded continuum of sounds with one type blending into another (Gurnell 1987).

Olfactory communication in fox and gray squirrels includes scent deposition via urination, cheek rubbing, face wiping, vaginal secretions, and anal dragging (Gurnell 1987). Scent marking at traditional marking points (Taylor 1977) occurs in both species, and is characterized by vigorous gnawing without ingestion of bark, and repeated cheek rubbing and chin rubbing against the gnawed area, followed occasionally by urination (Taylor 1977; Koprowski 1993a). Regularly used marking points become darkly stained by urine and other secretions (Gurnell 1987). Marking points are usually established on the underside of limbs, along exposed roots, or at the base of trees (Taylor 1968; Gurnell 1987; Koprowski 1991a). Face wiping is the alternate wiping, several times in succession, of both sides of the face on nontraditional marking points throughout an individual's home range (Benson 1980; Koprowski 1993a). Although the function of scent deposition is not fully understood, it may be used to signal occupation of an area by resident animals and serve as an indicator of social status, dominance, and reproductive condition (Taylor 1977; Benson 1980; Koprowski 1993a).

Breeding Behavior. Mating in fox and gray squirrels occurs in four phases: (1) prechase behavior, (2) mating chase, (3) copulation, and (4) postcoital behavior (McCloskey and Shaw 1977; Thompson 1977b; Benson 1980). Prechase or preliminary "following" occurs 1–5 days before females enter estrus. During this period, females are followed by males for up to 20 min. Females generally maintain a distance of 1–2 m from males, although closer contact has been observed (e.g., vaginal sniffing; Thompson 1977b). The mating chase begins with the aggregation of two or more males (up to 34; Goodrum 1961) following individual females during their 1 day of estrus in an attempt at copulation (Thompson 1977b; Gurnell 1987). Mixed-species mating chases have been reported in fox and gray squirrels (Moore 1968; Koprowski 1991b). A linear hierarchy is established in the group of pursuing males, with the dominant male positioning himself closest to the female and defending against approaching males (Thompson 1977b; Benson 1980). Once the female is receptive and assumes the copulatory position (Horwich 1972), copulation with the closest and presumed dominant (Gurnell 1987) male occurs immediately (Thompson 1977b). Dorsal copulation lasts <30 sec with locking, thrusting, and multiple intromissions (McCloskey and Shaw 1977; Koprowski 1993b). An estrous female may copulate several times with multiple males (Gurnell 1987; Koprowski 1994a, 1994b). Following copulation, the male and female separate and immediately groom their genitalia (Horwich 1972; McCloskey and Shaw 1977). During grooming, 50% of female fox and gray squirrels removed the copulatory plug from the vagina within 30 sec of copulation (Koprowski 1992). The function and significance of copulatory plugs and their removal by females, however, is uncertain (Koprowski 1992).

Dominance. Social organization in fox and gray squirrel populations is based on hierarchical relationships formed through agonistic encounters (Gurnell 1987). Neither species exhibits territorial behavior, with extensive overlap in home ranges observed (Don 1983). However, females may defend the area immediately surrounding their maternity den/nest (Bakken 1959; Havera and Nixon 1978). Agonistic behaviors include chases, jumping or running toward another individual, and tooth chattering (Thompson 1978a). Intense combat is rare, and most encounters are resolved with little aggression (Gurnell 1987). Under this hierarchical system, males dominate females, and social rank increases with age and size of the individual (Flyger 1956; Bakken 1959; Pack et al. 1967; Thompson 1978a; Benson 1980; Allen and Aspey 1986). Resident adults direct aggression toward immature animals and immigrants (Thompson 1978a, 1978b). Resident juveniles direct aggression toward other juveniles (Koprowski 1993c), and dominant juveniles are more successful in establishing residence within a local population (Pasitschniak-Arts and Bendell 1990). Moreover, kinship can influence agonistic and amicable interactions, spatial overlap, and recruitment success in gray squirrels (Thompson 1978a; Koprowski 1993c, 1996).

Communal Nesting. Both species exhibit communal (intraspecific) nesting characterized by nest sharing by two or more individuals (Bakken 1959; Christisen 1985; Weigl et al. 1989; Koprowski 1996; Layne 1998). Communal nesting is generally more common in gray squirrels and less frequently observed in fox squirrels (Christisen 1985; Koprowski 1996; J. Edwards, pers. obs.). Although nesting aggregations occur throughout the year, the prevalence during colder periods suggests a thermoregulatory function (Koprowski 1996).

Interspecific Competition. Fox and gray squirrels coexist over a large extent of their range (Flyger and Gates 1982; Gurnell 1987), despite their reported overlap in feeding habits (Barber 1954; Nixon et al. 1968; Smith and Follmer 1972; Webster et al. 1985), nest sites (Nixon and Hansen 1987; Weigl et al. 1989; Edwards and Guynn 1995), and habitat use (Flyger and Smith 1980; Flyger and Gates 1982). Sympatry in populations of fox and gray squirrels is believed to be maintained by differences in habitat preference (Taylor 1973; Flyger and Gates 1982; Weigl et al. 1989; Derge and Yahner 2000) and nest selection (Edwards and Guynn 1995).

Interspecific competition in midwestern populations of fox and gray squirrels is believed to be of limited importance in determining squirrel distributions (Brown and Yeager 1945; Armitage and Harris 1982; Brown and Batzli 1984, 1985). Changes in land-use patterns and silvicultural practices in the southeastern United States have altered or eliminated large portions of preferred fox squirrel habitat. Moreover, alteration of preferred habitats potentially has forced fox squirrels to use marginal habitats where they are at a competitive disadvantage with gray squirrels (Taylor 1973; Weigl et al. 1989). Edwards et al. (1998) examined niche breadth and niche overlap in habitat and nest characteristics among sympatric fox and gray squirrels in Georgia. On several dimensions, fox squirrels occupied narrower niches than gray squirrels. Intraspecific niche overlap between species was greater than interspecific overlap on 9 of 12 dimensions; intraspecific and interspecific niche overlap did not differ on overstory hardwood stems/hectare, overstory pine/hectare, and midstory tree species. Edwards et al. (1998) found little evidence of competitive interactions and suggested that coexistence between fox and gray squirrels was maintained, in part, through habitat partitioning.

Scatter-Hoarding. Fox and gray squirrels scatter-hoard seeds over much of their range (Gurnell 1987; Koprowski 1994a, 1994b). Loeb and Moncrief (1993), however, suggested that fox squirrels in the southeast may not cache food as readily as in other parts of the range. Caching behavior in fox squirrels was not observed in North Carolina (Weigl et al. 1989) or Georgia (J. Edwards, pers. obs.), but was reported in populations in Florida, although infrequently (Moore 1957; Wooding 1997). Squirrels scatter-hoard or store individual food items in separate cache sites at dispersed locations within their home ranges (Brown and Yeager 1945; Stapanian and Smith 1978). Gray and fox squirrels commonly cache black walnuts, hickory nuts, beechnuts, acorns, and possibly pine cones (Gurnell 1987; Vander Wall 1990). Individual nuts are carried in the jaws and buried <2 cm below the soil surface or covered with leaf litter (Cahalane 1942). Cached items are relocated over a period of a few days to several months by olfaction and memory (Thompson and Thompson 1980; Jacobs and Liman 1991). Cahalane (1942) reported fox squirrels recovered 83–99% of cached nuts in Michigan. Fox squirrels distribute caches at densities that minimize loss rates to seed competitors; memory of item placement gives the animal caching an advantage over naive competitors in retrieving the food (Stapanian and Smith 1978). Caching decisions may also be affected by predation risk (Lima and Valone 1986; Newman et al. 1988) and food storability (Smallwood and Peters 1986; Kotler et al. 1999).

Gray squirrels preferentially cache acorns of the red oak group (subgenus *Erythrobalanus*), which germinate in the spring (Fox 1982; Smallwood and Peters 1986). When caching acorns of the white oak group (subgenus *Leucobalanus*), which germinate in the fall, fox and gray squirrels often (>50%) excise the seed embryo before storing (Fox 1982). This practice prevents the transfer of up to 50% of the food material contained in the acorn (cotyledons) into the seedling taproot before consumption by squirrels (Fox 1982). Squirrels are not known to consume the taproot after germination (Smallwood and Peters 1986). This behavior may represent a countertactic evolved to negate autumn-germinating acorns and to enhance long-term storage of seeds (Smallwood and Peters 1986; Hadj-Chikh et al. 1996; Steele et al. 2001).

MORTALITY

Mortality Rates. Survivorship patterns in fox and gray squirrels are assumed to be similar (Gurnell 1987). Adult mortality in gray squirrels ranges from 30% to 57% (Mosby 1969; Barkalow et al. 1970; Thompson 1978b; Allen 1982; Gurnell 1987). In North Carolina, Barkalow et al. (1970) reported a mortality rate of 75% for gray squirrels during their first year in an unhunted population. Mean annual survival increased to 52% in adult age classes (>1 year old); adult females had a slightly higher annual survival rate than males, 59% versus 44%, respectively. Mean annual "disappearance" rates (mortality and emigration) ranged from 34% to 45% in unhunted populations of fox squirrels in Illinois (Hansen et al. 1986; Herkert et al. 1992). In Georgia, Conner (2001) reported similar annual survival rates for adult male (78%) and female (66%) fox squirrels; seasonal survival varied from 86% to 92%.

Survivorship in gray squirrel populations is dependent on mast production, with juvenile survival rates of 27–31% in years with abundant mast and 5% following a mast failure in North Carolina (Barkalow et al. 1970). Nixon et al. (1975) reported a significant correlation between adult gray squirrel survival and hickory nut production over a 10-year period in Ohio. Koprowski (1991c) found decreased juvenile survival in gray and fox squirrels following a late spring–early summer failure of mulberry (*Morus* spp.) mast and hackberry fruit in Kansas. Hansen et al. (1986), however, found no relation between mast production and fox squirrel recapture rates in Illinois. Whether supplemental feeding can enhance survival in fox and gray squirrel populations is uncertain (Havera and Nixon 1980; Brown and Batzli 1985).

Fox and gray squirrel longevity in the wild is similar (Flyger and Gates 1982). Longevity records for male and female fox and gray squirrels of ≥8 years and ≥12.5 years, respectively, were reported by Barkalow and Soots (1975) and Koprowski et al. (1988). In North Carolina, Barkalow et al. (1970) found mean life expectancy of gray squirrels at birth was approximately 1 year. Adult mean life expectancy increased from 1.8 years in the 1- to 2-year cohort to a high of 2.4 years in the 2- to 3-year cohort, and remained >1 year until the 7- to 8-year cohort, when it decreased to 0.50.

Predation. Predation is generally considered a minor factor in fox and gray squirrel population dynamics (Flyger and Gates 1982). Predators of adult and nestling gray and fox squirrels include rat snake (*Elaphe obsoleta*), pine snake (*Pituophis melanoleucus*), timber rattlesnake (*Crotalus horridus*), eastern diamondback rattlesnake (*C. adamanteus*), red-tailed hawk (*Buteo jamaicensis*), red-shouldered hawk (*B. lineatus*), rough-legged hawk (*B. lagopus*), ferruginous hawk (*B. regalis*), goshawk (*Accipiter gentilis*), Cooper's hawk (*A. cooperii*), bald eagle (*Haliaeetus leucocephalus*), golden eagle (*Aquila chrysaetos*), great horned owl (*Bubo virginianus*), barred owl (*Strix varia*), opossum (*Didelphis virginiana*), bobcat (*Lynx rufus*), wolf (*Canis lupus*), coyote (*C. latrans*), gray fox (*Urocyon cinereoargenteus*), red fox (*Vulpes vulpes*), raccoon (*Procyon lotor*), long-tailed weasel (*Mustela frenata*), mink (*M. vison*), dogs, and cats (Packard 1956; Moore 1957; Barkalow and Shorten 1973; Gurnell 1987; Weigl et al. 1989; U.S. Fish and Wildlife Service 1993; Koprowski 1994a, 1994b).

Parasites and Disease. Fox and gray squirrels are susceptible to numerous ecto- and endoparasites (Table 12.5). Mange mites cause hair loss and may result in considerable population decline during some winters (Allen 1943). Botflies (*Cuterebra emasculator*) parasitized 19% and 5% of gray and fox squirrels, respectively, in Mississippi (Jacobson et al. 1981). Fleas, ticks, lice, and chiggers are seasonally common parasites on squirrels (Flyger and Gates 1982). Ringworm (*Trichophyton mentagrophytes*) and other fungi from 19 genera were found on gray squirrels in Florida (Lewis et al. 1975). Viral and bacterial diseases reported in fox and gray squirrels include California encephalitis (Masterson et al. 1971; Moulton and Thompson 1971) and western equine encephalitis (Moulton and Thompson 1971), skin tumors caused by pox virus (Kilham 1954), leptospirosis (Shotts et al. 1975), tetanus (Wobeser 1969), tularemia (Davis et al. 1970), *Coxiella burneti* (Enright et al. 1971), cryptosporidiosis (Sundberg et al. 1982), and acute fatal toxoplasmosis (Roher et al. 1981). Rabies is rarely found in either species (Pritchett 1938; Capucci et al. 1972). Shivaprasad et al. (1984) reported carcinosarcoma mammary tumors in gray squirrels.

TABLE 12.5. Parasites of fox and gray squirrels

Parasite[a]	Fox	Gray	Source
Protozoa			
Eimeria ascotensis	X		Joseph 1975
E. confusa	X	X	Joseph 1972, 1975
E. kniplingi			Joseph 1975
E. lancasterensis	X	X	Joseph 1972, 1975
E. ontarioensis	X	X	Joseph 1972, 1975
Hepatozoon griseisciuri		X	Clark 1958
Toxoplasma gondii		X	Walton and Walls 1964
Cestodes			
Bothriocephalus sciuri	X		Self and Esslinger 1955
Catenotaenia pusilla		X	Rausch and Tiner 1948
Choanotaenia sciuricola	X		Harwood and Cooke 1949
Cittotaeiai pectinata		X	Rankin 1946
Taenia pisiformis	X		Baumgartner 1940
Hymenolepis diminuta	X	X	Rausch and Tiner 1948
H. nana		X	Olexik et al. 1969
Mesocestoides latus	X		Rausch and Tiner 1948
Raillietina bakeri	X	X	Chandler 1942, Moore 1957
Taenia crassiceps		X	Freeman 1962
T. hydatigena	X	X	Meggitt 1924, Baylis 1939
T. mustelae	X		Langham et al. 1990
T. pisiformis	X	X	Graham and Uhrich 1943, Brown and Yeager 1945
T. taeniaeformis	X		Harkema 1936
Nematodes			
Ascaris columnaris	X	X	Rausch and Tiner 1948, Tiner 1949
A. lumbricoides	X	X	Brown and Yeager 1945, Rausch and Tiner 1948
Bohmiella wilsoni	X	X	Lucker 1943, Rausch and Tiner 1948
Capillaria hepatica	X		McQuown 1954
Citellinema bifurcatum		X	Dikmans 1938, Parker 1968
Dipetalonema interstitium	X	X	Price 1962, Eckerlin 1993
Dirofilariaeformia pulmoni		X	Davidson 1975
Enterobius sciuri	X	X	Rausch and Tiner 1948, Parker and Holliman 1971
Gongylonema pulchrum	X	X	Parker and Holliman 1971, Coyner et al. 1996
Heligmodendrium hassalli	X	X	Rausch and Tiner 1948
Physaloptera sp.		X	Rausch and Tiner 1948
P. massino	X		Morgan 1943
Rictularia coloradensis		X	Parker 1968
Strongyloides papillosus		X	Reiber and Byrd 1942
S. robustus	X	X	Chandler 1942, Parker and Holliman 1971
Trichinella spiralis	X		Zimmerman and Hubbard 1969
Trichostrongylus calcaratus	X	X	Rausch and Tiner 1948, Eckerlin 1993
T. colubriformes		X	Baylis 1934
Acarina			
Amblyomma americanum	X	X	Morlan 1952, Moore 1957
A. maculatum	X	X	Hixon 1940, Moore 1957
A. tuberculatum	X		Moore 1957
Androlaelaps glasgowi		X	Parker and Holliman 1971
Atricholaelaps glasgowi	X		Uhrich and Graham 1941
A. megaventralis	X		Moore 1957
Bruelia rotundata		X	Parker 1968
Cheladonta micheneri		X	Lipovsky et al. 1955
Dermacentor variabilis	X	X	Tugwell and Lancaster 1962, Sonenshine and Stout 1971
Eulaelaps stabularis	X		Moore 1957
Eushoengastia diversa	X		Coyner et al. 1996
E. jonesi		X	Loomis 1956
E. setosa		X	Loomis 1956
Eutrombicula alfreddugesi	X		Jenkins 1948
Haemaphysalis leporispalustris	X	X	Bishopp and Trembley 1945
Haemogamasus ambulans		X	Clark 1958
Haemolaelaps megaventralis	X		Morlan 1952
Ixodes cookei	X	X	Bishopp and Trembley 1945
I. hearlei	X		Cooley and Kohls 1945
I. marxi		X	Bishopp and Trembley 1945
I. muris		X	Anastos 1947
I. scapularis	X		Moore 1957
Microtrombicula trisetica		X	Webb and Loomis 1971
Neotrombicula whartoni	X		Coyner et al. 1996
Notoedres douglasi	X		Kazacos et al. 1983
Ornithonyssus bacoti		X	Morlan 1952
Sarcoptes sp.		X	Chapman 1938
S. scabiei	X		Baumgartner 1940

TABLE 12.5—*Continued*

Parasite[a]	Fox	Gray	Source
Speleognathopsis sciuri		X	Clark 1960
Trombicula alfreddugesi	X	X	Jenkins 1948, Moore 1957
T. autumnalis		X	Loomis 1956
T. fitchi		X	Loomis 1954
T. gurneyi		X	Loomis 1955
T. splendens		X	Morlan 1952
T. sylvilagi		X	Kardos 1954
T. whartoni	X	X	Brennan and Wharton 1950, Moore 1957
Walchia americana		X	Loomis 1956
Anoplura			
Enderleinellus longiceps	X	X	Mathewson and Hyland 1962, Kim 1966a, 1966b
Hoplopleura hesperomydis		X	Mathewson and Hyland 1962
H. sciuricola	X	X	Morlan 1952
Neohaematopinus antennatus		X	Keegan 1943
N. sciuri		X	Mathewson and Hyland 1962
N. sciurinus	X	X	Harkema 1936, Johnson 1958
Polyplax spinulosa		X	Morlan 1952
Siphonaptera			
Cediopsylla simplex		X	Holland and Benton 1968
Ceratophyllus fasciatus	X		Baumgartner 1940
C. gallinae		X	Burbutis 1956
Conorhinopsylla stanfordi		X	I. Fox 1940
Ctenocephalides felis	X	X	Morlan 1952, Mathewson and Hyland 1964
Echidnophaga gallinacea	X	X	Morlan 1952, Ellis 1955
Epidedia wenmauni		X	Main 1970
Hoplopsyllus glacialis		X	Layne 1971
H. affinis	X		Graham and Uhrich 1943
Leptopsylla segnis	X		Morlan 1952
Megabothris asio		X	Mathewson and Hyland 1964
Monopsyllus vison		X	Woods and Larson 1969
Nosopsyllus fasciatus		X	Burbutis 1956
Opisocrostic bruneri		X	Fox 1940
Orchopeas howardi	X	X	Fox 1940, Holland and Benton 1968
Pulex irritans		X	Shaftesbury 1934
Spilopsyllus cuniculi		X	George 1954
Diptera			
Cuterebra sp.	X	X	Parker and Holliman 1971

SOURCE: Data from Davidson (1976), Flyger and Gates (1982), and Koprowski (1994a).
[a]Taxonomic name given as in original source.

AGE ESTIMATION

Several methods are used to estimate age in fox and gray squirrels (Dimmick and Pelton 1996). Tail pelage pattern and coloration of genitals and mammae are used in autumn to separate spring- from summer-born subadults and juveniles from adults (Allen 1943; Sharp 1958; Barrier and Barkalow 1967). McCloskey (1977) found nipple appearance and coloration in females and scrotal pigmentation in males the most accurate field criteria in fox squirrels. Reproductive condition of male gray squirrels only distinguishes breeding adults from nonbreeding animals, and this method is further compromised by seasonal peaks in breeding activity (Kirkpatrick and Barnett 1957). Chapman (1938) separated young from adult squirrels by weight. Body mass, however, correlates poorly with age, partially due to high seasonal variations (Kirkpatrick and Hoffman 1960; Dubock 1979).

Laboratory methods used to determine age of squirrels include histology of testes and accessary glands (Hoffman and Kirkpatrick 1959), baculum weight and length (Kirkpatrick and Barnett 1957), epiphyseal cartilage closure (Carson 1961), cementum annuli formation (Fogl and Mosby 1978; Edwards and Guynn 1993), toothwear and replacement (Hench et al. 1984), eye lens weight (Beale 1962; Fisher and Perry 1970), and development of suspensory tuberosities (Colburn 1986). Osteological measurements show a high degree of overlap among age classes and are therefore of limited use (Kirkpatrick and Barnett 1957). McCloskey (1977) found the presence or absence of an epiphyseal line a more accurate indicator of age than histological metrics. Dubock (1979) reported eye lens weight correlated well ($r^2 = 0.88$) with age. However, this relation was only reliable to 3 years of age. Fogl and Mosby (1978) found cementum annuli to more accurately determine age of young (<1 year) squirrels than the eye lens method. Edwards and Guynn (1993) reported cementum annuli underaged known-age gray and fox squirrels; however, the sample size was small ($n = 4$).

ECONOMIC STATUS AND MANAGEMENT

Damages and economic losses attributed to fox and gray squirrels are minor and localized (Gurnell 1987). Squirrels may cause damage by chewing on trees, electrical wires, or buildings (J. J. Jackson 1994) and can severely reduce production in commercial nut orchards (Huggins 1996). The most efficient method of control is to remove offending animals by shooting or trapping (Flyger and Gates 1982).

Harvest. Fox and gray squirrels are popular small-game animals. Annual harvest estimates exceed 1 million animals in Florida, Illinois, Indiana, Missouri, Ohio, and Pennsylvania. The total number of squirrels harvested annually in the United States is undetermined, but Flyger and Gates (1982) estimated the annual gray squirrel harvest at approximately 40 million. Many states are experiencing long-term declines in numbers of squirrel hunters and squirrels harvested (Table 12.6). Although squirrel populations in these states remain healthy, increased white-tailed deer (*Odocoileus virginianus*) season lengths and bag limits may be affecting hunter numbers and the amount of time they spend hunting squirrels. Annual expenditures by squirrel hunters represent

TABLE 12.6. Fox and gray squirrel harvests and number of squirrel hunters in Illinois, Pennsylvania, and Ohio, 1988–2000

	Illinois		Pennsylvania		Ohio	
Year	Hunters	Harvest	Hunters	Harvest	Hunters	Harvest
1988					379,849	2,279,094
1989	197,775	1,821,671			357,329	2,143,979
1990	189,700	1,718,101	369,848	2,044,264	355,376	2,132,254
1991	212,180	1,696,414	348,868	1,632,108	363,796	2,182,774
1992	188,949	1,591,189	329,726	1,761,285	340,637	2,043,820
1993	176,179	1,396,221	311,103	1,585,368	342,242	2,053,453
1994	184,710	1,497,257	326,271	1,826,618	335,972	2,015,831
1995	165,832	1,291,573	293,852	1,599,104	335,275	2,011,650
1996	157,083	1,322,138	279,259	1,442,560	325,481	1,952,888
1997	145,945	1,068,594	267,051	1,352,038	314,914	1,889,484
1998	143,680	1,327,876	252,738	1,331,051	298,884	1,793,304
1999			238,887	1,236,108	295,778	1,774,667
2000					287,963	1,727,778

SOURCE: Illinois, data from J. K. Garver, Illinois Department of Natural Resources, (pers. commun). Pennsylvania, data from Pennsylvania Game Commission. Ohio, data from D. Swanson, Ohio Department of Natural Resources (pers. commun.).

a substantial influx to state economies (Uhlig 1955; Flyger and Gates 1982). Additional revenues are generated from nonconsumptive users; squirrels ranked second to songbirds in value to nature watchers (Shaw and Mangun 1984).

Hunting mortality is considered compensatory to some extent, and is generally not thought of as a major factor controlling squirrel populations (Allen 1954; Uhlig 1956; Mosby 1969; Flyger and Gates 1982). Annual hunting mortality averaged 13–15% in gray squirrel populations in West Virginia (Uhlig 1956) and Louisiana (Dennett and Kidd 1960). In Maryland, Shugars (1986) reported annual hunting mortality in gray squirrels was relatively low at 7–9% when compared to overall mortality of 48–54%. Moreover, Mosby (1969) experimentally demonstrated that an average annual removal of 37% of the fall population did not adversely affect recruitment or annual mortality in a population of gray squirrels during a 6-year period in Virginia. Nixon et al. (1974), however, found a significant population decline in fox squirrels associated with an average annual hunting mortality of 48% in Ohio. Reproductive capacities of fox and gray squirrels are not able to compensate for hunting mortality rates of 40–45% (Rhodes 1989). Herkert et al. (1992) reported hunting mortality of 54–76% on a public hunting area in Illinois. In contrast to Nixon et al. (1974), they found that the successful recruitment of immigrants allowed this exploited population to recover quickly from high annual removals of squirrels. Gray squirrels generally inhabit more extensive areas relative to fox squirrels where hunting is more dispersed and greater opportunity exists for ingress. Intensively hunted populations of fox squirrels may be particularly vulnerable to overharvest depending on their level of isolation and potential for recolonization from nearby refuges or other lightly hunted areas (Allen 1943, 1954; Nixon et al. 1974; Herkert et al. 1992). Rhodes (1989) and Herkert et al. (1992) suggested that hunting mortality may be additive in some exploited populations of fox and gray squirrels.

Fox and gray squirrels are managed primarily through hunting regulations, which change little from year to year (Rhodes 1989), under the assumption that hunting intensity is self-regulating and will decrease proportionately with hunter success (Uhlig 1956; Jordan 1971b; Mosby et al. 1977). However, under some circumstances, this assumption may not be valid. For example, Nixon et al. (1975) found that hunting intensity on a small, public shooting area in Ohio remained high even as hunter success declined. Season starting date and length represent a compromise among several factors including sustainable harvest, number of recreation days, hunter satisfaction, acceptability of killing pregnant or lactating females, and botfly larvae infestation. Most states begin their squirrel hunting seasons between 1 September and 15 October (Fig. 12.4). Late summer/early fall start dates are preferred by sportsmen because squirrels are active gathering hard mast. This increased feeding in mast-bearing trees makes them more vulnerable during the early part of the season than after the mast crop has fallen to the ground later in the fall or early winter (Mosby et al. 1977). Early start dates, however, increase the likelihood of harvesting a greater proportion of pregnant or lactating females and squirrels infested with botfly larvae, which are often discarded by hunters because they are considered unfit

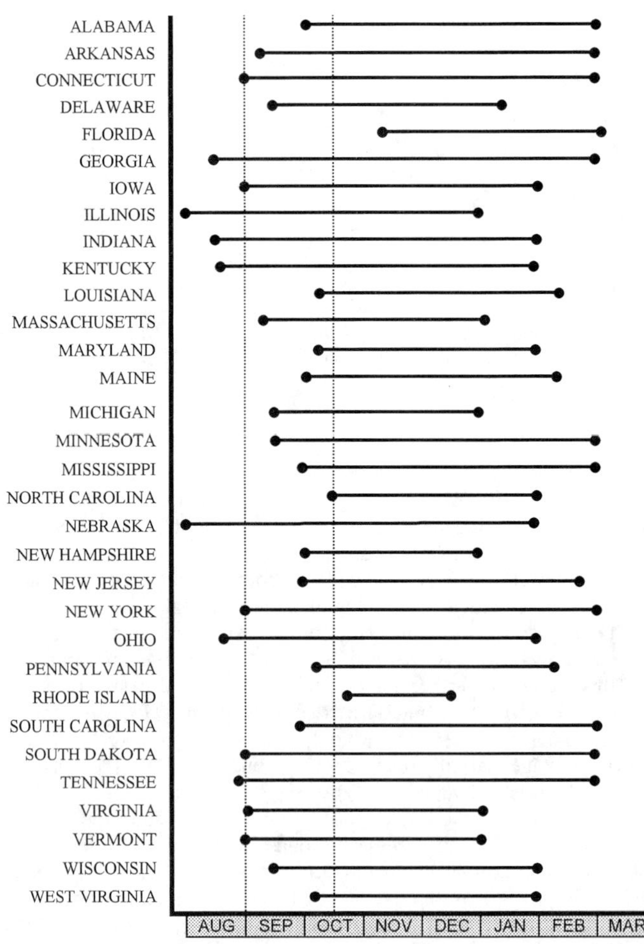

FIGURE 12.4. Legal squirrel hunting seasons in 32 states throughout the range. Seasons shown include the earliest and latest dates in each respective state; seasons may not be continuous for the entire period shown. Dashed lines indicate the period from 1 September to 15 October, during which most states open their squirrel season. State hunting seasons not shown include Kansas (1 June–28 February), Missouri (4 May–15 January), and Oklahoma (15 May–31 January).

as food (Jacobson et al. 1979). In most states, season length has little biological relevance because the majority of squirrel hunting occurs in the first weeks of the season (Uhlig 1956; Flyger and Gates 1982; Shugars 1986). In many states, no distinction is made between hunting regulations for fox and gray squirrels. Tappe and Guynn (1998) suggested that it may be more appropriate to manage southern fox squirrels as a relatively *K*-selected species compared to the gray squirrel or other small-game species. Fox squirrels in the southeast appear to be relatively long-lived, and have smaller and fewer litters and lower adult mortality. Ecological differences between gray and fox squirrels suggest the use of different strategies to manage harvest of each species in the southeast United States (Tappe and Guynn 1998; Conner 2001).

Habitat. Habitat improvement practices for fox and gray squirrels generally target forest structure and tree species composition. Changes in forest structure affect overstory and understory densities and tree cavity availability; tree species composition markedly affects mast production. Specific management practices for fox and gray squirrels vary between species and among regions, and may not be applicable throughout the range. Nixon and Hansen (1987) discussed silvicultural options and management recommendations regarding harvest, intermediate practices, stocking levels, and rotation lengths to maintain populations of gray and fox squirrels in the Midwest. The primary factor in maintaining squirrel populations in residual forest stands (i.e., postharvest) is sustaining adequate levels of winter-storable tree seeds (Nixon et al. 1975). Residual stands should have 1.3–1.5 m^2/ha basal area of mast-producing trees, including a 50:50 mix of fall/spring germinating oaks, and a minimum of 0.4–0.6 m^2/ha basal area of hickory (Nixon et al. 1975). Nixon and Hansen (1987) developed a stocking guide that managers can use to provide squirrels with 80% of their annual diet (30% oak, 40% hickory, 10% walnut). Nixon et al. (1980a) reported that removal of 37–55% of the basal area of a stand through single-tree or small group-selection harvests had no short-term (1–2 years after cutting) effect on fox and gray squirrel densities, breeding rates, or annual survival. Although selective harvests altered the physical structure of the stand by creating many small canopy gaps, tree species composition remained similar. In contrast, large-scale clearcutting of forested areas can decrease squirrel populations (Nixon et al. 1980b). However, use of small (<8 ha), narrow (<160 m), carefully located clearcuts in forests where 40–60% of stands are retained at a seed-producing age should not substantially reduce gray squirrel populations (Nixon et al. 1980b). Stream-side management zones also offer an option for providing high-quality habitat and travel corridors for fox and gray squirrels in areas affected by even-aged management. Practices such as heavy and continuous grazing and repeated burning, which reduce the number of tree cavities and understory density, favor fox squirrels over gray squirrels.

Whereas comprehensive management guidelines are available for fox and gray squirrels in the Midwest, recommendations from other parts of the range are based more on observational than experimental evidence. Results from habitat selection studies in the Southeast suggest that management practices that reduce dense understory vegetation and promote retention of mature mast-producing hardwoods will benefit fox squirrels. Such practices include use of prescribed fire, mowing, and retention of hardwood stringers in pine-dominated habitats (Lustig and Flyger 1975; Kantola 1986; Edwards et al. 1989; Weigl et al. 1989; U.S. Fish and Wildlife Service 1993; Lee 1999).

Fox and gray squirrels also benefit from promotion and retention of cavity trees and through addition of artificial cavities. Nixon and Hansen (1987) recommended 5–7 tree cavities/ha and 5–7 artificial cavities/ha if reasonably high squirrel densities are desired. The addition of artificial cavities increases carrying capacity for gray squirrels (Barkalow and Soots 1965; Burger 1969). Moreover, artificial cavities may allow gray squirrels to recolonize a young stand sooner (30–60 years) after clearcutting if food is not a limiting factor (Nixon and Donohoe 1979). Although of benefit to both species, cavities are a more important habitat component for gray squirrels than for fox squirrels (Sanderson et al. 1980; Flyger and Gates 1982; Edwards and Guynn 1995). Sanderson (1975) recognized the difficulty in providing tree cavities in intensively managed forests and suggested the following options: (1) retention of existing trees with cavities; (2) retention of potential cavity trees; (3) option 2, plus treatment (e.g., stub pruning) of selected trees to form cavities; and (4) option 2 and 3, plus provision of artificial cavities. In the midwestern and northeastern United States, canopy-reaching vines serve as important support structures for fox and gray squirrel leaf nests. Sanderson et al. (1980) and Nixon and Hansen (1987) recommended retention of ≥2–4 vines/ha for gray squirrels and ≥4–6 vines/ha for fox squirrels.

RESEARCH AND CONSERVATION NEEDS

Fox and gray squirrel research and management has made great strides in the nearly 20 years since the first edition of *Wild Mammals of North America* in 1982. Charles Nixon and colleagues, John Koprowski, and others have added tremendously to our knowledge of squirrel biology and ecology in the Midwest. This is not to say that further research is unnecessary, but that current research findings provide a substantial base from which to advance.

Although fox and gray squirrel populations remain stable in the Midwest, anthropogenic influences continue to fragment landscapes and alter the distribution and abundance of both species (Sheperd and Swihart 1995; Swihart and Nupp 1998). Conservation at the local level will continue to rely on habitat assessment, estimates of annual recruitment, and hunter harvest. Landscape-level conservation would benefit from long-term studies examining dispersal patterns, population genetics, and other facets of metapopulation theory. Such efforts could enable researchers and managers to further assess fragmentation and isolation impacts resulting from changes in land use/management practices on the conservation of fox and gray squirrels in the Midwest.

Research efforts in the Southeast lag behind those in the Midwest. This is somewhat disconcerting considering that four subspecies in the region are given species status. *S. n. cinereus* is listed as endangered by the U.S. Fish and Wildlife Service (1993), and *S. n. shermani*, *S. n. avicennia*, and *S. n. niger* are designated as Species of Concern in the states where they occur (Laerm et al. 2000). Although some progress has been made, well-designed quantitative studies examining anthropogenic-induced fragmentation, dispersal patterns, feeding habits, translocation success, hunting mortality, and population ecology are needed. Furthermore, much of the information on habitat selection by southeastern squirrels was gleaned from habitat use/availability studies. Although these provide important information, this research does not equate to manipulative treatments where habitats are changed and a corresponding species response is determined. The ultimate goal of management-oriented research should be to discover factors that enhance or limit the distribution and abundance of the species.

Changing land-use practices such as short-rotation forestry and large-scale agriculture combined with increasing development continue to affect fox and gray squirrels in the Southeast. Whereas local conservation can be accomplished through habitat management, population assessment, and regulated harvest, conservation of imperiled fox squirrel populations may require extensive research and management efforts. Landscape-scale research initiatives are needed to determine how best to mitigate habitat loss/modification and provide alternatives for ameliorating short- and long-term population-level effects. Although populations of threatened fox squirrels are believed to be stable, applied research is needed to ensure their conservation. Long-term stability will require compatible forest management and residential development guidelines. In addition to managing to conserve existing populations, our ultimate goal should be the increase and expansion of populations to prethreatened levels through habitat conservation plans, travel corridors, reintroductions, and other management options.

LITERATURE CITED

Adams, C. E. 1976. Measurement and characteristics of fox squirrel, *Sciurus niger rufiventer*, home ranges. American Midland Naturalist 95:211–15.

Allen, D. L. 1942. Populations and habits of the fox squirrel in Allegany County, Michigan. American Midland Naturalist 27:338–79.
Allen, D. L. 1943. Michigan fox squirrel management (Publication 100). Michigan Department of Conservation, Game Division.
Allen, D. S. 1982. Social status and survivorship in a population of eastern gray squirrels (*Sciurus carolinensis*). Ph.D. Dissertation, Ohio State University, Columbus.
Allen, D. S., and W. P. Aspey. 1986. Determinants of social dominance in eastern gray squirrels, (*Sciurus carolinensis*): A quantitative assessment. Animal Behaviour 43:81–89.
Allen, J. M. 1954. Gray and fox squirrel management in Indiana (P-R Bulletin 1). Indiana Department of Conservation.
Anastos, G. 1947. Hosts of certain New York ticks. Psyche 54:178–80.
Armitage, K. B., and K. S. Harris. 1982. Spatial patterning in sympatric populations of fox and gray squirrels. American Midland Naturalist 108:389–97.
Baker, R. H. 1946. An ecological study of tree squirrels in eastern Texas. Journal of Mammalogy 25:8–24.
Bakken, A. A. 1952. Interrelationships of *Sciurus carolinensis* (Gmelin) and *Sciurus niger* (Linnaeus) in mixed populations. Ph.D. Dissertation, University of Wisconsin, Madison.
Bakken, A. A. 1959. Behavior of gray squirrels. Proceedings of the Annual Conference of the Southeastern Association of Fish and Game Commissioners 13:393–407.
Barber, H. L. 1954. Gray and fox squirrel food habits investigations. Proceedings of the Annual Conference of the Southeastern Association of Fish and Game Commissioners 8:92–94.
Barkalow, F. S., Jr., and R. F. Soots, Jr. 1965. An analysis of the effect of artificial nest boxes on a gray squirrel population. Transactions of the North American Wildlife and Natural Resources Conference 30:349–60.
Barkalow, F. S., Jr., and R. F. Soots, Jr. 1975. Life span and reproductive longevity of the gray squirrel, *Sciurus c. carolinensis* Gmelin. Journal of Mammalogy 56:522–24.
Barkalow, F. S., Jr., and M. Shorten. 1973. The world of the gray squirrel. J. B. Lippincott, Philadelphia.
Barkalow, F. S., Jr., R. B. Hamilton, and R. F. Soots, Jr. 1970. The vital statistics of an unexploited gray squirrel population. Journal of Wildlife Management 34:489–500.
Barnett, R. J. 1977. Bergmann's rule and variation in structures related to feeding in the gray squirrel. Evolution 31:538–45.
Barrier, M. J., and F. S. Barkalow, Jr. 1967. A rapid technique for aging gray squirrels in winter pelage. Journal of Wildlife Management 31:715–19.
Baumgartner, L. L. 1939. Foods of the fox squirrel in Ohio. Transactions of the North American Wildlife Conference 4:579–84.
Baumgartner, L. L. 1940. The fox squirrel: Its life history, habits, and management in Ohio (Release 138). Ohio State University Wildlife Research Station.
Baumgartner, L. L. 1943a. Pelage studies of fox squirrels (*Sciurus niger rufiventer*). American Midland Naturalist 29:588–90.
Baumgartner, L. L. 1943b. Fox squirrels in Ohio. Journal of Wildlife Management 7:193–202.
Baylis, H. A. 1934. Miscellaneous notes on parasitic worms. Annual Magazine of Natural History 13:223–28.
Baylis, H. A. 1939. *Trichostrongylus colubriformis* from *S. carolinensis* in Oxfordshire, England. Annual Magazine of Natural History 18:796.
Beale, D. M. 1962. Growth of the eye lens in relation to age in fox squirrels. Journal of Wildlife Management 26:208–11.
Bendel, P. R., and G. D. Therres. 1994. Movements, site fidelity and survival of Delmarva fox squirrels following translocation. American Midland Naturalist 132:227–34.
Benson, B. N. 1980. Dominance relationships, mating behaviour and scent marking in fox squirrels (*Sciurus niger*). Mammalia 44:143–60.
Bishopp, F. C., and H. L. Trembley. 1945. Distribution and hosts of certain North American ticks. Journal of Parasitology 31:1–54.
Blair, W. F. 1935. The mammals of a Florida hammock. Journal of Mammalogy 16:271–77.
Bolls, N. J., and J. R. Perfect. 1972. Summer resting metabolic rate of the gray squirrel. Physiological Zoology 45:54–59.
Brennan, J. M., and G. W. Wharton. 1950. Studies on North American chiggers. No. 3, The subgenus *Neotrombicula*. American Midland Naturalist 44:153–97.
Brown, B. W., and G. O. Batzli. 1984. Habitat selection by fox and gray squirrels: A multivariate analysis. Journal of Wildlife Management 48:616–21.
Brown, B. W., and G. O. Batzli. 1985. Field manipulations of fox and gray squirrel populations: How important is interspecific competition? Canadian Journal of Zoology 63:2134–40.
Brown, L. G., and L. E. Yeager. 1945. Fox squirrels and gray squirrels in Illinois. Illinois Natural History Survey Bulletin 23:449–535.
Bugbee, R. E., and A. Riegel. 1945. Seasonal food choices of the fox squirrel in western Kansas. Transactions of the Kansas Academy of Science 48:199–203.
Burbutis, P. P. 1956. The Siphonaptera of New Jersey (Bulletin 782). New Jersey Agricultural Experiment Station.
Burger, G. V. 1969. Response of gray squirrels to nest boxes at Remington Farms, Maryland. Journal of Wildlife Management 33:796–801.
Burt, W. H. 1960. Bacula of North American mammals. Miscellaneous Publications of the Museum of Zoology, University of Michigan 113:1–75.
Cahalane, V. H. 1942. Caching and recovery of food by the western fox squirrel. Journal of Wildlife Management 6:338–52.
Cappucci, D. T., Jr., R. W. Emmons, and W. W. Sampson. 1972. Rabies in an eastern fox squirrel. Journal of Wildlife Diseases 8:340–42.
Carson, J. D. 1961. Epiphyseal cartilage as an age indicator in fox and gray squirrels. Journal of Wildlife Management 25:90–93.
Chandler, A. C. 1942. Helminths of tree squirrels in southeast Texas. Journal of Parasitology 28:135–40.
Chapman, F. B. 1938. Summary of the Ohio gray squirrel investigation. Transactions of the North American Wildlife Conference 3:677–84.
Chesemore, D. L. 1975. Ecology of fox and gray squirrels (*Sciurus niger* and *Sciurus carolinensis*) in Oklahoma. Ph.D. Dissertation, Oklahoma State University, Stillwater.
Christisen, D. M. 1985. Seasonal tenancy of artificial nest structures for tree squirrels. Transactions of the Missouri Academy of Science 19:41–48.
Clark, C. M. 1958. *Hepatozoon griseisciuri* n. sp., a new species of *Hepatozoon* from the grey squirrel (*Sciurus carolinensis* Gmelin, 1788) with studies on the life cycle. Journal of Parasitology 44:52–63.
Clark, C. M. 1960. Three new nasal mites (Acarina: Speleognathidae) from gray squirrel, the common grackle, and the meadowlark in the United States. Proceedings of the Helminthological Society of Washington 27:103–10.
Colburn, M. L. 1986. Suspensory tuberosities for aging and sexing squirrels. Journal of Wildlife Management 50:456–59.
Conner, L. M. 2001. Survival and cause-specific mortality of adult fox squirrels in southwestern Georgia. Journal of Wildlife Management 65:200–204.
Conner, L. M., L. J. Landers, and W. K. Michener. 1999. Fox squirrel density, habitat, and interspecific association with gray squirrels within minimally disturbed longleaf pine forests. Proceedings of the Annual Conference of the Southeastern Association of Fish and Wildlife Agencies 53:364–74.
Cooley, R. A., and G. M. Kohls. 1945. The genus *Ixodes* in North America (Bulletin 184). National Institutes of Health.
Cordes, C. L., and F. S. Barkalow, Jr. 1972. Home range and dispersal in a North Carolina gray squirrel population. Proceedings of the Annual Conference of the Southeastern Association of Fish and Game Commissioners 26:124–35.
Coyner, D. F., J. B. Wooding, and D. J. Forrester. 1996. A comparison of parasitic helminths and arthropods from two subspecies of fox squirrels (*Sciurus niger*) in Florida. Journal of Wildlife Diseases 32:492–97.
Dambach, C. A. 1942. Gray squirrel feeding on *Crataegus*. Journal of Mammalogy 23:337.
Davidson, W. R. 1975. *Dirofilariaeformia pulmoni* sp. n. (Nematoda: Onchocercidae) from the eastern gray squirrel (*Sciurus carolinensis* Gmelin). Journal of Parasitology 61:351–54.
Davidson, W. R. 1976. Endoparasites of selected populations of gray squirrels (*Sciurus carolinensis*) in the southeastern United States. Proceedings of the Helminthological Society of Washington 43:211–17.
Davis, J. W., L. H. Karstad, and D. L. Trainer. 1970. Infectious diseases of wild mammals. Iowa State University Press, Ames.
Dennett, D., and J. B. Kidd. 1960. An analysis by tag returns of three years of controlled squirrel hunting. Annual Conference of the Southeastern Game and Fish Commissioners 14:66–73.
Derge, K. L., and R. H. Yahner. 2000. Ecology of sympatric fox squirrels (*Sciurus niger*) and gray squirrels (*S. carolinensis*) at forest-farmland interfaces of Pennsylvania. American Midland Naturalist 143:355–60.
Dikmans, G. 1938. A consideration of the nematode genus *Citellinema* with a description of a new species, *Citellinema columbianum*. Proceedings of the Helminthological Society of Washington 5:55–58.
Dimmick, R. W., and M. R. Pelton. 1996. Criteria of sex and age. Pages 169–214 *in* T. A. Bookhout, ed. Research and management techniques for wildlife and habitats, 5th ed. The Wildlife Society, Bethesda, MD.
Doebel, J. H., and B. S. McGinnes. 1974. Home range and activity of a gray squirrel population. Journal of Wildlife Management 38:860–67.

Don, B. A. C. 1983. Home range characteristics and correlates in tree squirrels. Mammal Review 13:123–32.

Dubock, A. C. 1979. Methods of age determination in grey squirrels *Sciurus carolinensis* in Britain. Journal of Zoology (London) 188:27–40.

Ducharme, M. B., J. Larochelle, and D. Richard. 1989. Thermogenic capacity in gray and black morphs of the gray squirrel, *Sciurus carolinensis*. Physiological Zoology 62:1273–92.

Dudderar, G. R. 1967. A survey of the food habits of the gray squirrel (*Sciurus carolinensis*) in Montgomery County, Virginia. M.S. Thesis, Virginia Polytechnic Institute and State University, Blacksburg.

Dueser, R. D., J. L. Dooley, Jr., and G. J. Taylor. 1988. Habitat structure, forest composition and landscape dimensions as components of habitat suitability for the Delmarva fox squirrel. Pages 414–21 *in* R. C. Szaro, K. E. Severson, and D. R. Patton, eds. Management of amphibians, reptiles, and small mammals in North America (USDA Forest Service General Technical Report RM-166). Rocky Mountain Forest and Range Experiment Station.

Eason, P. K. 1998. Predation of a female house finch, *Carpodacus mexicanus*, by a gray squirrel, *Sciurus carolinensis*. Canadian Field-Naturalist 112:713–14.

Eckerlin, R. P. 1993. Helminth parasites from two populations of the Delmarva fox squirrel (*Sciurus niger cinereus*) in Maryland and Virginia. Pages 53–56 *in* N. D. Moncrief, J. W. Edwards, and P. A. Tappe, eds. Proceedings of the second symposium on southeastern fox squirrels, *Sciurus niger* (Special Publication No. 1). Virginia Museum of Natural History.

Edwards, J. W. 1986. Habitat utilization by southern fox squirrel in coastal South Carolina. M.S. Thesis, Clemson University, Clemson, SC.

Edwards, J. W. 1995. Resource use by sympatric populations of fox and gray squirrels. Ph.D. Dissertation, Clemson University, Clemson, SC.

Edwards, J. W., and D. C. Guynn, Jr. 1993. Assignment of age classes in fox squirrels (*Sciurus niger*) and gray squirrels (*S. carolinensis*) using cementum annuli. Pages 79–84 *in* N. D. Moncrief, J. W. Edwards, and P. A. Tappe, eds. Proceedings of the second symposium on southeastern fox squirrels, *Sciurus niger* (Special Publication No. 1). Virginia Museum of Natural History.

Edwards, J. W., and D. C. Guynn, Jr. 1995. Nest characteristics of sympatric populations of fox and gray squirrels. Journal of Wildlife Management 59:103–10.

Edwards, J. W., D. C. Guynn, Jr., and M. L. Lennartz. 1989. Habitat use by southern fox squirrel in coastal South Carolina. Proceedings of the Annual Conference of the Southeastern Association of Fish and Wildlife Agencies 43:337–45.

Edwards, J. W., D. G. Heckel, and D. C. Guynn, Jr. 1998. Niche overlap in sympatric populations of fox and gray squirrels. Journal of Wildlife Management 62:354–63.

Ellis, L. L. 1955. A survey of the ectoparasites of certain mammals in Oklahoma. Ecology 36:12–18.

Enright, J. B., C. E. Franti, D. E. Behymer, W. M. Longhurst, V. J. Dutson, and M. E. Wright. 1971. *Coxiella burneti* in a wildlife–livestock environment: Distribution of fever in wild animals. American Journal Epidemiology 94:79–90.

Faccio, S. D. 1996. Predation of an eastern chipmunk, *Tamias striatus*, by a gray squirrel, *Sciurus carolinensis*. Canadian Field-Naturalist 110:538.

Fischer, R. A., Jr. 1989. Habitat use and population densities of gray squirrels in south Alabama. M.S. Thesis, Auburn University, Auburn, AL.

Fischer, R. A., and N. R. Holler. 1991. Habitat use and relative abundance of gray squirrels in southern Alabama. Journal of Wildlife Management 55:52–60.

Fisher, E. W., and A. E. Perry. 1970. Estimating ages of gray squirrels by lens-weights. Journal of Wildlife Management 34:825–28.

Fitzwater, W. D., Jr., and W. J. Frank. 1944. Leaf nests of gray squirrels in Connecticut. Journal of Mammalogy 25:160–70.

Flyger, V. 1956. The social behavior and populations of the gray squirrel (*Sciurus carolinensis* Gmelin) in Maryland. Sc.D. Thesis, Johns Hopkins University, Baltimore.

Flyger, V. 1959. A comparison of methods for estimating squirrel populations. Journal of Wildlife Management 23:220–23.

Flyger, V. 1960. Movements and home range of the gray squirrel, *Sciurus carolinensis*, in two Maryland woodlots. Ecology 41:365–69.

Flyger, V. 1969. The 1968 squirrel 'migration' in the eastern United States. Transactions of the Northeast Section of the Wildlife Society 26:69–79.

Flyger, V., and J. E. Gates. 1982. Fox and gray squirrels. Pages 209–29 *in* J. A. Chapman and G. A. Feldhamer, eds. Wild mammals of North America. Johns Hopkins University Press, Baltimore.

Flyger, V., and D. A. Smith. 1980. A comparison of Delmarva fox squirrel and gray squirrel habitats and home range. Transactions of the Northeastern Section of the Wildlife Society 37:19–22.

Fogl, J. G., and H. S. Mosby. 1978. Ageing gray squirrels by cementum annuli in razor-sectioned teeth. Journal of Wildlife Management 42:444–48.

Fowler, L. J., and R. W. Dimmick. 1983. Wildlife use of nest boxes in eastern Tennessee. Wildlife Society Bulletin 11:178–81.

Fox, F. J. 1982. Adaptation of gray squirrel behaviour to autumn germination by white oak acorns. Evolution 36:800–809.

Fox, I. 1940. Fleas of the eastern United States. Iowa State College Press, Ames.

Freeman, R. S. 1962. Studies on the biology of *Taenia crassiceps* (Zeder, 1800) Rudolphi, 1910 (Cestoda). Canadian Journal of Zoology 40:969–90.

George, R. S. 1954. Siphonaptera from Gloucestershire. Entomological Gazette 5:85–94.

Gmelin, J. F. 1788. Caroli a Linne Systema naturae, 13th ed. George Emanuel Beer, Leipzig [Not seen, cited in Hall 1981].

Goodrum, P. D. 1937. Notes on the gray and fox squirrels of eastern Texas. Transactions of the North American Wildlife Conference 2:499–504.

Goodrum, P. D. 1961. The gray squirrel in Texas (Bulletin 42). Texas Parks and Wildlife Department.

Gouras, P. 1964. Duplex function in the gray squirrel's electroretinogram. Nature 203:767–68.

Graham, E., and J. Uhrich. 1943. Animal parasites of the fox squirrel in southeast Kansas. Journal of Parasitology 29:159–60.

Gurnell, J. C. 1987. The natural history of squirrels. Facts on File, New York.

Ha, J. C. 1983. Food supply and home range in the fox squirrel (*Sciurus niger*). M.S. Thesis, Wake Forest University, Winston Salem, NC.

Hadj-Chikh, L. Z., M. A. Steele, and P. D. Smallwood. 1996. Caching decisions by grey squirrels: A test of the handling-time and perishability hypotheses. Animal Behaviour 52:941–48.

Hall, E. R. 1981. The mammals of North America. John Wiley, New York.

Hall, E. R., and K. R. Kelson. 1959. The mammals of North America, Volume 1. Ronald Press, New York.

Hansen, L. P., C. M. Nixon, and S. P. Havera. 1986. Recapture rates and length of residence in an unexploited fox squirrel population. American Midland Naturalist 115:209–15.

Harkema, R. 1936. The parasites of some North Carolina rodents. Ecological Monographs 6:153–232.

Harnishfeger, R. L., J. L. Roseberry, and W. D. Klimstra. 1978. Reproductive levels in unexploited woodlot fox squirrels. Transactions of the Illinois State Academy of Science 71:342–55.

Harwood, P. D., and V. Cooke. 1949. The helminths from a heavily parasitized fox squirrel, *Sciurus niger.* Ohio Journal of Science 49:146–48.

Havera, S. P. 1977. Body composition and organ weights of fox squirrels. Transactions of the Illinois State Academy of Science 70:286–300.

Havera, S. P. 1979. Temperature variation in a fox squirrel nest box. Journal of Wildlife Management 43:251–53.

Havera, S. P., and C. M. Nixon. 1978. Interaction among adult female fox squirrels during their winter breeding season. Transactions of the Illinois State Academy of Science 71:24–38.

Havera, S. P., and C. M. Nixon. 1980. Winter feeding of fox and gray squirrel populations. Journal of Wildlife Management 44:41–55.

Havera, S. P., C. M. Nixon, and F. I. Collins. 1976. Fox squirrels feeding on Buckeye pith. American Midland Naturalist 95:462–64.

Hench, J. E., G. L. Kirkland, Jr., H. W. Setzer, and L. W. Douglass. 1984. Age classification for the gray squirrel based on eruption, replacement, and wear of molariform teeth. Journal of Wildlife Management 48:1409–14.

Herkert, J. R., C. M. Nixon, and L. P. Hansen. 1992. Dynamics of exploited and unexploited fox squirrel (*Sciurus niger*) populations in the midwestern United States. Pages 864–74 *in* D. R. McCullough and P. H. Barrett, eds. Wildlife 2001: Populations. Elsevier, New York.

Heuer, E. T., Jr., and H. R. Perry. 1976. Squirrel and rabbit abundances in the Atchafalaya Basin, Louisana. Proceedings of the Annual Conference of the Southeastern Association of Fish and Wildlife Agencies 30:552–59.

Hicks, E. A. 1949. Ecological factors affecting the activity of the western fox squirrel, *Sciurus niger rufiventer* (Geoffroy). Ecological Monographs 19:288–302.

Hilliard, T. H. 1979. Radio-telemetry of fox squirrels in the Georgia coastal plain. M.S. Thesis, University of Georgia, Athens.

Hixon, H. 1940. Field biology and environmental relationships of the Gulf Coast tick in southern Georgia. Journal of Economic Entomology 33:179–89.

Hoffman, R. A., and C. M. Kirkpatrick. 1959. Current knowledge of tree squirrel reproductive cycles and development. Proceedings of the Annual Conference of the Southeastern Association of Fish and Game Commissioners 13:363–67.

Holland, G. P., and A. H. Benton. 1968. Siphonaptera from Pennsylvania mammals. American Midland Naturalist 80:252–61.

Horwich, R. H. 1972. The ontogeny of social behavior in the gray squirrel (*Sciurus carolinensis*). Advances in Ethology 8:1–103.

Hougart, B., and V. Flyger. 1981. Activity patterns of radio-tracked squirrels. Northeast Wildlife Society 38:11–16.

Huggins, J. G. 1996. Economic effectiveness, efficiency, and selectivity of fox squirrel trapping in pecan groves. Pages 123–26 *in* R. M. Timm and A. C. Crabb, eds. Proceedings of the 17th vertebrate pest conference. University of California, Davis.

Husband, T. P. 1976. Energy metabolism and body composition of the fox squirrel. Journal of Wildlife Management 40:255–63.

Innes, S., and D. M. Lavigne. 1979. Comparative energetics of coat color polymorphs in the eastern gray squirrel *Sciurus carolinensis*. Canadian Journal of Zoology 57:585–92.

Ivey, T. L., and J. E. Frampton. 1987. Use of nest boxes by squirrels in the South Carolina Piedmont. Proceedings of the Annual Conference of the Southeastern Association of Fish and Wildlife Agencies 41:279–87.

Jackson, J. J. 1994. Tree squirrels. Pages B171–75 *in* S. E. Hygnstrom, R. M. Timm, and G. E. Larson, eds. Prevention and control of wildlife damage. University of Nebraska Cooperative Extension, Lincoln.

Jackson, L. L., H. E. Heffner, and R. S. Heffner. 1997. Audiogram of the fox squirrel (*Sciurus niger*). Journal of Comparative Psychology 111:100–104.

Jacobs, G. H. 1974. Scotopic and photopic visual capacities of an arboreal squirrel (*Sciurus niger*). Brain, Behavior, and Evolution 10:307–21.

Jacobs, L. F., and E. R. Liman. 1991. Grey squirrels remember the locations of buried nuts. Animal Behaviour 41:103–10.

Jacobson, H. A., D. C. Guynn, and E. J. Hackett. 1979. Impact of the botfly on squirrel hunting in Mississippi. Wildlife Society Bulletin 7:46–48.

Jacobson, H. A., M. S. Hetrick, and D. C. Guynn. 1981. Prevalence of *Cuterebra emasculator* in squirrels in Mississippi. Journal of Wildlife Diseases 17:79–87.

Jenkins, D. W. 1948. Trombiculid mites affecting man. I. Bionomics with reference to epidemiology in the United States. American Journal of Hygiene 48:22–35.

Jodice, P. G. R., and S. R. Humphrey. 1992. Activity and diet of an urban population of Big Cypress fox squirrels. Journal of Wildlife Management 56:685–93.

Johnson, P. T. 1958. Type specimens of lice (Order Anoplura) in the United States National Museum. Proceedings United States National Museum 108:39–49.

Jordan, J. S. 1971a. Dispersal period in a population of eastern fox squirrels (Research Paper NE-216). USDA Forest Service.

Jordan, J. S. 1971b. Yield from an intensively hunted population of eastern fox squirrels (Research Paper NE-186). USDA Forest Service.

Joseph, T. 1972. *Eimeria lancasterensis* Joseph, 1969 and *E. confusa* Joseph, 1969 from the grey squirrel, *Sciurus carolinensis*. Journal of Protozoology 19:143–50.

Joseph, T. 1975. Experimental transmission of *Eimeria confusa* Joseph 1969 to the fox squirrel. Journal of Wildlife Diseases 11:402–3.

Kantola, A. T. 1986. Fox squirrel home range and mast crops in Florida. M.S. Thesis, University of Florida, Gainesville.

Kantola, A. T., and S. R. Humphrey. 1990. Habitat use by Sherman's fox squirrel (*Sciurus niger shermani*) in Florida. Journal of Mammalogy 71:411–19.

Kardos, E. H. 1954. Biological and systematic studies on the subgenus *Neotrombicula* (genus *Trombicula*) in the central United States (Acarina, Trombiculidae). University of Kansas Science Bulletin 36:69–123.

Kazacos, E. A., K. R. Kazacos, and H. A. Demaree, Jr. 1983. Notoedric mange in two fox squirrels. Journal of American Veterinary Medical Association 183:1281–82.

Keegan, H. L. 1943. Some host records from the parasitological collection of the State University of Iowa. Bulletin of the Brooklyn Entomological Society 38:54–57.

Kenward, R. E. 1985. Ranging behaviour and population dynamics of grey squirrels. Pages 319–30 *in* R. M Sibly and R. H. Smith, eds. Behavioural ecology, Blackwell, Oxford.

Kilham, L. 1954. Metastasizing viral fibromas of gray squirrels: Pathogenesis and mosquito transmission. American Journal Hygiene 61:55–63.

Kiltie, R. A. 1989. Wildfire and the evolution of dorsal melanism in fox squirrels, *Sciurus niger.* Journal of Mammalogy 70:726–39.

Kim, K. C. 1966a. The nymphal stages of three North American species of the genus *Enderleinellus* Fahrenholz. Journal of Medical Entomology 2:327–30.

Kim, K. C. 1966b. The species of *Enderleinellus* (Anoplura, Hoplopleuridae) parasitic on the Sciurini and Tamiasciurini. Journal of Parasitology 52:988–1024.

Kirkpatrick, C. M. 1955. The testis of the fox squirrel in relation to age and seasons. American Journal of Anatomy 97:229–56.

Kirkpatrick, C. M., and E. M. Barnett. 1957. Age criteria in male gray squirrels. Journal of Wildlife Management 21:341–47.

Kirkpatrick, C. M., and R. A. Hoffman. 1960. Ages and reproductive cycles in a male gray squirrel population. Journal of Wildlife Management 24:218–21.

Koprowski, J. L. 1991a. Damage due to scent marking by eastern gray and fox squirrels. Proceedings of the Great Plains Wildlife Damage Control Workshop 10:101–5.

Koprowski, J. L. 1991b. Mixed-species mating chases of fox squirrels and gray squirrels. Canadian Field-Naturalist 105:117–18.

Koprowski, J. L. 1991c. Response of fox squirrels and gray squirrels to a late spring–early summer food shortage. Journal of Mammalogy 72:367–72.

Koprowski, J. L. 1992. Removal of copulatory plugs by female tree squirrels. Journal of Mammalogy 73:572–76.

Koprowski, J. L. 1993a. Sex and species biases in scent-marking by fox squirrels and eastern grey squirrels. Journal of Zoology (London) 230:319–56.

Koprowski, J. L. 1993b. Behavioral tactics, dominance, and copulatory success among male fox squirrels. Ethology, Ecology, and Evolution 5:169–76.

Koprowski, J. L. 1993c. The role of kinship in field interaction among juvenile gray squirrels (*Sciurus carolinensis.*) Canadian Journal of Zoology 71:224–26.

Koprowski, J. L. 1994a. *Sciurus niger*. Mammalian Species 479:1–9

Koprowski, J. L. 1994b. *Sciurus carolinensis*. Mammalian Species 480:1–9

Koprowski, J. L. 1996. Natal philopatry, communal nesting and kinship in fox squirrels and gray squirrels. Journal of Mammalogy 77:1006–17.

Koprowski, J. L., J. L. Roseberry, and W. D. Klimstra. 1988. Longevity records for the fox squirrel. Journal of Mammalogy 69:383–84.

Korschgen, L. 1981. Foods of fox and gray squirrels in Missouri. Journal of Wildlife Management 45:260–66.

Kotler, B. P., J. S. Brown, and M. Hickey. 1999. Food storability and the foraging behavior of fox squirrels (*Sciurus niger.*) American Midland Naturalist 142:77–87.

Laerm, J., W. M. Ford, and B. R. Chapman. 2000. Conservation status of terrestrial mammals in the southeastern United States. Pages 4–16 *in* B. R. Chapman and J. Laerm, eds. Fourth colloquium on conservation of mammals in the southeastern United States (Occasional Paper No. 12). North Carolina Museum of Natural Sciences and Biological Survey.

Langham, R. F., R. L. Rausch, and J. F. Williams. 1990. Cysticerci of *Taenia mustelae* in the fox squirrel. Journal of Wildlife Diseases 26:295–96.

Larson, B. J. 1990. Habitat utilization, population dynamics and long-term viability in an insular population of Delmarva fox squirrels (*Sciurus niger cinereus.*) M.S. Thesis, University of Virginia, Charlottsville.

Layne, J. N. 1971. Fleas (Siphonaptera) of Florida. Florida Entomologist 54:35–51.

Layne, J. N. 1998. Nest box use and reproduction in the gray squirrel (*Sciurus carolinensis*) in Florida. Pages 61–70 *in* M. A. Steele, J. F. Merritt, and D. A. Zegers, eds. Ecology and evolutionary biology of tree squirrels (Special Publication 6). Virginia Museum of Natural History.

Layne, J. N., and G. E. Woolfenden. 1958. Gray squirrels feeding on insects in car radiators. Journal of Mammalogy 39:595–96.

Lee, J. C. 1999. Ecology of the southern fox squirrel on Spring Island, South Carolina. M.S. Thesis, University of Georgia, Athens.

Lewis, E., G. L. Bigler, and M. B. Jefferies. 1975. Public health and the urban gray squirrel. Journal of Wildlife Diseases 11:502–4.

Lima, S. L., and and T. J. Valone. 1986. Influence of predation risk on diet selection: A simple example in the grey squirrel. Animal Behaviour 34:536–44.

Linnaeus, C. 1758. Systema naturae, Vol. 1, 10th ed., Laurentii Salvii, Stockholm [Not seen, cited in Hall 1981].

Lipovsky, L. J., D. A. Crossley, Jr., and R. B. Loomis. 1955. A new genus of chigger mites (Acarina, Trombiculidae). Journal of the Kansas Entomological Society 28:136–43.

Lishak, R. S. 1982a. Vocalization of nestling gray squirrels. Journal of Mammalogy 63:446–52.

Lishak, R. S. 1982b. Gray squirrel mating calls: A spectrographic and ontogenic analysis. Journal of Mammalogy 63:661–63.

Lishak, R. S. 1984. Alarm vocalizations of adult gray squirrels. Journal of Mammalogy 65:681–84.

Loeb, S. C., and M. L. Lennartz. 1989. The fox squirrel (*Sciurus niger*) in southeastern pine–hardwood forests. USDA Forest Service Southeastern Forest Experiment Station General Technical Report SE 58:142–48.

Loeb, S. C., and N. D. Moncrief. 1993. The biology of fox squirrels (*Sciurus niger*) in the Southeast: A review. Pages 1–19 *in* N. D. Moncrief, J. W. Edwards, and P. A. Tappe, eds. Proceedings of the second symposium on southeastern fox squirrels, *Sciurus niger* (Special Publication No. 1). Virginia Museum of Natural History.

Long, C. A., and T. Frank. 1968. Morphometric variation and function in the baculum, with comments on correlation of parts. Journal of Mammalogy 49:32–43.

Loomis, R. B. 1954. A new subgenus and six new species of chigger mites (genus *Trombicula*) from the central United States. University of Kansas Science Bulletin 36:919–41.

Loomis, R. B. 1955. *Trombicula gurneyi* Ewing and two new related chigger mites (Acarina, Trombiculidae). University of Kansas Science Bulletin 37:251–67.

Loomis, R. B. 1956. The chigger mites of Kansas (Acarina, Trombiculidae). University of Kansas Science Bulletin 37:1195–1443.

Lowery, G. H., Jr. 1974. The mammals of Louisana and its adjacent waters. Louisana State University Press, Baton Rouge.

Lucker, J. T. 1943. A new trichostrongylid nematode from the stomachs of American squirrels. Journal of the Washington Academy of Science 33:75–79.

Ludwick, R. J., J. P. Fontenot, and H. S. Mosby. 1969. Energy metabolism of the eastern gray squirrel. Journal of Wildlife Management 33:569–75.

Lustig, G. H., and V. Flyger. 1975. Observations and suggested management practices for the endangered Delmarva fox squirrel. Proceedings of the Annual Conference of the Southeastern Association of Game and Fish Commissioners 29:433–40.

MacDonald, I. M. V. 1992. Grey squirrels discriminate red from green in a foraging situation. Animal Behaviour 43:694–95.

Main, A. J., Jr. 1970. Distribution, seasonal abundance and host preference of fleas in New England. Proceedings of the Entomological Society of Washington 72:73–89.

Masterson, R. A., H. W. Stegmiller, M. A. Parsons, C. C. Croft, and C. B. Spencer. 1971. California encephalitis: An endemic puzzle in Ohio. Health Laboratory Science 8:89–96.

Mathewson, J. A., and K. E. Hyland. 1962. The ectoparasites of Rhode Island mammals. Part 2, A collection of Anoplura from nondomestic hosts. Journal of the New York Entomological Society 70:167–74.

Mathewson, J. A., and K. E. Hyland. 1964. The ectoparasites of Rhode Island mammals. Part 3, A collection of fleas from nondomestic hosts (Siphonaptera). Journal of the Kansas Entomological Society 37:157–63.

McCloskey, R. J. 1977. Accuracy of criteria used to determine age of fox squirrels. Proceedings of the Iowa Academy of Science 84:32–34.

McCloskey, R. J., and K. C. Shaw. 1977. Copulatory behavior of the fox squirrel. Journal of Mammalogy 58:663–65.

McCloskey, R. J., and P. A. Vohs, Jr. 1971. Chronology of reproduction of the fox squirrel in Iowa. Proceedings of the Iowa Academy of Science 78:12–15.

McComb, W. C., and R. E. Noble. 1981. Nest-box and natural-cavity use in three mid-south forest habitats. Journal of Wildlife Management 45:93–101.

McComb, W. C., S. A. Bonney, R. M. Sheffield, and N. D. Cost. 1986. Den tree characteristics and abundance in Florida and South Carolina. Journal of Wildlife Management 50:584–91.

McQuown, A. L. 1954. *Capillaria hepatica*. American Journal of Clinical Pathology 24:448–52.

Meggitt, F. J. 1924. The cestodes of mammals. Edward Goldston, London.

Merson, M. H., C. J. Cowles, and R. L. Kirkpatrick. 1978. Characteristics of captive gray squirrels exposed to cold and food deprivation. Journal of Wildlife Management 42:202–5.

Montgomery, S. D., J. B. Whelan, and H. S. Mosby. 1975. Bioenergetics of a woodlot gray squirrel population. Journal of Wildlife Management 39:709–17

Moore, J. C. 1957. The natural history of the fox squirrel, *Sciurus niger shermani*. Bulletin of the American Museum of Natural History 114:1–71.

Moore, J. C. 1968. Sympatric species of tree squirrels mix in mating chase. Journal of Mammalogy 49:531–33.

Morgan, B. B. 1943. The Physaloptera (Nematoda) of rodents. Wasmann Collector 5:99–107.

Morlan, H. B. 1952. Host relationships and seasonal abundance of some southwest Georgia ectoparasites. American Midland Naturalist 48:74–93.

Mosby, H. S. 1969. The influence of hunting on a woodlot gray squirrel population. Journal of Wildlife Management 33:59–73.

Mosby, H. S., R. L. Kirkpatrick, and J. O. Newell. 1977. Seasonal vulnerability of gray squirrels to hunting. Journal of Wildlife Management 41:284–89.

Mossman, H. W., R. A. Hoffman, and C. M. Kirkpatrick. 1955. The accessory genital glands of male gray and fox squirrels correlated with age and reproductive cycles. American Journal of Anatomy 97:257–302.

Moulton, D. W., and W. H. Thompson. 1971. California group virus infections in small, forest-dwelling mammals of Wisconsin: Some ecological considerations. American Journal of Tropical Medical Hygiene 20:474–82.

Muchlinski, A., and K. Shump. 1979. The sciurid tail: A possible thermoregulatory mechanism. Journal of Mammalogy 60:652–54.

Newman, J. A., G. M. Recer, S. M. Zwicker, and T. Caraco. 1988. Effects of predation hazard on foraging "constraints": Patch-use strategies in grey squirrels. Oikos 53:93–97.

Nichols, J. T. 1958. Food habits and behavior of the gray squirrel. Journal of Mammalogy 39:376–80.

Nixon, C. M., S. P. Havera, and R. E. Greenberg. 1978. Distribution and abundance of the gray squirrel in Illinois (Biological Note No. 105). Illinois Natural History Survey.

Nixon, C. M., S. P. Havera, and L. P. Hansen. 1980a. Initial response of squirrels to forest changes associated with selection cutting. Wildlife Society Bulletin 8:298–306.

Nixon, C. M., M. W. McClain, and R. W. Donohoe. 1980b. Effects of clear-cutting on gray squirrels. Journal of Wildlife Management 44:403–12.

Nixon, C. M., S. P. Havera, and L. P. Hansen. 1984. Effects of nest boxes on fox squirrel demography, condition, and shelter use. American Midland Naturalist 112:157–71.

Nixon, C. M., L. P. Hansen, and S. P. Havera. 1991. Growth patterns of fox squirrels in east-central Illinois. American Midland Naturalist 125:168–72.

Nixon, C. M., and R. W. Donohoe. 1979. Squirrel nest boxes: Are they effective in young hardwood stands? Wildlife Society Bulletin 7: 283–84.

Nixon, C.M., and L.P. Hansen. 1987. Managing forests to maintain populations of gray and fox squirrels (Technical Bulletin 5). Illinois Department of Conservation.

Nixon, C. M., and M. W. McClain. 1969. Squirrel population decline following a late spring frost. Journal of Wildlife Management 33:353–57.

Nixon, C. M., and M. W. McClain. 1975. Breeding seasons and fecundity of female gray squirrels in Ohio. Journal of Wildlife Management 39:426–38.

Nixon, C. M., D. M. Worley, and M. W. McClain. 1968. Food habits of squirrels in southeast Ohio. Journal of Wildlife Management 32:294–305.

Nixon, C.M., R W. Donohoe, and T. Nash. 1974. Overharvest of fox squirrels from two woodlots in western Ohio. Journal of Wildlife Management 38:67–80.

Nixon, C. M., M. W. McClain, and R. W. Donohoe. 1975. Effects of hunting and mast crops on a squirrel population. Journal of Wildlife Management 39:1–25.

Nupp, T. E. 1992. Nest box use and population densities of gray squirrels in southern Alabama. M.S. Thesis, Auburn University, Auburn, AL.

Olexik, W. A., A. E. Perry, and W. E. Wilhelm. 1969. Ectoparasites and helminth endoparasites of tree squirrels of southwest Tennessee. Journal of the Tennessee Academy of Science 44:4–6.

Pack, J. C., H. S. Mosby, and P. B. Siegel. 1967. Influence of social hierarchy on gray squirrel behavior. Journal of Wildlife Management 31:720–28.

Packard, R. L. 1956. The tree squirrels of Kansas. Museum of Natural History, State Biological Survey, University of Kansas, Lawrence.

Paglione, L. J. 1996. Population status and habitat management of Delmarva fox squirrels. M.S. Thesis, University of Massachusetts, Amherst.

Parker, J. C. 1968. Parasites of the gray squirrel in Virginia. Journal of Parasitology 54:633–34.

Parker, J. C., and R. B. Holliman. 1971. Observations on parasites of gray squirrels during the 1968 emigration in North Carolina. Journal of Mammalogy 52:437–41.

Pasitschniak-Arts, M., and J. F. Bendell. 1990. Behavioural differences between locally recruiting and dispersing gray squirrels, *Sciurus carolinensis*. Canadian Journal of Zoology 68:935–41.

Pitts, R. M., J. R. Choate, and V. L. Futrell. 1996. Predation on the house wren, *Troglodytes aedon*, by the eastern gray squirrel (*Sciurus carolinensis*). Transactions of the Missouri Academy of Science 30:70–71.

Powers, J. S. 1993. Fox squirrel home range and habitat use in the southeastern coastal plain. M.S. Thesis, Auburn University, Auburn, AL.

Price, D. L. 1962. Description of *Dipetalonema interstitium* n. sp. from the raccoon. Proceedings of the Helminthological Society of Washington 29:77–82.

Pritchett, H. D. 1938. Rabies in two gray squirrels. Journal of the American Veterinary Medical Association 45:563–64.

Rankin, J. S., Jr., 1946. Helminth parasites of birds and mammals in western Massachusetts. American Midland Naturalist 35:756–68.

Rausch, R., and J. D. Tiner. 1948. Studies on the parasitic helminths of the north central states. Part 1, Helminths of Sciuridae. American Midland Naturalist 39:728–47.

Reiber, R. J., and E. E. Byrd. 1942. Some nematodes from mammals of Reelfoot Lake in Tennessee. Journal of the Tennessee Academy of Science 17:78–89.

Reichard, T. A. 1976. Spring food habits and feeding behavior of fox squirrels and red squirrels. American Midland Naturalist 96:443–50.

Rhodes, M. N. 1989. Effects of exploitation on population parameters of fox squirrels and gray squirrels: A hunting experiment and computer model. Ph.D. Dissertation, University of Missouri, Columbia.

Riege, D. A. 1991. Habitat specialization and social factors in distribution of red and gray squirrels. Journal of Mammalogy 72:152–62.

Robb, J. R., M. S. Cramer, A. R. Parker, and R. P. Urbanek. 1996. Use of tree cavities by fox squirrels and raccoons in Indiana. Journal of Mammalogy 77:1017–28.

Roher, D. P., M. J. Ryan, S. W. Nielsen, and E. E. Roscoe. 1981. Acute fatal toxoplasmosis in squirrels. Journal of the American Veterinary Medical Association 179:1099–1101.

Ross, J. M. 1996. Habitat components and relative abundance of the fox squirrel in relation to forest management practices. M.S. Thesis, Mississippi State University, Starkville.

Sanderson, H. R. 1975. Den-tree management for gray squirrels. Wildlife Society Bulletin 3:125–31.

Sanderson, H. R., W. M. Healy, J. C. Pack, J. D. Gill, and J. W. Thomas. 1976. Gray squirrel habitat and nest tree preference. Proceedings of the Annual Conference of the Southeastern Association of Game and Fish Commissioners 30:609–16.

Sanderson, H. R., C. M. Nixon, R. W. Donohoe, and L. P. Hansen. 1980. Grapevines: An important component of gray and fox squirrel habitat. Wildlife Society Bulletin 8:307–10.

Schorger, A. W. 1949. Squirrels in early Wisconsin. Transactions of the Wisconsin Academy of Science, Arts, and Letters 39:195–247.

Self, J. T., and J. H. Esslinger. 1955. A new species of bothriocephalid cestode from the fox squirrel (*Sciurus niger* Linn.). Journal of Parasitology 41:256–58.

Seton, E. T. 1920. Migrations of the gray squirrel (*Sciurus carolinensis*). Journal of Mammalogy 1:53–58.

Shaftesbury, A. D. 1934. The Siphonaptera (fleas) of North Carolina with special reference to sex ratios. Journal of the Elisha Mitchell Scientific Society 49:247–63.

Sharp, W. M. 1958. Aging gray squirrels by use of tail-pelage characteristics. Journal of Wildlife Management 22:29–34.

Shaw, W. W., and W. R. Mangun. 1984. Nonconsumptive use of wildlife in the United States. (Resource Publication 154). U.S. Fish and Wildlife Service.

Shealer, D. A., J. P. Snyder, V. C. Dreisbach, D. F. Sunderlin, and J. A. Novak. 1999. Foraging patterns of eastern gray squirrels (*Sciurus carolinensis*) on goldenrod gall insects, a potentially important winter food resource. American Midland Naturalist 142:102–10.

Sheperd, B. F., and R. K. Swihart. 1995. Spatial dynamics of fox squirrels (*Sciurus niger*) in fragmented landscapes. Canadian Journal of Zoology 73:2098–2105.

Shivaprasad, H. L., J. P. Sundberg, and R. Ely. 1984. Malignant mixed (carcinosarcoma) mammary tumor in a gray squirrel. Veterinary Pathology 21:115–17.

Short, H. L., and W. B. Duke. 1971. Seasonal food consumption and body weights of captive tree squirrels. Journal of Wildlife Management 35:435–39.

Shorten, M. 1951. Some aspects of the biology of the grey squirrel (*Sciurus carolinensis*) in Great Britain. Proceedings of the Zoological Society of London 121:427–59.

Shotts, E. B., Jr., C. L. Andrews, and T. W. Harvey. 1975. Leptospirosis in selected wild mammals of the Florida panhandle and southwestern Georgia, U.S.A. Journal of the American Veterinary Medicine Association 167:587–89.

Shugars, J. C. 1986. Harvest analysis of a Maryland gray squirrel population. Proceedings of the Annual Conference of the Southeastern Association of Fish and Wildlife Agencies 40:382–88.

Smallwood, P. D., and W. D. Peters. 1986. Grey squirrel food preferences: The effects of tannin and fat concentration. Ecology 67:168–74.

Smith, C. C., and D. Follmer. 1972. Food preferences of squirrels. Ecology 53:82–91.

Smith, N. B. 1967. Some aspects of reproduction in female gray squirrels, *Sciurus carolinensis carolinensis* Gmelin, in Wake County, North Carolina. M.S. Thesis, North Carolina State University, Raleigh.

Smith, N. B., and F. S. Barkalow, Jr. 1967. Precocious breeding in the gray squirrel. Journal of Mammalogy 48:328–30.

Society of American Foresters. 1967. Forest cover types of North America (exclusive of Mexico). Society of American Foresters, Washington, DC.

Sonenshine, D. E., and T. J. Stout. 1971. Ticks infesting medium-sized wild mammals in two forest localities in Virginia (Acarina:Ixodidae). Journal of Medical Entomology 8:217–27.

Stapanian, M. A., and C. C. Smith. 1978. A model of seed scatterhoarding: Coevolution of fox squirrels and black walnuts. Ecology 59:884–96.

Steele, M. A., G. Turner, P. D. Smallwood, J. O. Wolff, and J. Radillo. 2001. Cache management by small mammals: Experimental evidence for the significance of acorn-embryo excision. Journal of Mammalogy 82:35–42.

Sundberg, J. P., D. Hill, and M. J. Ryan. 1982. Cryptosporidiosis in a gray squirrel. Journal of the American Veterinary Medicine Association 181:1420–22.

Swihart, R. K., and T. E. Nupp. 1998. Modeling population responses of north American tree squirrels to agriculturally induced fragmentation of forests. Pages 1–19 *in* M. A. Steele, J. F. Merritt, and D.A. Zegers eds. Ecology and evolutionary biology of tree squirrels (Special Publication 6). Virginia Museum of Natural History.

Tappe, P. A. 1991. Capture–recapture methods for estimating southern fox squirrel abundance. Ph.D. Dissertation, Clemson University, Clemson, SC.

Tappe, P. A., and D. C. Guynn, Jr. 1998. Southeastern fox squirrels: *r*- or *K*-selected? Implications for management. Pages 239–47 *in* M. A. Steele, J. F. Merritt, and D. A. Zegers, eds. Ecology and evolutionary biology of tree squirrels (Special Publication 6). Virginia Museum of Natural History.

Taylor, G. J. 1973. Present status and habitat survey of the Delmarva fox squirrel (*Sciurus niger cinereus*) with a discussion of reasons for its decline. Proceedings of the Annual Conference of the Southeastern Association of Game and Fish Commissioners 27:278–87.

Taylor, J. C. 1968. The use of marking points by grey squirrels. Journal of Zoology (London) 155:246–47.

Taylor, J. C. 1977. The frequency of grey squirrel (*Sciurus carolinensis*) communication by the use of scent marking points. Journal of Zoology (London) 183:543–45.

Thompson, D. C. 1977a. Diurnal and seasonal activity of the grey squirrel. Canadian Journal of Zoology 55:1185–89.

Thompson, D. C. 1977b. Reproductive behavior of the grey squirrel. Canadian Journal of Zoology 55:1176–84.

Thompson, D. C. 1978a. The social system of the grey squirrel. Behaviour 64:305–28.

Thompson, D. C. 1978b. Regulation of a northern grey squirrel (*Sciurus carolinensis*) population. Ecology 59:708–15.

Thompson, D. C., and P. S. Thompson. 1980. Food habits and caching behavior of urban grey squirrels. Canadian Journal of Zoology 58:701–10.

Tiner, J. D. 1949. Preliminary observations on the life history of *Ascaris columnaris*. Journal of Parasitology 35:13.

Tugwell, P., and J. L. Lancaster, Jr. 1962. Results of a tick-host study in northwest Arkansas. Journal of the Kansas Entomological Society 35:202–11.

Turner, D. A., and J. Laerm. 1993. Systematic relationships of populations of fox squirrel (*Sciurus niger*) in the southeastern United States. Pages 21–36 *in* N. D. Moncrief, J. W. Edwards, and P. A. Tappe, eds. Proceedings of the second symposium on southeastern fox squirrels, *Sciurus niger* (Special Publication 1). Virginia Museum of Natural History.

Uhlig, H. G. 1955. The gray squirrel: Its life history, ecology, and population characteristics in West Virginia (P-R Report, Project 31-R). West Virginia Conservation Commission.

Uhlig, H. G. 1956. Effect of legal restrictions and hunting on gray squirrel populations in West Virginia. Transactions of the North American Wildlife Conference 21:330–38.

Uhrich, J., and E. Graham. 1941. The animal parasites of the fox squirrel, *Sciurus niger rufiventer* and the gray squirrel, *Sciurus carolinensis carolinensis*. Anatomical Record 81:65.

U.S. Fish and Wildlife Service. 1993. Delmarva fox squirrel (*Sciurus niger cinereus*) recovery plan, 2nd rev. U.S. Fish and Wildlife Service, Northeast Region, Hadley, MA.

Vander Wall, S. B. 1990. Food hoarding in animals. University of Chicago Press, Chicago.

Wade, O., and P. T. Gilbert. 1942. The baculum of some Sciuridae and its significance in determining relationships. Journal of Mammalogy 23:52–63.

Walton, B. C., and K. W. Walls. 1964. Prevalence of toxoplasmosis in wild animals from Fort Stewart, Georgia, as indicated by serological tests and mouse inoculation. American Journal of Tropical Medicine and Hygiene 13:530–33.

Warren, R. C., and G. A. Hurst. 1980. Squirrel densities in pine–hardwood forests and streamside management zones. Proceedings of the Annual Conference of the Southeastern Association of Fish and Wildlife Agencies 34:492–98.

Webb, J. P., Jr., and R. B. Loomis. 1971. The subgenus *Scapuscutala* of the genus *Microtrombicula* (Acarina:Trombiculidae) from North America. Journal of Medical Entomology 8:319–29.

Webley, G. E., and E. Johnson. 1983. Reproductive physiology of the grey squirrel (*Sciurus carolinensis*). Mammal Review 13:149–54.

Webster, W. D., J. F. Parnell, and W. C. Bigg, Jr. 1985. Mammals of the Carolinas, Virginia, and Maryland. University of North Carolina Press, Chapel Hill.

Weigl, P. D., M. A. Steele, L. J. Sherman, J. C. Ha, and T. L. Sharpe. 1989. The ecology of the fox squirrel (*Sciurus niger*) in North Carolina: Implications for survival in the Southeast. Bulletin of Tall Timbers Research Station 24:1–93.

Weigl, P. D., L. J. Sherman, A. I. Williams, M. A. Steele, and D. S. Weaver. 1998. Geographic variation in the fox squirrel (*Sciurus niger*): A consideration of size clines, habitat vegetation, food habits and historical biogeography. Pages 171–84 *in* M. A. Steele, J. F. Merritt, and D. A. Zegers, eds. Ecology and evolutionary biology of tree squirrels (Special Publication 6). Virginia Museum of Natural History.

Williams, K. S., and S. R. Humphrey. 1979. Distribution and status of the endangered Big Cypress fox squirrel (*Sciurus niger avicennia*) in Florida. Florida Scientist 42:201–5.

Wobeser, G. 1969. Tetanus in a gray squirrel. Bulletin of the Wildlife Disease Association 5:18–19.

Wooding, J. B. 1997. Distribution and population ecology of the fox squirrel in Florida. Ph.D. Dissertation, University of Florida, Gainesville.

Woods, G. T. 1941. Mid-summer food of gray squirrels. Journal of Mammalogy 22:321–22.

Woods, C. E., and O. R. Larson. 1969. North Dakota fleas, Part 2, Records from man and other mammals. Proceedings of the North Dakota Academy of Science 23:28.

Yokoyama, S., and F. B. Radlwimmer. 1998. The "five-sites" rules and the evolution of red and green color vision in mammals. Molecular Biology and Evolution 15:560–67.

Yolton, R. L. 1975. The visual system of the western gray squirrel: Anatomical, electroretinographic and behavioral studies. Ph.D. Dissertation, University of Texas, Austin.

Zelley, R. A. 1971. The sounds of the fox squirrel, *Sciurus niger rufiventer*. Journal of Mammalogy 52:597–604.

Zielinski, D. L. 1996. Reproduction and environmental influences on energy regulation. Ph.D. Dissertation, Purdue University, West Lafayette, IN.

Zimmerman, W. J., and E. D. Hubbard. 1969. Trichiniasis in wildlife of Iowa. American Journal of Epidemiology 90:84–92.

John Edwards, Division of Forestry, West Virginia University, Morgantown, West Virginia 26505-6125. Email:jedwards@wvu.edu.

Mark Ford, USDA Forest Service, Northeastern Research Station, Parsons, West Virginia 26287. Email:mford@ff.fed.us.

David Guynn, Department of Forest Resources, Clemson University, Clemson, South Carolina 29634-1003. Email: dguynn@clemson.edu.

13

Pine Squirrels

Tamiasciurus hudsonicus and *T. douglasii*

Richard H. Yahner

NOMENCLATURE

COMMON NAMES. Red squirrel, Bang's red squirrel, boomer, chatterbox, chickaree, fairy diddle, piney, pine squirrel, rusty squirrel, spruce squirrel (Flyger and Gates 1982; Merritt 1987; Whitaker and Hamilton 1998)
SCIENTIFIC NAME. *Tamiasciurus hudsonicus*
SUBSPECIES. *T. h. abieticola, T. h. baileyi, T. h. columbiensis, T. h. dakotensis, T. h. dixiensis, T. h. fremonti, T. h. grahamensis, T. h. gymnicus, T. h. hudsonicus, T. h. kenaiensis, T. h. lanuginosus, T. h. laurentianus, T. h. loquax, T. h. lychnuchus, T. h. minnesota, T. h. mogollonensis, T. h. pallescens, T. h. petulans, T. h. picatus, T. h. preblei, T. h. regalis, T. h. richardsoni, T. h. streatori, T. h. ungavensis,* and *T. h. ventorum* (Hall 1981)

COMMON NAMES. Douglas' squirrel, chickaree, pine squirrel.
SCIENTIFIC NAME. *Tamiasciurus douglasii*
SUBSPECIES. *T. d. albolimbatus, T. d. douglasii, T. d. mearnsi,* and *T. d. mollipilosus* (Hall 1981)

DISTRIBUTION

The distribution of red squirrels is quite extensive, ranging from much of northern Canada and Alaska to the northeastern, upper midwestern, and western United States (Fig. 13.1). Populations extend southward along the mountainous regions to northern Georgia in the eastern United States and to southern Arizona and New Mexico in the western United States. Conversely, the distribution of Douglas' squirrels is more restricted, extending from only southwestern British Columbia to northern Baja California; the distribution of this sciurid does not overlap that of the red squirrel.

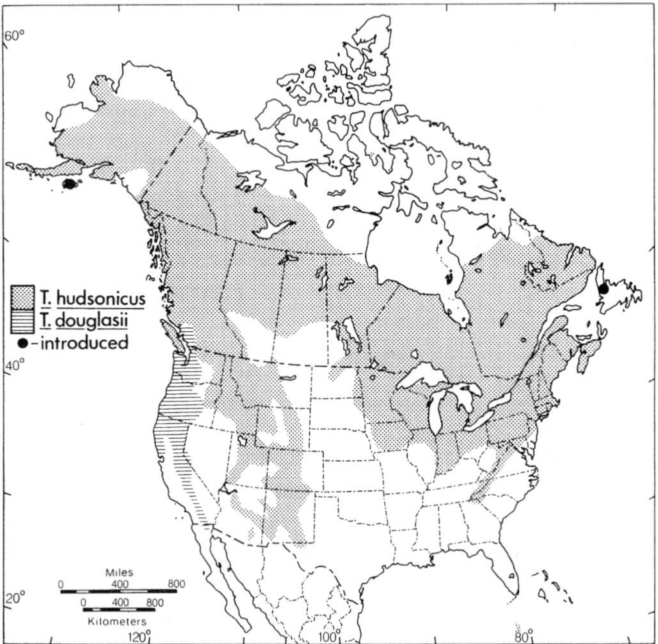

FIGURE 13.1. Distribution of the red squirrel (*Tamiasciurus hudsonicus*) and Douglas's squirrel (*Tamiasciurus douglasii*).

DESCRIPTION

Red squirrels are typically rusty reddish or reddish brown on the dorsal surface and white or grayish white on the ventral surface (Hall 1981; Flyger and Gates 1982). The tail is rusty to yellowish rusty, with white-to tawny-tipped hairs. Albinistic and melanistic red squirrels have been reported (Wood 1965; Mengel and Jenkinson 1971). By comparison, Douglas' squirrels are olive brown to grayish brown on the dorsal surface, with variation individually, geographically, and seasonally (Hall 1981). They are yellowish on the ventral surface and, like the red squirrel, have a blackish lateral line. The tail is black or blackish, and the hairs are white- or buffy-tipped.

The sexes in red and Douglas' squirrels are similar in coloration (Hall 1981; Flyger and Gates 1982). Each has a blackish lateral line, which is most distinct in summer, which separates the dorsal and ventral coloration. Tails of *Tamiasciurus* are shorter and flatter than tails of other tree squirrels (*Sciurus* spp.).

The body measurements of red and Douglas' squirrels, respectively, are as follows: total length, 270–380 and 270–345 mm; tail length, 92–158 and 102–156 mm; hind foot length, 35–57 and 45–55 mm (Hall 1981; Flyger and Gates 1982). The condylobasal length of the skull in both species is 42–48 mm, but tends to be longer in males than in females (Nellis 1969). The dental formula for *Tamiasciurus* is I 1/1, C 0/0, P 1/1, M 3/3 (Fig. 13.2).

Body weights of adult red squirrels and Douglas' squirrels vary from 198 to 250g and from 170 to 204g, respectively (Burt and Grossenheider 1976). Red and Douglas' squirrels are genetically similar in terms of chromosome number, with a diploid number (2*N*) of 46 (Nadler and Hoffman 1970).

ANATOMY AND PHYSIOLOGY

Red and Douglas' squirrels undergo two molts per year, and the winter pelage is more brightly colored than the summer pelage (Flyger and Gates 1982). The winter molt begins in the hindquarter region of the body and progresses anteriorly, whereas the summer molt begins in the nose region and moves posteriorly (Nelson 1945; Layne 1954).

Pine squirrels have sebaceous and sudoriferous glands near the oral angle (or cheek) region, as do many rodents (Quay 1965). At least in the red squirrel, both sexes use these glands when cheek rubbing (Ferron 1976). Squirrels often rub or gnaw objects, thereby leaving scent on objects along frequently used paths in their environment. Cheek rubbing probably helps squirrels to orient themselves and maintain familiarity with their home range (Ferron and Ouellet 1989). Cheek rubbing by

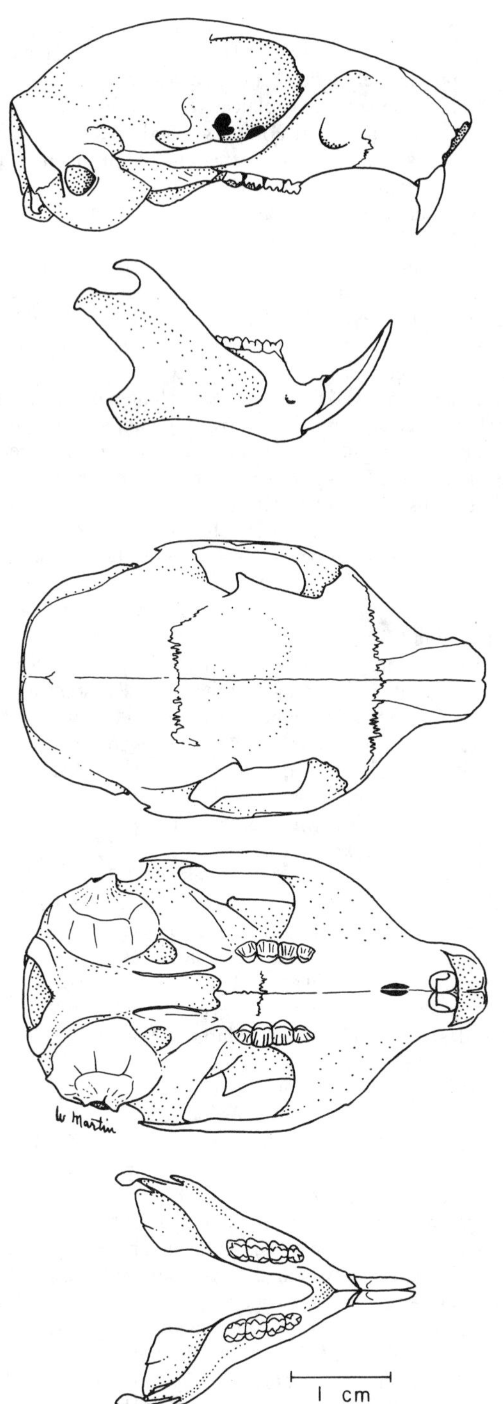

FIGURE 13.2. Skull of the red squirrel (*Tamiasciurus hudsonicus*), which is similar to Douglas' squirrel (*Tamiasciurus douglasii*). From top to bottom: lateral view of cranium, lateral view of mandible, dorsal view of cranium, ventral view of cranium, dorsal view of mandible.

a red squirrel may help advertise to neighboring conspecifics that its territory is occupied.

As with most squirrel species, red and Douglas' squirrels are diurnal, with eyes well adapted to seeing in the daytime (Flyger and Gates 1982). Nonetheless, the night vision of *Tamiasciurus* and *Sciurus* probably exceeds that of humans (Dippner and Armington 1971).

Pine squirrels have quite different reproductive anatomy from other tree squirrels (Flyger and Gates 1982; Whitaker and Hamilton 1998). For instance, male *Tamiasciurus* have a long, threadlike, symmetrical penis, no bulbar glands, and tiny Cowpers glands; in addition, the baculum is minute. Female *Tamiasciurus* have a uniquely coiled vagina. As in other male sciurids, the testes descend into the scrotum; scrota become pigmented and are hairless at about 10 months of age. The teats of females are pigmented at the first pregancy and thereafter for life.

The metabolic rate of pine squirrels is between 143 and 168 kcal/kg, which varies with season and ambient temperature (Grodzinski 1971). Red squirrels (and perhaps Douglas') have brown adipose (fat) tissue, which is mobilized faster in winter than summer (Aleksuik 1970, 1971) and quicker than in other rodents and mice (family Muridae) (Ferguson and Folk 1971).

REPRODUCTION AND DEVELOPMENT

Although two litters per year have been reported in southern populations of pine squirrels (Hamilton 1939; Layne 1954), a single litter is characteristic throughout the range (Kemp and Keith 1970; Dolbeer 1973). Estrus timing and hence breeding in pine squirrels are quite asynchronous compared to hibernating ground squirrels (e.g., *Spermophilus*) (Rusch and Reeder 1978). Estrus onset may vary by as much as 2 months among years (Rusch and Reeder 1978). Moreover, even during the same year, estrus timing may differ by 1–2 months among females in the same population, and individual squirrels tend to be consistent among years in timing of estrus (Becker 1993). Estrus timing appears to be a function of an interaction between an endogenous circannual reproductive cycle and net energy gain (e.g., food availability). In particular, availability of high-quality food rather than quantity of food may enable some pine squirrels to enter estrus and produce multiple litters in the same year (Kemp and Keith 1970; Sullivan 1990), thereby extending the length of the breeding season in a given squirrel population. Larsen et al. (1997) showed that size of litters, number of offspring emerging from nests, and number of offspring weaned in red squirrels are not affected by food supplementation experiments in the field.

Because pine squirrels are promiscuous (Smith 1968), the staggered occurrence of estrus among females in the same population may increase their opportunity to show preference for certain males with whom to mate (Becker 1993). In addition, asynchrony in the timing of estrus within a squirrel population will vary parturition and weaning dates, which may reduce competition among juveniles for limited territories at the same time of the year.

Estrus in an individual female pine squirrel lasts only a single day (Smith 1968), as in eastern chipmunks (*Tamias striatus*) (Yahner 1978). As is true of *T. striatus,* one to several males congregate on the estrous female's territory, and a conspicuous mating chase occurs. Furthermore, the female may mate with more than one male during the same chase (Lair 1985). Conception occurs immediately after mating because spontaneous ovulation occurs in *Tamiasciurus* (Layne 1954; Millar 1968; Zirul 1970; Modafferi 1972). Dominant males in a mating chase have a higher mating success than subordinant males; a dominant male usually stays near the female between copulations and chases other approaching males (Koford 1982).

The gestation period of pine squirrels is 33 ± 2 days (Lair 1985). Litter size ranges from one to seven young, with each young averaging about 7.5 g in body weight (Ferron and Prescott 1977). The mean numbers of placental scars and corpora lutea per female range from 3.6 to 4.0 and from 3.9 to 4.1, respectively (Layne 1954; Wood 1967; LaPierre 1986). The young are born during March–May (Davis 1969).

Young pine squirrels are born hairless and with their eyes closed (Flyger and Gates 1982). Only females care for the young (Smith 1968). At 6 days of age, the young change from a pink to a noticeably dark pigmentation (Flyger and Gates 1982). Juveniles develop the dark lateral line by 16 days and are well covered by dorsal hairs at 20 days. Their lower incisors erupt by 24 days and the upper incisors by 41 days. The eyes open at 27–35 days of age (Svihla 1930). Young are weaned by 7–8 weeks after birth (Ferron and Prescott 1977; Lair 1985). The age of first breeding is typically 10–12 months (Millar 1970).

ECOLOGY

Home Range and Territory. Home range size in pine squirrels may range from 0.29 to 1.5 ha, but it is often somewhat larger in males than in females (Davis 1969; Gurnell 1984). Size can change with food availability, and is smaller when food is abundant and larger when food is scarce (Sullivan 1990). All or part of the home range may be defended as a territory, consisting of 0.2–1.2 ha in size and averaging 0.65 ha or less (C. C. Smith 1968; Davis 1969; Gurnell 1984; Larsen and Boutin 1994, 1995). Territory size in Douglas' squirrel is typically less than 1 ha (Koford 1982). The shape of territories in pine squirrels is often circular or elliptical (C. C. Smith 1968; Larsen and Boutin 1994, 1995).

In highly suitable habitats of varying densities of conifer trees, territory size and hence habitat quality may not differ among individual squirrels (Larsen and Boutin 1995). Squirrels with territories in high-quality habitats typically remove only a small fraction of the staple food source, generally pine cones, from trees in a good cone-producing year. In less suitable habitats with few, if any, conifer trees, home ranges selected by squirrels include a high percentage of herbaceous cover (36–41%) and canopy cover (86–87%), with underground burrows and tree cavities in the immediate vicinity (Mahan and Yahner 1992).

Juveniles born in spring begin to make forays away from their natal home ranges that average 126 m (maximum = 323 m) in distance, beginning when they attain a body size of about 110–115 g (Larsen and Boutin 1994). These juveniles do not abandon their natal home range until they finally settle into their own territory. Most home ranges established by juveniles are on or adjacent to that of their mother's.

Population Density and Sex- and Age-Class Ratios. Red squirrel densities may be as high as 4.5 individuals/ha (Layne 1954). In unthinned and thinned stands of lodgepole pine (*Pinus contorta*) in British Columbia, densities were 1.2 and 0.22 squirrels/ha, respectively, in May–August, and increased to about 2 squirrels/ha in both stand types in September–October (Sullivan and Moses 1986). Other studies have documented densities of 1.3–7 individuals/ha (Sullivan and Sullivan 1982; Gurnell 1984). Densities of red squirrels were estimated at 0.4/ha in jack pine (*Pinus banksiana*)–black spruce (*Picea mariana*) stands compared to 2.3/ha in white spruce (*Picea glauca*) (Davis 1969). In the Adirondack Mountains of New York, populations of red squirrels apparently peaked every 3–4 years (Krull 1970).

Douglas' squirrel densities in California during the breeding season were 1.5 compared to 2.5 squirrels/ha during the postbreeding season (Koford 1982). Population densities of pine squirrels may be limited by food availability (Klenner and Krebs 1991; Sullivan and Klenner 1993; Ransome and Sullivan 1997).

Counts of vocalizations (rattle calls), middens, sightings of individual squirrels, and livetraps have been used to measure squirrel abundance (Koford 1982; Sullivan and Moses 1986; Sullivan 1990; Reinhart and Mattson 1990). Traps often are placed near logs, feeding sites, or middens (e.g., Yahner 1987; Mahan and Yahner 1992; Larsen and Boutin 1994). Handling of live squirrels can pose problems in the field because animals can die for unknown reasons from shock; the incidences of shock may vary among subspecies or populations (Yahner and Mahan 1992). Investigators should use a cone and mesh bag when handling live squirrels; if radio collars are applied to squirrels, consideration should be given to using remote-collaring techniques to minimize incidences of death (Yahner and Mahan 1992; Mahan et al. 1994).

Virtually all studies have shown a sex ratio bias in favor of males in pine squirrels (see review by Hurley 1987). This bias is not believed to be the result of survey or trapping methods used by investigators or differential production of sexes in litters, but probably is caused by higher mortality rates in females over 1 year of age. However, overall sex ratios in pine squirrels were 1 male:0.92 female in unthinned versus 1 male:1.78 females in thinned lodgepole stands in British Columbia (Sullivan and Moses 1986). Age ratios were 1 adult:0.27 juvenile and 1 adult:0.49 juvenile in these unthinned and thinned stands, respectively.

Survivorship. Pine squirrels may survive up to 9–10 years in the wild (Klugh 1927; Davis and Sealander 1971; Gurnell 1987). However, most squirrels (approximately 67%) presumably do not survive beyond 12 months after birth, and mortality plateaus at about 36 months; few squirrels live beyond 7 years of age (Halvorson and Engeman 1983). Thus, like other rodents, pine squirrels experience high juvenile mortality followed by a tapering of rates (Caughley 1966).

Survivorship in pine squirrels may be sex specific (Halvorson and Engeman 1983). The median longevity for males and females under 1 year of age is 3 and 6 months, respectively, which may be an adaptation in favor of females to ensure that they reach breeding age. Once squirrels are 1 year old, males tend to live longer than females, possibly because males do not have the stress associated with raising young (Smith 1968; Halvorson and Engeman 1983).

Habitat and food availability may affect longevity in pine squirrels. Survivorship in red squirrels, based on minimum survival (residency) rates/2 weeks, were higher in unthinned than in thinned lodgepole pine stands (Sullivan and Moses 1986). Squirrels born during years with high conifer seed crops of ponderosa pine (*Pinus ponderosa*) and Douglas fir (*Pseudotsuga menziesii*) lived to a median age of 7 months longer than squirrels born during years of low seed crops (Halvorson and Engeman 1983).

Habitat and Food Resources. Pine squirrels primarily occupy boreal coniferous forests throughout much of North America, with mature forest stands preferred over second-growth stands (Ransome and Sullivan 1997). Closed canopies in mature stands provide locations for home sites (tree nests) within the interlocking branches and produce important food resources (e.g., cone seeds) for pine squirrels (Vahle and Patton 1983). Pine squirrel populations occur in fir (*Abies* spp.), spruce, lodgepole pine, ponderosa pine, whitebark pine (*Pinus albicaulis*), and jack pine stands (Dice 1921; Bailey 1931; Harper 1932; Long 1940; Hatt 1943; Burnett and Dickerman 1956; Cecich and Rudolph 1982; Vahle and Patton 1983; Reinhart and Mattson 1990; A. A. Smith and Mannan 1994). In marginal portions of the range of red squirrel populations, stands with a mix of deciduous and coniferous trees, plantations of Norway spruce (*Picea abies*) and red pine (*Pinus resinosa*), fencerows, and farmstead shelterbelts may serve as habitat (Komarek and Komarek 1938; Petrides 1942; Odum 1949; Gurnell 1984; Yahner 1980, 1987; Mahan and Yahner 1992).

Pine squirrels are opportunistic feeders (Flyger and Gates 1982). Conifers are an important source of cone seeds for these sciurids, especially from pine (*Pinus* spp.) and spruce (*Picea* spp.) (Klugh 1927; Bailey 1936; Yeager 1937; Clarke 1939; Smith and Aldous 1947; Smith 1968; Smith 1968; Rusch and Reeder 1978; Sullivan et al. 1994). Pine squirrels also rely on buds, fungi, shoots, and cambium and phloem tissue as food. Consumption of cambium and phloem by pine squirrels causes debarking of trees, which results in injury to the trees (McKeever 1964). In some deciduous forest and farmstead shelterbelts, food of red squirrels can include a diversity of alternative foods, including apples and corn (Linduska 1942; Yahner 1980, 1987; Riege 1991). Pine squirrels also feed on sap from maple trees by biting holes in the xylem and harvesting the sap after water evaporates (Heinrich 1992) or from holes in birch (*Betula* spp.) made by foraging yellow-bellied sapsuckers (*Sphyrapicus varius*) (Kilham 1958).

A variety of other foods has been reported for pine squirrels. These items include nuts and seeds of deciduous trees and shrubs, including hazelnut (*Corylus* sp.), maple (*Acer* spp.), mockernut hickory (*Carya tomentosa*), witch hazel (*Hamamelis virginiana*), staghorn sumac (*Rhus typhina*), cotoneaster (*Cotoneaster* sp.), and basswood (*Tilia americana*) (Burton 1930; Mailliard 1931; Fox 1939; McClelland 1948; Yahner 1980), nestling songbirds (Adams 1939; King *et al.* 1998); young gray squirrels (*Sciurus carolinensis*) and eastern cottontails (*Sylvilagus floridanus*) (Hamilton 1934; Alkon 1962); forest insect larvae and pupae (Layne 1954; Jennings and Crawford 1989); beef (Howard 1935); and bones (Coventry 1940).

Conifer stands create a cool, moist microenvironment, which helps preserve cones stored in middens within the home ranges of pine

FIGURE 13.3. Midden of red squirrel (*Tamiasciurus hudsonicus*) on San Francisco Peaks, Arizona.

squirrels (Shaw 1936; Smith 1968; Rusch and Reeder 1978) (Fig. 13.3). The moisture in a midden or in wet areas, seepages, or springs helps keep cones from opening, thereby keeping the seeds in a cone viable for years (Shaw 1936; Flyger and Gates 1982).

A midden consists of large quantities of cone scales and fragments that accumulate on the ground beneath an elevated feeding perch of a squirrel (Gurnell 1984). Pine squirrels, which are central-place foragers (Elliott 1988), store freshly harvested cones in the midden, which are harvested in the surrounding territory. The size of middens (including the stored cones) often ranges from 50 to 154 m^2 and at least 40 cm deep (Gurnell 1984; Patton and Vahle 1986). Each midden may contain 2000 to over 3000 cones by mid-October, which may provide an energy supply for nearly 3 weeks for an individual squirrel (Gurnell 1984). As many as four middens may occur in the territory of an individual squirrel.

Middens typically are created by red squirrel populations in boreal forests of western North America, but for reasons that are unclear, are rarely in boreal forests of eastern North America (Hurley and Robertson 1990; Dempsey and Keppie 1993; Yahner 2001). In the eastern boreal forests, squirrels hoard individual conifer cones within their territory. In stands containing mixed conifers and deciduous trees in Pennsylvania, middens do not exist; instead, small piles of cone fragments occur immediately below feeding perches (Yahner 1987; Mahan and Yahner 1992).

Pine squirrels may also cache mushrooms in hollow trees or logs (Dice 1921; Cram 1924). Occasionally, a squirrel may place mushrooms along a branch to dry.

Home Sites. Nest sites commonly used by pine squirrels include leaf nests (Layne 1954; Fancy 1980; Vahle and Patton 1983) and occasionally cavities in snags, holes in rocks, root burrows (Gurnell 1984), abandoned burrow systems of 13-lined ground squirrels (*Spermophilus tridecemlineatus*) (Yahner 1980), and burrows under and within middens (Pruitt and Lucier 1958; Gurnell 1984; Larsen and Boutin 1995). Leaf nests consist of leaves and twigs placed in the branches of conifers about 5–6 m above ground (Rothwell 1979; Whitaker and Hamilton 1998). Tree nests seem to be preferred over tree cavities as home sites by pine squirrels (Flyger and Gates 1982). In winter, underground home sites can be an ideal source of shelter, particularly when habitats lack tree cavities or interlocking branches within which to construct tree nests (Yahner 1980).

BEHAVIOR

Territoriality. Adults pine squirrels defend individual territories from conspecifics (Smith 1968; Koford 1982). Territoriality in red squirrels probably reduces the recruitment of most breeding adults in spring and juvenile immigrants in autumn into the resident population, thereby limiting population densities (Klenner 1991).

In red squirrels, territorial behavior by females directed toward intruding males is relaxed only during the actual mating bout (Smith 1968). Conversely, in Douglas' squirrels, territorial behavior by females toward males is reduced during the entire breeding season (Koford 1982). This difference in territorial behavior by females between pine squirrel species may reflect that cache size in Douglas' squirrels is relatively small compared to that of red squirrels in western North America. Hence, at the onset of the breeding season, the cache (midden) of food is often depleted. Thus, the midden is negligible to the survival of Douglas' squirrels, making the benefit accrued by territoriality in this species less than in red squirrels.

Territories are defended by individual red squirrels rather than by a pair or small group of squirrels (Smith 1968; Yahner 2001). Apparently, the most efficient means of collecting, storing, and defending a midden of cones is for an individual squirrel to establish a circular territory and defend its centralized cache. In areas where middens are not created, as in the boreal forests of eastern North America, territorial defense is still exhibited by red squirrels (Hurley and Robertson 1990; Dempsey and Keppie 1993). In farmstead shelterbelts of southern Minnesota, no territorial defense was observed in red squirrels. In one shelterbelt, two adult squirrels were observed using the same abandoned ground squirrel burrow system as a home site (Yahner 1980).

Possession of a territory and a midden(s) is crucial to the survival of red squirrels in some regions of their geographic range (Larsen and Boutin 1994). A territory is most likely to evolve when resources critical to a territory holder can be in short supply (Brown 1964). In pine squirrels, a midden is often the critical resource positioned at the activity center of a squirrel's territory, which serves not only as a source of hoarded cones, but also as a winter retreat (Larsen and Boutin 1995). Because cone seeds may be a limited food resource to pine squirrels occupying boreal coniferous forests in years of cone failure, defense of whole cones stored in a midden from conspecifics helps to guarantee survival for the territory holder by providing a winter and early spring source of food (Smith 1968).

In western North America, middens are raided occasionally by other mammals (Kendall 1981; Reinhart and Mattson 1990; Mattson and Reinhart 1997). Cones of whitebark pine contain large edible seeds, which are raided by black (*Ursus americanus*) and grizzly bears (*U. arctos*) for food.

Vocalizations. Pine squirrels are very vocal compared to other tree squirrels (Flyger and Gates 1982). There is evidence that sympatric squirrels, including Douglas' squirrels, chipmunks (*Tamias* spp.), and ground squirrels (*Spermophilus* spp.), respond to alarm calls given by each other (Boyer 1943).

Pine squirrels have five types of vocalizations (Smith 1978) in addition to a scream, which is given when injured (Dice 1921). These five types include chirp, rattle, screech, growl, and buzz calls. Chirp calls are given in response to potential predators, such as goshawks (*Accipiter gentilis*), long-tailed weasels (*Mustela frenata*), and humans. Chirp calls may be given continuously during bouts lasting as long as 30 min; occasionally, these calls are given in response to vocalizations of sympatric sciurids, such as chipmunks.

Rattle calls are used by a resident pine squirrel when a conspecific enters its territory (Smith 1978). These calls may be a single call lasting 1 sec to a bout extending over 5 min. The presumed function of rattle calls is to facilitate spacing of individual squirrels on their territories or in relation to limited resources, such as estrous females or food, thereby minimizing time, energy, and risks of fighting (Smith 1968). The response of the intruding squirrel typically is to run from the caller.

Screech calls may be given alone or in association with rattle calls (Smith 1978). Like the rattle call, the screech call is used to defend a territory from an intruding squirrel, and the response of the intruder is to leave the area. Screech calls comprise bouts that are only a few seconds long and are essentially rattle calls whose notes are extended and vary in frequency.

Growl calls are given when two squirrels are fighting or in a close chase, when a female is trying to escape a male during copulation, or when an investigator approaches a squirrel caught in a livetrap (Smith 1978). Buzz calls generally are given by dominant and subordinate males during mating chases with females or when subordinant males approach each other.

Activity Patterns. Pine squirrels, like other tree squirrels except flying squirrels (*Glaucomys* spp.), are diurnal. They are very adept at climbing in trees while feeding or escaping danger, but may spend considerable time feeding or caching food on the ground (Whitaker and Hamilton 1998). During the midday heat of summer, pine squirrels are relatively inactive (Clarkson and Ferguson 1969). Pine squirrels do not hibernate in winter, but they may become very inactive during inclement weather, when they seek cover in subnivean or subterranean environments and are rarely seen in trees (Pruitt and Lucier 1958; Pruitt 1960). In severe weather, squirrels may remain inactive in nests for 1–2 days. In regions where winters are not severe, pine squirrels spend most activity on the ground or foraging trees.

Compared to gray squirrels, pine squirrels do not exhibit extensive fall movements (Flyger and Gates 1982). Emigrations by red squirrels, however, have been noted in the Adirondacks of New York (Klugh 1927).

Interspecific Competition. Pine squirrels experience little, if any, competition with sympatric tree and ground squirrels or chipmunks because of interspecific differences in habitats and food resources. Red squirrels have been known to rest simultaneously in the same nest box with gray squirrels (Ackerman and Weigl 1970). However, there is little credence to the anecdotal tale that a red squirrel may castrate a gray squirrel that intrudes on the red squirrel's territory (Whitaker and Hamilton 1998). The basis for this tale may be that red squirrels have been observed chasing gray squirrels during the nonbreeding season of gray squirrels when their testes are abdominal rather than scrotal.

MORTALITY

Predation. Known predators of pine squirrels include coyote (*Canis latrans*), red fox (*Vulpes vulpes*), American marten (*Martes americana*), fisher (*Martes pennanti*), ermine (*Mustela erminea*), long-tailed weasel, bobcat (*Lynx rufus*), Canada lynx (*Lynx canadensis*), great horned owl (*Bubo virginianus*), red-tailed hawk (*Buteo jamaicensis*), goshawk, Cooper's hawk (*Accipiter cooperii*), and bald eagle (*Haliaeetus leucocephalus*) (English 1934; Layne 1954; Van Zyll de Jong 1966; Nellis and Keith 1968; Johnson 1969; Luttich et al. 1970; Halvorson and Engeman 1983; Whitaker and Hamilton 1998). The American marten is particularly adept at chasing and capturing pine squirrels in trees (see Chapter 29).

Parasites and Diseases. Parasites of red squirrels have been summarized by Flyger and Gates (1982). Based on a more recent examination of 45 red squirrels in New Mexico, Patrick and Wilson (1995), found that 82.5% were affected by at least one species from the following groups of endoparasites and ectoparasites: Anoplura (sucking lice), Nematoda (nematodes), Cestoda (tapeworms), Coccidia, and Siphonaptera (fleas). Whitaker and Pascal (1979) examined 93 red squirrels in Indiana and found 19 species of Anoplura, Siphonaptera, Acarina (ticks and chiggers), Glycyphagidae (mites), and Laelapidae (mites).

Sarcocysts (*Sarcocystis* sp.) also are reported in red squirrels (Entzeroth et al. 1983). As in many rodents, pine squirrels may contain larvae of the botfly (*Cuterebra*) and some viruses (Whitaker and Hamilton 1998). Moreover, tularemia (Francis 1937; Burroughs et al. 1945), *Adiaspiromycosis* (Dvorak et al. 1965), *Haplosporanigium* (Dowding 1947), Silverwater virus (Hoff et al. 1971), California encephalitis virus (Masterson et al. 1971), and Powassan virus (McLean et al. 1968) have been noted in red squirrels.

AGE ESTIMATION

Age of pine squirrels is estimated primarily using eye lens weights from a sample of known-age, captive-reared squirrels or by the extent of closure of epiphyseal sutures on the femur or humerus (Davis and Sealander 1971; Halvorson and Engeman 1983). Adults can be distinguished from subadults by distinct pigmentation on the scrotum of adult males or teats of adult females (Flyger and Gates 1982).

ECONOMIC STATUS AND MANAGEMENT

Bark stripping by pine squirrels feeding on the cambium of conifers occurs mainly in spring and early summer when the bark is relatively easy to remove and the sap-sugar concentration is high (Kenward 1983; Sullivan and Sullivan 1982). Bark stripping is characterized by a lack of teeth marks on the stem and a pile of bark strips (3 × 8 cm) under an injured tree (Sullivan and Sullivan 1982). In contrast, bark stripping by porcupine (*Erethizon dorsatum*) is associated with broad incisor marks on the exposed stem.

Bark injury by pine squirrels has been well documented in various conifer stands (Pike 1934; Stillinger 1944; Schantz-Hansen 1945; Mitchell 1950; McKeever 1964), and the extent of damage caused by bark stripping has been given considerable attention (Sullivan and Sullivan 1982; Sullivan and Vyse 1987; Sullivan et al. 1994). In particular, lodgepole pine can be seriously damaged by red squirrels (Brockley and Sullivan 1988). Pine squirrels also are known to peel bark and cause the death of trees in paper birch (*Betula papyrifera*) stands (Lutz 1956). Damaged trees have open wounds, which can allow entry of fungi and bacteria (Shigo 1964).

Lodgepole pines most susceptible to stripping damage by pine squirrels are found in large stands with fire damage or affected by even-aged forest management (Sullivan et al. 1994). The percentage of damaged trees per stand may range from 0 to 96%. Squirrel damage may be especially high in managed stands with high shrub density, which presumably reduces predation risks for foraging squirrels. Larger pines (>6 cm diameter at breast height) are more likely to be attacked by feeding squirrels than are smaller pines (Sullivan and Vyse 1987). Although thinning of lodgepole pine stands can reduce the population densities of red squirrels, it may not reduce the tree damage to an acceptable level unless stem densities are quite low (<1000 stem/ha) (Sullivan and Moses 1986; Sullivan 1998). However, there is evidence that larger thinned stands are less susceptible to tree damage by pine squirrels than are smaller stands.

Another management practice for mitigating conifer damage by squirrels in thinned stands is to provide diversionary food (sunflower seeds) (Sullivan 1992, 1998). Diversionary food can be applied in a cost-effective manner (≤$45/ha) via aerial application. Even if used for a few years, this practice is worth the time and effort to protect commercially important timber resources (≥$2000/ha) (Sullivan 1992).

Red squirrels reportedly can damage orchards by feeding on cherry (*Prunus* sp.) blossoms and pear (*Pyrus* sp.) fruit (Storer 1875). Pine squirrels also are known to nip off the buds and ends of branches from a variety of coniferous trees (e.g., Hosley 1928; Cheyney 1929; Hart 1936; McCulloch 1937; Balch 1942; Schantz-Hansen 1945; Cook 1954). The extent of bud nipping by pine squirrels in northerly latitudes is probably greater with increased snow depth and during extended periods of snow cover (Hart 1936). If the canopies of trees are open, as in thinned conifer stands or orchards, sheet-metal sleeves can be installed on individual trees to prevent pine squirrels from climbing up and causing damage (Flyger and Gates 1982).

Squirrel removal programs probably are ineffective at mitigating damage to conifers (Sullivan and Moses 1986). Removal of all individuals in a population is nearly impossible, and suitable unoccupied habitat can quickly be colonized by squirrels, especially females, from nearby marginal habitats. Although adult squirrels are less likely to relocate to areas vacated by other squirrels, they may shift territorial boundaries in response to the absence of a neighbor (Larsen and Boutin 1995).

Pine squirrels are seldom hunted for sport (Flyger and Gates 1982). However, in Canada, pine squirrels can be an important furbearer from an economic perspective (Kemp and Keith 1970). Cones stored in middens can provide an excellent source of seeds for foresters, with those collected by foresters from middens having higher germination rates than those harvested directly from trees (Wagg 1964). In stands of ponderosa pine, jack pine, red pine, white pine, or black spruce, however, cone harvesting by red squirrels can potentially reduce seed production or reseeding of stands (Pulling 1924; Smith and Aldous 1947; Roe 1948; Schubert 1953; Finley 1969; Schmidt and Shearer 1971; Cecich and Rudolph 1982; West 1989).

In portions of the geographic range where middens are constructed by pine squirrels, sound habitat management practices should include protection of sites in the vicinity of middens (Smith and Mannan 1994). Human activities should be minimized near midden sites, especially those activities that remove large trees, snags, or logs or those that either open the forest canopy or create forest edge.

Populations of pine squirrels can be increased by planting conifers (Flyger and Gates 1982). Nest boxes can also be erected with dimensions measuring 15×15 cm, a depth of 30 cm, and a 4.5-cm hole placed near the top of the box. Boxes should be attached to trees at a height of about 6 m, and nesting material should be removed annually.

Nuisance pine squirrels (e.g., those that persist in entering buildings) can be shot or trapped (Flyger and Gates 1982). Live traps measuring 8 × 8 × 30 cm are effective when baited with peanut butter (Flyger and Gates 1982; Mahan and Yahner 1992).

CONSERVATION AND RESEARCH NEEDS

The abundance of pine squirrels throughout much of their geographic range is associated with the availability of cone seeds (e.g., Smith 1968; Kemp and Keith 1970; Sullivan 1990). Thus, more research is needed on the interaction among squirrel abundance, food availability, and feeding damage by squirrels to conifers (Sullivan and Vyse 1987).

The adaptiveness of the unique coiled vagina and tiny baculum in these squirrels deserves study (Whitaker and Hamilton 1998). Additional information also can be gathered on the invertebrate fauna associated with nests of pine squirrels.

Compared to red squirrels, much more research is necessary on the ecology and life history of the Douglas' squirrel (Flyger and Gates 1982; Verts and Carraway 1998). In addition, because pine squirrels often coexist with other sciurids (e.g., chipmunks, ground squirrels, and tree squirrels), investigations focusing on resource partitioning among these species are recommended.

The population of Mount Graham red squirrels (*T. h. grahamensis*), which is an endangered subspecies occupying the Mount Graham peak of the Pinaleño Mountains in southeastern Arizona, will continue to be the focus of future research (Allen et al. 1987; Istock and Hoffman 1995). This subspecies represents the southernmost population, which is geographically and genetically isolated from northern populations of red squirrels (Hall 1981; Sullivan and Yates 1995). Population estimates range between 200 and 500 animals occupying about 4800 ha of suitable habitat (Allen et al. 1987). The habitat of this subspecies has been altered by a variety of activities, including road building, timber harvesting, recreational development, and the development of a politically charged astronomical complex (Allen et al. 1987; Istock 1995). To date, however, Mount Graham red squirrels in the vicinity of the astronomical complex have not been negatively affected (Young 1995).

LITERATURE CITED

Ackerman, R., and P. D. Weigl. 1970. Dominance relationships of red and gray squirrels. Ecology 51:332–34.

Adams, L. 1939. Sierra chickaree eats young blue-fronted jays. Yosemite Nature Notes 18:93.

Aleksuik, M. 1970. The occurrence of brown adipose tissue in the adult red squirrel (*Tamiasciurus hudsonicus*). Canadian Journal of Zoology 48:188–89.

Aleksuik, M. 1971. Seasonal dynamics of brown adipose tissue function in the red squirrel (*Tamiasciurus hudsonicus*). Comparative Biochemistry 38:723–31.

Alkon, P. U. 1962. Red squirrel predation on nestling cottontail. New York Fish and Game Journal 9:142.

Allen, L. S., R. L. Wadleigh, P. Marshall, and R. B. Spicer. 1987. Mount Graham red squirrel: A biological assessment of impacts of the proposed Mt. Graham astrophysical project. Coronado National Forest. U.S. Forest Service, Tucson, AZ.

Bailey, V. 1931. Mammals of New Mexico. North American Fauna 53:1–412.

Bailey, V. 1936. Mammals and life zones of Oregon. North American Fauna 55:1–416.

Balch, R. E. 1942. A note on squirrel damage to conifers. Forestry Chronicle 18:42.

Becker, C. D. 1993. Environmental cues of estrus in the North American red squirrel (*Tamiasciurus hudsonicus* Bangs). Canadian Journal of Zoology 71:1326–33.

Boyer, R. H. 1943. Weasel versus squirrel in Sequoia National Park. Journal of Mammalogy 24:99–100.

Brockley, R. P., and T. P. Sullivan. 1988. Relationship of feeding damage by red squirrels to cultural treatments in young stands of lodgepole pine. Pages 322–29 *in* Proceedings of the future forests of the mountain West: A stand culture symposium (General Technical Report INT-243). USDA Forest Service, Intermountain Research Station.

Brown, J. L. 1964. The evolution of diversity in avian territorial systems. Wilson Bulletin 76:160–69.

Burnett, F. L., and R. W. Dickerman. 1956. Type locality of the Mogollon red squirrel, *Tamiasciurus hudsonicus mogollonensis*. Journal of Mammalogy 37:292–94.

Burroughs, A. L., R. Holdenreid, D. S. Longanecker, and K. F. Meyer. 1945. A field study of latent tularemia in rodents with a list of all known naturally infected vertebrates. Journal of Infectious Diseases 76:115–19.

Burt, W. H., and R. P. Grossenheider. 1976. A field guide to the mammals, 3rd ed. Houghton Mifflin, Boston.

Burton, S. S. 1930. A new diet for the red squirrel. Journal of Forestry 28:233.

Caughley, G. 1966. Mortality patterns in mammals. Ecology 47:906–18.

Cecich, R. A., and T. D. Rudolph. 1982. Time of jack pine seed maturity in Lake States provinces. Canadian Journal of Forest Research 12:368–73.

Cheyney, E. G. 1929. Damage to Norway and jack pine by red squirrels. Journal of Forestry 27:382–83.

Clarke, C. H. D. 1939. Some notes on hoarding and territorial behavior of the red squirrel, *Sciurus hudsonicus* (Erxleben). Canadian Field-Naturalist 53:42–43.

Clarkson, D. P., and H. J. Ferguson. 1969. Effect of temperature on activity in the red squirrel. American Zoologist 9:1110.

Cook, D. B. 1954. Susceptibility of larch to red squirrel damage. Journal of Forestry 52:491–92.

Coventry, A. F. 1940. The eating of bone by squirrels. Science 92:128.

Cram, W. E. 1924. The red squirrel. Journal of Mammalogy 5:37–41.

Davis, D. W. 1969. The behavior and population dynamics of the red squirrel, *Tamiasciurus hudsonicus,* in Saskatchewan. Ph.D. Dissertation, University of Arkansas, Fayetteville.

Davis, D. W., and J. A. Sealander. 1971. Sex ratio and age structure in two red squirrel populations in northern Saskatchewan. Canadian Field-Naturalist 85:303–8.

Dempsey, J. A., and D. M. Keppie. 1993. Foraging patterns of eastern red squirrels. Journal of Mammalogy 74:1007–13.

Dice, L. R. 1921. Notes on the mammals of interior Alaska. Journal of Mammalogy 2:20–38.

Dippner, R., and J. Armington. 1971. A behavioral measure of dark adaptation in the American red squirrel. Psychonometric Science 24:43–45.

Dolbeer, R. A. 1973. Reproduction in the red squirrel (*Tamiasciurus hudsonicus*) in Colorado. Journal of Mammalogy 54:536–40.

Dowding, E. S. 1947. *Haplosporangium* in Canadian rodents. Mycologia 39:372–73.

Dvorak, J., M. Otcenasek, and J. Prokopic. 1965. The distribution of adiaspiromycosis. Journal of Hygiene, Epidemiology, Microbiology, and Immunology 9:510–14.

Elliott, P. F. 1988. Foraging behavior of a central-place forager: Field tests of theoretical predictions. American Naturalist 131:159–74.

English, P. R. 1934. Some observations on a pair of red-tailed hawks. Wilson Bulletin 46:228–35.

Entzeroth, R., B. Chobotar, and E. Scholtyseck. 1983. Ultrastructure of a *Sarcocystis* species from the red squirrel (*Tamiasciurus hudsonicus*) in Michigan. Protistologica 19:91–94.

Fancy, S. G. 1980. Nest-tree selection by red squirrels in a boreal forest. Canadian Field-Naturalist 94:198.

Ferguson, J. H., and G. E. Folk, Jr. 1971. Effect of temperature and acclimation upon FFA levels in three separate species of rodents. Canadian Journal of Zoology 49:303–5.

Ferron, J. 1976. Comfort behaviour of the red squirrel (*Tamiasciurus hudsonicus*). Zeitschrift für Tierpsychologie 42:66–85.

Ferron, J., and J.-P. Ouellet. 1989. Behavioural context and possible function of scent marking by cheek rubbing in the red squirrel (*Tamiasciurus hudsonicus*). Canadian Journal of Zoology 67:1650–53.

Ferron, J., and J. Prescott. 1977. Gestation, litter size, and number of litters of the red squirrel (*Tamiasciurus hudsonicus*) in Quebec. Canadian Field-Naturalist 91:83–84.

Finley, R. B., Jr. 1969. Cone caches and middens of *Tamiasciurus* in the Rocky Mountain region. University of Kansas Museum of Natural History, Miscellaneous Publications 51:233–73.

Flyger, V., and J. E. Gates. 1982. Pine squirrels. Pages 230–38 *in* J. A. Chapman and G. A. Feldhamer, eds. Wild mammals of North America. Johns Hopkins University Press, Baltimore.

Fox, A. C. 1939. Red squirrels eat basswood and boxelder seeds. Journal of Mammalogy 20:257.

Francis, E. 1937. Sources of infection and seasonal incidence of tularemia in man. Public Health Report 52:103–13.

Grodzinski, W. 1971. Food consumption of small mammals in the Alaskan taiga forest. Annales Zoologici Fennici 8:133–36.

Gurnell, J. 1984. Home range, territoriality, caching behaviour and food supply of the red squirrel (*Tamiasciurus hudsonicus fremonti*) in a subalpine lodgepole pine forest. Animal Behaviour 32:1119–31.

Gurnell, J. 1987. The natural history of squirrels. Christopher Helm, London.

Hall, E. R. 1981. The mammals of North America, 2nd ed. John Wiley, New York.

Halvorson, C. H., and R. M. Engeman. 1983. Survival analysis for a red squirrel population. Journal of Mammalogy 64:332–36.

Hamilton, W. J., Jr. 1934. Red squirrel killing young cottontail and young gray squirrel. Journal of Mammalogy 15:322.

Hamilton, W. J., Jr. 1939. Observations on the life history of the red squirrel in New York. American Midland Naturalist 22:732–45.

Harper, F. 1932. Mammals of the Athabaska and Great Slave Lakes Region. Journal of Mammalogy 13:19–36.

Hart, A. C. 1936. Red squirrel damage to pine and spruce plantations. Journal of Forestry 34:729–30.

Hatt, R. T. 1943. The pine squirrel in Colorado. Journal of Mammalogy 24:311–45.

Heinrich, B. 1992. Maple sugaring by red squirrels. Journal of Mammalogy 73:51–54.

Hoff, G. L., T. M. Yuill, J. O. Iversen, and R. P. Hanson. 1971. Silverwater virus serology in snowshoe hares and other vertebrates. American Journal of Tropical Medicine and Hygiene 20:326–30.

Hosley, N. W. 1928. Red squirrel damage to coniferous plantations and its relations to changing food habits. Ecology 9:43–48.

Howard, W. J. 1935. Apparently neutral relations of weasel and squirrel. Journal of Mammalogy 16:322–26.

Hurley, T. A. 1987. Male-biased adult sex ratios in a red squirrel population. Canadian Journal of Zoology 65:1284–86.

Hurley, T. A., and R. J. Robertson. 1990. Variation in the food hoarding behavior of red squirrels. Behavioral Ecology and Sociobiology 26:91–97.

Istock, C. A. 1995. Telescopes, red squirrels, congress, courtrooms, and conservation. Pages 19–35 *in* C. A. Istock and R. S. Hoffman, eds. Storm over a mountain island: Conservation biology and the Mt. Graham affair. University of Arizona Press, Tucson.

Istock, C. A., and R. S. Hoffman, eds. 1995. Storm over a mountain island: Conservation biology and the Mt. Graham affair. University of Arizona Press, Tucson.

Jennings, D. T., and H. S. Crawford, Jr. 1989. Predation by red squirrels on the spruce budworm *Choristoneura fumiferana* (Clem.) (Lepidoptera: Tortricidae). Canadian Entomologist 121:827–28.

Johnson, W. J. 1969. Food habits of the Isle Royale red fox and population aspects of three of its principal prey species. Ph.D. Thesis, Purdue University, Lafayette, IN.

Kemp, G. A., and L. B. Keith. 1970. Dynamics and regulation of red squirrel (*Tamiasciurus hudsonicus*) populations. Ecology 51:763–79.

Kendall, K. C. 1981. Bear use of pine nuts. M.S. Thesis, Montana State University, Bozeman.

Kenward, R. E. 1983. The causes of damage by red and grey squirrels. Mammal Review 13:159–66.

Kilham, L. 1958. Red squirrels feeding at sapsucker holes. Journal of Mammalogy 39:596–97.

King, D. I., C. R. Griffin, and R. M. DeGraaf. 1998. Nest predator distribution among clearcut forest, forest edge and forest interior in an extensively forested landscape. Forest Ecology and Management 104:151–56.

Klenner, W. 1991. Red squirrel population dynamics. II. Settlement patterns and the response of removals. Journal of Animal Ecology 60:979–93.

Klenner, W., and C. J. Krebs. 1991. Red squirrel population dynamics. I. The effect of supplemental food on demography. Journal of Animal Ecology 60:961–78.

Klugh, A. B. 1927. Ecology of the red squirrel. Journal of Mammalogy 8:1–32.

Koford, R. R. 1982. Mating system of a territorial tree squirrel (*Tamiasciurus douglasii*) in California. Journal of Mammalogy 63:274–83.

Komarek, E. V., and R. Komarek. 1938. Mammals of the Great Smoky Mountains. Bulletin of the Chicago Academy of Sciences 5:137–62.

Krull, J. N. 1970. Response of chipmunks and red squirrels to commercial clearcut logging. New York Fish and Game Journal 17:58–59.

Lair, H. 1985. Length of gestation in the red squirrel, *Tamiasciurus hudsonicus*. Journal of Mammalogy 66:809–10.

LaPierre, L. 1986. Female red squirrel (*Tamiasciurus hudsonicus*) reproductive tracts from fenitrothion treated and untreated forest of southeastern New Brunswick. Forestry Chronicle 62:233–35.

Larsen, K. W. and S. Boutin. 1994. Movements, survival and settlement of red squirrel (*Tamiasciurus hudsonicus*) offspring. Ecology 75:214–23.

Larsen, K. W., and S. Boutin. 1995. Exploring territory quality in the North American red squirrel through removal experiments. Canadian Journal of Zoology 73:1115–22.

Larsen, K. W., C. D. Becker, S. Boutin, and M. Blower. 1997. Effects of hoard manipulations on life history and reproductive success of female red squirrels (*Tamiasciurus hudsonicus*). Journal of Mammalogy 78:192–203.

Layne, J. N. 1954. The biology of the red squirrel, *Tamiasciurus hudsonicus loquax* (Bangs), in central New York. Ecological Monographs 24:227–67.

Linduska, J. P. 1942. Winter rodent populations in field-shocked corn. Journal of Wildlife Management 6:353–63.

Long, W. S. 1940. Notes on the life histories of some Utah mammals. Journal of Mammalogy 21:170–80.

Luttich, S., D. H. Rusch, E. C. Meslow, and L. B. Keith. 1970. Ecology of red-tailed hawk predation in Alberta. Ecology 51:190–203.

Lutz, H. J. 1956. Damage to paper birch by red squirrels in Alaska. Journal of Forestry 54:31–33.

Mahan, C. G., and R. H. Yahner. 1992. Microhabitat use by red squirrels in central Pennsylvania. Northeast Wildlife 49:49–56.

Mahan, C. G., R. H. Yahner, and L. R. Stover. 1994. Development of remote-collaring techniques for red squirrels. Wildlife Society Bulletin 22:270–73.

Mailliard, J. 1931. Redwood chickaree testing and storing hazel nuts. Journal of Mammalogy 12:68–70.

Masterson, R. A., H. W. Stegmiller, M. A. Parsons, C. C. Croft, and C. B. Spencer. 1971. California encephalitis: An endemic puzzle in Ohio. Health Laboratory Science 8:89–96.

Mattson, D. J., and D. P. Reinhart. 1997. Excavation of red squirrel middens by grizzly bears in the whitebark pine zone. Journal of Applied Ecology 34:926–40.

McClean, D. W., S. R. Ladyman, and K. V. Purvingood. 1968. Westward extension of Powassan virus prevalence. Canadian Medical Association Journal 98:946–49.

McClelland, E. H. 1948. Notes on the red squirrel in Pittsburgh. Journal of Mammalogy 29:409–12.

McCulloch, W. F. 1937. Red squirrels attack Japanese larch. Journal of Forestry 35:692–93.

McKeever, S. 1964. Food habits of the pine squirrel in northeastern California. Journal of Wildlife Management 28:402–4.

Mengel, R. M., and M. A. Jenkinson. 1971. A melanistic specimen of the red squirrel. American Midland Naturalist 86:230–31.

Merritt, J. F. 1987. Guide to the mammals of Pennsylvania. University of Pittsburgh Press, Pittsburgh.

Millar, J. S. 1968. The reproductive biology of the western red squirrel. M.S. Thesis, University of British Columbia, Vancouver, Canada.

Millar, J. S. 1970. The breeding season and reproductive cycle of the western red squirrel. Canadian Journal of Zoology 48:471–73.

Mitchell, G. E. 1950. Wildlife-forest relationships in the Pacific Northwest. Journal of Forestry 48:26–30.

Modafferi, M. M. 1972. Aspects of the reproductive biology of the red squirrel (*Tamiasciurus hudsonicus*) in interior Alaska. M.S. Thesis, University of Alaska.

Nadler, C. F., and R. S. Hoffman. 1970. Chromosomes of some Asian and South American squirrels (Rodentia: Sciuridae). Experientia 26:1383–86.

Nellis, C. H. 1969. Sex and age variation in red squirrel skulls from Missoula County, Montana. Canadian Field-Naturalist 83:324–30.

Nellis, C. H., and L. B. Keith. 1968. Hunting activities and success of lynxes in Alberta. Journal of Wildlife Management 32:718–22.

Nelson, B. A. 1945. The spring molt of the northern red squirrel in Minnesota. Journal of Mammalogy 26:397–400.

Odum, E. P. 1949. Small mammals of the highlands (North Carolina) plateau. Journal of Mammalogy 30:179–92.

Patrick, M. J., and W. D. Wilson. 1995. Parasites of the Abert's squirrel (*Sciurus aberti*) and red squirrel (*Tamiasciurus hudsonicus*) of New Mexico. Journal of Parasitology 81:321–24.

Patton, D. R., and J. R. Vahle. 1986. Cache and nest characteristics of the red squirrel in an Arizona mixed-coniferous forest. Western Journal of Forestry 1:48–51.

Petrides, G. A. 1942. Relation of hedgerows in winter to wildlife in central New York. Journal of Wildlife Management 6:261–80.

Pike, G. W. 1934. Girdling of ponderosa pine by squirrels. Journal of Forestry 32:98–99.

Pruitt, W. O., Jr. 1960. Animals in the snow. Scientific American 202:60–68.

Pruitt, W. O., Jr. and C. V. Lucier. 1958. Winter activity of red squirrels in interior Alaska. Journal of Mammalogy 39:443–44.

Pulling, A. V. S. 1924. Small rodents and northeastern conifers. Journal of Forestry 22:813–14.

Quay, W. B. 1965. Comparative survey of the sebaceous and sudoriferous gland of the oral lips and angle in rodents. Journal of Mammalogy 46:23–37.

Ransome, D. B., and T. P. Sullivan. 1997. Food limitation and habitat preference of *Glaucomys sabrinus* and *Tamiasciurus hudsonicus*. Journal of Mammalogy 78:538–49.

Reinhart, D. P., and D. J. Mattson. 1990. Red squirrels in the whitebark zone. Pages 256–63 *in* W. C. Schmidt and K. J. MacDonald, eds. Symposium on whitebark pine ecosystems: Ecology and management of a high-mountain resource (General Technical Report INT 270). U. S. Forest Service, Bozeman, MT.

Riege, D. A. 1991. Habitat specialization and social factors in distribution of red and gray squirrels. Journal of Mammalogy 72:152–62.

Roe, E. I. 1948. Effects of red squirrels on red pine seed production in off years. Journal of Forestry 46:528–29.

Rothwell, R. 1979. Nest sites of red squirrels (*Tamiasciurus hudsonicus*) in the Laramie Range of southeastern Wyoming. Journal of Mammalogy 60:404–5.

Rusch, D. A., and W. G. Reeder. 1978. Population ecology of Alberta red squirrels. Ecology 59:400–420.

Schantz-Hansen, T. 1945. Red squirrel damage to mature red pine. Journal of Forestry 43:604–5.

Schmidt, W. C., and R. C. Shearer. 1971. Ponderosa pine seed: For animals or trees. U.S. Forest Service Research Paper INT-112:1–14.

Schubert, G. H. 1953. Ponderosa pine cone cutting by squirrels. Journal of Forestry 51:202.

Shaw, W. T. 1936. Moisture and its relation to the cone-storing habit of the western pine squirrel. Journal of Mammalogy 17:337–49.

Shigo, A. L. 1964. A canker on maple caused by fungi infecting wounds made by the red squirrel. Plant Diseases Report 48:794–96.

Smith, A. A., and R. W. Mannan. 1994. Distinguishing characteristics of Mount Graham red squirrel midden sites. Journal of Wildlife Management 58:437–45.

Smith, C. C. 1968. The adaptive nature of social organization in the genus of tree squirrels *Tamiasciurus*. Ecological Monographs 38:31–63.

Smith, C. C. 1978. Structure and function of the vocalizations of tree squirrels (*Tamiasciurus*). Journal of Mammalogy 59:793–808.

Smith, C. F., and S. E. Aldous., 1947. The influence of mammals and birds in retarding artificial and natural reseeding of coniferous forests in the United States. Journal of Forestry 45:361–69.

Smith, M. C. 1968. Red squirrel responses to spruce cone failure in interior Alaska. Journal of Wildlife Management 32:305–17.

Stillinger, C. R. 1944. Damage to conifers in northern Idaho by the Richardson red squirrel. Journal of Forestry 42:143–45.

Storer, F. H. 1875. Cherry blossoms destroyed by squirrels. Nature 13:26.

Sullivan, R. M. 1990. Responses of red squirrel (*Tamiasciurus hudsonicus*) populations to supplemental food. Journal of Mammalogy 71:579–90.

Sullivan, R. M., and T. L. Yates. 1995. Population genetics and conservation biology of relict populations of red squirrels. Pages 193–208 *in* C. A. Istock and R. S. Hoffman, eds. Storm over a mountain island: Conservation biology and the Mt. Graham affair. University of Arizona Press, Tucson.

Sullivan, T. P. 1992. Operational application of diversionary food in young lodgepole pine forests to reduce feeding damage by red squirrels. Pages 340–43 *in* J. E. Borrecco and R. E. Marsh, eds. Proceedings of the 15th Vertebrate pest conference. University of California, Davis.

Sullivan, T. P. 1998. Management of red squirrel feeding damage to lodgepole pine by stand density manipulation and diversionary food. Pages 196–202 *in* R. O. Baker and A. C. Crabb, eds. Proceedings of the 18th Vertebrate pest conference. University of California, Davis.

Sullivan, T. P., and W. Klenner. 1993. Influence of diversionary food on red squirrel populations and damage to crop trees in young lodgepole pine forest. Ecological Applications 3:708–18.

Sullivan, T. P., and R. A. Moses. 1986. Red squirrel populations in natural and managed stands of lodgepole pine. Journal of Wildlife Management 50:595–601.

Sullivan, T. P., and D. S. Sullivan. 1982. Barking damage by snowshoe hares and red squirrels in lodgepole pine stands in British Columbia. Canadian Journal of Forest Research 12:443–48.

Sullivan, T. P., and A. Vyse. 1987. Impact of red squirrel feeding damage on spaced stands of lodgepole pine in the Cariboo Region of British Columbia. Canadian Journal of Forest Research 17:666–74.

Sullivan, T. P., J. A. Krebs, and P. K. Diggle. 1994. Prediction of stand susceptibility to feeding damage by red squirrels in young lodgepole pine. Canadian Journal of Forest Research 24:14–20.

Svihla, R. D. 1930. Development of young red squirrels. Journal of Mammalogy 11:79–80.

Vahle, J. R., and D. R. Patton. 1983. Red squirrel cover requirements in Arizona mixed conifer forests. Journal of Forestry 81:14–15.

Van Zyll de Jong, C. G. 1966. Food habits of the lynx in Alberta and the Mackenzie District, N. W. T. Canadian Field-Naturalist 80:18–23.

Verts, B. J., and C. N. Carraway. 1998. Land mammals of Oregon. University of California Press, Los Angeles.

Wagg, J. W. B. 1964. Viability of white spruce seed from squirrel-cut cones. Forest Chronicle 40:98–110.

West, R. J. 1989. Cone depredations by the red squirrel in black spruce stands in Newfoundland: Implications for commercial cone collection. Canadian Journal of Forest Research 19:1207–10.

Whitaker, J. O., Jr., and D. D. Pascal, Jr. 1979. Ectoparasites of the red squirrel (*Tamiasciurus hudsonicus*) and the eastern chipmunk (*Tamias striatus*) from Indiana. Journal of Medical Entomology 16:350–51.

Whitaker, J. O., Jr., and W. J. Hamilton, Jr. 1998. Mammals of the eastern United States, 3rd ed. Cornell University Press, Ithaca, NY.

Wood, T. J. 1965. Albino red squirrel collected in Wood Buffalo Park. Blue Jay 23:90.

Wood, T. J. 1967. Ecology and population dynamics of the red squirrel *Tamiasciurus hudsonicus* in Wood Buffalo National Park. M.A. Thesis, University of Saskatchewan, Saskatoon.

Yahner, R. H. 1978. The adaptive nature of the social system and behavior in the eastern chipmunk, *Tamias striatus*. Behavioral Ecology and Sociobiology 3:397–427.

Yahner, R. H. 1980. Burrow system use by red squirrels. American Midland Naturalist 103:409–11.

Yahner, R. H. 1987. Feeding-site selection by red squirrels, *Tamiasciurus hudsonicus*, in a marginal habitat in Pennsylvania. Canadian Field-Naturalist 101:586–89.

Yahner, R. H. 2001. Fascinating mammals: Essays on conservation and ecology. University of Pittsburgh Press, Pittsburgh, PA.

Yahner, R. H., and C. G. Mahan. 1992. Use of a laboratory restraining device on wild red squirrels. Wildlife Society Bulletin 20:399–401.

Yeager, L. E. 1937. Cone-piling by Michigan red squirrels. Journal of Mammalogy 18:191–94.

Young, P. J. 1995. Monitoring the Mt. Graham red squirrel. Pages 226–46 *in* C. A. Istock and R. S. Hoffman, eds. Storm over a mountain island: Conservation biology and the Mt. Graham affair. University of Arizona Press, Tucson.

Zirul, D. L. 1970. Ecology of a northern population of the red squirrel, *Tamiasciurus hudsonicus preblei* (Howell). M.S. Thesis, University of Alberta, Edmonton, Canada.

Richard H. Yahner, School of Forest Resources,Pennsylvania State University, University Park, Pennsylvania 16802.Email: rhy@psu.edu.

14

Pocket Gophers

Geomyidae

Robert J. Baker
Robert D. Bradley
Lee R. McAliley, Jr.

NOMENCLATURE

Family Geomyidae. Species featured in this chapter are the following: *Thomomys bottae,* Botta's pocket gopher; *T. bulbivorus,* Camas pocket gopher; *T. clusius,* Wyoming pocket gopher; *T. idahoensis,* Idaho pocket gopher; *T. mazama,* western pocket gopher; *T. monticola,* mountain pocket gopher; *T. talpoides,* northern pocket gopher; *T. townsendii,* Townsend's pocket gopher; *T. umbrinus,* southern pocket gopher; *Geomys arenarius,* desert pocket gopher; *G. breviceps,* Baird's pocket gopher; *G. bursarius,* plains pocket gopher; *G. knoxjonesi,* Jones's pocket gopher; *G. personatus,* Texas pocket gopher; *G. pinetis,* southeastern pocket gopher; *G. texensis,* llano pocket gopher; *G. tropicalis,* tropical pocket gopher; and *Cratogeomys castanops,* yellow-faced pocket gopher.

Seven genera of pocket gopher (*Cratogeomys, Geomys, Heterogeomys, Orthogeomys, Pappogeomys, Thomomys,* and *Zygogeomys*) occur in North America. In this chapter, we will focus only on taxa occurring in the United States and northern Mexico (*Cratogeomys, Geomys,* and *Thomomys*). Taxa distributed in Mexico and Central America are lesser known. More than 35 species and 300 subspecies of gopher have been named (Hall 1981, Patton and Smith 1990, Wilson and Cole 2000) since the first taxonomic revision (Merriam 1895) of the family. Many of the named species have been recent elevations of previously named subspecies, primarily as a result of newly acquired molecular data. We anticipate that this trend will continue and that additional species will be recognized as more molecular data become available.

Pocket gophers are among the fauna of North America that possess fossorial adaptations. They are fusiform shaped; possess a short, usually naked tail; have short pelage, reduced ear pinnae, and small, but prominent eyes; and their forelimbs are modified for digging. All pocket gophers have fur-lined, external cheek pouches, which permit them to transport plant material through their burrow system. They also have skin that grows behind the incisors, which functions to limit soil from entering the mouth cavity. Although individuals may move aboveground to disperse to a new area, they generally do not come out of their burrow systems except to push dirt aboveground so the burrow systems can be expanded. Except during periods when the mother is nursing young, the burrow system is occupied by a single individual. The so-called mima mounds, which are the result of excavated dirt being deposited on the surface, are a characteristic of any area occupied by pocket gophers.

As described below, these rodents are an important component of North American fauna. First, the magnitude of genetic and morphological variation within the pocket gophers is important to biodiversity. Second, they are widely distributed and their burrowing activity has both a positive and a negative effect on the ecosystem. Third, their destruction of underground wires and pipes and consumption of crops, such as alfalfa, often makes control of populations necessary. Finally, their combination of unique biological characteristics is fascinating to study.

DISTRIBUTION

Pocket gophers are distributed from southern Canada to Panama. *Cratogeomys* occurs in the southern plains of the United States (Fig. 14.1),

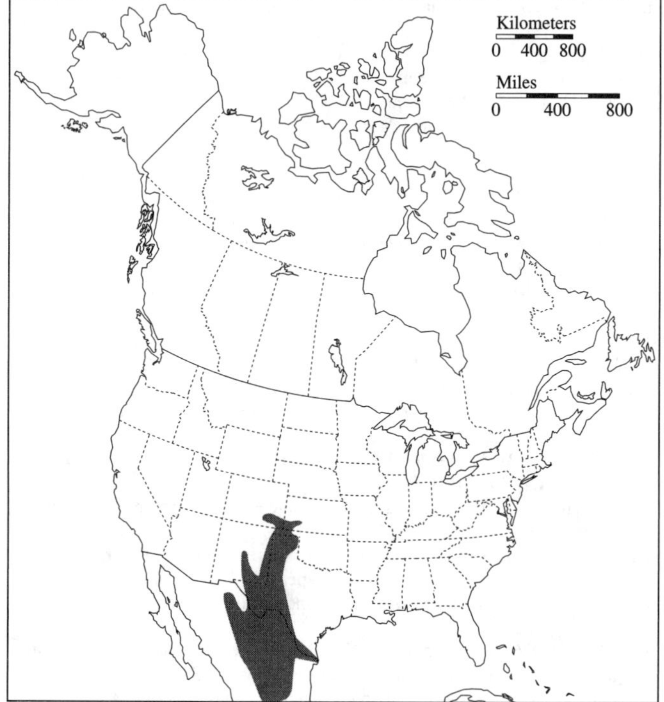

Figure 14.1. Distribution of the yellow-faced pocket gopher (*Cratogeomys castanops*).

Geomys in the central and southeastern United States and northeastern Mexico (Figs. 14.2 and 14.3), and *Thomomys* in the western United States and northern Mexico (Figs. 14.4–14.6). The distribution of pocket gophers typically is defined by soil type and, in some instances, elevation. All but two species (*G. pinetis* and *G. tropicalis*) occur in arid climates. *Cratogeomys* occupy chirty soils at low elevations; *Geomys* prefer sandy to loam soils at low elevations. *Thomomys* typically are found in rocky, gravelly soils at medium to high elevations; however, if no other species of pocket gopher are present, *Thomomys* may occupy more friable soils.

Typically, these three genera of pocket gopher are allopatric or at least parapatric in their distribution. However, several instances are known where *Geomys* and *Cratogeomys* are sympatric in west Texas and eastern New Mexico, and *Thomomys* and *Cratogeomys* are sympatric in the Davis Mountains in Texas. Although soil types may regulate these zones of contact, in the case of *Thomomys* and *Cratogeomys,* it has been hypothesized that the trend toward a more arid environment has favored the displacement of *Thomomys* by *Cratogeomys* (Reichman and Baker 1972; Williams and Baker 1976). The sympatry of *Geomys* and *Cratogeomys* may be the result of human activities, where roadbeds (chirty soil) have provided a mechanism for dispersal of *Cratogeomys* into traditional *Geomys* habitat.

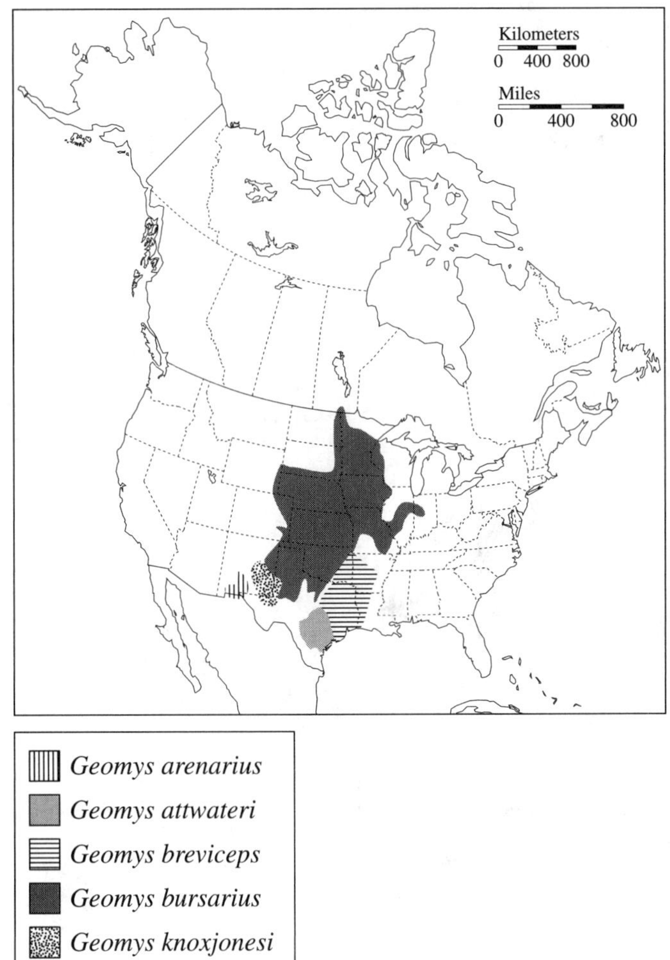

FIGURE 14.2. Distribution of five species of eastern pocket gopher: *Geomys arenarius, G. attwateri, G. breviceps, G. bursarius,* and *G. knoxjonesi.*

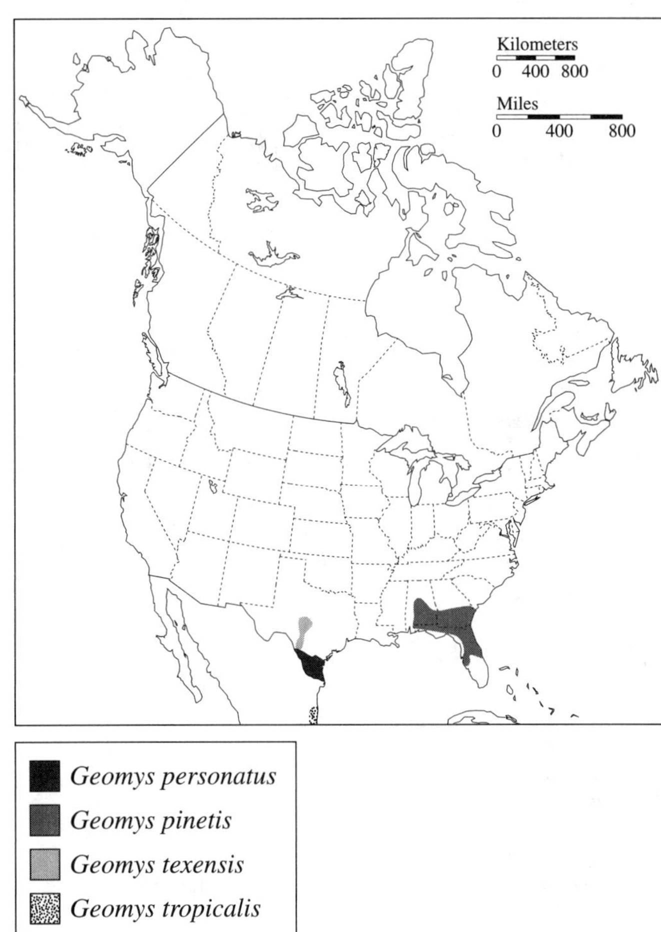

FIGURE 14.3. Distribution of four species of eastern pocket gopher: *Geomys personatus, G. pinetis, G. texensis,* and *G. tropicalis.*

Although most species within a genus exhibit a parapatric distribution as a result of soil preference (Miller 1964; Vaughan 1967; Thaeler 1968), narrow isolated contact zones may be common where different soil types meet or interdigitate. For example, at least three contact zones have been identified within *Geomys* (Kennerly 1959; Pembleton and Baker 1978; Honeycutt and Schmidly 1979; Tucker and Schmidly 1981; Heaney and Timm 1985; Baker et al. 1989; Dowler 1989; Bradley et al. 1991a, 1991c; Sudman et al. 1987; Burt and Dowler 1999) and several within *Thomomys* (Patton 1973; Thaeler 1974; Patton et al. 1979; Patton and Smith 1990, 1994; Ruedi et al. 1997).

Glaciation is thought to have had a major influence on the geographic distribution of *Geomys* (Russell 1968; Hart 1978). The diversity of species of *Geomys* in Texas supports the contention of Ice Age refugia and speciation as a result of pocket gophers from the central plains being forced farther south in response to changing climatic conditions. Physiography may have played a greater role in the divergence and speciation in *Thomomys* (Durrant 1952). For example, Dalquest and Kilpatrick (1973) described the separation in distribution between *Thomomys* and *Geomys* perhaps as recently as the last 4000 years.

DESCRIPTION

Pocket gophers are fossorial rodents and consequently have evolved several morphological adaptations to accommodate this lifestyle. The body is fusiform, with a small, somewhat flattened head. Eyes, ears, and tail are small to reduce exposure to soil while burrowing. The forelimbs are short, massive, muscular, and equipped with prominent claws. Specialization in the forelimbs and claws for digging is least developed in *Thomomys,* intermediate in *Cratogeomys,* and most highly developed in *Geomys*. Hind limbs generally are broad, flattened, and designed for moving soils loosened by the forelimbs. A defining characteristic associated with pocket gophers is the presence of fur-lined cheek pouches. Among North American mammals, only the pocket mice (subfamily Heteromyinae and Perognathinae) and kangaroo rats (subfamily Dipodomyinae) of the family Heteromyidae share this characteristic with pocket gophers. Size is variable within each genus, but typically *Thomomys* is the smallest (total length up to 300 mm; maximum body weight about 250 g), *Geomys* is intermediate (total length up to 330 mm; maximum body weight about 400 g), and *Cratogeomys* is the largest (total length up to 350 mm; maximum body weight about 420 g).

Pelage. The pelage is short, fine, and soft, probably to accommodate removal of soil during grooming. Coloration is highly variable and is closely associated with soil color. Color of fur is lighter on the venter than on the dorsum. Dorsal and lateral patches of white pelage are common. Historically, pelage color was an important characteristic in naming subspecies and species. However, recent studies have shown pelage color to be a highly variable characteristic and of little value in determining taxonomic status.

Dentition. Pocket gophers, like all rodents, have enlarged incisors (Fig. 14.7). The incisors are used for digging as well as for gnawing roots and tubers. The incisors protrude from the small mouth. This adaptation allows pocket gophers to use the incisors while the mouth is closed, thus preventing dirt from entering the oral cavity. Upper incisors have no

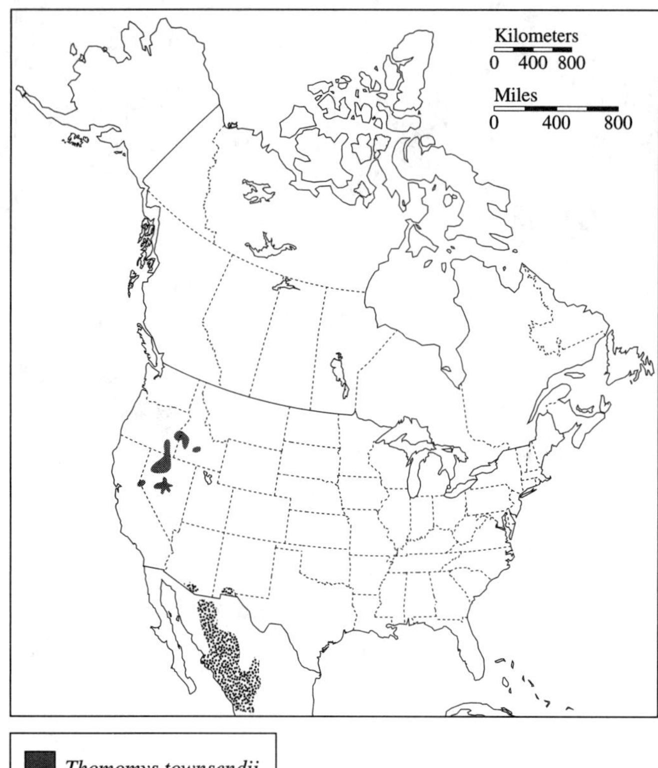

FIGURE 14.4. Distribution of two species of the western pocket gopher: *Thomomys townsendii* and *T. umbrinus.*

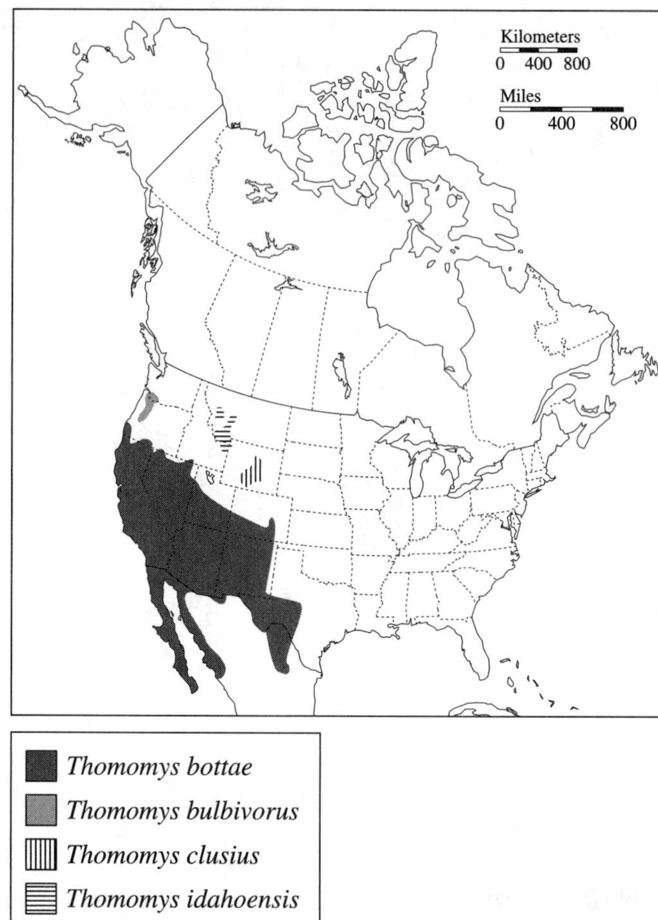

FIGURE 14.5. Distribution of four species of the western pocket gopher: *Thomomys bottae, T. bulbivorus, T. clusius,* and *T. idahoensis.*

grooves in *Thomomys* (Fig. 14.7, right), one groove in *Cratogeomys* (Fig. 14.7, left), and two grooves in *Geomys* (Fig. 14.7, middle). The dental formula in all pocket gophers is I 1/1, C 0/0, P 1/1, M 3/3. The incisors are open rooted and grow throughout life. The skull of Botta's pocket gopher is shown in Fig. 14.8.

Growth and Longevity. Smolen et al. (1980) reported that in *Cratogeomys,* females live longer than their male counterparts (females lived 56 weeks and males averaged 31 weeks). However, two females survived 105 weeks and one male was 76 weeks old and alive at the termination of the study. Male mortality increases during the peak of reproduction, whereas female mortality levels off from January to July or August. Mortality is especially high as subadults move to establish their own burrow system. Subjuveniles have good survivability, most likely due to the safety of the burrow system and aggressiveness/protectiveness of the mother.

Soil type (composition and texture) may play an important role in determination of body size. For example, Davis (1938), Kennerly (1954, 1959), Miller (1964), Hendrickson (1972), Smith and Patton (1988), and Wilkens and Swearingen (1990) suggested that cranial size increased in sandy soils, whereas gophers inhabiting heavier and more dense soils tend to be smaller. Together, soil type and other local environmental factors produce a large amount of geographic variation (both size and coloration) throughout the range of a species. Patton and Brylski (1987) and Smith and Patton (1988) found that pocket gopher size varies due to environmental factors as well as quality of the habitat. Over a 30-year period in which pocket gophers established populations in alfalfa fields, an increase in body size of males of 66.8% and in females of 53.6% was recorded (Patton and Brylski 1987).

GENETICS

Pocket gophers generally occur in small, isolated populations. Limited exchange of individuals between populations results in small effective population sizes (Patton and Feder 1981; Ruedi et al. 1997). The isolated nature of pocket gopher populations has produced low levels of intrapopulational and intraspecific genetic variation, but high levels among populations and species (Selander et al. 1975). For example, Patton and Sherwood (1982) and Sherwood and Patton (1982) demonstrated that the genome size in two species of *Thomomys* differed significantly. The observation that genome size is not correlated with genetic complexity is refered to as the *C*-value paradox (Swift 1950; Thomas 1971), and this paradox is explained in pocket gophers by variation in heterochromatin, which is usually inert relative to genetic complexity. Burton and Bickham (1989) and Bradley et al. (1991c) showed little variation in genome size in species of *Geomys.*

Heterozygosity is low compared to other rodent species (Patton et al. 1972; Selander et al. 1975; Penney and Zimmermann 1976; Avise et al. 1979; Zimmerman and Gayden 1981). Most species possess distinct karyotypes and allozymic markers. In fact, these genetic differences have been instrumental in recognizing and distinguishing among species (Patton and Dingman 1968; Patton et al. 1972; Tucker and Schmidly 1981; Baker et al. 1989, 1996; Block and Zimmerman 1991; Bradley et al. 1991a).

The fossorial behavior of pocket gophers, coupled with their solitary occupation of burrow systems and the magnitude of genetic plasticity, has resulted in a number of contact zones between populations that have not developed premating isolating mechanisms sufficient to prevent production of natural F_1 hybrids. As new genetic methods have been developed, studies of these natural hybrid zones have become more and more sophisticated (Patton and Dingman 1968; Baker et al. 1989, 1996; Bradley et al. 1991a, 1991b, 1991c). These studies

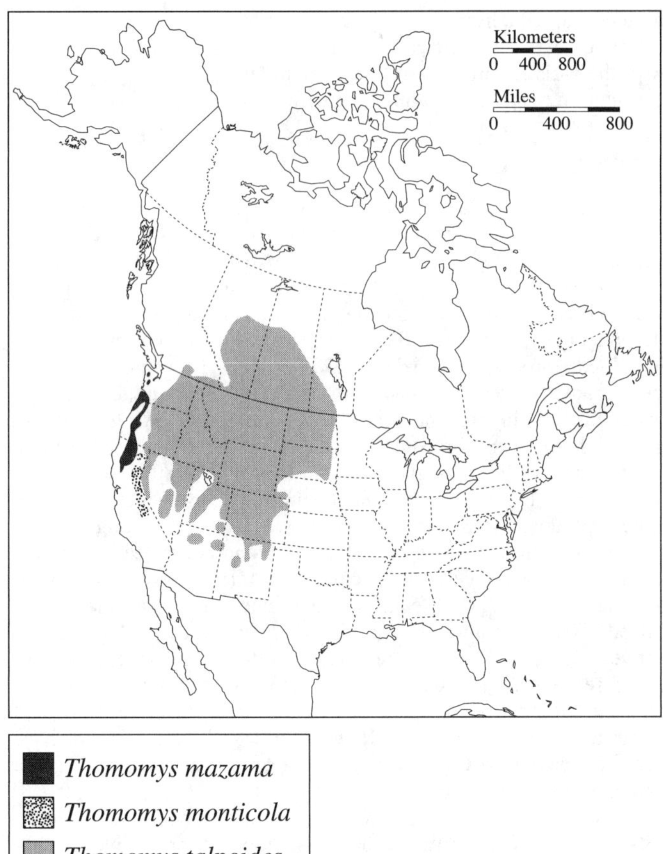

FIGURE 14.6. Distribution of three species of the western pocket gopher: *Thomomys mazama, T. monticola,* and *T. talpoides.*

have documented how complex are the stages necessary to complete reproductive isolation between species.

One of the first studies of hybrid zones involved populations that were chromosomally distinguishable (Patton and Dingman 1968). Previous to this publication, hybrids were known to occur in nature, but their frequency and relative numbers were underappreciated. Since the initial study by Patton and Dingman (1968), literature focusing on hybrid zones in pocket gophers has increased substantially (Pembleton and Baker 1978; Tucker and Schmidly 1981; Baker et al. 1989; Bradley et al. 1991a, 1991c; Chesser et al. 1996; Burt and Dowler 1999). These studies have generated an overall increase in our understanding of hybrid zones, population subdivisions, and speciation in mammals. For example, in a study of *G. bursarius* in eastern New Mexico (Baker et al. 1989; Bradley et al. 1991a), it was possible to document the frequency of F_1 and backcross individuals as well as the level of reciprocity in the production of F_1 individuals. This study used chromosomes, mitochondrial DNA, and nuclear DNA to establish complex genotypes for individuals. Ultimately, it was concluded that one cross, females of *G. bursarius* and males of *G. knoxjonesi,* did not occur, or had a very low frequency, in nature. The cross involving females of *G. knoxjonesi* and males of *G. bursarius* did occur, but F_1 males appeared to be sterile or at least possessed a low level of fertility. Although hybridization occurred to a point that it produced a substantial number of hybrid individuals, integration of the genes of one species into the other was restricted to a 200-m contact zone. Several relevant points can be made from these observations. First, pocket gophers provide an excellent model system for studying integration of genetic material between populations isolated by degrees of infertility in F_1 individuals. Some investigators interpret this as an opportunity to study how the speciation process occurs, whereas others argue that information from genetic hybrid zones has limited implications for mechanisms involved in the process of speciation. Second, this asymmetrical crossing provided support for evaluating the dynamics associated with mammalian hybrid zones (Bradley et al. 1991a; Patton and Smith 1994; Chesser et al. 1996) and the relevance to the Kaneshiro model (Kaneshiro 1976, 1987, 1989). Third, complex genotypes have allowed for testing of Haldane's theory (Haldane 1922) that the heterogametic sex is most likely to be sterile.

Even with the magnitude of molecular data and extent of hybridization documented in the *bursarius/knoxjonesi* complex, scientists have interpreted the implications of the data differently. For example, Baker et al. (1989) and Bradley et al. (1991a) concluded that *G. knoxjonesi* should be recognized as a species distinct from *G. bursarius*. Alternatively, Patton (1993) considered *G. knoxjonesi* to be a subspecies of *G. bursarius*. This disagreement lies in the difficulty of interpreting data within a theoretical framework as complicated as the species concept.

In addition to the wealth of information concerning patterns and processes of evolution obtained from the studies of pocket gopher hybrid zones, several lines of basic genetic information have been discovered. For example, Patton et al. (1979) and Baker et al. (1983) reported on a balanced chromosomal polymorphism present in a population of pocket gophers. This example is one of few cases in nature where a chromosomal polymorphism appears to provide a selective advantage. Furthermore, Bradley et al. (1993) provided the only description of a mechanism leading to the origin of a novel allozyme allele. Both of these studies contributed not only to our understanding of pocket gophers, but to an overall knowledge of mammalian genetics.

PHYSIOLOGY

Gettinger (1975) examined the thermoregulatory ability of two species of pocket gopher, *T. talpoides* and *T. umbrinus,* and concluded that *T. talpoides* had a more varied regulatory response that did *T. umbrinus.* Bradley et al. (1974) and McNabb (1966) related the distribution of

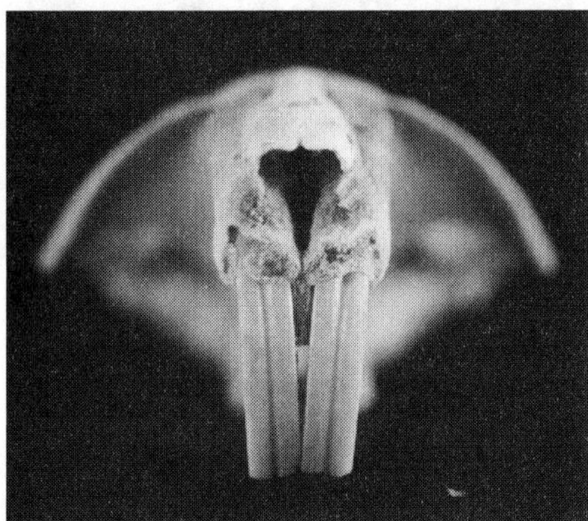
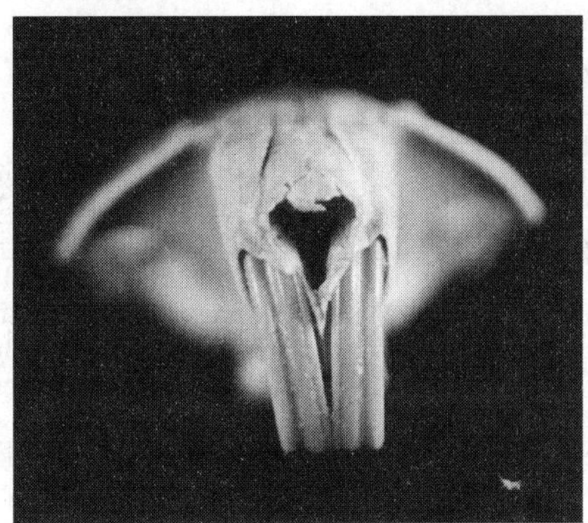
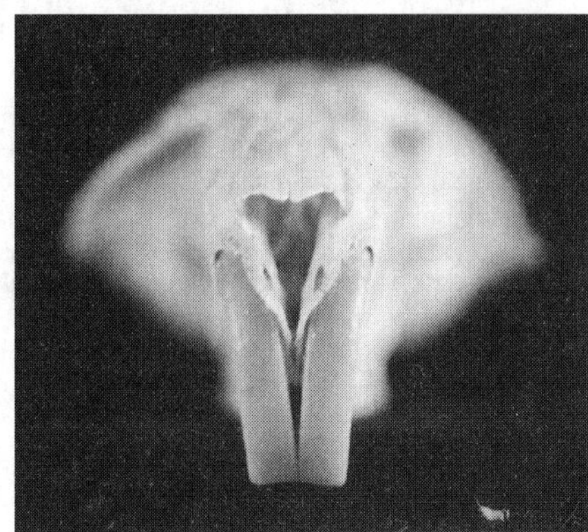

FIGURE 14.7. Incisors of (left) *Cratogeomys,* (middle) *Geomys,* and (right) *Thomomys* are diagnostic among the three genera. Incisors of *Cratogeomys* have only one groove, those of *Geomys* have two grooves, and those of *Thomomys* are smooth. SOURCE: L. R. McAliley.

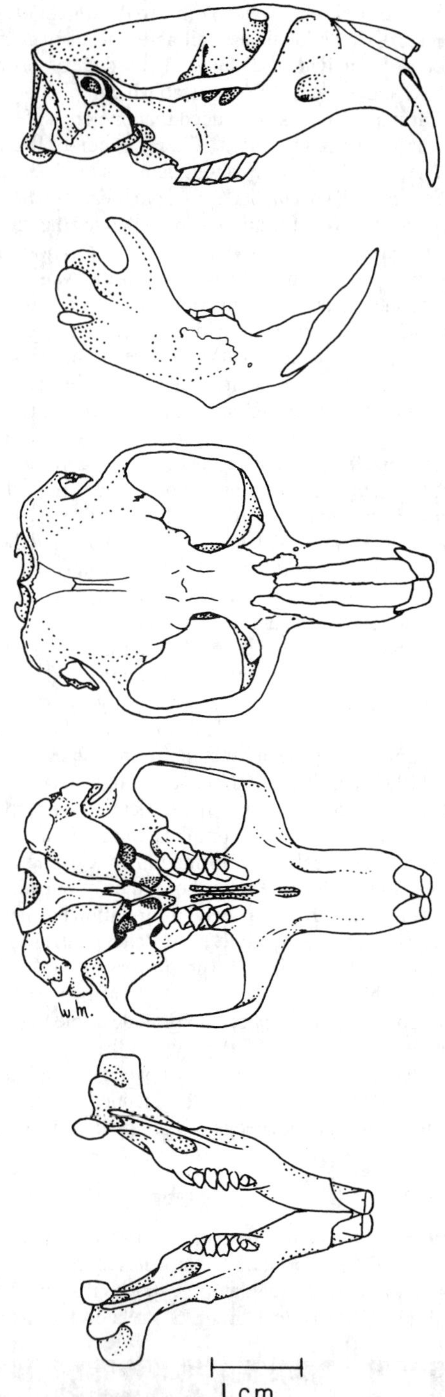

FIGURE 14.8. Skull of the Botta's pocket gopher *(Thomomys bottae)*. From top to bottom: lateral view of cranium, lateral view of mandible, dorsal view of cranium, ventral view of cranium, dorsal view of mandible.

some species of pocket gophers to their morphological and physiological adaptations to thermoregulation. Bradley et al. (1974) found that *T. talpoides* had larger bodies and shorter tails at higher altitudes than did members of this species located at lower elevations. Body temperatures of *T. talpoides* were maintained at 36.9°C when the ambient temperature was between 1°C and 30°C. As temperatures rise above 30°C, the ability of pocket gophers to effectively dissipitate heat is reduced. Pocket gophers exposed to temperatures at or above 38°C for short periods of time develop heat stress and die within an hour if not removed to a cooler area (Howard and Childs 1959). Pocket gophers have developed a highly vascularized naked tail to facilitate the dissipation of heat. This increased peripheral blood flow to the tail coupled with the variation in ratio of body size to tail length may explain the distribution of species within arid environments.

REPRODUCTION

Age, Season, and Gestation. Sexual maturity in females may be identified by examination of weight and pubic symphysis resorption (Smolen et al. 1980). In males, however, sexual maturity is much more difficult to estimate due to the lack of a visible external change associated with testes development. Age at sexual maturity appears to depend on time of birth for females. For example, Smolen et al. (1980) identified reproductive females of *Cratogeomys* that where first caught in livetraps between December and May; however young females first captured after May were not reproductive until the following season. In general, it seems that breeding begins near the end of the first year of a pocket gopher's life (Douglas 1969; Smolen et al. 1980).

Douglas (1969) reported *Thomomys* from Mesa Verde, Colorado, were reproductive only in the spring. Pregnant or lactating females where caught in April and early May. Cox and Hunt (1992) reported capturing reproductively active female *T. bottae* as early as November and the earliest pregnant females in December. Pregnant females continued to be caught through May. *Cratogeomys* were reported as being active as early as November (Smolen et al. 1980). Smolen et al. (1980) found reproduction reached its peak in April–May but continued to October. Most sources indicate that for the three genera, breeding and parturition take place in spring. However, in Texas, scrotal *Geomys* have been collected in November and early December, indicating that some breeding may occur in the fall.

Pocket gophers are typically characterized as being promiscuous (Eisenberg 1966). Patton and Feder (1981) found that reproductive contribution is skewed toward a small, select group of gophers in the population. Whereas there may be several reproductively active males in a group, only one appears to contribute to reproductive success. Patton and Feder (1981) found that sex ratios of offspring were biased toward females in both years of their study. Gestation period for *Thomomys* is 18–19 days (Schramm 1961; Douglas 1969). During reproduction, it is believed that males, being larger, must find females and entice them into the males's burrows. Asymmetrical crossing in hybrid zones suggests female choice (Bradley et al. 1991a; Patton and Smith 1994).

Litter Size and Number per Season. Most investigators report multiple litters during a breeding season. Smolen et al. (1980) identified female *Cratogeomys* with three or four litters. Contrary to this are reports that pocket gophers produce a single litter a year (Ellison and Aldous 1952; Howard and Childs 1959; Hansen 1960; Vaughan 1963). Others have suggested that it is more likely that pocket gophers produce more than one litter a year (Cahalane 1947; Hoffmeister 1971). Litter sizes in *T. bottae* average 4.6 pups (Douglas 1969). Estimates of litter sizes for *T. talpoides* range from 3.2 to 6.4 pups (Hansen 1960). Litter sizes for *Cratogeomys* are reported to average 2.0 (Ickenberry 1964).

SEX DETERMINATION

Typically, it is difficult to determine the sex of pocket gophers by external characteristics. Males lack a prominent penis, and the testes are generally internal, making males difficult to distinguish from females. Howard and Childs (1959) concluded size can sometimes be used to distinguish sexes. Males are bigger than females in all species: about 10% heavier and 3–4% greater in total length. Some evidence suggests that males may exhibit indeterminant growth (Baker and Genoways 1975; Williams and Genoways 1977, 1978, 1980, 1981). Howard and Childs (1959) determined the best method to identify the sex of immature pocket gophers was to apply pressure inward to cause the penis (if present) to extrude. We find the best way to sex adult live pocket gophers for mark–recapture studies is to digitally examine the ventral surface of the pelvic girdle. On females, the pubic symphysis has been reabsorbed, creating a narrow trough along the midline, whereas in males, the pubic symphysis is intact, forming an elevated peak along the midline. Although immature females may not have reabsorbed the pubic symphysis, these individuals have the pelage and external characteristics of an immature gopher and caution should be applied in determining the sex.

POPULATION DYNAMICS

Home Range. Pocket gophers are solitary individuals typically with one individual inhabiting a burrow system throughout most of the year. Because of this solitary lifestyle, pocket gophers partition habitat into territories or home ranges (Howard and Childs 1959). Pocket gophers rarely venture far from their territory, except during the reproductive season or to claim portions of abandoned tunnel systems where food availability and soils are more favorable (Howard and Childs 1959). Home range stability is vastly different and varies according to sex. Males are more likely to have plastic home ranges and females tend to have more rigid home ranges. Smolen et al. (1980) examined the home ranges of *Cratogeomys castanops* and found that those of males had an average linear measurement of 46.89 m, whereas those of females had an average linear measurement of 37.05 m. Howard and Childs (1959) examined home ranges of *T. bottae* and reported that average home range was 256 m^2 for males and 120 m^2 for females.

Reproductive status of pocket gophers appears to play an integral part in home range size (Reichman et al. 1982). Examination of burrow spacing of *T. bottae* in Arizona revealed that reproductively active males had larger home ranges, closer contact with neighboring pocket gophers, and an increased perimeter compared to nonreproductive males or females (Reichman et al. 1982). As male pocket gophers enter reproductive condition, they expand their home ranges in such a fashion as to increase their contact with potential mates (Reichman et al. 1982).

Availability and quality of food also may play a role in the size of pocket gopher home range. Rezsutek and Cameron (1998) found that removal of dicots from habitat reduced the occurrence of reproductively active females as well as the time gophers spent in the habitat. In arid environments, where available foods may be separated by greater distances, pocket gophers may be forced to increase home range sizes.

Movements. Williams and Baker (1976) described the movements of *Cratogeomys* and *Thomomys* in a zone of sympatry within the Davis Mountains of Texas. Movement of adult male *Cratogeomys* exceeding the burrow length was 135–595 m with an average of 391 m; movement by females was 60–100 m with an average of 85 m. Movement of male *Thomomys* exceeding the burrow length was 55–215 m with an average of 113 m, and females moved 45–120 m beyond the burrow with an average of 74 m. Smolen et al. (1980) studied movements of *Cratogeomys* in Lubbock County, Texas. Males moved 47–195 m, whereas females moved 37–194 m. The distances moved documented by Williams and Baker (1976) were substantially greater than those reported by Smolen et al. (1980). These differences probably reflect the nature of the gopher habitat available in the two studies. The Smolen et al. (1980) study was a rectangular area from which pocket gophers could immigrate in all directions, and trapping was limited to the rectangular area and the area immediately adjacent to the rectangle. Any gopher that moved beyond that area of study would not be observed. The area studied by Williams and Baker (1976) was a riparian area of favorable habitat (approximately 1800 m of river bank) flanked on both sides by steep canyon walls, which limited any movements to up and down the riverbank. Any area between the canyon walls which had any gopher activity was intensively sampled to identify the species and individuals that were present. Therefore, any gopher that dispersed and survived was likely to be observed. Although these are extreme conditions, they most likely provide some realistic insights into the dispersal potential of *Cratogeomys* and *Thomomys* under comparable conditions. Other studies indicate that in spite of movements there is considerable geographic isolation, resulting in population subdivision (Davis 1940; Kennerly 1954, 1959; Honeycutt and Schmidly 1979).

Density. Densities up to 37/ha have been reported (Butcher 1929; Kennerly 1954; Downhower and Hall 1966), but in some species it may be as low as 2.5/ha (*G. tropicalis*) (Selander et al. 1975). Density estimates for pocket gophers have long been considered to be influenced by such factors as local climate, soil suitability, body mass, and vegetation types. Smallwood and Morrison (1999a) reviewed 32 published studies on pocket gopher density and found that estimates may be a function of female body mass as much as any factor, with lower densities being associated with larger female body sizes and vice versa.

ECOLOGY

Habitat. Pocket gophers occur in a variety of habitats ranging from high mountain meadows to low-elevation plains and grasslands (Huntly and Inouye 1988). Pocket gophers generally are limited in their distribution by the quality and type of the soil. Soils that are fine sand to sandy loam, porous, well drained, and high in nutrient content are typically favored (Miller 1964; Hansen and Morris 1968; Pembleton and Baker 1978; Davis 1986; Huntly and Inouye 1988) over soils that are high in clay content, do not drain well, or are poor in nutrient quality (Huntly and Inouye 1988; Block and Zimmerman 1991). However, species that have a broad tolerance range in soil conditions, such as *T. talpoides* and *G. bursarius*, may be more widespread.

To avoid extreme fluctuations in surface and air temperatures, gophers tend to favor soils that are deeper, and avoid areas where the soil may be favorable but shallow (Huntly and Inouye 1988; Smallwood and Morrison 1999a). Deeper burrow systems not only allow gophers to avoid higher surface temperatures, but also may help to reduce collapse of tunnel systems and provide consistent atmosphere by controlling humidity and gas content. Deeper burrow systems also may allow pocket gophers to better avoid predators by providing better escape capabilities as well as making it difficult for predators to locate them. Relative humidity in burrow systems of pocket gophers is relatively constant because they are closed systems. In addition, humidity levels may be maintained by stems and leaves pulled into the burrows (Douglas 1969). These attributes allow pocket gophers to occupy and survive in arid environments (Douglas 1969).

Although nutrient content of soil affects pocket gopher distribution, this is ancillary. The effect of soil composition on distribution is due to the large amount of energy required for the fossorial lifestyle (Huntly and Inouye 1988). Because of high energy requirements, pocket gophers tend to be found in soils high in nitrogen content and primary productivity (Tilman 1983; Reichman and Smith 1985; Inouye et al. 1997). Soil nutrient content mainly affects pocket gopher distributions by influencing locations of food sources, thereby possibly eliminating what would otherwise be suitable soils as prime habitat (Turner et al. 1973; Huntly and Inouye 1988).

Burrows. Burrow systems of pocket gophers are complicated structures, which vary from species to species. However, there are basic characteristics associated with all types of systems. These characteristics include one or two main tunnels with multiple minor branches for foraging, as well as a chamber for depositing fecal matter, a nesting chamber, and a cache chamber for food.

Unlike most borrowing mammals, pocket gophers spend most of their time in their burrows performing daily routines such as foraging, waste disposal, and reproductive efforts. Depth of the burrow system appears to be directly correlated with body size (Anderson 1987; Roberts et al. 1997). However, there appears to be no correlation between body size and the extensiveness of the tunnel system (Roberts et al. 1997). The complexity and extent of burrow systems appears to be a function of food availability, food quality, quality of soil, and prolonged use.

Roberts et al. (1997) found spiral tunnel sections leading to the nesting chambers. Although these spiral sections appear to be rare, they seem to function in protecting the nesting chamber. Tunnels typically are linear with a diameter corresponding to the girth of the gopher. Branched tunnels radiate from the core tunnel, unlike those of other burrowing mammals, which construct tunnels without branching (Roberts et al. 1997). Tunnels vary in depth below soil surface, with the shallowest structures (feeding tunnels) located along the root zone. The deeper, general-use tunnels are located several centimeters below this depending on soil depth and type. Tunnels may range up to 150 m in overall length, with an average of 73 m, excluding the side tunnels (Smallwood and Morrison 1999a).

Excavation rates of pocket gophers vary according to body size and soil type. Burrowing is accomplished using the claws of the forefeet and the incisors (Lessa and Stein 1992). The hind legs are not used in digging, but to move the gopher forward during burrowing activities. Loose soil is pushed forward with the chin and forefeet and is carried to the surface, where it is deposited in mounds. Pocket gopher mounds in *G. bursarius* are crescent shaped and average 30–45 cm in radius and 8 cm in height. Smallwood and Morrison (1999a) found that each mound contained 0.01 m^3 of excavated soil. On average, gopher burrow systems were 0.59 m^3 in volume with an average excavation rate of 17.8 m^3/ha per year (Smallwood and Morrison 1999a). Pocket gophers have the ability to move a tremendous amount of soil; Grinnell (1923) estimated that pocket gophers in Yosemite National Park moved 8000 tons/year of soil to the surface.

During winter, when the ground is covered by snow, pocket gophers may extend their mounds into the snow pack, thereby increasing their foraging area. In addition, foraging in snow packs allows pocket gophers to forage in relative safety from predators. During spring, remnants of these tunnel systems are apparent by the cores of soil left as the snow melts.

Mima mounds found throughout the western portions of North America have long been attributed to the action of pocket gophers, especially in areas of poorly drained soils (Cox and Gakahu 1986; Cox 1990; Hunt 1992). These mounds are formed by the action of gophers moving soil to the surface in mounds as large as 2 m high and 25–50 m in diameter (Cox 1990; Hunt 1992). Pocket gophers typically deposit soils excavated during tunneling near the center of these mounds, maintaining the shape and size of the mound.

FEEDING HABITS

Because of the fossorial lifestyle maintained by pocket gophers, their daily caloric intake is greater than that of most other rodent species. Energy expenditure for burrowing and maintenance of tunnels is estimated to be 360–3400 times that required by mammals moving aboveground (Vleck 1979). Gettinger (1984) estimated that energy flow through pocket gopher populations in chaparral was 2200 MJ/ha per year. Anderson and MacMahon (1981) found that energy flow in pocket gopher populations in montane meadows was at least 1100 MJ/ha per year. These estimates far exceed values for any other rodents and are comparable to those expected in much larger herbivores. Due to this increased need for caloric intake, pocket gophers readily exploit a variety of herbaceous foods depending on season, species, and availability.

Pocket gophers are generalist herbivores whose diet may change seasonally and among species (Hunt 1992). Food items may include roots, tubers, corms, rhizomes, stolen, stems, or other plant parts collected either aboveground or below. Members of the genus *Geomys,* such as *G. attwateri* and *G. bursarius,* feed on subterranean plant tissues such as roots, corms, and tubers (Meyers and Vaughan 1964; Williams and Cameron 1986). In contrast, *T. mazama* consume mostly aboveground foods, including 40% forb shoots and 32% grass shoots (Burton and Black 1978).

As noted, season plays a role in the diets of pocket gophers. Burton and Black (1978) found *T. mazama* preferred forbs during the summer, but switched to roots and stems of grasses during November–May. Similarly, *T. bottae* in the San Jacinto Mountains of California preferentially foraged on herbaceous annuals during the summer, whereas their winter diets included *Pinus* roots, shrubs, and grasses (Gettinger 1984). This switch in diet from forbs may be due to the disappearance of the food source during fall and winter. In the grasslands of California, where forbs are more prominent during the wet winter months and grasses during the dry summer months, Hunt (1992) found that *T. talpoides* fed primarily on herbaceous annuals during the winter and grasses during the summer.

Rezsutek and Cameron (1998) studied the effects of dicots on the distribution of *G. attwateri* and found that removal of dicots reduced populations of pocket gophers. In that study, the diet of *G. attwateri* typically contained 29–54% dicots, depending on the age and the breeding condition of individual pocket gophers. *Geomys,* like *Thomomys,* will switch their diets during dry or cold periods when food sources are decreased. During these periods, pocket gophers will use plant materials cached during the growing season and/or switch to more monocots and less dicots.

BEHAVIOR

An important behavioral feature of pocket gophers is their response to any activity that opens the burrow. If they encounter an opening in their burrow, they will quickly seal it not only by moving dirt to the surface, but also by packing the burrow system some distance below the surface. It is thought that this behavior is adaptive in that it becomes difficult for snakes, badgers (*Taxidea taxus*) or other predators to have easy access to their burrow system. As noted below (Economic Status and Management), this commitment to sealing an open burrow is the key to successful trapping.

Pocket gophers are highly specialized for digging and moving soil. Their primary behavior is associated with excavating new burrows and pushing the excess dirt to the surface of the soil. Usually, when they push a load of dirt to the surface, they quickly deposit it without exposing their entire body. Often, the dirt can be seen being pushed from the burrow without actually seeing the gopher. Burrowing activity is both diurnal and nocturnal.

Construction of new burrows is strongly correlated with rainfall. In West Texas, during the summer, when there is rarely rain, no new burrows may be observed for 1 month or more. When the drought is over and a substantial rain occurs, burrowing activity becomes common and many new mounds are observed. In some mark–recapture studies (Smolen et al. 1980), severe drought restricted the ability to effectively capture animals. It is not clear whether periods of inactivation of mound building are associated with estivation or whether individuals simply are active but do not dig extensions of the burrow systems.

Vocalization. Literature on the vocalization of geomyid rodents is severely limited. Gophers grind their teeth, hiss, and make other chattering noises when trapped. There is, however, no published literature on vocalizations between pocket gophers. Wilkens et al. (1999), in their review of literature on hearing morphology in geomyids, noted the lack of published literature on vocalizations in gophers.

Aggression and Competition. The observation that different species of pocket gopher are almost never sympatric indicates that even species as divergent as *Cratogeomys* and *Thomomys* are intensive competitors for the same resources (Reichman and Baker 1972; Williams and Baker 1976). Similar patterns of contiguous allopatry have been observed for *Cratogeomys* and *Geomys*. Contact zones between congeneric species also document intense competition (Patton and Dingman 1968; Baker et al. 1989).

The behavior of *Cratogeomys, Geomys,* and *Thomomys* from the standpoint of being handled by humans or encountering other large mammals is highly variable and generically characterized. Individuals of *Cratogeomys* rarely are aggressive when being handled. They can be removed from a livetrap into the palm of one's hand and even gently stroked without offering any attempt to bite or defend themselves. The 94 individuals studied by Smolen et al. (1980) were all handled without gloves without any researcher being bitten. At the other extreme, individuals of *Geomys* are generally fiercely aggressive, even to the point of charging a human when they are released onto the surface of the ground. They almost always will bite any object that moves near to them and make aggressive gestures by clicking their teeth and lunging forward when approached. In our work on this genus, we always wear thick leather gloves. *Thomomys* is intermediate in their behavior, but often will bite when handled.

MORTALITY

Parasites. Considerable work has been performed on the parasites of pocket gophers. Douthitt (1915) described three cestodes (*Andrya macrocephala, Anoplocephaloides infrequens,* and *Anoplocephaloides variabilis*) from the subspecies *G. bursarius bursarius*. Timm and Price (1980) collected the pocket gopher louse *Geomydoecus geomydis*. Bartel and Gardner (2000) examined endo-and ectometazoan parasites from 144 pocket gophers in Minnesota and identified two species of flea (*Opisocrostis bruneri, Foxella ignota ignota*), one species of louse (*Geomydoecus geomydis geomydis*), one species of tick (*Dermacentor variabilis*), three species of nematodes (*Physaloptera limbata, Capallaria americana, Ransomus rodentorum*), four species of cestodes (*Anoplocephaloides infrequens, A. variabilis, Andrya macrocephala, Hymenolepis weldensis*), and one species of acanthocephalan (*Moniliformis clarki*). Bartel and Gardner (2000) reported that fleas infected 85% of the pocket gophers they examined; chewing lice, 79%; ticks, 0.8%; nematodes, 26.3%; cestodes, 46.3%; and acanthocephalans, 0.8%.

Chewing lice of the genera *Geomydoecus* and *Thomomydoecus* are found exclusively on pocket gophers and have been the subject of several studies of cospeciation (Demastes and Hafner 1993; Reed et al. 2000). Members of these two genera typically inhabit a single species of pocket gopher (Demastes and Hafner 1993; Reed et al. 2000).

Douglas (1969) reported larval botfly infestations in *Thomomys*. In some individuals, as many as three larvae were present. Incidence of botfly infestation was highest in September and absent in the spring.

Predators. The fossorial lifestyle of pocket gophers has resulted in a limited source of potential predators. Pocket gophers in burrows are safe from most predators with the exception of mammals such as badgers, coyotes (*Canis latrans*), weasels (*Mustela* spp.), and some snakes. Pocket gophers are most often preyed on during times of feeding at or near the surface or when moving soil and debris to the surface. During this vulnerable time, pocket gophers may be preyed on by hawks, owls, foxes, coyotes, bobcats (*Lynx rufus*), skunks, and house cats.

Avian predators appear to be the most efficient at capturing pocket gophers. Pocket gophers composed up to 7.4% of the diet of red-tailed hawks (*Buteo jamaicensis*) and 71.4% of the diet of barn owls (*Tyto alba*) (Douglas 1969). Coyotes in Mesa Verde were considered a major predator of pocket gophers; however, according to Douglas (1969), examination of 114 scat samples from coyotes found many rodents, but no pocket gophers. Predators do not appear to affect populations of gophers as much as habitat suitability, food availability, and other environmental factors (Hansen and Ward 1966; Douglas 1969). The inability of predators to control populations of pocket gophers have allowed large colonies to arise throughout their range, leading to increased economic problems.

RELATIONSHIP TO ENVIRONMENT

Pocket gophers drastically affect soil composition through burrowing, mound production, and tunneling. By moving soils loosened during tunneling to the surface, pocket gophers effect the mixing of geological materials, atmospheric chemicals, water, and humus (Smallwood and Morrison 1999b). Tunneling by pocket gophers increases the surface area of the soil and increases the effectiveness of the nitrification process (Reichman and Seabloom 2002). Through the action of mound formation and tunneling, pocket gophers are important contributors to the distribution of soil chemistry and the percolation of water through the system (Huntly and Inouye 1988; Inouye et al. 1997).

Along with benefits to soil composition, pocket gopher burrowing activities have several detrimental effects. Mounds produced by pocket gophers have patches of soil with little to no vegetation associated with them. Lack of vegetation and the presence of finer sand particles in mounds can lead to increased wind and water erosion. These mounds of soil left at the surface are bare patches, which may be higher or lower in nutrient content than intermound areas (Reichman and Seabloom 2002), allow for the exchange of gasses and the flow of water through the soils with greater ease. These patches are areas where succession must begin anew, increasing the diversity of plants in the area of disturbance.

Burrowing activities of pocket gophers have dramatic effects on the plant community. Feeding by pocket gophers can remove much of the root mass from plants, thus severely affecting plant survival. Reichman and Smith (1991) noted that removal of 25% of the root mass had more effect on goats beard (*Tragopogon dubias*) than did losing 75% of the leaves. In alpine meadows, members of the genus *Thomomys* can remove as much as 30% of the subterranean net primary productivity (Anderson and MacMahon 1981). Feeding tunnels of pocket gophers are concentrated around the root zone, thus increasing the chance that exposure of the roots to the interior of the tunnel may increase desiccation of the roots. Due to the herbivorous activity of pocket gophers, plant biomass may be greatly reduced in areas of continual occupation. Additionally, soils moved to the surface may bury or remove plants already in place. However, the long-term effects are one of increase (Reichman and Seabloom 2002).

ECONOMIC STATUS AND MANAGEMENT

Pocket gophers have interested scientists for many years due to their unique lifestyle and conflicts with humans. Pocket gophers historically have inhabited grasslands of the South and Midwest, areas desirable for their use as agricultural lands. Unfortunately, conflicts between pocket gophers and humans have arisen because of the desirability of these lands and the damage caused to agricultural products. As humans began moving westward and developing agriculture, pocket gophers were seen as threats to crops and livestock. Large numbers of pocket gophers in an area can drastically reduce the productivity of crops through burrowing and feeding on plants and causing damage to machinery and irrigation systems (Smallwood and Geng 1997; Smallwood and Morrison 1999b). Miller (1953) estimated pocket gophers reduced alfalfa production in the Sacramento Valley by as much as 25%, and Luce et al. (1981) estimated that pocket gophers damaged as much as 50% of the alfalfa crops throughout the Midwest.

Pocket gophers also damage buried electrical and telephone cables, causing millions of dollars worth of damage yearly (Connolly and Landstrom 1969; Smallwood and Geng 1997; Shumake et al. 1999). In the Northwest, logging companies have spent considerable time and effort on the control of pocket gophers. Pocket gophers feed on the roots of trees planted in regrowth areas, causing hundreds of thousands of dollars in damages yearly and killing many newly planted trees. Pocket gophers also have been implicated in damage to national monuments such as Mesa Verde National Monument in Colorado (Douglas 1969). These activities have contributed to the "eradication view" many take toward these fossorial rodents.

Due to these activities, means of pocket gopher control have been studied extensively. Through the years, there have been many chemical and mechanical means used in the control of pocket gophers. Chemicals such as strychnine, warfarin, capsaicin, denatonium benzoate, and many others (Smallwood and Geng 1997; Shumake et al. 1999) have been developed and used as a means of controlling populations of pocket gophers. Mechanical control methods include, for example, the use of plastic mesh and artificial burrow baiting systems, flooding with irrigation, and gasses (Kepner and Howard 1960; Smallwood and Geng 1997; Shumake et al. 1999).

Other means of controlling pocket gopher distribution include the use of herbicides to remove their food base (Keith et al. 1959; Tietjen et al. 1967). Whereas removing the food base for pocket gophers is effective for controlling populations, it is an impractical method due to

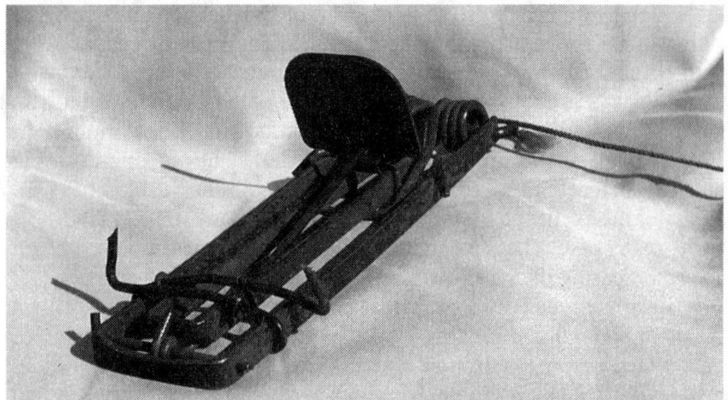

FIGURE 14.9. Macabee killtrap in the set position. The portion of the trap with the jaws is inserted into the burrow, and the trigger plate is pushed toward the burrow opening when the gopher responds to close the tunnel. This results in the jaws of the Macabee closing around the gopher, killing it. SOURCE: L. R. McAliley.

cost and time considerations. Killtraps and livetraps have been developed for pocket gophers. For a review of these and other methods, see Barnes (1973).

There are problems associated with the dissemination of data on means of controlling pocket gopher activity and population. These problems arise in part because research is performed through funding from private organizations, who wish to keep the information to themselves or do not develop the techniques for commercial application (Shumake et al. 1999).

Trapping. In addition to chemical methods of control, effective methods have been developed to trap pocket gophers. Two types of traps are available: one quickly kills the animal at the time it is trapped, and the other captures it alive, without injury. A variety of killtraps exists, but the most common type is referred to as a Macabee (Fig. 14.9). This small trap is easily set and remarkably effective. The amount of effort required to set the trap is small relative to that required to set a livetrap. The primary problem in the successful implementation of Macabees is locating an active burrow. This is done by observing the pattern of mounds on the surface of the ground and locating areas that have a high probability of an open burrow system below. The most effective method is to find a very fresh mound, moist and undisturbed, and follow the recently packed tunnel below it to a place where it interfaces with an open burrow system. Once an open burrow system is available, the Macabee trap is set on the floor of the burrow, about 15 cm beyond the burrow opening. The trap is more effective if a space in the floor of the burrow is excavated, which allows the trap base to be below the floor of the burrow. When the pocket gopher comes to close the open burrow system, it trips the jaws, which close around the body, usually around the chest cavity. In the case of removing a few pocket gophers from a lawn, garden, or area of concern, this trap may be the method of choice because no chemicals are introduced into the environment and the results are quick. Usually, within a few days to several weeks, another pocket gopher will move into the vacant burrow system, but this animal, too, can be quickly trapped. In West Texas, pocket gophers can be controlled in a yard or garden by removing two to five animals from a burrow system per year.

Livetrapping is much more labor-intensive and if there is no reason that live animals are desired, then the Macabee is the most efficient method. In the case of mark and recapture studies (Smolen et al. 1980) and trapping of individuals for laboratory studies, use of a livetrap is necessary. The most effective livetrap (Fig. 14.10) is described by Baker and Williams (1972). As far as we can ascertain, no one markets this trap, which means that one needs to build the number of traps desired. Instructions for building this trap are described by Baker and Williams (1972). As with the Macabee, this trap also is designed to take advantage of pocket gopher response to keep the burrow system closed. An open burrow is excavated and the livetrap is set as an extension to the open burrow system. It is necessary to have no light penetrate the plastic tube, which is the extension of the burrow. This is accomplished by covering the tube with dirt as well as the rat trap, which supplies the closing mechanism and trigger. A careful balance is necessary to cover the trap so that no light enters the tube, but the killing bar can efficiently advance on the trap. Trapping success is variable (Baker and Williams 1972), with *Thomomys* being the most trappable species, *Cratogeomys* being intermediate, and *Geomys* being the most difficult to trap.

CONSERVATION AND RESEARCH NEEDS

Only one species, *G. tropicalis,* is listed as endangered. This endangered status is the result of human encroachment on the restricted highly desirable habitat. This taxon is endemic to the coastal sand dunes of southeastern Tamalipas, and has a unique karyotype and a very reduced level of genetic variation (Selander et al. 1975). Monitoring of the population size is warranted.

The nature of pocket gopher population dynamics, however, often results in local extirpation and recolonization. The lack of sympatry by two species also results in local extirpation of competing populations (Williams and Baker 1976). In addition, glaciation events have produced populations with mixed genomes (Patton 1993; Jones et al. 1995). Recent molecular studies (Patton and Smith 1994; Jolly et al. 2000) document populations with small geographic ranges that are genetically distinct. These populations are either distinct species or near the completion of the speciation process. These two features (small range with unique genetic characters) pose a delimma for conservation efforts because within pocket gophers there may be many such examples, which would be expensive to protect. The problem is further exacerbated when these populations are on private property, where

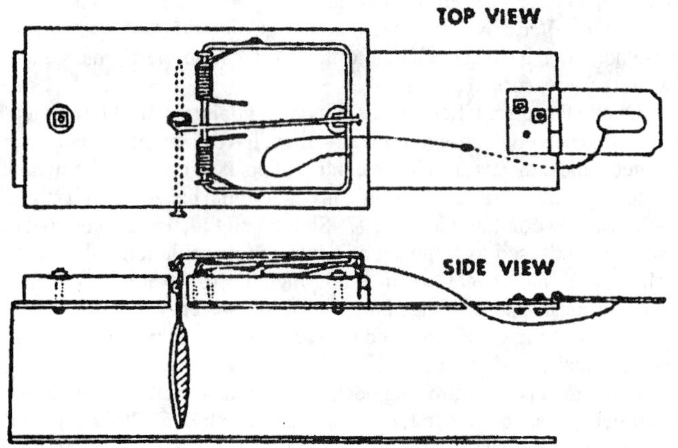

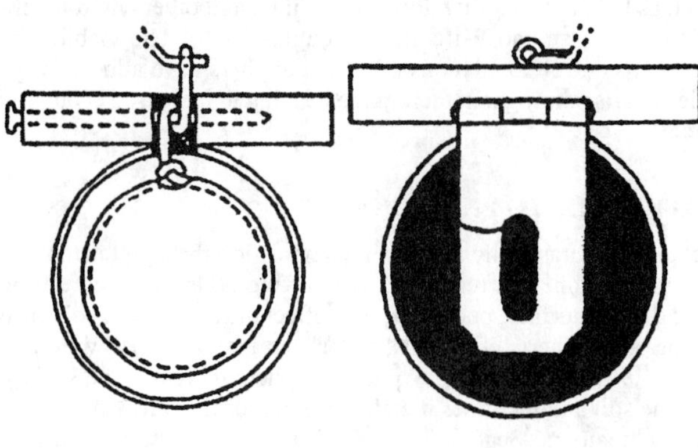

FIGURE 14.10. Design of a livetrap as described by Baker and Williams (1972).

pocket gophers are often viewed as a pest, which should be exterminated or at least intensively controlled.

The primary research need is similar to that summarized by Patton and Smith (1990), who provided a synthesis of 25 years of populations and species studied incorporating morphological, genetic, reproductive, geographic, and ecological data. Unfortunately, these types of studies require years of commitment and are labor-intensive, and it is expensive to identify metapopulations.

Another research need is the development of captive breeding programs to ensure conservation of unique genetic populations. We are unaware of any successful attempts to breed pocket gophers in captivity except for outdoor enclosures constructed in prime habitat that restrict movement. Obviously, basic systematic and taxonomic studies are needed. This is especially true for *Thomomys,* where more than 100 named taxa are represented. Molecular data are needed to resolve this issue. Finally, it would be appropriate to debate the needs for conservation efforts in a group of species and populations whose natural processes produce a high level of local extirpations associated with high levels of genetic variation and partitioning.

LITERATURE CITED

Anderson, S. 1987. *Geomys bursarius* burrowing patterns: Influence of season and food patch structure. Ecology 68:1306–18.

Anderson, D. C., and J. A. MacMahon. 1981. Population dynamics and bioenergetics of a fossorial herbivore, *Thomomys talpoides* (Rodentia: Geomyidae), in a spruce–fir sere. Ecological Monographs 51:179–202.

Avise, J. C., C. Giblin- Davidson, J. Laerm, J. C. Patton, and R. A. Lansman. 1979. Mitochondrial DNA clones and matriarchal phylogeny within and among geographic populations of the pocket gopher, *Geomys pinetis*. Proceedings of the National Academy of Science 76:6694–98.

Baker, R. J., and S. L. Williams. 1972. A live-trap for pocket gophers. Journal of Wildlife Management 36:1320–22.

Baker, R. J., and H. H. Genoways. 1975. A new subspecies of *Geomys bursarius* (Mammalia: Geomyidae) from Texas and New Mexico. Occasional Papers, the Museum, Texas Tech University 29:1–18.

Baker, R. J., R. K. Chesser, B. F. Koop, and R. A. Hoyt. 1983. Adaptive nature of chromosomal rearrangement: Differential fitness in pocket gophers. Genetica 61:161–64.

Baker, R. J., S. K. Davis, R. D. Bradley, M. J. Hamilton, and R. A. Van Den Bussche. 1989. Ribosomal-DNA, mitochondrial-DNA, chromosomal, and allozymic studies on a contact zone in the pocket gopher *Geomys*. Evolution 43:63–75.

Baker, R. J., A. D. Simmons, M. S. Powell, J. L. Longmire, and R. D. Bradley. 1996. Utility of a satellite DNA sequence as a genetic marker in hybrid zone of pocket gophers (genus *Geomys*). Pages 25–34 *in* H. H. Genoways and R. J. Baker, eds. Contributions in mammalogy: A memorial volume honoring Dr. Knox Jones Jr. Museum of Texas Tech University, Lubbock.

Barnes, V. G., Jr. 1973. Pocket gophers and reforestation in the Pacific northwest: A problem analysis (Special Science Report 155). U.S. Fish and Wildlife Service.

Bartel, M. H., and S. L. Gardner. 2000. Arthropod and helminth parasites from the plains pocket gopher, *Geomys bursarius bursarius* from the hosts' northern boundary range in Minnesota. Journal of Parasitology 86:153–56.

Block, S. B., and E. G. Zimmerman. 1991. Allozymic variation and systematics of plains pocket gophers (*Geomys*) of south central Texas. Southwestern Naturalist 36:29–36.

Bradley, W. G., J. S. Miller, and M. K. Yousef. 1974. Thermoregulatory pattern in pocket gophers: Desert and mountain. Physiological Zoology 47:172–79.

Bradley, R. D., S. K. Davis, and R. J. Baker. 1991a. Genetic control of premating-isolating behavior; Kanashiro's hypothesis and asymmetrical sexual selection in pocket gophers. Journal of Heredity 82:192–96.

Bradley, R. D., S. K. Davis, J. M. Bayouth, M. J. Hamilton, M. Maltbie, and R. J. Baker., 1991b. Chromosomal distribution of some repetitive DNA sequences in the pocket gopher (*Geomys, Cratogeomys, Thomomys*) as determined by in situ hybridization. Occasional Papers, the Museum, Texas Tech University 141:1–15.

Bradley, R. D., S. F. Lockwood, J. W. Bickham, and R. J. Baker. 1991c. Hybrid breakdown and cellular DNA content in a contact zone between two species of pocket gophers (*Geomys*). Journal of Mammalogy 72:697–705.

Bradley, R. D., J. J. Bull, A. D. Johnson, and D. M. Hillis. 1993. Origin of a novel allele in a mammalian hybrid zone. Evolution 90:8939–41.

Burt, M. S., and R. C. Dowler. 1999. Biochemical systematics of *Geomys breviceps* and two chromosomal races of *Geomys attwateri* in eastern Texas. Journal of Mammalogy 80:799–809.

Burton, D. W., and J. W. Bickham. 1989. Heterochromatin variation and DNA content conservatism in *Geomys attwateri* and *G. breviceps* (Rodentia: Geomyidae). Journal of Mammalogy 70:580–91.

Burton, D. H., and H. C. Black. 1978. Feeding habits of Mazama pocket gophers in south-central Oregon. Journal of Wildlife Management 42:383–90.

Butcher, F. D. 1929. Rodent pests of Iowa (Bulletin 153). Iowa State College Extension Service.

Cahalane, V. H. 1947. Mammals of North America. Macmillan, New York.

Chesser, R. K., R. D. Bradley, R. A. Van Den Busch, M. J. Hamilton, and R. J. Baker. 1996. Maintenance of a narrow hybrid zone in *Geomys:* Results from contiguous clustering analysis. Pages 35–45 *in* H. H. Genoways and R. J. Baker, eds. Contributions in mammalogy: A memorial volume honoring Dr. Knox Jones Jr. Museum of Texas Tech University, Lubbock.

Connolly, R. A., and R. E. Landstrom. 1969. Gopher damage to buried electric cable materials. American Society for Testing and Materials, Materials Research and Standards 9:13–18.

Cox, G. W. 1990. Soil mining by pocket gophers along topographic gradients in a mima mound-field. Ecology 71:837–43.

Cox, G. W., and C. G. Gakahu. 1986. A latitudinal test of the fossorial rodent hypothesis of mima mound origin. Zeitschrift für Geomorphologie 30:485–501.

Cox, G. W., and J. Hunt. 1992. Relation of seasonal activity patterns of valley pocket gophers to temperature, rainfall and food availability. Journal of Mammalogy 73:123–34.

Dalquest, W. W., and W. Kilpatrick. 1973. Dynamics of pocket gopher distribution on the Edwards Plateau of Texas. Southwestern Naturalist 18:1–9.

Davis, W. B. 1938. Relation of the size of pocket gophers to soil and altitude. Journal of Mammalogy 19:338–42.

Davis, W. B. 1940. Distribution and variation of pocket gophers (Genus *Geomys*) in the southwestern United States. Texas Agricultural Experiment Station.

Davis, S. K. 1986. Population structure and patterns of speciation in *Geomys* (Rodentia: Geomyidae): An analysis using mitochondrial and ribosomal DNA. Ph.D. Dissertation, Washington University, Saint Louis.

Demastes, J. W., and M. S. Hafner. 1993. Cospeciation of pocket gophers (*Geomys*) and their chewing lice (*Geomydoecus*). Journal of Mammalogy 74:521–30.

Douglas, C. L. 1969. Ecology of pocket gophers of Mesa Verde, Colorado. University of Kansas Museum of Natural History 51:147–75.

Douthitt, H. 1915. Studies on the Cestode family Anoplocephalidae. Illinois Biological Bulletin 1:355–446.

Dowler, R. C. 1989. Cytogenetic studies of three chromosomal races of pocket gophers (*Geomys bursarius* complex) at hybrid zones. Journal of Mammalogy 70:253–66.

Downhower, J. F., and E. R. Hall. 1966. The pocket gopher in Kansas. State Biological Survey, Kansas 44:1–32.

Durrant, S. D. 1952. The pocket gophers (genus *Thomomys*) of Utah. University of Kansas Publications, Museum of Natural History 1:1–82.

Eisenberg, J. F. 1966. The social organization of mammals. Handbook of Zoology 10:1–92.

Ellison, L., and C. M. Aldous. 1952. Influence of pocket gophers on vegetation of subalpine grassland in central Utah. Ecology 33:177–86.

Gettinger, R. D. 1975. Metabolism and thermoregulation of a fossorial rodent, the northern pocket gopher (*Thomomys talpoides*). Physiological Zoology 48:311–22.

Gettinger, R. D. 1984. Energy and water metabolism of free-ranging pocket gophers, *Thomomys bottae*. Ecology 65:740–51.

Grinnell, J. 1923. The burrowing rodents of California as agents in soil formation. Journal of Mammalogy 4:137–49.

Haldane, J. B. S. 1922. Sex-ratio and unisexual sterility in hybrid animals. Journal of Genetics 12:101–9.

Hall, E. R. 1981. The Mammals of North America, 2nd ed. John Wiley, New York.

Hansen, R. M. 1960. Age and reproductive characteristics of mountain pocket gophers in Colorado. Journal of Mammalogy 41:323–35.

Hansen, R. M., and M. J. Morris. 1968. Movement of rocks by northern pocket gophers. Journal of Mammalogy 49:391–99.

Hansen, R. M., and A. L. Ward. 1966. Some relations of pocket gophers to rangelands on Grand Mesa, Colorado. (Technical Bulletin 88). Colorado State University Agriculture Experimental Station.

Hart, E. B. 1978. Karyology and evolution of the plains pocket gopher, *Geomys bursarius*. Occasional Papers of the Museum of Natural History, University of Kansas, Lawrence 71:1–20.

Heaney, L. R., and R. M. Timm. 1985. Morphology, genetics and ecology of pocket gophers (genus *Geomys*) in a narrow hybrid zone. Biological Journal of the Linnean Society 25:301–17.

Hendrickson, R. L. 1972. Variation in the plains pocket gopher (*Geomys bursarius*) along a transect across Kansas and eastern Colorado. Transactions of the Kansas Academy of Science 75:322–68.

Hoffmeister, D. F. 1971. Mammals of the Grand Canyon. University of Illinois Press, Urbana.

Honeycutt, R. L., and D. J. Schmidly. 1979. Chromosomal and morphological variation in the Plains pocket gopher, *Geomys bursarius,* in Texas and adjacent states. Occasional Papers, the Museum, Texas Tech University 58:1–54.

Howard, W. E., and H. E. Childs, Jr. 1959. Ecology of pocket gophers with emphasis on *Thomoys bottae mewa*. Hilgardia 29:277–358.

Hunt, J. 1992. Feeding ecology of valley pocket gophers (*Thomomys bottae sactidiegi*) on a California coastal grassland. American Midland Naturalist 127:41–51.

Huntly, N., and R. Inouye. 1988. Pocket gophers in ecosystems: Patterns and mechanisms. Bioscience 38:786–93.

Ickenberry, R. D. 1964. Reproductive studies of the Mexican pocket gopher, *Cratogeomys castanops perplanus*. M.S. Thesis, Texas Tech University, Lubbock.

Inouye, R. S., N. Huntly, and G. A. Wasley. 1997. Effects of pocket gophers (*Geomys bursarius*) on microtopographic variation. Journal of Mammalogy 78:1144–48.

Jones, J. K., R. D. Bradley, and R. J. Baker. 1995. Hybrid pocket gophers and some thoughts on the relationships of natural hybrids to the rules of nomenclature and the endangered species act. Journal of Mammalogy 76:43–49.

Jolly, T. W., R. L. Honeycutt, and R. D. Bradley. 2000. Phylogenetic relationships of pocket gophers (genus *Geomys*) based on the mitochondrial 12s rRNA gene. Journal of Mammalogy 81:1025–34.

Kaneshiro, K. Y. 1976. Ethological isolation and phylogeny in the planitibia subgroup of Hawaiian *Drosophila*. Evolution 30:740–45.

Kaneshiro, K. Y. 1987. The dynamics in sexual selection and its pleiotropic effects. Behavioral Genetics 17:559–69.

Kaneshiro, K. Y. 1989. The dynamics of sexual selection and founder effects in species formation. Pages 279–96 *in* L. V. Giddings, K. Y. Kaneshiro, and W. W. Anderson, eds. Genetics, speciation, and the founder principle. Oxford University Press, Oxford.

Keith, J. J., R. M. Hansen, and A. L. Ward. 1959. Effect of 2,4D on abundance of foods of pocket gophers. Journal of Wildlife Management 23:137–45.

Kennerly, T. E., Jr. 1954. Local differentiation in the pocket gopher (*Geomys personatus*) in southern Texas. Texas Journal Science 6:297–329.

Kennerly, T. E., Jr. 1959. Contact between the ranges of two allopatric species of pocket gophers. Evolution 13:247–63.

Kepner, R. A., and W. E. Howard. 1960. Gopher-bait applicator. California Agriculture 14:7–14.

Lessa, E. P., and B. R. Stein. 1992. Morphological constraints in the digging apparatus of pocket gophers (Mammalia: Geomyidae). Biological Journal of the Linnean Society 47:439–53.

Luce, D.G., R. M. Case, and J. L. Stubbendieck. 1981. Damage to alfalfa fields by plains pocket gophers. Journal of Wildlife Management 45:258–60.

McNab, B. K. 1966. The metabolism of fossorial rodents: A study of convergence. Ecology 47:712–33.

Merriam, C. H. 1895. Monographic revision of the pocket gopher family Geomyidae (exclusive of the species of *Thomomys*). North American Fauna 8:1–258.

Meyers, G. T., and T. A. Vaughan. 1964. Food habits of the plains pocket gopher in eastern Colorado. Journal of Mammalogy 45:588–98.

Miller, M. A. 1953. Experimental studies on poisoning pocket gophers. Hilgardia 22:131–66.

Miller, R. S. 1964. Ecology and distribution of pocket gophers (Geomyidae) in Colorado. Ecology 45:256–72.

Patton, J. L. 1973. An analysis of natural hybridization between the pocket gophers, *Thomomys bottae* and *Thomomys umbrinus* in Arizona. Journal of Mammalogy 54:561–84.

Patton, J. L. 1993. Family Geomyidae. Pages 469–76 *in* D. E. Wilson and D. M. Reeder, eds. Mammal species of the world: A taxonomic and geographic reference, 2nd ed. Smithsonian Institution Press, Washington, DC.

Patton, J. L., and P. V. Brylski. 1987. Pocket gophers in alfalfa fields: Causes and consequences of habitat-related body size variation. American Naturalist 130:493–506.

Patton, J. L., and R. E. Dingman. 1968. Chromosome studies of pocket gophers, genus *Thomomys*. I. The specific status of *Thomomys umbrinus* (Richardson) in Arizona. Journal of Mammalogy 49:1–13.

Patton, J. L., and J. H. Feder. 1981. Microspatial genetic heterogeneity in pocket gophers: Non-random breeding and drift. Evolution 35:912–20.

Patton, J. L., and S. W. Sherwood. 1982. Genome evolution in pocket gophers (genus *Thomomys*): I. Heterochromatin variation and speciation potential. Chromosoma 85:149–62.

Patton, J. L., and M. F. Smith. 1990. The evolutionary dynamics of the pocket gopher *Thomomys bottae,* with emphasis on California populations. University of California Publications in Zoology 123:1–161.

Patton, J. L., and M. F. Smith. 1994. Paraphyly, polyphyly, and the nature of species boundaries in pocket gophers. Systematic Zoology 43:11–26.

Patton, J. L., R. K. Selander, and M. H. Smith. 1972. Genetic variations in hybridizing populations of gophers (genus *Thomomys*). Systematic Zoology 21:263–70.

Patton, J. L., J. C. Hafner, M. S. Hafner, and M. F. Smith. 1979. Hybrid zones in *Thomomys bottae* pocket gophers: Genetic, phenetic, and ecological concordance patterns. Evolution 33:860–76.

Pembleton, E. T., and R. J. Baker. 1978. Studies of a contact zone between chromosomally characterized populations of *Geomys bursarius*. Journal of Mammalogy 59:233–42.

Penney, D. F., and E. G. Zimmerman. 1976. Genic divergence and local population differentiation by random drift in the pocket gopher genus *Geomys*. Evolution 30:473–83.

Reed, D. L., M. S. Hafner, and S. K. Allen. 2000. Spatial partitioning of host habitat by chewing lice of the genera *Geomydoecus* and *Thomomydoecus* (Phthiraptera: Trichodetidae). Journal of Parasitology 86:951–55.

Reichman, O. J., and R. J. Baker. 1972. Distribution and, movements of two species of pocket gophers (Geomyidae) in an area of sympatry in the Davis Mountains, Texas. Journal of Mammalogy 53:21–33.

Reichman, O. J., and E. W. Seabloom. 2002. The role of pocket gophers as subterranean ecosystem engineers. Trends in Ecology and Evolution. 17:44–49.

Reichman, O. J., and S. C. Smith. 1985. Impact of pocket gopher burrows on overlying vegetation. Journal of Mammalogy 66:725–25.

Reichman, O. J., and S. C. Smith. 1991. Responses to simulated leaf and root herbivory by a biennial, *Tragopogon dubias*. Ecology 72:116–24.

Reichman, O. J., T. G. Whitham, and G. A. Ruffner. 1982. Adaptive geometry of burrow spacing in two pocket gopher populations. Ecology 63:687–95.

Rezsutek, M. J., and G. N. Cameron. 1998. Influence of resource removal on demography of Attwater's pocket gopher. Journal of Mammalogy 79:538–50.

Roberts, H. R., K. T. Wilkins, J. Flores, and A. Thompson-Gorozpe. 1997. Burrowing ecology of pocket gophers (Rodentia: Geomyidae) in Jalisco, Mexico. Southwestern Naturalist 42:323–27.

Ruedi, M., M. F. Smith, and J. L. Patton. 1997. Phylogenetic evidence of mitochondrial DNA introgression among pocket gophers in New Mexico (family Geomyidae). Molecular Ecology 6:453–62.

Russell, R. J. 1968. Evolution and classification of the pocket gophers of the subfamily Geomyinae. University of Kansas, Museum of Natural History Publications 16:473–579.

Schramm, P. 1961. Copulation and gestation in the pocket gopher. Journal of Mammalogy 42:167–70.

Selander, R. K., D. W. Kaufman, R. J. Baker, and S. L. Williams. 1975. Genetic and chromosomal differentiation in pocket gophers of the *Geomys bursarius* group. Evolution 28:557–64.

Sherwood, S. W., and J. L. Patton. 1982. Genome evolution in pocket gophers (genus *Thomomys*). II Variation in cellular DNA content. Chromosoma 85:163–79.

Shumake, S. A., R. T. Sterner, and S. E. Gaddis. 1999. Repellents to reduce gnawing by northern pocket gophers. Journal of Wildlife Management 63:1344–49.

Smallwood, K. S., and M. L. Morrison. 1999a. Spatial scaling of pocket gopher (Geomyidae) density. Southwestern Naturalist 44:73–82.

Smallwood, K. S., and M. L. Morrison. 1999b. Estimating burrow volume and excavation rate of pocket gophers (Geomyidae). Southwestern Naturalist 44:173–82.

Smallwood, S., and S. Geng. 1997. Multiscale influences of gophers on alfalfa yield and quality. Field Crops Research 49:59–168.

Smith, M. L., and J. L. Patton. 1988. Subspecies of pocket gophers: Causal bases for geographic differentiation in *Thomomys bottae*. Systematic Zoology 37:163–78.

Smolen, M. J., H. H. Genoways, and R. J. Baker. 1980. Demographic and reproductive parameters of the yellow-cheeked pocket gopher (*Pappogeomys castanops*). Journal of Mammalogy 61:224–36.

Sudman, P. D., J. R. Choate, and E. G. Zimmerman. 1987. Taxonomy of chromosomal races of *Geomys bursarius lutescens* Merriam. Journal of Mammalogy 68:526–43.
Swift, H. 1950. The constancy of deoxyribose nucleic acid in plant nuclei. Proceedings of the National Academy of Sciences of the USA 36:643–54.
Thaeler, C. S. 1968. An analysis of the distribution of pocket gopher species in northeastern California (genus *Thomomys*). University of California Publications in Zoology 86:1–46.
Thaeler, C. S. 1974. Four contacts between ranges of different chromosome forms of the *Thomomys talpoides* complex (Rodentia: Geomyidae). Systematic Zoology 23:343–54.
Thomas, C. A. 1971. The genetic organization of chromosomes. Annual Review of Genetics 5:237–56.
Tietjen, H. P., C. H. Halvoran, P. L. Hegdal, and A. M. Johnson. 1967. 2,4-D herbicide, vegetation and pocket gopher relationships; Black Mesa, Colorado. Ecology 48:634–43.
Tilman, D. 1983. Plant succession and gopher disturbance along an experimental gradient. Oecologia 60:285–92.
Timm, R. M., and R. D. Price. 1980. The taxonomy of *Geomydoecus* (Mallophaga: Trichodectidae) from the *Geomys bursarius* complex (Rodentia: Geomyidae). Journal of Medical Entomology 17:126–45.
Tucker, P. K., and D. J. Schmidly. 1981. Studies of a contact zone among three chromosomal races of *Geomys bursarius* in East Texas. Journal of Mammalogy 62:258–72.
Turner, G. T., R. M. Hansen, V. H. Reid, H. P. Tietjen, and A. L. Ward. 1973. Pocket gophers and Colorado mountain rangeland (Bulletin 554S). Colorado State University Experiment Station.
Vaughan, T. A. 1963. Movements made by two species of pocket gophers. American Midland Naturalist 69:367–73.
Vaughan, T. A. 1967. Two parapatric species of pocket gophers. Evolution 21:148–58.
Vleck, D. 1979. The energy cost of burrowing by the pocket gopher *Thomomys bottae*. Physiological Zoology 52:122–36.
Wilkens, K. T., and C. D. Swearingen. 1990. Factors affecting historical distribution and modern variation in the south Texas pocket gopher *Geomys personatus*. American Midland Naturalist 124:57–72.
Wilkens, K. T., J. C. Roberts, C. S. Roorda, and J. E. Hawkins. 1999. Morphometrics and functional morphology of middle ears of extant pocket gophers (Rodentia: Geomyidae). Journal of Mammalogy 80:180–98.
Williams, S. L., and R. J. Baker. 1976. Vagility and local movements of pocket gophers (Geomyidae: Rodentia). American Midland Naturalist 96:303–16.
Williams, S. L., and H. H. Genoways. 1977. Morphometric variation in the tropical pocket gopher (*Geomys tropicalis*). Annals of the Carnegie Museum 46:245–64.
Williams, S. L., and H. H. Genoways. 1978. Morphometric variation in the desert pocket gopher (*Geomys arenarius*). Annals of the Carnegie Museum 47:541–70.
Williams, S. L., and H. H. Genoways. 1980. Morphological variation in the southeastern pocket gopher (*Geomys pinetis*) (Mammalia: Rodentia). Annals of the Carnegie Museum 49:405–53.
Williams, S. L., and H. H. Genoways. 1981. Systematic review of the Texas pocket gopher, *Geomys personatus* (Mammalia: Rodentia). Annals of the Carnegie Museum 50:435–73.
Williams, L. R., and F. N. Cameron. 1986. Food habits and dietary preferences of Attwater's pocket gopher, *Geomys attwateri*. Journal of Mammalogy 67:489–96.
Wilson, D. E., and E. R. Cole. 2000. Common names of mammals of the world. Smithsonian Institution Press, Washington, DC.
Zimmerman, E. G., and N. A. Gayden. 1981. Analysis of genic heterozygosity among local populations of the pocket gopher, *Geomys bursarius*. Pages 272–78 *in* M. Smith and J. Joule, eds. Mammalian population genetics. University of Georgia Press, Athens.

Robert J. Baker, Department of Biological Sciences, Texas Tech University, Lubbock, Texas 79409-3131. Email: rjbaker@ttu.edu.

Robert D. Bradley, Department of Biological Sciences, Texas Tech University, Lubbock, Texas 79409-3131. Email: rbradley@ttacs.ttu.edu.

Lee R. McAliley, Jr., Department of Biological Sciences, Texas Tech University, Lubbock, Texas 79409-3131. Email: mcaliley1@cox.net.

15

Beaver

Castor canadensis

Bruce W. Baker
Edward P. Hill

NOMENCLATURE

COMMON NAMES. Beaver, North American beaver, Canadian beaver, American beaver, el Castor
SCIENTIFIC NAME. *Castor canadensis*
SUBSPECIES. *C. c. acadicus, C. c. baileyi, C. c. belugae, C. c. caecator, C. c. canadensis, C. c. carolinensis, C. c. concisor, C. c. duchesnei, C. c. frondator, C. c. idoneus, C. c. labradorensis, C. c. leucodontus, C. c. mexicanus, C. c. michiganensis, C. c. missouriensis, C. c. pallidus, C. c. phaeus, C. c. repentinus, C. c. rostralis, C. c. sagittatus, C. c. shastensis, C. c. subauratus, C. c. taylori,* and *C. c. texensis* (Hall 1981)

Castor canadensis (hereafter beaver) is endemic to North America and is one of two extant species in the genus *Castor. Castor fiber* (hereafter Eurasian beaver) is endemic to Europe and Asia, although its current range is severely reduced relative to its historical range. The general physical appearance of the two species is similar, but their karyotypes and several cranial and behavioral patterns are distinct (Lavrov and Orlov 1973). Multilocus allozyme electrophoresis can distinguish *C. canadensis* from *C. fiber* using tissue or blood samples from either live or dead animals, which makes the technique useful as a management tool for restoration of *C. fiber* in Europe (Sieber et al. 1999).

C. c. acadicus, C. c. canadensis, C. c. carolinensis, and *C. c. missouriensis* are the most widespread subspecies of beaver in North America (Hall 1981); however, reintroductions following extirpation have substantially altered pristine geographic variation among subspecies. The gene pools of some subspecies have been altered through introductions and subsequent mixing with other subspecies. Some subspecies may have disappeared entirely. Because subspecies are difficult to determine even with an animal in hand, subsequent discussions will be limited to species.

Fossil remains of a giant beaver, genus *Casteroides,* and a number of closely related prehistoric mammals also have been found in North America (Cahn 1932). The family Castoridae dates to the Oligocene and was highly diversified in the Tertiary period in North America (Kowalski 1976). The genus *Castor* dates to the Pleistocene (Garrison 1967) or late Tertiary (M. Schlosser 1902).

DISTRIBUTION

Historical Range. Seton (1929) estimated the beaver population at 60–400 million before European settlement of North America. Beaver occurred throughout the subarctic of mainland Canada below the northern tundra and the mouth of the MacKenzie River in the Northwest Territories (Novakowski 1965). They were widespread in Alaska, except along the Arctic Slope from Point Hope east to the Canadian border (Hakala 1952). Within the contiguous United States, they occupied suitable wetland and riparian habitat from coast to coast, even in the arid southwest. They were generally absent from the Florida peninsula and parts of southern California and southern Nevada. Although their original range in Mexico is difficult to determine, they were present in the Colorado River and Rio Grande River (Leopold 1959) as well as some coastal streams along the Gulf of Mexico.

Despite their legendary abundance, most beaver populations were decimated by fur trappers during the 1700s and 1800s, primarily to support the European fashion for felt hats (Bryce 1904). Large trading companies, such as the Hudson Bay Company, employed Europeans and Native Americans who supplied furs without regard for method or season of take. Because trappers continually moved to new territory, they likely were unaware of their cumulative effects on entire populations. In addition, intense harvest likely caused the local destruction of population structures, contributing to regional declines (Ingle-Sidorowicz 1982). Beaver populations in the eastern United States were largely extirpated by fur trappers before 1900.

Growing public concern over declines in beaver and other wildlife populations eventually led to regulations that controlled harvest through seasons and methods of take, initiating a continent-wide recovery of beaver populations. To supplement natural recovery, during the mid-1900s beaver were livetrapped and successfully reintroduced into much of their former range, a remarkable achievement of early wildlife managers. Although the area of pristine beaver habitat has been much reduced by human land-use practices, beaver have proved to be highly adaptable and occupy a variety of human-made habitats. In addition, beaver have been intentionally or accidentally introduced into areas outside their original range. Thus, the present range of beaver is a result of natural recovery and reintroduction to their original range, introduction and expansion into areas beyond their original range, the limits of native habitat as modified by human land uses, and adaptability to new human-made habitats such as urban areas, croplands, and areas with exotic vegetation.

Present Range. Beaver populations were estimated at 6–12 million by Naiman et al. (1988). Beaver now occupy much of their former range in North America, although habitat loss and other causes have severely restricted populations in many areas (Fig. 15.1) (Hall 1981; Larson and Gunson 1983). For example, since 1834, about 195,000–260,000 km^2 of wetlands has been converted to agricultural or other use in the United States, much of which was likely beaver habitat (Naiman et al. 1988). Nonetheless, beaver are remarkably adaptable. They can marginally subsist above timberline in mountainous areas; however, beaver have been unable to colonize Alaskan or Canadian arctic tundra, perhaps because tundra vegetation lacks essential woody plants for winter food and lodge construction or because thick ice limits surface access in winter. Although suitable beaver habitat in Canada has been reduced since pre-European settlement, fur harvest records indicate that beaver populations have fully recovered in many areas, perhaps a result of a return to earlier successional stages of forest cover (Ingle-Sidorowicz 1982). In the United States, beaver populations have continued to increase since major reintroductions ended in the 1950s. Populations in southeastern states have grown large enough to become a major nuisance to the timber industry and others (Larson and Gunson 1983). In the Far West, they have been reestablished in the Santa Ana and Colorado River systems of southern California. In Mexico, beaver may still subsist in some northern areas of Nuevo Leon and Chihuahua (Leopold 1959), although populations there likely are marginal (Landin 1980).

FIGURE 15.1. Distribution of the beaver (*Castor canadensis*). Modifications of the range map by Deems and Pursley (1978) include populations in Mexico, southern California, west-central Florida, and Delaware and the absence of beaver in the North Slope of Alaska.

Introductions of beaver in Finland (Lahti and Helminen 1974), Asian Kamchatka (Safonov 1979), Argentina (Lizarralde 1993), and other locations have resulted in the establishment of viable populations beyond their original range in North America. For example, 25 mated pairs of beaver were introduced (as a captive population) to Tierra del Fuego, Argentina, in 1946 to establish a fur industry. Animals that later escaped or were intentionally released resulted in a viable wild population. This population rapidly expanded in the absence of predators and other natural population controls, causing a substantial impact on native southern beech (*Nothofagus*) forests (Lizarralde 1993).

DESCRIPTION

Beaver are the largest rodents in North America. Most adults weigh 16–31.5 kg and attain a total length of up to 120 cm. They have heavily muscled bodies supported by large bones. Forelegs are shorter than hind legs, which results in greater height at the hips than at the shoulders. Viewed dorsally, beaver are short and thick, broadest just anterior to the hips, and taper gradually toward the nose; a short, thick neck appears almost continuous with the shoulders and head. Their most characteristic feature is a dorsoventrally flattened, paddle-like tail, the unfurred portion of which in most adults varies from 230 to 323 mm long and from 110 to 180 mm wide (Davis 1940). The distal three fourths of the tail is covered with black, leathery, uncornified scales (Kowalski 1976) containing a few scattered, coarse hairs. The caudal vertebrae are dorsoventrally flattened, with a complex arrangement of muscles and tendons to support the flat tail (Mahoney and Rosenburg 1981). Incisors are generally orangish in color, with the anterior surface in adults >5 mm wide, a feature that helps distinguish beaver damage from other rodent damage on the basis of toothmark width.

Beaver move with an awkward waddle on land, but can gallop if frightened. Adult beaver can walk upright in a bipedal fashion (partially supported by the tail) while carrying mud or sticks held against the chest with their chin and front legs. In water, they swim by alternate kicks of the hind legs, appearing graceful and efficient, though slow and deliberate. Beaver are shaped more like marine mammals than like other terrestrial mammals, with a fineness ratio (a hydrodynamic index of streamlining) of 4.8, a value similar to that for phocid seals (Reynolds 1993). In addition, the surface area of unfurred extremities (hind feet and scaly tail) is 30% of the total, perhaps a compromise between the need for propulsion and minimization of the area for heat exchange. Webbed toes on large hind feet (up to 200 mm long) facilitate swimming, and short, heavily clawed front feet facilitate digging. Great forepaw dexterity enables beaver to fold individual leaves into their mouth and to rotate small, pencil-sized stems as they gnaw off bark. The ears are rounded, short (30 mm), fleshy, and placed high on the rear of the head. The small eyes also are located high on the head, about midway between the nose and the base of the skull. Both these adaptations enable beaver to swim with minimum exposure above the water surface.

Pelt coloration is variable within and among populations, with reddish, chestnut, nearly black, and yellowish-brown specimens possible in the same watershed. Fur of the flanks, abdomen, and cheeks is usually shorter and lighter than back fur. Guard hairs are about 10 times the diameter of the hairs constituting the underfur, giving the pelt a coarse appearance. Guard hairs attain their greatest length (50 mm) and density along the back. Underfur is longest on the back (25 mm) and has wavy individual hairs, which give the pelt a downy softness. It may be dark gray to chestnut in color on the back and, like the guard hair, becomes lighter in color on the sides and ventral areas. Unlike the case in many furbearers, coloration of individual guard hairs is usually consistent throughout their length.

The two inside (medial) toes of each hind foot have movable, split nails, which beaver use as combs to groom their fur (Wilsson 1971). Beaver have closable nostrils, valvular ears, nictitating eye membranes, and lips that close behind large incisors, adaptions important to their semiaquatic existence. During periods of active lactation and when parturition is near, four pectoral mammae are discernible on the chest of the adult female. During pregnancy, beaver have a subplacenta located between the placenta and uterine tissues. Although its morphology has been well described, its function is unknown (Fischer 1985). The reproductive organs of both sexes are internal and lie anterior to a common anal cloaca containing the castor and anal glands (Svendsen 1978). A notable characteristic of beaver is the strong aroma from the paired castor glands. Contents of the castor glands (castoreum) and anal gland secretion may be deposited during scent marking. Castoreum has been used as a base aroma in perfume and in making trappers's lures.

Beaver skeletons are massive when compared to those of other mammals of similar length. The skull and the mandible are thick and heavy, providing a strong foundation for large incisors (Fig. 15.2). A less rugged skull would be unable to withstand the physical stress and strain of jaw muscle contractions of sufficient strength to cut hardwoods such as oak (*Quercus* spp.) and maple (*Acer* spp.). The braincase is narrow and there is a small infraorbital canal. A prominent rostrum is anterior to the massive zygomatic arch. Adult skulls are very large (120–148 mm condylobasal length), which minimizes the possibility of confusing them with other North American rodents. Juvenile skulls are smaller and may be similar in size to those of adult nutria (*Myocastor coypus*), porcupine (*Erethizon dorsatum*), or mountain beaver (*Aplodontia rufa*); however, differences other than size are apparent on close examination. As in other semiaquatic mammals, the acetabulum is shifted dorsally (Kowalski 1976). The male beaver has a baculum that generally enlarges with age (Friley 1949) and can be palpated as an aid in determining the sex of live beavers and unskinned carcasses (Denney 1952). Osteological changes during growth and development of beaver were described by Robertson and Shadle (1954). The dental formula is I 1/1, C 0/0, P 1/1, M 3/3. Incisors grow continuously and the chiseled edge is sharpened by grinding the uppers against the lowers (Wilsson 1971). The hard enameled front surface of incisors serves as the cutting edge to fell trees and peel bark. Cheek teeth are hypsodont and grow only through the deposition of cementum at the root base. Deciduous premolars are replaced at about 11 months of age by permanent premolars. Specializations such as large size; type and location of ears, eyes, and nose; size and function of front and hind legs; and a large, flattened tail appear to have individually and collectively enhanced the adaptability and survival of beaver in wetland environments.

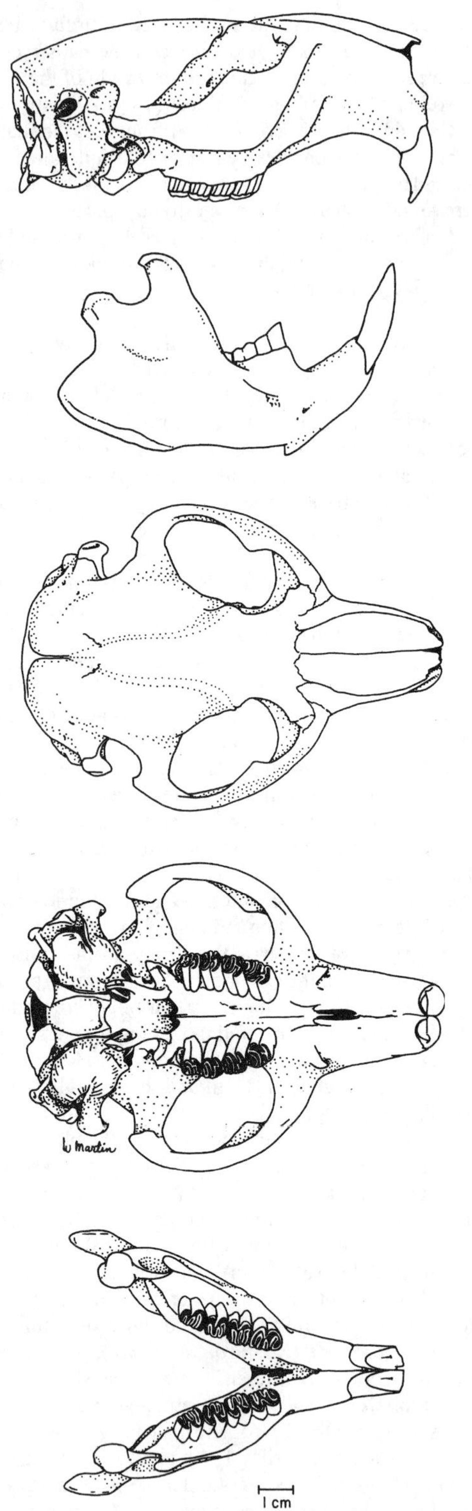

FIGURE 15.2. Skull of the beaver (*Castor canadensis*). From top to bottom: lateral view of cranium, lateral view of mandible, dorsal view of cranium, ventral view of cranium, dorsal view of mandible.

PHYSIOLOGY

Growth. Size of the adult beaver depends on latitude, climate, quality of available food, and extent of exploitation. In Alabama, a sample of 1450 beaver from an unexploited population showed mean body weight stabilized at 4 or 5 years of age and then diminished slightly after 9 years of age. Average weight of all specimens was 18.6 kg; maximum was 19.3 kg (review by Hill 1982). The relatively moderate climate of the midcontinent region may produce the largest beaver, where maximum weight can reach nearly 40 kg.

Growth of adults (body weight and tail size) occurs only in summer; however, kits (juveniles) continue to grow throughout their first winter (Novakowski 1965; Smith and Jenkins 1997). For northern beaver, winter ice formation on ponds and streams restricts or eliminates access to surface food, and adults and yearlings lose weight as fat stores are depleted. In southern beaver, adults and yearlings also lose weight in winter, even though their habitat typically remains ice free. Failure to maintain fat reserves during winter for beaver living in ice-free regions is likely not due to lack of adequate energy from available food, as it may be in the northern range, but instead may be associated with seasonal changes in physiology. Reduced food consumption, as described for captive beaver of a northern population, may also occur in southern beaver with the onset of warming trends in February and March, as beaver are frequently observed sunning themselves on lodges during clear sunny days of late winter, and early spring (review by Hill 1982).

In northern populations, Smith and Jenkins (1997) found that winter loss of body weight and tail size can vary among colonies by severity of winter, and sex and age composition of the colony. Beaver lost more body weight and tail size when winters were longer. Adults and yearlings that overwintered with young in the colony lost more weight than those without young. This supports Novakowski's (1965) conclusion that older members of the colony eat less stored food when young are present, and rely instead on other adaptations to survive the winter.

Thermoregulation. Northern populations of beaver in winter must contend with the thermoregulatory cost of foraging under the ice in near-freezing water and must subsist primarily on stored food and metabolized fat (Dyck and MacArthur 1992). Some mammals can conserve energy in winter by reducing their body temperature through seasonal torpor. Researchers have suspected torpor in beaver, but studies of change in body temperature in response to freezing ambient temperatures have been equivocal. Dyck and MacArthur (1992:1671) found the body temperature of free-ranging beaver averaged about 37° C throughout the year, with "no evidence of shallow torpor in either kits or adults." In contrast, D. W. Smith et al. (1991) found the mean daily body temperature of adult beaver declined by 1° C from fall to winter, but remained constant for kits. Body temperature can also vary by daily activity level. Before freeze-up, body temperature is higher during daylight hours, when beaver spend more time in the lodge, and lower at night, when they are away from the lodge (Dyck and MacArthur 1992). Thus, thermoregulation likely contributes to overwinter survival in beaver in combination with several other adaptations described in this chapter, including warmer winter fur, increased body fat, a stored food cache, a warmer microclimate in the lodge, huddling together in the lodge, and reduced activity in winter (D. W. Smith et al. 1991).

Digestion. Beaver are hind-gut fermenters. Digestion is enhanced by a prominent and unusual cardiogastric gland on the lesser curvature of the stomach (Vispo and Hume 1995), a glandular digestive area (Kowalski 1976), and a large trilobed cecum containing commensal microbiota. Beaver consume a high percentage of cellulose, but maximize the nutritional value of woody plants by eating only the bark. They can digest about 32% of available cellulose by microbial action in the cecum, which is similar to the case in some other mammals (review by Hill 1982, Buech 1984). Beaver have a relatively long small intestine, 70% longer than in the porcupine, which suggests a high absorptive capacity (Vispo and Hume 1995). Consumption of soft green excrement directly from the cloaca (coprophagy) occurs diurnally in the beaver (observed as early as 10 days of age; Buech 1984) as well as in the Eurasian beaver (Wilsson 1971), lagomorphs, and other rodents. Feces are reingested and chewed by the beaver and pass quickly through the digestive system (Buech 1984). In contrast, lagomorphs reingest and swallow mucous-covered entire pellets.

Circulation. Beaver heart weight averages 0.40% of body weight, which is consistent with heart ratios for other terrestrial mammals, but relatively small compared to fully aquatic mammals (Bisaillon 1982). The cardiac blood vessels are not specialized, but are typically mammalian and resemble those of both terrestrial and aquatic mammals (Bisaillon 1981). Beaver have no unusual oxygen storage capacity, but certain changes in blood parameters, heart rhythm, and circulation enable them to make dives lasting up to 15 min without asphyxiation (review by Hill 1982). Aleksiuk (1970a:145) noted that "minute blood vessels permeate the entire tail, and a countercurrent heat exchange system is present at the base." This specialized circulatory feature helps conserve heat energy in extremely cold water and radiate heat during hot weather.

REPRODUCTION AND DEVELOPMENT

Sexual Maturity. Beaver reach sexual maturity (defined as age at breeding that results in the first litter) at 1.5–3 years of age, although puberty may be reached several months before first breeding. Most studies have found at least some beaver had reached sexual maturity as yearlings (1.5–2.0 years old), although regional variation is evident. Gunson (1970) estimated that two thirds to three fourths of 2-year-old beaver produced young, and believed that early sexual maturity in Saskatchewan beaver was enhanced by high-quality habitat. Reproduction in yearlings may cease where >40% of suitable beaver habitat is occupied by established colonies. In Newfoundland, 24% of yearling females had bred (Payne 1975). In northern Canada, Novakowski (1965) found first pregnancies in 3 of 21 females that were approaching their third birthday, but no indications of conception in females that were almost 2 years old. In Alabama, in 2.5- to 3-year-old females, only 16 of 65 had ovulated, and there were no indications of ovulation or pregnancy in 50 yearlings. However, in Tennessee, Lizotte (1994) found sexual maturity occurred at 1.5–2.0 years of age, with a 25% pregnancy rate in this age class.

Breeding. Beaver are monogamous, described by Svendsen (1989:339) as "characterized by a single adult pair and young forming a family, a relatively long pair-bond where desertion of a mate is rare, and turnover of mates usually occurs after the death of one of the pair." Beaver typically breed in winter and give birth in late spring, producing only 1 litter/year. The potential breeding season is very long, with conception reported between November and March and parturition between February and November (review by Wigley et al. 1983). Latitude and climate can affect the breeding season, which is generally shorter in colder climates and longer in warmer climates (Hill 1982; Wigley et al. 1983). Breeding takes place in water (Kowalski 1976), bank dens, or lodges. Wilsson (1971) reported that *C. fiber* remains in estrus 10–12 hr and has a second estrus in 14 days if not fertilized. A gestation period of 100 days is typical for *C. canadensis* (Wigley et al. 1983), with a range of 98–111 days (review by Hill 1982).

Sex Ratios. Sex ratios in monogamous species are important because they can influence pregnancy rates. When averaged across age class, region, and harvest level, the sex ratio of beaver may be nearly even, but substantial variation among populations suggests caution in making this assumption. Sex ratios of trapped populations may reflect bias inherent in trapping methods, although results of different studies are inconsistent. For example, some studies found no difference in susceptibility of sexes to baited Conibear traps set under ice, but others suggested trapping was selective for adult females. Others have noted higher mortality from trapping and other causes for adult males (review by Hill 1982). In a review of 15 studies, Woodward (1977) found an average sex ratio of 98.5 males:100 females, but a ratio of 90.7:100 for adults and 111.4:100 for subadults. The combined average from studies in Saskatchewan, Newfoundland, Vermont, and Alabama was 105:100 (N = 4867) (review by Hill 1982).

Pregnancy Rates. Knowledge of pregnancy rates among age groups in monogamous species increases accuracy in computing estimates of reproductive performance. Pregnancy rates usually increase from 1.5 years of age until about age 4 years, remain high until old age, and then decrease (Lizotte 1994). Where populations are not exploited and suitable habitat is fully occupied, there likely is less dispersal and therefore less breeding among young adults, which remain in colonies containing older dominant pairs. Thus, both habitat quality and extent of exploitation should be considered when using pregnancy rates to calculate reproductive performance.

Reproductive Performance. Placental scar counts (Hodgdon 1949), counts of developing embryos, and, with some limitations, counts of corpora lutea and corpora albicantia are useful indices of reproductive performance in beaver (Provost 1958). In Mississippi, counts of fetuses, placental scars, and corpora lutea all yielded statistically similar estimates of litter size, despite pre- or postimplantation losses (Wigley et al. 1984). Preimplantation losses from unfertilized ova or failure of fertilized ova to implant and postimplantation losses resulting in resorptions account for differences between ovulation rates and litter size. Intrauterine mortality was 16% in 48 beaver from Ohio and almost 19% in 40 beaver from western Massachusetts (review by Hill 1982). Where rates of prenatal loss are high, correction factors should be developed to obtain more precise estimates of litter size and annual productivity (Wigley et al. 1984).

In areas where carcasses are available from fall or early winter trapping, counts of placental scars and persisting corpora albicantia, corrected for current resorption rates and prenatal loss, respectively, provide an index of litter size from the previous spring. Sources of error, such as regression or discoloration of implantation sites by some preservatives and degeneration of corpora albicantia with the onset of the breeding season (Provost 1962), make estimates of reproductive performance in fall less accurate than those made at other times. Where trapping seasons overlap the breeding season, a combination of placental scar counts corrected for the past season resorption rates and corpora lutea counts corrected for current resorption rates can provide information on litter size. However, early in gestation, it is difficult to distinguish between corpora lutea of ovulation and corpora lutea of pregnancy, which introduces a potential source of error in estimates of reproductive performance (Provost 1962). Also, embryo counts and ovulation rates of females trapped in January and February may not provide precise estimates of current-year breeding among yearlings and subadults. These age groups may breed later than adults (Grinnell et al. 1937), particularly at southern latitudes or if they had dispersed during the summer. Thus, winter samples may not reflect reproductive performance as well as those from May or June.

Where trapping seasons occur after the breeding season, counts of developing embryos, corrected for current resorption rates, provide an accurate index of litter size among age groups. Resorption is negligible if embryos are in an advanced stage of development. Where possible, delaying the trapping season until breeding has occurred may lower the incidence of unbred females whose mates were trapped from the population before breeding (Hodgdon and Hunt 1953). Such delay also facilitates measurement of current litter size through embryo counts.

The litter size of beaver is typically two to four, although local averages may be as high as six, and the number can vary from one to nine (reviews by Hill 1982; Wigley et al. 1983). These reviews suggest that beaver in the southeastern United States tend to have smaller litters, whereas northern and perhaps western beaver tend to have larger litters. Large litters may be associated with better quality habitats and increased weight of the mother. In Mississippi, age of mother was only weakly correlated with litter size, but weight of mother was strongly correlated. Litter size can also be reduced by lack of food (e.g., due to ice on ponds) or quality of food (e.g., limited supply of preferred plants) (Rutherford 1955). Because fewer yearlings breed in relatively dense populations and litter size may be inversely related to the number of beaver in the family, reproduction in beaver may be density dependent (Payne 1984). Some evidence also suggests that beaver may breed only during alternate years in very poor quality habitat, although this hypothesis needs further investigation (D. W. Smith, pers. commun., 2001).

Development of Young. Growth curves of the fetus were developed by Woodward (1977) for a 100-day gestation period. Curves may be useful for estimating peak periods of conception and parturition through extrapolation (Hodgdon and Hunt 1953).

Beaver kits are born precocial and fully furred, and weigh about 0.5 kg (review by Hill 1982). Lancia and Hodgdon (1983) studied the ontogeny of behavior in captive kits and found they were able to swim at 4 days and could dive and stay submerged at 2 months of age. Bipedal walking was noted at 1 month of age, and carrying construction materials while walking on the hind legs occurred at 90 days of age. Suckling peaked at 25 ml of milk/day at 1 month and decreased until weaned at 45–50 days. Zurowski et al. (1974) noted that the anterior nipples produced 50–75% less milk than the posterior nipples. Kits can take some solid food at 1–4 weeks of age and switch to mostly solid food by 1 month. However, they may suckle for up to 3 months even though they obtain little milk, perhaps to maintain the mother–infant bond. The fur of kits is not water repellent at birth, but after 3–4 weeks of age they begin to spread anal gland secretions on their fur, which creates water repellency by 5–8 weeks. Captive kits began to dive underwater in response to alarm at 8–10 days of age and initiated tail slapping in response to alarm at 3–4 weeks of age (Lancia and Hodgdon 1983). Rudimentary scent marking began at 13–14 days of age. Thus, very young kits express some adult behaviors, but require a long period in the family to develop their complex construction ability and other skills required for independent life.

BEHAVIOR

Social Organization. Individual beaver spend most of their lives in small, closed, extended-family units traditionally called colonies. Although the term "colony" is commonly used for beaver, its use has been questioned (Hodgdon and Lancia 1983) because a colony more often describes a spatially associated collection of individual families rather than a single family unit. For example, a family of prairie dogs (*Cynomys* spp.) living in the same burrow system is called a coterie and a group of families is called a colony. However, to maintain consistency with previous beaver literature, we use colony to represent an extended beaver family. Thus, a beaver colony typically contains the adult pair; young of the current year, or kits (<12 months old); and young of the previous year, or yearlings (12–24 months old). Sometimes older young may remain with the colony as subadults (>24 months old) before they disperse, especially if the available habitat is near carrying capacity (Busher 1987). A small percentage of colonies may contain more than one adult male or female (Busher 1983). Established colonies inhabit discrete and defended territories. Dispersing beaver of both sexes, also called floaters, remain transient until they settle with an unpaired beaver or they build dams or lodges, which may help attract a mate. Compared to many other mammals, especially other rodents, beaver populations are characterized by relatively low natality, low mortality of young, prolonged behavioral development, high parental care, and adult longevity (Hodgdon and Lancia 1983).

Social interactions involving close contact are fairly infrequent outside the lodge, perhaps an adaptation to minimize predation risk on land. The most common interaction among individual beaver concerns food items and usually involves kits begging for food from older siblings or adults (Busher 1983). Adults discourage yearlings from begging by snapping their head toward the yearling. Grooming fur to maintain water repellency is a common activity inside the lodge. Beaver groom themselves wherever they can reach, but rely on other family members to groom their back fur (Patenaude and Bovet 1984). This social grooming appears to be primarily to maintain a layer of air in the fur, as does self-grooming, rather than to maintain social bonds or as an appeasement gesture (Brady and Svendsen 1981). Aggressive interactions are rare among family members, with most aggression directed as threats that do not result in fights. Studies of dominance hierarchy systems in beaver have been equivocal. Hodgdon and Larson (1973) described dominance hierarchy as age class (older dominant over younger) and sexual (adult females dominant over adult males). Busher (1983), however, found only age-class hierarchy, and Brady and Svendsen (1981) found no clear patterns in any groups.

Vocalizations and Tail Slapping. Although seven vocal sounds have been described for beaver, most investigators recognize only three that are used outside the lodge: a whine, a hiss, and a growl (Hodgdon and Lancia 1983). The whine is the most frequent vocalization and can be repeated in rapid succession. Beaver of all ages whine, but kits account for two thirds of events, either when food is at risk of being taken away or when begging for food. Food begging by kits is usually effective, which provides kits with food without the risk of obtaining it from land (Brady and Svendsen 1981; Hodgdon and Lancia 1983). Vocalizations are also used to initiate grooming and play. Although beaver are typically docile with humans, they sometimes become aggressive, a behavior sometimes preceded by a hiss or a growl.

Probably the most familiar sound produced by beaver is the tail slap. The sound is made when a beaver forcefully strikes the water with its heavy paddle-like tail, a behavior that may precede diving underwater when alarmed (tail-slap dive). Tail slapping may function to (1) issue a warning signal to family members, which typically respond by moving to deep water or to the lodge (especially kits); (2) drive away potential predators; and (3) elicit a response from the source of disturbance (Brady and Svendsen 1981; Hodgdon and Lancia 1983). Tail slapping is used by all ages and both sexes, but studies of variation in frequency of use by sex and age have been equivocal. Hodgdon (1978) found older beaver slapped more often than younger ones, females were more easily provoked than males, and males slapped more times per event than females. Sudden alarm often elicits immediate tail slapping. However, if beaver are unsure, they often move to deep water and orient toward the disturbance with their nose in the air, a behavior that often precedes tail slapping. Smell, sound, sight, and movement are all important stimuli, either separately or in combination. The response of individual beaver to tail slapping also varies by age. Tail slapping by adults elicits the most response from all age classes, but adult beaver are the most responsive to tail slapping of other beaver. Kits are least likely to move or to elicit a response and yearlings are intermediate (Hodgdon and Lancia 1983).

Scent Marking. Scent marking is a highly developed communication method in beaver. Castor glands produce castoreum, a strong-smelling, urine-based brown paste containing phenolic, neutral, basic, and acidic compounds. Anal glands (also called oil glands) produce anal gland secretions consisting of waxy esters and fatty acids. Castoreum is likely derived from diet and thus subject to seasonal variation in odor; however, anal gland secretions are unique chemical identifiers of individual beaver (Sun and Muller-Schwarze 1998). Beaver use castoreum and anal gland secretions as scent marks, which they actively deposit on piles of mud and debris called scent mounds. Beaver deposit castoreum by rubbing it on scent mounds during and after construction; it is not clear how and when anal gland secretion is applied (Svendsen 1980a). Most scent mounds are constructed by adult males, who gather material in their forepaws and carry it to scent mounds in a bipedal fashion. Large numbers of scent mounds (>100) can be constructed within a territory, and they are usually placed on or near lodges, dams, and trails <1 m from water. Beaver of all ages place scent on mounds, but the frequency of marking increases with age. Males of all ages place the most scent marks (Hodgdon and Lancia 1983). In colder climates, construction and marking of scent mounds peaks soon after ice melts in the spring as beaver reoccupy their full territory and reapply scent that faded during the winter. In warmer, ice-free regions, scent marking can occur all year, but still may be more intense during the spring dispersal period. Scent marking has been observed in December and January in Alabama, where mounds can reach 35.5 cm in height.

The primary function of scent marking appears to be territorial. Scent marks may define the location and limits of the territory by creating a "scent fence" (Muller-Schwarze and Heckman 1980), which minimizes aggressive encounters with neighbors and discourages colonization by dispersing beaver. Beaver can distinguish the scent of castor fluid among family members, neighbors, and nonneighbors (beaver from beyond adjacent territories). They "overmark" the scents from strangers more often than scents from family members (Schulte 1998). Use of foreign castor scent (as in trapping) may elicit investigation, intense scent marking, or destruction of the foreign scent as well as hissing and tail slapping behavior. In Ohio, an adult female built over 70 mounds in 1 week, likely in response to the presence of castor bait applied by trappers (Brady and Svendsen 1981). In New York, beaver obliterated foreign scent by pawing the mud, overmarking foreign scent, and transferring the mud to their own scent mounds (Muller-Schwarze et al. 1983). In addition, a territorial function for scent mounds may be expressed as a change in the motivational state of beaver. Svendsen (1980a) suggested that scent may increase the confidence of resident beaver, which smell their own scent, and decrease the confidence of nonresident dispersing beaver, thus increasing the likelihood nonresidents will flee a territory "defended" by scent mounds. Experimental field studies support a territorial function for scent marking, as Schulte (1998) found that beaver could distinguish among the scents of adjacent neighbors, far neighbors, and family members. Beaver spent more time investigating and overmarking scent from unrelated beaver than from family members.

The ability of beaver to recognize relatives from nonrelatives via anal gland secretions (but not castoreum) may help prevent inbreeding as related individuals meet each other following dispersal outside their home territory (Sun and Muller-Schwarze 1997). Experimental comparisons of compounds in anal gland secretions have not yet clearly identified underlying mechanisms, but have shown that perhaps as few as two or three compounds may be important in communicating family membership (Sun and Muller-Schwarze 1998). In addition, scent marking may help orient beaver within their territory at night, although this has not been experimentally demonstrated.

Scent marking by beaver may be density dependent and vary by season and location. In Maine, beaver colonies with close neighbors had more scent mounds than isolated colonies (Muller-Schwarze and Heckman 1980). In South Carolina, scent marking was positively correlated with colony density. Peak scent marking occurred in fall and winter, with very little marking activity in summer (Davis et al. 1994). Southern beaver may increase marking activity in the fall in response to increased food competition among colonies (Davis et al. 1994). Thus, scent marking in southern beaver may differ from that in northern beaver, which exhibit a peak of marking in the spring.

Daily Activity Patterns. Beaver are crepuscular and nocturnal. In ice-free areas, they follow a normal 24-hr period yearlong, but in northern latitudes, they do so only during spring, summer, and fall. Winter activity periods of northern beaver commonly exhibit a free-running circadian rhythm of about 26–29 hr, likely because relatively constant light conditions preclude entrainment of a photoperiod (Lancia et al. 1982). To survive extreme cold in winter, beaver remain under the ice or inside a lodge, where temperatures are nearer a relatively moderate 0° C. When beaver venture above the ice, ambient air temperatures below about −10° C cause substantial energy deficits. During extreme cold, beaver may exhibit no detectable movement inside the lodge (Aleksuik and Cowan 1969; Lancia et al. 1982), an energy-saving mechanism that may reduce caloric needs by 20% (McKab 1963).

Dispersal and Other Movements. Bergerud and Miller (1977) classified the major movements of beaver as (1) movement of the entire colony between ponds within a territory; (2) short-term wandering of yearlings; (3) dispersal of beaver, usually at age 2 years, to establish new colonies; and (4) miscellaneous movement of adults, often following loss of a mate. Dispersal of 2-year-old beaver is the primary mechanism of population expansion. Dispersing individuals may return to the home colony for short periods of time, which suggests that dispersal is innate rather than learned from or encouraged by any aggressive behavior of parents (Hodgdon 1978). Dispersal of subadults often coincides with the birth of kits in the spring and/or high runoff, especially where ice in winter limits movements (Van Deelen and Pletscher 1996). Movements in ice-free areas are less restricted, as some dispersal occurs in late February and March, and scent marking and territorial defense may occur throughout the winter. Beaver in poor-quality habitat, or where trapping or other control measures have reduced populations below carrying capacity, may disperse at a higher rate than those in good-quality saturated habitat. Thus, habitat conditions may affect the length of time beaver remain in the family unit as subadults. However, beaver also may exhibit high dispersal in fully occupied habitat, which suggests dispersal patterns are inconsistent (Gunson 1970; Van Deelen and Pletscher 1996). Stochastic models of beaver population growth have assumed density-dependent dispersal rates (Molini et al. 1980). Distance of natural dispersal varies greatly, sometimes depending on the location of suitable but unoccupied habitat. Direction of dispersal can be either upstream or downstream within a watershed, or beaver can cross watersheds by overland travel of up to several kilometers. In a study of 46 dispersing beaver in New York, 74% initiated dispersal downstream, 35% moved to neighboring colonies, and females moved farther than males (Sun et al. 2000). In that study, 14% of dispersers were 1 year old, 64% were 2 years old, and 21% were 3 years old. If the entire colony moves to a new location, then movement usually occurs before parturition; single animals usually disperse before pairs (Hodgdon 1978). Distances moved and time of movement are important considerations in formulating management strategies.

Home Range. Home range of a beaver depends on sex, age, social organization of the family unit, type of occupied habitat, and seasonal constraints. During the summer, parental care for kits in and near the lodge or den can restrict the distance that adults can forage, with females staying closer to kits than do males. As young become more independent in the fall, the home range of adults may increase, although this is not always the case. In areas where ice confines movements in winter, home range is also constrained. For example, a radiotelemetry study of beaver in the taiga of southeastern Manitoba showed those in family units had smaller average summer home ranges (8 ha) than those in nonfamily units (18 ha) and that home ranges were larger in summer and fall than in winter (Wheatley 1997a, 1997b).

Habitat features, especially shoreline configuration, strongly affect home range shape and size. Home ranges tend to follow the irregular shoreline patterns of lakes, ponds, rivers, and streams. Small ponds may contain only a single family unit with a relatively circular home range, but in lakes, streams, and rivers, the home ranges of beaver are typically larger and more linear (Novak 1987; Wheatley 1997c). However, these habitat-related patterns may break down when beaver are not living in a sedentary family unit or during seasonal movements (Wheatley 1997c). In addition, intraspecific competition, or territoriality, is an important mechanism, which helps to regulate population density. Boyce (1981a) suggested that territoriality likely was responsible for a minimum intercolony distance of 1 km. Beyond that distance, availability of suitable sites for foraging, dams, and lodges more strongly influenced the distribution of colonies.

Population Density. Beaver population density varies spatially and temporally. Because the home ranges of adjacent beaver families are usually separated by unoccupied habitat, density estimates typically include some unoccupied habitat. Factors that contribute to variation in density of beaver populations include human exploitation (trapping), water quality, habitat suitability, area available for new colonization, length of habitation time relative to available resources, epizootic diseases, local predation events, and territoriality. There is a wide range in the density of beaver colonies, from near zero to at least 4.6/km^2 (reviews by Hill 1982; Novak 1987). Observers in different regions have attempted to estimate the maximum density or saturation point in local populations. Saturation has been reported to vary from 0.4 colony/km of stream in northern Alberta to 1.2 colonies/km of stream in New York and Utah (reviews by Hill 1982; Novak 1987). In the headwaters of four Alabama watersheds, saturation approached 1.9 colonies/km of stream (Hill 1976).

Trapping can suppress beaver populations below habitat-based carrying capacity and is an important consideration in understanding population dynamics. Trapping often removes a larger percentage of adult beaver than it does other age classes; thus, it can increase adult mortality and affect both the density and age structure of populations. Intense trapping over many years can entirely decimate populations, as it did during European settlement of North America. In previously unexploited populations, trapping can cause rapid population reductions. In a Wisconsin study, where trapping was resumed after 19 years of protection, beaver populations were reduced by 21% in the first year and 53% in the second year (Zeckmeister and Payne 1998). Trapping can also alter the age structure of populations as removal of adults from established territories frees suitable habitat, allowing beaver to disperse earlier from their natal colony and increasing their survival (Boyce 1981b). Comparing harvested and unharvested beaver populations in New York, Muller-Schwarze and Schulte (1999) found that in unharvested populations, beaver colonized steeper stream gradients, young remained longer in the natal colony, preferred forage species were depleted and less preferred species were used more often, and beaver foraged further from their pond, lodge, or den.

Density of beaver populations that occupy particular sites may also vary as a function of the length of time sites have been occupied. For example, after beaver returned to the Prescott Peninsula in Massachusetts following an absence of more than 200 years, the population showed slow growth the first 15 years, then 15 years of very rapid growth, and then a rapid decline in numbers until it stabilized at 23% of its peak. Populations in Sagehen Creek, California, also followed a pattern characterized first by slow growth, then rapid growth, then rapid decline to a level of relative stability (Busher and Lyons 1999).

Observers monitoring short-term trends in beaver populations should consider these and other intrinsic population regulation mechanisms (such as territoriality) as factors that might explain population change. These intrinsic factors can be important confounding variables when attempting to understand extrinsic factors affecting beaver populations, such as trapping, anthropogenic habitat alteration, and competition with other species.

Construction of Dams, Ponds, and Canals. Beaver are unique in their ability to create favorable aquatic habitat by building dams to restrict the flow of moving water, a behavior richly described in early literature on beaver and in many popular texts (Hilfiker 1991). The widely recognized beaver dam and beaver pond have made this seldom seen nocturnal rodent a familiar and well-studied mammal in North America.

Dams may be initiated by pushing sediment, rocks, or sticks into a ridge formed perpendicular to the flow of moving water or by locating sites to take advantage of existing substrate (Hodgdon and Lancia 1983). Structure is added by anchoring leafy branches, peeled branches, or other material to the substrate (stream bottom, stream banks, large rocks, or coarse, woody debris). Branches in the bulk of the dam may be anchored and intertwined perpendicular or parallel to the flow of water; however, material on the downstream side is usually placed with the cut end pushed into the stream bottom or bank and the branched end pointing upstream to support and stabilize the dam.

Beaver use woody vegetation (bark may be peeled and eaten before placement in dams) and many other materials in dams. Dams can include conifers, sagebrush (*Artemisia tridentata*), tamarisk (*Tamarix pentandra*), aquatic plants, corncobs, cornstalks, plastic, metal, or other debris. Interestingly, when preferred foods are limited and less-preferred foods are more abundant, beaver will select stems that are less palatable for dams and save the more palatable stems for food, especially for use in their winter food cache (Barnes and Mallik 1996; B. W. Baker, unpublished data). For example, Barnes and Mallik (1996) found that beaver preferred stems that were 1.5–3.5 cm in diameter and grew close to shore for dam-building material (mostly alder, *Alnus* spp.). They searched for and selected larger stems (>4.5 cm) that were further from the shore as food items. Thus, beaver increased risk of predation to obtain food but not dam-building material. Barnes and Mallik (1996) speculated that smaller stems were also better for construction of dams, as they might be easier to work with and provide a tighter seal against leaks. However, conventional wisdom suggests that larger material might make stronger dams in regions that experience high spring flows, although this hypothesis has not been tested experimentally. When woody material is in place, beaver seal the upstream side of the dam with mud and herbaceous vegetation (grass, leaves). They typically use mud from the stream bottom immediately upstream of the dam, making this area of the pond the deepest. If the pond overflows the channel as it develops behind the dam, then beaver will often extend the dam laterally by adding shallow wings. Often several dams are built in succession, with water from each pond backed up to the base of the upstream dam, creating a stair-step pattern of dams and ponds, which flattens the slope of the drainage.

The sound of running water is the primary cue for beaver to maintain and sometimes initiate dams (e.g., a noisy road culvert). Although beaver typically work on dams individually, sudden or loud sounds of running water may elicit cooperative behavior, especially to repair a breach in the dam (Aeschbacher and Pilleri 1983). Beaver of all ages inspect and repair dams, but adults perform most of the work. The literature is inconsistent about the relative efforts of males and females (Hodgdon 1978; Busher 1983). Beaver may initiate and maintain dams at any ice-free time of year; however, in many areas there is a peak of activity in the fall before freeze-up and again in the spring after high flows have subsided.

The size and number of dams in a colony and the surface area and volume of water in ponds vary greatly depending on duration of occupancy, topography, substrate, flow levels, available vegetation, and other factors. As water spreads from primary dams within main channels, beaver often build small dams on the surface of the floodplain to further spread and direct water. Thus, individual dams and ponds can be very large or very small, with area inundated generally increasing through the first few years of beaver occupancy.

Beaver often dig canals to facilitate movement of food and building material within and among their ponds or increase water depth for ice-free access to a lodge or food cache. The longer that beaver occupy a site, the more likely it is that they will build or extend canals to access new foraging areas. Canals built within the pond may not be visible unless the pond is drained, but canals built in the floodplain may become obvious features of the landscape. Some canals may contain burrows with an underwater entrance to provide a refuge from predators.

Beaver also create surface trails or "slides" as they transport woody material from their foraging area back to ponds and canals. These trails make it easier for beaver to drag material across the ground, permitting them to move material across greater distances. This is especially obvious in steep terrain, where gravity aids movement of material and can increase the effective foraging distance by several hundred meters.

Lodges and Bank Dens. Beaver construct bank dens and lodges, which are used for protection from predators and weather. Bank dens are often dug under a large tree or shrub on the stream bank to provide support for the roof of the den. They have a nest area above the water level, an underwater entrance, and small holes in the surface soil to permit air exchange. Where beaver live exclusively in rivers or deep lakes, bank dens are typically the only housing structures that are built. Even where beaver eventually build dams and lodges, they often live in a bank den until more permanent structures are completed. The only place that bank burrows are completely absent is where the substrate prohibits their construction, such as areas with very rocky soils or permafrost. In many areas, lodges and bank dens are used.

Lodges can be built in ponds or shallow lakes, where they are surrounded by water, or they can be built on the shore, often as an upward extension of a bank den. In this case, in which they often are called a bank lodge, beaver add sticks on top of the bank den and cut a hole to create a nest chamber. This process can be extended over several years if dam height and water level increase. Construction of a lodge in open water is similar, with sticks piled high enough to enable beaver to cut a nest chamber above the water surface. Mud is added to the surface of the lodge to provide a weather seal, but a portion of the top remains unsealed to allow air exchange. Beaver may have multiple active and inactive lodges within their territory. In addition to mud and freshly cut branches or dead sticks, beaver lodges may include some rocks or other material, although not as much as in dams. The presence of fresh mud or green branches on lodges is often used as an indicator of an active colony. As with dams, lodge construction is often most active in the fall immediately before freeze-up. In ice-free regions, construction of dams and lodges occurs all year, but is less active in the summer.

Food Caches. In regions where ponds or streams freeze during the winter, beaver build food caches, which they access from their lodge by swimming under the ice. The use of food caches is uncommon or absent where beaver inhabit ice-free regions. Beaver typically build a cache by first floating cut branches in a deep part of the pond and then adding new material under this raft. The branches eventually become water-logged and sink to the bottom, holding the cache in place. The upper layer of the cache, called a cap or raft, becomes frozen in ice and unavailable to beaver. Interestingly, beaver often use inedible or less-preferred species for the cap and place more-preferred food items deep enough in the cache to remain ice free and accessible throughout the winter (Slough 1978; B. W. Baker, unpublished data). Differential use of woody plants in food caches and dams can also occur. For example, beaver in Ontario preferentially used conifers and alder in dams and aspen (*Populus tremuloides*) and maple in food caches (Doucet et al. 1994a). Quality of food items in caches is especially important in colder climates, where gestation, parturition, and feeding of newborn young occur under the ice. Construction of a winter food cache usually occurs in late fall and is often initiated by the first hard frost. Beaver may build multiple food caches in a single colony and often do not consume the entire cache during the winter. In the spring, barked stems from the cache may be used to maintain the dam. During ice-free months, beaver sometimes forage by cutting stems on land and returning to a favored location at the edge of a pond to consume them in safety, leaving a pile of peeled stems suggestive of a winter food cache.

ECOLOGY

Diet. Beaver are choosy generalist herbivores, consuming a diet of herbaceous and woody plants, which varies considerably by region and season. The number of plant species in the diet is highest in the southern part of the range and decreases toward the northern and alpine limits of the range (Novak 1987). Herbaceous plants make up much of the diet when they are available and succulent (actively growing). In the central and southern United States, beaver eat a variety of aquatic and riparian forbs and grasses as well as cultivated row crops and grains. Roberts and Arner (1984) found that beaver in Mississippi depended on the bark of woody plants in late fall and winter, but abruptly shifted their diet to herbaceous species after spring greenup in March. Using stomach analysis, they identified 16 genera of herbaceous plants, 15 species of trees and shrubs, and four woody vines in the yearlong diet. Woody material constituted 53% of the annual diet (86% in winter, 16% in summer); grasses occurred in 25% of stomach samples, including some collected in midwinter. In an Ohio study, herbaceous plants accounted for 90% of the feeding time during summer and 40–50% during spring and fall (Svendsen 1980b). In the Mackenzie Delta, Northwest Territories, leaves and the growing tips of willow (*Salix* spp.) were the main foods in July and August. Bark of willow (76%), poplar (*Populus balsamifera*) (14%), and alder (*A. crispa*) (10%) made up the diet the rest of the year (Aleksiuk 1970b). The protein:calorie ratio was 40 mg/cal in summer and 8 mg/cal for the rest of the year, indicating that beaver in northern areas shift their diet to high-protein willow leaves whenever they are available. In northern latitudes, water lily (*Nymphaea, Nuphar*) is often the most important herbaceous component of the diet (Novak 1987). Its edible rhizomes remain succulent after cutting and are often stored in a food cache for winter use (Jenkins 1981). A variety of grasses, sedges (*Carex*), rushes (*Scirpus*), and cattails (*Typha*) is important in the West and Southwest.

Deciduous woody plants are usually the most important component of the diet of beaver and often are the primary limiting factor where ice forces subsistence on a winter food cache. Beaver eat the leaves, buds, twigs, noncorky bark, roots, and fruits of deciduous woody plants, as well as acorns when available (Grinnell et al. 1937; Novak 1987). There is wide regional variation in the number and composition of woody plant species used. As few as 3 species may be used by colonies in the northern range (Aleksiuk 1970), but in the southern range, 22 species were reported in Louisiana and 38 species in South Carolina (review by Hill 1982). In a review of regional food habits, Novak (1987) suggested that local populations of beaver in southern areas included more woody plant species in their diets than did northern populations, but at regional scales the number of woody species used was similar.

Conifers also are cut or gnawed by beaver and used for food or building material, although their value varies greatly by region and availability of preferred deciduous species. In the eastern United States, loblolly pine (*Pinus taeda*) and Virginia pine (*P. virginiana*) may make up over half of the woody material cut by beaver (Novak 1987). They may repeatedly gnaw the bark of pine trees to obtain sap (Svendsen 1980b) or sweetgum (*Liquidambar styraciflua*) trees to obtain storax, an aromatic balsam, which they lick from the injured site. In many areas, especially in their northern and western range, any substantial use of conifer is considered unusual and a sign that more-preferred species are lacking (Novak 1987). Dietary use of conifer also may be seasonal, as beaver in Massachusetts selected against pine during the fall, but not during the rest of the year (Jenkins 1979).

Food Preference. Preference for a particular food item indicates "it constitutes a significantly larger fraction of the diet than an unbiased sample of items of the various food types available" (Jenkins 1981:560). Thus, some foods may constitute a large percentage of the diet, but may not be preferred over less available, but more favored species.

Willow is often the most available and the most used woody riparian species in much of the beaver's range. In many areas of the far north, Rocky Mountains, and intermountain west, beaver may depend entirely on willow to supply winter forage and building material (Aleksiuk 1970b). Where aspen or poplar is available, it is usually more preferred than willow (Jenkins 1981). Cafeteria-style feeding experiments in Ontario showed the following preferences (in descending order): aspen, white water lily (*Nymphaea odorata*), raspberry (*Rubus idaeus*), speckled alder (*A. rugosa*), and red maple (*A. rubrum*). Similar experiments in Nevada showed that beaver preferred aspen and avoided Jeffery pine (*P. jeffreyi*) (Basey 1999).

Selection of forage items by beaver may be related to a variety of physical and chemical factors. Evidence suggests that beaver may select aspen resprouts based on their age-related growth form (Basey et al. 1988, 1990). Aspen reproduces asexually by resprouting within a clone. Aspen clones that have been repeatedly cut by beaver produce juvenile-form root sprouts (large leaves with an absence of lateral branching), which are avoided by beaver when compared to available adult-form root sprouts (small leaves with lateral branching). Although juvenile-form aspen sprouts have more protein and likely provide better nutrition, they contain secondary metabolites that apparently cause avoidance by beaver. The importance of secondary metabolites to selection was further demonstrated in experiments where leaf extracts from different deciduous and coniferous species were painted on aspen leaves and then presented to beaver. Selection favored aspen leaves painted with extracts from deciduous species more so than those painted with extracts from coniferous species (Basey 1999).

Retention time of forage passed through the digestive tract varies with diet composition (likely due to lignin and fiber content) and may also influence food selection by beaver. Experiments have shown that food preference and retention time are correlated; species with a shorter retention time, such as aspen, are more preferred than those with a longer retention time. Beaver "select a diet that maximizes long-term energy intake, subject to digestive limitations" (Doucet and Fryxell 1993:201). Thus, retention time may influence intake rates and energy gained from different forage species, indicating it may be an important factor in food selection by beaver (Fryxell et al. 1994).

Physical features of the food item may also influence selection. In an experimental study of foraging behavior, Doucet et al. (1994b) found that beaver could only distinguish differences in canopy biomass on a very coarse scale, which suggests they selected stems using diameter as an index of biomass. Beaver also select foods by taste, sometimes biting off small samples of bark before cutting down an entire tree. In feeding experiments, beaver avoided aspen that had been painted with an extract of red maple (Muller-Schwarze et al. 1994). In areas with a variety of trees available for food, red maple may be the only tree left standing at the edge of an older beaver pond. Odor may also affect selection. In a similar experiment using extracts of predator feces painted on aspen logs, there was a strong preference against the odors of coyote, lynx, and river otter. Thus, predator odors may be a useful management tool for preventing beaver damage (Engelhart and Muller-Schwarze 1995).

Central-Place Foraging. Beaver typically cut woody vegetation from terrestrial locations for food or construction material and bring it back to a central place, such as a pond, cache, aquatic feeding station, lodge, burrow, or dam. Because this behavior creates exposure to predation and has high energetic costs, beaver have been used to test general predictions of central-place foraging theory. These predictions suggest that beaver should modify their behavior to concentrate foraging near the central place and should increase their selectivity for size and species away from the central place (Fryxell 1992). Most studies have confirmed these general predictions, but exceptions to these patterns occur. For example, studies confirmed that beaver typically cut increasingly smaller stems (less provisioning time) further from the central place, as predicted by optimal foraging models (Jenkins 1980; Belovsky 1984; Fryxell and Doucet 1990). In contrast, where relatively small (1.5–30 mm) stems are the only woody plants available, beaver may select larger stems even when located further from the central place (McGinley and Whitman 1985). Selection for larger stems is particularly evident where beaver occupy shrub habitats, such as those containing only smaller species of willow and alder. In some cases, repeated cutting by beaver can cause trees to develop and maintain a shrubby growth form (e.g., Fremont cottonwood, *Populus fremontii;* McGinley and Whitman 1985).

The size–distance relation in food selection by beaver also may be affected by species preferences. For example, Jenkins (1980) found that beaver cut larger stems of preferred species, such as oak and cherry (*Prunus*), further from the central place than less-preferred species. In contrast, although Belovsky (1984) did find that beaver had strong food preferences, preferences did not change relative to distance from beaver ponds. In many cases, the interaction of species preferences with stem size and distance from the central place has been difficult to document in the field because depth to water and other plant-growing conditions preclude equal availability of stems to beaver. Modeling the diet of beaver is one way to overcome limitations of field experiments. For example, Belovsky (1984:220) found that beaver at Isle Royale selected their diet "consistent with an energy-maximizing solution to a linear-programming model." This energy maximization model was further confirmed by Fryxell (1992) with a second line of evidence, which found both density and distance were important predictors of food selection by beaver.

Chemical factors and stem size may also influence stem selection by beaver. If aspen responds to repeated beaver cutting by producing a juvenile growth form, then higher concentrations of phenolic compounds (secondary metabolites) may inhibit further cutting by beaver and influence the predictive value of optimal foraging models for beaver. Basey et al. (1988) found beaver avoided aspen stems <4.5 cm in diameter in favor of those >19.5 cm in diameter near a 20-year-old beaver pond where 51% of stems had been previously cut by beaver and the remaining 49% of stems were in juvenile form. In contrast, beaver at a newly occupied site selected smaller aspen stems and against larger ones. Taken together, these results suggest that phenolic compounds, or other factors in addition to size of stem, may influence selection by beaver.

Food Consumption and Production. Estimates of daily forage consumption rates (wet woody biomass) for beaver vary from 0.5 kg/day (Dyck and MacArthur 1993) to 2.0 kg/day (review by Stegeman 1954). In an interesting account of a Colorado beaver colony fed by a Forest Service contractor in the 1920s, it was estimated that each beaver consumed (although possible use in dams was not described) an average of about 900 kg of green aspen/year (Warren 1940). In a study of the energy content and digestibility of cached woody biomass, Dyck and MacArthur (1993) concluded that the total winter energy requirements could not be met from the submerged food cache in their study colony. Supporting research has shown that when food is limited, beaver may metabolize body tissue during winter.

Estimates of beaver food (twigs and bark) produced by trees and shrubs may be useful for predicting the carrying capacity of beaver where woody biomass limits population density. Beaver food estimates are derived from graphs or equations that model annual production (current annual growth) and total biomass based on measures of basal stem diameter. Estimates have been derived for aspen (Aldous 1938; Stegeman 1954), willow (Baker and Cade 1995), and five species of riparian shrubs in Minnesota (Buech and Rugg 1995). The maximum diameter of twigs that are entirely consumed by beaver is a critical consideration in deriving estimates of beaver food. The diameter used is often based on assumptions made from observing the size of peeled stems, but may vary greatly by species and region. For example, Aldous (1938) assumed beaver ate all stem portions <12.7 mm in diameter for aspen, Buech and Rugg (1995) assumed <5 mm for Minnesota shrubs, and Baker and Cade (1995) assumed <3 mm for coyote willow (*S. exigua*). Application of results among studies and different woody species requires caution. In addition, intense ungulate or livestock herbivory of the terminal leaders on shrubs may strongly reduce the biomass of beaver food relative to unbrowsed stems of equal diameter (B. W. Baker, pers. obs.).

Habitat Requirements. The ability of beaver to alter existing habitat conditions to meet their needs has allowed populations to inhabit a variety of natural and human-made habitats in North America. They have successfully colonized tundra and taiga in the far North, bottomland hardwood forests and marshes in the deep South, riparian areas in both cold and hot deserts, and elevations that vary from sea level to above 3400 m (reviews by Hill 1982; Novak 1987). Although beaver can occupy a wide variety of habitats, some generalizations are evident. A comprehensive evaluation of beaver habitat requirements in the Rocky Mountains showed they generally preferred wide valleys with a low (<6%) stream gradient, which offered relatively more food and reduced risk of severe floods (Retzer et al. 1956). Beaver typically inhabit streams with at least intermittent flow and lakes or ponds with standing water, but they can also inhabit bogs that lack open water. In Minnesota, they occur in sedge-moss and other bogs, where they can enlarge natural moats to create ponds of standing water and build floating lodges able to adjust to fluctuating water levels, and thus maintain protection from predators (Rebertus 1986).

Early studies of beaver formed the foundation for later mathematical models, which quantified habitat requirements and created a framework for making and testing predictions. Slough and Sadleir (1977) sampled colony density and associated habitat conditions at 136 lakes and 45 stream sites in Ontario and used regression analysis to develop a land classification system for beaver. Howard and Larson (1985) used principal component analysis of habitat variables to predict beaver colony density in Massachusetts. Percentage hardwood vegetation, watershed size, stream width, and stream gradient were important predictors in their classification system. Allen (1983) used existing literature and expert opinion to develop a general habitat suitability index (HSI) model for beaver, which used nine variables to rate habitat quality on a scale of 0.0–1.0. These variables included measures of canopy cover, height, stem diameter, species of trees and shrubs, stream gradient, water level fluctuation, and shoreline development (ratio of length and area). The HSI model assumed a minimum habitat area of 0.8 km of stream or 1.3 km^2 of lake or marsh as a prerequisite of suitable beaver habitat. This model has been widely used by environmental planners to quantify potential impacts from development projects and mitigate habitat loss.

Other researchers have developed alternative habitat models and modified existing models for different habitats and regions. For example, researchers in prairie regions found beaver selected riverine lodge-site locations in areas that had thick, concealing vegetation cover (which was often left uncut) and steep shoreline banks (Dieter and McCabe 1989). In the Truckee River basin of California, physical features of the stream, such as a lower gradient and a greater depth and width, were more important than vegetation in describing the location of lodge sites selected by beaver (Beier and Barrett 1987). In contrast, Barnes and Mallik (1997) found that concentration of woody plants 1.5–4.4 cm in diameter and size of the stream and its upstream watershed area were important predictors of dam-site location in northern Ontario. In Oregon streams, McComb et al. (1990) found that beaver selected dam sites where the substrate was less rocky, the water was shallower, the channel had a lower gradient, and woody vegetation (e.g., *Alnus*) had a greater canopy cover. They concluded that Allen's (1983) HSI model was useful in predicting habitat quality for beaver at their sites, but required some site-specific modifications.

In contrast, others have found that Allen's HSI model was a poor predictor of beaver habitat quality in their region. For example, a Kansas study suggested that water quality, river substrate type, and adjacent agricultural land-use practices are important predictors of riverine habitat quality for beaver in the central United States (Robel et al. 1993). In an Oregon study, many potential beaver sites were highly rated by the HSI model even though they were unoccupied by beaver at the time of study, which suggested poor model performance to the investigators (Suzuki and McComb 1998). However, low density or absence of beaver in apparently suitable habitat is not unusual and may be caused by many non-habitat-based factors, such as trapping, disease, territoriality, or simply the inherent variation of natural systems. At occupied sites, cutting of preferred species by beaver may alter the density and species composition of vegetation, and thus affect how the habitat might be rated by habitat models (Suzuki and McComb 1998).

MORTALITY

Predation. Predation by the timber wolf (*Canis lupus*) can be an important limiting factor of beaver populations where they occur together. Wolves prey on beaver during the ice-free period, when nearly half their diet may consist of beaver (Potvin et al. 1992). In Algonquin Park, Ontario, as white-tailed deer (*Odocoileus virginianus*) populations declined over a 9-year period, beaver gradually became the most important prey item: 55% of wolf scats had beaver remains (Voigt et al. 1976). In an experimental study in Quebec, density of beaver colonies increased 20% after 3 years of wolf control and then declined again after wolf control ceased, indicating wolf predation may have suppressed populations below the carrying capacity of the habitat (Potvin et al. 1992). Habitat conditions that force beaver to forage farther from water may increase predation rates by wolves. On Isle Royale, beaver foraged further (>100 m) from ponds when wolf populations were low and closer (<35 m) to ponds when wolf populations were high (D. W. Smith, unpublished data). Thus, the impact of wolf predation on beaver populations can be locally significant, but varies greatly depending on wolf density, alternative prey availability, and other factors.

Coyotes (*Canis latrans*) and mountain lions (*Puma concolor*) also prey on beaver, as do some other mammalian predators of generally minor importance, such as bears (*Ursus* spp.), wolverines (*Gulo gulo*), river otters (*Lontra canadensis*), lynx (*Lynx canadensis*), bobcat (*Lynx rufus*), and mink (*Mustela vison*) (review by Hill 1982). However, unusual circumstances can alter typical predation patterns. Black bear (*U. americanus*) predation strongly suppressed beaver populations on an island in Lake Superior (Smith et al. 1994). In this case, bears colonized one of two similar islands in the 1970s that beaver had colonized in the late 1940s or early 1950s. As the bear population grew, they focused predation on beaver, the only source of meat on the island, digging into 18 of 26 beaver lodges and causing surviving beaver to concentrate foraging on trees <30 m from water.

Disease. Water-borne tularemia is a zoonotic disease caused by the bacterium *Francisella tularensis* biovar *palaearctica* (type B), which commonly occurs in semiaquatic mammals such as beaver and muskrat (*Ondatra zibethicus*) and occasionally becomes epizootic. Type B tularemia is not fatal to humans and is responsible for only 5–10% of human tularemia infections in North America. Type A tularemia is responsible for the remaining human infections. It can be fatal to humans and has a terrestrial cycle in rabbits (*Sylvilagus* spp.) and ticks (Morner 1992). Tularemia infections in beaver are typically subclinical without noticeable effects on the individual or the population, but they can be fatal to beaver and cause mass mortality from local or regional epizootics. Tularemia in beaver sometimes can be traced to infections in terrestrial rodents that deposit urine or feces in water, or die in water, which then harbors *F. tularensis* bacteria. For example, an outbreak of tularemia in Montana during 1939–1940 caused widespread mortality of beaver (several hundred carcasses were found) and coincident infection and mortality of meadow voles (*Microtus pennsylvanicus*) that inhabited the grassy streambanks (Jellison et al. 1942). Interestingly, rabies has been documented in beaver, but little is known about its pathogenesis or epizootiology (J. Rupprecht, 47th Wildlife Disease Association Conference, oral commun., 1998).

Other Mortality. Starvation can be an important cause of mortality, especially at northern latitudes, when beaver are unable to construct a food cache large enough to sustain them through the winter (Gunson 1970; Bergerud and Miller 1977). Sudden snowmelts in midwinter or violent spring breakups can raise water levels in streams and may destroy lodges and occupants or drown large numbers of beaver under the ice (Hakala 1952).

BEAVER AS A KEYSTONE SPECIES AND AN ECOSYSTEM ENGINEER

A keystone species is one that greatly influences the species composition and physical appearance of ecosystems (Paine 1969) and whose effects on ecosystem structure and function are both large overall and disproportionately large relative to its abundance (Power et al. 1996). An ecosystem engineer is a species that directly or indirectly controls resource availability by causing "physical state changes in biotic or abiotic materials" (Jones et al. 1997:1946). The beaver is a definitive example of both a keystone species and an ecosystem engineer.

The dam-building, canal-building, and foraging activities of beaver have profound effects on ecosystem structure and function. Beaver dams slow current velocity, increase deposition and retention of sediment and organic matter in the pond, reduce turbidity downstream of the dam, increase the area of soil–water interface, elevate the water table, change the annual stream discharge rate by retaining precipitation runoff during high flows and slowly releasing it during low flows, alter stream gradients by creating a stair-step profile, and increase resistance to disturbance (reviews by Naiman et al. 1988; Gurnell 1998). Canals dug by beaver spread impounded water across a larger surface area, and thus magnify the effects of single dams. The foraging activity of beaver alters the species composition, density, growth form, and distribution of woody vegetation. These effects on vegetation and the physical characteristics of streams strongly alter the composition of the animal community. Because research did not begin until after beaver populations recovered following near extinction by presettlement trapping and extensive habitat loss, researchers have observed a human-impacted and likely conservative picture of how beaver have altered ecosystem structure and function (Naiman et al. 1988).

Effects of Beaver on Geological Processes. In 1938, researchers evaluating the formation of broad, flat alluvial valleys in the Catskill region of New York discovered that geological processes alone could not explain observed sediment deposition rates (Ruedemann and Schoonmaker 1938). They suggested that during the 25,000 years since the last glaciation, layer on layer of sediment-filled beaver ponds caused the "complete aggrading of valley floors, originally in small descending steps, which disappear in time and leave a gently graded, even valley plain horizontal from bank to bank" (Ruedemann and Schoonmaker 1938:525). Unfortunately, the theory of beaver-assisted alluvial valley formation remains largely unproven by rigorous research across varied landscapes. One exception is a study in Glacier National Park, Montana, where comparisons of sediment depth and pond age confirmed that beaver ponds gradually accumulated sediment as they aged (Meentemeyer and Butler 1999). Thus, unless dams fail, beaver ponds will eventually fill with sediment and become beaver meadows. This process of accelerated meadow development by beaver dams is likely more important in meandering, low-gradient, valley-bottom streams, where conditions favor stable dams that spread sediment-laden water over a large surface area, rather than in steep, V-shaped, high-energy streams, where beaver dams tend to fail at a greater rate.

Beaver dams also affect erosional processes within stream channels, typically increasing channel aggradation. As sediment-loaded water enters a beaver pond, it slows in velocity and drops sediment, increasing aggradation. However, water downstream of dams can be underloaded with sediment and increase erosional forces as the stream regains lost sediment (Meentemeyer and Butler 1999). This can lead to a localized increase in bank sloughing below beaver dams in areas with erosive soils. In most cases, the net effect of beaver dams is to decrease channel and streambank erosion and increase channel aggradation (Parker et al. 1985). Catastrophic beaver dam failures are rare events, but can occur following unusually large rainfalls or high spring runoffs. In 1994, a beaver dam in central Alberta failed and released 7500 m^3 of water, causing a flood wave that was 3.5 times the maximum recorded discharge for the stream (Hillman 1998). In this case, the flood wave was largely attenuated by downstream wetlands, including several beaver ponds.

Beaver dams may also act as a filter to decrease nonpoint-source water pollution. In a study of Currant Creek, Wyoming, a highly erosive second-order stream, concentrations of suspended solids, total phosphorus, sodium hydroxide-extractable phosphorus (biologically available P), and total Kjeldahl nitrogen were reduced in water flowing through beaver ponds during high spring runoff (Maret et al. 1987). The effect of beaver ponds on these parameters continued during summer low flows, but was of lower magnitude. Effects disappeared at about 1.6 km below the beaver dams, likely due to inputs from bank erosion (Parker et al. 1985; Maret et al. 1987). Beaver ponds may also influence water quality by affecting the number and composition of bacteria in the stream. In Wyoming, some species of bacteria apparently increased and others decreased as water flowed through beaver ponds; however, results were confounded by the effects of different livestock grazing systems (Skinner et al. 1984).

Effects of Beaver on Ecological Processes. In addition to geological affects, beaver alter the landscape by creating layers of spatially distinct volumetric units, or patch bodies, which include the bedrock, the water-saturated anerobic soil under the pond, the moist aerobic soil at the water's edge, the pond, the browse zone concentric to the pond or central place, and the overlying atmosphere (Johnston and Naiman 1987). These patch bodies create a shifting mosaic of conditions in the landscape, which varies spatially and temporally as beaver populations colonize new territory or abandon old sites (Naiman et al. 1988). The inherent habitat matrix strongly influences how these patch bodies affect ecosystem process. Effects are greatest where contrast is greatest. Beaver that create dams in existing wetlands have less effect than those that dam streams in upland forest or desert shrubland. Basin geomorphology and length of occupancy also affect the magnitude of influence. Beaver typically select the best pond sites first and move to less desirable sites as populations grow or as resources are depleted. Ponds built in better sites have greater longevity and affect the disturbance dynamics in the system. In a Minnesota study, a series of aerial photographs between 1940 and 1986 showed the rate of patch formation was much greater during the first two decades following colonization than during the second two decades, which suggests that geomorphology eventually limited the availability of pond sites after the initial better sites were occupied (Johnston and Naiman 1990a).

Beaver activity strongly alters the biogeochemical characteristics of watersheds through the accumulation, availability (standing stocks), and translocation of nutrients and ions (Naiman et al. 1994). In a comparison of stream riffle areas to beaver ponds, Naiman et al. (1988) showed riffles had only 48% of the carbon inputs, 5% of the carbon standing stock, and 6% of the carbon outputs as did beaver ponds. In addition, the turnover time of carbon in beaver ponds averaged 161 years, much slower than the 24 years found in riffles. Beaver ponds affect the amount and distribution of nitrogen in the system, as ponds create anerobic conditions, slow flows, and increase oxygen demand by retaining organic matter. In Quebec, a beaver-modified section of stream "accumulated 10^3 times more nitrogen than before alteration" (Naiman and Melillo 1984:150). Anerobic conditions caused by water-saturated soils in beaver ponds fundamentally alter biogeochemical pathways (Naiman et al. 1994). When beaver create ponds in forested uplands, most upland vegetation dies from inundation and woody material cut by beaver is moved to the stream for dams and lodges. The organic horizons of pond sediments accumulate chemical elements from formerly upland plants that become available for vegetative growth when the ponds fill with sediments or dams fail and abandoned sites become beaver meadows. Even small dams can accumulate a tremendous amount of sediment. For example, dams containing 4–18 m^3 of wood retained 2000–6500 m^3 of pond sediment in a boreal forest system in Quebec (Naiman et al. 1986). In a long-term Minnesota study, beaver activities increased the standing stock of chemical elements in the organic horizon of ponds by 20–295% (Naiman et al. 1994). Transport of water through beaver ponds can also neutralize acids, increase pH, and increase dissolved oxygen concentrations in acidic stream systems (M. E. Smith, et al. 1991). Thus, the activities of beaver can strongly modify the biogeochemical characteristics of stream systems and fundamentally influence forest ecosystem dynamics at landscape scales.

Effects of Beaver on Invertebrates. Beaver ponds affect the species composition and abundance of stream invertebrates as the community responds to increased sediment deposition and still water behind dams. Invertebrate taxa that prefer running water are replaced by pond taxa. Community function is changed as collectors and predators increase in importance and shredders and scrapers decrease in importance (McDowell and Naiman 1986; Naiman et al. 1988). In Quebec, density and biomass of invertebrates in pond sites were two to five times greater than in riffle sites in the spring and summer, but were similar in the fall (McDowell and Naiman 1986). The number of species in ponds was similar to that in the natural stream channel, but resembled that in slow-water habitats of larger order streams, indicating that invertebrates in the beaver ponds may not be unique in the watershed (Naiman et al. 1988). In a study that compared stream riffle sites above and below beaver dams in the Adirondack Mountains, sites immediately below dams had lower invertebrate richness and diversity, but higher total invertebrate, predator, and collector–gatherer densities (M. E. Smith et al. 1991).

Beaver may also affect the invertebrate community by changing the structure and chemistry of plant hosts. In a study of leaf beetles (*Chrysomela confluens*) and their cottonwood (*P. fremontii* × *P. angustifolia*) hosts, beetles were attracted to beaver-cut cottonwood regrowth even though it contained twice the level of defensive chemicals (to protect the plant from herbivory) as normal juvenile regrowth (Martinsen et al. 1998). In this case, beetles may have sequestered these chemicals for their own defense against mammalian predators. Beetles also may have obtained a nutritional benefit from beaver-cut regrowth because it contained more total nitrogen than nonresprout growth. Thus, beetles grew faster and were heavier at maturity.

Beaver impoundments also may affect local mosquito populations, but not necessarily as conventional wisdom might suggest. At a New York site, observers noted marked reductions in mosquito populations after beaver impounded an area of poorly drained forest (Butts 1992). Before impoundment, the area supported large larval populations of the *Aedes* mosquito, which is unable to breed in the permanent water developed by beaver.

Effects of Beaver on Fish. Dams, ponds, canals, and foraging of beaver may alter the density, distribution, species composition, and population characteristics of fish populations. Trout habitat in streams is usually improved by beaver where low flows or cold water temperatures limit trout distribution or production. However, trout may be harmed if water is warmed beyond their tolerance. Beaver benefit trout by creating deep pools, which resist freezing in winter (often capped with ice and snow, which prevents formation of anchor ice) and maintain cooler temperatures in summer. Beaver provide additional benefits by increasing the size of the wetland area, improving physical cover, and increasing the invertebrate forage base because of changes in the substrate and higher water temperatures in ponds (reviews by Hill 1982; Olson and Hubert 1994). In Wyoming, standing stock (kg/ha) and density of brook trout (*Salvelinus fontinalis*) in beaver ponds were correlated with surface area, mean water depth, water volume, discharge into pond, elevation, and a morphoedaphic index (Winkle et al. 1990). However, where trout populations are limited by high water temperature in the eastern United States, beaver ponds may increase temperatures beyond tolerable limits, and beaver dams are often removed to improve trout habitat. However, a study of the thermal effects of beaver dams in Wisconsin found no consistent relationship between size or number of beaver ponds and degree of downstream warming (McRae and Edwards 1994). They found that large ponds may act as thermal buffers, which dampen daily fluctuations in water temperature and only slightly raise downstream temperatures. Thus, large-scale beaver dam removal in headwater streams may have net negative consequences on trout populations as their positive effects on invertebrates and ecosystem processes are lost.

Beaver dams may be detrimental to populations of trout and salmon if they restrict or prevent fish passage. However, some studies have shown that trout can pass over dams during high water and may pass through newly constructed dams at any season (review by Olson and Hubert 1994). In the Pacific Northwest, many beaver dams are partially washed out each year by winter high water. However, dams can still benefit fish that use the remaining ends for cover, thus providing value as coarse, woody debris. During summer low flows in this area, beaver dams improved rearing habitat for coho salmon (*Oncorhynchus kisutch*), as beaver ponds were larger and contained more coho fry than pools without beaver (Leidholt-Bruner et al. 1992).

Beaver activities can have profound effects on fish community structure. In an experimental study of a headwater stream in northern Minnesota, beaver ponds appeared to act as reproductive "source" populations for fish, which dispersed to adjacent streams, which functioned as reproductive "sinks" in the landscape. Thus, the boundary influences of beaver ponds were "critical in controlling fish dispersal between ponds and streams and the subsequent abundance and composition of fish in lotic ecosystems" (I. J. Schlosser 1995:908). The age and size of beaver ponds also affect stream dynamics. In Georgia, fish species richness per pond increased until ponds were 9–17 years old and then decreased as ponds aged in headwater streams, but showed little change relative to pond age in downstream sections (Snodgrass and Meffe 1998). In larger ponds, fish species shifted from lotic to lentic species, and larger predators replaced small-bodied minnows in older ponds. The size of beaver ponds is also important. As expected from general species–area relationships, fish species richness increases with size of pond, but very small beaver ponds can have higher than expected richness compared to ponds of a similar size not impounded by beaver (Keast and Fox 1990). Thus, beaver have a strong effect on fish species richness, but the effect is dependent on the size and age of beaver ponds and how ponds are distributed within the landscape.

Effects of Beaver on Vegetation. Beaver affect vegetation by building dams and removing woody plants as food and building material. Beaver dams raise the water table by creating a pond area that may be inundated by several meters of water and an umbrella-shaped zone of influence that radiates out from the pond, creating a new water table gradient controlled by soil texture and other factors. The soil behind dams can act like a sponge, retaining water during wetter months and slowly releasing it during drier months. In areas of low or irregular precipitation, beaver dams may convert streams from intermittent flow to perennial flow. Changes in the amount, timing, or duration of available water can create a competitive advantage for many species of riparian–wetland plants, thus increasing their survival and dominance in the landscape. For example, in some western shrub–steppe ecosystems, beaver may help control invasive tamarisk by creating a competitive advantage for willow as dams alter hydrological and ecological conditions within incised stream channels (B. W. Baker, pers. obs.). Higher water tables caused by beaver ponds generally kill upland vegetation and promote establishment and growth of wetland vegetation.

Beaver can improve conditions for seedling establishment of willow, cottonwood, and other riparian species. Sediment deposited behind beaver dams creates an ideal moist soil substrate, which can become exposed as water levels in the pond decrease due to a dam washout or other cause. Beaver cuttings may be an important mechanism of plant establishment for some woody species such as willow (Cottrell 1995). Stems that are cut by beaver but not eaten can become imbedded in dams, lodges, or moist soil in or near ponds and sprout adventitious roots and new stem growth. Because a high percentage of stem segments is not consumed by beaver, they may substantially contribute to plant establishment, although the relative importance of this method is not well understood at the community level.

Cutting by beaver can stimulate vigorous resprouting, which may increase biomass production in many woody riparian species. Plants can sprout new shoots by activating dormant meristems located below the cut on the same stem or on below ground plant tissue, or from previously developed root suckers. In a study of red willow (*S. lasiandra*) in Oregon, trees that had a higher percentage of stems cut by beaver responded by producing a higher percentage of regrowth the following season (Kindschy 1985). Where stem cutting is concentrated in late fall to build dams or prepare a food cache, plants are dormant when cut and respond with new shoots in the next spring in an attempt to recover former root:shoot ratios, maximizing plant production and minimizing plant damage. Cutting by beaver can also stimulate plants to initiate growth earlier in the spring, further increasing stem production (Kindschy 1989). However, biomass of new shoots can be decreased if regrowth is browsed by native ungulates or livestock.

The interaction of stem cutting by beaver and intense browsing by livestock or native ungulates can strongly suppress regrowth and may result in declining riparian plant communities. Summer browsing by livestock that congregate along riparian areas can be particularly detrimental to recovery of beaver-cut willow, when new green shoots become highly preferred as grasses cure and become less palatable (Kindschy 1989). Intense browsing by native herbivores can also severely suppress regrowth of beaver-cut stems. When beaver cut tall stems, regrowth occurs at or near ground level, placing the apical portions of stems within easy reach of herbivores. Browsing the tips of stems (leaders) releases apical dominance and may cause plants to develop a short stature. In an experimental study at Rocky Mountain National Park, Colorado, intense elk (*Cervus elaphus*) browsing of willow (*S. monticola*) regrowth during 3 years following simulated beaver cutting produced plants of low vigor, which were small, short, and hedged with a high percentage of dead stems (B. W. Baker, unpublished data). In contrast, regrowth of unbrowsed plants was vigorous, large, tall, highly branched, and leafy with a low percentage of dead stems. Browsed and unbrowsed plants recovered more stems per plant (about 70 after 3 years) than were present before simulated beaver cutting, but elk browsing strongly suppressed recovery of plant biomass. After 2 years of regrowth, browsed plants had recovered only 6% of their precut biomass, whereas unbrowsed plants had recovered 84%. Thus, beaver cutting and elk browsing may interact to decrease woody plant height and biomass, eventually reducing beaver habitat suitability and the positive effects of beaver on community structure and function.

Foraging by beaver can significantly alter forest composition and plant succession. Besides humans, beaver are the only species in North America that can affect overstory vegetation by felling mature trees (Johnston and Naiman 1990b). In contrast to many other herbivores, beaver foraging is restricted to a central place within riparian communities where their habit of felling much more woody material than they actually consume dramatically increases the magnitude of their effects relative to other woody-plant foragers. In a Minnesota study, intense foraging by beaver decreased tree density and basal area by up to 43% near ponds, where individual beaver harvested an average of 1400 kg/ha/year of woody biomass over a 6-year period (Johnston and Naiman 1990b). Selective foraging decreased aspen and increased alder and conifer, with long-term effects on forest succession. In Wisconsin, beaver substantially reduced the density of preferred understory tree species (W. J. Barnes and Dibble 1988). Because understory species normally replace overstory species in areas without beaver, it was predicted that selective foraging by beaver strongly altered riparian forest succession. Donkor and Fryxell (1999) also found that the foraging activity of beaver altered forest succession by replacing deciduous species with conifers. They suggested this detracted from the keystone role of the beaver because the already dominant conifer community was increased.

Woody plant communities along major river systems can also be affected by the foraging behavior of beaver. In rivers too large for beaver to dam, bank-dwelling beaver can have significant effects on the structure of woody riparian vegetation. Regulation of rivers by major dams can modify the effects of beaver as peak flows are decreased and base flows are increased, which affects growth of woody vegetation and its use by beaver. In a comparative study of the unregulated Yampa River and the regulated Green River in northwestern Colorado, Breck (2001) found that beaver harvested a higher percentage of willow and cottonwood on the regulated river, where the constant flow regime caused vegetation patches to be more uniformly available to beaver. In the unregulated river, shifting channels constantly changed access to shoreline vegetation.

Abandonment of beaver ponds creates new habitat for plant establishment. Beaver meadows dominated by grasses and sedges often develop on the rich sediment that settles in beaver ponds. Meadows may eventually be replaced by forests, but in some cases succession is slower than expected. In northern forests, beaver meadows may resist invasion by conifers and persist for many decades as graminoid meadows (Terwilliger and Pastor 1999). Successful conifer (*Abies, Picea*) invasion requires the presence of ectomycorrhizal fungi, which form an obligate association with tree roots. Because past flooding would have killed these necessary fungi, beaver meadows require dispersal of fungi from nearby forest soils before conifers can invade. One possible mechanism of ectomycorrhizal fungi dispersal is via small-mammal feces. However, an evaluation of the red-backed vole (*Clethrionomys gapperi*) as a vector showed only about one third of plants grown in beaver meadow soils inoculated with vole feces had ectomycorrhizae present, compared to 100% of plants grown in forest soil. In addition, the potential of red-backed voles to facilitate dispersal of fungi is limited by their general avoidance of beaver meadows (Terwilliger and Pastor 1999). Thus, persistence of beaver meadows in forest communities may be controlled by complex ecological mechanisms.

Effects of Beaver on Mammals. A variety of mammals use the lush vegetation around beaver ponds as food and cover or rely on beaver to provide aquatic habitat (review by Olson and Hubert 1994). Beaver ponds are often important habitat for moose (*Alces alces*) because they increase production of woody plants and aquatic vegetation, which can contribute substantially to their total diet. In some cases, moose may compete with beaver for limited food supplies; however, the extent of their interaction and its effect on riparian vegetation are not well understood. Elk and deer can also be attracted to the increased abundance, palatability, or availability of woody and herbaceous forage at beaver ponds. Beaver ponds also create habitat for other semiaquatic mammals, such as river otter, mink, and muskrat, some of which may occur in a large percentage of active or abandoned ponds. In Idaho, the density and standing crop biomass of small-mammal populations was two to three times higher in willow-dominated beaver pond habitat than in adjacent riparian habitat (Medin and Clary 1991). Montane voles (*M. montanus*) and shrews (*Sorex* spp.) were the most abundant small mammals in the beaver pond habitat, and their populations accounted for most of the differences. However, neither species richness nor species diversity of small mammals was influenced by beaver in this study.

Effects of Beaver on Reptiles and Amphibians. Beaver ponds and associated riparian habitat likely alter the species composition and abundance of regional herpetofauna throughout the range of beaver. However, few studies have quantified these effects and some conclusions have been equivocal. For example, Russell et al. (1999) compared new (<5 years) and old (>10 years) beaver ponds to unimpounded streams in South Carolina. They found that richness and abundance did not differ among sites for amphibians, but both were higher for reptiles at the older beaver ponds than at new ponds or unimpounded streams. Reptile species diversity was highest at old ponds, intermediate at new ponds, and lowest at unimpounded sites. They suggested that beaver have generally benefited herpetofauna in the Piedmont of South Carolina, as the range of several species, such as common musk turtle (*Sternotherus odoratus*) and eastern mud turtle (*Kinosternon subrubrum*), has increased concurrent with the range expansion of beaver.

Effects of Beaver on Birds. Waterfowl use beaver ponds for nesting and brood-rearing habitat and as stopover sites during migration. Beaver ponds enhance vegetation growth, which improves dense nesting cover and enhances interspersion of cover and water, improving nest isolation for territorial waterfowl pairs and increasing nest density (review by McCall et al. 1996). When beaver ponds flood timber, dead trees provide nesting sites for waterfowl (cavity nesters) that are relatively safe from predators. Beaver ponds increase the production of invertebrates and aquatic vegetation, which improves brood-rearing habitat. The benefits of beaver ponds to waterfowl may be expressed by increased production of waterfowl as regional beaver populations increase. For example, when beaver populations in south-central Maine increased in the 3–4 years following a trapping closure, density of Canada geese (*Branta canadensis*), hooded mergansers (*Lophodytes cucullatus*), and mallards (*Anas platyrhynchos*) also increased, as did the number of wetlands used by these species (McCall et al. 1996). Waterfowl in Finland also responded positively to increasing beaver populations, but response varied by species. Teal (*Anas crecca,* the European race of the green-winged teal) increased to become the dominant species during the first 2 years following flooding by beaver, but populations of mallards and wigeon (*Anas penelope*) did not respond (Nummi and Poysa 1997).

Beaver ponds can be preferred to similar open-water wetlands by breeding waterfowl. In Ontario, open-water lakes made up 47% of the study area, but were avoided by blue-winged teal (*Anas discors*), mallards, and wood duck (*Aix sponsa*). However, beaver ponds constituted only 25% of the area and were preferred by wood ducks (Merendino et al. 1995). Preference for beaver ponds may be due to their increased production of invertebrates important to waterfowl broods. For example, populations of water fleas (*Cladocera* spp.), a small invertebrate preferred by young ducklings, were abundant in first-year beaver ponds, with populations of larger invertebrates more important in 2- and 3-year-old ponds (Nummi 1992). Beaver ponds substantially enhance waterfowl habitat in the western United States, where riparian areas constitute <2% of the landscape. Breeding waterfowl surveys in Wyoming found that streams with beaver ponds had 7.5 ducks/km, but similar streams without beaver ponds had only 0.1 duck/km (McKinstry et al. 2001). A survey of 125 land managers found that beaver had been removed from 23% of 28,297 km of Wyoming streams and that >3500 km of streams without beaver had potential for beaver reintroduction to improve waterfowl nesting habitat (McKinstry et al. 2001). Finally, beaver ponds can be highly important for migrating waterfowl throughout North America and for wintering waterfowl in ice-free southern regions (Ringelman 1991).

In addition to waterfowl, a variety of other bird species benefit from beaver activity. Beaver ponds can increase the area of open water, length of the shoreline, density of dead standing trees, and biomass, height, and canopy cover of shrubs. The abundance and diversity of terrestrial insects important to foraging birds are enhanced by changes in riparian and aquatic vegetation and availability of standing water for larval development. These habitat changes may dramatically increase bird species richness, diversity, and abundance (Krueger 1985; Medin and Clary 1990). Active beaver sites contained 92% of 106 bird species observed in 70 New York wetlands (Grover and Baldassarre 1995). Beaver sites also were important habitat for red-winged blackbirds (*Agelaius phoeniceus*), Brewer's blackbirds (*Euphagus cyanocephalus*), common snipe (*Gallinago gallinago*), and spotted sandpipers (*Actitis hypoleucos*) in Wyoming (Brown et al. 1996). In South Carolina, beaver ponds were an important avian habitat in all seasons, but reached maximum bird diversity in spring and summer and maximum number of individuals (all bird species combined) during fall migration (Reese and Hair 1976). Beaver ponds are especially important in arid environments, where riparian–wetland habitat provides an oasis for birds. Any increase in area or structural complexity of riparian vegetation usually benefits the avian community.

ECONOMIC STATUS, MANAGEMENT, AND CONSERVATION

Economic Value. The ecological role of beaver discussed in the previous section has tremendous indirect economic benefit through wetland creation, water storage, improved water quality, erosion control, sediment deposition, and recreation. These indirect benefits may far outweigh the direct monetary value obtained from their fur, and may offset any direct or indirect costs due to beaver damage. However, because the monetary value of these indirect ecological benefits has not been quantified, fur trapping and damage control are typically considered of primary economic importance.

Native Americans valued beaver for food and clothing and early Europeans valued the underfur of beaver pelts for the manufacture of felt hats. In 1610, France granted Samuel de Champlain the first fur-trading monopoly in North America, initiating 200 years of intensive beaver trapping, which continued until the early 1900s, when beaver populations had been nearly extirpated and demand for felt hats had declined (reviews by Hill 1982; Hill and Novakowski 1984; Novak 1987). Since then, strict harvest regulations, reintroduction programs, and cutting and burning of climax boreal forests (setting back succession) have greatly benefited beaver populations. Annual harvests during the 1980s sometimes exceeded 1 million beaver pelts, more than recorded at any other time (Novak 1987). Although beaver trapping is still an important part of the fur industry in Canada and the United States, erratic and relatively low pelt prices have caused the economic importance of trapping beaver (numbers marketed and value of pelts) to decline relative to furbearers such as red fox (*Vulpes vulpes*), mink, and several other species (Hill and Novakowski 1984; Novak 1987). In addition, the value of beaver pelts varies by region, with those from warmer southern climates valued lower and those of colder northern climates valued higher. Pelt prices are a primary incentive for most trappers. Prices influence annual harvest levels and the ability of resource managers to use recreational trapping as a tool to reduce beaver populations or remove unwanted individuals.

Strict trapping regulations or trapping bans can limit beaver trapping as a management tool. In recent years, recreational trapping has been the target of animal rights groups, which have successfully used the public initiative/ballot referendum process in some states (e.g., Amendment 14 in Colorado, November 1996) to eliminate or severely curtail trapping of furbearers. The long-term implications of trapping bans on beaver populations in these areas remain to be determined. One possible outcome may be that in some areas where private landowners once considered beaver a valuable fur resource they may instead be viewed as a pest species, with negative consequences to populations (Hill 1982). Regulations that control or ban recreational trapping in many jurisdictions do not apply to landowners who are protecting private property.

Beaver Damage. Expanding populations of humans and beaver inevitably lead to property damage. From the point of view of a landowner or other person experiencing damage, the cause is "nuisance beaver"; however, from the point of view of a beaver or some concerned citizens, the cause is "nuisance people." From either view, beaver damage can be prevalent, severe, and costly.

Beaver damage varies by type, magnitude, and region. Common complaints regarding beaver damage include flooding of roads (often by plugging culverts) and pastures, eating or flooding agricultural crops (e.g., corn), damage to timber by flooding and cutting (mainly in the relatively flat Southeast), burrowing damage to dikes, ditches, and dams (mainly the arid West), cutting or flooding ornamental plants around homes or businesses, flooding habitat of rare plants or animals, damaging wild trout habitat, damaging fish ponds by plugging the overflow pipe, and potentially increasing the risk of human infection via *Giardia lamblia,* although some data suggest this is unlikely (Hill 1982; Woodward et al. 1985; Hammerson 1994; Olson and Hubert 1994). Annual cost of damage can be very high. In the early 1980s, annual beaver damage in the United States was estimated at $75–100 million; in the Southeast, the 40-year cumulative damage was estimated at $4 billion (review by Novak 1987).

The importance of timber damage in the southeastern states was recognized soon after beaver were reintroduced; timber damage there far exceeds all other types of complaints (review by Hill 1982). Timber damage was reported by 67% of respondents to landowner questionnaires in Alabama (Hill 1976). In South Carolina, a 1984 survey showed beaver flooded >35,000 ha of timber, often flooding trees during the entire growing season with low dams <0.5 m high (Woodward et al. 1985). If the root systems remain inundated for one to two growing seasons, many species of tree usually will die. Beaver kill timber of all size classes, with loss of larger trees causing greater financial impact on producers. Beaver frequently gnaw bark from hardwoods, which increases the risk of disease and subsequent rotting.

Damage Control. Killing unwanted individual beavers and controlled harvests of beaver populations are the most common, and often the most effective, methods of reducing beaver damage, even though lethal control is becoming increasingly less acceptable to the public. In Mississippi, about 20,000 beaver are harvested annually, with 80% of the carcasses discarded without use of meat or fur (Schulte and Muller-Schwarze 1999). Nonetheless, annual harvest may be the most prudent approach to animal damage problems, particularly where such harvest can involve citizen participants with no resultant public expense (Hill 1974, 1976). Trapping techniques for taking beaver vary with trapper preference and climatic conditions, the greatest differences occurring between areas that have extremely thick ice and those that are ice free. Successful harvest techniques involve shooting, snaring, and trapping with either No. 3 or No. 4 leghold traps or No. 330 Conibear traps (Hill 1982; Miller and Yarrow 1994). Shooting beaver from boats or from land may or may not be an effective control method (Hill 1982; Olson and Hubert 1994) and raises significant safety concerns. Alligators (*Alligator mississippiensis*) have been evaluated as a control method in the Southeast, but are generally not effective (Hill 1982). Poison bait substances, such as strychnine alkaloid baits and compound 1080, have been evaluated as a lethal control method, but are not approved for this purpose. They also pose political and practical problems, which will likely preclude their development (Hill 1982; Hammerson 1994). Thus, if the decision is to kill problem beaver, then trapping likely remains the most effective method.

However, trapping beaver can be an ineffective control measure if landowners create a sink population by trapping beaver from their own land but fail to reach consensus for control among adjacent landowners containing a source population of beaver. Even when consensus for control is reached, migration of beaver from less-controlled sites to more-controlled sites imposes an external cost (negative diffusion externality) when landowners must incur the future cost of repeated trapping (Bhat et al. 1993). When attempting to minimize beaver damage to timber resources, the cost of trapping must be weighed against the benefit of increased timber production. Integrating a trapping plan with a timber plan via a cost-minimizing, area-wide bioeconomic model may help strike the optimal balance between timber damage and trapping costs and maintain a more even density and distribution of beaver in the management area (Bhat et al. 1993).

Timing of trapping is also important where inundation of timber is the primary problem. In this case, it may be most effective to eradicate problem beaver colonies by the end of May and continue to remove immigrants from June to September, or until dispersal rates are relatively low (Houston et al. 1995). In some cases, trees may tolerate relatively high levels of beaver damage without any major effect on the forest. In a study of beaver damage to bald cypress (*Taxodium distichum*) trees in Texas, 85% of the trees near lodges had their bark peeled by beaver with only minimal effects on tree survival (King et al. 1998). Thus, managing timber damage requires understanding the response of the particular tree species to foraging and flooding by beaver, gaining cooperation of all landowners within the affected area, and evaluating costs and benefits of beaver control relative to timber production.

Livetrapping and relocation of problem beaver is more acceptable to the general public and has become fairly popular in urban areas. However, this can be cost prohibitive without volunteer labor and is not an option if suitable relocation sites are not available (Hammerson 1994). Moreover, released beaver may suffer high mortality from predation or from their inability to successfully assimilate into established populations. Hancock traps are usually preferred to Bailey traps for removing problem beaver, although properly used snares may be more effective and less costly than either trap type (Hill 1982; Novak 1987; Hammerson 1994; McKinstry and Anderson 1998). In some situations, hand-held nets can help capture beaver on land or in shallow water from boats (Rosell and Hovde 2001). Chemosterilents and surgical sterilization also have been evaluated for beaver control, but remain impractical for treating wild populations (reviews by Hill 1982; Hammerson 1994).

Beaver repellents or wire fencing may be an effective means of protecting plants from cutting or discouraging beaver occupancy of selected sites. Tree trunks can be protected by hardware cloth, chain link, or similar wire mesh connected at the ends (leaving room to allow tree growth) and extending about 1 m above the ground (Hammerson 1994). A solution of 10% creosote and 90% diesel fuel sprayed or painted on tree trunks reduces gnawing damage by beaver, as does a mixture of acrylic paint and sand, which acts as an unpalatable abrasive. Commercial deer repellents (Thiram, Magic Circle) may also be effective chemical repellents (Hammerson 1994). Chemical extracts from native tree species that beaver avoid (Jeffery pine) also may be effective as a beaver repellent (Basey 1999). Marking areas with beaver scent (artificial scent mounds) may discourage occupancy in the short term, but this method is likely an ineffective long-term solution. Scent marking by beaver more likely acts as a form of territoriality rather than as a scent fence (Welsh and Muller-Schwarze 1989; Sun and Muller-Schwarze 1998; Schulte and Muller-Schwarze 1999).

Flooding by beaver can often be prevented or managed by installation of beaver-proof road culverts and other water control structures or by installation of flow control devices in existing beaver dams (Fig. 15.3). Damage occurs when beavers plug culverts or impound water against the beds of roads or railroads, causing flooding or washouts. Techniques for minimizing or preventing beaver from plugging road culverts include using oversized pipe-arch culverts (Fig. 15.3A), low-profile box culverts (Fig. 15.3B), and various designs of beaver exclosure fencing (Fig. 15.3C) which prevent beaver from building a dam at the upstream end of a road culvert (Buech 1985; Hammerson 1994; Olson and Hubert 1994; Jensen et al. 1999; S. Lisle, oral communication, Beaver and Common-Sense Conflict Solutions Conference, 1999). In most cases, preventing damage with proper culvert design is far more cost-effective than the repeated, labor-intensive unplugging of dammed culverts or removing unwanted beaver (Jensen et al. 1999). Flow control devices installed in existing beaver dams have been very effective at mitigating damage to roads or other developments while maintaining the ecological and esthetic values of beaver presence. For example, the Clemson beaver pond leveler (Fig. 15.3D) can control the water level in beaver ponds by using tubes or similar structures laid perpendicular to the dam with the upstream end porous and protected from plugging by beaver (Buech 1985; Hammerson 1994; Olson and Hubert 1994). The most widely used method to control flooding is to dynamite or otherwise remove problem beaver dams. However, this method is usually the least effective because beaver will often rebuild dams within a few days if building materials are still available. Altering habitat to make sites unsuitable to beaver may be possible in some cases, but may require habitat changes that are unacceptable to landowners.

Methods for controlling beaver or their damage often require special training, materials, and permits. State and local regulations should be consulted before attempting to control populations or to mitigate damage, such as by installing water control structures that affect in-stream flow.

Public Opinion. Public opinion is an important consideration when choosing methods to control beaver damage. Many people have an emotional attachment to beaver and are vehemently opposed to any control methods that may cause their pain, suffering, or death. However, attitudes toward beaver control methods vary greatly among individuals, communities, regions, and type of land tenure. Rural residents, especially in agricultural areas, may have less opposition to lethal control than urban residents. People who experience beaver problems may accept harsher control than those without problems. Cultural values attributed to the presence of beaver also vary by region. Beaver may be more socially and economically valued in regions of Canada and the northern United States than in the Southeast, where fur values are lower, damage to timber and development is more severe, and beaver have less historical and cultural value.

Social acceptability of beaver control depends on many factors. Residents in suburban areas of New York were willing to accept increasingly invasive control when they had increasingly severe concerns about problems (Loker et al. 1999). Contrary to predictions, residents were willing to use more severe control measures for nuisance and economic damage problems than for concerns about public health and safety (results combined for white-tailed deer, Canada geese, and beaver). In another New York survey, nearly 50% of stakeholders took lethal actions to solve beaver problems; highway superintendents were more likely than the general public to attempt to solve beaver problems and to use actions that were nonlethal but invasive (Enck et al. 1997). In Wyoming, managers of private and public lands were concerned that beaver caused problems when they blocked irrigation ditches, girdled timber, blocked culverts, and flooded pastures, roads, crops, and timber. However, 45% of private landowners with beaver on their land and all public land managers were interested in the proactive use of beaver for riparian management (McKinstry and Anderson 1999).

Population and Harvest Management. Do beaver populations "need" to be harvested so they do not "eat themselves out of house and home ?" This management philosophy is pervasive in the culture of laypeople and professional wildlife managers. But is it true ? Why would beaver need the help of humans to control their populations, whereas nongame mammals seem to manage just fine without it (aside from pervasive anthropogenic stressors) ? How does the absence of wolves or other predators affect beaver carrying capacity ? Clearly, answers to these questions are beyond the scope of this chapter, but recent evidence provides some interesting discussion points. For example, densities of beaver colonies did not differ in a comparison of unexploited, saturated populations and exploited, thinned-out populations. Beaver in the saturated populations, however, colonized steeper gradients, had families with a larger percentage of 3-year-olds, depleted preferred trees and fed on less palatable species, and extended trails to more distant foraging sites (Muller-Schwarze and Schulte 1999). Because the maximum density of colonies (usually a minimum distance of about 0.9 km between colonies) may be controlled by territorial behavior rather than by habitat conditions, beaver populations in optimum habitat may become saturated or self-limiting below habitat-based carrying capacity. Beaver typically maintain intercolony distances and body weight regardless of population density. In some areas, forage near the central place may become temporarily depleted or ponds may become silted-in, causing colonies to periodically move to former beaver habitat that has

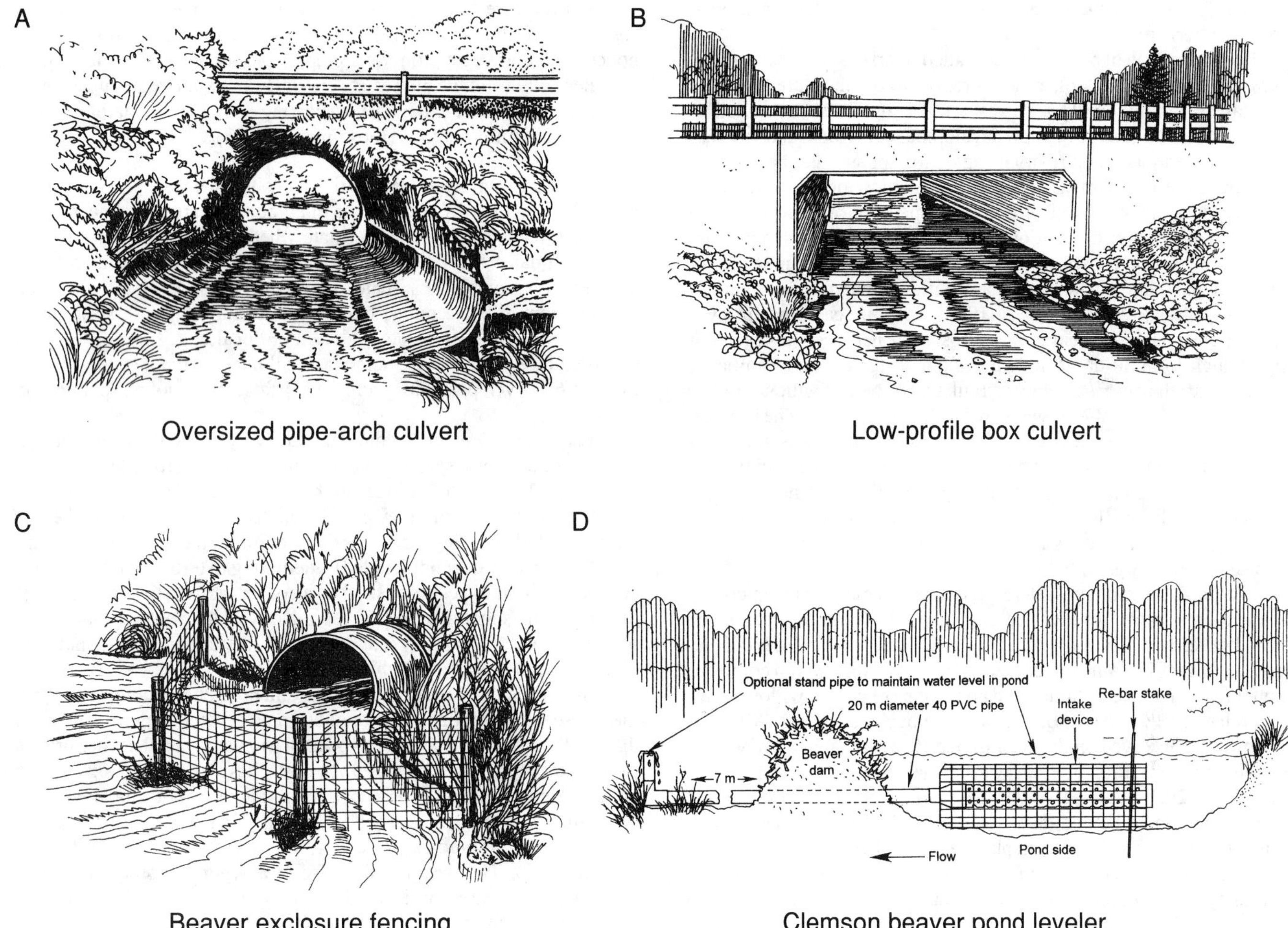

FIGURE 15.3. Examples of water control structures used to manage beaver (*Castor canadensis*) impacts. SOURCE: Adapted in part from Jensen et al. (1999).

recovered or to new habitat. In other areas, beaver may persist indefinitely within the same stream reach. In the Rocky Mountains, beaver may temporarily abandon mixed aspen–conifer sites after removing available aspen from nearby uplands, but may indefinitely occupy riparian–willow sites, where dams create moist, bare soil and a high water table, which helps establish and perpetuate the willow community. Despite lack of any inherent control needs, beaver harvest provides valuable recreational and income opportunities for trappers and is the primary means of reducing beaver damage to timber and other resources. Thus, justification for beaver harvest should be based on the economic value of fur trapping or "managerial decisions [that] address conflicts between beaver and humans, and not necessarily any requirements of the beaver themselves" (Muller-Schwarze and Schulte 1999:176).

Beaver populations are managed by state, provincial, and territorial wildlife agencies in the United States and Canada. Agencies are responsible for setting seasons, setting bag limits or area-specific quotas, licensing trappers, stamping or tagging pelts, licensing fur dealers and auction houses, and enforcing laws and regulations (Novak 1987). Trapper education is an important part of beaver management. Trappers in the field influence the age, sex, and distribution of animals removed as they make decisions about how, when, and where to set and check traps. Harvest rates, or quotas, are based on allowable harvest to maintain sustained yield. Rates suggested by different studies have varied from 10% to 70%, depending on habitat type, elevation, and region. The most typical recommended annual harvest rate is 20–30% of the population, which is about 1.0–1.5 beaver/colony/year. Beaver are highly vulnerable to overharvest, so managers must closely monitor populations. Their slow rate of reproduction and delayed sexual maturity preclude reproduction as a means to offset intensive annual harvest.

Regional differences among habitat types and land management prescriptions warrant regional beaver management plans (Snodgrass 1997). Managing beaver by managing their food supply may be possible in some regions. For example, prescribed burning may encourage aspen production in mixed conifer habitat, but may be of limited value where multiple forage species are available.

Using Beaver for Habitat Restoration and Improvement. The ability of beaver to store water, trap sediment, reduce channel erosion, and enhance establishment and production of riparian vegetation can be used as a proactive management tool to restore degraded riparian habitat. Beaver were abundant in forested, shrub–steppe, and some hot desert habitats in the western United States until fur trapping decimated populations. Ranchers followed trappers in settlement of the West, and immense numbers of sheep and cattle subjected newly beaver-free riparian areas to intense overgrazing. Overgrazing stripped streambanks of soil-binding vegetation, and channels responded with accelerated erosion and severe downcutting (see Elliott et al. [1999] for a discussion of possible mechanisms to explain observed channel incision). Mechanical restoration and revegetation (willow, cottonwood) of incised channels can be expensive, labor-intensive, and often unsuccessful. Thus, natural restoration of riparian systems can be an attractive alternative.

Reintroduction of beaver into degraded riparian systems has shown promise as a restoration tool, even where willow or other suitable winter food may be lacking (Fig. 15.4). Livestock grazing must be managed before beaver reintroduction to allow development of an adequate biomass of herbaceous aquatic and riparian vegetation for summer beaver food and to permit establishment and growth of willow or other woody riparian vegetation for winter beaver food. Aspen, cottonwood, or willow can be provided at reintroduction sites, or where beaver have initiated dam building on their own, to encourage beaver to remain at the site and to provide them with stronger dam-building material, which might otherwise be lacking (Apple et al. 1985). In some cases where overwinter food is lacking, beaver may subsist on herbaceous vegetation long enough to build dams and ponds. These features may initiate a positive riparian response, which stimulates growth of willow or other woody vegetation suitable as winter food. A winter food cache provides beaver with more permanent habitat and the ability to successfully raise young. Thus, beaver can create a positive feedback mechanism by temporarily expanding into marginal habitat (naturally or by introduction), improving conditions for the establishment and survival of woody riparian vegetation, and persisting long enough to raise young, which can disperse to new marginal habitat.

Beaver restoration in western riparian areas may also help control tamarisk, an invasive woody species (Fig. 15.4). In northwestern Colorado, beaver used tamarisk, sagebrush, and greasewood (*Sarcobatus vermiculatus*) as building material for a series of dams, which coincided with increase in the distribution and abundance of coyote willow (*S. exigua*) relative to tamarisk (B. W. Baker, unpublished data). A similar response was observed on the Zuni Indian Reservation in New Mexico following the relocation of 23 beaver to seven restoration sites (Albert and Trimble 2000). As beaver selectively cut vegetation and impound water and sediment behind dams, they alter conditions driving establishment and survival in riparian plant communities. Thus, beaver may create a competitive advantage for willow relative to tamarisk in some riparian systems, although specific mechanisms need further study at various spatial and temporal scales (B. W. Baker, unpublished data).

Beaver have also been useful as a timber management tool in the southeastern United States. Flooding by beaver ponds kills existing noncommercial vegetation, thus preparing sites for reforestation with commercially valuable timber. By draining ponds and removing beaver, landowners can reduce the cost of clearing land, often a major component of timber production (Houston et al. 1992).

Population Estimation. Population density of beaver is usually expressed as number of colonies or individuals in areal or linear units. Areal estimates are more appropriate where wetlands are diffuse and linear estimates are better where beaver habitat is limited to well-defined watercourses. Estimates of number of individuals are often derived from colony counts, but are not meaningful unless the mean number of beaver per colony is based on local data and not the general literature. Multiplying number of colonies by the general average of 5 or 6 beaver per colony only adds false precision to population estimates.

Estimates of mean colony size are very difficult to obtain and vary temporally and spatially, but are important in setting harvest quotas (Novak 1987). Colony size can be estimated by using night-vision scopes to count beaver as they move about their territory, driving beaver from their lodges using smoke or dogs, draining the pond and dismantling the lodge (not recommended), attempting to trap all the beaver in a colony (difficult to accomplish, thus a conservative estimate), and employing models that use age and reproductive data from populations of trapped beaver (Novak 1987) or data on the interactions among natality, mortality, and dispersal (Bishir et al. 1983).

Size of the food cache may be a useful predictor of colony size, but more research is needed to better understand relationships, as studies in different regions have been equivocal. In Montana, aerial cache surveys of prairie rivers located about 90% of caches, but colony size varied among years and areas. Thus, cache counts alone are not good predictors of population size or trend (Swenson et al. 1983). Estimates of cache size were a significant predictor of colony size in a Montana study (Easter-Pilcher 1990). This was not the case in a Wyoming study, however, perhaps due to variation in cache-building behavior among different age cohorts (Osmundson and Buskirk 1993).

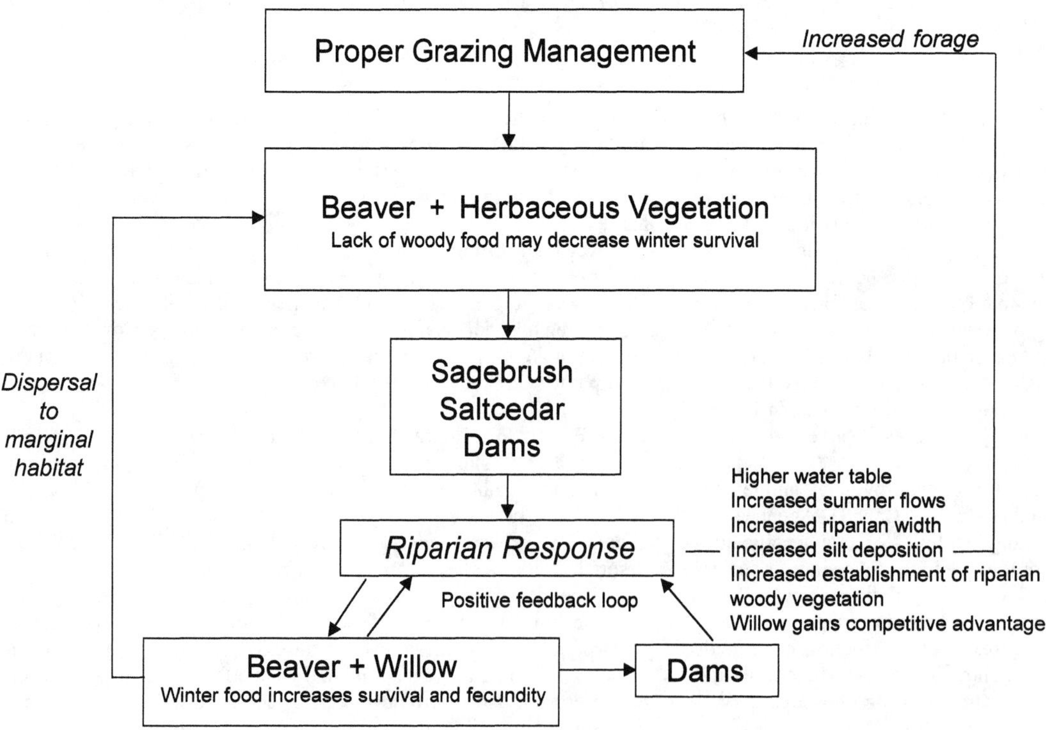

FIGURE 15.4. A conceptual model of the use of the beaver (*Castor canadensis*) activity as a possible mechanism for riparian restoration in shrub–steppe ecosystems.

Aerial surveys have been widely used to estimate the size and distribution of beaver populations, but evaluations of their efficiency, accuracy, and precision in terms of costs and benefits suggest their value is highly variable (Novak 1987). Surveys are usually conducted in the fall after leaf drop but before freeze-up. Results depend on search methods, topography, overstory vegetation, type of aircraft, and behavior of beaver (reviews by Hill 1982; Novak 1987). For example, aerial surveys only located 41 of 146 ground-located beaver colonies in Kansas riverine habitat, where beaver lived in bank dens and did not build food caches (Robel and Fox 1993).

Various combinations of aerial photography, aerial videography, and geographic information systems (GIS) can also be used to survey beaver populations and evaluate habitat conditions. Aerial photographs can be used as a reconnaissance aid to locate beaver habitat in remote locations for later ground or aerial survey; to map locations of dams, ponds, and lodges via photointerpretation methods; to find and/or plot colony locations during ground surveys (<1:3000-scale best); and to create overlays of beaver colony locations and habitat features that can be digitized into a GIS (Parsons and Brown 1978; Novak 1987; Baker et al. 1992; Werth and Boyd 1997). A GIS is essential for some types of landscape-level analysis of beaver habitat and may be better than manual methods for others, depending on costs and technical requirements (Johnston and Naiman 1990c). Global positioning systems (GPS) allow users to plot exact locations of active or inactive lodges and dens, which can be integrated with GIS systems to develop highly accurate beaver habitat models or monitoring programs. Aerial videography via a helicopter-mounted video camera may be useful for locating and monitoring beaver populations, especially in meandering riparian systems that are difficult to observe from low-level, fixed-winged aircraft (B. W. Baker, pers. obs.). Video has the advantage of instant availability on return from the flight and can be integrated with on-board GPS systems to create georeferenced data for computer analysis; however, videography is less useful than aerial photography for creating images for use in the field.

Sex Determination. Sex of beaver carcasses can be determined by the presence of the baculum and testes in males and the uterus in females, but the lack of any obvious external sex characteristics makes sex determination in live animals difficult (Novak 1987). The sex of live adult beaver can be determined by probing the cloaca or palpating the abdominal region for the baculum and testes, by noting the presence of four enlarged pectoral mammae in pregnant females (after 2 months), by giving beaver the anesthetic combelen (which causes the penis to lapse into the cloaca), by detecting Barr bodies in blood smears, and by using radiographs that reveal the presence of the baculum (review by Novak 1987). Each of these methods has some level of uncertainty or practical constraints for adult beaver and is especially difficult for young (reviews by Novak 1987; Schulte et al. 1995). A recently developed technique uses the color and viscosity of anal gland secretion to determine the sex of adult and juvenile beaver. This method produced relatively accurate and consistent results from three different regions in New York, which suggests it may have general field application for sex identification in beaver (Schulte et al. 1995).

Age Estimation. Characteristics of tooth eruption and annual cementum layers have been in standard use for age determination since the 1960s (reviews by Hill 1982; Novak 1987). A refinement in the cementum layer method used the independent variables cementum length and noncementum length to fairly accurately predict ages for a sample of 28 beaver (≥4 years old) with an exponential model, suggesting practical application in cases where more exact data are not necessary (Houston and Pelton 1995). Techniques for estimating age of live beaver are generally less accurate than the cementum method used for carcasses. The most common methods have used weight and skull measures, despite large error rates (Novak 1987; Hartman 1992). A refinement using regression models to estimate the age of live beaver based on live weight, tail width, and tail length was useful in identifying three age classes: kits, yearlings, and ≥ 2-year olds (Van Deelen 1994). A technique using dental radiographs to observe tooth root closure and annual deposition of cementum and dentine layers appears useful for estimating age of either live or dead beaver (Hartman 1992).

Livetrapping. Beaver can be captured alive using Hancock traps, Bailey traps, box traps, snares, or nets. Hancock traps are set on the sides of steep banks and Bailey traps are set in shallow water. Both are suitcase-type traps, which hold the beaver in a wire cage above water until released. Hancock traps are usually more effective than Bailey traps and are preferred in most situations (Novak 1987). Snares are cheap, light, and easy to set. They permit trappers to increase capture rates by saturating an area with traps, but have a greater mortality rate than Hancock traps due to increased risk of predation, entanglement leading to suffocation, and drowning (Hill 1982; McKinstry and Anderson 1998). Box traps baited with aspen and scent lure were effective for all age classes in spring and early summer when set along shoreline travel lanes in Massachusetts (C. Henner, pers. commun., 2001). In all cases, multiple traps per trapnight percolony will increase the likelihood of trapping all colony members or specific targeted individuals before colony members become trap shy. In some cases, beaver can be captured by nets or by hand after being flushed from lodges or dens by dogs or other means (Hill 1982; Rosell and Hovde 2001). In large river systems or lakes, surface-swimming beaver can be captured using a dip net. A dive net can be pushed down over the top of swimming beaver in shallow water, after locating them with the aid of spotlights and headlights from a motorboat (Rosell and Hovde 2001).

Tagging. Beaver can be individually marked with ear tags, neck collars, tail tags, and other methods (Novak 1987). Beaver may retain ear tags for several years, but retention rates have varied greatly among studies (review by Novak 1987). The tail can be marked with holes, notches, rivets, waterproof paint, branding, cryobranding with liquid nitrogen, and cattle ear tags (B. W. Baker, pers. obs.), and the fur can be bleached or dyed. Hind feet webs also can be tattooed or punched to individually mark beaver. Subcutaneously placed passive integrated transponder tags can positively identify recaptured beaver as well as free-ranging beaver, which can be scanned with a tag reader as they enter or exit burrows or lodges (although this method has not been field tested). These techniques all vary in their effectiveness and in trauma caused to beaver during and after marking, important considerations when selecting an appropriate marking technique.

Telemetry. Beaver are nocturnal and difficult to observe, so telemetry is often valuable in studies of their behavior and movements. However, transmitter attachment has been problematic. Neck collars often slip off because beaver have a thick neck relative to head size, and tail collars slip off if tail size decreases following release (B. W. Baker, pers. obs.). Free-floating intraperitoneal transmitter implants have proven successful in several studies, although their range can be relatively short (ground-to-ground signal of 0.1 km for beaver inside burrows and 2 km for active beaver; Davis et al. 1984). Implanted transmitters become tightly encapsulated with necrotic fibrous tissue and may be well tolerated by beaver (Guynn et al. 1987). Implants have been performed via dorsal (Davis et al. 1984) and ventrolateral (Wheatley 1997d) incisions. Although both techniques have been successful, the ventrolateral method appears to involve less risk to the animal.

Tail-mounted transmitters have shown variable success in Wyoming (Bothmeyer et al. 2002), Massachusetts (C. Henner, pers. commun., 2001), and Colorado (B. W. Baker, pers. obs.). This method uses a livestock ear tag transmitter attached to the dorsal surface of a beaver's tail. Dimensions and weight of tail-mounted transmitters can be designed to fit beaver tails of various sizes (kits, adults) and allow for variation in tail thickness. Durability and attachment effects (e.g., tearing the tail) of tail-mounted transmitters have yet to be determined, although preliminary evidence suggests short retention time for a relatively high percentage of individuals (B. W. Baker, pers. obs.).

Reintroduction. The first step in any reintroduction program should be to determine the purpose and feasibility of attempting to establish new beaver populations. This includes an understanding of why beaver are absent from the site, why they are important to the site, justification for relocation rather than natural dispersal, likelihood of movement beyond the intended relocation site, and potential conflicts or interaction with other resources or landowners. Relocation decisions and expectations are usually based on professional judgment after considering all available data, but also may be based on mathematical simulation models. South et al. (2000) developed a spatially explicit model to evaluate the proposed reintroduction of Eurasian beaver to Scotland. Their model simulated births and deaths of individuals and dispersal between habitat patches based on a land cover map and predicted that a reintroduction of 20 beaver would be sufficient to establish an initial population that would eventually expand and fill suitable habitat.

Most beaver do not stay at release sites and may move great distances following release. Average dispersal distances were 14.6 km in a North Dakota study, 16.7 km in a Colorado study, 11.7 km in a Maine study, 18 km in a Quebec study, and 7.4 km for beaver transplanted to streams and 3.2 km for beaver transplanted to lakes in a Wisconsin study (review by Novak 1987). Beaver can sometimes be encouraged to stay at specific release sites by providing aspen or other preferred species as food and building material or creating a base for a beaver dam by placing posts, wire fencing, silt-retaining fabric, rock, or other material (Apple et al. 1985; B. W. Baker, pers. obs.).

Survival of relocated beaver can depend on suitability of available habitat, timing of release (late summer in areas that freeze allows time to establish ponds and food cache), sex, age, composition, and number released (entire family units may improve establishment success), predation, and disease. In a study of 57 released Eurasian beaver in the Netherlands, 19 animals were found dead within the first year and 50% of these had died of infectious diseases (*Yersinia pseudotuberculosis, Y. enterocolitica,* Leptospirosis, and others) (Nolet et al. 1997). Stress during handling may increase risk of disease following release. Vaccination against likely disease agents and attention to animal hygiene before release may reduce mortality. Holding beaver in a cool, damp environment, providing adequate food and water, and minimizing length of captivity should improve survival following release (Novak 1987).

RESEARCH NEEDS

Beaver research has evolved from descriptions of beaver and their habitat to attempts to better understand mechanisms that explain behavior, population structure, and their keystone role in ecosystem structure and function. New knowledge about beaver will likely come from cross-discipline studies that stretch the boundaries of our already rich literature base on beaver. For example, a better understanding of the role of beaver as geological agents in the formation of alluvial floodplain valleys begs for the integration of the ecological sciences with new methods in geology and geomorphology, such as dating of buried organic layers that represent beaver dams and pond sediments. An important research need is to develop independent lines of evidence about how beaver affect ecosystem structure and function over the full range of ecological conditions inhabited by beaver, especially in the less well-known communities such as southeastern forests, western shrub–steppe, and desert grasslands. These studies should include the sequential events of development and abandonment that make up the life history of a beaver pond ecosystem. In South America and Europe, beaver are an invasive species with an ecological role that likely will require a different level of understanding than in their native North American habitat. The economic values of beaver wetlands in storing water, improving water quality, restoring wetland function, and mitigating development will be of critical importance as human populations expand and strain the land's ability to absorb their impact. The human dimensions of beaver management and control methods need improved understanding to better educate a public that is becoming more removed from the land and more inclined to use legislative or judicial means rather than the judgment of wildlife professionals to manage wildlife populations. Techniques need further refinement to more effectively estimate beaver populations, determine sex and age of live beaver in the field, and better use advanced telemetry as a research and management tool. For example, no radio-attachment technique has simultaneously solved problems of minimal effects to the animal, long retention time, long signal range, and long battery life.

The information presented in this chapter was based primarily on literature published since 1982. Readers are referred to the first edition of *Wild Mammals of North America* for a more thorough coverage of earlier work, and the following reviews and annotated bibliographies: Denney (1952), Yeager and Hay (1955), Jenkins and Busher (1979), Hodgdon and Larson (1980), Novak (1987), Medin and Torquemada (1988), and Olson and Hubert (1994).

ACKNOWLEDGMENTS

We are grateful to David Mitchell for compiling relevant literature, Dale Crawford for the preparing the artwork in Figures 15.3 and 15.4, and Stewart Breck, Doug Smith, and Butch Roelle for reviewing the manuscript. Funding was provided by the U.S. Geological Survey, Fort Collins Science Center, Fort Collins, Colorado.

LITERATURE CITED

Aeschbacher, A., and G. Pilleri. 1983. Observation on the building behaviour of the Canadian beaver (*Castor canadensis*) in captivity. Pages 83–98 *in* G. Pilleri, ed. Investigations on beaver, Vol. 1. Brain Anatomy Institute, Berne, Switzerland.

Albert, S., and T. Trimble. 2000. Beavers are partners in riparian restoration on the Zuni Indian Reservation. Ecological Restoration 18:87–92.

Aldous, S. E. 1938. Beaver food utilization studies. Journal of Wildlife Management 2:215–22.

Aleksuik, M. 1970a. The function of the tail as a fat storage depot in the beaver. (*Castor canadensis* Kuhl). Journal of Mammalogy 51:145–48.

Aleksiuk, M. 1970b. The seasonal food regime of arctic beaver. Ecology 51:264–70.

Aleksiuk, M., and I. Cowan. 1969. Aspects of seasonal energy expenditure in the beaver (*Castor canadensis*). Canadian Journal of Zoology 47:471–81.

Allen, A. W. 1983. Habitat suitability index models: Beaver. (FWS/OBS-82/10.30, rev.). U.S. Department of the Interior, Fish and Wildlife Service.

Apple, L. L., B. H. Smith, J. D. Dunder, and B. W. Baker. 1985. The use of beavers for riparian/aquatic habitat restoration of cold desert, gully-cut stream systems in southwestern Wyoming. Pages 123–30 *in* G. Pilleri, ed. Investigations on beaver, Vol. 4. Brain Anatomy Institute, Berne, Switzerland.

Baker, B. W., and B. S. Cade. 1995. Predicting biomass of beaver food from willow stem diameters. Journal of Range Management 48:322–26.

Baker, B. W., D. L. Hawksworth, and J. G. Graham. 1992. Wildlife habitat response to riparian restoration on the Douglas Creek watershed. Proceedings of the Annual Conference of the Colorado Riparian Association 4:62–80.

Barnes, D. M., and A. U. Mallik. 1996. Use of woody plants in construction of beaver dams in northern Ontario. Canadian Journal of Zoology 74:1781–86.

Barnes, D. M., and A. U. Mallik. 1997. Habitat factors influencing beaver dam establishment in a northern Ontario watershed. Journal of Wildlife Management 61:1371–77.

Barnes, W. J., and E. Dibble. 1988. The effects of beaver in riverbank forest succession. Canadian Journal of Botany 66:40–44.

Basey, J. M. 1999. Foraging behavior of beaver (*Castor canadensis*), plant secondary compounds, and management concerns. Pages 129–46 *in* P. E. Busher and R. M. Dzieciolowski, eds. Beaver protection, management, and utilization in Europe and North America. Kluwer Academic/Plenum, New York.

Basey, J. M., S. H. Jenkins, and P. E. Busher. 1988. Optimal central-place foraging by beaver: Tree size selection in relation to defensive chemicals of quaking aspen. Oecologia 76:278–82.

Basey, J. M., S. H. Jenkins, and G. C. Miller. 1990. Food selection by beavers in relation to inducible defenses of *Populus tremuloides*. Oikos 59:57–62.

Beier, P., and R. H. Barrett. 1987. Beaver habitat use and impact in Truckee River Basin, California. Journal of Wildlife Management 51:794–99.

Belovsky, G. E. 1984. Summer diet optimization by beaver. American Midland Naturalist 111:209–22.

Bergerud, A. T., and D. R. Miller. 1977. Population dynamics of Newfoundland beaver. Canadian Journal of Zoology 55:1480–92.

Bhat, M. G., R. G. Huffaker, and S. M. Lenhart. 1993. Controlling forest damage by dispersive beaver populations: Centralized optimal management strategy. Ecological Applications 3:518–30.

Bisaillon, A. 1981. Gross anatomy of the cardiac blood vessels in the North American beaver (*Castor canadensis*). Anatomischer Anzeiger 150:248–58.

Bisaillon, A. 1982. Anatomy of the heart in the North American beaver (*Castor canadensis*). Anatomischer Anzeiger 151:381–91.

Bishir, J., R. A. Lancia, and H. E. Hodgdon. 1983. Beaver family organization: Its implications for colony size. Pages 105–13 *in* G. Pilleri, ed. Investigations on beavers, Vol. 4, Brain Anatomy Institute, Berne, Switzerland.

Boyce, M. S. 1981a. Habitat ecology of an unexploited population of beaver in interior Alaska. Pages 155–86 *in* J. A. Chapman and D. Pursley, eds. Proceedings of the worldwide furbearer conference. Frostburg, MD.

Boyce, M. S. 1981b. Beaver life-history responses to exploitation. Journal of Applied Ecology 18:749–53.

Brady, C. A., and G. E. Svendsen. 1981. Social behavior in a family of beaver, *Castor canadensis*. Biology of Behaviour 6:99–114.

Breck, S. W. 2001. The effects of flow regulation on the population biology and ecology of beavers in northwestern Colorado. Ph.D. Dissertation, Colorado State University, Fort Collins.

Brown, D. J., W. A. Hubert, and S. H. Anderson. 1996. Beaver ponds create wetland habitat for birds in mountains of southeastern Wyoming. Wetlands 16:127–32.

Bryce, G. 1904. The remarkable history of the Hudson Bay Company. Reprint, 1968. Burt Franklin, New York.

Buech, R. R. 1984. Ontogeny and diurnal cycle of fecal reingestion in the North American beaver (*Castor canadensis*). Journal of Mammalogy 65:347–50.

Buech, R. R. 1985. Beaver in water impoundments: Understanding a problem of water-level management. Pages 95–105 *in* D. M. Knighton, Comp., Water impoundments for wildlife: A habitat management workshop (General Technical Report NC-100). U.S. Department of Agriculture, Forest Service.

Buech, R. R., and D. J. Rugg. 1995. Biomass of food available to beavers on five Minnesota shrubs (Research Paper NC-326). U.S. Department of Agriculture, Forest Service, North Central Forest Experiment Station.

Busher, P. E. 1983. Interactions between beavers in a montane population in California. Acta Zoologica Fennica 174:109–10.

Busher, P. E. 1987. Population parameters and family composition of beaver in California. Journal of Mammalogy 68:860–64.

Busher, P. E., and P. J. Lyons. 1999. Long-term population dynamics of the North American beaver, *Castor canadensis,* on Quabbin Reservation, Massachusetts, and Sagehen Creek, California. Pages 147–60 *in* P. E. Busher and R. M. Dzieciolowski, eds. Beaver protection, management, and utilization in Europe and North America. Kluwer Academic/Plenum, New York.

Butts, W. L. 1992. Changes in local mosquito fauna following beaver (*Castor canadensis*) activity: An update. Journal of the American Mosquito Control Association 8:331–32.

Cahn, A. R. 1932. Records and distribution of the fossil beaver, *Castoroides ohioensis*. Journal of Mammalogy 13:229–41.

Cottrell, T. R. 1995. Willow colonization of Rocky Mountain mires. Canadian Journal of Forest Research 25:215–22.

Davis, J. R., A. F. Von Recum, D. D. Smith, and D. C. Guynn, Jr. 1984. Implantable telemetry in beaver. Wildlife Society Bulletin 12:322–24.

Davis, J. R., D. C. Guynn, Jr., and G. W. Gatlin. 1994. Territorial behavior of beaver in the Piedmont of South Carolina. Proceedings of the Annual Conference of the Southeastern Association of Fish and Wildlife Agencies 48:152–61.

Davis, W. B. 1940. Critical notes on the Texas beaver. Journal of Mammalogy 21:84–86.

Deems, E. F., and D. Pursley. 1978. North American furbearers: Their management, research, and harvest status in 1976. University of Maryland Press, College Park, Maryland. 165 pp.

Denney, R. N. 1952. A summary of North American beaver management, 1946–1948 (Current Report 28). Colorado Game and Fish Department, Denver.

Dieter, C. D., and T. R. McCabe. 1989. Factors influencing beaver lodge-site selection on a prairie river. American Midland Naturalist 122:408–11.

Donkor, N. T., and J. M. Fryxell. 1999. Impact of beaver foraging on structure of lowland boreal forests of Algonquin Provincial Park, Ontario. Forest Ecology and Management 118:83–92.

Doucet, C. M., and J. M. Fryxell. 1993. The effect of nutritional quality on forage preference by beavers. Oikos 67:201–8.

Doucet, C. M., I. T. Adams, and J. M. Fryxell. 1994a. Beaver dam and cache composition: Are woody species used differently? Ecoscience 1:268–70.

Doucet, C. M., R. A. Walton, and J. M. Fryxell. 1994b. Perceptual cues used by beavers foraging on woody plants. Animal Behavior 47:1482–84.

Dyck, A. P., and R. A. MacArthur. 1992. Seasonal patterns of body temperature and activity in free-ranging beaver (*Castor canadensis*). Canadian Journal of Zoology 70:1668–72.

Dyck, A. P., and R. A. MacArthur. 1993. Daily energy requirements of beaver (*Castor canadensis*) in a simulated winter microhabitat. Canadian Journal of Zoology 71:2131–35.

Easter-Pilcher, A. 1990. Cache size as an index to beaver colony size in northwestern Montana. Wildlife Society Bulletin 18:110–13.

Elliott, J. G., A. C. Gellis, and S. B. Aby. 1999. Evolution of arroyos: Incised channels of the southwestern United States. Pages 153–85 *in* S. E. Darby and A. Simon, eds. Incised river channels: Processes, forms, engineering and management. John Wiley, New York.

Enck, J. W., N. A. Connelly, and T. L. Brown. 1997. Acceptance of beaver and actions to address nuisance beaver problems in New York. Human Dimensions of Wildlife 2:60–61.

Engelhart, A., and D. Muller-Schwarze. 1995. Responses of beaver (*Castor canadensis* Kuhl) to predator chemicals. Journal of Chemical Ecology 21:1349–64.

Fischer, T. V. 1985. The subplacenta of the beaver (*Castor canadensis*). Placenta 6:311–21.

Friley, C. E. Jr. 1949. Use of the baculum in age determination of Michigan beaver. Journal of Mammalogy 30:261–65.

Fryxell, J. M. 1992. Space use by beaver in relation to resource abundance. Oikos 64:474–78.

Fryxell, J. M., and C. M. Doucet. 1990. Provisioning time and central-place foraging in beaver. Canadian Journal of Zoology 69:1308–13.

Fryxell, J. M., S. M. Vamosi, R. A. Walton, and C. M. Doucet. 1994. Retention time and the functional response of beavers. Oikos 71:207–14.

Garrison, G. C. 1967. Pollen stratigraphy and age of an early postglacial beaver site near Columbus, Ohio. Ohio Journal of Science 67:96–105.

Grinnell, J., J. S. Dixon, and J. M. Linsdale. 1937. Fur bearing mammals of California, their natural history, systematic status, and relations to man. University of California Press, Berkeley.

Grover, A. M., and G. A. Baldassarre. 1995. Bird species richness within beaver ponds in south-central New York. Wetlands 15:108–18.

Gunson, J. R. 1970. Dynamics of the beaver of Saskatchewan's northern forest. M.S. Thesis, University of Alberta, Edmonton, Canada.

Gurnell, A. M. 1998. The hydrogeomorphological effects of beaver dam-building activity. Progress in Physical Geography 22:167–89.

Guynn, D. C., Jr., J. R. Davis, and A. F. Von Recum. 1987. Pathological potential of intraperitoneal transmitter implants in beaver. Journal of Wildlife Management 51:605–6.

Hakala, J. B. 1952. The life history and general ecology of the beaver (*Castor canadensis* Kuhl) in interior Alaska. M.S. Thesis, University of Alaska, Fairbanks.

Hall, E. R. 1981. The mammals of North America, Vol. 2, 2nd ed., John Wiley, New York.

Hammerson, G. A. 1994. Beaver (*Castor canadensis*): Ecosystem alterations, management, and monitoring. Natural Areas Journal 14:44–57.

Hartman, G. 1992. Age determination of live beaver by dental x-ray. Wildlife Society Bulletin 20:216–20.

Hilfiker, E. L. 1991. Beavers: Water, wildlife and history. Windswept Press, Interlaken, NJ.

Hill, E. P. 1974. Trapping beaver and processing their fur (Zoology and Entomology Department Series, Alabama Cooperative Wildlife Research Unit, No. 1). Agricultural Experiment Station, Auburn University, Auburn, AL.

Hill, E. P. 1976. Control methods for nuisance beaver in the southeastern United States. Pages 85–98 *in* Proceedings of the seventh vertebrate pest control conference.

Hill, E. P. 1982. Beaver (*Castor canadensis*). Pages 256–81 *in* J. A. Chapman and G. A. Feldhamer, eds. Wild mammals of North America. Johns Hopkins University Press, Baltimore.

Hill, E. P., and N. S. Novakowski. 1984. Beaver management and economics in North America. Acta Zoologica Fennica 172:259–62.

Hillman, G. R. 1998. Flood wave attenuation by a wetland following a beaver dam failure on a second order boreal stream. Wetlands 18:21–34.

Hodgdon, H. E. 1978. Social dynamics and behavior within an unexploited beaver (*Castor canadensis*) population. Ph.D. Dissertation, University of Massachusetts, Amherst.

Hodgdon, H. E., and R. A. Lancia. 1983. Behavior of the North American beaver, *Castor canadensis*. Acta Zoologica Fennica 174:99–103.

Hodgdon, H. E., and J. S. Larson. 1973. Some sexual differences in behavior within a colony of marked beaver (*Castor canadensis*). Animal Behavior 21:147–52.

Hodgdon, H. E., and J. S. Larson. 1980. A bibliography of the recent literature on beaver (Research Bulletin Number 665). Massachusetts Agricultural Experiment Station, University of Massachusetts, Amherst.

Hodgdon, K. W. 1949. Productivity data from placental scars in beaver. Journal of Wildlife Management 13:412–14.

Hodgdon, K. W., and J. J. Hunt. 1953. Beaver management in Maine (Maine Game Division Bulletin 3).

Houston, A. E., and M. R. Pelton. 1995. Premolar cementum and noncementum lengths as potential indicators of age for beavers, *Castor canadensis* (Rodentia: Castoridae). Brimleyana 22:67–72.

Houston, A. E., E. R. Buckner, and J. C. Rennie. 1992. Reforestation of drained beaver impoundments. Southern Journal of Applied Forestry 16:151–55.

Houston, A. E., M. R. Pelton, and R. Henry. 1995. Beaver immigration into a control area. Southern Journal of Applied Forestry 19:127–30.

Howard, R., and J. S. Larson. 1985. A stream habitat classification system for beaver. Journal of Wildlife Management 49:19–25.

Ingle-Sidorowicz, H. M. 1982. Beaver increase in Ontario. Result of changing environment. Mammalia 46:168–75.

Jellison, W. L., G. M. Kohls, W. J. Butler, and J. A. Weaver. 1942. Epizootic tularemia in the beaver, *Castor canadensis,* and the contamination of stream water with *Pasteurella tularensis*. American Journal of Hygiene 36:168–82.

Jenkins, S. H. 1979. Seasonal and year-to-year differences in food selection by beavers. Oecologia 44:112–16.

Jenkins, S. H. 1980. A size distance relation in food selection by beavers. Ecology 61:740–46.

Jenkins, S. H. 1981. Problems, progress, and prospects in studies of food selection by beaver. Pages 559–79 *in* J. A. Chapman and D. Pursley, eds. Proceedings of the worldwide furbearer conference. Frostburg, MD.

Jenkins, S. H., and P. E. Busher. 1979. *Castor canadensis*. Mammalian Species 120:1–8.

Jensen, P. G., P. D. Curtis, and D. L. Hamelin. 1999. Managing nuisance beaver along roadsides: A guide for highway departments. Media and Technology Services Resource Center, Cornell University, Ithaca, NY.

Johnston, C. A., and R. J. Naiman. 1987. Boundary dynamics at the aquatic–terrestrial interface: The influence of beaver and geomorphology. Landscape Ecology 1:47–57.

Johnston, C. A., and R. J. Naiman. 1990a. Aquatic patch creation in relation to beaver population trends. Ecology 71:1617–21.

Johnston, C. A., and R. J. Naiman. 1990b. Browse selection by beaver: Effects on riparian forest composition. Canadian Journal of Forest Research 20:1036–43.

Johnston, C. A., and R. J. Naiman. 1990c. The use of geographic information systems to analyze long term landscape alteration by beaver. Landscape Ecology 4:5–19.

Jones, C. G., J. H. Lawton, and M. Shachak. 1997. Positive and negative effects of organisms as physical ecosystem engineers. Ecology 78:1946–57.

Keast, A., and M. G. Fox. 1990. Fish community structure, spatial distribution and feeding ecology in a beaver pond. Environmental Biology of Fishes 27:201–14.

Kindschy, R. R. 1985. Response of red willow to beaver use in southeastern Oregon. Journal of Wildlife Management 49:26–28.

Kindschy, R. R. 1989. Regrowth of willow following simulated beaver cutting. Wildlife Society Bulletin 17:290–94.

King, S. L., B. D. Kneeland, and J. L. Moore. 1998. Beaver lodge distributions and damage assessments in a forested wetland ecosystem in the southern United States. Forest Ecology and Management 108:1–7.

Kowalski, K. 1976. Mammals, an outline of theriology. Polis, Warsaw.

Krueger, H. O. 1985. Avian response to mountainous shrub–willow riparian systems in southeastern Wyoming. Ph.D. Dissertation, University of Wyoming, Laramie.

Lahti, S., and M. Helminen. 1974. The beaver *Castor fiber* (L.) and *Castor canadensis* (Kuhl) in Finland. Acta Theriologica 19:177–89.

Lancia, R. A., W. E. Dodge, and J. S. Larson. 1982. Winter activity patterns of two radio-marked beaver colonies. Journal of Mammalogy 63:598–606.

Lancia, R. A., and H. E. Hodgdon. 1983. Observations on the ontogeny of behavior of hand-reared beaver (*Castor canadensis*). Acta Zoologica Fennica 174:117–19.

Landin, J. A. B. 1980. Estado actual del 'Castor' (*Castor canadensis mexicanus*) en el Estado de Nuevo Leon, Mexico. Pages 309–14 *in* Proceedings of the Inventarios de Recursos de Zonas Aridas. La Paz, B.C.S., Mexico [in Spanish with English summary].

Larson, J. S., and J. R. Gunson. 1983. Status of the beaver in North America. Acta Zoologica Fennica 174:91–93.

Lavrov, L. S., and V. N. Orlov. 1973. Karyotypes and taxonomy of modern beavers (*Castor,* Castoridae, Mammalia). Zoologicheskii Zhurnal 52:734–42.

Leidholt-Bruner, K., D. E. Hibbs, and W. C. McComb. 1992. Beaver dam locations and their effects on distribution and abundance of coho salmon fry in two coastal Oregon streams. Northwest Science 66:218–23.

Leopold, A. S. 1959. Wildlife of Mexico, the game birds and mammals. University of California Press, Berkeley.

Lizarralde, M. S. 1993. Current status of the introduced beaver (*Castor canadensis*) population in Tierra del Fuego, Argentina. Ambio 22:351–58.

Lizotte, R. E., Jr. 1994. Reproductive biology of beaver (*Castor canadensis*) at Old Hickory Lake in middle Tennessee. Journal of the Tennessee Academy of Science 69:23–26.

Loker, C. A., D. J. Decker, and S. J. Schwager. 1999. Social acceptability of wildlife management actions in suburban areas: 3 cases from New York. Wildlife Society Bulletin 27:152–59.

Mahoney, J. M., and H. I. Rosenberg. 1981. Anatomy of the tail of the beaver (*Castor canadensis*). Canadian Journal of Zoology 59:390–99.

Maret, T. J., M. Parker, and T. E. Fannin. 1987. The effect of beaver ponds on the nonpoint source water quality of a stream in southwestern Wyoming. Water Resources 21:263–68.

Martinsen, G. D., E. M. Driebe, and T. G. Whitman. 1998. Indirect interactions mediated by changing plant chemistry: Beaver browsing benefits beetles. Ecology 79:192–200.

McCall, T. C., T. P. Hodgman, D. R. Diefenbach, and R. B. Owen, Jr. 1996. Beaver populations and their relation to wetland habitat and breeding waterfowl in Maine. Wetlands 16:163–72.

McComb, W. C., J. R. Sedell, and T. D. Buchholz. 1990. Dam-site selection by beavers in an eastern Oregon basin. Great Basin Naturalist 50:273–81.

McDowell, D. M., and R. J. Naiman. 1986. Structure and function of a benthic invertebrate stream community as influenced by beaver (*Castor canadensis*). Oecologia 68:481–89.

McGinley, M. A., and T. G. Whitman. 1985. Central place foraging by beaver (*Castor canadensis*): A test of foraging predictions and the impact of selective feeding on the growth form of cottonwoods (*Populus fremontii*). Oecologia 66:558–62.

McKab, B. K. 1963. A model of the energy budget of a wild mouse. Ecology 44:521–32.

McKinstry, M. C., and S. H. Anderson. 1998. Using snares to live-capture beaver, *Castor canadensis*. Canadian Field-Naturalist 112:469–73.

McKinstry, M. C., and S. H. Anderson. 1999. Attitudes of private- and public-land managers in Wyoming, USA, toward beaver. Environmental Management 23:95–101.

McKinstry, M. C., P. Caffrey, and S. H. Anderson. 2001. The importance of beaver to wetland habitats and waterfowl in Wyoming. Journal of the American Water Resources Association 37:1571 77.

McRae, G., and C. J. Edwards. 1994. Thermal characteristics of Wisconsin headwater streams occupied by beaver: Implications for brook trout habitat. Transactions of the American Fisheries Society 123:641–56.

Medin, D. E., and W. P. Clary. 1990. Bird populations in and adjacent to a beaver pond ecosystem in Idaho (Research Paper INT-432). U.S. Department of Agriculture, Forest Service, Intermountain Research Station.

Medin, D. E., and W. P. Clary. 1991. Small mammals of a beaver pond ecosystem and adjacent riparian habitat in Idaho (Research Paper INT-445). U.S. Department of Agriculture, Forest Service, Intermountain Research Station.

Medin, D. E., and K. E. Torquemada. 1988. Beaver in western North America: An annotated bibliography, 1966 to 1986 (General Technical Report INT-242). U.S. Department of Agriculture, Forest Service, Intermountain Research Station.

Meentemeyer, R. K., and D. R. Butler. 1999. Hydrogeomorphic effects of beaver dams in Glacier National Park, Montana. Physical Geography 20:436–46.

Merendino, M. T., G. B. McCullough, and N. R. North. 1995. Wetland availability and use by breeding waterfowl in southern Ontario. Journal of Wildlife Management 59:527–32.

Miller, J. E., and G. K. Yarrow. 1994. Beaver. Pages B1–11 *in* S. E. Hygnstrom, R. M. Timm, and G. E. Larson, eds. Prevention and control of wildlife damage. University of Nebraska Press, Lincoln.

Molini, J. J., R. A. Lancia, J. Bishir, and H. E. Hodgdon. 1980. A stochastic model of beaver population growth. Pages 1215–45 *in* J. A. Chapman and D. Pursley, eds. Proceedings of the worldwide furbearer conference. Frostburg, MD.

Morner, T. 1992. The ecology of tularemia. Revue Scientifique et Technique O.I.E. (Office International des Epizooties) 11:1123–30.

Muller-Schwarze, D., and S. Heckman. 1980. The social role of scent marking in beaver (*Castor canadensis*). Journal of Chemical Ecology 6:81–95.

Muller-Schwarze, D., and B. A. Schulte. 1999. Behavioral and ecological characteristics of a "climax" population of beaver (*Castor canadensis*). Pages 161–77 *in* P. E. Busher and R. M. Dzieciolowski, eds. Beaver protection, management, and utilization in Europe and North America. Kluwer Academic/Plenum, New York.

Muller-Schwarze, D., S. Heckman, and B. Stagge. 1983. Behavior of free-ranging beaver (*Castor canadensis*) at scent marks. Acta Zoologica Fennica 174:111–13.

Muller-Schwarze, D., B. A. Schulte, L. Sun, A. Muller-Schwarze, and C. Muller-Schwarze. 1994. Red maple (*Acer rubrum*) inhibits feeding by beaver (*Castor canadensis*). Journal of Chemical Ecology 20:2021–34.

Naiman, R. J., and J. M. Melillo. 1984. Nitrogen budget of a subarctic stream altered by beaver (*Castor canadensis*). Oecologia 62:150–55.

Naiman, R. J., J. M. Melillo, and J. E. Hobbie. 1986. Ecosystem alteration of boreal forest streams by beaver (*Castor canadensis*). Ecology 67:1254–69.

Naiman, R. J., C. A. Johnston, and J. C. Kelley. 1988. Alteration of North American streams by beaver. BioScience 38:753–62.

Naiman, R. J., G. Pinay, C. A. Johnston, and J. Pastor. 1994. Beaver influences on the long-term biogeochemical characteristics of boreal forest drainage networks. Ecology 75:905–21.

Nolet, B. A., S. Broekhuizen, G. M. Dorrestein, and K. M. Rienks. 1997. Infectious diseases as main causes of mortality to beaver *Castor fiber* after translocation to the Netherlands. Journal of the Zoological Society of London 241:35–42.

Novak, M. 1987. Beaver. Pages 283–312 *in* M. Novak, J. A. Baker, M. E. Obbard, and B. Malloch, eds. Wild furbearer management and conservation in North America. Ontario Trappers Association and Ontario Ministry of Natural Resources.

Novakowski, N. S. 1965. Population dynamics of a beaver population in northern latitudes. Ph.D. Dissertation, University of Saskatchewan, Saskatoon, Canada.

Nummi, P. 1992. The importance of beaver ponds to waterfowl broods: An experiment and natural tests. Annales Zoologici Fennici 29:47–55.

Nummi, P., and H. Poysa. 1997. Population and community level responses in *Anas*-species to patch disturbance caused by an ecosystem engineer, the beaver. Ecography 20:580–84.

Olson, R., and W. A. Hubert. 1994. Beaver: Water resources and riparian habitat manager. University of Wyoming Press, Laramie.

Osmundson, C. L., and S. W. Buskirk. 1993. Size of food caches as a predictor of beaver colony size. Wildlife Society Bulletin 21:64–69.

Paine, R. T. 1969. A note on trophic complexity and community stability. American Naturalist 103:91–93.

Parker, M., F. J. Wood, Jr., B. H. Smith, and R. G. Elder. 1985. Erosional downcutting in lower order riparian ecosystems: Have historical changes been caused by removal of beaver? Proceedings of the North American Riparian Conference 1:35–38.

Parsons, G. R., and M. K. Brown. 1978. An assessment of aerial photograph interpretation for recognizing potential beaver colony sites. Transactions of the Northeast Fish and Wildlife Conference 35:181–84.

Patenaude, F., and J. Bovet. 1984. Self-grooming and social-grooming in the North American beaver, *Castor canadensis*. Canadian Journal of Zoology 62:1872–78.

Payne, N. F. 1975. Trapline management and population biology of Newfoundland beaver. Ph.D. Dissertation, Utah State University, Logan.

Payne, N. F. 1984. Population dynamics of beaver in North America. Acta Zoologica Fennica 172:263–66.

Potvin, F., L. Breton, C. Pilon, and M. Macquart. 1992. Impact of an experimental wolf reduction on beaver in Papineau-Labelle Reserve, Quebec. Canadian Journal of Zoology 70:180–83.

Power, M. E., D. Tilman, J. A. Estes, B. A. Menge, W. J. Bond, L. S. Mills, G. Daily, J. C. Castilla, J. Lubchenco, and R. T. Paine. 1996. Challenges in the quest for keystones. BioScience 46:609–20.

Provost, E. E. 1958. Studies on reproduction and population dynamics in beaver. Ph.D. Dissertation, State College of Washington, Pullman.

Provost, E. E. 1962. Morphological characteristics of the beaver ovary. Journal of Wildlife Management 26:272–78.

Rebertus, A. J. 1986. Bogs as beaver habitat in north-central Minnesota. American Midland Naturalist 116:240–45.

Reese, K. P., and J. D. Hair. 1976. Avian species diversity in relation to beaver pond habitats in the Piedmont Region of South Carolina. Proceedings of the Annual Conference of the Southeastern Association of Fish and Wildlife Agencies 30:437–47.

Retzer, J. L., H. W. Swope, J. D. Remington, and W. H. Rutherford. 1956. Suitability of physical factors for beaver management in the Rocky Mountains of Colorado (Technical Bulletin 2). Colorado Department of Game and Fish.

Reynolds, P. S. 1993. Size, shape, and surface area of beaver, *Castor canadensis,* a semiaquatic mammal. Canadian Journal of Zoology 71:876–82.

Ringelman, J. K. 1991. Managing beaver to benefit waterfowl (Fish and Wildlife Leaflet 13.4.7). U.S. Department of Interior, Fish and Wildlife Service, Washington, DC.

Robel, R. J., and L. B. Fox. 1993. Comparison of aerial and ground survey techniques to determine beaver colony densities in Kansas. Southwestern Naturalist 38:357–61.

Robel, R. J., L. B. Fox, and K. E. Kemp. 1993. Relationship between habitat suitability index values and ground counts of beaver colonies in Kansas. Wildlife Society Bulletin 21:415–21.

Roberts, T. H., and D. H. Arner. 1984. Food habits of beaver in east-central Mississippi. Journal of Wildlife Management 48:1414–19.

Robertson, R. A., and and A. R. Shadle., 1954. Osteologic criteria of age in beavers. Journal of Mammalogy 35:197–203.

Rosell, F., and B. Hovde. 2001. Methods of aquatic and terrestrial netting to capture Eurasian beavers. Wildlife Society Bulletin 29:269–74.

Rothmeyer, S. W., M. C. McKinstry, and S. H. Anderson. 2002. Tail attachment of modified ear-tag radio transmitters on beaver. Wildlife Society Bulletin 30:425–29.

Ruedemann, R., and W. J. Schoonmaker. 1938. Beaver-dams as geologic agents. Science 88:523–25.

Russell, K. R., C. E. Moorman, J. K. Edwards, B. S. Metts, and D. C. Guynn, Jr. 1999. Amphibian and reptile communities associated with beaver (*Castor canadensis*) ponds and unimpounded streams in the Piedmont of South Carolina. Journal of Freshwater Ecology 14:149–58.

Rutherford, W. H. 1955. Wildlife and environmental relationships of beaver in Colorado forests. Journal of Forestry 53:803–6.

Safonov, V. G. 1979. Experience of American beaver, *Castor canadensis,* introduction in Kamchatka. *In* Game management and fur farming USSR: international wildlife congress 14. All Union Research Institute [Abstract in English].

Schlosser, I. J. 1995. Dispersal, boundary processes, and trophic-level interactions in streams adjacent to beaver ponds. Ecology 76:908–25.

Schlosser, M. 1902. Extinct beaver (*Castor neglectus*) from Tertiary of South Germany. Palaeontoligische Abhandlungen 9:136.

Schulte, B. A. 1998. Scent marking and responses to male castor fluid by beavers. Journal of Mammalogy 79:191–203.

Schulte, B. A., and D. Muller-Schwarze. 1999. Understanding North American beaver behavior as an aid to management. Pages 109–128 *in* P. E. Busher and R. M. Dzieciolowski, eds. Beaver protection, management, and utilization in Europe and North America. Kluwer Academic/Plenum, New York.

Schulte, B. A., D. Muller-Schwarze, and L. Sun. 1995. Using anal gland secretion to determine sex in beaver. Journal of Wildlife Management 59:614–18.

Seton, E. T. 1929. Lives of game animals, Vol. 4, Part 2, Rodents, etc. Doubleday, Doran, Garden City, NY.

Sieber, J., F. Suchentrunk, and G. B. Hart. 1999. A biochemical–genetic discrimination method for the two beaver species, *Castor fiber* and *Castor canadensis,* as a tool for conservation. Pages 61–65 *in* P. E. Busher and R. M. Dzieciolowski, eds. Beaver protection, management, and utilization in Europe and North America. Kluwer Academic/Plenum, New York.

Skinner, Q. D., J. E. Speck, Jr., M. Smith, and J. C. Adams. 1984. Stream water quality as influenced by beaver within grazing systems in Wyoming. Journal of Range Management 37:142–46.

Slough, B. G. 1978. Beaver food cache structure and utilization. Journal of Wildlife Management 42:644–46.

Slough, B. G., and R. M. F. S. Sadleir. 1977. A land capability classification system for beaver (*Castor canadensis* Kuhl). Canadian Journal of Zoology 55:1324–35.

Smith, D. W., and S. H. Jenkins. 1997. Seasonal change in body mass and size of tail of northern beavers. Journal of Mammalogy 78:869–76.

Smith, D. W., R. O. Peterson, T. D. Drummer, and D. S. Sheputis. 1991. Overwinter activity and body temperature patterns in northern beavers. Canadian Journal of Zoology 69:2178–82.

Smith, D. W., D. R. Trauba, R. K. Anderson, and R. O. Peterson. 1994. Black bear predation on beavers on an island in Lake Superior. American Midland Naturalist 132:248–55.

Smith, M. E., C. T. Driscoll, B. J. Wyskowski, C. M. Brooks, and C. C. Cosentini. 1991. Modification of stream ecosystem structure and function by beaver

(*Castor canadensis*) in the Adirondack Mountains, New York. Canadian Journal of Zoology 69:55–61.

Snodgrass, J. W. 1997. Temporal and spatial dynamics of beaver-created patches as influenced by management practices in a south-eastern North American landscape. Journal of Applied Ecology 34:1043–56.

Snodgrass, J. W., and G. K. Meffe. 1998. Influence of beavers on stream fish assemblages: Effects of pond age and watershed position. Ecology 79:928–42.

South, A., S. Rushton, and D. Macdonald. 2000. Simulating the proposed reintroduction of the European beaver (*Castor fiber*) to Scotland. Biological Conservation 93:103–16.

Stegeman, L. C. 1954. The production of aspen and its utilization by beaver on the Huntington Forest. Journal of Wildlife Management 18:348–58.

Sun, L., and D. Muller-Schwarze. 1997. Sibling recognition in the beaver: A field test for phenotype matching. Animal Behavior 54:493–502.

Sun, L., and D. Muller-Schwarze. 1998. Anal gland secretion codes for family membership in the beaver. Behavioral Ecology and Sociobiology 44:199–208.

Sun, L., D. Muller-Schwarze, and B. A. Schulte. 2000. Dispersal pattern and effective population size of the beaver. Canadian Journal of Zoology 78:393–98.

Suzuki, N., and W. C. McComb. 1998. Habitat classification models for beaver (*Castor canadensis*) in the streams of the central Oregon coast range. Northwest Science 72:102–10.

Svendsen, G. E. 1978. Castor and anal glands of the beaver (*Castor canadensis*). Journal of Mammalogy 59:618–20.

Svendsen, G. E. 1980a. Patterns of scent-mounding in a population of beaver (*Castor canadensis*). Journal of Chemical Ecology 6:133–48.

Svendsen, G. E. 1980b. Seasonal change in feeding patterns of beaver in southeastern Ohio. Journal of Wildlife Management 44:285–90.

Svendsen, G. E. 1989. Pair formation, duration of pair-bonds, and mate replacement in a population of beavers (*Castor canadensis*). Canadian Journal of Zoology 67:336–40.

Swenson, J. E., S. J. Knapp, P. R. Martin, and T. C. Hinz. 1983. Reliability of aerial cache surveys to monitor beaver population trends on prairie rivers in Montana. Journal of Wildlife Management 47:697–703.

Terwilliger, J., and J. Pastor. 1999. Small mammals, ectomycorrhizae, and conifer succession in beaver meadows. Oikos 85:83–94.

Van Deelen, T. R. 1994. A field technique for aging live beavers, *Castor canadensis*. Canadian Field-Naturalist 108:361–63.

Van Deelen, T. R., and D. H. Pletscher. 1996. Dispersal characteristics of two-year-old beavers, *Castor canadensis,* in Western Montana. Canadian Field-Naturalist 110:318–21.

Vispo, C., and I. D. Hume. 1995. The digestive tract and digestive function in the North American beaver. Canadian Journal of Zoology 73:967–74.

Voight, D. R., G. B. Kolenosky, and D. H. Pimlott. 1976. Changes in summer foods of wolves in central Ontario. Journal of Wildlife Management 40:663–68.

Warren, E. R. 1940. A beaver's food requirements. Journal of Mammalogy 21:93.

Welsh, R. G., and D. Muller-Schwarze. 1989. Experimental habitat scenting inhibits colonization by beaver, *Castor canadensis*. Journal of Chemical Ecology 15:887–93.

Werth, L. F., and R. J. Boyd. 1997. Wildlife. Pages 495–516 *in* Manual of photographic interpretation, 2nd ed. American Society of Photogrammetry and Remote Sensing, Falls Church, VA.

Wheatley, M. 1997a. Beaver, *Castor canadensis,* home range size and patterns of use in the taiga of southeastern Manitoba: I. Seasonal variation. Canadian Field-Naturalist 111:204–10.

Wheatley, M. 1997b. Beaver, *Castor canadensis,* home range size and patterns of use in the taiga of southeastern Manitoba: II. Sex, age, and family status. Canadian Field-Naturalist 111:211–16.

Wheatley, M. 1997c. Beaver, *Castor canadensis,* home range size and patterns of use in the taiga of southeastern Manitoba: III. Habitat variation. Canadian Field-Naturalist 111:217–22.

Wheatley, M. 1997d. A new surgical technique for implanting radio transmitters in beaver, *Castor canadensis*. Canadian Field-Naturalist 111:601–6.

Wigley, T. B., T. H. Roberts, and D. H. Arner. 1983. Reproductive characteristics of beaver in Mississippi. Journal of Wildlife Management 47:1172–77.

Wigley, T. B., T. H. Roberts, and D. H. Arner. 1984. Methods of determining litter size in beaver. Proceedings of the Annual Conference of the Southeastern Association of Fish and Wildlife Agencies 38:197–200.

Wilsson, L. 1971. Observations and experiments on the ethology of the European beaver (*Castor fiber* L.). Viltrevy 8:115–266.

Winkle, P. L., W. A. Hubert, and F. J. Rahel. 1990. Relations between brook trout standing stocks and habitat features in beaver ponds in southwestern Wyoming. North American Journal of Fisheries Management 10:72–79.

Woodward, D. K. 1977. Status and ecology of the beaver (*Castor canadensis carolinensis*) in South Carolina with emphasis on the Piedmont. M.S. Thesis, Clemson University, Clemson, SC.

Woodward, D. K., R. B. Hazel, and B. P. Gaffney. 1985. Economic and environmental impacts of beavers in North Carolina. Proceedings of the Eastern Wildlife Damage Control Conference 2:89–96.

Yeager, L. E., and K. O. Hay. 1955. A contribution toward a bibliography on the beaver (Technical Bulletin 1). Colorado Department of Game and Fish.

Zeckmeister, M. T., and N. F. Payne. 1998. Effects of trapping on colony density, structure, and reproduction of a beaver population unexploited for 19 years. Transactions of the Wisconsin Academy of Sciences, Arts, and Letters 86:281–91.

Zurowski, W., J. Kisza, A. Kruk, and A. Roskosz. 1974. Lactation and chemical composition of milk of the European beaver (*Castor fiber* L.). Journal of Mammalogy 55:847–50.

Bruce W. Baker, U.S. Geological Survey, Fort Collins Science Center, 2150 Centre Avenue, Bldg. C., Fort Collins, Colorado 80526-8118. Email: Bruce_Baker@usgs.gov.

Edward P. Hill (retired), Mississippi Cooperative Fish and Wildlife Research Unit, Mississippi State University, Mississippi State, Mississippi 39762.

16

Muskrats

Ondatra zibethicus and *Neofiber alleni*

John Erb
H. Randolph Perry, Jr.

NOMENCLATURE

COMMON NAME. Muskrat
SCIENTIFIC NAME. *Ondatra zibethicus*
The muskrat is a semiaquatic gnawing mammal (Rodentia: Muridae). The rodents constitute the largest mammalian order in numbers of species and individuals.
SUBSPECIES Including the Newfoundland muskrat, formerly *Ondatra obscurus* (Pietsch 1970), there are 16 subspecies of *Ondatra zibethicus* in North America (Hall and Kelson 1959; Hall 1981): *O. z. zibethicus* (eastern United States, southeastern Canada), *O. z. albus* (Manitoba and adjacent central Canada), *O. z. aquilonius* (Labrador, adjacent Ungava and Quebec), *O. z. bernardi* (Colorado River areas of southeastern California, southern Nevada, western Arizona, and Mexico), *O. z. cinnamominus* (Great Plains region), *O. z. goldmani* (southwestern Utah, northwest Arizona, southeastern Nevada), *O. z. macrodon* (middle Atlantic coast), *O. z. mergens* (northern Nevada, parts of adjacent states), *O. z. obscurus* (Newfoundland), *O. z. occipitalis* (coastal Oregon and Washington), *O. z. osoyoosensis* (Rocky Mountains, southwestern Canada), *O. z. pallidus* (south-central Arizona, west-central New Mexico), *O. z. ripensis* (southwestern Texas, southeastern New Mexico), *O. z. rivalicius* (southern Louisiana, coasts of Mississippi, western Alabama, and eastern Texas), *O. z. spatulatus* (northwestern North America), and *O. z. zalophus* (southern Alaska); although morphologically similar, these subspecies vary widely as to population status, distribution, habits, and habitat; subspecific differences were discussed in detail by Errington (1963)

FIGURE 16.1. Distribution of the muskrat (*Ondatra zibethicus*) and round-tailed muskrat (*Neofiber alleni*).

DISTRIBUTION

Muskrats are indigenous, common, and widely distributed throughout most of North America. They occur throughout the greater part of the continent from northern Mexico to northern Alaska and northern Canada (Dozier 1953; Lowery 1974) (Fig. 16.1). Muskrats are absent from Florida, most of Mexico, and parts of extreme northern Alaska and Canada. Errington (1963) extensively discussed the range of the muskrat in North America.

Numerous authors described range extensions of the muskrat on the Pacific coast, especially into California and Oregon (Hansen 1965; Bleich 1974; Wood 1974). Twining and Hensley (1943) discussed the history of muskrats in California and listed localities inhabited.

Muskrat remains have been identified from Pleistocene fossils. For example, remains of *O. zibethicus* have been reported from Kansas (McMullen 1978), southern California (Langenwalter 1985), Louisiana (Arata 1964; Domning 1969), and South Carolina (Bentley et al. 1994). Bentley and Knight (1994) also made reference to Pleistocene remains from Iowa, Georgia, and Florida.

The muskrat is absent in some areas despite apparently suitable habitat, especially along the southeastern Atlantic coast (Dozier 1953; Wilson 1968). The absence of *O. zibethicus* from southern Georgia and Florida has been the subject of much discussion. Apparently, *O. zibethicus* did inhabit Florida during the Pleistocene (Sherman 1952; Bentley and Knight 1994). Errington (1963) noted that several authors believed the round-tailed muskrat (*Neofiber alleni*) replaced *O. zibethicus* in Florida. However, tidal fluctuations, habitat reduction by natural catastrophes, and scarcity of certain vegetative communities may limit muskrat distribution in the Southeast (Errington 1963). As discussed later, competition from introduced nutria (*Myocastor coypus*) may also limit muskrats in some areas.

Severe winters and lack of suitable habitat may prevent muskrats from inhabiting tundra. However, they have been harvested in the Northwest Territories at least 129 km beyond the tree line (Stewart et al. 1975). Similarly, Noble and Wright (1977) observed muskrats on an Alaskan tundra lake 160 km west of the tree line.

Muskrat translocations began after 1900. The early history of foreign and North American translocation efforts was described by Storer (1937) along with some of the resulting problems, especially in irrigated regions. No muskrats were introduced into the United States or Canada up to 1976 (Deems and Pursley 1978). They were, however, introduced into Europe in 1905 and today inhabit many parts of Eurasia (Danell 1978). We are unaware of any recent documented changes in the distribution of muskrats in North America.

DESCRIPTION

The muskrat is stocky, with a broad head, small ears, and small eyes; the neck is not distinctively restricted (Fig. 16.2). The legs are short and the tail is scaly, sparsely haired, and laterally flattened. Limbs are

FIGURE 16.2. The muskrat (*Ondatra zibethicus*) is a common inhabitant of many aquatic habitats. SOURCE: Photo by R. B. Forbes, Mammal Images Library, American Society of Mammalogists.

also modified for aquatic life. The front feet are unwebbed and have four toes, a rudimentary nailed thumb, and sharp claws. The five clawed toes of the hind feet are webbed proximally and have fringes of stiff hairs along the sides of the toes and the feet. Stein (1988) compared several aquatic adaptations of the muskrat with those of several other semiaquatic rodents; muskrats were the most aquatically adapted of those compared.

Scent Glands. The epithet "musky" refers to the typical odor during the breeding season. Paired glands, which enlarge during the mating period, emit a musky yellowish secretion through openings within the foreskin of the penis. This secretion mixes with the urine and is deposited throughout individual muskrat's territory on defecation posts, lodge bases, stations along travel routes, and during mating. Females have similar glands near their external genitalia, which are not as active as those of males (Schwartz and Schwartz 1959). Stevens and Erickson (1942) found that scent glands contain musk oil, a mixture of cyclopentadecanol and cycloheptadecanol and corresponding odiferous ketones. The scent remains viable long after exposure to air and serves as a means of advertisement (Errington 1963).

Size and Weight. Martin (1993) suggested that body mass of muskrats dramatically decreased 10,000 years ago. However, after analyzing additional data, Bentley and Knight (1994) concluded that Martin's results were likely an artifact of a small and geographically unrepresentative sample size. They concluded that further work is needed to elucidate any historic trends in body mass.

There is a small degree of sexual dimorphism in muskrats (Table 16.1), with males slightly heavier than females (Gould and Kreeger 1948; Boyce 1978). One notable reversal of the male:female weight relation was reported by Buss (1941): Based on 589 fall-trapped animals in Wisconsin, males averaged 42 g less than females.

Over large geographic areas, muskrats tend to be heavier in more northern latitudes (Boyce 1977, 1978). However, latitude may not be an exact correlate of body mass. Boyce (1978) proposed that small body size may be adaptive in highly seasonal environments and/or areas with diminished food quality or quantity. Presumably, smaller animals would have a lower absolute nutritional demand, thereby improving their ability to survive seasonal nutritional limitations or in nutritionally inferior habitats. Simpson and Boutin's (1993) results are consistent with this idea; muskrats in a highly nutritious cattail (*Typha* spp.) marsh in southern Ontario were larger than those in the more highly seasonal and nutritionally poorer habitat at Old Crow Flats, Yukon, near the northern boundary of muskrat distribution.

At the local scale, muskrat weights vary in the expected way with time and/or the quality of the habitat (Dozier and Allen 1942; Dozier 1945; Dozier et al. 1948; Scheffer 1955; Olsen 1959a; Neal 1968; Schacher and Pelton 1976; Virgil and Messier 1992a; Clark and Kroeker 1993; Clark 1994). Adequate and stable water levels, high-quality habitat, and a lack of population strife all result in greater body weight within a site.

Muskrats are 406–641 mm in total length; tail length ranges from 177 to 295 mm. Ranges of other body measurements include hind foot length, 63–92 mm; ear length, 15–25 mm; skull length, 60–70 mm; and skull width, 38–44 mm (Hall and Kelson 1959; Schwartz and Schwartz 1959; Lowery 1974).

Pelage. The pelage consists of a thick, waterproof, soft underfur overlain by long, glossy guard hairs. Johansen (1962) estimated the insulative value of muskrat fur. Dorsal coat color ranges from silvery to black. Ventrally, the coat is lighter in color, from whitish to brown. Dozier (1948a) described the genetic basis of color mutations in the muskrat. The common black agouti coat results from the joint action of three independent and dominant genes: *A*, a gene for the agouti, or wild-type, coat pattern; *B*, a gene that produces black pigment as the darkest element of the coat; and *C*, a gene for the development of full color. A recessive mutation of the *C* gene results in some type of albinism. Obbard (1987) provided color drawings of the various pelt colors and discussed the sections of North America commonly used in the fur trade for classifying muskrats based on various pelt characteristics. There are no apparent color differences between the sexes (Hall and Kelson 1959; Schwartz and Schwartz 1959).

Skull and Dentition. The skull of the muskrat (Fig. 16.3) was described in detail by Gould and Kreeger (1948). The dental formula is I 1/1, C 0/0, P 0/0, M 3/3. Galbreath (1954) described tooth growth and development in the muskrat. Occlusal surfaces of the molars characteristically form a pattern of angled enamel folds surrounded by dentine. Characteristic of rodents, the single pair of upper and lower incisors is chisel-like and separated from the cheek teeth by a diastema. The lips close behind the incisors, which allows underwater gnawing.

PHYSIOLOGY

Respiration, Circulation, and Thermoregulation. Smith (1938) stated that muskrats can suspend breathing when underwater for long periods; he described a muskrat that stayed submerged for 17 min, surfaced for about 3 sec, then submerged for another 10 min. Respiratory adaptations for diving include a high tolerance of carbon dioxide. Apparently, carbon dioxide does not normally elicit rapid internal responses for the escape from asphyxia during apnea generally characteristic of nondiving mammals (Errington 1963; MacArthur 1984). MacArthur (1986a) found that although daily patterns of oxygen consumption, abdominal body temperature, and foraging activity were minimally affected, metabolic rates were reduced by chronic CO_2 exposure. Also, muskrats breathing 9–10% CO_2 (the maximum reported from winter lodges; MacArthur 1984) made shorter voluntary dives, and the ability of muskrats to rewarm after cold water immersion was slightly depressed by hypercapnia. MacArthur (1986b) found that peak levels of CO_2 recorded in winter lodges are sufficient to impede restoration of acid–base balance following dives in cold water. Additionally, sudden and dramatic (i.e., 10–16%) changes in CO_2, though perhaps unrealistic in nature, did elicit apparent avoidance responses (e.g., exiting the lodge) by muskrats. Huenecke et al. (1958) reported a February peak of 5–7% CO_2 in muskrat houses in Minnesota.

MacArthur and Aleksiuk (1979) suggested feeding shelters and "push-ups" (described later) were spaced to allow access to all points within the home range by short excursions, thus minimizing exposure in cold water. Furthermore, based on aerobic-dive limits calculated previously (49–58 sec; MacArthur 1990), MacArthur (1992a) concluded that 80% of intershelter movements could be achieved aerobically by muskrats. Hence, aerobic-dive limits appear to be important, though not limiting (i.e., 20% of shelters were beyond the aerobic-dive limit), in the spacing of winter feeding shelters. MacArthur (1990) also noted a 42% increase in oxygen-storing capacity for muskrats in winter versus summer. This may be important because the metabolic costs of underwater movements in winter, in spite of a diving-induced reduction

TABLE 16.1. Selected adult muskrat weight data for North America

Location	Mean Weight (g) Males	Females	Source
Northwest Territories	1114	1010	Stevens 1953
Yukon	1048	939	Simpson and Boutin 1993
Alberta	1131	1053	Fuller 1951
New Brunswick	1365	1275	Parker and Maxwell 1984
Saskatchewan	926	913	Virgil and Messier 1992b
Ontario	1308	1305	Simpson and Boutin 1993
Maine	1093[a]	—	Clough 1987
Vermont	1207[a]	—	Seamans 1941 *in* Errington 1963
New York	1153	1181	Erickson 1963
	1370[b]	1323[b]	Erickson 1963
Michigan	1043[a]	—	Baumgartner and Bellrose 1943
Ohio	1299	1257	Donohoe 1966
Iowa	1100	—	Errington 1963
	1250[a,b]	—	Errington 1963
Illinois	1225[a]	—	Baumgartner and Bellrose 1943
Nebraska	1180	1089	Sather 1958

[a]Sexes combined.
[b]Second-year adults.

in peripheral circulation, are nearly as taxing as surface swimming (MacArthur and Krause 1989). Also, such increased capacity may permit greater spacing of feeding shelters, thereby improving access to more submerged forage in winter. MacArthur (1992b) found that by using gas bubbles previously released while diving under ice, muskrats were also able to increase cumulative immersion time and the duration of individual underwater excursions.

Fairbanks and Kilgore (1978) found that postdive excess oxygen consumption (volume of oxygen consumed above what would have been consumed if the animal had not dived) was always greater than the calculated oxygen debt. MacArthur (1984) found that postdive oxygen consumption increased with the time submerged and declining water temperatures. The postdive surge in metabolic rate likely reflects the combined costs of swimming, comfort movements in the chamber, grooming, and, at low water temperatures, rewarming.

Onset of diving bradycardia in muskrats occurs rapidly following submergence (Thornton et al. 1978; Jones et al. 1982; MacArthur and Karpan 1989), regardless of duration or the physical or behavioral circumstances surrounding the dive (MacArthur and Karpan 1989). However, forced submergence in the laboratory apparently reduces the diving response (Jones et al. 1982). During voluntary dives, heart rates appear approximately stable after submergence, with little evidence of postdive tachycardia or anticipatory changes in heart rate prior to dive termination (MacArthur and Karpan 1989). However, heart rates during dives are positively related to water temperature and core body temperature in muskrats (MacArthur and Karpan 1989). Bradycardia also appears most intense during escape dives, likely a result of reduced underwater movements and enhanced cardiovagal tone during escape dives (MacArthur and Karpan 1989). Estimates of postdive heart rates range from 17% to 50% of predive rates (Thornton et al. 1978; Drummond and Jones 1979; Gilbert and Gofton 1982; MacArthur and Karpan 1989). Data suggest that cardiac slowing during normal diving is largely triggered by nasal receptors (Drummond and Jones 1979; MacArthur and Karpan 1989), but that psychogenic influences may play a role in alarm bradycardia.

An increase in the metabolic rate and a decrease in whole-body insulation was observed by Fish (1979) for muskrats in water at 20–30°C, compared to those in air. He also reported that temperature of the appendages and whole-body insulation were inversely related. The higher thermal conductance of water was presumed responsible for the difference in physiological adjustment; apparently high pelage insulation (Johansen 1962) was insufficient to prevent general body cooling. Similarly, MacArthur (1984) found that core body temperature was dependent on both water temperature and dive duration, with core temperatures declining 0.15°C/min in 30°C water and 0.45°C/min in 3°C water. Body temperature continued to decline for 2–3 min after exiting the water, probably due to vasodilation of cooled peripheral tissues and a transient drop in blood temperature (MacArthur 1984). Subcutaneous cooling occurred at approximately twice the rate of abdominal cooling, but unlike core body temperature, peripheral rewarming began immediately. McEwan et al. (1974) and Sherer and Wunder (1979) estimated the thermoneutral zone of adult muskrats at 10–25°C, although the upper limit may be underestimated because tests were performed on animals with winter coats.

Although body cooling occurs during foraging activity in winter, it is retarded by periodic withdrawal from water (MacArthur 1979, 1984). Muskrats apparently reduce risks of hypothermia by elevating core body temperature before entering water and by periodic rewarming within feeding shelters. Harlow (1984) found that the Hardarian gland, situated behind the eye, is an important source of lipids for waterproofing the muskrat's fur. Time spent grooming while out of the water may play an important role in reducing heat loss when entering the water. Nonshivering thermogenesis also appears to be an important component of postdive rewarming (MacArthur 1986c). Also, huddling behavior may be an important thermoregulatory mechanism of muskrats in northern latitudes (MacArthur 1977; MacArthur and Aleksiuk 1979). Bazin and MacArthur (1992) estimated that resting metabolic rate for four huddled muskrats was 11–14% below that of single muskrats at temperatures encompassing the lowest recorded from winter shelters (−10°C to 0°C). However, despite benefits to resting metabolic rate, MacArthur et al. (1997) found that social grooming and aggregation behavior only slightly increased postimmersion rewarming rates of individuals and did not appear to extend foraging time in cold water.

Digestion. During early development, growth of all components of the digestive tract, including the cecum, gut, and small and large intestine, proceeds at a faster rate than growth in body mass (Virgil and Messier 1992b). By the time muskrats are weaned, they may attain 75% of adult length of large and small intestine. Such rapid growth likely prepares weaned muskrats for the transition from milk to the decreased digestibility and energy content of plant matter. In addition, such rapid development of the digestive tract may be a functional adaptation for rapid development of young muskrats.

The muskrat has a large, haustrated cecum, which empties into a complex, eight-spiral, proximal colon (Luppa 1957 in Campbell and MacArthur 1996a). Such complexity in the proximal colon may improve

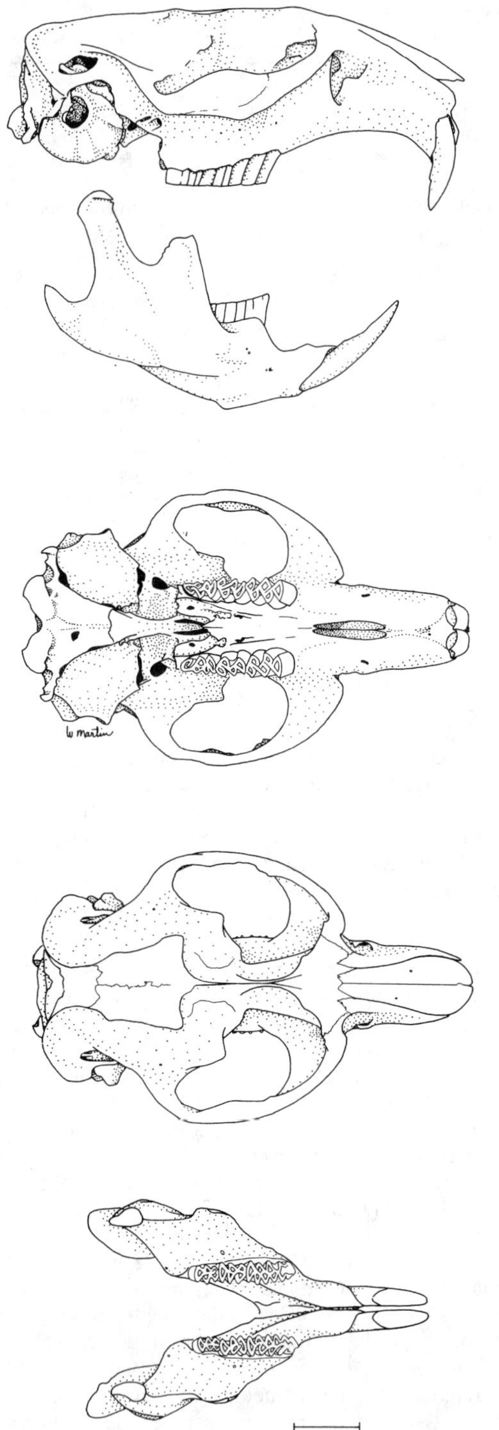

FIGURE 16.3. Skull of the muskrat (*Ondatra zibethicus*). From top to bottom: lateral view of cranium, lateral view of mandible, dorsal view of cranium, ventral view of cranium, dorsal view of mandible.

digestion of dry matter by providing a direct flow of bacteria and fine particles back toward the cecum, thereby selectively increasing fermentable surface area (i.e., fine particles) and supporting a higher concentration of bacteria.

Unlike many nonhibernating rodents, muskrats, particularly in marshes, appear to accrue substantial lipid stores during late fall and winter (Jelinski 1989; Virgil and Messier 1992a, 1992b, 1997; Campbell and MacArthur 1998). Some researchers have suggested this may result from a winter reduction in thyroid activity, lean body mass, and hence basal metabolic costs (Aleksiuk and Frohlinger 1971; Virgil and Messier 1992a). However, recent work in Canada has suggested that lipid accumulation may result primarily from increased intake of energy in winter (Campbell and MacArthur 1998; Campbell et al. 1998). Seasonal increase in cattail rhizome neutral detergent solubles and energy intake in muskrats was tracked by an increase in the mass of the stomach, cecum, large intestine, heart, and liver (Campbell et al. 1998). Such organ mass increases presumably are in response to the increased need to absorb, metabolize, and transport the additional nutrients ingested.

Campbell and MacArthur (1996a) noted that muskrats in Manitoba consuming natural vegetation (in a laboratory) had nitrogen excretion rates six times higher in spring and summer than in fall and winter, when dietary protein levels of aquatic vegetation appear lowest. Subsequently, Campbell and MacArthur (1997) documented that urea hydrolysis, a necessary step in urea recycling, increased in fall and winter, which supports the hypothesis that urea recycling is an important nitrogen-conserving tactic for muskrats on low-protein winter diets.

REPRODUCTION AND DEVELOPMENT

As early as the beginning of the twentieth century, published accounts of the breeding of muskrats varied widely (Lantz 1910). Reproductive data for the muskrat in North America are voluminous and are presented in practically every life history–ecological study published.

Anatomy. Superficially, the genitalia of male and female muskrats are similar. The prominent urinary papilla (clitoris) of the female lies anterior to the vaginal opening and resembles the penis of the male. However, the perineum is furred in males but naked near the urethral papilla in females. Immature females have a completely naked perineum. The anal to papilla distance is greater in males (Schwartz and Schwartz 1959; Taber 1969; Lowery 1974). A membrane covers the vaginal opening in unbred immature females. However, Chamberlain (1951) reported that in the fall and early winter a hymen may reform in adult females. Opening of the vaginal orifice of muskrats in southern and central Wisconsin occurred in March, with partial reclosure in about one third of the adults the last of October (Beer 1950).

Female muskrats have a duplex uterus, where the genital tract is double all the way to the urogenital sinus and the uteri are separate (Gilbert 1987). A mesenteric capsule, the bursa ovarii, encloses each ovary and the bursae have a small slit in one wall. In general, the function of the bursa is to isolate the ovaries from the general peritoneal cavity and help ensure that the ova pass into the oviducts (Gilbert 1987). Males have a baculum, a bony structure within the glans penis. The degree of ossification of the baculum has been used to estimate age of muskrats in Missouri (Elder and Shanks 1962).

Teats are absent on males. On females, three or four (rarely five) pairs of teats are present on the belly between the front and hind legs: two pairs are inguinal and two are pectoral (Lowery 1974). Dozier (1942, 1953) described a technique for determining sex by palpation. In males, the penis may be exposed or felt by stripping the urinary papilla posteriorly. Without palpation, females examined in the fall with vaginal openings sealed may be mistaken for males (Buss 1941). Gender determination of young muskrats was also described by Baumgartner and Bellrose (1943).

Sather (1958) found that gender determination of young muskrats by the presence or absence of hair on the perineum was subject to error. However, gender of very young muskrats could be determined by the presence or absence of mammary glands, and that of older animals by manipulation of the urinary sheath. Gender of dried muskrat pelts may be determined by the presence of mammae (Buss 1941).

Reproductive Cycle. Photoperiodism probably prepares muskrats for the breeding season, whereas the onset of mating may be triggered by a weather factor. Initiation of breeding and ambient temperature appear correlated (Olsen 1959a). Beer and Meyer (1951) evaluated seasonal changes in endocrine organs and reproductive tracts. Gonadotropic activity of the pituitary was highest during spring and early summer and lowest in late fall and early winter.

Follicle maturation began in January for adult females in east Tennessee; sperm were present in adult males throughout the year (Schacher and Pelton 1976). After the breeding season, adrenal weights of adult male muskrats were at a peak. However, adrenal weights of nonpregnant adult females were maximized during the early breeding season, coinciding with follicle maturation and early births. The maximum adrenal weight for pregnant females occurred in May (Schacher and Pelton 1976). Additional details of ovarian conditions during the reproductive cycle were presented by Enders (1939) and Forbes and Enders (1940).

Muskrats are polyestrous (Schwartz and Schwartz 1959) and are spontaneous ovulators (Gilbert 1987). Based on vaginal smears, Beer (1950) found the estrous cycle of southern and central Wisconsin muskrats averaged 28.7 days. Similarly, a 30-day estrous cycle (range = 24–34 days) was reported by Wilson (1955a). Initial research by McLeod and Bondar (1952) on Manitoba muskrats substantiated the finding by Beer (1950) that the estrous cycle was about 30 days. However, subsequent work by them produced quite variable results: Estrous cycles ranged from 2 to 22 days (the mean for 136 cycles was 6.1 days). They interpreted the longer cycles reported earlier as abnormal cycles or pseudopregnancies induced by the experimental technique and not estrous cycles.

Virgil and Messier (1992b) noted that there was a steady increase in juvenile testis mass from October to April in Canada, which generally tracked the increase in adult testis mass. The increase in juvenile testis mass was not due to a corresponding increase in body mass, because juveniles had stopped growing by that time. Laboratory findings of McLeod and Bondar (1952) indicated that females may go through three or four estrous periods before the male comes into sexual activity. Similarly, the progressive decline in number of litters after the first litter appeared to result from a progressive decline in the sexual activity of the male with the approach of warmer weather. Therefore, the reproductive rate of muskrat populations depends on the sexual activity of the male as well.

Errington (1937a) found that a postpartum estrus occurred in the muskrat. The interval was about 1 month between successive litters. The gestation period varied from an atypical minimum of 19 days to a more usual minimum of 22–23 days, with the ordinary period around 30 days. Similarly, Wilson (1955a) reported that normal gestation was 28–30 days.

Muskrats generally breed throughout the year in the southern United States (Svihla and Svihla 1931), whereas breeding in the far north is substantially restricted (mid-June to August) (Wilson 1955a; Ruttan and Wooley 1974; Boutin and Birkenholz 1987) (Table 16.2). Where ice cover is present in winter, breeding is usually initiated when waterways become ice free (Olsen 1959a; Boutin and Birkenholz 1987). In riverine environments, spring breeding and/or recruitment of individuals from early litters is influenced by the timing and magnitude of spring flooding (Clay and Clark 1985; Erb 1993).

Age at first breeding is variable. Gilbert (1987) lists age at first breeding for female muskrats at 5–6 months in the south and 12 months in the north, whereas males usually initiate breeding at 12 months of age regardless of locality. Schwartz and Schwartz (1959) stated that young born in early spring may breed in late summer, but that their first breeding period usually occurred during the spring following their birth. Some past estimates of precocial breeding rates may have been biased high due to underestimation of age from some age estimation methods (Moses and Boutin 1986, 1988). In general, precocial breeding in females appears rare (0–5%) in Canada and the northern United States (Errington 1963; Smith et. al 1981; Parker and Maxwell 1984; Proulx and Buckland 1985; Simpson and Boutin 1989). It may be higher in the central United States (3-year maximum of 14% in Missouri; Erb 1993), and appears most common in the southern United States (O'Neil and Linscombe 1976).

Mating Behavior and Parental Care. Muskrats are largely monogamous (Beer and Meyer 1951; Sather 1958; Errington 1963; Caley 1987; Marinelli and Messier 1993, 1995); however, polygynous strategies have been documented (Caley 1987; Marinelli and Messier 1993, 1995). Based on genetic analysis, Marinelli et al. (1997) detected no evidence of multiple paternity or cuckoldry in muskrats in a marsh in Saskatchewan. Furthermore, they concluded that paternity was consistent with that assumed by space-use patterns.

Up to weaning, muskrat offspring are believed to receive primarily maternal care (Errington 1963). However, following weaning, observed care appears largely paternal (Marinelli and Messier 1995) and includes such behaviors as food provisioning and vigilance at burrows. In a study in Saskatchewan, paternal care was highest for weaned offspring of monogamous pairs, intermediate for the weaned offspring of primary females of polygynous males, and lowest for weaned offspring from secondary females of polygynous males (Marinelli and Messier 1995; Marinelli et al. 1997). Secondary females of polygynous males may compensate by increasing maternal care to weaned offspring (Marinelli and Messier 1995). Higher paternal care of monogamous offspring, however, did not appear to confer any advantages to offspring survival, and polygynous males produce significantly more weaned offspring (Marinelli et al. 1997). Because these results are from a low-density, female-biased population, however, the generality of these conclusions is uncertain.

All copulations of penned muskrats observed by Svihla and Svihla (1931) occurred when they were partially submerged in water. Similarly, Salinger (1950) described copulation of a pair of muskrats in an Idaho stream; copulation lasted about 5 min.

Litters. Average adult litter sizes reportedly range from 3 in the southern United States to 9 or 10 in more northern areas. The number of litters produced exhibits the opposite trend, ranging from one to three, except in the southern United States, where breeding may occur throughout the year (Table 16.2; also see Boyce 1977). The largest single litter size—14 embryos—was reported from a muskrat in Labrador (Chubbs and Phillips 1993). Within location, litter sizes of females reproducing in the year they were born are smaller (3–5) (Sather 1958; Erb 1993). Proulx and Buckland (1985) documented larger litter sizes for three precocial breeders (7.7) compared to adults in other southern Ontario populations studied (6.3); however, there has been debate on the accuracy of age estimation in that study (Moses and Boutin 1988; Proulx and Gilbert 1988).

Boyce (1977) documented latitudinal differences in muskrat reproductive data from North America; litter size increased with latitude, but the number of litters decreased. The tradeoff likely reflects the shorter growing season at higher latitudes. In Europe, Danell (1978) noted a similar trend in the number of litters; however, litter size decreased with latitude. Simpson and Boutin (1993) did not find, as predicted, that litter sizes were larger for a northern population of muskrats in Canada compared to a southern population. To evaluate the discrepancy with Boyce's (1977) results, Simpson and Boutin (1993) analyzed 39 published estimates of litter size and found that a polynomial regression on latitude provided a better fit to the data, with litter size increasing up to 57°N latitude and declining at higher latitudes. Thus, an upper limit in litter size may be imposed by competing energetic demands for survival in the extremely seasonal environments at the northern extent of muskrat distribution.

Survival of young muskrats does not appear to be influenced by litter size (Boutin et al. 1988). However, up to weaning, survival of individuals within a litter does not appear to be independent; litters tend to disappear completely or have a higher proportion of individuals survive than expected (Dorney and Rusch 1953; Boutin et al. 1988). Following weaning, individuals within a litter tend to have independent survival probabilities.

In Mississippi, 13.3% fewer neonates occurred in the average litter than was indicated by embryo counts (Freeman 1945). Overall, embryo resorption in muskrats appears rare; Dozier (1947a) found only 1 incidence in over 15,000 female carcasses examined.

Young. Muskrats produce altricial young; neonates are blind, nearly helpless, and have hair so scant that they appear almost naked (Errington 1939). At birth they weigh about 15–21 g and are approximately 100 mm in total length. By 1 week of age, they are covered with coarse

TABLE 16.2. Selected reproductive data for muskrats in North America

Location	Habitat	Parturition Dates	Litter Size	Mean Number of Litters	Source
Yukon	Small lakes	Begins mid/late June	7.86	1, rarely 2	Simpson and Boutin 1993, Ruttan and Wooley 1974
Northwest Territories	—	June–Aug	—	—	Stevens 1953
Alberta	Streams	Begins late May	8.7	2	Fuller 1951
New Brunswick	—	Mid/late Apr–late Aug	8.4	2.4	Parker and Maxwell 1984
Saskatchewan	Marsh	Mid May–late July	6.4	2.6	Phillips 1979
	Marsh	Early May–mid-Aug	7.3	—	Virgil and Messier 1997
Manitoba	Marsh	Mid-May–late Aug	7.3	1.7	Olsen 1959a
Quebec	Ditches	Peak mid- to late Apr	6.3–6.6	2	Stewart and Bider 1974
Ontario	Creeks	—	6.0	2	Proulx and Buckland 1986
	Ditches/rivers	—	6.5	2	Proulx and Buckland 1986
	Ponds	—	6.7	2	Proulx and Buckland 1986
	Marshes	Begins late Apr–early May	6–7	2	Proulx and Gilbert 1983, Boutin et al. 1988
Maine	Marsh	Mar–July	5.4	2	Gashwiler 1948, 1950b
	Mostly upland	—	5.4	1.4	Clough 1987
Idaho	Marsh	May–Aug	7	1.6	Reeves and Williams 1956
Wisconsin	Marsh	Late Apr–Sep	7.3	—	Mathiak 1966, Dorney and Rusch 1953
	Marshes	—	7.4	2	Beer and Truax 1950
Minnesota	Multiple	Begins late Mar	6	2	McCann 1944
New York	Marsh	—	5.6	2	Alexander 1951
	Ponds	—	6.3	1.5	Erickson 1963
Massachusetts	Marsh	Mid-Mar–Sep	5	2.7	Chamberlain 1951
Iowa	Marsh	Apr–Aug	6–8	2	Errington 1937a, 1939, 1963
	River	Mid-Feb–Sep	7.1	1.5–2	Clay and Clark 1985
Nebraska	Marsh	Apr–Sep	6–7	2.6	Sather 1958
California	Marsh	—	4.7	—	Hensley and Twining 1946
Missouri	River	Feb–Sep	6.2	2.5	Erb 1993
	Pond	Midspring	—	1	Shanks and Arthur 1952
North Carolina	Marsh	Year-round except in cold winters; peaks Feb–May	3.7	3	Wilson 1954a
Tennessee	Streams	Late Feb–mid-Sep; peaks Apr–July	5.4	2.3	Schacher and Pelton 1975
Maryland	Marsh	Nearly year-round; peaks Mar–Sep	2.4–5	2–3	Smith 1938
	Marsh	Data collected Jan–March/ April–June, Aug–Sep	3.9/2.4	2	Harris 1952
Texas	Marsh	Nearly year-round	—	—	Lay 1945b
Mississippi	Marsh	Year-round; peaks winter/spring, low July/Aug	—	—	Freeman 1945
Louisiana		Year-round; peaks Nov/Mar, low July/Aug	4.0	5–6	Louisiana Department of Conservation 1931
			3–4	5–6	O'Neil and Linscombe 1976

gray-brown fur (Schwartz and Schwartz 1959). Eyes open between 14 and 16 days when young are about 80 g and 200 mm; after this, they can swim, dive, and climb (Errington 1939). Initially the young cling tightly to the mother's nipples. When they are older, she carries them by the skin of the belly. The female can dive while carrying young, but while swimming she holds them above the water. Generally, when the female leaves a nest containing naked young, they are well covered with shredded dry vegetation (Hensley and Twining 1946).

Young may suckle up to 21 days, though 15–16 days is more common (Smith 1938). Weaning usually occurs in the fourth week when young are about 180 g and 285 mm (Errington 1939; Virgil and Messier 1992b). By the time young are independent, their mother is usually ready to give birth again. The young may continue to use the nest or move to another area. If they remain and the female renests, she may construct a new chamber in the same house or bank (Schwartz and Schwartz 1959).

Young are referred to as kits when 70–90 days old (500–650 g, 400–480 mm). At 3.5 months of age, muskrats are similar in size and appearance to small–medium adults. By 1 year of age, they reach an average weight close to 1100 g and a total length of about 550 mm (Errington 1939). Virgil and Messier (1992b) reported that juvenile muskrats in Saskatchewan reached 95% of full skeletal size, but only 77% of full body weight, prior to their first breeding season. They also provided data on changes in body composition of juvenile muskrats in Saskatchewan.

ECOLOGY

Dwelling Structures. Muskrat dwellings are of two primary types: burrows excavated in shoreline banks and various types of lodges constructed predominantly with emergent marsh vegetation. Bank den entrances are generally underwater (Beshears and Haugen 1953; Earhart 1969; Willner et al. 1980) and are approximately 15 by 20 cm in diameter. Bank den entrances that are above water are often loosely plugged with vegetation. Burrow lengths may range from very shallow temporary burrows often used for feeding and/or constructed by young muskrats (Earhart 1969) to extensive burrow systems of >100 m (Errington 1963). Length and complexity of the burrows are often determined by soil type, bank slope, distance to water, and age of the burrow. An airshaft to the surface may occasionally be constructed for dens in dense soil (Schwartz and Schwartz 1959). In farm ponds of Alabama, the number of burrow entrances ranged from one to nine and averaged two. Typical burrows began about 15 cm beneath the water surface and nests were found in most of the major burrows excavated (Beshears 1951). Muskrats frequently construct new underwater burrow entrances as water levels decrease (Heatherly 1993). Brooks and Dodge (1986) estimated that an average family of muskrats on a river in Massachusetts used two separate burrows.

Water velocity, bank height and slope, soil type, width of the shoreline herbaceous belt, and presence of aquatic vegetation all play a role in the selection of sites for burrows (Brooks and Dodge 1981, 1986; Brooks 1985; Nadeau et al. 1995). Although muskrats may use a variety of sites for burrows, they prefer sites with low water velocity, high banks with steep slopes, clay–loam soils, and an abundance of shoreline and/or aquatic vegetation.

Fuller (1951) found that stream-dwelling muskrats on the Athabasca–Peace Delta, Canada, used bank burrows for shelter throughout the year, whereas slough-dwelling muskrats used burrows in summer but constructed houses in the fall. Use of burrows by muskrats at Delta Marsh, Manitoba, appeared greatest in summer (MacArthur and Aleksiuk 1979). MacArthur and Aleksiuk (1979) suggested that summer burrow use may be due to heat stress in summer lodges.

If suitable sites are available, muskrats appear to prefer denning in banks rather than houses (Dozier 1948b; Messier et al. 1990). Messier and Virgil (1992) concluded that preference for burrows was due to decreased costs of dwelling maintenance and possibly greater protection from predators throughout the year. Their findings did not support the suggestion by MacArthur and Aleksiuk (1979) that heat stress inside lodges may be responsible for preferential burrow use in summer. In spite of the frequent preference for burrows, lodges represent one of the most characteristic features of muskrat populations in many marshes and backwater sloughs. Proulx and Gilbert (1983) indicated that muskrats on a marsh in Ontario only used burrows when water depths and emergent vegetation were inadequate for house building.

Muskrats construct three main types of houses: primary houses, feeding huts, and push-ups. All are typically made of natural materials present in the marsh or pond and are conical in shape. Distinguishing among feeding huts, push-ups, and houses is frequently difficult. Houses or lodges are generally the largest (Fig. 16.4) and often vary in size and shape (Errington 1963). They are up to 2.5 m in diameter and vary in height from 0.4 to 1.8 m with a wall thickness of 0.1 to about 0.3 m (Dozier 1948b, 1953; O'Neil 1949; Schwartz and Schwartz 1959; Errington 1963). The feeding hut, or "feeder," is a place where muskrats bring food for consumption. It is generally smaller than a house, and provides a resting platform and protection from the elements and predators. Push-ups are generally associated with frozen marshes and snow. They are made by muskrats opening a 10- to 13-cm hole through thin ice and pushing up a 30- to 46-cm pile of fine fibrous roots, submerged vegetation, or other debris. The pile of debris, forming an enclosed cavity, rests on top of the ice. The cavity serves as a resting site and may be used as a feeding area in severe weather. Push-ups are temporary and collapse rapidly as the ice melts. On streams, houses can be distinguished from push-ups because houses are built on a firm base or are otherwise anchored, contain a well-defined chamber, are usually on the periphery of a slough, and are constructed largely of emergent rather than submergent plants (Fuller 1951).

MacArthur and Aleksiuk (1979) found that lodges of the Delta Marsh in Manitoba were usually constructed of cattails (*Typha latifolia*) or bulrush (*Scirpus* spp.) interspersed with pondweed (*Potamogeton* spp.) and bottom detritus. Muskrats on a California marsh used tule (*Scirpus* sp.) stalks and roots and waterweed (*Elodea* sp.) for house construction (Hensley and Twining 1946). They also constructed open "porch nests" on the side of the house for loafing and sleeping. Bellrose and Brown (1941) concluded that cattail communities supported more muskrat houses per unit area than other plant associations in the Illinois River valley. Other plants used for house construction included marsh smartweed (*Polygonum* sp.), hornwort or coontail (*Ceratophyllum demersum*), and American lotus (*Nelumbo lutea*). Muskrats generally use vegetation immediately surrounding the dwelling for construction purposes (Bellrose and Brown 1941). Muskrats appear to select house-building sites where water depths are a minimum of 10–15 cm (Bellrose and Brown 1941; Proulx and Gilbert 1983; Clark 1994), with averages in the range of 30–40 cm.

Inside primary houses, one or two internal nest chambers are constructed, with one or more plunge holes exiting the house. Individual

FIGURE 16.4. The muskrat house provides protection from predators and the elements as well as a suitable site for rearing young. SOURCE: Photo by Robert G. Linscombe.

family members may construct their own alcove, resulting in a multichambered house (Schwartz and Schwartz 1959; Errington 1963). Nests are located above water, and in coastal areas, they are above normal high tides (O'Neil 1949), with muskrats often constructing multiple nest chambers at different elevations to account for water level fluctuations (Kinler et al. 1990).

Houses provide dry nests and insulation from summer heat and winter cold. Although the maximum range in macroclimatic air temperature at Delta Marsh, Manitoba, Canada was –39°C to 34°C, microclimatic air temperature within lodges, burrows, and push-ups ranged from –9°C to 30°C (MacArthur 1977; MacArthur and Aleksiuk 1979). Mean temperatures within occupied dwelling lodges and burrows in summer and winter were close to the thermoneutral zone of single, adult muskrats (10–25°C). The insulative value of dwellings was highest in winter; during summer, nest temperatures were similar to air temperatures. However, Cook (1952) reported that temperatures in a house in New York composed of cattails remained moderate despite high ambient temperatures, similar to results from muskrat houses in Saskatchewan (Messier and Virgil 1992).

Although house building can occur throughout the year, primary activity begins in late summer in seasonal environments, and the most intensive house building is closely associated with the onset of cooler weather. Errington (1963) suggested that the first hard frost stimulated intensive house building in many areas. House building began in early October and peaked in the first week of November in Massachusetts (Chamberlain 1951) and began in August and September in an Ontario marsh (Proulx and Gilbert 1983). Nicholson and Davis (1957) found that the number of houses built per month in marshes on the eastern shore of Maryland was constant during fall, increased about fourfold in February, and then practically stopped. Houses built from September to December lasted about 7.7 months; houses built in February lasted only 3.5 months. Messier and Virgil (1992) estimated spring and summer subsidence rates for abandoned muskrat houses in a Saskatchewan marsh at 0.5–1.0 cm/day, which emphasizes the need for continual maintenance. None of the 53 lodges abandoned during winter lasted more than one summer. Proulx and Gilbert (1983) estimated the average number of houses used by a muskrat family in an Ontario marsh was 1.2–1.5 in early summer and 2.2–3.9 in late summer.

During periods of low water, canals are frequently constructed from dwellings to deeper water. Muskrats may construct an elaborate system of canals in the marsh. Canals are from 15 to 30 cm wide and vary from muddy surface trails to ditches ≥30 cm deep. Debris, which is usually piled up along the ditches, is excavated with the front feet and cast up with the hind feet (Schwartz and Schwartz 1959).

During winter in northern areas, feeding huts are also important for periodic rewarming and oxygen replenishment between under-ice forays. Further south, where weather is less severe, feeding huts are less frequently used and generally replaced by rather flimsy feeding shelters or feeding platforms (Louisiana Department of Conservation 1931; Glass 1952); muskrats in northern environments regularly use such platforms in summer. Feeding platforms or rafts are constructed by breaking down plants and/or piling up vegetation. Platforms are sometimes found on small, naturally elevated areas of the marsh.

A ratio of 2.7 feeding huts/house was recorded in a central New York marsh (Dozier 1948b). In Massachusetts, feeding huts were located at random and were not associated with lodges (Chamberlain 1951). However, aerobic-dive limits appear to be an important factor in the spacing of winter feeding shelters (and push-ups) in northern areas (MacArthur 1992a).

Muskrats may also use various other shelters, such as hollow logs, overhanging banks, or dens of other animals (MacArthur and Aleksiuk 1979; McKinstry et al. 1997). These retreats are often temporary and vary with local environmental conditions. For example, a floating hunting blind was used by muskrats in Manitoba to rear young (Anderson 1969).

Waterfowl and other birds, reptiles, amphibians, spiders, insects, and mammals frequently use muskrat houses as resting, basking, or residence sites. For example, Mathiak (1966) found 70 clutches of snapping turtle (*Chelydra serpentina*) eggs in muskrat houses on Horicon Marsh, Wisconsin, over a 2-year period. Svihla and Svihla (1931) reported that numerous animals occupied muskrat houses in Louisiana, including ants, beetles, ground skinks (*Leiolopisma laterale*), and many species of frogs, toads, and snakes. Muskrat houses in a California marsh were used as nesting sites by black terns (*Chlidonias niger*), Forster's terns (*Sterna forsteri*), Canada geese (*Branta canadensis*), ducks, and other birds (Hensley and Twining 1946).

Muskrat–Vegetation Interactions. Muskrats are a dominant herbivore in many aquatic ecosystems. It is through house building and foraging that muskrats are capable of altering the abundance, spatial distribution, diversity, and decomposition of aquatic vegetation.

Weller and co-workers (Weller and Spatcher 1965; Weller and Fredrickson 1973; Weller 1981) discussed muskrats as a key component of marsh habitat cycles. Many prairie marshes appear to proceed through five stages: (1) germination phase; (2) newly flooded, with sparse and often well-dispersed vegetation dominated by annuals and immature perennials; (3) flooded dense marsh dominated by perennials; (4) flooded, hemimarsh, an interspersion of roughly 50% dense emergents with numerous openings, many containing water, over the remaining 50%; and (5) deep, open marsh, rimmed by emergents. Although water levels strongly influence the timing and magnitude of such transitions (Murkin et al. 2000), muskrat foraging and house building are often important factors in the latter transitions (Weller and Spatcher 1965; Weller and Fredrickson 1973; Weller 1981). Muskrat populations appear to reach peak levels 3–5 years following reflooding (McLeod 1948; Weller and Spatcher 1965; Neal 1968; Weller and Fredrickson 1973; Bishop et al. 1979; Kroll and Meeks 1985).

In some cases, muskrat activity may become significant enough to cause severe "eat-outs." An eat-out is a condition that results when muskrats remove all or most of the existing emergent vegetation, often including the soil-binding root systems. Reports of severe eat-outs appear restricted to the southeastern United States (Pancoast 1937; Lynch et al. 1947; Dozier et al. 1948; O'Neil 1949; Wilson 1968; Lowery 1974). Lynch et al. (1947) found that significant marsh damage was inevitable when areas heavily populated by muskrats were undertrapped. However, Errington (1961) suggested that muskrat populations of the northern states and Canada are not as likely to completely destroy their habitats. We did not conduct a systematic survey of state and provincial biologists to assess the prevalence of eat-outs by muskrats. Nevertheless, we did not find any reports of such occurrences from northern regions, and Boutin and Birkenholz (1987) were unaware of any documented cases of severe eat-outs in Canada. Shorter growing seasons and the increasing influence of winter freeze-outs may reduce the likelihood that muskrat populations in northern regions can sustain high densities long enough to cause such drastic habitat alterations. Based in part on a long-term experiment in a Manitoba marsh, Clark (2000) concluded that although muskrats may have contributed to declines of emergent vegetation, they were not the principal cause of such declines as the term "eat-out" implies.

The hemimarsh condition is often a partial result of muskrat activity and appears to result in the greatest abundance and diversity of birds (Nelson 1954; Weller and Spatcher 1965; Weller and Fredrickson 1973; Bishop et al. 1979; Kaminski and Prince 1981), richer communities of invertebrates (Nelson and Kadlec 1984; Murkin 1989), and increased diversity of aquatic plants (Danell 1977). Hewitt and Miyanishi (1997) suggested that increased light availability was a significant factor in increasing plant species richness around openings created by muskrats. If muskrat activity becomes too extensive, however, marshes may be transformed into open-water habitats, and plant, invertebrate, and wildlife diversity may decline. Lynch et al. (1947) also suggested that wildlife that use eaten-out areas may be predisposed to diseases from stagnant water. Although high densities of muskrats are capable of substantially reducing plant biomass, such interactions should be viewed as natural processes in marsh environments. Depending on the severity of the eat-out, emergent vegetation may not recover until a drought or drawdown stimulates regermination of emergent vegetation. Hence, the duration of a marsh "cycle" may not be constant or predictable. High water and saltwater intrusion also appear to retard recovery in Gulf Coast marshes (Lynch et al. 1947).

Muskrat house building also influences localized decomposition rates of vegetation. Litter from muskrat mounds supported higher densities of microbes, possibly due to improved microbial substrates and more favorable microclimates for decomposition (Berg 1987; Wainscott et al. 1990).

Sex Ratios. The overall population sex ratio in the fall and winter tends to favor males (see Table 15.3 in Perry 1982). The degree of imbalance varies greatly, but ranges from about 1:1 to 2:1, depending on area, time of year, and method of evaluation. There is also evidence that the sex ratios may vary for different age classes of muskrats; overall sex ratios may thus vary depending on the status of the population.

Sex ratios of nestlings appear to be slightly to moderately biased toward males, with ratios ranging from 1.02:1 to 1.6:1 (Beer and Truax 1950; Gashwiler 1950a; Sather 1958; Olsen 1959a; Erickson 1963; Errington 1939, 1963). Although specific estimates may vary, Beer and Truax (1950) and Olsen (1959a) both found that the sex ratio shifted more toward males as nestlings became older, suggesting that female nestlings may experience higher mortality early in life. Olsen (1959a) also found that the sex ratio of nestlings tended more toward unity in litters born later in the breeding season.

Within a population, it appears that sex ratio bias toward males may continue to the subadult stage, but the sex ratio becomes more balanced or female biased from early spring to fall in adults (McCann 1944; Sooter 1946; Beer and Truax 1950; Sather 1958; Erickson 1963; Errington 1963). However, males outnumbered females in North Carolina in all age and weight groups, and a higher rate of mortality was indicated for adult females (Wilson 1956a). Muskrats trapped at southwest Lake Erie had a male:female ratio of 87:100 for animals <907 g; other weight groups showed a preponderance of males (Anderson 1947).

Clearly, overall population sex ratios based on samples of fall-trapped animals can be biased due to increased susceptibility of males. However, Beer and Truax (1950) concluded that there was not overwhelming support for such a conclusion in muskrats. In general, however, it does appear that neonatal sex ratios are slightly to moderately biased toward males. A mortality factor may affect a greater proportion of females from birth to fall, whereas males suffer greater mortality between their first and second winters. The latter is likely due to increased fighting and movements of males in the spring (and possibly greater trapping losses), which expose them to higher mortality. Such seasonal patterns, however, may be less distinct in the southern United States, where breeding may be more extended. An explanation for the suggested higher early mortality of female nestlings is less clear. Olsen (1959a) speculated that slower growth rates of female nestlings may render them more susceptible to various sources of mortality. Female muskrats in Louisiana had a parasite infestation rate nearly four times that of males: 15.6% versus 4.2%, respectively (O'Neil 1949).

Population Density and Dynamics. Population density varies widely, and depends on such factors as habitat type and condition, social interactions, water levels, competition, harvest, predation, and season. Density alone can often be a misleading metric for understanding wildlife populations. For example, Messier et al. (1990) and Clark (1994) found that muskrat density increased at a higher rate in less suitable portions of marshes than in higher quality portions, but overwinter survival in the less suitable portions appeared lower. In addition, methods for determining density frequently vary, and the reliability of estimates is often unknown. Because of these factors, interpreting causal differences in site-specific density estimates is fraught with problems, and broad generalizations regarding density are difficult. Nevertheless, a few patterns are worth noting.

Water levels appear to strongly influence population density. Other factors being equal, habitats with highly variable water levels tend to support lower densities of muskrats (Bellrose and Brown 1941; Bellrose and Low 1943; Errington 1939, 1963; Friend et al. 1964; Seabloom and Beer 1964; Donohoe 1966; Thurber et al. 1991; Proulx and Gilbert 1983; Virgil and Messier 1996). Low water levels directly affect muskrats by exposing them to increased predation, nutritional stress, and winter freeze-outs (Errington 1938b; Donohoe 1966; Proulx et al. 1987; Messier et al. 1990; Hjältén 1991; Clark and Kroeker 1993; Clark 1994; Virgil and Messier 1996). Similarly, abnormally high water levels can destroy muskrat dwellings, flood nest chambers, reduce emergent vegetation, and force increased movements, all of which can reduce muskrat survival and hence density (Bellrose and Low 1943; Errington 1963; Weller and Spatcher 1965; Donohoe 1966; Weller and Fredrickson 1973; Kinler et al. 1990).

In areas where burrows are the predominant dwelling structures, shoreline length is a more important factor than overall water body size in determining muskrat population levels. Other factors being equal, ponds or lakes with more edge tend to support proportionally larger populations (Glass 1952).

The density of muskrats that different habitats can support undoubtedly varies and is difficult to ascertain. Butler (1940) reported the carrying capacity near Mafeking, Manitoba, ranged from 7.4 muskrats/ha in sedge (*Carex* spp.) habitat to 64.2/ha in common reed (*Phragmites communis*) and sweetflag (*Acorus calamis*) habitats. Peak densities during studies of muskrats in cattail and bulrush habitats are often in the range of 20–40/ha (Errington 1948, 1963; O'Neil 1949; Smith et al. 1981; Proulx and Gilbert 1983; Clark and Kroeker 1993). However, Errington (1963) recorded a maximum of 86 muskrats/ha in *Typha* spp. marshes in Iowa and peaks approaching 50/ha in *Scirpus* spp. In addition, Lynch et al. (1947) and O'Neil (1949) suggested densities in coastal *Scirpus olneyi* marsh can exceed 100/ha. Brackish marshes appear to support higher densities of muskrats than freshwater marshes in coastal areas (Palmisano 1972b; O'Neil and Linscombe 1976). However, muskrat populations in brackish marshes often fluctuate drastically.

Maximum breeding densities of 5 pairs/ha were reported by Schwartz and Schwartz (1959). Breeding densities of 1 pair/46 m of shoreline and between 2.5 and 5 pairs/ha in deeper areas were indicated as most desirable for fur production in Iowa (Errington 1940). Breeding densities of 1 pair/183–457 m of shoreline were optimal for open lakes. Ditches and streams in agricultural landscapes may contain 6–8 pairs/1609 m; such watercourses in pasture may support only 3–4 breeding pairs/1609 m. Stewart and Bider (1974) found an average of 366 m of permanently watered collection ditch was needed to support each breeding female in Quebec.

A variety of other site-specific density estimates have been reported. They include estimates from northern Minnesota lakes (0.25–0.59 muskrats/ha of vegetation in fall; Thurber et al. 1991), Alabama farm ponds (2.8 muskrats/ha; Beshears 1951), backwaters of the upper Mississippi River (0.3–9.3 adults/ha in spring and 0.8–6.3 adults/ha in fall; Clay and Clark 1985), a Maine marsh (1.1–3.9 muskrats/ha of vegetation in fall; Gashwiler 1948), a Saskatchewan marsh (0.6–11.7 adults/ha in summer; Virgil and Messier 1996), several stretches on a Missouri river (0.58–4.97/km in summer and 0–4.65 in fall; Heatherly 1993), a Massachusetts River (49 muskrats/km in summer; Brooks 1985), and a Pennsylvania river (20 muskrats/km in summer; Brooks 1985).

Since muskrats are often aggressive toward conspecifics at certain times (Errington 1963) and that habitat selection in muskrats is density dependent (Errington 1963; Messier et al. 1990; Clark and Kroeker 1993; Clark 1994; Virgil and Messier 1996), it is reasonable to assume that density may have important consequences for the demographic rates of muskrat populations. Errington (1954a, 1961, 1963) provided some useful graphs from his Iowa studies depicting a reduction in spring-to-fall population growth rates with increasing muskrat density. Studies of muskrats in Manitoba (Clark and Kroeker 1993) and Wisconsin (Beer and Truax 1950) support this finding. Such reductions in population growth can be a result of intrinsically (socially induced) and/or extrinsically driven (e.g., habitat induced), density-dependent dispersal, survival, and/or reproduction.

Errington (1954a, 1961, 1963) noted that larger litter sizes (8.2 vs. 6.4) were associated with periods of minimal intraspecific aggression.

He suggested that psychological or behavioral responses to density may affect reproduction. Interestingly, however, the pattern he documented over the 20-year period in Iowa was a positive correlation between density and litter size (i.e., inverse density dependence), indicating that changes in habitat conditions were likely the ultimate driving factor (i.e., high density and reproduction corresponded to high habitat quality and vice versa). Results from Iowa (Errington 1963) and Nebraska (Sather 1958) indicate that, although still negligible, precocial breeding was also higher at peak densities. These results are not necessarily unexpected, as numerous authors have noted that habitat quality can influence muskrat reproduction to varying degrees (Errington 1941; Neal 1968; Danell 1978; Proulx and Gilbert 1983; Kroll and Meeks 1985; Proulx and Buckland 1986; Clark and Kroeker 1993). However, the population-level impacts of documented reproductive changes are not clearly understood.

Errington's (1954a, 1963) work suggested, however, that reproduction started to decline 1 year prior to the 1943 density peak, and may thus have been a density-dependent response (either socially or habitat driven). The subsequent 5-year declines in reproduction with declining density likely represent either delayed density-dependent responses due to habitat deterioration caused by previous high densities or responses to density-independent natural habitat deterioration. In other words, one would have predicted that reproduction would increase as density declined if in fact psychological or behavioral responses to immediate density were the ultimate determinant of observed changes in reproduction. Hence, support for such a psychologically or behaviorally driven reproductive response to density per se appears questionable, and Errington (1963) indicated that there were no apparent density-related trends in pregnancy rate or number of litters.

Olsen (1959a) observed apparent compensatory reproduction for muskrats in Manitoba. He noted that total natality within one summer increased in response to high newborn mortality from high water levels and increased in another summer due to a "psychological response to an uncrowded habitat." The former case, however, does not imply density dependence. Others either suggested slightly density-dependent reproduction (Clay and Clark 1985) or did not find increased litter size to be a significant population response to reduced densities (Smith et al. 1981; Parker and Maxwell 1984; Simpson and Boutin 1989). Compensatory reproduction in northern areas may be less likely due to the physiologically imposed upper limit on litter size and the restricted length of the breeding season (Simpson and Boutin 1989).

As noted, and as Clark (1990) discussed, Errington (1954a) was unable to draw any strong conclusions regarding a direct relationship between reproduction and density per se. The inconclusive and often contradictory statements regarding density-dependent reproduction need to be better studied through experimentation. Nevertheless, such uncertainty is not inherently contradictory to the well-supported observation that spring to fall recruitment is frequently density dependent (Beer and Truax 1950; Errington 1954a, 1963; Clark 1990; Clark and Kroeker 1993). These patterns are based largely on measures of population growth or recruitment, which confound the effects of reproduction, juvenile and/or adult survival, and immigration. Failure to clearly distinguish this appears to have created confusion in the literature.

There appears to be stronger support for the hypothesis of density-dependent survival in muskrat populations. For example, several reports indicate that summer survival of juveniles is likely density dependent (Errington 1961, 1963; Smith et al. 1981; Proulx and Gilbert 1983; Clay and Clark 1985; Clark 1987). Furthermore, Errington's (1963) belief that late-winter prebreeding densities were relatively constant regardless of fall densities led him to conclude that overwinter survival was, in part, inversely related to fall density. As Clark (1990) pointed out, however, Errington only partly recognized the contradiction between his belief in constant prebreeding densities and the previously discussed observation of density-dependent spring to fall recruitment, which requires spring densities to vary. Nevertheless, there has been some subsequent support for density-dependent overwinter survival as well (Messier et al. 1990; Clark 1990; Clark and Kroeker 1993). However, as Simpson and Boutin (1989) noted, in northern environments, overwinter survival can actually be reduced at low densities. A lower threshold density may be required to keep feeding huts and push-ups operable and facilitate thermoregulation.

Several authors have noted the importance of immigration in the establishment and buildup of low-density populations (Errington 1940, 1963; Simpson and Boutin 1989). At high densities, immigration may be blocked by established residents. Such a density-dependent response may, along with other factors, contribute to the higher population growth rates documented by Errington (1954a) for low-density breeding populations.

Population Cycles. The cyclic tendencies of many muskrat populations have been recognized for some time and have been presented in most detail from Canada (Elton 1924; Elton and Nicholson 1942; Erb et al. 2000), Iowa (Errington 1954a; 1963), and Louisiana (Lynch et al. 1947; O'Neil 1949; Lowery 1974). Factors responsible for cyclic changes in muskrat populations are still not entirely clear, but it appears that the relative importance of different factors may vary geographically.

O'Neil (1949) contended that food conditions formed the primary basis of population regulation in Louisiana and that muskrat population cycles were directly related to eat-outs. He did not believe that parasitism or predation were related to cyclic population fluctuations on the Louisiana coast.

The basic phases of muskrat cycles in Louisiana are low muskrat numbers, development of an abundant food supply, overpopulation, range damage, and then starvation (Lynch et al. 1947; Lowery 1974). Ten- to 14-year population cycles were reported on good *S. olneyi* marsh and longer cycles on less productive marsh types (O'Neil 1949). The cycles were local and independent and there was no set pattern for the entire brackish marsh (Lowery 1974). Palmisano (1972b) also reported that muskrat populations in some Louisiana coastal marshes frequently increased rapidly for 3–4 years and then declined precipitously within a few months.

As discussed earlier, strong muskrat–vegetation interactions have been observed for muskrat populations in the midwestern United States as well. However, Errington (1963) never observed such severe eats-outs as reported in Louisiana and believed that such severe overuse of habitat was of less significance in the northern United States. Boutin and Birkenholz (1987) were also unaware of eat-outs in Canada. Thus, although plant–herbivore interactions almost certainly influence muskrat numbers, shorter growing seasons and the increasing potential of winter freeze-outs to kill individuals may reduce the likelihood that muskrat populations in northern regions can build to and sustain densities high enough (relative to carrying capacity) to cause such drastic eat-outs.

Errington (1951, 1954a, 1963), after spending an estimated 32,000 hrs in the field observing and studying muskrats, was convinced that intrinsic factors, including social interactions, strongly influenced the pattern of muskrat population fluctuations. His studies in Iowa indicated that long-term muskrat population fluctuations roughly followed a 10-year cycle. He noted three peaks (1936, 1946, 1956) and two lows (1942 and 1952) in "restlessness and intolerances" in muskrats inhabiting marshes in central Iowa (Errington 1961). Similarly, he noted that litter sizes were higher (8.2 vs. 6.4) during the two periods of minimal fighting (in 1942 and 1952). Thus, he believed that social factors played a key role in the cyclic fluctuations. Increasing and decreasing muskrat populations in northeastern Iowa were also compared by Neal (1968). The decreasing population was associated with less dense vegetation and exhibited lighter average weight, a shorter breeding season, and fewer and smaller litters.

Another factor that can influence muskrat populations is disease. Tyzzer disease in muskrats (variously described as Errington or hemorrhagic disease) was first noted by Errington (1946a). Subsequently, Errington (1954a, 1963) documented numerous cases of severe die-offs of muskrats, which he attributed to disease. He did not believe that establishment of the disease occurred in a cyclic manner. However, he concluded that resistance of muskrats to the disease was approximately cyclic, with lowest resistance corresponding to the late peak and decline phases of the population cycle, when muskrat's strife was increasing and reproduction was decreasing. Tyzzer disease is not common in the southeastern portion of the United States, where muskrat populations also cycle, and is most common in more northern regions (Davidson and Nettles 1997). Interestingly, Davidson and Nettles (1997) noted that the bacterial spores that cause the disease can remain infectious in marshes for up to 5 years, approximately the length of the decline phase of 8- to10-year population cycles.

In summary, Errington believed that socially induced changes in reproduction and resistance to disease were key factors in the cyclic fluctuations of muskrat populations in Iowa. As discussed previously, however, it is important to distinguish between socially driven and externally driven forces. If social interactions were in fact the only ultimate force, then we would expect immediate density-dependent responses, with reproduction improving and fighting declining as density decreased. Such direct density-dependence would tend to stabilize densities rather than produce cyclical fluctuations. Hence, coincident changes in habitat quality or other external forces would also seem necessary to explain Errington's observations, and population cycles in general.

Population cycles of Canadian muskrats have been documented for some time (Elton 1924; Elton and Nicholson 1942). Interestingly, many researchers have reported cycles of different period length. For example, Elton (1924:147) suggested in his seminal paper on population fluctuations that the "musk-rat (*sic*) curve of skins (Canada) shows a short 3 or 4-year period mainly." It is unclear, however, to which populations he is referring, and he later (Elton and Nicholson 1942) makes reference only to 10-year cyclicity in muskrats. However, Elton's (1924) passing suggestion of 4-year cyclicity in muskrats was not unique. Work from northern Sweden (Danell 1978, 1985) and west-central Manitoba (McLeod 1948, 1950) indicated that the number of muskrat houses may often "cycle" with a periodicity of 4–5 years. Suggested causes of these periodic fluctuations were disease outbreaks (McLeod 1948, 1950) and a vole (*Microtus* and *Clethrionomys*)–red fox (*Vulpes vulpes*)–muskrat interaction (Danell 1978, 1985). Additional variation has been suggested by Butler (1962), who, based on a visual analysis of various muskrat harvest time series, presented evidence of a gradient in periodicity from 6 to 10 years moving through three habitat types from central to northern Saskatchewan. Periodic droughts were suggested as the likely cause of shorter periods in southern Saskatchewan (Butler 1962).

In an analysis of spatially extensive historic harvest data from Canada, Erb et al. (2000) concluded that average cycle lengths were approximately 7–8 years, except in northeastern Canada, where 4-year cycles were noted. Similar to conclusions regarding the 4-year muskrat population cycles in Sweden, they hypothesized that 4-year cycles in northeastern Canada may result from a vole/lemming–fox–muskrat interaction. Essentially, fox (*Vulpes vulpes* and/or *Alopex lagopus*) may be tracking (or causing) the 4-year vole/lemming cycles, with muskrats being an alternative prey taken in increasing numbers as other rodent numbers decline.

Other authors, noting the close relationship between mink (*Mustela vison*) and muskrat population cycles, have suggested that mink predation is a key cause of the longer muskrat population cycles in Canada (Butler 1953; Keith 1963; Bulmer 1974, 1975; Finerty 1980). Mink are important predators of muskrats (Seton 1909; Dearborn 1932; Errington 1943, 1954b, 1963; Hamilton 1959; Banfield 1974; Proulx et al. 1987), although the level of predation varies with habitat and population conditions (Errington 1943, 1963; Proulx et al. 1987).

The degree to which mink specialize on muskrats, however, is still equivocal, and mink are often characterized as generalist predators (Eagle and Whitman 1987). This is an important distinction because only predation by specialist predators is expected to produce cyclical fluctuations in their prey (Hanski et al. 1991). As shown for short-tailed weasels (*Mustela erminea*) in Europe (Hanski and Henttonen 1996), if the degree of prey specialization by mink is a function of the number of prey species available, then mink in more northern regions may specialize more on muskrats due to the lower number of prey species available (e.g., Kaufman 1995). Errington (1963) apparently realized this distinction, noting that the numerical dependence between mink and muskrat may possibly be stronger in the north. He did not, however, believe that such a strong and continual numerical dependence existed in Iowa and was furthermore convinced that most predation mortality was compensatory rather than additive. However, his observations have never been experimentally tested. For Canada, Erb et al. (2001) used statistical models to analyze the dependence between mink and muskrats as estimated from historic harvest data. They observed that mink population cycles lagged behind muskrats 2–3 years in northwestern Canada and 1 year in central Canada, with no distinct time lag in eastern Canada. In addition, the strength of the numerical dependence estimated from autoregressive predator–prey models was greatest in northwestern and north-central Canada and decreased moving through central and east/southeastern Canada. Combining these data, they interpreted this in part as a higher degree of prey specialization by mink in northwestern Canada and suggested, as others have, that mink are likely one factor important in explaining population cycles in Canada.

Movement and Home Range. Movements and home ranges of muskrats depend on the size, configuration, and diversity of aquatic habitat, social interactions, sex, age, season, and environmental conditions. Movements can occur within and between habitats. Movements often are an important mechanism for recovery of low-density muskrat populations reduced by excessive harvest or droughts (Errington 1963; Errington et al. 1963; Simpson and Boutin 1989). Beer and Meyer (1951) believed muskrat movements were ultimately caused by normal physiological cycles and emigrations from drought and high population density. Errington (1951) noted that under extreme population pressure, large numbers of muskrats may make long movements, sometimes as far as 34 km, into totally uninhabitable areas. Muskrats can home from distances of at least 1646 m (Erickson 1963).

Muskrats in Missouri ponds of size <0.4 ha were more likely to move (primarily from pond to pond) than animals in larger ponds (Shanks and Arthur 1951, 1952). Muskrats inhabiting Missouri streams showed no indication of moving to ponds; movements tended to be along the streams with localized concentrations.

In the warmer areas of the southern United States, pronounced movements generally occur at different times and/or in a less seasonal manner. Freeman (1945) found that a so-called spring emigration of muskrats in Mississippi begins in January and lasts for about 6 weeks, coinciding with what is considered the most active breeding period. In Texas, Lay (1945a) regularly found muskrats away from marshes during the winter months.

In more seasonal areas, dispersal movements are most common in spring and fall, with less frequent and more localized movements in summer and winter (Takos 1944; Shanks and Arthur 1951; 1952; Sprugel 1951; Mathiak and Linde 1954; Sather 1958; Erickson 1963; Errington 1963; MacArthur 1978; Proulx and Gilbert 1983). However, timing and frequency of movements may be altered depending on population density and habitat conditions.

Several investigators have suggested changing temperature, ice break-up, and onset of the breeding season as the proximate determinants of the timing of spring dispersal (Sprugel 1951; Mathiak and Linde 1954; Sather 1958; Erickson 1963; Errington 1963); spring flooding, particularly on rivers, may also play a proximate role in the timing, frequency, and distance of spring movements (Errington 1963; Clay and Clark 1985). Muskrats may disperse in spring, after which they become established in a new area, which, under normal conditions, is occupied until the following spring (Schwartz and Schwartz 1959). In Iowa, most spring dispersal activity was over within 6 weeks and territories were then established (Sprugel 1951).

Mean distance traveled during spring dispersal in New York ponds was 224 m (Erickson 1963). Almost all movement was from pond to pond, with a maximum distance traveled of 1600 m. Spring dispersal was biased toward males 3:1. The maximum distance muskrats moved between points of capture on an Alabama pond was about 457 m (Beshears 1951). Sather (1958) discussed cross-country emigration and within-marsh dispersal in spring; females tended to remain in their home territory, whereas the males dispersed. Within-marsh movements ranged from 4 m (i.e., no real dispersal) to 373 m. These estimates, however, were frequently based on animal recaptures 1.5 years apart, and do not necessarily represent single dispersal events. Heatherly (1993) reported long-distance movements by three male muskrats on a Missouri river, two downstream (7.2 and 27.6 river-km), and one upstream (19.6 river-km).

Fall dispersal also appears common in many areas and frequently is a general spatial rearrangement of young of the year into suitable overwintering areas. Errington (1963) suggested that fall dispersal was biased toward subadults. Similar results were noted by Erickson (1963), but noticeable age differences were not observed by Sather (1958) for summer and early fall movements. Shanks and Arthur (1951, 1952) reported that initial fall movements were often precipitated by population strife; they also suggested that pair formation may be involved. They found that by late fall most ponds contained a pair of muskrats. Erickson (1963) also indicated interspecific strife was the impetus for fall movements. Fall movements may also be a result of declining water levels (Arata 1959; Errington 1963) or natural exploratory drifting (Schwartz and Schwartz 1959; Errington 1963). Exploratory drifting was most common in August and September (Schwartz and Schwartz 1959).

Clearly, in habitats that freeze, winter movements will be restricted. MacArthur (1978) found that muskrats radio-tagged in winter were within 15 m of their primary dwelling lodge in 50% or more of the locations. Most foraging occurred within a 5- to 10-m radius of a lodge or push-up and few movements exceeded 150 m. Similarly, the majority of movements of muskrats in an Illinois lake were within 15 m of the home den (Coon 1965).

The percentage of animals moving to new areas can also vary. Only 5% of 1579 tagged muskrats at Horicon Marsh, Wisconsin, moved farther than the adjacent trapping unit (separated by dikes) (Mathiak 1966). Similarly, about 67% of the kits recovered by Dorney and Rusch (1953) on the Horicon Marsh had moved ≤91 m from the last site handled. In South Dakota, Aldous (1947) found that 54% of marked muskrats had not moved from the last release site; only 15% moved >156 m. Similarly, for a 120-acre marsh in central Maine, about 50% of the recaptures were within a 7.6-m radius of the original tagging site (Takos 1944). Less than 30% were beyond a 30.5-m radius and all immatures recaptured were within a 30.5-m radius. In a Manitoba marsh, Clark and Kroeker (1993) estimated that captures of nonresidents represented 7% of the total in summer through fall, but 41% from winter through spring. Virgil and Messier (1996) estimated monthly summer immigration rates for adult muskrats in a Saskatchewan marsh of 0–9%; monthly emigration rates of adults in summer ranged from 0 to 51%. In general, adult immigration was highest and emigration lowest in the higher quality portions of the marsh. Most of these studies were based on trapping results within a given marsh or marsh complex (i.e., longer movements outside the study area may not have been detected).

Although generalizations should be used with caution, animals that inhabit the center of marshes are more apt to have comparatively circular home ranges, whereas those that occupy edge habitat tend to have oblong ranges. Studies have indicated that summer home ranges of marsh muskrats may average around 45–65 m in diameter (Iowa, Errington 1963; Neal 1968; New York, Erickson 1963; Nebraska, Sather 1958), although the actual shapes may vary. Proulx and Gilbert (1983) suggested that those estimates would have overestimated the actual area used intensively by muskrats in an Ontario marsh. They found that areas used intensively by muskrats ranged from 17 to 33 m in diameter, similar to the 7- to 30-m range suggested by Takos (1944). They also noted that summer home ranges increased slightly in late summer. They also found that muskrats inhabiting ponds had summer home ranges that varied from 56 to 85 m in diameter. Marinelli and Messier (1993) reported larger summer home range estimates in a Saskatchewan marsh, with ranges of males (155–200 m in diameter) being larger than those of females (109–138 m in diameter).

River-dwelling muskrats in Missouri may range along the banks up to 183 m and across most rivers (Schwartz and Schwartz 1959). Brooks (1985) estimated home ranges on Pennsylvania and Massachusetts rivers to be 250–400 m long, similar to the results of Stewart and Bider (1974) for ditches in Quebec. Errington (1963) found that riverine muskrats in Iowa may use about 46 m of shoreline.

Muskrats are not the only major herbivore inhabiting some marshes. Nutrias coexist with muskrats in many Gulf Coast marshes, in scattered marshes throughout the western half of the United States, and on the Atlantic seaboard. Apparently, nutrias feed on coarser vegetation than do muskrats and competition for food may not be a serious problem. The two species have been shown to coexist in crowded cages (O'Neil 1949).

Gainey (1949) compared winter feeding of muskrats and nutrias in captivity and found considerable similarity in food preference. He stated that the degree of competition depends on whether nutrias concentrated in the brackish marshes generally preferred by muskrats. Later studies showed that nutrias generally preferred freshwater areas (Lowery 1974). Gainey (1949) indicated that, compared to muskrats, nutrias generally preferred the larger plants and ate more of the rougher, less palatable plant parts. On certain plants, such as alligatorweed (*Alternanthera philoxeroides*), nutrias ate the leaves, whereas muskrats ate the submerged part of the stem and rhizomes.

About 85% of the muskrats in Louisiana inhabit approximately 25% of the available marsh habitat—that portion dominated by *S. olneyi*—during peak density years (Lowery 1974). *S. olneyi* marsh has been gradually changing, and in places disappearing, since the 1940s from drought, hurricanes, industrialization, and eat-outs. The record high 8 million muskrats harvested in the 1945/1946 season declined to 200,000 in 1964/1965. Nutria catches increased from 436 in the 1943/1944 season to over 1.5 million pelts in 1964/1965. However, biologists found only circumstantial evidence that the muskrat decline was related to the nutria increase. Nevertheless, nutria removal from an area inhabited by both species often resulted in a rise in the muskrat population (Evans 1970). In addition to possible competition for food, muskrats may be harassed by nutrias, and they may compete for general living space and elevated areas for retreat during high water (Lowery 1974).

Although beavers and muskrats use many similar habitats (streams, ponds, ditches), we found no reported instances of apparent competition between these species in North America. Because feeding habits generally differ between these species, there is little likelihood of competition for food resources. If suitable bank denning areas (adequate slope, substrate, etc.) are limiting, it is possible that muskrats may be indirectly excluded from isolated areas if beavers occupy these den sites. It seems unlikely that such competition would be common, and certainly unlikely to influence muskrat numbers over large areas. Occasionally, beavers will dam outlets to marshes, thereby raising water levels and reducing emergent vegetation. In such cases, muskrat populations may be negatively affected. However, beaver dams often increase the amount of suitable habitat, including food resources, for muskrats by flooding new areas and slowing water in faster flowing streams. McKinstry et al. (1997) documented use of active beaver lodges by muskrats.

Generally, lack of competition between muskrat and beaver has also been noted in Russia: A brief review on muskrat and beaver interrelations in a lengthy Russian scientific work on muskrats (Sokolov and Lavrov 1993) indicated that the species usually do not compete. However, interestingly, Dyakov (1975) found that there may be some competition between muskrats and beavers for some common food items, mainly roots of aquatic and semiaquatic plants (reeds, cattail, water lily, etc.). Competition is most prevalent in autumn, winter, and early spring, when these food items are consumed intensively by both species. Such competition increased during periods when population density of both species is high.

FEEDING HABITS

Muskrats are chiefly herbivorous (Bailey 1937; Dozier 1953; O'Neil and Linscombe 1976; Perry 1982), although meat may represent a nutritionally important component of the diet for some muskrats

(Campbell and MacArthur 1996b). Muskrats primarily eat portions of aquatic plants (Table 16.3), especially shoots, roots, bulbs, tubers, stems, and leaves.

Muskrat foods and feeding habits vary with habitat, season, water levels, population density, and predator presence (Butler 1940; Errington 1941; Schwartz and Schwartz 1959; Perry 1982; Proulx and Gilbert 1983; Clough 1987; Jelinski 1989; Neves and Odom 1989; Lacki et al. 1990a; Campbell et al. 1998). Habitat type ultimately determines what foods are available for consumption. For example, Willner et al. (1975) studied muskrat feeding habits on one brackish and three freshwater areas in Maryland. More than 30 vascular plants were identified from fragments in the stomachs of muskrats inhabiting freshwater habitats. Green algae constituted the most important group of plants consumed. Only nine plant species were in the diet of muskrats inhabiting brackish water areas; cattail (*T. angustifolia*) was the most significant plant in their diet. Also, muskrat foods are often more diverse in stream or

TABLE 16.3. Foods of muskrats of North America

Northeastern United States[a]
- Cattail (*Typha latifolia, T. angustifolia*)
- Burreed (*Sparganium eurycarpum*)
- Bulrush (*Scirpus americanus, S. acutus, S. fluviatilis*)
- Arrowhead (*Saggitaria latifolia*)
- Sweetflag (*Acorus calamus*)
- Duckweed (*Wolffia* sp. and *Lemna minor*)
- Pondweed (*Potamogeton natans*)
- Sedges (*Dulichium arundinaceum, Carex rostrata, C. lacustris, C. lasiocarpa*)
- Bayonet rush (*Juncus militaris*)
- Pickerelweed (*Pontederia cordata*)
- Freshwater clams, carp, crayfish, turtles, snails

Mid-Atlantic United States[b]
- Bulrush (*Scirpus americanus, S. fluviatitis, S. olneyi, S. robustus*)
- Cattail (*Typha latifolia, T. angustifolia*)
- Cowlily (*Nuphar advena*)
- Pickerelweed (*Pontederia cordata*)
- Marshhay cordgrass (*Spartina patens*)
- Saltgrass (*Distichlis spicata*)
- Big cordgrass (*Spartina cynosuroides*)
- Common reed (*Phragmites communis*)
- Spikerush (*Eleocharis* sp.)
- Carp, mussels, turtles, blue crabs, dead birds, fish

Southeastern United States[c]
- Bulrush (*Scirpus americanus, S. robustus, S. olneyi*)
- Cattail (*Typha latifolia, T. angustifolia*)
- Arrowhead (*Sagittaria latifolia*)
- Waterlily (*Nymphaea odorata*)
- Big cordgrass (*Spartina cynosuroides*)
- Common reed (*Phragmites communis*)
- Crayfish, fish

North Central United States[d]
- Cattail (*Typha latifolia, T. angustifolia*)
- Bulrush (*Scirpus validus, S. fluviatilis, S. acutus*)
- Arrowhead (*Sagittaria latifolia*)
- Waterlily (*Nymphaea tuberosa*)
- Dry grasses (*Eragrostis, Panicum, Echinochloa, Bromus, Muhlenbergia, Agropyron, Elymus*)
- Corn (*Zea maize*)
- Common reed (*Phragmites communis*)
- Duckweed (*Lemna minor*)
- Burneed (*Sparganium eurycarpum*)
- Smartweed (*Polygonum* spp.)
- Willow (*Salix* spp.)
- Poplar (*Populus balsamifera*)
- Maple (*Acer* spp.)
- Shepherds purse (*Capsella bursa-pastoris*)
- Bluegrass (*Poa pratensis*)
- Goldenrod (*Solidago ulmifolia*)
- Barnyard grass (*Echinochloa crusgalli*)
- Beggartick (*Bidens* spp.)
- Hawthorn (*Crataegus* spp.)
- Love grass (*Eragrostis cilianensis*)
- Acorn (*Quercus* spp.)
- Aster (*Aster* spp.)
- Mint (Labiatae)
- Sweetclover (*Melilotus alba*)
- Giant ragweed (*Ambrosia trifida*)
- White clover (*Trifolium* sp.)
- Bluestem (*Andropogon virginicus*)
- White clover (*Trifolium* sp.)
- Bluestem (*Andropogon virginicus*)
- Stonewort (*Chara* spp.)
- Pondweed (*Potamogeton foliosus*)
- Sedge (*Carex crinita, C. rostrata*)
- Pickerelweed (*Pontederia cordata*)
- Crayfish, frogs, turtles, fish, freshwater clams, snails, birds (young)

South Central United States[e]
- Bulrush (*Scirpus olneyi, S. robustus*)
- Cattail (*Typha latifolia, T. angustifolia*)
- Spikerush (*Eleocharis quandrangulata*)
- Arrowhead (*Sagittaria* sp.)
- Maidencane (*Panicum hemitomon*)
- Rice (*Oryza* sp.)
- Rush (*Juncus roemerianus, J. effusus*)
- Pickerelweed (*Pontederia cordata, P. lanceolata*)
- Naiad (*Najas* sp.)
- Lotus (*Nelumbo* sp.)
- Johnsongrass (*Sorghum halepense*)
- Bermuda grass (*Cynodon dactylon*)
- Willow (*Salix* spp.)
- Waterlily (*Nymphaea* spp.)
- Pondweed (*Potamogeton* spp.)
- Lizard's tail (*Saururus cernuus*)
- Paspalum (*Paspalum* sp.)
- Burreed (*Sparganium* sp.)
- Smartweed (*Persicaria* spp.)
- Wild millet (*Echinochloa* spp.)
- Cordgrass (*Spartina* spp.)
- Switchgrass (*Panicum virgatum*)
- Alligatorweed (*Alternanthera philoxeroides*)
- Crayfish, mussels, fish, freshwater clams, turtles

Western United States[f]
- Cattail (*Typha latifolia, T. angustifolia*)
- Bulrush (*Scirpus acutus*)
- Clover (*Trifolium* sp.)
- Arrowhead (*Sagittaria* sp.)

Canada[g]
- Cattail (*Typha latifolia*)
- Bulrush (*Scirpus validus*)
- Pondweed (*Potamogeton* sp.)
- Water horsetail (*Equisetum fluviatile*)
- Reed (*Phragmites maximus*)

SOURCE: Adapted from Willner et al. 1975.

[a]From Johnson 1925; Enders 1931; Seamans 1941; Dozier 1945, 1950; Takos 1947; Gashwiler 1948; Bednarik 1956; Alexander 1956.

[b]From Lantz 1910; LeCompte 1928; Bailey 1937; Smith 1938; Dozier et al. 1948; Cofer 1950; Harris 1952; Wilson 1956a.

[c]From Svihla and Svihla 1931; Freeman 1945; O'Neil 1949.

[d]From Errington 1937c, 1939, 1941; Hamerstrom and Blake 1939; Butler 1940; Bellrose and Low 1943; McCann 1944; Aldous 1947; McLeod 1948; Bellrose 1950; Sprugel 1951; Sather 1958; Arata 1959; Schwartz and Schwartz 1959.

[e]From Louisiana Department of Conservation 1931; Svihla and Svihla 1931; Lay and O'Neil 1942; Freeman 1945; O'Neil 1949; Glass 1952.

[f]From Rawley et al. 1952; Borell and Ellis 1934.

[g]From Butler 1940; McLeod 1948; Fuller 1951.

canal systems. Lake, reservoir, and stream muskrats may feed more opportunistically and appear to feed more on animal matter than do marsh muskrats (O'Neil 1949; Schwartz and Schwartz 1959).

Rangewide, Willner et al. (1975) concluded that bulrush and cattail were the most important muskrat foods throughout the United States and that cattail was the most important in Canada. O'Neil (1949), for example, reported that 80% of the diet of muskrats in the brackish subdelta marshes of Louisiana was *S. olneyi*. Smith (1938) found that 80% of muskrat foods in Maryland marshes consisted of *S. olneyi, S. americanus, T. latifolia*, and *T. angustifolia*. In spite of the obvious importance of bulrush and cattail to muskrats, muskrats will often consume many other available species (Smith 1938; Takos 1947; Sather 1958; Proulx and Gilbert 1983).

Cultivated plants eaten by muskrats include carrots, corn, raw peanuts, clover, alfalfa, soybeans, and apples (Dozier 1953). MacArthur and Aleksiuk (1979) found that muskrats used field crops heavily. Muskrats in captivity have been successfully fed apples, sweet potatoes, rolled oats, corn, fresh lettuce, wheat, rice, and oats (Svihla and Svihla 1931; Bailey 1937).

Feeding habits may also vary seasonally in response to shifts in food availability, energy demands, or food quality. Food is generally scarcer in winter and is often restricted to underground plant parts or to what can be reached under ice (Fuller 1951; Dozier 1953). Parts of the lodge or bedding may be eaten in winter and spring (Errington 1941).

Food quality and palatability are factors influencing muskrat diet choice (Errington 1941; Takos 1947; Bellrose 1950). Campbell and MacArthur (1994), however, found minimal variation in the muskrat's ability to digest five common diets (sedge shoots, bulrush shoots, cattail shoots, cattail rhizomes, and mixed cattail shoots and rhizomes), although digestibility coefficients for dry matter and digestible and metabolizable energy were highest for cattail rhizomes. Lipid accumulation noted in fall and winter muskrats may result primarily from increased intake of energy in winter (Campbell and MacArthur 1998; Campbell et al. 1998). This increased intake of energy coincides with an increase in neutral detergent solubles in cattail rhizomes in late fall and winter (Campbell and MacArthur 1996a, 1998; Campbell et al. 1998).

Predation threat may also influence feeding habits, either directly through the immediate risk of predation while foraging (Clough 1987; Lacki et al. 1990a) or indirectly through the choice of lodge/burrow location (Messier and Virgil 1992). Clough (1987) found that an island population of muskrats foraged primarily in upland sites. He suggested that the near absence of predators on the island allowed muskrats to live and forage in areas that would have otherwise been uninhabitable due to predation. Lacki et al. (1990a) found that muskrats avoided high-quality food in close proximity to denning sites if aquatic escape routes were absent.

Although muskrats are largely herbivorous, in certain habitats or seasons they may consume meat (Lantz 1910; Bailey 1937; Errington 1941; Bellrose 1950; Convey et al. 1989). Stearns and Goodwin (1941) concluded that muskrats on a Delaware tidewater marsh consumed animal matter in an appreciable but undetermined amount during winter. Similarly, Schwartz and Schwartz (1959) indicated that Missouri stream muskrats feed on a variety of fauna, including clams, crustaceans, mussels, snails, and young birds. Sather (1958) suggested that consumption of animal matter may result from a shortage of preferred vegetation. However, Campbell and MacArthur (1994) suggested that carnivory (or scavenging) might serve a more adaptive purpose—maintaining nitrogen balance. They found that muskrats that were fed five common emergent plant diets tended to be in a negative nitrogen balance. This, combined with the high dietary crude protein requirement of muskrats on a diet of emergent vegetation, suggests that supplemental nitrogen may be important, especially in the summer diet. Furthermore, Campbell and MacArthur (1996b) found that muskrats offered mixed diets selected meat in higher proportion compared to its availability. They also found that muskrats were able to efficiently digest high levels of animal tissue without any loss in ability to digest fiber.

Muskrat predation may also play a role in structuring the community and population dynamics of certain prey items, particularly mussels. Such predation may influence the size, age, and species composition of mussels in certain areas (Convey et al. 1989; Hanson et al. 1989; Neves and Odom 1989). Furthermore, Jokela and Mutikainen (1995) noted that muskrat predation affected the spatial distribution of a freshwater clam.

Stomachs (with contents) of 2778 Delaware muskrats (body weight range of 450–1250 g) varied in volume from 2 to 130 cm^3; average volumes for males and females were 22.3 and 23.0 cm^3, respectively (Stearns and Goodwin 1941). Without contents, stomach weights ranged from 0.2 to 3.4 g, with an average of 1.5 g. Svihla (1931) found that Louisiana muskrats consume about 33% of mean body weight each day. At this consumption rate, one large muskrat can consume about 929 cm^2 daily of *S. olneyi* marsh, or about 1858 cm^2 of any other marsh type (O'Neil 1949). Based on research in Manitoba, Campbell et al. (1998) estimated that muskrats consume 750–1000 g of fresh vegetation/kg of body weight each day from spring until late fall. This is slightly higher than the estimate (734 g/kg) reported by Ching and Chih-Tang (1965) for muskrats in outdoor enclosures during spring and fall, but lower than the 1250 g/kg average reported by Akkermann (1975, in Campbell et al. 1998) for captive muskrats in summer. Campbell et al.'s (1998) estimate of daily winter consumption rate of 550–600 g/kg was similar to the finding in earlier work with captive winter-acclimated muskrats fed cattail rhizomes (Campbell and MacArthur 1996a).

The muskrat does not generally store large quantities of food (Smith 1938; Errington 1941; Schwartz and Schwartz 1959), although food caches of various amounts have been noted (Carter 1922; Errington 1963:22; Earhart 1969; Clough 1987). Also, MacArthur et al. (1997) found that, in addition to conveying thermal benefits in winter, social aggregation may influence food intake. Specifically, grouped animals in a laboratory ingested nearly three times as much food as animals feeding alone, which they concluded was a result of social facilitation rather than any specific gain in individual foraging efficiency arising from aggregation behavior out of water. However, the extent of such behavior in natural settings in unknown, and because aggregation may deplete local food supplies more rapidly in the restricted winter environment, it is unclear whether aggregation would convey any foraging-related advantages to winter survival of muskrats.

HABITAT

Muskrats generally prefer lentic or slightly lotic water containing vegetation. Such areas may be in coastal and inland marshes, lakes, ponds, sloughs, streams, and rivers. However, muskrats are adaptable, and they inhabit a wide range of community types, including strip-mined ponds, ditches, canals, and pits. They are usually absent from large open bodies of water due to excessive wave action (Errington 1963). In general, muskrats require water and some type of submergent, emergent, floating, or shoreline vegetation. However, they appear able to adapt to upland habitats in the absence of significant predation threats (Clough 1987). Errington (1963:Chapters 12–14) provides the most comprehensive, but general, overview of the habitats of the different subspecies of muskrats.

It is important to reiterate that habitat selection in muskrats is density dependent (Errington 1963; Messier et al. 1990; Clark and Kroeker 1993; Clark 1994; Virgil and Messier 1996). Because most aquatic habitats are heterogeneous, interpretation of habitat quality based on population densities can be misleading. High densities of animals may actually be indicative of lower quality habitat inhabited by subordinate animals. These animals may exhibit low survival and/or reproductive rates.

Research by biologists of the Louisiana Department of Conservation (1931) indicated the importance of humus in determining muskrat abundance. Areas that supported the greatest numbers of muskrats contained peaty humus several centimeters thick. This humus was about 90% vegetable matter. Humus was considered important to muskrats because it provided a medium from which roots and burrows could be dug with ease.

In Louisiana, highest muskrat populations occur in brackish areas where fresh and salt water mix (O'Neil and Linscombe 1976). Louisiana has >1.6 million ha of marsh, and most marsh types of the country are represented. Historically, >80% of the muskrats were produced on about 405,000 ha of brackish marsh where Olney bulrush (*S. olneyi*) was dominant or subdominant (O'Neil 1949). *S. olneyi* makes up 90% of the Gulf Coast muskrat's food supply. Advantages of *S. olneyi* marsh over other types of marsh are that it yields the greatest weight per square meter of edible material and it grows throughout the year (O'Neil 1949). The brackish marshes of the Texas coast, characterized by marshhay cordgrass or wiregrass (*Spartina patens*), saltgrass (*Distichlis spicata*), needle rush (*Juncus romenanus*), Olney bulrush, and saltmarsh bulrush (*S. robustus*), are also a very productive habitat for muskrats. The cordgrass is highly desirable for lodge building, but it must be controlled so it will not exclude food plants (Lay and O'Neil 1942).

Data presented by Lynch et al. (1947) and Palmisano (1972b) substantiate the importance of brackish marshes for muskrats of Gulf coastal marshes. Palmisano (1972b) reported that 72% of the muskrat houses counted on aerial surveys were in brackish marshes, although this habitat type accounted for only 37% of the habitat within the survey area. Less than 20% of the muskrat houses were recorded outside the brackish marsh in normal years. These data indicate that muskrat density in the brackish marsh was twice the average density for the entire coastal marsh area. Saline marshes represented 12% of the habitat and contained 14% of the houses, whereas fresh and intermediate marshes accounted for 40% of the survey area and contained only 13% of the houses observed. The maximum yield of muskrats in the brackish marsh was 10 times (16 pelts/ha) the maximum harvest in the fresh marshes. However, during periods of drought in the preferred brackish marshes, saline, intermediate, and fresh marshes increase in importance because they generally are not as affected by dry weather (Palmisano 1972b).

Scirpus spp. is also an important habitat component for muskrats in certain areas along the Atlantic coast. On Blackwater National Wildlife Refuge, Maryland, Dozier et al. (1948) reported that the fresh to slightly brackish *S. olneyi–Typha* spp. marshes adjacent to timbered areas were best for muskrat production (harvest). Similarly, on Montezuma National Wildlife Refuge, New York, the abundant supply of blue flag cattail (*T. glauca*) was the greatest factor contributing to the accelerated and continuous growth, large pelt size, and record weight of muskrats (Dozier 1950). Wilson (1949) listed needle rush, sawgrass (*Cladium jamaicense*), and big cordgrass (*Spartina cynosuroides*) as undesirable muskrat vegetation in coastal North Carolina; *Scirpus–Typha* marshes were considered desirable communities to support muskrats. *Scirpus–Typha* marshes are also among the most common plant associations frequented by muskrats of the inland portions of the United States and Canada (Sather 1958; Errington 1963; Proulx and Gilbert 1983; Kroll and Meeks 1985; Messier et al. 1990; Clark and Kroeker 1993; Clark 1994). Other plants of various importance often found in shallower areas include bur-reed (*Sparganium eurycarpum*), arrowhead (*Sagittaria latifolia*), whitetop (*Scolochloa festucacea*), reedgrass (*Phragmites australis*), sedges (*Carex* spp.), and horsetail (*Equisetum* sp.). Important submergents include pondweeds, milfoil (*Myriophyllum* spp.), and coontail. Lacki et al. (1990a) discussed plant associations for muskrats inhabiting a fen wetland in New York.

Hamerstrom and Blake (1939) found that Wisconsin ditches that were well shaped or had deep and swift water were practically unused by muskrats. Slow-running streams with numerous aquatic plants had the largest muskrat populations in northern Mississippi; old stream runs and lakes were preferred over flowing water areas (Freeman 1945). Based on his work in Iowa, Errington (1937c) felt that an evaluation of the suitability of stream habitat for muskrats should consider that (1) corn fields may enhance the attractiveness of stream environments to muskrats, (2) stream habitat for muskrats is dynamic, (3) semipermanent retreats that provide security are important, and (4) preferences for various aspects of stream habitat may vary greatly during the year. Overall, in stream environments where emergent vegetation is often absent, the presence of abundant submergent vegetation and/or herbaceous vegetation along the bank generally determines the suitability of the area for muskrats (Brooks and Dodge 1981, 1986; Brooks 1985; Nadeau et al. 1995). Low water velocity and steeper banks with good clay–loam soils were also very important. Arrowhead was the dominant habitat component for muskrats on the upper Mississippi River (Clay and Clark 1985).

Although the presence of vegetation must certainly be important, Greenwell (1948) and Shanks and Arthur (1952) found no correlation between muskrat populations in ponds and the type or kind of vegetation present. In northern Canada, muskrats commonly occupy ponds dominated by submergents (Ruttan and Wooley 1974), primarily pondweeds and milfoil.

BEHAVIOR

Activity Patterns. Muskrats are generally nocturnal, although they are sometimes out during the day, especially in spring and early summer. In Louisiana, about 80% of activities are nocturnal under normal conditions (O'Neil 1949). Most muskrats were in dwellings from daylight to midafternoon, after which the adults were seldom in the nest. Brooks (1985) reported that riverine muskrats in Pennsylvania and Massachusetts were active, primarily nocturnal, from 53% to 63% of each 24-hr period. They frequently returned to houses around midnight for 1–2 hr. Because survival depends largely on concealment, muskrats are wary and spend a large amount of time in tunnels and burrows or in their nests and feeding shelters.

Muskrat activity is apparently related to weather and environmental factors. Hensley and Twining (1946) indicated that muskrats in a California marsh were less active during periods with a full moon. However, Stewart and Bider (1977) found nocturnal light intensity (a combination of moon phase and cloud cover) was positively correlated with activity in Quebec. Stewart and Bider (1977) also found rainfall was the most important weather factor influencing muskrat activity in summer. Rainfall resulted in more frequent and longer movements, and diurnal rainfall resulted in more diurnal activity. On days without rain, mean daily temperature was the most important variable influencing summer activity, with activity greater following warmer days. However, summer activity declines if the weather becomes abnormally hot and dry (Lay 1945a; Stewart and Bider 1977). Svihla and Svihla (1931) reported that a very windy night or warm weather results in little muskrat activity. Rainfall and cool weather are the most favorable factors for muskrat activity in Louisiana (Svihla and Svihla 1931; Louisiana Department of Conservation 1931).

Feeding. Bailey (1937) described muskrats as "dainty feeders." The front feet are employed to gather food and transport it to the mouth. Frequently muskrats will rear on their hind legs, using the tail for support. Food is generally not eaten in the nest, but in some other area that affords protective cover. In winter in northern areas, food is brought into feeding huts and push-ups. Muskrats are sometimes active during the day, but most feeding is done at night (Freeman 1945; Errington 1963; Brooks 1985). Brooks (1985) noted that riverine muskrats in two areas of the northeastern United States generally made rapid (10–50 m/min) and direct movements to and from feeding areas.

When preying on mussels, muskrats dive and carry them to shore in their front feet (Convey et al. 1989; Hanson et al. 1989). They insert their lower incisors between the ventral edges of the valves, pry upward, and break the shell. The visceral mass is eaten, and the shells, with one side usually intact, are discarded in middens on shore or in shallow water near logs. Middens also occur near the entrances to feeding burrows (J. Erb, pers. obs.) on many rivers.

Territoriality. Muskrats are territorial (Errington 1943, 1963; Sather 1958; Willner et al. 1980; Messier et al. 1990; Hjälten 1991; Marinelli and Messier 1993), especially during the breeding season. Shanks and Arthur (1951) found the summer range of stream-dwelling muskrats was equal to about one half the distance between colonies. Similarly, Schwartz and Schwartz (1959) reported that houses are seldom closer than 8 m. Muskrats were described by Fuller (1951) as not being particularly tolerant of conspecifics, even though they are semicolonial.

Marinelli and Messier (1993) found summer home ranges exhibited very little intrasexual overlap; intersexual overlap generally involved males overlapping with one or more females, with females overlapping with only one male. During winter, muskrats are less territorial and more gregarious (Schwartz and Schwartz 1959; Errington 1963; Marinelli and Messier 1993). Through huddling and the maintenance of active feeding huts and push-ups, gregarious behavior may convey thermoregulatory and foraging advantages to muskrats in northern areas if the benefits exceed the costs of maintaining territories (MacArthur 1978; Bazin and MacArthur 1992).

Locomotion. On land, muskrats move with an amble or slow hop and, unless alarmed, they enter water slowly. They swim with their hind feet while holding the front feet against the chin. The tail is trailed and, although not generally used in surface swimming, may be used as a rudder in turning. Swimming speed is about 3–5 km/hr. In underwater swimming or in water with current, the tail is used more extensively. Fish (1982, 1984) concluded that, for water with current, the laterally flattened tail is important both for providing propulsion and preventing yawing.

Communication. Adults are usually silent (Smith 1938), but Svihla and Svihla (1931) and Schwartz and Schwartz (1959) reported that adult muskrats emit low squeaks, loud squeals, and snarls, and when cornered or fighting, they chatter. Both sexes may emit a high-pitched "n-n-n-n" during the breeding season. According to Schwartz and Schwartz (1959), muskrats may warn of impending danger by slapping their tail on the water. The young have a squeaky cry or squeal and are comparatively noisy (Smith 1938). When the young call, the mother attends to them rapidly. Frequently a mother may attempt to carry young away in her mouth. Svihla and Svihla (1931) stated that muskrats they observed did not indulge in play. They described a possible gesture of affection performed by males and females, which consisted in nibbling on each other's back and neck, especially during the breeding season.

Body Care. Muskrats are very clean and spend a great deal of time washing and combing their fur. Hensley and Twining (1946) described females using their front feet to groom and comb the backs and heads of suckling young. Excrement is usually deposited in water, and nests or houses are normally clean; pellets are sometimes found on a resting platform or feeding raft (Svihla and Svihla 1931). Although muskrats often defecate randomly throughout the marsh as they walk and swim, Smith (1938) described defecation "posts" where muskrats defecate repeatedly in definite spots. Most posts are on a slight elevation, such as the end of a log or piece of sod or pile of vegetation. Posts might be used by many animals and any suitable item in the muskrat's habitat may be used as a defecation post. Generally 3–12 black, oval droppings are deposited at one time (Schwartz and Schwartz 1959).

Aggression. Muskrats frequently jump when it is impossible to climb (Svihla and Svihla 1931). When caught or cornered, they fight desperately, striking at their opponents. Their long, sharp incisor teeth are capable of inflicting much damage. Muskrats encountered away from water sometimes are fierce and have been known to attack people without apparent provocation (Lantz 1910).

Muskrats readily fight each other, especially when food is scarce. Individuals are often more tolerant of their own than the opposite sex, with females being more tolerant (Beer and Meyer 1951; Schwartz and Schwartz 1959). Errington (1951) stated that population density was one of the most influential factors affecting muskrat behavior. During weaning, competition is keen when populations are dense, and older young may eat the newborn (Errington 1963; Neal 1968). Intraspecific strife among young muskrats in Nebraska was related to overpopulation (Sather 1958). A greater tolerance of crowding was noted during the "cyclic high," which was characterized by increased numbers of subadult and young-of-the-year breeders, and likely high habitat quality. However, Caley and Boutin (1985) estimated that infanticide occurred in 5% of the litters in an Ontario marsh. They did not find evidence that overcrowding, food limitation, or the presence of orphaned pups was responsible.

When muskrats are injured, they keep the wound clean by licking. They gnaw away infected flesh, and legs amputated by traps or other injuries reportedly heal rapidly; one-, two-, and three-legged muskrats have been captured (Svihla and Svihla 1931).

MORTALITY

Muskrats have a high reproductive potential and generally a short life span. However, one tagged Missouri muskrat reportedly lived 4 years in the wild (Schwartz and Schwartz 1959). Factors limiting muskrat populations include diseases, parasites, predators, accidents, climatic factors, food, intraspecific strife (fighting), and exploitation. These factors vary widely by area, season, and population level. Reported estimates of juvenile survival rates are shown in Table 16.4. In many instances, however, estimates are based on a comparison of placental scar counts with the age ratio in the harvest. Thus, many estimates may be high because they do not account for summer mortality of adult females. Furthermore, the possible influence of differential trap susceptibility on fall age ratios is unclear. Clearly, however, juveniles experience high losses during their first year of life.

Boutin et al. (1988) did not find that survival of young was influenced by litter size. Up to weaning, survival of individuals within a litter does not appear to be independent; litters tend to disappear completely or have a higher proportion of individuals survive than expected (Dorney and Rusch 1953; Boutin et al. 1988). Predation is often the suggested explanation. Following weaning, individuals within a litter tend to have independent survival probabilities.

Less information is available on summer and overwinter survival rates for adults. Virgil and Messier (1996) estimated summer survival rate ranged from 34% to 82% for adult muskrats in a Saskatchewan marsh. The lower survival estimate was for portions of the marsh considered poorer quality habitat. Clay and Clark (1985) estimated summer survival of 22–34% for adult muskrats on the upper Mississippi River. Using an estimated 37% overwinter survival for adults and subadults combined, they found an annual survival of adults on the Mississippi River of 13%. Subsequent analysis by Clark (1987) indicated that annual survival of adult muskrats on the Mississippi River was 6%. Simpson and Boutin (1993) estimated overwinter survival of 15% and 17% for adult muskrats in the Yukon and Ontario, respectively. However, muskrats were captured throughout the summer and early fall, and survival estimates may thus include some summer losses as well. Boyce (1977) suggested that adult survivorship and the percentage of adults in the population generally are inversely related to latitude.

Diseases and Parasites. Diseases of Maryland muskrats described by Smith (1938) include abscesses, septicemia (from *Chlamydia* sp.), coccidiosis (caused by a protozoan), leukemia, gallstones, and inflammation of the eye. Severe malocclusion of the incisors was described by Alexander and Dozier (1949). Mathiak (1966) reported pasteurellosis, hepatitis, uremia, pneumonia, and heart and liver degeneration in muskrats, but indicated that these conditions had an insignificant effect on local populations; tularemia, from *Francisella tularensis* (= *Pasteurella tularensis* = *Bacterium tularense*), and Errington disease (infection with *Bacillus piliformis*) were considered important diseases in muskrats.

Errington disease, or hemorrhagic disease, was first discussed in detail for muskrats by Errington (1946a). It is highly infectious and various studies have been performed to determine the causative agent. Lord et al. (1956a, 1956b) described the pathological changes and etiology of the condition and found the bacterium *Clostridium* sp. common to cultures from diseased animals. Wobeser et al. (1978) diagnosed Tyzzer disease in muskrats from a Saskatchewan marsh. Other investigators found similar infections of muskrats in other areas (Karstad et al. 1971; Chalmers and MacNeill 1977). Pathological and epizootiological similarities of Tyzzer and Errington diseases suggested to Wobeser et al. (1978) that these two diseases have a similar etiology, and Davidson and Nettles (1997) considered them to be the same. Tyzzer disease appears to be more common in the northern United States and Canada (Davidson and Nettles 1997).

TABLE 16.4. Estimated juvenile survival rates of muskrats in North America

Location	Survival rate (%)			Source
	Birth to Autumn	Autumn to Spring	Annual	
Marshes				
Iowa	32	—	—	Errington et al. 1963
	50	—	—	Errington 1963
Massachusetts	27–39	—	—	Chamberlain 1951
Connecticut	20–70	—	—	Smith et al. 1981
New York	38	—	—	Alexander 1951
Idaho	16-47	—	—	Reeves and Williams 1956
California	44	—	—	Sooter 1946
Ontario	31–37	32[a]	11	Proulx and Gilbert 1983
	—	—	9	Boutin et al. 1988
	—	22	—	Simpson and Boutin 1993
Maine	80	—	—	Clough 1987
Wisconsin	10–36	—	13	Mathiak 1966
	18	—	—	Dorney and Rusch 1953
Saskatchewan	—	—	10–35	Marinelli et al. 1997
	45	19	9	Phillips 1979
	59–87	—	—	Virgil and Messier 1997
Manitoba	54	—	—	Olsen 1959a
	—	9–42[a]	4–17	Clark and Kroeker 1993
Rivers/Creeks				
Alberta	85	68	58	Fuller 1951
Ontario	42–61	—	—	Proulx and Buckland 1986
Quebec	38	41	16	Stewart and Bider 1974
Iowa	34–55	37	15	Clay and Clark 1985
	—	—	16	Clark 1987
Lakes/Ponds				
Ontario	57	—	—	Proulx and Buckland 1986
New York	63	—	—	Erickson 1963
Yukon	—	13–43[a]	—	Simpson and Boutin 1989
	—	44	—	Simpson and Boutin 1993
Multiple habitats				
New Brunswick	65	—	—	Parker and Maxwell 1984
Missouri	33	—	—	Schwartz and Schwartz 1959
Minnesota	53	—	—	McCann 1944
Illinois/Michigan	30	—	—	Baumgartner and Bellrose 1943

[a] Adults and subadults combined.

Hockett (1968) isolated 19 genera of bacteria from muskrats. The most frequently found species (with percentage occurrence) were *Citrobacter freundii* (53%), *Enterobacter/Aerobacter* (57%), and *Proteus vulgaris* (40%). Cultural examinations of 200 muskrats by Friend and Muraschi (1963) yielded 28 isolates characteristic of *F. tularensis*.

A literature review by Jilek (1977) indicated 66 species of parasitic helminths in muskrats: 36 trematodes, 11 cestodes, 15 nematodes, and 4 acanthocephalans. General parasitological surveys of muskrats have been completed by Barker (1915), Chandler (1941), Ameel (1942), Penn (1942), Rausch (1946), Edwards (1949), Meyer and Reilly (1950), Fuller (1951), Knight (1951), Abram (1968b), Gash and Hanna (1972), and Rice and Heck (1975). A detailed review of the earlier literature on diseases and parasites was provided by Takos (1940). Generally, frequency occurrence in muskrats was 78–93% for trematodes, 0–41% for cestodes, and 8–25% for nematodes. However, muskrat parasite burdens are variable and are frequently related to habitat, sex, and age of the host (Anderson and Beaudoin 1966; Abram 1968a; Cromer 1968).

The major trematodes (≥35% frequency), including evaluations by Jilek (1977) and MacKinnon and Burt (1978), are *Echinochasmus schwartzi, Quinqueserialis quinqueserialis, Pseudodiscus zibethicus, Nudacotyle novica, Notocotylus filamentis, N. urbanensis, N. quinqueserialis, Echinostoma revolutum, Echinostomum coalitum,* and *Wardius zibethicus*. Important cestodes (≥10% frequency) reported, including work by Byrd (1953), are *Hymenolepis evaginata, Taenia tenuicollis, T. opaca,* and *T. taeniaeformis*. Frequent nematodes (≥10% frequency) are *Trichuris opaca, Rictularia ondatrae*, and *Capillaria hepatica* (see also Borucinska and Nielsen 1993).

Penn and Martin (1941) reported that about 9% of 1032 muskrats surveyed in Louisiana were infested with the pentastome nymph *Porocephalus crotali*. Other internal organisms that infect muskrats include the tapeworm *Taenia crassicollis*, the nematode *Dirofilaria* sp., the liver fluke *Parametorchis* sp. (Smith 1938), and two protozoans—*Giardia* sp. and probably *Trichomonas* sp. (Penn 1942). Uncommon parasites of muskrats are discussed by Beckett and Gallicchio (1967).

Errington (1942) described a dermatomycosis (fungal disease of the skin), chiefly attributed to *Trichophyton mentagrophytes*, which affects young muskrats. The condition was present in 9.6% of the litters examined and had an apparent mortality rate of 91.8% for infected individuals. This ringworm disease, which is reportedly of zoonotic importance, was also described in muskrats by Dozier (1943).

Ectoparasites of muskrats include the mites *Tetragonyssus spiniger*, *Ichoronyssus spiniger*, and *Laelaps multispinosus* and the larvae of the dipteran *Sarcophaga* sp. (Smith 1938; Penn 1942). Generally, acarina incidence is about 98%. Other maladies that may affect muskrats include yellow fat disease (nutritional panniculitis, steatitis), a dietary condition (Debbie 1968), and lumpy jaw (actinomycosis) (Dozier 1943).

Predation. Muskrats are heavily preyed on, but in better habitat their rate of production may be high enough to prevent population decline. For Louisiana, O'Neil (1949) listed 17 muskrat predators in descending order of effect: mink, raccoon (*Procyon lotor*), barn owl (*Tyto alba*), barred owl (*Strix varia*), alligator (*Alligator mississippiensis*), ant, northern harrier (marsh hawk; *Circus cyaneus*), eastern cottonmouth

(*Agkistrodon piscivorus*), bullfrog (*Rana catesbeiana*), garfish (*Lepisosteus* sp.), bowfin (*Amia calva*), snapping turtle (*Chelydra serpentina*), largemouth bass (*Micropterus salmoides*), crab (Decapoda), hog (*Sus scrofa*), house cat, and dog. The last 8 predators were identified from only limited scientific data. Although O'Neil (1949) included ants as a predator of muskrats, studies in Louisiana by Newsom et al. (1976) failed to reveal any significant deleterious effect of fire ants (*Solenopsis invicta*) on whether muskrat houses were active or contained young.

Marsh hawks and raccoons were listed by Lay (1945a) as common winter predators of muskrats in a Texas marsh. Hawks, owls, coyotes (*Canis latrans*), gray fox (*Urocyon cinereoargenteus*), red fox, dogs, raccoons, long-tailed weasels (*Mustela frenata*), snakes, snapping turtles, and certain fishes were also listed as predators of lesser importance.

Predators noted by Sather (1958) on muskrats in north-central Kansas were mink and coyote; badger (*Taxidea taxus*) and raccoon were also suspected predators in a few cases. Muskrat predators reported in Maryland marshes include bald eagle (*Haliaeetus leucocephalus*), marsh hawk, barred owl, and great horned owl (*Bubo virginianus*) (Smith 1938). Smith (1938) noted, however, that most birds of prey may have scavenged many of the muskrats they fed on. Additional predators of muskrats include weasels and pickerels (*Esox* sp.) (Lantz 1910; Sather 1958).

Wilson (1954b) found that muskrats were a minor fall–winter food of mink and an unimportant fall–winter food of otters (*Lontra canadensis*) in North Carolina. Similarly, Glass (1952) could not verify mink predation of muskrats in Oklahoma ponds. In Missouri, Schwartz and Schwartz (1959) reported mink and humans as the most important muskrat predators. The predatory relationship between mink and muskrat in Iowa was discussed by Errington (1943, 1951, 1954b). His 1943 analysis led to several conclusions: (1) environmental conditions, intraspecific tolerance, and drought (see also Proulx et al. 1987) are important in predisposing muskrats to mink predation, and, with few exceptions, mink had little influence on the net mortality of muskrats; (2) when mink predation is severe, losses of muskrats from other causes proportionally diminish (a decrease of mortality from predation was largely offset by increased killing of young by older muskrats or from increased miscellaneous losses); and (3) little increase in muskrat trapping revenues could be realized by mink repression in north-central areas where muskrat trapping occurs in fall and winter. His 1954 evaluation was based on analyses of 13,176 mink scats. Of the 2415 scats containing muskrats, 1600 were believed to have resulted from scavenging on muskrats with hemorrhagic disease, 100 from miscellaneous scavenging, and only 674 from predation. Therefore, mink predation pressure on muskrats was not considered great. However, Errington (1963) acknowledged that lower prey diversity in more northern areas may result in a stronger predator–prey relationship in those areas. The results of Erb et al. (2001) provide some support for the notion that predator–prey interactions between mink and muskrat in Canada may be stronger in areas of suspected lower prey diversity.

Wilson (1953) noted that raccoons were an important muskrat predator in the Currituck marshes of North Carolina. He documented extensive damage to muskrat houses by raccoons during the breeding season but negligible damage in December; apparently invasion of houses was influenced by the presence of young muskrats. Raccoons appeared to be an unimportant factor in the destruction of muskrat houses in the marshes of the Maryland eastern shore (Nicholson and Davis 1957). However, Harris (1952) found that 50% of 1892 muskrat structures in Maryland were disturbed by predators, largely raccoons. Muskrat remains occurred in 19.2% of 150 raccoon stomachs; 3% of 551 raccoon droppings collected in the marsh contained muskrat remains. Errington (1963) observed that raccoon predation on muskrats in Iowa marshes was primarily focused in shallow water areas where disease was impacting muskrats.

Errington and Scott (1945) documented heavy depredation of muskrats by red foxes under summer drought conditions in a marsh in north-central Iowa. Although the severe drought would have resulted in muskrat losses without fox depredation, the foxes seemed to have reduced muskrats in such a manner as to be noncompensatory. A possible net decrease of about 25% in the income of trappers was caused by fox depredation. Similarly, of 17 red fox stomachs from Maryland examined by Harris (1952), 59% had muskrat remains and 56% of 132 fox droppings contained muskrat. In Europe, Danell (1978, 1985) provided evidence that red fox may be the driving factor behind 4-year population cycles of muskrats in Sweden. Erb et al. (2000, 2001) suggested that red fox predation may be more important to muskrat populations in eastern Canada than in western Canada.

Clough (1987) described a muskrat population persisting primarily in upland habitat on an island off the coast of Maine. The near absence of mink (and other predators) has probably allowed such upland habitat use. Soper and Payne (1997) also suggested that, based on circumstantial evidence, the introduction of mink into Newfoundland had a depressing effect on muskrat populations. Predation has undoubtedly been an important evolutionary influence on muskrat life history.

Catastrophes and Accidents. Adverse or abnormal climate is an important factor that can limit muskrat population growth, as it can affect salinity, pH, dissolved oxygen, water tables, and food plants (Ferrigno 1967). Floods, hurricanes, and drought are natural factors that cause the largest degree of muskrat mortality and habitat destruction (Lantz 1910; Svihla and Svihla 1931; Errington 1963; Wilson 1968).

Muskrats in stream habitats seem especially vulnerable to sudden increases in water levels. Floods from heavy rains during late winter and spring were an important factor limiting muskrat populations in northern Mississippi; such floods cause stream rises, which fill muskrat burrows and drown the young (Freeman 1945). Errington (1937b) also noted drowning was an important mortality factor for young muskrats in Iowa. Bellrose and Low (1943) found that muskrats flooded in the Illinois River Valley primarily sought to remain on top of lodges, then on floating rafts of vegetative debris, and lastly on branches of willows (*Salix nigra*) and buttonbush (*Cephalanthus occidentalis*). During the first few days of displacement by flooding, muskrats were lethargic. Intraspecific strife, illustrated by fighting, occurred among muskrats most exposed, and death was frequently caused by muskrat-inflicted wounds and possibly drowning, disease, or exposure.

Equinoctial storms (i.e., rapid tidal influxes) are more damaging than rainwater floods in coastal marsh areas (Freeman 1945). O'Neil (1949) stated that Gulf Coast muskrats were safe from storms as long as the tides did not raise water levels over the tops of marsh grasses. Storms frequently cause high water to be blown into the marshes rapidly, giving muskrats little time to raise their houses or escape to higher ground. They may be buffeted by the rough water until they become exhausted and drown. Food is scarce or often damaged because of high and rough water, and muskrats driven from their homes are exposed to enemies (Freeman 1945). Floods in Maryland marshes have caused many muskrats to drown, but even more to succumb to disease, starvation, and lack of fresh water (because of the tide) (Smith 1938). The Louisiana Department of Conservation (1931) described how life rafts were constructed as refuges for muskrats during the 1928 flood.

Habitat deterioration from low precipitation and the associated salinity increase was identified as the causes of the decline of muskrats on the Atlantic coast during the early 1940s (Dozier 1947b). Drought may also concentrate muskrats and cause increased social pressures and higher rates of predation and disease. Errington (1938b) observed that a large proportion of muskrats in drought-inflicted habitats stayed in familiar range. Although they frequently suffered heavy mortality, they usually survived better than animals that attempted to disperse.

Muskrats subjected to low water conditions during summer may be less affected than during cold weather. Searching for food may become a special problem for muskrats during winter drought periods. Dry periods allow the lowered water to freeze deeper, thus sealing off food resources ordinarily available in unfrozen mud and water; animals may also become entrapped beneath the ice (Schwartz and Schwartz 1959). Snowless winters in northern regions may allow similar deep freezing of muskrat marshes. For example, heavy ice formation associated with lack of snow cover during winter resulted in a freeze-out and heavy mortality in muskrat populations in northeastern North Dakota (Seabloom and Beer 1964). Similar effects of severe winters on muskrats in New Jersey were described by Pancoast (1937).

Oil spills also are likely to have severe consequences for muskrats (see, e.g., Heatherly 1993). McEwan et al. (1974) found that heavy oiling increased the thermal conductance of muskrats by as much as 122%. To compensate for the loss of thermal insulation, oiled muskrats increased their dry-matter intake 2.5-fold. The investigators doubted that muskrats exposed to moderate quantities of oil could survive under natural conditions. Similarly, Wragg (1954) found that fuel oil had a persistent and cumulative wetting effect on muskrats.

Compensation. Although disease, predation, parasitism, climatic factors, harvest, and accidents act directly on individual muskrats, increasing one mortality factor may not have a net effect on annual survival if such mortality is compensatory. Certainly, for all wildlife populations there will be some level at which mortality becomes additive. However, because muskrats generally have low annual survival rates, the threshold level at which further mortality becomes additive in muskrats is likely to be comparatively high, that is, the opportunity for compensatory mortality is great. Density-dependent survival is most likely the underlying mechanism leading to compensatory mortality.

The concept of compensation was first derived by Errington (1946b, 1956, 1963), based in part on his extensive research on muskrats. His observation of relative constancy in spring breeding densities led him to believe that compensatory adjustments took place within the year. Subsequent research, discussed previously, supports density-dependent juvenile summer survival, and possibly density-dependent overwinter survival. In relation to harvest, Clark (1987) provided the most formal assessment of compensatory mortality in muskrats. He observed an increase in birth-to-trapping survival of juveniles that compensated for increases in previous harvest mortality.

In summarizing his work on muskrats, Errington (1963) largely concluded that predation did not have a depressing effect on muskrat populations. He felt that animals most susceptible to predation were those already in poor condition from poor habitat quality, fighting, and/or disease. However, there have been no formal experiments quantitatively evaluating the population impacts of predation.

Several miscellaneous examples of other apparent mechanisms for compensatory mortality have been noted. Errington (1937b, 1948) suggested that muskrat drowning rates tend to rise under natural conditions as populations become denser, and Sather (1958) indicated that hemorrhagic disease appeared associated with overpopulation. Penn and Martin (1941) related high parasitism to high muskrat populations in Louisiana. However, preliminary data of O'Neil (1949) indicated that an inverse relationship existed between degree of parasitic infestation and muskrat density: high muskrat populations (>25/ha) had a 5.7% infestation rate, whereas low densities (<25/ha) had an infestation rate of 13.8%. However, he pointed out that because high populations may decline very rapidly, a significant lag in degree of infestation may occur because of the time necessary for parasites to mature.

AGE ESTIMATION AND AGE RATIOS

Many techniques have been suggested for estimating age in muskrats: appearance of internal reproductive organs (Errington 1939; Alexander 1951), structure and weight of the baculum (Elder and Shanks 1962), size and structure of external genitalia (Baumgartner and Bellrose 1943), pattern of pelt primeness (Lay 1945a; Applegate and Predmore 1947; Shanks 1948; Linde 1963), length of the pelt (Lay 1945a), tail length (Dorney and Rusch 1953; LeBoulengé 1977), dental and skull characteristics (Brohn and Shanks 1948; Gould and Kreeger 1948; Alexander 1951; Galbreath 1954; Sather 1954, 1956; Doude van Troostwijk 1976), and eye lens weight (Le Boulengé 1977). The appropriateness of an age determination technique varies with time of year, status of the animals (dead/alive), accuracy level required, and availability of the necessary materials (pelts, skulls, etc.). Age determination has primarily focused on differentiating juvenile and adult muskrats or estimating the age (in days) of juveniles. We are unaware of reliable methods for separating year classes of adult muskrats.

Methods described by Baumgartner and Bellrose (1943) appear to be the most reliable for differentiating live juvenile and adult muskrats. The penis can be examined by palpating the urinary sheath. Adult males have a dark penis >5.1 mm in diameter with a blunt, round tip, whereas juvenile males have a lighter red penis <5.1 mm in diameter with a knob-shaped tip. Adult females have a thin or missing vaginal membrane, whereas immature females have a thick membrane. These methods have produced accurate results (95–100%) (Schofield 1955; see also Erb et al. 1999), although the method may be less reliable during the breeding season (Alexander 1951). For juveniles, growth curves have been formulated based on body mass (Iowa, Errington 1939; Wisconsin, Dorney and Rusch 1953; Nebraska: Sather 1958; Manitoba, Olsen 1959a; New York, Erickson 1963), total body length (Errington 1939; Sather 1958; Erickson 1963), tail length (Dorney and Rusch 1953; Sather 1958; Erickson 1963), and hind foot length (Erickson 1963). As with any method, however, caution should be exercised when estimating age of muskrats using growth curves developed in different areas.

A wide array of methods is available for assessing age of dead muskrats, although most of the methods were developed for use from late summer through late winter. The most commonly used methods for separating juveniles and adults have been the pelt primeness and molar fluting techniques (Moses and Boutin 1986).

The pelt primeness method is based on differences in the primeness patterns observed on skinned and fleshed muskrat pelts. Adults have a mottled, asymmetrical priming pattern, whereas juveniles have a bilaterally symmetrical pattern (Applegate and Predmore 1947; Shanks 1948; Linde 1963). The pelt primeness method is generally at least 95% accurate (Moses and Boutin 1986), although accuracy decreases during late winter and early spring (Linde 1963).

The molar fluting method is based on whether the fluting on the molars extends into the jaw bone (juveniles) or stops above the jaw (adults). Although early work found the molar fluting technique was accurate (Sather 1954), subsequent work suggested it was unreliable (Olsen 1959b; Elder and Shanks 1962; Moses and Boutin 1986, 1988). Proulx and Gilbert (1988) argued that the method may still be applicable in some situations, but accuracy is in doubt. Unfortunately, use of this method has been common (Proulx 1981; Erickson and Lindzey 1983; Parker and Maxwell 1984; Proulx and Buckland 1985) despite questionable accuracy.

Olsen (1959b) attempted to expand on the molar fluting technique by considering the fluting length and degree of root development. He found the method could be employed to classify animals as adult, subadult, or juvenile with a low degree of overlap. This approach of considering the size of different tooth segments is conceptually similar to more recent work. Erb et al. (1999), following the work of Doude van Troostwijk (1976) and Pankakoski (1980), found that molar (M^1) indices (crown length/total molar length) separated juvenile and adult muskrats in four populations in Wisconsin, Illinois, and Missouri with 100% accuracy during fall and early winter. However, the cutoff point for separating age classes was noticeably different for the Wisconsin muskrats (0.69 vs. 0.52–0.57 for the three Illinois and Missouri samples). They concluded that factors other than age at harvest, including genetics and feeding habits, were responsible for differences. The method appears accurate, but caution should be exercised when applying cutoff values to areas other than where they were developed. Also, the method may be less applicable in areas where muskrats breed throughout the year.

Numerous other methods have been explored to estimate age of muskrats. Incisor width appears to be a poor age criterion for muskrats (Brohn and Shanks 1948), although Sooter (1946) appeared to be more confident in the method.

Testes of adult males in late fall or early winter are usually ≥11 mm long, flattened, wrinkled, and discolored, whereas testes of animals that have not bred are smaller, turgid, and cream colored (Errington 1939; Schwartz and Schwartz 1959). Schofield (1955) found these criteria were 91% accurate for muskrats in Michigan. In females, uteri of adults are thickened and usually contain placental scars, whereas uteri from nulliparous subadults are thin, translucent, and lack placental scars (Errington 1939; Schwartz and Schwartz 1959). However, these methods are a function of age and breeding status and should be used with caution. Furthermore, the method appears unreliable just prior to and during the breeding season (Alexander 1951).

For males, Elder and Shanks (1962) found that the distal end of the baculum could be used to determine the age of Missouri muskrats with up to 97.6% accuracy during the harvest. In muskrats 5–8 months old, the distal processes of the baculum are cartilaginous and calcification has just begun. For adults, the baculum shaft is heavier and more rugose than in juveniles, and calcification is nearly complete. In juveniles, three to six ossification centers are present, with cartilaginous distal ends or digital processes exhibiting little calcification.

Zygomatic breadth of freshly skinned skulls was used by Alexander (1951, 1960a) to separate adults from subadults. Dividing points have ranged from 39.7 to 42.5 mm (Alexander 1951, 1960a; Schofield 1955). Schofield (1955) found the method was 89% accurate for muskrats in Michigan. Alexander (1960b) provided corrections to control for the effects of normal skull shrinkage and humidity on measurements. Male and female measurements appear to vary slightly (Sather 1956).

In Texas, Lay (1945a, 1945b) divided harvested muskrats into three age groups based on either pelt or total length measurements. However, this technique was unreliable for Missouri muskrats (Shanks 1948). Similarly, pelt length and/or total body length were of little value for estimating age of muskrats in Nebraska (Sather 1958) and New York (Alexander 1951). In another pelt-based method, Schwartz and Schwartz (1959) suggested that the teats on adult female pelts are usually blackened and 1.6–4.8 mm in diameter, whereas pelts of immature females usually have unpigmented teats <1.6 mm in diameter.

Le Boulengé (1977) also reported an age determination method for dead muskrats by using a reference growth curve of crystalline lens weight. The method appears reasonably reliable, although growth curves may differ geographically.

Age ratios of muskrats vary widely with measurement techniques, age, season, and geographic area. Because of this variation, Mathiak (1966) questioned the value of age ratios as production indicators. Reported ratios range from 7 to 650 young/100 adults (see Perry 1982). Large discrepancies in ratios between geographic areas may be partly due to differences in age determination techniques.

Lay (1945a) indicated that on a moderately trapped marsh in Texas, the standard age distribution was about 8% kits, 28% subadults, and 64% adults. Kits (≤624 g) constituted 6.6% of all trapped animals examined by Wilson (1956a) from Currituck County, North Carolina, fur sheds. However, age composition in the fall for muskrats on the Athabasca–Peace Delta may be as high as 7.3 young/adult (Fuller 1951). Beer and Truax (1950) found that the percentage of young in a Wisconsin marsh was inversely related to breeding population density, and that the best muskrat use of the habitat was obtained with a fall ratio of 4–5 young/adult.

ECONOMIC STATUS

Value. Muskrats have been of major economic importance in the North American fur trade since colonial times (Storer 1937) because of their wide distribution, abundance, and pelt value (O'Neil and Linscombe 1976). Historically, muskrats have been the most valuable fur animal in North America in number caught and total value (Dozier 1953; O'Neil and Linscombe 1976; Deems and Pursley 1978; Novak et al. 1987). Although the exact quality of the data is unknown, estimated harvests for all states and provinces have been summarized from 1931 to 1984 by Novak et al. (1987). Total estimated muskrat harvests for North America range from lows of around 1 million to highs exceeding 19 million (Novak et al. 1987; Figs. 16.5 and 16.6).

The quality of individual pelts varies according to size, color, primeness, damage, and, as a correlate of some of these factors,

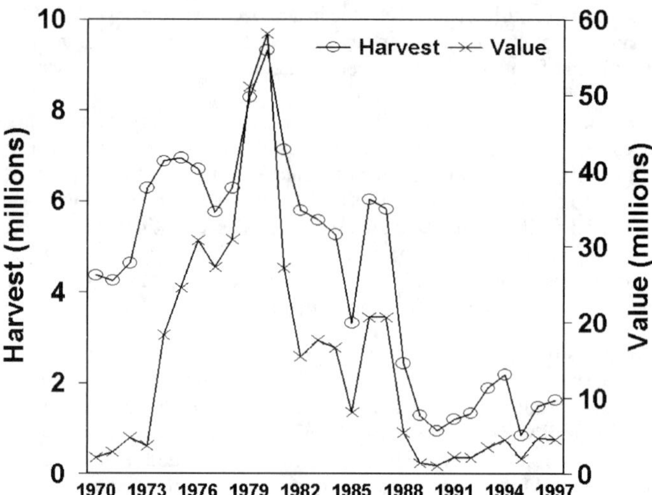

FIGURE 16.5. Muskrat harvests and fur value in the United States, 1970–1997. SOURCE: Data from International Association of Fish and Wildlife Agencies.

geographic region. Worthy et al. (1987) detailed the pelage priming process in furbearers, including how to visually identify pelt primeness. Generally, muskrats are prime from early December through March, with greatest primeness in March (Stains 1979). Obbard (1987) provided color drawings of the various pelt colors and discussed the sections of North America commonly used in the fur trade for classifying muskrats based on various pelt characteristics. Such quality differences arising from size, color, primeness, damage, and geographic region generally translate into different pelt values. However, actual and relative market values vary according to fashion trends, international economics, and normal supply and demand, and may change dramatically from year to year.

Muskrats as Food. Muskrat meat has been a food for stock poultry (Goff and Upp 1942) and for human and dog consumption (Gowanloch 1943; Novak 1975; Todd and Boggess 1987). The economic value of such consumption has rarely been quantified (Novak 1975; Monk 1981). Muskrats were esteemed as food by native peoples of North America and were eaten by early colonists, trappers, hunters, and voyageurs (Lantz 1910). The flesh has a gamey flavor and carcasses have been marketed in certain areas. Bailey (1937) stated that muskrats dressed to avoid a trace of musk and properly cooked and served have sweet, rich, and tender meat with the game flavor of wild duck. Detailed information on preparing muskrat meat for consumption is provided by Lantz (1910), Dozier (1943), and Dailey (1954). Muskrat was listed as a food

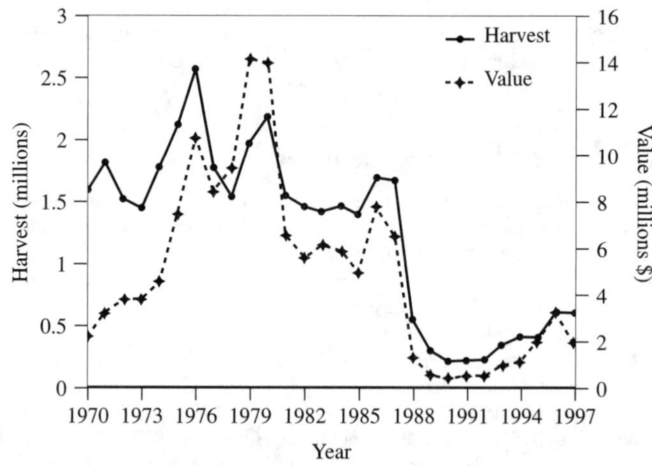

FIGURE 16.6. Muskrat harvests and fur value in Canada, 1970–1997. SOURCE: Data from Statistics Canada.

resource by Deems and Pursley (1978) in 18 states and 11 provinces or territories. Muskrat meat is often referred to as marsh rabbit, water squirrel, marsh hare, hare, or Chesapeake terrapin (Cook 1952, Indiana Department of Conservation 1963).

Carcasses taken during the breeding season are frequently impregnated with musk, which renders the meat offensive to humans (Schwartz and Schwartz 1959). A second factor that prevents wider acceptance of muskrats as food may be its designation as "rat." Dried musk was an important component in perfumes, especially before synthetics were developed (Erickson and Stevens 1944; Taylor 1980), and is used regularly in commercial trapping scents.

Farming. During 1925–1930, muskrat farming was promoted heavily in several states and many people were deceived into thinking such operations were highly profitable (Dozier 1953). Early muskrat ranching and farming operations have been described in detail (Louisiana Department of Conservation 1931). Field (1948) discussed muskrat farming and cited returns of 22–34% on investment, with annual yield varying with the quality of soil and skill of the farmer. The production of about 7 pelts/ha was considered good.

Dailey (1954) made the following points concerning muskrat farming: (1) most sections of America are acceptable for muskrat rearing; (2) muskrats develop quickly, and spring- and summer-born animals can be marketed the next spring (he advised against harvesting in fall and winter); (3) muskrat farming provides a quicker monetary return than the raising of any other furbearer because muskrats are prolific and can be cheaply maintained; (4) large numbers must be raised for farming to be successful, even though they are self-sustaining and less capital is needed than for some other aspects of the fur industry; and (5) muskrats should be raised under conditions as natural as possible.

In a farming situation, Dailey (1954) found 1 ha might support 124–185 pairs and their progeny each year. However, Bradley and Cook (1951) indicated that muskrats cannot be raised successfully in a seminatural environment where fencing restricts movement. They found that 75 of 86 muskrats released in fenced 0.3- to 1.2-ha enclosures died as a result of wounds from fighting. Similarly, Dozier (1953) stated that although muskrats can be raised on a limited scale in pens, it is not profitable because of irregular reproduction under restraint and losses from fighting, poor sanitation, and handling.

The greatest muskrat-producing areas of North America were Louisiana, Maryland, Delaware, New Jersey, Virginia, and parts of Canada (Dailey 1954). Today, muskrat farming in North America is limited, mainly because of low profitability.

Damage. Negative economic attributes of muskrats include burrowing-related damage to dikes, ditches, ponds, and levees, vegetation removal in wetlands used for wastewater treatment, and occasional losses of crops (Miller 1994). Glass (1952) described how muskrats damage ponds. Dams sodded with grass are frequently undercut by muskrats as much as 45–51 cm. When the water level rises or drops, muskrat holes are expanded to keep pace with the water line. Holes constructed during periods of low water are expanded upward when the water level rises. Erosion problems from muskrat activities are worsened by wave action caused by high winds. Excavation activities may seriously weaken a dam. Trees growing on the dam may exacerbate the problem because muskrats may tunnel along one or more of the roots. The most serious type of damage described is leakage, and damage is correlated with pond age and time inhabited by muskrats.

Muskrats occasionally damage portions of cornfields. Muskrat damage to farm crops in Iowa was reportedly confined mostly to cornfields close to watercourses (Errington 1938a). Lantz (1910) indicated that damage to crops by muskrats occurred only in limited areas, but noted muskrats were a pest in rice plantations in parts of Louisiana because of damage to embankments and plants. A 1969 survey indicated that damages by muskrats to Arkansas rice crops amounted to about $1 million annually in addition to damages to the levees and pond banks of fish farmers (Sealander 1979). A 1999 survey (Jack H. Berryman Institute, Utah State University, Logan Utah, unpublished data) estimated the total damage caused by muskrats annually in the United States at $6,027,000, averaging $123,000 per state. According to the same survey, an estimated 61,700 hr is spent annually by public employees responding to complaints regarding muskrats. Miller (1994) indicated that total monetary damage in four states (Arkansas, Louisiana, Mississippi, and California) exceeded that in all remaining states combined.

MANAGEMENT AND CONSERVATION

Management for muskrats centers around manipulation of the harvest and efforts to provide adequate food, water, and cover. Two books have been written and illustrated on muskrats and marsh management: O'Neil (1949) addressed southern coastal areas and Errington (1961) detailed the practical management of muskrats in northern areas.

A broad overview of earlier muskrat management in North America was presented by Deems and Pursley (1978). In the United States and Canada, no state, province, or territory granted total protection to the muskrat; 8 states had a muskrat hunting season; 47 of the 48 states where muskrats are present had a trapping season; and all of the 12 provinces or territories with muskrats had a trapping season. Arizona, Michigan, Texas, and Utah allowed year-round harvesting. Habitat management for the muskrat was being conducted in 15 states and 4 provinces or territories. Muskrat population inventories were conducted in 15 states and 7 provinces or territories.

Providing adequate water levels and desirable plant communities are key to managing for muskrats (see sections on Ecology, Habitat, and Feeding Habits). A thorough understanding of aquatic plant biology is necessary when attempting to manage for specific plant associations desirable for muskrat production. However, management philosophy, legal restrictions, and consideration of the needs of multiple species will ultimately guide management decisions. Even with sophisticated management practices, muskrat populations cannot be stockpiled from year to year (Dozier 1953). Investigations in Wisconsin by Mathiak and Linde (1954) and Mathiak (1966) indicated that closed seasons were unlikely to increase subsequent yields.

Habitat Manipulation. Burning can be an efficient tool for enhancing plant communities attractive to muskrats and has been used most commonly in southern coastal areas. Fire is a natural part of marshland ecology in many coastal areas and is beneficial for muskrats. Fire sets back succession, provides optimum growing conditions for Olney bulrush, prevents accumulation of roughs (dead, dry vegetation that may cause damaging wildfires), and controls undesirable plants (Freeman 1945; O'Neil 1949; Dozier 1953; Wilson 1968).

From a management standpoint, Olney bulrush is the most important food for muskrats in the Gulf coastal marshes. The greatest abundance of nonfoods consists of wiregrass or marshhay cordgrass, needle rush, sawgrass, giant southern wildrice (*Zizaniopsis miliacea*), and maidencane panicum (*Panicum hemitomon*). Olney bulrush is a subclimax species dependent on suppression of climax vegetation.

Several methods of eradicating undesirable climax marsh vegetation were investigated by Wilson (1949). Burning prevented marsh buildup, and in conjunction with flooding of needle rush–sawgrass marsh, resulted in 75–100% eradication of less desirable vegetation. Without proper burning, Olney bulrush and other food plants were invaded by needle rush, sawgrass, and wiregrass (Wilson 1968).

Lay (1945a) reported that Olney bulrush grows best during cool weather and grows very little in summer. Therefore, there appears to be an advantage to burning wiregrass in late summer to encourage Olney bulrush. Burning in the spring, on the other hand, encourages the grass rather than the preferred sedge. However, clean burning in summer seriously impaired muskrat marshes in Texas, especially when followed by drought or high tides (Lay and O'Neil 1942). Return of muskrats to the burned areas was apparently hampered by a lack of house-building materials.

Burning wiregrass and saltgrass during normal water periods was not very effective in enhancing *Scirpus* spp. in Louisiana (Palmisano 1967). However, burning such marshes during fall with low water levels and then immediately flooding with 25–38 cm of water until late spring was successful in controlling unwanted grasses and in enhancing salt-marsh bulrush (Babcock 1967; Palmisano 1967).

Specific burning dates recommended by other investigators vary by marsh type and geographic area. O'Neil (1949) stated that a normal three-cornered grass marsh in Louisiana should be burned between mid-October and early January with a 0- to 5-cm water level. The best time to burn Olney bulrush in Mississippi was in November after the vegetation had died down. Burns should be made the day after a rain or after a high tide to prevent injuring Olney bulrush roots or burning muskrat houses. Needle rush marsh that is suitable for Olney bulrush should be burned in late February or early March. The marsh should be dry enough so the fire will be hot enough to injure needle rush roots. With suitable marsh, switchgrass panicum (*Panicum virgatum*) and Olney bulrush will begin to establish in the second or third year (Freeman 1945). Wilson (1968) suggested burning between November and January in North Carolina. However, Dozier (1953) suggested waiting until late February to burn Atlantic coast marshes to ensure adequate muskrat cover during winter.

Specific burning techniques also vary by marsh type and area. For maximum production of muskrats, O'Neil (1949) recommended spot burning about 60% of Olney bulrush marshes in early October; then around 1 December approximately 20% of the remaining roughs should be burned. After the Olney bulrush gets a 20–25 cm start, the remaining 20% of the marsh should be burned. Burning at least every other year seemed necessary (Lay 1945a). A 3-year rotation, one third per year, was recommended for burning Olney bulrush in Mississippi, with needle rush areas being burned every other year, or 2 years in succession and then skipping a year (Freeman 1945). In Texas, Lay and O'Neil (1942) recommended burning alternate strips every other spring in marshes that do not have rank vegetation (e.g., common reed) growing profusely or with excessive vigor. Reported advantages were assurance of cover, discouragement of muskrat movements to other areas, conservation of marsh water supply, and provision of autumn vegetation too rank for geese. Proulx and Gilbert (1983) estimated that muskrat home ranges in an Ontario marsh were composed of 50% open water and 50% emergents, perhaps suggesting that any vegetation management practice should ensure adequate interspersion of emergents and open water areas. Marsh burning should not be done in a manner that would endanger spoil bank vegetation.

Water Control. Water level manipulation is a principal muskrat management technique, and the importance of adequate water levels should not be underestimated. For example, fluctuating water levels were found to be (1) more important than the type of marsh vegetation in determining muskrat population levels in the Illinois River valley (Bellrose and Brown 1941), (2) the principal factor limiting muskrat populations in the marshes of Currituck Sound, North Carolina (Wilson 1949), (3) adversely affecting muskrat populations in Illinois marshes (Bellrose 1950), and (4) preventing establishment of muskrat food plants in certain areas of Louisiana (Moody 1950).

Abnormally low water levels directly influence muskrats by exposing them to increased predation, nutritional stress, and/or winter freeze-outs (Errington 1938b; Friend et al. 1964; Donohoe 1966; Proulx et al. 1987; Messier et al. 1990; Hjältén 1991; Clark and Kroeker 1993; Clark 1994; Virgil and Messier 1996). Low water levels can affect impact muskrats indirectly by altering vegetative composition. For example, water-deficient marshes in Mississippi are frequently dominated by needle rush, and needle rush replaces the more desirable Olney bulrush when the water table is lowered (Freeman 1945). Water levels below the marsh floor also cause loss of peat through oxidation (Lay and O'Neil 1942; Lay 1945a).

Conversely, periodic droughts or drawdowns in many marshes can be important in re-establishing or increasing abundance of many emergents desirable to muskrats (Errington 1961; Weller and Fredrickson 1973). For many marshes in the north-central United States, Weller and Fredrickson (1973) believed that inundated basins with stable water levels remained productive for semiaquatic vertebrates for only 5–7 years. They noted that the usual product of long-term water level stability in prairie marshes is a central open-water body with a perimeter of dense cattail or other emergent vegetation. Such areas frequently exhibit low productivity for invertebrates and vertebrates. Where possible, Errington (1961) recommended rotational drawdowns of different sections of a marsh to continually revitalize areas without the complete (albeit temporary) elimination of muskrat habitat characteristic of whole-basin drawdowns.

Abnormally high water levels can also inhibit muskrat populations by destroying muskrat dwellings, flooding nest chambers, reducing and/or altering species composition of emergent vegetation, and forcing increased movements, all of which may reduce muskrat survival and hence density (Bellrose and Low 1943; Errington 1948, 1963; Weller and Fredrickson 1963; Weller and Spatcher 1965; Donohoe 1966; Kinler et al. 1990). In many prairie marshes of the north-central United States, prolonged high water levels may kill desirable muskrat plants such as common cattail (Errington 1948). In Gulf Coast marshes, excessive water depths may enhance less desirable plants such as cattail, bullwhip bulrush (*Scirpus californicus*), and southern wildrice (Lay and O'Neil 1942; Lay 1945a). However, water bodies with greater depth are not inherently synonymous with poor muskrat habitat. In deeper basins, submergent vegetation may often be abundant near shore and is regularly used by muskrats. In such settings, however, availability of house-building materials or banks for denning may be the limiting factor for muskrats.

Optimum water levels for muskrats vary geographically. Gulf Coast marshes should be maintained throughout the year between 2 and 31 cm, and preferably between 15 and 20 cm (Lay and O'Neil 1942; Lay 1945a). Water depth in Olney bulrush marshes in Louisiana averaged 5.8 cm in August (Palmisano 1970). Gashwiler (1948) recommended that water levels in Maine marshes should be maintained at 15–51 cm to improve habitat for muskrats. Such levels optimized muskrat use of the entire marsh; increased desirable food, cover, and breeding sites; and minimized the negative effects of destructive floods. Muskrats in other northern areas appear to select house-building sites with a minimum water depth of at least 10–15 cm (Illinois, Bellrose and Brown 1941; Ontario, Proulx and Gilbert 1983; Manitoba, Clark 1994). However, such minimum depths may be inadequate to ensure overwinter survival in areas where ice thickness exceeds this level. Proulx and Gilbert (1983) and Clark (1994) estimated average water depths around houses to be 54 and 38 cm in Ontario and Manitoba, respectively.

In coastal marshes, salinity is also a factor controlling plant associations and hence muskrat abundance (Lay and O'Neil 1942; Palmisano 1970). Marsh salinity, like water depth, is extremely variable, both seasonally and daily, and is affected by similar factors. Saline water blown over brackish and fresh marsh by storm tides will destroy stands of Olney bulrush and *Typha* spp. (Wilson 1968). A salt level of 1–1.5% in the soil water was recommended by Lay and O'Neil (1942) as most desirable for muskrats. Similarly, Palmisano (1967) found Olney bulrush growing in Louisiana where soluble salts in the soil ranged from 1.0% to 1.7% (salinity 10–17 parts per thousand) with a pH of 4.1–7.9. Later investigations by Palmisano (1970) revealed that the total soil salts in Louisiana marshes where Olney bulrush occurred averaged 0.9% in August. To enhance Olney bulrush, Palmisano (1967) indicated that salinity must be maintained between 0.5% and 2.0% salt. Interestingly, however, a 50% reduction in germination of Olney bulrush occurred at 0.4% salt.

Water control structures can help control salinity, maintain adequate water levels, and allow greater control of plant species composition. Even a small change in water levels can completely change marsh vegetation (Wilson 1968). Consequently, absence of water control structures will reduce management options and muskrat harvests (Wilson 1953, 1955b; Sather 1958; Donohoe 1961, 1966; Errington 1961). Water levels may be controlled by diking and impounding, construction of weirs, and pumping. Where water levels are affected by tides, dikes and weirs are especially useful to stabilize water levels, stop erosion, and control the supply of muskrat food plants.

Constructing ditches can also be a productive tool for improving muskrat habitat. The muskrat harvest from ditched plots in Wisconsin marsh was 4–10 times greater than the harvest in the surrounding bog (Mathiak and Linde 1956). Ditching was recommended by Mathiak and Linde (1956) as being especially useful for providing deeper water areas when flooding by dams or dikes is not possible. Ditches provide muskrats sufficient water to prevent water freezing solid in winter, access to food during the critical winter period, escape and feeding areas during spring floods, important muskrat food plants, denning and resting areas, and open water during summer drought periods (Lay 1945a; Anderson 1948; Mathiak and Linde 1956). Trapper access to the marsh is also enhanced by the ditches.

Ditches may be plowed, dynamited, or dug. Mathiak and Linde (1956) found that ditch dredging was superior to blasting because it was more economical, provided a spoil bank, and filled in less rapidly. Muskrats moved into ditched areas in the first few weeks following construction and the population increased for the next 3 years. Ditches should not be dredged in extremely long straight lines, so that boat travel will be safer during high winds. Spoil banks should be about 12 m long, staggered on alternate sides of the ditch to reduce the fire hazard, and seeded to grass soon after dredging. Mathiak and Linde (1956) observed that ditches spaced 15 m apart provided the greatest harvest (37 muskrats/ha/year); ditches spaced 122 m apart provided a harvest of only 14 muskrats/ha/yr. However, annualized returns on the ditching investment were highest with the 61- and 122-m ditch spacings. The 61-m spacing was recommended because it yielded more muskrats per hectare than the 122-m spacing.

For muskrat farms, Field (1948) recommended that ditches should be at least 1.2 m deep and not less than 2.4 m wide; when constructed in soft, floatable soil, ditches should be double-width to compensate for filling. Optimum ditch dimensions recommended by Mathiak and Linde (1956) were 1.5 m deep by 4–4.6 m wide. Ditches should remain productive for muskrats for about 6–10 years.

Miscellaneous Management Practices. In some coastal marshes, Olney bulrush has been introduced by planting seed or transplanting rhizomes or rooted plants. Areas to be planted should be pretreated with a hard burn, preferably in early spring (Freeman 1945), and flooded immediately after planting (Wilson 1968). Seed should be scattered when the area is wet or covered with water (Freeman 1945). Sometimes desired shifts in vegetation can be made quicker by sodding Olney bulrush and seeding *Typha* spp. (Wilson 1955c, 1968). Spring and summer were the best time to sod in North Carolina. Peat on the area planted should be at least 10 cm deep. Hand planting of marsh plants may be practical on small areas if circumstances justify the effort and expense. Temporary muskrat control may be necessary to allow new plantings to develop (Errington 1961).

Three techniques of site preparation were evaluated in Louisiana by Ross (1972): Tilling was best and burning was poorest. However, Olney bulrush survival in the burned area was almost twice that in the area with no site preparation. The combination of burning and tilling was recommended by Palmisano (1967) and Soileau (1968) for reducing competition from wiregrass and saltgrass in Louisiana marshes.

In Louisiana, Ross (1972) recommended planting Olney bulrush in December and January. Palmisano (1967) indicated that planting 8–9 rhizome nodes/m^2 in April will produce a dense stand by the end of the October. However, he stated that 2–3 nodes/m^2 is more than adequate for marsh revegetation; Ross (1972) suggested a node spacing of 1.8 m. Planting depth should be about 10–15 cm and water should be maintained 5–10 cm above the soil surface for 3–4 weeks after planting. Best survival and growth of plantings occurred at a salinity of 5 and 10 parts/1000 (Ross 1972). To be successful, rhizome predation from geese, nutrias, and muskrats must be prevented.

Mowing, disking, scalping, and herbicides are too expensive for large marshes but may be used on a small scale to enhance Olney bulrush and *Typha* spp. (Wilson 1968). A combination of burning and chemical treatment may be the only site preparation measure economically affordable for the average landowner (Ross 1972). Competition from wiregrass and saltgrass can be reduced significantly by the use of a soil sterilant (Soileau 1968), but use of such chemicals is regulated and has side effects. Wilson (1949) found that spraying with a sterilant resulted in vegetative mortality of 23–91% in unmowed and 12–70% in mowed plots. Mowing of needle rush–wiregrass–sawgrass areas followed by flooding also resulted in high mortality of climax vegetation.

Solid stands of Olney bulrush are more likely to sustain heavy herbivory than are mixed communities of climax plants (O'Neil 1949). Dozier (1953) suggested that vegetation in eaten-out areas could be rehabilitated by ditching. To assist vegetative growth, Donohoe (1961) suggested application of lime to bare areas of spoil banks, where acidity approaches a pH of 4.0–5.0. If the eat-out area is immediately drained and conditions are favorable for plant growth, the marsh may be revegetated the winter following the eat-out. With poor growing conditions, which is generally the case in denuded areas, the eat-out area will return to climax rough (wiregrass in brackish marsh) in 2–3 years. A subclimax marsh can be produced with proper management in 2–3 additional years (O'Neil 1949).

Because muskrats often prefer denning in banks when suitable sites are available (Dozier 1948b; Messier et al. 1990; Messier and Virgil 1992), managers may wish to consider constructing islands and/or improving existing shorelines along marshes or rivers to enhance muskrat use. Muskrats may use a variety of sites for burrows, but they prefer sites with low water velocity, high banks with steep slopes, clay–loam soils, and an abundance of shoreline and/or aquatic vegetation (Brooks and Dodge 1981, 1986; Brooks 1985; Nadeau et al. 1995). For streams, Errington (1961) suggested focusing shoreline management of slope and vegetation to areas near pools that contained slower water and adequate depth. Protecting stream banks from grazing or clearing is vitally important for preserving the integrity of stream and vegetative structure necessary to attract muskrats.

Important muskrat habitat and populations should be monitored to detect disease outbreaks early (Errington 1961). Managing disease outbreaks in muskrat should be dictated by type of disease; shape, size, and type area affected; importance of the population; likelihood of negatively influencing other animal populations; and potential for significant success. A multidisciplinary analysis of the need for ameliorative action and the best course of action, if any, should always be performed. In cases where muskrat populations need to be reduced because of disease, the best course may be to reduce the attractiveness of muskrat habitat (Errington 1948). However, the population may just shift if nearby habitat is suitable, thus potentially spreading the malady. The effectiveness of approaches to reducing muskrat disease outbreaks, including reducing or eliminating muskrats and/or their houses, is not well known.

Population Assessment. Evaluation of the effectiveness of management actions (or general population monitoring) usually requires sampling to assess changes in the population. The most common approaches for estimating size or trend in muskrat populations have been house counts and mark–recapture methods. Aerial surveys of key habitat variables along rivers have also been effectively used to evaluate populations (Nadeau et al. 1995).

Dozier (1948b, 1953) discussed the use of house counts as a population estimation method. Houses may be counted from the air, water, or ground over a specified area, band, or transect. Freeman (1945) suggested a survey of at least 20% of the area for occupied houses to determine suitable trapping intensity. Accuracy requires that the observer differentiate between unoccupied and occupied houses and between dwellings and feeding huts or shelters. Estimates of house detectability by each observer need to be formulated to statistically maximize the ability to develop unbiased population statistics (either trend or index data or population estimates). In addition, use of house counts as indicators of population change may, depending on objectives, require the assumption of constant (among years) number of houses per family, constant number of muskrats per house, and a constant proportion of inactive houses—assumptions that have rarely been formally evaluated.

Aerial censuses of muskrat houses permit coverage of a large area in a short time (Fig. 16.7). Dozier (1948b) stated that with good

FIGURE 16.7. During good weather conditions, trained observers can readily census muskrat houses from the air to provide a population or breeding index. SOURCE: Photo by Robert G. Linscombe.

visibility, larger houses are discernible from an altitude of 250–300 m within a 1.2-km radius. However, observations at 60 m or less may be necessary to accurately distinguish houses from other structures, depending on local conditions. For example, Proulx and Gilbert (1984) found that active and inactive houses in an Ontario marsh were of similar width, but active houses were on average 15 cm taller than inactive houses in midsummer. Such small vertical differences could be difficult to detect from the air. Low-level flights always carry high risk, and specialized training is required for the pilot and observer. Similarly, specialized protective equipment (e.g., fire-retardant clothing and shoes, helmet, and gloves) should be used by the observer and the aircrew, and the aircraft and pilot should be certificated for such flights. Furthermore, an aviation risk assessment should be completed for low-level flights and should include such things as development of a hazard map and securing flight-following services.

Counts of muskrat houses vary seasonally. In the marshes of Louisiana, they were highest in February, when water levels were high, temperatures low, and spring breeding was under way (Palmisano 1972a). Dozier (1948b) suggested that house counts should be taken in early fall to appropriately plan trapping strategies. However, Proulx and Gilbert (1984) found summer counts were a more reliable index of population size. If water levels decrease noticeably in a given fall, muskrats may abandon summer houses and construct new lodges, thereby inflating the number of houses in relation to muskrats.

To convert counts of houses to measures of population size or density, some measure of the number of muskrats per active house is needed. Earlier work, summarized by Dozier (1948b), gave 5 muskrats/large house as a conversion factor for estimating muskrat populations from house counts. However, he proposed using the average number of muskrats per litter as a conversion factor because during the fall and winter, the parent muskrats and their last litter frequently occupy the same dwelling until the spring breeding season. Others have reported conversion factors of 2.8 (Lay 1945a) and 3.5 (Parker and Maxwell 1980). In an Ontario marsh, Proulx and Gilbert (1984) estimated 1.9 and 4.5 muskrats/house in successive years, suggesting that even within a site, the number of muskrats per house cannot be assumed to be constant from year to year.

Counts of houses from the ground or air and the Lincoln index were of questionable value in estimating muskrat population abundance in north-central Kansas (Sather 1958). The most accurate population calculations were based on the number of young per adult female in the winter harvest and on the spring breeding-territory inventory. Winter muskrat populations were calculated by multiplying the number of breeding territories present during the previous breeding season and the number of young per adult female in the harvest (Errington 1943). Inventory of breeding territories, however, can be costly and time-consuming on larger marshes. Methods relying on house counts do not account for bank-dwelling muskrats and are thus not applicable to most riverine environments.

Intensive mark–recapture techniques have also been employed to estimate muskrat populations (e.g., Mathiak and Linde 1956). Estimates of population size with coefficients of variation <20% have been obtained under a variety of conditions (Clay and Clark 1985; Clark and Kroeker 1993). Marking large samples and getting high recapture rates are important in obtaining precise estimates. Segments of the muskrat population, such as different ages or genders, or animals that have previously been trapped, often have different capture probabilities (e.g., Clark and Kroeker 1993). Hence, population model assumptions must be verified and population estimation methods must be selected appropriately. Recaptures can be used to generate mark–recapture population estimates. Alternatively, trapper returns of tags, if close in time to tagging, can be used to obtain a Petersen population estimate. More details on population estimation are provided by Seber (1982) and Lancia et al. (1994).

Standard cage traps, generally 15 × 15 × 45–60 cm, can be used to capture muskrats. Lacki et al. (1990b) concluded that traps with spring-loaded doors were more effective than those with gravity-operated doors. Traps can be placed on houses, feeding platforms, banks, logs, toilet stations, or floating platforms. Double-door cage traps often work best when set on muskrat trails. To ensure traps are secure and cannot fall into the water or be moved, they should be anchored to the ground or platform; stiff wire may be used as support or traps may be wired to adjacent stakes. Family-type livetraps (Fig. 16.8) may also be used for capturing muskrats (Snead 1950). These traps have a base similar in design to a standard cage trap, except with one-way free-swinging doors and a hole cut in the top. The base is set at underwater den entrances, with a vertical chute (of appropriate depth) attached to allow muskrats to reach the water surface after being trapped by the swinging doors in the base. A holding pen, with or without a one-way swinging door, is then attached to the top of the vertical chute. In hot weather, wiring open the swinging door in the holding pen allows muskrats to reenter the water and can reduce heat stress (Heatherly 1993). For all livetraps, providing natural vegetation on top and within the trap reduces heat stress in summer, provides food, and generally increases capture success (Mathiak and Linde 1956; Erickson 1963; Heatherly 1993). Muskrats have also been successfully livecaptured by nightlighting and netting from boats. McCabe and Elison (1986) reported a capture rate of 10–12 muskrats/hr using this method.

Apples, carrots, cabbage, turnips, and potatoes have been used to attract muskrats into traps. Carrots and apples were superior baits in areas of central New York, with carrots more durable in warm months and slightly more effective (Erickson 1963). Use of commercially-prepared lures can increase trapping success (e.g., Williams 1951). Unbaited cage traps in natural trails can also be effective in catching muskrats (Robicheaux 1978).

Muskrats in livetraps may be carefully removed by the tail and placed head first into a wire cone for tagging, determination of age and sex, and measurement (Erickson 1963). A noose attached to a pole (Robicheaux 1978) can also be employed to remove and handle trapped muskrats safely. Though it is not necessary for tagging, muskrats have been anesthetized using inhalants such as halothane (Blanchette 1989), methoxyflurane (Lacki et al. 1989), and isoflurane (Belant 1995) or injectable anesthetics such as ketamine hydrochloride and sodium pentobarbital (MacArthur 1979; Dell et al. 1983). Ketamine–xylazine combinations have also been effectively used to anaesthetize muskrats (Belant 1996; Sleeman et al. 1997).

Serially numbered, size 1, Monel™ metal tags may be applied to the ears of muskrats for identification (Aldous 1946; Snead 1950). Tags should be inserted into the nonfleshy, dorsal portion of the ear to prevent infection (Erickson 1963). Tag sizes 1 and 3 may be similarly used on the webbing of the rear feet (Robicheaux 1978). A 5 × 24-mm aluminum tag inserted through two slits in the skin of the back was described by Errington and Errington (1937) and Errington (1944).

FIGURE 16.8. Multiple muskrats can be livetrapped using "family" traps constructed from welded mesh wire. SOURCE: Photo by John Erb.

Hensley and Twining (1946) described a waterproof plastic marking button, which can be threaded through the loose skin of the back with a flexible steel needle, and the use of a size 7 bird band around the base of the tail to mark adults. O'Neil (1949) recommended a band on the Achilles tendon. Evaluations by Takos (1943) of the back-slit, leg-ringing, and Achilles tendon methods of tagging muskrats indicated that the Achilles tendon (or Cook method) was best. Clark (1987) estimated that the probability of losing an individual ear- or leg tag was 0.14.

Muskrats have been toe clipped for identification purposes (Dorney and Rusch 1953). Concerns over use of toe clipping for identification include the following: (1) digits may be improperly severed, resulting in undue pain and producing partial digits that heal in various fashions, often resembling full toes; and (2) hair, skin, or other growth material may obscure a clipped digit, particularly if improperly removed. These factors introduce the likelihood that subsequent handlers, particularly if they are not well trained or very careful, will erroneously tally "clipped digits" as being present rather than clipped, corrupting the resulting data set. Most toe clipping has been done on the rear toes because the front feet are used extensively in feeding and burrowing. Biologists using toe clipping on kits up to 5 days of age clip a toe with cuticle clippers as close to the foot as possible; on older animals after toe elongation, a toe is clipped off at the first joint. Ferric chloride may be used to cauterize and sterilize the wound and prevent excessive blood loss, especially in small kits. Although toe clipping has been used successfully by qualified biologists on a variety of small mammals, it is viewed as inhumane by many, and unnecessary for larger animals by others, because of the availability of other reliable marking techniques.

Coon (1965) used radiotransmitters weighing about 43 g to monitor movements of muskrats. Several types of attachment collars and harnesses were tested, but a tight-fitting plastic neck collar was best. Intra-abdominal FM transmitters were used by MacArthur (1978) to monitor winter movements in Delta Marsh, Manitoba. Implanted transmitters have been used to monitor habitat use, home ranges, movements, activity patterns, survival, and physiological parameters (MacArthur 1978; Brooks 1985; MacArthur and Karpan 1989; Thurber et al. 1991; Marinelli and Messier 1993).

All methods of animal tagging and mutilation for identification purposes raise concerns about the humane treatment of animals used in studies. Such techniques should be carefully selected, vigorously justified, and used only by trained individuals. All animals should be handled in accordance with the spirit and intent of the Animal Welfare Act. All studies involving animal handling and marking should be reviewed by a credible animal care and use committee to ensure the highest standards of animal care and treatment.

Various reproductive parameters (pregnancy rate, litter size) for muskrats are most commonly obtained by counting placental scars (Gilbert 1987). Because muskrats usually produce multiple litters during one breeding season, placental scar counts from fall-trapped animals will provide estimates of total natality for the breeding season; counts at predetermined intervals during the breeding season could be used to estimate sizes of individual litters. The number of nestlings in a house has also been used to estimate the size of individual litters (Sather 1958; Olsen 1959a; Errington 1963). Counts of corpora lutea in the ovaries have been used to estimate ovulation rates (Forbes and Enders 1940). Juvenile-to-adult female ratios in the harvest have also been used as an index of reproductive output (Beer and Truax 1950; Mathiak 1966), although such measures are confound by the effects of reproduction and survival.

Survival can be monitored using mark–recapture methods (e.g., Clay and Clark 1985; Clark and Kroeker 1993) or radiotelemetry (e.g., Thurber et al. 1991). Juvenile survival rates can also be estimated by comparing placental scar counts to the juvenile-to-adult female ratio in the fall. Such an approach, however, does not account for summer mortality of adult females or immigration/emigration.

Assessment of body condition in muskrats also may provide information on the nutritional and demographic status of a population. Virgil and Messier (1993) found that body mass was a reliable predictor of protein mass, but was poorly correlated with fat mass and the energetic state of the animal. Water mass was a good predictor of protein and fat content, and they suggested that isotopic dilution space would provide a reliable and nondestructive method for estimating body composition in muskrats. Proulx (1997) discussed the estimation of whole-body weights from skinned carcasses in the fall.

Harvest. Regulation of the harvest is an important muskrat management tool. As with any species, a clear statement of objectives, such as a specified amount of recreational opportunity, maximum sustained yield, or population control, is important for establishing appropriate harvest regulations. Under some conditions, underharvesting may result in extensive eat-outs, which significantly reduce subsequent muskrat density (Dozier et al. 1948; Ferrigno 1967). Low populations of muskrats in some Louisiana marshes were attributed to such previous overpopulation caused by underharvest (O'Neil 1949). Hence, if prevention of eat-outs and/or maximum sustained yield are the stated objectives, harvest regulations and associated population monitoring may be different than if the management objective for a marsh is to allow natural processes to occur with only minimal concern for recreational or economic opportunity. Broadly, the length of the harvest season is constrained by pelt primeness in the fall and the breeding season in the spring.

Once management objectives are defined, a variety of factors may be considered in establishing muskrat seasons, including climatic conditions, trapper access to the marsh, timing of breeding, and pelt primeness. For example, a long trapping season has the advantage of allowing for variations in weather and for reducing overpopulation where desirable (Lay and O'Neil 1942).

Sather (1958) stated that the muskrat season should be (1) long enough to permit adequate harvesting, (2) during the time of year when pelts are of highest quality, and (3) closed during breeding seasons, including territory establishment. He suggested a 15 December–15 March season in Kansas. Highby (1941) suggested harvesting in early winter to reduce (1) muskrat mortality from freezeouts, food shortage, and disease; (2) loss by emigration, mink predation, and mink trapping; and (3) damage to pelts from muskrat fighting. Stewart and Bider (1974) suggested that a fall trapping season should be considered in southern Quebec. They noted that the number of muskrats available for harvest decreased as the season progressed: Young/breeding female declined from 5 in fall to 2.8 in spring.

Manipulating the timing of trapping seasons can be an especially effective means of managing the muskrat harvest (Erickson 1981; Clark 1986). Lay and O'Neil (1942) found a high degree of positive association between cold weather and muskrat trapping success. Catches on cold nights included approximately 6% more adults than those in warm weather (Lay 1945a). However, in northern areas, initiating seasons before inclement winter weather can increase harvest substantially. Inclement weather restricts trapper activity, and trapping muskrats is more difficult after freeze-up. Clark (1986) indicated that timing of the trapping season relative to weather conditions was the most influential factor influencing harvests on the upper Mississippi River. He noted that most muskrats were harvested in the first 3 weeks of the season.

Pelt quality, time of harvest, and geographic area are interrelated and should be considered jointly when setting harvest seasons. Worthy et al. (1987) discussed the pelt priming process in detail. Generally, the time of primeness for muskrats is from December through March (Stains 1979). The maximum percentage of furs that were graded "tops" for muskrats harvested in Texas occurred in January and the first half of February (Lay 1945a). Peak muskrat primeness was reached on Blackwater National Wildlife Refuge, Maryland, during the last half of February (Dozier et al. 1948). Although muskrats had reached their maximum size by this time, trapping yielded the smallest number of pelts.

Although no general rule is likely to apply over large spatial or temporal scales, several authors have reported maximum harvest rates for muskrat populations. It was reported that 50% of the muskrat population could have been harvested without jeopardizing the future of the population in Minnesota (McCann 1944), 60–65% in North Carolina (Wilson 1968), 66% along the Atlantic coast (Dozier 1953), 70% in New Jersey (Ferrigno 1967), and 75% in Wisconsin (Mathiak and Linde 1956). During years of high populations, 88–91% of the population from an area in Maine was trapped, whereas about 80% was trapped during years of low populations (Gashwiler 1948). Many of these estimates, however, should be viewed with caution because their accuracy is questionable.

Smith et al. (1981) used a harvest model and estimated that maximum sustained yield occurred between 70% and 75% for muskrats in a Connecticut estuary, with higher rates leading to reductions in subsequent harvests. Using tag recoveries to determine harvest rates, Clark (1987) estimated a maximum sustainable rate of 64% for muskrats on a pool in the upper Mississippi River. Parker and Maxwell (1984), using known harvests and mark–recapture estimates of population size, observed that harvest rates of 60% in either autumn or spring were sustainable; 60% harvests in both autumn and spring led to sharp declines in the population. Current information suggests, then, that the threshold where harvest becomes almost completely additive is approximately 65–75%. Wilson (1968) suggested that a 60–65% harvest might seldom be accomplished where muskrat populations exceed 74/ha in marsh habitat.

The actual number of muskrats trapped on a marsh is determined by weather, skill and economic needs of trappers, price of pelts, muskrat abundance, season length, and geographic area (Harris 1952; Erickson 1981; Clark 1986). Bailey (1937) indicated that yields of 7–10 muskrats/ha for Maryland marshes were good; yields of 15–20/ha were not unusual on good, well-managed marsh. During peak populations in Maryland, 12 muskrats/ha were trapped over extensive areas and up to 49 muskrats/ha have been recorded (Harris 1952). Wilson (1968) reported that productive Olney bulrush marshes in New Jersey and Maryland yielded catches of 7–25 muskrats/ha and occasionally 37/ha. On the better marshes of North Carolina, 10–15 muskrats/ha may be harvested, and catches exceeding 49/ha were reported in Louisiana (Wilson 1968). In South Dakota, Aldous (1947) reported that 7.9 muskrats were taken per hectare of good habitat.

Muskrat yield per house may be a more practical indicator of trapping efficiency than yield per hectare because the trapper can count houses while setting traps and can stop trapping when the desired catch per house is achieved. However, the number of harvestable muskrats represented by each house is extremely variable and is affected by both spatial and temporal factors, and no studies have rigorously estimated such a quantity. An average of 2.6–2.9 muskrats/house was taken in South Dakota (Aldous 1947). Palmisano (1972a) estimated an average of 3.2 harvestable muskrats/house in Louisiana but recommended a harvest rate of 2.5–3.0/house with a stable population. Potential catch estimates by other investigators range from a conservative 2/house (Lay 1945a) to 5/house (O'Neil 1949).

Age and sex ratios in the harvest have also been used as indicators of harvest intensity or population status. Based on age ratios obtained from trapped samples in Wisconsin, Beer and Truax (1950) concluded that there was a tendency for lower ratios (fewer young) after years of light trapping. Using 21 years of harvest data from a Wisconsin marsh, Mathiak (1966) also noted that the highest age ratio coincided with the year of highest harvest and vice versa. However, there was a very poor correlation during intermediate years. Similarly, Smith et al. (1981) noted that the number of young per adult female increased with increasing harvest, although harvest rates were implied from a harvest model. In southern areas where breeding may occur throughout the year, age groups become less distinct. Lay (1945a) suggested that a winter harvest ratio of 8% kits and "mice" (<459 mm total length), 28% subadults (460–509 mm), and 64% adults (≥510 mm) was indicative of moderate trapping; a 20:40:40 proportion was believed to represent overtrapping. Similarly, Freeman (1945) stated that good muskrat-producing marsh in Mississippi should yield a ratio of 30% juveniles to 70% adults (>12 weeks old). If the percentage of adults was higher, poor reproduction was indicated. If the number of kits and mice approaches the number of adults being caught, Freeman (1945) suggested stopping the trapping season. Unfortunately, reliable estimates of population size and harvest rates are frequently unavailable. Hence, the reliability of age ratios as harvest or population indicators remains uncertain.

Efforts are currently under way to develop species- and region-specific best management practices for trapping in the United States. Based on sound scientific research, they are designed to improve animal welfare in trapping programs while simultaneously considering factors such as trap selectivity, capture rate, and cost. Whether and how individual states adopt such recommendations (when complete) remain to be seen, but development of best management practices is independent of the political process. They are best viewed as a nonstatic educational tool for encouraging the use of only the best trapping systems available. In Canada, recent international agreements are in place that may determine what traps are approved for use on muskrats and other species.

Drop-door cage traps are perhaps the most humane trapping method for muskrats. However, when harvesting muskrats, cage traps still require the animal to be subsequently dispatched and may be too costly, bulky, and inefficient. Body-gripping traps, or foothold traps set to ensure drowning, are considered the most efficient muskrat trapping systems, and both have currently passed Canadian humane standards (see www.fur.ca for currently approved traps). "Submarine" traps (double- or single-door cage traps with one-way free-swinging doors) set underwater at trails or den entrances are effective multiple-catch traps, which can also be used to ensure drowning.

Body-gripping traps are designed to kill an animal by a blow to the neck or chest region and retain the entire animal, rather than just retaining an animal by its limb. However, Palmisano and Dupuie (1975) found that about 6% of muskrats were alive when removed from such traps. Replacing weak springs or using newer "magnum" versions (stronger springs and/or tighter closing jaws) of body-gripping traps should ensure death unless the animal is accidentally caught by the leg. Gilbert (1976) estimated impact energy thresholds necessary to cause death in muskrats at 58–63 cm kg for neck blows and 155 cm kg for chest strikes. Although "killing power" for body-gripping traps set on land is certainly tied to impact energy thresholds, ongoing trap testing in Canada uses time-to-death criteria for evaluating the acceptability of specific killing traps.

Effective sets for body-gripping traps can be made at den entrances, runways, or other natural trails. Bait can also be attached to the trigger wires. These traps may be used in water, under ice, or on the ground. Palmisano and Dupuie (1975), Linscombe (1976), and Parker (1983) compared the effectiveness of several trap types.

Overviews of trapping muskrats with foothold traps were provided by Freeman (1945), Lay (1945a), Cook (1952), Dozier (1953), and Wingard and Sharp (n.d.). Illustrations of various sets for trapping muskrats were presented by Errington (1961). In addition, many books have been written on trapping muskrats and are available in libraries and/or from trapping supply catalogs. Specific trap and set choices are influenced by factors such as habitat, season, trapper preference and goals, weather, and ultimately, legal restrictions.

When using foothold traps, "stop-loss" mechanisms are highly recommended for muskrats. Stop-loss traps have a spring arm that, after a muskrat is caught, helps prevent the muskrat from twisting at the wrist and breaking the limb. However, the trapping system employed should still ensure rapid drowning when foothold traps are employed. No matter what trapping method is used, traps should be checked frequently to prevent inhumane treatment and depredation and to ensure the highest-quality pelts.

When setting traps on muskrat lodges, care should be taken to reduce catches of waterfowl. If large numbers of waterfowl are in the area, use of body-gripping traps set underwater may help reduce nontarget catches (Parker 1983). Reported catches of waterfowl range from 1 duck per 11 muskrats to 1 duck per 37 muskrats (Gashwiler 1949; Coulter 1952; Parker 1983).

Predator and Competitor Control. Almost every vertebrate predator in the marsh will prey on muskrats (Wilson 1968). Lay (1945a) indicated that marsh intensively managed for muskrats may justify predator control. However, predator control for the enhancement of muskrats is not likely to be cost-effective under most situations (Errington 1961, 1963). Where attempted, legal restrictions must be considered before undertaking any control, and control programs should be monitored closely for effectiveness.

Techniques tested by Wilson (1956b) for the control of raccoon depredation on muskrats involved using corn, eggs, persimmons, prunes, and sardines treated with strychnine. Control effectiveness was influenced by the attractiveness and lethal effect of the bait, raccoon population density in the area and the surrounding habitat, and seasonal raccoon activity. Corn was the most effective bait and sardines were the least effective; however, 90% of the control effort was with corn. Control of raccoons was necessary from March to November to maximize muskrat production. Raccoons usually repopulated the treated areas from surrounding habitat in May and June.

Alligators are opportunistic feeders on muskrat. However, they may benefit muskrat populations by (1) forming holes that retain water during dry periods, (2) preventing pool water stagnation by their movements, and (3) scouring ditches, which reduces clogging by vegetation (Lay and O'Neil 1942).

Geese may feed on favorite muskrat foods such as Olney bulrush and saltmarsh bulrush. Use of burned areas by snow geese (*Chen caerulescens*) may convert vegetated areas of the marsh to mud, resembling muskrat eat-outs. Muskrats must emigrate when goose eat-outs occur (O'Neil 1949). Bare areas form ponds and may not recover for several years. To reduce competition between muskrats and geese, adjacent areas may be burned in autumn to produce succulent vegetation that attracts geese, and geese may be hunted on muskrat areas (Lay and O'Neil 1942).

Cattle in the marsh may disturb muskrats and their habitat by feeding on Olney bulrush and trampling beds and runs (Lay 1945a; Dozier 1953). Beds disturbed by cattle may not be rebuilt. In marsh heavily grazed by cattle, 5–10% of muskrat traps may be sprung by cattle (O'Neil 1949). Fencing to exclude cattle may be justified in the interest of muskrats and muskrat trapping (Lay and O'Neil 1942).

Damage Reduction. Muskrats can extensively damage ponds, impoundments, and ditches by tunneling into banks and dam areas. Burrowing activities may occur anytime but are often most pronounced from March through May (Erickson 1966). As with all wildlife damage management, prevention should be an integral part of any project. Integrated pest management should be the standard for all damage-reduction projects, recognizing that damage may be "managed" but not very well "controlled."

Ponds can be constructed and maintained to minimize muskrat damage, although it is difficult to manipulate habitat to prevent muskrats from occupying suitable ponds (Shanks and Arthur 1951, 1952). Specific measures for minimizing muskrat damage to impoundments include constructing or providing (1) a treeless dam of well-packed clay soil at least 6.1–7 m thick at the water line; (2) vertical concrete sheets or durable small-mesh fencing within the dam; (3) a 3:1 dam slope on the pond side, possibly with steeper (2:1) pond banks elsewhere to encourage burrowing away from the dam; (4) the dam and other burrow sites with a covering of 30 cm or more of sand to preclude effective construction of breeding burrows; (5) a good sod cover over the tops of embankments; (6) dam rip-rap to about 1 m below the lowest possible water level; (7) a sufficiently wide and sloped spillway to prevent water level increases (which encourage burrow expansion) of more than 15–20 cm during excessive rains; and (8) at least 0.9 m of dam freeboard, measured vertically from normal water level, to allow for denning activities (Nagel 1945; Glass 1952;Ueshears and Haugen 1953; Erickson 1956, 1966; Cook 1957; Brooks 1959; Schwartz and Schwartz 1959; Miller 1994). Exclusion of livestock and farm machinery from embankment tops can help prevent caving, and regular mowing of bank vegetation can increase the likelihood of discovering signs of burrowing

Trapping, poisoning, shooting, and water level manipulation have been used somewhat successfully to control muskrats in ponds. However, it is extremely difficult to prevent muskrats from occupying suitable habitat. Erickson (1966) indicated that poisoning yields the cheapest effective control, but use of chemicals for control of animals has legal limitations. The U.S. Fish and Wildlife Service (1973) described a zinc phosphide bait for muskrat (the only federally registered toxicant) for use in combination with cut carrots or sweet potatoes and corn oil.

Water level drawdowns can be used to concentrate muskrats for trapping, expose them to predators or winter freeze-outs, and/or cause them to disperse. Conversely, high water levels will drown nestlings during summer. Such modifications, unless regularly repeated, will not likely provide long-term population control. Trapping may be the most economical and efficient method to of reducing muskrat populations in many areas, particularly small ponds (Beshears 1951; Shanks and Arthur 1951; Beshears and Haugen 1953; Brooks 1959; Indiana Department of Natural Resources n.d.). Shanks and Arthur (1952) recommended conducting removal operations in ponds after spring dispersal but before litters are born.

Refuges and Restocking. Natural restocking of muskrat-vacant habitats is generally connected with spring breeding and dispersal but varies considerably. A system of refuges to ensure adequate maintenance of breeding densities was suggested by Errington (1940) for states having late-fall and winter trapping seasons. Suggested systems were 2.6-km^2 wintering refuges every 6–10 km for overtrapped streams and quarter-section refuges 2–3 km apart for extensive marsh areas. Highby (1941) also discussed a system of refuges for Minnesota whereby 0.8 km of stream would be selected every 10 km, and, for large marshes, a corner or a bay would be blocked off as a breeding ground. Muskrat populations in farm ponds may be increased by establishing impoundments of at least 0.8 ha as refuges (Shanks and Arthur 1952). However, Mathiak and Linde (1954) found that the muskrat population failed to increase in a centrally located 38-ha refuge in Horicon Marsh, Wisconsin. Muskrat movement from the refuge was insignificant.

Many muskrat restocking operations were successful, such as at Tule Lake, California, in 1930 and Moyock, North Carolina, in 1936. A Maryland firm sold 50,000 wild muskrats for restocking between 1915 and 1944. Stocking muskrats in bodies of water with repeated high water-level fluctuations is generally undesirable; stocking inland ponds having sufficient food plants may be practicable and desirable (Moody 1950). However, stocking may be unnecessary. Bradley and Cook (1951) found that muskrats moved from adjacent areas into newly created marsh developments in New York. Muskrats should be stocked only after careful study and all biological factors have been considered (Dozier 1953). The reason muskrats are not naturally colonizing the area to be stocked should be determined and carefully evaluated to ensure there is an impediment to natural recolonization. Similarly, a strong justification should be established if muskrats are to be introduced into an area outside their historical range.

RESEARCH AND MANAGEMENT NEEDS

Much of our knowledge of muskrats is highly descriptive. Such information has been, and remains, invaluable. This information, however, has limitations. Recently, wildlife managers have begun to more rigorously quantify their understanding of many aspects of muskrat ecology, but much work remains. Broadly, more formal experiments are needed to test many of the hypotheses developed over the years. Some specific informational needs on muskrats include a better understanding of (1) disease–population density relations, (2) muskrat–vegetation interactions, (3) cause-specific mortality rates, (4) the social organization of muskrats during the nonbreeding season, (5) density-dependent processes, (6) impacts of continued land-use changes, and (7) effects of environmental contaminants.

Although researchers have described the epizootiology of muskrat diseases, we have essentially no rigorous information on the role of density in disease outbreaks, nor the population-level impacts of disease. Knowledge of density–disease interactions, as well as other potential factors leading to disease outbreaks, would greatly advance our understanding of muskrat populations and would help wildlife managers determine whether manipulating the harvest has any impacts on disease outbreaks.

Muskrats are frequently the dominant herbivore in many aquatic systems. Clearly muskrats play a role in marsh vegetation dynamics. However, the exact role they play in influencing long-term vegetative patterns in marshes needs to be quantified. Additional information on muskrat food consumption rates and the effects of herbivory on shoot mortality and regrowth in natural settings will be vital. Large-scale vegetation exclosure experiments are necessary to separate the impacts of water levels and muskrat herbivory on short- and long-term marsh dynamics. Simultaneously monitoring the effects of herbivory on invertebrate communities and detrital pathways will provide a more complete look at the role of muskrat herbivory in marsh ecosystems.

Descriptive information on causal agents of muskrat mortality is voluminous. However, we also need reliable estimates of cause-specific mortality rates in muskrats. Additional intensive (and extensive) radiotelemetry studies, using sufficient sample sizes, are needed to provide such estimates. Frequent statements regarding the extent to which predators affect muskrat populations need to be formally evaluated. Combining such studies with mark–recapture efforts to monitor population size will help advance our understanding of season-specific density-dependent mortality. Ultimately, long-term experimental manipulation of density, while simultaneously monitoring reproduction and cause- and age-specific mortality rates, is the only approach that will yield holistic insight into muskrat population dynamics. Such experiments will be costly and time-consuming but will serve to substantiate or refute the many statements regarding density dependence and compensatory adjustments in muskrat populations. Muskrats lend themselves well to such work because of their abundance, ease of capture, widespread range, relative ease of delineating gender and age, and suitability for carrying tags and/or transmitters. Such studies will also improve our understanding of factors important in muskrat population cycles and generate data and conclusions that will be useful in making and testing important hypotheses about other species.

The social organization of muskrats has been discussed by many. Rarely, however, have the true relatedness of animals been confirmed. Additionally, most information on social organization applies to the breeding season, and little information is available regarding social structure in winter. Recent use of genetic evaluation to infer relatedness among individuals (Marinelli et al. 1997) shows promise for reliably understanding the social organization of muskrats. Advances in studies of other species have paved the way for nuclear minisatellite and nuclear and mitochondrial microsatellite DNA analysis in muskrats, polymorphism evaluations, development of species-specific DNA probes, and elucidation of the major histocompatability complex. This work will form the basis for identifying accurate familial relationships required to better understand the demographic consequences of social dynamics. More rigorous quantitative data are also needed to adequately determine the population-level importance of immigration and emigration. No experiments have been conducted to test the suggestions regarding the factors influencing the number and type of animals dispersing.

Land uses, and the various treatments applied to lands, have had major impacts on muskrat populations. The effects of direct wetland draining are obvious. Less obvious are the effects of practices that may more subtly alter the hydrological, and hence vegetative, patterns in rivers and marshes. Examples include tiling and ditching, channelization, sedimentation, flood control projects, and loss of ground cover, all of which influence infiltration rates. Experimental areas should be developed that study and illustrate how such practices influence the long-term hydrological and vegetative patterns in marshes and rivers. Developing solutions to land-use problems can never efficiently proceed until these effects have been quantified. Mitigation and synergistic strategies always need to be considered when management practices significantly alter habitat.

For instance, semidry marshes that are ditched properly can be efficiently managed for wildlife production rather than just to provide drainage for agricultural purposes (Mathiak and Linde 1956). Similarly, research on the Bombay Hook National Wildlife Refuge, Delaware, revealed that mosquito breeding on a tidal marsh may be effectively controlled by impoundment and proper ditching (Bourn and Cottam 1950), both potentially beneficial to muskrats. However, we need more studies to explore how detrimental effects on wildlife habitat can be minimized even when such projects are adequately constructed, operated, and maintained.

Relatively little or no information is available on the long- or short-term effects of herbicides, pesticides, and other contaminants on muskrats. Agricultural uses of chemicals remain diverse and widespread; some communities bordering marsh or swamp areas are routinely fogged for insect pests, especially mosquitoes. Ultimately, many contaminants enter the aquatic food chain and affect muskrats. Thus, muskrats should be further evaluated for their roles as bioindicators.

Erickson and Lindzey (1983) found that elevated levels of lead in cattail were tracked by elevated levels in muskrat tissues. However, more information is needed on agricultural chemicals, especially on the immediate toxic effects of pollutants on muskrats, the long-term effects on muskrat populations, and the potential implications for human consumption of muskrat meat. Physical condition and accumulation of contaminants were studied in muskrats by Halbrook et al. (1993), who noted 14 potential contaminants. Everett and Anthony (1976) documented similar heavy-metal accumulation in muskrats and related it to water quality. These and other data are compiled in the Contaminant Exposure and Effects—Terrestrial Vertebrates Database maintained by the U.S. Geological Survey Patuxent Wildlife Research Center (www.pwrc.usgs.gov), but additional research needs to be done to expand our contaminant knowledge base for muskrats. Data are especially needed on the immediate and long-term effects of commercial and industrial pollutants and on point and nonpoint eutrophic discharges from urban and agricultural areas.

ROUND-TAILED MUSKRAT

Neofiber alleni

NOMENCLATURE

COMMON NAME. Round-tailed muskrat
SCIENTIFIC NAME. *Neofiber alleni*

The round-tailed muskrat (*Neofiber alleni*) is a common microtine rodent frequently referred to as prairie rat, water rat, or muskrat (Harper 1927). All *Neofiber* occur in either Florida or southeastern Georgia (Paul 1967; Birkenholz 1972; Wassmer and Wolfe 1983; Lefebvre and Tilmant 1992) (Fig. 16.1), but its range formerly was more extensive (Schwartz 1953). Fossils of *Neofiber* have been described from the Pleistocene (Sherman 1952; Webb 1974). Climatic changes probably caused a range restriction from the late Pleistocene to the Recent (Birkenholz 1963; Frazier 1977).

The species was first discovered in 1883 (True 1884). Three subspecies of *Neofiber alleni* were recognized by Miller and Kellogg (1955), but five subspecies are listed in Schwartz (1953), Hall and Kelson (1959), and Hall (1981). These include *N. a. alleni* in Georgia and Brevard County, Florida; *N. a. apalachicolae* at Apalachicola, Franklin County, Florida; *N. a. exoristus* found southeast of Waycross, Ware County, Florida; *N. a. nigrescens* at Ritta, Palm Beach County, Florida; and *N. a. struix* found west of Miami, Dade County, Florida. Burt (1954) stated that subspecific designations of *Neofiber alleni* are arbitrary and should be discarded.

DESCRIPTION

The dorsal pelage of round-tailed muskrats is soft and lustrous; long guard hairs are glossy tipped and brown. Ventrally, the soft, dense, underfur is gray and brown tipped and overlain by scattered, pale guard hairs (Schwartz 1953). Juveniles are lead gray (Birkenholz 1972).

Measurements of the round-tailed muskrat include: total length, 285–381 mm; tail length, 99–168 mm; and hind foot length, 40–50 mm (Birkenholz 1962, 1963). Sexes are similar in size, but adult males are slightly heavier (average = 279 g) than females (average = 262 g). Adult weights generally range from 200 to 350 g. The skull is similar to that of *Ondatra*, but smaller, and the molars are rootless. The body form, fur, ears, and eyes resemble those of *Ondatra* with slight differences: Guard hairs produce a tuft above the tail, fringe hairs on hind foot margins are less developed, the hind feet are smaller in relation to the rest of the body, and the tail is terete and sparsely haired (Birkenholz 1972).

REPRODUCTION

Breeding in *Neofiber* is positively related to favorable water and cover conditions (Birkenholz 1962, 1972; Paul 1967). Peak reproduction often occurs in late fall and early winter, when emergent cover is most developed, but breeding occurs throughout the year (Harper 1927; Porter 1953). Birkenholz (1963) reported finding pregnant females in every month but March in central Florida.

Duration of the estrous cycle appears to be 15 days with a gestation period of 26–29 days (Hamilton 1956; Birkenholz 1963). Birkenholz (1962) found no indication of an immediate postpartum estrus, except in one case following a stillborn litter; breeding normally occurs 15 days after parturition.

The number of young per litter ranges from 1 to 4 (average = 2.3) (Birkenholz 1962). Four to six litters per female are produced each year, depending on conditions. Lefebvre (1982) reported an average litter size of 1.8 in winter.

Development of *Neofiber* young has been evaluated by Porter (1953), Hamilton (1956), and Birkenholz (1962). Average weight at birth is 12 g. At 2 days, they are covered by short, coarse, grayish to black fur. By 14–18 days, eyes are open and all teeth are erupted. By 3 weeks, the young are largely independent. They resemble adults by 1 month of age and they become sexually mature at 90–100 days of age (average weight = 275 g) (Birkenholz 1962, 1963).

ECOLOGY

The houses of round-tailed muskrats in Florida marshes are constructed of fine-textured grass and would not be very strong, durable, or protective in areas where the marsh freezes for long periods (Birkenholz 1962, 1972). They are dome shaped and nearly spherical, 18–61 cm in diameter. Houses are frequently constructed of tightly woven *Panicum hemitomon*, but *Pontederia* spp., *Sagittaria lancifolia*, *Ceratophyllum* spp., and *Polygonum* spp. are also used. The house base is usually on partly decayed vegetation and is partly supported by standing vegetation. The floor of the house is generally slightly above the water level (Birkenholz 1962, 1972).

Houses are constructed by bending vegetation over a platform and incorporating additional vegetation into the walls. Porter (1953) and Golley (1962) observed that shrubs were occasionally used for support. Interior chambers are approximately 10 cm in diameter and generally contain two plunge or exit holes. When young are not present, houses are usually occupied by only one individual, with an average of two houses for each animal (Birkenholz 1963). The chambers are lined with grasses when young are present (Birkenholz 1962, 1972). Houses may be used up to 6 months (Birkenholz 1962) or from year to year (Porter 1953).

Nests of *Neofiber* in Okefenokee Swamp are occasionally built at the bases of solitary bald cypress (*Taxodium distichum*) trees (Golley 1962) or clumps of bushes. Nest foundations are on sphagnum rather than in water. The nests are anchored with larger prairie plants such as *Nymphae macrophylla*, *Erianthus saccharoides*, *P. hemitomon*, and *Cephalanthus occidentalis*. Fibrous roots may also be used (Harper 1920, 1927).

In the Everglades region of Florida, *Mariscus jamaicense*, *C. occidentalis*, and *Salix amphibia* are used as nest foundations (Porter 1953). Nests of *Neofiber* were found by Chapman (1889) and Bangs (1899) in stumps at the bases of black mangroves (*Avicennia germinans*). As the water level rises, the floor of the nest is built higher and higher. If the level rises 0.9–1.2 m, new nests may be made using tree or bush limbs (Porter 1953). The houses of round-tailed muskrats may be used by turtles, frogs, snakes, other mammals, arthropods, and insects (Harper 1920, 1927; Porter 1953; Birkenholz 1962; Smith and Franz 1994).

Round-tailed muskrats construct feeding platforms and shelters for feeding and defecation. These consist of a pad of vegetation, about 10 × 15 cm, elevated slightly above the water. Feeding platforms contain one or two plunge holes. Feeding shelters are platforms that contain a roof and are more common in late winter when protective cover is scarcer. Platforms may also be converted to houses. Depending on population density, animals may share feeding platforms (Birkenholz 1962, 1972).

Grass structures that float on the surface of the water may also serve as feeding platforms (Golley 1962). During high water, platforms may be constructed on the limbs of trees. The size of the platform may vary with the abundance of food in the region (Chapman 1889).

Occasionally the round-tailed muskrat will burrow into banks and will live in muck, especially during periods of low water levels (Porter 1953; Paul 1967; Birkenholz 1972). Burrows and tunnels are usually filled with water and may be used for feeding. Extensive tunnel systems are often constructed in sugarcane fields, where water levels are maintained 0.6–0.75 m below the surface (Lefebvre and Tilmant 1992). Aboveground runways may also be constructed (Chapman 1889; Harper 1927; Golley 1962).

Reported densities in central Florida range from 100 to 300/ha (Birkenholz 1962). Densities fluctuate greatly and are regulated primarily by water level fluctuations. Density on a larger marsh area (1000 ha) averaged 50/ha. Adults constitute the greatest percentage of the population from May through August (Birkenholz 1962, 1972).

FEEDING HABITS

Round-tailed muskrats are herbivorous and mostly use grasses, including *P. hemitomon, Mariscus* sp., *Sporobolus* sp., *Leersia* spp., and *Echinochloa* sp. (Birkenholz 1963, 1972). Other plants eaten include the stems of *Sagittaria graminea* and *Brasenia* spp., seeds of *Peltandra* sp. and *Iris* spp., roots of *Anchistea virginica* (Harper 1927), *Sagittaria lancifolia, Pontederia* sp., and *Nymphaea lutea* (Porter 1953). In sugarcane fields, sugarcane, sedges (*Cyperus* spp.), and grasses (*Panicum* spp.) are important foods (Steffen 1978; Lefebvre 1982).

Golley (1962) stated that crayfish (Astacidae) and other invertebrates may be eaten by *Neofiber*, but crayfish were not found in analyses of stomach contents by Birkenholz (1963). Captive animals will eat lettuce, carrots, potatoes, and commercial rat pellets. Although they travel over a larger area, individuals feed in a circle about 9.1 m in diameter (Birkenholz 1962).

HABITAT

The niches of *Neofiber* and *Ondatra* differ relative to vegetation used. The round-tailed muskrat uses fine-textured grasses or sedges for house construction, whereas *Ondatra* prefer coarser vegetation, such as bulrushes (*Scirpus* spp.) (Birkenholz 1962). *Neofiber* occupy a variety of marsh habitats, but prefer shallow, grassy marshes and wet prairies (Chapman 1894; Rand and Host 1942; Porter 1953; Birkenholz 1962). Chapman (1889), Bangs (1899), and Tilmant (1975) considered the round-tailed muskrat common in salt savannas, around the edges of saltwater pools, and in saltwater and freshwater ponds and marshes. In Georgia, *Neofiber* often inhabit treeless wet areas and prefer bogs rather than more aquatic areas (Harper 1920, 1927; Golley 1962).

Population densities of round-tailed muskrats in central Florida are highest in areas with water depths of 15–46 cm, a sand substrate, and dense stands of *Panicum hemitomon* and *Leersia hexandra*. Slightly deeper areas that also contain *Pontederia lanceolata* are also used (Birkenholz 1972). Water areas nearly completely filled in with sphagnum and other aquatic vegetation are preferred in the Okefenokee Swamp (Harper 1927).

BEHAVIOR

Round-tailed muskrats are primarily nocturnal; peak activity occurs shortly after dark (Porter 1953; Birkenholz 1962, 1963; Webster et al. 1980). They are very shy. Birkenholz (1962) found no evidence of stress, conflict, or territoriality beyond the limits of the house. When aggression is initiated, animals frequently charge intruders with their front feet clenched on their chest and teeth chattering.

MORTALITY

Predators of *Neofiber* include cats and dogs (Porter 1953), barn owls (*Tyto alba*) (Schantz and Jenkins 1950), Northern harriers (*Circus cyaneus*), red-tailed hawks (*Buteo jamaicensis*), barred owls (*Strix varia*), great horned owls (*Bubo virginianus*), bobcats (*Lynx rufus*), and eastern cottonmouths (*Agkistrodon piscivorus*). Other reported predators include great blue herons (*Ardea herodius*) (Ehrhart 1984) and alligators (Kinsella 1982). They are normally preyed on while away from their houses, mostly in late winter and early spring (Birkenholz 1962, 1963). Changes in water level may be the most important factor contributing to mortality of *Neofiber*. Population reductions up to 85% were recorded by Birkenholz (1962) after flooding. Mortality resulted from predation and from animals being hit by automobiles.

Internal parasites of the round-tailed muskrat include the cestodes *Paranoplocephala neofibrinus, Anoplocephaloides neofibrinus, Cittotaenia praecoquis*, and *Taenia lyncis*; the trematode *Quinqueserialis floridensis*; and the nematodes *Carolinensis kinsellai, Longistriata adunca*, and *Litomosoides* sp. (Porter 1953; Birkenholz 1962, 1972; Forrester et al. 1987). Birkenholz (1962) reported various helminth infestation rates: tapeworms, 49.0%; flukes, 13.1%; and roundworms, 6.7%. Helminth prevalence varied from 16.7% for juveniles to 67.0% for adults.

Ectoparasites infesting *Neofiber* include *Laelaps evansi, Haemolaelaps glasglowi* (= *Androlaelaps fahrenholzi*), *Listrophorus caudatus, Listrophorus layni, Prolistrophorus birkenholzi, Garmania bulbicola, Macrocheles* sp., *Trombicula (Eutrombicula) splendens, Radfordia* spp., *Polygenis gwyni*, and *Tyrophagus linineri* (Porter 1953; Birkenholz 1972; Lefebvre 1982; Smith et al. 1988).

Porter (1953) described a condition of captive round-tailed muskrats in which slight breaks in the skin resulted in cysts swollen with pus. The condition was not observed in the wild and was attributed to a bacterial infection. Parasitism and disease do not appear to be important mortality factors for *Neofiber* (Birkenholz 1962).

MANAGEMENT AND CONSERVATION

Round-tailed muskrats are not harvested commercially. However, trapping may be conducted to reduce damage caused by this rodent. The burrowing and tunneling of the round-tailed muskrat may undermine canals and ditches and result in caving and bank sloughing. They may also damage sugarcane stalks and undermine the stools (stumps or rootstock, which produce shoots or suckers) so much that they are damaged by high winds (Porter 1953). Round-tailed muskrats occurred in 40.3% and 80.4% of the fields in two sugarcane areas evaluated by Steffen (1978). Damage was listed as probably economically significant. Fields harvested by hand had a higher occurrence rate of *Neofiber* than did fields harvested mechanically, probably because of increased cover in the hand-harvested fields. Potential control strategies include manipulation of field harvesting and husbandry practices (Steffen et al. 1981); such practices may form the basis of conservation and management strategies.

Number 110 body-gripping or No. 0 underspring steel traps have been used to capture round-tailed muskrats. Traps may be set in houses, covered feeding shelters, feeding pads, or runways. Animals may be captured alive using 10 × 10 × 20-cm hardware cloth traps with an inward-swinging wire door, particularly in house chambers. Standard cage traps, 15 × 15 × 45–60 cm and smaller, may also be used to capture *Neofiber*. Traps should be checked frequently and all trapping should be done in a way to prevent inhumane treatment of animals. All animals should be handled in accordance with the spirit and intent of the Animal Welfare Act. Studies involving animal handling and marking should be reviewed by a credible animal care and use committee to ensure the highest standards of animal care and treatment.

Birkenholz (1962) found that trapping success varied from 10% in warm weather and in areas of dense cover to 50% after water levels rose. Seasonally, the susceptibility of round-tailed muskrats to traps (captures/100 trap nights) varied from 1.5 in October to 7.6 in March

(Steffen 1978). *Neofiber* populations may be enumerated by counting houses, using the average of two houses per animal found by Birkenholz (1962), and/or examining houses for occupancy.

RESEARCH AND MANAGEMENT NEEDS

Compared to *Ondatra,* very little research has been conducted on *Neofiber*. There is virtually no aspect of their ecology and management that does not deserve additional attention. Given their limited distribution and uncommon status (Lefebvre and Tilmant 1992), research efforts to better document and understand this species are certainly warranted.

ACKNOWLEDGMENTS

This chapter could not have been completed without the assistance of others, to whom grateful appreciation is extended. The Minnesota Department of Natural Resources (MDNR) and the U.S. Geological Survey, Patuxent Wildlife Research Center (PWRC), provided us time and logistic support necessary for preparing this chapter. C. DePerno and R. Kimmel (MDNR) provided encouragement and work flexibility during revision of this chapter. B. Thompson, G. Feldhamer, and T. Carter provided helpful review suggestions. F. Messier and W. R. Clark offered suggestions on current research needs. L. J. Garrett and W. Manning, PWRC, provided assistance in obtaining needed references. L. G. Perry organized reference material and assisted with verification of questionable citations. E. Kuznetsov, M. Onyfrenya, and Y. Dgebuadze of Russia and P. W. Sumner, North Carolina Wildlife Resources Commission, provided helpful information on competition and muskrat ecology. D. H. Ellis assisted with Russian contacts and translation of Russian literature. J. A. Howell, J. A. Kushlan, G. W. Smith, M. A. Howe, and M. R. Whitehead, PWRC, cleared the bureaucratic way and provided encouragement throughout revision of this chapter.

LITERATURE CITED

Abram, J. B., Jr. 1968a. Ecological factors influencing gastrointestinal helminths of the Maryland muskrats. Ph.D. Dissertation, Oklahoma State University, Stillwater.

Abram, J. B., Jr. 1968b. Some gastrointestinal helminths of *Ondatra zibethica*, the muskrat, in Maryland. Proceedings of the Helminthological Society of Washington 36:93–95.

Aldous, S. E. 1946. Live trapping and tagging muskrats. Journal of Wildlife Management 10:42–44.

Aldous, S. E. 1947. Muskrat trapping on Sand Lake National Wildlife Refuge, South Dakota. Journal of Wildlife Management 11:77–90.

Aleksiuk, M., and A. Frohlinger. 1971. Seasonal metabolic organization in the muskrat (*Ondatra zibethica*). Part 1: Changes in growth, thyroid activity, brown adipose tissue, and organ weights in nature. Canadian Journal of Zoology 49:1143–54.

Alexander, M. M. 1951. The aging of muskrats on the Montezuma National Wildlife Refuge. Journal of Wildlife Management 15:175–86.

Alexander, M. M. 1956. The muskrat in New York State. Department of Forest Extension, State University of New York, Syracuse.

Alexander, M. M. 1960a. Dentition as an aid in understanding age composition of muskrat populations. Journal of Mammalogy 41:336–42.

Alexander, M. M. 1960b. Shrinkage of muskrat skulls in relation to aging. Journal of Wildlife Management 24:326–29.

Alexander, M. M., and H. L. Dozier. 1949. An extreme case of malocclusion in the muskrat. American Midland Naturalist 42:252–54.

Ameel, D. J. 1942. Two larval cestodes from the muskrat. Transactions of the American Microscopic Society 61:267–71.

Anderson, D. L. 1969. The persistent use of a floating blind by muskrats. American Midland Naturalist 81:601–2.

Anderson, D. R., and R. L. Beaudoin. 1966. Host habitat and age as factors in the prevalence of intestinal parasites of the muskrat. Bulletin of Wildlife Diseases 2:70–76.

Anderson, J. M. 1947. Sex ratio and weights of southwestern Lake Erie muskrats. Journal of Mammalogy 28:391–95.

Anderson, W. L. 1948. Level ditching to improve muskrat marshes. Journal of Wildlife Management 12:172–76.

Applegate, V. C., and H. E. Predmore, Jr. 1947. Age classes and patterns of primeness in a fall collection of muskrat pelts. Journal of Wildlife Management 11:324–30.

Arata, A. A. 1959. Ecology of muskrats in strip-mine ponds in southern Illinois. Journal of Wildlife Management 23:177–86.

Arata, A. A. 1964. Fossil vertebrates from Avery Island. Appendix pages 69–72 *in* S. M. Gagliana, ed. An archaeological survey of Avery Island. Coastal Studies Institute, Louisiana State University, Baton Rouge.

Babcock, K. M. 1967. The influence of water depth and salinity on wiregrass and saltmarsh grass. M.S. Thesis, Louisiana State University, Baton Rouge.

Bailey, V. 1937. The Maryland muskrat marshes. Journal of Mammalogy 18:350–54.

Banfield, A.W. F. 1974. The mammals of Canada. University of Toronto Press, Toronto.

Bangs, O. 1899. The land mammals of peninsular Florida and the coast region of Georgia. Proceedings of the Boston Society of Natural History 28:157–235.

Barker, F. D. 1915. Parasites of the American muskrat (*Fiber zibethicus*). Journal of Parasitology 1:184–97.

Baumgartner, L. L., and F. C. Bellrose. 1943. Determination of sex and age in muskrats. Journal of Wildlife Management 7:77–81.

Bazin, R. C., and R. A. MacArthur. 1992. Thermal benefits of huddling in the muskrat (*Ondatra zibethicus*). Journal of Mammalogy 73:559–64.

Beckett, J. V., and V. Gallicchio. 1967. A survey of helminths of the muskrat, *Ondatra zibethica*, in Portage County. Ohio Journal of Parasitology 53:1169–72.

Bednarik, K. 1956. Muskrat in Ohio Lake Erie marshes. Ohio Department of Natural Resources, Columbus.

Beer, J. R. 1950. The reproductive cycle of the muskrat in Wisconsin. Journal of Wildlife Management 14:151–56.

Beer, J. R., and R. K. Meyer. 1951. Seasonal changes in the endocrine organs and behavior patterns of the muskrat. Journal of Mammalogy 32:173–91.

Beer, J. R., and W. Truax. 1950. Sex and age ratios in Wisconsin muskrats. Journal of Wildlife Management 14:323–31.

Belant, J. L. 1995. Isoflurane as an inhalation anaesthetic for muskrats (*Ondatra zibethicus*). Journal of Wildlife Diseases 31:573–75.

Belant, J. L. 1996. Immobilization of muskrats (*Ondatra zibethicus*) with ketamine and xylazine. Journal of Wildlife Diseases 32:152–55.

Bellrose, F. C. 1950. The relationship of muskrat populations to various marsh and aquatic plants. Journal of Wildlife Management 14:299–315.

Bellrose, F. C., and L. G. Brown. 1941. The effect of fluctuating water levels on the muskrat population of the Illinois River Valley. Journal of Wildlife Management 5:206–12.

Bellrose, F. C., and J. B. Low. 1943. The influence of flood and low water levels on the survival of muskrats. Journal of Mammalogy 24:173–88.

Bentley, C. C., and J. L. Knight. 1994. Comments on the body mass trend of *Ondatra zibethicus* (Rodentia: Muridae) during the late Pleistocene. Brimleyana 21:37–43.

Bentley, C. C., J. L. Knight, and M. A. Knoll. 1994. The mammals of the Ardis local fauna (late Pleistocene), Harleyville, South Carolina. Brimleyana 21:1–35.

Berg, K. M. 1987. Effects of muskrat mounds on decomposition in a southeastern Michigan wetland. M. S. Thesis, Eastern Michigan University, Ypsilanti.

Beshears, W. W., Jr. 1951. Muskrats in relation to farm ponds. Proceedings of the Southeastern Association of Game and Fish Commissioners 5:1–8.

Beshears, W. W., Jr. and A. O. Haugen. 1953. Muskrats in your farm pond. Alabama Conservation 25:4–5, 22.

Birkenholz, D. E. 1962. A study of the life history and ecology of the round-tailed muskrat (*Neofiber alleni* True) in north-central Florida. Ph.D. Dissertation, University of Florida, Gainesville.

Birkenholz, D. E. 1963. A study of the life history and ecology of the round-tailed muskrat (*Neofiber alleni* True) in north-central Florida. Ecological Monographs 33:255–80.

Birkenholz, D. E. 1972. *Neofiber alleni*. Mammalian Species 15:1–4.

Bishop, R. A., R. D. Andrews, and R. J. Bridges. 1979. Marsh management and its relationship to vegetation, waterfowl, and muskrats. Proceedings of the Iowa Academy of Science 86:50–56.

Blanchette, P. 1989. Use of halothane to anaesthetize muskrats in the field. Journal of Wildlife Management 53:172–74.

Bleich, V. C. 1974. Muskrats (*Ondatra zibethicus*) in Amargosa Canyon, Inyo and San Bernardino Counties, California. Murrelet 55:7–8.

Borell, A. E., and R. Ellis. 1934. Mammals of the Ruby Mountains region of northeastern Nevada. Journal of Mammalogy 15:12–45.

Borucinska, J. D., and S. W. Nielsen. 1993. Hepatic capillariasis in muskrats (*Ondatra zibethicus*). Journal of Wildlife Diseases 29:518–20.

Bourn, W. S., and C. Cottam. 1950. Some biological effects of ditching tidewater marshes (Research Report 19). U.S. Fish and Wildlife Service.
Boutin, S., and D. E. Birkenholz. 1987. Muskrat and round-tailed muskrat. Pages 315–25 *in* M. Novak, J. A. Baker, M. E. Obbard, and B. Malloch, eds. Wild furbearer management and conservation in North America. Ontario Ministry of Natural Resources, Toronto.
Boutin, S., R. A. Moses, and M. J. Caley. 1988. The relationship between juvenile survival and litter size in wild muskrats (*Ondatra zibethicus*). Journal of Animal Ecology 57:455–62.
Boyce, M. S. 1977. Life histories in variable environments: applications to geographic variation in the muskrat (*Ondatra zibethicus*). Ph.D. Dissertation, Yale University, New Haven, CT.
Boyce, M. S. 1978. Climatic variability and body size variation in the muskrats (*Ondatra zibethicus*) of North America. Oecologia 36:1–19
Bradley, B. O., and A. H. Cook. 1951. Small marsh development in New York. Transactions of the North American Wildlife Conference 16:251–65.
Brohn, A., and C. F. Shanks. 1948. Incisor width as an age criterion in muskrats. Journal of Wildlife Management 12:437–39.
Brooks, D. M. 1959. Muskrat damage-prevention and control. Outdoor Indiana 2:11–15.
Brooks, D. M. 1985. Microenvironments and activity patterns of burrow-dwelling muskrats (*Ondatra zibethicus*) in rivers. Acta Zoologica Fennica 173:47–49.
Brooks, R. P., and W. E. Dodge. 1981. Identification of muskrat (*Ondatra zibethicus*) habitat in riverine environments. Pages 113–28 *in* J. A. Chapman, and D. Pursley, eds. Proceedings of the worldwide furbearer conference. Frostburg, MD.
Brooks, R. P., and W. E. Dodge. 1986. Estimation of habitat quality and summer population density for muskrats on a watershed basis. Journal of Wildlife Management 50:269–73.
Bulmer, M. G. 1974. A statistical analysis of the 10-year cycle in Canada. Journal of Animal Ecology 43:701–18.
Bulmer, M. G. 1975. Phase relations in the ten-year cycle. Journal of Animal Ecology 44:609–21.
Burt, W. H. 1954. The subspecies category in mammals. Systematic Zoology 3:99–104.
Buss, I. O. 1941. Sex ratios and weights of muskrats (*Ondatra zibethica*) from Wisconsin. Journal of Mammalogy 22:403–6.
Butler, L. 1940. A quantitative study of muskrat food. Canadian Field-Naturalist 54:37–40.
Butler, L. 1953. The nature of cycles in populations of Canadian mammals. Canadian Journal of Zoology 31:242–62.
Butler, L. 1962. Periodicities in the annual muskrat population figures for the province of Saskatchewan. Canadian Journal of Zoology 40:1277–86.
Byrd, M. A. 1953. High occurrence of *Taenia taeniaformis* in the muskrat. Journal of Wildlife Management 17:384–85.
Caley, M. J. 1987. Dispersal and inbreeding avoidance in muskrats. Animal Behavior 35:1225–33.
Caley, M. J., and S. Boutin. 1985. Infanticide in wild populations of *Ondatra zibethicus* and *Microtus pennsylvanicus*. Animal Behavior 33:1036–37.
Campbell, K. L., and R. A. MacArthur. 1994. Digestibility and assimilation of natural forages by muskrat. Journal of Wildlife Management 58:633–41.
Campbell, K. L., and R. A. MacArthur. 1996a. Seasonal changes in gut mass, forage digestibility, and nutrient selection of wild muskrats (*Ondatra zibethicus*). Physiological Zoology 69:1215–31.
Campbell, K. L., and R. A. MacArthur. 1996b. Digestibility of animal tissue by muskrats. Journal of Mammalogy 77:755–60.
Campbell, K. L., and R. A. MacArthur. 1997. Urea recycling in muskrats (*Ondatra zibethicus*): A potential nitrogen-conserving tactic? Physiological Zoology 70:222–29.
Campbell, K. L., and R. A. MacArthur. 1998. Nutrition and the energetic tactics of muskrats (*Ondatra zibethicus*): Morphological and metabolic adjustments to seasonal shifts in diet quality. Canadian Journal of Zoology 76:163–74.
Campbell, K. L., G. L. Weseen, and R. A. MacArthur. 1998. Seasonal changes in water flux, forage intake, and assimilated energy of free-ranging muskrats. Journal of Wildlife Management 62:292–99.
Carter, T. D. 1922. Notes on a Saskatchewan muskrat colony. Canadian Field-Naturalist 36:176.
Chalmers, G. A., and A. C. MacNeill. 1977. Tyzzer's disease in wild-trapped muskrats in British Columbia. Journal of Wildlife Diseases 13:114–16.
Chamberlain, J. L. 1951. The life history and management of the muskrat on Great Meadows Refuge. M.S. Thesis, University of Massachusetts, Amherst.
Chandler, A. C. 1941. Helminths of muskrats in southeast Texas. Journal of Parasitology 27:175–81.
Chapman, F. M. 1889. On the habits of the round-tailed muskrat (*Neofiber alleni* True). Bulletin of the American Museum of Natural History 11:119–22.
Chapman, F. M. 1894. Remarks on certain land mammals from Florida, with a list of the species known to occur in the state. Bulletin of the American Museum of Natural History 6:333–46.
Ching, C., and Y. Chih-Tang. 1965. Foods and food bases of the muskrat, *Ondatra zibethica* Linnaeus. Acta Zoologica Sinica 17:352–63.
Chubbs, T. E., and F. R. Phillips. 1993. Unusually high number of embryos in a muskrat, *Ondatra zibethicus*, from central Labrador. Canadian Field-Naturalist 107:363.
Clark, W. R. 1986. Influence of open season and weather on the harvest of muskrats. Wildlife Society Bulletin 14:376–80.
Clark, W. R. 1987. Effects of harvest on annual survival of muskrats. Journal of Wildlife Management 51:265–72.
Clark, W. R. 1990. Compensation in furbearer populations: Current data compared with a review of concepts. Transactions of the North American Wildlife and Natural Resources Conference 55:491–500.
Clark, W. R. 1994. Habitat selection by muskrats in experimental marshes undergoing succession. Canadian Journal of Zoology 72:675–80.
Clark, W. R. 2000. Ecology of muskrats in prairie wetlands. Pages 287–313 *in* H. R. Murkin, A. G. van der Valk, and W. R. Clark, eds. Prairie wetland ecology: The contribution of the Marsh Ecology Research Program. Iowa State University Press, Ames.
Clark, W. R., and D. W. Kroeker. 1993. Population dynamics of muskrats in experimental marshes at Delta, Manitoba. Canadian Journal of Zoology 71:1620–28.
Clay, R. T., and W. R. Clark. 1985. Demography of muskrats on the upper Mississippi River. Journal of Wildlife Management 49:883–90.
Clough, G. C. 1987. Ecology of island muskrats, *Ondatra zibethicus*, adapted to upland habitat. Canadian Field-Naturalist 101:63–69.
Cofer, H. P. 1950. Delaware furbearers. Delaware Game and Fish Commission, Dover.
Convey, L. E., J. M. Hanson, and W. C. MacKay. 1989. Size-selective predation of unionid clams by muskrats. Journal of Wildlife Management 53:654–57.
Cook, A. H. 1952. A study of the life history and management of the muskrat in New York State. Ph.D. Dissertation, Cornell University, Ithaca, NY.
Cook, A. H. 1957. Control of muskrat burrow damage in earthen dikes. New York Fish and Game Journal 4:213–18.
Coon, R. A. 1965. Daily movements and home range of the muskrat (*Ondatra zibethicus*). M.S. Thesis, Western Illinois University, Macomb.
Coulter, M. W. 1952. A comparison of spring and autumn muskrat trapping in Maine. Northeast Fish and Wildlife Conference 5:1–14.
Cromer, J. I. 1968. Helminths of the muskrat in the lower peninsula of Michigan. M.S. Thesis, University of Michigan, Ann Arbor.
Dailey, F. J. 1954. Practical muskrat raising. A. R. Harding, Columbus, OH.
Danell, K. 1977. Short-term plant successions following the colonization of a northern Swedish lake by the muskrat, *Ondatra zibethica* (L.) . Journal of Applied Ecology 14:933–47.
Danell, K. 1978. Population dynamics of the muskrat in a shallow Swedish lake. Journal of Animal Ecology 47:697–709.
Danell, K. 1985. Population fluctuations of the muskrat in coastal northern Sweden. Acta Theriologica 30:219–27.
Davidson, W. R., and V. F. Nettles. 1997. Field manual of wildlife diseases in the southeastern United States (Southeastern Cooperative Wildlife Disease Study). University of Georgia, Athens.
Dearborn, N. 1932. Foods of some predatory fur-bearing animals in Michigan (Bulletin No. 1). University of Michigan School of Forest Conservation.
Debbie, J. G. 1968. Yellow fat disease in muskrats. New York Fish and Game Journal 15:119–20.
Deems, F. F., Jr., and D. Pursley, eds. 1978. North American furbearers: Their management, research and harvest status in 1976. University of Maryland Press, College Park.
Dell, D. A., R. H. Chabreck, and R. G. Linscombe. 1983. Spring and summer movements of muskrats in a Louisiana coastal marsh. Proceedings of the Southeastern Association of Fish and Wildlife Agencies 37:210–18.
Domning, D. P. 1969. A list, bibliography, and index of the fossil vertebrates of Louisiana and Mississippi. Transactions of the Gulf Coast Association of Geological Societies 19:385–422.
Donohoe, R. W. 1961. Muskrat production in areas of controlled and uncontrolled water-level units. M.S. Thesis, Ohio State University, Columbus.
Donohoe, R. W. 1966. Muskrat reproduction in areas of controlled and uncontrolled water-level units. Journal of Wildlife Management 30:320–26.

Dorney, R. S., and A. J. Rusch. 1953. Muskrat growth and litter production (Technical Wildlife Bulletin 8). Wisconsin Conservation Department.

Doude Van Troostwijk, W. J. 1976. Age determination in musk-rats, *Ondatra zibethicus* (L.) , in the Netherlands. Lutra 18:33–43.

Dozier, H. L. 1942. Identification of sex in live muskrats. Journal of Wildlife Management 6:292–93.

Dozier, H. L. 1943. Occurrence of ringworm disease and lumpy jaw in the muskrat in Maryland. Journal of the American Veterinary Medical Association 102:451–53.

Dozier, H. L. 1945. Sex ratio and weights of muskrats from the Montezuma National Wildlife Refuge. Journal of Wildlife Management 9:232–37.

Dozier, H. L. 1947a. Resorption of embryos in the muskrat. Journal of Mammalogy 28:398–99.

Dozier, H. L. 1947b. Salinity as a factor in Atlantic coast tidewater muskrat production. Transactions of the North American Wildlife Conference 12:398–420.

Dozier, H. L. 1948a. Color mutations in the muskrat (*Ondatra macrodon*) and their inheritance. Journal of Mammalogy 29:393–405.

Dozier, H. L. 1948b. Estimating muskrat populations by house counts (Wildlife Leaflet 306). U.S. Fish and Wildlife Service.

Dozier, H. L. 1950. Muskrat trapping on the Montezuma National Wildlife Refuge, New York, 1943–1948. Journal of Wildlife Management 14:403–12.

Dozier, H. L. 1953. Muskrat production and management (Circular 18). U.S. Fish and Wildlife Service.

Dozier, H. L., and R. W. Allen. 1942. Color, sex ratios and weights of Maryland muskrats. Journal of Wildlife Management 6:294–300.

Dozier, H. L., M. H. Markley, and L. M. Llewellyn. 1948. Muskrat investigations on the Blackwater National Wildlife Refuge, Maryland, 1941–1945. Journal of Wildlife Management 12:177–90.

Drummond, P. C., and D. R. Jones. 1979. The initiation and maintenance of bradycardia in a diving mammal, the muskrat, *Ondatra zibethica*. Journal of Physiology 290:253–71.

Dyakov, Y. V. 1975. Beavers of the European part of the Soviet Union (morphology, ecology, ways and methods of use). Moskovsky Rabochy (Moscow Worker) Publishing House, Smolensk Branch, Moscow.

Eagle, T. C., and J. S. Whitman. 1987. Mink. Pages 615–24 *in* M. Novak, J. A. Baker, M. E. Obbard, and B. Malloch, eds. Wild furbearer management and conservation in North America. Ontario Ministry of Natural Resources, Toronto.

Earhart, C. M. 1969. The influence of soil texture on the structure, durability, and occupancy of muskrat burrows in farm ponds. California Fish and Game 55:179–96.

Edwards, R. L. 1949. Internal parasites of central New York muskrats (*Ondatra zibethica*). Journal of Parasitology 35:547–48.

Ehrhart, L. M. 1984. Some avian predators of the round-tailed muskrat. Florida Field Naturalist 12:98–99.

Elder, W. H., and C. E. Shanks. 1962. Age changes in tooth wear and morphology of the baculum in muskrats. Journal of Mammalogy 43:144–50.

Elton, C. S. 1924. Periodic fluctuations in the numbers of animals: Their causes and effects. British Journal of Experimental Biology 2:119–63.

Elton, C. S., and M. Nicholson. 1942. Fluctuations in numbers of the muskrat (*Ondatra zibethica*) in Canada. Journal of Animal Ecology 11:96–126.

Enders, R. K. 1931. Muskrat propagation in Ohio (Bulletin 19). Ohio Department of Agriculture.

Enders, R. K. 1939. The corpus luteum as an indicator of the breeding of muskrats. Transactions of the North American Wildlife Conference 4:631–34.

Erb, J. 1993. Age-specific reproduction of riverine muskrats. M.S. Thesis, University of Missouri, Columbia.

Erb, J., R. D. Bluett, E. K. Fritzell, and N. F. Payne. 1999. Aging muskrats using molar indices: A regional comparison. Wildlife Society Bulletin 27:628–35.

Erb, J., N. C. Stenseth, and M. S. Boyce. 2000. Geographic variation in population cycles of Canadian muskrats (*Ondatra zibethicus*). Canadian Journal of Zoology 78:1009–16.

Erb, J., M. S. Boyce, and N. C. Stenseth. 2001. Spatial variation in mink and muskrat interactions in Canada. Oikos 93:365–75.

Erickson, D. W. 1981. Furbearer harvest mechanics: An examination of variables influencing fur harvests in Missouri. Pages 1469–91 *in* J. A. Chapman, and D. Pursley, eds. Proceedings of the worldwide furbearer conference. Frostburg, MD.

Erickson, D. W., and J. S. Lindzey. 1983. Lead and cadmium in muskrat and cattail tissues. Journal of Wildlife Management 47:550–55.

Erickson, H. R. 1956. Muskrat damage to Pennsylvania farm ponds with emphasis on Indiana County. M.S. Thesis, Pennsylvania State University, University Park.

Erickson, H. R. 1963. Reproduction, growth, and movement of muskrats inhabiting small water areas in New York State. New York Fish and Game Journal 10:90–117.

Erickson, H. R. 1966. Muskrat burrowing damage and control procedures in New York, Pennsylvania and Maryland. New York Fish and Game Journal 13:176–87.

Erickson, J. L. E., and P. G. Stevens. 1944. American musk from muskrats used in perfume manufacture. Louisiana Conservation 2:3, 6, 8.

Errington, P. L. 1937a. The breeding season of the muskrat in northwest Iowa. Journal of Mammalogy 18:333–37.

Errington, P. L. 1937b. Drowning as a cause of mortality in musk-rats. Journal of Mammalogy 18:497–500.

Errington, P. L. 1937c. Habitat requirements of stream-dwelling muskrats. Transactions of the North American Wildlife Conference 2:411–16.

Errington, P. L. 1938a. Observations on muskrat damage to corn and other crops in central Iowa. Journal of Agricultural Research 57:415–21.

Errington, P. L. 1938b. Reaction of muskrat populations to drought. Ecology 20:168–86.

Errington, P. L. 1939. Observations on young muskrats in Iowa. Journal of Mammalogy 20:465–78.

Errington, P. L. 1940. Natural restocking of muskrat-vacant habitats. Journal of Wildlife Management 4:173–85.

Errington, P. L. 1941. Versatility in feeding and population maintenance of the muskrat. Journal of Wildlife Management 5:68–69.

Errington, P. L. 1942. Observations on a fungus skin disease of Iowa muskrats. American Journal of Veterinary Research 3:195–201.

Errington, P. L. 1943. An analysis of mink predation upon muskrats in north-central United States. Iowa State College Agricultural Experiment Station Research Bulletin 320:798–924.

Errington, P. L. 1944. Additional studies on tagged young muskrats. Journal of Wildlife Management 8:300–306.

Errington, P. L. 1946a. Special report on muskrat disease. Iowa Cooperative Wildlife Research Unit Quarterly Report 1946 (July–September):34–51.

Errington, P. L. 1946b. Predation and vertebrate populations. Quarterly Review of Biology 21:144–77, 221–45.

Errington, P. L. 1948. Environmental control for increasing muskrat production. Transactions of the North American Wildlife Conference 13:596–609.

Errington, P. L. 1951. Concerning fluctuations in populations of the prolific and widely distributed muskrat. American Naturalist 85:273–92.

Errington, P. L. 1954a. On the hazards of overemphasizing numerical fluctuations in studies of "cyclic" phenomena in muskrat populations. Journal of Wildlife Management 18:66–90.

Errington, P. L. 1954b. The special responsiveness of minks to epizootics in muskrat populations. Ecological Monographs 24:377–93.

Errington, P. L. 1956. Factors limiting higher vertebrate populations. Science 124:304–7.

Errington, P. L. 1961. Muskrats and marsh management. Stackpole, Harrisburg, PA.

Errington, P. L. 1963. Muskrat populations. Iowa State University Press, Ames.

Errington, P. L., and C. S. Errington. 1937. Experimental tagging of young muskrats for purposes of study. Journal of Wildlife Management 1:49–61.

Errington, P. L., and T. G. Scott. 1945. Reduction in productivity of muskrat pelts on an Iowa marsh through depredations of red foxes. Journal of Agricultural Research 71:137–48.

Errington, P. L., R. J. Siglin, and R. C. Clark. 1963. The decline of a muskrat population. Journal of Wildlife Management 27:1–8.

Evans, J. 1970. About nutria and their control (Publication 86). U.S. Department of Interior Bureau of Sport Fisheries and Wildlife Resources.

Everett, J. J., and R. G. Anthony. 1976. Heavy metal accumulation in muskrats in relation to water quality. Transactions of the Northeast Section of the Wildlife Society 33:105–18.

Fairbanks, E. S., and D. L. Kilgore, Jr. 1978. Post-dive oxygen consumption of restrained and unrestrained muskrats (*Ondatra zibethica*). Comparative Biochemical Physiology 59A:113–17.

Ferrigno, F. 1967. First in fur value: Muskrats and their management. Part 2: Research, management, and influences. New Jersey Outdoors 17:8, 13–19.

Field, W. H. 1948. Muskrats and muskrat farming. Wisconsin Conservation 1948 (August):9–12.

Finerty, J. P. 1980. The population ecology of cycles in small mammals. Yale University Press, New Haven, CT.

Fish, F. E. 1979. Thermoregulation in the muskrat (*Ondatra zibethicus*): The use of regional heterothermia. Comparative Biochemical Physiology 64A:391–97.

Fish, F. E. 1982. Function of the compressed tail of surface swimming muskrats (*Ondatra zibethicus*). Journal of Mammalogy 63:591–97.

Fish, F. E. 1984. Mechanics, power output and efficiency of the swimming muskrat (*Ondatra zibethicus*). Journal of Experimental Biology 110:183–201.

Forbes, T. R., and R. K. Enders. 1940. Observations on corpora lutea in the ovaries of Maryland muskrats collected during the winter months. Journal of Wildlife Management 4:169–72.

Forrester, D. J., D. B. Pence, A. O. Bush, D. M. Lee, and N. R. Holler. 1987. Ecological analysis of the helminths of round-tailed muskrats (*Neofiber alleni* True) in southern Florida. Canadian Journal of Zoology 65:2976–79.

Frazier, M. K. 1977. New records of *Neofiber lenardi* (Rodentia: Cricetidae) and the paleoecology of the genus. Journal of Mammalogy 58:368–73.

Freeman, R. M. 1945. Muskrats in Mississippi. Mississippi Game and Fish Commission, Jackson.

Friend, M., and T. F. Muraschi. 1963. Montezuma muskrats and tularemia: An investigation of the incidence of tularemia among muskrats at the Montezuma National Wildlife Refuge (Project Report W-35-R-18). New York State Division of Fish and Game.

Friend, M., G. E. Cummings, and J. S. Morse. 1964. Effect of changes in winter water levels on muskrat weights and harvest at the Montezuma National Wildlife Refuge. New York Fish and Game Journal 11:125–31.

Fuller, W. A. 1951. Natural history and economic importance of the muskrat in the Athabasca–Peace Delta, Wood Buffalo Park (Wildlife Management Bulletin Series 1, No. 2). Canadian Wildlife Service.

Gainey, L. F. 1949. Comparative winter food habits of the muskrat and nutria in captivity. M.S. Thesis, Louisiana State University, Baton Rouge.

Galbreath, E. C. 1954. Growth and development of teeth in the muskrat. Transactions of the Kansas Academy of Science 57:238–41.

Gash, S. L., and W. L. Hanna. 1972. Occurrence of some helminth parasites in the muskrats, *Ondatra zibethicus*, from Crawford County, Kansas. Transactions of the Kansas Academy of Science 75:251–54.

Gashwiler, J. S. 1948. Maine muskrat investigations. Maine Department of Inland Fisheries and Game, Augusta.

Gashwiler, J. S. 1949. The effect of spring muskrat trapping on waterfowl in Maine. Journal of Wildlife Management 13:183–88.

Gashwiler, J. S. 1950a. Sex ratios and age classes of Maine muskrats. Journal of Wildlife Management 14:384–98.

Gashwiler, J. S. 1950b. A study of the reproductive capacity of Maine muskrats. Journal of Mammalogy 31:180–85.

Gilbert, F. F. 1976. Impact energy thresholds for anesthetized raccoons, mink, muskrats, and beavers. Journal of Wildlife Management 40:669–76.

Gilbert, F. F. 1987. Methods for assessing reproductive characteristics of furbearers. Pages 180–90 *in* M. Novak, J. A. Baker, M. E. Obbard, and B. Malloch, eds. Wild furbearer management and conservation in North America. Ontario Ministry of Natural Resources, Toronto.

Gilbert, F. F., and N. Gofton. 1982. Heart rate values for beaver, mink and muskrat. Comparative Biochemical Physiology 73A:249–51.

Glass, B. P. 1952. Factors affecting the survival of the Plains muskrat *Ondatra zibethicus cinnamomina* in Oklahoma. Journal of Wildlife Management 16:484–91.

Goff, O. E., and C. W. Upp. 1942. Dried muskrat meal for poultry. Pages 100–101 *in* Louisiana Agricultural Experiment Station Annual Report, 1942–43. Baton Rouge, LA.

Golley, F. B. 1962. Mammals of Georgia. University of Georgia Press, Athens.

Gould, H. N., and N. H. Kreeger. 1948. The skull of the Louisiana muskrat (*Ondatra zibethica rivalicia* Bangs). Part 1: The skull in advanced age. Journal of Mammalogy 29:138–49.

Gowanloch, I. N. 1943. Louisiana muskrat industry as a source of human food. Transactions of the North American Wildlife Conference 8:213–17.

Greenwell, G. A. 1948. Wildlife values of Missouri farm ponds. Transactions of the North American Wildlife Conference 13:271–81.

Halbrook, R. S., R. L. Kirkpatrick, P. F. Scanlon, M. R. Vaughan, and H. P. Veit. 1993. Muskrat populations in Virginia's Elizabeth River: Physiological condition and accumulation of environmental contaminants. Archives of Environmental Contamination and Toxicology 25:438–45.

Hall, E. R. 1981. The mammals of North America. John Wiley, New York.

Hall, E. R., and K. R. Kelson. 1959. The mammals of North America. Ronald Press, New York.

Hamerstrom, F. N., Jr., and J. Blake. 1939. Central Wisconsin muskrat study. American Midland Naturalist 21:514–20.

Hamilton, W. J. 1956. The young of *Neofiber alleni*. Journal of Mammalogy 37:448–49.

Hamilton, W. J. 1959. Food habits of mink in New York. New York Fish and Game Journal 6:77–85.

Hansen, E. L. 1965. Muskrat distribution in south-central Oregon. Journal of Mammalogy 46:669–71.

Hanski, I., and H. Henttonen. 1996. Predation on competing rodent species: A simple explanation of complex patterns. Journal of Animal Ecology 65:220–32.

Hanski, I., L. Hansson, and H. Henttonen. 1991. Specialist predators, generalist predators, and the microtine rodent cycle. Journal of Animal Ecology 60:353–67.

Hanson, J. M., W. C. MacKay, and E. E. Prepas. 1989. Effect of size-selective predation by muskrats (*Ondatra zibethicus*) on a population of unionid clams (*Anodonta grandis simpsoniana*). Journal of Animal Ecology 58:15–28.

Harlow, H. J. 1984. The influence of hardarian gland removal and fur lipid removal on heat loss and water flux to and from the skin of muskrats (*Ondatra zibethicus*). Physiological Zoology 57:349–56.

Harper, F. 1920. The Florida water-rat (*Neofiber alleni*) in the Okefinokee Swamp, Georgia. Journal of Mammalogy 1:65–67.

Harper, F. 1927. The mammals of the Okefinokee Swamp region of Georgia. Proceedings of the Boston Society of Natural History 38:191–396.

Harris, V. T. 1952. Muskrats on tidal marshes of Dorchester County (Education Publication 91). Maryland Board of Natural Resources, Chesapeake Biological Laboratory.

Heatherly, W. G. 1993. Demographic characteristics of riverine muskrats after an oil spill. M.S. Thesis, University of Missouri, Columbia.

Hensley, A. L., and H. Twining. 1946. Some early summer observations on muskrats in a northeastern California marsh. California Fish and Game 32:171–81.

Hewitt, N., and K. Miyanishi. 1997. The role of mammals in maintaining plant species richness in a floating *Typha* marsh in southern Ontario. Biodiversity and Conservation 6:1085–102.

Highby, P. R. 1941. A management program for Minnesota muskrat. Proceedings of the Minnesota Academy of Science 9:30–34.

Hjälten, J. 1991. Muskrat (*Ondatra zibethica*) territoriality, and the impact of territorial choice on reproduction and predation risk. Annales Zoologica Fennici 28:15–21.

Hockett, R. N. 1968. Normal flora of the muskrat (*Ondatra zibethicus*) (Linnaeus), and attempts to isolate *Pasteurella pseudotuberculosis* and *Listeria monocytogenes*. M.S. Thesis, University of Michigan, Ann Arbor.

Huenecke, H. S., A. B. Erickson, and W. H. Marshall. 1958. Marsh gases in muskrat houses in winter. Journal of Wildlife Management 22:240–45.

Indiana Department of Conservation. 1963. Life series: The muskrat (Leaflet 5). Division of Fish and Game.

Indiana Department of Natural Resources. n.d. Muskrat control with Conibear traps (Leaflet 2). Division of Fish and Wildlife.

Jelinski, D. E. 1989. Seasonal differences in habitat use and fat reserves in an arctic muskrat population. Canadian Journal of Zoology 67:305–13.

Jilek, R. 1977. Trematode parasites of the muskrat, *Ondatra zibethicus*, in southern Illinois. Transactions of the Illinois State Academy of Science 70:105–7.

Johansen, K. 1962. Buoyancy and insulation in the muskrat. Journal of Mammalogy 43:64–68.

Johnson, C. E. 1925. The muskrat in New York. Roosevelt Wildlife Bulletin 3:193–230.

Jokela, J., and P. Mutikainen. 1995. Effect of size-dependent muskrat (*Ondatra zibethicus*) predation on the spatial distribution of a freshwater clam, *Anodonta piscinalis* Nilsson (Unionidae, Bivalvia). Canadian Journal of Zoology 73:1085–94.

Jones, D. R., N. H. West, O. S. Bamford, P. C. Drummond, and R. A. Lord. 1982. The effect of forcible submergence on the diving response in muskrats (*Ondatra zibethicus*). Canadian Journal of Zoology 60:187–93.

Kaminski, R. A., and H. H. Prince. 1981. Dabbling duck and aquatic macroinvertebrate responses to manipulated wetland habitat. Journal of Wildlife Management 45:1–15.

Karstad, L., P. Lusit, and D. Wright. 1971. Tyzzer's disease in muskrats. Journal of Wildlife Diseases 7:96–99.

Kaufman, D. M. 1995. Diversity of New World Mammals: Universality of the latitudinal gradients of species and bauplans. Journal of Mammalogy 76:322–34.

Keith, L. B. 1963. Wildlife's ten-year cycle. University of Wisconsin Press, Madison.

Kinler, Q. J., R. H. Chabreck, N. W. Kinler, and R. G. Linscombe. 1990. Effect of tidal flooding on mortality of juvenile muskrats. Estuaries 13:337–40.

Kinsella, J. M. 1982. Alligator predation on round-tailed muskrats. Florida Field Naturalist 10:79.

Knight, I. M. 1951. Diseases and parasites of the muskrat (*Ondatra zibethica*) in British Columbia. Canadian Journal of Zoology 29:188–214.

Kroll, R. W., and R. L. Meeks. 1985. Muskrat population recovery following habitat re-establishment near southwestern Lake Erie. Wildlife Society Bulletin 13:483–86.

Lacki, M. J., P. N. Smith, W. T. Peneston, and F. D. Vogt. 1989. Use of methoxyflurane to surgically implant transmitters in muskrats. Journal of Wildlife Management 53:331–33.

Lacki, M. J., W. T. Peneston, K. B. Adams, F. D. Vogt, and J. C. Houppert. 1990a. Summer foraging patterns and diet selection of muskrats inhabiting a fen wetland. Canadian Journal of Zoology 68:1163–67.

Lacki, M. J., W. T. Peneston, and F. D. Vogt. 1990b. A comparison of the efficacy of two types of live traps for capturing muskrats, *Ondatra zibethicus*. Canadian Field-Naturalist 104:594–96.

Lagenwalter, P. E. 1985. Indigenous muskrats, *Ondatra zibethicus*, in coastal southern California. California Fish and Game 72:121–22.

Lancia, R. A., J. D. Nichols, and K. H. Pollock. 1994. Estimating the number of animals in wildlife populations. Pages 215–43 *in* T. A. Bookhout, ed. Research and management techniques for wildlife and habitats. Wildlife Society, Bethesda, MD.

Lantz, D. E. 1910. The muskrat (Farmers' Bulletin 396). U.S. Department of Agriculture.

Lay, D. W. 1945a. Muskrat investigations in Texas. Journal of Wildlife Management 9:56–76.

Lay, D. W. 1945b. The problems of undertrapping in muskrat management. Transactions of the North American Wildlife Conference 10:75–78.

Lay, D. W., and T. O'Neil. 1942. Muskrats on the Texas coast. Journal of Wildlife Management 6:301–12.

Le Boulengé, E. 1977. Two ageing methods for muskrats: Live or dead animals. Acta Theriologica 22:509–20.

LeCompte, E. L. 1928. Muskrat industry of Maryland. Maryland Conservation Department, Annapolis.

Lefebvre, L. W. 1982. Population dynamics of the round-tailed muskrat (*Neofiber alleni*) in Florida sugarcane. Ph.D. Dissertation, University of Florida, Gainesville.

Lefebvre, L. W., and J. T. Tilmant. 1992. Round-tailed muskrat. Pages 277–86 *in* S. R. Humphrey, ed. Rare and endangered biota of Florida, Vol. I. University Press of Florida, Gainesville.

Linde, A. F. 1963. Muskrat pelt patterns and primeness (Technical Bulletin 29). Wisconsin Conservation Department.

Linscombe, G. 1976. An evaluation of the No. 2 Victor and 220 Conibear traps in coastal Louisiana. Proceedings of the Southeastern Association of Fish and Wildlife Agencies 30:560–68.

Lord, G. H., A. C. Todd, and C. Kabat. 1956a. Studies on Errington's disease in muskrats. Part 1: Pathological changes. American Journal of Veterinary Research 17:303–6.

Lord, G. H., A. C. Todd, and H. Mathiak. 1956b. Studies on Errington's disease in muskrats. Part 2: Etiology. American Journal of Veterinary Research 17:307–10.

Louisiana Department of Conservation. 1931. The fur animals of Louisiana (Bulletin 18, rev.). Louisiana Department of Conservation.

Lowery, G. H., Jr., 1974. The mammals of Louisiana and its adjacent waters. Louisiana State University Press, Baton Rouge.

Lynch, I. J., T. O'Neil, and D. W. Lay. 1947. Management significance of damage by geese and muskrats to Gulf Coast marshes. Journal of Wildlife Management 1:50–76.

MacArthur, R. A. 1977. Behavioral and physiological aspects of temperature regulation in the muskrat (*Ondatra zibethicus*). Ph.D. Dissertation, University of Manitoba, Winnipeg, Canada.

MacArthur, R. A. 1978. Winter movements and home range of the muskrat. Canadian Field-Naturalist 92:345–49.

MacArthur, R. A. 1979. Seasonal patterns of body temperature and activity in free-ranging muskrats (*Ondatra zibethicus*). Canadian Journal of Zoology 57:25–33.

MacArthur, R. A. 1984. Aquatic thermoregulation in the muskrat (*Ondatra zibethicus*): Energy demands of swimming and diving. Canadian Journal of Zoology 62:241–48.

MacArthur, R. A. 1986a. Metabolic and behavioral responses of muskrats (*Ondatra zibethicus*) to elevated CO_2 in a simulated winter microhabitat. Canadian Journal of Zoology 64:738–43.

MacArthur, R. A. 1986b. Effects of CO_2 inhalation on acid–base balance and thermal recovery following cold water dives by the muskrat (*Ondatra zibethicus*). Journal of Comparative Physiology B. Biochemical, Systematic, and Environmental Physiology 156:339–46.

MacArthur, R. A. 1986c. Brown fat and aquatic temperature regulation in muskrats, *Ondatra zibethicus*. Physiological Zoology 59:306–17.

MacArthur, R. A. 1990. Seasonal changes in the oxygen storing capacity and aerobic dive limits of the muskrat (*Ondatra zibethicus*). Journal of Comparative Physiology B. Biochemical, Systematic, and Environmental Physiology 160:593–99.

MacArthur, R. A. 1992a. Foraging range and aerobic endurance of muskrats diving under ice. Journal of Mammalogy 73:565–69.

MacArthur, R. A. 1992b. Gas bubble release by muskrats diving under ice: Lost gas or a potential oxygen pool? Journal of Zoology 226:151–64.

MacArthur, R. A., and M. Aleksiuk., 1979. Seasonal microenvironments of the muskrat (*Ondatra zibethicus*) in a northern marsh. Journal of Mammalogy 60:146–54.

MacArthur, R. A., and C. M. Karpan. 1989. Heart rates of muskrats diving under simulated field conditions: Persistence of the bradycardia response and factors modifying its expression. Canadian Journal of Zoology 67:1783–92.

MacArthur, R. A., and R. E. Krause. 1989. Energy requirements of freely diving muskrats (*Ondatra zibethicus*). Canadian Journal of Zoology 67:2194–200.

MacArthur, R. A., M. M. Humphries, and D. Jeske. 1997. Huddling behavior and the foraging efficiency of muskrats. Journal of Mammalogy 78:850–58.

MacKinnon, B. M., and M. D. B. Burt. 1978. Platyhelminth parasites of muskrats (*Ondatra zibethica*) in New Brunswick. Canadian Journal of Zoology 56:350–54.

Marinelli, L., and F. Messier. 1993. Space use and the social system of muskrats. Canadian Journal of Zoology 71:869–75.

Marinelli, L., and F. Messier. 1995. Parental-care strategies among muskrats in a female-biased population. Canadian Journal of Zoology 73:1503–10.

Marinelli, L., F. Messier, and Y. Plante. 1997. Consequences of following a mixed reproductive strategy in muskrats. Journal of Mammalogy 78:163–72.

Martin, R. A. 1993. Patterns of variation and speciation in Quarternary rodents. Pages 1–16 *in* R. A. Martin, and A. D. Barnosky, eds. Morphological change in Quarternary mammals of North America. Cambridge University Press, Cambridge.

Mathiak, H. A. 1966. Muskrat population studies at Horicon Marsh (Technical Bulletin 36). Wisconsin Conservation Department.

Mathiak, H. A., and A. F. Linde. 1954. Role of refuges in muskrat management (Technical Wildlife Bulletin 10). Wisconsin Conservation Department.

Mathiak, H. A., and A. F. Linde. 1956. Studies on level ditching for marsh management (Technical Wildlife Bulletin 12). Wisconsin Conservation Department.

McCabe, T. R., and G. Elison. 1986. An efficient live-capture technique for muskrats. Wildlife Society Bulletin 14:282–84.

McCann, L. J. 1944. Notes on growth, sex and age ratios, and suggested management of Minnesota muskrats. Journal of Mammalogy 25:59–63.

McEwan, L. H., N. Aitchison, and P. E. Whitehead. 1974. Energy metabolism of oiled muskrats. Canadian Journal of Zoology 52:1057–62.

McKinstry, M. C., R. R. Karhu, and S. H. Anderson. 1997. Use of active beaver, *Castor canadensis*, lodges by muskrats, *Ondatra zibethicus*, in Wyoming. Canadian Field-Naturalist 111:310–11.

McLeod, J. A. 1948. Preliminary studies on muskrat biology in Manitoba. Transactions of the Royal Society of Canada 42:81–95.

McLeod, J. A. 1950. A consideration of muskrat populations and population trends in Manitoba. Transactions of the Royal Society of Canada 44:69–79.

McLeod, J. A., and Bondar, G. F. 1952. Studies on the biology of the muskrat in Manitoba. Part 1: Oestrous cycle and breeding season. Canadian Journal of Zoology 30:243–53.

McMullen, T. L. 1978. Mammals of the Duck Creek local fauna, late Pleistocene of Kansas. Journal of Mammalogy 59:374–86.

Messier, F., and J. A. Virgil. 1992. Differential use of bank burrows and lodges by muskrats, *Ondatra zibethicus*, in a northern marsh environment. Canadian Journal of Zoology 70:1180–84.

Messier, F., J. A. Virgil, and L. Marinelli. 1990. Density-dependent habitat selection in muskrats: A test of the ideal free distribution model. Oecologia 84:380–85.

Meyer, M. C., and J. R. Reilly. 1950. Parasites of muskrats in Maine. American Midland Naturalist 44:467–77.

Miller, G. S., Jr., and R. Kellogg. 1955. List of North American Recent mammals (Bulletin 205). United States National Museum.

Miller, J. E. 1994. Muskrats. Pages B61–69 *in* S. E. Hygnstrom, R. M. Timm, and G. E. Larson, eds. Prevention and control of wildlife damage. University of Nebraska Press, Lincoln.

Monk, C. E. 1981. History and present status of fur management in Ontario. Pages 1501–23 *in* J. A. Chapman, and D. Pursley, eds. Proceedings of the worldwide furbearer conference. Frostburg, MD.

Moody, R. 1950. The possibilities of transplanting muskrats in East Baton Rouge, West Feliciana and Natchitoches Parishes, Louisiana. M.S. Thesis, Louisiana State University, Baton Rouge.

Moses, R. A., and S. Boutin. 1986. Molar fluting and pelt primeness techniques for distinguishing age classes of muskrats: A reevaluation. Wildlife Society Bulletin 14:403–6.

Moses, R. A., and S. Boutin. 1988. Problems associated with the use of the molar fluting technique: A response. Wildlife Society Bulletin 16:90–93.

Murkin, H. R. 1989. The basis for food chains in prairie wetlands. Pages 316–38 *in*A. G. van der Valk, ed. Northern prairie wetlands. Iowa State University Press, Ames.

Murkin, H. R., A. G. van der Valk, and W. R. Clark. 2000. Prairie wetland ecology: The contribution of the Marsh Ecology Research Program. Iowa State University Press, Ames.

Nadeau, S., R. Dé carie, D. Lambert, and M. St.-Georges. 1995. Nonlinear modeling of muskrat use of habitat. Journal of Wildlife Management 59:110–17.

Nagel, W. O. 1945. Controlling muskrat damage in ponds. Missouri Conservation 6:10–11.

Neal, T. J. 1968. A comparison of two muskrat populations. Iowa State Journal of Science 43:193–210.

Nelson, N. F. 1954. Factors in the development and restoration of waterfowl habitat at Ogden Bay Refuge, Weber County, Utah. Utah State Department of Fish and Game Publication 6:70–71.

Nelson, J. W., and J. A. Kadlec. 1984. A conceptual approach to relating habitat structure and macroinvertebrate production in freshwater wetlands. Transactions of the North American Wildlife and Natural Resources Conference 49:262–70.

Neves, R. J., and M. A. Odom. 1989. Muskrat predation on endangered freshwater mussels in Virginia. Journal of Wildlife Management 53:934–41.

Newsom, J. D., H. R. Perry, Jr., and P. E. Schilling. 1976. Fire ant–muskrat relationships in Louisiana coastal marshes. Proceedings of the Southeastern Association of Fish and Wildlife Commissioners 13:414–18.

Nicholson, W. R., and D. E. Davis. 1957. The duration of life of muskrat houses. Ecology 38:161–63.

Noble, R. E., and J. M. Wright. 1977. Nesting birds of the Shishmaref Inlet and Seward Peninsula, Alaska and the possible effects of reindeer herding and grazing on nesting birds (Progress Report). Alaska Cooperative Wildlife Research Unit.

Novak, M. 1975. The use of meat of furbearing animals. Ontario Ministry of Natural Resources, Toronto.

Novak, M., M. E. Obbard, J. G. Jones, R. Newman, A. Booth, A. J. Satterthwaite, and G. Linscombe. 1987. Furbearer harvests in North America, 1600–1984. Ontario Ministry of Natural Resources, Toronto.

Obbard, M. E. 1987. Fur grading and pelt identification. Pages 717–826 *in* M. Novak, J. A. Baker, M. E. Obbard, and B. Malloch, eds. Wild furbearer management and conservation in North America. Ontario Ministry of Natural Resources, Toronto.

Olsen, P. F. 1959a. Muskrat breeding biology at Delta, Manitoba. Journal of Wildlife Management 23:40–53.

Olsen, P. F. 1959b. Dental patterns as age indicators in musk-rats. Journal of Wildlife Management 23:228–31.

O'Neil, T. 1949. The muskrat in the Louisiana coastal marshes. Louisiana Department of Wildlife and Fisheries, New Orleans.

O'Neil, T., and G. Linscombe. 1976. The fur animals, the alligator, and the fur industry in Louisiana (Wildlife Education Bulletin 106). Louisiana Wildlife and Fisheries Commission.

Palmisano, A. W., Jr. 1967. Ecology of *Scirpus olneyi* and *Scirpus robustus* in Louisiana coastal marshes. M.S. Thesis, Louisiana State University, Baton Rouge.

Palmisano, A. W., Jr. 1970. Plant community–soil relationships in Louisiana coastal marshes. Ph.D. Dissertation, Louisiana State University, Baton Rouge.

Palmisano, A. W., Jr. 1972a. The distribution and abundance of muskrats (*Ondatra zibethicus*) in relation to vegetative types in Louisiana coastal marshes. Proceedings of the Southeastern Association of Fish and Wildlife Commissioners 26:160–77.

Palmisano, A. W., Jr. 1972b. Habitat preference of waterfowl and fur animals in the northern Gulf coast marshes. Pages 163–90 *in* R. H. Chabreck, ed. Proceedings of the marsh and estuary management symposium. Division of Continuing Education, Louisiana State University, Baton Rouge.

Palmisano, A. W., Jr. and H. H. Dupuie. 1975. An evaluation of steel traps for taking fur animals in coastal Louisiana. Proceedings of the Southeastern Association of Fish and Wildlife Commissioners 29:342–47.

Pancoast, J. M. 1937. Exhibit "A": Muskrat industry in southern New Jersey. Transactions of the North American Wildlife Conference 2:527–30.

Pankakoski, E. 1980. An improved method for age determination in the muskrat, *Ondatra zibethica* (L.) . Annales Zoologici Fennici 17:113–21.

Parker, G. R. 1983. An evaluation of trap types for harvesting muskrats in New Brunswick. Wildlife Society Bulletin 11:339–43.

Parker, G. R., and J. W. Maxwell. 1980. Characteristics of a population of muskrats (*Ondatra zibethicus*) in New Brunswick. Canadian Field-Naturalist 94:1–8.

Parker, G. R., and J. W. Maxwell. 1984. An evaluation of spring and autumn trapping seasons for muskrats, *Ondatra zibethicus*, in eastern Canada. Canadian Field-Naturalist 98:293–304.

Paul, J. R. 1967. Round-tailed muskrat in west central Florida. Quarterly Journal of the Florida Academy of Science 30:227–29.

Penn, G. H., Jr. 1942. Parasitological survey of Louisiana muskrats. Journal of Parasitology 28:348–49.

Penn, G. H., Jr. and E. C. Martin. 1941. The occurrence of porocephaliasis in the Louisiana muskrat. Journal of Wildlife Management 5:13–14.

Perry, H. R., Jr. 1982. Muskrats. Pages 282–325 *in* J. A. Chapman, and G. A. Feldhamer, eds. Wild mammals of North America. Johns Hopkins University Press, Baltimore.

Phillips, D. W. 1979. Muskrat population dynamics on a controlled wetland in southern Saskatchewan. M.S. Thesis, University of Regina, Regina, Saskatchewan, Canada.

Pietsch, M. 1970. Vergleichende untersuchungen an Schädeln nordamerikanisher und europäischer Bisamratten (*Ondatra zibethicus* L. 1766). Zeitshcrift für Saugetierkunde 35:257–88 [In German with English summary].

Porter, R. P. 1953. A contribution to the life history of the water rat, *Neofiber alleni*. M.S. Thesis, University of Miami, Coral Gables, FL.

Proulx, G. 1981. Relationship between muskrat populations, vegetation, and water level fluctuations and management considerations at Luther Marsh, Ontario. Ph.D. Dissertation, University of Guelph, Guelph, Ontario, Canada.

Proulx, G. 1997. Estimating fall whole-body weights of muskrats, *Ondatra zibethicus*, from skinned weights. Canadian Field-Naturalist 111:643–45.

Proulx, G., and M. L. Buckland. 1985. Precocial breeding in a southern Ontario muskrat, *Ondatra zibethicus*, population. Canadian Field-Naturalist 99:377–78.

Proulx, G., and M. L. Buckland. 1986. Productivity and mortality rates of southern Ontario pond- and stream-dwelling muskrat, *Ondatra zibethicus*, populations. Canadian Field-Naturalist 100:378–80.

Proulx, G., and F. F. Gilbert. 1983. The ecology of the muskrat, *Ondatra zibethicus*, at Luther Marsh, Ontario. Canadian Field-Naturalist 97:377–90.

Proulx, G., and F. F. Gilbert. 1984. Estimating muskrat population trends by house counts. Journal of Wildlife Management 48:917–22.

Proulx, G., and F. F. Gilbert. 1988. The molar fluting technique for aging muskrats: A critique. Wildlife Society Bulletin 16:88–89.

Proulx, G., J. A. McDonnell, and F. F. Gilbert. 1987. The effect of water level fluctuations on muskrat, *Ondatra zibethicus*, predation by mink, *Mustela vison*. Canadian Field-Naturalist 101:89–92.

Rand, A. L., and P. Host. 1942. Mammal notes from Highlands County, Florida: Results of the Archbold expeditions. Bulletin of the American Museum of Natural History 81:1–21.

Rausch, R. L. 1946. Parasites of Ohio muskrats. Journal of Wildlife Management 10:70.

Rawley, E., J. B. Low, and D. Sharp. 1952. The muskrat: A farm crop (Circular 168). Utah State Agriculture College Extension.

Reeves, H. M., and R. M. Williams. 1956. Reproduction, size, and mortality in the Rocky Mountain muskrat. Journal of Mammalogy 37:494–500.

Rice, E. W., and O. B. Heck. 1975. A survey of the gastrointestinal helminths of the muskrat, *Ondatra ziberhicus*, collected from two localities in Ohio. Ohio Journal of Science 75:263–64.

Robicheaux, B. L. 1978. Ecological implications of variably spaced ditches on nutria in a brackish marsh, Rockefeller Refuge, Louisiana. M.S. Thesis, Louisiana State University, Baton Rouge.

Ross, W. M. 1972. Methods of establishing natural and artificial stands of *Scirpus olneyi*. M.S. Thesis, Louisiana State University, Baton Rouge.

Ruttan, R. A., and D. R. Wooley. 1974. Studies of furbearers associated with proposed pipeline routes in the Yukon and Northwest Territories (Arctic Gas Biological Report Series, Vol. 9).

Salinger, H. E. 1950. Mating of muskrats. Journal of Mammalogy 31:97.

Sather, J. H. 1954. The dentition method of aging muskrats (Miscellaneous Publication 130). Chicago Academy of Science and Natural History.

Sather, J. H. 1956. Skull dimensions of the Great Plains muskrat, *Ondatra zibethicus cinnamominus*. Journal of Mammalogy 37:501–5.

Sather, J. H. 1958. Biology of the Great Plains muskrat in Nebraska. Wildlife Monographs 2:1–35.

Schacher, W. H., and M. R. Pelton. 1976. Sex ratios, morphology, and condition parameters of muskrats in East Tennessee. Proceedings of the Southeastern Association of Fish and Wildlife Agencies 30:660–66.

Schacher, W. H., and M. R. Pelton. 1975. Productivity of muskrats in East Tennessee. Proceedings of the Southeastern Association of Game and Fish Commissioners 29:594–608.

Schantz, V. S., and and J. H. Jenkins., 1950. Extension of range of the round-tailed muskrat, *Neofiber alleni*. Journal of Mammalogy 31:460–61.

Scheffer, V. B. 1955. Body size with relation to population density in mammals. Journal of Mammalogy 36:493–515.

Schofield, R. D. 1955. Analysis of muskrat age determination methods and their application in Michigan. Journal of Wildlife Management 19:463–66.

Schwartz, A. 1953. A systematic study of the water rat (*Neofiber alleni*) (Occasional Paper 547). University of Michigan Museum of Zoology.

Schwartz, C. W., and E. R. Schwartz. 1959. The wild mammals of Missouri. University of Missouri Press and Missouri Conservation Commission, Columbia.

Seabloom, R. W., and J. R. Beer. 1964. Observations of a muskrat (*Ondatra zibethica cinnamominus*) population decline in North Dakota. Proceedings of the North Dakota Academy of Science 17:66–70.

Sealander, J. A. 1979. A guide to Arkansas mammals. River Road Press, Conway, AK.

Seamans, R. 1941. Lake Champlain fur survey (State Bulletin 3–4). Vermont Fish and Game Service.

Seber, G. A. F. 1982. The estimation of animal abundance and related parameters. Charles Griffin, London.

Seton, E. T. 1909. Life histories of northern animals. Constable, London.

Shanks, C. E. 1948. The pelt-primeness method of aging muskrats. American Midland Naturalist 39:179–87.

Shanks, C. E., and G. C. Arthur. 1951. Movements and population dynamics of farm pond and stream muskrats in Missouri (Report W-13-R). Missouri Conservation Commission.

Shanks, C. E., and G. C. Arthur. 1952. Muskrat movements and population dynamics in Missouri farm ponds and streams. Journal of Wildlife Management 16:138–48.

Sherer, J., and B. A. Wunder. 1979. Thermoregulation of a semiaquatic mammal, the muskrat, in air and water. Acta Theriologica 24:249–56.

Sherman, H. B. 1952. A list and bibliography of the mammals of Florida, living and extinct. Quarterly Journal of the Florida Academy of Science 15:86–126.

Simpson, M. R., and S. Boutin. 1989. Muskrat, *Ondatra zibethicus*, population responses to harvest on the Old Crow Flats, Yukon Territory. Canadian Field-Naturalist 103:420–22.

Simpson, M. R., and S. Boutin. 1993. Muskrat life history: A comparison of a northern and southern population. Ecography 16:5–10.

Sleeman, J., R. Stevens, and E. Ramsey. 1997. Field immobilization of muskrats (*Ondatra zibethicus*) for minor surgical procedures. Journal of Wildlife Diseases 33:165–68.

Smith, F. R. 1938. Muskrat investigations in Dorchester County, Maryland, 1930–34 (Circular 474). U.S. Department of Agriculture.

Smith, L. L., and R. Franz. 1994. Use of Florida round-tailed muskrat houses by amphibians and reptiles. Florida Field Naturalist 22:69–96.

Smith, H. R., R. J. Sloan, and G. S. Walton. 1981. Some management implications between harvest rate and population resiliency of the muskrat (*Ondatra zibethicus*). Pages 425–42 *in* J. A. Chapman, and D. Pursley, eds. Proceedings of the worldwide furbearer conference, Frostburg, MD.

Smith, M. A., J. O. Whitaker, Jr., and J. N. Layne. 1988. Ectoparasites of the round-tailed muskrat (*Neofiber alleni*) with special emphasis on mites of the family Listrophoridae. American Midland Naturalist 120:268–75.

Snead, I. E. 1950. A family type live trap, handling cage, and associated techniques for muskrats. Journal of Wildlife Management 14:67–79.

Soileau, D. M. 1968. Vegetative reinvasion of experimentally treated plots in a brackish marsh. M.S. Thesis, Louisiana State University, Baton Rouge.

Sokolov, V. E., and N. P. Lavrov. 1993. The muskrat: Morphology, systematics, ecology. Nauka, Moscow.

Sooter, C. A. 1946. Muskrats of the Tule Lake Refuge, California. Journal of Wildlife Management 10:68–70.

Soper, L. R., and N. F. Payne. 1997. Relationship of introduced mink, an island race of muskrat, and marginal habitat. Annales Zoologica Fennici 34:251–58.

Sprugel, G., Jr. 1951. Spring dispersal and settling activities of central Iowa muskrats. Iowa State College of Science 26:71–84.

Stains, H. J. 1979. Primeness in North American furbearers. Wildlife Society Bulletin 7:120–25.

Stearns, L. A., and M. W. Goodwin. 1941. Notes on the winter feeding of the muskrats in Delaware. Journal of Wildlife Management 5:1–12.

Steffen, D. E. 1978. The occurrence of and damage by the Florida water rat in Florida sugar cane production areas. M.S. Thesis, Virginia Polytechnic Institute and State University, Blacksburg.

Steffen, D. E., N. R. Holler, L. W. Lefebvre, and P. F. Scanlon. 1981. Factors affecting the occurrence and distribution of Florida water rats in sugarcane fields. Proceedings of the American Society of Sugar Cane Technology 9:27–32.

Stein, B. R. 1988. Morphology and allometry in several genera of semiaquatic rodents (*Ondatra, Nectomys*, and *Oryzomys*). Journal of Mammalogy 69:500–511.

Stevens, P. G., and J. L. E. Erickson. 1942. The chemical constitution of the musk of the Louisiana muskrat. Journal of the American Chemical Society 64:144–47.

Stevens, W. E. 1953. The northwestern muskrat of the Mackenzie Delta, Northwest Territories, 1947–1948 (Wildlife Management Bulletin, Series 1, No. 8). Canadian Wildlife Service.

Stewart, R. E. A., J. R. Stephen, and R. J. Brooks. 1975. Occurrence of muskrat, *Ondatra zibethicus albus*, in the District of Keewatin, Northwest Territories. Journal of Mammalogy 56:507.

Stewart, R. W., and J. R. Bider. 1974. Reproduction and survival of ditch-dwelling muskrats in southern Quebec. Canadian Field-Naturalist 88:429–36.

Stewart, R. W., and J. R. Bider. 1977. Summer activity of muskrats in relation to weather. Journal of Wildlife Management 41:487–99.

Storer, T. I. 1937. The muskrat as native and alien. Journal of Mammalogy 18:443–60.

Svihla, A. 1931. The field biologist's report. Pages 278–87 *in* S. C. Arthur, comp. The fur animals of Louisiana (Bulletin 18, rev.). Louisiana Department of Conservation.

Svihla, A., and R. D. Svihla. 1931. The Louisiana muskrat. Journal of Mammalogy 12:12–28.

Taber, R. D. 1969. Criteria of sex and age. Pages 325–401 *in*R. H. Giles, Jr., ed. Wildlife management techniques. Wildlife Society, Washington, DC.

Takos, M. J. 1940. A review of the literature on diseases and parasites of the muskrat. Unpublished report, Maine Cooperative Wildlife Research Unit, Orono.

Takos, M. J. 1943. Trapping and banding muskrats. Journal of Wildlife Management 7:400–407.

Takos, M. J. 1944. Summer movements of banded muskrats. Journal of Wildlife Management 8:307–11.

Takos, M. J. 1947. A semi-quantitative study of muskrat food habits. Journal of Wildlife Management 11:331–39.

Taylor, M. 1980. Our valuable furbearers. Wildlife North Carolina 44:2–7.

Thornton, R., C. Gordon, and J. H. Ferguson. 1978. Role of thermal stimuli in the diving response of the muskrat (*Ondatra zibethica*). Comparative Biochemical Physiology 61A:369–70.

Thurber, J. M., R. O. Peterson, and T. D. Drummer. 1991. The effect of regulated lake levels on muskrats, *Ondatra zibethicus*, in Voyageurs National Park, Minnesota. Canadian Field-Naturalist 105:34–40.

Tilmant, J. T. 1975. Habitat utilization by round-tailed muskrats (*Neofiber alleni*) in Everglades National Park. M.S. Thesis, Humboldt State University, Arcadia, CA.

Todd, A. W., and E. K. Boggess. 1987. Characteristics, activities, lifestyles, and attitudes of trappers in North America. Pages 59–76 *in* M. Novak, J. A. Baker, M. E. Obbard, and B. Malloch, eds. Wild furbearer management and conservation in North America. Ontario Ministry of Natural Resources, Toronto.

True, F. W. 1884. A muskrat with a round tail. Science 4:34.

Twining, H., and A. L. Hensley. 1943. The distribution of muskrats in California. California Fish and Game 29:64–78.

U.S. Fish and Wildlife Service. 1973. Muskrat and nutria control with zinc phosphide (Wildlife Leaflet 504).

Virgil, J. A., and F. Messier. 1992a. Seasonal variation in body composition and morphology of adult muskrats in central Saskatchewan. Canadian Journal of Zoology 228:461–77.

Virgil, J. A., and F. Messier. 1992b. The ontogeny of body composition and gut morphology in free-ranging muskrats. Canadian Journal of Zoology 70:1381–88.

Virgil, J. A., and F. Messier. 1993. Evaluation of body size and body condition indices in muskrats. Journal of Wildlife Management 57:854–60.

Virgil, J. A., and F. Messier. 1996. Population structure, distribution, and demography of muskrats during the ice-free period under contrasting water fluctuations. Ecoscience 3:54–62.

Virgil, J. A., and F. Messier. 1997. Habitat suitability in muskrats: A test of the food limitation hypothesis. Journal of Zoology 243:237–53.

Wainscott, V. J., C. Bartley, and P. Kangas. 1990. Effect of muskrat mounds on microbial density on plant litter. American Midland Naturalist 123:399–401.

Wassmer, D. A., and J. L. Wolfe. 1983. New Florida localities for the round-tailed muskrat. Northeast Gulf Science 6:197–99.

Webb, S. D. 1974. Pleistocene mammals of Florida. University Press of Florida, Gainesville.

Webster, D. G., R. L. Evans, and D. A. Dewsbury. 1980. Behavioral patterns of round-tailed muskrats (*Neofiber alleni*). Florida Science 43:1–6.

Weller, M. W. 1981. Freshwater marshes. University of Minnesota Press, Minneapolis.

Weller, M. W., and L. H. Fredrickson. 1973. Avian ecology of a managed glacial marsh. Living Bird 12:269–91.

Weller, M. W., and C. E. Spatcher. 1965. Role of habitat in the distribution and abundance of marsh birds (Special Report Number 43). Iowa Agricultural and Home Economics Experiment Station.

Williams, R. M. 1951. The use of scent in live-trapping muskrats. Journal of Wildlife Management 15:117–18.

Willner, G. R., J. A. Chapman, and J. R. Goldsberry. 1975. A study and review of muskrat food habits with special reference to Maryland (Publication on Wildlife Ecology 1). Maryland Wildlife Administration.

Willner, G. R., G. A. Feldhamer, E. E. Zucker, and J. A. Chapman. 1980. *Ondatra zibethicus*. Mammalian Species 141:1–8.

Wilson, K. A. 1949. Investigations on the effects of controlled water levels upon muskrat production. Proceedings of the Southeastern Association of Game and Fish Commissioners 3:105–11.

Wilson, K. A. 1953. Raccoon predation on muskrats near Currituck, North Carolina. Journal of Wildlife Management 17:113–19.

Wilson, K. A. 1954a. Litter production of coastal North Carolina muskrats. Proceedings of the Southeastern Association of Game and Fish Commissioners 8:13–19.

Wilson, K. A. 1954b. The role of mink and river otter as muskrat predators in northeastern North Carolina. Journal of Wildlife Management 18:199–207.

Wilson, K. A. 1955a. A compendium of the principal data on muskrat reproduction (Federal Aid in Wildlife Restoration Project W-6-R). North Carolina Wildlife Resources Commission, Raleigh.

Wilson, K. A. 1955b. Effects of water level control on muskrat populations (Job Completion Report, Federal Aid in Wildlife Restoration Project W-6-R). North Carolina Wildlife Resources Commission, Raleigh.

Wilson, K. A. 1955c. Experimental marsh management near Currituck, North Carolina (Job Completion Report, Federal Aid in Wildlife Restoration Project W-6-R-15, Special Report 1). North Carolina Wildlife Resources Commission, Raleigh.

Wilson, K. A. 1956a. Color, sex ratios, and weights of North Carolina muskrats (Federal Aid in Wildlife Restoration Project W-6-R-16, Special Report 1). North Carolina Wildlife Resources Commission, Raleigh.

Wilson, K. A. 1956b. Control of raccoon predation on muskrats near Currituck, North Carolina. Proceedings of the Southeastern Association of Game and Fish Commissioners 10:221–33.

Wilson, K. A. 1968. Fur production on southeastern coastal marshes. Pages 149–62 *in* J. D. Newsom, ed. Proceedings of the marsh and estuary management symposium. Division of Continuing Education, Louisiana State University, Baton Rouge.

Wingard, R. G., and W. M. Sharp. n.d. Trapping and skinning muskrats (Circular 478). Pennsylvania State University College of Agriculture Extension Service.

Wobeser, G., D. B. Hunter, and P. Y. Daoust. 1978. Tyzzer's disease in muskrats: Occurrence in free-living animals. Journal of Wildlife Diseases 14:325–28.

Wood, W. 1974. Muskrat origin, distribution, and range extension through the coastal areas of Del Norte County California and Curry County Oregon. Murrelet 55:1–4.

Worthy, G. A. J., J. Rose, and F. Stormshak. 1987. Anatomy and physiology of fur growth: The pelage priming process. Pages 827–41 *in* M. Novak, J. A. Baker, M. E. Obbard, and B. Malloch, eds. Wild furbearer management and conservation in North America. Ontario Ministry of Natural Resources, Toronto.

Wragg, L. E. 1954. The effect of D. D.T. and oil on musk-rats. Canadian Field-Naturalist 68:11–13.

John Erb, Forest Wildlife Populations and Research Group, Minnesota Department of Natural Resources, Grand Rapids, Minnesota 55744. Email: john.erb@dnr.state.mn.us.

H. Randolph Perry, Jr. U.S. Geological Survey, Patuxent Wildlife Research Center, 12100 Beech Forest Road, Laurel, Maryland 20708-4039. Email: randy_perry@usgs.gov.

17

Voles

Microtus species

Stephen R. Pugh
Sherry Johnson
Robert H. Tamarin

NOMENCLATURE

COMMON NAMES. Voles, meadow mice, arvicolids, microtines; the various species may be referred to by specific common names (see species list below)
SCIENTIFIC NAME. Genus *Microtus*
SPECIES NORTH OF MEXICO. *Microtus abbreviatus,* insular vole (2 subspecies); *M. breweri,* beach vole; *M. californicus,* California vole (17 subspecies); *M. canicaudus,* gray-tailed vole; *M. chrotorrhinus,* rock vole (3 subspecies); *M. longicaudus,* long-tailed vole (15 subspecies); *M. miurus,* singing vole (4 subspecies); *M. mogollonensis,* Mogollon vole (4 subspecies); *M. montanus,* montane vole (15 subspecies); *M. ochrogaster,* prairie vole (7 subspecies); *M. oeconomus,* tundra vole (10 subspecies); *M. oregoni,* creeping vole, Oregon vole (4 subspecies); *M. pennsylvanicus,* meadow vole (26 subspecies); *M. pinetorum,* woodland vole, pine vole (7 subspecies); *M. richardsoni,* water vole, Richardson's vole (4 subspecies); *M. townsendii,* Townsend's vole (6 subspecies); *M. xanthognathus,* taiga vole, yellow-cheeked vole (Jones et al. 1997; Nowak 1999; Wilson and Ruff 1999)

Voles and lemmings were once included in a distinct family, Arvicolidae (Chaline et al. 1977; Johnson and Johnson 1982; Chaline and Graf 1988). Now the consensus is to include them as a subfamily (Arvicolinae) within the family Muridae, which includes rats and mice as well as voles and lemmings (Carleton and Musser 1984; Anderson 1985; Musser and Carleton 1993; Jones et al. 1997). Diversity within the genus *Microtus* is extensive, with 66 species worldwide. Although MacDonald and Cook (1996) and Nowak (1999) listed *M. coronarius* as a separate species, we follow the convention of Jones et al. (1997) and include it as a subspecies of *M. longicaudus. M. nesophilus,* from Gull Island in Long Island Sound, is probably extinct because of human activities (Hall 1981; Nowak 1999). Although it has been listed as a separate species (Nowak 1999), we include it as a subspecies of *M. pennsylvanicus* (Jones et al. 1986). *M. abbreviatus* and *M. miurus* usually are described as separate species (Musser and Carleton 1993; Nowak 1999; Wilson and Ruff 1999). However, Conroy and Cook (2000) suggested that because of similarities in cytochrome *b* sequences as well as chromosomal similarity (Rausch and Rausch 1968), they may be conspecific. *M. mogollonensis* has been listed as a subspecies of *M. mexicanus* (Johnson and Johnson 1982), but we consider it a separate species (Judd 1980; Frey 1989; Frey and LaRue 1993).

In addition to *Microtus,* other voles in North America include red-backed voles (*Clethrionomys californicus, C. gapperi,* and *C. rutilus*), the white-footed vole (*Arborimus albipes*), the red tree vole (*A. longicaudus*), the Sonoma tree vole (*A. pomo*), the western heather vole (*Phenacomys intermedius*), the eastern heather vole (*P. ungava*), and the sagebrush vole (*Lemmiscus curtatus*) (Jones et al. 1997).

Fossil remains of *Microtus* in North America range in age from the early Pleistocene to the Holocene. *Microtus* probably originated in Asia and migrated across the Bering Land Bridge into North America, where extensive and rapid diversification occurred (Zakrzewski 1985). The first invasion was about 2.1 million years ago (Conroy and Cook 1999, 2000). Analysis of mitochondrial cytochrome *b* gene sequences supports monophyly of the endemic North American species of *Microtus* and suggests that only two invasions occurred. The first resulted in the endemic North American species and the second was *M. oeconomus* (about 55,000 years ago) (Zakrzewski 1985; Lance and Cook 1998; Conroy and Cook 2000).

At the subgeneric level, well-supported sister groups include *M. californicus* and *M. mexicanus* (*mogollonensis*); *M. pinetorum* and *M. richardsoni; M. canicaudus* and *M. townsendii;* and *M. montanus* and *M. pennsylvanicus* (Conroy and Cook 2000). Other relationships include the *pennsylvanicus* clade uniting *M. pennsylvanicus* and *M. montanus* with *M. canicaudus* and *M. townsendii. M. longicaudus* and *M. oregoni* are both described as basal to the *pennsylvanicus* clade. *M. xanthognathus* is sister to the *M. miurus* and *M. abbreviatus* clade (Conroy and Cook 2000).

DISTRIBUTION

The genus *Microtus* has a circumboreal distribution. Many species occur throughout Europe and much of Asia (Musser and Carleton 1993; Nowak 1999). Of the 17 species listed above, 16 occur exclusively in North America north of Mexico. Five species are in Mexico and Central America, and 44 species are exclusively in the Old World. One species, *M. oeconomus,* occurs in both the Old and the New World. The 22 North American species are distributed from the northernmost reaches of the continent into Mexico and Guatemala. The highest diversity of vole species is in the temperate regions and in the western part of the continent. Generally, in the southern regions, the populations are isolated and occur only in higher elevations.

Microtus pennsylvanicus has the most extensive range, from coast to coast and north to above the Arctic Circle (Fig. 17.1). It is sympatric with several other species of *Microtus,* and is separated from these by differences in ecological niche, especially noted in zones of contact, and behavioral characteristics.

Other North American species have wide distributions. On a relative basis of size of range as compared to *M. pennsylvanicus,* these include *M. longicaudus,* 41%; *M. ochrogaster,* 30%; *M. pinetorum,* 24%; and *M. montanus,* 16%. These calculations do not take into account the frequently disjunct nature of the ranges of some species. They indicate a relative adaptive success within the North American ecosystems.

In extreme northern latitudes, four species exist: the circumboreal *Microtus oeconomus* in Alaska, Yukon Territory, and the Northwest Territories (Fig. 17.2); *M. miurus,* with a discontinuous range in much of Alaska, Yukon Territory, and the adjacent Northwest Territories; *M. abbreviatus,* confined to Hall and St. Matthew Islands in the Bering Sea (Fig. 17.3); and *M. xanthognathus* in Alaska, Yukon Territory, Northwest Territory, and the most northern portions of the Canadian prairie provinces (Manitoba, Saskatchewan, and Alberta; Fig. 17.4). *M. longicaudus* is distributed in the west from southern Alaska to the Great Plains and *M. pinetorum* is found in southeastern North America (Fig. 17.5). The greater number of species in the West probably reflects isolating mechanisms of mountain systems, climatologic differences, and available ecological niches. This is seen throughout the subfamily

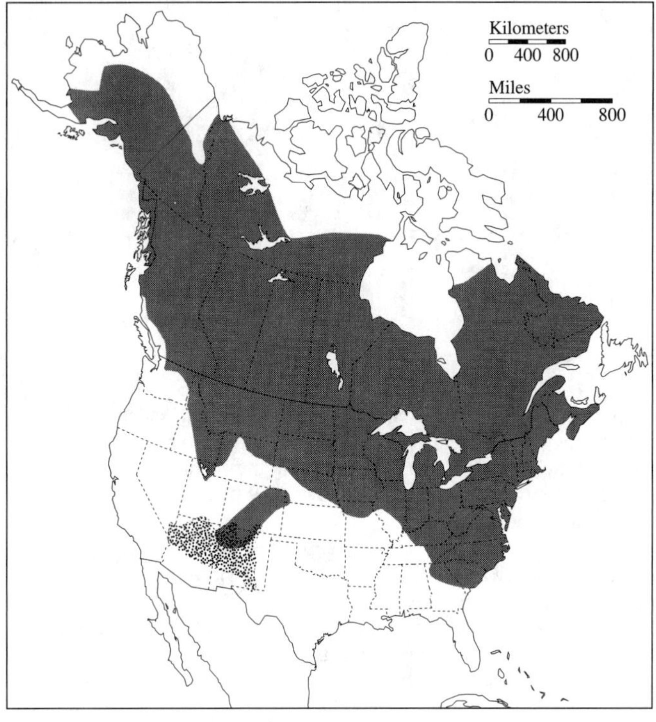

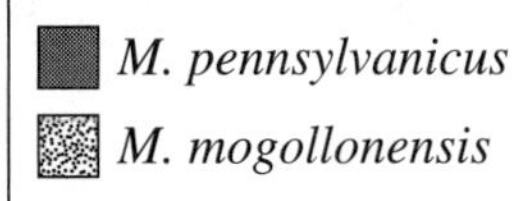

FIGURE 17.1. Distribution of the meadow vole (*Microtus pennsylvanicus*) and Mogollon vole (*M. mogollonensis*).

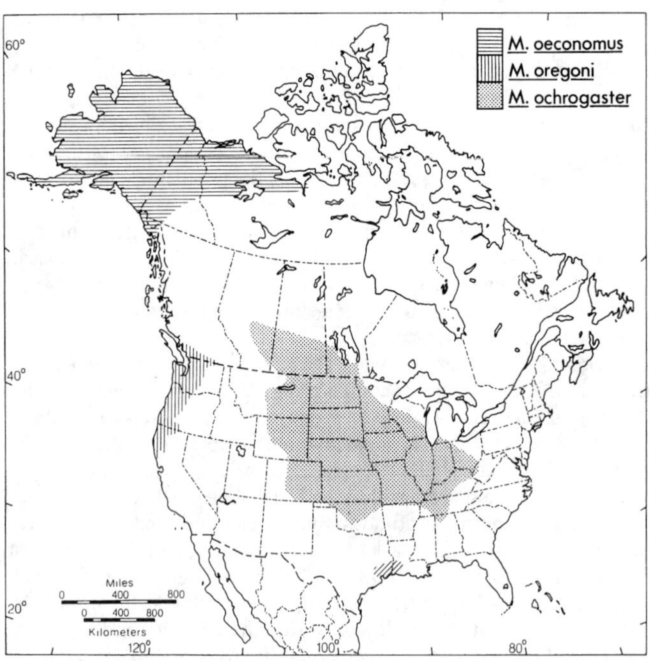

FIGURE 17.2. Distribution of the tundra vole (*Microtus oeconomus*), creeping vole (*M. oregoni*), and prairie voles (*M. ochrogaster*).

Arvicolinae as well as in the genus *Microtus*. In the East, the beach vole, *M. breweri,* is a recent offshoot of *M. pennsylvanicus*. It is found only on Muskeget Island, a small, sandy island off of Nantucket Island and Cape Cod, Massachusetts (Moyer et al. 1988).

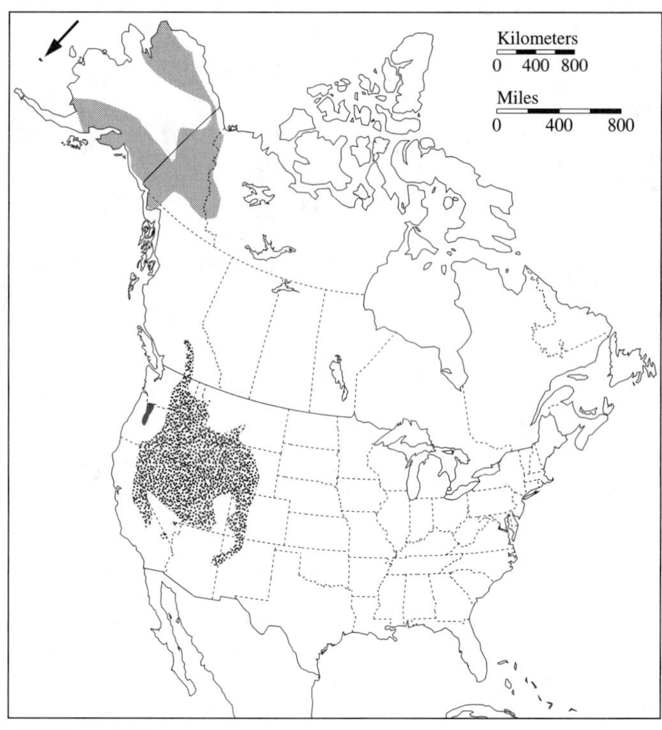

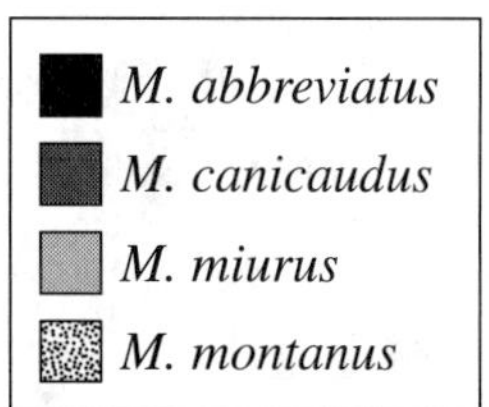

FIGURE 17.3. Distribution of the singing vole (*Microtus miurus*), gray-tailed vole (*M. canicaudus*), insular vole (*M. abbreviatus*), and montane vole (*M. montanus*).

In Mexico and Guatemala, the presumed result of long isolation shows in the spectrum of more differentiated forms of *Microtus*. Of the five species occurring in the south, four subgenera are represented. *M. mexicanus* is now considered to be confined to mountainous areas from north-central to southern Mexico. *M. mogollonensis,* once considered part of *M. mexicanus,* occurs in high elevations in New Mexico, Arizona, Colorado, Utah, and Texas (see Fig. 17.1). It is confined to habitat islands of high-elevation, moist, coniferous forests surrounded by inhospitable low-elevation desert. Because they have been isolated from each other since the last ice age (20,000 years), populations on different mountains tend to differ slightly from each other (Frey and La Rue 1993).

DESCRIPTION

General Morphology. Members of the North American genus *Microtus* are characterized by uniformity in morphology with little variation among species. They are generally adapted to subterranean, terrestrial, and in some cases, semiamphibious life.

The body is stocky and rounded. The nose is blunt, with vibrissae that are generally inconspicuous compared to those of many other rodents. The ears are small, rounded, a little longer than the body fur, and are usually more obvious in this genus than in some other arvicolids. The eyes are small; they are relatively smaller in the fossorial species, *M. oregoni* and *M. pinetorum*. The legs are short. The tail is scantily haired, longer than the hind foot, but shorter than the head and body length. The relative size of the tail varies among species. It is relatively shorter in juveniles, and achieves full length only in adults. The fur

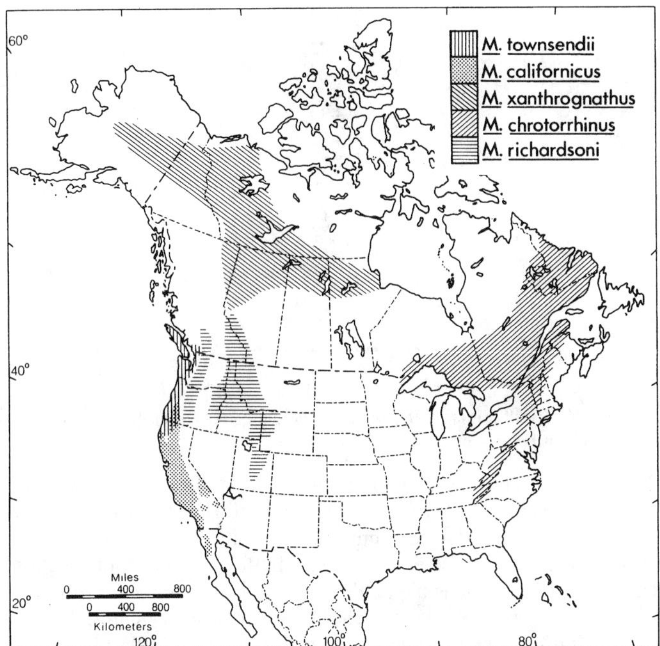

FIGURE 17.4. Distribution of the Townsend's vole (*Microtus townsendii*), California vole (*M. californicus*), yellow-cheeked vole (*M. xanthognathus*), rock vole (*M. crotorrhinus*), and water vole (*M. richardsoni*).

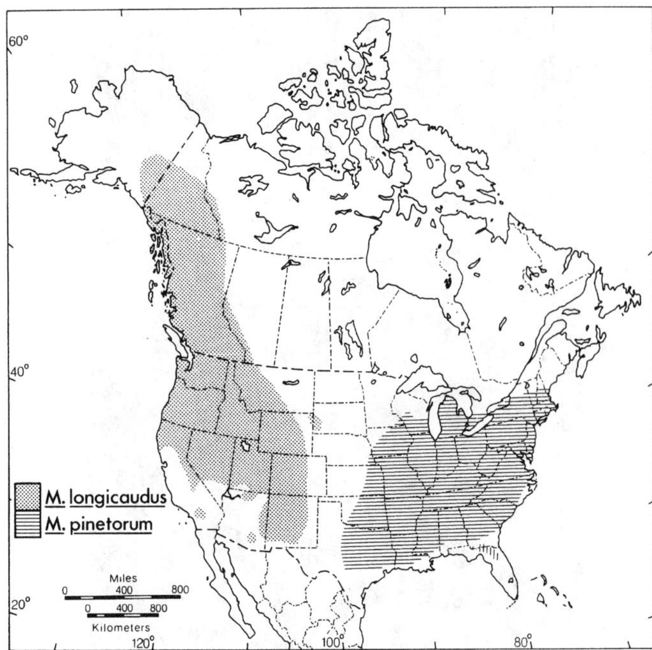

FIGURE 17.5. Distribution of the long-tailed vole (*Microtus longicaudus*) and pine vole (*M. pinetorum*).

appears moderately coarse and is various shades of brown to gray, lighter ventrally.

Size is variable within the genus. The largest two species are *M. richardsoni* and *M. xanthognathus,* both of which have adult weights that may exceed 120 g, comparable to the size of a half-grown Norway rat (*Rattus norvegicus*). Length of the head and body of these two species can exceed 180 mm. The smallest is *M. oregoni,* with a mean body weight usually <20g.

Pelage and Molts. Newborn voles have no external hair (except vibrissae) or dermal pigment. Hatfield (1935) found first dermal pigmentation at 8 hr of age in the California vole. This progresses from the top of the head posteriorly along the back. At 36 hr, pigment covers more than half the back. At 72 hr, the back is completely pigmented and the hair on the back is 2 mm long. By day 5, the hair over the rest of the body, exclusive of feet and tail, reaches 2 mm in length.

By 9 or 10 days of age, the skin pigmentation begins to disappear, leaving the skin pink under the fur until the third week. This fur is the juvenile pelage. At 3 weeks old, the juvenile molt begins, with dark pigmentation all over the skin and new fur growth all over the body. At 35 days of age, pigmentation is reduced to a dark dorsal streak. The adult pelage begins to grow at 8 weeks of age. In *M. californicus,* this begins on the belly, progressing to the mid-dorsal line. The molt line can be identified by this skin pigmentation, moving forward across the shoulders to the head region.

It is likely that the three molts—juvenile, subadult, and adult—occur in all species of *Microtus,* although there may be some variation in timing and pattern. The juvenile molt occurs between the juvenile and the subadult pelage. The subadult molt occurs between the subadult and the adult pelage. The adult molt tends to occur seasonally, in the spring and fall. In *M. breweri,* the juvenile molt occurs when individuals have attained a total length of 136–150 mm. This molt proceeds as a wave of hair replacement and pigmentation beginning on the ventral surface and proceeding laterally to the dorsal surface and terminating at the head. This pattern of molt is termed dorsad (Ling 1970). The subadult molt occurs at a length of 161–165 mm. Of all subadult molts described by Rowsemitt et al. (1975), 81% were of the dorsad type. The remainder were diffuse (Ling 1970), with a speckled or blotchy pattern of pigmentation and hair replacement that may or may not show bilateral symmetry. All molts in individuals >165 mm in length were adult molts. These were of either the dorsad or diffuse type and occurred seasonally, either in late spring to summer (June–August) or fall to winter (October–January) (Rowsemitt et al. 1975).

Adult molts are frequently seen in wild-caught and captive animals. Seasonal molts—winter and summer—appear to be of regular occurrence throughout the Arvicolinae. They are especially marked in the northern species. Hamilton (1938) described the summer molt of *M. pinetorum,* which replaced the dark winter pelage in May or June. The winter molt was noticeable in November or December.

Darkening of the skin is due to increased melanin content in the hair follicles, which actively produce hair. There is also probably increased vascularization during this time.

Skull and Dentition. Skulls are usually angular in adults and have a "standard" microtine appearance (Fig. 17.6). The teeth are characteristic in the genus *Microtus* (Fig. 17.7). Incisors are typically rodent, open rooted, and ever growing. The roots of the lower incisors, as in several genera, pass from the lingual to the labial side of molars between the bases of the roots of m2 and m3. They ascend behind the molars to terminate within or near the condylar process (Hall 1981). The molariform teeth are prismatic, hypsodont, and ever growing. The numbers and conformation of the prisms vary according to the species. As in other rodents of the suborder Myomorpha, the medial portion of the masseter muscle originating from the maxilla sends only a small slip of muscle through the infraorbital canal to attach on the mandible. The dental formula is I 1/1, C 0/0, P 0/0, M 3/3.

GENETICS

Chromosomes. Chromosome number in *Microtus* is characterized by interspecific and intraspecific variation. The diploid number of chromosomes ($2n$) and the fundamental number (FN = the number of autosomal arms) are reported in Table 17.1. The number of chromosomes varies from as few as 17–18 in *M. oregoni* to as many as 70 in *M. longicaudus*. Fundamental number ranges from 32 in *M. oregoni* to 108 in *M. longicaudus*.

The greatest intraspecific variation occurs in *M. longicaudus.* Matthey (1955) was the first to report a diploid number of 56, for a

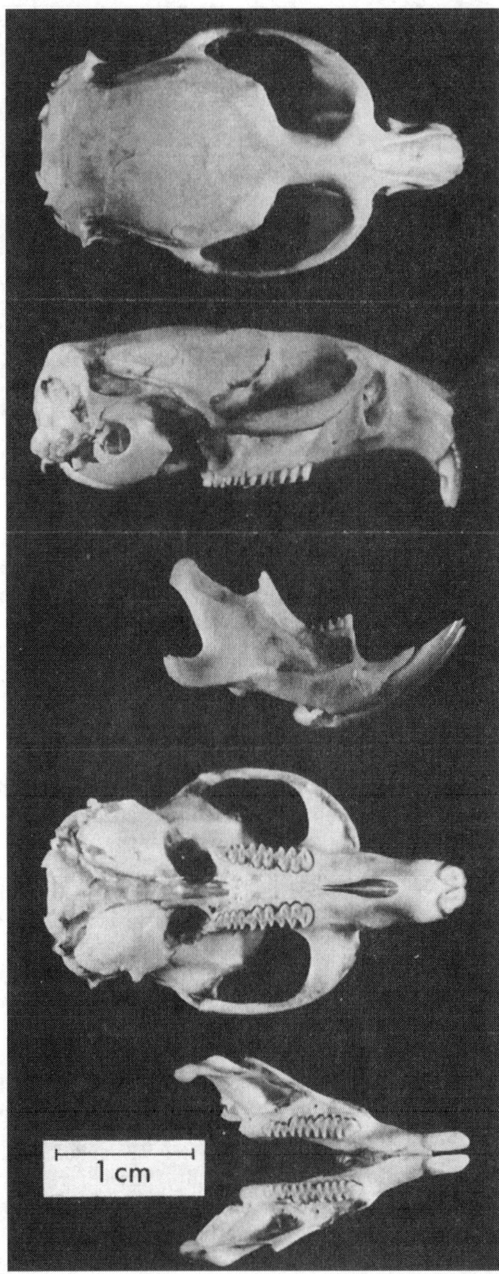

FIGURE 17.6. Skull of the meadow vole (*Microtus pennsylvanicus*). From top to bottom: dorsal view of cranium, lateral view of cranium, lateral view of mandible, ventral view of cranium, dorsal view of mandible.

specimen of *M. longicaudus mordax* from Whitman County, Washington. This number was confirmed by Hsu and Benirschke (1967–1971) in a sample from the same subspecies collected in Oregon. Judd and Cross (1980) found that specimens of *M. longicaudus* from Arizona, Colorado, and New Mexico were similar ($2n = 56$). However, they also observed six other chromosomal forms ($2n = 57, 58, 59, 62, 66,$ and 70) from southern Oregon and northern California. The differences among these seven karyotypes are thought to be due to pericentric inversion in 1 pair of acrocentric chromosomes and variations in the number of supernumerary (minute) chromosomes. Specimens in which $2n = 56$ had 10 pairs of medium to large biarmed chromosomes, 12 pairs of acrocentric chromosomes, and 5 pairs of small metacentric chromosomes. Specimens from northern California and southern Oregon had 6 pairs of small metacentric chromosomes and only 11 pairs of acrocentric chromosomes (Judd and Cross 1980). Whereas supernumerary

FIGURE 17.7. Right upper toothrow of the montane vole (*Microtus montanus*), showing characteristic prisms.

chromosomes have been found in other rodents, for example, western harvest mice (*Reithrodontomys megalotis*) and Bailey's pocket mouse (*Chaetodipus baileyi*) (Shellhammer 1969; Patton 1977), they are not known for any other species of *Microtus*. Differences in nondisjunction are thought to be responsible for the variation in the number of supernumerary chromosomes. The presence of supernumerary chromosomes is of evolutionary significance because they provide for an increase in the number of centromeres and could give rise to larger chromosomes by reciprocal or nonreciprocal translocation (Judd and Cross 1980).

Several other species have exhibited variation in chromosome number. Gill (1982) reported a polymorphism in chromosome number in populations of *M. californicus* with diploid numbers of 52, 53, or 54. Fundamental numbers were 60 or 62. Gill (1982) suggested that speciation may be occurring. Judd et al. (1980) reported chromosomal variation in *M. montanus*. Specimens from Arizona and New Mexico had $2n = 24$ and $FN = 44$. Specimens from Oregon had $2n = 22$ and $FN = 40$. The most obvious difference between these two karyotypes was the absence of chromosome pair 8, the smallest metacentric pair, in the 22-chromosomal form. Judd et al. (1980) suggested that the two forms of *M. montanus* might be specifically distinct.

Microtus oregoni is unique among voles in that the two sexes have different diploid numbers of chromosomes. Male *M. oregoni* have 18 chromosomes including the X and Y chromosomes. Females have 17 chromosomes with only one X chromosome. Male-producing sperm contain a single Y chromosome and female-producing sperm contain no sex chromosome (Pinkel et al. 1982).

Staining techniques have also been used to compare chromosomal structure in different species and populations of voles. The most common banding techniques used are C bands (constitutive heterochromatin) and G bands (Giemsa stain). In addition to the differences in chromosome number among populations of *M. montanus* described above, Judd et al. (1980) also documented chromosomal differences among these populations with respect to C-band and G-band patterns. The Oregon populations of *M. montanus* ($2n = 22$) have smaller C bands and less centromeric heterochromatin on chromosome pair 5. The G-banding patterns of the two populations are almost identical with the exception of chromosome pair 2, which has additional G bands at the terminal end of the short arm. Chromosome 2 is also slightly larger in the $2n = 22$ form. Judd et al. (1980) suggested that nonreciprocal translocation of chromosome 8 in the $2n = 22$ form could account for the additional length of chromosome 2.

Modi (1987a) used G-banding patterns to document chromosomal rearrangements such as centric fusion, tandem fusion, and pericentric inversion. These karyotypic features were used as characters in a cladistic analysis and phylogenetic trees were produced. The trees generated were similar but not identical to trees produced using more traditional techniques.

Allozymic Variation. The earliest attempts to analyze the genetic structure of vole populations at the molecular level involved the use of protein electrophoresis to study allozymic variation. Because most

TABLE 17.1. Diploid number of chromosomes (2*n*) and fundamental number (FN) of North American species of *Microtus*

Species	2*n*	FN	Source
M. abbreviatus	54	—	Rausch and Rausch 1968
M. breweri	46	50	Modi 1986, 1987a; Fivush et al. 1975
M. californicus	52, 53, 54	60, 61, 62	Modi 1987a, 1987b; Gill 1982
M. canicaudus	24	44	Hsu and Benirschke 1967–1971; Modi 1986
M. chrotorrhinus	60	64	Meylan 1967
M. longicaudus	56–70	84–108	Hsu and Benirschke 1967–1971; Judd and Cross 1980
M. miurus	54	—	Rausch and Rausch 1968
M. mogollonensis	44	54	Judd 1980
M. montanus	22, 24	40, 44	Hsu and Benirschke 1967–1971; Modi 1986; Judd et al. 1980
M. ochrogaster	54	64	Hsu and Benirschke 1967–1971; Modi 1986, 1987a, 1987b
M. oeconomus	30	54	Hsu and Benirschke 1967–1971; Modi 1986, 1987a, 1987b
M. oregoni	17 female, 18 male	32	Hsu and Benirschke 1967–1971; Modi 1986, 1987a, 1987b
M. pennsylvanicus	46	50	Hsu and Benirschke 1967–1971; Modi 1986
M. pinetorum	62	62, 64	Modi 1986, 1987a, 1987b; Beck and Mahan 1978
M. richardsoni	56	56–58	Modi 1986, 1987a, 1987b
M. townsendii	50	48	Hsu and Benirschke 1967–1971; Modi 1987a, 1987b
M. xanthognathus	54	62	Rausch and Rausch 1974

of these studies were long term and based on natural populations in which animals were not sacrificed, researchers relied on blood and saliva samples. Therefore, the number of proteins examined is rather limited. However, the level of genetic variability, measured as the degree of heterozygosity, was similar to that of other mammals, and ranged from near 0 to 0.488 (Gaines 1985). Several studies have been conducted to attempt to quantify fitness differences between genotypes and the intensity and direction of selection on genotypes. Authors of these studies suggest that at least for some loci, natural selection may have a role in maintaining polymorphisms (Gaines 1985).

Molecular Genetics. More recently, molecular genetic techniques have been used to analyze the genetic structure of vole populations. Southern blotting and various staining techniques such as C banding, G banding, and fluorescence hybridization techniques have been used to assess chromosomal variation within the genomes of vole species. Libbus and Johnson (1988) described populations of *M. oregoni* in Oregon in which the X chromosome was 39% longer than that of populations from Washington. They explained that this was due to the presence of major blocks of constitutive heterochromatin. Modi (1993) found extensive variation in the amount of constitutive heterochromatin among eight species examined. The sex chromosomes of *M. chrotorrhinus* are also unusually large and contain substantial quantities of constitutive heterochromatin (Ivanov and Modi 1996). Three highly repeated satellite DNA families, MSAT-21, MSAT-160, and MSAT-2570 (the number refers to the size of the repeat), have been localized to these heterochromatic regions of the sex chromosomes. Sequence analysis suggests that these satellite families may have a retroviral origin (Ivanov and Modi 1996).

Molecular techniques have been used recently to assess genetic variability among individuals and patterns of relatedness in vole populations. DNA fingerprinting using Southern blot techniques has been used to assess genetic variability and patterns of relatedness in the prairie vole, although the challenge of obtaining sufficient quantities of DNA from animals without sacrificing them may limit its usefulness (Hoagland et al. 1991). Moncrief et al. (1997) identified 18 primers for the polymerase chain reaction that successfully amplified *M. pennsylvanicus* DNA. Of the 18 microsatellite loci examined, 6 were polymorphic in the population studied, with the number of alleles per locus ranging from 2 to 10. Polymerase chain reaction shows great promise for evaluating genetic diversity and studying phenomena such as dispersal, relatedness patterns, immigration, and emigration (Moncrief et al. 1997). Gordon et al. (1998) used randomly amplified polymorphic DNA to determine genetic relatedness of *M. canicaudus*. They found that DNA profiles of wild-caught voles were less similar than the DNA profiles of inbred laboratory voles. They also determined paternity based on a comparison of bands between progeny and putative fathers (Gordon et al. 1998).

ANATOMY

Skeletal System. The skeleton shows no remarkable evolutionary changes. It consists of 7 cervical, 13 dorsal, and 6 lumber vertebrae; the number of caudal vertebrae varies with the length of the tail. The hyoid bones are fused into a single, curved, unattached bone. There are well-developed clavicles joining the sternum with the acromion processes, helping to stabilize the shoulder joints. There are four functional digits on the forelimb, five on the hind limb. The radius and ulna are not fused; a supracondylar foramen is not present. The distal fibula is solidly fused with the tibia.

Other Anatomy. There usually are eight mammary glands: two pairs of pectoral and two pairs of inguinal glands (Carleton 1985). This number may be reduced in some species. In younger animals, the pectoral mammae especially may not be fully developed, even during pregnancy (Howell 1924). Niethammer (1972) suggested a positive correlation between the number of mammae and litter size among different species. Posterolateral (flank) glands and anal glands are usually present and are particularly noticeable in adult males.

The gastrointestinal system is similar to that of most other rodents (Vorontsov 1962; Carleton 1985). There is a simple stomach, which is mostly lined by squamous epithelium, with a small pouch on the greater curvature, lined with glandular epithelium. The cecum is very large. This is apparently the site for symbiotic bacterial digestion, for no mammal has developed a cellulase (Grassé 1973; Hume 1978). The colon is small and forms the typical fecal pellets, with some storage enlargement in the rectum.

The kidneys of *Microtus* do not show elongated renal papillae. It is presumed that there is no particular adaptation for water conservation.

Testes are in a shallow scrotum in a fully mature animal. There is a characteristic baculum composed of a shaft and three separate digitate processes, which become fully ossified at maturity. The uterus is bicornate.

PHYSIOLOGY

Temperature Regulation. Voles are homeotherms. Despite their small size and the fact that they live in cool environments and feed primarily on vegetative plant parts, which have a high fiber content and low digestibility (due to plant secondary compounds), no vole has ever demonstrated any kind of seasonal hibernation, daily torpor, or prolonged fasting (Wunder 1985).

Body temperatures are controlled within a narrow range by metabolic activity, regulating external heat loss or gain by various external maneuvers, and anatomical characteristics. One strategy that voles use for winter acclimatization is nonshivering thermogenesis, which allows for an increase in metabolism and heat production. Alternatively, an increase in basal metabolic rate would result in an increase in energetic costs at warmer temperatures. The advantage of nonshivering thermogenesis is that it is used only when needed at cold temperatures (Wunder 1985). Bradley (1976) found that several species of voles were capable of maintaining a constant body temperature when the ambient temperature was as low as 2°C. Another important factor in the ability of voles to withstand extremely cold temperatures is their tendency to avoid the cold through the use of burrows, runways, nests, and snowcover (subnivian habitat). Adjustment of diurnal and nocturnal activity and winter communal nesting also contribute to winter survival. A short, globular body shape with short ears and limbs, a relatively short tail, and a good insulating pelage, especially in northern and high-altitude species, also contribute to efficient thermoregulation.

Water Regulation. Arvicolids are not known to have any particular physiological mechanism for water conservation, nor are they considered to be desert-adapted animals. However, certain species occur, occasionally in great numbers, in regions considered to be arid (Krebs 1966; Rose and Gaines 1978). These include the sagebrush (*Artemisia* spp.) ecosystem (*M. montanus*) and the southwestern grasslands ecosystem (*M. californicus*). However, optimal habitat contains stable amounts of moisture and moisture-laden plant food, influenced by permanent water sources including the water table or by elevation and microclimate effects of geomorphology. From these stable ecosystems, populations increase at irregular intervals and seasons as the plant communities expand from increased moisture.

Digestion. The most abundant source of plant energy is in the cell wall (Parra 1978). This is a polysaccharide comprising 3000 or more glucose units. Mammals break down the fiber directly by hydrolysis of the 1,4 glycosidic bonds of cellulose (Hume 1978) and with the help of symbionts, bacteria and protozoa within the intestinal tract of the fiber-eating animals (fibrivores). These symbionts live in the microtine gut, where they are provided a supply of finely ground foodstuffs.

In the vole, compared to other fibrivores, considerably lower amounts of fermentation products have been found within the gastrointestinal tract (McBee 1970); fermentation contents were 6.0% of body weight, calculated at 25 g/kg. This may be compared to cattle and camels, in which >17% fermentation contents occur, calculated at 728 g/kg (Parra 1978). Although the microbiology of the intestinal tract and use of the fermentation products of *Microtus* have been little studied, hind gut microbial populations apparently are under similar microconditions as those in the rumen of the larger herbivores, about which more is known (Parra 1978). Herbivores evidently contain comparable numbers of bacteria and a spectrum of morphologically similar bacteria (*Streptococcus, Bacteroides,* gram-negative spore-forming bacilli, and unidentified cellulytic bacteria) in their expanded gut compartments. In the cecum of the vole, culture counts of about 300×10^9 bacteria/g of dry matter were obtained (McBee 1970). However, symbiotic digestion may not be as efficient in the voles as in the larger herbivores (Keys and Van Soest 1970).

REPRODUCTION AND DEVELOPMENT

Voles in general have a potentially high rate of reproduction. This provides the potential for the rapidly developing population irruptions found in most species studied.

Anatomy. Voles can be identified as female externally by the close proximity of the urethral orifice to the anus. The vaginal introitus is in between. In young animals that have recently copulated, the vagina is visibly perforate. Slight traction anteriorly will usually reveal a vaginal mucous membrane. Immediately postpartum, the vagina will be dilated and sometimes blood stained. Lactating individuals can usually be identified by the presence of inflated lactation tissue surrounding each nipple.

As noted, the uterus is bicornate (Raynaud 1969). There are two functional horns, which open into a common lower segment, then through a single cervix into the vagina. Pregnancies occur simultaneously and equally in both horns. In *M. longicaudus, M. californicus,* and *M. montanus,* a tiny os clitoridis has been found, although its occurrence is sporadic (Ziegler 1961).

Externally, males can be identified by an anogenital distance of >8 mm, and in adults by the shallow scrotum and descended testes. The tissue micromorphology of the penis has some value as a taxonomic character. The baculum comprises a main shaft and three distal digital processes (Burt 1960; Wiseman 1967). There is extensive interspecific variation in the structure of lateral bacular processes. They are conspicuous in most species to diminutive in *M. richardsoni, M. ochrogaster, M. pinetorum, M. miurus,* and *M. abbreviatus,* and are absent in *M. californicus* and *M. mogollonensis* (Carleton 1985). Accessory glands consist of the paired seminal vesicles, urethral glands, paired bulbourethral glands, paired preputial glands, and the prostate. The prostate is a complex gland constituted of a pair of large anterior lobes ("coagulation glands") associated with the seminal vesicles, and paired dorsal, lateral, and ventral lobes (Raynaud 1969). Sperm morphology, as related to taxonomy, has been analyzed in *M. mexicanus* by Wilhelm (1977). Some variation among populations was found; differences might exist at the species level comparable with other groups of animals. This is a little-studied subject, but has promise for clarifying relationships.

Reproductive Cycle. There is a remarkable uniformity of the reproductive patterns of voles. Female voles reach reproductive maturity earlier than males. They may ovulate as early as 3 weeks of age and become pregnant. In most species examined, ovulation appears to be induced, either by copulation or hormonal stimulation (Seabloom 1985). Males, however, require >6–8 weeks before spermatogenesis produces mature sperm, and, equally important, before hormonal influences stimulate appropriate mating behavior. As noted by Hamilton (1941), this discrepancy helps to prevent inbreeding, for the juvenile females are pregnant before their sibling brothers are in breeding condition. These early reproductive activities occur well before adult size is attained. Early reproductive maturity allows an early beginning of population increase during an environmental cycle favorable to the vole. A mild mitigation of the effect of early breeding is the lower number of offspring in the first or second pregnancy (Negus et al. 1977).

The short gestation period of 20–23 days is another important factor in producing a high population level of *Microtus*. In a theoretical calculation, Negus et al. (1977) showed that a beginning population of 100 pairs of voles in April could potentially create a population of 8900 by September.

Reproductive Behavior and Copulation. Mating behavior occurs when a female is receptive to a male; at other times, agonistic behaviors may occur. During the receptive period, a female may actively repulse and severely wound certain males and accept others. Males vary in their aggressiveness for mating, and, in general, it is the female that determines mating success. After a few exploratory contacts between male and female, the female will pursue the male, smelling his genital area and then in turn be followed by the male.

Odor, primarily from gonadal hormones, is the predominant cue used to determine reproductive status and receptivity. In odor preference tests, males and females tend to prefer opposite-sex odors during the breeding season. During the nonbreeding season, females prefer the odors of females, whereas males show no preference (Ferkin and Seamon 1987; Ferkin et al. 1992a; Ferkin and Johnston 1993). Odor-based preferences can be affected by reproductive status. Males showed no attraction to the odors of either pregnant or lactating females. However, males were attracted to females around the time of parturition, when females had entered postpartum estrus. Females, on the other hand, maintained a preference for males during pregnancy, lactation, and postpartum estrus (Ferkin and Johnston 1995).

During copulation, the male mounts the female from behind, grasping her sides with his forepaws. The female assumes a lordotic posture in which she lowers her head and arches her back. The male performs a series of rapid pelvic thrusts, which are followed by intromission and ejaculation. The number and timing of mounting, thrusting, intromission, and ejaculation varies vary among species (Dewsbury and Hartung 1982; Wolff 1985). The Coolidge effect, rapid rejuvenation of satiated males when exposed to an unfamiliar female, has been documented in *M. pennsylvanicus* and *M. montanus,* but not *M. ochrogaster* (Wolff 1985). A copulation plug consisting of a congealed mucoid substance forms within the vagina and lasts about 2 days (Hamilton 1941).

Mating Systems. Mating systems in voles are characterized by interspecific variation. Mating systems have been observed that range from monogamy (*M. ochrogaster*) to monogamy or facultative polygyny (*M. californicus, M. montanus*) to polygyny (*M. xanthognathus*) to promiscuity *(M. pennsylvanicus, M. richardsoni*) (Wolff 1985). Intraspecific variation also appears to be common. For example, Fitzgerald and Madison (1981, 1983) observed that *M. pinetorum* lived in discrete family units with little overlap with other family groups. The principal mating system in the populations they studied was monogamy, but polygyny and promiscuity also occurred (Fitzgerald and Madison 1983). Although monogamy may be the principal mating system in *M. pinetorum,* polygyny or promiscuity may occur depending on the size and degree of relatedness among group members (Marfori et al. 1997; Solomon et al. 1998). Ostfeld and Klosterman (1990) suggested that mating systems of voles are not as rigid as they may seem and researchers should be cautious in placing a mating system label on a particular species.

Mating systems in voles appear to be related to social structure, space use, and ecological factors. Wolff (1985) placed vole social systems into four categories: (1) Males defend exclusive home ranges and females mate promiscuously or polygynously in home ranges that overlap with those of males and other females, (2) females defend exclusive territories and males mate promiscuously in home ranges that overlap with those of females and other males, (3) males and females are territorial, and (4) more than 1 male and female occupy exclusive communal groups. Ostfeld (1985) suggested that ecological factors, especially the distribution of resources, were the principal factors that determined social structure and therefore mating systems. He proposed that in voles, female reproductive success is dependent on acquiring the necessary energy resources to produce offspring. If food resources are clumped, they represent a defendable resource and females should defend territories. If food resources are evenly distributed, they are undefendable and females should not be territorial. Male reproductive success, on the other hand, depends on access to females. If females are clumped in space (i.e., nonterritorial), they represent a defendable resource and males should defend territories containing two or more females. If females are not clumped (i.e., territorial), males should not defend territories. This hypothesis is generally supported. Vole populations in which one sex is territorial and the other is not tend to have polygynous or promiscuous mating systems. Only in *M. ochrogaster* are both sexes territorial and the mating system in this species is monogamous (Carter and Getz 1993).

Microtus ochrogaster has been the subject of intensive research because, unlike most other species of voles (and mammals in general), it tends to form long-lasting pair bonds and males and females demonstrate parental care (Carter and Getz 1993). Pair bonding and parental care may be under hormonal influence. Vasopressin and oxytocin promoted pair formation and parental care in male and female *M. ochrogaster,* but not in polygynous *M. montanus* (Wang et al. 1994; Insel and Hulihan 1995; Cho et al. 1999; Young et al. 1999). Young et al. (1999) also documented changes in the structure of the vasopressin receptor gene (V_{1a}) that may contribute to differences in vasopressin expression. In addition, dopamine D2-like receptors are involved in social attachment in M. *ochrogaster* (Wang et al. 1999; Gingrich et al. 2000).

The most common type of social group in *M. ochrogaster* is the communal group (Getz et al. 1993; Getz and Carter 1996). Male–female pairs and single females also occur. Male–female pairs tend to remain together until one member dies. If the male dies, the female rarely acquires a new mate and lives as a single female. If the female dies, the male wanders (Getz and Carter 1996). Communal groups form primarily in the fall, not in response to the environmental stress of winter as in *M. pennsylvanicus* and *M. xanthognathus* (Wolff 1980; Wolff and Lidicker 1980; Madison and McShea 1987), but by the addition of philopatric offspring to male–female pairs. High juvenile mortality, primarily by snakes, limits the formation of communal groups in spring and summer (Getz et al. 1993; Getz and Carter 1996).

Pregnancy and Parturition. Implantation occurs on one or both horns of the bicornate uterus. Preimplantation mortality may occur, possibly from genetic or nutritional causes. Absorption of implanted embryos regularly occurs; embryo counts done early in pregnancy are higher than the number of young born. It is assumed that at least some of this mortality is due to defective embryos. The preimplantation losses in various species may vary from 5% to 10%; postimplantation losses are 1.5–7% (Rose and Gaines 1978).

Hamilton (1941) described parturition as beginning usually in the morning in captive *M. pennsylvanicus*. The vole assumes a sitting posture and licks her external genitalia. She moves about in a slow and "seemingly laborious" fashion. Contractions are evident and presentation is either breech or cephalic. The birth is frequently aided by the female with gentle traction by her teeth, and may take only 4–5 sec. Intervals between births are usually a few minutes, but may be as long as 7 hr. Little discomfort is evident.

Death of the mother or fetus occasionally occurs in later pregnancy or during parturition. Hybridization between sibling species of the red tree vole (*Arborimus*) causes this type of mortality, a phenomenon assumed to be part of the reproductive barrier between similar species.

The placenta is expelled after each young and is usually eaten by the mother. The umbilical cord breaks spontaneously; the young is cleaned by maternal licking immediately after birth. Nursing may begin before the entire litter is delivered.

Birth weight in *M. pennsylvanicus* ranged from 1.6 to 3.0 g/offspring, averaging 2.1 g (Hamilton 1941). Average birth weights in other species ranged from a low of 1.7 g in *M. oregoni* to a high of 5.4 g in *M. townsendii* (Nadeau 1985). Litter size data were summarized by Keller (1985) and showed significant interspecific and intraspecific variation. *M. pinetorum* had the smallest average litter size (1.9–2.5); *M. xanthognathus* had the highest (8.0–8.8). In most species, average litter size ranged from 3 to 6. Litter size is inversely correlated with birth weight (Nadeau 1985). Sex ratios in most cases approached unity. Deviations from unity seem to have little effect on demography (Keller 1985; Nadeau 1985).

Early Development. Young are born in a nest. They are pink and have closed eyes and ears. The only external hair is the mystacial bristles. In *M. pennsylvanicus,* growth rate is fairly constant for the first 21 days with a mean of 0.38 g/day (Innes and Millar 1979). In *M. californicus,* growth rate in early development is faster than later in development. Growth rates average 1.07 g/day from parturition to day 7, and 0.76 g/day from day 8 to day 21 (Hatfield 1935). Brown fur appears between 3 and 5 days of age, incisors emerge between days 1 and 7, neonates first crawl between days 3 and 9, eyes open between days 5 and 12, and weaning occurs between days 8 and 17 (Nadeau 1985).

Reproductive Potential. Bailey (1924) reported a captive female *M. pennsylvanicus* that produced 17 consecutive litters in 1 year (25 May–20 May), for a total of 83 young. One of her young from the 25 May litter produced 13 litters, totaling 78 young, before she was 1 year old. Thus, *Microtus* are very prolific mammals. There are species differences, and *M. pinetorum* "stands out for its conservation of reproductiveness" (Schadler and Butterstein 1979).

ECOLOGY

The ecology of the genus *Microtus* is intimately tied to grasslands. Humans are invading many natural ecosystems, and in creating their own agroecologic system, humans are in a competitive situation with *Microtus*.

Evolution of the Grasses and Herbivores. During the Tertiary, and especially following the Eocene, grasses and mammals underwent rapid evolution and diversification (Dawson 1967; Raven and Axelrod 1974). As the plants provided an abundant energy source, herbivores expanded in number and evolved into the new ecological niches. The Perissodactyla and Artiodactyla both evolved in the Eocene, about 40–55 million years ago. The grass-eating marsupials probably arose 25–38 million years ago (Hume 1978). The Holarctic microtine rodents probably have a history no older than 10 million years, with rapid evolution and diversification in the past 6 million years (Repenning 1980).

Grasses became the most important of all flowering plants in a modern, economic sense because of their nutritious grains and soil-forming function. Thus, they developed into the principal sustenance of humans, domestic grazing livestock, and wild herbivores, including *Microtus*. Two major adaptive characteristics (the masticatory system and the digestive system) have evolved in *Microtus* that place this genus in competition with humans. Both adaptations are related to the ability to eat grasses, including cereals, on which humans and voles subsist.

The first adaptive modification for a fibrivore was the ability to bite and masticate the tough celluloses. The toughness of grass is compounded into abrasiveness by the presence of silica cells or bodies in the leaf epidermis interspersed among other cells. Abrasiveness is particularly critical in small mammals, especially in *Microtus,* which have tiny teeth. The solution to this ecological problem was ever-growing teeth: As noted, incisors and molariform teeth are ever growing in *Microtus*. As they are abraded and worn down, they grow, providing continuous biting and grinding surfaces. The wearing pattern provides both sharpness for the cutting function of the incisors and a ridged rasping surface for the grinding molars. The specialization of ever-growing teeth evolved around the early Pleistocene (Phillips and Oxberry 1972).

The second major adaptive modification was the physiological need to digest complex carbohydrates (the celluloses and hemicelluloses) contained in the fibrous portion of grasses. Most omnivores and carnivores are unable to digest these celluloses (Grassé 1973). Herbivores have evolved strategies that allow them to use this part of the energy store.

The central theme of this critical cellulose digestion is gut fermentation by microorganisms. This may be further divided into foregut fermentation (with the anatomical evolution of large sacculated chambers, the rumen, the reticulum, and the omasum) and hind gut fermentation (also with the development of large chambers, the cecum and the proximal colon) (Hume 1978). The artiodactyls and marsupials, in general, possess the foregut adaptation. As a rule, the Perissodactyla, Lagomorpha, and Rodentia have the cecum-proximal colon systems. Experimental data on many species are lacking, but correlation with the better-studied domestic animals provides clear evidence for these relationships (Vorontsov 1962, 1967; Eisenberg 1978).

Ecosystems and Niches. There are numerous well-recognized ecosystems in North America, and *Microtus* occur in most of them. Several species of *Microtus* may exist in a single ecosystem and, when this occurs, partitioning of niches can usually be recognized. Frequently, partitioning separates sympatric species; there have been questions of particular adaptations of these populations to certain microenvironments or the possibility of exclusion by dominance of one species over another. Trapping experience has indicated considerable overlap, with more than one species being collected at a single trap station. The truth of the local situation can be found only by extensive study, with saturation trapping and sampling of various niches within the ecosystem. Even with this information, if the investigation is only during a single season, the conclusions may be confounded in subsequent years by population peaks, asynchronous cycles among species, or subtle or drastic climatic changes. Also, other vertebrate species may iufluence competitive outcomes.

Therefore, the following should be undertaken before making firm conclusions of the ecological relationships of *Microtus* in any particular environment: (1) Adequate samples should be obtained from a sufficiently wide area to examine all niches and (2) the data base should extend over each season during several consecutive years.

Niche preferences are known for various species of North American *Microtus* (Table 17.2). The niche preferences also will indicate those species that are potential or actual pests, and a "pest index" (from 1 to 3, where is most harmful) can be applied to each species.

Runways, Nests, and Burrows. Most species of *Microtus* construct networks of well-kept runways aboveground or tunnels or burrows belowground. *M. pennsylvanicus* and *M. ochrogaster* construct runways through litter and vegetation, whereas *M. pinetorum* live almost exclusively in underground tunnels. *M. xanthognathus* and *M. richardsoni* partition their time between aboveground runways and belowground tunnels (Wolff and Lidicker 1980; Ludwig 1981). Fresh cuttings in runways inditite the species and parts of the plants that are being eaten. Areas with a tamped-down appearance of the ground show that extra time is spent in certain spots. Concentration of scats usually indicates some degree of sanitation. In a high-density area, however, fecal pellets may be found scattered throughout. Some reingestion occurs, but this appears minimal under most circumstances. Exploratory activity away from the obvious trail often can be identified.

Nest building is well developed. The nests are used as nurseries, resting areas, and protection against extreme environmental conditions. They are constructed with dried grass and plant fibers and can be either aboveground or belowground. Nests of *M. ochrogaster* lie about 12 cm below the surface and are about 10–15 cm in diameter (Getz and Carter 1996). Nests of *M. richardsoni* are 10–24 cm in diameter (Ludwig 1981). Nests occur under rocks and logs, near boards or fence posts, in or under hay bales, or in brush piles. Subnivian nests are frequently under snow cover, usually placed against some natural bulwark.

Competition. Interspecific competition is probably very common in microtines either because various species of *Microtus* have near-identical niches or there is a broad overlap of two or more species of *Microtus* (see Table 17.2) In many cases where several species of *Microtus* occur in overlapping areas, a definite niche difference can be demonstrated. These are presumed to be situations of coevolution with dynamic competition continuously at work.

Several studies have shown the importance of intraspecific competition in voles, especially with regard to population regulation. This competition is demonstrated by density-dependent changes in population characteristics, which can be either direct or with some sort of time lag between cause and effect. Ostfeld et al. (1993) and Ostfeld and Canham (1995) demonstrated direct density dependence in *M. pennsylvanicus*. Meadow voles maintained in fenced populations showed a reduction of breeding effort, reduction in individual growth rates, increase in size at sexual maturity, and reduction in movement distances when at high densities. May (1976) developed a model in which density dependence with a lag of about 9 months could result in multiannual cycles in density (see section on cycles). Lagged density dependence could be mediated through predator–prey interactions, herbivore–resource interactions, disease, maternal effects, or behavioral polymorphism. Hörnfeldt (1994) demonstrated delayed density dependence with a lag of about 9 months in several species of European voles, but did not provide support for any biological mechanism. Hanski et al. (1993) concluded that multiannual fluctuations in Fennoscandian rodents are due to delayed density dependence imposed by mustelid predators. Ostfeld et al. (1993) and Pugh and Ostfeld (1998) found no evidence of herbivore–resource-mediated lag in population growth rate or space use in *M. pennsylvanicus*.

TABLE 17.2. Niche preference of *Microtus* as it relates to competitive and noncompetitive overlap and pest index

Reference Number	*Microtus* Species	Niche Preference	Competitive Overlap[a]	Noncompetitive Overlap[a]	Pest Index[b]	Reference and Remarks
1	*abbreviatus*	Arctic or alpine tundra, shrubby areas, burns	—	—	3	Rausch 1964
2	*breweri*	Beach grass and poison ivy	—	—	3	Tamarin and Kunz 1974
3	*californicus*	Annual grasslands of California	6	12, 16	1	Bailey 1936; Krebs 1966
4	*canicaudus*	Grass-dominated meadows, pastures	6	12	2	Bailey 1936
5	*chrotorrhinus*	Rocky outcrops in deciduous forests, disturbed areas	13	3, 14	3	Timm et al. 1977; Kirkland 1977
6	*longicaudus*	Rock, grass-dominated wide range	3, 4, 7, 8,13, 15, 16	12	3	Overlap agricultural areas but no reports of damage
7	*mogollonensis*	Pinyon–juniper zone, yellow pine-forested highlands, grasses	6, 8	—	2	Findley and Jones 1962
8	*montanus*	Grass-dominated meadows, agroecosystem, wide range	3?, 6, 12	12?	1	Severe irruptions reported (Piper 1909a; Oregon State College 1959)
9	*miurus*	Arctic or alpine tundra, shrub areas, burns	11	17	3	Rausch 1964
10	*ochrogaster*	Wet and dry grasslands, agroecosystem	13	14	1	Martin 1956
11	*oeconomus*	Grass-dominated tundra, Arctic and subarctic	6, 13, 17	—	1	May affect productivity of tundra ecosystem; carries *Echinococcus* (Rausch 1975)
12	*oregoni*	Grass-dominated meadow and forest, burns	—	3, 4, 5, 8, 15, 16	3	Gashwiler 1972
13	*pennsylvanicus*	Grass-dominated old field, marsh, bog mats	5, 8, 11?	14	2	Getz 1961
14	*pinetorum*	Dense grass, forbs, and brush; orchards, forests	5, 10, 13		1	Major damage to fruit orchards (Goertz 1971)
15	*richardsoni*	Grass-dominated waterside, subalpine	6, 8	12	2	Overlaps agroecosystem and affects subalpine ecosystem (Racey 1960); affects alpine and subalpine ecosystem (Johnson and Johnson 1982).
16	*townsendii*	Grass-dominated meadow marshes, agroecosystem	4, 6, 8	3, 12	2	Overlaps agroecosystem; carrier of giardiasis (Sheffield 1979)
17	*xanthognathus*	Grass-dominated taiga with microrelief, waterside	11	8	3	Douglas 1977

[a]By reference number in this table.
[b]Most harmful is 1.

Home Range and Territoriality. Home range has been defined as "that area traversed by the individual in its normal activities of food gathering, mating, and caring for young" (Burt 1943). It may or may not overlap with the home range of another individual. A territory is a defended area from which conspecifics are excluded. Individuals may defend territories against all conspecifics or only against members of the same sex. A territory may include the entire home range or only a smaller portion of it (Madison 1985).

There is extensive interspecific and intraspecific variation in the use of home ranges and territories among voles. Conclusions are further complicated by the many different techniques that have been used to study space use in voles. Although complications have made it difficult to make any general statements, we do have a basic understanding of how several vole species use space.

During the breeding season, *M. pennsylvanicus* and *M. richardsoni* females defend territories, whereas males do not. Home ranges of males are larger than the territories of females and the home ranges of males overlap with those of other males and one or more females (Madison 1980; Ludwig 1981). In *M. californicus, M. montanus,* and *M. townsendii,* males and females are territorial. There is extensive overlap between males and females, and territories of males tend to be larger than those of females (Lidicker 1980; Jannett 1982; Lambin and Krebs 1991). In *M. xanthognathus,* males defend large territories, whereas females do not. The home ranges of two to four females overlap with that of a single male (Wolff 1980). In *M. ochrogaster* and *M. pinetorum,* males and females defend shared territories and extended family units are produced because offspring tend to remain in the natal territory (Fitzgerald and Madison 1983; Getz et al. 1993).

Within a single population, there can be variation in space use patterns. Ostfeld et al. (1988) found that reproductively successful females had significantly smaller home ranges than unsuccessful females. This relationship did not occur among males. Ostfeld et al. (1988) speculated that successful females occupied higher quality territories and were able to acquire the resources necessary for reproduction in a smaller area. Taitt and Krebs (1981) and Taitt et al. (1981) described a similar pattern in *M. townsendii,* in which home range size increased with decreasing food resources.

Patterns of space use can also change seasonally. For instance, in *M. pennsylyanicus,* territoriality among females tends to break down in the winter and voles often live in large communal groups of males and females. *M. pennsylvanicus* also exhibit day-to-day changes in space use in response to weather conditions (Madison et al. 1984). Madison (1978) also showed that, depending on their reproductive cycle, territories of female meadow voles fluctuated in size. Territories became smaller during early lactation, but returned to a more typical size at weaning of the litter.

Dispersal. Dispersal is defined as movements of voles away from their home range. There can be a variety of motivations and purposes for these movements. Lidicker (1985) proposed a classification system in which he organized dispersal movements into five categories: saturation dispersal and four types of presaturation dispersal—seasonal, ontogenetic, colonizing, and interference. Saturation dispersal occurs when a population is at or near its carrying capacity. Competition for limited resources forces individuals to leave their home ranges. Presaturation dispersal occurs when the population is below its carrying capacity. Although these categories are not necessarily discrete or mutually exclusive, they do provide a framework in which to consider the evolutionary implications of dispersal on vole populations.

In spite of the variety of motivations for dispersal, some general comments can be made about the characteristics of dispersers. Males tend to predominate slightly, especially among subadults, but saturation dispersal does not seem to have a sex bias. Dispersers tend to be reproductively active if adults, or, if subadults, to reach reproductive status at a younger age than nondispersers. Dispersers tend to be less aggressive then nondispersers (Turner and Iverson 1973; Krebs et al. 1978) and they may be genetically distinct (Myers and Krebs 1971; Keith and Tamarin 1981; Baird and Birney 1982). Individuals that successfully disperse are as reproductively successful as residents (Pugh and Tamarin 1991).

Dispersal and spacing behavior are an essential feature of vole demography and contribute to the multiannual fluctuations characteristic of many species of *Microtus* (see next section). Two pieces of evidence point to this conclusion. The first is what happens to vole populations when dispersal is prevented. When this is done in fenced populations, densities tend to get abnormally high, the voles overexploit their resources, and the population eventually crashes. This sequence has been termed the "fence effect" (Krebs et al. 1969). Although the universality of this phenomenon has been questioned, especially in productive habitats (Ostfeld 1994), and its influence on demography may be limited to peak densities (Boonstra and Krebs 1977), the occurrence of the fence effect does indicate that dispersal plays an important role in regulating density. The second is the observation that populations restricted to small islands, where dispersal is impossible, tend to have stable population densities (Tamarin et al. 1987).

Cycles. One of the most interesting characteristics of many populations of *Microtus* is that they undergo multiannual fluctuations in density, or population cycles. In cyclic populations, peaks occur at intervals of about 4–5 years, with densities changing by several orders of magnitude from near zero at low density to as high as 427/ha (Myers and Krebs 1974; Taitt and Krebs 1985). This phenomenon has been studied extensively since it was first described by Charles Elton (1942), yet an understanding of the factors that regulate population density and drive population cycles remains elusive (Taitt and Krebs 1985; Lidicker 1988; Krebs 1996). Not all species of *Microtus* undergo cycles and not all populations within a cyclic species undergo cycles on a regular basis (Taitt and Krebs 1985). A given population can show multiannual fluctuations in some years and only annual fluctuations in others (Mihok 1984). A complete explanation for population cycles must also explain why some populations do not cycle.

Microtus townsendii (Beachem 1980), *M. pennsylvanicus* (Mihok 1984; Mihok et al. 1985), *M. ochrogaster* (Gaines and Rose 1976), and *M. californicus* (Krebs 1966) exhibit multiannual fluctuations. *M. oregoni* (Sullivan and Krebs 1981), *M. breweri* (Tamarin et al. 1987), *M. longicaudus* (Van Horne 1982), *M. mexicanus* (Conley 1976), and *M. xanthognathus* (Wolff and Lidicker 1980) exhibit only annual fluctuations. Circumpolar *M. oeconomus* exhibit multiannual fluctuations in Europe (Taitt and Krebs 1985), but evidence for cycles is equivocal in North America (Whitney 1976).

Many hypotheses have been proposed to explain this phenomenon. Krebs (1996) grouped these into five classes: (1) food supply, (2) predation, (3) interaction of food supply and predation, (4) qualitative changes in individuals, and (5) multiple factors. Krebs (1996:11) asserted that "changes in food supply have never been shown to be necessary or by itself sufficient to cause cycles." He discounted the food hypothesis for lack of evidence. Support for the predation hypothesis has been found primarily in Fennoscandian populations (Hanski et al. 1993). However, it has received criticism for relying on correlational evidence and not documenting cause and effect (Krebs 1996). The predation and food hypothesis suggests that predation may interact with changes in food supply to generate cycles.

The individual differences hypothesis suggests that changes in genetic structure, physiology, or spacing behavior within a population can drive cyclic changes in density. Several different mechanisms have been proposed. The Chitty or polymorphic behavior hypothesis (Chitty 1967; Krebs 1978) suggests that spacing behavior and aggression are heritable and that changes in gene frequency can result in changes in density. In an indirect test of this hypothesis, Boonstra and Boag (1987) did not find evidence of high heritability for life history traits in *M. pennsylvanicus*. The Charnov and Finerty (1980) sociobiological hypothesis suggests that patterns of individual relatedness change within a cycle. These patterns can affect aggression and can precipitate cyclic fluctuations in density. This hypothesis has been rejected by Ims (1989), Kawata (1990), and Pugh and Tamarin (1990). Ferkin (1990) and Ferkin et al. (1992b) argued that this hypothesis was problematic in *M. pennsylvanicus* because individual meadow voles determine relatedness based on familiarity and association. For this hypothesis to work, individuals must be able to determine the level of relatedness of unfamiliar as well as familiar individuals. The stress hypothesis (Christian 1950) describes physiological changes that may occur as density increases that can decrease survivorship or reproduction. However, little or no evidence has been documented to support this hypothesis. Lastly, the multiple-factor hypothesis (Lidicker 1988) proposes that as many as eight intrinsic and extrinsic factors may be necessary to explain vole cycles. For *M. californicus,* these include productivity of vegetation, dry season resources, vegetation mosaic, predation, dispersal, fecundity, physiological responses to stress, and social interactions.

Krebs (1996) concluded that vole cycles are probably the result of a combination of factors including intrinsic processes such as spacing behavior and dispersal and extrinsic processes such as food and predation. Specific and well-designed experiments involving relatedness, dispersal, predators, food, and the fragmentation of habitat are needed to answer the many questions that remain.

FEEDING HABITS

Several methods of studying feeding habits have been applied to voles. Direct examination of cuttings and food caches, stomach and intestinal contents, and fecal pellets has been used. In reviewing many of these reports, it is apparent that (1) agricultural products used for food by humans and domestic animals are also prime food for *Microtus,* and hence they are a pest; (2) there is a large spectrum of food items used by most species of *Microtus;* (3) food items may change in relative proportions and species from season to season; (4) food items may change from year to year; (5) food items may differ from one part of the range to another; and (6) there may.be subtle differences (niche differences ?) between sympatric species.

Voles are herbivores. Generally, <10% of the diet is from animal sources (Batzli 1985). The diet of *M. californicus* changes seasonally. Seeds and fruits are the principal summer diet, whereas the leaves, stems, and roots of dicots are the principal winter food (Batzli and Pitelka 1971). The diet of *M. pennsylvanicus* shows regional variation. In Indiana, meadow voles ate primarily monocots, whereas in prairie habitats in Illinois, dicots were more prevalent in the diet. Lichens and fungi also made up a significant portion of the diet (Zimmerman 1965; Lindroth and Batzli 1984a). *M. ochrogaster* consumed leaves, roots, and stems, with monocots preferred in winter and dicots in summer. They also consumed large amounts of seeds (as high as 35% of total volume; Cole and Batzli 1979). *M. xanthognathus* and *M. oeconomus* both used horsetails *(Equisetum* spp.) to a greater extent than other vole species (Tast 1974; Wolff and Lidicker 1980).

Why is there such an extreme degree of variability in diet among a group of similar herbivores? In his review of vole nutrition, Batzli (1985) suggested that three main factors were involved: (1) diet choice may simply reflect availability of food items in the local environment, (2) there may be different nutritional adaptations among different microtine species, and (3) microtines may show strong food preferences due to the quality of a particular food item. Clearly, availability has a significant impact in some species. *M. breweri* lives only on Muskeget Island off the coast of Nantucket. Its diet consists almost entirely of beach grass (*Ammophila breviligulata*) and bayberry (*Myrica pennsylvanica*) because that is the extent of the limited diversity of plants on the island (Rothstein and Tamarin 1977). However, when vegetational diversity exists, preferences are often established. For instance, cotton grass (*Eriophorum*) was five times more prevalent in the diet of voles in Alaska than it was in the environment (Batzli and Jung 1980). Palatability is determined by nutritional content (especially protein content), which has a positive effect, and fiber content and the presence of plant secondary compounds, which have a negative effect (Lindroth and Batzli 1984b; Bergeron and Jodoin 1987, 1989; Marquis and Batzli 1989).

Agricultural Relationships. Depending on season, food availability, and population density, microtines can be agricultural pests. A general vegetable and cultivated plant classification indicates which types of crops tend to be damaged by voles (Johnson and Johnson 1982):

1. Root vegetables (carrots, beets, turnips, sweet potatoes). These are frequently used by voles in small home and truck gardens, especially when easy access is provided by mole (*Scapanus* spp.) tunnels.
2. Stem vegetables (asparagus and kohlrabi). Not usually included in the vole diet.
3. Tuber vegetables (potato, yam). As item 1, these may become an item of vole food.
4. Leaf and leafstalk vegetables (cabbage, lettuce, celery, spinach). These are naturally exploited items, and are frequently fed to captive voles.
5. Immature inflorescent vegetables (artichoke, broccoli, cauliflower). Not usually included in the vole diet.
6. Fruits as vegetables: (a) immature (bean, okra, sweet corn), (b) mature (melon, pumpkin, tomato). Although these are not usually included in the vole diet, under the stress of high populations some items may be eaten.
7. Fruit trees (bark and root systems). May be used, even killed, by overwintering populations, especially *M. pinetorum, M ochrogaster,* and *M. montanus*.
8. Pasture, grassland, hay crops, and grain. These plant species may become the primary food item for *Microtus*, especially during population highs, and may be dramatically destroyed during population irruptions.

Underground Storage. In 1804, Lewis and Clark (cited by Bailey 1920) found and recorded the Native American custom of raiding the underground stores of native rodents for food. Bailey (1920) reported that Native Americans and settlers greatly prized these stores. He identified the foods as, among others, the beans produced on the underground shoots of a trifoliate bean vine (*Falcata comosa*) and the tubers of a wild sunflower (*Helianthus tuberosa*). The rodent was identified as *Microtus pennsylvanicus*. However, recent work suggests that seed-eating rodents, such as *Peromyscus* spp., are more active at food storage than grazing voles (Barry 1976). With the exception of *M. xanthognathus*, which has been reported to store rhizomes (Wolff and Lidicker 1980), and the suggestion by Wunder (1978) that voles may clip grass for winter storage, there is little documentation of food storage among voles (Wunder 1985).

Reingestion of Feces. Coprophagy occurs in many species of mammals. It allows vitamins produced in the gastrointestinal tract to be used, and increases the efficiency of nutrient use after partial digestion by symbiotic organisms in the cecum and lower bowel. Coprophagy has been documented in many species of rodents, including voles, and may be an adaptation for a high-fiber diet (Kenagy and Hoyt 1980). *M. pennsylvanicus* and *M. pinetorum* showed a reduction in body mass when prevention of coprophagy was combined with a low-quality diet (Cranford and Johnson 1989). *M. californicus* reingests about 25% of its feces in a series of rhythmic, short-term alternations of 1 to several hours duration between reingestion and nonreingestion. This pattern correlates with the day-and-night foraging pattern observed in *M. californicus* (Kenagy and Hoyt 1980). *M. pennsylvanicus* has also been shown to reingest feces in a similar rhythmic pattern (Ouellette and Heisinger 1980).

In general, (1) *Microtus* species feed primarily on green succulent vegetation; (2) they may be selective plant feeders (Zimmerman 1965) or general plant feeders (Whitaker and Martin 1977); (3) they frequently eat items other than green plants, including roots, bark, fungi, and, rarely, insects and meat; (4) there are species differences in feeding habits; and (5) coprophagy occurs as part of the feeding behavior of most voles.

BEHAVIOR

An understanding of the evolutionary significance of the life history patterns of voles requires an understanding of behavior patterns. Also, knowledge of the field behavior of voles can help in protecting agricultural lands.

Activity Patterns. *Microtus* are active at all times of the year and may be out day or night. This generalization may be modified by differences in species, season, habitat, cover, temperature, and other factors. There is typically intermittent activity. Voles must eat at frequent intervals, for the food volume needed is large in proportion to the energy budget realized. It is usual to see voles during the daytime. Trapping data, radiotelemetry, radioisotopes, photography, and activity wheels have shown *M. pennsylvanicus* to be nocturnal, diurnal, and crepuscular (Hamilton 1937; Heidt 1971; Dewsbury 1980; Madison 1985). *M. ochrogaster* have variation in peak periods of activity (Calhoun 1945; Carley et al. 1970; Baumgardner et al. 1980; Dewsbury 1980; Glass and Slade 1980). *M. canicaudus, M. murius, M. montanus,* and *M. pinetorum* are primarily nocturnal (Swade and Pittendrigh 1967; Gettle 1975; Dewsbury 1980; Rowsemitt 1981). *M. californicus* are crepuscular or nocturnal (Pearson 1960; Shields 1976).

Locomotion. These animals generally move at a run, which may be intermittent and appears to be related to a need for some protection, as they have spurts of speed over open areas and may pause in sheltered spots. There is none of the bounding, jumping progression that most other mammals exhibit. In captivity, there is little of the stereotyped captive pacing. As long as they have food, water, and space, they are restful, although some species and individuals (as with other animals) are hyperactive. Most voles do not usually climb, as they inhabit a two-dimensional grassland habitat (Wolff 1985). However, some species, such as *M. longicaudus*, climb trees and bushes with no difficulty.

Most voles swim well. Evans et al. (1978) documented that *M. pennsylvanicus, M. montanus,* and *M. californicus* were all capable of swimming, with *M. ochrogaster* described as the best swimmer. Species that live along stream banks, such as *M. richardsoni* and *M. xanthognathus*, have been described as good swimmers (Ludwig 1981; Wolff 1985). Digging and burrowing behavior is well developed in voles, as demonstrated by their habit of creating extensive networks of runways and tunnels.

Senses. Of all the senses, olfaction is probably the best-developed in voles, which seem to live in a world of odors. They use odors to mediate individual identity, including age, sex, reproductive condition, diet, social status, and degree of relatedness (e.g., parent–offspring discrimination and sibling recognition) (Ferkin and Seamon 1987; Drickamer 1989; Ferkin et al. 1992a; Fortier et al. 1996; Phillips and Tang-Martinez 1998). Scents are produced by glands on the flanks or hips and are distributed by scent-marking behaviors such as scratching or rubbing (Quay 1968). Scents can also be distributed through urination and scat pile construction. *M. xanthognathus* produces scat piles at the junctions of runways, which seem to have communication significance (Wolff 1980). Studies indicate that *M. pennsylvanicus* is even capable of discriminating between the top and bottom scent when two individuals deposit scent in the same location (Ferkin et al. 1999).

Hearing is also well developed in voles. A variety of vocalizations has been described. These are usually high-pitched, squeaky sounds associated with agonistic behavior (Househecht 1968; Colvin 1973; Jannett 1981). Wolff (1980) documented high-pitched alarm calls in *M. xanthognathus*. Youngman (1975) described a similar call in *M. miurus*. Several species of voles use ultrasonic vocalizations, primarily among neonates to elicit parental care (Colvin 1973).

Vision is less well developed and probably only has significance over short distances and in response to shadows and movements. Touch is well developed, even in neonates; contact with a warm object stimulates nuzzling in newborns (Hatfield 1935).

MORTALITY

Avian Predators. The genus *Microtus* occurs in a variety of habitats, most of which are relatively open and grassy. These areas are ideal for hunting by avian predators. Many feeding habit studies of predatory birds have been done, and *Microtus* are usually the most common food item reported. In southern Idaho, 66.3% of the total prey and 80.0% of the prey biomass of the barn owl (*Tyto alba*) was *M. montanus* (Roth and Powers 1979); for the great horned owl (*Bubo virginianus*), the figures were 52.9% and 48.2%, respectively. In the same region, Sonnenberg and Powers (1976) found that *M. montanus* was also the preferred prey species (53.7%) of the long-eared owl (*Asio otus*). In central Oregon, on a yearly basis, *M. montanus* made up 13.4% of the prey species of the great horned owl, 3.6% of that of the long-eared owl, and 9.8% of that of the short-eared owl (Maser et al. 1970). This was in an area of low populations of *Microtus*. Fitzner and Fitzner (1975) reported that *M. montanus* made up 65.5% and *M. longicaudus* 8.0% of winter food items of the short-eared owl in the Palouse prairies of eastern Washington. Seidensticker (1970) reported 8% occurrence of *Microtus* sp. among 28 kinds of food items during the nesting season of red-tailed hawks (*Buteo jamaicensis*) in Montana. Rageot (1957) found that the barred owl (*Strix varia*) took *M. pennsylvanicus* and *M. pinetorum* in Virginia.

Bent (1937, 1938) cited numerous examples of hawks and owls feeding on *Microtus*. Almost all species of raptor take *Microtus*. Other birds not usually considered predators of mice may actively catch and eat *Microtus*, especially during times of "mouse plagues." Sea gulls may be attracted in great numbers. Piper (1909a) showed a photograph of gulls hunting voles in an alfalfa field during a mouse plague in Nevada. Craighead (1959) identified excessively high counts of California and ring-billed gulls (*Larus californicus* and *L. delawarensis*) during an Oregon vole irruption. Pellet analysis indicated that the diet of the gulls was 99% *M. montanus*.

Other species of birds prey on *Microtus*, especially during periods of irruptions. This list includes the northern shrike (*Lanius excubitor*) and magpie (*Pica pica;* Craighead 1959), crow (*Corvus brachyrhynchos;* Bent 1946), white-tailed kite (*Elanus leucurus;* Waian and Stendell 1970), great blue heron (*Ardea herodias*), and bittern (*Botaurus lentiginosus;* Piper 1909b; Jewett et al. 1953).

Mammalian Predators. Natural mammalian enemies of *Microtus* include the full spectrum of predators (Maser and Storm 1970). Those that eat voles as a major food item are northern short-tailed shrews (*Blarina brevicauda;* Eadie 1952), badgers (*Taxidea taxus*), coyotes (*Canis latrans;* Murie 1940), red foxes (*Vulpes vulpes;* Murie 1936), skunks (*Mephitis* spp.), weasels (*Mustela* spp.; Fitzgerald 1977), wolves (*Canis lupus;* Murie 1944), lynx (*Lynx canadensis;* Saunders 1963), bobcats (*Lynx rufus;* Piper 1909b), and domestic cats (Christian 1975), as well as other carnivores. As a general rule, predators are opportunists; during periods of low populations, they will shift their diet to other items. During high populations, they eat predominately the readily available *Microtus*.

Other Predators. Other vertebrate predators often take *Microtus*, either in accidental encounters or, as in the case of snakes, by deliberate hunting strategies. The following predators have been observed to take or contain ingested voles: trout (*Salmo* sp.), garter snake (*Thamnophis* sp.), western yellow-bellied racer (*Coluber constrictor*), gopher snake (*Pituophis melanoleucus*), rattlesnake (*Crotalus viridis*), and rubber boa (*Charina bottae*) (Johnson and Johnson 1982). Other predators include the Pacific giant salamander (*Dicamptodon ensatus;* Bury 1971), bullfrogs (Korschgen and Baskett 1963), copperheads (*Agkistrodon contortrix;* Fitch 1960), and black pilot snakes (*Elaphe obsoleta;* Madison 1978).

Ectoparasites. The literature on parasites of New World *Microtus* was reviewed by Timm (1985). Fleas (*Siphonaptera*) are well-known ectoparasites and occur on a variety of animals; numerous species have been reported from *Microtus*. Fleas tend to be host specific, at least to the genus level of the host. However, because they are so mobile, they frequently occur on other genera. If an animal dies, the fleas soon leave and may temporarily engage the next mammal that comes along. Because *Microtus* runways are used by many other small mammals, there also is a good chance of temporary exchanges of fleas between species using these runways. For this reason, predators often have "rodent fleas" on them.

According to Hubbard (1947), many more fleas may be found in nests than on animals. However, compared to those of other small mammals, *Microtus* nests seldom have many fleas in them; the voles, too, often are entirely without fleas.

Other ectoparasites (mites, ticks, lice, and others) are less well known on voles. Reports are fragmentary and scattered and there are many lapses of information related to their biology. As part of the ecological community in which *Microtus* exists, there is much to be learned about them.

Endoparasites. Helminths characteristically found in *Microtus* belong to genera that are represented in rodents of various families. In members of the subfamily Arvicolinae, helminths frequently exhibit little host specificity at the generic level, and some species may occur in two or more rodent genera. Qualitative differences in the helminth faunas of *Microtus* usually are attributable to ecological factors that influence the distribution of various parasite–host assemblages. Nonetheless, voles of some species differ in degree of susceptibility to infection by certain helminths.

As indicated by both diversity of species and numbers of individuals, cestodes and nematodes predominate among helminths occurring in *Microtus*. Trematodes are less common. Acanthocephalans are generally rare and probably accidental in occurrence, although Benton (1955) reported *Moniliformes clarki* as common in *M. pinetorum* in New York. Most of the helminths of voles are the adult (sexually reproducing) stage, which inhabit the lumen of the alimentary canal; larvae usually localize in the liver or in other organs.

Helminths for which voles serve as the final host may have direct cycles (some nematodes), or intermediate hosts may be involved (indirect cycles of cestodes, trematodes, and some nematodes). Voles serve as the intermediate host of some helminths, the most important of which are cestodes, which occur in the strobilar stage either in carnivores (canids or mustelids) or hawks or owls. Completion of the cycles of these cestodes is favored by the predator–prey relationship existing between the respective final and intermediate hosts.

The most common cestodes in *Microtus* represent the subfamily Anoplocephalinae, family Anoplocephalidae. As far as is known, oribatid mites serve exclusively as intermediate hosts of these cestodes (Denegri et al. 1998). Voles become infected by the incidental ingestion of the mites while feeding. The common anoplocephaline cestodes of voles represent the genera *Anoplocephaloides, Paranoplocephala,* and *Andrya* (Timm 1985; Frey and Patrick 1995). Although most species are not host specific, *P. omphalodes* may be specific to *M. miurus* (Haukisalmi et al. 1995). Also relatively common are cestodes (genus *Hymenolepis*) of the family Hymenolepididae, of which insects serve as intermediate hosts (Timm 1985).

Voles and other small rodents serve as intermediate hosts for several cestodes of the family Taeniidae, genera *Taenia* and *Echinococcus*. All of the species considered here are Holarctic, and some occur most commonly in the Arctic and subarctic. *T. crassiceps* and *T. polyacantha* are parasites of foxes (*Vulpes* and *Alopex*). Their larval stages, which reproduce asexually in the intermediate host, occur in liver tissue and peritoneal cavities of voles (Rausch and Fay 1989). *T. mustelae* and *T. martis* are host-specific parasites in the strobilar stage of mustelids. The larvae are in the liver of the intermediate host, but other tissues or organs may be involved. In inhabited areas, the larval *T. taeniaeformis* often occurs in *Microtus,* in which it forms a large cyst at the hepatic surface. Larvae of other taeniid species may sometimes occur in voles, but are not common. Additional information concerning the taxonomy of *Taenia* sp. is available in Verster (1969).

Microtus, as well as other rodents, serve as the intermediate host of *Echinococcus multilocularis,* of which the strobilar stage typically occurs in canids, including foxes, coyotes, and domestic dogs (Marquardt et al. 2000). The larval cestode develops in the liver of the intermediate host, often causing extreme enlargement of the organ. With massive infections, movement of the rodent is hampered, probably making it more vulnerable to predation. Humans can be infected by inhaling or ingesting eggs shed in the feces of the definitive host. Fur trappers can become infected when pelting animals and pet owners can become infected from contact with dog feces. The larval stage develops in the human liver, causing alveolar hydatid disease, which can be fatal in 75% of cases if untreated (Marquardt et al. 2000). *E. multilocularis* was first observed in the contiguous United States in 1964, when it was found in foxes in North Dakota. Since that time, the cestode has been found in rodents or foxes in northern plains states and provinces of the United States and Canada (Rausch 1975; Marquardt et al. 2000). The first human case was reported in a resident of Minnesota (Gamble et al. 1979). Because of the general presence of suitable hosts, continuing spread of this cestode in the United States is possible. However, Hildreth et al. (2000) conducted a serological survey of 115 trappers from South Dakota who were at high risk for exposure. In spite of a high incidence in foxes and coyotes in the area, none of the trappers showed antibody evidence for the presence of *E. multilocularis*.

The families Paruterinidae and Dilepididae include cestodes whose larval stages occur in meadow voles (Timm 1985). Certain species of *Paruterina* are host-specific parasites of owls, whereas cestodes of the dilepidid genus *Cladotaenia* occur only in birds of the order Falconiformes. Both have cysticercoid-like larvae, which are found in the liver or other organs of voles and other small rodents that are preyed on by hawks and owls.

The helminth fauna of *Microtus* includes a large component of nematodes, of which relatively few occur commonly in these rodents in North America. Nematodes of the genus *Heligmosomoides,* family Heligmosomidae, inhabit the small intestine of voles throughout the Holarctic (Durette-Desset 1971). The life cycle is direct, involving the ingestion of the embryonated egg, with subsequent development taking place in the small intestine. *Heligmosomoides polygyrus* (formerly known as *Nematospiroides dubius*), has been intensively studied under laboratory conditions. Sato and Kamiya (1992) reported that a syngamid nematode (*Syngamus* sp.) was found in the respiratory tract of 29% of tundra voles (*M. oeconomus*) from St. Lawrence Island, Alaska.

Nematodes of the oxyurid genus *Syphacia* (pinworms) inhabit the cecum of many species of voles (Timm 1985). Quentin (1971) concluded that all species of *Syphacia* parasitizing North American *Microtus* are *S. nigeriana* and that *S. obvelata* is restricted to *Rattus* and other murids. However, Frey and Patrick (1995) reported that *S. obvelata* was the most common helminth in the Hualapai vole, *M. mogollonensis hualpaiensis,* in northwestern Arizona.

More common nematodes that require intermediate hosts represent the genera *Pterygodermatities* (Rictulariidae) and *Mastophorus* (Spiruridae). *P. microti* occurs in the small intestine of *Microtus* in northwestern North America. *M. muris* is a Holarctic species and occurs widely in rodents. It is a large nematode and typically inhabits the stomach of the host. Insects serve as intermediate hosts for both of these nematodes (Timm 1985). *Capillaria* sp. occurs in many species of rodents, showing high organ specificity but low host specificity (Dunaway et al. 1968). However, Childs et al. (1989) documented a high incidence of *C. hepatica* in Norway rats (*Rattus norvegicus*) near Baltimore, where meadow voles were free from infection.

As noted earlier, *Microtus* harbor comparatively few trematodes. A netocotylid, *Quinqueserialis quinqueserialis,* is common in voles in wet habitat, where the metacercariae encyst on emergent vegetation. The trematodes inhabit the cecum of the final host and may be very numerous. *Mediogonimus ovilacus* (Plagiorchiidae) is a parasite of *M. pennsylvanicus,* where it occurs in the bile duct (Timm 1985). Voles occupying wet habitat may also become infected by a schistosome, *Schistosomatium douthitti,* the adults of which occur in the veins of the portal circulation. Snails are the intermediate host. This trematode is of unusual interest because its cercarial stage is a cause of dermatitis (swimmer's itch) in humans (Zajic and Williams 1980, 1981).

Diseases. One important deleterious effect of *Microtus* on human beings is disease transmission. Some of these are fatal to humans, *Microtus,* and other native fauna. The etiology of some diseases are little known and their potential effects on rodents and people are unknown. Others are better known, including tularemia and plague.

Tularemia, caused by the bacterium *Francisella tularensis,* occurs throughout the Northern Hemisphere. It has been reported from >100 mammalian species including several species of lagomorphs and rodents, including *Microtus,* especially during periods of high vole population density (Gage et al. 1995). This is a serious illness in human beings (7% fatality rate in untreated cases). Infection can occur by direct contact with an infected animal (e.g., skinning an infected animal), bites from infected arthropods such as ticks, or contact with water contaminated with urine from infected animals (Gage et al. 1995). Kartman et al. (1959) reported an infection rate of 43 of 126 *M. montanus* collected in Oregon. The role this disease has as an effective agent of vole mortality during irruptions is not clear, but dead voles were positive for *F. tularensis*. Live voles given known dosages of *F. tularensis* showed a high susceptibility to the disease. However, the oral route of infection, assumed to be the method of natural infection, required 1000–100,000 times as many organisms as a subcutaneous injection. The suspected route of infection during the winter (when there are few ticks) was cannibalism.

Rausch et al. (1968) isolated *F. tularensis* from *M. oeconomus*. The isolate resembled the less virulent Eurasian strains. They suggested that this strain of organism was responsible for a relatively high rate of subclinical tularemia in humans in northern and western Alaska, as indicated by serological tests.

Isolates of plague bacillus, *Yersinia pestis,* from *M. californicus* have been found repeatedly in California (Hubbard 1947). *M. californicus* has been implicated in enzootic (maintenance) transmission during interepizootic periods (Gage et al. 1995). Kartman et al. (1959) reported isolation of the plague bacillus from *M. montanus* in eight instances during the Oregon vole irruption of 1958. No epizootic or human infection resulted. The greatest risk of human infection occurs when humans live or travel in areas with plague epizootics. Although there has been no documented case of plague among North American mammalogists, individuals who handle infected animals or animals infested with infected ticks should consider themselves at risk (Gage et al. 1995).

Other disease-causing organisms occur in *Microtus*. In general, these reports are not the result of long continued research, and in-depth quantitative data are not available. Most, but not all, organisms are assumed to be symbiotic in nature and of little significance to the hosts or to the public health of humans and animals.

Several parasites occur in the blood of *Microtus*. In Alaska, Fay and Rausch (1969) reported a *Grahamella*-like organism (Bartonellaceae); a trypanosome, *T. microti* (Trypanosomatidae); and two morphologically similar but biologically different strains of piroplasm (Theileriidae) in *M. oeconomus*. A piroplasm was also found in

M. pennsylvanicus. These parasites were found during a period of high population density.

Babesia (*Babesia microti*), a widespread piroplasm, has been diagnosed in humans in New England (Healy 1979). The dominant vector is the deer tick, *Ixodes scapularis* (*dammini*), which is an indiscriminant feeder; larvae and nymph stages may bite and inject mice, deer, or humans. This disease has a potential of being very widespread, related to its occurrence in smaller mammals, including *Microtus*. In Colorado, 13 of 15 *M. ochrogaster* tested positive for *B. microti* (Burkot et al. 2000). In Grand Teton National Park, Wyoming, 103 of 257 *M. montanus* and 5 of 12 *M. pennsylvanicus* were infected (Watkins et al. 1991). Serological examination in New England revealed that 11 of 577 (1.9%) humans tested had a positive titer, but 10 of 133 (7.5%) humans with a tick bite or a fever of at least 3 days were positive (Healy 1979).

Lyme disease is caused by a treponema-like spirochete, *Borrelia burgdorferi,* and is transmitted by the ticks *Ixodes scapularis* and *I. pacificus*. The white-footed mouse, *Peromyscus leucopus,* is the most common host in eastern and central North America. Although the bank vole, *Clethrionomys glareolus,* is a reservoir in Europe, North American voles have not been identified as hosts (Gage et al. 1995).

Some flagellated protozoans are intestinal parasites. One, *Giardia* sp., indistinguishable from human *G. lamblia,* has been recovered from intestines and fecal pellets of *Microtus townsendii* (Sheffield 1979). *G. lamblia* is the infecting organism of giardiasis, a serious worldwide diarrheal disease, often called "beaver fever" or "backpacker's diarrhea." Analysis of the ecological niches and behavior of several species of *Microtus* suggests positive correlation to giardiasis. *Giardia* sp. have been found in *Microtus californicus* and *M. pennsylvanicus* in the United States, and *Microtus* (*Pitymys*) *savii* in Europe, not associated with disease outbreak (Sheffield 1979).

No viral diseases have been associated with North American *Microtus* with certainty. There has, however, been little investigation in this field. In Europe, there have been reports of rabies virus from *Microtus arvalis, M. agrestis, Clethrionomys glareolus,* and *Apodemus flavicollis,* all common species (Sodja et al. 1971, 1973; Arata 1975). It has been hypothesized that microtines may serve as an asymptomatic reservoir for rabies, which could then be transmitted to predators by ingestion (Mallory and Dieterich 1985). In 1993, an outbreak of hantavirus pulmonary syndrome was attributed to strains of hantavirus isolated from deer mice (*Peromyscus maniculatus*) in the southwestern United States. Although strains of hantavirus have been identified from *M. californicus* and *M. pennsylvanicus,* these have not been identified as pathogenic in humans (Childs et al. 1995). Other viral pathogens, such as eastern equine encephalitis, western equine encephalitis, and Colorado tick fever virus, could potentially use microtines as a reservoir because the viruses have been isolated from ticks and mosquitoes known to feed on voles (Mallory and Dieterich 1985). However, no definitive link between these diseases and voles has been documented.

AGE ESTIMATION

Research involving anatomy, physiology, population structure, and reproduction generally requires that the age of individuals be estimated. Many evaluations require only that a general age status of "immature" or "mature" be obtained, but because of the great influence of age on many functions, more accurate determinations have been sought (Hoffmeister and Getz 1968).

All vole species achieve reproductive maturity at the early age of a few weeks. However, many have a growth period that proceeds over a period of many months. In general, maximum size is reached between 2 and 10 months of age; few voles survive more than 1 year.

With experience, or a large number of specimens, it is possible to segregate (1) the *juvenile,* by pelage (duller and darker), size, and virgin reproductive condition; (2) the *subadult,* which is larger and shows evidence of breeding (perforated vagina, corpora lutea of pregnancy, enlarged testes); (3) the *mature adult,* which has achieved a general maximum body size by weight and measurement; and (4) the *old adult,* which may show scruffy tail and pelage, deteriorating conditions of bones and teeth, and frequently a weight reduction.

Size. Size can be used only after sufficient validation. In large series of animals, weight alone or length of head and body can be suitably accurate. Livetrapped animals may have abnormally low weights unless traps are examined frequently. Captive animals frequently are overweight. Under some natural conditions, seasonal or individual obesity may invalidate weight criteria. Body weight alone has been used routinely in field studies (Krebs et al. 1969).

Reproductive System. Examination of the reproductive system gives information primarily regarding sexual maturity, not age, but is useful. In the female, gross examination may give evidence of perforation of the vagina, recent parturition, or pregnancy; autopsy may reveal corpora lutea or uterine scars. The simultaneous presence of uterine scars and an active pregnancy indicates age beyond the first pregnancy (generally 3 weeks to first conception and 3 weeks of gestation period, or a minimum age of 6 weeks). Evidence of active lactation or enlargement of the teats postweaning can be found. In the male, the best external evidence is descent and enlargement of the testes. In older males, darkening and some hair loss of the scrotal skin occurs. Internally, enlargement of the testes and the turgid enlargement of the seminal vesicles indicate full maturity; measurement of the testis may be used to indicate this relative physiological age. Microscopic examination of the testis and epididymis and of the ovary will show mature sperm or ova, respectively, accurately timing sexual maturity (Didow and Hayward 1969). The baculum of the male can be cleared and stained, and can accurately determine age of juveniles and younger males by the progressive ossification of the digital processes as well as by size (Sozen et al. 1999).

Skeleton. In addition to skeletal growth, there is one marked qualitative change at sexual maturity of females. This is the opening of the symphysis pubis, which occurs during the development of the first pregnancy. This is of limited use, but can be helpful in certain studies. Closure of epiphyses is another method useful within certain age groups and requiring known-age controls. The epiphysis of various bones may close at different ages.

Myers (1978) used a regression model with known age as the dependent variable and several skeletal measurements obtained from; X rays of *M. californicus* as independent variables. All regressions were significant ($p < .0001$; $R^2 = .855$) and Myers (1978) successfully predicted the age of another known-age sample.

Skull. The general size and conformation of the skull is helpful in age determination (see Fig. 17.6). The rounded, softer skull of the juvenile becomes angular and hard. There is gradual closing of the suture lines of the skull bones; several remain open during the lifetime of the vole, but there is a reasonable progressive closure, different for each suture. Due to the wide range of times of closure, status of the sutures is not useful in age determination.

Irruption of the molariform teeth occurs at an early age, before weaning; it is useful only at this restricted time (see Fig. 17.7). Because all the teeth are ever growing, wear patterns are not useful in age determination; the crown pattern is usually as good in an old vole as in a juvenile. Howell (1924) provided an excellent, detailed anatomical study of age variation in *M. montanus*.

Eye Lens and Lens Proteins. Eye lens parameters are widely used for age determination in small mammals, with apparent good results (Birney et al. 1975). Total lens weight and determination of the-soluble lens protein appear to be best for the first 4 months, after which examination of the insoluble proteins is better.

CAPTIVE MAINTENANCE

Most species of *Microtus* adapt very well to captivity (Mallory and Dieterich 1985). They appear to do equally well in small standard laboratory cages or other modified caging. Maintenance of health, increased longevity, breeding in captivity, and rearing of young are the prime criteria for determining the adaptability of a species. By these criteria, *Microtus* does well. They are hardy, their food is readily available, they are prolific breeders, and they are easily transported. For these reasons, *Microtus* has been well studied for their own merits or as a laboratory animal to investigate other biological problems. Over much of North America (see Distribution), one or several species of *Microtus* are readily available.

There are a few disadvantages. Compared to laboratory mice or rats, voles do not handle easily; they can be aggressive and bite readily.

Food. Most species do best if they are given root vegetables, especially carrots, greens (such as lettuce and sprouts), seeds, and fruit daily. Lab Chow (Purina) and other laboratory food commercially available can be a part of the standard fare. Rabbit pellets, alfalfa, and other hay are good food, and hay makes an excellent nesting material. Food is best supplied in some form of manger up off the floor level, which conserves food and prevents waste and soiling. Water should be present at all times.

Social Structure. Colonies may be maintained with several breeding animals of either sex present or a harem arrangement if the enclosure is large enough. This arrangement may have some advantages under certain conditions. However, controlled breeding not only identifies the genetic background, but in general allows for better survival. It also demands better record keeping. Breeders should generally be housed as single pairs. Males can be removed 7 days postcoitum or after the birth of the litter to take advantage of postpartum estrus. On weaning at 3 weeks postpartum, immature males should be housed individually and immature females should be housed with unfamiliar males to maximize colony productivity (Mallory and Dieterich 1985).

Miscellaneous. Ambient room temperature is usually satisfactory. Colonies may be kept out of doors in all but the coldest months. Nest boxes may be helpful in handling and providing a secure microenvironment. Long-day photoperiod (14–18 hr of light/day) can enhance reproduction in breeding colonies.

Litter must be changed at 2- to 6-week intervals to avoid parasites and disease problems. Especially in wild-caught animals, buildups of potentially noxious organisms can be prevented by good sanitation.

Anesthesia is frequently needed for certain manipulations, such as blood drawing or tissue biopsy. In general, ether works well and is safe for the animal, but it is explosive. Nembutal and other agents can be used for longer sessions. Blood samples may be obtained by heart puncture, lower cervical vessel puncture, or ocular plexus puncture.

ECONOMIC STATUS

Within the temperate and boreal zones of Europe, Asia, and North America, humans have evolved with an extensive supportive agricultural base in areas where voles, *Microtus,* have prospered. In North America, as in the Old World, the ecological communities existed with *Microtus* as an integral part; the vole has continued to compete with humans for food (Lantz 1907; Bailey 1924; Elton 1942; Myllymäki 1979). Continuous habitat was fragmented by human activity into islands of lesser size, but this was not all bad for the voles, for irrigation and forest clearing opened up large new areas. Food for domestic grazing animals provided a source for *Microtus,* too. New species of grass and other forage were introduced. Fence rows and highway rights-of-way provided access for the migration of small mammals, particularly *Microtus* (Schmidly and Wilkins 1977; Getz et al. 1978).

A new habitat, the agroecosystem, is now recognized (Loucks 1977). In the modern context of economic accountability and systems analysis, a different kind of evaluation of the effects of *Microtus* on the bioenergetics of agriculture is possible.

Beyond the consideration of humans and domestic animals, voles compete with larger grazing herbivores, which are often managed. During periods of vole irruptions, however, the biomass of voles may become extremely large. This deleterious effect of voles on large herbivores has not been well documented. Competition between voles and small mammals of other species has not been critically studied.

Voles are responsible for severe crop damage by directly consuming bulbs, vegetables, grain, sagebrush, alfalfa, pasture land, and hay. Voles can also kill plants such as blackberry, raspberry, blueberry, strawberry, melon, and ornamental plants by girdling the roots and stems. This can result in serious economic losses (Lantz 1907; Batzli and Pitelka 1970; Byers 1985).

Damage to Trees. Severe and at times devastating harm is done to nursery stock, fruit orchards, and, in some cases, trees in the eastern deciduous forests (Lantz 1907; Byers 1984, 1985). Economic damage has been reported as exceeding $50 million annually (Merwin et al. 1999). Tree damage consists in debarking under the cover of snow. Destruction of roots is associated whenever extensive damage has been reported during an irruption of rodent populations (Littlefield et al. 1946). The extent of damage also has a positive correlation with long, hard winters.

The species of *Microtus* involved depends on the region. The pine vole, *M. pinetorum;* the meadow vole, *M. pennsylvanicus;* and the montane vole, *M. montanus,* are the species most often involved in tree damage reports (Piper 1909b; Garlough and Spencer 1944; Sartz 1970; Baumgartner 1978; Myllymäki 1979). Tree species include orchard trees, shade trees, common deciduous forest trees in eastern forests, and several coniferous species in young plantations of silvicultural operations. In this latter group, the preferred trees were Scotch pine (*Pinus sylvestris*), Douglas fir (*Pseudotsuga menziesii*), Austrian pine (*Pinus nigra*), Norway spruce (*Picea abies*), and European larch (*Larix decidua*). Red (*Pinus resinosa*), white (*P. strobus*), and jack (*P. banksiana*) pine, white cedar (*Thuja occidentalis*), balsam fir (*Abies balsamea*), and several larch (*Larix*) species were secondarily attacked (Littlefield et al. 1946). White spruce (*Picea glauca*) was avoided except by *M. ochrogaster* (Sartz 1970). In eastern North American forests, locust (*Robinia pseudoacacia*), black cherry (*Prunus serotina*), oak (*Quercus* sp.), chestnut (*Castanea dentata*), and hickory (*Carya* sp.) plantations have been particularly at risk (Lantz 1907).

Tree damage is not necessarily complete. Root damage may be hidden underground and go unrecognized. Aboveground debarking is more noticeable; it may completely encircle the tree and cause death, or it may be partial and the tree may survive. Such partial damage results in lowered production and promotes secondary decay and insect depredation.

Population Irruptions. Population irruptions are variously referred to as "mouse years" and "mouse plagues." They have been well documented in the Old World for centuries (Bailey 1924), but were not officially recognized as a problem in North America until 1907–1908.

An irruption of *Microtus montanus* in 1907–1908 in Nevada, Utah, and northeast California was the most serious recorded in the United States up to that time (Piper 1909a, 1909b). Numbers of mice were estimated (probably overestimated) at 20,000–30,000/ha. Fields were honeycombed by up to 60,000 holes/ha. During the summer growing season, one third of the area's alfalfa crops, three fourths of the potatoes, and much of the root crops of beets and carrots were ruined. In the fall, green food disappeared and the roots of alfalfa and orchard and shade trees were extensively damaged until the dramatic abatement occurred by the following summer.

In 1948–1949, an irruption of the montane vole occurred in eastern Washington and was investigated on the experimental agricultural plot at Orondo. As the snow cover left, examination of the experimental stone fruit tree plots showed that in three plots containing no ground cover, there was no mouse damage; three plots where some cover was maintained suffered 15.8% losses. Of special interest was one plot where 38 of 294 trees were girdled where ground cover was present, and only 1 of 182 trees in the adjoining clean-cultivated section was girdled.

An irruption of *Microtus montanus* was documented in eastern Oregon in 1957–1958 (Oregon State College 1959). This was widespread in adjoining northern California, Nevada, and Idaho. The 1957/58 winter was mild; population numbers rose to a peak at the end of the year and began to decrease in February. Maximum counts of vole populations in the agricultural areas ranged from 500 to 10,000/ha. Severe damage resulted in many fields at a level of 500–1200 voles/ha. There was an irregular die-off; in some areas, complete disappearance occurred, whereas in others, a buildup continued into July and August 1958.

Alfalfa, grain fields, and grass fields were damaged. Some fields harvested in 1957 were not worth cutting in 1958. Hundreds of hectares of ordinarily high-yield crops were not harvested. Red clover seed and grain in some areas were not harvested because the voles had eaten the seed heads. Potato fields and cellars suffered damage up to 50%. Cattlemen had to move cattle from natural meadows many weeks before the usual time.

Grass had better ability to regrow in areas of severe infestation, but much clover was removed. In late December and January, there was an apparent movement of voles into brushland-type habitat. Serious damage in rangeland occurred in bitterbrush (*Purshia tridentata*), rabbitbrush (*Chrysothamnus viscidiflorus*), wild cherry (*Prunus emarginata*), wild plum (*Prunus subcordata*), and greasewood (*Sarcobatus vermiculatus*).

Additional potential damage by tunneling in irrigation ditches was averted by heavy poisoning. Other damage occurred in baled hay by tunneling, chewing, and fouling. In numerous cases, 5- to 15-year old apple and other fruit trees were completely girdled and their supporting roots cut.

MANAGEMENT AND CONSERVATION

A most effective means of controlling voles in agricultural areas is eradication of local populations of pest animals. This means primarily control of roadside areas, fence lines, and ditches. Such control entails either frequent cultivation, application of herbicides, or grass cutting in the agricultural areas and in the surrounding borders. This may be a severe economic drain on the landholder and farmer. Because control requires working a large amount of unproductive land, it may not be economically feasible. Furthermore, the word *control* must be clearly understood. To people involved in vertebrate pest control, the aim is not to exterminate all members of a species, but "to reduce the numbers of these animals to tolerable densities only in areas where they are pests" (California Vertebrate Pest Control Committee 1964:2).

In most situations of mouse plagues, it has been documented that there is a buildup of population numbers the year before the irruption. Thus, control measures may be taken before the population irruption. Once the population cycle is in its ascent, it has been impossible to control the development of these populations.

On a year-to-year basis, to minimize the vole problem, farmers can (1) manage ground cover ("clean" farming), (2) protect natural enemies (see Mortality), and (3) control directly. In initiating control, the farmer should be certain that action is necessary. The following should be evaluated: (1) how much damage has already occurred, (2) how much damage is anticipated without control, (3) what the economic factors are of cost versus control and its benefits, and (4) the effect of a control program on the ecosystems and nontarget animals.

Ground cover can be managed by close mowing, using herbicides, mulching, or rototilling. Fruit yields in apple orchards were maximized by mulching and using herbicides. Tilling and close mowing were less effective (Byers 1985; Merwin and Stiles 1994). However, there may be a compensatory response of the small-mammal community to herbicide application. Deer mice and chipmunks (*Tamias* spp.) increased in abundance after treatment with the herbicide glyphosate, whereas montane vole populations declined (Sullivan et al. 1998). Merwin et al. (1999) concluded that a combination of close mowing, trunk protection with mesh guards, contiguous habitat for vole predators, and herbicide application within tree rows provided effective control without the use of rodenticides. Alternatively, broadcast burning (Sullivan and Boateng 1996) and trapping (Niemeyer 1997) have shown some success, but movement of individuals into areas where animals have been removed may limit the efficacy of trapping (Lapasha and Powell 1994). Repellents have not yet proven to be effective (Myllymäki 1979).

Poisons. Use of rodenticides has evolved as one of the most efficient (and most destructive) management tools. Bromadiolone, difethialone (Pascal et al. 1989), zinc phosphide (Sterner et al. 1996; Hygnstrom et al. 2000), sodium fluoroacetate, fluoroacetamide (Moran 1995), and Guthion 2S-R (Edge et al. 1996) have all been evaluated as rodenticides and can be reasonably effective. However, carcass residues can persist (Sterner and Mauldin 1995), although in one study there was no evidence of exposure of the principal predator (barn owls) to rodenticide residue in prey (Eadsforth et al. 1996). Alternatives to rodenticides such as repellents (Wager and Mason 1996; Zimmerling and Zimmerling 1996), alternative foods (Bergeron et al. 1998), and physical barriers (Zimmerling and Zimmerling 1998) have been effective. The best results (and least harm) occur when the application is done by an experienced person.

As a final note, it cannot be stressed too heavily that the total ecosystem must be considered. The risk of all of the poisoning methods to nontarget species should be evaluated.

Threatened or Endangered. No species of *Microtus* north of Mexico is listed by the U.S. Fish and Wildlife Service as threatened or endangered. Three subspecies of voles are listed as endangered. The Florida salt marsh vole, *M. pennsylvanicus dukecampbelli,* was listed as endangered in 1991. It is restricted to salt marsh habitats on Waccasassa Bay in Levy County, Florida (Department of the Interior 1991). The amargosa vole, *M. californicus scirpensis,* inhabits isolated wetlands of the Central Mojave Desert in Inyo County, California, and was listed as threatened in 1984 (Department of the Interior 1984). The Hualapai vole, *M. mogollonensis hualpaiensis,* was listed as endangered in 1987. It inhabits isolated alpine regions in northern Arizona (Department of the Interior 1987). All three subspecies are designated as vulnerable by the International Union for Conservation of Nature and Natural Resources (IUCN; Hilton-Taylor 2000).

Other subspecies, although not listed by the Fish and Wildlife Service, may be at risk because of small or declining population numbers, fragmented populations, or loss of habitat. *M. californicus mohavensis, M. montanus fucosis,* and *M. montanus nevadensis* are designated as vulnerable by the IUCN (Hilton-Taylor 2000).

RESEARCH NEEDS

In spite of more than nine decades of research and thousands of publications, there is still much to learn about the biology of voles. Based on chromosomal evidence, there are unresolved issues about the taxonomic status of several species including *M. abbreviatus, M. miurus, M. californicus, M. longicaudus,* and *M. montanus*.

There are numerous gaps in the information about the nutritional system of *Microtus,* leaving several unanswered questions:

1. What microorganisms are found in the intestinal tracts of various species of *Microtus?*
2. What role do microorganisms in general play in vole digestion?
3. What role do flagellated protozoans play in vole digestion? Are they parasites, which at times may produce harmful effects, or are they true symbionts? Do they aid in digestion and produce byproducts used by voles, or are they themselves eventually digested by the vole's system?
4. What degree of coprophagy exists in nature that allows *Microtus* to use partially digested cellulose?
5. Is there any significant biochemical breakdown of the celluloses in the upper gastrointestinal tract, and, if so, what microorganisms are involved?
6. What production of vitamins (especially B complex and K) takes place in the gastrointestinal tract? Is this internal production required by the vole?

7. Are there any valid biological comparisons among *Microtus,* other herbivores, and humans? Are there breakdown products of certain ingested items that are potentially toxic to the host?

Forty years of intensive research into the processes of population regulation have not yielded satisfactory answers to questions about the cause(s) of the multiannual fluctuations in density that occur in some species of voles (Krebs 1996). The exact relationship between density dependence (direct or lagged) and population regulation requires further research. Although we have learned much about the patterns of space use in voles, there are unanswered questions concerning the factors that contribute to the variation in space use patterns among individual voles and its role in population regulation. The relationship between voles and disease, especially diseases that can be transmitted to humans, is still not completely understood, and the potential effect of disease on behavior and population regulation has received little attention (Childs 1995).

Although no species of vole is listed as threatened or endangered, several subspecies are so listed and require attention. Other species, such as the island species *M. breweri* and *M. abbreviatus,* may be at risk in the future because of their restricted distribution. Of greater concern is the potential impact of irruptions in vole density and their economic impact on agriculture. Satisfactory mechanisms for monitoring, and, when needed, controlling vole populations must be developed that minimize damage to other species and ecosystems as a whole.

LITERATURE CITED

Anderson, S. 1985. Taxonomy and systematics. Pages 52–83 *in* R. H. Tamarin, ed. Biology of New World *Microtus* (Special Publication No. 8). American Society of Mammalogists.

Arata, A. A. 1975. The importance of small mammals in public health. Pages 349–59 *in* F. B. Golley, K. Petrusewicz, and L. Ryszkowski, eds. Small mammals: Their productivity and population dynamics. Cambridge University Press, Cambridge.

Bailey, V. 1920. Identity of the bean mouse of Lewis and Clark. Journal of Mammalogy 1:70–72.

Bailey, V. 1924. Breeding, feeding and other life habits of meadow mice (*Microtus*). Journal of Agricultural Research 27:523–36.

Bailey, V. 1936. The mammals and life zones of Oregon. North American Fauna 55:1–416.

Baird, D. D., and E. C. Birney. 1982. Characteristics of dispersing meadow voles *Microtus pennsylvanicus*. American Midland Naturalist 107:262–83.

Barry, W. J. 1976. Environmental effects on food hoarding in deermice (*Peromyscus*). Journal of Mammalogy 57:731–46.

Batzli, G. O. 1985. Nutrition. Pages 779–811 *in* R. H. Tamarin, ed. Biology of New World *Microtus* (Special Publication No. 8). American Society of Mammalogists.

Batzli, G. O., and H. G. Jung. 1980. Nutritional ecology of microtine rodents: Resource utilization near Atkasook, Alaska. Arctic Alpine Research 12:483–99.

Batzli, G. O., and F. A. Pitelka. 1970. Influence of meadow mouse populations on California grasslands. Ecology 51:1027–39.

Batzli, G. O., and F. A. Pitelka. 1971. Condition and diet of cycling populations of the California vole, *Microtus californicus*. Journal of Mammalogy 52:141–63.

Baumgardner, D. J., S. E. Ward, and D. A. Dewsbury. 1980. Diurnal patterning of eight activities in 14 species of muroid rodents. Animal Learning and Behavior 8:322–30.

Baumgartner, D. M. 1978. Animal damage control in the Pacific Northwest (EM 3908). Cooperative Extension Service, Washington State University, Pullman.

Beachem, T. D. 1980. Growth rates of the vole *Microtus townsendii* during a population cycle. Oikos 35:99–106.

Beck, M. L., and J. T. Mahan. 1978. The chromosomes of *Microtus pinetorum*. Journal of Heredity 69:343–44.

Bent, A. C. 1937. Life histories of North American birds of prey. Part 1: Order Falconiformes. Smithsonian Institution Press, Washington, DC.

Bent, A. C. 1938. Life histories of North American birds of prey. Part 2: Order Falconiformes and Strigiformes. Smithsonian Institution Press, Washington, DC.

Bent, A. C. 1946. Life histories of North American jays, crows, and titmice. Smithsonian Institution Press, Washington, DC.

Benton, A. H. 1955. Observations on the life history of the northern pine mouse. Journal of Mammalogy 36:52–62.

Bergeron, J. M., and L. Jodoin. 1987. Defining "high quality" food resources of herbivores: The case for meadow voles (*Microtus pennsylvanicus*). Oecologia 71:510–17.

Bergeron, J. M., and L. Jodoin. 1989. Patterns of resource use, food quality, and health status of voles *(Microtus pennsylvanicus*) trapped from fluctuating populations. Oecologia 79:306–14.

Bergeron, J. M., R. Goulet, and V. A. Gonzalez. 1998. The use of coniferous seedlings as alternative food to protect red oak (*Quercus rubra*) from vole girdling. Scandinavian Journal of Forest Research 13:50–53.

Birney, E. C., R. Jennes, and D. D. Baird. 1975. Eye lens proteins as criteria of age in cotton rats. Journal of Wildlife Management 39:718–28.

Boonstra, R., and P. T. Boag. 1987. A test of the Chitty hypothesis: Inheritance of life history traits in meadow voles *Microtus pennsylvanicus*. Evolution 41:929–47.

Boonstra, R., and C. J. Krebs. 1977. A fencing experiment on a high-density population of *Microtus townsendii*. Canadian Journal of Zoology 55:1166–75.

Bradley, S. R. 1976. Temperature regulation and bioenergetics of some microtine rodents. Ph.D. Dissertation, Cornell University, Ithaca, NY.

Burkot, T. R., B. S. Schneider, N. J. Pieniazek, C. M. Happ, J. S. Rutherford, S. B. Slemenda, E. Hoffmeister, G. O. Maupin, and N. S. Zeidner. 2000. *Babesia microti* and *Borrelia bissettii* transmission by *Ixodes spinipalpis* ticks among prairie voles, *Microtus ochrogaster,* in Colorado. Parasitology 121:595–99.

Burt, W. H. 1943. Territoriality and home range concepts as applied to mammals. Journal of Mammalogy 24:346–52.

Burt, W. H. 1960. Bacula of North American mammals. Miscellaneous Publications of the Museum of Zoology, University of Michigan 113:1–76.

Bury, R. B. 1971. Small mammals and other prey in the diet of the Pacific giant salamander (*Dicamptodon ensatus*). American Midland Naturalist 87:524–26.

Byers, R. E. 1984. Control and management of vertebrate pests in deciduous orchards. HortScience 10:391–92.

Byers, R. E. 1985. Management and control. Pages 621–46 *in* R. H. Tamarin, ed. Biology of New World *Microtus* (Special Publication No. 8). American Society of Mammalogists.

Calhoun, J. B. 1945. Diel activity rhythms of the rodents *Microtus ochrogaster* and *Sigmodon hispidus*. Ecology 26:251–73.

California Vertebrate Pest Control Committee, ed. 1964. Proceedings of second vertebrate pest control conference. California College of Agriculture, University of California, Davis.

Carleton, M. D. 1985. Macroanatomy. Pages 116–75 *in* R. H. Tamarin, ed. Biology of New World *Microtus* (Special Publication No. 8). American Society of Mammalogists.

Carleton, M. D., and G. G. Musser. 1984. Muroid rodents. Pages 289–379 *in* S. Anderson, and J. K. Jones, eds. Orders and families of recent mammals of the world. John Wiley, New York.

Carley, C. J., E. D. Fleharty, and M. A. Mares. 1970. Occurrence and activity of *Reithrodontomys megalotis, Microtus ochrogaster,* and *Peromyscus maniculatus* as recorded by a photographic device. Southwestern Naturalist 15:209–16.

Carter, C. S., and L. L. Getz. 1993. Monogamy and the prairie vole. Scientific American 268(6):100–106.

Chaline, J., and J. D. Graf. 1988. Phylogeny of the Arvicolidae (Rodentia): Biochemical and paleontological evidence. Journal of Mammalogy 69:22–33.

Chaline, J., P. Mein, and F. Petter. 1977. Les grandes lignes d'une classification évolution des Muroidea. Mammalia 41:245–52.

Charnov, E. L., and J. P. Finerty. 1980. Vole population cycles: A case for kin-selection? Oecologia 45:1–2.

Childs, J. E. 1995. Special feature: Zoonoses. Journal of Mammalogy 76:663.

Childs, J. E., G. E. Glass, and G. W. Korch, Jr. 1989. The comparitive epizootiology of *Capillaria hepatica* (Nematoda) in urban rodents from different habitats of Baltimore, Maryland (USA). Canadian Journal of Zoology 66:2769–75.

Childs, J. E., J. N. Mills, and G. E. Glass. 1995. Rodent-borne hemorrhagic fever viruses: A special risk for mammalogists? Journal of Mammalogy 76:664–80.

Chitty, D. 1967. The natural selection of self-regulatory behaviour in animal populations. Proceedings of the Ecological Society of Australia 2:51–78.

Cho, M. M., A. C. DeVries, J. R. Williams, and C. S. Carter. 1999. The effects of oxytocin and vasopressin on partner preferences in male and female prairie voles (*Microtus ochrogaster*). Behavioral Neurosciences 113:1071–79.

Christian, D. P. 1975. Vulnerability of meadow voles, *Microtus pennsylvanicus,* to predation by domestic cats. American Midland Naturalist 93:498–502.
Christian, J. J. 1950. The adreno-pituitary system and population cycles in mammals. Journal of Mammalogy 31:247–59.
Cole, F. R., and G. O. Batzli. 1979. Nutrition and population dynamics of the prairie vole, *Microtus ochrogaster,* in central Illinois. Journal of Animal Ecology 48:455–70.
Colvin, M. A. 1973. Analysis of acoustic structure and function in ultrasounds of neonatal *Microtus*. Behaviour 44:234–63.
Conley, W. 1976. Competition between *Microtus:* A behavioral hypothesis. Ecology 57:224–37.
Conroy, C. J., and J. A. Cook. 1999. MtDNA evidence for repeated pulses of speciation within arvicoline and murid rodents. Journal of Mammalian Evolution 6:221–45.
Conroy, C. J., and J. A. Cook. 2000. Molecular systematics of a holarctic rodent (*Microtus:* Muridae). Journal of Mammalogy 81:344–59.
Craighead, J. J. 1959. Predation by hawks, owls, and gulls. Pages 35–42 *in* Oregon State College, ed., The Oregon meadow mouse irruption of 1957–1958. Federal Cooperative Extension Service, Corvallis, OR.
Cranford, J. A., and E. O. Johnson. 1989. Effects of coprophagy and diet quality on two microtine rodents (*Microtus pennsylvanicus* and *Microtus pinetorum*). Journal of Mammalogy 70:494–502.
Dawson, M. R. 1967. Fossil history of the families of recent mammals. Pages 12–53 *in* S. Anderson and J. K. Jones, eds. Recent mammals of the world. Ronald Press, New York.
Denegri, G., W. Bernadina, S. J. Perez, and C. F. Rodriquez. 1998. Anoplocephalid cestodes of veterinary and medical significance: A review. Folia Parasitologica Ceske Budejovice 45:1–8.
Department of the Interior. 1984. Determination of endangered status and critical habitat for the amargosa vole. Federal Register 49:45160–64.
Department of the Interior. 1987. Determination of endangered status for the Hualapai vole. Federal Register 52:36776–80.
Department of the Interior. 1991. Endangered status for the Florida salt marsh vole. Federal Register 56:1457–59.
Dewsbury, D. A. 1980. Wheel-running behavior in 12 species of muroid rodents. Behavioural Processes 5:271–82.
Dewsbury, D. A., and T. G. Hartung. 1982. Copulatory behavior of three species of *Microtus*. Journal of Mammalogy 63:306–9.
Didow, L. A., and J. S. Hayward. 1969. Seasonal variations in the mass and composition of brown adipose tissue in the meadow vole, *Microtus pennsylvanicus*. Canadian Journal of Zoology 47:547–55.
Douglas, R. J. 1977. Population dynamics, home range and habitat association of the yellow-cheeked vole, *Microtus xanthognathus,* in the Northwest Territories. Canadian Field-Naturalist 91:237–47.
Drickamer, L. C. 1989. Pheromones: Behavioral and biomedical aspects. Pages 269–348 *in* J. Balthazart, ed. Advances in comparative and environmental physiology. Springer-Verlag, Berlin.
Dunaway, P. B., G. E. Cosgrove, and J. D. Story. 1968. *Capillaria* and *Trypanosoma* infestations in *Microtus ochrogaster*. Journal of Wildlife Diseases 4:18–20.
Durette-Desset, M. C. 1971. Essai de classification des nématodes héligmosomes: Corrélations avec la paléobiogéographie des hôtes. Memoires, Museum National d' Histoire, Naturelle, Paris, Series A, Zoologie 69:1–126.
Eadie, W. R. 1952. Shrew predation and vole populations on a localized area. Journal of Mammalogy 33:185–89.
Eadsforth, C. V., A. Gray, and E. G. Harrison. 1996. Monitoring the exposure of barn owls to second-generation rodenticides in southern Eire. Pesticide Science 47:225–33.
Edge, W. D., R. L. Carey, J. O. Wolff, L. M. Ganio, and T. Manning. 1996. Effects of Guthion 2S-R on *Microtus canicaudus:* A risk assessment validation. Journal of Applied Ecology 33:269–78.
Eisenberg, J. F. 1978. The evolution of arboreal herbivores in the class Mammalia. Pages 135–52 *in* G. G. Montgomery, ed. The ecology of arboreal folivores. Smithsonian Institution Press, Washington, DC.
Elton, C. 1942. Voles, mice, and lemmings. Clarendon Press, Oxford.
Evans, R. L., E. M. Katz, N. L. Olson, and D. A. Dewsbury. 1978. A comparative study of swimming behavior in eight species of muroid rodents. Bulletin of the Psychonomic Society 11:168–70.
Fay, F. H., and R. L. Rausch. 1969. Parasitic organisms in the blood of Arvicoline rodents in Alaska. Journal of Parasitology 55:1258–65.
Ferkin, M. H. 1990. Kin recognition and social behavior in microtine rodents. Pages 11–24 *in* R. H. Tamarin, R. S. Ostfeld, S. R. Pugh, and G. Bujalska, eds. Social systems and population cycles in voles. Birkhäuser, Basel, Switzerland.
Ferkin, M. H., and R. E. Johnston. 1993. Roles of gonadal hormones in control of five sexually attractive odors of meadow voles (*Microtus pennsylvanicus*). Hormones and Behavior 27:523–38.
Ferkin, M. H., and R. E. Johnston. 1995. Effects of pregnancy, lactation and postpartum oestrus on odour signals and the attraction to odours in female meadow voles, *Microtus pennsylvanicus*. Animal Behaviour 49:1211–17.
Ferkin, M. H., and J. O. Seamon. 1987. Odor preference and social behavior in meadow voles, *Microtus pennsylvanicus*. Canadian Journal of Zoology. 65:2931–37.
Ferkin, M. H., M. R. Gorman, and I. Zucker. 1992a. Influence of gonadal hormones on odours emitted by male meadow voles (*Microtus pennsylvanicus*). Journal of Reproduction and Fertility 95:729–36.
Ferkin, M. H., R. H. Tamarin, and S. R. Pugh. 1992b. Cryptic relatedness and the opportunity for kin recognition in microtine rodents. Oikos 63:328–32.
Ferkin, M. H., J. Dunsavage, and R. E. Johnston. 1999. What kind of information do meadow voles (*Microtus pennsylvanicus*) use to distinguish between the top and bottom scent of an over-mark? Journal of Comparative Psychology 113:43–51.
Findley, J. S., and C. J. Jones., 1962. Distribution and variation of voles of the genus *Microtus* in New Mexico and adjacent areas. Journal of Mammalogy 43:154–66.
Fitch, H. S. 1960. Autecology of the copperhead. University of Kansas Museum of Natural History Publications 13:85–288.
Fitzgerald, B. M. 1977. Weasel predation on a cyclic population of the montane vole (*Microtus montanus*) in California. Journal of Animal Ecology 46:367–97.
Fitzgerald, R. W., and D. M. Madison. 1981. Spacing, movements, and social organization of a free-ranging population of pine voles *Microtus pinetorum*. Pages 54–59 *in* R. E. Byers, ed. Proceedings of the fifth eastern pine and meadow vole symposium, Gettysburg, PA.
Fitzgerald, R. W., and D. M. Madison. 1983. Social organization of a free-ranging population of pine voles, *Microtus pinetorum*. Behavioral Ecology and Sociobiology 13:183–87.
Fitzner, R. E., and J. N. Fitzner. 1975. Winter food habits of short-eared owls in the Palouse prairie. Murrelet 56:2–4.
Fivush, B., R. Parker, and R. H. Tamarin. 1975. Karyotype of the beach vole, *Microtus breweri,* an endemic island species. Journal of Mammalogy 56:272–73.
Fortier, G. M., M. S. Erskin, and R. H. Tamarin. 1996. Female familiarity influences odor preferences and plasma estradiol levels in the meadow vole, *Microtus pennsylvanicus*. Physiology and Behavior 59:205–8.
Frey, J. K. 1989. Morphologic variation in the Mexican vole (*Microtus mexicanus*). M.S. Thesis, Emporia State University, Emporia, KS.
Frey, J. K., and C. T. LaRue. 1993. Notes on the distribution of the Mogollon vole (*Microtus mogollonensis*) in New Mexico and Arizona. Southwestern Naturalist 38:176–78.
Frey, J. K., and M. J. Patrick. 1995. Gastrointestinal helminths from the endangered hualapal vole, *Microtus mogollonensis hualpaiensis* (Rodentia: Cricetidae). Journal of Parasitology 81:641–43.
Gage, K. L., R. S. Ostfeld, and J. G. Olson. 1995. Nonviral vector-borne zoonoses associated with mammals in the United States. Journal of Mammalogy 76:695–715.
Gaines, M. S. 1985. Genetics. Pages 845–83 *in* R. H. Tamarin, ed. Biology of New World *Microtus* (Special Publication No. 8). American Society of Mammalogists.
Gaines, M. S., and R. K. Rose. 1976. Population dynamics of *Microtus ochrogaster* in eastern Kansas. Ecology 57:1145–61.
Gamble, W. B., M. Segal, P. M. Schantz, and R. L. Rausch. 1979. Alveolar hydatid disease in Minnesota: First human case acquired in the contiguous United States. Journal of the American Medical Association 241:904–7.
Garlough, F. E., and D. A. Spencer. 1944. Control of destructive mice. U. S. Fish and Wildlife Service Conservation Bulletin 36:1–37.
Gashwiler, J. S. 1972. Life history notes on the Oregon vole, *Microtus oregoni*. Journal of Mammalogy 53:558–69.
Gettle, A. S. 1975. Densities, movements and activities of pine voles (*Microtus pinetorum*) in Pennsylvania. M.S. Thesis, Pennsylvania State University, State College.
Getz, L. L. 1961. Factors influencing the local distribution of *Microtus* and *Synaptomys* in Southern Michigan. Ecology 42:110–19.
Getz, L. L., and C. S. Carter. 1996. Prairie-vole partnerships. American Scientist 84:56–62.
Getz, L. L., F. R. Cole, and D. L. Gates. 1978. Interstate roadsides as dispersal routes for *Microtus pennsylvanicus*. Journal of Mammalogy 59:208–12.

Getz, L. L., B. McGuire, J. Hofmann, T. Pizzuto, and B. Frase. 1993. Social organization of the prairie vole (*Microtus ochrogaster*). Journal of Mammalogy 74:44–58.

Gill, A. E. 1982. Variability in the karyotype of the California vole, *Microtus californicus*. Mammalian Chromosomes Newsletter 23:18.

Gingrich, B., Y. Liu, C. Cascio, Z. Wang, and T. R. Insel. 2000. Dopamine D2 receptors in the nucleus accumbens are important for social attachment in female prairie voles (*Microtus ochrogaster*). Behavioral Neurosciences 114:173–83.

Glass, G. E., and N. A. Slade. 1980. The effect of *Sigmodon hispidus* on spatial and temporal activity of *Microtus ochrogaster:* Evidence for competition. Ecology 60:358–70.

Goertz, J. W. 1971. An ecological study of *Microtus pinetorum* in Oklahoma. American Midland Naturalist 86:1–12.

Gordon, D. A., D. L. Lattier, R. N. Silbiger, J. Torsella, J. O. Wolff, and M. K. Smith. 1998 Determination of genetic diversity and paternity in the gray-tailed vole (*Microtus canicaudus*) by RAPD PCR. Journal of Mammalogy 79:604–11.

Grassé, P. P. 1973. Traité de zoologie, Tome 16: Manniferes, Vol. 5, Fasc. 1. Splanchnologie. Paris.

Hall, E. R. 1981. The mammals of North America. John Wiley, New York.

Hamilton, W. J., Jr. 1937. Activity and home range of the field mouse, *Microtus pennsylvanicus pennsylvanicus*. Ecology 18:255–63.

Hamilton, W. J., Jr. 1938. Life history notes on the northern pine mouse. Journal of Mammalogy 19:163–70.

Hamilton, W. J., Jr. 1941. Reproduction of the field mouse *Microtus pennsylvanicus* (Ord) (Memoir 237). Cornell University Agricultural Experiment Station.

Hanski, I., P. Turchin, E. Korpimäki, and H. Henttonen. 1993. Population oscillations of boreal rodents: Regulation by mustelid predators leads to chaos. Nature 364:232–35.

Hatfield, D. M. 1935. A natural history study of *Microtus californicus*. Journal of Mammalogy 16:261–71.

Haukisalmi, V., H. Henttonen, and G. O. Batzli. 1995. Helminth parasitism in the voles *Microtus oeconomus* and *M. miurus* on the North Slope of Alaska: Host specificity and the effects of host sex, age and breeding status. Annales Zoologici Fennici 32:193–201.

Healy, G. R. 1979. *Babesia* infections in man. Hospital Practice 1979 (June):107–16.

Heidt, G. A. 1971. Daily summer activity of the meadow vole, *Microtus pennsylvanicus*. Michigan Academician 3:31–39.

Hildreth, M. B., S. Sriram, B. Gottstein, M. Wilson, and P. M. Schantz., 2000. Failure to identify alveolar echinococcosis in trappers from South Dakota in spite of high prevalence of *Echinococcus multilocularis* in wild canids. Journal of Parasitology 86:75–77.

Hilton-Taylor, C., comp. 2000. 2000 IUCN red list of threatened species. IUCN, Gland, Switzerland.

Hoagland, D. B., N. Tilakaratne, R. F. Weaver, and M. S. Gaines. 1991. "DNA fingerprinting" of prairie voles (*Microtus ochrogaster*). Journal of Mammalogy 72:422–426.

Hoffmeister, D. F., and L. L. Getz. 1968. Growth and age classes in the prairie vole, *Microtus ochrogaster*. Growth 32:57–69.

Hörnfeldt, B. 1994. Delayed density dependence as a determinant of vole cycles. Ecology 75:791–806.

Househecht, C. R. 1968. Sonographic analysis of vocalizations of three species of mice. Journal of Mammalogy 49:555–60.

Howell, A. B. 1924. Individual and age variations in *Microtus montanus yosemite*. Journal of Agricultural Research 28:977–1015.

Hubbard, C. A. 1947. Fleas of western North America. Iowa State College Press, Ames.

Hume, I. D. 1978. Evolution of the Macropodidae digestive system. Australian Mammalogy 2: 37–42.

Hsu, T. C., and K. Benirschke. 1967–1971. An atlas of mammalian chromosomes, Springer-Verlag, New York.

Hygnstrom, S. E., K. C. VerCauteren, R. A. Hines, and C. W. Mansfield. 2000. Efficacy of in-furrow zinc phosphide pellets for controlling rodent damage in no-till corn. International Biodeterioration and Biodegradation 45:215–22.

Ims, R. A. 1989. Kinship and origin effects on dispersal and space sharing in *Clethrionomys rufocanus*. Ecology 70:607–16.

Innes, D. G. L., and J. S. Millar. 1979. Growth of *Clethrionomys gapperi* and *Microtus pennsylvanicus* in captivity. Growth 43:208–17.

Insel, T. R., and T. J. Hulihan. 1995. A gender-specific mechanism for pair bonding: Oxytocin and partner preference formation in monogamous voles. Behavioral Neurosciences 109:782–89.

Ivanov, S. V., and W. S. Modi. 1996. Molecular characterization of the complex sex-chromosome heterochromatin in the rodent *Microtus chrotorrhinus*. Cytogenetics and Cell Genetics 75:49–56.

Jannett, F. J. 1981. Scent mediation of intraspecific, interspecific, and intergeneric agonistic behavior among sympatric species of voles (Microtinae). Behavioral Ecology and Sociobiology 8:293–96.

Jannett, F. J. 1982. Social dynamics of the montane vole, *Microtus montanus,* as a paradigm. Biologist 62:3–19.

Jewett, S. G., W. P. Taylor, W. T. Shaw, and J. W. Aldrich. 1953. Birds of Washington state. University of Washington Press, Seattle.

Johnson, M. L., and S. Johnson. 1982. Voles. Pages 326–54 *in* J. Chapman, and G. Feldhamer, eds., Wild mammals of North America: Biology, management, and economics. Johns Hopkins University Press, Baltimore.

Jones, C., R. S. Hoffmann, D. W. Rice, M. D. Engstrom, R. D. Bradley, D. J. Schmidly, C. A. Jones, and R. J. Baker. 1997. Revised checklist of North American mammals north of Mexico. Occasional Papers, the Museum, Texas Tech University 173:1–19.

Jones, J. K., Jr., D. C. Carter, H. H. Genoways, R. S. Hoffmann, D. W. Rice, and C. Jones. 1986. Revised checklist of North American mammals north of Mexico. Occasional Papers, the Museum, Texas Tech University 107:1–22.

Judd, S. R. 1980. Observations of the chromosome variation in *Microtus mexicanus* (Rodentia:Microtinae). Mammalian Chromosome Newsletter 21:110–13.

Judd, S. R., and S. P. Cross. 1980. Chromosomal variation in *Microtus longicaudus*. Murrelet 61:2–5.

Judd, S. R., S. P. Cross, and S. Pathek. 1980. Non-Robertsonian chromosomal variation in *Microtus montanus*. Journal of Mammalogy 61:109–13.

Kartman, L., F. M. Prince, and S. F. Quan. 1959. Pages 43–54 *in* Oregon State College, ed. The Oregon meadow mouse irruption of 1957–1958. Federal Cooperative Extension Service, Corvallis, OR.

Kawata, M. 1990. Fluctuating populations and kin interaction in mammals. Trends in Ecology and Evolution 5:17–20.

Keith, T. P., and R. H. Tamarin. 1981. Genetic and demographic differences between dispersers and residents in cycling and noncycling vole populations. Journal of Mammalogy 62:713–25.

Keller, B. L. 1985. Reproductive patterns. Pages 725–78 *in* R. H. Tamarin, ed. Biology of New World *Microtus* (Special Publication No. 8). American Society of Mammalogists.

Kenagy, G. J., and D. F. Hoyt. 1980. Reingestion of feces in rodents and its daily rhythmicity. Oecologia 44:403–9.

Keys, J. E., and P. J. Van Soest. 1970. Digestibility of forages by the meadow vole *(Microtus pennsylvanicus*). Journal of Dairy Science 53:1502–8.

Kirkland, G. L., Jr. 1977. The rock vole *Microtus chrotorrhinus* (Miller) in West Virginia. Annals of the Carnegie Museum 46:45–53.

Korschgen, L. J., and T. S. Baskett. 1963. Food of impoundment- and stream-dwelling bullfrogs in Missouri. Herpetologica 19:89–99.

Krebs, C. J. 1966. Demographic changes in fluctuation populations of *Microtus californicus*. Ecological Monographs 36:239–73.

Krebs, C. J. 1978. A review of the Chitty hypothesis of population regulation. Canadian Journal of Zoology 56:2463–80.

Krebs, C. J. 1996. Population cycles revisited. Journal of Mammalogy 77:8–24.

Krebs, C. J., B. L. Keller, and R. H. Tamarin. 1969. *Microtus* population biology: Demographic changes in fluctuating populations of *M. ochrogaster* and *M. pennsylvanicus* in southern Indiana. Ecology 50:587–607.

Krebs, C. J., J. A. Redfield, and M. J. Taitt. 1978. A pulsed-removal experiment on the vole *Microtus townsendii*. Canadian Journal of Zoology 56:2253–62.

Lambin, X., and C. J. Krebs. 1991. Spatial organization and mating system of *Microtus townsendii*. Behavioral Ecology and Sociobiology 28:353–64.

Lance, E. W., and J. A. Cook. 1998. Biogeography of tundra voles (*Microtus oeconomus*) of Beringia and the southern coast of Alaska. Journal of Mammalogy 79:53–65.

Lantz, D. E. 1907. An economic study of field mice (genus *Microtus*) (Bulletin 31). U. S. Department of Agriculture Biological Survey.

Lapasha, D. G., and R. A. Powell. 1994. Pine vole (*Microtus pinetorum*) movement toward areas in apple orchards with reduced populations. Journal of Horticultural Science 69:1077–82.

Libbus, B. L., and L. A. Johnson. 1988. The creeping vole, *Microtus oregoni:* Karyotype and sex-chromosome differences between two geographical populations. Cytogenetics and Cell Genetics 47:181–84.

Lidicker, W. Z., Jr. 1980. The social biology of the California vole. Biologist 62:46–55.

Lidicker, W. Z., Jr. 1985. Dispersal. Pages 420–54 *in* R. H. Tamarin, ed. Biology of New World *Microtus* (Special Publication No. 8). American Society of Mammalogists.

Lidicker, W. Z., Jr. 1988. Solving the enigma of microtine "cycles." Journal of Mammalogy 69:225–35.
Lindroth, R. L., and G. O. Batzli. 1984a. Food habits of the meadow vole (*Microtus pennsylvanicus*) in bluegrass and prairie habitats. Journal of Mammalogy 65:600–606.
Lindroth, R. L., and G. O. Batzli. 1984b. Plant phenolics as chemical defenses: The effects of natural phenolics on survival and growth of prairie voles. Journal of Chemical Ecology 10:229–44.
Ling, J. K. 1970. Pelage and molting in wild mammals with special reference to aquatic forms. Quarterly Review of Biology 45:16–54.
Littlefield, E. W., W. J. Schoomaker, and D. B. Cook. 1946. Field mouse damage to coniferous plantations. Journal of Forestry 44:756–49.
Loucks, O. L. 1977. Emergence of research on agroecosystems. Annual Review of Ecology and Systematics 8:173–92.
Ludwig, D. R. 1981. The population biology and life history of the water vole, *Microtus richardsoni*. Ph.D. Dissertation, University of Calgary, Calgary, Alberta, Canada.
MacDonald, S. O., and J. A. Cook. 1996. The land mammal fauna of southeast Alaska. Canadian Field-Naturalist 110:571–98.
Madison, D. M. 1978. Movement indicators of reproductive events among female meadow voles as revealed by radiotelemetry. Journal of Mammalogy 59:835–43.
Madison, D. M. 1980. Space use and social structure in meadow voles, *Microtus pennsylvanicus*. Behavioral Ecology and Sociobiology 7:65–71.
Madison, D. M. 1985. Activity rhythms and spacing. Pages 373–419 *in* R. H. Tamarin, ed. Biology of New World *Microtus* (Special Publication No. 8). American Society of Mammalogists.
Madison, D. M., and W. J. McShea. 1987. Seasonal changes in reproductive tolerance, spacing, and social organization in meadow voles: A microtine model. American Zoologist 27:899–908.
Madison, D. M., R. W. Fitzgerald, and W. J. McShea. 1984. Dynamics of social nesting in overwintering meadow voles (*Microtus pennsylvanicus*). Behavioral Ecology and Sociobiology 15:9–17.
Mallory, F. F., and R. A. Dieterich. 1985. Laboratory management and pathology. Pages 647–84 *in* R. H. Tamarin, ed. Biology of New World *Microtus* (Special Publication No. 8). American Society of Mammalogists.
Marfori, M. A., P. G. Parker, T. G. Gregg, J. G. Vandenbergh, and N. G. Solomon. 1997. Using DNA fingerprinting to estimate relatedness within social groups of pine voles, *Microtus pinetorum*. Journal of Mammalogy 78:715–24.
Marquardt, W. C., R. S. Demaree, and R. B. Grieve. 2000. Parasitology and vector biology, 2nd ed. Academic Press, San Diego, CA.
Marquis, R. J., and G. O. Batzli. 1989. Influence of chemical factors on palatability of forage to voles. Journal of Mammalogy 70:503–11.
Martin, E. P. 1956. A population study of the prairie vole (*Microtus ochrogaster*) in northeastern Kansas. University of Kansas Museum of Natural History Publications 8:361–416.
Maser, C., and R. M. Storm. 1970. A key to Microtinae of the Pacific Northwest (Oregon, Washington, Idaho). Oregon State University Book Stores, Corvallis.
Maser, C., E. W. Hammer, and S. Anderson. 1970. Comparative food habits of three owl species in central Oregon. Murrelet 51:29–33.
Matthey, R. 1955. Nouveaux documents sur les chromosomes des Muridae. Review Suisse Zoology 62:163–206.
May, R. M. 1976. Models for single populations. Pages 4–25 *in* R. M. May, ed. Theoretical ecology: Principles and applications. W. B. Saunders, Philadelphia.
McBee, R. H. 1970. Metabolic contributions of the cecal flora. American Journal of Clinical Nutrition 23:1514–18.
Merwin, I. A., and W. C. Stiles. 1994. Orchard groundcover management impacts on apple tree growth and yield, and nutrient availability and uptake. Journal of the American Society for Horticultural Science 119:209–15.
Merwin, I. A., J. A. Ray, and P. D. Curtis. 1999. Orchard groundcover management systems affect meadow vole populations and damage to apple trees. HortScience 34:271–74.
Meylan, A. 1967. Karyotype and giant sex chromosomes of *Microtus chrotorrhinus* (Miller) (Mammalia: Rodentia). Canadian Journal of Genetics and Cytology 9:700–703.
Mihok, S. 1984. Life history profiles of boreal meadow voles (*Microtus pennsylvanicus*). Pages 91–102 *in* J. F. Merritt, ed. Winter ecology of small mammals (Special Publication No. 10). Carnegie Museum of Natural History, Pittsburgh.
Mihok, S., B. N. Turner, and S. L. Iverson. 1985. The characterization of vole (*Microtus pennsylvanicus*) population dynamics. Ecological Monographs 55:399–420.
Modi, W. S. 1986. Karyotypic differentiation among two sibling species pairs of New World microtine rodents. Journal of Mammalogy 67:159–65.
Modi, W. S. 1987a. Phylogenetic analyses of chromosomal banding patterns among the Nearctic Arvicolidae (Mammalia: Rodentia). Systematic Zoology 36:109–36.
Modi, W. S. 1987b. C-banding analysis and the evolution of heterochromatin among arvicolid rodents. Journal of Mammalogy 68:704–13.
Modi, W. S. 1993. Comparative analyses of heterochromatin in *Microtus:* Sequence heterogeneity and localized expansion and contraction of satellite DNA arrays. Cytogenetics and Cell Genetics 62:142–48.
Moncrief, N. D., N. E. Cockett, A. D. Neff, W. L. Thomas, and R. D. Dueser. 1997. Polymorphic microsatellites in the meadow vole, *Microtus pennsylvanicus:* Conservation of loci across species of rodents. Molecular Ecology 6:299–301.
Moran, S. 1995. Reducing sodium fluoroacetate and fluoroacetamide concentrations in field rodent baits. Phytoparasitica 23:195–203.
Moyer, C. A., G. H. Adler, and R. H. Tamarin. 1988. Systematics of New England *Microtus,* with emphasis on *Microtus breweri*. Journal of Mammalogy 69:782–94.
Murie, A. 1936. Following fox trails. Miscellaneous Publications of the Museum of Zoology, University of Michigan 32:1–45.
Murie, A. 1940. Ecology of the coyote in the Yellowstone. (Fauna of the National Parks of the United States, No. 4). U. S. Department of the Interior, Washigton, DC.
Murie, A. 1944. The wolves of Mount McKinley (Fauna of the National Parks of the United States, No. 5). U. S. Department of the Interior, Washington, DC.
Musser, G. G., and M. D. Carleton. 1993. Family Muridae. Pages 501–756 *in* D. E. Wilson and D. M. Reeder, eds. Mammal species of the world: A taxonomic and geographic reference. Smithsonian Institution Press, Washington, DC.
Myers, J. H., and C. J. Krebs. 1971. Genetic, behavioral, and reproductive attributes of dispersing field voles *Microtus pennsylvanicus* and *Microtus ochrogaster*. Ecological Monographs 41:53–78.
Myers, J. H., and C. J. Krebs. 1974. Population cycles in small mammals. Advances in Ecological Research 8:267–399.
Myers, P. 1978. A method for determining the age of living small mammals. Journal of the Zoological Society of London 186:551–56.
Myllymäki, A. 1979. Importance of small mammals as pests in agriculture and stored products. Pages 239–79 *in* D. M. Stoddard, ed. Ecology of small mammals. Chapman and Hall, London.
Nadeau, J. H. 1985. Ontogeny. Pages 254–85 *in* R. H. Tamarin, ed. Biology of New World *Microtus* (Special Publication No. 8). American Society of Mammalogists.
Negus, N. C., P. J. Berger, and L. Forsland. 1977. Reproductive strategy of *Microtus montanus*. Journal of Mammalogy 58:347–53.
Niemeyer, H. 1997. Field tests on rodenticide-free control of voles in forestry. Anzeiger für Schädlingskunde Pflanzenschutz Umweltschutz 70:25–29.
Niethammer, J. 1972. Die Zahl der Mammae bei *Pitymys* und bei den Microten. Bonner Zoologische Bieträge 23:49–60.
Nowak, R. M. 1999. Walker's mammals of the world, Vol. 2, 6th ed. Johns Hopkins University Press, Baltimore.
Oregon State College, ed. 1959. The Oregon meadow mouse irruption of 1957–1958. Federal Cooperative Extension Service, Corvallis, OR.
Ostfeld, R. S. 1985. Limiting resources and territoriality in microtine rodents. American Naturalist 126:1–15.
Ostfeld, R. S. 1994. The fence effect reconsidered. Oikos 70:340–48.
Ostfeld, R. S., and C. D. Canham. 1995. Density-dependent processes in meadow voles: An experimental approach. Ecology 76:521–32.
Ostfeld, R. S., and L. L. Klosterman. 1990. Microtine social systems, adaptation, and the comparative method. Pages 35–44 *in* R. H. Tamarin, R. S. Ostfeld, S. R. Pugh, and G. Bujalska, eds. Social systems and population cycles in voles. Birkhäuser, Basel, Switzerland.
Ostfeld, R. S., S. R. Pugh, J. O. Seamon, and R. H. Tamarin. 1988. Space use and reproductive success in a population of meadow voles. Journal of Animal Ecology 57:385–94.
Ostfeld, R. S., C. D. Canham, and S. R. Pugh. 1993. Intrinsic density-dependent regulation of vole populations. Nature 366:259–61.
Ouellette, D. E., and J. F. Heisinger. 1980. Reingestion of feces by *Microtus pennsylvanicus*. Journal of Mammalogy 61:366–68.
Parra, R. 1978. Comparison of foregut and hindgut fermentation in herbivores. Pages 205–29 *in* G. G. Montgomery, ed. The ecology of arboreal folivores. Smithsonian Institution Press, Washington, DC.
Pascal, M., B. Pradier, and M. Habert. 1989. Comparative efficacy of two rodenticides, bromadiolone and difethialone, used against the fossorial form of the water vole (*Arvicola terrestris scherman* (Shaw, 1801)): Efficacy tests

of two rodenticides in the field: Methodological approach. Acta Oecologica Oecologia Applicata 9:371–84.

Patton, J. L. 1977. B-chromosome systems in the pocket mouse, *Perognathus baileyi:* Meiosis and C-band studies. Chromosoma 60:1–14.

Pearson, O. P. 1960. Habits of *Microtus californicus* revealed by automatic photographic records. Ecological Monographs 30:231–49.

Phillips, C. J., and B. Oxberry. 1972. Comparative histology of molar dentitions of *Microtus* and *Clethrionomys* with comments on dental evolution in microtine rodents. Journal of Mammalogy 53:1–20.

Phillips, M. L., and Z. Tang-Martinez. 1998. Parent–offspring discrimination in the prairie vole and the effects of odors and diet. Canadian Journal of Zoology 76:711–16.

Pinkel, D., B. L. Gledhill, S. Lake, D. Stephenson, and M. A. Van Dilla. 1982. Sex preselection in mammals? Separation of sperm bearing Y and "O" chromosomes in the vole *Microtus oregoni*. Science 218:904–6.

Piper, S. E. 1909a. Mouse plagues, their control and prevention. Pages 301–10 *in* Yearbook of the Department of Agriculture (1908). U. S. Department of Agriculture, Washington, DC.

Piper, S. E. 1909b. The Nevada mouse plague of 1907–1908. United States Department of Agriculture Farmers Bulletin 352:1–23.

Pugh, S. R., and R. S. Ostfeld. 1998. Effects of prior population density on use of space by meadow voles, *Microtus pennsylvanicus*. Journal of Mammalogy 79:551–57.

Pugh, S. R., and R. H. Tamarin. 1990. A test of Charnov and Finerty hypothesis of population regulation in meadow voles. Pages 111–20 *in* R. H. Tamarin, R. S. Ostfeld, S. R. Pugh, and G. Bujalska, eds. Social systems and population cycles in voles. Birkhäuser, Basel, Switzerland.

Pugh, S. R., and R. H. Tamarin. 1991. A comparison of population characteristics and reproductive success of resident and immigrant meadow voles. Canadian Journal of Zoology 69:2638–43.

Quay, W. B. 1968. The specialized posterolateral sebaceous glandular regions in microtine rodents. Journal of Mammalogy 49:427–45.

Quentin, J. C. 1971. Morphologie comparée des structures céphaliques et génitales des Oxyures du genre *Syphacia*. Annales de Parasitogie Humaine et Comparée 46:15–60.

Racey, K. 1960. Notes relative to the fluctuation in numbers of *Microtus richardsoni richardsoni* about Alta Lake and Pemberton Valley, B.C. Murrelet 41:13–14.

Rageot, R. H. 1957. Predation on small mammals in the Dismal Swamp, Virginia. Journal of Mammalogy 38:281.

Rausch, R. L. 1964. The specific status of the narrow-skulled vole (subgenus *Stenocranius* Kashchenko) in North America. Zeitschrift für Säugertierkunde 29:343–58.

Rausch, R. L. 1975. Taeniidae. Pages 678–707 *in* W. T. Hubbert, W. F. McCulloch, and P. R. Schnurrenberger, eds. Diseases transmitted from animals to man, 6th ed. Charles C. Thomas, Springfield, IL.

Rausch, R. L., and F. H. Fay. 1989. Postoncospheral development and cycle of *Taenia polycantha* Leuckart, 1856 (Cestoda: Taeniidae). Annales de Parasitologie Humaine et Comparée 63:263–77.

Rausch, R. L., and V. R. Rausch. 1968. On the biology and systematic position of *Microtus abbreviatus* Miller, a vole endemic to the St. Matthew Islands, Bering Sea. Zeitschrift für Säugertierkunde 33:65–99.

Rausch, R. L., and V. R. Rausch. 1974. The chromosomal complement of the yellow-cheeked vole, *Microtus xanthognathus* (Leach). Canadian Journal of Genetics and Cytology 16:267–72.

Rausch, R. L., B. E. Huntley, and J. G. Bridgens. 1968. Notes on *Pastuerella tularensis* isolated from a vole, *Microtus oeconomus* Pallas, in Alaska. Canadian Journal of Microbiology 15:47–55.

Raven, P. H., and D. I. Axelrod. 1974. Angiosperm biogeography and past continental movements. Annals of the Missouri Botanical Gardens 61:539–637.

Raynaud, A. 1969. Les organes génitaux de mammiféres. Pages 149–636 *in* P. P. Grassé, ed. Traité de zoologie. Tome 16: Manniferes, Vol. 6, Mammelles, appareil génital, gamétogenése, fécondation, gestation. Masson, Paris.

Repenning, C. A. 1980. Faunal exchanges between Siberia and North America. Canadian Journal of Anthropology 1:37–44.

Rose, R. K., and M. S. Gaines. 1978. The reproductive cycle of *Microtus ochrogaster* in eastern Kansas. Ecological Monographs 48:21–42.

Roth, D., and L. R. Powers. 1979. Comparative feeding and roosting habits of three sympatric owls in southwestern Idaho. Murrelet 60:12–15.

Rothstein, B. E., and R. H. Tamarin. 1977. Feeding behavior of the insular beach vole, *Microtus breweri*. Journal of Mammalogy 58:84–85.

Rowsemitt, C. N. 1981. Hormonal regulation of activity patterns in *Microtus montanus,* the montane vole. Ph.D. Dissertation, University of Utah, Salt Lake City.

Rowsemitt, C., T. H. Kunz, and R. H. Tamarin. 1975. The timing and patterns of molt in *Microtus breweri*. Occasional Papers of the Museum of Natural History, University of Kansas 34:1–11.

Sartz, R. S. 1970. Mouse damage to young plantations in southwestern Wisconsin. Journal of Forestry 68:88–89.

Sato, H., and M. Kamiya. 1992. Occurrence of *Syngamus* sp. in tundra voles (*Microtus oeconomus*) collected on St. Lawrence Island, Bering Sea. Journal of Wildlife Diseases 28:134–37.

Saunders, J. K. 1963. Food habits of the lynx in Newfoundland. Journal of Wildlife Management 27:384–90.

Schadler, M. H., and G. M. Butterstein. 1979. Reproduction in the pine vole, *Microtus pinetorum*. Journal of Mammalogy 60:841–44.

Schmidly, D. J., and K. T. Wilkins. 1977. Composition of small mammal populations on highway rights-of-way in east Texas (Report of Study 2-8-76-197; NITS PB-275089). U.S. Federal Highway Administration and Texas State Department of Highways and Public Transportation.

Seabloom, R. W. 1985. Endocrinology. Pages 685–724 *in* R. H. Tamarin, ed. Biology of New World *Microtus* (Special Publication No. 8). American Society of Mammalogists.

Seidensticker, J. C. 1970. Food of nesting red-tailed hawks in south-central Montana. Murrelet 51:38–40.

Sheffield, S. 1979. Intestinal sarcomastigophorans in *Peromyscus maniculatus* from southwestern Washington. M.S. Thesis, University of Puget Sound, Tacoma, WA.

Shellhammer, H. S. 1969. Supernumerary chromosomes of the harvest mouse, *Reithrodontomys megalotis*. Chromosoma 27:102–8.

Shields, L. J. 1976. Telemetric determination of the activity of free-ranging rodents; the fine structure of *Microtus californicus* activity patterns. Ph.D. Dissertation, University of California, Los Angeles.

Sodja, I., D. Lim, and O. Matouch. 1971. Isolation of rabies virus from small wild rodents. Journal of Hygiene Epidemiology, Microbiology, and Immunology 15:271–77.

Sodja, I., D. Lim, and O. Matouch. 1973. Isolation of rabies strains from small wild rodents and their biological properties. Folia Microbiologica 18:182.

Solomon, N. G., J. G. Vandenbergh, and W. T. Sullivan. 1998. Social influences on intergroup transfer by pine voles (*Microtus pinetorum*). Canadian Journal of Zoology 76:2131–36.

Sonnenberg, E. L., and L. R. Powers. 1976. Notes on the food habits of long-eared owls in southwestern Idaho. Murrelet 57:63–64.

Sozen, M., E. Colak, N. Yigit, and S. Ozkurt. 1999. Age variations in *Microtus guntheri* Danford and Alston, 1880 (Mammalia: Rodentia) in Turkey. Turkish Journal of Zoology 23:145–55.

Sterner, R. T., and R. E. Mauldin. 1995. Regressors of whole-carcass zinc phosphide/phosphine residues in voles: Indirect evidence of low hazards to predators/scavengers. Archives of Environmental Contamination and Toxicology 28:519–23.

Sterner, R. T., C. A. Ramey, W. D. Edge, T. Manning, J. O. Wolff, and K. A. Fagerstone. 1996. Efficacy of zinc phosphide baits to control voles in alfalfa: An enclosure study. Crop Protection 15:727–34.

Sullivan, T. P., and J. O. Boateng. 1996. Comparison of small mammal community responses to broadcast burning and herbicide application in cutover forest habitats. Canadian Journal of Forest Research 26:462–73.

Sullivan, T. P., and C. J. Krebs. 1981. *Microtus* population biology: Demography of *M. oregoni* in southwestern British Columbia. Canadian Journal of Zoology 59:2092–2102.

Sullivan, T. P., D. S. Sullivan, E. J. Hogue, R. A. Lautenschlager, and R. G. Wagner. 1998. Population dynamics of small mammals in relation to vegetation management in orchard agroecosystems: Compensatory responses in abundance and biomass. Crop Protection 17:1–11.

Swade, R. H., and C. S. Pittendrigh. 1967. Circadian locomotor rhythms of rodents in the Arctic. American Naturalist 101:431–66.

Taitt, M. J., and C. J. Krebs. 1981. The effect of extra food on small rodent populations: II. Voles (*Microtus townsendii*). Journal of Animal Ecology 50:125–37.

Taitt, M. J., and C. J. Krebs. 1985. Population dynamics and cycles. Pages 567–620 *in* R. H. Tamarin, ed. Biology of New World *Microtus* (Special Publication No. 8). American Society of Mammalogists.

Taitt, M. J., J. H. W. Gipps, and C. J. Krebs. 1981. The effect of extra food and cover on declining populations of *Microtus townsendii*. Canadian Journal of Zoology 59:1593–99.

Tamarin, R. H., and T. H. Kunz. 1974. *Microtus breweri*. Mammalian Species 45:1–3.

Tamarin, R. H., G. H. Adler, M. Sheridan, and K. Zwicker. 1987. Similarity of spring population densities of the island beach vole (*Microtus breweri*), 1972–1986. Canadian Journal of Zoology 65:2039–41.

Tast, J. 1974. The food and feeding habits of the root vole, *Microtus oeconomus,* in Finnish Lapland. Aquilo Serie Zoologica 15:25–32.

Timm, R. M. 1985. Parasites. Pages 455–534 *in* R. H. Tamarin, ed. Biology of New World *Microtus*) (Special Publication No. 8). American Society of Mammalogists.

Timm, R. M., L. M. Heaney, and D. D. Baird. 1977. Natural history of rock voles *(Microtus chrotorrhinus*) in Minnesota. Canadian Field-Naturalist 91:177–81.

Turner, B. N., and S. L. Iverson. 1973. The annual cycle of aggression in male *Microtus pennsylvanicus,* and its relation to population parameters. Ecology 54:967–81.

Van Horne, B. 1982. Demography of the longtailed *Microtus longicaudus* in seral stages of coastal coniferous forest, southeast Alaska. Canadian Journal of Zoology 60:1690–1709.

Verster, A. 1969. A taxonomic revision of the genus *Taenia* Linnaeus 1758. Journal of Veterinary Research 36:3–58.

Vorontsov, N. N. 1962. The ways of food specialization and evolution of the alimentary tract. Pages 360–77 *in* J. Kratochvbil, and J. Pelikban, eds. Symposium theriologicum. Czechoslovak Academy of Science, Prague.

Vorontsov, N. N. 1967. Evolution of the alimentary system of myomorph rodents. Nauka, Novosibirsk, Russia [In Russian, with English summary and title].

Wager, P. S., and J. R. Mason. 1996. Ortho-aminoacetophenone, a non-lethal repellent: The effect of volatile cues vs. direct contact on avoidance behavior by rodents and birds. Pesticide Science 46:55–60.

Waian, L. B., and R. Stendell. 1970. The white-tailed kite in California with observation of the Santa Barbara population. California Fish and Game 56:188–98.

Wang, Z., C. F. Ferris, and G. J. DeVries. 1994. Role of septal vasopressin innervation in paternal behavior in prairie voles (*Microtus ochrogaster*). Proceedings of the National Academy of Science of the USA 91:400–404.

Wang, Z., G. Yu, C. Cascio, Y. Liu, B. Gingrich, and T. R. Insel. 1999. Dopamine D2 receptor-mediated regulation of partner preferences in female prairie voles (*Microtus ochrogaster*): A mechanism for pair bonding? Behavioral Neurosciences 113:602–11.

Watkins, R. A., S. E. Moshier, W. D. O'Dell, and A. J. Pinter. 1991. Splenomegaly and reticulocytosis caused by *Babesia microti* infections in natural populations of the montane vole, *Microtus montanus*. Journal of Protozoology 38:573–76.

Whitaker, J. O., Jr., and R. L. Martin. 1977. Food habits of *Microtus chrotorrhinus* from New Hampshire, New York, Labrador, and Quebec. Journal of Mammalogy 58:99–100.

Whitney, P. 1976. Population ecology of two sympatric species of subarctic microtine rodents. Ecological Monographs 46:85–104.

Wilhelm, D. E. 1977. Zoogeographic and evolutionary relationships of selected populations of *Microtus mexicanus*. Ph.D. Dissertation, Texas Tech University, Lubbock.

Wilson, D. E., and S. Ruff, eds. 1999. The Smithsonian book of North American mammals. Smithsonian Institution Press, Washington, DC.

Wiseman, V. S. 1967. A morphological study of the baculum in some microtine rodents. M.S. Thesis, University of Puget Sound, Tacoma, WA.

Wolff, J. O. 1980. Social organization of the taiga vole (*Microtus xanthognathus*). Biologist 62:34–45.

Wolff, J. O. 1985. Behavior. Pages 340–72 *in* R. H. Tamarin, ed. Biology of New World *Microtus* (Special Publication No. 8). American Society of Mammalogists.

Wolff, J. O., and W. Z. Lidicker. 1980. Population ecology of the taiga vole, *Microtus xanthognathus,* in interior Alaska. Canadian Journal of Zoology 48:1800–1812.

Wunder, B. A. 1978. Implications of a conceptual model for the allocation of energy resources by small mammals. Pages 68–75 *in* D. P. Snyder, ed. Populations of small mammals under natural conditions. Pymatuning Laboratory of Ecology, University of Pittsburgh, Pittsburgh.

Wunder, B. A. 1985. Energetics and thermoregulation. Pages 812–44 *in* R. H. Tamarin, ed. Biology of New World *Microtus* (Special Publication No. 8). American Society of Mammalogists.

Young, L. J., R. Nilsen, K. G. Waymire, G. R. MacGregor, and T. R. Insel. 1999. Increased affiliative response to vasopressin in mice expressing the V1a receptor from a monogamous vole. Nature 400:766–68.

Youngman, P. M. 1975. Mammals of the Yukon Territory. National Museum of Natural Science, Ottawa.

Zajic, A. M., and J. F. Williams. 1980. Infection with *Schistosomatium douthitti* (Schistosomitidae) in the meadow vole (*Microtus pennsylvanicus*) in Michigan. Journal of Parasitology 66:366–77.

Zajic, A. M., and J. F. Williams. 1981. The pathology of infection with *Schistosomatium douthitti* in the laboratory mouse and the meadow vole, *Microtus pennsylvanicus*. Journal of Comparative Pathology 91:1–10.

Zakrzewski, R. J. 1985. The fossil record. Pages 1–51 *in* R. H. Tamarin, ed. Biology of New World *Microtus* (Special Publication No. 8). American Society of Mammalogists.

Ziegler, A. C. 1961. Occurrence of os clitoridis in *Microtus*. Journal of Mammalogy 42:101–3.

Zimmerling, T. N., and L. M. Zimmerling. 1996. A comparison of the effectiveness of predator odor and plant antifeedant in deterring small mammal feeding damage on lodgepole pine seedlings. Journal of Chemical Ecology 22:2123–32.

Zimmerling, T. N., and L. M. Zimmerling. 1998. Effectiveness of a physical barrier in deterring vole and snowshoe hare feeding damage to lodgepole pine seedlings. Western Journal of Applied Forestry 13:12–14.

Zimmerman, E. G. 1965. A comparison of habitat and food of two species of *Microtus*. Journal of Mammalogy 46:605–12.

Stephen R. Pugh, University of New Hampshire, Manchester, New Hampshire 03101. Email: spugh@cisunix.unh.edu.

Sherry Johnson, affiliated with the Puqet Sound Muscum of Natural History, University of Puqet Sound, Taroma, WA.

Robert H. Tamarin, College of Arts and Sciences, University of Massachusetts, Lowell, Massachusetts 01854-2882. Email: robert_tamarin@uml.edu.

18

Porcupine

Erethizon dorsatum

Uldis Roze
Linda M. Ilse

NOMENCLATURE

Common Name. Porcupine
Scientific Name. *Erethizon dorsatum*
Subspecies *E. d. dorsatum* (Canada, east of Yukon Territory and British Columbia except Labrador and Newfoundland, and northeastern and north-central United States), *E. d. myops* (Alaska and Yukon Territory), *E. d. nigrescens* (British Columbia, northern Washington), *E. d. bruneri* (Great Plains east of Rocky Mountains), *E. d. epixanthum* (central Washington, California, Idaho, Nevada, Utah, western Colorado, Wyoming, western Montana), *E. d. couesi* (western Texas, New Mexico, central Arizona), and *E. d. picinum* (Labrador) (Woods 1999)

The porcupine is a large, atypical member of the order Rodentia, suborder Hystricognatha, and one of four genera of the New World family Erethizontidae. The stump-tailed porcupine (*Echinoproctus rufescens*; Amazon); prehensile-tailed porcupines (*Coendou* spp.), and hairy dwarf porcupines (*Sphiggurus* spp.; Mexico and South America) constitute the remaining extant genera of erethizontids (Roze 1989). However, Nowak (1991) describes *Sphiggurus* as a subgenus of *Coendou* because of limited differentiation between these genera. *Erethizon* is derived from *Coendou* and shares many anatomical and behavioral characteristics despite differences in ecological strategies and distribution (Dodge 1982; Roze 1989).

PHYLOGENY

Phylogenetic relationships among porcupines, particularly Old and New World genera, have been controversial, in part due to incomplete fossil records. Rodents are believed to have first appeared in the Paleocene and radiated during the Tertiary. The suborder Hystricognatha appeared in North America during the later Tertiary, and extant species are found in North America, Europe, the Near East, Africa, Sri Lanka, Java, the Philippines, China, and India (Nowak 1991).

The relationship between Hystricidae (Old World porcupines) and Erethizontidae (New World porcupines) was debated by researchers for many years. Wood (1950) asserted the relationship between these families exemplified convergence or parallel evolution. Landry (1957) provided the earliest challenge to Woods, contending that relatedness was based on similarity of a variety of morphological features and was not the result of convergence. The consensus favoring parallelism was seriously challenged in 1985 at the NATO conference on evolutionary relationships among rodents (Luckett and Hartenberger 1985). Significant evidence was presented at that time supporting Landry. Reproductive and chromosomal data (George 1985), dental homologies and placental characteristics (Luckett 1985), and musculoskeletal features (Woods and Hermanson 1985) indicate a common evolutionary origin of the hystricomorphs. Landry (1999) provided further evidence of a monophyletic relationship based on morphology, parasitology, nature of the quills, and preliminary molecular data.

DISTRIBUTION

Distribution of the North American porcupine is widespread and includes most of the western, north-northwestern, and northeastern United States, northeastern Mexico, and most of Canada (Fig. 18.1). Despite lack of extant populations in plains, coastal regions, and southeastern coastal plains of the United States, fossils indicate that *Erethizon* formerly occurred in Alabama (Barkalow 1971), Florida (White 1970), Missouri (Parmalee 1971), plains of Nebraska (Voorhies 1981), and West Virginia (Guilday and Hamilton 1973). However, northward expansion of the porcupine has been reported in Quebec at tree line along Hudson Bay (Payette 1987) and in the Thelon River Valley of the Northwest Territories (Norment et al. 1999). Similarly, southward expansions into southern and eastern portions of Texas have been reported by Davis and Schmidly (1994) and Ilse and Hellgren (2001). However, the species is classified as endangered in northern Mexico (List et al. 1999).

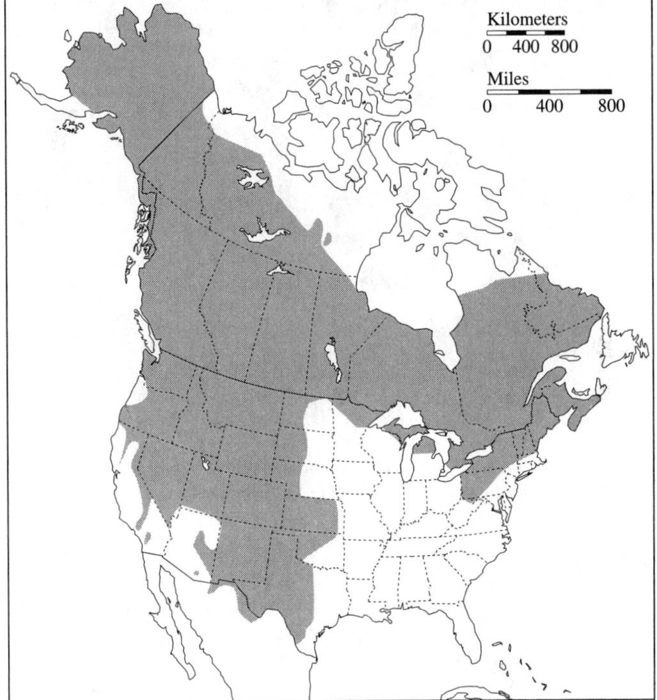

Figure 18.1. Distribution of the North American porcupine (*Erethizon dorsatum*).

DESCRIPTION

The porcupine characterizes North America as strongly as do the bison (*Bison bison*) and the wild turkey (*Meleagris gallopavo*). Its body outline, often irregular, is defined by its quill complement. When hunched up in a tree, the animal may look like a dark knot on a branch. When threatened, it presents a bristling back of white quill bases. Quills cover

FIGURE 18.2. Porcupine (*Erethizon dorsatum*) climbing a tree, showing the typical pattern of quills. SOURCE: Photo by Uldis Roze.

all parts of the body except the face, the belly, the inner limbs, and the underside of the tail (Fig. 18.2).

Limbs are relatively short and locomotion is plantigrade, which render the animal a slow runner but an accomplished tree climber. External adaptations for tree climbing include naked and textured foot soles, long claws (four on the forefeet, five on hind feet), concealed genitals in both sexes, and a muscular tail with stiff bristles on the underside.

The claws of the front feet are adapted for grasping and holding objects such as apples or small branches by flexion of the claws against the footpads. Footpad size is proportional to body weight and is typically larger in males than females (Sweitzer and Berger 1997).

The porcupine typically has four nipples (although two, three, and five have been observed). With time, the process of lactation stains fur posterior to each nipple a contrasting brown color.

Weight. Adult body weights vary by habitat (Table 18.1). Peak body weights are reached in the fall. Porcupines lose 17–31% of their weight in winter (Roze 1985; Sweitzer and Berger 1993). Weights also decline with advancing age (Earle and Kramm 1980).

Pelage and Quills. The body is covered by a variety of modified hairs, including vibrissae for sensory information; guard hairs with sensory, insulation, and rain-shedding functions; underfur for insulation; tail bristles for climbing support; and quills for defense and aposematic (warning) signaling. Only the underfur is shed in a regular spring molt, which coincides with the switch from low-protein winter foods to high-protein spring foods (Roze 1989). Porcupines through most of their range are dark colored with white highlights, but the western subspecies (*E. d. epixanthum*) appears yellowish because of its yellow-gold guard hairs.

The porcupine is the only North American mammal with body hairs modified as quills. Adult quills typically range between 2 and 10 cm in length. They consist of a conical follicle, a thin neck, a broad shoulder leading to a stout, sponge-filled midsection, and a tapering, barb-covered tip.

Quills are driven into would-be predators either by passive collision with erected quills or an active tail-slap. Separation of quills from the porcupine is enhanced by a facilitated quill–release mechanism (Roze 2002). The tension required to extract a quill from a porcupine is reduced by 35–40% if the quill is first driven back into the body of the porcupine, as would follow a tail-slap or other violent contact with an antagonist. A spool of connective tissue in the dermis of the porcupine arrests downward movement of the quill at its shoulder, thus preventing back-stabbing (Chapman and Roze 1997). Structure and function of quills differ depending on their location on the body (see Behavior).

Once the quill tip penetrates the flesh of the antagonist, it may be pulled deeper by ratchet action of the one-way barbs (Fig. 18.3) against muscle. In extreme cases, a quill or quill fragment may disappear completely below the skin surface, then travel through the body until it either emerges at a point distant from its origin or is stopped against bone (Shadle and Po-Chedley 1949; Maser and Rohweder 1983; Roze 1989).

Traveling quills seldom cause infections, perhaps because they are coated with an antibiotic layer of free fatty acids (Roze et al. 1990). Major quill fatty acids are C-16 to C-18 saturated and monounsaturated fatty acids, all of which show antibiotic activity against gram-positive microorganisms. The antibiotic coating benefits the porcupine, which frequently suffers from self-quilling as a result of tree-climbing accidents. Accidents are fairly common; healed fractures were evident in 35.1% of 37 museum specimens examined (Roze et al. 1990).

Skull and Dentition. As a hystricomorph rodent, the porcupine has a large infraorbital foramen, through which passes part of the masseter muscle. The dental formula is I 1/1, C 0/0, P 1/1, M 3/3 (Fig. 18.4). The ever-growing incisors are stained orange on the anterior surface by iron salts, which function as hardening agents. More than 50 museum specimens and more than 500 live captures have presented no instance of malocclusion of the incisors, as is relatively common in woodchucks (*Marmota monax*) (U. Roze, pers. obs.).

At birth, the animal has four incisors and four fully erupted deciduous premolars. During the first 4 months, the jaw elongates to

TABLE 18.1. Representative mean body weights of male and female porcupines in various habitats

Source	Habitat	Weight (kg)	
		Males	Females
Sweitzer and Berger 1993	Nevada desert	11.8 (45)	8.7 (47)
Roze 1989	New York mixed deciduous forest	6.1 (15)	5.1 (29)
Dodge 1967	Massachusetts mixed deciduous forest	6.0 (20)	5.2 (21)
Ilse and Hellgren 2001	Texas pinyon–juniper	10.1 (5)	6.5 (17)
Fournier and Thomas 1999	Quebec mixed deciduous forest	6.6 (5)	5.6 (5)
Craig and Keller 1986	Idaho desert	11.5 (5)	

NOTE: Sample sizes are in parentheses.

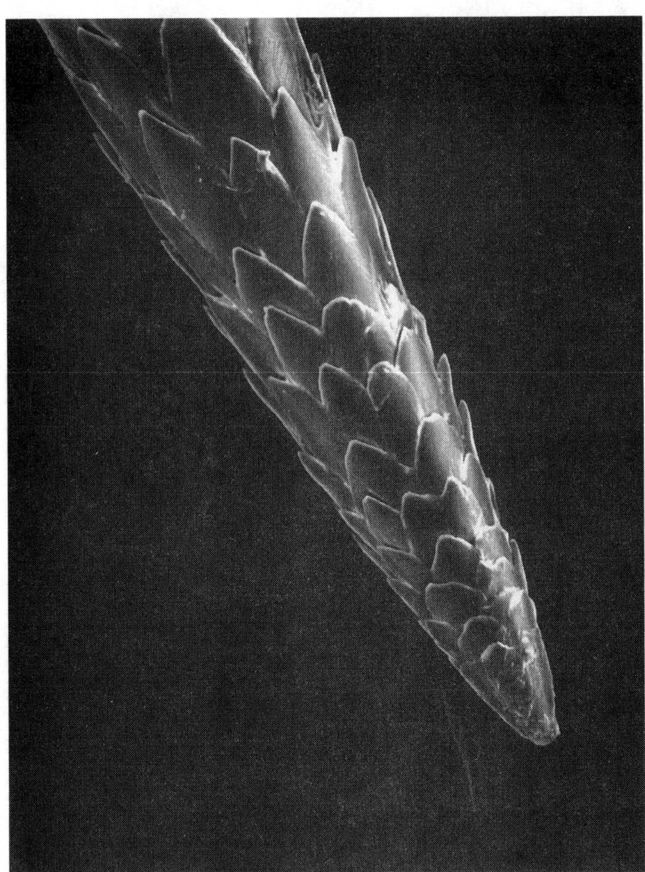

FIGURE 18.3. Photomicrograph of the tip of a porcupine quill. SOURCE: Photo by Uldis Roze.

accommodate the first and then the second molars. Third molars erupt at the end of the first year. At age 2 years, the deciduous premolars are replaced by permanent premolars. Because the latter are the largest teeth in the jaw, they are easily identified and indicate a minimum age of 2 years (Dodge 1967). Beyond age 2 years, accurate ages can be obtained by tooth sectioning and staining for cementum annuli (Earle and Kramm 1980) or by knowledge of past history of tagged individuals.

Digestive Tract. A porcupine's mouth has two chambers. The incisors are located in the outer chamber (vestibule). The lips meet across the diastema, posterior to the incisors, conserving heat and allowing the animal to rasp away unwanted outer bark of trees without opening its mouth. The cheek teeth are located in the inner chamber. Their occlusal grinding surfaces work in an anterior-to-posterior direction to macerate ingested food. This food consists largely of the leaves and inner bark of trees, a diet rich in fiber and secondary compounds (tannins, terpenes, phenols, flavonoids, etc.).

Digestive function is enhanced by cellulolytic bacteria in the digestive tract (Balows and Jennison 1949) and a large fermenting cecum (Johnson and McBee 1967). The cecum makes up 34% by weight of the digestive tract and contributes significantly to the animal's energy requirements (Johnson and McBee 1967; Vispo and Hume 1995).

The porcupine's cecum also functions as a sodium reservoir, comparable to the reticulorumen of moose (*Alces alces*) (Belovsky and Jordan 1981; Roze 1989; Vispo and Hume 1995). There is evidence of sodium resorption in the colon. Sodium concentrations decrease threefold between proximal and distal colon, and relatively little sodium is lost in the feces under normal conditions. Along with sodium, water is strongly resorbed in the colon, and distal colon pellets contain 40% dry matter (Vispo and Hume 1995).

Scent Glands. The porcupine has two major skin scent glands: the rosette skin, which generates a warning odor, and paired perineal glands

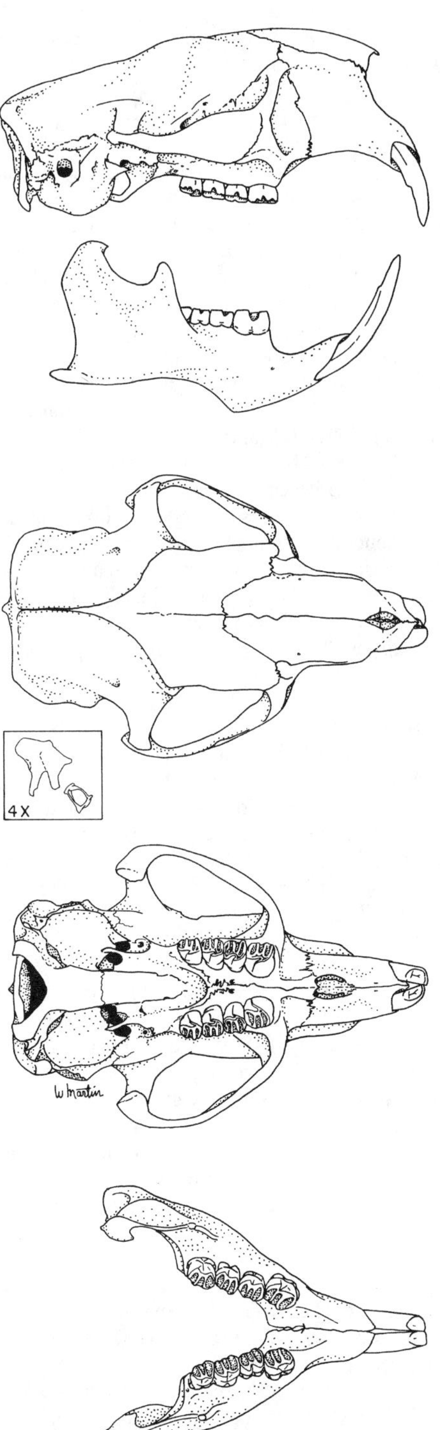

FIGURE 18.4. Skull of the porcupine (*Erethizon dorsatum*). From top to bottom: lateral view of cranium, lateral view of mandible, dorsal view of cranium, ventral view of cranium, dorsal view of mandible. Inset: middle ear ossicles.

in the anogenital region. The perineal glands, present in males and females, consist of glandular skin with shallow, paired, sexually dimorphic pockets. Pocket openings are elongated anteroposteriorly in females, laterally in males. The pockets house a seasonal bacterial flora and contain emergent hairs with osmetrichial specializations (D. M. Chapman and U. Roze, unpublished obs.). As in other hystricomorphs (Beruter et al. 1974), secretions of the perineal glands are disseminated by anal dragging. In the North American porcupine, as in *Coendou*, anal dragging is practiced primarily by males (Shadle et al. 1946;

Roberts et al. 1985). The social and ecological context of this behavior in *Erethizon* has not been determined.

REPRODUCTION

Male Reproductive System. The penis is normally retracted within a penile sheath beneath the perineum, where it lies flexed with the urethra pointed backward (Mirand and Shadle 1953). Erection is accomplished by a combination of rapid muscular contraction, which everts the penis, followed by engorgement of the *corpora cavernosa*. The penis can be readily everted manually in an anesthetized porcupine. Without manual eversion, the sexes are not easily differentiated.

The glans penis is covered by spines approximately 0.3 mm thick and 1 mm long. In addition to the urethral opening, the tip of the penis contains a pocket, the *sacculus urethralis*, which houses larger spines everted during copulation (Hooper 1961).

Testes are abdominal in juvenile males and descend around 2 years of age. Testes are also descended in newborns, but retract in 8 days (Mirand and Shadle 1953). In some sexually dominant males, testes may remain descended throughout the year (U. Roze, pers. obs.). More typically, testes descend in midsummer and remain descended through the breeding season. They become abdominal again when, presumably, the testosterone levels decline. Accessory organs include a seminal vesicle, an exceptionally large prostate, and Cowper's glands (Dodge 1982).

Female Reproductive System. Females typically begin breeding in their second autumn of life, though in exceptional cases, breeding may take place during the first autumn (Roze 1989; Sweitzer and Holcombe 1993). The vagina is normally closed by a membrane that forms part of the perineal skin. The vaginal closure membrane dissolves at estrus and before parturition. Dodge (1982) stated that ovulation in the porcupine is unilateral from the right ovary, with the embryo implanting in the right horn of the uterus. The left side of the abdominal cavity accommodates the large cecum.

Breeding and Pregnancy. In the Catskill Mountains of New York, 22 of 33 observed breeding encounters took place between 15 September and 15 October (Roze 1989). In the Great Basin of Nevada, for 20 breeding events observed, the breeding period lasted from 30 September to 15 October in 1990 and from 13 October to 2 November in 1991 (Sweitzer and Holcombe 1993).

Onset of the breeding season in participating males is marked by an expansion of their seasonal home range (Roze 1989). When a male encounters a receptive female at this time, a guarding episode begins. The female sits in a tree branch, the male typically in a lower branch in the same tree. At this time, the male may intermittently spray high-velocity urine drops at the female (U. Roze, pers. obs.).

During the guarding episode, which may last for several days, other wandering males may discover the receptive female and challenge the resident male. Vicious battles between males may result, with use of quills and teeth, and combatants may sustain serious injuries (Roze 1989; Sweitzer and Berger 1997). The highest incidence of injuries is observed in older males with high body mass (Sweitzer and Berger 1997). Under captive conditions, where flight may not be possible, fatalities have been observed (Dodge 1982).

Male reproductive success, measured by number of guarding episodes, is highly variable and directly correlates with nonwinter home range and body mass (Roze 1989; Sweitzer and Berger 1997; Sweitzer 2003). In contrast, adult female reproductive success is high and shows little variability. Roze (1989) found 89% of adult females pregnant ($n = 18$); Sweitzer and Holcombe 1993 found an adult pregnancy rate of 92% ($n = 50$).

As the female approaches estrus and the vaginal closure membrane dissolves, the male urine salvos increase (Shadle et al. 1946; Dathe 1963). Although the physiological reason for urine sprays has not been established, male urine spraying is the norm in hystricomorph rodents (Macdonald 1985). It may serve the same function as in mice, where male urine accelerates estrus in the female (Marsden and Bronson 1964; Bronson and Whitten 1968).

Copulation usually occurs on the ground, with the female pressing her quills against the body, elevating her hindquarters, and arching her tail over her back, providing the male with a safe surface for chest and forepaws. After copulation, a vaginal plug forms and the female becomes nonreceptive.

If copulation fails to establish pregnancy, the female may recycle on a 28-day schedule, possibly extending the breeding period into November or December (Burge 1966; Dodge 1982; Roze 1989). The 210-day gestation period is supported by accessory corpora lutea and is atypical for most rodent species (Shadle 1948; Mossman and Judeas 1949). Pregnancy is characterized by a rise in plasma progesterone from a nonpregnancy level of 1.1 ng/ml to a mean of 13.6 ± 6.5 ng/ml (Sweizer and Holcombe 1993). Estrogen and progesterone levels also rise in feces of pregnant individuals, offering a potentially noninvasive test of pregnancy. However, preliminary results failed to correctly characterize all pregnancies (Bogdan and Monfort 2001).

In New York, young are typically born in late April or early May. In the wild, several thousand observations have failed to document any but single births (Shadle 1950). The neonate is delivered wrapped in its amnion, which the mother licks off and consumes. The newborn weighs about 450 g, is fully quilled and open-eyed, and defends itself with vigorous tail flicks. For the first 2 weeks, it is completely dependent on its mother for nutrition. It spends daylight hours hidden in some crevice near the mother's resting tree; the two reunite at night (Roze 1989). Though the juvenile begins feeding on green food at about 2 weeks of age, it continues nursing for 127 days in New York (Roze 1989) and 124–128 days in Nevada (Sweitzer and Holcombe 1993). As a result, females may spend 11 months a year in reproductive processes. The consequent drain on maternal resources may contribute to the male–female dimorphism in weight.

ECOLOGY

Habitat. Though the bulk of their species range coincides with coniferous forest, porcupines inhabit a variety of other habitats including mixed forests (Dodge 1967; Roze 1989; Griesemer et al. 1998), desert shrub (Reynolds 1957), and juniper (*Juniperus* sp.) and sagebrush (*Artemisia* sp.) (Sweitzer 1990). Habitat use generally reflects seasonal and vegetation changes, and also may be dictated by activities including foraging, predator avoidance, and resting or sleeping. Although the porcupine is primarily arboreal (Roze 1989; Griesemer et al. 1998), seasonal ground dwelling as a response to feeding opportunities has been reported (Taylor 1935; Curtis 1941; Dodge 1967). Sweitzer and Berger (1992) reported that feeding forays occurred in open areas where predation risk was heightened, whereas sleeping animals were observed only in more sheltered grove areas. Similarly, younger and smaller animals used more protected areas while foraging than did older or larger animals. In the montane forests of southwestern Alberta, Harder (1980) concluded that habitat use was influenced by density of porcupines, availability and age of tree species, and prevailing temporal and climatic conditions.

Longevity. In upper Michigan, Earle and Kramm (1980) used 74 skulls and carcasses from a single population to study population age structure. Teeth were sectioned to estimate age of individuals. The median age of the population was 5 years; the oldest animal was 18 years of age. This is not the upper limit for longevity of porcupines in the wild. One radiotelemetered porcupine died at age 21years and weighed 7.1 kg at death. A female captured and released in 1988 may have been 25–30 years old, based on comparison with the 21-year old individual (U. Roze, unpublished obs.).

Population Density. Population densities are difficult to estimate and highly variable across the geographic ranges of the porcupine. Density estimates may lack accuracy because of methods used and the reclusivity of the porcupine (Hale and Fuller 1999) and may be affected by predators (Earle and Kramm 1982; Powell 1993). Densities range from 1.9–3.3 animals/km^2 in pinyon (*Pinus remota*)–juniper woodlands of Texas using visual observations, radiotelemetry, and known home ranges (Ilse and Hellgren 2001), to 10–42 porcupines/km^2 in central Massachusetts using radiotelemetry and mark–resight techniques (Hale and Fuller 1999). Brander (1973) reported densities of 9.3–16.6 animals/km^2 in mixed hardwood–hemlock (*Tsuga canadensis*) forests in Michigan during winter. Roze (1984) counted feeding trails in snow to derive an estimate of 10.7 animals/km^2 in the Catskill Mountains of New York. Shapiro (1949), using a line strip census in the Adirondack Mountains of New York, reported an estimate of 4.8 animals/km^2. Hendricks and Allard (1988) used sighting surveys to estimate a density of 6 animals/km^2 in prairie habitat in Montana. In a mixed pine (*P. ponderosa*)–fir (*Abies grandis*) forest in Oregon, census data and estimated home ranges were used to estimate a density of 12.6 animals/km^2 (Smith 1977). Curtis (1944) described a density of 7–9 animals/km^2 in mixed hemlock–hardwood forests of Maine using census data. In a long-term study in upper Michigan, Powell (1993) reported a decline in porcupines from 11.8 to 0.4 animals/km^2 following the establishment of fishers (*Martes pennanti*). Similarly, Earle and Kramm (1982) estimated 3.5 animals/km^2 in areas without fishers and 0.4 animals/km^2 in areas with established populations of fishers.

FEEDING HABITS

Despite its widespread geographic distribution, few studies have thoroughly explored the feeding habits of the porcupine. Strict herbivores, porcupines generally feed on inner bark of trees and coniferous foliage during winter and increase consumption of deciduous leaves, forbs, and herbs during spring and summer. Roze (1989) provided the most thorough analysis of feeding habits throughout the year and reported dramatic seasonal shifts in diet influenced by changing plant chemistry and nutrient availability. Generally, the winter diet is low in nitrogen and high in fiber. It is not uncommon for the porcupine to lose body weight during this time of nutritional stress (Roze 1989; Sweitzer and Berger 1993). High-nitrogen, low-fiber foods may become more available in some buds and young leaves during spring, and animals begin to gain body weight lost during winter (Roze 1989). During summer, porcupines begin to consume relatively undefended leaves of linden (*Tilia americana*) and aspen (*Populus* spp.), avoiding sugar maple (*Acer saccharum*), which has higher tannin levels (Roze 1989). Fall diets are characterized by the inclusion of high-calorie, low-nitrogen items such as acorns (Roze 1989). Although these seasonal characterizations are most applicable to porcupine activity in temperate mixed deciduous forests, they provide a possible framework for examination of porcupine feeding habits in other regions. However, seasonal shifts may not be as dramatic in regions with more moderate temperature regimes (L. M. Ilse, pers. obs.).

Regional use of feeding trees varies with forest composition. Eastern hemlock (*Tsuga canadensis*) is a preferred winter feeding tree in the northeastern and Great Lakes regions (Dodge 1967; Brander 1973; Griesemer et al. 1998), whereas western hemlock (*T. heterophylla*), Douglas fir (*Pseudotsuga menziesii*), and ponderosa pine (*Pinus ponderosa*) constitute preferred feeding trees in the Pacific northwest (Dodge and Canutt 1969; Dodge 1970). Lodgepole pine (*P. contorta*), western white pine (*P. monticola*), eastern white pine (*P. strobus*), white cedar (*Thuja occidentalis*), limber pine (*P. flexilis*), and pinyon pines (*P. edulis, P. monophylla, P. remota*) also are commonly used conifers (Taylor 1935; Speer and Dilworth 1978; Harder 1979; Dodge 1982; Ilse and Hellgren 2001). White spruce (*Picea glauca*) is used in northern Quebec (Payette 1987) and Alaska (Murie 1926).

Porcupines are not limited to use of coniferous species for winter feeding. Deciduous species, including sugar maple, striped maple (*A. pennsylvanicum*), beech (*Fagus grandifolia*), birches (*Betula* spp.), aspens, oaks (*Quercus* spp.), and linden (*Tilia americana*), also are favored for feeding (Dodge 1967; Roze 1984; Griesemer et al. 1998). Orchard trees including apple (*Malus* spp.) and cherry (*Prunus* spp.) also are favored when available (Dodge 1982). Stricklan et al. (1995) concluded that Gambel oak (*Quercus gambelii*) was selected over coniferous species in Montana during winter as a result of its higher protein content.

In addition to trees, porcupines feed on a variety of woody shrubs including elderberry (*Sambucus* spp; Dodge 1970), spiny buffaloberry (*Shepherdia argentea*; Taylor 1935), and evergreen sumac (*Rhus virens*; L. M. Ilse, pers. obs.). In Arizona, they have been observed feeding on dwarf mistletoe (*Arceuthobium* spp.; Taylor 1935).

Less attention has been given to summer and spring diets of porcupines. In some regions, porcupines increase ground feeding during spring and summer months, when succulent, nutritious vegetation is abundant (Craig and Keller 1986; Sweitzer and Berger 1992). Taylor (1935) concluded that ground vegetation was a principal component of the diet during these periods in Arizona, constituting 85% of stomach contents during summer and 40% in spring. Identified items included *Taraxacum, Solidago, Trifolium, Carex, and Rubus*. Only 10% of the feeding observations by Griesemer et al. (1998) occurred on the ground, but these included consumption of grasses (*Graminae*), violets (*Viola* spp.), milkweed (*Asclepias* spp.), and grapes (*Vitis* spp.). Although less conspicuous in the diet, arrowhead (*Sagittaria* spp.), pondweeds (*Potamogeton* spp.), water lilies (*Nymphaea* spp.), and aquatic liverwort (*Riccia* spp.) are favored by animals in riparian and wetland habitats (Taylor 1935; Dodge 1967, 1982). The consumption of the latter aquatic species is evidence of the female-biased springtime sodium drive described by Roze (1989). Porcupines are not averse to swimming and may feed in shallow water (Dodge 1982) or swim to retrieve food items which are then brought to shore for consumption (Dean 1950). In arid regions, they may feed on ocotillo (*Fouqueria splendens*; Reynolds 1957) or cacti (*Opuntia* spp.; L. M. Ilse 2001). Additionally, feeding on inner bark, a habit most often associated with the winter diet, occurs throughout the year in the South and Southwest (Taylor 1935; Ilse and Hellgren 2001). The catholic diet of the porcupine combined with its ability to extract energy and protein from nutrient-poor forage (Felicetti et al. 2000) contributes to its widespread geographic distribution.

Digestive Physiology. The porcupine subsists on a diet typically high in fiber, low in nitrogen, and containing significant amounts of tannins and terpenoids. Yet the species is able to meet its energy and nitrogen needs on this unpromising diet while extending its geographic range northward to near the subarctic tree line (Payette 1987). Studies of porcupine digestive physiology indicate some of the reasons for this success.

Porcupines digest 62–96% of dry matter for natural foods (Felicetti et al. 2000) and 81–98% for pelleted chow (Fournier and Thomas 1997). They digest highly lignified fiber better than many large hindgut fermenters and ruminants. In this, the relatively large, fermenting cecum contributes 16% of the basal metabolic energy requirements in the form of low molecular weight straight-chain and branched-chain fatty acids (Johnson and McBee 1967). In addition, porcupines have a distal colon four times larger than that of similarly sized beavers (*Castor canadensis*) (Vispo and Hume 1995). Moreover, porcupines have a long retention time for particles in their digestive tract: 38.4 hr versus 21 hr expected in a typical 11-kg hindgut fermenter (Felicetti et al. 2000). The extended retention time allows more time for microbial and chemical breakdown of plant fiber.

Porcupines also have a highly efficient nitrogen metabolism. On a nontannin diet, they achieve nitrogen balance at a relatively low level of nitrogen intake, about 346–389 mg N/$W^{0.75}$/day, where W is weight in kg (Fournier and Thomas 1997; Felicetti et al. 2000). They lose nitrogen by two pathways: as urinary endogenous nitrogen, estimated as 205–223 mg N/$W^{0.75}$/day, and by the more important pathway of fecal nitrogen loss (0.92–2.8 mg N/g dry matter intake). Although the rate of urinary nitrogen loss is significantly higher than the mean of 160 mg N/$W^{0.75}$/day for eutherians, this is offset by the extremely low rate of fecal nitrogen loss, a quantitatively more important pathway. At 0.92 mg N/g dry matter intake, this is only 11–29% of the value reported for other rodents (Fournier and Thomas 1997).

Dietary tannins reduce the porcupine's ability to digest protein. However, Felicetti et al. (2000) found that tannin assays via bovine serum albumin precipitation failed to predict the degree of reduction of protein digestibility. They postulate that the unpredictability of this assay may be caused by highly specific salivary binding proteins in the porcupine or by its ability to break down specific tannin-protein complexes.

BEHAVIOR

Tree Climbing. Tree climbing typically involves use of the claws as anchors in bark crevices. The ventral surface of the tail is pressed against the tree trunk to check backsliding. When tree trunks or branches become too thin to climb with claws, the animal can climb by using its footpads alone. Powerful clenching thigh muscles allow a porcupine to clasp small trunks or branches by sole friction, leaving the front limbs free for food handling or exploration (Woods 1972; McEvoy 1982).

Quill Function. Though its quill armament gives a porcupine excellent protection against most predators, it is in the porcupine's interest to avoid an attack in the first place. The porcupine warns potential predators in three sensory modes. Tooth clacking provides an auditory warning. The quills deliver both visual and olfactory warnings.

The visual warning (aposematic coloration) is effective at night, when the porcupine and its predators are active. It consists of regions of black-and-white contrast on the porcupine's body, generated by black and white sections of the quills. A black-bordered white chevron travels down the lower back and tail. A white tuft (more pronounced in males than females) highlights the top of the head. Baby porcupines, though fully quilled, lack the white highlights and appear dark. The white portions of porcupine quills fluoresce under long-wave ultraviolet excitation. The fluorescence is due to brighteners that enhance the quill whiteness. The chemical nature of the brighteners is under study (U. Roze and D. C. Locke, unpublished obs.).

The quills also disseminate an olfactory warning (olfactory aposematism). Aroused porcupines erect their quills and release a pungent, porcupine-specific warning odor. The active ingredient is R-delta-decalactone, an optically active 10-carbon molecule (Li et al. 1997). The unique odor is generated by skin of the lower back (the rosette). This region of the porcupine has good quill cover but sparse underfur, allowing quills to come in contact with naked skin. The skin is richly supplied with sebaceous glands, the source of the warning odor. On quill erection, the odor-coated quills greatly increase the surface area of expressed odorant and thus help disseminate the odor. The rosette quills therefore function as classic osmetrichia. In contrast with quills from the thoracic region, the rosette quills are barbed along a greater percentage of their length, the barbs show greater overlap, and the quills carry greater lipid loads.

Quills serve a number of additional functions in the porcupine. Their watertight, sponge-filled interiors aid in flotation, enhancing the porcupine's swimming capabilities. Furthermore, quills, in combination with guard hairs and fur, provide the porcupine with a continuously variable and efficient insulation layer. The pelage of the porcupine provides highly effective insulation (DeMatteo and Harlow 1997; Fournier and Thomas 1999). Although the basis of this thermoconservation may involve physiological as well as anatomical factors, a part of the cold response involves pressing the quills against the body while elevating fur and guard hairs.

Salt Drive. As noted, in sodium-depleted environments, porcupines exhibit a strong seasonal salt drive (Roze 1989). Salt seeking often leads the animals to anthropogenic salt sources, including residual road salt. The salt drive peaks in May and June, with a minor peak in late summer. It essentially ends during winter, when the animals are less mobile while denned and are dependent on endogenous sodium stores from the cecum.

Females predominate at salt sources because salt craving is stimulated by the hormones of pregnancy and lactation (Denton 1982). The specific ions sought are sodium, that is, sodium salts are sought without regard for the anion (Roze 1989). Materials such as plywood and rubber are attacked because they contain sodium ions derived from the manufacturing process. In seeking sodium, porcupines avoid potassium ions and try to maximize the sodium/potassium ratio of ingested material. In the absence of an anthropogenic supply, porcupines exploit natural sodium sources such as aquatic vegetation and carrion.

Activity Cycle. Brander (1973) reported that Michigan porcupines were strongly diurnal, apparently during summer months. Dodge and Barnes (1975) reported porcupine activity in Washington as primarily nocturnal, with most movement occurring between 2100 and 0700 hr. These observations are based on six 24-hr records from five individuals. Roze and Marcus (unpublished obs.) found porcupines in New York were primarily nocturnal, as indicated by 900 hr of monitoring three radiotelemetered porcupines equipped with activity sensors. In January and March, activity cycles showed a bimodal nocturnal pattern, with a first peak lasting from sundown to 2300 hr and a second peak from 0300 to 0600 hr. In May, irregular bouts of diurnal activity were added to the bimodal nocturnal pattern. In June, activity was irregular and evenly distributed across the diel cycle, with little day–night differentiation. In September, animals returned to their bimodal nocturnal activity patterns. Total hours of daily activity were significantly less in winter than in spring and summer.

Den Use. Porcupines may enter dens episodically during nonwinter months for reasons ranging from blackfly avoidance to shelter from rain. In New York, porcupines enter permanent dens in late October and remain until April, emerging only for nightly foraging sessions (Roze 1989). The animals return year after year to preferred dens in deep rock crevices, which offer better thermal protection than dens such as tree holes, hollow logs, outbuildings, and, rarely, the thick vegetation of hemlocks (*Tsuga canadensis*). No insulation or nest material is carried into the den. Among New York porcupines, den sharing was observed only 12% of the time (Roze 1987). In habitats where suitable den sites are limited, porcupines may routinely share dens, although there is evidence of strife (Dodge 1967; Griesemer et al. 1996).

Social Structure. The North American porcupine has an exceptionally large geographic range, ranking in the upper 12% of North American mammals (Burt and Grossenheider 1976). The species range encompasses habitat types that include boreal forest, montane coniferous forest, mixed deciduous forest, and desert scrub. As might be expected, feeding behavior and social structure show variation across this broad habitat spectrum.

One constant appears to be a solitary lifestyle. Predictable associations between conspecifics are limited to nocturnal meetings between juveniles and their lactating mothers and between adult males and females during mating season.

In the mixed deciduous forest of New York, porcupines show strong female territoriality, whereas male territories overlap those of females as well as other males (Roze 1989). Females show year-to-year fidelity to core territorial locations, although territories may expand or contract with changes in population density. Minimum annual female territory size is 23 ha. Male nonwinter territory sizes do not differ significantly between years, but differ significantly between individual males. Territory size correlates positively with male mating success. Adult male nonwinter territories average 73 ha. For males and females, winter territories are a small fraction of the size of nonwinter territories.

Young of the year spend their first summer in the mother's territory. At weaning, young-of-the-year females undergo juvenile dispersal, whereas weaned males remain loosely associated with their mothers. Adults may also disperse, with males the predominant dispersing sex. There is limited evidence that such dispersal occurs in response to poor reproductive success in the home territory. In the desert scrub habitat of Nevada, adult females had nonwinter territories of 8.2 ha, adult males 15.3 ha. Female territories showed a high degree of female–female overlap. Core home ranges (60% minimum convex polygon) overlapped significantly less among males (9.4%) than among females (27%) (Sweitzer 2003). The difference between porcupines from Nevada and New York may reflect greater clumping of resources in the desert habitat, forcing the animals to share. However, both populations showed female-biased juvenile dispersal (Sweitzer and Berger 1998).

MORTALITY

Predation. Despite its formidable quill-based defense, the porcupine is prey to several carnivores. The fisher is the most effective of these predators and this medium-sized mustelid may significantly reduce porcupine numbers (Earle and Kramm 1982; Powell 1993). Mountain lions (*Puma concolor*) also pose a potential threat. Sweitzer et al. (1997) attributed near extirpation of porcupines in their study area in the Great Basin Desert of Nevada to activity of this predator. True effects of predation may be confounded by possible natural population cycles with peaks between 12 and 20 years (Spencer 1964).

Other, less significant predators on the porcupine include the lynx (*Lynx canadensis*), the bobcat (*Lynx rufus*), the coyote (*Canis latrans*), the wolf (*Canis lupus*), the wolverine (*Gulo gulo*), and the great horned owl (*Bubo virginianus*) (Dodge 1967; Woods 1973; Roze 1989). However, humans may pose the most serious threat to porcupines. The concept of porcupines being a pest species has resulted in them being shot or poisoned in encounters with people (List et al. 1999). Highway fatalities are common, particularly in regions where highways are salted to enhance snow removal.

Parasites and Diseases. Porcupines harbor heavy loads of ecto- and endoparasites, but rarely exhibit overt adverse reactions to these conditions. *Monoecocestus americanus* and *M. variabilis* are the most commonly occurring cestodes, whereas *Dipetalonema arbuta* (*Molinema arbuta*), and *Wellcomia evaginata* constitute the most common nematodes associated with porcupines (Olsen and Tollman 1951; Bartlett and Anderson 1985; Roze 1989). Major ectoparasites associated with porcupines include fleas (*Ceratophyllus wickhami*), lice (*Trichodectes setosus*), and ticks (*Ixodes cookei, Dermacentor andersoni, D. variabilis*) (Dodds et al. 1969; Woods 1973; Dodge 1982). Although *Ixodes* may be more prevalent on porcupines, Jellison (1933) and McLean et al. (1993) reported occurrence of *Dermacentor* in Montana and Colorado, respectively, and Dodds et al. 1969 in Nova Scotia. However, mites (*Sarcoptes scabiei*) may pose the most serious threat to porcupines (Woods 1973; Roze 1989). Mange or scabies is characterized by thick yellow-white encrustations, which may occur in patches or cover much of the body. Encrustations may thicken to the point of incapacitation. Fissures may occur, causing the skin underneath to split and expose subcutaneous tissue (Payne and O'Meara 1958; Roze 1989). Dodge (1967) described the disease as episodic, but epidemic.

In addition, a variety of viral and bacterial diseases has been reported in porcupine populations. Dodge (1967) described the occurrence of colds with typical symptoms including coughing, sneezing, and mucosal discharge. Papillomas, small, black, wartlike lesions resulting from viral infections, also are common among adults. These pose no apparent harm to the animal with the possible exception of a mild decrease in thermoregulatory ability in the affected area (Dodge 1967). An unusual case of tularemia transmitted from porcupines to two boys and their dogs was reported by Kuhn et al. (1953). Sweitzer et al. (1997) reported infections of the brain, which appeared to be initiated by ear infection.

The basic ecology of the porcupine is not conducive to major disease propagation, and may explain why parasites are not a major cause of mortality in porcupine populations. Generally solitary, porcupines engage in gregarious behavior primarily during the breeding season and during occasional communal denning in winter and periods of inclement weather (Dodge 1967; Roze 1987; Griesemer et al. 1996). A common feature of den sites is the massive amount of fecal matter in and surrounding the den. These sites are likely the sources of parasite and disease transmission among conspecifics and other species using these dens, including raccoons (*Procyon lotor*), ringtails (*Bassariscus astutis*), and red foxes (*Vulpes vulpes*).

Affinity of the porcupine for resting and feeding in trees limits their access to many ground-dwelling disease vectors and intermediate hosts. Their arboreal habits do, however, expose them to seasonal abundance of mosquito vectors. During spring and summer, when animals, primarily in the Northwest, switch from feeding on cambium and phloem of trees to feeding on herbaceous vegetation, more time is spent on the ground, which increases their exposure to mites and ticks. Porcupines in the northeastern United States generally remain arboreal at this time. However, young are born on the ground and hidden under roots and leaf litter. This exposes them to a variety of parasites, and explains why young exhibit high parasite loads (Dodge 1967; Roze 1989). Because females spend more time on the ground nursing these young, they may be more vulnerable than males to higher infestation evident during spring and summer (Roze 1989).

ECONOMIC STATUS AND MANAGEMENT

The porcupine is often noted for its deleterious effects in forests. Its feeding activities can result in diminished radial growth, reduction of merchantable trees, and increase in structural damage (Krefting et al. 1962). However, trees are seldom killed by porcupine activity (Curtis 1941; Curtis and Wilson 1953; Sullivan et al. 1986) and the often exaggerated claims of economic loss and damage have likely prejudiced people against this species (Curtis 1941). The lack of current documentation of economic loss may be a testament to this situation. Furthermore, difficulty in accounting for loss to insects, browsing and feeding by other animal species, and human and environmental impacts confounds the ability to accurately estimate damage attributable to porcupines. Economic loss as the result of porcupine activity may in fact be inconsequential compared to the volume of timber lost to fire, fungi, and insects (Curtis and Kozicky 1944). Tenneson and Oring (1985) reported that porcupine damage was most prevalent on mature trees, and that decreased regeneration of trees may pose a more serious threat to some wooded systems than porcupine damage. A better approach to determining economic status of the porcupine might be to examine damage in the context of its basic ecology. An understanding of the selection of trees, characteristics of damage, and response of the tree may provide the best approach to determining the impact of the porcupine on forests.

Selection of tree species by porcupines varies across their distribution by season and availability. Generally, winter feeding on inner bark poses a more serious threat than summer feeding, which in many areas is limited to feeding on foliage. Economic impact is also largely dependent on the commercial significance of the targeted species. In central Massachusetts, oak (*Quercus* spp.), quaking aspen (*Populus tremuloides*), and ash (*Fraxinus americana*) are preferred trees for summer feeding, whereas hemlock is selected in winter (Griesemer et al. 1998). Hemlock also is a preferred species in Maine, New York, and British Columbia (Curtis and Kozicky 1944; Shapiro 1949; Sullivan et al. 1986). In the northern Catskills of New York, linden and aspens are preferred species in summer (Roze 1989). Beech (*Fagus grandifolia*) and sugar maple are preferred species in winter (Roze 1989). In New Brunswick, white pine (*Pinus strobus*), eastern larch (*Larix laricina*), and gray birch (*Betula populifolia*) are targeted in winter (Speer and Dilworth 1978). In Arizona and Texas, pinyon pines (Taylor 1935; Ilse 2001) are frequently targeted and bark feeding occurs throughout the year.

Individual tree selection may be influenced by concentrations of nitrogen (Roze 1989) and phosphorus and fiber (Stricklan et al. 1995). Complete girdling of trees generally is rare (Curtis 1941; Curtis and Wilson 1953); however, Sullivan et al (1986) reported 30.9% mortality resulting from severe porcupine damage in a 1296-ha study area in British Columbia. Although porcupines do girdle trees, patchy, ovate bark removal is more typical. The animal rests comfortably on a branch where bark can be easily grasped and removed (Spencer 1964). Feeding is most prevalent on distal ends of branches and the bole within the upper crown (Sullivan et al. 1986; Roze 1989) of vigorous, generally larger trees (Harder 1979). The crown, an area of greater foliage and annual incremental phloem, provides the porcupine with food of higher nutrient content and less cork than other regions of the bole. Van Deusen and Myers (1962) reported that only 13% of trees damaged by porcupines were unsuitable for 16-ft sawlogs because of damage from spike-tops resulting from feeding in the upper portions of trees. In addition, bark-feeding activity is often clustered around precipitous rocky ledges surrounding dens, where timber is likely to be low grade and difficult to harvest (Dodge 1982). Use of larger diameter trees is common (Curtis and Wilson 1953; Van Deusen and Myers 1962; Harder 1980; Ilse 2001) and, combined with other nutritional and morphological characteristics of trees, reflects foraging and energetic optimization by porcupines. In trees that have been scarred but not girdled, deleterious effects are most apparent in the first 2 years following injury, but by the sixth year following injury these effects are absent (Storm and Halvorson 1967).

In addition to individual tree characteristics, levels of porcupine damage may be influenced by stand characteristics. Taylor (1935) and Van Deusen and Myers (1962) reported that damage was most significant in trees near openings and speculated that this provided easier access to herbaceous plant matter used by porcupines in these regions. High-density stands were targeted in Idaho (Curtis and Wilson 1953); however, Harder (1980) reported greater use of low-density stands in Alberta, except when porcupine abundance was high. Monotypic stands are often favored over mixed stands (Harder 1980), and Roze (1984) found that locally abundant species were preferred.

Research on physiological effects to trees resulting from porcupine feeding has been limited. In a simulated study of girdling effects on photosynthates in regulation of tree defense, Dunn and Lorio (1992) reported declines in carbohydrate supply and resin flow and an increase in successful beetle colonization below the affected area. Ilse (2001) detected an association between porcupine feeding activity on pinyon pines (*P. remota*) and pine engraver (Coleoptera: Scolytidae) infestation. Accumulation of sugars above the porcupine feeding scars may attract additional porcupine attack and precipitate bark beetle attacks. However, beetle infestations also can be precipitated by a variety of environmental factors, and the multistemmed structure of the pinyon pine minimizes the potential impact of beetle and porcupine damage (Ilse 2001).

Damage resulting from porcupines is not restricted to forest, crops, and other vegetation. Gnawing on inanimate objects of economic value is a problem in some regions (Dodge 1982). Buildings and signs, particularly plywood, are subject to porcupine damage. Automotive equipment, including tires, and fuel and hydraulic lines have been targeted by porcupines. Canoes, paddles, and tools bearing human perspiration may be gnawed due to the salt drive of the porcupine (Dodge 1982; Roze 1989). Sap lines, sugar houses, and communication cables may be chewed or gnawed, and power outages have been attributed to porcupines chewing on electrical and power lines and cables (Dodge 1982). Damage to these human artifacts is further evidence of the sodium drive of porcupines.

Areas with significant timber and forestry industry have attempted control of porcupine activity. Bounties were common, particularly in New England, but like most bounty systems, were unsuccessful. Most of these efforts were discontinued in the 1920s, but New Hampshire, the last state with subsidized porcupine control, did not repeal bounties until 1979 (Dodge 1982; Roze 1989).

Strychnine salt blocks (Spencer 1950) and sodium arsenite gels (Faulkner and Dodge 1962) have been used to control porcupines, but these methods have generally been discontinued by legislation (Dodge 1982). Mechanical barriers, including aluminum flashing and electric fences, have been used with limited success for individual ornamental and fruit trees; however, their utility on a large scale is economically prohibitive (Dodge 1982).

Recent efforts have focused on natural means of control that capitalize on the ecology of the porcupine. Reintroduction of the fisher into areas where it has been extirpated by trapping has proven successful in Michigan, Wisconsin, California, and Montana (Cook and Hamilton 1957; Dodge 1982; Earle and Kramm 1982; Powell 1993) and may be the most effective means of control in other areas where distributions of these species overlap.

CONSERVATION AND RESEARCH NEEDS

In terms of conservation, the porcupine is not endangered in any part of its range except northern Mexico, where an unorganized but widespread extermination policy is directed against it. Change in this policy awaits a change in cultural attitudes.

With respect to research needs, the biggest gap in biological knowledge about the porcupine exists in the heart of its geographic range, the boreal coniferous forest. Issues such as year-round diet (with nutritional analysis), social structure, and territorial relationships have not been systematically studied in this habitat. A related question is what limits the species's southward expansion in the eastern deciduous forest—why do they not extend through Virginia and Tennessee into the southern Appalachians? In addition, attention should be focused on areas where the porcupine is extending its range, such as the pinyon–juniper woodlands of Texas. What are the required elements of ecological success in these peripheral areas?

At the population level, long-term studies of population trends would be highly desirable. Do porcupine populations display long-term cycles? How do such cycles vary with latitude and habitat? At the physiological level, how is the cecal sodium pool maintained and regulated? What is the role of male urine showers during precopulatory behavior? What factors account for the high efficiency of water conservation? All of these questions could be profitably studied in a comparative perspective by comparing the North American porcupine with *Coendou*. In addition, the physiological effects of porcupine feeding on trees warrants study both in terms of porcupine feeding ecology and in terms of controlling damage to economically significant tree species. Do trees produce specific secondary compounds following feeding attack? Thus, at every level of biology, interesting questions await the future porcupine researcher.

LITERATURE CITED

Balows, A., and M. W. Jennison. 1949. Thermophilic, cellulose-decomposing bacteria from the porcupine. Journal of Bacteriology 57:135.

Barkalow, F. F., Jr. 1971. The porcupine and fisher in Alabama archaeological sites. Journal of Mammalogy 52:835.

Bartlett, C. M., and R. C. Anderson. 1985. The third-stage larva of *Molinema arbuta* (Highby 1943) (Nematoda) and development of the parasite in the porcupine (*Erethizon dorsatum*). Annals of Parasitology 60:703–08.

Belovsky, G. E., and P. A. Jordan. 1981. Sodium dynamics and adaptations of a moose population. Journal of Mammalogy 62:613–21.

Beruter, J., G. K. Beauchamp, and E. L. Muetterties. 1974. Mammalian chemical communication: Perineal gland secretions of the guinea pig. Physiological Zoology 47:130–36.

Bogdan, D., and S. L. Montfort. 2001. Fecal estrogen and progesterone profiles in breeding and non-breeding female North American porcupines (*Erethizon dorsatum*). Mammalia 65:73–81.

Brander, R. B. 1973. Life history notes on the porcupine in a hardwood–hemlock forest in upper Michigan. Michigan Academy of Science 5:425–33.

Bronson, F. H., and W. K. Whitten. 1968. Estrus-accelerating pheromone of mice: Assay, androgen dependency and presence in bladder urine. Journal of Reproductive Fertility 15:131–34.

Burge, B. L. 1966. Vaginal casts passed by captive porcupine. Journal of Mammalogy 47:713–14.

Burt, W. H., and R. P. Grossenheider. 1976. A field guide to the mammals, 3rd ed. Houghton Mifflin, Boston.

Chapman, D. M., and U. Roze. 1997. Functional histology of quill erection in the porcupine, *Erethizon dorsatum*. Canadian Journal of Zoology 75:1–10.

Cook, D. B., and W. J. Hamilton, Jr. 1957. The forest, the fisher, and the porcupine. Journal of Forestry 55:719–22.

Craig, E. H., and B. L. Keller. 1986. Movements and home range of porcupines, *Erethizon dorsatum*, in Idaho shrub desert. Canadian Field-Naturalist 100:167–73.

Curtis, J. 1941. The silvicultural significance of porcupines. Journal of Forestry 39:583–94.

Curtis, J. 1944. Appraisal of porcupine damage. Journal of Wildlife Management 8:88–91.
Curtis, J., and E. L. Kozicky. 1944. Observations on the eastern porcupine. Journal of Mammalogy 25:137–46.
Curtis, J., and A. K. Wilson. 1953. Porcupine feeding on ponderosa pine in central Idaho. Journal of Forestry 51:339–41.
Dathe, H. 1963. Vom Harnspritzen des Ursons (*Erethizon dorsatus*) (*sic*). Zeitschrift für Saugetierkunde 28:369–75.
Davis, W. R., and D. J. Schmidly. 1994. Mammals of Texas, 5th rev. Texas Parks and Wildlife Department, Austin.
Dean, H. J. 1950. Porcupine swims for food. Journal of Mammalogy 31:94.
DeMatteo, K. E., and H. J. Harlow. 1997. Thermoregulatory responses of the North American porcupine (*Erethizon dorsatum bruneri*) to decreasing ambient temperature and increasing wind speed. Comparative Biochemical Physiology 116B: 339–46.
Denton, D. 1982. The hunger for salt. Springer-Verlag, New York.
Dodds, D. G., A. M. Martell, and R. Y. Yescott. 1969. Ecology of the American dog tick, *Dermacentor variabilis* (Say) in Nova Scotia. Canadian Journal of Zoology 47:171–81.
Dodge, W. E. 1967. The biology and life history of the porcupine *(Erethizon dorsatum*) in western Massachusetts. Ph.D. Dissertation, University of Massachusetts, Amherst.
Dodge, W. E. 1970. The porcupine in western Washington. Washington Forest Protection Association, Seattle.
Dodge, W. E. 1982. Porcupine. Pages 355–66 *in* J. A. Chapman, and G. A. Feldhamer, eds. Wild mammals of North America: Biology, management, and economics. Johns Hopkins University Press, Baltimore.
Dodge, W. E., and V. G. Barnes. 1975. Movements, home range, and control of porcupines in western Washington (Wildlife Leaflet No. 507). U.S. Department of Interior, Fish and Wildlife Service.
Dodge, W. E., and P. R. Canutt. 1969. A review of the status of the porcupine (*Erethizon dorsatum epixanthum*) in western Oregon. Bureau of Sport Fisheries and Wildlife and U.S. Forest Service, Portland, OR.
Dunn, J. P., and P. L. Lorio, Jr. 1992. Effects of bark girdling on carbohydrate supply and resistance of loblolly pine to southern pine beetle (*Dendroctonus frontalis*, Zimm) attack. Forest Ecology and Management 50:317–30.
Earle, R. D., and K. R. Kramm. 1980. Techniques for age determination in the Canadian porcupine. Journal of Wildlife Management 44:413–419.
Earle, R. D., and K. R. Kramm. 1982. Correlation between fisher and porcupine abundance in upper Michigan. American Midland Naturalist 107:244–49.
Faulkner, C. E., and W. E. Dodge. 1962. Control of the porcupine in New England. Journal of Forestry 60:36–37.
Felicetti, L. A., L. A. Shipley, G. W. Witmer, and C. T. Robbins. 2000. Digestibility, nitrogen excretion, and mean retention time by North American porcupines (*Erethizon dorsatum*) consuming natural forages. Physiological and Biochemical Zoology 73:772–80.
Fournier, F., and D. W. Thomas. 1997. Nitrogen and energy requirements of the North American porcupine (*Erethizon dorsatum*). Physiological Zoology 70:615–20.
Fournier, F., and D. W. Thomas. 1999. Thermoregulation and repeatability of oxygen-consumption measurements in winter-acclimatized North American porcupines (*Erethizon dorsatum*). Canadian Journal of Zoology 767:194–202.
George, W. 1985. Reproductive and chromosomal characters of ctenodactylids as a key to their evolutionary relationships. Pages 453–74 *in* W. P. Luckett, and J. L. Hartenberger, eds. Evolutionary relationships among rodents. Plenum Press, New York.
Griesemer, S. J., T. K. Fuller, and R. M. DeGraaf. 1996. Denning patterns of porcupines, *Erethizon dorsatum*. Canadian Field-Naturalist 110:634–37.
Griesemer, S. J., T. K. Fuller, and R. M. DeGraaf. 1998. Habitat use by porcupines (*Erethizon dorsatum*) in central Massachusetts: Effects of topography and forest composition. American Midland Naturalist 140:271–79.
Guilday, J. E., and H. W. Hamilton. 1973. The late Pleistocene small mammals of Eagle Cave, Pendleton County, West Virginia. Annals of the Carnegie Museum 44:45–58.
Hale, M. O., and T. K. Fuller. 1999. Estimating porcupine (*Erethizon dorsatum* Linnaeus, 1758) density using radiotelemetry and replicated mark–resight techniques. Zeitschrift für Saugetierkunde 64:85–90.
Harder, L. D. 1979. Winter feeding by porcupines in montane forests of southwestern Alberta. Canadian Field-Naturalist 93:405–10.
Harder, L. D. 1980. Winter use of montane forests by porcupines in southwestern Alberta: Preferences, density effects, and temporal changes. Canadian Journal of Zoology 58:13–19.
Hendricks, P., and H. F. Allard. 1988. Winter food habits of porcupines in Montana. Prairie Naturalist 21:1–6.
Hooper, E. T. 1961. The glans penis in *Proechimys* and other caviomorph rodents. Occasional Papers of the Museum of Zoology of the University of Michigan 623:1–18.
Ilse, L. M. 2001. Porcupines, pinyon pines, and pine engravers: Multitrophic interactions in pinyon–juniper woodlands of Texas. Ph.D. Dissertation, Oklahoma State University, Stillwater.
Ilse, L. M., and E. C. Hellgren. 2001. Demographic and behavioral characteristics of North American porcupines (*Erethizon dorsatum*) in pinyon–juniper woodlands of Texas. American Midland Naturalist 146:329–38.
Jellison, W. L. 1933. Parasites of the genus *Erethizon* (Rodentia). Transactions of the American Microscopical Society 52:42–47.
Johnson, J. L., and R. H. McBee. 1967. The porcupine caecal fermentation. Journal of Nutrition 91:540–46.
Krefting, L. W., J. H. Stoeckeler, B. J. Bradle, and W. D. Fitzwater. 1962. Porcupine–timber relationships in the Lake states. Journal of Forestry 60:325–30.
Kuhn, E., C. S. Houtz, and A. Axley. 1953. Tularemia from a porcupine. Rocky Mountain Medical Journal 50:736.
Landry, S., Jr. 1957. The interrelationships of the Old and New World hystricomorph rodents. University of California Publications in Zoology 56:1–118.
Landry, S., Jr. 1999. A proposal for a new classification and nomenclature for the Glires (Lagomorpha and Rodentia). Mitteilungen aus dem Museum für Naturkunde in Berlin: Zoologische Reihe 75:283–316.
Li, G., U. Roze, and D. C. Locke. 1997. Warning odor of the North American porcupine (*Erethizon dorsatum*). Journal of Chemical Ecology 23:2737–54.
List, R., G. Ceballos, and J. Pacheco. 1999. Status of the North American porcupine (*Erethizon dorsatum*) in Mexico. Southwestern Naturalist 44:400–404.
Luckett, W. P. 1985. Superordinal and infraordinal affinities of rodents: Developmental evidence from the dentition and placentation. Pages 227–76 *in* W. P. Luckett, and J. L. Hartenberger, eds. Evolutionary relationships among rodents. Plenum Press, New York.
Luckett, W. P., and J. L. Hartenberger, eds. 1985. Evolutionary relationships among rodents. Plenum Press, New York.
Macdonald, D. W. 1985. The rodents IV: Suborder Hystricomorpha. Pages 480–506 *in* R. E. Brown, and D. W. Macdonald, eds. Social odors in mammals, Vol., 1. Clarendon Press, Oxford.
Marsden, H. M., and F. H. Bronson. 1964. Estrous asynchrony in mice: Alteration by exposure to male urine. Science 144:1469.
Maser, C., and R. S. Rohweder. 1983. Winter food habits of cougars from northeastern Oregon. Great Basin Naturalist 43:425–28.
McEvoy, J. S. 1982. Comparative myology of the pectoral and pelvic appendages of the North American porcupine (*Erethizon dorsatum*) and the prehensile-tailed porcupine (*Coendou prehensilis*). Bulletin of the American Museum of Natural History 173:337–421.
McLean, R. G., A. B. Carey, L. J. Kirk, and D. B. Francy. 1993. Ecology of porcupines (*Erethizon dorsatum*) and Colorado tick fever virus in Rocky Mountain National Park, 1975–1977. Journal of Medical Entomology 30:236–38.
Mirand, E. A., and A. R. Shadle. 1953. Gross anatomy of the male reproductive system of the porcupine. Journal of Mammalogy 34:210–20.
Mossman, H. W., and I. Judeas. 1949. Accessory corpora lutea, lutein cell origin, and the ovarian cycle in the Canadian porcupine. American Journal of Anatomy 85:1–40.
Murie, O. J. 1926. The porcupine in northern Alaska. Journal of Mammalogy 7:109–13.
Norment, C. J., A. Hall, and P. Hendricks. 1999. Important bird and mammal records in the Thelon River Valley, Northwest Territories: Range expansions and possible causes. Canadian Field-Naturalist 113:375–85.
Nowak, R. M. 1991. Walker's mammals of the world; Vol. II, 5th ed. Johns Hopkins University Press, Baltimore.
Olsen, O. W., and C. D. Tollman. 1951. *Wellcomia evaginata* (Smitt 1908) (Oxyuridae, Nematoda) of porcupines in mule deer, *Odocoileus hemionus*, in Colorado. Proceedings of the Helminthological Society of Washington 18:120–22.
Parmalee, P. W. 1971. Fisher and porcupine remains from cave deposits in Missouri. Transactions of the Illinois State Academy of Science 64:225–29.
Payette, S. 1987. Recent porcupine expansion at tree line: A dendroecological analysis. Canadian Journal of Zoology 65:551–57.
Payne, D. D., and D. C. O'Meara. 1958. *Sarcoptes scabiei* infestation of a porcupine. Journal of Wildlife Management 22:321–22.
Powell, R. A. 1993. The fisher: Life history, ecology and behavior, 2nd ed. University of Minnesota Press, Minneapolis.
Reynolds, H. G. 1957. Porcupine behavior in the desert–shrub type of Arizona. Journal of Mammalogy 38:418–19.

Roberts, M., S. Brand, and E. Maliniak. 1985. The biology of captive prehensile-tailed porcupines, *Coendou prehensilis*. Journal of Mammalogy 66:476–82.

Roze, U. 1984. Winter foraging by individual porcupines. Canadian Journal of Zoology 62:2425–28.

Roze, U. 1985. How to select, climb, and eat a tree. Natural History 94 (5):62–68.

Roze, U. 1987. Denning and winter range of porcupines. Canadian Journal of Zoology 65:981–86.

Roze, U. 1989. The North American porcupine. Smithsonian Institution Press, Washington, DC.

Roze, U. 2002. A facilitated release mechanism for quills of the North American porcupine (*Erethizon dorsatum.*) Journal of Mammalogy. 83 (2):381–85.

Roze, U., D. C. Locke, and N. Vatakis. 1990. Antibiotic properties of porcupine quills. Journal of Chemical Ecology 16:725–34.

Shadle, A. R. 1948. Gestation period in the porcupine, *Erethizon dorsatum*. Journal of Mammalogy 29:162–64.

Shadle, A. R. 1950. The North American porcupine up to date. Ward's Natural Science Bulletin 24:5–11.

Shadle, A. R., and D. Po-Chedley. 1949. Rate of penetration of a porcupine spine. Journal of Mammalogy 30:172–73.

Shadle, A. R., M. Smelzer, and M. Metz. 1946. The sex reactions of the porcupine, *Erethizon d. dorsatum*, before and after copulation. Journal of Mammalogy 27:116–21.

Shapiro, J. 1949. Ecological and life history notes on the porcupine in the Adirondacks. Journal of Mammalogy 30:247–57.

Smith, G. W. 1977. Population characteristics of the porcupine in northeastern Oregon. Journal of Mammalogy 58:674–76.

Speer, R. J., and T. G. Dilworth. 1978. Porcupine winter foods and utilization in central New Brunswick. Canadian Field-Naturalist 92:271–74.

Spencer, D. A. 1950. The porcupine: Its economic status and control (Wildlife Leaflet 328). U.S. Department of the Interior, Fish and Wildlife Service.

Spencer, D. A. 1964. Porcupine fluctuations in past centuries revealed by dendrochronology. Journal of Applied Ecology 1:127–49.

Storm, G. L., and C. H. Halvorson. 1967. Effect of injury by porcupines on radial growth of ponderosa pine. Journal of Forestry 65:740–43.

Stricklan, D., J. T. Flinders, and R. G. Cates. 1995. Factors affecting selection of winter food and roosting resources by porcupines in Utah. Great Basin Naturalist 55:29–36.

Sullivan, T. P., W. T. Jackson, J. Pojar, and A. Banner. 1986. Impact of feeding damage by the porcupine on western hemlock-Sitka spruce forests of north-coastal British Columbia. Canadian Journal of Forest Research 16:642–47.

Sweitzer, R. A. 1990. Winter ecology and predator avoidance in porcupines (*Erethizon dorsatum*) in the Great Basin Desert. M.S. Thesis. University of Nevada, Reno.

Sweitzer, R. A. 2003. Breeding movements and reproductive activities of porcupines in the Great Basin desert. Western North American Naturalist. 63 (1):1–11.

Sweitzer, R. A., and J. Berger. 1992. Size-related effects of predation on habitat use and behavior of porcupines (*Erethizon dorsatum*). Ecology 73:867–75.

Sweitzer, R. A., and J. Berger. 1993. Seasonal dynamics of mass and body condition in Great Basin porcupines (*Erethizon dorsatum*). Journal of Mammalogy 74:198–203.

Sweitzer, R. A., and J. Berger. 1997. Sexual dimorphism and evidence for intrasexual selection from quill impalements, injuries, and mate guarding in porcupines (*Erethizon dorsatum*). Canadian Journal of Zoology 75:847–54.

Sweitzer, R. A., and J. Berger. 1998. Evidence for female-biased dispersal in North American porcupines (*Erethizon dorsatum*). Journal of Zoology (London) 244:159–66.

Sweitzer, R. A., and D. H. Holcombe. 1993. Serum-progesterone levels and pregnancy rates in Great Basin porcupines (*Erethizon dorsatum*). Journal of Mammalogy 74:769–76.

Sweitzer, R. A., S. H. Jenkins, and J. Berger. 1997. Near extinction of porcupines by mountain lions and consequences of ecosystem change in the Great Basin Desert. Conservation Biology 11:1407–17.

Taylor, W. P. 1935. Ecology and life history of the porcupine (*Erethizon epixanthum*) as related to the forests of Arizona in the southwestern United States. University of Arizona Bulletin 6:1–177.

Tenneson, C., and L. W. Oring. 1985. Winter food preferences of porcupines. Journal of Wildlife Management 49:28–33.

Van Deusen, J. L., and C. A. Myers. 1962. Porcupine damage in immature stands of ponderosa pine in the Black Hills. Journal of Forestry 60:811–13.

Vispo, C., and I. D. Hume. 1995. The digestive tract and digestive function in the North American porcupine and beaver. Canadian Journal of Zoology 73:967–74.

Voorhies, M. R. 1981. A fossil record of the porcupine (*Erethizon dorsatum*) from the Great Plains. Journal of Mammalogy 62:835–37.

White, J. A. 1970. Late Cenozoic porcupines (Mammalia: Erethizontidae) of North America. American Museum Novitates 2421:1–15.

Wood, A. E. 1950. Porcupines, paleogeography, and parallelism. Evolution 4:87–98.

Woods, C. A. 1972. Comparative myology of jaw, hyoid, and pectoral appendicular regions of New World and Old World hystricomorph rodents. Bulletin of the American Museum of Natural History 147:117–98.

Woods, C. A. 1973. *Erethizon dorsatum*. Mammalian Species 29:1–6.

Woods, C. A. 1999. North American porcupine, *Erethizon dorsatum*. Pages 671–73 *in* D. Wilson, and S. Ruff, eds. Smithsonian book of North American mammals. Smithsonian Institution Press, Washington, DC.

Woods, C. A., and J. W. Hermanson. 1985. Myology of hystricognath rodents: An analysis of form, function, and phylogeny. Pages 515–48 *in* W. P. Luckett, and J. L. Hartenberger, eds. Evolutionary relationships among rodents. Plenum Press, New York.

Uldis Roze, Department of Biology, Queens College, Flushing, New York 11367. Email: uldis_roze@qc.edu.

Linda M. Ilse, Biology Department, Southwestern University, Georgetown, Texas 78626. Email: ilsel@southwestern.edu.

19

Eastern Woodrat

Neotoma floridana and Allies

Anne-Marie Monty
Robert E. Emerson

NOMENCLATURE

Common Names. Eastern woodrat, wood rat, pack rat, trade rat, Florida wood rat, bush rat, brush rat, cave rat, mountain rat
Scientific Name. *Neotoma floridana*
Subspecies. *N. f. attwateri* (Mearns 1897), *N. f. baileyi* (Merriam 1894), *N. f. campestris* (Allen 1894), *N. f. floridana* (Ord 1818), *N. f. haematoreia* (Howell 1934), *N. f. illinoensis* (Howell 1910), *N. f. rubida* (Bangs 1898), and *N. f. smalli* (Sherman 1955); Hayes (1990) recommended that *N. f. haematoreia* be subsumed under *N. f. floridana*

The eastern woodrat was collected and described by Ord (1818) from a location near the St. Johns River, Florida. According to Wilson and Reeder (1993), the genus *Neotoma* contains 20 species. Hayes (1990) suggested that the Allegheny woodrat, *N. magister* (formerly a subspecies of *N. floridana*), be promoted to the species level. The recommended change would result in 10 species living within the United States and 11 species endemic from Mexico south to Nicaragua.

DISTRIBUTION

The eastern woodrat occurs across the southeastern United States south of the Tennessee River to central Florida. The species range continues west to central Texas, Oklahoma, and Colorado. The northern portion of the range includes Kansas, central Missouri, southern Illinois, and western Kentucky and Tennessee. Disjunct populations occur in Nebraska and on Key Largo, Florida (Fig. 19.1).

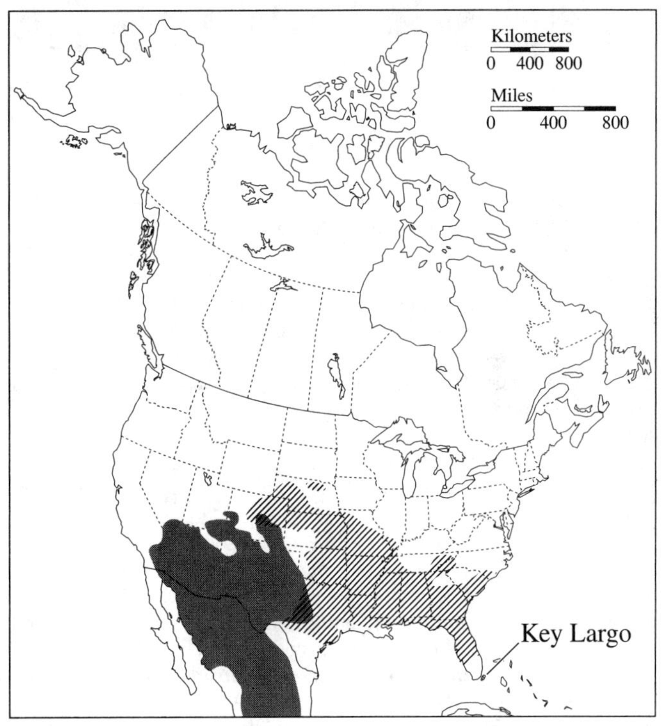

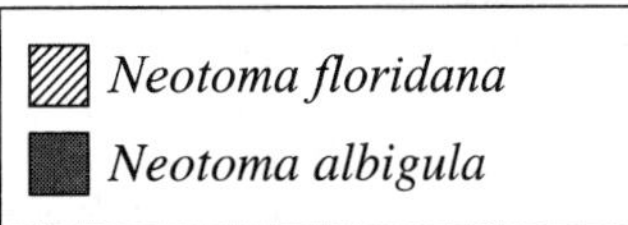

Figure 19.1. Distribution of the eastern woodrat (*Neotoma floridana*) and the white-throated woodrat (*Neotoma albigula*).

DESCRIPTION

The eastern woodrat is a member of the order Rodentia and the family Muridae. In appearance (Fig. 19.2) it looks like a large white-footed mouse. Body measurements are as follows: total length, 305–450 mm; tail length, 130–180 mm; hind foot length, 35–42 mm; and ear length, 24–29 mm (Rainey 1956). They have four clawed digits and a rudimentary thumb on the forelimbs and five clawed digits on the hind limbs. Woodrats possess long, curving vibrissae, large naked ears, and prominent eyes, which allow nocturnal activity. Females have two pairs of inguinal mammae (Finley 1958). Mean weight of 21 adult males and 14 nonpregnant females in Kansas was 299 and 216 g, respectively (Rainey 1956). In Illinois, the mean adult weight of 68 males and 77 females (including pregnant individuals) was 284 and 250 g, respectively (Monty 1997). Maximum weights reported by Goertz (1970) were 425 g for 142 males and 350 g for 139 females.

Pelage. The sides and the dorsal surface of adult eastern woodrats are gray-brown to bright cinnamon-orange with black-tipped hairs. The ventral surface of adults is white or creamy white. Juveniles are gray with a white ventral surface. The feet are white in all age groups. The tail is almost as long as the body, has short hairs, and is bicolored: dusky gray above and white below. Southern populations may have a unicolor tail (Schwartz and Odum 1957). A brown stain on the midventral pelage results from secretions of a ventral abdominal gland, which is present in both sexes, but increases in size in adult males during the breeding season. This gland is believed to play a role in scent communication (Poole 1940) and mother–litter recognition (Clarke 1973).

Skull and Dentition. The average greatest length of the skull ranged from 49.3 to 50.3 mm for 131 skulls representing five subspecies (Schwartz and Odum 1957). In the past, the presence of a forked anterior palatal spine was used to distinguish *N. floridana* from adjacent species of *Neotoma* (Wiley 1980). With the return of *N. magister* to the species level, a reliable skull characteristic is necessary to distinguish between *N. floridana* and the Allegheny woodrat. The maxillovomerine notch is present in *N. magister* but not in *N. floridana*. The forked anterior palatal spine is less reliable than the maxillovomerine notch, because only 71% of 386 *N. floridana* skulls examined had a notched or forked anterior palatal spine (Hayes 1990). The dental formula for *N. floridana* is I 1/1, C 0/0, P 0/0, M 3/3 (Fig. 19.3). Molars are moderately high crowned and prismatic (Hoffmeister 1989).

FIGURE 19.2. Eastern woodrat (*Neotoma floridana*). SOURCE: Photo by Joseph Whittaker.

GENETICS

Karyotypic analyses of the genus *Neotoma* were conducted by Baker and Mascarello (1969). Eastern woodrats possess a diploid number of 52. The species has one large and two small pairs of autosomal biarmed elements and 22 pairs of graded acrocentric chromosomes. The X sex chromosome is a large submetacentric and the Y is a medium subtelocentric. Birney (1973) reported that the large acrocentrics possess chromatin visible beyond the centromere and opposite the arm, which if counted would increase the fundamental number above 56. Eastern woodrats are polymorphic for the number of large biarmed chromosomes. In *N. f. baileyi,* two biarmed autosomes may be replaced by acrocentric chromosomes. A submetacentric Y was found in populations of *N. f. baileyi, N. f. attwateri,* and *N. f. campestris* (Birney 1973).

Several genetic techniques have been used to study the evolutionary relationships between members of the genus *Neotoma*. Electrophoresis of multiple hemoglobins of *N. floridana,* the southern plains woodrat (*N. micropus),* and their hybrids were reported by Birney and Perez (1971). Zimmerman and Nejtek (1975, 1977) studied hemoglobin and albumin variation among *N. floridana* and three other species. Shipley et al. (1990) reported that immunoelectrophoresis using tissue homogenates supported systematic relationships proposed by other techniques. Hayes and Harrison (1992) investigated variation in mitochondrial DNA restriction sites in *N. floridana* subspecies and *N. magister*. DNA fingerprinting has been conducted on the bushy-tailed woodrat (*N. cinerea*) to determine mating patterns and reproductive success (Topping and Millar 1998). Woodrat-specific microsatellite DNA primers have been developed to study genetic structure and gene flow among populations (Castleberry et al. 2002).

PHYSIOLOGY

Molts. Finley (1958) described the molt cycle of woodrats as involving two or three molts during the first year; thereafter, molts into fresh winter pelage occur annually. Winter pelage is gray-brown; worn fur is brown to rusty brown (Goldman 1910). The pattern of molting is juvenile pelage, postjuvenile molt, subadult pelage, second molt, first autumn pelage, third molt, first winter pelage, annual molt. The first molt takes place at 5–6 weeks of age and begins on the abdomen, chest, and throat, and then progresses dorsally (Rainey 1956). The annual molt may be delayed 1–3 months in females until the breeding season ends. Southern populations may not exhibit well-synchronized molting cycles, but the annual molt usually occurs during June–October (Birney 1973). In Illinois, 21 adult woodrats were observed molting from February to June in 1994–1996 (A.-M. Monty, pers. obs.), possibly exhibiting what Birney (1973) described as a "vernal molt" to maintain the pelage.

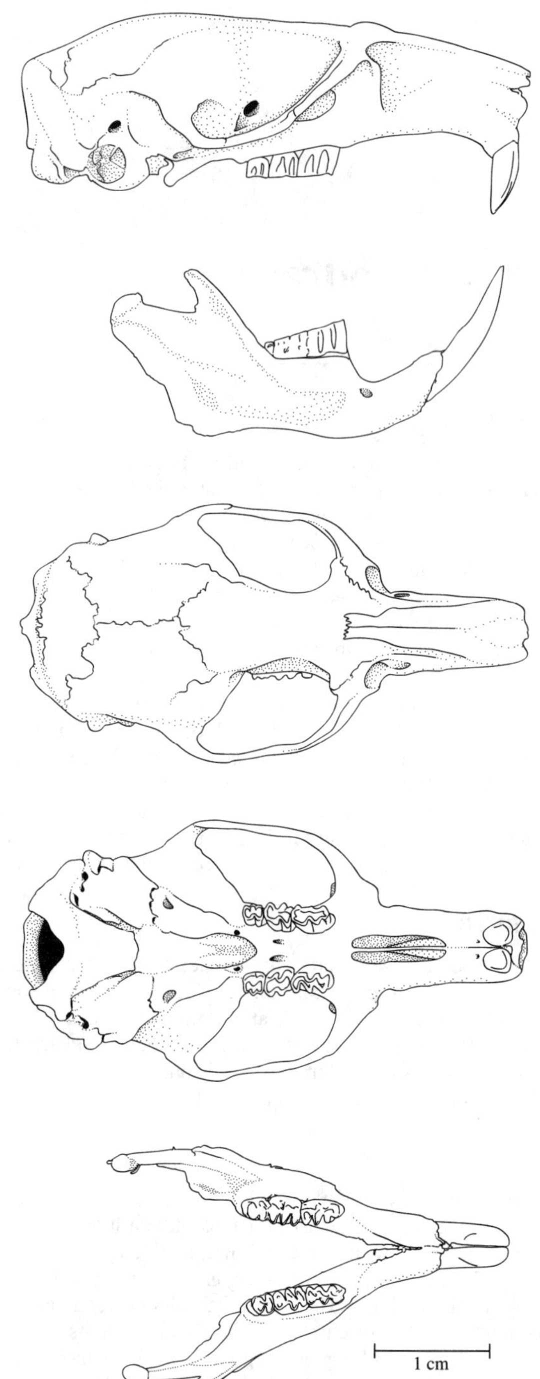

FIGURE 19.3. Skull of the eastern woodrat (*Neotoma floridana*). From top to bottom: lateral view of cranium, lateral view of mandible, dorsal view of cranium, ventral view of cranium, dorsal view of mandible.

Body Weight. Adult female woodrats vary in weight throughout much of the year because of pregnancy. A sharp decline in weight may indicate parturition. Rainey (1956) noted that adult male woodrats tend to reach maximum weight in February and March at the beginning of the breeding season and then decrease to their annual minimum in May. Their weight recovers in June, but declines again in July during hot weather. In September and October, weight declines because of increased food-collecting activity. Weight increases in late fall and winter until the maximum is again reached.

AGE ESTIMATION

Rainey (1956) classified eastern woodrats into three age groups based on pelage color, weight, reproductive condition, and time of year when first captured. Juvenile woodrats had soft gray fur, had not completed their first molt, and were usually less than 150 g. The pelage of subadults was not as bright as that of adults, and body weight was usually less than 250 g for males and 200 g for females. Woodrats were classified as adults at the first signs of sexual maturity. If an individual was captured outside of the breeding season, the adult status was determined based on size, pelage, and body weight. Hamilton (1953) documented the pattern of tooth eruption and skull measurements in juveniles of known age. Birney (1973) classified eight age groups based on the degree of eruption of the upper molars and their wear pattern. In adult *N. floridana,* the proximal portion of the baculum is more triangular shaped laterally than in the young (Burt and Barkalow 1942). In most other species of woodrats, the baculum of immatures is a miniature of the adult form.

REPRODUCTION AND DEVELOPMENT

Anatomy. Rodents have a duplex uterus consisting of two uteri, two cervixes, and a single vagina (Feldhamer et al. 1999). Uterine length reported for three species of *Neotoma* ranges from 30 to 43 mm (Howell 1926). During sexual inactivity, the vagina is closed, the nipples are small, and hair covers the abdomen (Rainey 1956). Adult male *Neotoma* have paired testes, which descend into the scrotum during the breeding season (Howell 1926). The baculum of eastern woodrats has a broad proximal end with upturned lateral projections, resulting in a U-shaped cross section. From the proximal end, the baculum gradually narrows to a rounded shaft (Burt and Barkalow 1942).

Breeding. Duration of the breeding season varies geographically. Woodrats in Florida and southern Georgia appear to breed throughout the year (Harper 1927; Pearson 1952; Hamilton 1953). Breeding occurs from March through October in Oklahoma (Goertz 1970) and from February through August in Kansas (Rainey 1956). Goertz (1970) captured males with fully scrotal testes 15–40 mm long from April to November.

Eastern woodrats are polyestrous, with estrous cycles usually lasting 4–6 days (range = 3–8) (Asdell 1964). During estrus, the vagina is perforate, the clitoris is swollen, and the uteri and ovaries enlarge (Rainey 1956). Females reenter estrus soon after parturition. In the laboratory, Pearson (1952) observed a mating in the presence of the female's 7-day-old litter.

Females reach sexual maturity at 5–6 months of age or when they weigh about 160 g, unless this time period occurs during winter in northern populations (Fitch and Rainey 1956). Females born in the early spring exhibit reproductive condition during their first autumn, but only occasionally produce a litter that season. Birney (1973) captured a subadult in late August that was nursing young less than 2 weeks old. In males and females, sexual maturity usually is reached the year following their birth (Rainey 1956).

Parturition. In the late stages of gestation and during lactation, the nipples enlarge and soften and hair loss occurs in the inguinal region (Rainey 1956). Parturition occurs after gestation of 32–38 days, with a mode of 35 days (Birney 1973). If the female is still nursing her previous litter, then delayed implantation may lengthen gestation. Oswald and McClure (1990) reported that a second pregnancy was delayed 3 days when the female was suckling one pup and up to 14 days when suckling four pups. The delay allows the energetically expensive late stages of gestation to occur after the previous litter is weaned.

Murphy (1952) observed the birth of the final three young in a litter of four pups. The female displayed muscular contractions of the abdominal region and held her back in an arched position; her hind feet were extended forward and parted slightly. Her front feet were held off the floor so the tail and hind legs supported the body. Her head was bent down and her eyes were closed. Following each birth, the mother licked the young and consumed the placenta. Each young then attached to a nipple. The litter was born in about 17 min.

Newborn. Females in Kansas have two to four litters per year (Rainey 1956). Litter size varies from one to six in Kansas (Kellogg 1915) and Missouri (Schwartz and Schwartz 1959). Two to seven ($\bar{X} = 3.2$) embryos or placental scars were found in 50 females in Oklahoma (Goertz 1970). In Florida, only one to four young were born per litter (Worth 1950). Birney (1973) speculated that northern subspecies tend to have larger litter size because they produce fewer litters each year.

Newborn woodrats are altricial, sparsely haired, and slate gray on the back, and have a pink muzzle (Rainey 1956). Vibrissae and claws are present. The eyes and pinnae are closed. The young can roll and make uncoordinated movements of the limbs. The upper and lower incisors have erupted, forming a six-sided opening for attachment to the mother's nipple (Hamilton 1953). Females remove nursing young by turning in circles and pushing with the forefeet and nose. Measurements of 21 neonates were as follows: total length, 87–96 mm; length of tail, 24–27 mm; length of hind foot, 13.5–14.8 mm; and weight, 11.8–14.1 g (Hamilton 1953). At 9 days of age, the ears begin to open and the entire dorsum is furred. Around 15 days, the eyes open, and the young are well furred and can run (Rainey 1956). By 20 days of age, solid food can be eaten. Hamilton (1953) assumed that in the wild the young would be weaned when 1 month old, although they will nurse longer in captivity. By 70 days of age, the young may leave the nest permanently. Fitch and Rainey (1956) found that woodrats increase in weight about 1.5 g/day until 2 months of age or until they reach 100 g. While 100–149 g, they increased 0.71–0.92 g/day and when greater than 150 g, they increased 0.68–0.83 g/day. Adult body weight is reached in about 8 months (Rainey 1956).

Sex Ratio. The sex ratio at birth is 1:1 (Birney 1973). Goertz (1970) reported a nearly 1:1 ratio across all age groups in 281 eastern woodrats captured in Oklahoma. Rainey (1956) captured 55.3% males and 44.7% females among 105 woodrats in Kansas. Monty (1997) captured 41.7% males and 58.3% females among 283 individuals (all ages) in Illinois.

ECOLOGY

Eastern woodrats live in a variety of habitats throughout their range. They occur in upland woods, brushy riparian woods (Goertz 1970), wooded bottomland (Neal 1967), swamps (Hamilton 1953), and rocky bluffs with caves and crevices (Nawrot and Klimstra 1976). In grassland habitat, woodrats are associated with hedgerows or rock outcrops. Rainey (1956) believed the species range was limited in Kansas by the lack of overhead cover and outcroppings with suitable crevices. In Colorado, grassland alone was insufficient for eastern woodrats (Finley 1958). If no trees were available, woodrats were associated with shrubs, yucca (*Yucca* spp.), or tree cactus (*Opuntia* spp.), which were used for shelter and food. A structure of sticks or other materials, referred to as a house or den, is built for shelter.

Activity. Woodrats become active about 30 min before complete darkness and remain active until 30 min before sunrise (Wiley 1971). Most activity occurs between 2030 and 2230 hr and is highest on nights of new or quarter moon phases and lowest on nights of full moon. Murphy (1952) found woodrat activity decreased during extremely cold or rainy weather, although Rainey (1956) stated that woodrats are more active on dark and rainy nights than on clear nights.

Home Range and Population Density. Goertz (1970) estimated the average home range was 0.26 ha for males and 0.17 ha for females captured 5–24 times. Tate (1970) reported an average home range for eight woodrats was 0.03–0.2 ha depending on the calculation method. Subadults have larger home ranges than adults. Adult males have larger home ranges than adult females because they tend to make longer movements in search of mates.

Estimates of population density vary widely by location and census method. In Louisiana, Neal (1967) estimated woodrat density using the strip census method. Houses that appeared active were counted and assumed to house one woodrat. This resulted in estimates of 0.82 woodrat/ha in 1965 and 0.2/ha in 1966 following a population decline. Fitch and Rainey (1956) used trapping data to estimate a density of about 4.25/ha after a decline. Humphrey (1988) used capture histories and probability models from the computer program CAPTURE to estimate a mean density of 7.6/ha on Key Largo. He concluded that past estimates from Key Largo were unreliable because they measured abundance of sign (houses, latrine sites, burrows), which is not well correlated with population size. The most recent survey of Key Largo woodrats by Frank et al. (1997) found that small sample size and low number of sequential nights trapped prohibited the use of program CAPTURE. An enumeration estimation technique yielded a mean density of 0.9 woodrat/ha.

Movements. Pearson (1952) reported that in Florida the mean maximum distance between points of recapture was 54 m, with an extreme of 165 m. Eastern woodrats may tend to make shorter movements in more homogeneous forest habitats because resources are more concentrated (Wiley 1971). In grassland with rock outcrop habitat, such as the western portion of the species range, eastern woodrats may travel greater distances to locate sufficient food, shelter, and mates. Fitch and Rainey (1956) recorded a mean maximum distance of 105 m for 27 adult males and 44 m for 39 subadult and adult females. Maximum distances moved were 329 m for males and 198 m for females. Adult males in reproductive condition moved the greatest distance. Wiley (1971) reported a male that moved 302 m in 4.5 hr. From livetrapping data, Fitch and Rainey (1956) estimated that woodrats usually forage within 23 m of their house.

FEEDING HABITS

Woodrats are food generalists and consume primarily mast and herbaceous materials. Individuals that live near agricultural fields collected corn, wheat, and sorghum (Murphy 1952; Rainey 1956), but did not significantly damage the crop. Remains of invertebrates have been found in stomach contents: beetles (Pearson 1952), cicada (Rainey 1956), grasshopper, scorpion, and snail shell (Murphy 1952). Diet varies among habitats and within populations. Each individual collects food within several meters of the house, which results in variation of diet because of availability.

Rainey (1956) believed that quantity of food and availability of house sites were more important limiting factors than the presence specific food plants in Kansas. Osage orange (*Maclura pomifera*) leaves and seeds were the most used food source, although many others were used in lower frequency (Table 19.1). Eastern woodrats in Colorado consume highly variable diets depending on species availability (Table 19.1): joints and fruits of cactus (*Opuntia* spp.), fruit and seeds of *Yucca glauca,* skunkbush (*Rhus trilobata*), greasewood (*Sarcobatus vermiculatus*), saltbush (*Atriplex canescens*), and chokecherry (*Prunus virginiana*) (Finley 1958). In different parts of Texas, *N. floridana* consume the leaves and nuts of walnuts (*Juglans nigra*), cactus, acorns (*Quercus* spp.), juniper berries (*Juniperus* spp.), wild grapes, mushrooms, Mexican buckeyes (*Ungnadia speciosa*) (Bailey 1905), and pecans (*Carya illinoensis*) (Strecker 1929). Murphy (1952) found grass in the diet during early summer in Oklahoma. During the rest of the year acorns; the bark, fruit, and seeds of sumac (*Rhus* spp.); fruit of poison ivy (*Rhus radicans*) and dogwood (*Cornus* spp.); and the seeds of redbud (*Cercis canadensis*) and Kentucky coffee tree (*Gymnocladus dioica*) were consumed. Palmetto seeds (*Sabal* sp.) and acorns were preferred by woodrats in Louisiana (Lowery 1974). Pearson (1952) found that captive *N. floridana* ate 91% of herbaceous plants, 57% of shrubs, and 37% of tree species presented (Table 19.1).

Analysis of woodrat fecal pellets in Illinois revealed a diet of 61–67% mast throughout the year, with the remaining portion consisting of Virginia creeper (*Parthenocissus quinquefolia*), sedge (*Carex* sp.), hickory nut leaves (*Carya* spp.), and the fruit and leaves of spicebush (*Lindera benzoin*) (Wagle and Feldhamer 1997). Little mast was available on the ground during spring and summer, which indicated that the source was caches. Stored mast was rationed to ensure availability throughout the year.

Nutrient content of plant parts varies seasonally and after caching by woodrats (Post 1992). To assess the value of available foods, woodrats are able to detect differences in nutrient concentrations (Post

TABLE 19.1. Food sources for *Neotoma floridana*

Reference	Location	Food Source
Pearson 1952	Florida	Bead grass (*Paspalum* sp.)
		Big leaf snowbell bush (*Styrax grandifolia*)
		Chickasaw plum (*Prunus angustifolia*)
		Coneflower (*Rudbeckia* sp.)
		Elephant's foot (*Elephantopus* sp.)
		Elm (*Ulmus floridana*)
		Farkleberry (*Vaccinium arboreum*)
		Fireweed (*Erechtities* sp.)
		Goldenrod (*Solidago* sp.)
		Groundsel-tree (*Baccharis halimifolia*)
		Horse nettles (*Solanum* spp.)
		Ironweed (*Veronia* sp.)
		Marsh fleabane (*Pluchea longifolia*)
		Panic grass (*Panicum* sp.)
		Viburnum spp.
		Wax myrtle (*Myrica cerifera*)
Rainey 1956	Kansas	Bittersweet (*Celastrus scandens*)
		Black nightshade (*Solanum nigrum*)
		Black walnut (*Juglans nigra*)
		Bur oak (*Quercus macrocarpa*)
		Chestnut oak (*Q. muehlenbergii*)
		Climbing false buckwheat (*Polygonum scandens*)
		Coralberry (*Symphoricarpos orbiculatus*)
		Corn (*Zea mays*)
		Dogwood (*Cornus drummondi*)
		Goldenrod (*Solidago* sp.)
		Gooseberry (*Ribes missouriense*)
		Green briar (*Smilax tamnoides*)
		Hackberry (*Celtis occidentalis*)
		Honey locust (*Gleditsia triacanthos*)
		Horse-nettle (*Solanum carolinense*)
		Kentucky coffee tree (*Gymnocladus dioica*)
		Osage orange tree (*Maclura pomifera*)
		Poison ivy (*Rhus radicans*)
		Poison oak (*Rhus toxicodendron*)
		Red haw (*Crataegus mollis*)
		Smooth sumac (*Rhus glabra*)
		Sorghum (*Sorghum* sp.)
		Thistle (*Cirsium undulatum*)
		Wheat (*Triticum aestivum*)
		Wild plum (*Prunus americana*)
		Winter grape (*Vitis vulpina*)
Finley 1958	Colorado	Cactus (*Opuntia* spp.)
		Chokecherry (*Prunus virginiana*)
		Cottonwood (*Populus* sp.)
		Evening star (*Mentzelia* spp.)
		Golden currant (*Ribes aureum*)
		Greasewood (*Sarcobatus vermiculatus*)
		Kansas sunflower (*Helianthus annuus*)
		Mugwort (*Artemisia ludoviciana*)
		Nootka rose (*Rosa nutkana*)
		Poison ivy (*Rhus radicans*)
		Ragweed (*Ambrosia* sp.)
		Rubber rabbitbrush (*Chrysothamnus nauseosus*)
		Russian thistle (*Salsola kali*)
		Saltbush (*Atriplex* spp.)
		Scurfpea (*Psoralea linearifolia*)
		Skunkbush (*Rhus trilobata*)
		Snakeweed (*Gutierrezia* sp.)
		Snowberry (*Symphoricarpos* sp.)
		Wild gourd (*Curcurbita foetidissima*)
		Wild grape (*Vitis* sp.)
		Yucca (*Yucca glauca*)

1993). Captive individuals consumed more standard diet food than foods reduced in protein, lipids, and energy.

Eastern woodrats seem able to obtain sufficient water from dew, rain, vegetation, and metabolic processes, so that open water is not necessary (Murphy 1952). Woodrats are able to survive in hot, arid environments by consuming succulent vegetation, exhibiting nocturnal activity, and remaining in the house for protection from daytime heat (Lee 1963). Laboratory experiments indicate that if vegetation is not provided, woodrats will drink water, using a lapping motion (Murphy 1952).

BEHAVIOR

Mating Behavior. During laboratory interactions, woodrats sniff the ventral abdominal gland and genitalia on introduction. Sparring and fighting between the pair may follow. Birney (1973) found that if the female established dominance, she might bite or kill the male if he was not removed. If the male established dominance, then mating might occur following a short fight. Hamilton (1953) noted that the male followed the female, drummed his hind feet, and sniffed or licked at her perineal region. The male mounted the female from behind, displayed rapid shallow thrusts, and penetrated, and intromission was indicated on a final deep thrust (Dewsbury 1974). The female next often tried to move away, but the pair remained joined in a copulatory lock.

Aggressive Behavior. Eastern woodrats are solitary animals (Murphy 1952; Layne 1958) and both sexes are highly antagonistic toward other individuals. Rainey (1956) observed two adult males placed in the same cage. They approached, touched noses, and then began fighting. Both reared up on the hind legs and bit while lashing with the front feet. The vibrissae move rapidly and the tail is flicked from side to side. A captive male acted aggressively toward a female through a partition and later forced entry into her pen and killed her.

Eastern woodrats are territorial and will defend their house and possibly the surrounding area (Fitch and Rainey 1956). Adults are dominant over subadults and will chase them away from food sources (Wiley 1971). Kinsey (1977) observed territorial behavior in a confined population of Allegheny woodrats. Females defended a territory during spring and summer, whereas males were aggressive and territorial in fall and winter.

Worth (1950) and Pearson (1952) described captive *N. floridana* living in mixed groups in apparent harmony. These observations and their patchy distribution in the wild led to early speculation that they might be colonial in the source populations in Florida.

Communication. Eastern woodrats make loud squeaking noises when fighting or angry and may grate their teeth to make a chattering noise (Svihla and Svihla 1933). Young squeak when in distress. Foot stamping when alarmed, angry, or surprised has been reported in individuals as young as 26 days (Hamilton 1953). The tail may be vibrated rapidly to produce a buzzing noise or slapped against the substrate (Rainey 1956).

Escape Behavior. Adults learn the available escape routes from the house so they can respond quickly when in danger. If the house is located at the base of a tree, the woodrat will climb part way up, assess the situation, and may proceed high into the branches. Nursing young are dragged from dangerous situations while clinging to the mother's nipples. If a pup becomes detached, the mother may pick it up and carry it in her mouth (Rainey 1956).

House Construction. Eastern woodrats construct one or more houses (also called dens) to protect an inner nest and food caches, moderate temperature fluctuations, provide shelter from precipitation, and provide protection from predators (Fig. 19.4). Depending on the available habitat, houses may be built in diverse locations. In Kansas, woodrats build at the base of trees in forests or osage orange hedges, in blackberry thickets, in hilltop limestone outcrops, inside standing hollow trees, under root tangles along gullies, and in abandoned buildings (Fitch and Rainey 1956). Eastern woodrats may live among brush piles and refuse heaps in Texas (Schmidly 1983) and Missouri (A.-M. Monty, personal observation). Houses have been observed up to 8 m above ground among tree branches or vines (Lowery 1974). In swamps, houses are built at the base of trees on high ground (Hamilton 1953). In Colorado, houses are often built in the branches of tree cactus (Finley 1958). Subterranean chambers, either naturally occurring (Pearson 1952), dug by another species, or dug by the woodrat (Finley 1958), may be used as house sites. Habitat in Illinois with rough, rocky terrain was preferred over smooth areas due to the prevalence of ledges and caves, which provide cover for houses (Layne 1958). When located in crevices, houses may appear as sticks and rubbish filling gaps between rocks.

FIGURE 19.4. House of an eastern woodrat built in a cave. SOURCE: Photo by Robert E. Emerson.

Houses built away from rocks tend to exhibit a dome or pyramid shape (Fitch and Rainey 1956) typically 0.5–1 m in diameter by >1 m high (Schwartz and Schwartz 1959). Larger woodrats tend to live in larger houses (Horne et al. 1998). Some houses reach large dimensions as they are used over many generations: Neal (1967) reported a house 2 m high and 1 m in diameter, and Murphy (1952) observed a linear house built along a fallen tree that measured 4 m long by 2 m wide by 1 m high. Houses are built with available materials: twigs, plant stems (Lowery 1974), bark, leaves, stones (Murphy 1952), and cactus joints (Finley 1958). Individual items are usually picked up in the mouth and carried to the site. Larger items may be dragged (Svihla and Svihla 1933). Middens are collections of plant materials, fecal pellets, pollen, bones, and remains of invertebrates within or outside of the house. Over time a midden may become embedded in crystallized urine (amberat). The materials may be preserved for thousands of years if located in a dry, protected environment (Wells 1976; Betancourt et al. 1990).

One to three spherical or cup-shaped nests are built within the house either aboveground or belowground (Finley 1958). Nests are up to 20 cm in diameter and may be built of shredded bark, leaves, grass (Schwartz and Schwartz 1959), yucca fibers, or feathers (Finley 1958). Houses built outside of caves tend to have more than one entrance leading to the nest, providing multiple escape routes in times of danger (Murphy 1952).

Caching. Woodrats have earned the nickname pack rats because of their tendency to collect not only food and building materials, but also any other materials that happen to be present for placement on the house: bones, dry dung, broken glass, cans, empty shotgun shells, bits of paper (Rainey 1956), corn cobs, and rusty nails (Nawrot and Klimstra 1976). In the caches of females, Horne et al. (1998) found bones, possibly indicating an increased mineral requirement for reproduction. Eastern woodrats begin caching food in September or October, but do not cache much during spring or summer (Rainey 1956). Among captive woodrats, young model their caching choices after their mother's caching patterns, which should lead to collection of appropriate foods (Post et al. 1998). Woodrats discriminate between foods based on perishability and then decide whether to consume or cache accordingly (Reichman 1988; Post and Reichman 1991). Both adults and young cache dry foods but consume moist foods (Reichman 1988; Post et al. 1998). In Kansas, larger individuals had more days of energy stored in the cache than smaller individuals (range = 25–271 days). When smaller individuals deplete their cache, they must forage outside the house and become more vulnerable to predation. The mean energy content of caches was 3682 kcal (*SE* = 53 kcal) in October and 2369 kcal (*SE* = 22 kcal) in December (Post et al. 1993).

Other Behavior. Woodrats groom their pelage by licking and separating the fur with the front feet. Woodrat skin is relatively loose. Areas that are not easily groomed are pulled to the mouth. The face is cleaned by licking the forefeet and rubbing them in a circular motion. Dust bathing areas have been found near houses in Kansas (Rainey 1956).

Defecation and urination are usually restricted to latrine sites away from the house. Scat piles 450 mm by 250 mm and up to 50 mm deep have been recorded (Schwartz and Schwartz 1959). Urination sites 150–200 mm in diameter are indicated by a dark stain on rocks.

MORTALITY

Of 27 individuals caught as juveniles in Kansas, only 6 survived to adult size and only 3 survived long enough to reproduce (Rainey 1956). Goertz (1970) reported that males disappeared from the study area sooner than females, possibly because of exposure to predation while traveling longer distances. Eastern woodrats live longer than most other small mammals. Fitch and Rainey (1956) reported an adult male that was recaptured 827 days later. An adult female was captured throughout a 1089-day period (Monty 1997). About 5% of woodrats were recaptured after 1 year in Oklahoma during a 2-year study in prairie, savanna-edge, upland woods, and brushy riparian woods (Goertz 1970). In an Illinois upland woods and bluff habitat, 23% of woodrats remained for at least 1 year in a 3-year study (Monty 1997). In captivity, woodrats tend to live about 2 years (Birney 1973), but have lived as long as 4 years (Schwartz and Schwartz 1959).

Rainey (1956) believed that predators could keep woodrat numbers suppressed at low population levels because woodrats have relatively low reproductive potential. Possible woodrat predators include spotted skunk (*Spilogale putorius*), long-tailed weasel (*Mustela frenata*), black rat snake (*Elaphe obsoleta*), great horned owl (*Bubo virginianus*), timber rattlesnake (*Crotalus horridus*) (Fitch and Rainey 1956), red fox (*Vulpes vulpes*), gray fox (*Urocyon cinereoargenteus*), raccoon (*Procyon lotor*), opossum (*Didelphis virginiana),* cottonmouth (*Agkistrodon piscivorous*), and copperhead (*A. contortrix*) (Crim 1961). Juveniles are highly susceptible to predation by snakes while in the den before weaning.

Fitch and Rainey (1956) attributed the decline of woodrat populations in Kansas to unusually cold winters with heavy accumulations of snow and ice during two consecutive years. Extreme cold in March 1948 could have caused the death of the season's first litter. In January 1949, there was ice cover for at least 21 days. Lack of available food could have led to starvation of some individuals. Nawrot and Klimstra (1976) attributed the decline in Illinois woodrat populations to severe winters during 1912 and 1918, in conjunction with some formerly occupied sites being in marginal habitat. Neal (1967) concluded that a decline in woodrat population density in 1966 in the lower Mississippi River Basin was caused by poor acorn crops during 1964/65.

Parasites. Eastern woodrats are hosts to many kinds of parasites (Table 19.2), but the infestations are rarely a serious threat to a population's survival. Some ectoparasites are seasonal in abundance. Rainey (1956) found that 16 of 105 individuals were infested with a subcutaneous bot fly larva (*Cuterebra beameri*). One individual possessed four larvae. *Cuterebra* larvae were observed from May to November, with peak occurrence in July and August. Infestations are most severe during dry years. Ticks of the genera *Dermacentor* (Rainey 1956) and *Ixodes* (Durden et al. 1997) also occur on eastern woodrats. Nymphs and larvae of *Dermacentor* were collected from the pinnae or muzzle from May through September (Rainey 1956). Chiggers were the most numerous ectoparasite of woodrats in Kansas and often were found in clusters in the ears or scattered over the body. They were found on individuals all months of the year except April and May. Rainey (1956) collected 10 species of chiggers including the genera *Trombicula, Euschöngastia,* and *Pseudoschöngastia*. Fleas were present on woodrats in Kansas year-round (6–54% infestation rate). Three species of fleas were reported on woodrats in Kansas: *Conorhinopsylla nidicola, Epitedia wenmanni,*

TABLE 19.2. Parasites of *Neotoma floridana*

Reference	Location	Parasite
Hall 1916	Nebraska	*Nematodirus neotoma*
Stiles 1932	Florida	*Moniliformis* sp.
Murphy 1952	Oklahoma	*Longistriata neotoma*
		Böhmiella wilsoni
		Trichuris muris
		Taenia taeniaeformis
		Andrya sp.
		Cuterebra sp.
		Ixodes sp.
		Eutrombicula sp.
Rainey 1956	Kansas	*Cuterebra beameri*
		Conorhinopsylla nidicola
		Epitedia wenmanni
		Orchopeas sexdentatus
		Trombicula lipovskyi
		Trombicula sylvilagi
		Trombicula alfreddugèsi
		Trombicula lipovskyana
		Trombicula trisetica
		Trombicula cynos
		Euschöngastia peromysci
		Euschöngastia diversa
		Euschöngastia setosa
		Pseudoschöngastia farneri
		Pseudoschöngastia hungerfordi
		Dermacentor variabilis
Finley 1958	Colorado	*Brevisterna utahensis*
		Haemolaelaps glasgowi
		Euschöngastia criceticola
		Neohaematopinus neotomae
		Stenistomera alpina
		Anomiopsyllus sp.
		Orchopeas sexdentatus
		Malaraeus sp.
Wheat and Ernst 1974	Alabama	*Eimeria glauceae*
		Eimeria dusii
Durden et al. 1997	South Carolina	*Amblyomma maculatum*
		Dermacentor variabilis
		Ixodes affinis
		Ixodes minor
		Ixodes scapularis
		Androlaelaps fahrenholzi
		Eutrombicula alfreddugesi
		Eutrombicula splendens
		Listrophorus neotomae
		Ornithonyssus bacoti
		Orchopeas sexdentatus
		Polygenis gwyni
Durden et al. 1997	Georgia	*Dermacentor variabilis*
		Ixodes minor
		Androlaelaps fahrenholzi
		Androlaelaps casalis
		Eulaelaps stabularis
		Euschoengastia peromysci
		Eutrombicula alfreddugesi
		Eutrombicula batatas
		Listrophorus neotomae
		Miiyatrombicula jonesi
		Myocoptes neotomae
		Ornithonyssus bacoti
		Orchopeas howardi
		Polygenis gwyni
Lewis 1998	Illinois	*Orchopeas illinoiensis*

and the most common, *Orchopeas sexdentatus*. A recently described species of flea, *Orchopeas illinoiensis,* was collected from woodrats in Illinois (Lewis 1998). One species of sucking louse, found only on mammals, was collected from *N. floridana* in Colorado (Finley 1958).

Only 2 species of ectoparasite, the fleas *Epitedia cavernicola* and *E. neotomae,* have been described as host specific to *N. floridana* (Durden et al. 1997). In contrast, western woodrat species have at least 71 species or subspecies of host-specific ectoparasites. This discrepancy may be explained by historical woodrat dispersal and divergence events. Eastern and Allegheny woodrats are the most recent descendants of ancestral *Neotoma* forms. Therefore, it is possible that some ectoparasites of western ancestors were unable to move to the eastern United States with the dispersers and there has not been sufficient time for many host-specific ectoparasites to coevolve with the eastern species.

Endoparasites of *N. floridana* are less studied than ectoparasites. Worms from three phyla have been collected from eastern woodrats. An acanthocephalan, *Moniliformis* sp., was identified in woodrats from Florida, which represents the only member of the phylum found in *Neotoma* (Stiles 1932). Four nematode species were collected (with infestation rates in Oklahoma from Murphy [1952] in parentheses): *Nematodirus neotoma* (Hall 1916), *Longistriata neotoma* (20%), *Böhmiella wilsoni* (34%), and *Trichuris muris* (34%). Infection with *Baylisascaris procyonis,* an intestinal roundworm of raccoons, has not been reported in eastern woodrats, although it is a mortality factor in Allegheny woodrats (see Management and Conservation Needs). Two cestodes also were collected by Murphy (1952): *Taenia taeniaeformis* and *Andrya* sp. Two protists in the genus *Eimeria* were identified in woodrats in Alabama (Wheat and Ernst 1974). Confirmed reports of naturally acquired rabies in rodents are extremely rare. There is a record of an eastern woodrat captured in South Carolina that tested positive for rabies. The woodrat was acting inappropriately and having difficulty walking (Dowda et al. 1981).

Western woodrats also seem to be hosts for more microscopic endoparasites than eastern woodrats. Five western species are hosts to one or more arenaviruses (Kosoy et al. 1996). A serologic study of southwestern rodents identified hantavirus antibody in one Mexican woodrat (Hjelle et al. 1994). Dusky-footed woodrats (*N. fuscipes*) infected with *Ehrlichia* sp. were captured in northern California (Nicholson et al. 1999). *N. fuscipes* (Brown and Lane 1994), *N. mexicana* (Maupin et al. 1994), and *N. floridana* (Oliver et al. 2000) may become infected with the bacterium *Borrelia burgdorferi,* the causative agent of Lyme disease. *N. fuscipes* in southern Oregon (Hammer and Maser 1973) and *N. cinerea* in northern California may become infected with *Yersinia pestis,* the causative agent of bubonic plague, which leads to sporadic local extirpations (Nelson and Smith 1976). In Texas, the southern plains woodrat (*N. micropus*) is a reservoir of *Leishmania mexicana* (McHugh et al. 1996), which causes leishmaniasis in humans, and *Trypanosoma cruzi,* which causes Chagas disease in humans (Packchanian 1942). Forty-two percent of white-throated woodrats (*N. albigula*) collected in New Mexico had oocysts of two species of *Eimeria* in their feces (Reduker and Duszynski 1985).

OTHER WOODRATS

Allegheny Woodrat. The Allegheny woodrat (*Neotoma magister*) ranged historically from southern New York, western Connecticut, and northern New Jersey south through much of Pennsylvania, western Maryland, Virginia, North Carolina, and southern Ohio and Indiana. The Tennessee River limits the southern extent of the range. Populations have been declining during the last 20 years across the northeastern portion of the range (Fig. 19.5). New Jersey and Ohio each support only one known population (Woodrat Recovery Group 1993). Three principal causes for declines have been suggested: changes in the landscape (Balcom and Yahner 1996), reduced availability of acorns (McManus and McIntyre 1981) and American chestnuts (*Castanea dentata*) (Woods and Shanks 1959), and a nematode parasite, *Baylisascaris procyonis,* carried by raccoons (LoGiudice 2000). Woodrats are an intermediate host for *B. procyonis* and may ingest the nematode

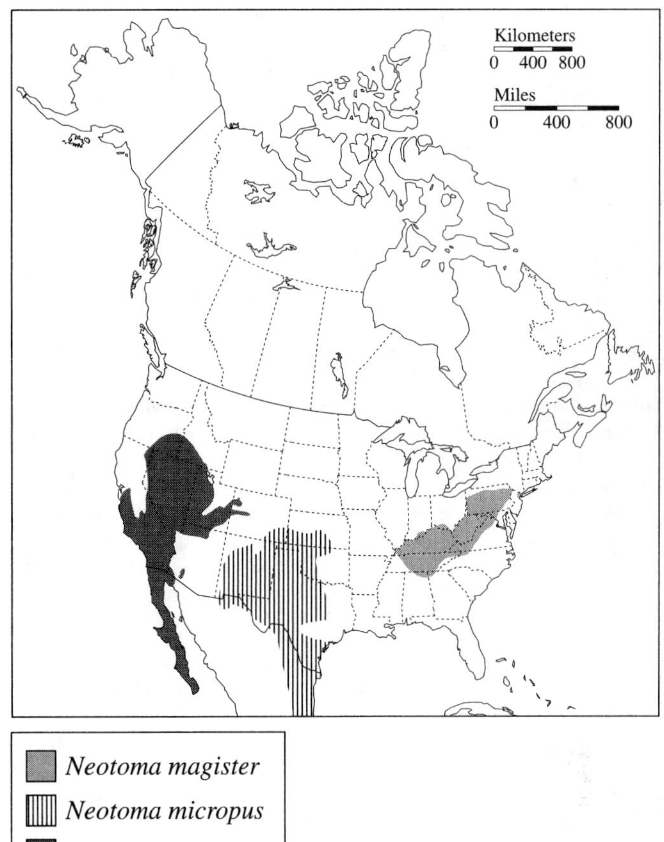

FIGURE 19.5. Distribution of the Allegheny woodrat (*Neotoma magister*), the southern plains woodrat (*Neotoma micropus*), and the desert woodrat (*Neotoma lepida*).

eggs while collecting animal feces. Infection leads to lethargy, loss of muscle control, and eventual death (Kazacos et al. 1981; Kazacos and Boyce 1989). In experimental populations, LoGiudice (2000) found that woodrats in sites with higher levels of *B. procyonis* contamination had lower survival than those in lower contamination sites.

Since the 1950s, the Allegheny woodrat has been considered a subspecies of *N. floridana* (Schwartz and Odum 1957). An investigation of genetic and morphological characteristics by Hayes (1990) indicated that the group should be returned to the species level. Allegheny woodrats usually inhabit crevices in rocky bluffs, talus slopes, and caves (Howell 1921). The propensity to live in protective rock shelters may explain why they may not cover the nest proper with a stick house, as would an eastern woodrat. Large middens of collected food and rubbish may be constructed near the nest or at the cave's mouth (Newcombe 1930). Average litter size is two, and two or three litters may be produced annually (Poole 1940).

Southern Plains Woodrat. The eastern range of *Neotoma micropus* is adjacent to the western range of *N. floridana* (see Fig. 19.5). Several characteristics distinguish the two species. The southern plains woodrat is more gray, most individuals possess a straight anterior palatal spine, and they prefer open habitat of grass (Birney 1973) and cactus (*Opuntia* spp.) rather than woodland (Box 1959). Houses built within dense cactus patches provide protection from avian and mammalian predators (Schmidly 1983). Where cacti are available, the flesh and fruits are used as a source of water during drought (Finley 1958), and cactus pads may constitute over half of house construction materials (Thies et al. 1996). Two to three young are born/litter and females may have up to five litters/year in southern populations (Birney 1973). Southern plains woodrats may hybridize with white-throated woodrats (*N. albigula*) where they are sympatric. Hybridization with *N. floridana* has produced fertile offspring in captivity (Birney 1973).

Southern plains woodrats exhibit a higher level of resistance to western diamondback rattlesnake (*Crotalus atrox*) venom than other small mammals (Perez et al. 1978a, 1978b). A serum protein antihemorrhagic factor, which has been purified and described (Garcia and Perez 1984), seems to allow woodrats to survive attacks of this common predator.

White-Throated Woodrat. White-throated woodrats (*Neotoma albigula*) are small to medium in size, which is adaptive for heat dissipation in the warm climate (Brown and Lee 1969) of their range in the southwestern United States (see Fig. 19.1). In New Mexico, average adult weights were 224 g for males and 188 g for females (Bailey 1931). They are usually associated with arid, moderately rocky slopes, clumps of prickly pear (*Opuntia* spp.), or pinyon–juniper woodland (Finley 1958). Cactus (Vorhies and Taylor 1940) and juniper (Finley 1958) are important sources of food and building materials. Houses in rock are preferred over houses built under trees or cacti (Olsen 1973). The dorsal pelage is brown. The throat and pectoral regions are white to the base of the hairs. Melanistic populations occur on lava beds (Blair 1954). Litter size ranges from one to three, and two or more litters may be produced each year (Finley 1958).

Desert Woodrat. The desert woodrat (*Neotoma lepida*) is a small ($\bar{X} =$ 122 and 107 g for adult males and females, respectively), buffy-colored species (Finley 1958), which uses diverse habitats throughout its range (see Fig. 19.5). In western Colorado, *N. lepida* prefers rocky habitat. Houses were constructed of sticks, cactus spines, and cactus joints in horizontal crevices. Foliage of shrubs, forbs, and small trees and the joints of cacti were used for food and water (Finley 1958). A population in Utah lived in a juniper–sagebrush community (Stones and Hayward 1968). Stick houses were built at the base of junipers. The foliage and berries of juniper were most prevalent in food caches. In the eastern Mojave Desert in California, *N. lepida* built primarily around Mojave yucca (*Yucca schidigera*) and buckhorn cholla (*Opuntia acanthocarpa*) (Smith 1995). Foods used in the Mojave included turbinella oak (*Quercus turbinella*) cuttings and yucca shoots (Cameron and Rainey 1972). Burt (1934) reported that at least two litters of four young were produced annually.

Arizona Woodrat. Previously considered a subspecies of the desert woodrat, *N. devia* was recommended for promotion to the species level by Mascarello (1978) and Koop et al. (1985). Arizona woodrats build small houses in rock fissures or areas with little cover throughout their range (Fig. 19.6). The dorsal pelage is pale gold to buffy, the venter is white, and the throat has dark gray fur (Hoffmeister 1986). Preferred foods are cholla cactus (*Opuntia* spp.), Mormon tea (*Ephedra viridis*), wolfberry (*Lycium* sp.), creosote bush (*Larrea tridentata*), and mesquite (*Prosopis glandulosa*). Mean litter size is 2.1 and young may be born throughout the year.

Stephens' Woodrat. *Neotoma stephensi* is limited to rocky habitat in northern Arizona, southern Utah, and western New Mexico (Fig. 19.7). They specialize in using junipers for building materials, shelter, food, and water (Vaughan 1982). Harris (1963) observed Stephens' woodrat houses in vertical crevices, at the base of junipers, and 1 m above the ground in the branches of junipers. The dorsal pelage is yellow to gray-buff and the ventral pelage is creamy or white (Hall 1981). The tail is semibushy. The litter size usually is one. Females may lose up to 40% of their body weight during lactation and the young grow more slowly than those of other *Neotoma* species. Most females had only a few litters annually, but some were estimated to have up to five litters/year (Vaughan and Czaplewski 1985).

Mexican Woodrat. The Mexican woodrat, *N. mexicana,* occurs from Colorado south to El Salvador (see Fig. 19.6). The dorsal pelage ranges from gray to bright rufous and the ventral fur is white to yellow (Hall 1981). Melanistic individuals have been collected on dark lava beds in New Mexico (Blair 1941). The species is primarily montane and tends

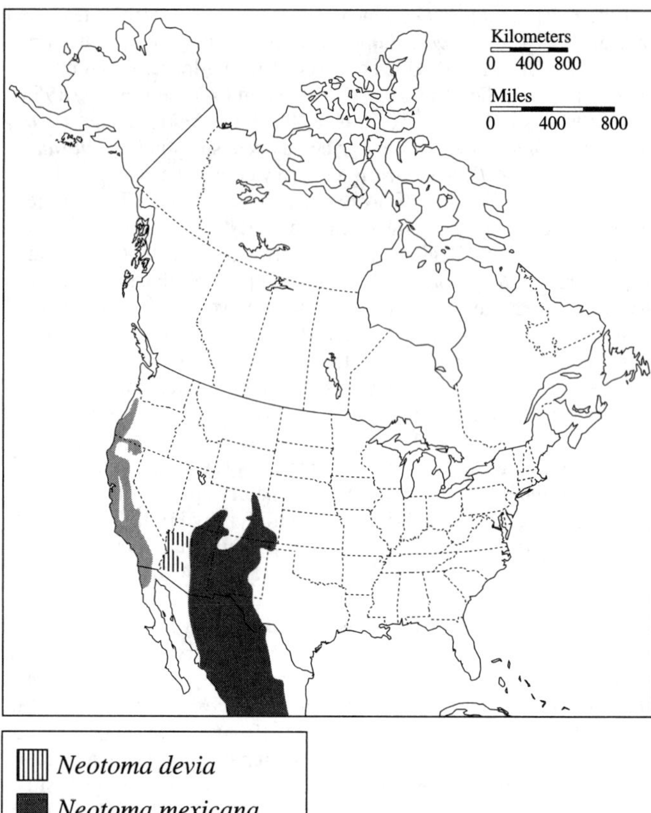

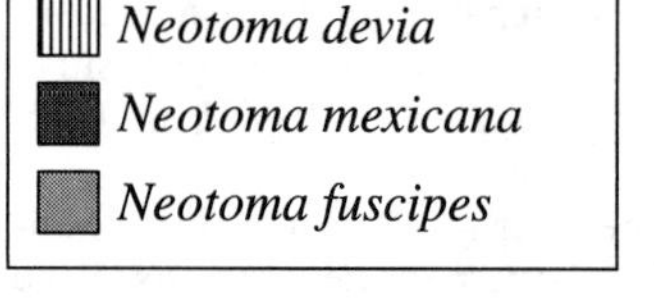

FIGURE 19.6. Distribution of the Arizona woodrat (*Neotoma devia*), the Mexican woodrat (*Neotoma mexicana*), and the dusky-footed woodrat (*Neotoma fuscipes*).

to use rock outcrops, cliffs (Finley 1958), and boulder fields (Findley et al. 1975). In Mexico, they are located in riparian forests (Hooper 1955) and tropical thorn forests (Birney and Jones 1971). Stick houses typical of other species of *Neotoma* are less commonly built by Mexican woodrats. Nests often are built in rock crevices, tree cavities, or abandoned buildings. They forage on common plants including scrub oak (*Quercus gambellii*), skunkbush, and juniper (Finley 1958). In Colorado, litter size was two to five, with two litters/year. Females born in April or May matured rapidly and produced a litter in June or July (Brown 1969).

Bushy-Tailed Woodrat. The bushy-tailed woodrat (*N. cinerea*) is the largest member of the genus. Adult males may weigh more than 500–600 g (Escherich 1981) in the northern portion of their range (see Fig. 19.7). The dorsal pelage varies from pale gray to brown-black and the ventral fur is white to buff. The tail and hind foot sole (from heel to posterior tubercle) are fully furred (Finley 1958). Bushy-tailed woodrats live in mountainous habitat among Douglas fir (*Pseudotsuga menziesii*), spruce (*Picea* spp.), aspen (*Populus tremuloides*), ponderosa pine (*Pinus ponderosa*), or pinyon–juniper woodland. Free-standing houses are seldom constructed. Crevices and caves are preferred house sites, which are supplemented with dry vegetation and sticks. Human-made structures are also used for shelter (Findley et al. 1975). The leaves of a variety of shrubs and trees may be dried before caching (Finley 1958). Following investigations of different populations, researchers suggested that bushy-tailed woodrats exhibit a polygynous (Escherich 1981) or promiscuous (Topping and Millar 1996, 1998) mating system. Females exhibit philopatry and interact amicably with kin (Moses and Millar 1992). In most areas females have one or two litters/year and litter size ranges from one to six (Egoscue 1962).

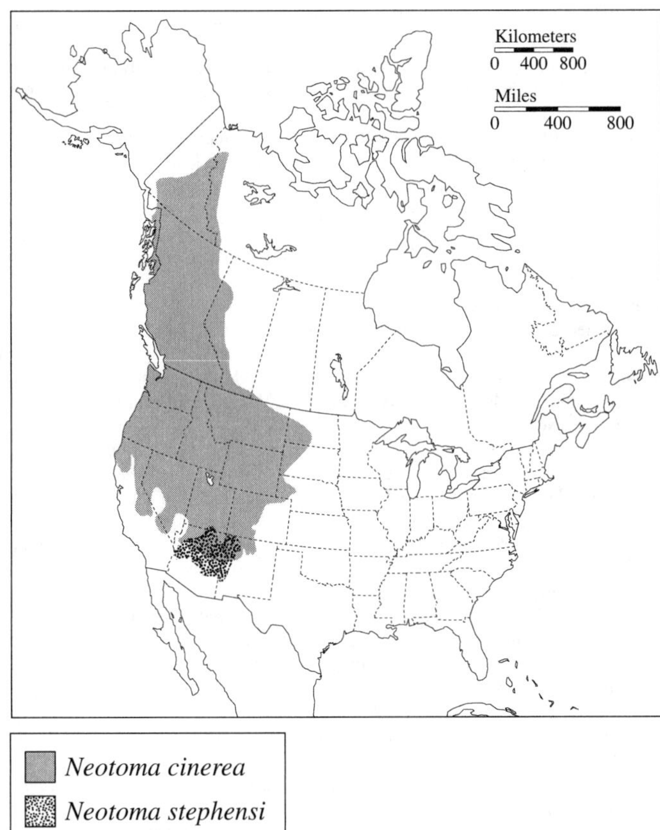

FIGURE 19.7. Distribution of the bushy-tailed woodrat (*Neotoma cinerea*) and the Stephens' woodrat (*Neotoma stephensi*).

Dusky-Footed Woodrat. The dusky-footed woodrat (*N. fuscipes*) is found from western Oregon south into northern Baja California (see Fig. 19.6). The dorsal pelage is gray-brown, the ventral pelage is pale or white, and the top of the hind foot is covered in sooty-colored fur (Howell 1926). Large conical houses are built of sticks and plant cuttings in or under trees, on brushy hillsides (English 1923), or below rocky bluffs (Emerson and Howard 1978). The dusky-footed woodrat is a specialist in scrub and woodland communities (Meserve 1974). Population density decreases if underbrush is removed from the habitat (Linsdale and Tevis 1956). Cuttings of shrubs and other vegetation, fruits, seeds, and acorns are consumed and cached (English 1923). Oak cuttings are preferred when available (Atsatt and Ingram 1983). One litter is born annually (Vestal 1938), with a mean litter size of 2.8 (English 1923). Populations are most abundant in brushy areas (Sakai and Noon 1997) and forest edge (Ward et al. 1998). While foraging in old-growth forest at night, they serve as an important food source for Northern spotted owls (*Strix occidentalis*) (Sakai and Noon 1997; Ward et al. 1998). Dusky-footed woodrats are considered the primary reservoir of *Borrelia burgdorferi* in the western United States (Gage et al. 1995).

ECONOMIC STATUS

Although woodrats were regarded as agricultural pests in the past, more recently they have not been considered to be of economic importance (Rainey 1956). Where woodrats live close to agricultural fields or storage buildings, some crop depredation may occur (Goldman 1910; Howell 1921; Murphy 1952). One cache in Texas held almost a bushel of pecans (Strecker 1929). Shrubs and cacti may be damaged or killed due to girdling or overuse for food, water, and building materials in desert regions (Goldman 1910). Woodrats may live in parked automobiles or chew on wires and other vehicle parts (Salmon and Gorenzel 1994). An Allegheny woodrat was found electrocuted after gnawing on wires in an Indiana cave (Mumford and Whitaker 1982). Substantial damage may be inflicted on buildings if woodrats take up residency. Upholstered furniture and mattresses may be shredded for nest materials (Salmon and Gorenzel 1994). Woodrats are hosts of parasites that can transmit zoonoses, but this rarely occurs because of minimal contact between humans and woodrats. On the positive side, woodrats increase soil fertility by depositing vegetation and fecal material into houses and facilitate seed dispersal for many species. Humans have used woodrats as food in the northeastern states (Newcombe 1930), the Southwest (Finley 1958), and Mexico (Goldman 1910; Rainey 1956).

MANAGEMENT AND CONSERVATION NEEDS

Eastern Woodrat Management. In many states, eastern woodrats are considered to be relatively common, so that few resources have been invested in population assessments. Lack of studies of long-term population dynamics or habitat requirements results in few management plans for the species. In regions experiencing population declines, monitoring programs and habitat protection are taking place. To ensure long-term survival of the species, suitable habitat must remain undeveloped by humans. As development encroaches on more remote areas, local populations may become threatened. Woodrats that settle in human structures may be perceived as a nuisance and removed by trapping. Populations in isolated habitat may be extirpated because of stochastic factors (Soulé 1987). Such areas are unlikely to naturally repopulate, due to human-influenced changes in landscapes, including agricultural use, housing developments, and roads. Reintroduction into previously inhabited sites may aid in local recovery projects. For example, it was recommended that formerly inhabited sites in southeastern Illinois should be repopulated by translocations of woodrats from southwestern Illinois (Monty 1997). Research on the effects of timber harvests near eastern woodrat colonies should be conducted (see Threatened Woodrat Species). Removal of mature mast trees would reduce food availability, although brush piles left behind could provide shelter for houses.

Researchers face many challenges while collecting basic data. Woodrat behavior is difficult to observe because of their nocturnal activity and use of houses or caves. Photography and direct observation using red light have revealed activity outside of shelters (Finley 1959; Wiley 1971). Mate choice, breeding success, parentage, and genetic diversity studies will be facilitated as genetic techniques are developed and the technology becomes more widely available. Radiotelemetry projects could investigate dispersal patterns of subadults, although cliff habitat makes location of individuals difficult. Periodic monitoring of endangered populations is necessary, but livetrapping may pose a threat to captured woodrats. Mortality may result from disturbance by predators, attack on neonates by ants (Stones and Hayward 1968), and accidental trap deaths. Little information exists on the prevalence of *Baylisascaris procyonis* in raccoons living in the vicinity of *N. floridana* populations. Birch et al. (1994) found only a 5% occurrence of *B. procyonis* among raccoons sampled in southern Illinois. Prevalence of *B. procyonis* seems to decrease in southern latitudes (Harkema and Miller 1964), which may protect eastern woodrats from the extirpations experienced by Allegheny woodrats.

The Key Largo woodrat (*N. floridana smalli*) is listed as endangered by both the Florida Fish and Wildlife Conservation Commission and the U.S. Fish and Wildlife Service. The most recent population assessment was conducted in 1995, 3 years after Hurricane Andrew (Frank et al. 1997). The resulting density estimates were much lower than estimates from the 1980s and indicated a significant decline in numbers at the southern end of North Key Largo. To protect the remaining tracts of hammock, land has been acquired by state and federal agencies to create preserves for native species. The following management recommendations and research areas have been identified: semiannual livetrapping for monitoring presence/absence and long-term population dynamics; determination of structural requirements of woodrat habitat, the effects of exotic species, human development, and habitat fragmentation; and investigation of the possibility of future habitat restoration efforts. The effect of fire ants (*Solenopsis invicta*) on woodrat survival has not been studied. Livetrapped small mammals may be killed by fire ants, and similarly, helpless woodrat pups in terrestrial houses could experience increased mortality.

Threatened Woodrat Species. More research must be conducted to determine the causes of declines in northeastern populations of Allegheny woodrats. The nematode parasite *Baylisascaris procyonis* has been identified as a mortality factor in New York (McGowan 1993a), New Jersey (LoGiudice 2000), and Indiana (Johnson 2002). LoGiudice (1995) developed and tested a raccoon bait dosed with piperazine, an anthelmintic, for short-term, local control of *B. procyonis*. Periodic worming of resident raccoons could increase survival for small woodrat populations of special concern. Reintroduction of Allegheny woodrats into historical sites could occur if the causes of the extirpations are determined and rectified. Historical sites and currently used habitat should be monitored for raccoon latrine sites and *B. procyonis* (McGowan 1993b). Allegheny woodrat population monitoring programs have been implemented to detect trends in population size and survival. In West Virginia, timber harvests near woodrat colonies had minimal impact as long as intact forest remained nearby (Castleberry et al. 2001).

The Coronados Island woodrat (*N. bunkeri*), endemic to Coronados Island in the Gulf of California, may be at extremely low numbers or extinct because of introduced feral cats and the destruction of vegetation for firewood (Smith et al. 1993). In Zacatecas, Mexico, four species of *Neotoma* face drastic alterations to the landscape due to agriculture. Grassland has been planted to crops and stream valleys have been cleared of tropical vegetation for sugar cane (Matson and Baker 1986). Goldman's woodrats (*N. goldmani*), despite living in rocky, desert habitat, have experienced a population decline possibly related to human influences (Matson and Baker 1986).

Woodrats as Environmental Indicators. In the western United States, middens of *N. lepida, N. mexicana, N. albigula,* and *N. cinerea* have been studied to conduct paleoenvironmental reconstruction (Betancourt et al. 1990). Plant macrofossils present reflect the local flora within the woodrat's home range, whereas airborne pollen deposits represent the regional flora. Crystallized urine forms a hard coating over the deposited materials and stops decay of organic matter. Upon soaking in water, the urine dissolves and macrofossils, pollen, bones, and invertebrate remains can be identified and dated (Wells 1976). Middens dating from recent to >40,000 radiocarbon years before present reveal changing vegetation communities because of climatic shifts during the late Pleistocene and Holocene (Wells 1983). Between 18,000 and 12,000 years ago, the Mojave Desert supported juniper woodland below 1000 m, pinyon–juniper woodland at 1000–1800 m, and cold-tolerant trees at higher elevations (Spaulding 1990). Pollen and macrofossils from the present-day Sonoran Desert reveal a mixture of woodland and desert species during the late Pleistocene (King and Van Devender 1977). If removed from their dry shelters by vandals, middens break down quickly and the information contained is lost.

Smith and Betancourt (1998) reported that woodrat body size is a precise paleothermometer. Size of *N. cinerea* was estimated from fecal pellets recovered from middens. Body size decreased as temperature increased following the last glacial maximum about 21,000 years ago. The inverse relationship between temperature and body size in *Neotoma* was further supported by a study in New Mexico that found *N. albigula* body size decreased from 1989 through 1996 when the temperature increased 2–3°C (Smith et al. 1998). Woodrats may be used as indicators for assessing the effects of global warming on animal morphology, physiology, life histories, abundance, distribution, and availability of food sources.

LITERATURE CITED

Allen, J. A. 1894. Descriptions of ten new North American mammals, and remarks on others. Bulletin of the American Museum of Natural History 6:317–32.

Asdell, S. A. 1964. Patterns of mammalian reproduction. Cornell University Press, Ithaca, NY.

Atsatt, P. R., and T. Ingram. 1983. Adaptation to oak and other fibrous, phenolic-rich foliage by a small mammal, *Neotoma fuscipes*. Oecologia (Berlin) 60:135–42.

Bailey, V. 1905. Biological survey of Texas. North American Fauna 25:1–222.

Bailey, V. 1931. Mammals of New Mexico. North American Fauna 53:1–412.

Baker, R. J., and J. T. Mascarello. 1969. Karyotypic analyses of the genus *Neotoma* (Cricetidae, Rodentia). Cytogenetics 8:187–98.

Balcom, B. J., and R. H. Yahner. 1996. Microhabitat and landscape characteristics associated with the threatened Allegheny woodrat. Conservation Biology 10:515–25.

Bangs, O. 1898. The land mammals of peninsular Florida and the coast region of Georgia. Proceedings of the Boston Society of Natural History 28:157–235.

Betancourt, J. L., T. R. Van Devender, and P. S. Martin, eds. 1990. Packrat middens: The last 40,000 years of biotic change. University of Arizona Press, Tucson.

Birch, G. L., G. A. Feldhamer, and W. G. Dyer. 1994. Helminths of the gastrointestinal tract of raccoons in southern Illinois with management implications of *Baylisascaris procyonis* occurrence. Transactions of the Illinois State Academy of Science 87:165–70.

Birney, E. C. 1973. Systematics of three species of woodrats (genus *Neotoma*) in central North America. University of Kansas Museum of Natural History Miscellaneous Publications 58:1–173.

Birney, E. C., and J. K. Jones, Jr. 1971. Woodrats (genus *Neotoma*) of Sinaloa, Mexico. Transactions of the Kansas Academy of Science 74:197–211.

Birney, E. C., and J. E. Perez. 1971. Inheritance of multiple hemoglobins in two species of woodrats, genus *Neotoma* (Rodentia, Cricetidae). University of Kansas Science Bulletin 49:345–56.

Blair, F. W. 1941. Annotated list of mammals of the Tularosa Basin, New Mexico. American Midland Naturalist 26:218–29.

Blair, F. W. 1954. A melanistic race of the white-throated packrat (*Neotoma albigula*) in Texas. Journal of Mammalogy 35:239–42.

Box, T. W. 1959. Density of plains wood rat dens on four plant communities in south Texas. Ecology 40:715–16.

Brown, J. H., and A. K. Lee. 1969. Bergmann's rule and climatic adaptation in woodrats (*Neotoma*). Evolution 23:329–38.

Brown, L. N. 1969. Reproductive characteristics of the Mexican woodrat at the northern limit of its range in Colorado. Journal of Mammalogy 50:536–41.

Brown, R. N., and R. S. Lane. 1994. Natural and experimental *Borrelia burgdorferi* infections in woodrats and deer mice from California. Journal of Wildlife Diseases 30:389–98.

Burt, W. H. 1934. The mammals of southern Nevada. Transactions of the San Diego Society of Natural History 7:375–427.

Burt, W. H., and F. S. Barkalow, Jr. 1942. A comparative study of the bacula of wood rats (subfamily Neotominae). Journal of Mammalogy 23:287–97.

Cameron, G. N., and D. G. Rainey. 1972. Habitat utilization by *Neotoma lepida* in the Mojave Desert. Journal of Mammalogy 53:251–66.

Castleberry, S. B., W. M. Ford, P. B. Wood, N. L. Castleberry, and M. T. Mengak. 2001. Movements of Allegheny woodrats in relation to timber harvesting. Journal of Wildlife Management 65:148–56.

Castleberry, S. B., T. L. King, P. B. Wood, and W. M. Ford. 2002. Microsatellite DNA analysis of population structure in Allegheny woodrats (*Neotoma magister*). Journal of Mammalogy 83:1058–70.

Clarke, J. W. 1973. The specialized midventral gland of the eastern woodrat, *Neotoma floridana osagensis*. M.S. Thesis, Kansas State Teachers College, Emporia.

Crim, J. A. 1961. The habitat of the woodrat in southern Illinois. M.S. Thesis, Southern Illinois University, Carbondale.

Dewsbury, D. A. 1974. Copulatory behavior of *Neotoma floridana*. Journal of Mammalogy 55:864–66.

Dowda, H., A. F. DiSalvo, and S. Redden. 1981. Naturally acquired rabies in an eastern wood rat (*Neotoma floridana*). Journal of Clinical Microbiology 13:238–39.

Durden, L. A., C. W. Banks, K. L. Clark, B. V. Belbey, and J. H. Oliver, Jr. 1997. Ectoparasite fauna of the eastern woodrat, *Neotoma floridana:* Composition, origin, and comparison with ectoparasite faunas of western woodrat species. Journal of Parasitology 83:374–81.

Egoscue, H. J. 1962. The bushy-tailed woodrat: A laboratory colony. Journal of Mammalogy 43:328–37.

Emerson, D. O., and W. E. Howard. 1978. Mineralogy of woodrat, *Neotoma cinerea,* urine deposits from northeastern California. Journal of Mammalogy 59:424–25.

English, P. F. 1923. The dusky-footed wood rat (*Neotoma fuscipes*). Journal of Mammalogy 4:1–9.

Escherich, P. C. 1981. Social biology of the bushy-tailed woodrat, *Neotoma cinerea*. University of California Publications in Zoology 110:1–132.

Feldhamer, G. A., L. C. Drickamer, S. H. Vessey, and J. F. Merritt. 1999. Mammalogy: Adaptation, diversity, and ecology. WCB/ McGraw-Hill, Boston.

Findley, J. S., A. H. Harris, D. E. Wilson, and C. Jones. 1975. Mammals of New Mexico. University of New Mexico Press, Albuquerque.

Finley, R. B., Jr. 1958. The wood rats of Colorado: Distribution and ecology. University of Kansas Publications of the Museum of Natural History 10:213–552.

Finley, R. B., Jr. 1959. Observations of nocturnal animals by red light. Journal of Mammalogy 40:591–94.

Fitch, H. S., and D. G. Rainey. 1956. Ecological observations on the woodrat, *Neotoma floridana*. University of Kansas Museum of Natural History Publications 8:499–533.

Frank, P., F. Percival, and B. Keith. 1997. A status survey for the Key Largo woodrat (*Neotoma floridana smalli*) and Key Largo cotton mouse (*Peromyscus gossypinus allapaticola*) on North Key Largo, Monroe County, Florida. Unpublished report to the U.S. Fish and Wildlife Service.

Gage, K. L., R. S. Ostfeld, and J. G. Olson. 1995. Nonviral vector-borne zoonoses associated with mammals in the United States. Journal of Mammalogy 76:695–715.

Garcia, V. E., and J. C. Perez. 1984. The purification and characterization of an antihemorrhagic factor in woodrat (*Neotoma micropus*) serum. Toxicon 22:129–38.

Goertz, J. W. 1970. An ecological study of *Neotoma floridana* in Oklahoma. Journal of Mammalogy 51:94–104.

Goldman, E. A. 1910. Revision of the wood rats of the genus *Neotoma*. North American Fauna 31:1–124.

Hall, E. R. 1981. The mammals of North America, 2nd ed. John Wiley, New York.

Hall, M. C. 1916. Nematode parasites of mammals of the orders Rodentia, Lagomorpha, and Hyracoidea. Proceedings of the U.S. National Museum 50:1–258.

Hamilton, W. J., Jr. 1953. Reproduction and young of the Florida wood rat, *Neotoma f. floridana* (Ord). Journal of Mammalogy 34:180–89.

Hammer, E. W., and C. Maser. 1973. Distribution of the dusky-footed woodrat, *Neotoma fuscipes* Baird, in Klamath and Lake counties, Oregon. Northwest Science 47:123–27.

Harkema, R., and G. C. Miller. 1964. Helminth parasites of the raccoon, *Procyon lotor* in the southeastern United States. Journal of Parasitology 50:60–66.

Harper, F. 1927. The mammals of Okefenokee Swamp region of Georgia. Proceedings of the Boston Society of Natural History 38:191–396.

Harris, A. 1963. Ecological distribution of some vertebrates in the San Juan Basin, New Mexico. Museum of New Mexico Papers in Anthropology 8:1–63.

Hayes, J. P. 1990. Biogeographic, systematic, and conservation implications of geographic variation in woodrats of the eastern United States. Ph.D. Dissertation, Cornell University, Ithaca, NY.

Hayes, J. P., and R. G. Harrison. 1992. Variation in mitochondrial DNA and the biogeographic history of woodrats (*Neotoma*) of the eastern United States. Systematic Biology 41:331–44.

Hjelle, B., F. Chavez-Giles, N. Torrez-Martinez, T. Yates, J. Sarisky, J. Webb, and M. Ascher. 1994. Genetic identification of a novel hantavirus of the harvest mouse *Reithrodontomys megalotis*. Journal of Virology 68:6751–54.

Hoffmeister, D. F. 1986. Mammals of Arizona. University of Arizona Press and Arizona Game and Fish Department, Tucson.

Hoffmeister, D. F. 1989. The mammals of Illinois. University of Illinois Press, Urbana.

Hooper, E. T. 1955. Notes on mammals of western Mexico. Occasional Papers of the Museum of Zoology, University of Michigan 565:1–26.

Horne, E. A., M. McDonald, and O. J. Reichman. 1998. Changes in cache contents over winter in artificial dens of the eastern woodrat (*Neotoma floridana*). Journal of Mammalogy 79:898–905.

Howell, A. B. 1926. Anatomy of the woodrat. Williams and Wilkins, Baltimore.

Howell, A. H. 1910. Notes on mammals of the middle Mississippi Valley, with descriptions of a new woodrat. Proceedings of the Biological Society of Washington 12:23–34.

Howell, A. H. 1921. A biological survey of Alabama. North American Fauna. 45:1–88.

Howell, A. H. 1934. Description of a new subspecies of the Florida woodrat. Proceedings of the Academy of Natural Sciences of Philadelphia 86:403–4.

Humphrey, S. R. 1988. Density estimates of the endangered Key Largo woodrat and cotton mouse (*Neotoma floridana smalli* and *Peromyscus gossypinus allapaticola*) using the nested grid approach. Journal of Mammalogy 69:524–31.

Johnson, S. A. 2002. Reassessment of the Allegheny woodrat (*Neotoma magister*) in Indiana. Proceedings of the Indiana Academy of Science 1:56–66.

Kazacos, K. R., and W. M. Boyce. 1989. *Baylisascaris* larva migrans. Journal of the American Veterinary Medicine Association 195:894–903.

Kazacos, K. R., W. L. Wirtz, and P. P. Burger. 1981. Raccoon ascarid larvae as a cause of fatal central nervous system disease in subhuman primates. Journal of the American Veterinary Medicine Association 179:1089–94.

Kellogg, R. 1915. The mammals of Kansas with notes on their distribution, life histories and economic importance. M.S. Thesis, University of Kansas, Lawrence.

King, J. E., and T. R. Van Devender. 1977. Pollen analysis of fossil packrat middens from the Sonoran Desert. Quaternary Research 8:191–204.

Kinsey, K. P. 1977. Agonistic behavior and social organization in a reproductive population of Allegheny woodrats, *Neotoma floridana magister*. Journal of Mammalogy 58:417–19.

Koop, B. F., R. J. Baker, and J. T. Mascarello. 1985. Cladistical analysis of chromosomal evolution within the genus *Neotoma*. Occasional Papers, the Museum, Texas Tech University 96:1–9.

Kosoy, M. L., L. H. Elliott, T. G. Ksiazek, C. F. Fulhorst, P. E. Rollin, J. E. Childs, J. N. Mills, G. O. Maupin, and C. J. Peters. 1996. Prevalence of antibodies to arenaviruses in rodents from the southern and western United States: Evidence for an arenavirus associated with the genus *Neotoma*. American Journal of Tropical Medicine and Hygiene 54:570–76.

Layne, J. N. 1958. Notes on mammals in southern Illinois. American Midland Naturalist 60:219–54.

Lee, A. K. 1963. The adaptations to arid environments in wood rats of the genus *Neotoma*. University of California Publications in Zoology 64:57–96.

Lewis, R. E. 1998. A new species of *Orchopeas,* Jordan, 1933 from the midwestern United States (Siphonaptera: Ceratophyllidae). Journal of Medical Entomology 35:399–403.

Linsdale, J. M., and L. P. Tevis, Jr. 1956. A five-year change in an assemblage of wood rat houses. Journal of Mammalogy 37:371–74.

LoGiudice, K. 1995. Control of *Baylisascaris procyonis* (Nematoda) in raccoons (*Procyon lotor*) through the use of anthelmintic baits: A potential method for reducing mortality in the Allegheny woodrat (*Neotoma floridana magister*). M.S. Thesis, Rutgers University, New Brunswick, NJ.

LoGiudice, K. 2000. *Baylisascaris procyonis* and the decline of the Allegheny woodrat (*Neotoma magister*). Ph.D. Dissertation, Rutgers University, New Brunswick, NJ.

Lowery, G. H. 1974. The mammals of Louisiana and its adjacent waters. Louisiana State University Press, Baton Rouge.

Mascarello, J. T. 1978. Chromosomal, biochemical, mensural, penile, and cranial variation in desert woodrats (*Neotoma lepida*). Journal of Mammalogy 59:477–95.

Matson, J. O., and R. H. Baker. 1986. Mammals of Zacatecas. Texas Tech University Museum Special Publications 24:1–88.

Maupin, G. O., K. L. Gage, J. Piesman, J. Montenieri, S. L. Sviat, L. VanderZanden, C. M. Happ, M. Dolan, and B. J. B. Johnson. 1994. Discovery of an enzootic cycle of *Borrelia burgdorferi* in *Neotoma mexicana* and *Ixodes spinipalpis* from northern Colorado, an area where Lyme disease is nonendemic. Journal of Infectious Disease 170:636–43.

McGowan, E. 1993a. Experimental release and fate study of the Allegheny woodrat (*Neotoma magister*). Unpublished report. New York State Department of Environmental Conservation, Endangered Species Unit.

McGowan, E. 1993b. Procedure for assessing *Baylisascaris procyonis* contamination in Allegheny woodrat habitat. Unpublished report. New York State Department of Environmental Conservation, Endangered Species Unit.

McHugh, C. P., P. C. Melby, and S. G. LaFon. 1996. Leishmaniasis in Texas: Epidemiology and clinical aspects of human cases. American Journal of Tropical Medicine and Hygiene 55:547–55.

McManus, M. L., and T. McIntyre. 1981. Introduction. Pages 1–32 *in* C. C. Doane, and M. L. McManus, eds. The gypsy moth: Research toward integrated pest management (Technical Bulletin 1584). USDA Forest Service Science and Education Agency Animal and Plant Health Inspection Service.

Mearns, E. A. 1897. Descriptions of six new mammals from North America. Proceedings of the U.S. National Museum 19:719–24.

Merriam, C. H. 1894. Abstract of a study of the American wood rats, with descriptions of fourteen new species and subspecies of the genus *Neotoma*. Proceedings of the Biological Society of Washington 9:117–28.

Meserve, P. L. 1974. Ecological relationships of two sympatric woodrats in a California coastal sage scrub community. Journal of Mammalogy 55:442–47.

Monty, A.-M. 1997. The eastern woodrat (*Neotoma floridana*) in southern Illinois: Population assessment and genetic variation. Ph.D. Dissertation, Southern Illinois University, Carbondale.

Moses, R. A., and J. S. Millar. 1992. Behavioural asymmetries and cohesive mother–offspring sociality in bushy-tailed wood rats. Canadian Journal of Zoology 70:597–604.

Mumford, R. E., and J. O. Whitaker, Jr. 1982. Mammals of Indiana. Indiana University Press, Bloomington.

Murphy, M. F. 1952. Ecology and helminths of the Osage wood rat, *Neotoma floridana osagensis,* including description of *Longistriata neotoma* n. sp. (Trichostrongylidae). American Midland Naturalist 48:204–18.

Nawrot, J. R., and W. D. Klimstra. 1976. Present and past distribution of the endangered southern Illinois woodrat (*Neotoma floridana illinoensis*). (Natural History Miscellanea No. 196). Chicago Academy of Sciences.

Neal, W. A. 1967. A study of the ecology of the woodrat in the hardwood forests of the lower Mississippi River Basin. M.S. Thesis, Louisiana State University, Baton Rouge.

Nelson, B. C., and C. R. Smith. 1976. Ecological effects of a plague epizootic on the activities of rodents inhabiting caves at Lava Beds National Monument, California. Journal of Medical Entomology 13:51–61.

Newcombe, C. L. 1930. An ecological study of the Allegheny cliff rat (*Neotoma pennsylvanica* Stone). Journal of Mammalogy 11:204–11.

Nicholson, W. L., M. B. Castro, V. L. Kramer, J. W. Sumner, and J. E. Childs. 1999. Dusky-footed wood rats (*Neotoma fuscipes*) as reservoirs of granulocytic ehrlichiae (Rickettsiales: Ehrlichieae) in northern California. Journal of Clinical Microbiology 37:3323–27.

Oliver, J. H., Jr., K. L. Clark, F. W. Chandler, Jr., L. Tao, A. M. James, C. W. Banks, L. O. Huey, A. R. Banks, D. C. Williams, and L. A. Durden. 2000. Isolation, cultivation, and characterization of *Borrelia burgdorferi* from rodents and ticks in the Charleston area of South Carolina. Journal of Clinical Microbiology 38:120–24.

Olsen, R. W. 1973. Shelter-site selection in the white-throated woodrat, *Neotoma albigula*. Journal of Mammalogy 54:594–610.

Ord, G. 1818. *In* H. De Blainville, Sur une nouvelle espèce de rongeur de la Floride, par M. Ord de Philadelphie. Bulletin des Sciences (Société Philomatique de Paris) 1818:181–82.

Oswald, C., and P. A. McClure. 1990. Energetics of concurrent pregnancy and lactation in cotton rats and woodrats. Journal of Mammalogy 71:500–9.

Packchanian, A. 1942. Reservoir hosts of Chagas disease in the state of Texas. American Journal of Tropical Medicine 22:623.

Pearson, P. G. 1952. Observations concerning the life history and ecology of the wood rat *Neotoma floridana floridana* (Ord). Journal of Mammalogy 33:459–63.

Perez, J. C., W. C. Haws, V. E. Garcia, and B. M. Jennings, III. 1978a. Resistance of warm-blooded animals to snake venoms. Toxicon 16:375–83.

Perez, J. C., W. C. Haws, and C. H. Hatch. 1978b. Resistance of woodrats (*Neotoma micropus*) to *Crotalus atrox* venom. Toxicon 16:198 200.

Poole, E. L. 1940. A life history sketch of the Allegheny woodrat. Journal of Mammalogy 21:249–70.

Post, D. M. 1992. Change in nutrient content of foods stored by eastern woodrats (*Neotoma floridana*). Journal of Mammalogy 73:835–39.

Post, D. M. 1993. Detection of differences in nutrient concentrations by eastern woodrats (*Neotoma floridana*). Journal of Mammalogy 74:493–97.

Post, D. M., and O. J. Reichman. 1991. Effects of food perishability, distance, and competitors on caching behavior by eastern woodrats. Journal of Mammalogy 72:513–17.

Post, D. M., O. J. Reichman, and D. E. Wooster. 1993. Characteristics and significance of the caches of eastern woodrats (*Neotoma floridana*). Journal of Mammalogy 74:666–92.

Post, D. M., M. W. McDonald, and O. J. Reichman. 1998. Influence of maternal diet and perishability on caching and consumption behavior of juvenile eastern woodrats. Journal of Mammalogy 79:156–62.

Rainey, D. G. 1956. Eastern woodrat, *Neotoma floridana:* Life history and ecology. University of Kansas Publications of the Museum of Natural History 8:535–646.

Reduker, D. W., and D. W. Duszynski. 1985. *Eimeria ladronensis* n. sp. and *E. albigulae* (Apicomplexa: Eimeriidae) from the woodrat, *Neotoma albigula* (Rodentia: Cricetidae). Journal of Protozoology 32:548–50.

Reichman, O. J. 1988. Caching behavior by eastern woodrats, *Neotoma floridana,* in relation to food perishability. Animal Behaviour 36:1525–32.

Sakai, H. F., and B. R. Noon. 1997. Between-habitat movement of dusky-footed woodrats and vulnerability to predation. Journal of Wildlife Management 61:343–50.

Salmon, T. P., and W. P. Gorenzel. 1994. Pages B-133–36 *in* S. E. Hygnstrom, R. M. Timm, and G. E. Larson, eds. Prevention and control of wildlife damage. University of Nebraska Cooperative Extension, USDA, and Great Plains Agricultural Council, Lincoln, NB.

Schmidly, D. J. 1983. Texas mammals east of the Balcones Fault Zone. Texas A&M University Press, College Station.

Schwartz, A., and E. P. Odum. 1957. The woodrats of the eastern United States. Journal of Mammalogy 38:197–206.

Schwartz, C. W., and E. R. Schwartz. 1959. The wild mammals of Missouri. University of Missouri Press, Columbia.

Sherman, H. B. 1955. Description of a new race of woodrat from Key Largo, Florida. Journal of Mammalogy 36:113–20.

Shipley, M. M., F. B. Stangl, Jr., R. L. Cate, and C. S. Hood. 1990. Immunoelectrophoretic relationships among four species of woodrats (Cricetidae: *Neotoma*). Southwestern Naturalist 35:173–76.

Smith, F. A. 1995. Den characteristics and survivorship of woodrats (*Neotoma lepida*) in the eastern Mojave Desert. Southwestern Naturalist 40:366–72.

Smith, F. A., and J. L. Betancourt. 1998. Response of bushy-tailed woodrats (*Neotoma cinerea*) to late Quaternary climatic change in the Colorado Plateau. Quaternary Research 50:1–11.

Smith, F. A., B. T. Bestelmeyer, J. Biardi, and M. Strong. 1993. Anthropogenic extinction of the endemic woodrat, *Neotoma bunkeri* Burt. Biodiversity Letters 1:149–55.

Smith, F. A., H. Browning, and U. L. Shepherd. 1998. The influence of climate change on the body mass of woodrats *Neotoma* in an arid region of New Mexico, USA. Ecography 21:140–48.

Soulé, M. E., ed. 1987. Viable populations for conservation. Cambridge University Press, Cambridge.

Spaulding, W. G. 1990. Vegetational and climatic development of the Mojave Desert: The last glacial maximum to the present. Pages 166–99 *in* J. L. Betancourt, T. R. Van Devender, and P. S. Martin, eds. Packrat middens: The last 40,000 years of biotic change. University of Arizona Press, Tucson.

Stiles, C. W. 1932. Helminthological notes. Journal of Parasitology 19:90.

Stones, R. C., and C. L. Hayward. 1968. Natural history of the desert woodrat, *Neotoma lepida*. American Midland Naturalist 80:458–76.

Strecker, J. K. 1929. Notes on the Texan cotton and Attwater wood rats in Texas. Journal of Mammalogy 10:216–20.

Svihla, A., and R. D. Svihla. 1933. Notes on the life history of the woodrat, *Neotoma floridana rubida* Bangs. Journal of Mammalogy 14:73–75.

Tate, W. H., Jr. 1970. Movements of *Neotoma floridana attwateri* in Brazos County, Texas. M.S. Thesis, Texas A&M University, College Station.

Thies, K. M., M. L. Thies, and W. Caire. 1996. House construction by the southern plains woodrat (*Neotoma micropus*) in southwestern Oklahoma. Southwestern Naturalist 41:116–22.

Topping, M. G., and J. S. Millar. 1996. Spatial distribution in the bushy-tailed wood rat (*Neotoma cinerea*) and its implications for the mating system. Canadian Journal of Zoology 74:565–69.

Topping, M. G., and J. S. Millar. 1998. Mating patterns and reproductive success in the bushy-tailed woodrat (*Neotoma cinerea*), as revealed by DNA fingerprinting. Behavioral Ecology and Sociobiology 43:115–24.

Vaughan, T. A. 1982. Stephens' woodrat, a dietary specialist. Journal of Mammalogy 63:53–62.

Vaughan, T. A., and N. J. Czaplewski. 1985. Reproduction in Stephens' woodrat: The wages of folivory. Journal of Mammalogy 66:429–43.

Vestal, E. H. 1938. Biotic relations of the wood rat (*Neotoma fuscipes*) in the Berkeley Hills. Journal of Mammalogy 19:1–36.

Vorhies, C. T., and W. P. Taylor. 1940. Life history and ecology of the white-throated wood rat, *Neotoma albigula albigula* Hartley, in relation to grazing in Arizona. University of Arizona, College of Agriculture Technical Bulletin 86:454–529.

Wagle, E. R., and G. A. Feldhamer. 1997. Feeding habits of the eastern woodrat (*Neotoma floridana*) in southern Illinois. Transactions of the Illinois State Academy of Science 90:171–77.

Ward, J. P., Jr., R. J. Gutierrez, and B. R. Noon. 1998. Habitat selection by northern spotted owls: The consequences of prey selection and distribution. Condor 100:79–92.

Wells, P. V. 1976. Macrofossil analysis of wood rat (*Neotoma*) middens as a key to the Quaternary vegetational history of arid America. Quaternary Research 6:223–48.

Wells, P. V. 1983. Paleobiogeography of montane islands in the Great Basin since the last glaciopluvial. Ecological Monographs 53:341–82.

Wheat, B. E., and J. V. Ernst. 1974. *Eimeria glauceae* sp. n. and *Eimeria dusii* sp. n. (Protozoa: Eimeriidae) from the eastern woodrat, *Neotoma floridana,* from Alabama. Journal of Parasitology 60:403–5.

Wiley, R. W. 1971. Activity periods and movements of the eastern woodrat. Southwestern Naturalist 16:43–54.

Wiley, R. W. 1980. *Neotoma floridana*. Mammalian Species 139:1–7.
Wilson, D. E., and D. M. Reeder, eds. 1993. Mammal species of the world, 2nd ed. Smithsonian Institution Press, Washington, DC.
Woodrat Recovery Group. 1993. The Allegheny woodrat: Where now and where next? Presented at conference held 19–20 March, Dickinson College/Pennsylvania Biological Survey.
Woods, F. W., and R. E. Shanks. 1959. Natural replacement of chestnut by other species in the Great Smoky Mountains National Park. Ecology 13:349–61.
Worth, C. B. 1950. Observations on the behavior and breeding of captive rice rats and woodrats. Journal of Mammalogy 31:421–26.
Zimmerman, E. G., and M. E. Nejtek. 1975. The hemoglobins and serum albumins of three species of woodrats (*Neotoma* Say and Ord). Comparative Biochemistry and Physiology 50B:275–78.
Zimmerman, E. G., and M. E. Nejtek. 1977. Genetics and speciation of three semispecies of *Neotoma*. Journal of Mammalogy 58:391–402.

Anne-Marie Monty, Biology Department, Indiana University Purdue University Indianapolis, Indianapolis, Indiana 46202. Email: monty@indy.net.

Robert E. Emerson, Department of Pathology and Laboratory Medicine, Indiana University School of Medicine, Indianapolis, Indiana 46202. Email: reemerso@iupui.edu.

IV

Toothed Whales and Baleen Whales

20

Bottlenose Dolphin, Harbor Porpoise, Sperm Whale and Other Toothed Cetaceans

Tursiops truncatus, Phocoena phocoena, and *Physeter macrocephalus*

Randall R. Reeves
Andrew J. Read

NOMENCLATURE

COMMON NAMES. Bottlenose(d) dolphin or porpoise, Tursiops (especially by people who work with cetaceans in captivity); tonina, Tursión, bufeo, delfín naríz de botella (Spanish); Tursion, dauphin à gros nez (French)
SCIENTIFIC NAME. *Tursiops truncatus* (Montagu, 1821)

Systematics. Odontoceti, a suborder of Cetacea, contains more than 70 species worldwide, of which well over half occur in waters bordering the North American continent (Rice 1998). The odontocetes, or toothed whales, are subdivided into at least 10 well-defined families and about 34 genera. The bottlenose dolphins belong to the most speciose family, the Delphinidae, which includes more than 30 extant species in 17 genera. A phylogenetic study using cytochrome *b* sequencing (LeDuc et al. 1999) demonstrated that the Delphinidae are badly in need of systematic revision.

The taxonomy of the genus *Tursiops* is muddled (Rice 1998), and the genus is likely polyphyletic (LeDuc et al. 1999). Two nominal species are widely recognized: *T. truncatus,* the common bottlenose dolphin; and *T. aduncus,* the Indo-Pacific bottlenose dolphin. Only the former occurs in North American waters and is discussed in this chapter. In some areas where bottlenose dolphins have been studied closely, two morphologically and ecologically separate forms of *T. truncatus* have been described: a large-bodied form and a smaller, more attenuate form (Walker 1981; Hersh and Duffield 1990; Mead and Potter 1990, 1995). Analyses of mitochondrial and nuclear sequences indicate fixed differences between these two forms, so they may warrant differentiation at the species level (LeDuc and Curry 1997; Hoelzel et al. 1998).

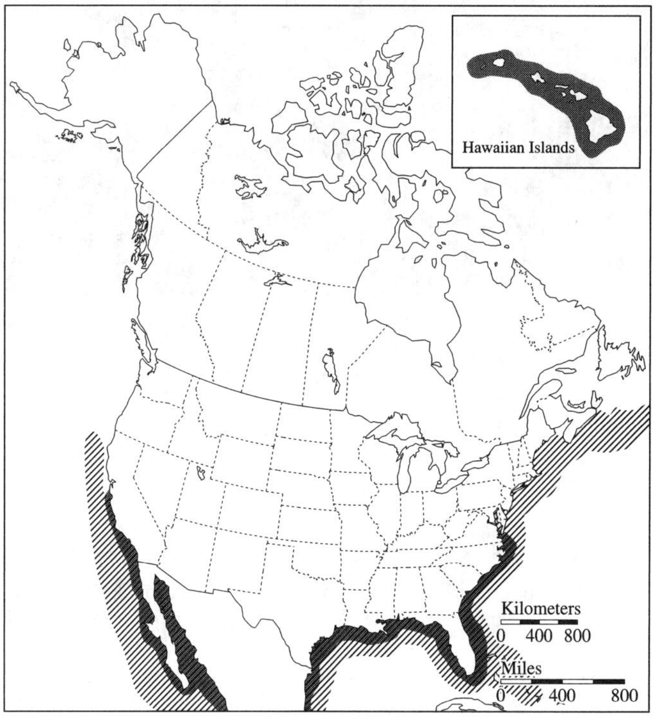

Hawaiian Islands
Kilometers
0 400 800
Miles
0 400 800

FIGURE 20.1. Distribution of the bottlenose dolphin (*Tursiops truncatus*).

DISTRIBUTION

The global distribution of bottlenose dolphins spans all tropical and temperate marine waters (Wells and Scott 1999). Along the western margins of the North Atlantic, the coastal ecotype is found close to shore (usually <5 km from land) and in bays, sounds, and estuaries (Mead and Potter 1990) (Fig. 20.1). The offshore ecotype occurs primarily in deeper waters along the shelf break (200–2000 m) north and east to Georges Bank (Kenney 1990) and the Scotian Shelf (Gowans and Whitehead 1995). The two ecotypes overlap in some areas, and mixed schools have been reported (Torres et al. 2003). The year-round distribution of coastal bottlenose dolphins is continuous from North Carolina to Texas; they are common north of Cape Hatteras only during summer (Mead and Potter 1990). In many areas of the southeastern United States, coastal bottlenose dolphins occur as a series of resident "communities" defined by patterns of association and individual home ranges (Scott et al. 1990; Scott 1990; Wells and Scott 1999).

In the eastern Pacific, bottlenose dolphins occur in the neritic zone as far north as the Southern California Bight. During the El Niño event in 1982–1983, they extended their range north of Point Conception, and individual dolphins from San Diego were photographed in Monterey Bay (Wells et al. 1990). During the 1990s, bottlenose dolphins continued to be present along the central California coast, expanding their normal range to at least as far north as San Francisco (Feinholz 1995; Carretta et al. 2002). An animal judged to be of the coastal form stranded about 100 km north of Seattle, Washington, in 1988 (Ferrero and Tsunoda 1989), and bottlenose dolphin remains found in an Indian midden in southern Washington suggest past incursions that far north (Baird et al. 1993). Although coastal animals in the eastern Pacific are distinguishable morphologically from offshore animals (Walker 1981), genetic separation between inshore and offshore populations has not been confirmed (Curry and Smith 1997). Because the coastal population is almost entirely confined to waters within 1 km of shore, dolphins seen a few kilometers or farther from shore off Baja California, California, Oregon, or Washington are likely offshore animals. The normal range of offshore dolphins extends at least as far north as 41°N (Forney et al. 2000).

The tuna purse seine fishery in the eastern tropical Pacific (see Gosliner [1999] for a thorough review) has made it possible to plot the extensive occurrence of bottlenose dolphins in this vast area (Scott and Chivers 1990). They are present in pelagic waters, frequently associated

FIGURE 20.2. Bottlenose dolphins are robust and demonstrative, occasionally leaping high above the surface. SOURCE: Photo by Kim W. Urian.

with other delphinids (see below). Communities of bottlenose dolphins are often associated with volcanic island groups such as the Hawaiian Islands, the Marquesas, and the Tuamotus.

DESCRIPTION

The head is robust and the rostrum relatively short, particularly in comparison with many tropical delphinids. The dorsal fin is moderately falcate and located near the midpoint of the back. The body tapers to a keel behind the dorsal fin. The flippers are moderately long and convex, tapering to a rounded point at the tips. The flukes are curved along the rear margin and separated by a distinct notch (Fig. 20.2). A moderate degree of sexual dimorphism can be present in some forms. For example, in the western North Atlantic coastal form, the dorsal fins of adult males are larger and more scarred than those of adult females (Tolley et al. 1995) (Fig. 20.3).

As is true of their morphology, the coloration of bottlenose dolphins is that of a generalized delphinid (Mitchell 1970). They are slate to charcoal gray on the dorsal surfaces, although the appearance of animals in the field can vary with lighting conditions. Subtle but complex patterns of shading occur over the head, including a dark stripe from the eye to the insertion of the flipper and a well-defined dark "bridle" (Mitchell 1970). A darker cape demarcates the dorsal field from the lighter lateral field. The abdomen is light gray and can be clearly demarcated from the lateral surfaces. The bellies of dolphins that are excited or exercising can be pink or pink tinged. The coloration of individual dolphins varies considerably.

In coastal waters of the Gulf of Mexico, male and female bottlenose dolphins are approximately 110 cm long at birth (Fernandez and Hohn 1998). In this region, males and females reach total body lengths of approximately 265 and 250 cm, respectively (Read et al. 1993; Fernandez and Hohn 1998). Although not pronounced, this dimorphism in length is significant (Tolley et al. 1995). Adult males of this form are approximately 25% heavier than females, and grow to greater than 250 kg. They achieve this greater size by continuing to add mass after females begin to invest energy in reproduction (Read et al. 1993). In other areas, bottlenose dolphins have been reported to reach almost 4 m in length, although such large specimens are uncommon. Body size can be used to distinguish populations in some areas. For example, in the western North Atlantic, offshore specimens are, on average, approximately 15% larger than their coastal counterparts (Mead and Potter 1995).

The skull of the bottlenose dolphin, like that of all odontocetes, is telescoped and asymmetrical (Howell 1930) (Fig. 20.4). The rostrum is

FIGURE 20.3. Individual recognition based on high-resolution photographs of the dorsal fin has been the principal research tool in many studies of coastal bottlenose dolphin populations. The notching pattern along the trailing edge of the fin provides a unique identifier. SOURCE: Photo by Kim W. Urian.

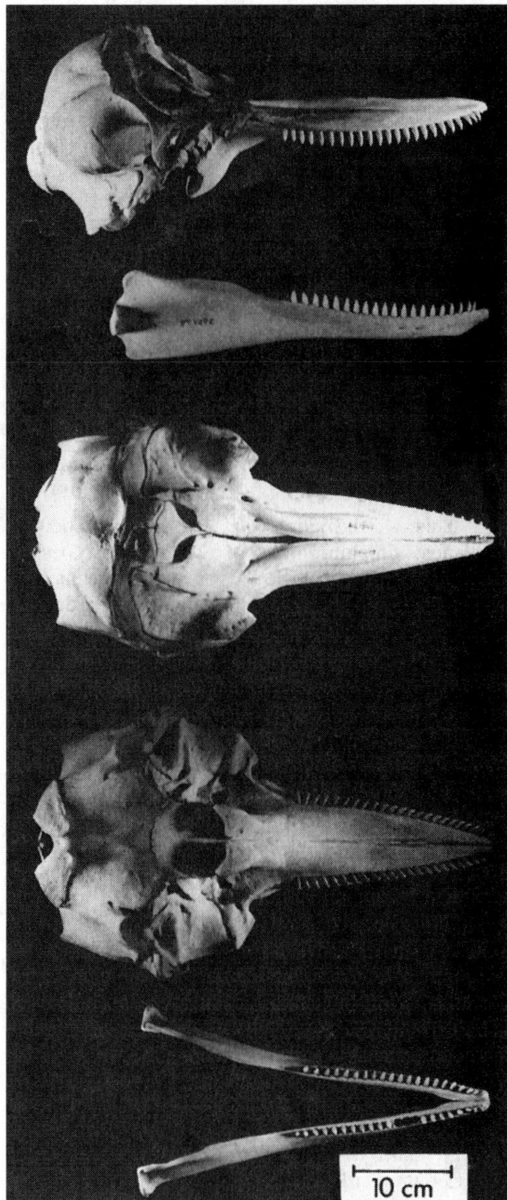

FIGURE 20.4. Skull of the bottlenose dolphin (*Tursiops truncatus*). From top to bottom: lateral view of cranium, lateral view of mandible, dorsal view of cranium, ventral view of cranium, dorsal view of mandible.

relatively short, narrow, and somewhat blunt, especially in older specimens. The mandibular symphysis is relatively short and does not exceed 10–15% of the length of the mandible. The mandibles are hollow and filled with lipid-rich adipose tissue, which is used to conduct sound energy. There are 18–24 simple, robust conical teeth in each row. In the western North Atlantic, the relative diameter of the internal nares of offshore dolphins is greater than that of coastal dolphins, and the two forms can be separated unequivocally using this feature (Mead and Potter 1995).

Seven cervical vertebrae are present; the first two are completely fused. Some of the other cervical vertebrae may be ankylosed to the atlas and axis, although this is uncommon. There are 12–14 thoracic vertebrae, 15–19 lumbar vertebrae, and 23–30 caudal vertebrae (Rommel 1990; Wells and Scott 1999). Twelve to 14 vertebral ribs are present, the first 4 or 5 of which are double headed. A pair of floating ribs is also present. The skull and postcranial skeleton are described fully by Rommel (1990), the axial musculature and connective tissues by Pabst (1990), and the central nervous system by Ridgway (1990).

REPRODUCTION

Female bottlenose dolphins ovulate spontaneously and may cycle several times in a season (Kirby and Ridgway 1984; Schroeder 1990). Both ovaries are functional, although there is a preponderance of ovulatory activity in the left organ (Harrison 1977). Gestation lasts for 12 months and a single calf is born, usually tail first. The seasonality of reproduction varies from region to region (Urian et al. 1996). In some areas, such as the Gulf coast of Texas, reproduction is synchronous, with a single seasonal peak of births (Fernandez and Hohn 1998). Other populations exhibit bimodal birth peaks or diffuse reproductive seasonality. Female dolphins captured in the wild and held in captivity tend to maintain the reproductive seasonality of their population of origin (Urian et al. 1996). Testes size and sperm production also vary seasonally (Schroeder and Keller 1989).

Newborn calves are precocious and are seldom separated from their mothers in the first few months of life (Mann and Smuts 1998). The calf will be nursed by its mother for several years, although it may begin to forage independently during its first or second year of life (Wells et al. 1987). The period of dependency varies considerably and may be influenced by the age, nutritive condition, and reproductive status of the mother (Whitehead and Mann 2000). In general, calves are weaned completely during their mother's next pregnancy (Mann et al. 2000). Typical interbirth intervals range from 2 to 6 years (Connor et al. 2000).

After a few years of juvenile independence, female bottlenose dolphins generally attain sexual maturity between the ages of 6 and 10 years (Connor et al. 2000). The age at attainment of sexual maturity in males has not been well documented because the process is more gradual and less apparent than in females.

AGE ESTIMATION

Bottlenose dolphins are long-lived mammals and some females are estimated to have lived for more than 50 years (Hohn et al. 1989; Wells and Scott 1999). Ages are estimated from field observations of known-age animals (Wells et al. 1987) or from counts of dentinal and cemental "growth layer groups" in thin, decalcified sections of teeth (Hohn et al. 1989). Annuli have been calibrated by examining tooth sections from known-age individuals.

ECOLOGY

Home Ranges and Movement. The habitat of bottlenose dolphins varies from brackish, semienclosed estuaries to the pelagic ocean (Fig. 20.5). Coastal dolphins probably have smaller ranges than pelagic animals, but this supposition has not been tested (Fig. 20.6). It is clear, however, that coastal bottlenose dolphins exhibit considerable plasticity in their movement patterns. In California, dolphins are relatively mobile, with individuals traveling as much 286 km along the coast in 14 days (Defran et al. 1999). Bottlenose dolphins along the Atlantic coast may have the largest home ranges: They migrate annually from North Carolina in the winter to as far north as New York in the summer (Mead and Potter 1990; McLellan et al. 2002). In contrast, the dolphin community in Sarasota Bay, Florida, occupies a home range of only about 125 km^2, primarily in bays that are protected by barrier islands (Wells et al. 1987; Scott et al. 1990). The reasons for this regional variation in ranging patterns are not fully understood, but may be related to seasonal variation in prey availability, water temperature, and dispersion of prey patches.

Within an area, there may be significant individual variation in movement patterns. In Sarasota Bay, for example, adult males travel more widely than adult females (Urian 2002), perhaps because males are roving in search of receptive females (Connor et al. 2000). There is no evidence that coastal dolphins of either sex disperse far from their natal waters. For example, rates of immigration into and emigration out of the Sarasota community are <2–3% per year (Wells and Scott 1990). As more researchers adopt standard approaches to measuring

FIGURE 20.5. Bottlenose dolphins are often seen in the surf zones along the southeastern and southwestern coasts of North America. Their large, falcate dorsal fin can arouse shark fears in casual observers. SOURCE: Photo by Danielle Waples.

home range, it will be possible to test hypotheses regarding variation in the movements of dolphins at the individual and population levels.

In the southeastern United States, bottlenose dolphins are often found in or near seagrass beds (Shane 1990). Some authors have suggested that this association can be explained by the diverse assemblage of potential prey in such areas (Barros and Wells 1998). Allen et al. (2001) tested the hypothesis that bottlenose dolphins prefer to forage in seagrass habitats in the eastern Gulf of Mexico. In this area at least, dolphins preferred to forage in other types of habitat where the prey density was low but the mean prey size was greater than in seagrasses. Seagrass beds are, nonetheless, critical to the health of bottlenose dolphin populations because they provide important "nursery" habitat for the juvenile stages of many important prey species (Allen et al. 2001).

The open nature of their habitat suggests that offshore bottlenose dolphins have a high degree of mobility. Suggestive evidence is provided by the satellite tracks of two stranded dolphins that were

FIGURE 20.6. True (1891) noticed that northward-migrating dolphins taken in the Cape Hatteras fishery often had stalked barnacles (*Xenobalanus*) on the trailing edges of their fins, while south-bound migrants in the autumn and winter did not. The fishermen called the former "tassel-fins." SOURCE: Photo by Kim W. Urian.

rehabilitated and released in Florida (Wells et al. 1999). One of them, classified as "intermediate" between the coastal and offshore forms based on hematocrit and red blood cell count, moved from Clearwater in the Gulf of Mexico to the Atlantic Ocean off North Carolina, covering at least 2050 km in 43 days after release. The other dolphin, judged an offshore animal from blood values and body measurements, moved from a release site off Cape Canaveral to the Virgin Islands, covering at least 4200 km in the 47 days following release. Both dolphins remained in waters between 23°C and 26°C. The depth preferences exhibited by these two animals were consistent with their "ecotype" classifications—the one classified as intermediate remained in shelf waters and the one classified as offshore moved into very deep water seaward of the shelf edge (Wells et al. 1999). There is no other direct evidence of movements and ranging patterns of individual offshore bottlenose dolphins.

ABUNDANCE

Mitchell (1975b) made a cumulative catch estimate of about 14,000 for the stock(s) of bottlenose dolphins available to the Cape Hatteras dolphin fishery in the early 1880s (see below). Recent estimates of abundance have been derived from direct counts, line-transect surveys, and photographic mark–recapture studies. Estimates are available for all stocks of bottlenose dolphins in the United States, as mandated by the Marine Mammal Protection Act (MMPA). However, the accuracy and precision of these estimates vary. For example, relatively precise estimates and trend data exist for many of the more than 20 bottlenose dolphin communities ("stocks") identified in coastal waters of the Gulf of Mexico. Minimum total estimates for the Gulf of Mexico in the mid-1990s were somewhat more than 43,000 on the outer continental shelf, about 4500 on the shelf edge and slope, and 19,000 near the coast (Waring et al. 2002). Attempts to estimate abundance of coastal dolphins along the east coast of the United States have been hampered because their distribution overlaps with that of the offshore form and because the population structure of the coastal dolphins themselves is complex. Thus, estimates used in management prior to 2002 were simply uncorrected counts made in a 1-km strip adjacent to the coast. The current approach recognizes smaller management units within the "Western North Atlantic coastal stock," with estimates of overall abundance ranging between about 9200 from summer surveys to 19,500 from winter surveys (Waring et al. 2002). An estimate of the California coastal population is 250–300, with no evidence of trend (Defran and Weller 1999).

Abundance of offshore bottlenose dolphins generally is estimated using shipboard line transect surveys. For example, based on surveys in 1998, the estimate for offshore dolphins in the western North Atlantic is 30,633 (coefficient of variation [CV] = 0.25) (Waring et al. 2002). The most recent (1991–1996) estimate for U.S. waters of the eastern North Pacific is 956 (CV = 0.14) (Carretta et al. 2002).

BEHAVIOR

Bottlenose dolphins live in fission–fusion societies, in which group composition changes frequently as individuals join and leave groups (Connor et al. 2000). Within dolphin societies, however, some stable, long-term bonds may exist. Calves stay with their mothers for several years until they are weaned. In areas such as Sarasota, pairs of adult males form close bonds that last for decades, perhaps to increase their individual reproductive success and better defend themselves against predators (Wells et al. 1987). Male pairs are not found in all populations, however. In Moray Firth, Scotland, for example, no evidence has been found of such pairing behavior (Wilson 1995).

The mating system of bottlenose dolphins is the subject of much current research. In Sarasota, solitary males and pairs of males associate with receptive females, sometimes for extended periods (Wells et al. 1987). It is not clear whether these associations are coerced, as they are in the Indo-Pacific bottlenose dolphins in Shark Bay, Australia (Connor et al. 1992). It is also not clear which of the males in a pair, or both, mate with a female during these extended "consortships." The sexual dimorphism in size and greater frequency of scarring in adult males could mean that they compete for females (Tolley et al. 1995).

The size of bottlenose dolphin groups varies with habitat, region, and season. In many areas, it is difficult to define a "group" of dolphins in the field, and researchers have adopted various definitions, making comparisons among studies difficult. In odontocetes generally, there is a trend toward larger groups as one moves farther offshore from coastal to pelagic habitats (Norris and Dohl 1980). Bottlenose dolphins can occur in very large groups in pelagic environments (Scott and Chivers 1990), but similarly large groups are sometimes observed in coastal habitats.

FEEDING HABITS

The diet of bottlenose dolphins has been studied extensively from stomach contents (e.g., Leatherwood et al. 1978; Barros and Odell 1990; Barros and Wells 1998) and by watching animals in the wild (e.g., Leatherwood 1975; Shane 1990; Rossbach and Herzing 1997). Like many other aspects of their biology, the diet and feeding behavior of bottlenose dolphins is remarkably variable, and they have adapted to a variety of ecological conditions.

Bottlenose dolphins eat fish and cephalopods, but there is considerable variation in diet among individuals, regions, and populations. For example, there is a distinct difference in feeding habits between coastal and offshore animals in the western North Atlantic; individuals can be assigned to an ecotype based on their stomach contents alone (Mead and Potter 1990, 1995). In this region, coastal dolphins feed primarily on demersal fish of the family Sciaenidae and offshore dolphins consume mesopelagic fish (family Myctophidae) and squid (Barros and Odell 1990; Mead and Potter 1990, 1995; Walker et al. 1999). Many coastal bottlenose dolphins prey on soniferous (sound-producing) fishes, such as croakers (Sciaenidae), grunts (Haemulidae), and toadfish (Batrachoididae). They may locate these prey by passively listening (Barros and Myrberg 1987).

In several areas, coastal bottlenose dolphins have developed unusual methods of prey capture; unfortunately, very little is known of the predatory behavior of the offshore ecotype. In the Bahamas, dolphins bury their entire heads in the sandy substrate while pursuing infaunal prey (Rossbach and Herzing 1997). In the salt marshes of Georgia and South Carolina, small groups feed cooperatively by driving schools of prey onto mud banks and then following them *out of the water* and capturing individual fish on the banks (Hoese 1971) (Fig. 20.7). Near Sarasota, individual dolphins propel fish many meters in the air by powerful blows of their flukes, a behavior referred to as "fish-whacking" (Wells et al. 1987). This list of feeding behaviors is by no means complete. The cooperative aspects of prey capture in the "strand-feeding" behavior are unusual, and, contrary to popular belief, most coastal bottlenose dolphins pursue and capture prey as individuals.

Bottlenose dolphins use natural and anthropogenic features of their environment to enhance prey capture. They often herd fish against barriers, such as sand bars or sea walls (Reynolds et al. 2000). Dolphins have also learned to feed around commercial fishing vessels, particularly shrimp trawlers in the Gulf of Mexico and off the southeastern United States (Gunter 1942; Leatherwood 1975). Dolphins may feed on organisms that have been concentrated or disturbed by fishing gear or on fish that are discarded by fishermen.

MORTALITY

Bottlenose dolphins are large, upper-trophic-level animals, but in most areas their lives are not free from predators. Various elasmobranchs are known to prey on dolphins (Wood et al. 1970). Evidence comes from direct observations, scars indicating unsuccessful attacks, and the stomach contents of sharks. In Sarasota Bay, for example, >30% of bottlenose dolphins carry the scars of unsuccessful predation attempts (Urian et al. 1998); most shark attacks focus on calves and juveniles (Corkeron et al. 1987). Captive dolphins trained to attack smaller sharks became agitated when bull sharks (*Carcharhinus leucas*), known

FIGURE 20.7. Some bottlenose dolphins in coastal areas pursue fish schools up against, and even onto, the muddy banks of salt marshes. These animals managed to wriggle back into the water after being photographed while partially stranded in a marsh at Hilton Head, South Carolina. SOURCE: Photo by Kim W. Urian.

predators of bottlenose dolphins, were introduced into their tanks (Irvine et al. 1973). Killer whales (*Orcinus orca*) also prey on bottlenose dolphins in some areas (Würsig and Würsig 1979).

Bottlenose dolphins sometimes die from injuries caused by stingray (Dasyatidae) barbs (Walsh et al. 1988). The barbs migrate through body tissues and may eventually reach critical organ systems, killing the dolphin months or perhaps years after its encounter with the stingray.

Large numbers of bottlenose dolphins have died from epizootics in several areas of North America (Lipscomb et al. 1996). Most notable was the event of 1987–1988, in which more than 740 dolphin carcasses were recovered from beaches along the eastern seaboard (Scott et al. 1988). The die-off was initially attributed to exposure to brevitoxins associated with a red tide event, but later determined to have been caused by a morbillivirus (Lipscomb et al. 1994; Duignan et al. 1996). Federal resource managers estimated that >50% of the bottlenose dolphins along the U.S. east coast may have died (Scott et al. 1988).

ECONOMIC STATUS

Historical. Direct fisheries for bottlenose dolphins have existed in many parts of the world (e.g., Mitchell 1975b; Wells and Scott 1999). Among the most intensive of these were operations in North Carolina (from 1797, off and on until 1929) and New Jersey (1884–1885) (Mead 1975; Leatherwood and Reeves 1982).

According to Mead (1975; citing Stick 1958), the Hatteras "porpoise" fishery (for bottlenose dolphins) began at Ocracoke Inlet in 1797. Davis et al. (1982:37) claimed that in 1800 a "short lived" seine fishery began near Cape Lookout. During a survey of the coast between Cape Fear and Cape Hatteras in 1806, Tatham (1806) noted that the people of Beaufort, North Carolina, "carry on the Porpoise and other fisheries, jointly with the People of the Straights of Core Sound, at Cape Look-out, where about two hundred barrels of oil were produced last year." This amount of oil, if taken only from bottlenose dolphins (at 7 gal per animal and 31.5 gal per barrel), would represent a catch of about 900 animals. Some of the oil, however, may have come from whales (Reeves and Mitchell 1988a). Oil from "domestic fisheries" was expected to supply the Cape Lookout lighthouses in those days (Tatham 1806) (Fig. 20.8).

The fishing equipment and methods were described by Copley Amory in a 1928 manuscript prepared on behalf of the William F. Nye Company, New Bedford, Massachusetts, a major supplier of fine oils for watches and other instruments. Amory visited the Outer Banks to investigate the "porpoise" fisheries there. The outer net, which served as a fence, was in four sections, each about 40 fathoms long (1 fathom = 6 ft or approximately 1.83 m) and two fathoms deep, with 5.5-in. mesh. The inner or sweep net was 120 fathoms long, in two sections. The seaward end of each section was 5 fathoms deep, with 4-in. mesh; the shore ends tapered to 1 fathom deep, with 5.25-in. mesh. These nets were deployed by four dory crews, each consisting of two oarsmen, a net handler, and a steersman. At least two "spare" men remained on the beach. The dolphins were herded toward shore and killed in the shallows with long knives. The dories were 21 ft long, 6 ft across, and 22 in. deep. On shore, there was a central camp and two signal stations, 1.25 miles north and south of the main camp, respectively. Each camp and signal station had a signal pole and "waif" (a bundle of dry brush used in place of a flag) for alerting the camps when a school of dolphins was sighted. The investment in boats amounted to about $1000, and in nets about $1500.

Carcasses were processed in a machine "splitter," which separated the body blubber from the hides. The blubber was boiled in open kettles, then strained after settling, and the oil was put in barrels for marketing. The average yield of body oil per dolphin varied from about 8 gal in the winter to only 3–4 gal later in the season (True 1891). In the early twentieth century, this oil was worth about 35 cents/gal, the same as the oil from baleen whales (Stevenson 1904). The most valuable product was the "fine" or "head" oil, obtained from the jaws and melon. Each dolphin produced about 1 quart of this oil, according to C. Amory (see above). The jaws, according to Stevenson (1904), produced about $\frac{1}{2}$ pint of high-quality oil, worth $5–10/gal. The flensed carcasses were transported to a boiling house, where they were steamed to separate the bones from the flesh. Oil was extracted from the soft tissues, while the bones were ground and mixed with other materials to produce 200-pound bags of phosphate fertilizer (news article from the Philadelphia *Times,* reprinted in the Grand Rapids, Michigan, *Democrat,* 4 June 1899).

Leather was the principal incentive for the fishery during the nineteenth century. The carcass was skinned in halves after cutting along

A

B

FIGURE 20.8. A: A haul of bottlenose dolphins at Cape Hatteras in November 1913. Some of the animals were kept alive in a saltwater pond near the beach for about 24 hr, recaptured and placed in shipping tanks, carried to Norfolk on a schooner, then to New York on a steamer, and finally introduced to the New York Aquarium, where some of them lived for more than half a year. (B) Carcasses were processed for oil. Each dolphin produced about 3–8 gal of body oil and a quart of head oil. SOURCE: Townsend 1914.

the dorsal and ventral midlines. Each half-hide was worth about $2 "green" and $10–12 tanned (Stevenson 1904). "Porpoise leather" was noted for its tractility and durability as well as its ability to absorb water, making it especially popular for outdoor footwear. Beluga leather, made from the skin of white whales (*Delphinapterus leucas;* see Other Toothed Cetaceans), was even more valuable because of its greater tensile strength and softness. Because it came in large pieces, beluga leather was especially well adapted for machinery belts (Stevenson 1904). Beluga shoelaces were popular in Great Britain and Europe, selling for $8–10/gross, compared with $1.25/gross for calfskin laces (Stevenson 1904). C. Amory reported in 1928 that dolphin hides were no longer valuable enough to bother curing, and they were being dry salted at the time of his visit to the Outer Banks (see above).

Mead (1975) inferred from a collection of skulls in the U.S. National Museum that a *Tursiops* fishery was active at Cape Lookout in 1871. Stephens (1984:33) claimed that a "porpoise camp" was established at Rice Path on Bogue Banks (west of Cape Lookout) in the 1880s. He noted, referring to this camp, that although oil was the main product, the meat (i.e., "the lean part that came out the back of the neck") was consumed locally. Some of the oil was burned in lamps, the rest sold. In the late 1880s, R. G. Salomen, of Newark, New Jersey, published a notice in a local Beaufort, North Carolina, newspaper (*Weekly Record,* 27 January 1887) soliciting salted hides (at 15¢/lb, "free of blubber and cork") to be shipped by steamer to New York. Three months later, it was reported that the fishery on Bogue Banks (one of three active in Carteret County at the time) had taken more than 600 "porpoises" thus far in

TABLE 20.1. Monthly catch data from the "porpoise" fishery at Cape Hatteras, North Carolina

Month	1884–85 (True)	1885–86 (Stick)	1885–86 (JWR)	1886–87 (JWR)	1886–87 (Fulcher)	1886–87 (Stow)	1887–88 (JWR)	1888–89 (JWR)	1890–91 (JWR)	1891–92 (JWR)
January	36	165	165	25	0	19	335	119	122	—
February	111	210	210	145	0	85	273	146	0	—
March	219	205	205	304	23	80	185	132	52	—
April	264	282	282	43	62	60	33	0	225	—
May	303	262	262	0	2	46	137	—	43	—
June	—	—	—	—	—	—	—	—	78	—
July	—	—	—	—	—	—	—	—	—	—
August	—	—	—	—	—	—	—	—	—	—
September	—	—	—	—	—	—	—	—	—	—
October	—	—	—	—	—	—	—	—	—	—
November	246	—	4	618	262	—	—	0	—	—
December	89	171[a]	167	178	117	—	—	182	—	57[b]
Total	1268	1295	1295[c]	1313[d]	466	290	963[e]	579	620	57
Season began			21 Nov	1 Nov	Nov	25 Jan	4 Jan	12 Nov	12 Jan	25 Nov
Season ended			21 May	May	May	May	10 May	Apr	3 June	?

SOURCE: Data attributed to True (1891) and Stick (1958) are taken from Mead (1975:Table 1); the rest (JWR) are from Rolinson (1845–1905).

[a]"Combined catch for November to end of December" (Mead 1975:Table 1).

[b]The entire catch was made on 14 "smooth" days between 25 November and 31 December.

[c]In addition, 754 were taken by Tom Fulcher for Wainwright at "Trent," giving the 2049 total for the year as shown in Mead (1975:Table 2).

[d]In addition, 466 were taken by Fulcher and 290 by James M. Stow on behalf of Zimmerman, giving a total of 2069, which differs from Mead's (1975:Table 2) figure of "?6450 (2293)" for 1886–87 (no source given).

[e]In addition, J. W. Austin caught 852 dolphins from 15 October 1887 to 10 May 1888, which gives a total of 1815 in two fisheries for the year 1887–88.

that season (*Weekly Record,* 21 April 1887). Several weeks later, 219 dolphins were taken in a single haul of the net at Rice Path. Described as the largest haul ever made, this catch was reportedly worth about $1000 (*Weekly Record,* 26 May 1887). Stick (1958) reported single-day catches of 142 (23 April) and 170 (1 November) at Cape Hatteras in 1886.

A diary kept by John W. Rolinson (1845–1905), superintendent of the Hatteras dolphin fishery from 1885 to 1891, is the primary source for much of the catch data tabulated by Mead (1975) (Table 20.1). The fishery continued at least intermittently until 1929, when the nets, boats, and other equipment were liquidated (Angell 1981). The shore fisheries for bottlenose dolphins along the U.S. east coast must have had access to a total dolphin population of well over 10,000 animals in the late nineteenth century. Mitchell's (1975b) estimate mentioned above was based on North Carolina catches reported by True (1891) for 1884–1885 (1268 dolphins), Kellogg (1926–1927) for either 1885 or 1886 ("?2000"), and Stevenson (1904) for 1887 (6450), 1889 (2283), and 1890 (1747). His figure of about 14,000 for the dolphin population in the early 1880s is probably an underestimate because of the short period of coverage and the fact that substantial catches continued after 1890.

Modern. Bottlenose dolphins are no longer hunted, but they are taken as a by-catch in many fisheries of the United States, including gill net, purse seine, longline, pelagic trawl, and crab pot gear. In some cases, captures may occur as dolphins attempt to investigate or take fish from the gear. Up to 115 offshore bottlenose dolphins were taken in one year (1990) in the pelagic drift net fishery for swordfish (*Xiphias gladius*) off the U.S. east coast during 1989–1998 (Waring et al. 2002). Virtually all of the dolphins that become entangled in drift nets die. In contrast, a few of the offshore bottlenose dolphins taken in the pelagic longline fishery in the western North Atlantic are released alive (Waring et al. 2002). A large but undetermined number of coastal bottlenose dolphins are killed each year in small-scale gill net fisheries from New Jersey to North Carolina, as shown by the frequent appearance of carcasses on the beach bearing evidence of entanglement (Waring et al. 2002). In some southeastern states, coastal gill net fishing has been banned.

Although most of the attention given to fishery conflicts has been directed at commercial fisheries, there is evidence from the Gulf of Mexico that at least some of the serious injury and mortality experienced by bottlenose dolphins is related to recreational fisheries. Monofilament line is a particular hazard, whether encountered via ingestion or entanglement (Gorzelany 1998; Wells et al. 1998b).

Bottlenose dolphins have always been the species of choice in cetacean display facilities and in research holdings throughout the world. More than 1500 were taken from the wild from the late 1930s through the early 1980s, the vast majority of them from the southeastern United States (Leatherwood and Reeves 1982; Reeves and Leatherwood 1984; Wells et al. 1998a).

Bottlenose dolphins appear to be well adapted to living in coastal waters, where they are routinely exposed to a large number of diverse human activities. As upper-trophic-level predators, they accumulate a wide variety of contaminants, although the effects of these compounds are poorly understood. Dolphins in coastal waters of eastern North America have high tissue concentrations of polychlorinated biphenyls, polybrominated biphenyls, and various other toxic compounds; some of these are known to be strong immunosuppressive agents (Kuehl et al. 1991; Lahvis et al. 1995). Particular concern has been expressed over the apparent increase in disease outbreaks, sometimes associated with high levels of contaminant exposure (O'Shea et al. 1999).

MANAGEMENT AND CONSERVATION

Like all marine mammals in the United States, bottlenose dolphins are subject to management under the MMPA of 1972, which prohibits *takes* of marine mammals, except under specific circumstances, where *to take* is defined as "*to harass, hunt, capture or kill or attempt to harass, hunt, capture or kill*" (Baur et al. 1999). The primary goal of the act is to maintain populations at or above the optimum sustainable population level. This level is defined by the National Marine Fisheries Service as a population size exceeding the maximum net productivity level (MNPL). In turn, the MNPL is defined as the population size that yields the greatest net increment from additions to the population due to reproduction and growth, minus losses due to natural mortality. MNPL is generally assumed to occur between 50% and 70% of a stock's carrying capacity or historic population size (Taylor and DeMaster 1993). Stocks below this level are considered *depleted.*

In 1994, the MMPA was amended to deal specifically with the by-catch of marine mammals in commercial fisheries. Under the 1994 amendments, the National Marine Fisheries Service and U.S. Fish and Wildlife Service must prepare assessment reports for each of more than 150 stocks of marine mammals in the United States. These reports must contain information on stock structure, abundance, trends, and sources and magnitude of anthropogenic mortality and an evaluation of whether this mortality exceeds threshold levels specified by the Act (Read and Wade 2000). A maximum allowable level of anthropogenic mortality is determined for each stock of marine mammals. This potential biological removal (PBR) level is defined as PBR $= N_{min} \times 0.5 R_{max} F_R$, where N_{min} is the minimum population estimate for the stock, R_{max} is the maximum theoretical or estimated net productivity of the stock at a small population size, and F_R is a recovery factor, which is set between 0.1 and 1.0 (Wade 1998). The recovery factor is automatically set at 0.1 for endangered stocks and at 0.65 for depleted or threatened stocks of cetaceans. Higher values are only assigned to stocks known to be at or above the optimum sustainable population level or known to be increasing in spite of takes greater than the calculated PBR (Barlow et al. 1995). The intent of this scheme is to provide a conservative removal level that will allow populations to recover to or remain above the MNPL. Simulation models indicate that the approach is robust to biases associated with estimates of mortality, abundance, stock structure, and other parameters (Wade 1998). As long as the magnitude of anthropogenic removals, such as by-catches, are below PBR, a stock of marine mammals should equilibrate above the MNPL.

If the magnitude of by-catch or other anthropogenic mortality exceeds PBR, the stock is deemed to be *strategic*. In such cases, the MMPA requires that a *take-reduction plan* be developed. The plan must include regulatory and/or voluntary measures that will reduce anthropogenic mortality and serious injury to below PBR within 6 months of their implementation. The take-reduction plans are developed by teams of stakeholders, including representatives from government agencies, academic and scientific organizations, environmental groups, and commercial and recreational fishing groups. These stakeholders go through a process of negotiated rule making in which they work with a federally appointed mediator to develop the plan.

An analysis of strandings along the U.S. east coast indicated that PBR was exceeded for the coastal migratory stock of bottlenose dolphins (Waring et al. 2000). Thus, the National Marine Fisheries Service convened a take-reduction team in 2001 to reduce the by-catch of bottlenose dolphins in this region to below PBR. No other stocks of bottlenose dolphins in the United States are considered strategic (Fig. 20.9).

HARBOR PORPOISE

NOMENCLATURE

Common Names. Harbor or harbour porpoise, common porpoise (UK), puffing pig (Newfoundland and Labrador), marsouin (French, in Quebec)
Scientific Name. *Phocoena phocoena* (Linnaeus 1758)

Systematics. True porpoises, family Phocoenidae, are distinguished from other odontocetes by their spatulate (rather than conical) teeth and lack of an external rostrum ("beak"). Three genera are recognized: *Neophocaena* Palmer, 1899, which includes the finless porpoise of the Indo-West Pacific rim; *Phocoenoides* Andrews, 1911, which includes the Dall's porpoise (*P. dalli dalli*) and True's porpoise (*P. d. truei*) of the North Pacific; and *Phocoena* G. Cuvier, 1816, which includes four well-defined species worldwide (Rice 1998). Geographic variation in mitochondrial DNA and morphology supports designation of several subspecies of *Phocoena phocoena,* although there is disagreement as to which of these are valid (compare Rice 1998 and Read 1999). The subspecies from the Atlantic coast of North America is *P. p. phocoena;* that from the Pacific coast is *P. p. vomerina*.

Figure 20.9. Interactions with fishing gear, including gill nets, are a common feature in the lives of coastal bottlenose dolphins. Although many individuals appear to be adept at feeding in close proximity to the gear with no ill effect, the risk of entanglement and death is always present. Source: Photo by Andrew J. Read.

DISTRIBUTION

Harbor porpoises are widely distributed in shelf waters of the temperate Northern Hemisphere (Read 1999) (Fig. 20.10). Limits of their range in the western Atlantic are northern Florida in the south and Upernavik, West Greenland, in the north. Analyses of mitochondrial DNA indicate four populations in the northwestern Atlantic: western Greenland, Newfoundland–Labrador, Gulf of St. Lawrence, and Bay of Fundy–Gulf of Maine (Rosel et al. 1999). Individuals from these populations make fairly extensive seasonal movements, and some animals from different populations may mix in waters off the central east coast of the United States during winter (Rosel et al. 1999).

In the northeastern Pacific, harbor porpoises are found more or less continuously from Point Conception, California, north to Alaska and east to the Mackenzie Delta. Movements of individual porpoises along the U.S. west coast appear to be more restricted than those of individuals in the northwestern Atlantic and the resultant population structure is considerably more complex (Chivers et al. 2002). For example, molecular analyses and density discontinuities indicate the existence of at least nine distinct populations from Morro Bay, California, to the Bering Sea (Angliss and Lodge 2002).

DESCRIPTION

Harbor porpoises are small animals with rounded heads, which lack an external rostrum or beak. Their stocky bodies are covered in a 1.5- to 2-cm-thick blubber layer and taper to form a laterally flattened keel just anterior to the flukes. A small, triangular dorsal fin is located at

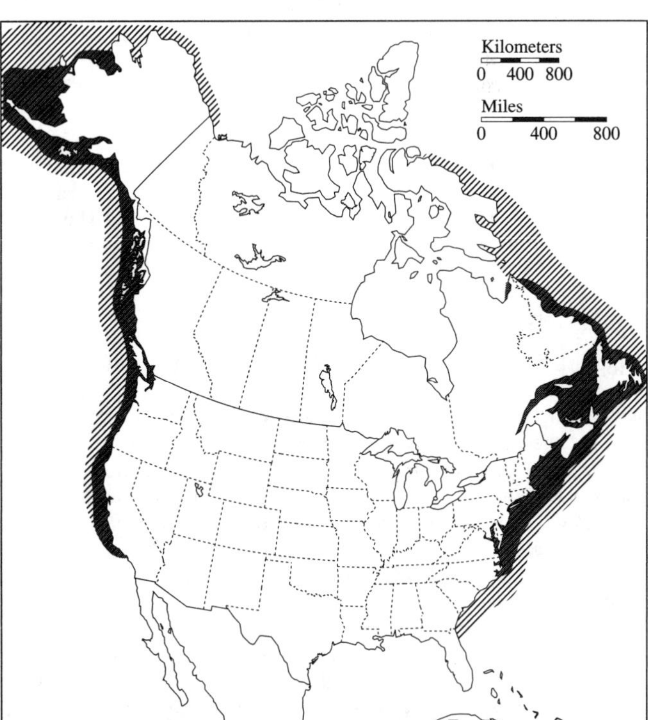

Figure 20.10. Distribution of the harbor porpoise (*Phocoena phocoena*).

FIGURE 20.11. Although harbor porpoises usually appear evenly gray when sighted at sea, their color pattern is fairly complex, shading from dark gray dorsally to white ventrally. This animal in the lower Bay of Fundy, southeastern Canada, was rescued from a herring weir and released in nearby waters. SOURCE: Photo by John Y. Wang.

approximately the midpoint of the back. The relatively small, pointed flippers are located behind and below the angle of the mouth.

A black cape extends over the dorsal and lateral surfaces, although its extent varies considerably among individuals and populations. The flanks are a mottled grayish white and fade to an almost white ventral surface. Individuals may exhibit dark eye, chin, and lip patches. Single or multiple dark stripes may extend from the angle of the mouth to the anterior insertion of the flippers. Koopman and Gaskin (1994) provide detailed descriptions of the highly variable pigmentation patterns of this species (Fig. 20.11).

Harbor porpoises are among the smallest cetaceans, and few individuals of any population exceed 2 m in total length. At birth, porpoise calves are approximately 75 cm long and weigh about 6 kg (Read 1999). The calves grow rapidly and triple their body mass by 3 months of age (Read 2001). The species is sexually dimorphic, with females attaining greater lengths and body masses than males. In the Bay of Fundy, for example, females reach approximately 160 cm and 65 kg, compared with 145 cm and 50 kg for males (Read and Tolley 1997).

The skull is characterized by a large, rounded cranium and a short rostrum. As in other odontocetes, the skull is asymmetrical, with the center of the nares located on the left side of the midline (Yurick and Gaskin 1987). There are 21–29 spade-shaped teeth in each maxillary tooth row and 20–29 in each mandibular row. There are 7 cervical, 12–14 thoracic, 13–18 lumbar, and 27–32 caudal vertebrae (Gaskin et al. 1974).

LIFE HISTORY

Reproduction in all populations studied is seasonal, with ovulation and conception limited to a few weeks in late spring or early summer (Read 1990a). Ovulation is exclusively sinistral (inclined to the left) and the right ovary retains an immature appearance throughout life. Gestation lasts for 10–11 months, followed by a lactation period of at least 8 months. Females may exhibit postpartum estrus and, in some populations, most female porpoises spend the majority of their adult lives simultaneously pregnant and lactating (Read 1990b).

Males exhibit pronounced seasonal variation in testicular size and activity, with peak sperm production occurring around the period of ovulation (Neimanis et al. 2000). The testes are large, reaching 4% of body mass during the peak breeding season, which suggests that male porpoises are sperm competitors (Connor et al. 2000). Little is known about mating behavior, however, and few direct observations of reproductive events have been made in the wild.

The juvenile period is short and females typically mature in their third year of life, although age at sexual maturation is variable within populations (Read and Gaskin 1990). The species is relatively short lived compared to other odontocetes and few individuals live past their teens. The maximum reported life span is 24 years, derived from counts of dentinal growth layers in thin, decalcified, and stained sections of teeth (Lockyer 1995).

Considering life history patterns of odontocetes, the harbor porpoise has a very short life span compared to the sperm whale (Read and Hohn 1995). Thus, it is likely that porpoises have a relatively high potential rate of increase compared to many of the larger toothed whales. Attempts to estimate this rate have been thwarted by a lack of information on survival or mortality rates (see discussion in Caswell et al. 1998).

ECOLOGY

Porpoises are difficult to study in the wild because they are small, unobtrusive, wary of human observers, and possess few marks that allow for individual recognition. Thus, knowledge of their ecology and behavior is limited. Individuals equipped with satellite-linked radio transmitters in the Bay of Fundy and Gulf of Maine have traveled more than 50 km in a single day. Home ranges may encompass the entire Gulf of Maine, an area of thousands of square kilometers (Read and Westgate 1997). As noted above, porpoises along the Pacific coast may be less mobile and have smaller home ranges.

There are no reliable records of coordinated migratory movements by harbor porpoises in North America, although their distribution in some areas changes seasonally. Along the central east coast of the United States, for example, harbor porpoises are present only during winter and move northward during the early spring as water temperatures rise. In northern portions of their range, harbor porpoises are probably excluded by ice cover during the winter.

Harbor porpoises occur primarily over the temperate continental shelves, although individuals are occasionally found in deeper waters. The species, true to its name, is sometimes found in bays and harbors, particularly during the summer, and occasionally penetrates far into estuaries. It is well adapted to cold water and is seldom found in water warmer than 17°C (Gaskin et al. 1993).

ABUNDANCE

The abundance of harbor porpoises is estimated using line transect techniques, corrected for submerged animals, from aerial or shipboard surveys. Estimates of abundance are available for most stocks in U.S. waters (Carretta et al. 2002; Angliss and Lodge 2002; Waring et al. 2002). The size of these stocks varies widely. In the Gulf of Maine and Bay of Fundy, for example, abundance is estimated at 89,700 (CV = 0.22) (Waring et al. 2002). In contrast, only about 9000 porpoises are present in central California, an area that may contain three genetically distinct stocks (Carretta et al. 2002). Estimates are also available from portions of the species range in eastern Canadian waters (Kingsley and Reeves 1998). Trends in abundance are available only for central California, where sighting rates from 1986 to 1995 were negatively correlated with sea surface temperature (Forney 1999). In this area at least, interannual variation in abundance is driven by changes in oceanic conditions, likely mediated through the movements of prey species.

BEHAVIOR

Little is known about the social behavior of harbor porpoises because it is difficult to identify individuals in the field. Observations of a small number of naturally marked females in the Bay of Fundy indicated that their social groupings are fluid, much like those of other odontocetes, and that individual porpoises may use the same areas in successive years (Watson 1976). Porpoises tagged together in the Bay of Fundy did not remain together after release, which supports the idea that this species exhibits a fission–fusion pattern of social behavior. Harbor porpoises are usually observed in small groups of a few individuals or alone, although larger aggregations of several hundred animals have been reported (Hoek 1992). As noted above, the mating system of this species likely involves sperm competition, and frequent sexual behavior has been documented in captive specimens (Andersen and Dziedzic 1964).

FEEDING HABITS

Information on the diet of harbor porpoises comes almost exclusively from examination of prey remains in the stomachs of by-caught and stranded animals. The diet includes a variety of small fishes and cephalopods, usually <30 cm in length. Energy-rich species, such as herring (*Clupea harengus*) and capelin (*Mallotus villosus*), constitute the bulk of the diet in many areas (Fontaine et al. 1994; Gannon et al. 1998). At least some prey items are demersal, living on or near the sea floor; porpoises feeding on such items are at risk of entanglement in bottom-set gill nets (see below). They are capable of diving to depths of >200 m (Westgate et al. 1995). Harbor porpoises forage independently, although many individuals may aggregate to feed in areas of prey concentrations. In the Bay of Fundy, porpoises sometimes exploit schools of prey brought to the surface by tidally driven convergences and upwellings.

MORTALITY

Harbor porpoises are preyed on by white sharks (*Carcharodon carcharias*) (Arnold 1972) and killer whales (Jefferson et al. 1991). There are no estimates of the numbers of porpoises consumed by these predators, nor are there estimates of the rates of natural mortality for any population. In a few areas, including the Moray Firth, Scotland, and the coast of Virginia, porpoises are sometimes killed by bottlenose dolphins (Ross and Wilson 1996). The reasons behind this unusual behavior are unclear, but may be related to infanticide practiced by some bottlenose dolphins. Adult harbor porpoises are approximately the same size as bottlenose dolphin calves, and the dolphins may be practicing their infanticidal skills on the porpoises (Patterson et al. 1998). Little is known about the role of disease in the natural mortality of harbor porpoises. Each spring, however, many emaciated juveniles are found stranded along the Atlantic coast between New York and North Carolina, apparently having starved to death.

FIGURE 20.12. Monofilament gill nets kill many hundreds, if not thousands, of harbor porpoises in North American waters each year. Bottom-set, or "sink," gill nets take the largest toll, but porpoises die in surface drift nets as well. Because of the porpoise by-catch, fishery management is key to conservation of harbor porpoise populations in most areas. SOURCE: Photo by Bill Eppridge.

ECONOMIC STATUS

In the past, harbor porpoises were taken in directed fisheries in many areas of their range. The best-documented hunt occurred near the entrance of the Baltic Sea, where Danish fishermen captured large numbers of porpoises each autumn as they moved toward the Kattegat. This fishery was practiced from the fourteenth century until World War II, when the numbers of porpoises decreased and the demand for oil, which was obtained by rendering the blubber, also declined. Detailed catch statistics are available for this fishery (Kinze 1995).

Harbor porpoises were also hunted by aboriginal people on both coasts of North America in the nineteenth and early twentieth centuries (Scammon 1874; Leighton 1937). The number of animals taken was not recorded, but in the Bay of Fundy, several hundred porpoises were likely taken each year. Hunters worked from canoes on calm days, when it was possible to follow and approach porpoises; shotguns were used to wound or kill the animals. The blubber and mandibular fat pads were rendered for oil and the meat was used for human consumption (Leighton 1937). About 2–3 gal of body oil was rendered from each porpoise (up to 5–6 gal from large individuals) (Stevenson 1904). This was sold for 60–80 cents/gal pure, or for less when adulterated with seal oil. It was used mainly for lighting and machine lubrication. The superior oil from the mandibles, obtained by hanging jaws and allowing the oil to drip into cans, sold at a very high price. It was used for lubricating clocks, watches, and other precision instruments. Only about $\frac{1}{2}$ pint of this "jaw oil" was secured from each porpoise (Stevenson 1904). Archaeological examination of coastal middens has provided no evidence that porpoises were hunted before the introduction of firearms by Europeans. A small hunt by members of the Passamaquoddy tribe in Maine continued sporadically into the late twentieth century, with the last animals taken in 1997 (Waring et al. 2000).

Harbor porpoises are no longer hunted in North America (except possibly to a minor extent in Labrador, Newfoundland, and Quebec), but are taken frequently in commercial fisheries as a by-catch (Fig. 20.12). The primary fishing gear responsible for this by-catch are bottom-set gill nets, which are used to capture demersal fish species such as cod (*Gadus morhua*). This by-catch has existed since gill nets were first introduced into North American fisheries in 1880 by Spencer Baird, then U.S. Commissioner of Fish and Fisheries. In the first report of the efficacy of these nets, Collins (1886) noted that "in addition to the various species of Gadidae which have been taken, porpoises (locally called 'puffers')... have been caught." In the 1990s, the by-catch of harbor porpoises in the bottom-set gill net fishery in the Gulf of Maine was the largest take of marine mammals in any domestic U.S. fishery (Read and Wade 2000). The estimated annual by-catch of porpoises in this fishery between 1990 and 2000 ranged from 270 to 2900 (Waring et al. 2002). These estimates were derived by placing observers onboard commercial fishing vessels to generate by-catch rates and then using some measure of overall fishing effort to extrapolate. By-catches of harbor porpoises also occur in gill net fisheries along the coast southward from New Jersey in the winter (Waring et al. 2002), in Monterey Bay,

FIGURE 20.13. Entrapment of harbor porpoises in herring weirs in the lower Bay of Fundy has had one important scientific benefit: It has provided researchers with opportunities to handle and instrument healthy porpoises before releasing them back into the wild. Much of what is known about porpoise diving behavior and movements has come from the porpoise rescue program based on Grand Manan Island, New Brunswick. SOURCE: Photo by John Y. Wang.

California (Carretta et al. 2002), and in Washington State (Gearin et al. 2000). Small numbers of porpoises are taken in herring weirs, large fish traps placed along the shores of the Bay of Fundy, but most of these individuals are released alive (Waring et al. 2002) (Fig. 20.13).

Harbor porpoises do not adapt readily to a captive environment and are seldom kept in oceanaria. Several stranded, rehabilitated juveniles, however, have been maintained for years in captivity, and observations of these individuals have provided considerable insight into the biology of the species (Read et al. 1997). Some stranded juveniles have been released after periods of rehabilitation that lasted for months or years (Westgate et al. 1998).

Harbor porpoises are shy animals, and intensive human activities in coastal waters may have adverse effects on their populations. One important, but poorly understood concern is the effect of noise. For example, devices that produce high-intensity sound are frequently used to deter pinnipeds from approaching mariculture operations. These loud sounds, designed to cause pain to seals and sea lions, are also effective in deterring harbor porpoises (Olesiuk et al. 2002). Controlled playback studies in the Bay of Fundy indicate that porpoises can be displaced several kilometers away from the source of these sounds (Johnston, 2002). Proliferation of mariculture sites in many coastal areas could reduce the amount of habitat available to harbor porpoises.

MANAGEMENT

In the United States, harbor porpoises are managed under the MMPA, as previously described for bottlenose dolphins. Two take-reduction teams have been formed to reduce the by-catches of harbor porpoises in commercial fisheries, one for the Gulf of Maine and one for the "Mid-Atlantic" (meaning waters from New York to North Carolina). Both of these teams produced plans designed to reduce by-catches to below PBR, estimated recently at 747 individuals per year (Waring et al. 2002). Among the measures included in these plans are use of acoustic alarms (pingers) designed to keep porpoises away from gill nets, time–area fishery closures, and modifications to fishing gear. Acoustic alarms have been shown experimentally to reduce by-catch of harbor porpoises in the Gulf of Maine (Kraus et al. 1997), Bay of Fundy (Trippel et al. 1999), and Washington State (Gearin et al. 2000). Take-reduction plans came into force in January 1999, and by-catches of porpoises in the Gulf of Maine and Mid-Atlantic declined to below PBR in the first year of implementation (Waring et al. 2000). Simultaneous management measures were adopted in these areas to address overfishing, however, and it is not clear whether these measures or those designed to reduce the porpoise by-catch were responsible for this successful outcome.

By-catches also occur in Canadian commercial fisheries. Canada has no legislative counterpart to the MMPA. In Canadian waters, porpoises are managed under poorly defined provisions of the federal Fisheries Act. The United States and Canada have cooperated in developing a management framework for the harbor porpoise in the Bay of Fundy, however, and Canada has agreed to limit by-catches to approximately 100/year in this area. Overfishing in the Bay of Fundy during the past two decades has resulted in the collapse of the sink gill net fishery in this area, precluding the need for management intervention to reduce porpoise by-catches. As noted above, harbor porpoises are also taken in sink gill net fisheries in California and Washington State. Recent estimates of porpoise mortality in the set gill net fishery in Monterey Bay (80 animals/year) exceeds the PBR (11 porpoises/year), so this stock is considered "strategic" under the MMPA (Carretta et al. 2002).

SPERM WHALE

DESCRIPTION

The sperm whale (*Physeter macrocephalus*) is, by far, the largest of the toothed cetaceans (Rice 1989). Males grow to 18.3 m in length and can weigh more than 50 metric tons. Females are considerably smaller, reaching only 12.5 m and 24 metric tons. Length at birth is about 4 m.

The body is dominated by a disproportionately large, barrel-shaped head, which is one fourth to one third of the total body length. Although the skin may appear brown in bright sunlight, the basic color is dark gray. The upper lips and lingual portion of the narrow, rod-shaped, underslung lower jaw are white. There are usually about 20–26 pairs of large, well-developed mandibular teeth. Teeth are functionally absent (i.e., vestigial and unerupted) in the upper jaw. Unlike any other cetacean, the sperm whale's blowhole is situated at the front and top of the head, strongly skewed to the left of the midline. This causes the blow to be angled forward, often making it possible to identify sperm whales at a considerable distance (Fig. 20.14). The sperm whale's dorsal fin is low, thick, and rounded or obtuse. Photographs of distinctive markings on the dorsal fins and flukes of sperm whales are used in studies of life history and behavior (Whitehead and Gordon 1986; Whitehead 1990).

The sperm whale's extremely thick skin and blubber (nearly 40 cm in a fat male; Berzin 1972) and, most significantly, the large reservoir of liquid wax, or spermaceti, in the supracranial basin were responsible for its historical economic value. The spermaceti organ, or "case" as the whalers called it, occupies much of the huge head (Cranford 1999). It is separated from the rostral portion of the skull by a mass of whitish tissue, called the "junk," also saturated with spermaceti. The musculature is exceptionally high in myoglobin—up to 44 g/kg in the body muscle and 6.7 g/kg in the heart (Rice 1989).

FIGURE 20.14. In this aerial view of a sperm whale off the east coast of the United States, the blowhole is visible at the front of the head, left of the midline. The animal has just exhaled, and its bushy forward-oriented blow is easily visible. SOURCE: Photo by Robert D. Kenney, courtesy of Cetacean and Turtle Assessment Program, University of Rhode Island.

DISTRIBUTION AND MIGRATION

The sperm whale has the most extensive distribution of any marine mammal except the killer whale (see below) (Rice 1989). It occurs in all deep, ice-free marine regions from equatorial waters to the edges of polar pack ice. Its presence in certain warm-water areas throughout the year suggests that there may be "resident" populations (e.g., Watkins et al. 1985; Gordon et al. 1998). Its aggregate distribution is certainly influenced by the patchiness of global marine productivity (Jaquet and Whitehead 1996). No physical barriers, apart from land masses, appear to obstruct its dispersal (Berzin 1972; Jaquet 1996). There is a marked difference in migratory behavior between adult males and females. Only adult males move into high latitudes, whereas all age classes and both sexes range throughout tropical and temperate seas (Fig. 20.15). The combination of wide dispersal by males, ontogenetic changes in association patterns, and female pod fidelity and cohesion complicates any evaluation of population structure. In general, groups of females and immatures remain for periods of at least a decade or so within areas about 1000 km wide, whereas adult males range over much larger distances, especially latitudinally (Dufault et al. 1999). Movements by males, in particular, across and between ocean basins have resulted in a relatively high degree of genetic uniformity among sperm whales worldwide (e.g., Dillon 1996; Lyrholm et al. 1999).

Two major nineteenth-century whaling grounds for sperm whales, the Southern Ground and the Charleston Ground, were situated directly off the eastern United States (Townsend 1935). In the late 1960s and early 1970s, high densities of sperm whales were found in the "North Sargasso Sea Region" and the "Gulf Stream Region" (Mitchell 1972). The Gulf Stream has an important influence on the distribution of sperm whales (Townsend 1935; Waring et al. 1993). They occur in relatively high densities in canyon waters, around the edges of banks, and along the shelf break northward to Nova Scotia (Mitchell 1975a; CETAP 1982; Waring et al. 2002). Occasionally, they move onto the continental shelf in waters <100 m deep on the southern Scotian Shelf and south of New England, particularly between late spring and autumn (Whitehead et al. 1992; Scott and Sadove 1997; Waring et al. 2000).

Sperm whales are the most common large cetaceans in the northern Gulf of Mexico, where they occur in greatest density along and seaward of the 1000-m contour (Mullin et al. 1991, 1994; Jefferson and Schiro 1997; Davis et al. 1998). In the north-central Gulf near the Mississippi Canyon, sperm whales are especially common and present year-round (Davis et al. 1998; Weller et al. 2000). Although they are not common near DeSoto Canyon to the east in the Gulf, their total range includes much of the wider Caribbean region (Townsend 1935; Watkins and Moore 1982).

FIGURE 20.15. Female sperm whales and their young calves are present along the coasts of the United States, but do not venture into the cooler northern latitudes, where only adult males are found. Sperm whales are slow to reproduce and make large investments in their single young, which are born at intervals of at least 4 years. SOURCE: Photo by Robert D. Kenney, courtesy of Cetacean and Turtle Assessment Program, University of Rhode Island.

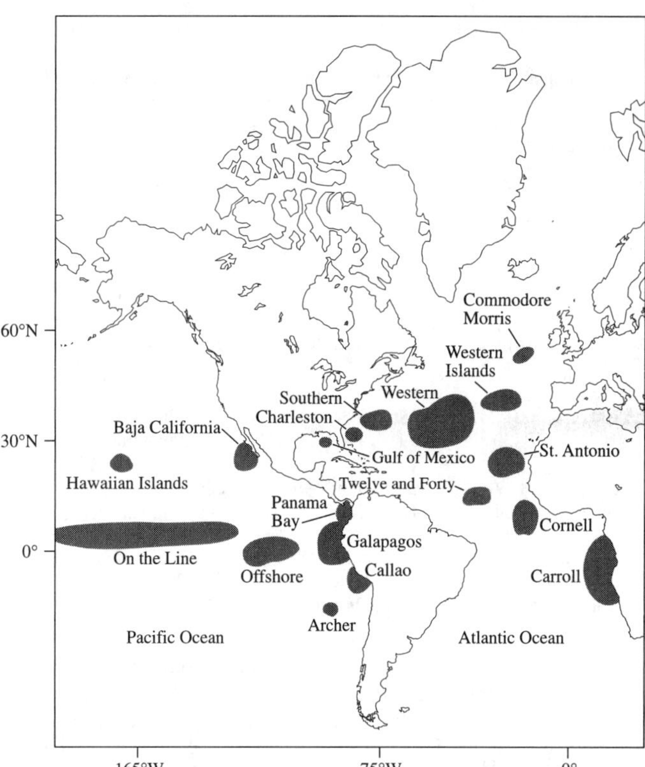

FIGURE 20.16. Chief nineteenth-century sperm whaling grounds in the Atlantic and eastern Pacific Oceans.

In the Pacific, the chief nineteenth-century whaling grounds for sperm whales were the Panama, Galápagos, and Offshore Grounds; the On-the-Line Ground; the Hawaiian Ground; areas off Baja California and mainland Mexico; and the Japan, Coast of Japan, and Bonin Islands Grounds (Townsend 1935) (Fig. 20.16). The more northern areas were visited mainly in summer and fall. Adult males generally go no farther north than a line from Cape Navarin, Russia, to the Pribilof Islands in the northeastern Bering Sea (Berzin and Rovnin 1966). Females and young whales usually go no farther north than about 50–51°N in the southern Gulf of Alaska (Berzin and Rovnin 1966). Sperm whales are present all year off California (Dohl et al. 1983; Barlow 1995; Forney et al. 1995), but reach peak abundance from April through mid-June and again from the end of August through mid-November (Rice 1974). They are also present in all seasons except midwinter (December–February) off Oregon and Washington (Green et al. 1992). Sperm whales, including females and young males, were abundant on the whaling grounds up to 200 miles offshore from Vancouver Island and the Queen Charlotte Islands, British Columbia, from spring through fall (Pike and MacAskie 1969). Gregr et al. (2000) interpreted catch data to mean that sperm whales mate (in April–May) and calve 15 months later (in July–August) in temperate waters off Vancouver Island.

Although Townsend's (1935) charts show little evidence of sperm whales in the Gulf of Alaska and around the Aleutians, modern shore and pelagic whalers took adult males regularly in summer in deep offshore waters of the eastern Aleutians and Kodiak Island (Reeves et al. 1985). Large concentrations of breeding schools were reported by modern pelagic whalers.

ABUNDANCE

No attempt has been made to estimate the total abundance of sperm whales in either the North Atlantic Ocean or the North Pacific Ocean. Instead, a patchwork of estimates is available, most of them referring to relatively small portions of the species's range (Table 20.2).

All estimates from sighting surveys are negatively biased because of the long submergence times of sperm whales, but the bias associated

TABLE 20.2. Abundance estimates of sperm whales in the North Atlantic and North Pacific Oceans

Number	Survey Dates	Survey Method	Area Sampled	Source
Atlantic				
22,000 (no CV)[a]	1966–1969	Ship; strip census	10–70°N, 20–80°W	Mitchell 1972
2,848 (CV = 0.49)[b]	6 July–6 Sept 1998	Combined ship/aerial; line transect	U.S. EEZ waters north from 38°N	Waring et al. 2002
1,854 (CV = 0.53)[b]	8 July–17 Aug 1998	Ship; line transect	U.S. EEZ waters south from 38°N	Waring et al. 2002
530 (CV = 0.31)	1991–1994	Ship; line transect	U.S. EEZ waters of northern Gulf of Mexico	Waring et al. 2002
2,548 (CV = 0.27)	4–28 July 1988	Ship; line transect	Offshore, northern Norway	Øien 1990
1,234 (CV = 0.16)	June–July 1987	Ship; line transect	Iceland	Gunnlaugsson and Sigurjónsson 1990
308 (CV = 0.38)	June–Aug 1987	Ship; line transect	Faroes	Gunnlaugsson and Sigurjónsson 1990
Pacific				
22,700 (CV = 0.224)	1986–1990	Ship; line transect	Eastern tropical Pacific	Wade and Gerrodette 1993
1,191 (CV = 0.22)	1991–1996	Ship; line transect	Waters within 300 nmi of the coasts of California, Oregon, and Washington	Barlow 1997
39,200 (CV = 0.60)	March–June 1997	Ship; line transect (acoustic)	Area from U.S. west coast to Hawaii	Barlow and Taylor 1998
24,000 (CV = 0.46)	March–June 1997	Ship; line transect (visual)	Area from U.S. west coast to Hawaii	Barlow and Taylor 1998

NOTE: CV, Coefficient of variation; EEZ, exclusive economic zone.
[a]Positively biased according to author.
[b]Negatively biased according to authors.

with a given estimate can be highly variable, depending on the survey platform. Availability bias for ship surveys may be relatively small because of the substantial distance at which sperm whale blows can be detected (Barlow and Sexton 1996). Abundance estimates from aerial surveys, in contrast, could be low by a factor of three to eight (Barlow 1994).

BEHAVIOR AND SOCIAL ORGANIZATION

The fascinating subjects of sperm whale behavior and social organization are discussed here mainly in relation to interactions with human activities. Readers interested in more details about sperm whale behavior and social patterns are referred to Caldwell et al. (1966), Waters and Whitehead (1990), Gordon et al. (1998), Watkins et al. (1999), Whitehead and Weilgart (2000), and their contained references.

Sperm whales are acoustically active and presumably depend on echolocation as well as passive listening to navigate and find food in the dark ocean depths. Echolocation involves the production of sounds (usually called "clicks") followed by the detection of echoes that bounce off distant objects. This highly specialized form of biological sonar has been studied experimentally with small odontocetes, but only inferred for sperm whales (Tyack 2000). Clicks produced by sperm whales (and presumably heard by them) are in the range of <100 Hz to as high as 30 kHz, often with most of the energy in the 2- to 4-kHz range (Watkins 1980). The whales respond, sometimes dramatically, to unfamiliar noise. On one occasion, whales exposed to pinger sounds in the frequency range of 6–13 kHz temporarily fell silent, apparently out of curiosity rather than fear (Watkins and Schevill 1975). A stronger response was observed in sperm whales exposed to intense sonar signaling and ship propeller noise from military operations during the U.S. invasion of Grenada in 1983. The whales fell silent, changed their activities, scattered, and moved away from the sound sources (Watkins et al. 1985). They also showed longer term responses by becoming quieter and seemingly more wary of a research vessel that had visited the same area in previous years (Watkins et al. 1985). Sperm whales in the Indian Ocean ceased vocalizing in response to air-gun pulses from a seismic vessel more than 300 km away and to other low-frequency sounds transmitted during a scientific experiment (Bowles et al. 1994). The results of playback experiments using aversive sounds to drive sperm whales away from a ferry route were interpreted as indicating that these animals have a high tolerance for certain kinds of noise (André et al. 1997).

Entire schools of sperm whales occasionally strand, but the causes of this phenomenon are uncertain (Rice 1989). Among the possible causal factors cited in the literature on mass strandings by cetaceans are (1) local coastal conditions forming "whale traps" (e.g., broad tidal flats, unusual currents, extreme tidal volume), (2) impaired echolocation in shallow waters, (3) geomagnetic anomalies or disturbances, and (4) the natural need for a "population regulation" mechanism. However, the only feature that seems to be common to most mass strandings by odontocete cetaceans is social cohesion—"a force strong enough to ensure that when a single animal comes ashore, for whatever reason, others in the pod are likely to follow" (Geraci et al. 1999; also see Geraci and Lounsbury 1993). Social cohesion is a major feature of sperm whale societies (Whitehead and Weilgart 2000; Pitman et al. 2001), although fighting occurs between adult males (Caldwell et al. 1966; Best 1979; Kato 1984; Clarke and Paliza 1988; Whitehead 1993).

Based on whaling statistics and observations by whalers, it was formerly assumed that the sperm whale had a "harem mode of life" (Berzin 1972), that is, social groups were led or otherwise dominated by a single mature bull. This model has been effectively supplanted by a more complex one based on matrilineal cohesion and male roving (Whitehead and Weilgart 2000). Stable, long-term associations among related and unrelated females form the core units of sperm whale societies (Whitehead et al. 1991; Christal et al. 1998). Up to about a dozen females usually live in such groups, accompanied by their female and young male offspring. Males start leaving these family groups at about 6 years of age, after which they live in "bachelor schools." Cohesion among males within a bachelor school declines as the animals age. During their breeding prime and old age, male sperm whales are essentially solitary (but see Christal and Whitehead 1997).

REPRODUCTION

Sperm whales mature slowly. Females usually begin ovulating at 7–13 years of age. Maturation in males usually begins in this same age interval, but most individuals do not become fully mature until their 20s. Prime bulls, in their late 20s and older, rove among groups of females on the tropical breeding grounds (Whitehead and Weilgart 2000). A male's association with a female group can be as brief as several hours. Because females within a group often come into estrus synchronously, the male need not remain with them for an entire season to achieve maximal breeding success (Best and Butterworth 1980).

The peak period of conceptions is March–May, with some mating activity spanning the entire period from midwinter to midsummer (December–August). Gestation lasts well over 1 year, with credible estimates of the normal duration ranging from 15 months to >18 months. Lactation lasts at least 2 years. The interval between births is 4–6 years for prime-aged females and apparently much longer for females >40 years old. Male juveniles up to 13 and females up to 7.5 years old have tested positive for lactose in their stomachs (Best et al. 1984).

At least two aspects of the sperm whale's reproductive biology are relevant to management. First, the maximal rate of population increase is very low, perhaps no more than 1–2%/year. Second, selective killing of large males by whalers could have had the residual effect of reducing reproductive rates (Whitehead et al. 1997).

FEEDING HABITS

Sperm whales feed mainly on large and medium-sized squids, although they also consume octopuses and large and medium-sized rays, sharks, and teleosts (Berzin 1972; Clarke 1977, 1980; Rice 1989). A large proportion of the diet consists of low-fat, ammoniacal (using ammonium chloride for buoyancy control), luminescent squids (Clarke 1980, 1996; Martin and Clarke 1986). One sperm whale's stomach contained a giant squid (*Architeuthis* sp.) 12 m long, which weighed 200 kg (Berzin 1972).

The diet of large males in some areas, especially in high northern latitudes, is dominated by fish (Rice 1989). Lumpsuckers (*Cyclopterus lumpus*), for example, are frequently taken in Denmark Strait (Martin and Clarke 1986). In some areas of the North Atlantic, males prey heavily on the oil-rich squid *Gonatus fabricii,* a species frequently also eaten by bottlenose whales (*Hyperoodon ampullatus*) (Clarke 1997).

Sperm whales are deep and prolonged divers and can therefore use the entire water column even in very deep areas. Dives lasting longer than 1 hr are unexceptional, and a single dive of 2 hr, 18 min has been reported (Watkins et al. 1985). Sperm whales certainly dive to depths of at least 2000 m (Watkins et al. 1993) and perhaps much deeper. In some areas, they seem to forage mainly on or near the bottom, often ingesting stones, sand, sponges, and other nonfood items (Rice 1989). In other areas, they may dive only to depths of a few hundred meters and remain submerged for about 40 min (Papastavrou et al. 1989). As far as is known, sperm whales feed regularly throughout the year. Lockyer (1981) estimated that they consumed about 3.0–3.5% of their body weight per day.

Whitehead et al. (1989) and Smith and Whitehead (1993) used observed defecation rate as an index of "feeding success" in sperm whales near the Galápagos and related this index to oceanographic conditions. In a separate study, Jaquet et al. (1996) compared nineteenth-century sperm whale distribution in the tropical Pacific (based on whaling catch positions) to the distribution of phytoplankton pigment concentrations derived from satellite color observations. They found that over large spatial scales, and when the data were averaged over large temporal scales, chlorophyll concentration was a good indicator of sperm whale distribution.

Sperm whales are known to interact with longline fisheries in many parts of the world (e.g., around South Georgia; Ashford et al. 1996), including the Gulf of Alaska, where they reportedly take fish from gear set to catch sablefish (*Anoplopoma fimbria*) (Rice 1989) and halibut (*Hippoglossus stenolepis*) (Ferrero et al. 2000; Angliss and Lodge 2002).

MORTALITY AND MORBIDITY

Although sperm whales can live >60 years (Rice 1989), they face a number of hazards apart from whaling (which is not currently a major threat in either the eastern North Pacific or western North Atlantic; see below). Calves are subject to predation by killer whales (Arnbom et al. 1987) and possibly large sharks (Best et al. 1984). Adult sperm whales have long been thought to be almost free of natural predation (Rice 1989; Dufault and Whitehead 1995), but recent observations off California indicate that attacks by killer whales on groups of adult females can result in severe wounding and death (Pitman et al. 2001). Sperm whales are also "harassed" by pilot whales (*Globicephala* spp.) and false killer whales (*Pseudorca crassidens*), but most "attacks" by these species are probably unsuccessful (Palacios and Mate 1996; Weller et al. 1996).

Very little is known about the role of disease in the natural mortality of sperm whales. Only two potentially lethal diseases have been identified in sperm whales: myocardial infarction associated with coronary atherosclerosis, and gastric ulceration associated with nematode infection (Lambertsen 1997).

The bottom-feeding habit of sperm whales, which might involve a suction mechanism, means that they often ingest marine debris (Lambertsen 1997). The consequences can be debilitating and even fatal. Of 32 sperm whales examined for pathology in Iceland, 1 had a lethal disease thought to have been caused by the complete obstruction of the gut with plastic marine debris (Lambertsen 1990).

A dramatic increase in the rate of sperm whale strandings in western Europe since the early 1980s raised concerns about the potential effects of pollution. Many of the whales had high tissue levels of mercury, cadmium, and certain organochlorines (Law et al. 1996, 1997; Bouquegneau et al. 1997). However, detailed pathological examinations carried out on some of the animals revealed no clear link between contamination and stranding (Jacques and Lambertsen 1997).

Levels of organochlorine contaminants in sperm whales killed off northwestern Spain were intermediate between the levels found in fin whales (*Balaenoptera physalus*) and small odontocetes in the same region (Aguilar 1983). Also, levels in females were consistently higher than those in males, a finding contrary to the usual situation in cetaceans. Placental and milk transfer from mothers to their young normally results in a net lowering of contaminant burdens in adult females. Since male and female sperm whales are geographically separated during much of the year, it is possible that males feed in less polluted waters or perhaps on less contaminated prey than do females.

ECONOMIC STATUS AND HISTORY OF WHALING

Sperm whales were subject to commercial hunting for more than 250 years. The principal products driving the commercial hunt were sperm oil and spermaceti (Fig. 20.17). The latter is a semiliquid waxy oil found only in the head of the sperm and some other toothed whales (Norris and Harvey 1972; Rice 1989). Sperm oil obtained from the bodies of sperm whales has special lubricant properties, and the spermaceti was formerly prized for use in candle making and illumination. Ambergris (a perfume fixative found occasionally in the lower intestines) (Gilmore 1969; Berzin 1972) and tooth ivory were valuable byproducts of sperm whale hunting. The skin was used as low-grade leather in the Soviet Union (Berzin 1972). Having declined in the late nineteenth and early twentieth centuries because of the increasing availability of petroleum for lubrication and lamp fuel, the demand for sperm oil and spermaceti greatly expanded after World War II (Berzin 1972; Rice 1989). It was used in cosmetics and soaps and as machine oil. Some sperm oil was used in submarines. Only in a few areas, notably some parts of Japan, the West Indies, and Indonesia, has the meat of sperm whales been used regularly for human consumption.

On a global scale, the exploitation of sperm whales fell into two main eras: the open-boat, sailing-vessel, hand-harpooning period from about 1715 to 1925 (Starbuck 1878; Hegarty 1959; Stackpole 1972) and the modern period from about 1910 to the early 1980s (Best et al. 1984). The total take, worldwide, between 1800 and 1909 may have been close to 700,000; that between 1910 and 1973, close to 605,000 (Best et al. 1984). For North Atlantic data, see Clarke (1954), Mitchell (1974, 1975a), Jonsgård (1977), Kapel (1979), Anonymous (1981), Martin and Ávila de Melo (1983), Avila de Melo and Martin (1985), Price (1985), and Sanpera and Aguilar (1992). For North Pacific data, see Scheffer and Slipp (1948), Pike and MacAskie (1969), Berzin (1972), Rice (1974), Ohsumi (1980), Best (1983), Reeves et al. (1985), Clapham et al. (1997), and Gregr et al. (2000).

Although minimum size limits of 38 ft for male and 35 ft for female sperm whales were included in the first regulations of the International Whaling Commission (IWC), sperm whaling was essentially unregulated until 1970, when quotas were introduced in the North Pacific Ocean (Allen 1980). Quotas were introduced in the Southern

FIGURE 20.17. The American sperm whale fishery was a huge industry, which delivered tens of thousands of barrels of sperm oil and spermaceti to American ports each year throughout the nineteenth century. An estimated total of 164,073,918 gal (621,019,780 L) of sperm oil (including spermaceti) was landed at U.S. ports between 1804 and 1925 (Rice 2002). The spermaceti was favored for candle making, whereas the body oil was used mainly as lamp fuel. SOURCE: Photo courtesy of the New Bedford Whaling Museum.

TABLE 20.3. North American stocks of white whales and narwhals

Stock or "Aggregation"	Center(s) of Distribution	Abundance	Exploitation	Other Threats	Conservation Status
White Whales					
South Greenland	Qaqortoq to Maniitsoq (winter; Oct–June)	—	Disappeared after intensive hunting in first third of 20th century (Heide-Jørgensen 1994)	—	Extirpated
Southwest Greenland	Maniitsoq to Disko Bay (winter)	6722 (3562–12,688) in 1998	>400/year in Greenland	—	Declining due to overhunting
Northwest Greenland	Migrating southward in autumn	A few 1000s counted in 1998	>100/year in Greenland	—	Declining due to overhunting
North Water	Polynya in northern Baffin Bay (winter)	—	<50/year in Canada and Greenland	—	Uncertain
Canadian High Arctic	Prince Regent Inlet, Barrow Strait, Peel Sound (summer)	28,499 (13,886–58,491) in 1996	<50/year in Canada	—	Portions of the stock are overhunted in Greenland in autumn and winter
Southeastern Baffin Island (Pangnirtung)	Cumberland Sound (summer)	484 counted in 1986 (Richard et al. 1990)	35/year, under quota in Canada	—	Small population, at risk from continued hunting
Southeastern Baffin Island (Iqaluit)	Frobisher Bay (summer)	—	ca. 25/year in Canada	—	Uncertain
Southeastern Baffin Island (Kimmirit)	North coast of Hudson Strait, spring and autumn migration	—	<30/year in Canada	—	Uncertain
St. Lawrence River	Lower river year-round	1238 ($SE = 119$) in 1997	Not hunted	Pollution, disturbance by vessel traffic	Small population with restricted distribution, probably increasing slowly
Ungava Bay	River mouths (summer)	20 counted in 1993 (Kingsley 2000)	Hunted opportunistically	—	Critically depleted with little prospect of recovery
Northern Hudson Bay	Coasts of Southampton Island (summer)	1000 (600–1600) in 1990 (partial survey; Richard et al. 1990)	80–140/year	—	Uncertain
Eastern Hudson Bay coastal	Remnant in Little Whale and Nastapoka Rivers (summer)	1014 ($SE = 421$) plus 115–148 (Kingsley 2000)	>50/year	Upstream diversion and impoundment of rivers for hydroelectric power; boat traffic in estuaries	Depleted and probably continuing to decline due to overhunting
Belcher Islands	Offshore of eastern Hudson Bay (winter and summer)	—	ca. 30/year	—	Uncertain
Hudson Strait	Spring and autumn migration	—	150–200/year	—	Uncertain
Western Hudson Bay	Churchill and Nelson River estuaries (summer)	25,100 (18,300–52,800) in 1987 (partial survey; Richard et al. 1990)	130–200/year (also hunted in nonsummer months in Hudson Strait)	Upstream diversion and impoundment of rivers for hydroelectric power; boat traffic in estuaries	Offtake may be sustainable
Southern Hudson Bay	Severn and Winisk River estuaries (summer)	1299 counted in 1987 (partial survey; Richard et al. 1990)	Generally not hunted in summer, but may be hunted in nonsummer months in other areas	—	Uncertain
James Bay	Scattered throughout bay, mainly in summer	3141 ($SE = 787$) in 1993 (Kingsley 2000)	Generally not hunted	Upstream diversion and impoundment of rivers for hydroelectric power	Probably stable
Foxe Basin	Unpredictable summer and autumn occurrence	—	<100/year	—	Uncertain
Beaufort Sea	Scattered widely, Mackenzie River Delta (summer)	39,258 in 1992 (Angliss and Lodge 2002)	ca. 120/year in Canada (also hunted in nonsummer months in Alaska [ca. 61/year] and Russia)	Oil and gas development, commercial fishing	Probably stable

(*Continues*)

TABLE 20.3—*Continued*

Stock or "Aggregation"	Center(s) of Distribution	Abundance	Exploitation	Other Threats	Conservation Status
Eastern Chukchi Sea	Kotzebue Sound, off Kasegaluk Lagoon (summer)	3710 in 1989–91 (Angliss and Lodge 2002)	70/year in Alaska	Coastal mining and development; vessel traffic	Probably stable
Eastern Bering Sea	Norton Sound and Yukon Delta (summer)	17,675 (9056–34,515 95% CI) in 1996–98	130/year in Alaska	Interactions with salmon stocks; vessel traffic	Uncertain
Bristol Bay	Inner Bristol Bay, Alaska (summer)	At least 1000 in early 1990s (Frost and Lowry 1990)	13/year	Interactions with salmon stocks	Stable or increasing
Cook Inlet	Northern Cook Inlet, Alaska (summer); Cook Inlet and Gulf of Alaska (winter)	347 (CV = 0.29) in 1998	65–75/year	Oil and gas development, tourism, shipping, commercial fishing, sewage from Anchorage	Seriously depleted and declining
Narwhals					
West Greenland–Disko Bay	Wintering area for whales from several different summering areas	5210 (1285–21,115) (partial survey)	140–205/year	—	Variable, depending on the summering stock
West Greenland–Uummannaq	November aggregation (also hunted in other areas)	—	210–400/year	—	Uncertain, but could be overhunted
West Greenland–Upernavik–Melville Bay	Occupied in summer, spring, and autumn, involving more than one stock	No estimates, but local summer population small	60–70/year (also hunted in Disko Bay)	—	Uncertain
West Greenland–Avanersuaq	Inglefield Bay (summer)	3539 (95% CI 1869–6722) in 1986 (partial survey; Born et al. 1994)	ca. 140–150/year (also hunted in other areas)	—	Uncertain
Eclipse Sound	Pond Inlet, Koluktoo Bay, Eclipse Sound (summer)	543 (90% CI 166–1784) plus 675 in 1984 (partial survey; Richard et al. 1994)	80–100/year (also hunted in other areas)	—	Uncertain but of concern
Admiralty Inlet	Summer aggregation and migration route	5556 (90% CI 3759–8213) in 1984 (Richard et al. 1994)	ca. 80/year (also hunted in other areas)	—	Considered stable
Prince Regent Inlet	Summer aggregation	9754 (90% CI 6057–15,816) in 1984 (Richard et al. 1994)	ca. 25/year (also hunted in other areas)	—	Considered stable
Peel Sound	Summer aggregation	1701 (90% CI 1029–2816) in 1984 (Richard et al. 1994)	<10/year (also hunted in other areas)	—	Uncertain
Jones Sound	Summer aggregation and migration route	—	<20/year (also hunted in other areas)	—	Uncertain
Canadian High Arctic	Aggregate of the foregoing five summering groups	14,240 (95% CI 6658–30,931) in 1996 (partial survey)	Combined annual catch in various hunting areas ca. 280/year	—	—
Baffin Bay	Pack ice wintering area for animals from many of the foregoing summering groups	34,363 (*SE* = 8282) in 1979 (partial survey; Koski and Davis 1994)	Combined annual catch including Greenland and Canada ca. 885/year	—	Variable, depending on the summering stock
Northern Hudson Bay	Summer aggregation of animals that winter in Hudson Strait	1355 (90% CI 1000–1900) in 1984 (partial survey; Richard 1991)	20–25/year	—	Considered stable

SOURCES: Data from IWC (2000) and NAMMCO (2000). Other sources as indicated.
NOTE: There is some overlap among stock units, so numbers are not necessarily additive within Abundance and Exploitation columns. Also, abundance estimates are not standardized for bias, and catch levels generally do not account for hunting loss. SE, Standard Error; CV, Coefficient of variation; CI, confidence interval.

Hemisphere in 1971. Soon thereafter, catch limits were set separately for males and females. No catch limits were placed on sperm whales in the North Atlantic until 1977. The IWC accorded sperm whales complete protection from commercial whaling (by member states) beginning with the 1982 pelagic season and the 1986 coastal season (IWC 1982:19–20). Japan formally objected to this whaling ban and continued its shore-based hunt through the 1987 season, after which its objection was withdrawn (IWC 1989:1). Although commercial in nature, sperm whale hunting in the Azores and Madeira in the North Atlantic was exempt from IWC regulation because Portugal, which owned these islands, was not a member of the IWC. Shore-based whaling continued at the Azores until the 1980s.

Data for Soviet pelagic whaling in the Southern Hemisphere and North Pacific were falsified on a large scale (Yablokov 1994; Zemsky

FIGURE 20.18. Pilot whales were driven ashore in large numbers at Cape Cod during the nineteenth century. This drive fishery took advantage of the fact that pilot whales (or blackfish, as they were known to the fishermen) occur in close-knit groups, which remain together in the face of danger. The head and jaw oil of pilot whales was sold under the name of "porpoise-jaw oil" and was highly valued by watch and instrument makers (Clark 1887). SOURCE: Photo courtesy of the New Bedford Whaling Museum.

et al. 1995; Brownell et al. 1998), as were the official catch statistics reported by Japan for their North Pacific land stations (Kasuya 1991, 1999). The latter included not only falsification of body lengths of undersized whales and underreporting of catches, but also deliberate listing of females as males.

MANAGEMENT AND CONSERVATION

Impacts of whaling (e.g., selective removal and effective elimination of large breeding males; see Whitehead et al. 1997) and the complex social structure and reproductive behavior of sperm whales (Whitehead and Weilgart 2000) have confounded stock assessment. Also, catch records are poor for some areas and periods, and gross underreporting or intentional misreporting of modern catch data has taken place on a large scale. The wide-ranging, generally offshore distribution of sperm whales and their long submergence times complicate efforts to estimate abundance. Although the aggregate abundance worldwide is probably >100,000 individuals, the extent of depletion and degree of recovery of populations are uncertain. It has been suggested that the large twentieth-century catches of sperm whales in the North Pacific not only further reduced the population below its pre-exploitation level and (possibly) reduced pregnancy rates by reducing the number of breeding males (Whitehead 1987), but also may have (1) increased mortality within family units because key individuals were lost, making groups less able to defend themselves against killer whales and less adept at tracking resources; and (2) affected social structure, forcing depleted or fragmented pods to coalesce and form groups of mixed maternal lineages (Richard et al. 1996).

There is always a possibility of illegal whaling or a resumption of legal commercial whaling for sperm whales. Also, sperm whales are susceptible to entanglement in fishing gear and collisions with ships. Gill nets set in deep water for pelagic fish (e.g., sharks, billfish, and tuna) are especially hazardous for sperm whales (e.g., Di Natale and Notarbartolo di Sciara 1994; Haase and Félix 1994; Barlow et al. 1994; Palacios and Gerrodette 1996; Félix et al. 1997; Carretta et al. 2002; Waring et al. 2002). Sperm whales are also killed after becoming trapped in tuna purse-seine nets (Haase and Félix 1994) or entangled in longlines (Waring et al. 2002; Angliss and Lodge 2002). They spend long periods (typically up to 10 min; e.g., Jaquet et al. 2000) "rafting" at the surface between deep dives, making them vulnerable to ship strikes. There are reports of sperm whales being struck by passenger ships, tug boats, and high-speed ferries, which sometimes results in serious injuries to passengers as well as the whales (Berzin 1972; André et al. 1997; Laist et al. 2001). Incidental mortality should be viewed as equivalent to an ongoing whaling enterprise, with the removal rate dependent on, among other factors, fishing or shipping intensity and the density of whales in the area. A major difference, however, is that data on fishery by-catch and ship strikes are generally not as reliable as whaling catch data.

Their demonstrated responsiveness to loud, unfamiliar underwater sounds makes it likely that sperm whales are adversely affected, at least transiently, by artificial noise in the marine environment. Moreover, there is some evidence from Pacific equatorial waters that sperm whale feeding success and, in turn, calf production rates are negatively affected by increases in sea surface temperature (Smith and Whitehead 1993; Whitehead 1997). This could mean that global warming (regardless of whether it is driven primarily by natural or anthropogenic processes) will reduce the productivity of at least some sperm whale populations (Whitehead 1997).

In U.S. waters, sperm whales are protected under both the Endangered Species Act and the Marine Mammal Protection Act. They are also listed in Appendix I of the Convention on International Trade in Endangered Species of Wild Fauna and Flora, which means that commercial trade in their products is prohibited. Sperm whales are not currently hunted commercially in the North Atlantic or North Pacific. Although Canada and Iceland withdrew from the IWC in 1982 and 1992, respectively, only Iceland has expressed an intention to resume commercial whaling (e.g., Sigurjónsson 1989). There is no evidence that whaling will resume in the near future in the Azores or Madeira. In the North Pacific, Japan has expressed a strong, ongoing interest in maintaining a whaling industry. Japan authorized a "scientific catch" of five sperm whales in the western North Pacific in 2000 (Normile 2000).

OTHER TOOTHED CETACEANS

Small Arctic Whales: Narwhal and White Whale. Two odontocetes are adapted to living year-round in the Arctic. Both the narwhal (*Monodon monoceros*) and the white whale, or beluga (*Delphinapterus leucas*), lack a dorsal fin, which probably makes it easier for them to surface and maneuver in sea ice. They also have thick skin and blubber, the latter providing insulation and lipid storage. The two species are similar in body shape and size, neither growing longer than about 4.7 m or heavier than 1600 kg, with males considerably larger than females (Hay and Mansfield 1989; Stewart and Stewart 1989). Narwhals change in color from gray at birth to solid dark gray or black as juveniles. Adults are white ventrally, spotted or mottled on the sides, and dark gray to black dorsally. Old males can be almost entirely white. Belugas are also gray at birth and remain so for the first few years of life. They lighten gradually and become essentially all white after sexual maturity. White whales have 8–11 pairs of conical teeth in both jaws. Narwhals are functionally toothless, with no erupted teeth inside the mouth. In males, and occasionally females, a horizontally oriented tooth (possibly analogous to an upper incisor) rooted along the left maxillary–premaxillary suture erupts through the upper lip at about the time of sexual maturation. This uncurved but leftward-spiraling tooth continues to grow and, if unbroken, can reach an erupted length of well over 2 m (Hay and Mansfield 1989).

TABLE 20.4. Status of selected odontocetes in the Western North Atlantic and Gulf of Mexico

Species	Stock	Abundance (CV)	Annual By-Catch	Status
Grampus griseus	WNA	29,110 (0.29)	51	None
Lagenorhynchus acutus	WNA[a]	51,640 (0.38)	118	None
Delphinus delphis	WNA	30,768 (0.32)	375	Strategic
Stenella frontalis	WNA	52,279 (0.87)	8[b]	None
Stenella frontalis	GOM	3,213 (0.44)	2[b]	None
Stenella attenuata	WNA	13,117 (0.56)	0[b]	None
Stenella attenuata	GOM	31,320 (0.20)	2[b]	None
Stenella coeruleoalba	WNA	61,546 (0.40)	7	None
Stenella clymene	GOM	5,571 (0.37)	0	None
Steno bredanensis	GOM	852 (0.31)	0	None
Orcinus orca	GOM	277 (0.42)	0	None
Globicephala spp.[c]	WNA	14,524 (0.30)	199	Strategic
Kogia spp.[d]	WNA	536 (0.45)	6	Strategic (*K. breviceps*)
Beaked whales[e]	WNA	3,196 (0.34)	0	Strategic

SOURCE: Data from Waring et al. (2002).

NOTE: Stock refers to either Western North Atlantic (WNA) or Gulf of Mexico (GOM). Status refers to listings under the U.S. Marine Mammal Protection Act. Some closely related species are lumped due to the difficulty of identifying individuals at sea. CV, Coefficient of variation.

[a]Gulf of Maine component only.

[b]Includes *Stenella attenuata* and *Stenella frontalis*.

[c]Includes *Globicephala melas* and *G. macrorhynchus*.

[d]Includes *Kogia breviceps* and *K. sima*.

[e]Includes four species in the genus *Mesoplodon* and *Ziphius cavirostris*.

TABLE 20.5. Status of selected odontocetes in coastal waters of California, Oregon, and Washington

Species	Abundance (CV)	Annual By-Catch	Status
Phocoenoides dalli	117,545 (0.45)	12	None
Lagenorhynchus obliquidens	25,825 (0.49)	7	None
Grampus griseus	16,483 (0.28)	6	None
Stenella coeruleoalba	20,235 (0.14)	0	None
Delphinus delphis	373,573 (0.19)	79	None
Delphinus capensis	32,239 (0.18)	14	None
Lissodelphis borealis	13,705 (0.38)	15	None
Globicephala macrorhynchus	970 (0.37)	3	None
Berardius bairdii	379 (0.23)	0	None
Ziphius cavirostris	5,870 (0.38)	0	None
Mesoplodon beaked whales[a]	4,098 (0.50)	0	None
Kogia spp.[b]	4,746 (0.67)	0	None

SOURCE: Data from Carretta et al. 2002.

NOTE: Status refers to listings under the U.S. Marine Mammal Protection Act. Some closely related species are lumped due to problems of identification at sea. CV, Coefficient of variation.

[a]Includes five species in the genus *Mesoplodon*.

[b]Includes *Kogia breviceps* and *K. sima* but primarily the former.

White whales have a nearly circumarctic distribution in waters north of about 60°N, and there are populations in several lower-latitude sites as well, notably in the southern Sea of Okhotsk, Cook Inlet (Alaska), James Bay, southern Hudson Bay, Ungava Bay, and the lower St. Lawrence River (Canada) (IWC 2000). Numerous stocks, with varying degrees of genetic and demographic distinctness, occur in North American waters (Table 20.3). The distribution of narwhals is much less extensive. Most of the world's narwhals live in Arctic waters adjacent to the North Atlantic Ocean, with centers of abundance in northern Hudson Bay–Hudson Strait, Baffin Bay–Lancaster Sound, and the Greenland Sea (IWC 2000; NAMMCO 2000) (Table 20.3). The estimated worldwide abundance of white whales is about 150,000; that of narwhals, 50,000 (IWC 2000; NAMMCO 2000).

Although the two species are partially sympatric, especially in the winter, there are clear ecological differences between them. White whales tend to overwinter in open or partially open water along the margins of the pack ice. Narwhals are generally more dispersed within the heavy pack ice. In spring and early summer, both species migrate into areas that were largely inaccessible during winter because of ice cover (Reeves and St. Aubin 2001). White whales congregate in estuaries, apparently at least in part to facilitate molting (St. Aubin et al. 1990), whereas large herds of narwhals occur in deep fiord waters during the open-water period. Both species are subject to predation by killer whales, and predator avoidance may at least partly account for their movements (Reeves and Mitchell 1988b). In the winter, whales trapped in small areas of open water are vulnerable to predation by polar bears (*Ursus maritimus*) or human hunters or to starvation and suffocation (Lowry et al. 1987; Siegstad and Heide-Jørgensen 1994).

Serious conservation problems exist in a number of areas where white whales were historically depleted by commercial whaling and subsequently by "subsistence" whaling. Several populations continue (or at least continued until recently) to be overhunted by local residents, for example, in Cook Inlet (Alaska), West Greenland, and eastern Canada. The population in Canada's St. Lawrence River is threatened primarily by pollution and disturbance from vessel traffic, having been reduced historically by overhunting. Narwhals are hunted intensively in the eastern Canadian Arctic and Greenland, largely for food, but also for their tusk ivory, which enters international markets as a novelty item (Reeves 1993). Consumption of meat and skin from white whales and narwhals has serious health implications for Native people in the north, some of them good (e.g., intake of vitamin C, retinol, and zinc; Geraci and Smith 1979; Kinloch et al. 1992) and some of them bad (e.g., exposure to organochlorine contaminants; Krahn et al. 1999).

Killer Whale. As mentioned above, the killer whale (*Orcinus orca*), in recent years often called the orca, is the most widely distributed marine mammal. It is truly cosmopolitan in every sense, occurring from the tropics and as far poleward in both hemispheres as ice conditions permit (Dahlheim and Heyning 1999). The species is best known for its habit of preying on fellow homeotherms, including cetaceans of all sizes, seals, large tunas, seabirds, and marine turtles (Jefferson et al. 1991). There is evidence that killer whales even consume terrestrial mammals (e.g., moose, *Alces alces;* other deer, *Odocoileus* sp.; river otters, *Lontra canadensis*) swimming in marine waters. Remarkably, there is no authenticated record of killer whales attacking human swimmers or divers.

Killer whales have been closely studied in inshore waters of Washington and British Columbia (Bigg et al. 1990; Olesiuk et al. 1990) and also in Prince William Sound and the eastern Gulf of Alaska (Matkin and Saulitis 1994; Matkin et al. 1999). In these areas, two sympatric forms have been distinguished on the basis of association and dispersal patterns, shape of the dorsal fin, pigmentation, genetics, and acoustics (Baird 2000). One form (so-called "residents") prey on fish, the other ("transients") on marine mammals (Ford et al. 1998).

Killer whales were hunted commercially by Norway in the North Atlantic until the late 1960s and to some extent by Japan in the western North Pacific, but since then, most direct removals in the Northern Hemisphere have been either opportunistic kills by Native hunters (Heide-Jørgensen 1988; Reeves and Mitchell 1988b) or livecaptures for oceanaria (Bigg and Wolman 1975; Sigurjónsson and Leatherwood 1988). With changes in North American attitudes toward cetaceans in captivity (Lavigne et al. 1999; Reeves and Mead 1999), livecapture of killer whales has essentially stopped. In the absence of commercial whaling and a significant livecapture fishery, conservation concerns now relate to the high levels of toxic contaminants found in killer whale tissues, the potential for disturbance by boat-based whale watching in

TABLE 20.6. Status of selected odontocetes in Alaska

Species	Stock	Abundance (CV)	Annual Harvest	Annual By-Catch	Status
Delphinapterus leucas	Beaufort Sea	39,258 (NA)	177	0	None
Delphinapterus leucas	Eastern Chukchi Sea	3,710 (NA)	60	0	None
Delphinapterus leucas	Eastern Bering Sea	18, 142 (0.24)	164	0	None
Delphinapterus leucas	Bristol Bay	1,888 (NA)	15	1	None
Delphinapterus leucas	Cook Inlet	435 (0.23)	0[a]	0	Depleted
Lagenorhynchus obliquidens	North Pacific (Gulf of Alaska)	26,880 (NA)	0	4	None
Phocoenoides dalli	Alaska	83,400 (0.10)	0	42	None

SOURCE: Data from Angliss and Lodge (2002).

NOTE: Status refers to listings under the U.S. Marine Mammal Protection Act or Endangered Species Act. NA signifies that coefficient of variation (CV) is not available.

[a]The average subsistence take from 1995 to 1997 was 87 whales; a government-imposed moratorium on hunting was responsible for zero catches in 1999 and 2000.

British Columbia and Washington, and the depletion of salmonid populations, which are a major component of the diet of "resident" killer whales (Baird 2000).

Pilot Whales, Beaked Whales, and Others. Numerous other cetacean species occur in North American coastal and shelf waters (Fig. 20.18). Some basic information on a selection of these is given in Tables 20.4–20.6.

LITERATURE CITED

Aguilar, A. 1983. Organochlorine pollution in sperm whales, *Physeter macrocephalus,* from the temperate waters of the eastern North Atlantic. Marine Pollution Bulletin 14:349–52.

Allen, K. R. 1980. Conservation and management of whales. University of Washington Press, Seattle.

Allen, M. C., A. J. Read, J. Gaudet, and L. S. Sayigh. 2001. Fine-scale habitat selection of foraging bottlenose dolphins (*Tursiops truncatus*) near Clearwater, Florida. Marine Ecology Progress Series 222:253–264.

Andersen, S., and A. Dziedzic. 1964. Behaviour patterns of captive harbour porpoise *Phocaena phocaena* (L.) Bulletin de l'Institut Océanographique (Monaco) 63:1–20.

André, M., M. Terada, and Y. Watanabe. 1997. Sperm whale (*Physeter macrocephalus*) behavioural response after the playback of artificial sounds. Report of the International Whaling Commission 47:499–504.

Angell, L. 1981. The porpoise factory: A little known business that prospered on Hatteras Island for over 100 years, and made a living for many families. The State: Down Home in North Carolina (Raleigh, NC) 49(5):20–23, 39.

Angliss, R. P., and K. L. Lodge, eds. 2002. Alaska marine mammal stock assessments, 2002 (NOAA Technical Memorandum NMFS-AFSC-133). Alaska Fisheries Science Center, Seattle, WA.

Anonymous., 1981. Annex K. Catches of sperm whales in the North Atlantic in the twentieth century. Report of the International Whaling Commission 31:703–5.

Arnbom, T., V. Papastavrou, L. S. Weilgart, and H. Whitehead. 1987. Sperm whales react to an attack by killer whales. Journal of Mammalogy 68:450–53.

Arnold, P. W. 1972. Predation on harbour porpoise, *Phocoena phocoena,* by a white shark, *Carcharodon carcharias*. Journal of the Fisheries Research Board of Canada 29:1213–14.

Ashford, J. R., P. S. Rubilar, and A. R. Martin. 1996. Interactions between cetaceans and longline fishery operations around South Georgia. Marine Mammal Science 12:452–57.

Avila de Melo, A. M., and A. R. Martin. 1985. A study of male sperm whale length data from the Azorean and Madeiran catches, 1947–82. Report of the International Whaling Commission 35:209–15.

Baird, R. W. 2000. The killer whale: Foraging specializations and group hunting. Pages 127–53 *in* J. Mann, R. C. Connor, P. L. Tyack, and H. Whitehead, eds. Cetacean societies: Field studies of dolphins and whales. University of Chicago Press, Chicago.

Baird, R. W., E. L. Walters, and P. J. Stacey. 1993. Status of the bottlenose dolphin, *Tursiops truncatus,* with special reference to Canada. Canadian Field-Naturalist 107:466–80.

Barlow, J. 1994. Recent information on the status of large whales in California waters (NOAA Technical Memorandum NMFS, NOAA-TM-NMFS-SWFSC-203). National Marine Fisheries Service, La Jolla, CA.

Barlow, J. 1995. The abundance of cetaceans in California waters. Part I. Ship surveys in summer and fall of 1991. Fishery Bulletin 93:1–14.

Barlow, J. 1997. Preliminary estimates of cetacean abundance off California, Oregon, and Washington based on a 1996 ship survey and comparisons of passing and closing modes (Administrative Report LJ-97-11). NMFS, Southwest Fisheries Science Center, La Jolla, CA.

Barlow, J., and S. Sexton. 1996. The effect of diving and searching behavior on the probability of detecting track-line groups, g_0, of long-diving whales during line-transect surveys (Administrative Report LJ-96-14). Southwest Fisheries Science Center, National Marine Fisheries Service, La Jolla, CA.

Barlow, J., and B. L. Taylor. 1998. Preliminary abundance of sperm whales in the northeastern temperate Pacific estimated from a combined visual and acoustic survey (Document SC/50/CAWS 20). International Whaling Commission, Scientific Committee. Cambridge, UK.

Barlow, J., S. Swartz, T. C. Eagle, and P. R. Wade. 1995. U.S. marine mammal stock assessments: Guidelines for preparation, background, and a summary of the 1995 assessments (NOAA Technical Memorandum NMFS-OPR-6). U.S. Department of Commerce, NOAA National Marine Fisheries Service. Office of Protected Resources, Silver Spring, MD.

Barlow, J., R. W. Baird, J. E. Heyning, K. Wynne, A. M. Manville, L. F. Lowry, D. Hanan, J. Sease, and V. N. Burkanov. 1994. A review of cetacean and pinniped mortality in coastal fisheries along the west coast of the USA and Canada and the east coast of the Russian Federation. Report of the International Whaling Commission (Special Issue) 15:405–26.

Barros, N. B., and A. A. Myrberg, Jr. 1987. Prey detection by means of passive listening in bottlenose dolphins (*Tursiops truncatus*). Journal of the Acoustical Society of America 62(Supplement 1):S97.

Barros, N. B., and D. K. Odell. 1990. Food habits of bottlenose dolphins in the southeastern United States. Pages 309–28 *in* S. Leatherwood and R. R. Reeves, eds. The bottlenose dolphin. Academic Press, San Diego, CA.

Barros, N. B., and R. S. Wells. 1998. Prey and feeding patterns of resident bottlenose dolphins (*Tursiops truncatus*) in Sarasota Bay, Florida. Journal of Mammalogy 79:1045–59.

Baur, D. C., M. J. Bean, and M. L. Gosliner. 1999. The laws governing marine mammal conservation in the United States. Pages 48–86 *in* J. R. Twiss, Jr., and R. R. Reeves, eds. Conservation and management of marine mammals. Smithsonian Institution Press, Washington, DC.

Berzin, A. A. 1972. The sperm whale (kashalot). Pacific Scientific Research Institute of Fisheries and Oceanography, Moscow.

Berzin, A. A., and A. A. Rovnin. 1966. Distribution and migration of whales in the northeastern part of the Pacific Ocean, Bering and Chukchi seas. Pages 103–6 *in* K. I. Panin, ed. Soviet research on marine mammals of the Far East. U.S. Department of the Interior, Bureau of Commercial Fisheries, Seattle, WA.

Best, P. B. 1979. Social organization in sperm whales, *Physeter macrocephalus*. Pages 227–89 *in* H. E. Winn and B. L. Olla, eds. Behavior of marine animals, Vol. 3. Plenum Press, New York.

Best, P. B. 1983. Sperm whale stock assessments and the relevance of historical whaling records. Report of the International Whaling Commission (Special Issue) 5:41–55.

Best, P. B., and D. S. Butterworth. 1980. Timing of oestrus within sperm whale schools. Report of the International Whaling Commission (Special Issue) 2:137–40.

Best, P. B., P. A. S. Canham, and N. Macleod. 1984. Patterns of reproduction in sperm whales, *Physeter macrocephalus*. Report of the International Whaling Commission (Special Issue) 8:51–79.

Bigg, M. A., and A. A. Wolman. 1975. Live-capture killer whale (*Orcinus orca*) fishery, British Columbia and Washington, 1962–73. Journal of the Fisheries Research Board of Canada 32:1213–21.

Bigg, M. A., P. F. Olesiuk, G. M. Ellis, J. K. B. Ford, and K. C. Balcomb III. 1990. Social organization and genealogy of resident killer whales (*Orcinus orca*) in the coastal waters of British Columbia and Washington State. Report of the International Whaling Commission (Special Issue) 12:383–405.

Born, E. W., M. P. Heide-Jørgensen, F. Larsen, and A. R. Martin. 1994. Abundance and stock composition of narwhals (*Monodon monoceros*) in Inglefield Bredning (NW Greenland). Meddelelser om Grønland, Bioscience 39:51–68.

Bouquegneau, J. M., V. Debacker, S. Govert, and J. P. Nellissen. 1997. Toxicological investigations on four sperm whales stranded on the Belgian coast: Inorganic contaminants. Bulletin de l'Institut Royal des Sciences Naturelles de Belgique, Biologie 67 (Supplement):75–8.

Bowles, A. E., M. Smultea, B. Würsig, D. P. DeMaster, and D. Palka. 1994. Relative abundance and behavior of marine mammals exposed to transmissions from the Heard Island feasibility test. Journal of the Acoustical Society of America 96:2469–84.

Brownell, R. L., Jr., A. V. Yablokov, and V. A. Zemsky. 1998. USSR pelagic catches of North Pacific sperm whales, 1949–1979: Conservation implications (Document SC/50/CAWS 27). International Whaling Commission, Scientific Committee. Cambridge, UK.

Caldwell, D. K., M. C. Caldwell, and D. W. Rice. 1966. Behavior of the sperm whale *Physeter catodon* L. Pages 677–717 *in* K. S. Norris, ed. Whales, dolphins, and porpoises. University of California Press, Berkeley.

Carretta, J. V., M. M. Muto, J. Barlow, J. Baker, K. A. Forney, and M. Lowry. 2002. U.S. Pacific marine mammal stock assessments: 2002 (NOAA Technical Memorandum NMFS-SWFSC-346). Southwest Fisheries Science Center, La Jolla, CA.

Caswell, H., S. Brault, A. J. Read, and T. D. Smith. 1998. Harbor porpoise and fisheries: An uncertainty analysis of incidental mortality. Ecological Applications 8:226–38.

CETAP. 1982. A characterization of marine mammals and turtles in the mid- and north Atlantic areas of the U.S. outer continental shelf (Final Report to Bureau of Land Management, Contract AA551-CTB-48). Cetacean and Turtle Assessment Program, University of Rhode Island.

Chivers, S. J., A. E. Dizon, P. J. Gearin, and K. M. Robertson. 2002. Small-scale population structure of eastern North Pacific harbour porpoises (*Phocoena phocoena*) indicated by molecular genetic analyses. Journal of Cetacean Research and Management 4:111–122.

Christal, J., and H. Whitehead. 1997. Aggregations of mature male sperm whales on the Galápagos Islands breeding ground. Marine Mammal Science 13:59–69.

Christal, J., H. Whitehead, and E. Lettevall. 1998. Sperm whale social units: Variation and change. Canadian Journal of Zoology 76:1431–40.

Clapham, P. J., S. Leatherwood, I. Szczepaniak, and R. L. Brownell, Jr. 1997. Catches of humpback and other whales from shore stations at Moss Landing and Trinidad, California. Marine Mammal Science 13:368–94.

Clark, A. H. 1887. The blackfish and porpoise fisheries. Pages 295–310 *in* G. B. Goode, ed. The fisheries and fishery industries of the United States. Section V. History and methods of the fisheries, Vol. II. U.S. Government Printing Office, Washington, DC.

Clarke, M. R. 1977. Beaks, nets and numbers. Symposium of the Zoological Society of London 38:89–126.

Clarke, M. R. 1980. Cephalopods in the diet of sperm whales of the Southern Hemisphere and their bearing on sperm whale biology. Discovery Reports 37:1–324.

Clarke, M. R. 1996. Cephalopods as prey. III. Cetaceans. Philosophical Transactions of the Royal Society of London B 351:1053–65.

Clarke, M. R. 1997. Cephalopods in the stomach of a sperm whale stranded between the islands of Terschelling and Ameland, southern North Sea. Bulletin de l'Institut Royal des Sciences Naturelles de Belgique, Biologie 67 (Supplement):53–5.

Clarke, R. 1954. Open boat whaling in the Azores. The history and present methods of a relic industry. Discovery Reports 26:281–354.

Clarke, R., and O. Paliza. 1988. Intraspecific fighting in sperm whales. Report of the International Whaling Commission 38:235–41.

Collins, J. W. 1886. Gill-nets in the cod fishery: A description of Norwegian cod-nets, etc., and a history of their use in the United States. Pages 265–85 *in* Report of the U.S. Fisheries Commission for 1884, Part XII.

Connor, R. C., R. A. Smolker, and A. F. Richards. 1992. Dolphin alliances and coalitions. Pages 415–43 *in* A. H. Harcourt and F. B. M. de Waal, eds. Coalitions and alliances in humans and other animals. Oxford University Press, Oxford.

Connor, R. C., R. S. Wells, J. Mann, and A. J. Read. 2000. The bottlenose dolphin: Social relationships in a fission–fusion society. Pages 91–126 *in* J. Mann, R. C. Connor, P. L. Tyack, and H. Whitehead, eds. Cetacean societies: Field studies of dolphins and whales. University of Chicago Press, Chicago.

Corkeron, P. J., R. J. Morris, and M. M. Bryden. 1987. Interactions between bottlenose dolphins and sharks in Moreton Bay, Queensland. Aquatic Mammals 13:109–13.

Cranford, T. W. 1999. The sperm whale's nose: Sexual selection on a grand scale. Marine Mammal Science 15:1133–57.

Curry, B. E., and J. Smith. 1997. Phylogeographic structure of the bottlenose dolphin (*Tursiops truncatus*): Stock identification and implications for management. Pages 227–47 *in* A. E. Dizon, S. J. Chivers, and W. F. Perrin, eds. Molecular genetics of marine mammals (Special Publication No. 3). Society for Marine Mammalogy.

Dahlheim, M. J., and J. E. Heyning. 1999. Killer whale *Orcinus orca* (Linnaeus, 1758). Pages 281–322 *in* S. H. Ridgway and R. Harrison, eds. Handbook of marine mammals. Vol. 6: The second book of dolphins and the porpoises. Academic Press, San Diego, CA.

Davis, P. D., K. H. Hamilton, and K. S. Hewitt, eds. 1982. The Heritage of Carteret County, North Carolina, Vol. 1. Carteret Historical Research Association, Beaufort, NC.

Davis, R. W., G. S. Fargion, N. May, T. D. Leming, M. Baumgartner, W. E. Evans, L. J. Hansen, and K. Mullin. 1998. Physical habitat of cetaceans along the continental slope in the north-central and western Gulf of Mexico. Marine Mammal Science 14:490–507.

Defran, R. H., and D. W. Weller. 1999. Occurrence, distribution, site fidelity, and school size of bottlenose dolphins (*Tursiops truncatus*) off San Diego, California. Marine Mammal Science 15:366–80.

Defran, R. H., D. W. Weller, D. L. Kelly, and M. A. Espinosa. 1999. Range characteristics of Pacific coast bottlenose dolphins (*Tursiops truncatus*) in the Southern California Bight. Marine Mammal Science 15:381–93.

Dillon, M. C. 1996. Genetic structure of sperm whale populations assessed by mitochondrial DNA sequence variation. Ph.D. Dissertation, Dalhousie University, Halifax, Nova Scotia, Canada.

Di Natale, A., and G. Notarbartolo di Sciara. 1994. A review of the passive fishing nets and trap fisheries in the Mediterranean Sea and of the cetacean bycatch. Report of the International Whaling Commission (Special Issue) 15:189–202.

Dohl, T. P., R. C. Guess, D. L. Duman, and R. C. Helm. 1983. Cetaceans of central and northern California, 1980–1983: Status, abundance and distribution (OCS Study MMS 84-0045, Contract 14-12-0001–29090). Minerals Management Service.

Dufault, S., and H. Whitehead. 1995. An encounter with recently wounded sperm whales (*Physeter macrocephalus*). Marine Mammal Science 11:560–63.

Dufault, S., H. Whitehead, and M. Dillon. 1999. An examination of the current knowledge on the stock structure of sperm whales (*Physeter macrocephalus*) worldwide. Journal of Cetacean Research and Management 1:1–10.

Duignan, P. J., C. House, D. K. Odell, R. S. Wells, L. J. Hansen, M. T. Walsh, D. J. St. Aubin, B. K. Rima, and J. R. Geraci. 1996. Morbillivirus infection in bottlenose dolphins: Evidence for recurrent epizootics in the western Atlantic and Gulf of Mexico. Marine Mammal Science 12:499–515.

Feinholz, D. M. 1995. Northern range extension, abundance, and distribution of Pacific coastal bottlenose dolphins (*Tursiops truncatus gilli*) in Monterey Bay, California. Page 35 *in* Abstracts, 11th biennial conference on the biology of marine mammals, Orlando, FL.

Félix, F., B. Haase, J. W. Davis, D. Chiluiza, and P. Amador. 1997. A note on recent strandings and bycatches of sperm whales (*Physeter macrocephalus*) and humpback whales (*Megaptera novaeangliae*) in Ecuador. Report of the International Whaling Commission 47:917–19.

Fernandez, S., and A. A. Hohn. 1998. Age, growth, and calving season of bottlenose dolphins, *Tursiops truncatus,* off coastal Texas. Fishery Bulletin 96:357–65.

Ferrero, R. C., and L. M. Tsunoda. 1989. First record of a bottlenose dolphin (*Tursiops truncatus*) in Washington State. Marine Mammal Science 5:302–5.

Ferrero, R. C., D. P. DeMaster, P. S. Hill, M. M. Muto, and A. L. Lopez. 2000. Alaska marine mammal stock assessments, 2000 (NOAA Technical Memorandum NMFS-AFSC-119) Alaska Fisheries Science Center, Seattle, WA.

Fontaine, P.-M., M. O. Hammill, C. Barrette, and M. C. S. Kingsley. 1994. Summer diet of the harbour porpoise (*Phocoena phocoena*) in the estuary and the northern Gulf of St. Lawrence. Canadian Journal of Fisheries and Aquatic Sciences 51:172–78.

Forney, K. A. 1999. Trends in harbour porpoise abundance off central California, 1986–95: Evidence for interannual changes in distribution? Journal of Cetacean Research and Management 1:73–80.

Ford, J. K. B., G. M. Ellis, L. G. Barrett-Lennard, A. B. Morton, R. S. Palm, and K. C. Balcomb. 1998. Dietary specialization in two sympatric populations of killer whales (*Orcinus orca*) in coastal British Columbia and adjacent waters. Canadian Journal of Zoology 76:1456–71.

Forney, K. A., J. Barlow, and J. V. Carretta. 1995. The abundance of cetaceans in California waters. Part II: Aerial surveys in winter and spring of 1991 and 1992. Fishery Bulletin 93:15–26.

Forney, K. A., J. Barlow, M. M. Muto, M. Lowry, J. Baker, G. Cameron, J. Mobley, C. Stinchcomb, and J. V. Carretta. 2000. U.S. Pacific marine mammal stock assessments: 2000 (NOAA Technical Memorandum NMFS-SWFSC-300). Southwest Fisheries Science Center, La Jolla, CA.

Frost, K. J., and L. F. Lowry. 1990. Distribution, abundance, and movements of beluga whales, *Delphinapterus leucas,* in coastal waters of western Alaska. Canadian Bulletin of Fisheries and Aquatic Sciences 224:39–57.

Gannon, D. P., J. E. Craddock, and A. J. Read. 1998. Autumn food habits of harbor porpoises, *Phocoena phocoena,* in the Gulf of Maine. Fishery Bulletin 96:428–37.

Gaskin, D. E., P. W. Arnold, and B. A. Blair. 1974. *Phocoena phocoena*. Mammalian Species 42:1–8.

Gaskin, D. E., S. Yamamoto, and A. Kawamura. 1993. Harbor porpoise, *Phocoena phocoena* (L.), in the coastal waters of northern Japan. Fishery Bulletin 91:440–54.

Gearin, P. J., M. E. Gosho, J. L. Laake, L. Cooke, R. L. DeLong, and K. M. Hughes. 2000. Experimental testing of acoustic alarms (pingers) to reduce bycatch of harbour porpoise, *Phocoena phocoena,* in the state of Washington. Journal of Cetacean Research and Management 2:1–9.

Geraci, J. R., and V. J. Lounsbury. 1993. Marine mammals ashore: A field guide for strandings. Texas A&M University Sea Grant College Program, Galveston.

Geraci, J. R., and T. G. Smith. 1979. Vitamin C in the diet of Inuit hunters from Holman, Northwest Territories. Arctic 32:135–39.

Geraci, J. R., J. Harwood, and V. J. Lounsbury. 1999. Marine mammal die-offs: Causes, investigations, and issues. Pages 367–95 *in* J. R. Twiss, Jr., and

R. R. Reeves, eds. Conservation and management of marine mammals. Smithsonian Institution Press, Washington, DC.

Gilmore, R. M. 1969. Ambergris. Page 23 *in* F. E. Firth, ed. The encyclopedia of marine resources. Van Nostrand Reinhold, New York.

Gordon, J., A. Moscrop, C. Carlson, S. Ingram, R. Leaper, J. Matthews, and K. Young. 1998. Distribution, movements and residency of sperm whales off the Commonwealth of Dominica, eastern Caribbean: Implications for the development and regulation of the local whalewatching industry. Report of the International Whaling Commission 48:551–57.

Gorzelany, J. F. 1998. Unusual deaths of two free-ranging Atlantic bottlenose dolphins (*Tursiops truncatus*) related to ingestion of recreational fishing gear. Marine Mammal Science 14:614–17.

Gosliner, M. L. 1999. The tuna–dolphin controversy. Pages 120–55 *in* J. R. Twiss, Jr., and R. R. Reeves, eds. Conservation and management of marine mammals. Smithsonian Institution Press, Washington, DC.

Gowans, S., and H. Whitehead. 1995. Distribution and habitat partitioning by small odontocetes in the Gully, a submarine canyon on the Scotian Shelf. Canadian Journal of Zoology 73:1599–1608.

Green, G. A., J. J. Brueggeman, R. A. Grotefendt, C. E. Bowlby, M. L. Bonnell, and K. C. Balcomb III. 1992. Cetacean distribution and abundance off Oregon and Washington, 1989–1990. Chapter 1 *in* J. J. Brueggeman, ed. Oregon and Washington marine mammal and seabird surveys (Contract Report 14-12-0001-30426 prepared for the Pacific OCS Region). Minerals Management Service.

Gregr, E. J., L. Nichol, J. K. B. Ford, G. Ellis, and A. W. Trites. 2000. Migration and population structure of northeastern Pacific whales off coastal British Columbia: An analysis of commercial whaling records from 1908–1967. Marine Mammal Science 16:699–727.

Gunnlaugsson, T., and J. Sigurjónsson. 1990. NASS-87: Estimation of whale abundance based on observations made onboard Icelandic and Faroese survey vessels. Report of the International Whaling Commission 40: 571–80.

Gunter, G. 1942. Contributions to the natural history of the bottle-nose dolphin, *Tursiops truncatus* (Montague), on the Texas coast, with particular reference to food habits. Journal of Mammalogy 23:267–76.

Haase, B., and F. Félix. 1994. A note on the incidental mortality of sperm whales (*Physeter macrocephalus*) in Ecuador. Report of the International Whaling Commission (Special Issue) 15:481–83.

Harrison, R. J. 1977. Ovarian appearances and histology in *Tursiops truncatus*. Pages 195–204 *in* S. H. Ridgway and K. Benirschke, eds. Breeding dolphins: Present status and suggestions for the future (Marine Mammal Commission Report No. MMC-76/07, NTIS No. PB-273673). Available from National Technical Information Service, Arlington, VA.

Hay, K. A., and A. W. Mansfield. 1989. Narwhal *Monodon monoceros* Linnaeus, 1758. Pages 145–76 *in* S. H. Ridgway and R. Harrison, eds. Handbook of marine mammals. Vol. 4: The river dolphins and the larger toothed whales. Academic Press, London.

Hegarty, R. B. 1959. Returns of whaling vessels sailing from American ports. A continuation of Alexander Starbuck's "History of the American Whale Fishery" 1876–1928. Old Dartmouth Historical Society and Whaling Museum, New Bedford, MA.

Heide-Jørgensen, M.-P. 1988. Occurrence and hunting of killer whales in Greenland. Rit Fiskideildar 11:115–35.

Heide-Jørgensen, M. P. 1994. Distribution, exploitation and population status of white whales (*Delphinapterus leucas*) and narwhals (*Monodon monoceros*) in West Greenland. Meddelelser om Grønland, Bioscience 39:135–49.

Hersh, S. L., and D. A. Duffield. 1990. Distinction between northwest Atlantic offshore and coastal bottlenose dolphins based on hemoglobin profile and morphometry. Pages 129–39 *in* S. Leatherwood and R. R. Reeves, eds. The bottlenose dolphin. Academic Press, San Diego, CA.

Hoek, W. 1992. An unusual aggregation of harbor porpoises (*Phocoena phocoena*). Marine Mammal Science 8:152–55.

Hoelzel, A. R., C. W. Potter, and P. B. Best. 1998. Genetic differentiation between parapatric 'nearshore' and 'offshore' populations of the bottlenose dolphin. Proceedings of the Royal Society of London B 265:1177–83.

Hoese, H. D. 1971. Dolphins feeding out of water in a salt marsh. Journal of Mammalogy 52:222–23.

Hohn, A. A., M. D. Scott, R. S. Wells, J. C. Sweeney, and A. B. Irvine. 1989. Growth layers in teeth from known-age, free-ranging bottlenose dolphins. Marine Mammal Science 5:315–42.

Howell, A. B. 1930. Aquatic mammals: Their adaptations to life in the water. Charles C. Thomas, Baltimore.

International Whaling Commission. 1982. Chairman's report of the thirty-third annual meeting. Report of the International Whaling Commission 32: 17–42.

International Whaling Commission. 1989. International Whaling Commission report 1987–88. Report of the International Whaling Commission 39:1–9.

International Whaling Commission. 2000. Annex I. Report of the Subcommittee on Small Cetaceans. Journal of Cetacean Research and Management 2(Supplement):235–63.

Irvine, A. B., R. S. Wells, and P. W. Gilbert. 1973. Conditioning an Atlantic bottle-nosed dolphin, *Tursiops truncatus,* to repel various species of sharks. Journal of Mammalogy 54:503–5.

Jacques, T. G., and R. H. Lambertsen, eds. 1997. Sperm whale deaths in the North Sea: Science and management. Bulletin de l'Institut Royal des Sciences Naturelles de Belgique, Biologie 67(Supplement):1–133.

Jaquet, N., S. M. Dawson, and E. Slooten. 2000. Seasonal distribution and diving behaviour of male sperm whales off Kaikoura: foraging implications. Canadian Journal of Zoology 78:407–19.

Jaquet, N. 1996. How spatial and temporal scales influence understanding of sperm whale distribution: A review. Mammal Review 26:51–65.

Jaquet, N., and H. Whitehead. 1996. Scale-dependent correlation of sperm whale distribution with environmental features and productivity in the South Pacific. Marine Ecology Progress Series 135:1–9.

Jaquet, N., H. Whitehead, and M. Lewis. 1996. Coherence between 19th century sperm whale distributions and satellite-derived pigments in the tropical Pacific. Marine Ecology Progress Series 145:1–10.

Jefferson, T. A., and A. J. Schiro. 1997. Distribution of cetaceans in the offshore Gulf of Mexico. Mammal Review 27:27–50.

Jefferson, T. A., P. A. Stacey, and R. W. Baird. 1991. A review of killer whale interactions with other marine mammals: Predation to co-existence. Mammal Review 21:151–80.

Johnston, D. W. 2002. The effect of acoustic harassment devices on harbour porpoises (*Phocoena phocoena*) in the Bay of Fundy, Canada. Biological Conservation 108:113–118.

Jonsgård, Å. 1977. Tables showing the catch of small whales (including minke whales) caught by Norwegians in the period 1938–75, and large whales caught in different North Atlantic waters in the period 1868–1974. Report of the International Whaling Commission 27:413–26.

Kapel, F. O. 1979. Exploitation of large whales in West Greenland in the twentieth century. Report of the International Whaling Commission 29:197–214.

Kasuya, T. 1991. Density dependent growth in North Pacific sperm whales. Marine Mammal Science 7:230–57.

Kasuya, T. 1999. Examination of the reliability of catch statistics in the Japanese coastal sperm whale fishery. Journal of Cetacean Research and Management 1:109–22.

Kato, H. 1984. Observation of tooth scar on the head of male sperm whale, as an indication of inter-sexual fightings. Scientific Reports of the Whales Research Institute (Tokyo) 35:39–46.

Kellogg, R. 1926–1927. Report on researches by Remington Kellogg. Carnegie Institute Year Book 26:366 (Cited from Mitchell 1975b).

Kenney, R. D. 1990. Bottlenose dolphins off the northeastern United States. Pages 369–86 *in* S. Leatherwood and R. R. Reeves, eds. The bottlenose dolphin. Academic Press, San Diego, CA.

Kingsley, M. C. S. 2000. Numbers and distribution of beluga whales, *Delphinapterus leucas,* in James Bay, eastern Hudson Bay, and Ungava Bay in Canada during the summer of 1993. Fishery Bulletin 98:736–47.

Kingsley, M. C. S., and R. R. Reeves. 1998. Aerial surveys of cetaceans in the Gulf of St. Lawrence in 1995 and 1996. Canadian Journal of Zoology 76:1529–50.

Kinloch, D., H. Kuhnlein, and D. C. G. Muir. 1992. Inuit foods and diet: A preliminary assessment of benefits and risks. Science of the Total Environment 122:247–78.

Kinze, C. C. 1995. Exploitation of harbour porpoises (*Phocoena phocoena*) in Danish waters: A historical review. Report of the International Whaling Commission (Special Issue) 16:141–53.

Kirby, V. L., and S. H. Ridgway. 1984. Hormonal evidence of spontaneous ovulation in captive dolphins, *Tursiops truncatus* and *Delphinus delphis*. Report of the International Whaling Commission (Special Issue) 6:459–64.

Koopman, H., and D. E. Gaskin. 1994. Individual and geographic variation in pigmentation patterns of the harbour porpoise, *Phocoena phocoena* (L). Canadian Journal of Zoology 72:135–43.

Koski, W. R., and R. A. Davis. 1994. Distribution and numbers of narwhals (*Monodon monoceros*) in Baffin Bay and Davis Strait. Meddelelser om Grønland, Bioscience 39:15–40.

Krahn, M. M., D. G. Burrows, J. E. Stein, P. R. Becker, M. M. Schantz, D. C. G. Muir, T. M. O'Hara, and T. Rowles. 1999. White whales (*Delphinapterus leucas*) from three Alaskan stocks: concentrations and patterns of persistent organochlorine contaminants in blubber. Journal of Cetacean Research and Management 1:239–49.

Kraus, S. D., A. J. Read, A. Solow, K. Baldwin, T. Spradlin, E. Anderson, and J. Williamson. 1997. Acoustic alarms reduce porpoise mortality. Nature 388:525.

Kuehl, D. W., R. Haebler, and C. Potter. 1991. Chemical residues in dolphins from the U.S. Atlantic coast including Atlantic bottlenose obtained during the 1987/88 mass mortality. Chemosphere 22:1071–84.

Lahvis, G. P., R. S. Wells, D. W. Kuehl, J. L. Stewart, H. L. Rhinehart, and C. S. Via. 1995. Decreased lymphocyte responses in free-ranging bottlenose dolphins (*Tursiops truncatus*) are associated with increased concentrations of PCBs and DDT in peripheral blood. Environmental Health Perspectives 103(Supplement 4):67–72.

Laist, D. W., A. R. Knowlton, J. G. Mead, A. S. Collet, and M. Podesta. 2001. Collisions between ships and whales. Marine Mammal Science 17:35–75.

Lambertsen, R. H. 1990. Disease biomarkers in large whales of the North Atlantic and other oceans. Pages 395–417 *in* J. F. McCarthy and L. R. Shugart, eds. Biomarkers of environmental contamination. CRC Press, Boca Raton, FL.

Lambertsen, R. H. 1997. Natural disease problems of the sperm whale. Bulletin de l'Institut Royal des Sciences Naturelles de Belgique, Biologie 67 (Supplement):105–12.

Lavigne, D. M., V. B. Scheffer, and S. R. Kellert. 1999. The evolution of North American attitudes toward marine mammals. Pages 10–47 *in* J. R. Twiss, Jr., and R. R. Reeves, eds. Conservation and management of marine mammals. Smithsonian Institution Press, Washington, DC.

Law, R. J., R. L. Stringer, C. R. Allchin, and B. R. Jones. 1996. Metals and organochlorines in sperm whales (*Physeter macrocephalus*) stranded around the North Sea during the 1994/1995 winter. Marine Pollution Bulletin 32:72–77.

Law, R. J., R. J. Morris, C. R. Allchin, and B. R. Jones. 1997. Metals and chlorobiphenyls in tissues of sperm whales (*Physeter macrocephalus*) and other cetacean species exploiting similar diets. Bulletin de l'Institut Royal des Sciences Naturelles de Belgique, Biologie 67 (Supplement):79–89.

Leatherwood, S. 1975. Some observations of feeding behavior of bottle-nosed dolphins (*Tursiops truncatus*) in the northern Gulf of Mexico and (*Tursiops* cf *gilli*) off southern California, Baja California, and Nayarit, Mexico. Marine Fisheries Review 37(9):10–16.

Leatherwood, S., and R. R. Reeves. 1982. Bottlenose dolphin (*Tursiops truncatus*) and other toothed cetaceans. Pages 369–414 *in* J. A. Chapman and G. A. Feldhamer, eds. Wild mammals of North America: Biology, management, and economics. Johns Hopkins University Press, Baltimore.

Leatherwood, S., M. W. Deerman, and C. W. Potter. 1978. Food and reproductive status of nine *Tursiops truncatus* from the northeastern United States coast. Cetology 28:1–6.

LeDuc, R. G., and B. E. Curry. 1997. Mitochondrial DNA sequence analysis indicates need for revision of the genus *Tursiops* (Document SC/48/SM 27). International Whaling Commission, Scientific Committee. Cambridge, UK.

LeDuc, R. G., W. F. Perrin, and A. E. Dizon. 1999. Phylogenetic relationships among the delphinid cetaceans based on full cytochrome *b* sequences. Marine Mammal Science 15:619–48.

Leighton, A. H. 1937. The twilight of the Indian porpoise hunters. Natural History 40:410–16, 458.

Lipscomb, T. P., F. Y. Schulman, D. Moffett, and S. Kennedy. 1994. Morbilliviral disease in Atlantic bottlenose dolphins (*Tursiops truncatus*) from the 1987–1988 epizootic. Journal of Wildlife Diseases 30:567–71.

Lipscomb, T. P., S. Kennedy, D. Moffett, A. Krafft, B. A. Klaunberg, J. H. Lichy, G. T. Regan, G. A. J. Worthy, and J. K. Taubenberger. 1996. Morbilliviral epizootic in bottlenose dolphins in the Gulf of Mexico. Journal of Veterinary Diagnostic Investigations 8:283–90.

Lockyer, C. 1981. Estimates of growth and energy budget for the sperm whale, *Physeter catodon*. FAO Fisheries Series 5:489–504.

Lockyer, C. 1995. Investigation of aspects of the life history of the harbour porpoise, *Phocoena phocoena,* in British waters. Report of the International Whaling Commission (Special Issue) 16:189–209.

Lowry, L. F., J. J. Burns, and R. R. Nelson. 1987. Polar bear, *Ursus maritimus,* predation on belugas, *Delphinapterus leucas,* in the Bering and Chukchi seas. Canadian Field-Naturalist 101:141–46.

Lyrholm, T., O. Leimar, B. Johanneson, and U. Gyllensten. 1999. Sex-biased dispersal in sperm whales: Contrasting mitochondrial and nuclear genetic structure of global populations. Proceedings of the Royal Society of London B 266:347–54.

Mann, J., and B. B. Smuts. 1998. Natal attraction: Allomaternal care and mother–infant separations in wild bottlenose dolphins. Animal Behaviour 55:1097–1113.

Mann, J., R. C. Connor, L. M. Barre, and M. R. Heithaus. 2000. Female reproductive success in wild bottlenose dolphins (*Tursiops* sp.): Life history, habitat, provisioning, and group size effects. Behavioral Ecology 11: 210–19.

Martin, A. R., and M. R. Clarke. 1986. The diet of sperm whales (*Physeter macrocephalus*) between Iceland and Greenland. Journal of the Marine Biological Association of the United Kingdom 66:779–90.

Martin, A. R., and A. M. Ávila de Melo. 1983. The Azorean sperm whale fishery: A relic industry in decline. Report of the International Whaling Commission 33:283–86.

Matkin, C. O., and E. L. Saulitis. 1994. Killer whale (*Orcinus orca*) biology and management in Alaska (Contract T75135023). Report to Marine Mammal Commission, Washington, DC.

Matkin, C. O., G. M. Ellis, P. Olesiuk, and E. L. Saulitis. 1999. Association patterns and genealogies of resident killer whales (*Orcinus orca*) in Prince William Sound, Alaska. Fishery Bulletin 97:900–19.

McLellan, W. A., A. S. Friedlaender, J. G. Mead, C. W. Potter, and D. A. Pabst. 2002. Analysing 25 years of bottlenose dolphin (*Tursiops truncatus*) strandings along the Atlantic coast of the USA: Do historic records support the coastal migratory stock hypothesis? Journal of Cetacean Research and Management 4:297–304.

Mead, J. G. 1975. Preliminary report on the former net fisheries for *Tursiops truncatus* in the western North Atlantic. Journal of the Fisheries Research Board of Canada 32:1155–62.

Mead, J. G., and C. W. Potter. 1990. Natural history of bottlenose dolphins along the central Atlantic coast of the United States. Pages 165–95 *in* S. Leatherwood and R. R. Reeves, eds. The bottlenose dolphin. Academic Press, San Diego, CA.

Mead, J. G., and C. W. Potter. 1995. Recognizing two populations of the bottlenose dolphin (*Tursiops truncatus*) off the Atlantic coast of North America: Morphologic and ecologic considerations. International Marine Biological Institute, Kamogawa, Japan, IBI Report 5:31–44.

Mitchell, E. 1970. Pigmentation pattern evolution in delphinid cetaceans: An essay in adaptive coloration. Canadian Journal of Zoology 48:717–40.

Mitchell, E. 1972. Estimates of stock size of the sperm whale (*Physeter catodon*) in the central and western North Atlantic from shipboard census data (Document SP72/11). Unpublished. International Whaling Commission, Cambridge, UK.

Mitchell, E. 1974. Present status of Northwest Atlantic fin and other whale stocks. Pages 108–69 *in* W. E. Shevill, ed. The whale problem: A status report. Harvard University Press, Cambridge, MA.

Mitchell, E. 1975a. Preliminary report on Nova Scotian fishery for sperm whales (*Physeter catodon*). Report of the International Whaling Commission 25:226–35.

Mitchell, E. 1975b. Porpoise, dolphin and small whale fisheries of the world: Status and problems (IUCN Monograph 3). International Union for the Conservation of Nature and Natural Resources, Morges, Switzerland.

Mullin, K., W. Hoggard, C. Roden, R. Lohoefener, C. Rogers, and B. Taggart. 1991. Cetaceans on the upper continental slope in the north-central Gulf of Mexico (OCS Study/MMS 91–0027). U.S. Department of the Interior, Minerals Management Service, Gulf of Mexico OCS Regional Office, New Orleans, LA.

Mullin, K. D., W. Hoggard, C. L. Roden, R. R. Lohoefener, C. M. Rogers, and B. Taggart. 1994. Cetaceans on the upper continental slope in the north-central Gulf of Mexico. Fishery Bulletin 92:773–86.

NAMMCO. 2000. Report of the NAMMCO Scientific Committee working group on the population status of beluga and narwhal in the North Atlantic. North Atlantic Marine Mammal Commission, Annual Report 1999:153–88.

Neimanis, A. S., A. J. Read, R. A. Foster, and D. E. Gaskin. 2000. Seasonal regression in testicular size and histology of harbour porpoises (*Phocoena phocoena,* L.) from the Bay of Fundy and Gulf of Maine. Journal of Zoology (London) 250:221–29.

Normile, D. 2000. Japan's whaling program carries heavy baggage. Science 289:2264–65.

Norris, K. S., and T. P. Dohl. 1980. The structure and functions of cetacean schools. Pages 211–61 *in* L. M. Herman, ed. Cetacean behavior: Mechanisms and functions. Robert E. Kreiger, Malabar, FL.

Norris, K. S., and G. W. Harvey. 1972. A theory for the function of the spermaceti organ of the sperm whale (*Physeter catodon* L.), Pages 397–417 *in* S. R. Galler et al., ed. Animal orientation and navigation. National Aeronautics and Space Administration, Washington, DC.

Ohsumi, S. 1980. Catches of sperm whales by modern whaling in the North Pacific. Report of the International Whaling Commission (Special Issue) 2:11–18.

Øien, N. 1990. Sightings surveys in the northeast Atlantic in July 1988: Distribution and abundance of cetaceans. Report of the International Whaling Commission 40:499–511.

Olesiuk, P. F., M. A. Bigg, and G. M. Ellis. 1990. Life history and population dynamics of resident killer whales (*Orcinus orca*) in the coastal waters of British Columbia and Washington State. Report of the International Whaling Commission (Special Issue) 12:209–43.

Olesiuk, P. F., L. M. Nichol, M. J. Sowden, and J. K. B. Ford. 2002. Effect of the sound generated by an acoustic harassment device on the relative abundance and distribution of harbor porpoises (*Phocoena phocoena*) in Retreat Passage, British Columbia. Marine Mammal Science 18:843–862.

O'Shea, T. J., R. R. Reeves, and A. K. Long, eds. 1999. Marine mammals and persistent ocean contaminants: Proceedings of the Marine Mammal Commission workshop, Keystone, Colorado, 12–15 October 1998. Marine Mammal Commission, Bethesda, MD.

Pabst, D. A. 1990. Axial muscles and connective tissues of the bottlenose dolphin. Pages 51–67 *in* S. Leatherwood and R. R. Reeves, eds. The bottlenose dolphin. Academic Press, San Diego, CA.

Palacios, D. M., and T. Gerrodette. 1996. Potential impact of artisanal gillnet fisheries on small cetacean populations in the eastern tropical Pacific (Administrative Report LJ-96-11). National Marine Fisheries Service, Southwest Fisheries Science Center, La Jolla, CA.

Palacios, D. M., and B. R. Mate. 1996. Attack by false killer whales (*Pseudorca crassidens*) on sperm whales (*Physeter macrocephalus*) in the Galápagos Islands. Marine Mammal Science 12:582–87.

Papastavrou, V., S. C. Smith, and H. Whitehead. 1989. Diving behaviour of the sperm whale, *Physeter macrocephalus,* off the Galapagos Islands. Canadian Journal of Zoology 67:839–46.

Patterson, I. A. P., R. J. Reid, B. Wilson, K. Grellier, H. M. Ross, and P. M. Thompson. 1998. Evidence for infanticide in bottlenose dolphins: An explanation for violent interactions with harbour porpoises? Proceedings of the Royal Society of London B 265:1167–70.

Pike, G. C., and I. B. MacAskie. 1969. Marine mammals of British Columbia. Bulletin of the Fisheries Research Board of Canada 171:1–54.

Pitman, R. L., L. T. Ballance, S. L. Mesnick, and S. J. Chivers. 2001. Killer whale predation on sperm whales: Observations and implications. Marine Mammal Science 17:494–507.

Price, W. S. 1985. Whaling in the Caribbean: Historical perspective and update. Report of the International Whaling Commission 35:413–20.

Read, A. J. 1990a. Reproductive seasonality in harbour porpoises, *Phocoena phocoena,* from the Bay of Fundy. Canadian Journal of Zoology 68:284–88.

Read, A. J. 1990b. Age at sexual maturity and pregnancy rates of harbour porpoises *Phocoena phocoena* from the Bay of Fundy. Canadian Journal of Fisheries and Aquatic Sciences 47:561–65.

Read, A. J. 1999. Harbour porpoise *Phocoena phocoena* (Linnaeus, 1758). Pages 323–55 *in* S. H. Ridgway and R. Harrison, eds. Handbook of marine mammals. Vol. 6: The second book of dolphins and the porpoises. Academic Press, San Diego, CA.

Read, A. J. 2001. Trends in the maternal investment of harbour porpoises are uncoupled from the dynamics of their primary prey. Proceedings of the Royal Society of London B 268:573–77.

Read, A. J., and D. E. Gaskin. 1990. Changes in growth and reproduction of harbour porpoises, *Phocoena phocoena,* from the Bay of Fundy. Canadian Journal of Fisheries and Aquatic Sciences 47:2158–63.

Read, A. J., and A. A. Hohn. 1995. Life in the fast lane: The life history of harbor porpoises from the Gulf of Maine. Marine Mammal Science 11:423–40.

Read, A. J., and K. A. Tolley. 1997. Postnatal growth and allometry of harbour porpoises from the Bay of Fundy. Canadian Journal of Zoology 75:122–30.

Read, A. J., and P. R. Wade. 2000. Status of marine mammals in the United States. Conservation Biology 14:929–40.

Read, A. J., and A. J. Westgate. 1997. Monitoring the movements of harbour porpoises (*Phocoena phocoena*) with satellite telemetry. Marine Biology 130:315–22.

Read, A. J., R. S. Wells, A. A. Hohn, and M. D. Scott. 1993. Patterns of growth in wild bottlenose dolphins, *Tursiops truncatus*. Journal of Zoology (London) 231:107–23.

Read, A. J., P. R. Wiepkema, and P. E. Nachtigall, eds. 1997. The biology of the harbour porpoise. De Spil, Woerden, The Netherlands.

Reeves, R. R. 1993. Domestic and international trade in narwhal products. TRAFFIC Bulletin 14 (1):13–20.

Reeves, R. R., and S. Leatherwood. 1984. Live-capture fisheries for cetaceans in USA and Canadian waters, 1973–1982. Report of the International Whaling Commission 34:497–507.

Reeves, R. R., and J. G. Mead. 1999. Marine mammals in captivity. Pages 412–36 *in* J. R. Twiss, Jr., and R. R. Reeves, eds. Conservation and management of marine mammals. Smithsonian Institution Press, Washington, DC.

Reeves, R. R., and E. Mitchell. 1988a. History of whaling in and near North Carolina. NOAA Technical Report NMFS 65:1–28.

Reeves, R. R., and E. Mitchell. 1988b. Distribution and seasonality of killer whales in the eastern Canadian Arctic. Rit Fiskideildar 11:136–60.

Reeves, R. R., and D. J. St. Aubin, eds. 2001. Belugas and narwhals: Application of new technology to whale science in the Arctic. Arctic (Special Issue) 54 (3):i–vi+207–341.

Reeves, R. R., S. Leatherwood, S. A. Karl, and E. R. Yohe. 1985. Whaling results at Akutan (1912–39) and Port Hobron (1926–37), Alaska. Report of the International Whaling Commission 35:441–57.

Reynolds, J. E., III, R. S. Wells, and S. D. Eide. 2000. The bottlenose dolphin: Biology and conservation. University of Florida Press, Gainesville.

Rice, D. W. 1974. Whales and whale research in the eastern North Pacific. Pages 170–95 *in* W. E. Schevill, ed. The whale problem: A status report. Harvard University Press, Cambridge, MA.

Rice, D. W. 1989. Sperm whale *Physeter macrocephalus* Linnaeus, 1758. Pages 177–233 *in* S. H. Ridgway and R. Harrison, eds. Handbook of marine mammals. Vol. 4: River dolphins and the larger toothed whales. Academic Press, London.

Rice, D. W. 1998. Marine mammals of the world. Systematics and distribution (Special Publication No. 4). Society for Marine Mammalogy, Lawrence, KS.

Rice, D. W. 2002. Spermaceti. Pages 1163–65 *in* W. F. Perrin, B. Würsig, and J. G. M. Thewissen, eds. Encyclopedia of marine mammals. Academic Press, San Diego, CA.

Richard, K. R., M. C. Dillon, H. Whitehead, and J. M. Wright. 1996. Patterns of kinship in groups of free-living sperm whales (*Physter macrocephalus*) revealed by multiple molecular genetic analyses. Proceedings of the National Academy of Sciences of the USA 93:8792–95.

Richard, P. R. 1991. Abundance and distribution of narwhals (*Monodon monoceros*) in northern Hudson Bay. Canadian Journal of Fisheries and Aquatic Sciences 48:276–83.

Richard, P. R., J. R. Orr, and D. G. Barber. 1990. The distribution and abundance of belugas, *Delphinapterus leucas,* in eastern Canadian Subarctic waters: a review and update. Canadian Bulletin of Fisheries and Aquatic Sciences 224:23–38.

Richard, P., P. Weaver, L. Dueck, and D. Barber. 1994. Distribution and numbers of Canadian High Arctic narwhals (*Monodon monoceros*) in August 1984. Meddelelser om Grønland, Bioscience 39:41–50.

Ridgway, S. H. 1990. The central nervous system of the bottlenose dolphin. Pages 69–97 *in* S. Leatherwood and R. R. Reeves, eds. The bottlenose dolphin. Academic Press, San Diego, CA.

Rolinson, J. W. 1845–1905. [Rolinson's diary.] Microfilm, Southern Historical Collection, Chapel Hill, NC. Accession No. M-3122.

Rommel, S. 1990. Osteology of the bottlenose dolphin. Pages 29–49 *in* S. Leatherwood and R. R. Reeves, eds. The bottlenose dolphin. Academic Press, San Diego, CA.

Rosel, P. E., R. Tiedemann, and M. Walton. 1999. Genetic evidence for limited trans-Atlantic movements of the harbor porpoise *Phocoena phocoena*. Marine Biology 133:583–91.

Ross, H. M., and B. Wilson. 1996. Violent interactions between bottlenose dolphins and harbour porpoises. Proceedings of the Royal Society of London B 263:283–86.

Rossbach, K. A., and D. L. Herzing. 1997. Underwater observations of benthic-feeding bottlenose dolphins (*Tursiops truncatus*) near Grand Bahama Island, Bahamas. Marine Mammal Science 13:498–504.

Sanpera, C., and A. Aguilar. 1992. Modern whaling off the Iberian Peninsula during the twentieth century. Report of the International Whaling Commission 42:723–30.

Scammon, C. M. 1874. The marine mammals of the north-western coast of North America, described and illustrated; together with an account of the American whale-fishery. Carmany & Co., San Francisco.

Scheffer, V. B., and J. W. Slipp. 1948. The whales and dolphins of Washington State with a key to the cetaceans of the west coast of North America. American Midland Naturalist 39:257–337.

Schroeder, J. P. 1990. Breeding bottlenose dolphins in captivity. Pages 435–46 *in* S. Leatherwood and R. R. Reeves, eds. The bottlenose dolphin. Academic Press, San Diego, CA.

Schroeder, J. P., and K. V. Keller. 1989. Seasonality of testosterone and sperm density in *Tursiops truncatus*. Journal of Experimental Zoology 249: 316–26.

Scott, G. P. 1990. Management-oriented research on bottlenose dolphins by the Southeast Fisheries Center. Pages 632–39 *in* S. Leatherwood and R. R. Reeves, eds. The bottlenose dolphin. Academic Press, San Diego, CA.

Scott, G. P., D. M. Burn, and L. J. Hansen. 1988. The dolphin dieoff: Long-term effects and recovery of the population. Pages 819–23 *in* Proceedings of the Oceans '88 Conference, Baltimore.

Scott, M. D., and S. J. Chivers. 1990. Distribution and herd structure of bottlenose dolphins in the eastern tropical Pacific Ocean. Pages 387–402 *in* S. Leatherwood and R. R. Reeves, eds. The bottlenose dolphin. Academic Press, San Diego, CA.

Scott, M. D., R. S. Wells, and A. B. Irvine. 1990. A long-term study of bottlenose dolphins on the west coast of Florida. Pages 235–44 *in* S. Leatherwood, and R. R. Reeves, eds. The bottlenose dolphin. Academic Press, San Diego, CA.

Scott, T. M., and S. S. Sadove. 1997. Sperm whale, *Physeter macrocephalus,* sightings in the shallow shelf waters off Long Island, New York. Marine Mammal Science 13:317–21.

Shane, S. H. 1990. Behavior and ecology of the bottlenose dolphin at Sanibel Island, Florida. Pages 245–65 *in* S. Leatherwood and R. R. Reeves, eds. The bottlenose dolphin. Academic Press, San Diego, CA.

Siegstad, H., and M. P. Heide-Jørgensen. 1994. Ice entrapments of narwhals (*Monodon monoceros*) and white whales (*Delphinapterus leucas*) in Greenland. Meddelelser om Grønland, Bioscience 39:151–60.

Sigurjónsson, J. 1989. To Icelanders, whaling is a godsend. Oceanus 32 (1):29–36.

Sigurjónsson, J., and S. Leatherwood. 1988. The Icelandic live-capture fishery for killer whales, 1976–1988. Rit Fiskideildar 11:307–16.

Smith, S. C., and H. Whitehead. 1993. Variations in the feeding success and behaviour of Galápagos sperm whales (*Physeter macrocephalus*) as they relate to oceanographic conditions. Canadian Journal of Zoology 71:1991–96.

Stackpole, E. A. 1972. Whales and destiny: The rivalry between America, France, and Britain for control of the southern whale fishery, 1785–1825. University of Massachusetts Press, Amherst.

Starbuck, A. 1878. History of the American whale fishery; from its earliest inception to the year 1876 (Report of the U.S. Commissioner of Fish and Fisheries, Part 4).

St. Aubin, D. J., T. G. Smith, and J. R. Geraci. 1990. Seasonal epidermal moult in beluga whalees, *Delphinapterus leucas*. Canadian Journal of Zoology 68:359–67.

Stephens, K. H. R. 1984. Judgment Land: The story of Salter Path, Book 1. The Print Shop, Havelock, NC.

Stevenson, C. H. 1904. Aquatic mammals in arts and industries. Fish oils, fats, and waxes. Fertilizers from aquatic products. Pages 177–279 *in* Report of the Commissioner for the year ending June 30, 1902, Part XXVIII. U.S. Commission of Fish and Fisheries, Government Printing Office, Washington, DC.

Stewart, B. E., and R. E. A. Stewart. 1989. *Delphinapterus leucas*. Mammalian Species 336:1–8.

Stick, D. 1958. The Outer Banks of North Carolina, 1584–1958. University of North Carolina Press, Chapel Hill.

Tatham, W. 1806. Original report of William Tatham on the survey of the coast of North Carolina from Cape Fear to Cape Hatteras. U.S. National Archives, Washington, DC.

Taylor, B. L., and D. P. DeMaster. 1993. Implications of non-linear density dependence. Marine Mammal Science 9:360–71.

Tolley, K. A., A. J. Read, R. S. Wells, K. W. Urian, M. D. Scott, A. B. Irvine, and A. A. Hohn. 1995 Sexual dimorphism in wild bottlenose dolphins (*Tursiops truncatus*) from Sarasota, Florida. Journal of Mammalogy 76:1190–98.

Torres, L. G., P. E. Rosel, C. D'Agrosa, and A. J. Read. 2003. Improving management of overlapping bottlenose dolphin ecotypes through spatial analysis and genetics. Marine Mammal Science 19.

Townsend, C. H. 1914. The porpoise in captivity. Zoologica 1:289–99.

Townsend, C. H. 1935. The distribution of certain whales as shown by logbook records of American whaleships. Zoologica 19:1–50.

Trippel, E. A., M. B. Strong, J. M. Terhune, and J. D. Conway. 1999. Mitigation of harbour porpoise (*Phocoena phocoena*) by-catch in the gillnet fishery in the lower Bay of Fundy. Canadian Journal of Fisheries and Aquatic Sciences 56:113–23.

True, F. W. 1891. Observations on the life history of the bottlenose porpoise. Proceedings of the United States National Museum 13:197–203.

Tyack, P. L. 2000. Functional aspects of cetacean communication. Pages 270–307 *in* J. Mann, R. C. Connor, P. L. Tyack, and H. Whitehead, eds. Cetacean societies: Field studies of dolphins and whales. University of Chicago Press, Chicago.

Urian, K. W. 2002. Community structure of bottlenose dolphins (*Tursiops truncatus*) in Tampa Bay, Florida, USA. M.Sc. thesis, University of North Carolina, Wilmington, 26 pp.

Urian, K. W., D. A. Duffield, A. J. Read, R. S. Wells, and E. D. Shell. 1996. Seasonality of reproduction in bottlenose dolphins, *Tursiops truncatus*. Journal of Mammalogy 77:394–403.

Urian, K. W., R. S. Wells, M. D. Scott, A. B. Irvine, A. J. Read, and A. A. Hohn. 1998. When the shark bites: An analysis of shark bite scars on wild bottlenose dolphins (*Tursiops truncatus*) from Sarasota, Florida. Page 139 *in* Abstracts, World marine mammal science conference. Monaco.

Wade, P. R. 1998. Calculating limits to the allowable human-caused mortality of cetaceans and pinnipeds. Marine Mammal Science 14:1–37.

Wade, P. R., and T. Gerrodette. 1993. Estimates of cetacean abundance and distribution in the eastern tropical Pacific. Report of the International Whaling Commission 43:477–93.

Walker, J. L., C. W. Potter, and S. A. Macko. 1999. The diets of modern and historic bottlenose dolphin populations reflected through stable isotopes. Marine Mammal Science 15:335–50.

Walker, W. A. 1981. Geographical variation in morphology and biology of bottlenose dolphins (*Tursiops*) in the eastern North Pacific (Administrative Report LJ-81-03C). National Marine Fisheries Service, Southwest Fisheries Center, La Jolla, CA.

Walsh, M. T., D. Beusse, G. D. Bossart, W. G. Young, D. K. Odell, and G. W. Patton. 1988. Ray encounters as a mortality factor in Atlantic bottlenose dolphins (*Tursiops truncatus*). Marine Mammal Science 4:154–62.

Waring, G. T., C. P. Fairfield, C. M. Ruhsam, and M. Sano. 1993. Sperm whales associated with Gulf Stream features off the north-eastern USA shelf. Fisheries and Oceanography 2:101–5.

Waring, G. T., J. M. Quintal, and S. L. Swartz, eds. 2000. U.S. Atlantic and Gulf of Mexico marine mammal stock assessments, 2000 (NOAA Technical Memorandum NMFS-NE-162). Northeast Fisheries Science Center, Woods Hole, MA.

Waring, G. T., J. M. Quintal, and C. F. Fairfield, eds. 2002. U.S. Atlantic and Gulf of Mexico marine mammal stock assessments—2002 (NOAA Technical Memorandum NMFS-NE-169). Northeast Fisheries Science Center, Woods Hole, MA.

Waters, S., and H. Whitehead. 1990. Aerial behaviour of sperm whales. Canadian Journal of Zoology 68:2076–82.

Watkins, W. A. 1980. Acoustics and the behavior of sperm whales. Pages 283–90 *in* R.-G. Busnel and J. F. Fish, eds. Animal sonar systems. Plenum Press, New York.

Watkins, W. A., and K. E. Moore. 1982. An underwater acoustic survey for sperm whales (*Physeter catodon*) and other cetaceans in the southeast Caribbean. Cetology 46:1–7.

Watkins, W. A., and W. E. Schevill. 1975. Sperm whales (*Physeter catodon*) react to pingers. Deep-Sea Research 22:123–29.

Watkins, W. A., K. E. Moore, and P. Tyack. 1985. Sperm whale acoustic behaviors in the southeast Caribbean. Cetology 49:1–15.

Watkins, W. A., M. A. Daher, K. M. Fristrup, T. J. Howard, and G. Notarbartolo di Sciara. 1993. Sperm whales tagged with transponders and tracked underwater by sonar. Marine Mammal Science 9:55–67.

Watkins, W. A., M. A. Daher, N. A. DiMarzio, A. Samuels, D. Wartzok, K. M. Fristrup, D. P. Gannon, P. W. Howey, R. R. Maiefski, and T. R. Spradlin. 1999. Sperm whale surface activity from tracking by radio and satellite tags. Marine Mammal Science 15:1158–80.

Watson, A. P. 1976. The diurnal behaviour of the harbour porpoise (*Phocoena phocoena* L.) in the coastal waters of the western Bay of Fundy. M.Sc. Thesis, University of Guelph, Guelph, Ontario, Canada.

Weller, D. W., B. Würsig, H. Whitehead, J. C. Norris, S. K. Lynn, R. W. Davis, N. Clauss, and P. Brown. 1996. Observations of an interaction between sperm whales and short-finned pilot whales in the Gulf of Mexico. Marine Mammal Science 12:588–93.

Weller, D. W., B. Würsig, S. K. Lynn, and A. J. Schiro. 2000. Preliminary findings on the occurrence and site fidelity of photo-identified sperm whales (*Physeter macrocephalus*) in the northern Gulf of Mexico. Gulf of Mexico Science 2000:35–39.

Wells, R. S., and M. D. Scott. 1990. Estimating bottlenose dolphin population parameters from individual identification and capture-release techniques. Report of the International Whaling Commission (Special Issue) 12:407–15.

Wells, R. S., and M. D. Scott. 1999. Bottlenose dolphin *Tursiops truncatus* (Montagu, 1821). Pages 137–82 *in* S. H. Ridgway and R. Harrison, eds. Handbook of marine mammals. Vol. 6: The second book of dolphins and the porpoises. Academic Press, San Diego, CA.

Wells, R. S., M. D. Scott, and A. B. Irvine. 1987. The social structure of free-ranging bottlenose dolphins. Pages 247–305 *in* H. H. Genoways, ed. Current Mammalogy 1. Plenum Press, New York.

Wells, R. S., L. J. Hansen, A. Baldridge, T. P. Dohl, D. L. Kelly, and R. H. Defran. 1990. Northward extension of the range of bottlenose dolphins along the California coast. Pages 421–31 *in* S. Leatherwood and R. R. Reeves, eds. The bottlenose dolphin. Academic Press, San Diego, CA.

Wells, R. S., K. Bassos-Hall, and K. S. Norris. 1998a. Experimental return to the wild of two bottlenose dolphins. Marine Mammal Science 14:51–71.

Wells, R. S., S. Hofmann, and T. L. Moors. 1998b. Entanglement and mortality of bottlenose dolphins, *Tursiops truncatus,* in recreational fishing gear in Florida. Fishery Bulletin 96:647–50.

Wells, R. S., H. L. Rhinehart, P. Cunningham, J. Whaley, M. Baran, C. Koberna, and D. P. Costa. 1999. Long distance offshore movements of bottlenose dolphins. Marine Mammal Science 15:1098–1114.

Westgate, A. J., A. J. Read, P. Berggren, H. N. Koopman, and D. E. Gaskin. 1995. Diving behaviour of harbour porpoises, *Phocoena phocoena.* Canadian Journal of Fisheries and Aquatic Sciences 52:1064–73.

Westgate, A. J., A. J. Read, T. M. Cox, T. D. Schofield, B. R. Whitaker, and K. E. Anderson. 1998. Monitoring a rehabilitated harbor porpoise using satellite telemetry. Marine Mammal Science 14:599–604.

Whitehead, H. 1987. Social organization of sperm whales off the Galapagos: Implications for management and conservation. Report of the International Whaling Commission 37:195–99.

Whitehead, H. 1990. Assessing sperm whale populations using natural markings: Recent progress. Report of the International Whaling Commission (Special Issue) 12:377–82.

Whitehead, H. 1993. The behaviour of mature male sperm whales on the Galápagos Islands breeding grounds. Canadian Journal of Zoology 71:689–99.

Whitehead, H. 1997. Sea surface temperature and the abundance of sperm whale calves off the Galápagos Islands: Implications for the effects of global warming. Report of the International Whaling Commission 47:941–44.

Whitehead, H., and J. Gordon. 1986. Methods of obtaining data for assessing and modelling sperm whale populations which do not depend on catches. Report of the International Whaling Commission (Special Issue) 8:149–65.

Whitehead, H., and J. Mann. 2000. Female reproductive strategies of cetaceans: Life histories and calf care. Pages 219–46 *in* J. Mann, R. C. Connor, P. L. Tyack, and H. Whitehead, eds. Cetacean societies: Field studies of dolphins and whales. University of Chicago Press, Chicago.

Whitehead, H., and L. Weilgart. 2000. The sperm whale: Social females and roving males. Pages 154–72 *in* J. Mann, R. C. Connor, P. L. Tyack, and H. Whitehead, eds. Cetacean societies: Field studies of dolphins and whales. University of Chicago Press, Chicago.

Whitehead, H., V. Papastavrou, and S. C. Smith. 1989. Feeding success of sperm whales and sea-surface temperature off the Galápagos Islands. Marine Ecology Progress Series 53:201–3.

Whitehead, H., S. Waters, and T. Lyrholm. 1991. Social organization in female sperm whales and their offspring: Constant companions and casual acquaintances. Behavioral Ecology and Sociobiology 29:385–89.

Whitehead, H., S. Brennan, and D. Grover. 1992. Distribution and behaviour of male sperm whales on the Scotian Shelf, Canada. Canadian Journal of Zoology 70:912–18.

Whitehead, H., J. Christal, and S. Dufault. 1997. Past and distant whaling and the rapid decline of sperm whales off the Galápagos Islands. Conservation Biology 11:1387–96.

Wilson, B. 1995. The ecology of bottlenose dolphins in the Moray Firth, Scotland: A population at the northern extreme of the species' range. Ph.D. Dissertation, University of Aberdeen, Scotland.

Wood, F. G., Jr., D. K. Caldwell, and M. C. Caldwell. 1970. Behavioral interaction between porpoises and sharks. Investigations on Cetacea 2:264–77.

Würsig, B., and M. Würsig. 1979. Behavior and ecology of the bottlenose dolphin, *Tursiops truncatus,* in the South Atlantic. Fishery Bulletin 77:399–412.

Yablokov, A. V. 1994. Validity of whaling data. Nature 367:108.

Yurick, D. B., and D. E. Gaskin. 1987. Morphometric and meristic comparisons of skulls of harbour porpoise *Phocoena phocoena* (L.) from the North Atlantic and North Pacific. Ophelia 27:53–75.

Zemsky, V. A., A. A. Berzin, Yu. A. Mikhalyev, and D. D. Tormosov. 1995. Soviet Antarctic whaling data (1947–1972). Center for Russian Environmental Policy, Moscow.

Randall R. Reeves, Okapi Wildlife Associates, 27 Chandler Lane, Hudson, Quebec, Canada J0P 1H0. Email: rrreeves@total.net.

Andrew J. Read, Nicholas School of Environment and Earth Sciences, Duke University Marine Laboratory, Beaufort, North Carolina 28516. Email: aread@duke.edu.

21

Baleen Whales: Right Whales and Allies

Eubalaena spp.

Randall R. Reeves
Robert D. Kenney

This chapter follows the approach of Reeves and Brownell (1982), emphasizing the right whales (*Eubalaena* spp.) and providing limited discussion of the other baleen whale species found in waters bordering the North American continent—bowhead whale (*Balaena mysticetus*), blue whale (*Balaenoptera musculus*), fin whale (*Balaenoptera physalus*), sei whale (*Balaenoptera borealis*), Bryde's whale (*Balaenoptera brydei*), minke whale (*Balaenoptera acutorostrata*), humpback whale (*Megaptera novaeangliae*), and gray whale (*Eschrichtius robustus*). In the two decades since the corresponding chapter in the previous edition of *Wild Mammals of North America,* public awareness and concern about the precarious status of North Atlantic and North Pacific right whales have grown enormously. In response, U.S. and Canadian government agencies as well as the International Whaling Commission (IWC) and many nongovernmental organizations have invested large amounts of resources in right whale research and management. This chapter reflects the tremendous amount of work that has been accomplished during the past 20 years.

NOMENCLATURE

Common Names. It is often stated that right whales were so named because they swam slowly and thus were approachable for killing, occurred near shore and thus were relatively accessible, usually floated after being killed, and produced large amounts of valuable oil and baleen (known commercially as "whalebone"). These characteristics made them the "right" whales to hunt. In addition, however, many early writers (e.g., Eschricht and Reinhardt 1866) considered "right" to connote "true" or "proper," meaning typical of the group.

In modern usage, the name *right whale* has been applied only to the animals formerly known as Nördkaper (a Norwegian term, in reference to North Cape, Norway), Biscay(an) whale (referring to the Bay of Biscay off Spain, where Basque whaling originated), sarde or sarda (origin uncertain), and sletbag (from the Old Icelandic *slätbag,* meaning "smooth-back"). In the past, the name *right whale* also was used for what is now known as the bowhead whale. The bowhead was called the Greenland or arctic right whale, as distinct from the so-called black right whales of temperate regions. Most recent literature refers to the temperate-region right whales in the North Atlantic and North Pacific as northern right whales to distinguish them from the right whales of the Southern Ocean, called southern right whales (see Systematics).

Systematics. The nomenclature of balaenids is somewhat controversial. Some authorities, including Árnason and Gullberg (1994) and Rice (1998), have insisted that there is no credible scientific basis for recognizing the genus *Eubalaena* and that all balaenids should be subsumed under *Balaena*. Others, notably Schevill (1986), Bannister et al. (1999), and Rosenbaum et al. (2000), have argued that the generic distinction between the temperate-region right whales and the bowhead is justified by "customary usage of right whale biologists over many years" (Bannister et al. 1999: footnote 2). The argument for maintaining separate generic names for right whales and bowheads for purposes of nomenclatural stability, however, may not be a strictly valid application of the International Code of Zoological Nomenclature, which applies only to questions of nomenclature, not of taxonomy (Ride et al. 1985).

There has also been debate about how many species of temperate-region right whales should be recognized. For many years, the custom was to recognize two species, one in the Southern Hemisphere and one in the Northern Hemisphere, even though some authors argued for a single species due to the lack of consistently demonstrable morphological differences (Tomilin 1967; Reeves and Brownell 1982; Rice 1998). Molecular genetic analyses have now provided convincing evidence of species-level differences among right whales in the North Atlantic, the North Pacific, and the Southern Hemisphere (Rosenbaum et al. 2000). The appropriate names are *E. glacialis* (Müller 1776) for the North Atlantic species, *E. japonica* (Lacépède 1818) for the North Pacific species, and *E. australis* (Desmoulins 1822) for the southern species.

Other Baleen Whales. The Mysticeti, or baleen whales, constitute a suborder of the Cetacea: whales, dolphins, and porpoises. They are readily distinguished from odontocetes (suborder Odontoceti), or toothed whales, by having baleen instead of teeth (see below), two blowholes (external nostrils) rather than one, and a variety of cranial and postcranial skeletal differences.

Four living families of Mysticeti are recognized globally, of which three are represented along North American shores (Fig. 21.1). The Balaenidae, or right whales, consist of the three temperate-region species of *Eubalaena* and the bowhead whale, an arctic endemic (see later). Most molecular phylogenies show the balaenids to be the most primitive clade of extant mysticetes (Árnason et al. 1992; 1993; Árnason and Gullberg 1994; Milinkovitch et al. 1994; Nikaido et al. 2001). The families Neobalaenidae and Eschrichtiidae both contain single species. These are, respectively, the pygmy right whale (*Caperea marginata*) of the cold temperate Southern Ocean and the gray whale of the Northern Hemisphere (see later). Morphological analyses have classified Neobalaenidae as a sister taxon of Balaenidae (Mitchell 1989; McLeod et al. 1993; Fordyce and Barnes 1994), whereas some molecular phylogenies have indicated that it is a primitive sister of the combined Balaenopteridae and Eschrichtiidae (Árnason and Best 1991; Árnason et al. 1992, 1993). Balaenopteridae is the most diverse family of mysticetes, with two morphologically well-defined genera (*Balaenoptera* and *Megaptera*) and at least eight species (blue, fin, sei, two minke, two Bryde's, and humpback whales; see later). The balaenopterids, or rorquals as they are commonly called (including the humpback whale), are distinguished by having a dorsal fin and a series of "ventral grooves"—long, parallel, distensible grooves or pleats—on the throat and chest. Although the genus- and family-level classifications of *Eschrichtius, Balaenoptera,* and *Megaptera* are well supported by anatomical characters (Rice 1998), a variety of molecular studies calls this taxonomy into question, as they show both the gray whale and the humpback whale to be nested within *Balaenoptera* (Árnason and Best 1991; Árnason et al. 1992, 1993; Árnason and Gullberg 1994, 1996; Nikaido et al. 2001).

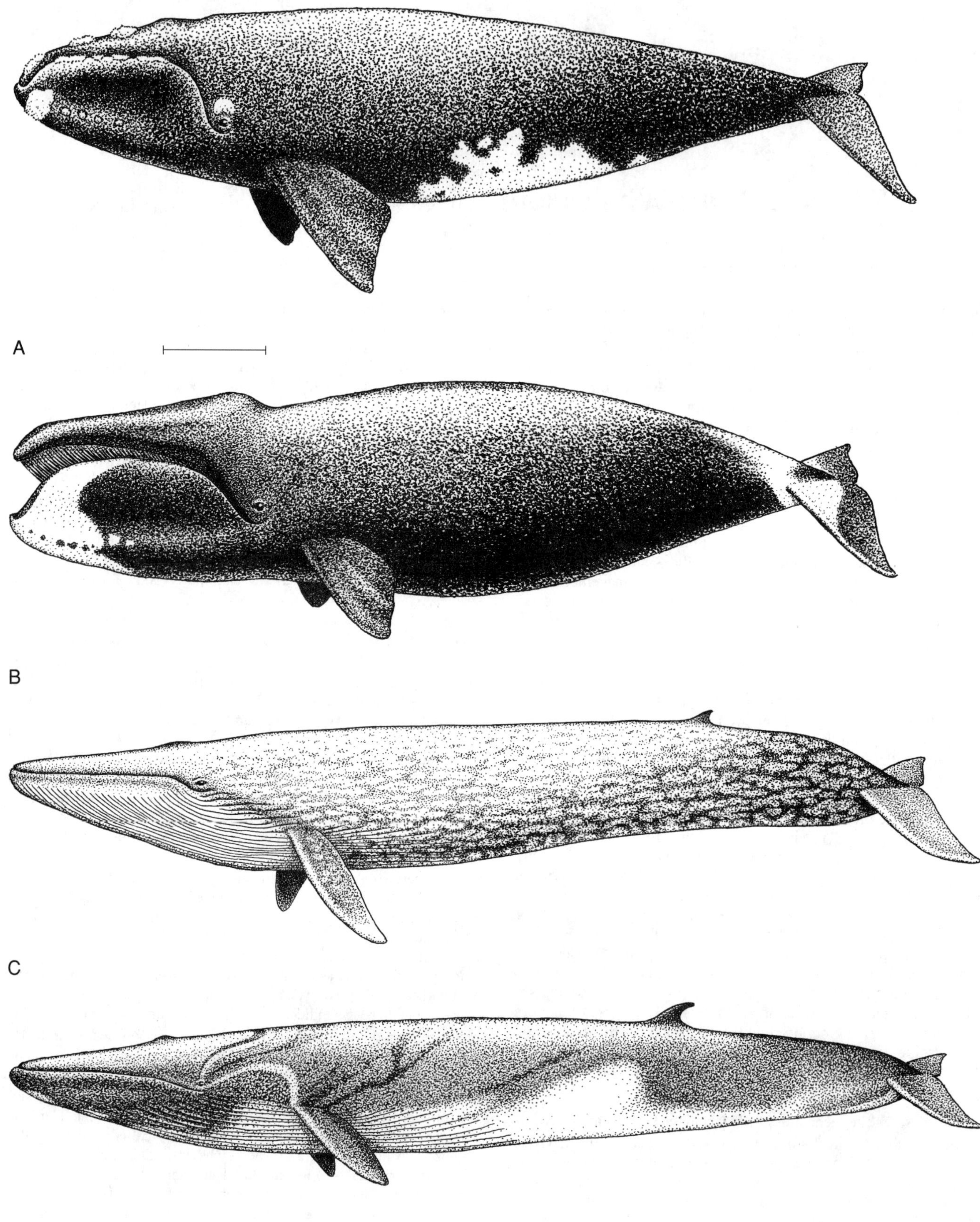

FIGURE 21.1. (A) Right whale (*Eubalaena* spp.). (B) Bowhead whale (*Balaena mysticetus*). (C) Blue whale (*Balaenoptera musculus*). (D) Fin whale (*Balaenoptera physalus*).

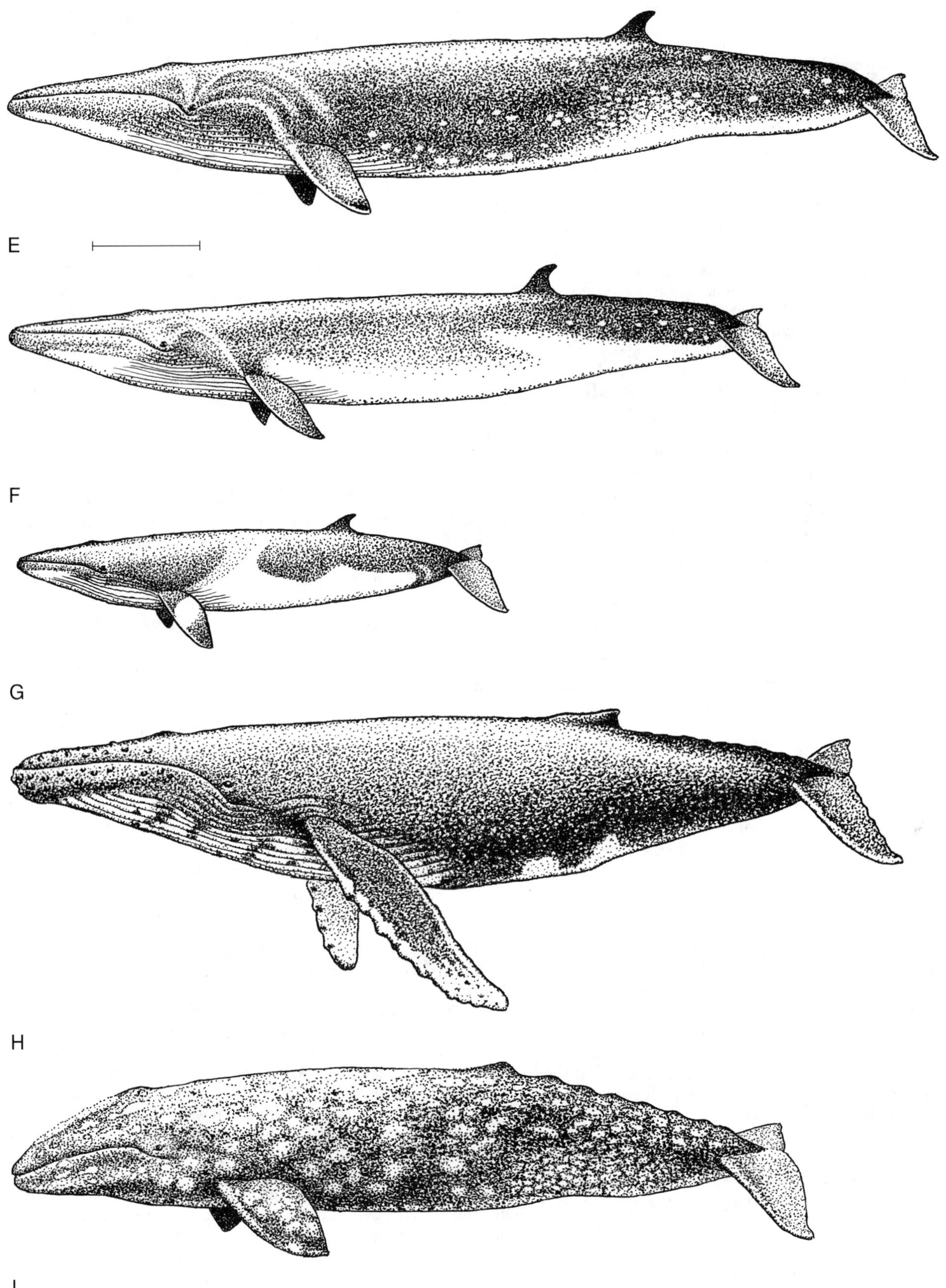

FIGURE 21.1. *Continued*. (E) Sei whale (*Balaenoptera borealis*). (F) Bryde's whale (*Balaenoptera brydei*). (G) Minke whale (*Balaenoptera acutorostrata*). (H) Humpback whale (*Megaptera novaeangliae*). (I) Gray whale (*Eschrichtius robustus*). (Scale bar = 2m.)

ORIGINS

The fossil record of cetaceans dates to the early Eocene, some 50 million years ago (mya) (Fordyce and Barnes 1994; Gingerich 1998; Thewissen 1998; Williams 1998). They most likely diverged from a terrestrial mammal lineage around 55 mya (Montgelard et al. 1997; Bajpai and Gingerich 1998; Gingerich 1998; Berta and Sumich 1999; Árnason et al. 2000; Lum et al. 2000). Their highly derived body plans, modified and specialized for an aquatic existence, have made it very difficult to understand their phylogenetic affinities with other mammalian orders (Simpson 1945; Milinkovitch 1995, 1997). Flower (1883) was the first to propose, based on anatomical characteristics, that cetaceans were related to the ungulates, but there has been some controversy about whether their affinities are closer to artiodactyls (Slijper 1962; Gingerich et al. 1990; Thewissen et al. 2001) or perissodactyls (McKenna 1987; Prothero et al. 1988; Novacek 1989; Thewissen 1994; Heyning 1997). Van Valen (1966) hypothesized that cetaceans were derived from Mesonychia, an extinct ungulate group. A cetacean–mesonychid link, based on morphological and especially dental characters, has been the consensus of most morphologists and paleontologists (McKenna 1975; Prothero et al. 1988; Thewissen 1994; Zhou et al. 1995; Gingerich 1998; O'Leary 1998). A growing consensus based on molecular studies, however, is that the Cetacea are nested within the Artiodactyla. Most investigators have concluded that cetaceans and hippopotamuses are sister groups (Graur and Higgins 1994; Árnason and Gullberg 1996; Gatesy et al. 1996, 1999; Smith et al. 1996; Gatesy 1997, 1998; Montgelard et al. 1997; Shimamura et al. 1997, 1999; Milinkovitch et al. 1998; Ursing and Árnason 1998; Liu and Miyamoto 1999; Nikaido et al. 1999; Árnason et al. 2000; Lum et al. 2000). Recognizing the divergent conclusions between the morphological and molecular studies, Thewissen (1998:453) wrote, "The presence of a trochleated astragalus in early whales would probably lead most morphologists to accept the molecular phylogeny and consider cetaceans as artiodactyls." Gingerich et al. (2001) reported two new species of 47-million-year-old fossil protocetid whales from Pakistan, both with a characteristic artiodactylan astragalus in the ankle joint, which decisively supports a cetacean–artiodactyl relationship.

The mysticetes arose and diversified during the Oligocene, between 35 and 23 mya (Fordyce and Barnes 1994; Árnason et al. 2000). Among their diagnostic features are the loss of teeth before birth, the development of large body sizes, including large heads, and a shortening of the intertemporal and neck regions. In mysticetes, the skull is uniquely "telescoped," with the maxilla extending posteriorly under the orbit to form a platelike infraorbital process (Miller 1923). It is from the epithelium on the sides of this expanded maxilla that the baleen plates grow (see Baleen).

The balaenids are the most ancient living mysticetes, having first appeared in the fossil record in the earliest Miocene at least 23 mya (McLeod et al. 1993). Fossil balaenids were widespread throughout the Northern Hemisphere during the late Miocene (10 to 5 mya) and Pliocene (5 to 2 mya) (Barnes and McLeod 1984; McLeod et al. 1993). The balaenopterids were present by the late Miocene (16 to 10 mya) and possibly arose in the middle Miocene (Fordyce and Barnes 1994). The fossil records of both Eschrichtiidae and Neobalaenidae are extremely sparse (Barnes and McLeod 1984; Fordyce and Barnes 1994).

DISTRIBUTION AND MIGRATION

Historically, right whales inhabited temperate marine waters worldwide (Perry et al. 1999) (see Fig. 21.2 for their distribution off North America). They were extirpated or severely depleted in all areas by commercial whaling. Steady recovery has occurred in parts of the Southern Hemisphere since the early 1970s, when whaling ended there (Tormosov et al. 1998), and southern right whales appear to be re-establishing much of their historic range (International Whaling Commission [IWC] 2001b). The situation is much different in the Northern Hemisphere, where intensive and prolonged whaling left vast portions of the North Atlantic and North Pacific essentially devoid of right whales. Historical and present-day distribution of right whales along the North American coasts of these two oceans is summarized briefly below.

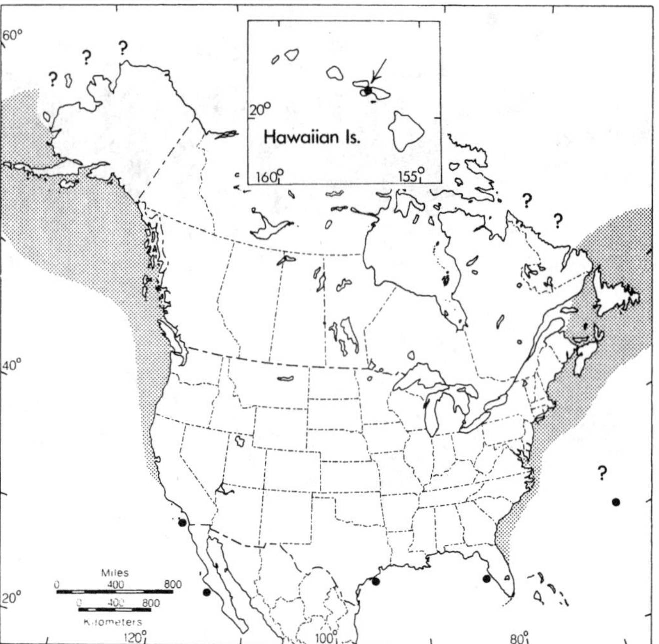

FIGURE 21.2. Distribution of right whales (*Eubalaena glacialis* and *E. japonica*). Dots indicate records outside "normal" range; question marks indicate uncertainty about occurrence in the area so marked. SOURCE: Data from published accounts (see text).

Eastern North Pacific. The historical limits of distribution in the eastern North Pacific are the southern Bering Sea in the north and the Pacific coast of Baja California in the south (Scarff 1986a, 1991). However, large concentrations of right whales are only known to have occurred during the summer months (June–September) in the northern parts of this range, particularly in the Gulf of Alaska (between about 50–58°N and 140–152°W), along the south side of the Aleutian chain, and in the southeastern Bering Sea. Unlike in other parts of the world, where areas of winter concentration near continental or island coasts are well known, in the North Pacific, no such winter concentration areas are known. In the absence of evidence for a distinct winter calving ground on either the east or the west side of the North Pacific, it is possible that North Pacific right whales winter and calve far offshore (Scarff 1986a, 1991).

In recent years, small groups of right whales have been seen between April and September in the southeastern Bering Sea, particularly in July and August, in a specific area just outside the mouth of Bristol Bay (Goddard and Rugh 1998; Brownell et al. 2001; LeDuc et al. 2001; Moore et al. 2000; McDonald and Moore 2002; Tynan et al. 2001). These sightings have occurred in midshelf waters about 50–70 m deep, unlike the historical sightings in the same region, which were in deeper waters (>200 m) (Tynan et al. 2001). Other sightings have been scattered along the coast from Washington State southward to the outer coast of Baja California, Mexico, usually involving solitary individuals. All biopsies and photographs of the animals in the southeastern Bering Sea prior to 2002 showed them to be males in the size range of 14.7–17.6 m, which raised concerns that little or no reproduction was occurring in the eastern North Pacific population (LeDuc et al. 2001). A mother and calf were observed and sampled on 24 August 2002 (Ballance 2002; National Oceanic and Atmospheric Administration 2002).

The migratory habits of right whales in the North Pacific are poorly known. There is no direct photographic matching or telemetric evidence of movements, so migration can only be inferred from trends in catch and sighting data, which are inevitably subject to effort bias. Judging by the fact that almost all records of right whales in relatively low latitudes (e.g., off Baja California, California, and Hawaii) are from winter or spring months (Rice and Fiscus 1968; Herman et al. 1980; Rowntree et al. 1980; Woodhouse and Strickley 1982; Carretta et al. 1994; Rowlett et al. 1994; Gendron et al. 1999; Salden and Mickelsen 1999; see Brownell et al. 2001 for a complete review of twentieth-century records), it is likely that North Pacific right whales migrate. However, it would be premature to speculate on the exact nature or route of their migration. In addition, some animals apparently either remain at relatively high latitudes through the winter or return northward in the early spring (e.g., an April sighting in Unimak Pass in the eastern Aleutians) (Goddard and Rugh 1998). An interesting feature of recent sightings in Hawaii is that they usually have involved solitary right whales in proximity to, and interacting with, groups of humpback whales. The Hawaiian Islands are a regular winter destination and calving area for North Pacific humpbacks (see later).

Western North Atlantic. The historical distribution of right whales included continental shelf waters from the Labrador Sea south to northern Florida. They were formerly common in summer along the east coast of Labrador, in the Strait of Belle Isle and Gulf of St. Lawrence, and far offshore along the eastern edges of the Grand Banks (Aguilar 1986; Reeves and Mitchell 1986b; Reeves 2001). The sparsity of records from these areas in recent years may be an artifact of insufficient search effort and reporting, as well as low abundance, rather than an indication that the range has been reduced. Rosenbaum et al. (2000) found that the mitochondrial DNA of several museum specimens of baleen from Iceland and Scotland matched the most common haplotype from the western North Atlantic, and they inferred, therefore, that the right whales across the rim of the North Atlantic may have formed a single biological population. The sample size was small, however, and further analyses are needed to test such a hypothesis.

The known present-day spring–fall distribution of North Atlantic right whales is centered in the Gulf of Maine and adjacent waters (Winn et al. 1986). Areas of significant, consistent aggregations include Cape Cod Bay and Massachusetts Bay in late winter and early spring (Schevill et al. 1986; Hamilton and Mayo 1990; Kraus and Kenney 1991), the Great South Channel east of Cape Cod in late spring (Kenney et al. 1995), and the lower Bay of Fundy and Scotian Shelf in summer and fall (Kraus et al. 1988; Brown et al. 1994). Sightings on Jeffreys Ledge (east and northeast of Cape Ann, Massachusetts) during the summer and fall suggest that this area may be an important feeding area as well (Weinrich et al. 2000). Recently, an increase in right whale aggregations has been observed in the central Gulf of Maine in early summer (Waring et al. 2002). In some years, a few animals also are sighted in the northwestern Gulf of St. Lawrence and in coastal waters of eastern and southern Newfoundland (Lien et al. 1989; Knowlton et al. 1992; R. Sears, Mingan Island Cetacean Study, St. Lambert, Quebec, pers. commun., 1996). Individuals from the western North Atlantic population sometimes wander northeastward to waters off southern Greenland, Iceland, and northern Norway (Knowlton et al. 1992; N. Øien and M. K. Marx, pers. commun., in IWC 2001c). Occasional sightings in the eastern North Atlantic, including a mother and a small calf off southwestern Portugal in February 1995 (Martin and Walker 1997), indicate that portions of the species's range in the eastern North Atlantic are still used by at least a few animals.

Recent photographic matching and the tracking of individual right whales with satellite-linked radio transmitters have provided a basis for revising and elaborating on the traditional model of right whale migration outlined by Winn et al. (1986). These authors proposed a six-phase model, as follows:

Phase 1. Parturient females and their calves overwintering in shallow, near-shore waters of the southeastern United States, with the rest of the population scattered widely along the eastern seaboard to at least as far north as Cape Cod.

Phase 2. Initiation of a northward migration in late winter and early spring, with some animals moving along shore (passing, e.g., Cape Hatteras and Long Island) and others remaining in deep water offshore, possibly in the Gulf Stream.

Phase 3. Whales stopping to feed during March–June in food-rich areas, including Great South Channel, Cape Cod Bay, and Massachusetts Bay.

Phase 4. Arrival on the northern feeding grounds (e.g., the lower Bay of Fundy, northern Gulf of Maine, and southern Scotian Shelf) in July.

Phase 5. Intensive feeding during August–September on the summering grounds.

Phase 6. Diffuse southward migration from October to January, with some animals passing Cape Cod and others possibly following the eastern edge of Georges Bank.

Female right whales tend to take their calves to the same summer feeding (nursery) areas to which they were taken by their own mothers (Schaeff et al. 1993).

Several radio-tracked whales initially tagged in the Bay of Fundy moved long distances during the late summer and autumn, including excursions south along the U.S. coast and back, east and north over the Scotian Shelf, and directly offshore into deep water (Mate et al. 1997; Slay and Kraus 1998). This evidence of high mobility during the feeding season, taken together with the photographic matching evidence of long-distance movements (Knowlton et al. 1992; see above), demonstrates that the population's total summer and autumn distribution extends far beyond the core areas in the Bay of Fundy and southern Scotian Shelf (Reeves 2001).

Moreover, major changes in the movements and distribution of the population have been observed since the mid-1980s. For example, the intensive use of the Great South Channel as a feeding ground during spring and early summer (April–early July) appeared to be a stable feature of the annual cycle from 1979 to 1991, but during the 1990s this pattern became much less consistent, unfortunately confounded by changes in surveys at the same time. Right whales abandoned the Great South Channel in 1992, which was apparently a 1-year anomaly (Kenney 2001). Surveys ceased during the mid-1990s. When they resumed in the late 1990s, the distribution of the whales seemed to be concentrated farther east (Waring et al. 2002), although additional data are necessary to determine whether this is a long-term or short-term phenomenon. Shifts in zooplankton communities, themselves likely related to large-scale climatic variability, have been cited as the probable cause (Kenney et al. 1995; Kenney 1998, 2001). The large aggregations of right whales seen in Roseway Basin on the Scotian Shelf in August and September, typical of the 1980s (Stone et al. 1988), were not seen between 1993 and 1999 (Brown et al. 2001). Instead, a much higher proportion of the population congregated and remained in the lower Bay of Fundy throughout the summer and the early fall (IWC 2001c).

DESCRIPTION

All baleen whales have a pair of external blowholes situated posteriorly on the top of the head. The forelimbs are paddle-like flippers, and the pelvic girdle and hind limbs are reduced to small bones resting in the ventral musculature. Genitalia and mammary glands are concealed inside folds on the abdomen. The tail is well developed and muscular, and ends in a pair of horizontally flattened flukes, supported entirely by connective tissue with no bony skeleton. When oscillated vertically, these flukes propel the body forward. Body shape varies from long, slender, and streamlined in some of the rorquals, to bulky and rotund in the balaenids (see Fig. 21.1). For summaries of descriptive characters and identification, see Leatherwood et al. (1976, 1988), Jefferson et al. (1993), Katona et al. (1993), Wynne (1997), Wynne and Schwartz (1999), and Reeves et al. (2002).

External Appearance. Right whales are stout compared to the balaenopterids and the gray whale (Figs. 21.1 and 21.3). Their axillary girth is often more than half their body length (Thompson 1918; Omura et al. 1969). The mean ratio of maximal girth to total body length in 10 North Pacific right whales was 0.73, with a high of 0.89 in a large

FIGURE 21.3. A 41-ft (12.5-m) North Pacific right whale (*Eubalaena japonica*) at the Coal Harbour, British Columbia, whaling station, May 1951. Note the long baleen, fleshy tongue, serrated upper margin of the lower lip, and white patch on the side behind the flipper. The eye is clearly seen just above and slightly behind the corner of the mouth. The chunky appearance, absence of a dorsal fin, and broad flipper are all characteristic of the right whales.

(17.1 m in length) male (Omura et al. 1969). Right whales are the heaviest whale species relative to body length, with the possible exception of the closely related bowhead (Lockyer 1976).

The head of a right whale occupies about one fourth to one third of its body length, the proportion increasing with age (Allen 1908). Head shape is distinctive. The rostrum is narrow and arched in lateral view. Viewed from above, it widens anteriorly and appears pinched or enfolded from the sides by the massive lower lips. The upper margins of the lips are usually scalloped. The eyes are set midway down the sides of the head, just above the corners of the mouth. There are no external ear pinnae. The paired blowholes, tightly closed to slits except during a breath, are set on the highest point on the posterior part of the head. They are directed somewhat laterally, so the blow (spout) appears V-shaped when viewed from ahead or behind.

The back is broad and smooth, with neither a dorsal fin as in the balaenopterids nor a dorsal ridge as in the gray whale. Ventral grooves also are absent. Flippers are up to 1.7 m long, broad, and squarish, longer on the anterior than the posterior edge. Flukes are broad (up to 6 m tip to tip), tapered to points, with a smooth trailing edge and a deep central notch.

Size. Females grow somewhat larger than males, which is the general pattern in baleen whales. The longest specimen reported from the North Atlantic was 18 m (Norman and Fraser 1937); from the North Pacific, 18.3 m (Klumov 1962). Right whales in the North Pacific are thought to grow larger than those in the North Atlantic and the Southern Hemisphere (Omura 1958); at least their yield of oil and baleen was consistently greater (Best 1987; see later). A 17.4-m female from the North Pacific weighed 106.5 metric tons (Klumov 1962); a 16.4-m male, 78.5 metric tons (Omura et al. 1969). In western North Atlantic stranding records since 1970, the longest reliably measured individual was a 15.5-m female, although there was a male estimated at 16.8 m (Knowlton and Kraus 2001). Body mass can be estimated with the formula $W = 0.0132L^{3.06}$, where W is weight in metric tons and L is length in meters (Lockyer 1976). A 13.7-m female stranded on Cape Cod, Massachusetts, in 1999 weighed approximately 52 metric tons, compared with 39.7 metric tons predicted by the formula. This was minus blood and fluid loss, but included an unknown quantity of sand scooped up with sections of the carcass (M. W. Brown, Center for Coastal Studies, Provincetown, pers. commun., 2002). At birth, right whale calves are 4–4.5 m long.

Coloration. Overall body color of right whales is black, although many individuals also have irregular white patches on the ventral surfaces. Of 50 right whales examined in the Hebrides, Scotland, during the early twentieth century, 10 had white ventral markings (Collett 1909). Of 13 animals examined in the North Pacific, only 2 were entirely black, and white ventral markings were extensive on 4 of the others (Omura et al. 1969). About one third of the current population of North Atlantic right whales have white markings—belly patches, chin patches, or both (Schaeff and Hamilton 1999). White ventral markings represent the recessive genotype and have no obvious selective value (Schaeff and Hamilton 1999). Many individuals also may appear mottled, with irregular gray markings caused by uneven sloughing of epidermis.

Callosities. Right whales have callosities—raised, thickened patches of cornified epidermal tissue—on their heads (Ridewood 1901; Payne and Dorsey 1983). Callosity tissue is relatively smooth surfaced on fetuses and calves, but becomes roughened with age. Its true color is primarily gray or black, but at a distance it usually appears white, and on close inspection often appears yellow, orange, or pink. The whiteness and brighter colors are caused by dense clusters of cyamids, or "whale lice," which live on the callosities. Cyamidae is a highly specialized family of amphipod crustaceans, related to the skeleton shrimp (Caprellidae) that live on seaweeds, bryozoans, and hydroids (Rowntree 1983, 1996; Pfeiffer 2002). There are at least 23 cyamid species, and they are known exclusively from their occurrence on the skin of whales. Whale lice tend to accumulate at sites where water flow is reduced, such as in folds of skin around the flippers, eyes, and blowholes; in ventral slits or grooves, lip margins, and wounds; and around barnacles on gray whales. They feed on whale skin (Schell et al. 2000). Right whale callosities seem to provide ideal substrates for these organisms. Some cyamid species are host specific, and some individual whales are infested simultaneously by more than one species.

Placement, size, and configuration of callosities on the heads of right whales are individually distinctive and thus provide a means of identifying individuals through photographs (Fig. 21.4) (Payne et al. 1983; Kraus et al. 1986a). Right whales usually have a large oval or kidney-shaped callosity on at least the anterior half of the rostrum, called the "bonnet." This callosity can be either continuous along almost the entire length of the rostrum or broken at about midlength, with smaller callosities ("rostral islands") variably arranged behind the bonnet. There usually is a large callosity over each eye, one on each side and toward

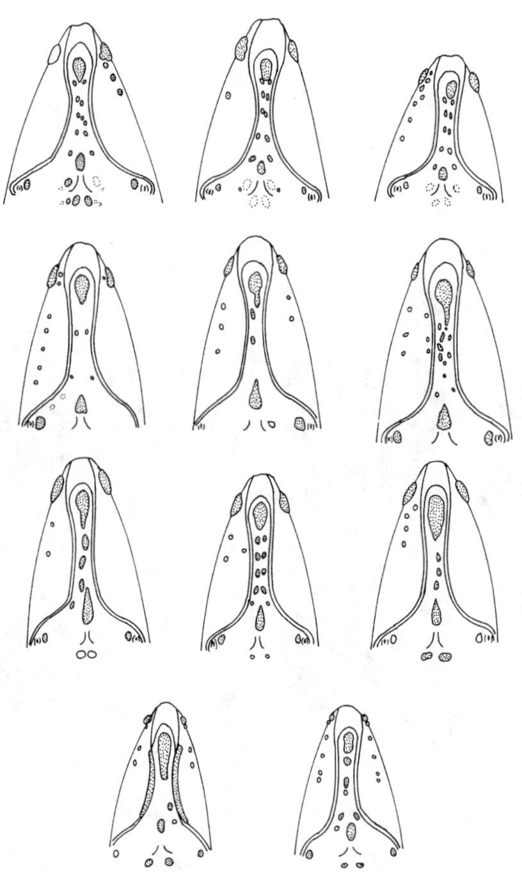

FIGURE 21.4. Dorsal views of the heads of 11 right whales (*Eubalaena japonica*) from the North Pacific, showing variability and distinctiveness of callosity patterns. SOURCE: After Omura et al. (1969), Fig. 3.

the front of the lower jaw and one immediately in front of the blowholes (the "coaming" or splash barrier). Some animals have callosities along the scalloped upper edges of the lower lips.

The function of callosities is not entirely clear, although it has been suggested that their horny texture makes them potentially useful in aggressive interactions with conspecifics or predators. In a population of southern right whales, males had more and larger callosities than females. The higher incidence of scrape marks on the bodies of males was interpreted as supporting the hypothesis that callosities are used by males to bump and scrape each other (Payne and Dorsey 1983).

Hair. Baleen whales are hairless except for superficially inconspicuous bristles on various parts of the head. On the bowhead, for example, many such hairs occur on the chin, lower and upper lips, and rostrum and in association with the blowholes (Haldiman and Tarpley 1993). These hairs arise from specialized follicles, which extend about 4 cm into the blubber. Large clusters of short (0.5–1.0 cm), grayish hairs usually are present on at least the tip of the rostrum and on the chin of right whales (Andrews 1908; Allen 1916; Matthews 1938; Omura 1958; see photo in Payne 1976). Smaller clumps sometimes are present near the blowholes (Omura et al. 1969). The sites of hairs generally match the positions of callosities. These hairs are richly enervated and likely function as tactile sensory organs similar to the vibrissae of terrestrial mammals (Ling 1977; Haldiman and Tarpley 1993). Kenney et al. (2001) hypothesized that the hairs are important in assessing the microscale quality of food patches.

Baleen. The main feeding apparatus of all mysticetes is baleen, a series of brushlike structures, which grows as two rows of keratinous plates suspended from the palate (Figs. 21.3 and 21.5) (Pivorunas 1976; St. Aubin et al. 1984; Haldiman and Tarpley 1993; McLeod et al. 1993).

FIGURE 21.5. Baleen from a male right whale (*Eubalaena japonica*) killed in the North Pacific. The photo shows the baleen in place, rooted in the upper jaw. SOURCE: Courtesy of H. Omura, Whales Research Institute, Tokyo.

Each plate is oriented perpendicular to the longitudinal axis of the head, and there is one row on each side of the mouth. These plates emerge as fibers from the soft epidermal tissue. The matted, brushy fringes along the inner edges of the baleen plates strain food organisms from seawater as it is expelled from the mouth.

Compared to other mysticetes, balaenids have supple, finely fringed baleen (Nemoto 1970). The bristles of right whale baleen are almost as fine as silk (Collett 1909). The baleen plates of balaenids are also much longer than those of other mysticetes, and the two rows, or racks, one on each side of the head, do not meet at the front of the rostrum. In balaenopterids, the two rows are connected by a series of hairlike structures (Watkins and Schevill 1976). Balaenids are regarded as true filter feeders, or "skimmers," in contrast to the balaenopterids, which normally gulp or swallow rather than skim their prey (Nemoto 1970; Pivorunas 1979). Experimental work with right whale baleen (Mayo et al. 2001) has shown it to be well adapted for capturing large, energy-rich zooplankton organisms, such as the late juvenile and adult stages of larger calanoid copepods (see Feeding Habits), but not particularly efficient at capturing smaller prey.

There is an average of about 250 plates in each row in North Atlantic right whales (Collett 1909; Allen 1916) and about 223 in North Pacific right whales (Omura et al. 1969). Bowheads generally have about 330 per side (Scammon 1874). The longest plates are situated near the middle of the mouth. The maximum length of a single plate in right whales is about 2.7 m, measured from the gum line along the lateral edge to the tip. Reports of 11-ft (3.4-m) baleen in North Pacific right whales (Clark 1887) probably refer to cleaned plates that include the portion buried inside the gum. Bowhead baleen can be 3.5–4.6 m long. Color of the baleen varies among species. Right whales and bowheads have mainly dark gray to black baleen, although some of the anteriormost plates can be white (Collett 1909).

Skull and Skeleton. The most conspicuous features of the balaenid skull are its long, outwardly bowed mandibles and the almost equally long and strongly arched maxillary and premaxillary bones (True 1904; Allen 1908) (Fig. 21.6). The shape of the skull is centered around supporting the massive baleen apparatus, housing the huge tongue, and providing space to accommodate large quantities of zooplankton (Haldiman and Tarpley 1993). Other skeletal features that distinguish balaenids from balaenopterids include the presence of five fingers in the hand instead of four; a greater number of ribs (14–15), which have a double articulation with the vertebrae; the complete or nearly complete

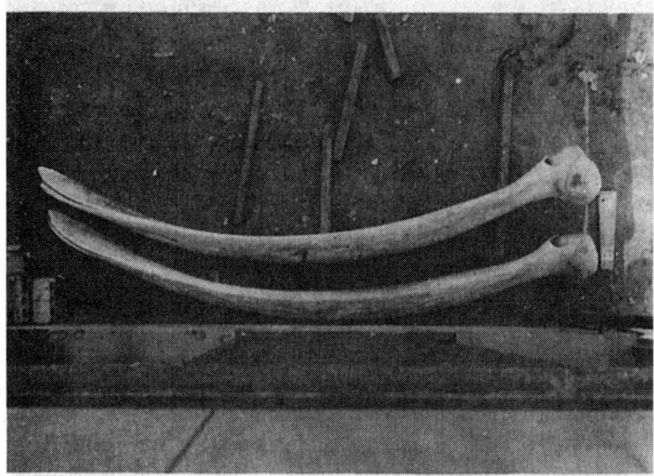

FIGURE 21.6. Skull of a North Pacific right whale (*Eubalaena japonica*). From top to bottom: lateral view of cranium, view of two left mandibles, view of occipital area. SOURCE: Courtesy of H. Omura, Whales Research Institute, Tokyo.

fusion and anteroposterior compression of the cervical vertebrae (see Fig. 1 in Omura et al. 1971); the relatively large vestigial femurs; and the absence of a distinct coronoid process of the mandible.

REPRODUCTION

Anatomy. As in all cetaceans (see Slijper 1966), the male sex organs of right whales are permanently abdominal, and there is no scrotum or baculum. The enormous size of the right whale's testes is exceptional. One individual from the North Pacific had a testis 201 cm long, 78 cm in diameter, and 525 kg in mass (Omura et al. 1969). The combined testicular mass of nearly a metric ton or more is about six times larger than would be predicted on the basis of body mass (Brownell and Ralls 1986). The darkly pigmented penis is long, slender, and retractile, reaching a length of 215–270 cm with a girth at the base of 90–110 cm (Omura et al. 1969) (Fig. 21.7). The glans penis is nearly rectangular in cross section rather than round as in the balaenopterids. In a 17-m-long specimen from the North Pacific, the center of the genital opening was 1.6 m posterior to the umbilicus and 2.4 m anterior to the anus (Omura et al. 1969).

FIGURE 21.7. Penis of a 17.1-m right whale (*Eubalaena japonica*) taken in the North Pacific. SOURCE: Courtesy of H. Omura, Whales Research Institute, Tokyo.

The female's ovaries are in approximately the same position as the male's testes. In mature female mysticetes, follicles in different stages of development protrude from various points along the surface of the ovaries, and resemble a bunch of grapes (Slijper 1966). The largest ovary measured from a right whale weighed 6.3 kg (Omura et al. 1969). It is generally assumed that most but not all ovulations result in the formation of a corpus luteum. The corpora lutea eventually degenerate to permanent corpora albicantia, visible as scars on the surface of the ovary. Distinguishing corpora of simple ovulation from those of pregnancy can be difficult (Perrin and Donovan 1984).

The cetacean uterus is bicornuate. In most mysticetes, the fetus can develop in either horn (Slijper 1966), and this presumably is the case for right whales. Females, unlike males, have a single groove on the abdomen, which contains both the genital and anal openings. Toward the front of this groove, a unilobular clitoris projects into the genital slit. A single, short cleft housing a single mammary teat is located on each side of and about halfway along the genital groove (Fig. 21.8).

FIGURE 21.8. Genital groove and mammary slits of a 16.1-m right whale (*Eubalaena japonica*) taken in the North Pacific. SOURCE: Courtesy of H. Omura, Whales Research Institute, Tokyo.

Reproductive Cycle. Most evidence suggests that mysticetes are spontaneous ovulators, with sexual activity determined by photoperiod (Slijper 1966). Ovulation rates have been little studied in right whales, but considerable variability has been found in studies of other mysticetes (Perrin and Donovan 1984). Some evidence suggests that right and gray whales are polyestrous (Matthews 1938; Rice and Wolman 1971). This would be of selective advantage to species that produce no more than one offspring every 2 or 3 years, do not form long-term pair bonds, and are sufficiently dispersed that a male might not be available when the female first comes into estrus. It is uncertain whether spermatogenic activity in male right whales is seasonal. However, male gray whales exhibit strong seasonality in sperm production, with the peak occurring in late autumn and early winter (Rice and Wolman 1971). Given the strong seasonality of calving in right whales (Kraus et al. 1986b; Best 1994), it is likely that sperm production is also seasonal.

The gestation period of southern right whales is between 357 and 396 days, or roughly 12–13 months (Best 1994), and there is no reason to believe that Northern Hemisphere right whales differ in any major way (see Breeding Season). Southern right whales give birth at intervals of 3–4 years (Best 1990; R. Payne et al. 1990; Burnell 2001; Cooke et al. 2001), and this was also true of North Atlantic right whales until the early 1990s (Knowlton et al. 1994). However, during the 1990s, the average birth interval in the North Atlantic population increased significantly, to >5 years by 2000 (Kraus et al. 2001).

Breeding Season. Most right whale births in the western North Atlantic take place in December and January, with the peak in early January (Kraus et al. 1993). Given the expected 12- to 13-month gestation, mating likely occurs in November–January. The whereabouts of most adult males and noncalving adult females at that time are largely unknown (Winn et al. 1986; Brown et al. 2001). Courtship activity occurs in all seasons in the calving ground and feeding grounds (see Mating Behavior and Table 21.1), but of the photoidentified individuals observed in courtship groups during winter in the calving area off Florida and Georgia, most were juveniles (60%), whereas 27% were adults and 13% were of unknown age (Kraus et al. 1993; North Atlantic Right Whale Consortium, unpublished data).

Birth. Direct observations of births of mysticetes are extremely rare. Caudal presentation is normal, judging by the few whale births that have been witnessed and by the positioning of the fetus within the uterus (Slijper 1966). During a right whale birth observed in South Africa, the mother rolled onto her back and slapped the water with her flippers; soon thereafter a small tail was seen protruding from her genital slit. From that time onward, a small calf was seen in close association with the adult (Best 1970).

There is no well-documented record of multiple births of right whales, although a female was reported as being accompanied by two young off Spain in the mid-nineteenth century (Markham 1881), and Scammon (1874) claimed that twins had been observed. Multiple embryos have been documented occasionally in other mysticetes (Harrison 1969).

Lactation and Weaning. Duration of lactation in right whales appears to vary from about 1/2 year to more than 1 year, with most calves probably weaned toward the end of their first year of life (Hamilton et al. 1995; Burnell 2001). Yearlings sometimes accompany their mothers well into their second year, but are by then almost certainly nutritionally independent (Thomas and Taber 1984). Estimated lactation periods of balaenopterids and the gray whale range from 4 to 10 months (Lockyer 1984), and thus the right whale may be near the high end of the range for mysticetes.

DEVELOPMENT

Growth. From a birth length of about 4.5 m and a mass of 800 kg, North Atlantic right whales almost double their length and increase their mass by more than six times by the time they are weaned (Brown

TABLE 21.1. Differential habitat use patterns among sex, age, and reproductive categories of identified right whales in the five known high-use habitats in the western North Atlantic

	Percentage of Noncalf Sightings Within a Habitat				
	SEUS	MB	GSC	BOF	RB
Mothers	45	19	7	10	1
Males (all ages)	20	29	49	53	69
Females (all ages)	77	60	34	40	16
Unknown sex (all ages)	3	11	17	7	16
Adults	65	52	31	53	38
Adult males	5	10	10	29	28
Adult females	60	39	18	22	5
Adults, unknown sex	1	3	3	3	4
In SAGs[a]	36	20	19	28	67

SOURCE: Data from the North Atlantic Right Whale Consortium photoidentification catalog.

NOTE: Calves were removed before analysis. SEUS, Southeastern United States; MB, Massachusetts and Cape Cod Bays; GSC, Great South Channel; BOF, Bay of Fundy; RB, Roseway Basin; SEUS is a calving ground; the other four are feeding grounds.

[a]Surface active groups (Kraus and Hatch 2001).

et al. 2001). A postweaning hiatus in body growth has been inferred from analyses of stable isotope traces in the baleen of southern right whales (Best and Schell 1996). This diapause may last for several years. Growth presumably accelerates at 4–5 years of age, by which time the animal's baleen and mouth are large enough for effective filter feeding (Best and Schell 1996).

Maturity. The mean age at first parturition for right whales throughout their range is about 9–10 years (Best et al. 2001; Cooke et al. 2001; Kraus et al. 2001). Therefore, females probably become sexually mature at a mean age of about 8–9 years. Interestingly, the youngest mature female in the western North Atlantic population, who first calved at age 5 years, apparently was not weaned until well into her second year (Hamilton et al. 1995). Therefore, maturity in female right whales may be related to the attainment of a certain size rather than a specific age. There are no data on age at sexual maturity in males.

Longevity. Mean or maximum longevity in right whales is not known. Age estimation is difficult at best (see Age Estimation), and there is no current whaling to provide samples. A number of individuals in the western North Atlantic population have sighting histories of 20 years or more, but reliable estimation of longevity in such a long-lived species requires a much longer study period. One female identified from a photograph taken in Florida in 1935 when her calf was shot by fishermen (therefore, when she was likely at least 5–10 years old) was last photographed in 1995 with severe injuries from a probable ship strike, at an age of at least 65–70 years (Hamilton et al. 1998a; Kenney 2002). Recent findings for bowhead whales suggest that they can live to remarkably great ages (see Age Estimation).

FIGURE 21.9. This extraordinary view of a surface-feeding North Atlantic right whale (*Eubalaena glacialis*), near Cape Cod, Massachusetts, shows clearly the arrangement of the long baleen plates. The animal is swimming from right to left in the photo, with its mouth open. SOURCE: Photo by William A. Watkins.

FEEDING HABITS

Right whales in the western North Atlantic feed on a variety of organisms, but seem to depend most heavily on the later developmental stages of the copepod *Calanus finmarchicus*. This dependence is evidenced by the remains of copepod hard parts found in fecal material (Kraus and Prescott 1982; Murison 1986; Kraus and Stone 1995) and the high-density copepod aggregations found in the immediate vicinity, or exactly on the paths, of feeding right whales (Murison and Gaskin 1989; Mayo and Marx 1990). Also, spring, summer, and autumn aggregations of right whales occur primarily in areas with high densities of these copepods (Kenney et al. 1986, 1995; Wishner et al. 1988, 1995; Kenney and Wishner 1995). In Cape Cod Bay, not all plankton samples collected in the feeding paths of right whales have been dominated by *Calanus*; many are dominated by smaller calanoids (*Pseudocalanus, Centropages),* and >20% of the samples have been dominated by zooplankters other than copepods, including cyprids (planktonic larvae of barnacles) (Mayo and Marx 1990). Such observations suggest that a variety of taxa are at least occasionally important in the nutrition of these whales. Watkins and Schevill (1976) reported finding *C. finmarchicus* and juvenile euphausiids "in approximately equal amounts" in plankton tows near feeding right whales, and Collett (1909) found only euphausiids, about 1.25 cm long, in the stomachs of right whales killed off Iceland and the Hebrides. Therefore, euphausiids apparently contribute to the diet of right whales, at least in some areas at some times. In the North Pacific, calanoid copepods (including *Neocalanus plumchrus, N. cristatus,* and *Calanus marshallae*) and euphausiids are known prey (Omura 1958; Omura et al. 1969; Tynan et al. 2001).

Feeding behavior has never been observed in the winter calving ground off Florida and Georgia (Kraus and Kenney 1991). It is not known whether the whales that go to other wintering habitats feed there. Although it is often assumed that right whales fast during migration and on the winter range, they may feed opportunistically whenever they encounter appropriate prey concentrations. Annual patterns in stable isotope ratios in South African right whale baleen support the hypothesis that feeding is suspended at the start of the fall migration and does not resume until the spring migration (Best and Schell 1996).

Right whales can sometimes be observed skim feeding at or just below the surface (Watkins and Schevill 1976, 1979; Hamner et al. 1988; Mayo and Marx 1990) (Fig. 21.9). In the western North Atlantic, the animals typically feed continuously for hours, maintaining a steady swimming speed of about 3 knots, but frequently change course to keep within the densest patches of plankton. The mouth remains open except when, about once per hour, it is closed abruptly and water is expelled ("flushing") (Mayo and Marx 1990). Most feeding appears to occur at depth and out of sight of surface observers (Murison and Gaskin 1989; Kenney et al. 1995; Winn et al. 1995; Goodyear 1996; Novacek et al. 2001).

Although their filtration of zooplankton may appear indiscriminate, the foraging behavior of right whales is in fact highly selective and adapted for energy efficiency. They concentrate on patches of zooplankton that are unusually dense and compressed into layers $\geq$1 m thick (Watkins and Schevill 1979). In one study, it was estimated that they need to find patches with concentrations of tens to hundreds of thousands of copepods/m^3 to achieve a long-term net energetic benefit from feeding (Kenney et al. 1986). Some of the highest recorded copepod densities in the North Atlantic have been measured near feeding right whales (Kenney et al. 1986; Murison 1986; Murison and Gaskin 1989; Mayo and Marx 1990; Mayo and Goldman 1992, 1998; Macaulay et al. 1995; Wishner et al. 1995; Beardsley et al. 1996; Goodyear 1996). As summarized by Mayo et al. (2001:228),

> "Restricted to the capture of large zooplankton by the baleen's filtering characteristics, yet limited in their ability to harvest larger and more mobile nektonic prey by the ponderous filtering apparatus, the right whale's trophic position is narrowly focused on the most productive but labile part of the marine food web. The success of the right whale, as for all filter feeders, is thus determined by the availability of a limited variety of prey organisms whose occurrence may vary widely and unpredictably in the natural system."

Successful foraging requires an ability to detect and track prey patches, which are highly dynamic and multidimensional. The swimming patterns of right whales are analogous to the area-restricted foraging patterns documented in many terrestrial species. Their movements appear to be cued by the density characteristics of the zooplankton patch (Mayo and Marx 1990). The sensory mechanisms involved in this precise tracking of prey probably include at least sight and touch, if not also sound and possibly taste (Kenney et al. 2001).

Right whales track zooplankton over tidal cycles (Murison 1986) and regularly move 10–15 km on large spring tides in the Bay of Fundy. Adverse weather can disrupt the feeding pattern, usually by increasing

FIGURE 21.10. This North Atlantic right whale has rolled onto its side, extending one of its broad, spatulate flippers (sometimes called "pectoral fins") clear of the sea surface. The flipper shape distinguishes right whales from most other temperate-region baleen whales, although the flippers of gray whales are also wide in comparison to those of the balaenopterids. SOURCE: Photo courtesy of the New England Aquarium.

the amount of time spent feeding during the day. Because many zooplankton make vertical migrations and are nearer the surface at night, the influence of increased surface wave action on prey patch stability would be greater at night than during the day.

BEHAVIOR

Right whales are generally slower and less predictable in their movements than the more streamlined rorquals. Watching right whales can feel like watching slow-motion film or video. They often linger at the surface with the back exposed, sometimes remaining almost stationary for long periods. They are tactile and rub against each other or slap each other with their large flippers. Sometimes a whale raises its head or tail vertically high above the water, or rolls onto its side with a flipper extended into the air (Fig. 21.10). Occasionally an individual breaches repeatedly, bursting high above the water surface and falling back onto its side and causing a loud report and impressive splash (Fig. 21.11). Successive breaches are generally less high and energetic than the preceding ones.

There is no reason to believe that males participate in calf rearing or in the defense of females or calves. Females, in contrast, are extremely solicitous of their young offspring. They regularly position themselves between a calf and an approaching vessel or aircraft (Schevill et al. 1986). Whalers took advantage of this care-giving behavior by first harpooning the calf so that the mother would not leave her offspring before being harpooned herself.

Diving. Diving behavior in right whales is similar to that seen in other large whales—a series of blows and short submergences followed by a longer dive (Slijper 1962; Winn et al. 1995). In right whales, as in humpback, bowhead, gray, and sperm (*Physeter macrocephalus*) whales, the flukes are frequently raised above the water's surface during the final or "terminal" dive in a surfacing sequence (Leatherwood et al. 1976, 1988). The regular occurrence of raised-fluke dives in some species but not others may be related to buoyancy, with those animals raising flukes being more buoyant (Kenney and Winn 1987). Whether this behavior has functional significance or is simply an unavoidable consequence of buoyancy, however, is not known.

Although generally not regarded as deep or prolonged divers, right whales are capable of remaining submerged for 15–20 min (Watkins and Schevill 1982, 1983; Goodyear 1993; Winn et al. 1995). Their diving behavior greatly depends on activity state, which in turn is influenced by season, prey distribution, and other factors. In Cape Cod Bay, whales that have been remaining near the surface and making only brief submergences can suddenly change their behavior and begin making long dives with short surfacing intervals (Watkins and Schevill 1982). On one occasion off Cape Cod, three feeding whales alternated at the surface, each on a 13.5- to 15-min dive cycle, while swimming at about 6 km/h (Watkins and Schevill 1983). In the Great South Channel, Winn et al. (1995) showed that dive times shifted from longer dives during daylight hours to shorter dives at night, corresponding to diel vertical migration by the zooplankton layer on which the whales were likely feeding.

In the Bay of Fundy, whales monitored with combination sonic/radio tags (Goodyear 1993) made long, stereotyped "bottom excursion" dives (Goodyear 1996). The animals were solitary, with no evident opportunities for social interaction. Their swimming speed was similar to that of surface-feeding whales in Cape Cod Bay (Mayo and Marx 1990). Moreover, sampling with a remotely operated vehicle, sonar, and conical net tows revealed that the greatest biomass of calanoid copepods was at the same depths as reached by the whales in their bottom excursion dives (Goodyear 1996).

Novacek et al. (2001) tagged right whales in the Bay of Fundy with digital recording tags, which measured depth, water temperature, heading, speed, and acceleration in all three dimensions. The pitch-axis accelerometer was sufficiently sensitive to record individual fluke strokes. The dive profiles were essentially the same as shown by Goodyear (1993, 1996), with a rapid descent and ascent phase and a prolonged excursion at a relatively constant depth. Descents were at very steep angles and driven by the most powerful fluke strokes observed. Ascent rates were the same as or greater than descent rates, but 15–60% of the time was spent gliding at relatively shallow angles, driven by positive buoyancy with no fluke movement.

Group Size and Composition. The term *surface active group* (SAG) was coined to describe the boisterous interactions of adult right whales that appear to be engaged in courtship (Kraus and Hatch 2001; see Mating). When not engaged in SAGs, right whales appear to be at most only moderately social. Concentrations of a few tens of individuals occur on the feeding grounds, but these are probably adventitious rather than socially significant. Other than SAGs, the animals recorded as a "group" in one sighting are typically only loosely associated within a localized area and not tightly associated into a "pod" or herd as might be expected with most odontocetes (Tyack 1986). Solitary animals and small groups of up to about eight are typical of nonbreeding right whales (Watkins and Schevill 1982; Hamilton and Mayo 1990). The most frequently observed group size is a solitary individual; of more than 20,000 sighting records in the North Atlantic Right Whale Consortium (NARWC) database, 71% represent single animals, and 96% of sightings are of four or fewer whales (NARWC, unpublished data; see Ongoing Research Programs). Most eastern North Pacific sightings are similarly of single animals, with a sighting in the Bering Sea in July 1997 of five to seven whales the largest group seen (LeDuc et al. 2001; Tynan et al. 2001). This mild sociality is typical for mysticetes in general (Clapham 2000) and distinguishes them from the many species of strongly social and gregarious odontocetes (Tyack 1986; Connor 2000).

Segregation. Adult female right whales generally do not visit the calving ground off the southeastern United States (denoted as SEUS in Table 21.1) during their noncalving years, that is, while they are pregnant or "resting" between pregnancies (Brown et al. 2001). The calving ground is populated primarily by cow–calf pairs and juveniles, and sightings of known adult males there are infrequent (Kraus and Kenney 1991; IWC 2001b). Noncalving (pregnant and resting) adult females also are seen less often than juveniles, adult males, and cow–calf pairs in the spring and summer feeding areas off New England and southeastern Canada. Their diet may differ from those of other age, sex, and reproductive classes, or they may be actively avoiding concentrations of adult males (Brown et al. 2001). In general, there are consistent differences among the four principal feeding grounds in their occupancy by different age and sex classes (Brown et al. 2001) (Table 21.1). Cow–calf pairs are most common in Massachusetts Bay and quite rare in Roseway Basin. Massachusetts Bay and the Bay of Fundy contain a higher proportion of adults than the Great South Channel or Roseway Basin. Females (all ages or only adults) are most common in Massachusetts

FIGURE 21.11. The energy in a right whale breach is both impressive and intimidating. It has been estimated that to achieve a full breach, a humpback whale needs to break the water surface at a speed of about 15 knots (roughly 8 m/sec), expending about 1% of its daily resting metabolic budget (Whitehead 2002). The purpose of breaching has been the subject of extensive speculation by scientists, with suggestions ranging from removal of ectoparasites, to acoustic signaling, to a form of play. SOURCE: Photo by Stephanie Martin, courtesy of the New England Aquarium.

Bay and rarest in Roseway Basin. Males of all ages outnumber females in the Great South Channel, Bay of Fundy, and Roseway Basin, but adult males are more numerous than adult females only in the latter two, the summer/fall habitats.

Habitat Preferences. Feeding habitat for right whales is defined by the presence of sufficiently high densities of prey, especially calanoid copepods in the North Atlantic and the North Pacific. Development of those patches is essentially a function of oceanic conditions, such as temperature, stratification, bottom topography, and currents, that concentrate zooplankton, and concentration is probably enhanced by the behavior of the organisms themselves (Murison and Gaskin 1989; Kenney and Wishner 1995; Beardsley et al. 1996; Epstein and Beardsley 2001). To the extent that annual fluctuations in oceanic conditions cause the formation of zooplankton concentrations to vary temporally and spatially, locations of preferred feeding habitat may change. For example, dramatically "anomalous" years have been described in which concentrations of right whales either failed to appear in the usual feeding areas (Kenney 2001) or remained through the feeding season in areas that they usually leave for the summer and autumn (Hamilton and Mayo 1990; P. M. Payne et al. 1990). The near absence of right whales on their spring and early summer feeding ground in the Great South Channel in 1992 was attributed to a lack of sufficiently dense patches of *Calanus finmarchicus,* probably caused by an anomalous influx of cold Scotian Shelf water, which began in the late winter and resulted in below-average temperatures over much of Georges Bank through the spring (Kenney 2001). Tynan et al. (2001) attributed the shift in Bering Sea right whale occurrences from deep waters in the mid-twentieth century to the mid-shelf region in the late 1990s to changes in the availability of optimal zooplankton patches, possibly related to climatic forcing (variability in oceanic conditions caused by changes in atmospheric patterns).

In the Florida–Georgia calving and nursery ground, right whales occur most often in very shallow nearshore waters, commonly in depths less than the body length of an adult right whale (Kraus and Kenney 1991; Kraus et al. 1993). The calving ground is located where the very shallow zone is broadest, probably providing some protection from heavy seas. In addition, the whales most often occur in the coolest waters available, inshore of a persistent midshelf front (Ward 1999). The adaptive significance of their selection of such shallow, relatively cool water is not clear, but it may be related to avoidance of predation on neonates by sharks. Sharks are more common in warmer waters, and white sharks (*Carcharodon carcharias*) are known to attack their prey from below (Klimley 1994). Further development or rejection of this preliminary hypothesis might be accomplished by comparing the physical features of other known right whale calving areas to see whether the pattern is consistent.

Response to Humans. Responses of right whales to approaching vessels can be highly variable. Animals engaged in social interaction or feeding appear oblivious (unresponsive) to vessel approaches. They tend to move slowly but consistently away from passing ships, or dive abruptly, often without lifting their flukes, when approached rapidly and incautiously (Watkins 1986). Researchers attempting to obtain biopsies of right whales need to close within 5–20 m of the animal before firing the dart. In general, whales visibly respond to darting about 20% of the time, sometimes with as little as a twitch of the skin surface, and at other times by arching the back and swimming away or by "lobtailing" (slapping the surface with the flukes) (Brown et al. 1991). Whales that are "logging" (motionless at the surface) when approached and darted are more likely to respond, perhaps because they are awakened from a state of rest.

Mating. Courtship (i.e., SAG) behavior in right whales has been described primarily from observations on the summer feeding grounds in the Bay of Fundy (Kraus and Hatch 2001). SAGs in the western North Atlantic can consist of as many as 35 whales, although the mean group size is about 5. Typically, there is one focal female and the rest of the animals are males. A second female is present occasionally. Males converge from considerable distances (at least several kilometers) to participate in what is clearly sexual activity. They jostle, bump, and slap one another as they compete for position near the focal female. The female may incite intermale competition by calling continuously while she attempts to maintain an inverted posture, with her genital area exposed above the water surface. Although males attempt to copulate even when the female is on her back, most successful copulations probably occur underwater as she rolls into an upright posture to breathe. SAGs usually stay together for at least 1 hr, during which time the female swims slowly in circles near the surface.

SAGs have been observed in all seasons and in all of the known high-use right whale habitats in the western North Atlantic (Table 21.1), although involving mainly juveniles in the winter on the calving ground (Schevill et al. 1986; Kraus et al. 1988, 1993; Hamilton and Mayo 1990; Kraus and Kenney 1991; Kraus and Hatch 2001). The frequency of occurrence of SAGs increases in summer and fall, and is highest in the Roseway Basin habitat (Table 21.1). In contrast to the winter calving ground, animals in SAGs in Roseway Basin were 40% adults, 38% unknown age, and only 21% juveniles. The occurrence of apparent sexual activity throughout the year, which also occurs in bowheads (Würsig and Clark 1993), seems inconsistent with the assumption of a 12- to 13-month gestation period. There is no evidence of delayed implantation in cetaceans, so such an explanation for this inconsistency is unlikely. Females do not appear to actively select males, but apparently create conditions that lead to male–male competition; perhaps they are assessing male quality for future mating (Kraus and Hatch 2001). In addition to competition for opportunities to copulate, sperm competition is likely (Brownell and Ralls 1986).

Vocalization. Vocalizations by right whales consist of a variety of low-frequency moans and groans as well as amplitude-modulated pulsive calls (Clark 1982, 1983; McDonald and Moore 2002). They range from 30 Hz to 2.2 kHz, with most energy at 50–500 Hz. See Thompson et al. (1979), Watkins and Wartzok (1985), Thomson and Richardson (1995), Wartzok and Ketten (1999), and Frankel (2002) for reviews of cetacean sound production.

ABUNDANCE AND TRENDS

There are no good estimates of overall abundance in the North Pacific (Table 21.2). Ship line-transect surveys in the Okhotsk Sea during 1989–1992 produced an estimate of 922 right whales (coefficient of variation [CV] = 0.433; 95% confidence interval [CI] 404–2108) (IWC 2001b). Elsewhere in the North Pacific, numbers are much lower and, as indicated earlier, most survivors appear to be mature males. Only one mother-calf pair has been observed in recent times.

Nearly the entire western North Atlantic population has been photoidentified. Only about three "new" animals, other than first-year calves, have been photoidentified each year since 1989 (Brown et al. 2001). With known or assumed deaths taken into account, the total living population is thought to consist of <350, and probably not many more than about 300 individuals (IWC 2001c) (Table 21.2). Although Knowlton et al. (1994) estimated that the population was growing at a rate of about 2.5% ($SE = 0.3\%$) per year in the late 1980s and early 1990s, subsequent analyses suggested that by the late 1990s the rate of increase was lower, possibly close to zero or declining at 2–3% (Caswell et al. 1999; Fujiwara and Caswell 2001; IWC 2001c). In contrast, several Southern Hemisphere right whale populations are growing at rates of 7% or higher (IWC 2001b).

Reproductive histories of individual North Atlantic right whales, based on photoidentification data, indicate that there are only about 75–85 "reproductively active" females, that is, those that have given birth at least once (Knowlton et al. 1994; Kraus et al. 2001; A. R. Knowlton, New England Aquarium, pers. commun., 2003). Annual per-capita calf production has been highly variable. During the 1990s, the total number of calves observed (believed to reflect the true total in the population) was about half of what would have been expected from comparison with Southern Hemisphere females, whose calving intervals are slightly greater than 3 years (Best et al. 2001; Cooke et al. 2001). From 1998 to

TABLE 21.2. Abundance and status of selected stocks of baleen whales in the North Atlantic and North Pacific, including all stocks covered by annual MMPA stock assessment reports as well as other stocks of particular interest

Species	Ocean	Total Abundance[a]	Management Stock	Stock Abundance[a]	Human-Caused Mortality[b]	Status[c]			
						ESA	MMPA	IUCN	CITES
Right	Atlantic	Unknown	Western N. Atlantic	~300	1.8	E	S	E	I
			Eastern N. Atlantic	Very small	?	"	na	"	"
Right	Pacific	Unknown	Eastern N. Pacific	Very small	<1	E	S	E	I
			Western N. Pacific	922 (0.43)	?	"	na	"	"
Bowhead	Atlantic	Unknown	Spitsbergen/Barents	10s	?	E	na	CE	I
			Baffin Bay/Davis Strait	350	?	"	na	E	"
			Hudson Bay/Foxe Basin	100+	?	"	na	V	"
Bowhead	Pacific	~8,400	Bering/Chukchi/Beaufort	8,200 (0.07)	53	"	S	LR/cd	"
			Okhotsk	150–200+	?	"	na	E	"
Blue	Atlantic	Unknown	Gulf of St. Lawrence	320	<1	E	S	V	I
Blue	Pacific	3,300	California	1,940 (0.15)	<1	"	S	LR/cd	"
Fin	Atlantic	50–60,000	Western N. Atlantic	2,814 (0.21)	1.6	E	S	E	I
Fin	Pacific	14–19,000	California/Oregon/Washington	1,851 (0.19)	1.9	"	S	"	"
			Northeast Pacific	Unknown	<1	"	S	"	"
Sei	Atlantic	Unknown	Nova Scotia	1,400–2,200	<1	E	S	E	I
Sei	Pacific	Unknown	Eastern N. Pacific	Unknown	<1	"	S	"	"
Bryde's	Atlantic	Unknown	Gulf of Mexico	218	<1	U	N	D	I
Bryde's	Pacific	20–30,000	Eastern tropical Pacific	13,000 (0.20)	<1	"	N	"	"
			Gulf of California	160	<1	"	na	"	"
Minke	Atlantic	Unknown	Canadian east coast	4,108 (0.16)	2.6	U	N	LR/nt	I
			West Greenland	Unknown	159	"	na	"	II
Minke	Pacific	Unknown	California/Oregon/Washington	631 (0.45)	<1	"	N	"	I
			Alaska	Unknown	<1	"	N	"	"
Humpback	Atlantic	10,600 (0.07)	Gulf of Maine	902 (0.41)	3.0	E	S	V	I
Humpback	Pacific	6–8,000	Eastern N. Pacific	856 (0.12)	<1	"	S	"	"
			Western N. Pacific	394 (0.08)	>0.6	"	S	"	"
			Central N. Pacific	4,005 (0.10)	4.7	"	S	"	"
Gray	Pacific	Unknown	California	26,635 (0.10)	107	U	N	LR/cd	I
			Korea	~100	?	E	na	CE	I

SOURCE: Stock assessment reports: Waring et al. 2002; Carretta et al. 2002; Angliss and Lodge 2002. Some details from a variety of additional sources (see text for each species).

[a]The numbers in parentheses after some abundance estimates are coefficients of variation (CV).

[b]Human-caused mortality is the 1996–2000 five-year mean reported in the year 2002 stock assessment reports, which includes subsistence whaling in the case of Bering/Chukchi/Beaufort bowhead whales and California gray whales. Since West Greenland minke whales are not covered by these reports, the number here is the mean subsistence harvest for the same 5-year period from Table 21.3.

[c]The Status columns show the classification for the species, population, or stock under the Endangered Species Act (ESA) (E, endangered; U, unlisted), Marine Mammal Protection Act (MMPA) (S, strategic; N, not strategic; na, not applicable because the stock occurs entirely outside U.S. waters; strategic stocks are those where total human-caused mortality exceeds a threshold based on estimated stock abundance, with ESA-listed species automatically defined as strategic), International Union for the Conservation of Nature (IUCN) Red List (CE, critically endangered; E, endangered; V, vulnerable; LR/nt, lower risk/near-threatened; LR/cd, lower risk/conservation dependent; D, data deficient), and Convention on International Trade in Endangered Species of Wild Flora and Fauna (CITES) appendices. Ditto marks indicate that the population or stock is not considered separately under the relevant listing. (See text for additional details and sources.)

2000, annual calf production in North Atlantic right whales was lower in absolute terms (seven, four, and one, respectively) than in all but 1 of the preceding 17 years. However, in 2001, 31 new calves were documented, the highest 1-year total since 1980, when regular surveys of the calving grounds began. In spite of this pulse in calf production, the population's long-term prospects continue to appear bleak because of the continuing conflicts with human activities (see Mortality).

AGE ESTIMATION

Age estimation of balaenopterids is most reliably accomplished by counting laminae in wax ear plugs from the auditory meatus (Purves 1955). Although present in right whales, these plugs are soft and difficult to remove without distortion and breakage (Omura et al. 1969). Another approach to age estimation is counting corpora on the ovaries of adult females (Laws 1961), but this method assumes that age at sexual maturity and the ovulation rate are known with certainty. A third method is to count the transverse ridges on the surface of baleen plates (Ruud 1940, 1945), and a fourth is to trace annual fluctuations in stable isotope ratios in the baleen (Schell et al. 1989). However, even if these patterns were calibrated, they would be of limited use because the distal ends of the baleen plates abrade with age. All of these methods of age estimation require adequate samples of dead whales for meaningful analyses, which is problematic in species with small populations and no harvests. The method applied to bowhead whales of quantifying isomers of aspartic acid in the eye lens (George et al. 1999; see later) could be useful for right whales, but with the same sample-size caveat.

The method of choice for studying age-related parameters in Northern Hemisphere right whales is to monitor living individuals from year to year using photoidentification. Unique morphological features, such as callosity patterns, scarring, and lip crenulations, have made it possible to identify virtually every whale older than about 1 year in the western North Atlantic population (Hamilton et al. 1998b) (see Longevity).

MORTALITY

Conflicts with Human Activities. The potential importance of ship collisions and entanglement in fishing gear as causes of morbidity and mortality of right whales has long been recognized (Reeves et al. 1978), but Kraus's (1990) empirical analysis was the first demonstration of the conservation implications of these events. He found that at least half of the documented nonneonatal mortality in the western North Atlantic right whale population between 1970 and 1989 had been caused by ship strikes or gear entanglement. Of the deaths attributed to human causes, well over half were from ship strikes. In addition to the outright mortality, about 7% of the living population had "major wounds" on the back or tail peduncle caused by ship propellers. Nine additional deaths were documented from 1990 to June 1998; of these, eight resulted from ship strikes (Kenney and Kraus 1993; Knowlton and Kraus 2001). The rate of documented mortality caused by ship strikes has increased since 1990, from about 20% of documented mortality through 1990 to 47% since then (Knowlton and Kraus 2001). There are at least three possible explanations of this trend, and they are not mutually exclusive: (1) there has been increased effort to retrieve carcasses and perform rigorous necropsies on them, (2) the rate has increased over the past decade, and (3) the rate was underestimated by Kraus (1990). Since not all deaths are documented and the condition of some carcasses is too poor to determine the cause of death, the true effect of ship strikes and entanglements is underestimated by these statistics. Moreover, they fail to fully reflect the morbidity and lowered productivity and longevity of animals with injuries classified as "nonfatal" or "possibly fatal," such as propeller cuts and deep gashes, or of animals bearing entangled fishing gear for long periods (Knowlton and Kraus 2001). In the North Pacific, 1 of 13 reported strandings of right whales in the twentieth century involved gillnet entanglement, and Brownell et al. (2001:284) considered entanglement a potentially significant threat to North Pacific right whales, "particularly given the present operation of Japanese salmon driftnet fisheries within the Russian EEZ [exclusive economic zone] inside the Okhotsk Sea."

Approximately 60% of living North Atlantic right whales bear scars caused by gear entanglement, whereas only about 6% have scars (mainly propeller wounds) from vessel strikes (Hamilton et al. 1998b). Clearly, not all entanglements result in death. However, a relatively benign entanglement, by compromising locomotion, may make the whale more susceptible to being struck by a vessel. Entanglement deaths may be underestimated because attached gear prevents the carcass from floating to the surface. Also, animals that are towing gear may become emaciated because of impaired feeding or the increased energy expenditure needed to swim and dive. When they die, their carcasses would be less buoyant and more likely to sink (Knowlton and Kraus 2001).

Predation. The only observed natural predator of right whales is the killer whale (*Orcinus orca*), although large sharks are certainly capable of taking prey the size of a right whale calf (Klimley 1994). Descriptions of attacks on right whales and other large whales emphasize that small groups of killer whales cooperate in harassing their prey, with some of them biting the flukes and flippers to impair mobility, and others leaping onto the target animal's back and blowholes to prevent it from surfacing and respiring normally (Jefferson et al. 1991). Killer whales show a consistent preference for eating the tongues and lips of the whales that they kill.

Disease and Parasites. The incidence of disease and parasites is best known in those marine mammal species that are (1) commonly kept in captivity, (2) subject to intensive hunting, or (3) frequently stranded (see reviews in Dailey and Brownell 1972; Britt and Howard 1983; Howard et al. 1983a, 1983b, 1983c; Migaki and Jones 1983; Dailey 1985; Dierauf 1990, Dunn 1990; Kennedy-Stoskopf 1990; Haebler and Moeller 1993). Since right whales are not among those species, there is limited information on infections in them. In general, the incidence of viral and bacterial disease in baleen whales appears to be low, possibly because of their generally low degree of sociality, unlike the more social or gregarious odontocetes and pinnipeds. Moreover, neoplasms (tumors) have not been reported in right whales, although they have been observed in blue, fin, and bowhead whales.

Lesions in the muscle of a stranded southern right whale calf contained a zygomycetous fungus of uncertain origin (Best and McCully 1979). The lesions were reminiscent of the fibrous, nonneoplastic "husks" commonly found in the skeletal muscle, and less often on the pleural and peritoneal surfaces of the viscera and in the mammary glands, of large whales. Howard et al. (1983a) suggested that the husks were caused by *Crassicauda* nematode infections.

Presumably because their diet is composed primarily of low-trophic-level organisms, right whales are relatively free of endoparasites (Klumov 1963). At least 10 different species have been documented in right whales, most of which also occur in other mysticetes. These include two cestodes (*Priapocephalus* and *Phyllobothrium*), an acanthocephalan (*Bolbosoma*), and a nematode (*Crassicauda*) (Dailey and Brownell 1972; Yablokov et al. 1974). *Phyllobothrium, Bolbosoma,* and *Crassicauda* all are very common in other mysticetes, as are *Lecithodesmus* (trematode), *Tetrabothrius* (cestode), and *Anisakis* (nematode).

Right whales also are relatively free of ectoparasites, apart from abundant cyamids, which infest the callosities, ventral slits, eyes, blowholes, mouth parts, and even the flukes of some individuals (Omura et al. 1969) (see Callosities). These might more properly be called ectocommensals rather than ectoparasites (Kenney 2002; however, see Pfeiffer 2002). Patches of diatoms (*Cocconeis ceticola*) are sometimes present on the skin (Omura et al. 1969). Reeves and Brownell (1982) cited evidence that the barnacle *Tubicinella major* is endemic to right whales, but only or primarily in the Southern Hemisphere, and concluded that there was no good evidence of barnacles on right whales in the Northern Hemisphere. A right whale photographed off central California in March 1982, however, had about 300 barnacles on its head and flukes. These represented at least two species, probably *Coronula diadema* and *C. reginae,* possibly also *Cetopirus complanatus* (Scarff 1986b). *Coronula* barnacles are common on humpback whales, and it is likely that this right whale became infested while swimming near or associating with humpbacks. Parasitic copepods (*Penella*) are known from a variety of mysticete whales. Lampreys (family Petromyzontidae) and remoras (family Echenedidae) often are included in lists of baleen whale parasites; however, the former is more a specialized predator than a parasite, whereas the latter is an ectocommensal (Fertl and Landry 2002).

ECONOMIC STATUS

History of Whaling. Basque whalers hunted right whales in the Bay of Biscay from as early as the eleventh century and progressively extended their activities along the rim of the North Atlantic, reaching the shores of Newfoundland and Labrador by perhaps the 1530s (Aguilar 1986). They killed hundreds of right whales in the western North Atlantic annually (Barkham 1984). A rough estimate of the cumulative catch in Newfoundland–Labrador waters from 1530 to 1610 is 25,000–40,000 (Aguilar 1986). Some (possibly a high proportion; B. N. White, Trent University, pers. commun., 2003) of the animals taken in the Strait of Belle Isle were bowheads rather than right whales (Cumbaa 1986). The Basque whale hunt in the New World had become less profitable by the mid-1600s. In the American colonies to the south, however, shore whaling enterprises were established during the early to mid-seventeenth century, and these persisted until the early 1900s (Edwards and Rattray 1932; Reeves and Mitchell 1986a, 1988; Reeves et al. 1999). At least 2000 right whales were killed by the shore whalers in New England, New York, and Pennsylvania between 1696 and 1734, but thereafter the industry in the North Atlantic had to depend on other whale species to remain profitable. The American offshore (pelagic) whaling fleet killed right whales opportunistically in the North Atlantic from the mid-1700s until the late 1800s (Reeves and Mitchell 1986b; Reeves 2001). Modern Norwegian whalers took fewer than 150 right whales at stations in Iceland, the Faroe Islands, western Norway, the Shetland Islands, the Hebrides, and Ireland between 1900 and 1937 (Brown 1986). The last known kill in the eastern United States was a calf killed in Florida in 1935 (Moore 1953). Two right whales were killed in Newfoundland in 1937 and 1951, respectively (Mitchell et al. 1986), and three in Madeira (one in 1959, a mother and calf in 1967; Brown 1986).

The commercial hunt for right whales in the North Pacific did not begin in earnest until the 1830s, when American pelagic whalers arrived in the Gulf of Alaska (Kugler 1986; Webb 1988). The so-called Northwest Coast whaling ground became a major theater of right whaling, but only for less than two decades. The American fleet whaling north of 50°N in the North Pacific increased from 2 ships in 1839 to 108 in 1843 and 292 in 1846 (Scarff 1986a). With the discovery of bowhead whales in the Bering Strait region in 1848, however, the bulk of American whaling effort shifted to the north. The term "Northwest Coast" has never been precisely defined, but it certainly encompassed the Gulf of Alaska north of 50°N (Clark 1887). In the 1840s, American whalers began referring to the northern portion of the Gulf of Alaska and waters westward along the Alaskan Peninsula as the Kodiak Ground (Webb 1988), and they also recognized Bristol Bay as a summer whaling ground for right whales (Bockstoce 1986).

Best (1987) estimated that American vessels took somewhat more than 70,000 right whales worldwide (20% of them in the North Pacific) between 1805 and 1909. The total number of right whales removed from the North Pacific by commercial whalers (American, French, British, German, and Hawaiian) in 40 years (approximately 1830s–1870s) was probably much larger than 14,000, however. Scarff (2001) adjusted catch data to account for whales that were struck but not secured, and concluded that 26,500–37,000 right whales were killed in the North Pacific, Okhotsk Sea, and Bering Sea from 1839 to 1909. Of that number, he estimated that 21,000–30,000 were killed in the single decade of 1840–1849. It is probably true, as suggested by Scarff (2001), that no other large population of whales was depleted so quickly as this one in the days before modern steam-powered catcher boats and explosive harpoons.

Brownell et al. (2001) compiled records of about 740 right whales taken in the North Pacific during the twentieth century, more than half of them in the western sector. The Soviet Union took 10 off the Kurile Islands under a scientific research permit in 1955 (Klumov 1962), and Japan took 13 more under scientific permits—2 off eastern Japan in 1956 (Omura 1958), 9 off Kodiak Island and in the southeastern Bering Sea in 1961–1963 (Omura et al. 1969), and 2 southeast of Sakhalin Island in the Okhotsk Sea in 1968 (Omura et al. 1969). About 28 were taken by shore stations in Alaska and British Columbia between 1911 and 1938 (Reeves et al. 1985; Brueggeman et al. 1986). Most importantly, Soviet whalers killed about 200 right whales in the southeastern Bering Sea and Gulf of Alaska in 1964 (Brownell et al. 2001). This illegal kill, which was not acknowledged until recently, either severely retarded the recovery of eastern North Pacific right whales (at best) or rendered them incapable of recovery (at worst). Similar illegal killing took place in the western North Pacific. For example, 126 were killed by Soviet whalers off Sakhalin Island in 1967 (Brownell et al. 2001), although the population in the Sea of Okhotsk may be in the middle to high hundreds and therefore capable of recovery (IWC 2001a).

Products. The principal products from the right whale fishery were oil and baleen (Clark 1887). The oil was rendered primarily from the thick blubber layer and from other fatty body parts, such as the lower lips and tongue. Its principal use was for illumination [see Stevenson (1904), Brandt (1940), and Ross (1993) for reviews of the uses of whale oil]. The baleen (referred to as "whalebone" or simply "bone") was used mainly to stiffen and support fashionable women's clothing and in the manufacture of items requiring a combination of strength and flexibility (Ross 1993) (Fig. 21.12). Use of whales for human food was not of particular interest to North American commercial whalers. In some other countries, however, notably Japan and Norway, the meat of baleen whales has always been in demand (Foote 1975; Food and Agriculture Organization 1978).

North Pacific right whales are much larger than those from the Southern Hemisphere and the North Atlantic. Whaling records indicate that North Pacific right whales yielded about twice as much oil (average about 125–130 barrels [bbl] compared with no more than about 60 bbl) and baleen (1000 lb [455 kg] vs. 600 lb [273 kg] per 100 bbl of oil) as whales in other areas (Scammon 1874; Clark 1887). Approximately 250–330 kg of baleen could be obtained from a mature adult North Atlantic right whale (Collett 1909), whereas a North Pacific right whale could yield up to 682 kg (Scammon 1874). Calculated averages from whaling statistics include 122 bbl of oil and 1126–1189 lb of baleen (512–540 kg) for North Pacific whales (Best 1987), 67 bbl and 563 lb (256 kg) for whales from outside the North Pacific (Best 1987), and 44 bbl and 647 lb (294 kg) for North Atlantic whales (Reeves et al. 1999).

The blubber of right whales varies in color from pale pink to pure white (Collett 1909). The amount of oil obtained from individual whales varied considerably, depending on season, life stage, and location. The Long Island shore fishery, for example, took whales mainly during the northbound spring migration, after a winter of fasting or much reduced feeding, with generally lower oil yields (Reeves and Mitchell 1986a). The majority of the whales landed were juveniles, calves, and lactating females. The oil yield of a calf, which would be only a few months old during the spring, could be as high as 50 bbl, and was often higher than the mother's. Lactating females were referred to by the whalers as "dry-skins" because of their low oil yield.

Whale Watching. Whale watching has become a popular branch of tourism in North America and indeed in many parts of the world where whales are predictably accessible within a short distance of shore (Hoyt 2002). Although whale-watching enterprises centered on southern right whales have flourished for a considerable time in Argentina (Rivarola et al. 2001; Rowntree et al. 2001) and South Africa (Best 1995; Findlay 1998), right whale watching in the North Atlantic has been controversial. In part out of concern that close approaches by whale-watching vessels could be a source of disturbance to the animals, the U.S. National Marine Fisheries Service issued regulations in 1997 prohibiting boats and aircraft from approaching any right whale closer than about 500 m (Silber and Payne 1998). This effectively precludes commercial whale watching focused on right whales. In Canada, where right whales are readily accessible on their summer feeding ground in the lower Bay of Fundy, they are a major tourist attraction (Fig. 21.13).

Indirect Impact of Human Activities. The evident decline in reproductive success of North Atlantic right whales in the 1990s prompted investigations of possible causes, including genetic factors, nutritional problems, chemical contaminants, biotoxins, and disease (Reeves et al. 2001a). None of these could be ruled out, but no specific cause-and-effect link has been established. In retrospect, given the high interannual variability in calf production and the large number of calves produced in 2001, it seems likely that transient fluctuations in prey availability in the western North Atlantic feeding areas were at least a contributory factor to the poor reproductive performance during the late 1990s. Studies of tissue contaminant levels and biomarkers in right whales and most other marine mammals have been largely descriptive and are likely to remain so, as it is impossible to conduct anything resembling a controlled experiment with these animals (Reijnders 1988; O'Shea 1999; O'Shea et al. 1999). Total polychlorinated biphenyl fractions in the blubber are an order of magnitude lower than those in seals and toothed cetaceans known or thought to have experienced toxic effects (Woodley et al. 1991; Weisbrod et al. 2000). Concentrations of organochlorine compounds in the tissues of baleen whales are typically lower than those demonstrated to have reproductive effects on other mammals (O'Shea and Brownell 1994; O'Shea 1999). Seasonal trends suggest that there is some release of lipid-stored organochlorines during the winter, when a whale's food intake is reduced and body fat becomes depleted. In addition to being exposed to the more conventional chemicals of concern, mainly the persistent organic pollutants, North Atlantic right whales may be exposed to certain "nontraditional" contaminants. This is because of their relatively low position in the food chain, presence of regional sources near their feeding grounds, introduction of new compounds or new uses of old ones, and their tendency to feed at least occasionally in the sea surface microlayer, a concentrated source of lipophilic contaminants (Reeves et al. 2001a). Potentially harmful levels of paralytic shellfish poisoning toxins were detected recently in copepods (*Calanus finmarchicus*) from the Bay of Fundy (Durbin et al. 2002).

Although the possibility has been raised that the North Atlantic right whale population is experiencing "inbreeding depression," manifest as reduced fertility, fecundity, and juvenile survival, a direct link

ARTICLES MADE OF WHALEBONE

BY

J. A. SEVEY,

40 Essex Street - - - - Boston.

No.		No.	
1.	Whip Makers' Bone.	27.	Fishing Rod Tips.
2.	Whip Makers' Bone (Patent).	28.	Landing Rods for Nets.
3.	Parasol Makers' Bone.	29.	Divining Rods.
4.	Umbrella Makers' Bone.	30.	Drill Bows.
5.	Gross Dress Bone.	31.	Ferules.
6.	Samples to sell by.	32.	Busks.
7.	Dress Bone in Boxes.	33.	Back Supporters.
8.	Round Dress Bone.	34.	Fore Arm Bones.
9.	White Dress Bone.	35.	Probangs.
10.	Corset Makers' Bone.	36.	Tongue Scrapers.
11.	Corset, showing Place of Bone.	37.	Pen Holders.
12.	Corset Clasps.	38.	Plait Raisers.
13.	Hip Busk Bone.	39.	Paper Folder and Cutters.
14.	Cap Makers' Bone.	40.	Painters' Graining Combs.
15.	Hat Makers' Bone.	41.	Boot Shanks.
16.	Whalebone Hats.	42.	Shoe Horns.
17.	Ribbon Weavers' Bone.	43.	Bone Fibre Shoe Brushes.
18.	Suspender Makers' Bone.	44.	Fibre of Whalebone.
19.	Neck Stock Makers' Bone.	45.	Fibre Curled for Beds.
20.	Bonnet Makers' Shurs.	46.	Brush Makers' Bone.
21.	Whalebone Canes.	47.	Flue Brushes.
22.	Whalebone Riding Whips.	48.	Caterpillar Brushes.
23	Whalebone Rosettes.	49.	Whalebone Shavings for Beds.
24.	Ribbon Bone.	50.	Policeman's Clubs.
25.	Billiard Cushion Springs.	51.	All Bone Whips.
26.	Fishing Rod (New Style).	52.	Twisted Whip Handles.

53. Rio Corset Bone, all lengths and sizes.

Copyright, by J. A. Sevey.

FIGURE 21.12. The uses of baleen, or whalebone, were diverse. Fringe hair and bristles were trimmed off and used for upholstery stuffing and light brushes. The main slab was then cut into parallel strips of varying width (e.g., very narrow strips for umbrella tines and wider strips for corset stays) (Bockstoce, 1986). SOURCE: Photo courtesy of the New Bedford Whaling Museum.

FIGURE 21.13. Whale watching has become firmly established among the array of tourist options at several seaside communities in the northeastern United States and southeastern Canada. In fact, this "nonconsumptive use" of whales as resources has been touted by some conservationists as preferable, both ecologically and economically, to commercial whaling. Debate continues, however, about the extent to which whale watching itself has a negative effect on whales and needs to be tightly regulated to minimize disturbance. This photo shows a right whale and a whale-watching vessel in the lower Bay of Fundy near Grand Manan Island, New Brunswick. SOURCE: Photo by Laurie Murison, Grand Manan Whale and Seabird Research Station.

between indicative symptoms and the loss of genetic variability is difficult to verify (Schaeff et al. 1997; IWC 2001c). Another possible explanation for the population's failure to recover more quickly is depensation, or the so-called Allee effect (Fowler and Baker 1991; also see Best 1993; Clapham et al. 1999; Wade 2002). Genetic analyses and information on whaling do, in fact, indicate that the population was reduced to only a very few individuals (tens) by the early 1900s (Schaeff et al. 1997; Reeves 2001), so if depensation occurs in baleen whales at very low population levels, it could help explain the continued precarious status of right whales throughout the North Atlantic and at least the eastern sector of the North Pacific.

MANAGEMENT AND CONSERVATION

North Atlantic and North Pacific right whales are classified as endangered under the U.S. Endangered Species Act of 1973 (ESA) and in the International Union for the Conservation of Nature Red List of Threatened Species (Table 21.2). They also are listed in Appendix I of the Convention on International Trade in Endangered Species of Wild Fauna and Flora, which means that all international trade is prohibited except under exceptional circumstances, for example, the scientific exchange of research specimens. Protection from deliberate killing has been key to preventing the extinction of right whales and other commercially valuable species. Since 1935, when the first International Convention for the Regulation of Whaling came into force, commercial whaling for right whales has been prohibited. This ban certainly had an effect on whaling operations (e.g., Reeves et al. 1985) but neither it nor its successor, the 1946 convention, which remains in force, has been entirely successful in preventing continued killing (Best 1988; Tormosov et al. 1998; Katona and Kraus 1999). In North American waters, the most recent reported kills were in 1951—one in Newfoundland and one in British Columbia. In addition, hundreds of right whales were killed in the North Pacific by Soviet and Japanese whalers between 1955 and the early 1970s (Brownell et al. 2001). For the last few decades, the right whales in the eastern North Pacific and the western North Atlantic have been well protected from whaling—there is no existing or proposed "aboriginal subsistence" whaling on these species.

The National Marine Fisheries Service (NMFS), a branch of the U.S. Department of Commerce's National Oceanic and Atmospheric Administration, is the lead agency responsible for whale management and conservation under both the Marine Mammal Protection Act of 1972 (MMPA) and the ESA (Baur et al. 1999). The Marine Mammal Commission and its Committee of Scientific Advisors on Marine Mammals provide executive oversight for agency implementation of the MMPA, and they have played important roles in guiding research and management action with respect to right whales (Katona and Kraus 1999; also see the Marine Mammal Commission Annual Reports to Congress, available from the Commission at 4340 East-West Highway, Room 905, Bethesda, MD 20814).

Although other potential threats are being investigated (see above), it is generally understood that the clearest and most immediate detriment to North American right whale populations is the mortality caused by ship collisions and entanglement in fishing gear. Therefore, NMFS and the Canadian Department of Fisheries and Oceans are under pressure to manage ship traffic and fishing activities in ways that will reduce the risk of killing right whales. Both countries have "recovery plans" in place for right whales (NMFS 1991; Right Whale Recovery Team 2000), and a revised U.S. recovery plan for the North Atlantic was recently distributed for public comment. A separate U.S. plan is to be developed for North Pacific right whales. Implementation teams in both countries meet regularly to review progress and evaluate new information relevant to right whale conservation. A newsletter, *Right Whale News*, is published quarterly and distributed free of charge on behalf of these teams (available on-line at http://www.graysreef.nos.noaa.gov/rightwhalenews.html and www.gepinstitute.com).

The ESA requires that the status of all listed species be reviewed at 5-year intervals, although this requirement is not achieved in practice (Shelden and Rugh 1995; Perry et al. 1999). Moreover, under the 1994 amendments to the MMPA, NMFS is obliged to publish annual stock assessment reports that review the status of all marine mammal stocks in U.S. waters. Assessments must be reviewed and updated annually for "strategic" stocks (ESA listed or for which human-caused mortality is estimated to exceed a defined threshold; see Chapter 20), and at least every 3 years for others. ESA status reviews, MMPA stock assessments, updates of take-reduction plans, and other documentation of federal marine mammal management are accessible on-line at http://www.nmfs.noaa.gov/prot_res/.

In the United States, "critical habitat" for North Atlantic right whales was designated in 1994 under Section 7 of the ESA, and includes the following three areas: the Georgia–Florida calving ground, the Great South Channel spring feeding ground, and the Cape Cod–Massachusetts Bay late winter feeding/nursery area (Katona and Kraus 1999). This designation requires that federal agencies consult with NMFS before approving or conducting potentially harmful activities in the designated areas, such as dredging, waste disposal, mining or drilling, and discharges of explosives. For example, several major ports in northern Florida and Georgia require periodic dredging. Until the early 1980s, when it became known that the area serves as an important right whale calving ground, the dredging was confined to winter months to protect endangered sea turtles, which are vulnerable from late spring to autumn. In response to concerns about collisions with right whales, the Corps of Engineers was forced to develop a series of operating standards in consultation with NMFS. Aerial surveys are conducted daily (weather permitting) throughout the season to locate right whales and advise the dredge operators. Dredging vessels must carry whale observers while transiting between channels and the dump sites offshore. The dredges are required to maintain slow speeds as a precaution during periods of poor visibility, at night, and whenever a right whale has been sighted within 16 km of the transit and dumping zones (Katona and Kraus 1999). In Canada, right whale conservation areas were designated by the federal government in 1993 in the lower Bay of Fundy and in Roseway Basin between Browns and Baccaro Banks on the western Scotian Shelf. These designations have no regulatory significance, but are marked on nautical charts to alert mariners to the risk of whale collisions (Colborn et al. 1998).

An analysis of ship noise in the Bay of Fundy in relation to the presumed hearing abilities of right whales led Terhune and Verboom (1999) to conclude that although the whales may hear approaching ships, they are unlikely to react in ways that preclude collisions. The onus, therefore, must be on ship operators to detect and avoid the whales. Laist et al. (2001) inferred that most whales hit by ships are not detected beforehand or are detected too late to divert the vessel's course, and also that serious injuries are infrequent when the vessel is traveling at speeds of <14 knots and are rare at speeds of <10 knots. These inferences suggest that management strategies that depend on the vessel operator's detection and avoidance of whales are largely ineffective for large ships with limited maneuverability. A more effective strategy might be to minimize travel distances and reduce speeds of vessels traversing areas of high whale densities (Laist et al. 2001). A report recommending a suite of region-specific vessel-traffic management actions (Russell et al. 2001) was developed through a lengthy process of stakeholder workshops and negotiations involving two regional recovery implementation teams. This report has been submitted to NMFS for review and potential implementation.

Although the shipping industry has been hopeful that technological solutions could be found, such as passive acoustic "whale detection" systems or "whale-avoidance" sonars, these approaches have yet to be tested and shown to be effective. Exclusion of traffic, or at least enforced vessel-speed reduction, in high-risk areas appears to be the most practical option available. A mandatory ship-reporting system came into effect in 1999 for right whale critical habitat along the east coast of the United States. Under this system, commercial vessels larger than 300 tons that enter specified right whale areas must report their location, course, speed, destination, and route to a clearinghouse on shore, which in turn gives advice on right whale occurrence. The system is active throughout the year in Massachusetts and from 15 November to 15 April in Georgia and northern Florida. Various other early-warning alert measures have been in place for a number of years in both the United States and Canada. These programs involve dissemination of real-time information on right whale distribution to mariners as they enter areas seasonally occupied by right whales (Colborn et al. 1998; Silber and Payne 1998).

In addition to measures intended to reduce the incidence of vessel collisions, much attention has been devoted to entanglement reduction. At least portions of some areas designated as critical habitat are permanently closed to fixed-gear fishing. Both dynamic area management and seasonal area management have been used to some extent off the eastern United States to reduce entanglement risk, but these approaches are unpopular with fishermen, and dynamic management in particular requires costly monitoring and enforcement effort. Dynamic area management means that once a certain number of whales have been observed in an area, a short-term restricted-fishing zone is established to clear it of high-risk gear. Seasonal area management involves closures of areas known to be used by whales.

An alternative approach has been to seek ways of modifying fishing gear to reduce the frequency of entanglements and the seriousness of those that occur. Effort has focused on vertical lines, floating groundline, and line that does not break in time to prevent serious injury or death to a whale that hits it. For example, floating polypropylene line in gillnet and lobster-trap arrays might be replaced with neutrally buoyant or sinking line.

An important adjunct to efforts to reduce the incidence of collisions and entanglements is a right whale rescue, or disentanglement, program. Begun in 1984 at the Center for Coastal Studies in Provincetown, Massachusetts, the Atlantic Large Whale Disentanglement Network has become recognized as a major component of the overall recovery effort for right whales. Since the program's inception, 56 large whales, including 17 right whales, have been disentangled or released from fishing gear (Mayo et al. 1998; D. K. Mattila and B. Bowman, Center for Coastal Studies, pers. commun., 2002). The rescue program should not be viewed, however, as a solution to the entanglement problem. In 2000, for example, the network responded to 26 calls for assistance, but the team managed to free only 3 whales. Three other attempts were unsuccessful and the whales probably died. The rest of the whales either were never relocated, judged not to be in need of disentanglement, or escaped from the gear on their own. There is, of course, no way to determine how many other entanglements were never observed or reported. The specially trained disentanglement team at the Center for Coastal Studies works in close cooperation with the U.S. Coast Guard, federal fishery agencies in both Canada and the United States, and the research team at the New England Aquarium in Boston. Timely reporting of entanglements is obviously key, as is the ability to transport personnel and equipment to the entangled whale's locality. If possible, a satellite-tag buoy is attached to the gear trailing from an entangled whale so that the animal can be tracked through the night and until weather conditions and logistics are amenable to a hands-on release effort.

ONGOING RESEARCH PROGRAMS

In 1982, when the first edition of *Wild Mammals of North America* was published, the entire annual NMFS budget for marine mammal research on the U.S. east coast was about $500,000. More than ten times that amount has been committed to right whale research alone under special congressional appropriations in each of the last several years. Numerous governmental agencies and private foundations also fund research on right whales in both the United States and Canada.

Much of the research on North Atlantic right whales is coordinated through the North Atlantic Right Whale Consortium, which includes as dues-paying members individuals who are active or interested in right whale research and conservation. This unusual, ad hoc, nonprofit organization developed as a response to the need for a coordinating body to provide advice to funding agencies and a forum for information exchange among right whale researchers. The consortium is governed by an elected steering committee and meets annually to review research progress and needs. One of its key functions is to control access to and use of the very large databases of photographs, sightings, strandings, tissue samples, etc., that have accumulated over the past 20-plus years on the western North Atlantic whale population. The two main repositories of data are the New England Aquarium in Boston and the University of Rhode Island in Narragansett. Although these data are not externally accessible, they are provided for research and management projects on request (through the Consortium secretary, Marilyn Marx, New England Aquarium, mmarx@neaq.org; see for details http://www.rightwhaleweb.org).

Recent and ongoing research on right whales in the eastern North Pacific has consisted primarily of shipboard and aerial surveys in the southeastern Bering Sea to obtain photographs, biopsies, and plankton samples (LeDuc et al. 2001; Tynan et al. 2001) and compilation of data from "platforms of opportunity" in the Gulf of Alaska and Bering Sea (Goddard and Rugh 1998). The former have been conducted mainly by the NMFS Southwest Fisheries Science Center in La Jolla, California, and the latter by the National Marine Mammal Laboratory of the NMFS Alaska Fisheries Science Center in Seattle. Funding has come principally from the Office of Protected Resources, National Oceanic and Atmospheric Administration, Recover Protected Species Program.

Research and Management Needs. It has been stated repeatedly in workshop and meeting reports that the highest priority for right whale conservation in the western North Atlantic is to reduce human-caused mortality toward zero as quickly as possible, without waiting for further research results (IWC 2001b, 2001c; Reeves et al. 2001a). Thus, the most urgent types of research are those related to reducing the risks of ship strikes and entanglement in fishing gear. Continuation of the long-term monitoring program, with standard protocols for obtaining photographs and biopsies, is essential for evaluation of trends that in turn could influence management efforts. In the eastern North Pacific, knowledge about right whale phenology is far inferior to that in the western North Atlantic, so the immediate need is for better information on the distribution and movements of the few remaining animals. Only after such information is available will it be possible to develop useful management strategies in U.S. and Canadian west coast waters. It goes without saying that the long-term survival of right whales in the Northern Hemisphere requires their continued full protection from whaling for the foreseeable future.

Numerous research and development projects have been identified within synoptic reviews of status and trends in North Atlantic right

TABLE 21.3. Reported North Pacific and North Atlantic whaling kills since the beginning of the worldwide moratorium on commercial whaling in 1986

Stock	1986	1987	1988	1989	1990	1991	1992	1993	1994	1995	1996	1997	1998	1999	2000
North Pacific															
Bowhead, Alaska	28	31	29	26	44	47[a]	50	52	46	57	45[a]	66	54	48[b]	54[b]
Bryde's, Japan	317	317	0	0	0	0	0	0	0	0	0	0	0	0	43
Minke, Japan	311	304	0	0	0	0	0	0	21	100	77	100	100	100	40
Minke, Korea	69	0	0	0	0	0	0	0	0	0	0	0	0	0	0
Minke, Alaska	0	0	0	2	0	0	0	0	0	0	0	0	0	0	0
Gray, Chukotka	171[c]	158	151[d]	180[d]	162	169	0	0	44	92[c]	43	79	122	126[e]	115
North Atlantic															
Bowhead, Canada	0	0	0	0	0	0	0	0	1	0	1	0	1	0	1
Fin, Iceland	76	80	68	68	0	0	0	0	0	0	0	0	0	0	0
Fin, W Greenland	9[f]	9	9	10	19	16	22	13	20[g]	12	19	12	11[f]	11	7
Sei, Iceland	40	20	10	0	0	0	0	0	0	0	0	0	0	0	0
Bryde's, Bequia	0	0	0	0	0	0	0	0	0	0	0	0	0	1	0
Minke, Norway	379	375	29	17	5	1	95	226	280	218	388	503	625	589	487
Minke, W Greenland	145	86	109	63	89	93	103	107	104	156	164	148	166	170	145
Minke, E Greenland	2	4	10	10	6	7	11	9	8	7	12	14	10	14	10
Humpback, Bequia	2	2	1	0	0	1	2	2	0	0	1	0	2	2	2
Humpback, Greenland	0	0	1	2	1	1	1	0	0	0	0	0	0	0	0

SOURCE: Data from the annual reports of the International Whaling Commission's Scientific Committee (IWC 1988, 1989, 1990, 1991, 1992, 1993, 1994, 1995, 1996, 1997, 1998, 1999, 2000, 2001a, 2002a), Rugh et al. (1999), and Angliss and Lodge (2002).

NOTE: Numbers include whales landed and struck-but-lost.

[a]Includes 1 taken in western Canada.
[b]Includes 1 taken in Chukotka.
[c]Includes 2 taken in Alaska.
[d]Includes 1 taken in Alaska.
[e]Includes 1 taken on the U.S. west coast.
[f]Includes 1 possibly misidentified sei whale.
[g]Mistakenly identified in IWC (1996) as sei whales, apparently a tabulation error.

whales (IWC 2001c; Reeves et al. 2001a). Among the high-priority items are the following:

- Find "missing" habitats, that is, where most of the population spends the winter, and the travel routes and destinations of animals that take their calves somewhere other than the Bay of Fundy or Scotian Shelf in summer.
- Develop and test "whale-safe" fishing gear.
- Develop ways of managing ship traffic that minimize risks of collisions with right whales.
- Develop ways of managing fishing activity that minimize risks of right whale entanglement.
- Determine the cause or causes of the reduced, or highly variable, reproductive success.
- Improve understanding of habitat requirements, both to remediate or mitigate human-caused degradation and to predict whale occurrence in time and space.

OTHER BALEEN WHALES

Eight other species of baleen whales are found in North American waters, seven in both the Atlantic and the Pacific (bowhead, blue, fin, sei, Bryde's, minke, and humpback) and one (gray) only in the Pacific, although it occurred in the North Atlantic historically (see Fig. 21.1). All of these species have been hunted commercially at some time in the past (Table 21.3). In the following abbreviated species accounts, the topics included are past and present status (Table 21.2), description, North Atlantic and North Pacific distribution, feeding habits, life history, and acoustic and other behavior.

BOWHEAD WHALE

The bowhead whale (*Balaena mysticetus*), a circumarctic balaenid, is the closest living relative of the right whales. Bowhead whales that stranded or became trapped in sea ice provided windfalls of meat, oil, and construction materials for early human inhabitants of high northern latitudes. Beginning at least 2000 years ago, Eskimos began hunting these whales from skin boats using hand-thrown harpoons and lances (Stoker and Krupnik 1993). North Atlantic commercial whaling on bowheads began in the early sixteenth century in Labrador (Barkham 1984) and in the early seventeenth century at Svalbard (Scoresby 1820; see review in Ross 1993). As the bowheads in the latter region became depleted, the Dutch, Danish, Basque, and British fleets extended their operations into Davis Strait and northward into Baffin Bay. American and Scottish whalers hunted bowheads intensively in northwestern Hudson Bay from 1860 to 1915. By the early twentieth century, all Atlantic bowhead populations had been severely depleted and commercial whaling ceased; total takes are estimated to have been at least 90,000 from the Svalbard stock, 29,000 from the Davis Strait stock, and perhaps 600 from the Hudson Bay stock (Ross 1993). In the North Pacific, whaleships entered the Sea of Okhotsk and the Bering Sea and began exploiting bowheads there in the 1840s, likely reducing the stocks by 60% within two decades (Braham 1984b; Bockstoce and Burns 1993). By the early twentieth century, when commercial whaling ended in the western Arctic, about 22,000 bowheads had been taken from the Bering/Chukchi/Beaufort Seas stock (Fig. 21.14). Whaling by Eskimos for subsistence has continued in the western Arctic, where the annual take since 1990 has ranged from 44 to 66 whales (including struck and lost) (Table 21.3). Inuit in Canada's eastern Arctic also have resumed a hunt for bowheads (<1 whale/year thus far), a right entrenched in their land-claim agreement. Quotas for hunting in Alaska are decided under a cooperative agreement between NMFS and the Alaska Eskimo Whaling Commission, which incorporates advice from the IWC (Huntington 1992), whereas in Canada (a non-IWC member), bowhead whaling is "co-managed" by the local resource agencies and the federal Department of Fisheries and Oceans (DFO 1999). The total number of bowheads around the Arctic prior to whaling was likely on the order of 50,000–60,000 (Woodby and Botkin 1993); it may now be no more than about 9000 (Zeh et al. 1993, 1995; DFO 1999) (Table 21.2). The population off Alaska is thought to be increasing at about 3% per year despite the continuing harvest (Raftery et al. 1995).

In appearance, bowheads closely resemble right whales (Reeves and Leatherwood 1985; Reeves et al. 2002) (Fig. 21.1B). However, they

FIGURE 21.14. A “forest” of baleen (whalebone) at the Arctic Oil Works yard in New Bedford, Massachusetts, during the heyday of Yankee whaling for bowhead whales in the western Arctic. The baleen was scraped, washed, dried, and bundled while at sea, and further cleaning was undertaken in the Oil Works yard to combat foul odors. The plates were then sorted according to their length—6 ft or longer was “size bone,” and shorter was “undersize bone.” Baleen from right whales was kept separate, as it was coarser and less pliable than the “arctic bone” from bowheads (Bockstoce, 1986). SOURCE: Photo courtesy of the New Bedford Whaling Museum.

lack the callosities on the head. Most individuals have a white chin, and some also have an area of light gray to white around the caudal peduncle, just anterior to the flukes. In addition, bowheads have a more prominent crest on top of the head, at the site of the blowholes, thus the name "steepletop" sometimes used by American commercial whalers. Their baleen is longer than that of right whales, with a maximum length of 4.6 m. Although the maximum body length of 17 m is similar to the right whale's, a large bowhead generally yields more oil and baleen than a right whale of comparable size. The bowhead's extremely thick blubber layer provides efficient insulation, theoretically enough to maintain a 200°C temperature gradient between the skin and muscle and "good enough to enable it to swim in liquid oxygen" (Hokkanen 1990:469).

Bowhead whales occur in the Arctic and subarctic between about 55°N and 85°N (Moore and Reeves 1993), and they are closely associated with sea ice during most of the year (Moore and DeMaster 1997). Annual patterns of distribution and migration follow the advance and retreat of the sea ice. More detailed information on distribution and movements is available for the North Pacific, where current abundance is higher. The Bering/Chukchi/Beaufort (or western Arctic) stock winters in the northern Bering Sea in November–March, moves through the Chukchi Sea following leads in the ice in March–June, summers (mid-May–September) in the Beaufort Sea, and migrates back through the Chukchi Sea to the Bering Sea in October–November. The fall migration passes through more open water than the spring migration, and extends west along the coast of Chukotka.

Bowheads are skim feeders like right whales, and similarly possess long, narrow baleen plates with very fine fringing hairs (Nemoto 1970; Pivorunas 1976, 1979; Würsig et al. 1985; Lowry 1993). Their principal prey are copepods and euphausiids, 3–30 mm long, and a variety of other prey of the same size range are taken, including amphipods, mysids, other crustaceans, pteropods, and fishes (Lowry 1993). Feeding near the surface can be observed easily, but the whales typically feed well below the surface, sometimes apparently at the bottom, as evidenced by visible mud plumes (Würsig and Clark 1993). Feeding has been observed in spring, summer, and fall (Würsig and Clark 1993), and may also occur in winter, judging by stable isotope levels in baleen (Schell et al. 1989; Schell and Saupe 1993).

The bowhead's life history, like its anatomy, is adapted for survival in a frigid habitat with low energy availability (George et al. 1999; see review by Koski et al. 1993). Females mature at body lengths of 13–14 m and ages of perhaps 12–18 years (Givens et al. 1995). Conception occurs in late winter or early spring, although sexual activity is observed throughout the year (Würsig and Clark 1993). Calves, 4–4.5 m long, are born in April–early June, following a 13- to 14-month gestation. They grow only to 6–6.5 m by the end of summer, a significantly slower growth rate than right whales (Schell and Saupe 1993). Weaning is at 9–15 months and body lengths of 6.6–9.4 m. Growth after weaning also is slow (Koski et al. 1992). Bowheads may be the longest lived mammals. George et al. (1999) estimated ages of 42 bowheads from the Alaska hunt using the relative proportions of D and L enantiomers ("right-handed" and "left-handed" mirror-image isomers of the same molecule) of aspartic acid in the eye lens. Four animals, all males, were estimated to have been >100 years old, with the oldest estimated to be 211 years old ($SE = 35$). The age–length curves constructed from the data confirmed the existence of slow growth rates, and suggested an age at maturity of around 25 years. These remarkable results are supported by the fact that six different whales taken during 1981–2001 had ancient, traditional harpoon points (four stone, one metal, and one ivory with a metal tip) embedded in their tissues.

Vocalizations produced by bowheads include a variety of low-frequency sounds, with most energy below 400 Hz (Würsig and Clark 1993). Bowheads produce long, patterned sequences, which have been termed "songs." They also produce a variety of amplitude-modulated and pulsed calls up to about 5 kHz. A sharp, pulsed sound has been likened to a gunshot, and has also been heard from right whales. Not only is the function of this sound unknown, but it is also uncertain how and where in the animal it is produced.

BLUE WHALE

Blue whales (*Balaenoptera musculus*) occur from the polar ice zones to the tropics in all the world's oceans (Yochem and Leatherwood 1985; Rice 1998; Perry et al. 1999). Populations in both the North Atlantic and the North Pacific have experienced periods of heavy commercial exploitation. After the development of steam-powered catcher boats and cannon-fired explosive harpoons in the late nineteenth century, the blue whale, because of its huge size and large oil yield, became the whale of choice within the industry (Mackintosh 1965), and stocks were depleted rapidly. At least 11,000 blue whales were taken by commercial whalers in the North Atlantic from the late nineteenth century to mid-twentieth century before protection came in 1960 (Sigurjónsson and Gunnlaugsson 1990). In the North Pacific, about 9500 blue whales were taken between 1910 and 1965, including more than 1300 by factory ships off California and Baja California between 1913 and 1937 (Reeves et al. 1998).

Although the IWC recognizes a single stock of blue whales in the North Atlantic (Donovan 1991), there is considerable evidence in the patterns of whaling catches and sighting trends to suggest the existence of discrete "feeding stocks" (Sigurjónsson and Gunnlaugsson 1990). There may have been close to 15,000 blue whales in the North Atlantic prior to whaling (Klinowska 1991; Sigurjónsson 1995), whereas today their numbers are probably only in the hundreds or low thousands. No current abundance estimate is available for the ocean basin as a whole (Perry et al. 1999), although >320 different animals have been photographically identified in the Gulf of St. Lawrence since 1979 (Sears et al. 1987, 1990; Hammond et al. 1990; Reeves et al. 1998). The number of blue whales visiting Icelandic waters in summer was estimated to be increasing at about 5% per year from 1979 to 1988 (Sigurjónsson and Gunnlaugsson 1990).

In the North Pacific, there is evidence for as many as five blue whale stocks (Reeves et al. 1998; Stafford et al. 2001). Catches peaked in British Columbia in August (Gregr et al. 2000) and in Alaska in June (Brueggeman et al. 1985), which can be interpreted as either two populations feeding on opposite sides of the Gulf of Alaska or a single population moving from Alaska in early summer to British Columbia in late summer. Estimates of prewhaling abundance in the North Pacific are only in the mid-thousands (Rice 1974; Gambell 1976). There may be as many as about 3500 today in the eastern North Pacific, including Californian, Mexican, and offshore tropical waters (Reeves et al. 1998). Anthropogenic mortality is relatively low in both the North Pacific and the North Atlantic (Carretta et al. 2002; Waring et al. 2002).

Blue whales, reaching >30 m in length and >150,000 kg in weight, are the largest animals known to have lived on earth (Yochem and Leatherwood 1985). Northern Hemisphere adults are 22–27 m in length (Leatherwood et al. 1976; Yochem and Leatherwood 1985). Like other *Balaenoptera* spp., the body is long and streamlined, unlike the more robust balaenids and humpbacks (Fig. 21.1C). The body is blue-gray with distinctive light mottling except on the head, and with a lighter colored belly. A yellowish film of diatoms, often present on the ventral surfaces, prompted the whaler's name of "sulfurbottom" for the species. The rostrum of a blue whale is broad, U-shaped in dorsal profile, and flat on top. The flippers are long, smooth, and tapered, and, as in all balaenopterids, contain only four digits. The dorsal fin is relatively small and located on the posterior third of the back.

In the North Atlantic, blue whales most commonly occur from Nova Scotia north to Baffin Bay, in waters around Iceland and Svalbard, and in the Barents Sea (Jonsgård 1966; Sutcliffe and Brodie 1977; Yochem and Leatherwood 1985; Sigurjónsson and Gunnlaugsson 1990; Rice 1998). The population that summers along the north shore of the Gulf of St. Lawrence is relatively well studied (Sears et al. 1990). Blue whales occur only sporadically in the Gulf of Maine (Wenzel et al. 1988). One photoidentified animal was seen during August 1984 and 1985 in the Gulf of St. Lawrence, then off southwestern Greenland during August 1988 and 1989, and back in the St. Lawrence during July 1991, October 1992, and August 1994 (Sears and Larsen 2002). Sparse stranding and sighting data suggest that the historical range extended south to Florida, the Gulf of Mexico, and the Caribbean Sea in the western Atlantic and to the Cape Verde Islands off West Africa in the eastern Atlantic (Harmer 1923; Jonsgård 1966; Leatherwood et al. 1976; Yochem and Leatherwood 1985). Recent passive acoustic studies using Navy hydrophones found that blue whales are relatively abundant off the Newfoundland Grand Banks south to 35°N and west of the British Isles (Clark 1995). They also were detected throughout much of the deeper pelagic western North Atlantic. One individual tracked in waters around Bermuda in February and March 1993 traveled over 1700 miles in 43 days. Although acoustic detections in offshore waters peaked from September to February (Clark 1995), it is uncertain whether this pattern represents a seasonal trend in whale abundance or simply variation in acoustic activity by the whales.

In the eastern North Pacific, blue whales are observed from Costa Rica north to Alaska (Yochem and Leatherwood 1985; Rice 1998). Off western North America, they occupy a feeding range off California from June to November, and migrate south off Mexico and into the Gulf of California in winter and spring (Carretta et al. 2002). Blue whales occur throughout the year in the area of the Costa Rica Dome. Passive acoustic detection of blue whales is common across the North Pacific from Mexico to Hawaii to Wake Island and north to the Aleutians, but sightings are rare. The acoustic data show a consistent pattern of occurrence at high latitudes in the summer and fall and low latitudes in the winter and spring (Stafford et al. 1999, 2001; Watkins et al. 2000). Many blue whales may inhabit pelagic waters far from land throughout the year.

Feeding in blue whales, as in most other rorquals, is by "gulping" (Nemoto 1970; see Fig. 6 in Pivorunas 1979). The whale engulfs a large volume of seawater and prey, greatly distending the ventral grooves, before closing the mouth and squeezing the water out through the baleen (Pivorunas 1979; Lambertsen 1983). Blue whales have relatively short, coarsely fringed baleen, and they prey almost exclusively on euphausiids, which are shrimplike crustaceans commonly called krill (Nemoto 1970).

Female blue whales mature at 5–15 years of age (Yochem and Leatherwood 1985). Females are about 1/2 m longer than males at maturity, possibly an adaptation to facilitate carrying a large fetus or for storing extra lipid in the blubber for pregnancy and lactation (Boyd et al. 1999). The 7- to 8-m-long calves are born after a gestation of 10–12 months and are weaned at 6 or 7 months and about 16 m in length (Mizroch et al. 1984; Yochem and Leatherwood 1985). The usual interval between calves is 2 years. Calving appears to take place mainly during the winter, probably well offshore and in relatively low latitudes.

Blue whale sounds are elongate, low-frequency, downswept tonal calls lasting up to 36 sec. Their frequency range is 12–400 Hz, with dominant energy in the infrasonic range at 12–25 Hz. They occur in long repetitive sequences repeated at 3- to 10-min intervals, often over several hours (Watkins et al. 2000, and references therein).

FIN WHALE

Fin whales (*Balaenoptera physalus*) are broadly distributed throughout the world's oceans, in continental shelf and pelagic waters (Gambell 1985b; Rice 1998). They began to be targeted after the depletion of blue whale stocks early in the modern whaling era from the late nineteenth to the early twentieth centuries (Tønnessen and Johnsen 1982). Tens of thousand of fin whales were taken during the twentieth century in the North Atlantic and the North Pacific. Many of these were delivered to Pacific shore stations in Alaska (Reeves et al. 1985), British Columbia (Gregr et al. 2000), and California (Clapham et al. 1997) or Atlantic shore stations in Newfoundland and Nova Scotia (Mitchell 1974). Shore whaling in Canada continued until the early 1970s. At its annual meeting in July 1982, the International Whaling Commission approved, by a vote of 25 to 7 with 5 abstentions, a measure stating that "catch limits for the killing for commercial purposes of whales from all stocks for the 1986 coastal and the 1985/1986 pelagic seasons and thereafter shall be zero" (IWC 1983; Gambell 1999). This effectively created a worldwide moratorium on commercial whaling from 1986 onward. Legal whaling since then has been conducted only under (1) the exception for "aboriginal subsistence" whaling (see Reeves 2002b), (2) scientific research permits, or (3) objection (nations that formally object to specific IWC regulations are not bound by them). Iceland took 292 fin whales from 1986 to 1989 under a research permit (see Table 21.3). The "aboriginal subsistence" hunt in West Greenland has an annual quota of 19 fin whales (IWC 1998).

There are no reliable estimates of the abundance of the entire fin whale populations in either the North Atlantic or the North Pacific, although there may be 50,000–60,000 in the North Atlantic and 14,000–19,000 in the North Pacific (Perry et al. 1999). A recent estimate of about 2800 in the U.S. Atlantic (Waring et al. 2002) was negatively biased because it lacked a correction for submerged animals that were missed during surveys. A more realistic estimate would be on the order of 5000–6000 (Cetacean and Turtle Assessment Program, University of Rhode Island [CETAP] 1982; Hain et al. 1992; Kenney et al. 1997). Recent estimates for portions of the North Pacific include about 2000 in the stock that feeds off California, Oregon, and Washington (Barlow and Taylor 2001); about 5000 in the central Bering Sea (Moore et al. 2000); and at least 150 in the Gulf of California (Tershy et al. 1990). Some human-caused mortality is known to occur in North American waters, mainly from ship collisions and entanglement in fishing gear, but it is not believed to be great enough to affect recovery (Carretta et al. 2002; Waring et al. 2002).

The fin whale is the second-largest species of whale, with adults reaching body lengths of 17–24 m (Gambell 1985a; Reeves et al. 2002). Their bodies are sleek and streamlined, and the flattened rostrum is V-shaped, with one central ridge (Fig. 21.1D). There is a distinct ridge along the back from the dorsal fin to the tail, and the dorsal fin is usually prominent, falcate, and pointed. The color ranges from gray to brownish, with a much lighter belly. The head is asymmetrically colored: The lower jaw is white on the right and dark on the left. There is a pale, V-shaped chevron on the back, pointing forward, as well as light swirls on the sides above the flippers, especially the right.

The overall range of fin whales in the North Atlantic extends from the Gulf of Mexico/Caribbean Sea and the Mediterranean Sea north to Greenland, Iceland, and Norway (Jonsgård 1966; Gambell 1985a). Seven stocks have been defined (Donovan 1991; Bérubé et al. 1998): Nova Scotia, Newfoundland/Labrador, West Greenland, East Greenland/Iceland, and three in the eastern North Atlantic. Along the North American coast, fin whales are the most commonly sighted large whale in continental shelf waters from Virginia north to eastern Canada (Sergeant 1977; Sutcliffe and Brodie 1977; Hain et al. 1992; Waring et al. 2002). There have been scattered strandings and sightings in the southeastern United States, mainly from late fall through mid-spring (Schmidly 1981; Waring et al. 2002). Fin whale abundance in U.S. Atlantic shelf waters declines in winter to about 20–25% of spring–fall levels (Hain et al. 1992). Fin whale calls were by far the most abundant cetacean vocalizations detected throughout the deep North Atlantic, with a peak in winter. The acoustic data showed a definite southward movement of fin whales to south of Bermuda and into the West Indies during fall and a northward trend in spring, mainly toward Newfoundland and Labrador.

In the North Pacific, fin whales are widespread from about 20°N to the Bering Sea (Gambell 1985a; Rice 1998). The IWC has defined a single widespread fin whale stock in the North Pacific, with a second smaller stock in the East China Sea (Donovan 1991). However, there is some evidence of finer scale stock structuring, perhaps related to matrilineal habitat fidelity as in humpback whales (Clapham 1996). For management purposes, NMFS has defined separate stocks for Alaska, Hawaii, and California/Oregon/Washington (Carretta et al. 2002). Fin whales occur throughout the year off California (Barlow 1995; Forney et al. 1995) and in the Gulf of California (Tershy et al. 1993; Silber et al. 1994). Fin whales also have been observed in summer in the central Bering Sea (Moore et al. 2000). Wide-area passive acoustic monitoring has revealed fin whale calls off Hawaii in winter, off the U.S. Pacific coast from late summer through winter, around the Aleutian Islands mainly in late spring and summer, but also occasionally in fall and winter, and in the northwestern Pacific near the Emperor Seamounts throughout the year, with a peak in fall (Moore et al. 1998; McDonald and Fox 1999; Watkins et al. 2000).

Like blue whales, fin whales are gulp feeders (Nemoto 1970; Pivorunas 1979). They prey on a variety of small, schooling prey, including many small fishes, squids, and crustaceans such as krill and copepods (review in Kenney et al. 1985; Gambell 1985a). Off the U.S. Atlantic coast, fin whales are frequently observed in large feeding aggregations with humpback whales, minke whales, and Atlantic white-sided dolphins (CETAP 1982).

Female fin whales mature at 6–10 or more years of age and about 17 m in total length (Lockyer 1972; Gambell 1985a). Peak calving is in winter, with some variability (Mitchell 1974; Haug 1981; Hain et al. 1992). Gestation probably lasts 10–12 months, and weaning occurs at 6–11 months of age (Best 1966; Haug 1981; Gambell 1985a). Calves are about 6 m long at birth and 12 m at weaning. The interval between calves is 2–3 years (Christensen et al. 1992; Agler et al. 1993).

The most typical fin whale sound is a 20-Hz infrasonic pulse (actually a frequency-modulated sweep from about 23 to 18 Hz) with a duration of about 1 sec and a source level of about 170 dB re 1 μPa (micropascal) at 1 m (Charif et al. 2002). These calls are produced throughout the year, singly or in groups of up to five. However, they are produced in long, patterned series during the reproductive season from late fall to early spring, which indicates that they could be displays associated with breeding, similar to humpback whale songs (Watkins et al. 1987). Fin whales produce a variety of other sounds, with a frequency range up to 750 Hz.

SEI WHALE

Sei whales (*Balaenoptera borealis*) occur in all oceans of the world, but they tend to move less toward the poles and tropics than the other rorquals (Gambell 1985b; Horwood 1987; Rice 1998). The common name is from the Norwegian *sejhval*, referring to the fact that sei whales arrived in local waters at the same time as the *seje* (pollack, *Pollachius virens*) (Horwood 1987). In those cultures where whale meat is eaten, sei whales are frequently the preferred catch. In the North Atlantic, sei whales have been hunted in the waters off Norway, the British Isles, Iceland, Greenland, and Canada, with total takes of more than 14,000 whales (Horwood 1987). The total take in the North Pacific in the twentieth century was at least 74,000 (Horwood 1987). Seventy sei whales were taken by Iceland in 1986–1988 under a scientific research permit (Table 21.3). Since 1988, sei whales have not been hunted in either the North Atlantic or the North Pacific, and mortality from vessel collisions or entanglement in fishing gear appears to be infrequent (Carretta et al. 2002, Waring et al. 2002).

There are no reliable estimates of total abundance of sei whales in either the North Atlantic or the North Pacific (Perry et al. 1999). Sightings during recent assessment surveys in U.S. waters have been too infrequent to support meaningful estimates (Carretta et al. 2002; Waring et al. 2002). Mitchell and Chapman (1977) estimated the Nova Scotia stock at 1400–2200 whales, which is similar to the estimate of about 2200 for the U.S. Atlantic from the CETAP (1982) surveys if corrected for diving behavior using the correction factor derived for fin whales (Kenney et al. 1997).

Adult sei whales are generally 12–17 m long, with a very sleek appearance (Fig. 21.1E). They are dark gray or brown to almost black, with a lighter belly (Leatherwood et al. 1976; Gambell 1985b; Horwood 1987; Reeves et al. 2002). They often have pale mottling or scars on the body. The rostrum is sharply pointed with a single longitudinal ridge, and curves noticeably downward toward the sides and tip. The dorsal fin is erect and falcate, and meets the back at a sharp, almost square, angle. An impediment to understanding the distribution of sei whales is that they are similar in appearance to Bryde's whales. The two species are difficult to differentiate at sea and often are combined in sighting data. Sei whales also are known for their sporadic occurrence in areas where they are not regularly seen (Gambell 1985b; Horwood 1987; P. M. Payne et al. 1990; Schilling et al. 1992; Clapham et al. 1997).

The IWC recognizes three separate sei whale stocks in the North Atlantic—Nova Scotia, Iceland–Denmark Strait, and northeastern Atlantic (Donovan 1991). In the western North Atlantic, sei whales occur from Georges Bank to Davis Strait, with perhaps separate stocks off the U.S./Nova Scotia and Newfoundland/Labrador coasts (Mitchell 1974; Mitchell and Chapman 1977; Gambell 1985b; Horwood 1987). Most sightings are along the continental shelf edge and slope (Mitchell 1975; Martin 1983; Hain et al. 1985). The winter range is poorly known, but there are scattered sighting and stranding records from the southeastern United States, Gulf of Mexico, Bay of Campeche, and northern Caribbean Sea (Leatherwood et al. 1976; Mead 1977; Schmidly 1981; Gambell 1985b; Rice 1998). The Nova Scotia stock apparently moves from spring feeding grounds on or near Georges Bank, especially along the southern edge and Northeast Peak, to the Scotian Shelf in June and July, eastward to perhaps Newfoundland and the Grand Banks in late summer, back to the Scotian Shelf in fall, and offshore and south in winter (Mitchell 1975; Mitchell and Chapman 1977; CETAP 1982).

In the North Pacific, the IWC recognizes a single sei whale stock (Donovan 1991), but there may be separate eastern and western (Horwood 1987) or eastern, central, and western stocks (Masaki 1977). The species seems to be broadly distributed across the temperate North Pacific north of 40°N and south of the Aleutians, but sei whales generally are rare near continental coasts and in marginal seas, gulfs, and bays (Horwood 1987). Japanese whaling records suggest that sei whales occur almost continuously across the North Pacific basin between 45°N and 55°N (Masaki 1977). A generalized north–south seasonal pattern can be discerned, with most records north of 40°N in summer and south of 35°N in winter (Horwood 1987).

Sei whales are flexible in their feeding behavior, sometimes gulping like blue, fin, or humpback whales and sometimes skimming like right or bowhead whales (Ingebrigtsen 1929; Nemoto 1970; Pivorunas 1979; Watkins and Schevill 1979). Their principal prey are copepods and euphausiids (Kawamura 1974; Mitchell 1975; Jonsgård and Darling 1977; Mitchell et al. 1986; Christensen et al. 1992; Schilling et al. 1992), although North Pacific sei whales are more euryphagous. Their diet includes larger prey, such as cephalopods, sardines (*Sardinops*), anchovies (*Engraulis*), and other schooling fishes up to the size of mackerels (*Scomber*) (Gill and Hughes 1971; Kawamura 1982; Gambell 1985b; Clapham et al. 1997).

Sei whales typically mature at 5–15 years of age (peak at 8–10 years) and about 13 m long, and follow a 2- to 3-year reproductive cycle (Mitchell and Chapman 1977; Rice 1977; Lockyer and Martin 1983; Gambell 1985b; Horwood 1987; Boyd et al. 1999). Calves are 4.4–4.5 m long at birth, and are born mainly during the winter after a gestation of 10–12 months. Weaning takes place at an age of 6–9 months, when body length is about 9 m.

Sei whale vocalizations have been recorded only on a few occasions (Thompson et al. 1979; Knowlton et al. 1991). They consist of paired sequences (0.5–0.8 sec, separated by 0.4–1.0 sec) of 7–20 short (30–40 msec) frequency-modulated sweeps between 1.5 and 3.5 kHz.

BRYDE'S WHALE

Bryde's whales occur in tropical and subtropical waters worldwide (Rice 1998). The common name honors Johan Bryde, a Norwegian consul who built the first whaling stations in South Africa (Cummings 1985). Two forms of Bryde's whale were long known to exist in tropical and warm temperate regions (Kawamura and Satake 1976; Best 1977), but only in recent years has it become clear that more than one species is involved (e.g., Wada and Numachi 1991; Pastene et al. 1997). Preliminary mitochondrial DNA analyses indicate that the "standard-form" Bryde's whale is more closely related to the sei whale than to the so-called pygmy Bryde's whale (Dizon et al. 1997). According to Rice (1998), the appropriate names for the two species are *B. brydei* for the larger form and *B. edeni* for the smaller form, the latter restricted to coastal waters of the tropical Indian and southwestern Pacific Oceans.

The IWC's Working Group on Nomenclature acknowledged that there was more than one species, but decided to continue using *B. edeni* for all Bryde's whales until remaining nomenclatural questions have been resolved (IWC 2001a). Here, we treat all Bryde's whales in North American waters as *B. brydei,* following Rice (1998).

Bryde's whales have never been exploited heavily, although some were mistakenly listed as "sei whales" in whaling records. Commercial removals under IWC quotas took place in the western North Pacific beginning in the mid-1970s. After the commercial whaling moratorium began in 1986, Japan lodged an objection and took 317 Bryde's whales in both 1986 and 1987. The objection was withdrawn after the 1987 season. In 2000, Japan began whaling for Bryde's whales in the North Pacific under a national scientific research permit and took 43 whales (IWC 2002a). Artisanal whalers in the West Indies very occasionally take a Bryde's whale (e.g., in 1999 at Bequia) (Table 21.3).

Best (1975) estimated that there were 20,000–30,000 Bryde's whales in the entire North Pacific, and there are a number of highly variable estimates for the western North Pacific. Wade and Gerrodette (1993) estimated that there were about 13,000 in the eastern tropical North Pacific, and Tershy et al. (1990) identified 160 different individuals in the Gulf of California. Bryde's whales occur in Hawaiian waters, but sightings are rare (Carretta et al. 2002). Recent estimates of Bryde's whale abundance in the North Atlantic are completely lacking. Three sightings during a survey in the Gulf of Mexico in 1991 resulted in an imprecise estimate of about 200 (Waring et al. 2002). No Bryde's whales have been reported as having died in recent years from ship strikes or fishery by-catch in U.S. waters (Waring et al. 2002, Carretta et al. 2002).

Adult Bryde's whales are 13–15 m long and closely resemble sei whales (Reeves et al. 2002) (Fig. 21.1F). They are generally dark gray with a pale ventral surface. The key differentiating characteristic of a Bryde's whale is the presence of three longitudinal ridges on the rostrum—a central ridge and one lateral ridge on each side. The dorsal fin is tall, pointed, and strongly falcate.

Bryde's whales apparently are widely distributed in pelagic waters between 40°N and 40°S (Mead 1977; Kawamura and Satake 1976; Rice 1998). Their northern distributional boundary in the eastern North Pacific is approximately the California–Mexico border. In the western North Atlantic, it is Virginia, based largely on stranding records. Although the migratory behavior of Bryde's whales is poorly known, animals tagged in equatorial waters of the western Pacific were later killed on feeding grounds east of southern Japan at 25–30°N (Ohsumi 1978, 1979, 1980).

Bryde's whale baleen is relatively coarse (Kawamura and Satake 1976). Reported prey species include krill, other crustaceans, and a variety of pelagic fishes and squids (Best 1977; Kawamura 1980).

Bryde's whales may reach maturity at 9–13 years of age and 12–13 m in length (Best 1977). Calves are about 4 m long at birth. Gestation period and birth interval are poorly known, but probably are about 1 and 2 years, respectively.

Bryde's whales produce low-frequency moans, with most energy between about 90 and 500 Hz (Cummings et al. 1986; Edds et al. 1993). Frequency modulations within a call vary from 10 to 65 Hz. Sounds recorded from a live-stranded juvenile that was held in a pool at Sea World of Florida for several weeks, then released back into the Gulf of Mexico, included pulsed moans in the 200- to 900-Hz range (Edds et al. 1993).

MINKE WHALE

The minke whales are the smallest of the balaenopterids. Consequently, they were of little commercial significance until the 1970s, by which time whaling for the larger, more valuable species was either prohibited or greatly scaled back. The common name comes from a Norwegian whale spotter named Meincke, who was derided by other whalemen for having mistaken a minke for a blue whale, thus the smaller species became known as "Meincke's" whale (Stewart and Leatherwood 1985; Horwood 1990). Although two morphs, one with and one without white flipper patches, had long been recognized, the differences were not clearly specified until the 1980s (Best 1985), and it was only in the 1990s that the species-level distinction became widely accepted (Rice 1998; Nikaido et al. 1999, 2001; Perrin and Brownell 2002). The "white-shouldered" northern species (*B. acutorostrata*) is found throughout the North Atlantic (nominally *B. a. acutorostrata*) and the North Pacific (*B. a. scammoni*), whereas the "dark-shouldered" Antarctic species (*B. bonaerensis*) is widely distributed in the Southern Ocean. Still unresolved is the taxonomic status of a "dwarf" form of white-shouldered minke whale found in the Southern Hemisphere (Best 1985; Arnold et al. 1987). It is provisionally viewed as an unnamed subspecies of *B. acutorostrata* because it is genetically and morphologically closer to the northern species (Rice 1998).

Northern minke whales were hunted on a relatively small scale by shore-based whalers in both the North Atlantic and the North Pacific (Stewart and Leatherwood 1985; Horwood 1990). Minke whaling in the fjords of Norway dates to at least the Middle Ages. During the twentieth century, >100,000 minke whales were taken from the North Atlantic stocks, most of them by Norwegian pelagic whalers. In contrast, only somewhat greater than 1000 were taken by shore whalers along the Canadian east coast and nearly 8000 off West Greenland. In the North Pacific, substantial catches were made only in the west, with more than 31,000 taken primarily by Japan and South Korea. Since the moratorium began in 1986, minke whaling has continued in both the North Pacific and the North Atlantic (Table 21.3). Japan took >300 in the western North Pacific in 1986 and again in 1987 under objection, discontinued the hunt from 1988 to 1993, and then began taking 100 annually under a national scientific research permit in 1994. South Korea took 69 under a research permit in 1986. In the North Atlantic, Norway took >300 minke whales in both 1986 and 1987 under objection, then smaller numbers from 1988 to 1994 under a research permit. Norway resumed commercial whaling under objection in 1993, and catches increased to several hundred per year in the late 1990s. Greenlanders continue to hunt minke whales under the exemption for aboriginal subsistence whaling, taking >150/year.

Minke whales are difficult to survey because they are inconspicuous sighting targets and tend to be solitary; they can be detected consistently only in calm sea conditions. Most abundance estimates are imprecise and negatively biased (Kenney et al. 1997). Estimates do not exist for the entire North Atlantic or the North Pacific, but considerable effort has been invested in surveys in the whaling areas of the central and eastern North Atlantic and the western Pacific (Horwood 1990). The Atlantic stocks (not including Canadian and U.S. waters) total about 150,000; the western Pacific stocks (including the Okhotsk Sea), about 25,000 (Gambell 1999). Waring et al. (2000) estimated about 4000 between Virginia and the Gulf of St. Lawrence, and Kingsley and Reeves (1998) estimated about 1000 in the Gulf. Moore et al. (2000) estimated about 1000 in the central Bering Sea; Carretta et al. (2002), about 650 off California/Oregon/Washington. As mentioned above, these may be underestimates. Minke whales die relatively often in fishing traps or nets, and they are at least occasionally struck by vessels. The levels of such mortality in U.S. waters are presumed to be low, however, with little effect on the populations (Angliss and Lodge 2002; Carretta et al. 2002; Waring et al. 2002).

Minke whales grow to about 10 m and perhaps 8000 kg (an 8.3-m female weighed 7650 kg; Tomilin 1967). They are generally 6–9 m long, and are somewhat more robustly shaped than the larger *Balaenoptera* spp. (Stewart and Leatherwood 1985; Reeves et al. 2002) (Fig. 21.1G). The body is dark gray to black above and pale below, and often lighter gray on the sides behind the head. The head is relatively short and sharply pointed. The flippers are narrow and pointed, with a diagnostic white band across the middle. The dorsal fin is tall, falcate, and located about two thirds of the way back on the body.

Four North Atlantic stocks are recognized by the IWC (Donovan 1991): Canadian east coast (including U.S. waters), West Greenland, central, and northeastern. Minke whales are widespread and can be encountered from the Gulf of Mexico and Caribbean Sea, Azores, and Mediterranean Sea north to the Arctic (Stewart and Leatherwood 1985;

Horwood 1990; Rice 1998). They can be present virtually anywhere on the continental shelf, including inshore bays and estuaries, except when the latter are ice covered. Minke whales apparently move offshore and southward in winter (Mitchell 1991; Mellinger et al. 2000). In the North Pacific, the IWC has defined three stocks (Donovan 1991)—Sea of Japan/East China Sea, western Pacific, and "remainder." The stock centered in the Sea of Japan (J stock) is genetically differentiated from that centered off the east coast of Japan and in the Okhotsk Sea (O stock) (Goto and Pastene 1997). According to IWC (2001a:41), the mitochondrial DNA differences between O and J stocks "are dramatic and arguably as high as anything observed between sympatric or partially sympatric stocks within the same hemisphere and ocean basin." Minke whales are present in the North Pacific from near the Equator to the Arctic (Horwood 1990). They are relatively common in the Bering Sea, the Chukchi Sea, and the Gulf of Alaska. Minke whales in the northern regions probably are migratory, whereas at least some of those along the U.S. west coast seem to be resident (Angliss and Lodge 2002; Forney et al. 2002).

Minke whales take a variety of prey, including copepods, krill, pteropods, squids, and many kinds of small and medium-sized fishes (review in Horwood 1990). In the northeastern North Atlantic, where stomach contents have been studied extensively, krill and herring (*Clupea harengus*) are the principal prey, followed by several gadoids, including cod (*Gadus morhua*), haddock (*Melanogrammus aeglefinus*), and pollack (*Pollachius virens*), and capelin (*Mallotus villosus*) (Folkow et al. 2000).

Minke whales mature at 5–10 years old and about 7 m long (Stewart and Leatherwood 1985; Horwood 1990). Births are concentrated in winter, and neonates are 2.4–2.8 m long. Gestation period is 10–11 months, and most prime-aged females probably give birth annually. Calves are weaned after 4–6 months. In some parts of their range, minke whales are regularly preyed on by killer whales (e.g., Wenzel and Sears 1988).

Minke whales produce a variety of sounds, including low-frequency grunts (80–140 Hz, 165–320 msec duration), frequency-modulated tonal sounds, which sweep down 10–70 Hz from 80–200 Hz in 100–600 msec; 100- to 200-Hz thumps (50–70 msec duration), which are sometimes produced in long "thump trains"; and 3- to $\geq$12-kHz pings or clicks (Schevill and Watkins 1972; Beamish and Mitchell 1973; Winn and Perkins 1976; Thompson et al. 1979; Thomson and Richardson 1995; Mellinger et al. 2000).

HUMPBACK WHALE

Humpback whales (*Megaptera novaeangliae*) occur in all of the world's oceans. They are strongly migratory, and are among the few baleen whale species for which good information is available on winter distribution (Kellogg 1929; Jonsgård 1966; Winn and Reichley 1985; Rice 1998). There is a long history of whaling for humpbacks in the North Atlantic, beginning in the 1600s in Bermuda. American open-boat whalers hunted humpbacks on the wintering grounds between sperm whaling seasons, and their technology and practices were adopted by local people on islands in the West Indies and Cape Verdes (Mitchell and Reeves 1983). Many thousands of humpbacks were taken by nineteenth-century whalers, and thousands more in the twentieth century by shore stations in southeastern Canada, West Greenland, Iceland, the Faroe Islands, the British Isles, and Norway (Smith and Reeves 2002; Reeves and Smith 2002). People from Bequia, a small island in St. Vincent and the Grenadines, continue to hunt humpbacks using open-boat, hand-harpoon techniques, but the annual take does not exceed two whales (Table 21.3).

In the North Pacific, large catches at shore stations appear to have depleted local humpback populations. In British Columbia, >5600 were killed, the vast majority between 1908 and 1917 (Gregr et al. 2000). More than 1300 were delivered to the station at Bay City, Washington, between 1911 and 1919; nearly 2500 more to the Bay City, Trinidad (California), and Moss Landing (California) stations between 1919 and 1926 (Scheffer and Slipp 1948; Clapham et al. 1997); and >1500 to stations at Akutan and Port Hobron, Alaska, between 1912 and 1939 (Reeves et al. 1985). Clapham et al. (1997) tabulated close to 5100 humpbacks taken along the west coast of North America, from Baja California to Alaska, in the 8-year period 1919–1926.

Recent mark–recapture estimates of 10,000–12,000 humpbacks in the North Atlantic basin were obtained using photographs of flukes and DNA fingerprints (Palsbøll et al. 1997; Smith et al. 1999; Stevick et al. 2001). The population in the Gulf of Maine appears to be increasing at about 6.5% annually (Barlow and Clapham 1997) and the aggregate population in the West Indies at about 3% (Stevick et al. 2001). There are few statistically reliable estimates for the regional feeding stocks along the east coast of North America, although the Gulf of Maine stock includes about 800–900 animals (Waring et al. 2002; Clapham et al. 2003). The total number of humpbacks in the North Pacific is approximately 6000–8000 (Perry et al. 1999). Current estimates of individual stocks are about 400 for the western stock, 4000 for the central stock, and 1000 for the eastern stock (Angliss and Lodge 2002; Carretta et al. 2002), although these are negatively biased by incomplete samples across the full range of each stock. An average of at least 9–10/year are killed in U.S. waters (6 Atlantic, 3–4 Pacific) by entanglements in a variety of fixed fishing gear or collisions with ships (Angliss and Lodge 2002; Carretta et al. 2002; Waring et al. 2002). Additional mortality in cod traps along the Newfoundland shore was a concern during the 1970s and 1980s, with around 50 entanglements and a 20% mortality rate each year (Lien 1994; Volgenau et al. 1995). Closure of the Newfoundland cod fishery since 1992 has greatly reduced the incidence of humpback entanglements.

Humpbacks reach body lengths of 11–16 m, with a record length of 18 m (Winn and Reichley 1985). They are more robust than the other balaenopterids, but less so than the balaenids (Fig. 21.1H). The color is generally black, with areas of white on the ventral surface. Their most conspicuous characteristics are the extremely long, frequently white flippers, about one third the total body length, and rounded knobs on the surface of the head. The flipper skeleton includes only four digits, as in *Balaenoptera*. The dorsal fin varies from low and rounded to erect and falcate, and is preceded by a rounded hump and followed by a series of dorsal crenulations along the tail stock. The flukes are deeply notched and irregularly serrate on the trailing edge, and the pigmentation pattern on their ventral surface (ranging from nearly all black to nearly all white) is useful for identification of individuals (Katona et al. 1979).

In the North Atlantic, humpbacks occur from the Caribbean Sea and Cape Verde Islands in the south to Greenland, Iceland, Svalbard, and the Barents Sea in the north (Jonsgård 1966; Winn et al. 1975; Winn and Winn 1978; Whitehead and Moore 1982; Martin et al. 1984; Winn and Reichley 1985; Katona and Beard 1990; Clapham et al. 1992, 1993a, 1993b; Clapham 1996; Palsbøll et al. 1997; Rice 1998; Stevick et al. 1998; Smith et al. 1999). There is a strongly seasonal component to their occurrence in any particular habitat, and these whales undertake some of the longest migrations known for any mammal (Kellogg 1929). The vast majority of sightings are in nearshore and continental shelf waters, but the whales also migrate across deep oceanic regions. During the winter, humpbacks from all North Atlantic feeding grounds migrate south to calving and breeding grounds on shallow banks in the West Indies/Caribbean region or around the Cape Verde Islands (IWC 2002b). The peak season in these areas is January–March, with some animals arriving as early as December and a few stragglers not leaving until June (Reeves et al. 2001b). The habitat requirements for humpbacks in winter appear to be determined by optimal calving and breeding conditions: shallow water with little bottom relief in the temperature range of 24–28°C, often protected from heavy seas by the presence of reefs (Winn et al. 1975; Balcomb and Nichols 1978; Whitehead and Moore 1982; Mattila et al. 1989, 1994; Mignucci-Giannoni 1998).

Not all humpback whales migrate every year to the tropical wintering grounds. Some individuals remain at high latitudes, but they likely represent a small fraction of the total population (Whitehead 1982; Swingle et al. 1993; Clapham et al. 1993a). North Atlantic humpbacks occur from spring through fall on feeding grounds in continental shelf waters from south of New England north to Davis Strait and from the British Isles north to Iceland, Svalbard, and Finnmark, Norway (Martin et al. 1984; Katona and Beard 1990; Sigurjónsson and Gunnlaugsson 1990; Clapham et al. 1992; Clapham 1996; Palsbøll et al. 1997; Rice 1998; Stevick et al. 1998). At least six separate feeding stocks are recognized—Gulf of Maine/Nova Scotia, Gulf of St. Lawrence, Newfoundland/Labrador, West Greenland, Iceland/Denmark Strait, and Norway. There appears to be little exchange between these feeding areas (Palsbøll et al. 2001).

Stock structure in the North Pacific is extremely complex and somewhat different than in the North Atlantic, as there are three well-known calving areas: Mexico, Hawaii, and Japan (Baker et al. 1998; Calambokidis et al. 2001). Eastern North Pacific humpbacks winter mainly in Mexican and Central American waters and feed off the North American coast from California north to Alaska. The animals that winter in the Hawaiian Islands summer mainly in Alaska and British Columbia. Finally, the Asian stock winters around the Ryukyu (Okinawa) and Bonin (Ogasawara) Islands south of Japan, with movements to feeding areas all the way east to British Columbia and the Gulf of Alaska. Matrilineal fidelity to feeding grounds appears to be important in structuring humpback populations (Baker et al. 1994), and indeed, strong fidelity to feeding areas has been observed in the North Pacific (Calambokidis et al. 2001). There also is finer structuring within these "stocks." For example, Calambokidis et al. (1996) reported only limited exchange between California/Oregon/Washington and British Columbia. Some exchange occurs, however, between the three main "stocks" (Darling and McSweeney 1985; Baker et al. 1986; Darling and Cerchio 1993; Darling et al. 1996; Calambokidis et al. 2001). The feeding group of about 600 whales that ranges between southern California and northern Washington during summer and autumn and migrates to Baja California, mainland Mexico, and Costa Rica for the winter appears to have a lower reproductive rate than other groups of humpbacks (Steiger and Calambokidis 2000).

Humpbacks are gulpers like the other rorquals (Nemoto 1970; Pivorunas 1979), but they exhibit a wider variety of feeding behaviors (Ingebrigtsen 1929; Jurasz and Jurasz 1979; Hain et al. 1982; 1995; Weinrich et al. 1992; Swingle et al. 1993). They may lunge dramatically with the mouth open, or surface open-mouthed very slowly and smoothly. They also employ bubbles in feeding—either columns of large bubbles in lines or partial or complete circles ("bubble-nets") or large clouds of tiny bubbles ("bubble clouds"). Some whales add tail slaps or other vigorous splashing to their feeding repertoire. Humpbacks feed on a variety of small, schooling prey, including krill and fish (Watkins and Schevill 1979; Winn and Reichley 1985; Clapham 1996). Humpback feeding distributions change over time, tracking shifts in the distribution or relative abundance of prey species (Lien et al. 1979; Bryant et al. 1981; Whitehead and Lien 1982; P. M. Payne et al. 1986, 1990; Kenney et al. 1996; Weinrich et al. 1997).

Female humpbacks mature at 4–9 years of age (Clapham and Mayo 1987, 1990; Clapham 1992, 1996; Craig and Herman 2000). Calves are born from January to March after a gestation period of about 1 year (Rice 1967; Johnson and Wolman 1984) and are weaned by 1 year of age (Clapham 1992). Calving intervals are usually 2–3 years, although females occasionally give birth in successive years (Clapham and Mayo 1990; Clapham 1996; Steiger and Calambokidis 2000). Reproductive behavior in humpback whales includes aggressive encounters and vocal displays (Winn and Reichley 1985; Clapham 1996; Boyd et al. 1999; Tyack 1999). Males compete for access to receptive females by aggressive, sometimes violent, interactions (Tyack 1981; Clapham 1996; Tyack 1999). They also produce long, stereotyped "songs," involving multiple repeats of a limited number of phrases (Payne and McVay 1971; Winn and Winn 1978).

Humpback whales produce three classes of vocalization: calls made at high latitudes on the feeding grounds, social calls produced in the groups on the breeding grounds where males are aggressively competing for access to females, and songs produced by solitary males mainly during the breeding season, but also to some extent during migration. "Feeding" calls are 20 Hz to 2 kHz, <1 sec in duration. Social calls are from 50 Hz to >10 kHz, with dominant energy below 3 kHz. Components of the song range from <20 Hz to occasionally 8 kHz. Passive acoustic monitoring, using bottom-mounted hydrophones, has been used to infer movements (Charif et al. 2001).

GRAY WHALE

The gray whale (*Eschrichtius robustus*) is a Northern Hemisphere endemic with a primarily coastal distribution (Wolman 1985). It was extirpated in the North Atlantic in early historical times (review in Mead and Mitchell 1984), although it apparently survived in Icelandic waters into the early eighteenth century (Lindquist 2000). Dudley (1725) described the "scrag" whale of early New England whalers, which in fact was the basis for the first scientific description of the gray whale under the name *Balaena gibbosa,* although that name has been dropped in favor of the junior synonym (based on a North Atlantic gray whale, a subfossil skeleton from Sweden; Barnes and McLeod 1984), due to uncertainty about the identity of Dudley's whale. Gray whales survive in two geographically separate populations in the North Pacific. One of these, the eastern or "California" population, migrates annually between the warm lagoon and bay waters of Baja California, Mexico, and the productive shelf waters of the Bering and Chukchi Seas. The other, the western or "Korean" population, migrates between the South China Sea in winter and the Okhotsk Sea in summer.

Gray whales were hunted by the aboriginal inhabitants of North America and Asia for thousands of years (Krupnik 1984; O'Leary 1984; Wolman 1985). Shore whaling in Japan began by the sixteenth century, and possibly by the tenth century (Omura 1984). Commercial whaling using nets began in Japan in the seventeenth century; records are sketchy, but at least 2000 gray whales were taken by the end of the nineteenth century (Omura 1984). Gray whales in Baja California were first subjected to whaling in 1846, and American whalers had taken nearly 11,000 there by 1874 (Scammon 1874; Henderson 1984). The whales also were hunted from shore stations as they migrated along the coast of California, mainly from the 1850s to 1880s (Sayers 1984). Modern Norwegian pelagic whaling started in 1914 and took about 1000 gray whales between then and 1946 (Reeves 1984).

Beginning in 1969, a modern catcher boat replaced the traditional open boats used by the aboriginal gray whale hunters on Russia's Chukotka Peninsula. Whales were delivered to villages for use as human and animal food (including for ranched foxes; Freeman et al. 1998). Takes averaged more than 160 California gray whales per year through 1991 (Table 21.3), during which time the population increased. No catcher boat was available and whaling ceased in 1992 and 1993, but in 1994 it resumed, with the whalers using open boats and firearms. The IWC quota for all California gray whale takes for 1998–2002 is 620, with a maximum of 140 in any 1 year (IWC 1998). Most of the take is by Chukotkan whalers, as Alaskan Native hunters rarely shoot gray whales. However, the Makah tribe in Washington State recently established their right to take five gray whales per year for "subsistence," and they took their first whale in 1999. Since then, the Makah hunt has been suspended because of legal action brought by a number of environmental groups. Aboriginal groups on Vancouver Island, British Columbia, also have expressed interest in establishing their right to hunt whales under land-claim agreements, and gray whales would be the most likely targets (Reeves 2002b).

Abundance of the California stock was estimated at 26,600 in 1997–1998 (Hobbs and Rugh 1999; Rugh et al. 1999), and the population has been increasing in recent years at 2.4–3.3%/year (Buckland et al. 1993; Wade and DeMaster 1996; Rugh et al. 1999). The California gray whale stock was removed from the U.S. endangered species list under the ESA in 1994 (Perry et al. 1999; Rugh et al. 1999). The Korean stock was almost exterminated by 1933. It was thought to be extirpated until Brownell and Chun (1977) revealed that 67 were killed between 1948 and 1967 by whalers from the Republic of Korea (South Korea). Only about 100 gray whales are known to survive in a relict group that feeds in summer near Sakhalin Island in the Okhotsk Sea

(Weller et al. 1999). This population is classified by the International Union for the Conservation of Nature as critically endangered.

The gray whale was once thought to be the closest living relative of the extinct cetotheres (e.g., Gaskin 1982; Wolman 1985). However, phylogenies based on both morphological and molecular characters show it to be closer to modern mysticete families (Barnes and McLeod 1984; Nikaido et al. 2001). The gray whale's body is more robust than those of *Balaenoptera* spp., but less than the humpback's (Wolman 1985; Reeves et al. 2002) (Fig. 21.1). Adults range from 11 to 15 m in length, with females slightly larger than males; neonates are 4.6–4.9 m long. The color is gray with irregular mottling, and large numbers of barnacles and whale lice are typical. The head is short relative to body length. The rostrum is convex, but less so than in balaenids, and tapered anteriorly. There is no dorsal fin, but there is a rounded hump followed by a ridge with irregular crenulations. There are two to four ventral grooves about 2 m in length. The flippers are narrow, with only four digits as in the balaenopterids.

California gray whales feed in the northern Bering Sea, the Chukchi Sea, and the western Beaufort Sea and in coastal waters off southeastern Alaska, British Columbia, Washington, Oregon, and California (Swartz 1986). Southbound migration is in October–December. The whales funnel through Unimak Pass in the eastern Aleutians and travel close to shore thereafter (Braham 1984a). The population segregates during migration—pregnant females first, followed by other adult females, immature females and adult males, and finally immature males. Most of the whales winter along the western side of the Baja California peninsula, but a few enter the Gulf of California. The northbound migration, also near shore, is in February–May. The order is newly pregnant females, other adult females, adult males, immatures, and then mothers and calves.

Gray whale baleen is short and coarse. These whales are mainly benthic feeders (Nemoto 1970; Nerini 1984). Their primary prey are benthic, infaunal gammaridean amphipods (Ampeliscidae), which live in mats of tubes in the sediment. A whale swims to the bottom and rolls to one side, placing the side of its head against the bottom, and sucks up a volume of sediment, water, and any prey included, probably using the muscular tongue in a piston-like fashion. A benthic-feeding gray whale leaves behind characteristic oval pits, which can be mapped using side-scan sonar (Johnson and Nelson 1984; Kvitek and Oliver 1986; Nelson and Johnson 1987). Gray whales also feed on pelagic prey in the water column, including krill, other crustaceans, small fishes, and squid. More than 80 species have been documented as gray whale prey (Jones and Swartz 2002).

Gray whales mature at 5–11 years, with a mean of about 8 years, and at a body length of 11–12 m (Wolman 1985). Calves are born between late December and early March (median late January) after a gestation of about 13 months (Rice 1983; Jones and Swartz 2002). Breeding therefore occurs from late November until early February, with conceptions occurring either on the wintering grounds or in the southern portions of the migration (Jones and Swartz 2002). Weaning occurs at about 7–8 months of age. The usual birth interval for prime-aged females is probably about 2 years. Calf production indices from 1994 to 2000, ranging between 1.1% and 5.8%, were positively correlated with the length of time that prime feeding areas in the Bering and Chukchi Seas were ice free during the previous year (Perryman et al. 2002). This was interpreted to mean that shortened feeding seasons caused pregnant females to be in suboptimal nutritive condition when they migrated south, an inference consistent with the findings of LeBoeuf et al. (2000) in relation to high observed mortality and low recruitment in 1999 (see IWC 2001a, 2002a).

Gray whale vocalizations include a variety of low-frequency, broad-band signals described as "rasps, croaks, snorts, moans, groans, grunts, pops, roars, quick series of clicks, belches, and metallic knocks and bongs" (Jones and Swartz 2002:531). The frequency range of these sounds is mainly from 100 Hz to about 4 kHz, occasionally up to 12 kHz. The most common sounds on the feeding and breeding grounds are bursts of pulsive metallic knocks, whereas migrating whales mainly produce tonal moans.

MANAGEMENT AND CONSERVATION OF THE OTHER BALEEN WHALES

Because of their migratory nature, and in view of the fact that much of their economic value has traditionally been realized through international trade, most baleen whale stocks need to be conserved and managed at an appropriate international scale. The most effective instrument in this regard has been the International Whaling Commission (IWC) (Gambell 1999; Donovan 2002). Although Mexico, the United States, and Canada were among the founding signatories of the 1946 convention establishing the IWC, Canada withdrew from membership in 1982 on the grounds that it no longer had an interest in commercial whaling (Lavigne et al. 1999). This withdrawal has acquired particular significance in recent years as Inuit in northern Canada have resumed their hunting of bowhead whales and indigenous people on Vancouver Island, British Columbia, have expressed their intention to resume whaling for gray and/or humpback whales (Reeves 2002b). From a strictly North American perspective, aboriginal whaling for "subsistence" is likely to be a more relevant, and contentious, issue than commercial whaling for the foreseeable future (MacLean et al. 2002). Beyond whaling, however, there are serious ongoing concerns about incidental mortality from ship strikes and entanglement in fishing gear [see the stock assessment reports (Angliss and Lodge 2002; Carretta et al. 2002; Waring et al. 2002); annual updates available on-line at http://www.nmfs.noaa.gov/prot_res/], as well as the potential effects of toxic chemicals, prey depletion, noise disturbance, and climate change (Evans and Raga 2002; Reeves 2002a). These issues apply most immediately and most critically to right whales, as discussed earlier in this chapter, but they are also relevant to other baleen whale populations.

ACKNOWLEDGMENTS

The North Atlantic Right Whale Consortium granted permission for the use of unpublished data.

LITERATURE CITED

Agler, B. A., R. L. Schooley, S. E. Frohock, S. K. Katona, and I. E. Seipt. 1993. Reproduction of photographically identified fin whales, *Balaenoptera physalus,* from the Gulf of Maine. Journal of Mammalogy 74:577–87.

Aguilar, A. 1986. A review of old Basque whaling and its effect on the right whales (*Eubalaena glacialis*) of the North Atlantic. Report of the International Whaling Commission (Special Issue) 10:191 99.

Allen, G. M. 1916. The whalebone whales of New England. Memoirs of the Boston Society of Natural History 8(2):107–322.

Allen, J. A. 1908. The North Atlantic right whale and its near allies. Bulletin of the American Museum of Natural History 24:277–329.

Andrews, R. C. 1908. Notes upon the external and internal anatomy of *Balaena glacialis* Bonn. Bulletin of the American Museum of Natural History 24:171–82.

Angliss, R. P., and K. L. Lodge. 2002. Alaska marine mammal stock assessments, 2002 (NOAA Technical Memorandum NMFS-AFSC-133). National Marine Fisheries Service, Seattle, WA.

Árnason, Ú., and P. B. Best. 1991. Phylogenetic relationships within the Mysticeti (whalebone whales) based upon studies of highly repetitive DNA in all extant species. Hereditas 114:263–69.

Árnason, Ú., and A. Gullberg. 1994. Relationship of baleen whales established by cytochrome *b* gene sequence comparison. Nature 367:726–28.

Árnason, Ú., and A. Gullberg. 1996. Cytochrome *b* nucleotide sequences and the identification of five primary lineages of extant cetaceans. Molecular Biology and Evolution 13:407–17.

Árnason, Ú., S. Grétarsdóttir, and B. Widegren. 1992. Mysticete (baleen whale) relationships based upon the sequence of the common cetacean DNA satellite. Molecular Biology and Evolution 9:1018–28.

Árnason, Ú., S. Grétarsdóttir, and B. Widegren. 1993. Cetacean mitochondrial DNA control region: Sequences of all extant baleen whales and two sperm whale species. Molecular Biology and Evolution 10:960–70.

Árnason, Ú., A. Gullberg, S. Grétarsdóttir, B. Ursing, and A. Janke. 2000. The mitochondrial genome of the sperm whale and a new molecular reference for estimating eutherian divergence dates. Journal of Molecular Evolution 50:569–78.

Arnold, P., H. Marsh, and G. Heinsohn. 1987. The occurrence of two forms of minke whales in east Australian waters with a description of external characters and skeleton of the diminutive or dwarf form. Scientific Reports of the Whales Research Institute (Tokyo) 38:1–46.

Bajpai, S., and P. D. Gingerich. 1998. A new Eocene archaeocete (Mammalia, Cetacea) from India and the time of origin of whales. Proceedings of the National Academy of Sciences of the USA 95:15464–68.

Baker, C. S., L. M. Herman, A. Perry, W. S. Lawton, J. M. Straley, A. A. Wolman, G. D. Kaufman, H. E. Winn, J. D. Hall, J. M. Reinke, and J. Östman. 1986. Migratory movement and population structure of humpback whales (*Megaptera novaeangliae*) in the central and eastern North Pacific. Marine Ecology Progress Series 31:105–19.

Baker, C. S., R. W. Slade, J. L. Bannister, R. B. Abernethy, M. T. Weinrich, J. Lien, J. Urban, P. Corkeron, J. Calambokidis, O. Vazquez, and S. R. Palumbi. 1994. The hierarchical structure of mitochondrial DNA gene flow among humpback whales worldwide. Molecular Ecology 3:313–27.

Baker, C. S., L. Medrano-Gonzalez, J. Calambokidis, A. Perry, F. Pichler, H. Rosenbaum, J. M. Straley, J. Urban-Ramirez, M. Yamaguchi, and O. von Ziegesar. 1998. Population structure of nuclear and mitochondrial DNA variation among humpback whales in the North Pacific. Molecular Ecology 7:695–707.

Balcomb, K. C., and G. Nichols. 1978. Western North Atlantic humpback whales. Report of the International Whaling Commission 28:159–64.

Ballance, L. 2002. North Pacific right whale survey 2002, NOAA ship McArthur, weekly report. http:// swfsc. nmfs. noaa. gov/prd/2002% 20Cruises/NorthPac/WeeklyReports/weekly8. pdf.

Bannister, J. L., L. A. Pastene, and S. R. Burnell. 1999. First record of movement of a southern right whale (*Eubalaena australis*) between warm water breeding grounds and the Antarctic Ocean, south of 60°S. Marine Mammal Science 15:1337–42.

Barkham, S. H. 1984. The Basque whaling establishments in Labrador, 1536–1632: A summary. Arctic 37:515–19.

Barlow, J. 1995. The abundance of cetaceans in California waters. Part 1. Ship surveys in summer and fall of 1991. Fishery Bulletin 93:1–14.

Barlow, J., and P. J. Clapham. 1997. A new birth-interval approach to estimating demographic parameters of humpback whales. Ecology 78:535–46.

Barlow, J., and B. L. Taylor. 2001. Estimates of large whale abundance off California, Oregon, Washington, and Baja California based on 1993 and 1996 ship surveys (Administrative Report LJ-01-03). Southwest Fisheries Science Center, La Jolla, CA.

Barnes, L. G., and S. A. McLeod. 1984. The fossil record and phyletic relationships of gray whales. Pages 3–32 *in* M. L. Jones, S. L. Swartz, and S. Leatherwood, eds. The gray whale, *Eschrichtius robustus*. Academic Press, Orlando, FL.

Baur, D. C., M. J. Bean, and M. L. Gosliner. 1999. The laws governing marine mammal conservation in the United States. Pages 48–86 *in* J. R. Twiss, Jr., and R. R. Reeves, eds. Conservation and management of marine mammals. Smithsonian Institution Press, Washington, DC.

Beamish, P., and E. Mitchell. 1973. Short pulse length audio frequency sounds recorded in the presence of a minke whale (*Balaenoptera acutorostrata*). Deep-Sea Research 20:375–86.

Beardsley, R. C., A. W. Epstein, C. Chen, K. F. Wishner, M. C. Macaulay, and R. D. Kenney. 1996. Spatial variability in zooplankton abundance near feeding right whales in the Great South Channel. Deep-Sea Research 43:1601–25.

Berta, A., and J. L. Sumich. 1999. Marine mammals: Evolutionary biology. Academic Press, San Diego, CA.

Bérubé, M., A. Aguilar, D. Dendanto, F. Larsen, G. Notarbartolo di Sciara, R. Sears, J. Sigurjónsson, J. Urbán-R., and P. J. Palsbøll. 1998. Population genetic structure of North Atlantic, Mediterranean, and Sea of Cortez fin whales, *Balaenoptera physalus* (Linnaeus, 1758): Analysis of mitochondrial and nuclear loci. Molecular Ecology 15:585–99.

Best, P. B. 1966. A case for prolonged lactation in the fin whale. Norsk Hvalfangst-tidende 55:118–22.

Best, P. B. 1970. Exploitation and recovery of right whales *Eubalaena australis* off the Cape Province, South Africa (Investigational Report No. 80). Division of Sea Fisheries, Department of Industries, Republic of South Africa.

Best, P. B. 1977. Two allopatric forms of Bryde's whale off South Africa. Report of the International Whaling Commission (Special Issue) 1:10–38.

Best, P. B. 1985. External characters of southern minke whales and the existence of a diminutive form. Scientific Reports of the Whales Research Institute (Tokyo) 36:1–33.

Best, P. B. 1987. Estimates of the landed catch of right (and other whalebone) whales in the American fishery, 1805–1909. Fishery Bulletin 85:403–18.

Best, P. B. 1988. Right whales (*Eubalaena australis*) at Tristan da Cunha—A clue to the 'non-recovery' of depleted stocks? Biological Conservation 46:23–51.

Best, P. B. 1990. Natural markings and their use in determining calving intervals in right whales off South Africa. South African Journal of Zoology 25:114–23.

Best, P. B. 1993. Increase rates in severely depleted stocks of baleen whales. ICES Journal of Marine Science 50:169–86.

Best, P. B. 1994. Seasonality of reproduction and the length of gestation in southern right whales *Eubalaena australis*. Journal of Zoology (London) 232:175–89.

Best, P. B. 1995. Whale watching in South Africa. The southern right whale. Mammal Research Institute, University of Pretoria, Pretoria, South Africa.

Best, P. B., and R. M. McCully. 1979. Zygomycosis (phycomycosis) in a right whale (*Eubalaena australis*). Journal of Comparative Pathology 89:341–48.

Best, P. B., and D. M. Schell. 1996. Stable isotopes in southern right whale (*Eubalaena australis*) baleen as indicators of seasonal movements, feeding and growth. Marine Biology 124:483–94.

Best, P. B., A. Brandão, and D. S. Butterworth. 2001. Demographic parameters of southern right whales off South Africa. Journal of Cetacean Research and Management (Special Issue) 2:161–69.

Bockstoce, J. R. 1986. Whales, ice, and men: The history of whaling in the western Arctic. University of Washington Press, Seattle.

Bockstoce, J. R., and J. J. Burns. 1993. Commercial whaling in the North Pacific sector. Pages 563–77 *in* J. J. Burns, J. J. Montague, and C. J. Cowles, eds. The bowhead whale (Special Publication No. 2). Society for Marine Mammalogy, Lawrence, KS.

Boyd, I. L., C. Lockyer, and H. D. Marsh. 1999. Reproduction in marine mammals. Pages 218–86 *in* J. E. Reynolds III and S. A. Rommel, eds. Biology of marine mammals. Smithsonian Institution Press, Washington, DC.

Braham, H. W. 1984a. Distribution and migration of gray whales in Alaska. Pages 249–66 *in* M. L. Jones, S. L. Swartz, and S. Leatherwood, eds. The gray whale *Eschrichtius robustus*. Academic Press, Orlando, FL.

Braham, H. 1984b. The status of endangered whales: An overview. Marine Fisheries Review 46(4):2–6.

Brandt, K. 1940. Whale oil: An economic analysis. Food Research Institute, Stanford University, Stanford, CA.

Britt, J. O., Jr., and E. B. Howard. 1983. Virus disease. Pages 47–67 *in* E. B. Howard, ed. Pathobiology of marine mammal diseases, Vol. I. CRC Press, Boca Raton, FL.

Brown, M. W., S. D. Kraus, and D. E. Gaskin. 1991. Reaction of North Atlantic right whales (*Eubalaena glacialis*) to skin biopsy sampling for genetic and pollutant analysis. Report of the International Whaling Commission (Special Issue) 13:81–89.

Brown, M. W., S. D. Kraus, D. E. Gaskin, and B. N. White. 1994. Sexual composition and analysis of reproductive females in the North Atlantic right whale (*Eubalaena glacialis*) population. Marine Mammal Science 10:253–65.

Brown, M. W., S. Brault, P. K. Hamilton, R. D. Kenney, A. R. Knowlton, M. K. Marx, C. A. Mayo, C. K. Slay, and S. D. Kraus. 2001. Sighting heterogeneity of right whales in the western North Atlantic: 1980–1992. Journal of Cetacean Research and Management (Special Issue) 2:245–50.

Brown, S. G. 1986. Twentieth-century records of right whales (*Eubalaena glacialis*) in the northeast Atlantic Ocean. Report of the International Whaling Commission (Special Issue) 10:121–27.

Brownell, R. L., Jr., and C.-I. Chun. 1977. Probable existence of the Korean stock of the gray whale (*Eschrichtius robustus*). Journal of Mammalogy 58:237–39.

Brownell, R. L., Jr., and K. Ralls. 1986. Potential for sperm competition in baleen whales. Report of the International Whaling Commission (Special Issue) 8:97–112.

Brownell, R. L., Jr., P. J. Clapham, T. Miyashita, and T. Kasuya. 2001. Conservation status of North Pacific right whales. Journal of Cetacean Research and Management (Special Issue) 2:269–86.

Brueggeman, J. J., T. C. Newby, and R. A. Grotefendt. 1985. Seasonal abundance, distribution and population characteristics of blue whales reported in the 1917 to 1939 catch records of two Alaska whaling stations. Report of the International Whaling Commission 35:405–11.

Brueggeman, J. J., T. Newby, and R. A. Grotefendt. 1986. Catch records of twenty North Pacific right whales from two Alaska whaling stations, 1917–1939. Arctic 39:43–46.

Bryant, P. J., G. Nichols, T. B. Bryant, and K. Miller. 1981. Krill availability and the distribution of humpback whales in southeastern Alaska. Journal of Mammalogy 62:427–30.

Buckland, S. T., J. M. Breiwick, K. L. Cattanach, and J. L. Laake. 1993. Estimated population size of the California gray whale. Marine Mammal Science 9:235–49.

Burnell, S. R. 2001. Aspects of the reproductive biology, movements and site fidelity of right whales off Australia. Journal of Cetacean Research and Management (Special Issue) 2:89–102.

Calambokidis, J., G. H. Steiger, J. R. Evenson, K. R. Flynn, K. C. Balcomb, D. E. Claridge, P. Bloedel, J. M. Straley, C. S. Baker, O. von Ziegesar, M. E. Dahlheim, J. M. Waite, J. D. Darling, G. Ellis, and G. A. Green. 1996 Interchange and isolation of humpback whales in California and other North Pacific feeding grounds. Marine Mammal Science 12:215–26.

Calambokidis, J., G. H. Steiger, J. M. Straley, L. M. Herman, S. Cerchio, D. R. Salden, J. Urbán R., J. K. Jacobsen, O. von Ziegesar, K. C. Balcomb, C. M. Gabriele, M. E. Dahlheim, S. Uchida, G. Ellis, Y. Miyamura, P. Ladrón de Guevara P., M. Yamaguchi, F. Sato, S. A. Mizroch, L. Schlender, K. Rasmussen, J. Barlow, and T. J. Quinn II. 2001. Movements and population structure of humpback whales in the North Pacific. Marine Mammal Science 17:769–94.

Carretta, J. V., M. S. Lynn, and C. A. LeDuc. 1994. Right whale (*Eubalaena glacialis*) sighting off San Clemente Island, California. Marine Mammal Science 10:101–5.

Carretta, J. V., M. M. Muto, J. Barlow, J. Baker, K. A. Forney, and M. Lowry. 2002. U.S. Pacific marine mammal stock assessments, 2002. (NOAA Technical Memorandum NMFS-SWFSC-346) National Marine Fisheries Service, La Jolla, CA.

Caswell, H., M. Fujiwara, and S. Brault. 1999. Declining survival probability threatens the North Atlantic right whale. Proceedings of the National Academy of Sciences of the USA 96:3308–13.

Cetacean and Turtle Assessment Program, University of Rhode Island. 1982. A characterization of marine mammals and turtles in the Mid- and North Atlantic areas of the U.S. outer continental shelf, final report (Contract AA551-CT8-48). Bureau of Land Management, Washington, DC.

Charif, R. A., P. J. Clapham, and C. W. Clark. 2001. Acoustic detections of singing humpback whales in deep waters off the British Isles. Marine Mammal Science 17:751–68.

Charif, R. A., D. K. Mellinger, K. J. Dunsmore, K. M. Fristrup, and C. W. Clark. 2002. Estimated source levels of fin whale (*Balaenoptera physalus*) vocalizations: Adjustments for surface interference. Marine Mammal Science 18:81–98.

Christensen, I., T. Haug, and N. Øien. 1992. A review of feeding and reproduction in large baleen whales (Mysticeti) and sperm whales *Physeter macrocephalus* in Norwegian and adjacent waters. Fauna Norvegica, Series *A* 13:39–48.

Clapham, P. J. 1992. Age at attainment of sexual maturity in humpback whales, *Megaptera novaeangliae*. Canadian Journal of Zoology 70:1470–72.

Clapham, P. J. 1996. The social and reproductive biology of humpback whales: An ecological perspective. Mammal Review 26:27–49.

Clapham, P. J. 2000. The humpback whale: Seasonal feeding and breeding in a baleen whale. Pages 173–96 *in* J. Mann, R. C. Connor, P. L. Tyack, and H. Whitehead, eds. Cetacean societies: Field studies of dolphins and whales. University of Chicago Press, Chicago.

Clapham, P. J., and C. A. Mayo. 1987. Reproduction and recruitment of individually identified humpback whales, *Megaptera novaeangliae*, observed in Massachusetts Bay, 1979–1985. Canadian Journal of Zoology 65:2853–63.

Clapham, P. J., and C. A. Mayo. 1990. Reproduction of humpback whales (*Megaptera novaeangliae*) in the Gulf of Maine. Report of the International Whaling Commission (Special Issue) 12:171–75.

Clapham, P. J., P. J. Palsbøll, D. K. Mattila, and O. Vásquez. 1992. Composition and dynamics of humpback whale competitive groups in the West Indies. Behaviour 122:182–94.

Clapham, P. J., L. S. Baraff, C. A. Carlson, M. A. Christian, D. K. Mattila, C. A. Mayo, M. A. Murphy, and S. Pittman. 1993a. Seasonal occurrence and annual return of humpback whales, *Megaptera novaeangliae*, in the southern Gulf of Maine. Canadian Journal of Zoology 71:440–43.

Clapham, P. J., D. K. Mattila, and P. J. Palsbøll. 1993b. High-latitude-area composition of humpback whale competitive groups in Samana Bay: Further evidence for panmixis in the North Atlantic population. Canadian Journal of Zoology 71:440–43.

Clapham, P. J., S. Leatherwood, I. Szczepaniak, and R. L. Brownell, Jr. 1997. Catches of humpback and other whales from shore stations at Moss Landing and Trinidad, California, 1919–1926. Marine Mammal Science 13:368–94.

Clapham, P. J., S. B. Young, and R. L. Brownell, Jr. 1999. Baleen whales: Conservation issues and the status of the most endangered populations. Mammal Review 29:35–60.

Clapham, P. J., J. Barlow, M. Bessinger, T. Cole, D. Mattila, R. Pace, D. Palka, J. Robbins, and R. Seton. 2003 Abundance and demographic parameters of humpback whales from the Gulf of Maine, and stock definition relative to the Scotian Shelf. Journal of Cetacean Research and Management. 5:13–22.

Clark, A. H. 1887. History and present condition of the fishery. Pages 3–218 *in* G. B. Goode, ed. The fisheries and fishery industries of the United States. Section 5: History and methods of the fisheries. Vol. 2, Part 15. The whale fishery. U.S. Government Printing Office, Washington, DC.

Clark, C. W. 1982. The acoustic repertoire of the southern right whale, a quantitative analysis. Animal Behavior 30:1060–71.

Clark, C. W. 1983. Acoustic communication and behavior of the southern right whale (*Eubalaena australis*). Pages 163–98 *in* R. Payne, ed. Communication and behavior of whales (AAAS Selected Symposium 76). Westview Press, Boulder, CO.

Clark, C. W. 1995. Annex M. Matters arising out of the discussion of blue whales; Annex M1. Application of the US Navy underwater hydrophone arrays for scientific research on whales. Report of the International Whaling Commission 45:210–12.

Colborn, K., G. Silber, and C. Slay. 1998. Avoiding collisions with right whales. Professional Mariner 35:24–26.

Collett, R. 1909. A few notes on the whale *Balaena glacialis* and its capture in recent years in the North Atlantic by Norwegian whalers. Proceedings of the Zoological Society of London 1909:91–98.

Connor, R. C. 2000. Group living in whales and dolphins. Pages 199–218 *in* J. Mann, R. C. Connor, P. L. Tyack, and H. Whitehead, eds. Cetacean societies: Field studies of dolphins and whales. University of Chicago Press, Chicago.

Cooke, J., V. J. Rowntree, and R. Payne. 2001. Estimates of demographic parameters for southern right whales (*Eubalaena australis*) observed off Península Valdés, Argentina. Journal of Cetacean Research and Management (Special Issue) 2:125–32.

Craig, A., and L. M. Herman. 2000. Habitat preferences of female humpback whales *Megaptera novaeangliae* in the Hawaiian Islands are associated with reproductive status. Marine Ecology Progress Series 193:209–16.

Cumbaa, S. L. 1986. Archaeological evidence of the 16th century Basque right whale fishery in Labrador. Report of the International Whaling Commission (Special Issue) 10:187–90.

Cummings, W. C. 1985. Bryde's whale *Balaenoptera edeni* Anderson, 1878. Pages 137–54 *in* S. H. Ridgway and R. Harrison, eds. Handbook of marine mammals. Vol. 3: The sirenians and baleen whales. Academic Press, London.

Cummings, W. C., P. O. Thompson, and S. J. Ha. 1986. Sounds from Bryde, *Balaenoptera edeni*, and finback, *B. physalus*, whales in the Gulf of California. Fishery Bulletin 84:359–70.

Dailey, M. D. 1985. Diseases of Mammalia: Cetacea. Pages 805–47 *in* O. Kinne, ed. Diseases of marine animals. Vol. 4, Part 2: Introduction, Reptilia, Aves, Mammalia. Biologische Anstalt Helgoland, Hamburg, Germany.

Dailey, M. D., and R. L. Brownell, Jr. 1972. A checklist of marine mammal parasites. Pages 528–89 *in* S. H. Ridgway, ed. Mammals in the sea: Biology and medicine. Charles C. Thomas, Springfield, IL.

Darling, J. D., and S. Cerchio. 1993. Movement of a humpback whale (*Megaptera novaeangliae*) between Japan and Hawaii. Marine Mammal Science 9:84–89.

Darling, J. D., and D. J. McSweeney. 1985. Observations on the migrations of North Pacific humpback whales (*Megaptera novaeangliae*). Canadian Journal of Zoology 63:308–14.

Darling, J. D., J. Calambokidis, K. C. Balcomb, P. Bloedel, K. Flynn, A. Mochizuki, K. Mori, F. Sato, and M. Yamaguchi. 1996. Movement of a humpback whale (*Megaptera novaeangliae*) from Japan to British Columbia and return. Marine Mammal Science 12:281–87.

Department of Fisheries and Oceans. 1999. Hudson Bay/Foxe Basin bowhead whale (Stock Status Report E5-52). Canadian Department of Fisheries and Oceans, Winnipeg, Manitoba, Canada.

Dierauf, L. A. 1990. Marine mammal parasitology. Pages 89–96 *in* L. A. Dierauf, ed. CRC handbook of marine mammal medicine: Health, disease, and rehabilitation. CRC Press, Boca Raton, FL.

Dizon, A., C. A. Lux, R. G. LeDuc, J. Urbán R., M. Henshaw, C. S. Baker, F. Cipriano, and R. L. Brownell, Jr. 1997. Molecular phylogeny of the Bryde's whale/sei whale complex: Separate species status for the pygmy Bryde's form? Report of the International Whaling Commission 47:398.

Donovan, G. P. 1991. A review of IWC stock boundaries. Report of the International Whaling Commission (Special Issue) 13:39–68.

Donovan, G. P. 2002. International Whaling Commission. Pages 637–41 *in* W. F. Perrin, B. Würsig, and J. G. M. Thewissen, eds. Encyclopedia of marine mammals. Academic Press, San Diego, CA.

Dudley, P. 1725. An essay upon the natural history of whales, with a particular account of the ambergris found in the sperma ceti whale. Philosophical Transactions of the Royal Society of London 33:256–69.

Dunn, J. L. 1990. Bacterial and mycotic diseases of cetaceans and pinnipeds. Pages 73–87 *in* L. A. Dierauf, ed. CRC handbook of marine mammal medicine: Health, disease, and rehabilitation. CRC Press, Boca Raton, FL.

Durbin, E., G. Teegarden, R. Campbell, A. Cembella, M. F. Baumgartner, and B. R. Mate. 2002. North Atlantic right whales, *Eubalaena glacialis,* exposed to paralytic shellfish poisoning (PSP) toxins via a zooplankton vector, *Calanus finmarchicus*. Harmful Algae 1:243–51.

Edds, P. L., D. K. Odell, and B. R. Tershy. 1993. Vocalizations of a captive juvenile and free-ranging adult-calf pairs of Bryde's whales, *Balaenoptera edeni*. Marine Mammal Science 9:269–84.

Edwards, E. J., and J. E. Rattray. 1932. 'Whale off!' The story of American shore whaling. Frederick A. Stokes, New York.

Epstein, A. W., and R. C. Beardsley. 2001. Flow-induced aggregation of plankton at a front: A 2-D Eulerian study. Deep-Sea Research II 48:395–418.

Eschricht, D. F., and J. Reinhardt. 1866. On the Greenland right whale (*Balaena mysticetus* Linn.); with especial reference to its geographical distribution and migrations in times past and present, and to its external and internal characteristics. Pages 1–150 and plates 1–6 *in* W. H. Flower, ed. Recent memoirs on the Cetacea by Professors Eschricht, Reinhardt, and Lilljeborg. Robert Hardwicke, London.

Evans, P. G. H., and J. A. Raga, eds. 2002. Marine mammals: Biology and conservation. Plenum Press, New York.

Fertl, D., and A. M. Landry, Jr. 2002. Remoras. Pages 1013–15 *in* W. F. Perrin, B. Würsig, and J. G. M. Thewissen, eds. Encyclopedia of marine mammals. Academic Press, San Diego, CA.

Findlay, K. 1998. Aspects of whale watching of right whales off South Africa. (Document SC/M98/RW17). Unpublished, International Whaling Commission, Cambridge, UK.

Flower, W. H. 1883. On whales, past and present, and their probable origin. Notices of the Proceedings at the Meetings of the Members of the Royal Institution, with Abstracts of the Discourses delivered at the Evening Meetings 10:360–76.

Folkow, L. P., T. Haug, K. T. Nilssen, and E. S. Nordøy. 2000. Estimated food consumption of minke whales *Balaenoptera acutorostrata* in northeast Atlantic waters in 1992–1995. North Atlantic Marine Mammal Commission Scientific Publications 2:65–80.

Food and Agriculture Organization. 1978. Mammals in the seas. Vol. I: Report of the FAO ACMRR Working Party on Marine Mammals. Advisory Committee on Marine Resources Research, Food and Agriculture Organization of the United Nations, Rome.

Foote, D. C. 1975. Investigations of small whale hunting in northern Norway. Journal of the Fisheries Research Board of Canada 32:1163–89.

Fordyce, R. E., and L. G. Barnes. 1994. The evolutionary history of whales and dolphins. Annual Review of Earth and Planetary Sciences 22:419–55.

Forney, K. A., J. Barlow, and J. V. Carretta. 1995. The abundance of cetaceans in California waters. Part II. Aerial surveys in winter and spring of 1991 and 1992. Fishery Bulletin 93:15–26.

Forney, K. A., J. Barlow, M. M. Muto, M. Lowry, J. Baker, G. Cameron, J. Mobley, C. Stinchcomb, and J. V. Carretta. 2000. U.S. Pacific marine mammal stock assessments 2000. (NOAA Technical Memorandum NMFS-SWFSC-300).

Fowler, C. W., and J. D. Baker. 1991. A review of animal population dynamics at extremely reduced population levels. Report of the International Whaling Commission 41:545–54.

Frankel, A. S. 2002. Sound production. Pages 1126–38 *in* W. F. Perrin, B. Würsig, and J. G. M. Thewissen, eds. Encyclopedia of marine mammals. Academic Press, San Diego, CA.

Freeman, M. M. R., L. Bogoslovskaya, R. A. Caulfield, I. Egede, I. I. Krupnik, and M. G. Stevenson. 1998. Inuit, whaling, and sustainability. AltaMira Press, Walnut Creek, CA.

Fujiwara, M., and H. Caswell. 2001. Demography of the endangered North Atlantic right whale. Nature 414:537–41.

Gambell, R. 1976. World whale stocks. Mammal Review 61:41–53.

Gambell, R. 1985a. Fin whale *Balaenoptera physalus* (Linnaeus, 1758). Pages 171–92 *in* S. H. Ridgway and R. Harrison, eds. Handbook of marine mammals. Vol. 3: The sirenians and baleen whales. Academic Press, London.

Gambell, R. 1985b. Sei whale *Balaenoptera borealis* Lesson, 1828. Pages 155–70 *in* S. H. Ridgway and R. Harrison, eds. Handbook of marine mammals. Vol. 3: The sirenians and baleen whales. Academic Press, London.

Gambell, R. 1999. The International Whaling Commission and the contemporary whaling debate. Pages 179–98 *in* J. R. Twiss, Jr., and R. R. Reeves, eds. Conservation and management of marine mammals. Smithsonian Institution Press, Washington, DC.

Gaskin, D. E. 1982. The ecology of whales and dolphins. Heinemann, London.

Gatesy, J. 1997. More DNA support for a Cetacea/Hippopotamidae clade: The blood-clotting protein gene γ-fibrinogen. Molecular Biology and Evolution 14:537–43.

Gatesy, J. 1998. Molecular evidence for the phylogenetic affinities of Cetacea. Pages 63–111 *in* J. Thewissen, ed. The emergence of whales: Evolutionary patterns in the origin of Cetacea. Plenum Press, New York.

Gatesy, J., C. Hayashi, M. A. Cronin, and P. Arctander. 1996. Evidence from milk casein genes that cetaceans are close relatives of hippopotamid artiodactyls. Molecular Biology and Evolution 13:954–63.

Gatesy, J., M. Milinkovitch, V. Waddell, and M. Stanhope. 1999. Stability of cladistic relationships between Cetacea and higher-level artiodactyl taxa. Systematic Biology 48:6–20.

Gendron, D., S. Lanham, and M. Carwardine. 1999. North Pacific right whale (*Eubalaena glacialis*) sighting south of Baja California. Aquatic Mammals 25:31–34.

George, J. C., J. Bada, J. Zeh, L. Scott, S. E. Brown, T. O'Hara, and R. Suydam. 1999. Age and growth estimates of bowhead whales (*Balaena mysticetus*) via aspartic acid racemization. Canadian Journal of Zoology 77:571–80.

Gill, C. D. and S. E. Hughes. 1971. A sei whale, *Balaenoptera borealis*, feeding on Pacific saury, *Cololabis saira*. California Fish and Game 57:218–19.

Gingerich, P. D. 1998. Paleobiological perspectives on Mesonychia, Archaeoceti, and the origin of whales. Pages 423–49 *in* J. G. M. Thewissen, ed. The emergence of whales: Evolutionary patterns in the origin of Cetacea. Plenum Press, New York.

Gingerich, P. D., B. H. Smith, and E. L. Simons. 1990. Hind limbs of *Basilosaurus isis*: Evidence of feet in whales. Science 229:154–57.

Gingerich, P. D., M. ul Haq, I. S. Zalmout, I. H. Khan, and M. S. Malkani. 2001. Origin of whales from early artiodactyls: Hands and feet of Eocene Protocetidae from Pakistan. Science 293:2239–42.

Givens, G. H., J. E. Zeh, and A. E. Raftery. 1995. Assessment of the Bering–Chukchi–Beaufort Seas stock of bowhead whales using the BALEEN II model in a Bayesian synthesis framework. Report of the International Whaling Commission 45:345–64.

Goddard, P. D., and D. J. Rugh. 1998. A group of right whales seen in the Bering Sea in July 1996. Marine Mammal Science 14:344–49.

Goodyear, J. D. 1993. A sonic/radio tag for monitoring dive depths and underwater movements of whales. Journal of Wildlife Management 57:503–15.

Goodyear, J. D. 1996. Significance of feeding habitats of North Atlantic right whales based on studies of diel behaviour, diving, food ingestion rates, and prey. Ph.D. Dissertation, University of Guelph, Guelph, Ontario, Canada.

Goto, M., and L. A. Pastene. 1997. Population structure of the western North Pacific minke whale based on an RFLP analysis of the mtDNA control region. Report of the International Whaling Commission 47:531–37.

Graur, D., and D. G. Higgins. 1994. Molecular evidence for the inclusion of cetaceans within the order Artiodactyla. Molecular Biology and Evolution 11:357–64.

Gregr, E. J., L. Nichol, J. K. B. Ford, G. Ellis, and A. W. Trites. 2000. Migration and population structure of northeastern Pacific whales off coastal British Columbia: An analysis of commercial whaling records from 1908–1967. Marine Mammal Science 16:699–727.

Haebler, R., and R. B. Moeller, Jr. 1993. Pathobiology of selected marine mammal diseases. Pages 217–44 *in* J. A. Couch and J. W. Fournie, eds. Pathobiology of marine and estuarine organisms. CRC Press, Boca Raton, FL.

Hain, J. H. W., G. R. Carter, S. D. Kraus, C. A. Mayo, and H. E. Winn. 1982. Feeding behavior of the humpback whale, *Megaptera novaeangliae*, in the western North Atlantic. Fishery Bulletin 80:259–68.

Hain, J. H. W., M. A. M. Hyman, R. D. Kenney, and H. E. Winn. 1985. The role of cetaceans in the shelf-edge region of the northeastern United States. Marine Fisheries Review 47 (1):13–17.

Hain, J. H. W., M. J. Ratnaswamy, R. D. Kenney, and H. E. Winn. 1992. The fin whale, *Balaenoptera physalus*, in waters of the northeastern United States continental shelf. Report of the International Whaling Commission 42:653–69.

Hain, J. H. W., S. L. Ellis, R. D. Kenney, P. J. Clapham, B. K. Gray, M. T. Weinrich, and I. G. Babb. 1995. Apparent bottom feeding by humpback whales on Stellwagen Bank. Marine Mammal Science 11:464–79.

Haldiman, J. T., and R. J. Tarpley. 1993. Anatomy and physiology. Pages 71–156 *in* J. J. Burns, J. J. Montague, and C. J. Cowles, eds. The bowhead whale (Special Publication No. 2). Society for Marine Mammalogy, Lawrence, KS.

Hamilton, P. K., and C. A. Mayo. 1990. Population characteristics of right whales (*Eubalaena glacialis*) observed in Cape Cod and Massachusetts Bays, 1978–1986. Report of the International Whaling Commission (Special Issue) 12:203–8.

Hamilton, P. K., M. K. Marx, and S. D. Kraus. 1995. Weaning in North Atlantic right whales. Marine Mammal Science 11:386–90.

Hamilton, P. K., A. R. Knowlton, M. K. Marx, and S. D. Kraus. 1998a. Age structure and longevity in North Atlantic right whales *Eubalaena glacialis* and their relation to reproduction. Marine Ecology Progress Series 171:285–92.

Hamilton, P. K., M. K. Marx, and S. D. Kraus. 1998b. Scarification analysis of North Atlantic right whales (*Eubalaena glacialis*) as a method of assessing human impacts (Document SC/M98/RW28). Unpublished, International Whaling Commission, Cambridge, UK.

Hammond, P. S., R. Sears, and M. Bérubé. 1990. A note on problems in estimating the numbers of blue whales in the Gulf of St. Lawrence from photoidentification data. Report of the International Whaling Commission (Special Issue) 12:141–42.

Hamner, W. M., G. S. Stone, and B. S. Obst. 1988. Behavior of southern right whales, *Eubalaena australis*, feeding on the antarctic krill, *Euphausia superba*. Fishery Bulletin 86:143–50.

Harmer, S. F. 1923. Cervical vertebrae of a giant blue whale from Panama. Proceedings of the Zoological Society of London 1923:1085–89.

Harrison, R. J. 1969. Reproduction and reproductive organs. Pages 253–348 *in* H. T. Andersen, ed. The biology of marine mammals. Academic Press, New York.

Haug, T. 1981. On some reproduction parameters in fin whales *Balaenoptera physalus* (L.) caught off Norway. Report of the International Whaling Commission 31:373–78.

Henderson, D. A. 1984. Nineteenth century gray whaling: Grounds, catches and kills, practices and depletion of the whale population. Pages 159–86 *in* M. L. Jones, S. L. Swartz, and S. Leatherwood, eds. The gray whale *Eschrichtius robustus*. Academic Press, Orlando, FL.

Herman, L. M., C. S. Baker, P. H. Forestell, and R. C. Antinoja. 1980. Right whale *Balaena glacialis* sightings near Hawaii: A clue to the wintering grounds? Marine Ecology Progress Series 2:271–75.

Heyning, J. E. 1997. Sperm whale phylogeny revisited: Analysis of the morphological evidence. Marine Mammal Science 13:596–613.

Hobbs, R. C., and D. J. Rugh. 1999. The abundance of gray whales in the 1997/98 southbound migration in the eastern North Pacific (Document SC/51/AS10). Unpublished, International Whaling Commission, Cambridge, UK.

Hokkanen, J. E. I. 1990. Temperature regulation of marine mammals. Journal of Theoretical Biology 145:465–85.

Horwood, J. 1987. The sei whale. Population biology, ecology and management. Croom Helm, New York.

Horwood, J. 1990. Biology and exploitation of the minke whale. CRC Press, Boca Raton, FL.

Howard, E. B., J. O. Britt, Jr., and G. K. Matsumoto. 1983a. Parasitic diseases. Pages 119–232 *in* E. B. Howard, ed. Pathobiology of marine mammal diseases, Vol. I. CRC Press, Boca Raton, FL.

Howard, E. B., J. O. Britt, Jr., G. K. Matsumoto, R. Itehara, and C. N. Nagano. 1983b. Bacterial diseases. Pages 69–118 *in* E. B. Howard, ed. Pathobiology of marine mammal diseases, Vol. I. CRC Press, Boca Raton, FL.

Howard, E. B., J. O. Britt, Jr., and J. G. Simpson. 1983c. Neoplasms in marine mammals. Pages 95–163 *in* E. B. Howard, ed. Pathobiology of marine mammal diseases, Vol. II. CRC Press, Boca Raton, FL.

Hoyt, E. 2002. Whale watching. Pages 1305–10 *in* W. F. Perrin, B. Würsig, and J. G. M. Thewissen, eds. Encyclopedia of marine mammals. Academic Press, San Diego, CA.

Huntington, H. P. 1992. The Alaska Eskimo Whaling Commission and other cooperative marine mammal management organizations in northern Alaska. Polar Record 28:119–26.

Ingebrigtsen, A. 1929. Whales caught in the North Atlantic and other seas. Rapports et Procès-verbaux des Réunions, Conseil Permanent International pour L'exploration de la Mer 56 (2):1–26.

International Whaling Commission. 1983. Chairman's report of the thirty-fourth annual meeting. Report of the International Whaling Commission 33:20–42.

International Whaling Commission. 1988. International Whaling Commission report 1986–87. Report of the International Whaling Commission 38:1–231.

International Whaling Commission. 1989. International Whaling Commission report 1987–88. Report of the International Whaling Commission 39:1–204.

International Whaling Commission. 1990. International Whaling Commission report 1988–89. Report of the International Whaling Commission 40:1–222.

International Whaling Commission. 1991. International Whaling Commission report 1989–90. Report of the International Whaling Commission 41:1–269.

International Whaling Commission. 1992. International Whaling Commission report 1990–91. Report of the International Whaling Commission 42:1–376.

International Whaling Commission. 1993. International Whaling Commission report 1991–92. Report of the International Whaling Commission 43:1–307.

International Whaling Commission. 1994. International Whaling Commission report 1992–93. Report of the International Whaling Commission 44:1–251.

International Whaling Commission. 1995. International Whaling Commission report 1993–94. Report of the International Whaling Commission 45:1–267.

International Whaling Commission. 1996. International Whaling Commission report 1994–95. Report of the International Whaling Commission 46:1–290.

International Whaling Commission. 1997. International Whaling Commission report 1995–96. Report of the International Whaling Commission 47:1–376.

International Whaling Commission. 1998. International Whaling Commission report 1996–97. Report of the International Whaling Commission 48:1–374.

International Whaling Commission. 1999. Report of the Scientific Committee. Journal of Cetacean Research and Management 1(Supplement):1–284.

International Whaling Commission. 2000. Report of the Scientific Committee. Journal of Cetacean Research and Management 2(Supplement):1–318.

International Whaling Commission. 2001a. Report of the Scientific Committee. Journal of Cetacean Research and Management 3(Supplement):1–374.

International Whaling Commission. 2001b. Report of the workshop on the comprehensive assessment of right whales: A worldwide comparison. Journal of Cetacean Research and Management (Special Issue) 2:1–60.

International Whaling Commission. 2001c. Report of the workshop on the status and trends of western North Atlantic right whales. Journal of Cetacean Research and Management (Special Issue) 2:61–87.

International Whaling Commission. 2002a. Report of the Scientific Committee. Journal of Cetacean Research and Management 4(Supplement):1–78.

International Whaling Commission. 2002b. Report of the sub-committee on the comprehensive assessment of North Atlantic humpback whales. Journal of Cetacean Research and Management 4(Supplement): 230–60.

Jefferson, T. A., P. J. Stacey, and R. W. Baird. 1991. A review of killer whale interactions with other marine mammals: Predation to co-existence. Mammal Review 21:151–80.

Jefferson, T. A., S. Leatherwood, and M. A. Webber. 1993. FAO species identification guide; marine mammals of the world. United Nations Environment Programme, Food and Agriculture Organization of the United Nations, Rome.

Johnson, J. H., and A. A. Wolman. 1984. The humpback whale (*Megaptera novaeangliae*). Marine Fisheries Review 46 (4):30–37.

Johnson, K. R., and C. H. Nelson. 1984. Side-scan sonar assessment of gray whale feeding in the Bering Sea. Science 225:1150–52.

Jones, M. L., and S. L. Swartz. 2002. Gray whale *Eschrichtius robustus*. Pages 524–36 *in* W. F. Perrin, B. Würsig, and J. G. M. Thewissen, eds. Encyclopedia of marine mammals. Academic Press, San Diego, CA.

Jonsgård, Å. 1966. The distribution of Balaenopteridae in the North Atlantic Ocean. Pages 114–24 *in* K. S. Norris, ed. Whales, dolphins, and porpoises. University of California Press, Berkeley.

Jonsgård, Å., and K. Darling. 1977. On the biology of the eastern North Atlantic sei whale, *Balaenoptera borealis* Lesson. Report of the International Whaling Commission (Special Issue) 1:124–29.

Jurasz, C. M., and V. P. Jurasz. 1979. Feeding modes of the humpback whale, *Megaptera novaeangliae,* in southeast Alaska. Scientific Reports of the Whales Research Institute (Tokyo) 31:69–83.

Katona, S. K., and J. A. Beard. 1990. Population size, migrations and feeding aggregations of the humpback whale (*Megaptera novaeangliae*) in the western North Atlantic Ocean. Report of the International Whaling Commission (Special Issue) 12:295–305.

Katona, S. K., and S. D. Kraus. 1999. Efforts to conserve the North Atlantic right whale. Pages 311–31 *in* J. R. Twiss, Jr., and R. R. Reeves, eds. Conservation and management of marine mammals. Smithsonian Institution Press, Washington, DC.

Katona, S., B. Baxter, O. Brazier, S. Kraus, J. Perkins, and H. Whitehead. 1979. Identification of humpback whales by fluke photographs. Pages 33–44 *in* H. E. Winn and B. L. Olla, eds. Behavior of marine animals: Current perspectives in research. Vol. 3: Cetaceans. Plenum Press, New York.

Katona, S. K., V. Rough, and D. T. Richardson. 1993. A field guide to whales, porpoises, and seals from Cape Cod to Newfoundland, 4th ed., rev. Smithsonian Institution Press, Washington, DC.

Kawamura, A. 1974. Food and feeding ecology in the southern sei whale. Scientific Reports of the Whales Research Insititute (Tokyo) 26:25–144.

Kawamura, A. 1980. Food habits of the Bryde's whales taken in the South Pacific and Indian Oceans. Scientific Reports of the Whales Research Institute (Tokyo) 32:1–23.

Kawamura, A. 1982. Food habits and prey distributions of three rorqual species in the North Pacific Ocean. Scientific Reports of the Whales Research Institute (Tokyo) 34:59–91.

Kawamura, A. and Y. Satake. 1976. Preliminary report on the geographical distribution of the Bryde's whale in the North Pacific with special reference to the structure of filtering apparatus. Scientific Reports of the Whales Research Institute (Tokyo) 28:1–35.

Kellogg, R. 1929. What is known of the migration of some of the whalebone whales. Annual Report of the Smithsonian Institution 1928:467–94.

Kennedy-Stoskopf, S. 1990. Viral diseases in marine mammals. Pages 97–113 *in* L. A. Dierauf, ed. CRC handbook of marine mammal medicine: Health, disease, and rehabilitation. CRC Press, Boca Raton, FL.

Kenney, R. D. 1998. Global climate change and whales: Western North Atlantic right whale calving rate correlates with the Southern Oscillation Index (Document SC/M98/RW29). Unpublished. International Whaling Commission, Cambridge, UK.

Kenney, R. D. 2001. Anomalous 1992 spring and summer right whale (*Eubalaena glacialis*) distributions in the Gulf of Maine. Journal of Cetacean Research and Management (Special Issue) 2:209–23.

Kenney, R. D. 2002. North Atlantic, North Pacific, and southern right whales *Eubalaena glacialis*, *E. japonica*, and *E. australis*. Pages 806–13 *in* W. F. Perrin, B. Würsig, and J. G. M. Thewissen, eds. Encyclopedia of marine mammals. Academic Press, San Diego, CA.

Kenney, R. D., and S. D. Kraus. 1993. Right whale mortality: A correction and an update. Marine Mammal Science 9:445–46.

Kenney, R. D., and H. E. Winn. 1987. Why some whales fluke: Suggesting a novel hypothesis. CETUS 7 (2):15–19.

Kenney, R. D., and K. F. Wishner. 1995. The South Channel Ocean Productivity Experiment. Continental Shelf Research 15:373–84.

Kenney, R. D., M. A. M. Hyman, and H. E. Winn. 1985. Calculation of standing stocks and energetic requirements of the cetaceans of the northeast United States outer continental shelf (NOAA Technical Memorandum NMFS-F/NEC-41). National Marine Fisheries Service, Woods Hole, MA.

Kenney, R. D., M. A. M. Hyman, R. E. Owen, G. P. Scott, and H. E. Winn. 1986. Estimation of prey densities required by western North Atlantic right whales. Marine Mammal Science 2:1–13.

Kenney, R. D., H. E. Winn, and M. C. Macaulay. 1995. Cetaceans in the Great South Channel, 1979–1989: Right whale (*Eubalaena glacialis*). Continental Shelf Research 15:385–414.

Kenney, R. D., P. M. Payne, D. J. Heinemann, and H. E. Winn. 1996. Shifts in Northeast shelf cetacean distributions relative to trends in Gulf of Maine/Georges Bank finfish abundance. Pages 169–96 *in* K. Sherman, N. A. Jaworski, and T. J. Smayda, eds. The Northeast Shelf ecosystem: Assessment, sustainability, and management. Blackwell Science, Boston.

Kenney, R. D., G. P. Scott, T. J. Thompson, and H. E. Winn. 1997. Estimates of prey consumption and trophic impacts of cetaceans in the USA northeast continental shelf ecosystem. Journal of Northwest Atlantic Fisheries Science 22:155–71.

Kenney, R. D., C. A. Mayo, and H. E. Winn. 2001. Migration and foraging strategies at varying spatial scales in western North Atlantic right whales: A review of hypotheses. Journal of Cetacean Research and Management (Special Issue) 2:251–60.

Kingsley, M. C. S., and R. R. Reeves. 1998. Aerial surveys of cetaceans in the Gulf of St. Lawrence in 1995 and 1996. Canadian Journal of Zoology 76:1529–50.

Klimley, A. P. 1994. The predatory behavior of the white shark. American Scientist 82:122–33.

Klinowska, M. 1991. Dolphins, porpoises and whales of the world. The IUCN red data book. International Union for the Conservation of Nature, Gland, Switzerland.

Klumov, S. K. 1962. Gladkiye (Yaponskiye) kity Tikhogo Okeana [The right whales in the Pacific Ocean]. Trudy Instituta Okeanologii Akademii Nauk SSSR 58:202–97 (In Russian with English summary).

Klumov, S. K. 1963. Pitanie I gelmintofauna usatykh kitov (Mystacoceti) v osnovnykh promyslovykh raionakh morivogo oceana [Food and helminth fauna of whalebone whales (Mystacoceti) in the main whaling regions of the world ocean]. Trudy Instituta Okeanologii Akademii Nauk SSSR 71:94–194 (English translation: Fisheries Research Board of Canada Translation Series No. 589, 1965).

Knowlton, A. R., and S. D. Kraus. 2001. Mortality and serious injury of northern right whales (*Eubalaena glacialis)* in the western North Atlantic Ocean. Journal of Cetacean Research and Management (Special Issue) 2:193–208.

Knowlton, A. R., C. W. Clark, and S. D. Kraus. 1991. Sounds recorded in the presence of sei whales, *Balaenoptera borealis*. Page 40 *in* Abstracts, ninth biennial conference on the biology of marine mammals. Chicago, IL.

Knowlton, A. R., J. Sigurjónsson, J. N. Ciano, and S. D. Kraus. 1992. Long-distance movements of North Atlantic right whales (*Eubalaena glacialis*). Marine Mammal Science 8:397–405.

Knowlton, A. R., S. D. Kraus, and R. D. Kenney. 1994. Reproduction in North Atlantic right whales (*Eubalaena glacialis*). Canadian Journal of Zoology 72:1297–305.

Koski, W. R., R. A. Davis, and G. W. Miller. 1992. Growth rates of bowhead whales as determined from low-level aerial photography. Report of the International Whaling Commission 42:491–99.

Koski, W. R., R. A. Davis, G. W. Miller, and D. E. Withrow. 1993. Reproduction. Pages 239–74 *in* J. J. Burns, J. J. Montague, and C. J. Cowles, eds. The bowhead whale (Special Publication No. 2). Society for Marine Mammalogy, Lawrence, KS.

Kraus, S. D. 1990. Rates and potential causes of mortality in North Atlantic right whales (*Eubalaena glacialis*). Marine Mammal Science 6:278–91.

Kraus, S. D., and J. J. Hatch. 2001. Mating strategies in the North Atlantic right whale (*Eubalaena glacialis*). Journal of Cetacean Research and Management (Special Issue) 2:237–44.

Kraus, S. D., and R. D. Kenney. 1991. Information on right whales (*Eubalaena glacialis*) in three proposed critical habitats in United States waters of the western North Atlantic Ocean (Final report, Contracts T-75133740 and T-75133753; National Technical Information Service PB91-194431). Marine Mammal Commission, Washington, DC.

Kraus, S. D., and J. H. Prescott. 1982. The North Atlantic right whale (*Eubalaena glacialis*) in the Bay of Fundy, 1981, with notes on distribution, abundance, biology and behavior (Annual report, contract number NA-81-FA-C-00030). National Marine Fisheries Service, Northeast Fisheries Center, Woods Hole, MA.

Kraus, S. D., and G. S. Stone. 1995. Coprophagy by Wilson's storm petrels, *Oceanites oceanicus,* on North Atlantic right whale, *Eubalaena glacialis*, faeces. Canadian Field-Naturalist 109:443–44.

Kraus, S. D., J. H. Prescott, A. R. Knowlton, and G. S. Stone. 1986a. Migration and calving of right whales (*Eubalaena glacialis*) in the western North Atlantic. Report of the International Whaling Commission (Special Issue) 10:139–44.

Kraus, S. D., K. E. Moore, C. A. Price, M. J. Crone, W. A. Watkins, H. E. Winn, and J. H. Prescott. 1986b. The use of photographs to identify individual North Atlantic right whales (*Eubalaena glacialis*). Report of the International Whaling Commission (Special Issue) 10:145–51.

Kraus, S. D., M. J. Crone, and A. R. Knowlton. 1988. The North Atlantic right whale. Pages 684–98 *in* W. J. Chandler, ed. Audubon wildlife report 1988/1989. Academic Press, San Diego, CA.

Kraus, S. D., R. D. Kenney, A. R. Knowlton, and J. N. Ciano. 1993. Endangered right whales of the southwestern North Atlantic (Final report, Contract No. 14-35-0001-30486). U.S. Department of the Interior, Minerals Management Service, Herndon, VA.

Kraus, S. D., P. K. Hamilton, R. D. Kenney, A. R. Knowlton, and C. K. Slay. 2001. Reproductive parameters of the North Atlantic right whale. Journal of Cetacean Research and Management (Special Issue) 2:231–36.

Krupnik, I. I. 1984. Gray whales and the aborigines of the Pacific Northwest: The history of aboriginal whaling. Pages 103–20 *in* M. L. Jones, S. L. Swartz, and S. Leatherwood, eds. The gray whale *Eschrichtius robustus*. Academic Press, Orlando, FL.

Kugler, R. C. 1986. Random notes on the history of right whaling on the 'Northwest Coast'. Report of the International Whaling Commission (Special Issue) 10:17–19.

Kvitek, R. G., and J. S. Oliver. 1986. Side-scan sonar estimates of the utilization of gray whale feeding grounds along Vancouver Island, Canada. Continental Shelf Research 6:639–54.

Laist, D. W., A. R. Knowlton, J. G. Mead, A. S. Collett, and M. Podesta. 2001. Collisions between ships and whales. Marine Mammal Science 17:35–75.

Lambertsen, R. H. 1983. The internal mechanism of rorqual feeding. Journal of Mammalogy 64:76–88.

Lavigne, D. M., V. B. Scheffer, and S. R. Kellert. 1999. The evolution of North American attitudes toward marine mammals. Pages 10–47 *in* J. R. Twiss, Jr., and R. R. Reeves, eds. Conservation and management of marine mammals. Smithsonian Institution Press, Washington, DC.

Laws, R. M. 1961. Reproduction, growth, and age of southern fin whales. Discovery Reports 21:327–486.

Leatherwood, S., D. K. Caldwell, and H. E. Winn. 1976. Whales, dolphins, and porpoises of the western North Atlantic: A guide to their identification (NOAA Technical Report NMFS CIRC-396). National Marine Fisheries Service, Seattle, WA.

Leatherwood, S., R. R. Reeves, W. F. Perrin, and W. E. Evans. 1988. Whales, dolphins, and porpoises of the eastern North Pacific and adjacent Arctic waters: A guide to their identification, rev. ed. Dover, New York.

LeBoeuf, B. J., H. Pérez-Cortés M., J. Urbán R., B. R. Mate, and F. Ollervides U. 2000. High gray whale mortality and low recruitment in 1999: Potential causes and implications. Journal of Cetacean Research and Management 2:85–99.

LeDuc, R. G., W. L. Perryman, J. W. Gilpatrick, Jr., J. Hyde, C. Stinchcomb, J. V. Carretta, and R. L. Brownell, Jr. 2001. A note on recent surveys for right whales in the southeastern Bering Sea. Journal of Cetacean Research and Management (Special Issue) 2:287–89.

Lien, J. 1994. Entrapments of large cetaceans in passive inshore fishing gear in Newfoundland and Labrador (1979–1990). Report of the International Whaling Commission (Special Issue) 15:149–57.

Lien, J., S. Johnson, and B. Merdsoy. 1979. Whale distribution in Newfoundland during 1979. Osprey 11(2):21–32.

Lien, J., R. Sears, G. B. Stenson, P. W. Jones, and I.-H. Ni. 1989. Right whale, *Eubalaena glacialis*, sightings in waters off Newfoundland and Labrador and the Gulf of St. Lawrence, 1978–1987. Canadian Field-Naturalist 103:91–93.

Lindquist, O. 2000. The North Atlantic gray whale (*Eschrichtius robustus*): An historical outline based on Icelandic, Danish-Icelandic, English and Swedish sources dating from ca 1000 AD to 1792. (Occasional Papers, No. 1). Centre for Environmental History and Policy, Universities of St. Andrews and Stirling, Scotland.

Ling, J. K. 1977. Vibrissae of marine mammals. Pages 387–415 *in* R. J. Harrison, ed. Functional anatomy of marine mammals, Vol. 3. Academic Press, London.

Liu, F.-G. R., and M. M. Miyamoto. 1999. Phylogenetic assessment of molecular and morphological data for eutherian mammals. Systematic Biology 48:54–64.

Lockyer, C. 1972. The age at sexual maturity of the southern fin whale (*Balaenoptera physalus*) using the annual layer counts in the ear plug. Journal du Conseil International pour l'Exploration de la Mer 34:276–94.

Lockyer, C. 1976. Body weights of some species of large whales. Journal du Conseil International pour l'Exploration de la Mer 36:259–73.

Lockyer, C. 1984. Review of baleen whale (Mysticeti) reproduction and implications for management. Report of the International Whaling Commission (Special Issue) 6:27–50.

Lockyer, C., and A. R. Martin. 1983. The sei whale off western Iceland: II. Age, growth and reproduction. Report of the International Whaling Commission 33:465–76.

Lowry, L. F. 1993. Foods and feeding ecology. Pages 201–38 *in* J. J. Burns, J. J. Montague, and C. J. Cowles, eds. The bowhead whale (Special Publication No. 2). Society for Marine Mammalogy, Lawrence, KS.

Lum, J. K., M. Nikaido, M. Shimamura, H. Shimodaira, A. M. Shedlock, N. Okada, and M. Hasegawa. 2000. Consistency of SINE insertion topology and flanking sequence tree: Quantifying relationships among cetartiodactyls. Molecular Biology and Evolution 17:1417–24.

Macaulay, M. C., K. F. Wishner, and K. L. Daly. 1995. Acoustic scattering from zooplankton and micronekton in relation to a whale feeding site near Georges Bank and Cape Cod. Continental Shelf Research 15:509–37.

Mackintosh, N. A. 1965. The stocks of whales. Fishing News, London.

MacLean, S. A., G. W. Sheehan, and A. M. Jensen. 2002. Inuit and marine mammals. Pages 641–52 *in* W. F. Perrin, B. Würsig, and J. G. M. Thewissen, eds. Encyclopedia of marine mammals. Academic Press, San Diego, CA.

Markham, C. R. 1881. On the whale-fishery of the Basque provinces of Spain. Proceedings of the Zoological Society of London 1881:969–76.

Martin, A. R. 1983. The sei whale off western Iceland. I. Size, distribution and abundance. Report of the International Whaling Commission 33:457–63.

Martin, A. R., and F. J. Walker. 1997. Sighting of a right whale (*Eubalaena glacialis*) with calf off S. W. Portugal. Marine Mammal Science 13:139–40.

Martin, A. R., S. K. Katona, D. Mattila, D. Hembree, and T. D. Waters. 1984. Migration of humpback whales between the Caribbean and Iceland. Journal of Mammalogy 65:330–33.

Masaki, Y. 1977. The separation of the stock units of sei whales in the North Pacific. Report of the International Whaling Commission (Special Issue) 1:71–77.

Mate, B. R., S. L. Nieukirk, and S. D. Kraus. 1997. Satellite-monitored movements of the northern right whale. Journal of Wildlife Management 61:1393–1405.

Matthews, L. H. 1938. Notes on the southern right whale, *Eubalaena australis*. Discovery Reports 17:169–82.

Mattila, D. K., P. J. Clapham, S. K. Katona, and G. S. Stone. 1989. Population composition of humpback whales on Silver Bank. Canadian Journal of Zoology 67:281–85.

Mattila, D. K., P. J. Clapham, O. Vásquez, and R. Bowman. 1994. Occurrence, population composition, and habitat use of humpback whales in Samana Bay, Dominican Republic. Canadian Journal of Zoology 72:1898–1907.

Mayo, C. A., and L. Goldman. 1992. Right whale foraging and the plankton resources in Cape Cod and Massachusetts Bays. Pages 43–44 *in* J. Hain, ed. The right whale in the western North Atlantic: A science and management workshop (Northeast Fisheries Science Center Reference Document 92–05). National Marine Fisheries Service, Woods Hole, MA.

Mayo, C. A., and L. Goldman. 1998. Fine scale characteristics of copepod patches foraged by the northern right whale (*Eubalaena glacialis*) (Document SC/M98/RW7). Unpublished. International Whaling Commission, Cambridge, UK.

Mayo, C. A., and M. K. Marx. 1990. Surface foraging behaviour of the North Atlantic right whale, *Eubalaena glacialis*, and associated zooplankton characteristics. Canadian Journal of Zoology 68:2214–20.

Mayo, C. A., E. Lyman, and D. K. Mattila. 1998. Disentanglement of northern right whales: A model for immediate response (Document SC/M98/RW47). Unpublished. International Whaling Commission, Cambridge, UK.

Mayo, C. A., B. H. Letcher, and S. Scott. 2001. Zooplankton filtering efficiency of the baleen of a North Atlantic right whale, *Eubalaena glacialis*. Journal of Cetacean Research and Management (Special Issue) 2:225–29.

McDonald, M. A., and C. G. Fox. 1999. Passive acoustic methods applied to fin whale population density estimation. Journal of the Acoustical Society of America 105:2643–51.

McDonald, M. A., and S. E. Moore. 2002. Calls recorded from North Pacific right whales (*Eubalaena japonica*) in the eastern Bering Sea. Journal of Cetacean Research and Management 4:261–66.

McKenna, M. C. 1975. Toward a phylogenetic classification of the Mammalia. Pages 21–46 *in* W. D. Luckett and F. S. Szalay, eds. Phylogeny of the primates. Plenum Press, New York.

McKenna, M. C. 1987. Molecular and morphological analysis of high-level mammalian interrelationships. Pages 55–93 *in* C. Patterson, ed. Molecules and morphology in evolution: Conflict or compromise? Cambridge University Press, Cambridge.

McLeod, S. A., F. C. Whitmore, Jr., and L. G. Barnes. 1993. Evolutionary relationships and classification. Pages 45–70 *in* J. J. Burns, J. J. Montague, and C. J. Cowles, eds. The bowhead whale (Special Publication No. 2). Society for Marine Mammalogy, Lawrence, KS.

Mead, J. G. 1977. Records of sei and Bryde's whales from the Atlantic coast of the United States, Gulf of Mexico and the Caribbean. Report of the International Whaling Commission (Special Issue) 1:113–16.

Mead, J. G., and E. D. Mitchell. 1984. Atlantic gray whales. Pages 33–53 *in* M. L. Jones, S. L. Swartz, and S. Leatherwood, eds. The gray whale *Eschrichtius robustus*. Academic Press, Orlando, FL.

Mellinger, D. K., C. D. Carson, and C. W. Clark. 2000. Characteristics of minke whale (*Balaenoptera acutorostrata*) pulse trains recorded near Puerto Rico. Marine Mammal Science 16:739–56.

Migaki, G., and S. R. Jones. 1983. Mycotic diseases. Pages 1–27 *in* E. B. Howard, ed. Pathobiology of marine mammal diseases, Vol. II. CRC Press, Boca Raton, FL.

Mignucci-Giannoni, A. A. 1998. Zoogeography of cetaceans off Puerto Rico and the Virgin Islands. Caribbean Journal of Science 34:173–90.

Milinkovitch, M. C. 1995. Molecular phylogeny of cetaceans prompts revision of morphological transformations. Trends in Ecology and Evolution 10:328–34.

Milinkovitch, M. C. 1997. The phylogeny of whales: A molecular approach. Pages 317–38 *in* A. E. Dizon, S. J. Chivers, and W. F. Perrin, eds. Molecular genetics of marine mammals. Society for Marine Mammalogy, Lawrence, KS.

Milinkovitch, M. C., A. Meyer, and J. R. Powell. 1994. Phylogeny of all major groups of cetaceans based on DNA sequences from three mitochondrial genes. Molecular Biology and Evolution 11:939–48.

Milinkovitch, M. C., M.Bérubé, and P. J. Palsbøll. 1998. Cetaceans are highly derived artiodactyls. Pages 113–31 *in* J. G. M. Thewissen, ed. The emergence of whales: Evolutionary patterns in the origin of Cetacea. Plenum Press, New York.

Miller, G. S., Jr. 1923. The telescoping of the cetacean skull. Smithsonian Miscellaneous Collections 76:1–71.

Mitchell, E. 1974. Present status of northwest Atlantic fin and other whale stocks. Pages 108–69 *in* W. E. Schevill, ed. The whale problem, a status report. Harvard University Press, Cambridge, MA.

Mitchell, E. 1975. Preliminary report on Nova Scotia fishery for sei whales (*Balaenoptera borealis*). Report of the International Whaling Commission 25:218–25.

Mitchell, E. 1989. A new cetacean from the late Eocene La Meseta Formation, Seymour Island, Antarctic Peninsula. Canadian Journal of Fisheries and Aquatic Sciences 46:2219–35.

Mitchell, E. D., Jr., 1991. Winter records of the minke whale (*Balaenoptera acutorostrata acutorostrata* Lacépède 1804) in the southern North Atlantic. Report of the International Whaling Commission 41:455–57.

Mitchell, E., and D. G. Chapman. 1977. Preliminary assessment of stocks of northwest Atlantic sei whales (*Balaenoptera borealis*). Report of the International Whaling Commission (Special Issue) 1:117–20.

Mitchell, E., and R. R. Reeves. 1983. Catch history, abundance, and present status of northwest Atlantic humpback whales. Report of the International Whaling Commission (Special Issue) 5:153–212.

Mitchell, E., V. M. Kozicki, and R. R. Reeves. 1986. Sightings of right whales, *Eubalaena glacialis*, on the Scotian Shelf, 1966–72. Report of the International Whaling Commission (Special Issue) 10:83–107.

Mizroch, S. A., D. W. Rice, and J. M. Breiwick. 1984. The blue whale, *Balaenoptera musculus*. Marine Fisheries Review 46 (4):15–19.

Montgelard, C., F. M. Catzeflis, and E. Douzery. 1997. Phylogenetic relationships of artiodactyls and cetaceans as deduced from the comparison of cytochrome *b* and 12S rRNA mitochondrial sequences. Molecular Biology and Evolution 14:550–59.

Moore, J. C. 1953. Distribution of marine mammals to Florida waters. American Midland Naturalist 49:117–58.

Moore, S. E., and D. P. DeMaster. 1997. Cetacean habitats in the Alaskan Arctic. Journal of Northwest Atlantic Fishery Science 22:55–69.

Moore, S. E., and R. R. Reeves. 1993. Distribution and movement. Pages 313–86 *in* J. J. Burns, J. J. Montague, and C. J. Cowles, eds. The bowhead whale (Special Publication No. 2). Society for Marine Mammalogy, Lawrence, KS.

Moore, S. E., K. M. Stafford, M. E. Dahlheim, C. G. Fox, H. W. Braham, J. J. Polovina, and D. E. Bain. 1998. Seasonal variation in reception of fin whale calls at five geographic areas in the North Pacific. Marine Mammal Science 14:617–27.

Moore, S. E., J. M. Waite, L. L. Mazzuca, and R. C. Hobbs. 2000. Mysticete whale abundance and observations of prey associations on the central Bering Sea shelf. Journal of Cetacean Research and Management 2:227–34.

Murison, L. D. 1986. Zooplankton distributions and feeding ecology of right whales (*Eubalaena glacialis glacialis*) in the outer Bay of Fundy, Canada. M. Sc. thesis, University of Guelph, Guelph, Ontario, Canada.

Murison, L. D., and D. E. Gaskin. 1989. The distribution of right whales and zooplankton in the Bay of Fundy, Canada. Canadian Journal of Zoology 67:1411–20.

National Oceanic and Atmospheric Administration. 2002. Right whale calf sighted in Bering Sea (press release NOAA 2002-212). http://www.publicaffairs.noaa.gov/releases2002/sep02/noaa02122.html.

National Marine Fisheries Service. 1991. Final recovery plan for the northern right whale (*Eubalaena glacialis*). U.S. Department of Commerce, National Oceanic and Atmospheric Administration, National Marine Fisheries Service, Office of Protected Resources, Silver Spring, MD.

Nelson, C. H., and K. R. Johnson. 1987. Whales and walruses as tillers of the sea floor. Scientific American 255:112–17.

Nemoto, T. 1970. Feeding patterns of baleen whales in the ocean. Pages 241–52 *in* J. H. Steele, ed. Marine food chains. University of California Press, Berkeley.

Nerini, M. 1984. A review of gray whale feeding ecology. Pages 423–50 *in* M. L. Jones, S. L. Swartz, and S. Leatherwood, eds. The gray whale *Eschrichtius robustus*. Academic Press, Orlando, FL.

Nikaido, M., A. P. Rooney, and N. Okada. 1999. Phylogenetic relationships among cetartiodactyls based on insertions of short and long interspersed elements: Hippopotamuses are the closest extant relatives of whales. Proceedings of the National Academy of Sciences of the USA 96:10261–66.

Nikaido, M., F. Matsumo, H. Hamilton, R. L. Brownell, Jr., Y. Cao, W. Ding, Z. Zuoyan, A. M. Shedlock, R. E. Fordyce, M. Hasegawa, and N. Okada. 2001. Retroposon analysis of major cetacean lineages: The monophyly of toothed whales and paraphyly of river dolphins. Proceedings of the National Academy of Sciences of the USA 98:7384–89.

Norman, J. R., and F. C. Fraser. 1937. Giant fishes, whales and dolphins. Putnam, London.

Novacek, D. P., M. P. Johnson, P. L. Tyack, K. A. Shorter, W. A. McLellan, and D. A. Pabst. 2001. Buoyant balaenids: The ups and downs of buoyancy in right whales. Proceedings of the Royal Society of London B 268:1811–16.

Novacek, M. J. 1989. Higher mammal phylogeny: The morphological–molecular synthesis. Pages 421–35 *in* B. Fernholm, K. Bremer, and H. Jornvall, eds. The hierarchy of life. Elsevier, Amsterdam.

Ohsumi, S. 1978. Bryde's whales in the North Pacific in 1976. Report of the International Whaling Commission 28:277–80.

Ohsumi, S. 1979. Bryde's whales in the North Pacific in 1977. Report of the International Whaling Commission 29:265–66.

Ohsumi, S. 1980. Bryde's whales in the North Pacific in 1978. Report of the International Whaling Commission 30:315–18.

O'Leary, B. L. 1984. Aboriginal whaling from the Aleutian Islands to Washington State. Pages 79–102 *in* M. L. Jones, S. L. Swartz, and S. Leatherwood, eds. The gray whale *Eschrichtius robustus*. Academic Press, Orlando, FL.

O'Leary, M. A. 1998. Phylogenetic and morphometric reassessment of the dental evidence for a mesonychian and cetacean clade. Pages 133–61 *in* J. G. M. Thewissen, ed. The emergence of whales: Evolutionary patterns in the origin of Cetacea. Plenum Press, New York.

Omura, H. 1958. North Pacific right whale. Scientific Reports of the Whales Research Institute (Tokyo) 13:1–52.

Omura, H. 1984. History of gray whales in Japan. Pages 57–77 *in* M. L. Jones, S. L. Swartz, and S. Leatherwood, eds. The gray whale *Eschrichtius robustus*. Academic Press, Orlando, FL.

Omura, H., S. Ohsumi, T. Nemoto, K. Nasu, and T. Kasuya. 1969. Black right whales in the North Pacific. Scientific Reports of the Whales Research Institute (Tokyo) 21:1–78.

Omura, H., M. Nishiwaki, and T. Kasuya. 1971. Further studies on two skeletons of the black right whale in the North Pacific. Scientific Reports of the Whales Research Institute (Tokyo) 23:71–81.

O'Shea, T. J. 1999. Environmental contaminants and marine mammals. Pages 485–563 *in* J. E. Reynolds III and S. A. Rommel, eds. Biology of marine mammals. Smithsonian Institution Press, Washington, DC.

O'Shea, T. J., and R. L. Brownell, Jr. 1994. Organochlorine and metal contaminants in baleen whales: A review and evaluation of conservation implications. Science of the Total Environment 154:179–200.

O'Shea, T. J., R. R. Reeves, and A. K. Long, eds. 1999. Marine mammals and persistent ocean contaminants: Proceedings of the Marine Mammal Commission workshop, Keystone, Colorado, 12–15 October 1998. Marine Mammal Commission, Washington, DC.

Palsbøll, P. J., J. Allen, M. Bérubé, P. J. Clapham, T. P. Feddersen, P. Hammond, H. Jørgensen, S. Katona, A. H. Larsen, F. Larsen, J. Lien, D. K. Mattila, J. Sigurjónsson, R. Sears, T. Smith, R. Sponer, P. Stevick, and N. Øien. 1997. Genetic tagging of humpback whales. Nature 388:767–69.

Palsbøll, P. J., J. Allen, T. H. Anderson, M. Bérubé, P. J. Clapham, T. P. Feddersen, N. Friday, P. Hammond, H. Jørgensen, S. K. Katona, A. H. Larsen, F. Larsen, J. Lien, D. K. Mattila, F. B. Nygaard, J. Robbins, R. Sponer, R. Sears, J. Sigurjónsson, T. D. Smith, P. T. Stevick, G. Vikingsson, and N. Øien. 2001. Stock structure and composition of the North Atlantic humpback whale, *Megaptera novaeangliae* (Document SC/53/NAH11). Unpublished. International Whaling Commission, Cambridge, UK.

Pastene, L. A., M. Goto, S. Itoh, S. Wada, and H. Kato. 1997. Intra and interoceanic patterns of mitochondrial DNA variations in the Bryde's whale *Balaenoptera edeni*. Report of the International Whaling Commission 47:569–74.

Payne, P. M., J. R. Nicolas, L. O'Brien, and K. D. Powers. 1986. The distribution of the humpback whale, *Megaptera novaeangliae*, on Georges Bank and in the Gulf of Maine in relation to densities of the sand eel, *Ammodytes americanus*. Fishery Bulletin 84:271–77.

Payne, P. M., D. N. Wiley, S. B. Young, S. Pittman, P. J. Clapham, and J. W. Jossi. 1990. Recent fluctuations in the abundance of baleen whales in the southern Gulf of Maine in relation to changes in selected prey. Fishery Bulletin 88:687–96.

Payne, R. 1976. At home with right whales. National Geographic 140:322–39.

Payne, R., and E. M. Dorsey. 1983. Sexual dimorphism and aggressive use of callosities in right whales (*Eubalaena australis*). Pages 295–329 *in* R. Payne, ed. Communication and behavior of whales (AAAS Selected Symposium 76). Westview Press, Boulder, CO.

Payne, R. S., and S. McVay. 1971. Songs of humpback whales. Science 173:587–97.

Payne, R., O. Brazier, E. M. Dorsey, J. S. Perkins, V. J. Rowntree, and A. Titus. 1983. External features in southern right whales (*Eubalaena australis*) and their use in identifying individuals. Pages 371–445 *in* R. Payne, ed. Communication and behavior of whales (AAAS Selected Symposium 76). Westview Press, Boulder, CO.

Payne, R., V. Rowntree, J. S. Perkins, J. G. Cooke, and K. Lankester. 1990. Population size, trends and reproductive parameters of right whales (*Eubalaena australis*) off Peninsula Valdes, Argentina. Report of the International Whaling Commission (Special Issue) 12:271–78.

Perrin, W. F., and R. L. Brownell, Jr. 2002. Minke whales *Balaenoptera acutorostrata* and *B. bonaerensis*. Pages 750–54 *in* W. F. Perrin, B. Würsig, and J. G. M. Thewissen, eds. Encyclopedia of marine mammals. Academic Press, San Diego, CA.

Perrin, W. F., and G. P. Donovan, eds. 1984. Report of the workshop. Report of the International Whaling Commission (Special Issue) 6:1–24.

Perry, S. L., D. P. DeMaster, and G. K. Silber. 1999. The great whales: History and status of six species listed as endangered under the U.S. Endangered Species Act of 1973. Marine Fisheries Review 61 (1):1–74.

Perryman, W. L., M. A. Donahue, P. C. Perkins, and S. B. Reilly. 2002. Gray whale calf production 1994–2000: Are observed fluctuations related to changes in seasonal ice cover? Marine Mammal Science 18:121–44.

Pfeiffer, C. J. 2002. Whale lice. Pages 1302–5 *in* W. F. Perrin, B. Würsig, and J. G. M. Thewissen, eds. Encyclopedia of marine mammals. Academic Press, San Diego, CA.

Pivorunas, A. 1976. A mathematical consideration on the function of baleen plates and their fringes. Scientific Reports of the Whales Research Institute (Tokyo) 28:37–55.

Pivorunas, A. 1979. The feeding mechanisms of baleen whales. American Scientist 67:432–40.

Prothero, D. R., E. M. Manning, and M. Fischer. 1988. The phylogeny of the ungulates. Pages 201–34 *in* M. J. Benton, ed. The phylogeny and classification of the tetrapods, Vol., 2. Clarendon Press, Oxford.

Purves, P. L. 1955. The wax plug in the external auditory meatus of the Mysticeti. Discovery Reports 27:239–302.

Raftery, A., G. H. Givens, and J. E. Zeh. 1995. Inference from a deterministic population dynamics model for bowhead whales. Journal of the American Statistical Association 90:402–30.

Reeves, R. R. 1984. Modern commercial pelagic whaling for gray whales. Pages 187–200 *in* M. L. Jones, S. L. Swartz, and S. Leatherwood, eds. The gray whale *Eschrichtius robustus*. Academic Press, Orlando, FL.

Reeves, R. R. 2001. Overview of catch history, historic abundance and distribution of right whales in the western North Atlantic and in Cintra Bay, West Africa. Journal of Cetacean Research and Management (Special Issue) 2:187–92.

Reeves, R. R. 2002a. Conservation efforts. Pages 276–97 *in* W. F. Perrin, B. Würsig, and J. G. M. Thewissen, eds. Encyclopedia of marine mammals. Academic Press, San Diego, CA.

Reeves, R. R. 2002b. The origins and character of 'aboriginal subsistence' whaling: A global review. Mammal Review 32:71–106.

Reeves, R. R., and R. L. Brownell, Jr. 1982. Baleen whales (*Eubalaena glacialis* and allies). Pages 415–44 *in* J. A. Chapman and G. A. Feldhamer, eds. Wild mammals of North America: Biology, management, and economics. Johns Hopkins University Press, Baltimore.

Reeves, R. R., and S. Leatherwood. 1985. Bowhead whale *Balaena mysticetus* Linnaeus, 1758. Pages 305–44 *in* S. H. Ridgway and R. Harrison, eds. Handbook of marine mammals. Vol. 3: The sirenians and baleen whales. Academic Press, London.

Reeves, R. R., and E. Mitchell. 1986a. The Long Island, New York, right whale fishery: 1650–1924. Report of the International Whaling Commission (Special Issue) 10:201–20.

Reeves, R. R., and E. Mitchell. 1986b. American pelagic whaling for right whales in the North Atlantic. Report of the International Whaling Commission (Special Issue) 10:221–54.

Reeves, R. R., and E. Mitchell. 1988. History of whaling in and near North Carolina (NOAA Technical Report NMFS-65). National Marine Fisheries Service, Seattle, WA.

Reeves, R. R., and T. D. Smith. 2002. Historical catches of humpback whales in the North Atlantic Ocean: An overview of sources. Journal of Cetacean Research and Management 4:219–34.

Reeves, R. R., J. G. Mead, and S. Katona. 1978. The right whale, *Eubalaena glacialis*, in the western North Atlantic. Report of the International Whaling Commission 28:303–12.

Reeves, R. R., S. Leatherwood, S. A. Karl, and E. R. Yohe. 1985. Whaling results at Akutan (1912–39) and Port Hobron (1926–37), Alaska. Report of the International Whaling Commission 35:441–57.

Reeves, R. R., P. J. Clapham, R. L. Brownell, Jr., and G. K. Silber. 1998. Recovery plan for the blue whale (*Balaenoptera musculus*). U.S. Department of Commerce, National Oceanic and Atmospheric Administration, National Marine Fisheries Service, Office of Protected Resources, Silver Spring, MD.

Reeves, R. R., J. M. Breiwick, and E. D. Mitchell. 1999. History of whaling and estimated kill of right whales, *Balaena glacialis*, in the northeastern United States, 1620–1924. Marine Fisheries Review 61(3):1–36.

Reeves, R. R., R. Rolland, and P. J. Clapham, eds. 2001a. Report of the workshop on the causes of reproductive failure in North Atlantic right whales: New avenues of research (Reference Document 01–16). Northeast Fisheries Science Center, National Marine Fisheries Service, Woods Hole, MA.

Reeves, R. R., S. L. Swartz, S. E. Wetmore, and P. J. Clapham. 2001b. Historical occurrence and distribution of humpback whales in the eastern and southern Caribbean Sea, based on data from American whaling logbooks. Journal of Cetacean Research and Management 3:117–29.

Reeves, R. R., B. S. Stewart, P. J. Clapham, and J. A. Powell. 2002. National Audubon Society guide to marine mammals of the world. Knopf, New York.

Reijnders, P. J. H. 1988. Ecotoxicological perspectives in marine mammalogy: Research principles and goals for a conservation policy. Marine Mammal Science 4:91–102.

Rice, D. W. 1967. Cetaceans. Pages 291–324 *in* S. Anderson and J. K. Jones, eds. Recent mammals of the world: A synopsis of families. Ronald Press, New York.

Rice, D. W. 1974. Whales and whale research in the eastern North Pacific. Pages 170–95 *in* W. E. Schevill, ed. The whale problem, a status report. Harvard University Press, Cambridge, MA.

Rice, D. W. 1977. Synopsis of biological data on the sei whale and Bryde's whale in the eastern North Pacific. Report of the International Whaling Commission (Special Issue) 1:92–97.

Rice, D. W. 1983. Gestation period and fetal growth of the gray whale. Report of the International Whaling Commission 33:539–44.

Rice, D. W. 1998. Marine mammals of the world: Systematics and distribution (Special Publication No. 4). Society for Marine Mammalogy, Lawrence, KS.

Rice, D. W., and C. H. Fiscus. 1968. Right whales in the southeastern North Pacific. Norsk Hvalfangst-tidende 57(5):105–7.

Rice, D. W., and A. A. Wolman. 1971. The life history and ecology of the gray whale (*Eschrichtius robustus*) (Special Publication No. 3). American Society of Mammalogists, Lawrence, KS.

Ride, W. D. L., C. W. Sabrosky, G. Bernardi, R. V. Melville, J. O. Corliss, J. Forest, K. H. L. Key, and C. W. Wright, eds. 1985. International Code of Zoological Nomenclature, 3rd ed. University of California Press, Berkeley.

Ridewood, W. G. 1901. On the structure of the horny excrescence, known as the "bonnet," of the southern right whale (*Balaena australis*). Proceedings of the Zoological Society of London 1901:44–47.

Right Whale Recovery Team. 2000. Canadian North Atlantic right whale recovery plan. World Wildlife Fund Canada, Toronto, Ontario, and Department of Fisheries and Oceans, Dartmouth, Nova Scotia, Canada.

Rivarola, M., C. Campagna, and A. Tagliorette. 2001. Demand-driven commercial whalewatching in Península Valdés (Patagonia): Conservation implications for right whales. Journal of Cetacean Research and Management (Special Issue) 2:145–51.

Rosenbaum, H. C., R. L. Brownell, Jr., M. W. Brown, C. Schaeff, V. Portway, B. N. White, S. Malik, L. A. Pastene, N. J. Patenaude, C. S. Baker, M. Goto, P. B. Best, P. J. Clapham, P. Hamilton, M. Moore, R. Payne, V. Rowntree, C. T. Tynan, J. L. Bannister, and R. DeSalle. 2000. World-wide genetic differentiation of *Eubalaena*: Questioning the number of right whale species. Molecular Ecology 9:1793–802.

Ross, W. G. 1993. Commercial whaling in the North Atlantic sector. Pages 511–61 *in* J. J. Burns, J. J. Montague, and C. J. Cowles, eds. The bowhead whale (Special Publication No. 2). Society for Marine Mammalogy, Lawrence, KS.

Rowlett, R. A., G. A. Green, C. E. Bowlby, and M. A. Smultea. 1994. The first photographic documentation of a northern right whale off Washington State. Northwestern Naturalist 75:102–4.

Rowntree, V. 1983. Cyamids: The louse that moored. Whalewatcher 17(4):14–17.

Rowntree, V. 1996. Feeding, distribution, and reproductive behavior of cyamids (Crustacea: Amphipoda) living on humpback and right whales. Canadian Journal of Zoology 74:103–9.

Rowntree, V., J. Darling, G. Silber, and M. Ferrari. 1980. Rare sighting of a right whale (*Eubalaena glacialis*) in Hawaii. Canadian Journal of Zoology 58:309–12.

Rowntree, V. J., R. S. Payne, and D. M. Schell. 2001. Changing patterns of habitat use by southern right whales (*Eubalaena australis*) on their nursery ground at Península Valdés, Argentina, and in their long-range movements. Journal of Cetacean Research and Management (Special Issue) 2:133–43.

Rugh, D. J., M. M. Muto, S. E. Moore, and D. P. DeMaster. 1999. Status review of the eastern North Pacific stock of gray whales (NOAA Technical Memorandum NMFS-AFSC-103). National Marine Fisheries Service, Seattle, WA.

Russell, B. A., A. R. Knowlton, and B. Zoodsma. 2001. Recommended measures to reduce ship strikes of North Atlantic right whales (Report submitted to National Marine Fisheries Service, Northeast Implementation Team for the Recovery of the Northern Right Whale and Humpback Whale, and Southeast Implementation Team for the Recovery of the Northern Right Whale). J S & A Environmental Services, Chevy Chase, MD.

Ruud, J. T. 1940. The surface structure of the baleen plates as a possible clue to age in whales. Hvalrådets Skrifter 23:1–24.

Ruud, J. T. 1945. Further studies of the structure of baleen plates and their application to age determination. Hvalrådets Skrifter 29:1–69.

Salden, D. R., and J. Mickelsen. 1999. Rare sighting of a North Pacific right whale (*Eubalaena glacialis*) in Hawaii. Pacific Science 53:341–45.

Sayers, H. 1984. Shore whaling for gray whales along the coast of the Californias. Pages 121–57 *in* M. L. Jones, S. L. Swartz, and S. Leatherwood, eds. The gray whale *Eschrichtius robustus*. Academic Press, Orlando, FL.

Scammon, C. M. 1874. The marine mammals of the north-western coast of North America, described and illustrated; together with an account of the American whale-fishery. Carmany, San Francisco.

Scarff, J. E. 1986a. Historic and present distribution of the right whale (*Eubalaena glacialis*) in the eastern North Pacific south of 50°N and east of 180°W. Report of the International Whaling Commission (Special Issue) 10:43–63.

Scarff, J. E. 1986b. Occurrence of the barnacles *Coronula diadema*, *C. reginae* and *Cetopirus complanatus* (Cirripedia) on right whales. Scientific Reports of the Whales Research Institute (Tokyo) 37:129–53.

Scarff, J. E. 1991. Historic distribution and abundance of the right whale (*Eubalaena glacialis*) in the North Pacific, Bering Sea, Sea of Okhotsk and Sea of Japan from the Maury whale charts. Report of the International Whaling Commission 41:467–89.

Scarff, J. E. 2001. Preliminary estimates of whaling-induced mortality in the 19th century North Pacific right whale (*Eubalaena japonicus*) fishery, adjusting for struck-but-lost whales and non-American whaling. Journal of Cetacean Research and Management (Special Issue) 2:261–68.

Schaeff, C. M., and P. K. Hamilton. 1999. Genetic basis and evolutionary significance of ventral skin color markings in North Atlantic right whales (*Eubalaena glacialis*). Marine Mammal Science 15:701–11.

Schaeff, C. M., S. D. Kraus, M. W. Brown, and B. N. White. 1993. Assessment of the population structure of western North Atlantic right whales (*Eubalaena glacialis*) based on sighting and mtDNA data. Canadian Journal of Zoology 71:339–45.

Schaeff, C. M., S. D. Kraus, M. W. Brown, J. S. Perkins, R. Payne, and B. N. White. 1997. Comparison of genetic variability of North Atlantic right whales (*Eubalaena glacialis*) using DNA fingerprinting. Canadian Journal of Zoology 75:1073–80.

Scheffer, V. B., and J. W. Slipp. 1948. The whales and dolphins of Washington State with a key to the cetaceans of the west coast of North America. American Midland Naturalist 39:257–337.

Schell, D. M., and S. M. Saupe. 1993. Feeding and growth as indicated by stable isotopes. Pages 491–509 *in* J. J. Burns, J. J. Montague, and C. J. Cowles, eds. The bowhead whale (Special Publication No. 2). Society for Marine Mammalogy, Lawrence, KS.

Schell, D. M., S. M. Saupe, and N. Haubenstock. 1989. Bowhead whale (*Balaena mysticetus*) growth and feeding as estimated by delta-13C techniques. Marine Biology 103:433–43.

Schell, D. M., V. J. Rowntree, and C. J. Pfeiffer. 2000. Stable-isotope and electron-microscopic evidence that cyamids (Crustacea: Amphipoda) feed on whale skin. Canadian Journal of Zoology 78:721–27.

Schevill, W. E. 1986. Right whale nomenclature. Report of the International Whaling Commission (Special Issue) 10:19.

Schevill, W. E., and W. A. Watkins. 1972. Intense low-frequency sounds from an Antarctic minke whale, *Balaenoptera acutorostrata*. Breviora 388:1–7.

Schevill, W. E., W. A. Watkins, and K. E. Moore. 1986. Status of *Eubalaena glacialis* off Cape Cod. Report of the International Whaling Commission (Special Issue) 10:79–82.

Schilling, M. R., I. Seipt, M. T. Weinrich, S. E. Frohock, A. E. Kuhlberg, and P. J. Clapham. 1992. Behavior of individually-identified sei whales *Balaenoptera borealis* during an episodic influx into the southern Gulf of Maine in 1986. Fishery Bulletin 90:749–55.

Schmidly, D. J. 1981. Marine mammals of the southeastern United States coast and the Gulf of Mexico (Biological Services Program Report FWS/OBS-80/41). U.S. Fish and Wildlife Service, Washington, DC.

Scoresby, W., Jr., 1820. An account of the arctic regions, with a history and description of the northern whale fishery. Constable, Edinburgh.

Sears, R., and F. Larsen. 2002. Long range movements of a blue whale (*Balaenoptera musculus*) between the Gulf of St. Lawrence and West Greenland. Marine Mammal Science 18:281–85.

Sears, R., F. W. Wenzel, and J. M. Williamson. 1987. The blue whale: A catalog of individuals from the western North Atlantic (Gulf of St. Lawrence). Mingan Island Cetacean Study, St. Lambert, Quebec, Canada.

Sears, R., J. M. Williamson, F. W. Wenzel, M. Bérubé, D. Gendron, and P. Jones. 1990. Photographic identification of the blue whale (*Balaenoptera musculus*) in the Gulf of St. Lawrence, Canada. Report of the International Whaling Commission (Special Issue) 12:335–42.

Sergeant, D. E. 1977. Stocks of fin whales *Balaenoptera physalus* L. in the North Atlantic Ocean. Report of the International Whaling Commission 27:460–73.

Shelden, K. E. W., and D. J. Rugh. 1995. The bowhead whale, *Balaena mysticetus*: Its historic and current status. Marine Fisheries Review 57(3–4):1–20.

Shimamura, M., H. Yasue, K. Ohshima, H. Abe, H. Kato, T. Kishiro, M. Goto, I. Munechika, and N. Okada. 1997. Molecular evidence from retroposons that whales form a clade within even-toed ungulates. Nature 388:666–70.

Shimamura, M., H. Abe, M. Nikaido, K. Ohshima, and N. Okada. 1999. Genealogy of families of SINEs in cetaceans and artiodactyls: The presence of a huge superfamily of $tRNA^{Glu}$-derived families of SINEs. Molecular Biology and Evolution 16:1046–60.

Sigurjónsson, J. 1995. On the life history and autecology of North Atlantic rorquals. Pages 425–41 *in* A. S. Blix, L. Walløe, and Ø. Ulltang, eds. Whales, seals, fish and man. Elsevier Science, Amsterdam.

Sigurjónsson, J., and T. Gunnlaugsson. 1990. Recent trends in abundance of blue (*Balaenoptera musculus*) and humpback whales (*Megaptera novaeangliae*) off west and southwest Iceland, with a note on occurrence of other cetacean species. Report of the International Whaling Commission 40:537–51.

Silber, G. K., and P. M. Payne. 1998. Implementation of the northern right whale recovery plan (Document SC/M98/RW9). Unpublished. International Whaling Commission, Cambridge, UK.

Silber, G. K., M. W. Newcomer, P. C. Silber, H. Pérez-Cortés M., and G. M. Ellis. 1994. Cetaceans of the northern Gulf of California: Distribution, occurrence, and relative abundance. Marine Mammal Science 10:283–98.

Simpson, G. G. 1945. The principles of classification and a classification of the mammals. Bulletin of the American Museum of Natural History 85:1–350.

Slay, C. K., and S. D. Kraus. 1998. Right whale tagging in the North Atlantic. Marine Technology Society Journal 32:102–3.

Slijper, E. J. 1962. Whales. Hutchinson, London.

Slijper, E. J. 1966. Functional morphology of the reproductive system in Cetacea. Pages 277–319 *in* K. S. Norris, ed. Whales, dolphins, and porpoises. University of California Press, Berkeley.

Smith, M. R., M. S. Shivji, V. G. Waddell, and M. J. Stanhope. 1996. Phylogenetic evidence from the IRBP gene for the paraphyly of toothed whales, with mixed support of Cetacea as a suborder of Artiodactyla. Molecular Biology and Evolution 13:918–22.

Smith, T. D., and R. R. Reeves. 2002. Estimating historical humpback whale removals from the North Atlantic. Journal of Cetacean Research and Management 4 (Supplement):242–55.

Smith, T. D., J. Allen, P. J. Clapham, P. S. Hammond, S. Katona, F. Larsen, J. Lien, D. Mattila, P. J. Palsbøll, J. Sigurjónsson, P. T. Stevick, and N. Øien. 1999. An ocean-basin-wide mark–recapture study of the North Atlantic humpback whale (*Megaptera novaeangliae*). Marine Mammal Science 15:1–32.

Stafford, K. M., S. L. Nieukirk, and C. G. Fox. 1999. An acoustic link between blue whales in the eastern tropical Pacific and the Northeast Pacific. Marine Mammal Science 15:1258–68.

Stafford, K. M., S. L. Nieukirk, and C. G. Fox. 2001. Geographic and seasonal variation of blue whale calls in the North Pacific. Journal of Cetacean Research and Management 3:65–76.

St. Aubin, D. J., R. H. Stinson, and J. R. Geraci. 1984. Aspects of the structure and composition of baleen, and some effects of exposure to petroleum hydrocarbons. Canadian Journal of Zoology 62:193–98.

Steiger, G. H., and J. Calambokidis. 2000. Reproductive rates of humpback whales off California. Marine Mammal Science 16:220–39.

Stevenson, C. H. 1904. Aquatic products in arts and industries: fish oils, fats, and waxes; fertilizers from aquatic products. Pages 177–279 and plates 10–25 *in* Report of the Commissioner for 1902, Part 28. U.S. Commission on Fish and Fisheries, Washington, DC.

Stevick, P. T., N. Øien, and D. Mattila. 1998. Migration of a humpback whale (*Megaptera novaeangliae*) between Norway and the West Indies. Marine Mammal Science 14:162–66.

Stevick, P. T., J. Allen, P. J. Clapham, N. Friday, S. K. Katona, F. Larsen, J. Lien, D. K. Mattila, P. J. Palsbøll, R. Sears, J. Sigurjónsson, T. D. Smith, G. Vikingsson, N. Øien, and P. S. Hammond. 2001. Trends in abundance of North Atlantic humpback whales, 1979–1993 (Document SC/53/NAH2). Unpublished. International Whaling Commission, Cambridge, UK.

Stewart, B. S., and S. Leatherwood. 1985. Minke whale *Balaenoptera acutorostrata* Lacépède, 1804. Pages 91–136 *in* S. H. Ridgway and R. Harrison, eds. Handbook of marine mammals. Vol. 3: The sirenians and baleen whales. Academic Press, London.

Stoker, S. W., and I. I. Krupnik. 1993. Subsistence whaling. Pages 579–629 *in* J. J. Burns, J. J. Montague, and C. J. Cowles, eds. The bowhead whale (Special Publication No. 2). Society for Marine Mammalogy, Lawrence, KS.

Stone, G. S., S. D. Kraus, J. H. Prescott, and K. W. Hazard. 1988. Significant aggregations of the endangered right whale, *Eubalaena glacialis*, on the continental shelf of Nova Scotia. Canadian Field-Naturalist 102:471–74.

Sutcliffe, W. H., and P. F. Brodie. 1977. Whale distributions in Nova Scotia waters (Technical Report 722). Fisheries and Marine Service, Bedford Institute of Oceanography, Dartmouth, Nova Scotia, Canada.

Swartz, S. L. 1986. Gray whale migratory, social and breeding behavior. Report of the International Whaling Commission (Special Issue) 8:207–29.

Swingle, W. M., S. G. Barco, T. D. Pitchford, W. A. McLellan, and D. A. Pabst. 1993. Appearance of juvenile humpback whales feeding in the nearshore waters of Virginia. Marine Mammal Science 9:309–15.

Terhune, J. M., and W. C. Verboom. 1999. Right whales and ship noises. Marine Mammal Science 15:256–58.

Tershy, B. R., D. Breese, and C. S. Strong. 1990. Abundance, seasonal distribution, and population composition of balaenopterid whales in the Canal de Ballenas, Gulf of California, Mexico. Report of the International Whaling Commission (Special Issue) 12:369–75.

Tershy, B. R., J. Urbán-R., D. Breese, L. Rojas-B., and L. T. Findley. 1993. Are fin whales resident to the Gulf of California? Revista de Investigación Científica de la Universidad Autónoma de Baja California Sur 1:69–71.

Thewissen, J. G. M. 1994. Phylogenetic aspects of cetacean origins: A morphological perspective. Journal of Mammalian Evolution 2:157–84.

Thewissen, J. G. M. 1998. Cetacean origins: Evolutionary turmoil during the invasion of the seas. Pages 451–64 *in* J. G. M. Thewissen, ed. The emergence of whales: Evolutionary patterns in the origin of Cetacea. Plenum Press, New York.

Thewissen, J. G. M., E. M. Williams, L. J. Roe, and S. T. Hussain. 2001. Skeletons of terrestrial cetaceans and the relationship of whales to artiodactyls. Nature 413:277–81.

Thomas, P. O., and S. M. Taber. 1984. Mother–infant interaction and behavioral development in southern right whales, *Eubalaena australis*. Behaviour 88:41–60.

Thompson, D. W. 1918. On whales landed at the Scottish whaling stations, especially during the years 1908–1914. Scottish Naturalist 81:197–208.

Thompson, T. J., H. E. Winn, and P. J. Perkins. 1979. Mysticete sounds. Pages 403–31 *in* H. E. Winn and B. L. Olla, eds. Behavior of marine animals: Current perspectives in research. Vol. 3: Cetaceans. Plenum Press, New York.

Thomson, D. H., and W. J. Richardson. 1995. Marine mammal sounds. Pages 159–204 *in* W. J. Richardson, C. R. Greene, C. I. Malme, and D. H. Thomson, eds. Marine mammals and noise. Academic Press, San Diego, CA.

Tomilin, A. G. 1967. Mammals of the U.S.S.R. and adjacent countries. Vol. 9: Cetacea. Israel Program for Scientific Translations, Jerusalem.

Tønnessen, J. N., and A. O. Johnsen. 1982. The history of modern whaling. University of California Press, Berkeley.

Tormosov, D. D., Y. A. Mikhaliev, P. B. Best, V. A. Zemsky, K. Sekiguchi, and R. L. Brownell, Jr. 1998. Soviet catches of southern right whales *Eubalaena australis*, 1951–1971. Biological data and conservation implications. Biological Conservation 86:185–97.

True, F. W. 1904. The whalebone whales of the western North Atlantic, compared with those occurring in European waters, with some observations on the species of the North Pacific. Smithsonian Contributions to Knowledge 33:1–331.

Tyack, P. 1981. Interactions between singing Hawaiian humpback whales and conspecifics nearby. Behavioral Ecology and Sociobiology 8:105–16.

Tyack, P. 1986. Population biology, social behavior and communication in whales and dolphins. Trends in Ecology and Evolution 1:144–50.

Tyack, P. L. 1999. Communication and cognition. Pages 287–323 *in* J. E. Reynolds III and S. A. Rommel, eds. Biology of marine mammals. Smithsonian Institution Press, Washington, DC.

Tynan, C. T., D. P. DeMaster, and W. T. Peterson. 2001. Endangered right whales on the southeastern Bering Sea shelf. Science 294:1894.

Ursing, B. M., and Ú. Árnason. 1998. Analyses of mitochondrial genomes strongly support a hippopotamus–whale clade. Proceedings of the Royal Society of London B 265:2251–55.

Van Valen, L. 1966. Deltatheridia, a new order of mammals. Bulletin of the American Museum of Natural History 132:1–126.

Volgenau, L., S. D. Kraus, and J. Lien. 1995. The impact of entanglement on two substocks of the western North Atlantic humpback whale, *Megaptera novaeangliae*. Canadian Journal of Zoology 73:1689–98.

Wada, S., and K. Numachi. 1991. Allozyme analyses of genetic differentiation among the populations and species of *Balaenoptera*. Report of the International Whaling Commission (Special Issue) 13:125–54.

Wade, P. R. 2002. Population dynamics. Pages 974–79 *in* W. F. Perrin, B. Würsig, and J. G. M. Thewissen, eds. Encyclopedia of marine mammals. Academic Press, San Diego, CA.

Wade, P. R., and D. P. DeMaster. 1996. A Bayesian analysis of eastern Pacific gray whale population dynamics (Document SC/48/AS3). Unpublished. International Whaling Commission, Cambridge, UK.

Wade, P. R., and T. Gerrodette. 1993. Estimates of cetacean abundance and distribution in the eastern tropical Pacific. Report of the International Whaling Commission 43:477–93.

Ward, J. A. 1999. Right whale (*Balaena glacialis*) South Atlantic Bight habitat characterization and prediction using remotely sensed oceanographic data. M. Sc. Thesis, University of Rhode Island, Kingston, RI.

Waring, G. T., J. M. Quintal, and C. P. Fairfield, eds. 2002. U.S. Atlantic and Gulf of Mexico marine mammal stock assessments—2002 (NOAA Technical Memorandum NMFS-NE-169). National Marine Fisheries Service, Woods Hole, MA.

Wartzok, D., and D. R. Ketten. 1999. Marine mammal sensory systems. Pages 117–75 *in* J. E. Reynolds III and S. A. Rommel, eds. Biology of marine mammals. Smithsonian Institution Press, Washington, DC.

Watkins, W. A. 1986. Whale reactions to human activities in Cape Cod waters. Marine Mammal Science 2:251–62.

Watkins, W. A., and W. E. Schevill. 1976. Right whale feeding and baleen rattle. Journal of Mammalogy 57:58–66.

Watkins, W. A., and W. E. Schevill. 1979. Aerial observations of feeding behavior in four baleen whales: *Eubalaena glacialis*, *Balaenoptera borealis*, *Megaptera novaeangliae*, and *Balaenoptera physalus*. Journal of Mammalogy 60:155–63.

Watkins, W. A., and W. E. Schevill. 1982. Observations of right whales, *Eubalaena glacialis*, in Cape Cod waters. Fishery Bulletin 80:875–80.

Watkins, W. A., and W. E. Schevill. 1983. Three right whales (*Eubalaena glacialis*) alternating at the surface. Journal of Mammalogy 64:506–8.

Watkins, W. A., and D. Wartzok. 1985. Sensory biophysics of marine mammals. Marine Mammal Science 1:219–60.

Watkins, W. A., P. Tyack, K. E. Moore, and J. E. Bird. 1987. The 20-Hz signals of finback whales (*Balaenoptera physalus*). Journal of the Acoustical Society of America 82:1901–12.

Watkins, W. A., M. A. Daher, G. M. Reppucci, J. E. George, D. L. Martin, N. A. DiMarzio, and D. P. Gannon. 2000. Seasonality and distribution of whale calls in the North Pacific. Oceanography 13:62–67.

Webb, R. L. 1988. On the Northwest: Commercial whaling in the Pacific Northwest 1790–1967. University of British Columbia Press, Vancouver, Canada.

Weinrich, M. T., M. R. Schilling, and C. R. Belt. 1992. Evidence for acquisition of a novel feeding behavior: Lobtail feeding in humpback whales, *Megaptera novaeangliae*. Animal Behavior 44:1059–72.

Weinrich, M. T., M. Martin, R. Griffiths, J. Bove, and M. Schilling. 1997. A shift in distribution of humpback whales, *Megaptera novaeangliae*, in response to prey in the southern Gulf of Maine. Fishery Bulletin 95:826–36.

Weinrich, M. T., R. D. Kenney, and P. K. Hamilton. 2000. Right whales (*Eubalaena glacialis*) on Jeffreys Ledge: A habitat of unrecognized importance? Marine Mammal Science 16:326–37.

Weisbrod, A. V., D. Shea, M. J. Moore, and J. J. Stegeman. 2000. Organochlorine exposure and bioaccumulation in the endangered northwest Atlantic right whale (*Eubalaena glacialis*) population. Environmental Toxicology and Chemistry 19:654–66.

Weller, D. W., B. Würsig, A. L. Bradford, A. M. Burdin, S. A. Blokhin, H. Minakuchi, and R. L. Brownell, Jr. 1999. Gray whales (*Eschrichtius robustus*) off Sakhalin Island, Russia: Seasonal and annual patterns of occurrence. Marine Mammal Science 15:1208–27.

Wenzel, F. W., and R. Sears. 1988. A note on killer whales in the Gulf of St. Lawrence, including an account of an attack on a minke whale. Rit Fiskideildar 11:202–4.

Wenzel, F. W., D. K. Mattila, and P. J. Clapham. 1988. *Balaenoptera musculus* in the Gulf of Maine. Marine Mammal Science 4:172–75.

Whitehead, H. 1982. Populations of humpback whales in the western North Atlantic. Report of the International Whaling Commission 32:345–53.

Whitehead, H. 2002. Breaching. Pages 162–64 *in* W. F. Perrin, B. Würsig, and J. G. M. Thewissen, eds. Encyclopedia of marine mammals. Academic Press, San Diego, CA.

Whitehead, H., and J. Lien. 1982. Changes in the abundance of whales, and whale damage, along the Newfoundland coast 1973–1981 (Document SC/34/O1). Unpublished. International Whaling Commission, Cambridge, UK.

Whitehead, H., and M. J. Moore. 1982. Distribution and movements of West Indian humpback whales in winter. Canadian Journal of Zoology 60:2203–11.

Williams, E. M. 1998. Synopsis of the earliest cetaceans. Pages 1–28 *in* J. G. M. Thewissen, ed. The emergence of whales: Evolutionary patterns in the origin of Cetacea. Plenum Press, New York.

Winn, H. E., and P. J. Perkins. 1976. Distribution and sounds of the minke whale, with a review of mysticete sounds. Cetology 19:1–12.

Winn, H. E., and N. E. Reichley. 1985. Humpback whale (*Megaptera novaeangliae*). Pages 241–73 *in* S. H. Ridgway and R. Harrison, eds. Handbook of marine mammals. Vol. 3: The sirenians and baleen whales. Academic Press, London.

Winn, H. E., and L. K. Winn. 1978. The song of the humpback whale (*Megaptera novaeangliae*) in the West Indies. Marine Biology 47:97–114.

Winn, H. E., R. K. Edel, and A. G. Taruski. 1975. Population estimate of the humpback whale (*Megaptera novaeangliae*) in the West Indies by visual and acoustic techniques. Journal of the Fisheries Research Board of Canada 32:499–506.

Winn, H. E., C. A. Price, and P. W. Sorensen. 1986. The distributional ecology of the right whale *Eubalaena glacialis* in the western North Atlantic. Report of the International Whaling Commission (Special Issue) 10:129–38.

Winn, H. E., J. D. Goodyear, R. D. Kenney, and R. O. Petricig. 1995. Dive patterns of tagged right whales in the Great South Channel. Continental Shelf Research 15:593–611.

Wishner, K., E. Durbin, A. Durbin, M. Macaulay, H. Winn, and R. Kenney. 1988. Copepod patches and right whales in the Great South Channel off New England. Bulletin of Marine Science 43:825–44.

Wishner, K., J. R. Schoenherr, R. Beardsley, and C. Chen. 1995. Abundance, distribution and population structure of the copepod *Calanus finmarchicus* in a springtime right whale feeding area in the southwestern Gulf of Maine. Continental Shelf Research 15:475–507.

Wolman, A. A. 1985. Gray whale *Eschrichtius robustus* (Lilljeborg, 1861). Pages 67–90 *in* S. H. Ridgway and R. Harrison, eds. Handbook of marine mammals. Vol. 3: The sirenians and baleen whales. Academic Press, London.

Woodby, D. A., and D. B. Botkin. 1993. Stock sizes prior to commercial whaling. Pages 387–407 *in* J. J. Burns, J. J. Montague, and C. J. Cowles, eds. The bowhead whale (Special Publication No. 2). Society for Marine Mammalogy, Lawrence, KS.

Woodhouse, C. D., Jr., and J. Strickley. 1982. Sighting of a northern right whale (*Eubalaena glacialis*) in the Santa Barbara Channel. Journal of Mammalogy 63:701–2.

Woodley, T. H., M. W. Brown, S. D. Kraus, and D. E. Gaskin. 1991. Organochlorine levels in North Atlantic right whale (*Eubalaena glacialis*) blubber. Archives of Environmental Contaminants and Toxicology 21:141–45.

Würsig, B., and C. W. Clark. 1993. Behavior. Pages 157–99 *in* J. J. Burns, J. J. Montague, and C. J. Cowles, eds. The bowhead whale (Special Publication No. 2). Society for Marine Mammalogy, Lawrence, KS.

Würsig, B., E. M. Dorsey, M. A. Fraker, R. S. Payne, and W. J. Richardson. 1985. Behavior of bowhead whales, *Balaena mysticetus*, summering in the Beaufort Sea. Fishery Bulletin 83:357–77.

Wynne, K. 1997. Guide to marine mammals of Alaska; 2nd ed. Alaska Sea Grant College Program, University of Alaska, Fairbanks.

Wynne, K., and M. Schwartz. 1999. Guide to marine mammals and turtles of the U.S. Atlantic and Gulf of Mexico. Rhode Island Sea Grant, Narragansett.

Yablokov, A. V., V. M. Bel'kovich, and V. I. Borisov. 1974. Whales and dolphins. Publications Research Service, Arlington, VA.

Yochem, P. K., and S. Leatherwood. 1985. Blue whale *Balaenoptera musculus* (Linnaeus, 1758). Pages 193–240 *in* S. H. Ridgway and R. Harrison, eds. Handbook of marine mammals. Vol. 3: The sirenians and baleen whales. Academic Press, London.

Zeh, J. E., C. W. Clark, J. C. George, D. Withrow, G. M. Carroll, and W. R. Koski. 1993. Current population size and dynamics. Pages 409–89 *in* J. J. Burns, J. J. Montague, and C. J. Cowles, eds. The bowhead whale (Special Publication No. 2). Society for Marine Mammalogy, Lawrence, KS.

Zeh, J. E., A. E. Raftery, and A. A. Schaffner. 1995. Revised estimates of bowhead population size and rate of increase (Document SC/47/AS10). Unpublished. International Whaling Commission, Cambridge, UK.

Zhou, X., R. Zhai, P. D. Gingerich, and L. Chen. 1995. Skull of a new mesonychid (Mammalia, Mesonychia) from the late Paleocene of China. Journal of Vertebrate Paleontology 15:387–400.

Randall R. Reeves, Okapi Wildlife Associates, 27 Chandler Lane, Hudson, Quebec, Canada J0P 1H0. Email: rrreeves@total.net.

Robert D. Kenney, Graduate School of Oceanography, University of Rhode Island, Narragansett, Rhode Island 02882-1197. Email: rkenney@gso.uri.edu.

V

Carnivores

22

Coyote

Canis latrans

Marc Bekoff
Eric M. Gese

NOMENCLATURE

COMMON NAMES. Coyote, brush wolf, prairie wolf, Heul wolf, Steppenwolf, lobo, American jackal; the word *coyote* means "barking dog" and is taken from the Aztec word *coyotl*
SCIENTIFIC NAME. *Canis latrans*

The coyote is one of eight recognized species in the genus *Canis*. By the late Pliocene, the ancestral coyote, *Canis lepophagus*, was widespread in North America. The Eastern coyote, formerly referred to as the New England canid, appears to be a recent immigrant, having predominantly coyote ancestry with some introgression of wolf (*C. lupus*) and dog (*C. familiaris*) genes (Lawrence and Bossert 1969, 1975; Silver and Silver 1969; Bekoff et al. 1975; Hilton 1978; Wayne and Lehman 1992). Lehman et al. (1991) and Wayne and Lehman (1992) presented evidence that coyotes have interbred with wolves in areas where wolves are rare and conspecific mates may not be readily available.

Subspecies. There are 19 recognized subspecies of *C. latrans*. However, because of the mobility of coyotes, the integrity of individual subspecies and their taxonomic utility are questionable (Nowak 1978). In chronological order of being named, the subspecies are *C. l. latrans, C. l. ochropus, C. l. cagottis, C. l. frustror, C. l. lestes, C. l. mearnsi, C. l. microdon, C. l. peninsulae, C. l. vigilis, C. l. clepticus, C. l. impavidus, C. l. goldmani, C. l. texensis, C. l. jamesi, C. l. dickeyi, C. l. incolatus, C. l. hondurensis, C. l. thamnos,* and *C. l. umpquensis* (Jackson 1951).

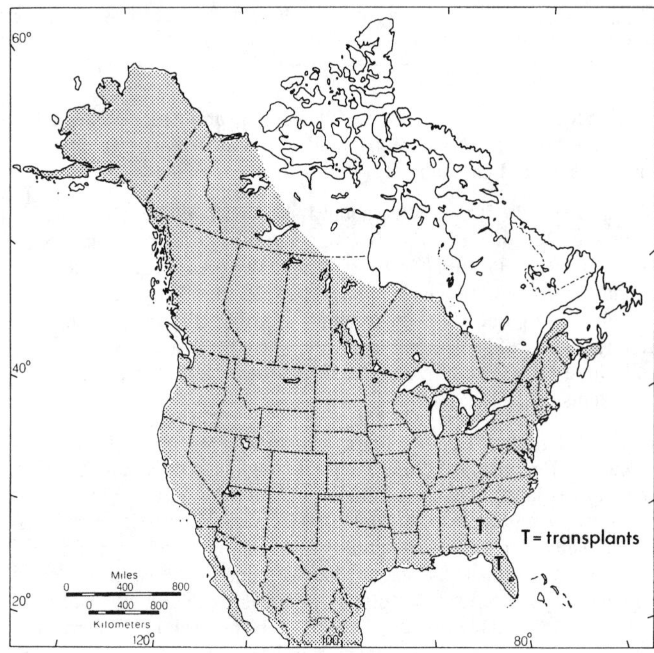

FIGURE 22.1. Distribution of the coyote (*Canis latrans*).

DISTRIBUTION

Coyotes are Nearctic canids. They occupy many diverse habitats between about 10°N latitude (Costa Rica) and 70°N latitude (northern Alaska). They are found throughout the continental United States and in many areas of Canada (Fig. 22.1). In some southeastern states, such as Florida (Cunningham and Dunford 1970) and Georgia (Fisher 1975), and several eastern states, it appears that coyotes were transplanted, introduced, or liberated by humans (Schultz 1955; Hill et al. 1987). See Moore and Parker (1992) for the history of coyote expansion and changes in distribution. With its ability to adapt to an environment modified by humans, coyotes are now observed in large cities (Shargo 1988; Quinn 1997a, 1997b; Grinder and Krausman 1998; Finkel 1999).

DESCRIPTION

Coyotes are often confused with other canids, such as gray wolves, red wolves (*C. rufus*), and domestic dogs. Coyotes can successfully interbreed and produce fertile hybrids with all these species (Dice 1942; Kennelly and Roberts 1969; Kolenosky 1971). However, coyotes can usually be differentiated (although overlap and hybridization can occur) using serologic parameters, dental characteristics, cranial measurements, neuroanatomical features, diameter of the nose pad, diameter of the hind foot pad, ear length, track size, stride length, pelage, behavior, and genetics (for reviews, see Lawrence and Bossert 1967; Bekoff 1977a; Elder and Hayden 1977). For example, coyotes are typically smaller than gray wolves (Table 22.1), thus the nose pad (about 25 mm in diameter) and hind foot pads (less than 32 mm) are correspondingly smaller.

Coyotes may be differentiated from dogs using the ratio of palatal width (distance between the inner margins of the alveoli of the upper first molars) to the length of the upper molar toothrow (from the anterior margin of the alveolus of the first premolar to the posterior margin of the last molar alveolus) (Howard 1949). If the toothrow is 3.1 times the palatal width, the specimen is a coyote; if the ratio is less than 2.7, the specimen is a dog (this method is about 95% reliable).

The coyote has a relatively larger braincase than *C. lupus* (Mech 1974). The coyote brain is anatomically different from that of gray wolves (Radinsky 1973; Atkins 1978). The wolf has a dimple in the middle of the coronal gyrus, whereas the coyote does not (see also Atkins and Dillon 1971). There is no overlap when comparing large coyotes to small wolves in zygomatic breadth (greatest distance across zygomata), greatest length of the skull, or bite ratio (width across the outer edges of the alveoli of the anterior lobes of the upper carnassials divided by the length of the upper molar toothrow) (Paradiso and Nowak 1971). *C. latrans* is usually smaller than *C. rufus* and there is almost no overlap between them in greatest length of the skull. Red wolves also have a more pronounced sagittal crest than coyotes. Multivariate techniques have clearly shown that coyotes, wolves, and dogs can be differentiated anatomically (Lawrence and Bossert 1967; Elder and Hayden 1977) and behaviorally (Bekoff et al. 1975; Bekoff 1978) and

TABLE 22.1. Representative mean coyote weights from a variety of locales

Source	Adults (kg)		Juveniles (kg)		State/Province
	Males	Females	Males	Females	
Gier 1968	14.1	11.8	—	—	Kansas
Hawthorne 1971	11.2	9.8	—	—	NE California
Richens and Hugie 1974	15.8	13.7	—	—	Maine
Andrews and Boggess 1978	13.4	11.4	—	—	Iowa
Berg and Chesness 1978	12–13	11–12	10–11	10	Minnesota
Boggess and Henderson 1978	13.1	11.0	—	—	Kansas
Bowen 1978	12.1	11.5	—	—	Alberta
Litvaitis 1978	14.7	12.1	—	—	Oklahoma
Murray and Boutin 1991	10.3	8.0	—	—	Yukon
Thurber and Peterson 1991	12.9	11.1	—	—	Alaska
Windberg et al. 1991	10.6–11.4	9.1–9.6	10.0–10.8	8.6–8.9	Texas
Poulle et al. 1995	12.5–16.0	11.0–14.2	—	—	Quebec
Windberg 1995	10.9–11.0	—	8.8–9.2	8.0–8.4	Texas

provide for more rigorous analyses than do univariate methods. Further refinement of genetic techniques will also assist in differentiation of the canids (e.g., Lehman et al. 1991; Wilson et al. 2000).

Size and Weight. Coyotes are about 1–1.5 m in body length; the tail is about 400 mm long. Size varies with geographic locale and subspecies (Jackson 1951; Hall and Kelson 1959). Adult males are usually heavier and larger than adult females (Table 22.1). Temporal changes in coyote morphology may be occurring in some parts of North America (Schmitz and Lavigne 1987; Thurber and Peterson 1991), but see Larivière and Crête (1993) and Peterson and Thurber (1993) for reviews and comments.

Pelage. The banded nature of coyote hair is responsible for the appearance of the blended color, gray mixed with a reddish tint. Coyotes show great variation in color, ranging from almost pure gray to rufous. Melanistic coyotes are rare (Young 1951; Van Wormer 1964; Gipson 1976; Mahan 1978). Texture and color of the fur also vary geographically. In northern subspecies, the hair is longer and coarser. In desert habitats, coyotes tend to be fulvous, whereas those at higher latitudes are more gray and black (Jackson 1951). The belly and throat are paler than the rest of the body. Course guard hairs are about 50–90 mm long; in the mane, they tend to be 80–110 mm. The fine underfur (up to 50 mm long) has coronal-shaped cuticular scales (Adorjan and Kolenosky 1969; Ogle and Farris 1973). The summer coat is shorter than the winter coat. Coyote hair may be differentiated from hair of dogs and red foxes (*Vulpes vulpes*) by the number, order, and color of the bands, the cross-sectional translucence and shape, and the coronal scale pattern (Hilton and Kutscha 1978). Coyote hairs typically are coarser, longer, larger in diameter, and rougher and stiffer.

The coyote's fur is similar in insulative value to that of the gray wolf (Ogle and Farris 1973). The critical temperature of *C. latrans* is −10°C (Shield 1972). When wearing the shorter summer coat, there is a decrease of about 87% in thermal conductivity (Ogle and Farris 1973). There is usually one main molt between late spring and autumn. About 50 mm down from the base of the tail there is an oval tail gland (Hildebrand 1952).

Skull and Dentition. Among adult coyotes, males show greater development of the sagittal crest than females. The dental formula is I 3/3, C 1/1, P 4/4, M 2/3 (Fig. 22.2). The skull of a mature male is about 180–205 mm long from the tip of the premaxilla to the posterior rim of the coronal crest (Gier 1968) and weighs between 170 and 210 g.

PHYSIOLOGY

Central Nervous System. Although the cerebrum and cerebellum share many common features with other canids, there are some interspecific differences (for reviews see Atkins and Dillon 1971; Atkins 1978). With respect to cerebellar morphology, coyotes may be distinguished from all other canids as follows: The anterior lobe is more than one half the total width, the parafloccular process is relatively prominent, the vermian lobule reaches its greatest size, there are fewer and larger posterior hemispheric folia, the posterior ventral parafloccular limb is reduced in size, and there is a broad vermian twist (Atkins 1978). The remainder of the central nervous system, the brain stem and spinal cord, is similar to that of the domestic dog.

Adrenals. Coyote adrenals are similar in structure to those of most other canids (Heinrich 1972). In males and females, the left adrenal is heavier than the right, and the adrenals of females tend to be heavier than those of males.

Audition and Vision. The region of maximal sensitivity to auditory stimuli is 100–30,000 Hz with a limit of approximately 80 kHz (Petersen et al. 1969). The retina is duplex and has a preponderance of rods. The absolute scotopic (rod) threshold is about 1.4 ft-candles and the adaptation curve shows distinct rod–cone breaks (Horn and Lehner 1975).

REPRODUCTION

Genetics and Hybridization. The coyote has 38 pairs of chromosomes (Wurster and Benirschke 1968). The autosomes are acrocentric or telocentric and the sex chromosomes are submetacentric (Mech 1974). Fertile hybrids have been produced by matings of coyotes with domestic dogs (Dice 1942; Young 1951; Kennelly and Roberts 1969; Silver and Silver 1969; Mengel 1971), red and gray wolves (Young 1951; Kolenosky 1971; Paradiso and Nowak 1971; Riley and McBride 1975), and golden jackals (*C. aureus;* Seitz 1965). Coyote–dog hybrids exhibit decreased fecundity (Mengel 1971; Gipson et al. 1975). Hybridization between coyotes and red wolves is becoming problematic for red wolf recovery in the southeast United States.

Anatomy and Physiology. There are no detailed reports of the gross or microscopic anatomy of coyote reproductive systems. Patterns of reproductive hormones and reproductive behaviors in coyotes have been described (Bekoff and Diamond 1976; Hodges 1990; Parrish 1994). With the exception of the seasonality of breeding and associated changes in reproductive anatomy and physiology, there appear to be only minor differences, if any, between coyotes and domestic dogs (Kennelly 1978). Berg and Chesness (1978) found no correlation between carcass weight and ovarian weight ($n = 105, r = .218, p > .05$).

Kennelly (1972, 1978) documented spermatogenesis and the estrous cycle. Proestrus lasts about 2–3 months and estrus up to 10 days, depending on locale (Hamlett 1939; Kennelly 1978). Copulation ends with the copulatory "tie," during which time (up to 25 min) the male's penis is locked in the female's vagina (Grandage 1972). Juvenile males and females are able to breed (see below), although juvenile females may ovulate less than adult females.

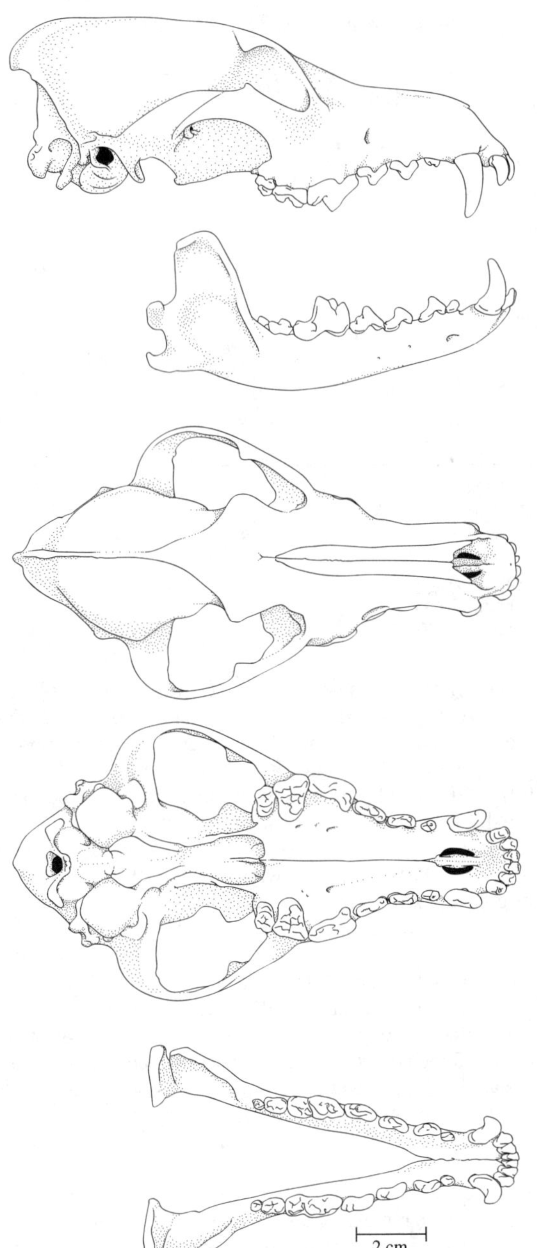

FIGURE 22.2. Skull of the coyote (*Canis latrans*). From top to bottom: lateral view of cranium, lateral view of mandible, dorsal view of cranium, ventral view of cranium, dorsal view of mandible.

Males and females show annual cyclic changes in reproductive anatomy and physiology (Kennelly 1978). Females are seasonally monestrus, showing one period of "heat" per year, usually during January and March, depending on geographic locale (Hamlett 1939; Gier 1968; Kennelly 1978).

Pair Bonding. The dynamics of heterosexual pair bonding are not known for wild coyotes in any detail. Data on captive individuals do not appear to differ significantly from those for wild coyotes. Courtship may begin as long as 2–3 months before successful copulation (Bekoff and Diamond 1976). Associated changes in behavior have been described, especially increases in scent marking and howling observed at the beginning of the breeding season (Bekoff and Diamond 1976; Wells and Bekoff 1981; Hodges 1990; Gese and Ruff 1997, 1998). During early stages of courtship, the male becomes increasingly attracted to the female's urine and feces. When the female is ready to copulate, she will tolerate mounting attempts by the male and will flag her tail to one side. When tying, the male steps over the female's back and the couple remain locked at 180° for up to 25 min. The same pair may breed from year to year, but not necessarily for life. Adult coyotes may maintain pair bonds and whelp or sire pups even when 10–12 years of age (Gese 1990).

Pregnancy Rate, Gestation, and Litter Size. The percentage of females that breed in a given year varies with local conditions (Gier 1968; Knowlton 1972; Gipson et al. 1975; Gese et al. 1989a; Knowlton and Gese 1995). Food supply is usually the prime factor; in good years, more females, especially yearlings, breed (Gier 1968; Knowlton and Gese 1995). Usually, about 60–90% of adult females and 0–70% of female yearlings will produce litters (Knowlton 1972; Gese et al. 1989a; Knowlton et al. 1999). The greatest annual variation in the number of breeding females is related to the number of juveniles that become sexually mature (Knowlton 1972; Kennelly 1978; Gese et al. 1989a). Adults may show more placental scars than yearlings; however, Nellis and Keith (1976) reported that the difference was not statistically significant. Gier (1975) estimated that the number of young born was about 80% of the ova shed, and Knowlton (1972) estimated that about 87% of implants were represented by viable young. Gipson et al. (1975) reported the mean number of ova per breeding female was 6.2, with 4.5 (73%) becoming implanted. Hamlett (1939) and Asdell (1964) reported that 85–92% of embryos develop into viable young.

Gestation lasts approximately 63 days. Average litter size is 6, but it is known that litter size is affected by population density and food availability the previous winter (Knowlton 1972; Gese et al. 1996a, 1996b; Knowlton et al. 1999). Knowlton (1972) reported average litter sizes of 4.3 at high population densities and 6.9 at low densities. Gier (1968) reported the effects of food on litter size. During years of high rodent density, mean litter size is higher (5.8–6.2) than in years of reduced densities of rodents (4.4–5.1). In northern climes, coyote litter size changes in response to cycles of snowshoe hares (*Lepus americanus*) (Todd et al. 1981; Todd and Keith 1983; O'Donoghue et al. 1997). Gese et al. (1996a) found an increase in coyote litter size after cold, snowy winters in Yellowstone National Park. Winters with harsh temperatures and deep snow increased the number of ungulate carcasses (because of coyote predation and weather) available to ovulating females. The sex ratio of litters is generally 1:1, though it may be skewed toward females in areas of high exploitation (Knowlton 1972; Berg and Chesness 1978; Kleiman and Brady 1978; Gese et al. 1989a).

Development. Young are born blind and helpless in a den. Birth weight is about 240–275 g and the length of the body from the tip of the head to the base of the tail is about 160 mm (Gier 1968; Bekoff and Jamieson 1975). Eyes open at about 14 days. Teeth erupt on the average as follows: upper canines, day 14; lower canines and upper incisors, day 15; and lower incisors, day 16 (Bekoff and Jamieson 1975). The young are able to urinate and defecate on their own by 2–3 weeks. They emerge from the den at about 3 weeks. Young are cared for by the mother and by other "helpers," usually siblings from a previous year (Bekoff and Wells 1980; Hatier 1995). Food provisioning of the female may occur during the nursing period (Hatier 1995). The alpha male and other associates will also help to rear the young by providing food as the young grow older (Hatier 1995). Young are weaned at about 5–7 weeks (Snow 1967). They begin to eat solid food at about 3 weeks, when the caregivers regurgitate semisolid food. Between birth and week 8, average weight increase is about 310 g/week. The pups reach adult weight at about 9 months.

ECOLOGY

More is known about the ecology of coyotes than perhaps any other carnivore (Bekoff 2001a). Coyotes occupy a variety of habitats, including grasslands, deserts, and mountains. They do not compete well with larger carnivores and may be killed by them or avoid areas and habitats occupied by these species. Studies have documented direct and indirect competition with larger carnivores, such as wolves (Mech

1966, 1974; Krefting 1969; Fuller and Keith 1981; Thurber et al. 1992; Peterson 1995; Arjo and Pletscher 1999; Crabtree and Sheldon 1999) and cougars, *Puma concolor* (Young 1951; Boyd and O'Gara 1985; Koehler and Hornocker 1991; Murphy 1998). Arjo and Pletscher (1999) documented behavioral changes in coyotes after several years of increased wolf abundance in northwestern Montana. Paquet (1991, 1992) reported that coyotes did not spatially avoid wolves, but actually followed their trails and scavenged wolf-killed ungulates.

Interspecific killing appears to be common in carnivore communities (Peterson 1995; Palomares and Caro 1999). In Yellowstone National Park, Crabtree and Sheldon (1999) reported that wolves killing coyotes during the winters of 1997 and 1998 reduced coyote numbers; average pack size decreased from 6 coyotes to 4 (a 33% decline). However, Gese et al. (1996a, 1996b) documented similar annual variations in coyote pack sizes in the same area in the absence of wolves. Average pack size changed from 4.6 to 6.8 coyotes during three winters (1990–93) several years before wolf reintroduction, an increase of 32%. Thus, annual variation in coyote pack size and population size attributed to wolf reintroduction should be viewed in the context of baseline population data that documented these same fluctuations had occurred before wolf reintroduction. Changes in the abundance of food resources, mainly cyclic lagomorph populations, causes even greater annual variations (3–10 times) in coyote populations than populations exposed to wolves (Todd et al. 1981; Knowlton and Stoddart 1992; O'Donoghue et al. 1997). Direct predation and competition for food and space with wolves may limit coyote numbers in some areas and under certain environmental conditions (Peterson 1995; Arjo and Pletscher 1999).

Coyotes may not tolerate red foxes in some areas (Voigt and Earle 1983; Major and Sherburne 1987; Sargeant et al. 1987; Harrison et al. 1989; Sargeant and Allen 1989), but appear to be more tolerant of red foxes when food is abundant (Gese et al. 1996d). Coyotes also will kill many of the small canids, particularly swift (*Vulpes velox*), kit (*V. macrotis*), and gray foxes (*Urocyon cinereoargenteus*). Coexistence between coyotes and these small canids is typically mediated by resource partitioning (Ralls and White 1995; White et al. 1995; Cypher and Spencer 1998; Kitchen et al. 1999). Bobcats (*Lynx rufus*) also may not be well tolerated by coyotes (Young 1951), but Major and Sherburne (1987) found no evidence of interference competition between bobcats and coyotes. Soulé et al. (1988) and Crooks and Soulé (1999) reported that coyotes in southern California apparently control the abundance and distribution of smaller predators. In the absence of coyotes, these mesopredators (i.e., foxes and feral cats) increase in density. They prey on native bird species and negatively impact the avifaunal community. Henke and Bryant (1999) documented that coyotes were considered a keystone predator shaping faunal community structure in west Texas.

Population Regulation. Population demographics of coyotes have been studied throughout North America (e.g., Gier 1968; Knowlton 1972; Todd and Keith 1983; Windberg 1995). They exhibit a land-tenure system of exclusive territories (Camenzind 1978; Bowen 1981, 1982; Messier and Barrette 1982; Windberg and Knowlton 1988; Knowlton and Gese 1995), and within resident packs, they display a dominance hierarchy similar to that of wolves (Camenzind 1978; Bowen 1978; Bekoff and Wells 1986; Gese et al. 1996a). The social organization and land-tenure system mediate the regulation of coyote numbers as packs space themselves across the landscape in relation to available food and habitat (Knowlton and Stoddart 1983; Bekoff and Wells 1986; Gese et al. 1988a; Knowlton and Gese 1995; Knowlton et al. 1999). The social hierarchy and dominance structure among members of resident packs also influence accessibility to food resources (Gese et al. 1996a, 1996b). Older, experienced pack members are more successful hunters of large prey (Gese and Grothe 1995), have greater access to ungulate carcasses (Gese et al. 1996b), and are more proficient hunters of small mammals (Gese et al. 1996c). Transient or nomadic coyotes also exist across the landscape (Camenzind 1978; Bowen 1982; Bekoff and Wells 1986; Gese et al. 1988a) and may move into territories whenever vacancies occur.

TABLE 22.2. Coyote densities in different geographic areas and seasons

Source	Location	Density (individuals/km^2)
Gier 1968	Kansas	0.8[a]
Knowlton 1972	Texas	0.9[a]; 1.5–2.3[b]
Chesness and Bremicker 1974	Minnesota	0.2–0.4[a]
Nellis and Keith 1976	Alberta	0.1–0.6[c]
Bowen 1978	Alberta	0.46[a]; 0.35[d]
Todd et al. 1981	Alberta	0.08–0.44[c]
Pyrah 1984	Montana	0.15[e]; 0.39[f]
Andelt 1985	Texas	0.9[g]
Gese et al. 1989a	Colorado	0.26–0.33[g]
Babb and Kennedy 1989	Tennessee	0.35[g]
O'Donoghue et al. 1997	Yukon	0.01–0.09[c]
Henke and Bryant 1999	Texas	0.12–0.14[g]

[a]Postwhelping.
[b]Fall.
[c]Winter.
[d]Late winter.
[e]Spring.
[f]Summer.
[g]Prewhelping.

Coyote density varies geographically and seasonally (Table 22.2) in response to changing food resources. Available food, whether rabbits, rodents, or ungulates, is the major factor regulating coyote populations (Gier 1968; Clark 1972; Knowlton and Stoddart 1992; O'Donoghue et al. 1997). Nutritional pressures for limited food resources are mediated through social dominance and territoriality (Knight 1978; Davison 1980; Knowlton and Stoddart 1983; Gese et al. 1989a; Knowlton and Gese 1995; Windberg 1995; Knowlton et al. 1999). Food abundance regulates coyote numbers by influencing pack size, reproductive rates, survival, dispersal, and space-use patterns (Gier 1968; Knowlton 1972; Todd et al. 1981; Todd and Keith 1983; Todd 1985; Bekoff and Wells 1986; Mills and Knowlton 1991; Harrison 1992; Windberg 1995; Gese et al. 1996a; O'Donoghue et al. 1997; Knowlton et al. 1999). In areas with hard winters, when carrion biomass is low, coyote pack sizes remain small and the core social unit subsists on small mammals (Gese et al. 1996a, 1996b). In contrast, during winters when carrion biomass is greater, more coyotes remain in their social groups and pack size increases (Gese et al. 1996a, 1996b). In addition, dominant individuals in resident packs have greater access to carcasses and are thereby less likely to disperse (Gese et al. 1996a, 1996b). Thus, dominance plays an important role in regulating coyote numbers. Knowlton et al. (1999) reported that the acquisition of a territory is also important because resident coyotes are more apt to survive, have more breeding opportunities, and are more likely to have access to carcasses in winter than are transient individuals (Andelt 1985; Bekoff and Wells 1986; Gese et al. 1989a, 1996a, 1996b).

Within resident coyote packs, size and structure change seasonally (Knowlton 1972; Davison 1980; Bekoff and Wells 1986; Gese et al. 1996a). Pack size increases during the whelping season (April), followed by a gradual decline as pups die or disperse and associated pack members disperse during winter as food resources become more limited (Knowlton 1972; Davison 1980; Gese et al. 1996a; Knowlton et al. 1999). If more food resources are available over winter, pack size may increase (Gese et al. 1996a, 1996b).

Effects of Exploitation. There have been few studies of unexploited coyote populations (Andelt 1985; Bekoff and Wells 1986; Crabtree 1988; Gese et al. 1989a, 1996a; Windberg 1995). These populations generally differ from exploited populations by having an older age structure, higher survival rates, lower reproductive rates, larger pack size, and lower recruitment into the adult population (Andelt 1985; Windberg et al. 1985; Gese et al. 1989a, 1996a, 1996b; Windberg 1995; Knowlton et al. 1999). Under high levels of exploitation, coyote populations usually have a younger age structure, lower survival, increased numbers of yearlings reproducing, increased litter size, and relatively small packs (Gier 1968; Knowlton 1972; Berg and Chesness 1978; Davison 1980; Andelt 1987; Knowlton et al. 1999). Litter size may increase due to reduced coyote density, likely in response to reduced competition for food (Andelt 1987, 1996) or breeding among younger females (Knowlton et al. 1999).

Life Expectancy. Coyotes in captivity may live as long as 18 years (Young 1951), but in wild populations, life expectancy is considerably shorter. Maximum ages reported for wild coyotes are 13.5 (Nellis and Keith 1976), 14.5 (Knowlton 1972), and 15.5 years (Gese 1990).

Dens. Coyotes den in a variety of places, including brush-covered slopes, steep banks, under rock ledges, thickets, and hollow logs. Dens of other animals, including badgers (*Taxidea taxus*), are frequently used. Dens may have more than one entrance (Harrison and Gilbert 1985) and often there are many interconnecting tunnels. Dens may be oriented to the south to maximize solar radiation (Gier 1968; Hallett 1977; Harrison and Gilbert 1985). The same den may be used from year to year. Den sharing occurs only rarely (Nellis and Keith 1976; Camenzind 1978). Movement of pups among dens is very common (Harrison and Gilbert 1985). Reasons for these moves are not known, but disturbance (Harrison and Gilbert 1985) and possibly infestation by parasites may be factors. Most moves are relatively short; however, moves of >2 km are common (Harrison and Gilbert 1985). Van Wormer (1964) reported that a male coyote moved four pups, individually, a distance of 8.0 km. The den and pup-rearing activities are the focal point for coyote families for several months until pups are large and mobile (Andelt et al. 1979; Harrison and Gilbert 1985; Bekoff and Wells 1986; Hatier 1995).

Activity and Movements. Coyotes are active throughout the day, but tend to be more active during the early morning and around sunset (Woodruff and Keller 1982; Andelt 1985; Gese et al. 1989b). Gipson and Sealander (1972) showed a principal activity peak at sunset and a minor peak at daybreak in Arkansas. Activity patterns may change seasonally (Holzman et al. 1992a) or in response to human disturbance and persecution (Gese et al. 1989b; Kitchen et al. 2000a). Seasonal activity patterns also are obvious, especially during winter months, when there is a change in food base (Bekoff and Wells 1986; Gese et al. 1996a). For example, coyotes living in northern climates rest more during winter months, when they are dependent primarily on ungulate carrion for food, than during other seasons, when they feed mainly on small rodents (Bekoff and Wells 1981; Gese et al. 1996a). Hunting attempts also decrease during winter months, when rodents are relatively inaccessible because of snow-covered ground, one of the major rodent food items (Uinta ground squirrels, *Spermophilus armatus*) are hibernating, and ungulate vulnerability to predation increases with snow accumulation (Bekoff and Wells 1986; Gese and Grothe 1995; Gese et al. 1996a). In contrast, Shivik et al. (1997) reported a change in coyote activity levels during winter that was not dependent on a change in prey base. They theorized that the reduced activity levels helped reduce energy expenditures in an area with a limited food base. Coyotes living in packs rest more and travel less during winter months than do coyotes living as mated pairs or alone (Bekoff and Wells 1981, 1986; but see Gese et al. 1996a). Larger energy savings have been shown for a pregnant female living in a pack than for a female living only with her mate (Bekoff and Wells 1981). However, it is unknown whether females living in packs are better off reproductively because of the energy saved during pregnancy.

Coyotes, similar to wolves, are typically territorial, and primarily remain within their territory. However, they also may make extraterritorial movements into neighboring territories. Dispersal from the natal site may be into a vacant or occupied territory in an adjacent area or they may disperse long distances (Harrison et al. 1991). In general, it is the pups or subordinate yearlings that typically disperse (Knowlton 1972; Nellis and Keith 1976; Berg and Chesness 1978; Gese et al. 1989a, 1996a). Dispersal appears to be voluntary as social and nutritional pressures intensify during the winter and availability and access to food becomes more limited (Gese et al. 1996a; Bekoff 2001a). There are no consistent sex differences in dispersal distance (Gese et al. 1989a; Harrison et al. 1991), although in one study, female pups moved farther than male pups (Nellis and Keith 1976). Dispersal by juveniles usually occurs during autumn and early winter, though some individuals do not disperse during the first year and remain to provide care for future siblings (Hatier 1995; Gese et al. 1996a). Dispersal direction appears to be random, and pups will move 80–160 km (Bowen 1982; Gese et al. 1989a), with a record dispersal of 544 km (Carbyn and Paquet 1986). Berg and Chesness (1978) reported mean dispersal distances of 48 km, which occurred at a mean rate of about 11 km/week. Forays or exploratory movements also may occur before dispersal (Harrison et al. 1991). Increased mortality may be associated with dispersal as animals move into unfamiliar areas and low-security habitats (Tzilkowski 1980; Pyrah 1984; Bekoff and Wells 1986; Gese et al. 1989a). Observations of coyote dispersal are summarized in Table 22.3.

Home Range and Territory. Home range size of coyotes has been studied throughout their range. Home range size varies geographically and seasonally (Gipson and Sealander 1972; Bekoff and Wells 1980; Laundré and Keller 1984) and within a population (Springer 1977; Bowen 1982; Gese et al. 1988a) (Table 22.4). Variation in home range size among resident coyotes within a population depends on energetic requirements, season, sex, physiographic makeup, habitat, and food distribution (Laundré and Keller 1984; Gese et al. 1988a). Home range size also is influenced by social organization. Transient individuals

TABLE 22.3. Average distances of coyote dispersal movements from established home ranges

Source	Adults (km)		Juveniles (km)		
	Males	Females	Males	Females	Both Sexes
Robinson and Cummings 1951	12.6	17.8	—	—	16.8
Young 1951	36.2	—	—	—	—
Young 1951	45.3	40.0	—	—	—
Robinson and Grand 1958	45.6	34.2	—	—	40.6
Hawthorne 1971	6.4	7.6	5.2	6.4	—
Gipson and Sealander 1972	20.5	8.2	—	7.4	—
Nellis and Keith 1976	5.5	6.6	28.2 (5.8)[a]	31.5 (23.5)[b]	—
Andrews and Boggess 1978	30.2	31.2[c]	—	—	—
Bowen 1982	19.3	24.6	17.1	51.4	—
Gese et al. 1989a	—	—	—	—	59.0
Harrison 1992	—	—	113	94	—

[a]Male pups.
[b]Female pups.
[c]Excludes data for one female that moved a record distance of 323.2 km.

TABLE 22.4. Mean coyote home range sizes from some representative studies

Study	Adults (km²)		Juveniles and Pups (km²)		State/Province
	Males	Females	Males	Females	
Gipson and Sealander 1972	32.8	13.1	—	11.8	Arkansas
Chesness and Bremicker 1974	41.9	10.0	—	—	Minnesota
Berg and Chesness 1978	68.0	16.0	—	—	Minnesota
Hibler 1977	17.8	20.2	—	—	Utah
Litvaitis 1978	15.0	27.9	—	21.3	Oklahoma
Bowen 1982	13.9[a]	10.3[a]	—	—	Alberta
	14.9[b]	16.3[b]	—	—	
Pyrah 1984	9.7	7.8	—	—	Montana
Andelt 1985	2.7[c]	2.6[c]	—	—	Texas
	3.0[d]	2.1[d]	—	—	
	2.8–3.3[e]	2.3–2.5[e]	—	—	
	3.5[f]	2.7[f]	—	—	
Gese et al. 1988a	8.5[c]	8.7[c]	—	—	Colorado
	10.0[d]	7.7[d]	—	—	
	11.9[e]	9.1[e]	—	—	
	9.3[f]	8.8[f]	—	—	
Windberg and Knowlton 1988	—	2.8[g]	—	—	Texas
	—	2.3[h]	—	—	
Atkinson and Shackleton 1991	7.7	17.0	—	—	British Columbia
Person and Hirth 1991	10.4[a]	6.6[a]	—	—	Vermont
	18.8[b]	11.8[b]	—	—	
Holzman et al. 1992a	6.0[c]	18.9[c]	2.6[c]	4.5[c]	Georgia
	3.3[d]	9.8[d]	2.2[d]	2.4[d]	
	4.3[e]	21.9[e]	2.9[e]	4.6[e]	
	6.5[f]	24.8[f]	5.1[f]	6.5[f]	
Windberg et al. 1997a	5.9	5.4	—	—	New Mexico

[a]Summer.
[b]Winter.
[c]Breeding.
[d]Gestation.
[e]Pup rearing.
[f]Dispersal.
[g]Fall.
[h]Spring.

range over large areas, whereas residents occupy distinct territories (Bowen 1982; Bekoff and Wells 1986; Gese et al. 1988a). Coyotes living in packs who defend ungulate carrion during the winter have much smaller, compressed home ranges than coyotes living in pairs or alone (Bekoff and Wells 1980). Coyotes actively defend well-defined territorial boundaries using direct confrontation and indirect means involving scent marking and howling (Camenzind 1978; Bekoff and Wells 1980, 1986; Wells and Bekoff 1981; Gese and Ruff 1997, 1998; Gese 2001). Typically, only pack members maintain and defend territories; solitary individuals do not (Bekoff and Wells 1980; Gese and Ruff 1997, 1998; Gese 2001). Fidelity to the home range area is very high among resident animals, may persist for many years (Kitchen et al. 2000b), and may be passed to successive generations. Shifts in territorial boundaries may occur in response to loss of one or both of the alpha pair (Camenzind 1978; Gese 1998).

FEEDING HABITS

Coyotes are opportunistic, generalist predators and eat a variety of food items in relation to changes in availability (Van Vuren and Thompson 1982; Todd and Keith 1983; Andelt et al. 1987; Windberg and Mitchell 1990). They will consume food items ranging in size from fruit and insects to large ungulates and livestock (Fig. 22.3). MacCracken and Hansen (1987) suggested that coyotes may select prey as predicted by optimal foraging models, but see Boutin and Cluff (1989) and MacCracken (1989) for review and comments.

Gipson (1974) found regional and seasonal differences in feeding habits of coyotes living in Arkansas. The most common food items from 168 coyote stomachs were poultry (34%), persimmons (23%), insects (11%), rodents (9%), songbirds (8%), cattle (7%), rabbits (7%), deer (5%), woodchucks (4%), goats (4%), and watermelon (4%). Korschgen (1973), in an analysis of coyote feeding habits in Missouri, also found seasonal differences. For example, rabbits were found in 57% of coyote stomachs in winter versus 14.8% in spring. Carrion was found in 15.6% of stomachs in winter versus 37.0% in spring. Livestock and wild ungulates often may be found in coyote stomachs and scats as carrion (Murie 1935, 1951; Ozoga and Harger 1966; Korschgen 1973; Weaver

FIGURE 22.3. Coyote feeding on elk calf carcass, Yellowstone National Park, Wyoming. SOURCE: Photo by E. Gese.

1977; Bekoff and Wells 1980), but actual predation on large ungulates, both native and domestic, does occur (Gier 1968; Andelt 1987; Gese and Grothe 1995). In Wyoming, the percentage of coyote stomachs containing ungulate meat from scavenging carrion may increase as much as threefold in winter compared to summer (Weaver 1977, 1979; Houston 1978). Carrion availability during winter plays a large role in coyote foraging and population ecology in northern regions (Weaver 1979; Todd 1985; Gese et al. 1996a, 1996b). Reproductive status may also influence coyote feeding habits. Coyotes that are provisioning pups may switch to larger, more energetically "profitable" prey items as compared to nonreproductive coyotes (Till and Knowlton 1983; Harrison and Harrison 1984; Bromley 2000). Coyotes in suburban areas are equally adept at exploiting human-made food resources and will readily consume dog food or other human-related items (Shargo 1988; Atkinson and Shackleton 1991; McClure et al. 1995). As a caveat for food habit analysis, whenever scat analysis data are used to determine feeding habits, the inherent problems of prey digestibility and recovery of food items in relation to actual prey consumption can confound results and should be considered (Weaver and Hoffman 1979; Kelly and Garton 1997).

BEHAVIOR

Direct observation of coyote behavior in the wild is difficult because of the elusive nature of the species (Kleiman and Brady 1978; Bekoff 2001a). However, observations of coyotes in national parks have provided insight into many aspects of the behavioral ecology of the species (Camenzind 1978; Bekoff and Wells 1980, 1981, 1986; Wells and Bekoff 1981; Gese et al. 1996a, 1996b, 1996c; Gese and Ruff 1997, 1998; Allen et al. 1999). Data on behavioral development are presented in Bekoff (1972, 1974, 1977b, 1978) and Bekoff and Dugatkin (2000). Coyotes show early development of aggressive behavior compared to wolves and most domestic dogs; they will engage in serious fights when they are only 19–24 days old (Knight 1978; Bekoff et al. 1981; Bekoff and Dugatkin 2000). The early development of rank relationships within litters appears to last up to 4.5 months (Knight 1978), but the dominance hierarchy later in life may not reflect the hierarchy observed within the litter (Bekoff 1977b; Knight 1978). General behavioral patterns (postures, gestures, tail movements, facial expressions, vocalizations) are described and discussed in Kleiman (1966), Fox (1970), Bekoff (1972, 1974, 1978), Knight (1978), Lehner (1978a, 1978b), and Wells and Bekoff (1981). Gait, stance, ear and tail position, and retraction of the lips to expose the teeth are all very important in social communication and may vary independently or together, depending on individual "mood." Comparative reviews are in Kleiman and Eisenberg (1973) and Kleiman and Brady (1978).

Social Organization. Generally, coyotes are considered less social than wolves (but see Gese et al. 1996a, 1996b). In Yellowstone National Park, high prey biomass, high survival rates, and lack of persecution has allowed formation of large packs numbering up to 10 individuals in winter (Gese et al. 1996a, 1996b). The basic social unit, even in large packs, is the dominant adult heterosexual pair, often referred to as the "alpha" pair (Bowen 1978; Bekoff and Wells 1980, 1986; Gese et al. 1996a). Associate animals will remain in the pack and possibly inherit or displace a member of the breeding pair and become an alpha themselves (Gese et al. 1996a). These associates (or beta animals) participate in territorial maintenance and pup rearing (Hatier 1995), but not to the extent of the alpha pair (Gese and Ruff 1997, 1998; Gese 2001). Other coyotes exist on the landscape outside of the resident packs and are considered transient or nomadic individuals (Camenzind 1978; Bowen 1982; Gese et al. 1988a). These transients usually travel alone over a large area, do not breed, and do not maintain a territory (Bekoff and Wells 1986; Gese and Ruff 1997, 1998; Gese 2001). Coyote packs resemble wolf packs (Bowen 1978; Bekoff and Wells 1980; Gese et al. 1996a) and it appears that differences between coyotes and wolves are quantitative rather than qualitative.

Just as with all other aspects of coyote biology, there is considerable variability in observed social organization (Bowen 1978; Bekoff and Wells 1980, 1981, 1986). In many areas, solitary individuals are frequently observed (Berg and Chesness 1978) outside of the breeding season. In other areas, such as Jackson Hole, Wyoming (Camenzind 1978; Bekoff and Wells 1980, 1981), and Jasper, Alberta (Bowen 1978), groups of coyotes are commonly observed. Prey size may be important in affecting coyote sociality (Bowen 1978; Bekoff and Wells 1980). In populations where the major prey items throughout the year are small rodents, coyotes tend to be in pairs or trios (Bekoff and Wells 1986; Gese et al. 1989a). In populations where large animals are available (e.g., elk, deer) either as live individuals or as carrion, coyotes form large groups (Bekoff and Wells 1986; Gese et al. 1996a, 1996b). However, coyote groups are not necessarily formed to capture large prey (Gese et al. 1988b), though coyotes occasionally hunt as a pair or a group (Hamlin and Schweitzer 1979; Bowyer 1987; Gese and Grothe 1995). Instead, cooperative group defense appears to be the major selective force favoring increased sociality (Berger 1978; Bowen 1978; Lamprecht 1978; Bekoff and Wells 1980). However, even within a resident pack, dominance plays a large role in access to feeding on carcasses (Gese et al. 1996b) and influences the ability of associates to remain in the pack (Gese et al. 1996a). Importantly, coyotes tend to be more social during winter, when carrion is a very important food resource (Bowen 1978; Camenzind 1978; Bekoff and Wells 1980). In areas without large ungulate carrion, increased pack size also occurs in the breeding season, which then facilitates capture of large prey (Gese et al. 1988b).

Predatory Behavior. There have been a few detailed studies of the predatory behavior of wild coyotes on small mammals (Wells and Bekoff 1982; Bekoff and Wells 1986; Gese et al. 1996b, 1996c). Predatory sequences may be divided into at least six components: search, orientation, stalk, pounce, head thrust/close search into ground cover, and rush. Pouncing is used mostly for capturing small microtine rodents, whereas the rush is used most frequently on larger animals, such as ground squirrels (Bekoff and Wells 1986; Gese et al. 1996c). Age of the animal, wind, habitat, and snow conditions (depth and hardness) all influence a coyote's ability to detect and capture small mammals (Wells and Bekoff 1982; Murray and Boutin 1991; Gese et al. 1996c). Coyotes generally hunt small mammals alone, even though the pack size may be quite large (Gese et al. 1996b, 1996c). Cooperative hunting for small mammals would be inefficient because only one individual can consume the prey item (Bowen 1981; Andelt 1985; Gese et al. 1988b). It has been shown experimentally that coyotes depend on various senses to locate prey. In order of decreasing importance, they are vision, audition, and olfaction (Wells and Lehner 1978), though these "priorities" may change depending on environmental conditions (Wells 1978).

Predatory behavior and habits of wild coyotes preying on domestic livestock have received increased attention (e.g., Henne 1977; Shivik et al. 1996; Windberg et al. 1997b; Neale et al. 1998; Sacks et al. 1999). Connolly et al. (1976) found that captive coyotes killed sheep (in confinement) only in 20 of 38 tests (52.6%); defensive behavior by sheep deterred coyotes only 31.6% of the time. Mean latency to attack was very long (47 min), with considerably variability (standard deviation = 48 min) among attacks. Mean killing time, likewise, was rather long (13 min), and considerably variable. In most instances, coyotes attacked sheep by biting the throat, and the sheep died of suffocation. Various factors influence coyote depredation rates on sheep, including breed of sheep, sheep management practices, coyote behavior, environmental factors, and depredation management programs. In general, changes in animal size and behavior, differences in group cohesiveness, sociality, grazing dispersion, attentiveness, and maternal protection all affect vulnerability to coyote predation (see review by Knowlton et al. 1999). Windberg et al. (1997b) and Bromley (2000) documented that coyotes generally killed the smallest or lightest individuals in the flock.

Cooperative hunting of adult ungulates (i.e., deer, elk, antelope) by coyotes rarely has been documented (Cahalane 1947; Murie 1951; Bruns 1970; Bowyer 1987). When coyotes attack large wild ungulates, environmental factors such as snow depth and hardness of the crust are important to the success or failure of the attempt (Gese and Grothe 1995). Presence of the alpha pair also appears to be important in the outcome of the attack. Generally, younger animals do not participate in the attack sequence. It appears that the number of coyotes in the pack is not as important as which coyotes are involved in the attack. Even in a pack of seven coyotes, only the alpha pair was involved in attacks on large ungulates (Gese and Grothe 1995). Also, the ability of ungulates to escape into water, defensive abilities of individuals and other members of the herd or the parent, and nutritional state of the individual under attack will contribute to the outcome (Ozoga and Harger 1966; Bowyer 1987; Gese and Grothe 1995).

Communication. Coyotes communicate using auditory, visual, olfactory, and tactile cues (Lehner 1978a, 1978b). Communication by howling or vocalization is very common among coyotes (Gier 1975; Lehner 1978a, 1978b). Studies of wild coyotes have identified many different types of vocalizations (McCarley 1975; Lehner 1978a, 1978b). Seasonal and diel patterns (Laundré 1981; Walsh and Inglis 1989), lunar phase (Bender et al. 1996), and social status in the pack (Gese and Ruff 1998) also influence coyote vocalization rates. Howling among coyotes plays a role in territorial maintenance and pack spacing. It signals territorial boundaries and the presence of alpha animals, who will confront intruders and defend the territory (Gese and Ruff 1998).

Studies on olfactory communication (i.e., scent marking) among coyotes also have been conducted (Barrette and Messier 1980; Bowen and Cowan 1980; Wells and Bekoff 1981; Bekoff and Wells 1986; Gese and Ruff 1997; Allen et al. 1999). Alpha coyotes perform the majority of scent-marking duties. Rates of scent marking vary seasonally, pack size does not influence scent-marking rates, and scent marks are located more than expected along the periphery of the territory and likely contribute to territory maintenance (Bowen and Cowan 1980; Bekoff and Wells 1986; Gese and Ruff 1997). Scent marking may be a mechanism for sex recognition (Bekoff 1979a) and serve as an indicator of sexual condition, maturity, or synchrony (Bekoff and Diamond 1976). Internal information to orient members of the resident pack (Wells and Bekoff 1981) and alert dispersing animals of occupied territories also may be communicated via scent marks.

MORTALITY

Mortality Rates. Different-aged coyotes have different mortality rates. Mortality rates depend on the level of control to which a population is exposed and levels of food availability (Knowlton 1972; Todd et al. 1981; Todd and Keith 1983; Todd 1985; Gese et al. 1989a; Knowlton et al. 1999). Pups (<1 year old) and yearlings (individuals 1–2 years old) tend to have the highest mortality rates (Nellis and Keith 1976; Gese et al. 1989a). For individuals >1 year old, mortality rates vary geographically (Knowlton 1972; Nellis and Keith 1976; Bowen 1978; Andrews and Boggess 1978; Gese et al. 1989a). Mathwig (1973) found greatest life expectancy for coyotes in Iowa at 1.5 years of age and lowest life expectancy at 5.5 years. Knowlton (1972), Crabtree (1988), and Gese et al. (1989a) reported relatively high survival in 4- to 8-year-old coyotes. About 70–75% of coyote populations are 1–4 years old (see Knowlton et al. [1999] for a summary of population studies). To maintain population stability, net survival of about 33–38% seems to be necessary (Knowlton 1972; Nellis and Keith 1976).

Causes of Mortality. Most studies indicate that human activity is involved in a high proportion of the deaths of adult coyotes (Davison 1980; Tzilkowski 1980; Windberg et al. 1985; Gese et al. 1989a; Windberg 1995), although predation by large carnivores (Peterson 1995; Arjo 1998; Crabtree and Sheldon 1999) and starvation during crashes of food resources also may be substantial mortality factors. Disease also can be a factor, especially among young of the year; parvovirus killed several radio-marked pups in Yellowstone (Gese et al. 1997). Even in lightly exploited populations, most mortality is attributable to human causes. Human exploitation can be substantial in some coyote populations (Knowlton 1972; Knowlton et al. 1999).

Diseases and Parasites. Serological analyses for antibodies in coyotes have shown that they have been exposed to many diseases. In Yellowstone National Park, prevalence of antibodies against canine parvovirus (CPV) was 100% for adults, yearlings, and old pups (4–12 months old), but 0% for young pups (<3 months old) (Gese et al. 1997 [$n = 110$]). In Texas, Utah, Idaho, and Colorado, >70% of the coyotes had antibodies against CPV (Thomas et al. 1984 [$n = 1184$]; Gese et al. 1991 [$n = 72$]). In Georgia, 65% of 17 coyotes had antibodies against CPV (Holzman et al. 1992b). High prevalence of antibodies is often associated with a highly contagious, but nonfatal infection because prevalence is measured among survivors (Thomas et al. 1984). However, of 21 coyote pups implanted with radios in 1992 in Yellowstone, 8 of 14 deaths were from CPV infection (Gese et al. 1997).

Presence of antibodies against canine distemper virus was 88%, 54%, 23%, and 0% prevalence among adults, yearlings, old pups, and young pups, respectively, in Yellowstone National Park (Gese et al. 1997). In Texas, Trainer and Knowlton (1968) found 37% of 33 coyotes had antibodies to the virus; Guo et al. (1986) reported 56% of 228. Williams et al. (1988) reported that 50% of 10 coyotes in Wyoming tested positive for this virus. In Georgia, no coyotes were found to have been exposed (Holzman et al. 1992b).

Prevalence of canine infectious hepatitis virus antibodies was 97%, 82%, 54%, and 33% for adults, yearlings, old pups, and young pups, respectively, in Yellowstone National Park (Gese et al. 1997). Coyotes in Texas and Georgia had a lower prevalence (41–51%) of virus exposure (Trainer and Knowlton 1968; Holzman et al. 1992b). The degree to which this virus affects coyote populations is unknown.

In Yellowstone National Park, the prevalence of antibodies against the plague bacterium *Yersinia pestis* was 86%, 33%, 80%, and 7% for adults, yearlings, old pups, and young pups, respectively (Gese et al. 1997). This high prevalence was similar to results in other western states (Barnes 1982 [$n = 12{,}405$]). In contrast, coyotes in California had very low antibody prevalence (<6% of 338 coyotes sampled) (Thomas and Hughes 1992). Coyotes may become infected with *Y. pestis* by being bitten by fleas or possibly by ingesting infected rodents (Thomas et al. 1989). Infected coyotes usually do not develop clinical signs, but develop antibody titers that last about 6–8 months, making coyotes an indicator species for plague (Barnes 1982). Changes in prevalence of plague in the coyote population likely reflect prevalence of plague in the small-mammal prey base.

Serological prevalence of antibodies against *Francisella tularensis* (tularemia) was found in a coyote population in Yellowstone, but at relatively low levels (<25%) (Gese et al. 1997). In Texas, Trainer and Knowlton (1968) found no serological evidence of tularemia. In contrast, 88% of the coyotes in Idaho were seropositive (Gier et al. 1978). Coyotes may contract tularemia, but are relatively unsusceptible and will likely recover (Gier and Ameel 1959). No coyotes in Yellowstone had serological evidence of exposure to brucellosis, either *Brucella abortus* or *B. canis* (Gese et al. 1997). Similarly, coyotes in Texas and Georgia had not been exposed to brucellosis (Trainer and Knowlton 1968, Holzman et al. 1992b). Coyotes do not appear to act as significant hosts for brucellosis.

Prevalence of antibodies against leptospirosis in the coyote population in Yellowstone was low (Gese et al. 1997), similar to results for coyotes in Texas (Trainer and Knowlton 1968) and Georgia (Holzman et al. 1992b). In contrast, four of nine coyotes tested in Arizona were seropositive for leptospirosis (Drewek et al. 1981). The impact of leptospirosis on coyotes is unknown, but infected animals may survive and remain carriers for a short time.

Coyotes also carry a variety of parasites. Ectoparasites commonly found among coyotes include fleas (the most common external parasite), various ticks, mites, and lice; see Gier and Ameel (1959) and Gier et al. (1978) for species identification. Internal parasites include several species of flukes (trematodes), tapeworms (cestodes), intestinal worms (nematodes, ascarids), hookworms (ancylostomids), heartworms (filaroids), esophageal worms (spiruroids), lungworms (trichinellids), kidney worms (dioctophymoides), spiny-headed worms (acanthocephalids), protozoans, and coccidia fungus (Eads 1948; Gier and Ameel 1959; Hirsch and Gier 1974; Mitchell and Beasom 1974; Thornton et al. 1974; Conder and Loveless 1978; Gier et al. 1978;

Mercer et al. 1988; Wixsom et al. 1991; Holzman et al. 1992b). Coyotes also may carry rabies and may suffer from cardiovascular diseases, aortic aneurysms (Thornton et al. 1974), mange (Pence et al. 1983; Pence and Windberg 1994), and cancer. For reviews and taxonomic analyses of external and internal parasites, see Bekoff (1977a) and (Gier et al. 1978).

AGE ESTIMATION

Coyote age can be estimated by counting dental cementum annuli (Linhart and Knowlton 1967; Nellis et al. 1978) and coarsely estimated with tooth wear (Gier 1968). Before tooth sectioning for cementum analysis, age class (juveniles and adults) can be distinguished based on the relative pulp cavity size using radiography (Knowlton and Whittemore 2001). Roberts (1978) pointed out that there is variation in age determination from different teeth and suggested using canines in age determination. Eye lens weight, baculum weight, and thermal contraction of long tendons can also be used to estimate age accurately. Ages of young coyotes can be estimated from body mass, body length, and length of the hind foot (Gier 1968; Bekoff and Jamieson 1975; Barnum et al. 1979). The regression equation for the body mass of hand-reared pups (0–30 days of age) is $y = 0.2685 + 0.197x$; for coyotes 31–154 days of age, the equation is $y = 0.5049 + 0.0469x$ (Barnum et al. 1979). From hand-raised coyotes, the regression equations for predicting the weight of known-age coyotes are $y = -13.57 + 50.59x$ (0–30 days) and $y = 11.386 + 21.11x$ (31–154 days). The correlations between weight and age for 0–30 days and 31–154 days are 0.999 and 0.995, respectively (Barnum et al. 1979).

ECONOMIC STATUS AND MANAGEMENT

Coyotes are victims of success. By taking advantage of poorly devised domestication practices that left most livestock virtually defenseless against predation, coyotes have established a reputation for efficient and effective predation. Sheep have been selectively bred to produce animals that are suited to particular husbandry practices, regions, conditions, and cultural tastes as well as to provide food and fiber (Knowlton et al. 1999). Differences in group cohesiveness, sociality, grazing dispersion, location, attentiveness, mobility, behavior, and maternal protection all affect their vulnerability to predation (Gluesing et al. 1980; Blakesley and McGrew 1984; Knowlton et al. 1999).

Importantly, not all coyotes kill sheep. Some are never exposed to sheep (Wagner 1988), whereas others do not develop sheep-killing tendencies (U.S. Fish and Wildlife Service 1978; Timm and Connolly 1980). Shivik et al. (1996), Conner et al. (1998), and Sacks et al. (1999) described radio-collared coyotes near sheep with few lambs being killed; losses were usually attributed to the breeding, territorial coyotes. In contrast, Connolly (1988) and Windberg et al. (1997b) reported a high number of coyotes that killed and consumed livestock. In the intermountain West, Till and Knowlton (1983) and Bromley (2000) demonstrated that adult coyotes with pups were more likely to kill lambs than adults without pups. In contrast, territorial coyotes in north-coastal California kill lambs as soon as they become available in December and January, outside the pup-rearing season (Conner 1995; Sacks 1996).

Economics of Coyote Predation. The amount of livestock losses that producers attribute to coyotes is a contentious issue. Wildlife protection advocates claim that few or no depredations actually occur, whereas producers indicate losses from coyotes threaten their economic livelihood (Caine et al. 1972; Gee et al. 1977; U.S. Fish and Wildlife Service 1978; Wagner 1988; Knowlton et al. 1999). In situations where there was predator control, losses of sheep to coyotes were 1.0–6.0% for lambs and 0.1–2.0% for ewes (U.S. Fish and Wildlife Service 1978). In situations where there was no predator control, losses to coyotes ranged from 12% to 29% of lambs and from 1% to 8% of ewes (Delorenzo and Howard 1976; Brawley 1977; Henne 1977; Munoz 1977; McAdoo and Klebenow 1978; O'Gara et al. 1983). In 1994, the economic value of domestic sheep lost to predators was reported as $17.7 million (U.S. Department of Agriculture 1995). Sterner and Shumake (1978) and U.S. Department of Agriculture (1995) reported that coyotes were responsible for the majority of all sheep lost to predators. However, it is important to consider that coyote predation is not the major cause of losses (U.S. Fish and Wildlife Service 1978; Bekoff 1979b). Early et al. (1974) found that 77% of the total economic loss of sheep in 1970–1971 was due to disease (43%), unspecified causes (31%), and predators (23%).

Depredation Management. Protecting livestock or wildlife species from coyotes requires consideration of legal, social, economic, biological, technical, and ethical factors (Sterner and Shumake 1978; Wade 1978; Bekoff 1998, 2001b; Knowlton et al. 1999). Successful resolution of conflicts should consider the efficacy, selectivity, and efficiency of various procedures. Several techniques are available for reducing coyote depredations on livestock (Sterner and Shumake 1978; Wade 1978; Fall 1990; U.S. Department of Agriculture 1994). Many techniques are within the operational purview of the producers, such as nonlethal techniques. Others involve lethal control in either a preventative or corrective context. Procedures that are more benign in their effects on natural systems are typically preferred. Fall (1990), Andelt (1996), and Knowlton et al. (1999) reviewed many of the techniques for reducing coyote depredations.

Nonremoval Techniques. Several techniques have been used to deter coyotes from attacking livestock (Linhart 1984a; Wagner 1988; Fall 1990; Green et al. 1994). Whereas some methods have been successful, others are complete failures. Knowlton et al. (1999) suggested several husbandry practices that may reduce depredations, including confining or concentrating flocks during periods of vulnerability, using herders, shed lambing, removing livestock carrion from pastures, synchronizing birthing, and keeping young animals in areas with little cover and/or nearest to human activity (Robel et al. 1981; Wagner 1988).

Several types of protective fencing can exclude or reduce coyote use of an area (De Calesta and Cropsey 1978; Linhart et al. 1982; Shelton 1984, 1987; Nass and Theade 1988), but few are "coyote proof." Frightening devices that emit bursts of light or sound have deterred coyote predation on sheep in fenced pastures (Linhart et al. 1982, 1984; Linhart 1984b) and on open range (Linhart et al. 1992); however, coyotes often become habituated to such devices (Knowlton et al. 1999). Use of guard animals may reduce damage by coyotes. Dogs, llamas, and donkeys or mules are used as livestock guards (Linhart et al. 1979; Coppinger et al. 1983; Green and Woodruff 1983, 1987; Black and Green 1984; Green et al. 1984; Timm and Schmidt 1989; Powell 1993; Conner 1995).

There are no repellents and learned aversions that will deter coyote predation. A variety of gustatory, olfactory, and irritating products have been tested. Food consumption may be reduced (Hoover 1996), but predation is not (Lehner 1987; Burns and Mason 1997). Investigations of conditioned taste aversion using lithium chloride showed mixed results. Some researchers reported success at reducing food consumption (e.g., Gustavson et al. 1974, 1982; Ellins and Martin 1981; Forthman-Quick et al. 1985a, 1985b), whereas others reported no reduction of predation by coyotes (Bekoff 1975; Conover et al. 1977; Burns 1980, 1983; Burns and Connolly 1980, 1985; Bourne and Dorrance 1982).

Attempts to reduce coyote reproductive rates and population levels have been investigated, with the assumption that fewer coyotes will result in fewer depredations (Knowlton et al. 1999). Coyote reproduction was reduced using diethylstilbesterol (Balser 1964; Linhart et al. 1968), but was impractical without effective bait delivery systems. Alternatively, reproductive interference using sterilization of territorial, breeding coyotes may be an effective way to reduce depredations by coyotes (Bromley 2000) by changing the predatory behavior of adults when provisioning young (Till and Knowlton 1983). With respect to chemosterilants, both regulation (federal and state) and distribution are problematic (Stellflug et al. 1978).

Coyote Removal Techniques. When nonremoval techniques do not stop depredations, removing one or more coyotes may achieve management objectives (Knowlton et al. 1999), especially if the removed animals are the "problem individuals" (Linnell et al. 1999). Removing one or two individuals in a small area (i.e., corrective control) may stop the problem, whereas in other cases, population reduction may be warranted. Knowlton et al. (1999) suggested that selection of the appropriate method should consider the nature of the problem, presence of a historical pattern, size of the area, season of the year, timing, efficacy, selectivity, efficiency, and animal welfare considerations (Bekoff 1998, 2001b). On small areas where specific coyotes pose the most immediate risk, calling and shooting can selectively remove coyotes killing livestock (Coolahan 1990), but requires correctly identifying the areas used by the killers.

One method that may be the most selective for removing coyotes responsible for depredations is the use of the livestock protection collar (McBride 1974, 1982; Burns et al. 1988, 1996; Connolly and Burns 1990; Connolly 1993; Rollins 1995). These devices release a toxic chemical into a coyote's mouth when the coyote attacks and punctures the collar on a sheep's neck. They are registered by the Environmental Protection Agency in seven states and require approved training and accountability programs (Moore 1985; Knowlton et al. 1999). Coyotes causing depredations also can be removed with traps, snares, and M-44 devices, but the selectivity for the offending coyote(s) is lower (Brand et al. 1995). Most depredations are attributed to territorial, dominant coyotes (Till and Knowlton 1983; Sacks 1996). Within these territories, coyotes are less vulnerable to capture devices because of their familiarity of the area (Harris 1983; Windberg and Knowlton 1990; Windberg 1996), which makes removal of offending individuals difficult (Conner et al. 1998; Sacks et al. 1999). Aerial hunting in winter is efficient and effective (Wagner and Conover 1999). However, concerns regarding the safety of pilots and gunners, selectivity of animals, and economics (but see Wagner and Conover 1999) are a matter of public debate (e.g., Finkel 1999).

Coyote Population Reduction. There may be situations in which a reduction in the number of coyotes requires serious consideration, including instances where coyotes pose a risk to other wildlife species, the spread of infectious diseases needs to be curtailed, or predation on livestock needs to be prevented when more benign techniques have been ineffective (Knowlton et al. 1999). Population reduction programs are most effective when conducted after the dominance and territorial patterns are established for the coming breeding period and before whelping, to prevent other breeding pairs from becoming established and producing offspring that season (Knowlton 1972; Connolly 1978; Knowlton et al. 1999). As noted previously, before any population reduction measures are implemented, the technical, legal, social, biological, economic, and ethical considerations as well as the efficiency, effectiveness, and selectivity of the technique should be considered.

LITERATURE CITED

Adorjan, A. S., and G. B. Kolenosky. 1969. A manual for the identification of hairs of selected Ontario mammals (Research Report 90). Ontario Department of Lands and Forests.

Allen, J., M. Bekoff, and R. Crabtree. 1999. An observational study of coyote (*Canis latrans*) scent-marking and territoriality in Yellowstone National Park. Ethology 105:289–302.

Andelt, W. F. 1985. Behavioral ecology of coyotes in south Texas. Wildlife Monographs 94:1–45.

Andelt, W. F. 1987. Coyote predation. Pages 128–40 *in* M. Novak, J. A. Baker, M. E. Obbard, and B. Malloch, eds. Wild furbearer management and conservation in North America. Ontario Ministry of Natural Resources and the Ontario Trappers Association.

Andelt, W. F. 1996. Carnivores. Pages 133–55 *in* P. R. Krausman, ed. Rangeland wildlife. Society for Range Management, Denver, CO.

Andelt, W. F., D. P. Althoff, and P. S. Gipson. 1979. Movements of breeding coyotes with emphasis on den site relationships. Journal of Mammalogy 60:568–75.

Andelt, W. F., J. G. Kie, F. F. Knowlton, and K. Cardwell. 1987. Variation in coyote diets associated with season and successional changes in vegetation. Journal of Wildlife Management 51:273–77.

Andrews, R. D., and E. K. Boggess. 1978. Ecology of coyotes in Iowa. Pages 249–65 *in* M. Bekoff, ed. Coyotes: Biology, behavior, and management. Academic Press, New York.

Arjo, W. M. 1998. The effects of recolonizing wolves on coyote populations, movements, behaviors, and food habits. Dissertation, University of Montana, Missoula.

Arjo, W. M., and D. H. Pletscher. 1999. Behavioral responses of coyotes to wolf recolonization in northwestern Montana. Canadian Journal of Zoology 77:1919–27.

Asdell, S. A. 1964. Patterns of mammalian reproduction. Cornell University Press, Ithaca, NY.

Atkins, D. L. 1978. Evolution and morphology of the coyote brain. Pages 17–35 *in* M. Bekoff, ed. Coyotes: Biology, behavior and management. Academic Press, New York.

Atkins, D. L., and L. S. Dillon. 1971. Evolution of the cerebellum in the genus *Canis*. Journal of Mammalogy 52:96–107.

Atkinson, K. T., and D. M. Shackleton. 1991. Coyote, *Canis latrans*, ecology in a rural–urban environment. Canadian Field-Naturalist 105:49–54.

Babb, J. G., and M. L. Kennedy. 1989. An estimate of minimum density for coyotes in western Tennessee. Journal of Wildlife Management 53:186–88.

Balser, D. S. 1964. Management of predator populations with antifertility agents. Journal of Wildlife Management 28:352–58.

Barnes, A. M. 1982. Surveillance and control of bubonic plague in the United States. Symposia of the Zoological Society (London) 50:237–70.

Barnum, D. A., J. S. Green, J. T. Flinders, and N. L. Gates. 1979. Nutritional levels and growth rates of hand-reared coyote pups. Journal of Mammalogy 60:820–23.

Barrette, C., and F. Messier. 1980. Scent-marking in free-ranging coyotes, *Canis latrans*. Animal Behaviour 28:814–19.

Bekoff, M. 1972. An ethological study of the development of social interaction in the genus *Canis:* A dyadic analysis. Dissertation, Washington University, St. Louis, MO.

Bekoff, M. 1974. Social play and play-soliciting by infant canids. American Zoologist 14:323–41.

Bekoff, M. 1975. Predation and aversive conditioning in coyotes. Science 187:1096.

Bekoff, M. 1977a. *Canis latrans*. Mammalian Species 79:1–9.

Bekoff, M. 1977b. Mammalian dispersal and ontogeny of individual behavioral phenotypes. American Naturalist 111:715–32.

Bekoff, M. 1978. Behavioral development in coyotes and eastern coyotes. Pages 97–126 *in* M. Bekoff, ed. Coyotes: Biology, behavior and management. Academic Press, New York.

Bekoff, M. 1979a. Ground scratching by male domestic dogs: A composite signal. Journal of Mammalogy 60:847–48.

Bekoff, M. 1979b. Coyote damage assessment in the west: Review of a report. BioScience 29:754.

Bekoff, M. 1998. Encyclopedia of animal rights and animal welfare. Greenwood, Westport, CT.

Bekoff, M. 2001a. Cunning coyotes: Tireless tricksters, protean predators. Pages 381–407 *in* L. Dugatkin, ed. Model systems in behavioral ecology. Princeton University Press, Princeton, NJ.

Bekoff, M. 2001b. Human–carnivore interactions: Adopting proactive strategies for complex problems. Pages 179–195 *in* J. L. Gittleman, S. M. Funk, D. W. Macdonald, and R. K. Wayne, eds. Carnivore conservation. Cambridge University Press, London.

Bekoff, M., and J. Diamond. 1976. Precopulatory and copulatory behavior in coyotes. Journal of Mammalogy 57:372–75.

Bekoff, M., and L. A. Dugatkin. 2000. Winner and loser effects and the development of dominance in young coyotes: An integration of data and theory. Evolutionary Ecology Research 2:871–83.

Bekoff, M., and R. Jamieson. 1975. Physical development in coyotes (*Canis latrans*) with a comparison to other canids. Journal of Mammalogy 56:685–92.

Bekoff, M., and M. C. Wells. 1980. Social ecology and behavior of coyotes. Scientific American 242:130–48.

Bekoff, M., and M. C. Wells. 1981. Behavioural budgeting by wild coyotes: The influence of food resources and social organization. Animal Behaviour 29:794–801.

Bekoff, M., and M. C. Wells. 1986. Social ecology and behavior of coyotes. Advances in the Study of Behavior 16:251–338.

Bekoff, M., H. L. Hill, and J. B. Mitton. 1975. Behavioral taxonomy in canids by discriminant function analysis. Science 190:1223–25.

Bekoff, M., M. Tyrrell M, V. E. Lipetz, and R. A. Jamieson. 1981. Fighting patterns in young coyotes: Initiation, escalation, and assessment. Aggressive Behavior 7:225–44.

Bender, D. J., E. M. Bayne, and R. M. Brigham. 1996. Lunar condition influences coyote (*Canis latrans*) howling. American Midland Naturalist 136:413–17.

Berg, W. E., and R. A. Chesness. 1978. Ecology of coyotes in northern Minnesota. Pages 229–47 *in* M. Bekoff, ed. Coyotes: Biology, behavior and management. Academic Press, New York.

Berger, J. 1978. "Predator harassment" as a defensive strategy. American Midland Naturalist 102:197–99.

Black, H. L., and J. S. Green. 1984. Navajo use of mixed-breed dogs for management of predators. Journal of Range Management 38:11–45.

Blakesley, C. S. and J. C. McGrew. 1984. Differential vulnerability of lambs to coyote predation. Applied Animal Behavior Science 12:349–61.

Boggess, E. K., and F. R. Henderson. 1978. Regional weights of Kansas coyotes. Transactions of the Kansas Academy of Science 80:79–80.

Bourne, J., and M. J. Dorrance. 1982. A field test of lithium chloride aversion to reduce coyote predation on domestic sheep. Journal of Wildlife Management 46:235–39.

Boutin, S., and H. D. Cluff. 1989. Coyote prey choice: Optimal or opportunistic foraging? A comment. Journal of Wildlife Management 53:663–66.

Bowen, W. D. 1978. Social organization of the coyote in relation to prey size. Dissertation, University of British Columbia, Vancouver, Canada.

Bowen, W. D. 1981. Variation in coyote social organization: The influence of prey size. Canadian Journal of Zoology 59:639–52.

Bowen, W. D. 1982. Home range and spatial organization of coyotes in Jasper National Park, Alberta. Journal of Wildlife Management 46:201–15.

Bowen, W. D., and I. M. Cowan. 1980. Scent marking in coyotes. Canadian Journal of Zoology 58:473–80.

Bowyer, R. T. 1987. Coyote group size relative to predation on mule deer. Mammalia 51:515–26.

Boyd, D., and B. W. O'Gara. 1985. Cougar predation on coyotes. Murrelet 66:17.

Brand, D. J., N. Fairall, and W. M. Scott. 1995. The influence of regular removal of black-backed jackals on the efficiency of coyote getters. South African Journal of Wildlife Research 25:44–48.

Brawley, K. C. 1977. Domestic sheep mortality during and after tests of several predator control methods. Thesis, University of Montana, Missoula.

Bromley, C. 2000. Coyote sterilization as a means of reducing predation on domestic lambs. Thesis, Utah State University, Logan.

Bruns, E. H. 1970. Winter predation of golden eagles and coyotes on pronghorn antelopes. Canadian Field-Naturalist 84:301–4.

Burns, R. J. 1980. Evaluation of conditioned predation aversion for controlling coyote predation. Journal of Wildlife Management 44:938–42.

Burns, R. J. 1983. Microencapsulated lithium chloride bait aversion did not stop coyote predation on sheep. Journal of Wildlife Management 47:1010–17.

Burns, R. J., and G. E. Connolly. 1980. Lithium chloride bait aversion did not influence prey killing by coyotes. Proceedings of the Vertebrate Pest Conference 9:200–204.

Burns, R. J., and G. E. Connolly. 1985. A comment on "Coyote control and taste aversion." Appetite 6:276–81.

Burns, R. J., and J. R. Mason. 1997. Effectiveness of Vichos non-lethal collars in deterring coyote attacks on sheep. Proceedings of the Vertebrate Pest Conference 17:204–6.

Burns, R. J., G. E. Connolly, and P. J. Savarie. 1988. Large livestock protection collars effective against coyotes. Proceedings of the Vertebrate Pest Conference 13:215–19.

Burns, R. J., D. E. Zemlicka, and P. J. Savarie. 1996. Effectiveness of large livestock protection collars against depredating coyotes. Wildlife Society Bulletin 24:123–27.

Cahalane, V. H. 1947. A deer–coyote episode. Journal of Mammalogy 28:36–39.

Caine, S. A., J. A. Kadlec, D. L. Allen, R. A. Cooley, M. C. Hornocker, A. S. Leopold, and F. H. Wagner. 1972. Predator control—1971—Report to the Council on Environmental Quality and the Department of the Interior by the Advisory Committee on Predator Control. Council on Environmental Quality and U.S. Department of the Interior, Washington, DC.

Camenzind, F. J. 1978. Behavioral ecology of coyotes on the National Elk Refuge, Jackson, Wyoming. Pages 267–94 *in* M. Bekoff, ed. Coyotes: biology, Behavior, and management. Academic Press, New York.

Carbyn, L. N., and P. C. Paquet. 1986. Long distance movement of a coyote from Riding Mountain National Park. Journal of Wildlife Management 50:89.

Chesness, R. A., and T. P. Bremicker. 1974. Home range, territoriality, and sociability of coyotes in north-central Minnesota. Coyote Research Workshop, Denver, CO.

Clark, F. W. 1972. Influence of jackrabbit density on coyote population change. Journal of Wildlife Management 36:343–56.

Conder, G. A., and R. M. Loveless. 1978. Parasites of the coyote (*Canis latrans*) in central Utah. Journal of Wildlife Diseases 14:247–49.

Conner, M. M. 1995. Identifying patterns of coyote predation on sheep on a northern California ranch. Thesis, University of California, Berkeley.

Conner, M. M., M. M. Jaeger, T. J. Weller, and D. R. McCullough. 1998. Effect of coyote removal on sheep depredation in northern California. Journal of Wildlife Management 62:690–99.

Connolly, G. E. 1978. Predator control and coyote populations: A review of simulation models. Pages 327–45 *in* M. Bekoff, ed. Coyotes: Biology, behavior, and management. Academic Press, New York.

Connolly, G. E. 1988. Aerial hunting takes sheep-killing coyotes in western Montana. U.S. Forest Service, General Technical Report RM 154:184–88.

Connolly, G. E. 1993. Livestock protection collars in the United States, 1988–1993. Proceedings of the Great Plains Wildlife Damage Control Workshop 11:25–33.

Connolly, G. E., and R. J. Burns. 1990. Efficacy of compound 1080 livestock protection collars for killing coyotes that attack sheep. Proceedings of the Vertebrate Pest Conference 14:269–76.

Connolly, G. E., R. M. Timm, W. E. Howard, and W. M. Longhurst. 1976. Sheep killing behavior of captive coyotes. Journal of Wildlife Management 40:400–407.

Conover, M. R., J. G. Francik, and D. E. Miller. 1977. An experimental evaluation of using taste aversion to control sheep loss due to coyote predation. Journal of Wildlife Management 41:775–79.

Coolahan, C. 1990. The use of dogs and calls to take coyotes around dens and resting areas. Proceedings of the Vertebrate Pest Conference 14:260–62.

Coppinger, R., J. Lorenz, and L. Coppinger. 1983. Introducing livestock guarding dogs to sheep and goat producers. Proceedings of the Eastern Wildlife Damage Control Conference 1:129–32.

Crabtree, R. L. 1988. Sociodemography of an unexploited coyote population. Dissertation, University of Idaho, Moscow.

Crabtree, R. L., and J. W. Sheldon. 1999. Coyotes and canid coexistence in Yellowstone. Pages 127–63 *in* T. W. Clark, A. P. Curlee, S. C. Minta, and P. M. Kareiva, eds. Carnivores in ecosystems. Yale University Press, New Haven, CT.

Crooks, K. R., and M. E. Soulé. 1999. Mesopredator release and avifaunal extinctions in a fragmented system. Nature 400:563–66.

Cunningham, V. D., and R. D. Dunford. 1970. Recent coyote record from Florida. Quarterly Journal of the Florida Academy of Science 33:279–80.

Cypher, B. L., and K. A. Spencer. 1998. Competitive interactions between coyotes and San Joaquin kit foxes. Journal of Mammalogy 79:204–214.

Davison, R. P. 1980. The effect of exploitation on some parameters of coyote populations. Dissertation, Utah State University, Logan.

De Calesta, D. S., and M. G. Cropsey. 1978. Field test of a coyote-proof fence. Wildlife Society Bulletin 6:256–59.

DeLorenzo, D. G., and V. W. Howard, Jr. 1976. Evaluation of sheep losses on a range lambing operation without predator control in southeastern New Mexico. Final report to the U.S. Fish and Wildlife Service, Denver Wildlife Research Center, New Mexico State University, Las Cruces.

Dice, L. R. 1942. A family of dog–coyote hybrids. Journal of Mammalogy 23:186–92.

Drewek, J., Jr., T. H. Noon, R. J. Trautman, and E. J. Bicknell. 1981. Serologic evidence of leptospirosis in a southern Arizona coyote population. Journal of Wildlife Diseases 17:33–37.

Eads, R. B. 1948. Ectoparasites from a series of Texas coyotes. Journal of Mammalogy 29:268–71.

Early, J. O., J. C. Roetheli, and O. R. Brewer. 1974. An economic study of predation in the Idaho range sheep industry, 1970–71 production cycle (Progress Report No. 182). Idaho Agricultural Research, University of Idaho, Moscow.

Elder, W. H., and C. M. Hayden. 1977. Use of discriminant function in taxonomic determination of canids from Missouri. Journal of Mammalogy 21:17–24.

Ellins, S. R. and G. C. Martin. 1981. Olfactory discrimination of lithium chloride by the coyote (*Canis latrans*). Behavior and Neurological Biology 31:214–24.

Fall, M. W. 1990. Control of coyote depredation on livestock—Progress in research and development. Proceedings of the Vertebrate Pest Conference 14:245–51.

Finkel, M. 1999. The ultimate survivor. Audubon 101:52–59.

Fisher, J. 1975. The plains dog moves east. National Wildlife 13:14–16.

Forthman-Quick, D. L., C. R. Gustavson, and K. W. Rusiniak. 1985a. Coyote control and taste aversion. Appetite 6:253–64.

Forthman-Quick, D. L., C. R. Gustavson, and K. W. Rusiniak. 1985b. Coyotes and taste aversion: The authors' reply. Appetite 6:284–90.

Fox, M. W. 1970. A comparative study of the development of facial expressions in canids: Wolf, coyote, foxes. Behavior 36:49–73.

Fuller, T. K., and L. B. Keith. 1981. Non-overlapping ranges of coyotes and wolves in northeastern Alberta. Journal of Mammalogy 62:403–5.

Gee, C. K., R. S. Magleby, D. B. Nielsen, and D. M. Stevens. 1977. Factors in the decline of the western sheep industry (Agriculture Economics Report No. 377). U.S. Department of Agriculture, Economic Research Service.

Gese, E. M. 1990. Reproductive activity in an old-age coyote in southeastern Colorado. Southwestern Naturalist 35:101–2.

Gese, E. M. 1998. Response of neighboring coyotes (*Canis latrans*) to social disruption in an adjacent pack. Canadian Journal of Zoology 76:1960–63.

Gese, E. M. 2001. Territorial defense by coyotes (*Canis latrans*) in Yellowstone National Park, Wyoming: Who, how, where, when, and why. Canadian Journal of Zoology 79:980–87.

Gese, E. M., and S. Grothe. 1995. Analysis of coyote predation on deer and elk during winter in Yellowstone National Park, Wyoming. American Midland Naturalist 133:36–43.

Gese, E. M., and Ruff, R. L. 1997. Scent-marking by coyotes, *Canis latrans:* The influence of social and ecological factors. Animal Behaviour 54:1155–66.

Gese, E. M., and R. L. Ruff. 1998. Howling by coyotes (*Canis latrans*): Variation among social classes, seasons, and pack sizes. Canadian Journal of Zoology 76:1037–43.

Gese, E. M., O. J. Rongstad, and W. R. Mytton. 1988a. Home range and habitat use of coyotes in southeastern Colorado. Journal of Wildlife Management 52:640–46.

Gese, E. M., O. J. Rongstad, and W. R. Mytton. 1988b. Relationship between coyote group size and diet in southeastern Colorado. Journal of Wildlife Management 52:647–53.

Gese, E. M., O. J. Rongstad, and W. R. Mytton. 1989a. Population dynamics of coyotes in southeastern Colorado. Journal of Wildlife Management 53:174–81.

Gese, E. M., O. J. Rongstad, and W. R. Mytton. 1989b. Changes in coyote movements due to military activity. Journal of Wildlife Management 53:334–39.

Gese, E. M., R. D. Schlutz, O. J. Rongstad, and D. E. Andersen. 1991. Prevalence of antibodies against canine parvovirus and canine distemper virus in wild coyotes in southeastern Colorado. Journal of Wildlife Diseases 27: 320–23.

Gese, E. M., R. L. Ruff, and R. L. Crabtree. 1996a. Social and nutritional factors influencing dispersal of resident coyotes. Animal Behaviour 52:1025–43.

Gese, E. M., R. L. Ruff, and R. L. Crabtree. 1996b. Foraging ecology of coyotes (*Canis latrans*): The influence of extrinsic factors and a dominance hierarchy. Canadian Journal of Zoology 74:769–83.

Gese, E. M., R. L. Ruff, and R. L. Crabtree. 1996c. Intrinsic and extrinsic factors influencing coyote predation of small mammals in Yellowstone National Park. Canadian Journal of Zoology 74:784–97.

Gese, E. M., T. E. Stotts, and S. Grothe. 1996d. Interactions between coyotes and red foxes in Yellowstone National Park, Wyoming. Journal of Mammalogy 77:377–82.

Gese, E. M., R. D. Schlutz, M. R. Johnson, E. S. Williams, R. L. Crabtree, and R. L. Ruff. 1997. Serological survey for diseases in free-ranging coyotes (*Canis latrans*) in Yellowstone National Park, Wyoming. Journal of Wildlife Diseases 33:47–56.

Gier, H. T. 1968. Coyotes in Kansas, rev. Kansas Agricultural Experiment Station, Kansas State University, Manhattan.

Gier, H. T. 1975. Ecology and social behavior of the coyote. Pages 247–62 *in* M. W. Fox, ed. The wild canids. Van Nostrand Reinhold, New York.

Gier, H. T., and D. J. Ameel. 1959. Parasites and diseases of Kansas coyotes. Kansas State University Agricultural Experiment Station, Technical Bulletin 91:1–34.

Gier, H. T., S. M. Kruckenburg, and R. J. Marler. 1978. Parasites and diseases of coyotes. Pages 37–71 *in* M. Bekoff, ed. Coyotes: Biology, behavior, and management. Academic Press, New York.

Gipson, P. S. 1974. Food habits of coyotes in Arkansas. Journal of Wildlife Management 38:848–53.

Gipson, P. S. . 1976. Melanistic *Canis* in Arkansas. Southwestern Naturalist 21:124–26.

Gipson, P. S., and J. A. Sealander. 1972. Home range and activity of the coyote (*Canis latrans frustror*) in Arkansas. Proceedings of the Annual Conference of Southeastern Association of Game and Fish Commissioners 26:82–95.

Gipson, P. S., I. K. Gipson, and J. A. Sealander. 1975. Reproductive biology of wild *Canis* in Arkansas. Journal of Mammalogy 56:605–12.

Gluesing, E. A., D. F. Balph, and F. F. Knowlton. 1980. Behavioral patterns of domestic sheep and their relationship to coyote predation. Applied Animal Ethology 6:315–30.

Grandage, J. 1972. The erect dog penis: A paradox of flexible rigidity. Veterinary Record 91:141–47.

Green, J. S., and R. A. Woodruff. 1983. The use of three breeds of dog to protect rangeland sheep from predators. Applied Animal Ethology 11:141–61.

Green, J. S., and R. A. Woodruff. 1987. Livestock-guarding dogs for predator control. Pages 62–68 *in* J. S. Green, ed. Protecting livestock from coyotes. U.S. Department of Agriculture, Agricultural Research Service, U.S. Sheep Experiment Station, Dubois, ID.

Green, J. S., R. A. Woodruff, and T. T. Tueller. 1984. Livestock-guarding dogs for predator control: Costs, benefits and practicality. Wildlife Society Bulletin 12:44–50.

Green, J. S., F. R. Henderson, and M. D. Collinge. 1994. Coyotes. Pages C51–76 *in* S. E. Hygnstrom, R. M. Timm, and G. E. Larson, eds. Prevention and control of wildlife damage, Vol., 1. University of Nebraska Press, Lincoln.

Grinder, M. I., and P. R. Krausman. 1998. Ecology and management of coyotes in Tucson, Arizona. Proceedings of the Vertebrate Pest Conference 18:293–98.

Guo, W., J. F. Evermann, W. J. Foreyt, F. F. Knowlton, and L. A. Windberg. 1986. Canine distemper virus in coyotes: A serologic survey. Journal of the American Veterinary Medical Association 189:1099–1100.

Gustavson, C. R., J. Garcia, W. G. Hankins, and K. W. Rusiniak. 1974. Coyote predation control by aversive conditioning. Science 184:581–83.

Gustavson, C. R., J. R. Jowsey, and D. N. Milligan. 1982. A 3-year evaluation of taste aversion coyote control in Saskatchewan. Journal of Range Management 35:57–59.

Hall, E. R., and K. R. Kelson. 1959. Mammals of North America. Ronald Press, New York.

Hallett, D. L. 1977. Post-natal mortality, movements, and den sites of Missouri coyotes. Thesis, University of Missouri, Columbia.

Hamlett, G. W. D. 1939. The reproductive cycle of the coyote. U.S. Department of Agriculture Technical Bulletin 616:1–11.

Hamlin, K. L., and L. L. Schweitzer. 1979. Cooperation by coyote pairs attacking mule deer fawns. Journal of Mammalogy 60:849–50.

Harris, C. E. 1983. Differential behavior of coyotes with regard to home range limits. Dissertation, Utah State University, Logan.

Harrison, D. J. 1992. Dispersal characteristics of juvenile coyotes in Maine. Journal of Wildlife Management 56:128–38.

Harrison, D. J., and J. R. Gilbert. 1985. Denning ecology and movements of coyotes in Maine during pup rearing. Journal of Mammalogy 66:712–19.

Harrison, D. J., and J. A. Harrison. 1984. Foods of adult Maine coyotes and their known-aged pups. Journal of Wildlife Management 48:922–26.

Harrison, D. J., J. A. Bissonette, and J. A. Sherburne. 1989. Spatial relationships between coyotes and red foxes in eastern Maine. Journal of Wildlife Management 53:181–85.

Harrison, D. J., J. A. Harrison, and M. O'Donoghue. 1991. Predispersal movements of coyote (*Canis latrans*) pups in eastern Maine. Journal of Mammalogy 72:756–63.

Hatier, K. G. 1995. Effects of helping behaviors on coyote packs in Yellowstone National Park, Wyoming. Thesis, Montana State University, Bozeman.

Hawthorne, V. M. 1971. Coyote movements in Sagehen Creek Basin, northeastern California. California Fish and Game 57:154–61.

Heinrich, D. 1972. Vergleichende Untersuchungen an Nebennieren einger Arten der Familie Canidae Gray 1821. Zoologische Wissenschaftliche. (Leipzig) 185:122–92.

Henke, S. E., and F. C. Bryant. 1999. Effects of coyote removal on the faunal community in western Texas. Journal of Wildlife Management 63:1066–81.

Henne, F. R. 1977. Domestic sheep mortality on a western Montana ranch. Pages 133–46 *in* R. L. Phillips and C. Jonkel, eds. Proceedings of the 1975 predator symposium. Montana Forest and Conservation Experiment Station, School of Forestry, University of Montana, Missoula.

Hibler, S. J. 1977. Coyote movement patterns with emphasis on home range characteristics. Thesis, Utah State University, Logan.

Hildebrand, M. 1952. The integument in Canidae. Journal of Mammalogy 33:419–28.

Hill, E. P., P. W. Sumner, and J. B. Wooding. 1987. Human influences on range expansion of coyotes in the southeast. Wildlife Society Bulletin 15:521–24.

Hilton, H. 1978. Systematics and ecology of the eastern coyote. Pages 209–28 *in* M. Bekoff, ed. Coyotes: Biology, behavior and management. Academic Press, New York.

Hilton, H., and N. P. Kutscha. 1978. Distinguishing characteristics of the hairs of eastern coyote, domestic dog, red fox and bobcat in Maine. American Midland Naturalist 100:223–27.

Hirsch, R. P., and H. T. Gier. 1974. Multiple-species infections of intestinal helminths in Kansas coyotes. Journal of Parasitology 60:650–53.

Hodges, C. M. 1990. The reproductive biology of the coyote (*Canis latrans*). Dissertation, Texas A&M University, College Station.

Holzman, S., M. J. Conroy, and J. Pickering. 1992a. Home range, movements, and habitat use of coyotes in southcentral Georgia. Journal of Wildlife Management 56:139–46.

Holzman, S., M. J. Conroy, and W. R. Davidson. 1992b. Diseases, parasites and survival of coyotes in south-central Georgia. Journal of Wildlife Diseases 28:572–80.

Hoover, S. 1996. Effectiveness of volatile trigeminal irritants at reducing egg consumption by mammalian predators: An experimental analysis. Thesis, Utah State University, Logan.

Horn, S. W., and P. N. Lehner. 1975. Scotopic sensitivity in the coyote (*Canis latrans*). Journal of Comparative Physiology and Psychology 89:1070–76.

Houston, D. B. 1978. Elk as winter–spring food for carnivores in northern Yellowstone National Park. Journal of Applied Ecology 15:653–61.

Howard, W. E. 1949. A means to distinguish skulls of coyotes and domestic dogs. Journal of Mammalogy 30:169–71.

Jackson, H. H. T. 1951. Part 2, Classification of the races of coyotes. Pages 227–341 *in* S. P. Young and H. H. T. Jackson. eds. The clever coyote. Wildlife Management Institute, Washington, DC.

Kelly, B. T., and E. O. Garton. 1997. Effects of prey size, meal size, meal composition, and daily frequency of feeding on the recovery of rodent remains from carnivore scats. Canadian Journal of Zoology 75:1811–17.

Kennelly, J. J. 1972. Coyote reproduction. Part 1, The duration of the spermatogenic cycle and epididymal sperm transport. Journal of Reproductive Fertility 31:163–70.

Kennelly, J. J. 1978. Coyote reproduction. Pages 73–93 *in* M. Bekoff, ed. Coyotes: Biology, behavior and management. Academic Press, New York.

Kennelly, J. J., and J. D. Roberts. 1969. Fertility of coyote–dog hybrids. Journal of Mammalogy 50:830–31.

Kitchen, A. M., E. M. Gese, and E. R. Schauster. 1999. Resource partitioning between coyotes and swift foxes: Space, time, and diet. Canadian Journal of Zoology 77:1645–56.

Kitchen, A. M., E. M. Gese, and E. R. Schauster. 2000a. Changes in coyote activity patterns due to reduced exposure to human persecution. Canadian Journal of Zoology 78:853–57.

Kitchen, A. M., E. M. Gese, and E. R. Schauster. 2000b. Long-term spatial stability of coyote (*Canis latrans*) home ranges in southeastern Colorado. Canadian Journal of Zoology 78:458–64.

Kleiman, D. G. 1966. Scent marking in Canidae. Symposium of the Zoological Society of London 18:167–77.

Kleiman, D. G., and C. A. Brady. 1978. Coyote behavior in the context of recent canid research: Problems and perspectives. Pages 163–88 *in* M. Bekoff, ed. Coyotes: Biology, behavior and management. Academic Press, New York.

Kleiman, D. G., and J. F. Eisenberg. 1973. Comparison of canid and felid social systems from an evolutionary perspective. Animal Behaviour 21:637–59.

Knight, S. W. 1978. Dominance hierarchies of captive coyote litters. Thesis, Utah State University, Logan.

Knowlton, F. F. 1972. Preliminary interpretations of coyote population mechanics with some management implications. Journal of Wildlife Management 36:369–82.

Knowlton, F. F., and E. M. Gese. 1995. Coyote population processes revisited. Pages 1–6 *in* D. Rollins, C. Richardson, T. Blankenship, K. Canon, and S. Henke, eds. Coyotes in the southwest: A compendium of our knowledge. Texas Parks and Wildlife Department, Austin.

Knowlton, F. F., and L. C. Stoddart. 1983. Coyote population mechanics: Another look. Pages 93–111 *in* F. L. Bunnell, D. S. Eastman, and J. M. Peek, eds. Symposium on natural regulation of wildlife populations. Forest, Wildlife, and Range Experiment Station, University of Idaho, Moscow.

Knowlton, F. F., and L. C. Stoddart. 1992. Some observations from two coyote–prey studies. Pages 101–21 *in* A. H. Boer, ed. Ecology and management of the eastern coyote. Wildlife Research Unit, University of New Brunswick, Fredericton, Canada.

Knowlton, F. F., and S. L. Whittemore. 2001. Pulp cavity-tooth width ratios from known-age and wild-caught coyotes determined by radiography. Wildlife Society Bulletin 29:239–244.

Knowlton, F. F., E. M. Gese, and M. M. Jaeger. 1999. Coyote depredation control: An interface between biology and management. Journal of Range Management 52:398–412.

Koehler, G. M., and M. G. Hornocker. 1991. Seasonal resource use among mountain lions, bobcats, and coyotes. Journal of Mammalogy 72:391–96.

Kolenosky, G. B. 1971. Hybridization between wolf and coyote. Journal of Mammalogy 52:446–49.

Korschgen, L. J. 1973. Food habits of coyotes in north-central Missouri (Federal Aid Project No. W-13-R-27). Missouri Department of Conservation.

Krefting, L. W. 1969. The rise and fall of the coyote on Isle Royale. Naturalist 20:24–32.

Lamprecht, J. 1978. The relationship between food competition and foraging group size in some larger carnivores. Zeitschrift für Tierpsychologie 46:337–43.

Larivière, S., and M. Crête. 1993. The size of eastern coyotes (*Canis latrans*): A comment. Journal of Mammalogy 74:1072–74.

Laundré, J. W. 1981. Temporal variation in coyote vocalization rates. Journal of Wildlife Management 45:767–69.

Laundré, J. W., and B. L. Keller. 1984. Home-range size of coyotes: A critical review. Journal of Wildlife Management 48:127–39.

Lawrence, B., and W. H. Bossert. 1967. Multiple character analysis of *Canis lupus, latrans* and *familiaris,* with a discussion of the relationships of *Canis niger*. American Zoologist 7:223–32.

Lawrence, B., and W. H. Bossert. 1969. The cranial evidence of hybridization in New England *Canis*. Breviora 330:1–13.

Lawrence, B., and W. H. Bossert. 1975. Relationships of North American *Canis* shown by a multiple character analysis of selected populations. Pages 73–96 *in* M. W. Fox, ed. The wild canids. Van Nostrand Reinhold, New York.

Lehman, N., A. Eisenhawer, K. Hansen, L. D. Mech, R. O. Peterson, P. J. P. Gogan, and R. K. Wayne. 1991. Introgression of coyote mitochondrial DNA into sympatric North American gray wolf populations. Evolution 45:104–19.

Lehner, P. N. 1978a. Coyote communication. Pages 127–62 *in* M. Bekoff, ed. Coyotes: Biology, behavior, and management. Academic Press, New York.

Lehner, P. N. 1978b. Coyote vocalizations: A lexicon and comparisons with other canids. Animal Behaviour 26:712–22.

Lehner, P. N. 1987. Repellents and conditioned avoidance. Pages 56–61 *in* J. S. Green, ed. Protecting livestock from coyotes. U.S. Department of Agriculture, Agricultural Research Service, U.S. Sheep Experiment Station, Dubois, ID.

Linhart, S. B. 1984a. Managing coyote damage problems with nonlethal techniques: Recent advances in research. Proceedings of the Eastern Wildlife Damage Conference 1:105–18.

Linhart, S. B. 1984b. Strobe-light and siren devices for protecting fenced-pasture and range sheep from coyote predation. Proceedings of the Vertebrate Pest Conference 11:154–56.

Linhart, S. B., and F. F. Knowlton. 1967. Determining age of coyotes by tooth cementum layers. Journal of Wildlife Management 31:362–65.

Linhart, S. B., H. H. Brusman, and D. S. Balser. 1968. Field evaluation of an antifertility agent, stilbesterol, for inhibiting coyote reproduction. Transactions of the North American Wildlife Conference 33:316–26.

Linhart, S. B., R. T. Sterner, T. C. Carrigan, and D. R. Henne. 1979. Komondor guard dogs reduce sheep losses to coyotes: A preliminary evaluation. Journal of Range Management 32:238–41.

Linhart, S. B., J. D. Roberts, and G. J. Dasch. 1982. Electric fencing reduces coyote predation on pastured sheep. Journal of Range Management 35:276–81.

Linhart, S. B., R. T. Sterner, G. J. Dasch, and J. W. Theade. 1984. Efficacy of light and sound stimuli for reducing coyote predation upon pastured sheep. Protection Ecology 6:75–84.

Linhart, S. B., G. J. Dasch, R. R. Johnson, J. D. Roberts, and C. J. Packham. 1992. Electronic frightening devices for reducing coyote depredation on domestic sheep: Efficacy under range conditions and operational use. Proceedings of the Vertebrate Pest Conference 15:386–92.

Linnell, J. D. C., J. Odden, M. E. Smith, R. Aanes, and J. E. Swenson. 1999. Large carnivores that kill livestock: Do "problem individuals" really exist? Wildlife Society Bulletin 27:698–705.

Litvaitis, J. A. 1978. Movements and habitat use of coyotes on the Wichita Mountains National Wildlife Refuge. Thesis, Oklahoma State University, Stillwater.

MacCracken, J. G. 1989. Coyote prey choice: A reply. Journal of Wildlife Management 53:666–67.

MacCracken, J. G., and R. M. Hansen. 1987. Coyote feeding strategies in southeastern Idaho: Optimal foraging by an opportunistic predator? Journal of Wildlife Management 51:278–85.

Mahan, B. R. 1978. Occurrence of melanistic canids in Nebraska. Proceedings of the Nebraska Academy of Science 5:121–22.

Major, J. T., and J. A. Sherburne. 1987. Interspecific relationships of coyotes, bobcats, and red foxes in western Maine. Journal of Wildlife Management 51:606–16.

Mathwig, H. J. 1973. Food and population characteristics of Iowa coyotes. Iowa State Journal of Research 47:167–89.

McAdoo, J. K., and D. A. Klebenow. 1978. Predation on range sheep with no predator control. Journal of Range Management 31:111–14.

McBride, R. T. 1974. Predator protection collar for livestock. Patent No. 3,842,806, registered October 22, 1974. U.S. Patent Office, Washington, DC.

McBride, R. T. 1982. Predator control toxic collar. Patent No. 4,338,886, registered July 13, 1982. U.S. Patent Office, Washington, DC.

McCarley, H. 1975. Long-distance vocalizations of coyotes (*Canis latrans*). Journal of Mammalogy 56:847–56.

McClure, M. F., N. S. Smith, and W. W. Shaw. 1995. Diets of coyotes near the boundary of Saguaro National Monument and Tucson, Arizona. Southwestern Naturalist 40:101–4.

Mech, L. D. 1966. The wolves of Isle Royale. U.S. National Park Service Fauna Series 7:1–210.
Mech, L. D. 1974. *Canis lupus*. Mammalian Species 37:1–6.
Mengel, R. M. 1971. A study of coyote–dog hybrids and implications concerning hybridization in *Canis*. Journal of Mammalogy 52:316–36.
Mercer, S. H., L. P. Jones, J. H. Rappole, D. Twedt, L. L. Laack, and T. M. Craig. 1988. *Hepatozoon* sp. in wild carnivores in Texas. Journal of Wildlife Diseases 24:574–76.
Messier, F., and C. Barrette. 1982. The social system of the coyote (*Canis latrans*) in a forested habitat. Canadian Journal of Zoology 60:1743–53.
Mills, L. S., and F. F. Knowlton. 1991. Coyote space use in relation to prey abundance. Canadian Journal of Zoology 69:1516–21.
Mitchell, R. L., and S. L. Beasom. 1974. Hookworms in south Texas coyotes and bobcats. Journal of Wildlife Management 38:455–58.
Moore, G. C., and G. R. Parker. 1992. Colonization by the eastern coyote (*Canis latrans*). Pages 23–37 *in* A. Boer, ed. Ecology and management of the eastern coyote. Wildlife Research Unit, University of New Brunswick, Fredericton, Canada.
Moore, J. A. 1985. Registration of compound 1080. Environmental Protection Agency notice. Federal Register 50:28986.
Munoz, J. R. 1977. Causes of sheep mortality at the Cook Ranch, Florence, Montana, 1975–76. Thesis, University of Montana, Missoula.
Murie, O. J. 1935. Food habits of the coyote in Jackson Hole (Circular 362). U.S. Department of Agriculture.
Murie, O. J. 1951. The elk of North America. Wildlife Management Institute, Washington, DC.
Murie, O. J. 1954. A field guide to animal tracks. Houghton Mifflin, Boston.
Murphy, K. M. 1998. The ecology of the cougar (*Puma concolor*) in the northern Yellowstone ecosystem: Interactions with prey, bears, and humans. Dissertation, University of Idaho, Moscow.
Murray, D. L., and S. Boutin. 1991. The influence of snow on lynx and coyote movements: Does morphology affect behavior? Oecologia 88:463–69.
Nass, R. D., and J. Theade. 1988. Electric fences for reducing sheep losses to predators. Journal of Range Management 41:251–52.
Neale, J. C. C., B. N. Sacks, M. M. Jaeger, and D. R. McCullough. 1998. A comparison of bobcat and coyote predation on lambs in north-coastal California. Journal of Wildlife Management 62:700–706.
Nellis, C. H., and L. B. Keith. 1976. Population dynamics of coyotes in central Alberta, 1964–68. Journal of Wildlife Management 40:380–99.
Nellis, C. H., S. P. Wetmore, and L. B. Keith. 1978. Age-related characteristics of coyote canines. Journal of Wildlife Management 42:680–83.
Nowak, R. 1978. Evolution and taxonomy of coyotes and related *Canis*. Pages 3–16 *in* M. Bekoff, ed. Coyotes: Biology, behavior and management. Academic Press, New York.
O'Donoghue, M., S. Boutin, C. J. Krebs, and E. J. Hofer. 1997. Numerical responses of coyotes and lynx to the snowshoe hare cycle. Oikos 80:150–62.
O'Gara, B. W., K. C. Brawley, J. R. Munoz, and D. R. Henne. 1983. Predation on domestic sheep on a western Montana ranch. Wildlife Society Bulletin 11:253–64.
Ogle, T. F., and E. M. Farris. 1973. Aspects of cold adaptation in the coyote. Northwest Science 47:70–74.
Ozoga, J. J., and E. M. Harger. 1966. Winter activities and feeding habits of northern Michigan coyotes. Journal of Wildlife Management 30:809–18.
Palomares, F., and T. M. Caro. 1999. Interspecific killing among mammalian carnivores. American Naturalist 153:492–508.
Paquet, P. C. 1991. Winter spatial relationships of wolves and coyotes in Riding Mountain National Park, Manitoba. Journal of Mammalogy 72:397–401.
Paquet, P. C. 1992. Prey use strategies of sympatric wolves and coyotes in Riding Mountain National Park, Manitoba. Journal of Mammalogy 73:337–43.
Paradiso, J. L., and R. Nowak. 1971. A report on the taxonomic status and distribution of the red wolf (Special Science Report 145). U.S. Fish and Wildlife Service.
Parrish, I. R. 1994. Development and statistical validation of radioimmunoassays for measuring six reproductive hormones in coyote sera. Thesis, Utah State University, Logan.
Pence, D. B., and L. A. Windberg. 1994. Impact of a sarcoptic mange epizootic on a coyote population. Journal of Wildlife Management 58:624–33.
Pence, D. B., L. A. Windberg, B. C. Pence, and R. Sprowls. 1983. The epizootiology and pathology of sarcoptic mange in coyotes, *Canis latrans,* from south Texas. Journal of Parasitology 69:1100–1115.
Person, D. K., and D. H. Hirth. 1991. Home range and habitat use of coyotes in a farm region of Vermont. Journal of Wildlife Management 55:433–41.
Petersen, E. A., W. C. Heaton, and S. D. Wruble. 1969. Levels of auditory response in fissiped carnivores. Journal of Mammalogy 50:566–87.
Peterson, R. O. 1995. Wolves as interspecific competitors in canid ecology. Pages 315–23 *in* L. N. Carbyn, S. H. Fritts, and D. R. Seip, eds. Ecology and conservation of wolves in a changing world. Canadian Circumpolar Institute, University of Alberta, Edmonton.
Peterson, R. O., and J. M. Thurber. 1993. The size of eastern coyotes (*Canis latrans*): A rebuttal. Journal of Mammalogy 74:1075–76.
Poulle, M. L., M. Crête, and J. Huot. 1995. Seasonal variation in body mass and composition of eastern coyotes. Canadian Journal of Zoology 73:1625–33.
Powell, K. J. 1993. The use of guard llamas to protect sheep from coyote depredation. Thesis, Iowa State University, Ames.
Pyrah, D. 1984. Social distribution and population estimates of coyotes in north-central Montana. Journal of Wildlife Management 48:679–90.
Quinn, T. 1997a. Coyote (*Canis latrans*) habitat selection in urban areas of western Washington via analysis of routine movements. Northwest Science 71:289–97.
Quinn, T. 1997b. Coyote food habits in three urban habitat types of western Washington. Northwest Science 71:1–5.
Radinsky, L. B. 1973. Evolution of the canid brain. Brain, Behavior, and Evolution 7:169–202.
Ralls, K., and P. J. White. 1995. Predation on San Joaquin kit foxes by larger canids. Journal of Mammalogy 76:723–29.
Richens, V. B., and R. D. Hugie. 1974. Distribution, taxonomic status, and characteristics of coyotes in Maine. Journal of Wildlife Management 38:447–54.
Riley, G. A., and R. T. McBride. 1975. A survey of the red wolf (*Canis rufus*). Pages 263–77 *in* M. W. Fox, ed. The wild canids. Van Nostrand Reinhold, New York.
Robel, R. J., A. D. Dayton, F. R. Henderson, R. L. Meduna, and C. W. Spaeth. 1981. Relationships between husbandry methods and sheep losses to canine predators. Journal of Wildlife Management 45:894–911.
Roberts, J. D. 1978. Variation in coyote age determination from annuli in different teeth. Journal of Wildlife Management 42:454–56.
Robinson, W. B., and M. W. Cummings. 1951. Movements of coyotes from and to Yellowstone National Park (Special Scientific Report Wildlife No. 11). U.S. Fish and Wildlife Service.
Robinson, W. B., and E. F. Grand. 1958. Comparative movements of bobcats and coyotes as disclosed by tagging. Journal of Wildlife Management 22:117–22.
Rollins, D. 1995. The livestock protection collar for removing depredating coyotes: A search for perfect justice? Pages 168–71 *in* D. Rollins, C. Richardson, T. Blankenship, K. Canon, and S. Henke, eds. Coyotes in the Southwest: A compendium of our knowledge. Texas Parks and Wildlife Department, Austin.
Sacks, B. N. 1996. Ecology and behavior of coyotes in relation to depredation and control on a California sheep ranch. Thesis, University of California, Berkeley.
Sacks, B. N., M. M. Jaeger, J. C. C. Neale, and D. R. McCullough. 1999. Territoriality and breeding status of coyotes relative to sheep predation. Journal of Wildlife Management 63:593–605.
Sargeant, A. B., and S. H. Allen. 1989. Observed interactions between coyotes and red foxes. Journal of Mammalogy 70:631–33.
Sargeant, A. B., S. H. Allen, and J. O. Hastings. 1987. Spatial relations between sympatric coyotes and red foxes in North Dakota. Journal of Wildlife Management 51:285–93.
Schmitz, O. J., and D. M. Lavigne. 1987. Factors affecting body size in sympatric Ontario *Canis*. Journal of Mammalogy 68:92–99.
Schultz, V. 1955. Status of the coyote and related forms in Tennessee. Journal of the Tennessee Academy of Science 30:44–46.
Seitz, A. 1965. Früchtbare Kreuzungen Goldshakal × Coyote und veziprok Coyote × Goldshakal; erste früchtabre Ruckkreuzung. Zoologische Garten 31:174–83.
Shargo, E. S. 1988. Home range, movements, and activity patterns of coyotes (*Canis latrans*) in Los Angeles suburbs. Dissertation, University of California, Los Angeles.
Shelton, M. 1984. The use of conventional and electric fencing to reduce coyote predation on sheep and goats (MP 1556). Texas Agricultural Experiment Station.
Shelton, M. 1987. Antipredator fencing. Pages 30–37 *in* J. S. Green, ed. Protecting livestock from coyotes. U.S. Department of Agriculture, Agricultural Research Service, U.S. Sheep Experiment Station, Dubois, ID.
Shield, J. 1972. Acclimation and energy metabolism of the dingo, *Canis dingo,* and the coyote, *Canis latrans*. Journal of Zoology (London) 168:483–501.
Shivik, J. A., M. M. Jaeger, and R. H. Barrett. 1996. Coyote movements in relation to the spatial distribution of sheep. Journal of Wildlife Management 60:422–30.
Shivik, J. A., M. M. Jaeger, and R. H. Barrett. 1997. Coyote activity patterns in the Sierra Nevada. Great Basin Naturalist 57:355–58.
Silver, H., and W. T. Silver. 1969. Growth and behavior of the coyote-like canid

of northern New England with observations on canid hybrids. Wildlife Monograph 17:1–41.

Snow, C. J. 1967. Some observations on the behavioral and morphological development of coyote pups. American Zoologist 7:353–55.

Soulé, M. E., D. T. Bolger, A. C. Alberts, J. Wright, M. Sorice, and S. Hill. 1988. Reconstructed dynamics of rapid extinctions of chaparrel-requiring birds in urban habitat islands. Conservation Biology 2:75–92.

Springer, J. T. 1977. Movement patterns of coyotes in south-central Washington as determined by radio telemetry. Dissertation, Washington State University, Pullman.

Stellflug, J. N., N. L. Gates, and R. G. Sasser. 1978. Reproductive inhibitors for coyote population control: Developments and current status. Proceedings of the Vertebrate Pest Conference 8:185–89.

Sterner, R. T., and S. A. Shumake. 1978. Coyote damage-control research: A review and analysis. Pages 297–325 *in* M. Bekoff, ed. Coyotes: Biology, behavior, and management. Academic Press, New York.

Thomas, C. U., and P. E. Hughes. 1992. Plague surveillance by serological testing of coyotes (*Canis latrans*) in Los Angeles County, California. Journal of Wildlife Diseases 28:610–13.

Thomas, N. J., W. J. Foreyt, J. F. Evermann, L. A. Windberg, and F. F. Knowlton. 1984. Seroprevalence of canine parvovirus in wild coyotes from Texas, Utah, and Idaho (1972–1983). Journal of the American Veterinary Medical Association 185:1283–87.

Thomas, R. E., M. L. Beard, T. J. Quan, L. G. Carter, A. M. Barnes, and C. E. Hopla. 1989. Experimentally induced plague infection in the northern grasshopper mouse (*Onychomys leucogaster*) acquired by consumption of infected prey. Journal of Wildlife Diseases 25:477–80.

Thornton, J. E., R. R. Bell, and M. J. Reardon. 1974. Internal parasites of coyotes in southern Texas. Journal of Wildlife Diseases 10:232–36.

Thurber, J. M., and R. O. Peterson. 1991. Changes in body size associated with range expansion in the coyote (*Canis latrans*). Journal of Mammalogy 72:750–55.

Thurber, J. M., R. O. Peterson, J. D. Woolington, and J. A. Vucetich. 1992. Coyote coexistence with wolves on the Kenai Peninsula, Alaska. Canadian Journal of Zoology 70:2494–98.

Till, J. A., and F. F. Knowlton. 1983. Efficacy of denning in alleviating coyote depredations upon domestic sheep. Journal of Wildlife Management 47:1018–25.

Timm, R. M. and G. E. Connolly. 1980. How coyotes kill sheep. National Wool Grower 70:14–15.

Timm, R. M., and R. H. Schmidt. 1989. Management problems encountered with livestock guarding dogs on the University of California, Hopland Field Station. Pages 54–58 *in* reat Plains wildlife damage control workshop (Forest Service General Technical Report RM-171). U.S. Department of Agriculture.

Todd, A. W. 1985. Demographic and dietary comparisons of forest and farmland coyote, *Canis latrans,* populations in Alberta. Canadian Field-Naturalist 99:163–71.

Todd, A. W., and L. B. Keith. 1983. Coyote demography during a snowshoe hare decline in Alberta. Journal of Wildlife Management 47:394–404.

Todd, A. W., L. B. Keith, and C. A. Fischer. 1981. Population ecology of coyotes during a fluctuation of snowshoe hares. Journal of Wildlife Management 45:629–40.

Trainer, D. O., and F. F. Knowlton. 1968. Serologic evidence of diseases in Texas coyotes. Journal of Wildlife Management 32:981–83.

Tzilkowski, W. M. 1980. Mortality patterns of radio-marked coyotes in Jackson Hole, Wyoming. Dissertation, University of Massachusetts, Amherst.

U.S. Department of Agriculture. 1994. Final environmental impact statement. Animal Damage Control Program, Vol., 2. Animal and Plant Health Inspection Service, Washington, DC.

U.S. Department of Agriculture. 1995. Sheep and goat predator loss. U.S. Department of Agriculture, National Agricultural Statistics Board, Washington, DC.

U.S. Fish and Wildlife Service. 1978. Predator damage in the west: A study of coyote management alternatives. U.S. Fish and Wildlife Service, Washington, DC.

Van Vuren, D., and S. E. Thompson, Jr. 1982. Opportunistic feeding by coyotes. Northwestern Science 56:131–35.

Van Wormer, J. 1964. The world of the coyote. Lippincott, Philadelphia.

Voigt, D. R., and B. D. Earle. 1983. Avoidance of coyotes by red fox families. Journal of Wildlife Management 47:852–57.

Wade, D. H. 1978. Coyote damage: A survey of its nature and scope, control measures and their application. Pages 347–68 *in* M. Bekoff, ed. Coyotes: Biology, behavior, and management. Academic Press, New York.

Wagner, F. H. 1988. Predator control and the sheep industry. Regina Books, Claremont, CA.

Wagner, K. K., and M. R. Conover. 1999. Effect of preventative coyote hunting on sheep losses to coyote predation. Journal of Wildlife Management 63:606–12.

Walsh, P. B., and J. M. Inglis. 1989. Seasonal and diel rate of spontaneous vocalization in coyotes in south Texas. Journal of Mammalogy 70:169–71.

Wayne, R. K., and N. Lehman. 1992. Mitochondrial DNA analysis of the eastern coyote: Origins and hybridization. Pages 9–22 *in* A. Boer, ed. Ecology and management of the eastern coyote. Wildlife Research Unit, University of New Brunswick, Fredericton, Canada.

Weaver, J. L. 1977. Coyote–food base relationships in Jackson Hole, Wyoming. Thesis, Utah State University, Logan.

Weaver, J. L. 1979. Influence of elk carrion upon coyote populations in Jackson Hole, Wyoming. Pages 152–57 *in* M. S. Boyce and L. D. Hayden-Wing, eds. North American elk: Ecology, behavior and management. University of Wyoming, Laramie.

Weaver, J. L., and S. W. Hoffman. 1979. Differential detectability of rodents in gray wolf scats. Journal of Wildlife Management 43:783–86.

Wells, M. C. 1978. Coyote senses in predation: Environmental influences on their relative use. Behavioral Proceedings 3:149–58.

Wells, M. C., and P. N. Lehner. 1978. The relative importance of the distance senses in coyote predatory behaviour. Animal Behaviour 26:251–58.

Wells, M. C., and M. Bekoff. 1981. An observational study of scent marking by wild coyotes. Animal Behaviour 9:332–50.

Wells, M. C., and M. Bekoff. 1982. Predation by wild coyotes: Behavioral and ecological analyses. Journal of Mammalogy 63:118–27.

White, P. J., K. Ralls, and C. A. Vanderbilt White. 1995. Overlap in habitat and food use between coyotes and San Joaquin kit foxes. Southwestern Naturalist 40:342–49.

Williams, E. S., E. T. Thorne, M. J. G. Appel, and D. W. Belitsky. 1988. Canine distemper in black-footed ferrets (*Mustela nigripes*) from Wyoming. Journal of Wildlife Diseases 24:385–98.

Wilson, P. J., S. Grewal, I. D. Lawford, J. N. M. Heal, A. G. Granacki, D. Pennock, J. B. Theberge, M. T. Theberge, D. R. Voigt, W. Waddell, R. E. Chambers, P. C. Paquet, G. Goulet, D. Cluff, and B. N. White. 2000. DNA profiles of the eastern Canadian wolf and the red wolf provide evidence for a common evolutionary history independent of the gray wolf. Canadian Journal of Zoology 78:2156–66.

Windberg, L. A. 1995. Demography of a high-density coyote population. Canadian Journal of Zoology 73:942–54.

Windberg, L. A. 1996. Coyote responses to visual and olfactory stimuli related to familiarity with an area. Canadian Journal of Zoology 74:2248–54.

Windberg, L. A., and F. F. Knowlton. 1988. Management implications of coyote spacing patterns in southern Texas. Journal of Wildlife Management 52:632–40.

Windberg, L. A., and F. F. Knowlton. 1990. Relative vulnerability of coyotes to some capture procedures. Wildlife Society Bulletin 18:282–90.

Windberg, L. A., and C. D. Mitchell. 1990. Winter diets of coyotes in relation to prey abundance in southern Texas. Journal of Mammalogy 71:439–47.

Windberg, L. A., H. L. Anderson, and R. M. Engeman. 1985. Survival of coyotes in southern Texas. Journal of Wildlife Management 49:301–7.

Windberg, L. A., R. M. Engeman, and J. F. Bromaghin. 1991. Body size and condition of coyotes in southern Texas. Journal of Wildlife Diseases 27:47–52.

Windberg, L. A., S. M. Ebbert, and B. T. Kelly. 1997a. Population characteristics of coyotes (*Canis latrans*) in the northern Chihuahuan desert of New Mexico. American Midland Naturalist 138:197–207.

Windberg, L. A., F. F. Knowlton, S. M. Ebbert, and B. T. Kelly. 1997b. Aspects of coyote predation on Angora goats. Journal of Range Management 50:226–30.

Wixsom, M. J., S. P. Green, R. M. Corwin, and E. K. Fritzell. 1991. *Dirofilaria immitis* in coyotes and foxes in Missouri. Journal of Wildlife Diseases 27:166–69.

Woodruff, R. A., and B. L. Keller. 1982. Dispersal, daily activity, and home range of coyotes in southeastern Idaho. Northwest Science 56:199–207.

Wurster, D. H., and K. Benirschke, K. 1968. Comparative cytogenetic studies in the order Carnivora. Chromosoma 24:336–82.

Young, S. P. 1951. Part 1, Its history, life habits, economic status, and control. Pages 1–226 *in* S. P. Young and H. H. T. Jackson. The clever coyote. Wildlife Management Institute, Washington, DC.

Marc Bekoff, Department of Environmental, Population, and Organismic Biology, University of Colorado, Boulder, Colorado 80309-0334. Email: marc.bekoff@colorado.edu.

Eric M. Gese, National Wildlife Research Center, Department of Forest, Range, and Wildlife Services, Utah State University, Logan, Utah 84322-5295. Email: egese@cc.usu.edu.

23

Gray Wolf

Canis lupus and Allies

Paul C. Paquet
Ludwig N. Carbyn

NOMENCLATURE

COMMON NAMES. Gray wolf, timber wolf, tundra wolf, plains wolf; Spanish, lobo; French, loup; Inuktituk, amaruq; here, we use the common name "gray" wolf for the description of all subpopulations, although gray may not be the predominant color phase over large regional areas (Wilson and Reeder 1993)
SCIENTIFIC NAMES. *Canis lupus*

Subspecies. As expected in a widely distributed species, considerable variation occurs across the vast range of the gray wolf. The designation of wolf subspecies according to recently formulated criteria (e.g., Avise and Ball 1990; O'Brien and Mayr 1991), which place subspecies on a more empirical foundation, is not feasible at this time. The subspecific taxonomy of the gray wolf in North America has not been adequately analyzed by modern techniques and is unsatisfactory. A comprehensive molecular taxonomic revision of the species, using the much larger and geographically diverse collections now available, has yet to be done.

Relationships within the gray wolf are complicated by the possible introgression of genes from domestic dogs (Grace 1976; Miller 1978, 1993; Friis 1985; Wayne et al. 1992; Clutton-Brock et al. 1994) and by introgressive hybridization with coyotes (*C. latrans*) and red wolves (*C. rufus*) (Pilgrim et al. 1998; Wilson et al. 2000). Geographic variation also may have resulted from human-induced ecological disruption (Friis 1985; Lehman et al. 1991), including extirpation of local populations and their subsequent replacement by wolves of different subspecies from neighboring ranges (Nowak 1983).

Until recently, 24 subspecies of the gray wolf were recognized for North America (Hall and Kelson 1952, 1959; Hall 1981). These were based on a revision of Goldman (1944), who used cranial features, external measurements, and pelage characteristics, but did not use statistical analyses to evaluate his results. Many studies, however, cast doubt on the validity of Hall's (1981) taxonomic arrangement and suggest there are fewer subspecies (Jolicoeur 1959, 1975; Nowak 1973, 1979; Skeel and Carbyn 1977; Kolenosky and Standfield 1975; Pedersen 1982; Bogan and Mehlhop 1983; Friis 1985; Nowak 1995). Based on a cursory examination of specimens and speculation on the species's Pleistocene biogeography, Nowak (1983) suggested dividing *C. lupus* into five subspecific North American groups: northern (*occidentalis*), presumed to have originated in Beringia; southern (*nubilus*), which originated on the central plains south of the ice sheet; arctic (*arctos*), which originated in the Pearyland Refugium; eastern (*lycaon*), with origins in a southeastern refugium; and *baileyi,* a small form from the Southwest.

A more recent revision by Nowak (1995), using quantitative evaluation of cranial morphology, retained his original five subspecific designations, but revised the groupings. Mulders (1997), however, identified three subspecies of wolves in the Canadian North using statistical analyses of skull measurements. He separated mainland tundra wolves from central boreal forest wolves while retaining arctic wolves as described by Nowak (1983). In addition, new genetic analyses suggest that the eastern timber wolf (*C. lupus lycaon*) and red wolf (*C. rufus*) of the eastern United States are the same species (Wilson et al. 2000). Accordingly, the researchers proposed changing the scientific name of both to *Canis lycaon,* with a common name of red wolf. The recommended name is based on historical precedence (Brewster and Fritts 1995). This proposal raises numerous legal, policy, and management questions.

As an alternative to conventional taxonomic classifications, Theberge (1991) proposed an ecological classification of wolf subspecies. Using major prey species and vegetation zones as the principal criteria, his classification comprises 10 categories or "ecotypes." These ecological criteria may or may not be correlated with morphological and genetic differences.

EVOLUTION

The gray wolf is a member of the Canidae, or dog family, which is part of the order Carnivora. Although a distinct taxon, it is closely related to coyotes and Simien jackals (*C. simensis*) (Wayne et al. 1995). The closest relative is the domestic dog (*C. familiaris*). Generally considered among the most morphologically primitive of the living carnivores, along with the coyote, the gray wolf is usually placed at the beginning of the systematic treatments of the order. The genus *Canis* seems to have originated from foxlike ancestors in the early to middle Pliocene (Wayne et al. 1995).

Wayne et al. (1995) suggest several wolflike species evolved from a common ancestor. *Canis lupus* first appeared in Eurasia during the Pleistocene period, about 1 million years ago. The dire wolf (*Canis dirus*) is thought to be a descendant of *Canis lupus,* which migrated to North America around 750,000 years ago. The two species seem to have coexisted for about 400,000 years. As prey began to vanish due to climate changes, the dire wolf gradually became extinct, vanishing completely about 7000 years ago.

Much recent debate has centered on the evolution and relatedness of gray wolves, red wolves, and coyotes (Wayne and Jenks 1991; Nowak and Federoff 1998; Wayne et al. 1998; Wilson et al. 2000). In theory, the wolf and the coyote are sister species, which diverged during the late Pliocene (3 million years before the present) (Nowak 1979; Wayne and O'Brien 1987). The relationship is evident from the ability of the two species to produce fertile hybrids (Kolenosky 1971). Yet, the tendency for eastern gray wolves to hybridize with coyotes is not observed in western gray wolves (Pilgrim et al. 1998; Wilson et al. 2000). In addition, red and eastern wolves share morphological characteristics not observed in western gray wolves, such as smaller size (Goldman 1944). Wilson et al. (2000) maintained that red and eastern gray wolves are more closely related to each other than either is to western gray wolves. Moreover, they believed the red wolf and eastern wolf have a common North American origin separate from that of the western gray wolf.

According to Wilson et al. (2000), North America was inhabited by a common canid ancestor 1–2 million years ago. Some of these animals traveled to Eurasia over the Bering land bridge and evolved into the gray wolf. The remaining canids evolved wholly in North America. Between 150,000 and 300,000 years ago, they diverged into the coyote, which adapted to preying on smaller mammals in the arid southwest, and the eastern/red wolf, which adapted to preying on white-tailed deer

(*Odocoileus virginianus*) in eastern forests. Gray wolves returned to the North American continent approximately 300,000 years ago, and adapted to preying on large ungulates throughout the western United States and Canada. According to this hypothesis, coyotes are more closely related to the eastern/red wolf than to the western gray wolf, therefore the propensity for interbreeding (Wilson et al. 2000).

Domestic dogs are believed to be recent derivatives of the wolf (Scott 1968; Epstein 1971; Turnbull and Reed 1974; Olsen 1985). Domesticated canids are clearly distinguishable from wolves by starch gel electrophoresis of RBC acid phosphatase (Elliot and Wong 1972). Mitochondrial DNA sequences from dogs and wolves show considerable diversity and support the hypothesis that wolves were the ancestors of dogs (Vilà and Wayne 1999). Most dog sequences belonged to a divergent monophyletic clade sharing no sequences with wolves. Contrary to earlier speculation that the domestic dog originated 10,000–15,000 years ago, the sequence divergence within this clade suggests that dogs originated more than 100,000 years before the present. Associations of dog haplotypes with other wolf lineages indicate episodes of admixture between wolves and dogs. Repeated genetic exchange between dog and wolf populations may have been an important source of variation for artificial selection. Although researchers from Israel (Mendelssohn 1982), Finland (Pulliainen 1982) and Greenland (Vibe 1981; Maargaard and Graugaard 1994) have observed matings between domestic dogs and wild wolves, none have been reported from North America. Vilà and Wayne (1999) concluded that hybridization may not be an important conservation concern even in small, endangered wolf populations near human settlements. Behavioral and physiological differences between domestic dogs and gray wolves may be sufficiently great that mating is unlikely and hybrid offspring rarely survive to reproduce in the wild.

DISTRIBUTION

The gray wolf has one of the most extensive distributional ranges of any mammal (Nowak 1983), being circumpolar throughout the northern hemisphere north of 15–20°N latitude. Except for humans the only mammalian species that has ever had a more extensive natural range is the lion (*Panthera leo*). The historical range included nearly all Eurasia and North America. Present distributions are much restricted. In recent times, the species has been extirpated from large portions of its former range and is now found mostly in remote and undeveloped areas with sparse human populations.

The gray wolf originally occupied all habitats in North America north of about 20°N latitude (Fig. 23.1). On the mainland, it was found everywhere except the southeastern United States, California west of the Sierra Nevada, and the tropical and subtropical parts of Mexico. In the southeastern United States, the red wolf replaced the gray wolf (Fig. 23.2). The species also occurred on large continental islands, such as Newfoundland, Vancouver Island, the islands off the coast of southeast Alaska, and throughout the Arctic Archipelago and Greenland, but was absent from Prince Edward Island, Anticosti, and the Queen Charlotte Islands.

An increase in the human population in North America and the expansion of agriculture initiated a general decline in the distribution and abundance of the gray wolf. At the turn of the twentieth century, wolves had nearly vanished from the eastern United States, except for some areas in the Appalachians and the northwestern part of the Great Lakes Region (Young 1944; Nowak 1983). In Canada, the species was exterminated in New Brunswick and Nova Scotia between 1870 and 1921, and in Newfoundland around 1911 (Ganong 1908; Allen and Barbour 1937; Cameron 1958; Lohr and Ballard 1996). They disappeared from the southern parts of Quebec and Ontario between 1850 and 1900 (Peterson 1966). In the prairies, the decline of the species began with the extirpation of the bison (*Bison bison*) in the 1860s and 1870s. Overhunting of other ungulate prey contributed further to the decline of gray wolves. Subsequently, in the period 1900–1930, intensive predator control aimed at the eradication of wolves virtually eliminated the species from the western United States and adjoining parts of Canada. By 1960,

FIGURE 23.1. Past and present distribution of the gray wolf (*Canis lupus*).

the wolf was exterminated by federal and state governments from all of the United States except Alaska and northern Minnesota.

In the 1930s to the early 1950s, the decline in distribution and abundance was reversed, particularly in southwestern Canada (Nowak 1983). This recovery was the result of expanding ungulate populations—following improved regulation of big game hunting—and a moderation in predator control programs (Gunson 1995). The increase in the number of wolves triggered the resumption of wolf control in western and northern Canada, which resulted in the killing of thousands of wolves from the early 1950s to the early 1960s, mostly by poisoning (Heard 1983; Stardom 1983; Hayes and Gunson 1995). Recovery followed the cessation of indiscriminate control and by the mid-1970s wolf populations had increased.

The distribution of the gray wolf in North America is now confined primarily to the northern half the continent (i.e., Alaska and Canada). In the conterminous United States, populations exist in northern Minnesota, northern Wisconsin and Michigan's Upper Peninsula, and parts of Washington, Idaho, and Montana. A program to reintroduce wolves from Alberta and northeastern British Columbia to Yellowstone National Park and Idaho was carried out in 1995 and 1996, respectively. In the Southwest and Mexico, the Mexican wolf (*C. l. baileyi*) is effectively extinct in the wild, but a small number survives in captivity. Reintroductions were begun in Arizona and New Mexico in 1998.

In Canada, the gray wolf is still found throughout most of its historical range including coastal islands (Miller and Reintjes 1995; Paquet and Darimont 2002).The species is completely gone from insular Newfoundland, Nova Scotia, and New Brunswick and is absent or rare in the densely populated and developed parts of the other provinces. Overall distribution of the species in Canada has not changed substantially in the last 40 years and still constitutes approximately 80% of the historical range (Carbyn 1983a). In many areas within its world wide range, wolf populations have been decimated or completely extirpated, which makes Canada an important stronghold of the species.

DESCRIPTION

The gray wolf is the largest living canid. Externally, the gray wolf resembles a large domestic dog, such as a husky, but usually differs in having proportionally longer legs, larger feet, and a narrower chest (Banfield 1974). The wolf's face can be distinguished by its wide tufts of hair, which project down and outward from below the ears (Mech

FIGURE 23.2. History of the distribution of the red wolf (*Canis rufus*).

1970). A wolf's tail is straight and does not curl up posteriorly as with some domestic dogs. Adult wolves, except melanistic individuals, have white fur around the mouth, but dogs usually have black fur in this area.

Mature males weigh from 20 to 80 kg, depending on subspecies, and vary in total length from 1.27 to 1.64 m. Shoulder height varies from 66 to 81 cm. Adult females are usually smaller, weighing from 16 to 55 kg, and are 1.37 to 1.52 m in total length (Young and Goldman 1944; Mech 1970, 1974). The overall size and weight of wolves increase from southern to northern latitudes. For more discussion on weight in *C. lupus,* see Mech (1970, 1974) and Young and Goldman (1944).

Wolves are digitigrade, walking so only the toes contact the ground. The front foot has five toes; the first is rudimentary and does not reach the ground, but has a well-developed dew claw. The hind foot has four toes. The claws are not retractable, are blunt, and are nearly straight. Depending on weight, adult wolves have a foot load of 89–114 g/cm^2 (Formozov 1946; Nasimovich 1955). Young (1944) reported that wolf tracks in the Rocky Mountains averaged 90 mm in length and 70 mm in width for the front foot, and 82 mm in length and 64 mm in width for the hind foot. In comparison with most dogs, tracks of wolves are more elongated, have the front two toe prints closer together, and show the marks of the front two toenails more prominently.

The red wolf resembles the gray wolf in most respects but is smaller in average size. Total length is usually about 1300–1600 mm and weight usually 20–35 kg for males and 16–25 kg for females. The red wolf has longer legs, larger ears, and shorter fur. The color is not really red, as in a red fox (*Vulpes vulpes*), but much like that of most *C. lupus,* with a stronger reddish tinge to the flanks and limbs. Some gray wolves, however, also are reddish, especially on the west coast of British Columbia and Alaska. A dark-colored or black phase of *C. rufus* apparently was locally common in the heavily forested parts of the range of the species.

Pelage. Pelage of wolves consists of long, coarse guard hair, 60–100 mm long, and much shorter, thicker, and softer underfur (Young and Goldman 1944; Mech 1974). The fur is much longer and darker in northern populations. Dorsal hairs are longer and darker than ventral pelage. The longest hairs of all, 120–150 mm, are in the mane, a special erectile part of the pelage, which extends along the center of the back from the neck to behind the shoulders. Wolves usually have one long annual molt, which begins in late spring when the old coat is shed. Simultaneously, the new, short summer coat develops, and grows through fall and winter.

Coloration of wolves is highly variable and usually of little use in ascertaining the geographic origin of specimens, although arctic populations are predominantly white. Over most of the range, "gray" wolves vary from a pure white to coal black. The usual color is light tan or cream mixed with brown, black, and white. Much of the black is concentrated on the back, the forehead tends to be brown, and the lower parts of the head and body are whitish. Dark or all-black wolves are more common in the interior of western Canada and Alaska than the conterminous United States. Standfield (1970) stated that, in Ontario, wolves to the north of Lake Superior varied in pelage color from white to black, but those east and southeast of Lakes Superior and Huron were invariably gray-brown.

Certain specialized hairs are present in the pelage of wolves. Elongated whiskers, or vibrissae, on the muzzle are tactile organs. A group of stiff hairs surrounds the precaudal gland on the back about 70 mm above the base of the tail. These hairs are usually tipped with black, even in animals that are otherwise white (Mech 1970).

Skull and Dentition. The skull of a gray wolf usually has a greatest length of 230–290 mm and a zygomatic width of 120–150 mm. The largest skulls of *C. lupus* on record (one was 305 mm in greatest length) are from Alberta (Gunson and Nowak 1979). A wolf skull has an elongated rostrum, a broadly spreading zygomata, a heavily ossified braincase, and usually a pronounced sagittal crest (Fig. 23.3). A skull of *C. familiaris* of equivalent size can usually be distinguished by a much more massive, steeply rising frontal region (a usual result of which is a higher orbital angle; see Mech 1970) and its relatively smaller teeth.

Teeth of wolves are designed to tear and cut large chunks of meat and to crush and crack bone. The normal dental formula for all members of the genus *Canis* is I 3/3, C 1/1, P 4/4, M 2/3. Incisors are relatively small, and the canines are large with an exposed dorsoventral length of about 26 mm in *C. lupus*. The fourth upper premolar and first lower molar form the carnassials. Molars of wolves retain a flattened or chewing surface, but not to the same extent as in the coyote, which depends more on vegetable matter in its diet.

GENETICS

The karyotype of the wolf appears to be the same as that of the domestic dog (diploid number $2n = 78$). The autosomal complement is 38 acrocentric chromosomes in decreasing order of size. The X is the largest submetacentric chromosome and Y the smallest meta- or submetacentric. Thus, the fundamental number (n.f. after Stains 1975) is 80 (Chiarelli 1975). Since the early 1970s (Seal et al. 1975), many molecular studies have been carried out to resolve taxonomic and conservation problems related to *Canis lupus* and other canids (Kennedy et al. 1991; Lehman et al. 1991, 1992; Wayne et al. 1991, 1992; Roy et al. 1994, 1996; Meier et al. 1995; Forbes and Boyd 1996, 1997; García-Moreno et al. 1996; Vilà and Wayne 1999; Wilson et al. 2001).

The pattern of genetic diversity in sexually reproducing species is principally due to genetic drift, gene flow, and natural selection (Allendorf 1983). Of these three factors, genetic drift is probably not important with respect to the pattern of variation observed in the gray wolf, as it would be if it occurred in small, isolated populations. The second factor, gene flow, is related to the dispersal of reproductive individuals among demes. Because the wolf is a highly vagile species, considerable gene flow is expected among demes (Forbes and Boyd 1996). This would help maintain genetic similarity among demes. Distance alone, however, may act to reduce gene flow (Chesser 1983). Whereas gene flow tends to reduce variation among populations, differential directional natural selection tends to act in the opposite direction and maintain divergence among demes. Gene flow and natural selection are probably the principal factors underlying the observed within-species diversity in wolves.

An alternative view (Shields 1983) holds that, without disruptions, philopatry would limit dispersal and promote breeding within packs, thus maintaining variation among demes. Studies of the genetic relationships of wild populations (Kennedy et al. 1991; Lehman et al. 1992; Wilson et al. 2000) suggest that significant outbreeding takes

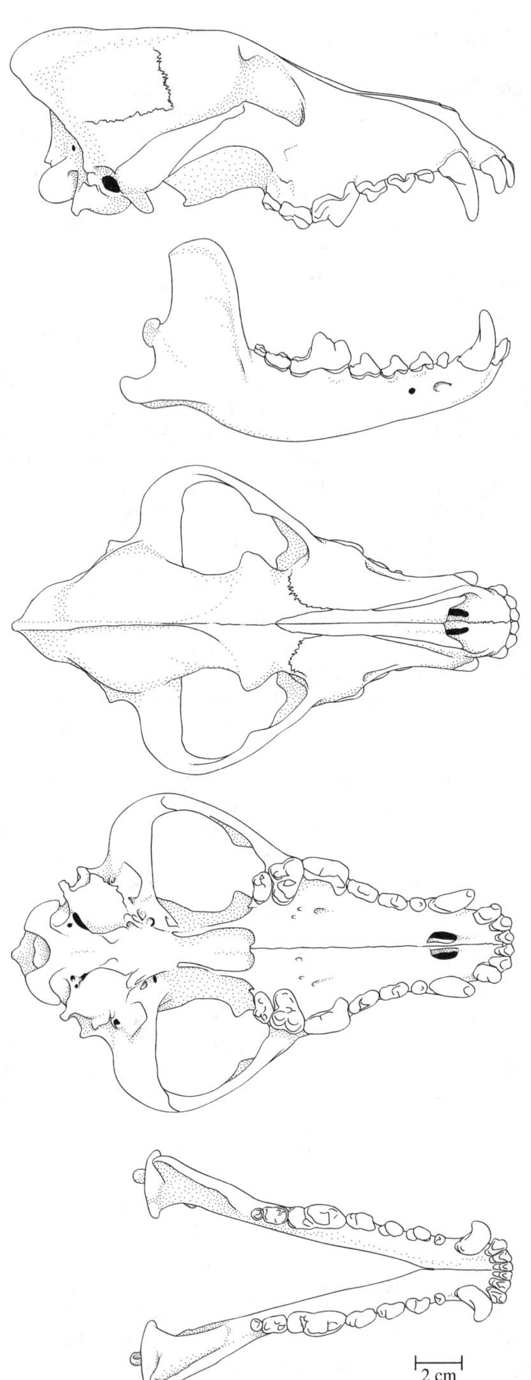

FIGURE 23.3. Skull of the gray wolf (*Canis lupus*). From top to bottom: lateral view of cranium, lateral view of mandible, dorsal view of cranium, ventral view of cranium, dorsal view of mandible.

place. Little regional genetic differentiation may occur (Wayne et al. 1992). Hybridization appears to occur readily in the genus *Canis*. Viable hybrids have been reported between *C. lupus* and *C. familiaris, C. lupus* and *C. latrans,* and *C. rufus* and *C. latrans* (Gray 1972; Nowak 1979).

Lehman et al. (1991) suggested coyote genes have introgressed into sympatric North American gray wolf populations. One explanation of this relationship was that "hybridizing wolves" contained similar amounts of coyote genetic material to the red wolf (Roy et al. 1994). The similarity between the eastern wolf and the red wolf has been noted previously. Both species were described as small eastern wolves long before the eastward expansion of coyotes occurred (Brewster and Fritts 1995). Recent studies by Wilson et al. (2000) may have clarified the relationship. Their results suggest that the DNA of the eastern Canadian wolf and the red wolf is not of gray wolf origin, but is similar to that of coyotes because of their recent divergence from a common ancestor (see Evolution).

PHYSIOLOGY

The internal anatomy of the gray wolf is not known to differ substantially from that of domestic dogs as described by Miller et al. (1964). The digestive system of the gray wolf was discussed in detail by Mech (1970), who commented on its efficiency in absorbing large amounts of meat while ridding itself of indigestible matter such as hair and bone. He also observed that malnutrition generally is not a direct threat to the survival of individuals.

REPRODUCTION AND DEVELOPMENT

Mech (1974) summarized the breeding data on wolves, and much of the following is based on his discussion. A more comprehensive review is provided by Asa and Valdespino (1998). The wolf has a high reproductive potential, although the rate of reproduction may be adjusted to the carrying capacity of the environment (Mech 1970). Wolves mate from January to April, depending on latitude. Courtship takes place between pack members or between lone wolves that pair during mating season. The female is in estrus for 5–7 days and blood may flow from the vulva for a few days to a few weeks before estrus (Carbyn 1987). Not all females in a wolf pack produce pups (Harrington et al. 1982). The dominant pair within a pack usually breeds, with subdominant females under behaviorally induced reproductive suppression (Packard et al. 1985). Subordinates occasionally reproduce successfully (Paquet et al. 1982; Harrington et al. 1982; Van Ballenberghe 1983b; Packard 1985). Reproduction of female wolves may be regulated by aggression among females, direct interference during copulation attempts, or deferred reproduction by younger females (Packard et al. 1983; Asa et al. 1990). Copulation is in typical canine fashion, with the bulbous base of the male's penis locking into the female's vaginal sphincter, the tie lasting up to 30 min.

The young are born in spring after a gestation period of 62–63 days. Birth dates extend from early April to early June. Birth usually occurs in a sheltered place in a hole, rock crevice, hollow log, or overturned stump. Duration of lactation is 8–10 weeks. Litter size averages 6 and ranges from 1 to 11. Young are blind and deaf at birth and weigh an average of 450 g. They become sexually mature between 9 and 46 months (Lentfer and Sanders 1973; Medjo and Mech 1976), but generally do not mate until they are at least 3 years old because of the social structure of the pack. Average age of sexual maturity is 22 months. Longevity is up to 13 years in the wild (Mech 1988) and 16 years in captivity. Mech (1975) found, by examining wolf litters, that males were more common in high-density populations. Kuyt (1972) also found male pups were predominant in the Northwest Territories.

A pack generally has one to three breeding females, but other adults in the pack help in rearing the young. Behavioral limitations on mating, including mate preferences, may hold the productivity of wolves considerably below the theoretical maximum, and often only one litter of pups is born, even in large packs. The instinct to raise the young appears strong among most pack members. Two lone-female wolves and a lone-male wolf each successfully raised litters of young in northwestern Montana and southeastern British Columbia. This unusual reproductive behavior occurred in a low-density population of wolves colonizing an area containing a relatively dense population of ungulates (Boyd and Jimenez 1994). A surrogate female adopted four orphaned pups in Banff National Park. She successfully reared the litter and was able to lactate, although never pregnant (P. Paquet, unpublished data).

In an unexploited population, an estimated 60% of adult females breed, compared with 90% in populations exploited by humans (Rausch 1967). The ability of wolves to respond to increased mortality is also reflected by the percentages of pups in unexploited versus exploited populations. In Wood Buffalo National Park, the percentage of pups in a population under natural control escalated from 20% to 35% one year after wolf control was initiated and to 55% two years later (Fuller and Novakowski 1955). Similar results were obtained in the Great Slave Lake area. There, wolf control resulted in an increased percentage of pups, 46% and 73% one and five years later, compared with 13% under natural control (Kelsall 1968). Studies in Alaska (Rausch 1967) showed no decline in an exploited population where the pup–adult ratio remained at around 44:56. On the other hand, the wolf population in the Great Slave Lake area, with a pup–adult ratio of 73:27, was declining. Mech (1970) concluded from these findings that wolves can compensate, by increased reproduction, for annual losses of at least 50% of animals 5 months or older. We believe this figure is high if recruitment through immigration is not included. Our calculations suggest that 30–40% would be more realistic, whereas Keith (1983) established a value of about 30% of the fall population.

The 36% reproductive success is the lowest on record. Reproductive success from other studies has ranged between 62% and 93% (Messier 1985b; Potvin 1987). Ideally, it is necessary to measure the percent of pups in a fall or winter population (annual recruitment) because this reflects both natality and survival of juveniles through the summer relative to adults (Peterson 1977; Fritts and Mech 1981).

Development of Pups. Hillis and Mallory (1996a), working in the Northwest Territories, examined 73 fetuses from 16 wolves. Fetuses grew at a mean rate of 5.17 g/day between day 32 after coitus and parturition. During the same period, fetuses increased in length at a mean rate of 0.204 cm/day. No significant sexual dimorphism in body mass or other morphological features was found at this stage of development. Hillis and Mallory (1996b) also compared 22 skeletal, visceral, and adipose characteristics of 425 adult wolves. Fifteen parameters differed significantly by sex. Males were usually larger than females. The degree of sexual dimorphism, however, varied with structure. The authors concluded that sexual dimorphism in wolves has evolved primarily as a foraging strategy, owing to division of labor between the sexes, and males are more highly specialized for capturing and killing large ungulate prey; females are more specialized for a nurtural role.

As with other altricial mammals, wolves are born helpless. Neonates are blind and deaf, have little ability to thermoregulate, and receive assistance from the mother to eliminate wastes (Mech 1970). Van Ballenberghe and Mech (1975), working in northern Minnesota, obtained weights, growth rates, canine tooth lengths, and survival data from 73 wild wolf pups. Relative weights of wild pups were expressed as percentages of a standard weight curve based on data from captive pups of similar age. These relative weights varied greatly within litters, among litters, and among years; extremes of 31–144% of the standard were observed. Growth rates ranged from 0.05 to 0.23 kg/day, and similar variations in general development and in replacement and growth of canine teeth were noted. Survival data based on radio-tracking and tag returns indicated that pups with relative weights <65% of standard had a poor chance of survival, whereas pups of at least 80% of standard weight had high survival.

Home Sites, Dens, and Rendezvous. Wolf home sites are important and comparatively small areas where reproductive activities take place. Pups are born, fed, raised, and protected in the natal and secondary den sites, a series of rendezvous sites, and surrounding areas. Research on wolf dens has been conducted in Alaska (Rausch 1969; Stephenson 1974; Chapman 1977; Ballard and Dau 1983; Lawhead 1983), Canada (Criddle 1947; Mech and Packard 1990; Coscia 1993; Bloch in Paquet 1993), and Minnesota (Fuller 1989; Ciucci and Mech 1992). In other studies, wolf den characteristics and den-site selection have been topics peripheral to the main objectives of research (Murie 1944; Haber 1968, 1977 [Alaska]; Joslin 1966, 1967 [Ontario]; Banfield 1954 and Clark 1971; Walton 2000 [Canadian Arctic]; and Carbyn 1975a [Canadian Rockies]). Several studies have investigated habitat features associated with den-site selection and construction (Joslin 1966; Clark 1971; Carbyn 1975a; Stephenson 1974; Ryon 1977; Ballard and Dau 1983; Ciucci and Mech 1992; Heard and Williams 1992; Matteson 1992; Walton 2000). For example, investigators have identified secondary dens, and dens are sometimes found at rendezvous sites (Haber 1968; Clark 1971; Chapman 1977).

A den is an underground burrow or other sheltered place used by wolves (Young and Goldman 1944; Banfield 1954; Lawhead 1983). Wolves visit and prepare one to several den sites within their home range as much as 4–5 weeks before giving birth (Jordan et al. 1967; Haber 1968; Clark 1971; Stephenson and Johnson 1973). As parturition nears, the pregnant female remains near the selected den (Young and Goldman 1944; Mech 1970). Wolf dens may be burrow systems, hollow logs, spaces between roots of trees, caves or crevices in rocks, abandoned beaver (*Castor canadensis*) lodges, or expanded mammal burrows. Pups also have been born in excavations in snow (Kelsall 1960), on surface beds at the base of spruce trees (Soper 1942), and in very shallow surface dens (Mech 1970). Most dens face south or near south; their exposure to sun and wind is such that the den area is often snow free at the onset of denning (Clark 1971; Stephenson 1974). Banfield (1954), however, noted one den where the burrow length included 1 m of snow. Generally, dens are near a source of water (Joslin 1967; Haber 1968; Clark 1971; Voigt 1973; Stephenson 1974; Carbyn 1975a). Rendezvous sites are characteristically centered near open, grassy areas bordered by trees or thickets with water within 50 m of the site (Joslin 1967; Pimlott et al. 1969; Van Ballenberghe et al. 1975).

The natal den serves a brief but important purpose by providing protection from the elements and potential predators for the first few weeks of life. Temperature and humidity in the den are generally moderate and stable compared with the outside environment. Even after pups emerge from the den and begin to eat semisolid food regurgitated by adults, at 3–4 weeks (Mech 1970), wolf dens temporarily remain the center of activity, the point from which adults go out to hunt and to which they return with food for the young.

Dens may be found at locations other than where pups are born. Multiple den sites within a given pack's home range may be concentrated in a small area. For example, five dens were found within a 15-km^2 area in Jasper National Park (Carbyn 1975a). Two or more dens may be within a few hundred meters of each other, with other dens several kilometers away (Banfield 1954; Clark 1971). In Mount McKinley National Park, pairs of dens were approximately 0.2, 0.3, and 0.4 km apart (Chapman 1977). In those few reported cases where single packs had two litters of pups, the dens were 6.4 km (Murie 1944) and 3.2 km (Clark 1971) apart. Average distance between dens of neighboring packs in Alaska was 45 km (Ballard et al. 1987). Denning areas of three neighboring packs occupying islands in coastal British Columbia were separated by <500 m of water (C. Darimont, University of Victoria, Victoria, British Columbia, pers. commun., 2001).

Rendezvous sites are areas where pups are left, usually with an adult, while pack members forage. Rest and play dominate the activities at rendezvous sites (Theberge 1969). They are characteristically centered near open, grassy areas that are bordered by trees or thickets with sources of water within 50 m (Joslin 1967; Pimlott et al. 1969; Van Ballenberghe et al. 1975). In coastal areas, estuaries provide ideal habitat for coastal rendezvous sites (Darimont and Paquet 2002). Abandonment of rendezvous sites appears to occur during September or October at all latitudes (Pimlott et al. 1969; Clark 1971; Voigt 1973; Van Ballenberghe et al. 1975). In coastal British Columbia, rendezvous sites near salmon streams often are used until runs are exhausted in late October and beyond (C. Darimont and P. Paquet, pers. obs.).

Repeated use of established natal dens and rendezvous sites has occurred in as many as 15 consecutive years (P. Paquet, pers. obs.). Several authors report natal den and rendezvous sites being used 4 consecutive years (Clark 1971; Voigt 1973; Peterson 1974; Carbyn 1975a). In Minnesota, den use was traditional in 86% of the denning alpha females studied for >1 year (Ciucci and Mech 1992). Voigt (1973) found that one rendezvous site was used at least five times, and four others at least three times, during a period of 9 years. Rendezvous sites may be occupied several times within the same year (Pimlott et al. 1969, P. Paquet, pers. obs.).

Availability of stable sources of food and water, suitable physical characteristics for den construction, location of neighboring packs, and security from human activity may influence den location. As discussed in Van Ballenberghe and Mech (1975:59), "The quality and quantity of prey eaten and the frequency of its composition probably influence the growth of wild wolf pups more than any other single factor." Thus, to the extent that the den's location helps, or impedes, swift, easy access to prey, placement of the den plays a role in the health of the pups until the time of its abandonment. Den selection in the arctic and subarctic was related to prey availability (Chapman 1977) and habitat characteristics (Heard and Williams 1992; Walton 2000). Within the tree-line zone of the taiga, wolves preferred to den where the roots of trees and shrubs provided structural support for their tunnels (Heard and Williams 1992). The denning area of a wolf pack in Jasper National Park was near elk (*Cervus elaphus*) calving grounds and several mineral licks (Carbyn 1975b). Kuyt (1972) and Clark (1971) noted that migratory wolves often move to denning areas before caribou (*Rangifer tarandus*) begin to migrate through. They also found that wolves usually denned along the major caribou migration routes. Heard and Williams (1992) reported that most wolves on migratory caribou ranges in the Northwest Territories denned near the northern limit of tree growth. The density of dens in the forest was lower than expected if dens were randomly dispersed. Within the tundra zone, wolves did not show any preference for denning near caribou calving grounds.

Wolves in northern Montana used dens that were the greatest distance possible from human disturbance (D. Boyd, University of Montana, Missoula, pers. common., 1992). Wolf dens within the central 60% of winter territories in northeastern Minnesota were randomly located compared with territory centers (Ciucci and Mech 1992). Only 10.5% of the dens were within a 1-km-wide strip inside the territory boundaries, suggesting possible avoidance of neighboring packs. A negative relationship between territory size and den distance from territory centers also was found, suggesting that in large territories wolves might select the denning site that reduces travel distance from and to the den. Lack of suitable denning sites can be a limiting factor in poorly drained tundra (Fleck and Gunn 1982).

Pups are usually born during the first 2 weeks of May in Mount McKinley National Park (Haber 1968). In the Arctic, most pups are whelped between mid-May and early June (Kelsall 1968; Clark 1971; Stephenson 1974), during late April to early May in southeast Alaska (Garceau 1960), early to mid-April in Wood Buffalo National Park (Soper 1942), early to mid-May in Algonquin Provincial Park (Rutter and Pimlott 1968), mid-April to mid-May in Jasper National Park (Carbyn 1975a), and mid-March to late April in Isle Royale National Park and in Minnesota (Van Ballenberghe and Mech 1975).

The age at which pups are carried or led from the natal den to another den or rendezvous site varies considerably. Joslin (1966) reported that one pack moved its litter to a new den when the pups were <3 weeks old. The usual time for pups to leave the natal den seems closer to that reported by Murie (1944), who observed packs abandoning natal dens when the pups were 8–10 weeks old. In Mount McKinley National Park, most recorded dates for the movement of pups to a secondary den or a rendezvous site were between early June and early July, when pups are 4–8 weeks old (Haber 1968). In 1976, however, a natal den was used until approximately 24 August (Matteson 1992). Arctic wolves usually leave natal dens in July, but moves have been reported in early June and in August (Stephenson and Johnson 1972). On Baffin Island, abandonment of natal dens occurred between early July and early August, when pups were 4–9 weeks old (Clark 1971). In southern Canada and in Minnesota, it is probably unusual for pups to remain at natal dens beyond 1 July (Mech 1970). In Jasper National Park, pups were moved between late May and mid-June, when they were 3–6 weeks old (Carbyn 1975a). In Alaska, natal dens were usually abandoned between late June and late July (Ballard et al. 1987). Earlier abandonment resulted from human disturbance.

Reported movements of pups from natal dens to secondary dens averaged 3 km and ranged from 0.3 to 11.2 km (Chapman 1977). Duration of occupancy of secondary home sites is quite variable and has ranged from 2 to 90 days. Termination of rendezvous site use apparently occurs during September or October at all latitudes (Pimlott et al. 1969; Clark 1971; Voigt 1973; Peterson 1974; Van Ballenberghe et al. 1975).

ECOLOGY

Habitat Use. The gray wolf once occurred in all major habitats including deserts, grasslands, forests, and arctic tundra. Because wolves are not highly habitat specific, move long distances, and require large home ranges, the species is regarded as a habitat generalist (Mech 1970; Fuller et al. 1992; Mladenoff et al. 1995). Populations, however, are adapted to local conditions and specialized concerning den-site use, foraging habitats, physiography, and prey selection (Fritts et al. 1995; Mladenoff et al. 1995, 1997, 1999; Paquet et al. 1996; Alexander et al. 1996, 1997; Haight et al. 1998; Mladenoff and Sickley 1998; Callaghan 2002). Thus, wolves are better characterized as ecosystem generalists that are idiosyncratic concerning the surroundings in which they live. Habitat use by wolves is strongly influenced by availability and abundance of prey (Carbyn 1974, 1975a; Keith 1983; Fuller 1989; Huggard 1991, 1993a; Weaver 1994; Paquet et al. 1996), snow conditions (Nelson and Mech 1986a, 1986b; Fuller 1991a, 1991b; Paquet et al. 1996), protected and public lands (Woodroffe 2000), absence or low occurrence of livestock (Bangs and Fritts 1996), road density (Thiel 1985; Jensen et al. 1986; Mech 1988; Fuller 1989; Thurber et al. 1994; Mladenoff et al. 1995, 1997, 1999; Alexander et al. 1996), human presence (Mladenoff et al. 1995; Paquet et al. 1996; Callaghan 2002), and topography (Paquet et al. 1996; Callaghan 2002). Although wolves continue to occupy diverse regions of North America, the species is no longer present in areas with dense human populations or those under intense cultivation (Mladenoff et al. 1995, 1997; Paquet et al. 1996; Mladenoff and Sickley 1998; Haight et al. 1998; Callaghan 2002). Protected and public lands likely encourage wolf presence because of fewer lethal encounters with humans (Mech 1995). Due to conflicts with ranchers, wolves are prone to local extirpation in areas with high densities of livestock (Bangs and Fritts 1996; Bangs et al. 1998, 2001).

Home Range. Many researchers have reported that packs of wolves occupy stable home ranges that are exclusive territories (Mech 1970, 1974; Peterson et al. 1984; Messier 1985a, 1985b). Territorial behavior is thought to be a spacing mechanism, which adjusts wolf densities to their food level. In some circumstances, however, home ranges are dynamic and nonexclusive. Reasons for this instability are not well understood, but likely relate to availability of food. Forshner (2000) reported home ranges of wolves in Ontario overlapped extensively. Others found the areal extent and geographic location of home ranges changed among years (Carbyn 1981, 1982b; Potvin 1987; Mech et al. 1995a).

Biologists usually define the home range of a wolf as an area within which it can meet all of its annual biological requirements. Seasonal feeding habitat, thermal and security needs, travel, denning, and the bearing and raising of young, are all essential life requirements. The manner in which habitats for these requirements are used and distributed influences home range size and local and regional population distributions. Generally, wolves locate their home ranges in areas where adequate prey are available and human disturbance minimal (Mladenoff et al. 1995, 1997; Mladenoff and Sickley 1998) and seem to cognitively map their territories (Peters 1978, 1979). Wolves use areas within those home ranges in ways that maximize encounters with prey (Huggard 1993a, 1993b). In mountainous areas, selection of home ranges and travel routes is influenced by topographic complexity (Paquet et al. 1996). Wolf use of valley bottoms and lower slopes corresponds to the presence of wintering ungulate prey and snow depth in these areas (Singer 1979; Jenkins and Wright 1988; Paquet et al. 1996). In areas of higher prey density, pack sizes increase (Messier 1985b) and home range size is closely correlated with pack size (Peterson et al. 1984; Messier 1985b). Mech (1970) and Ballard et al. (1997) suggested that wolves denning on the tundra and relying on migratory caribou range over larger areas than wolves occupying forested areas and relying on resident prey.

The size of a pack's home range varies considerably from area to area, depending principally on the type and density of prey and season. Territory and home range sizes are more closely correlated with pack size than with prey density (Peterson et al. 1984; Messier 1985b). In areas of higher prey density, pack sizes increase. Messier (1985b) concluded that between 0.2 and 0.4 moose/km^2, the territory area per wolf is independent of moose abundance. A colonizing pack might have a larger, more fluid home range than a pack surrounded by other wolf packs (Boyd et al. 1995; Boyd and Pletscher 1999). Territories tend to be smaller in summer, when packs are tied to dens and home sites (Mech 1977a). No such restrictions exist in winter, which allows wolves to roam more freely. In Algonquin Park, where the principal prey is deer, territories ranged from 104 to 311 km^2 (Pimlott et al. 1969). In the boreal areas, sizes of home ranges reported and the predominant prey are as follows: 283 km^2 (deer, *Odocoileus virginianus;* moose, *Alces alces;* elk) in central Manitoba (Carbyn 1981); 568 km^2 (195–629 km^2 in summer, 357–1779 km^2 in winter) (moose) in northern Alberta (Fuller and Keith 1980); and 1250 km^2 (bison) in Wood Buffalo National Park (Oosenbrug and Carbyn 1982). In the mountain zone, home ranges in the Rocky Mountains range from 1058 to 3374 km^2 (Paquet 1993) and in the Yukon from 583 to 794 km^2 (Hayes et al. 1991). In the arctic region, home ranges on Ellesmere Island were >2500 km^2 (Mech 1987, 1988). Winter-territory size in Minnesota averaged 78–153 km^2 (Fuller 1989).

Population Densities. Keith (1983) and Fuller (1989) reviewed numerous studies of North American wolf populations and concluded that average wolf densities are correlated with the biomass of ungulates available per wolf. Densities of wolves are highest where prey biomass is highest (Keith 1983; Fuller 1989; Fuller and Murray 1998). In North America, reported wolf densities range from 1/2 km^2 to 1/3274 km^2 for stable populations (Table 23.1) Average annual wolf densities do not often exceed about 1 wolf /24 km^2 and are usually far lower. Pimlott (1967) suggested that 1 wolf/20–25 km^2 was the maximum density tolerable by a natural wolf society. During certain periods of exceptionally high prey concentrations, however, the density of wolves may increase dramatically (Mech 1974). For example, Kuyt (1972) reported that in some parts of the Northwest Territories (Mackenzie), winter wolf densities can be compressed to 1 wolf/10 km^2 as a response to high concentrations of migrating caribou. In 1998, at least 633 wolves were killed in an area of about 8000 km^2 in the Rennie Lake region of the Northwest Territories, or about 1 wolf /13 km^2 (D. Cluff, Government of Northwest Territories, Yellowknife, Northwest Territories, pers. commun., 2001). Van Ballenberghe et al. (1975), working in Superior National Forest, found a 550-km^2 area in which the density of wolves reached an average of about 1/14 km^2.

The highest density of wolves ever recorded was 1/2 km^2 at a winter deer yard near Algonquin, Ontario (Forbes and Theberge 1995), reflecting a seasonal concentration of wolf packs. Wolves reintroduced to Coronation Island, Alaska, reached a density of about 1/8 km^2 before dying out (Merriam 1964; Klein 1995). Sustained densities of about 1 wolf/13–29 km^2 have been reported on coastal islands of Alaska (Person 2001) and British Columbia (Darimont and Paquet 2000). The lowest reported density is 1 wolf /3274 km^2, on Ellesmere Island in the Canadian Arctic. In the central Canadian Rocky Mountains, density was 1 wolf /250–333 km^2 over 10 years (Paquet et al. 1996; Callaghan 2002). This is the lowest reported density in North America for a stable population. The lowest reported density in forested, nonmountainous regions is 1 wolf /260–500 km^2 in Ontario (Pimlott et al. 1969).

Movements. Movement patterns of gray wolves have been studied in much of their current range in North America (Fritts and Mech 1981, Messier 1985a, 1985b; Potvin 1988; Ballard et al. 1997). Most studies were of territorial wolves that prey on ungulates including deer, elk, moose, and sheep (*Ovis* spp.). Although some of these ungulates may undergo seasonal migrations, they are of lesser magnitude than the migrations of barren-ground caribou (*Rangifer tarandus groenlandicus*). Consequently, most studies have concluded that wolves maintain relatively stable annual territories. Migratory wolves, however, have been documented in wolf–caribou systems in northern Canada and Alaska (Parker 1973; Stephenson and James 1982; Ballard et al. 1997; Walton 2000; Walton et al. 2001) and wolf–bison ecosystems in Wood Buffalo National Park, Northwest Territories (Carbyn et al. 1993).

Movements by wolves can be divided into migrational movements following prey, movements within territories, and dispersal (Mech 1974). Travel patterns of wolves are influenced by elevation, topography, distribution of important prey, and seasonal changes of climate. Wolves prefer the easiest possible traveling and therefore make use of logging roads, survey lines, trails, lake shores, and passes between hills. Although wolves appear to select areas of low road density (see below), travel routes are generally close to trails and roads (Gehring 1995; Singleton 1995; Boyd-Heger 1997; Callaghan 2002). Ski

TABLE 23.1. Winter density for various North American wolf populations within broadly defined regions

Region	Density (wolf/km2)	Reference
Great Lakes/St. Lawrence		
Quebec	1/20–50	Banville 1983
	1/50–100	
	1/100–1000	
Ontario Algonquin Park	1/26	Pimlott et al. 1969
	1/2	Forbes and Theberge 1995
Ontario	1/60–130	Kolenosky 1983
Ontario Pukaskwa N.P.	1/104–139	P. Paquet, pers. obs.
Michigan Isle Royale N.P.	1/20–25	Jordan et al. 1967
	1/11–33	Peterson and Page 1988
Boreal		
Quebec	1/50–100	Banville 1983
	1/100–1000	
	1/71–125	Messier 1985b
Ontario	1/130–260	Kolenosky 1983
Minnesota	1/32	Mech 1973
	1/24	Van Ballenberghe et al. 1975
	1/17–35	Fuller 1989
Manitoba wolf range	1/500	R. Stardom, pers. commun.
Riding Mountain N.P.	1/30–109	P. Paquet, pers. obs.
Prince Albert N.P.	1/74–85	Parks Canada, pers. commun.
Alberta wolf range	1/55–78	J. Gunson, pers. commun.
Alberta Oil Sands	1/128	J. Gunson, pers. commun.
Swan Hills	1/90	Fuller and Keith 1980
Fort McMurray	1/158	Fuller and Keith 1980
Wood Buffalo N.P.	1/89	Oosenburg and Carbyn 1982
	1/83	Gunson 1995
Simonette River	1/40–66	Gunson 1995
Nordegg	1/42	Gunson 1995
Mountain		
Alberta, Jasper N.P.	1/250–333	Gunson 1995
	1/111–143	
	1/250–500	
Alberta, Banff N.P.	1/250–333	Paquet et al. 1996
British Columbia	1/70–171	Tompa 1983
	1/100–110	Bergerud and Elliot 1986
Arizona	1/109	Theberge 1991
Pacific		
Vancouver Island	1/12–17	Tompa 1983
Arctic		
Barren Grounds	1/154–307	Kelsall 1957
	1/588	Kelsall 1968
	1/500	Parker 1972
	1/11–13	Parker 1973
Baffin Island	1/255	Miller 1993
Southern tier of islands	1/329	Miller 1993
Queen Elizabeth Islands	1/2026	Miller 1993
Ellesmere Island	1/3274	Miller 1993
	1/900	Riewe 1975
Alaska	1/71	Boertje et al. 1996
Alaska Kenai	1/50–91	Peterson et al. 1984
Alaska south central	1/97–385	Ballard et al. 1987
	1/227–667	Ballard et al. 1997

trails, snowmobile trails, ploughed roads, and snow-packed roads can enhance the range and efficiency of winter forays (Singleton 1995; Paquet et al. 1996). Highways, other human structures, and human activities may impede or alter use of travel routes (Paquet and Callaghan 1996; Paquet et al. 1996; Boyd-Heger 1997). Elevation can also govern seasonal movements of wolves. In mountainous areas with high snowfall, use of low-elevation valleys increases during winter, where frozen rivers and lakes, shorelines, and ridges are preferred because of ease of travel. Singleton (1995) suggested that variation in pack size, variation in home range size, and interactions with sympatric predators may also influence habitat use and travel patterns. Musiani et al. (1998) reported an average travel speed of 3.78 km/hr for wolves in Bialowieza Primeval Forest, Poland. Wolves moved faster on forest trails, roads, and frozen rivers than in the forest. In addition, individuals traveling with other pack members moved faster than those walking alone. On Ellesmere Island, Northwest Territories, mean travel speed of wolves was measured during summer on barren ground at 8.7 km/hr during regular travel and 10.0 km/hr when returning to a den (Mech 1994b).

Daily distances traveled in a pack's territory can range from a few kilometers up to 200 km (Mech 1970). On Isle Royale, a pack of 16 wolves travelled 443 kilometers over 31 days, an average of 14.3 km/day (Mech 1966). In Alaska, Burkholder (1959) followed a pack that averaged 24 km/day for 15 days. Mech (1970) reported daily movements of 1.6–4.6 km for wolves preying on deer. Kolenosky (1972) found that a pack of eight wolves in Ontario traveled 327 km over 46 days, averaging 7.1 km/day. Peterson (1977) calculated the average daily distance traveled by packs on Isle Royal between 1971 and 1974 was 11.1 km. Oosenbrug and Carbyn (1982) reported daily movements of 2.3–18.7 km for wolves preying on bison. Five wolves traveled about 32 km along the coast of northeastern Bathurst Island in about 5 hr (F. Miller, Canadian Wildlife Service, Edmonton, Alberta, pers. commun., 1999).

Migration. Migration involves the seasonal movement of a pack between widely separated geographic locations. Migratory wolf populations are those that depend on caribou, especially migratory barren-ground caribou, but also woodland caribou. Movements of migratory wolves are generally long term, involve entire packs, and can be related to the availability of preferred food on the resident territory. Both Stephenson and James (1982) and Ballard et al. (1997) suggested that migration of wolves following caribou is not always an annual event. If wolves have sufficient access to prey on resident territories, migration probably represents a historical pattern that has evolved. When prey such as caribou are highly aggregated and predictable, strict territoriality may be abandoned and the social system swings toward "group" nomadism.

In the western Arctic of Alaska, wolf packs usually did not follow migratory caribou, but maintained year-round resident territories, which averaged 1868 km^2 (Ballard et al. 1997). Wolves only migrated with the western Arctic caribou herd in years when alternate ungulate prey densities were too low to sustain wolf packs (Ballard et al. 1997). Pack areas and territories did not normally overlap, but when overlaps occurred, packs were separated temporally. Wolves in south-central Alaska did not follow migratory movements of moose or caribou (*Rangifer tarandus granti*) outside their pack areas, but did follow elevational movements of moose within their areas (Ballard et al. 1987). Differing migratory strategies of Alaskan wolves may be related to the availability of moose as an alternative prey (Stephenson and James 1982; Ballard et al. 1997).

Radio-collared wolves in the Northwest Territories and western Nunavut showed a distinct migratory pattern associated with movements of caribou. Packs left tundra denning areas in autumn and moved over large areas throughout the winter before returning to the tundra to give birth in early spring. Returning wolves began to restrict movements around den sites on the tundra by late April. Thus, they did not exhibit territorial behavior typical of other wolf populations in North America (Walton et al. 2001). Annual home range sizes averaged 63,058 km^2 for males and 44,936 km^2 for females. Home range in summer averaged 2022 km^2 for males and 1130 km^2 for females. Straight-line distances from the most-distant location on the winter range to the den site averaged 508 and 265 km, respectively.

Dispersal. Dispersal involves the movement of an individual away from the territory of its birth and its pack. Dispersal movements are important for gene flow among demes and aid in the establishment of new packs. Wolves may increase their reproductive fitness by dispersing as yearlings, but remaining in packs as older adults. Dispersal in wolves appears as a gradual and dynamic dissociation process. As offspring begin to mature, they disperse from the pack as young as 9 months of age (Fritts and Mech 1981; Messier 1985a; Fuller 1989; Gese and Mech 1991). Separation from the pack may extend from a few months to several years (Messier 1985a). Most wolves disperse when 1–2 years old, and few remain with the pack beyond 3 years of age (Mech et al. 1998). Thus, young members constitute a temporary portion of most packs and the only long-term members are the breeding pair (Mech 1999b).

Dispersal may be directional long-distance travel or nomadic (Carbyn 1987; Boyd and Pletscher 1999). Some evidence suggests that wolf packs colonize areas that were first "pioneered" by dispersing lone wolves (Ream et al. 1991; Pletscher et al. 1991, 1997). Dispersing wolves are often deposed alpha animals or younger, low-ranking pack members (Zimen 1975, 1976) driven away by aggression over food or mates. In the Yukon, dispersal rate was density independent and related to mean pack size and prey biomass–wolf index (Hayes and Harestad 2000a). Yearling and pup dispersal rates in Minnesota were highest when the wolf population was increasing or decreasing and low when the population was stable (Gese and Mech 1991). Potvin (1987) concluded dispersal in Quebec resulted from the onset of sexual maturity and, possibly, from social stress. A study in Minnesota recorded up to six exploratory moves were made before dispersal (Fuller 1989). Conversely, Boyd and Pletscher (1999) found most dispersing wolves left their natal home range quickly after separating from the pack.

Wolves that disperse frequently try to establish new packs. Most new packs are likely formed by dispersers (Rothman and Mech 1979; Fritts and Mech 1981; Fuller 1989; Boyd and Pletscher 1999). Pack fission is another mechanism (Mech 1966; Meier et al. 1995). Dispersing wolves typically establish territories or join packs within 50–100 km of their natal pack (Fritts and Mech 1981; Fuller 1989; Gese and Mech 1991; Boyd et al. 1995; Wydeven et al. 1995). Some wolves, however, move longer distances. Some reported dispersal distances are 206 km in 2 months (Mech 1974), 670 km in 81 days (Van Camp and Gluckie 1979), and 20–390 km for five dispersing wolves (Fritts and Mech 1981). In Alaska, several wolves from the same pack traveled 732 km in a 9-month period (Ballard et al. 1983). Fuller (1989) reported that dispersing wolves traveled 5–100 km during periods of 1–265 days. Dispersing wolves in Quebec traveled an average distance of 40 km (Potvin 1987). Fritts (1983) reported a record dispersal distance for a male wolf of 886 km. A yearling female dispersed a record 840 km from the Rocky Mountains in Montana north into British Columbia (Boyd et al. 1995).

Wolves may disperse at any age but young wolves do so more frequently. In north Minnesota, most dispersers left when they were 11–12 months old, and only a few wolves dispersed as adults (Gese and Mech 1991). In Papineau–Labelle Reserve, most wolves dispersed when they were 10–20 months old (Potvin 1987). Several studies found both sexes disperse equally (Fuller 1989; Gese and Mech 1991; Boyd et al. 1995; Boyd and Pletscher 1999). In Alaska, however, 74% of dispersers were males (Ballard et al. 1987). Rates of dispersal appear to vary with age and environmental conditions. Annual dispersal rates in northern Minnesota were about 17% for adults, 49% for yearlings, and 10% for pups (Fuller 1989). Of 316 wolves monitored in Superior National Forest, Minnesota, 75 were dispersers. Eight percent of adults, 75% of yearlings, and 16% of pups dispersed (Gese and Mech 1991). In Alaska, 28% of 135 wolves dispersed from their original area (Ballard et al. 1987).

The times of reported dispersals vary, although January–February seems consistent among studies. In north-central Minnesota, pups left natal packs during January–March and older wolves left frequently during September–April (Fuller 1989). Dispersal occurred mainly in February–April and October–November in another Minnesota study (Gese and Mech 1991). Dispersal in Alaska occurred mainly during April–June and October–November (Ballard et al. 1987). January–February and May–June were peak periods for wolf dispersal in the northern Rockies (Boyd and Pletscher 1999).

The fate of dispersing wolves is probably related to their age, the density of a wolf population, availability of prey, availability of unoccupied habitat, and presence of humans (Fuller 1989; Gese and Mech 1991; Boyd et al. 1995). In northern Minnesota, adults dispersed short distances into nearby territories, but yearlings and pups dispersed short or long distances. Adults had the highest pairing and denning success, yearlings had a moderate pairing and low denning success, and pups had low pairing and denning success. Yearlings and pups that dispersed a short distance had a higher success of settling in a new territory, which likely reflected available vacancies in nearby territories (Gese and Mech 1991). Fuller (1989) found only 1 disperser joined an established pack, but 16 others formed new packs. In Alaska, 28% of 135 wolves dispersed from their original area. Twenty-two were accepted into existing packs (Ballard et al. 1987). Dispersers in a newly established population produced more litters than biders (philopatric wolves) (Boyd and Pletscher 1999). Annual survival rate for dispersers and biders did not differ. Proportionately more dispersers (90%) than biders (60%) died of human causes.

Interactions with Other Carnivores. As summit predators, gray wolves likely have a profound influence on other top carnivores (Hairston et al. 1960). However, except for coyotes (Fuller and Keith 1981; Carbyn 1982b; Schmitz and Kolenosky 1985a, 1985b; Meleshko 1986; Paquet 1991a, 1991b; 1992; Thurber et al. 1992; Peterson 1995; Arjo and Pletscher 1999) and red foxes (Peterson 1995), competition between wolves and other carnivores has been the focus of relatively few studies. Interference competition is the best-known mechanism by which wolves subordinate other carnivores (Ballard 1982; Gehring 1993) or are themselves displaced (Ballard 1982; Hornbeck and Horejsi 1986). Interference occurs when competitively subordinate species are aggressively displaced, killed, driven away, or choose to avoid more dominant wolves. Resource competition and exploitation competition, which are indirect and difficult to demonstrate, have not been documented.

Wolves can exclude coyotes from individual pack territories to entire regions. However, coexistence also is common (Paquet 1991a, 1991b, 1992; Thurber et al. 1992) and the outcome of competition may be influenced in subtle ways by topography, snow cover, seasons, food abundance, niche overlap, population characteristics, and the overriding influence of humans (Peterson 1995). In theory, wolves might affect other predators by reducing availability of ungulates, or conversely, by increasing availability of carrion. Less obvious influences of wolves might include modified community relationships. By providing carcasses for scavenging, wolves might affect interactions of sympatric canids, ursids, felids, mustelids, and avian scavengers such as ravens (*Corvus* spp.) and bald eagles (*Haliaeetus leucocephalus*). Ravens, for example, appear to fly toward howling, especially in winter when wolves are killing large animals and carrion may be available (Harrington 1978). Wolves might affect brown bears (*Ursus arctos*) by reducing the availability of a limiting resource (possibly an ungulate) or, conversely, by increasing the carrion available to bears (Ballard 1982). In complex systems with multiple prey and predators, all of these relationships would become increasingly more complicated.

Fatal encounters have been documented between wolves and cougars (*Puma concolor*) (Schmidt and Gunson 1985; White and Boyd 1989; Boyd and Neale 1992), wolves and coyotes (Seton 1925; Young and Goldman 1944; Munro 1947; Stenlund 1955; Berg and Chesness 1978; Carbyn 1982a; Fuller and Keith 1981; Paquet 1991b), wolves and red foxes (Stenlund 1955; Mech 1970; Banfield 1974; Allen 1979; Peterson 1995), wolves and black bears (*U. americanus*) (Rogers and Mech 1981; Ramsay and Sterling 1984; Horejsi et al. 1984; Paquet and Carbyn 1986), wolves and grizzly bears (P. Paquet, unpublished data), wolves and polar bears (*U. maritimus*) (Ramsay and Sterling 1984), wolves and wolverines (*Gulo gulo*) (Burkholder 1962), and wolves and river otters (*Lontra canadensis*) (Route and Peterson 1991; Kohira and Rexstad 1995). Depredation is not necessarily motivated by food attainment, as carcasses of wolf-killed carnivores are often not consumed (Carbyn 1982a; Paquet 1991b). Characteristically, smaller canids (Wobeser 1992) and cougars (White and Boyd 1989) killed by wolf aggression are not eaten, but left with fatal bites in the head and neck and frequent puncture wounds through the torso.

FEEDING HABITS

Wolves are obligate carnivores whose use of prey depends largely on the availability and vulnerability of ungulates (Weaver 1994). Beavers, hares (*Lepus americanus*), other smaller mammals, and scavenging (Forbes and Theberge 1992) supplement the diet, particularly during wolf denning and rendezvous site activities. In North America, important ungulate prey include deer (see citations throughout chapter), moose (Atwell 1964; Frenzel 1974; Peterson 1977; Bergerud et al. 1983; Messier 1984; Potvin et al. 1988; Forbes and Theberge 1996b; Ballard and Van Ballenberghe 1997), caribou (Banfield 1954; Kuyt 1972; Bergerud 1974; Seip 1992; Dale et al. 1994; 1995; Adams et al. 1995; Boertje et al. 1996; Ballard et al. 1997), elk (Cowan 1947; Carbyn 1983b; Paquet 1992; Huggard 1993b, 1993c; Larter et al. 1994; Kunkel et al. 1999), bison (Oosenbrug and Carbyn 1982; Carbyn and Trottier 1987, 1988; Joly and Messier 2000; Smith et al. 2000), muskoxen (*Ovibos moschatus*) (Gray 1970, 1983; Heard 1992; Mech 1999a), mountain goats (*Oreamnos americanus*) (Smith 1983; Fox and Streveler 1986; Festa-Bianchet et al. 1994), and mountain sheep (Murie 1944; Haber 1977; Gasaway et al. 1983; Hoefs et al. 1986; Paquet et al. 1996). When two or more ungulate species inhabit the same area, wolves usually concentrate on the smallest or easiest to catch (Mech 1970; Paquet 1992; Weaver 1994; Paquet et al. 1996). Most ungulate prey are young, weakened, debilitated, or older animals (Fuller and Keith 1980; Carbyn 1983b; Paquet 1992), although wolves do kill healthy adult animals. The proportion of debilitated prey may be higher than reported, as wolves are keen observers of behavior, able to detect subtle susceptibilities not evident to humans (Frenzel 1974).

Stephenson and James (1982) reported that caribou constituted 97% and 96% of the biomass consumed by two radiocollared wolf packs during winter. Calves constituted 20% and 6% of caribou killed by the same two packs during summer. In the Nelchina Basin of south-central Alaska, 72% of 330 kills located during winter were moose. Fox and Streveler (1986) found that 62% of wolf scats in the northern coast ranges of southeastern Alaska contained mountain goat remains. Kuyt (1972) found caribou were the staple diet of wolves on caribou winter ranges. Fuller (1962), working in Wood Buffalo National Park, recorded bison remains in 39 of 95 stomachs of poisoned wolves. Oosenbrug and Carbyn (1982) and Joly and Messier (2000), also working in Wood Buffalo National Park, confirmed that bison were the primary prey for wolves. The major prey of two packs of Vancouver Island wolves were black-tailed deer (*Odocoileus hemionous*), elk, and beaver, respectively. In the southern Rocky Mountains of Canada and northern Rockies of the United States, elk were the most common prey species (Cowan 1947; Huggard 1993b; Boyd et al. 1994; Weaver 1994; Paquet et al. 1996; Kunkel et al. 1999), followed by deer and mountain sheep. Moose predominated in the diet of wolves in north-central Alberta (Fuller and Keith 1980). Elk, moose, and white-tailed deer were the major prey species in Riding Mountain National Park (Carbyn 1983b). In the Superior National Forest of northeastern Minnesota, white-tailed deer constituted 80% of dietary occurrences during winters of 1946–1948 (Stenlund 1955) and 56% and 66% of summer and winter occurrences, respectively, in scats during 1969–1971 (Van Ballenberghe et al. 1975). In north-central Minnesota, deer were the primary prey in winter and spring, but beavers were an important secondary prey (20–47% of items in scats) during April–May. Neonatal deer fawns occurred in 25–60% of scats during June–July, whereas the occurrence of beavers declined markedly. Overall, deer provided 79–98% of biomass consumed each month. Adult wolves consumed an estimated 19 deer/year, of which 11 were fawns (Fuller 1989).

Packs have increased energetic demands when raising pups (Mech 1970). Those demands can be met seasonally by higher rates of kill, more complete use of carcasses, and increased consumption of small mammals such as beavers (Meleshko 1986; Potvin et al. 1992a). Nevertheless, most studies of the summer diet of gray wolves show ungulates are the predominant source of biomass consumed. Use of fawns and calves increases, apparently related to availability and not local abundance (Murie 1944; Cowan 1947; Mech 1966; Pimlott et al. 1969; Clark 1971; Kuyt 1972; Carbyn 1975a; Van Ballenberghe et al. 1975; Voigt et al. 1976; Peterson 1977; Scott and Shackleton 1980; Peterson et al. 1984; Messier and Crête 1985; Meleshko 1986; Ballard et al. 1987; Potvin et al. 1988; Fuller 1989; Thurber and Peterson 1993). In the Nelchina Basin of south-central Alaska, analysis of wolf scats collected at den and rendezvous sites indicated moose represented 53% and caribou 7% of the summer diet by occurrence (Van Ballenberghe 1991). Kuyt (1972) found caribou constituted 47% of food items during spring and summer in mainland Northwest Territories. The type and quantity of food consumed by wolves while attending den and rendezvous sites is of interest to researchers. Theberge et al. (1978), and Scott and Shackelton (1980) found scat collected on trails and at rendezvous sites differed significantly in content. Carbyn (1983b) reported that wolf scats collected along trails were the same in content as those collected at rendezvous sites. Fuller and Keith (1980) also reported similar results.

Beavers, lagomorphs, microtine rodents, and a variety of birds (especially waterfowl) (Kuyt et al. 1981) and their eggs supplement the diet of wolves. Fish, berries, and carrion are consumed seasonally where available (Young and Goldman 1944; Bromley 1973; Meleshko 1986; Kohira and Rexstad 1995; Darimont et al. 2002). Coastal wolves also feed on marine mammal carcasses, crabs, mussels, and even barnacles (Darimont and Paquet 2000). Wolves occasionally scavenge at refuse dumps, rubbish bins, and bone yards even when wild prey are available (Fuller and Keith 1980; Krizan 1997). On occasion, wolves kill and consume other carnivores such as bears (Rogers and Mech 1981; Ballard 1982; Ramsay and Stirling 1984; Horejsi et al. 1984; Paquet and Carbyn 1986) and river otters (Route and Peterson 1991; Kohira and Rexstad 1995; Paquet pers. obs.). These minor food items may be sustaining between periods of ungulate kills, during declines in ungulate populations, or while wolves are denning and using rendezvous sites. For example, wolves denning in the Canadian Arctic may subsist on small mammals, birds, and fish when their primary prey, caribou, migrate to summer range (Kuyt 1972; Williams and Heard 1993; Williams 1995). Tener (1954), working on Ellesmere Island, recorded remains of arctic hare (*Lepus arcticus*) in 83% of 70 summer and winter scats.

Peterson (1977) reported beaver remains in 76% of the scats collected on Isle Royale. Voigt et al. (1976), in central Ontario, found beavers represented 55–75% of occurrences in summer scats. Frequent occurrence of beavers throughout the year has been reported for Vancouver Island (Scott and Shackleton 1980; Milne et al. 1989) and southeastern Alaska (Kohira and Rexstad 1995). Milne et al. (1989) attributed this year-round use to the mild winter of the region. Although beavers are common food items in some regions of North America, they do not control the distribution or abundance of wolves (National Research Council 1997).

As mentioned above, wolves occasionally consume fish. Francis (1960) reported six wolves feeding on concentrations of minnows and water bugs in a hole in the frozen Torch River, Saskatchewan. Young and Goldman (1944) record instances of wolves catching spawning salmon (*Oncorhyncus* spp.). A wolf pack on midcoast British Columbia was documented catching and partially consuming >200 salmon during one night (Darimont et al. 2002). Adult wolves in the same pack were observed catching >20 salmon/hr. Often, wolves consume only the head, leaving the rest for a diversity of scavengers (Paquet pers. obs.). Wolves in areas of low ungulate densities use more alternative foods, especially during summer (Voigt et al. 1976; Peterson 1977; Messier and Crête 1985; Ballard et al. 1987). On Isle Royale, the incidence of beaver in scats increased from 13–15% to 76% during a time of high beaver populations and low moose productivity. Seasonal declines in ungulate density occur in areas where caribou migrate to calving grounds in early spring. Thus, many wolf packs that prey on caribou during winter become separated from caribou during summer (Kuyt 1972; Williams and Heard 1993; Williams 1995; Walton et al. 2001). Banfield (1954) suggested that small, nonungulate prey might be critically important in the summer diet of wolves denning in areas where caribou are migratory. Pimlott (1967) believed that wolves would prey on low numbers of caribou when other ungulate prey were not available. Subsequent studies have supported his argument that caribou or other ungulates are the primary source of food for wolves in summer and winter (Clark 1971; Kuyt 1972; James 1983; Meleshko 1986; Ballard et al. 1987; Dale et al. 1994; Spaulding et al. 1998). Dale et al. (1994) found no evidence of prey switching owing to changes in ungulate abundance. Wolves continued to prey on caribou even when moose were twice as abundant. Spaulding et al. (1998) reported that wolves switched to preying on moose during winter in years when caribou numbers were low and moose were more vulnerable because of snow conditions, but by June of each year caribou were again the predominant prey item. Wolves in the Yukon did not show a strong switching response away from moose as the ratio of caribou to moose increased in winter (Hayes et al. 2000).

Kill and Consumption Rates. The reported rates of kill and food consumption by wolves vary considerably. Hebblewhite (2000) concluded the wide array of methods used to estimate kill rates confounds attempts to compare studies. Much of the variation likely reflects the remarkable ability of wolves to adjust to seasonal and annual availability, vulnerability, distribution, and abundance of ungulate prey. Regional differences in size and energy requirements of wolves might also explain some differences in rates of predation (Oosenbrug and Carbyn 1982). For example, wolves in Wood Buffalo National Park are substantially larger than wolves in eastern North America. In the early 1970s, Isle Royale wolves exhibited an immediate response to high moose vulnerability by increasing their kill rate. As kill rate went up, wolf use of carcasses declined. During these years, killing another moose was easier for wolves than digging out frozen remains of an old carcass (Peterson and Allen 1974). Throughout most North America, the condition of winter snow, particularly the interaction of depth and hardness, could be an important determinant of prey susceptibility and rates of predation (Kolenosky 1972; Peterson and Allen 1974; Haber 1977; Peterson 1977; Carbyn 1983b). Kill rates often increase as the depth of snow increases (Peterson and Allen 1974; Mech and Nelson 1986; Huggard 1993a, 1993b; Paquet et al. 1996; DelGuidice 1998). Fuller (1991a) reported wolves in north-central Minnesota changed winter activity, movement patterns, sociality, and feeding behavior in response to snow-induced changes in deer distribution and mobility.

Surplus or excessive killing by wolves of caribou (Miller et al. 1985) and white-tailed deer (DelGuidice 1998) has been reported. In Minnesota, excessive killing by wolves related to poor physical condition of deer owing to effects of a severe winter. The authors predicted excessive killing will occur when snow depth exceeds 70 cm for 4–8 weeks. In the Northwest Territories, Miller et al. (1985) found 34 newborn caribou calves killed by wolves. The calves were killed within minutes of each other and clumped in a 3-km^2 area. Wolves did not feed on 17 of the carcasses and only partially ate the other 17. Miller et al. (1985) reported that a single wolf killed three calves on one occasion and three and four calves on a second occasion at average kill rates of 1 calf/min, 1 calf/8 min, and 1 calf/6 min, respectively, between first and last deaths. They attributed surplus killing of newborn caribou calves to their high densities and vulnerability on calving grounds.

A pack of five wolves in Riding Mountain National Park, Manitoba, under extreme late winter conditions killed at the rate of 1 elk, white-tailed deer, or moose every 2.7 days (Carbyn 1983b). Fuller and Keith (1980) found a summer kill rate of 1 moose every 4.7 days for a pack of 10 wolves in Alberta. The mid- to late winter predation rate of wolves in Wood Buffalo National Park was 1 bison/7.8 days (Oosenbrug and Carbyn 1982). In Alaska, wolf predation rates during summer were 1 kill/7–16 days, whereas winter rates were 1 kill/5–11 days. Large packs killed ungulates more frequently than did smaller packs. Kill rates per wolf, however, were greater for smaller packs (Ballard et al. 1987). Kill rates by wolves on a rapidly growing moose population in the east-central Yukon reached 2.4 moose/wolf/100 days (Hayes and Harestad 2000a). Kill rates by individual wolves were inversely related to pack size and unrelated to prey density or snow depth (Hayes et al. 2000). In Banff National Park, where seven species of ungulate prey are sympatric with wolves, the predation rate of wolf packs averaged 1 kill/3 days. This was composed of 1 elk/4.4 days, 1 mule deer/25 days, 1 white-tailed deer/46 days, 1 bighorn sheep/59 days, and 1 moose/67 days (Hebblewhite 2000).

Most estimates of food consumption by wolves (summarized by Mech 1970 and Schmidt and Mech 1997) are made by calculating the weight of edible material available from a carcass and dividing that by the number of wolves and days. The method, however, does not account for underuse of carcasses by wolves (Peterson 1977; Paquet 1992), food lost to scavengers (Carbyn 1983b; Messier and Crête 1985; Ballard et al. 1987; Fuller 1989; Hayes et al. 1991; Paquet 1992; Thurber and Peterson 1993; Dale et al. 1995), and undetected caching (Murie 1944; Cowan 1947; Mech 1988; Mech et al. 1998; Mech and Adams 1999). In Riding Mountain National Park, 91% of wolf-killed elk and 86% of wolf-killed moose were abandoned before all edible portions had been consumed. Large wolf packs consumed a higher proportion than small wolf packs (Paquet 1992). Promenberger (1992) found that large groups of juvenile ravens removed up to 37 kg of food/day from fresh ungulate carcasses. He suggested these flocks were more important competitors with small than large wolf packs because fewer wolves consume kills more slowly than larger packs. Hayes et al. (2000), studying a recovering wolf population in the Yukon, confirmed that scavenging by ravens decreased the amount of prey biomass available for wolves to consume, especially for wolves in smaller packs.

As noted above, wolves can adjust to a wide variation in amount of food available, and will eat as much as four times their daily maintenance requirement of 1.7 kg/wolf (Mech 1970). A mean daily rate of >3.2 kg/wolf is required for successful reproduction (Mech 1977a). On Isle Royale, Mech (1966) estimated daily consumption of moose at 4.4–6.3 kg/wolf. Peterson (1977), also working on Isle Royale, calculated daily winter food consumption of 6.21–10.0 kg/wolf during 1971–1973 and 4.4–5.0 kg/wolf during 1974. In Ontario, Pimlott et al. (1969) and Kolenosky (1972) estimated a daily rate of consumption of 3.8 and 2.9 kg/wolf, respectively, composed primarily of white-tailed deer. Mech and Frenzel (1971) calculated a daily rate of consumption in northeastern Minnesota of 2.7 kg/wolf. Daily winter consumption averaged 2.0 kg deer/wolf in north-central Minnesota (Fuller 1989). In Riding Mountain National Park, wolves consumed about 8 kg of prey/day under unusual winter conditions and high elk densities (Carbyn 1983b). Daily per capita consumption of wolves in northeastern Alberta varied between 0.12 and 0.15 kg prey/kg wolf (Fuller and Keith 1980). The daily consumption of bison in Wood Buffalo National Park was 5.3 kg/wolf. A pack of eight wolves rearing five pups required 2526 kg of edible meat between 1 May and 1 October (Fuller and Keith 1980).

Predator–Prey Relations. Wolves specialize on vulnerable individuals of large prey (e.g., elk and moose), yet readily generalize to common prey such as deer and beavers. As a species, wolves display remarkable behavioral plasticity in using different prey and habitats (Mech 1991; Weaver et al. 1996). Although wolf predation has been investigated in more than 35 locales (Fuller 1989; Gasaway et al. 1992; Messier 1994), most studies have involved one or two ungulate species. Potvin et al. (1992b) studied the effects of wolf predation on beaver in Papineau-Labelle Reserve, Québec. Studies of wolf predation amid high ungulate diversity are limited (Cowan 1947; Carbyn 1974, 1975a, 1983b; Paquet 1992; Huggard 1993a, 1993b, 1993c; Boyd et al. 1994; Kunkel et al. 1999; Hebblewhite 2000). This is due, in part, to wolf extirpation where multiple prey species are common (Young and Goldman 1944).

Keith (1983) and Fuller (1989) found ungulate biomass per wolf is highest for heavily exploited (Ballard et al. 1987) or newly protected (Fritts and Mech 1981) wolf populations and lowest for unexploited wolf populations (Oosenbrug and Carbyn 1982; Mech 1986) or those where ungulates are heavily harvested (Kolenosky 1972). Wolf densities lower than predicted by Keith (1983) and Fuller (1989) were charcterized by high ungulate diversity in which at least one prey species occurred in large groups (Weaver 1994). Crête and Manseau (1996) compared the biomass of ungulates with primary productivity along a 1000-km latitudinal gradient on the Québec–Labrador peninsula, and Crête (1999) did the same over North America. For the same latitude, ungulate biomass was five to seven times higher in areas where wolves were absent than where wolves were present. In areas of former wolf range, but where no wolves currently exist, a regression of ungulate biomass to primary productivity produced a positive slope (Crête 1999). Thus, elimination of wolves from an area adapted to an evolutionary history of strong predator–prey interactions may have a severe impact through a trophic cascade (Terborgh et al. 1999).

Interactions of ungulates and their predators may overshadow habitat capability as a controlling factor for ungulate populations. Many researchers have reported wolf predation decreases survival and population growth rates of ungulate populations (Gauthier and Theberge 1986; Gasaway et al. 1992; Potvin et al. 1992a; Hatter and Janz 1994; Boertje et al. 1996; Jedrzejewska et al. 1997; Bergerud and Elliot 1998; Kunkel and Pletscher 1999; Hayes and Harestad 2000a). Despite difficulties in applying rigorous experimental design to wolf–prey studies (Boutin 1992; Orians et al. 1997; Minta et al. 1999), many researchers have concluded that wolf predation can limit, and possibly regulate, populations of moose, caribou, and white-tailed deer (Bergerud et al. 1983; Messier and Crête 1985; Gauthier and Theberge 1986; Messier 1991, 1994, 1995b; Skogland 1991; Gasaway et al. 1992; Seip 1992, Van Ballenberghe and Ballard 1994; Boertje et al. 1996; Ballard et al. 1997; Eberhardt 1997; Messier and Joly 2000; Hayes and Harestad 2000a; Hayes et al. 2000; but see Boutin 1992; Theberge 1990; Theberge and Gauthier 1985; Thompson and Peterson 1988). Potvin et al. (1988) suggested that under certain conditions, wolf predation can have antiregulatory effects on white-tailed deer.

Group size, landscape structure, and winter severity may influence whether wolf predation is density dependent or density independent. The functional and numerical responses of wolves to changing prey density likely vary with prey species, availability of alternative prey, presence of other predators (Messier 1994; Eberhardt 1997; Eberhardt and Peterson 1999), habitat overlap, herd sizes, and herd behavior (Huggard 1993b, 1993c; Weaver 1994; Hebblewhite 2000). In diverse prey systems, wolf predation may shift among species depending on annual fluctuations in winter severity (Nelson and Mech 1986a; Paquet 1992; Huggard 1993a; Post et al. 1999) or landscape changes (Weaver et al. 1996). Where wolf predation is a factor, ungulates may exist at levels well below carrying capacity for relatively long periods. Unusually mild or severe winter weather can result in ungulate populations that are temporarily higher or lower than predicted habitat potential (which reflects long-term average maximum). Wolf packs may react to changing conditions in varying ways, depending on the location of their territories in relation to other packs and prey distribution (Nelson and Mech 1986b; Mech 1994a). If packs have lower prey densities within their territories, they may exploit territories more intensely. This may be achieved by (1) persevering in each attack, (2) using carcasses thoroughly, (3) feeding on alternative and possibly second-choice food resources such as beaver (Messier and Crête 1985), and (4) patrolling their territory more intensely (Messier 1985b).

Messier (1984, 1991, 1994, 1995b) and Messier and Crête (1985) proposed that two-state equilibrium models best described the dynamics of wolf predation on ungulates. Eberhardt (1998) and Eberhardt and Peterson (1999), however, concluded that single-state models were adequate. In multiprey systems, the stability, or equilibrium, of ungulate prey and wolf populations seems to depend on a variety of factors, including wolf predation rate, number and age of ungulates killed by hunters, ratio of ungulates to wolves, and population growth rate of different ungulate species (Cowan 1947; Carbyn 1983b; Huggard 1991; Weaver 1994; Kunkel and Pletscher 1999). Elk are the primary prey of wolves in many multiple-prey systems, and are often the preferred prey when available (Carbyn 1983b; Paquet 1992; Huggard 1993b; Weaver 1994; but see Kunkel et al. 1999). The consequences of wolf preference for elk on population dynamics are complex due to prey switching (Oaten and Murdoch 1975), alternate prey increasing predator density at low primary prey density (Messier 1994, 1995), spatial distribution of multiple prey species (Iwasa et al. 1981), and differential encounter rates across species (Messier 1994, 1995b).

Most studies emphasize the direct effects (e.g., prey mortaliy) wolves have on the population dynamics of their ungulate prey (Carbyn 1975a; Mech and Karns 1977; Carbyn 1983b; Gasaway et al. 1983; Gunson 1983; Peterson et al. 1984; Messier and Crête 1985; Ballard et al. 1987; Boutin 1992; Messier 1994; Hayes and Harestad 2000a). However, predation can also profoundly affect the behavior of prey, including use of habitat, time of activity, foraging mode, diet, mating systems, and life histories (Sih et al. 1985). Accordingly, several studies describe the influence wolves have on movements, distribution, and habitat selection of caribou, moose, and white-tailed deer (Carbyn 1975b; Mech 1977c; Rogers et al. 1980; Nelson and Mech 1981; Bergerud et al. 1984; Stephens and Peterson 1984; Messier and Barrette 1985; Ballard et al. 1987; Messier 1994). A failed wolf attack on barren-ground caribou changed the movement pattern of a postcalving herd. Three attacking wolves caused the caribou to reverse their direction of travel, recross a river, and return to the direction from which they had come (Scotter 1995). Wolves can increase the rate at which they accrue resources by seeking out areas with dense concentrations of prey (Huggard 1991; Weaver 1994; Hebblewhite 2000). Prey, in turn, can lower their expected mortality rate by preferentially residing in areas with few or no wolves. Several studies have suggested that ungulate prey seek out predator-free refugia to avoid predation by wolves (Mech 1977c; Holt 1987; Paquet 1993; Hebblewhite 2000). Wolf predation in the Superior National Forest of northern Minnesota affected deer distributions within wolf territories (Mech 1977c). Densities were greater along edges of territories where predation was thought to be less. Nelson and Mech (2000), however, observed female white-tailed deer remained on their traditional home ranges despite proximity to wolf home sites and did not attempt to reduce exposure to wolves by moving away.

Recent evidence suggests the importance of cascading trophic interactions on terrestrial ecosystem function and processes (Glanz 1982; Emmons 1984; Terborgh 1988; Terborgh et al. 1999). Accordingly, system-wide effects of wolf predation may be more profound than previously expected. For example, on Isle Royale, Michigan, wolf predation on moose positively influenced biomass production in trees of boreal forest (McLaren and Peterson 1994). Growth rates of balsam fir (*Abies balsamea*) were regulated by moose density, which in turn was controlled by wolf predation (McLaren and Peterson 1994). When the wolf population declined for any reason, moose reached high densities and suppressed fir growth. Research elsewhere suggests elk populations not regulated by large predators negatively affect the growth of aspen (*Populus tremuloides*) (Kay 1990; Kay and Wagner 1994; Kay 1997; but see Barnett and Stohlgren 2001). Wolves are a significant predator of elk, and wolves may positively influence aspen overstory through a trophic cascade caused by reducing elk numbers, modifying their movement, and changing elk browsing patterns on aspen (White et al. 1998). Ripple and Larson (2000) reported that aspen overstory recruitment ceased when wolves disappeared from Yellowstone National Park.

Climatic patterns, such as El Niño or La Niña, affect the relationship of wolves and ungulate prey (Ballard and Van Ballenberghe 1997; Post et al. 1999). In years that the North Atlantic Oscillation produced deep snow cover, moose were more vulnerable to wolf predation (Ballard and Van Ballenberghe 1997; Post et al. 1999). Thus, the fir forest of Isle Royale was released from heavy browsing, more seedlings were established, more saplings survived, and litter production and nutrient dynamics were affected (Post et al. 1999).

POPULATION DYNAMICS

Many processes are involved in the dynamics of wolf populations. Included are limitations of habitat, environmental variation that causes regular or episodic fluctuations in reproduction; dispersal; intrinsic processes such as demographic stochasticity, effects of age structure, and social system (Vucetich et al. 1997); and genetics (Peterson et al. 1998). The effect of food on wolf populations is mediated by intrinsic social factors, including pack formation, territorial behavior, exclusive breeding, deferred reproduction, intraspecific aggression, dispersal, and primary-prey shifts (Packard and Mech 1980). Population dynamics, however, are primarily dictated by per capita amount of ungulate prey (Keith 1983; Messier and Crête 1985; Fuller 1989; Messier 1994; Eberhardt 1997, 1998; Eberhardt and Peterson 1999) and secondarily by the vulnerability of ungulates to predation (Boertje and Stephenson 1992), disease (see below), level of human-caused mortality (Keith 1983; Fuller 1989), and human activities that displace wolves and their prey from critical habitats (Thiel 1985; Fuller et al. 1992; Paquet et al. 1996; Mladenoff et al. 1995). Any one of these factors, however, can predominate at different times and may interact synergistically with other influences. In addition, the weakening of one factor may enhance another.

Boertje and Stephenson (1992) concluded that wolf productivity declines as prey availability per wolf declines. However, significant declines in reproductive potential only occur when ungulate biomass per wolf declines below threshold levels. Eberhardt and Peterson (1999) estimated an equilibrium ratio for wolves and ungulates of 122 deer-equivalents per wolf. Messier (1987) concluded that in wolf–moose ecosystems with <0.4 moose/km^2, wolves would struggle to subsist, the habitat would not be saturated, and pack sizes would be relatively small with few or no extra adults physiologically capable of breeding. Messier (1985b) estimated that, with moose as the sole ungulate present, minimum prey biomass for a pack to maintain itself is equivalent to a density of 0.2 moose/km^2. In the Superior National Forest, where prey were scarce and the wolf population was declining from high densities, litter size and pack size were inversely related. In the Beltrami Island State Forest, where prey were relatively abundant and the wolf population was increasing, pack size and litter size were positively correlated (Harrington et al. 1983).

Limiting and Regulatory Factors. Limiting factors are density independent, such as the effects of climate on growth rates, whereas regulatory factors are density independent, such as density-induced starvation (Sinclair 1989). Several factors have been reported to limit or regulate growth of wolf populations. These include ungulate biomass (Carbyn 1974; Van Ballenberghe and Mech 1975; Fuller and Keith 1980; Packard and Mech 1980; Keith 1983; Messier 1985a, 1985b, 1987; Peterson and Page 1988), disease (see below), and human-caused mortality (Gasaway et al. 1983; Keith 1983; Peterson et al. 1984; Fuller 1989; Ballard et al. 1987, 1997; Van Ballenberghe 1991; Paquet et al. 1996; Noss et al. 1996).

Quantifying the importance of food in limiting population growth based on cause of death alone is difficult. Results vary among studies. On Isle Royale, annual mortality from starvation and intraspecific strife (both related to low food availability) ranged from 18% to 57% during a 20-year period (Peterson and Page 1988). In populations where some human-caused mortality occurs and thus compensates for natural mortality (starvation, accidents, disease, and intraspecific strife), about 8% of individuals >6 months of age can be lost each year (Ballard et al. 1987; Fuller 1989). Some researchers have accepted this variability and decided any sign of starvation among adult wolves means food is limiting population growth (Fritts and Mech 1981; Ballard et al. 1997). This assumption is reasonable, since adults typically are the last members of the population affected by food shortage (Eberhardt 1977). As such, they may be the most sensitive indicators of food shortages.

Several researchers have suggested that wolf numbers stabilize at a limiting population density. Saturation is thought to occur at about 1 wolf/26 km^2 (Pimlott et al. 1969; Mech 1970), where social behavior is believed to make space a limiting factor and thereby regulate wolf populations at levels below the level that would adversely affect food resources. Accordingly, territoriality would regulate the number of breeding units and social dominance would limit the number of breeders within each unit. Another theory is that wolf numbers are not limited by territoriality, but exceed the supposed saturation level of 1 wolf/26 km^2, and wolves continue to prey on available biomass to a point of diminishing returns (Packard and Mech 1980). Wolf numbers at one point reach high densities, and wolves continue to exert heavy losses on prey; this, together with other mortality factors, inevitably results in prey declines, and local extinctions are thought to result (Mech and Karns 1977). With the loss of food, wolf numbers also decline and under favorable conditions prey numbers increase, followed by increases in wolves, and the cycle repeats. Because of lag phases, the wolf–prey systems are perceived to stabilize at different equilibria and follow cyclical predator–prey oscillations. This view seems to have support in the technical literature. It is likely that neither view is applicable in all situations and combinations may occur. Dynamics of multiple-prey systems are further complicated because of the potential for prey switching and opportunistic predation (Seip 1992).

The rate of annual mortality that controls growth of wolf populations is unknown. Mech (1970) concluded an annual harvest (hunting or trapping) of 50% or more was necessary to control wolf populations based on pup–adult ratios, but did not distinguish between harvest and natural mortality. However, Keith (1983) and Fuller (1989) reviewed numerous wolf studies across North America and concluded that harvests exceeding 28–30% of fall populations resulted in declines. Fuller (1989) further concluded that populations would stabilize with an overall annual mortality rate of 35%. Peterson et al. (1984) and Fuller (1989) found evidence that harvest effects vary with time and population structure. For instance, if productivity is high, and consequently so is the ratio of pups to adults, the population can withstand a higher overall mortality because pups (nonproducers) make up a disproportionately larger amount of the harvest (Fuller 1989). Furthermore, net immigration or emigration may mitigate the effects of harvest (Fuller 1989). Multiple denning within individual packs (Harrington et al. 1982; Ballard et al. 1987) can also have a significant influence on rates of increase and sustainable mortality rates.

Ballard et al. (1997) reviewed three factors (pack size, number of pups, and multiple denning) that they felt determined the amount of mortality populations of wolves can withstand. Wolf populations comprising small wolf packs can withstand higher levels of exploitation than those comprising large packs, provided that reproductively active females are not killed. Populations with high proportions of pups can also withstand heavier exploitation than populations comprising large proportions of adults because pups are more vulnerable to exploitation and in populations with fewer pups, adults may make up a larger proportion of the harvest (Fuller 1989).

Rates of Growth. Three factors dominate wolf population dynamics: wolf density, ungulate density, and human exploitation. These are linked through wolf predation, social behavior, and functional and numerical responses. Rate of increase and densities in wolf populations are primarily a function of ungulate biomass and secondarily age structure of the population and human-caused mortality. In most cases, stable wolf populations are in equilibrium with their ungulate prey (Keith 1983). Reported rates of growth in wild wolf populations have varied between 0.93 and 2.40 (Fuller and Keith 1980; Fritts and Mech 1981; Ballard et al. 1987; Fuller 1989; Hayes et al. 1991; Messier 1991; Pletscher et al. 1997). Few, if any, populations achieve a theoretical exponential rate of 0.833 ($\lambda = 2.30$) given maximum reproduction (Rausch 1967), a stable age distribution, and no deaths. Keith (1983), calculated a maximum rate of increase ($r = 0.304$, $\lambda = 1.36$) based on the highest reproductive and survival rates reported from studies on wild wolves. The rate of increase of different populations likely varies with environmental and ecological factors.

Wolf populations can compensate demographically for excessive mortality. Under certain circumstances, this compensation enables wolves to respond to increased rates of juvenile or adult mortality with increased reproduction and/or survival, thereby mitigating demographic fluctuations (Weaver et al. 1996). Dominant wolves can reproduce at a very young age and usually reproduce every year after that (Weaver et al. 1996). Age at reproductive senescence has not been well documented, but few females survive to reproduce past the age of nine years (Mech 1988). Wolves also display remarkable ability to recover from exploitation. Human-induced mortality in wolf populations tends to reduce wolf densities, which alters predator–prey ratios (i.e., more prey per predator), which in turn raises rates of increase.

MORTALITY

Causes of Mortality. Wolves die because of accidents (Fuller and Keith 1980; Boyd et al. 1992), starvation (Mech 1972; Seal et al. 1975; Van Ballenberghe and Mech 1975; Mech 1977b; Fuller and Keith 1980), intraspecific strife (Mech 1972; Van Ballenberghe et al. 1975; Messier 1985a, 1985b; Ballard et al. 1987), interspecific conflicts (Stanwell-Fletcher 1942; Frijlink 1977; Ballard 1980, 1982; Nelson and Mech 1985; Mech and Nelson 1990; Weaver 1992), disease (see above), and human-related causes. Causes of human-related mortality includes legal harvest (Fuller and Keith 1980; Gasaway et al. 1983; Keith 1983; Peterson et al. 1984; Messier 1985a; Ballard et al. 1987, 1997; Potvin 1987; Bjorge and Gunson 1989; Fuller 1989; Hayes et al. 1991; Pletscher et al. 1997), illegal harvest (Fritts and Mech 1981; Fuller 1989; Pletscher et al. 1997), vehicles on highways (Berg and Kuehn 1982; Potvin 1987; Fuller 1989; Forbes and Theberge 1995; Thiel and Valen 1995; Bangs and Fritts 1996; Paquet and Callaghan 1996; Paquet et al. 1996), trains (Paquet 1993; Paquet et al. 1996; Krizan 1997; Forshner 2000), and introduced diseases (see below).

Researchers have noted that starvation and intraspecific aggression are more common when wolves are faced with a low density of prey (Van Ballenberghe and Erickson 1973). For instance, Messier (1985b) reported that in southwestern Quebec, the mortality rate of wolves living in an area with low densities of prey was higher than the rate for wolves living in an area of high densities of prey. This was because wolves with less prey incurred more deaths from starvation and intraspecific aggression. Similarly, Mech (1977a) found that intraspecific aggression increased as prey availability declined in Minnesota. However, Mech (1977a) found that only pups seemed to starve. Other investigators have also reported that a shortage of prey increases natural mortality (Van Ballenberghe and Mech 1975; Mech 1977a; Messier 1985b).

Wolves risk injury (Rausch 1967; Phillips 1984; Pasitschniak-Arts et al. 1988) and death in attempting to kill large prey (Mech 1970). Healthy, vigorous prey often escape wolf predation by fighting back or fleeing (Mech 1984; Nelson and Mech 1993; Stephenson and Van Ballenberghe 1995). Weaver (1992) noted that about 25% of 1450 wolves killed by humans in control programs in Alaska showed traumatic skull injuries, presumably inflicted by moose and other large prey. On occasion, moose, bison, elk, and deer can gain the upper hand and kill attacking wolves (Stanwell-Fletcher 1942; Frijlink 1977; Nelson and Mech 1985; Mech and Nelson 1990; Weaver 1992). The risk to wolves appears to increase with size of prey. Several wolf–moose studies have shown that wolves often search for less risky opportunities rather than attack such dangerous prey (Stephens and Krebs 1986; Forbes 1989). Avalanche-caused wolf mortality has been reported in Alberta (Boyd et al. 1992) and Alaska (Mech et al. 1992 in Boyd et al. 1992). Wolves also are killed occasionally by other predators (see below).

Diseases. Many authors have concluded that food abundance by itself, or in combination with social stress, is the main regulatory factor of wolf populations. Diseases, however, can be important modulators of the many processes that determine population dynamics of wolves (Carbyn 1982a; Bailey et al. 1995). A wolf pack may be affected by disease through loss of experienced adults, reduced recruitment of young, and decimation or disruption of pack social structure. Disease has not been linked to low food availability, but the relationship makes sense intuitively. A population of wolves that suffers from lack of food should be more vulnerable to disease than a population with more food available. Furthermore, food shortage leading to nutritional stress may combine with disease factors to increase the significance of otherwise innocuous or sublethal conditions (Brand et al. 1995).

Diseases of wolves have been summarized (Mech 1970; Custer and Pence 1981; Brand et al. 1995), but the affects of epizootics and enzootics on the dynamics of wolf populations have not been well documented. Most studies report no disease-related or disease-caused deaths of wolves (Van Ballenberghe et al. 1975; Mech 1977a; Fritts and Mech 1981; Messier 1985b; Potvin 1987; Ballard et al. 1987; Hayes et al. 1991; Meier et al. 1995; Pletscher et al. 1997). Where documented, from 2% to 21% of wolf mortality has been attributed to disease (Carbyn 1982a; Peterson et al. 1984; Fuller 1989; Ballard et al. 1997). The transmission of disease, such as parvovirus, from domestic dogs to wild wolves is a serious conservation concern (Bailey et al. 1995).

Rabies is an important zoonosis, but is infrequently reported in wolves (Cowan 1949; Rausch 1958; Chapman 1978; Theberge et al. 1994). Ballard et al. (1997) concluded that rabies was a significant factor in a decline of wolves from Alaska. In that study, rabies-caused mortality was 21%. Little is known about the effects of bacterial zoonoses on wolf populations. Wolves are definitive hosts for various protozoan infections. The most notable are toxoplasmosis and sarcosporidosis. Although numerous arthropod parasites are known, only sarcoptic mange is an epizootic disease of significance. There is little evidence that noninfectious deseases are serious problems in the morbidity and mortality of wolves. In contrast, many infectious disease agents have been reported. Some of these reach epizootic proportions and occasionally affect their populations. Important viral infections thought to be significant are distemper and canine hepatitis (Custer and Spence 1981; Brand et al. 1995). Other diseases that occur in wolf populations are canine parvovirus (Mech and Goyal 1993, 1995; Johnson et al. 1994; Bailey et al. 1995; Mech et al. 1997), Lyme disease (Kazmierczak et al. 1988; Thieking et al. 1992), heartworm (*Dirofilaria immitis;* Mech and Fritts 1987), leptospiroseus, blastomycosis (Thiel et al. 1987; Krizan 2000; Paquet et al. 2001a), tuberculosis (Carbyn 1982a), and coccidiosis (Mech and Kurtz 1999). The effects of these diseases are largely unknown.

Murie (1944) discussed mange, canine distemper, and rabies as possible regulating factors in Alaskan wolf populations. Distemper and mange may have reduced wolf populations in Jasper National Park during the 1940s (Carbyn 1982a). Sarcoptic mange is an important, sometimes common, but rarely reported disease of wolves (Todd et al. 1981), caused by a mite, *Sarcoptes scabiei*. Based largely on circumstantial evidence, several researchers believed that mange is an important regulating factor in wild canid populations (Pike 1892; Murie 1944; Cowan 1951; Green 1951; Todd et al. 1981). For example, during a 10-year period in Alberta, mange was present in wolves each year, but the number of cases increased when wolf densities increased, and the number of pups surviving decreased as prevalence of mange increased (Todd et al. 1981). Todd et al. (1981) found that weights of xiphoid-process fat globules of mangy coyotes and wolves were lower than weights from nonmangy animals although many animals were lightly infested. In addition, body weights of mangy wolves averaged 4–10% less than weights of nonmangy wolves.

BEHAVIOR

Social Behavior and the Pack. Although some wolves are solitary, most are highly gregarious and live in packs with complex social structures. Pack size is largest in fall and early winter when pups are integrated into the pack. Reductions in pack size by late winter are typically due to pup and adult mortality as well as dispersal of younger wolves. Wolf packs are usually made up of 5–12 individuals, although larger packs may be found (Mech 1974). For example, Mech (2000) documented a pack of 22–23 wolves in central Minnesota. Because the pack preyed primarily on white-tailed deer, he cited this as evidence that prey size and pack size are not tightly linked (see discussion below). Packs in the Northwest Territories of Canada occasionally coalesce into groups of 20–30 animals when hunting bison (Carbyn et al. 1993). A pack of 28 individuals in spring and a pack of 40 in autumn were observed in association with migratory barren-ground caribou (F. Miller, Canadian Wildlife Service, Edmonton, Alberta, pers. commun., 1999).

The proximal mechanism underlying the regulation of pack size is complex and imperfectly understood (Zimen 1976, 1982; Packard and Mech 1980; Haber 1996a). An increase in prey abundance seems to produce a direct increment in the in-group recruitment and survival rates (Keith 1983). As prey abundance increases, the territory mosaic becomes progressively saturated (lower territory vacancy), which could provoke delayed dispersal (Packard and Mech 1980; Messier and Barrette 1982; Messier 1985a). Nudds (1978) detected an apparent relationship between wolf pack size and food acquired per wolf, which implied that packs smaller than optimal size acquired substantially less food per wolf than those of optimal size. He also speculated that there were different optimal pack sizes for wolves preying on moose and other large prey than for those preying principally on deer. However, Thurber and Peterson (1993) demonstrated that for wolves preying primarily on moose on Isle Royale, food acquisition per wolf decreased with increased pack size. Hayes (1995) reported the same for wolves preying on moose and caribou. Dale et al. (1995) concluded that the larger packs killing moose, caribou, and Dall sheep (*Ovis dalli*) acquired no more food per wolf than smaller packs. A comprehensive review of prey use by wolves in several geographic areas showed a negative relationship between pack size and food acquisition per wolf (Schmidt and Mech 1997). Taken together, these studies provide strong evidence against the idea that wolves live in packs to facilitate predation on large prey.

The pack is usually a breeding pair and their offspring of the previous 1–3 years, or sometimes two or three such families (Murie 1944; Young and Goldman 1944; Mech 1970 1988; Clark 1971; Haber 1977; Mech and Nelson 1989). Most offspring disperse near 3 years of age (Fritts and Mech 1981; Peterson et al. 1984; Ballard et al. 1987; Fuller 1989; Gese and Mech 1991; Boyd and Pletscher 1999). Although female wolves in captivity have bred successfully at 10 months of age (Medjo and Mech 1976), wild wolves typically do not breed until at least 22 months (Rausch 1967; Mech 1970). Occasionally, an unrelated wolf is adopted into a pack (Van Ballenberghe 1983a; Lehman et al. 1992; Mech et al. 1998) or a relative of a breeder is included (Haber 1977; Mech and Nelson 1990; Mech et al. 1998), or a dead parent is replaced by an outside wolf (Rothman and Mech 1979; Fritts and Mech 1981; D. Smith, National Park Service, Yellowstone National Park, Wyoming, pers. commun., 2000). In the latter case, an offspring of opposite sex from the newcomer may replace its parent and breed with the stepparent (Fritts and Mech 1981; Mech and Hertel 1983).

Communication is through postures (Schenkel 1947, 1967; Crisler 1958; Fox 1971, 1973, 1975; Zimen 1976, 1982; Fox and Cohen 1977), vocalizations (Harrington and Mech 1978a, 1978b, 1979, 1983; Harrington 1986, 1989; Coscia et. al. 1991; Coscia 1995), and scents (Kleiman 1966; Theberge and Falls 1967; Peters 1978; Harrington 1981; Asa et al. 1985a, 1985b; Merti-Millhollen et al. 1986; Paquet 1989; Asa 1997; Asa and Valdespino 1998). All wolves exhibit similar behaviors, although a degree of individual variability exists (Fox 1975; Zimen 1982). Fixed patterns of behavior express the inner state of a wolf to which other wolves respond. As with domestic dogs, an elevated tail and erect ears convey alertness and sometimes aggression. Facial expressions, emphasized by the position of the lips and display of the teeth, are the most dramatic form of communication. Scent from urine and possibly fecal matter is used to express social status and breeding condition and advertise territorial occupancy (Peters and Mech 1975, 1978; Asa et al. 1985a). A gland at the ventral base of the tail and anal glands may also exude chemicals used in communication (Asa et al. 1985b).

Wolves organize themselves into strict dominance hierarchies where individual position reflects status and privilege. Details of social structure vary with the number, sex, age, and reproductive structure of the group. In large packs, males and females have separate linear hierarchies in which each animal knows its position (Zimen 1976, 1982). At the top of these hierarchies are the highest ranking male and female, one of which serves as alpha wolf or pack leader. Females are as likely to lead the pack as males. Interactions between the sexes are more complex because of breeding relationships. Rank positions are not permanent and agonistic contests are most intense during the winter breeding period (Peterson 1979).

A few people have observed the social behavior of wild wolves around dens. Murie (1944) gave an anecdotal account. Clark (1971) presented a quantified summary of the pack social relationships. Haber (1977) described his interpretation of a pack's social hierarchy without supporting evidence. Based on summer observations of wolves on Ellesmere Island, Mech (1999b) described the wolf pack social order, the alpha concept, and social dominance and submission. He concluded that adult parents guide the activities of the pack in a division of labor system. The female predominates primarily in such activities as pup care and defense, and the male primarily during foraging and food provisioning.

The complex social organization found within and between wolf packs may have subtle influences on physiology and behavior that are of regulatory importance (Haber 1996a). Social relationships within the pack may also be sensitive to food supply and thus influence size of a pack. The most common explanation for the highly evolved social behavior of wolves is the need for cooperation in hunting large prey (Murie 1944; Mech 1970; Zimen 1976; Peterson 1977; Nudds 1978; Pulliam and Caraco 1978; Rodman 1981), although even single wolves can kill prey the size of moose (Thurber and Peterson 1993) and bison (Carbyn and Trottier 1988; Carbyn et al. 1993). An alternative view proposes that wolves live in packs so adult pairs can efficiently share with their offspring surplus food resulting from the pair's predation on large mammals (Schoener 1971; Rodman 1981; Hayes 1995; Schmidt and Mech 1997). A social capacity limit independent of food supply may also influence group size (Mech 1970; Zimen 1976; Packard 1980; Packard and Mech 1980; Packard et al. 1983, 1985).

Chemical and Vocal Communication. Scent markings and vocalizations are used by wolves to maintain territories and communicate among themselves. For territorial advertisement, vocalizations are thought to be less important than scent marking (Harrington and Mech 1978b, 1983). Scent marking is long term and site specific, whereas howling is immediate and long range (Harrington and Mech 1979). Distinctive howling (Joslin 1967, Harrington and Mech 1978b) and the presence of recent urine marks may limit direct aggression between packs by encouraging wolves to avoid alien territories (Peters and Mech 1975). Scent marking involves urine (Raymer et al. 1984, 1986), feces, and anal scent glands (Kleiman 1966; Ewer 1973; Raymer et al. 1985; Asa et al. 1985a, 1985b), often in conjunction with scratch marks (Paquet 1991a). Scent rubbing may also be involved in communication (Harrington et al. 1986). The raised-leg urination of wolves is generally accepted as a form of scent marking (Kleiman 1966; Ewer 1973). Scent marking may play a role in intrapack communication by expressing sex, reproductive state (Ryon and Brown 1990), and dominance (Macdonald 1985). Establishment and maintenance of pair bonds may also involve scent marking (Rothman and Mech 1979).

Urine Marking. Time spent in each part of a territory influences the frequency of urine marking (Paquet and Fuller 1990; P. Paquet, unpublished data). Accordingly, the number of marks could be elevated in areas of high prey density as a reflection of foraging behavior by wolves (Paquet and Fuller 1990). Physiography also influences the frequency and distribution of scent marking (Peters and Mech 1975; Paquet 1991a). In winter, established roads and trails exhibit the highest number of marks and frozen water bodies the lowest. Therefore, territories encompassing large bodies of water and/or unusual configuration of roads and trails might show an uneven representation of scent marks.

Wolves increase their rate of scent marking when they encounter scent marks from members of other packs (Peters and Mech 1975). The fresher the scent mark, the more likely it is to elicit another mark (Paquet 1989). Dominant wolves mark more than subdominants, and female wolves mark more than males (Haber 1977; Asa et al. 1990; Ryon and Brown 1990). Nonbreeding wolves seldom scent mark (Rothman and Mech 1979), whereas newly formed pairs mark the most (Rothman and Mech 1979). Lone wolves rarely mark (Rothman and Mech 1979; Paquet 1991a). Urine marks of female wolves are responded to more frequently than those of males. Visual display, used in conjunction with urine marking, may play a role in interpack communication by signaling dominance. Marking frequency of female wolves increases during courtship and breeding season, and the response of other wolves to the marks of females also increases during these periods (Ryon and Brown 1990). Because the volatile chemical constituents of the urine of male and female wolves change seasonally (Raymer et al. 1984, 1986), urine marks may also provide information on reproductive status. Urine is also used to mark caches of food (Harrington 1981).

Peters and Mech (1975) concluded that wolf-marking sites are more numerous along territorial borders, and proposed as a model an "olfactory bowl" of scent, in which the number of marks decreases from the edge to the middle of a territory. In theory, this higher density of scent marks enables packs to recognize the periphery and keep from trespassing into more dangerous areas beyond. Barrette and Messier (1980), however, questioned the appropriateness of the model because it was based on marks/km rather than marks/km^2. Moreover, wolf packs in Riding Mountain National Park showed no difference between peripheral and interior marking rates, although scent marks were more abundant in some areas than others (Paquet and Fuller 1990).

Howling. Howls can be heard for several kilometers under certain conditions. Joslin (1967) reported that howling could advertise the presence of wolves over a 130-km^2 area. Spontaneous and elicited howls are influenced by time of year and social circumstances (Harrington and Mech 1978a, 1978b). Rate of howling by two wolf packs in Minnesota increased throughout the pup-rearing season (Harrington and Mech 1978a). Wolves characteristically respond to human disturbance near their pups by barking and howling (Chapman 1977). A midwinter increase in howling is associated with reproductive behavior, especially for groups containing breeding animals. Through the year, the rate of elicited howling is higher among packs and lone wolves attending kills, The more food remaining at a kill, the higher the rate of reply. Larger packs reply more often than smaller packs. Lone wolves rarely reply, reflecting the low-profile behavior expected of surplus animals in territorial populations. The responsiveness of lone wolves depends on the status of the wolf before becoming a loner and amount of time since it left the pack. For wolves separated from their pack, the howling reply rate depends on their age and social role. Dominant adults are more likely to howl than subordinate younger animals. Howling may reflect the status and motivational state of wolves. Harrington (1987) reported that howls of antagonistic wolves were deeper in pitch than those of passive animals. He speculated that during aggressive encounters, use of low-frequency, harsh sound expresses body size, which is a primary determinant in the outcome of aggressive interactions. Although animals of larger size can produce sounds of lower pitch and harsher tonal quality, size can be exaggerated by vocal manipulation.

Howling may be involved in coordination of pack activities (Harrington and Mech 1978b). Harrington (1975) reported that howling is important in maintaining pack structure in populations of high mortality, helping to assemble individuals after they have been separated. On Isle Royale, howling was important in coordinating moves of a large pack (Peterson 1977). Most howling was heard at night when adults were hunting and spatially separated. Such howling may help wolves synchronize hunting efforts. Carbyn (1975a) recorded crepuscular peaks in howling at wolf rendezvous sites in Jasper National Park, Alberta. Increased howling at dawn and dusk may be associated with departures and arrivals of adults at rendezvous areas (Harrington and Mech 1978a, 1978b).

ECONOMIC STATUS, MANAGEMENT, AND CONSERVATION

Viable, well-distributed wolf populations depend on abundant, available, and stable ungulate populations. Flexible feeding habits, high annual productivity, and dispersal capabilities enable wolves to respond to natural and human-induced disturbances (Weaver et al. 1996). Though evidence is lacking, movement among many North American subpopulations of wolves appears relatively unimpeded and, for the most part, the current rate of mortality sustainable. The fate of wolves ultimately depends on our ability to coexist with them at a local level. Therefore, successful management and conservation of wolves depends as much on social acceptance as on protecting the species's biological requisites.

Controlling wolves to protect livestock, enhance ungulate populations, and protect endangered species such as mountain caribou and whooping cranes (Kuyt et al. 1981; Edmonds 1988) for the benefit of humans remains controversial (Archibald et al. 1991; Cluff and Murray 1995; Buss and de Almeida 1998; Haber 1996; Thompson et al. 2000). Most of the debate focuses on the relative contributions of overhunting, industrial development, recreational development, and wolf predation to the decline of ungulates. Currently, wolves are controlled by shooting, poisoning, trapping, and sterilization (Haight and Mech 1997). Wolf control programs have been demonstrated to increase ungulate numbers (Gasaway et al. 1983; Bergerud and Elliott 1986, 1998; Farnell and McDonald 1988), but due to negative public reaction such programs have been delivered at substantial costs to the agencies involved. As noted above, some research suggests wolf predation can have a regulatory affect on ungulates (e.g., Mech and Karns 1977; Gasaway et al. 1983; Larsen et al. 1989; Bergerud and Elliot 1998; Hayes and Harestad 2000b). Less clear, however, is the role of predation in initiating such declines (Gauthier and Theberge 1987).

For the most part, management of wolf populations outside protected areas could be improved (Mech 1995; Thompson et al. 2000). Livestock and wolves need to be managed in areas of conflict. As wolf populations continue to grow in newly colonized or reestablishment areas, there may be an increasing need for control of those wolves preying on livestock (Fritts 1993). Because the public has so strongly supported wolf recovery and reintroduction, understanding the need for control may be difficult for many. Thus, strong efforts at public education will be required. Social approval for protection of livestock and enhancement of subsistence hunting may be higher than for furtherance of sport hunting. In addition, public acceptance of methods used to reduce wolf populations varies regionally and culturally.

In most of North America, regional wolf populations require core wilderness areas to persist (Mladenoff et al. 1995, 1999; Paquet et al. 1996; Woodroffe 2000; Carroll et al. 2001; Callaghan 2002). In this human-dominated world, however, requirements of the wolf are quickly becoming rare commodities. Even the largest North American parks and reserves are inadequate in area to fully protect wolves (Woodroffe and Ginsberg 1998). Packs living in highly productive environments such as Yellowstone National Park require about 150–300 km^2. In mountainous areas, annual home ranges can be as large as 3000 km^2 (Paquet et al. 1996). Wolves living in the Arctic, which depend on caribou, may use areas of 60,000 km^2 or larger (Walton et al. 2001). The number of protected areas should be increased in some areas and the effectiveness of existing reserves that are too small, or have unsuitable configurations, could be improved by the creation of buffer zones.

In many parts of North America, wolves live in networks of disjunct populations, many of which are close to human settlement. Wolves can survive in disjunct populations if movement between populations is unobstructed, human persecution is not excessive, and prey is abundant (Haight et al. 1998). Wolves do move throughout human-occupied landscape, across many unfavorable areas, but establishment success is restricted to higher quality habitat characterized by low human presence.

Population Status. The status of the gray wolf in its global range (North America, Eurasia, and the Middle East) is listed as vulnerable (Hilton-Taylor 2000). The North American gray wolf was added to Appendix II of the Convention on the International Trade in Endangered Species of Wild Flora and Fauna (CITES) in 1977. That agreement regulates international trade of animals and plants when that trade (1) threatens the species survival or the survival of a geographic population of that species or (2) the species looks like a threatened species. The North American gray wolf was listed under CITES to help control trade of endangered gray wolf populations in other parts of the world, not because the species is threatened or endangered globally.

TABLE 23.2. Estimates of adult wolf populations in the United States and Canada

State/Province/Territory	Trend	Population Estimate
Newfoundland/Labrador	Increasing	1,500
Quebec	Stable	4,000
Ontario	Stable	8,000–10,000
Manitoba	Stable	4,000–6,000
Saskatchewan	Undetermined	2,000–4,500
Alberta	Decreasing	3,000–5,000
British Columbia	Stable	4,000–8,000
Northwest Territories and Nunavut	Stable, declining on Queen Elizabeth Islands	10,000
Yukon	Stable	4,500
Michigan (Upper)	Increasing	100
Michigan (Isle Royale)	Oscillating	10–25
Wisconsin	Increasing	100
Minnesota	Stable	2,500
North Dakota	Dispersers, no resident population	<10
Montana (northern)	Stable	63
Wyoming/Montana (Greater Yellowstone)	Increasing	170
Arizona/New Mexico	?	30–40
Idaho	Increasing	192
Oregon	Dispersers, no resident population	<10
Washington	?	<10
Alaska	Stable	6,000–7,500
Total		50,000–55,000

SOURCE: United States, Bangs et al. (2001). Canada, Van Zyll De Jong and Carbyn (1999).

Because of diversity in climate, topography, vegetation, human settlement, and development, wolf populations in various parts of the original North American range vary from extirpated to pristine. As of March 2002, the species is listed as threatened in Minnesota and endangered (U.S. Endangered Species Act) in states other than Alaska. Review of the current status of all wolf populations in contiguous states is under way (U.S. Fish and Wildlife Service 2000). All Canadian populations are unlisted (Matthews and Moseley 1990). Note, however, that Canada currently lacks endangered species legislation at the federal level. Newly proposed taxonomic classifications and the success of reintroduction programs may require a reassessment of North American wolf populations (Wilson et al. 2000).

The estimated population of gray wolves in North America is about 50,000 (Table 23.2). An estimated 3000 wolves occupy the lower 48 states and approximately 6500 wolves inhabit Alaska. All wolf populations in the contiguous 48 states are increasing, whereas populations in Alaska are stable. Several new populations have become established in Montana, Wyoming, and Idaho through natural recolonization and reintroduction of wild-caught Canadian wolves (U.S. Fish and Wildlife Service 1987; Bangs and Fritts 1996; Bangs et al. 1998, 2001). Natural expansion (Licht and Fritts 1994) into Washington, Oregon, and California is probable (Carroll et al. 2001). Reintroduction of captive-bred Mexican wolves into Arizona is currently under way (U.S. Fish and Wildlife Service 1982; García-Moreno et al. 1996; Parsons 1998; Kalinowski et al. 1999; Brown and Parsons 2001; Paquet et al. 2001c). Additional reintroductions are being considered for northeastern United States (U.S. Fish and Wildlife Service 1987; Wydeven et al. 1998; Harrison and Chapin 1998; Mladenoff and Sickley 1998; Paquet et al.

2001b) and Colorado. In Minnesota, wolves occupy all suitable areas and have begun to colonize agricultural regions (U.S. Fish and Wildlife Service 1992). The occupied range of wolf populations and numbers of wolves in Wisconsin and Michigan are expanding. The number of gray wolves in Canada is estimated at 50,000–60,000 (Theberge 1991; Carbyn 1994), occupying 80% of their former range. The status of populations in the High Arctic is unknown because inventory and survey records are not available. Moreover, the area is huge and largely uninhabited by people. Canadian wolf populations most sensitive to human activities and/or natural events are the eastern wolf (*C. l. lycaon*), high arctic wolf (*C. l. arctos*) (van Zyll De Jong and Carbyn 1999), and all wolves inhabiting coastal islands.

Economic Status. Systematic economic assessments involving wild wolves are rare (Duffield and Neher 1996; Rasker and Hackman 1996). Most government agencies responsible for management of wolves have not conducted such analyses. Incomplete information is occasionally buried in reports but lacks the economic context necessary for evaluation. The emphasis is usually on financial costs incurred in lethal management of wolves to reduce predation on livestock and wild ungulates (Wagner et al. 1997). Costs and benefits associated with the sale of wolf fur are seldom available. Potential biological costs and ecological benefits of managing wolves are seldom quantified. Mech (1998), for example, evaluated the financial costs of wolf management options, but did not consider possible monetary benefits or other less tangible values. Clearly, comprehensive and rigorous economic evaluations are necessary for informed management decisions.

Good economic information is also needed to counter market forces that encourage destruction of formerly secure wolf habitats. Over the last 200 years, the North American landscape has been modified by an economy that ignores the environment or views it as an obstacle to overcome. Within this context, the decline of wolves has been considered a measure of the success of an enterprising economy. This attitude continues to prevail because only monetary benefits and costs associated with resource products are recognized in conventional marketplace transactions. Whereas conservation and restoration efforts are directed at improving current and future conditions, market interests usually discount future benefits and costs in favor of present consumption. Because information about the future is limited, a premium is placed on the present. Accordingly, short-term profits are usually favored over the uncertain profits of the future. Wolves can help society come to terms with the total value of biological diversity because their presence causes us to consider the comparative value of tangible and intangible aspects of the things that make up our lives (Pimlott et al. 1969).

Response of Wolves to Humans. To assess the effects of human influence accurately, we must know the uninfluenced norms and ranges. Because such information is lacking, the degree of influence by humans is imperfectly understood. Interpretation of the wolf–human interaction is confounded by multiple factors that influence how wolves use the landscape and respond to people (Mladenoff et al. 1995; Fritts and Carbyn 1995; Paquet et al. 1996; Carroll et al. 2001; Duke et al. 2001). The extent and intensity of response appear to vary with environmental conditions, social context, and disturbance history. Disturbance history is a critical concept in understanding the behavior of long-lived animals that learn through social transmission (Curatolo and Murphy 1986). Given the wolf's inherent behavioral variability, it is unlikely that all individuals, packs, or populations react equally to human-induced change or humans. Because researchers have developed no reasonable expression of individual differences, our understanding of wolf/human interactions is limited to the pack and population levels (Mladenoff et al. 1995, 1997, 1999; Paquet et al. 1996; Boyd-Heger 1997; Mladenoff and Sickley 1998; Carroll et al. 2001; Callaghan 2002).

Specific conditions that impair the distribution, movements, survival, or fecundity of wolves are believed to be highly variable. Although wolves are sensitive to human predation and harassment (Thiel 1985; Jensen et al. 1986; Mech et al. 1988; Fuller 1989; Mech 1989, 1993, 1995; Fuller et al. 1992; Thurber et al. 1994; Mladenoff et al. 1995, 1999; Paquet et al. 1996), we have limited empirical information on tolerance to human disturbance. Reactions of wolves to people likely depend on the type of human activity, where the interaction occurs, the distance between the activity (person) and the animal, cover, a wolf's experiences, inherited tolerance, and age/sex class. Wolves can habituate to human activities, at least partially, provided these activities are repetitive and innocuous. A wolf's experience with humans is important because habituation affects the wolf's sense of security. Studies have shown wolves avoid humans in time (Boitani 1982) and space (Mladenoff et al. 1995; Paquet et al. 1996) or are displaced via human-induced mortality (Paquet et al. 1996; Duke et al. 2001). Although human activities influence the distribution (Thiel 1985; Fuller et al. 1992; Paquet et al. 1996; Mladenoff et al. 1995) and survival of wolves (Mech et al. 1995b; Mladenoff et al. 1995; Fritts and Carbyn 1995; Paquet et al. 1996), human-caused mortality is consistently cited as the major cause of displacement (Fuller et al. 1992; Mech and Goyal 1993; Fritts and Carbyn 1995; Bangs et al. 2001).

Recent reports suggest wolves in Minnesota tolerate higher levels of disturbance than previously thought possible. Wolves, for example, are now occupying ranges formerly assumed to be marginal because of prohibitive road densities and high human populations (Mech 1993, 1995). Legal protection and changing human attitudes are cited as the critical factors in the wolf's ability to use areas that have not been wolf habitat for decades. Several studies suggest adequate prey density is the main factor limiting wolves where they are present and tolerated by humans (Keith 1983; Fuller et al. 1992; Mech 1993, 1996; Fuller and Murray 1998). If wolves are not persecuted, they seem able to occupy areas of greater human activity than previously assumed (Fuller et al. 1992; Mech 1993; Fritts et al. 1994; Fritts and Carbyn 1995). Based on these observations, Mech (1995) commented that misconceptions about the gray wolf's inherent ability to tolerate human activity encourage unwarranted protectionism.

Nonetheless, wolves continue to occur most often where road density and human population are low (Fuller et al. 1992; Mladenoff et al. 1995; Paquet et al. 1996; Callaghan 2002). Gray wolves from the Great Lakes region of the United States and Canada may have hybridized with coyotes (Kolenosky 1971; Schmitz and Kolenosky 1985a, 1985b; Schmitz and Lavigne 1987; Wayne et al. 1991, 1992; Lehman et al. 1991; M. T. Theberge et al. 1996) and red wolves (Wilson et al. 2000; Fascione et al. 2001), which may affect their behavior (Fox 1971, 1975) and their relationship with humans. Consequently, extrapolating information from Minnesota, Michigan, Minnesota, and Ontario may be inappropriate for other parts of North America. For example, wolves in the Rocky Mountains show no introgression of coyote genes (Arjo and Pletscher 1999). Moreover, the fact that wolves are using areas of greater human activity suggests dispersers or marginalized individuals and packs are being pushed into lower quality habitats (Mladinoff et al. 1999). This suggests that wolves occupy habitat closer to humans only if necessary to acquire life requisites (Paquet et al. 1996; Woodroffe 2000). A similar phenomenon has been observed with grizzly bears (Mattson et al. 1987). An alternative explanation is that the social flexibility of wolves allows the species to adapt their behavior to survive in human-altered habitats. For example, Eurasian wolves have become more secretive where they coexist with people, adopting a strategy of nocturnal scavenging (Boitani 1992).

Use of Habitats. Essential to any evaluation of the relationship between wolves and humans is an understanding of which habitats are inherently attractive to wolves. A general assumption is that habitat use is strongly related to availability of ungulate prey (Huggard 1991; Keith 1983), ease of travel (Cowan 1947; Mech 1970; Peek et al. 1991; Paquet et al. 1996), availability of den sites (Chapman 1977), and availability of rendezvous areas (Theberge 1969; Carbyn 1975a; Mech 1970). Evidence from field studies strongly suggests that when habitat is very attractive, wolves move closer to human activities because of the benefit. Conversely, when quality of habitat is low, displacement is greater because security risks outweigh advantages (Boitani 1982; Paquet et al. 1996).

In recent years, researchers have used geographic information systems and spatial radio-collar and wolf occurrence data to assess the importance of landscape-scale factors in defining favorable wolf habitat (Mladenoff et al. 1995, 1999; Paquet et al. 1996; Meriggi et al. 1996; Massolo and Meriggi 1998; Corsi et al. 1999; Callaghan 2002). These and earlier studies (e.g., Meriggi et al. 1991; Fritts et al. 1994; Bangs et al. 1998) agree that an adequate prey base, existence of sufficient protected areas, and absence or low occurrence of livestock are necessary to maintain wolf populations south of the treeline. Wolf presence also depends on areas with forest cover, few roads, and low human density. Wolves thrive in areas with high ungulate densities, but tend toward extirpation in areas with high densities of livestock because of conflicts with ranchers (Bangs et al. 1998). Protected and public lands encourage wolf presence, likely because there are fewer lethal encounters with humans (Forbes and Theberge 1996a). Some authors (e.g., Mech 1995) maintain that such areas are the least accessible to humans, and that the lack of human presence is the most important variable in predicting wolf viability.

We are aware of only five studies that have systematically and explicitly examined the landscape relationship of wolves and humans (Mladenoff et al. 1995, 1999; Paquet et al. 1996; Theberge et al. 1996; Corsi et al. 1999; Callaghan 2002). Observed patterns of displacement suggest the presence of humans repulses wolves, although a strong attraction to highly preferred habitats increases a wolf's tolerance for disturbance. As conditions become less favorable, the quality of habitat likely takes on greater importance. In the northern Great Lakes (Mladenoff et al. 1995), human population density was much lower in pack territories than in nonpack areas. Wolf pack territories also had more public land, forested areas with at least some evergreens, and lower proportions of agricultural land. Notably, no difference was detected between white-tailed deer densities in pack territories and nonpack areas. Overall, wolves selected areas most remote from human influence, with <1.54 humans/km^2 and <0.15 km roads/km^2. Most wolves in Minnesota (88%) were in townships with <0.70 km roads/km^2 and <4 humans/km^2 or with <0.50 km^2 and <8 humans/km^2. In Italy, absence of wolves was related to human density, road density, urban areas, cultivated areas, and cattle and pig density. However, because human density, road density, and urbanized areas were highly intercorrelated, no specific human effect was established (Corsi et al. 1999). In the Bow River Valley, Alberta, the selection or avoidance of particular habitat types was related to human use levels and habitat potential (Paquet et al. 1996). Changes in patterns of habitat use were evident when human activity exceeded 100 people/month. Nearly complete alienation of wolves occurred when >10,000 people/month used an area, regardless of habitat suitability. In portions of the Bow Valley where high elk abundance was associated with high road and/or human population density, wolves were completely absent. Several studies have suggested that ungulate prey seek out predator-free refugia to avoid predation by wolves (Mech 1977c; Holt 1987; Hebblewhite 2000). These changes can lead to different intrinsic rates of growth for ungulates using different habitat patches. Species other than wolves and their prey also may be affected by these human-induced changes in predator–prey relationships.

Use of Den and Rendezvous Sites. Wolves are usually intolerant of humans near dens and pups (Chapman 1977). Researchers, however, have successfully observed den and rendezvous sites without apparent disturbance to the wolves (Joslin 1966, 1967; Theberge 1969; Carbyn 1975a; Mech 1987; 1988). Wolves characteristically respond to human presence near their pups by barking and howling, leaving the area, moving the pups, or deserting the home site (Chapman 1977). Most pups in the presence of humans retreat to the den. Severity of disturbance is the most critical factor influencing desertion of home sites. In only 1 of 51 den site disturbances examined by Chapman (1977) were pups abandoned by the pack. Low-intensity disturbance at den sites seems unlikely to affect the fitness of a wolf population. The seriousness of human disturbance, however, is ultimately a human judgment. Consequently, any alteration of the normal activities of wolves at home sites may be judged by some to be undesirable.

Wolf dens within 1.0 km of established centers of human activity were usually permanently abandoned. Dens within 2.4 km of roads or campgrounds, however, were frequently used by wolves (Chapman 1977). Wolves pups in Banff National Park detected road construction activity from >4 km away and remained in their den until the construction stopped (P. Paquet, unpublished data). Avoidance of human activity seems reduced where artificial sources of food such as garbage dumps are present (Chapman 1977; Paquet et al. 1996; Krizan 1997) or where substantial innocuous human activity occurs. Mortality of pups because of human disturbance has not been reported. Ballard et al. (1987) suggested pup survival is not decreased by den site disturbances. In some areas, wolves may be adapting to human activity and disturbances (Mech 1995). Thiel et al. (1998) reported wolves tolerating human activity near dens and rendezvous sites with pups. These include moss harvesting in Wisconsin and military maneuvers and road construction in Minnesota. In Montana, a pack of wolves kept its pups in a rendezvous site 0.8 km from a helicopter logging operation during summer 1994 (Jimenez 1995).

Chapman (1977) concluded that human disturbance of wolves at levels characteristic of National Parks does not significantly affect survival of wolf pups or seriously alter ecological relationships between wolves and their prey. Denali National Park, Alaska, maintains closures around wolf dens and rendezvous sites, including some rendezvous sites that have not been used in many years. Regulations allowing wolf reintroduction in Yellowstone National Park allow closing areas to human visitation for 1.6 km around active dens from 15 March to 1 July (Fritts et al. 1994). The Wisconsin Department of Natural Resources recommends closing areas within 100 m of dens and restricting use from 100 to 800 m from dens from 1 March to 31 July (A. P. Wydeven, Wisconsin Department of Natural Resources, Park Falls, pers. commun., 2000). Banff National Park, Alberta, protects dens by closing entire watersheds to humans from 1 May to 1 August. Peter Lougheed Provincial Park, Alberta, and Riding Mountain National Park, Manitoba, restrict human activities within 1.6 km of known dens. Based on a comprehensive study of North American Parks, Chapman (1977) recommended a protective buffer around den and rendezvous sites of 2.4 km radius in open country.

Influence of Linear Developments. Ensured connectivity of quality habitats is important for survival of large carnivores (Beier 1993; Doak 1995; Noss et al. 1996), especially for those that face a high risk of mortality from humans or vehicles when traveling across settled landscapes (Noss 1992; Beier 1993). Many human activities associated with linear corridors (highways, secondary roads, railways, power line corridors, gas lines, and seismic lines) fragment wolf ranges and result in the death of wolves (De Vos 1949; Fuller 1989; Paquet et al. 1996; Krizan 1997; Callaghan 2002). Such developments also may be physical and/or psychological impediments to wolf movement (Paquet and Callaghan 1996; Paquet et al. 1996; Duke et al. 2001).

Conversely, linear developments may enhance movements of wolves. Thurber et al. (1994) speculated that roads with low human activity provide easy travel corridors for wolves. Specifically, they serve as conduits or travel corridors for wolves (Paquet et al. 1996; Paquet and Callaghan 1996). The provision of artificial travel corridors, however, should not be construed as a positive development. The overwhelming effects of vehicle collisions and other human-caused mortality factors resulting from increased access, including poaching, hunting, and trapping, outweigh the benefits (Jalkotzy et al. 1997). Moreover, use by wolves of linear corridors increases predation pressure on woodland caribou (James and Stuart-Smith 2000), elk, deer, and coyotes (Paquet 1989).

Studies in Wisconsin, Michigan, Ontario, and Minnesota have shown a strong relationship between road density and the absence of wolves (Thiel 1985; Jensen et al. 1986; Mech et al. 1988; Fuller 1989). Persistent occupancy of wolves is usually assured at mean road densities below 0.6–0.70 km/km^2 (Thiel 1985; Jensen et al. 1986; Mech et al. 1988; Fuller 1989; Mech 1989; Fuller et al. 1992; Shelley and Anderson 1995; Boyd-Heger 1997; Frair 1999; but see Merrill 2000).

These thresholds, however, probably do not apply to areas where public access is restricted and activities are regulated (Merrill 2000). To a point, road density may be less important than the mortality of wolves caused by humans using roads. Overall, lethality of a road is a function of frequency of use, traffic speed, and the attitude/motivation of drivers (Merrill 2000). Mech (1989) reported wolves using an area with a road density of 0.76 km/km^2, but it was next to a large, roadless area. He speculated that excessive mortality experienced by wolves in the roaded area was compensated for by individuals that dispersed from the adjacent roadless area. Wolves on Prince of Wales Island, Alaska, used areas with road densities >0.58 km/km^2. Core areas, however, were mostly in the least densely roaded areas of the home range. In addition, wolf activity that does occur in densely roaded areas occurs primarily at night.

A study in Alaska concluded that wolves avoid heavily used roads and areas inhabited by humans, despite low human-caused wolf mortality (Thurber et al. 1994). Landscape level analysis in northern Great Lakes region found mean road density was much lower in pack territories (0.23 km/km^2 in 80% use area) than in random nonpack areas (0.74 km/km^2) or the region overall (0.71 km/km^2). Few areas used by wolves had a road density of >0.45 km/km^2 (Mladenoff et al. 1995). In the Rocky Mountains, wolves killed by humans died closer to roads than wolves that died of other causes (Boyd and Pletscher 1999). However, the relationship of road density and wolf distribution is not well understood in mountainous topography (Singleton 1995; Paquet et al. 1996; Boyd-Heger 1997; Callaghan 2002). In complex mountain terrain, wolves must use valley bottoms, where roads converge with high-quality habitat.

There are several plausible explanations for the absence of wolves in densely roaded areas. Wolves may behaviorally avoid these areas depending on the type of use the road receives (Thurber et al. 1994; Person 2001). In other instances, their absence may be a direct result of higher mortality in areas with greater road density (Van Ballenberghe et al. 1975; Mech 1977a; Berg and Kuehn 1982). Roads and other linear developments provide people access to remote regions, which allows them to deliberately, accidentally, or incidentally kill wolves (Van Ballenberghe et al. 1975; Mech 1977a; Berg and Kuehn 1982; Fuller 1989; Mech 1989). On Prince of Wales Island, Alaska, researchers reported a significant increase in wolf mortality in areas where road density was >0.25 km/km^2 (Person 2001).

RESEARCH NEEDS

The future of wild wolves in an increasingly human-dominated world depends on informed science-based management and decisions (Mech et al. 1997; Theberge et al. 1996). Much of the essential science remains ambiguous however, because most field research is outside the domain of reproducibility and control. Moreover, the impossibility of accounting for unknown processes and variables adds to scientific uncertainty. Wolf researchers, therefore, may need to acknowledge ignorance and emphasize uncovering the limits to reliable knowledge rather than proving existing knowledge to be correct.

In North America, extensive field research on wolf biology and ecology has been carried out in the last 30 years. Information regarding social behavior and physiology of wolves, however, has come largely from captive studies, supplemented by incidental observations in the wild (Haber 1996). Most field studies have emphasized wolf–ungulate interactions primarily as related to humans. Largely overlooked has been the natural role of the wolf as a summit predator (McLaren and Peterson 1994; Terborgh et al. 1999), especially in complex multipredator and multiprey systems. The relationship of wolves with domestic livestock has had only limited scientific inquiry. In addition, a preponderance of quality research from Alaska and the Great Lakes region of the United States and Canada has slanted our understanding of wolves to those environments. Because wolves from the Great Lakes region may be hybrids of gray wolf, red wolf, and coyotes, we should be cautious about generalizing to other regions of North America. Further genetic investigations are needed to more clearly understand the history of hybridization and the implications for gray wolf recovery efforts.

Future research efforts should focus on coexistence of wolves and humans and the ecological processes that sustain them. Because of public interest in the species, refinements in wolf population estimates (Créte and Messier 1987; Fuller and Sampson 1988; Fuller and Snow 1988) and better documentation of numbers of wolves killed annually are required. True sustained-yield management requires more emphasis on qualitative biological features to determine the extent to which wolves and other species with evolutionary histories as predators rather than as prey should be harvested. Research is needed on viability of small wolf populations in human-dominated landscapes, spatial assessments of source–sink populations, and human dimensions of wolf management. The latter should address social issues, economics, traditional knowledge, hunting, trapping, and the potential for aggressive interactions of wolves with humans. More behavioral and ecological research needs to be carried out in agricultural areas, mountainous topography, coastal and island environments, and arid locations such as the southwestern United States and the Arctic. Finally, the role of the wolf in the ecological community needs to be clarified. Specifically, interactions with other carnivores and top-down effects on ungulates and vegetation should be studied. Ironically, the species once regarded as a threat to our survival is turning out to be a test of how likely we are to achieve sustainability and coexistence with the elements that sustain us.

LITERATURE CITED

Adams, L. G., B. W. Dale, and L. D. Mech. 1995. Wolf predation on caribou calves in Denali National Park, Alaska. Pages 245–60 *in* L. N. Carbyn, S. H. Fritts, and D. R. Seip, eds. Ecology and conservation of wolves in a changing world. Canadian Circumpolar Institute, Edmonton, Alberta.

Alexander, S., C. Callaghan, P. C. Paquet, and N. Waters. 1996. GIS Predictive Model of Habitat Use by Wolves (*Canis lupus*). CD Rom publication, GIS '96: Ten years of Excellence, GIS World 1996 Conference Proceedings. Vancouver, BC.

Alexander, S., P. C. Paquet, and N. Waters. 1991. Playing God with GIS: Uncertainty in wolf habitat suitability models. Pages 449–53 *in* GIS '97 Integrating Spatial Information Technology for Tomorrow. Eleventh Annual GIS World Conference Proceedings. Vancouver, BC.

Allen, D. 1979. The wolves of Minong: Their vital role in a wild community. Houghton Mifflin, Boston.

Allen, G. M., and T. Barbour. 1937. The Newfoundland wolf. Journal of Mammalogy 18:229–34.

Allendorf, F. W. 1983. Isolation, gene flow, and genetic differentiation among populations. Pages 51–65 *in* C. M. Schonewald-Cox, ed. Genetics and conservation. Benjamin Cummings, London.

Archibald, W. R., D. Janz, and K. Atkinson. 1991. Wolf control: A management dilemma. Transactions of the North American Wildlife and Natural Resources Conference 56:497–511.

Arjo, W. M., and D. H. Pletscher. 1999. Behavioral responses of coyotes to wolf recolonization in northwestern Montana. Canadian Journal of Zoology 77:1919–27.

Asa, C. S. 1997. Hormonal and experiential factors in the expression of social and parental behavior in canids. Pages 129–49 *in* N. G. Solomon and J. A. French, eds. Cooperative breeding in mammals. Cambridge University Press, Cambridge.

Asa, C. S., and C. Valdespino. 1998. Canid reproductive biology: An integration of proximate mechanisms and ultimate causes. American Zoologist 38:251–59.

Asa, C. S., L. D. Mech, and U. S. Seal. 1985a. The use of urine, faeces and anal gland secretions in scent marking by a captive wolf (*Canis lupus*) pack. Animal Behavior 33:1034–36.

Asa, C. S., E. K. Peterson, U. S. Seal, and L. D. Mech. 1985b. Deposition of anal-sac secretions by captive wolves (*Canis lupus*). Journal of Mammalogy 66:89–93.

Asa, C. S., L. D. Mech, U. S. Seal, and E. D. Plotka. 1990. The influence of social and endocrine factors on urine-marking by captive wolves (*Canis lupus*). Hormones and Behavior 24:497–509.

Atwell, G. 1964. Wolf predation on calf moose. Journal of Mammalogy 45:313–14.

Avise, J. C., and R. M. Ball, Jr. 1990. Principles of genealogical concordance in species concepts and biological taxonomy. Pages 43–67 *in* D. Futuyma and J. Antonovics, eds. Oxford surveys in evolutionary biology. Oxford University Press, Oxford.

Bailey, T. N., E. E. Bangs, and R. O. Peterson. 1995. Exposure of wolves to canine parvovirus and distemper on the Keni National Wildlife Refuge, Kenai Peninsula, Alaska 1976–1988. Pages 441–46 *in* L. N. Carbyn, S. H. Fritts, and D. R. Seip, eds. Ecology and conservation of wolves in a changing world (Occasional Publication No. 35). Canadian Circumpolar Institute, Edmonton, Alberta.

Ballard, W. B. 1980. Brown bear kills gray wolf. Canadian Field–Naturalist 94:91.

Ballard, W. B. 1982. Gray wolf–brown bear relationships in the Nelchina Basin of south-central Alaska. Pages 71–80 *in* F. H. Harrington and P. C. Paquet, eds. Wolves of the world: Perspectives of behavior, ecology, and conservation. Noyes, Park Ridge, NJ.

Ballard, W. B., and J. R. Dau. 1983. Characteristics of gray wolf, *Canis lupus*, den and rendezvous sites in southcentral Alaska. Canadian Field-Naturalist 97:299–302.

Ballard, W. B., and V. Van Ballenberghe. 1997. Predator/prey relationships. Pages 247–73 *in* A. W. Franzman and C. C. Schwartz, eds. Ecology and management of the North American moose. Smithsonian Institution Press, Washington, DC.

Ballard, W. B., R. Farnell, and R. O. Stephenson. 1983. Long distance movement by gray wolves, *Canis lupus*. Canadian Field-Naturalist 97:333.

Ballard, W. B., J. S. Whitman, and C. L. Gardner. 1987. Ecology of an exploited wolf population in south-central Alaska. Wildlife Monographs 98:1–54.

Ballard, W. B., L. A. Ayres, P. R. Krausman, D. J. Reed, and S. G. Fancy. 1997. Ecology of wolves in relation to a migratory caribou herd in northwest Alaska. Wildlife Monographs 135:1–47.

Banfield, A. W. F. 1954. Preliminary investigation of the barren ground caribou. Part 2. Life history, ecology and utilization (Wildlife Management Bulletin, Series 1, No. 10B). Canadian Wildlife Service, Ottawa.

Banfield, A. W. F. 1974. The mammals of Canada. University of Toronto Press, Toronto.

Bangs, E. E., and S. H. Fritts. 1996. Reintroducing the gray wolf into central Idaho and Yellowstone National Park. Wildlife Society Bulletin 24:402–13.

Bangs, E. E., S. H. Fritts, J. A. Fontaine, D. W. Smith, K. M. Murphy, C. M. Mack, and C. C. Niemeyer. 1998. Status of gray wolf restoration in Montana, Idaho, and Wyoming. Wildlife Society Bulletin 26:785–98.

Bangs, E., J. Fontaine, M. Jimenez, T. Meier, C. Niemeyer, D. Smith, K. Murphy, D. Guernsey, L. Handegard, M. Collinge, R. Krischke, J. Shivik, C. Mack, I. Babcock, V. Asher, and D. Domenici. 2001. Gray wolf restoration in the northwestern United States. Endangered Species Update 18:147–52.

Banville, D. 1983. Status and management of wolves in Quebec. Pages 41–43 *in* L. N. Carbyn, ed. Wolves in Canada and Alaska: Their status, biology, and management (Report Series 45). Canadian Wildlife Service, Ottawa.

Barnett, D. T., and T. J. Stohlgren. 2001. Aspen persistence near the National Elk Refuge and Gros Ventre Valley elk feedgrounds of Wyoming, USA. Landscape Ecology 16:569–80.

Barrette, C., and F. Messier. 1980. Scent-marking in free ranging coyotes, *Canis latrans*. Animal Behavior 28:814–19.

Beier, P. 1993. Determining minimum habitat areas and habitat corridors for cougars. Conservation Biology 7:94–108.

Berg, W. E., and R. A. Chesness. 1978. Ecology of coyotes in northern Minnesota. Pages 229–47 *in* M. Bekoff, ed. Coyotes: Biology, behavior, and management. Academic Press, New York.

Berg, W. E., and D. W. Kuehn. 1982. Ecology of wolves in north-central Minnesota. Pages 4–11 *in* F. H. Harrington and P. C. Paquet, eds. Wolves of the world: Perspectives of behavior, ecology, and conservation. Noyes, Park Ridge, NJ.

Bergerud, A. T. 1974. Decline of caribou in North America following settlement. Journal of Wildlife Management 38:757–70.

Bergerud, A. T., and J. P. Elliot. 1986. Dynamics of caribou and wolves in northern British Columbia. Canadian Journal of Zoology 64:1515–29.

Bergerud, A. T., and J. P. Elliott. 1998. Wolf predation in a multiple-ungulate system in northern British Columbia. Canadian Journal of Zoology 76:1551–69.

Bergerud, A. T., W. Wyett, and B. Snider. 1983. The role of wolf predation in limiting a moose population. Journal of Wildlife Management 47:977–88.

Bergerud, A. T., H. E. Butler, and D. R. Miller. 1984. Antipredator tactics of calving caribou: Dispersion in mountains. Canadian Journal of Zoology 62:1566–75.

Bjorge, R. R., and J. R. Gunson. 1989. Wolf, *Canis lupus*, population characteristics and prey relationships near Simonette River, Alberta. Canadian Field-Naturalist 103:327–34.

Boertje, R. D., and R. O. Stephenson. 1992. Effects of ungulate availability on wolf reproduction potential in Alaska. Canadian Journal of Zoology 70:441–43.

Boertje, R. D., P. Valkenburg, and M. E. McNay. 1996. Increases in moose, caribou, and wolves following wolf control in Alaska. Journal of Wildlife Management 60:474–89.

Bogan, M. A., and P. Mehlhop. 1983. Systematic relationships of gray wolves (*Canis lupus*) in southwestern North America (Occasional Papers, No. 1). Museum of Southwestern Biology, Albuquerque, NM.

Boitani, L. 1982. Wolf management in intensively used areas of Italy. Pages 158–72 *in* F. H. Harrington and P. C. Paquet, eds. Wolves of the world: Perspectives of behavior, ecology, and conservation. Noyes, Park Ridge, NJ.

Boitani, L. 1992. Wolf research and conservation in Italy. Biological Conservation 61:125–32.

Boutin, S. 1992. Predation and moose population dynamics: A critique. Journal of Wildlife Management 56:116–27.

Boyd, D. K., and M. D. Jimenez. 1994. Successful rearing of young by wild wolves without mates. Journal of Mammalogy 75:14–17.

Boyd, D. K., and G. K. Neale. 1992. An adult cougar, *Felis concolor*, killed by gray wolves, *Canis lupus*, in Glacier National Park, Montana. Canadian Field-Naturalist 106:524–25.

Boyd, D. K., L. B. Secrest, and D. H. Pletscher. 1992. A wolf, *Canis lupus*, killed in an avalanche in southwestern Alberta. Canadian Field-Naturalist 106:526.

Boyd, D. K., D. H. Pletscher, R. R. Ream, and M. W. Fairchild. 1994. Prey characteristics of colonizing wolves and hunters in the Glacier National Park area. Journal of Wildlife Management 58:289–95.

Boyd, D. K., P. C. Paquet, S. Donelon, R. R. Ream, D. H. Pletscher, and C. C. White. 1995. Transboundary movements of a recolonizing wolf population in the Rocky Mountains. Pages 135–41 *in* L. N. Carbyn, S. H. Fritts, and D. R. Seip, eds. Ecology and conservation of wolves in a changing world (Occasional Publication No. 35). Canadian Circumpolar Institute, Edmonton, Alberta.

Boyd, D. K., and Pletscher, D. H. 1999. Characteristics of dispersal in a colonizing wolf population in the central Rocky Mountains. Journal of Wildlife Management 63:1094–1108.

Boyd-Heger, D. K. 1997. Dispersal, genetic relationships, and landscape use by colonizing wolves in the Central Rocky Mountains. Ph.D. Dissertation, University of Montana, Missoula.

Brand, C. J., M. J. Pybus, W. B. Ballard, and R. O. Peterson. 1995. Infectious and parasitic diseases of the gray wolf and their potential effects on wolf populations in North America. Pages 419–29 *in* L. N. Carbyn, S. H. Fritts, and D. R. Seip, eds. Ecology and conservation of wolves in a changing world (Occasional Publication No. 35). Canadian Circumpolar Institute, Edmonton, Alberta.

Brewster, W. G., and S. H. Fritts. 1995. Taxonomy and genetics of the gray wolf in western North America. Pages 353–74 *in* L. N. Carbyn, S. H. Fritts, and D. R. Seip, eds. Ecology and conservation of wolves in a changing world (Occasional Publication No. 35). Canadian Circumpolar Institute, Edmonton, Alberta.

Bromley, R. G. 1973. Fishing behavior of a wolf on the Taltson River, Northwest Territories. Canadian Field-Naturalist 87:301–3.

Brown, W. M., and D. R. Parsons. 2001. Restoring the Mexican gray wolf to the mountains of the Southwest. Pages 169–86 *in* D. S. Maehr, R. F. Noss, and J. L. Larkin, eds. Large mammal restoration: Ecological and sociological challenges in the 21st century. Island Press, Covelo, CA.

Burkholder, B. L. 1959. Movements and behavior of a wolf pack in Alaska. Journal of Wildlife Management 23:1–11.

Burkholder, B. L. 1962. Observations concerning wolverine. Journal of Mammalogy 43:263–64.

Buss, M., and M. de Almeida. 1998. A review of wolf and coyote status and policy in Ontario, December 1997. Ontario Ministry of Natural Resources, Fish and Wildlife Branch, Queens Printer for Ontario, Toronto.

Callaghan, C. J. 2002. The ecology of gray wolf (*Canis lupus*) habitat use, survival, and persistence of gray wolves in the central Rocky Mountains. Ph.D. Dissertation, University of Guelph, Guelph, Ontario, Canada.

Cameron, A. W. 1958. Mammals of the islands in the Gulf of St. Lawrence (Bulletin 154). National Museum of Canada, Ottawa.

Carbyn, L. N. 1974. Wolf population fluctuations in Jasper National Park, Canada. Biological Conservation 6:94–101.

Carbyn, L. N. 1975a. Wolf predation and behavioral interactions with elk and other ungulates in an area of high prey density. Ph.D. Dissertation, University of Toronto, Toronto.

Carbyn, L. N. 1975b. Factors influencing activity patterns of ungulates at mineral licks. Canadian Journal of Zoology 53:377–84.

Carbyn, L. N. 1981. Territory displacement in a wolf population with abundant prey. Journal of Mammalogy 62:193–95.

Carbyn, L. N. 1982a. Incidence of disease and its potential role in the population dynamics of wolves in Riding Mountain National Park, Manitoba. Pages 106–16 *in* F. H. Harrington and P. C. Paquet, eds. Wolves of the world: Perspectives of behaviour, ecology, and conservation. Noyes, Park Ridge, NJ.

Carbyn, L. N. 1982b. Coyote population fluctuations and spatial distribution in relation to wolf territories in Riding Mountain National Park, Manitoba. Canadian Field-Naturalist 96:176–83.

Carbyn, L. N. 1983a. Management of non-endangered wolf populations in Canada. Acta Zoologica Fennica 174:239–43.

Carbyn, L. N. 1983b. Wolf predation on elk in Riding Mountain National Park, Manitoba. Journal of Wildlife Management 47:963–76.

Carbyn, L. N. 1987. Gray wolf and red wolf. Pages 359–76 *in* M. Novak, J. A. Baker, M. E. Obbard, and B. Malloch, eds. Wild furbearer management and conservation in North America. Ontario Trappers Association and Ministry of Natural Resources, Toronto.

Carbyn, L. N. 1994. Canada's 50,000 wolves. International Wolf 4:3–8.

Carbyn, L. N., and T. Trottier. 1987. Responses of bison on their calving grounds to predation by wolves in Wood Buffalo National Park. Canadian Journal of Zoology 65:2072–78.

Carbyn, L. N., and T. Trottier. 1988. Descriptions of wolf attacks on bison calves in Wood Buffalo National Park. Arctic 41:297–302.

Carbyn, L. N., S. M. Oosenbrug, and D. W. Anions. 1993. Wolves, bison and the dynamics related to the Peace–Athabasca Delta in Canada's Wood Buffalo National Park (Circumpolar Research Series No. 4). Canadian Circumpolar Institute, Edmonton, Alberta.

Carroll, C., R. F. Noss, N. H Schumaker, and P. C. Paquet. 2001. Is the return of wolf, wolverine, and grizzly bear to Oregon and California biologically feasible? Pages 25–46 *in* D. Maehr, R. F. Noss, and J. Larkin, eds. Large mammal restoration: Ecological and sociological implications. Island Press, Washington, DC.

Chapman, R. C. 1977. The effects of human disturbance on wolves (*Canis lupus*). M.S. Thesis, University of Alaska, Fairbanks.

Chapman, R. C. 1978. Rabies: Decimation of a wolf pack in arctic Alaska. Science 201:365–67.

Chesser, R. K. 1983. Isolation by distance: Relationship to the management of genetic resources. Pages 66–77 *in* C. M. Schonewald-Cox, ed. Genetics and conservation. Benjamin Cummings, London.

Chiarelli, A. B. 1975. The chromosomes of the canidae. Pages 40–53 *in* M. W. Fox, ed. The wild canids: Their systematics, behavioral ecology and evolution. Van Nostrand Reinhold, New York.

Ciucci, P., and L. D. Mech. 1992. Selection of wolf dens in relation to winter territories in northeastern Minnesota. Journal of Mammalogy 73:899–905.

Clark, K. R. F. 1971. Food habits and behaviour of the tundra wolf on central Baffin Island. Ph.D. Dissertation, University of Toronto, Toronto.

Cluff, H. D., and D. L. Murray. 1995. Review of wolf control methods in North America. Pages 491–504 *in* L. N. Carbyn, S. H. Fritts, and D. R. Seip, eds. Ecology and conservation of wolves in a changing world (Occasional Publication No. 35). Canadian Circumpolar Institute, Edmonton, Alberta.

Clutton-Brock, J., A. C. Kitchener, and J. H. Lynch. 1994. Changes in the skull morphology of the arctic wolf, *Canis lupus arctos,* during the twentieth century. Journal of Zoology (London) 233:19–36.

Corsi, F., E.Dupré , and L. Boitani. 1999. A large-scale model of wolf distribution in Italy for conservation planning. Conservation Biology 13:150–59.

Coscia, E. M. 1993. Swimming and aquatic play by timber wolf, *Canis lupus,* pups. Canadian Field-Naturalist 107:361–62.

Coscia, E. M. 1995. Ontogeny of timber wolf vocalizations: Acoustic properties and behavioral contexts. Ph. D. Dissertation, Dalhousie University, Halifax, Nova Scotia, Canada.

Coscia, E. M., D. P. Phillips, and J. C. Fentress. 1991. Spectral analysis of neonatal wolf *Canis lupus* vocalizations. Bioacoustics 3:275–93.

Cowan, I. McT. 1947. The timber wolf in the Rocky Mountain National Parks of Canada. Canadian Journal of Research 25:139–74.

Cowan, I. McT. 1949. Rabies as a possible population control of arctic Canidae. Journal of Mammalogy 30:396–98.

Cowan, I. McT. 1951. The diseases and parasites of big game mammals of western Canada. Proceedings of the Annual Game Convention 5:37–64.

Cowan, I. McT. 1954. The occurrence of the Pleistocene wolf *Canis dirus* in the Rocky Mountains of central Alberta. Canadian Field-Naturalist 68:44.

Crête, M. 1999. The distribution of deer biomass in North America supports the hypothesis of exploitation ecosystems. Ecology Letters 2:223–27.

Crête, M., and M. Manseau. 1996. Natural regulation of cervidae along a 1000 km latitudinal gradient: Change in trophic dominance. Evolutionary Ecology 10:51–62.

Crète, M., and F. Messier. 1987. Evaluation of indices of gray wolf, *Canis lupus* density in hardwood–conifer forests of southwestern Québec. Canadian Field-Naturalist 101:147–52.

Criddle, S. 1947. Timber wolf dens and pups. Canadian Field-Naturalist 61:115.

Crisler, L. 1958. Arctic wild. Harper and Bros., New York.

Curatolo, J. A., and S. M. Murphy. 1986. The effects of pipelines, roads, and traffic on the movements of caribou, *Rangifer tarandus*. Canadian Field-Naturalist 100:218–24.

Custer, J. W., and D. B. Pence. 1981. Host–parasite relationships in the wild canids of North America. II. Pathology of infections diseases in the genus *Canis*. Pages 760–845 *in* J. A. Chapman and D. Pursley, eds. Proceedings of the worldwide furbearer conference. Frostburg, MD.

Dale, B. W., L. G. Adams, and R. T. Bowyer. 1994. Functional response of wolves preying on barren-ground caribou in a multiple-prey ecosystem. Journal of Animal Ecology 63:644–52.

Dale, B. W., L. G. Adams, and R. T. Bowyer. 1995. Winter wolf predation in a multiple ungulate prey system, Gates of the Arctic National Park, Alaska. Pages 223–30 *in* L. N. Carbyn, S. H. Fritts, and D. R. Seip, eds. Ecology and conservation of wolves in a changing world (Occasional Publication No. 35). Canadian Circumpolar Institute, Edmonton, Alberta.

Darimont, C. T., and P. C. Paquet. 2000. The gray wolves (*Canis lupus*) of British Columbia's coastal rainforests: Findings from year 2000 pilot study and conservation assessment. Prepared for Raincoast Conservation Society, Victoria, British Columbia, Canada.

Darimont, C. T., and P. C. Paquet. 2002. The Gray Wolves, *Canis lupus,* of British Columbia's Central and North Coast: Distribution and conservation assessment. Canadian Field-Naturalist 116:416–422.

Darimont C. T., T. E. Reimchen, and P. C. Paquet. 2003. Foraging behaviour by gray wolves on salmon streams in coastal British Columbia. Canadian Journal of Zoology 81:349–353.

DelGuidice, G. D. 1998. Surplus killing of white-tailed deer by wolves in north-central Minnesota. Journal of Mammalogy 79:227–35.

De Vos, A. 1949. Timber wolves (*Canis lupus lycaon*) killed by cars on Ontario highways. Journal of Mammalogy 30:197.

Doak, D. F. 1995. Source–sink models and the problem of habitat degradation: General models and applications to the Yellowstone grizzly. Conservation Biology 9:1370–79.

Duffield, J. W., and C. J. Neher. 1996. Economics of wolf recovery in Yellowstone National Park. Transactions of the North American Wildlife and Natural Resources Conference 61:285–92.

Duke, D. L., M. Hebblewhite, P. C. Paquet, C. Callaghan, and M. Percy. 2001. Restoration of a large carnivore corridor in Banff National Park, Alberta. Pages 261–76 *in* D. Maehr, R. F. Noss, and J. Larkin, eds. Large mammal restoration: Ecological and sociological implications. Island Press, Washington, DC.

Eberhardt, L. L. 1977. Optimal policies for conservation of large mammals, with special reference to marine ecosystems. Environmental Conservation 4:205–12.

Eberhardt, L. L. 1997. Is wolf predation ratio-dependent? Canadian Journal of Zoology 75:1940–44.

Eberhardt, L. L. 1998. Applying difference equations to wolf predation. Canadian Journal of Zoology 76:380–86.

Eberhardt, L. L., and R. O. Peterson. 1999. Predicting the wolf–prey equilibrium point. Canadian Journal of Zoology 77:494–98.

Edmonds, E. J. 1988. Population status, distribution, and movements of woodland caribou in west central Alberta. Canadian Journal of Zoology 66:817–26.

Elliot, D. G., and M. Wong. 1972. Acid phosphatase, handy enzyme that separates the dog from the wolf. Acta Biologica et Medica Germanica 28:957–62.

Emmons, L. E. 1984. Geographic variation in densities and diversities of non-flying mammals in Amazonia. Biotropica 16:210–22.

Epstein, H. 1971. The origins of the domestic animals of Africa, Vol. 1. Africans, New York.

Ewer, R. F. 1973. The carnivores. Cornell University Press, Ithaca, NY.

Farnell, R., and J. McDonald. 1988. The influence of wolf predation on caribou mortality in Yukon's Finlayson caribou herd. Proceedings of the North American Caribou Workshop 3:52–70.

Fascione, N., L. G. L. Osborn, S. R. Kendrot, and P. C. Paquet. 2001. *Canis soupus*: Eastern wolf genetics and its implications for wolf recovery in the northeast United States. Endangered Species Update 18:159–63.

Festa-Bianchet, M., M. Urquhart, and K. G. Smith. 1994. Mountain goat recruitment: Kid production and survival to breeding age. Canadian Journal of Zoology 72:22–27.

Fleck, E. S., and A. Gunn. 1982. Characteristics of three barren-ground caribou calving grounds in the Northwest Territories (Report 7). Northwest Territories Wildlife Service.

Forbes, G. J., and J. B. Theberge. 1992. Importance of scavenging on moose by wolves in Algonquin Park, Ontario. Alces 28:235–41.

Forbes, G. J., and J. B. Theberge. 1995. Influences of a migratory deer herd on wolf movements and mortality in and near Algonquin Park, Ontario. Pages 303–14 *in* L. N. Carbyn, S. H. Fritts, and D. R. Seip, eds. Ecology and conservation of wolves in a changing world (Occasional Publication No. 35). Canadian Circumpolar Institute, Edmonton, Alberta.

Forbes, G. J., and J. B. Theberge. 1996a. Cross-boundary management of Algonquin Park wolves. Conservation Biology 10:1091–97.

Forbes, G. J., and J. B. Theberge. 1996b. Response by wolves to prey variation in central Ontario. Canadian Journal of Zoology 74:1511–20.

Forbes, L. S. 1989. Prey defences and predator handling behavior: The dangerous prey hypothesis. Oikos 55:155–58.

Forbes, S. H., and D. K. Boyd. 1996. Genetic variation of naturally colonizing wolves in the central Rocky Mountains. Conservation Biology 10:1082–90.

Forbes, S. H., and D. K. Boyd. 1997. Genetic structure and migration in native and reintroduced Rocky Mountain wolf populations. Conservation Biology 11:1226–34.

Formozov, A. N. 1946. Snow cover as an integral factor of the environment and its importance in the ecology of mammals and birds (Occasional Publication No. 1). Boreal Institute, University of Alberta, Edmonton, Canada.

Forshner, S. A. 2000. Population dynamics and limitation of wolves (*Canis lupus*) in the Greater Pukaskwa Ecosystem, Ontario. M.Sc. Thesis, University of Alberta, Edmonton, Canada.

Fox, M. W. 1971. Behaviour of wolves, dogs and related canids. Harper and Row, New York.

Fox, M. W. 1973. Social dynamics of three captive wolf packs. Behavior 47:290–301.

Fox, M. W. 1975. Evolution of social behavior in canids. Pages 429–59 *in* M. W. Fox, ed. The wild canids: Their systematics, behavioral ecology, and evolution. Van Nostrand Reinhold, New York.

Fox, M. W., and J. A. Cohen. 1977. Canid communication. Pages 728–48 *in* T. A. Sebeok, ed. How animals communicate. Indiana University Press, Bloomington.

Fox, J. L., and G. P. Streveler. 1986. Wolf predation on mountain goats in southeastern Alaska. Journal of Mammalogy 67:192–95.

Frair, J. L. 1999. Crosing paths: Gray wolves and highway in the Minnesota–Wisconsin border region. M.S. Thesis, University of Wisconsin, Stevens Point.

Francis, C. S. 1960. Wolves feeding on water bugs and minnows. Blue Jay 18:139.

Frenzel, L. D. 1974. Occurrence of moose in food of wolves as revealed by scat analyses: A review of North American studies. Naturaliste Canadien 101:467–79.

Friis, L. K. 1985. An investigation of subspecific relationships of the grey wolf, *Canis lupus,* in British Columbia. M.Sc. Thesis, University of Victoria, Victoria, British Columbia, Canada.

Frijlink, J. H. 1977. Patterns of wolf pack movements prior to kills as read from tracks in Algonquin Park, Ontario, Canada. Bijdragen Tot De Dierkunde 47:131–37.

Fritts, S. H. 1983. Record dispersal by a wolf from Minnesota. Journal of Mammalogy 64:166–67.

Fritts, S. H. 1993. The downside of wolf recovery. International Wolf 3:24–26.

Fritts, S. H., and L. N. Carbyn. 1995. Population viability, nature reserves, and the outlook for gray wolves conservation in North America. Restoration Ecology 3:26–38.

Fritts, S. H., and L. D. Mech. 1981. Dynamics, movements and feeding ecology of a newly protected wolf population in northwestern Minnesota. Wildlife Monographs 80:1–79.

Fritts, S. H., E. E. Bangs, and J. F. Gore. 1994. The relationship of wolf recovery to habitat conservation and biodiversity in the northwestern United States. Landscape and Urban Planning 28:23–32.

Fritts, S. H., E. E. Bangs, J. A. Fontaine, W. G. Brewster, and J. F. Gore. 1995. Restoring wolves to the northern Rocky Mountains of the United States. Pages 107–26 *in* L. N. Carbyn, S. H. Fritts, and D. R. Seip, eds. Ecology and conservation of wolves in a changing world (Occasional Publication No. 35). Canadian Circumpolar Institute, Edmonton, Alberta.

Fuller, T. K. 1989. Population dynamics of wolves in north-central Minnesota. Wildlife Monographs 105:1–41.

Fuller, T. K. 1990. Dynamics of a declining white-tailed deer population in north-central Minnesota. Wildlife Monographs 110:1–37.

Fuller, T. K. 1991a. Effect of snow depth on wolf activity and prey selection in north central Minnesota. Canadian Journal of Zoology 69:283–87.

Fuller, T. K. 1991b. Erratum: Effect of snow depth on wolf activity and prey selection in north central Minnesota. Canadian Journal of Zoology 69:821.

Fuller, T. K., and L. B. Keith. 1980. Wolf population dynamics and prey relationships in northeastern Alberta. Journal of Wildlife Management 44:583–602.

Fuller, T. K., and L. B. Keith. 1981. Non-overlapping ranges of coyotes and wolves in northeastern Alberta. Journal of Mammalogy 62:403–5.

Fuller, T. K., and D. L. Murray. 1998. Biological and logistical explanations of variation in wolf population density. Animal Conservation 1:153–57.

Fuller, T. K., and B. A. Sampson. 1988. Evaluation of a simulated howling survey for wolves. Journal of Wildlife Management 52:60–63.

Fuller, T. K., and W. J. Snow. 1988. Estimating winter wolf densities using radiotelemetry data. Wildlife Society Bulletin 16:367–70.

Fuller, T. K., W. E. Berg, G. L. Radde, M. S. Lenarz, and G. B. Joselyn. 1992. A history and current estimate of wolf distribution and numbers in Minnesota. Wildlife Society Bulletin 20:42–55.

Fuller, W. A. 1962. The biology and management of bison of Wood Buffalo National Park. Canadian Wildlife Service Wildlife Management Bulletin Series 1:1–52.

Fuller, W. A., and N. S. Novakowski. 1955. Wolf control operations, Wood Buffalo National Park, 1951–1952 (Management Bulletin Series 1, No. 11). Canadian Wildlife Service, Ottawa.

Ganong, W. F. 1908. On the occurrence of the wolf in New Brunswick. Bulletin Natural History Society, New Brunswick 26:30–35.

Garceau, P. 1960. Reproduction, growth and mortality of wolves. Alaska Department of Fish and Game. Annual Report Progress, Investigation Project 1:458–83.

García-Moreno, J., M. D. Matocq, M. S. Roy, E. Geffen, and R. K. Wayne. 1996. Relationships and genetic purity of the endangered Mexican wolf based on analysis of microsatellite loci. Conservation Biology 10:376–89.

Gasaway, W. C., R. O. Stephenson, J. L. Davis, P. E. K. Shepherd, and O. E. Burris. 1983. Interrelationships of wolves, prey, and man in interior Alaska. Wildlife Monographs 84:1–50.

Gasaway, W. C., R. D. Boertje, D. V. Grangaard, D. G. Kellyhouse, R. O. Stephenson, and D. G. Larsen. 1992. The role of predation in limiting moose at low densities in Alaska and Yukon and implications for conservation. Wildlife Monographs 120:1–59.

Gauthier, D. A., and J. B. Theberge. 1986. Wolf predation in the Burwash caribou herd, southwest Yukon. Proceedings of the International Reindeer/Caribou Symposium 4:137–44.

Gauthier, D. A., and J. B. Theberge. 1987. Wolf predation. Pages 119–27 *in* M. Novak, J. A. Baker, M. E. Obbard, and B. Malloch, eds. Wild furbearer management and conservation in North America. Ontario Ministry of Natural Resources, Toronto.

Gehring, T. M. 1993. Adult black bear, *Ursus americanus,* displaced from a kill by a wolf, *Canis lupus,* pack. Canadian Field-Naturalist 107:373–74.

Gehring, T. M. 1995. Winter wolf movements in northwestern Wisconsin and east-central Minnesota: Aquantitative approach. M.S. Thesis, University of Wisconsin, Stevens Point.

Gese, E. M., and L. D. Mech. 1991. Dispersal of wolves (*Canis lupus*) in northeastern Minnesota, 1969–1989. Canadian Journal of Zoology 69:2946–55.

Glanz, W. E. 1982. The terrestrial mammal fauna of Barro Colorado Island. Census and long-term changes. Pages 239 *in* E. G. Leigh, Jr., A. S. Rand, and D. M. Windsor, eds. The ecology of a tropical forest: Seasonal rhythms and long-term changes. Smithsonian Institution Press, Washington, DC.

Goldman, E. A. 1944. The wolves of North America, Part 2. Classification of wolves. Pages 387–507 *in* S. P. Young and E. A. Goldman, eds. The wolves of North America. Dover, New York.

Grace, E. S. 1976. Interactions between men and wolves at an Arctic outpost on Ellesmere Island. Canadian Field-Naturalist 90:149–56.

Gray, A. P. 1972. Mammalian hybrids. Commonwealth Agricultural Bureau, Slough, UK.

Gray, D. R. 1970. The killing of a bull muskox by a single wolf. Arctic 23:197–99.

Gray, D. R. 1983. Interactions between wolves and muskoxen on Bathurst Island, Northwest Territories, Canada. Acta Zoologica Fennica 174:255–57.

Green, H. U. 1951. The wolves of Banff National Park. Canadian Department of Natural Resources and Development, National Parks Branch, Ottawa.

Gunson, J. R. 1983. Status and management of wolves in Alberta. Pages 25–29 *in* L. N. Carbyn, ed. Wolves in Canada and Alaska: Their status, biology, and management (Report Series 45). Canadian Wildlife Service, Ottawa.

Gunson, J. R. 1995. Wolves: Their characteristics, history, prey relationships and management in Alberta. Alberta Environmental Protection.

Gunson, J. R., and R. M. Nowak. 1979. Largest gray wolf skulls found in Alberta. Canadian Field-Naturalist 93:308–9.

Haber, G. C. 1968. The social structure and behavior of an Alaskan wolf population. M.A. Thesis, Northern Michigan University, Marquette.

Haber, G. C. 1977. Socio-ecological dynamics of wolves and prey in a subarctic ecosystem. Ph.D. Dissertation, University of British Columbia, Vancouver, Canada.

Haber, G. C. 1996. Biological, conservation and ethical implications of exploiting and controlling wolves. Conservation Biology 10:1068–81.

Haight, R. G., and L. D. Mech. 1997. Computer simulation of vasectomy for wolf control. Journal of Wildlife Management 61:1023–31.

Haight, R. G., D. J. Mladenoff, and A. P. Wydeven. 1998. Modeling disjunct gray wolf populations in semi-wild landscapes. Conservation Biology 12:879–88.

Hairston, N. G., F. E. Smith, and L. B. Slobodkin. 1960. Community structure, population control, and competition. American Naturalist 94:421–25.

Hall, E. R. 1981. The mammals of North America, Vol. II, 2nd ed. John Wiley, New York.

Hall, E. R., and K. R. Kelson. 1952. Comments on the taxonomy and geographic distribution of some North American marsupials, insectivores, and carnivores. University of Kansas Museum of Natural History Publications 5:319–41.

Hall, E. R., and K. R. Kelson. 1959. The mammals of North America, Vol. 2. Ronald Press, New York.

Harrington, F. H. 1975. Response parameters of elicited wolf howling. Ph.D. Dissertation, State University of New York, Stony Brook.

Harrington, F. H. 1978. Ravens attracted to wolf howling. Condor 80:236–37.

Harrington, F. H. 1981. Urine-marking and caching behavior in the wolf. Behaviour 76:280–88.

Harrington, F. H. 1986. Timber wolf howling playback studies: Discrimination of pup from adult howls. Animal Behavior 34:1575–77.

Harrington, F. H. 1987. Aggressive howling in wolves. Animal Behavior 35:7–12.

Harrington, F. H. 1989. Chorus howling by wolves: Acoustic structure, pack size, and the Beau Geste effect. Bioacoustics 2:117–36.

Harrington, F. H., and L. D. Mech. 1978a. Howling at two Minnesota wolf pack summer homesites. Canadian Journal of Zoology 56:2024–28.

Harrington, F. H., and L. D. Mech. 1978b. Wolf vocalization. Pages 109–32 *in* R. L. Hall and H. S. Sharp, eds. Wolf and man: Evolution in parallel. Academic Press, New York.

Harrington, F. H., and L. D. Mech. 1979. Wolf howling and its role in territory maintenance. Behaviour 68:207–49.

Harrington, F. H., and L. D. Mech. 1983. Wolf pack spacing: Howling as a territory-independent spacing mechanism in a territorial population. Behavioral Ecology and Sociobiology 12:161–68.

Harrington, F. H., P. C. Paquet, J. Ryon, and J. C. Fentress. 1982. Monogamy in wolves: A review of the evidence. Pages 209–22 *in* F. H. Harrington and P. C. Paquet, eds. Wolves of the world: Perspectives of behavior, ecology, and conservation. Noyes, Park Ridge, NJ.

Harrington, F. H., L. D. Mech, and S. H. Fritts. 1983. Pack size and wolf pup survival: Their relationship under varying ecological conditions. Behavioral Ecology and Sociobiology 13:19–26.

Harrington, F. H., J. Ryon, J. C. Fentress, and S. Bragdon. 1986. Scent rubbing in wolves (*Canis lupus*): The effect of novelty. Canadian Journal of Zoology 64:573.

Harrison, D. J., and T. G. Chapin. 1998. Extent and connectivity of habitat for wolves in eastern North America. Wildlife Society Bulletin 26:767–75.

Hatter, I. W., and D. W. Janz. 1994. Apparent demographic changes in black-tailed deer associated with wolf control on northern Vancouver Island. Canadian Journal of Zoology 72:878–84.

Hayes, R. D. 1995. Numerical and functional responses of wolves and regulation of moose in the Yukon. M.S. Thesis, Simon Fraser University, Burnaby, British Columbia, Canada.

Hayes, R., and G. Gunson. 1995. Status and management of wolves in Canada. Pages 21–35 *in* L. N. Carbyn, S. H. Fritts, and D. R. Seip, eds. Ecology and conservation of wolves in a changing world (Occasional Publication No. 35). Canadian Circumpolar Institute, Edmonton, Alberta, Canada.

Hayes, R. D., and A. Harestad. 2000a. Wolf functional response and regulation of moose in the Yukon. Canadian Journal of Zoology 78:60–66.

Hayes, R. D., and A. Harestad. 2000b. Demography of a recovering wolf population in the Yukon. Canadian Journal of Zoology 78:36–48.

Hayes, R. D., A. M. Baer, and D. G. Larsen. 1991. Population dynamics and prey relationships of an exploited and recovering wolf population in the southern Yukon (Final Report TR-91-1). Yukon Fish and Wildlife Branch.

Hayes, R. D., A. M. Baer, U. Wotschikowsky, and A. S. Harestad. 2000. Kill rate by wolves on moose in the Yukon. Canadian Journal of Zoology 78:49–59.

Heard, D. C. 1983. Historical and present status of wolves in the Northwest Territories. Pages 44–47 *in* L. N. Carbyn, ed. Wolves in Canada and Alaska: Their status, biology, and management (Report Series 45). Canadian Wildlife Service, Ottawa.

Heard, D. C. 1992. The effect of wolf predation and snow cover on musk-ox group size. American Naturalist 139:190–204.

Heard, D. C., and T. M. Williams. 1992. Distribution of wolf dens on migratory caribou ranges in the Northwest Territories, Canada. Canadian Journal of Zoology 70:1504–10.

Hebblewhite, M. 2000. Wolf and elk predator–prey dynamics in Banff National Park. M.S. Thesis, University of Montana, Missoula.

Hillis, T. L., and F. F. Mallory. 1996a. Fetal development in wolves, *Canis lupus*, of the Keewatin District, Northwest Territories, Canada. Canadian Journal of Zoology 74:2211–18.

Hillis, T. L., and F. F. Mallory. 1996b. Sexual dimorphism in wolves (*Canis lupus*) of the Keewatin District, Northwest Territories, Canada. Canadian Journal of Zoology 74:721–25.

Hilton-Taylor, comp. 2000. Red list of threatened species. International Union for the Conservation of Nature, Species Survival Commission, Gland, Switzerland.

Hoefs, M., H. Hoefs, and D. Burles. 1986. Observations on Dall sheep (*Ovis dalli*)–grey wolf, (*Canis lupus pambasileus*) in the Kluane. Canadian Field-Naturalist 100:78–84.

Holt, R. D. 1987. Prey communities in patchy environments. Oikos 50:276–90.

Horejsi, B. L., G. E. Hornbeck, and R. M. Raine. 1984. Wolves, *Canis lupus*, kill female black bear, *Ursus americanus*, in Alberta. Canadian Field-Naturalist 98:368–69.

Hornbeck, G. E., and B. L. Horejsi. 1986. Grizzly bear, *Ursus arctos*, usurps wolf, *Canis lupus*, kill. Canadian Field-Naturalist 100:259–60.

Huggard, D. J. 1991. Prey selectivity of wolves in Banff National Park. M.Sc. Thesis, University of British Columbia, Vancouver, Canada.

Huggard, D. J. 1993a. Effect of snow depth on predation and scavenging by gray wolves. Journal of Wildlife Management 57:382–88.

Huggard, D. J. 1993b. Prey selectivity of wolves in Banff National Park. I. Prey species. Canadian Journal of Zoology 71:130–39.

Huggard, D. J. 1993c. Prey selectivity of wolves in Banff National Park, II. Age, sex, and condition of elk. Canadian Journal of Zoology 71:140–47.

Iwasa, Y., M. Higashi, and N. Yamamura. 1981. Prey distribution as a factor determinging the choice of optimal foraging strategy. American Naturalist 117:710–23.

Jalkotzy, M. G., P. I. Ross, and M. D. Nasserden. 1997. The effects of linear developments on wildlife: A review of selected scientific literature. Prepared for Canadian Association of Petroleum Producers, Calgary, Alberta.

James, D. D. 1983. Seasonal movements, summer food habits, and summer predation rates of wolves in northwest Alaska. M. S. Thesis, University of Alaska, Fairbanks.

James, A. R. C., and A. K. Stuart-Smith. 2000. Distribution of caribou and wolves in relation to linear corridors. Journal of Wildlife Management 64:154–59.

Jedrzejewska, B., W. Jedrzejewski, A. N. Bunevich, L. Milkowski, and Z. Krasinski. 1997. Factors shaping population densities and increase rates of ungulates in Bialoweiza Primeval Forest (Poland and Belarus) in the 19th and 20th centuries. Acta Theriologica 42:399–451.

Jenkins, K. J., and R. G. Wright. 1988. Resource partitioning and competition among cervids in the northern Rocky Mountains. Journal of Applied Ecology 25:11–24.

Jensen, W. F., T. K. Fuller, and W. L. Robinson. 1986. Wolf, *Canis lupus*, distribution on the Ontario–Michigan border near Sault Ste. Marie. Canadian Field-Naturalist 100:363–66.

Jimenez, M. D. 1995. Tolerance and respect help nine-mile wolves recover. International Wolf 5:18–19.

Johnson, M. R., D. K. Boyd, and D. H. Pletscher. 1994. Serology of canine parvovirus and canine distemper in relation to wolf (*Canis lupus*) pup mortalities. Journal of Wildlife Diseases 30:270–73.

Jolicoeur, P. 1959. Multivariate geographical variation in the wolf *Canis lupus* L. Evolution 13:283–99.

Jolicoeur, P. 1975. Sexual dimorphism and geographical distance as factors of skull variation in the wolf *Canis lupus* L. Pages 54–61 *in* M. W. Fox, ed. The wild canids: Their systematics, behavioral ecology and evolution. Van Nostrand Reinhold, New York.

Joly, D. O., and F. Messier. 2000. A numerical response of wolves to bison abundance in Wood Buffalo National Park, Canada. Canadian Journal of Zoology 78:1101–4.

Jordan, P. A., P. C. Shelton, and D. L. Allen. 1967. Numbers, turnover, and social structure of the Isle Royale wolf population. American Zoologist 7:233–52.

Joslin, P. W. 1966. Summer activities of two timber wolf (*Canis lupus*) packs in Algonquin Park. M.S. Thesis, University of Toronto, Toronto.

Joslin, P. W. 1967. Movements and home sites of timber wolves in Algonquin Provincial Park. American Zoologist 7:279–88.

Kalinowski, S. T., P. W. Hedrick, and P. S. Miller. 1999. No inbreeding depression observed in Mexican and red wolf captive breeding programs. Conservation Biology 13:1371–77.

Kay, C. E. 1990. Yellowstone's northern elk herd: A critical evaluation of the "natural regulation" paradigm. Ph.D. Dissertation, Utah State University, Logan.

Kay, C. E. 1997. Is aspen doomed? Journal of Forestry 95:4–11.

Kay, C. E., and F. H. Wagner. 1994. Historic condition of woody vegetation on Yellowstone's "natural regulation" paradigm. Pages 159–69 *in* D. Despain, ed. Plants and their environment. Proceedings first biennial conference. Greater Yellowstone Ecosystem. U.S. National Park Service.

Kazmierczak, J. J., E. C. Burgess, and T. E. Amundson. 1988. Susceptibility of the gray wolf *Canis lupus* to infection with the Lyme disease agent *Borrelia burgdorferi*. Journal of Wildlife Diseases 24:522–27.

Keith, L. B. 1983. Population dynamics of wolves. Pages 66–77 *in* L. N. Carbyn, ed. Wolves in Canada and Alaska: Their status, biology, and management (Report Series 45). Canadian Wildlife Service, Ottawa.

Kelsall, J. P. 1957. Continued barren-ground caribou studies (Wildlife Management Bulletin Series 1, No. 12). Canadian Wildlife Service, Ottawa.

Kelsall, J. P. 1960. Cooperative studies of barren-ground caribou. (Wildlife Management Bulletin Series 1, No. 15). Canadian Wildlife Service, Ottawa.

Kelsall, J. P. 1968. The migratory caribou of Canada. Canadian Wildlife Service Monograph 3:1–340.

Kennedy, P. K., M. L. Kennedy, P. L. Clarkson, and I. S. Liepins. 1991. Genetic variability in natural populations of the gray wolf, *Canis lupus*. Canadian Journal of Zoology 69:1183–88.

Kleiman, D. G. 1966. Scent marking in the Canidae. Symposium of the Zoological Society of London 18:167–78.

Klein, D. R. 1995. The introduction, increase, and demise of wolves on Coronation Island, Alaska. Pages 275–82 *in* L. N. Carbyn, S. H. Fritts, and D. R. Seip, eds. Ecology and conservation of wolves in a changing world (Occasional Publication No. 35). Canadian Circumpolar Institute, Edmonton, Alberta.

Kohira, M., and E. A. Rexstad. 1995. Diets of wolves, *Canis lupus,* in logged and unlogged forests of southeastern Alaska. Canadian Field-Naturalist 111:429–35.

Kolenosky, G. B. 1971. Hybridization between wolf and coyote. Journal of Mammalogy 52:446–49.

Kolenosky, G. B. 1972. Wolf predation on wintering deer in east-central Ontario. Journal of Wildlife Management 36:357–69.

Kolenosky, G. B. 1983. Status and management of wolves in Ontario. Pages 35–40 *in* L. N. Carbyn, ed. Wolves in Canada and Alaska: Their status, biology, and management (Report Series 45). Canadian Wildlife Service, Ottawa.

Kolenosky, G. B., and D. H. Johnston. 1967. Radio-tracking timber wolves in Ontario. American Zoologist 7:289–303.

Kolenosky, G. B., and R. O. Standfield. 1975. Morphological and ecological variation among gray wolves (*Canis lupus*) of Ontario, Canada. Pages 62–72 *in* M. W. Fox, ed. The wild canids. Van Nostrand Reinhold, New York.

Krizan, P. 1997. The effects of human development, landscape features, and prey density on the spatial use of wolves (*Canis lupus*) on the north shore of Lake Superior. M.S. Thesis, Acadia University, Wolfville, Nova Scotia, Canada.

Krizan, P. 2000. Blastomycosis in a free ranging wolf, *Canis lupus,* on the north shore of Lake Superior, Ontario. Canadian Field-Naturalist 114:491–93.

Kunkel, K. E., and D. H. Pletscher. 1999. Species-specific population dynamics of cervids in a multipredator ecosystem. Journal of Wildlife Management 63:1082–93.

Kunkel, K. E., T. K. Ruth, D. H. Pletscher, and M. G. Hornocker. 1999. Winter prey selection by wolves and cougars in and near Glacier National Park, Montana. Journal of Wildlife Management 63:901–10.

Kuyt, E. 1972. Food habits and ecology of wolves on barren-ground caribou range in the Northwest Territories (Report Series 21). Canadian Wildlife Service, Ottawa.

Kuyt, E., B. E. Johnson, and R. C. Drewien. 1981. A wolf kills a juvenile whooping crane. Blue Jay 392:116–19.

Larsen, D. G., D. A. Gauthier, and R. L. Markel. 1989. Causes and rate of moose mortality in the southwest Yukon. Journal of Wildlife Management 53:548–57.

Larter, N. C., A. R. E. Sinclair, and C. C. Gates. 1994. The response of predators to an erupting bison, *Bison bison athabascae,* population. Canadian Field-Naturalist 108:318–27.

Lawhead, B. E. 1983. Wolf den site characteristics in the Nelchina Basin, Alaska. M.S. Thesis, University of Alaska, Fairbanks.

Lehman, N., A. Eisenhauer, K. Hansen, L. D. Mech, R. O. Peterson, P. J. P. Gogan, and R. K. Wayne. 1991. Introgression of coyote mitochondrial DNA into sympatric North American gray wolf populations. Evolution 45:104–19.

Lehman, N., P. Clarkson, L. D. Mech, Y. Meier, and R. K. Wayne. 1992. A study of the genetic relationship within and among wolf packs using DNA fingerprinting and mitochondrial DNA. Behavioral Ecology and Sociobiology 30:83–94.

Lentfer, J. W., and D. K. Sanders. 1973. Notes on the captive wolf (*Canis lupus*) colony, Barrow, Alaska. Canadian Journal of Zoology 51:623–27.

Licht, D. S., and S. H. Fritts. 1994. Gray wolf (*Canis lupus*) occurrences in the Dakotas. American Midland Naturalist 132:74–81.

Lohr, C., and W. B. Ballard. 1996. Historical occurrence of wolves, *Canis lupus,* in the Maritime Provinces. Canadian Field-Naturalist 10:607–10.

Maargaard, L., and J. Graugaard. 1994. Female arctic wolf, *Canis lupus arctos,* mating with domestic dogs, *Canis familiaris,* in northeast Greenland. Canadian Field-Naturalist 108:374–75.

Macdonald, D. W. 1985. The carnivores: Order Carnivora. Pages 619–722 *in* R. E. Brown and D. W. Macdonald, eds. Social odours in mammals. Clarendon Press, Oxford.

Massolo, A., and A. Meriggi. 1998. Factors affecting habitat occupancy by wolves in northern Apennines (northern Italy): A model of habitat suitability. Ecography 21:97–107.

Matteson, M. Y. 1992. Denning ecology of wolves in northwest Montana and southern Canadian Rockies. M.S. Thesis, University of Montana, Missoula.

Matthews, J. R., and C. J. Moseley, eds. 1990. The official World Wildlife Fund guide to endangered species of North America, Vol. 1. Beacham, Washington, DC.

Mattson, D. J., R. R. Knight, and B. M. Blanchard. 1987. The effects of developments and primary roads on grizzly bear habitat use in Yellowstone National Park, Wyoming. International Conference on Bear Research and Management 7:259–73.

McLaren, B. E., and R. O. Peterson. 1994. Wolves, moose, and tree rings on Isle Royale. Science 266:1555–58.

Mech, L. D. 1966. The wolves of Isle Royale (Fauna Series Number 7). U.S. National Park Service.

Mech, L. D. 1970. The wolf: The ecology and behavior of an endangered species. Natural History Press, Garden City, NY.

Mech, L. D. 1972. Spacing and possible mechanisms of population regulation in wolves. American Zoologist 4:642.

Mech, L. D. 1973. Wolf numbers in the Superior National Forest of Minnesota. (USDA Forest Service Research Paper NC-97). North Central Forest Experiment Station, St. Paul, MN.

Mech, L. D. 1974. *Canis lupus*. Mammalian Species 37:1–6.

Mech, L. D. 1975. Disproportionate sex ratios of wolf pups. Journal of Wildlife Management 39:737–40.

Mech, L. D. 1977a. Productivity, mortality, and population trends of wolves in northeastern Minnesota. Journal of Mammalogy 58:559–74.

Mech, L. D. 1977b. Population trend and winter deer consumption in a Minnesota wolf pack. Pages 55–83 *in* R. L. Phillips and C. Jonkel, eds. Proceedings of the 1975 predator symposium. Montana Forest and Conservation Experiment Station, University of Montana, Missoula.

Mech, L. D. 1988. Wolf-pack buffer zones as prey reservoirs. Science 198:320–21

Mech, L. D. 1984.

Mech, L. D. 1986. Wolf numbers and population trend in central Superior National Forest, 1967–1985 (Research Paper No. NC-270). U. S. Department of Agriculture Forest Service.

Mech, L. D. 1987. At home with the arctic wolf. National Geographic 171:562–94.

Mech, L. D. 1988. The arctic wolf: Living with the pack. Voyageur Press, Stillwater, MN.

Mech, L. D. 1989. Wolf population survival in an area of high road density. American Midland Naturalist 121:387–89.

Mech, L. D. 1991. The way of the wolf. Voyageur Press, Stillwater, MN.

Mech, L. D. 1993. Updating our thinking on the role of human activity in wolf recovery (Research Information Bulletin 57). U.S. Fish and Wildlife Service, St. Paul, MN.

Mech, L. D. 1994a. Buffer zones of territories of gray wolves as regions of intraspecific strife. Journal of Mammalogy 75:199–202.

Mech, L. D. 1994b. Regular and homeward travel speeds of arctic wolves. Journal of Mammalogy 75:741–42.

Mech, L. D. 1995. The challenge and opportunity of recovering wolf populations. Conservation Biology 9:270–78.

Mech, L. D. 1996. A new era for carnivore conservation. Wildlife Society Bulletin 24:397–401.

Mech, L. D. 1998. Estimated costs of maintaining a recovered wolf population in agricultural regions of Minnesota. Wildlife Society Bulletin 26:817–22.

Mech, L. D. 1999a. Killing of a muskox, *Ovibos moschatus,* by two wolves, *Canis lupus,* and subsequent caching. Canadian Field-Naturalist 113:673–74.

Mech, L. D. 1999b. Alpha status, dominance, and division of labor in wolf packs. Canadian Journal of Zoology 77:1196–1203.

Mech, L. D. 2000. A record large wolf, *Canis lupus,* pack in Minnesota. Canadian Field-Naturalist 114:504–5.

Mech, L. D., and L. G. Adams. 1999. [Review of] The wolves of Denali. Journal of Wildlife Management 63:412–14.

Mech, L. D., and L. D. Frenzel, eds. 1971. Ecological studies of the timber wolf in northeastern Minnesota (Research Paper NC-52), USDA Forest Service, North Central Forest Experiment Station, St. Paul, MN.

Mech, L. D., and S. H. Fritts. 1987. Parvovirus and heartworm found in Minnesota wolves. Endangered Species Technical Bulletin 12:5–6.

Mech, L. D., and S. M. Goyal. 1993. Canine parvovirus effect on wolf population change and pup survival. Journal of Wildlife Diseases 29:330–33.

Mech, L. D., and S. M. Goyal. 1995. Effects of canine parvovirus on a wolf population in Minnesota. Journal of Wildlife Management 59:565–70.

Mech, L. D., and H. H. Hertel. 1983. An eight-year demography of a Minnesota wolf pack. Acta Zoologica Fennica 174:249–50.

Mech, L. D., and P. D. Karns. 1977. Role of the wolf in a deer decline in the Superior National Forest (Research Paper NC-148). U.S. Forest Service.

Mech, L. D., and H. J. Kurtz. 1999. First record of coccidiosis in wolves, *Canis lupus*. Canadian Field-Naturalist 113:305–6.

Mech, L. D., and M. E. Nelson. 1986. Relationship between snow depth and gray wolf predation on white-tailed deer. Journal of Wildlife Management 50:471–74.

Mech, L. D., and M. E. Nelson. 1989. Polygyny in a wild wolf pack. Journal of Mammalogy 70:675–76.

Mech, L. D., and M. E. Nelson. 1990. Evidence of prey-caused mortality in three wolves. American Midland Naturalist 123:207–8.

Mech, L. D., and J. M. Packard. 1990. Possible use of wolf den over several centuries. Canadian Field Naturalist 104:484–85.

Mech, L. D., S. H. Fritts, G. L. Radde, and W. J. Paul. 1988. Wolf distribution and road density in Minnesota. Wildlife Society Bulletin 16:85–87.

Mech, L. D., S. H. Fritts, and D. Wagner. 1995a. Minnesota wolf dispersal to Wisconsin and Michigan. American Midland Naturalist 133:368–70.

Mech, L. D., T. J. Meier, J. W. Burch, and L. G. Adams. 1995b. Patterns of prey selection by wolves in Denali National Park, Alaska. Pages 231–43 *in* L. N. Carbyn, S. H. Fritts, and D. R. Seip, eds. Ecology and conservation of wolves in a changing world (Occasional Publication No. 35). Canadian Circumpolar Institute, Edmonton, Alberta.

Mech, L. D., S. H. Fritts, and M. E. Nelson. 1996. Wolf management in the 21st century: From public input to sterilization. Journal of Wildlife Diseases 1:195–98.

Mech, L. D., H. J. Kurtz, and S. Goyal. 1997. Death of wild wolf from canine parvoviral enteritis. Journal of Wildlife Diseases 33:321–32.

Mech, L. D., L. G. Adams, T. J. Meier, J. W. Burch, and B. W. Dale. 1998. The wolves of Denali. University of Minnesota Press, Minneapolis.

Medjo, D. C., and L. D. Mech. 1976. Reproductive activity in nine- and ten-month-old wolves. Journal of Mammalogy 57:406–8.

Meier, T. J., J. W. Burch, L. D. Mech, and L. G. Adams. 1995. Pack structure and genetic relatedness among wolf packs in a naturally-regulated population. Pages 293–302 *in* L. N. Carbyn, S. H. Fritts, and D. R. Seip, eds. Ecology and conservation of wolves in a changing world (Occasional Publication No. 35). Canadian Circumpolar Institute, Edmonton, Alberta.

Meleshko, D. 1986. Feeding habits of sympatric canids in an area of moderate ungulate density. M.S. Thesis, University of Alberta, Edmonton, Canada.

Mendelssohn, H. 1982. Wolves in Israel. Pages 173–95 *in* F. H. Harrington and P. C. Paquet, eds. Wolves of the world: Perspectives of behavior, ecology, and conservation. Noyes, Park Ridge, NJ.

Meriggi, A., P. Rosa, A. Brangi, and C. Matteucci. 1991. Habitat use and diet of the wolf in northern Italy. Acta Theriologica 36:141–51.

Meriggi, A., A. Brangi, C. Matteucci, and O. Sacchi. 1996. The feeding habits of wolves in relation to large prey availability in northern Italy. Ecography 19:287–95.

Merriam, H. R. 1964. The wolves of Coronation Island. Page 27 *in* G. Dahlgren, ed. Proceedings of the 15th Alaskan science conference. Alaska Division of the American Association for the Advancement of Science, College, AK.

Merrill, S. B. 2000. Road densities and wolf, *Canis lupus,* habitat suitability: An exception. Canadian Field-Naturalist 114:312–14.

Merti-Millhollen, A. S., P. A. Goodmann, and E. Klinghammer. 1986. Wolf scent marking with raised-leg urination. Zoo Biology 5:7–20.

Messier, F. 1984. Moose–wolf dynamics and the natural regulation of moose populations. Ph.D. Dissertation, University of British Columbia, Vancouver, Canada.

Messier, F. 1985a. Solitary living and extraterritorial movements of wolves in relation to social status and prey abundance. Canadian Journal of Zoology 63:239–45.

Messier, F. 1985b. Social organization, spatial distribution, and population density of wolves in relation to moose density. Canadian Journal of Zoology 63:1068–77.

Messier, F. 1987. Physical condition and blood physiology of wolves in relation to moose density. Canadian Journal of Zoology 65:91–95.

Messier, F. 1991. The significance of limiting and regulating factors on the demography of moose and white-tailed deer. Journal of Animal Ecology 60:377–93.

Messier, F. 1994. Ungulate population models with predation: A case study with the North American moose. Ecology 75:478–88.

Messier, F. 1995a. Is there evidence for a cumulative effect of snow depth on moose and deer populations? Journal of Animal Ecology 64:136–40.

Messier, F. 1995b. On the functional and numeric responses of wolves to changing prey density. Pages 187–97 *in* L. N. Carbyn, S. H. Fritts, and D. R. Seip, eds. Ecology and conservation of wolves in a changing world (Occasional Publication No. 35). Canadian Circumpolar Institute, Edmonton, Alberta.

Messier, F., and C. Barrette. 1985. The efficiency of yarding behavior by white-tailed deer as an antipredator strategy. Canadian Journal of Zoology 63:785–89.

Messier, F., and M. Crête. 1985. Moose–wolf dynamics and the natural regulation of moose populations. Oecologia 65:503–12.

Messier, F., and D. O. Joly. 2000. Comment: Regulation of moose populations by wolf predation. Canadian Journal of Zoology 78:506–10.

Miller, F. L. 1978. Interactions between men, dogs and wolves on western Queen Elizabeth Islands, Northwest Territories, Canada. Musk-ox 22:70–72.

Miller, F. L. 1993. Status of wolves in the Canadian Arctic Archipelago (Technical Report Series No. 173). Canadian Wildlife Service, Western and Northern Region, Alberta.

Miller, F. L., and F. D. Reintjes. 1995. Wolf-sightings on the Canadian arctic islands. Arctic 48:313–23.

Miller, F. L., A. Gunn, and E. Broughton. 1985. Surplus killing as exemplified by wolf predation on newborn caribou. Canadian Journal of Zoology 63:295–300.

Miller, M. E., G. C. Christensen, and H. E. Evans. 1964. Anatomy of the dog. W. B. Saunders, Philadelphia.

Milne, D. G., A. S. Harestad, and K. Atkinson. 1989. Diets of wolves on northern Vancouver Island. Northwest Science 63:83–86.

Minta, S. C., P. M. Kareiva, and A. P. Curlee. 1999. Carnivore research and conservation: Learning from history and theory. Pages 323–404 *in* T. K. Clark, A. P. Curlee, S. C. Minta, and P. M. Kareiva, eds. Carnivores in eocsystems: The Yellowstone experience. Yale University Press, New Haven, CT.

Mladenoff, D. J., and T. A. Sickley. 1998. Assessing potential gray wolf restoration in the northeastern United States: A spatial prediction of favorable habitat and potential population levels. Journal of Wildlife Management 62:1–10.

Mladenoff, D. J., T. A. Sickley, R. G. Haight, and A. P. Wydeven. 1995. A regional landscape analysis and prediction of favorable gray wolf habitat in the northern Great Lakes region. Conservation Biology 9:279–94.

Mladenoff, D. J., R. G. Haight, T. A. Sickley, and A. P. Wydeven. 1997. Causes and implications of species restoration in altered ecosystems: A spatial landscape projection of wolf population recovery. Bioscience 47:21–31.

Mladenoff, D. J., T. A. Sickley, and A. P. Wydeven. 1999. Predicting gray wolf landscape recolonization: Logistic regression models vs. new field data. Ecological Applications 9:37–44.

Mulders, R. 1997. Geographic variation in the cranial morphology of the wolf (*Canis lupus*) in northern Canada. M.Sc. Thesis, Laurentian University, Sudbury, Ontario, Canada.

Munro, J. A. 1947. Observations of birds and mammals in central British Columbia (Occasional Paper No. 6). British Columbia Provincial Museum.

Murie, A. 1944. The wolves of Mount McKinley (Fauna of the national parks of the United States, Fauna Series No. 5). U.S. National Park Service, Washington, DC.

Musiani, M., H. Okarma, and W. Jedrzewski. 1998. Speed and actual distance traveled by radiocollared wolves in Bialowieza Primeval Forest (Poland). Acta Theriologica 43:409–16.

Nasimovich, A. A. 1955. The role of the regime of snow cover in the life of ungulates in the USSR. Canadian Wildlife Service, Ottawa.

National Research Council. 1997. Wolves, bears, and their prey in Alaska: Biological and social challenges in wildlife management. National Academy Press, Washington, DC.

Nelson, M. E., and L. D. Mech. 1981. Deer social organization and wolf predation in northeastern Minnesota. Wildlife Monograph 77:1–53.

Nelson, M. E., and L. D. Mech. 1985. Observations of a wolf killed by a deer. Journal of Mammalogy 66:187–88.

Nelson, M. E., and L. D. Mech. 1986a. Relationship between snow depth and gray wolf predation on white-tailed deer. Journal of Wildlife Management 50:471–74.

Nelson, M. E., and L. D. Mech. 1986b. Wolf predation risk associated with white-tailed deer movements. Canadian Journal of Zoology 69:2696–99.

Nelson, M. E., and L. D. Mech. 1993. Prey escaping wolves despite close proximity. Canadian Field-Naturalist 107:245–46.

Nelson, M. E., and L. D. Mech. 2000. Proximity of white-tailed deer, *Odocoileus virginianus,* ranges to wolf, *Canis lupus,* pack homesites. Canadian Field-Naturalist 114:504–5.

Noss, R. F. 1992. The Wildlands Project land conservation strategy. Wild Earth (Special issue: The Wildlands Project. Plotting a North American wilderness recovery strategy) 1992:10–25.

Noss, R. F., H. B. Quigley, M. G. Hornocker, T. Merrill, and P. C. Paquet. 1996. Conservation biology and carnivore conservation in the Rocky Mountains. Conservation Biology 10:949–63.

Nowak, R. M. 1973. North American quaternary *Canis*. Ph.D. Dissertation, University of Kansas, Lawrence.

Nowak, R. M. 1979. North American quaternary *Canis* (Monograph No. 6). Museum of Natural History, University of Kansas, Lawrence.

Nowak, R. M. 1983. A perspective on the taxonomy of wolves in North America. Pages 10–19 *in* L. N. Carbyn, ed. Wolves in Canada and Alaska: Their status, biology, and management (Report Series 45). Canadian Wildlife Service, Ottawa.

Nowak, R. M. 1995. Another look at wolf taxonomy. Pages 375–99 *in* L. N. Carbyn, S. H. Fritts, and D. R. Seip, eds. Ecology and conservation of wolves in a changing world (Occasional Publication No. 35). Canadian Circumpolar Institute, Edmonton, Alberta.

Nowak, R. M., and N. E. Federoff. 1998. Validity of the red wolf: Response to Roy et al. Conservation Biology 12:722–25.

Nudds, T. D. 1978. Convergence of group size strategies by mammalian social carnivores. American Naturalist 112:957–60.

Oaten, A., and W. W. Murdoch. 1975. Functional response and stability in predator–prey systems. American Naturalist 109:289–98.

O'Brien, S. J., and E. Mayr. 1991. Bureaucratic mischief: Recognizing endangered species and subspecies. Science 231:1187–88.

Olsen, S. J. 1985. Origins of the domestic dog: The fossil record. University of Arizona Press, Tucson.

Oosenbrug, S. H., and L. N. Carbyn. 1982. Winter predation on bison and activity patterns of a wolf pack in Wood Buffalo National Park. Pages 43–53 *in* F. H. Harrington and P. C. Paquet, eds. Wolves of the world: Perspectives of behavior, ecology, and conservation. Noyes, Park Ridge, N J.

Orians, G., P. A. Cochran, J. W. Duffield, T. K. Fuller, R. J. Gutierrez, W. M. Henemann, F. C. James, P. Kareiva, S. R. Kellert, D. Klein, B. N. McLellan, P. D. Olson, and G. Yaska. 1997. Wolves, bears, and their prey in Alaska. National Academy Press, Washington, DC.

Packard, J. M. 1980. Deferred reproduction in wolves (*Canis lupus*). Ph.D. Dissertation, University of Minnesota, Minneapolis.

Packard, J. M., and L. D. Mech. 1980. Population regulation in wolves. Pages 135–50 *in* M. N. Cohen, R. S. Malpass, and H. G. Klein, eds. Biosocial mechanisms of population regulation. Yale University Press, New Haven, CT.

Packard, J. M., L. D. Mech, and U. S. Seal. 1983. Social influences on reproduction in wolves. Pages 78–85 *in* L. N. Carbyn, ed. Wolves of Canada and Alaska, their status, biology and management. (Report Series, No. 45). Canadian Wildlife Service, Ottawa.

Packard, J. M., U. S. Seal, L. D. Mech, and E. D. Plotka. 1985. Causes of reproductive failure in two family groups of wolves (*Canis lupus*). Zeitschrift für Tierpsychologie 68:24–40.

Paquet, P. C. 1989. Behavioral ecology of sympatric canids in an area of moderate ungulate density. Ph.D. Dissertation, University of Alberta, Edmonton, Canada.

Paquet, P. C. 1991a. Scent-marking behavior of sympatric wolves (*Canis lupus*) and coyotes (*C. latrans*) in Riding Mountain National Park. Canadian Journal of Zoology 69:1721–27.

Paquet, P. C. 1991b. Winter spatial relationships of wolves and coyotes in Riding Mountain National Park, Manitoba. Journal of Mammalogy 72:397–401.

Paquet, P. C. 1992. Prey use strategies of sympatric wolves and coyotes in Riding Mountain National Park, Manitoba. Journal of Mammalogy 73:337–43.

Paquet, P. C. 1993. Summary reference document: Ecological studies of recolonizing wolves in the Central Canadian Rocky Mountains. Canadian Parks Service, Banff, Alberta.

Paquet, P. C., and L. N. Carbyn. 1986. Wolves, *Canis lupus,* killing denning black bears, *Ursus americanus,* in the Riding Mountain National Park area. Canadian Field-Naturalist 100:371–72.

Paquet, P. C., and C. Callaghan. 1996. Effects of linear developments on winter movements of gray wolves in the Bow River Valley of Banff National Park, Alberta. Pages 1–21 *in* G. L. Evink, D. Zeigler, and J. Berry, eds. Trends in addressing transportation related wildlife mortality. Florida Department of Transportation, Orlando.

Paquet, P. C., and W. A. Fuller. 1990. Scent marking and territoriality in wolves of Riding Mountain National Park. Pages 394–400 *in* D. W. Macdonald, D. Muller-Schwarze, and S. E. Natynczuk, eds. Chemical signals in vertebrates 5. Oxford University Press, New York.

Paquet, P. C., S. Bragdon, and S. McCusker. 1982. Cooperative rearing of simultaneous litters in captive wolves. Pages 223–37 *in*F. H. Harrington and P. C. Paquet, eds. Wolves of the world: Perspectives of behavior, ecology, and conservation. Noyes, Park Ridge, NJ.

Paquet, P. C., J. Wierzchowski, and C. Callaghan. 1996. Summary report on the effects of human activity on gray wolves in the Bow River Valley, Banff National Park, Alberta. Chapter 7 *in* J. Green, C. Pacas, S. Bayley, and L. Cornwell, eds. A cumulative effects assessment and futures outlook for the Banff Bow Valley. Prepared for the Banff Bow Valley Study, Department of Canadian Heritage, Ottawa.

Paquet, P. C., J. R. Strittholt, N. L. Stauss, P. J. Wilson, S. Grewel, and B. N. White. 2001a. Feasibility of timber wolf reintroduction in Adirondack Park, New York. Pages 47–64 *in* D. Maehr, R. F. Noss, and J. Larkin, eds. Large mammal restoration: Ecological and sociological implications. Island Press, Washington, DC.

Paquet, P. C., J. A. Vucetich, M. K. Phillips, and L. M. Vucetich. 2001b. Mexican wolf recovery: Three-year program review and assessment. Prepared by the Conservation Breeding Specialist Group Apple Valley, MN, for the U.S. Fish and Wildlife Service, Albuquerque, NM.

Paquet, P. C., F. Burrows, A. Forshner, G. Neale, and K. Wade. 2001c. Blastomycosis in a free ranging lone wolf, *Canis lupus,* on the North Shore of Lake Superior, Ontario: A response to P. Krizan. Canadian Field-Naturalist 115:185.

Parker, G. R. 1972. Biology of the Kaminuriak population of barren-ground caribou. Part I: Total number, mortality, recruitment, and seasonal distribution (Report Series, 20). Canadian Wildlife Service, Ottawa.

Parker, G. R. 1973. Distribution and densities of wolves within barren-ground caribou range in northern mainland Canada. Journal of Mammalogy 54:341–48.

Parsons, D. R. 1998. 'Green fire' returns to the Southwest: Reintroduction of the Mexican wolf. Wildlife Society Bulletin 26:799–807.

Pasitschniak-Arts, M., M. E. Taylor, and L. D. Mech. 1988. Skeletal injuries in an adult arctic wolf. Arctic Alpine Research 20:360–65.

Pedersen, S. 1982. Geographical variation in Alaskan wolves. Pages 345–61 *in* F. H. Harrington and P. C. Paquet, eds. Wolves of the world: Perspectives of behavior, ecology, and conservation. Noyes, Park Ridge, NJ.

Peek, J. M., D. E. Brown, S. R. Kellert, L. D. Mech, J. D. Shaw, and V. Van Ballenberghe. 1991. Restoration of wolves in North America. Wildlife Society Technical Review 91-1:1–21.

Person, D. K. 2001. Wolves, deer and logging: Population viability and predator–prey dynamics in a disturbed insular landscape. Ph.D. Dissertation, University of Alaska, Fairbanks.

Peters, R. P. 1978. Communication, cognitive mapping, and strategy in wolves and hominids. Pages 95–107 *in* R. L Hall and H. S. Sharp, eds. Wolf and man: Evolution in parallel. Academic Press, New York.

Peters, R. P. 1979. Mental maps in wolf territory. Pages 119–52 in E. Klinghammer, ed. The behavior and ecology of wolves. Garland STPM Press, New York.

Peters, R. P., and L. D. Mech. 1975. Scent-marking in wolves. American Scientist 63:628–37.

Peters, R. P., and L. D. Mech. 1978. Scent-marking in wolves. Pages 133–47 *in* R. L. Hall and H. S. Sharp, eds. Wolf and man: Evolution in parallel. Academic Press, New York.

Peterson, R. L. 1966. The mammals of eastern Canada. Oxford University Press, Toronto.

Peterson, R. O. 1974. Wolf ecology and prey relationships on Isle Royale. Ph.D. Dissertation, Purdue University, West Lafayette, IN.

Peterson, R. O. 1976. The role of wolf predation in a moose population decline. Pages 329–33 *in* Proceedings of the first conference on scientific research in the national parks (Transactions and Proceedings Series No. 5). U.S. Department of the Interior, National Park Service.

Peterson, R. O. 1977. Wolf ecology and prey relationships on Isle Royale (Fauna Series 11). U.S. National Park Service, Washington, DC.

Peterson, R. O. 1979. Social rejection following mating of a subordinate wolf. Journal of Mammalogy 60:219–21.

Peterson, R. O. 1995. Wolves as interspecific competitors in canid ecology. Pages 315–24 *in* L. N. Carbyn, S. H. Fritts, and D. R. Seip, eds. Ecology and conservation of wolves in a changing world (Occasional Publication No. 35). Canadian Circumpolar Institute, Edmonton, Alberta.

Peterson, R. O., and D. L. Allen. 1974. Snow conditions as a parameter in moose–wolf relationships. Naturaliste Canadien 101:481–92.

Peterson, R. O., and R. E. Page. 1983. Cyclic fluctuations of wolves and moose at Isle Royale National Park, U.S.A. Acta Zoologica Fennica 174:252–54.

Peterson, R. O., and R. E. Page. 1988. The rise and fall of Isle Royale wolves, 1975–1986. Journal of Mammalogy 69:89–99.

Peterson, R. O., J. D. Woolington, and T. N. Bailey. 1984. Wolves of the Kenai Peninsula, Alaska. Wildlife Monographs 88:1–52.

Peterson, R. O., N. J. Thomas, J. M. Thurber, J. A. Vucetich, and T. A. Waite. 1998. Population limitation and the wolves of Isle Royale. Journal of Mammalogy 79:828–41.

Phillips, M. K. 1984. The cost to wolves of preying on ungulates. Australian Mammalogy 8:99.

Pike, W. M. 1892. The barren ground caribou of northern Canada. Macmillan, London, Ontario, Canada.

Pilgrim, K. L., D. K. Boyd, and S. H. Forbes. 1998. Testing for wolf–coyote hybridization in the Rocky Mountains using mitochondrial DNA. Journal of Wildlife Management 62:683–89.

Pimlott, D. 1967. Wolf predation and ungulate populations. American Zoologist 7:267–78.

Pimlott, D. H., J. A. Shannon, and G. B. Kolenosky. 1969. The ecology of the timber wolf in Algonquin Provincial Park (Research Report, Wildlife No. 87). Ontario Department of Lands and Forests, Toronto.

Pletscher, D. H., R. R. Ream, R. Demarchi, W. G. Brewster, and E. E. Bangs. 1991. Managing wolf and ungulate populations in an international ecosystem. Transactions of the North American Wildlife and Natural Resources Conference 56:539–49.

Pletscher, D. H., R. R. Ream, D. K. Boyd, M. W. Fairchild, and K. E. Kunkel. 1997. Population dynamics of a recolonizing wolf population. Journal of Wildlife Management 61:459–65.

Post, E., R. O. Peterson, N. C. Stenseth, and B. E. McLaren. 1999. Ecosystem consequences of wolf behavioural response to climate. Nature 401:905–7.

Potvin, F. 1987. Wolf movements and population dynamics in Papineau–Labelle Reserve, Quebec. Canadian Journal of Zoology 66:1266–73.

Potvin, F. 1988. Wolf movements and population dynamics in Papineau–Labelle Reserve, Quebec. Canadian Journal of Zoology 66:1266–73.

Potvin, F., H. Jolicoeur, and J. Huot. 1988. Wolf diet and prey selectivity during two periods for deer in Québec: Decline versus expansion. Canadian Journal of Zoology 66:1274–79.

Potvin, F., H. Jolicoeur, L. Breton, and R. Lemieux. 1992a. Evaluation of an experimental wolf reduction and its impact on deer in Papineau–Labelle Reserve, Québec. Canadian Journal of Zoology 70:1595–1603.

Potvin, F., L. Breton, C. Pilon, and M. Macquart. 1992b. Impact of an experimental wolf reduction on beaver in Papineau–Labelle Reserve, Quebec. Canadian Journal of Zoology 70:180–83.

Promenberger, C. 1992. Woelfe und scavenger [wolves and scavengers]. Diplomarbeit, Ludwig Maximillians Universität, Munich, Germany.

Pulliainen, E. 1982. Behavior and structure of an expanding wolf population in Karelia, northern Europe. Pages 134–45 *in* F. H. Harrington and P. C. Paquet, eds. Wolves of the world: Perspectives of behavior, ecology, and conservation. Noyes, Park Ridge, NJ.

Pulliam, H. R., and T. Caraco. 1978. Living in groups: Is there an optimal group size? Pages 122–47 *in* J. T. Krebs and N. B. Davies, eds. Behavioural ecology: An evolutionary approach. Sinauer, Sunderland, MA.

Ramsay, M. A., and I. Stirling. 1984. Interactions of wolves and polar bears in northern Manitoba. Journal of Mammalogy 65:693–94.

Rasker, R., and A. Hackman. 1996. Economic development and the conservation of large carnivores. Conservation Biology 10:991–1002.

Rausch, R. 1958. Some observations on rabies in Alaska, with special reference to wild canidae. Journal of Wildlife Management 22:246–60.

Rausch, R. A. 1967. Some aspects of the population ecology of wolves, Alaska. American Zoologist 7:253–65.

Rausch, R. A. 1969. Wolf and wolverine–wolf summer food habits and den studies (Project No. W-017-R-01/ Wk.PL.O/Job 03). Alaska Department Fish and Game.

Raymer, J., D. Weisler, M. Novotny, C. Asa, U. S. Seal, and L. D. Mech. 1984. Volatile constinuents of wolf (*Canis lupus*) urine as related to gender and season. Experientia 40:707–9.

Raymer, J., D. Wiesler, M. Movotny, C. Asa, U. S. Seal, and L. D. Mech. 1985. Chemical investigations of wolf (*Canis lupus*) anal-sac secretion in relation to breeding season. Journal of Chemical Ecology 11:593–608.

Raymer, J., D. Wiesler, M. Novotny, C. Asa, U. S. Seal, and L. D. Mech. 1986. Chemical scent constituents in urine of wolf *Canis lupus* and their dependence on reproductive hormones. Journal of Chemical Ecology 12:297–314.

Ream, R. R., M. W. Fairchild, D. K. Boyd, and D. H. Pletscher. 1991. Pages 349–66 *in* R. B. Keiter and M. S. Boyce, eds. The Greater Yellowstone Ecosystem: Redefining America's wilderness heritage. Yale University Press, New Haven, CT.

Riewe, R. 1975. The high arctic wolf in the Jones Sound Region. Arctic 28:209–12.

Ripple, W. J., and E. J. Larsen. 2000. Historic aspen recruitment, elk, and wolves in northern Yellowstone National Park. Biological Conservation 95:361–70.

Rodman, P. S. 1981. Inclusive fitness and group size with a reconsideration of group size in lions and wolves. American Naturalist 118:275.

Rogers, L. L., and L. D. Mech. 1981. Interactions of wolves and black bears in northeastern Minnesota. Journal of Mammalogy 62:434–36.

Rogers, L. L., L. D. Mech, D. K. Dawson, J. M. Peek, and M. Korb. 1980. Deer distribution in relation to wolf pack territory edges. Journal of Wildlife Management 44:253–58.

Rothman, R. V., and L. D. Mech. 1979. Scent-marking in lone wolves and newly formed pairs. Animal Behavior 27:750–60.

Route, W. T., and R. O. Peterson. 1991. An incident of wolf, *Canis lupus,* predation on a river otter, *Lutra canadensis,* in Minnesota. Canadian Field-Naturalist 105:567–68.

Roy, M. S., E. Geffen, D. Smith, E. A. Ostrander, and R. K. Wayne. 1994. Pattern of differentiation and hybridization in North American wolflike canids, revealed by analysis of microsatellite loci. Molecular Biology and Evolution 1:553–70.

Roy, M. S., E. Geffen, D. Smith, and R. K. Wayne. 1996. Molecular genetics of pre-1940 red wolves. Conservation Biology 10:1413–24.

Rutter, R. J., and D. H. Pimlott. 1968. The world of the wolf. J. B. Lippincott, Philadelphia.

Ryon, C. J. 1977. Den digging and related behavior in a captive timber wolf pack. Journal of Mammalogy 58:87–89.

Ryon, J., and R. E. Brown. 1990. Urine-marking in female wolves (*Canis lupus*): An indicator of dominance status and reproductive state. Pages 346–51 *in* D. W. Macdonald, D. Muller-Schwarze, and S. E. Natynczuk, eds. Chemical signals in vertebrates 5. Oxford University Press, New York.

Schenkel, R. 1947. Expression studies of wolves. Behaviour 1:81–129.

Schenkel, R. 1967. Submission: Its features and function in the wolf and dog. American Zoologist 7:319–29.

Schmidt, K. P., and J. R. Gunson. 1985. Evaluation of wolf–ungulate predation near Nordegg, Alberta: Second year progress report, 1984–85. Alberta Energy and Natural Resources Fish and Wildlife Division, Edmonton, Alberta, Canada.

Schmidt, P. A., and L. D. Mech. 1997. Wolf pack size and food acquisition. American Naturalist 150:513–17.

Schmitz, O. J., and G. B. Kolenosky. 1985a. Hybridization between wolf and coyote in captivity. Journal of Mammalogy 66:402–5.

Schmitz, O. J., and G. B. Kolenosky. 1985b. Wolves and coyotes in Ontario: Morphological relationships and origins. Canadian Journal of Zoology 63:1130–37.

Schmitz, O. J., and D. M. Lavigne. 1987. Factors affecting body size in sympatric Ontario *Canis*. Journal of Mammalogy 68:92–99.

Schoener, T. 1971. Theory of feeding strategies. Annual Review of Ecology and Systematics 2:369–404.
Scott, B. M. V., and D. M. Shackleton. 1980. Food habits of two Vancouver Island wolf packs: A preliminary study. Canadian Journal of Zoology 58:1203–7.
Scott, J. P. 1968. Evolution and domestication of the dog. Evolutionary Biology 2:243–75.
Scotter, G. W. 1995. Influence of harassment by wolves, *Canis lupus,* on barren-ground caribou, *Rangifer tarandus groenlandicus,* movements near the Burnside River, Northwest Territories. Canadian Field-Naturalist 109:452–53.
Seal, U. S., L. D. Mech, and V. Van Ballenberghe. 1975. Blood analyses of wolf pups and their ecological and metabolic interpretation. Journal of Mammalogy 56:64–75.
Seip, D. R. 1992. Factors limiting woodland caribou populations and their interrelationships with wolves and moose in southeastern British Columbia. Canadian Journal of Zoology 70:1494–1503.
Seton, E. T. 1925. Lives of game animals. Vol. 1, Parts 1 and 2. Doubleday/Doran, Garden City, NY.
Shelley, D. P., and E. M. Anderson. 1995. Impact of U. S. Highway 53 expansion on timber wolves—baseline data. Final Report. University of Wisconsin, Stevens Point.
Shields, W. M. 1983. Genetic considerations in the management of the wolf and other large vertebrates: An alternative view. Pages 90–92 *in* L. N. Carbyn, ed. Wolves in Canada and Alaska: Their status, biology, and management (Report Series 45). Canadian Wildlife Service, Ottawa.
Sih, A., P. Crowley, M. McPeek, J. Petranka, and K. Strohmeier. 1985. Predation, competition, and prey communities: A review of field experiments. Annual Review of Ecology and Systematics 16:269–311.
Sinclair, A. R. E. 1989. Population regulation in animals. Pages 197–241 *in* J. M. Cherrett, ed. Ecological concepts. Blackwell, Oxford, UK.
Singer, F. J. 1979. Status and history of timber wolves in Glacier National Park, Montana. Pages 19–42 *in* E. Klinghammer, ed. The behavior and ecology of wolves. Garland STPM Press, New York
Singleton, P. H. 1995. Winter habitat selection by wolves in the North Fork of the Flathead River Basin, Montana and British Columbia. M.Sc. Thesis, University of Montana, Missoula.
Skeel, M. A., and L. N. Carbyn. 1977. The morphological relationship of gray wolves (*Canis lupus*) in national parks of central Canada. Canadian Journal of Zoology 55:737–47.
Skogland, T. 1991. What are the effects of predators on large ungulate populations? Oikos 61:401–11.
Smith, C. A. 1983. Responses of two groups of mountain goats, *Oreamnos americanus,* to a wolf, *Canis lupus*. Canadian Field-Naturalist 97:110.
Smith, D. W., W. E. Clark, M. K. Phillips, J. A. Mack, L. D. Mech, M. Meagher, and R. Jaffe. 2000. Wolf–bison interactions in Yellowstone National Park. Journal of Mammalogy 81:1128–35.
Soper, J. D. 1942. Mammals of Wood Buffalo Park, northern Alberta, and District of Mackenzie. Journal of Mammalogy 23:119–45.
Spaulding R. L., P. R. Krausman, and W. B. Ballard. 1998. Summer diet of gray wolves, *Canis lupus,* in Northwestern Alaska. Canadian Field-Naturalist 112:262–66.
Stains, H. J. 1975. Distribution and taxonomy of the Canidae. Pages 429–59 *in* M. W. Fox, ed. The wild canids: Their systematics, behavioral ecology, and evolution. Van Nostrand Reinhold, New York.
Standfield, R. O. 1970. Some consideration on the taxonomy of wolves in Ontario. Pages 32–38 *in* S. E. Jorgensen, C. E. Faulkner, and L. D. Mech, eds. Proceedings of a symposium on wolf management in selected areas of North America. U. S. Fish and Wildlife Service, Bureau of Sport Fishing and Wildlife, Region 3, Twin Cities, MN.
Stanwell-Fletcher, J. F. 1942. Three years in the wolves' wilderness. Natural History 49:136–47.
Stardom, R. R. P. 1983. Status and management of wolves in Manitoba. Pages 30–34 *in* L. N. Carbyn, ed. Wolves in Canada and Alaska: Their status, biology, and management (Report Series 45). Canadian Wildlife Service, Ottawa.
Stenlund, M. H. 1955. A field study of the timber wolf (*Canis lupus*) on the Superior National Forest (Technical Bulletin No 4). Minnesota Department of Conservation, St. Paul.
Stephens, D. W., and J. R. Krebs. 1986. Foraging theory. Princeton University Press, Princeton, NJ.
Stephens, P. W., and R. O. Peterson. 1984. Wolf-avoidance strategies of moose. Holarctic Ecology 7:239–44.
Stephenson, R. O. 1974. Characteristics of wolf den sites (Project W-17-2, W-17-3, W-17-4, W-17-5, and W-17-6; Job 14.6 R). Alaska Department Fish and Game.
Stephenson, R. O., and D. James. 1982. Wolf movements and food habits in northwest Alaska. Pages 26–41 *in* F. H. Harrington and P. C. Paquet, eds. Wolves of the world: perspectives of behavior, ecology, and conservation. Noyes, Park Ridge, NJ.
Stephenson, R. O., and L. J. Johnson. 1972. Wolf report (Federal Aid in Wildlife Restoration Program Report, Project W-17-3, Vol. X). Alaska Department of Fish and Game.
Stephenson, R. O., and L. J. Johnson. 1973. Wolf report (Federal Aid in Wildlife Restoration Program Report, Project W-17-4, Vol. X). Alaska Department of Fish and Game.
Stephenson, T. R., and V. Van Ballenberghe. 1995. Defense of one twin calf against wolves, *Canis lupus,* by a female moose, *Alces alces*. Canadian Field-Naturalist 109:251–53.
Tener, J. S. 1954. A preliminary study of muskoxen on Fosheim Peninsula, Ellesmere Island, Northwest Territories (Wildlife Management Bulletin, Series 1, No. 9). Canadian Wildlife Service, Ottawa.
Terborgh, J. 1988. The big things that run the world: A sequel to E. O. Wilson. Conservation Biology 2: 402–3.
Terborgh, J., J. A. Estes, P. C. Paquet, K. Ralls, D. Boyd-Heger, B. J. Miller, and R. F. Noss. 1999. The role of top carnivores in regulating terrestrial ecosystems. Pages 39–64 *in* Continental Conservation, eds. M. E. Soule and J. Terborgh. Island Press, Washington, DC. 227pp.
Theberge, J. B. 1969. Observations of wolves at a rendezvous site in Algonquin Park. Canadian Field-Naturalist 83:122–28.
Theberge, J. B. 1990. Potentials for misinterpreting impacts of wolf predation through prey:predator ratios. Wildlife Society Bulletin 18:188–92.
Theberge, J. B. 1991. Ecological classification, status, and management of the gray wolf, *Canis lupus,* in Canada. Canadian Field-Naturalist 105:459–63.
Theberge, J. B., and J. B. Falls. 1967. Howling as a means of communication in timber wolves. American Zoologist 7:331–38.
Theberge, J. B., and D. A. Gauthier. 1985. Models of wolf–ungulate relationships: When is wolf control justified. Wildlife Society Bulletin 13:449–58.
Theberge, J. B., S. M. Oosenbrug, and D. H. Pimlott. 1978. Site and seasonal variations in food of wolves in Algonquin Park, Ontario. Canadian Field-Naturalist 92:91–94.
Theberge, J. B., G. J. Forbes, I. K. Barker, and T. Bollinger. 1994. Rabies in wolves of the Great Lakes Region. Journal of Wildlife Diseases 30:563–66.
Theberge, J. B., M. T. Theberge, and G. J. Forbes. 1996. What Algonquin Park wolf research has to instruct about recovery in northeastern United States. Pages 34–40 *in* Wolves of America conference proceedings. Defenders of Wildlife, Washington, DC.
Theberge, M. T., J. B. Theberge, G. J. Forbes, and S. Stewart. 1996. Is the Algonquin canid a wolf or a coyote? Pages 208–11 *in* Wolves of America conference proceedings. Defenders of Wildlife, Washington, DC.
Thieking, A., S. M. Goyal, R. F. Berg, K. L. Loken, L. D. Mech, and R. P. Thiel. 1992. Seroprevalence of Lyme disease in Minnesota and Wisconsin wolves. Journal of Wildlife Diseases 28:177–82.
Thiel, R. P. 1985. Relationship between road densities and wolf habitat suitability in Wisconsin. American Midland Naturalist 113:404–7.
Thiel, R. P., and J. Valen. 1995. Developing a state timber wolf recovery plan with public input: The Wisconsin experience. Pages 169–75 *in* L. N. Carbyn, S. H. Fritts, and D. R. Seip, eds. Ecology and conservation of wolves in a changing world (Occasional Publication No. 35). Canadian Circumpolar Institute, Edmonton, Alberta.
Thiel, R. P., L. D. Mech, G. R. Ruth, J. R. Archer, and L. Kaufman. 1987. Blastomycosis in wild wolves. Journal of Wildlife Diseases 23:321–23.
Thiel, R. P., S. Merrill, and L. D. Mech. 1998. Tolerance by denning wolves, *Canis lupus,* to human disturbance. Canadian Field-Naturalist 112:340–42.
Thompson, B. C., J. S. Prior-Magee, M. L. Munson-McGee, W. Brown, D. Parsons, and L. Moore. 2000. Beyond release: Incorporating diverse publics in setting research priorities for the Mexican wolf recovery program. Transactions of North American Wildlife and Natural Resources Conference 65:278–91.
Thompson, I. D., and R. O. Peterson. 1988. Does wolf predation alone limit the moose population in Pukaskwa Park, Ontario, Canada: A comment. Journal of Wildlife Management 52:556–59.
Thurber, J. M., and R. O. Peterson. 1993. Effects of population density and pack size on the foraging ecology of gray wolves. Journal of Mammalogy 74:879–89.
Thurber, J. M., R. O. Peterson, J. D. Woolington, and J. A. Vucetich. 1992. Coyote coexistence with wolves on the Kenai Peninsula, Alaska. Canadian Journal of Zoology 70:2494–98.
Thurber, J. M., R. O. Peterson, T. D. Drummer, and S. A. Thomasma. 1994. Gray wolf response to refuge boundaries and roads in Alaska. Wildlife Society Bulletin 22:61–68.

Todd, A. W., J. R. Gunson, and W. B. Samuel. 1981. Sarcoptic mange: An important disease of coyotes and wolves in Alberta, Canada. Pages 706–29 *in* J. A. Chapman and D. Pursely, eds. Worldwide furbearer conference. Frostburg, MD.

Tompa, F. S. 1983. Status and management of wolves in British Columbia. Pages 20–23 *in* L. N. Carbyn, ed. Wolves in Canada and Alaska: Their status, biology, and management (Report Series 45). Canadian Wildlife Service, Ottawa.

Turnbull, P. F., and C. A. Reed. 1974. The fauna from the terminal Pleistocene of Palegawra cave. Fieldiana Anthropology 63:81–146.

U.S. Fish and Wildlife Service., 1982. Mexican wolf recovery plan. U.S. Fish and Wildlife Service, Albuquerque, NM.

U.S. Fish and Wildlife Service. 1987. Northern Rocky Mountain wolf recovery plan. U.S. Fish and Wildlife Service, Denver, CO.

U.S. Fish and Wildlife Service. 1992. Recovery plan for the eastern timber wolf. U.S. Fish and Wildlife Service, Twin Cities, MN.

U.S. Fish and Wildlife Service. 2000. Endangered and threatened wildlife and plants; proposal to reclassify and remove the gray wolf from the list of endangered and threatened wildlife in portions of the conterminous United States; proposals to establish three special regulations for threatened gray wolf wolves; proposed rule. Federal Register 65:43450–96.

Van Ballenberghe, V. 1983a. Extraterritorial movements and dispersal of wolves in southcentral Alaska. Journal of Mammalogy 64:168–71.

Van Ballenberghe, V. 1983b. Two litters raised in one year by a wolf pack. Journal of Mammalogy 64:171–72.

Van Ballenberghe, V. 1991. Forty years of wolf management in the Nelchina basin, southcentral Alaska: A critical review. Transactions of the North American Wildlife and Natural Resource Conference 56:561–56.

Van Ballenberghe, V., and W. B. Ballard. 1994. Limitation and regulation of moose populations: The role of predation. Canadian Journal of Zoology 72:2071–77.

Van Ballenberghe, V., and A. W. Erickson. 1973. A wolf pack kills another wolf. American Midland Naturalist 90:490–93.

Van Ballenberghe, V., and L. D. Mech. 1975. Weights, growth, and survival of timber wolf pups in Minnesota. Journal of Mammalogy 56:44–63.

Van Ballenberghe, V., A. W. Erickson, and D. Byman. 1975. Ecology of the timber wolf in northeastern Minnesota. Wildlife Monograph 43:1–43.

Van Camp, J., and R. Gluckie. 1979. A record long-distance movement by a wolf (*Canis lupus*). Journal of Mammalogy 60:236–37.

Van Zyll De Jong, C. G., and L. N. Carbyn. 1999. COSEWIC—Status report on the gray wolf (*canis lupus*) in Canada. Committee on the Status of Endangered Wildlife in Canada, Ottawa.

Vibe, C. 1981. Pattedyr (Mammalia). Pages 363–459 *in* F. Salomonsen, ed. Gronlands Fauna. Nordisk Forlag, Copenhagen.

Vilà, C., and R. K. Wayne. 1999. Hybridization between wolves and dogs. Conservation Biology 13:195–98

Voigt, D. R. 1973. Summer food habits and movements of wolves (*Canis lupus*) in central Ontario. M.S. Thesis, University of Guelph, Ontario, Canada.

Voigt, D. R., G. B. Kolenosky, and D. H. Pimlott. 1976. Changes in summer food of wolves in central Ontario, Canada. Journal of Wildlife Management 40:663–68.

Vucetich, J. A., R. O. Peterson, and T. A. Waite. 1997. Effects of social structure and prey dynamics on extinction risk in gray wolves. Conservation Biology 11:957–65.

Wagner, K. K., R. H. Schmidt, and M. R. Conover. 1997. Compensation programs for wildlife damage in North America. Wildlife Society Bulletin 25:312–19.

Walton, L. R. 2000. Investigation into the movements of migratory wolves in the central Canadian Arctic. M.S. Thesis, University of Saskatchewan, Saskatoon, Canada.

Walton, L. R., H. D. Cluff, P. C. Paquet, and M. A. Ramsay. 2001. Movement patterns of barren-ground wolves in the central Canadian arctic. Journal of Mammalogy 82:867–76.

Wayne, R. K., and S. Jenks. 1991. Mitochondrial DNA analysis implying extensive hybridization of the endangered red wolf, *Canis rufus*. Nature 351:565–68.

Wayne, R. K., and S. J. O'Brien. 1987. Allozyme divergence within the Canidae. Systematic Zoology 36:339–55.

Wayne, R. K., D. A. Gilbert, N. Lehman, K. Hansen, A. Eisenhawer, D. Girman, L. D. Mech, P. J. P. Gogan, U. S. Seal, and R. J. Krumenaker. 1991. Conservation genetics of the endangered Isle Royale gray wolf. Conservation Biology 5:41–51.

Wayne, R. K., N. Lehman, M. W. Allard, and R. L. Honeycutt. 1992. Mitochondrial DNA variability of the gray wolf: Genetic consequences of population decline and habitat fragmentation. Conservation Biology 6:499–569.

Wayne, R. K., N. Lehman, and T. K. Fuller. 1995. Conservation genetics of the gray wolf. Pages 399–408 *in* L. N. Carbyn, S. H. Fritts, and D. R. Seip, eds. Ecology and conservation of wolves in a changing world (Occasional Publication No. 35). Canadian Circumpolar Institute, Edmonton, Alberta.

Wayne, R. K., M. S. Roy, and J. L. Gittleman. 1998. Origin of the red wolf: Response to Nowak and Federoff and Gardener. Conservation Biology 12:726–29.

Weaver, J. L. 1992. Two wolves, *Canis lupus,* killed by a moose, *Alces alces,* in Jasper National Park, Alberta. Canadian Field-Naturalist 106:126–27.

Weaver, J. L. 1994. Ecology of wolf predation amidst high ungulate diversity in Jasper National Park, Alberta. Ph.D. Dissertation, University of Montana, Missoula.

Weaver, J. L., P. C. Paquet, and L. F. Ruggiero. 1996. Resilience and conservation of large carnivores in the Rocky Mountains. Conservation Biology 10:964–76.

White, P. A., and D. K. Boyd. 1989. A cougar, *Felis concolor,* killed and eaten by gray wolves, *Canis lupus,* in Glacier National Park, Montana. Canadian Field-Naturalist 103:408–9.

White, C. A., C. E. Olmsted, and C. E. Kay. 1998. Aspen, elk, and fire in the Rocky Mountain national parks of North America. Wildlife Society Bulletin 26:449–62.

Williams, T. M. 1995. Pup survival and home range use of wolves associated with dens on the range of the Bathurst caribou herd, June–August 1993. Government of the Northwest Territories, Department of Renewable Resources, Yellowknife, Canada.

Williams, T. M., and D. C. Heard. 1993. Predation rates, movement and activity patterns, pup survival and diet of wolves associated with dens on the range of the Bathurst caribou herd, June–September 1992. Government of the Northwest Territories, Department of Renewable Resources, Yellowknife, Canada.

Wilson, D. E., and D. M. Reeder. 1993. Mammal species of the world: A taxonomic and geographic reference, 2nd ed. Smithsonian Institution Press, Washington, DC.

Wilson, P. J., S. Grewal, I. D. Lawford, J. N. M. Heal, A. G. Granacki, D. Pennock, J. B. Theberge, M. T. Theberge, D. R. Voigt, W. Waddell, R. E. Chambers, P. C. Paquet, G. Goulet, D. Cluff, and B. N. White. 2000. DNA profiles of the eastern Canadian wolf and the red wolf provide evidence for a common evolutionary history independent of the gray wolf. Canadian Journal of Zoology 78:1–11.

Wobeser, G. 1992. Traumatic, degenerative, and developmental lesions in wolves and coyotes from Saskatchewan. Journal of Wildlife Diseases 28:268–75.

Woodroffe, R. 2000. Predators and people: Using human densities to interpret declines of large carnivores. Animal Conservation 3:165–73.

Woodroffe, R., and J. R. Ginsberg. 1998. Edge effects and the extinction of populations inside protected areas. Science 280:2126–28.

Wydeven, A. P., R. N. Schultz, and R. P. Thiel. 1995. Monitoring of a recovering gray wolf population in Wisconsin, 1979–1991. Pages 147–56 *in* L. N. Carbyn, S. H. Fritts, and D. R. Seip, eds. Ecology and conservation of wolves in a changing world (Occasional Publication No. 35). Canadian Circumpolar Institute, Edmonton, Alberta.

Wydeven, A. P., T. K. Fuller, W. Weber, and K. MacDonald. 1998. The potential for wolf recovery in the northeastern United States via dispersal from southeastern Canada. Wildlife Society Bulletin 26:776–84.

Young, S. P. 1944. The wolves of North America. Part 1. Their history, life habits, economic status, and control. Pages 1–385 *in* S. P. Young and E. A. Goldman, eds. The wolves of North America. Dover, New York.

Young, S. P., and E. A. Goldman 1944. The wolves of North America. Dover, New York, and American Wildlife Institute, Washington, DC.

Zimen, E. 1975. Social dynamics of the wolf pack. Pages 336–62 *in* M. W. Fox, ed. The wild canids: Their systematics, behavioral ecology, and evolution. Van Nostrand Reinhold, New York.

Zimen, E. 1976. On the regulation of pack size in wolves. Zeitschrift für Tierpsychologie 40:300–341.

Zimen, E. 1982. A wolf pack sociogram. Pages 282–322 *in* F. H. Harrington and P. C. Paquet, eds. Wolves of the world. Noyes, Park Ridge, NJ.

Paul C. Paquet, World Wildlife Fund Canada, Meacham, Saskatchewan, Canada S0K 2V0. Email: ppaquet@sasktel.net.

Ludwig N. Carbyn, Canadian Wildlife Service, Edmonton, Alberta, Canada T6B 2X3. Email: lu.carbyn@ec.gc.ca.

24

Foxes

Vulpes species, *Urocyon* species, and *Alopex lagopus*

Brian L. Cypher

NOMENCLATURE

Common Names (Moore and Collins 1995; Macdonald 1999). *Alopex lagopus*: arctic fox, white fox, polar fox, blue fox; *Urocyon cinereoargenteus*: gray fox, tree fox; *Urocyon littoralis*: island fox, Channel Island fox, island gray fox, coast fox, short-tailed fox, insular gray fox; *Vulpes macrotis*: kit fox, desert fox; *Vulpes velox*: swift fox, plains fox; *Vulpes vulpes*: red fox, black fox, cross fox, silver fox
Scientific Name. The six North American fox species are *Alopex lagopus, Urocyon cinereoargenteus, U. littoralis, Vulpes macrotis, V. velox,* and *V. vulpes*

TAXONOMY

Foxes are in the family Canidae in the order Carnivora. The taxonomy of foxes has been dynamic, with frequent revisions occurring at the genus, species, and subspecies levels. The taxonomic status of the arctic fox is unsettled. Based on its very distinctive morphology, the arctic fox (Fig. 24.1) typically has been placed in the monotypic genus *Alopex* (Clutton-Brock et al. 1976; Van Gelder 1978). However, recent genetic analyses consistently indicate that *Alopex* and *Vulpes* are closely related and that species in these genera are congeneric (Wayne et al. 1989; Geffen et al. 1992). In particular, arctic foxes are very closely related to kit foxes and swift foxes (Wayne and O'Brien 1987; Mercure et al. 1993). Indeed, arctic foxes appear to be as closely related to swift foxes as are subspecies of swift foxes to each other (Mercure et al. 1993). However, the reduction of *Alopex* to a species within *Vulpes* has not yet gained wide usage. Within *Alopex,* five North American subspecies are recognized (Hall 1981): *A. l. lagopus* (northern Alaska and Canada northwest of Hudson Bay), *A. l. groenlandicus* (Greenland), *A. l. hallensis* (St. Matthew Island in the Bering Sea), *A. l. pribilofensis* (Pribilof Islands in the Bering Sea), and *A. l. ungava* (Canada northeast of Hudson Bay). Additional subspecies occur in northern Europe and Asia.

Gray foxes (Fig. 24.2) were commonly considered to be distinct from vulpine foxes, thereby warranting placement in the genus *Urocyon* (Geffen et al. 1992). Clutton-Brock et al. (1976) and Van Gelder (1978) proposed reclassifying gray foxes as *Vulpes*. However, more recent genetic analyses indicate that gray foxes represent a distinct evolutionary lineage and warrant generic status separate from the vulpine foxes (Geffen et al. 1992). Seven subspecies are recognized (Hall 1981): *U. c. borealis* (New England), *U. c. cinereoargenteus* (eastern United States), *U. c. floridanus* (Gulf states), *U. c. ocythous* (central plains states), *U. c. scottii* (southwestern United States, northern Mexico), *U. c. townsendi* (California, Oregon), and *U. c. californicus* (southern California). Eight additional subspecies occur in Central and South America.

Gray foxes may have colonized the northern Channel Islands off southern California during the mid- or late Pleistocene, where they evolved in isolation. These island foxes (Fig. 24.3) warrant recognition as a separate species from mainland gray foxes based on morphological and genetic differences (Collins 1982; Moore and Collins 1995). Biological and archaeological evidence indicates that foxes from the northern Channel Islands were transported to the southern Channel Islands by early Native Americans more than 2000 years ago (Collins 1991a). Foxes now occur on the six largest Channel Islands and each of these populations is recognized as a separate subspecies (Grinnell et al. 1937): *U. l. santacruzae* (Santa Cruz Island), *U. l. catalinae* (Catalina Island), *U. l. clementae* (San Clemente Island), *U. l. littoralis* (San Miguel Island), *U. l. dickeyi* (San Nicolas Island), and *U. l. santarosae* (Santa Rosa Island).

The taxonomy of arid land foxes also has been problematic. Kit foxes (Fig. 24.4) and swift foxes (Fig. 24.5) initially were recognized as distinct species, but this has occasionally been questioned (Egoscue 1979). A relatively narrow (ca. 100 km) and historically stable contact zone between these two species occurs in eastern New Mexico and western Texas. Although some hybridization occurs, Rohwer and Kilgore (1973) concluded that selection favored parental forms over hybrids. This and evidence by others (Packard and Bowers 1970; Thornton and Creel 1975) supported species recognition for both kit foxes and swift foxes. More recently, mitochondrial DNA analyses further supported retaining separate species (Mercure et al. 1993).

Historically, eight subspecies of kit fox were recognized (Hall and Kelson 1959): *V. m. arsipus* (southeastern California, Arizona), *V. m. devia* (southern Baja California, Mexico), *V. m. macrotis* (southwestern California—extinct), *V. m. mutica* (San Joaquin Valley, California), *V. n. neomexicana* (New Mexico, Mexico), *V. n. nevadensis* (Great Basin), *V. m. tenuirostris* (northern Baja California, Mexico), and *V. m. zinseri* (northern Mexico). Two subspecies of swift fox were recognized (Merriam 1902): *V. v. hebes* (northern prairie regions), and *V. v. velox* (southern prairie regions). However, Stromberg and Boyce (1986) concluded that subspecific division was not warranted based on morphological analyses.

North American and Eurasian red foxes (Fig. 24.6) were once considered distinct species (*V. fulva* and *V. vulpes,* respectively). However, all red foxes generally are now considered to be conspecific within *V. vulpes*. Ten subspecies have been identified in North America (Hall 1981): *V. v. fulva* (eastern United States), *V. v. rubricosa* (southern Quebec, Nova Scotia), *V. v. regalis* (prairie regions of United States and Canada), *V. v. macroura* (Rocky Mountains), *V. v. necator* (Sierra Nevada), *V. v. abietorum* (western Canada), *V. v. alascensis* (Alaska), *V. v. harrimani* (Kodiak Island), *V. v. kenaiensis* (Kenai Peninsula), and *V. v. cascadensis* (northwestern coastal United States). Despite numerous recognized subspecies, substantial mixing of gene pools has occurred. Red foxes from Europe were introduced into portions of the United States in the 1700s (Kamler and Ballard 2002). Also, foxes have been transported among regions for hunting and fur farming, and some of these animals have escaped and interbred with local populations.

DISTRIBUTION

Arctic foxes have a circumpolar distribution in North America, Europe, and Asia (Fig. 24.7). They primarily inhabit tundra and pack ice, but occasionally have been observed in boreal forests in southeastern Canada

FIGURE 24.1. Arctic fox (*Alopex lagopus*) in summer pelage. SOURCE: Photo by B. Moose Peterson.

FIGURE 24.2. Gray fox (*Urocyon cinereoargenteus*). SOURCE: Photo by B. Moose Peterson.

and the Kenai Peninsula of Alaska (Underwood and Mosher 1982; Garrott and Eberhardt 1987). Arctic foxes also have been introduced onto >450 Aleutian Islands (Buskirk and Gipson 1981; Tietjen et al. 1988; Bailey 1992).

Gray foxes occur in woodland and shrubland habitats from southern Canada into northern South America. Their distribution did not historically extend into boreal regions, but gray foxes have expanded into these regions in the northeastern United States and southeastern Canada (Fig. 24.8), probably due to the creation of favorable habitat conditions by agricultural activity (Fritzell and Haroldson 1982). Gray foxes also have expanded into the Great Plains, probably as a result of increased woodland cover resulting from fire suppression and tree planting (Fritzell 1987). Island foxes occur in the brushlands and grasslands on the six largest Channel Islands (Moore and Collins 1995).

Kit foxes and swift foxes are considered arid land species, and are adapted to habitats with relatively open vegetation structures. Kit foxes (Fig. 24.9) occur primarily in desert and semiarid shrublands and grasslands in the southwestern United States and northwestern Mexico (McGrew 1979). Swift foxes (Fig. 24.9) occur primarily in short-grass prairie regions of the west-central United States, with populations extending into southern Canada (Egoscue 1979).

The red fox is native to Europe and Asia, but its origin in North America has been disputed. Although some authorities suggest that red foxes may have been introduced into North American from Europe, evidence suggests that they probably were native to North America but limited to boreal and mixed hardwood habitats north of 40–45°N (Gilmore 1946; Churcher 1959; Kamler and Ballard 2002). In the 1700s, red foxes from England were introduced into the southeastern United States (Churcher 1959) and the New England region (Gilmore 1946; Waters 1964) for hunting.

The red fox is the most widely distributed of the North American fox species (Hall 1981), and indeed, has the most extensive natural distribution of any extant terrestrial mammal except humans (Nowak and Paradiso 1983). The current distribution of the red fox (Fig. 24.10) appears to be a function of several factors including interspecific

FIGURE 24.3. Island fox (*Urocyon littoralis*). SOURCE: Photo by B. Moose Peterson.

FIGURE 24.4. Kit fox (*Vulpes macrotis*). SOURCE: Photo by B. Moose Peterson.

FIGURE 24.5. Swift fox (*Vulpes velox*). SOURCE: Photo by Cynthia Moehrenschlager.

FIGURE 24.6. Red fox (*Vulpes vulpes*). SOURCE: Photo by B. Moose Peterson.

competition, adaptability, habitat modification, and human influences (Sargeant 1982). Red foxes apparently increased in the southeastern United States following the reduction of gray wolves (*Canis lupus*) and red wolves (*C. rufus*) and the clearing of forests (Churcher 1959; Godin 1977). Red foxes may have colonized prairie regions following the elimination of wolves and the significant reduction of coyotes (*C. latrans*) by humans (Janes and Gier 1966; Sargeant 1982). Expansion of red foxes into arid areas of the southwestern United States has probably been facilitated by irrigated agriculture. In California, introduced eastern red foxes have expanded rapidly throughout low-elevation areas, and may be interbreeding with the native subspecies, *V. v. necator* (Jurek 1992). Red foxes also have expanded their range further into arctic regions (Voight 1987), and have been introduced onto some Aleutian islands (Hersteinsson and Macdonald 1992). The northern limit of red foxes may be determined by the availability of suitable prey (Hersteinsson and Macdonald 1992).

DESCRIPTION

North American foxes are small to medium-sized canids. All six species share certain diagnostic characteristics (Hall 1981). All are doglike in appearance with five toes on each forefoot (the "dew claw" is higher up on the leg than other toes) and four toes on each hind foot. Typical of canids, all species have long, nonretractable claws, and all have a subcaudal gland on the dorsal surface of the tail. In all species, there is substantial regional and individual variation in external measurements and body mass (hence, ranges of averages are given below). Males are generally 5–15% larger than females.

Arctic foxes are characterized by seasonally dimorphic pelage; relatively short appendages and rostrum, short, rounded ears; and thickly furred feet (Underwood and Mosher 1982). The arctic fox is one of the larger North American fox species (Table 24.1). Some individuals reach 10–11 kg (Chesemore 1967; Garrott and Eberhardt 1987). The coat of arctic foxes is soft and thick with dense underfur and long guard hairs. Two main color phases occur among arctic foxes (Underwood and Mosher 1982). In one phase, winter pelage is almost pure white except for the black nose and yellowish eyes. In the "blue" phase, winter pelage is dark to dull slate gray. In both phases, summer pelage

FIGURE 24.7. Distribution of the arctic fox (*Alopex lagopus*).

FIGURE 24.9. Distribution of the kit fox (*Vulpes macrotis*) and swift fox (*Vulpes velox*).

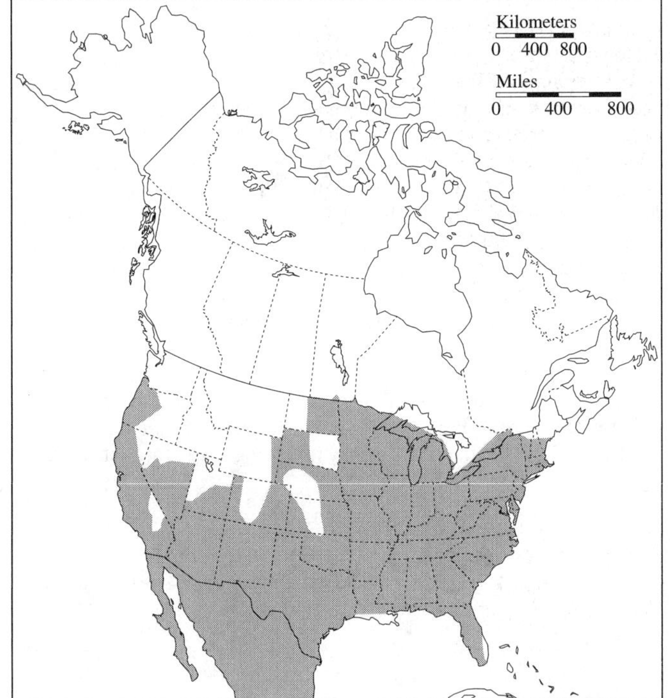

FIGURE 24.8. Distribution of the gray fox (*Urocyon cinereoargenteus*).

FIGURE 24.10. Distribution of the red fox (*Vulpes vulpes*).

consists of much shorter fur, which is gray along the back and sides, whereas the underside and flanks are cream to fawn colored. There are two molts/year, which occur in spring and autumn (Chesemore 1975).

Gray foxes have a generally stocky appearance and relatively short legs compared to other foxes. The gray fox is one of the larger North American fox species (Table 24.1). The pelage is grizzled gray with rusty highlights on the neck, sides, and legs. The ventral surface is buffy to white (Fritzell and Haroldson 1982). Gray foxes have a diagnostic ridge of coarse black guard hairs on the dorsal surface of the tail, and the tip of the tail is black. The fur is relatively coarse. The long caudal gland extends one third to one half the length of the tail (Seton 1923; Hildebrand 1952a).

Island foxes are the smallest of the six North American fox species (Table 24.1). They are very similar in appearance to gray foxes, but are

TABLE 24.1. Ranges of average morphological measurements for the six North American fox species

Species	Head–Body Length (mm)	Tail Length (mm)	Hind Foot Length (mm)	Mass (kg)	Source
	Range of Average Measurements				
Arctic fox	650–700	260–340	120–150	2.5–6.7	Chesemore 1967; Garrott and Eberhardt 1987
Gray fox	525–682	275–443	100–150	3.0–7.0	Hall 1981; Fritzell and Haroldson 1982
Island fox	480–497	110–290	98–157	1.0–2.7	Hall 1981; Collins 1982
Kit fox	375–517	225–323	111–137	1.7–3.0	Grinnell et al. 1937; Hall 1981
Swift fox	462–530	240–350	113–135	1.9–2.4	Kilgore 1969; Thornton and Creel 1975; Sheldon 1992
Red fox	536–642	291–455	124–182	3.0–8.0	Storm et al. 1976; Larivière and Pasitshniak-Arts 1996

25–50% smaller (Crooks 1994; Moore and Collins 1995). The largest individuals are on Santa Catalina Island, whereas the smallest occur on Santa Cruz Island (Collins 1982). Pelage is somewhat darker than that of the gray fox, but it is otherwise very similar, including the ridge of coarse black hairs down the dorsal surface of the tail (Grinnell et al. 1937; Collins 1982).

Kit foxes are relatively small (Table 24.1) and slender in appearance, with "delicate" features. Pelage is grizzled to tawny gray with buffy highlights on the neck, sides, and legs. The ventral surface is white to tawny. The tip of the tail is black and there is no dark ridge of hairs down the dorsal side of the tail as in gray foxes. The underfur is dense and the summer coat is coarse. The feet have considerable hair between the pads, and the pinnae have a thick border of hairs on the forward inner edge and inner base (McGrew 1979).

Swift foxes are very similar in appearance and size to kit foxes (Table 24.1), but there are some characteristics that differentiate the two species (Thornton and Creel 1975, McGrew 1979). The ears of kit foxes are longer (>75 mm from notch) and more pointed, whereas those of swift foxes are shorter (<75 mm) and more rounded. In frontal view, the base of the ears of kit foxes are set closer to the midline of the skull compared to swift foxes, where the ears are spaced more widely. The head of kit foxes is broader between the eyes and the snout is narrower compared to swift foxes. Also, the tail of kit foxes is longer relative to body length than in swift foxes.

The pelage of swift foxes is grizzled to yellowish gray with buffy highlights on the neck, chest, sides and legs. The ventral surface is light buff to white. The underfur is dense and the summer coat is coarse. The tail is black tipped and the feet are well furred between the pads (Egoscue 1979).

Red foxes are the largest of the North American foxes (Table 24.1) and have a distinctive appearance. They are generally slender and relatively long legged. Red foxes exhibit at least four pelage variations. The most common phase is the distinctive bright red to yellowish-red coat. The ventral surface is white, the ears are trimmed with black, and the lower legs and feet are black. The white-tipped tail is a diagnostic characteristic. The other color phases are more common in colder regions (Samuel and Nelson 1982; Voigt 1987). In the black or silver phase, the guard hairs are black with a silver frosting on the tips. In the cross phase, the pelage is grayish-brown with light buffy patches on the legs, shoulders, and hips, and there is a dark cross pattern on the shoulders (Voigt 1987). In some areas, 10% of individuals exhibit the black/silver phase and 25% the cross phase (Nowak 1991). In the bastard phase, the pelage is bluish gray (Samuel and Nelson 1982).

Skull and Dentition. North American foxes all share certain skull and dental characteristics (Ewer 1973; Hall 1981), including an elongated rostrum. The dorsal surface of the postorbital process is concave, and the sagittal crest is not prominent. All also exhibit the typical canid dental formula, I 3/3, C 1/1, P 4/4, M 2/3. The incisors are moderately large and slightly curved. Also, the canines are relatively large but not particularly sharp, and are slightly recurved. The lower canines are slightly anterior to the upper canines. A small diastema in front of the upper canines and behind the lowers allows the teeth to interlock when the mouth is closed. These incisor and canine characteristics facilitate gripping and tearing. Also characteristic of canids and many other carnivores, the last upper premolar and the first lower molar constitute the carnassial teeth. The scissor-like action of these teeth allows foxes to shear flesh and cut through tendons, ligaments, other tough tissues, including some smaller bones. The molars are adapted for crushing flesh and vegetable foods. Canids are diphyodonts, and deciduous and permanent teeth have closed roots.

Representative skull measurements are presented in Table 24.2. Compared to the other North American foxes, arctic foxes have a shorter and broader rostrum (Fig. 24.11), and the interorbital region is elevated (Underwood and Mosher 1982). Gray foxes (Fig. 24.12) and island foxes

TABLE 24.2. Selected skull measurements (mm) for five North American species of foxes

Skull Measurement	Gray Fox M	Gray Fox F	Island Fox M	Island Fox F	Kit Fox	Swift Fox M	Swift Fox F	Red Fox M	Red Fox F
Condylobasal length	121.0	117.4	97.7	95.1	114.4	111.0	113.3	144.0	137.0
	(20)	(14)	(64)	(64)	(35)	(3)	(1)	(16)	(11)
Zygomatic breadth	67.9	64.9	56.7	55.2	62.1	64.1	63.0	76.7	75.2
	(20)	(14)	(64)	(64)	(35)	(3)	(1)	(16)	(11)
Interorbital width	25.6	24.4	31.3	30.8	23.1	24.1	24.0	—	—
	(20)	(14)	(64)	(64)	(35)	(3)	(1)		
Postorbital constriction	—	—	27.3	27.3	21.4	23.2	22.5	23.0	23.1
			(64)	(64)	(35)	(3)	(1)	(16)	(11)
Maxillary toothrow length	—	—	42.8	41.8	42.4	52.5	52.7	54.6	52.3
			(64)	(64)	(185)	(3)	(1)	(16)	(11)

SOURCE: Gray fox, Grinnell et al. 1937. Island fox, Moore and Collins 1995. Kit fox, McGrew 1979; Dragoo et al. 1990. Swift fox, Egoscue 1979. Red fox, Storm et al. 1976.

NOTE: Samples sizes are given in parentheses below each mean measurement.

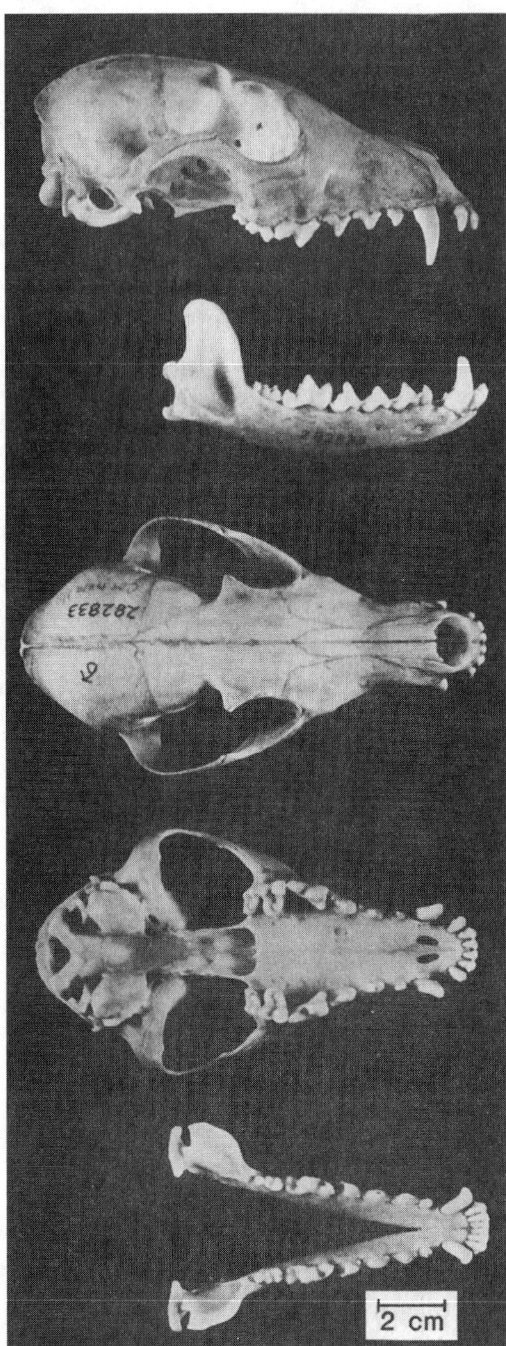

FIGURE 24.11. Skull of the arctic fox (*Alopex lagopus*). From top to bottom: lateral view of cranium, lateral view of mandible, dorsal view of cranium, ventral view of cranium, dorsal view of mandible.

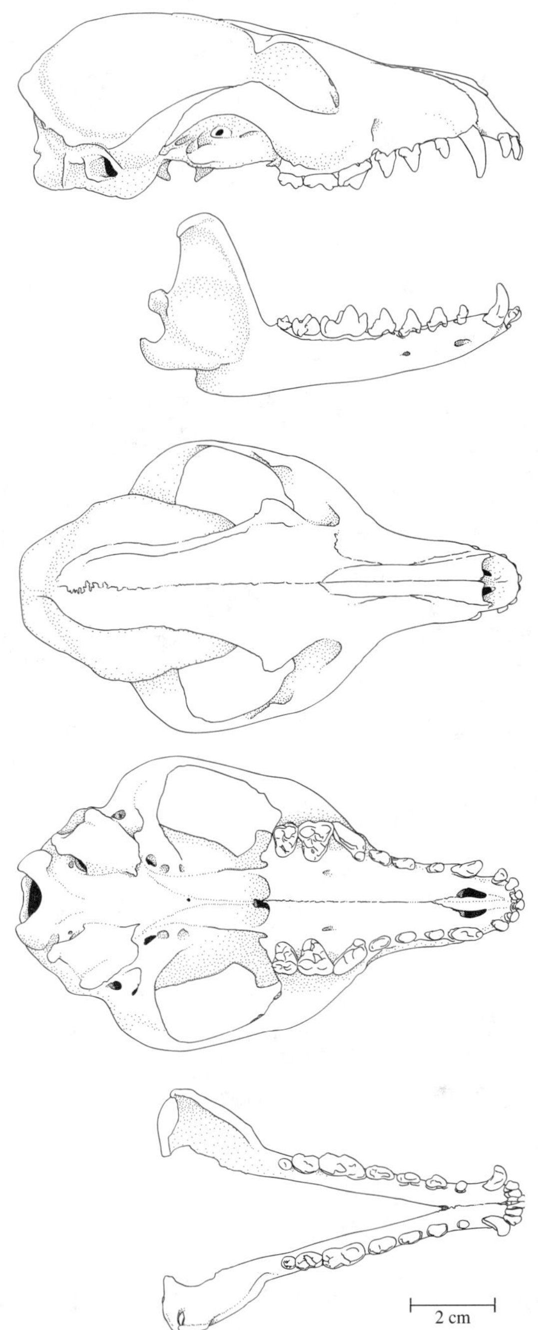

FIGURE 24.12. Skull of the gray fox (*Urocyon cinereoargenteus*). From top to bottom: lateral view of cranium, lateral view of mandible, dorsal view of cranium, ventral view of cranium, dorsal view of mandible.

exhibit generally similar skull and dental morphology, although island fox measurements are approximately 17% smaller (Collins 1982). In both species, the posterior ventral border of the dentary has a prominent notch or "step" (Hall 1981). The temporal ridges are separated anteriorly, but connect posteriorly to form a distinctive U shape (Fig. 24.12). The temporal bones are roughened below the temporal ridges. Collins (1982) summarized some subtle differences in cranial characteristics between gray foxes and island foxes.

Kit foxes and swift foxes also exhibit similar skull and dental characteristics, reflecting the close evolutionary relationship between these species. The skull is narrow and delicate with a long, slender rostrum. The auditory bullae are relatively large and are well inflated (Stains 1975). Kit fox and swift fox skulls are morphologically similar to but smaller than red fox skulls. In red foxes, the temporal ridges are separated anteriorly, but connect posteriorly to form a distinctive V shape (Fig. 24.13). The rostrum is long and narrow, and the interorbital and frontal regions are relatively flat (Samuel and Nelson 1982; Larivière and Pasitschniak-Arts 1996).

GENETICS

Arctic foxes have a diploid chromosome number ($2n$) of 48–50 and a fundamental number (NF) of either 88 (Gustavsson and Sundt 1965) or 94 (Wurster and Benirschke 1968). Chiarelli (1975) mentioned that

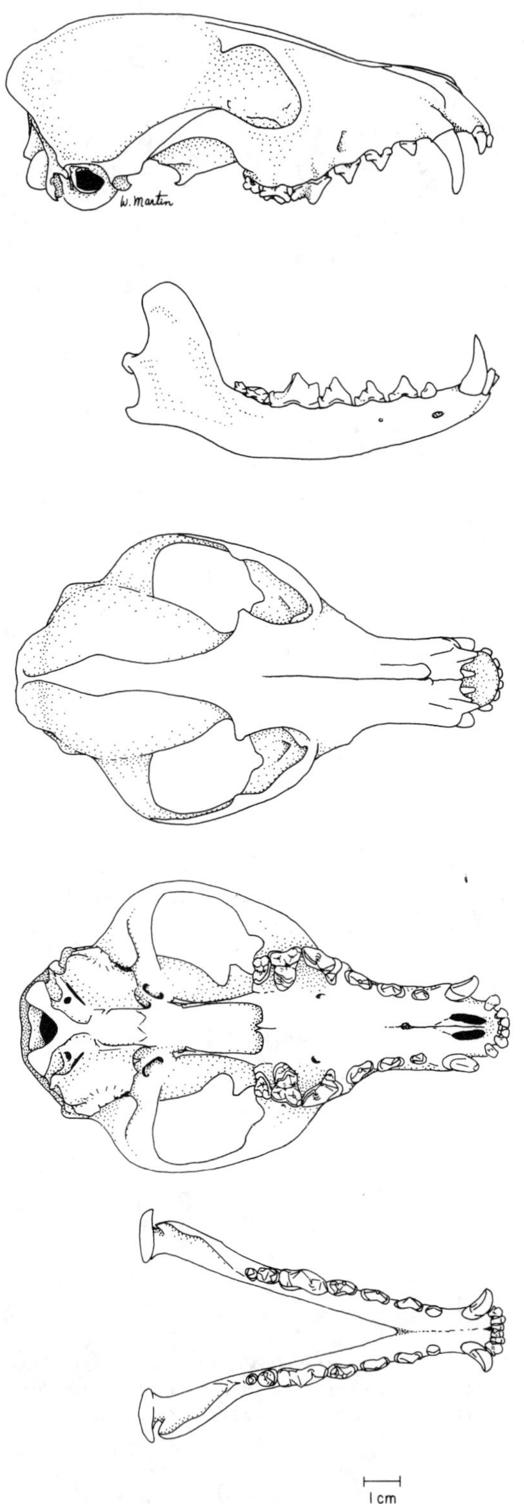

FIGURE 24.13. Skull of the red fox (*Vulpes vulpes*). From top to bottom: lateral view of cranium, lateral view of mandible, dorsal view of cranium, ventral view of cranium, dorsal view of mandible.

arctic fox–red fox crosses will produce fertile hybrids, further supporting the close evolutionary relationship between *Alopex* and *Vulpes*. Pelage coloration in the arctic fox is determined by a single gene locus (Johansson 1960; Banfield 1974). Homozygous recessive alleles produce white coloration, whereas the homozygous dominant produces a dark "blue" coloration. A heterozygous genotype produces a lighter blue coloration. The presence of fewer blue-phase individuals may be a function of differences in litter size, breeding season, timing of molt, and body size between the two color phases (Chiarelli 1975).

In gray foxes, $2n = 66$ and NF = 70 (Wurster and Benirschke 1968). Individuals exhibiting a "Samson" condition (a genetically based absence of guard hairs) have been reported (Grinnell et al. 1937; Root 1981).

The $2n$ number in island foxes also is 66 (Moore and Collins 1995) and NF = 70. The genetics of island foxes has been extensively investigated because they are rare and insular and present an opportunity to examine evolutionary processes. Analyses of mitochondrial and nuclear DNA from the six island populations revealed that genetic variability is greater among populations than it is within any single population (Gilbert et al. 1990). Allozyme heterozygosity is lower among island foxes than it is for closely related gray foxes on the nearby mainland. Also, no allozyme variation was evident among foxes on San Nicolas and Santa Catalina, and variation was low for foxes on San Miguel and San Clemente (Wayne et al. 1991a, 1991b). Foxes on San Clemente, San Miguel, San Nicolas, and Santa Cruz can be distinguished by the presence of unique hypervariable DNA restriction fragments (Gilbert et al. 1990; Wayne et al. 1991b). Unique mitochondrial genotypes are found in fox populations on San Nicolas and Santa Catalina, and foxes on San Nicolas, San Miguel, and San Clemente exhibit only one mitochondrial genotype (Wayne et al. 1991a, 1991b). Foxes on San Nicolas exhibit evidence of extensive inbreeding or genetic bottlenecking (Gilbert et al. 1990; George and Wayne 1991).

Kit foxes have $2n = 50$, as do swift foxes (Creel and Thornton 1974). Consistent with their close evolutionary relationship, the karyotypes of kit foxes and swift foxes are similar. As noted, kit foxes and swift foxes readily hybridize along a narrow ($\leq$100 km) contact zone in eastern New Mexico, and hybrid genotypes do not appear to occur outside of this zone (Rohwer and Kilgore 1973). Red fox–kit fox hybrids also have been reported from Texas (Thornton et al. 1971; Creel and Thornton 1974).

The $2n = 34$–38 in red foxes and NF = 72 (Chiarelli 1975). The polychromatism in red foxes is genetically based and is determined by three alleles. Adalsteinsson et al. (1987) describe the combinations of dominant and recessive alleles that produce the three color phases (red, silver/black, and cross) observed in red foxes. All three color phases can occur in the same litter (Murie 1944). Also, Keeler (1975) described behavioral traits that appeared to be consistently associated with each color phase. For example, red-phase individuals appear to be more wary than silver/black- or cross-phase individuals. Individuals with the Samson condition also occur among red foxes (Samuel and Nelson 1982).

MORPHOLOGICAL AND PHYSIOLOGICAL ADAPTATIONS

Like most canids, foxes are morphological and physiological generalists, which in part contributes to their ability to exploit a variety of widely distributed habitats. However, North American foxes do have some specialized adaptations. Arctic foxes exhibit adaptations to withstand extremely cold temperatures in their arctic environment (Chesemore 1975), including relatively short limbs, muzzle, and ears, which reduce heat loss. Another adaptation is long, dense winter pelage with excellent insulative qualities, which allows arctic foxes to maintain body-to-ambient air temperature gradients of up to 100°C for many weeks (Scholander et al. 1950). This fur may have the best insulation value of any mammal pelage (Ewer 1973; Chesemore 1975; Clutton-Brock et al. 1976).

Other adaptations of arctic foxes include their characteristic dichromatism. White winter pelage and dark summer pelage provide cryptic coloration, which may facilitate stalking prey. Compared to the winter pelage, the summer pelage is much shorter and less dense, thereby facilitating heat loss during the warm season (Underwood 1971). Arctic foxes also have densely furred soles of the feet, which provide traction and warmth on snow and ice (Bee and Hall 1956). Arctic foxes can maintain foot temperatures at just above 0°C, which keeps the feet from freezing while minimizing heat loss (Henshaw et al. 1972). These foxes also have an extremely acute olfactory sense, which they use to detect rodent nests under several centimeters of snow (Banfield 1974).

Gray foxes and island foxes exhibit adaptations that facilitate tree climbing. These adaptations include a stocky build and relatively short legs with long, sharp recurved claws. Gray foxes in the forested eastern regions of Mexico had sharper and more recurved claws then individuals in more arid western regions (Goldman 1938). Gray foxes and island foxes also are able to rotate their foreleg more than any other canid, which may help them to get a better grip when climbing tree trunks (Hildebrand 1952b, 1954; Ewer 1973).

Kit foxes exhibit adaptations for living in hot, arid environments. Stiff tufts of hair between the toe pads ("fimbriation") on the soles of the feet improve traction on loose, sandy surfaces and may help protect the feet from hot sand (Grinnell et al. 1937). As is common among desert animals, kit foxes have large ears, which facilitate heat radiation. Also, long, dense hairs along the inner edge and base of the pinnae completely cover the auricle opening to exclude dust and sand (McGrew 1979). Through panting, canids use evaporative cooling to lose excess heat. Many canids pant at the resonant frequency of the thorax, but kit foxes pant at a lower rate, which increases with the ambient temperature until reaching the resonant frequency (the maximum recorded is 576 cycles/min). This is energetically more costly, but reduces water loss (Denver Wildlife Research Center 1975).

A significant challenge for kit foxes is obtaining adequate water in an environment where free water is sparse or absent. Kit foxes will drink free water if it is available (Egoscue 1956; B. L. Cypher, pers. obs.). However, they are almost exclusively carnivorous and are able to obtain adequate moisture from their prey (Egoscue 1962; Morrell 1972). Metabolic water from the digestion and assimilation of prey provides approximately 18% of the total daily water requirement for kit foxes. The remainder of this requirement is met by preformed water from prey (Golightly and Ohmart 1984). To acquire adequate preformed water, kit foxes must consume substantially more prey than is required to meet daily energetic demands. Golightly and Ohmart (1984) estimated that kit foxes need to consume 175 g of prey daily to meet the water requirements. Disposition of the excess energy consumed is unclear.

Another significant challenge for kit foxes is dissipating excess heat in a hot, arid environment. Due to the scarcity of water in their environment, kit foxes rely minimally on evaporative heat loss. Instead, they exhibit a high thermal conductance and rely primarily on passive heat dissipation (Golightly and Ohmart 1983). Behavioral adaptations such as daily diurnal den use and nocturnal activity also assist significantly in reducing heat loads and conserving water. Indeed, kit foxes are poorly adapted physiologically to cope with high temperatures. They are not tolerant of high ambient temperatures, and their body temperature can quickly rise to lethal levels during prolonged exposure to ambient temperatures exceeding 35°C.

Klir and Heath (1992) examined the relationship between body morphology and heat loss in arctic foxes, kit foxes, and red foxes. The face region, nose, dorsal head area, pinna, lower legs, and paws were important thermoregulatory areas in kit foxes and red foxes. The face region, nose, front of pinna, lower legs, and paws were important thermoregulatory areas in arctic foxes. The lower legs and paws are the regions where heat loss is most efficient. All of these body regions have short fur year-round. Excessive heat loss is avoided during cool temperatures by reducing blood flow to these regions through vasoconstriction. To eliminate excess heat during warm temperatures, blood flow to these regions is increased through vasodilation. Also, because these regions are not as well insulated, vasodilation also increases blood flow to these body areas to prevent freezing during extremely cold temperatures. The amount of the total body surface occupied by these regions is about 38% in kit foxes, 33% in red foxes, and 22% in arctic foxes, and reflects the environment of each species.

No morphological or physiological specializations are evident among red foxes. Their adaptations are very generalized, which facilitates ecological plasticity and widespread distribution. Red foxes have a slender build and relatively long legs, characteristic of other highly mobile canids such as gray wolves and coyotes. Although red foxes usually are associated with more mesic environments, they may not require free water, but can obtain adequate water from food (Sargeant 1978).

REPRODUCTION

North American foxes share a number of reproductive attributes, many of which also are common to most canids. All six species are basically monogamous, although some polygamy may occur (see Social Ecology). Foxes generally form pair bonds, and mates will commonly remain together until one dies. All species also are monestrous. Typically, unpaired individuals locate mates in mid- to late fall. Pairs begin courtship in late fall to early winter. Breeding usually occurs during winter, and young are born from late winter into early summer. The chronology of breeding events varies among and within species on a latitudinal gradient, with events occurring earlier in the south and later in the north. During mating, a copulatory tie occurs in which the bulbus glandis near the proximal end of the penis enlarges, preventing withdrawal for some period of time (Ewer 1973). Pregnancy rates and reproductive success are lower among first-year females than older females. Gestation lasts 50–55 days. Parturition is commonly timed such that energetically costly lactation and provisioning of newly weaned pups coincide with peak abundance of primary prey. Reproductive success and litter size are strongly influenced by food availability.

Young are born in a sheltered location such as a den or hollow log. Pups are born with fur, but the eyes do not open for 1–2 weeks following birth. Young nurse for 5–10 weeks. The adult males assist with pup rearing by provisioning the adult females during lactation and then by provisioning the pups after weaning. Juveniles become independent during the summer following their birth and then disperse between late summer and early winter, although some young may delay dispersal (see Social Ecology).

In arctic foxes, pair formation occurs during February and March (Eberhardt 1977), and most mating occurs in March and April (Fay 1973). Females are estrus for 12–14 days (Dementyeff 1958, cited in Garrott and Eberhardt 1987). Although females are sexually mature at 9–10 months, relatively few breed in their first or second year (Banfield 1974). MacPherson (1969) reported that 34% of 1- and 2-year-old females had placental scars, whereas 85% of 33-year-old females had scars. Gestation averages about 52 days (Barabash-Nikiforov 1938; Johnson 1946). Parturition occurs from April to July; most litters are born in May or June (Garrott and Eberhardt 1987; Sheldon 1992).

Litter size in arctic foxes ranks among the largest in the Canidae (Tannerfeldt and Angerbjörn 1998). Arctic foxes are morphologically adapted to produce larger litters (Moehlman 1989) in that they have more teats (12–14) than the other North American foxes (8) (Ewer 1973). Litter size among arctic foxes varies considerably depending on annual food availability (Eberhardt et al. 1982; Tannerfeldt and Angerbjörn 1998) and can change two- to threefold between successive years (MacPherson 1969; Eberhardt et al. 1983a). Large litters appear to be an adaptation to exploit periodic superabundant food resources (Tannerfeldt and Angerbjörn 1998; Strand et al. 2000; Loison et al. 2001). Speller (1972) suggested that low food availability may lead to abortion or resorbtion of embryos. However, MacPherson (1969) found little annual variation in average placental counts and suggested that higher postnatal pup mortality may be occurring in years of low food availability. Tannerfeldt and Angerbjörn (1998) reviewed litter sizes reported in 23 studies. Mean litter size ranged from 2.8 on Rat Island, Alaska, during low food availability (Berns 1969) to 11.2 in Norway (Frafjord 1992). Maximum values included 18 (Dorogoi 1987, cited in Tannerfeldt and Angerbjörn 1998) and an anecdotal report of 22 (Chirkova et al. 1959, cited in Tannerfeldt and Angerbjörn 1998). Average litter size generally is three to six in coastal areas and six to nine in inland areas (Garrott and Eberhardt 1987). This may be due to the greater preponderance of blue-phase individuals in coastal areas. Braestrup (1941, cited in Chesemore 1975) suggested that blue-phase arctic foxes may be genetically predisposed to produce smaller litters.

Arctic fox pups weigh 60–90 g at birth, open their eyes at about 16 days, emerge from their den at 3–4 weeks, and are weaned at 6–7 weeks. Parents may switch dens during pup rearing (Underwood 1975; Eberhardt 1977), and a litter may be split between two dens (Eberhardt et al. 1983b). Anthony (1996) reported a single litter split among four dens. Juveniles develop adult-colored pelage at about 8 weeks (Novikov 1962, cited in Garrott and Eberhardt 1987), and are essentially independent from adults by about 12–14 weeks (Garrott et al. 1984). They attain adult size and weight by 6–7 months of age (Prestrud and Nilssen 1995). Juveniles may overwinter in their natal ranges if food is abundant (Eberhardt et al. 1983a).

Gray foxes breed from January to April (Sheldon 1949; Sullivan 1956). Both sexes are sexually mature at about 10 months of age (Follman 1978; Root 1981). Wood (1958) reported that 8% of <1-year-old foxes did not reproduce. Layne (1958) suggested that some first-year foxes may reach maturity too late to breed successfully. Among females of all ages, the percentage of nonreproducing females observed included 3.3% in New York (Sheldon 1949), 6.4% in Georgia and Florida (Wood 1958), and 6.5% in Alabama (Sullivan 1956). Pregnancy rates among gray foxes in Wisconsin were 82% for juveniles (<1 year), 96% for yearlings (1.5 years), and 100% for adults (>1.5 years) (Root and Payne 1985). Reported estimates of gestation range from 53 days (Sheldon 1949; Wood 1958) to 63 days (Grinnell et al. 1937; Asdell 1964), but these estimates are not precise or are based on other species (Trapp and Hallberg 1975).

Fritzell (1987) summarized litter size data based on embryo and placental scar counts from eight studies. Litter sizes ranged from 1 to 10, with 4 being the most commonly reported number. Root (1981) reported that litter size did not differ between 1.5-year-old and 4.5-year-old females. Gray fox pups weigh about 86 g at birth (Wood 1958) and are nearly hairless. Eyes open at 10–12 days (Samuel and Nelson 1982). Pups accompany adults on foraging expeditions at 3 months and forage independently at 4 months (Trapp and Hallberg 1975; Nicholson et al. 1985). Juveniles reach adult size and weight at about 210 days (Wood 1958).

Island foxes engage in pair formation and courtship in January–March, and most breeding occurs from late February to early March, (Laughrin 1977). Females can breed at 10 months of age (Laughrin 1977). A pregnancy rate of 31% was reported for females on Santa Cruz Island (Moore and Collins 1995), and few females ≥36 years old were in breeding condition. On San Nicolas Island, 79% of females ≥31 year old had litters in 1980. Gestation is 50–53 days, and parturition occurs from late April to early May (Laughrin 1977).

Litter size averages 2–3 with a range of 1–5. An average of 2.17 pups was observed at 24 dens on Santa Cruz Island (Moore and Collins 1995). Pups emerge from dens at 3–4 weeks, and begin foraging with their parents in mid- to late June. Juveniles remain with the parents through summer and resemble the parents in size and pelage by fall (Laughrin 1977). Juveniles achieve adult weight by their first winter (Crooks 1994).

Kit foxes pair during October and November, and breed in December and January (Egoscue 1956). Reproductive success is strongly influenced by food availability (Egoscue 1975; White and Garrott 1997; Cypher et al. 2000). In central California, annual reproductive success during a 16-year period varied from 20% to 100% for females >1 year old, with a mean of 61%. Mean success for 1-year-old females was 18%, with no evidence of reproduction by yearlings in many years (Cypher et al. 2000). Spiegel (1996) reported a success rate of 64% for adults and 39% for yearling foxes. Gestation is 49–55 days (Egoscue 1962; Zoellick et al. 1987), and parturition occurs from late January to March (Egoscue 1956; Zoellick et al. 1987; Spiegel 1996).

In central California, litter size for adult (>1 year old) kit foxes averaged 3.8, with a range of 1–6 and a mode of 4. For yearling females, the mean was 2.5 and the range was 1–4 (Cypher et al. 2000). O'Neal et al. (1987) reported a mean litter size of 4.6 for kit foxes in Utah, and White and Ralls (1993) reported a mean of 2.0 in California during a period of low food availability. Pups emerge from the den after 4 weeks (Egoscue 1956). Juveniles begin foraging with their parents at 3–4 months of age, exhibit adult pelage at 4–5 months (Morrell 1972), and achieve 90% of adult mass by 10 months (Warrick and Cypher 1999).

Swift foxes breed from late December into January (Kilgore 1969). Both sexes can breed in their first year (Scott-Brown et al. 1986). Gestation averages 51 days (Schroeder 1985), and parturition occurs from late March to May (Kilgore 1969; Schroeder 1985). Litter size based on counts of pups at dens averages four to five, with a range of one to eight (Kilgore 1969, Hillman and Sharps 1978). Pups open their eyes at 10–15 days and are weaned at 6–7 weeks (Kilgore 1969). Juveniles exhibit adult pelage at 2–3 months and achieve adult size by August (Loy and Fitzgerald 1980). Juveniles remain with the adults into late summer and begin dispersing in August and September (Kilgore 1969).

Red foxes breed from December to April, depending on latitude, with most matings occurring during January and February (Ables 1975; Storm et al. 1976; Allen 1984). Estrus in females lasts 1–6 days (Asdell 1964). Copulation averages 26 min, with a range of 1–67 min (Pearson and Bassett 1946). Both sexes are capable of breeding in their first year (Storm et al. 1976; Allen 1984). Storm et al. (1976) reported that 5% of females were unsuccessful in breeding, and Layne and McKeon (1956c) reported that 11.5% of females were barren. As with the other fox species, pregnancy rates and reproductive success are highly variable. They are strongly influenced by food availability and possibly by fox density as well (Harris 1979; Englund 1980). Most yearlings do not breed during vole (*Microtus* spp. and *Clethrionomys* spp.) declines in Sweden (von Schantz 1981a, 1981b; Lindström 1982). In areas of low fox density in Canada, 80–90% of yearlings and 95% of adults may breed successfully (Voigt and Macdonald 1984). Conversely, in urban areas of England, where fox density is high and mortality is low, relatively low fecundity is observed even though food availability is relatively high (Voigt and Macdonald 1984). Gestation is 51–54 days (Sheldon 1949), and parturition occurs from March to May (Voigt 1987).

As with pregnancy rates, litter size among red foxes is influenced by food availability (Englund 1980, Kolb and Hewson 1980; Lindström 1989). Estimates of litter size also are influenced by methodology. Average reported litter sizes based on embryo counts range from 3.7 (Sheldon 1949) to 6.8 (Hoffman and Kirkpatrick 1954; Storm et al. 1976). Average reported litter sizes based on placental scar counts range from 3.7 (Ryan 1976) to 7.1 (Storm et al. 1976). In Ontario, where red foxes were experiencing high mortality rates from trapping, hunting, and rabies, a 1-year-old vixen was found with 14 placental scars (Voigt 1987). Average reported litter sizes based on pups in dens range from 3.8 (McIntosh 1963) to 5.6 (Pils and Martin 1978). Some ova may fail to implant, and 5–8% of fetuses may be resorbed (Layne and McKeon 1956b; Ryan 1976). Lloyd (1980) reported prenatal losses as high as 22%. Absolute litter size based on pups at dens ranges from 1–12, but most often is 3–6 (Allen 1984; Vos 1994). Holcomb (1965) reported observing 17 pups from a single litter. Communal denning has been reported among red foxes (Tullar et al. 1976; Tullar and Berchielli 1980), and this may explain some unusually large litter sizes (Holcomb 1965).

Red fox pups weigh about 71–119 g at birth (Storm and Ables 1966) and have grayish-brown, silky hair (Linhart 1968). Parents may move pups to new dens three times in the first 6 weeks and may even split a litter between two dens (Scott 1943, Sheldon 1949; Storm et al. 1976). Eyes open in 8–9 days and the pups begin to walk at 3 weeks (Linhart 1968). Pups are weaned in 8–10 weeks (Ables 1975), and adult pelage appears in 9–14 weeks (Linhart 1968; Sargeant et al. 1981). Juveniles attain adult size in about 6 months and begin dispersing in early fall (Storm et al. 1976).

ECOLOGY

Ecological Role. North American foxes are considered to be "mesopredators" because they are not the top terrestrial predators in the ecosystems in which they occur. Larger predators such as wolves, coyotes, and bears (*Ursus* spp.) almost always co-occur with foxes and occasionally kill them (see Mortality). Island foxes are the largest mammalian predators in their ecosystem, but they are occasionally preyed on by large avian predators such as bald eagles (*Haliaeetus leucocephalus*) and golden eagles (*Aquila chrysaetos*). Although foxes are occasionally preyed on by other species, they are not the primary prey of any other species.

Trophically, North American foxes are primary, secondary, and tertiary consumers, and also are scavengers (see Feeding Habits). All six species are primarily carnivorous and consume a diversity of prey species. Degree of omnivory varies considerably among species. Gray foxes and red foxes are highly frugivorous and consume a variety of wild fruits. Thus, these species also function as seed dispersers in their ecosystems. Conversely, kit foxes and swift foxes rarely consume fruits. Most of the species also are seasonally insectivorous and may consume large quantities of insects. All of the species exhibit extremely flexible foraging patterns, and thus trophic interactions can be quite complex.

Predators that can switch to alternate food resources sometimes can limit abundance of primary prey species. Foxes do not appear to regulate; primary prey populations. To the contrary, fox populations more likely are regulated by primary prey abundance (see Population Dynamics). However, red foxes in conjunction with other mesopredators such as skunks (*Mephitis* spp.) and raccoons (*Procyon lotor*) may limit waterfowl productivity in prairie regions (Sargeant 1972 Cowardin et al. 1985; Sargeant et al. 1993). Also, foxes may include uncommon species as secondary prey, thereby contributing to their rarity. This is particularly problematic where foxes have been introduced into previously unoccupied areas (see Economic Status and Management).

Dens. All six North American fox species use earthen dens, some of which they dig themselves. These dens may be used by other species either concurrently or after being vacated by the foxes. This is particularly true of kit foxes and swift foxes because they use or excavate numerous dens, most of which are vacant at any given time. Although distinctive vegetation changes may be associated with older arctic fox and kit fox dens (Egoscue 1962; Chesemore 1975), these changes are very localized and widely scattered, and therefore probably have little effect on ecosystem functions.

Most North American fox species use dens primarily for bearing and rearing pups, but kit and swift foxes use dens virtually every day of their lives. All species are capable of excavating their own dens, but also will readily use dens excavated by other species. Some dens are "traditional" in that they are used over multiple years. Some of these traditional dens increase in size over time and can become rather extensive, with numerous entrances.

Arctic foxes primarily use dens during the pup-rearing period, but occasionally will use them at other times of year during inclement weather. In winter, they may dig burrows in the snow for shelter during blizzards (Chesemore 1975; Eberhardt 1977). Arctic fox dens can be found in open tundra, sand dunes, talus slopes, alpine fellfields, rock outcrops, and under boulder piles (Murie 1959; Stephenson 1970; Speller 1972; Fay 1973). MacPherson (1969) suggested that optimal den sites were locations with reduced accumulations of winter snow, good exposure to spring sun, protection from severe summer winds, and sufficient elevation above water and permafrost. In tundra, den construction is limited to sites where soils thaw to at least 1 m above the permafrost. Therefore, dens are frequently on mounds 1–4 m high or on ridges (Chesemore 1967; Underwood 1975; Eberhardt 1977; Anthony 1996). Most dens occur in sandy soils, where digging is easier, and have a south-facing exposure (Chesemore 1969; Underwood 1975). Optimal sites may be somewhat limited, resulting in multiple dens being clustered on favorable sites and dens being used repeatedly (Chesemore 1969). MacPherson (1969) suggested that some dens may have been in use for several centuries.

Arctic fox dens commonly are expanded each spring before parturition. Thus, size and number of entrances increase with time. Den complexes are roughly circular in shape and may cover from 3 to 250 m^2 (Chesemore 1967). The number of entrances can range from 1 to >100 (Chesemore 1969; MacPherson 1969; Underwood 1975; Eberhardt 1977; Garrott et al. 1983a). Older, repeatedly used dens commonly develop characteristic vegetation due to soil aeration and the addition of organic material (Chesemore 1969). This diagnostic vegetation consists of lush mats of plants dominated by grasses. MacPherson (1969) classified arctic fox dens as follows: "youthful" dens have few entrances and no diagnostic vegetation, "mature" dens have good mats of diagnostic vegetation and no collapsing entrances, "old" dens have large mats of diagnostic vegetation dominated by grasses and obvious collapsing, and "senile" dens are inactive, with extensive collapsing.

Gray foxes and island foxes also use dens primarily during parturition and pup rearing. Gray foxes may occasionally use dens at other times to avoid predators (Fritzell 1987). Earthen dens are usually modified burrows of other species (Schmeltz and Whitaker 1977) or are dug by the foxes. Dens dug by island foxes tend to be simple tunnels (Laughrin 1977). Both species also will den in wood piles, brush piles, rock crevices, hollow logs and trees, hollows under shrubs, and under abandoned houses (Latham 1943; Laughrin 1973, 1977; Trapp and Hallberg 1975). Gray foxes have been found denning in hollow trees 7.6 m (Grinnell et al. 1937) to 9.1 m (Davis 1960) above the ground. In eastern deciduous forests, gray fox dens commonly have east-, southeast-, or south-facing exposures (Richards and Hine 1953) and are in brushy or wooded areas, where they are less conspicuous than dens of co-occurring red foxes (Layne and McKeon 1956a; Sullivan 1956; Nicholson et al. 1985). Island foxes will reuse dens in successive years (Laughrin 1977). Daytime resting sites of gray foxes commonly are located in dense vegetation (Hallberg and Trapp 1984; Nicholson et al. 1985), but foxes also have been observed resting in the shade of boulders (Trapp 1978) and sunning on top of large rocks and in trees (Grinnell et al. 1937; Yeager 1938).

Kit foxes and swift foxes use dens year-round, and dens are a critical component of the life history of both species. Seton (1929) characterized swift foxes as the most subterranean of North American foxes, although den use by kit foxes is equally extensive. Kit foxes and swift foxes spend most days in dens, although occasionally they will rest or bask aboveground outside of den entrances (Egoscue 1962; Kilgore 1969; Morrell 1972; Kitchen et al. 1999). Dens are used to bear and rear young, escape predators, avoid extreme temperatures, conserve water, and as daytime resting sites. Both species will enlarge badger (*Taxidea taxus*) digs or enlarge burrows of other species. Kit foxes will enlarge burrows of California ground squirrels (*Spermophilus beechyi*) and kangaroo rats (*Dipodomys* spp.); kit foxes and swift foxes will enlarge burrows of prairie dogs (*Cynomys* spp.) (Kilgore 1969; Morrell 1972; Herrero et al. 1986; List 1997). Entrances to the dens of both species are usually sufficiently narrow to restrict entry by larger predators such as coyotes and badgers (Egoscue 1962; Kilgore 1969). Dens dug by kit foxes and swift foxes are usually in loose-textured, well-drained soils (Egoscue 1956; Cutter 1958b; Kilgore 1969; Morrell 1972). Multiple dens are located within the home ranges of both species, and are primarily used by the members of resident family groups (Egoscue 1956; Kilgore 1969; Morrell 1972; Koopman et al. 1998). Many dens are used repeatedly over multiple generations (Egoscue 1956, 1962, 1979).

Kit fox dens have from 1 to 24 entrances (Egoscue 1962), although most have 2–7 (Berry et al. 1987; Reese et al. 1992). Entrances are approximately 20–25 cm high and <20 cm wide (Egoscue 1962), and can have a characteristic "keyhole" shape (Fig. 24.14). Den tunnels average 2–6 m in length (McGrew 1979) and can be up to 3 m deep (Seton 1929). Escape and daytime resting dens are more likely to have single entrances and to be relatively simple. Natal dens, where pups are born and raised, typically have multiple entrances and are more complex, with multiple tunnels and one or more chambers (Egoscue 1962). Natal dens also can be identified by trampled vegetation up to 10 m around the den and the presence of pup-sized scats and prey remains (Egoscue 1962; O'Farrell and Gilbertson 1986). Kit foxes also may den in human-made structures such as culverts or oilfield pipes or under buildings (Egoscue 1956; Morrell 1972; Berry et al. 1987; Reese et al. 1992), but young are almost always born in earthen dens (Zoellick et al. 1987; Spiegel 1996).

Kit fox dens exhibit a clumped distribution with as many as 8–10 per 1–2 ha (Egoscue 1956). This may be because females will

FIGURE 24.14. Den of a kit fox (*Vulpes macrotis*). Note the "keyhole" shape of the entrance. SOURCE: Photo by B. Cypher.

frequently move pups to new dens. The relatively short distance between dens suggests that these switches may not be for predator avoidance, but instead may be a response to flea infestations in dens (Egoscue 1956). Estimates of the average number of dens used by individual kit foxes include 15.5 (Reese et al. 1992) and 11.8 (Koopman et al. 1998), and some individuals may use as many as 49 dens in a year (Reese et al. 1992).

Swift fox dens occur in ridges, slopes, hill tops, pastures, roadside ditches, fence rows, and cultivated fields. Dens typically are on higher ground with well-drained soils and relatively short, sparse vegetation (Cutter 1958b; Kilgore 1969; Pruss 1999). The short vegetative structure around dens probably increases visibility (Cameron 1984). Dens also may be located in rangeland as well as cultivated and fallow cropland (Jackson and Choate 2000). Dens may be relatively close (<100 m) to human habitations (Cutter 1958a) and swift foxes occasionally den in human-made structures such as culverts (Kilgore 1969). Older dens and natal dens typically have two or more entrances (Cutter 1958a; Kilgore 1969; Hillman and Sharps 1978). Den chambers may be up to 100 cm deep (Egoscue 1979).

Red foxes use dens primarily for bearing and rearing young. They may dig their own den or modify and use those abandoned by other species such as woodchucks (*Marmota monax*) or badgers (Samuel and Nelson 1982). Dens dug by foxes are usually in sandy soils. Entrances may be 40 cm high and tunnels can be up to 22.5 m long (Sheldon 1950). Dens may have several entrances; Ables (1975) reported a den with 19 entrances. Some dens may be used over multiple generations (Stanley 1963) and may be enlarged each year (Pils and Martin 1978).

Activity Patterns. All six North American fox species are primarily nocturnal with peaks in activity common during crepuscular periods. Kit foxes and swift foxes are probably the most nocturnal, and island foxes exhibit the greatest diurnal activity. Arctic foxes generally are active during the same hours in winter and summer despite extreme differences in day length between these two seasons (Speller 1972; Eberhardt et al. 1982; Burgess 1984; Garrott et al. 1984). Gray foxes were active in 25–54% of diurnal telemetry locations and in 77–87% of nocturnal locations (Yearsley and Samuel 1980; Haroldson and Fritzell 1984). Greater diurnal activity characteristic of island foxes may be a function of the absence of large predators on the Channel Islands and relative freedom from human harassment (Laughrin 1977). Island foxes on some islands became more nocturnal following high predation rates by colonizing golden eagles (G. Roemer, University of California at Los Angeles, pers. commun., 2001); whether this was a result of adaptation or selection is unknown. Activity patterns of island foxes also may vary with temperature. Nocturnal activity increases during warm seasons and diurnal activity increases during cool seasons (Fausett 1982). Activity levels did not differ between males and females nor between females with and without young (Fausett 1982). Red foxes exhibit increased diurnal activity when pups are young, in winter (Ables 1969; 1975; Henry 1986), and in far northern areas (Murie 1944). Red fox activity periods may coincide with those of primary prey species (Ables 1969; Lovari et al. 1994). Climatological factors influence red fox activity. Foxes were more active on cooler and windier days and nights in summer. Conversely, deep snows and strong winds in winter cause foxes to curtail activity and seek sheltered areas (Ables 1969).

Home Range and Movements. Home range size in North American foxes is not static, but instead varies temporally and spatially. Factors that influence home range size include food abundance, presence of young, habitat quality, presence of natural physical barriers such as rivers and lakes, and intra- and interspecific competition. Therefore, seasonal, annual, and regional differences are observed among all species. Also, most home range estimates are derived in the absence of data on some or all of the factors above and use different techniques. Thus, caution should be exercised when comparing estimates.

Home range size estimates for arctic foxes vary considerably (Table 24.3). Anthony (1997) reported relatively small home ranges (Table 24.3) in an area with a high density of goose nests. Arctic foxes occupy particular ranges during spring and summer while bearing and raising young. Many foxes abandon these ranges in winter and move to areas of greater food abundance. In some coastal areas, foxes may continue to use productive beaches year-round (West and Rudd 1983). In some tundra areas, arctic foxes also may remain in summer ranges over winter (Eberhardt et al. 1983b), but most individuals abandon these areas in late summer or early fall (Garrott and Eberhardt 1987). In years of low food availability, many individuals may adopt a more transient behavior and range extensively (MacPherson 1968).

Daily movements primarily involve searches for food, and males tend to travel further than females (MacPherson 1969). Chesemore (1969, 1975) described two major seasonal movements among arctic foxes in Alaska: seaward in fall and early winter and landward in late winter and early spring. After leaving land-based home ranges in fall and early winter, foxes move out onto sea ice, where food presumably is more plentiful. Movements during this period can be extensive, with some individuals averaging 24 km/day. Arctic foxes have been observed on pack ice 800 km from the nearest land and within 150 km of the true north pole (Wrigley and Hatch 1976). Eberhardt and Hanson (1978) reported seven long-distance movements, which ranged from 128 to 945 km. MacPherson (1968) reported that one fox moved a straight-line distance of 1120 km during a 2-year period. Garrott and Eberhardt (1987) reported that an adult male tagged in northern Alaska moved a straight-line distance of 2300 km before being recovered in northern Canada 3 years later. Occasionally, arctic foxes move southward from

TABLE 24.3. Home range estimates for the six North American fox species

Species	Location	Sex[a]	*n*	Mean Home Range Size (ha)	Source
Arctic fox	Northwest Territories	C	4	290	Speller 1972
	Alaska	C	4	2080	Eberhardt et al. 1982
	Alaska	M	11	1022	Anthony 1997
		F	11	457	
Gray fox	West Virginia	M	3	97	Yearsley and Samuel 1980
		F	1	75	
	Illinois	M	2	136	Follman 1973
		F	4	107	
	Alabama	M	5	653	Nicholson and Hill 1984
		F	15	626	
	Utah	M	4	102	Trapp 1978
		F	4	113	
	Utah	C	4	129	Hallberg and Trapp 1984
	California	F	4	122	Fuller 1978
	Florida	C	3	550	Sunquist 1989
	Missouri	C	7	676	Haroldson and Fritzell 1984
	Alabama and Mississippi	C	11	493	Wooding 1984
Island fox	Santa Cruz Island	M	6	34	Crooks and Van Vuren 1996
		F	6	33	
	Santa Cruz Is.	C	18	55	Roemer et al. 2001
Kit fox	California	C	13	251	Knapp 1978
	California	C	8	434	Koopman 1995
	California	C	21	462	Zoellick et al. 2002
	California	C	21	1160	White and Ralls 1993
	Nevada	C	11	974	Golightly and Hardenbrook 1986
	Arizona	C	7	1120	Zoellick and Smith 1992
	Utah	C	11	310	O'Neal et al. 1987
Swift fox	Colorado	M	3	191	Fitzgerald et al. 1981, cited in Scott-Brown et al. 1987
		F	1	86	
	Colorado	C	54	760	Kitchen et al. 1999
	Nebraska	M	4	1730	Hines 1980
		F	3	1240	
Red fox	North Dakota	C	19	1190	Ables 1969
	Maine	C	4	1990	Major and Sherburne 1987
	Maine	C	6	1470	Harrison et al. 1989
	Ontario	C	6	900	Voigt and Tinline 1980
	British Columbia	C	6	1610	Jones and Theberge
	Illinois	M	2	975	Storm 1965
	Minnesota	M	3	699	Sargeant 1972

[a]M, Males; F, Females; C, both sexes combined.

tundra regions into boreal forest areas. These movements seem to involve large numbers of individuals and may be related to population crashes among rodents, particularly lemmings (*Dicrostonyx* spp. and *Lemmus* spp.), in tundra areas (Elton 1931; Chesemore 1975; Wrigley and Hatch 1976).

Among gray foxes, home range size varies immensely (Table 24.3) with habitat quality and the distribution of food resources (Fuller 1978). Home range size also may vary with sex. Male ranges generally are larger, although Chamberlain and Leopold (2000) reported no differences between sexes in Mississippi. Reported home range sizes vary from 13 ha (Richards and Hine 1953) to 2755 ha (Nicholson 1982).

Gray fox home ranges are generally elongate in shape (Follman 1973; Montague 1975; Trapp 1978), and home ranges seem to be larger in eastern compared to western North America (Fritzell 1987). Home range size of gray foxes varies seasonally. Ranges for both sexes increase in late fall and winter, possibly in response to breeding activity or declining food availability (Follman 1973; Nicholson 1982; Chamberlain and Leopold 2000). When pups are young, female ranges decrease by 80% (Nicholson et al. 1985), but male ranges may increase, probably because they are foraging for food for the females and pups as well as themselves (Follman 1973; Nicholson et al. 1985). In Utah, Trapp (1973) estimated average nightly movements of 600.5 m for four females and 475.5 m for five males. In Mississippi, movement rates for both sexes were greatest during pup rearing and in winter and lowest during breeding (Chamberlain and Leopold 2000).

Home range size also can vary with sex and season among island foxes (Table 24.3). Laughrin (1977) and Fausett (1982) reported that female home ranges were similar in all seasons. During spring and summer, male home ranges approximated those of their female mates. However, during fall and winter, males increased their ranges and overlapped extensively with ranges of other males, and also encompassed the ranges of two to three females. Daily movements do not differ significantly between sexes or age classes. Among 5- to 6-year-old foxes, daily distances of 1.9–2.7 km were observed (Laughrin 1977).

Among kit foxes, home range size usually does not differ between sexes, and most estimates reported are for both sexes combined (Table 24.3). Differences in home range size among areas are probably related to differences in food availability (Spiegel 1996; Zoellick et al. 2002). The relatively large home ranges (Table 24.3) reported by White and Ralls (1993) were recorded during a period of low food availability. In Utah, Daneke et al. (1984) reported home range sizes of 190 ha for males and 140 ha for females in summer and 420 ha for males and 190 ha for females in winter; thus, sex and season differences were observed.

Nightly movements of kit foxes may vary seasonally. In California, Zoellick et al. (2002) reported nightly movement distances of 10.7 km during the pup-rearing season (mid-February–May), 9.4 km during the pup-dispersal season (May–September), and 14.6 km during the breeding season (December–mid-February). In Arizona, kit foxes exhibited more extensive nightly movements during the breeding season and more restricted movements during gestation and pup rearing (Zoellick et al. 1989). Males exhibited higher nightly movements (14.3 km) than females (11.8 km) in Arizona (Zoellick et al. 1989), as also was observed in California (Koopman 1995).

Home range size of swift foxes is similar to that of kit foxes (Table 24.3). Covell et al. (1996) estimated nightly movements of 13.0 km in winter and 5.7 km in summer. Kitchen et al. (1999) found that movement rates were similar between sexes, but varied seasonally with rates of 0.4 km/hr during the pup-rearing season and 0.74 km/hr during the dispersal season.

Home range sizes reported for red foxes are extremely variable (Table 24.3). Some of this variation is attributable again to differences in methodologies, but much of it is a reflection of the diversity of environments inhabited by red foxes and the variation in attributes among these environments. Home ranges in tundra regions tend to be relatively larger due to a lower abundance of prey suitable for red foxes (Jones and Theberge 1982). Conversely, in some urban environments, where food availability can be quite high, red fox home ranges may be as small as 10 ha (Hersteinsson and Macdonald 1982). Seasonal variation also occurs, with ranges being larger in winter, probably in response to lower food availability (Sheldon 1950).

Daily movements of red foxes may cover 10 km (Storm 1965; Macdonald 1981; Voigt 1987). For a few weeks after pups are born, the adults may remain within 0.8 km of the den (Scott 1943). Movement patterns within home ranges are strongly influenced by the distribution of food resources (Ables 1975).

Habitat. Virtually all terrestrial habitats in North America are occupied by one or more fox species. Arctic foxes primarily occur in arctic habitats, although they occasionally range southward into subarctic regions. Garrott and Eberhardt (1987) defined four major categories of arctic fox habitat: coastal, inland, alpine, and marine. Coastal habitat is usually associated with islands in the Bering Sea and North Atlantic Ocean and portions of the Greenland coast. This habitat occurs on the coastline, where arctic foxes forage on animals and carrion in beach and intertidal zones, and on nesting seabirds on rocky outcrops and cliffs. Inland habitat occurs in continental areas and possibly the interior of larger islands. It is primarily tundra habitat, characterized by continuous permafrost just below a shallow active-soil layer, lack of trees, and a prevalence of low-growing shrubs, herbs, and grasses. Primary food resources are rodents and ptarmigans (*Lagopus* spp.). Alpine habitat is characterized by alpine tundra and is similar to inland tundra habitat; alpine habitat is used by arctic foxes in Europe, but not North America. Marine habitat consists of pack ice in the Arctic Ocean. In North America, most breeding occurs in inland and coastal habitats (MacPherson 1969; Chesemore 1975; Eberhardt 1977), and marine habitat is used in winter. The occurrence of arctic foxes in boreal forest areas appears to be associated with food shortages in tundra regions (Elton 1931; Chesemore 1975; Wrigley and Hatch 1976). Arctic foxes also are occasionally observed around villages and other human habitations, where they are attracted by anthropogenic food sources (Eberhardt 1976).

Gray foxes were considered to be primarily a woodland species closely associated with deciduous hardwood forests of eastern North America (Hall 1981). Indeed, gray foxes use a greater proportion of wooded habitats than any of the other North American fox species. However, they also thrive in mixed agricultural/woodland landscapes and a diversity of western North American habits, although brush and tree cover are critical habitat components. In eastern North America, gray foxes prefer a mix of fields and woods rather than large tracts of homogeneous forest (Wood et al. 1958). Areas well interspersed with forest and farmland are high-quality gray fox habitat in Wisconsin (Richards and Hine 1953; Petersen et al. 1977), Georgia (Wood et al. 1958), Virginia (Carey et al. 1978), and Texas (Wood 1952). In Illinois, gray foxes use early old fields more than expected based on availability, agricultural and brushy habitats less than expected, and woodlands in proportion to their availability (Follman 1973). Old fields appear to be used for foraging (Follman 1973; Yearsley and Samuel 1980), whereas woodlands, particularly dense or brushy stands, commonly are used for daytime cover (Follman 1973; Yearsley and Samuel 1980; Hallberg and Trapp 1984; Haroldson and Fritzell 1984). In Mississippi, gray foxes use mature pine (*Pinus* spp.) more than expected, possibly because small mammals are more abundant in these habitats (Chamberlain and Leopold 2000). Habitats used in Florida included scrubby woodlands, old fields, and citrus groves (Wasmer 1984). Habitats used in Texas included post oak (*Quercus stellata*) woodlands, pinon–juniper (*Pinus–Juniperus*) woodlands, and wooded areas within short-grass prairie (Trapp and Hallberg 1975).

In western North America, habitats used by gray foxes include oak woodlands, chaparral, pinon–juniper woodlands, brushy washes and meadows, and riparian forests. In California, gray foxes used riparian and old field habitats more than expected, and agricultural habitats less than expected (Fuller 1978). Old fields also were used more than expected in Utah (Trapp 1978). In California, gray foxes were most abundant in oak woodlands at elevations of 1150–1525 m; lower elevation habitats generally are too arid for gray foxes, and dense conifer forests occur at higher elevations (Grinnell et al. 1937; Johnson et al. 1948). Gray foxes also use reclaimed surface mines with heterogeneous vegetative cover (Yearsley and Samuel 1980) and do well on urban fringes (Harrison 1997).

Island foxes occur in all natural habitats on the six occupied Channel Islands (Laughrin 1977, 1980). These include valley and foothill grasslands, dunes, sage and cactus scrub, chaparral, oak woodlands, riparian woodlands, pine forests, and coastal marsh. Woodlands are preferred over grassland habitats (Laughrin 1977).

Kit foxes are adapted to desert and semiarid habitats that primarily include desert shrub and shrub–grass communities. These communities are dominated by saltbush (*Atriplex* spp.), creosote bush (*Larrea tridentata*), greasewood (*Sarcobatus vermiculatus*), and sagebrush (*Artemisia tridentata*) (McGrew 1979). Kit foxes prefer areas with sparse ground cover. In Utah, 75% of kit fox sightings were in areas with <20% ground cover (McGrew 1977). Areas with loose-textured soils are preferred (Egoscue 1962; Laughrin 1970; McGrew 1977). Kit foxes also prefer flat or gentle terrain (slopes <5%) over more rugged terrain, where predation risk may be greater (Warrick and Cypher 1998). Kit foxes occur in areas adjacent to irrigated cropland (Swick 1973; Morrell 1975), and occasionally will use nut and fruit orchards (G. Warrick, Center for Natural Lands Management, pers. commun, 2001). Kit foxes also occur near and within some urban areas (Jensen 1972; Cypher and Warrick 1993).

Swift foxes primarily occur in short-grass and mixed-grass prairies (Egoscue 1979). Historically, they also may have occupied drier, well-drained portions of tall grass prairies in western Iowa (Allen 1870, cited in McGrew 1979) and possibly western Minnesota (Swanson et al. 1945). Optimal vegetation structure is sparse, short ground cover with shrubs absent or sparsely distributed (Scott-Brown et al. 1987). Much of the remaining natural habitat is grazed (Egoscue 1979; Uresk and Sharps 1986). In addition to using rangeland, swift foxes also are found in cultivated and fallow cropland (Cutter 1958b; Jackson and Choate 2000; Sovada et al. 2001).

Consistent with their immense ecological plasticity, red foxes inhabit diverse habitats including arctic tundra, boreal forest, deciduous forest, prairie and grasslands, shrublands, semiarid desert, agricultural landscapes, and urban environments. Heterogeneous and fragmented landscapes characterized by high habitat diversity and interspersion are preferred (Ables 1975; Lloyd 1980; Catling and Burt 1995). Such areas include landscapes where woodlots are interspersed with croplands and pastures (Voigt 1987), and red foxes also can use intensively cultivated landscapes (Ables 1975; Sargeant et al. 1975; Hewson 1986). Dense forests seem to constitute lower quality habitat (Lloyd 1975). Follman (1973) reported that wooded areas were used more in winter by red foxes in Illinois, but others have reported that forest stands are avoided in winter because of greater snow accumulations (Halpin and Bissonette 1988; Theberge and Wedeles 1989). Food availability may be the most important factor determining habitat use (Jones and Theberge 1982; Halpin and Bissonette 1988; Phillips and Catling 1991), and areas with higher habitat diversity may provide more food in all seasons. Voigt (1987) suggested that a lack of suitable foods in some seasons may limit red fox abundance in arctic tundra regions. Presence of competitors, particularly other larger canids such as coyotes, also may limit abundance in some habitats (Voigt and Earle 1983; Sargeant et al. 1987). Red foxes also thrive near and in some urban areas, and urban red fox populations in England are infamous (Harris 1977; Macdonald and Newdick 1982; Harris and Rayner 1986). Red foxes also occur in some North American urban areas such as Orange County, California (Lewis et al. 1993), and Toronto (Rosatte et al. 1991). Food is abundant in urban environments, and daytime resting sites may be the primary limiting factor (Harris 1977).

Population Densities. Relatively few density estimates are available for North American foxes. Because of large ranges, secretive habits, and capture difficulty, density estimates are difficult to obtain. Also, fox abundance, and therefore density, is highly variable both temporally (annually and seasonally) and spatially (within and among habitats). Factors that affect density include habitat quality, food availability, interspecific competition, disease, and human exploitation. Also, a variety of methods have been used to estimate density. Thus, density estimates should be considered location and time specific, and caution should be exercised when comparing estimates derived using different methods.

Arctic fox density is particularly difficult to estimate because foxes may occupy distinct ranges in summer during breeding and pup rearing, and then in winter may abandon these summer areas and range widely. Following a population crash in Iceland, arctic fox density was estimated at 0.086/km^2 (Hersteinsson and Macdonald 1982). For gray foxes, Grinnell et al. (1937) estimated a density of 0.4/km^2. Lord (1961) reported a density of 1.2–1.5/km^2 in Florida. Errington (1933) reported a density of 10.4/km^2 in Wisconsin, but this is considered to be highly unusual (Trapp and Hallberg 1975).

Possibly because of limited predators and dispersal opportunities on the Channel Islands, island fox densities are generally higher than those of mainland gray foxes (Roemer et al. 1994; Roemer et al. 2001). Estimates of island fox density based on line transect methods include 0.3/km^2 on Santa Catalina Island, 1.2/km^2 on San Nicolas Island, 2.7/km^2 on San Miguel Island, 4.2/km^2 on both San Clemente Island and Santa Rosa Island, and 7.9/km^2 on Santa Cruz Island (Laughrin 1980). More recent density estimates include 7/km^2 on Santa Cruz Island, 4.8–8.0/km^2 on San Clemente Island, and from near zero to 15.9/km^2 on San Miguel Island (Roemer et al. 1994; Coonan et al. 1998). Densities on San Clemente Island have been reported to be relatively stable, whereas densities on other islands have fluctuated markedly (Coonan 2001). Island fox densities are higher in woodland habitats because of greater food availability (Laughrin 1977, 1980). On Santa Rosa Island, Laughrin (1977) estimated densities of 5.8/km^2 in woodland habitats and 2.4/km^2 in grassland habitats. In general, island foxes are less dense and more evenly distributed on islands with low topographic relief, reduced habitat diversity, and limited woody vegetation (Laughrin 1977).

In Utah, kit fox density was estimated at 1 pair/9.2 km^2 (0.2 foxes/km^2), with 1 pair/2.6 km^2 (0.8 foxes/km^2) in optimal habitat (Egoscue 1956). During a 3-year population decline in Utah, kit fox density declined from 0.2/km^2 to 0.1/km^2 (Egoscue 1975). A number of density estimates have been derived for endangered San Joaquin kit foxes in central California. Grinnell et al. (1937) estimated that density before 1925 was about 0.4/km^2. Laughrin (1970) estimated rangewide density in 1969 at 0.2–0.5/km^2. On a small study site (2.6 km^2), Morrell (1972) estimated a density of 2.3/km^2, but this likely included the resident pair, young of thc ycar, and some neighboring foxes. A range-wide density of 0.6/km^2 was estimated in 1974–1975 (Morrell 1975). More recently, White et al. (1996) reported that kit fox density declined from 0.24/km^2 in 1989 to 0.15/km^2 in 1991 during drought conditions. At another study site during the drought, density declined from 1.0/km^2 in 1988 to 0.5/km^2 in 1990 (Berry and Standley 1992). After the drought, Spiegel (1996) estimated density at 1.2/km^2. At a study site where the kit fox population was monitored from 1981 to 1995, annual density based on mark–recapture ranged from 0.2 to 1.7/km^2 (Cypher et al. 2000). Abundance also varied spatially, with kit foxes being more abundant in flat or gentle terrain and less abundant in more rugged terrain (Warrick and Cypher 1998; Cypher et al. 2000).

For swift foxes, Banfield (1974) estimated a density of 0.2/km^2 on "ideal range." For red foxes, a density in southern Ontario was estimated to be 1.0/km^2 in spring when pups were present (Voigt 1987). In northern boreal forests and arctic tundra, density is much lower and was estimated to be 0.1/km^2 (Voigt 1987). Density for an urban fox population in Toronto was estimated at 1.0–1.3/km^2 (Rosatte et al. 1991). In urban areas of England, where food availability is high, red fox densities have been estimated to be 2.1–30/km^2 (Harris 1977; Page 1981; Macdonald and Newdick 1982; Harris and Rayner 1986).

Population Dynamics. All North American fox species exhibit annual fluctuations in abundance, some of which can be dramatic. The causal factor driving these population fluctuations usually is food availability. However, in some instances disease (e.g., rabies) and human exploitation can cause marked changes in fox abundance. In food-related fluctuations, the causal mechanism typically is reduced reproductive success and lower juvenile survival, particularly among young pups. Adult survival also may be affected as individuals forage further and longer, thus increasing the risk of mortality from predation and other sources. In fluctuations related to disease and exploitation, the causal mechanism is increased direct mortality among all age classes.

Arctic fox populations exhibit distinct cycles with peak abundance occurring every 3–5 years. The causal factor in these cycles is the abundance of small mammals, particularly lemmings (Elton 1942; Chitty 1950p; Chesemore 1967). A lag effect sometimes is evident in that peak fox abundance may follow peak rodent abundance by 1–2 years (Chitty 1950; MacPherson 1969; Angerbjörn et al. 1999). In Iceland, Hersteinsson et al. (1989) reported fluctuations in fox abundance related to 10-year cycles in ptarmigan abundance. Availability of other foods, such as seal carcasses, also may affect annual fox abundance (Hiruki and Stirling 1989). When food availability decreases, arctic fox numbers can decline by over 80% (Strand et al. 2000). During declines in arctic fox populations, fewer and smaller litters are observed (Hiruki and Stirling 1989). Also, aggression among siblings may intensify and smaller, weaker pups may be killed by larger, stronger pups (MacPherson 1969). Disease, parasites, predation, competition, human activities, and weather are secondary factors in arctic fox population fluctuations and primarily affect local abundance (Tchirkova 1958a, cited in Underwood and Mosher 1982).

Few data are available on gray fox population dynamics. Richmond (1952) reported that abundance of gray foxes in Pennsylvania was highest in years in which the period of January–March was wetter and warmer than average and lowest in years in which this period was colder and drier. Also, relatively little is known about island fox population dynamics, although anecdotal information suggests that abundance historically may have fluctuated widely (Laughrin 1980). Prolonged drought may cause population declines (Laughrin 1977; Crooks 1994). During a 5-year study in chaparral–woodland habitat on Santa Cruz Island, density ranged from 5.9 to 9.8/km^2. From 1971 to 1977 on San Nicolas Island, density declined from 2.7 to 0.12/km^2 (Laughrin 1980). More recently, marked population declines have been attributed to two factors. Golden eagles have emigrated to several islands and have preyed heavily on island foxes. On San Miguel Island, island fox density declined from 15/km^2 in 1994 to 1/km^2 in 1999 (David 2001). On Santa Catalina Island, an anthropogenic introduction of distemper resulted in an epidemic among foxes and drastically reduced their numbers (Coonan 2001).

Typical of many species in arid environments, kit foxes also exhibit marked population fluctuations. During a 15-year study at a site in California, the highest and lowest kit fox numbers were observed within a 3-year period (Cypher et al. 2000). These fluctuations appear to be driven by food availability, which in turn is strongly influenced by annual precipitation levels (Ralls and Eberhardt 1997; White and Garrott 1997; Cypher et al. 2000). In Utah, kit fox abundance declined as black-tailed jackrabbit (*Lepus californicus*) abundance declined (Egoscue 1975). During this decline, the number of nonbreeding female foxes increased and litter size decreased among foxes that did breed. In California, kit fox abundance declined markedly during the late 1980s as food availability decreased in response to drought conditions. Precipitation affects primary productivity, which in turn affects prey populations. As discussed previously, reproductive success among foxes is strongly influenced by food availability. Consistent with this, reduced reproductive success occurred among kit foxes during the drought (White and Ralls 1993; Spiegel 1996; Cypher et al. 2000). At least a 1-year lag effect is evident in the response of kit foxes to environmental conditions because effects of annual precipitation on primary productivity and prey abundance occur after the kit fox reproductive season (Egoscue 1975; Ralls and Eberhardt 1997). Furthermore, this relation appears to be even more complex in that precipitation from the previous 3 years may have a cumulative effect on fox abundance (Cypher et al. 2000).

Kit foxes are subject to intense interference and exploitative competition from other species, particularly coyotes (White et al. 1995; Cypher and Spencer 1998). However, this competition does not drive kit fox population dynamics. Instead, competitors influence local abundance and distribution of kit foxes (Warrick and Cypher 1998) and may affect the amplitude of population fluctuations (Cypher et al. 2000). Disease is generally not an important factor in kit fox population dynamics. However, White et al. (2000) offered evidence that rabies may have played a role in the decline of a kit fox population in California. Few data are available on swift fox population dynamics, but the factors that influence population dynamics probably are similar to those for kit foxes.

As with the other species, red foxes also exhibit population fluctuations that are a function of food-mediated variation in reproductive success. In northern Sweden, red fox numbers peak every 3–4 years in response to vole abundance (Lindström 1980). Similar cycles are not observed in southern Sweden, however, where vole populations do not fluctuate greatly and where alternate prey are available (von Schantz 1981a). In Illinois, red fox numbers declined in response to decreasing food availability during years of low rainfall (Scott and Klimstra 1955). Some population fluctuations among red foxes in North America are caused by rabies epizootics (Johnston and Beauregard 1969). In eastern Canada, rabies outbreaks would reduce red fox numbers locally, and in 2–3 years, when foxes had increased sufficiently, another epizootic would occur, resulting in cyclic-appearing dynamics.

Sex and Age Structure. Sex and age ratios are dynamic and vary temporally and spatially, and "characteristic" ratios may be difficult to define. In addition to biological and environmental influences, anthropogenic factors, particularly harvests, can alter sex and age ratios. Sex and age ratios are frequently estimated from livetrapping or harvests. However, capture vulnerability commonly varies with age class (e.g., juveniles are more vulnerable) and sometimes with sex. Therefore, sex and age ratios should be considered specific to a given time and location.

For arctic foxes, adult sex ratios near 1:1 are typical (Petrides 1950; Richards and Hine 1953; Layne 1958; Wood 1958; Carey et al. 1978; Root 1981; Wigal and Chapman 1983; Root and Payne 1985). Hiruki and Stirling (1989) also reported a 1:1 adult sex ratio. They also reported that the proportion of juveniles (<1 year of age) was 8% in a poor reproduction year, but was 71–86% in years of higher productivity. For gray foxes, fetal and neonatal sex ratios generally are not significantly different from 1:1 (Layne 1958). Of 102 fetuses examined in three studies, 54 were male and 48 were female (Sheldon 1949; Layne and McKeon 1956b; Wood 1958). Layne and McKeon (1956b) reported male-biased ratios of 1.17:1 to 1.87:1 among 1132 adult gray foxes in New York, but these values were based on trapping data and may have reflected greater male vulnerability to capture. Davis and Wood (1959) reported that 48–61% of gray fox populations consisted of individuals <1 year old. Over a 3-year period in Wisconsin, 60–72% of females examined were <1 year old (Richards and Hine 1953), whereas just 7% of gray foxes examined in Alabama were juveniles (Sullivan and Haugen 1956). Of 636 gray foxes harvested in Wisconsin, 66% were juveniles (<1 year), 16% were yearlings (1.5 years), and 15% were adults (>1.5 years) (Root and Payne 1985).

For adult island foxes, Crooks (1994) reported a male:female ratio of 1:1.2 ($n = 128$), which did not differ significantly from 1:1. Similarly, ratios of 1.6:1 for adults ($n = 42$) and 0.75:1 for pups ($n = 28$) did not differ significantly from 1:1 (Roemer et al. 2001). Laughrin (1977) reported a sex ratio of 1:1 among pups at 24 dens. Compared to mainland gray foxes, a higher proportion of individuals 4–6 years old was observed among island foxes. This older age structure was attributed to lower mortality among island foxes (Laughrin 1977, 1980). A mean adult/juvenile ratio of 3.85 was reported across all occupied islands (Laughrin 1977). Crooks (1994) reported that >75% of individuals captured were juveniles on Santa Cruz Island in 1991–1992, and speculated that the population was in a growth phase following 6 years of drought.

For kit foxes, Egoscue (1962) reported that sex ratios among adults and pups were slightly male biased in Utah in the 1950s. During a 4-year period in Utah when food availability was initially declining and then began increasing, the percentage of males among adults was 46% in 1966, 62% in 1967, 56% in 1968, and 50% in 1969. Males among pups went from 67% in 1968 to 27% in 1969. Egoscue (1975) suggested that the proportion of males increases as kit fox abundance increases relative to food availability. Furthermore, Egoscue (1962, 1975) reported a slight male bias among resident adults, but a strong male bias among unpaired and transient kit foxes. In California, annual sex ratios among adult kit foxes usually approximated 1:1 during 1981–1995, but a strong male bias (up to 1.62:1) was observed during 1992–1994, when the population was growing rapidly. A similar male bias was observed among pups during this same period (Cypher et al. 2000). Spiegel (1996) reported equal or female-biased sex ratios at a site in California during a period when kit fox density was relatively low while food availability was increasing rapidly. During a 4-year period in Utah, the average age of kit foxes was about 2 years (Egoscue 1975). Adult/juvenile ratios in California averaged 1.65 over 15 years, but ranged from 0.83 to 8.38. The higher adult ratios were observed during years of low food availability, when reproductive success also was low (Cypher et al. 2000). During a period when kit fox numbers were increasing rapidly in California, Spiegel (1996) reported an adult:juvenile ratio of 1:2.

For swift foxes, sex ratio at birth for 76 pups in captive populations approximated 1:1 (Schroeder 1985). Of 66 adult and juvenile swift foxes captured in Colorado, 53% were male and 47% were female (Fitzgerald et al. 1981, cited in Scott-Brown et al. 1987). Also in Colorado, Cameron (1984) reported sex ratios of 1.3 males per 1 female for adults and 1.1 females per 1 male for juveniles, neither of which was significantly different from 1:1. Among red foxes, relatively equal fetal sex ratios were observed by Sheldon (1949) and Storm et al. (1976), but Layne and McKeon (1956b) reported that 61% of fetuses were male. Storm et al. (1976) reported a significantly higher proportion of male pups in dens.

Interspecific Interactions. Island foxes, because of their insular and isolated distribution, are the only species not subjected to interspecific interactions with other fox species. In the narrow contact zone in eastern New Mexico and western Texas where kit foxes and swift foxes overlap, interbreeding commonly occurs between these closely related species (Thornton et al. 1971; Rohwer and Kilgore 1973; Thornton and Creel 1975). However, all other interactions between sympatric foxes are competitive, and this competition can be intense, with larger species dominating smaller species. These interactions include interference competition (e.g., direct mortality, spatial exclusion) and exploitative competition (e.g., competition for food, dens, and other resources).

Red foxes, because of their wide distribution across diverse habitats, interact with several other fox species. In arctic regions, they compete with arctic foxes. Red foxes are more aggressive than arctic foxes (Rudzinski et al. 1982). Arctic foxes avoid red foxes (Schamel and

Tracy 1986) and are occasionally killed by them (Frafjord et al. 1989; Bailey 1992). The two species also compete for food and den sites (Hersteinsson et al. 1989). Hersteinsson and Macdonald (1992) suggested that red foxes may limit the southern distribution of arctic foxes. As red foxes are expanding their distribution northward in North America, they may be displacing arctic foxes (Chesemore 1967; MacPherson 1969). Introduced red foxes have eliminated arctic foxes from some Aleutian Islands. They have been proposed as a biological control strategy to eliminate introduced arctic foxes from additional islands, where they adversely impact indigenous bird species (West and Rudd 1983; Bailey 1992).

Red foxes also are dominant over kit foxes and swift foxes. For the most part, habitats occupied by kit foxes are too arid for red foxes. However, red foxes have been increasing in some portions of the kit fox range, particularly California, and this increase has been facilitated by anthropogenic landscape changes such as irrigated agriculture and urban development (Jurek 1992; Lewis et al. 1993). Red foxes will kill kit foxes (Ralls and White 1995; Clark 2001), and also compete with kit foxes for food and den sites (Clark 2001; Cypher et al. 2001). Red foxes may have similar detrimental impacts on swift foxes (A. Moehrenschlager, Calgary Zoo, pers. commun., 2000).

Interactions between red foxes and gray foxes are unclear. Historically, overlap between these species may not have been extensive, with red foxes primarily using more open and mesic habitats and gray foxes using more wooded as well as some semiarid habitats. Forest clearing and other anthropogenic changes may have produced favorable habitat conditions for red foxes, which allowed them to increase in areas primarily occupied by gray foxes (Churcher 1959; Godin 1977). In Illinois, red fox and gray fox home ranges overlapped considerably (Follman 1973). Saggese and Tullar (1974) reported possible predation on a red fox pup by a gray fox. Hockman and Chapman (1983) suggested that gray foxes might have a competitive advantage in altered landscapes because of a higher degree of omnivory. Rue (1969) provided anecdotal data indicating that gray foxes are dominant over red foxes, and Richmond (1952) concluded that gray foxes could displace red foxes. However, Trapp and Hallberg (1975) suggested that gray fox distribution may be limited by competition from red foxes. In general, habitat and food partitioning may reduce competition between these two species (Cypher 1993). Sargeant (1982) suggested that where red foxes and gray foxes were sympatric, the combined density of both species approximated that which would have been achieved by one species in the absence of the other. The affinity of gray foxes for brushy or woody cover and the aversion of kit foxes and swift foxes for such habitat limits interactions between gray foxes and these species.

North American foxes also have positive and negative interactions with coyotes and wolves. Coyotes engage in interference and exploitative competition with all fox species except island foxes. Coyotes will kill foxes, exclude them from some areas, and compete for some foods. Fox mortality from coyotes appears to be due to competition rather than predation because the fox carcasses frequently are not consumed (Spiegel 1996; Cypher and Spencer 1998). This competition can reduce or even significantly limit fox populations in some areas. Negative impacts to red foxes attributable to coyotes have been documented in a number of locations. Coyotes will opportunistically kill red foxes (Sargeant and Allen 1989). Red foxes will avoid areas and habitats heavily used by coyotes (Dekker 1983, 1989; Major and Sherburne 1987) and red fox home ranges frequently are located on the periphery of and between coyote home ranges (Voigt and Earle 1983; Sargeant et al. 1987). In Ontario, red foxes and coyotes used some areas in common, but at different times (Voigt and Earle 1983). In Yukon Territory, red foxes and coyotes also exhibited overlap in space as well as food use, but differences in habitat preferences provided some spatial segregation and facilitated sympatry (Theberge and Wedeles 1989). Interestingly, Gese et al. (1996) reported that coyotes and red foxes overlapped spatially in Wyoming and coyotes appeared to tolerate the presence of red foxes, but this may have been a function of abundant food resources. Coyote were reported to limit red fox abundance in North Dakota (Sargeant et al. 1987) and Maine (Harrison et al. 1989), and fox numbers increased in North Dakota when coyote numbers declined (Johnson and Sargeant 1977).

Coyotes also will opportunistically kill gray foxes (Wooding 1984; B. L. Cypher unpublished data) and may limit their abundance in some areas. Coyote and gray fox abundance were inversely related in California (Crooks and Soulé 1999), and gray fox numbers increased following coyote removal in Texas (Henke and Bryant 1999). However, habitat partitioning, food partitioning, and tree climbing may facilitate coexistence between gray foxes and coyotes (Wooding 1984; Cypher 1993).

Coyotes also will kill kit foxes (Cypher et al. 2000) and swift foxes (Carbyn 1998), and generally are the primary source of mortality for these species. Thus, coyotes can have a significant impact on the populations of these species, and at times may limit their abundance. Increased swift fox abundance was reported in an area of intensive coyote control (Linhart and Robinson 1972). However, coyotes naturally occur everywhere that kit foxes and swift foxes occur, and therefore some degree of coevolution has occurred. Both fox species have adaptive strategies that help reduce competition and facilitate coexistence. Some resource partitioning occurs relative to food availability, kit foxes may exhibit greater dietary breadth, and year-round den use by foxes and the presence of multiple dens within home ranges allow foxes to avoid coyotes (Kilgore 1969; White et al. 1994, 1995; Cypher and Spencer 1998; Kitchen et al. 1999). Also, in an interesting interaction among species, the presence of coyotes potentially provides a benefit to kit foxes and swift foxes by limiting red fox abundance (Cypher et al. 2001). Red foxes historically were uncommon or absent in habitats occupied by kit foxes and swift foxes, and therefore these species have not evolved strategies for mitigating competition from red foxes. Because of anthropogenic landscape modifications, red foxes are increasing within the range of kit foxes and swift foxes and pose a significant threat to these species (U.S. Fish and Wildlife Service 1998). Interspecific aggression from coyotes appears to limit red fox abundance in kit fox and swift fox habitat.

Wolves have been reported to occasionally kill foxes (Mech 1966; Chesemore 1975; Larivière and Pasitschniak-Arts 1996). However, this is not common, nor does it affect fox population dynamics. In general, wolves likely confer a benefit on foxes by providing carcasses on which foxes can scavenge. Arctic foxes, swift foxes, and red foxes all have been observed scavenging on the carcasses of large animals killed by wolves (MacPherson 1969; Garrott et al. 1983b; Hersteinsson et al. 1989). Indeed, arctic foxes may be detrimentally affected by elimination of wolves (Hersteinsson et al. 1989). In the interesting dynamics of canid interactions, wolves may provide another benefit to foxes by reducing coyote abundance. Through interspecific competition, wolves may reduce coyote abundance, which relieves the competitive pressure exerted on foxes by coyotes (Peterson 1977).

Foxes also interact with a number of other species (see Mortality for other species that kill foxes, and Feeding Habits for species preyed on by foxes). Arctic foxes frequently scavenge polar bear (*Ursus maritimus*) kills (Seton 1929; Chesemore 1975; Hiruki and Stirling 1989). Latham (1952) and Richmond (1952) observed that weasel (*Mustela* spp.) populations in Pennsylvania were inversely related to the abundance of red and gray foxes. Weasel abundance increased markedly in the vicinity of a poultry farm where gray foxes had been intensively trapped, and then declined again as gray fox abundance increased (Hensley and Fisher 1975). Scott (1943) reported that weasels were frequently killed but not eaten by red foxes, indicating that this may have been a competitive interaction. Primary competitors for island foxes on San Clemente and Santa Catalina Islands were feral cats (*Felis cattus*), which competed with foxes for food and excluded them from certain areas (Laughrin 1973). Spotted skunks (*Spilogale gracilis*) and even some rodents may compete with island foxes for food (Laughrin 1973).

Kit foxes and swift foxes benefit from the presence of badgers in that both species will enlarge badger diggings into new dens (Kilgore 1969; Morrell 1972). Kit foxes also will enlarge kangaroo rat and ground squirrel burrows, and kit foxes and swift foxes will enlarge prairie dog burrows into dens. Also, the vacant dens of both fox species are readily used by a variety of other species including burrowing owls (*Athene cunicularia*), antelope squirrels (*Ammospermophilus leucurus*), striped skunks, deer mice (*Peromyscus maniculatus*), side-blotched lizards (*Uta stansburiana*), Great Plains toads (*Bufo cognatus*), prairie rattlesnakes (*Crotalus viridis*), and a variety of invertebrates (Egoscue 1956; Kilgore 1969). Merriam (1966) reported that red foxes and woodchucks occasionally have been found in the same den.

FEEDING HABITS

Items Used. As is common among most canids, all North American fox species are opportunistic foragers and consume a variety of foods. Feeding habits have been well documented for all species. Small rodents, leporids, birds, insects, fruit, carrion, and anthropogenic foods (e.g., food refuse, pet food) are the primary items in the diet of North American foxes. Although some plant material such as fruits and nuts is purposely consumed, trace amounts of grass and twigs occur frequently in food habit samples. This material likely is incidentally consumed. Foxes also will cache surplus food, which may be consumed within 1–2 days or saved and consumed weeks or even months later during periods of food scarcity. Foraging patterns are strongly influenced by availability of items; foods consumed and their relative contribution to the diet vary by season, year, and location for each species. This variation should be taken into account when examining feeding habit results or comparing results among studies. Also, a variety of techniques has been used to determine fox feeding habits including examining scat contents, stomach contents, prey remains at dens, prey remains along trails, and direct observations. These techniques vary in their ability to detect certain food items, and again caution should be exercised when comparing results derived using different techniques.

Arctic foxes in tundra regions rely heavily on lemmings as a food source (Angerbjörn et al. 1999). Lemmings are particularly prevalent in the diet in summer, but may be used year-round by foxes that do not move into coastal areas or onto pack ice in the winter (Anthony et al. 2000). Where abundant, other rodents such as voles may be important dietary constituents. Other important foods include various birds, bird eggs, arctic ground squirrels (*Spermophilus parryii*), arctic hares (*Lepus arcticus*), insects, snails, fish, berries, and carrion of caribou (*Rangifer tarandus*) (Chesemore 1968; MacPherson 1969; Stephenson 1970; Speller 1972; Eberhardt 1977; Garrott et al. 1983b). Fewer data are available on winter diets in tundra areas, although dietary diversity likely is lower, as fewer items are available. Chesemore (1968) reported that 90% of winter scats contained lemming; other items included birds, eggs, and carrion from caribou and marine mammals. Arctic foxes also follow caribou herds and attack weak fawns (Chesemore 1975).

In some coastal areas, arctic foxes feed extensively in summer on nesting colonies of seabirds (Murie 1959; Stephenson 1970) and waterfowl (Stickney 1991). They also forage along beaches, where they consume carrion and various marine invertebrates including amphipods, isopods, and sea urchins (West and Rudd 1983). On pack ice in winter, marine mammals appear to be primary foods. Arctic foxes commonly scavenge seal carcasses left by polar bears (Elton 1949), and may even routinely follow polar bears (Seton 1929; Chesemore 1975; Garrott et al. 1983b) as well as wolves (MacPherson 1969). Arctic foxes also scavenge walrus (*Odobenus rosmarus*) carcasses (Schiller 1954; Chesemore 1975), but likely feed on any available carrion. Arctic foxes also prey on ringed seal (*Phoca hispida*) pups, which they excavate from subnivean birthing dens (Smith 1976; Riewe 1977). Smith (1976) estimated that 58% of ringed seal pups in the Amundsen Gulf may have been killed by arctic foxes.

Arctic foxes commonly cache food, particularly when foraging on birds and eggs in seabird colonies. Foxes foraging in waterfowl nesting areas may cache over 80% of the eggs they find (Stickney 1991), and Murie (1959) reported an arctic fox cache containing over 135 birds. Anthony et al. (2000) found that cached eggs were consumed throughout the year. Other items found in caches included rodents and carrion (Seton 1929; MacPherson 1969; Stephenson 1970). Arctic foxes also will scavenge on refuse in human-occupied areas, and may become quite dependent on such food when natural foods are scarce (Urquhart 1973; Eberhardt 1977; Fine 1980; Eberhardt et al. 1982, 1983b; Garrott et al. 1983b). In Iceland, arctic foxes occasionally prey on domestic sheep (Hersteinsson et al. 1989). When food availability is low, arctic foxes also may kill and cannibalize adults and pups (Chesemore 1975).

Gray foxes consume primarily mammalian prey in winter. Cottontails (*Sylvilagus* spp.) and various rodents are the primary items. Diets are more diverse in spring and summer as other items become available. In summer, gray foxes are highly insectivorous and particularly consume large quantities of orthopterans such as grasshoppers. As natural fruits and nuts ripen, they are consumed in increasing quantities, and by fall constitute as much as 70% by volume of food habit samples (Pils and Klimstra 1975). Gray foxes have been described as the most omnivorous of the North American fox species (Fritzell and Haroldson 1982).

Mammalian prey consumed by gray foxes includes rabbits, voles, mice (*Peromyscus* spp.), woodrats (*Neotoma* spp.), cotton rats (*Sigmodon* spp.), pocket gophers (*Thomomys* spp.), squirrels (*Sciurus* spp. and *Tamiasciurus* spp.), opossums (*Didelphis virginiana*), and carrion of deer (*Odocoileus* spp.). Birds consumed include pheasants (*Phasianus colchicus*), ducks, and a variety of passerines. Insects consumed include orthopterans, coleopterans, and lepidopterans. Fruits used by gray foxes include persimmons (*Diospyros virginiana*), grapes (*Vitis* spp.), huckleberries (*Gaylussacia* spp.), hawthorns (*Crataegus* spp.), elderberries, (*Sambucus* spp.), apples (*Malus* spp.), juniper berries (*Juniperus* spp.), prickly pears (*Opuntia* spp.), mesquite beans (*Prosopis* spp.), corn, and various grains. Wild nuts eaten include hickory nuts (*Carya* spp.), acorns (*Quercus* spp.), beechnuts (*Fagus americanus*), and peanuts. Sources for gray fox food habit data include Kozicky (1943), Glover (1949), Scott (1955), Korschgen (1957), Wood et al. (1958), Turkowski (1969), Pils and Klimstra (1975), Trapp and Hallberg (1975), and Trapp (1978).

Like gray foxes, island foxes also are highly omnivorous. Among all of the North American foxes, island foxes consume the least amount of vertebrate prey, primarily because of depauperate vertebrate fauna on the Channel Islands (Wenner and Johnson 1980). Consequently, island foxes feed extensively on insects and fruits. Island fox diets vary seasonally and among islands (Laughrin 1977). The primary vertebrate prey consumed are deer mice. Laughrin (1977) reported that deer mice were present in 4–13% of fecal samples on Santa Cruz Island. Collins (1980) found that deer mice made up 53% by volume of winter feces of foxes on San Miguel Island. Other mammals consumed include California ground squirrels, harvest mice (*Reithrodontomys megalotis*), California voles (*Microtus californicus*), rats (*Rattus rattus*), and house mice (*Mus musculus*) (von Bloeker 1967; Laughrin 1973). Remains of ground-nesting birds were present in 22% of fox feces during spring on San Miguel Island (Collins 1980) and 3–6.2% of feces year-round on Santa Cruz Island (Laughrin 1977). The most frequently occurring birds include horned larks (*Eremophila alpestris*), western meadowlarks (*Sturnella neglecta*), and chukars (*Alectoris chukar*) (Laughrin 1977; Collins 1980; Moore and Collins 1995). Lizards are occasionally consumed, and the Pacific treefrog (*Pseudacris regilla*) is the only amphibian known to be eaten by island foxes (Laughrin 1977).

Insects are a mainstay in the diet of island foxes in all seasons. Orthopterans such as crickets and grasshoppers are particularly important foods (Laughrin 1977). Jerusalem crickets (*Stenopelmatus* spp.) are found year-round in scats. Grasshoppers are frequently eaten in summer and fall. Beetles and lepidopteran larvae are commonly consumed in spring and fall (Laughrin 1977; Collins 1980; David 2001). Doyen (1974) found nine different species of beetles in fox scats collected in spring on San Clemente Island. The most frequently occurring were *Trigonoscuta* sp., june beetles (Scarabaeidae), and two tenebrionids (*Coelus remotus, Eusattus robustus*). Laughrin (1977) also reported that land snails (*Helminthoglypta* sp.) were eaten occasionally.

Plant material, especially fruits, also are primary items in the diet of island foxes. The types and quantities of fruits consumed vary with island and season (Laughrin 1977). Frequently consumed fruits include those of *Arctostaphylos, Comarostaphylis, Heteromeles, Mesembryenthemum, Opuntia,* and *Prunus*. Other species consumed include fruits and seeds of *Atriplex, Carpobrotus, Ficus, Rhamnus, Rhus, Rosa, Schinus, Solanum,* and *Vaccinium* (Laughrin 1973, 1977; Collins 1980; David 2001). In addition to the foods above, island foxes also consume human refuse and carrion of pigs, sheep, cattle, and marine mammals (Laughrin 1973, 1977).

Kit foxes primarily consume animal prey. Primary items in the diet are usually rodents or leporids, depending on local availability. Black-tailed jackrabbits and desert cottontails (*Sylvilagus audubonii*) were the primary items consumed in Utah (Egoscue 1962, 1975) and California (Knapp 1978). Kangaroo rats were the primary items at locations in California (Laughrin 1970; Morrell 1972; Spiegel 1996). California ground squirrels were the primary items eaten at other locations in California (Logan et al. 1992; Cypher and Warrick 1993), whereas California pocket mice (*Chaetodipus californicus*) were the primary items consumed at another location (White et al. 1996). Prairie dogs (*Cynomys ludovicianus*) were the main prey item in Mexico (List 1997). Kit fox food habits also can vary markedly with annual availability. At a site in California where kit fox feeding habits were monitored during 1980–1995, leporids were present in 44% of scats. Kangaroo rats were present in 31% (Cypher et al. 2000). However, annual use of leporids varied from 7% to 94% and annual use of kangaroo rats varied from 3% to 84%. Use of leporids was high when kangaroo rat abundance was low, but foxes switched to kangaroo rats when they became more abundant. A preference for kangaroo rats has been noted in other studies (Laughrin 1970; Morrell 1972; Fisher 1981; Koopman 1995). Also, the distribution of kit foxes may be closely associated with the presence of kangaroo rats (Benson 1938; Grinnell et al. 1937; Laughrin 1970).

Other items commonly consumed by kit foxes include deer mice, pocket mice (*Perognathus* spp.), antelope squirrels, gophers, various birds, snakes, lizards, Orthopteran and Coleopteran insects, and sheep carrion (Egoscue 1962, 1975; Morrell 1972; White et al. 1996; Cypher et al. 2000). Spiegel (1996) reported that insects were important food items when rodent availability was low. Fleshy fruits are usually not abundant in habitats used by kit foxes, but they eat cactus fruits (Egoscue 1956). Kit foxes on the edge of agricultural areas in California ate cotton seeds, almonds, and tomatoes (G. Warrick, Center for Natural Lands Management, pers. commun., 2000). Kit foxes also consume anthropogenic foods (Cypher and Warrick 1993) and will cache food (Morrell 1972).

Similar to kit foxes, swift foxes primarily consume animals, with leporids and rodents the most frequent prey. Leporids were taken most frequently in Texas (Cutter 1958a), Oklahoma (Kilgore 1969), Colorado (Cameron 1984), and Kansas (Scott-Brown et al. 1987). Prairie dogs were the primary prey in South Dakota (Hillman and Sharps 1978; Uresk and Sharps 1981, cited in Scott-Brown et al. 1987). Other items commonly consumed include 13-lined ground squirrels (*Spermophilus tridecemlineatus*), deer mice, voles, pocket mice, pocket gophers, grasshopper mice (*Onychomys leucogaster*), harvest mice, mourning doves (*Zenaida macroura*), western meadowlarks, and occasionally reptiles and amphibians (Kilgore 1969; Hillman and Sharps 1978; Zumbaugh et al. 1985; Uresk and Sharps 1981, cited in Scott-Brown et al. 1987). Insects also are frequently consumed, particularly Orthopterans and Coleopterans. In Texas, insects, primarily grasshoppers and beetles, were found in 29% of swift fox stomachs and constituted 55% of scats by volume (Cutter 1958b). Carrion may be an important food source, particularly in winter (Zumbaugh et al. 1985). Swift foxes do not consume much plant material, but readily eat commercially grown sunflower seeds in agricultural areas (Sovada et al. 2001). As with other fox species, swift foxes will cache food (Scott-Brown et al. 1987).

Feeding habits of red foxes have been documented through numerous investigations conducted throughout the range of the species (Sheldon 1992). Red foxes, because of their ubiquitous distribution, consume a diversity of items. However, rodents and leporids commonly are the primary dietary items. Depending on location, the most frequently consumed rodents include voles, woodchucks, pocket gophers, and deer mice. The most frequently consumed leporids include cottontails, black-tailed jackrabbits, snowshoe hares (*Lepus americanus*), and European hares (*L. europaeus*). Other mammalian prey include squirrels (Sciuridae), mice (Muridae), muskrats (*Ondatra zibethicus*), porcupines (*Erethizon dorsatum*), raccoons, and opossums. Moles (Talpidae) and shrews (Soricidae) are occasionally taken, but are not preferred items (Scott and Klimstra 1955). Deer carrion can be an important food source in some locations (Schofield 1960; Ozoga et al. 1982; Cypher and Yahner 1996), particularly in winter.

Birds also are consumed by red foxes. Species taken include ring-necked pheasants, grouse (Tetraonidae), woodcock (*Scolopax minor*), quail (Phasionidae), turkeys (*Meleagris gallopavo*), ducks and geese, various seabirds, and passerines. Bird eggs are consumed as well. In the prairie pothole region, red foxes can be the primary cause of mortality for ducks and can significantly affect duck production through predation on adults and eggs (Sargeant 1972, 1978; Sargeant et al. 1984; Cowardin et al. 1985). Red foxes also eat turtles and their eggs, other reptiles and amphibians, fish, and a variety of insects. Although not documented in North America, red foxes in Europe can feed extensively on earthworms (Macdonald 1980a).

Similar to other fox species, red foxes feed extensively on fruits when they are available, and in some locations fruit may constitute 100% of the diet in autumn. Fruits commonly consumed include blueberries, blackberries and raspberries (*Rubus* spp.), cherries, persimmons, mulberries (*Morus rubra*), apples, plums, grapes, and acorns. Other plant foods consumed include grasses, sedges, and tubers. In an unusual occurrence, Sklepkovych (1994) observed red foxes climbing balsam fir trees (*Abies balsamea*) in Newfoundland and feeding on fir cones. Red foxes also feed on some crops including corn, wheat, and other grains, and prey on livestock. Sargeant et al. (1986) reported that commercial sunflower seeds left over after harvest were important winter foods for red foxes in North Dakota. Red foxes also may prey on lambs, goat kids, piglets, chickens, and turkeys, but such depredations are usually localized. Not surprisingly, red foxes also readily consume human refuse (Doncaster et al. 1990; Cypher and Yahner 1996).

Red fox feeding habits vary markedly with annual and seasonal availability of food items (e.g., Scott and Klimstra 1955). Red foxes also will cache food (Dekker 1983; Henry 1986). Murie (1936) reported that red foxes in Michigan cached 40% of food items. Sargeant and Eberhardt (1975) reported that red foxes cached eggs from duck nests and even cached ducks that were still alive.

Energy Requirements. Relatively little information is available on the energetic requirements of North American foxes. Most species may experience a negative energy balance during breeding and pup rearing, resulting in mass loss. Recovery of mass occurs in late summer and early fall (Kolb and Hewson 1980; Prestrud and Nilssen 1992; Winstanley et al. 1999). For arctic foxes, Speller (1972) estimated that an average litter of 10.6 pups required 60 lemmings/day. Working with captive arctic foxes, Riewe (1977) determined that adults required 125 kcal/kg/day to maintain weight. Underwood (1981) reported that energy requirements of arctic foxes vary seasonally. Captive foxes consumed 240 kcal/kg/day in summer, but only 85 kcal/kg/day in winter. This difference was assumed to be an adaptation to higher food availability in summer along with increased energetic requirements for reproduction. Underwood (1971) also found that arctic foxes will accumulate considerable body fat when food is plentiful. This fat primarily is deposited around visceral organs and along the dorsal and ventral surfaces of the trunk. These deposits can be up to 1 cm thick and weigh 1.8 kg/fox with an estimated caloric value of 16,200 kcal.

Captive gray foxes maintained weight on natural and prepared diets that by weight equaled about 3.8% of body mass (Fritzell and Haroldson 1982). Captive kit foxes consumed an average 175 g of fresh meat/day (Egoscue 1962). Egoscue (1962) also estimated that the food requirements for a family of seven kit foxes (two adults and five pups) for the first 64 days following birth of young was 44,605 g. Golightly (1981) estimated that kit foxes required 101 g of prey in summer and 115 g in winter to meet energy needs. Golightly and Ohmart (1984) further estimated that kit foxes without access to free water would need to consume approximately 40% more prey to meet their water requirements.

Kit foxes lose mass between winter and summer (Warrick and Cypher 1999), possibly associated with reproduction. Kilgore (1969) estimated that the average food consumption of swift foxes was about 200 g/day.

Sargeant (1978) examined food requirements for captive red foxes of different ages. Pups 5–8 weeks old required about 1.4 kg of prey/week; pups 9–12 weeks old required 1.9 kg of prey/week. Pups older than 12 weeks required 2.5 kg of prey/week, and adults required about 2.3 kg of prey/week. Based on this, it was estimated that a family would require 18.5 kg of prey/km^2 during the denning season, but only 10–15% of that amount after the denning season. In Europe, Haltenorth and Roth (1968, cited in Sheldon 1992) estimated that adult red foxes required 0.5–1.0 kg of food/day.

BEHAVIOR

Social Ecology. North American foxes share some similarities with respect to their social ecology. All species exhibit some flexibility in that mating systems, group composition, and other social behaviors can vary with factors such as population density and the abundance and dispersion of food (e.g., Macdonald 1983; Lindström 1986). As with other small canids, North American foxes are primarily monogamous, with occasional polygyny. Moehlman (1989) summarized an evolutionary scenario that results in polygyny. In small canids, females produce relatively fewer and larger young, which potentially require less paternal postpartum investment. If females can potentially raise a litter with minimal or no assistance from males, then males can potentially breed with more than one female. Also, again depending on population density and food availability, one or more young from the current year's litter may remain in the parental home range through the next breeding season and therefore more than two adults may be present. These philopatric individuals commonly are females. These younger females may assist the parents in raising their new litter of pups through provisioning and protection, and may even continue raising the pups if the mother dies. These younger individuals also benefit by remaining in a known territory, which they may eventually inherit. Less commonly, the younger females will produce their own litters, resulting in two litters of pups in a given home range. In most species, mated pairs remain together until either pairmate dies. Nonbreeding individuals tend to be solitary and somewhat transient.

Consistent with the patterns above, arctic foxes are primarily monogamous and reportedly mate for life (Chesemore 1975; Hersteinsson and Macdonald 1982). However, members of a pair may only be together during the breeding and pup-rearing season and are usually solitary the rest of the year. During the breeding season, the basic social unit is the mated pair and their pups. On occasion, one or two additional adults, typically female young from a previous litter, may be present and may assist in raising the current year's litter (Hersteinsson and Macdonald 1982; Eberhardt et al. 1983a), although Strand et al. (2000) reported that these extra adults do not engage in significant food provisioning. The bulk of food provisioning for the pups is conducted by the mother with some assistance from the male (Garrott et al. 1984). During breeding and pup rearing, arctic foxes are territorial and will actively repel and expel intruding conspecifics (Chesemore 1967; Speller 1972; Eberhardt et al. 1982; Hersteinsson and Macdonald 1982).

In Iceland, Hersteinsson and Macdonald (1982) reported that social ecology of arctic foxes varied with the spatiotemporal availability of food. Along the coast, foods primarily consisted of seal carcasses, seabirds, fish, invertebrates, and berries, and dispersion was clumped. Fox density was higher, territories were smaller, litter sizes were larger (mean = 4.5), and helper foxes (additional adults) were common. Inland, foods consisted of ptarmigan and sheep carcasses in winter and migrant birds in summer, and availability was more evenly distributed. Fox density was lower, territories were larger, litter sizes were smaller (mean = 4.0), and helper foxes were rare.

Fewer data are available on the social ecology of gray foxes. Monogamy with occasional polygyny probably is typical (Trapp and Hallberg 1975). The basic social unit is the mated pair and their pups. There are no reports of alloparental care by helper foxes, although Banfield (1974) reported that two females may rear pups in a communal den. Evidence suggests that females primarily provision pups (Nicholson et al. 1985). Gray foxes do not appear to be strongly territorial (Hallberg and Trapp 1984; Haroldson and Fritzell 1984). Although home ranges of adjacent family groups may overlap, core areas appear to be used exclusively by resident pairs (Chamberlain and Leopold 2000). It is unknown whether breeding pairs mate in subsequent years or find new partners (Banfield 1974).

Island foxes are solitary from September to December, mated pairs are together from January to May, and family groups are observed from May to September (Laughrin 1977). Like other fox species, island foxes are primarily monogamous (Laughrin 1977; Fausett 1982; Crooks and Van Vuren 1996), with occasional instances of extrapair fertilizations by males (Roemer et al. 2001). Island foxes are territorial, with little overlap among neighbors (Roemer et al. 1994).

Closely related kit foxes and swift foxes have virtually identical social ecologies. Both species apparently mate for life (Grinnell et al. 1937; Egoscue 1956; Kilgore 1969), and are primarily monogamous with occasional instances of polygyny (Egoscue 1962; Kilgore 1969; Ralls et al. 2001). Additional adults, typically young from previous years, also may be present and assist in pup rearing (Covell 1992; Ralls et al. 2001). These extra individuals are usually females (Covell 1992; Koopman et al. 2000; Ralls et al. 2001), but extra males also occur (Spencer et al. 1992). Ralls et al. (2001) found that among kit foxes, mated males and females are not closely related, but that females in adjacent home ranges commonly were related. When pups are very young, the female rarely leaves the den and the male provides food for her. As the pups get older, both the male and the female bring food to the den (Egoscue 1962). Among kit foxes, yearling females assisting with pup rearing have adopted and raised litters following the death of the mother (Spiegel 1996; Koopman et al. 2000). Neither kit foxes nor swift foxes appear to be strongly territorial, even during the breeding season (Morrell 1972; Cameron 1984). Home range overlap can be extensive, although core areas usually are used exclusively by resident family groups (Ralls and White 1995; Zoellick et al. 2002). Among kit foxes, trespassing into adjacent territories is occasionally observed during the breeding season, and these individuals are usually males that may be attempting to mate with additional females (Ralls et al. 2001).

The social ecology of red foxes appears to be the most complex and variable. This variability and complexity is more often observed among red foxes in Europe, and to an extent results from suburban and urban situations where food resources are superabundant and fox densities can be extremely high. In such situations, social units frequently consist of an adult male; two or more adult females, which probably are related (as many as five have been observed); and young from the current year. Usually, only the dominant female produces pups; reproduction by subordinates is behaviorally or endocrinologically inhibited. However, polygyny is common, and the resulting litters are sometimes raised in communal dens (Macdonald 1979, 1980b, 1981; Niewold 1980; Macdonald and Newdick 1982). Macdonald (1981) proposed the resource dispersion hypothesis to explain territory and group size among red foxes. According to this hypothesis, territory size is determined by the area necessary to reliably support a mated pair. In years when food resources in this territory exceed that required by the pair, additional adults (usually related to the pair) are tolerated within the territory and may confer some benefits to the pair, such as alloparental care.

In North America, mated pairs and their young of the year are the most commonly observed social unit. Additional related adults that function as helpers sometimes occur and can increase reproductive success (Zabel and Taggart 1989). Occasionally, two reproductive females have been observed in the home range of a male. Such polygyny may be more frequent when food is abundant (Zabel and Taggart 1989), consistent with the resource dispersion hypothesis. Pups from two litters are sometimes raised in communal dens, where all young may be nursed by both females (Tullar et al. 1976; Zabel and Taggart 1989). In Wisconsin, Pils and Martin (1978) found five communal dens (11% of all dens in their sample). Particularly during the breeding season, territories are actively defended against intrusion by conspecifics (Preston 1975; Storm and Montgomery 1975; Voigt and Macdonald 1984). Trespassing into adjacent territories is occasionally observed during the breeding season, and these individuals may be attempting to mate with additional females (Voigt and Macdonald 1984).

Dispersal. Dispersal consists of permanently leaving a natal home range with the intention of establishing a home range elsewhere, and most commonly involves juveniles. Among small canids, including North American foxes, dispersal is typically male biased. The proportion of individuals dispersing, onset of dispersal, and dispersal distances all vary spatially and temporally with population density, mortality rates among parents, and annual food availability. As mentioned previously, some individuals, particularly females, may delay dispersal until their second year and may inherit natal ranges on the death of parents (Moehlman 1989).

Among arctic foxes, dispersal by young primarily occurs in late summer and early fall (Chesemore 1967). Eberhardt et al. (1983b) reported that when food was abundant, many juveniles remain in natal home ranges over winter and then disperse in late winter or early spring. Among gray foxes, juveniles remain in natal ranges until January or February. Of five juveniles in Illinois (two males, three females), none exhibited long-range dispersal movements (Follman 1973). Of nine juveniles in Alabama, all three males dispersed an average distance of 15 km. The six females made exploratory movements of up to 3 km, but always returned to their natal ranges (Nicholson et al. 1985). In New York, 63% of juvenile female and 73% of juvenile male gray foxes dispersed (Tullar and Berchielli 1982). Dispersal distances of 83 km (Sullivan 1956), 84 km (Sheldon 1953), and 135 km (Banfield 1974) have been recorded.

Few data are available on dispersal in island foxes, although dispersal potential is somewhat limited by the insular situation. Roemer et al. (2001) reported that dispersal rates were low, with an average distance of 1.4 km. Laughrin (1977) and Fausett (1982) reported that the parents actually disperse from natal ranges in September and that juveniles remain in these ranges until at least December.

Of 209 juvenile kit foxes monitored in California during 1980–1996, 33% dispersed (Koopman et al. 2000), 49% of males and 24% of females. However, annual dispersal rates varied from 0 to 79% for males, 0 to 50% for females, and 0 to 52% for all juveniles. Male dispersal was weakly correlated to mean annual litter size, whereas female dispersal rates were weakly and inversely correlated to small-mammal abundance. Dispersal began in June and peaked in July, but individuals dispersed in almost every month. The mean age at dispersal was 8 months, and 87% of juveniles dispersed in their first year. Survival increased with age of dispersal, but 65% of individuals died within 10 days of dispersing. Predators were the primary cause of mortality. Survival was similar between dispersing males and females, but philopatric males had lower survival rates than philopatric females. Survival of philopatric males was lower than for dispersing males, but was similar between dispersing and philopatric females. O'Neal et al. (1987) also reported lower survival among dispersing kit foxes in Utah. Warrick et al. (1999) found that juvenile dispersal rates were lower among supplementally fed family groups than among nonfed groups. In California, the mean dispersal distance of juveniles was 7.8 km and did not differ between males and females (Scrivner et al. 1987). Some philopatric juveniles engage in helping behavior (Spiegel 1996; B. L. Cypher unpublished data). Also, some adults disperse on occasion (O'Neal et al. 1987; Koopman et al. 2000).

Few data are available on dispersal among swift foxes, but patterns may be similar to those for kit foxes. Covell (1992) reported mean dispersal distances of 9.4 km for males and 2.1 km for females in Colorado.

As with other species, dispersal rates among juvenile red foxes are higher among males than females (Storm et al. 1976; Voigt and Macdonald 1984; Allen and Sargeant 1993). Dispersal occurs from August to March, but peaks during fall (Phillips et al. 1972; Storm et al. 1976; Pils and Martin 1978). Most individuals of both sexes disperse in their first year (Phillips et al. 1972; Storm et al. 1976). In Canada, Voigt and Macdonald (1984) reported that dispersal rates increased when food availability decreased. Dispersal usually is preceded by exploratory movements (Voigt and Macdonald 1984), and males usually disperse earlier and further than females (Storm et al. 1976). Mean dispersal distances reported for males include 26 km (Ables 1968; Phillips 1970), 29 km (Phillips et al. 1972), 30 km (Voigt 1987), 31 km (Storm et al. 1976), and 43 km (Arnold 1956). Mean dispersal distances reported for females include 7 km (Phillips 1970), 8 km (Arnold 1956; Voigt 1987), 11 km (Storm et al. 1976), 13 km (Ables 1968), and 16 km (Phillips et al. 1972). Some long-distance movements include 203 km (Longley 1962), 257 km (Errington and Berry 1937), 302 km (Allen and Sargeant 1993), and 394 km (Ables 1965). Dispersal routes are influenced by the presence of cities, highways, lakes, and rivers (Storm et al. 1976; Allen and Sargeant 1993).

Adult red foxes also may disperse, although they usually travel shorter distances than juveniles (Storm et al. 1976; Voigt and Macdonald 1984). Of 30 adults monitored in Ontario, 3 males and 1 female dispersed after raising litters of pups (Voigt and Macdonald 1984). Pils and Martin (1978) reported that 83% of adult males and 56% of adult females dispersed.

Other Behaviors. All North American fox species routinely scent mark using urine and feces. This may be a particularly important activity among arctic foxes and red foxes, which apparently are more territorial than the other species (Macdonald 1977). Marking is particularly important along home range boundaries, although all species also mark within their home ranges. Urine and feces typically are deposited on conspicuous and prominent objects, novel objects, prey remains, rocks, cow dung, and along travel routes (Grinnell et al. 1937; Egoscue 1962; Laughrin 1977; Henry 1986). Island foxes were reported to scent mark about every 6–9 m (Laughrin 1977). Use of latrine sites, where many animals deposit feces, has been observed among gray foxes (Trapp 1978), island foxes (Laughrin 1977), and kit foxes (B. L. Cypher, pers. obs.). The other species likely engage in this behavior as well. Red foxes also may use anal-sac secretions to mark territories (White et al. 1989).

North American foxes also communicate using a variety of vocalizations. Growls and barks are reportedly used as alarm and threat vocalizations by arctic foxes, gray foxes, red foxes (Cohen and Fox 1976), island foxes (Fausett 1982), kit foxes (Egoscue 1962), and swift foxes (Cameron 1984) in agonistic interactions with conspecifics. Screams are used during aggressive interactions or when in distress (Cohen and Fox 1976). Greeting and contact-solicitation vocalizations have been described as "coos" in arctic foxes, gray foxes, and red foxes (Cohen and Fox 1976) and "mewing" in gray foxes and red foxes (Tate 1931; Cohen and Fox 1976). Other vocalizations include whines and whimpers, chirps, shrill yaps, purrs, hisses, and rasping cries. When distressed, kit foxes make a distinctive "roop" noise with their mouths closed, which also has been described as a croaking noise (Egoscue 1962) or the sound of a "perking coffee pot" (Morrell 1972). North American foxes also communicate using a variety of visual signals including facial expressions, body postures, and actions (Fox 1970; Laughrin 1977; Henry 1986; Wakely and Mallory 1988). Gray foxes and island foxes also allogroom; paired adults will groom each other and adults groom juveniles (Fox 1970; Laughrin 1977).

North American foxes vary in their innate response to humans. Arctic foxes (Pederson 1975; Eberhardt 1976), island foxes (Grinnell et al. 1937; Laughrin 1977), kit foxes (Egoscue 1956; Laughrin 1970) and swift foxes (Cutter 1958a) all are relatively unwary around humans and may even be attracted to human activity. Wrigley and Hatch (1976) reported that arctic foxes will follow biologists in the field. Gray foxes and red foxes usually are more wary and are more apt to avoid humans. However, all species, if not harassed, will acclimate to the presence of humans. They may inhabit areas of human activities and habitations and can become nuisances. Arctic foxes can become "camp thieves" (Eberhardt 1976) and are frequently observed around oil field development camps and communities (Eberhardt 1977). Populations of gray foxes (Harrison 1997), kit foxes (Cypher et al. 2001), and red foxes

(Lewis et al. 1993) occur in suburban and urban environments in North America. Red foxes inhabiting many British cities are infamous.

Hunting behaviors of foxes include pouncing (i.e., characteristic "mousing" behavior), stalking, and chasing. Arctic foxes carefully locate lemmings under the snow by sound or smell, and then suddenly pounce and dig rapidly to capture the prey (Banfield 1974). When searching for food, kit foxes systematically meander in open areas and circle clumps of brush. When a prey item is located, kit foxes make a stealthy approach and then use a swift rush to capture it (Egoscue 1962). Red foxes use the "mousing" technique to capture small mammals. After prey is located by sound, foxes leap into the air, pin it with their forefeet, and then bite it (Seton 1929; Scott 1943). Red foxes capture rapid terrestrial prey like rabbits by stalking followed by a quick pursuit (Murie 1936; Scott 1947).

Gray foxes are noteworthy for their tree-climbing behavior. They can climb vertical, branchless tree trunks to heights of 18 m. They do this by clasping the trunk with their forelegs and pushing upward with their hind legs (Seton 1929; Terres 1939; Taylor 1943). They also can climb by jumping among branches (Seton 1929; Grinnell et al. 1937). Gray foxes climb trees and shrubs to forage, rest, and escape predators (Grinnell et al. 1937; Yeager 1938; Terres 1939). They descend by backing down vertical trunks (Seton 1929) or running head first down slanted trees (Yeager 1938). Island foxes also are agile climbers (Laughrin 1973). All North American foxes likely can swim, and this behavior has been observed among arctic foxes (Murie 1959) and kit foxes (Reeder 1949).

MORTALITY

Sources and Rates. Sources and rates of mortality for North American foxes are not static, but instead are spatially and temporally dynamic. For a given location and time period, mortality is dependent on environmental and anthropogenic factors. For example, starvation may be more important in some years and areas when food availability is critically low. Likewise, disease may be more important during epizootics. Much of the early data on fox mortality was collected in the mid-1900s, when trapping and hunting for fur harvest and predator control programs were the primary sources of mortality in many areas. Harvesting foxes for fur is declining rapidly and has ceased entirely in many areas, and predator control efforts have been significantly reduced and rarely target foxes. Thus, mortality factors for foxes have changed over time, and mortality data should be considered specific to a given area and time period. Another generality is that mortality rates for juvenile foxes are typically higher than rates for adults. Juveniles are less experienced at avoiding dangers, less skilled at foraging, and more likely to disperse into unfamiliar territory, all of which puts them at greater risk.

Arctic foxes reportedly have been killed by wolves, grizzly bears, red foxes, wolverines (*Gulo gulo*), snowy owls (*Nyctea scandiaea*), large hawks, eagles, and jaegers (*Stercorarius* spp.) (Chesemore 1967, 1975; Berns 1969). Avian predators primarily take pups. Domestic dogs occasionally kill arctic foxes near villages. Disease also can be a significant source of mortality, and disease risk may increase during periods of food scarcity (Speller 1972). Rabies is enzootic in arctic fox populations (Crandell 1975; Secord et al. 1980), and periodic epizootics are common, particularly when fox densities are high (Elton 1931; Rausch 1958; Fay 1973). Other natural sources of mortality include starvation when food is scarce (MacPherson 1969). During such periods, sibling aggression increases, as does abandonment of natal dens by parents. Both of these factors lead to high pup mortality. MacPherson (1969) reported that annual den abandonment ranged from 0 to 57% during a 4-year period in Canada. Hiruki and Stirling (1989) estimated mean annual survival rates of 58% for adults and 23% for juveniles. Juvenile survival is strongly influenced by food availability (Fay and Rausch 1992).

Arctic foxes are still hunted and trapped throughout much of their range. Harvesting by humans still is a significant, and sometimes is the primary, cause of mortality. Arctic foxes usually do not survive for more than 3–4 years, and on rare occasions individuals may reach 9–10 years of age (MacPherson 1969; Hammill 1983). In the Northwest Territories, Canada, maximum age was 6 years for males and 7 years for females (Hiruki and Stirling 1989).

Gray foxes are killed by golden eagles, bobcats (*Lynx rufus*), mountain lions (*Puma concolor*), and coyotes, and pups may be taken by large raptors (Grinnell et al. 1937; Mollhagen et al. 1972). Disease also can be a significant source of mortality, and local population reductions have occurred during distemper (Nicholson and Hill 1984) and rabies (Trapp and Hallberg 1975; Steelman et al. 2000) epizootics. Gray foxes are harvested throughout most of their range, and in many areas this is the primary cause of mortality. As with other fox species, juveniles usually are most vulnerable to trapping and may constitute 80% of the harvest (Carey et al. 1978). In Florida, Lord (1961) estimated that 43–47% of gray foxes die during their first 7 months, and that mortality rates for adults were 61–64%. In southern Georgia, 50% of juveniles died by the end of their first summer and 90% did not survive their first winter. Annual adult mortality rates were 50% (Wood 1958). Most gray foxes do not live past 4–5 years of age (Wood 1958; Lord 1961), although Seton (1929) claimed that some individuals may reach 14–15 years of age in the wild.

Island foxes have no natural terrestrial predators. Red-tailed hawks (*Buteo jamaicensis*) occasionally prey on pups, and Laughrin (1973, 1977) reported that domestic dogs, ravens (*Corvus corax*), bald eagles, and golden eagles were potential predators. In recent years, golden eagles increased in abundance on San Miguel, Santa Rosa, and Santa Cruz Islands and predation by these birds caused significant declines in fox abundance on all three islands. Vehicles are an important source of mortality on San Nicolas and San Clemente Islands (Moore and Collins 1995). Before 1994, mean annual probability of survival of all age classes ranged from 0.75 to 1.0. By 1998, it averaged <0.4 (Roemer 1999). Island foxes are relatively free of diseases (Laughrin 1977), probably because of their insular situation and the lack of interaction with other fox or carnivore populations. Harvests probably were never an important source of mortality, and all island fox populations currently are fully protected. Island foxes live an average of 4–6 years in the wild (Collins 1982) and may live 8 or more years in captivity (Laughrin 1973).

Predators known to kill kit foxes include bobcats, coyotes, red foxes, badgers, feral dogs, and golden eagles. Predators, particularly coyotes, are the primary source of mortality for kit foxes in many locations. In four studies in California, predators were responsible for 75–89% of kit fox mortalities for which the cause of death could be identified (Standley et al. 1992; Ralls and White 1995; Spiegel 1996; Cypher et al. 2000). Disease does not appear to be a significant source of mortality for kit foxes. O'Farrell and Gilbertson (1986) reported that starvation may occur among kit foxes during periods of low food availability.

Anthropogenic sources of mortality for kit foxes have decreased in importance, but still can be important in some locations. Kit foxes previously were harvested throughout most of their range (Grinnell et al. 1937; Egoscue 1956, 1962), but harvest has declined, and in California and Oregon is no longer permitted. Historically, many kit foxes died from toxicants distributed for other predators, primarily wolves and coyotes (Grinnell et al. 1937; Egoscue 1956). Rodent control programs can result in the primary or secondary poisoning of kit foxes, but this source of mortality is now infrequent (Snow 1973). Vehicles have been and continue to be an important source of kit fox mortality (Egoscue 1962), and in some locations are responsible for over 10% of kit fox mortalities (Cypher et al. 2000). Other anthropogenic sources of mortality include illegal shooting (Laughrin 1970; Morrell 1972; Cypher et al. 2000) and accidental death associated with agricultural and urban development (Knapp 1978; B. L. Cypher unpublished data).

Annual survival rates of kit foxes exhibit extreme annual variation in response to yearly fluctuations in environmental conditions. Recent estimates of the mean annual probability of survival for adult kit foxes in California ranged from 0.44 (Cypher et al. 2000) to 0.60 (Ralls and White 1995). Annual estimates ranged from 0.36 to 0.71 over 4 years at one location (Spiegel 1996) and from 0.20 to 0.81 over 16 years at another location (Cypher et al. 2000). Estimates of the mean annual probability of survival for juvenile kit foxes in California ranged from 0.14 (Cypher et al. 2000) to 0.21 (Ralls and White 1995). Spiegel (1996) reported that 85% of foxes died by the end of their second year. Kit foxes can live to 7 years in the wild (Egoscue 1975; Cypher et al. 2000), and 10–12 years in captivity (Mann 1930, Crandell 1964).

Swift foxes are killed by wolves and coyotes, and possibly also by bobcats, badgers, and golden eagles (Kilgore 1969). Coyotes are the most significant source of mortality in many locations (Covell 1992; Kahn et al. 1997; Sovada et al. 1998; Kitchen et al. 1999; Matlack et al. 2000), and coyote predation has been a problem in attempts to reintroduce swift foxes where they were extirpated (Scott-Brown and Reynolds 1984; Scott-Brown et al. 1986; Carbyn et al. 1994). Similar to kit foxes, disease does not appear to be a significant mortality factor for swift foxes. Harvests used to be the primary source of mortality in most locations (Kilgore 1969; Hillman and Sharps 1978). Swift foxes are still harvested in some states, but not to the extent they were historically; they now are fully protected throughout most of their historical range. Inadvertent poisoning and trapping during predator control programs aimed at wolves and coyotes historically was an important source of mortality (Seton 1929; Robinson 1953; Kilgore 1969). Vehicles and farm equipment also are important sources of mortality in some locations (Kilgore 1969; Hillman and Sharps 1978; Sovada et al. 1998; Matlack et al. 2000). Sovada et al. (1998) estimated mortality rates of 0.55 for adults and 0.67 for juveniles in Kansas. Estimated survival rates at a site in Colorado include 0.52 (Rongstad et al. 1989), 0.53 (Covell 1992), and 0.64 (Kitchen et al. 1999). Swift foxes have lived for 10–12 years in captivity (Mann 1930, Crandall 1964).

Red foxes are killed by wolves, coyotes, mountain lions, lynx (*Lynx canadensis*), bobcats, and dogs (Larivière and Pasitschniak-Arts 1996). Disease also can be a significant source of mortality. Rabies is enzootic in many populations, and occasional epizootics can result in marked reductions in local and regional red fox abundance (Johnston and Beauregard 1969; Chomel 1993). Sarcoptic mange also commonly occurs in red fox populations and can cause local reductions in abundance (Trainer and Hale 1969; Stone et al. 1972). Epizootics of rabies and mange are most common when red fox densities are high. Harvests have historically been a primary source of mortality and still are in many locations. Vehicles also can be an important source of mortality. Storm et al. (1976) reported that 80% of tagged red foxes were killed by shooting, trapping, or vehicles in the midwestern United States. Juveniles are particularly susceptible to being struck by vehicles (Storm et al. 1976). Voigt (1987) found that 25% of juveniles in Ontario were killed by vehicles and 60% were taken by hunters and trappers. Red foxes also are killed by farm machinery (Storm et al. 1976). Most red foxes do not live past 5 years of age in the wild (Storm et al. 1976), but some may live more than 8 years (Tullar 1983; Allen and Sargeant 1993).

Diseases and Parasites. North American foxes are afflicted by a number of diseases and parasites. However, not all are fatal or even debilitating, and only a relatively small number produce population-level impacts. The ones most likely to affect populations include rabies, canine distemper, and sarcoptic mange.

Rabies is a virus that is usually fatal once contracted by a fox. It can be endemic in fox populations, or introduced into a population either by a conspecific or by another species. Rabies appears to be density dependent in that epizootics occur most frequently when fox populations attain high densities. Some minimal density appears to be necessary to sustain transmission rates required to produce an epizootic. Extensive research has been conducted on the epidemiology of rabies because of its importance to animal populations and humans (Baer 1975; MacInnes 1987). Annual costs associated with rabies epidemics and protecting public health in North America are substantial (MacInnes and LeBer 2000).

Rabies has been documented in arctic foxes, gray foxes, kit foxes, and red foxes, and likely occurs occasionally in swift foxes. Island foxes are protected from natural transmission pathways because they are insular. Among arctic foxes, rabies is endemic and epizootics have been observed in most populations (Elton 1931; Rausch 1958; Crandell 1975). During population lows, 0.7–1.0% of arctic foxes may be infected in a given area, but this figure may rise to 20% when densities are high (Syuzyumova 1968, cited in Underwood and Mosher 1982). Young foxes are particularly susceptible, and prevalence of infected individuals may be higher during colder months (Crandell 1975). Arctic foxes may be the main vector for transmission of rabies to sled dogs (Crandell 1975). Rabies epizootics also have been reported among gray foxes and have resulted in local population reductions (Jennings et al. 1960; Trapp and Hallberg 1975; Steelman et al. 2000). Rabies epizootics have not been documented among kit foxes. However, kit fox deaths attributable to rabies have been reported (Standley et al. 1992), and in California, a population decline was concurrent with a rabies epizootic among skunks (White et al. 2000).

Red foxes have been described as the most widespread natural reservoir of rabies in the world (Chomel 1993). Epizootics have been documented in numerous locations, but geographic foci where epizootics are recurring include central Europe (Anderson et al. 1981) and the region encompassing southeastern Canada and northeastern United States (Johnston and Beauregard 1969; MacInnes 1987; Chomel 1993). Rabies was first reported in North America in 1753, and epizootics were documented as early as 1812 (Carey 1982). Rabies may be enzootic among fox populations in eastern North America, primarily in the Appalachian and Ozark Mountains (McLean 1970), and southern Ontario (Voigt 1987).

Canine distemper also is a viral disease. Mortality rates are unknown among foxes in the genus *Vulpes*. However, antibodies to distemper are commonly detected in kit foxes, swift foxes, and red foxes, which suggests that most animals survive exposures to the virus. Estimates of antibody prevalence among kit foxes include 0–25% (Miller et al. 2000), 14% (McCue and O'Farrell 1988), and 20% (Standley and McCue 1997). Prevalence among swift fox populations ranges from 18% to 100% (Miller et al. 2000). During a distemper epizootic among red foxes in New York, 31% of 133 foxes tested positive for antibodies (Monson and Stone 1976).

Distemper causes significant mortality among foxes of the genus *Urocyon*, where it almost always is fatal (Nicholson and Hill 1984), although individuals apparently can recover (Tullar 1979, cited in Samuel and Nelson 1982). During a distemper epizootic in New York, 64% of 131 gray foxes tested positive (Monson and Stone 1976). Local population reductions associated with distemper epizootics have been documented among gray foxes (Nicholson and Hill 1984). Davidson et al. (1992b) considered distemper to be a significant mortality factor among gray foxes in the southeastern United States. Distemper recently was introduced onto Santa Catalina Island, apparently by people or their pets, and quickly spread among island foxes. That fox population declined rapidly by about 90%, and intervention (e.g., captive breeding) has been initiated in an attempt to avoid extirpation (Coonan 2001).

A variety of other diseases have been reported among foxes (Table 24.4), but significant mortality associated with these diseases has not been documented. In many cases, the presence of infectious agents is documented through the presence of antibodies, which suggests that foxes have been exposed to the disease but have survived.

A multitude of external and internal parasites are found among North American foxes. External parasites include fleas, lice, ticks, chiggers, and mites. Internal parasites include protozoans, trematodes (flukes), cestodes (tapeworms), nematodes (roundworms, hookworms), and acanthocephalans. Many have no measurable effects on individuals or populations, and therefore no attempt is made to list all of the parasites. However, noteworthy parasites are discussed below.

Sarcoptic mange can cause reductions in fox abundance. Sarcoptic mange is caused by a mite (*Sarcoptes scabei*). The mites irritate the skin, causing flaking and cracking, lesions, and hair loss. This leads to other complications, particularly vulnerability to cold temperatures due to the loss of insulating hair, and frequently results in death (Trainer and Hale 1969; Stone et al. 1972). Mange seems to affect red foxes more than other species (Pryor 1956; Trainer and Hale 1969). Tullar

TABLE 24.4. Infectious diseases other than rabies and canine distemper detected in the six North American fox species

Disease	Arctic Fox	Gray Fox	Island Fox	Kit Fox	Swift Fox	Red Fox
Brucellosis	*			*		*
Cache Valley virus				*	*	
Canine adenovirus		*	*	*		*
Canine coronavirus			*			
Canine herpes virus			*			*
Canine parvovirus			*	*	*	*
Coccidiomycosis				*		
Colorado tick fever				*		
Histoplasmosis		*				
Jamestown Canyon virus				*	*	
Leptospirosis	*	*	*	*		*
Listeriosis		*				
Parainfluenza virus						*
Q fever	*					
Rotavirus						*
San Miguel sea lion virus			*			
St. Louis encephalitis		*				*
Staphylococcosis						*
Toxoplasmosis		*	*	*		
Tularemia		*		*		*
Tyzzer disease		*				
Vesicular stomatitus				*	*	
Western equine encephalitis				*		*

SOURCE: Arctic fox, Underwood and Mosher 1982. Gray fox, Fritzell 1987. Island fox, Prato et al. 1977; Garcelon et al. 1992. Kit fox, Vest et al. 1965; McCue and O'Farrell 1988; Standley and McCue 1997; Miller et al. 2000. Swift fox, Miller et al. 2000. Red fox, Ross and Fairley 1969; Quinn et al. 1976; Barker et al. 1983; Evans 1984; Davidson et al. 1992a.

(1979, cited in Samuel and Nelson 1982) reported mange among 4.5% of red foxes in New York. Smith (1978) found mange mites on 67% of red foxes in New Brunswick and Nova Scotia. Pils and Martin (1978) reported that 7% of red fox mortality in Wisconsin was attributable to mange, but suggested that this estimate was low due to the difficulty in finding carcasses of foxes killed by mange. Trainer and Hale (1969) also reported a decrease in red fox numbers in Wisconsin associated with a mange outbreak. Gray foxes are reported to be resistant to mange mites (Stone et al. 1972), and "mangey" gray foxes are rarely observed (Stone et al. 1982). Mange also has been observed among island foxes (Moore and Collins 1995). Arctic fox declines on some Russian islands have been attributed to mange (Ginsberg and Macdonald 1990).

Heartworm (*Dirofilaria immitis*) is found in all North American foxes and can cause serious debilitation among individuals. Prevalence among wild foxes generally is low (Miller et al. 1998). Stuht and Youatt (1972) reported heartworm in 11 of 39 red foxes from an area of Michigan where this parasite was endemic in dogs. Gray foxes seem to have some resistance to heartworm infection (Monson et al. 1973; Simmons et al. 1980). However, recent prevalence in four island fox populations ranged from 58% to 100% (Roemer et al. 2000).

The hydatid tapeworm (*Echinococcus multilocularis*) has minimal effects on foxes, but can cause severe debilitation and death in humans. Therefore, its distribution is of interest from a public health perspective. This tapeworm first was reported in North America among arctic foxes and red foxes in Alaska (Rausch 1956). Subsequent surveys revealed that the parasite occurred in arctic foxes throughout tundra habitat (Rausch 1967, cited in Underwood and Mosher 1982) with occurrence rates of 40–100% (Fay 1973). The range of this parasite in the upper latitudes of North America appears to correspond to the range of arctic foxes (Fay 1973). This tapeworm also has been reported from arctic foxes and red foxes in Canada (Choquett et al. 1962; Hnatiuk 1969) and red foxes in a number of states, primarily in the northern prairie region (Dyer and Klimstra 1980). Seese et al. (1993) found this parasite in 16.5% of 532 red foxes examined in Montana and concluded that the tapeworm was endemic among foxes in the state. Hydatid tapeworms also are prevalent in North Dakota, where they were found in 16–70% of red foxes examined (Rausch and Richards 1971; Kritsky and Leiby 1978).

Several species of fleas have been collected from kit foxes (Egoscue 1962, 1985), and infestations can be substantial both on animals and in dens. Egoscue (1962) suggested that den switching by kit foxes, particularly the frequent switching of dens when pups are present, may be a function of flea infestations.

AGE ESTIMATION

A variety of techniques have been used to estimate age in canids, including foxes. Some can be used only on dead animals, as the examination of internal structures is required. Others can be used on live animals or on samples collected from live animals. Techniques used on dead animals include using eye lens weight, cranial suture closure, and baculum and skull morphology. Techniques that can be used on live animals as well as dead ones include use of body morphology, body mass, testicular development, epiphyseal closure, tooth root canal characteristics, cementum annuli counts, and tooth eruption and wear patterns. Age estimation usually is not possible from a distance, except for young individuals. Small size and mass, juvenile features (e.g., relatively short rostrum), and pelage characteristics generally can be used to distinguish juvenile foxes until about 6 months old. After that, age estimations based on observations at a distance become very unreliable.

Skull, body, and baculum measurements, body mass, and testicular development are all reasonably straightforward techniques where values are compared to ranges for juveniles and adults to estimate age. These techniques may only permit differentiation into two age classes: juveniles and adults. Sutures are evident between cranial bones in juvenile canids before the bones fuse. Harris (1978) suggested that the basioccipital–basisphenoid suture was the most useful and reliable to use for age estimation. This technique also only differentiates juveniles and adults, as does epiphyseal closure. In juveniles, epiphyseal cartilage is present at the distal ends of the limb bones. As growth slows toward the end of the first year, the epiphyses ossify, resulting in "closure." Radiography (x-ray) can be used to examine limbs for epiphyseal closure (Sullivan and Haugen 1956).

The eye lens in vertebrates grows continuously for the life of the animal and does not shed cells (Bloemendal 1977). Thus, the lens can be used as an indicator of age. However, variation in lens weights can be high, and this technique may be most reliable for distinguishing juveniles from adults. Also, the lens must be collected soon after death, and special preservation and drying are required before weighing. See Friend (1967) and Dimmick and Pelton (1994) for additional information on this technique.

The root canal or pulp cavity within canine teeth also can be used to distinguish juvenile foxes from adults. The canal is slowly occluded and generally is fully closed by age 1 year; open canals indicate individuals <1 year old. Radiography can be used to examine the canals (Dimmick and Pelton 1994). Similarly, presence of an apical root foramen on the canine indicates a juvenile (Root and Payne 1984).

A commonly used estimation technique is to count cementum annuli. Layers of cementum, which consist of a dark and a light region, are deposited on the roots of teeth annually. Teeth can be removed from dead animals, or an incisor or premolar can be removed from a live animal (Allen 1974; Johnston and Watt 1981). Canine teeth afford the greatest accuracy, but can only be taken from dead animals. The accuracy of this technique decreases with increasing age (Dimmick and Pelton 1994).

Tooth eruption patterns can be used to determine age of juveniles (in months) until all of the deciduous teeth are replaced by permanent teeth. Tooth wear patterns are commonly used to distinguish juveniles from adults and absolute age in adults. For some species (e.g., island foxes; Collins 1982), wear patterns have been carefully documented, and this technique can be fairly accurate for individuals >2 years old. However, wear patterns vary considerably among individuals and populations depending on such factors as diet and oral injuries.

ECONOMIC STATUS AND MANAGEMENT

Human perspectives and attitudes regarding North American foxes have evolved with time and most likely will continue to do so for the foreseeable future. Perspectives and attitudes are influenced by culture, economic conditions, and social values. Humans have viewed foxes as competitors for game, depredators of livestock and crops, disease carriers, valuable fur resources, recreational opportunities, important components of ecosystems, photography subjects, threats to endangered species, and endangered species themselves.

Human views of foxes vary spatially and temporally for a given species. Up through the early 1900s, foxes routinely were killed any time the opportunity presented itself. Even as recently as the mid-1900s, foxes were still considered to be "vermin" in some areas and were the subject of predator control efforts and bounties. State-wide bounties on foxes persisted in Wisconsin until 1963 (Pils and Martin 1978), and some counties still had bounties into the 1980s (Samuel and Nelson 1982). Foxes were not considered a threat to humans, but were considered competitors with humans for game and also occasionally depredated livestock and crops. Concurrently, foxes also were recognized as a valuable resource because of their fur. Fur harvesting and trading was practiced by native peoples in North America long before the arrival of Europeans. The harvesting and trading of fur at times has constituted an industry on which many people depended as a supplemental and even primary source of income. Currently, relatively few people depend on fur harvests as a significant source of income, and most fur harvesting is recreational in nature.

Social changes also have significantly affected the number of people that hunt and trap foxes. As human populations have shifted from being more rural oriented to being more urban oriented, fewer people now are interested in hunting and trapping (U.S. Fish and Wildlife Service 1997). Armstrong and Rossi (2000) estimated that only about 145,000 licenses for recreational trapping were sold in the United States in 1999. Also, a growing proportion of the public is opposed to hunting and trapping on moral or other grounds. This has led to efforts to restrict trapping in many areas, including legislative bans in entire states including Arizona, California, Colorado, Florida, Massachusetts, and New Jersey (Armstrong and Rossi 2000). Concurrent with this shift in attitudes and social values has been a growing recognition and appreciation of the role and importance of predators in ecosystem function.

Historically, the primary management issues for foxes were regulating harvests and reducing depredation. Contemporary management issues include conflicts with humans in urban–suburban areas, transmission of zoonotic diseases, and predation on rare species. As growing human populations encroach on natural habitats, foxes increasingly are in close association with people and pets. Although this helps maintain higher numbers of foxes, potential detriments include disease transmission to pets and humans and nuisance problems. This increasing contact between humans and foxes increases the potential for transmission of zoonotic diseases. Rabies and alveolar hydatid disease resulting from the tapeworm *Echinococcus multilocularis* are the two diseases of greatest concern. Some species of foxes prey on rare species, thereby further endangering them. Measures taken to protect rare species from foxes include exclusion, relocation, and population control.

Arctic Foxes. Arctic foxes have long been of great economic importance to peoples in arctic regions. Income from their pelts still provides income for many individuals in these regions, although the economic importance of furs has declined (Chesemore 1975). Around the middle of the 1900s, 10,000–80,000 arctic foxes were harvested annually in Canada, and in the early 1980s, arctic fox pelts were worth as much as $75 (Underwood and Mosher 1982). The worldwide harvest is probably around 100,000 foxes annually (Garrott and Eberhardt 1987). During 1994–1995, 149 arctic foxes were harvested in Alaska, and the average pelt price was $22.00. Total value of the harvest was $3278 (International Association of Fish and Wildlife Agencies, unpublished data).

Management of arctic foxes in North America historically has consisted primarily of regulating the harvest. More recently, increased petroleum exploration and production has created another management concern. Arctic foxes have been reported to chew through seismic exploration cables (Urquhart 1973). More importantly, they frequently concentrate around petroleum facilities to feed on garbage, litter, and handouts. This increases local fox density and the potential for disease epidemics. Also, it increases the potential exposure of humans to rabies and alveolar hydatid disease (Eberhardt et al. 1982, 1983b). In some instances, petroleum companies have sponsored trapping efforts to reduce fox densities around facilities (Garrott and Eberhard 1987).

In an effort to increase fur harvest opportunities, arctic foxes were introduced onto most Aleutian Islands beginning in 1886 (West and Rudd 1983; Tietjen et al. 1988). However, these foxes significantly affected populations of native animals, which had evolved in the absence of terrestrial predators and therefore lacked defenses. In particular, breeding colonies of marine birds were heavily predated by foxes, resulting in the elimination of some colonies. Species most seriously affected included ancient murrelets (*Synthliberamphus antiquus*), whiskered aulets (*Aethia pygmaea*), Cassin's auklet (*Ptychoramphus aleuticus*), storm petrels (*Oceanodroma* spp.), and Aleutian Canada geese (*Branta canadensis leucopareia*), which are threatened with extinction (Springer et al. 1978). Efforts to eradicate introduced foxes on these islands have included poisoning, trapping, and shooting. Also, sterilized red foxes were introduced onto some islands to competitively eliminate arctic foxes, and this strategy appeared to be successful (Bailey 1992). In coastal mainland areas, arctic foxes also have significantly affected bird populations including the black brant (*Branta bernicla nigricans*) (Raveling 1989; Anthony et al. 1991) and various waterfowl (Stickney 1991). Experimental removals of arctic foxes from black brant colonies resulted in significantly higher nesting success (Anthony et al. 1991).

Gray Foxes. Gray foxes are harvested in most states as well as in Ontario and Manitoba. During 1994–1995, 82,514 gray foxes were harvested in 40 states. The states with the highest harvests were Pennsylvania (33,387), Texas (8066), New York (5008), and Georgia (3558). Average pelt price was $9.56 and the total value of the harvest was $788,990 (International Association of Fish and Wildlife Agencies, unpublished data). In general, gray fox fur is not as desirable as the fur of other species. Most gray foxes harvested are captured incidental to trapping efforts directed at other species. Hunting gray foxes with hounds is a cultural tradition in the southeastern United States (Fritzell 1987).

Management of gray foxes has primarily consisted of regulation of fur harvests and hunting. However, Hall (1983) reported that gray foxes once were fed at feeders in the southeastern United States to increase numbers. In research on nonlethal control measures, Oleyar and McGinnes (1974) reported some success in suppressing reproduction using diethylstilbestrol delivered in baits. Baits for delivering rabies vaccine to gray foxes have been tested in Texas (Steelman et al. 2000).

Island Foxes. Island foxes were never considered to be economically important, primarily because of their limited distribution. They were harvested by Native Americans and played prominent roles in religious and ceremonial practices (Collins 1991b). Harvests by nonnative peoples probably were minimal.

Island foxes are declining on most islands. However, an interesting management situation exists on San Clemente Island, where foxes are abundant. They are preying on San Clemente loggerhead shrikes (*Lanius ludovicianus mearnsi*), a subspecies classified as "Endangered." Breeding and foraging habitat for the shrikes was severely degraded by feral goats, sheep grazing, and fires. Shrikes, their eggs, and young also are preyed on by feral cats, introduced black rats, ravens, and various raptors. Removal of feral animals has resulted in the initiation of habitat rehabilitation and has reduced predation pressure on shrikes. To further reduce predation pressure, control measures have been applied to island foxes and have included euthanasia, relocation, and retention in a holding facility (Coonan 2001). A shock-collar system has been used with some success (U.S. Department of Defense 1998).

Kit Foxes. Kit foxes also have never been economically important. Their fur has low market value, and most trapped kit foxes are incidental captures. The total harvest in 1987 was 5400–7800 pelts, and as much as $23 has been paid for their pelts (O'Farrell 1987). Kit foxes are relatively unwary and easy to trap or shoot. A trapper in the 1930s once captured 100 kit foxes in a week along a 20-mile trapline in California (Grinnell et al. 1937). Kit foxes are harvested in Arizona, Colorado, Nevada, New Mexico, Texas, Utah, and Mexico. Harvests are prohibited in California, Oregon, and Idaho. During 1994–95, 247 kit foxes were harvested in Nevada and 531 in Utah. Kit fox and swift fox harvest totals were combined for other states and include 279 in Colorado, 273 in New Mexico, and 363 in Texas. Average pelt price was $5.46 (International Association of Fish and Wildlife Agencies, unpublished data).

Swift Foxes. Similar to kit foxes, swift foxes never were economically important, although 117,025 swift foxes were harvested in Canada between 1853 and 1877 (Herrero et al. 1986). Most swift foxes taken are incidental captures. Swift foxes are harvested in Kansas, New Mexico, Texas, and Wyoming. Harvests are not permitted in Colorado, Montana, Nebraska, North Dakota, Oklahoma, and South Dakota (Kahn et al. 1997). During 1994–1995, only 11 swift foxes were harvested in Wyoming and 11 in Kansas. Average pelt price was $5.46 (International Association of Fish and Wildlife Agencies, unpublished data).

Red Foxes. Red foxes are important economically as furbearers, because of the costs associated with controlling them for rabies suppression, and because of the effects they have on game species and rare species. During 1994–1995, 186,344 red foxes were harvested in 38 states. Red fox harvests were highest in Minnesota, Pennsylvania, New York, and Iowa. Average pelt price was $13.48 and the total value of the U.S. harvest was $2.5 million (International Association of Fish and Wildlife Agencies, unpublished data).

A significant cost as well as management challenge associated with red foxes is the control of rabies. Rabies epizootics are common among red fox populations, and the potential transmission of rabies to humans and livestock is a concern. Voigt (1987) reported that approximately $20 million is spent annually in Ontario on rabies control. Strategies for controlling this threat include oral vaccination and reducing fox numbers. Testing of both strategies has been conducted in Ontario, where rabies epizootics among red foxes occur frequently (Johnston and Voigt 1982; Rosatte et al. 1992). Methods for reducing red fox numbers have included large-scale poisoning, trapping, hunting, and gassing. Bounty payments also have been offered to encourage the killing of red foxes. These strategies have produced variable success, although sustained intensive harvesting appears to help reduce the frequency of epizootics (Voigt and Tinline 1982).

Vaccination of red foxes also has been attempted. Baits laced with oral vaccines are distributed across large areas to try to vaccinate a significant portion of the red fox population. In general, eradication or at least control of rabies appears possible if 60–70% of the fox population can be immunized (Rosatte et al. 1992). This strategy has proven promising in Europe (Steck et al. 1982; Schneider et al. 1988; Ginsberg and Macdonald 1990). Rosatte et al. (1992) reported an immunization rate of 46–80% among red foxes in Toronto, after which only one case of rabies was reported. Rabies appears to have been eliminated in a 3000-km^2 area of Ontario using oral vaccination (MacInnes and LeBer 2000).

Historically, red foxes were considered to significantly reduce game bird populations, although few data were available. However, significant impacts on nesting duck populations in the prairie pothole region of North America have been demonstrated. Red foxes depredate duck nests and also prey on hens incubating eggs (Sargeant 1972; Cowardin et al. 1985; Sargeant et al. 1993). Red fox abundance in this region may have increased following the reduction of wolves and coyotes, thus exacerbating the problem (Johnson and Sargeant 1977). Strategies proposed to help reduce red fox effects on breeding ducks have included increased harvests and increasing the abundance of coyotes, which competitively exclude red foxes (Sargeant and Arnold 1984). Also, reproductive suppression using diethylstilbesterol delivered in baits has had some success (Allen 1982).

An interesting situation and management/conservation challenge exists in California. Native red foxes are restricted to the Sierra Nevada and Cascade ranges. However, red foxes began appearing in low elevations of the state in late 1800s. These individuals are thought to be eastern red foxes that escaped from fur farms or were released for hunting (Jurek 1992; Lewis et al. 1993). Consistent with the immense adaptability of this species, these introduced foxes are spreading rapidly throughout California and are increasing in abundance. They present a significant threat to a variety of rare species in California. One concern is that they will interbreed with and swamp the gene pool of the native Sierra Nevada red fox (*V. v. necator*), which is rare. Another concern is that they will negatively effect endangered San Joaquin kit foxes through interference and exploitative competition. At least three kit foxes have been killed by red foxes (Ralls and White 1995; Clark 2001), and competition for resources has been documented (Clark 2001; Cypher et al. 2001). Similarly, recent increased abundance of red foxes in southern Alberta and Saskatchewan is considered a potential threat to reintroduced swift foxes (Carbyn 1998). Red foxes also prey on a variety of rare species, including endangered California least terns (*Sterna antillarum browni*), California light-footed clapper rails (*Rallus longirostris*), and Pacific pocket mice (*Perognathus longimembris pacificus*) (Jurek 1992). Close proximity to human habitations and a ban on the use of many types of traps in California limit options for controlling red foxes. Coyotes may constitute a potential strategy for reducing numbers of nonnative red foxes in some areas (Cypher et al. 2001).

CONSERVATION AND RESEARCH NEEDS

Conservation. Foxes are rare or extirpated in some areas because of habitat loss, predator control programs, past exploitation, and other factors. At least one subspecies of kit fox is extinct and another is endangered, and swift foxes have been extirpated over large areas of their range. Paradoxically, some fox species are harvested intensely in some parts of their range and are fully protected with harvests prohibited in other parts.

Arctic fox populations appear secure throughout their range in North America. They now are considered to be rare in Finland, Sweden, Norway, and some parts of Russia, and conservation efforts have been implemented (Garrott and Eberhardt 1987; Ginsberg and Macdonald 1990). Gray foxes are relatively widespread and common and are not considered to be at risk in any portion of their range.

Island foxes present a continuing conservation challenge. Habitat on all six Channel Islands occupied by island foxes is protected to some extent. However, this insular situation results in limited geographic distribution, small population sizes, low genetic variability, low food diversity, and lack of previous exposure to potentially lethal canine diseases (Gilbert et al. 1990; Garcelon et al. 1992; Roemer et al. 2000). Also, as an adaptation to their insular environment, island foxes exhibit relatively low reproductive rates and high population densities. Consequently, populations always will be vulnerable to stochastic as well as anthropogenic events. Island foxes were classified as "threatened" in California in 1971.

The vulnerability of island foxes was emphasized by dramatic declines in populations on four of islands in the 1990s. These declines were attributable to anthropogenic factors. Bioaccumulation of pesticides caused the disappearance of bald eagles (primarily piscivorous) in the 1960s and 1970s, and this coupled with the presence of feral animals facilitated colonization by golden eagles (primarily terrestrial feeders) in the 1980s and 1990s. In addition to feeding on piglets, goat kids, and other feral animals, golden eagles preyed extensively on island foxes, causing population declines of >90% on Santa Cruz, Santa Rosa, and San Miguel Islands (Roemer 1999). A program to capture and relocate golden eagles has been implemented. Also, most of the remaining foxes on Santa Rosa and San Miguel Islands have been captured and placed in captive breeding colonies (Coonan 2001). On Santa Catalina Island, an epizootic of distemper has reduced the fox population by over 90% (Coonan 2001). Distemper likely was introduced by people or their pets. A captive breeding colony also has been established on this island. A recovery plan has been prepared for island foxes on the northern Channel Islands (Coonan 2001).

Kit foxes are classified as "endangered" in Oregon. They may always have been uncommon in Oregon and Idaho, where they are on the periphery of their range. The San Joaquin kit fox (*V. m. mutica*) occurs in the San Joaquin Valley and some adjacent valleys in California. It is classified as "endangered" federally and as "threatened" in California, primarily because of habitat loss from agricultural, industrial, and urban development (U.S. Fish and Wildlife Service 1998). A recovery plan was completed in 1983, and an updated multispecies recovery plan that includes the San Joaquin kit fox was approved in 1998. Ongoing conservation efforts include habitat preservation and restoration, prohibitions on killing foxes, ecological research, and research on conservation strategies (U.S. Fish and Wildlife Service 1998). Kit foxes also are classified as "endangered" in Mexico (Maldonado et al. 1997), where they appear to be strongly associated with prairie dog colonies (List 1997).

Swift fox abundance in many portions of their historical range has been reduced significantly. They have disappeared from some regions because of habitat loss, intensive harvests, and predator and rodent control programs (Hillman and Sharps 1978; Herrero et al. 1986). Northern populations of swift foxes briefly were classified as "endangered" in the United States, but then protection was retracted based on taxonomic questions. Swift foxes again were proposed for federal protection, but protection has been deemed unnecessary (U.S. Fish and Wildlife Service 2001) because of a coordinated effort by agencies and groups to conserve them, including the preparation of a conservation plan (Kahn et al. 1997). Swift foxes are classified as "endangered" in Nebraska and "threatened" in South Dakota. Numbers in the United States appear to be increasing (Stromberg and Boyce 1986; Kahn et al. 1997).

In Canada, swift foxes declined and the last free-ranging fox was observed in 1938. The species was officially declared "extirpated" in 1978 (Herrero et al. 1986). However, an extensive reintroduction effort was initiated in 1983 (Carbyn and Schroeder 1987), and since then nearly 1000 individuals have been released in Alberta and Saskatchewan (Carbyn 1998). Animals used in this effort primarily were from captive breeding facilities supplemented with wild foxes from the United States. A population of about 290 foxes was estimated in Alberta and Saskatchewan in 1998, and swift foxes are now classified as "endangered" in Canada (Carbyn 1998). A recovery plan for Canadian swift foxes has been completed (Brechtel et al. 1996). Also, some of these Canadian individuals have dispersed southward into Montana (Kahn et al. 1997). Swift foxes also have been reintroduced into Montana on Blackfeet tribal lands (C. Smeeton, Cochrane Ecological Institute, pers. commun., 2000). Another reintroduction effort is planned in South Dakota (K. Kunkel, Turner Endangered Species Fund, pers. commun., 2000), and consideration also has been given to reintroducing swift foxes into North Dakota (Scott-Brown et al. 1987).

Red fox populations are relatively secure in most regions. One exception is the subspecies *V. v. necator,* which is restricted to the Sierra Nevada and Cascade ranges of California. This subspecies may never have been very abundant, and harvests were discontinued in 1974 (Gray 1975). The subspecies was listed as "threatened" in California in 1980.

Research Needs. North American foxes are among the more thoroughly studied mammal species. Despite this, questions remain regarding their biology, ecology, management, and conservation. Needs common to multiple species are discussed first followed by needs specific to each species.

A significant research need is to study the ecology and demography of all of the fox species in anthropogenically altered landscapes. Most of the remaining natural habitats have been altered to some degree by people. These alterations have affected foxes, in some cases in positive ways. Foxes either persist in or have colonized radically altered landscapes such as irrigated croplands and urban and suburban environments. Data are needed on the ecological and demographic patterns of fox populations in these environments to develop effective and cost-efficient management strategies to reduce nuisance or damage problems. A particularly important issue is the protection of people and domestic animals from diseases potentially carried by foxes. Conversely, ecological and demographic data on these fox populations also are necessary to conserve foxes in these altered environments. The populations of many wildlife species decline as their habitats are altered. However, largely because of the adaptability of foxes, they may persist and thrive in such landscapes. This presents an opportunity to maintain populations and offset reductions in historical abundance associated with habitat loss. Conservation of foxes in altered habitats is particularly important for populations such as endangered San Joaquin kit foxes.

As humans continue to encroach on natural habitats, and given the colonization of urban habitats by foxes, an extremely important issue is the control of zoonotic diseases. Rabies is the most important zoonotic disease carried by foxes. Further research is warranted both to understand the epidemiology of rabies and to develop effective strategies for controlling epizootics and protecting public health. Population monitoring programs could be established in areas subject to chronic rabies epizootics. Considerable research has been conducted on methods for administering oral vaccines to wild foxes, and further research is warranted. Research also is needed on effective population control measures, both lethal and nonlethal. Control of fox populations may be necessary when epizootics are imminent or in progress. A significant challenge with lethal control measures is determining how to safely administer such measures in areas where people and pets are present. Research also is warranted on methods of nonlethal population control such as reproductive suppression.

Alveolar hydatid disease resulting from the tapeworm *Echinococcus multilocularis* is another potential threat to humans, mostly in more northern regions of North America. As contact between foxes and humans increases in these regions, strategies for protecting humans from this disease may need to be developed. Such strategies may be as simple as public education and proper sanitation methods. However, fox population control potentially may be necessary and, as with rabies control, both lethal and nonlethal strategies could be developed and implemented.

Interspecific interactions among fox species warrant further investigation. Of particular importance are data on the effects of red foxes on other species. Available information suggests that red foxes may outcompete and even competitively exclude other foxes. Recent range expansions by red foxes into habitats occupied by arctic foxes, kit foxes, and swift foxes could negatively affect these species. The increase of nonnative red foxes into areas occupied by endangered San Joaquin kit foxes is a particular concern. Interactions between species should be investigated to determine whether interference or exploitative competition is occurring, to identify those factors favoring one species over another, and to identify potential strategies for mitigating impacts on vulnerable species.

Arctic foxes are subject to increasing human presence within their range. In particular, tourism and oil exploration have been increasing in recent years. Of concern are the effects of these activities on arctic fox populations, and so are the potential threats posed to humans by rabies and hydatid tapeworms, both of which are carried by arctic foxes. The attraction of arctic foxes to anthropogenic food sources (e.g., garbage dumps) is a particular concern, as the potential for disease transmission to humans and among foxes increases in such situations. Strategies to reduce impacts of arctic foxes on nesting-bird colonies, particularly rare species, also need further investigation. Lethal control measures may be effective in some circumstances, although the reproductive capacity of arctic foxes allows them to rapidly increase following population suppression. In some situations, nonlethal measures may be effective. For example, fox-proof fencing has been used to effectively protect bird colonies in Iceland (Ginsberg and Macdonald 1990).

Gray foxes are relatively common and of little economic value, and therefore comparatively less research has been conducted on them. Consequently, basic ecological and demographic information is needed on gray foxes in each of the habitats they occupy. Also, data are needed on factors driving population dynamics. Such information will be necessary to develop management strategies if gray foxes ever present a nuisance problem, and also to develop conservation strategies if human activities cause gray fox populations to become rare in some areas.

For reasons discussed previously, island foxes will always be vulnerable. Therefore, permanent population monitoring programs should be established on all islands. This would facilitate identification of population declines and provide opportunities to implement protective measures before fox numbers reach population-threatening levels. Additional investigation of basic biology, ecology, and population demography of island foxes is warranted on all islands. These data will facilitate the development of management and conservation strategies, and also can be used in population modeling and viability analyses, which will further contribute to conservation efforts. This information also may help determine whether additional protection, including further restrictions on human access or federal protection under the U.S. Endangered Species Act, is warranted. Ongoing research to develop strategies for vaccination against introduced disease also should continue. Paradoxically, management alternatives are needed on San Clemente Island, where foxes are affecting endangered shrikes. Ongoing research on lethal and nonlethal alternatives should continue.

For kit foxes, habitat-specific ecological and demographic data would be useful to develop future management and conservation strategies, should they become necessary. Long-term population monitoring throughout the range of the endangered San Joaquin kit fox and additional demographic studies are needed to provide information necessary to conserve this subspecies. The San Joaquin kit fox currently exists in a metapopulation structure. Demographic data are needed to conduct viability analyses and determine optimal preserve design. Also, additional habitat needs to be protected to ensure the long-term viability of this subspecies. Furthermore, this habitat may need to be actively managed to counter range-wide reductions in habitat quality associated with the invasion of exotic grasses (U.S. Fish and Wildlife Service 1998). Further investigation is warranted on kit fox–red fox interactions. Nonnative red foxes pose a potential threat to San Joaquin kit foxes; control strategies may be needed to reduce red foxes without adversely affecting kit foxes. Additionally, investigations of basic biology, ecology, and population demography are needed on kit fox populations in Mexico so that effective conservation strategies can be developed for these populations.

Swift foxes are repopulating some regions within their historical range (e.g., southern Montana) and have been reintroduced into others (e.g., Alberta, northern Montana). Recolonizing populations should be protected and encouraged, and additional reintroduction efforts should be considered. Investigations of the ecology and demography of these populations are necessary to facilitate their conservation. Monitoring programs also are needed throughout the range of the swift fox to assess population status. In some areas, populations may always be small because of limited habitat availability. Monitoring is especially important for these populations to determine whether additional conservation measures are warranted. Further investigation of habitat relationships is warranted to identify suitable habitat and target critical areas for protection. The recently developed conservation plan for swift foxes established a goal of expanding the current distribution of swift foxes to occupy at least 50% of the existing suitable habitat (Kahn et al. 1997).

Red foxes have been extensively studied in North America. They are still extensively harvested, therefore, healthy populations of this species are desirable. Red foxes may not need much assistance in this regard, other than ensuring that adequate habitat remains available. Conversely, red fox abundance may need to be controlled in areas where high numbers pose a threat to human interests and domestic animals, especially where chronic rabies epidemics occur. Lethal and nonlethal control strategies have been the subject of numerous investigations and additional research is warranted. Also, range extensions and introductions of red foxes into new areas potentially pose a threat to indigenous wildlife in those areas. Of particular concern are threats to a number of rare species; strategies to mitigate these threats are needed. Additional research and conservation efforts also are needed for the rare subspecies in the Sierra Nevada of California. Investigations of the distribution, ecology, and demography of this subspecies are ongoing so that effective conservation strategies can be developed.

LITERATURE CITED

Ables, E. D. 1965. An exceptional fox movement. Journal of Mammalogy 46:102.

Ables, E. D. 1968. Ecological studies on red foxes in southern Wisconsin. Ph.D. Dissertation, University of Wisconsin, Madison.

Ables, E. D. 1969. Activity studies of red foxes in southern Wisconsin. Journal of Wildlife Management 33:145–53.

Ables, E. D. 1975. Ecology of the red fox in America. Pages 216–36 *in* M. W. Fox, ed. The wild Canids. Van Nostrand Reinhold, New York.

Adalsteinsson, S., P. Hersteinsson, and E. Gunnarsson. 1987. Fox colors in relation to colors in mice and sheep. Journal of Heredity 78:235–37.

Allen, S. H. 1974. Modified techniques for aging red fox using canine teeth. Journal of Wildlife Management 38:152–54.

Allen, S. H. 1982. Bait consumption and diethylstilbesterol influence on North Dakota red fox reproductive performance. Wildlife Society Bulletin 10:370–74.

Allen, S. H. 1984. Some aspects of reproductive performance in female red fox in North Dakota. Journal of Mammalogy 65:246–55.

Allen, S. H., and A. B. Sargeant. 1993. Dispersal patterns of red foxes relative to population density. Journal of Wildlife Management 57:526–33.

Anderson, R. M., H. C. Jackson, R. M. May, and A. M. Smith. 1981. Population dynamics of fox rabies in Europe. Nature 289:765–71.

Angerbjörn, A., M. Tannerfeldt, and S. Erlinge. 1999. Predator–prey relationships: Arctic foxes and lemmings. Journal of Animal Ecology 68:34–49.

Anthony, R. M. 1996. Den use by arctic foxes (*Alopex lagopus*) in a subarctic region of western Alaska. Canadian Journal of Zoology 74:627–31.

Anthony, R. M. 1997. Home range and movements of arctic fox (*Alopex lagopus*) in western Alaska. Arctic 50:147–57.

Anthony, R. M., P. L. Flint, and J. S. Sedinger. 1991. Arctic fox removal improves nest success of black brant. Wildlife Society Bulletin 19:176–84.

Anthony, R. M., N. L. Barten, and P. E. Seiser. 2000. Foods of arctic foxes (*Alopex lagopus*) during winter and spring in western Alaska. Journal of Mammalogy 81:820–28.

Armstrong, J. B., and A. N. Rossi. 2000. Status of avocational trapping based on the perspectives of state furbearer biologists. Wildlife Society Bulletin 28:825–32.

Arnold, D. A. 1956. Red foxes of Michigan. Michigan Conservation Department, Lansing.

Asdell, S. A. 1964. Patterns of mammalian reproduction, 2nd ed. Cornell University Press, Ithaca, NY.

Baer, G. 1975. The natural history of rabies, Vol. 2. Academic Press, New York.

Bailey, E. P. 1992. Red foxes, *Vulpes vulpes,* as biological control agents for introduced arctic foxes, *Alopex lagopus,* on Alaskan Islands. Canadian Field-Naturalist 106:200–205.

Banfield, A. W. F. 1974. The mammals of Canada. University of Toronto Press, Toronto.

Barabash-Nikiforov, I. 1938. Mammals of the Commander Islands and the surrounding sea. Journal of Mammalogy 19:423–29.

Barker, I. K., R. C. Povey, and D. R. Voigt. 1983. Response of mink, skunk, red fox and raccoon to inoculation with mink virus enteritis, feline panleukopenia and canine parvovirus and prevalence of antibody to parvovirus in wild carnivores in Ontario. Canadian Journal of Comparative Medicine 47:188–97.

Bee, J. W., and E. R. Hall. 1956. Mammals of northern Alaska on the arctic slope (Miscellaneous Publication 8). University of Kansas Museum of Natural History, Lewrence.

Benson, S. B. 1938. Notes on kit foxes (*Vulpes macrotis*) from Mexico. Proceedings of the Biological Society of Washington 51:17–24.

Berns, V. D. 1969. Notes on the blue fox of Rat Island, Alaska. Canadian Field-Naturalist 83:404–5.

Berry,W. H., and W. G.Standley. 1992. Population trends of San Joaquin kit fox at Camp Roberts Army National Guard Training Site, California (Topical Report No. EGG 10617–2155). U.S. Department of Energy, Washington, DC.

Berry, W. H.,T. P. O'Farrell,T. T.Kato, and P. M.McCue. 1987. Characteristics of dens used by radiocollared San Joaquin kit fox (*Vulpes macrotis mutica*), Naval Petroleum Reserve #1, Kern County California (Topical Report EGG 10282–2177). U.S. Department of Energy, Washington, DC.

Bloemendal, H. 1977. The vertebrate eye lens. Science 197:127–38.

Brechtel, S., L. Carbyn, G. Erickson, D. Hjertaas, C. Mamo, and P. McDougall. 1996. National recovery plan for the swift fox (Report No. 15). Recovery of Nationally Endangered Wildlife Committee, Ottawa.

Burgess, R. M. 1984. Investigations of patterns of vegetation, distribution and abundance of small mammals and nesting birds, and behavioral ecology of arctic foxes at Demarcation Bay, Alaska. M.S. Thesis, University of Alaska, Fairbanks.

Buskirk, S. W., and P. S. Gipson. 1981. Zoogeography of arctic foxes (*Alopex lagopus*) on the Aleutian Islands. Pages 38–54 *in* J. A. Chapman and D. Pursley, eds. Proceedings of the worldwide furbearer conference, Frostburg, MD.

Cameron, M. W. 1984. The swift fox (*Vulpes velox*) on the Pawnee National Grassland: Its food habits, population dynamics and ecology. M.S. Thesis, University of Northern Colorado, Greeley.

Carbyn, L. N. 1998. Update COSEWIC status report on swift fox (*Vulpes velox*). Canadian Wildlife Service, Ottawa.

Carbyn, L. N., and C. Schroeder. 1987. Preliminary recovery plan proposal for the swift fox programme in Canada. Canadian Wildlife Service, Edmonton, Alberta, Canada.

Carbyn, L. N., H. J. Armbruster, and C. Mamo. 1994. The swift fox reintroduction program in Canada from 1983 to 1992. Pages 247–71 *in* M. L. Bowles and C. J. Whelan, eds. Restoration of endangered species: Conceptual issues, planning and implementation. Cambridge University Press, Cambridge.

Carey, A. B. 1982. The ecology of red foxes, gray foxes, and rabies in the eastern United States. Wildlife Society Bulletin 10:18–26.

Carey, A. B., R. H. Giles, Jr., and R. G. McLean. 1978. The landscape epidemiology of rabies in Virginia. American Journal of Tropical Medicine and Hygiene 27:573–80.

Catling, P. C., and R. J. Burt. 1995. Why are red foxes absent from some eucalypt forests in eastern New South Wales? Wildlife Research 22:535–46.

Chamberlain, M. J., and B. D. Leopold. 2000. Spatial use patterns, seasonal habitat selection, and interactions among adult gray foxes in Mississippi. Journal of Wildlife Management 64:742–51.

Chesemore, D. L. 1967. Ecology of the arctic fox in northern and western Alaska. M.S. Thesis, University of Alaska, Fairbanks.

Chesemore, D. L. 1968. Notes on the food habits of arctic foxes in northern Alaska. Canadian Journal of Zoology 46:1127–30.

Chesemore, D. L. 1969. Den ecology of the arctic fox in northern Alaska. Canadian Journal of Zoology 47:121–29.

Chesemore, D. L. 1975. Ecology of the arctic fox (*Alopex lagopus*) in North America: A review. Pages 143–63 *in* M. W. Fox, ed. The wild canids. Van Nostrand Reinhold, New York.

Chiarelli, A. B. 1975. The chromosomes of the Canidae. Pages 40–53 *in* M. W. Fox, ed. The wild canids. Van Nostrand Reinhold, New York.

Chitty, H. 1950. Canadian arctic wildlife enquiry, 1943–49: With a summary of results since 1933. Journal of Animal Ecology 19:180–93.

Chomel, B. B. 1993. The modern epidemiological aspects of rabies in the world. Comparative Immunology and Microbiology of Infectious Diseases 16:11–20.

Choquett, L. P. E., A. H. Macpherson, and J. G. Cousineau. 1962. Note on the occurrence of *Echinococcus multilocularis* Leuckart, 1863 in the arctic fox in Canada. Canadian Journal of Zoology 40:1167.

Churcher, C. S. 1959. The specific status of the new world red fox. Journal of Mammalogy 40:513–20.

Clark, H. O., Jr. 2001. Endangered San Joaquin kit fox and non-native red fox interspecific interactions. M.S. Thesis, California State University, Fresno.

Clutton-Brock, J., G. B. Corbert, and M. Hills. 1976. A review of the family Canidae, with a classification by numerical methods. Bulletin of the British Museum of Natural History, Zoology 29:117–99.

Cohen, J. A., and M. W. Fox. 1976. Vocalizations in wild canids and possible effects of domestication. Behavior Proceedings 1:77–92.

Collins, P. W. 1980. Food habits of the island fox (*Urocyon littoralis littoralis*) on San Miguel Island, California. Pages 152–64 *in* Proceedings of the second conference on scientific research in the national parks. U.S. Department of the Interior, National Park Service, Washington, DC.

Collins, P. W. 1982. Origin and differentiation of the island fox: A study of evolution in insular populations. M.S. Thesis, University of California, Santa Barbara.

Collins, P. W. 1991a. Interaction between island foxes (*Urocyon littoralis*) and Indians on islands off the coast of southern California: I. Morphologic and archaeological evidence of human assisted dispersal. Journal of Ethnobiology 11:51–81.

Collins, P. W. 1991b. Interaction between island foxes (*Urocyon littoralis*) and Native Americans on islands off the coast of southern California: II. Ethnographic, archaeological, and historical evidence. Journal of Ethnobiology 11:205–29.

Coonan, T. 2001. Recovery plan for island foxes (*Urocyon littoralis*) on the northern Channel Islands. Channel Islands National Park, Ventura, CA.

Coonan, T. J., G. Austin, and C. Schwemm. 1998. Status and trend of island fox, San Miguel Island, Channel Islands National Park (Technical Report 98–01). Channel Islands National Park, Ventura, CA.

Covell, D. F. 1992. Ecology of the swift fox (*Vulpes velox*) in southwestern Colorado. M.S. Thesis, University of Wisconsin, Madison.

Covell, D. F., D. S. Miller, and W. H. Karasov. 1996. Cost of locomotion and daily energy expenditure by free-living swift foxes (*Vulpes velox*): A seasonal comparison. Canadian Journal of Zoology 74:283–90.

Cowardin, L. M., D. S. Gilmer, and C. W. Shaiffer. 1985. Mallard recruitment in the agricultural environment of North Dakota. Wildlife Monographs 92:1–37.

Crandall, L. S. 1964. The management of wild animals in captivity. University of Chicago Press, Chicago.

Crandell, R. A. 1975. Arctic fox rabies. Pages 23–40 *in* G. Baer, ed. The natural history of rabies, Vol. 2. Academic Press, New York.

Creel, G. C., and W. A. Thornton. 1974. Comparative study of a *Vulpes fulva–Vulpes macrotis* hybrid fox karyotype. Southwestern Naturalist 18:465–68.

Crooks, K. R. 1994. Demography and status of the island fox and the island spotted skunk on Santa Cruz Island, California. Southwestern Naturalist 39:257–62.

Crooks, K. R., and M. E. Soulé. 1999. Mesopredator release and avifaunal extinctions in a fragmented system. Nature 400:563–66.

Crooks, K. R., and D. Van Vuren. 1996. Spatial organization of the island fox (*Urocyon littoralis*) on Santa Cruz Island, California. Journal of Mammalogy 77:801–6.

Cutter, W. L. 1958a. Food habits of the swift fox in northern Texas. Journal of Mammalogy 39:527–32.

Cutter, W. L. 1958b. Denning of the swift fox in northern Texas. Journal of Mammalogy 39:70–74.

Cypher, B. L. 1993. Food item use by three sympatric canids in southern Illinois. Transactions of the Illinois State Academy of Science 86:139–44.

Cypher, B. L., and K. A. Spencer. 1998. Competitive interactions between coyotes and San Joaquin kit fox. Journal of Mammalogy 79:204–14.

Cypher, B. L., and G. D. Warrick. 1993. Use of human-derived food items by urban kit foxes. Transactions of the Western Section of the Wildlife Society 29:34–37.

Cypher, B. L., and R. H. Yahner. 1996. Foods used by red foxes at Valley Forge National Historical Park. Northeast Wildlife 53:19–24.

Cypher, B. L., G. D. Warrick, M. R. M. Otten, T. P. O'Farrell, W. H. Berry, C. E. Harris, T. T. Kato, P. M. McCue, J. H. Scrivner, and B. W. Zoellick. 2000. Population dynamics of San Joaquin kit foxes at the Naval Petroleum Reserves in California. Wildlife Monographs 145:1–43.

Cypher, B. L., H. O. Clark, Jr., P. A. Kelly, C. Van Horn Job, G. D. Warrick, and D. F. Williams. 2001. Interspecific interactions among mammalian predators: Implications for the conservation of endangered San Joaquin kit foxes. Endangered Species Update 18:171–74.

Daneke, D., M. Sunquist, and S. Berwick. 1984. Notes on kit fox biology in Utah. Southwestern Naturalist 29:361–62.

David, H. E. 2001. Food habits and prey availability of the threatened San Miguel island fox, *Urocyon littoralis littoralis,* during a decline in population, 1993–1998. M.S. Thesis, California Polytechnic State University, San Luis Obispo.

Davidson, W. R., M. J. Appel, G. L. Doster, O. E. Baker, and J. F. Brown. 1992a. Diseases and parasites of red foxes, gray foxes, and coyotes from commercial sources selling to fox-chasing enclosures. Journal of Wildlife Diseases 28:581–89.

Davidson, W. R., V. F. Nettles, L. E. Hayes, E. W. Howerth, and C. E. Couvillion. 1992b. Diseases diagnosed in gray foxes (*Urocyon cinereoargenteus*) from the southeastern United States. Journal of Wildlife Diseases 28:28–33.

Davis, D. E., and J. E. Wood. 1959. Ecology of foxes and rabies control. Public Health Report 74:115–18.

Davis, W. B. 1960. The mammals of Texas (Bulletin 41) Texas Game and Fish Commission, Austin.

Dekker, D. 1983. Denning and foraging habits of red foxes, *Vulpes vulpes,* and their interaction with coyotes, *Canis latrans,* in central Alberta, 1972–1981. Canadian Field-Naturalist 97:303–6.

Dekker, D. 1989. Population fluctuations and spatial relationships among wolves, *Canis lupus,* coyotes, *Canis latrans,* and red foxes, *Vulpes vulpes,* in Jasper National Park, Alberta. Canadian Field-Naturalist 103:261–64.

Denver Wildlife Research Center. 1975. Annual report—Research highlights. U.S. Department of Agriculture, Washington, DC.

Dimmick, R. W., and M. R. Pelton. 1994. Criteria of sex and age. Pages 169–214 *in* T. A. Bookhout, ed. Research and management techniques for wildlife and habitats. Wildlife Society, Bethesda, MD.

Doncaster, C. P., C. R. Dickman, and D. W. Macdonald. 1990. Feeding ecology of red foxes (*Vulpes vulpes*) in the city of Oxford, England. Journal of Mammalogy 71:188–94.

Doyen, J. T. 1974. Differential predation of darkling ground beetles (Coleoptera: Tenebrionidae) by channel islands fox. Pan-Pacific Entomologist 50:86–87.

Dragoo, J. W., J. R. Choate, T. L. Yates, and T. P. O'Farrell. 1990. Evolutionary and taxonomic relationships among North American arid-land foxes. Journal of Mammalogy 71:318–32.

Dyer, W. G., and W. D. Klimstra. 1980. A survey of grey foxes (*Urocyon cinereoargenteus*) for *Echinococcus multilocularis* in southern Illinois. Transactions of the Illinois State Academy of Science 73:72–74.

Eberhardt, L. E., and W. C. Hanson. 1978. Long-distance movements of arctic foxes tagged in northern Alaska. Canadian Field-Naturalist 92:386–89.

Eberhardt, L. E., W. C. Hanson, J. L. Bengston, R. A. Garrott, and E. E. Hanson. 1982. Arctic fox home range characteristics in an oil-development area. Journal of Wildlife Management 46:183–90.

Eberhardt, L. E., R. A. Garrott, and W. C. Hanson. 1983a. Den use by arctic foxes in northern Alaska. Journal of Mammalogy 64:97–102.

Eberhardt, L. E., R. A. Garrott, and W. C. Hanson. 1983b. Winter movements of arctic foxes, *Alopex lagopus,* in a petroleum development area. Canadian Field-Naturalist 97:66–77.

Eberhardt, W. L. 1976. The biology of arctic and red foxes on the North Slope. Pages 238–39 *in* G. C. West, ed. Proceedings of the 27th Alaska science conference. Alaska Division of the American Association for the Advancement of Science, Fairbanks.

Eberhardt, W. L. 1977. The biology of arctic and red foxes on the north slope. M.S. Thesis, University of Alaska, Fairbanks.

Egoscue, H. J. 1956. Preliminary studies of the kit fox in Utah. Journal of Mammalogy 37:351–57.

Egoscue, H. J. 1962. Ecology and life history of the kit fox in Tooele County, Utah. Ecology 43:481–97.

Egoscue, H. J. 1975. Population dynamics of the kit fox in western Utah. Bulletin of the Southern California Academy of Science 74:122–27.

Egoscue, H. J. 1979. *Vulpes velox*. Mammalian Species 122:1–5.

Egoscue, H. J. 1985. Kit fox flea relationships on the Naval Petroleum Reserves, Kern County, California. Bulletin of the Southern California Academy of Sciences 84:127–32.

Elton, C. 1931. Epidemics among sled dogs in the Canadian arctic and their relation to disease in the arctic fox. Canadian Research 5:673–92.

Elton, C. 1942. Voles, mice and lemmings: Problems in population dynamics. Oxford University Press, Oxford.

Elton, C. 1949. Movements of arctic fox populations in the region of Baffin Bay and Smith Sound. Polar Research 3:296–305.

Englund, J. 1980. Yearly variations of recovery and dispersal rates of fox cubs tagged in Swedish coniferous forests. Pages 195–207 *in* E. Zimen, ed. Biogeographica. Vol. 18: The red fox symposium on behavior and ecology. W. Junk, The Hague, the Netherlands.

Errington, P. L. 1933. Bobwhite winter survival in an area heavily populated with gray foxes. Iowa State College Journal of Science 8:127–30.

Errington, P. L., and R. M. Berry. 1937. Tagging studies of red foxes. Journal of Mammalogy 18:203–5.

Evans, R. H. 1984. Rotavirus-associated diarrhea in young raccoons (*Procyon lotor*), striped skunks (*Mephitis mephitis*) and red foxes (*Vulpes vulpes*). Journal of Wildlife Diseases 20:79–85.

Ewer, R. F. 1973. The carnivores. Cornell University Press, Ithaca, NY.

Fausett, L. L. 1982. Activity and movement patterns of the island fox, *Urocyon littoralis,* Baird 1857 (Carnivora: Canidae). Ph.D. Dissertation, University of California, Los Angeles.

Fay, F. H. 1973. The ecology of *Echinococcus multilocularis* Leuckart, 1863 (Cestoda: Taeniidae), on St. Lawrence Island, Alaska. Annales of Parasitology 48:523–42.

Fay, F. H., and R. L. Rausch. 1992. Dynamics of the arctic fox population on St. Lawrence Island, Bering Sea. Arctic 45:393–97.

Fine, H. 1980. Ecology of arctic foxes at Prudhoe Bay, Alaska. M.S. Thesis, University of Alaska, Fairbanks.

Fisher, J. L. 1981. Kit fox diet in south-central Arizona. M.S. Thesis, University of Arizona, Tucson.

Follman, E. H. 1973. Comparative ecology and behavior of red and gray foxes. Ph.D. Dissertation, Southern Illinois University, Carbondale.

Follman, E. H. 1978. Annual reproductive cycle of the male gray fox. Transactions of the Illinois State Academy of Science 71:304–11.

Fox, M. W. 1970. A comparative study of the development of facial expressions in canids; wolf, coyote and foxes. Behaviour 36:49–73.

Frafjord, K. 1992. Denning behaviour and activity of arctic fox *Alopex lagopus* pups: Implications of food availability. Polar Biology 12:707–12.

Frafjord, K., D. Becker, and A. Angerbjörn. 1989. Interactions between arctic and red foxes in Scandinavia—Predation and aggression. Arctic 42:354–56.

Friend, M. 1967. A review of research concerning eye-lens weight as a criterion of age in animals. New York Fish and Game Journal 14:152–65.

Fritzell, E. K. 1987. Gray fox and island gray fox. Pages 408–20 *in* M. Novak, J. A. Baker, M. E. Obbard, and D. Malloch, eds. Wild furbearer management and conservation in North America. Ontario Trappers Association and Ontario Ministry of Natural Resources, Toronto.

Fritzell, E. K., and K. J. Haroldson. 1982. *Urocyon cinereoargenteus*. Mammalian Species 189:1–8.

Fuller, T. K. 1978. Variable home range sizes of female gray foxes. Journal of Mammalogy 59:446–49.

Garcelon, D. K., R. K. Wayne, and B. J. Gonzales. 1992. A serologic survey of the island fox (*Urocyon littoralis*) on the Channel Islands, California. Journal of Wildlife Diseases 28:223–29.

Garrott, R. A., and L. E. Eberhardt. 1987. Arctic fox. Pages 394–406 *in* M. Novak, J. A. Baker, M. E. Obbard, and B. Malloch, eds. Wild furbearer management and conservation in North America. Ontario Trappers Association and Ontario Ministry of Natural Resources, Toronto.

Garrott, R. A., L. E. Eberhardt, and W. C. Hanson. 1983a. Arctic fox den identification and characteristics in northern Alaska. Canadian Journal of Zoology 61:423–26.

Garrott, R. A., L. E. Eberhardt, and W. C. Hanson. 1983b. Summer food habits of juvenile arctic foxes in northern Alaska. Journal of Wildlife Management 47:540–45.

Garrott, R. A., L. E. Eberhardt, and W. C. Hanson. 1984. Arctic fox denning behavior in northern Alaska. Canadian Journal of Zoology 62:1636–40.

Geffen, E., A. Mercure, D. J. Girman, D. W. Macdonald, and R. K. Wayne. 1992. Phylogenetic relationships of the fox-like canids: Mitochondrial DNA restriction fragment, site and cytochrome *b* sequence analyses. Journal of Zoology (London) 228:27–39.

George, S. B., and R. K. Wayne. 1991. Island foxes: A model for conservation genetics. Terra 30:18–23.

Gese, E. M., T. E. Stotts, and S. Grothe. 1996. Interactions between coyotes and red foxes in Yellowstone National Park, Wyoming. Journal of Mammalogy 77:377–82.

Gilbert, D. A., N. Lehman, S. J. O'Brien, and R. K. Wayne. 1990. Genetic fingerprinting reflects population differentiation in the California Channel Island fox. Nature 344:764–67.

Gilmore, R. M. 1946. Mammals in archaeological collections from southwestern Pennsylvania. Journal of Mammalogy 27:227–35.

Ginsberg, J. R., and D. W. Macdonald. 1990. Foxes, wolves, jackals, and dogs: An action plan for the conservation of canids. International Union for the Conservation of Nature and Natural Resources, Gland, Switzerland.

Glover, F. A. 1949. Fox foods on West Virginia wild turkey range. Journal of Mammalogy 30:78–79.

Godin, A. J. 1977. Wild mammals of New England. Johns Hopkins University Press, Baltimore.

Goldman, E. A. 1938. List of the gray foxes of Mexico. Journal of the Washington Academy of Science 28:494–98.

Golightly, Jr., R. T. 1981. Comparative energetics of two desert canids: The coyote (*Canis latrans*) and the kit fox (*Vulpes macrotis*). Ph.D. Dissertation, Arizona State University, Tempe.

Golightly, Jr., R. T., and D. B. Hardenbrook. 1986. Habitat usage and den sites of kit foxes in Nevada (Final Report, Federal Aid Project W-48-R-15, Study R-XIX) Nevada Department of Wildlife, Reno.

Golightly, Jr., R. T., and R. D. Ohmart. 1983. Metabolism and body temperature of two desert canids: Coyotes and kit foxes. Journal of Mammalogy 64:624–35.

Golightly, Jr., R. T., and R. D. Ohmart. 1984. Water economy of two desert canids: Coyote and kit fox. Journal of Mammalogy 65:51–58.

Gray, R. L. 1975. Sacramento Valley red fox survey, 1975 (Progress Report W-54-R, Job II-1.2). California Department of Fish and Game, Sacramento.

Grinnell, J., D. S. Dixon, and J. M. Linsdale. 1937. Fur-bearing mammals of California, Vol. 2. University of California Press, Berkeley.

Gustavsson, I., and C. O. Sundt. 1965. Chromosome complex of the family of Canidae. Hereditas 54:249–54.

Hall, E. R. 1981. The mammals of North America, 2nd ed. John Wiley, New York.

Hall, E. R., and K. R. Kelson. 1959. The mammals of North America, Vol. 2. Ronald Press, New York.

Hall, M. W. 1983. Artificial dens and feeders in the stocking and management of gray foxes for sport hunting. M.S. Thesis, University of Georgia, Athens.

Hallberg, D. L., and G. R. Trapp. 1984. Gray fox temporal and spatial activity in a riparian–agricultural zone in California's Central Valley. Pages 920–28 *in* R. W. Warner and K. M. Hendrix, eds. Proceedings of the California riparian systems conference. University of California Press, Berkeley.

Halpin, M. A., and J. A. Bissonette. 1988. Influence of snow depth on prey availability and habitat use by red fox. Canadian Journal of Zoology 66:587–92.

Hammill, M. O. 1983. The arctic fox *Alopex lagopus* as a marine mammal: Physical condition and population age structure. M.S. Thesis, McGill University, Montreal.

Haroldson, K. J., and E. K. Fritzell. 1984. Home ranges, activity, and habitat use by gray foxes in an oak–hickory forest. Journal of Wildlife Management 48:222–27.

Harris, S. 1977. Distribution, habitat utilization and age structure of a suburban fox (*Vulpes vulpes*) population. Mammal Review 7:25–39.

Harris, S. 1978. Age determination in the red fox (*Vulpes vulpes*): An evaluation of technique efficiency as applied to a sample of suburban foxes. Journal of the Zoological Society of London 184:91–117.

Harris, S. 1979. Age-related fertility and productivity in red foxes, *Vulpes vulpes,* in suburban London. Journal of Zoology (London) 187:195–99.

Harris, S., and J. M. V. Rayner. 1986. Urban fox (*Vulpes vulpes*) population estimates and habitat requirements in several British cities. Journal of Animal Ecology 55:575–91.

Harrison, D. J., J. A. Bissonette, and J. A. Sherburne. 1989. Spatial relationships between coyotes and red foxes in eastern Maine. Journal of Wildlife Management 53:181–85.

Harrison, R. L. 1997. A comparison of gray fox ecology between residential and undeveloped rural landscapes. Journal of Wildlife Management 61:112–21.

Henke, S. E., and F. C. Bryant. 1999. Effects of coyote removal on the faunal community in western Texas. Journal of Wildlife Management 63:1066–81.

Henry, J. D. 1986. Red fox: The catlike canine. Smithsonian Institution Press, Washington, DC.

Henshaw, R. E., L. S. Underwood, and T. M. Casey. 1972. Peripheral thermoregulation: Foot temperature in two arctic canines. Science 175:988–90.

Hensley, M. S., and J. E. Fisher. 1975. Effects of intensive gray fox control on population dynamics of rodents and sympatric carnivores. Southeastern Association of Game and Fish Commissioners Annual Conference Proceedings 29:694–705.

Herrero, S., C. Schroeder, and S. Scott-Brown. 1986. Are Canadian foxes swift enough? Biological Conservation 36:159–67.

Hersteinsson, P., and D. W. Macdonald. 1982. Some comparisons between red and arctic foxes, *Vulpes vulpes* and *Alopex lagopus,* as revealed by radio tracking. Symposium of the Zoological Society of London 49:259–89.

Hersteinsson, P., and D. W. Macdonald. 1992. Interspecific competition and the geographical distribution of red and arctic foxes *Vulpes vulpes* and *Alopex lagopus*. Oikos 64:505–15.

Hersteinsson, P., A. Angerbjörn, K. Frafjord, and A. Kaikusalo. 1989. The arctic fox in Fennoscandia and Iceland: Management problems. Biological Conservation 49:67–81.

Hewson, R. 1986. Distribution and density of fox breeding dens and the effects of management. Journal of Applied Ecology 23:531–38.

Hildebrand, M. 1952a. The integument in Canidae. Journal of Mammalogy 33:419–28.

Hildebrand, M. 1952b. An analysis of body proportions in the Canidae. American Journal of Anatomy 90:217–56.

Hildebrand, M. 1954. Comparative morphology of the body skeleton in recent Canidae. University of California Publication in Zoology 52:399–470.

Hillman, C. N., and J. C. Sharps. 1978. Return of swift fox to northern great plains. Proceedings of the South Dakota Academy of Science 57:154–62.

Hines, T. 1980. An ecological study of *Vulpes velox* in Nebraska. M.S. Thesis, University of Nebraska, Lincoln.

Hiruki, L. M., and I. Stirling. 1989. Population dynamics of the arctic fox, *Alopex lagopus,* on Banks Island, Northwest Territories. Canadian Field-Naturalist 103:380–87.

Hnatiuk, J. M. 1969. Occurrence of *Echinococcus multilocularis* Leuckart, 1863 in *Vulpes fulva* in Saskatchewan. Canadian Journal of Zoology 47:264.

Hockman, J. G., and J. A. Chapman. 1983. Comparative feeding habits of red foxes (*Vulpes vulpes*) and gray foxes (*Urocyon cinereoargenteus*) in Maryland. American Midland Naturalist 110:276–85.

Hoffman, R. A., and C. M. Kirkpatrick. 1954. Red fox weights and reproduction in Tippecanoe County, Indiana. Journal of Mammalogy 55:504–9.

Holcomb, L. C. 1965. Large litter size of red fox. Journal of Mammalogy 46:530.

Jackson, V. L., and J. R. Choate. 2000. Dens and den sites of the swift fox, *Vulpes velox*. Southwestern Naturalist 45:212–20.

Janes, D. W., and H. T. Gier. 1966. Distribution, numbers, and hunting of foxes in Kansas. Transactions of the Kansas Academy of Science 69:23–31.

Jennings, W. L., N. J. Schneider, A. L. Lewis, and J. E. Scatterday. 1960. Fox rabies in Florida. Journal of Wildlife Management 24:171–79.

Jensen, C. C. 1972. San Joaquin kit fox distribution. Bureau of Sport Fisheries and Wildlife, Division of Wildlife Services, Sacramento, CA.

Johansson, I. 1960. Inheritance of the color phases in ranch bred blue foxes. Hereditas 46:753–66.

Johnson, D. H., and A. B. Sargeant. 1977. Impact of red fox predations on the sex ratio of prairie mallards (Wildlife Research Report 6). U.S. Fish and Wildlife Service.

Johnson, D. H., M. D. Bryant, and H. H. Miller. 1948. Vertebrate animals of the Providence Mountains area of California. University of California Publications in Zoology 48:221–375.

Johnson, S. T. 1946. Breeding blue foxes for profit. Black Fox Magazine 30:24.

Johnston, D. H., and M. Beauregard. 1969. Rabies epidemiology in Ontario. Bulletin of the Wildlife Disease Association 5:357–70.

Johnston, D. H., and D. R. Voigt. 1982. A baiting system for the oral rabies vaccination of wild foxes and skunks. Comparative Immunology, Microbiology, and Infectious Diseases 5:185–86.

Johnston, D. H., and I. D. Watt. 1981. A rapid method for sectioning undecalcified carnivore teeth for aging. Pages 407–22 *in* J. A. Chapman and D. Pursley, eds. Proceedings of the worldwide furbearer conference. Frostburg, MD.

Jones, D. B., and J. B. Theberge. 1982. Summer home range and habitat utilization of the red fox (*Vulpes vulpes*) in a tundra habitat, northwest British Columbia. Canadian Journal of Zoology 60:807–12.

Jurek, R. M. 1992. Nonnative red foxes in California (Nongame Bird and Mammal Section Report 92–04). California Department of Fish and Game, Sacramento.

Kahn, R., L. Fox, P. Horner, G. Giddings, and C. Roy. 1997. Conservation assessment and conservation strategy for swift fox in the United States. South Dakota Department of Game, Fish and Parks, Pierre.

Kamler, J. F., and W. B. Ballard. 2002. A review of native and non-native red foxes in North America. Wildlife Society Bulletin 30:370–379.

Keeler, C. 1975. Genetics of behavior variations in color phases of the red fox. Pages 40–53 *in* M. W. Fox, ed. The wild canids. Van Nostrand Reinhold, New York.

Kilgore, D. L., Jr. 1969. An ecological study of the swift fox (*Vulpes velox*) in the Oklahoma Panhandle. American Midland Naturalist 81:512–34.

Kitchen, A. M., E. M. Gese, and E. R. Schauster. 1999. Resource partitioning between coyotes and swift foxes: Space, time, and diet. Canadian Journal of Zoology 77:1645–56.

Klir, J. J., and J. E. Heath. 1992. An infrared thermographic study of surface temperature in relation to external thermal stress in three species of foxes: The red fox (*Vulpes vulpes*), arctic fox (*Alopex lagopus*), and kit fox (*Vulpes macrotis*). Physiological Zoology 65:1011–21.

Knapp, D. K. 1978. Effects of agricultural development in Kern County, California, on the San Joaquin kit fox in 1977 (Final Report, Project E-1-1, Job V-1. 21). California Department of Fish and Game, Non-game Wildlife Investigations, Sacramento.

Kolb, H. H., and R. Hewson. 1980. A study of fox populations in Scotland from 1971 to 1976. Journal of Applied Ecology 17:7–19.

Koopman, M. E. 1995. Food habits, space use, and movements of the San Joaquin kit fox on the Elk Hills Naval Petroleum Reserves in California. M.S. Thesis, University of California, Berkeley.

Koopman, M. E., J. H. Scrivner, and T. T. Kato. 1998. Patterns of den use by San Joaquin kit foxes. Journal of Wildlife Management 62:373–79.

Koopman, M. E., B. L. Cypher, and J. H. Scrivner. 2000. Dispersal patterns of San Joaquin kit foxes (*Vulpes macrotis mutica*). Journal of Mammalogy 81:213–22.

Korschgen, L. J. 1957. Food habits of coyotes, foxes, house cats and bobcats in Missouri (Pittman-Robertson Series 15). Missouri Conservation Commission.

Kozicky, E. L. 1943. Food habits of foxes in wild turkey territory. Pennsylvania Game News 14:8–9, 28.

Kritsky, D. C., and P. D. Leiby. 1978. Studies on sylvatic echinococcosis. V. Factors influencing prevalence of *Echinococcus multilocularis* Leuckart 1863, in red foxes from North Dakota, 1965–1972. Journal of Parasitology 64:625–34.

Larivière, S., and M. Pasitschniak-Arts. 1996. *Vulpes vulpes*. Mammalian Species 537:1–11.

Latham, R. M. 1943. An ecological study of the red and gray foxes in Southeastern Pennsylvania (Pittman-Robertson Report 1-R). Pennsylvania Game Commission, Harrisburg.

Latham, R. M. 1952. The fox as a factor in the control of weasel populations. Journal of Wildlife Management 16:516–17.

Laughrin, L. L. 1970. San Joaquin kit fox, its distribution and abundance (Administrative Report 70–2). California Department of Fish and Game, Wildlife Management Branch, Sacramento.

Laughrin, L. L. 1973. The island fox. Environment Southwest 458:6–9.

Laughrin, L. L. 1977. The island fox: A field study of its behavior and ecology. Ph.D. Dissertation, University of California, Santa Barbara.

Laughrin, L. L. 1980. Populations and status of the island fox. Pages 745–49 *in* D. M. Power, ed. The California islands: Proceedings of a multidisciplinary symposium. Santa Barbara Museum of Natural History, Santa Barbara, CA.

Layne, J. N. 1958. Reproductive characteristics of the gray fox in southern Illinois. Journal of Wildlife Management 22:157–63.

Layne, J. N., and W. H. McKeon. 1956a. Notes on red and gray fox den sites in New York. New York Fish and Game Journal 3:248–49.

Layne, J. N., and W. H. McKeon. 1956b. Notes on the development of the red fox fetus. New York Fish and Game Journal 3:120–28.

Layne, J. N., and W. H. McKeon. 1956c. Some aspects of red fox and gray fox reproduction in New York. New York Fish and Game Journal 3:44–74.

Lewis, J. C., K. L. Sallee, and R. T. Golightly, Jr. 1993. Introduced red fox in California (Report 93–10). California Department of Fish and Game, Nongame Bird and Mammal Section, Sacramento.

Lindström, E. 1980. The red fox in a small game community of the south taiga region in Sweden. Pages 177–84 *in* E. Zimen, ed. Biogeographica. Vol. 18: The red fox. W. Junk, The Hague, the Netherlands.

Lindström, E. 1982. Population ecology of the red fox (*Vulpes vulpes* L.) in relation to food supply. Ph.D. Dissertation, University of Stockholm, Stockholm.

Lindström, E. 1986. Territory inheritance and the evolution of group living in carnivores. Animal Behavior 34:1825–35.

Lindström, E. 1989. Food limitation and social regulation in a red fox population. Holarctic Ecology 12:70–79.

Linhart, S. B. 1968. Dentition and pelage in the juvenile red fox (*Vulpes vulpes*). Journal of Mammalogy 49:526–28.

Linhart, S. B., and W. B. Robinson. 1972. Some relative carnivore densities in areas under sustained coyote control. Journal of Mammalogy 53:880–84.

List, R. 1997. Ecology of the kit fox (*Vulpes macrotis*) and coyote (*Canis latrans*) and the conservation of the prairie dog ecosystem in northern Mexico. Ph.D. Dissertation, University of Oxford, Oxford.

Lloyd, H. G. 1975. The red fox in Britain. Pages 207–15 *in* M. W. Fox, ed. The wild canids. Van Nostrand Reinhold, New York.

Lloyd, H. G. 1980. The red fox. B. T. Batsford, London.

Logan, C. G., W. H. Berry, W. G. Standley, and T. T. Kato. 1992. Prey abundance and food habits of San Joaquin kit fox at Camp Roberts Army National Guard Training Site, California (Topical Report No. EGG 10617–2158). U.S. Department of Energy, Washington, DC.

Loison, A., O. Strand, and J. D. C. Linnell. 2001. Effect of temporal variation in reproduction on models of population viability: A case study for remnant arctic fox (*Alopex lagopus*) populations in Scandinavia. Biological Conservation 97:347–59.

Longley, W. H. 1962. Movements of red fox. Journal of Mammalogy 43:107.

Lord, R. D. 1961. A population study of the gray fox. American Midland Naturalist 66:87–109.

Lovari, S., P. Valier, and M. Ricci Lucchi. 1994. Ranging behaviour and activity of red foxes (Vulpes vulpes: Mammalia) in relation to environmental variables, in a Mediterranean mixed pinewood. Journal of Zoology (London) 232:323–39.

Loy, R., and J. P. Fitzgerald. 1980. Status of the swift fox (*Vulpes velox*) on the Pawnee National Grasslands, Colorado. Journal of the Colorado–Wyoming Academy of Science 12:43.

Macdonald, D. W. 1977. The behavioural ecology of the red fox, *Vulpes vulpes*: A study of social organization and resource exploitation. Ph.D. Dissertation, University of Oxford, Oxford.

Macdonald, D. W. 1979. Helpers in fox society. Nature 282:69–71.

Macdonald, D. W. 1980a. The red fox, *Vulpes vulpes,* as a predator upon earthworms, *Lumbricus terrestris*. Zeitschriftfür Tierpsychologie 52:171–200.

Macdonald, D. W. 1980b. Social factors affecting reproduction amongst red foxes. Pages 123–75 *in* E. Zimen, ed. Biogeographica. Vol. 18: The red fox. W. Junk, The Hague, the Netherlands.

Macdonald, D. W. 1981. Resource dispersion and the social organization of the red fox (*Vulpes vulpes*). Pages 918–49 *in* J. A. Chapman and D. Pursley, eds. Proceedings of the worldwide furbearer conference. Frostburg, MD.

Macdonald, D. W. 1983. The ecology of carnivore social behavior. Nature 301:379–84.

Macdonald, D., ed. 1999. The encyclopedia of mammals. Facts on File, New York.

Macdonald, D. W., and M. T. Newdick. 1982. The distribution and ecology of foxes, *Vulpes vulpes* (L.), in urban areas. Pages 123–35 *in* R. Bornkamm, J. A. Lee, and M. R. D. Seaward, eds. Urban ecology. Blackwell, Oxford.

MacInnes, C. D. 1987. Rabies. Pages 910–29 *in* M. Novak, J. A. Baker, M. E. Obbard, and D. Malloch, eds. Wild furbearer management and conservation in North America. Ontario Trappers Association and Ontario Ministry of Natural Resources, Toronto.

MacInnes, C. D., and C. A. LeBer. 2000. Wildlife management agencies should participate in rabies control. Wildlife Society Bulletin 28:1156–67.

MacPherson, A. H. 1968. Apparent recovery of translocated arctic fox. Canadian Field-Naturalist 82:287–89.

MacPherson, A. H. 1969. The dynamics of Canadian arctic fox populations. Canadian Wildlife Service Report Series 8:1–49.

Major, J. T., and J. A. Sherburne. 1987. Interspecific relationships of coyotes, bobcats, and red foxes in western Maine. Journal of Wildlife Management 51:606–16.

Maldonado, J. E., M. Cotera, E. Geffen, and R. K. Wayne. 1997. Relationships of the endangered Mexican kit fox (*Vulpes macrotis zinseri*) to North American arid-land foxes based on mitochondrial DNA sequence data. Southwestern Naturalist 42:460–70.

Mann, W. M. 1930. Wild animals in and out of zoos. Smithsonian Science Series 6:1–362.

Matlack, R. S., P. S. Gipson, and D. W. Kaufman. 2000. The swift fox in rangeland and cropland in western Kansas: Relative abundance, mortality, and body size. Southwestern Naturalist 45:221–25.

McCue, P. M., and T. P. O'Farrell. 1988. Serological survey for selected diseases in the endangered San Joaquin kit fox (*Vulpes macrotis mutica*). Journal of Wildlife Diseases 24:274–81.

McGrew, J. C. 1977. Distribution and habitat characteristics of the kit fox (*Vulpes macrotis*) in Utah. M.S. Thesis, Utah State University, Logan.

McGrew, J. C. 1979. *Vulpes macrotis*. Mammalian Species 123:1–6.

McIntosh, D. L. 1963. Reproduction and growth of the red fox in the Canberra district. Wildlife Research 8:132–41.

McLean, R. G. 1970. Wildlife rabies in the United States: Recent history and current concepts. Journal of Wildlife Diseases 6:229–35.

Mech, L. D. 1966. The wolves of Isle Royale (National Parks Fauna Series No. 7). U.S. Government Printing Office, Washington, DC.

Mercure, A., K. Ralls, K. P. Koepeli, and R. K. Wayne. 1993. Genetic subdivisions among small canids: Mitrochondrial DNA differentiation of swift, kit, and arctic foxes. Evolution 47:1313–28.

Merriam, C. H. 1902. Three new foxes of the kit and desert fox groups. Proceedings of the Biological Society of Washington 15:73–74.

Merriam, H. G. 1966. Temporal distribution of woodchuck interburrow movements. Journal of Mammalogy 47:103–10.

Miller, D. S., B. G. Campbell, R. G. McLean, E. Campos, and D. F. Covell. 1998. Parasites of swift fox (*Vulpes velox*) from southeastern Colorado. Southwestern Naturalist 43:476–79.

Miller, D. S., D. F. Covell, R. G. McLean, W. J. Adrian, M. Niezgoda, J. M. Gustafson, O. J. Rongstad, R. D. Schultz, L. J. Kirk, and T. J. Quan. 2000. Serologic survey for selected infectious disease agents in swift and kit foxes from the western United States. Journal of Wildlife Diseases 36:798–805.

Moehlman, P. D. 1989. Intraspecific variation in canid social systems. Pages 164–82 *in* J. L. Gittleman, ed. Carnivore behavior, ecology, and evolution. Cornell University Press, Ithaca, NY.

Mollhagen, T. R., R. W. Wiley, and R. L. Packard. 1972. Prey remains in golden eagle nests: Texas and New Mexico. Journal of Wildlife Management 36:784–92.

Monson, R. A., and W. B. Stone. 1976. Canine distemper in wild carnivores in New York. New York Fish and Game Journal 23:149–54.

Monson, R. A., W. B. Stone, and E. Parks. 1973. Aging red foxes (*Vulpes fulva*) by counting the annular cementum rings of their teeth. New York Fish and Game Journal 20:54–61.

Montague, F. H., Jr. 1975. The ecology and recreational value of the red fox in Indiana. Ph.D. Dissertation, Purdue University, Lafayette, IN.

Moore, B. M., and P. W. Collins. 1995. *Urocyon littoralis*. Mammalian Species 489:1–7.

Morrell, S. 1972. Life history of the San Joaquin kit fox. California Fish and Game 8:162–74.

Morrell, S. 1975. San Joaquin kit fox distribution and abundance in 1975 (Administrative Report 75–3). California Department of Fish and Game, Wildlife Management Branch, Sacramento.

Murie, A. 1936. Following fox trails (Miscellaneous Publication 32). University of Michigan Museum of Zoology.

Murie, A. 1944. The wolves of Mount McKinley (U.S. National Parks Fauna Series 5). U.S. Government Printing Office, Washington, DC.

Murie, O. J. 1959. Fauna of the Aleutian Islands and Alaska Peninsula. North American Fauna 61:1–406.

Nicholson, W. S. 1982. An ecological study of the gray fox in east central Alabama. M.S. Thesis, Auburn University, Auburn, AL.

Nicholson, W. S., and E. P. Hill. 1984. Mortality in gray foxes from east-central Alabama. Journal of Wildlife Management 48:1429–32.

Nicholson, W. S., E. P. Hill, and D. Briggs. 1985. Denning, pup-rearing, and dispersal in the gray fox in east-central Alabama. Journal of Wildlife Management 49:33–37.

Niewold, F. J. J. 1980. Aspects of the social structure of red fox populations: A summary. Pages 185–94 *in* E. Zimen, ed. Biogeographica. Vol. 18: The red fox. W. Junk, The Hague, the Netherlands.

Nowak, R. M. 1991. Walker's mammals of the world, 5th ed. Johns Hopkins University Press, Baltimore.

Nowak, R. M., and J. L. Paradiso. 1983. Walker's mammals of the world, Vol. II, 4th ed. Johns Hopkins University Press, Baltimore.

O'Farrell, T. P. 1987. Kit fox. Pages 423–31 *in* M. Novak, J. A. Baker, M. E. Obbard, and B. Malloch, eds. Wild furbearer management and conservation in North America. Ontario Trappers Association and Ontario Ministry of Natural Resources, Toronto.

O'Farrell, T. P., and L. Gilbertson. 1986. Ecology of the desert kit fox, *Vulpes macrotis arsipus,* in the Mojave Desert of southern California. Bulletin of the Southern California Academy of Science 85:1–15.

Oleyar, C. M., and B. S. McGinnes. 1974. Field evaluation of diethylstilbestrol for suppressing reproduction in foxes. Journal of Wildlife Management 38:101–6.

O'Neal, G. T., J. T. Flinders, and W. P. Clary. 1987. Behavioral ecology of the Nevada kit fox (*Vulpes macrotis nevadensis*) on a managed desert rangeland. Pages 443–81 *in* H. H. Genoways, ed. Current Mammalogy. Plenum Press, New York.

Ozoga, J. J., C. S. Bienz, and L. J. Verme. 1982. Red fox feeding habits in relation to fawn mortality. Journal of Wildlife Management 46:242–43.

Packard, R. L., and J. H. Bowers. 1970. Distributional notes on some foxes from western Texas and eastern new Mexico. Southwestern Naturalist 14:450–51.

Page, R. J. C. 1981. Dispersal and population density of the fox (*Vulpes vulpes*) in an area of London. Journal of Zoology (London) 194:485–91.

Pearson, O. P., and C. F. Basssett. 1946. Certain aspects of reproduction in a herd of silver foxes. American Naturalist 80:45–67.

Petersen, L. R., M. A. Martin, and C. M. Pils. 1977. Status of gray foxes in Wisconsin, 1975. Wisconsin Department of Natural Resources Report 94:1–18.

Peterson, R. O. 1977. Wolf ecology and prey relationships on Isle Royale (Scientific Monograph Series, No. 11). U.S. National Park Service.

Petrides, G. A. 1950. The determination of sex and age ratios in fur animals. American Midland Naturalist 43:355–82.

Phillips, M., and P. C. Catling. 1991. Home range and activity patterns of red foxes in Nadgee nature reserve. Wildlife Research 18:677–86.

Phillips, R. L. 1970. Age ratios of Iowa foxes. Journal of Wildlife Management 34:52–56.

Phillips, R. L., R. D. Andrews, G. L. Storm, and R. A. Bishop. 1972. Dispersal and mortality of red foxes. Journal of Wildlife Management 36:237–48.

Pils, C. M., and W. D. Klimstra. 1975. Late fall foods of the gray fox in southern Illinois. Transactions of the Illinois State Academy of Science 68:255–62.

Pils, C. M., and M. A. Martin. 1978. Population dynamics, predator–prey relationships and management of the red fox in Wisconsin (Technical Bulletin 105). Wisconsin Department of Natural Resources.

Prato, C. M., T. G. Akers, and A. W. Smith. 1977. Calicivirus antibodies in wild fox populations. Journal of Wildlife Diseases 13:448–50.

Preston, E. M. 1975. Home range defense in the red fox (*Vulpes vulpes* L.) . Journal of Mammalogy 56:645–52.

Prestrud, P., and K. Nilssen. 1992. Fat deposition and seasonal variation in body composition of arctic foxes in Svalbard. Journal of Wildlife Management 56:221–33.

Prestrud, P., and K. Nilssen. 1995. Growth, size, and sexual dimorphism in arctic foxes. Journal of Mammalogy 76:522–30.

Pruss, S. D. 1999. Selection of natal dens by the swift fox (*Vulpes velox*) on the Canadian prairies. Canadian Journal of Zoology 77:646–52.

Pryor, L. 1956. Sarcoptic mange in wild foxes in Pennsylvania. Journal of Mammalogy 37:90–93.

Quinn, P., R. O. Ramdsen, and D. H. Johnston. 1976. Toxoplasmosis: A serological survey in Ontario wildlife. Journal of Wildlife Diseases 12:504–10.

Ralls, K., and L. L. Eberhardt. 1997. Assessment of abundance of San Joaquin kit foxes by spotlight surveys. Journal of Mammalogy 78:65–73.

Ralls, K., and P. J. White. 1995. Predation on San Joaquin kit foxes by larger canids. Journal of Mammalogy 76:723–29.

Ralls, K., K. Pilgrim, P. J. White, E. E. Paxinos, and R. C. Fleischer. 2001. Kinship, social relationships, and den-sharing in kit foxes. Journal of Mammalogy 82:858–866.

Rausch, R. L. 1956. Studies on the helminth fauna of Alaska—The occurrence of *Echinococcus multilocularis* Leuckart, 1863, on the mainland of Alaska. American Journal of Tropical Medicine and Hygiene 5:1086–92.

Rausch, R. 1958. Some observations on rabies in Alaska, with special reference to wild Canidae. Journal of Wildlife Management 22:246–60.

Rausch, R. L., and S. H. Richards. 1971. Observations on parasite–host relationships in *Echinococcus multilocularis* Leuckart, 1863, in North Dakota. Canadian Journal of Zoology 49:1317–30.

Raveling, D. G. 1989. Nest-predation rates in relation to colony size of black brant. Journal of Wildlife Management 53:87–90.

Reeder, W. G. 1949. Aquatic activity of a desert kit fox. Journal of Mammalogy 30:196.

Reese, E. A., W. G. Standley, and W. H. Berry. 1992. Habitat, soils, and den use of San Joaquin kit fox (*Vulpes velox macrotis*) at Camp Roberts Army National Guard Training Site, California (Topical Report EGG 10617–2156). U.S. Department of Energy, Washington, DC.

Richards, S. W., and R. L. Hine. 1953. Wisconsin fox populations (Technical Bulletin 6). Wisconsin Department of Conservation.

Richmond, N. D. 1952. Fluctuations in gray fox populations in Pennsylvania and their relationship to precipitation. Journal of Wildlife Management 16:198–208.

Riewe, R. R. 1977. Mammalian carnivores utilizing Truelove Lowland. Pages 493–501 *in* L. C. Bliss, ed. Truelove Lowland, Devon Island, Canada: A high arctic ecosystem. University of Alberta Press, Edmonton, Alberta, Canada.

Robinson, W. B. 1953. Population trends of predators and fur animals in 1080 station areas. Journal of Mammalogy 34:220–27.

Roemer, G. W. 1999. The ecology and conservation of the island fox. Ph.D. Dissertation, University of California, Los Angeles.

Roemer, G. W., D. K. Garcelon, T. J. Coonan, and C. Schwemm. 1994. The use of capture–recapture methods for estimating, monitoring, and conserving island fox populations. Pages 387–400 *in* W. L. Halvorson and G. J. Maender, eds. The fourth California islands symposium: Update on the status of resources, Santa Barbara, California. Santa Barbara Museum of Natural History, Santa Barbara, CA.

Roemer, G. W., T. J. Coonan, D. K. Garcelon, C. H. Starbird, and J. W. McCall. 2000. Spatial and temporal variation in the seroprevalence of canine heartworm antigen in the island fox. Journal of Wildlife Diseases 36:723–28.

Roemer, G. W., D. A. Smith, D. K. Garcelon, and R. K. Wayne. 2001. The behavioural ecology of the island fox (*Urocyon littoralis*). Journal of Zoology 255:1–14.

Rohwer, S. A., and D. L. Kilgore, Jr. 1973. Interbreeding in the aridland foxes, *Vulpes velox* and *V. macrotis*. Systematic Zoology 22:157–65.

Rongstad, O. J., T. R. Laurion, and D. E. Andersen. 1989. Ecology of swift fox on the Pinon Canyon Maneuver Site, Colorado (Final Report). Directorate of Engineering and Housing, Fort Carson, CO.

Root, D. A. 1981. Productivity and mortality of gray foxes and raccoons in southwestern Wisconsin. M.S. Thesis, University of Wisconsin, Stevens Point.

Root, D. A., and N. F. Payne. 1984. Evaluation of techniques for aging gray fox. Journal of Wildlife Management 48:926–33.

Root, D. A., and N. F. Payne. 1985. Age-specific reproduction of gray foxes in Wisconsin. Journal of Wildlife Management 49:890–92.

Rosatte, R. C., M. J. Power, and C. D. MacInnes. 1991. Ecology of urban skunks, raccoons, and foxes in metropolitan Toronto. Pages 31–38 *in* L. W. Adams and D. L. Leedy, eds. Wildlife conservation in metropolitan environments. National Institute for Urban Wildlife, Columbia, MD.

Rosatte, R. C., M. J. Power, C. D. MacInnes, and J. B. Campbell. 1992. Trap–vaccinate–release and oral vaccination for rabies control in urban skunks, raccoons and foxes. Journal of Wildlife Diseases 28:562–71.

Ross, J. G., and J. S. Fairley. 1969. Studies of disease in the red fox (*Vulpes vulpes*) in northern Ireland. Journal of Zoology 157:375–81.

Rudzinski, D. R., H. B. Graves, A. B. Sargeant, and G. L. Storm. 1982. Behavioral interactions of penned red and arctic foxes. Journal of Wildlife Management 46:877–84.

Rue, L. L., III. 1969. The world of the red fox. J. B. Lippincott, New York.

Ryan, G. E. 1976. Observations on the reproduction and age structure of the fox, *Vulpes vulpes* L. in New South Wales. Australian Wildlife Research 3:11–20.

Saggese, E. P., and B. F. Tullar, Jr. 1974. Possible predation of a gray fox on a red fox pup. New York Fish and Game Journal 21:86.

Samuel, D. E., and B. B. Nelson. 1982. Foxes. Pages 475–90 *in* J. A. Chapman and G. A. Feldhamer, eds. Wild mammals of North America. Johns Hopkins University Press, Baltimore.

Sargeant, A. B. 1972. Red fox spatial characteristics in relation to waterfowl predation. Journal of Wildlife Management 36:225–36.

Sargeant, A. B. 1978. Red fox prey demands and implications to prairie duck production. Journal of Wildlife Management 42:520–27.

Sargeant, A. B. 1982. A case history of a dynamic resource—The red fox. Pages 121–37 *in* G. C. Sanderson, ed. Midwest furbearer management. Proceedings of the 43rd Midwest fish and wildlife conference. Wichita, KS.

Sargeant, A. B., and S. H. Allen. 1989. Observed interactions between coyotes and red foxes. Journal of Mammalogy 70:631–33.

Sargeant, A. B., and P. M. Arnold. 1984. Predator management for ducks on waterfowl production areas in the northern plains. Pages 161–67 *in* Proceedings of the 11th vertebrate pest control onference. University of California, Davis.

Sargeant, A. B., and L. E. Eberhardt. 1975. Death feigning by ducks in response to predation by red foxes (*Vulpes fulva*). American Midland Naturalist 94:108–19.

Sargeant, A. B., S. H. Allen, and D. H. Johnson. 1981. Determination of age and whelping dates of live red fox pups. Journal of Wildlife Management 45:760–65.

Sargeant, A. B., W. K. Pfeifer, and S. H. Allen. 1975. A spring aerial census of red foxes in North Dakota. Journal of Wildlife Management 39:30–39.

Sargeant, A. B., S. H. Allen, and R. T. Eberhardt. 1984. Red fox predation on breeding ducks in mid-continent North America. Wildlife Monographs 89:1–41.

Sargeant, A. B., S. H. Allen, and J. P. Fleskes. 1986. Commercial sunflowers: Food for red foxes in North Dakota. Prairie Naturalist 18:91–94.

Sargeant, A. B., S. H. Allen, and J. O. Hastings. 1987. Spatial relations between sympatric coyotes and red foxes in North Dakota. Journal of Wildlife Management 51:285–93.

Sargeant, A. B., R. J. Greenwood, M. A. Sovada, and T. L. Shaffer. 1993. Distribution and abundance of predators that affect duck production—Prairie Pothole Region (Resource Publication 194). U.S. Fish and Wildlife Service.

Schamel, D., and D. M. Tracy. 1986. Encounters between arctic foxes, *Alopex lagopus*, and red foxes, *Vulpes vulpes*. Canadian Field-Naturalist 100:562–63.

Schiller, E. L. 1954. Unusual walrus mortality on St. Lawrence Island, Alaska. Journal of Mammalogy 35:203–9.

Schmeltz, L. L., and J. O. Whitaker, Jr. 1977. Use of woodchuck burrows by woodchucks and other mammals. Transactions of the Kentucky Academy of Science 38:79–82.

Schofield, R. D. 1960. A thousand miles of fox trails in Michigan's ruffed grouse range. Journal of Wildlife Management 24:432–34.

Scholander, P. R., V. Walters, R. Hoek, and L. Irving. 1950. Body insulation of arctic and tropical mammals and birds in relation to body temperature, insulation and basal metabolic rate. Biological Bulletin 99:225–36.

Schneider, L. G., J. H. Cox, W. W. Muller, and K. P. Hohnsbeen. 1988. Current oral rabies vaccination in Europe: An interim balance. Review of Infectious Diseases 10:654–59.

Schroeder, C. 1985. A preliminary management plan for securing swift fox for reintroduction into Canada. University of Calgary, Calgary, Alberta, Canada.

Scott, T. G. 1943. Some food coactions of the northern plains red fox. Ecological Monographs 13:427–73.

Scott, T. G. 1947. Comparative analysis of red fox feeding trends on two central Iowa areas. Iowa Agricultural Experiment Station Research Bulletin 353:427–87.

Scott, T. G. 1955. Dietary patterns of red and gray foxes. Ecology 36:366–67.

Scott, T. G., and W. D. Klimstra. 1955. Red foxes and a declining prey population (Monograph Series No. 1). Southern Illinois University, Carbondale.

Scott-Brown, J. M., and J. Reynolds. 1984. Monitoring of released swift foxes in southern Alberta. Canadian Wildlife Service, Edmonton, Alberta, Canada.

Scott-Brown, J. M., S. Herrero, and C. Mamo. 1986. Monitoring of released swift foxes in Alberta and Saskatchewan—Final report 1986. Canadian Wildlife Service Report, Edmonton, Alberta.

Scott-Brown, J. M., S. Herrero, and J. Reynolds. 1987. Swift fox. Pages 432–41 *in* M. Novak, J. A. Baker, M. E. Obbard, and B. Malloch, eds. Wild furbearer management and conservation in North America. Ontario Trappers Association and Ontario Ministry of Natural Resources, Toronto.

Scrivner, J. H., T. P. O'Farrell, and T. T. Kato. 1987. Dispersal of San Joaquin kit foxes, *Vulpes macrotis mutica,* on Naval Petroleum Reserve #1, Kern County, California (Topical Report EGG 10282–2190). U.S. Department of Energy, Washington, DC.

Secord, D. C., J. A. Bradley, R. D. Eaton, and D. Mitchell. 1980. Prevalence of rabies virus in foxes trapped in the Canadian Arctic. Canadian Veterinary Journal 21:297–300.

Seese, F. M., M. C. Sterner, and D. E. Worley. 1993. *Echinococcus multilocularis* (Cestoda: Taeniidae) in Montana: Additional locality records in foxes and coyotes. Proceedings of the Montana Academy of Sciences 53:9–14.

Seton, E. T. 1923. The mane on the tail of the gray-fox. Journal of Mammalogy 4:180–82.

Seton, E. T. 1929. Lives of game animals. Doubleday, Garden City, NY.

Sheldon, J. W. 1992. Wild dogs. Academic Press, San Diego, CA.

Sheldon, W. G. 1949. Reproductive behavior of foxes in New York state. Journal of Mammalogy 30:236–46.

Sheldon, W. G. 1950. Denning habits and home range of red foxes in New York state. Journal of Wildlife Management 14:33–42.

Sheldon, W. G. 1953. Returns on banded red and gray foxes in New York state. Journal of Mammalogy 34:125.

Simmons, J. M., W. S. Nicholson, E. P. Hill, and D. B. Briggs. 1980. Occurrence of *Dirofilaria immitis* in gray fox (*Urocyon cinereoargenteus*) in Alabama and Georgia. Journal of Wildlife Diseases 16:225–28.

Sklepkovych, B. 1994. Arboreal foraging by red foxes, *Vulpes vulpes,* during winter food shortage. Canadian Field-Naturalist 108:479–81.

Smith, H. J. 1978. Parasites of red foxes in New Brunswick and Nova Scotia. Journal of Wildlife Diseases 14:366–77.

Smith, T. G. 1976. Predation of ringed seal pups (*Phoca hispida*) by the arctic fox (*Alopex lagopus*). Canadian Journal of Zoology 54:1610–16.

Snow, C. 1973. San Joaquin kit fox (Habitat Management Series for Endangered Species, Report No. 6). U.S. Bureau of Land Management, Denver Service Center, Denver, CO.

Sovada, M. A., C. C. Roy, J. B. Bright, and J. R. Gillis. 1998. Causes and rates of mortality of swift foxes in western Kansas. Journal of Wildlife Management 62:1300–1306.

Sovada, M. A., C. C. Roy, and D. J. Telesco. 2001. Seasonal food habits of swift fox (*Vulpes velox*) in cropland and rangeland landscapes in western Kansas. American Midland Naturalist 145:101–11.

Speller, S. W. 1972. Food ecology and feeding behavior of denning arctic foxes at Aberdeen Lake, Northwest Territories. Ph.D. Dissertation, University of Saskatchewan, Saskatoon, Saskatchewan, Canada.

Spencer, K. A., W. H. Berry, W. G. Standley, and T. P. O'Farrell. 1992. Reproduction of the San Joaquin kit fox on Camp Roberts Army National Guard Training Site, California (Topical Report EGG 10617–2154). U.S. Department of Energy, Washington, DC.

Spiegel, L. K., ed. 1996. Studies of San Joaquin kit fox in undeveloped and oil-developed areas. California Energy Commission, Sacramento.

Springer, P. F., G. V. Byrd, and D. W. Woolington. 1978. Reestablishing Aleutian Canada geese. Pages 311–38 *in* S. A. Temple, ed. Endangered birds: Management techniques for preserving threatened species. University of Wisconsin Press, Madison.

Stains, H. J. 1975. Distribution and taxonomy of the Canidae. Pages 3–26 *in* M. W. Fox, ed. The wild canids. Van Nostrand Reinhold, New York.

Standley, W. G., and P. M. McCue. 1997. Prevalence of antibodies against selected diseases in San Joaquin kit foxes at Camp Roberts, California. California Fish and Game 83:30–37.

Standley, W. G., W. H. Berry, T. P. O'Farrell, and T. T. Kato. 1992. Mortality of San Joaquin kit fox at Camp Roberts Army National Guard Training Site, California (Topical Report No. EGG 10627–2157). U.S. Department of Energy, Washington, DC.

Stanley, W. C. 1963. Habits of the red fox in northeastern Kansas. University of Kansas Museum of Natural History Miscellaneous Publications 34:1–31.

Steck, F., A. Wandeler, P. Bichsel, S. Capt, U. Hafliger, and L. Schneider. 1982. Oral immunization of foxes against rabies: Laboratory and field studies. Comparative Immunology, Microbiology, and Infectious Diseases 5:165–71.

Steelman, H. G., S. E. Henke, and G. M. Moore. 2000. Bait delivery for oral rabies vaccine to gray foxes. Journal of Wildlife Diseases 36:744–51.

Stephenson, R. O. 1970. A study of the summer food habits of the arctic fox on St. Lawrence Island, Alaska. M.S. Thesis, University of Alaska, Fairbanks.

Stickney, A. 1991. Seasonal patterns of prey availability and the foraging behavior of arctic foxes (*Alopex lagopus*) in a waterfowl nesting area. Canadian Journal of Zoology 69:2853–59.

Stone, W. B., E. Parks, B. L. Wever, and F. J. Parks. 1972. Experimental transfer of sarcoptic mange from red foxes and wild canids to captive wildlife and domestic animals. New York Fish and Game Journal 19:1–11.

Stone, W. B., E. Parks, P. Bichsel, S. Capt, U. Hafliger, and L. Schneider. 1982. Oral immunization of foxes against rabies: Laboratory and field studies. Comparative Immunology, Microbiology, and Infectious Disease 5:165–71.

Storm, G. L. 1965. Movements and activities of foxes as determined by radio-tracking. Journal of Wildlife Management 29:1–13.

Storm, G. L., and E. D. Ables. 1966. Notes on newborn and full-term wild red foxes. Journal of Mammalogy 47:116–18.

Storm, G. L., and G. B. Montgomery. 1975. Dispersal and social contact among red foxes: Results from telemetry and computer simulation. Pages 237–46 *in* M. W. Fox, ed. The wild canids. Van Nostrand Reinhold, New York.

Storm, G. L., R. D. Andrews, R. L. Phillips, R. A. Bishop, D. B. Siniff, and J. R. Tester. 1976. Morphology, reproduction, dispersal, and mortality of Midwestern red fox populations. Wildlife Monographs 49:1–82.

Strand, O., A. Landa, J. D. C. Linnell, B. Zimmermann, and T. Skogland. 2000. Social organization and parental behavior in the arctic fox. Journal of Mammalogy 81:223–33.

Stromberg, M. R., and M. S. Boyce. 1986. Systematics and conservation of the swift fox, *Vulpes velox*, in North America. Biological Conservation 35:97–110.

Stuht, J. N., and W. G. Youatt. 1972. Heartworms and lung flukes from red foxes in Michigan. Journal of Wildlife Management 36:166–70.

Sullivan, E. G. 1956. Gray fox reproduction, denning, range, and weights in Alabama. Journal of Mammalogy 37:346–51.

Sullivan, E. G., and A. O. Haugen. 1956. Age determination of foxes by x-ray of forefeet. Journal of Wildlife Management 20:210–12.

Sunquist, M. E. 1989. Comparison of spatial and temporal activity of red foxes and gray foxes in north-central Florida. Florida Field Naturalist 17:11–18.

Swanson, G. A., T. Surber, and T. S. Roberts. 1945. The mammals of Minnesota (Technical Bulletin 2). Minnesota Department of Conservation Division of Game and Fish.

Swick, C. D. 1973. Determination of San Joaquin kit fox range in Contra Costa, Alameda, San Joaquin, and Tulare Counties, 1973. (Special Wildlife Investigations Program Report W-54-R4). California Department of Fish and Game, Sacramento.

Tannerfeldt, M., and A. Angerbjörn 1998. Fluctuating resources and the evolution of litter size in the arctic fox. Oikos 83:545–59.

Tate, G. H. H. 1931. Random observations on habits of South American mammals. Journal of Mammalogy 12:248–56.

Taylor, W. P. 1943. The gray fox in captivity. Texas Game and Fish 1:12–13, 19.

Terres, J. K. 1939. Tree-climbing technique of a gray fox. Journal of Mammalogy 20:256.

Theberge, J. B., and C. H. R. Wedeles. 1989. Prey selection and habitat partitioning in sympatric coyote and red fox populations, southwest Yukon. Canadian Journal of Zoology 67:1285–90.

Thornton, W. A., and G. C. Creel. 1975. The taxonomic status of kit foxes. Texas Journal of Science 26:127–36.

Thornton, W. A., G. C. Creel, and R. E. Trimble. 1971. Hybridization in the fox genus *Vulpes* in west Texas. Southwestern Naturalist 15:473–84.

Tietjen, H., F. Deines, and W. Stephensen. 1988. Sodium monofluoroacetate (1080): A study of residues in arctic fox muscle tissue. Bulletin of Environmental Contamination and Toxicology 40:707–10.

Trainer, D. O., and J. B. Hale. 1969. Sarcoptic mange in red foxes and coyotes of Wisconsin. Bulletin of the Wildlife Disease Association 5:387–91.

Trapp, G. R. 1973. Comparative behavioral ecology of two southwest Utah carnivores: *Bassariscus astutus* and *Urocyon cinereoargenteus*. Ph.D. Dissertation, University of Wisconsin, Madison.

Trapp, G. R. 1978. Comparative behavioral ecology of the ringtail (*Bassariscus astutus*) and gray fox (*Urocyon cinereoargenteus*) in southwestern Utah. Carnivore 1:3–32.

Trapp, G. R., and D. L. Hallberg. 1975. Ecology of the gray fox (*Urocyon cinereoargenteus*): A review. Pages 164–78 *in* M. W. Fox, ed. The wild canids. Van Nostrand Reinhold, New York.

Tullar, B. F., Jr. 1983. An unusually long-lived red fox. New York Fish and Game Journal 30:227.

Tullar, B. F., Jr., and L. T. Berchielli, Jr. 1980. Movement of the red fox in central New York. New York Fish and Game Journal 27:179–204.

Tullar, B. F., Jr., and L. T. Berchielli, Jr. 1982. Comparison of red foxes and gray foxes in central New York with respect to certain features of behavior, movement and mortality. New York Fish and Game Journal 29:127–33.

Tullar, B. F., Jr., L. T. Berchielli, Jr., and E. P. Saggese. 1976. Some implications of communal denning and pup adoption among red foxes in New York. New York Fish and Game Journal 23:92–96.

Turkowski, F. J. 1969. Food habits and behavior of the gray fox (*Urocyon cinereoargenteus*) in the lower and upper Sonoran life zones of the southwestern United States. Ph.D. Dissertation, Arizona State University, Tempe.

Underwood, L. S. 1971. The bioenergetics of the arctic fox (*Alopex lagopus* L.). Ph.D. Dissertation, Pennsylvania State University, University Park.

Underwood, L. S. 1975. Notes on the arctic fox (*Alopex lagopus*) in the Prudhoe Bay area of Alaska. Pages 145–49 *in* J. Brown, ed. Ecological investigations of the tundra biome in the Prudhoe Bay region, Alaska (Biological Papers Special Report No. 2). University of Alaska, Fairbanks.

Underwood, L. S. 1981. Seasonal energy requirements of the Arctic fox (*Alopex lagopus*). Pages 368–85 *in* J. A. Chapman and D. Pursley, eds. Proceedings of the worldwide furbearer conference. Frostburg, MD.

Underwood, L. S., and J. A. Mosher. 1982. Arctic fox. Pages 491–503 *in* J. A. Chapman and G. A. Feldhamer, eds. Wild mammals of North America. Johns Hopkins University Press, Baltimore.

Uresk, D. W., and J. C. Sharps. 1986. Denning habitat and diet of the swift fox in western South Dakota. Great Basin Naturalist 46:249–53.

Urquhardt, D. R. 1973. Oil exploration and Banks Island wildlife: A guideline for the preservation of caribou, muskox, and arctic fox populations on Banks Island. Northwest Territories Government, Game Management Division, Yellowknife, Northwest Territories, Canada.

U.S. Department of Defense. 1998. Final environmental assessment: Predator damage management to protect the federally endangered San Clemente Loggerhead shrike on San Clemente Island, California. Department of the Navy, San Diego, CA.

U.S. Fish and Wildlife Service. 1997. 1996 national survey of hunting, fishing, and wildlife-associated recreation. U.S. Government Printing Office, Washington, DC.

U.S. Fish and Wildlife Service. 1998. Recovery plan for upland species of the San Joaquin Valley, California. U.S. Fish and Wildlife Service, Region 1, Portland, OR.

U.S. Fish and Wildlife Service. 2001. Endangered and threatened wildlife and plants; annual notice of findings on recycled petitions. Federal Register 66:1295–1300.

Van Gelder, R. G. 1978. A review of canid classification. American Museum Novitates 2646:1–10.

Vest, E. D., D. Lundgren, D. D. Parker, D. E. Johnson, E. L. Morse, J. B. Bushman, R. W. Sidwell, and B. D. Thorpe. 1965. Results of a five-year survey for certain enzootic diseases in the fauna of western Utah. American Journal of Tropical Medicine and Hygiene 14:124–35.

Voigt, D. R. 1987. Red fox. Pages 379–392 *in* M. Novak, J. A. Baker, M. E. Obbard, and B. Malloch, eds. Wild furbearer management and conservation in North America. Ontario Trappers Association and Ontario Ministry of Natural Resources, Toronto.

Voigt, D. R., and B. D. Earle. 1983. Avoidance of coyotes by red fox families. Journal of Wildlife Management 47:852–57.

Voigt, D. R., and D. W. Macdonald. 1984. Variation in the spatial and social behaviour of the red fox, *Vulpes vulpes*. Acta Zoologica Fennica 171:261–65.

Voigt, D. R., and R. L. Tinline. 1980. Strategies for analyzing radio tracking data. Pages 387–404 *in* C. J. Amlaner Jr., and D. W. Macdonald, eds. A handbook on biotelemetry and radio tracking. Pergamon Press, Oxford.

Voigt, D. R., and R. L. Tinline. 1982. Fox rabies and trapping: A study of disease and fur harvest interaction. Pages 139–56 *in* G. C. Sanderson, ed. Midwest furbearer management. Proceedings of a symposium at the 43rd Midwest fish and wildlife conference, Wichita, KS.

von Bloeker, J. C., Jr. 1967. The land mammals of the Southern California Islands. Pages 245–63 *in* R. N. Philbrick, ed. Proceedings of the symposium on the biology of the California Islands. Santa Barbara Botanic Garden, Santa Barbara, CA.

von Schantz, T. 1981a. Evolution of group living, and the importance of food and social organization in population regulation: A study on the red fox (*Vulpes vulpes*). Ph.D. Dissertation, Lund University, Lund, Sweden.

von Schantz, T. 1981b. Female cooperation, male competition, and dispersal in the red fox *Vulpes vulpes*. Oikos 37:63–68.

Vos, A. C. 1994. Reproductive performance of the red fox, *Vulpes vulpes,* in Garmisch-Partenkirchen, Germany, 1987–1992. Zeitschriftfür Säugetierkunde 59:326–31.

Wakely, L. G., and F. F. Mallory. 1988. Hierarchical development, agonistic behaviours, and growth rates in captive arctic fox. Canadian Journal of Zoology 66:1672–78.

Warrick, G. D., and B. L. Cypher. 1998. Factors affecting the spatial distribution of a kit fox population. Journal of Wildlife Management 62:707–17.

Warrick, G. D., and B. L. Cypher. 1999. Variation in body mass of San Joaquin kit foxes. Journal of Mammalogy 80:972–79.

Warrick, G. D., J. H. Scrivner, and T. P. O'Farrell. 1999. Demographic responses of kit foxes to supplemental feeding. Southwestern Naturalist 44:367–74.

Wasmer, D. A. 1984. Movements and activity patterns of a gray fox in south-central Florida. Florida Science 47:76–77.

Waters, J. H. 1964. Red and gray fox from New England archeological sites. Journal of Mammalogy 45:307–8.

Wayne, R. K., and S. J. O'Brien. 1987. Allozyme divergence within the canidae. Systematic Zoology 36:339–55.

Wayne, R. K., R. E. Benveniste, D. N. Janczewski, and S. J. O'Brien. 1989. Molecular and biochemical evolution of the carnivores. Pages 465–94 *in* J. L. Gittleman, ed. Carnivore behaviour, ecology and evolution. Cornell University Press, Ithaca, NY.

Wayne, R. K., S. B. George, D. Gilbert, and P. W. Collins. 1991a. The channel island fox (*Urocyon littoralis*) as a model of genetic change in small populations. Pages 639–49 *in* E. C. Dudley, ed. The unity of evolutionary biology: Proceedings of the fourth international congress of systematic and evolutionary biology, Vol. II. Dioscorides Press, Portland, Oregon.

Wayne, R. K., S. B. George, D. Gilbert, P. W. Collins, S. D. Kovach, D. Girman, and N. Lehman. 1991b. A morphologic and genetic study of the island fox, *Urocyon littoralis*. Evolution 45:1849–68.

Wenner, A. M., and D. L. Johnson. 1980. Land vertebrates on the California Channel Islands: sweepstakes or bridges? Pages 497–530 *in* D. M. Power, ed. The California Islands: Proceedings of a multidisciplinary symposium. Santa Barbara Museum of Natural History, Santa Barbara, CA.

West, E. W., and R. L. Rudd. 1983. Biological control of Aleutian Island arctic fox: A preliminary strategy. International Journal of Animal Problems 4:305–11.

White, P. J., and K. Ralls. 1993. Reproduction and spacing patterns of kit foxes relative to changing prey availability. Journal of Wildlife Management 57:861–67.

White, P. J., and R. A. Garrott. 1997. Factors regulating kit fox populations. Canadian Journal of Zoology 75:1982–88.

White, P. J., T. J. Kreeger, J. R. Tester, and U.S. Seal. 1989. Anal-sac secretions deposited with feces by captive red foxes (*Vulpes vulpes*). Journal of Mammalogy 70:814–16.

White, P. J., K. Ralls, and R. A. Garrott. 1994. Coyote-kit fox interactions as revealed by telemetry. Canadian Journal of Zoology 72:1831–36.

White, P. J., K. Ralls, and C. A. Vanderbilt White. 1995. Overlap in habitat and food use between coyotes and San Joaquin kit foxes. Southwestern Naturalist 40:342–49.

White, P. J., C. A. Vanderbilt White, and K. Ralls. 1996. Functional and numerical responses of kit foxes to a short-term decline in mammalian prey. Journal of Mammalogy 77:370–76.

White, P. J., W. H. Berry, J. J. Eliason, and M. T. Hanson. 2000. Catastrophic decrease in an isolated population of kit foxes. Southwestern Naturalist 45:204–11.

Wigal, R. A., and J. A. Chapman. 1983. Age determination, reproduction, and mortality of the gray fox (*Urocyon cinereoargenteus*) in Maryland, U.S.A. Zeitschriftfür Säugetierkunde 48:226–45.

Winstanley, R. K., W. A Buttemer, and G. Saunders. 1999. Fat deposition and seasonal variation in body composition of red foxes (*Vulpes vulpes*) in Australia. Canadian Journal of Zoology 77:406–12.

Wood, J. E. 1952. The effects of agriculture (ranching and farming) on the habitat and food supply of furbearers in the post oak region of Texas. Transactions of the North American Wildlife Conference 7:427–27.

Wood, J. E. 1958. Age structure and productivity of a gray fox population. Journal of Mammalogy 39:74–86.

Wood, J. E., D. E. Davis, and E. V. Komarek. 1958. The distribution of fox populations in relation to vegetation in southern Georgia. Ecology 39:160–62.

Wooding, J. B. 1984. Coyote food habits and the spatial relationship of coyotes and foxes in Mississippi and Alabama. M.S. Thesis, Mississippi State University, State College.

Wrigley, R. E., and D. R. M. Hatch. 1976. Arctic fox migrations in Manitoba. Arctic 29:147–58.

Wurster, D. H., and K. Benirschke. 1968. Comparative cytogenetic studies in the order Carnivora. Chromosoma 24:336–82.

Yeager, L. E. 1938. Tree-climbing by a gray fox. Journal of Mammalogy 19:376.

Yearsley, E. F., and D. E. Samuel. 1980. Use of reclaimed surface mines by foxes in West Virginia. Journal of Wildlife Management 44:729–34.

Zabel, C. J., and S. J. Taggart. 1989. Shift in red fox, *Vulpes vulpes,* mating system associated with El Niño in the Bering Sea. Animal Behaviour 38:830–38.

Zoellick, B. W., and N. S. Smith. 1992. Size and spatial organization of home ranges of kit foxes in Arizona. Journal of Mammalogy 73:83–88.

Zoellick, B. W., T. P. O'Farrell, P. M. McCue, C. E. Harris, and T. T. Kato. 1987. Reproduction of the San Joaquin kit fox on Naval Petroleum Reserve #1, Elk Hills, California, 1980–1985 (Topical Report No. EGG 10282–2144). U.S. Department of Energy, Washington, DC.

Zoellick, B. W., N. S. Smith, and R. S. Henry. 1989. Habitat use and movements of desert kit foxes in western Arizona. Journal of Wildlife Management 53:955–61.

Zoellick, B. W., C. E. Harris, B. T. Kelly, T. P. O'Farrell, T. T. Kato, and M. E. Koopman. 2002. Movements and home ranges of San Joaquin kit foxes relative to oil-field development. Western North American Naturalist 62:151–159.

Zumbaugh, D. M., J. R. Choate, and L. B. Fox. 1985. Winter food habits of the swift fox on the central high plains. Prairie Naturalist 17:41–47.

Brian L. Cypher, Endangered Species Recovery Program, P. O. Box 9622, Bakersfield, California 93389. Email: bcypher@esrp.org.

25

Black Bear

Ursus americanus

Michael R. Pelton

NOMENCLATURE

COMMON NAME. Black bear
SCIENTIFIC NAME. *Ursus americanus*
SUBSPECIES. *U. a. altifrontalis, U. a. amblyceps, U. a. americanus, U. a. californiensis, U. a. carlottae, U. a. cinnamomum, U. a. emmonsii, U. a. eremicus, U. a. floridanus, U. a. hamiltoni, U. a. kermodei, U. a. luteolus, U. a. machetes, U. a. perniger, U. a. pugnax,* and *U. a. vancouveri* (Hall 1981; Lariviere 2001).

DISTRIBUTION

Black bears are the most common and widely distributed of the three ursids in North America. The primitive range of *U. americanus* covered the forested areas of North America, including Mexico. Black bears are now found primarily in less settled, forested regions in 39 states, 11 Canadian provinces and territories, and possibly 12 Mexican states (Fig. 25.1). The current range represents approximately 62% of the historical range (Pelton and Vanmanen 1994). Their status and density vary considerably within the existing range. In some western states and many Canadian provinces, the species is almost relegated to pest status, with thousands being harvested annually. In contrast, states such as Alabama, Louisiana, Mississippi, Kentucky, Ohio, Missouri, Oklahoma, and Texas report only remnant numbers surviving in enclaves of relatively inaccessible habitat. The species has been extirpated from some midwestern states, for example, Illinois and Indiana. More than 40,000 black bears are harvested each year in North America (Pelton et al. 1999).

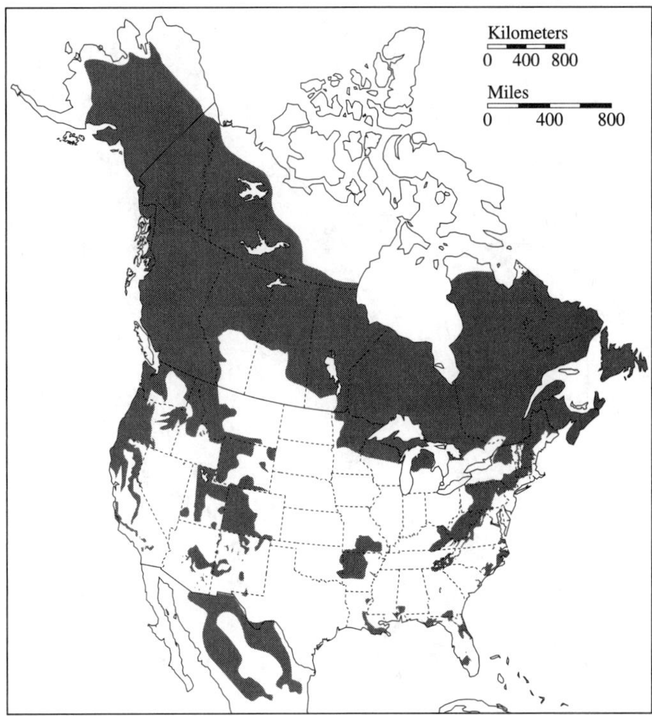

FIGURE 25.1. Distribution of the black bear (*Ursus americanus*).

DESCRIPTION

Black bears are plantigrade and pentadactyl, and have short, curved, nonretractable claws. Average weights range from 40 to 70 kg for adult females and from 60 to 140 kg for adult males. An occasional adult male will exceed 250–300 kg. Body length ranges from 1 to 2 m. Black bears are generally full-grown at 4 years of age. The eyes are small; the ears are small, rounded, and erect; the tail is short. Fur is uniform in color except for a brown muzzle; an occasional white blaze occurs on the chest. The black color phase is most prevalent in the East and the brown phase most prevalent in the West (Rounds 1987). The color phase of an individual can change between molts (Rogers 1980). Unique white and bluish phases occur on the Pacific coast in the Northwest. The skull has a large cranium with prominent sagittal crest and zygomatic arches. The dental formula is I 3/3, C 1/1, P 4/4, M 2/3. The first three upper and lower premolars are usually rudimentary (Fig. 25.2). Dentition is bunodont and reflects the black bear's omnivorous food habits. See Lariviere (2001) for details.

REPRODUCTION

Female black bears become sexually mature between 2 and 7 years of age, normally at 3–4 years. Rausch (1961) suggested that age of reproductive maturity may increase with latitude. However, nutrition has a dramatic effect on age of reproductive maturity and subsequent fecundity of females (Rogers 1976; Hamilton 1978). Years of poor berry or acorn production result in delayed first estrus, decreased litter sizes, and increased incidence of barren females (Eiler et al. 1989; Elowe and Dodge 1989; Noyce and Garshelis 1994).

Breeding occurs in summer and peaks in the latter part of June and July. However, females in estrus have been observed as early as late May and as late as mid-August (Jonkel and Cowan 1971). Black bears have a seasonally constant estrus; females remain in estrus until bred or until the ovarian follicles begin to degenerate. Within the breeding season, multiple matings may occur for both sexes (Schenk and Kovacs 1995). Black bears are induced ovulators (Wimsatt 1963; Erickson and Nellor 1964; Boone et al. 1998); ovulation occurs only as a result of coital stimulation. Poor nutrition may result in no implantation of the blastocysts, resorption of the implanted fetuses, or early death of the neonates (Hellgren et al. 1990).

The gestation period for black bears is 7–8 months. However, most fetal development occurs only during the last 6–8 weeks. Delayed implantation occurs in the black bear; blastocysts float free in the uterus and do not implant until late November or early December. Only minimal cell differentiation occurs before implantation (Wimsatt 1963).

Cubs are born in winter dens at the end of January or the beginning of February. Young are altricial—helpless, hairless, and eyes closed at

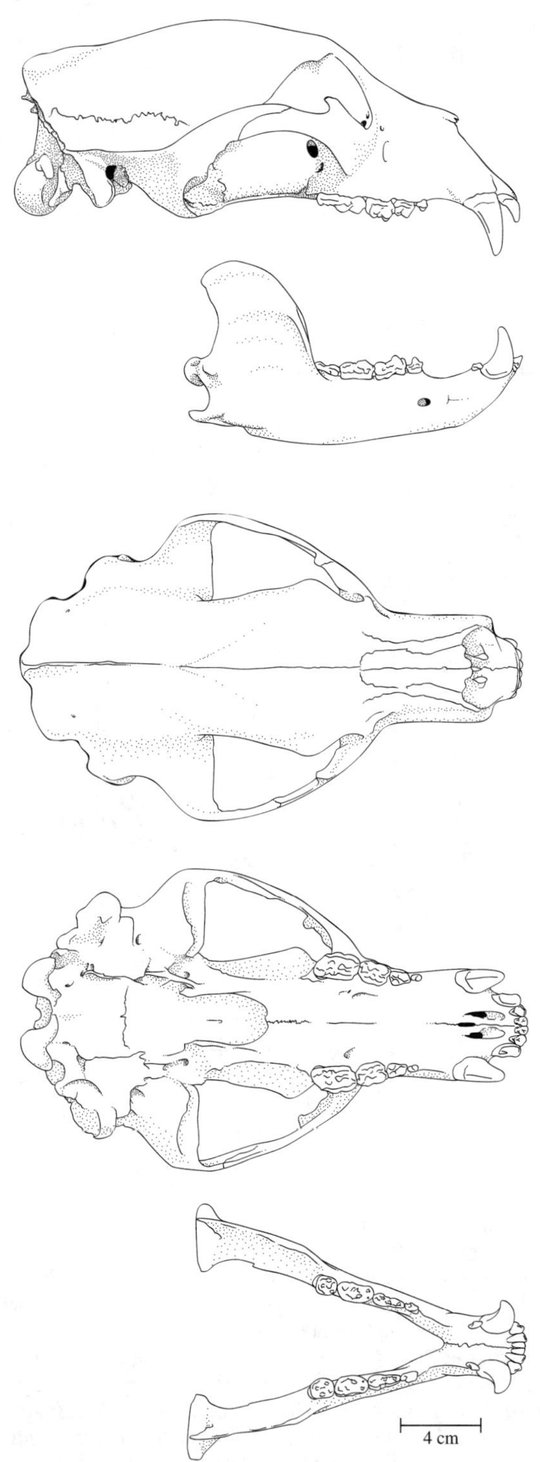

FIGURE 25.2. Skull of the black bear (*Ursus americanus*). From top to bottom: lateral view of cranium, lateral view of mandible, dorsal view of cranium, ventral view of cranium, dorsal view of mandible.

birth—and about 200 mm in length. Normal litter size is two, but three and four cubs are not uncommon and litters of five are occasionally reported (Jonkel and Cowan 1971; Alt 1981). Parturient females emerge from dens from late March to early May. Cubs stay with their mother through summer and fall and den with them the second winter. Lactation suppresses estrus, but females occasionally mate while raising cubs (Lecount 1983). Young disperse as yearlings (~16 months) in spring or summer, before the female's period of estrus. Thus, the adult female normally breeds every other year. Females will occasionally skip a year between reproductive cycles (Jonkel and Cowan 1971). Years with poor food can cause reproductive failure, thus initiating population-wide breeding synchrony (McLaughlin et al. 1994). Because of the relatively low biological potential of the species, changes in the various factors influencing reproduction—age of reproductive maturity, number of cubs produced—can significantly alter population stability. Thus, low biological potential becomes an important consideration for management.

ECOLOGY

Population Density. Because of their generally sparse numbers, characteristic shy and secretive nature, and inaccessible habitat, black bears are often difficult to census. Historically, researchers have used a variety of methods relying on intensive capture, tag, and recapture efforts. Several early studies used multiple recapture methods to estimate black bear density. Kemp (1972) used the Lincoln and Schnabel methods to estimate population density in Canada; the size of the study area was small and this enabled him to engage in a very intensive trapping program. Hornocker (1962) and Troyer and Hensel (1964) correlated estimates obtained by the Schnabel method with direct counts of grizzly bears (*U. arctos*) in the Northwest. However, too often the results based on capture–recapture data are not reliable because samples are too small and the data are biased due to nonrandom sampling, loss of marks, and unequal vulnerability of the individuals in the population. Pelton and Marcum (1977) reported on the use of radioisotope feces tags as a technique for estimating population density of black bears using the Schnabel method. They felt that several of the biases inherent in the technique were removed or lessened. Piekielek and Burton (1975) used the Lincoln index as well as the technique derived by Edwards and Eberhardt (1967) to estimate population densities of black bears in California. Most of the above efforts were hampered by a failure of one or more of the basic assumptions necessary for reliable estimates and an inability to delineate the area inhabited by the estimated population (Miller et al. 1997). More recent reported densities range from <1/10 km^2 (Clark 1991; Miller et al. 1997) to nearly 1/km^2 (Hellgren and Vaughan 1989). Both habitat quality and the variety of procedures used to derive the estimates likely explain this wide variation. Recent advances in more powerful statistical programs, better telemetry techniques, use of remote cameras, and DNA technology should help resolve some of the sampling bias and provide more reliable population estimates (Eberhardt 1990; Miller 1990; Garshelis 1992). Other, less costly, less time-consuming, and less intrusive—but also less reliable—census methods are population indices such as the incidence of scats, tracks, mark trees, and use of bait stations (Spencer 1955; Pelton 1972). However, only bait station indices have been shown to be a reliable indicator of population trends and relative population size.

Removal techniques are not applicable in areas where populations are not hunted or sample sizes are too small. Direct counts of bears also have been used to determine density. Barnes and Bray (1967) made direct counts of black bears along roadsides and backcountry areas in Yellowstone National Park. Hornocker (1962) believed that direct counts of bears were reliable where there were adequate open areas for observations to be made. However, in many parts of North America where the vegetation is dense and often continuous, the technique is not feasible. The roadside census technique often is biased because of human traffic and activities (Hayne 1949). Mark–reobserve estimates in park areas are often biased because of the influence of the differential observability of "panhandler" versus wild bears (Marcum 1974).

Harvest data have been used in a variety of ways to estimate population size. Poelker and Hartwell (1973) in Washington, Carpenter (1973) in Virginia, and Spencer (1955) in Maine, in computing statewide estimates of black bears, used methods based on mean annual harvest figures. These authors assumed that the total harvest is a known percentage of the population. Carpenter (1973) assumed that the annual kill in Virginia was 20% of the population. Spencer (1955) arrived at an annual harvest percentage based on the total harvest and the percentage of young animals in the harvest. Erickson and Petrides (1964) used the ratio of marked to unmarked individuals in the harvest of Michigan black bears to determine estimates using the Lincoln index. Pennsylvania uses this method statewide on an annual basis to determine the size of its black bear population (Alt and Carr 1984). Many state wildlife agencies continue to rely on analysis and interpretation of harvest data as the preferred method of population assessment and status.

Recent advances in population modeling enable biologists to predict harvest levels and population trends by using data on natality, mortality, and sex and age ratios. However, sex and age composition of the harvest and hunting success have proven to be unreliable indicators of bear population status (Bunnell and Tait 1980; Miller 1990; Garshelis 1991). Hunting methods, season timing, numbers of hunters, and their individual selectivity often change (Kolenosky 1986; Litvaitis and Kane 1994). Also, black bear populations that exhibit similar sex and age compositions may have different growth rates (Caughley 1974; Harris and Metzgar 1987). Finally, vulnerability to harvest can be affected by annual changes in weather and food availability (Lindzey et al. 1983; Noyce and Garshelis 1997). Although black bears enjoy a higher reproductive rate than the other two species of bears in North America, they are nonetheless a *K*-selected species. As such, the margin of error for making miscalculations in management (such as too much harvest pressure) is narrow. Consequently, the need to arrive at more accurate estimates of population density is important. Because of the number of variables affecting any attempt at determining population density or size, resource agencies should use as many estimators and indices as possible.

Sex Ratios. The neonatal sex ratio of black bears is essentially 50:50 (Kolenosky 1990). However, sex ratios determined from older animals are biased strongly toward males. Most of this bias results from the sampling procedures used. Males tend to be more aggressive and bold than females, thus increasing their chances of contact with people. Males also exhibit greater mobility and home range size, and thus greater vulnerability, than females. As a result, sex ratios based on hunter harvest are predominantly male. Nuisance bears trapped at campgrounds and garbage dumps exhibit an even greater predominance of males than females. Consequently, males have a higher mortality rate and the actual sex ratio of the adult population is skewed toward females. Because one adult male is capable of breeding with a number of females, a greater mortality of males normally would not detrimentally affect a population.

Age Ratios. Ages of black bears are most accurately estimated using the tooth section, cementum annuli technique (Willey 1974); with this method, animals are placed in 1-year age classes. The average age of males in a healthy black bear population ranges from 3 to 5 years, whereas females average 5–8 years old (Fig. 25.3). The more restricted movements, smaller home range, and less aggressive behavior of females contribute to their longer life span. Few black bears in a population ever reach 10 years of age. However, an occasional animal will survive to 20–25 years. Changes in age ratios in a population may reflect the relative health of that population if other parameters such as population density are also known. As with sex ratios, constructing age

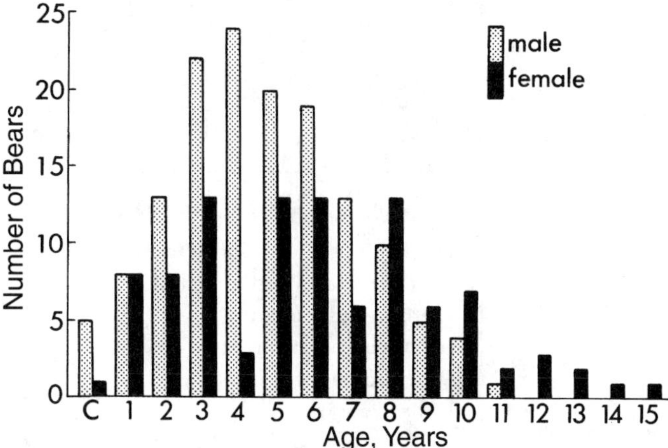

FIGURE 25.3. Typical age structure of captured male and female black bears. SOURCE: Johnson and Pelton 1980b.

TABLE 25.1. Home range sizes of black bears derived from annual 100% convex polygons

Location	Home Range Size (km^2)		Source
	Females	Males	
Arkansas	34.7	89.7	Clark 1991
California	17.1	22.4	Novick and Stewart 1982
Idaho	48.9	112.1	Amstrup and Beecham 1976
Massachusetts	28.0	318.0	Elowe 1984
Tennessee	6.9	51.2	Vanmanen 1994
Virginia	27.0	111.7	Hellgren and Vaughan 1990

structures from harvested bears can be biased and may contribute to a misinterpretation of the actual age structure. Factors such as selectivity by hunters, laws prohibiting shooting of smaller bears, differential dispersal of young animals, and timing of hunting seasons all affect the availability to hunters of various sex and age classes of bears.

Home Range and Movements. The size and shape of a black bear's home range is determined by the capability of an area to provide the animal's annual needs (Hamilton 1978; Garshelis et al. 1983). Home ranges also may vary considerably depending on such factors as sex, age, season, and population density. Because these factors change, so does the home range size for an individual. For example, black bears can respond dramatically to changing food resources; individuals have moved more than 160 km to take advantage of isolated pockets of available food (Rogers 1977).

Concentrations of soft mast (Piekielek and Burton 1975), hard mast (Sauer et al. 1969), or artificial food sources (Rogers 1977) provide the stimulus for extensive movements and consequent range expansion, at least temporarily. These movements occur particularly during late summer and fall when foraging activities increase. There are conflicting reports of average home range sizes of various sex and age groups of bears (Table 25.1); some of these differences can be attributed to the variables noted above. However, methods of data collection and analysis also contribute to the differences in home range sizes. Consistent among past studies, though, is that the home range size of adult male bears is typically three to eight times larger than that of adult females. Contributing to these movements is the social pressure exerted by larger, older adult males on younger animals, causing them to disperse from the areas occupied by older animals.

The great mobility of black bears, particularly males, also results in a significantly increased mortality rate. Their movements often take them beyond the confines of familiar areas and, in many instances, into closer contact with people and their activities, thus making them more vulnerable.

Habitat Requirements. Black bears are very adaptable. Populations are maintained surprisingly well in the presence of humans if the bears are not overharvested. However, in most instances, if habitat areas that provide refuge are not available, local populations succumb to intolerances of humans. Throughout its geographic range, prime black bear habitat is characterized by relatively inaccessible terrain, thick understory vegetation, and abundant sources of food in the form of shrub or tree-borne soft or hard mast. As the pressures of an expanding human population decrease existing habitat, quality food and cover become more critical. If bears are forced to leave relatively protected enclaves to forage on less protected sites, increased rates of mortality result in a declining population.

In the Southwest, prime black bear habitat is restricted to vegetated, mountainous areas ranging from 900 to 3000 m in elevation. Habitats consist mostly of chaparral and pinyon–juniper (*Pinus–Juniperus*) woodland sites (Waddell 1979). Bears occasionally move out of the chaparral into more open sites and feed on prickly pear cactus (*Opuntia* spp.).

There are two distinct prime habitat types in the Southeast: mountains and coastal plain. In the Appalachian, Ozark, and Ouachita Mountains, black bears survive in a predominantly oak–hickory (*Quercus–Carya*) and mixed mesophytic forest. Understory consists of such food plants as blueberry (*Vaccinium* sp.), huckleberry (*Gaylussacia* sp.), raspberry, and blackberry (*Rubus* sp.). Laurel (*Kalmia latifolia*) and rhododendron (*Rhododendron* sp.) provide additional thick cover. In the coastal plain, bears inhabit a mixture of flatwoods, bays, and swampy hardwood sites; black gum (*Nyssa sylvatica*) and cypress (*Taxodium distichum*) are common overstory species on wetter sites. On drier sites, pine and oak predominate. The understory consists primarily of dense thickets of evergreen, woody species, and greenbriar (*Smilax* sp.). Holly (*Ilex* sp.), greenbriar, gallberry (*Ilex coriacea*), arrow-arum (*Peltandra virginica*), and huckleberry are representative of the understory food species of this habitat. Black gum seed and some acorns provide additional food (Hamilton 1978).

In the Northeast, prime habitat consists of a forest canopy of hardwoods such as beech (*Fagus* spp.), maple (*Acer* spp.), and birch (*Betula* spp.) and coniferous species including red spruce (*Picea rubens*) and balsam fir (*Abies balsamea*). Swampy habitat areas are mainly white cedar (*Thuja occidentalis*). Raspberries and blueberries are common understory food plants (Hugie 1974). Abandoned apple orchards also play an important role as a source of food. Corn crops and oak–hickory mast are also common sources of food in some sections of the Northeast; small, thick swampy areas provide excellent refuge cover.

Along the Pacific coast, redwood (*Sequoia sempervirens*), sitka spruce (*Picea sitchensis*), and hemlocks (*Tsuga* spp.) predominate as overstory cover. Ponderosa pine (*Pinus ponderosa*), lodgepole pine (*P. contorta*), and Douglas-fir (*Pseudotsuga menziesii*) are common on the drier sites. Within these forest types are early successional areas important for black bears, such as brushfields, wet and dry meadows, high tidelands, riparian areas, and a variety of mast-producing hardwood species (Lawrence 1979).

Spruce–fir forest dominates much of the range of the black bear in the Rockies. Important nonforested areas are wet meadows, riparian areas, avalanche chutes, roadsides, burns, sidehill parks, and subalpine ridgetops (Kemp 1979).

Winter Denning. Winter dormancy is an interesting and important aspect of the biology of bears. Black bears, the least predacious of the North American carnivores, circumvent food shortages and severe weather conditions by becoming dormant. They apparently exhibit a dormant period even in the southern extremities of their range in Florida and Arizona. Bears in the northern part of their range take advantage of heavy snowfall for concealment and insulation; bears in the Southeast may select cavities high above ground in large trees (Johnson and Pelton 1981) or simply construct a crude nest on the ground in dense thickets of briars and/or low shrubs (Hamilton and Marchinton 1977). Secure dens are most important because birth and early maternal care of cubs is limited to winter dens. In more southern latitudes, black bears appear to be less lethargic and can be easily aroused from their dens. However, this may be related more to den selection and climate relationships than to differences in physiology and denning behavior. After increased movements and intense foraging in the fall, black bears usually return to their spring and summer home ranges to den (Garshelis and Pelton 1980). Denning is preceded and followed by a period of decreased movements and activities (Johnson and Pelton 1979). This transition into and out of denning, which may last as long as 1 month, may be a physiological and behavioral adaptation through which bears adjust their digestive system to a lengthy period of quiescence. Bears may enter dens between October and early January depending on latitude, available food, sex and age, and local weather conditions. Adult females generally den first, followed by subadults and adult males (Johnson and Pelton 1980a). Unless disturbed because of weather, people, or other animals, they tend to stay in one den. Foot pads may be shed during denning, and new pads formed during the postdenning recovery period. Also, on emergence, black bears generally defecate a fecal plug that has blocked the gastrointestinal tract since they entered the den in the fall.

Unique physiological adaptations of black bears to winter dormancy have resulted in a variety of classifications of this behavior. Body temperature is reduced only 7–8°C, metabolism is reduced 50–60%, heart rate decreases from 40–50 bpm to 8–19 bpm, and a weight loss of 20–27% (adipose tissue only) occurs; the animals do not eat, drink, defecate, or urinate during the entire denning period (Hock 1960; Folk et al. 1972; Nelson et al. 1973). A typical mammalian hibernator reduces its body temperature, heart rate, and metabolism to within 1°C of the ambient temperature. A regular awakening every few days to eat, drink, defecate, and urinate is also typical, but when in hibernation the animal can be handled and even removed from the den without awakening (Hock 1960; Folk et al. 1972). On the contrary, black bears can be easily aroused and will react to a disturbance (Jonkel and Cowan 1971; Folk et al. 1972).

Deviation from the norm established for "true" hibernators has resulted in such phrases as "dormancy" (Matson 1946), "ecological hibernators" (Morrison 1960), and "carnivorean lethargy" (Hock 1960) to describe the denning behavior of bears. Further physiological evidence, particularly concerning the electrocardiogram of bears (Folk et al. 1972) and their metabolic and excretory mechanisms (Lundberg et al. 1976), indicates that the "hibernator" designation applies to bears. The winter weight loss of bears (20–27%) versus that of smaller hibernators (25–30%) (Hock 1960; Kayser 1961) indicates that the adaptations of bears are equivalent or even superior to those of other hibernators because a relatively high body temperature enables black bears to remain somewhat alert and care for their young in the winter den. This ability to react to disturbances is important to an animal as large as a black bear because complete concealment in a den is not usually possible.

FEEDING HABITS

Throughout their range in North America, black bears consume primarily grasses and forbs in spring, soft mast in the form of shrub and tree-borne fruits in summer, and a mixture of hard and soft mast in fall. However, availability of different food types varies regionally. Only a small portion of their diet consists of animal matter, and then primarily in the form of colonial insects and beetles (Fig. 25.4).

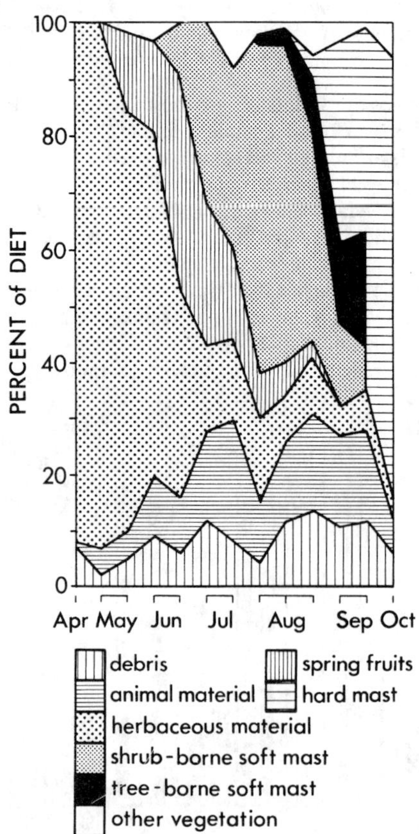

FIGURE 25.4. Typical food habits for black bears, illustrating the high percentage of plant material in their diet. SOURCE: Eagle and Pelton 1983.

Most vertebrates are consumed in the form of carrion. Black bears are not active predators and feed on vertebrates only if the opportunity arises. The diet of black bears is high in carbohydrates and low in proteins and fats. Consequently, they generally prefer foods with high protein or fat content, thus their propensity for the food and garbage of people. Bears feeding on a protein-rich food source show significant weight gains and enhanced fecundity (McLean and Pelton 1990).

Spring, after emergence from winter dens, is a period of relative food scarcity. Bears tend to lose weight during this period and continue to subsist partly off of body fat stored during the preceding fall. They take advantage of any succulent and protein-rich foods available; however, these are often not in sufficient quantity to maintain body weight.

As summer approaches, a variety of berry crops becomes available. Summer is generally a period of abundant and diverse foods for black bears, enabling them to recover from the energy deficits of winter and spring. Late summer and fall are critical periods for black bears to increase fat stores. Extensive foraging may occur and animals travel great distances to take advantage of available foods. Weight gains of >1 kg/day have been recorded during this period (Jonkel and Cowan 1971). However, it is also a period of greater vulnerability to mortality. The extensive movements increase the chances of traveling into unfamiliar habitats and subsequent contact with people. During years of abundant fall foods, bears tend to forage longer before returning to their spring–summer ranges and entering winter dens.

BEHAVIOR

Black bears are generally crepuscular, although breeding and feeding activities may alter this pattern seasonally (Fig. 25.5) (Garshelis and Pelton 1980). Where human food or garbage is available, individuals may become distinctly diurnal (on roadsides) or nocturnal (in campgrounds). Nuisance activities are usually associated with sources of artificial food and the very opportunistic feeding behaviors of black bears. Activities are depressed at temperatures above 25°C or below freezing. As with many other mammalian species, they tend to be most active after the passage of a low-pressure weather front (Garshelis and Pelton 1980). During periods of inactivity, black bears use bed sites in forest habitat; these sites generally consist of a simple shallow depression in the forest leaf litter.

Feeding activities intensify as the summer progresses. Bears tend to leave abundant sign of their presence in an area. Scats on game trails, rotten logs broken open, large rocks overturned, opened yellowjacket nests, matted paths through berry patches, and broken limbs and climb marks on a variety of mast-bearing plants are all signs of their intensified summer and fall feeding activities. The amount and ease of observation of such signs indicate their potential as an index of bear abundance or activity.

"Mark trees" are also evident in parts of a black bear's range (Fig. 25.6). Why bears mark objects is still open to question. Marks are usually made by biting or clawing at 1.5–2.0 m above the ground on conifer or deciduous trees (Burst and Pelton 1980). Marked trees are usually located along animal trails, ridge tops, abandoned roads, or hiking trails. Although the function of marking in the natural history of black bears is unknown, the incidence of marking peaks during the breeding season (midsummer), indicating a possible relationship to breeding behavior. The ritualistic nature of the behavior, its intensity, and its defined location suggest that it is associated with an important aspect of the social structure of the population.

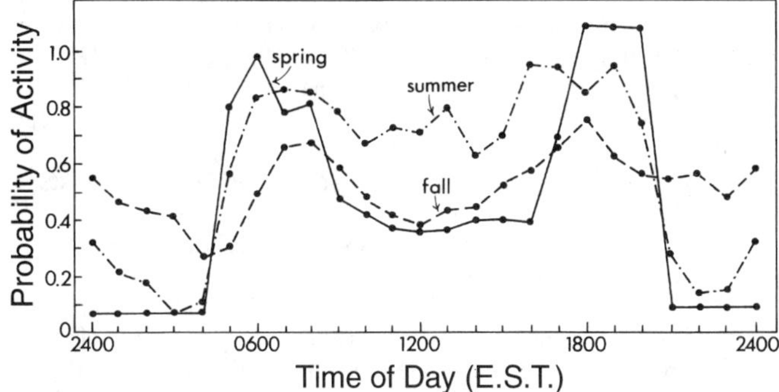

FIGURE 25.5. Seasonal differences in the daily activity patterns of black bears. SOURCE: Garshelis and Pelton 1980.

FIGURE 25.6. Mark trees are indicators of the presence of black bears. SOURCE: Burst and Pelton 1983.

Black bears are normally solitary except for female groups (adult female and cubs), breeding pairs in summer, and congregations at feeding sites. Adult females establish territories during the summer (Rogers 1977). Temporal spacing is exhibited by individuals at other times of the year and is likely maintained through a dominance hierarchy system.

The highly evolved family behavioral relationships probably are the result of the slow maturation of cubs and the high degree of learning associated with the family organization. Black bears possess a high level of intelligence and exhibit a high degree of curiosity and exploratory behaviors (Bacon and Burghardt 1976; Pruitt 1976). Black bears are generally characterized as shy and secretive toward humans. However, they have a much wider array of intraspecific and interspecific behaviors than originally thought, and exhibit agonistic behavior toward conspecifics as well as humans. Threat behaviors are well defined by such characteristics as ear position, body postures, and vocalizations. Tate and Pelton (1980) delineated seven types of aggression by black bears toward humans; these include noncontact behaviors such as a low moan vocalization, blow vocalizations coupled with lip extensions and head oriented downward, and charge. Quadripedal and bipedal swat and snapping or biting are two types of aggressive contact behavior. During the above study, 624 aggressive acts of bears were recorded toward visitors to Great Smoky Mountains National Park; only 37 (5.9%) resulted in contact. Thus, from the standpoint of actual physical contact, black bears are less aggressive than the other North American ursids. Defensively, black bears have had the security of a dense habitat or trees in which to escape. Conversely, grizzly bears and polar bears (*U. maritimus*) evolved on relatively open plains or tundra habitat and were forced to "stand and defend" against antagonists (Herrero 1979).

TABLE 25.2. Major diseases of black bears

Type	Disease	Source
Neoplastic	Liposarcoma and unidentified tumors	King et al. 1960; Rausch 1961
Rickettsial	Elokomin fluke fever	Farrell et al. 1973
Viral	Rabies (reported in Ursidae)	Schoening 1956
Bacterial	Bronchiectasis	King et al. 1960
	Bronchopneumonia	Rausch 1961
	Dental caries	Rausch 1961
	Osteomyelitis	Rausch 1961
	Periodontal disease	Erickson 1965

MORTALITY

Much of the mortality in black bear populations is human related and includes hunting, poaching, roadkills, and depredation control. During years of severe natural food shortages, young bears may starve and older bears are killed as they move greater distances in search of food. Some very old animals (>15 years old) may suffer from serious dental problems (worn teeth, abscesses) and are rendered incapable of feeding. Also, some mortality is attributed to infanticide by males toward younger bears (LeCount 1987). Other forms of incidental natural mortality include flooding of winter ground den, lightning strike, snake bite, and falling from a tree. Mortality rates are highest among young bears after they leave their mother; this rate likely exceeds 35% (Kolenosky 1986). Mortality rate of males is twice that of females (Bunnell and Tait 1985). Although a number of neoplastic, rickettsial, viral, bacterial, and traumatic diseases have been reported for black bears, none appears to contribute greatly to the natural regulation of black bear populations (Table 25.2). Black bears also exhibit a relatively low prevalence and diversity of parasites compared to other mammalian species. There are literature reports of 25 genera and 37 species of endoparasite and 8 genera and 12 species of ectoparasite in black bears (Hamilton 1978). As in other mammals, a greater prevalence and variety of parasites have been detected in the southeastern United States, where combined higher temperatures and humidity provide a more favorable environment. In this region, 22 different species of parasites have been recovered from black bears (Crum 1977).

MANAGEMENT

How resource management agencies deal with black bears is as varied as the status of the species, which ranges from pest to threatened in North America. Where there are large expanses of forested areas and relatively sparse human populations, such as in Canada (including British Columbia, Alberta, and Ontario) and the northwestern United States, harvest regulations are liberal, sometimes amounting to a nearly year-round season. However, small residual enclaves may receive total protection or a very restricted harvest. For example, there is special concern for some populations in the southeastern coastal plain. The Louisiana black bear (*U. a. luteolus*) was listed as threatened under the federal Endangered Species Act in 1992 (Weaver 1992). Recovery of this subspecies has been orchestrated by the Black Bear Conservation Committee, a broad coalition of landowners, state and federal agencies, private conservation groups, forest industry, agricultural interests, and the academic community.

Management of black bears has primarily consisted of harvest regulations to sustain populations at certain levels. To achieve this, hunting and/or trapping seasons, bag limits, and the relative amount of protection are adjusted in response to the assessed status of these populations based on harvest trends and/or research results. For example, an adjustment made by many resource agencies is to move the bear hunting season back to late fall. This change occurred after telemetry studies revealed that females were less vulnerable to hunting the closer it was to the denning period (Johnson and Pelton 1979). In some cases, the amount of protection is mandated by the land rights where the bear population exists. For instance, due to the policy of no hunting in national parks in the United States, bears are afforded complete protection under the law. Regulations and special protective measures have not had to be initiated or altered due to the commercial trade in bear parts (gall bladders, paws) that has had such a detrimental impact on four species of bears in parts of Asia (Mills and Servheen 1994).

There are eight basic steps to good black bear management:

1. Using spatial management by controlling human access through such actions as establishing no-hunting zones and gating roads.
2. Managing habitat for soft and hard mast production, denning sites, and adequate escape cover.
3. Protecting populations through laws, regulations, and enforcement.
4. Developing and implementing proactive nuisance control policies.
5. Initiating a comprehensive information and education program, particularly in dealing with nuisance situations.
6. Harvesting the population using proven and acceptable methods.
7. Monitoring the population using as many habitat and population parameters and indicators as possible.
8. Undertaking a continuous, consistent, and long-term research program.

Wherever black bears exist, if allowed to do so, the species has a tendency to adapt to the presence of humans. It is by far the most adaptable of the eight bear species of the world. Management strategies designed to deal with human–bear interactions are relatively new. The comparatively high intelligence of bears and the emotions the species evokes in people combine to present a singular dilemma for responsible resource agencies. Regulations dealing with separating people and their food or garbage from bears are often difficult to enforce, particularly where high densities of both may interact. Translocation efforts offer only a partial solution because a high percentage of the moves result in bears either returning to their original capture area or being killed soon after release in unfamiliar habitat. A homing success rate of >50% was achieved by translocated bears in Yosemite National Park (Harms 1977). Homing success is negatively correlated with the distance bears are moved (i.e., the further the animal is moved, the less chance there is that it will return). Furthermore, intensive trapping in areas near release sites has resulted in very few recaptures of the relocated bears. This suggests that a large portion of the bears that are not successful in homing likely succumb to some form of mortality. However, most managers still prefer this method to the more drastic option of destroying problem bears.

Much hope has been placed in the future use of aversive stimuli to deal with problem bears rather than shooting or translocating the animals. Aversive conditioning employs any of a number of materials that will evoke a negative response in the animal when it is involved in conflict with humans. For instance, lithium chloride, a strong emetic, placed in a specific food may make a bear sick and keep it from eating that particular type of food again. Shock from an electric fence may keep a bear from trying to get to beehives. However, neither of the above examples has been completely effective in separating bears from human food or garbage. Research is being undertaken to assess the effectiveness of these and other aversive stimuli in dealing with human–bear conflicts (Hunt 1985; Gillin et al. 1994). There is recent interest in the potential use of "capture and release on site" as a more economical and humane method of dealing with nuisance bears (Shull 1994; Clark et al. 2003). Ultimately, however, the most effective management is aimed at the human side of the problem: making unnatural food sources unattainable to the black bear. Given this species's strength and intelligence, this is a formidable task, but it offers a better long-range solution to a persistent problem.

Often, wild or nuisance bears cross political boundaries, which necessitates multiagency cooperation such as the Southern Appalachian Black Bear Study Group (Clark and Pelton 1998). Such groups are able to approach management questions at the ecosystem or landscape level and resolve problems that are most effectively addressed at that level.

Development and/or improvement of several techniques in the past 30 years have enabled biologists to collect and analyze data on bear populations to an extent never before realized. Aldrich spring-activated snares now make it possible to capture bears efficiently, humanely, and in relatively inaccessible locations (Johnson and Pelton 1980b). New immobilization drugs (Ketaset, ketamine hydrochloride; Rompun, xylazine hydrochloride; and Telazol, tiletamine hydrochloride and zolazepam hydrochloride) allow for greater safety of bears and biologists (Kreeger 1996). Refinements in the tooth section method provide a high degree of accuracy in age estimation (Harshyne et al. 1998). New telemetry systems allow researchers to monitor movements and activities of large numbers of animals more frequently and for longer periods of time (McLellan and Hovey 2001). More sophisticated and powerful statistical programs allow for data analysis on smaller and larger, complex data sets (Bozdogan 1987; Burnham and Anderson 1998). Identification of individual animals using DNA from hair or scat samples has the potential for revolutionizing our ability to delve into the population dynamics and social structure of bears as never before (Mowat and Strobeck 2000). Use of geographic information system programs to develop habitat and metapopulation models allow us to see how bears best fit on the landscape or how the landscape best fits the bears (Vanmanen and Pelton 1997). All of the above techniques provide the means by which wildlife biologists can piece together comprehensive life history information to develop better conservation strategies for black bears.

RESEARCH NEEDS

Although the development of new techniques and refinements of existing procedures have allowed researchers to accumulate much valuable information on black bear populations, there are still many significant gaps in our knowledge about this species. We are only beginning to understand the social structure of bear populations, the mechanisms controlling certain behaviors, and the implications of such behaviors on population regulation. We are aware only in general terms of the possible factors affecting natality and mortality. Furthermore, we must begin to document and sort out the long-term natural (forest succession, weather) and anthropogenic (urban sprawl, exotic pests, land-use policies) impacts of a wide array of interacting variables on the population dynamics of black bears. Few studies have measured the before, during, and after impacts of a particular parameter or set of parameters and the resulting response of a bear population. Therefore, more emphasis should be placed on long-term studies (Pelton and Vanmanen 1996). Through long-term (>10 years) intensive field efforts, many parameters not obtainable in short-term (1–5 years) efforts, such as age-specific natality and mortality rates, can be derived. The methods of censusing used by agencies responsible for managing black bears often are crude at best. As future pressures demand that we know with greater certainty the number of animals per unit area that we are trying to manage, we will have to intensify our efforts toward developing and using more reliable population estimators. If bears and people are to coexist in the future, we must also develop ways to decrease or eliminate the negative interactions that occur. We know very little about the dynamics of that segment of a bear population that interacts with people, or human food and/or garbage. More research is needed in the area of proper handling of translocated bears, better techniques for dealing with troublesome bears, and especially more innovative methods of separating bears from sources of human food and/or garbage. Land-use practices will ultimately decide the fate of black bears throughout their range. Results of research on black bear populations must be integrated with land-use agencies (for example, U.S. Forest Service, Bureau of Land Management, National Park Service, and private timber companies) to ensure survival of the species. The black bear is in competition with other natural resources such as timber, mining, and commercial development. Black bears have important economic, esthetic, recreational, social, and scientific values. Black bear management must be based on sound research so this resource can compete with other demands on the land. This can be accomplished only through interagency cooperative team research in a long-term commitment.

LITERATURE CITED

Alt, G. L. 1981. Reproductive biology of black bears of Northeastern Pennsylvania. Transactions of the Northeast Fish and Wildlife Conference 38:88–89.

Alt, G. L., and P. C. Carr. 1984. Pennsylvania status report. Proceedings of the Eastern Workshop on Black Bear Research and Management 7:27–29.

Amstrup, S. C., and J. Beecham. 1976. Activity patterns of radiocollared black bears in Idaho. Journal of Wildlife Management 40:340–48.

Bacon, E. S., and G. M. Burghardt. 1976. Learning and color discrimination in the American black bear. Pages 27–36 *in* M. R. Pelton, J. W. Lentfer, and G. E. Folk, Jr., eds. Bears: Their biology and management. Third international conference on bear research and management (Publication New Series No. 40). International Union for the Conservation of Nature, Morges, Switzerland.

Barnes, V. G., and O. E. Bray. 1967. Population characteristics and activities of black bears in Yellowstone National Park. Colorado Cooperative Wildlife Research Unit and Colorado State University, Fort Collins.

Boone, W. R., J. C. Catlin, K. G. Casey, E. T. Boone, P. S. Dye, R. J. Schuett, J. O. Rosenburg, T. Tsubota, and J. M. Bahr. 1998. Bears as induced ovulators—A preliminary study. Ursus 10:503–5.

Bozdogan, H. 1987. Model selection and Akaike's information criterion (AIC): The general theory and its analytical extensions. Psychometrika 52:345–70.

Bunnell, F. L., and D. E. N. Tait. 1980. Bears in models and in reality—Implications to management. International Conference on Bear Research and Management 4:15–23.

Bunnell, F. L., and D. E. N. Tait. 1985. Mortality rates of North American bears. Arctic 38:316–25

Burnham, K. P., and D. R. Anderson. 1998. Model selection and inference: A practical information theoretic approach. Springer-Verlag, New York.

Burst, T. L., and M. R. Pelton. 1983. Black bear mark trees in the Smoky Mountains. International Conference on Bear Research and Management 5:45–53.

Carpenter, M. 1973. The black bear in Virginia. Virginia Commission Game and Inland Fisheries, Richmond.

Caughley, G. 1974. Interpretation of age ratios. Journal of Wildlife Management 38:557–62.

Clark, J. D. 1991. Ecology of two black bear (*Ursus americanus*) populations in the interior highlands of Arkansas. Ph. D. Dissertation, University of Arkansas, Fayetteville.

Clark, J. D., and M. R. Pelton. 1998. Management of a large carnivore: Black bear. Pages 209–23 *in* J. D. Peine, ed. Ecosystem management for sustainability. Lewis, Boca Raton, FL.

Clark, J. E., F. T. Vanmanen, and M. R. Pelton. 2002. Correlates of success for on-site releases of nuisance black bears in Great Smoky Mountains National Park. Wildlife Society Bulletin. 30:104–111.

Crum, J. M. 1977. Some parasites of black bears (*Ursus americanus*) in the southeastern United States. M. S. Thesis, University of Georgia, Athens.

Eagle, T. C., and M. R. Pelton. 1983. Seasonal nutrition of black bears in the Great Smoky Mountains National Park. International Conference on Bear Research and Management 5:94–104.

Eberhardt, L. L. 1990. Using radio-telemetry for mark–recapture studies with edge effects. Journal of Applied Ecology 27:259–71.

Edwards, W. R., and L. L. Eberhardt. 1967. Estimating cottontail abundance from live trapping data. Journal of Wildlife Management 31:87–96.

Eiler, J. H., W. G. Wathen, and M. R. Pelton. 1989. Reproduction in black bears in the southern Appalachian Mountains. Journal of Wildlife Management 53:353–60.

Elowe, K. D. 1984. Home range, movements, and habitat preferences of black bear (*Ursus americanus*) in western Massachusetts. M. S. Thesis, University of Massachusetts, Amherst.

Elowe, K. D., and W. E. Dodge. 1989. Factors affecting black bear reproductive success and cub survival. Journal of Wildlife Management 53:962–68.

Erickson, A. W. 1965. The black bear in Alaska: Its ecology and management (Federal Aid in Wildlife Restoration, W-6-R-5, Work Plan F). Alaska Department of Fish and Game.

Erickson, A. W., and J. E. Nellor. 1964. Breeding biology of the black bear. Part 1. Pages 1–45 *in* A. W. Erickson, J. Nellor, and G. A. Petrides, eds. The black bear in Michigan (Research Bulletin Number 4). Michigan State Agricultural Experiment Station.

Erickson, A. W., and G. A. Petrides. 1964. Population structure, movements, and mortality of tagged black bears in Michigan. Part 2. Pages 46–67 *in* A. W. Erickson, J. Nellor, and G. A. Petrides, eds. The black bear in Michigan (Research Bulletin Number 4). Michigan State Agricultural Experiment Station.

Farrell, R. K., R. W. Leader, and S. O. Johnston. 1973. Differentiation of salmon poisoning disease and Elokomin fluke fever: Studies with the black bear (*Ursus americanus*). American Journal of Veterinary Research 34:919–22.

Folk, G. E., M. A. Folk, and J. J. Minor. 1972. Physiological condition of three species of bears in winter dens. Pages 107–24 *in* S. Herrero, ed. Bears: Their biology and management (New Series Publication No. 23). International Union for the Conservation of Nature, Morges, Switzerland.

Garshelis, D. L. 1991. Monitoring effects of harvest on black bear populations in North America: A review and evaluation of techniques. Eastern Workshop Black Bear Research and Management 10:120–44.

Garshelis, D. L. 1992. Mark–recapture estimation for animals with large home ranges. Pages 1098–11 *in* D. R. McCullough and R. H. Barrett, eds. Wildlife 2001: Populations. Elsevier Applied Science, London.

Garshelis, D. L., and M. R. Pelton. 1980. Activity of black bears in the Great Smoky Mountains National Park. Journal of Mammalogy 61:8–19.

Garshelis, D. L., H. B. Quigley, C. R. Vilarrubia, and M. R. Pelton. 1983. Diel movements of black bears in the southern Appalachians. International Conference on Bear Research and Management 5:11–19.

Gillin, C. M., F. M. Hammond, and C. M. Peterson. 1994. Evaluation of an aversive conditioning technique used on female grizzly bears in the Yellowstone Ecosystem. International Conference on Bear Research and Management 9:503–12.

Hall, E. R. 1981. The mammals of North America, 2nd ed. John Wiley, New York.

Hamilton, R. J. 1978. Ecology of the black bear in southeastern North Carolina. M.S. Thesis, University of Georgia, Athens.

Hamilton, R. J., and R. L. Marchinton. 1977. Denning and related activities of black bears in the Coastal Plain of North Carolina. International Conference on Bear Research and Management 4:121–26.

Harms, D. 1977. Black bear management in Yosemite. International Conference on Bear Research and Management 4:205–12.

Harris, R. B., and L. H. Metzgar. 1987. Harvest age structures as indicators of decline in small populations of grizzly bears. International Conference on Bear Research and Management 7:109–16.

Harshyne, W. A., D. R. Diefenbach, G. L. Alt, and G. M. Matson. 1998. Analysis of error from cementum-annuli age estimates of known-age Pennsylvania black bears. Journal of Wildlife Management 62:1281–1991.

Hayne, D. W. 1949. An examination of the strip census method for estimating animal populations. Journal of Wildlife Management 13:145–57.

Hellgren, E. C., and M. R. Vaughan. 1989. Demographic analysis of a black bear population in the Great Dismal Swamp. Journal of Wildlife Management 53:969–77.

Hellgren, E. C., and M. R. Vaughan. 1990. Range dynamics of black bears in Great Dismal Swamp, Virginia–North Carolina. Proceedings of the Annual Conference of Southeastern Association of Fish and Wildlife Agencies 44:268–78.

Hellgren, E. C., M. R. Vaughan, F. C. Gwazkauska, B. Williams, P. F. Scanlon, and R. L. Kirkpatrick. 1990. Endocrine and electrophoretic profiles during pregnancy and nonpregnancy in captive female black bears. Canadian Journal of Zoology 69:892–98.

Herrero, S. M. 1979. Black bears: The grizzly's replacement. Pages 179–95 *in* D. Burk, ed. The black bear in modern North America. Amwell Press, Clinton, NY.

Hock, R. J. 1960. Seasonal variations in physiologic functions of arctic ground squirrels and black bears. Harvard University Museum of Comparative Zoology Bulletin 124:155–73.

Hornocker, M. G. 1962. Population characteristics and social reproductive behavior of the grizzly bear in Yellowstone National Park. M.S. Thesis, University of Montana, Missoula.

Hugie, R. H. 1974. Habitat of the black bear in Maine. Eastern Workshop on Black Bear Management and Research 2:151–57.

Hunt, C. L. 1985. Descriptions of five promising deterrent and repellent products for use on bears. Final Report to U.S. Fish and Wildlife Service, Grizzly Bear Recovery Coordinator.

Johnson, K. G., and M. R. Pelton. 1979. Denning behavior of black bears in the Great Smoky Mountains National Park. Proceedings of the Annual Conference of Southeastern Association of Fish and Wildlife Agencies 33:239–49.

Johnson, K. G., and M. R. Pelton. 1980a. Environmental relationships and the denning period of black bears in Tennessee. Journal of Mammalogy 61:653–60.

Johnson, K. G., and M. R. Pelton. 1980b. Prebaiting and snaring techniques for black bears. Wildlife Society Bulletin 8:46–54.

Johnson, K. G., and M. R. Pelton. 1981. Selection and availability of dens for black bears in Tennessee. Journal of Wildlife Management 45:111–19.

Jonkel, C. J., and I. Mct. Cowan. 1971. The black bear in the spruce–fir forest. Wildlife Monographs 27:1–57.

Kayser, C. 1961. The physiology of natural hibernation. Pergamon Press, New York.

Kemp, G. A. 1972. Black bear population dynamics at Cold Lake, Alberta, 1968–1970. Pages 26–31 *in* S. Herrero, ed. Bears: Their biology and management. Second international conference on bear research and management (New Series Publication No. 23). International Union for the Conservation of Nature, Morges, Switzerland.

Kemp, G. A. 1979. The Rocky Mountain working group. Pages 217–36 *in* D. Burk, ed. The black bear in modern North America. Amwell Press, Clinton, NY.

King, J. M., H. C. Black, and O. M. Hewitt. 1960. Pathology, parasitology, and hematology of the black bear in New York. New York Fish and Game Journal 7:99–111.

Kolenosky, G. B. 1986. The effects of hunting on an Ontario black bear population. International Conference on Bear Research and Management 6:45–55.

Kolenosky, G. B. 1990. Reproductive biology of black bears in east-central Ontario. International Conference on Bear Research and Management 8:385–92.

Kreeger, T. J. 1996. Handbook of wildlife chemical immobilization. Wildlife Pharmaceuticals, Laramie, WY.

Lariviere, S. 2001. *Ursus americanus*. Mammalian Species 647:1–11.

Lawrence, W. 1979. Working group reports: Pacific working group: Habitat management and land use practices. Part 3. Pages 196–201 *in* D. Burk, ed. The black bear in modern North America. Amwell Press, Clinton, NY.

Lecount, A. L. 1983. Evidence of wild black bears breeding while raising cubs. Journal of Wildlife Management 47:264–68.

Lecount, A. L. 1987. Causes of black bear cub mortality. International Conference on Bear Research and Management 7:75–82.

Lindzey, J. S., G. L. Alt, C. R. Mc Laughlin, and W. S. Kordek. 1983. Population response of Pennsylvania black bears to hunting. International Conference on Bear Research and Management 5:34–39.

Litvaitis, J. A., and D. M. Kane. 1994. Relationship of hunting techniques and hunter selectivity to composition of black bear harvest. Wildlife Society Bulletin 22:604–6.

Lundberg, D. A., R. A. Nelson, H. W. Wahner, and J. P. Jones. 1976. Protein metabolism in the black bear before and during hibernation. Mayo Clinic Proceedings 51:716–22.

Marcum, L. C. 1974. An evaluation of radioactive feces-tagging as a technique for determining population densities of the black bear in the Great Smoky Mountains National Park. M.S. Thesis, University of Tennessee, Knoxville.

Matson, J. R. 1946. Notes on the dormancy of the black bear. Journal of Mammalogy 27:203–12.

McLaughlin, C. R., G. J. Matula, Jr., and R. J. O'Connor. 1994. Synchronous reproduction by Maine black bears. International Conference on Bear Research and Management 9:471–79.

McLean, P. K., and M. R. Pelton. 1990. Some demographic comparisons of wild and panhandler bears in the Smoky Mountains. International Conference on Bear Research and Management 8:105–12.

McLellan, B. N., and F. W. Hovey. 2001. Habitat selected by grizzly bears in a multiple use landscape. Journal of Wildlife Management 65:92–99.

Miller, S. D. 1990. Population management of bears in North America. International Conference on Bear Research and Management 8:357–73.

Miller, S. D., G. C. White, R. A. Sellers, H. V. Reynolds, J. W. Schoen, K. Titus, V. G. Barnes, Jr., R. B. Smith, R. R. Nelson, W. B. Ballard, and C. C. Schwartz. 1997. Brown and black bear density estimation in Alaska using radiotelemetry and replicated mark-resight techniques. Wildlife Monographs 133:1–55.

Mills, J., and C. Servheen. 1994. The Asian trade in bears and bear parts: Impacts and conservation recommendations. International Conference on Bear Research and Management 9:161–67.

Morrison, P. 1960. Some interrelations between weight and hibernation function. Harvard University Museum of Comparative Zoology Bulletin 124:75–91.

Mowat, G., and C. Strobeck. 2000. Estimating population size of grizzly bears using hair capture, DNA profiling, and mark–recapture analysis. Journal of Wildlife Management 64:183–93.

Nelson, R. A., H. W. Wahner, J. D. Jones, R. D. Ellefson, and P. E. Zollman. 1973. Metabolism of bears before, during, and after winter sleep. American Journal of Physiology 224:491–96.

Novick, H. J., and G. R. Stewart. 1982. Home range and habitat preferences of black bears in the San Bernardino Mountains of southern California. California Fish and Game Journal 67:21–35.

Noyce, K. V., and D. L. Garshelis. 1994. Body size and blood characteristics as indicators of condition and reproductive performance in black bears. International Conference on Bear Research and Management 9:481–96.

Noyce, K. V., and D. L. Garshelis. 1997. Influence of natural food abundance on black bear harvests in Minnesota. Journal of Wildlife Management 61:1067–74.

Pelton, M. R. 1972. Use of foot trail travellers in the Great Smoky Mountains National Park to estimate black bear (*Ursus americanus*) activity. Pages 36–43 *in*S. Herrero, ed. Bears: Their biology and management. International conference on bear research and management (New Series Publication No. 23). International Union for the Conservation of Nature, Morges, Switzerland.

Pelton, M. R., and L. C. Marcum. 1977. The potential use of radio-isotopes for determining densities of black bears and other carnivores. Pages 221–36 *in* R. L. Phillips and C. Jonkel, eds. Proceedings of the 1975 predator Symposium. Montana Forest and Conservation Experiment Station, University of Montana, Missoula.

Pelton, M. R., and F. T. Vanmanen. 1994. Distribution of black bears in North America. Proceedings of the Eastern Workshop on Black Bear Research and Management 12:133–38.

Pelton, M. R., and F. T. Vanmanen. 1996. Benefits and pitfalls of long-term research: A case study of black bears in Great Smoky Mountains National Park. Wildlife Society Bulletin 24:443–50.

Pelton, M. R., A. B. Coley, T. H. Eason, D. L. D. Martinez, J. A. Pederson, F. T. Vanmanen, and K. M. Weaver. 1999. American black bear conservation action plan. Pages 144–56 *in* C. Servheen, S. Herrero, and B. Peyton, eds. Bears: Status survey and conservation action plan. International Union for the Conservation of Nature, Species Survival Commission Bear Specialist Group, Gland, Switzerland, and Cambridge, UK.

Piekielek, W., and T. S. Burton. 1975. A black bear population study in northern California. California Fish and Game Journal 61:4–25.

Poelker, R. J., and H. D. Hartwell. 1973. Black bear of Washington (Biological Bulletin No. 18). Washington State Game Department.

Pruitt, C. H., 1976. Play and agonistic behavior in captive black bears. Pages 79–86 *in* M. R. Pelton, J. W. Lentfer, and G. E. Folk, Jr., eds. Bears: Their biology and management. Third international conference on bear management and research (New Series Publication No. 40). International Union for the Conservation of Nature, Morges, Switzerland.

Rausch, R. L. 1961. Notes on the black bear, *Ursus americanus,* in Alaska, with particular reference to dentition and growth. Zeitschrift für Saugertierkunde 26:77–107.

Rogers, L. L. 1976. Effects of mast and berry crop failures on survival, growth, and reproductive success of black bears. Transaction of the North American Wildlife and Natural Resources Conference 41:432–38.

Rogers, L. L. 1977. Movements and social relationships of black bears in northeastern Minnesota. Ph.D. Dissertation, University of Minnesota, St. Paul.

Rogers, L. L. 1980. Inheritance of coat color and changes in pelage coloration in black bears in northeastern Minnesota. Journal of Mammalogy 61:324–27.

Rounds, R. C. 1987. Distribution and analysis of colour morphs of the black bear (*Ursus americanus*). Journal of Biogeography 14:521–38.

Sauer, P. R., S. L. Free, and S. D. Browne. 1969. Movement of tagged black bears in the Adirondacks. New York Fish and Game Journal 16:205–23.

Schenk, A., and K. M. Kovacs. 1995. Multiple mating between black bears revealed by DNA fingerprinting. Animal Behavior 50:1483–90.

Schoening, H. W. 1956. Rabies. Pages 195–202 *in*A. Stefferud, ed. Animal diseases: United States Department of Agriculture yearbook. U.S. Government Printing Office, Washington, DC.

Shull, S. D. 1994. Management of nuisance black bears in the interior highlands of Arkansas. M. S. Thesis, University of Arkansas, Fayetteville.

Spencer, H. E. 1955. The black bear and its status in Maine (Game Division Bulletin Number 4). Maine Department of Inland Fisheries and Game.

Tate, J., and M. R. Pelton. 1980. Human–bear interactions in Great Smoky Mountains National Park. International Conference on Bear Research and Management 5:312–21.

Troyer, W. A., and R. J. Hensel. 1964. Structure and distribution of a Kodiak bear population. Journal of Wildlife Management 28:769–72.

Vanmanen, F. T. 1994. Black bear habitat use in Great Smoky Mountains National Park. Ph.D. Dissertation, University of Tennessee, Knoxville.

Vanmanen, F. T., and M. R. Pelton. 1997. A GIS model to predict black bear habitat use. Journal of Forestry 95:6–12.

Waddell, T. 1979. State and provincial status reports: Arizona. Pages 33–37 *in*A. LeCount, ed. First western black bear workshop. Tempe, AZ.

Weaver, K. M. 1992. Louisiana status report. Proceedings of Eastern Black Bear Workshop 11:16–21.

Willey, C. H. 1974. Aging black bears from first premolar tooth sections. Journal of Wildlife Management 38:97–100.

Wimsatt, W. A. 1963. Delayed implantation in the Ursidae, with particular reference to the black bear. Pages 49–76 *in* A. C. Ender, ed. Delayed implantation. University of Chicago Press, Chicago.

Michael R. Pelton, Department of Forestry, Wildlife, and Fisheries, University of Tennessee, Knoxville, Tennessee 37996-4563. Email: mpelton @utk.edu.

26

Grizzly Bear

Ursus arctos

Charles C. Schwartz
Sterling D. Miller
Mark A. Haroldson

NOMENCLATURE

Common Names. Brown bear, grizzly bear, Kodiak bear
Scientific Name. *Ursus arctos* Linnaeus

The grizzly bear inspires fear, awe, and respect in humans to a degree unmatched by any other North American wild mammal. Like other bear species, it can inflict serious injury and death on humans and sometimes does. Unlike the polar bear (*Ursus maritimus*) of the sparsely inhabited northern arctic, however, grizzly bears still live in areas visited by crowds of people, where presence of the grizzly remains physically real and emotionally dominant. A hike in the wilderness that includes grizzly bears is different from a stroll in a forest from which grizzly bears have been purged; nighttime conversations around the campfire and dreams in the tent reflect the presence of the great bear. Contributing to the aura of the grizzly bear is the mixture of myth and reality about its ferocity, unpredictable disposition, large size, strength, huge canines, long claws, keen senses, swiftness, and playfulness. They share characteristics with humans such as generalist life history strategies, extended periods of maternal care, and omnivorous diets. These factors capture the human imagination in ways distinct from other North American mammals. Precontact Native American legends reflected the same fascination with the grizzly bear as modern stories and legends (Rockwell 1991).

Dominance of the grizzly in human imagination has played a significant role in the demise of the species. Conquest of the western wilderness seemed synonymous with destruction of the great bear. The challenge of the twenty-first century is to avoid repeating and attempt to correct the errors of the nineteenth and twentieth centuries.

Ursus arctos is widely distributed throughout the Palearctic (Europe and Asia) and Nearctic (North America) faunal regions. In the Palearctic, *U. arctos* is commonly referred to as the brown bear, whereas in North America it is called the grizzly bear in the lower 48 states and most of Alaska. Typically only the coastal populations of Alaska or those in Canada are referred to as brown bears. Here, we use the terms interchangeably recognizing that there is only one species with different common names. The grizzly bear is one of eight species of bears distributed worldwide, and one of six members of the genus *Ursus*. The brown/grizzly bear occupies a diverse array of habitats, from arctic tundra, to boreal and coastal forests, to the mountain forest/grassland ecotone. Classified as an omnivorous carnivore, its diet varies widely over its North American range. To a large degree, abundance of high-quality foods dictates body size, reproductive rates, and population density. Human influences on the landscape continue to alter once pristine habitats to the detriment of grizzly bears. Habitat degradation and losses coupled with human-caused mortality are the major conservation issues the species has faced historically and continues to face today.

Subspecies. By necessity, early classification relied heavily on paleontological and morphological data, but such classifications of ursids were inconclusive at best (Kurtén 1968; Kitchener 1994; Waits et al. 1999). Merriam (1918) proposed over 90 subspecies that described the geographic variants of *U. arctos,* but this classification is considered obsolete (Waits et al. 1998a). As summarized by Craighead and Mitchell (1982) and Waits et al. (1998a), Rausch (1963) identified two extant subspecies of brown bears in North America primarily from skull measurements. He classified bears from the mainland as *U. arctos horribilis* Ord and those from the Kodiak Island archipelago as *U. a. middendorffi* Merriam. Rausch (1963) reconsidered his earlier classification (Rausch 1953) of the bears from the Alaska Peninsula as being a distinct subspecies (*U. a. gyas* Merriam). Kurtén (1973) used skull measurements from Rausch (1963) to propose three North American subspecies, *U. a. middendorffi* from Kodiak Island archipelago, *U. a. dalli* Merriam of southern coastal regions of the Alaska panhandle, including the islands of Admiralty, Baranof, and Chichagof (ABC), and *U. a. horribilis* for all other brown bears. Finally, Hall (1984) used cranial and dentition dimensions to propose seven North American subspecies. Five were restricted to Alaska: (1) *U. a. middendorffi* (Kodiak islands), (2) *U. a. gyas* (Kenai Peninsula), (3) *U. a. dalli* (northwest panhandle), (4) *U. a. sitkensis* Merriam (southeast Alaska including ABC islands), and (5) *U. a. alascensis* Merriam (the remaining mainland). The subspecies *U. a. stikeenensis* Merriam was restricted to coastal British Columbia, Washington, and Oregon, and *U. a. horribilis* included all inland brown bear populations in Canada and the lower 48 states. The generally accepted current classification is that proposed by Rausch (1963), but this is likely to change based on DNA analysis.

With the advent of DNA analysis and the technological advancements in this field, we now know considerably more about evolution of ursids and subspecific classification within species (Waits et al. 1999). Using mitochondrial DNA (mtDNA) of brown bears across their geographic range, several researchers have defined five mtDNA lineage groups defined as clades (Cronin et al. 1991; Taberlet and Bouvet 1994; Kohn et al. 1995; Randi et al. 1995; Taberlet et al. 1995; Talbot and Shields 1996; Waits et al. 1998a). Clade I brown bears are from southern Scandinavia and southern Europe; Clade II are from the ABC islands; Clade III are from eastern Europe, Asia, and western Alaska; Clade IV are from southern Canada and the lower 48 states; and Clade V are from eastern Alaska and northern Canada (Fig. 26.1).

The mtDNA phylogeny does not support any of the historic taxonomic classifications (Waits et al. 1998a). There is no support for *U. a. middendorffi, U. a. horribilis,* or *U. a. gyas.* The classification by Kurtén (1968) and Hall (1984) of bears from the ABC islands and adjacent mainland probably is incorrect. Brown bears from the ABC islands constitute the oldest and most genetically unique mtDNA clade in the New World and are a sister taxa to the polar bear (Talbot and Shields 1996; Shields et al. 2000). However, as stated by Waits et al. (1998a:415), "a revision of the taxonomy of North American brown bears in accordance with the phylogenetic species concept (Cracraft 1983) would result in drastic changes in the current classification. The most frequently recognized subspecies, *U. a. middendorffi,* would be abolished, and 4 new subspecific distributions would be added. But it seems unreasonable to dramatically alter the current taxonomy based on the results from a single mtDNA region." Additional research using

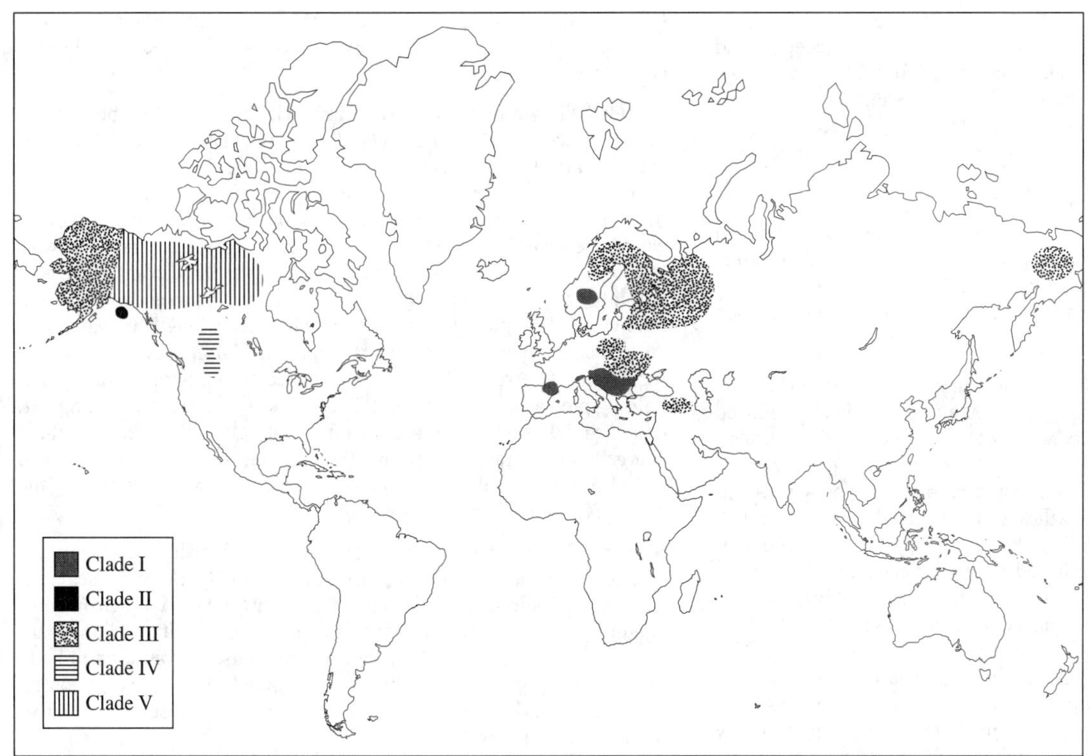

FIGURE 26.1. Worldwide geographic distribution of the five mitochondrial phylogenetic clades in the brown/grizzly bear. SOURCE: After Waits et al. (1999).

additional genes, particularly from the Y chromosome, is needed for taxonomic clarification (Waits et al. 1998a).

Evolution. Thenius (1959) and Kurtén (1968) provided much of the original paleontological work on the Ursinae. Herrero (1972), Martinka (1974), and Craighead and Mitchell (1982) provided excellent summaries, which we paraphrase here. All living and fossil bears of the genus *Ursus* descended from *U. minimus,* a small, forest-dwelling bear of the Pliocene. The grizzly bear differentiated from the Etruscan bear (*Ursus etruscus*) in Asia during the middle Pleistocene (2–3 million years ago). The earliest records of *U. arctos* are from about 500,000 years ago from Choukoutien, China. The species entered Europe some 250,000 years ago during formation of glacial land bridges (Pasitschniak-Arts 1993). Speciation occurred during a period of extensive glaciation in northern continental areas. Forests were replaced with tundra, and adaptation to these open habitats was a key element associated with genetic separation of the grizzly from its forest-dwelling ancestors (Herrero 1972). Steppe and tundra forms dominated late dispersal, and it appears that the grizzly did not successfully colonize Alaska until the Wisconsin glacial period (Herrero 1972). Recession of the continental ice sheets allowed expansion into most of North America by the early Holocene (Martinka 1974).

DISTRIBUTION

Historical Range. Following recession of the ice sheets, *Ursus arctos* was widely distributed across North America (Fig. 26.2). Distribution expanded eastward to Ontario (Peterson 1965) and Ohio and Kentucky (Guilday 1968), and southward to Mexico (Storer and Tevis 1955). The range possibly extended northeast as far as Labrador (Speiss 1976; Speiss and Cox 1977). Distribution apparently receded following this eastward expansion in response to unfavorable environmental conditions (Guilday 1968).

Before European settlement of the North American continent, the brown bear had a wide distribution (Roosevelt 1907; Wright 1909; Dobie 1950; Storer and Tevis 1955; Herrero 1972; Stebler 1972;

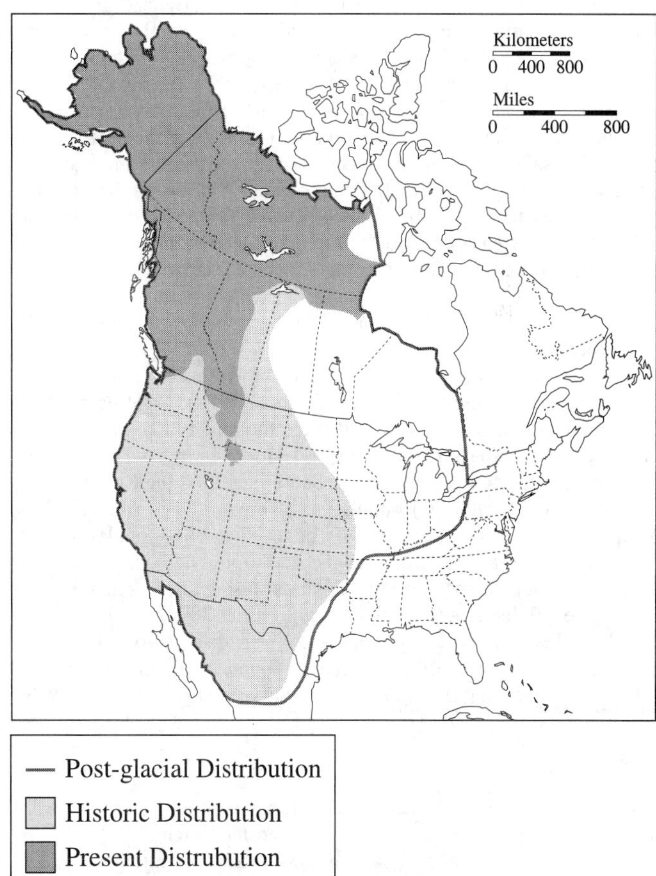

FIGURE 26.2. Postglacial, historical, and current distribution of the brown/grizzly bear. SOURCE: Data from Rausch (1963), Martinka (1976), Servheen et al. (1999).

Schneider 1977; Craighead and Mitchell 1982). The distribution provided by Rausch (1963) seems to be the most commonly accepted and we use it here (Fig. 26.2) with additional detail from Herrero (1972). According to MacPherson (1965), the eastern extent of the barren-grounds grizzly was near the Thelon, Back, Dubawnt, and Kazan Rivers in Northwest Territories. In the southwestern extreme of the range in the grassland and chaparral of California, the grizzly probably was numerous, extending to the coast (Herrero 1972). In interior North America, grizzlies used riparian bottoms of the Great Plains. The range extended eastward to the great bend of the Missouri River in North Dakota, southward to the Moreau River in South Dakota, and possibly to the Red River region of Texas (Stebler 1972). The southern extent reached into Mexico.

Current Range. With the advent of European settlement of the North American continent, grizzly bear range was drastically reduced (Mattson et al. 1995). These reductions were greatest in the southern and eastern parts of the range. In the lower 48 states, grizzly bears were eliminated from 98% of their historical range during a 100-year period (Mattson et al. 1995). Grizzlies were extirpated throughout much of their range in the 1920s and 1930s. Thirty-one of 37 bear populations present in 1922 were eliminated by 1975 (Servheen 1999), when the species was declared threatened (U.S. Fish and Wildlife Service 1993). Currently there are 5 recognized populations south of 49°N, but 3 populations contain <35 individuals each (Servheen 1999). The Yellowstone and Northern Continental Divide populations each contain >350 individuals. Servheen (1999) estimated 800–1020 bears reside in the United States south of Canada. Some people contend that grizzly bears still occupy the Bitterroot Mountains of central Idaho and western Montana, but there is no evidence supporting this belief (Melquist 1985; Groves 1987; Servheen et al. 1990; Kunkel et al. 1991).

Macey (1979) and more recently Banci (1991), Banci et al. (1994), and McLellan and Banci (1999) reviewed the status of the brown bear in Canada. Macey (1979) concluded that brown bears were not endangered in Canada, but were extremely vulnerable in some areas. Brown bears have been extirpated from part of their historical range in Manitoba, Saskatchewan, and Alberta, primarily in the prairies and boreal plains. Densities are most depressed in the southern portions of Canada, particularly in British Columbia and Alberta. In the hot, dry plateaus of British Columbia, brown bears are considered threatened. In inhospitable areas of the north or in the rugged mountains, there are limited human settlements and brown bears are relatively numerous for the habitat, whereas brown bears are rare where people have settled (McLellan and Banci 1999). Banci (1991) estimated that about 25,000 brown bears live in Canada.

Alaska has the largest population of brown bears of any state or province in North America (Miller and Schoen 1999). Their distribution has remained relatively unchanged since the mid-1700s, and populations are considered stable (Miller 1993). However, in November 1998, the Alaska Department of Fish and Game identified the Kenai Peninsula (south of Anchorage) population of brown bears as a "Species of Special Concern" because it was believed to be a population that "is vulnerable to a significant decline due to low numbers, restricted distribution, dependence on limited habitat resources, or sensitivity to environmental disturbance." (Schoen and Miller 2002).

Excellent reviews of brown/grizzly bear distribution in Alaska, Canada, and the lower 48 states can be found in Miller and Schoen (1999), McLellan and Banci (1999), and Servheen (1999), respectively.

DESCRIPTION

Early accounts and descriptions of the grizzly bear are superficial, subjective, and often sensationalized (Craighead and Mitchell 1982). Many focused on natural history, behavior, and hunting techniques, emphasizing extremes in body size and strength, with little attention to the typical. The earliest scientific descriptions of the grizzly bear based on adequate sample size were those of Swainson, Baird, and Elliot from the arctic, western United States, and British Columbia, respectively (Storer and Tevis 1955). With the advent of immobilization drugs, our knowledge of individual bear populations has advanced considerably (LeFranc et al. 1987).

General Morphology and Structure. The brown/grizzly bear varies greatly in size and shape throughout the range in North America. However, certain characteristics are consistent. The skeletal structure of the brown bear is larger and heavier than that of most other ursids, but the axial and appendicular skeleton is similar to that of the American black bear (*Ursus americanus*). Brown bears are tetrapedal, with legs of approximately equal length, tapering to large plantigrade feet (Craighead and Mitchell 1982). Each foot has five toes ending with a relatively long claw. Foreclaws can reach 8 cm in length and are much larger than on black or polar bears. Claws of *U. arctos* evolved as tools for digging (Herrero 1972) rather than tree climbing or capturing and holding prey as in *U. maritimus.* They walk with a heavy shuffling gait (Pasitschniak-Arts 1993). Features that distinguish the species include a large hump of muscle overlying the scapulae, characteristic skull and dental structure, and, in some individuals, color and appearance of the pelage (Craighead and Mitchell 1982).

Size and Weight. Size varies greatly across the North American range, among sex and age classes of bears, and seasonally. Body masses from various populations are reviewed in the Interagency Grizzly Bear Compendium (LeFranc et al. 1987) and supplemented with additional information (McLellan 1994). These records illustrate variation in body mass among populations (Table 26.1). Brown bears occupying coastal habitats of Alaska and British Columbia are the largest representative of the species in North America.

Bears from coastal Alaska with access to salmon are the heaviest. For example, males from the Alaska Peninsula (Miller and Sellers 1992) average 357 kg, whereas males from the Yukon (Pearson 1975) average 145 kg. Females from the same areas average 226 and 98 kg, respectively.

Popular literature often sensationalizes the "1000-pound bear." Although brown bears have been documented to reach and exceed this weight (Craighead and Mitchell 1982), most are smaller. Mass in bears is related to diet (Hilderbrand et al. 1999a). Bear populations with better nutrition from consuming large quantities of animal flesh (salmon and ungulates) tend to be larger (Fig. 26.3). Bears consuming principally vegetal diets are smaller.

Brown bears are sexually dimorphic, with males about 1.2–2.2 times larger than females (LeFranc et al. 1987; Stringham 1990; Hilderbrand et al. 1999a). Differences in body mass between males and females are influenced by age at sexual maturity, samples from within the population, season of sampling, reproductive status, and differential mortality. Any or all of these factors can contribute to a slightly different ratio. Dimorphism begins early in life and is apparent between ages 2 and 4 years (Troyer and Hensel 1969; Pearson 1975; Blanchard 1987). Dimorphism is believed to be related to dominance competition among males during the breeding season.

Body mass is dynamic in brown bears. During late summer and fall, brown bears gain weight rapidly, primarily as fat (Troyer and Hensel 1969; Pearson 1975; Craighead and Mitchell 1982; Kingsley et al. 1983; Nagy et al. 1983a, 1983b; Blanchard 1987; Hilderbrand et al. 2000) when they feed intensively before denning (Nelson 1980; Nelson et al. 1983a). Because bears rely solely on their stored energy reserves during hibernation, this predenning weight gain is essential for reproduction and survival. Peak body mass generally occurs in fall just before hibernation. Bears metabolize fat and muscle during the denning period (Hellgren 1998; Hilderbrand et al. 2000).

Weight loss during the denning season depends on condition of the bear when entering the den (Atkinson and Ramsay 1995; Atkinson et al. 1996; Hilderbrand et al. 1999a), length of the denning season, and reproductive status (Hilderbrand et al. 1999a). Bears in poor body condition use more muscle mass relative to fat compared to fatter individuals (Figure 1 in Hilderbrand et al. 2000). Daily loss of mass in six adult female Alaskan brown bears was 352 ±136 g/day over a 208 ± 19-day period (Hilderbrand et al. 2000). Over the course of the winter,

TABLE 26.1. Estimated characteristics of grizzly bear populations in North America, with sample sizes in parentheses

Study Area	Density (Bears/100 km²)	Litter Size	Reproductive Interval[a]	Age at First Litter (years)	Weight (kg) Adult Male	Weight (kg) Adult Female	Cub Mortality Rate	Percent Adult Male[a]	Hunted?
Interior population									
East Front Montana	0.7	2.2 (41)	2.6 (11)	6.0 (4)	—	125	—	54	Yes
Flathead	8.0	2.2 (26)	3.1 (17)	6.1 (7)	176 (22)	114 (16)	0.18	37	Yes
Eastern Brooks	0.4	1.8 (13)	—	—	179 (26)	108 (31)	—	49	Yes
Alaska Range	1.5[b]	2.2 (36)	4.2[c] (38)	7.6[c] (8)	224[d] (24)	135[d] (32)	0.29	33	Yes
Nelchina	1.0	2.1 (64)	3.8[c] (44)	5.6[c] (24)	269[d] (12)	144[d] (21)	0.30	27[e]	Yes
Tuktoyaktuk	0.4	2.3 (18)	3.3[c] (8)	6.4[c] (10)	195 (16)	124 (36)	—	33	Yes
MacKenzie Mountains	1.2	1.8 (6)	3.8 (5)	—	148 (20)	110 (28)	—	—	Yes
Glacier National Park	4.7	1.7 (35)	—	—	—	—	—	—	No
Yellowstone 1959–1970	—	2.2 (173)	3.2 (68)	5.7 (16)	245 (33)	152 (72)	0.26	46	No
Yellowstone 1975–1989	—	1.9 (232)	2.6 (20)	5.7 (23)	193 (65)	134 (63)	0.15	55	No
Western Brooks	2.4	2.0 (6)	4.1[c] (16)	7.9[c] (14)	182 (26)	117 (35)	0.44	42	No
Kluane Park	3.7	1.7 (11)	—	7.7 (7)	145 (26)	98 (16)	—	—	No
Northern Yukon	2.8	2.0 (6)	4.0[c] (4)	7.0[c] (3)	173 (59)	116 (35)	—	51	No
Coastal population									
Kodiak Island	28.0	2.5 (29)	4.6[c] (41)	6.7[c] (12)	312 (10)	202 (16)	0.37	38	Yes
Alaska Peninsula	18.4	2.3 (200)	3.0 (81)	4.4 (9)	357 (21)	226 (63)	0.40	28	Yes
Admiralty Island	40.0	1.8 (32)	3.9 (7)	8.1 (7)	260 (10)	169 (18)	0.20	—	Yes
McNeil Sanctuary	—	2.2 (137)	3.8 (37)	6.5 (11)	257	160	0.31	55	No

SOURCE: Adapted from McLellan (1994).
NOTE: Cautious interpretation is necessary because variables have been collected in different ways among studies.
[a]Due to a variety of methods used in their derivation, comparisons must be done cautiously.
[b]The original estimate was for bears >1 year old. This value was adjusted to all bears by multiplying by 1.3 (Miller 1988).
[c]Includes incomplete intervals and births.
[d]Spring-only weights and adjusted by 1.28 for females and 1.24 for males.
[e]Adult sex ratio changed from 53% to 27% male during the study period due to intensive harvests.

these bears lost about 32% (±10%) of their body weight. These results are comparable to those in a study of captive nonlactating (335 g/day) and lactating (490 g/day) brown bears of similar mass (Farley and Robbins 1995). Weight loss during winter was highly variable for grizzly bears in a northern Northwest Territories study (Nagy et al. 1983b). Total weight loss during the denning period averaged 190 g/day during a 256-day period for two adult males (24% of body mass), but only 20 g/day for the same time period for a subadult male (5% of body mass). Five adult females lost on average 180 g/day over a 249-day period (30% of body mass) and subadult females lost 100 g/day (34% of body mass). Pearson (1975) documented an average of 200 g/day loss over a 220-day period in four grizzly bears in the southern Yukon. These animals lost 28–43% of their fall mass during the denning period.

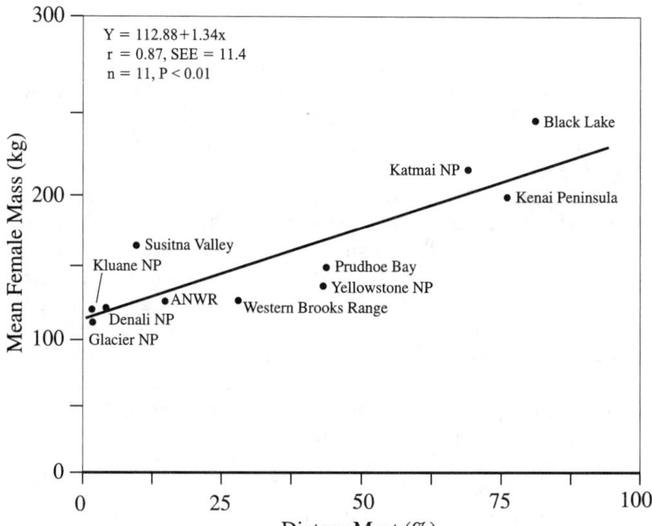

FIGURE 26.3. Body mass of adult female brown bears and the amount of meat in the diet. SOURCE: Data from Hilderbrand et al. (1999a).

Depending on availability and quality of spring forage, bears can continue to lose body mass until resources improve (Troyer and Hensel 1969; Craighead and Mitchell 1982; Blanchard 1987). Weight increases rapidly in the fall. Pearson (1975) measured a 410 g/day gain for a 126-day period for an adult male and a 640 g/day gain over 16 days for a subadult female in August.

Bears continue to grow throughout their life, but the sexes grow differently. Kingsley et al. (1983) fitted growth curves to age-specific data from spring and fall captures from northern Canada (Fig. 26.4). Although the curves are specific to that area, they illustrate general patterns in grizzly bear growth, seasonal weight dynamics, and the magnitude and allocation of weight gain through life. In that area, males took nearly 14 years to reach 95% of their maximum weight, whereas females required 9 years. The maximum rate of increase in basal weight was similar for the two sexes. When females divert resources into reproduction, they stop growing. Fall weight of males is related to spring weight, but increases approximately 28%; winter maintenance is remarkably constant at 22% (Fig. 26.5). Weight gain more than triples during the first summer, and declines continuously thereafter. Mature females cycle more weight than males, both relatively (Fig. 26.5) and absolutely (Fig. 26.6). Gain and loss in females continues to increase through maturity, until the oldest females cycle 70% of their spring weight. The relative gain and loss of weight in females exceeds that for males from the age of first reproduction onward. Females have greater weight fluctuations than males because females must expend energy in gestation and lactation (Kingsley et al. 1983). Several researchers (Troyer and Hensel 1969; Glenn 1980; Nagy et al. 1984; Blanchard 1987; Kingsley et al. 1988) established similar relationships of various body measurements with age and weight.

Pelage. Throughout their range, brown bears vary from light blond to black (LeFranc et al. 1987). Many specimens have silver or cream tipping on the guard hairs, creating a grizzled appearance; hence the origin of the name grizzly bear. Cubs in their first year are typically

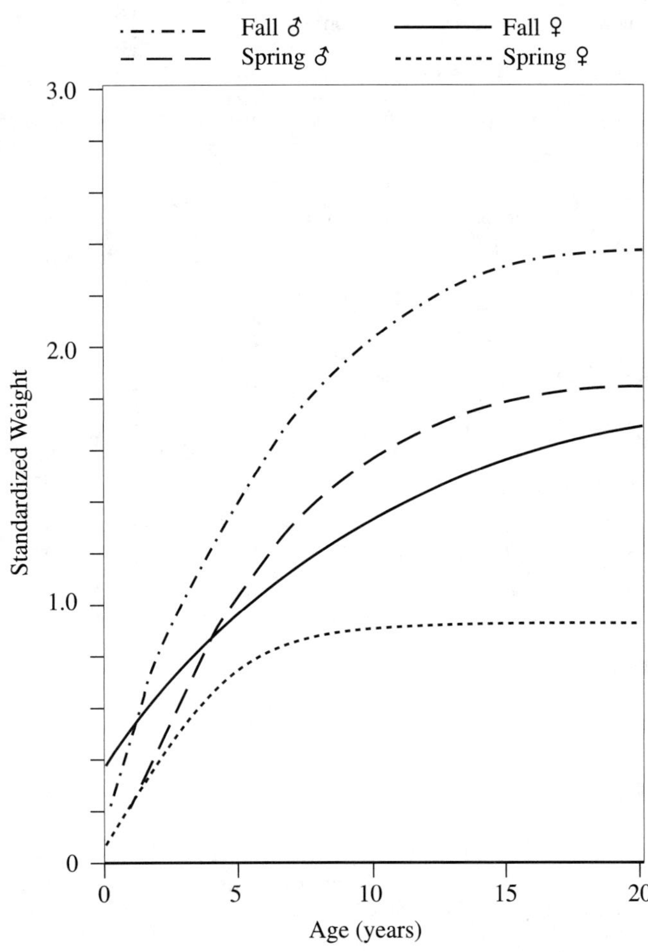

FIGURE 26.4. Growth curves depicting typical growth patterns in brown bears. Curves fitted to spring and fall weights; difference between spring and fall represents summer weight gain. SOURCE: Data from Kingsley et al. (1988) for northern Canadian grizzly bears.

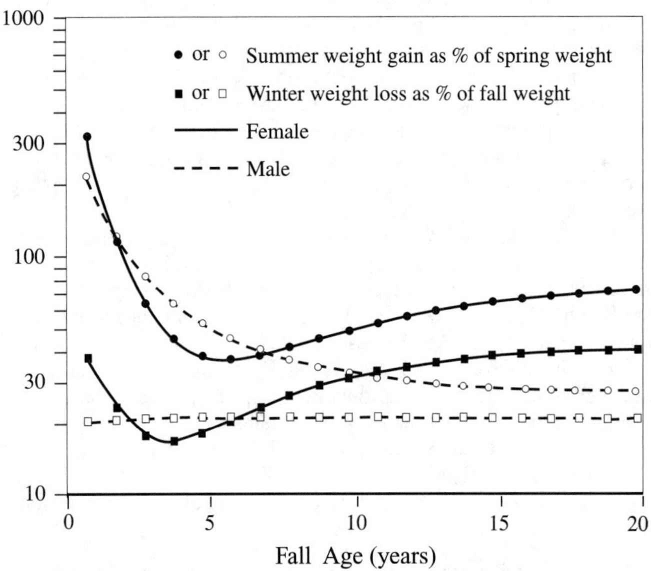

FIGURE 26.5. Annual weight gain and loss in grizzly bears as percentages (y axis), respectively, of spring and fall weight with respect to bear age. SOURCE: After Kingsley et al. (1988).

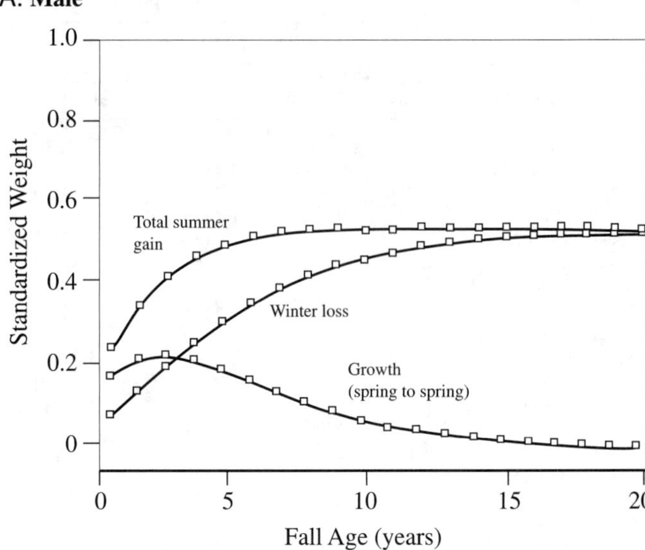

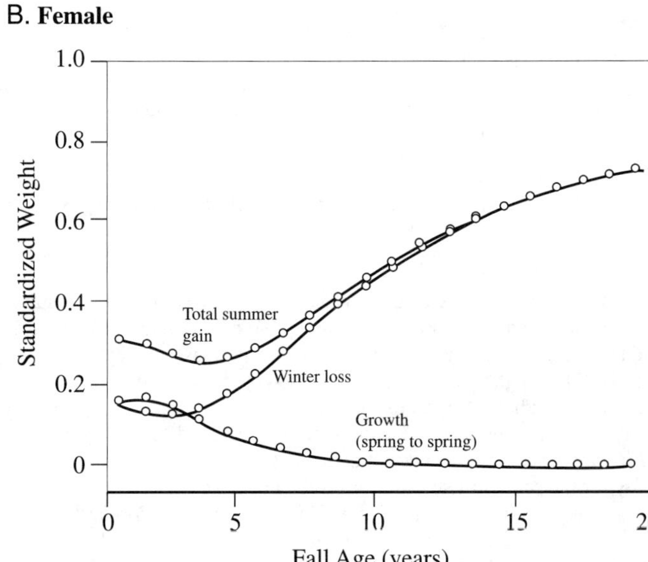

FIGURE 26.6. Magnitude and allocation of summer weight gain in (A) grizzly bear males and (B) females. SOURCE: After Kingsley et al. (1988).

brown with a natal ring of whitish hair around the neck and on the chest. The shade of color of a bear varies according to the direction from which the light strikes it relative to the position of the viewer. The bear appears darker when facing away from the light because of reduced reflection (Murie 1981).

Reynolds (1987) reported that color in Alaskan brown bears varies regionally and may be related to habitat use. Although all colors from blond to dark brown were found in most populations, lighter colors were more prevalent in open tundra habitats of the Arctic and interior Alaska. In areas where bears used darker, more forested habitats, grizzled, brown, and dark brown pelage was more frequent. Rausch (1953), Erickson (1965), and Quimby and Snarski (1974) supported his generalization. In central and arctic areas of Alaska, the general pattern is a light-colored head and shoulders; dark back, sides, and belly; and darker legs and feet. This color phase is sometimes called Toklat, after the Toklat River in Denali National Park. Overall, color varies from pale yellow to black. In coastal areas, such as the Alaska Peninsula and Kodiak Island, most bears are uniformly dark brown but exceptions are common.

In the boreal forests of Canada, LeFranc et al. (1987) noted that northern Alberta bears are mostly brown. Some have brown underfur with blond to white guard hairs on the head, shoulders, and back; legs are darker. Russell et al. (1978) considered about half the grizzlies in Jasper National Park as brown, with some yellowish tinge on the sides and back. The rest have prominent blond, yellow, or silver-tipped guard hairs on the sides, back, and neck. Heads are brown to yellow, the hump is darker than the head, and the legs are darker yet. In the Yukon, Pearson (1975) classed about 75% of the brown bears as brown, mostly chocolate, with grizzled silver or yellow guard hairs. The rest are blond to yellow, usually with a dorsal stripe along the back, and darker legs.

Knight et al. (1981) reported five major color patterns in pelt characteristics of Yellowstone grizzlies. The most prevalent have medium to dark brown underfur; brown legs, hump, and underparts; light to medium grizzling on the head and part of the back; and a light-colored girth band or patch behind the forelegs. Other patterns include (1) an overall gold or silver appearance and brown underparts, with an occasional dark back stripe; (2) no distinct silver tipping, giving a general black or brown appearance; and (3) medium to dark brown underfur, rump, legs, and hump, with medium to heavily grizzled forequarters and face. Subadults often appear multicolored with various shadings of red, blond, brown and great variation in silver tipping. Light-colored "yolks" on the chest and dark stripes on the back are common. These patterns fade as the bear matures into one of the four patterns described in adults.

Molt. Latitude, sex, and age influence molting of hair in the brown bear. Brown bears replace their hair annually. In general, adult males begin to molt first, followed by young males and other lone individuals; females with dependent young molt last (Pearson 1965, 1975; Quimby and Snarski 1974; Nagy et al. 1983a). Molt is generally complete by late July or August. Color, color pattern, and general appearance change markedly over time (Pearson 1965, 1975; Quimby and Snarski 1974; Nagy et al. 1983a). Quimby and Snarski (1974) found that dark-colored bears predominated in spring and fall, whereas lighter colors predominate during summer. They attributed these trends to differences in timing of emergence, sex-specific differences in color, bleaching, and observability. Rausch (1953) and Troyer and Hensel (1969) examined spring hides with rub marks, suggesting that molting may begin at emergence from dens; they noted substantially less rubbing in the fall.

Skull and Dentition. The skull of the brown/grizzly bear is highly variable across the North American range. It is stout and heavy (Fig. 26.7) and sexually dimorphic (Merriam 1918; Rausch 1953, 1963; Kurtén 1973; Craighead and Mitchell 1982; Pasitschniak-Arts 1993). Records of skull measurements from dead bears (Byers and Bettas 1999) provide potential maxima for the species. The largest skull length and width recorded for the Alaskan brown bear are 45.56 and 32.54 cm, respectively.

The skull grows and changes in dimension throughout life. Cubs have an oval-shaped skull, which lengthens during the active growth phase and reaches standard configuration at sexual maturity (Zavatsky 1976). Condylobasal length and zygomatic width are frequently measured skull characteristics, with the later continuing to increase after length is attained (Rausch 1963). Rausch (1963) presented the most comprehensive study comparing skull morphology throughout North America. Variation in mean condylobasal length is clinal with an increasing gradient along the coastal zone from British Columbia to the end of the Alaska Peninsula. A similar gradient was evident along the Arctic Coast. Bears from the interior are smaller.

The dental formula is I 3/3, C 1/1, P 4/4, M 2/3 (LeFranc et al. 1987; Pasitschniak-Arts 1993), however, some premolars can be missing (Glass 1974). Craighead and Mitchell (1982) incorrectly reported the dental formula for molars as 3/2. The skull of *U. arctos* can be distinguished from that of *U. americanus* based on molar measurements. The most accurate method (Gordon 1977) separates brown bears from black bears based on the first mandibular molar (M_1). A crown length greater than 20.4 mm or width greater than 10.5 mm indicates *U. arctos;* smaller measurements indicate *U. americanus.* This method showed no overlap for the two species in a sample of 128 skulls of all ages and both sexes. Grinnell et al. (1937) and Storer and Tevis (1955) separated the species based on the greatest crown length of maxillary molar 2 (M_2); it is seldom <38 mm in *U. arctos,* and seldom >31 mm in *U. americanus.*

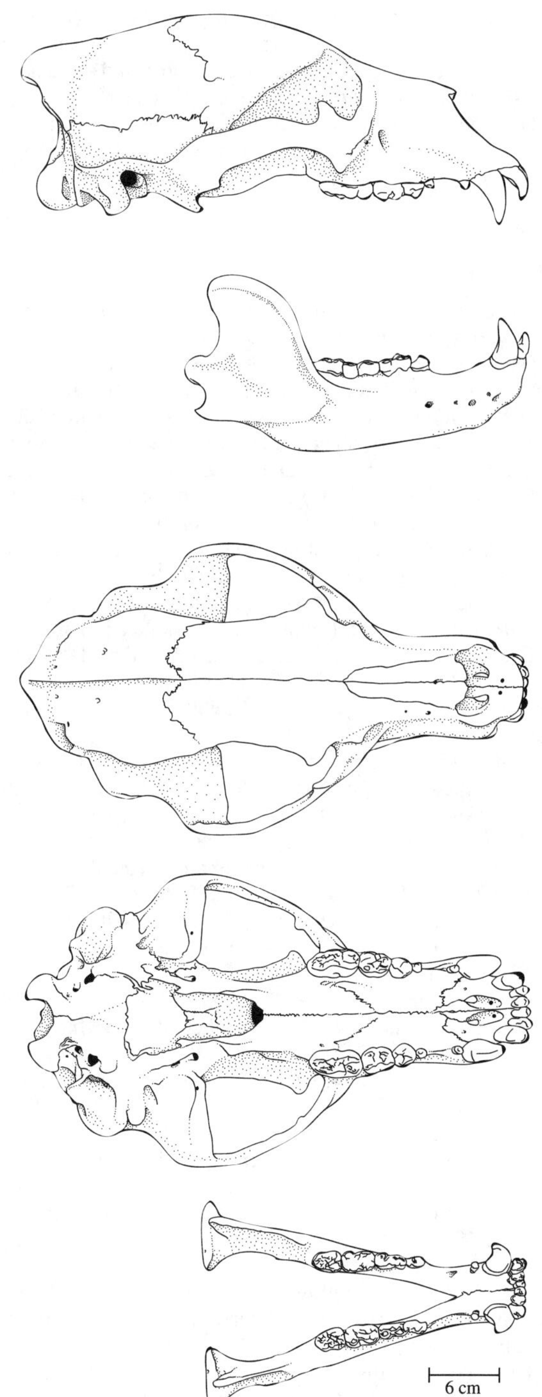

FIGURE 26.7. Skull of the brown/grizzly bear (*Ursus arctos*). From top to bottom: lateral view of cranium, lateral view of mandible, dorsal view of cranium, ventral view of cranium, dorsal view of mandible.

GENETICS

The six ursine bears (sun bear, *Helarctos malayanus;* American black bear; Asiatic black bear, *Ursus thibetanus;* brown bear; polar bear; and sloth bear, *Melursus ursinus*) have a nearly identical karyotype with

74 diploid chromosomes (Ewer 1973; Waits et al. 1999). These consist of 72 autosomes (60 acrocentric and 12 metacentric or submetacentric) and 2 sex chromosomes, a large metacentric X and small acrocentric Y (Pasitschniak-Arts 1993). According to Pasitschniak-Arts (1993) crosses between *U. arctos* and *U. maritimus* in zoos have produced fertile offspring (Davis 1950); hybrids of brown and black bears have been recorded, but the young died when just a few weeks old (Gray 1954).

PHYSIOLOGY

In general, bears exhibit the basic systemic physiology common to other carnivorous mammals (Bielanska-Osuchowska and Szankowska 1970). The digestive system of the grizzly bear is similar in form and function to that of canids and other ursids (Davis 1964). They possess a single stomach (monogastric), which constitutes about half of the total digestive capacity (Jaczewski et al. 1960). The intestines are short and nondifferentiated and are 5.1–7.7 times the total body length (Jaczewski et al. 1960; Bielanska-Osuchowska and Szankowska 1970). The cecum is either absent (Ewer 1973; Mealey 1975) or quite small (Jaczewski et al. 1960). Because bears are noncecal monogastrics, they cannot digest fiber efficiently (Bunnell and Hamilton 1983). Likewise, they cannot significantly increase fat reserves on foliage alone (Poelker and Hartwell 1973; Bunnell and Hamilton 1983; Eagle and Pelton 1983). Highly digestible, high-calorie foods are essential to their diet (Pritchard and Robbins 1990; Welch et al. 1997; Hilderbrand et al. 1999a). Foods pass through the digestive system rapidly, with meat, for example, taking about 13 hr and clover (*Trifolium* spp.) only 7 hr (Pritchard and Robbins 1990). Brown bears are unlikely to attain large body size consuming vegetable diets (Welch et al. 1997; Hilderbrand et al. 1999a; Jacoby et al. 1999; Rode 1999).

We know very little about the sensory system of the brown bear. The most acute sense is smell, but hearing and eyesight facilitate foraging and predatory behaviors (Pruitt and Burghardt 1977; LeFranc et al. 1987). The popular myth that bears do not see well is contradicted by our personal observations of grizzly bears observing other bears or humans from distances of 1–2 km.

Body temperature in the brown bear ranges from 36.5°C to 38.5°C when active, but declines 4–5°C during hibernation (Irving and Krog 1954; Folk et al. 1968, 1972, 1976; Nelson 1973; Follman et al. 1979; LeFranc et al. 1987). Resting heart rate is 40–50 beats/min in summer and declines to 8–12 beats/min during hibernation (Folk 1967; Folk et al. 1972, 1976). Metabolic rate (51 kcal/kg$^{0.75}$/day) (Farley and Robbins 1995) during the denning period is approximately 68–73% (Watts and Cuyler 1988; Watts and Jonkel 1988; Farley and Robbins 1995) of the interspecific basal rate for active mammals (70 kcal/kg$^{0.75}$/day) (Kleiber 1947); no measurements of metabolic rates are available for the nondenning season.

The most noticeable physiological difference between bears and other carnivores is their state of winter dormancy. Based on nearly 30 years of research by G. E. Folk, Jr., R. A. Nelson, and others, investigators have unequivocally stated that hibernation is the fitting term for the dormant or torpid state of bears during denning (Hellgren 1998). Nelson (1980) argued that bear hibernation represents the most refined response to starvation of any mammal. Bears exhibit continuous dormancy for up to 7 months without eating, drinking, defecating, or urinating (Craighead and Craighead 1972; Folk et al. 1972). However, others (Watts et al. 1981; Lyman et al. 1982; Pasitschniak-Arts 1993) do not consider winter denning in the bear to be deep hibernation because body temperature does not go below 15°C. Body temperature in small mammals considered deep hibernators (see Lyman et al. [1982:2] for definitions) decreases from around 39°C to below 10°C (Lyman et al. 1982), whereas the bear's body temperature only declines to 31–35°C. However, as demonstrated in simulations by Guppy (1986), the difference may be related to surface area:volume ratios. Bears depress metabolic rate to the same level as ground squirrels, but are not faced with problems of hypothermia. Bears likely have a more efficient torpor metabolism than the ground squirrel (Guppy 1986).

Female bears produce young during the denning period and face additional energetic costs of gestation and lactation during their winter fast. Ramsay and Dunbrack (1986:735) proposed that bears produce small neonates relative to their body size when compared to other large mammals to conserve maternal proteins. Brown bear cubs are born during January–March; they are altricial and generally weigh about 0.5 kg (Pasitschniak-Arts 1993).

Lactation is the most energetically costly mammalian process (Thompson 1992; Robbins 1993). Though lactation in many species occurs when food resources are abundant, brown bear cubs are born in winter when the female is fasting. Farley and Robbins (1995) examined milk composition, lactation characteristics, cub growth, and maternal mass changes for grizzly bears during the denning season. Composition of various constituents in milk varies through time, but when averaged over the lactation period, grizzly bear milk contains 1.3% ash, 33% dry matter, 18% lipids, and 2.3 kcal/g of energy. Grizzly bear milk is more concentrated than that of most terrestrial carnivores. It is similar in protein content to that of the polar bear and the black bear. Brown bear milk contains about half the fat and total energy of polar bear milk. Cubs consume relatively small amounts of milk (353 g/day) during the denning period; milk consumption increases rapidly after den emergence, peaks at midsummer (1350 g/day), and ceases by hibernation. The mass of milk consumed throughout lactation averages 224 kg/cub (Farley and Robbins 1995).

Lactating females lose body mass throughout hibernation. Mass loss for lactating females averages about 500 g/day, and is about 95% higher than for nonlactating grizzly bears of the same mass. Each kilogram of tissue lost by the lactating mother above normal hibernating costs results in 0.7 kg gained by the cub (Farley and Robbins 1995).

Most of the physiological studies of hibernation have been conducted with black bears. Where comparable data are available, it appears that the mechanisms are similar in the grizzly. For a detailed review, see Hellgren (1998).

Nelson et al. (1983a) described four behavioral and biochemical patterns in bears. Stage I, hibernation, has been described above. Stage II, walking hibernation, occurs after den emergence and lasts 10–14 days in the brown bear. During this period, bears are active, yet anorexic, with low intake of water and limited urine output, suggesting the biochemical stage of hibernation persists in part or in full after denning. Stage III, normal activity, lasts from May to September (this may be shorter for some populations; see Mattson et al. 1991a, 1994; Mattson 1997). During this period, bears cannot duplicate the hibernation phase. If deprived of food or water, they burn muscle tissue, suffer dehydration, and become uremic. Body mass increases during this phase, with most (78% in adult female Alaskan bears) as lean tissue (Hilderbrand et al. 1999a). Stage IV, hyperphagia, is the period of fat accumulation. Food intake rates increase and animals gain significant body mass, primarily as fat (81% in adult female Alaskan bears) (Hilderbrand et al. 1999a).

REPRODUCTION

Reproductive biology of the brown/grizzly bear is similar to that of the black bear (J. J. Craighead et al. 1995). Breeding occurs in late spring. The fertilized ova develop to the blastocyst stage and then arrest development. Implantation occurs in late November, followed by a 6- to 8-week gestation period and birth (Pasitschniak-Arts 1993). On average, females reach sexual maturity sometime between 4 and 7 years of age, and give birth to one to three cubs about every 3 years (Craighead and Mitchell 1982). Offspring remain with the female for 2–4 years before weaning.

There is some confusion in the technical literature regarding the term *cub.* Some use the term broadly to refer to all dependent young, whereas others use it narrowly to refer only to offspring <1 year old. Here we use the term *cub(s)* in the narrow sense, with age calculated from an assumed February birth date. Bears >1 but <2 years old are yearlings; bears >2 but <3 years old are referred to as 2-year-olds. Age at first reproduction, litter size, and interbirth interval vary among populations. These factors are linked to body size, which depends on nutrition (Stringham 1990; Hilderbrand et al. 1999a). The brown bear has a low reproductive rate relative to other mammals, a trait that critically affects survival in the presence of humans (Pasitschniak-Arts 1993; J. J. Craighead et al. 1995).

Early research into the reproductive biology of the species was based on field observation and examination of reproductive tracts from dead specimens (Craighead and Mitchell 1982). With the advent of radiotelemetry, biologists have been able to follow individual females through several breeding cycles. Such studies have provided more accurate insight into reproduction of the species and inherent variation among populations. Estimates of male reproductive success are possible with the development of DNA fingerprinting techniques (F. L. Craighead et al. 1995).

Breeding Season. The breeding season is narrowly defined as that period when copulation occurs, or more inclusively the period of male–female consorting, plus pre- and postcopulatory behavior (LeFranc et al. 1987). Variations among populations in breeding season chronologies are influenced by definition, length of study, numbers of observations, habitats, and biological differences among areas. However, it is nearly impossible to determine the exact date of conception under natural conditions, so no studies provide such detailed information. Data compiled from 20 different study sites across North America suggest that, on average, the breeding season (broadly defined) begins around mid-May and ends in early-July (Fig. 26.8).

J. J. Craighead et al. (1995) provided detailed breeding data from Yellowstone National Park during an 8-year period. Earliest date of observed copulation was 18 May and latest was 11 July, a period of 55 days. The period of observed copulation in any given year averaged 29 days with a range of 17–45 days. They predicted a mating season of approximately 63 days. Dittrich and Kronberger (1963) reported a mating season of approximately 72 days from captive brown bears. The earliest recorded date from the 20 North American studies (Fig. 26.8) was 21 April (courtship association), whereas the latest recorded was early August (breeding pairs). Average time between recorded start and end dates for the 20 reported studies was 49 days, with a minimum and maximum time for any one study of 25 and 92 days, respectively.

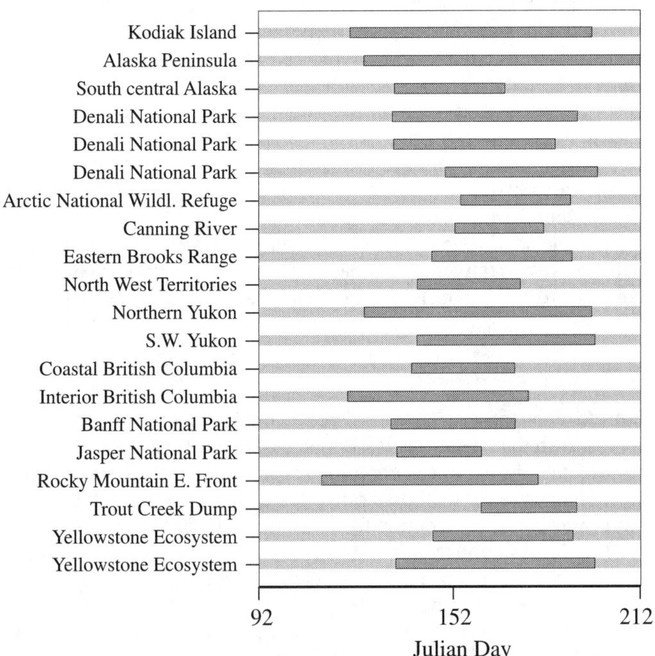

FIGURE 26.8. The period of male–female association, breeding plus postcopulatory association, for grizzly bear populations. Julian days 92 and 213 are the first day of April and last day of July, respectively. SOURCE: Data from LeFranc et al. (1987), Table 7.

Copulation. Copulation by grizzly bears is vigorous and prolonged (Craighead and Mitchell 1982). Vigor of the male, receptivity of the female, and privacy of the event (J. J. Craighead et al. 1995) influence duration. Probably the best data set available on observed copulation comes from Yellowstone National Park during the mid-1960s, when the open-pit garbage dumps were still operating (J. J. Craighead et al. 1969, 1995). The mean duration of 64 successful copulations ($\geq$10 min) was 24.3 min, with more than half being <24 min; the longest observed was 60 min.

Brown bears are promiscuous. Females mate with multiple males and may have a litter with offspring sired by different males; males can sire litters with multiple females in a breeding season (F. L. Craighead et al. 1995, 1998). Dominant males attempt to sequester a receptive female during her estrous period (Hornocker 1962; Herrero and Hamer 1977; Hamer and Herrero 1990; Brady and Hamer 1992). Plasticity is associated with this reproductive behavior. Mating can occur at concentrated food sources (Glenn et al. 1974; J. J. Craighead et al. 1995) or in poor-quality foraging sites (Herrero and Hamer 1977; Hamer and Herrero 1990; Brady and Hamer 1992). Pair bonds can last several weeks (Murie 1944; Herrero and Hamer 1977; Hamer and Herrero 1990) or may last only a few hours (Craighead et al. 1969). Females may enter estrus (defined here as the period of sexual receptivity) more than once (Dittrich and Kronberger 1963; Reynolds and Hechtel 1984; Reynolds 1989, 1992; J. J. Craighead et al. 1995). Not all breeding results in cub production the following spring, particularly in subadult females (Craighead and Mitchell 1982; J. J. Craighead et al. 1995).

Age at Puberty. Age of first litter production in brown bears varies widely geographically (LeFranc et al. 1987; Blanchard 1987; Stringham 1990; McLellan 1994), and is related to age at maturation and body size (Blanchard 1987; Stringham 1990), which is positively related to diet quality (Hilderbrand et al. 1999a) (Table 26.1). Nagy and Haroldson (1990), however, cautioned against interpreting body size–habitat relationships in the absence of information on population density. Age at first conception is estimated in brown bears by following subadult females through their first litter production. Age is generally determined from tooth sectioning, but in some cases can include known-age animals. However, the conventional method for calculating age of first conception, using only bears whose first litters are observed, gives a low-biased estimate (Garshelis et al. 1998). Cub production can be detected by actual observation of offspring or indirectly by examination of condition of mammae or ovarian structures (Stringham 1990).

Female brown bears do not reach sexual maturity until 3.5 years old (Hensel et al. 1969; Ballard et al. 1982; Craighead and Mitchell 1982; Aune et al. 1994), with some females producing first litters at age 4. In Yellowstone National Park, for example, from a sample of 15 females observed long enough to produce their first litters, 7, 5, 2, 0, and 1 produced first litters at age 5, 6, 7, 8, and 9 years of age, respectively. Mean age of first litter production from this sample was 5.9 years (J. J. Craighead et al. 1995). Mean age at first litter production varies from as low as 4.4 years for a growing population on the Alaska Peninsula (Miller and Sellers 1992) to as high as 8.1 years on Admiralty Island (Schoen and Beier 1990) (Table 26.1).

Litter Size. The number of cubs varies among individuals and populations but is typically one to three/litter. Litters of four are rare (Onoyama and Haga 1982; Bunnell and Tait 1985; Wilk et al. 1988; Sellers and Aumiller 1994; Case and Buckland 1998), but litters as large as six (Wilk et al. 1988) have been documented. However, adoption and/or exchange of cubs among different maternal females has been observed (Erickson and Miller 1963; Glenn et al. 1974; Barnes and Smith 1993), making empirical documentation based on field observations difficult. Mean litter size has been correlated with adult female body mass; intake of dietary meat, primarily salmon and ungulates (Bunnell and Tait 1981; Stringham 1990; McLellan 1994; Hilderbrand et al. 1999a); and garbage (Stringham 1986). Litter size also has been related to latitude (Bunnell and Tait 1981; Stringham 1984), climate, and a climate–carrion index (Picton 1978; Picton and Knight 1986); there are exceptions (Wielgus and Bunnell 2000). Litter size also is age related, with young and old females producing fewer cubs per litter than prime-age adults (Craighead et al. 1974, 1995; Sellers and Aumiller 1994). Reported mean litter sizes (Table 26.1) range from 1.7 to 2.5 cubs/litter across North America (Blanchard 1987; LeFranc et al. 1987; Stringham 1990; McLellan 1994), although smaller or larger means have been reported within a year or age class (Sellers and Aumiller 1994; Pac and Dood 1998). Litter size typically averages close to 2, and has less demographic significance than age at first parturition, interbirth interval, and cub survivorship.

Interbirth Interval. The interval between production of cubs is related to maternal nutrition and litter loss before weaning, but is generally ≥3 years in North America (Table 26.1). Females that lose either cub or yearling litters can have shorter interbirth intervals, but from a demographic standpoint this statistic may be misleading. Females that wean offspring as yearlings (Craighead and Mitchell 1982) can rebreed and produce their next litters at a shorter interval (2 years), which is demographically meaningful if the offspring survive to adulthood. For most females that successfully rear their offspring to weaning, the interval between litters is 3 years, but can extend up to 6 years (Craighead and Mitchell 1982; Stringham 1990; McLellan 1994; Miller 1997). Most cubs remain with their mother for 2.5 years, but some are weaned at 1.5 years (this is rare in North America) or remain as long 4.5 years (Reynolds 1976; McLellan 1994).

Method of calculation can influence the reproductive interval statistic. Inclusion of only females completing a reproductive cycle (litter to litter) results in an estimate that is biased low. This is because longer cycles are more likely to be missed because some animals die or have radiocollar failure before their next successful litter production. Interweaning interval, the time between successful weaning of offspring, is a more informative statistic from a demographic perspective, but is very difficult to quantify because of the number of years required to monitor each female. Inclusion or exclusion of incomplete intervals also influences calculation of this statistic. Miller (1997) calculated interweaning interval in Alaska at 3.2 years when including only complete intervals. The same statistic including only incomplete or both complete and incomplete was 5.0 and 3.5 years, respectively.

Depending on maternal body condition, age, and other factors, adult females may or may not produce their next cub litter the year following loss or weaning of offspring. In rich environments, females may be more likely to produce a litter in the year following litter loss or weaning, whereas in poor environments, they may require additional year(s) to replenish body reserves (Reynolds 1976; Reynolds and Hechtel 1980, 1984; Nagy et al. 1983b; Case and Buckland 1998).

Reproductive Rates. Natality can be expressed in various ways, but for large mammals it is common to express it as birth rate per unit of time or per female per unit time (Odum 1959). Natality can be used synonymously with maternity, although the latter generally expresses the number of offspring produced by an adult female in a given breeding season (Akcakaya et al. 1999). Natality is derived from litter size and interbirth interval, and so it varies among brown bear populations. Natality (cubs/female/year) varies from a low of 0.42 to as high as 1.07 (Stringham 1990), although the upper value was based on only four litters. A more recent review (Case and Buckland 1998) presents an upper value of 0.87.

Population ecologists are interested in the number of female offspring produced per reproductive female per year. Most assume an equal sex ratio at birth (Eberhardt 1990), although there are data suggesting a slight predominance of male cubs (55–59%) born in some populations (Craighead et al. 1974; Craighead and Mitchell 1982; Knight and Eberhardt 1985, 1987). The sex for 1326 cubs born in zoos was 51% male (U.S. Fish and Wildlife Service 1993:Appendix C).

Reproductive Longevity. Craighead and Mitchell (1982) and Pasitschniak-Arts (1993) indicated that reproductive longevity approximates physical longevity. A recent study, which compiled data from 20 geographically distinct areas across North America and Sweden, clearly demonstrates that reproductive senescence in brown bears occurs before physical senescence (Schwartz et al. 2003). Maximum per capita litter production occured at age 8.7 years and reproductive performance remained relatively high between about 8 and 25 years of age. Thereafter, productivity declined rapidly, with the rate of decline peaking around age 28 years.

Delayed Implantation. Grizzly bears exhibit obligate delayed implantation or embryonic diapause (Renfree and Calaby 1981). Studies in brown bears have described the presence of unimplanted embryos in the uterus several months after mating season (Craighead et al. 1969; Tsubota and Kanagawa 1993). Implantation is assumed to occur in late November to early December as based on changes in serum progesterone concentrations and fetal growth (Tsubota et al. 1987). For more information on delayed implantation in bears, see Chapter 25.

Male Reproductive Characteristics. White et al. (1998) provided an excellent review of the reproductive characteristics of male grizzly bears. Based on presence or absence of spermatozoa in the lumen of the seminiferous tubules, mean age of sexual maturity in a sample 20 grizzly bears from the continental Unites States was 5.5 years (White et al. 1998). The youngest bear with fully formed spermatozoa was 3.5 years old and killed in July. Only 1 of 11 bears that were ≤4.5 years of age had spermatozoa, whereas 8 of 9 bears ≥5.5 years of age did (White et al. 1998). Erickson et al. (1968) reported spermatozoa in seminiferous tubules of brown bears in Alaska at 4.5 years of age, whereas Pearson (1975) reported them in bears 5–7 years of age. In Hokkaido, Tsubota and Kanagawa (1991) concluded that sexual maturity in captive brown bears was reached between 2 and 5 years of age. As with females, sexual maturity in males is probably related to nutrition; variation among populations is expected.

Testicular mass is greatest during the breeding season, regresses by September, and is smallest by mid- to late October. By late September, sperm are no longer produced, although they may be present in the epididymides (Erickson et al. 1964, 1968; Pearson 1975). By mid- to late October or early November, testicular regression is nearly complete; by mid-November, the testicles are infiltrated by adipose tissue and loose fibrous connective tissue (Erickson et al. 1964; Pearson 1975; Tsubota and Kanagawa 1989). Testicular weights in early winter are the lowest found in mature bears through the year (Erickson et al. 1964, 1968; Pearson 1975).

Testicular recrudescence begins before den emergence, when seminiferous tubules enlarge and Leydig cell activity increases (Erickson et al. 1964). Spermatogenesis, with spermatozoa in the epididymides, occurs in bears in late May to the middle of July. Sperm are present at least 1 month before and several months after the breeding season (Dittrich and Kronberger 1963; Erickson et al. 1968; Tsubota and Kanagawa 1989).

Testicular growth is linearly related to age (Tsubota and Kanagawa 1991; White et al. 1998); seminiferous tubule diameter is curvilinearly related to age. Mean testicular mass, volume, and seminiferous tubule diameter are smaller in immature bears than in mature bears (White et al. 1998).

ECOLOGY

Habitat. Johnson (1980) considered habitat selection a hierarchical process, with four spatial scales, defined as orders. First-order selection included the physical or geographic range of a species; second-order selection operates at the home range scale within a geographic range. Third-order selection occurs at feeding sites within the home range, and fourth-order selection refers to specific foraging decisions. Most studies of brown bear habitat use focus on second- and third-order selection.

Brown bears occupy a variety of primary habitats (first-order selection) throughout North America, indicating relatively broad environmental limits (Craighead 1998). Their ability to effectively use vastly different landscapes can be attributed to their omnivorous generalist lifestyle and intelligence, which in effect translate to adaptability. Because the active season for brown bears is compressed to 5–7 months, during which bears must gain sufficient weight to supply their energetic needs for the next denning cycle, they tend to concentrate their activity seasonally in the most productive habitats available.

On the north slope of Alaska and the barren grounds of northern Canada, brown bears occupy a treeless landscape. In the central arctic, esker complexes and riparian tall shrub habitats were preferred by bears throughout the year (McLoughlin 2000). Bears in these regions rely extensively on herbaceous plants, roots, and berries when seasonally available (Gebhard 1982; Hechtel 1985; Phillips 1987). Meat from scavenging or predation on caribou (*Rangifer tarandus*), ground squirrels, and microtines also is seasonally important (Nagy et al. 1983b; Hechtel 1985; Phillips 1987; Gau 1998).

In Alaska and British Columbia, bears use a variety of habitats including old-growth forests, coastal sedge meadows, and south-facing avalanche slopes. During early summer, most bears use alpine and subalpine meadows. From midsummer through early fall, they move to coastal habitats and concentrate along streams to feed on spawning salmon (LeFranc et al. 1987; Schoen et al. 1994). Not all bears follow this typical pattern of habitat use; some do not visit salmon streams (Schoen et al. 1986), but remain in high-elevation habitats throughout the year. Mace and Waller (1997) observed that habitat selection often varies among individuals, even in an environment that appears consistently similar to humans. During late fall, bears alternately fish or use berry-producing habitats (LeFranc et al. 1987; Schoen et al. 1994).

Grizzly bears in the northern Rocky Mountains rely on a fairly predictable sequence of habitats that provide seasonally available forage. Seasonal habitats are often separated into (1) a spring/early-summer preberry period, when bears forage on a variety of locally available graminoids, forbs, and roots; and (2) a summer/early-fall berry-producing period when bears fatten on locally available berry corps (LeFranc et al. 1987; Mace and Waller 1997; Herrero et al. 2000). During spring, bears are generally in lower elevation habitats eating emergent vegetation and winter-killed ungulates. During late spring, they move to higher elevations following the phenological advance of vegetal foods. During summer, bears move to lower sites to exploit habitats with early-ripening berry crops. They repeat their altitudinal movements, following the ripening fruits to higher elevations during early fall (Darling 1987; Hamer and Herrero 1987; Mace and Waller 1997).

In the Greater Yellowstone Ecosystem (GYE), the pattern of seasonal elevation use is similar to that found for other populations occupying interior western mountains (Mealey 1980). During the spring, grizzly bear use of ungulates, both scavenged and as neonate prey, is extensive (French and French 1990; Gunther and Renkin 1990; Green 1994). The annual percentage of energy obtained from ungulate meat is considerably higher in the GYE than for other interior populations (Hilderbrand et al. 1999a). Use of ungulates abates during summer as bears use habitats that supply a variety of graminoids, forbs, and root crops (Mattson et al. 1991a). Yellowstone lacks significant berry-producing habitats. Consequently, bears use high-elevation sites to feed on whitebark pine (*Pinus albicaulis*) nuts (Blanchard and Knight 1991; Mattson et al. 1991a) and army cutworm moths (*Euxoa auxiliaris*) at insect aggregation sites (Mattson et al. 1991b; French et al. 1994).

In much of Alaska and northern Canada, habitats occupied by the grizzly bear are not significantly altered by humans. However, in the contiguous 48 states and some portions of southern Canada, most of the productive lands are dominated by humans. As a result, grizzly bear populations are relegated to "what's left," which usually constitutes the most remote and rugged mountainous areas; these may not represent what historically were "the best" habitats (Craighead and Mitchell 1982; Gibeau 1998). For bear populations in these areas, human settlement and alteration of the landscape limits habitat choices.

Home Range and Movements. Since 1970, movements and patterns of landscape use by brown bears have been investigated throughout North America (LeFranc et al. 1987). Movement patterns can be extremely variable within and among populations of brown bears. Movements are influenced by many factors, including key food items, breeding, reproductive and individual status (i.e., dominance), security, and human disturbance. Such factors dictate the pattern and extent of the landscape used throughout a season, a year, and the life of an individual, and define its home range (Burt 1943). It is generally believed that animals establish home ranges because it is more efficient to exploit familiar rather than unfamiliar areas (McLellan 1985).

Boulanger and White (1990) observed that use of different home range estimators could produce confusion in interpretation due to differences among the estimators themselves and not the behavior of the animal being studied. For brown bears, differences may also be influenced by sample size, which is typically small for wide-ranging bears (Nagy and Haroldson 1990). Most authors reporting brown bear home ranges used Mohr's (1947) minimum convex polygon method (Table 26.2); some lack sufficient locations to accurately estimate true home range size because the polygon method is sensitive to sample size (Gustafson and Fox 1983; Bekoff and Mech 1984).

More recently, kernel estimators (Worton 1989) have been employed to estimate home range extent for grizzly bears, with more attention paid to the adequacy of sample sizes (Blanchard and Knight 1991; Holms 1998; McLoughlin 2000). With the application of global positioning system technology, future knowledge of movements and range extent for brown bears will improve (Arthur and Schwartz 1999; Schwartz and Arthur 1999).

Though direct comparisons of home range statistics are difficult, several consistent patterns of grizzly bear home range size are evident. Craighead and Mitchell (1982) suggested that movements and range use by brown bears could be separated into two distinct patterns based on whether or not the population had access to high-quality food resources that concentrated individuals. Where brown bear populations have access to dependable, high-quality food resources, traditional patterns of movement to exploit them are well established. Average seasonal, annual, and life ranges for bears in these populations are typically smaller than those reported for populations that do not rely on dependable concentrated foods. For example, brown bear populations with access to rich salmon fisheries on the coast of Alaska have some of the smallest annual ranges observed in North America (Table 26.2). In contrast, annual ranges for brown bear populations in interior Alaska that do not use salmon were much larger. In the GYE, range sizes reported during years when bears were feeding extensively in open garbage dumps (Craighead 1976) were significantly smaller than those reported after dumps were closed (Blanchard and Knight 1991).

Differences in annual range size observed among study areas have generally been attributed to differences in habitat quality and distribution (Blanchard and Knight 1991). In support of this, McLoughlin et al. (1999) found a significant negative correlation between an index of primary productivity and grizzly bear home range size. However, Nagy and Haroldson (1990) speculated that social factors such as kinship, density, and population structure, all of which are influenced by turnover rates (human-caused or natural), may also affect range size observed in different regions.

Another consistent finding is that adult male bears typically have annual ranges that are several times larger than those observed for adult females (Table 26.2). This pattern usually is attributed to breeding activity of males (Blanchard and Knight 1991) or increased energy demand due to larger body size (Harested and Bunnell 1979; McLoughlin et al. 1999). Ranges of adult males overlap those of several females. During the 13-year study conducted by Blanchard and Knight (1991), multiannual or life ranges for most adult male bears did not plateau over time, but increased annually with additional radiotracking. Multiannual ranges of females were more likely to plateau at some maximum size (Blanchard and Knight 1991).

Seasonal ranges for specific sex and age classes of bears can be very restricted. Spring and early-summer ranges of females with cubs are often the smallest (Pearson 1975; Russell et al. 1979; Aune and Kasworm 1989; Blanchard and Knight 1991). This is attributed to the lack of mobility of young cubs and/or the need for security of cubs to reduce intraspecific predation. Sizes of late-summer and fall ranges, which coincide with the hyperphagic period of intense foraging (Nelson et al. 1983b), are usually more variable where key fall foraging opportunities are temporally and spatially unpredictable.

TABLE 26.2. Estimated mean home ranges of grizzly bears in North America

Study Area	Females		Males	
	Range (km²)	n	Range (km²)	n
Admiralty Island (Hawk Inlet), Alaska	24	12	115	6
Khutzeymateen River Valley, BC[a]	52	13	130	4
Kodiak Island, Alaska	71	33	185	6
Kluane National Park, Yukon	86	8	287	5
Revelstoke, BC	89	14	318	23
South Fork Flathead, Montana	99	2	286	5
Alaska Range	132	11	710	6
Mission Mountains, Montana	133	2	1398	3
Ivvavik National Park, Yukon[a]	149	15	447	8
Copper River Delta, Alaska[b]	174	4	295	2
Kananaskis, Alberta	179	5	1198	4
Akamina-Kishinena/Flathead, BC	200	5	446	5
Northern Yukon[c]	210	8	645	6
Western Brooks Range, Alaska	225	35	872	14
East Front Montana	226	3	747	5
Eastern Brooks Range, Alaska[c,d]	230	8	702	5
MacKenzie Mountains, NWT	265	6	—	—
Yellowstone National Park, Wyoming	281	48	874	28
Alaska Peninsula	293	30	262	4
Jasper National Park, Alberta[a]	331	6	948	6
West-central Alberta[e]	364	—	1918	17
Selkirk Mountains, Idaho	402	2	—	—
Upper Susitna River Basin, Alaska[c]	408	13	769	10
Tuktoyaktuk Peninsula, NWT[e]	670	—	1154	7
Noatak River, Alaska	993	33	1437	15
Anderson-Horton Rivers, NWT	1182	14	433	7
Central Northwest Territories	2434	35	8171	19

SOURCE: After McLoughlin et al. (1999).

NOTE: Ranges are primarily adult annual home ranges calculated using the minimum convex polygon approach unless otherwise indicated; weighted means were calculated if ranges were estimated with small or variable numbers of locations. Ordered on female home range size.

[a]Weighted means calculated from data presented.

[b]Cited in LeFranc et al. (1987:28–30).

[c]Estimate contains some multiannual ranges (Woods et al. 1997).

[d]Ranges calculated using the modified exclusive boundary technique.

[e]Weighted means cited in Nagy and Haroldson (1990). For females, data are presented as the midpoint between the mean for females with and without young, except for the northern Yukon, where the mean is only for females without young.

Except for subadult bears that may not have established permanent home ranges, female brown bears exhibit a high degree of range fidelity, especially during spring (Nagy et al. 1983a, 1983b; Aune and Kasworm 1989; Blanchard and Knight 1991). Fidelity to fall ranges is more variable due to unpredictability in abundance and location of fall foods. If key fall foods fail in areas where traditional use has occurred, bears must search out alternative food. During failure of key natural food items, the search for alternative foods often results in an increased number of bear–human conflicts and an increase in human-caused bear mortality (Blanchard 1990; Riley et al. 1994; Blanchard and Knight 1995).

At natural feeding sites (salmon streams) and unnatural sites (garbage dumps) where bears congregate, spacing is effected through intraspecific aggression and formation of dominance hierarchies (Hornocker 1962; Stonorov and Stokes 1972). Aggression arguably forms the basis for social organization in all bears and also probably functions to affect spacing among individuals not aggregated at concentrated food resources (Lindzey and Meslow 1977). Classical territorialism (Burt 1943; Brown and Orians 1970) has not been found in brown bear populations (Mace and Waller 1997), but this may be a difference in degree rather than kind of sociality, as both hierarchies and territories are manifestations of aggressiveness within a species (Fisler 1969).

Natal philopatry (Waser and Jones 1983) may be viewed as an extension of maternal care past the age of independence. Rogers (1977) postulated that by residing within maternal ranges, yearling black bears are buffered from social conflicts while they continue to mature. The same may be true for newly independent grizzly bears that continue to reside within their maternal range. Philopatry beyond 3 years of age in brown bears is sexually biased toward females. Dispersal of subadult males is common, whereas female dispersal is rare in these age classes (Glenn and Miller 1980; Blanchard and Knight 1991). Waser and Jones (1983) commented that sex-biased philopatry tends to be stronger in long-lived species whose adults are iteroparous. This trend is consistent with the views that sex biases reflect selection against inbreeding or that reproductive competition with parents discourages philopatry in one sex or the other (Waser and Jones 1983). Both views appear valid for differential dispersal among subadult brown bears. However, intrinsic rather than extrinsic factors probably influence male dispersal. Philopatric female offspring have the selective advantage of range familiarity (Waser and Jones 1983) and can occupy vacancies that may occur in adjacent habitats. Having close kin as neighbors may also decrease the cost of mutual tolerance (Waser and Jones 1983) and account in part for the considerable home range overlap observed among females (Mace and Waller 1997; Holms 1998).

Denning. Denning behavior in bears has been described as an elaborate bedding process that probably evolved as a result of adverse environmental conditions, primarily seasonal lack of food and unfavorable weather (Mystrud 1983). Nelson and Beck (1984) separated the physiological from the behavioral aspects of denning in black bears. They characterized denning as the physical act of reducing mobility

TABLE 26.3. Chronology of denning for brown bears in North America

Location	Latitude (°N)	Who Dens?	Denning Period[a]									
			Sep	Oct	Nov	Dec	Jan	Feb	Mar	Apr	May	June
NW Alaska	68	All		ooo•	oooo	++++	++++	++++	++++	++oo	o•o	
Central Alaska	62	All	o	o•oo	o+++	++++	++++	++++	++++	+ooo	•ooo	o
SE Alaska	57	All		oo•	oooo	oo++	++++	++++	+++o	oooo	•ooo	
NE Kodiak Island	57	Not all adult males		oo	•ooo	oooo	++++	++++	++oo	oooo	•ooo	oooo
SW Kodiak Island	57	All		oo	ooo•	oooo	++++	++++	oooo	oooo	•ooo	oooo
Banff NP, Alberta	52	—			oo++	++++	++++	++++	+++o	oo		
NW Montana	48	All			o•o	++++	++++	++++	+++o	o•o		
NW Montana	48	All		ooo	•ooo	o+++	++++	++++	+ooo	•ooo	oo	
Yellowstone NP	44	All		o	o•++	++++	++++	++++	++++	oo		
Yellowstone NP	44	All	o	oooo	o•oo	ooo+	++++	++oo	ooo•	oo		

SOURCE: Adapted from Linnell et al. (2000).
[a]Each month is divided into four quarters. Shown are (•) the quarters containing the average entrance and emergence dates, (o) the range of quarters in which bears began to den or emerge, and (+) the quarters during which all bears were denned.

and presumably conserving energy by entering a constructed or natural cavity, and hibernation as physiological adaptations that allow bears to survive for several months without food or water. This same distinction can logically apply to brown bears. Thus, as a necessary prerequisite to the behavioral aspects of denning, brown bears must first attain a hibernating physiology.

Physiologically, North American black, grizzly, and polar bears are true hibernators (Folk et al. 1976; Hellgren 1998). This condition allows bears to go up to 7 months without eating, drinking, defecating, or urinating (Folk et al. 1976; Nelson 1980). Yet female bears can support fetal development and lactation, as young are born in midwinter during the denning period (Nelson 1973). Unlike other true hibernators such as ground squirrels, bears can be aroused almost instantly for defense (Nelson 1973). Nelson et al. (1983a) reported that the physiological condition is not readily and/or intermittently attained in response to fluctuating weather and suggested that a neurocircumannual cycle is involved. Bears are generally thought to be in a physiological state of hibernation well before they enter dens in the fall. This is indicated by the predenning lethargy described by Craighead and Craighead (1972) and for the period (stage II, walking hibernation) after emerging from dens in the spring (Nelson et al. 1983b). Hellgren (1998) provided a good review of literature pertaining to the physiology of hibernation in black, brown, and polar bears.

A comprehensive summary (Table 26.3) of denning chronology for brown bear populations worldwide was compiled by Linnell et al. (2000). They reported that almost all brown bear populations studied in North America exhibit denning behavior. An exception occurs on a portion of Kodiak Island, Alaska, where >25% of radiocollared male bears remained active through at least one winter of a 6-year study (Van Daele et al. 1990). These males reportedly spent much of their time bedded, intermittently traveling short distances, and appeared to be in a state of "walking hibernation" (Nelson et al. 1983b).

Food availability and weather conditions are proximal factors that influence timing of den entry among most brown bears (Craighead and Craighead 1972; Van Daele et al. 1990). Den entry and duration also are somewhat correlated with latitude; brown bears in northern latitudes enter dens earlier and remain longer than bears at more southerly latitudes (Fig. 26.9). Pregnant females generally enter dens earlier and emerge later than other sex and age classes. Males are typically the last class of brown bear to enter dens in the fall and the first to emerge in the late winter or early spring (Linnell et al. 2000). Duration of denning may be as short as several weeks for adult males or as long as 7 months for females that emerge from dens with cubs. Females that emerge from dens with cubs may loiter near the den for several weeks (Craighead and Craighead 1972; Vroom et al. 1977).

Linnell et al. (2000) also summarized den and den site characteristics for brown bear populations worldwide (Table 26.4). The typical den documented for North America brown bears is excavated (Linnell et al. 2000), often under trees where root systems provide stability for the roof. Use of natural cavities or caves as dens has been observed less frequently, but is typical in study areas where natural structures are available, such as southeastern Alaska (Schoen et al. 1987). Van Daele et al. (1990:265) stated that "suitable den sites were those that remained dry throughout the denning period, and provided adequate soil depth and stability for excavation of a den or a suitable natural cavity." Thus, suitable den sites are probably not limiting in most populations of brown bears in North America; however, local exceptions may occur. Linnell et al. (2000) concluded that natural cavities were reused more often than excavated dens. Reynolds et al. (1977) and Miller (1990a) found that excavated dens in Alaska did not persist long enough for reuse to occur.

Specific sites and habitats chosen for dens are highly variable both within and among study areas, and show the considerable behavioral plasticity with regard to environmental condition exhibited by bears. Van Daele et al. (1990) concluded that brown bears likely used the most suitable denning habitat within their home range and local tradition plays a role in selection and construction. Habitats used for denning vary from open tundra to forested sites, depending on availability to local populations (Harding 1976; Vroom et al. 1977; Judd et al. 1986; Schoen et al. 1987; Van Daele et al. 1990). Selection of den sites with a

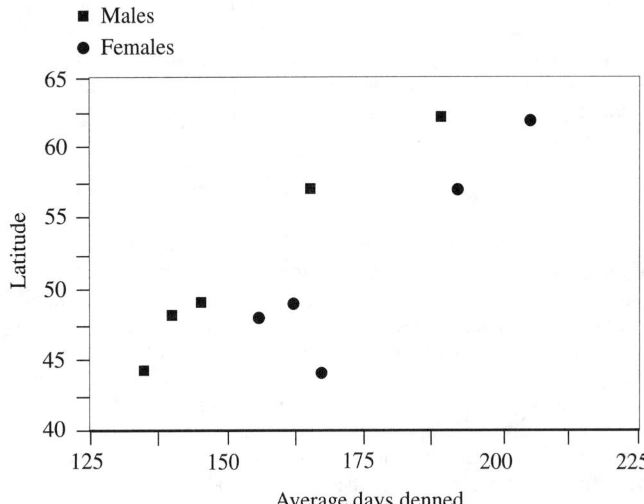

FIGURE 26.9. Average days denned relative to latitude for five interior brown bear study areas. SOURCE: Data from Aune et al. (1986), Judd et al. (1986), Schoen et al. (1987), Miller (1990a), Mace and Waller (1997).

TABLE 26.4. Topographic and habitat characteristics [mean (range)] and den construction type of brown bear dens in North America

Location	°N	Slope Use (deg)	Elevation (m)		Den Construction (%)[a]				Habitat Type (%)[b]				
			Used	Available	*n*	Exc	Cave	Sn	*n*	Tun	For	Eco	Alp
N Yukon	70	40 (20–80)	—	—	24	96	—	4	24	100	—	—	—
N Alaska	70	54	816	—	29	93	7	—	29	—	—	10	90
N Alaska	70	—	1063	—	49	70	30	—	—	—	—	—	—
NW Alaska	68	>30	500	0–1200	86	100	—	—	—	—	—	—	—
N Alaska	68	— (20–35)	—	—	52	75	25	—	—	—	—	—	—
Central Alaska	62	32 (11–60)	1200 (320–1626)	—	96	99	1	—	102	—	—	48	52
SW Yukon	62	35 (30–40)	1250	—	10	100	—	—	—	—	—	—	—
SE Alaska[c]	57	35 (5–75)	640 (6–1190)	1–1400	38	29	63	8	38	—	52	22	13
Kodiak Island (2 sites)	57	>45	450 (128–915) 665 (91–1189)	0–1000 0–1300	135	82	13	5	320	—	1	43	56
SW Alaska	57	40 (0–60)	450 (30–1000)	0–1200	30	96	—	4	30	—	—	50	50
Rocky Mts.	53	57	2147	—	60	93	7	—	60	—	—	100	—
Rocky Mts.	53	26	2057	—	—	—	—	—	24	—	42	8	50
Jasper NP	53	27 (15–40)	2236	—	10	90	10	—	—	—	—	—	—
Banff NP	53	33 (21–35)	2200 (2050–2300)	1300–3500	47	100	—	—	38	—	71	29	—
NW Montana	48	30 (21–35)	2124 (2050–2500)	850–3000	15	100	—	—	—	—	—	—	—
NW Montana	48	57 (51–62)	2166	1280–2800	—	—	—	—	—	—	—	—	—
Yellowstone NP	44	45 (30–60)	2470 (2000–3050)	1500–4200	33	91	6	3	55	—	100	—	—
Yellowstone NP	44	—	—	—	11	100	—	—	—	—	—	—	—

SOURCE: adapted from Linnell et al. 2000

[a]Exc, Excavated; Cave, natural cave or cavity; Sn, snow den.

[b]Tun, Tundra/muskeg; For, forest/swamp forest and shrub; Eco, forest/alpine ecotone; Alp, alpine meadows.

[c]Habitat type does not include rock.

steep slope relative to available slopes was a consistent pattern among studied populations. Aspect and elevation of den sites were much more variable within and among study areas. Female brown bears generally exhibit greater fidelity to den areas than do males (Linnell et al. 2000).

Security at den sites appears to be an important factor, especially if human disturbance occurs near the time of den entry. Craighead and Craighead (1972) observed that bears disturbed by their approach were more likely to abandon dens shortly after entry than they were during midwinter. Reynolds et al. (1986) observed increases in activity and heart rate and one instance of possible den abandonment by brown bears that were likely due to activities of seismic crews working near den sites. However, they concluded that the effects of these activities on denned bears were probably minimal. Mace and Waller (1997) did not observe any overt effects of snowmobiles within 2 km of dens in western Montana. They believed the greatest potential impact on bears was during spring when females with cubs were still confined to the vicinity of the den, and also after bears had moved to gentler terrain more suitable to use by snow machines. Predictable denning chronology and the behavioral plasticity bears exhibit toward den and den site characteristics suggest that potential human impacts on denning brown bears may be mitigated by careful consideration when implementing strategies for human activity.

FEEDING HABITS

As a group, bear species deviate from most other meat-eating members of the Carnivora by the volume and variety of vegetative foods in their diets. Comparing the three North American bear species, feeding habits of brown bears fall somewhere between those of the largely herbivorous black bear and the primarily carnivorous polar bear. Brown bears are opportunistic omnivores; few taxa, from insects to vertebrates and fungi to angiosperms, are overlooked as potential foods. Evolutionarily, brown bears have developed several adaptations for herbivory, including expansion of molar chewing surfaces and longer claws for digging. Nevertheless, they have maintained an unspecialized digestive system capable of digesting protein with efficiency equal to that of obligate carnivores (Bunnell and Hamilton 1983).

Most commonly, brown bear feeding habits have been quantified by analysis of scat contents. However, because of the differential digestibility of foods, contents of fecal residue are rarely equivalent to amounts of foods ingested by bears. The resulting underestimation of highly digestible foods is most pronounced for meat and fish diets (Hewitt and Robbins 1996). Fecal correction factors have been developed to convert results of scat analyses to actual volume of foods consumed; however, high variability in residues relative to methods of feeding reduces their utility (Hewitt and Robbins 1996). Recently, analyses of stable isotopes in hair and bone samples have been used to assess the relative importance of terrestrial animal, marine fish, and plant matter to brown bear populations (Hilderbrand et al. 1999a; Jacoby et al. 1999).

Major foods consumed by brown bears (LeFranc et al. 1987) can be grouped a variety of ways. Major categories, characterized by taxonomic group and method of acquisition, include (1) vegetative matter readily available by grazing, including graminoids, horsetails (*Equisetum* spp.), and forbs; (2) roots, corms, and bulbs acquired by digging, including hedysarum (*Hedysarum* spp.), biscuitroot (*Lomatium* spp.), glacier lily (*Erythronium* spp.), and yampa (*Perideridia* spp.); (3) fruits harvested from shrubs, including huckleberries/blueberries (*Vaccinium* spp.), buffaloberry (*Shepherdia* spp.), bearberry (*Arctostaphylos* spp.), and American devil's club (*Oplopanax horridus*); (4) whitebark pine nuts excavated from red squirrel (*Tamiasciurus hudsonicus*) middens; (5) insects harvested from nests or aggregation sites, including ants (Formicidae), wasps (Vespidae), army cutworm moths, and ladybird beetles (*Hippodamia casey*); (6) mammals and birds, acquired through predation or scavenging, including ungulates and rodents; and (7) fish acquired through predation or scavenging, including salmon and trout (*Oncorhynchus* spp.).

Grizzly bears commonly consume herbaceous vegetation during spring and early summer in many ecosystems. Even in areas with abundant meat or fish resources, grasses, forbs, and sedges can make up the majority of the diet in spring and early summer (LeFranc et al. 1987). In the northern Rocky Mountains (McLellan and Hovey 1995) and in captive feeding studies (Rode et al. 2001), brown bears selected forbs over grasses. This is likely because forbs retain more of their nutritional value longer than grasses with advancing phenology (McLellan and Hovey 1995). In captive feeding trials, small brown bears met their nutritional needs and gained weight on herbaceous diets. Very large bears had difficulty meeting their energy requirements on vegetation diets because of the combination of their absolute energy requirements and relatively small mouth. In the wild, male bears are more carnivorous than females (Jacoby et al. 1999). Meat eating by adult males provides the necessary calories to maintain a large body size, which leads to sexual dimorphism (Hilderbrand et al. 1999a).

Due to high digestibility and energy content, animal matter is arguably a highly valuable bear food (Welch et al. 1997; Hilderbrand et al. 1999a). However, a bear's ability to acquire these foods may be compromised by its size, its status in the social order, or the needs of its dependent offspring. Bears are most successful when feeding on animals that are abundant and vulnerable to their predatory skills. Bears inhabiting the coastal regions of Alaska and British Columbia commonly feed on spawning salmon, often centering their activities at falls, where upstream movement of fish is impeded. Under these circumstances, bears can be quite efficient predators. Many bears have access to this high-quality food for nearly the entire active season because of extended availability afforded by sequential runs of several salmon species. At Karluk Lake, Alaska, brown bears killed up to 79% of salmon migrating upstream (Gard 1971). In other coastal areas, bears may feed on postspawning salmon with little impact on the salmon run (Clark 1957). For some interior bear populations, trout provide a high-quality seasonal food. In the GYE, an estimated 30–50 grizzly bears forage annually on spawning cutthroat trout (*Oncorhynchus clarki*) in tributary streams of Yellowstone Lake (Reinhart and Mattson 1990).

In contrast to coastal environments with anadromous fish, meat is much less available and more difficult to obtain for interior brown bear populations. Use of ungulates as prey and carrion is common and seasonally important. Following spring emergence, brown bears feed on winter-starved ungulates including caribou, moose (*Alces alces*), elk (*Cervus elaphus*), and bison (*Bison bison*). Bears can also be effective predators. In early summer, neonates are actively hunted. Moose, caribou, and elk calves are seasonally important foods (Ballard et al. 1981; Larsen et al. 1989; Gunther and Renkin 1990; Hamer and Herrero 1991; Green et al. 1997; Mattson 1997; Gau 1998). Marine mammals, rodents, and ground-nesting birds and their eggs are eaten when available (Nagy et al. 1983b; LeFranc et al. 1987).

In the southern Rocky Mountains, army cutworm moths and ladybird beetles are valuable seasonal foods (Klaver et al. 1986; Mattson et al. 1991b; White 1996). Bears forage on moths in the talus where they are vulnerable to predation. Studies from Glacier National Park (White et al. 1999) indicate that a foraging bear can consume as many as 40,000 moths/day, ingesting approximately 20,000 kcal. These insects are high in lipid content (Kevan and Kendall 1997) and represent one of the most calorie-rich foods consumed by bears (White et al. 1999). Cutworm moth aggregation sites can attract large numbers of bears (French et al. 1994), but are geographically limited in North America.

Fruits of blueberries, huckleberries, buffaloberry, devil's club, bearberry, and other species are seasonally important foods for bears throughout much of their range in North America. High carbohydrate content makes berries important summer and fall foods. When available, bears spend up to 50% of the day foraging on berries; foraging efficiency is related to fruit abundance, size, and distribution (Welch et al. 1997).

Roots, corms, and bulbs are commonly used by bears in the Rocky Mountains and interior Alaska. Roots of hedysarum (*Hedysarum* spp.) are dug in all mountainous and arctic habits of Canada and Alaska, but are not a major diet item south of Canada (LeFranc et al. 1987). Here, biscuitroot, glacier lily, and yampa are seasonally important. These foods are typically higher in starch and digestible energy than herbaceous foods. They can serve as alternate fall foods during years when berry crops fail.

Whitebark pine nuts are an important fall food wherever the species is abundant in the contiguous United States (Mattson et al. 1991a; Mattson and Reinhart 1997). Almost all seeds consumed by bears are excavated from the middens of red squirrels (Mattson and Reinhart 1997). Pine nuts are high in fat and one of the most energy-rich foods consumed by bears. When abundant, they use pine nuts to the exclusion of most other foods. Unfortunately, whitebark pine has been eliminated or significantly reduced over much of its former range by an exotic fungus, white pine blister rust (*Cronartium ribicola*) (Kendall and Arno 1990). Most stands persist in the GYE where the climate is dryer. However, even there, rust is present and spreading (Smith and Hoffman 1998).

Geophagy, the purposeful consumption of soils, has been documented in the GYE (Mattson et al. 1999a). Soils consumed were high in potassium, magnesium, and sulfur. This behavior peaked primarily during March–May and secondarily during August–October and occurred during peak consumption of ungulate meat and mushrooms. Mattson et al. (1999a) speculated that bears were consuming soils to remedy potassium deficiencies incurred during hibernation, stimulate motility, and reduce parasites and harmful bacteria in the intestines.

Anthropogenic foods (i.e., garbage, livestock feed, pet food, bird seed, human foods, garden crops, honey) are used by brown bears wherever humans and bears coexist (Herrero 1985). Open garbage dumps can be a source of highly nutritious foods when available. Use of dumps can lead to food conditioning, habituation, and increases in property damage and human-caused bear mortality. In the GYE, considerable effort has gone into eliminating availability of anthropogenic foods (Meagher and Phillips 1983). These efforts have been largely successful in reducing incidents of bear–human conflicts. Here and in other regions where bears and people live in close proximity to one another, most conflicts occur during years when important natural foods fail (Blanchard 1990; Riley et al. 1994; Blanchard and Knight 1995).

DEMOGRAPHICS

Sex and Age Composition. Constructing the sex and age composition of a grizzly bear population is difficult. Sample method, sample size, number of years of study, sightability, natural variation during the study period, human harvest, age of adults, and other factors all influence estimates. Capture records and visual observations are biased by differential capture and sighting probabilities. Harvest records can be biased by selective harvest regulations (protection of females with offspring) and differential vulnerability of different gender and age classes to harvest. With ground-based trapping operations and helicopter capture, potential biases exist due to heterogeneity of capture for certain age–gender classes of bears (Miller et al. 1997). For example, adult females with cubs tend to be underrepresented in samples because of their secretive nature (Miller et al. 1987, 1997). Aerial observations can be subject to error and misclassification, with certain groups of bears underrepresented and others overrepresented (Erickson and Siniff 1963; Dean 1987; O'Brien and Lindzey 1998). At best, reconstruction of sex and age composition for grizzly bear populations based on field observations and capture records is an approximation. LeFranc et al. (1987) provides a summary of gender and age composition from several populations in North America. Many studies are constrained by small sample sizes. Miller (1997) provided a "weighted snapshot" estimate of population composition designed to reduce bias associated with different rates of movements between males and females.

A sample of multiple-year studies and relatively large sample sizes suggests ratios among adults, subadults, yearlings, and cubs vary widely (Table 26.5). The proportion of cubs in any population is a reflection of reproductive performance and early mortality, and should in general be higher for more fecund populations. Cub production varies yearly (Craighead et al. 1974), so as multiple-year sampling increases, a more accurate picture of age structure emerges. As expected, yearlings usually make up a smaller proportion of the population than cubs due to mortality during the first year of life. The proportion of adults, particularly adult males from populations that are harvested, tends to be lower than from unharvested populations (Miller 1990b). Ages of males and females in harvested populations are younger and older, respectively, with intensive harvest (Miller 1990b), although not in all cases (Miller 1997).

The sex ratio in bear populations tends to be skewed toward females, particularly in harvested populations (Table 26.5). Although sex ratio at birth can favor males (see Reproductive Rates), in general,

TABLE 26.5. Age and sex composition of selected grizzly bear populations in North America with multiple years of study

Location	Years of Study	Cubs (%)	Yearlings (%)	Subadults (%)	Adults (%) (age, years)	M:F Adult Sex Ratio	Hunted?	Reference
Alaska Peninsula	5	25.0	15.0	34.3	25.6 (>4)	15:85	Yes	Glenn 1975
Northwest Territories	3	15.0	8.0	18.0	59.0 (>4)	37:63	Yes	Clarkson and Liepins 1994
Northwest Territories	4	15.6	13.3	34.6	36.5 (>6)	31:69	Yes	Nagy et al. 1983b
Southeast British Columbia	8	21.5	17.5	26.5	34.5 (>5)	38:62	Yes	McLellan 1989a
Swan Mountains, Montana	7	16.9	10.5	27.2	48.2 (>4)	32:68	No	Mace and Waller 1998
Yellowstone National Park	9	18.7	13	25.1	43.2 (>4)	46:54	No	J. J. Craighead et al. 1995

males have a lower survival rate. Differential mortality is apparent between the genders following weaning. These differences are due to gender-linked behavioral characteristics including dispersal, denning chronology, home range size, and vulnerability to harvest (Bunnell and Tait 1980). Sex ratio of the adult population is skewed toward females (Tables 26.1 and 26.5) in heavily harvested populations such as the Black Lake area on the Alaska Peninsula (Glenn 1975) and nearly equal in unharvested populations as in Yellowstone National Park (J. J. Craighead et al. 1995). Variation in sex ratio and age structure among populations is primarily driven by differential mortality among the various gender–age classes and is discussed below.

Survival. Survival in bears is estimated in various ways. In some studies, a simple ratio of animals dead to alive at the end of a study or year is used. If a sample of animals is marked at the start of a period of interest, calculation of survival and cause-specific mortality rates as simple percentages is appropriate (Hessler et al. 1970). More often, however, animals are radio-marked at different times or even tracked during periods for which survival rates differ; simple ratios applied to such sampling lead to serious biases (Heisey and Fuller 1985). As a consequence, more sophisticated analytical techniques have been developed to correct such biases (White and Garrott 1990). One commonly applied method is the Kaplan–Meier or product limit estimator (Kaplan and Meier 1958). The Kaplan–Meier approach is simple and flexible and allows for staggered entry of newly tagged animals. Although this approach is widely used to estimate survival, it is not uniquely the best in all circumstances (Pollock et al. 1989).

Many estimates of survival rates in grizzly bears are generated with the Kaplan–Meier approach, making it easier for comparisons among populations. A sample of recent studies (Table 26.6) indicates that annual survival of adult females is usually quite high (≥0.90%). Survival of adult males varies among populations, but is generally lower in hunted populations (Tables 26.1 and 26.6). Subadult female survival is also high, normally equal to or slightly less than adult female survival. Subadult male survival can be quite variable, but tends to be lower than that of the other independent bears. Survival of dependent young is lower than adults; yearling survival is usually greater than that of cubs.

Cub survival is generally estimated by tracking the fate of cubs of radio-marked females. Because cubs are seldom collared, actual causes of mortality are difficult to document. Cub mortality is highly variable (Tables 26.1 and 26.6) and can be as low as 13% or as high as 44%. Modafferi (1984) reported 31% mortality for cubs on the Alaskan Peninsula between the ages of 0.5 and 1.5 years. From 1978 to 1991, 31% of observed cubs at McNeil River Bear Sanctuary disappeared between 0.5 and 1.5 years of age (Sellers and Aumiller 1994). Where litters were typically observed shortly after den emergence in Alaska, reported cub mortality was 33% in south-central Alaska (Miller 1988), 29% in the north-central Alaska Range (Reynolds 1993), and 37% on Kodiak Island (Smith and Van Daele 1991). Cub mortality was estimated at 9% in the Northern Continental Divide ecosystem (Aune et al. 1994).

Management agencies often resort to translocation to reduce human-caused mortality associated with problem bears. Relocating grizzly bears from human–bear conflict situations is often a short-term solution to an immediate crisis because many bears return to the conflict site (Judd and Knight 1980; Miller and Ballard 1982). Blanchard and Knight (1995) found that survival rate of transported bears was

TABLE 26.6. Kaplan–Meier survival estimates for several North American grizzly bear populations

Location	Adult Male	Adult Female	Subadult Male	Subadult Female	Yearling	Cub	Age of Adults (years)	Reference
Noatak, AK	0.91	0.94	—	0.89[a]	0.89[a]	0.87	≥5	Ballard et al. 1991
Nelchina, AK	0.80	0.92	—	—	—	0.69	≥5	Miller 1990c, 1997
McNeil River, AK	0.94	0.93	—	—	0.89[b]	0.67[b]	≥5	Sellers and Aumiller 1994
Mountain Parks, Canada[c]	0.89	0.91	0.74	0.95	—	—	≥6	McLellan et al. 1999
Flathead River, British Columbia	0.92	0.94	0.91	0.94	0.88	0.82	>5	McLellan 1989b
Blackfeet–Waterton[d]	0.62	0.92	0.80	0.86	—	—	≥6	McLellan et al. 1999
Selkirk–Yaak[e]	0.84	0.95	0.81	0.93	—	—	≥6	McLellan et al. 1999
North Fork Flathead, MT	0.89	0.96	0.78	0.94	—	—	≥6	McLellan et al. 1999
South Fork Flathead, MT	0.89	0.89	0.78	0.87	—	—	≥6	McLellan et al. 1999
Swan Mountains, MT	0.87	0.90	0.83	0.83	0.90	0.77	≥5	Mace and Waller 1998
Rocky Mountain East Front, MT	0.811	0.94	0.66	0.92	0.82[f]	0.82[f]	≥5	Aune and Kasworm 1989
Greater Yellowstone Ecosystem	—	0.94	—	0.80[g]	—[g]	0.84	≥5	Eberhardt 1995

NOTE: Survival generally was estimated from radio-collared bears; cub and yearling survival was estimated for most studies by observing marked females with offspring.

[a]Calculations of survival for yearlings combined males and females.

[b]Early survival prior to arrival at the sanctuary not recorded. Estimate is high relative to survival estimated from den emergence to den entrance in other studies.

[c]Includes Jasper, Cascade Valley, Eastern Slope, Upper Columbia, Yoho-Kootenay, and Kananaskis areas that are Canadian national and provincial parks.

[d]Waterton and the Blackfeet Indian Reservation are adjacent and some bears moved between study areas; they were pooled.

[e]The Cabinet–Yaak Ecosystem and Selkirk Mountains encompass ecosystems in both the continental United States and Canada. Although geographically distinct, their management goals were similar and were combined to improve sample size.

[f]Cubs included with yearlings.

[g]Yearlings included with subadults.

TABLE 26.7. Cause-specific mortality (%) from a sample of grizzly bear studies in North America

Number of Deaths	Natural	Hunter Harvest[a]	Citizen Killing[b]	Management Control[c]	Accident[d]	Unknown	Location	Reference
22	4.8	81.0	—	—	14.2	—	Noatak, AK	Ballard et al. 1991
14	28.6	64.3	7.1	—	—	—	Northwest Territories	Clarkson and Liepins 1994
10	—	60.0	30.0	—	10.0	—	Kananaskis Country, AB	Carr 1989
83	16.9	19.3	36.2	12.0	2.4	13.2	Interior mountains of Canada and United States	McLellan et al. 1999
38	15.8	50.0	26.3	2.6	5.3	—	Flathead River, BC	McLellan 1989b
35	28.5	2.8	34.3	17.1	2.8	14.3	Swan Mountains, MT	Mace and Waller 1998
43	11.6	25.6	27.9	32.5	2.3	—	Rocky Mountain front, MT	Aune and Kasworm 1989:213
365	1.6	29.3[e]	19.7	39.2	3.0	7.1	Yellowstone Ecosystem (1959–1972)	Craighead et al. 1988
145	13.8	8.3[f]	42.8	24.8	6.9	3.4	Yellowstone Ecosystem (1973–1985)	Knight et al. 1988

NOTE: Data include known and probable deaths, except in the Greater Yellowstone Ecosystem, which includes possible deaths.
[a]Hunter harvest includes only bears harvested legally during a sport-hunting season.
[b]Citizen killing includes defense-of-life or property killing, poaching, mistaken identification, and malicious killing. In some cases, killing of bear for defense of life or property is legal.
[c]Management control represents removal of problem bears by agency staff.
[d]Accident includes train and automobile kills, electrocution, and research deaths.
[e]Legal hunting ended in Montana and Wyoming in 1973 and 1974, respectively. Management control includes humane removals and trap casualty.
[f]Data span 1973–1985. Legal hunting occurred in 1973 only.

83%; survival for nontransported bears was 89%. Survival was largely affected by whether the bear returned to the capture site; return rates were most affected by distance transported and age and gender of the bear. Return rates decreased at distances of ≥75 km, and subadult females returned the least. Because of low survival and high return rates, transporting grizzly bears should be considered a final action to eliminate a conflict situation. However, transporting females must be considered a viable technique because some translocated females have contributed to the population through successful reproduction.

Causes of Mortality. Bears die for a number of reasons, primarily human related (Table 26.7). Natural mortality can result from old age, intra- and interspecific killing, starvation, rock or snow avalanche, den collapse, or unknown reasons. Natural mortality constitutes a greater proportion of total mortality for dependent young (Nagy et al. 1983b). Cubs and yearlings are killed by conspecifics, although the cause of mortality in dependent young is often unknown because few are radio-collared; loss of dependent young from marked mothers is generally considered mortality. McLellan et al. (1999) found different mortality rates due to natural causes among gender–age classes, with adult females having a higher rate than adult or subadult males. Work by Mace and Waller (1998) supports this.

Hunting, management removal, and defense of life and property by citizens can constitute as much as 90% of all recorded mortalities for adult bears (Table 26.7). Even in areas with no hunting, human-caused mortality dominates. Deer and elk hunters killing grizzly bears in self-defense, hunters mistaking a grizzly bear for a black bear, and malicious killing are major causes of bear deaths in Montana (Craighead et al. 1988; McLellan et al. 1999). Agency removal of problem bears either by euthanasia or relocating to zoos and shooting by citizens protecting livestock, homes, and campsites constitute a major mortality factor in many areas (Table 26.7).

Most bears die during the nondenning season. Although an occasional mortality is documented during winter (McLellan et al. 1999), most deaths occur when bears are active. Aune and Kasworm (1989) and Mace and Waller (1998) found that most grizzly bears in Montana died during autumn. Natural mortality was prominent during spring and summer, whereas management removal was the primary cause of loss during autumn. Mortality due to mistaken identification by black bear hunters was the leading cause of subadult female mortality. Adult males were most likely to die during ungulate hunting season in defense-of-life killings by hunters. Subadult males were equally susceptible to malicious killing and mistaken identification (Mace and Waller 1998).

Because most bears are killed by humans, proximity of kills to human facilities and access routes (roads, trails, back country sites) are common. Aune and Kasworm (1989) found that of 43 grizzly bear mortalities on the Rocky Mountain front, 63% occurred within 1 km of the nearest road. Knight et al. (1988) found that the majority of grizzly bear deaths in the GYE were clustered near foci within and on the periphery of Yellowstone National Park. Major population sinks included communities such as West Yellowstone, Cooke City, and Gardiner, Montana; recreational developments, sheep grazing allotments, and various other human concentration areas. Also, diverse attractants such as apple orchards, outfitter camps, and locations where people have persistently fed individual bears or unlawfully disposed of garbage enticed bears into conflict situations, especially during periods of natural food shortage. Hunter harvest also tends to be greater in areas with enhanced human access (Miller 1990b). On Chichagof Island in southeastern Alaska, increased cumulative miles of road construction was strongly correlated with fall brown bear harvests from 1978 to 1989 (Titus and Schoen 1992). This happened even after closure of hunting seasons, because of defense-of-life and property kills and illegal kills (Titus and Beier 1991; Schoen et al. 1994).

Grizzly bears, like most other animals, are afflicted with an array of parasites and diseases (LeFranc et al. 1987). Occasionally a bear succumbs to such ailments, but documenting cause of death is difficult, particularly under natural conditions. Animals carrying a heavy load of parasites can die from starvation, malnutrition, or in a conflict situation. The parasite may ultimately be the cause of their demise, but the proximal cause may differ. We are unaware of a documented major die-off in a grizzly bear population linked either to parasites or diseases.

Intraspecific Killing. On occasion, grizzly bears kill one another. Adult males have been implicated as the killers in nearly 78% of the 27 documented cases where the age and gender of the killer is known (McLellan 1994). Of 57 cases of intraspecific killing, cubs of the year are the greatest victims (44%, $n = 25$), but adult females are also killed (18%, $n = 10$). Some adult female victims are protecting their cubs. Victims are of all age and sex classes, indicating that intraspecific killing is not limited to infanticide (McLellan 1994). Adult females have also been implicated in killing cubs (Hessing and Aumiller 1994). In 10 cases where age and gender of the killer were known, adult females were implicated in 5 (McLellan 1994).

There are two competing theories on the impacts of intraspecific killing in bear populations (Miller 1990c, 1990d). One suggests that greater mortality of adult bears will result in increased survival of young bears, particularly cubs. Although some studies have demonstrated a negative relationship between recruitment of subadults and number of adult male bears (McCullough 1981, 1986; Stringham 1983), Stringham (1983) and others (Miller 1990c; Garshelis 1994; McLellan

1994) caution against density-dependent interpretation until the effects of nutrition and other confounding factors can be distinguished.

The second theory proposes that conspecific killing of unrelated cubs by adult male bears may increase male fitness if females that lose their offspring are subsequently impregnated by the male doing the killing (Hausfater and Hrdy 1984). The sexually selective infanticide hypothesis predicts that survival of cubs would decline after a resident adult male was killed due to immigration of nonresident males (Swenson et al. 1997). No study has been specifically designed to test this theory. However, data by Swenson et al. (1997) support the theory, but are inadequate to draw strong conclusions.

Janson and Van Schaik (2000) and Boyce et al. (1999) cited Swenson et al. (1997) as an example illustrating that an increased rate of infanticide might be a consequence of male based hunting in mammal populations. Boyce et al. (2001) cited the studies in Scandinavia (Swenson et al. 1997) and southern Canada (Wielgus 1993) studies as illustrating possible relationships meriting consideration in management of bear hunting. In contrast, a panel of 6 scientists reviewed brown bear hunt management in British Columbia and concluded that presently available data on the effects of selective removal of males by hunting are equivocal, and therefore hunting-related changes in density or social structure should not be incorporated into the British Columbia harvest management program (Peek et al. 2003).

Hunting. Legal hunting seasons for brown/grizzly bears exist in Alaska and all Canadian provinces that have grizzly bears. In the United States south of Canada, there are no hunting seasons; the species is protected as "threatened" under the Endangered Species Act.

During 1989/90–1998/99, an average of 1600 bears were annually harvested by recreational hunters in North America (Table 26.8). Most of these were taken in Alaska (73%) and British Columbia (18%), with the remaining 9% from the more eastern Canadian provinces, where bear densities are lower (Table 26.8). An average annual total of 1825 known human-caused moralities occurred in North America (Table 26.8). This documented total included control actions, illegal kills, and defense-of-life and property kills of bears. Such nonsport kills ranged from a low of 5% of total human-caused mortalities in Alaska (Miller and Tutterrow 1999) to 48–50% in areas of northern Canada, where bears are sparse and hunting quotas are low (Table 26.8). In the contiguous United States, all human-caused mortalities were from defense-of-life, incidental, and control kills (Table 26.8).

Grizzly bear hunting is highly valued by participants. In an Alaska study, nonresident bear hunters reported trip expenditures of $10,677 compared to $1247 for resident hunters (Miller et al. 1998). The estimated cumulative annual value of bear-viewing trips ($29.1 million) was higher than for bear-hunting trips taken by nonresidents ($17.05 million) or resident hunters ($4.15 million) (Miller et al. 1998).

Access to brown/grizzly bear hunting opportunities varies in different geographic regions. Where bears are relatively abundant, as in Alaska, residents and nonresidents can hunt them in some areas, and hunter participation is unlimited. Lotteries are used to limit hunting intensity in areas where bear populations are sparse (such as in northern Canada and Alberta) or where bears are abundant but potential hunting intensity is especially high (such as on Kodiak Island, Alaska, and in British Columbia). On the Alaska Peninsula, where brown bear hunting is very popular, hunter participation is limited by closing the season in alternate years rather than by limiting the number of hunters through a lottery.

Like any renewable resource, brown bear populations can sustain a certain level of mortality without declining. Sustainable harvest in most areas is derived from estimates of population size and reproduction data (Miller 1990e). Because brown bears can sustain only very low mortality rates (a maximum of 5.7% was estimated by Miller [1990e]), most managers adopt conservative regulations to avoid overharvests. The Canadian provinces and settlement areas have quotas on total human-caused mortality designed to avoid population declines. In different areas, these quotas are 2–6% of conservative estimates of population size. The 2% figure applies to areas with lower reproductive rates, the 6% quota to areas with higher reproductive rates. In British Columbia, conservative estimates of population size are assured by subtracting one standard deviation from point estimates of population size before calculating quota size (Province of British Columbia 1999). In Canada, as well as on the Kenai Peninsula in Alaska, mortalities by control actions or defense-of-life or property kills are the first to be counted against the quota, with any remainder available to be taken by hunters. In Yukon Territory, resident sport hunters receive a priority in harvest allocations over nonresidents guided by big game outfitters. On the Kenai Peninsula in Alaska, brown bear numbers have been reduced by large numbers of control, defense-of-life and property kills combined with habitat deterioration from road building and increased human presence. This situation on the Kenai Peninsula demonstrates that even Alaska is not immune from the decimating factors that have caused dramatic declines in grizzly bear abundance in southern Canada and the lower 48 states.

In areas of North America where bears are hunted, the principle of "sustainable yield" is practiced except in portions of interior Alaska. In 1994, the Alaska legislature passed an "intensive management" law intended to assure maintenance of high levels of human harvest of moose and caribou through control of predators like bears and wolves (Alaska Statutes 16.05.255). In several portions of interior Alaska, this law has been implemented in attempts to reduce grizzly bear numbers by killing in excess of sustainable rates. This was done regardless of analyses indicating that increased hunting of grizzly bears did not benefit moose or caribou in one of these areas (Miller and Ballard 1992) and the absence

TABLE 26.8. Number of brown bears taken in sport harvests and defense-of-life or property (DLP) circumstances in North America

Location	Period	Hunter-Killed				Annual Average Number of Control and DLP Kills
		Male	Female	Sex Unknown	Annual Average	
Alaska	1989/1990–1998/1999	7883	3872	119	1187	92.1
Yukon	1989–1998	522	289	0	81	15
Northwest Territories and Nunavut	1989–1998	81	17	10	11	10
Gwich'in Settlement Area	1990–1999	35	6	12	5	2.6a
Inuvialuit Settlement Area	1990–1999	176	46	29	25	5.7a
British Columbia	1990–1999	1878	1018	12	291	53
Alberta	1990–1999	109	52	7	17	6.7b
Northern Continental Divide Ecosystem	1990–1999	—	—	—	0	13
Greater Yellowstone Ecosystem	1992–1998	—	—	—	0	6
Total		10,684	5300	189	1617	204.1

NOTE: Ordered north to south.
[a]These kills are included in the hunter-killed data.
[b]"Nonhunter" kills

of data indicating it would be beneficial in other areas where intensive bear management was adopted.

Hunting affects population composition in different ways, and regulations can affect the composition of harvests (Miller 1990e; Van Daele et al. 1990). Because bears are promiscuous, regulations that direct harvests toward males and away from adult females permit higher hunter quotas (Taylor et al. 1987). In early spring, hunters kill primarily males because they are the first to emerge from dens. Females accompanied by newborn cubs are the last to emerge from dens. Similarly, males are the last to enter dens in the fall, so late fall seasons have higher proportions of males. In central Alaska, females constituted 18% of the spring season hunter kill before 1 May, but >40% of the harvest after the third week in May (Miller 1990a). In the fall, females represented 53% of the kill during the first week of September, but <43% of the kill during October (Miller 1990a). Bears enter dens later on northern Kodiak Island and are more vulnerable to hunters during fall seasons than on southwestern Kodiak Island (Van Daele et al. 1990). In Alaska and Canada, regulations prohibit shooting females accompanied by cub-of-year or yearling offspring, which contributes to a male bias in hunter harvests. In the Yukon, a point system is used that provides incentives for outfitters to avoid harvesting females (Yukon Renewable Resources 1997). It is difficult for hunters to distinguish between males and female bears unless the female is accompanied by offspring or the male is exceptionally large. Regardless of regulations, male bears are more vulnerable to hunters than female bears because they range more widely and are more likely to encounter areas frequented by hunters (Bunnell and Tait 1980). Correspondingly, across North America, males constitute between 64% (Yukon) and 85% (northern Canada) of hunter harvests (Table 26.8).

Hunting regulations can influence the composition of hunted populations of bears (Reynolds 1993; Miller 1997). In an extremely heavily hunted population in south-central Alaska that included spring and early-fall seasons, population composition (bears ≥2 years) shifted from 70 males/100 females to 21 males/100 females over a 10-year period. For bears >5 years old, sex ratio shifted from 53 males/100 females to 26 males/100 females. In this area, 58% of the bears harvested during this period were males (Miller 1997). Percentage males in the harvest is a potentially misleading statistic to use in evaluating harvest level because as the proportion of males in the population declines, the proportion of females in the harvest will increase (Frasier et al. 1982). Populations in which hunter effort is not uniformly distributed will also frequently show a prevalence of males in hunter harvest greater than in the population because males have larger home ranges and a correspondingly higher chance of encountering hunters (Bunnell and Tait 1980). In a heavily hunted area of Alaska, there was no significant change in the age of males or females in the population, although there was a tendency for both sexes to be older following the period of heavy hunter kills (Miller 1997). In spite of these changes in population composition in this area, grizzly bear density was not significantly changed (Miller 1995a). In another portion of Alaska, heavy hunting pressure caused a decline in grizzly bear density (Reynolds 1990).

Reporting Rate. Not all bear deaths are detected and recorded. Miller (1990b) indicated that unreported sport or nuisance kills and wounding losses could represent significant sources of mortality that managers should consider. Studies by McLellan et al. (1999), for example, show that without the aid of radiotelemetry, management agencies would have been aware of only 46–51% of grizzly bear deaths and 54–66% of human-caused deaths. Large portions of radio-collared grizzly bear deaths in British Columbia are legal, reported sport kills. However, even in British Columbia, the management agency would have only recorded 53–59% of the mortalities and 67–83% of the human-caused deaths. In rural northwestern Alaska, less than half the grizzly bear sport and subsistence harvest is reported (Miller 1990b). In Montana, where hunting is illegal, agencies would have recorded only 38–41% of deaths and 44–55% of human-caused deaths (McLellan et al. 1999). In the GYE, Knight et al. (1988) suggested that the overall fraction of recorded deaths of grizzly bears ranges from 40% to 60%. They concluded that most deaths due to legal hunting, removal by management agencies, and road kills were confirmed, whereas 32 of 73 (44%) of deaths associated with illegal activities were not confirmed. In a subsequent analysis of the Yellowstone data, Mattson (1998) concluded that there was a high prevalence (60–76%) of radio-marked bears among recorded deaths, and different causes of mortality were not reported equally. He cautioned against use of a simple correction for unknown, unreported mortality.

Density. For brown/grizzly bears, like most species, density (number/unit area) is a key population parameter. High-density bear populations can exist in areas with abundant and uniformly distributed food resources. Low-density bear populations exist in areas where food resources are sparse and/or patchy with long distances between patches (or where there has been excessive human killing of bears). The highest documented grizzly bear density in North America is about 140 times greater than in low-density areas (Table 26.9).

The greatest brown bear densities in North America occur in coastal areas of Alaska, where bears thrive on summer and fall runs of salmon. Coastal maritime climate leads to longer growing seasons, which also benefit bears. Documented densities in these areas are 175–550 bears (all ages)/1000 km^2 (Miller et al. 1997) (Table 26.9). Salmon import energy from rich marine systems into frequently nutrient-impoverished terrestrial systems. Because of this importation of energy, bears living in salmon-rich areas not only have more dense populations, but they are 1.5–3 times larger in body mass (Glenn 1980; Hilderbrand et al. 1999a). Populations with the lowest densities occur in the extreme northern part of North America, between the Alaska Range and the Beaufort Sea in Alaska, and in northern Yukon and Northwest Territories in Canada (Kingsley et al. 1988). Densities in these areas are typically <10 bears/1000 km^2 (Table 26.9). Higher densities can be maintained even in these northern environments in areas where caribou are abundant (Reynolds and Garner 1987). Migratory caribou, like anadromous salmon, are net importers of energy into these energetically impoverished northern systems. Nutrients from salmon that are imported into forest ecosystems and distributed as bear feces may be important for forests growing as far inland as Idaho (Hilderbrand et al. 1999b).

Techniques for estimating bear density are not standardized; consequently, density estimates presented in Table 26.9 are not directly comparable. In Alaska, however, 19 brown bear density estimates were obtained using the same techniques in different habitats; all are directly comparable and have measures of precision (Miller 1995b; Miller et al. 1997; Testa et al. 1998). These techniques required the use of radiocollars, which largely eliminate the problem of geographic closure common to other density estimation techniques.

Radio-marking techniques are not broadly applied outside of Alaska because of expense, need to capture bears to apply radiocollars, and low sightability of bears in heavily forested habitats. Instead, many researchers in Canada and the United States have focused on the development of techniques to estimate number of bears and density employing hair-snaring methods. With this procedure, bears are attracted to sampling stations with a scent lure. At each sampling station, barbed wire is strung between trees, and when the bear passes under the wire, a small tuft of hair is snagged in the barb of the wire (Woods et al. 1996, 1999). The follicles from these hair samples contain DNA, which can be used to identify individual animals. This technique is conceptually similar to techniques developed to identify bears based on photos taken when bears trip cameras (Mace et al. 1994). Advantages of these DNA and camera techniques include reduced need to mark bears or see them from aircraft. However, these techniques are labor intensive and expensive, and typically have problems identifying the area inhabited by the estimated population. This closure problem creates difficulties in estimating density. So far, the DNA and camera techniques are not standardized for design or data analysis, hence results from different areas may not be comparable. In Glacier National Park, U.S. Geological Survey researcher Kate Kendall has conducted the most extensive effort to estimate grizzly bear abundance using hair-snaring and

TABLE 26.9. Density estimates (bears/1000 km^2) of brown/grizzly bear populations based on bears of all ages in different North American study areas

Study Area	Density	Reference
Interior populations		
Tuktoyaktuk Peninsula and northern Yukon[a]	3–4	Nagy et al. 1983a, 1983b
Arctic National Wildlife Refuge coastal plain, Alaska	4	Reynolds 1976
West-central Alberta	4–5	Nagy and Haroldson 1990
Eastern Brooks Range, Alaska	7	Reynolds and Garner 1987
East Front, Montana[b]	7	Aune and Kasworm 1989; Aunu and Brannon 1987
Jasper National Park, Alberta	10–12	Nagy and Haroldson 1990
South-central Alaska Range	10–15[b]	Miller et al. 1997
South-central Alaska	11–41[b,c]	Miller et al. 1987, 1997; Miller 1995a; Testa et al. 1998
MacKenzie Mountains	12	Miller et al. 1982
Yellowstone Ecosystem	14–18	Calculated based on data in Servheen 1999
Southwest Alberta (Waterton Lakes)	15[b]	Mowat and Strobeck 2000
Arctic National Wildlife Refuge, Alaska	16	Reynolds and Garner 1987
East-central Alaska Range	16	Boertje et al. 1987; Gasaway et al. 1992
Northern Continental Divide Ecosystem, Montana	17–22	Calculated based on data in Servheen 1999
Seward Peninsula, Alaska	18[b]	Miller et al. 1997
Northern British Columbia, Prophet River	21[b]	Boulanger and McLellan 2001
Northern Yukon Territory[a]	26–30	Nagy and Haroldson 1990
Southeastern British Columbia (Selkirks)	27[b]	Mowat and Strobeck 2000
Western Brooks Range, Alaska	30[b]	Miller et al. 1997
Denali National Park, Alaska	34[b]	Dean 1987
Kluane National Park, Yukon Territory	37	Pearson 1975
Glacier National Park and adjacent National Forest, Montana	47[b]	K. Kendall, U.S. Geological Survey, pers. commun., 1998, hair snare results
Glacier National Park, Montana	47	Martinka 1974
Glacier National Park, Montana	79[b]	K. Kendall, U.S. Geological Survey, pers. commun., 1998, hair snare results
Flathead River, Montana	80	McLellan 1989a, 1989b, 1989c; British Columbia Forest Service, unpublished data
Coastal populations		
Alaska Peninsula, Black Lake	191[b]	Miller and Sellers 1992; Miller et al. 1997
Chichagof Island, SE Alaska	318[b]	Miller et al. 1997
Kodiak Island, Alaska	323–342[b,c]	Miller et al. 1997
Admiralty Island	399–440[b,c]	Schoen and Beier 1990; Miller et al. 1997
Alaska Peninsula, Katmai National Park	551[b]	Miller et al. 1997

SOURCE: Adapted from McLellan (1994).
NOTE: Ordered by increasing density
[a]Currently Inuvialuit Settlement Region.
[b]Technique used included estimate of precision; other approaches had no estimates of precision, and due to a variety of methods used in their derivation, comparisons must be done cautiously.
[c]Range reflects different study areas or different times in the same study area.

DNA analysis. Although her research is in progress, she has identified a minimum number of different individuals (>200) in Glacier National Park and vicinity that is larger than previously suspected (K. Kendall, pers. commun., 2000).

Estimates of density frequently have problems associated with differential inclusion of age or gender groups. Because newborn cubs have high mortality rates, estimates made early in the year will be larger than estimates made later in the year for the same population. Closure problems may result in overestimation of males, the more mobile sex, in a density estimation area. With DNA hair-snaring techniques, efforts are made to exclude cubs by setting the barbed wire too high to snag their hair. Nonetheless, some cubs leave hair samples behind and some bears >1 year old may be able to go under the barbed wire without leaving hair. The age of a bear is not revealed by DNA analyses. The Alaska capture–mark–resight technique avoids most of these problems, but estimates of precision may be exaggerated by tabulating each member of a family group as a separate individual (Miller et al. 1997). Biologists attempting to estimate bear density need to be aware of these sources of potential bias and specify which sex and age groups occur in their density estimates.

Demographic Modeling. Models are useful tools in evaluating hypotheses about grizzly bears because they integrate large amounts of information. They are also useful when incorporating uncertainties in available data by bounding input parameters within feasible ranges. Demographic models are used to guide the management decision process for wild populations of grizzly bears.

Although modeling efforts can take various forms, a common application to threatened or endangered populations is termed population viability analysis (PVA). A PVA estimates the likelihood of persistence of a population over time and is most frequently employed in endangered species or small population management. There are many different concepts of what composes a PVA, from simple, deterministic models for estimating population change to complex, spatially explicit individual-based models of landscape and population dynamics (Beissinger and Westphal 1998). Single deterministic models are among the simplest analyses and demand the least amount of data (Beissinger and Westphal 1998). Demographic vigor of a population can be measured by its survival-fecundity rate of increase (Caughley 1977). This intrinsic rate of increase (r_s) is the exponential rate at which a population with a stable age distribution changes when resources are not limiting.

As reviewed by Hovey and McLellan (1996), several researchers have estimated r_s or its antilogarithm, the finite rate of increase ($\lambda = e^{r_s}$), to assess status of grizzly bears. A $\lambda > 1.0$ indicates an increasing population, whereas $\lambda < 1.0$ indicates a declining population; $\lambda = 1.0$ suggests a stable population. Most published estimates of λ for grizzly bear populations are derived with the Lotka model (Lotka 1907) as proposed by Eberhardt (1985); many lack confidence intervals. The highest published rate of increase (1.085 ± 0.026) was derived by Hovey and McLellan (1996) in the North Fork of the Flathead River in

British Columbia and Montana. The λ value for the GYE was 0.97–1.12 (Eberhardt 1995). Stable population growth was estimated for grizzlies in the Kananaskis area of southwestern Alberta ($\lambda = 0.99$–1.01; Wielgus and Bunnell 1994) and the Selkirk Mountains of British Columbia and Idaho ($\lambda = 1.00$; Wielgus et al. 1994). A declining population was estimated for the Swan Mountains of Montana ($\lambda = 0.977$, 95% confidence interval [CI] = 0.875–1.046; Mace and Waller 1998). Some of these rates are point estimates based on small sample sizes. For nearly all estimates, the 95% CI bounds 1.0, making it impossible to determine true population trajectory. For a slowly reproducing species like grizzly bears, in which even a maximum lambda will always be close to 1.0, it will seldom be possible to have a 95% CI that does not overlap 1.0. Uncertainty primarily associated with subadult and adult female survival explains most of the variance associated with these estimates (Eberhardt et al. 1994; Hovey and McLellan 1996; Mace and Waller 1998).

Shaffer (1978, 1983) was the first to use stochastic models to help guide grizzly bear management in Yellowstone National Park. This pioneering work was the first PVA for any species. His model estimated a minimum viable population, or the smallest population size necessary with a 95% chance of remaining extant after 100 years. Initial simulations indicated that a population of 35 grizzly bears might be expected to survive 100 years. Because of uncertainty associated with his original estimate, Shaffer (1983) later suggested that this value should be increased to 50–90 bears. Later Suchy et al. (1985) updated these estimates to 40–125 or 50–225 bears depending on a low versus high mortality schedule. To be conservative, Suchy et al. (1985) recommended a population >125 be maintained to ensure a high probability of persistence for at least 100 years. Soulé (1987) and Shaffer (1992) expressed concern that targeting a minimum population level is inadequate for sound conservation and that larger populations are necessary to ensure long-term persistence of the species. More recent reviews of PVA (Boyce 1992; Boyce et al. 2001) have pointed out that traditional PVA models are demographically based; they lack a link to habitat, particularly habitat changes. Most PVAs do not consider genetic effects, including inbreeding depression, loss of evolutionary potential, and accumulation of harmful mutations (Allendorf and Ryman 2002).

AGE ESTIMATION

Assessing growth annuli in teeth is the most accurate means of age determination for many mammalian species (Thomas 1977; Fancy 1980). The technique has been applied to the canine (Rausch 1969), the lower third molar (Mundy and Fuller 1964), and the first upper and lower premolars (Matson et al. 1993) of brown bears. Because of the ease in collection, its vestigial nature, and small root size, the premolar is the tooth most commonly extracted from live bears. Eruption of permanent premolars occurs before denning in grizzly bears at about 8 months of age (Pearson 1975). The first annulus is formed during the denning season around the time the bear has its first birthday. By spring, the premolar of a yearling brown bear has the light cementum of the previous summer/fall and a single annulus of the winter just past (Matson et al. 1993). Each denning season, a new annulus is formed. Accuracy of the technique is dependent on tooth quality, experience of the technician, and age of the bear. For older bears (>9 years), errors for exact age can be as high as 70%; errors decline to 40% if accuracy within 1 year is acceptable (Matson et al. 1993). A thin annual layer of light cementum has been correlated with successful cub rearing in some female black bears (Carrel 1992; Coy and Garshelis 1992), but has proven unreliable in brown bears.

MANAGEMENT AND CONSERVATION

Conservation. As noted, grizzly bears south of Canada have been dramatically reduced in abundance and distribution to perhaps as few as 1000 individuals in mountainous areas of the northern Rocky Mountains near Canada and in the Yellowstone Ecosystem (Servheen 1999). Additionally, there may be a small transitory population in the North Cascades near the border with British Columbia (Servheen 1998, 1999). Grizzlies occupy only 1–2% of their historical range south of Canada. In the United States, healthy populations of brown bears remain only in Alaska, where some 31,700 individuals are estimated to live (Miller 1993; Miller and Schoen 1999). Even in Alaska, however, there are areas such as the Kenai Peninsula where the same decimating factors of excessive mortality and habitat destruction that reduced the population south of Canada have placed the persistence of brown bears at risk.

Habitat loss and human-caused mortality operate in southern Canada. Brown bears have been exterminated from the open plains in the provinces of Manitoba, Saskatchewan, and western Alberta (McLellan and Banci 1999). An estimated 25,300 brown bears remain in Canada (Banci et al. 1994; McLellan and Banci 1999). The most secure of these populations are in the high-density zones along the Pacific coast, but even here they were listed as threatened or vulnerable by McLellan and Banci (1999). Banci et al. (1994:140) noted, "It can no longer be assumed that there will be some areas of [Canada] that will be left natural and untrammeled and that can serve as refugia for grizzly bears and other wildlife that require large areas of relative solitude." In the far north of Canada, populations exist at densities that are close to prehistoric conditions, but because densities are naturally very low in northern Canada (Nagy and Haroldson 1990), these populations are inherently vulnerable.

Six and one-half years following listing under the Endangered Species Act, a recovery plan for grizzly bears was published (U.S. Fish and Wildlife Service 1982). The plan was revised in 1993. This revised plan presents recovery targets for grizzly bears in the Northern Continental Divide Ecosystem (Glacier National Park and vicinity), the GYE including Yellowstone National Park, the Cabinet/Yaak Ecosystem, and the Selkirk Ecosystem. The recovery plan also identified an objective of reestablishing a grizzly bear population in the Bitterroot Ecosystem and mentioned the need to develop a plan for the North Cascades of Washington State. When the targets identified in the recovery plan are reached, the U.S. Fish and Wildlife Service will propose removing grizzly bears from protection under the Endangered Species Act and returning management authority to state agencies responsible for wildlife management.

In the Northern Continental Divide Ecosystem (NCDE) and GYE recovery targets are based on numerical, distributional, and mortality objectives. Because of the difficulty of accurately estimating abundance, numerical targets in the recovery plan are based on counts of unduplicated females with newborn cubs (Knight et al. 1995). This segment of the population is most recognizable. Females accompanied by newborn cubs observed by qualified personnel are tallied based on location, date, pelage color, size, and number of cubs. Because cub production varies yearly, trends are based on a 6-year running average count. The target running average is 15 females with newborn cubs in the GYE, 22 in the NCDE (10 within Glacier National Park and 12 outside the park), 6 in the Cabinet/Yaak Ecosystem, and 6 in the Selkirk Ecosystem. This numerical target is conservative because not all females with newborn cubs are observed and there is a conservative protocol for excluding possible duplicate counts.

Recovery cannot occur unless recruitment exceeds mortality. Under the Recovery Plan (U.S. Fish and Wildlife Service 1993) for all recovery areas, maximum mortality level is set at $\leq 4\%$ of the population estimate/year during any two consecutive years. The population estimate derives from an extrapolation on the number of females with newborn cubs based on a conservative set of assumptions about the proportion of the population constituted by females with newborn cubs. In addition to this numerical mortality limit, the plan specifies that no more than 30% of the mortality can be females. The female quota recognizes that a population of bears can persist with higher rates of mortality to males than to females. Mortality quotas established in the recovery plan have proven more difficult to achieve than numerical goals.

The recovery plan also recognizes that bears must be well distributed in each ecosystem. Recovery zones have been subdivided into bear management units (BMUs). Recovery targets identify the proportion of BMUs within which females with offspring must be observed (70%, 82%, 89%, and 91% in the Selkirk, Cabinet/Yaak, GYE, and NCDE, respectively). These occupancy rates must be obtained during a 6-year sum of sightings during which no two adjacent BMUs can remain unoccupied.

The U.S. Grizzly Bear Recovery Plan also recognizes that there are areas of acceptable habitat where grizzly bears have been eliminated and should be restored. Foremost among these is the Selway–Bitterroot in central Idaho and western Montana, where two designated wilderness areas of 14,983 km^2 (5785 mi^2) form the core recovery area for a 65,100-km^2 (25,140-mi^2) experimental population area that includes the wilderness areas and surrounding national forests. The plan for restoring grizzly bears was developed by a coalition of advocates for the grizzly bear (the National Wildlife Federation and the Defenders of Wildlife) and representatives of timber and labor groups (Resource Organization on Timber Supply and the Intermountain Forest Association). Their plan was adopted by the U.S. Fish and Wildlife Service as its preferred alternative for restoring grizzly bears into the largest area of acceptable habitat (without grizzly bears) south of Canada. In this area, grizzly bears appear to have been extirpated >50 years earlier (U.S. Fish and Wildlife Service 2000).

A key element of the plan was that bears would be reintroduced as an "experimental nonessential" population under Section 10(j) of the Endangered Species Act. The designation as experimental allows more flexibility in managing the reintroduced population and the "nonessential" designation means that the population is unnecessary to avoid extinction of the species. Another element of the plan is that the reintroduced population would be managed by a local citizens management committee (CMC). The CMC is composed of nominees from the governors of Idaho and Montana plus representatives from the U.S. Fish and Wildlife Service, the U.S. Forest Service, and the Nez Perce Tribe. The CMC is an effort to reassure local citizens who are currently using these areas that their concerns about the way the reintroduced bears are managed will be heard. If the decisions of the CMC do not lead toward recovery goals, then the plan calls for mediation. If this is unsuccessful, it allows the U.S. Secretary of Interior to assume management authority over the reintroduced population. Estimates of carrying capacity in the core recovery area suggest the wilderness areas alone can support nearly 250 bears; more can be supported in the national forest lands surrounding the wilderness areas (Boyce and Waller 2000).

The Record of Decision to implement the Bitterroot reintroduction was published in the Federal Register in November, 2000. Shortly thereafter, the State of Idaho filed suit to block implementation and the Secretary of Interior of the newly elected Bush Administration entered into negotiations. In February 2001, the Secretary of Interior proposed to substitute a "no action" alternative for the published Record of Decision and announced a 60-day comment period. During this period, 98% of the public comments received opposed substituting a "no action" alternative for the previously adopted reintroduction decision; opposition included 98% of comments from Idaho and 93% from Montana (Schoen and Miller 2002). In addition, 8 prominent professional organizations of scientists and biologists wrote letters in opposition to the proposed change (Schoen and Miller 2003). Perhaps because of this negative response, no new record of decision has been announced and the existing decision remains in place although no funding to implement it has been provided by the Department of Interior putting the reintroduction into limbo.

Reasons for the decline of brown/grizzly bears in North America are excessive human-caused mortality and habitat loss (Storer and Tevis 1955; Brown 1985; Servheen 1999). Habitat loss results from conversion of native vegetation to agriculture, depletion of preferred food resources (i.e., salmon and whitebark pine), disturbance, displacement from human developments and activities (roads, mines, subdivisions), and fragmentation of habitat into increasingly smaller blocks inadequate to maintain viable populations. Until the 1950s, brown/grizzly bears were considered dangerous to humans and livestock, and consequently, excessive mortality was intentional and sanctioned by government agencies. Currently, in most areas where human-caused mortality is excessive, it results from the same motives despite government efforts to limit mortality.

Management. Radiotelemetry studies have identified roads as significant factors in habitat deterioration and increased mortality of brown/grizzly bears (Archibald et al. 1987; Mattson et al. 1987; Peek et al. 1987; McLellan and Shackleton 1988; Schoen 1990; Mace et al. 1996, 1999; Mace and Waller 1997; Claar et al. 1999). Areas of adult female displacement by roads and developments totaled about 16% of available habitat in Yellowstone National Park (Mattson et al. 1987). In southeastern British Columbia, McLellan and Shackleton (1988) estimated that brown bear habitat loss from roads totaled 8.7% of their entire study area. The percentage of habitat loss as a consequence of behavioral displacement from roads is a function of road density. Percentage is higher in areas having higher road density regardless of the distance at which roads affect bear behavior.

The distance at which bears appear to be displaced by roads varies in different areas and seasons. In Yellowstone National Park, bears avoided areas within 500 m from roads during spring and summer. During summer, daytime disruption of foraging was observed out to 2 km from roads (Mattson et al. 1987). In southeastern British Columbia, bears used areas within 100 m of roads significantly less than expected (McLellan and Shackleton 1988). Habitat use was 58% less than expected within 100 m from roads and 7% less than expected 101–250 m from roads (McLellan and Shackleton 1988). Displacement of grizzlies from habitats near roads may extend up to 3 km for primary roads and 1.5 km for secondary roads (Kasworm and Manley 1990). Roads are typically constructed along streams in riparian areas most intensively used by bears early in the spring following emergence from dens. Correspondingly, the impact of roads on displacement from preferred habitats is greatest in spring (Mace et al. 1996, 1999). During fall, bears tend to move to higher elevations to forage. At this time, they select habitats that are typically more distant from existing roads. Consequently, the importance of disturbance displacement by roads is less evident during fall than during spring (Mace et al. 1996, 1999).

Level of traffic appears to influence degree of bear avoidance of roads. Buffer zones of 500 m around roads with no traffic or with use by ≤10 vehicles/day received positive or neutral selection. In contrast, bears avoided buffers surrounding roads having >10 vehicles/day (Mace et al. 1996). In another analysis, radio-marked bears significantly avoided roads with even low levels of human use (<1 vehicle/day) as well as roads with moderate (1–10) and high (>10) levels of use (Mace et al. 1999).

In southeastern British Columbia, areas near roads were used less than expected by adult males and more than expected by weaned yearlings and adult females (McLellan and Shackleton 1988). This pattern may result from selection by female and subadults, the classes most vulnerable to intraspecific predation, of the roadside areas avoided by adult males (Mattson et al. 1987; McLellan and Shackleton 1988; Mattson 1990). Bears living near roads have higher probability of human-caused mortality as a consequence of being mistakenly identified as a black bear by hunters, illegal shooting, and control actions influenced by attraction to unnatural food sources (Mattson 1990; McLellan 1990; Mace et al. 1996).

In Montana's Flathead National Forest, Amendment 19 to the forest plan recognizes the impact of roads to grizzly bear habitat security based on research in the northern Swan Mountains of Montana (Mace and Manley 1993; Mace et al. 1996; Mace and Waller 1997). This plan divided the grizzly bear habitat into three categories and specified the maximum road density in each category for each BMU:

1. At least 68% had to be roadless core area (classified as >0.5 km from any roads) at least 1012 ha (2500 acres) in size.
2. No more than 19% of the grizzly bear habitat in each BMU could have a road density >2.0 mi/mi^2 ("roads" were defined as any driveable road or trail).
3. No more than 19% of the grizzly bear habitat could have >1 mi/mi^2 of open roads (defined as roads or trails used more than six times/week by motorized vehicles).

Road density was measured based on moving window analyses. This technique involves randomly moving a window of 1 mi^2 across a map and summing the lengths of segments of roads and trails within the window (EPPL7, Minnesota Land Management Information Center, 658 Cedar Street, St. Paul, MN 55155). This technique precludes direct conversion of the above standards to metric equivalents.

These guidelines acknowledge the importance of large core areas with no roads and only low road density in the remaining grizzly bear habitat. However, these guidelines have been difficult to implement because of opposition to the extensive road closures required to meet them. Biologists also recognized that there was a seasonal component to grizzly bear use of the habitats influenced by roads. Acceptable levels of habitat security for bears require seasonal closures of roads and trails during periods when they are most commonly in these areas. Consequently, the guidelines listed above are under review in an effort to close roads in areas that will result in maximum benefit to bear habitat security. Although there is clear evidence of the detrimental impact of roads in grizzly bear habitat, the threshold of road density may vary among areas. J. J. Craighead et al. (1995) suggested that road density >1 km/6.4 km (0.25 mi/mi^2) has detrimental impacts on bear use of landscapes.

In addition to habitat loss by disturbance displacement, roads facilitate killing of bears by humans via hunting and control actions. These are the greatest sources of mortality to adult bears in the GYE (Weaver et al. 1986; Mattson et al. 1987). Risk of mortality was estimated as five times higher near roads (Doak 1995). On Chichagof Island in southeastern Alaska, where brown bears are legally hunted, there was a direct positive correlation between bear kill by hunters and cumulative kilometers of constructed roads (Titus and Beier 1991).

Habitat evaluation for grizzly bears requires knowledge about the abundance and distribution of food and shelter patches as well as knowledge about human influences that may make bears avoid some areas or expose them to higher risk of mortality. Excellent early work on evaluation of bear habitat concentrated on vegetation analyses was presented by Craighead (1977) and Craighead et al. (1982). More recent efforts to evaluate habitat incorporate similar vegetative analyses with those of mortality risks to brown bears and likelihood of disturbance avoidance of preferred habitats. Risks to bears come from many sources, and managers use cumulative effects models (CEM) to assess habitat values (Weaver et al. 1986; Schoen et al. 1994; Suring et al. 1998; Mattson et al. 1999b). CEM models assign qualitative importance scores to different components of the habitat and then sum these scores for all factors to obtain a measure of habitat value and overall risk. The first CEM for grizzly bears included measures of human-induced risk of mortality, habitat alteration, and displacement from habitat (Weaver et al. 1986). Each of these parameters incorporated numerous coefficients (Mattson et al. 1999b). The value for mortality was derived from indices of habitat quality and type and intensity of human activities. The value for habitat displacement included components of distance to cover and nature and intensity of bear activity in that habitat (Weaver et al. 1986).

A more recent approach integrates empirical information from telemetry studies into models to derive resource selection functions (RSF) (Schoen et al. 1994; Mace et al. 1996, 1999; Boyce and McDonald 1999; Merrill et al. 1999; see Carroll et al. 1999 for a review). RSFs are proportional to the probability of an area being used by an animal. The key to this approach is to correctly identify the important parameters; some include satellite imagery (greenness), elevation, human activity points, roads, and trails (Mace et al. 1999) or human numbers, human distribution, and abundance and quality of bear foods (Merrill et al. 1999).

South of Canada, grizzly bears once occupied the landscape continuously from mountain tops to valley bottoms and plains. With ever-increasing human presence in the valley bottoms and plains, bears have become isolated in islands of remaining mountainous and forested habitat surrounded by a threatening sea of subdivisions, agricultural fields, and pastures. Bears that venture beyond the borders of these remaining islands venture into areas described as "mortality sinks," which can drain the island population and threaten its viability (Knight et al. 1988). Identification of zones of connectivity or linkages between the islands is an essential element of habitat analyses, and numerous approaches have been described. Linkage zone models predict relative probability of grizzly bear movements through an area as a function of factors such as visual cover, riparian corridors, and anthropogenic features (Gibeau 1993; Servheen et al. 2001; Gibeau et al. 1996; Apps 1997). A simulation model was used to predict dispersal routes for grizzly bears based on permeability of different habitat types (Boone and Hunter 1996). Based on a literature review, whitebark pine/lodgepole pine (*Pinus contorta*) habitats were assigned high permeability values, whereas clear-cut and early-seral stage forests had low permeability. Walker and Craighead (1997) combined the permeability data and dispersal mortality risk to map potential dispersal routes for grizzly bears in the northern Rocky Mountains. These models can be used to plot the "least-cost path" for bears moving between ecosystems. These approaches recognize that the correct paradigm for a linkage zone is not a corridor that bears use to move between ecosystems, but rather an area of habitat between ecosystems that bears can safely occupy at low densities with acceptable levels of mortality risk.

Bears are archetypal flagship species—species so charismatic that they symbolize an entire conservation program (Simberloff 1999). The grizzly bear and other large carnivores are a flagship for the Yellowstone-to-Yukon Biodiversity Strategy (Y2Y), a broad program to maintain and restore natural diversity and ecological health of the Rocky Mountains. The Y2Y mission is to establish an interconnected system of core protected areas and wildlife movement corridors that extend from the GYE to the Yukon's Mackenzie Mountains (Tabor 1996; Tabor and Soulé 1999). The concept is premised on protecting existing core areas within existing national parks and preserves, state and provincial parks, and wilderness areas. Core areas will be interconnected with corridors, allowing migration of wildlife among them. Conservation benefits of Y2Y encompass more than the grizzly bear in the United States, and include large carnivores and other wildlife species. Such broad thinking is heretofore unheard of in North America and generally beyond traditional agency thinking or mandates. The vision has inspired over 200 conservation groups to work together beyond international boundaries. The research, planning, and implementation that has gone into the Y2Y effort will benefit the management of grizzly bears and other wildlife in North America that require large interconnected landscapes to support healthy populations.

RESEARCH NEEDS

One of the earliest studies of grizzly bears in North America was in Denali National Park (formerly Mount McKinley National Park) by the great naturalist Adolph Murie (Murie 1944, 1981). Using observational techniques, Murie discovered much about grizzly bear ecology before the development of effective means of immobilizing bears and tracking them with radiotelemetry. However, through observation he was unable to quantify some important parameters such as population density, reproductive and mortality rates, distances moved by individual animals, and characteristics of denning locations. Information on these awaited the pioneering studies of grizzly bears in Yellowstone National Park by the Craighead brothers using radiotelemetry techniques they developed (Craighead et al. 1963). Later, they pioneered the use of satellite monitoring of bears (Craighead et al. 1971).

Although the importance of telemetry techniques is well recognized, Craighead and Mitchell (1982) expressed concern that some biologists are overusing telemetry. They note that capture and handling imposes unnecessary stress on the animals, particularly from populations inhabiting similar environments. We agree with this concern and suggest that radiotelemetry is only one tool available to biologists. Application of such a tool must be employed when the technique is applicable to address a specific objective and answer certain questions. Telemetry studies must be well designed, adequately reviewed, and competently conducted. Scientific ethics requires that agencies proposing to conduct telemetry studies on grizzly bears or other rare animals must seek and incorporate review of their proposed research from biologists outside their agency before studies are initiated.

However, properly executed telemetry studies remain critical to improving our ability to manage grizzly bear populations and habitats. With modern drugs, competently conducted capture efforts should have a mortality rate of handled bears of $<1\%$. Only collars that are designed to drop off without the need to recapture an individual should be used on bears. This reduces the risk that collars will injure growing animals or that animals will wear the collar longer than necessary to complete the project.

More recent advances in satellite telemetry employ global positioning system technology (Schwartz and Arthur 1999) and are providing insight into bear movements, home range analyses, habitat use, and other spatial statistics not obtainable via conventional telemetry (Arthur and Schwartz 1999). Satellite telemetry enables the collection of thousands of locations during the life of a transmitter, thus reducing the need to put collars on numerous individuals. It also allows for continuous 24-hr sampling, which is independent of personnel, weather, terrain, and other factors that limit conventional tracking. In designing telemetry studies, it is essential that the information obtained result in sufficient gains in improved management to offset the costs to bear populations associated with handling.

Humans have the greatest influence on brown/grizzly bear distribution and abundance in North America. Understanding, mitigating, and managing human impacts on bears and their habitats are the greatest challenges facing resource managers, wildlife and conservation biologists, politicians, and the general public. Much of the research needs now and in the future must address these issues.

Improved population abundance estimation and trend assessment remain critical research needs. Today's techniques are expensive and labor intensive. Also, some population estimation techniques are subjective, have no estimate of precision, and cannot be replicated in a systematic manner. Some techniques require radio-marking large numbers of individuals (i.e., Reynolds and Boudreau 1992; Miller et al. 1997), which may not be feasible in some environments. These techniques also typically provide density estimates in only small portions of the area inhabited by the entire population of interest to bear managers. Hair snaring and DNA analysis techniques (Woods et al. 1996, 1999; Mowat and Strobeck 2000; K. Kendall, unpublished data) have the potential to become a standardized and objective approach to abundance estimation without the need to handle and radio-collar individual bears. However, these techniques are currently expensive and have problems with demographic and geographic closure, potential capture biases, and standardization of experimental design. Design issues include grid size and scent lure rotation frequency, sample collection frequency, and mathematical techniques for data analysis. On Kodiak Island, Alaska, the bear population may be too homozygous (Waits et al. 1998a, 1998b) to distinguish between individuals based on DNA. Techniques based on visual observations of unduplicated adult females accompanied by newborn cubs (Knight et al. 1995) have been used to estimate minimum population size and establish mortality quotas for bears in the Yellowstone Ecosystem but extrapolation to a total population number or population density remains problematic (Boyce et al. 2001). Observational techniques using double-count procedures are under investigation in Alaska (E. Becker, Alaska Department of Fish and Game, pers. commun., 2000).

Brown bears are hunted in all jurisdictions north of the border between the United States and Canada. However, methods of assessing the impacts of hunting on populations are poorly developed (Miller 1990e). Better means of assessing impacts of hunting are needed because brown/grizzly bears have the lowest reproductive rates among North American game mammals. Without such techniques, appropriate hunting opportunities may be needlessly curtailed or populations may be overharvested.

Especially in the lower 48 states, better means of assessing the biological carrying capacity of actual or potential grizzly bear habitats are needed. Early assessment efforts were pioneered by Craighead et al. (1982). Boyce and McDonald (1999), Hogg et al. (1999), and Merrill et al. (1999) have proposed other approaches. Such assessments are important to ensure restoration efforts for grizzly bears are successful in areas where they have been extirpated or to adapt management policy to environmental change to ensure long-term persistence in extant habitats.

Further research is also needed on the importance of anthropogenic impacts on bear habitats. As documented elsewhere in this section, roads, commercial activities (mining, logging), livestock grazing, suburban sprawl, and recreational uses affect (i.e., through snow machining or off-road vehicles) the ability of bear populations to persist in an area (e.g., Mattson et al. 1987; McLellan and Shackleton 1988; McLellan 1989a, 1990; Mace et al. 1996, 1999). More intensive research is needed on threshold levels at which these impacts become significant and possible ways to mitigate adverse human impacts on brown/grizzly bear populations. Similarly, it is important to find ways to identify threshold levels of tolerance for adverse impacts of grizzly bears on humans.

Efforts to restore grizzly bears also require better information on economic costs and benefits of bears and social attitudes toward bears. Swanson et al. (1994), Bath (1998), Duda et al. (1998), and Miller et al. (1998) presented such analyses. Among other reasons, such information is needed to demonstrate the value of preserving wildlife corridors among fragmented islands of habitat that characterize grizzly bear habitat in the continental United States and southern Canada (Noss and Harris 1986; Craighead and Vyse 1996). Additional research is required to document the impacts of introduced exotics on grizzly bears including blister rust, lake trout (*Salvelinus namaycush*), and brucellosis, to name a few (Reinhart et al. 2001).

There are still some basic biological issues still unresolved. Hellgren (1998) identified aspects of the hibernation physiology in bears where our knowledge is lacking. We do not clearly understand the role of intraspecific killing as it relates to bear behavior, male density, and hunting. Furthermore, our understanding of density dependence is still unclear.

ACKNOWLEDGMENTS

Numerous people contributed to this chapter. We especially wish to thank Maureen Hartmann for efforts in searching out and providing pertinent literature. We thank Karrie West and Cecily Costello for providing numerous editorial comments. We also thank Marsha Branigan, Matt Austin, Eldon Bruns, John Hechtel, Mike McDonald, Ray Case, and Alistair Bath.

LITERATURE CITED

Akcakaya, H. R., M. A. Burgman, and L. R. Ginzburg. 1999. Applied population ecology: Principles and computer exercises using RAMAS EcoLab 2.0. Sinauer Associates, Sunderland, MA.

Allendorf, F. W., and N. Ryman. 2002. The role of genetics in population viability analysis. Pages 50–85 *in* S. R. Beissinger and D. R. McCullough, eds. Population viability analysis. University of Chicago Press, Chicago.

Apps, C. D. 1997. Identification of grizzly bear linkage zone along the Highway 3 corridor of southeast British Columbia and southwest Alberta (Prepared for British Columbia Ministry of Environment, Lands and Parks and World Wildlife Fund-Canada). Aspen Wildlife Research, Calgary, Alberta, Canada.

Archibald, W. R., R. Ellis, and A. N. Hamilton. 1987. Responses of grizzly bears to logging truck traffic in the Kimsquit River Valley, British Columbia. International Conference on Bear Research and Management 7:251–57.

Arthur, S. M., and C. C. Schwartz. 1999. Effects of sample size and sampling frequency on studies of brown bear home ranges and habitat use. Ursus 11:139–48.

Atkinson, S. N., and M. A. Ramsay. 1995. The effects of prolonged fasting on the body composition and reproductive success of female polar bears (*Ursus maritimus*). Functional Ecology 9:559–67.

Atkinson, S. N., R. A. Nelson, and M. A. Ramsay. 1996. Changes in body composition of fasting polar bears (*Ursus maritimus*): The effect of relative fatness on protein conservation. Physiological Zoology 69:304–16.

Aune, K., and B. Brannon. 1987. East Front grizzly studies. Montana Department of Fish, Wildlife and Parks, Helena.

Aune, K., and W. Kasworm. 1989. Final report. East Front grizzly studies. Montana Department of Fish, Wildlife and Parks, Helena.

Aune, K., M. Madel, and C. Hunt. 1986. Rocky Mountain East Front grizzly bear monitoring and investigations. Unpublished data. Montana Department of Fish, Wildlife and Parks.

Aune, K. E., R. D. Mace, and D. W. Carney. 1994. The reproductive biology of female grizzly bears in the Northern Continental Divide Ecosystem with supplemental data from the Yellowstone Ecosystem. International Conference on Bear Research and Management 9(1):451–58.

Ballard, W. B., T. H. Spraker, and K. P. Taylor. 1981. Causes of neonatal moose calf mortality in south-central Alaska. Journal of Wildlife Management 45:335–42.

Ballard, W. B., S. D. Miller, and T. H. Spraker. 1982. Home range, daily movements, and reproductive biology of the brown bear in southcentral Alaska. Canadian Field-Naturalist 96:1–5.

Ballard, W. B., L. A. Ayres, K. E. Roney, D. J. Reed, and S. G. Fancy. 1991. Demography of Noatak grizzly bears in relation to human exploitation and mining development (Federal Aid in Wildlife Restoration Final Report W-22-5, W-22-6, W-23-1, W-23-2, and W23-3, Study 4.20), Alaska Department of Fish and Game, Juneau.

Banci, V. 1991. The status of the grizzly bear in Canada in 1990 (Status Report). Committee on the Status of Endangered Wildlife in Canada.

Banci, V., D. A. Demarchi, and W. R. Archibald. 1994. Evaluation of the population status of grizzly bears in Canada. International Conference on Bear Research and Management 9(1):129–42.

Barnes, V. G., Jr., and R. B. Smith. 1993. Cub adoption by brown bears, *Ursus arctos middendorffi,* on Kodiak Island, Alaska. Canadian Field, Naturalist 107:365–67.

Bath, A. J. 1998. The role of human dimensions in wildlife resource research in wildlife management. Ursus 10:343–48.

Beissinger, S. R., and M. I. Westphal. 1998. On the use of demographic models of population viability in endangered species management. Journal of Wildlife Management 62:821–41.

Bekoff, M., and L. D. Mech. 1984. Simulation analyses of space use: Home range estimates, variability, and sample size. Behavioral, Research Methods, Instruction and Computation 16:32–37.

Bielanska-Osuchowska, Z., and S. Szankowska. 1970. Histological and histochemical studies on the alimentary tract in the brown bear. Acta Theriologica 15:303–42.

Blanchard, B. M. 1987. Size and growth patterns of the Yellowstone grizzly bear. International Conference on Bear Research and Management 7:99–107.

Blanchard, B. M. 1990. Relationship between whitebark pine cone production and fall grizzly bear movements. Pages 362–63 *in* W. C. Schmidt and K. J. McDonald, comps. Proceedings of symposium on whitebark pine ecosystems: Ecology and management of a high-mountain resource (U.S. Forest Service General Technical Report INT-270). U.S. Department of Agriculture, Forest Service, Ogden, UT.

Blanchard, B. M., and R. R. Knight. 1991. Movements of Yellowstone grizzly bears, 1975–87. Biological Conservation 58:41–67.

Blanchard, B. M., and R. R. Knight. 1995. Biological consequences of relocating grizzly bears in the Yellowstone Ecosystem. Journal of Wildlife Management 59:560–65.

Boertje, R. D., W. C. Gasaway, D. V. Grangaard, D. G. Kelleyhouse, and R. O. Stephenson. 1987. Factors limiting moose population growth in Subunit 20E (Federal Aid in Wildlife Restoration Progress Report Project W-22-5, Job 1.37R). Alaska Department of Fish and Game, Juneau.

Boone, R. B., and M. L. Hunter. 1996. Using diffusion models to simulate the effects of land use on grizzly bear dispersal in the Rocky Mountains. Landscape Ecology 11:51–64.

Boulanger, J. G., and B. McLellan. 2001. Estimating closure violation bias in DNA based mark–recapture censuses of grizzly bears. Canadian Journal of Zoology 79:642–651.

Boulanger, J. G., and G. C. White. 1990. A comparison of home-range estimators using Monte Carlo simulation. Journal of Wildlife Management 54:310–15.

Boyce, M. S. 1992. Population viability analysis. Annual Review of Ecology and Systematics 23:481–506.

Boyce, M. S., and L. L. McDonald. 1999. Relating populations to habitats using resource selection functions. Trends in Ecology and Evolution 14:268–72.

Boyce, M. S., A. R. E. Sinclair, and G. C. White. 1999. Seasonal compensation of predation and harvesting. Oikos 87:419–26.

Boyce, M. S., and J. Waller. 2000. The application of resource selection functions analysis to estimate the number of grizzly bears that could be supported by habitats in the Bitterroot Ecosystem. Appendix 21B. Pages 231–46 *in* Grizzly bear recovery in the Bitterroot Ecosystem, final environmental impact statement. U.S. Fish and Wildlife Service.

Boyce, M. S., B. M. Blanchard, R. R. Knight, and C. Servheen. 2001. Population viability for grizzly bears: A critical review (Monograph Series No. 4). International Association for Bear Research and Management.

Brady, K. S., and D. Hamer. 1992. Use of summit mating area by a pair of courting grizzly bears, *Ursus arctos,* in Waterton Lakes National Park, Alberta. Canadian Field-Naturalist 106:519–20.

Brown, D. E. 1985. The grizzly in the southwest, documentary of an extinction. University of Oklahoma Press, Norman.

Brown, J. L., and G. H. Orians. 1970. Spacing patterns in mobile animals. Annual Review of Ecological Systematics 1:239–62.

Bunnell, F. L., and T. Hamilton. 1983. Forage digestibility and fitness in grizzly bears. International Conference on Bear Research and Management 5:179–85.

Bunnell, F. L., and D. E. N. Tait. 1980. Bears in models and reality: Implications to management. International Conference on Bear Research and Management 3:15–23.

Bunnell, F. L., and D. E. N. Tait. 1981. Population dynamics of bears: Implications. Pages 75–98 *in* C. W. Fowler and T. D. Smith, eds. Dynamics of large mammal populations. John Wiley, New York.

Bunnell, F. L., and D. E. N. Tait. 1985. Mortality rates of North American bears. Arctic 38:316–23.

Burt, W. H. 1943. Territoriality and home range concepts as applied to mammals. Journal of Mammalogy 24:346–352.

Byers, C. R., and G. A. Bettas. 1999. Records of North American big game, 11th ed. Boone and Crockett Club, Missoula, MT.

Carr, H. D. 1989. Distribution, numbers, and mortality of grizzly bears in and around Kananaskis Country, Alberta (Wildlife Research Series No. 3). Alberta Forest Lands and Wildlife.

Carrel, W. K. 1992. Reproductive history in dental cementum of female black bears. International Conference on Bear Research and Management 9(1):205–12.

Carroll, C., P. C. Paquet, and R. F. Noss. 1999. Modeling carnivore habitat in the Rocky Mountain region: A literature review and suggested strategy. World Wildlife Fund-Canada, Toronto.

Case, R. L., and L. Buckland. 1998. Reproductive characteristics of grizzly bears in the Kugluktuk area, Northwest Territories, Canada. Ursus 10:41–47.

Caughley, G. 1977. Analysis of vertebrate populations. John Wiley, New York.

Claar, J. J., N. Anderson, D. Boyd, B. Conard, G. Hickman, R. Hompesch, G. Olson, H. I. Pac, T. Wittinger, and H. Youmans. 1999. Carnivores. Pages 7.1–7.61 *in* G. Joslin and H. Youmans, coords. Effects of recreation on Rocky Mountain wildlife: A review for Montana. Committee on Effects of Recreation on Wildlife, Montana Chapter of the Wildlife Society, Missoula.

Clark, W. K. 1957. Seasonal food habits of the Kodiak bear. Transactions of the North American Wildlife Conference 22:145–51.

Clarkson, P. L., and I. S. Liepins. 1994. Grizzly bear population estimate and characteristics in the Anderson and Horton Rivers area, Northwest Territories. International Conference on Bear Research and Management 9(1):213–21.

Coy, P. L., and D. L. Garshelis. 1992. Reconstructing reproductive histories of black bears from the incremental layering in dental cementum. Canadian Journal of Zoology 70:2150–60.

Cracraft, J. 1983. Species concepts and speciation analysis. Current Ornithology 1:159–87.

Craighead, D. J. 1998. An integrated satellite technique to evaluate grizzly bear habitat use. Ursus 10:187–201.

Craighead, F. C., Jr. 1976. Grizzly bear ranges and movement as determined by radiotracking. International Conference on Bear Research and Management 3:97–109.

Craighead, F. C., Jr., and J. J. Craighead. 1972. Grizzly bear prehibernation and denning activities as determined by radiotracking. Wildlife Monograph 32:1–35.

Craighead, F. L., and E. Vyse. 1996. Brown/grizzly bear metapopulations. Pages 325–51 *in* D. McCullough, ed. Metapopulations and wildlife conservation management. Island Press, Washington, DC.

Craighead, F. C., Jr., J. J. Craighead, and R. S. Davies. 1963. Radiotracking of grizzly bears. Pages 133–48 *in* L. E. Slater, ed. Biotelemetry: The use of telemetry in animal behavior and physiology in relation to ecological problems. Macmillan, New York.

Craighead, F. L., D. Paetkau, H. V. Reynolds, E. R. Vyse, and C. Strobeck. 1995. Microsatellite analysis of paternity and reproduction in arctic grizzly bears. Journal of Heredity 86:255–61.

Craighead, F. L., D. Paetkau, H. V. Reynolds, C. Strobeck, and E. R. Vyse. 1998. Use of microsatellite DNA analysis to infer breeding behavior and demographic processes in an arctic grizzly bear population. Ursus 10:323–27.

Craighead, J. J. 1977. A proposed delineation of critical grizzly bear habitat in the Yellowstone Region (Monograph Series No. 1). International Association of Bear Research and Management.

Craighead, J. J., and J. A. Mitchell. 1982. Grizzly bear. Pages 515–56 *in* J. A. Chapman and G. A. Feldhamer, eds. Wild mammals of North America. Johns Hopkins University Press, Baltimore.

Craighead, J. J., M. G. Hornocker, and F. C. Craighead. 1969. Reproductive biology of young female grizzly bears. Journal of Reproductive Fertility (Supplement) 6:447–75.

Craighead, J. J., F. C. Craighead, Jr., J. R. Varney, and C. E. Cote. 1971. Satellite monitoring of black bear. Bioscience 21:1206–12.

Craighead, J. J., J. R. Varney, and F. C. Craighead. Jr. 1974. A population analysis of the Yellowstone grizzly bears (Montana Forest and Conservation Experiment Station Bulletin 40). University of Montana, Missoula.

Craighead, J. J., J. S. Sumner, and G. B. Scaggs. 1982. A definitive system for analysis of grizzly bear habitat and other wilderness resources (Wildlife-Wildlands Institute Monograph 1). University of Montana Foundation, Missoula.

Craighead, J. J., K. R. Greer, R. R. Knight, and H. I. Pac. 1988. Grizzly bear mortalities in the Yellowstone Ecosystem, 1959–1987. Montana Department of Fish, Wildlife and Parks, Craighead Wildlife Institute, Interagency Grizzly Bear Study Team, and National Fish and Wildlife Foundation.

Craighead, J. J., J. S. Sumner, and J. A. Mitchell. 1995. The grizzly bears of Yellowstone: Their ecology in the Yellowstone Ecosystem, 1959–1992. Island Press, Washington, DC.

Cronin, M. A., S. C. Amstrup, G. W. Garner, and E. R. Vyse. 1991. Interspecific and intraspecific mitochondrial DNA variation in North American bears (*Ursus*). Canadian Journal of Zoology 69:2985–92.

Darling, L. M. 1987. Habitat use by grizzly bear family groups in interior Alaska. International Conference on Bear Research and Management 7:169–78.

Davis, D. D. 1950. Hybrids of the polar and Kodiak bear. Journal of Mammalogy 31:449–50.

Davis, D. D. 1964. The giant panda: A morphological study of evolutionary mechanisms. Fieldiana Zoology Memoirs 3:1–339.

Dean, F. C. 1987. Brown bear density, Denali National Park, Alaska, and sighting efficiency adjustment. International Conference on Bear Research and Management 7:37–43.

Dittrich, L., and H. Kronberger. 1963. Biological–anatomical research concerning the biology of reproduction of the brown bear (*Ursus arctos* L.) and other bears in captivity. Zeitschrift für Säugetierkunde 28:129.

Doak, D. 1995. Source–sink models and the problem of habitat degradation: General models and applications to the Yellowstone grizzly. Conservation Biology 9:1370–79.

Dobie, J. F. 1950. The Ben Lilly legend. Little, Brown, Boston.

Duda, M. D., S. J. Bissell, and K. C. Young. 1998. Wildlife and the American mind. Public opinion on and attitudes toward fish and wildlife management. Responsive Management, Harrisonburg, VA.

Eagle, T. C., and M. R. Pelton. 1983. Seasonal nutrition of black bears in the Great Smoky Mountains National Park. International Conference on Bear Research and Management 5:94–101.

Eberhardt, L. L. 1985. Assessing the dynamics of wild populations. Journal of Wildlife Management 49:997–1012.

Eberhardt, L. L. 1990. Survival rates required to sustain bear populations. Journal Wildlife Management 54:587–90.

Eberhardt, L. L. 1995. Population trend estimates from reproductive and survival data. Pages 13–19 *in* R. R. Knight and B. M. Blanchard, eds. Yellowstone grizzly bear investigations: Report of the Interagency Study Team, 1994. National Biological Service, Bozeman, MT.

Eberhardt, L. L., B. M. Blanchard, and R. R. Knight. 1994. Population trend of the Yellowstone grizzly bear as estimated from reproductive and survival rates. Canadian Journal of Zoology 72:360–63.

Erickson, A. W. 1965. The brown-grizzly bear in Alaska: Its ecology and management (Federal Aid in Wildlife Restoration, Vol. V, Project W-6-R-5, Work Plan F, 2nd producing February 1967). Alaska Department of Fish and Game, Juneau.

Erickson, A. W., and L. H. Miller. 1963. Cub adoption in the brown bear. Journal of Mammalogy 44:584–85.

Erickson, A. W., and D. B. Siniff. 1963. A statistical evaluation of factors influencing aerial survey results on brown bears. Transactions of the North American Wildlife Conference 28:391–408.

Erickson, A. W., J. Nellor, and G. A. Petrides. 1964. The black bear in Michigan (Research Bulletin No. 4). Michigan State University Agricultural Experiment Station, East Lansing.

Erickson, A. W., H. W. Moosman, R. J. Hensel, and W. A. Troyer. 1968. The breeding biology of the male brown bear (*Ursus arctos*). Zoologica 53:85–105.

Ewer, R. F. 1973. The carnivores. Cornell University Press, Ithaca, NY.

Fancy, S. G. 1980. Preparation of mammalian teeth for age determination by cementum layers: A review. Wildlife Society Bulletin 8:242–48.

Farley, S. D., and C. T. Robbins. 1995. Lactation, hibernation, and mass dynamics of American black bears and grizzly bears. Canadian Journal of Zoology 73:2216–22.

Fisler, G. E. 1969. Mammalian organizational systems. Contributions in Science of the Los Angeles County Museum 167:1–32.

Folk, G. E., Jr. 1967. Physiological observations of subarctic bears under winter den conditions. Pages 75–85 *in* K. Fisher, ed. Mammalian hibernation. American Elsevier, New York.

Folk, G. E., Jr., R. C. Simmonds, M. C. Brewer, and M. A. Folk. 1968. Physiology of winter denning of polar and grizzly bears. Proceedings of the Alaska Science Conference 9:26–27.

Folk, G. E., Jr., M. A. Folk, and J. J. Minor. 1972. Physiological condition of three species of bears in winter dens. International Conference on Bear Research and Management 2:107–24.

Folk, G. E., Jr., A. Larson, and M. A. Folk. 1976. Physiology of hibernating bears. International Conference on Bear Research and Management 3:373–80.

Follman, E. H., L. M. Philo, and H. V. Reynolds. 1979. Annual variations in body temperature of grizzly bears. Proceedings of the Alaska Science Conference 29:647.

Frasier, D. J., F. Gardner, G. B. Kolenosky, and S. Strathern. 1982. Estimation of harvest rate of black bears from age and sex data. Wildlife Society Bulletin 10:53–57.

French, S. P., and M. G. French. 1990. Predatory behavior of grizzly bears feeding on elk calves in Yellowstone National Park, 1986–88. International Conference on Bear Research and Management 8:335–41.

French, S. P., M. G. French, and R. R. Knight. 1994. Grizzly bear use of army cutworm moths in the Yellowstone Ecosystem. International Conference on Bear Research and Management 9:389–99

Gard, R. 1971. Brown bear predation on sockeye salmon at Karluk Lake, Alaska. Journal of Wildlife Management 35:193–204.

Garshelis, D. L. 1994. Density-dependent population regulation of black bears. Pages 3–14 *in* M. Taylor, ed. Density-dependent population regulation in black, brown, and polar bears (Monograph Series No. 3). International Conference on Bear Research and Management.

Garshelis, D. L., K. V. Noyce, and P. L. Coy. 1998. Calculating average age of first reproduction free of the biases prevalent in bear studies. Ursus 10:437–47.

Gasaway, W. C., R. D. Boertje, D. V. Grangaard, D. G. Kelleyhouse, R. O. Stephenson, and D. G. Larson. 1992. The role of predation in limiting moose at low densities in Alaska and Yukon and implications for conservation. Wildlife Monograph 120:1–59.

Gau, R. J. 1998. Food habitats, body condition, and habitat of the barren-ground grizzly bear. M.S. Thesis, University of Saskatchewan, Saskatoon, Canada.

Gebhard, J. G. 1982. Annual activities and behavior of a grizzly bear (*Ursus arctos*) family in northern Alaska. M.S. Thesis, University of Alaska, Fairbanks.

Gibeau, M. L. 1993. Grizzly bear habitat effectiveness model for Banff, Yoho, and Kootenay National Parks. Parks Canada Report, Banff National Park, Banff, Alberta.

Gibeau, M. L. 1998. Grizzly bear habitat effectiveness model for Banff, Yoho and Kootenay National Parks, Canada. Ursus 10:235–41.

Gibeau, M. L., S. Herrero, J. L. Kansas, and B. Benn. 1996. Grizzly bear population and habitat status in Banff National Park: A report to the Banff Bow Valley Task Force by the Eastern Slopes Grizzly Bear Project. University of Calgary, Calgary, Alberta, Canada.

Glass, B. P. 1974. A key to the skulls of North American mammals, 2nd ed. Author, Stillwater, OK.

Glenn, L. P. 1975. Report on 1974 brown bear studies: Distribution and movement of Alaska Peninsula brown bears (Federal Aid in Wildlife Restoration Project W-17-6 and W17-7, Jobs 4.4R and 4.6R). Alaska Department of Fish and Game, Juneau.

Glenn, L. P. 1980. Morphometric characteristics of brown bears on the central Alaska Peninsula. International Conference on Bear Research and Management 4:313–19.

Glenn, L. P., and L. E. Miller. 1980. Seasonal movements of an Alaska Peninsula brown bear population. International Conference on Bear Research and Management 4:307–12

Glenn, L. P., J. W. Lentfer, J. B. Faro, and L. H. Miller. 1974. Reproductive biology of female brown bears (*Ursus arctos*), McNeil River, Alaska. International Conference on Bear Research and Management 3:381–90.

Gordon, K. R. 1977. Molar measurements as a taxonomic tool in *Ursus*. Journal of Mammalogy 58:247–48.

Gray, A. P. 1954. Mammalian hybrids. A check list with bibliography. Commonwealth Bureau of Animal Breeding and Genetics, Edinburgh Technical Communications 10:1–144.

Green, G. 1994. Use of spring carrion by bears in Yellowstone National Park. M.S. Thesis, University of Idaho, Moscow.

Green, G. I., D. J. Mattson, and J. M. Peek. 1997. Spring feeding on ungulate carcasses by grizzly bears in Yellowstone National Park. Journal of Wildlife Management 6:1040–55.

Grinnell, J., J. S. Dixon, and J. Linsdale. 1937. Fur-bearing mammals of California, their natural history, systematic status, and relations to man. University of California Press, Berkeley.

Groves, C. 1987. A compilation of grizzly bear reports from central and northern Idaho (Endangered Species Projects E-III, E-IV). Idaho Department of Fish and Game, Boise.

Guilday, J. E. 1968. Grizzly bears from eastern North America. American Midland Naturalist 79:247–50.

Gunther, K. A., and R. A. Renkin. 1990. Grizzly bear predation on elk calves and other fauna of Yellowstone National Park. International Conference on Bear Research and Management 8:329–34.

Guppy, M. 1986. The hibernating bear: Why is it so hot, and why does it cycle urea through the gut? Trends in Biochemical Science 11:274–76.

Gustafson, K. A., and L. B. Fox. 1983. A comprehensive interactive computer program for calculating and plotting home ranges and distributions. Pages 299–317 *in* D. G. Pincock, ed. Fourth international conference on wildlife biotelemetry. Applied Microelectronics Institute and Technology, University of Nova Scotia, Halifax, Canada.

Hall, E. R. 1984. Geographic variation among brown and grizzly bears (*Ursus arctos*) in North America (Special Publication 13). Museum of Natural History, University of Kansas, Lawrence.

Hamer, D., and S. Herrero. 1987. Grizzly bear food and habitat in the Front Ranges of Banff National Park, Alberta. International Conference on Bear Research and Management 7:199–213.

Hamer, D., and S. Herrero. 1990. Courtship and use of mating areas by grizzly bears in the Front Ranges of Banff National Park, Alberta. Canadian Journal of Zoology 68:2695–97.

Hamer, D., and S. Herrero. 1991. Elk, *Cervus elaphus,* calves as food for grizzly bears, *Ursus arctos,* in Banff National Park, Alberta. Canadian Field-Naturalist 105:101–3.

Harding, L. 1976. Den-site selection of arctic coastal grizzly bears (*Ursus arctos* L.) on Richards Island, Northwest Territories, Canada. Canadian Journal of Zoology 54:1357–63.

Harestad, A. S., and F. L. Bunnell. 1979. Home range and body weight: A reevaluation. Ecology 60:389–402.

Hausfater, G., and S. B. Hrdy, eds. 1984. Infanticide: Comparative and evolutionary perspectives. Aldine, New York.

Hechtel, J. L. 1985. Activity and food habits of barren-ground grizzly bears in Arctic Alaska. M.S. Thesis, University of Montana, Missoula.

Heisey, D. M., and T. K. Fuller. 1985. Evaluation of survival and cause-specific mortality rates using telemetry data. Journal of Wildlife Management 49:668–74.

Hellgren, E. C. 1998. Physiology of hibernation in bears. Ursus 10:467–77.

Hensel, R. J., W. A. Troyer, and A. W. Erickson. 1969. Reproduction in the female brown bear. Journal of Wildlife Management 33:357–65.

Herrero, S. 1972. Aspects of evolution and adaptation in American black bears (*Ursus americanus* Pallas) and brown and grizzly bears (*Ursus arctos* Linné) of North America. International Conference on Bear Research and Management 2:221–31.

Herrero, S. 1985. Bear attacks: Their causes and avoidance. Lyons and Burford, New York.

Herrero, S., and D. Hamer. 1977. Courtship and copulation of a pair of grizzly bears, with comments on reproductive plasticity and strategy. Journal of Mammalogy 58:441–44.

Herrero, S., P. S. Miller, and U. S. Seal, eds. 2000. Population and habitat viability assessment for the grizzly bear of the Central Rockies Ecosystem (*Ursus arctos*). Eastern Slopes Grizzly Bear Project, University of Calgary, Calgary, Canada, and Conservation Breeding Specialist Group, Apple Valley, MN.

Hessing, P., and L. Aumiller. 1994. Observations of conspecific predation by brown bears, *Ursus arctos,* in Alaska. Canadian Field-Naturalist 108:332–36.

Hessler, E., J. R. Tester, D. B. Siniff, and M. M. Nelson. 1970. A biotelemetry study of survival of pen-reared pheasants released in selected habitats. Journal of Wildlife Management 34:267–74.

Hewitt, D. G., and C. T. Robbins. 1996. Estimating grizzly bear food habits from fecal analysis. Wildlife Society Bulletin 24:547–50.

Hilderbrand, G. V., C. C. Schwartz, C. T. Robbins, M. E. Jacoby, T. A. Hanley, S. M. Arthur, and C. Servheen. 1999a. The importance of meat, particularly salmon, to body size, population productivity, and conservation of North American brown bears. Canadian Journal of Zoology 77:132–38.

Hilderbrand, G. V., T. A. Hanley, C. T. Robbins, and C. C. Schwartz. 1999b. Role of brown bears (*Ursus arctos*) in the flow of marine nitrogen into a terrestrial ecosystem. Oecologia 121:546–50.

Hilderbrand, G. V., C. C. Schwartz, C. T. Robbins, and T. A. Hanley. 2000. Effect of hibernation and reproductive status on body mass and condition of coastal brown bears. Journal of Wildlife Management 64:178–83.

Hogg, J. T., N. S. Weaver, J. J. Craighead, M. L. Pokorny, B. M. Steele, R. L. Redmond, and F. B. Fisher. 1999. Abundance and spatial distribution of grizzly bear plant-food groups in the Salmon–Selway Ecosystem: A preliminary analysis and report. Craighead Wildlife-Wildlands Institute, Missoula, MT [Excerpt in Appendix 21D, Pages 6-247–6-270 *in* Grizzly bear recovery in the Bitterroot Ecosystem (Final Environmental Impact Statement). U.S. Department of the Interior, U.S. Fish and Wildlife Service (March 2000)].

Holms, G. W. 1998. Interactions of sympatric black and grizzly bears in northwest Wyoming. M.S. Thesis, University of Wyoming, Laramie.

Hornocker, M. G. 1962. Population characteristics and social and reproductive behavior of the grizzly bear in Yellowstone National Park. M.S. Thesis, University of Montana, Missoula.

Hovey, F. W., and B. N. McLellan. 1996. Estimating population growth of grizzly bears from the Flathead River drainage using computer simulations of reproduction and survival rates. Canadian Journal of Zoology 74:1409–16.

Irving, L., and J. Krog. 1954. Body temperatures of arctic and subarctic birds and mammals. Journal of Applied Physiology 6:667–80.

Jacoby, M. E., G. V. Hilderbrand, C. Servheen, C. C. Schwartz, S. M. Arthur, T. A. Hanley, C. C. Robbins, and R. Michener. 1999. Trophic relations of brown and black bears in several western North American ecosystems. Journal of Wildlife Management 63:921–29.

Jaczewski, Z., J. Gill, and S. Kozniewski. 1960. Capacity of the different parts of the digestive tract in the brown bear. Transactions of the International Union of Game Biologists 4:146–54.

Janson, C. H., and C. P. Van Schaik. 2000. The behavioral ecology of infanticide by males. Pages 469–94 *in* C. P. Van Schaik and C. H. Janson, eds. Infanticide by males and its implications. Cambridge University Press, New York, New York. USA.

Johnson, D. H. 1980. The comparisons of usage and availability measurements for evaluations of resource preference. Ecology 61:65–71.

Judd, S., and R. R. Knight. 1980. Movements of radio-instrumented grizzly bears within the Yellowstone transport area. International Conference on Bear Research and Management 4:359–67.

Judd, S. L., R. R. Knight, and B. M. Blanchard. 1986. Denning of grizzly bear in the Yellowstone National Park Area. International Conference on Bear Research and Management 6:111–17.

Kaplan, E. L., and P. Meier. 1958. Nonparametric estimation from incomplete observations. Journal of the American Statistical Association 53:457–81.

Kasworm, W. F., and T. L. Manley. 1990. Road and trail influences on grizzly bears and black bears in northwest Montana. International Conference on Bear Research and Management 8:79–84.

Kendall, K. C., and S. F. Arno. 1990. Whitebark pine—An important but endangered wildlife resource. Pages 264–73 *in* W. C. Schmidt and K. J. McDonald, comps. Proceedings of symposium on whitebark pine ecosystems: Ecology and management of a high mountain resource (General Technical Report INT-270). U.S. Department of Agriculture, Forest Service, Ogden, UT.

Kevan, P. G., and D. M. Kendall. 1997. Liquid assets for fat bankers: Summer nectarivory by migratory moths in the Rocky Mountains, Colorado, U.S.A. Arctic and Alpine Research 29:478–82.

Kingsley, M. C. S., J. A. Nagy, and R. H. Russell. 1983. Patterns of weight gain and loss for grizzly bears in northern Canada. International Conference on Bear Research and Management 5:174–78.

Kingsley, M. C. S., J. A. Nagy, and H. V. Reynolds. 1988. Growth in length and weight of northern brown bears: Differences between sexes and populations. Canadian Journal of Zoology 66:981–86.

Kitchener, A. C. 1994. A review of the evolution, systematics, functional morphology, distribution, and status of the Ursidae. International Zoo News 242:4–24.

Klaver, R. W., J. J. Claar, D. B. Rockwell, H. R. Mays, and C. F. Acevedo. 1986. Grizzly bears, insects, and people: Bear management in the McDonald Peak region, Montana. Pages 204–11 *in* G. Contreras and K. Evans, comps.

Proceedings of the grizzly habitat symposium (U.S. Forest Service General Technical Report INT-207). U.S. Forest Service Intermountain Research Station, Ogden, UT.

Kleiber, M. 1947. Body size and metabolic rate. Physiological Review 27:511–41.

Knight, R. R., and L. L. Eberhardt. 1985. Population dynamics of Yellowstone grizzly bears. Ecology 66:323–34.

Knight, R. R., and L. L. Eberhardt. 1987. Prospects for Yellowstone grizzlies. International Conference on Bear Research and Management 7:45–50.

Knight, R. R., B. Blanchard, and K. Kendall. 1981. Yellowstone grizzly bear investigations: Annual report of the Interagency Study Team, 1980. U.S. Department of the Interior, Interagency Grizzly Bear Study Team, Bozeman, MT.

Knight, R. R., B. M. Blanchard, and L. L. Eberhardt. 1988. Mortality patterns and population sinks for Yellowstone grizzly bears, 1973–1985. Wildlife Society Bulletin 16:121–25.

Knight, R. R., B. M. Blanchard, and L. L. Eberhardt. 1995. Appraising status of the Yellowstone grizzly bear population by counting females with cubs-of-the-year. Wildlife Society Bulletin 23:245–48.

Kohn, M., F. Knauer, A. Stoffella, W. Schroder, and S. Paabo. 1995. Conservation genetics of European brown bear: A study using excremental PCR. Molecular Ecology 4:95–103.

Kunkel, K., W. Clark, and C. Servheen. 1991. A remote camera survey for grizzly bears in low human areas of the Bitterroot grizzly bear evaluation area. Unpublished report. Idaho Department of Fish and Game, Boise.

Kurtén, B. 1968. Pleistocene mammals of Europe. Weidenfeld & Nicolson, London.

Kurtén, B. 1973. Transberingian relationships of *Ursus arctos* Linné (brown and grizzly bears). Commentationes Biologicae 65:1–10.

Larsen, D. G., D. A. Gauthier, and R. L. Markel. 1989. Causes and rate of moose mortality in the southwest Yukon. Journal of Wildlife Management 53:548–57.

LeFranc, M. N., Jr., M. B. Moss, K. A. Patnode, and W. C. Sugg, eds. 1987. Grizzly bear compendium. Interagency Grizzly Bear Committee, Washington, DC.

Lindzey, F. G., and E. C. Meslow. 1977. Home range and habitat use by black bear in southwestern Washington. Journal of Wildlife Management 41:413–25.

Linnell, J. D. C., J. E. Swenson, R. Anderson, and B. Barnes. 2000. How vulnerable are denning bears to disturbance? Wildlife Society Bulletin 28:400–413.

Lotka, A. J. 1907. Relation between birth rates and death rates. Science 26:21–22.

Lyman, C. P., J. Willis, A. Malan, and L. Wang. 1982. Hibernation and torpor in mammals and birds. Academic Press, New York.

Mace, R. D., and T. L. Manley. 1993. South Fork Flathead River grizzly bear project (Progress Report 1992). Montana Department of Fish, Wildlife and Parks, Helena.

Mace, R. D., and J. S. Waller. 1997. Final report: Grizzly bear ecology in the Swan Mountains. Montana Department of Fish, Wildlife and Parks, Helena.

Mace, R. D., and J. S. Waller. 1998. Demography and population trend of grizzly bears in the Swan Mountains, Montana. Conservation Biology 12:1005–16.

Mace, R. D., S. C. Minta, T. L. Manley, and K. E. Aune. 1994. Estimating grizzly bear population size using camera sightings. Wildlife Society Bulletin 22:74–83.

Mace, R. D., J. S. Waller, T. L. Manley, L. J. Lyon, and H. Zuuring. 1996. Relationships among grizzly bears, roads, and habitat in the Swan Mountains, Montana. Journal of Applied Ecology 33:1395–1404.

Mace, R. D., J. S. Waller, T. L. Manley, K. Ake, and W. T. Wittinger. 1999. Landscape evaluation of grizzly bear habitat in western Montana. Conservation Biology 13:367–77.

Macey, A. 1979. The status of the grizzly bear in Canada. National Museum of Natural Science, Ottawa.

MacPherson, A. H. 1965. The barren-ground grizzly bear and its survival in Canada. Canadian Audubon 27:2–8.

Martinka, C. J. 1974. Population characteristics of grizzly bears in Glacier National Park, Montana. Journal of Mammalogy 55:21–29.

Martinka, C. J. 1976. Ecological role and management of grizzly bears in Glacier National Park, Montana. International Conference on Bear Research and Management 3:147–56.

Matson, G. M., L. Van Daele, E. Goodwin, L. Aumiller, H. Reynolds, and H. Hristienko. 1993. A laboratory manual for cementum age determination of Alaskan brown bear first premolar teeth. Matson's Laboratory, Milltown, MT.

Mattson, D. J. 1990. Human impacts on bear habitat use. International Conference on Bear Research and Management 8:33–56.

Mattson, D. J. 1997. Use of ungulates by Yellowstone grizzly bears *Ursus arctos*. Biological Conservation 81:161–77.

Mattson, D. J. 1998. Changes in mortality of Yellowstone's grizzly bears. Ursus 10:129–38.

Mattson, D. J., and D. P. Reinhart. 1997. Excavation of red squirrel middens by grizzly bears in the whitebark pine zone. Journal of Applied Ecology 34:926–40.

Mattson, D. J., R. R. Knight, and B. M. Blanchard. 1987. The effects of developments and primary roads on grizzly bear habitat use in Yellowstone National Park, Wyoming. International Conference on Bear Research and Management 7:259–73.

Mattson, D. J., B. M. Blanchard, and R. R. Knight. 1991a. Food habits of Yellowstone grizzly bears, 1977–1987. Canadian Journal of Zoology 69:1619–29.

Mattson, D. J., C. M. Gillin, S. A. Benson, and R. R. Knight. 1991b. Bear use of alpine insect aggregations in the Yellowstone ecosystem. Canadian Journal of Zoology 69:2430–35.

Mattson, D. J., D. P. Reinhart, and B. M. Blanchard. 1994. Variation in production and bear use of whitebark pine seeds in the Yellowstone area. Pages 205–20 *in* D. G. Despain, ed. Plants and their environments: Proceedings of the first biennial science conference on the Greater Yellowstone Ecosystem (Technical Report NPS/NRYELL/NRTR). U.S. National Park Service.

Mattson, D. J., R. G. Wright, K. C. Kendall, and C. J. Martinka. 1995. Grizzly bears. Pages 103–5 *in* E. T. LaRoe, G. S. Farris, C. E. Puckett, P. D. Doran, and M. J. Mac, eds. Our living resources: A report to the nation on the distribution, abundance, and health of U.S. plants, animals, and ecosystems. U.S. Department of the Interior, National Biological Service, Washington, DC.

Mattson, D. J., G. I. Green, and R. Swalley. 1999a. Geophagy by Yellowstone grizzly bears. Ursus 11:109–16.

Mattson, D. J., K. Barber, R. Maw, and R. Renkin. 1999b. Coefficients of productivity for Yellowstone's grizzly bear habitat (Technical Report). U.S. Geological Survey Forest and Rangeland Ecosystem Science Center, Corvallis, OR.

McCullough, D. R. 1981. Population dynamics of the Yellowstone grizzly bear. Pages 173–96 *in* C. W. Fowler and T. D. Smith, eds. Dynamics of large mammal populations. John Wiley, New York.

McCullough, D. R. 1986. The Craigheads' data on Yellowstone grizzly bear populations: Relevance to current research and management. International Conference on Bear Research and Management 6:21–32.

McLellan, B. N. 1985. Use availability analysis and timber selection by grizzly bears. Pages 163–66 *in* G. Contreras and K. Evans, comps. Proceedings of the grizzly habitat symposium, Missoula, Montana (U.S. Forest Service General Technical Report INT-207). U.S. Forest Service Intermountain Research Station, Ogden, UT.

McLellan, B. N. 1989a. Dynamics of a grizzly bear population during a period of industrial resource extraction. I. Density and age–sex composition. Canadian Journal of Zoology 67:1856–60.

McLellan, B. N. 1989b. Dynamics of a grizzly bear population during a period of industrial resource extraction. II. Mortality rates and causes of death. Canadian Journal of Zoology 67:1861–64.

McLellan, B. N. 1989c. Population dynamics of grizzly bears during a period of resource extraction development. III. Natality and rate of change. Canadian Journal of Zoology 67:1865–68.

McLellan, B. N. 1990. Relationships between human industrial activity and grizzly bears. International Conference on Bear Research and Management 8:57–64.

McLellan, B. N. 1994. Density-dependent population regulation of brown bears. Pages 15–24 *in* M. Taylor, ed. Density-dependent population regulation of black, brown, and polar bears. (Monograph Series 3). International Conference on Bear Research and Management.

McLellan, B. N., and V. Banci. 1999. Status and management of the brown bear in Canada. Pages 46–50 *in* C. Servheen, S. Herrero, and B. Peyton, comps. Bears: Status survey and conservation action plan. IUCN/SSC Bear and Polar Bear Specialist Groups, IUCN, Gland, Switzerland, and Cambridge, UK.

McLellan, B. N., and F. W. Hovey. 1995. The diet of grizzly bears in the Flathead River drainage of southeastern British Columbia. Canadian Journal of Zoology 73:704–12.

McLellan, B. N., and D. M. Shackleton. 1988. Grizzly bears and resource extraction industries: Effects of roads on behavior, habitat use and demography. Journal of Applied Ecology 25:451–60.

McLellan, B. N., F. W. Hovey, R. D. Mace, J. G. Woods, D. W. Carney, M. L. Gibeau, W. L. Wakkinen, and W. F. Kasworm. 1999. Rates and causes of grizzly bear mortality in the interior mountains of British Columbia,

Alberta, Montana, Washington, and Idaho. Journal of Wildlife Management 63:911–20.

McLoughlin, P. D. 2000. The spatial organization and habitat selection patterns of barren-ground grizzly bears in the central Arctic. Ph.D. Dissertation, University of Saskatchewan, Saskatoon, Canada.

McLoughlin, P. D., R. L. Case, R. J. Gau, S. H. Ferguson, and F. Messier. 1999. Annual and seasonal movement patterns of barren-ground grizzly bears in the central Northwest Territories. Ursus 11:79–86.

Meagher, M., and J. R. Phillips. 1983. Restoration of natural populations of grizzly and black bear in Yellowstone National Park. International Conference on Bear Research and Management 5:152–58.

Mealey, S. P. 1975. The natural food habits of free ranging grizzly bears in Yellowstone National Park, 1973–1974. M.S. Thesis, Montana State University, Bozeman.

Mealey, S. P. 1980. The natural food habits of grizzly bears in Yellowstone National Park, 1973–74. International Conference on Bear Research and Management 4:281–92.

Melquist, W. 1985. A preliminary survey to determine the status of grizzly bears (*Ursus arctos horribilis*) in the Clearwater National Forest of Idaho. Idaho Cooperative Fish and Wildlife Research Unit, University of Idaho, Moscow.

Merriam, C. H. 1918. Review of the grizzly and big brown bears of North America (genus *Ursus*) with the description of a new genus, *Vetularctos*. North American Fauna 41:1–136.

Merrill, T., D. J. Mattson, R. G. Wright, and H. B. Quigley. 1999. Defining landscapes suitable for restoration of grizzly bears *Ursus arctos* in Idaho. Biological Conservation 87:231–48.

Miller, S. D. 1988. Impacts of increased hunting pressure on the density, structure, and dynamics of brown bear populations in Alaska's Management Unit 13 (Federal Aid in Wildlife Restoration Final Report, Project W-22-6, Job 4.21). Alaska Department of Fish and Game, Juneau.

Miller, S. D. 1990a. Denning ecology of brown bears in southcentral Alaska and comparisons with sympatric black bear populations. International Conference on Bear Research and Management 8:279–87.

Miller, S. D. 1990b. Detection of differences in brown bear density and population composition caused by hunting. International Conference on Bear Research and Management 8:393–404.

Miller, S. D. 1990c. Impact of increased bear hunting on survivorship of young bears. Wildlife Society Bulletin 18:462–67.

Miller, S. D. 1990d. Impacts of increased hunting pressure on the density, structure, and dynamics of brown bear populations in Alaska's Game Management Unit 13 (Project. W-23-3). Alaska Department of Fish and Game, Juneau.

Miller, S. D. 1990e. Population management of bears in North America. International Conference on Bear Research and Management 8:357–73.

Miller, S. D. 1993. Brown bears in Alaska: A statewide management overview (Wildlife Technical Bulletin 11). Alaska Department of Fish and Game, Juneau.

Miller, S. D. 1995a. Impacts of heavy hunting pressure on the density and demographics of brown bear populations in southcentral Alaska (Federal Aid in Wildlife Restoration Progress Report W-24-3, Study 4.26). Alaska Department of Fish and Game, Juneau.

Miller, S. D. 1995b. Impacts of increased hunting pressure on the density, structure, and dynamics of brown bear populations in Alaska's Game Management Unit 13 (Federal Aid in Wildlife Restoration Progress Report W-24-3, Study 4.21). Alaska Department Fish and Game, Juneau.

Miller, S. D. 1997. Impacts of heavy hunting pressure on the density and demographics of brown bear populations in southcentral Alaska (Federal Aid in Wildlife Restoration, Final Report W-24-2, W-24-3, and W-24-4, Study 4.26). Alaska Department of Fish and Game, Juneau.

Miller, S. D., and W. B. Ballard. 1982. Homing of transported Alaskan brown bears. Journal of Wildlife Management 46:869–76.

Miller, S. D., and W. B. Ballard. 1992. Analysis of an effort to increase moose calf survivorship by increased hunting of brown bears in southcentral Alaska. Wildlife Society Bulletin 20:445–54.

Miller, S. D., and J. Schoen. 1999. Status and management of the brown bear in Alaska. Pages 40–46 *in* C. Servheen, S. Herrero, and B. Peyton, comps. Bears: Status survey and conservation action Plan. IUCN/SSC Bear and Polar Bear Specialist Groups, IUCN, Gland, Switzerland, and Cambridge, UK.

Miller, S. D., and R. A. Sellers. 1992. Brown bear density on the Alaska Peninsula at Black Lake, Alaska. Final report on completion of density estimation objectives of cooperative interagency brown bear studies on the Alaska Peninsula. Alaska Department of Fish and Game, Juneau.

Miller, S. D., and V. L. Tutterrow. 1999. Characteristics of nonsport mortalities to brown and black bears and human injuries from bears in Alaska. Ursus 11:235–48.

Miller, S. D., E. F. Becker, and W. B. Ballard. 1987. Black and brown bear density estimates using modified capture–recapture techniques in Alaska. International Conference on Bear Research and Management 7:23–35.

Miller, S. D., G. C. White, R. A. Sellers, H. V. Reynolds, J. W. Schoen, K. Titus, V. G. Barnes, Jr., R. B. Smith, R. R. Nelson, W. B. Ballard, and C. C. Schwartz. 1997. Brown and black bear density estimation in Alaska using radiotelemetry and replicated mark-resight techniques. Wildlife Monograph 133:1–55.

Miller, S. J., N. Barichello, and D. Tait. 1982. The grizzly bears of the MacKenzie Mountains, Northwest Territories (Completion Report 3). Northwest Territories Wildlife Service, Yellowknife, Canada.

Miller, S. M., S. D. Miller, and D. W. McCollum. 1998. Attitudes toward and relative value of Alaskan brown and black bears to resident voters, resident hunters, and nonresident hunters. Ursus 10:357–76.

Modafferi, R. D. 1984. Review of Alaska Peninsula brown bear investigations (Federal Aid in Wildlife Restoration Final Report Project W-17-10, W17-11, W-21-1, W-21-2, and W-22-1, Job 4.12R). Alaska Department of Fish and Game, Juneau.

Mohr, C. O. 1947. Table of equivalent populations of North American small mammals. American Midland Naturalist 37:223–49.

Mowat, G., and C. Strobeck. 2000. Estimating population size of grizzly bears using hair capture, DNA profiling, and mark–recapture analysis. Journal of Wildlife Management 64:183–93.

Mundy, K. R. D., and W. A. Fuller. 1964. Age determination in the grizzly bear. Journal of Wildlife Management 28:863–66.

Murie, A. 1944. The wolves of Mount McKinley (National Park Service, Fauna Series No. 5). U.S. Government Printing Office, Washington, DC.

Murie, A. 1981. The grizzlies of Mount McKinley (Scientific Monograph Series No. 14). U.S. Department of the Interior, National Park Service, Washington, DC.

Mystrud, I. 1983. Characteristics of summer beds of European brown bears in Norway. International Conference on Bear Research and Management 5:208–22.

Nagy, J. A., and M. A. Haroldson. 1990. Comparisons of some home range and population parameters among four grizzly bear populations in Canada. International Conference on Bear Research and Management 8:227–35.

Nagy, J. A., R. H. Russell, A. M. Pearson, M. C. Kingsley, and B. C. Goski. 1983a. Ecological studies of the grizzly bear in arctic mountains, northern Yukon Territories, 1972 to 1975. Canadian Wildlife Service, Edmonton, Alberta.

Nagy, J. A., R. H. Russell, A. M. Pearson, M. C. Kingsley, and C. B. Larsen. 1983b. A study of grizzly bears on the barren grounds of Tuktoyaktuk Peninsula and Richards Island, Northwest Territories, 1974–1978. Canadian Wildlife Service Report, Edmonton, Alberta.

Nagy, J. A., M. C. S. Kingsley, R. H. Russell, and A. M. Pearson. 1984. Relationship of weight to chest girth in the grizzly bear. Journal of Wildlife Management 48:1439–40.

Nelson, R. A. 1973. Winter sleep in the black bear: A physiologic and metabolic marvel. Mayo Clinic Proceedings 48:733–37.

Nelson, R. A. 1980. Protein and fat metabolism in hibernating bears. Federal Proceedings 39:2955–58.

Nelson, R. A., and T. D. I. Beck. 1984. Hibernation adaptation in the black bear. Implications for management. Proceedings Eastern Workshop on Black Bear Research and Management 7:48–53.

Nelson, R. A., D. L. Steiger, and T. D. I. Beck. 1983a. Neuroendrocine and metabolic interactions in the hibernating black bear. Acta Zoologica Fennica 174:137–41.

Nelson, R. A., G. E. Folk, Jr., E. W. Pfeiffer, J. J. Craighead, C. J. Jonkel, and D. L. Steiger. 1983b. Behavior, biochemistry, and hibernation in black, grizzly, and polar bears. International Conference on Bear Research and Management 5:284–90.

Noss, R. F., and L. D. Harris. 1986. Nodes, networks, and MUMs: Preserving diversity at all scales. Environmental Management 10:299–309.

O'Brien, S. L., and F. G. Lindzey. 1998. Aerial sightability and classification of grizzly bears at moth aggregation sites in the Absaroka Mountain, Wyoming. International Conference on Bear Research and Management 10:427–35.

Odum, E. P. 1959. Fundamentals of ecology, 2nd ed. W. B. Saunders, Philadelphia, PA.

Onoyama, K., and R. Haga. 1982. New record of four fetuses in a litter of Yeso brown bears (*Ursus arctos yesoensis*) Lydekker with mention of prenatal growth and development. Journal of the Mammal Society of Japan 9:1–8.

Pac, H. I., and A. R. Dood. 1998. Five year update of the programmatic environmental impact statement: The grizzly bear in Northwestern Montana, 1991–1995. Montana Department of Fish, Wildlife and Parks, Helena.

Pasitschniak-Arts, M. 1993. Mammalian species: *Ursus arctos*. American Society of Mammalogy 439:1–10.

Peek, J. M., M. R. Pelton, H. D. Picton, J. W. Schoen, and P. Zager. 1987. Grizzly bear conservation and management: A review. Wildlife Society Bulletin 15:160–69.

Peek, J., J. Beecham, D. Garshelis, F. Messier, S. Miller, and D. Strickland. 2003. Management of grizzly bears in British Columbia: A review by an independent scientific panel. Unpublished report submitted to Minister of Water, Land and Air Protection, Government of British Columbia, Victoria, British Columbia.

Pearson, A. M. 1965. Study of the grizzly bear in the Yukon (Progress Report for 1964, Project No. M5-1-2). Canadian Wildlife Service.

Pearson, A. M. 1975. The northern grizzly bear *Ursus arctos* L. (Report Series No. 34). Canadian Wildlife Service, Ottawa.

Peterson, R. L. 1965. A well-preserved grizzly bear skull recovered from a late glacial deposit near Lake Simco, Ontario. Nature 208:1233–34.

Phillips, M. K. 1987. Behavior and habitat use of grizzly bears in northeastern Alaska. International Conference on Bear Research and Management 7:159–67.

Picton, H. D. 1978. Climate and the reproduction of grizzly bears in Yellowstone National Park. Nature 274:888–89.

Picton, H. D., and R. R. Knight. 1986. Using climate data to predict grizzly bear litter size. International Conference on Bear Research and Management 6:41–44.

Poelker, R. J., and H. D. Hartwell. 1973. Black bear of Washington: Its biology, natural history, and relationship to forest regeneration (Biological Bulletin No. 14). Washington State Game Department, Olympia.

Pollock, K. H., S. R. Winterstein, C. M. Bunck, and P. D. Curtis. 1989. Survival analysis in telemetry studies: The staggered entry design. Journal of Wildlife Management 53:7–15.

Pritchard, G. T., and C. T. Robbins. 1990. Digestive and metabolic efficiencies of grizzly and black bears. Canadian Journal of Zoology 68:1645–51.

Province of British Columbia. 1999. Grizzly bear harvest management. Procedure manual, Vol. 4, Sect. 7, Subsect. 04.04.

Pruitt, C. H., and G. M. Burghardt. 1977. Communication in terrestrial carnivores: Mustelidae, Procyonidae, and Ursidae. Pages 767–93 *in* T. A. Sebeok, ed. How animals communicate, Vol. 2. Indiana University Press, Bloomington.

Quimby, R., and D. J. Snarski. 1974. A study of fur-bearing mammals associated with gas pipeline routes in Alaska. Pages 11–36 *in* R. D. Jakimchuk, ed. Distribution of moose, sheep, muskox and fur-bearing mammals in northeastern Alaska (Arctic Gas Biological Report Series, Vol. 6).

Ramsay, M. A., and R. L. Dunbrack. 1986. Physiological constraints on life history phenomena: The example of small bear cubs at birth. American Naturalist 127:735–43.

Randi, E., L. Gentile, G. Boscagli, D. Huber, and H. U. Roth. 1995. Mitrochondrial DNA sequence divergence among some west European brown bear (*Ursus arctos*) populations: Lessons for conservation. Heredity 73:480–89.

Rausch, R. L. 1953. On the status of some arctic mammals. Arctic 6:91–148.

Rausch, R. L. 1963. Geographic variation in size of North American brown bears, *Ursus arctos* L, as indicated by condylobasal length. Canadian Journal of Zoology 41:33–45.

Rausch, R. L. 1969. Morphogenesis and age-related structure of permanent canine teeth in the brown bear, *Ursus arctos* L. in Arctic Alaska. Zeitschrift für Morphologie der Tiere 66:167–88.

Reinhart, D. P., and D. J. Mattson. 1990. Bear use of cutthroat trout spawning streams in Yellowstone National Park. International Conference on Bear Research and Management 8:343–50.

Reinhart, D. P., M. A. Haroldson, D. J. Mattson, and K. A. Gunther. 2001. Effects of exotic species on Yellowstone's grizzly bears. Western North American Naturalist 61:277–288.

Renfree, M. B., and J. H. Calaby. 1981. Background to delayed implantation and embryonic diapause. Journal of Reproductive Fertility (Supplement) 29:1–9.

Reynolds, H. V. 1976. North Slope grizzly bear studies (Federal Aid in Wildlife Restoration, Final Report Project W-17-6 and W-17-7, Jobs 4.8R, 4.9R, 4.10R, and 4.11R). Alaska Department of Fish and Game, Juneau.

Reynolds, H. V. 1987. The brown/grizzly bear. Pages 41–42 *in* J. Rennicke, ed. Bears of Alaska in life and legend. Roberts Rinehart, Boulder, CO.

Reynolds, H. V. 1989. Grizzly bear population ecology in the Western Brooks Range, Alaska (Progress Report 1988). Alaska Department of Fish and Game and U.S. National Park Service, Fairbanks.

Reynolds, H. V. 1990. Population dynamics of a hunted grizzly bear population in the northcentral Alaska Range (Federal Aid in Wildlife Restoration Progress Report Project W-23-2, Study 4.19). Alaska Department of Fish and Game, Juneau.

Reynolds, H. V. 1992. Grizzly bear population ecology in the western Brooks Range, Alaska (Progress Report, 1990 and 1991). Alaska Department of Fish and Game and U.S. National Park Service, Fairbanks.

Reynolds, H. 1993. Effects of harvest on grizzly bear population dynamics in the northcentral Alaska Range (Federal Aid in Wildlife Restoration Final Report Project W-23-1, W-23-2, W-23-3, W-23-4, and W-23-5). Alaska Department of Fish and Game, Juneau.

Reynolds, H. V., and T. A. Boudreau. 1992. Effects of harvest rates on grizzly bear population dynamics in the northcentral Alaska Range (Federal Aid in Wildlife Restoration Progress Report Project W-23-5, Study 4.19). Alaska Department of Fish and Game, Juneau.

Reynolds, H. V., and G. Garner. 1987. Patterns of grizzly bear predation on caribou in northern Alaska. International Conference on Bear Research and Management 7:59–67.

Reynolds, H. V., and J. L. Hechtel. 1980. Big game investigations: Structure, status, reproductive biology, movements, distribution and habitat utilization of a grizzly bear population (Federal Aid in Wildlife Restoration Progress Report, Project W-17-11). Alaska Department of Fish and Game, Juneau.

Reynolds, H. V., and J. Hechtel. 1984. Structure, status, reproductive biology, movement, distribution, and habitat utilization of a grizzly bear population (Federal Aid in Wildlife Restoration Final Project Report W-21-1, W-21-2, W-22-1, and W-22-2, Job 4.14R). Alaska Department of Fish and Game, Juneau.

Reynolds, H. V., J. A. Curatolo, and R. Quimby. 1977. Denning ecology of grizzly bears in northeastern Alaska. International Conference on Bear Research and Management 3:403–9.

Reynolds, P. E., H. V. Reynolds, and E. H. Follmann. 1986. Response of grizzly bears to seismic surveys in northern Alaska. International Conference on Bear Research and Management 6:169–75.

Riley, S. J., K. Aune, R. D. Mace, and M. J. Madel. 1994. Translocation of nuisance grizzly bears in northwestern Montana. International Conference on Bear Research and Management 9(1):567–73.

Robbins, C. T. 1993. Wildlife feeding and nutrition, 2nd ed. Academic Press, San Diego, CA.

Rockwell, D. 1991. Giving voice to bear: North American Indian rituals, myths, and images of the bear. Roberts Rinehart, Niwot, CO.

Rode, K. D. 1999. Nutritional limitations on the consumption of plant foods by grizzly bears. M.S. Thesis, Washington State University, Pullman.

Rode, K. D., C. T. Robbins, and L. A. Shipley. 2001. Constraints on herbivory by grizzly bears. Oecologia. 128:62–71.

Rogers, L. L. 1977. Social relationships, movements and population dynamics of black bears in northeastern Minnesota. Ph.D. Dissertation, University of Minnesota, Minneapolis.

Roosevelt, T. 1907. Hunting the grizzly and other sketches. Part 2. The wilderness hunter. Current Literature, New York.

Russell, R. H., J. W. Nolan, N. G. Woody, G. H. Anderson, and A. M. Pearson. 1978. A study of the grizzly bear (*Ursus arctos*) in Jasper National Park: A progress report, 1976–1977. Canadian Wildlife Service, Edmonton, Alberta.

Russell, R. H., J. W. Nolan, N. G. Woody, and G. H. Anderson. 1979. A study of the grizzly bear (*Ursus arctos*) in Jasper National Park, 1975 to 1978. Prepared for Parks Canada by Canadian Wildlife Service, Edmonton, Alberta.

Schneider, B. 1977. Where the grizzly walks. Mountain Press, Missoula, MT.

Schoen, J. W. 1990. Bear habitat management: A review and future perspective. International Conference on Bear Research and Management 8:143–54.

Schoen, J., and L. Beier 1990. Brown bear habitat preferences and brown bear logging and mining relationships in southeast Alaska (Federal Aid in Wildlife Restoration Research Final Report, W-22-1-6 and W-23-1-3, Study 4.17). Alaska Department of Fish and Game, Juneau.

Schoen, J. W., J. W. Lentfer, and L. R. Beier. 1986. Differential distribution of brown bears on Admiralty Island, Southeastern Alaska: A preliminary assessment. International Conference on Bear Research and Management 6:1–5.

Schoen, J. W., and S. D. Miller. 2002. New strategies for bear conservation: Collaboration between Resource Agencies and Environmental Organizations. Ursus 13:316–67.

Schoen, J. W., L. I. Beier, J. W. Lentfer, and L. J. Johnson. 1987. Denning ecology of brown bears on Admiralty and Chichagof Islands. International Conference on Bear Research and Management 7:293–304.

Schoen, J. W., R. W. Flynn, L. H. Suring, K. Titus, and L. R. Beier. 1994. Habitat-capability model for brown bear in southeast Alaska. International Conference on Bear Research and Management 9(1):327–37.

Schwartz, C. C., and S. M. Arthur. 1999. Radio-tracking large wilderness mammals: Integration of GPS and Argos technology. Ursus 11:261–74.

Schwartz, C. C., K. A. Keating, H. V. Reynolds, III., V. G. Barnes, Jr., R. A. Sellers, J. E. Swenson, S. D. Miller, B. N. McLellan, J. Keay, R. McCann, M. Gibeau, W. F. Wakkenin, R. D. Mace, W. Kasworm, and R. Smith. 2003. Reproductive maturation and senescence in the female brown/grizzly bear. Ursus 14(2) in press.

Sellers, R. D., and L. D. Aumiller. 1994. Brown bear population characteristics at McNeil River, Alaska. International Conference on Bear Research and Management 9(1):283–93.

Servheen, C. 1998. The grizzly bear recovery program: Current status and future considerations. Ursus 10:591–96.

Servheen, C. 1999. Status and management of the grizzly bear in the lower 48 United States. Pages 50–54 *in* C. Servheen, S. Herrero, and B. Peyton, comps. Bears: Status survey and conservation action plan. IUCN/SSC Bear and Polar Bear Specialist Groups, IUCN, Gland, Switzerland, and Cambridge, UK.

Servheen, C., J. Waller, and P. Sandstrom. 2001. Identification and Management of linkage zones for grizzly bears between the large blocks of public land in the Northern Rocky Mountains. Pages 161–98 *in* Proceedings of the 2001. International Conference on Ecology and transportation. September, 2001.

Servheen, G., A. Hamilton, R. Knight, and B. McLellan. 1990. Report of the technical review team: Evaluation of the Bitterroot and North Cascades to sustain viable grizzly bear populations (Report to the Interagency Grizzly Bear Committee). U.S. Fish and Wildlife Service, Boise, ID.

Servheen, C., S. Herrero, and B. Peyton, comps. 1999. Bears: Status survey and conservation action Plan. IUCN/SSC Bear and Polar Bear Specialist Groups, IUCN, Gland, Switzerland, and Cambridge, UK.

Shaffer, M. L. 1978. Determining minimum viable population size: A case study of the grizzly bear (*Ursus arctos* L.). Ph.D. Dissertation, Duke University, Durham, NC.

Shaffer, M. L. 1983. Determining minimum viable population size for the grizzly bear. International Conference on Bear Research and Management 5:133–39.

Shaffer, M. L. 1992. Keeping the grizzly bear in the American West: A strategy for real recovery. Wilderness Society, Washington DC.

Shields, G. F., D. Adams, G. Garner, M. Labelle, J. Pietsch, M. Ramsay, C. Schwartz, K. Titus, and S. Williamson. 2000. Phylogeography of mitochondrial DNA variation in brown bears and polar bears. Molecular Phylogenetics and Evolution 15:319–26.

Simberloff, D. 1999. Biodiversity and bears: A conservation paradigm shift. Ursus 11:25–31.

Smith, J., and J. Hoffman. 1998. Status of white pine blister rust in Intermountain Region white pines (Forest Health Protection Report No. R4-98-02). U.S. Forest Service Intermountain Region, State and Private Forestry.

Smith, R. B., and L. J. Van Daele. 1991. Terror Lake hydroelectric project, Kodiak Island, Alaska. Final report 1. Brown bear studies (1982–86). Alaska Department of Fish and Game and Alaska Power Authority.

Soulé, M. E., 1987. Viable populations for conservation. Cambridge University Press, Cambridge.

Speiss, A. 1976. Labrador grizzly (*Ursus arctos* L.): First skeletal evidence. Journal of Mammalogy 57:787–90.

Speiss, A., and S. Cox. 1977. Discovery of the skull of a grizzly bear in Labrador. Arctic 29:194–200.

Stebler, A. M. 1972. Conservation of the grizzly: Ecologic and cultural considerations. International Conference on Bear Research and Management 2:297–303.

Stonorov, D., and A. W. Stokes. 1972. Social behavior of the Alaska brown bears. International Conference on Bear Research and Management 2:232–42.

Storer, T. I., and L. P. Tevis. 1955. California grizzly. University of California Press, Berkeley.

Stringham, S. F. 1983. Roles of adult males in grizzly bear population biology. International Conference on Bear Research and Management 5:140–51.

Stringham, S. F. 1984. Responses by grizzly bear population dynamics to certain environmental and biosocial factors. Ph.D. Dissertation, University of Tennessee, Knoxville.

Stringham, S. F. 1986. Effects of climate, dump closure, and other factors on Yellowstone grizzly bear litter size. International Conference on Bear Research and Management 6:33–39.

Stringham, S. F. 1990. Grizzly bear reproductive rate relative to body size. International Conference on Bear Research and Management 8:433–43.

Suchy, W., L. L. McDonald, M. D. Strickland, and S. H. Anderson. 1985. New estimates of minimum viable population size for grizzly bears of the Yellowstone ecosystem. Wildlife Society Bulletin 13:223–28.

Suring, L. H., K. R. Barber, C. C. Schwartz, T. N. Bailey, W. C. Shuster, and M. D. Tetreau. 1998. Analysis of cumulative effects on brown bears on the Kenai Peninsula, southcentral Alaska. Ursus 10:107–17.

Swanson, C. S., D. W. McCollum, and M. Maj. 1994. Insights into the economic value of grizzly bears in the Yellowstone recovery area. International Conference on Bear Research and Management 9:575–82.

Swenson, J. E., F. Sandegren, A. Söderberg, A. Bjärvall, R. Franzén, and P. Wabakken. 1997. Infanticide caused by hunting of male bears. Nature 386:450–52.

Taberlet, P., and J. Bouvet. 1994. Mitrochondrial DNA polymorphism phylogeography, and conservation genetics of the brown bear *Ursus arctos* in Europe. Proceedings of the Royal Society London, Series B 255:195–200.

Taberlet, P., J. E. Swenson, F. Sandegren, and A. Bjarvall. 1995. Localization of a contact zone between two highly divergent mitrochondrial DNA lineages of brown bear (*Ursus arctos*) in Scandinavia. Conservation Biology 9:1255–61.

Tabor, G. M. 1996. Yellowstone to Yukon (Y2Y). Report to the Kendall Foundation, Boston, MA.

Tabor, G. M., and M. E. Soulé. 1999. Yellowstone to Yukon (Y2Y). Report to the Wilburforce Foundation, Seattle, WA.

Talbot, S. L., and G. F. Shields. 1996. A phylogeny of the bears (Ursidae) inferred from complete sequences of three mitochondria genes. Molecular Phylogenetics and Evolution 5:567–75.

Taylor, M. K., D. DeMaster, F. L. Bunnell, and R. Schweinsburg. 1987. Modeling the sustainable harvest of female polar bears. Journal of Wildlife Management 51:811–20.

Testa, J. W., W. P. Taylor, and S. D. Miller. 1998. Impacts of increased hunting pressure on the density and dynamics of brown bear populations in southcentral Alaska (Federal Aid in Wildlife Restoration Progress Report, W-27-1, Study 4.28). Alaska Department of Fish and Game, Juneau.

Thenius, E. 1959. Ursidenphylogenese und biostratigraphie. Zeitschrift für Säugertierkunde 24:78–84.

Thomas, D. C. 1977. Metachromatic staining of dental cementum for mammalian age determination. Journal of Wildlife Management 41:207–10.

Thompson, S. D. 1992. Energetics of gestation and lactation in small mammals: Basal metabolic rate and the limits of energy use and reproduction. Pages 213–25 *in* T. Tomasi and T. Horton, eds. Mammalian energetics: Interdisciplinary views of metabolism and reproduction. Cornell University Press, Ithaca, NY.

Titus, K., and L. Beier. 1991. Population and habitat ecology of brown bears on Admiralty and Chichagof Islands (Federal Aid in Wildlife Restoration Project Progress Report W-23-4). Alaska Department of Fish and Game, Juneau.

Titus, K., and J. W. Schoen. 1992. A plan for maintaining viable and well-distributed brown bear populations in southeast Alaska. Pages 171–205 *in* L. H. Suring, D. C. Crocker- Bedford, R. W. Flynn, C. S. Hale, G. C. Iverson, M. D. Kirchhoff, T. E. Schenck, II, L. C. Shea, and K. Titus, eds. A proposed strategy for maintaining well-distributed, viable populations of wildlife associated with old-growth forests in southeast Alaska. Unpublished report. U.S. Department Agriculture Forest Service, Alaska Region, Juneau.

Troyer, W. A., and R. J. Hensel. 1969. The brown bear of Kodiak Island. U.S. Department of the Interior, Bureau of Sport Fish and Wildlife, Branch of Wildlife Refuges, Kodiak, AK.

Tsubota, T., and H. Kanagawa. 1989. Annual changes in serum testosterone levels and spermatogenesis in Hokkaido brown bear, *Ursus arctos yesoensis.* Journal of the Mammalogy Society of Japan 14:11–17.

Tsubota, T., and H. Kanagawa. 1991. Sexual maturity of the Hokkaido brown bear. Asiatic Bear Conference 1:1–9.

Tsubota, T., and H. Kanagawa. 1993. Morphological characteristics of the ovary, uterus and embryo during the delayed implantation period in the Hokkaido brown bear (*Ursus arctos yesoensis*). Journal of Reproductive Development 39:325–31.

Tsubota, T., Y. Takahashi, and H. Kanagawa. 1987. Changes in serum progesterone levels and growth of fetuses in Hokkaido brown bear, *Ursus arctos yesoensis.* International Conference on Bear Research and Management 7:355–58.

U.S. Fish and Wildlife Service. 1982. Grizzly bear recovery plan. U.S. Fish and Wildlife Service, Denver, CO.

U.S. Fish and Wildlife Service. 1993. Grizzly bear recovery plan. Missoula, MT.

U.S. Fish and Wildlife Service. 2000. Grizzly bear recovery in the Bitterroot Ecosystem. Final environmental impact statement. U.S. Department of the Interior, Washington DC.

Van Daele, L. J., V. G. Barnes, and R. B. Smith. 1990. Denning characteristics of brown bears on Kodiak Island, Alaska. International Conference on Bear Research and Management 8:257–67.

Vroom, G. W., S. Herrero, and T. T. Oglive. 1977. The ecology of winter den sites of grizzly bears in Banff National Park, Alberta. International Conference on Bear Research and Management 3:321–30.

Waits, L. P., S. L. Talbot, R. H. Ward, and G. F. Shields. 1998a. Mitochondrial DNA phylogeography of the North American brown bear and implications for conservation. Conservation Biology 12:408–17.

Waits, L. P., D. Paetkau, C. Strobeck, and R. H. Ward. 1998b. A comparison of genetic variability in brown bears (*Ursus arctos*) from Alaska, Canada, and the lower 48 states. Ursus 10:307–14.

Waits, L., D. Paetkau, and C. Strobeck. 1999. Genetics of the bears of the world. Pages 25–32 *in* C. Servheen, S. Herrero, and B. Pelton, comps. Bears: Status survey and conservation action plan. IUCN/SSC Bear and Polar Bear Specialist Groups, IUCN, Gland, Switzerland, and Cambridge, UK.

Walker, R., and L. Craighead. 1997. Corridors: Key to wildlife from Yellowstone to Yukon. Pages 113–23 *in* L. Willcox, B. Robinson, and A. Harvey, eds. A sense of place: Issues, attitudes, and resources in the Yellowstone to Yukon Ecoregion. Yellowstone to Yukon Conservation Initiative, Canmore, Alberta, Canada.

Waser, P. M., and W. T. Jones. 1983. Natal philopatry among solitary mammals. Quarterly Review of Biology 58:355–90.

Watts, P. D., and C. Cuyler. 1988. Metabolism of the black bear under simulated denning conditions. Acta Physiology Scandinavia 134:149–52.

Watts, P. D., and C. Jonkel. 1988. Energetic cost of winter dormancy in grizzly bear. Journal of Wildlife Management 52:654–56.

Watts, P. D., C. Jonkel, and K. Ronald. 1981. Mammalian hibernation and the oxygen consumption of a denning black bear (*Ursus americanus*). Comparative Biochemistry and Physiology 69A:121–23.

Weaver, J., R. Escano, D. Mattson, T. Puchlerz, and D. Despain. 1986. A cumulative effects model for grizzly bear management in the Yellowstone Ecosystem. Pages 234–46 *in* G. P. Contreras and K. E. Evans, comps. Proceedings of the grizzly bear habitat symposium (U.S. Forest Service General Technical Report INT-207). U.S. Forest Service Intermountain Research Station, Ogden, UT.

Welch, C. A., J. Keay, K. C. Kendall, and C. T. Robbins. 1997. Constraints on frugivory by bears. Ecology 78:1105–19.

White, D., Jr. 1996. Two grizzly bear studies: Moth feeding ecology and male reproductive biology. Ph.D. Dissertation, Montana State University, Bozeman.

White, D., Jr., J. G. Berardinelli, and K. Aune. 1998. Reproductive characteristics of the male grizzly bear in the continental United States. Ursus 10:497–501.

White, D., Jr., K. C. Kendall, and H. D. Picton. 1999. Potential energetic effects of mountain climbers on foraging grizzly bears. Wildlife Society Bulletin 27:146–51.

White, G. C., and R. A. Garrott. 1990. Analysis of wildlife radio-tracking data. Academic Press, San Diego, CA.

Wielgus, R. B. 1993. Causes and consequences of sexual habitat segregation in grizzly bears. Ph.D. Dissertation, University of British Columbia, Vancouver, British Columbia, Canada.

Wielgus, R. B., and F. L. Bunnell. 1994. Dynamics of a small, hunted brown bear *Ursus arctos* population in southwestern Alberta, Canada. Biological Conservation 67:161–66.

Wielgus, R. B., and F. L. Bunnell. 2000. Possible negative effects of adult male mortality on female grizzly bear reproduction. Biological Conservation 93:145–54.

Wielgus, R. B., F. L. Bunnell, W. L. Wakkinen, and P. E. Zagar. 1994. Population dynamics of Selkirk Mountain grizzly bears. Journal of Wildlife Management 58:266–72.

Wilk, R. J., J. W. Solberg, V. D. Berns, and R. A. Sellers. 1988. Brown bear, *Ursus arctos,* with six young. Canadian Field-Naturalist 102:541–43.

Woods, J. G., B. N. McLellan, D. Paetkau, M. Proctor, P. Ott, and C. Strobeck. 1996. DNA fingerprinting applied to mark–recapture bear studies. International Bear News 5:9–10.

Woods, J. G., B. N. McLellan, D. Paetkau, M. Proctor, and C. Strobeck. 1997. West Slopes bear research project: Second progress report. Parks Canada, Revelstroke, British Columbia.

Woods, J. G., D. Paetkau, D. Lewis, B. N. McLellan, M. Proctor, and C. Strobek. 1999. Genetic tagging free ranging black and brown bears. Wildlife Society Bulletin 27:616–27.

Worton, B. J. 1989. Kernel methods for estimating the utilization distribution in home-range studies. Ecology 70:164–68.

Wright, W. H. 1909. The grizzly bear. University of Nebraska Press, Lincoln.

Yukon Renewable Resources. 1997. Grizzly bear management guidelines. Whitehorse, Yukon Territory, Canada.

Zavatsky, B. P. 1976. The use of the skull in age determination of the brown bear. International Conference on Bear Research and Management 4:275–79.

Charles C. Schwartz, U.S. Geological Survey, Forestry Sciences Lab., Montana State University, Bozeman, Montana 59717. Email: chuck_schwartz@usgs.gov

Sterling D. Miller, National Wildlife Federation, 240 N. Higgins #2, Missoula, Montana 59802. Email: millers@nwf.org

Mark A. Haroldson, U.S. Geological Survey, Forestry Sciences Laboratory, Montana State University, Bozeman, Montana 59717. Email: mark_haroldson@usgs.gov

27

Polar Bear

Ursus maritimus

Steven C. Amstrup

NOMENCLATURE

COMMON NAMES. Polar bear, nanook, nanuq, nanuk, ice bear, sea bear, eisbär, isbjørn, white bear
SCIENTIFIC NAME. *Ursus maritimus*

Phipps (1774) first described the polar bear as a species distinct from other bears and gave the name *Ursus maritimus*. Subsequently, alternative generic names including *Thalassarctos, Thalarctos,* and *Thalatarctos* were suggested. Erdbrink (1953) and Thenius (1953) settled on *Ursus* (*Thalarctos*) *maritimus,* citing interbreeding between brown bears (*Ursus arctos*) and polar bears in zoos. Kurtén (1964) described the evolution of polar bears based on the fossil record and recommended the name *Ursus maritimus* as adopted by Phipps (1774). Harington (1966), Manning (1971), and Wilson (1976) subsequently promoted use of the name *Ursus maritimus,* and it has predominated ever since.

DISTRIBUTION

Polar bears occur only in the Northern Hemisphere. Their range is limited to areas in which the sea is ice covered for much of the year. Over most of their range, polar bears remain on the sea-ice year-round or visit land only for short periods. Polar bears are common in the Chukchi and Beaufort Seas north of Alaska. They occur throughout the East Siberian, Laptev, and Kara Seas of Russia and the Barent's Sea of northern Europe. They are found in the northern part of the Greenland Sea, and are common in Baffin Bay, which separates Canada and Greenland, as well as through most of the Canadian Arctic Archipelago (Fig. 27.1). Because their principal habitat is the sea-ice surface rather than adjacent land masses, they are classified as marine mammals. In most areas, pregnant females come ashore to create a den in which to give birth to young. Even then, however, they are quick to return to the sea ice as soon as cubs are able. In some areas, notably the Beaufort and Chukchi Seas of the polar basin, many females den and give birth to their young on drifting pack ice (Amstrup and Gardner 1994).

Polar bears are most abundant in shallow-water areas near shore and in other areas where currents and upwellings increase productivity and keep the ice cover from becoming too solidified in winter (Stirling and Smith 1975; Stirling et al. 1981; Amstrup and DeMaster 1988; Stirling 1990; Stirling and Øritsland 1995; Stirling and Lunn 1997; Amstrup et al. 2000). Despite apparent preferences for the more productive waters near shorelines and polynyas (areas of persistent open water), polar bears occur throughout the polar basin including latitudes >88°N (Stefansson 1921; Papanin 1939; Durner and Amstrup 1995).

Because they derive their sustenance from the sea, the distribution of polar bears in most areas changes with the seasonal extent of sea-ice cover. In winter, for example, sea-ice extends as much as 400 km south of the Bering Strait, which separates Asia from North America, and polar bears extend their range to the southernmost extreme of the ice (Ray 1971). Sea-ice disappears from most of the Bering and Chukchi Seas in summer, and polar bears occupying these areas may migrate as much as 1000 km to stay with the southern edge of the pack ice (Garner et al. 1990, 1994). Throughout the polar basin, polar bears spend their summers concentrated along the edge of the persistent pack ice. Significant northerly and southerly movements appear to be dependent on seasonal melting and refreezing of ice near shore (Amstrup et al. 2000). In other areas, for example, Hudson Bay, James Bay, and portions of the Canadian High Arctic, when the sea-ice melts, polar bears are forced onto land for up to several months while they wait for winter and new ice (Jonkel et al. 1976; Schweinsburg 1979; Prevett and Kolenosky 1982; Schweinsburg and Lee 1982; Ferguson et al. 1997; Lunn et al. 1997).

FIGURE 27.1. Approximate worldwide winter distribution of polar bears (light gray). Polar bears are distributed throughout most ice-covered seas of the Northern Hemisphere. Hatched areas indicate known coastal regions preferred for maternal denning. Only in the Beaufort Sea adjacent to Alaska is denning in the pack-ice thought to be common.

Until the 1960s, the prevalent belief was that polar bears wandered throughout the Arctic. Some naturalists felt that individual polar bears were carried passively with the predominant currents of the polar basin (Pedersen 1945). Researchers have known for some time that is not the case (Stirling et al. 1980, 1984). However the advent of radiotelemetry (Amstrup et al. 1986), including the use of satellites (Fancy et al. 1988; Harris et al. 1990; Messier et al. 1992; Amstrup et al. 2000), detailed knowledge of polar bear movements was not available.

DESCRIPTION

Size and Weight. The polar bear is the largest of the extant bears (DeMaster and Stirling 1981). In Hudson Bay, the mean scale weight of 94 males >5 years of age was 489 kg. The largest bear in that group was a 13-year-old, which weighed 654 kg (Kolenosky et al. 1992). The heaviest bear we have weighed in Alaska was 610 kg, and several animals were heavy enough that we could not raise them with our helicopter or weighing tripod. Some animals too heavy to lift have been estimated to weigh 800 kg (DeMaster and Stirling 1981). Females are smaller, with peak weights usually not exceeding 400 kg. Total lengths of males in the Beaufort Sea of Alaska ranged up to 285 cm. Such an animal may reach nearly 4 m when standing on its hind legs and is 1.7 m shoulder height when standing on all four legs. Chest girth for large males is close to 200 cm. Although smaller, females in the Beaufort sea were as long as 247 cm with chest girths up to 175 cm. Only prehistoric polar bears and the giant short-faced bear (*Arctodus* spp.) of the Pleistocene were of greater stature than today's polar bears (Kurtén 1964; Stirling and Derocher 1990).

Manning (1971) suggested there is a cline in size of polar bears across the Arctic. Size increases, he suggested, with distance from east Greenland across the Nearctic to the Chukchi Sea between Alaska and Russia. Manning (1971) also suggested that polar bears from Svalbard may be larger than those from east Greenland. A cline in size across the Palearctic also might occur, but samples from the Russian Arctic are inadequate to confirm it (Manning 1971).

The hypothesized cline was based on measurements made from skulls housed in museums around the world. Unfortunately, the sources of skulls in the various collections were not similar. Of particular note was that many of the skulls originating in the Chukchi Sea may have been donated by trophy hunters. These hunters worked over the ice in teams of aircraft (Tovey and Scott 1957) and were quite effective in killing a great number of the largest polar bears (Amstrup et al. 1986). Another potential problem is that ages of bears in the sample were estimated only by class or life stage. Hence, older bears from one locale might have been compared to younger bears (of the same age class) in another.

Potentially nonstandardized collection methods prevent any meaningful conclusions about relative sizes of polar bears from different locales. Also, if there is a cline in skull sizes around the world, it appears that body sizes and weights of polar bears do not follow a similar cline. The largest bears for which actual scale weights are known have come from the Hudson and James Bay areas of Canada and from the Beaufort Sea of Alaska, not from the Chukchi Sea. That observation, too, may be subject to some bias, as the most prolonged and intensive polar bear studies have been conducted in Hudson Bay and the Beaufort Sea. Greater numbers of captures in those locations may have increased the probability that very large bears were included in the sample.

Despite their large adult sizes, the young of polar bears are among the most altricial (undeveloped) of eutherian mammals (Ramsay and Dunbrack 1986). Newborn polar bears weigh only 600–700 g. They are blind, only lightly furred, and totally helpless (Blix and Lentfer 1979). Mother polar bears when giving birth commonly weigh over 300 kg, and can weigh 400 kg (Ramsay 1986). If only a single cub is born, the ratio of maternal to neonate weights could be between 400 and 500 to 1. Even with the more common two-cub litter, the ratio of maternal to neonate mass is extraordinarily large (Ramsay and Dunbrack 1986). Cubs grow very fast after birth. In Alaska, they average 13 kg on emergence from the den in late March or early April, with maximum weights of 22 kg. Cubs continue to grow rapidly through their first summer on the sea-ice and some weigh over 100 kg as they approach 1 year of age.

Pelage. Polar bears are completely furred except for the tip of the nose. Pelage density is more even than in other ursids, which are often more sparsely furred ventrally and in axillary and groin areas. Even the pads of the feet of polar bears may be covered with hair, especially in late winter (Fig. 27.2). Furred foot pads may provide a more secure purchase on the slippery sea ice surface and add another layer of insulation between the bear's foot and the substrate of ice and snow. Under the fur, pads of the feet of polar bears are made up of the same cornified epidermis characteristic of the pads of other bears (Storer and Tevis 1955; Ewer 1973).

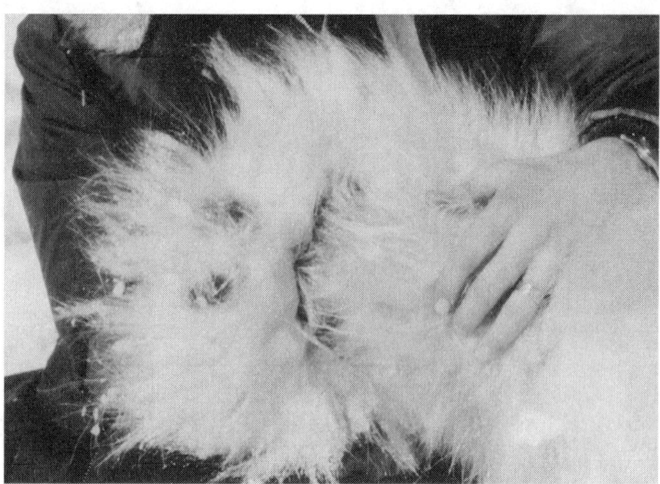

FIGURE 27.2. In winter, polar bear foot pads may be densely furred. This may provide a better purchase on the slippery ice surface than naked pads. SOURCE: Photo by Steven C. Amstrup.

The skin of polar bears is uniformly black. Hence, if polar bears lose hair due to physical trauma or disease, they appear from a distance to have black patches on their bodies. Polar bear fur appears white when it is clean and in even sunlight. Because it actually is without pigment, however (Øritsland and Ronald 1978; Grojean et al. 1980), bears may take on the yellow-orange hues of the setting and rising sun and the blue of sunlight filtered through clouds and fog. They appear the whitest right after molting. In spring and late winter, however, many polar bears are "off-white" or yellowish because of oils from their prey and other impurities that have attached to and been incorporated into their hair.

The molt appears to be somewhat variable, but begins by late April and May. The molt appears to be complete by late summer, and bears captured in autumn have notably shorter coats than those captured in spring. The pelt is thick with a dense underfur and guard hairs of various lengths. Polar bear fur may have a high propensity to take on the colors of environmental impurities because the guard hairs have a hollow medulla (or core) where impurities may lodge. In zoo environments, some species of algae can enter the hollow cores of guard hairs and result in a pronounced "greening" of the fur (Lewin and Robinson 1979).

Lavigne and Oritsland (1974) noted that polar bears effectively absorb ultraviolet (UV) light, and suggested that could be useful in remote-sensing surveys to enumerate them. The discovery that polar bears appear to absorb UV light led to much speculation about their ability to capture the energy in that light. Popular and scientific reports claimed that the ability to absorb energy in the UV spectrum was an adaptation to help maintain body heat in the rigorous Arctic environment (Anonymous 1978; Grojean et al. 1980; Lopez 1986; Mirsky 1988). Suddenly, the hollow hairs of polar bears, adept at catching algae and other contaminants (Lewin and Robinson 1979) also were endowed with the powers of optic fibers to funnel UV light to the skin. According to this theory, the skin was black to better absorb such energy without damage. Capturing this high-frequency electromagnetic energy would be a great adaptation for polar bears. This ability has attained the status of an Arctic legend, and contributed to the mystique surrounding the great white bears of the north. Unfortunately, this supposed adaptation has no basis in fact. Lavigne (1988) and Koon (1998) established unequivocally that the hair of polar bears, although transparent in the visible spectrum, absorbs UV light. If the hair of polar bears absorbs UV light, it does not efficiently transmit UV light. As UV light moves down the shaft of the hair, its energy is absorbed, preventing significant energy from being transmitted to the skin.

Claws. The claws of polar bears are shorter and more strongly curved than those of brown bears. They also are larger and heavier than those of black bears (*Ursus americanus*). They appear to be very well adapted to clambering over blocks of ice and snow and especially to securely gripping prey animals. The claws are normally black (Fig. 27.3), but rarely may, like polar bear fur, lack pigment (Fig. 27.4).

Skull and Dentition. Polar bears share the general ursid dental formula : I 3/3, C 1/1, P 4/4, M 2/3. The first premolars are vestigial and occur in a long diastema or gap between the functional canine and molariform teeth. That gap allows the powerful canines to penetrate deeply into the bodies of seals and other prey without interference from adjacent cheek teeth. Although polar bears apparently evolved from brown bears <250,000 years ago, their teeth have changed significantly from the brown bear form. The cheek teeth are greatly reduced in size and surface area, and the carnassials are more pronounced than in brown bears, reflecting the predatory lifestyle. The teeth of polar bears are well suited to the tasks of grabbing and holding prey and shearing meat and hide. They no longer are as suited to grinding grasses and other vegetation as are those of brown bears. The canine teeth of males are larger and heavier, relative to the size of the jaw, than those of females (Kurtén 1955), and the molar arcade of males is longer than in females (Larsen 1971). The proportionately larger canines coincide with the pronouced sexual dimorphism which is more accentuated in polar bears than it is in any other ursid (Stirling and Derocher 1990).

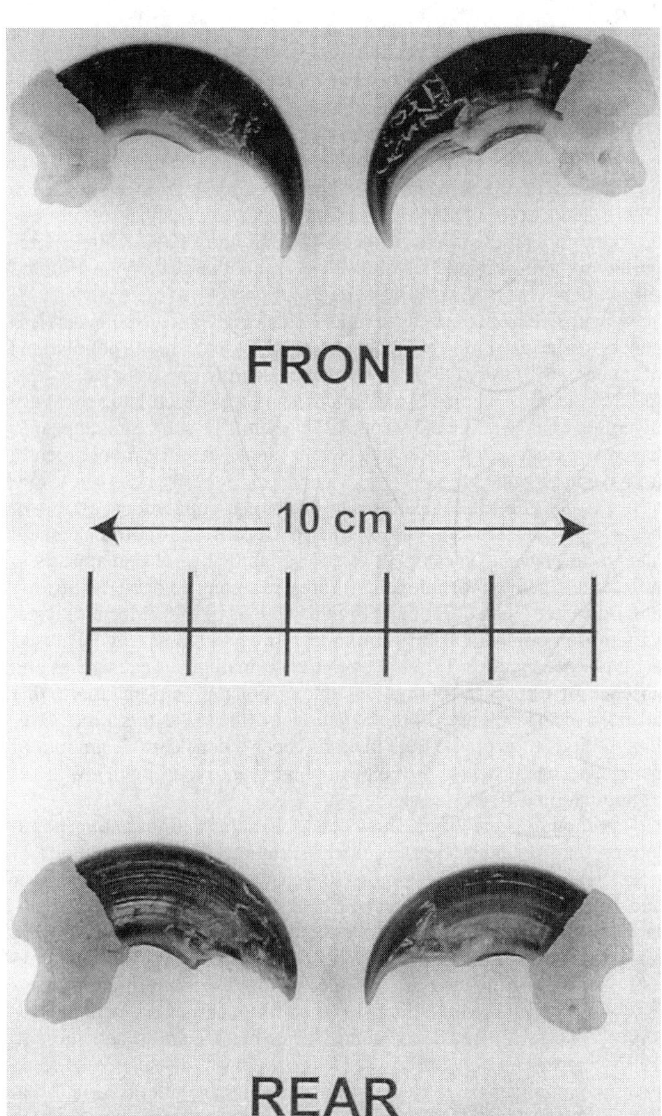

FIGURE 27.3. Normal front and rear claws of a female polar bear from the Beaufort Sea. Note strong curve and sharp points for clinging to blocks of ice and for capturing prey. SOURCE: Photo by Steven C. Amstrup.

FIGURE 27.4. Rare unpigmented polar bear claws, on a polar bear captured in the southern Beaufort Sea. SOURCE: Photo by Steven C. Amstrup.

The skull of the polar bear shares the principal characteristics of the skulls of other ursids. The largest brown bear skulls are larger than the largest polar bear skulls. Polar bear skulls are proportionately narrower across the palate between second molars than skulls of brown bears (Kurtén 1964). The ratio of condylobasal length to zygomatic width (L/W) also is larger in the polar bear, accentuating the narrower skull. The L/W for 279 brown bears taken by hunters was 1.59, whereas the L/W for 150 polar bears was 1.63 (calculated from Nesbitt and Parker 1977). The difference in actual measurements is not as pronounced as the visual impression suggests. This is because of the more strongly developed and overhanging occiput and significantly greater height in skulls of brown bears (Kurtén 1964). In lateral view, the lower height, combined with absence of the pronounced brow ridge that tends to give brown bears a "dish-faced" appearance, yields a smooth curve from canines across the maxillary bones to the cranium (Fig. 27.5). These features combine to give the polar bear a "Roman nose" appearance.

GENETICS

Despite the evidence of population segregation from marking, survey, and radiotelemetry data, initial evaluations using genetic techniques suggested small differences among polar bears in different geographic regions. Such small differences might be expected under Pedersen's (1945) hypothesis of a globally wandering panmictic polar bear population, but not in light of current knowledge of movements. Using

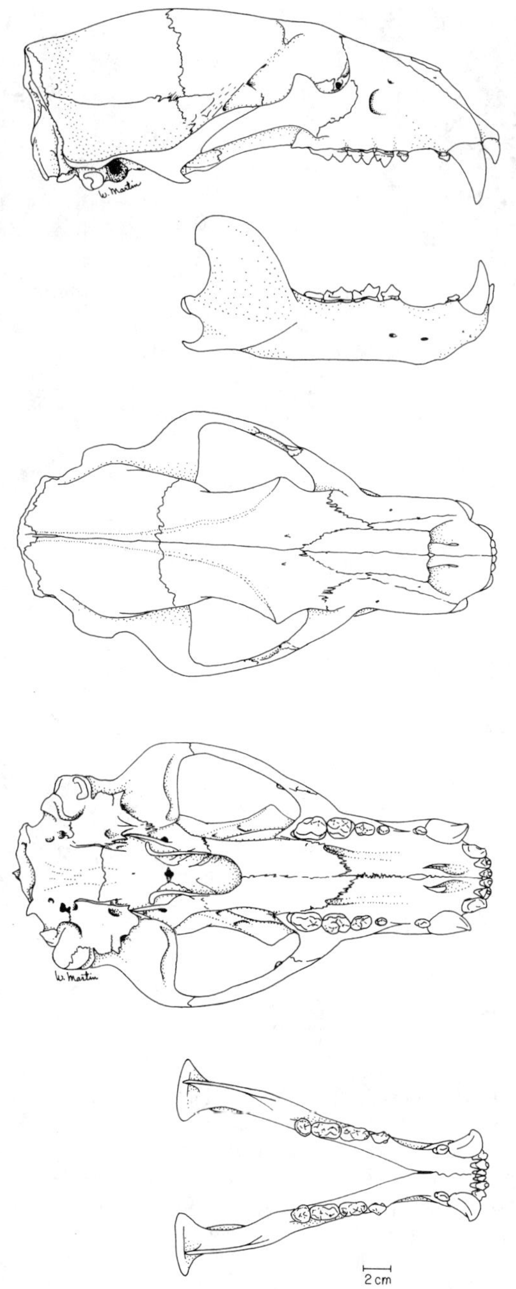

FIGURE 27.5. Skull of the polar bear (*Ursus maritimus*). From top to bottom: lateral view of cranium, lateral view of mandible, dorsal view of cranium, ventral view of cranium, dorsal view of mandible.

protein electrophoresis, Larsen et al. (1983b) found little variation in allozymes among polar bears. They concluded there was no reason to reject a hypothesis of one common polar bear population worldwide. Larsen et al. (1983b) assumed that high gene flow across the Arctic and strong common selective pressures reduced variation among polar bears. Supporting that concept, Durner and Amstrup (1995) recorded the movement of a radio-collared polar bear from near Prudhoe Bay, Alaska, to Greenland. That bear was only 1 of 106 equipped with platform transmitter terminals or satellite radio-collars between 1985 and 1992. Relatively few such movements, however, would be required to genetically homogenize polar bear populations worldwide (Paetkau et al. 1995).

Corroborating the work of Larsen et al. (1983b), Cronin et al. (1991) found little mitochondrial DNA (mtDNA) differentiation among bears of the Beaufort and Chukchi seas. Mitochondrial DNA and protein electrophoresis, however, may have a relatively low ability to resolve genetic variation among populations (Scribner et al. 1997). Therefore, the absence of differences in these markers may not be surprising.

More recent studies using highly variable nuclear genetic markers called microsatellites have resolved differences among polar bears living in different areas. Paetkau et al. (1995) found significant differences in genetic distances among members of four Canadian populations of polar bears. They concluded that the long-distance movements of polar bears have not resulted in complete genetic mixing of populations. Furthermore, Paetkau et al. (1999) reported significant correlations among movement data and genetic data from polar bear populations worldwide. They found greater differences among populations in the Canadian Arctic than among populations surrounding the polar basin. Such contrasts fit well with observed differences in movement patterns in these areas (Amstrup et al. 1986, 2000; Messier et al. 1992; Amstrup and Gardner 1994; Bethke et al. 1996; Scribner et al. 1997).

Genetic management units that correspond with boundaries defined by radiotelemetry have now been identified for most polar bear populations (Paetkau et al. 1999). The correspondence between movement data and recent genetic data allows managers to make better decisions about harvest or other human activities that could have population-level impacts.

Comparisons of the relative genetic variability among putative populations of different bear species are difficult because literature on brown, black, and polar bears has not presented easily comparable or consistent measures of interpopulation genetic variation. Nonetheless, microsatellite data that can be compared suggest there may be less genetic variation among populations of polar bears than among populations of black bears and brown bears (Paetkau et al. 1995, 1999). Paetkau et al. (1999) also found genetic distances among polar bear populations were at the lower extreme of the distances reported for the gray wolf (*Canus lupus*), another widely distributed carnivore.

Evidence from patterns in mtDNA also may hint at somewhat less genetic variation among polar bear populations than among populations of other bears. Cronin et al. (1991) reported only one basic polar bear mtDNA lineage, whereas black and brown bears each have two very divergent lineages. The older species (black and brown bears) appear to have more genetic variation across their ranges than the more recently derived polar bears.

Greater morphological variation among populations of brown bears (e.g., very large individuals, such as those living on Kodiak Island and coastal Alaska, vs. smaller interior or arctic bears) also appears to reflect more genetic variation than is present among polar bears (Stirling and Derocher 1990; Talbot and Shields 1996a, 1996b). Morphological variation among polar bears is minimal throughout their range. Paetkau et al. (1999) concluded from the relatively small genetic distances and absence of major discontinuities among polar bear populations that all polar bears belong to one evolutionary lineage at this time. Over long periods of geologic time there has been a considerable amount of genetic exchange across the range of polar bears, resulting in low levels of population differentiation.

Although polar bears show less genetic variation *among* populations than do other bears, genetic variation *within* populations of polar bears appears to be similar to that within populations of black and brown bears. Paetkau and Strobeck (1998) concluded that polar bear populations were less variable than brown bears, which were less variable than black bears. When levels were averaged over several putative populations of each species, however, microsatellite heterozygosity within populations was 0.68 for polar bears (Paetkau et al. 1999), 0.66 for brown bears, and 0.72 for black bears (Paetkau et al. 1997), suggesting little difference. This pattern was also observed when two functional genes were compared. Considerable allelic variation in DNA sequences at the κ-casein and major histocompatability complex (MHC) DQβ genes was observed in polar, brown, and black bears (M. A. Cronin and S. C. Amstrup, unpublished data) and no species appeared more variable than the others. It is thought that genes for

κ-casein and the MHC are highly conserved because they influence milk quality and production and disease resistance. The functional importance of these genes may have led polar bears to retain their ancestral variability.

EVOLUTION

The polar bear appears to share a common ancestor with the present-day brown bear. It apparently branched off the brown bear lineage during the late Pleistocene. Kurtén (1964) suggested that ancestors of the modern polar bear were "gigantic." Although still the largest of the extant bears, the polar bear, like many other mammals, has decreased in size since the Pleistocene. Also, significant morphological changes have continued within the last 20,000–40,000 years, perhaps through the present (Kurtén 1964). Stanley (1979) described the many recently derived traits of polar bears as an example of "quantum speciation."

Evidence of polar bear evolution contained in the sparse samples of fossils has been strengthened recently by molecular genetics. Whereas traits of fossil teeth and bones from polar bears clearly indicate their brown bear origins, fossil remains include only a handful of specimens (Kurtén 1964). Genetic data from extant bears can provide phylogenetic information unavailable in the fossil record. Shields and Kocher (1991) first analyzed mtDNA sequences and showed a close relationship between brown bears and polar bears. Cronin et al. (1991) then discovered that mtDNA of brown bears is paraphyletic with respect to polar bears. That is, the mtDNA of brown bears of the Alexander Archipelago in southeastern Alaska is more closely related to the mtDNA of polar bears than it is to the mtDNA of other brown bears. Cronin et al. (1991) reported that mtDNA sequence divergence between Alexander Archipelago brown bears and polar bears is only about 1%, whereas a divergence of about 2.6% separates polar bears from brown bears occurring elsewhere. Cronin et al. (1991) and Cronin (1993) emphasized that mtDNA sequence divergence trees are not species trees and that mtDNA is not, by itself, a good measure of overall genetic differentiation. Nonetheless, these relationships provide a compelling argument regarding the origin and evolution of polar bears.

Following the discovery of Cronin et al. (1991), others corroborated the finding of paraphyletic mtDNA in brown bears and polar bears. Talbot and Shields (1996a, 1996b) suggested that the Alexander Archipelago brown bears represent descendents of ancestral stock that gave rise to polar bears. This stock may have survived Pleistocene glaciers in an ice-free refugium in southeastern Alaska, isolated from brown bears in other Pleistocene refugia (Heaton et al. 1996). This island-dwelling ancestral stock apparently has remained isolated from the more recent mainland bears by broad ocean passages.

Talbot and Shields (1996b) found mtDNA sequence divergence rates similar to those reported by Cronin et al. (1991), and proposed that ancestors of the Alexander Archipelago brown bears diverged from the other mtDNA lineages of brown bears 550,000–700,000 years ago. The mtDNA sequence divergences also suggested that polar bears branched from the Alexander Archepelago ancestral stock of brown bears about 200,000–250,000 years ago, a date closely corresponding with that suggested in the fossil record (Thenius 1953; Kurtén 1964). Shields and Kocher (1991) and Cronin et al. (1991) reported that the mtDNA nucleotide sequence divergence between brown and polar bears (grouped together) and black bears was 7–9%. Applying the substitution rate (6%/million years) for mtDNA genes reported by Talbot and Shields (1996a) to the sequence divergence reported by Cronin et al. (1991) suggests that brown bear ancestral stock diverged from that of black bears approximately 1.2–1.5 million years ago. This "molecular clock" estimate may be low. The fossil record suggests black bears diverged from the brown bear lineage 1.5–2.5 million years ago.

Cronin (1993) cautioned that mutation rates vary among genes as well as among taxa, and that conclusions based on "molecular clocks" must be viewed with caution and in the context of other evidence. For example, DNA sequences for two functional nuclear genes, κ-casein and the DQβ gene of the major histocompatability complex, show polyphyletic relationships among the three species of bears (M. Cronin and S. Amstrup, unpublished data). That is, the DNA sequences do not resolve the relationships among the species. These functional genes are presumably under strong selection and do not diverge as rapidly as mtDNA. Nonetheless, the mtDNA analyses indicate that Alexander Archipelogo brown bears derive from more ancient stocks and are more closely related to polar bears than are other members of the brown bear clan. These conclusions also corroborate the recent appearance of the polar bear in the fossil record and the more ancient roots of the black bear (Thenius 1953; Kurtén 1964). All DNA evidence, regardless of some areas of uncertainty, corroborate conclusions from the fossil record that the polar bear is a recently derived species and is undergoing rapid evolution. The extreme arctic marine environment is undoubtedly exerting strong selection pressures for rapid adaptation.

FIGURE 27.6. Ringed seal (*Phoca hispida*), named for the ringlike pattern in the fur. Ringed seals, which weigh <100 kg, make up the greatest portion of the polar bear diet worldwide. SOURCE: Photo by Steven C. Amstrup.

FEEDING HABITS

The polar bear is more predatory than other bears and is the apical predator of the arctic marine ecosystem. Polar bears prey heavily throughout their range on ringed seals (*Phoca hispida*) (Fig. 27.6) and, to a lesser extent, bearded seals (*Erignathus barbatus*) (Fig. 27.7). Ringed seals

FIGURE 27.7. Bearded seals (*Erignathus barbatus*) are much larger than ringed seals, with adults weighing 350 kg. They are the second-most-important prey species for polar bears in many regions of the world. SOURCE: Photo by Steven C. Amstrup.

apparently have been a principal food of polar bears for a significant portion of their coevolutionary history and ringed seal behaviors appear to be oriented around avoidance of polar bear predation. Stirling (1977) contrasted the behavioral ecology of ringed seals and Weddell seals (*Leptonychotes weddelli*). Steady predation pressure from polar bears may have led ringed seals to use subnivian birthing lairs and to interrupt spring and summer basking with frequent periods of scanning their surroundings for predators. Weddell seals, on the other hand, evolved in the Antarctic system, where surface predators are absent. They give birth unsheltered on the surface of the sea ice, and they are so ambivalent about activities on the ice surface that human researchers often can walk right up to them for study purposes (Stirling 1977).

Although seals are their primary prey, polar bears also have been known to kill much larger animals such as walruses (*Odobenus rosmarus*) and belugas (*Delphinapterus leucas*) (Stirling and Archibald 1977; Kiliaan et al. 1978; Smith 1980, 1985; Lowry et al. 1987; Calvert and Stirling 1990). The heaviest prey may be taken mainly by large male polar bears (Stirling and Derocher 1990), and unusual circumstances may be required. Nonetheless, in some areas and under some conditions, alternate prey may be quite important to polar bear sustenance. Stirling and Øritsland (1995) suggested that in areas where the estimated numbers of ringed seals are proportionately reduced relative to numbers of polar bears, other prey species were being substituted.

Overall, polar bears are most effective predators of young ringed seals, perhaps because they are naive with regard to predator avoidance. In spring, polar bears may concentrate their predatory efforts on capture of new-born ringed seal pups (Smith and Stirling 1975; Smith 1980). In some areas, predation on pups is extensive. Hammill and Smith (1991) estimated that polar bears annually kill up to 44% of new born seal pups if conditions are right. Throughout the rest of the year, polar bears take seals predominantly from the first two year classes (Stirling et al. 1977a; Smith 1980). Whereas abundance of ringed seals may regulate density of polar bears in some areas, polar bear predation may regulate density and reproductive success of ringed seals in other areas (Hammill and Smith 1991; Stirling and Øritsland 1995).

Polar bears apparently digest fat more easily than protein (Best 1984). They seem to prefer the fatty portions of seals (and presumably other animals) to muscle and other tissues. Stirling (1974) reported that polar bears often remove the fat layer from beneath the skin of freshly killed seals and consume it immediately. Because over half of the calories in a whole seal carcass may be located in the layer of fat between the skin and underlying muscle (Stirling and McEwan 1975), a bear that quickly consumes most of the fat available has maximized its caloric return in the minimal amount of time possible. This may be important to all but the largest polar bears because there is considerable competition for kills. Younger and smaller bears often are driven away from their kills by larger bears.

A high-fat and low-protein diet apparently serves polar bears physiologically as well. They are very efficient at recycling nitrogenous products of catabolism, and can use metabolic water released from fat metabolism (Nelson et al. 1983). Digestion of protein requires water, whereas digestion of fat releases water. In a cold environment, free water is available only at the energetic cost of melting ice and snow. The lipophilic habits of the polar bear minimize energy expended to obtain water in winter (Nelson 1981).

Polar bears tend not to cache prey animals they have killed like grizzly bears do (Stirling 1974; DeMaster and Stirling 1981; Stirling and Derocher 1990). This may be another reason why they consume the highest reward portion of their prey first. Although they have not been observed to cache, polar bears are surplus killers. Stirling and Derocher (1990) reported seeing a polar bear kill two seals within an hour of feeding extensively on another seal. Neither of the latter two seals killed was eaten. Stirling and Øritsland (1995) also have reported surplus killing in polar bears. I once observed a young male polar bear still-hunting at a breathing hole on new autumn ice. There was a partially consumed seal nearby, and between that feeding site and where he was still-hunting were three freshly killed ringed seals stacked like cordwood. When my helicopter approached the bear to capture him, he abandoned his still-hunting site, ran to the pile of dead seals, and covered them with his body as if to protect his stash. This bear apparently had eaten his fill from the first seal but was continuing to hunt, catch, and stack seals despite a low probability that he would consume much of them.

An interesting adaptation to the carnivorous diet, and a difference between polar bears and other temperate and arctic bears, is that only the pregnant females enter dens for the entire winter. Other members of the population continue to hunt seals on the sea-ice throughout the winter. The year-around availability of seals allows denning in polar bears to be strictly a reproductive strategy (affording an acceptable environment for neonates), whereas in most bears it is largely a foraging strategy (avoiding the winter period of food unavailability).

Like other ursids, polar bears will eat human refuse (Lunn and Stirling 1985), and when trapped on land for long periods they will consume coastal marine and terrestrial plants and other terrestrial foods (Derocher et al. 1993). The significance of other foods to polar bears may be limited, however (Lunn and Stirling 1985; Derocher et al. 1993). Over most of their range, polar bears have little opportunity to take foods of shoreline or terrestrial origin. Derocher et al. (1993) found that 31% of pregnant polar bears in the Hudson Bay area fed on berries before denning in autumn. The significance of this to their productivity was not known. Ramsay and Hobson (1991) and Hobson and Stirling (1997) differed in opinions of the value of supplemental terrestrial food. In general, the significance of terrestrial foraging to polar bears is poorly understood.

Clearly the value of alternate foods for polar bears depends on their richness and digestability. Polar bears are poorly equipped to consume and digest most plant parts (Bunnell and Hamilton 1983), and it seems likely that except for fruiting bodies, plants will contribute little to their energy balance. Lunn and Stirling (1985) found that polar bears using human refuse at a dump maintained their weight or lost less weight than bears not using anthropogenic foods. Some bears using the dump even gained weight, but the supplemental food did not appear to confer a reproductive advantage (Lunn and Stirling 1985). Derocher et al. (2000) reported that some polar bears in Svalbard have become adept at catching reindeer (*Rangifer tarandus*). Considering the high digestibility of meat, it seems plausible that if readily available, reindeer could be an important alternate food of polar bears. Likewise, in the Beaufort Sea, dozens of polar bears each year have developed a habit of gathering at the butchering sites of bowhead whales (*Balaena mysticetus*) that are killed by local Native people. The value of this alternate food is apparently great, as nearly every bear seen near whale carcasses in autumn is obese.

MOVEMENTS

Data collected from radio-collared polar bears have confirmed their close ties to the ice. For example, between May 1985 and April 2001, we obtained 34,034 high-quality satellite radio-locations of polar bears in the Chukchi and Beaufort Sea areas of Alaska and northwestern Canada. Some collars had duty cycles that allowed them to transmit more frequently than other collars. When duty cycles were standardized so that each bear contributed one relocation per week, only 975 (7%) of 14,622 weekly locations were on land (Amstrup et al. 2000; Amstrup unpubl. data). Most of those were bears occupying maternal dens for the winter. In the polar basin area, polar bears truly are pelagic organisms (Garner et al. 1994)!

Telemetry data also have proven that polar bears do not wander aimlessly on the ice, nor are they carried passively with the ocean currents as previously thought (Pedersen 1945). Rather, they occupy multiannual activity areas outside of which they seldom venture. Annual activity areas of female polar bears monitored by radiotelemetry for multiyear periods varied among years. Collared animals, however, seemed to use seasonally preferred or "core" regions every year despite variation in annual activity area boundaries (Amstrup et al. 2000). This suggests that activity areas of polar bears, when viewed over multiyear periods, might be called home ranges. All areas of the home range, however, will not be used each year. Sea-ice habitat quality varies temporally as well as geographically (Stirling and Smith 1975; Ferguson et al. 1997, 1998, 2000a, 2000b; DeMaster et al. 1980; Amstrup et al. 2000). In areas of volatile ice, a large multiannual home range of which only a portion is used in any one season or year is an important part of the polar bear life history strategy.

Linear movements and activity areas are very large compared to those of most terrestrial mammals, and they vary in different regions of the globe, presumably because of variation in patterns of productivity and other sea-ice characteristics. In the Beaufort Sea, where polar bears have been followed by radiotelemetry for 20 years (Amstrup et al. 2000), total annual movements, calculated as the sum of straight-line distances separating consecutive weekly relocations, averaged 3415 km and ranged up to 6200 km. Movement rates of >4 km/hr were sometimes sustained for long periods, and movements of >50 km/day were observed. Annual activity areas of 75 radio-collared female polar bears in the Beaufort Sea region averaged 149,000 km^2. The smallest annual activity area was nearly 13,000 km^2, whereas the largest was 597,000 km^2 (Amstrup et al. 2000).

Whereas movements of polar bears in the Beaufort Sea are impressive in their magnitude, movements of bears in areas of more dynamic ice may be even greater. The mean activity area size for six bears followed by satellite telemetry in the Chukchi Sea was 244,463 km^2 (Garner et al. 1990). The mean annual distance moved by those bears was 5542 km. The potential mobility of polar bears in regions of volatile ice was illustrated by a mean rate of northerly spring movement of 14.1 km/day at a time when ice was moving as much as 15.5 km/day in the opposite direction (Garner et al. 1990).

In contrast, Schweinsburg and Lee (1982) reported maximum activity areas of <23,000 km^2 in the Canadian Arctic Archipelago. Ferguson et al. (1999) also reported very large-scale movements for polar bears in the volatile sea-ice conditions of Davis Strait and Baffin Bay, and much smaller movements for bears in the interior of the Canadian Arctic Archipelago. The sea-ice of the Chukchi and Beaufort Seas and Baffin Bay is more dynamic and unpredictable than the ice in much of the Canadian Arctic Archipelago. The mobility of polar bears appears to be directly related to that variability (Garner et al. 1990, 1994; Gloersen et al. 1992; Messier et al. 1992; Ferguson et al. 2001).

Seasonal movement patterns of polar bears serve to emphasize the role of sea-ice in their life cycle. In the Beaufort Sea, the largest monthly activity areas were in June–July and November–December. These also were the months of highest movement rate. This matches the patterns of ice ablation and formation observed in the area (Gloersen et al. 1992). Polar bears catch seals mainly by still-hunting (Stirling and LaTour 1978). The volatile summer and autumn ice must minimize predictability of seal hunting opportunity. That unpredictability could require longer movements and larger activity areas during seasons of freeze-up and break-up. From May through August, measured net monthly movements of polar bears in the Beaufort Sea were significantly to the north for all bears. In October bears moved back to the south (Stirling 1990; Amstrup et al. 2000). Those movements appeared to be correlated with general patterns of ice formation and ablation. Between May and August, the ice of the southern Beaufort Sea is degrading (Gloersen et al. 1992). October is usually the month of freeze-up in the southern Beaufort Sea and may be the first time in months when ice is available over the shallow water near-shore. Polar bears summering on the persistent pack ice quickly move into shallow-water areas as soon as new annual ice forms in autumn, and they disperse easterly and westerly as ice solidifies through winter.

In contrast to polar bears of the Beaufort Sea region, Messier et al. (1992) reported that peak movement rates of instrumented polar bears in Viscount Melville Sound within the Canadian High Arctic Archipelago occurred from May to July. Movements, although increasing after January, were less from October through March. Ferguson et al. (2001) reported high movement rates in spring and summer in the High Arctic, and Messier et al. (1992) reported increasing mobility from January through spring in the Canadian Arctic. Polar bears in the Beaufort Sea also demonstrate high summer movement rates apparently because of rapidly changing ice conditions. In the southern and northern Beaufort Sea areas, movement rates remained high in November and December and low in May. The lower level of winter movement among polar bears of Viscount Melville Sound may be a consequence of the year-round abundance of multiyear ice (Gloersen et al. 1992; Messier et al. 1992; Ferguson et al. 2001). The density of ringed seals is lower there than in most other areas of polar bear habitat from Alaska through to West Greenland (Stirling and Øritsland 1995), and seals that are present in Viscount Melville Sound tend to be more concentrated along tidal cracks and pressure ridges that parallel the island coastlines (Kingsley et al. 1985). By comparison, the annual ice that predominates in most of the southern Beaufort Sea is more dynamic, and allows a greater amount of sunlight into the water column to support primary productivity. This facilitates easier access to air for seals to breath, and supports higher densities and numbers of ringed seals and polar bears (Stirling et al. 1982; Kingsley et al. 1985; Stirling and Øritsland 1995).

Polar bears in the Beaufort Sea may spend more time in winter actively foraging, and those in the Viscount Melville Sound area may spend more time resting and conserving energy. Messier et al. (1992) reported that long periods of "sheltering" were common among bears wintering in Viscount Melville Sound, and attributed this behavior to the poor foraging conditions there. Another factor may be the greater predictability of the foraging conditions in the stable ice of the High Arctic. With less change in the character of the sea-ice after freeze-up, polar bears may be able to determine the profitable hunting areas in early winter. Predictable sea-ice conditions could help bears minimize midwinter searching for good hunting areas and maximize benefits of sheltering. The constantly changing sea-ice in places like the Beaufort Sea or Baffin Bay, however, may require major modifications of foraging strategy from month to month or even day to day during break-up, freeze-up, or periods of strong winds. Polar bears are adaptable enough to modify their foraging patterns for the extreme range of sea-ice scenarios (Ferguson et al. 2001).

Just as the labile nature of the sea-ice results in annual variability in the distribution of suitable habitat for polar bears, it also eliminates any benefit to polar bears of defending territories. The location of resources is less predictable than resources on which terrestrial predators depend. Seals tend to be distributed over very large areas at low densities (Bunnell and Hamilton 1983). Furthermore, their distribution, density, and productivity are extremely variable among years (DeMaster et al. 1980; Stirling et al. 1982; Stirling and Øritsland 1995). As radiotelemetry studies have shown, female polar bears show only general fidelity to seasonal feeding areas (Ferguson et al. 1997; Amstrup et al. 2000). Absence of strict fidelity, especially during breeding and denning seasons (Garner et al. 1994; Amstrup and Gardner 1994), essentially prohibits defendable territories. Males similarly must be free of the need to defend territories if they are to maximize their potential for finding mates each year (Ramsay and Stirling 1986b).

Although there may be limited spatial segregation among individual polar bears, telemetry studies have demonstrated spatial segregation among groups or stocks of polar bears in different regions (Schweinsburg and Lee 1982; Amstrup et al. 1986, 2000; Garner et al. 1990, 1994; Messier et al. 1992; Amstrup and Gardner 1994; Bethke et al. 1996; Ferguson et al. 1999). Patterns in spatial segregation suggested by telemetry data, survey and reconnaissance, marking and tagging studies, and traditional knowledge resulted in recognition of 19 partially discrete polar bear groups (Lunn et al. 2002:21–35). There is considerable overlap in areas occupied by members of these groups, and boundaries separating the groups have been adjusted as new data were collected. Nonetheless, these boundaries are thought to be ecologically meaningful, and the units they describe are managed as populations.

A 20th polar bear population may occur in the central polar basin (Table 27.1). It is unclear whether bears that occur in this region are simply visitors from populations nearer to islands and continental shorelines or whether there are animals that spend all of their time in these high-latitude regions far from any land. The frequency of recent observations deep in the polar basin, however, mandates recognition that a separate stock could occur there (Fig. 27.8).

TABLE 27.1. Summary of polar bear population status as determined by both historical harvest (1995–96 to 1999–00) levels and current management practices. Abundance estimates are based on the best available data for each population, which ranges from little or no information to detailed inventory studies. The percent females statistic excludes bears of unknown sex, and natural deaths are not included

Population	Abundance Estimate	Certainty of Estimate	Monitoring of Harvest and Other Removals	% Females in Kill	Sustainable Kill[a]	Mean Annual Kill	Environmental Concerns[b]	Status[c]
East Greenland	2000	poor (1997)	fair	38	unknown	80	P, W	?
Barents Sea	2000–5000	poor (1982)	Norway – good Russia – poor		na	Norway – 2 Russia – ?	P, W	?
Kara Sea	unknown	unknown	poor		na	unknown	P, I	?
Laptev Sea	800–1200	poor (1993)	poor		na	unknown	P	?
Chukchi Sea	2000+	poor (1997)	US – good Russia – poor	US – 35 Russia – ?	86+	US – 76 Russia – ?	W, I	S?
Southern Beaufort Sea	1800	good (2001)	good	33	81	50	W, I	I
Northern Beaufort Sea	1200	good (1987)	good	33	54	32	W	I
Queen Elizabeth	200	poor (1995)			9?	0	P	S?
Viscount Melville Sound	230	fair (1992)	good	25	4	4		S
Norwegian Bay	100	fair (1979)	good	32	4	4	W	S[a]
Lancaster Sound	1700	fair (1996)	good	25	77	76	W	S[a]
M'Clintock Channel	350	fair (2001)	good	26	11	24	W	S?
Gulf of Boothia	900	poor (1986)	good	40	34	37		S[a]
Foxe Basin	2300	good (1996)	good	36	97	90	W	S[a]
Western Hudson Bay	1200	good (1997)	good	35	52	49	W	S[a]
Southern Hudson Bay	1000	fair (1986)	good	36	41	45		S
Kane Basin	200	fair (1996)	fair	32	9	10		S
Baffin Bay	2200	fair (1996)	fair	36	93	139		D
Davis Strait	1400	fair (1996)	fair	38	56	63	W	D?
Arctic Basin	unknown	unknown	none		na			?
Total estimate for world abundance: 21,500–25,000								

SOURCE: Lunn et al. 2002:22.

[a]Except for Viscount Melville Sound, sustainable harvest is based on population estimate (N), estimated rates of birth and death, and harvest sex ratio (Taylor et al. 1987):

$$\text{Sustainable harvest} = \frac{N \times 0.015}{\text{Proportion of harvest that was female}}$$

Proportion of harvest that was female is the greater of the actual value or 0.33. Unpublished modeling indicates a sex ratio of 2 males: 1 female is sustainable, although mean age and abundance of males will be reduced at maximum sustainable yeild. Harvest data (Lee and Taylor, 1994) indicate that selection of males can be achieved

[b]I – industrial development current or proposed; P – evidence of pollutants in bear tissues; W – evidence global warming effects on sea ice or populations

[c]D – decreasing; I – increasing; S – stationary; S[a] – stationary, population managed with a flexible quota system in which any over-harvest in one year results in a fully compensatory reduction to the following year's quota; ? – indicated trend uncertain

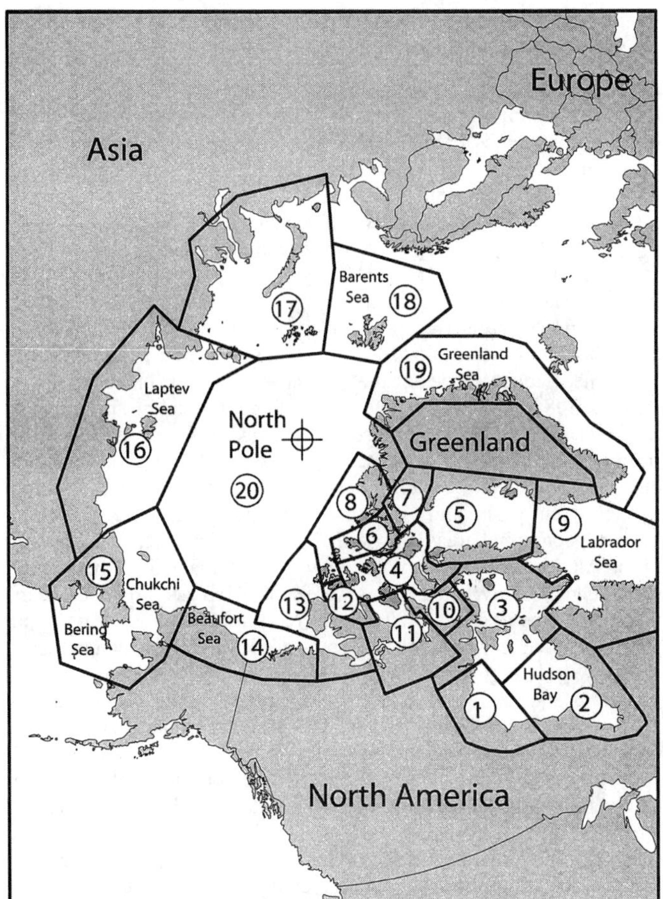

FIGURE 27.8. The circumpolar range of polar bears is subdivided, according to observed movement patterns, into 20 hypothesized populations or stocks. 1, Western Hudson Bay; 2, southern Hudson Bay; 3, Foxe Basin; 4, Lancaster Sound; 5, Baffin Bay; 6, Norwegian Bay; 7, Kane Basin; 8, Queen Elizabeth Islands; 9, Davis Strait; 10, Gulf of Boothia; 11, M'Clintock Channel; 12, Viscount Melville Sound; 13, northern Beaufort Sea; 14, southern Beaufort Sea; 15, Chukchi Sea; 16, Laptev Sea; 17, Novaya Zemlya; 18, Svalbard; 19, East Greenland; 20, Arctic basin. Boundaries are constantly being adjusted as new data and ecological insights are obtained. SOURCE: Adapted from Lunn et al. (2003:23).

DENNING

Across most of their range, pregnant female polar bears excavate dens in snow and ice in early winter (Harington 1968; Lentfer and Hensel 1980; Ramsay and Stirling 1990; Amstrup and Gardner 1994). They give birth in those dens during midwinter (Kostyan 1954; Harington 1968; Ramsay and Dunbrack 1986) (see section on reproduction), and emerge from dens when cubs are approximately 3 months old. Because neonates are so altricial, the period of denning is essential to their early survival. Recognizing it as a critical phase in the polar bear life cycle, scientists have devoted much attention to aspects of maternal denning.

Distribution of Denning. Historically, polar bear dens were thought to represent the "core areas" of their ranges (Harington 1968). In those core areas, large numbers of polar bears repeatedly and predictably concentrated their denning within relatively small geographic regions (see Fig. 27.1). Examples of concentrated denning areas include particular islands of the Svalbard Archipelago north of Norway (Lønø 1970; Larsen 1985); Franz Josef Land, Novaya Zemlya, and Wrangel Island in Russia (Uspenski and Chernyavski 1965; Uspenski and Kistchinski 1972); and the west coast of Hudson Bay in Canada (Harington 1968; Jonkel et al. 1975; Stirling et al. 1977b; Ramsay and Andriashek 1986; Ramsay and Stirling 1990).

Concentration or core areas were easily detected by aerial or ground surveys, and often had been known from reports of early explorers or Native people residing in the area. Early knowledge of concentrated denning led to the view that all polar bears must use such core areas, and that areas without them did not significantly contribute to polar bear reproduction. Harington (1968 :8) implied as much when he stated, "No major denning areas seem to exist in Alaska." It was assumed geographic regions without such areas mainly were populated by visiting polar bears. That concept led Native people of northern Canada to believe that harvests in Alaska were being sustained by polar bears produced in Canada (Stirling and Andriashek 1992). Over much of their range, we now know, polar bears den in a more diffuse pattern where individual dens are scattered over broad reaches of habitat at low density (Lentfer and Hensel 1980; Stirling and Andriashek 1992; Amstrup 1993; Amstrup and Gardner 1994; Messier et al. 1994; Ferguson et al. 2000a; Durner et al. 2001, 2003).

Successful denning by polar bears requires accumulation of sufficient snow that a pregnant female can create a snow cave early in winter and subsequently be covered over. A variety of weather and topographic conditions meet that requirement. Bears denning in the concentration areas of the Svalbard Islands or the large islands north of the Russian coast meet their needs in rugged mountains and fjordlands (Uspenski and Chernyavski 1965; Lønø 1970; Uspenski and Kistchinski 1972; Larsen 1985). One famous concentration area, however, is in the relatively flat tundra along the west coast of Hudson Bay in Canada (Ramsay and Andriashek 1986; Ramsay and Stirling 1990). There, dens are concentrated along relatively low banks and ridges supporting small trees. In the regions where denning is dispersed, the topography ranges from mountainous to essentially flat (Harington 1968; Lentfer and Hensel 1980; Stirling and Andriashek 1992; Amstrup 1993; Amstrup and Gardner 1994; Messier et al. 1994; Ferguson et al. 2000a; Durner et al. 2001, 2003).

Research on Wrangel Island (Belikov 1976) and Hudson Bay (Kolenosky and Prevett 1983) suggested that polar bears select den sites based on specific topography and habitat. Whether the habitats are scattered or concentrated, however, all denning areas have micro- or macrohabitats that predictably catch snow in the autumn and early winter (Durner et al. 2003). In the mountains, snow-catching features are obvious. The snow-catching ability of the very flat terrain in coastal Alaska, where until recently denning was presumed to be insignificant, is not so obvious. There, the most frequently used denning habitats are along coastal and river banks. Although the mean bank height where female bears den is 5.4 m ($SD = 7.4$ m), banks as low as 1.3 m provided sufficient snow depth for successful denning (Durner et al. 2001). The Alaskan northern coast gets relatively little snow. However, the landscape is so flat that what snow there is, is blown incessantly across the plain throughout the winter. Any areas of relief in the otherwise flat terrain are filled solidly with snow from the very early winter. Banks used for denning in Alaska most commonly had water or level ground below the slope and relatively level ground above, enhancing the chance for sufficient snow build-up for denning (Durner et al. 2001).

Across the range of polar bears, most denning, whether in concentration areas or dispersed, occurs relatively near the coast. In early visual surveys, Harington (1968) found that 61% of dens located over broad regions of the Canadian Arctic were within 8 km and 81% were within 16 km of the coast. All dens seen by Stirling and Andriashek (1992) in the Canadian Beaufort Sea were along the coast. Likewise, Stishov (1991) reported that 83% of dens found during surveys of northern Russia were in snow banks formed under shore slopes and precipices. Most polar bear dens were within 3 km of the coast of Svalbard (Larsen 1985). Even on Wrangel Island, where bears move inland to den in high mountains, most are located within 8 km of the coast. The main exception to coastal denning appears to be the Hudson Bay area, where females moved from 29 to 118 km inland to traditional denning areas (Kolenosky and Prevett 1983; Stirling and Ramsay 1986).

The trend toward denning very near the coast has now been confirmed in most regions by radiotelemetry studies. Telemetry allows investigators to locate dens anywhere bears establish them. More than 80% of maternal dens found on land by radiotelemetry in the Alaskan Beaufort Sea were within 10 km of the coast and over 60% are right on the coast or on coastal barrier islands (S. C. Amstrup, unpublished data). Messier et al. (1994) used satellite telemetry to learn that maternal dens in the Canadian High Arctic were widely scattered in coastal areas, and averaged 8.6 km from the coast. Ferguson et al. (2000a) reported that most dens found using telemetry in the Canadian High Arctic and Baffin Bay areas were within 20 km of the coast.

Denning on the Pack Ice. Although most maternal denning appears to occur on coastlines of mainlands and islands, Amstrup and Gardner (1994) discovered that 53% of the dens of polar bears radio-collared between 1981 and 1991 were on drifting pack ice. They also found that 4% were on land-fast ice adjacent to shore. Lentfer and Hensel (1980) recognized the occurrence of dens on pack ice, but suggested that it was limited to bears that could not make it to shore to den. Harington (1968) concluded that denning on ice was not preferred, and Messier et al. (1994) reported no maternal denning on pack ice, although some "shelter denning" on pack ice was observed. The discovery that half of the bears in the Beaufort Sea may den on drifting sea ice, therefore, was not expected.

Bears that den on pack ice potentially are subject to a number of disruptions that could not affect bears denning on land. First, the sea-ice changes throughout the year. It shifts, breaks up, and refreezes. Ice floes can turn over or have other floes rafted onto them. Therefore, a maternal den could be overturned, buried, or otherwise compromised any time in the denning cycle. Amstrup and Gardner (1994) reported observations of six polar bears in pack-ice dens that were swept past Point Barrow and southwest into the Chukchi Sea due to unusually unstable ice. Two of those females were observed after their dens had been destroyed by rafting action of sea-ice in mid-February. The females carried tiny cubs in their mouths, probably in desperate attempts to relocate to a new den site. When observed later that spring, however, neither of these bears was accompanied by young. Only one of six females swept into the Chukchi Sea that year had cubs when reobserved later. Hence, there are risks involved in denning on sea-ice.

On emergence of the female and her new cubs, the predictability of available resources may be limited even if a pack-ice den remains intact. Bears that den on pack ice may drift up to 1000 km during the winter (Amstrup and Gardner 1994). Despite observed and hypothetical risks, production of cubs from dens at sea was not significantly different than that from dens on land (Amstrup and Gardner 1994), and sea-ice denning has obviously been maintained as a successful reproductive strategy in the Beaufort Sea region.

Despite the absence of conclusive reports, sea-ice denning probably occurs at some level in other areas. When engaged in polar bear aerial surveys on the high pack ice northeast of Greenland and north of Svalbard, Larsen et al. (1983a) observed numerous tracks of females with cubs of the year near 84°N latitude. These animals were moving predominantly in a southeasterly direction, toward Svalbard. The distance of these sightings from land and the time of the year in which they were recorded suggested those cubs were born on the pack ice. Little significance was attached to those footprints at that time. However, that observation takes on a greater significance in light of the confirmed frequency of sea-ice denning in the Beaufort Sea (Amstrup and Gardner 1994; Amstrup et al. 2000) and the recent recognition of a possible polar basin stock of polar bears. Because other recent studies using satellite radiotelemetry have not revealed significant amounts of sea-ice denning, it seems reasonable to assume that its overall frequency is low. The linear coastline of central Arctic Russia may be more similar to the Beaufort Sea than other areas, and hence may be another area where sea-ice denning is common. Satellite data from that region, however, are too few to test that hypothesis.

In addition to questions about security of animals while in dens, the phenomenon of pack-ice denning also raises questions about navigation capabilities of polar bears. No other vertebrate is passively transported this far "in the blind." Thus, not only do polar bears range far and wide, they are able to determine where they are and return to previously used areas after long distances of passive transport. How polar bears accomplish this is unknown.

Fidelity to Denning Locales. Although there are no historical data regarding denning fidelity, it has logically been assumed that concentrated denning areas are maintained by fidelity of individual females to those sites (Uspenski and Chernyavski 1965; Lønø 1970; Uspenski and Kistchinski 1972; Larsen 1985). Pregnant females return, it is assumed, to areas where they have successfully denned in the past.

The greatest number of records of den-site fidelity derives from the Beaufort Sea. There, 27 polar bears were followed to more than one suspected or confirmed maternity den (Amstrup and Gardner 1994). One radio-collared polar bear was followed to four maternal denning sites, 7 were followed to three dens each, and 19 to two dens. Confirmed sequential dens were separated from their precursors by a mean of 308 km ($SD = 262$, $n = 30$), and the minimum distance was 23 km. Distances separating sequential land dens were not different from those separating sequential pack-ice dens. Bears that denned once on pack ice were more likely to den on pack ice than on land in subsequent years, and vice versa. Similarly, bears were faithful to general geographic areas. Those that denned once in the eastern half of the Alaskan coast were more likely to den there than to the west in subsequent years. When all years were considered, denning polar bears preferred some areas, but no areas were used by collared bears in all years. Weather, ice conditions, and prey availability, all of which varied annually, probably determined where bears denned. Those annual variations and the long-distance movements of polar bears (Amstrup et al. 1986, 2000; Garner et al. 1990) make seasonal recurrence at exactly the same location unlikely.

Although Beaufort Sea polar bears were not faithful to particular denning sites, data on den distribution and fidelity of females to denning areas indicated there are both "pack-ice" and "land-denning" bears. Den substrate switching appeared to be limited. This segregation may have begun when some females were prevented from reaching land in the fall.

The only other region where data are available on fidelity to denning areas is Hudson Bay. There, pregnant females initiate their overwinter denning period in earthen dens they occupy in summer. During winter, they burrow into adjacent snow drifts (Watts and Hansen 1987). The presence of hundreds of earthen dens in the region suggests a long tradition of use. In three instances, cavities were reused by different bears, but no observations of reuse of a cavity by the same bear were reported (Ramsay and Stirling 1990). On average, bears followed to second dens chose locations 27 km (4–52 km) from previous attempts (Ramsay and Stirling 1990). Hence, there was greater fidelity to local areas than in the Beaufort Sea, but site-specific philopatry was not apparent.

Despite general fidelity to local areas, the overall distribution of denning along the west coast of Hudson Bay shifted markedly over a 20-year period (Ramsay and Stirling 1990). Because bears have the navigational skills to return to the same area, the reason for the shift is not clear. A similar shift appears to be occurring in the Beaufort Sea region as well, however. During the 1990s, more females appeared to choose den locations in the central and western portions of the northern Alaska shore than during the 1980s (S. C. Amstrup, unpublished data). Such shifts must reflect changing ice formation and ablation patterns, food availability, or other unidentified ecological factors. Harington (1968), Larsen (1985), and Lønø (1970) concluded that variation in the local pattern of sea-ice movements during the preceding summer and autumn accounts for annual changes in the distribution of winter dens. Multiple-year trends in sea-ice patterns could clearly alter denning and other behavioral patterns.

Denning Chronology. Pregnant female polar bears enter their dens in the autumn after drifts large enough to excavate a snow cave are formed. Because polar bears (in most areas) den only in ice and snow rather than in the soil under the snow, the annually variable snow and ice conditions mediate when and where bears enter their dens each autumn. Polar bears depart dens in the spring when their cubs are able to survive in the outside climate. Until the advent of effective radiotelemetry, little was known about the chronology of denning. Larsen (1985) reported that most dens on Svalbard were opened in late March and vacated by mid-April. Lentfer and Hensel (1980) reported Alaskan polar bears came ashore to den in late October and early November and left their dens in late March and early April. Lønø (1970) concluded dens on Svalbard were entered in November and December and abandoned between 10 and 25 April. At the far north of Svalbard, he speculated that bears entered dens as early as late October. Observations in other High Arctic areas suggest abandonment between mid-March and mid-April (Uspenski and Chernyavski 1965; Kistchinski 1969; Belikov et al. 1977). Hansson and Thomassen (1983) suggested the first dens were opened in the first week of March and most were abandoned by mid-April. Kolenosky and Prevett (1983) and Ramsay and Andriashek (1986) reported emergence from dens in the Hudson Bay area in late February and early March. Polar bears are largely food deprived while on land in the ice-free period. During this time, they survive by mobilizing stored fat. Pregnant females that spend the late summer on land and then go right into dens may not feed for 8 months (Watts and Hansen 1987; Ramsay and Stirling 1988). This may be the longest period of food deprivation of any mammal, and it occurs at a time when the female must give birth and nourishment to her new cubs.

Satellite telemetry has now confirmed that the chronology of denning varies somewhat around the world. In the Beaufort Sea, mean dates of den entry were 11 and 22 November for land ($n = 20$) and pack-ice ($n = 16$) dens, respectively (Amstrup and Gardner 1994). Female bears continued foraging right up to the time of den entry. Then they denned near where they happened to be foraging. On average, Beaufort Sea polar bears emerged from their dens with new cubs on 26 March if they were on the pack ice ($n = 10$) and 5 April if they were on land ($n = 18$). Dates of entry and exit varied somewhat among years depending on sea-ice, snow, and weather conditions.

Messier et al. (1994) reported the mean entry into maternal dens in the Canadian Arctic was 17 September (*SE* = 3 days; range 27 August–12 October) and mean emergence was 21 March (*SE* = 3 days; range 4 March–7 April). Females and their cubs remained near dens for a mean 13 (*SE* = 3) days in the spring before leaving the denning area. Those data may indicate an earlier and more protracted denning period at higher latitudes than in the Beaufort Sea. Ferguson et al. (2000a), on the other hand, observed that bears denning at higher latitudes entered their dens a bit later than those to the south, but that exit times did not differ by latitude. They reported a mean den entry of 15 September (1 September–7 October), a mean exit of 20 March (15–28 March), and a mean 180 days in dens (163–200 days).

As noted, initiation of denning depends on sufficient snow accumulation to allow excavation of a den cavity. For bears denning on sea-ice or moving from sea-ice to land denning habitat, timing of sea-ice consolidation can alter the onset of denning. Sea-ice dens must be in ice stable enough to stay intact for up to 164 days while being pushed by currents for hundreds of kilometers.

Whereas only pregnant female polar bears enter dens for the entire winter, any bear may enter shelters for shorter periods to avoid storms, extreme cold or heat, or periods of poor hunting. Sheltering is best known along the west coast of Hudson Bay. Because the ice in Hudson and James Bays disappears entirely, the whole population there is forced onto land in summer. Feeding opportunities are minimal, and many animals take shelter in earthen dens, where it is cooler and they minimize insect harassment. When the ice forms in fall, most of the bears in earthen shelters go out on the new ice to hunt. Pregnant females, however, remain in the dens and eventually move into snow that drifts over their earthen structures (Stirling et al. 1977b; Derocher and Stirling 1990).

Use of shelter dens also occurs at higher latitudes. Messier et al. (1992) reported that long periods of "sheltering" were common among all classes of female bears (except those in maternal dens) wintering in Viscount Melville Sound. Females entered shelters on average on 18 December (*SD* = 7 days) and stayed an average of 53 days (*SD* = 9 days). The duration of sheltering ranged from 25 to150 days. Messier et al. (1992) attributed this behavior to the poor foraging conditions in the Viscount Melville Sound region. In Baffin Bay and the eastern Canadian High Arctic, Ferguson et al. (2000a) reported a bimodal incidence of sheltering. Autumn sheltering occurred from mid-September to early November, whereas winter shelters were occupied mainly from late December to March. Autumn shelters were occupied for a mean 56 days (range = 50–70), whereas winter shelters were occupied for a mean 65 days (range = 35–86). At higher latitudes, the frequency of winter sheltering increased and the frequency of autumn sheltering decreased.

Although more female polar bears have been followed for longer periods there than anywhere, sheltering for protracted periods has not commonly been observed in the Beaufort Sea region (Amstrup and Gardner 1994; S. C. Amstrup, unpublished data). Yet, the latitude of the Beaufort Sea is in the middle of the range reported by Ferguson et al. (2000a). Clearly, use of sheltering in the eastern Canadian High Arctic and Baffin Bay is not so much a function of latitude as of sea-ice and other ecological conditions. Sea-ice formation and ablation, weather, and prey availability, although influenced by latitude, are as much controlled by the shapes of coastlines, presence of islands, water depths, currents and other factors. In areas of Baffin Bay and Davis Strait, seasonal absence of sea-ice forces polar bears onto land in autumn as it does in Hudson Bay. While on land, the best strategy is to conserve energy (Nelson et al. 1973; Guppy 1986). Conversely, ice is available in the High Arctic year-round, and autumn sheltering is less necessary. The winter increase in sheltering at higher latitudes is probably an adaptation to avoid the harshest of winter weather and heavy ice. In the Beaufort Sea, the sea-ice substrate is available year-round, but winter ice conditions may not be as harsh as in the eastern High Arctic. Those realities may ameliorate the need for sheltering. Messier et al. (1994) also concluded that variations in the availability of satisfactory hunting conditions may encourage a facultative approach to use of shelters.

Human Influences on Denning. The time spent in maternal dens is the only period in their life cycle during which polar bears are unable to simply move away from a potential disturbance. Premature exposure of altricial neonates to the outside arctic environment can be fatal (Amstrup and Gardner 1994). Therefore, disturbance of denning bears could result in reproductive failures. The only quantitative data on sensitivity of denning polar bears to human disturbances are from the Beaufort Sea region. Amstrup (1993) reported considerable tolerance of human activities near dens. Subsequent observations (S. C. Amstrup, unpublished data) corroborate those early records. Polar bears seem secure in their dens, and appear very tolerant of aerial and ground traffic very near maternal dens in winter and spring. These observations corroborate the results of Blix and Lentfer (1992), who observed that only seismic testing <100 m from a den and a helicopter taking off at a distance of 3 m produced noises inside the dens that were notably above background levels. They also concluded that a polar bear in its den is unlikely to feel vibrations unless the source is very close. Preliminary analyses of more recent work (MacGillivray et al. 2002) confirm that sound penetration into dens is greatly ameliorated. They also suggested, however, that helicopters and some ground vehicles may be detectable in dens at much greater distances than suggested by Blix and Lentfer (1992).

Observations of grizzly bears also suggested substantial tolerance of such activities. McLellan and Shackleton (1989) found that grizzly bears in summer were not displaced from the immediate vicinity of seismic testing supported by helicopters. Reynolds et al. (1986) reported some movements and possible increased heart rates when denned grizzly bears were exposed to seismic testing activities. However, they also observed that similar movements and heart rate patterns sometimes occurred in absence of human activities, and they concluded that "effects on the bears were probably minimal" (Reynolds et al. 1986 :174). Although the observations appear compelling, the sample sizes reported by Amstrup (1993) are small. The tests of Blix and Lentfer (1992) also were relatively limited in scope, and the degree to which information from grizzly bears applies to polar bears is uncertain. We know polar bear behaviors are highly variable among individuals. This variability means that additional data will be necessary to quantify the possible effects human activities may have on denning female polar bears. Fortunately, prudent spatial and temporal management of human activities using the best available information, can prevent most potentially disruptive activities from overlapping with polar bear denning.

Delayed implantation and birth of altricial young mean that early in pregnancy, parental investment is low. Female polar bears have less to lose by leaving a den in the fall than they do by leaving after parturition. Perhaps that explains why polar bears appear to be more willing to abandon dens in fall than later in the denning period (Amstrup 1993; S. C. Amstrup, unpublished data). Belikov (1976) also reported that polar bears were more easily displaced from their dens in the fall. Five Alaskan polar bears thought to abandon dens because of human interference in the autumn were successful in redenning elsewhere (Amstrup 1993, S. C. Amstrup, unpublished data). Likewise, three bears disturbed from dens near Hudson Bay relocated to other den sites (Ramsay and Stirling 1986a). The relative resilience of denning bears to disruptions in spring and their plasticity regarding den selection in autumn have significant management ramifications.

In the Beaufort Sea, individual polar bears have strong ties only to general denning areas and substrata. Denning habitat there is widely scattered across broad areas and is not limiting. Where such circumstances prevail, temporal and spatial management of human activities should eliminate most conflicts between those activities and maternal denning (Amstrup 1993; Amstrup and Gardner 1994). For example, proposed human activities can be directed around most of the narrow bands of habitat that are suitable for denning (Durner et al. 2001). Furthermore, initiation of intense human activities in autumn would give bears enroute to land dens the opportunity to den in less disturbed areas. If a bear encountered activities it didn't like, if could move up or down the coast to a place where it is comfortable. Also, bears already in dens could relocate more easily in autumn than after parental investment increases.

Much relocation of denning appears to occur naturally in autumn before bears finally settle down for the winter. Natural fluctuations in areas used for denning (Uspenski and Chernyavski 1965; Uspenski and Kistchinski 1972; Ramsay and Stirling 1986a, 1990) suggest that, even in some concentrated or core denning areas, alternate den sites may be available to bears if they are disturbed. However, data regarding fidelity to denning locales and responses to disturbances of dens are largely unavailable outside of the Beaufort Sea. More importantly, a human activity with potential to disturb denning bears will affect more individuals where dens are geographically clustered than where they are widely scattered. In the Beaufort Sea region for example, even expansive human activities, such as some related to oil exploration, would likely overlap with only a very small number of dens in any given year. In the concentrated denning areas of the world, a similar activity could overlap with dozens of dens. The risks that human activities could have population level effects are greater where dens are geographically clusterred. Clearly, human activities around all maternal denning areas must be managed with utmost caution. Each development scenario must be approached with full understanding of the ecological and behavioral situation, and wherever human activities proceed, outcomes must be carefully monitored so that management can be adjusted as needed.

PHYSIOLOGY

Liver Toxicity. Polar bear liver can contain very high levels of vitamin A (Rodahl and Moore 1943; Lewis and Lentfer 1967; Russell 1967). The concentration varies greatly among individuals, but does not seem to be age dependent. The liver is toxic to humans if eaten. Rodahl and Moore (1943) summarized the variety of human health effects reported by Arctic explorers who had eaten polar bear liver. Effects ranged from drowsiness, headache, and general irritability to large-scale peeling of the skin. Peeling was often localized, but sometimes covered victims from head to foot. Variation in the quantity of liver consumed and the vitamin A concentrations within each liver probably accounted for the diversity of reported symptoms.

Thermoregulation. Polar bears appear to be highly specialized for life in the arctic marine environment. However, Scholander et al. (1950) reported relatively high thermal conductivity for polar bear fur in both air and water. Likewise, Øritsland (1970) concluded that polar bears depend on a combination of fur, fat, and subdermal vascularization to maintain their body temperature. Øritsland (1970) and Best (1982) showed that polar bears can increase effective peripheral insulation with vasomotor controls. Such controls could be most effective during water immersion. All adaptations, however, were inadequate to contend with either exceptionally cold or hot air temperatures; polar bears may depend on postural and behavioral mechanisms during extremes of air temperature.

Newborn cubs have short, thin hair and no subcutaneous fat (Blix and Lentfer 1979). Therefore, they are poorly equipped for survival outside of the maternal dens in which they are born. On emergence from the den, however, cubs are much better equipped for outside exposure. Blix and Lentfer (1979) reported a lower critical temperature for a 12.5-kg cub of –30°C. At –45°C, the cub's oxygen consumption increased only 33%, and there was no decrease in core temperature. Immersion of this cub into ice water resulted in a precipitous and immediate drop of body temperature. Despite the small size and minimal subcutaneous fat, it appears that cubs are ready to face the outside world at the time of den departure. They are, however, not ready for immersion.

Locomotion. Øritsland et al. (1976), Hurst et al. (1982a, 1982b), and Best (1982) concluded that polar bears are relatively inefficient walkers. Measurements were made from two polar bears walking on treadmills. Oxygen consumption and heat storage were higher than might have been predicted for other mammals of comparable size. Inefficient walking was attributed to aspects of polar bear morphology, specifically the massive forelimbs evolved for capture of prey (Øritsland et al. 1976; Hurst et al. 1982a, 1982b). Economy of transport, they suggested, was compromised by considerations of thermoregulation and hunting strategy. Nonetheless, the typical daily, seasonal, and annual movements of polar bears place them among the most mobile of all quadrupeds (Amstrup et al. 2000). Locomotion in polar bears is clearly an area where additional research is in order.

Hibernation. Like other ursids, polar bears have evolved a very specialized winter dormancy. Females occupy maternal dens of ice and snow for periods of 4–8 months. During that time, they neither eat nor drink and they do not urinate or defecate (Nelson et al. 1973; Folk and Nelson 1981; Nelson 1987; Watts and Hansen 1987; Ramsay and Stirling 1988). In hibernating bears, normal mineral levels are maintained, lean body mass is constant, blood electrolyte balance is preserved, and levels of blood metabolites are largely unchanged despite loss of nearly half of their total body mass after den entry (Nelson et al. 1973; Folk and Nelson 1981; Guppy 1986; Nelson 1987; Atkinson and Ramsay 1995). They appear able to maintain constant fluid levels by using metabolic water produced from fat catabolism (Guppy 1986; Nelson 1987).

Polar bears may be even more highly evolved with regard to their ability to survive food deprivation than the other ursids. Behavior and physiology of polar bears are well adapted to a feast-and-famine feeding regimen (Lunn and Stirling 1985; Watts and Hansen 1987; Ramsay and Stirling 1988; Derocher and Stirling 1990; Derocher et al. 1990). It now appears that they can alter their metabolism during periods of food deprivation at any time of the year (Nelson et al. 1983). Atkinson and Ramsay (1995) and Derocher et al. (1990) demonstrated that polar bears, unlike other bears, can shift as needed into a hibernation-like metabolic pattern when confronted by a period of food shortage. Facultative changes into and out of a hibernation-like state would magnify the value of summer and winter shelter denning described by Messier et al. (1994) and Ferguson et al. (2000a). This ability could make polar bears the most advanced of all mammals when it comes to dealing with food and water deprivation (Nelson 1987).

REPRODUCTION

Reproduction in the female polar bear is similar to that in other ursids. They enter a prolonged estrus between March and June. In the polar basin, the peak of estrus as evidenced by turgidity of the vulva and vaginal discharge seems to be in late April and early May. Ovulation is thought to be induced by coitus (Wimsatt 1963; Ramsay and Dunbrack 1986; Derocher and Stirling 1992). Implantation is delayed until autumn, and total gestation is 195–265 days (Uspenski 1977), although during most of this time, active development of the conceptus is arrested. Young are born by early January (see below), but stay within the shelter of the den until March or early April (Amstrup and Gardner 1994). Litters of two cubs are most common over most of the polar bear range. Litters of three cubs are seen sporadically across the Arctic, and were most commonly reported in the Hudson Bay region (Stirling et al. 1977b; Ramsay and Stirling 1988; Derocher and Stirling 1992). Young bears will stay with their mothers until weaning most commonly in early spring when the cubs are 2.3 years of age. Female polar bears undergo a lactational anestrus and are available to breed again after weaning. Therefore, in most areas, the minimum successful reproductive interval for polar bears is 3 years (see below).

Newborn polar bears have hair, but are blind and weigh only 0.6 kg (Blix and Lentfer 1979). The growth of cubs is very rapid, and they may weigh 10–12 kg by the time they emerge from the den in the spring. After leaving the den, the rapid growth continues, and cubs may increase their weight by an order of magnitude between den exit and their first birthday (S. C. Amstrup, unpublished data). Cubs can double their weight between their first and second birthdays. Cubs receive an especially rich milk from their mothers. The milk of polar bears typically has a higher fat content than that of other bears, and in general the milk of bears is richer in fat and protein than the milk of other carnivores (Jenness et al. 1972). Polar bear milk is more similar to that of pinnipeds than it is to milk of most terrestrial mammals (Jenness et al. 1972; Ramsay and Dunbrack 1986). Although polar bears may nurse cubs through their second birthday, some females apparently stop allowing cubs to suckle sometime after their first birthday. The contribution to growth from milk during the second year of life is much lower than during the first year (Arnould and Ramsay 1994). Arnould and Ramsay (1994) noted that fat content begins to decline fairly early in lactation, but the biggest differences are between the first and the second year of the cubs's lives. Mean fat content of milk provided to cubs of the year was 31.2 $\pm$ 1.6%, whereas the fat content of milk fed to yearlings was 18.3 $\pm$ 2.4%. The energy contribution from milk is a significant contributor to the observed rapid growth of cubs and comes at a significant cost to mother bears (Arnould and Ramsay 1994).

The exact timing of birth may vary across the range of polar bears. Harington (1968) reported births as early as 30 November with a median date of 2 December. Derocher et al. (1992), reported, based on progesterone spikes in the blood of pregnant bears and the implied date of implantation, that births of Hudson Bay bears probably occur from mid-November through mid-December. Messier et al. (1994) suggested that polar bears give birth by 15 December. In contrast, many pregnant female polar bears in the Beaufort Sea did not enter dens until late November or early December (Amstrup and Gardner 1994; S. C. Amstrup, unpublished data). Unless those bears were giving birth immediately on den entry, a later date of birth can be assumed. One captive female in Barrow, Alaska gave birth on 27 December, corroborating that assumption (Blix and Lentfer 1979). Similarly, Lønø (1972) reported that implantation of the conceptus into the uterus of the polar bear began in November, around the peak of den entry in the Beaufort Sea. The timing of implantation, and hence that of birth, is likely dependent on body condition of the female. Condition of the female, in turn, depends on a variety of environmental factors. The interaction between environmental and physiological factors that control births is clearly an area in need of further research.

Testes of male polar bears reside in the abdomen for most of the year. They descend into the scrotum in late winter, and remain there through May. Descent of the testes permits spermatogenesis, which is thought to occur from February to May (Erickson 1962; Lentfer and Miller 1969; Lønø 1970). Lønø (1970) reported that male/female pairs were observed as early as 8 March and as late as 20 June. According to histological examination of testes and ovaries, Lønø (1970) further concluded breeding could last into July. Deteriorating ice conditions preclude scientific observations in most polar bear habitats by June, so the frequency of summer breeding cannot be easily documented.

Lentfer and Miller (1969) concluded, from presence of mature spermatozoa in epididymides, that male polar bears in Alaska may be able to breed as early as 3 years of age. Presence of sperm also guaranteed reproductive capability until at least age 19 years (Lentfer and Miller 1969). A recent study in Greenland found that 2 of 7 two-year-old males, 5 of 10 three-year-olds, and 4 of 9 four-year-olds had some spermatazoa in epididymides (Rosing-Asvid et al. 2002). Although spermatazoa occurred at low density in the younger bears, all bears $\geq$5 years old, except for one very thin individual, had produced abundant spermatozoa and appeared capable of breeding. Lentfer et al. (1980) observed males 3–11 years old in consort with estrous females, confirming at least the age of earliest breeding ability for male polar bears. It should be noted, however, that excessive hunting in Alaska just before and during the time those observations were made had all but eliminated prime males (aged >10 years) from the population (Amstrup et al. 1986). Subsequently, few male bears that young have been observed with females. Since 1980, the proportion of prime males in Alaskan waters has been high (Amstrup 1995). Presently, large males weighing 400–500+ kg are abundant in this region. Three- and 4-year-old bears typically weigh $\leq$250 kg, and would not be able to compete successfully for mates with the now-abundant large males. Currently, young males must have very low reproductive output despite their apparent reproductive potential.

Productivity of polar bear populations appears to be largely dependent on numbers and productivity of ringed seals. For example, in the Beaufort Sea, ringed seal densities are lower than in some areas of the Canadian High Arctic or Hudson Bay. As a possible consequence, female polar bears in the Beaufort Sea usually do not breed for the first time until they are 5 years of age (Stirling et al. 1976; Lentfer and Hensel 1980). This means they give birth for the first time at age 6. In contrast, across many areas of Canada, females reach maturity at age 4 and produce their first young at age 5 (Stirling et al. 1977b, 1980, 1984; Ramsay and Stirling 1982, 1988; Furnell and Schweinsburg 1984).

Craighead and Mitchell (1982:527) reported that in grizzly bears "reproductive longevity approximates physical longevity." Female polar bears, on the other hand, may show a reproductive senescence long before the end of their lives. Derocher et al. (1992) calculated an average age of first breeding in the Hudson Bay area of 4.1 years. Productivity, assessed by estimated pregnancy rates, remained high between 5 and 20 years of age and declined thereafter (Derocher et al. 1992). Unfortunately, long-term monitoring of individual polar bears is uncommon and data addressing senescence are few. One 32-year-old female in the Beaufort Sea was monitored for the last 25 years of her life and seen annually during her last 10 years. This bear was in extraordinary condition nearly every autumn. Although she was not recaptured during the autumn of her 30th year, she was observed standing next to a 400-kg female that was captured that season. The 30-year-old female appeared larger, but still did not enter a den that autumn. Despite her apparent excellent physical condition, she last produced cubs at age 22, suggesting a prolonged reproductive senescence. Some contrary evidence also is available. One 29-year-old female in the Beaufort Sea was clearly in estrus (based on turgidity of the vulva) and traveling with an adult male in the spring of 2001. Derocher et al. (1992) also indicated that some females retained reproductive competency throughout life. The reproductive longevity of brown bears and polar bears appears to be fertile ground for further research.

Derocher and Stirling (1994) noted that litter size varied with maternal age, increasing until age 14 years, after which it declined. Heavy hunting reduces numbers of prime-age and older polar bears of both sexes (Amstrup et al. 1986). If such changes occurred without density-dependent increases in reproductive performance for young animals, overharvesting could have the additional population-depressing effect of actually reducing reproduction at low population densities rather than increasing it. Polar bears in the Hudson Bay area were heavily harvested into the 1970s, but numbers there appear to have increased since then (Prevett and Kolenosky 1982; Derocher et al. 1997). Litter size, litter production rate, and other reproductive factors can be expected to change with population size relative to carrying capacity. It also changes in a response to hunting pressure and other population perturbations. Hence, comparisons among populations or within populations over time must take into account the status of the population relative to natural and anthropogenic features of the environment.

In most parts of the Arctic, female polar bears cannot complete a reproductive cycle more frequently than every 3 years. The interbirth interval is determined by the length of time cubs are attended by their mothers, which most commonly is 2.3 years (Stirling et al. 1976, 1980; Lentfer et al. 1980; Amstrup et al. 1986; Amstrup and Durner 1995) (Fig. 27.9). Lønø (1970) concluded that in the Svalbard area, most cubs were weaned by about 17 months of age. Likewise, Ramsay and Stirling (1988) reported that during the 1970s and early 1980s, a significant proportion of female polar bears in the Hudson Bay region weaned their cubs at about 1.3 years of age. After weaning her cubs in the spring of their second year (at age 1.3 years), a female bear could breed again that same spring and achieve a 2-year reproductive interval.

The historically shorter reproductive interval of polar bears living in Hudson Bay (Stirling et al. 1977b) meant that they were more prolific than most other populations of polar bears. Captures of many hundreds of female polar bears and their young in Alaska, Canada, and Svalbard have suggested geographic differences in litter size, litter production, onset of maturity, and reproductive interval. For example, mean litter sizes of cubs and yearlings in Alaska were 1.63 and 1.49, respectively (Amstrup 1995). In Svalbard, these values were 1.81 and 1.32, respectively, whereas litter sizes of polar bears in Hudson Bay during the early 1980s were 1.9 and 1.7 for cubs and yearlings, respectively (Ramsay and Stirling 1988, Derocher and Stirling 1992). Annual litter production rates as high as 0.45 litters/female have been reported for polar bears in the Hudson Bay area (Derocher and Stirling 1992). Nearly half of the females in that population were annually producing a litter of cubs at that time. By comparison, only one fourth of the female polar bears in the Beaufort Sea produce a litter of cubs each year (a litter production rate of 0.25) (Amstrup 1995). That is, in Hudson Bay, each female had a litter nearly every other year, but in the Beaufort Sea, each female produced a litter only every fourth year. Because polar bears in Hudson Bay also produced larger litter sizes, these differences in litter production rates translated into a much higher overall reproductive rate there than in the Beaufort Sea. Female polar bears in the Beaufort Sea produced only ~0.40 cub/year, whereas in the Hudson Bay area they produced up to 0.90 cub/year at the time those studies were conducted (Derocher and Stirling 1992). Reproductive rates in most other areas appear to be more similar to those in the Beaufort Sea than in Hudson Bay.

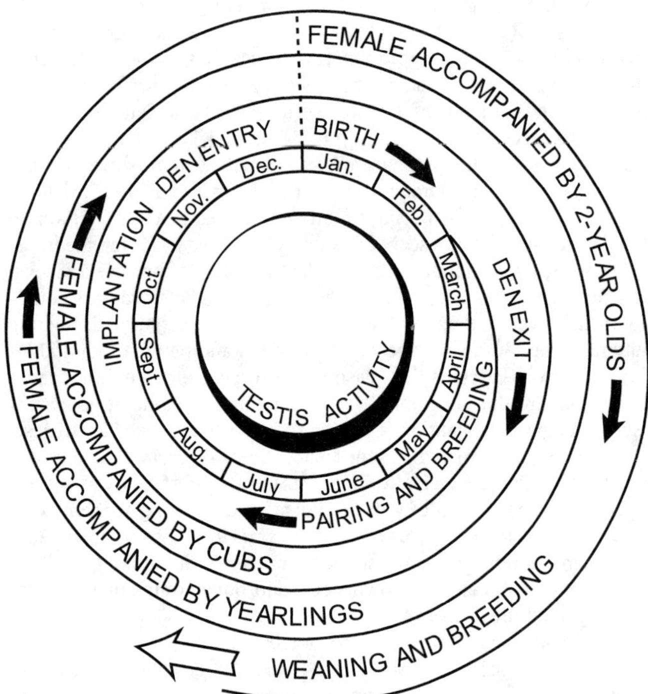

FIGURE 27.9. The 3-year reproductive cycle typical of polar bears throughout most of their range. Exact timing of birth, shown here as 1 January, is not well known and may vary geographically from November to January. Within geographic regions, the timing of birth also may depend on the condition of the female on den entry. SOURCE: Modified from Lønø (1970).

In assessing reproductive intervals, it is critical to confirm weaning, as opposed to mortality of cubs. Many polar bear cubs die in their first year of life (Amstrup and Durner 1995). Those females can breed again in the year of the loss (if it occurs early enough in the spring) or the next year. The breeding frequency, by itself, might suggest a short reproductive interval when it is actually prolonged by poor cub survival. In addition to documenting that tagged females were no longer accompanied by yearling cubs in the spring, Ramsay and Stirling (1988) also captured many weaned yearlings in the autumn of their second year (approximately 1.8 years of age), confirming that many females in the Hudson Bay region actually did have a 2-year reproductive interval.

Lønø (1970), Stirling et al. (1977b), and Ramsay and Stirling (1988) reported on populations that may have been well below carrying capacity due to unregulated hunting (Stirling et al. 1977b; Larsen 1986; Derocher and Stirling 1992, 1995a). Likewise, breeding intervals in the Hudson Bay area have increased, possibly in response to increased relative density of bears in the area (Derocher and Stirling 1992, 1995b). Annual litter production rate in the Hudson Bay region declined from 0.45 litter/female in the period from 1965–1979 to 0.35 during 1985–1990 (Derocher and Stirling 1992). A higher proportion produced cubs every 3 years in the latter period. The inverse of the litter production rate is the interbirth interval. That increased from 2.22 years in 1965–1979 to 2.86 years in 1985–1990. Simultaneously, cub mortality from spring to autumn was significantly higher in the latter period (Derocher and Stirling 1992). The proximate factor associated with all of these trends was the declining weight of adult females during this 25-year period (Derocher and Stirling 1992). Age of first successful reproductive effort increased, although pregnancy rates did not change noticeably. An increasing age of maturation may indicate that a population is approaching carrying capacity. Age of maturation in mammals is often associated with attainment of a threshold body mass (Sadleir 1969) which could be more difficult to attain as competition for resources increases.

A delay in reaching that threshold mass may signal density-dependent influences on the population. Such influences, however, also could result from environmental changes that reduce carrying capacity rather than from increases in polar bear numbers. The documented declines in body weights of females, declines in numbers of independent yearlings, and protracted reproductive intervals appear to be closely related to earlier deterioration of the sea-ice of Hudson Bay (Stirling et al. 1999). The sea-ice extent in the Arctic has been declining throughout the past two decades (Gloersen and Campbell 1991; Vinnikov et al. 1999). Declining Arctic sea-ice cover by itself is difficult to link with polar bear reproductive performance. The timing of melt of the sea-ice in Hudson Bay, however, is more easily connected. Polar bears there, especially pregnant females, depend heavily on the spring and early summer foraging for seals to carry them through the ice-free period (late summer to autumn). Pregnant females, unlike other polar bears in Hudson Bay, remain ashore in autumn when ice returns, and may be food deprived for up to 8 months. Those females must secure sufficient fat stores during the spring and summer to see them through that long period of food deprivation (Stirling 1977; Derocher and Stirling 1992). The mean date of sea-ice break-up in the late 1990s was >2 weeks earlier than it was in the 1970s and early 1980s (Stirling et al. 1999). Earlier break-up and the shortened foraging period accompanying it may mean a significant reduction in the fat stores female polar bears can accumulate before denning. This hypothesis is strengthened by the observation of a transient increase in condition of females coming ashore during the early 1990s when cooler than normal temperatures resulted in later break-ups (Stirling et al. 1999).

Evidence of the critical link between availability of seal prey and reproduction in polar bears is also available in more northerly parts of the range. Weights of females and their reproductive output in the Beaufort Sea decreased markedly in the mid-1970s following a decline in ringed and bearded seal populations (Stirling et al. 1976, 1977b; Kingsley 1979; DeMaster et al. 1980; Stirling et al. 1982; Amstrup et al. 1986). The strength and longevity of declines in reproductive parameters varied both geographically and temporally with the severity of ice conditions that reduced numbers and productivity of seals (Amstrup et al. 1986).

SURVIVAL

The very low reproductive rate of polar bears means that there must be a high rate of survival to maintain population levels. In fact, polar bears "defer" reproduction in favor of survival when foraging conditions are difficult (Derocher et al. 1992). A complete reproductive effort is energetically expensive for polar bears. So, when energetically stressed, female polar bears will forgo reproduction rather than increase risks by incurring the energetic costs of the reproductive process. The reproductive cycle lends itself to convenient early termination if that is appropriate (Ramsay and Dunbrack 1986; Derocher and Stirling 1992). Many radio-collared female polar bears in the Beaufort Sea region entered dens and then abandoned them early without cubs (Amstrup and Gardner 1994). Others lost cubs shortly after emerging from their den and bred again that same spring. Bears leaving dens early may have resorbed their fetuses or may have experienced a pseudopregnancy (Derocher et al. 1992). In any event, they did not complete a full reproductive cycle.

Breeding takes place in the early spring, long before the female can assess whether she will secure sufficient resources to bring her pregnancy to fruition. After fertilization, if she has been able to secure sufficient reserves, birth and rearing can follow pregnancy with some reasonable probability of success. Polar bears, however, also are equipped to abandon a reproductive effort if reserves are insufficient. Because implantation is delayed many months (Wimsatt 1963), and because neonates are so undeveloped (Blix and Lentfer 1979; Ramsay and Dunbrack 1986), early stages of reproduction are relatively inexpensive. Termination of the reproductive process, through abortion or resorbtion of the fetus or failure to nurse after birth, costs a female relatively little (Derocher et al. 1992). The biggest maternal investment begins with postpartum lactation (Ramsay and Dunbrack 1986). Even after emergence from the den, however, it is not unusual for females in poor condition to lose their cubs (Amstrup and Gardner 1994; Amstrup and Durner 1995). Polar bears that terminate a pregnancy and leave their dens early or lose their cubs in early spring usually breed again, preparing them for an opportunity to successfully rear cubs the following year if conditions improve.

In the Hudson Bay region during the 1980s, the survival rate of more than 200 cubs from spring through the ice-free period of autumn was 44% (Derocher and Stirling 1996). Although less mortality was thought to occur after the ice returned in autumn, first-year survival clearly was lower than the 48% reported by Larsen (1985). The body mass of cubs was a significant determinant of survival during early life that included the ice-free period of food deprivation. The mass of cubs, of course, is at least partly dependent on the mass of their mother. Survival of Hudson Bay cubs ($N > 400$) from their first to their second autumn was 35% (Derocher and Stirling 1996). Annual survival of yearlings ranged from 43% to 53%. The survival estimates Derocher and Stirling (1996) calculated for cubs >1 year old were derived from bears that were actually captured. Because many in that age class were independent of their radio-collared mothers, they were not recaptured or reobserved, and their fate was not known. Hence, these must be considered minimum survival values, and likely are below the actual values.

In the Beaufort Sea, survival of cubs was approximately 65% from den exit to the end of their first year of life. Yearlings fared much better, with 86% surviving to weaning (Amstrup and Durner 1995). Observations of the young of radio-collared females substantiate the observation from Hudson Bay that most cub mortality comes early in the period after emergence from the den (Amstrup and Durner 1995; Derocher and Stirling 1996), but depart radically from the very minimal yearling survivals observed there. Derocher and Stirling (1996) suggested that a heavy harvest accounted for much of the yearling mortality in Hudson Bay. Nonetheless, only 15% (44% × 35%) of the cubs produced were confirmed to survive through their second autumn. This contrasts with the survival of 56% (65% × 86%) of Beaufort Sea bears surviving until weaning, 5 months beyond their second autumn. If actual values are close to the minimums reported there, the differences in survival could more than compensate for the apparent reproductive differences between bears in the Beaufort Sea and Hudson Bay.

Tait (1980) hypothesized that brown bear females may choose to abandon a single cub on the chance that they might enhance fitness by breeding again and giving birth to twin or larger litters. That concept has resulted in much discussion and debate about parental care and investment in young. Whether or not it makes sense mathematically, such a strategy apparently does not prevail among polar bears. In the Hudson Bay area, single cubs may actually survive at a somewhat higher rate than cubs from larger litters (Derocher and Stirling 1996). Furthermore, deaths of dependent young in the Beaufort Sea were independent of litter size, and cubs were lost at similar rates whether as whole litters or portions of litters (Amstrup and Durner 1995). Parental investment in single polar bear cubs is not different from investment in litters of two or more. Single cubs are often much heavier than twin cubs (S. C. Amstrup, unpublished data), and survival of cubs appears to be heavily dependent on their weight (Derocher and Stirling 1992).

Estimating survival rates of independent polar bears has been an even greater challenge than estimating survival of dependent young. Eberhardt (1985) argued that survival of adult marine mammals must be in the high 90% range for their populations to be sustaining. However, early estimates of survival in polar bears derived by mark and recapture methods were much lower (Amstrup et al. 1986). More recent estimates derived from Hudson Bay, where the intensity of marking exceeds all other study areas, still have ranged between 0.86 and 0.90 (Derocher and Stirling 1995a; Lunn et al. 1997). Only by relying on radiotelemetry monitoring of individual animals have estimates in line with Eberhardt's (1985) theory been developed. Amstrup and Durner (1995) estimated that survival of adult females in the prime age groups may exceed 96%. Although that estimate fits well with population dynamics theory, the fact that it is much higher than estimates derived by other methods suggests added work on polar bear survival is necessary.

Causes of natural mortalities among polar bears are largely unknown. Because polar bears spend most of their time on drifting sea-ice, dead animals are likely to go undiscovered and cause of death for animals that are discovered is seldom discernible. Therefore, we are forced to extrapolate from a very few observations to understand natural mortality patterns and causes. Accidents involving unskilled young must be a common cause of natural death in the harsh arctic environment (Derocher and Stirling 1996). Starvation of independent young as well as very old animals must account for much of the natural mortality among polar bears. Age-specific differences in hunting success rates have been reported by Stirling and Latour (1978) for polar bears of the central Canadian High Arctic. Cubs of any age spent little time hunting, and were not effective at taking seals in the spring of the year. During summer, the success rate of 2-year-olds was similar to that of adults, although they spent much less time hunting. Young of the year and yearlings were less successful than adults. Cubs abandoned prior to the normal weaning age of 2.5 years likely have poor survival (Stirling and LaTour 1978). That conclusion is corroborated by the dearth of observations of independent bears <2 years old in all populations except Hudson Bay. Also, age structure data show that subadults aged 2–5 years survive at lower rates than adults (Amstrup 1995), probably because they are still learning hunting and survival skills. I once observed a 3-year-old subadult that weighed only 70 kg in November. This was near the end of the autumn period in which Beaufort Sea bears reach their peak weights (Durner and Amstrup 1996), and his cohorts at that time weighed in excess of 200 kg. This young animal apparently had not learned the skills needed to survive and was starving to death. As they age, polar bears that avoid serious injury must simply get too old and feeble to catch food, and thus literally die of old age. Local and widespread climatic phenomena that make seals less abundant or less available also can significantly affect polar bear populations (Stirling et al. 1976; Kingsley 1979; DeMaster et al. 1980; Amstrup et al. 1986).

Injuries sustained in fights over mates or in predation attempts also may lead to natural mortalities of polar bears. Some injuries are immediately fatal. I have seen three instances where a bear has killed another and consumed it. Broken teeth and even broken jaws may frequently result from fighting and failed predation attempts. In brown and black bears, such injuries commonly are not life-threatening. L. Aumiller (Alaska Department of Fish and Game, pers. commun.) has observed several brown bears at Alaska's McNeil River Sanctuary with jaws that had broken and healed in a variety of distorted conformations. D. Garshelis (Minnesota Department of Natural Resources, pers. commun.) captured a 2-year-old black bear with a missing lower jaw. The jaw and all lower teeth were destroyed by gunshot wounds that had largely healed when Garshelis examined the bear in its winter den. The bear was radio-tracked through the following spring and summer and killed by a hunter the following autumn as a normal-size 3-year-old. Brown bears and black bears often survive on a diet including plant parts, fish, insects, small animals, and carrion. A videotape made by the hunter revealed how ingenious the young Minnesota black bear was in feeding without a lower jaw. These and other observations of injured brown and black bears (D. Moody, Wyoming Game and Fish Department, Pers. Commun.; M. Haroldson, USGS Interagency Grizzly Bear Study Team, pers. commun.) suggest they regularly survive with severely damaged mouth-parts, perhaps because of their great adaptability and the small particle size of most of their foods.

Injuries to polar bear's mouth parts also may not be immediately fatal, but they probably are deadly in the long run. Despite capture of thousands of polar bears worldwide, confirmed observations of mended jaws or survival of polar bears with broken jaws are rare or lacking. The long penetrating canine teeth are the polar bear's most important trophic appendage and are critical to holding and killing large prey. Polar bears usually cannot switch to a diet of smaller food particles, and a broken jaw may simply reduce hunting efficiency below the survival threshold. I captured an emaciated but very large male polar bear one autumn when he should have been near his maximum weight. His weight was less than half that of similar-size males at that time. He seemed to be fit and his teeth were in excellent shape. On examination, however, we discovered that his maxilla was broken through (Fig. 27.10), and there was a pronounced gap in his palate. The front portion of his upper jaw was attached only by the skin and musculature of his lips. His ability to bite and hold large prey was seriously compromised. How this injury was sustained is not clear. He has not been recaptured, and given the bear's lean state just before the harshest season of the year, I suspect he did not survive the winter.

In addition to trauma of various kinds, an array of maladies occurs at low frequencies in polar bears just as they do with other wild and domestic mammals including humans. For example, a very large male in the Beaufort Sea died of gastric dilatation and volvulus (Amstrup and Nielsen 1989). This is a condition in which the alimentary organs, including the stomach and much of the intestine, rotate around the mesentaries that support those structures in the abdominal cavity. Blood supply is cut off, resulting in edema, shock, and rapid death. This is a phenomenon common in large, deep-chested dogs and in bears in zoos. Another bear apparently died as a result of occlusion of the bile duct by numerous large gall-stones (S. C. Amstrup unpublished data).

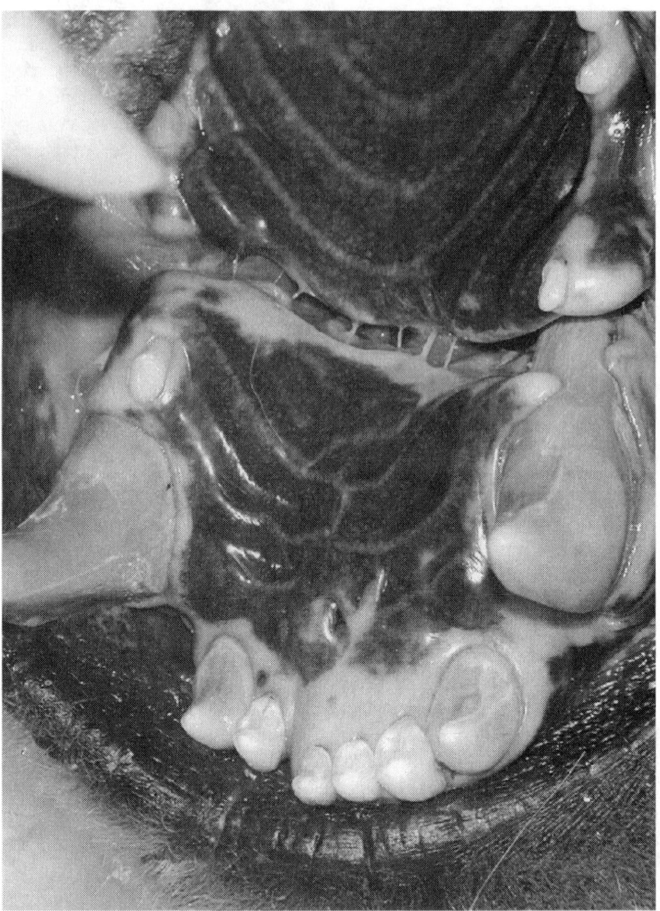

FIGURE 27.10. Broken upper jaw of a large male polar bear captured in the Beaufort Sea in autumn 1999. Because they rely on strong jaws to catch and hold large prey animals, such injuries are probably fatal to polar bears. SOURCE: Photo by Steven C. Amstrup.

Reported diseases and parasites of polar bears are few. In 21 years of research in Alaska, I have not seen any evidence of ectoparasites. In an extensive review of ursid parasites, Rogers and Rogers (1976) found that seven endoparasites had been reported in polar bears. Only *Trichinella* spp., however, had been observed in wild animals. The three species of nematode and three species of cestodes that had been reported in captive polar bears had not occurred in the wild. *Trichinella* can be quite common in polar bears and has been observed throughout their range. Concentrations of this parasite in some tissues can be high, but infections are not normally fatal (Rausch 1970; Dick and Belosevic 1978; Larsen and Kjos-Hanssen 1983; Taylor et al. 1985). Arctic foxes (*Alopex lagopus*) are common carriers of rabies and they routinely interact with polar bears. However, only one instance of rabies has been confirmed in a polar bear (Taylor et al. 1991). Although polar bears are not immune to diseases and parasites, they seem to be plagued by fewer of these problems than most terrestrial mammals.

Male polar bears, like males of other ursids, will kill and eat dependent cubs (Hansson and Thomassen 1983; Larsen 1985; Taylor et al. 1985; Derocher and Wiig 1999). Although this activity does not account for a large percentage of the mortality, it is a curious cause of death in young bears. A male bear that kills cubs fathered by another probably confers some survival advantage to cubs he fathered by eliminating possible competitors for resources. Also, female bears undergo a lactational anestrus. By killing her cubs, a male interrupts that anestrus, and theoretically could breed with the female, inducing her to have his cubs rather than the cubs of some other male. Infanticide, therefore, is a mechanism by which males can increase their relative fitness.

To increase his fitness in this manner assumes that male bears recognize their own cubs. Clearly, with all of the risks to a conceptus that occur between breeding and emergence of cubs onto the sea-ice in spring, there is no selective advantage to a male if he kills cubs he fathered a year before. For the benefits of infanticide to be maximized, the male also must have some reasonable assurance of being around when the female comes back into estrus. In terrestrial bears with limited home range sizes and the ability to defend definable territories, it may be reasonable for a male bear to keep track of a female during the several days between loss of her cubs and onset of estrus. For polar bears, with no territories or other restrictions on movements, the likelihood of a male remaining with a female for that period seems small. In two cases of infanticide I observed in the Beaufort Sea, the male and female involved were already separated by dozens of kilometers the day after the cubs were killed, and they were going in opposite directions. In one case, the male and female were >200 km apart 2 weeks after the male killed her cubs. At least in that case of infanticide, it seems very unlikely that breeding was the goal of the male.

Polar bears will eat the flesh of their own kind, and often a bear that kills another will eat it. The killing of young cubs is probably not motivated by predatory instincts. Small cubs provide a very limited amount of energy, especially considering the risk of injury to a predatory male imposed by the defending female. Males that kill cubs may not even consume them (Derocher and Wiig 1999; S.C. Amstrup, unpublished data), perhaps due to their limited energy value. In terrestrial bears, harassment, or infanticide by large males may be a mechanism of density-dependent population regulation (McCullough 1981; Young and Ruff 1982; Stringham 1983). Derocher and Wiig (1999) also speculated that infanticide may be a density-dependent phenomenon, increasing in frequency with population size. Harassment of subadults by adult males at scavenging sites (Smith 1980) also may be an important regulating factor among polar bears. Infanticide has been detected more often in the Svalbard area than in other parts of the polar bear range where relative densities may be lower (Taylor et al. 1985). In all areas, however, frequencies of infanticide and cannibalism appear to be low enough that understanding their significance to population regulation is difficult. Infanticide in polar bears may be nothing more than an atavistic trait carried over from their terrestrial ancestors, and quantitative effects male polar bears have on their population are unknown.

AGE ESTIMATION

Polar bears can be assigned to age classes based on examination of the skull and dentition. As with other mammals, progressive closure of skull sutures is adequate to separate young and adult polar bears (Manning 1964). Hensel and Sorenson (1980) assigned living polar bears to approximate age classes based on reproductive status, physical measurements, and tooth replacement and wear. Ages of cub-of-the year and yearling polar bears can be assigned without error by observation of dental eruption despite huge variation in size and weight in these age classes. In the autumn and early winter of their first year, permanent canines of young polar bears appear as small cones barely longer than the incisors. Early in the second year of life, canines have grown, but still appear conical in overall cross section. By autumn and early winter, as the second birthday approaches, canines have taken on the shape of mature teeth with a distinctive base and crown. More precise assignment of polar bear ages can be made, as in other ursids, by counting cementum annuli in microscopic cross sections of tooth roots (Stoneberg and Jonkel 1966; Hensel and Sorenson 1980; Calvert and Ramsay 1998).

MANAGEMENT AND CONSERVATION

Hunting. Early Eurasian explorers viewed polar bears as fearless marauders (Larsen 1978). They killed polar bears in self-defense, before they could become a threat, or just because they could. For centuries, Arctic travelers killed as many polar bears as possible (Seton 1929). In Alaska, explorers of the late 1800s and local residents both affected polar bears. Historically, polar bears occupied St. Matthew Island, which lies over 350 km south of the Bering Strait. Unlike polar bears of the Beaufort Sea and other parts of the polar basin, many of these individuals spent the summer on land instead of remaining with the sea-ice as it retreated to the north. Commercial hunters eliminated polar bears from St. Matthew Island by the early 1900s (Hanna 1920). Likewise, overwintering commercial whalers, along with local residents, may have nearly eliminated the bears that once denned along the north coast of Alaska (Leffingwell 1919).

Although the wanton destruction of polar bears by Arctic explorers decreased during the 1900s, polar bears continued to be harvested in large numbers through the middle of the century. In recognition of the polar bear's increasing vulnerability to human activities, the five nations (the Soviet Union, Canada, Denmark, Norway, and the United States) with jurisdiction over polar bear habitats negotiated the International Agreement on Conservation of Polar Bears (Treaty). The Treaty, negotiated in 1973 and ratified in 1976, prohibited the taking of polar bears from aircraft or large motor vessels or in areas where they were not taken by traditional means in the past (Prestrud and Stirling 1994). This prohibition created a *de facto* sanctuary in the central polar basin. A resolution appended to the Treaty requested governments to prohibit the taking of cubs or females with cubs, and hunting in denning areas during periods when pregnant females are moving into them or are denning. Another resolution requested governments to establish an international system to identify and control the trafficking of hides (Lentfer 1974; Stirling 1986; Prestrud and Stirling 1994). Finally, the Treaty required each signatory nation to conduct research and cooperate in management and research of populations that overlap jurisdictional boundaries.

Subsequent to the Treaty, most polar bear populations continue to be hunted. Hunting is not allowed in Svalbard, although a limited number of polar bears there are killed each year in defense of life and property. Hunting in the other jurisdictions, with two exceptions (see below), is limited to Native people continuing a centuries-long tradition. Modern Native hunters, in most areas, use snow-mobiles and high-powered rifles and can be more effective in harvesting polar bears than ever before. Shooting by local hunters accounted for 85% of the deaths of adult female polar bears documented in the Beaufort Sea during the 1980s and early 1990s (Amstrup and Durner 1995). Despite the effects of technology on abilities of hunters to kill polar bears, a combination of government regulations and user's agreements has kept hunting by Native people in balance with the allowable yields of most populations.

In Canada, where the largest numbers of polar bears are harvested, the take is regulated by a quota system and, with specific exceptions, is limited to Native people (Prestrud and Stirling 1994). Quotas are developed through the best available scientific information and enforced through a system of harvest tags distributed by the local hunters and trappers organizations. Representatives of the hunters groups work with scientists to help set quotas throughout Canada, and are fundamental participants in all managment processes. In some parts of Canada, Native hunters use a fraction of their allotted quota tags to guide sport hunters, who may be non-Native (Prestrud and Stirling 1994). Such hunts generate significant financial returns for small northern communities. Sport hunts also have resulted in smaller harvests and a higher proportion of males in the harvests. Sport hunters are not as efficient as local people in catching polar bears, and they strive for larger bears, which are most commonly males. Both the shifted composition of the harvest and the smaller total take resulting from the use of quota tags for sport hunts benefit the hunted populations. The Canadian quota system, whether tags are used for sport hunts or not, has resulted in strict controls on the size of the harvest and exceptionally high quality reporting of harvest composition. Although the Canadian polar bear harvest is larger than that in any other nation, that harvest also is the most scientifically controlled and the best managed. Vigorous efforts are being made to improve harvest monitoring where it is weak, and to implement quotas in the provinces of Quebec and Ontario, where they are not at this writing, in place. Most Canadian management units are being harvested at levels below maximum sustainable yield, and the status of most stocks is thought to be good (Prestrud and Stirling 1994 Lunn et al. 2002).

Hunting in Greenland is limited to "professional" hunters who derive all of their income and sustenance from hunting and fishing. Theoretically, non-Native people could become professional hunters if they established long-term residence in remote areas and gave up all other income sources. In practice, however, this requirement has limited polar bear hunting to Native Greenlanders. The size of the harvest in Greenland is limited only by availability (which is controlled by weather and ice conditions) and the requirement that polar bear hunters use dog teams rather than snow-mobiles to pursue polar bears. In 2000 and 2001, the Greenland Home Rule Government began intensive work with hunters to improve harvest monitoring and work toward harvest limits and a quota system (Amalie Jessen, Vice Director, Greenland Home Rule, Department of Industry, pers. commun. June 2001).

In northern Alaska, the harvest is regulated by agreements among the local users and international agreements between Inupiat hunters in Alaska and Inuvialuit hunters in Canada (Treseder and Carpenter 1989). Current population estimates (Amstrup et al. 1986, 2001) suggest the harvest in northern Alaska is under the maximum sustainable yield. There currently are no restrictions on the harvest by Native Alaskans in western portions of the state, and harvests there largely have been regulated by availability of bears to hunters. A new agreement between Russia and Alaska, however, will soon bring more control and better monitoring to this region (see below).

The biggest threat from hunting may be in Russia. That conclusion is ironic because under Soviet management, hunting was banned in 1956. After the fall of the Soviet regime, however, management of hunting and other uses of renewable resources has suffered from lack of funding and absence of enforcement. So, although shooting polar bears is still illegal, there is nothing to stop such activity when it occurs. This is cause for concern because the economic gains from organized hunting are potentially great and other opportunities to generate income in northern Russia are extremely limited. Also, Russia controls nearly half of the Arctic and much of the world's polar bear habitat. A potentially uncontrolled harvest over such a broad area could be a problem with far-reaching implications.

Risks of excessive take in Russia have been partially addressed by the Agreement between the Government of the United States of America and the Government of the Russian Federation on the Conservation and Management of the Alaska–Chukotka Polar Bear Population. This agreement, signed on 16 October 2000, followed more than 8 years of discussions and negotiations between the local Native people and government representatives of both countries. Although still awaiting congressional ratification, this agreement recognizes the needs of Native people to harvest polar bears for subsistence purposes and includes provisions for developing binding harvest limits, allocation of the harvest between jurisdictions, compliance, and enforcement. Each jurisdiction is entitled to up to one half of the harvest limit. The agreement reiterates requirements of the 1973 multilateral Treaty and includes restrictions on harvesting denning bears, females with cubs, or cubs <1 year old, and prohibitions on the use of aircraft, large motorized vessels, and snares or poison for hunting polar bears. The agreement does not allow hunting for commercial purposes or commercial use of polar bears or their parts. It also commits the parties to the conservation of ecosystems and important habitats, with a focus on conserving polar bear habitats such as feeding, congregating, and denning areas. As with the agreement between the Inuvialuit of northwest Canada and the Inupiat of Alaska (Treseder and Carpenter 1989), the commitment of the users seems likely to assure that in western Alaska and in the eastern Russian Arctic, harvests will be kept within sustainable limits. The remainder of the Russian Arctic is still of concern.

In sum, although hunting can still pose a threat to the welfare of polar bears, it is maintained at sustainable levels in most jurisdictions by a combination of regulations and user's agreements. In the jurisdictions where formal agreements or rules are lacking, the need for limits on the take are increasingly obvious and gaining acceptance.

Impact of Arctic Industrialization. Human activities and habitat alterations associated with industrial development could interfere with movement, feeding, and breeding patterns and could result in exposure to contaminants (Amstrup et al. 1989; Amstrup 1995). In the Beaufort Sea of Alaska, polar bears have been exposed to activities related to hydrocarbon exploration and development for >30 years. Throughout those same 30 years, the population appears to have grown (Amstrup et al. 2001). The population growth during industrialization of the Beaufort Sea coast suggests that management of potential negative interactions between polar bears and humans has been successful. Proposed activities indicate increased vigilance may be necessary to continue those management successes. Oil development activities currently span >160 km of the Beaufort Sea coastal area, but proposed developments would more than double that. The potential for direct and indirect interactions between polar bears and humans only can increase with greater numbers of people and more activities, and more area under development.

If management is to respond properly to the added perturbations that could result from more expansive developments, the processes bringing about natural changes in polar bear populations must be understood. For example, responses to perturbations vary with population size if density plays an important role in population regulation. This means that when numbers are near carrying capacity, populations regulated largely by density-dependent factors may compensate for increased levels of mortality by increasing recruitment (McCullough 1979; Clutton-Brock et al. 1982). Such compensation would result in greater resilience of the population to perturbations than could be predicted from recruitment and survival rates measured prior to the perturbation. Accurate prediction of effects of human activities could only be made, therefore, if population-regulating factors and mechanisms are understood.

The potential for disturbances of denning polar bears may be especially great because undeveloped young must remain in a maternal den that protects them from the rigors of the arctic winter until they are approximately 3 months old (Lentfer and Hensel 1980; Stirling 1990; Stirling and Andriashek 1992; Amstrup 1993; Amstrup and Gardner 1994). Although polar bears may be less sensitive to activities near their dens than previously assumed (Amstrup 1993), the potential for disruptions can only increase as human activities in the Arctic escalate. The general distribution of dens is now known worldwide, but it is still largely unexplained. Except for some critical habitat requirements that can be defined, why bears chose to den at particular sites is unknown. The influences of slope, aspect, and elevation are beginning to be understood (Durner et al. 2001, 2003), although we do know that some bears make mistakes in their choices (Clarkson and Irish 1991). We have made much progress in describing the kinds of habitats polar bears prefer for denning. If we are to adequately manage human activities that could interfere with denning, however, we must either know how to predict which specific sites polar bears will use or learn how to detect newly established dens under the snow.

Global Warming. With reasonable management flexibility, the future of the polar bear as it relates to interactions with humans appears bright. Even in areas of intense industrial activity, such as the Alaskan Beaufort Sea, polar bear populations have flourished. Given adequate vigilance, humans and polar bears should be able to continue to coexist for the foreseeable future.

Significant larger concerns loom on the horizon, however. Evidence that the average temperatures of the globe are increasing continues to mount (Stirling and Derocher 1993). Along with increasing temperatures, the extent of the sea-ice in the Northern Hemisphere has declined from about 12.5 to 11.5 million km^2 during the past 25 years (Gloersen and Campbell 1991; Vinnikov et al. 1999). Reductions in the amount of time polar bears of southern regions can spend hunting on the sea-ice already may have resulted in significant declines in productivity. Body weights of female polar bears and numbers of independent yearlings have declined, while reproductive intervals have increased at the same time the sea-ice of Hudson Bay has been melting earlier in summer (Stirling et al. 1999). Because Hudson Bay is at the southern extreme of polar bear range (see Figs. 27.1 and 27.8), the effects of warmer temperatures and earlier ice melt are likely to be felt there sooner than in more northerly parts of the polar bear range. In fact, in areas where the heavy sea-ice limits access to food (Messier et al. 1992; Ferguson et al. 2000b) and where productivity traditionally has been lower than Hudson Bay (Amstrup et al. 1986, 2000), a milder climate may actually benefit polar bears in the short run. Thinner ice cover and shorter ice seasons of time could enhance primary productivity in northerly areas and such increases could be passed through the food chain. Ultimately, however, if sea-ice cover continues to decline, polar bears worldwide will suffer. Polar bears are so closely tied to the presence of the sea-ice platform from which to hunt, mate, and carry on other life functions that continuing extensive declines in ice coverage will restrict their productivity and could ultimately threaten their survival.

Contaminants. Throughout the 1900s, numerous organic compounds were released into the global environment. Organochlorine pesticides like chlordane, dichlorodiphenyltrichloroethane (DDT), hexachlorocyclohexanes (HCHs), and hexachlorobenzene (HCB) have been used in agricultural areas worldwide and in many undeveloped areas to control mosquitoes and other insects that pose a threat to human health. Many of these compounds are resistant to physical as well as biological degradation and persist in the environment for extended periods. Their persistence allows these compounds to be spread by atmospheric and oceanic circulation, and many have concentrated in the Arctic.

The position of polar bears at the top of the Arctic food chain means they are likely to build up high body burdens of these compounds. Recent studies have verified that likelihood (Norstrom et al. 1998) and raised concerns for health effects of such contamination. The highest concentrations of many chlorinated hydrocarbon compounds have occurred in polar bears of western Russia, Svalbard, and portions of the Canadian High Arctic (Bernhoft et al. 1997; Norstrom et al. 1998). Wiig et al. (1998) hypothesized that environmental contaminants could have caused pseudohermaphroditism by disrupting normal endocrine function of Svalbard polar bears. Although linking contaminants to pseudohermaphroditism remains hypothetical, more recent studies have explicitly linked contaminants to polar bear health. Bernhoft et al. (2000) found significant negative correlations between immunoglobulin (IgG) levels and polychlorinated biphenyl (PCB) and HCB levels in the plasma of Svalbard Polar bears. PCBs and HCB also were negatively correlated with plasma levels of retinal and thyroid hormones in Svalbard polar bears (Skaare et al. 2001). Although specific maladies were not identified, IgG and thyroid hormone levels are associated with a broad variety of health-related systems. Their depression by external contaminants must be of concern. A study nearing completion in summer 2001 provided the strongest evidence yet that pollutants can reduce biological function in polar bears. Wild free-ranging polar bears in Canada, where contamination levels are relatively low, and in Svalbard were captured and vaccinated with harmless viruses. Blood samples collected 5 weeks after vaccination revealed reduced immune system response in the heavily polluted Svalbard polar bears (Skaare et al. 2002).

Although population-level effects of reduced ability to build immunity to disease have yet to be observed, it is clear that organochlorine contamination places polar bears at greater risk to a variety of possible environmental challenges. Use of some of these compounds has been dramatically reduced in recent years, and those reductions may already be reflected in polar bear tissues (Norstrom et al. 1998). Many other chemicals have been introduced in recent years, however, and risks they may cause will continue to create polar bear management uncertainties for decades to come.

RESEARCH NEEDS

Understanding Movement Patterns. With modern radiotelemetry techniques, we have gained greater understanding of movements and distribution patterns of polar bears (Amstrup et al. 1986, 2000; Garner et al. 1990; Ferguson et al. 1997, 1998, 2000a, 2000b). Seasonal movement patterns of polar bears emphasize the role of sea-ice in their life cycle (Garner et al. 1990, 1994; Gloersen et al. 1992; Messier et al. 1992; Amstrup et al. 2000; Ferguson et al. 2001). Just as clearly, distribution and availability of prey are important in movement patterns of polar bears (Stirling and Øritsland 1995). The links between sea-ice, prey, and polar bears, however, are still poorly understood. If we are to explain the movements and activities of polar bears, we need to understand the ecological and energetic components of the predator–prey interactions (Lunn et al. 1997). Also, we need to understand explicitly how that interaction is mediated by the volatile sea-ice platform on which seals and bears depend. At present, descriptions of sea-ice patterns in the Beaufort Sea are too general to provide needed explanations (Stirling et al. 1981; Gloersen et al. 1992). Many logistical obstacles will make understanding seals, sea-ice, and the activities of polar bears a formidable task. Given that polar bears may be important indicators of the health of the arctic marine ecosystem (Stirling and Derocher 1993), overcoming such obstacles is necessary.

Recent measurements, derived from many observations of numerous individuals, are unequivocal and indicate polar bears are among the most mobile of all quadrupeds (Amstrup et al. 2000). However, physiological evaluations suggest that walking polar bears are energetically inefficient (Best 1982; Hurst et al. 1982a, 1982b). The physiology of locomotion in polar bears clearly needs to be reevaluated in light of their known extensive travel. Furthermore, the cues polar bears use to navigate during long movements need to be understood. No other animal is transported as far "in the blind" as are female polar bears that den on drifting pack ice. Going to sleep in one location and waking up, months later, 1000 km from that location must challenge abilities to return, but somehow polar bears are able to do so.

An obvious shortcoming of the data on movements of polar bears is that they were collected with satellite telemetry. Building platform satellite radiotransmitters (platform transmitter terminals or PTTs) into neck collars and attaching them to polar bears has provided previously unobtainable insights into polar bear movements and behaviors (Amstrup et al. 1986, 2000; Messier et al. 1992; Amstrup and Gardner 1994; Bethke et al. 1996). However, the necks of male polar bears are larger than their heads, and radio-collaring does not work for them. Neck collar radios also cannot be fitted to subadults, for fear of injury that could result as they grow and the collar does not. Hence, inferences regarding the movements of all members of the population must be extrapolated from the movements of females only. Males and subadults constitute a large portion of the population, and are often the most likely to be harvested or otherwise interact with humans. Failing to understand what they do is a significant limitation. Males are not only necessary to maintain the population, they also may play a role in limiting population size (McCullough 1981; Young and Ruff 1982; Stringham 1983). Polar bear populations can sustain higher harvests of males than of females (Taylor et al. 1987), but males also appear to be more vulnerable to human hunters. If male polar bears move in patterns that are significantly different than those of females, adjustments to management plans that are currently based on telemetry results from females (Treseder and Carpenter 1989; Nageak et al. 1991) might be required.

There is no satisfactory method of long-term attachment of transmitters to males, and information on movements or activities of male bears is minimal. A limited number of satellite radiotelemetry observations from seven male polar bears suggested that movement rates, distances moved, and areas occupied do not differ greatly between males and females (Amstrup et al. 2001). Those observations corroborated the tag and recovery findings of Lentfer (1983), Schweinsburg et al. (1981) and Stirling et al. (1980, 1984) as well as the correlations between genetic and telemetric population groupings (Paetkau et al. 1999). Clearly, as management needs intensify, better knowledge of the movements of all components of the population will be necessary. Nonetheless, an understanding of the movements of males comparable to that now available for females remains a challenge.

Typically, radiotelemetry data provide retrospective views of the movements of wild animals. With these data, we can outline areas occupied by instrumented animals during specified times. We can measure the rates of movement, total distances moved, and net movements for a period of time. We often lack, however, an understanding of why animals select particular locales or habitats. Visits to land in the Beaufort Sea region, for example, appear to be increasing. Time on land and near shore exposes bears to anthropogenic risk factors and increases the probability of humans being injured by bears. Direct interactions between people and bears have increased markedly in recent years, and that trend can be expected to continue. If those interactions are to be handled properly, managers must know why bears are in areas frequented by people, which bears they are, and how to minimize prospects of bears and people ending up at the same places at the same time.

Research also is needed to understand how to predict probabilities of polar bear occurrence at various locations. When trying to understand the potential of a harvest or other human perturbation on highly mobile species that occur in overlapping management units, we need a way to convert retrospective telemetry data into predictions of the probability of occurrence at locales of interest. As the need for intensive management grows with increasing human presence in the Arctic, predicting where bears will be at particular times and understanding why they are there will be essential.

Maternal Denning. Worldwide, most available denning information has been derived from visual surveys. Except in Alaska and Canada, few animals have been followed to dens by radiotelemetry. Outside Alaska, most radio-collared polar bears have not been resighted on emergence to confirm denning outcome. Still fewer have been monitored long enough to assess patterns of fidelity to den sites or habitat types. Hence, data that might provide an understanding of the significant differences in denning patterns worldwide are not available. Why polar bears in different regions choose to den where they do is largely unknown.

The absence of comparative data among different geographic regions minimizes our understanding of the influence of different sea-ice, climate, and biotic conditions on the chronology and geographic distribution of denning. Polar bears in dens are more vulnerable to anthropogenic as well as natural disturbances than at other times in their life cycle. Therefore, denning data that are comparable over broad geographic regions are essential to interpretation of proposed increases in human uses of the Arctic.

Estimating Numbers. Worldwide, 20 populations of polar bears currently are recognized (see Fig. 27.8). The total combined number of bears in those populations is thought to be between 21,500 and 25,000 (Lunn et al. 2002, Table 1). Population estimates vary in quality from educated guesses in some of the populations to rigorous values complete with confidence intervals in others. Because most of the populations are hunted and many are subject to other potential human perturbations, better estimates of abundance are needed. Optimistic population estimates in the 1950s and 1960s resulted in excessive harvest and declines in Beaufort Sea polar bear numbers (Amstrup et al. 1986). Other populations also were suffering from overharvest at that time (Prestrud and Stirling 1994). Managers need reliable estimates of population size and trend to prevent overharvests from recurring. Despite over 30 years of polar bear capture, however, estimates of population size and trend have been elusive. Failures of past capture–recapture efforts to provide reliable estimates apparently resulted from biases caused by heterogeneity in capture or survival probabilities (Seber 1982; Hwang and Chao 1995). Amstrup et al. (2001) and McDonald and Amstrup (2001) improved estimates of numbers of polar bears by modeling heterogeneity in capture probability with covariates. Their analyses also pointed out many shortcomings of past efforts as well as weaknesses in available data. Estimates of population size based on mark and recapture data are dependent on the number of animals marked and the number of occasions during which marking occurs. Larger numbers of marked animals increase probabilities of capture and reduce the variance of the estimated population size. Increased numbers of occasions allow selection of covariates that help to compensate for heterogeneity. The lessons learned from past efforts must be applied to future studies if more accurate and precise estimates are to be obtained.

Understanding Reproduction. Delayed implantation is known mainly from animals that are tractable in the laboratory. How it works in ursids and how they use it to their advantage in highly variable environments is poorly understood. The timing of implantation, and hence that of birth, appears to vary geographically and also among individual females. It is likely dependent on body condition of the female, which in turn depends on a variety of environmental factors. The interaction between environmental and physiological factors that controls timing of birth and onset of lactation is simply not understood. Further research is necessary to understand how global warming and other broad-scale changes in Arctic conditions may affect reproductive processes.

Estimating Survival. Because polar bears are harvested in most areas and also increasingly exposed to other human perturbations, understanding the processes of reproduction and survival is essential. Estimating the numbers of cubs produced in polar bear populations can be logistically difficult, but is technically straightforward. Estimating survival, on the other hand, is both logistically and technically challenging. Making reasonable estimates of survival continues to provide a stumbling block for researchers and managers alike. Although estimates of survival derived from radiotelemetry studies are consistent with ecological theory (Amstrup and Durner 1995; Eberhardt 1985), the lack of concordance among estimates derived by different methods is troubling. The best estimates derived without the aid of telemetry (Derocher and Stirling 1995a; Lunn et al. 1997) are not in the range thought to be necessary to sustain populations (Eberhardt 1985). Also, telemetry estimates can be derived only for females because male polar bears will not retain radio-collars. As management needs intensify, precise estimates of survival of independent juvenile polar bears as well as adult males will need to be developed while survival patterns of females and dependent young are reevaluated.

Jurisdictional Inequalities. Most current information about polar bears has been derived from studies in Alaska, Canada, and Svalbard. Large portions of the Arctic either have not been studied or have been the site of less consistent efforts. These inequalities create difficulties when we attempt to manage shared populations or wish to draw conclusions applicable over large regions of the Arctic. The greatest information gap is in the Russian Arctic. Russia controls nearly half of the world's Arctic habitat and undoubtedly many of the world's polar bears. However, few detailed studies have occurred or are planned. Lack of funding has environmental research over most of Russia at a standstill. Limited availability of logistics support, concerns over personal safety, and political restrictions on activities in many geographic regions further inhibit studies that might be accomplished in Russia by researchers from other countries.

The inability to study polar bears in Russia and limitations in information from previous work are significant. As indicated under Management and Conservation Issues, the future of polar bears in Russia may be less secure than in any other jurisdiction. Contamination of bears in portions of the Russian Arctic is high. Poaching is thought to be on the increase, and dramatic increases in harvest could result from the current economic unrest and absence of regulatory capacity. Some of the highest density maternal denning areas in the world are within Russian boundaries. The earliest work on maternal denning of polar bears was done by Soviet scientists. All of that work preceded the development of modern telemetry and other new technologies. Hence, those findings are dated and not directly comparable with current studies in other jurisdictions. The limited telemetry data from Russia has precluded discovery of new denning areas there. Large concentration areas on a few offshore islands in Russia have been known by conventional observations for decades. Much of the Russian Arctic, however, has a long linear coastline similar to that of Alaska. Scattered denning undoubtedly occurs there. Also, the long linear coast, as in Alaska, may encourage denning in the heavy offshore pack-ice. No denning studies are being conducted. Although specific local threats to polar bears in Russia are not known, uncertainties with regard to population status there are clearly significant and beg for better understanding.

LITERATURE CITED

Amstrup, S. C. 1993. Human disturbances of denning polar bears in Alaska. Arctic 46:246–50.

Amstrup, S. C. 1995. Movements, distribution, and population dynamics of polar bears in the Beaufort Sea. Ph.D. Dissertation, University of Alaska Fairbanks, Fairbanks.

Amstrup, S. C., and D. P. DeMaster. 1988. Polar bear—*Ursus maritimus*. Pages 39–56 *in* J. W. Lentfer, ed. Selected marine mammals of Alaska: Species accounts with research and management recommendations. Marine Mammal Commission, Washington, DC.

Amstrup, S. C., and G. M. Durner. 1995. Survival rates of radio-collared female polar bears and their dependent young. Canadian Journal of Zoology 73:1312–22.

Amstrup, S. C., and C. Gardner. 1994. Polar bear maternity denning in the Beaufort Sea. Journal of Wildlife Management 58:1–10.

Amstrup, S. C., and C. A. Nielsen. 1989. Acute gastric dilatation and volvulus in a free-living polar bear. Journal of Wildlife Diseases 25:601–4.

Amstrup, S. C., I. Stirling, and J. W. Lentfer. 1986. Past and present status of polar bears in Alaska. Wildlife Society Bulletin 14:241–54.

Amstrup, S. C., C. Gardner, K. C. Myers, and F. W. Oehme. 1989. Ethylene glycol (antifreeze) poisoning in a free-ranging polar bear. Veterinary and Human Toxicology 31:317–19.

Amstrup, S. C., G. Durner, I. Stirling, N. J. Lunn, and F. Messier. 2000. Movements and distribution of polar bears in the Beaufort Sea. Canadian Journal of Zoology 78:948–66.

Amstrup, S. C., T. L. McDonald, and I. Stirling. 2001. Polar bears in the Beaufort Sea: A 30 year mark–recapture case history. Journal of Agricultural, Biological, and Environmental Statistics 6:221–34.

Amstrup, S. C., G. M. Durner, T. L. McDonald, D. M. Mulcahy, and G. W. Garner. 2001. Comparing movement patterns of satellite-tagged male and female polar bears. Canadian Journal of Zoology 79:2147–2158.

Anonymous. 1978. Furry funnels. Time 112(23):82–84.

Arnould, J. P. Y., and M. A. Ramsay. 1994. Milk production and milk consumption in polar bears during the ice-free period in western Hudson Bay. Canadian Journal of Zoology 72:1365–70.

Atkinson, S. N., and M. A. Ramsay. 1995. The effects of prolonged fasting on the body composition and reproductive success of female polar bears. Functional Ecology 9:559–67.

Belikov, S. E. 1976. Behavioral aspects of the polar bear, *Ursus maritimus*. International Conference on Bear Research and Management 3:37–40.

Belikov, S. E., S. M. Uspenski, and A. G. Kuprijanov. 1977. Ecology of the polar bear on Wrangell Island in the denning period. Pages 7–18 *in* S. E. Belikov and S. M. Uspenski, eds. The polar bear and its conservation in the Soviet Arctic. Centralnaja Laboratorija Ochrany Prirody, Moscow.

Bernhoft, A., Ø. Wiig, and J. U. Skaare. 1997. Organochlorines in polar bears (*Ursus maritimus*) at Svalbard. Environmental Pollution 95:159–75.

Bernhoft, A., J. U. Skaare, Ø. Wiig, A. E. Derocher, and H. J. S. Larsen. 2000. Possible immunotoxic effects of organochlorines in polar bears (*Ursus maritimus*) at Svalbard. Journal of Toxicology and Environmental Health, Part A 59:561–74.

Best, R. C. 1982. Thermoregulation in resting and active polar bears. Journal of Comparative Physiology B 146:63–73.

Best, R. C. 1984. Digestibility of ringed seals by the polar bear. Canadian Journal of Zoology 63:1033–36.

Bethke, R., M. Taylor, S. Amstrup, and F. Messier. 1996. Population delineation of polar bears using satellite collar data. Ecological Applications 6:311–17.

Blix, A. S., and J. W. Lentfer. 1979. Modes of thermal protection in polar bear cubs: at birth and on emergence from the den. American Journal of Physiology 236:R67–74.

Blix, A. S., and J. W. Lentfer. 1992. Noise and vibration levels in artificial polar bear dens as related to selected petroleum exploration and developmental activities. Arctic 45:20–24.

Bunnell, F. L., and T. Hamilton. 1983. Forage digestibility and fitness in grizzly bears. International Conference for Bear Research and Management 5:179–85.

Calvert, W., and M. A. Ramsay. 1998. Evaluation of age determination of polar bears by counts of cementum growth layer groups. Ursus 10:449–53.

Calvert, W., and I. Stirling. 1990. Interactions between polar bears and overwintering walruses in the central Canadian High Arctic. International Conference on Bear Research and Management 8:351–56.

Clarkson, P. L., and D. Irish. 1991. Den collapse kills female polar bear and two newborn cubs. Arctic 44:83–84.

Clutton-Brock, T. H., F. E. Guinness, and S. D. Alban. 1982. Red deer: Behavior and ecology of two sexes. University of Chicago Press, Chicago.

Craighead, J. J., and J. A. Mitchell. 1982. Grizzly bear. Pages 515–56 *in* J. Chapman and G. Feldhamer, eds. Wild mammals of North America: Biology, management, and economics. Johns Hopkins University Press, Baltimore.

Cronin, M. A. 1993. Mitochondrial DNA in wildlife taxonomy and conservation biology: Cautionary notes. Wildlife Society Bulletin 21:339–48.

Cronin, M. A., S. C. Amstrup, G. W. Garner, and E. R. Vyse. 1991. Interspecific and intraspecific mitochondrial DNA variation in North American bears (*Ursus*). Canadian Journal of Zoology 69:2985–92.

DeMaster, D. P., and I. Stirling. 1981. *Ursus maritimus*. Polar bear. Mammalian Species 145:1–7.

DeMaster, D. P., M. C. S. Kingsley, and I. Stirling. 1980. A multiple mark and recapture estimate applied to polar bears. Canadian Journal of Zoology 58:633–38.

Derocher, A. E., and I. Stirling. 1990. Distribution of polar bears (*Ursus maritimus*) during the ice-free period in western Hudson Bay. Canadian Journal of Zoology 68:1395–1403.

Derocher, A. E., and I. Stirling. 1992. The population dynamics of polar bears in western Hudson Bay. Pages 1150–59 *in* D. R. McCullough and R. H. Barrett, eds. Wildlife 2001: Populations. Elsevier, Amsterdam.

Derocher A. E., and I. Stirling. 1994. Age-specific reproductive performance of female polar bears (*Ursus maritimus*). Journal of Zoology (London) 234:527–36.

Derocher, A. E., and I. Stirling. 1995a. Estimation of polar bear population size and survival in western Hudson Bay. Journal of Wildlife Management 59:215–21.

Derocher, A. E., and I. Stirling. 1995b. Temporal variation in reproduction and body mass of polar bears in western Hudson Bay. Canadian Journal of Zoology 73:1657–65.

Derocher, A. E., and I. Stirling. 1996. Aspects of survival in juvenile polar bears. Canadian Journal of Zoology 74:1246–52.

Derocher, A. E., and Ø. Wiig. 1999. Infanticide and cannibalism of juvenile polar bears (*Ursus maritimus*) in Svalbard. Arctic 52:307–10.

Derocher, A. E., R. A. Nelson, I. Stirling, and M. A. Ramsay. 1990. Effects of fasting and feeding on serum urea and serum creatinine levels in polar bears. Marine Mammal Science 6:196–203.

Derocher, A. E., I. Stirling, and D. Andriashek. 1992. Pregnancy rates and serum progesterone levels of polar bears in western Hudson Bay. Canadian Journal of Zoology 70:561–66.

Derocher, A. E., D. Andriashek, and I. Stirling. 1993. Terrestrial foraging by polar bears during the ice–free period in western Hudson Bay. Arctic 46:251–54.

Derocher, A. E., I. Stirling, and W. Calvert. 1997. Male-biased harvesting of polar bears in western Hudson Bay. Journal of Wildlife Management 61:1075–82.

Derocher, A. E., Ø. Wiig, and G. Bangjord. 2000. Predation of Svalbard reindeer by polar bears. Polar Biology 23:675–78.

Dick, T. A., and M. Belosevic. 1978. Observations on a *Trichinella spiralis* isolate from a polar bear. Journal of Parasitology 64:1143–45.

Durner, G. M., and S. C. Amstrup. 1995. Movements of a polar bear from northern Alaska to northern Greenland. Arctic 48:338–41.

Durner, G. M., and S. C. Amstrup. 1996. Mass and body-dimension relationships of polar bears in northern Alaska. Wildlife Society Bulletin 24:480–84.

Durner, G. M., S. C. Amstrup, and K. J. Ambrosius. 2001. Remote identification of polar bear maternal den habitat in northern Alaska. Arctic 54:115–21.

Durner, G. M., S. C. Amstrup, and A. S. Fischbach. 2003. Habitat characteristics of polar bear terrestrial maternal den sites in northen Alaska. Arctic 56(1):55–62.

Eberhardt, L. L. 1985. Assessing the dynamics of wild populations. Journal of Wildlife Management 49:997–1012.

Erdbrink, D. 1953. A review of fossil and recent bears of the Old World with remarks on their phylogeny based upon their dentition. Ph.D. Dissertation, University of Utrecht, Utrecht, Netherlands.

Erickson, A. W. 1962. Bear investigations. Breeding biology and productivity. Pages 1–8 *in* Alaska wildlife investigations. Job completion report, Vol. 3 (Federal Aid in Wildlife Restoration, Project W-6-R-3, Work Plan F, Job 4). Alaska Department of Fish and Game.

Ewer, R. R. 1973. The carnivores. Cornell University Press, Ithaca, NY.

Fancy, S. G., L. F. Pank, D. C. Douglas, C. H. Curby, G. W. Garner, S. C. Amstrup, and W. L. Regelin. 1988. Satellite telemetry: A new tool for wildlife research and management (Resource Publication 172). U.S. Fish and Wildlife Service.

Ferguson, S. H., M. K. Taylor, and F. Messier. 1997. Space use by polar bears in and around Auyuittuq National Park, Northwest Territories, during the ice-free period. Canadian Journal of Zoology 75:1585–94.

Ferguson, S. H., M. K. Taylor, E. W. Born, and F. Messier. 1998. Fractals, sea-ice landscape and spatial patterns of polar bears. Journal of Biogeography 25:1081–92.

Ferguson, S. H., M. K. Taylor, E. W. Born, A. Rosing-Asvid, and F. Messier. 1999. Determinants of home range size for polar bears (*Ursus maritimus*). Ecology Letters 2:311–18.

Ferguson, S. H., M. K. Taylor, E. W. Born, A. Rosing-Asvid, and F. Messier. 2001. Activity and movement patterns of polar bears inhabiting consolidated versus active pack ice. Arctic 54:49–54.

Ferguson, S. H., M. K. Taylor, A. Rosing-Asvid, E. W. Born, and F. Messier. 2000a. Relationships between denning of polar bears and conditions of sea ice. Journal of Mammalogy 81:1118–27.

Ferguson, S. H., M. K. Taylor, and F. Messier. 2000b. Influence of sea ice dynamics on habitat selection by polar bears. Ecology 81:761–72.

Folk, G. E., Jr., and R. A. Nelson. 1981. The hibernation of polar bears: A model for the study of human starvation. Proceedings of the International Symposium on Circumpolar Health 5:617–19 (Nordic Council for Arctic Medical Research Report Series 33).

Furnell, D. J., and R. E. Schweinsburg. 1984. Population dynamics of central Canadian Arctic island polar bears. Journal of Wildlife Management 48:722–28.

Garner, G. W., S. T. Knick, and D. C. Douglas. 1990. Seasonal movements of adult female polar bears in the Bering and Chukchi Seas. International Conference on Bear Research and Management 8:219–26.

Garner, G. W., S. C. Amstrup, I. Stirling, and S. E. Belikov. 1994. Habitat considerations for polar bears in the North Pacific Rim. Transactions of the North American Wildlife and Natural Resources Conference 59:111–20.

Gloersen, P., and W. J. Campbell. 1991. Recent variations in Arctic and Antarctic sea-ice covers. Nature 352:33–36.

Gloersen, P., W. J. Campbell, D. J. Cavalieri, J. C. Comiso, C. L. Parkinson, and H. J. Zwally. 1992. Arctic and Antarctic sea ice, 1978–1987: Satellite passive-microwave observations and analysis (Special Publication SP-511). National Aeronautics and Space Administration.

Grojean, R. E., J. A. Sousa, and M. C. Henry. 1980. Utilization of solar radiation by polar animals: An optical model for pelts. Applied Optics 19:339–46.

Guppy, M. 1986. The hibernating bear: Why is it so hot, and why does it cycle urea through the gut? Trends in Biochemical Sciences 11:274–76.

Hammill, M. O., and T. G. Smith. 1991. The role of predation in the ecology of the ringed seal in Barrow Strait, Northwest Territories, Canada. Marine Mammal Science 7:123–35.

Hanna, G. D. 1920. Mammals of the St. Matthew Islands, Bering Sea. Journal of Mammalogy 1:118–22.

Hansson, R., and J. Thomassen. 1983. Behavior of polar bears with cubs of the year in the denning area. International Conference on Bear Research and Management 5:246–54.

Harington, C. R. 1966. A polar bear's life. Pages 3–7 *in* Report on Polar Bears. Lectures presented at the Eighth Annual Meeting of the Washington Area Associates. Arctic Institute of North America Research Paper No. 34.

Harington, C. R. 1968. Denning habits of the polar bear (*Ursus maritimus* Phipps) (Report Series 5). Canadian Wildlife Service, Ottowa.

Harris, R. B., S. G. Fancy, D. C. Douglas, G. W. Garner, S. C. Amstrup, T. R. McCabe, and L. F. Pank. 1990. Tracking wildlife by satellite: Current systems and performance. (Technical Report 30). U.S. Fish and Wildlife Service.

Heaton, T. H., S. L. Talbot, and G. F. Shields. 1996. An ice age refugium for large mammals in the Alexander Archipelago, southeastern Alaska. Quaternary Research 46:189–192.

Hensel, R. J., and F. E. Sorenson, Jr. 1980. Age determination of live polar bears. International Conference on Bear Research and Management 4:93–100.

Hobson, K. A., and I. Stirling. 1997. Low variation in blood δ^{13}C among Hudson Bay polar bears: Implications for metabolism and tracing terrestrial foraging. Marine Mammal Science 13:359–67.

Hurst, R. J., M. L. Leonard, P. D. Watts, P. Beckerton, and N. A. Øritsland. 1982a. Polar bear locomotion: Body temperature and energetic cost. Canadian Journal of Zoology 60:40–44.

Hurst, R. J., N. A. Øritsland, and P. D. Watts. 1982b. Body mass, temperature and cost of walking in polar bears. Acta Physiologica Scandinavica 115:391–95.

Hwang, W. D., and A. Chao. 1995. Quantifying the effects of unequal catchabilities on Jolly–Seber estimators via sample coverage. Biometrics 51:128–41.

Jenness, R., A. W. Erickson, and J. J. Craighead. 1972. Some comparative aspects of milk from four species of bears. Journal of Mammalogy 53:34–47.

Jonkel, C. J., I. Stirling, G. B. Kolenosky, S. Miller, and R. Robertson. 1975. Polar bear research in Canada, 1972–74. International Union for the Conservation of Nature and Natural Resources Publications, New Series Supplemental Paper 42:23–36.

Jonkel, C., P. Smith, I. Stirling, and G. B. Kolenosky. 1976. The present status of the polar bear in the James Bay and Belcher Islands area (Occasional Paper No. 26). Canadian Wildlife Service, Ottawa.

Kiliaan, H. P. L., I. Stirling, and C. J. Jonkel. 1978. Polar bears in the area of Jones Sound and Norwegian Bay (Progress Notes No. 88). Canadian Wildlife Service, Ottawa.

Kingsley, M. C. S. 1979. Fitting the von Bertalanffy growth equation to polar bear age–weight data. Canadian Journal of Zoology 57:1020–25.

Kingsley, M. C. S., I. Stirling, and W. Calvert. 1985. The distribution and abundance of seals in the Canadian High Arctic, 1980–1982. Canadian Journal of Fisheries and Aquatic Sciences 42:1189–1210.

Kistchinski, A. A. 1969. The polar bear on the Novosibirsk Islands. Pages 103–13 *in* A. G. Bannikov and A. A. Kistchinski, eds. The polar bear and its conservation in the Soviet Arctic. Gidrometerologiydet, Leningrad.

Kolenosky, G. B., and J. P. Prevett. 1983. Productivity and maternity denning of polar bears in Ontario. International Conference on Bear Research and Management 5:238–45.

Kolenosky, G. B., K. F. Abraham, and C. J. Greenwood. 1992. Polar bears of southern Hudson Bay. Polar Bear Project, 1984–88, final report. Unpublished report. Ontario Ministry of Natural Resources, Maple, Ontario, Canada.

Koon, D. W. 1998. Is polar bear hair fiber optic? Applied Optics 37:3198–3200.

Kostyan, E. Y. 1954. New data on the reproduction of the polar bear. Zoologicheskii Zhurnal 33:207–15.

Kurtén, B. 1955. Sex dimorphism and size trends in the cave bear, *Ursus spelaeus,* Rosenmüller and Heinroth. Acta Zoologica Fennica 90:4–48.

Kurtén, B. 1964. The evolution of the polar bear, *Ursus maritimus* Phipps. Acta Zoologica Fennica 108:1–30.

Larsen, T. 1971. Sexual dimorphism in the molar rows of the polar bear. Journal of Wildlife Management 35:374–77.

Larsen, T. 1978. The world of the polar bear. Hamlyn, London.

Larsen, T. 1985. Polar bear denning and cub production in Svalbard, Norway. Journal of Wildlife Management 49:320–26.

Larsen, T. 1986. Population biology of the polar bear (*Ursus maritimus*) in the Svalbard area (Norsk Polarinstitutt Skrifter 184). Oslo, Norway.

Larsen, T., and B. Kjos-Hanssen. 1983. *Trichinella* sp. in polar bears from Svalbard, in relation to hide length and age. Polar Research 1:89–96.

Larsen, T., C. Jonkel, and C. Vibe. 1983a. Satellite radio-tracking of polar bears between Svalbard and Greenland. International Conference on Bear Research and Management 5:230–37.

Larsen, T., H. Tegelström, R. K. Juneja, and M. K. Taylor. 1983b. Low protein variability and genetic similarity between populations of the polar bear (*Ursus maritimus*). Polar Research 1:97–105.

Lavigne, D. M. 1988. Letters to the Editor Re: Solar polar bears. Scientific American 259:8.

Lavigne, D. M., and N. A. Øritsland. 1974. Ultraviolet photography: A new application for remote sensing of mammals. Canadian Journal of Zoology 52:939–41.

Leffingwell, D. K. E. 1919. The Canning River region, northern Alaska (U.S. Geological Survey Professional Paper 109). U.S. Government Printing Office, Washington, DC.

Lentfer, J. 1974. Agreement on conservation of polar bears. Polar Record 17:327–30.

Lentfer, J. W. 1983. Alaskan polar bear movements from mark and recovery. Arctic 36:282–88.

Lentfer, J. W., and R. J. Hensel. 1980. Alaskan polar bear denning. International Conference on Bear Research and Management 4:101–8.

Lentfer, J. W., and L. H. Miller. 1969. Big game investigations: Polar bear studies (Project Segment Report, Vol. 10 (Federal Aid in Wildlife Restoration, Projects W-15-R-3 and W-17-1, Work Plans M and R). Alaska Department of Fish and Game, Division of Game, Juneau.

Lentfer, J. W., J. R. Hensel, J. R. Gilbert, and F. E. Sorensen. 1980. Population characteristics of Alaskan polar bears. International Conference on Bear Research and Management 4:109–15.

Lewin, R. A., and P. T. Robinson. 1979. The greening of polar bears in zoos. Nature 278:445–47.

Lewis, R. W., and J. W. Lentfer. 1967. The vitamin A content of polar bear liver: Range and variability. Comparative Biochemistry and Physiology 22:923–26.

Lønø, O. 1970. The polar bear (*Ursus maritimus* Phipps) in the Svalbard area (Norsk Polarinstitutt Skrifter 149). Oslo, Norway.

Lønø, O. 1972. Polar bear fetuses found in Svalbard. Norsk Polarinstitutt-Arbok 1970:294–98.

Lopez, B. 1986. Arctic dreams. Scribner, New York.

Lowry, L. F., J. J. Burns, and R. R. Nelson. 1987. Polar bear, *Ursus maritimus*, predation on belugas, *Delphinapterus leucas*, in the Bering and Chukchi Seas. Canadian Journal of Zoology 101:141–46.

Lunn, N. J., and I. Stirling. 1985. The significance of supplemental food to polar bears during the ice-free period of Hudson Bay. Canadian Journal of Zoology 63:2291–97.

Lunn, N. J., I. Stirling, D. Andriashek, and G. B. Kolenosky. 1997. Re-estimating the size of the polar bear population in western Hudson Bay. Arctic 50:234–40.

Lunn, N. J., S. Schliebe, and E. W. Born, eds. 2002. Polar Bears. Proceedings of the 13th Working Meeting of the IUCN/SSC Polar Bear Specialist Group. Occasional Paper of the International Union for Conservation of Nature Species Survival Commission No. 26. Gland.

MacGillivray, A. O., D. E. Hannay, R. G. Racca, C. J. Perham, S. A. Maclean, and M. T. Williams. 2002. Assessment of industrial sounds and vibrations received in artificial polar bear dens, Flaxman Island, Alaska. ExxonMobil Production Co. Draft Report by LGL Alaska Research Associates, Inc., Anchorage, AK, and JASCO Research Ltd., Victoria, BC.

Manning, T. H. 1964. Age determination in the polar bear *Ursus maritimus* Phipps (Occasional Paper No. 5). Canadian Wildlife Service, Department of Northern Affairs and National Resources, National Parks Branch.

Manning, T. H. 1971. Geographical variation in the polar bear *Ursus maritimus* Phipps (Report Series No. 13). Canadian Wildlife Service, Ottawa.

McCullough, D. R. 1979. The George Reserve deer herd: Population ecology of a *K*-selected species. University of Michigan Press, Ann Arbor.

McCullough, D. R. 1981. Population dynamics of the Yellowstone grizzly bear. Pages 173–96 *in* C. W. Fowler, ed. Dynamics of large mammal populations. John Wiley, New York.

McDonald, T. L., and S. C. Amstrup. 2001. Estimation of population size using open capture–recapture models. Journal of Agricultural, Biological, and Environmental Statistics 6:206–20.

McLellan, B. N., and D. M. Shackleton. 1989. Grizzly bears and resource-extraction industries: Habitat displacement in response to seismic exploration, timber harvesting and road maintenance. Journal of Applied Ecology 26:371–80.

Messier, F., M. K. Taylor, and M. A. Ramsay. 1992. Seasonal activity patterns of female polar bears (*Ursus maritimus*) in the Canadian Arctic as revealed by satellite telemetry. Journal of Zoology (London) 226:219–29.

Messier, F., M. K. Taylor, and M. A. Ramsay. 1994. Denning ecology of polar bears in the Canadian Arctic Archipelago. Journal of Mammalogy 75:420–30.

Mirsky, S. D. 1988. Solar polar bears. Scientific American 258:24–26.

Nageak, B. P., C. D. Brower, and S. L. Schliebe. 1991. Polar bear management in the southern Beaufort Sea: An agreement between the Inuvialuit Game Council and North Slope Borough Fish and Game Committee. Transactions of the North American Wildlife and Natural Resources Conference 56:337–43.

Nelson, R. A. 1981. Polar bears: Active yet hibernating? Pages 244–47 *in* G. B. Stone, ed. Spirit of enterprise—The 1981 Rolex awards. W. H. Freeman, San Francisco.

Nelson, R. A. 1987. Black bears and polar bears—Still metabolic marvels. Mayo Clinic Proceedings 62:850–53.

Nelson, R. A., H. W. Wahner, J. D. Jones, R. D. Ellefson, and P. E. Zollman. 1973. Metabolism of bears before, during, and after winter sleep. American Journal of Physiology 224:491–96.

Nelson, R. A., G. E. Folk, Jr., E. W. Pfeiffer, J. J. Craighead, C. J. Jonkel, and D. L. Steiger. 1983. Behavior, biochemistry, and hibernation in black, grizzly, and polar bears. International Conference on Bear Research and Management 5:284–90.

Nesbitt, W. H., and J. S. Parker. 1977. North American big game. R. R. Donnelley, Chicago.

Norstrom, R. J., S. E. Belikov, E. W. Born, G. W. Garner, B. Malone, S. Olpinski, M. A. Ramsay, S. Schliebe, I. Stirling, M. S. Stishov, M. K. Taylor, and Ø. Wiig. 1998. Chlorinated hydrocarbon contaminants in polar bears from eastern Russia, North America, Greenland, and Svalbard: Biomonitoring of Arctic pollution. Archives of Environmental Contamination and Toxicology 35:354–67.

Øritsland, N. A. 1970. Temperature regulation of the polar bear (*Thalarctos maritimus*). Comparative Biochemistry and Physiology 37:225–33.

Øritsland, N. A., and K. Ronald. 1978. Solar heating of mammals: Observations of hair transmittance. Journal of Biometeorology 22:197–201.

Øritsland, N. A., C. Jonkel, and K. Ronald. 1976. A respiration chamber for exercising polar bears. Norwegian Journal of Zoology 24:65–67.

Paetkau, D., and C. Strobeck. 1998. Ecological genetic studies of bears using microsatellite analysis. Ursus 10:299–306.

Paetkau, D., W. Calvert, I. Stirling, and C. Strobeck. 1995. Microsatellite analysis of population structure in Canadian polar bears. Molecular Ecology 4:347–54.

Paetkau, D., L. P. Waits, P. L. Clarkson, L. Craighead, and C. Strobeck. 1997. An empirical evaluation of genetic distance statistics using microsatellite data from bear (Ursidae) populations. Genetics 147:1943–57.

Paetkau, D., S. C. Amstrup, E. W. Born, W. Calvert, A. E. Derocher, G. W. Garner, F. Messier, I. Stirling, M. K. Taylor, Ø. Wiig, and C. Strobeck. 1999. Genetic structure of the world's polar bear populations. Molecular Ecology 8:1571–84.

Papanin, I. 1939. Life on an ice-floe. Hutchinson, London.

Pedersen, A. 1945. The polar bear: Its distribution and way of life. Aktieselskabet E. Bruun, Copenhagen.

Phipps, C. J. 1774. A voyage towards the North Pole. J. Nourse, London.

Prestrud, P., and I. Stirling. 1994. The international polar bear agreement and the current status of polar bear conservation. Aquatic Mammals 20:113–24.

Prevett, J. P., and G. B. Kolenosky. 1982. The status of polar bears in Ontario. Naturaliste Canadien 109:933–39.

Ramsay, M. A. 1986. The reproductive biology of the polar bear: A large, solitary carnivorous mammal. Ph.D. Dissertation, University of Alberta, Edmonton. Canada.

Ramsay, M. A., and D. S. Andriashek. 1986. Long distance route orientation of female polar bears (*Ursus maritimus*) in spring. Journal of Zoology A (London) 208:63–72.

Ramsay, M. A., and R. L. Dunbrack. 1986. Physiological constraints on life history phenomena: The example of small bear cubs at birth. American Naturalist 127:735–43.

Ramsay, M. A., and K. A. Hobson. 1991. Polar bears make little use of terrestrial food webs: Evidence from stable-carbon isotope analysis. Oecologia 86:598–600.

Ramsay, M. A., and I. Stirling. 1982. Reproductive biology and ecology of female polar bears in western Hudson Bay. Naturaliste Canadien 109:941–46.

Ramsay, M. A., and I. Stirling. 1986a. Long–term effects of drugging and handling free-ranging polar bears. Journal of Wildlife Management 50:619–26.

Ramsay, M. A., and I. Stirling. 1986b. On the mating system of polar bears. Canadian Journal of Zoology 64:2142–51.

Ramsay, M. A., and I. Stirling. 1988. Reproductive biology and ecology of female polar bears (*Ursus maritimus*). Journal of Zoology (London) 214:601–34.

Ramsay, M. A., and I. Stirling. 1990. Fidelity of female polar bears to winter-den sites. Journal of Mammalogy 71:233–36.

Rausch, R. L. 1970. Trichinosis in the Arctic. Pages 348–73 *in* S. E. Gould, ed. Trichinosis in man and animals. C. C. Thomas, Springfield, IL.

Ray, C. E. 1971. Polar bear and mammoth on the Pribilof Islands. Arctic 24:9–19.

Reynolds, P. E., H. V. Reynolds, and E. H. Follmann. 1986. Responses of grizzly bears to seismic surveys in northern Alaska. International Conference on Bear Research and Management 6:169–75.

Rodahl, K., and T. Moore. 1943. The vitamin A content and toxicity of bear and seal liver. Biochemical Journal 37:166–68.

Rogers, L. L., and S. M. Rogers. 1976. Parasites of bears: A review. International Conference on Bear Research and Management 3:411–30.

Rosing-Asvid, A., E. W. Born, and M. C. S. Kingsley. 2002. Age at sexual maturity of males and timing of the mating season of polar bears (*Ursus maritimus*) in Greenland. Polar Biology 25:878–883.

Russell, F. E. 1967. Vitamin A content of polar bear liver. Toxicon 5:61–62.

Sadleir, R. M. F. S. 1969. The ecology of reproduction in wild and domestic mammals. Methuen, London.

Scholander, P. F., R. Hock, V. Walters, and L. Irving. 1950. Adaptation to cold in arctic and tropical mammals and birds in relation to body temperature, insulation, and basal metabolic rate. Biological Bulletin (Woods Hole) 99:259–71.

Schweinsburg, R. E. 1979. Summer snow dens used by polar bears in the Canadian High Arctic. Arctic 32:165–69.

Schweinsburg, R. E., and L. J. Lee. 1982. Movement of four satellite-monitored polar bears in Lancaster Sound, Northwest Territories. Arctic 35:504–11.

Schweinsburg, R. E., D. J. Furnell, and S. J. Miller. 1981. Abundance, distribution, and population structure of the polar bears in the lower central Arctic islands (Northwest Territories Wildlife Service Completion Report No. 2).

Scribner, K. T., G. W. Garner, S. C. Amstrup, and M. A. Cronin. 1997. Population genetic studies of the polar bear (*Ursus maritimus*): A summary of available data and interpretation of results. Pages 185–96 *in* A. E. Dizon, S. J. Chivers, and W. F. Perrin, eds. Molecular genetics of marine mammals (Special Publication No. 3). Society for Marine Mammalogy.

Seber, G. A. F. 1982. The estimation of animal abundance and related parameters, 2nd ed. Charles Griffin, London.
Seton, E. T. 1929. Lives of game animals. Vol. II, Part I. Doubleday, Doran, New York.
Shields, G. F., and T. D. Kocher. 1991. Phylogenetic relationships of North American ursids based on analysis of mitochondrial DNA. Evolution 45:218–21.
Skaare, U. J., A. Bernhoft, Ø. Wiig, K. R. Norum, E. Haug, M. D. Eide, and A. E. Derocher. 2001. Relationships between plasma levels of organochlorines, retinol and thyroid hormones from polar bears (*Ursus maritimus*) at Svalbard. Journal of Toxicology and Environmental Health, Part A 62:227–41.
Skaare, J. U., H. J. Larsen, E. Lie, A. Bernhoft, A. E. Derocher, R. Norstrom, E. Ropstad, N. F. Lunn, and O. Wiig. 2002. Ecological risk assessment of persistent organic pollutants in the arctic. Toxicology 181–182:193–197.
Smith, T. G. 1980. Polar bear predation of ringed and bearded seals in the land-fast sea ice habitat. Canadian Journal of Zoology 58:2201–9.
Smith, T. G. 1985. Polar bears, *Ursus maritimus*, as predators of belugas, *Delphinapterus leucas*. Canadian Field-Naturalist 99:71–75.
Smith, T. G., and I. Stirling. 1975. The breeding habitat of the ringed seal (*Phoca hispida*). The birth lair and associated structures. Canadian Journal of Zoology 53:1297–1305.
Stanley, S. M. 1979. Macroevolution, pattern and process. W. H. Freeman, San Francisco.
Stefansson, V. 1921. The friendly Arctic. Macmillan, New York.
Stirling, I. 1974. Midsummer observations on the behavior of wild polar bears (*Ursus maritimus*). Canadian Journal of Zoology 52:1191–98.
Stirling, I. 1977. Adaptations of Weddell and ringed seals to exploit the polar fast ice habitat in the absence or presence of surface predators. Pages 741–48 *in* G. A. Llano, ed. Adaptations within Antarctic ecosystems. Gulf, Houston, TX.
Stirling, I. 1986. Research and management of polar bears, *Ursus maritimus*. Polar Record 23:167–76.
Stirling, I. 1990. Polar bears and oil: Ecological perspectives. Pages 223–34 *in* J. R. Geraci and D. J. St. Aubin, eds. Sea mammals and oil: Confronting the risks. Academic Press, San Diego, CA.
Stirling, I., and D. Andriashek. 1992. Terrestrial maternity denning of polar bears in the eastern Beaufort Sea area. Arctic 45:363–66.
Stirling, I., and W. R. Archibald. 1977. Aspects of predation of seals by polar bears. Journal of the Fisheries Research Board of Canada 34:1126–29.
Stirling, I., and A. E. Derocher. 1990. Factors affecting the evolution and behavioral ecology of the modern bears. International Conference on Bear Research and Management 8:189–204.
Stirling, I., and A. E. Derocher. 1993. Possible impacts of climatic warming on polar bears. Arctic 46:240–45.
Stirling, I., and P. B. LaTour. 1978. Comparative hunting abilities of polar bear cubs of different ages. Canadian Journal of Zoology 56:1768–72.
Stirling, I., and N. J. Lunn. 1997. Environmental fluctuations in Arctic marine ecosystems as reflected by variability in reproduction of polar bears and ringed seals. Pages 167–81 *in* S. J. Woodin and M. Marquiss, eds. Ecology of arctic environments. Blackwell, Oxford.
Stirling, I., and E. H. McEwan. 1975. The caloric value of whole ringed seals (*Phoca hispida*) in relation to polar bear (*Ursus maritimus*) ecology and hunting behavior. Canadian Journal of Zoology 53:1021–27.
Stirling, I., and N. A. Øritsland. 1995. Relationships between estimates of ringed seal (*Phoca hispida*) and polar bear (*Ursus maritimus*) populations in the Canadian Arctic. Canadian Journal of Fisheries and Aquatic Sciences 52:2594–2612.
Stirling, I., and M. A. Ramsay. 1986. Polar bears in Hudson Bay and Fox Basin: Present knowledge and research opportunities. Pages 341–54 *in* I. P. Martini, ed. Canadian inland seas. Elsevier, Amsterdam.
Stirling, I., and T. G. Smith. 1975. Interrelationships of Arctic Ocean mammals in the sea ice habitat. Circumpolar Conference on Northern Ecology 2:129–36.
Stirling, I., A. M. Pearson, and F. L. Bunnell. 1976. Population ecology studies of polar and grizzly bears in northern Canada. Transactions of the North American Wildlife and Natural Resources Conference 41:421–30.
Stirling, I., W. R. Archibald, and D. DeMaster. 1977a. Distribution and abundance of seals in the eastern Beaufort Sea. Journal of the Fisheries Research Board of Canada 34:976–88.
Stirling, I., C. Jonkel, P. Smith, R. Robertson, and D. Cross. 1977b. The ecology of the polar bear (*Ursus maritimus*) along the western coast of Hudson Bay (Occasional Paper No. 33). Canadian Wildlife Service, Ottawa.
Stirling, I., W. Calvert, and D. Andriashek. 1980. Population ecology studies of the polar bear in the area of southeastern Baffin Island (Occasional Paper No. 44). Canadian Wildlife Service, Ottawa.
Stirling, I., H. Cleator, and T. G. Smith. 1981. Marine mammals. Pages 44–58 *in* I. Stirling and H. Cleator, eds. Polynyas in the Canadian Arctic (Occasional Paper No. 45). Canadian Wildlife Service, Ottawa.
Stirling, I., M. Kingsley, and W. Calvert. 1982. The distribution and abundance of seals in the eastern Beaufort Sea, 1974–79 (Occasional Paper No. 47). Canadian Wildlife Service, Ottawa.
Stirling, I., W. Calvert, and D. Andriashek. 1984. Polar bear (*Ursus maritimus*) ecology and environmental considerations in the Canadian High Arctic. Pages 201–22 *in* R. Olsen, F. Geddes, and R. Hastings, eds. Northern ecology and resource management. University of Alberta Press, Edmonton, Canada.
Stirling, I., N. J. Lunn, and J. Iacozza. 1999. Long-term trends in the population ecology of polar bears in western Hudson Bay in relation to climatic change. Arctic 52:294–306.
Stishov, M. S. 1991. Results of aerial counts of the polar bear dens on the arctic coasts of the extreme northeast Asia. Polar Bears. Proceedings of the Working Meeting of the IUCN/SSC Polar Bear Specialist Group. Occasional Paper of the International Union for Conservation of Nature Species Survival Commission No. 7. 10:90–92.
Stoneberg, R. P., and C. J. Jonkel. 1966. Age determination of black bears by cementum layers. Journal of Wildlife Management 30:411–14.
Storer, T. I., and L. P. Tevis. 1955. California grizzly. University of Nebraska Press, Lincoln.
Stringham, S. F. 1983. Roles of adult males in grizzly bear population biology. International Conference on Bear Research and Management 5:140–51.
Tait, D. E. N. 1980. Abandonment as a reproductive tactic: The example of grizzly bears. American Naturalist 115:800–808.
Talbot, S. L., and G. F. Shields. 1996a. A phylogeny of the bears (Ursidae) inferred from complete sequences of three mitochondrial genes. Molecular Phylogenetics and Evolution 5:567–75.
Talbot, S. L., and G. F. Shields. 1996b. Phylogeography of brown bears (*Ursus arctos*) of Alaska and paraphyly within the Ursidae. Molecular Phylogenetics and Evolution 5:477–94.
Taylor, M. K., T. Larsen, and R. E. Schweinsburg. 1985. Observations of intraspecific aggression and cannibalism in polar bears (*Ursus maritimus*). Arctic 38:303–9.
Taylor, M. K., D. P. DeMaster, F. L. Bunnell, and R. E. Schweinsburg. 1987. Modeling the sustainable harvest of female polar bears. Journal of Wildlife Management 51:811–20.
Taylor, M., B. Elkin, N. Maier, and M. Bradley. 1991. Observation of a polar bear with rabies. Journal of Wildlife Diseases 27:337–39.
Thenius, E. 1953. Concerning the analysis of the teeth of polar bears. Mammalogical Bulletin 1:14–20.
Tovey, P. E., and R. E. Scott. 1957. A preliminary report on the status of polar bear in Alaska. Presented at the eighth Alaska science conference, Anchorage.
Treseder, L., and A. Carpenter. 1989. Polar bear management in the southern Beaufort Sea. Information North 15(4):2–4.
Uspenski, S. M., ed. 1977. The polar bear and its conservation in the Soviet Arctic. A collection of scientific papers. Central Laboratory of Nature Conservation, Moscow.
Uspenski, S. M., and F. B. Chernyavski. 1965. Maternity home of the polar bear. Priroda 4:81–86.
Uspenski, S. M., and A. A. Kistchinski. 1972. New data on the winter ecology of the polar bear (*Ursus maritimus*) on Wrangell Island. International Conference on Bear Research and Management 2:181–97.
Vinnikov, K. Y., A. Robock, R. J. Stouffer, J. E. Walsh, C. L. Parkinson, D. J. Cavalieri, J. F. B. Mitchell, D. Garrett, and V. F. Zakharov. 1999. Global warming and northern hemisphere sea ice extent. Science 286:1934–36.
Watts, P. D., and S. E. Hansen. 1987. Cyclic starvation as a reproductive strategy in the polar bear. Symposium of the Zoological Society of London 57:306–18.
Wiig, Ø., A. E. Derocher, M. M. Cronin, and J. U. Skaare. 1998. Female pseudohermaphrodite polar bears at Svalbard. Journal of Wildlife Diseases 34:792–96.
Wilson, D. E. 1976. Cranial variation in polar bears. International Conference on Bear Research and Management 3:447–53.
Wimsatt, W. A. 1963. Delayed implantation in the Ursidae, with particular reference to the black bear (*Ursus americanus* Pallas). Pages 49–76 *in* A. C. Enders, ed. Delayed implantation. University of Chicago Press, Chicago.
Young, B. F., and R. L. Ruff. 1982. Population dynamics and movements of black bears in east central Alberta. Journal of Wildlife Management 46:845–60.

Steven C. Amstrup, U.S. Geological Survey, Alaska Science Center, Biological Science Office, 1011 E. Tudor Road, Anchorage, Alaska 99503-6199. Email: steven_amstrup@usgs.gov.

28

Raccoon

Procyon lotor and Allies

Stanley D. Gehrt

NOMENCLATURE

COMMON NAMES. Northern, Common Raccoon, Raccoon, coon
SCIENTIFIC NAME. *Procyon lotor*
SUBSPECIES. *Procyon lotor lotor, P. l. maritimus, P. l. solutus, P. l. litoreus, P. l. elucus, P. l. marinus, P. l. inesperatus, P. l. auspicatus, P. l. incautus, P. l. varius, P. l. megalodous, P. l. fuscipes, P. l. hirtus, P. l. excelsus, P. l. vancouverensis, P. l. pacificus, P. l. psora, P. l. pallidus, P. l. mexicanus, P. l. grinnelli, P. l. hernandezii, P. l. shufeldti, P. l. dickeyi, P. l. crassidens, P. l. pumilus* (Hall 1981).

The center of procyonid radiation is thought to have been Central America, although most of the fossil record has come from North America (Baskin 1982). Forests or woodlands, the presumed habitats used by prehistoric procyonids, are not conducive to fossilization and may explain the relative paucity of fossil record for this family. Nevertheless, *Procyon* first appeared in the fossil record in the Miocene. Fossil records for the genus occur in North America by the late Miocene, and are well represented in the Pliocene (Martin 1989). Among procyonids, the raccoon has the most widespread distribution in North America. Their success is probably due to a combination of a generalist strategy with associated behavioral plasticity, intelligence, a relatively high metabolic rate, and favorable landscape changes such as agriculture and urbanization.

Other Procyonids. *Procyon lotor* is 1 of 18 species recognized for the family Procyonidae (Wozencraft 1993), which includes the following genera: olingos (*Bassaricyon* sp.) and kinkajous (*Potos* sp.) of Central and South America, ringtails (*Bassariscus* sp.) and raccoons (*Procyon* sp.) of North and Central America, and coatis (*Nasua* sp., *Nasuella* sp.) of North, Central, and South America. A current phylogeny of recent Procyonidae was provided by Decker and Wozencraft (1991). Seven species are recognized for the genus *Procyon,* although five of these (*P. minor, P. maynardi, P. insularis, P. gloveralleni,* and *P. pygmaeus*) are insular and may simply be variations of *P. lotor* (Corbet and Hill 1986) or may be extinct and replaced by *P. lotor* (Lotze and Anderson 1979; Hall 1981). For example, raccoons on one of the Bahamian islands may be introduced *P. lotor,* rather than *P. maynardi* (Koopman et al. 1957; McKinley 1959). Recent genetic analysis indicated the Guadeloupean raccoon, *P. minor,* is a variation of *P. lotor* and not a separate species (Pons et al. 1999). The other continental congeneric is the crab-eating raccoon, *Procyon cancrivorous,* which occurs in South and Central America.

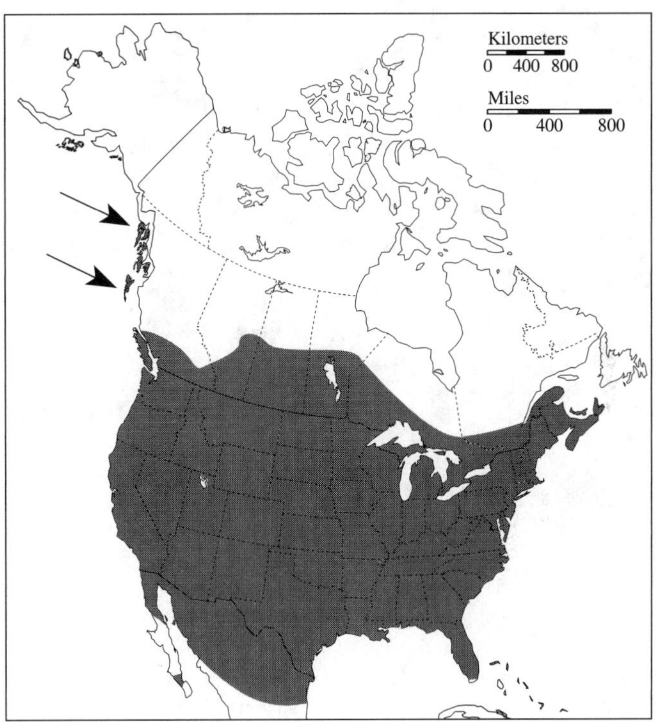

FIGURE 28.1. Distribution of the raccoon (*Procyon lotor*).

DISTRIBUTION

Raccoons have a widespread distribution across North America, extending from Mexico to parts of Canada and almost continuously between both coasts (Fig. 28.1). Historically, raccoons were rare in southern Canada and in parts of the Rocky Mountains and western deserts. Raccoons dramatically increased in density beginning in the 1940s, with a concomitant increase in distribution. Today they occur in the mountains and deserts where they were previously rare or absent, and have extended their range in Canada. Some range extensions were the result of introductions, such as islands off the coast of southeastern Alaska (Scheffer 1947), the Queen Charlotte Islands of British Columbia (Hartman and Eastman 1999), and some Bahamian Islands (Sherman 1954). From 1920 to 1987, the total estimated geographic range in North America increased from 6.6 million km^2 to 8.8 million km^2 (Sanderson 1987). There have been occasional records of raccoons occurring at extreme northern latitudes in Canada (Soper 1942; Sutton 1964; Lynch 1971), but these probably do not represent established populations (Kaufmann 1982). MacDonald and Cook (1996) considered the raccoons introduced to Alaskan islands to be extirpated. Sanderson (1987) estimated there were 15–20 times as many raccoons in North America during the 1980s as there were in the 1930s. With recent declines in pelt value and a concomitant reduction in harvest, raccoons may have experienced another significant population increase during the 1990s.

Raccoons were introduced to Russia in the 1930s for fur production. They have become established in Eastern Europe in the south of Belorussia and along the Caucasian chain, although at low numbers. There are currently three population centers in Western Europe, all the result of introductions; two foci are located in Germany (Lutz 1984, 1995) and another is located in eastern France (Leger 1999).

FIGURE 28.2. The black mask of the raccoon (*Procyon lotor*) is familiar to most people. SOURCE: Photo by Stan Gehrt.

DESCRIPTION

The raccoon is a medium-sized carnivore, with a stocky torso and short limbs. The pelage coloration, with the striking black mask (Fig. 28.2) and ringed tail, makes the raccoon one of the most recognizable mammals in North America. The spine is curved posteriorly, giving the animal a roundish appearance similar to a bear. The feet are plantigrade with naked footpads. Front and back feet are pentadactyl with no webbing between toes, and each toe has a sharp, curved, nonretractile claw. Tracks resemble human prints, with the long toes of the front foot resembling fingers and the shorter toes of the back foot, with its longer foot pad, resembling human toes. The pinnae are relatively small compared to those of canids or felids, but larger than those of mustelids. Similarly, the snout is medium length and pointed. Cheek tufts that flare to the sides give the impression of a broad head when seen from the front. There are no pelage differences between males and females. Males are larger than females in body measurements and weight.

The largest raccoons apparently occur in the northwestern United States and the smallest in the Southeast (Goldman 1950; Ritke and Kennedy 1988). Adult body weights usually range from 4 to 9 kg, depending on geographic location, season, and sex. The smallest raccoons occur in south Florida (Goldman 1950); mean winter body weights are 2.4 and 2.0 kg for adult males and females, respectively. Mean body weights for adult raccoons in Minnesota or Michigan range from 6.9 to 10.4 kg during the winter. Variation in body size may conform to a latitudinal gradient in the eastern United States (Johnson 1970; Mugaas and Seidensticker 1993; Sanderson 1987); however, gradients may not occur elsewhere. Adult body weights in South Texas averaged 9.0 and 6.7 kg for males and females, respectively (Gehrt and Fritzell 1999a). These are similar to weights at more northern latitudes and significantly greater than weights from Florida raccoons at a similar latitude (Goldman 1950). Local factors such as soil quality may indirectly affect body size of raccoons. In statewide comparisons in Kansas and Missouri, raccoons from locations with higher quality soils were larger than those from poor soils (Crawford 1950; Stains 1956). The largest recorded body weights are 25.4 and 28.3 kg (Wood 1922; Scott 1951), but raccoons have rarely been observed over 18 kg (Kaufmann 1982). Sanderson (1987) recorded body weights during the harvest season for 29 years in Illinois. The range for 2115 adult males was 7.0–8.3 kg and for 2089 adult females was 5.1–7.1 kg.

Pelage. The pelage consists primarily of guard hairs and short, thick underfur. The black mask usually extends from slightly above the eyes to the base of the snout and flares out along the cheeks. It is accentuated by white lines immediately above and below the mask. As a raccoon ages, the mask tends to fade. The anterior side of the ear is entirely white; the posterior side has a black base, which extends to nearly half the ear. The top of the nose and forehead are grayish or reddish brown. Pelage color on the shoulders and back is dominated by reddish-brown guard hairs. However, variations on this color scheme are common, including nearly black, orange, or cinnamon variants, and geographic variation in pelage color occurs (Goldman 1950; Stains 1956; Johnson 1970; Lotze and Anderson 1979). The ventral side is lighter, usually a light brown, with sparse guard hairs. The hair on each foot is short and whitish-gray. Four to seven blackish rings and a dark tip characterize the tail. Albinism occurs in free-ranging raccoons, and it appears to be the product of a recessive gene (Whitney and Underwood 1952; Allen and Neill 1956).

The previous winter's coat is shed during the spring, either gradually all over the body or in patches. Summer pelage is thin, short hairs.

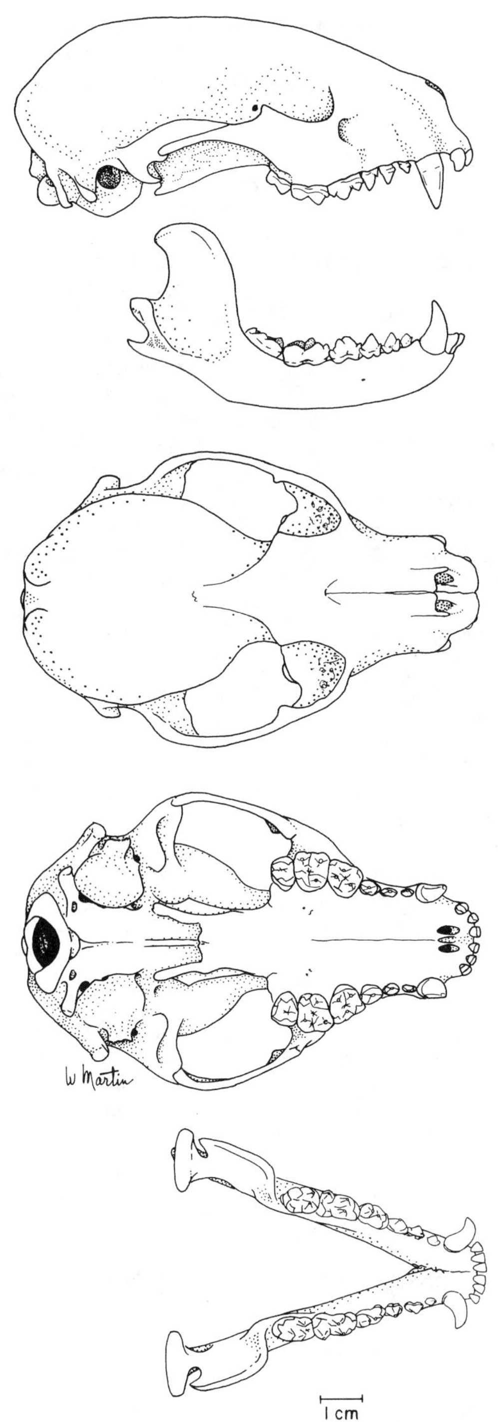

FIGURE 28.3. Skull of the raccoon (*Procyon lotor*). From top to bottom: lateral view of cranium, lateral view of mandible, dorsal view of cranium, ventral view of cranium, dorsal view of mandible.

In the autumn, thick underfur and new guard hairs are produced during a 6-week process in preparation for winter (Sanderson 1987). The fur in most temperate areas becomes prime in December, but some raccoons may not become prime at all, particularly at southern latitudes.

Skull and Dentition. Adult raccoons have 40 teeth, with a dental formula I 3/3, C 1/1, P 4/4, M 2/2 (Fig. 28.3). The dentition reflects the omnivorous habits of this species, with sharp incisors and canines suitable for cutting or tearing, contrasting with square-shaped molars and fourth upper premolars with small, rounded cusps (bunodont), suited for grinding plant material. The sagittal crest becomes more prominent in older individuals (Kaufmann 1982).

Kennedy and Lindsay (1984) examined relationships between a suite of environmental variables and cranial measurements of raccoons across the southeastern United States. They found a clinal gradient for skull morphometrics, although they discounted a simple climatic explanation for their results. A similar study that included cranial measurements of raccoons recorded over much of the United States failed to find consistent clinal variations in dimensions (Ritke and Kennedy 1988).

Skulls are sexually dimorphic for multiple characteristics (Grau et al. 1970; Ritke and Kennedy 1993). There is little geographic variation in the sexual dimorphism of skull characteristics, which suggests that selection pressures responsible for dimorphism are consistent throughout the United States (Ritke and Kennedy 1993). Intersexual differences in canine width vary to a greater degree than other skull characteristics. Canines of males are larger than those of females, possibly as a result of intrasexual selection (Ritke 1990a).

GENETICS

The diploid number of chromosomes is 38 (Kaufmann 1982); however, $2n = 42$ has been reported for some raccoons of unknown origin (Lotze and Anderson 1979).

Protein electrophoresis analyses have revealed genetic differentiation among polymorphic loci between geographic regions in North America (Hamilton and Kennedy 1987), but little variation within geographic regions, particularly the Southeast (Beck and Kennedy 1980; Dew and Kennedy 1980). Overall genic heterogeneity among geographic regions was 0.9%, which is similar to reported levels for other carnivores. However, heterogeneity was occasionally much higher within regional populations. Dew and Kennedy (1980) suggested that artificial dispersal (translocations for management purposes) might have influenced regional genetic patterns, although Hamilton and Kennedy (1987) questioned its role in gene flow in the southeast.

Temporal genetic variation during a 9-year period was described for a hunted population of raccoons at Reelfoot National Wildlife Refuge in Tennessee and Kentucky (White et al. 1998). Average heterozygosity was 0.042 for six polymorphic loci and ranged from 0.033 to 0.046 during the period, but did not differ among years. There were annual differences in allele frequencies, although variation was low and was probably the result of stochastic events (White et al. 1998).

PHYSIOLOGY

Raccoons have a higher mass-specific basal metabolic rate than other procyonids. This may partially explain why this species has a more widespread distribution than other procyonids (Mugaas et al. 1993). The basal metabolism (ml O_2/g/hr) of raccoons in Virginia during the summer was 0.46 for captive males and 0.42 for captive females. During the winter it was 0.47 and 0.46 for males and females, respectively. Body temperatures for males during active and rest phases of the daily cycle were 38.4°C and 37.5°C in summer and 38.6°C and 37.6°C in winter. Female results were similar, with respective active and resting body temperatures of 38.2°C and 37.6°C during summer and 38.3°C and 37.3°C in winter. Thus, raccoon metabolic rates and body temperatures do not vary among seasons; apparently pelage molt is sufficient to manage the potential for summer heat stress (Mugaas et al. 1993). For Virginia raccoons, minimum thermal conductance was 49% higher during summer than winter, which was due to the difference in insulation between summer and winter pelts.

Adults of both sexes experience notable weight gains during autumn and weight loss during winter and early spring (Mech et al. 1968; Johnson 1970; Moore and Kennedy 1985). Timing and magnitude of these fluctuations change with latitude. Raccoons in northern latitudes lose >50% of their body weight during winter where extreme winter conditions occur over extended periods (Stuewer 1943a; Mech et al. 1968). During winter, raccoons in temperate climates may enter a dormant period and remain inactive in dens for weeks or months. The heart rate does not decrease during winter dormancy, and may even increase during prolonged winter denning (Folk et al. 1968). Similarly, body temperature does not decrease during winter denning. Because raccoons do not decrease their metabolic rate during winter denning, they metabolize fat reserves gained during autumn. Consequently, the length of time a raccoon can remain inactive will be affected primarily by the amount of fat deposited during the previous autumn. Seasonal weight fluctuation appears to scale with climate. At southern latitudes, raccoons only add 10–30% of their lean body mass during autumn (Johnson 1970; Moore and Kennedy 1985; Gehrt and Fritzell 1999a), whereas raccoons at northern latitudes where winters are harsh may add up to 50% of their lean body mass (Stuewer 1943a; Mech et al. 1968). Thus, Mugaas and Seidensticker (1993) suggested that climate may select for body size in raccoons. An increase in lean body mass results in an increase in potential fat deposition, which in turn affects the length of time raccoons can remain inactive during the winter. Longer winters apparently select for larger body size.

Thorkelson and Maxwell (1974) illustrated the importance of tree cavities for winter thermoregulation. They found 65% of the resistance to heat transfer came from the pelt and 35% from characteristics associated with the den, including conductance of the air in the den and the den walls. Raccoons could further reduce the cost of thermoregulation during winter by denning communally, which may be particularly important for juveniles at northern latitudes (Mugaas et al. 1993). This is a period of particular hardship for juveniles, which usually lack abundant fat reserves during their first winter (Mech et al. 1968).

Weight loss in mild or subtropical climates may be a physiological response to the need to lose heat during warm weather rather than maintain endothermy during the winter. Goldman (1950) suggested that raccoons at subtropical climates may not exhibit seasonal weight fluctuations because they remain active year-round. Raccoons in mild climates exhibit seasonal variation in body weight, but the magnitude and timing differ from those in northern populations (Gehrt and Fritzell 1999a). In South Texas, raccoons gain weight in autumn as do northern populations, and subsequently lose 27% of their body weight in the following seasons. However, in contrast to northern populations, most of their weight loss occurs during the spring and early summer, and may be a response to a need for evaporative cooling during summer rather than energetic costs during winter (Gehrt and Fritzell 1999a).

Capacity of the alimentary canal may change with altered energetic requirements. In Kentucky, lactating females have greater masses of stomach, colon, and gastrointestinal tract than do nonreproductive females and males (Derting 1996). This difference suggests energetic demands of reproduction and lactation stimulate a physiological response in the alimentary canal. However, Derting (1996) reported no seasonal differences in gut capacity between summer and winter for males.

The nervous system is consistent with a highly developed tactile sense. Highly innervated forefeet and developments in the cerebral neocortex are associated with forepaw sensitivity (summarized by Kaufmann 1982).

REPRODUCTION AND DEVELOPMENT

Anatomy. The reproductive morphology of raccoons has been described in detail by Sanderson and Nalbandov (1973). Males lack seminal vesicles and Cowper's glands, but the reproductive system is otherwise typical of other carnivores. Males possess a well-defined, bicurved baculum, which can reach a length of 92–111 mm (Kaufmann 1982). The baculum is curved downward and the anterior end is bifurcated. This bone is frequently broken, particularly among juveniles, when it is still relatively undeveloped (Sanderson and Nalbandov 1973).

The uterus is intermediate between bicornuate and bipartite, with a single cervix and two distinct horns. An *os clitoridis* is rarely found in raccoons (Rinker 1944), is much smaller than a baculum, and its frequency of occurrence may vary geographically (Sanderson and Nalbandov 1973). The placenta is deciduous, and placental scars remain for months following parturition. Because each scar corresponds to an implantation site for an embryo, scars provide a means for estimating the litter size of free-ranging raccoons. Raccoons have six mammae, although eight have been observed for some individuals (Sanderson 1987).

Physiology. Ovulation in raccoons is spontaneous (Sanderson and Nalbandov 1973; Sanderson 1987), although raccoons previously were described as induced ovulators. Kaufmann (1982) may have resolved the discrepancy by pointing out that induced ovulation has been observed for species that are characterized as spontaneous ovulators. Perhaps raccoon ovulation, although typically spontaneous, may be triggered by stimuli from mating activity (Kaufmann 1982). It has been shown experimentally that increasing photoperiod is a stimulant for ovulation and sperm production in raccoons, whereas temperature or snow amount had no effect (Bissonnette and Csech 1938). However, winter weather can inhibit movements and activity and thereby affect the timing of successful matings. Corpora lutea are formed following pregnancy or pseudopregnancy and disappear 14–16 days after parturition (Sanderson and Nalbandov 1973). Nipples become enlarged and, in many cases, pigmented on the formation of corpora lutea. The amount of pigment varies among individuals, with some nipples heavily blackened, whereas others may remain pink. The condition of the nipples is a useful identifier of reproductive females because the nipples never completely return to the preovulatory condition (Sanderson and Nalbandov 1973; Sanderson 1987). Transuterine migration of ova has been reported in the species (Llewellyn and Enders 1954; Dunn and Chapman 1983).

Juvenile females may ovulate during their first mating season and produce litters as yearlings, although the actual percentage that does so varies greatly among years and among populations (Table 28.1). In Illinois and Missouri, 38–77% of yearling females were parous each year (Fritzell et al. 1985). For other populations, the pregnancy rate for yearlings has ranged from 0% (Scheffer 1950) to 73% (Junge and Sanderson 1982). Pregnancy rates for adults have ranged from 68% to 100% in various studies (Table 28.1).

Adult testis size fluctuates seasonally and, for Illinois raccoons, there is a period of 3–4 months in summer and autumn in which sperm is absent from the epididymides (Sanderson and Nalbandov 1973). Thus, adult males may not be sexually active year-round (Dunn and Chapman 1983; Sanderson 1987). In Maryland, adult males are probably in breeding condition from October to May (Dunn and Chapman 1983). Testes reach maximum size about 2 months before the mating season, indicating that males are in prime reproductive condition when the first females become reproductive (Johnson 1970; Sanderson and Nalbandov 1973; Dunn and Chapman 1983). Fluctuations in testes size are not as dramatic for southern populations as in northern populations (Johnson

TABLE 28.1. Percentage of parous raccoons in approximate relation to latitude

	Yearling		Adult		
Location	%	*n*	%	*n*	Reference
Manitoba	26	27	86	52	Cowan 1973
North Dakota	14	14	93	28	Fritzell 1978a
Washington	0	17	74	23	Scheffer 1950
Michigan	54	28	96	27	Stuewer 1943b
Oregon	32	129	92	184	Fiero and Verts 1986a
Iowa	45–70	—	88–100	—	Clark et al. 1989
Maryland	9	43	14	77	Dunn and Chapman 1983
Illinois	73	27	95	32	Junge and Sanderson 1982
Illinois	66	176	94	182	Fritzell et al. 1985
Missouri	49	136	84	159	Fritzell et al. 1985
Alabama	9	38	95	104	Johnson 1970
Texas	40[a]	—	—	—	Wood 1955

[a]Yearling sample size not given; total sample for females was 37.

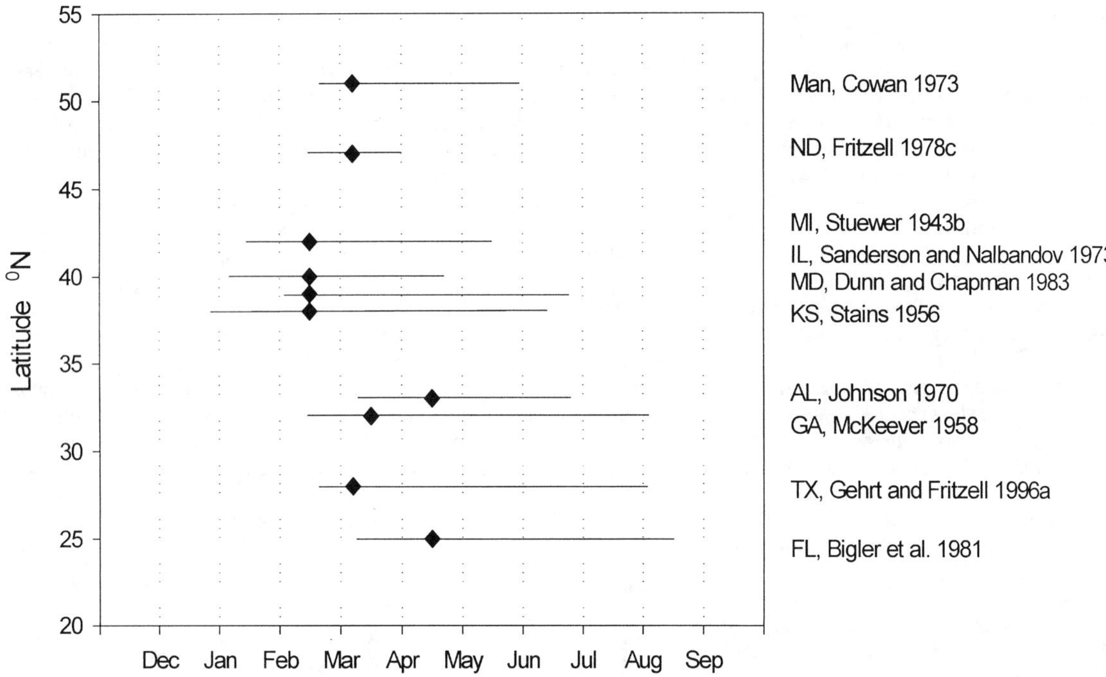

FIGURE 28.4. Latitudinal distribution of peak mating periods (diamonds) and ranges (lines) reported for raccoons.

1970; S. D. Gehrt, unpublished data), which might be a product of prolonged mating activity. Juvenile males do not become sexually mature until 3–4 months following the primary peak of mating, and therefore do not participate in mating activities in most seasons. However, because of their delayed development, yearling males may have sperm present at a time of year when adult males are not reproductively active, leading to the suggestion by Sanderson and Nalbandov (1973) that yearling males may father the majority of late litters. However, yearling males were never observed consorting with estrous females during late mating in South Texas (Gehrt 1994). Additionally, yearling males in North Dakota were not sexually mature until the following mating season (Fritzell 1978a). Captive-reared juvenile males may sire litters in captivity (Pope 1944).

Mating. Most mating occurs during February and March throughout much of North America and somewhat later at more southern latitudes (Fig. 28.4). However, reports of litters conceived outside the normal mating period are common, and range from January to August (Berard 1952; Dorney 1953; McKeever 1958; Lehman 1968). In Maryland, most mating occurs in February and March, but can extend to late June (Dunn and Chapman 1983). In North Dakota and Manitoba, mating takes place later than in other northern regions, suggesting the extreme winter season may delay mating (Cowan 1973; Fritzell 1978a). In Texas, a discrete, bimodal distribution of mating occurred with a primary peak in February and March, followed by a less synchronized, secondary peak during May and June (Gehrt and Fritzell 1996a). This bimodal pattern is probably typical of most natural populations, although the timing will vary. Onset of the first mating period is probably stimulated by lengthening photoperiod (Bissonnette and Csech 1938). The second mating period involves females coming into a second estrus after losing a litter or failing to produce a litter from the first mating period. The second period is less synchronized than the first because litters are not lost simultaneously. In Texas, all females that mated during the second mating period also were in estrus during the first mating period. In northern populations, extreme winter conditions may inhibit movements and prevent matings with estrous females or possibly postpone estrus until later in the year (Dorney 1953; Sanderson and Nalbandov 1973; Fritzell 1978a).

In general, the only association between adult males and females occurs briefly during the mating season. Consortships between adult males and estrous females have been observed for captive (Stuewer 1943b) and free-ranging raccoons (Gehrt and Fritzell 1996a, 1999b). In both cases, consortships usually last 3 days, although for free-ranging pairs the range was 1–3 days. Only one male consorts with a female at a time, but some females will sequentially consort with multiple males. In South Texas, 62% of the time females consorted with only one male during estrus; in other cases, females sequentially consorted with two to six males during an estrus (Gehrt and Fritzell 1999b). Tevis (1947) described an apparent consort between a free-ranging adult male and female in which the male actively defended her from other males. It is likely some matings occur without consortships. Pair bonds lasting months before mating (Whitney and Underwood 1952) have not been observed for free-ranging raccoons and probably do not occur.

Free-ranging raccoons have rarely been observed copulating, and only two detailed descriptions of this behavior are published (Goldman 1950; Stains 1956). In each case, the male was mounted on top of the female in a semicrouched position and the pair was coupled for nearly 1 hr. Stains (1956) reported the female uttered shrill cries with the thrusting motions and often attempted to bite the male. Copulatory movements alternated between rhythmic, slow thrusting and quick outward motion and sideways movements, and little thrusting at all. Eventually the male slipped from the female and each left without further interactions. Goldman (1950) described a similar position and thrusting, but there was little animosity from the female toward the male until copulation was terminated. Again, both participants left separately. These observations suggest that raccoon copulation may last 1 hr or more and may vary in the level of aggression between the participants.

There is no evidence that females come into a second estrus in the same year after successfully rearing a litter, and no third periods of estrus in one year have been observed. A third estrus in one year was not possible even under experimental conditions (Bissonnette and Csech 1938). Although Kaufmann (1982) speculated that raccoons may breed throughout the year in the Southeast, data from southern latitudes suggest seasonal breeding, albeit a possibly extensive season, throughout North America (Bigler et al. 1981; Gehrt and Fritzell 1996a).

Litters. Gestation is about 63 days (Kaufmann 1982; Sanderson 1987); however, a range of 54–70 days has been reported (Gander 1928). Pseudopregnancy occurred in about 2.5% of the raccoons in Illinois (Sanderson 1987). Prenatal mortality is probably negligible in most populations (Johnson 1970; Dunn and Chapman 1983). However, in Manitoba, prenatal mortality rates were 67% for yearling females and 4% for adult females (Cowan 1973).

Parturition typically occurs in April at northern latitudes and sometime later at southern latitudes. Mean date of birth was 18 June in Alabama (Johnson 1970) and 18 April in Illinois (Sanderson 1987), and there was a peak parturition period in May in Manitoba (Cowan 1973). Median parturition dates ranged from 30 April to 28 May in South Texas (Gehrt and Fritzell 1998a). Extreme records include February births reported for Louisiana (Arthur 1928, as cited in Cagle 1949), September births for Indiana (Lehman 1968), and parturition dates from April to October in Georgia (McKeever 1958) and April to September in Texas (Gehrt and Fritzell 1998a). Cowan (1973) reported 14% of births occurred as late as early September in Manitoba.

The mother reduces her activity and movements during the week of parturition (Montgomery 1969; Schneider et al. 1971) and becomes increasingly intolerant of conspecifics (Bissonnette and Csech 1938). Tree cavities are often preferred for natal dens, although litters have been reared in a variety of den types (Kaufmann 1982). The mother makes few modifications to the den before parturition. Males play no direct role in rearing litters, and females may actively exclude them from the immediate area.

Number of young per litter is usually 3–4; however, this varies by region and latitude. Larger litters are produced at northern latitudes, probably because of the larger size of northern raccoons (Ritke 1990b). Mean litter size, reported for samples sizes of greater than 10, has ranged from 2.5 in Alabama (Johnson 1970) to 4.8 in North Dakota and Manitoba (Cowan 1973; Fritzell 1978a). Sanderson (1987) reported a 29-year average of 3.6 for litter size in Illinois raccoons. A geographic range of published litter sizes is presented in Ritke (1990b). Litter size also varies with age of the female. Yearling females average smaller litters than adults, and there may be a small decline in fecundity with the oldest age groups (Fiero and Verts 1986a; Clark et al. 1989). For example, mean litter size was 3.3 for yearlings and 3.6 for adults in Missouri and Illinois (Fritzell et al. 1985). Cowan (1973) reported a mean litter size of 0.8 for yearlings and 4.1 for adults in Manitoba.

Newborns are sparsely furred, and eyes and ear canals remain closed for about 3 weeks (Montgomery 1969; Johnson 1970). They cannot walk until about 4 weeks of age. Observations of captive raccoons in Illinois revealed that juveniles began eating solid food at 9 weeks, although they continued to nurse until about 16 weeks of age (Montgomery 1969). Growth rates increased dramatically when they began eating solid food. Young began moving out of their nest boxes at 6–7 weeks of age (Montgomery 1969). Nestlings in Alabama had slower growth rates while in the den than did northern raccoons (Johnson 1970). Deciduous incisors, canines, and premolars begin to erupt soon after parturition, and most of the permanent dentition eruptes by 4 months (Montgomery 1964).

In some areas, females will move young from the tree den to a ground rest site at about 6–9 weeks of age (Schneider et al. 1971). In other cases, females move young to other dens soon after parturition (Bissonnette and Csech 1938). Gehrt and Fritzell (1998a) reported litters remained in natal dens an average of 72 days (range 51–104) postpartum.

Postden Development. Growth rates of young raccoons are largely unknown for free-ranging populations. Juvenile males in South Texas grew more quickly than juvenile females; however, despite the faster growth rate, males did not complete their growth earlier than females (Gehrt and Fritzell 1999a). Females attained adult size before their second winter, whereas males attained adult size following their second winter. Young raccoons of both sexes attained adult size during their second autumn in Alabama (Johnson 1970). Because juvenile raccoons at southern latitudes experience relatively smaller weight loss during winter than northern raccoons (Stuewer 1943a; Mech et al. 1968), they may attain adult size earlier than northern populations (Gehrt and Fritzell 1999a).

Few studies have monitored the dissolution of familial bonds. Three raccoon families were intensively monitored in Minnesota (Schneider et al. 1971) and 11 families in South Texas (Gehrt and Fritzell 1998a). Both studies found that cubs begin traveling from the natal den with the mother on foraging excursions at 10–12 weeks of age. In Minnesota, raccoons continue traveling and resting as a family until the cubs reach 17–18 weeks of age. After 18 weeks, cubs begin traveling independently of family members during the autumn. However, as winter approaches, family bonds strengthen and the family begins denning together through the winter. Family break-up occurs during the following spring. Juvenile raccoons from other regions become independent during autumn (Stuewer 1943a; Sharp and Sharp 1956; Urban 1970; Fritzell 1977); however, investigators did not monitor specific family groups. In South Texas, the age at which cubs first rested independently from the mother ranged from 5 to 9 months; however, familial bonds remained strong until the mother came into estrus. In a few instances, juveniles reestablished bonds with the mother following her estrus, and continued to den and travel with her until parturition. There was no difference between juvenile males and females in the amount of time spent with family members before male dispersal.

ECOLOGY

Habitat. Raccoons exploit a variety of habitats; however, those associated with water are always important (Kaufmann 1982). Raccoons use wetlands heavily during the summer in North Dakota (Fritzell 1978b), and bottomland forests have higher densities of raccoons than do uplands (Yeager and Rennels 1943; Glueck 1985; Leberg and Kennedy 1988). Woodlots are also used heavily by raccoons in all landscapes (e.g., Stuewer 1943a; Twichell and Dill 1949; Ellis 1964). Raccoons tend to avoid uplands such as pastures, grasslands, and croplands, but will use linear features such as fence rows, shelter belts, and roads as travel lanes (Fritzell 1978b; Glueck et al. 1988). In Louisiana, raccoons congregated in marshes and used levees for foraging and traveling (Cagle 1949).

Pedlar et al. (1997) examined habitat use at different spatial scales in Ontario and found that many features important at the microspatial scale were repeated at the macrospatial scale. Raccoons frequented woodlands at the microscale, and extensive agricultural edge and woodlands with extensive corn cover at the macroscale. In the urban landscape of Toronto, Ontario, habitat was the most important determinant for raccoon density, with the highest densities occurring in woodland parks (Rosatte et al. 1992a). The lowest densities occurred in areas dominated by industry and fields.

Raccoons' affinity for riparian habitat occasionally exposes them to flooding, during which their arboreal adaptations serve them well. In a flood pool of a reservoir, raccoons subjected to extensive flooding that lasted weeks remained within their home ranges, even when these were completely inundated, and used treetops until the water subsided (Gehrt et al. 1993). Similarly, raccoons in coastal salt marshes foraged in areas exposed by low tide, and remained in these flooded areas during high tide (Ivey 1948) using floating beds of vegetation.

Population Demographics. Raccoon densities have been estimated in a wide range of studies using a variety of methods. Thus, caution should be used when comparing estimates among studies. Also, it is important to note that raccoon populations are subject to dynamic changes between seasons or years (Lotze and Anderson 1979), often as the result of epizootics or anthropogenic effects such as harvest. Nevertheless, in a general way densities appear to reflect habitat quality, with respect to the distribution and abundance of resources. Summaries of some density estimates for raccoons were provided by Kaufmann (1982) and Riley et al. (1998). Typical rural densities have ranged from 1 to 27 raccoons/km^2 (Moore and Kennedy 1985; Kennedy et al. 1986; Gehrt 1988; Seidensticker et al. 1988). Lowest densities (0.5–1/km^2) usually occur near the northern limits of their distribution, particularly in prairies (Cowan 1973; Fritzell 1978c). However, Bigler et al. (1981) reported average densities of only 3.9/km^2 for raccoons in the

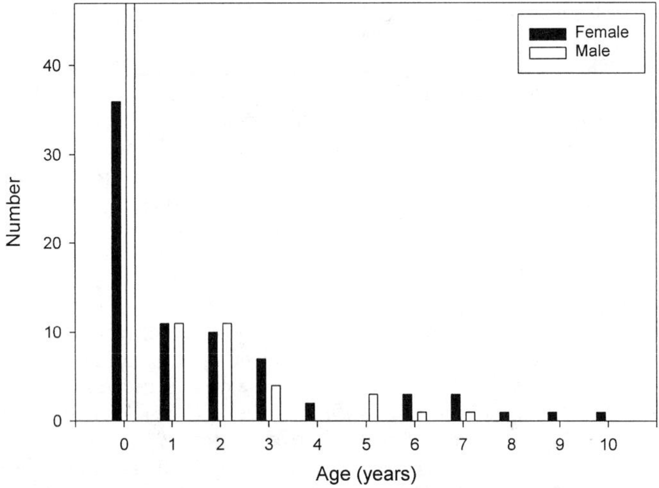

FIGURE 28.5. Age distribution, determined from cementum annuli of incisors, of raccoons ($n = 153$) recorded from the Max McGraw Wildlife Foundation, Illinois, 1995–1998. Ages are calculated based on birth in April; age 0 is <1 year old.

Keys region of Florida, but this population regularly experienced rabies outbreaks.

The highest density of raccoon populations usually occurs in suburban/urban areas, although some patches of woodland or swamp within agricultural landscapes also may have large concentrations. Riley et al. (1998) estimated raccoon densities of 67–333/km^2, with a mean of 125/km^2, for a population in an urban park in Washington D.C. These are the highest estimates reported from mark–recapture data, although some of their estimates are probably inflated because the trapping areas were small portions of the park where raccoons concentrated their activity. Similarly, Twichell and Dill (1949) reported an exceptionally high density of 244/km^2 for a population of raccoons in Missouri. This estimate was determined by intensively searching den trees in a small woodlot; thus, movements of raccoons were not considered in the density estimate. Other high densities reported for urbanized areas include 66.7 raccoons/km^2 in a suburban area in Cincinnati (Hoffmann and Gottschang 1977), 111/km^2 in Cincinnati (Schinner and Cauley 1974), and 55.6/km^2 in Toronto (Rosatte et al. 1990).

In terms of age distribution, raccoon populations are usually composed of a majority of individuals <2 years old. However, Johnson (1970) and Bigler et al. (1981) provided evidence that populations in the southern United States are often dominated by older age classes. Harvest, other anthropogenic factors, or high adult survival combined with a low reproductive rate may affect the age distribution of a population. Figure 28.5 displays the age distribution of a monitored raccoon population from the Max McGraw Wildlife Foundation, Illinois, during 1995–1998. Data are pooled across years, and were obtained from resident mortalities, primarily road-kills, collected from the study area. This age distribution is probably typical of most raccoon populations located in temperate climates that are protected from harvest. Age distributions are similar between the sexes, but there are slightly more females in the older age categories. There is a dramatic decrease in numbers between 1 year of age and older age classes.

Most published sex ratios have been male biased (Lotze and Anderson 1979; Kaufmann 1982), although usually there have not been significant differences from an even ratio. During 29 years, the sex ratio of harvested raccoons in Illinois ranged from 49% to 58% males (Sanderson 1987). Reported sex ratios may reflect biases in sampling techniques. Many estimates have been derived from harvest or live-trapping data, and males may be more prone to capture than females (Sanderson 1987; Gehrt and Fritzell 1996b). It is important to note bias exists even when estimated sex ratios are near parity. Sex ratios determined from other techniques have been female biased. Twichell and Dill (1949) captured raccoons by hand from trees in a Missouri refuge and recorded 60 females:40 males. Similarly, sex ratios were biased toward females during a study in Texas in which nearly all residents were marked (Gehrt and Fritzell 1996a, 1998b). Thus, most raccoon populations are probably slightly female biased or near parity, which is consistent with a promiscuous or mildly polygynous mating system.

Activity. Raccoons are primarily nocturnal (Lotze and Anderson 1979; Kaufmann 1982). Exceptions to this pattern may occur during winter when daytime temperatures are relatively warm, and raccoons may begin foraging before dusk. Also, recently parturient females may forage during the day (Schneider et al. 1971; S. D. Gehrt, unpublished data), especially when their activity periods are abbreviated because of the care of young. Ladine (1997), using timers on traps in Tennessee, reported most raccoon captures occurred before midnight. Only 10% of the captures occurred during daytime hours.

At night, raccoon activity can be partitioned into local foraging, running, and walking (Greenwood 1982). During the spring and summer in North Dakota, raccoons spent most of their time walking (49%) and in local movements (38%) and relatively little time running (13%); however, activity patterns vary with season and habitat (Greenwood 1982). Females and yearlings exhibited similar activity patterns, whereas adult males differed from those groups in that they traveled farther, moved locally half as often, and ran twice as often. In most areas, raccoons probably exhibit a similar general pattern. On leaving daytime retreats, they move quickly to foraging habitats, such as wetlands, and spend much of their nightly activity foraging in local areas (Johnson 1970; Fritzell 1978b; Seidensticker et al. 1988). On concluding their nightly activity, they may move quickly to a particular den for the day or use an available den nearest their last foraging site. Variations on this theme depend on season, sex, and age group. Females reduce their activity during the week of parturition (Montgomery 1969) and for various periods following parturition.

Shelter. Given the widespread distribution and flexible behavior of raccoons, it is not surprising they have been observed using a variety of types of diurnal resting sites (reviewed in Kaufmann 1982). Raccoons usually do not create their own dens, but rather use structures already present. Tree cavities are frequently used as den sites, although their use may vary seasonally. In addition to cavities, raccoons may use squirrel nests in trees or simply lie on limbs or vines in the canopy. Females generally prefer tree cavities for parturition and rearing of young, probably more for protection than for den temperatures (Rabinowitz and Pelton 1986). If rock crevices are available, raccoons often use these, particularly during the winter, when the dens afford more thermal insulation and opportunities for communal denning than tree cavities. Where trees are rare, raccoons will use muskrat (*Ondatra zibethicus*) houses or levees, or construct ground nests from vegetation. Raccoons also use artificial structures such as buildings, mines, sewers, and brush piles. On the northern prairies where trees are limited, old farmsteads and other buildings are important den types, and snow cavities have been used during the winter (Lynch 1974; Fritzell 1978b). For raccoons in Manitoba near the limit of their distribution, den selection during winter is crucial to their survival (Cowan 1973). Communal denning has often been reported for raccoons during the winter, particularly at northern latitudes, where it may be an important behavior for survival in addition to den type selection. An extreme example was the presence of 23 raccoons in one winter den in Minnesota (Mech and Turkowski 1966).

Comparisons between sex–age groups are undoubtedly affected by high individual variation in preferences for den types and den site fidelity (Shirer and Fitch 1970; Rabinowitz and Pelton 1986; Hadidian et al. 1991; Endres and Smith 1993). Hadidian et al. (1991) described daytime rest site use in an urban park. They observed almost exclusive use of residential buildings for rest sites for some raccoons, whereas most others restricted their use to dens within the park. Some studies have reported raccoons frequently shifting daytime rest sites, particularly when they are using ground rest sites during the summer and autumn (Mech et al. 1966; Shirer and Fitch 1970). Fritzell (1978b) observed raccoons in North Dakota rarely reusing daytime rest sites on successive days. The lack of day site fidelity may partially be a consequence of home-range size and extent of nightly movements; as nightly travels increase in scope, it may become increasingly difficult to return to a previous rest site. Alternatively, if raccoons must return to a particular rest site each day, as may be the case for females with litters, they may have to reduce their movements at night.

However, other studies have reported greater degrees of den site fidelity. Endres and Smith (1993) monitored denning habits of raccoons in Tennessee and reported a range of 22–51 different dens used by individual raccoons over a year. Some individuals repeatedly used a few dens during a season, whereas others used a greater variety of dens. This variation was not associated with particular sex or age classes. At another study area in Tennessee, raccoons of both sexes often reused dens on consecutive days (Rabinowitz and Pelton 1986). Some raccoons will select den sites near their current foraging areas, but others will travel extensive distances each morning to return to particular dens.

Shirer and Fitch (1970) were the first to describe certain den sites that receive use by multiple individuals as the "communal property of the population." These dens can be rock crevices (Shirer and Fitch 1970), tree cavities (Gehrt et al. 1990; Hadidian et al. 1991), or structures such as brush piles (Gehrt and Fritzell 1998b) or buildings. Multiple raccoons use such den sites either sequentially or simultaneously. Such use is not always a result of limited availability of den sites or communal denning during winter. Some of these, such as rock crevices or brush piles, may have multiple entrances or chambers, which facilitate multiple users. On an area with few tree cavities but numerous ground dens, two brush piles situated near a permanent water source were used repeatedly by every member of the local population during the year, regardless of age or sex (Gehrt and Fritzell 1998b). Additionally, certain tree cavities receive similar use, where as many as seven raccoons use the same dens, sometimes simultaneously (Gehrt et al. 1990; Hadidian et al. 1991). These communal dens may be located in opportune locations, possess favorable microclimates, or reflect limited availability for that den type. Also, social factors may influence their use.

Home Range and Movement. Most estimates of home range sizes are from 50 to 300 ha, despite wide variation in sample sizes and monitoring effort. In most populations, average home ranges of males are larger than those of females, although at times individual variation can be substantial. Sizes of home range have a negative relationship with population densities (Ellis 1964; Sherfy and Chapman 1980); thus, both measures reflect the distribution and quality of local resources (see Table 28.2). The largest estimates were summer home ranges reported for raccoons in North Dakota, with an average of 2560 ha for males and 806 ha for females (Fritzell 1978b). The smallest home ranges occur in urban/suburban populations, and include estimates for residential areas in Cincinnati (16 ha for males, 5 ha for females; Hoffmann and Gottschang 1977) and New Jersey (62 ha for males, 94 ha for females; Slate 1985).

Glueck et al. (1988) reported that sizes of home range of raccoons in Iowa decreased during the winter, and that harvest activities probably did not affect movements of either males or females. A similar pattern occurred in Kansas (Gehrt 1988). In South Texas, individual seasonal home ranges varied from 204 to 681 ha for males and from 15 to 315 ha for females, with no difference in average size of home range among seasons (Gehrt and Fritzell 1997). In that study, home ranges of males were larger than predicted based on energetic requirements alone, suggesting that male social behavior and female distribution, in addition to resource distribution, influence the size of home ranges.

Nocturnal movements of raccoons are typically characterized by lengthy periods of small, localized movements at foraging sites interspersed with short periods of extensive movements between sites and dens (Fritzell 1978b; Greenwood 1982). Fritzell (1978b) reported adult males in North Dakota had a mean travel rate of 0.93 km/hr during the summer, which was greater than the 0.39 km/hr for parous females and the 0.36 km/hr for nulliparous females. Similarly, in South Texas, movement rates of males were greater than those of females during spring and autumn (Gehrt and Fritzell 1996b). Maximum distance traveled in one night by a raccoon in North Dakota was 14.5 km by an adult male (Greenwood 1982).

Raccoons forage opportunistically, and distribution of resources affects movement patterns and spatial distribution of the local population (Johnson 1970). It is not uncommon for raccoons to occur in feeding aggregations to exploit locally abundant resources such as berries or other preferred food items. During a study of free-ranging raccoons in Virginia (Seidensticker et al. 1988), an artificial rich food patch was established in a 0.5-ha area to determine how many raccoons would find the patch and how often they would visit it. Twenty-one of 23 radio-collared raccoons inhabiting the study area eventually used the artificial food patch during the 27-day period. The average number of visitors during 3-day trials steadily increased from 7 during the beginning of the period to 14 on the last trial.

Raccoons easily cross water and probably do so regularly in some areas. They are occasionally documented traveling between islands or colonizing new islands. In the Florida Keys, livetrapping revealed movements of 165–645 m across water between islands (Bigler et al. 1981). During extensive flooding in a floodplain in Kansas, raccoons continued to move within their home ranges, even when they were completely inundated with water (Gehrt et al. 1993). The raccoons resided in treetops for at least 2 weeks, until flood waters receded below tree cavities. One exception was a female that swam at least 1.2 km to higher ground.

Some adult raccoons of both sexes exhibit strong spatial fidelity for multiple years, whereas the home ranges of others may slowly drift over the years. I have observed both patterns for males and females. Some adult raccoons in Texas and Illinois continued to use the same

TABLE 28.2. Population densities and mean home range estimates for raccoons in the United States

Location	Density[a] (raccoons/km^2)	Mean Home Range (ha)[b] Females	Males	Reference
North Dakota	0.5–1	806	2560	Fritzell 1978c
Ontario	55.6	—	42	Rosatte et al. 1990
New Jersey	56	94	62	Slate 1985
Ohio	66.7	4	16	Hoffmann and Gottschang 1977
Ohio	17.5	—	88	Urban 1970
Indiana	11.8	264	486	Lehman 1984
Indiana	12.6	99	102	Lehman 1980
Iowa	17.5	79	131	Hasbrouck 1991
Kansas	28.2	107	156	Gehrt 1988
Alabama	11.8	—	93	Johnson 1970
Texas	8.5	79	339	Gehrt 1994, Gehrt and Fritzell 1997

[a]Average densities were determined for multiyear studies.

[b]Home range estimates vary in season and monitoring protocol; most estimates are minimum convex polygons or variations thereof.

dens and spatial areas for at least 3 years. Long-term spatial fidelity may be characteristic of populations in habitats where resources such as food and dens are predictable between years, and annual drift may occur when those distributions change.

Dispersal and Philopatry. Natal dispersal usually takes place during the autumn by raccoons at 0.5 or 1.5 years of age, and sometimes during the spring at approximately 1 year of age (Stuewer 1943a; Butterfield 1944; Urban 1970; Gehrt and Fritzell 1998a). At more northern latitudes, natal dispersal takes place after the first winter (Schneider et al. 1971; Fritzell 1977). However, at southern latitudes, dispersal may take place at any time and individual differences in sexual maturation may explain variations in the timing of dispersal (Gehrt and Fritzell 1998a). Male-biased dispersal has been reported for most studies (Stuewer 1943a; Urban 1970; Fritzell 1977, 1978a; Clark et al. 1989; Gehrt and Fritzell 1998a), although female dispersal also may occur occasionally.

Dispersal distances are difficult to determine and may be underestimated in many studies. Distances up to 33 km have been reported from multiple mark–recapture and radiotelemetry studies (Stuewer 1943a; Fritzell 1977, 1978c; Clark et al. 1989; Gehrt and Fritzell 1998a). Maximum dispersal distance was 10.9 km during a radiotelemetry study in Virginia (Seidensticker et al. 1988). An exceptional distance of 275 km was reported for two males on separate occasions (Lynch 1967; Priewert 1961).

Natal philopatry is female biased and conforms to philopatry by parental consent (Waser and Jones 1983), where the mother shares all or some of her home range with her adult daughters. This spatial tolerance between related females may extend to communal denning and foraging. Fritzell (1978c) reported female philopatry in North Dakota, and for females in South Texas, distances between birth sites and locations 2 years later ranged from 0.11 to 1.38 km (Gehrt and Fritzell 1998a). In populations with high survivorship, natal philopatry will produce matrilines where segments of the local population comprise related females.

FEEDING HABITS

Perhaps no other aspect of raccoon ecology is as well documented as dietary habits. Many studies from throughout the range have reported on the diversity of food items. For example, the annual diet of raccoons in Kansas was conservatively determined at 43 animal species and 33 plant species (Stains 1956). Nearly all studies have suggested that raccoon diets reflect availability of food items, and that raccoons are highly opportunistic at exploiting foods they prefer (Yeager and Elder 1945; Stains 1956; Lotze and Anderson 1979; Greenwood 1981).

In most systems, plant foods constitute the majority of the diet in all seasons, although the relative importance varies seasonally (Kaufmann 1982). During Stuewer's (1943a) study in Michigan, acorns represented the single dominant food item in all seasons, and overall plants constituted the most common food item in all seasons except spring. Animals associated with riparian areas such as crayfish, amphibians, and clams were more common in spring. In landscapes dominated by agriculture, grain crops, especially corn, often replace acorns as important food items. In Kansas, corn is the most important food item during autumn, winter, and spring, and fruits such as grapes (*Vitis* sp.), mulberries (*Morus rubra*), and gooseberries (*Ribes missouriense*) are exploited heavily during the summer (Stains 1956). Animal prey increase in importance during late winter and spring, when plants are relatively limited. A similar pattern was observed in Alabama, where insects increased in importance during the winter and early spring (Johnson 1970).

Although plant foods are important in most systems, in some locations or seasons animal prey are used almost exclusively. In southwest Washington, summer diets of raccoons were exclusively animals, primarily invertebrates (e.g., mollusks, shrimp, crabs) and fish (Tyson 1950). Dorney (1954) observed raccoons regularly preying on muskrats in marshes. Raccoons on an Illinois goose refuge exploited large numbers of waterfowl crippled due to fall hunting; 65% of the posthunting diet was Canada geese (*Branta canadensis;* Yeager and Elder 1945). Curiously, frogs are rarely reported as food items, despite the presumed availability of this potential prey item (Johnson 1970; Kaufmann 1982). In urban areas, raccoons exploit anthropogenic resources such as pet food, bird feed, and human refuse (Kaufmann 1982).

When nesting prey species are concentrated in space and time, raccoons can be significant predators on eggs or nestlings. Examples include seabirds (Kadlec 1971; Hartman et al. 1997), waterfowl (Urban 1970; Greenwood 1981), songbirds (Robinson et al. 1995), and reptiles (Cagle 1949; Burger 1977; Davis and Whiting 1977; Hopkins et al. 1978; Christiansen and Gallaway 1984). Raccoon predation can be particularly acute on islands where they have been introduced. On islands where seabirds form colonies and nest in burrows, raccoons can depredate birds in all stages of life history (Hartman and Eastman 1999).

Most dietary studies of free-ranging raccoons have obtained data from fecal analyses of scats deposited by unidentified raccoons or stomach analyses of dead raccoons. Thus, despite the abundant literature on raccoon feeding habits, certain aspects such as age and sex differences or variation among individuals (Johnson 1970) are poorly understood.

BEHAVIOR

Senses and Intelligence. The well-developed senses and intelligence of raccoons inspired stories among Native Americans and European pioneers (Whitney and Underwood 1952) and have been the subject of scientific queries for nearly a century (summarized by Kaufmann 1982; Sanderson 1987). Raccoons have good hearing capabilities (Cole 1912), with a maximum limit of 85 kHz and high sensitivity range of 50 kHz (Peterson et al. 1960). As is typical of many nocturnal mammals, raccoons are keen to strange noises (Tevis 1947; Sharp and Sharp 1956; Kaufmann 1982). Raccoons are color-blind, but have excellent nighttime vision (Davis 1907; Johnson 1957; Michels et al. 1960). The characteristic eyeshine at night, produced by light reflected by the tapetum behind the retina, and the raccoon's habit of looking at light sources enhance spotlight census surveys.

Olfactory senses are adequate, but tactile senses are what set raccoons apart from most other carnivores. Their sense of touch and manipulative abilities have received considerable study (Davis 1907; Thorgeson 1958; Iwaniuk and Whishaw 1999). The forepaws have an exceptional sense of touch, which is used to identify food items more so than other senses (Cole 1912). The tactile sense, a product of highly developed nerves in the pads (Sanderson 1987) and manipulative abilities of raccoons are more advanced than those of most carnivores. Whitney and Underwood (1952) and Kaufmann (1982) summarized highlights from studies of tactile ability.

Some measures of intelligence have placed raccoons above cats but below primates in their ability to discriminate objects (Cole 1907). Davis (1907) observed that raccoons had quick learning abilities with regard to opening containers, and they retained this knowledge for at least 1 year without reinforcement. Similarly, raccoons were better at learning certain formations than all other tested species except higher primates (Johnson 1957).

Locomotion. The typical gait of raccoons is a lateral sequence, lateral couplet walk, resulting in a characteristic lateral swaying (McClearn 1992). Hind feet are usually plantigrade, whereas front feet are semidigitigrade. At faster speeds, raccoons will use a pacing gait with swing phase, and at top speeds they use a full gallop for short distances. Raccoons are adept at climbing, and are well suited for an arboreal lifestyle. A modified ankle joint enables the hind feet to be reversed for head-first descents from trees. While investigating an object or area, raccoons often stand in a bipedal position. This posture frees the front feet for manipulation of objects and food handling.

The front limb movement of raccoons is more flexible than that of most carnivores (Iwaniuk and Whishaw 1999). Their manipulative ability is more advanced than that of the ringtail and coati, but less than that of the kinkajou and olingo (Iwaniuk and Whishaw 1999). Patting and probing with the front feet are common behaviors. The front feet are highly sensitive and are typically used to locate food. They are not capable of complete flexion, however. If an object is grasped by one hand, it is usually with a scissor grasp of the second and third digits, which is common in mammals (Iwaniuk and Whishaw 1999). Although the bipedal posture is most typical while manipulating an object, raccoons also have been observed sitting with the back feet extended in front of the body, lying on the back with all limbs off the ground, and lying flat on the stomach with all limbs extended on the ground (Iwaniuk and Whishaw 1999).

Feeding and Defecation. Raccoons use olfaction, vision, and especially hapsis to locate, identify, and capture food (Iwaniuk and Whishaw 1999). As noted, the highly sensitive front feet have limited grasping capabilities, and most objects are held or rolled between the two front feet. This is evidenced by the balls of mud often found in a box trap with a captured raccoon. The head is often turned away while using the front feet during foraging (Tevis 1947), which illustrates how important the tactile sense is for locating and identifying food. Ewer (1973) suggested that species with such a well-developed tactile sense and dexterity of the front feet may have evolved from ancestors that foraged on crustaceans or in aquatic areas at night.

Raccoons are capable of defecating anywhere, but they often deposit multiple scats at one location called a latrine. Stains (1956) reported that raccoons are more likely to deposit feces in latrines during late summer, autumn, and winter, but defecate randomly during spring. Latrines were most common during October. However, it is unclear whether changes in food habits or weather conditions affect detectability of latrines. Latrines are usually located near den sites, particularly structures such as barns or attics, fallen logs, or the base of trees (Yeager and Rennels 1943; Kennedy et al. 1991; Page et al. 1998). Latrines also have been found on tree limbs as high as 12 m above the ground. Especially popular locations are fallen trees that serve as bridges across drainages (Stains 1956).

It is unknown what function, if any, latrines may serve, or if they are simply a result of loafing or resting. Most latrines are small and are largely produced by a single individual during a season. However, some latrines become quite large, are probably created by multiple individuals, and are maintained in successive years. Stains (1956) collected over 700 scats from five latrines during a 2-year period, and Page et al. (1998) observed at least three raccoons using the same latrine. Given that defecation may be an important marking behavior in raccoons, communal latrines may serve as information centers for the population (Eisenberg and Kleiman 1972; Sneddon 1991). Page et al. (1999) monitored vertebrate use of raccoon latrines in Indiana, and observed 14 mammal and 15 avian species foraging on undigested seeds within the feces. Small granivorous rodents were especially frequent visitors to the latrines.

Communication. Sieber (1984) recognized 13 different call types from the species' vocalizations, of which 7 involved mothers and young. Some of the calls were given exclusively in the natal den. These included the "purr" uttered by the mother while nursing, a "churr" emitted by young that are nibbled by the mother; a "squeal," which served as a distress call by young; a "grunt," which served as a warning from the mother; and a "screech" produced by all raccoons that were physically threatened. In addition, a "chitter" emitted by the mother and the cub's "whistle" were used to maintain contact after the young become mobile. Mothers did not emit chitters until after parturition, and the frequency of these calls peaked during the first month following excursions of the cubs. Then vocalizations decreased until termination during the third month following the first excursions of the cubs (Seiber 1984).

Analyses of chitters and whistles revealed individually distinct calls, and playback experiments suggested a potential for individual identification. However, Sieber (1984) thought it unlikely that in natural populations different litters might be in close proximity to each other; thus, selection for individual recognition of vocalizations might not occur. Actually, there may be numerous instances where different litters may be in close proximity to each other in natural systems. Multiple litters were born and reared simultaneously in the same brush pile in South Texas (Gehrt and Fritzell 1998a), and multiple families forage in the same areas in urban areas (S. D. Gehrt, unpublished data). Where cubs from multiple litters are in close proximity, individual identification might be important. Since raccoons are more social than previously thought, further analyses of raccoon vocalization may yield other modes of acoustic communication.

Other vocalizations are associated with agonistic interactions, including screams, growls, hissing, and snarls (Sanderson 1987). During threatening situations, raccoons also employ visual displays of dominance such as arching the back, drawing back the lips to bare teeth, laying back the ears, raising and flicking the tail, and raising the hackles (Tevis 1947; Barash 1974; Kaufmann 1982). Kaufmann (1982) described a sideways movement in a series of hops when faced with a threat. Subordinate behavior is exhibited by lowering the chin, ventral body surface, and tail (Barash 1974).

Captive raccoons scent mark (Ough 1982), although the frequency and function of this behavior are essentially unknown. Anal rubbing is one of two types of observed scent marking, and appears to be common. The other type is neck rubbing, where a raccoon tucks its head between the front legs and rubs the dorsal part of the neck and shoulders against an object. Scent marking also may be accomplished through urination and defecation. Social status (Ough 1982), age, and sex may affect the types and frequency of scent marking.

Social Organization. Social behavior of raccoons has remained enigmatic because it is difficult to observe carnivores at night. Descriptions of interactions between raccoons have been largely anecdotal (Tevis 1947) or limited to captive individuals (Barash 1974) and feeding stations (Sharp and Sharp 1956). Most studies of the spacing patterns of adults have reported overlapping home ranges with both sexes and little interaction between individuals. Exceptions occur during feeding aggregations and winter denning. However, there are overt and subtle geographic variations to this theme, especially for males. Whereas females always exhibited some degree of spatial overlap with other females, male spatial patterns varied geographically. On the northern prairies of North Dakota where raccoon densities are very low, males are territorial and maintain intrasexually exclusive home ranges that overlap two to three female home ranges (Fritzell 1978c). Males actively defend these territories from other males. At more hospitable locations with higher population densities, males have overlapping home ranges but apparently exhibit little interaction with conspecifics (Stuewer 1943a; Johnson 1970; Urban 1970; Hoffmann and Gottschang 1977; Seidensticker et al. 1988). The lack of interaction between adjacent raccoons despite extensive spatial overlap, and anecdotal reports of agonistic behavior (Tevis 1947; Sharp and Sharp 1956), led to the concept of local territorialism for males (Johnson 1970; Lotze and Anderson 1979; Kaufmann 1982), where males are intolerant of each other. However, many of these studies were of limited scale and scope, which may have affected subsequent descriptions of behavior (Gehrt 1994). The degree of sociality, or interactions between raccoons, may vary among populations. In Virginia, despite relatively small numbers of radio-collared raccoons, evidence indicated that raccoons were often in close proximity to each other, and concurrent den use was documented for intra- and intersexual dyads (Seidensticker et al. 1988).

On the coastal plains of South Texas, most adult males occurred in spatially distinct groups made up of three to four individuals, in which members frequently denned and traveled together (Gehrt and Fritzell 1998b). These male bonds were stable and group memberships, often made up of different-aged individuals, continued annually. When a group member was lost to mortality or dispersal, an immigrant would take his place. Home ranges of a group usually overlapped multiple female home ranges, and there was little overlap with neighboring groups. Solitary males inhabited areas in which females were absent. Variations of this social system may occur elsewhere, and suggest more social complexity for raccoons than previously reported. Social bonds in which adult males denned and traveled together, but not spatial groups, also were observed for raccoons in Kansas (Gehrt 1988; Gehrt et al. 1990). Similarly, neighbor recognition and dominance systems have been observed for captive males (Barash 1974; Ough 1982) and free-ranging raccoons at a winter feeding station (Sharp and Sharp 1956).

These patterns of intraspecific variation in social systems indicate male social behavior is determined by the spatial orientation of females and their densities. Female distributions reflect the distribution of resources such as food, water, and den sites (Gehrt and Fritzell 1998b). Given the diverse conditions in which raccoons occur, the possibility exists that the full range of intraspecific variation in social structure has yet to be described.

The common social unit among all raccoon populations is the family unit of mothers and young, and the lengths of time families maintain bonds are highly variable within and among populations. There has been speculation that the long winters at northern latitudes and long periods of winter denning prolong family bonds (Johnson 1970; Schneider et al. 1971; Fritzell 1977). However, in South Texas, all families maintained their bonds through most of the winter, and sometimes until parturition (Gehrt and Fritzell 1998a). There may be geographic differences in the duration of familial bonds, but the importance of familial relationships in raccoon populations is probably underestimated. Because of strong natal philopatry, closely related females share the same areas. These matrilineal relationships may explain why some females share den structures and occasionally associate together. Gehrt and Fritzell (1998a) reported three generations of the same matriline simultaneously reared litters in the same brush pile, and rare instances of positive interactions occurred among adult females.

Johnson (1970) suggested that cultural inheritance might explain how raccoons seemingly exploited resources so efficiently. The long-term association with family members and female philopatry may provide the mechanism for cultural transmission. Family bonds persist for months following weaning, during which offspring are undoubtedly learning to exploit ever-changing resources. Philopatry ensures that those individuals will remain and often continue using preferred den sites. Thus, cultural inheritance may explain certain foraging patterns and repeated den use exhibited by local raccoons.

Mating System. By quantifying consortships for a free-ranging population of raccoons in Texas, Gehrt and Fritzell (1999b) determined the mating system is on a continuum between polygyny and promiscuity. Variance in consortship success and seasonal incidence of wounding suggested a male polygynous system based on a dominance hierarchy, where males competed for access to estrous females. Consortship success, the number of estrous females consorted by a male during a mating season, ranged from zero to six females/male. Females consorted with one to four males during their estrous period; however, 62% of the females consorted with only one male during estrus. Body weight was correlated with consortship success for males, and canine width may also be important in determining mating success among males (Ritke 1990a). Canines may be important for fighting and dominance, and wider canines may be less likely to break. The degree of polygyny, however, may have been influenced by the distribution of females and synchrony of estrus. When estrus was highly synchronized, the mating system shifted toward promiscuity, as male variance in consortship success declined and the frequency of multiple consorting of males by females increased. During a radiotelemetry study in Virginia, the mating system resembled promiscuity as every radiotagged male was found in a den with every radiotagged female at least once during the mating season (Seidensticker et al. 1988). Although Fritzell (1978c) did not monitor raccoons during the peak mating season, the territorial spacing system strongly suggested a polygynous mating system, as each male territory exclusively overlapped two to three female home ranges.

MORTALITY

Longevity. Throughout most of their range, the primary mortality factors for raccoons are related to human activities (Lotze and Anderson 1979,; Kaufmann 1982), which are usually harvest, collisions with vehicles, and nuisance control. Other mortality factors include predation, disease, starvation, accidents, and possibly infanticide. Raccoons have died while stuck in crevices or tree cavities and are occasionally bitten by poisonous snakes (personal observation). The oldest estimated age for free-ranging raccoons is 12 years, 7 months (Haugen 1954) from livetrapping, and 13–16 years from tooth analysis (Johnson 1970). The longevity record for captive-reared raccoons is 17 years (Garrett and Goertz 1975). Estimates of average life span include 1.8 years in Missouri (Sanderson 1951) and Manitoba (Cowan 1973) and 3.1 years in Alabama (Johnson 1970). Lotze and Anderson (1979) stated most raccoons live <5 years. However, longevity is largely dependent on harvest intensity or other anthropogenic factors and climate.

Data on juvenile survival are limited and results are variable. Gehrt and Fritzell (1999c) estimated survival rate from 0.52 to 0.65 during the nestling stage for a population in South Texas, where natal den type was important for survival of litters. In Iowa, where most of the natal dens were tree cavities, nestling survival apparently was high (Judson et al. 1994). Postemergent survival of juveniles is highly variable. Judson et al. (1994) reported a 0.65 survival rate of juveniles during postemergence to September in Iowa. In southern populations, postemergence survival is relatively high (Johnson 1970), for example, 0.90 (Gehrt and Fritzell 1999c). In populations with intensive harvest, the juvenile class experiences low (<0.60) survival during autumn and winter (Clark et al. 1989; Hasbrouck et al. 1992). At more northern latitudes, juvenile survival decreases during late winter and early spring as a result of extreme winter conditions (Mech et al. 1968; Cowan 1973). Once young raccoons become yearlings, survival is similar to that of adults except possibly for dispersers.

For adult raccoons in Mississippi, the primary mortality factors were harvest and canine distemper. Annual survival was 0.63 for males and 0.50 for females (Chamberlain et al. 1999). Adult and yearling survival was 0.93 and 0.81, respectively, during the spring and summer in North Dakota (Fritzell and Greenwood 1984). However, survival likely is relatively low during the winter and early spring at northern latitudes. In a harvested population in Iowa, annual survival for adults ranged from 0.45 to 0.69 and juvenile survival ranged from 0.17 to 0.63 during the 6 years of the study (Hasbrouck et al. 1992).

In populations exposed to minimal or no harvest pressure and few roads, adult survival is generally high. Disease is usually the most common cause of natural mortality, and predation, especially for adults, is relatively minor. Average annual survival of adults in a protected population in South Texas was 0.84, with no difference between sexes (Gehrt and Fritzell 1999c). This agrees with Johnson's (1970) assessment of Alabama raccoons that natural mortality factors, excluding disease, have little effect. In urban areas, harvest may be replaced by increased mortality from vehicle collisions, attacks from domestic animals, and disease. The most important causes of mortality in an urban population in Cincinnati were attacks from dogs (33%) and canine distemper (29%) (Schinner and Cauley 1974).

Predators. Species reported as potential predators of raccoons include coyotes (*Canis latrans*), bobcats (*Lynx rufus*), red foxes (*Vulpes vulpes*), owls (multiple species), and alligators (*Alligator mississippiensis*) (Stains 1956; Johnson 1970; Kaufmann 1982; Shoop and Ruckdeschel 1990). Predation by red foxes and owls is probably restricted to young raccoons. Whitney and Underwood (1952) described a mother raccoon successfully defending an offspring from a red fox. Predation is probably not an important cause of mortality in most raccoon populations (Johnson 1970; Kaufmann 1982), especially for adults. Predation was minor for radiomarked raccoons in Iowa and Mississippi (Hasbrouck et al. 1992; Chamberlain et al. 1999), and despite relatively high predator densities, few adult raccoons succumbed to predation in South Texas (Gehrt and Fritzell 1999c). One exception to this pattern was an insular population of coyotes in which raccoons were an important food item (O'Connell et al. 1992).

Parasites and Diseases. Raccoons host a variety of ecto- and endoparasites, and their relative importance in the ecology of raccoons probably depends on geographic location, season, and age and sex of the host. At least 10 species of flea, 1 species of sucking louse, 6 species of chewing louse, and 14 species of tick and mite have been reported for raccoons (summarized in Stains 1956). Raccoons may be reservoirs for Lyme disease through their role as host for the immature tick, *Ixodes scapularis* (Fish and Daniels 1990; Magnarelli et al. 1991).

At least 56 species of endoparasites have been identified in raccoons (Stains 1956; Johnson 1970), including protozoans (4 species), flatworms (22 species), nematodes (27 species), and spiny-headed worms (3 species).

The raccoon roundworm, *Baylisascaris procyonis,* is an important pathogen because it is common in raccoon populations, infects a wide range of intermediate hosts, and is a potentially devastating zoonosis. Raccoons are the host of the adult form, which resides in the gastrointestinal tract and usually is relatively benign to the host. However, *Baylisascaris* larvae, which are transmitted via eggs shed in the feces of infected raccoons, can be pathogenic to a wide variety of intermediate hosts, including over 50 species of mammals and birds (Sheppard and Kazacos 1997). The parasite is common in raccoon populations in many parts of North America (Bafundo et al. 1980; Jones and McGinnes 1983; Snyder and Fitzgerald 1985; Ermer and Fodge 1986), and infection rates are highly variable among populations and seasons. There appears to be a latitudinal gradient to infection rates in raccoons, with higher rates at northern versus southern latitudes (Kazacos and Boyce 1989). Infection rates from populations in the Midwest and Northeast can be as high as 68–82% (Kazacos and Boyce 1989). Infection rates of juvenile raccoons by *B. procyonis* can be over 90% and can be pathogenic.

Because of the accumulation of scats, raccoon latrines may serve as the primary mode of transmission of *B. procyonis* to intermediate hosts, including humans (Page et al. 1999). Especially susceptible to infection are granivorous rodents and birds, which consume the undigested seeds in raccoon feces (Page et al. 1999). Tiner (1953, 1954) estimated that *B. procyonis* infection was responsible for 5% of the natural mortality of *Peromyscus leucopus* in Illinois woodlots. Intermediate host infection by *B. procyonis* from raccoon scats has been suggested as a cause of the decline of woodrats (*Neotoma* sp.) in the eastern and midwestern United States (Birch et al. 1994; also see Chapter 19).

Although *B. procyonis* has not often been diagnosed in humans, it is of concern because of its pathogenicity (Kazacos and Boyce 1989). Among humans, children are most at risk of infection because they are most likely to ingest eggs and they are most susceptible to disease. Prevention is the best precaution against *B. procyonis,* as postinfection treatment is limited.

The list of microparasites associated with raccoons is long and continues to grow. Stains (1956) and Johnson (1970) provided summaries of diseases in raccoons, including tularemia, *Trypanosoma cruzi* (Chagas disease), pseudorabies, *Histoplasma capsulatum* (histoplasmosis), and feline distemper. The following pathogens are particularly notable.

Canine distemper is an enzootic disease in many raccoon populations (Roscoe 1993; Mitchell et al. 1999). The disease can be an important cause of mortality (Mech et al. 1968; Johnson 1970), and epizootics may occur in 4-year cycles (Hoff et al. 1974; Roscoe 1993). There appears to be a seasonal pattern to infection rates, with highest rates in late winter or early spring and lowest rates during summer (Roscoe 1993). Although epizootics usually reduce raccoon abundance, a canine distemper epizootic in an urban area did not appear to negatively affect raccoon populations (Schubert et al. 1998). Hamir et al. (1992) observed testicular lesions in raccoons infected with canine distemper and suggested the disease may cause infertility in survivors.

Leptospirosis is another enzootic disease in many raccoon populations, and seroprevalence for multiple variants has been found, most recently *L. interrogans grippotyphosa* (Mitchell et al. 1999). Raccoons have been implicated in human outbreaks of *Leptospirosis interrogans* (Jackson et al. 1993), *L. pomona* (Galton 1959), and *L. autumnalis* (Galton et al. 1959). Additional pathogens frequently associated with raccoons that have important implications for people and domestic animals include pseudorabies (Wright and Thawley 1980) and toxoplasmosis (Dubey et al. 1992; Hill et al. 1998; Mitchell et al. 1999).

Rabies. Since 1980, raccoon rabies has become one of the most important wildlife diseases in the eastern United States. Historically, raccoons have been the primary terrestrial reservoir for rabies in the southeastern United States. However, the disease has spread dramatically across the eastern coast and now appears to be headed westward. During the 1950s and 1960s, rabies was endemic to raccoons in Florida and Georgia (McLean 1970; Bigler et al. 1973), and raccoon rabies appeared to be restricted to this region. This changed dramatically in 1980 when raccoon rabies was reported in Virginia, apparently the result of the legal translocation of more than 3500 raccoons from Florida for hunting purposes (Jenkins et al. 1988). Rabies cases increased dramatically following the translocation and by 1984 the outbreak had spread to West Virginia, Maryland, Pennsylvania, and the District of Columbia (Beck et al. 1987; Fischman et al. 1992). The front of the epizootic continued to move at an average rate of 40 km/year up the Atlantic coast (Fig. 28.6). The epizootic has reached New York, begun spreading westward into Pennsylvania and Ohio (Uhaa et al. 1992), and has crossed the border into Ontario. In some cases, the spread was unintentionally augmented by trash trucks carrying infected raccoons (Wilson et al. 1997).

During raccoon rabies outbreaks, spillover to other wildlife species can occur, but apparently the raccoon strain is not perpetuated by this spillover. Other wildlife species that have been infected with raccoon rabies include foxes, striped skunks (*Mephitis mephitis*), and other improbable species such as rabbits (*Sylvilagus* sp.), woodchucks (*Marmota monax*), and opossums (*Didelphis virginiana;* Beck et al. 1987). Of most concern to humans are transmissions to domestic pets, especially cats. In Maryland, rabid raccoons exposed an average of 1.6 animals, whereas an average of 8 animals were exposed to each cat that became infected (Fischman et al. 1992). Raccoon rabies is easily spread to domestic cats because many cats are allowed to roam and vaccination rates usually are low. More than 80% of human exposure cases resulted from indirect contact with infected animals, usually through domestic pets, during the raccoon rabies epizootic in Connecticut (Wilson et al. 1997). The story of raccoon rabies illustrates the repercussions that can occur following wildlife translocations, and it appears the final chapter is not yet complete.

AGE ESTIMATION

Age of raccoons has been estimated from fusion of cranial sutures (Junge and Hoffmeister 1980), weight of eye lens (Sanderson 1961a), epiphyscal plate closure, reproductive condition, baculum size (Sanderson 1961b), tooth measurements and wear (Grau et al. 1970), and tooth eruption (Montgomery 1964). Fiero and Verts (1986b) evaluated six age-determination techniques for raccoons in Oregon. They found that dental annuli were not present until after 2 years of age, and this technique also underestimated older ages. In contrast, baculum mass and length, epiphyseal closure, and dental attrition were useful techniques for discriminating age class I (0–14 months) from older age classes. Cranial suture and dental annuli techniques were most useful in classifying older age classes. The pattern of ossification of cranial sutures in Oregon differed from that reported in Illinois. Body weight and external sexual characteristics are useful for discriminating juveniles from adults during the summer and early autumn, but they are not useful later in the year.

ECONOMIC STATUS AND MANAGEMENT

Economic Status. Historically raccoons have been one of the most economically important mammals in North America through their role as furbearer or game species (Sanderson 1987). They are trapped during an autumn–winter harvest season and may be pursued by hunters with dogs in some states. In a few states in the Southeast, pursuit with dogs is permitted during the summer. Because of the large number of raccoons taken annually, the species generates a large revenue through the sale of pelts. For much of the 1900s, raccoon pelts generated the largest revenue among furbearers in North America (Shieff and Baker 1987).

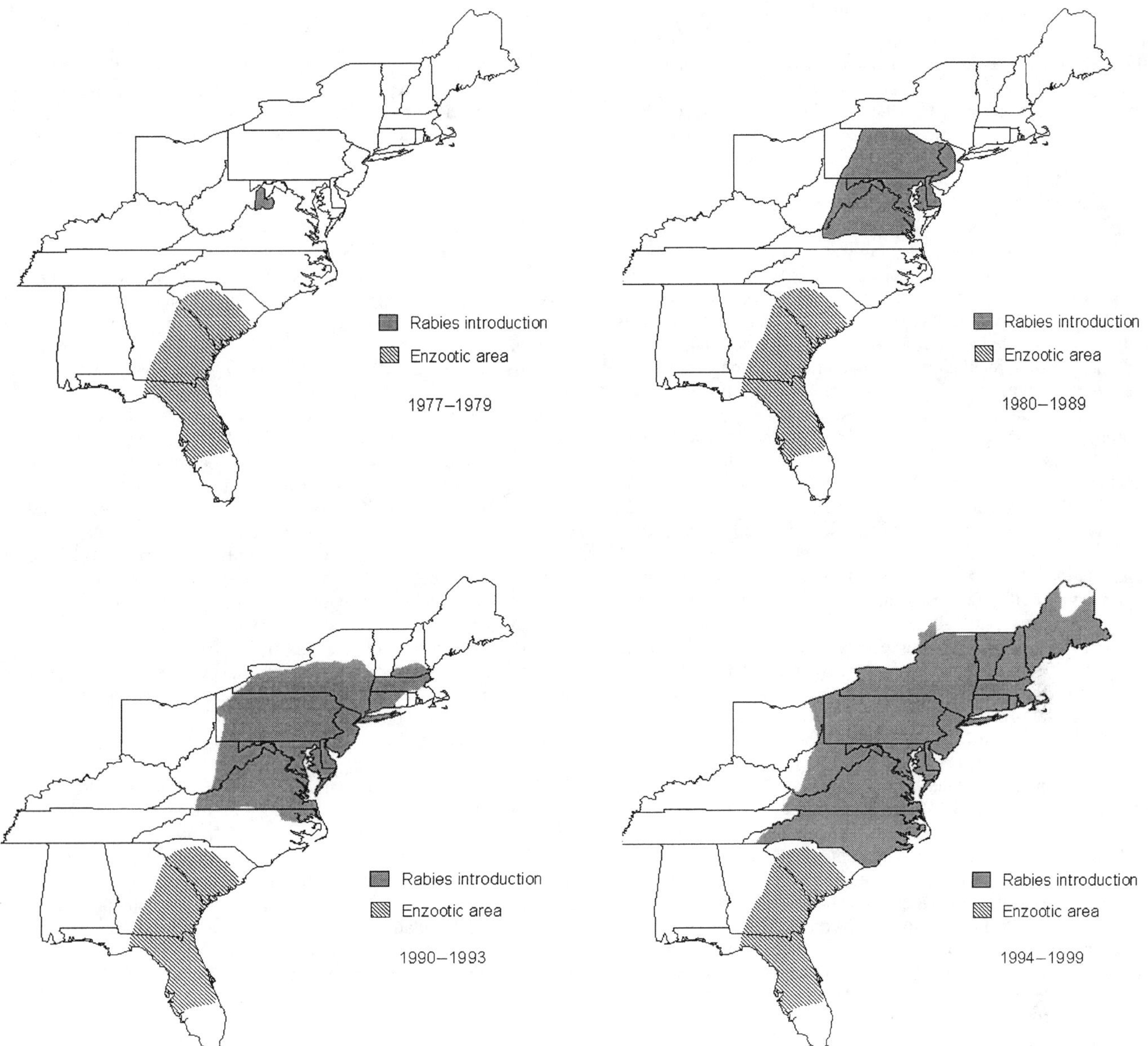

FIGURE 28.6. Geographic distribution of raccoon rabies by year. SOURCE: Data from Centers for Disease Control (2000).

During the late 1980s and 1990s, there was a shift economically from raccoons being an important resource to them causing significant costs in some areas of North America. Raccoons are still economically important in many states, but are an economic drawback as a host of disease, particularly rabies, and as a nuisance species. Because of their ubiquitous distribution and adaptability to human-dominated landscapes, raccoons are a frequently reported nuisance species in urbanized areas (de Almeida 1987). For example, they were the most commonly reported nuisance wildlife species in Illinois during the 1990s. More specifically, over 18,000 raccoons were handled as nuisances in the Chicago region during 1999 (Bluett 2000), at a conservative cost of over $1 million (approximately $60.00 per raccoon) to the public and municipalities.

The raccoon rabies outbreak in the eastern United States has had substantial economic consequences. For example, costs associated with raccoon rabies in two counties in New Jersey were $405,565/100,000 human population during the pre-epizootic period and $979,027/100,000 population during the epizootic period (Uhaa et al. 1992). Cost of rabies control in these counties increased $1.2 million from 1988 (before raccoon rabies) through 1990 when raccoon rabies was introduced to the area. In Connecticut, the estimated number of individuals receiving a postexposure prophylaxis following exposure to the virus increased from 41 in 1990 to 887 in 1994. The median cost of these treatments was $1500 per person (Centers for Disease Control 1996). Similarly, expenditures associated with raccoon rabies in New York increased 4000% during the outbreak from 1989 to 1993 (Wyatt et al. 1999).

Management. Raccoons are an important furbearer in North America based on volume and economic value of annual harvests (Sanderson 1987). As such, harvest has historically been the focus of management of raccoons. Harvest, or the hunting and trapping of raccoons for pelts or sport, is a primary cause of mortality throughout large parts of their range. Harvest intensity is largely determined by pelt prices and market demand. Erickson (1981) compared the effects of various factors on raccoon harvest in Missouri and found that pelt price had the greatest influence on harvest rate. Annual raccoon harvests in the United States from the 1930s through the 1960s fluctuated between 400,000 and

2 million (Sanderson 1987), and pelt prices were usually under $5.00. During the 1970s, pelt prices dramatically increased to a peak of nearly $30.00/pelt, with a concomitant increase in harvest, which reached an all-time high of 5.1 million in 1979–1980 (Sanderson 1987). Although pelt prices and harvests during the 1980s declined from the record high numbers of the previous decade, they remained relatively high until 1988 and 1989. During most of the 1990s, pelt prices decreased to <$10.00/pelt, and harvest also declined accordingly.

There is little evidence that, even during the years of high pelt prices and intensive harvest, harvest has a long-term detrimental impact on raccoon populations. There may be local populations that decline under prolonged intense hunting and trapping, but on a geographic scale harvest does not seem to irretrievably reduce raccoon populations. Indeed, Sanderson (1987) observed that numbers of harvested raccoons have fluctuated over the years as a response to pelt price and weather, and that management regulations have had little effect on populations. Harvest regulations are even less influential when pelt prices are low and there is little hunting or trapping activity.

Although raccoon populations appear to be capable of withstanding varied harvest intensity, the actual effects on populations have been difficult to elucidate. By varying the harvest rate for an Iowa population, Clark et al. (1989) and Hasbrouck et al. (1992) determined that effects varied by sex and age as harvest rates increased. Hasbrouck et al. (1992) found that harvest acts as an additive mortality factor when at least 40% of the population is harvested each year, otherwise it acts as a sequential compensatory mortality factor. Similarly, Sanderson (1987) presented a model that estimated that 49–59% of the total raccoon population in west-central Illinois could be harvested annually without decreasing the population. From a behavioral perspective, harvest activities seem to have little effect on movement patterns or habitat use of raccoons (Glueck et al. 1988; Hodges et al. 2000).

In addition to harvest regulations, translocations have been used to augment populations for sporting purposes or to remove nuisance animals. Raccoons translocated to new areas rarely remain near the release spot (Butterfield 1944; Ellis 1964) and they exhibit extensive movements (Wright 1977; Rosatte and MacInnes 1989; Tabatabai and Kennedy 1989). Thus, attempts to augment local raccoon populations with translocations are generally unsuccessful.

In recent years, the emphasis of raccoon management seems to have shifted from that of propagation and exploitation as a resource to reduction as a predator or nuisance. This is partially because of related effects of increasing urbanization and a decrease in society's interest in harvest activities, particularly trapping (Andelt et al. 1999; Manfredo et al. 1999).

Because of their success at adapting to urbanized landscapes, raccoons are a commonly reported nuisance species in urban and suburban areas (de Almeida 1987). Management of nuisance raccoons in urban areas often involves capturing the offending individuals and relocating them to rural areas or parks. However, monitoring of relocated nuisance raccoons revealed that all but 3 of 16 rural raccoons and 3 of 17 nuisance urban raccoons translocated to a rural forest preserve dispersed within 50 days after release (Mosillo et al. 1999). By comparison, 18 of 20 resident raccoons were still present during the same period. In most cases, the translocated raccoons dispersed within the first few days following release. Three translocated raccoons moved extensive distances of 24, 25, and 60 km from the release site. Some of the translocated raccoons settled near human residences and subsequently were captured as nuisances. Consequently, translocation does not necessarily remedy nuisance problems, but simply transfers the problem elsewhere. Additionally, translocation may help to spread certain diseases or disrupt the dynamics of the local, resident population (Mosillo et al. 1999).

To reduce nuisance problems, prevention should be encouraged through public education programs. Human–raccoon conflicts can be minimized by removing food sources (such as trash) and preventing access to buildings and homes. When problem raccoons need to be removed, trapping should be specific for the problem individuals to reduce nontarget captures. In areas where large numbers of raccoons are captured as nuisances each year, the most effective and safe management is humane euthanasia of problem animals.

The raccoon's role as a predator of nests has received attention from resource managers. To control predation of waterfowl nests by raccoons and other predators, managers have used electrified exclosures, toxicants, and predator removal (Greenwood et al. 1990). However, each technique has limitations, and their efficacy has been less than conclusive (Lagrange et al. 1995; Beauchamp et al. 1996; Cowardin et al. 1998). Similar techniques also have been implemented for conservation of sea turtle nests (Ratnaswamy et al. 1997), where raccoon predation can be particularly acute.

Conditioned taste aversion as a management strategy usually involves lacing artificial eggs with a chemical, such as estrogen, to inhibit raccoon predation (Conover 1989, 1990). However, this technique is not always feasible and has met with limited success. Another concern is the possible negative effects of this technique on nontarget wildlife. Ratnaswamy et al. (1997) found that screening individual turtle nests, though relatively more expensive, was more effective than removal or conditioned taste aversion. However, removal was only conducted for 2 years, and prolonged removal eventually may be effective. The most effective deterrent for raccoon predation may vary from site to site. Ratnaswamy and Warren (1998) pointed out that lethal removal should be carefully considered because it may have additional, unintended ecological consequences. Lethal removal, either by trapping or shooting from boats, has been used to reduce raccoon predation on islands or beaches (Hartman and Eastman 1999). However, effectiveness of this technique has been variable or not assessed at all.

Although removal of raccoons to reduce predation may have intuitive appeal, in most systems predation is a complex ecological process involving multiple species with intricate relationships, so that removal of one species may not have the desired result. It may be that in many areas predation is a symptom of larger problems such as habitat fragmentation or loss, and that removal simply provides a short-term solution but does not address the larger problem. Conditions where removal is warranted are when the viability of a prey species is endangered and predation is a direct threat (e.g., sea turtle nests) or it has been determined that predator abundance has significantly increased and led to unprecedented levels of predation. Removal is a short-term, temporary solution, and every effort should be made to understand perturbations in predator–prey relationships and develop more long-term strategies.

Because of their omnivorous diet and preference for foraging in aquatic or riparian areas, raccoons are often used as bioindicators of environmental pollutants, such as heavy metals, chemicals, and radioactivity (Gaines et al. 2000).

Rabies Management. To prevent the spread of rabies, four strategies are possible: (1) reduction of the host population, (2) use of oral baits laced with rabies vaccines, (3) adoption of trap–vaccinate–release programs, and (4) making no intervention and allowing the disease to run its course.

Large-scale population reduction appears to have been successful for a rabies outbreak in striped skunks in Alberta (Rosatte et al. 1986). However, a generalized removal of raccoons often is not possible and may not be effective, especially in urbanized areas.

Oral vaccination programs are being used to control spread of the virus (Robbins et al. 1998). A vaccinia–rabies glycoprotein recombinant vaccine contained in oral baits is distributed by air over large areas at the edge of the outbreak. A raccoon may become immunized after chewing or ingesting the bait–vaccine. After a rabid raccoon was documented in Ohio near the Pennsylvania border in 1996, a task force made up of federal, state, and local officials began deploying vaccine baits over a 16-km wide buffer zone. By 1998, 355,000 baits containing the vaccine had been disbursed over four counties by the Ohio Department of Health (Farello 2000). However, many factors are involved in the efficacy of this strategy, including bait placement and density, season, and attractiveness of the bait to raccoons.

In areas where oral baiting is not acceptable or where the potential spread of the disease is in a relatively small area, trap–vaccinate–release programs are employed (Rosatte et al. 1992b; Rosatte et al. 1997). This entails trapping resident raccoons, administering a vaccination, and releasing them at the capture sites. In Ontario, this program is used in association with a general reduction of raccoons in a removal zone (Rosatte et al. 1997). Despite these efforts to prevent the spread of infection, raccoon rabies was detected in Ontario in 1999.

Culling and vaccination as control measures for rabies are expensive and may not be effective in North America. Bruggemann (1992) argued the only management necessary is public education and vaccination of pets. Despite high densities of humans and raccoons in the areas of rabies epizootics, there have been no documented cases of humans contracting rabies from raccoons. Public notices and education regarding the importance of reporting sick raccoons, seeking help immediately after exposure to a sick raccoon, and vaccinating pets have effectively prevented human tragedy during rabies outbreaks. A postexposure prophylaxis is highly effective in treating human victims shortly after a bite. In addition, a pre-exposure vaccine series is available for individuals with a high potential for exposure to the virus.

RESEARCH AND MANAGEMENT NEEDS

Most life history characteristics are well described for raccoons in the eastern and midwestern United States. Comparatively little basic information is available for raccoons in the western United States. Large-scale patterns in morphological traits differ between these macroregions, which suggests that selective pressures may differ between these areas. We have little idea of how raccoons cope with the variety of ecosystems in western North America.

Most studies reporting on the social organization of raccoons were not designed for that purpose, which may have influenced subsequent observations or patterns. The full variety of sociospatial systems exhibited by raccoons may not be described. In addition, molecular studies of paternity and mating success have yet to be conducted on free-ranging raccoons.

As the influence of harvest continues to decline over much of North America, it is becoming increasingly important to determine the ecological and economic consequences of the burgeoning raccoon population and possibly develop alternative ways to manage them, particularly in urbanized areas. The efficacy and ecological consequences of current management techniques should continue to be evaluated.

Raccoons have a multifaceted role in most ecological communities that relate to other wildlife, including predator, competitor, and host for a variety of micro and macroparasites. Historically, most research on raccoons has been species specific, owing to its importance as a furbearer, and despite numerous observations on the natural history of raccoons we still know very little about their relationships with other wildlife species.

RINGTAIL

The ringtail, *Bassariscus astutus,* is one of two members of the genus *Bassariscus,* and 14 subspecies are recognized (Hall 1981). Common names include miner's cat, ringtailed cat, and civet. The genus appeared in the Miocene with fossil records in various locations in North America. Current distribution of ringtails extends from Mexico up the west coast to southwest Oregon, eastward through parts of Nevada, Utah, Colorado, and possibly parts of Kansas; and south to western portions of Arkansas and Louisiana (Fig. 28.7). In some of these states, the species is considered rare or uncommon.

Ringtails are the smallest procyonids in North America. The most striking physical characteristic is a long tail, approximately the same length as the head and body, with seven to eight alternating white and dark rings. The distal end of the tail is black tipped, and a white streak breaks the dark rings on the ventral side. The dorsal pelage is gray, brown, or tan and varies geographically (Poglayen-Neuwall and Toweill 1988), whereas the ventral pelage is usually white or gray. The pinnae are large and thin, and the muzzle is small and pointed with prominent vibrissae. The eyes are large and ringed with dark hair, which

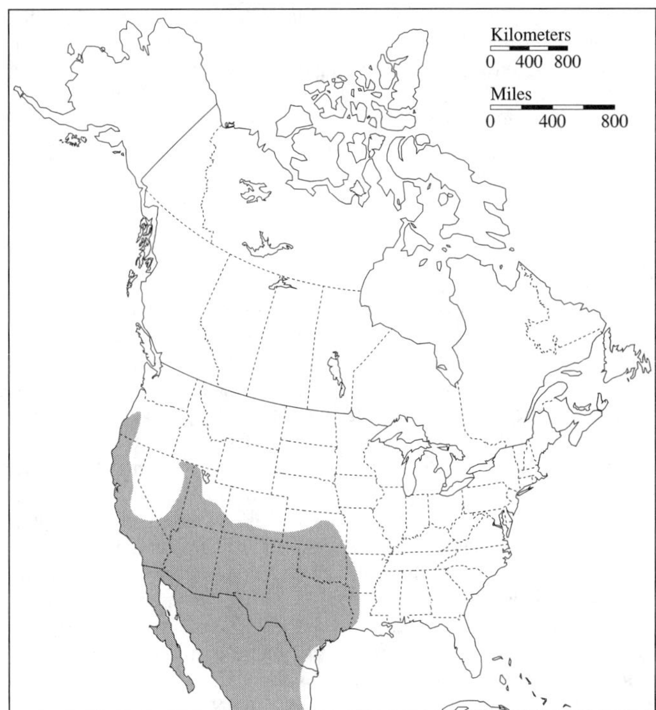

FIGURE 28.7. Distribution of the ringtail (*Bassariscus astutus*).

is surrounded by striking white patches. The body is elongated with relatively short legs and small feet. Feet are semiplantigrade and pentadactyl, with dense hair surrounding the footpad and toes. Tactile sensory hairs on the foreleg are used to detect and assist with the capture of prey (Toweill and Toweill 1978). Claws are short and semiretractile (Hall 1981). Seasonal molt begins in summer and is completed by autumn. Paired glands near the anus (Pocock 1921) can discharge pungent secretions when the ringtail is alarmed (Toweill and Toweill 1978).

Skulls appear to be sexually dimorphic in size and shape (Grinnell et al. 1937). Published dental formulas are inconsistent; Poglayen-Neuwall and Toweill (1988) gave a formula of I 3/3, C 1/1, P 3/4, M 3/2, whereas Kaufmann (1987) described the dentition with P 4/4 and M 2/2. Body size varies geographically, with larger specimens recorded from Mexico and smaller specimens from the southwestern United States (Poglayen-Neuwall and Toweill 1988). The ranges of external measurements (mm) are as follows: total length, 616–811; tail length, 310–438, hind foot, 57–78; and ear, 44–50. Body mass is 870–1100 g, with males slightly larger than females.

Male reproductive systems include paired scrotal testes, epididymides, *vasa deferentia,* and a baculum (Poglayen-Neuwall and Toweill 1988). The female system includes paired ovaries, ovarian bursae, oviducts, a bicornuate uterus, and an *os clitoris* located within the *glans clitoris* (Arata 1965). Typically there are two pairs of mammae, although five and six mammae may occur (Poglayen-Neuwall and Toweill 1988).

Although ringtails depend on open water (Grinnell et al. 1937), the kidneys are modified for water conservation and allow ringtails to metabolize water from succulent fruits and other dietary items (Chevalier 1984). When ringtails are water stressed, they can produce the most concentrated urine of any carnivores (Richards 1976). The normal body temperature is 37.6°C during activity and 23.0°C while inactive (Chevalier 1984; Mugaas et al. 1993). Basal metabolism is 0.43 ml O_2/g/hr, which is relatively low compared to that of raccoons. Mugaas et al. (1993) suggested that the insulative properties of the pelt, in addition to a low metabolism, allow the ringtail to use less energy than expected for thermoregulation at low temperatures. Thus, they can exploit temperate climates.

The mating peak occurs during March and April, but some mating may occur from February into May (Bailey 1974; Poglayen-Neuwall and Poglayen-Neuwall 1980; Poglayen-Neuwall and Toweill 1988). One estrus is typical, although a second estrus may occur if a litter is lost soon after parturition (Kaufmann 1982). The female is receptive to the male for 24–36 hr (Bailey 1974; Poglayen-Neuwall and Toweill 1988). During estrus, the male chases and mounts the female several times per hour. Copulation is frequent and consists of a series of quick thrustings separated by pauses. Each copulation event lasts 1–2 min (Poglayen-Neuwall and Toweill 1988). Pairs may stay together following mating until parturition (Poglayen-Neuwall and Toweill 1988), although to what extent this occurs in natural populations is unknown.

Parturition usually occurs during May or June after a gestation of 51–54 days. As parturition approaches, the female becomes antagonistic toward the male and drives him away (Richardson 1942). A detailed description of parturition is provided by Poglayen-Neuwall and Poglayen-Neuwall (1980). Litter sizes range from one to five, but most are four or less. The young are born with closed eyes and ear canals, limited locomotion, and sparse hair (Poglayen-Neuwall and Toweill 1988). Deciduous teeth emerge at 3–4 weeks of age, and permanent dentition is fully emerged at 17–20 weeks (Toweill and Toweill 1978). At 10 days, rear legs are able to assist with locomotion and the forepaws can grasp objects (Richardson 1942). By 45 days postpartum, nestlings can hold their bodies off the ground and move short distances. The mother consumes the excrement of the young until they begin eating solid food, and by 120 days they develop toilet areas (Richardson 1942; Bailey 1974; Poglayen-Neuwall and Toweill 1988). At 6 weeks, the young begin to resemble adults. They are fully furred and pinnae are erect (Richardson 1942; Toweill and Toweill 1978), and they can walk. Weaning occurs at about 10 weeks, and testes descend at 16 weeks. Young approach adult size at about 30 weeks of age (Richardson 1942; Toweill and Toweill 1978). Vocalizations of the young gradually change from metallic squeaks to a bark at about 10 weeks (Toweill and Toweill 1978). Radiotelemetry studies indicate that only the female rears the young (Trapp 1972; Toweill and Toweill 1978), but Fry (1926) reported the male also assists with provisioning the young. During the weaning process, the young may lick saliva from inside the female's mouth as a means of maintaining fluids as the milk supply diminishes (Poglayen-Neuwall and Poglayen-Neuwall 1980; Chevalier 1984). Young leave the den and begin foraging with the mother between 60 and 100 days postpartum (Fry 1926; Toweill and Toweill 1978; Kaufmann 1982). Ringtails of both sexes typically achieve sexual maturity during their second year; however, some successful matings have been reported for yearlings (Poglayen-Neuwall and Poglayen-Neuwall 1980).

Ringtails occur in a variety of habitats, but are most often associated with rocky outcroppings in semiarid systems and riparian woodlands. Throughout their range, they use woodlands typical of the Southwest, including oak (*Quercus* sp.), pinyon pine (*Pinus* sp.), juniper (*Juniperus* sp.), montane forests, and canyons of chaparral and desert communities (Poglayen-Neuwall and Toweill 1988). They are rarely found far from water (Kaufmann 1982). Ringtails occupy a variety of habitats in California, particularly in riparian areas (Orloff 1988). The only habitats they do not use in California are the northern juniper woodlands and the agricultural portions of the San Joaquin Valley. In Texas, ringtails occur in cedar–oak habitat, live oak (*Quercus virginiana*) savannas, and mesquite. They will occupy buildings and use caves or mines as shelter. They have been observed at high elevations between 1327–2804 m. Water availability and den sites may be critical features of habitat use (Richards 1976).

Ringtails do not construct their own dens, but use those already available. Common den types include rock crevices or boulder piles, but less frequently ringtails will use tree cavities or ground burrows created by other animals. They also will use anthropogenic structures such as brush piles and buildings. They rarely reuse dens except during extreme weather (Toweill and Teer 1980).

Food habits have been reported for ringtails from a variety of areas. In most studies, vertebrates are a consistently high proportion of the diet (Poglayen-Neuwall and Toweill 1988), at least relative to other procyonids. Vertebrate prey include numerous small mammals, lizards, and birds. Other food items include insects, fruit, and carrion (Davis 1960; Taylor 1954; Trapp 1973; Toweill and Teer 1977). In Oregon, plant material was found in 93% of ringtail feces and animal matter in 91% (Alexander et al. 1994). Mammalian remains were found in 66% of the samples, which is higher than frequencies reported for other areas (Toweill and Teer 1977; Trapp 1978).

Population dynamics are poorly understood. Predators include owls, coyotes, raccoons, bobcats, snakes, and domestic dogs and cats (Kaufmann 1982). Ringtails also are vulnerable to vehicles and trappers. The pelt is thin and of poor quality, and thus ringtails are usually captured in traps set for other furbearers. Ringtails are directly or indirectly poisoned through animal damage control efforts.

Ringtails are susceptible to diseases such as feline and canine panleucopenia, rabies, and canine distemper. At least 24 species of ectoparasites have been reported for ringtails (Toweill and Price 1976; Custer and Pence 1979), including fleas (seven species), mites (eight species), lice (one species), and ticks (eight species). Endoparasites include the cestodes *Taenia martis* and *Mesocestoides* and the nemotodes *Pneumospirura bassarisci, Physaloptera, Uncinaria lotoris,* and *Macracantherhynchus ingens* (Price 1928; Pence and Stone 1977; Pence and Willis 1978). Longevity in the wild is unknown; captive animals can live as long as 16.5 years (Poglayen-Neuwall and Toweill 1988).

Most reported population densities have ranged from 1 to 4 individuals/km^2, and appear to vary with habitat and method of estimation. Densities ranged from 0.08 to 2.3/km^2 in the chaparral country in the Sierra Nevada, California (Grinnell et al. 1937). Studies using data from radiotelemetry yielded similar density estimates, including 1.5–2.9/km^2 for pinyon pine/juniper woodland habitat in Utah (Trapp 1978) and 2.2–4.2/km^2 for juniper/oak woodland habitat on the Edwards Plateau, Texas (Toweill and Teer 1980). However, studies in the Central Valley in California using mark–recapture reported densities of 10.5–20.5/km^2 (Belluomini 1983; Belluomini and Trapp 1984). A radiotelemetry study in the same area in California reported estimates of 7–20 individuals/km^2 along shorelines of ponds and sloughs (Lacy 1983). Riparian areas apparently have the highest densities of ringtails in Texas (Toweill and Teer 1980).

Few radiotelemetry studies have been conducted on ringtails, and even fewer collected long-term data that extend across seasons and years. Limited information suggests size and location of ringtail home ranges vary by season and sex. In Texas, home range sizes averaged 43.4 ha for males and 20.3 ha for females (Toweill and Teer 1980). Home ranges averaged 136 ha for males and females combined for a population in Utah (Trapp 1978), a larger value than estimates for Texas ringtails despite shorter monitoring periods. Home ranges of ringtails using riparian habitat in California ranged from 5.0 to 13.8 ha during the summer (Lacy 1983); however, a conservative estimator was used in this study and may have minimized estimated area used by individuals. Ringtails shift their movements with the seasons, and there may be little or no overlap between successive seasonal home ranges. These studies did not report intrasexual overlap of home ranges, although it is unknown whether territoriality is responsible for this spatial system. Home range overlap can be extensive between sexes.

Descriptions of social organization are varied; contradictions may be due to differences in behavior between free-ranging and captive individuals (e.g., Kaufmann and Kaufmann 1963). Free-ranging individuals are largely solitary and rarely share dens (Grinnell et al. 1937), whereas captives of both sexes will share dens at any time of the year (Poglayen-Neuwall and Toweill 1988). Poglayen-Neuwall and Toweill (1988) described the social system, where males and females share space, as based on land tenure as opposed to some bond between them.

Autogrooming consists of licking the forepaws and wiping from behind the ears, over the head, to the muzzle. This behavior may be accompanied by scratching with front or rear feet or grooming with claws or teeth (Poglayen-Neuwall and Toweill 1988).

The vocal repertoire of ringtails includes squeaks, metallic chirps, and whimpers, which are associated with juveniles; chitters, which are associated with stress; chucking and barks, which occur with alarm and defensive threats; and hisses, grunts, growls, and ululations, which are associated with aggression (Bailey 1974; Willey and Richards 1981).

Scent apparently is important for communication. It is distributed through urine rubbed on the ground and on objects (Kaufmann 1982) and accumulations of fecal deposits in latrines (Toweill 1976; Lemoine 1977; Trapp 1978). Urine and fecal deposits may be used as home range markers (Poglayen-Neuwall and Toweill 1988), and their frequency of occurrence increases just before and following the mating season (Bailey 1974).

Locomotion is primarily a steady walk over and around objects, with the tail carried directly behind and slightly off the ground (Grinnell et al. 1937; Trapp 1978). Two types of walks have been described: a high walk with straight legs, and a crouchlike low walk when the animal is cautious or stressed. Other types of locomotion include ricocheting, chimney stemming, and power leaps (Trapp 1972), which are used to maneuver in rocky, steep terrain. Like other procyonids, ringtails are capable of descending head first by rotating the hind foot 180°(Trapp 1972).

Ringtails are managed as furbearers in some states within their range, particularly Texas, New Mexico, and Arizona (Kaufmann 1987). Although their pelts are usually not economically important, they are occasionally caught in traps set for other furbearers. Ringtails are killed incidentally by traps and poison baits used for animal damage control (Kaufmann 1987). Ringtail populations were declining in the 1970s in Colorado (Willey and Richards 1974). Elsewhere, ringtail populations appear to be stable, but little information exists to confirm this. Kaufmann (1987) advocated complete protection in New Mexico and Arizona in the absence of information on population status. Little research has been conducted on ringtails since Kaufmann's (1982) summary, therefore many aspects of the ecology and behavior of free-ranging ringtails remain enigmatic.

COATI

The white-nosed coati (*Nasua narica*) is one of two species of the genus *Nasua,* and comprises four subspecies (Hall 1981; Decker 1991). Common names include chulo, chulo bears, coati, and coatimundi (Gompper 1995). They occur from Central America to southeastern Arizona, southwestern New Mexico, and extreme southern Texas (Fig. 28.8). The species is uncommon or at low abundance throughout its U. S. distribution and especially Texas, where it is rare. Persistent populations occur from the Animas Mountains in southwestern New Mexico to the Baboquivari Mountains in Arizona, with the Gila River representing the northern distributional limit (Kaufmann 1982, 1987). Coatis may occur north of the Gila River, but are probably wanderers or dispersers (Gompper 1995).

White-nosed coatis are intermediate in size between raccoons and ringtails, with adult weights ranging from 4 to 6 kg. Males are slightly larger than females. Characteristic features include a relatively elongated snout and long tails, which are usually held vertical when individuals are on the ground. The ears are small and round with white tips. Feet are plantigrade with nonretractile claws that are longer on the front feet than the rear. Pelage over most of the body is typically a dark brown, but inter- and intrapopulation coloration varies from a sandy brown to almost black (Kaufmann 1987). The tail may have dark rings, or the rings may be indistinct or absent. A splash of white encircles the snout and two streaks continue up to between the eyes, which are accentuated by white eye rings. Molts occur during June–August for males and nonbreeding females in Arizona; reproductive females may shed until November (Gompper 1995). Males possess a baculum (Pocock 1921; Burt 1960; Decker 1991) and females an *os clitoris* (Layne 1954).

The dental formula is the same as for raccoons, but the canines are relatively narrow and the premolars and molars are small with sharp cusps (Kaufmann 1982). The brain differs from that of other procyonids. It has a large sensory cortex region for reception from the snout and a forepaw reception area to receive joint movement rather than tactile stimuli (Gompper 1995). In some ways, the rhinarium of the coati is analogous to the forepaws of the raccoon, where each possess

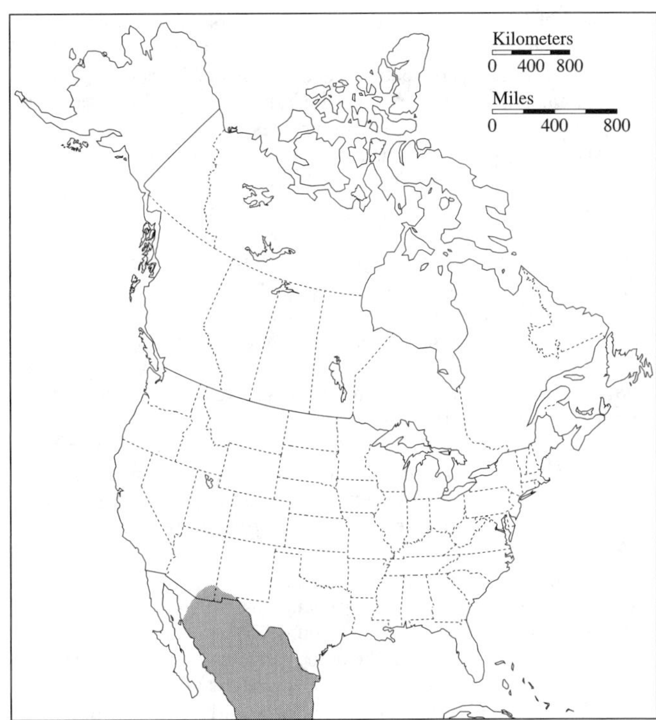

FIGURE 28.8. Distribution of the coati (*Nasua narica*).

a strong tactile sense used to find food. The coati rhinarium has a variety of anatomical modifications that improve its function as a sensory organ (Pocock 1921; Compton 1966, 1973; Barker and Welker 1969; Gompper 1995), including considerable variety in movements and an abundance of sensory receptors at the rostral surface. Coatis often inspect holes or crevices with their snouts and may use the snout to shred litter or old logs (McClearn 1992).

The limbs of coatis are relatively shorter, and front claws relatively larger, than those of raccoons. They reflect adaptations for digging rather than the climbing or manipulation characteristic of raccoons (McClearn 1992). Coatis are capable climbers, but they have difficulties scaling large trees with smooth bark. Locomotion is similar to that of raccoons, with a lateral sequence, lateral couplets gait (McClearn 1992). Coatis may gallop for short distances. As with raccoons and ringtails, coatis usually descend trees head first by rotating the rear ankle joints, although the physical mechanism by which this is accomplished differs among the species (Kaufmann 1982). The normal body temperature is 38.9°C when active and 37.4°C when inactive (Chevalier 1984; Mugaas et al. 1993). Measurements of basal metabolism of white-nosed coatis have ranged from 0.25 to 0.42 ml O_2/g/hr (Mugaas et al. 1993).

Coatis are found in a variety of woodland habitats and riparian areas, often with oaks and sycamores (*Platanus* sp.) dominating. They may also occur in canyons with a mixture of oaks and pines. In Arizona, coatis inhabit piñon–oak–juniper woodlands associated with riparian areas (Kaufmann et al. 1976; Lanning 1976). Water is important to their distribution, and coatis are usually found near streams or creeks.

In contrast to most procyonids, coatis are strongly diurnal (Kaufmann 1982), although males are more nocturnal than females (Valenzuela and Ceballos 2000). Most of their foraging is on the ground, with occasional forays into trees. The diet is omnivorous, although mainly made up of invertebrates and fruit. In addition to insects, coatis use reptiles such as lizards and snakes, rodents, carrion, nuts, and fruits (Ingles 1957; Risser 1963; Kaufmann et al. 1976; Kaufmann 1982, 1987; Russell 1982; Delibes et al. 1989). In tropical areas, the seasonality of fruit abundance appears to be important to coatis because parturition occurs during the peak of fruit production (Smythe 1970; Gompper 1995).

The dichotomous social organization of coatis is unique among carnivores and consequently has received considerable attention from researchers, especially for coatis in Panama. All females and their young ≤2 years old form social groups called "bands," which can range in size from 6 to 38 individuals. Adult males are solitary most of the year, except during the brief mating period, when they temporarily join bands (Kaufmann 1962; Gompper 1995). Bands are largely composed of closely related females, although a few unrelated individuals may be present (Gompper et al. 1997). Within bands are patterns of complex cooperative behaviors such as cooperative grooming, nursing, vigilance, antipredator behavior, and coalition support (Kaufmann 1962; Gompper 1995), which are influenced by degree of relatedness and possibly reciprocal altruism (Russell 1979, 1983; Gompper et al. 1997). Benefits accruing to group members include increased access to food patches, protection from predation, and reduced harassment from males (Russell 1981, 1983; Gompper 1996).

Usually, solitary adult males share the same area with a band and other males. A male avoids interactions or is chased away by band members. Occasionally, there are exceptions where adult males assimilate in groups outside the mating season for short periods of time (Gompper and Krinsley 1992).

Females are philopatric and remain within their natal bands except when bands dissolve or band fission occurs (Kaufmann 1962; Gompper 1997; Gompper et al. 1997). Interestingly, males also are philopatric, so that solitary males sharing the home range with a band are related to band members. They are more closely related to neighboring males than more distant males (Gompper et al. 1998). Adult males temporarily disperse from their home range to pursue matings in other bands and then return to their home ranges following mating (Gompper et al. 1997).

Mating is highly synchronized during a 2- to 4-week period. It usually occurs during April in Arizona (Risser 1963; Kaufmann et al. 1976), which is later than the January–February mating period in Panama (Kaufmann 1962; Smythe 1970; Gompper 1995). Copulation takes place on the ground or in trees. Multiple males may mate within a band in a given year, and consorting males for particular bands may vary among years (Gompper 1995).

Two types of copulatory behavior have been observed in free-ranging coatis. A male and a female may copulate through multiple mounts of short duration (1–6 sec) (Risser 1963; Smith 1980), or copulation may be a prolonged mount for 1 hr or more with periodic thrusting (Hass and Roback 2000). Hass and Roback (2000) suggested variation in types of copulation may be related to social status and different reproductive strategies. Prolonged copulations may be typical of satellite males and females on the preiphery of a group, whereas short, repeated copulations occur with resident males and females within the group.

The gestation period ranges from 70 to 77 days (Gompper 1995), which is longer than that for either raccoons or ringtails. Just before parturition, a female leaves the band and bears the litter alone in a tree or den. The peak parturition period is June in Arizona, April in Panama, and June–August in various locations in Mexico (Gompper 1995). Litter sizes range from one to six (Kaufmann 1962; Risser 1963; Smith 1980). Captive coatis weighed 180 g at birth, with a head–body length of 155–165 mm (Kaufmann 1962). Eyes and ear canals are closed at birth. Eyes open at 4–11 days, deciduous dentition begins to erupt at 15 days, and young begin walking and holding the tail erect at 11 days postpartum (Kaufmann 1962; Risser 1963). Deciduous dentition is complete by 2 months of age. By 9 months of age, males are larger than females and permanent dentition begins to erupt.

Females and their young rejoin the band at about 40 days postpartum. At this point, the band apparently restricts its movements, as the young are small and have difficulty keeping up (Gompper 1995). Females continue to nurse young for up to 3 months following band reaggregation, and some adult females have been observed nursing young other than their own (Russell 1979).

Sexual development differs between males and females. Testes descend at 24 or 25 months of age, when males become solitary. Although they are sexually mature and may compete for breeding opportunities by the time they are 34 months old, matings for these males may be rare because of high intrasexual competition (Gompper 1995). Females first become sexually active at 22 months of age; however, some may not breed until 46 months of age (Kaufmann 1962; Gompper 1995). Females may become postreproductive after 7 years of age, but continue normal band activities despite this senescence (Kaufmann 1962; Russell 1982).

Coatis exhibit "urine rubbing" and penile dragging (Kaufmann 1962), which suggests that such marking is important for social structure. Many types of vocalizations have been described for coatis, including those associated with aggression, alarm, appeasement, and maintaining contact with band members (Kaufmann 1962, 1982; Gompper 1995). Analyses of vocalizations have indicated distinctions between individuals (Krinsley 1989), which may allow individual identification, particularly for contact calls. Visual communication involves nose-up, head-down, and tail-switching displays (Kaufmann 1962, Krinsley 1989). Some behaviors may be learned and culturally transmitted, such as auto- and allogrooming with the resin of *Trattinnickia aspera* (Burseraceae), which may have pharmaceutical properties for coatis (Gompper and Hoylman 1993).

Population densities are highly variable within years, but are lower in the southwestern United States than in the tropics. Reported densities in Arizona have ranged from 1.2 to 2 individuals/km^2 (Risser 1963; Kaufmann et al. 1976; Lanning 1976). The high variation within and among years in Arizona suggested to Kaufmann et al. (1976) that populations in that region may be nomadic or migratory. Estimated densities in tropical areas have ranged from 15 to 70 individuals/km^2 (Kaufmann 1962; Glanz 1991; Gompper 1995). On Barro Colorado Island, Panama, Kaufmann (1962) estimated densities ranging from 26 to 42 individuals/km^2, with a 50% decrease occurring within 8 months. Gompper (1997) estimated the density for the island at 51.5/km^2.

Home ranges vary in size depending on, among other things, geographic location, season, and sex. Size of coati home ranges appears to follow a latitudinal gradient, with the largest home ranges to the north and smaller home ranges at southern latitudes (Table 28.3). On Barro Colorado Island, home range size varied seasonally, with the smallest sizes during the nesting season (Russell 1982). Although band home ranges frequently overlap, the core areas within each home range exhibit little overlap (Kaufmann 1962). Movements of solitary, reproductive females in Arizona included an abrupt shift at the time of parturition from their prepartum home ranges (Ratnayeke et al. 1994). Mean home range size during the postpartum phase was 175 ha.

It is important to note that descriptions of the social organization and genetic analyses of coatis have come almost entirely from the population on Barro Colorado Island, Panama. Geographic differences in population densities and movement patterns suggest coati social structure also may vary geographically, particularly with regard to patterns of dispersal.

Predation appears to be a major limiting factor for coati populations. Documented predators of coatis include felids such as jaguars (*Panthero onca*) (Currier 1983; Rabinowitz and Nottingham 1986; Valenzuela and Ceballos 2000); raptors, including golden

TABLE 28.3. Average home range estimates for coatis

	Bands		Males		
Location	*n*	Range (ha)	*n*	Range (ha)	Reference
Arizona	?	200–300[a]	4	178	Kaufmann et al. 1976; Lanning 1976
Arizona	1	341	1	661	Ratnayeke et al. 1994
Mexico	10	381	6	386	Valenzuela and Ceballos 2000
Panama	3	40	2	41	Kaufmann 1962
Panama	5	36	7	37	Gompper 1997

[a]Individuals were not monitored, a general range was given without sample sizes.

eagles (*Aquila chrysaetos*) and red-tailed hawks (*Buteo jamaicensis*) (Kaufmann et al. 1976; Glanz 1991; Jorgenson and Redford 1993); and in tropical areas, snakes (Janzen 1970) and primates (Fedigan 1990). Infanticide by adult males has been reported (Russell 1981). Mortality rates are particularly high among juveniles, which may be due primarily to predation (Russell 1982). In a tropical population, 20 of 22 juveniles disappeared during May–August (Russell 1982).

Coatis can host numerous endo- and ectoparasites (summarized by Gompper 1995). Coatis are susceptible to rabies and canine distemper virus (Risser 1963; Kaufmann et al. 1976). Other diseases include tuberculosis and salmonellosis (Risser 1963). A canine distemper outbreak may have been responsible for a population crash in Arizona during 1960–1961 (Risser 1963). A major threat to coati populations in North America is indiscriminant killing associated with predator control and harvest (Kaufmann 1987).

In Arizona, coatis were first designated as furbearers in 1948–1949, and until 1969 an unlimited number could be harvested throughout the year. In 1980–1981, coatis were reclassified as nongame animals and harvest was prohibited. This designation continued until 1986–1987, when a season was established and they were listed among "predatory, furbearing, and other mammals."

The coati was listed as a U.S. Forest Service sensitive species in 1990. In New Mexico, it was listed as state endangered in 1979, but was delisted by 1990. In Arizona, little protection has been afforded the coati. The species was listed as state endangered in Texas in 1993. Coati populations have declined dramatically in northern Mexico because of subsistence hunting, consequently coati populations in the United States may be genetically isolated (Kaufmann 1987). Given the tenuous status of coatis in the United States, Kaufmann's (1987) recommendation of complete legal protection for coatis seems warranted. Perhaps the recent restrictions on trapping in Arizona (Andelt et al. 1999) will reduce incidental take of coatis.

LITERATURE CITED

Allen, E. R., and W. T. Neill. 1956. Albinistic sibling raccoons from Florida. Journal of Mammalogy 37:120.

Alexander, L. F., B. J. Verts, and T. P. Farrell. 1994. Diet of ringtails (*Bassariscus astutus*) in Oregon. Northwestern Naturalist 75:97–101.

Andelt, W. F., R. L. Phillips, R. H. Schmidt, and R. B. Gill. 1999. Trapping furbearers: An overview of the biological and social issues surrounding a public policy controversy. Wildlife Society Bulletin 27:53–64.

Arata, A. A. 1965. The *os clitoris* of *Bassariscus*. Journal of Mammalogy 46:523.

Bafundo, K. W., W. E. Wihelm, and M. L. Kennedy. 1980. Geographic variation in helminth parasites from the digestive tract of Tennessee raccoons, *Procyon lotor*. Journal of Parasitology 66:134–39.

Bailey, E. P. 1974. Notes on the development, mating behavior, and vocalization of captive ringtails. Southwestern Naturalist 19:117–19.

Barash, D. P. 1974. Neighbor recognition in two "solitary" carnivores: The raccoon (*Procyon lotor*) and the red fox (*Vulpes vulpes*). Science 185:794–96.

Barker, D. J., and W. I. Welker. 1969. Receptive fields of first-order somatic sensory neurons innervating rhinarium in coati and raccoon. Brain Research 14:367–86.

Baskin, J. A. 1982. Tertiary Procyoninae (Mammalia: Carnivora) of North America. Journal of Vertebrate Paleontology 2:71–93.

Beauchamp, W. D., T. D. Nudds, and R. G. Clark. 1996. Duck nest success declines with and without predator management. Journal of Wildlife Management 60:258–64.

Beck, M. L., and M. L. Kennedy. 1980. Biochemical genetics of the raccoon, *Procyon lotor*. Genetica 54:127–32.

Beck, A. B., S. R. Felser, and L. T. Glickman. 1987. An epizootic of rabies in Maryland, 1982–1984. American Journal of Public Health 77:42–44.

Belluomini, L. A. 1983. Ringtail (*Bassariscus astutus*) distribution and abundance in the Central Valley of California. M.S. Thesis, California State University, Sacramento.

Belluomini, L. A., and G. R. Trapp. 1984. Ringtail distribution and abundance in the Central Valley of California. Pages 906–14 *in* R. E. Warner and K. M. Hendrix, eds. California riparian systems conference. University of California Press, Berkeley.

Berard, E. V. 1952. Evidence of a late birth for the raccoon. Journal of Mammalogy 33:247–48.

Bigler, W. J., R. G. McLean, and H. A. Trevino. 1973. Epizootic aspects of raccoon rabies in Florida. American Journal of Epidemiology 98:326–55.

Bigler, W. J., G. L. Hoff, and A. S. Johnson. 1981. Population characteristics of *Procyon lotor marinus* in estuarine mangrove swamps of southern Florida. Florida Scientist 44:151–57.

Birch, G. L., G. A. Feldhamer, and W. G. Dyer. 1994. Helminths of the gastrointestinal tract of raccoons in southern Illinois with management implications of *Baylisascaris procyonis* occurrence. Transactions of the Illinois Academy of Science 87:165–70.

Bissonnette, T. H., and A. G. Csech. 1938. Sexual photoperiodicity of raccoons on low protein diet and second litters in the same breeding season. Journal of Mammalogy 19:342–48.

Bluett, R. D. 2000. Nuisance wildlife control in Illinois: 1999 summary (Note 00–3). Illinois Department of Natural Resources, Furbearer Program Management.

Bruggemann, E. P. 1992. Rabies in the mid-Atlantic states: Should raccoons be vaccinated? Bioscience 42:694–99.

Burger, J. 1977. Determinants of hatching success in diamondback terrapin, *Malaclemys terrapin*. American Midland Naturalist 97:444–64.

Burt, W. H. 1960. Bacula of North American mammals. Miscellaneous Publications, Museum of Zoology, University of Michigan 113:1–75.

Butterfield, R. T. 1944. Populations, hunting pressure, and movement of Ohio raccoons. Transactions of the North American Wildlife Conference 9:337–44.

Cagle, F. R. 1949. Notes on the raccoon, *Procyon lotor megalodous* Lowery. Journal of Mammalogy 30:45–47.

Centers for Disease Control. 1996. Rabies postexposure prophylaxis, Connecticut, 1990–1994. Morbidity and Mortality Weekly Report 45:232–34.

Centers for Disease Control. 2000. Update: Raccoon rabies epizootic, United States and Canada, 1999. Morbidity and Mortality Weekly Report 49:31–35.

Chamberlain, M. J., K. M. Hodges, B. D. Leopold, and T. S. Wilson. 1999. Survival and cause-specific mortality of adult raccoons in central Mississippi. Journal of Wildlife Management 63:880–88.

Chevalier, C. D. 1984. Water requirements of free-ranging and captive ringtail cats (*B. astutus*) in the Sonoran Desert. M.S. Thesis, Arizona State University, Tempe.

Christiansen, J. L., and B. J. Gallaway. 1984. Raccoon removal, nesting success, and hatchling emergence in Iowa turtles with special reference to *Kinosternon flavescens* (Kinosternidae). Southwestern Naturalist 29:343–48.

Clark, W. R., J. J. Hasbrouck, J. M. Kienzler, and T. F. Glueck. 1989. Vital statistics and harvest of an Iowa raccoon population. Journal of Wildlife Management 53:982–90.

Cole, L. W. 1907. Concerning the intelligence of raccoons. Journal of Comparative Neurological Psychology 17:211–61.

Cole, L. W. 1912. Observations of the senses and instincts of the raccoon. Journal of Animal Behavior 2:299–309.

Compton, R. W. 1966. The skeletal musculature innervated by the facial nerve in the coati (*Nasua*). American Zoologist 6:540(A).

Compton, R. W. 1973. Morphological, physiological, and behavioral studies of the facial musculature of the coati (*Nasua*). Brain, Behavior and Evolution 7:85–126.

Conover, M. R. 1989. Potential compounds for establishing conditioned food aversions in raccoons. Wildlife Society Bulletin 17:430–35.

Conover, M. R. 1990. Reducing mammalian predation on eggs by using a conditioned taste aversion to deceive predators. Journal of Wildlife Management 54:360–65.

Corbet, G. B., and J. E. Hill. 1986. A world list of mammalian species, 2nd ed. British Museum (Natural History), London.

Cowan, W. F. 1973. Ecology and life history of the raccoon (*Procyon lotor hirtus* Nelson and Goldman) in the northern part of its range. Ph.D. Dissertation, University of North Dakota, Grand Forks.

Cowardin, L. M., P. J. Pietz, J. T. Lokemoen, H. T. Sklebar, and G. A. Sargeant. 1998. Response of nesting ducks to predator exclosures and water conditions during drought. Journal of Wildlife Management 62:152–63.

Crawford, B. T. 1950. Some specific relationships between soils and wildlife. Journal of Wildlife Management 14:115–23.

Currier, M. J. P. 1983. *Felis concolor*. Mammalian Species 200:1–7.

Custer, J. W., and D. B. Pence. 1979. Ectoparasites of the ringtail, *Bassariscus astutus*, from West Texas. Journal of Medical Entomology 15:132–33.

Davis, G. E., and M. C. Whiting. 1977. Loggerhead sea turtle nesting in Everglades National Park, Florida, USA. Herpetologica 33:18–28.

Davis, H. B. 1907. The raccoon: A study in intelligence. American Journal of Psychology 18:447–89.

Davis, W. B. 1960. The mammals of Texas. Bulletin of the Texas Game and Fish Commission 41:1–267.

de Almeida, M. H. 1987. Nuisance furbearer damage control in urban and suburban areas. Pages 996–1006 *in* M. Novak, J. A. Baker, M. E. Obbard, and B. Malloch, eds. Wild furbearer management and conservation in North America. Ontario Trappers Association, North Bay, Canada.

Decker, D. M. 1991. Systematics of the coatis, genus *Nasua* (Mammalia: Procyonidae). Proceedings of the Biological Society of Washington 104:370–86.

Decker, D. M., and W. C. Wozencraft. 1991. Phylogenetic analysis of Recent procyonid genera. Journal of Mammalogy 72:42–55.

Delibes, M., L. Hernandez, and F. Hiraldo. 1989. Comparative food habits of three carnivores in Western Sierra Madre, Mexico. Zeitschrift für Säugetierkunde 54:107–10.

Derting, T. L. 1996. Changes in gastrointestinal characteristics of an omnivorous species, the raccoon, with lactation and season. Journal of Mammalogy 77:440–48.

Dew, R. D., and M. L. Kennedy. 1980. Genic variation in the raccoon *Procyon lotor*. Journal of Mammalogy 61:697–702.

Dorney, R. S. 1953. Some unusual juvenile raccoon weights. Journal of Mammalogy 34:122–23.

Dorney, R. S. 1954. Ecology of marsh raccoons. Journal of Wildlife Management 18:217–25.

Dubey, J. P., A. N. Hamir, C. A. Hanlon, and C. E. Rupprecht. 1992. Prevalence of *Toxoplasma gondii* infection in raccoons. Journal of the American Veterinary Medical Association 200:534–36.

Dunn, J. P., and J. A. Chapman. 1983. Reproduction, physiological responses, age structure, and food habits of raccoon in Maryland, USA. Zeitschrift für Säugetierkunde 48:161–75.

Eisenberg, J. F., and D. G. Kleiman. 1972. Olfactory communication in mammals. Annual Review of Ecology and Systematics 3:1–32.

Ellis, R. J. 1964. Tracking raccoons by radio. Journal of Wildlife Management 28:363–68.

Endres, K. M., and W. P. Smith. 1993. Influence of age, sex, season and availability on den selection by raccoons within the Central Basin of Tennessee. American Midland Naturalist 129:116–31.

Erickson, D. W. 1981. Furbearer harvest mechanics: An examination of variables influencing fur harvests in Missouri. Pages 1469–91 *in* J. A. Chapman and D. Pursely, eds. Worldwide furbearer conference proceedings. Wildlife Administration, Annapolis, MD.

Ermer, E. M., and J. A. Fodge. 1986. Occurrence of the raccoon roundworm in raccoons in western New York. New York Fish and Game Journal 33:58–61.

Ewer, R. F. 1973. The carnivores. Cornell University Press, Ithaca, NY.

Farello, C. A. 2000. Prediction of rabies spread in urban Illinois raccoons and evaluation of control strategies. M.S. thesis, University of Illinois, Urbana-Champaign.

Fedigan, L. M. 1990. Vertebrate predation in *Cebus capucinus:* Meat eating in a neotropical monkey. Folia Primatologica 54:196–205.

Fiero, B. C., and B. J. Verts. 1986a. Age-specific reproduction in raccoons in northwestern Oregon. Journal of Mammalogy 67:169–72.

Fiero, B. C., and B. J. Verts. 1986b. Comparison of techniques for estimating age in raccoons. Journal of Mammalogy 67:392–95.

Fischman, H. R., J. T. Horman, and E. Israel. 1992. Epizootic of rabies in raccoons in Maryland. Journal of the American Veterinary Medical Association 201:1883–86.

Fish, D., and T. J. Daniels. 1990. The role of medium-sized mammals as reservoirs of *Borrelia burgdorferi* in southern New York. Journal of Wildlife Diseases 26:339–45.

Folk, G. E., Jr., K. B. Coady, and M. A. Folk. 1968. Physiological observations on raccoons in winter. Proceedings of the Iowa Academy of Science 75:301–5.

Fritzell, E. K. 1977. Dissolution of raccoon sibling bonds. Journal of Mammalogy 58:427–28.

Fritzell, E. K. 1978a. Reproduction of raccoons (*Procyon lotor*) in North Dakota. American Midland Naturalist 100:253–56.

Fritzell, E. K. 1978b. Habitat use by prairie raccoons during the waterfowl breeding season. Journal of Wildlife Management 42:118–27.

Fritzell, E. K. 1978c. Aspects of raccoon (*Procyon lotor*) social organization. Canadian Journal of Zoology 56:260–71.

Fritzell, E. K., and R. J. Greenwood. 1984. Mortality of raccoons in North Dakota. Prairie Naturalist 16:1–4.

Fritzell, E. K., G. F. Hubert, Jr., B. E. Meyen, and G. C. Sanderson. 1985. Age-specific reproduction in Illinois and Missouri raccoons. Journal of Wildlife Management 49:901–5.

Fry, W. 1926. The California ring-tailed cat. California Game and Fish 12:77–78.

Gaines, K. F., C. G. Lord, C. S. Boring, I. L. Brisbin, Jr., M. Gochfeld, and J. Burger. 2000. Raccoons as potential vectors of radionuclide contamination to human food chains from a nuclear industrial site. Journal of Wildlife Management 64:199–208.

Galton, M. M. 1959. The epidemiology of Leptospirosis in the United States. Public Health Report 74:141–48.

Galton, M. M., N. Hirschberg, R. W. Menges, M. P. Hines, and R. Habermann. 1959. An investigation of possible wild animal hosts of leptospires in the area of the "Fort Bragg Fever" outbreaks. American Journal of Public Health 49:1343–48.

Gander, F. F. 1928. Period of gestation in some American mammals. Journal of Mammalogy 9:75.

Garrett, H., and J. Goertz. 1975. Longevity record of a captive raccoon (*Procyon lotor*). Transactions of the Missouri Academy of Science 9:44–45.

Gehrt, S. D. 1988. Movement patterns and related behavior of the raccoon, *Procyon lotor,* in east-central Kansas. M.S. Thesis, Emporia State University, Emporia, KS.

Gehrt, S. D. 1994. Raccoon social organization in south Texas. Ph.D. Dissertation, University of Missouri, Columbia.

Gehrt, S. D., and E. K. Fritzell. 1996a. Second estrus and late litters in raccoons. Journal of Mammalogy 77:388–93.

Gehrt, S. D., and E. K. Fritzell. 1996b. Sex-biased response of raccoons (*Procyon lotor*) to live traps. American Midland Naturalist 135:23–32.

Gehrt, S. D., and E. K. Fritzell. 1997. Sexual differences in home ranges of raccoons. Journal of Mammalogy 78:921–31.

Gehrt, S. D., and E. K. Fritzell. 1998a. Duration of familial bonds and dispersal patterns for raccoons in South Texas. Journal of Mammalogy 79:859–72.

Gehrt, S. D., and E. K. Fritzell. 1998b. Resource distribution, female home range dispersion and male spatial interactions: Group structure in a solitary carnivore. Animal Behaviour 55:1211–27.

Gehrt, S. D., and E. K. Fritzell. 1999a. Growth rates and intraspecific variation in body weights of raccoons (*Procyon lotor*) in southern Texas. American Midland Naturalist 141:19–27.

Gehrt, S. D., and E. K. Fritzell. 1999b. Behavioural aspects of the raccoon mating system: Determinants of consortship success. Animal Behaviour 57:593–601.

Gehrt, S. D., and E. K. Fritzell. 1999c. Survivorship of a nonharvested raccoon population in south Texas. Journal of Wildlife Management 63:889–94.

Gehrt, S. D., D. L. Spencer, and L. B. Fox. 1990. Raccoon denning behavior in eastern Kansas as determined from radio-telemetry. Transactions of the Kansas Academy of Sciences 93:71–78.

Gehrt, S. D., L. B. Fox, and D. L. Spencer. 1993. Locations of raccoons during flooding in eastern Kansas. Southwestern Naturalist, 38:404–6.

Glanz, W. E. 1991. Mammalian densities at protected versus hunted sites in Central Panama. Pages 163–173 *in* J. G. Robinson and K. H. Redford, eds. Neotropical wildlife use and conservation. University of Chicago Press, Illinois.

Glueck, T. F. 1985. Demography of an exploited raccoon population in Iowa. M.S. Thesis, Iowa State University, Ames.

Glueck, T. F., W. R. Clark, and R. D. Andrews. 1988. Raccoon movement and habitat use during the fur harvest season. Wildlife Society Bulletin 16:6–11.

Goldman, E. A. 1950. Raccoons of North and Middle America. North America Fauna 60:1–153.

Gompper, M. E. 1995. *Nasua narica*. Mammalian Species 487:1–10.

Gompper, M. E. 1996. Sociality and asociality in white-nosed coatis (*Nasua narica*): Foraging costs and benefits. Behavioral Ecology 7:254–63.

Gompper, M. E. 1997. Population ecology of the white-nosed coati (*Nasua narica*) on Barro Colorado Island, Panama. Journal of Zoology (London) 241:441–55.

Gompper, M. E., and A. M. Hoylman. 1993. Grooming with *Trattinnickia* resin: Possible pharmaceutical plant use by coatis in Panama. Journal of Tropical Ecology 9:533–40.

Gompper, M. E., and J. S. Krinsley. 1992. Variation in social behavior of adult male coatis (*Nasua narica*) from Panama. Biotropica 24:216–19.

Gompper, M. E., J. L. Gittleman, and R. K. Wayne. 1997. Genetic relatedness, coalitions, and social behaviour of white-nosed coatis (*Nasua narica*). Animal Behaviour 53:781–97.

Gompper, M. E., J. L. Gittleman, and R. K. Wayne. 1998. Dispersal, philopatry, and genetic relatedness in a social carnivore: Comparing males and females. Molecular Ecology 7:157–63.

Grau, G. A., G. C. Sanderson, and J. P. Rogers. 1970. Age determination of raccoons. Journal of Wildlife Management 34:364–72.

Greenwood, R. J. 1981. Foods of prairie raccoons during the waterfowl nesting season. Journal of Wildlife Management 45:754–60.

Greenwood, R. J. 1982. Nocturnal activity and foraging of prairie raccoons (*Procyon lotor*) in North Dakota. American Midland Naturalist 107:238–43.

Greenwood, R. J., P. M. Arnold, and B. G. McGuire. 1990. Protecting duck nests from mammalian predators with fences, traps, and a toxicant. Wildlife Society Bulletin 18:75–82.

Grinnell, J. J., J. B. Dixon, and J. M. Linsdale. 1937. Fur-bearing mammals of California. University of California Press, Berkeley.

Hadidian, J., D. A. Manski, and S. Riley. 1991. Daytime resting site selection in an urban raccoon population. Pages 39–45 *in* L. W. Adams and D. L. Leedy, eds. Proceedings of the 2nd national symposium on urban wildlife, Cedar Rapids, Iowa. National Institute for Urban Wildlife, Columbia, MD.

Hall, E. R. 1981. The mammals of North America, 2nd ed. John Wiley, New York.

Hamilton, M. J., and M. L. Kennedy. 1987. Genic variability in the raccoon *Procyon lotor*. American Midland Naturalist, 118:266–74.

Hamir, A. N., N. Raju, C. Hable, and C. E. Rupprecht. 1992. Retrospective study of testicular degeneration in raccoons with canine distemper infection. Journal of Veterinary Diagnostic Investigations 4:159–63.

Hartman, L. H., and D. S. Eastman. 1999. Distribution of introduced raccoons *Procyon lotor* on the Queen Charlotte Islands: Implications for burrow-nesting seabirds. Biological Conservation 88:1–13.

Hartman, L. H., A. J. Gaston, and D. S. Eastman. 1997. Raccoon predation on ancient murrelets on East Limestone Island, British Columbia. Journal of Wildlife Management 61:377–88.

Hasbrouck, J. J. 1991. Demographic responses of raccoons to varying exploitation rates. Ph.D. Dissertation, Iowa State University, Ames.

Hasbrouck, J. J., W. R. Clark, and R. D. Andrews. 1992. Factors associated with raccoon mortality in Iowa. Journal of Wildlife Management 56:693–99.

Hass, C. C., and J. F. Roback. 2000. Copulatory behavior of white-nosed coatis. Southwestern Naturalist 45:329–31.

Haugen, O. L. 1954. Longevity of the raccoon in the wild. Journal of Mammalogy 35:439.

Hill, R. E., J. J. Zimmerman, R. W. Wills, S. Patton, and W. R. Clark. 1998. Seroprevalence of antibodies against *Toxoplasma gondii* in free-ranging mammals in Iowa. Journal of Wildlife Diseases 34:811–15.

Hodges, K. M., M. J. Chamberlain, and B. D. Leopold. 2000. Effects of summer hunting on ranging behavior of adult raccoons in central Mississippi. Journal of Wildlife Management 64:194–98.

Hoff, G. L., W. J. Bigler, S. J. Proctor, and L. P. Stallings. 1974. Epizootic of canine distemper virus infection among urban raccoons and gray foxes. Journal of Wildlife Diseases 10:423–28.

Hoffmann, C. O., and J. L. Gottschang. 1977. Numbers, distribution, and movements of a raccoon population in a suburban residential community. Journal of Mammalogy 58:623–36.

Hopkins, S. R., T. M. Murphy, Jr., K. B. Stansell, and P. M. Wilkinson. 1978. Biotic and abiotic factors affecting nest mortality in the Atlantic loggerhead turtle. Proceedings of the Annual Conference of the Southeastern Association of Fish and Wildlife Agencies 32:213–23.

Ingles, L. G. 1957. Observations on behavior of the coatimundi. Journal of Mammalogy 38:263–64.

Ivey, R. D. 1948. The raccoon in the salt marshes of northeastern Florida. Journal of Mammalogy 29:290–91.

Iwaniuk, A. N., and I. Q. Whishaw. 1999. How skilled are the skilled limb movements of the raccoon (*Procyon lotor*)? Behavioural Brain Research 99:35–44.

Jackson, L. A., A. F. Kaufman, W. G. Adams, M. B. Phelps, C. Andreasen, C. W. Langkop, B. J. Francis, and J. D. Wenger. 1993. Outbreak of leptospirosis associated with swimming. Pediatric Infectious Disease Journal 12:48–54.

Janzen, D. H. 1970. Altruism by coatis in the face of predation by *Boa constrictor*. Journal of Mammalogy 51:387–89.

Jenkins, S. R., B. D. Perry, and W. G. Winkler. 1988. Ecology and epidemiology of raccoon rabies. Reviews of Infectious Diseases 10:620–25.

Johnson, A. S. 1970. Biology of the raccoon (*Procyon lotor varius* Nelson and Goldman) in Alabama (Bulletin 402). Auburn University Agricultural Experiment Station.

Johnson, J. I., Jr. 1957. Studies of visual discrimination by raccoons. Ph.D. Dissertation, Purdue University, West Lafayette, IN.

Jones, E. J., and B. S. McGinnes. 1983. Distribution of adult *Baylisascaris procyonis* in raccoons from Virginia. Journal of Parasitology 69:653.

Jorgenson, J. P., and K. H. Redford. 1993. Humans and big cats as predators in the Neotropics. Symposia of the Zoological Society of London 65:367–90.

Judson, J. J., W. R. Clark, and R. D. Andrews. 1994. Post-natal survival of raccoons in relation to female age and denning behavior. Journal of Iowa Academy of Sciences 101:24–27.

Junge, R., and D. F. Hoffmeister. 1980. Age determination in raccoons from cranial suture obliteration. Journal of Wildlife Management 44:725–29.

Junge, R. E., and G. C. Sanderson. 1982. Age related reproductive success of female raccoons. Journal of Wildlife Management 46:527–29.

Kadlec, J. A. 1971. Effects of introducing foxes and raccoons on herring gull colonies. Journal of Wildlife Management 35:625–36.

Kaufmann, J. H. 1962. Ecology and social behavior of the coati *Nasua narica*, on Barro Colorado Island, Panama. University of California Publications of Zoology 60:95–222.

Kaufmann, J. H. 1982. Raccoon and allies. Pages 567–85 *in* J. A. Chapman and G. A. Feldhamer, eds. Wild mammals of North America: Biology, management, and economics. Johns Hopkins University Press, Baltimore.

Kaufmann, J. H. 1987. Ringtail and coati. Pages 501–8 *in* M. Novak, J. A. Baker, M. E. Obbard, and B. Malloch, eds. Wild furbearer management and conservation in North America. Ontario Ministry of Natural Resources, Toronto.

Kaufmann, J. H., and A. Kaufmann. 1963. Some comments on the relationships between field and laboratory studies of behaviour, with special reference to coatis. Animal Behaviour 11:464–69.

Kaufmann, J. H., D. V. Lanning, and S. E. Poole. 1976. Current status and distribution of the coati in the United States. Journal of Mammalogy 57:621–37.

Kazacos, K. R., and W. M. Boyce. 1989. *Baylisascaris* larva migrans. Journal of the American Veterinary Medical Association 195:894–903.

Kennedy, M. L., and S. L. Lindsay. 1984. Morphologic variation in the raccoon *Procyon lotor*, and its relationship to genic and environmental variation. Journal of Mammalogy 65:195–205.

Kennedy, M. L., G. D. Baumgardner, M. E. Cope, F. R. Tabatabai, and O. S. Fuller. 1986. Raccoon (*Procyon lotor*) density as estimated by the census-assessment line technique. Journal of Mammalogy 67:166–68.

Kennedy, M. L., J. P. Nelson, Jr., F. W. Weckerly, D. W. Sugg, and J. C. Stroh. 1991. An assessment of selected forest factors and lake level in raccoon management. Wildlife Society Bulletin 19:151–54.

Koopman, K. F., M. K. Hecht, and E. Ledecky-Janecek. 1957. Notes on the mammals of the Bahamas with special reference to the bats. Journal of Mammalogy 38:164–74.

Krinsley, J. S. 1989. An ethogram of the white-nosed coati (*Nasua nasua narica*). M.S. Thesis, University of Wisconsin, Madison.

Lacy, M. K. 1983. Home range size, intraspecific spacing and habitat preference of ringtails (*Bassariscus astutus*) in a riparian forest in California. M.S. Thesis, University of California, Sacramento.

Ladine, T. A. 1997. Activity patterns of co-occurring populations of Virginia opossums (*Didelphis virginiana*) and raccoons (*Procyon lotor*). Mammalia 61:345–54.

Lagrange, T. G., J. L. Hansen, R. D. Andrews, A. W. Hancock, and J. M. Kienzler. 1995. Electric fence predator exclosure to enhance duck nesting: A long-term case study in Iowa. Wildlife Society Bulletin 23:261–66.

Lanning, D. V. 1976. Density and movements of the coati in Arizona. Journal of Mammalogy 57:609–11.

Layne, J. N. 1954. The *os clitoridis* of some North American Sciuridae. Journal of Mammalogy 35:357–67.

Leberg, P. L., and M. L. Kennedy. 1988. Demography and habitat relationships of raccoons in western Tennessee. Proceedings of the Annual Conference of Southeastern Association of Fish and Wildlife Agencies 42:272–82.

Leger, F. 1999. Le raton-laveur en France. Bulletin Mensuel de l'Office National de la Chasse 241:16–37.

Lehman, L. E. 1968. September birth of raccoons in Indiana. Journal of Mammalogy 49:126–27.

Lehman, L. E. 1980. Raccoon population ecology on the Brownstone Ranger District, Hoosier National Forest (Pittman-Robertson Bulletin No. 11). Indiana Department of Natural Resources, Indianapolis.

Lehman, L. E. 1984. Raccoon density, home range, and habitat use on south-central Indiana farmland (Pittman-Robertson Bulletin No. 15). Indiana Department of Natural Resources, Indianapolis.

Lemoine, J. 1977. Some aspects of ecology and behavior of ringtails (*Bassariscus astutus*) in St. Helena, California. M.S. Thesis, Antioch College, Yellow Springs, OH.

Llewellyn, L. M., and R. K. Enders. 1954. Trans-uterine migration in the raccoon. Journal of Mammalogy 35:439.

Lotze, J., and S. Anderson. 1979. *Procyon lotor*. Mammalian Species 119:1–8.

Lutz, W. 1984. Die Verbreitung des Waschbaeren (*Procyon lotor*, L. 1758) im mitteleuropaeischen Raum. Zeitschrift für Jagdwissenschaften 30:218–28.

Lutz, W. 1995. Occurrence and morphometrics of the raccoon *Procyon lotor* L. in Germany. Annales Zoologici Fennici 32:15–20.

Lynch, G. M. 1967. Long-range movement of a raccoon in Manitoba. Journal of Mammalogy 48:659–60.

Lynch, G. M. 1971. Raccoons increasing in Manitoba. Journal of Mammalogy 52:621–22.

Lynch, G. M. 1974. Some den sites of Manitoba raccoons. Canadian Field-Naturalist 88:494–95.

MacDonald, S. O., and J. A. Cook. 1996. The land mammal fauna of southeast Alaska. Canadian Field-Naturalist 110:571–98.

Magnarelli, L. A., J. H. Oliver, Jr., H. J. Hutcheson, and J. F. Anderson. 1991. Antibodies to *Borrelia burgdorferi* in deer and raccoons. Journal of Wildlife Diseases 27:562–68.

Manfredo, M. J., C. L. Pierce, D. Fulton, J. Pate, and B. R. Gill. 1999. Public acceptance of wildlife trapping in Colorado. Wildlife Society Bulletin 27:499–508.

Martin, L. D. 1989. Fossil history of the terrestrial Carnivora. Pages 536–68 *in* J. L. Gittleman, ed. Carnivore behavior, ecology, and evolution. Cornell University Press, Ithaca, NY.

McClearn, D. 1992. Locomotion, posture, and feeding behavior of kinkajous, coatis, and raccoons. Journal of Mammalogy, 73:245–61.

McKeever, S. 1958. Reproduction in the raccoon in the southeastern United States. Journal of Wildlife Management 22:211.

McKinley, D. 1959. Historical note on the Bahama raccoon. Journal of Mammalogy 40:248–49.

McLean, R. G. 1970. Wildlife rabies in the United States: Recent history and current concepts. Journal of Wildlife Diseases 6:229–35.

Mech, L. D., and F. J. Turkowski. 1966. Twenty-three raccoons in one winter den. Journal of Mammalogy 47:529–30.

Mech, L. D., J. R. Tester, and D. W. Warner. 1966. Fall daytime resting habits of raccoons as determined by telemetry. Journal of Mammalogy 47:450–66.

Mech, L. D., D. M. Barnes, and J. R. Tester. 1968. Seasonal weight changes, mortality, and population structure of raccoons in Minnesota. Journal of Mammalogy 49:63–73.

Michels, K. M., B. E. Fischer, and J. I. Johnson, Jr. 1960. Raccoon performance on color discrimination problems. Journal of Comparative Physiological Psychology 53:379–80.

Mitchell, M. A., L. L. Hungerford, C. Nixon, T. Esker, J. Sullivan, R. Koerkenmeier, and J. P. Dubey. 1999. Serologic survey for selected infectious disease agents in raccoons from Illinois. Journal of Wildlife Diseases 35:347–55.

Montgomery, G. G. 1964. Tooth eruption in preweaned raccoons. Journal of Wildlife Management 28:582–84.

Montgomery, G. G. 1969. Weaning of captive raccoons. Journal of Wildlife Management 33:154–59.

Moore, D. W., and M. L. Kennedy. 1985. Weight changes and population structure of raccoons in western Tennessee. Journal of Wildlife Management 49:906–9.

Mosillo, M., E. J. Heske, and J. D. Thompson. 1999. Survival and movements of translocated raccoons in northcentral Illinois. Journal of Wildlife Management 63:278–86.

Mugaas, J. N., and J. Seidensticker. 1993. Geographic variation of lean body mass and a model of its effect on the capacity of the raccoon to fatten and fast. Bulletin of the Florida Museum of Natural History 36:85–107.

Mugaas, J. N., J. Seidensticker, and K. P. Mahlke-Johnson. 1993. Metabolic adaptation to climate and distribution of the raccoon *Procyon lotor* and other Procyonidae. Smithsonian Contributions to Zoology 542:1–34.

O'Connell, A. F., D. J. Harrison, B. Connery, and K. N. Anderson. 1992. Food use by an insular population of coyotes. Northeast Wildlife 49:36–42.

Orloff, S. 1988. Present distribution of ringtails in California. California Fish and Game 74:196–202.

Ough, W. D. 1982. Scent marking by captive raccoons. Journal of Mammalogy 63:318–19.

Page, L. K., R. K. Swihart, and K. R. Kazacos. 1998. Raccoon latrine structure and its potential role in transmission of *Baylisascaris procyonis* to small vertebrates. American Midland Naturalist 140:180–85.

Page, L. K., R. K. Swihart, and K. R. Kazacos. 1999. Implications of raccoon latrines in the epizootiology of *Baylisascaris*. Journal of Wildlife Diseases 35:474–80.

Pedlar, J. H., L. Fahrig, and H. G. Merriam. 1997. Raccoon habitat use at 2 spatial scales. Journal of Wildlife Management 61:102–12.

Pence, D. B., and J. E. Stone. 1977. Lungworms (Nematoda: Pneumospiruridae) from West Texas carnivores. Journal of Parasitology 63:979–91.

Pence, D. B., and K. D. Willis. 1978. Helminths of the ringtails, *Bassariscus astutus,* from West Texas. Journal of Parasitology 64:568–69.

Peterson, F. A., W. C. Heaton, and S. D. Wruble. 1960. Levels of auditory response in fissiped carnivores. Journal of Mammalogy 50:566–78.

Pocock, R. I. 1921. On the external characters and classification of the Procyonidae. Proceedings of the Zoological Society of London 1921:389–422.

Poglayen-Neuwall, I., and I. Poglayen-Neuwall. 1980. Gestation period and parturition of the ringtail, *Bassariscus astutus* (Lichtenstein, 1830). Zeitschrift für Säugetierkunde 45:73–81.

Poglayen-Neuwall, I., and D. E. Toweill. 1988. *Bassariscus astutus*. Mammalian Species 327:1–8.

Pons, J.-M., V. Volobouev, J.-F. Ducroz, A. Tilllier, and D. Reudet. 1999. Is the Guadeloupean racoon (*Procyon minor*) really an endemic species? New insights from molecular and chromosomal analyses. Journal of Zoological Systematics and Evolutionary Research 37:101–8.

Pope, C. H. 1944. Attainment of sexual maturity in raccoons. Journal of Mammalogy 25:91.

Price, W. W. 1928. The civet, *Bassariscus astutus flavus,* a new host for *Uncinaria lotoris*. Journal of Parasitology 14:197.

Priewert, F. W. 1961. Record of an extensive movement by a raccoon. Journal of Mammalogy 42:113.

Rabinowitz, A. R., and B. G. Nottingham, Jr. 1986. Ecology and behaviour of the jaguar (*Panthera onca*) in Belize, Central America. Journal of Zoology (London) 210:149–59.

Rabinowitz, A. R., and M. R. Pelton. 1986. Day-bed use by raccoons. Journal of Mammalogy 67:766–69.

Ratnaswamy, M. J., and R. J. Warren. 1998. Removing raccoons to protect sea turtle nests: Are there implications for ecosystem management? Wildlife Society Bulletin 26:846–50.

Ratnaswamy, M. J., R. J. Warren, M. T. Kramer, and M. D. Adam. 1997. Comparisons of lethal and nonlethal techniques to reduce raccoon depredation of sea turtle nests. Journal of Wildlife Management 61:368–76.

Ratnayeke, S., A. Bixler, and J. L. Gittleman. 1994. Home range movements of solitary, reproductive female coatis, *Nasua narica,* in south-eastern Arizona. Journal of Zoology (London) 233:322–26.

Richards, R. E. 1976. The distribution, water balance, and vocalization of the ringtail, *Bassariscus astutus*. Ph.D. Dissertation, University of Northern Colorado, Greeley.

Richardson, W. B. 1942. Ring-tailed cats (*Bassariscus astutus*): Their growth and development. Journal of Mammalogy 23:17–26.

Riley, S. P. D., J. Hadidian, and D. A. Manski. 1998. Population density, survival, and rabies in raccoons in an urban national park. Canadian Journal of Zoology 76:1153–64.

Rinker, G. C. 1944. *Os clitoridis* from the raccoon. Journal of Mammalogy 25:91–92.

Risser, A. C., Jr. 1963. A study of the coatimundi (*Nasua narica*) in southern Arizona. M.S. Thesis, University of Arizona, Tucson.

Ritke, M. E. 1990a. Sexual dimorphism in the raccoon (*Procyon lotor*): Morphological evidence for intrasexual selection. American Midland Naturalist 124:342–51.

Ritke, M. E. 1990b. Quantitative assessment of variation in litter size of the raccoon *Procyon lotor*. American Midland Naturalist 123:390–98.

Ritke, M. E., and M. L. Kennedy. 1988. Intraspecific morphologic variation in the raccoon (*Procyon lotor*) and its relationship to selected environmental variables. Southwestern Naturalist 33:295–314.

Ritke, M. E., and M. L. Kennedy. 1993. Geographic variation of sexual dimorphism in the raccoon *Procyon lotor*. American Midland Naturalist 129:257–65.

Robbins, A. H., M. D. Borden, B. S. Windmiller, M. Niezgoda, L. C. Marcus, S. M. O'Brien, S. M. Kreindel, M. W. McGuill, A. Demaria, Jr., C. E. Rupprecht, and S. Rowell. 1998. Prevention of the spread of rabies to wildlife by oral vaccination of raccoons in Massachusetts. Journal of the American Veterinary Medical Association. 213:1407–12.

Robinson, S. K., F. R. Thompson, III., T. M. Donovan, D. R. Whitehead, and J. Faaborg. 1995. Regional forest fragmentation and the nesting success of migratory birds. Science 267:1987–90.

Rosatte, R. C., and C. D. MacInnes. 1989. Relocation of city raccoons. Proceedings of the Great Plains Wildlife Damage Control Workshop 9:87–92.

Rosatte, R. C., and M. J. Pyrus, and J. R. Gunson. 1986. Population reduction as a factor in the control of skunk rabies in Alberta. Journal of Wildlife Diseases 22:459–67.

Rosatte, R., M. J. Power, C. D. MacInnes, and K. F. Lawson. 1990. Rabies control for urban foxes, skunks, and raccoons. Pages 160–67 *in* L. R. Davis and R. E. Marsh, eds. Proceedings of the 14th vertebrate pest conference, Sacramento, California. University of California, Davis.

Rosatte, R. C., M. J. Power, and C. D. MacInnes. 1992a. Density, dispersion, movements and habitat of skunks (*Mephitis mephitis*) and raccoons (*Procyon lotor*) in metropolitan Toronto. Pages 932–44 *in* D. E. McCullough and R. E. Barrett, eds. Wildlife 2001: Populations. Elsevier, Essex, UK.

Rosatte, R. C., M. J. Power, C. D. MacInnes, and J. B. Campbell. 1992b. Trap–vaccinate–release and oral vaccination for rabies control in urban skunks, raccoons, and foxes. Journal of Wildlife Diseases 28:562–71.

Rosatte, R. C., C. D. MacInnes, W. R. Taylor, and W. Owen. 1997. A proactive prevention strategy for raccoon rabies in Ontario, Canada. Wildlife Society Bulletin 25:110–16.

Roscoe, D. E. 1993. Epizootiology of canine distemper in New Jersey raccoons. Journal of Wildlife Diseases 29:390–95.

Russell, J. K. 1979. Reciprocity in the social behavior of coatis (*Nasua narica*). Ph.D. Dissertation, University of North Carolina, Chapel Hill.

Russell, J. K. 1981. Exclusion of adult male coatis from social groups: Protection from predation 62:206–8.

Russell, J. K. 1982. Timing of reproduction by coatis (*Nasua narica*) in relation to fluctuations in food resources. Pages 413–31 *in* E. G. Leigh, Jr., A. S. Rand, and D. M. Windsor, eds. The ecology of a tropical forest. Smithsonian Institution Press, Washington, DC.

Russell, J. K. 1983. Altruism in coati bands: Nepotism or reciprocity? Pages 263–90 *in* S. K. Wasser, ed. Social behavior of female vertebrates. Academic Press, New York.

Sanderson, G. C. 1951. Breeding habits and a history of the Missouri raccoon population from 1941 to 1948. Transactions of the North American Wildlife Conference 16:445–60.

Sanderson, G. C. 1961a. The lens as an indicator of age in the raccoon. American Midland Naturalist 65:481–85.

Sanderson, G. C. 1961b. Techniques for determining age of raccoons. Illinois Natural History Survey, Biological Notes 45:1–16.

Sanderson, G. C. 1987. Raccoon. Pages 487–99 *in* M. Novak, J. A. Baker, M. E. Obbard, and B. Malloch, eds. Wild furbearer management and conservation in North America. Ontario Trappers Association, North Bay, Canada.

Sanderson, G. C., and A. V. Nalbandov. 1973. The reproductive cycle of the raccoon in Illinois. Illinois Natural History Survey Bulletin 31:29–85.

Scheffer, V. B. 1947. Raccoons transplanted in Alaska. Journal of Wildlife Management, 11:350–51.

Scheffer, V. B. 1950. Notes on the raccoon in southwest Washington. Journal of Mammalogy 31:444–48.

Schinner, R., and D. Cauley. 1974. The ecology of urban raccoons in Cincinnati, Ohio. Pages 125–30 *in* J. H. Noyes and D. R. Progulske, eds. Wildlife in an urbanizing environment. University of Massachusetts, Springfield.

Schneider, D. G., L. D. Mech, and J. R. Tester. 1971. Movements of female raccoons and their young as determined by radio-tracking. Animal Behaviour Monograph 4:1–43.

Schubert, C. A., I. K. Barker, R. C. Rosatte, C. D. MacInnes, and T. D. Nudds. 1998. Effect of canine distemper on an urban raccoon population: An experiment. Ecological Applications 8:379–87.

Scott, W. E. 1951. Wisconsin's first prairie spotted skunk, and other notes. Journal of Mammalogy 32:363.

Seidensticker, J., A. J. T. Johnsingh, R. Ross, G. Sanders, and M. B. Webb. 1988. Raccoons and rabies in Appalachian mountain hollows. National Geographic Research 4:359–70.

Sharp, W. M., and L. H. Sharp. 1956. Nocturnal movements and behavior of wild raccoons at a winter feeding station. Journal of Mammalogy 37:170–77.

Sheppard, C. H., and K. R. Kazacos. 1997. Susceptibility of *Peromyscus leucopus* and *Mus musculus* to infection with *Baylisascaris procyonis*. Journal of Parasitology 83:1104–11.

Sherfy, F. C., and J. A. Chapman. 1980. Seasonal home range and habitat utilization of raccoons in Maryland. Carnivore 3:8–18.

Sherman, H. B. 1954. Raccoons of the Bahama Islands. Journal of Mammalogy 35:126.

Shieff, A., and J. A. Baker. 1987. Marketing and international fur markets. Pages 862–77 *in* M. Novak, J. A. Baker, M. E. Obbard, and B. Malloch, eds. Wild furbearer management and conservation in North America. Ontario Trappers Association, North Bay, Canada.

Shirer, H. W., and H. S. Fitch. 1970. Comparison from radiotracking of movements and denning habits of the raccoon, striped skunk, and opossum in northeastern Kansas. Journal of Mammalogy 51:491–503.

Shoop, C. R., and C. A. Ruckdeschel. 1990. Alligators as predators on terrestrial mammals. American Midland Naturalist 124:407–12.

Sieber, O. J. 1984. Vocal communication in raccoons (*Procyon lotor*). Behaviour 90:80–113.

Slate, D. 1985. Movement, activity and home range patterns among members of a high density suburban raccoon population. Ph.D. Dissertation, Rutgers University, New Brunswick, NJ.

Smith, H. J. 1980. Behavior of the coati (*Nasua narica*) in captivity. Carnivore 2:88–136.

Smythe, N. 1970. The adaptive value of the social organization of the coati (*Nasua narica*). Journal of Mammalogy 51:818–20.

Sneddon, I. A. 1991. Latrine use by the European rabbit (*Oryctolagus cuniculus*). Journal of Mammalogy 72:769–75.

Snyder, D. E., and P. R. Fitzgerald. 1985. The relationship of *Baylisascaris procyonis* to Illinois raccoons (*Procyon lotor*). Journal of Parasitology 71:596–98.

Soper, J. D. 1942. Mammals of Wood Buffalo Park, northern Alberta and District of Mackenzie. Journal of Mammalogy 23:119–45.

Stains, H. J. 1956. The raccoon in Kansas: Natural history, management, and economic importance (Miscellaneous Publications 10). University of Kansas Museum of Natural History and State Biological Survey.

Stuewer, F. W. 1943a. Raccoons: Their habits and management in Michigan. Ecological Monographs 13:203–57.

Stuewer, F. W. 1943b. Reproduction of raccoons in Michigan. Journal of Wildlife Management 7:60–73.

Sutton, R. W. 1964. Range extension of raccoon in Manitoba. Journal of Mammalogy 45:311–12.

Tabatabai, F. R., and M. L. Kennedy. 1989. Movements of relocated raccoons (*Procyon lotor*) in western Tennessee. Journal of the Tennessee Academy of Science 64:221–24.

Taylor, W. P. 1954. Food habits and notes on life history of the ring-tailed cat in Texas. Journal of Mammalogy 35:55–63.

Tevis, L., Jr. 1947. Summer activities of California raccoons. Journal of Mammalogy 28:323–32.

Thorgeson, H. L. 1958. Studies of tactual discrimination by raccoons. Ph.D. Dissertation, Purdue University, West Lafayette, IN.

Thorkelson, J., and R. K. Maxwell. 1974. Design and testing of a heat transfer model of a raccoon (*Procyon lotor*) in a closed tree den. Ecology 55:29–39.

Tiner, J. D. 1953. Fatalities in rodents caused by larval *Ascaris* in the central nervous system. Journal of Mammalogy 34:153–67.

Tiner, J. D. 1954. The fraction of *Peromyscus leucopus* fatalities caused by raccoon ascarid larvae. Journal of Mammalogy 35:589–92.

Toweill, D. E. 1976. Movements of ringtails in Texas' Edwards Plateau region. M.S. Thesis, Texas A & M University, College Station.

Toweill, D. E., and M. E. Price. 1976. Ectoparasites of ringtails collected from Kerr Co., Texas. Southwestern Entomology 1:20.

Toweill, D. E., and J. G. Teer. 1977. Food habits of ringtails in the Edwards Plateau region of Texas. Journal of Mammalogy 58:660–63.

Toweill, D. E., and J. G. Teer. 1980. Home range and den habits of Texas ringtails (*Bassariscus astutus flavus*). Pages 1103–20 *in* J. A. Chapman and D. Pursley, eds. Proceedings of the worldwide furbearers conference. Frostburg, MD.

Toweill, D. E., and D. B. Toweill. 1978. Growth and development of captive ringtails (*Bassariscus astutus flavus*). Carnivore 1:46–53.

Trapp, G. R. 1972. Some anatomical and behavioral adaptations of ringtails, *Bassariscus astutus*. Journal of Mammalogy 53:549–57.

Trapp, G. R. 1973. Comparative behavioral ecology of two southwestern Utah carnivores: *Bassariscus astutus* and *Urocyon cinereoargenteus*. Ph.D. Dissertation. University of Wisconsin, Madison.

Trapp, G. R. 1978. Comparative behavioral ecology of the ringtail and gray fox in southwestern Utah. Carnivore 1:3–32.

Twichell, A. R., and H. H. Dill. 1949. One hundred raccoons from one hundred and two acres. Journal of Mammalogy 30:130–33.

Tyson, E. L. 1950. Summer food habits of the raccoon in southwest Washington. Journal of Mammalogy 31:448–49.

Uhaa, I. J., V. M. Dato, F. E. Sorhage, J. W. Berkley, D. E. Roscoe, R. D. Gorsky, and D. B. Fishein. 1992. Benefits and costs of using an orally absorbed vaccine to control rabies in raccoons. Journal of the American Veterinary Medical Association 201:1873–82.

Urban, D. 1970. Raccoon populations, movement patterns, and predation on managed waterfowl marsh. Journal of Wildlife Management 34:372–82.

Valenzuela, D., and G. Ceballos. 2000. Habitat selection, home range, and activity of the white-nosed coati (*Nasua narica*) in a Mexican tropical dry forest. Journal of Mammalogy 81:810–19.

Waser, P. M., and W. T. Jones. 1983. Natal philopatry among solitary mammals. Quarterly Review of Biology 58:355–90.

White, S. E., P. K. Kennedy, and M. L. Kennedy. 1998. Temporal genetic variation in the raccoon, *Procyon lotor*. Journal of Mammalogy 79:747–54.

Whitney, L. F., and A. B. Underwood. 1952. The raccoon. Practical Science, Orange, CT.

Willey, R. B., and R. E. Richards. 1974. The ringtail (*Bassariscus astutus*): Vocal repertoire and Colorado distribution. Journal of the Colorado–Wyoming Academy of Science 7:58.

Willey, R. B., and R. E. Richards. 1981. Vocalization of the ringtail (*Bassariscus astutus*). Southwestern Naturalist 26:23–30.

Wilson, M. L., P. M. Bretsky, G. H. Cooper, Jr., S. H. Egbertson, H. J. Van Kruiningen, and M. L. Cartter. 1997. Emergence of raccoon rabies in Connecticut, 1991–1994: Spatial and temporal characteristics of animal infection and human contact. American Journal of Tropical Medicine and Hygiene 57:457–63.

Wood, J. E. 1955. Notes on reproduction and rate of increase of raccoons in the Post Oak region of Texas. Journal of Wildlife Management 19:409–10.

Wood, N. A. 1922. The mammals of Washtenaw, County, Michigan. University of Michigan Museum of Zoology Occasional Papers 123:1–23.

Wozencraft, W. C. 1993. Order Carnivora. Pages 279–348 *in* D. E. Wilson and D. M. Reeder, eds. Mammal species of the world: A taxonomic and geographic reference, 2nd ed. Smithsonian Institution Press, Washington, DC.

Wright, G. A. 1977. Dispersal and survival of translocated raccoons in Kentucky. Proceedings of the Annual Conference of the Southeastern Association of Fish and Wildlife Agencies 31:285–94.

Wright, J. C., and D. G. Thawley. 1980. Role of the raccoon in transmission of pseudorabies: A field and laboratory investigation. American Journal of Veterinary Research 41:581–83.

Wyatt, J. D., W. H. Barker, N. M. Bennett, and C. A. Hanlon. 1999. Human rabies postexposure prophylaxis during a raccoon rabies epizootic in New York, 1993–1994. Emerging Infectious Diseases 5:415–22.

Yeager, L. E., and W. H. Elder. 1945. Pre- and post-hunting season foods of raccoons on an Illinois goose refuge. Journal of Wildlife Management 9:48–56.

Yeager, L. E., and R. G. Rennels. 1943. Fur yield and autumn foods of the raccoon in Illinois river bottom lands. Journal of Wildlife Management 7:45–60.

Stanley D. Gehrt, Max McGraw Wildlife Foundation, P.O. Box 9, Dundee, Illinois 60118. Email: sgehrt@mcgrawwildlife.org.

29

Fisher and Marten

Martes pennanti and *Martes americana*

Roger A. Powell
Steven W. Buskirk
William J. Zielinski

NOMENCLATURE

FISHER

Common Names. Fisher, wejack, fisher cat, black cat, Pennant's marten, marten, pekan; only fisher is used commonly today, and locally fisher cat in some places in New England; the name fisher may have come from early immigrants, who noted the fisher's similarity to the dark European polecat (*Mustela putorius*), names for which included fitchet, fitche, and fitchew (Brander and Books 1973)
Scientific Name. *Martes pennanti*
Subspecies. *M. p. pennanti, M. p. pacifica,* and *M. p. columbiana*. Hagmeier (1959) concluded that none of the subspecies is separable on the basis of pelage or skull characteristics, and questioned the classification. As have most authors since Hagmeier, we ignore subspecies taxonomy except when discussing certain conservation issues, which are often organized around designated subspecies rather than around detailed ranges.

MARTEN

Common Names. Marten, American marten, and pine marten are the only names commonly used today; Marten cat, American sable, Canadian sable, saple, and Hudson Bay sable have been used in the past and appear in old literature
Scientific Name. *Martes americana*
Subspecies. The subspecies have been divided into *americana* and *caurina* subspecies groups (Hagmeier 1961). The *americana* group includes *M. a. americana, M. a. brumalis, M. a. atrata, M. a. abieticola, M. a. actuosa, M. a. kenaiensis,* and *M. a. abietinoides*. The *caurina* group includes *M. a. caurina, M. a. humboldtensis, M. a. vancouverensis, M. a. nesophila, M. a. vulpina, M. a. origenes,* and *M. a. sierrae*. Although Hagmeier concluded that the subspecies differed too little to warrant continued recognition, Carr and Hicks (1997) argued from mitochondrial DNA data that the *americana* and *caurina* groups constitute two separate species, *M. americana* and *M. caurina,* as originally described by Merriam (1890). The groups intergrade morphologically and genetically, indicating hybridization (Wright 1953; Stone 2000), and although the groups differ in mitochondrial DNA genotypes by six nucleotide substitutions, no two specimens within a group differ by more than two substitutions. Thus, Carr and Hicks (1997) concluded that the putative *M. americana* and *M. caurina* differ to a similar degree as do the Eurasian pine marten, *M. martes,* and the sable, *M. zibellina*. From mitochondrial and nuclear DNA, Stone (2000) concluded that the *americana* and *caurina* groups have descended from a single colonization of North America and not two colonizations as previously hypothesized, and are indeed more closely related than are the Eurasian pine marten and sable. Whether the groups of American martens will be determined to be distinct species is not yet known, and Hagmeier (1961) and Anderson (1970) suggested that the American marten, Eurasian pine marten, sable, and Japanese marten, *M. melampus,* are conspecific. Because evidence is not yet compelling, we shall treat the *caurina* group as *M. americana* and consider the entire *M. americana* to be a good species. As with fishers, we ignore subspecies taxonomy except when discussing certain conservation issues.

DISTRIBUTION

The genus *Martes* is circumboreal in distribution, with extensions into southern (*M. gwatkinsii*) and southeast Asia as far as 7°S latitude (*M. flavigula*; Anderson 1970). The fisher (subgenus *Pekania*) is endemic to the New World and restricted to mesic coniferous forest of the boreal zone and its southern peninsular extensions (Hagmeier 1956; Gibilisco 1994). The boreal forest martens, comprising four sibling species (*M. martes, M. zibellina, M. melampus, M. americana*), are distributed parapatrically or allopatrically across the boreal zone from Ireland eastward across Eurasia to North America as far east as Newfoundland Island (Buskirk 1994). The American marten (Fig. 29.1) is distributed similarly to the fisher (Fig. 29.2) in the southern parts of its range, but is found farther north, to the northern limit of trees, than the fisher. In the Rocky Mountains, the range of the marten extends much farther south (to New Mexico) than does that of the fisher (to Montana and Idaho). In the Pacific states, both species occur as far south as the southern Sierra Nevada.

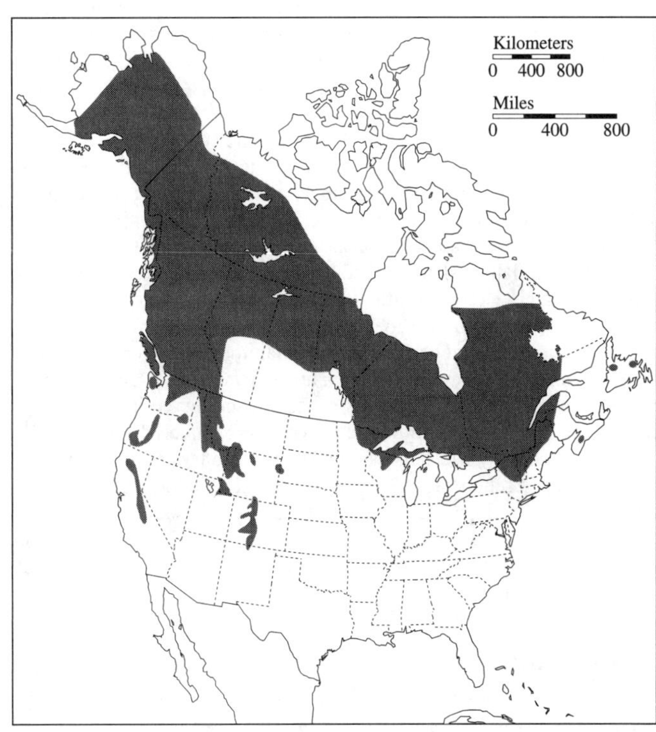

Figure 29.1. Distribution of the marten (*Martes americana*).

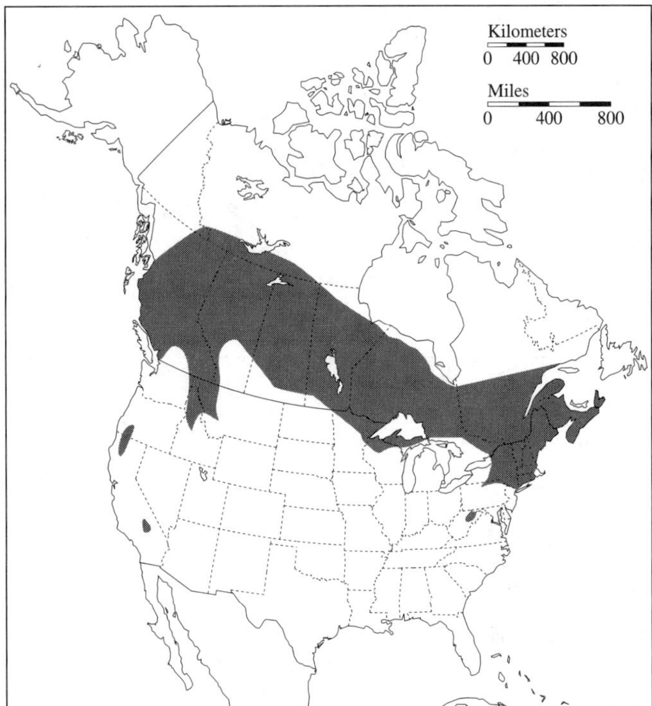

FIGURE 29.2. Distribution of the fisher (*Martes pennanti*).

Distributional changes during historical times, comprising regional extirpations, natural colonization and recolonization, and translocations, summarized by Berg (1982), Powell (1993), and Strickland et al. (1982a, 1982b), have affected the fisher more than the American marten. For the fisher, major distributional changes from presettlement times include losses from much of its previous range in the Pacific states (Gibilisco 1994; Zielinski et al. 1995), including apparently complete extirpation from Washington and northern Oregon. The fisher has undergone a slight range expansion to northwestern British Columbia and southeastern Alaska, losses from central Alberta and Saskatchewan, and losses from the southern Great Lakes and Ohio Valley regions (Gibilisco 1994). Parts of Quebec and Labrador also appear to have dropped from the fisher's range during the historical period. During the last two decades of the twentieth century, the fisher expanded its range southward into southern New England (Gibilisco 1994).

Salient changes in the distribution of the American marten since European settlement include loss of much of its range in the coast ranges of California and Oregon (Zielinski et al. 2001). The entire range of *M. a. humboldtensis* is nearly vacant and may be represented by only one small population. The geographic range in the Rocky Mountains is approximately as in presettlement times, with possible losses from some isolated areas (Gibilisco 1994). The marten is absent from much of its presettlement range in southern Manitoba, the southern Great Lakes region, the upper Ohio Valley, and southern New England. The range of the marten, however, has expanded southward into the latter area and into the north-central states since approximately 1980 (Buskirk and Ruggiero 1994). Martens have been introduced to several islands of the Alexander Archepelago of southeast Alaska (MacDonald and Cook 1996) and reintroduced to Nova Scotia and the Black Hills of South Dakota. The species is now extirpated on Prince Edward Island and very restricted in distribution on Cape Breton Island and Newfoundland (Buskirk and Ruggiero 1994).

DESCRIPTION

Fishers and martens are medium-sized carnivores with a general weasel shape but lacking the extreme elongation of the weasels. They have well-furred bodies and full tails. Their tails constitute about one third of total body length. Their faces are triangular, with muzzles less pointed than those of foxes. The fisher's ears are wide and rounded, whereas the marten's ears are pointed. Males are considerably larger than females in both species.

Fishers and martens are digitigrade with five toes on each large, well-furred paw. Claws are sharp, curved, and semiretractable but not sheathed.

Size and Weight. The fisher is the largest member of the genus *Martes*. Adult males generally weigh 3.5–5.5 kg, though larger males are not uncommon in parts of the species' range, and an exceptional male weighed over 9 kg (Blanchard 1964). Females generally weigh 2.0–2.5 kg. Males are also longer than females; total length is 90–120 cm versus 75–95 cm (Powell 1993). Although martens vary considerably in size across their range (Hagmeier 1961), they are nowhere as large as fishers. Like fishers, male martens are larger than females; males generally weigh 0.6–1.0 kg, but can be up to 1.4 kg, whereas females generally weigh 0.4–0.7 kg, but can be up to 1 kg (Buskirk and McDonald 1989). Males generally are 50–70 cm in total length, whereas females are 45–60 cm.

Juvenile martens reach adult length, but not weight, by 3 months of age, whereas fishers reach adult length, but not weight, at about 6 months of age. By 7 months, the epiphyses of female fishers long bones have ossified, indicating they have reached full size, but males do not reach this stage until 10 months of age (Dagg et al. 1975).

Pelage. On both species, the fur on the body and head is lighter in color than on the tail, legs, and shoulders. Fishers have dark-brown bodies with black legs and tails; their heads have a gold or silver, hoary appearance created by tricolored guard hairs (Coulter 1966). Many fishers have irregular white or cream patches on their chests or groins; the chest patches never become so large as to resemble a chin or throat patch. The pelage color of martens varies more across the specie's range than does that of fishers. Martens may have light-brown to dark brown to gold bodies with dark-brown tails and legs. They have irregular, often large patches of cream to gold on their chins and throats. The color of these patches varies with season, and tends to be darkest in autumn.

Skeletal Morphology. Fishers and martens have generalized skeletons capable of a wide variety of movements and locomotion, including running on the ground, running along tree limbs, and climbing trees (Leach 1977a, 1977b; Leach and de Kleer 1978; Holmes 1980). Members of both species can rotate their hind feet, allowing them to descend trees head first. The scapulae of fishers each have enlarged postscapular fossae, which probably accommodate large muscles used in climbing trees (Powell 1993).

Skull and Dentition. Skulls of male fishers generally are 110–130 mm long, whereas those of females are 95–105 mm long. Skull width for males is 62–84 mm, whereas that for females is 52–61 mm (Peterson 1966). For martens, skulls generally are 80–95 and 69–80 mm in length and 46–53 and 38–46 mm in width for males and females, respectively. The dental formula for both species is I 3/3, C 1/1, P 4/4, M 1/2. Among mustelids, only fishers, martens, and wolverines *Gulo gulo*) have four upper and lower premolars. Fishers' skulls (Fig. 29.3) can be distinguished from those of wolverines by their generally shorter length (<130 mm) and from those of martens (Fig. 29.4) by their longer length (>95 mm). In addition, fisher and marten skulls arch less than do wolverine skulls. Skulls of fully adult male fishers have a large sagittal crest that usually exceeds 1 cm in height and is an exceptional feature of the skull (Figs. 29.5 and 29.6). Skulls of small female fishers can be distinguished from those of large male martens by the exposed lateral root of the fourth upper premolar (Anderson 1970), which is diagnostic for fisher skulls in general. Skulls of martens can be distinguished from those of minks (*Mustela vison*) and skunks (*Mephitis mephitis, Spilogale* spp.) by having four premolars and from minks by their larger size and taller cranium.

Scent Glands. Fishers and American martens mark with several scents. Urine and feces are presumed to be scent marks (Buskirk 1994) because fishers and martens often place them on stumps or other prominent

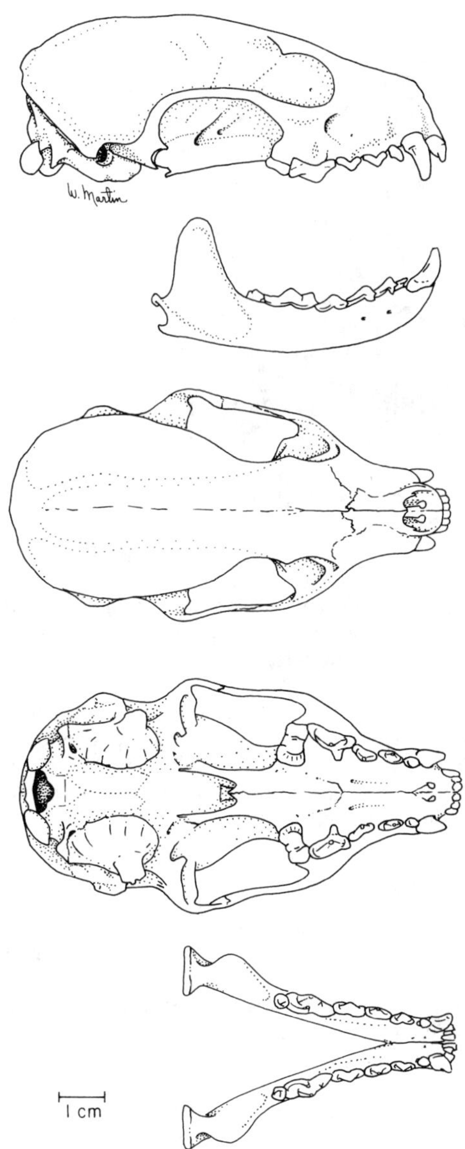

FIGURE 29.3. Skull of the fisher (*Martes pennanti*). From top to bottom: lateral view of cranium, lateral view of mandible, dorsal view of cranium, ventral view of cranium, dorsal view of mandible.

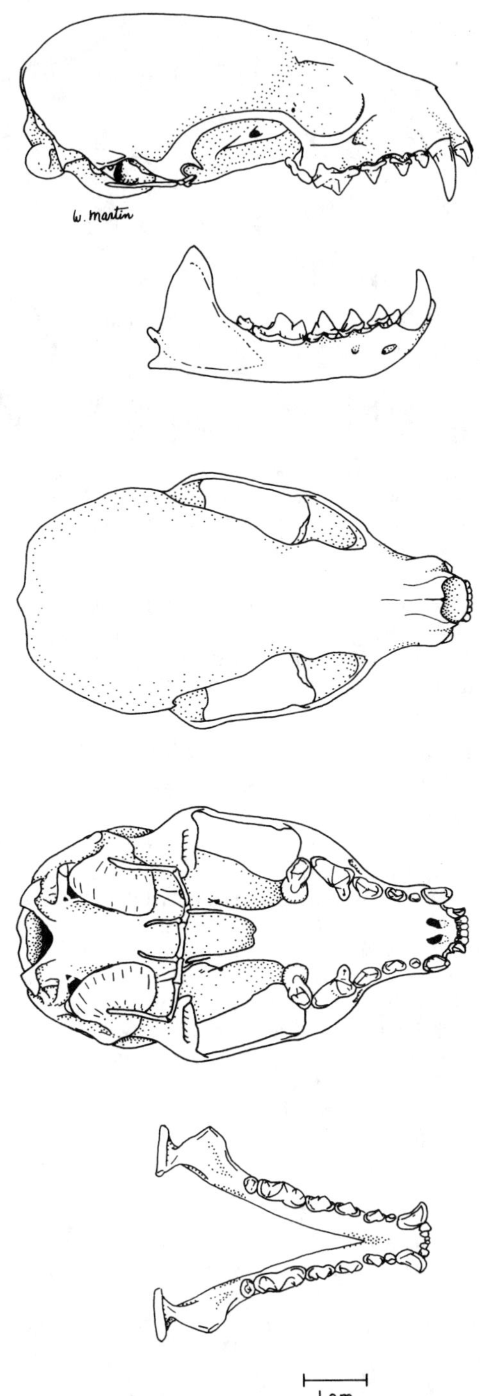

FIGURE 29.4. Skull of the marten (*Martes americana*). From top to bottom: lateral view of cranium, lateral view of mandible, dorsal view of cranium, ventral view of cranium, dorsal view of mandible.

structures. In addition, both species have abdominal glands (Hall 1926), plantar glands (Buskirk et al. 1986; Powell 1982), and anal sacs, the latter typical of most fissipede carnivores. The scents from the latter are not as perceptible by humans as are those of other mustelids (Buskirk 1994). Plantar glands in male and female fishers undergo a large annual cycle in size (Frost et al. 1997), peaking in size at the time of mating. Ecological and behavioral consequences of scent marking are poorly known.

Body Composition and Fat Deposits. Body composition and fat depots have been described for martens in Alaska, Ontario, and Wyoming, and in relation to sex, age, season, and habitat (Buskirk and Harlow 1989; Cobb 2000), but fat depots of fishers are less well described (Leonard 1980). Martens and fishers are very lean for their body size, with only 2.4–4.6% of the body mass of martens as extractable lipid. Fat depots are located as in other carnivores (i.e., omental, perirenal), but only exceptionally do fat animals exhibit obvious subcutaneous fat. Absolute and relative dry mass of the omental and perirenal fat are the best indicators of percentage body fat for both species (Leonard 1980; Cobb 2000) and account for up to 65% of the pooled variation in total body fat for both sexes of martens (Cobb 2000). The effect of age on body fat differs between sexes in martens (Cobb 2000), but body fat does not vary over winter (Buskirk and Harlow 1989). Body fat of martens in Ontario varies with habitat. Animals living in boreal (conifer-dominated) forests have more fat than those from "mixed forests" having a strong broad-leaved tree component (Cobb 2000). For fishers in Manitoba, body fat is directly related to the abundance of snowshoe hares (*Lepus americanus*) in their dict (Leonard 1980). Protein constitutes about 17.5% and water about 70% of the bodies of martens, values similar to those of other lean mammals. Leanness is usually seen as an adaptation to highly athletic movement styles, including foraging in small spaces beneath the snow. Although protein serves as an important secondary

FIGURE 29.5. Skulls and bacula of male fishers (*Martes pennanti*), showing variation with age. Note the development of the sagittal crest and baculum and the fusion of the sutures with age. Numerals denote age estimated from cementum annuli. The juvenile at the extreme left was captured in November; the one adjacent was captured in February. Contrast the skull of the adult male at the extreme right with the drawing of the skull of a juvenile female in Fig. 29.3.

energy store to fat (Harlow and Buskirk 1991), martens have limited fasting endurance. They must balance their energy budgets over brief periods and must gear their activity and foraging to balance short-term energy budgets.

AGE ESTIMATION AND SEX DETERMINATION

Poole et al. (1994) reviewed methods of determining sex and estimating age of fishers and martens for protected populations and populations that are trapped for fur. Arthur et al. (1992) reviewed use of cementum annuli to estimate ages of fishers.

Cementum analysis is the only technique that provides an estimate of the ages of adult animals (Poole et al. 1994). Livetrapped animals can be aged if a tooth is extracted under anesthetic. A first premolar is usually extracted, with probably no functional effect on the animal. Canine teeth are often used to estimate age of animals trapped for fur. Age estimates made by different technicians or using different teeth agree 80–95% of the time (Poole et al. 1994), but disagreements highlight that uncontrollable factors affect these estimates. This technique often underestimates ages of very old animals (Arthur et al. 1992; Poole et al. 1994).

Radiographs of canine teeth, which show an open or closed pulp cavity, correctly distinguish juveniles from adults over 90% of the time (Poole et al. 1994). Estimating the length of the sagittal crest from coalescence of the temporal muscles on carcasses of martens trapped for fur allowed Poole et al. (1994) to distinguish juvenile from adult martens trapped in Alaska and the Northwest Territories, but many adult females from Ontario were incorrectly classed as juveniles. A number of measurements taken from teeth and skulls correctly identify sex of fishers and martens when skulls only are available. Length and width of lower canines distinguish the sexes of fishers 100% of the time, and skull length and zygomatic width distinguish the sexes with 98–100% accuracy. Upper or lower canine root width, lower canine root length, skull length, and zygomatic width distinguish the sexes of martens 92–100% of the time (Poole et al. 1994).

REPRODUCTION

Anatomy. Ovaries in both species are completely encapsulated by a bursa and the uterus has two horns with a common corpus uteri, allowing migration of blastocysts between the horns (Strickland et al. 1982a, 1982b). Males have bacula (Fig. 29.5). Testes increase in size and weight before the breeding season, which occurs in late March through early May for fishers and July and August for martens.

Reproductive Cycle and Delayed Implantation. Both sexes of both species reach sexual maturity by 1 year of age, but effective breeding may not occur before 2 years of age. One-year-old male fishers produce sperm, but their bacula have not reached adult size and shape (Coulter 1966; Wright and Coulter 1967; Frost et al. 1997). When bred, 1-year old males fail to produce offspring (Douglas 1943). Female fishers generally breed for the first time when 1 year old (Wright and Coulter

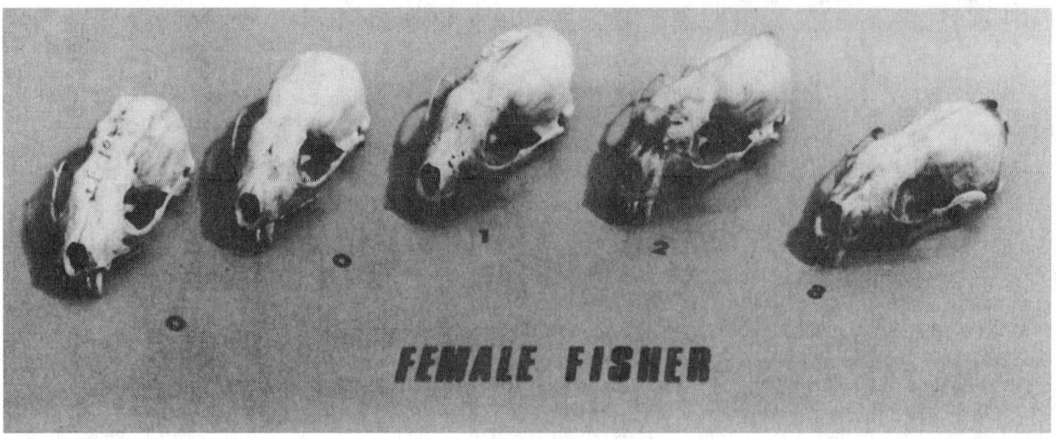

FIGURE 29.6. Skulls of female fishers (*Martes pennanti*), showing variation with age. Note the sutures and minimal development of the sagittal crest. The juvenile at the extreme left was captured in November; the one adjacent was captured in February.

1967; Strickland et al. 1982a). Literature concerning age of maturity for martens is confusing, in part because early researchers had difficulty estimating the ages of martens (Strickland et al. 1982b). Although most early literature reported that martens did not breed before 2 years of age, Strickland et al. (1982b) reported that 80% of 18- to 20-month-old female martens trapped in southern Ontario had corpora lutea, indicating pregnancy.

Female fishers and martens both give birth in late March or April (Fig. 29.7). Female fishers then enter estrus and breed approximately 7–10 days later (Hall 1942; Asdell 1946; Eadie and Hamilton 1958; Coulter 1966; Wright and Coulter 1967; Strickland 1982a). Female martens enter estrus in July or August (Strickland et al. 1982b). Ovulation in many *Mustela* and *Martes* species is induced by copulation, and shape and size of the baculum may be critical for proper stimulation of the vagina (Mead 1994). Cherepak and Connor (1992), however, recovered an unfertilized ovum in August from the reproductive tract of a female fisher who had not bred. As Ewer (1973) noted, induced ovulation may be a matter of degree, and vaginal stimulation during copulation may induce ovulation in females who might have ovulated spontaneously at a later date without breeding.

Both martens and fishers delay implantation. A fertilized zygote develops to a blastocycst and then becomes inactive in the uterus, its metabolic rate falls, and cell division ceases (Ewer 1973). The blastocyst remains dormant until late winter, when change in day length induces implantation and active gestation (Enders and Pearson 1943; Frost et al. 1997). Consequently, adult female fishers are pregnant nearly all year, except for 7–10 days following parturition. Adult female martens are pregnant for 7–8 months. Active gestation is approximately 50 days (<10 days before to delay, 40 days postimplantation) in fishers (Frost et al. 1997). Active gestation for martens may be somewhat shorter. Jonkel and Weckwerth (1963) reported that the postimplantation period for martens was 27–28 days, but their ability to detect implantation time was limited. The postimplantation period for Eurasian pine martens is 30–35 days (Mead 1994).

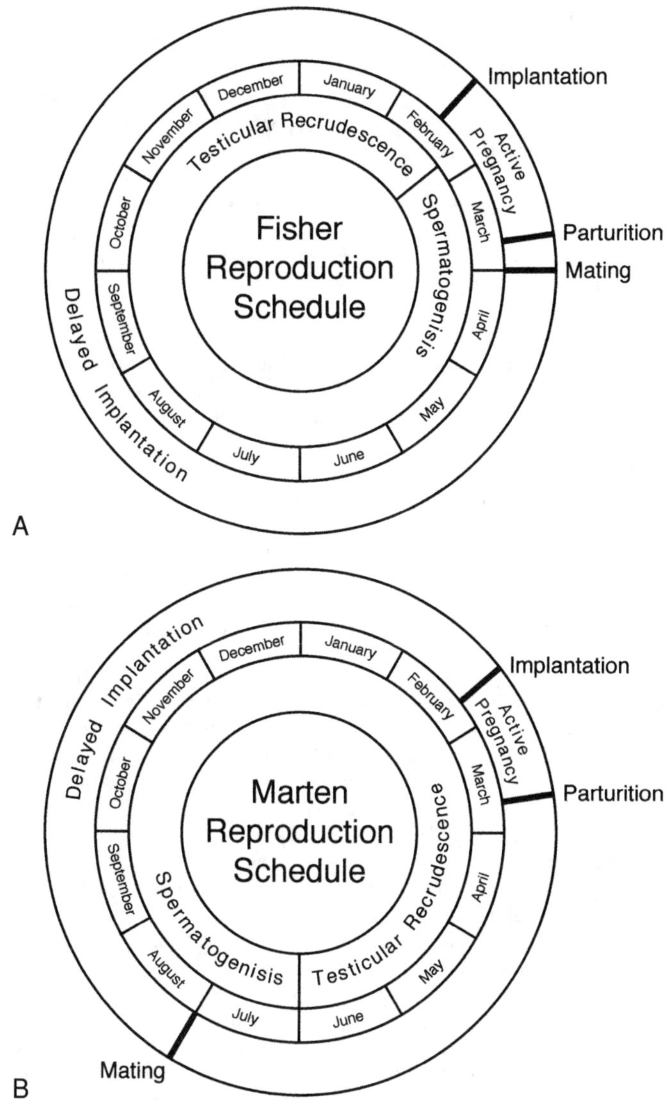

FIGURE 29.7. Reproduction schedules of (A) fishers (*Martes pennanti*) and (B) martens (*Martes americana*). SOURCE: Figure drawn by C. B. Powell.

Changes in soft body parts signal reproductive condition. Plantar glands on the hind feet of female fishers increase in size from <10 cm^2 in December to >10 cm^2 in January, remain enlarged through May, and then regress to <10 cm^2 again by June (Frost et al. 1997). For female fishers who produce offspring in a given year, nipples begin to enlarge in February–March, reach a peak in size in August–September, then decrease in size through November (Frost et al. 1999). Width times height of anterior nipples is an index of nipple size that differs significantly between breeding and nonbreeding females (Frost et al. 1999).

In male fishers, concentrations of testosterone begin to rise in December in adults and in January in juveniles, followed by an increase in testicular size, which becomes maximum in March in adults and April in juveniles (Frost et al. 1997). Production of sperm is maximal in March–May for both age classes. Concentration of testosterone begins to decrease in April and testes are fully regressed by June. Plantar glands on the hind feet of male fishers begin to increase in size from <15 cm^2 in December to >30 cm^2 in May, then regress rapidly by June (Frost et al. 1997).

Breeding Behavior. Male fishers and martens are undoubtedly polygynous and females may well be both polyandrous and selective. Large sexual dimorphism in body size, as found in martens and fishers, is strongly correlated with polygynous mating patterns in mammals (Kleiman 1977). Courtship of fishers is similar to that of other mustelines and may be prolonged and vigorous (Hodgson 1937; Laberee 1941; Enders and Enders 1963; Heidt et al. 1968; C. Kline and M. Don Carlos, Minnesota Zoological Society, unpublished records). Copulations resulting in pregnancy have been reported to last from 20 min to 7 hr (Hodgson 1937; Laberee 1941; C. Kline and M. Don Carlos, Minnesota Zoological Society, unpublished records). Such long copulations are not needed to induce ovulation, yet when copulation is interupted after 5–12 min in ferrets (*Mustela furo*) and minks (*M. vison*), no fertilization occurs (Mead 1994).

Parturition and Litter Size. Parturition dates for fishers occur from late February through early May, but most litters are born in March through early April (Hodgson 1937; Laberee 1941; Hall 1942; Douglas 1943; Hamilton and Cook 1955; Coulter 1966; Wright and Coulter 1967; Leonard 1980; Paragi 1990; Frost and Krohn 1994; Frost et al. 1997; reviewed by Powell 1993). Recorded parturition dates for martens show less variation than do those for fishers (reviewed by Strickland and Douglas 1987) and range from mid-March through the end of April. C. Kline and M. Don Carlos (Minnesota Zoological Society, unpublished records) recorded parturition and breeding dates for three captive-bred litters of fishers at the Minnesota Zoo. One female bred on 16 April 1989, gave birth on 31 March 1990, and bred again on 7–9 April 1990. Another female bred on 22 April 1990, gave birth on 2 April 1991, and bred again on 11–12 April. This second female gave birth again on 10 April 1992. The consistency of the parturition and breeding dates for these females suggests that individual females may implant around the same date each year and consequently give birth and breed at consistent times. This pattern, in turn, suggests that the variability of parturition dates reported in the literature stems from variation among females for implantation dates and not variation from year to year by individual females.

Reported litter sizes for fishers range from one to six and the means for all studies are between two and three inclusive (reviewed by Powell

1993; also Frost and Krohn 1994; Frost et al. 1997), whereas reported litter sizes for martens range from one to five (reviewed by Strickland and Douglas 1987). Mean numbers of corpora lutea for fishers ranged from 2.7 to 3.9 across several studies (reviewed by Powell 1993). Strickland and Douglas (1987) reported a mean of 3.5 for 880 female martens studied in Ontario between 1973 and 1985. Reported mean numbers of implanted blastocysts, implanted embryos, placental scars, and litter sizes show a slightly decreasing pattern in fishers, indicating small loss of young in utero through pregnancy (Powell 1993).

Pregnancy rates for fishers and martens are generally calculated as the proportion of adult females whose ovaries contain corpora lutea in carcasses turned in by trappers (Shea et al. 1985; Douglas and Strickland 1987; Crowley et al. 1990). Although placental scars should, in theory, document birth and litter sizes more accurately then corpora lutea, Frost et al. (1999) showed that placental scars sometimes fail to identify reproduction by female fishers. Four of 13 female fishers who produced young in their study did not have identifiable placental scars.

Corpora lutea generally indicate ovulation rates of 95% or more for fishers (Shea et al. 1985; Douglas and Strickland 1987; Crowley et al. 1990), but Arthur and Krohn (1991) and Paragi (1990) reported a denning rate of about 65% for the fishers they studied in Maine. Strickland and Douglas (1987) reported a pregnancy rate of 87% (90% if $1\frac{1}{2}$-year-old females are excluded) for martens in Ontario. Strickland and Douglas (1987) also found greater variation in annual pregnancy rates for martens than for fishers. They attributed the variation in pregnancy rates to be food-related in martens, but did not speculate on why martens show greater variation than fishers. In the southern Sierra Nevada of California, 50–60% of 26 female fishers captured in spring were lactating each year from 1994 through 1996 (Truex et al. 1998). Further north in California, Truex et al. (1998) found greater variation in lactation rates for a smaller sample of fishers (18 over 2 years).

Bull and Heater (1995, 2001) documented 13 reproductive efforts by martens in Oregon. Of these, 4 females weaned at least one kit, 8 females failed to raise any kits, and the outcome for 1 female was unknown.

DEVELOPMENT

Martens and fishers, similar to other mustelines, are altricial and are born completely helpless with their eyes and ears closed (Hodgson 1937; Coulter 1966; LaBarge et al. 1990). They are partially covered with fine hair.

Development of fishers has been documented better than that of martens. Fishers remain helpless for weeks following birth and do not begin to crawl until $\geq$3 weeks old (Hodgson 1937; Coulter 1966; Powell 1993). By 3 months, they climb well (Grinnell et al. 1937; Powell 1993). Somewhere between 6 and 8 weeks, kits open their eyes (Hodgson 1937; Coulter 1966; LaBarge et al. 1990; Powell 1993). Deciduous teeth begin to erupt when kits are about 6 weeks old and canines erupt by 7–9 weeks (Coulter 1966; LaBarge et al. 1990; Powell 1993).

By 2 weeks of age, fisher kits are covered with light silver-gray fur (LaBarge et al. 1990). Around 3–4 weeks of age, they begin to change to a chocolate-brown color. By 10–12 weeks, most kits are completely chocolate-brown. From this age on, the tricolored guard hairs characteristically found on the head, neck, and shoulders of adults can be seen, but with a restricted distribution. Thus, through the summer and the early autumn, young fishers are the same general color as adults, but are more uniform in color.

Fishers weigh well under 50 g at birth, reach about 500 g by 40–50 days, and thereafter begin to exhibit sexual dimorphism in size. By late summer or early fall, they approach adult size and sexual dimorphism is pronounced (Powell 1993). Martens reach adult body weight around 3 months of age (Markley and Bassett 1942).

Fisher kits are completely dependent on milk until they are 8–10 weeks old (Coulter 1966; LaBarge 1990; Powell 1993). A litter raised by its mother in captivity was weaned by her at approximately 4 months of age (Coulter 1966).

When about $7\frac{1}{2}$ weeks old, fisher kits may begin shaking pieces of bedding or other objects in play (Powell 1993), and kits 3 months old may play with prey before eating (Coulter 1966). At about 4 months, they may attack the head and neck region of small prey. Young American martens are able to kill prey proficiently by $2\frac{1}{2}$ months of age (Remington 1952), a considerably younger age than fishers. Marten and fisher mothers tend to spend extensive time with their kits during the days shortly after birth and then spend progressively less time, in shorter periods, with them (Leonard 1980; Henry et al. 1997), although not all females show this pattern (Paragi 1990).

Intraspecific aggression in captive fisher kits appears when they are around 3 months old and kits are intolerant of each other by $5\frac{1}{2}$ months (Hodgson 1937; Powell 1993). A mother fisher observed by Coulter (1966) became increasingly hostile toward her two kits beginning late in their fourth month. At age $5\frac{1}{2}$ months, one kit was killed and the other injured by the mother. Wild kits followed by Paragi (1990) using radiotelemetry remained in their mothers' territories into the winter. The kits, now juveniles, tended to avoid areas used most by their mothers. By age 1 year, juveniles have established their own home ranges.

ECOLOGY

Population Dynamics. Unharvested populations of fishers and martens exhibit marked fluctuations in size, sometimes in excess of an order of magnitude, in response to fluctuations in prey populations (Powell 1994a). Weckwerth and Hawley (1962) and Thompson and Colgan (1987) reported fourfold and sixfold changes in population densities of martens. Where fishers and martens prey heavily on snowshoe hares, their populations fluctuate in response to the roughly 10-year cycle in hare density (Bulmer 1974, 1975). Fryxell et al. (1999) suggested that a harvested population of martens in southern Ontario was relatively stable because the martens switch to alternate prey as prey populations change. Nonetheless, Fryxell et al. (1999) reported that the rate of increase for their marten population changed significantly with abundances of major prey. Douglas and Strickland (1987) questioned whether many fisher populations fluctuate in response to snowshoe hare cycles.

Age structures of natural fisher and marten populations change with population density and therefore are never stable (Powell 1994a). When prey populations increase, the juvenile cohort constitutes a larger than average proportion of a population, and when prey populations decline, older cohorts predominate (Strickland and Douglas 1987; Thompson and Colgan 1987). When prey populations remain low for extended periods, young animals again make up a larger proportion of a population because old animals finally die and total population size decreases. Age structure of harvested populations differs from that of unharvested populations because few individuals reach old ages in harvested populations, especially males (Douglas and Strickland 1987; Strickland and Douglas 1987).

Density, Spatial Organization, and Home Range. Original estimates of densities of fishers came from tracking studies (deVos 1952; Hamilton and Cook 1955; Coulter 1966) and yielded densities as high as 1 fisher/2.6 km^2. Recent research using live-trapping and radiotelemetry have yielded consistently lower estimates of density (Table 29.1). Douglas and Strickland (1987) estimated preharvest densities in southern Ontario were about 1 fisher/6.5 km^2, whereas estimates from live-trapping data have ranged from 1 fisher/2.6 km^2 to 1 fisher/20 km^2, depending on habitat and season.

Because martens are smaller than fishers, their densities tend to be considerably higher than those of fishers. Francis and Stephenson (1972) estimated a density of about 1.5 martens/km^2 in southern Ontario. Soutiere (1979) estimated a similar density for a population in relatively undisturbed forest in Maine, but only 0.4/km^2 in commercially clearcut forests. Archibald and Jessup (1984) estimated similar densities for martens in Yukon Territory in both fall and winter.

Home range sizes for fishers and martens vary across their ranges, probably depending on densities of prey (Powell 1994a). Buskirk and

TABLE 29.1. Recent density estimates for fishers and martens in North America

Place	Density		Technique	Source
	Animals/km^2	km^2/Animal		
Fisher				
Ontario	0.15	6.5	Harvest returns	Douglas and Strickland 1987
New Hampshire	0.11	8.9–9.2	Livetrapping	Kelly 1977
New Hampshire (suitable habitat)	0.13–0.26	3.9–7.5	Livetrapping	Kelly 1977
Maine (summer)	0.09–0.34	2.8–10.5	Livetrapping	Arthur et al. 1989
Maine (winter)	0.05–0.12	8.3–20.0	Livetrapping	Arthur et al. 1989
Maine (suitable habitat)	0.09–0.38	2.6–11.7	Livetrapping	Coulter 1966
Upper Peninsula Michigan	0.08	12–13	Trapper survey	Peterson et al. 1977
Upper Peninsula Michigan	0.08	12–13	Livetrapping	Powell 1977
California	0.31	3.2	Livetrapping	Buck et al. 1983
Marten				
Yukon Territory (fall)	0.6	1.7	Livetrapping	Archibald and Jessup 1984
Yukon Territory (winter)	0.4	2.5	Livetrapping	Archibald and Jessup 1984
Southern Ontario	1.5	0.67	Livetrapping	Francis and Stephenson 1972
Maine (undisturbed forest)	1.2	0.83	Livetrapping	Soutiere 1979
Maine (disturbed forest)	0.4	2.5	Livetrapping	Soutiere 1979

NOTE: References to "suitable habitat" refer to authors's evaluations of habitat.

McDonald (1989) found no geographic pattern to the variation in home range sizes for martens, but Thompson and Colgan (1987) found that home range sizes of martens were smaller when prey density was high. From 17 studies, Powell (1994a) calculated a mean home range size of 8.1 km^2 for male martens and 2.3 km^2 for females. Means for individual studies ranged from 2.0 ± 2.6 km^2 ($\pm$ *SD*) for males ($n = 5$) and 0.6 km^2 for females ($n = 1$) in Montana (Burnett 1981) to 27 and 17 km^2 ($n = 1$ for each sex) in Newfoundland (Bateman 1986; Bissonette et al. 1988). From six studies, Powell (1994a) calculated a mean home range size of 38 km^2 for male fishers and 15 km^2 for females. Means for individual studies ranged from 19 ± 17 km^2 ($\pm$ *SD*) for adult males ($n = 3$) and 15 ± 5 km^2 for females ($n = 5$) in New Hampshire (Kelly 1977) to 79 ± 35 km^2 for males ($n = 6$) and 32 ± 23 km^2 for females ($n = 4$) in Idaho (Jones 1991).

Fishers and martens exhibit intrasexual territoriality consistently (Powell 1994a). Intrasexual territoriality allows the home ranges of a male and a female to overlap, although the animals may compete for limiting resources in their area of overlap. For all mustelids, and for fishers specifically, Powell (1979a, 1993) found that intrasexual territoriality, large sexual dimorphism in body size, elongate shape, and high degree of carnivory are correlated. Intrasexual territoriality may be possible in fishers and martens because their large sexual dimorphism in body size might allow members of the two sexes to have different diets. This hypothesis, however, is unsupported (Coulter 1966; Clem 1977; Powell 1979a, 1993; Holmes and Powell 1994). Spacing behavior of a species often varies across its range and through time, and such variation has been documented for fishers and martens (reviewed by Powell 1994a). Powell (1994a) proposed that intrasexual territoriality is part of a continuum, from transiency through exclusive territories to intrasexual territories to extensive home range overlap, that depends on the distribution and availability of limiting resources. For mustelids, he proposed that patchy distribution of prey and temporary resource depression of prey availability for foraging predators allow two individuals to overlap territories with minor cost. Females gain no benefit from this overlap, but males minimize their chances of reproductive failure. Thus, intrasexual territoriality is imposed on females by males, who are larger than, and dominant to, females.

Habitat. In the broadest sense, fishers and American martens occupy mesic, conifer-dominated forests with abundant physical structure near the ground. Both species avoid areas lacking overhead cover (Buskirk and Powell 1994). Sometimes, talus or boulders, subterranean lava tubes, or shrubs provide suitable overhead structure in otherwise open areas. How dependent martens and fishers are on late-successional forests with their associated physical complexity is a matter of considerable debate. In western North America, the need for old growth forest is fairly clear. Xeric forest types are subject to episodic fire, which removes woody structure near the ground, whereas mesic types burn less often and retain woody structures near the ground (Thomas et al. 1988). Here, fishers are closely associated with riparian areas, and martens consistently select mesic, late-successional stands. By contrast, in the temperate East, a stronger deciduous component is typical of late seres, spruce budworm outbreaks occur fairly often, and old-growth conditions are less typical of old stands, particularly in black spruce *Picea mariana*). Also in the East, balsam fir (*Abies balsamifera*) can produce complex structure near the ground after only a few decades of succession. In addition, snowshoe hares, which are more important prey for martens in the East than the West, associate with early deciduous seres in the East (Potvin et al. 2000).

Exceptions to these patterns provide insight into important aspects of habitat for martens. For example, martens use young conifer forests regenerating from clearcuts on Vancouver Island, but seek large stumps that predate the cutting for resting sites (Baker 1992). Thus, the part of the early-successional forest that is actually used by martens is a remnant of the old-growth forest and not an element related to cutting. Fishers and martens need physical structure near the ground in winter for access to subnivean spaces in which to forage and to rest (Buskirk and Ruggiero 1994).

Most studies of habitat selection have been conducted at the scale of the stand or microsite and have found that fishers prefer sites dominated by mid- to late-successional stands of conifers, although they will use partially or entirely deciduous stands. Riparian stands dominated by Douglas fir (*Pseudotsuga menziesii*) are important to fishers in the West. Martens use stands dominated by conifers, but use those with a greater deciduous component in the East than the West. In the West, stand types preferred by martens are moist-site associations like Pacific silver fir (*Abies amabilis*)–western hemlock (*Tsuga heterophylla*), Engelmann spruce (*Picea engelmannii*)–subalpine fir (*Abies lasiocarpa*), and white spruce (*Picea glauca*)–black spruce (reviewed by Buskirk and Ruggiero 1994). In the East, balsam fir and black spruce tend to be preferred. Although Chapin et al. (1997) found little selection by martens for coniferous overstory among stands in Maine, Cobb (2000) found that martens in conifer habitats in Ontario had more fat than martens living in mixed-hardwood habitats.

In Maine, marten populations declined precipitously when timber harvest led to >60% of the forest in early-successional stages, and they selected home ranges with <40% early-successional forest (Chapin et al. 1998; Payer 1999). Partial harvesting of timber reduces habitat quality, in part through reduction of prey and possibly through increased risk of predation (Fuller 1999), and industrial forest management may reduce carrying capacity for martens by half (Payer 1999). Hargis and Bissonette (1997) and Hargis et al. (1999) found similar responses of martens to habitat fragmentation in the intermountain West. In addition, natural and trapping mortalities are both lower in areas not logged (Thompson 1994).

Powell (1994b) considered landscape-scale selection by fishers, and found that, at a coarse scale, fishers selected for pine and lowland conifer habitat. For resting sites, however, fishers were more selective of local cover types, strongly preferring lowland conifer. Kelly (1977), Arthur et al. (1989), and Jones and Garton (1994) all found similar results. Carroll et al. (1999) and Kelly (1977) both found that habitat selection by fishers appears to be dominated by factors acting at the home range scale and above. Weir and Harestad (1997), however, found no landscape-level trends in habitat selection by fishers. For martens, Chapin et al. (1998) in Maine, Hargis et al. (1999) in Utah, and Potvin et al. (2000) in Quebec examined selection at landscape scales. All three studies found a fairly consistent upper limit to the amount of openings in the forest (including clear-cutting and natural openings) tolerated by martens: 25–30% of a marten's home range. Potvin et al. (2000) showed a strong, negative, linear relationship ($r = -0.78$) between the size of the core area of the home range of martens and the proportion of it that was uncut forest. Wisz (1999) found that the areas of mountain ranges and their distances from either the main Rocky Mountains or the nearest other isolated mountain range predicted well the distribution of martens in insular mountain ranges of Montana. Also, nonforested habitats below the elevational limit of trees were effective barriers to travel by martens, with as little as 6.5 km of nonforest habitat precluding colonization of suitable habitat (Wisz 1999). For areas of continuous forest, Minta et al. (1999) hypothesized that martens are most selective at the finest scales (scale of foraging and resting sites) and coarsest scales (population selection of landscape types), but relatively less selective at intermediate scales (patch and home range).

For fishers, seasonal differences in habitat use are not well documented, but Kelly (1977) and Arthur et al. (1989) noted potentially less habitat selection in summer than in winter. Fishers avoid habitats with deep, soft snow because of their heavy foot loadings (Powell and Zielinski 1994; Krohn et al. 1995, 1997). Seasonally, martens show their strongest habitat specialization in winter (reviewed by Buskirk and Ruggiero 1994), when habitat specialization is most closely tied to structurally complex forests. In summer, martens venture into more diverse cover types, including alpine areas above elevational treeline in western mountains (Streeter and Braun 1968), but all studies that have examined seasonal use of nonforested areas by martens have shown less use in winter (Koehler and Hornocker 1977; Soutiere 1979; Spencer et al. 1983). In alpine areas in summer, boulders fill the need of martens for overhead structure. Within seasons, martens vary their habitat selection depending on weather. Buskirk et al. (1989) and Wilbert et al. (2000) showed that, during winter, martens associated more closely with coniferous forest with old-growth characteristics during cold periods than at other times. This pattern of use likely provides martens with warm microsites in severe weather.

Young martens (<1 year old) use a broader range of habitats than do older individuals (Burnett 1981; Buskirk et al. 1989). Older animals tend to use more physically complex forest types than juveniles. Whether inexperienced martens are less selective or are excluded from the best habitats by territory-holding adults is not clear.

Foraging, resting, and Denning. Habitat use has also been studied in relation to specific behaviors. For fishers, Powell (1994b) showed that habitat selection for resting sites was stronger than for foraging, and that resting sites tended to be in closed habitats. For martens, a gradient in the strength of habitat selection has been proposed (Burnett 1981; Schumacher 1999): foraging (weakest selection), resting, maternal denning (females with older kits), and natal denning (females with young kits, strongest selection). For members of both species, this gradient of selection tends to place the most structurally complex stands at the preferred end of the scale. In the West, the most structurally complex stands are old growth coniferous forest.

Natal dens are used by mothers and neonatal young, and typically are in cavities in very large logs, snags, or live trees (Ruggiero et al. 1998; Schumacher 1999). Maternal dens are used by mothers and older, but still dependent young, and tend to be in less-specialized structures, more like resting sites (Ruggiero et al. 1998). Nearly all dens used to raise young fishers have been high in hollow trees. Thirty-eight natal dens used by fishers outfitted with transmitter collars by Leonard (1980), Arthur (1987), Paragi (1990), and Truex et al. (1998) were in large trees and 1 (in Montana) was in a hollow log (Roy 1991). Fifty-six natal and maternal dens of fishers found by Powell et al. (1997a) were also in large trees. In northern studies, over half the natal dens were in aspens *Populus* spp.). Powell et al. (1997a) found that choice of habitat for natal and maternal dens did not differ from that of the landscape, but female fishers did choose large trees. Female fishers will use one to three dens to raise a litter and these are also invariably in large trees (Paragi 1990; Powell et al. 1997a; Truex et al. 1998). The natal den observed by Leonard (1980) was approximately 7 m high in a hollow, living quaking aspen (*P. tremuloides*) and had two entrances: an entrance too small for the mother to use at the bottom of the hollow and a larger hole approximately 10 m above the ground from which the mother descended to her kits. The nest area was on flat, rough wood with no nesting material and was extremely neat after the kits left, with no sign that fishers had been raised there.

Energetics. For martens, energy is believed to be a currency closely linked to fitness in winter (Wilbert et al. 2000), and several aspects of their life history contribute to limitation of energy used. They have very limited body-fat reserves (Buskirk and Harlow 1989; Cobb 2000), have limited fasting endurance (Harlow and Buskirk 1991), and have very large home ranges (Buskirk and McDonald 1989). The lower critical temperature (T_{lc}) for animals at rest in winter is 16°C (Buskirk et al. 1988), which is well above virtually all winter temperatures experienced by martens. This high T_{lc} is due to small body size and to only moderately efficient fur, and leads to high mass-specific heat loss while resting (Buskirk et al. 1988). To reduce this heat loss, martens enter shallow torpor daily in winter (Buskirk et al. 1988). In addition, heat generated by martens while foraging should lower T_{lc}, as it does for weasels (Sandell 1989), and may put martens in thermal neutrality. In winter, martens thermoregulate behaviorally (Buskirk et al. 1989; Taylor and Buskirk 1994), alternating among positions relative to the snow surface and among forest types to minimize thermal losses while resting. Martens are thought to maximize foraging efficiency in winter by investigating sites where coarse woody debris penetrates the snow surface, providing olfactory information about prey and access to subnivean spaces.

Powell (1979b, 1981) modeled energy budgets of fishers, estimated parameter values using captive fishers trained to run on a treadmill while connected to an oxygen analyzer, and tested the model using field data. The fishers he studied in Michigan's Upper Peninsula appeared to be on a positive energy budget preying predominantly on snowshoe hares and porcupines (*Erethizon dorsatum*) and scavenging dead white-tailed deer (*Odocoileus virginianus*). Patterns of sexual dimorphism in that fisher population were consistent with Moors's (1974) hypothesis that females are small to minimize energy expenditure for maintenance during reproduction, whereas sexual selection selects for large males. In addition, daily energy expenditure of a female with kits is predicted to exceed that of a large male (Powell and Leonard 1983), but lifetime energy expenditure is more equal. Thus, large sexual dimorphism may equalize lifetime reproductive costs for males and females.

BEHAVIOR

The time that an animal spends active and the timing of that activity are influenced by proximate environmental events, such as a recent meal and weather, but the animal's endogenous circadian pacemaker favors activity at times of day (i.e., night, day, dawn, and dusk) that are favored by natural selection (Daan and Aschoff 1982; Zielinski 2000). Martens have been estimated to be active as little as 16% of the day in the winter to 60% in the summer (Thompson and Colgan 1994). Powell (1979b, 1981) estimated that fishers are active about 30% of the day in winter. The marten's patterns of activity vary with the activity budgets of its dominant prey, which vary seasonally and geographically (Hauptman 1979; Zielinski et al. 1983; Zielinski 2000). Consequently, martens are diurnal in winter and crepuscular in summer (More 1978), or largely diurnal in summer and nocturnal in winter (Zielinski et al. 1983), or diurnal in summer (Martin 1987), or arythmic in summer and diurnal in winter (Thompson and Colgan 1994). Fishers are characterized as generally crepuscular, though their activity patterns are variable (Kelly 1977; Powell 1979b, 1981; Johnson 1984; Arthur and Krohn 1991). Activity in male fishers increases during the mating season (Arthur and Krohn 1991). Activity of females is least during pregnancy and increases with the age of their kits (Leonard 1980).

FOOD HABITS

Foods. Diets of martens and fishers are similar in that both eat primarily rodents, lagomorphs, birds, and sometimes insects, fruits, and carrion when available (Martin 1994). Individual studies reveal diverse diets, suggesting that these mammals are opportunistic predators influenced by local abundance and availability of potential prey (Ben-David et al. 1997; Powell and Zielinski 1994; Powell et al. 1997b). This flexibility explains why diets are more diverse in summer than in winter (Buskirk and Ruggierro 1994; Martin 1994; Powell and Zielinski 1994) and in southern and eastern than in northern and western regions (Zielinski et al. 1983, 1999; Buskirk and MacDonald 1984; Martin 1994). Although members of both species eat a large variety of foods, only a few items dominate and the abundance of these items in particular may directly affect the abundance of martens (Weckwerth and Hawley 1962) and fishers (Powell 1993).

Voles (*Clethrionomys* spp., *Microtus* spp., and *Phenacomys* spp.) are the dominant prey of martens across their geographic range (Buskirk and Ruggierro 1994; Martin 1994). Because food preferences can be determined only by comparing what is available with what is consumed, few studies can distinguish foods that are consumed from those that are actually preferred. *Microtus* spp., however, appear to be preferred prey because, unlike red-backed voles (*Clethrionomys* spp.), deer mice (*Peromyscus* spp.) and shrews (*Sorex* spp., *Blarina* spp.), *Microtus* are consumed in a proportion greater than that of their availability (Weckwerth and Hawley 1962; Buskirk and MacDonald 1984). Prevalence of microtines in the diet may increase with latitude; voles constitute the greatest proportion of the diet of martens in Alaska and the Yukon (Douglas et al. 1983; Buskirk and MacDonald 1984).

Ben-David et al. (1997) suggested that preference for small mammals as prey appears to increase when they are least abundant. Thompson and Colgan (1990) and Poole and Graf (1996), in contrast, believed that martens primarily forage for large prey (e.g., snowshoe hares) and that they capture small mammals proficiently, but incidentally. Other researchers have found that large prey constitute more of the winter than the summer diet (Zielinski et al. 1983; Thompson 1986), but this pattern may be a product of prey availability rather than preference. Fluctuations in the commercial harvest of marten pelts were correlated with the snowshoe hare cycle until around 1920 (Bulmer 1974, 1975). Martens have important ecological relationships with red squirrels (*Tamiasciurus hudsonicus*) and Douglas's squirrels (*T. douglasii*) (Buskirk 1984; Corn and Raphael 1992; Sherburne and Bissonnette 1993) and in some studies these species constitute a significant part of the diet (Martin 1994). In coastal areas, fish and birds appear to substitute for small mammal prey (Nagorsen et al. 1989, 1991).

Fishers eat snowshoe hares and porcupines in most places where the diet has been studied (Powell 1993; Martin 1994). The fisher–porcupine predator–prey system has been the subject of considerable study (reviewed by Powell 1993) and the importance of porcupines as prey for fishers is reflected in the evolution of unique hunting and killing behaviors (Powell and Brander 1977; Powell 1978). That snowshoe hares are important prey is suggested by the correlation between the number of commercially harvested fisher pelts and the 10-year cycle of hare abundance (Bulmer 1974, 1975). Where snowshoe hares and porcupines are uncommon, the diets of fishers become more diverse and can include significant quantities of other mammals, reptiles, insects, and fungi (Zielinski et al. 1999). Small mammals are frequent components of the fisher's diet, but, unlike the diet of the marten, voles are not always dominant in this category (Martin 1994). Diets of fishers are consistent with optimal diet choices (Powell 1993).

Larger size of fishers generally makes more prey species potentially available to them, thus their diets are usually more diverse than those of martens (Martin 1994). Males of both species also appear to have a greater range of prey sizes available to them (Martin 1994; Poole and Graf 1996), but despite suggestions to the contrary (Rosenzweig 1966), no consistent differences in the diet exist between the sexes in either species (Thompson and Colgan 1990; Powell 1993) nor does evidence exist for morphological change induced by resource partitioning (Holmes and Powell 1994). Diversity of the diets of both species decreases when large prey (e.g., snowshoe hares) are consumed, which typically occurs in the winter (Buskirk and Ruggiero 1994; Martin 1994) and in northern portions of the ranges (Nagorsen et al. 1989).

Both species eat fruits and disperse seeds (Martin 1994; Hickey et al. 1999). Fruits of shrubs and trees can constitute as much as 30% of the summer diet (Stevens 1968), and two California studies reported evidence for the consumption by fishers of the fruiting bodies of false truffles (*Rhizopogon* spp.) (Grenfell and Fasenfest 1979; Zielinski et al. 1999). Davison (1975) could not induce captive fishers to eat fruit, which suggests that they do so only as a last resort. Martens and fishers scavenge carrion readily (e.g., Powell 1979b, 1981; Martin 1994; Ben-David et al. 1997) and are easily lured to traps, track plates, or camera stations using meat as bait (Strickland and Douglas 1987; Zielinski and Kucera 1995).

Foraging. Martens and fishers are active year-round, have a demanding metabolism (Davison et al. 1978; Buskirk et al. 1988; Powell 1993; Harlow 1994), and, especially martens, store very little energy as fat (Buskirk and Harlow 1989). Thus, selection is strong for efficient foraging behavior, particularly in winter. Both species use a variety of hunting methods to maximize rates of prey capture. They respond to variation in forest patch structure, physical structure near the ground, prey abundance, and probably prey behaviors (Buskirk and Powell 1994). They search for prey where prey are abundant and available (More 1978; Powell 1981; Spencer et al. 1983; Buskirk and MacDonald 1984; Arthur et al. 1989), but it is not clear whether they search for prey that are easy to catch or instead search habitats in which prey are vulnerable to capture (Buskirk and Powell 1994). Both species climb and move through tree canopies proficiently, but forage predominantly on the ground (Clark and Campbell 1976; Powell 1980; Raine 1981; Zielinski et al. 1983). During winter, martens, more than fishers, forage beneath the snow for small mammals. Unlike weasels, however, martens and fishers are too large to pursue microtine rodents in their burrows. Subnivean access points used by martens are frequently associated with the middens of red squirrels (Sherburne and Bissonette 1993). Fishers rarely forage under the snow, undoubtedly because their large body size makes travel through snow and in subnivean spaces difficult (Raine 1983; Krohn et al. 1995).

Fishers and martens appear to share two types of foraging: area-restricted search and directional search. The former is used opportunistically to surprise prey in temporary refuges and is characterized by "zigzag" search typical of the mustelines (Powell 1978). The latter is used by martens to investigate tree squirrel activity centers (Corn and Raphael 1992; Sherburne and Bissonette 1993) and by fishers to investigate snowshoe hare habitat (Powell 1978, 1993). Both species minimize the amount of time spent foraging in forest openings (Buskirk and Powell 1994; Buskirk and Ruggierro 1994; Powell and Zielinski 1994). Martens and fishers do not generally pursue prey long distances (Powell 1993), although Raine (1981) documented two instances of a fisher chasing a snowshoe hare over 1 km. Neither fishers nor martens use the "mouse pounce" hunting behavior that is so successful for canids (Buskirk and Powell 1994). Instead, evidence from following tracks in snow indicates that prey are surprised in refuges, sometimes after being tracked in the snow, and are captured only if they are overtaken quickly. Martens use a variety of additional techniques, including ambush, excavation, and hunting perches (Spencer and Zielinski 1983).

MORTALITY

Fishers in captivity have lived longer than 10 years (Bronx Zoo, New York, unpublished records), and Arthur et al. (1992) livetrapped a wild fisher estimated to be $10^{1}/_{2}$ years old. Other studies have found no fishers older than 7 years (Weckwerth and Wright 1968; Kelly 1977).

Natural causes of mortality for fishers and martens are poorly known. Fishers exhibit a low incidence of disease, although enough fishers have been livetrapped to document sarcoptic mange (O'Meara et al. 1960; Coulter 1966), Aleutian disease, leptospirosis, toxoplasmosis, and trichinosis (Douglas and Strickland 1987). Fishers also exhibit low levels of parasitism (Hamilton and Cook 1955), and those animals with parasites have low infestation (Coulter 1966). Fourteen genera of nematodes, two of cestodes, two of trematodes, and a protozoan have been documented in fishers (Powell 1993). Fishers in Ontario carried detectable levels of DDT, chlordane, dieldrin, mirex, and polychlorinated biphenyls (Douglas and Strickland 1987). Five genera of fleas were collected from martens in California, and 4 of 18 individual martens were positive for plague antibodies (Zielinski 1984). Helminth parasites are common, at least in western populations of martens (Holmes 1963; Hoberg et al. 1990; Foreyt and Lagerquist 1993).

Little evidence exists that healthy fishers are subject to predation by other predators, though occasional mortality from other predators has been noted (Buck et al. 1983; Douglas and Strickland 1987; Krohn et al. 1994). Roy (1991), however, reported that a reintroduction of fishers to the Cabinet Mountains of Montana was hindered by predation on fishers by predators they had not experienced in their original range, including mountain lion (*Puma concolor*), coyote (*Canis latrans*), wolverine (*Gulo gulo*), golden eagle (*Aquila chrysaetos*), and lynx (*Lynx canadensis*).

Martens may be more subject to predation than fishers. Eight of 15 martens studied by Hodgman et al. (1997) that died in an area without trapping were killed by mammalian and avian predators. Similarly, 22 of 35 martens studied by Bull and Heater (1995, 2001) in Oregon died, and 18 of these were killed by bobcats (*Lynx rufus*), raptors, and other martens. Bull and Heater also found that the probability of survival of martens >9 months old was 0.56 for 1 year, 0.38 for 2 years, 0.22 for 3 years, and 0.16 for 4 years, which yields an average survival of approximately 0.65 per year over 4 years.

MANAGEMENT AND CONSERVATION

Conservation of fishers and martens has two areas of emphasis: (1) concern about populations that are vulnerable to extirpation and (2) interest in the sustainable harvest of animals from healthy populations for their fur. The particular emphasis depends on the size of the population and the historical use of, and current interest in, fur as a commodity in a particular region. In some instances, the status of a species can change from rare and protected to abundant and commercially harvested over a relatively short period (e.g., fisher: Coulter 1966; Kohn et al. 1993; Powell 1993). These recoveries occur when animals recolonize improved habitat, as in the northeastern and midwestern United States, where fishers have responded to successional change (Balser and Longley 1966; Brander and Books 1973; Powell 1993), and as a result of reintroductions of animals into favorable habitat in unoccupied portions of their range (Powell 1993; Slough 1994). Martens are trapped for their fur in all but a few states and provinces in the United States and Canada, whereas fishers are protected throughout most of the western states (but not provinces) and legally trapped throughout Canada and the midwestern and eastern United States (Ruggiero et al. 1994; Ray 2000). In the past, management of fishers included using them to control porcupine populations to reduce timber damage (Powell 1993).

Protected Populations. When populations become protected from trapping for fur, important information about their distribution and demography is no longer available. A number of nonlethal approaches, such as track plates and camera stations, have substituted for trapping as a means of determining the status of martens and fishers (e.g., Raphael 1994; Zielinski and Kucera 1995; Foran et al. 1997). These methods have been used to map the current distribution in regions where commercial trapping no longer occurs (e.g. Kucera et al. 1995; Zielinski et al. 1995) and to produce habitat models (Carroll et al. 1999) that can be used to assess and monitor change in habitat suitability over time.

The concept of a metapopulation of populations (Levins 1968; Wright 1978; Harrison and Taylor 1997) linked by occasional dispersal may be especially applicable to conservation of fishers and martens in the fragmented habitat in the western United States. The discontinous pattern of habitat for fishers in the West (e.g., Weir and Harestad 1997; Carroll et al. 1999) and the pattern of marten occurrence across western mountain ranges (Wisz 1999) are examples of this phenomenon operating at very large spatial and temporal scales. On a smaller scale, evidence suggests that martens abandon, or fail to colonize, home range-sized landscapes with less than about 60% mature forest cover (Bissonette et al. 1997; Chapin et al. 1998; Hargis et al. 1999; Payer 1999; Potvin et al. 2000). This reinforces earlier studies showing that martens avoid regenerating clearcuts for several decades (Campbell 1979; Thompson and Harestad 1994). Fishers demonstrate a similar avoidance of clearcut areas, especially in the West (Harris et al. 1982; Buck et al. 1994), and are associated with unfragmented forest (Rosenberg and Raphael 1986; Carroll et al. 1999). Martens may exhibit the greatest selection of habitat at the smallest (microhabitat) and the largest (landscape) scales (Minta et al. 1999), which suggests that managers should provide adequate densities of snags, large trees, and logs and also be conscious of the arrangement, composition, and connectivity of blocks of habitat (Bissonette et al. 1989). Conservation of marten and fisher populations at risk will require planning to protect large blocks of mature forest and the connections between them (Bissonette and Broekhuizen 1995).

The following subspecies are attracting conservation concern due either to their small population sizes or to their vulnerability to human activities: *M. a. humboldtensis* in northwestern California (Zielinski et al. 2001), *M. a. atrata* in Newfoundland (Burnett et al. 1989), *M. a. americana* in eastern Canada (Thompson 1991), and *M. p. pacifica* in the western United States (Powell and Zielinski 1994; Zielinski et al. 1995). *M. a. humboldtensis* was exploited heavily in the early 1900s (Twining and Hensley 1947) and occurs in a region where much of the original redwood forest has been harvested and is managed by private timber companies under short-rotation harvests (Thornburgh et al. 2000; Zielinski et al. 2001). *M. a. atrata* and *M. a. americana* in eastern Canada have suffered from timber harvest and trapping, and *M. a. atrata* is also frequently killed in snares set for snowshoe hares in Newfoundland. *M. p. pacifica* has been extirpated from most of Washington and Oregon (Aubry and Houston 1992; Drew et al. 2003; Stinson and Lewis 1998) and from the northern Sierra Nevada (Zielinski et al. 1995), presumably because of the historical effect of trapping and the widespread effects of clearcut timber harvest. Reintroductions of fishers into apparently suitable habitats in the western United States have not resulted in the success achieved in the eastern United States (Irvine et al. 1964; Roy 1991; Hienemeyer 1993; Powell 1993; Aubry and Lewis 2003).

Trapped Populations. Trapping of martens and fishers is controlled differently in the jurisdictions across the species's ranges using various combinations of limiting the number of licensed trappers, varying the time and length of trapping seasons, and setting quotas. Trapping may be limited to November and December, when pelts are prime. Such timing of trapping also targets young animals and may remove many animals that would die of natural causes later in the winter (Strickland et al. 1982a, 1982b; Krohn et al. 1994; Strickland 1994).

Fishers and martens are easily overharvested, even in a short, early trapping season, especially when trapping intensity is heavy. Such overharvesting led, at least in part, to the reductions of the ranges of both fishers and martens through the early 1900s. Where trapping is heavy, limiting the number of licensed trappers and setting quotas is necessary (Strickland et al. 1982a, 1982b).

Determining when trapping restrictions are necessary is not easy. Strickland (1994) reported that ratios of juveniles to adult females, juveniles to adults, and females to males had been used successfully to estimate the percentage of the populations being harvested. The first two ratios decrease and the third increases as greater percentages of populations are harvested. One must establish these relationships, however, before this approach can be used to regulate harvest, and some populations appear not to maintain stable relationships (Fortin and Cantin 1994). Fryxell et al. (2001) used age distribution in the harvest to back-calculate minimum number of martens alive in previous years and thereby set harvest quotas. Ratios of juveniles to adults and age distributions depend on reproduction in the preceding spring and survival of juveniles, both of which depend on prey abundance. Prey abundance varies by orders of magnitude from year to year for fishers and martens (reviewed by Powell 1993, 1994a), varies with changes in habitat quality, and affects sex ratio of martens (Payer 1999).

Recent research indicates that habitat quality interacts with trapping and that habitat quality must be considered whenever fishers and martens are trapped, especially martens. Extensive timber harvest leads to reduced marten populations and may reduce carrying capacity by half (Fuller 1999; Payer 1999). In addition, trapping mortality appears to be additive to natural mortality in industrial forests (Payer 1999). Responses of martens to habitat change and to forest fragmentation may not be linear (Hargis and Bissonette 1997). Consequently, managers must be alert to and avoid habitat and mortality thresholds when establishing trapping quotas. In some forests, even very low levels of trapping may limit marten (Schneider 1997) and fisher (Powell 1979c) populations.

A proposal by deVos (1951) to manage fisher and marten populations using refuges has received considerable discussion (Archibald and Jessup 1984; Thompson and Colgan 1987; Buskirk 1993) and the strategy has been proposed for managing martens in areas where management is highly contended (e.g., Tongass National Forest; USDA. Forest Service 1993). Refuges of appropriate size can function to maintain viable populations of martens and fishers from which juveniles disperse, thereby providing colonizers for refuges with low populations and supplementing trapped populations between refuges (Archibald and Jessup 1984; Thompson and Colgan 1987; Hodgman et al. 1994). If refuges are established, however, but populations not monitored, managers may assume that populations are stable within refuges when they are not and that animals can travel between refuges when they cannot (Buskirk 1993; Hodgman et al. 1997). Sizes of refuges, habitat quality, variability among and within refuges, and genetic variability of the target species affect the source–sink dynamics (Pulliam 1988) of a refuge system of management and must be considered in any such management plan for fishers and martens. In areas with high road densities and high trapping pressure, a refuge system alone probably will not suffice (Hodgman et al. 1997).

RESEARCH NEEDS

As is the case with most mammals discussed in this volume, martens and fishers have much to gain from research that (1) builds on inductions, deductions, and solidly grounded hunches resulting from field research and (2) tests ecological, genetic, developmental, and evolutionary theory on martens and fishers, which inevitably differ from the animals used to develop the theory and from the animals usually used to test it. Intriguing theory relative to martens and fishers deals with the relationship between habitat and fitness, especially related to multiple habitat scales (microhabitat, home range, landscape, region).

Some of the most compelling new knowledge about carnivore ecology involves competitive interactions. Competition, particularly interference competition, appears to affect carnivore communities in powerful ways, yet the ways in which martens and fishers participate in these interactions is poorly understood. In addition, carnivores may have powerful top-down effects on communities and ecosystems, effects that may even affect primary production through trophic cascades.

Management agencies are particularly interested in understanding habitat requirements of martens and fishers. Specifically, region-specific responses of martens and fishers and their prey to novel patterns of landscape alteration, timber harvest, and fire management must be learned. Understanding how martens and fishers view habitat from the framework of source–sink theory may fundamentally alter our understanding of how martens and fishers coexist with managed forests.

LITERATURE CITED

Anderson, E. 1970. Quaternary evolution of the genus *Martes* (Carnivora: Mustelidae). Acta Zoologica Fennica 130:1–133.

Archibald, W. R., and R. H. Jessup. 1984. Population dynamics of the pine marten (*Martes americana*) in the Yukon Territory. Pages 81–97 *in* R. Olson, R. Hastings, and F. Geddes, eds. Northern ecology and resource management. University of Alberta Press, Edmonton, Canada.

Arthur, S. M. 1987. Ecology of fishers in south-central Maine. Ph.D. Dissertation, University of Maine, Orono.

Arthur, S. M., and W. B. Krohn. 1991. Activity patterns, movements, and reproductive ecology of fishers in southcentral Maine. Journal of Mammalogy 72:379–85.

Arthur, S. M., W. B. Krohn, and J. R. Gilbert. 1989. Habitat use and diet of fishers. Journal of Wildlife Management 53:680–88.

Arthur, S. M., R. A. Cross, T. F. Paragi, and W. B. Krohn. 1992. Precision and utility of using cementum annuli for determining ages of fishers. Wildlife Society Bulletin 20:402–5.

Asdell, S. A. 1946. Patterns of mammalian reproduction. Comstock Press, Ithaca, NY.

Aubry, K. B., and D. B. Houston. 1992. Distribution of the fisher (*Martes pennanti*) in Washington. Northwestern Naturalist 73:69–79.

Aubry, K. B., and J. C. Lewis. 2003. Extirpation and reintroduction of fishers (*Martes pennanti*) in Oregon: Implications for their conservation in the Pacific states. Biological Conservation in press.

Baker, J. M. 1992. Habitat use and spatial organization of pine marten on southern Vancouver Island, British Columbia. M.S. Thesis, Simon Fraser University, Burnaby, British Columbia, Canada.

Balser, D. S., and W. H. Longley. 1966. Increase of the fisher in Minnesota. Journal of Mammalogy 47:547–50.

Bateman, M. C. 1986. Winter habitat use, food habits and home range size of the marten, *Martes americana,* in western Newfoundland. Canadian Field-Naturalist 100:58–62.

Ben-David, M., R. W. Flynn, and D. M. Schell. 1997. Annual and seasonal changes in diets of martens: Evidence from stable isotope analysis. Oecologia 111:280–91.

Berg, W. E. 1982. Reintroduction of fisher, pine marten, and river otter. Pages 159–73 *in* G. C. Sanderson, ed. Midwest furbearer management. Proceedings of the 43rd Midwest fish and wildlife conference. Wichita, KS.

Bissonette, J. A., and S. Broekhuizen. 1995. *Martes* populations as indicators of habitat spatial patterns: The need for a multiscale approach. Pages 95–121 *in* W. Z. Lidicker, Jr., ed. Landscape approaches in mammalian ecology and conservation. University of Minnesota Press, Minneapolis.

Bissonette, J. A., R. J. Frederickson, and B. J. Tucker. 1988. The effects of forest harvesting on marten and small mammals in western Newfoundland. Newfoundland and Laborador Wildlife Division and Corner Brook Pulp and Paper, Ltd. Unpublished report. Utah State University, Logan.

Bissonette, J. A., R. J. Frederickson, and B. J. Tucker. 1989. Pine marten: A case for landscape level management. Transactions of the North American Wildlife and Natural Resources Conference 54:89–101.

Bissonette, J. A., D. J. Harrison, C. D. Hargis, and T. G. Chapin. 1997. The influence of spatial scale and scale-sensitive properties on habitat selection by American marten. Pages 368–85 *in* J. A. Bissonette, ed. Wildlife and landscape ecology. Springer-Verlag, New York.

Blanchard, H. 1964. Weight of a large fisher. Journal of Mammalogy 45:487–88.

Brander, R. B., and D. J. Books. 1973. Return of the fisher. Natural History 82:52–57.

Buck S., C. Mullis, and A. Mossman. 1983. Final report: Coral Bottom-Hayfork Bally fisher study. Unpublished report. USDA Forest Service and Humboldt State University, Arcata, CA.

Buck, S., C. Mullis, and A. Mossman. 1994. Habitat use by fishers in adjoining heavily and lightly harvested forest. Pages 368–76 *in* S. W. Buskirk, A. Harestad, and M. Raphael, eds. Martens, sables and fishers: Biology and conservation. Cornell University Press, Ithaca, NY.

Bull, E. L., and T. W. Heater. 1995. Intraspecific predation on American marten. Northwestern Naturalist 76:132–34.

Bull, E. L., and T. W. Heater. 2001. Survival, causes of mortality, and reproduction of the American marten in northeastern Oregon. Northwestern Naturalist. 82:1–6.

Bulmer, M. G. 1974. A statistical analysis of the 10-year cycle in Canada. Journal of Animal Ecology 43:701–18.

Bulmer, M. G. 1975. Phase relations in the ten-year cycle. Journal of Animal Ecology 44:609–22.

Burnett, G. W. 1981. Movements and habitat use of American marten in Glacier National Park, Montana. M.S. Thesis, University of Montana, Missoula.

Burnett, J. A., C. T. Dauphine, S. H. McCrindle, and T. Mosquin. 1989. On the brink: Endangered species in Canada. Western Producer Prairie Books, Saskatoon, Saskatchewan, Canada.

Buskirk, S. W. 1984. Seasonal use of resting sites by martens in south-central Alaska. Journal of Wildlife Management 48:950–53.

Buskirk, S. W. 1993. The refugium concept and the conservation of forest carnivores. Proceedings of the International Union of Game Biologists 21:242–45.

Buskirk, S. W. 1994. Introduction to the genus *Martes*. Pages 1–10 *in* S. W. Buskirk, A. S. Harestad, M. G. Raphael, and R. A. Powell, eds. Martens, sables, and fishers: Biology and conservation. Cornell University Press, Ithaca, NY.

Buskirk, S. W. 1999. Mesocarnivores of Yellowstone. Pages 165–87 *in* T. W. Clark, S. C. Minta, P. K. Karieva, and A. P. Curlee, eds. Carnivores in ecosystems: The Yellowstone experience. Yale University Press, New Haven, CT.

Buskirk, S. W., and H. J. Harlow. 1989. Body-fat dynamics of the American marten (*Martes americana*) in winter. Journal of Mammalogy 70:191–93.

Buskirk, S. W., and L. L. McDonald. 1989. Analysis of variability in home-range size of the American marten. Journal of Wildlife Management 53:997–1004.

Buskirk, S. W., and S. O. MacDonald. 1984. Seasonal food habits of marten in southcentral Alaska. Canadian Journal of Zoology 62:944–50.

Buskirk, S. W., and R. A. Powell. 1994. Habitat ecology of fishers and American martens. Pages 283–96 *in* S. W. Buskirk, A. S. Harestad, M. G. Raphael, and R. A. Powell, eds. Martens, sables, and fishers: Biology and conservation. Cornell University Press, Ithaca, NY.

Buskirk, S. W., and L. F. Ruggierro. 1994. The American marten. Pages 7–37 *in* L. F. Ruggierro, K. B. Aubry, S. W. Buskirk, L. J. Lyon, and W. J. Zielinski, eds. The scientific basis for conservation of forest carnivores: American marten, fisher, lynx, and wolverine in the western United States (General Technical Report RM-254). U.S. Forest Service.

Buskirk, S. W., P. F. A. Maderson, and R. M. O'Conner. 1986. Plantar glands in North American Mustelidae. Pages 617–22 *in* D. Duvall, D. Muller-Schwarze, and R. M. Silverstein, eds. Chemical signals in vertebrates, Vol. 4. Plenum Press, New York.

Buskirk, S. W., H. J. Harlow, and S. C. Forrest. 1988. Temperature regulation in American marten (*Martes americana*) in winter. National Geographic Research 4:208–18.

Buskirk, S. W., S. C. Forrest, M. G. Raphael, and H. J. Harlow. 1989. Winter resting site ecology of marten in the central Rocky Mountains. Journal of Wildlife Management 53:191–96.

Campbell, T. M. 1979. Short-term effects of timber harvests on pine marten ecology. M.S. Thesis, Colorado State University, Fort Collins.

Carr, S. M., and S. A. Hicks. 1997. Are there two species of marten in North America? Genetic and evolutionary relationships within *Martes*. Pages 15–28 *in* G. Proulx, H. N. Bryant, and P. M. Woodard, eds. *Martes:* Taxonomy, ecology, techniques, and management. Provincial Museum of Alberta, Edmonton, Canada.

Carroll, C., W. J. Zielinski, and R. F. Noss. 1999. Using presence–absence data to build and test spatial habitat models for the fisher in the Klamath Region, USA. Conservation Biology 13:1259–1344.

Chapin, T. G., D. M. Phillips, D. J. Harrison, and E. C. York. 1997. Seasonal selection of habitats by resting martens in Maine. Pages 166–81 *in* G. Proulx, H. N. Bryant, and P. M. Woodard, eds. *Martes:* Taxonomy, ecology, techniques, and management. Provincial Museum of Alberta, Edmonton, Canada.

Chapin, T. G., D. J. Harrison, and D. D. Katnik. 1998. Influence of landscape pattern on habitat use by American marten in an industrial forest. Conservation Biology 12:1327–37.

Cherepak, R. B., and M. L. Connor. 1992. Constantly pregnant ... well almost. Reproductive hormone levels of the fisher (*Martes pennanti*), a delayed implanter. Norwegian Journal of Agriculture Sciences (Supplement) 9:150–54.

Clark, T. W., and T. M. Campbell III. 1976. Population organization and regulatory mechanisms of pine martens in Grand Teton National Park, Wyoming. Pages 293–95 *in* R. M. Linn, ed. Conference on scientific research in national parks, Vol., 1. Washington, DC: National Park Service.

Clem, M. K. 1977. Food habits, weight changes and habitat selection of fisher during winter. M.S. Thesis, University of Guelph, Guelph, Ontario, Canada.

Cobb, E. W. 2000. Physical condition of American martens, *Martes americana,* from two forest regions in northeastern Ontario. M.S. Thesis, Laurentian University Sudbury, Ontario, Canada.

Coffin, K. W., Q. J. Kujala, R. J. Douglass, and L. R. Irby. 1997. Interactions among marten prey availability, vulnerability and habitat structure. Pages 199–210 *in* G. Proulx, H. N. Bryant, and P. M. Woodard, eds. *Martes:* Taxonomy, ecology, techniques and management. Provincial Museum of Alberta, Edmonton, Canada.

Corn, J. G., and M. G. Raphael. 1992. Habitat characteristics at marten subnivean access sites. Journal of Wildlife Management 56:442–48.

Coulter, M. W. 1966. Ecology and management of fishers in Maine. Ph.D. Dissertation, State University College of Forestry, Syracuse University, Syracuse, NY.

Crowley, S. K., W. B. Krohn, and T. F. Paragi. 1990. A comparison of fisher reproductive estimates. Transactions of the Northeastern Section of the Wildlife Society 47:36–42.

Daan, S., and J. Aschoff. 1982. Circadian contributions to survival. Pages 305–21 *in* J. Aschoff, S. Daan, and G. Groos, eds. Vertebrate circadian systems. Springer-Verlag, Berlin.

Dagg, A. I., D. Leach, and G. Sumner-Smith. 1975. Fusion of the distal femoral epiphyses in male and female marten and fisher. Canadian Journal of Zoology 53:1514–18.

Davison, R. P. 1975. The efficiency of food utilization and energy requirements of captive fishers. M.S. Thesis, University of New Hampshire, Concord.

Davison, R. P., W. W. Mautz, H. H. Hayes, and J. B. Holter. 1978. The efficiency of food utilization and energy requirements of captive female fishers. Journal of Wildlife Management 42:811–21.

deVos, A. 1951. Overflow and dispersal of marten and fisher from wildlife refuges. Journal of Wildlife Management 15:164–75.

deVos, A. 1952. Ecology and management of fisher and marten in Ontario (Technical Bulletin). Ontario Department of Lands and Forests, Toronto.

Douglas, C. W., and M. A. Strickland. 1987. Fisher. Pages 511–30 *in* M. Novak, J. A. Baker, M. E. Obbard, and B. Malloch, eds. Wild furbearer management and conservation in North America. Ontario Ministry of Natural Resources, Toronto.

Douglas, R. J., L. G. Fisher, and M. Mair. 1983. Habitat selection and food habitas of marten, *Martes americana,* in the Northwest Territories. Canadian Field-Naturalist 97:71–74.

Douglas, W. O. 1943. Fisher farming has arrived. American Fur Breeder 16:18–20.

Drew, R. E., J. G. Hallett, K. B. Aubry, K. W. Cullings, M. Koepfs, and W. J. Zielinski. 2003. Conservation genetics of the fisher (*Martes pennanti*) based on mitochondrian DNA sequencing. Molecular Ecology 12:51–62.

Eadie, W. R., and W. J. Hamilton, Jr. 1958. Reproduction of the fisher in New York. New York Fish and Game Journal 5:77–83.

Enders, R. K., and A. C. Enders. 1963. Morphology of the female reproductive tract during delayed implantation in the mink. Pages 129–40 *in* A. C. Enders, ed. Delayed implantation. University of Chicago Press, Chicago.

Enders, R. K., and O. P. Pearson. 1943. The blastocyst of the fisher. Anatomical Review 85:285–87.

Ewer, R. F. 1973. The carnivores. Cornell University Press, Ithaca, NY.

Foran, D. R., K. R. Crooks, and S. C. Minta. 1997. Species identification from scat: An unambiguous method. Wildlife Society Bulletin 25:835–39.

Foreyt, W. J., and J. E. Lagerquist. 1993. Internal parasites from the marten (*Martes americana*) in eastern Washington. Journal of Helminthology 60:72–75.

Fortin, C., and M. Cantin. 1994. Effects of trapping on a newly exploited American marten population. Pages 179–92 *in* S. W. Buskirk, A. S. Harestad, M. G. Raphael, and R. A. Powell, eds. Martens, sables, and fishers: Biology and conservation. Cornell University Press, Ithaca, NY.

Francis, G. R., and A. B. Stephenson. 1972. Marten ranges and food habits in Algonquin Provincial Park, Ontario (Research Report No. 91). Ontario Ministry of Natural Resources, Toronto.

Frost, H. C., and W. B. Krohn. 1994. Capture, care, and handling of fishers (*Martes pennanti*) (Technical Bulletin No. 157). Maine Agricultural and Forest Experiment Station.

Frost, H. C., W. B. Krohn, and C. R. Wallace. 1997. Age-specific reproductive characteristics of fishers. Journal of Mammalogy 78:598–612.

Frost, H. C., E. C. York, W. B. Krohn, K. D. Elow, T. A. Decker, S. M. Powell, and T. K. Fuller. 1999. An evaluation of parturition indices in fishers. Wildlife Society Bulletin 27:221–30.

Fryxell, J., J. B. Falls, E. A. Falls, R. J. Brooks, L. Dix, and M. Strickland. 2001. Harvest dynamics of mustelid carnivores in Ontario, Canada. Wildlife Biology 7:151–159.

Fryxell, J. M., J. B. Falls, and M. A. Strickland. 1999. Density dependence, prey dependence, and population dynamics of martens in Ontario. Ecology 80:1311–21.

Fuller, A. K. 1999. Influence of partial timber harvesting on American marten and their primary prey in northcentral Maine. M.S. Thesis, University of Maine, Orono.

Gibilisco, C. J. 1994. Distributional dynamics of modern *Martes* in North America. Pages 59–71 *in* S. W. Buskirk, A. S. Harestad, M. G. Raphael, and R. A. Powell, eds. Martens, sables, and fishers: Biology and conservation. Cornell University Press, Ithaca, NY.

Grenfell, W. E., and M. Fasenfest. 1979. Winter food habits of fishers, *Martes pennanti,* in northwestern California. California Fish and Game 65:186–89.

Grinnell, J., J. S. Dixon, and L. M. Linsdale. 1937. Fur-bearing mammals of california: Their natural history, systematic status and relations to man, Vol. 1. University of California Press, Berkeley.

Hagmeier, E. M. 1956. Distribution of marten and fisher in North America. Canadian Field-Naturalist 70:149–68.

Hagmeier, E. M. 1959. A re-evaluation of the subspecies of fisher. Canadian Field-Naturalist 73:185–97.

Hagmeier, E. M. 1961. Variation and relationships in North American marten. Canadian Field-Naturalist 75:122–38.

Hall, E. R. 1926. The abdominal skin gland of *Martes*. Journal of Mammalogy 7:227–29.

Hall, E. R. 1942. Gestation period of the fisher with recommendation for the animals' protection in California. California Fish and Game Journal 28:143–47.

Hamilton, W. J., and A. H. Cook. 1955. The biology and management of the fisher in New York. New York Fish and Game Journal 2:13–35.

Hargis, C. D., and J. A. Bissonette. 1997. Effects of forest fragmentation on populations of American marten in the intermountain West. Pages 437–51 *in* G. Proulx, H. N. Bryant, and P. M. Woodard, eds. *Martes:* Taxonomy, ecology, techniques, and management. Provincial Museum of Alberta, Edmonton, Canada.

Hargis, C. D., J. A. Bissonette, and D. L. Turner. 1999. The influence of forest fragmentation and landscape pattern on American martens. Journal of Applied Ecology 36:157–72.

Harlow, H. 1994. Trade-offs associated with size and shape of American martens. Pages 391–403 *in* S. W. Buskirk, A. S. Harestad, M. G. Raphael, and R. A. Powell, eds. Martens, sables, and fishers: Biology and conservation. Cornell University Press, Ithaca, NY.

Harlow, H. J., and S. W. Buskirk. 1991. Comparative plasma and urine chemistry of fasting white-tailed prairie dogs (*Cynomys leucurus)* and American marten (*Martes americana*): Representative fat- and lean-bodied animals. Physiological Zoology 64:1262–78.

Harris, L. D., C. Maser, and A. McKee. 1982. Patterns of old-growth harvest and implications for Cascades wildlife. Transactions of the North American Wildlife and Natural Resources Conference 47:374–92.

Harrison, S., and A. D. Taylor. 1997. Empirical evidence for metapopulation dynamics. Pages 27–42 *in* I. Hanski and M. E. Gilpin, eds. Metapopulation biology: Ecology, genetics, and evolution. Academic Press, San Diego, CA.

Hauptman, T. N. 1979. Spatial and temporal distribution and feeding ecology of the pine marten. M.S. Thesis, Idaho State University, Pocatello.

Heidt, G. A., M. K. Petersen, and G. L. Kirkland, Jr. 1968. Mating behavior and development of least weasels (*Mustela nivalis*) in captivity. Journal of Mammalogy 49:413–19.

Henry, S. E., E. C. O'Doherty, L. F. Ruggiero, and W. C. van Sickle. 1997. Maternal den attendance patterns of female American martens. Pages 78–85 *in* G. Proulx, H. N. Bryant, and P. M. Woodard, eds. *Martes:* Taxonomy, ecology, techniques, and management. Provincial Museum of Alberta, Edmonton, Canada.

Hickey, J. R., R. W. Flynn, S. W. Buskirk, K. G. Gerow, and M. F. Willson. 1999. An evaluation of a mammalian predator, *Martes americana,* as a disperser of seeds. Oikos 87:499–508.

Hienemeyer, K. S. 1993. Temporal dynamics in the movements, habitat use, activity, and spacing of reintroduced fishers in northwestern Montana. M.S. Thesis, University of Montana, Missoula.

Hoberg, E. P., K. A. Aubry, and J. D. Brittell. 1990. Helminth parasites in martens (*Martes americana*) and ermines (*Mustela erminea*) from Washington, with comments on the distribution of *Trichinella spiralis*. Journal of Wildlife Diseases 26:447–52.

Hodgman, T. P., D. J. Harrison, D. D. Katnik, and K. D. Elowe. 1994. Survival in an intensively trapped marten population in Maine. Journal of Wildlife Management 58:593–600.

Hodgman, T. P., D. J. Harrison, D. M. Phillips, and K. D. Elowe. 1997. Survival of American marten in an untrapped forest preserve in Maine. Pages 86–99 *in* G. Proulx, H. N. Bryant, and P. M. Woodard, eds. *Martes:* Taxonomy, ecology, techniques, and management. Provincial Museum of Alberta, Edmonton, Canada.

Hodgson, R. G. 1937. Fisher farming. Fur Trade Journal of Canada (Toronto).

Holmes, J. C. 1963. Helminth parasites of pine marten, *Martes americana,* from the district of MacKenzie. Canadian Journal of Zoology 41:333.

Holmes, T. 1980. Locomotor adaptations in the limb skeletons of North American mustelids. M.A. Thesis, Humboldt State University, Arcata, CA.

Holmes, T., and R. A. Powell. 1994. Morphology, ecology, and the evolution of sexual dimorphism in North American *Martes*. Pages 72–84 *in* S. W. Buskirk, A. S. Harestad, M. G. Raphael, and R. A. Powell, eds. Martens, sables, and fishers: Biology and conservation. Cornell University Press, Ithaca, NY.

Irvine, G. W., L. T. Magnus, and B. J. Bradle. 1964. The restocking of fishers in lake states forests. Transaction of the North American Wildlife and Natural Resource Conference 29:307–15.

Johnson, S. A. 1984. Home range, movements, and habitat use of fishers in Wisconsin. M.S. Thesis, University of Wisconsin, Stevens Point.

Jones, J. L. 1991. Habitat use of fishers in northcentral Idaho. M.S. Thesis, University of Idaho, Moscow.

Jones, J. L., and E. O. Garton. 1994. Selection of successional stages by fishers in north-central Idaho. Pages 377–88 *in* S. W. Buskirk, A. S. Harestad, M. G. Raphael, and R. A. Powell, eds. Martens, sables, and fishers: Biology and conservation. Cornell University Press, Ithaca, NY.

Jonkel, C. J., and R. P. Weckwerth. 1963. Sexual maturity and implantation of blastocysts in wild pine marten. Journal of Wildlife Management 27:93–98.

Kelly, G. M. 1977. Fisher (*Martes pennanti*) biology in the White Mountain National Forest and adjacent areas. Ph.D. Dissertation, University of Massachusetts, Amherst.

Kleiman, D. 1977. Monogamy in mammals. Quarterly Review of Biology 52:39–69.

Koehler, G. H., and M. G. Hornocker. 1977. Fire effects on marten habitat in the Selway–Bitterroot Wilderness. Journal of Wildlife Management 41:500–505.

Kohn, B. E., N. F. Payne, J. E. Ashbrenner, and W. A. Creed. 1993. The fisher in Wisconsin (Technical Bulletin No. 183). Wisconsin Department of Natural Resources, Madison.

Krohn, W. B., S. M. Arthur, and T. F. Paragi. 1994. Mortality and vulnerability of a heavily trapped fisher population. Pages 137–46 *in* S. W. Buskirk, A. S. Harestad, M. G. Raphael, and R. A. Powell, eds. Martens, sables, and fishers: Biology and conservation. Cornell University Press, Ithaca, NY.

Krohn, W. B., K. D. Elowe, and R. B. Boone. 1995. Relations among fishers, snow, and martens: Development and evaluation of two hypotheses. Forestry Chronicle 71:97–105.

Krohn, W. B., W. J. Zielinski, and R. B. Boone. 1997. Relations among fishers, snow, and martens in California: Results from small-scale spatial comparisons. Pages 211–32 *in* G. Proulx, H. N. Bryant, and P. M. Woodard, eds. *Martes:* Taxonomy, ecology, techniques, and management. Provincial Museum of Alberta, Edmonton, Canada.

Kucera, T. E., W. J. Zielinski, and R. H. Barrett. 1995. Current distribution of the American marten, *Martes americana,* in California. California Fish and Game 81:96–103.

LaBarge, T., A. Baker, and D. Moore. 1990. Fisher (*Martes pennanti*): Birth, growth and development in captivity. Mustelid and Viverrid Conservation 2:1–3.

Laberee, E. E. 1941. Breeding and reproduction in fur bearing animals. Fur Trade Journal of Canada (Toronto).

Leach, D. 1977a. The forelimb musculature of marten (*Martes americana* Turton) and fisher (*Martes pennanti* Erxleben). Canadian Journal of Zoology 55:31–41.

Leach, D. 1977b. The description and comparative postcranial osteology of marten (*Martes americana* Turton) and fisher (*Martes pennanti* Erxleben): The appendicular skeleton. Canadian Journal of Zoology 55:199–214.

Leach, D., and V. S. de Kleer. 1978. The description and postcranial osteology of the marten (*Martes americana* Turton) and fisher (*Martes pennanti* Erxleben): The axial skeleton. Canadian Journal of Zoology 56:1180–91.

Leonard, R. D. 1980. Winter activity and movements, winter diet and breeding biology of the fisher in southeast Manitoba. M.S. Thesis, University of Manitoba, Winnipeg, Canada.

Levins, R. 1968. Evolution in changing environments. Princeton University Press, Princeton, NJ.

MacDonald, S. O., and J. A. Cook. 1996. The land mammal fauna of southeast Alaska. Canadian Field-Naturalist 110:571–98.

Markley, M. H., and C. F. Bassett. 1942. Habits of captive marten. American Midland Naturalist 28:604–16.

Martin, S. K. 1987. Ecology of the pine marten (*Martes americana*) at Sagehen Creek, California. Ph.D. Dissertation, University of California, Berkeley.

Martin, S. K. 1994. Feeding ecology of American martens and fishers. Pages 297–315 *in* S. W. Buskirk, A. Harestad, and M. Raphael, eds. Martens, sables and fishers: Biology and conservation. Cornell University Press, Ithaca, NY.

Mead, R. A. 1994. Reproduction in *Martes*. Pages 404–22 *in* S. W. Buskirk, A. S. Harestad, M. G. Raphael, and R. A. Powell, eds. Martens, sables, and fishers: Biology and conservation. Cornell University Press, Ithaca, NY.

Merriam, C. H. 1890. Descriptions of 26 new species of North Americam mammals. North American Fauna 4:1–55.

Minta, S. C., P. M. Kareiva, and A. P. Curlee. 1999. Carnivore research and conservation: Learning from history and theory. Pages 323–404 *in* T. W. Clark, A. P. Curlee, S. C. Minta, and P. M. Kareiva, eds. Carnivores in ecosystems: The Yellowstone experience. Yale University Press, New Haven, CT.

Moors, P. J. 1974. The annual energy budget of a weasel (*Mustela nivalis* L.) in farmland. Ph.D. Dissertation, University of Aberdeen, Aberdeen, Scotland.

More, G. 1978. Ecological aspects of food selection in pine marten (*Martes americana*). M.S. Thesis, University of Alberta, Edmonton, Canada.

Nagorsen, D. W., K. F. Morrison, and J. E. Forsberg. 1989. Winter diet of Vancouver Island marten (*Martes americana*). Canadian Journal of Zoology 67:1394–1400.

Nagorsen, D. W., R. W. Campbell, and G. R. Giannico. 1991. Winter food habits of marten, *Martes americana,* on the Queen Charlotte Islands. Canadian Field-Naturalist 105:55–59.

O'Meara, D. C., D. D. Payne, and J. F. Witter. 1960. *Sarcoptes* infestation of a fisher. Journal of Wildlife Management 24:339.

Paragi, T. F. 1990. Reproductive biology of female fishers in southcentral Maine. M.S. Thesis, University of Maine, Orono.

Payer, D. C. 1999. Influences of timber harvesting and trapping on habitat selection and demography of marten. Ph.D. Dissertation, University of Maine, Orono.

Petersen, L. R., M. A. Martin, and C. M. Pils. 1977. Status of fishers in Wisconsin, 1975 (Report No. 92).Wisconsin Department of Natural Resources, Madison.

Peterson, R. L. 1966. The mammals of eastern Canada. Oxford University Press, Toronto.

Poole, K. G., and R. P. Graf. 1996. Winter diet of marten during a snowshoe hare decline. Canadian Journal of Zoology 74:456–66.

Poole, K. G., G. M. Matson, M. A. Strickland, A. J. Magoun, R. P. Graf, and L. M. Dix. 1994. Age and sex determination for American martens and fishers. Pages 204–23 *in* S. W. Buskirk, A. S. Harestad, M. G. Raphael, and R. A. Powell, eds. Martens, sables, and fishers: Biology and conservation. Cornell University Press, Ithaca, NY.

Potvin, F., L. Belanger, and K. Lowell. 2000. Marten habitat selection in a clearcut boreal landscape. Conservation Biology 14:844–57.

Powell, R. A. 1977. Hunting behavior, ecological energetics and predator–prey community stability of the fisher (*Martes pennanti*). Ph.D. Dissertation, University of Chicago, Chicago.

Powell, R. A. 1978. A comparison of fisher and weasel hunting behavior. Carnivore 1:28–34.

Powell, R. A. 1979a. Mustelid spacing patterns: Variations on a theme by *Mustela*. Zeitschrift für Tierpsychologie 50:153–65.

Powell, R. A. 1979b. Ecological energetics and foraging strategies of the fisher (*Martes pennanti*). Journal of Animal Ecology 48:195–212.

Powell, R. A. 1979c. Fishers, population models and trapping. Wildlife Society Bulletin 7:149–54.

Powell, R. A. 1980. Fisher arboreal activity. Canadian Field-Naturalist 94:90.

Powell, R. A. 1981. Fisher food requirements and hunting behavior. Pages 883–917 *in* J. A. Chapman and D. Pursley, eds. Proceedings of the first worldwide furbearer conference. Frostburg, MD.

Powell, R. A. 1982. The fisher: Life history, ecology and behavior. University of Minnesota Press, Minneapolis.

Powell, R. A. 1993. The fisher: Life history, ecology and behavior, 2nd ed. University of Minnesota Press, Minneapolis.

Powell, R. A. 1994a. Structure and spacing in *Martes* populations. Pages 101–21 *in* S. W. Buskirk, A. S. Harestad, M. G. Raphael, and R. A. Powell, eds. Martens, sables, and fishers: Biology and conservation. Cornell University Press, Ithaca, NY.

Powell, R. A. 1994b. Effects of scale on habitat selection and foraging behavior of fishers in winter. Journal of Mammalogy 75:349–56.

Powell, R. A., and R. B. Brander. 1977. Adaptations of fishers and porcupines to their predator–prey system. Pages 45–53 *in* R. L. Phillips and C. Jonkel, eds. Proceedings of 1975 predator symposium. Montana Forest Conservation Experiment Station, University of Montana, Missoula.

Powell, R. A., and R. D. Leonard. 1983. Sexual dimorphism and energy expenditure for reproduction in female fisher *Martes pennanti*. Oikos 40:166–74.

Powell, R. A., and W. J. Zielinski. 1994. The fisher. Pages 38–73 *in* L. F. Ruggiero, K. B. Aubry, S. W. Buskirk, L. J. Lyon, and W. J. Zielinski, eds. The scientific basis for conserving forest carnivores: American marten, fisher, lynx, and wolverine in the western United States (General Technical Report RM-254). USDA Forest Service, Rocky Mountain Forest and Range Experiment Station, Fort Collins, CO.

Powell, S. M., E. C. York, J. J. Scanlon, and T. K. Fuller. 1997a. Fisher maternal den sites in central New England. Pages 265–78 *in* G. Proulx, H. N. Bryant, and P. M. Woodard, eds. *Martes:* Taxonomy, ecology, techniques, and management. Provincial Museum of Alberta, Edmonton, Canada.

Powell, S. M., E. C. York, and T. K. Fuller. 1997b. Seasonal food habits of fishers in central New England. Pages 279–305 *in* G. Proulx, H. N. Bryant, and P. M. Woodard, eds. *Martes:* Taxonomy, ecology, techniques and management. Provincial Museum of Alberta, Edmonton, Canada.

Pulliam, H. R. 1988. Sources, sinks, and population regulation. American Naturalist 132:652–61.

Raine, R. M. 1981. Winter food habits, responses to snow cover and movements of fisher (*Martes pennanti*) and marten (*Martes americana*) in southeastern Manitoba. M.S. Thesis, University of Manitoba, Winnipeg, Canada.

Raine, R. M. 1983. Winter habitat use and responses to snow cover of fisher (*Martes pennanti*) and marten (*Martes americana*) in southeastern Manitoba. Canadian Journal of Zoology 61:25–34.

Raphael, M. G. 1994. Techniques for monitoring populations of fishers and American martens. Pages 224–40 *in* S. W. Buskirk, A. S. Harestad, M. G. Raphael, and R. A. Powell, eds. Martens, sables, and fishers: Biology and conservation. Cornell University Press, Ithaca, NY.

Ray, J. C. 2000. Mesocarnivores of northeastern North America: Status and conservation issues (Working Paper No. 15). Wildlife Conservation Society, New York.

Remington, J. D. 1952. Food habits, growth, and behavior of two captive pine martens. Journal of Mammalogy 33:66–70.

Rosenberg, K. V., and and R. G. Raphael. 1986. Effects of forest fragmentation on vertebrates in Douglas-fir forests. Pages 263–72 *in* J. Verner, M. L. Morrison, and C. J. Ralph, eds. Wildlife 2000: Modeling habitat relationships of terrestrial vertebrates. University of Wisconsin Press, Madison.

Rosenzweig, M. L. 1966. Community structure in sympatric Carnivora. Journal of Mammalogy 47:602–12.

Roy, K. D. 1991. Ecology of reintroduced fishers in the Cabinet Mountains of northwest Montana. Thesis, University of Montana, Missoula.

Ruggiero, L. F., K. B. Aubry, S. W. Buskirk, L. J. Lyon, and W. J. Zielinski. 1994. American marten, fisher, lynx and wolverine in the western United States (General Technical Report RM-254). USDA Forest Service, Rocky Mountain Forest and Range Experiment Station, Fort Collins, CO.

Ruggiero, L. F., D. E. Pearson, and S. E. Henry. 1998. Characteristics of American marten den sites in Wyoming. Journal of Wildlife Management 62:663–73.

Sandell, M. 1989. Ecological energetics, optimal body size and sexual dimorphism: A model applied to the stoat, *Mustela erminea* L. Functional Ecolology 3:315–24.

Schneider, R. 1997. Simulated spatial dynamics of martens in response to habitat succession in the Western Newfoundland Model Forest. Pages 419–36 *in* G. Proulx, H. N. Bryant, and P. M. Woodard, eds. *Martes:* Taxonomy, ecology, techniques, and management. Provincial Museum of Alberta, Edmonton, Canada.

Schumacher, T. V. 1999. A multi-scale analysis of habitat selection at dens and resting sites of American martens in southeast Alaska. M.S. Thesis, University of Wyoming, Laramie.

Shea, M. E., N. L. Rollins, R. T. Bowyer, and A. G. Clark. 1985. Corpora lutea number as related to fisher age and distribution in Maine. Journal of Wildlife Management 49:37–40.

Sherburne, S. S., and J. A. Bissonette. 1993. Squirrel middens influence marten (*Martes americana*) use of subnivean access points. American Midland Naturalist 129:204–7.

Slough, B. G. 1994. Transplants of American martens: An evaluation of factors in success. Pages 165–78 *in* S. W. Buskirk, A. S. Harestad, and A. S. Raphael, eds. Martens, sables, and fishers: Biology and conservation. Cornell University Press, Ithaca, NY.

Soutiere, E. C. 1979. Effects of timber harvesting on marten in Maine. Journal of Wildlife Management 43:850–60.

Spencer, W. D., and W. J. Zielinski. 1983. Predatory behavior of pine martens. Journal of Mammalogy 64:715–17.

Spencer, W. D., R. H. Barrett, and W. J. Zielinski. 1983. Marten habitat preferences in the northern Sierra Nevada. Journal of Wildlife Management 47:1181–86.

Stevens, C. L. 1968. The food of fisher in New Hampshire. Unpublished report. New Hampshire Department of Fish and Game, Concord.

Stinson, D. W., and J. C. Lewis. 1998. Washington state status report for the fisher. Washington Department of Fish and Wildlife, Wildlife Management Program, Olympia.

Stone, K. D. 2000. Molecular evolution of martens (genus *Martes*). Ph.D. Dissertation, University of Alaska, Fairbanks.

Streeter, R. G., and C. E. Braun. 1968. Occurrence of pine marten, *Martes americana* (Carnivora: Mustelidae) in Colorado alpine areas. Southwestern Naturalist 13:449–51.

Strickland, M. A. 1994. Harvest management of fishers and American martens. Pages 149–64 *in* S. W. Buskirk, A. S. Harestad, M. G. Raphael, and R. A. Powell, eds. Martens, sables, and fishers: Biology and conservation. Cornell University Press, Ithaca, NY.

Strickland, M. A., and C. W. Douglas. 1987. Marten. Pages 531–46 *in* M. Novak, J. A. Baker, M. E. Obbard, and B. Malloch, eds. Wild furbearer management and conservation in North America. Ontario Ministry of Natural Resources, Toronto.

Strickland, M. A., C. W. Douglas, M. Novak, and N. P. Hunziger. 1982a. Fisher. Pages 586–98 *in* J. A. Chapman and G. A. Feldhamer, eds. Wild mammals of North America. Johns Hopkins University Press, Baltimore.

Strickland, M. A., C. W. Douglas, M. Novak, and N. P. Hunziger. 1982b. Marten. Pages 599–612 *in*J. A. Chapman and G. A. Feldhamer, eds. Wild mammals of North America. Johns Hopkins University Press, Baltimore.

Taylor, S. L., and S. W. Buskirk. 1994. Forest microenvironments and resting energetics of the American marten *Martes americana*. Ecography 17:249–56.

Thomas, J. W., L. F. Ruggiero, R. W. Mannan, J. W. Schoen, and R. A. Lancia. 1988. Management and conservation of old-growth forests in the United States. Wildlife Society Bulletin 16:252–62.

Thompson, I. D. 1986. Diet choice, hunting behavior, activity patterns, and ecological energetics of marten in natural and logged areas. Ph.D. Dissertation, Queen's University, Kingston, Ontario, Canada.

Thompson, I. D. 1991. Will marten become the spotted owl of the east? Forestry Chronicle 67:136–40.

Thompson, I. D. 1994. Marten populations in uncut and logged boreal forests in Ontario. Journal of Wildlife Management 58:272–80.

Thompson, I. D., and P. W. Colgan. 1987. Numerical responses of martens to a food shortage in northcentral Ontario. Journal of Wildlife Management 51:824–35.

Thompson, I. D., and P. W. Colgan. 1990. Prey choice by marten during a decline in prey abundance. Oecologia 83:443–51.

Thompson, I. D., and P. W. Colgan. 1994. Marten activity in uncut and logged boreal forests in Ontario. Journal of Wildlife Management 58:280–88.

Thompson, I. D., and A. S. Harestad. 1994. Effects of logging on American martens, and models for habitat management. Pages 355–67 *in* S. W. Buskirk, A. S. Harestad, M. G. Raphael, and R. A. Powell, eds. Martens, sables, and fishers: Biology and conservation. Cornell University Press, Ithaca, NY.

Thornburgh, D. A., R. F. Noss, D. P. Angelides, C. M. Olson, F. Euphrat, and H. H. Welsh, Jr. 2000. Managing redwoods. Pages 229–61 *in* R. F. Noss, ed. The redwood forest. Island Press, Covelo, CA.

Truex, R. L., W. J. Zielinski, R. T. Gollightly, R. H. Barrett, and S. M. Wisely. 1998. A meta-analysis of regional variation in fisher morphology, demography, and habitat ecology in California (Draft Report). California Department of Fish and Game, Wildlife Management Division, Nongame Bird and Mammal Section.

Twining, H., and A. Hensley. 1947. The status of pine martens in California. California Fish and Game Journal 33:133–37.

USDA Forest Service. 1993. Tongass Lands management plan. Tongass National Forest, Alaska.

Weckwerth, R. P., and V. D. Hawley. 1962. Marten food habits and population fluctuations in Montana. Journal of Wildlife Management 26:55–74.

Weckwerth, R. P., and P. L. Wright. 1968. Results of transplanting fishers in Montana. Journal of Wildlife Management. 32:977–80.

Weir, R. D., and A. S. Harestad. 1997. Landscape-level selectivity by fishers in south-central British Columbia. Pages 252–64 *in* G. Proulx, H. N. Bryant, and P. M. Woodard, eds. *Martes:* Taxonomy, ecology, techniques, and management. Provincial Museum of Alberta, Edmonton.

Wilbert, C. J., S. W. Buskirk, and K. G. Gerow. 2000. Effects of weather and snow on habitat selection by American martens (*Martes americana*). Canadian Journal of Zoology 78:1691–96.

Wisz, M. S. 1999. Islands in the Big Sky: Equilibrium biogeography of isolated mountain ranges in Montana. M.A. Thesis, University of Colorado, Boulder.

Wright, P. L. 1953. Intergradation between *Martes americana* and *Martes caurina* in western Montana. Journal of Mammalogy 34:74–86.

Wright, P. L., and M. W. Coulter. 1967. Reproduction and growth in Maine fishers. Journal of Wildlife Management 31:70–87.

Wright, S. 1978. Evolution and the genetics of populations. Vol. 4: Variability within and among natural populations. University of Chicago Press, Chicago.

Zielinski, W. J. 1984. Plague in martens and the fleas associated with its occurrence. Great Basin Naturalist 44:170–75.

Zielinski, W. J. 2000. Weasels and martens: Carnivores in northern latitudes. Pages 94–118 *in*S. Halle and N. C. Stenseth, eds. Activity patterns in small mammals: An ecological approach. Springer-Verlag, Berlin.

Zielinski, W. J., and T. E. Kucera. 1995. American marten, fisher, lynx, and wolverine: Survey methods for their detection (General Technical Report PSW-GTR-157). U.S. Forest Service.

Zielinski, W. J., W. D. Spencer, and R. D. Barrett. 1983. Relationship between food habits and activity patterns of pine martens. Journal of Mammalogy 64:387–96.

Zielinski, W. J., T. E. Kucera, and R. H. Barrett. 1995. The current distribution of fisher, *Martes pennanti,* in California. California Fish and Game 81:104–12.

Zielinski, W. J., N. P. Duncan, E. C. Farmer, R. L. Truex, A. P. Clevenger, and R. H. Barrett. 1999. Diet of fishers (*Martes pennanti*) at the southernmost extent of their range. Journal of Mammalogy 80:961–71.

Zielinski, W. J., K. M. Slauson, C. R. Carroll, C. J. Kent, and D. G. Kudrna. 2001. The status of American martens in coastal forests of the Pacific states. Journal of Mammalogy. 82:478–490.

ROGER A. POWELL, Department of Zoology, North Carolina State University, Raleigh, North Carolina 27695-7617. Email: newf@ncsu.edu.

STEVEN W. BUSKIRK, Department of Zoology and Physiology, University of Wyoming, Laramie, Wyoming 82071. Email: marten@uwyo.edu.

WILLIAM J. ZIELINSKI, Redwood Sciences Laboratory, Pacific Southwest Research Station, U.S. Forest Service., 1700 Bayview Drive, Arcata, California 95521. Email: bzielinski@fs.fed.us.

30

Weasels and Black-footed Ferret

Mustela species

Gerald E. Svendsen

NOMENCLATURE

Common Names. Weasel, ermine, stoat, short-tailed weasel, long-tailed weasel, least weasel, pygmy weasel, black-footed ferret
Scientific Name. Black-footed ferret (*Mustela nigripes*), ermine, short-tailed weasel, or stoat (*Mustela erminea*); long-tailed weasel (*Mustela frenata*); and least weasel (*Mustela nivalis*)

Weasels and the black-footed ferret are small, terrestrial carnivores in the family Mustelidae. The family is large and contains 25 genera and nearly 70 species, which occupy terrestrial, freshwater, and marine habitats throughout the world except for Australia, the Antarctic, and most oceanic islands. The genus *Mustela* contains 16 living species (Wilson and Reeder 1993) of weasels, polecats, ferrets, and mink. Three species of weasels occur in North America. The least weasel and ermine occur in both the New and Old World (Allen 1933; Hall 1944). The long-tailed weasel is limited to the New World (Hall 1951). Hall (1981) recognized 20 subspecies of *M. erminea,* 4 subspecies of *M. nivalis,* and 35 subspecies of *M. frenata* in North America.

There is a single species of black-footed ferret, *M. nigripes* (Hall 1981). The black-footed ferret once was abundant in prairies from Canada to Mexico (Hillman and Clark 1980; Anderson et al. 1986). Today it is endangered and its fate remains uncertain and tied to a reintroduction program that began in 1991 (Miller et al. 1994, 1996; Biggins and Godbey 1995).

Figure 30.1. Distribution of the ermine (*Mustela erminea*).

DISTRIBUTION

The New World *Mustela* occur from the Arctic to the Tropics. *M. erminea* is a circumboreal species and has the most widespread distribution of any mustelid. It is found in the Nearctic from the Arctic south to northern California, Nevada, New Mexico, and Colorado in the West, to the upper Great Lakes states in the Midwest, and to Pennsylvania and Maryland in the East (Fig. 30.1). The species occurs in a variety of habitats, from agricultural lowlands, woodlands, and meadows to montane habitats at elevations of 3000–4000 m. It avoids dense coniferous forests and deserts.

The least weasel also is circumboreal in distribution. In North America, it is found from Alaska southward to Montana and Nebraska in the West; through Minnesota, Wisconsin, Michigan, Illinois, Indiana, and Ohio in the Midwest; and south into western Pennsylvania and through the Appalachians into North Carolina in the East (Fig. 30.2). It is absent along the east coast from New England north through the Maritime Provinces. The distribution of *M. nivalis* is sporadic or disjunct throughout much of its range (Hatt 1940). It inhabits marshes, meadows, cultivated fields, brushy areas, and open woods (Jackson 1961). In Indiana, it occurs primarily in open, cultivated fields (Mumford 1969), whereas in Wisconsin it prefers marshes (Beer 1950).

The long-tailed weasel is distributed from southern Canada throughout all of the United States (Fig. 30.3) and Mexico, southward through all of Central America and into northern South America. It inhabits a variety of habitats from alpine–arctic to tropical, but does not inhabit arid habitats or deserts. The availability of water in summer appears to limit its distribution (Hall 1951). Favored habitats include brushland and open timber, brushy borders to fields, grasslands along creeks and lakes, and swamps.

The black-footed ferret once was distributed over the prairies from southern Alberta and Saskatchewan south to Texas and Arizona, but was never abundant (Fig. 30.4). The historical geographic range of the black-footed ferret coincided with the range of prairie dogs (*Cynomys* sp.) (Hillman et al. 1979; Hillman and Clark 1980; Powell 1982; Anderson et al. 1986). This secretive mustelid spends most of its life underground in the burrows of prairie dogs, with which it shares a close predator–prey association. This may not have always been the case, however. A middle Pleistocene record dating back 750,000–850,000 years was not associated with prairie dogs. This suggests that the predator–prey relationship between prairie dogs and black-footed ferrets was not always as close as it is today. Black-footed ferrets may have had food and habitat requirements much closer to those of their sister taxon the Siberian polecat (*M. eversmanni*), and the predator–prey relationship between black-footed ferrets and prairie dogs may have been a secondary effect of colonization of North America (Owen et al. 2000).

DESCRIPTION

A long, slender body and short legs characterize weasels and the black-footed ferret. The feet have five digits with nonretractible, curved claws. Ears are short and rounded. Pelage of weasels living in northern latitudes

FIGURE 30.2. Distribution of the least weasel (*Mustela nivalis*).

FIGURE 30.3. Distribution of the long-tailed weasel (*Mustela frenata*).

turns white during the winter, but that of weasels living in southern latitudes may remain brown or become a mixture of brown and white. Males are generally larger than females.

Size. The largest weasel is *Mustela frenata,* with a total length of 300–350 mm. The tail is 40–70% of the head and body length. Males are about 10–15% larger than the females in most populations. The dorsal pelage is brown in summer. The ventral pelage is whitish, tinged with yellowish or buffy brown from the chin to the inguinal region. The tail is uniformly brown except for the distinct black tip. Populations in Florida and the southwestern United States sometimes have white or yellowish facial markings. The winter pelage in northern populations is normally entirely white, sometimes tinged with yellow, except for the black tip of the tail. In regions where the two species are sympatric, *M. frenata*

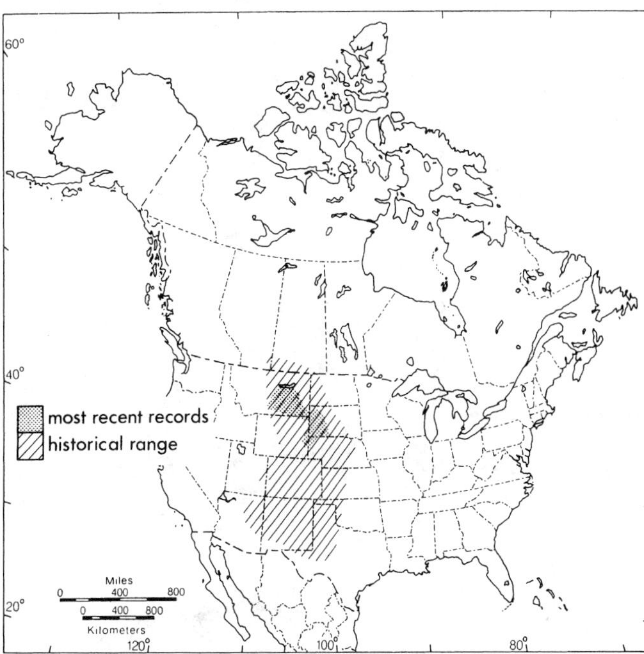

FIGURE 30.4. Distribution of the black-footed ferret (*Mustela nigripes*).

is separable from *M. erminea* by size. The tail is >44% of the head and body length in *M. frenata,* whereas it is about 40% in *M. erminea.* When the sex is known, the greatest breadth of skull will distinguish the two species; *M. frenata* is always larger than *M. erminea.*

M. erminea is a medium-sized to small weasel. The total length is 225–340 mm in males and 190–290 mm in females. The tail is 30–45% of the head and body length and has a distinct black tip. In summer pelage, the dorsal surface is brown and the venter is whitish from the chin to the inguinal region. In some specimens, the white ventral pelage is interrupted by the brown of the upper parts encircling the abdominal region. The winter pelage is white, again with a black-tipped tail. Soles of the feet are densely haired in winter and only a small area of pad is exposed in summer. Males are about 30% larger and heavier than females.

Least weasels are small. The total length is <250 mm in males and 225 mm in females. The tail length is ≤25% of the head and body length, and the tail lacks a distinct black tip. The summer dorsal pelage is a rich chocolate-brown extending well down on each side onto the ventral parts, sometimes meeting midventrally. Underparts are white, or rarely the same brown as the upper parts. The normal winter pelage is entirely white; in rare cases the summer pelage is maintained throughout winter. The small size, length of tail, and absence of a black tip to the tail readily distinguish *M. nivalis* from *M. erminea* and *M. frenata*. Also, the fur of *M. nivalis* will fluoresce under ultraviolet light, producing a lavender color, whereas the fur of the other two species remains dull brown under ultraviolet light (Latham 1953).

The black-footed ferret is larger than any North American weasel. Total length is 500–600 mm, with the tail about one third the length of the head and body. The ears are relatively large and rounded. Males are 10% larger and heavier than females. The pelage is buffy, becoming lighter on the face, throat, and ventrum and brownish over the middle of the back. The feet and the end of the tail are black and a black mask occurs across the face and eyes. The pelage is slightly paler in winter. The skull is large and massive, broad between the orbits, and constricted behind the postorbital processes.

Skull. The skull shape of weasels and the black-footed ferret has a slight facial angle, a long braincase, a short rostrum, and greatly inflated tympanic bullae with the paraoccipital processes closely appressed to the bullae. The skull of adult *M. frenata* (Fig. 30.5) is more angular and the postorbital processes are more pointed than in the skull of

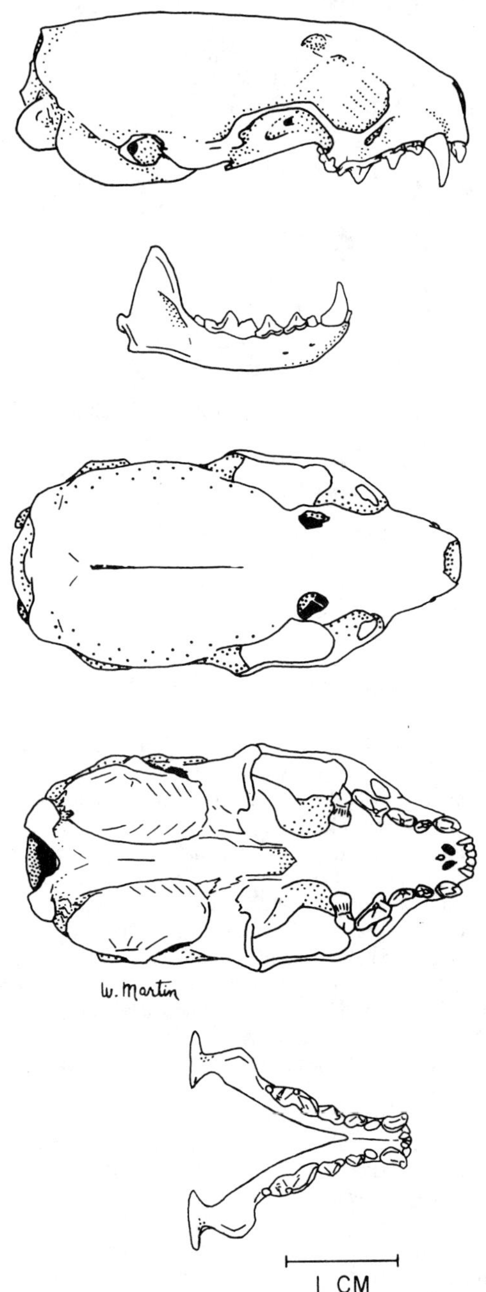

FIGURE 30.5. Skull of the long-tailed weasel (*Mustela frenata*). From top to bottom: lateral view of cranium, lateral view of mandible, dorsal view of cranium, ventral view of cranium, dorsal view of mandible.

M. erminea. The skull of *M. erminea* is elongated with a smoothly rounded braincase. The sagittal crest is weakly developed and the frontal region is flat. The zygomata are slender, weak, and moderately arched. The rostrum is short, the auditory bullae are large and flattened, and there is no auditory meatus. The skull of *M. nivalis* is a miniature of that of *M. erminea*. The braincase is narrower through the mastoid process and has a more distinct sagittal ridge and a more prominent postorbital process. The skull of *M. nigripes* is large and massive compared to the that of the weasels, with the mastoid process projecting and angular (Hillman and Clark 1980; Hall 1981).

Dentition. The dental formula for all species is I 3/3, C 1/1, P 3/3, M 1/2. The permanent dentition is second only to that of the Felidae in the degree of specialization for a diet of flesh. The metaconid of the first lower molar is completely suppressed and the talonid forms an elevated blade for cutting food rather than a basin for crushing. The upper blade of the last upper premolar and the first lower molar (carnassials) form a shearing surface. Other modifications of the dentition for carnivory include reduction in size of the second lower molar and the absence of the last two upper molars and third lower molar. The teeth vary little among the species except in size (Hall 1981).

Specialization for a diet of flesh is equally evident in deciduous and permanent dentition. The upper carnassial of the deciduous dentition is more highly sectorial than the permanent tooth, suggesting that the deciduous dentition is more specialized for carnivory than the permanent dentition. Milk teeth are used for eating solid food as soon as the carnassials are in place, at about 3 weeks after birth (Hamilton 1933). The deciduous teeth are used for almost 2 months before being replaced by permanent teeth in *M. frenata* (Hall 1951).

ANATOMY

A long, slender body that enables them to enter the hiding places of rodents, their main prey, characterizes weasels and the black-footed ferret. There is no elongation of the vertebral column compared to mammals with a body shape like a raccoon (*Procyon lotor*) (Hall 1951). The different species of weasels do not vary in the number of vertebrae except for the caudal vertebrae, which are fewest in the short-tailed species and most numerous in the long-tailed species. There is also some geographic variation in the number of caudal vertebrae within each species. The long, thin shape has disadvantages in cold climates because it exposes a large surface area to cold air. As a result, cold-stressed weasels have metabolic rates 50–100% higher than less slender mammals of similar size (Brown and Lasiewski 1972; King 1989, 1990; McNab 1989).

The shape of the baculum varies among the species of weasels and black-footed ferret. It is a useful characteristic for distinguishing *M. nivalis* from *M. erminea,* and indicates maturity in *M. frenata*. Male *M. frenata* reach adult body size in 3–4 months, yet do not reach sexual maturity until they are about 1 year old. The baculum remains small during the year of immaturity, but develops to adult proportions when the animals become sexually active (Wright 1951).

External characters vary according to geographic distribution, season, age, and gender. Measurements of body size and skull vary with age (Hamilton 1933), and the skull of the female is lighter and more slender than that of the male (Hall 1951). Individual variation occurs in molt patterns, skull measurements, and dentition. Pelage differences occur seasonally and among populations at different latitudes. Body size and shape of the skull vary geographically (Ralls and Harvey 1985; Dayan et al. 1989; Reig 1997). Rosenzweig (1966) statistically analyzed variation in the body size of mammalian carnivores including weasels with respect to latitude, temperature, community production, competition, and size of available prey. He concluded that temperature and latitude appeared to be likely causes of change in body size. Many endotherms follow Bergmann's rule in that races of the same species from cooler climates tend to be larger than those in warmer climates (Mayr 1966). The usual explanation for this trend is that large animals expend less energy for thermoregulation because of the small surface-to-volume ratio, thus it is physiologically more economical for them to live in cold climates. Scholander (1955) criticized this interpretation, and McNab (1971) found no consistent pattern among length, latitude, and feeding habits for either gender of black-footed ferret or weasels. Only least weasels and ermine exhibited increased body length at high latitudes. Variation in the size of weasels and the ferret is believed to be due to the size of available prey and the interaction among sympatric species of weasel, a conclusion also reached by Erlinge (1979) and Powell and King (1997).

PHYSIOLOGY

Metabolism. The elongated body shape of weasels and the ferret results in a higher surface-to-volume ratio than that of standard-shaped mammals of the same weight. The metabolic rates of adult male

M. frenata are greater than those of adult females. At 32°C, the thermoneutral zone, the metabolic rate of resting, fasting postabsortive males, is 136 ± 0.12 kcal/hr. Metabolism was inversely related to ambient temperature below 35°C, and mean body temperature did not vary with ambient temperature. Weasels lost almost twice as much heat to the environment as comparably sized woodrats (*Neotoma* sp.) Evaporative heat loss accounted for about 14% of the minimal metabolic coefficient. The higher heat loss in weasels was attributed largely to their greater surface area, shorter pelage, and inability to achieve a spherical resting posture. Weasels are unable to minimize their heat loss at low ambient temperatures by rolling into a nearly round ball assumed by most mammals. Instead, they coil their elongated bodies into a flattened disc, a shape with considerably more surface area exposed than a sphere. Compensation of heat loss through the evolution of greater insulation qualities of the pelage has not occurred (Brown and Lasiewski 1972).

Least weasels in the Arctic have a metabolic rate two to three times higher than expected from the values for basal metabolic rate and body size derived from the standard "mouse to elephant curve" (McNab 1989). In experiments where ambient temperature was low, weasels were able to mobilize heat production nearly four times greater than the basal rate. This is as high a metabolic rate as a mammal can maintain for long periods of time in a small cage. Compared to arctic weasels, weasels in Wisconsin have a lower metabolic rate, indicating that arctic weasels have an increased resting metabolic rate compared with the southern form (Scholander et al. 1950). Parer and Metcalfe (1967) postulated that in animals weighing <1 kg there is an increased oxygen need by the tissues. This is reflected in an inverse relationship between capillary density and body weight (Schmidt-Nielsen and Pennycuik 1961). Iversen (1972) repudiated this and suggested that the high basal metabolic rate observed in weasels is due to a physiological adjustment to high metabolic rates in the smaller species of mustelids.

Molt. The summer pelage of all North American weasels is brown with white or whitish underparts. Pelage dimorphism occurs in the northern ranges. Increasing day length initiates molting and growth of brown hair and decreasing day length causes molt and growth of white hair (Bissonnette and Bailey 1944). The white pelage is termed *ermine* in the fur trade. Weasels in the southern part of their range in North America retain the brown pelage during both molts. Intermediate populations may have a mottling of white and brown.

The pineal gland produces melatonin, which initiates changes in the central nervous system and causes molt and growth of the white winter pelage (Rust and Meyer 1969). The seasonal pelage cycle is controlled by light (Bissonnette and Bailey 1944), which acts to mediate the activity of the pituitary gland (Lyman 1942; Rust 1965; Rust and Meyer 1968).

Onset of molt to the brown pelage was delayed in a cold room at −7°C, whereas at 21°C brown hairs appeared as soon as the white hairs were shed (Rust 1962). Temperature was a modifying factor affecting the speed and nature of the spring pelage change, but it was not the major factor responsible. Weasels in the north have thicker pelage than those in the south. The soles of the feet are least hairy in regions of high average temperature and hairiest in regions of low average temperature. The degree of hairiness is determined by the diameter and length of individual hairs and the number of hairs/dermal surface area (Hall 1951).

M. nigripes also molts twice a year, but dichromism is not evident. Summer pelage is sleek and short, whereas winter pelage is only slightly longer and darker. A captive ferret was shedding in June, but no abrupt seasonal change in pelage followed (Progulske 1969).

Fall molt of weasels begins on the ventral surface and moves along the flanks, encroaching on the facial region and limbs. The last area to whiten is the dorsum; a few patches of old brown fur on the shoulders, neck, and head are the last to remain. The molt is completed in a little over 3 weeks (Hamilton 1933). Spring molt begins on the dorsum and head and spreads over the body and legs; it is not necessarily symmetrical. Small patches of white in front of the ears may be the last remnants of the winter pelage. Each individual weasel has a slightly different pattern of molt, and the pattern will vary yearly in the same weasel.

REPRODUCTION AND DEVELOPMENT

Delayed Implantation. Many mustelids exhibit an unusually long period between conception and parturition because of delayed implantation (Hamlett 1935; Wright 1942; Sandell 1984). Both *M. frenata* and *M. erminea* have delayed implantation, whereas *M. nivalis* and *M. nigripes* do not (Miller et al. 1996).

Litter Size. Weasels have large litters typical of mustelids in general (Erlinge 1974; King and McMillan 1982; King 1983), whereas black-footed ferrets have smaller litters. Litter size for *M. erminea, M. nivalis,* and *M. frenata* ranges from 4 to 9 with an average of about 6. Mean litter size for *M. nigripes* was 3.3 in Wyoming (Forrest et al. 1988) and 3.4 in South Dakota (Linder et al. 1972).

Mating System. The mating system of weasels and black-footed ferret is polygyny. Territorial males enlarge their home range at the beginning of the mating season and visit females occupying territories within their territory. Males that lack a territory are probably excluded from mating (Sandell 1986).

Breeding and Gestation. Male *M. frenata* become sexually mature at about 1 year of age. The bacula of sexually immature *M. frenata* weigh 14–29 mg. Those of sexually mature individuals weigh 53–100 mg (Petrides 1950). The earliest sign of spermatogenic activity occurs in March, coincident with the beginning of the spring molt. Testes reach a maximum volume seven to eight times larger than the inactive size, and do not regress until August or September. Testes are inactive throughout the winter (Wright 1948a). Female *M. frenata* are fully grown and sexually mature at 3–4 months of age. Sanderson (1949) reported that a female came into heat (estrus) at the age of 9 weeks. Adult and young-of-the-year females come into heat from June through August. A swollen, doughnut-shaped vulva indicates heat, and females are receptive for 3–4 days from the time of first mating. The swollen vulva of successfully mated females shows regression within 4–5 days, whereas the vulva of unmated females remains swollen for several weeks. Estrus does not occur during lactation; females that nurse young come into heat 55–104 days after parturition. Females that give birth and do not nurse come into heat 36–71 days after parturition (Wright 1948a).

Breeding occurs during early summer in *M. erminea* when females are receptive to courting males. Breeding females include the adults as well as young only 60–70 days old but sufficiently sexually mature to mate (Simms 1979). Males do not attain sexual maturity until the following spring. Ermine breed once per year and females may survive for two breeding seasons.

Unlike the ermine and long-tailed weasels, *M. nivalis* does not have delayed implantation and prolonged gestation (Heidt et al. 1968). Males are sexually active in all months but December and January (Baker 1983), and litters have been reported during most months (Banfield 1974). Females reach sexual maturity at 4 months and males at 8 months of age. Females that are born early in the year will breed and produce young that same year.

The breeding season for wild black-footed ferrets probably occurs in March and April (Richardson et al. 1987), corresponding to a shift of activity comparable to that seen in weasels (Erlinge and Sandell 1986) and Siberian polecats (Stroganov 1962). In captivity, the main breeding season for black-footed ferrets is March and April (Carpenter 1985; Miller et al. 1996; Wolf et al. 2000a, 2000b).

Courtship and Copulation. Copulation is prolonged and rough in weasels and the black-footed ferret. Copulation may last >3 hr and occurs repeatedly over several days. There is little difference in mating behavior among the species. Although a female in heat allows a male to approach, vigorous struggling and fighting before copulation is typical. However, the brief struggle and resistance to the advances of the male ends abruptly when he grabs the female by the scruff of the neck with his teeth. If the female continues to struggle, the male holds the female down by the neck until she is subdued, then clasps her lower abdomen with the forelegs and arches his back over her posterior. Copulation occurs in the upright position or from the side. The female remains rather passive during coitus. Bursts of pelvic thrusts alternate with periods of rest. At the end of coitus, the female breaks away (Wright 1948a; Hartman 1964; Heidt et al. 1968). Weasels and black-footed ferret do not exhibit induced ovulation where prolonged copulation is required to initiate release of mature ova. Instead, prolonged copulation may serve to increase sperm transport and subsequent fertilization. This is the case in domestic ferrets (Miller and Anderson 1989) and sables (*Martes zibellina*) (Reed 1946).

Development. The first signs of pregnancy in *M. frenata* occur 50–80 hr after copulation. The two-cell stage is reached by 70–85 hr, the four-cell stage by 81–99 hr, the eight-cell stage by 4–8 days, and morula at 11 days. The early blastula is positioned in the upper end of the uterus. The full blastula with early stages of cell differentiation remains quiescent until spring, then implantation occurs and further development takes place. Active but not implanted blastocysts were recovered 251 days after mating and 27 days before birth (Wright 1948b). Parturition occurs in April and May and a single litter is produced each year.

Male *M. erminea* reach 85% of their adult weight by July, but do not become sexually mature until the following May or June. Spermatogenesis begins with initiation of the spring molt. Testes enlarge to about 13 times the resting size by April and then begin to regress after July (Deanesly 1944). Treatment with melatonin causes regression of testes, indicating that the pineal gland is involved in the regulation of seasonal reproduction (Rust and Meyer 1969). Female *M. erminea* also reach 85% of their adult weight by 4 months of age, but, unlike males, they are sexually mature during their first summer of life. Spontaneous ovulation may begin in June and repeat monthly unless the female is bred. Nine to 10 ova are shed each ovulation (Hamilton 1933). Delayed implantation accounts for all but 21–28 days of the 10-month gestation period.

The least weasel typically has two litters/year and delayed implantation does not occur. Pregnancies are recorded in most months, but peak numbers of pregnancies occur in spring and midwinter. Males from spring litters grow quickly and produce active sperm by fall, but most males do not reproduce until the following year. As noted, spring-born females breed in the year of their birth. Males and females of the fall litter do not become sexually mature until the following spring (Southern 1964). The gestation period for least weasels is about 35 days (Heidt et al. 1968).

The captive breeding program for the endangered black-footed ferret has generated information on reproduction in captivity, but the details of reproduction in the wild are not as well known (Wolf et al. 2000a, 2000b). There are many similarities to reproduction of the domestic ferret (*Mustela putorus*) and the black-footed ferret. Delayed implantation does not occur in black-footed ferrets. Mating takes place in early spring and the young are born about 6 weeks after copulation. This reproductive pattern resembles that of the domestic ferret and the closely related Siberian polecat. Young black-footed ferrets were observed in early July in South Dakota and attained adult size in August (Hillman 1968a). Parturition apparently occurred in May.

Neonates. Weasels and black-footed ferrets are blind and helpless at birth, but postnatal development is rapid. Weasels attain adult size in 4–6 weeks, depending on the species. Least weasels are the smallest and attain adult size and pelage earliest; they become self sufficient in 4–5 weeks. The eyes and ears of young *M. nivalis* open at 26–29 days, whereas those of *M. erminea* and *M. frenata* do not open until 5 weeks of age. Permanent dentition is complete by 40–42 days in *M. nivalis,* but not until 75 days in *M. frenata* (Hamilton 1933). *M. nivalis* young can kill prey by 40 days of age. In general, self-sufficiency of the two larger species of weasel is attained 2–3 weeks later than in *M. nivalis*.

Young *M. erminea* can be distinguished from the other species by a prominent dark-colored mane extending from the forehead to the shoulders, which is evident shortly after birth and persists up to 6 weeks of age. Both *M. erminea* and *M. frenata* exhibit sexual dimorphism in body size as adults. Dimorphism is evident by 2 weeks of age in *M. frenata,* but not until 4 weeks of age in *M. erminea.*

Black-footed ferret kits are born with a sparse coat of white hair, closed eyes, no teeth, flattened ears, and an average birth weight of 7–10 g (Hillman and Carpenter 1983). At 16 days of age, kits have deciduous canines and some premolars. They begin to eat solid foods at about 30 days of age. Eyes open at 34–37 days, and young begin walking when 40 days old. Pigmentation is evident at 16–18 days after birth, with the adult color pattern of dark legs and mask evident at 33 days of age. Kits are weaned at 42 days and appear aboveground when about 60 days old (Vargas 1994; Vargas and Anderson 1998b).

Predatory skills of captive black-footed ferrets improved as the young became older and gained experience. Black-footed ferrets exposed to a more complex environment and greater prey-search opportunities were more likely to make a successful kill than ferrets in a deprived environment (Vargas and Anderson 1999). Experience and witnessing the mother kill a prairie dog increased efficiency. Placement of the bite is innate, and both handling and killing are enhanced by experience (Vargas and Anderson 1998a).

ECOLOGY

Activity. Weasels and the black-footed ferret do not hibernate and are active in winter and summer. Weasels are commonly thought to be nocturnal, but evidence does not support this. In Pennsylvania, peak activity was at dusk in winter (Glover 1942). In Iowa, peak winter activity occurred at sunrise (Polder 1968), and Fitzgerald (1977) observed both hunting and traveling during the day. In a montane habitat in Colorado, *M. frenata* was active throughout the day during the summer, with highest activity occurring in the late afternoon and morning. *M. erminea* alternates periods of activity and rest over a 24-hr period, but appears to be more diurnal in summer and more nocturnal in winter. *M. nivalis* remains active throughout 24-hr periods (Ognev 1935/1962). King (1975) caught more *M. nivalis* during the day than at night, and 98% of all *M. frenata* livetrapped in Colorado were caught between dawn and dusk.

The black-footed ferret is mainly nocturnal (Clark 1976) and appears aboveground for only a few minutes every few days (Hillman 1968b). Both season and weather influenced activity of a captive black-footed ferret. This ferret fed and dug tunnels mostly at night. During the days, it retreated to a hollow tile pipe. During cold and snow, it emerged daily to feed and romp, but spent most of the time in the nest box (Progulske 1969). Most aboveground activity by black-footed ferrets is recorded during late summer, when the young are old enough to be moving about on their own (Hillman 1968b).

In laboratory experiments, both *M. frenata* and *M. nivalis* preferred the brightest illumination available (500–900 lux) (Kavanau 1969). This preference was related to the visual needs of these carnivores for greater color, contrast, pattern, and intensity discrimination. In later studies, it was determined that the visual systems were best adapted to dim light, but suitable for daylight. Complete darkness inhibited running activity (Kavanau and Ramos 1975).

Movement. Movement is related primarily to hunting and seeking mates. Glover (1942) found that male *M. frenata* traveled farther at night than did females in Pennsylvania. Males traveled an average distance of 214 m (18–775), whereas females traveled 105 m (6–430) per night. Weasels in open timber traveled farther per trip than did those in brushland and dense stands of trees. Trails lead from one rodent den to another; dense vegetation with cover was used more than open areas and the same route was not used two nights in a row. In Iowa fields, the average cruising radius was 125 m from a central den; the maximum was 225 m (Polderboer et al. 1941). In Idaho, *M. frenata* and *M. erminea* followed circuits and did not use central dens. *M. frenata* followed 7- to 12-day circuits and the smaller *M. erminea* 10- to 15-day circuits. The same general route was followed on each circuit (Musgrove 1951). Where snow is deep, weasels burrow beneath the snow to hunt (Fitzgerald 1977). They also will use snowshoe hare

(*Lepus americanus*) runways (Keith and Meslow 1966) and pocket gopher (Geomyidae) burrows as foraging routes (Vaughan 1961).

A large male *M. erminea* holds the record for the longest known distance traveled by a weasel, crossing 35 linear km in 7 months (Burns 1964). Peak long-distance movements seem to occur in spring, coincident with mating (Erlinge 1977b), and primarily involve males. Knowledge of movements of *M. nigripes* is limited. Nightly movements are restricted to their home range and consist of forays from their home burrow to surrounding prairie dog burrows; these movements are influenced by the size of the prairie dog town.

Home Range. Home range size varies with habitat, population density, season, gender, food availability, and species. In general, home ranges of males are larger and overlap with several smaller home ranges of females. Spacing results from intrasexual exclusion (Powell 1979). *M. nivalis* is the smallest weasel and it has the smallest mean home range; *M. frenata* is the largest weasel and it has the largest mean home range. Male *M. nivalis* use 7–15 ha, whereas mean home range of females is 1–4 ha. Home ranges are smaller in young pine plantations (1–5 ha) than in mature deciduous woodland (7–15 ha), and home ranges generally are 65–85% smaller in habitats with the most food (King 1975).

Because *M. erminea* is larger than *M. nivalis,* it as a larger mean home range. Males use an average of 34 ha and females 7.4 ha based on snow tracking (Nyholm 1959). Using radio-collared animals, Erlinge (1977b) determined home ranges to be 2–3 ha for females and 8–13 ha for males in late autumn. In late summer, females used 4–10 ha and males 15 ha. Males increased the area of use considerably in spring, coincident with mating. When small rodents were scarce, home ranges were two to three times larger than during periods with food abundance. The largest weasel, *M. frenata,* has home ranges of 12–16 ha, and males have larger home ranges than do females in summer.

Female *M. nigripes* have smaller home ranges than do males. The larger home range of males overlaps those of several females (Biggins et al. 1985; Richardson et al. 1987). Intrasexual exclusion is typical. Burrow structure and density of prairie dogs within a colony influence the spacing of black-footed ferrets.

Population Dynamics. Weasel population densities vary with season, food availability, and species. In favorable habitat, maximum densities of *M. nivalis* may reach 25/km^2; *M. erminea,* 8/km^2; and *M. frenata,* 6–7/km^2 (Glover 1942; Quick 1944; Jackson 1961; King 1975). Population density fluctuated yearly in relation to small-mammal abundance (MacLean et al. 1974; Fitzgerald 1977) and there are great differences in densities in different habitats and parts of the geographic distribution (Polderboer et al. 1941; Glover 1942; King 1975). In central and western New York, *M. frenata* and *M. erminea* occurred in equal numbers, in northern New York *M. erminea* was the most common, and in southern New York *M. frenata* was most common (Hamilton 1933). Estimated average density of *M. nigripes* is 1/50 ha of white-tailed prairie dog colony (Miller et al. 1996).

Sex Ratios. King (1975) reviewed the sex ratios of trapped *M. nivalis* from 12 studies. Males usually outnumbered females by about 3 to 1, a ratio similar to those reported for trapped *M. frenata* and *M. erminea* (Hamilton 1933; Jackson 1961). The sex ratio at birth is 1:1 and King (1975) found no evidence of differential mortality. Thus, the observed imbalance probably was due to sampling bias. For *M. nigripes,* the ratio of juvenile males to females was 1:1 in Wyoming (Forrest et al. 1988).

Sociality. The social organization of weasel populations is that of a solitary existence except during the breeding season. Residents maintain intrasexual territories and transients do not. Males defend territories against males, and females defend against females, but intersexual territory overlap occurs (Lockie 1966; Erlinge 1977a). Male *M. nivalis* and *M. erminea* have nearly nonoverlapping interspecific territories in winter; overlap was greatest during the summer, but even while there was overlap, exclusive areas of the territory were maintained. Females of both species occupy smaller territories than do males. Transients regularly moved through the area occupied by residents. Even though a female territory was completely within the boundaries of a male territory, the male seldom ventured into the female territory except during the breeding season (Erlinge 1977a). When a male did enter a female territory during the nonbreeding season, he used different parts of it (Erlinge 1975). In the region of overlap between two male weasels, avoidance characterized the behavior of residents toward each other; each occupied a different portion of the area at any given time (King 1975).

Dominance relationships play an important part in maintaining territories in Sweden. Male *M. erminea* and females exhibited avoidance and defensive actions, including mutual avoidance, retreat, flee, escape, and submissive behaviors, more frequently than offensive actions such as approach, threat and threat vocalizations, chase, nest occupation, and prey robbery in paired encounters. Adult males were dominant to juvenile males and nonbreeding females. Pregnant females were the most aggressive and dominant, or of equal dominance, to adult males. Adult males were ranked in order of dominance based on body size, age, and weight of anal glands. Adult females did not exhibit this relationship. Established males and females were dominant to introduced individuals, but established nonbreeding females and juvenile males were not able to maintain dominance over an introduced male. Established animals exhibited threat behavior, causing the introduced animal to flee. Important factors in the spatial organization of *M. erminea* are solitary habits maintained by avoidance and threat display (Erlinge 1977a).

Territories of *M. nivalis* have defined boundaries. Animals holding territories come into breeding condition from March to August. Animals not holding territories are variable as to when they enter breeding condition and sometimes they do not breed. According to Lockie (1966), *M. nivalis* females in Scotland are always subdominant to males and the females live on the male's territory at his pleasure. When food in scarce, resident males treat females like trespassing males. This results in poor breeding and high female mortality. Lockie (1966) felt that before a territorial system could be established, a certain minimum threshold density is necessary. Erlinge (1974) failed to find a minimum density necessary for territory formation, but did not detect seasonal changes in the pattern of territoriality. In summer and autumn, males were territorial, but in spring, the males moved about and stayed in one area for only a short time. Prey density and individual qualities were important in territory formation. Some individuals failed to maintain a territory at any time of the year.

The size of territories varies greatly among individuals and with the season (Erlinge 1974, 1977a; King 1975). Vacated territories are likely to be occupied by adjacent residents even though nonresidents are present. Not all available habitat is occupied. Those habitats lacking food and cover are avoided (Moors 1975). During the breeding season, males travel widely and well-defined male territories are not maintained (Lockie 1966; Erlinge 1974, 1977a; Moors 1975). However, females maintain well-defined territories during the breeding season (Erlinge 1974).

The social organization of *M. nigripes* appears to be much the same as that of weasels. Adult ferrets of the same sex have not been reported to inhabit the same prairie dog town. Adults of the opposite sex are found in the same town, but not in the same burrow system (Hillman 1968a, 1968b).

FEEDING HABITS AND HUNTING BEHAVIOR

Between 50% and 80% of the yearly food intake of weasels consists of small mammals, mainly rodents. Other foods vary in proportion depending on the season, availability, and individual preferences. Largest weasels generally take larger-sized prey than do smaller weasels. Based on scat analysis, Erlinge (1974) determined that *M. nivalis* in Sweden used rodents as the staple food item; they represented 80% of the total prey eaten. The vole *Microtus agrestis* was the primary prey and other small rodents and hares (*Lepus* spp.) were of secondary importance. Shrews (Soricidae), birds, and lizards were taken only rarely. In autumn and winter, rodents represented 94% of all food items. However, in spring and summer, as rodent populations declined, the proportion of rodents taken as food also declined and use of lagomorphs increased.

TABLE 30.1. Food items reported for North American weasels

M. erminea
10–36% of items in scat and stomach contents
Voles (*Microtus* spp)
Northern short-tailed shrew (*Blarina brevicauda*)
Deer mice (*Peromyscus maniculatus*)
Cottontail rabbits (*Sylvilagus floridanus*)
Meadow jumping mice (*Zapus hudsonius*)
< 10% of items in scat and stomach contents
Rats (*Rattus* ssp.)
Eastern chipmunk (*Tamias striatus*)
Vagrant shrew (*Sorex vagrans*)
Songbirds
Grasshoppers and crickets
Frogs
M. frenata
10–35% of items in scat and stomach contents
Voles
Cottontail rabbit
Unidentified mice
Rats
Harvest mice (*Reithrodontomys megalotis*)
<10% of items in scat and stomach contents
Northern short-tailed shrew
Squirrels (*Sciurus* spp.)
Eastern chipmunk
Star-nosed mole (*Condylura cristata*)
Muskrat (*Ondatra zibethicus*)

SOURCE: Data from Hamilton (1933) and Northcott (1971).

Few shrews were taken in captive studies, and weasels showed little interest in them when hunting.

Male *M. nivalis* are twice as large as females and there is evidence of food segregation between them. During food scarcity, males were able to shift to larger prey, including lagomorphs and the larger water vole (*Arvicola terrestris*), whereas the smaller vole, *M. agrestis,* remained the primary prey of females. Differences in hunting behavior were also seen. Smaller females were better suited for entering rodent burrows. They spent more time hunting in rodent tunnels, whereas males spent more time hunting aboveground. Overall, small rodent populations were exploited in relation to their numbers. The low rate of predation on birds was attributed to the low density of breeding birds in the study area. Brugge (1977) also reported that male *M. erminea* and *M. nivalis* preyed on larger animals than did females.

Small mammals were predominant as prey items of *M. erminea* in New York (Table 30.1). Winter food items determined from stomach analysis in the Northwest Territories and Alberta, Canada, included 55% mammals, 13% fish, 7% amphibia, 4% birds, 4% insects, and 14% vegetation (Table 30.1). At both sites, the jumping mouse (*Zapus hudsonius*) appeared in the diets at rates higher than anticipated, especially because these mice were hibernating at the time of the study. Weasels live under snow when the air temperature falls below $-13°C$ and they probably entered the hibernacula of the jumping mice and preyed on them during these periods of subnivean existence (Northcott 1971).

Rodents, shrews, and young cottontail rabbits (*Sylvilagus* spp.) were major prey items for *M. frenata* in New York State (Table 30.1). Of 163 scats analyzed, 51% of the prey consisted of voles (*Microtus* spp.) and cottontail rabbits (Hamilton 1933). Voles were most common in diets of *M. frenata* in Colorado, with deer mice (*Peromyscus maniculatus*) and chipmunks (*Tamias* spp.) next in order of abundance (Quick 1951). Moles (Talpidae), deer mice, and harvest mice (*Reithrodontomys* spp.) made up 75% of the prey of *M. frenata* in Iowa (Polderboer et al. 1941). Rabbits were taken primarily when they were a few weeks old. Songbirds were rarely taken by *M. frenata.* Errington (1936) reported that 13-lined ground squirrels (*Spermophilus tridecemlineatus*) bore the brunt of the predation pressure by a weasel family, but eastern cottontail rabbits (*Sylvilagus floridanus*), voles, deer mice, birds, and insects were also taken. Among the insects preyed on were ground beetles, ants, grasshoppers, and blowflies.

Loss of three blue-winged teal (*Anus discors*) nests was attributed to *M. frenata* in Manitoba. Captive weasels made small, paired conical punctures in eggs that were identical to those found in the eggs of plundered teal nests (Teer 1964). Stoddard (1931) included weasels as predators on bobwhite quail (*Colinus virginianus*) eggs based on the fact that they feed on them in captivity. Bump et al. (1947) ranked weasels second to red foxes (*Vulpes vulpes*) and gray foxes (*Urocyon cinereoargenteus*) as predators on ruffed grouse (*Bonasa umbellus*) eggs, estimating that weasels destroyed about 10% of all grouse nests.

During periods of rodent scarcity, *M. frenata* and *M. erminea* may shift to alternative prey that includes poultry and other domestic fowl (Jackson 1961). *M. frenata* have preyed on 3-day-old pigs (Polderbeer 1948). *M. nivalis* are not known to prey on domestic animals, however. At normal prey densities, weasels occupy the same buildings as poultry, do not molest them, and effectively reduce the numbers of rats and mice.

Shrews were part of the normal diet of weasels in some studies (Hamilton 1933; Aldous and Manweiler 1942; Jackson 1961; Ognev 1962), but were uncommon in other studies (Day 1968; Erlinge 1974). Individual preferences may reflect these differences to some degree. A single captive female *M. nivalis* preferred shrews to voles when offered both (Erlinge 1974). Jackson (1961) reported that *M. erminea* preyed on shrews mainly in winter.

MacLean et al. (1974) and Fitzgerald (1977) found that predation by weasels had a great effect on the population dynamics of microtines. Predation by *M. nivalis* during the declining phase of the population cycle of a collared lemming (*Dicrostonyx groenlandicus*) population was important in sustaining the amplitude of the cycle. In a California study on the impact of weasels on the vole cycle, predation was related to population density of the voles. *M. frenata* and *M. erminea* preyed on voles under the snow and then used the vole nests for dens. Both species of weasel varied in population density over the course of the study. *M. erminea* was most common. In winter, both species preyed entirely on the voles, and winter food habits did not change with changing vole density. The percentage of vole nests occupied by weasels varied from 5% to 54%. At low vole densities, all losses between autumn and spring were accounted for by predation, whereas at high vole densities, there was less mortality because of predation (Fitzgerald 1977). These studies support the hypothesis of Pearson (1971) that predation during and after the crash of microtine populations is responsible for timing and amplitude of microtine cycles.

It is generally assumed that weasels use living prey and do not scavenge. However, *M. nivalis* scavenged on frozen carcasses of brown lemmings (*Lemmus sibiricus*) that were systematically placed under the snow in various tundra habitats (Mullen and Pitelka 1972).

Weasels have evolved a highly successful prey capture and killing behavior consisting of rapidly dashing to the prey, biting them at the base of the neck, and entwining them with the body and legs. In this manner, prey are securely held even though the bite may be changed to a better vantage point for the kill. Weasels do not suck blood, nor do they subsist on blood. After the prey is dead, weasels may devour the entire body or select certain parts and leave the rest. It is common for a well-fed weasel to devour only the viscera, muscle, and brains and leave the feet, tail, and skin.

Heidt (1972) detailed the killing and feeding behaviors of *M. nivalis.* The killing behavior is stereotyped. Weasels seize prey at the nape of the neck and bite through the base of the skull and/or the throat. The initial bite may be on any part of the prey's body to gain a hold and leverage for the neck bite. Weasels manipulate and position the prey with their feet, and wrap their body about the prey to provide further leverage. Mice were commonly dropped after they had quit struggling. If more than one mouse was present, the weasel dropped the first after it quit struggling, then caught and killed the second before returning to the first. The time elapsed per kill ranged from 10 to 60 sec.

Killing appears to be innate (Heidt 1970). Young *M. nivalis* that were separated from their mother before their eyes had opened could kill mice when the weasels were 50–60 days of age without having previous experience. However, training by the mother caused young weasels to be more efficient in capturing and killing mice at 40–45 days of age (Heidt et al. 1968).

Weasels hunt by traveling through the habitat in a "random search" manner (Ewer 1973), investigating tunnels, nests, and potential hiding places of rodents as they encounter them. Sight, sound, and odors are all potential cues for prey detection. *M. erminea* was observed pursuing the flight sounds of grasshoppers and responding to them in a manner that indicated that it was hunting them (Willey 1970). The visual clues of a moving prey elicit an immediate response and attack in *M. nivalis* (Heidt 1972) and *M. frenata,* but stationary prey are not seen readily. In confined conditions, weasels pass within centimeters of still prey and do not notice them, indicating less reliance on the sense of smell in searching for prey than on other cues.

Prairie dogs are the principal food of the black-footed ferret. Scats recovered from burrows occupied by a mother and four young ferrets contained 18% mouse remains and 82% prairie dog remains (Sheets and Linder 1969). A captive ferret readily ate freshly shot prairie dogs, ground squirrels (*Spermophilus* spp.), small rodents, birds, and cottontail rabbits. It also ate fresh fish, calf liver, hamburger, fat pork, milk, and bread (Aldous 1940). A black-footed ferret lived for several days under a wooden sidewalk in Hays, Kansas, where it killed rats.

Like other members of the weasel family, the black-footed ferret kills prey by attacking the neck and base of the skull. A caged black-footed ferret stalked individual prairie dogs and approached them from the rear. The ferret lunged for the hind end of the animal and then struck the base of the skull. When the black-footed ferret bit, a fierce fight followed until the prairie dog was killed. Prairie dogs will kick loose soil at the ferret to deter an attack. Under natural conditions, prairie dogs are probably pursued into their burrows, killed within the confines of the burrow or nest, and devoured. Like weasels, the black-footed ferret evidently has poor distance vision and it needs to move close to a prairie dog or other small mammal prey before noticing it (Progulske 1969).

Caching food occurs among all weasels. When weasels encounter a local abundance of food in excess of what they can consume, they will store the unused food and return to eat it later. The cache site may be a side burrow off the main home burrow or a site located near the kill. A female *M. frenata* was observed hunting in an alpine meadow in Colorado, where it found two nests of 1- to 2-week-old golden-mantled ground squirrels (*Spermophilus lateralis*). It killed the young ground squirrels in the burrow and then carried them to an old pocket gopher burrow, where they were cached. There were nine young ground squirrels cached in all. After removing all the young squirrels, the weasel returned to the burrow where they were cached and remained for 30 min. It then emerged and crossed the meadow to a creek and then returned to the cache. This was repeated 50 min later. The next morning it returned to the food cache and was livetrapped and marked. The weasel was lactating and had a burrow containing young about 175 m away. Over the next 2 weeks, it cached in the same pocket gopher burrow on three occasions. Prey caching likely is beneficial when prey are abundant. Prey are killed quickly and used later, thus conserving energy used in foraging. Other accounts of caching by weasels include immense numbers of rats dragged together and piled in a compact heap beneath the floorboards of a barn. Piles of 100 or more rats and mice are reported (Seton 1929).

Weasels require a constant supply of drinking water. They take only a little at a time, but drink frequently. *M. frenata* drinks about 25 cm^3 daily (Hamilton 1933).

BEHAVIOR

Weasels have the reputation of being objectionable, bloodthirsty, wandering demons of carnage. This view is expressed by Dr. Elliot Coues: "A glance at the physiognomy of the weasel would suffice to betray their character. The teeth are almost the highest known raptorial character; the jaws are worked by enormous masses of muscles covering all sides of the skull. The forehead is low, and the nose is sharp; the eyes are small, penetrating, cunning, and glitter with an angry green light. There is something peculiar, moreover, in the way that this fierce face surmounts a body extraordinarily wiry, lithe, and muscular. It ends a remarkable long and slender neck, in such a way that it may be held at right angles with the axis of the latter. When the creature is glancing forward, swaying from one side to the other, we catch the likeness in a moment—it is the image of a serpent" (Seton 1929:599).

Seton goes on to single out weasels alone among the animals that seem to revel in slaughter for its own sake and to find unholy joy in the horrors of a dying squeak, final quiver, and wholesale extermination.

These anthropomorphisms are of little value in understanding weasel behavior. Instead, weasel behavior should be interpreted as a highly specialized and adaptive carnivorous way of life that is a result of a long and successful evolutionary process. Weasels are not angry, cunning, or wanton killers. Instead, they are efficient predators with behavioral, anatomical, physiological, and sensory adaptations that allow them to survive as small carnivores. Their quick actions and curious nature are tools with which to hunt and find small and agile prey. These same features may also help avoid being taken as prey. Their fearless and pugnacious nature is necessary to allow the weasel to attack, capture, and kill prey that may be much larger. The "wanton slaughter" of prey is an efficient way of exploiting a locally abundant food source that is easily captured. Once prey are killed, some are eaten immediately and the rest cached for future use.

The long, wiry, serpentlike body enables weasels to enter burrows and hiding places of prey and aids in capture and killing. Weasels are not bloodsuckers, but instead are like most carnivores that lap up the blood flowing from the wounds of freshly killed prey as part of the feeding pattern. This enables them to use all of the nutrients in prey.

Locomotion. The traveling gait of weasels and the black-footed ferret is a slow gallop or series of jumps; frequently they walk. *M. nivalis* moves at a slow pace, about 8–10 km/hr; the larger species are capable of moving faster when pressed, but probably travel at about the same pace (Jackson 1961). Weasels are capable of climbing trees in pursuit of prey, but are not adept in doing so. They also enter water, but are slow swimmers. There is no information indicating that weasels dig, whereas black-footed ferrets do.

Black-footed ferrets must be good burrowers. When hunting prairie dogs, ferrets remove dirt plugs that prairie dogs construct to block the tunnels. Loose dirt that ferrets pile at the entrance of prairie dog tunnels is a good indication that the area is occupied by ferrets (Sheets et al. 1971; Richardson et al. 1987).

Interactions Among Individuals. Young weasels are playful and spend a great deal of time in play fighting and in reproductive behavior (Hamilton 1933; Heidt et al. 1968). However, adults live a solitary existence. Interactions among adults occur primarily during mating, otherwise individuals avoid each other. Complex social behaviors characterizing the social carnivores are not well developed in this group. When two strange *M. erminea* were placed together simultaneously, the result was avoidance. When one animal was established and then a second animal was added, the intruder was met with threat behavior that caused the intruder to escape. Threat display through scent marking and visual and acoustic signals were involved in territorial defense (Erlinge 1977c).

Like other mustelids, weasels and the black-footed ferret produce a pungent odor from their anal glands. When irritated, they discharge the odor, which can be detected at some distance. *M. frenata* has been observed rubbing and dragging its body over surfaces, possibly to leave scent from its anal glands (Jackson 1961), and a specific anal drag behavior has been observed in all species. Male *M. erminea* and *M. nigripes* scent mark throughout the year and increase the frequency of scent marking during breeding season. Dominant male ermine scent mark three times more frequently than do subordinates (Erlinge et al. 1982; Richardson et al. 1987). As in other mammals, odor probably serves to identify individuals just as visual features do in humans. However, knowledge of scent communication in weasels and black-footed ferrets is not sufficient to make succinct statements of either the role or the information content of the scent.

Vocalizations. Vocalizations of weasels can be grouped into three types: a trill, a screech, and a squeal. The trill is a common vocalization accompanying several behaviors. It occurs when animals are investigating their surroundings, playing, in conjunction with mating, and when hunting. The trill is composed of a series of short-duration, rapidly occurring, low-frequency calls. The smallest weasel, *M. nivalis,* has the highest frequency trill; the larger weasels, *M. erminea* and *M. frenata,* produce trills at a lower frequency. When a weasel is suddenly disturbed, the screech is elicited. An open-mouthed gape and a sudden lunge accompany this vocalization. The probable function is a startle effect in the defensive or threat display. The screech is composed of complex harmonics and white noise caused by forcefully expelling air. The squeal is elicited under stress, possibly accompanying pain (Heidt et al. 1968; Huff and Price 1968; Svendsen 1976).

Four distinct vocalizations have been identified for the black-footed ferret: hiss, bark, scream, and chatter. A prolonged hiss is elicited when threatened. If the threat is intensified, a ferret gives a single bark, which leads to a series of quick barks forming a chatter when the ferret is cornered and contact is inevitable. The scream is heard during fights (Miller et al. 1996). A chattering scold was made by an adult female toward men trying to capture one of its young (Aldous 1940). When people came near a captive female, it chattered constantly in a staccato voice, and between bursts of six to seven loud chirps it emitted low, hissing sounds. Chatter was also related to excitement and was made in response to unusual conditions, such as a bird flying overhead or people nearby (Progulske 1969).

Behavior Toward Other Animals. Weasels are well known for their boldness and courage. It is said that a weasel will face any animal, including humans. There are several reports of weasels attacking people; almost always this occurred when someone was trying to take freshly killed prey away from a weasel (Wight 1932; Oehler 1944). On occasions when someone has tried to grab a live weasel, the power and tenacious grip of the jaws were quickly evident in the bite. A bleached skull of a weasel was found with a "death grip" on the throat of a shot bald eagle (*Haliaeetus leucocephalus*). It was surmised that the eagle had captured the weasel and in its struggle the weasel had latched onto the eagle and died (Seton 1929).

MORTALITY

Longevity in the wild is not well documented for weasels or the black-footed ferret. Jackson (1961) reported the life span of *M. erminea* to be 4–6 years in the wild, potentially up to 10 years. King (1975) determined that a resident *M. nivalis* lived on her study area about 1 year and the mean age of death of 171 trapped weasels was 11 months. Marked adult *M. frenata* were resident on a Colorado site for up to 3 years (G. E. Svendsen, unpublished data). Natural mortality of weasels is a result of several factors interacting simultaneously: disease, parasites, nutrition, population stress, and predation.

Predation. Black-footed ferrets have many potential predators, including badgers (*Taxidea taxus*), coyotes (*Canis latrans*), bobcats (*Lynx rufus*), rattlesnakes (*Crotalus* spp.), eagles, hawks, and owls. There are reports of predation or pursuit of ferrets by badgers, great horned owls (*Bubo virginianus*), domestic dogs, domestic cats, golden eagles (*Aquila chrysaetos*), prairie falcons (*Falco mexicanus*), ferruginous hawks (*Buteo regalis*), and coyotes (Henderson et al. 1974; Forrest et al. 1988). A great horned owl was observed diving at an adult ferret (Hillman 1968a). Sperry (1941) reported the remains of black-footed ferrets in three coyote stomachs. The underground tactics when hunting prairie dogs offers some protection against surface predators.

The reliance of black-footed ferrets on prairie dogs for food is central to understanding their endangered status. The population decline of black-footed ferrets is due primarily to loss of habitat caused by eradication of prairie dog towns from disease and pest control. Prairie dogs are susceptible to plague, a bacterial disease caused by *Yersinia pestis,* which was introduced from Asia around 1900. Plague can completely eradicate a prairie dog town or it can reduce the number of individuals to a point that the town will not support a population of black-footed ferrets (Miller et al. 1996).

The total area covered by prairie dog towns once was between 40 and 100 million ha. By 1960, prairie dogs occupied 600,000 ha, or 2% of the area they occupied in the late 1880s (Miller et al. 1996). The few prairie dog towns that remained were small and geographically isolated. At the root of the demise of prairie dog towns was the widespread belief that they were pests that competed with livestock for forage. This belief led to reducing the numbers of prairie dogs to improve the range for cattle. Poisoning began by individual ranchers in the late 1880s, and in 1915 the federal government began paying for prairie dog control. Strychnine, sodium cyanide, and sodium monofluoracetate (1080) all were used extensively in the poisoning program. Not only was the food base for black-footed ferrets eradicated in town after town, but ferrets were also killed as they fed on poisoned prairie dogs. Fewer and fewer black-footed ferrets were seen and by 1964 the federal government was ready to declare the ferret extinct. In that year, a small population of black-footed ferrets was found in Mellette County, South Dakota, and they were declared endangered in 1966 (Miller et al. 1996).

Unlike the black-footed ferret, none of the species of weasels are endangered. However, land use practices undoubtedly have resulted in local extirpation and low density. Powell (1973) suggested that weasel population density is controlled by predation. There are records of weasels being eaten by rattlesnakes, blacksnakes (*Elaphe obsoleta*), snowy owls (*Nyctea scandiaca*), great horned owls, barred owls (*Strix varia*), rough-legged hawks (*Buteo lagopus*), goshawks (*Accipiter gentilis*), red foxes, gray foxes, and domestic cats (Hamilton 1933; Handley 1949; Errington et al. 1940; Jackson 1961). The small *M. nivalis* is sometimes prey for the larger *M. frenata* (Polderboer et al. 1941). The erratic hunting pattern and black tip of the tail have been postulated as adaptations to reduce losses to other predators. Powell (1973) suggested that a black tip to the tail of the very short-tailed *M. nivalis* would improve the hawk's chances of catching this small weasel because the black tip would direct the predator's attention to the short tail and body. Hence, *M. nivalis* has no black tip to its tail. Conversely, the black tip on the longer-tailed *M. erminea* and *M. frenata* reduces the chances of being captured by a hawk. The predator's attention is directed toward the black tip and not at the body. This is especially true when the weasels are in their white winter coat on a background of snow.

Diseases and Parasites. The incidence of disease and parasites of weasels and the black-footed ferret are poorly known. External parasites of the weasels include ticks (*Dermacentor variabilis, Ixodes cookei*), fleas (*Ceratophyllus vison, C. fasciatus, Nearctopsylla brooksi, Neotrichodectes mephitidis*), the biting louse (*Stacheilla kingi*), the sucking louse (*Hoplopleura erratica*), chigger mites (*Euschoengastia peromysci, Eutrombicula alfreddugesi*), and mites (*Lutrilichus canadensis, Myocoptes japonensis, Androlaelaps fahrenholzi, Listrophorus mexicanus, Zibethacarus ondatrae,* and *Laelaps alaskensis*). Reported internal parasites include nematodes (*Trichinella spiralis, Dracunculus medinensis, Molineus* spp., *Physaloptera maxillaris, Filaroides martis, Capillaria mustelorum, Skjabingylus nasicola*), the cestode *Taenia taeniaformis,* and trematodes (*Alaria mustelae, A. taxideae*) (Goble 1942; Jackson 1961; Schmidt 1965; Soleman and Warner 1969; Fain et al. 1974; Mumford and Whitaker 1982). The nematode *Dracunculus medinensis* invades the frontal sinuses, causing swelling and perforations of the skull in the supraorbital region (Dougherty and Hall 1955; van Soest et al. 1972). Canine distemper was identified as the primary cause of the black-footed ferret population decline at Meeteetze, Wyoming (Williams et al. 1988). Parker (1934) reported that the black-footed ferret and weasels were highly susceptible to tularemia (*Francisella tularensis*).

ECONOMIC STATUS AND MANAGEMENT

Overall, weasels are more of an asset than a liability to people. They destroy rats and mice that otherwise would eat and damage crops and produce. This asset is partially counterbalanced by the fact that weasels occasionally kill beneficial animals and game species. The killing of domestic poultry may come only after the rat population around a farmyard is diminished. In fact, rats may have destroyed more poultry than the weasel. In most cases, a farmer can live in harmony with weasels on the farm for years without realizing that they are even there until they kill a chicken. Quick (1951) estimated that 8000 weasels inhabited Gunnison County, Colorado, an area of 10,240 km^2. These were killing >30,000 small mammals/day or 10,000,000/year in that county alone. In New York State, a conservative estimate of the number of rats and mice killed per year by weasels was 60,000,000 mice and several million rats (Hamilton 1933).

The white winter pelage of *M. erminea* and *M. frenata* is valued in the fur trade as "ermine." Select winter pelts of *M. frenata* bring more than smaller pelts of *M. erminea*. Pelts of *M. nivalis* have practically no commercial value. Prices for pelts vary greatly. The total number of weasel pelts sold in the United States and Canada between 1970 and 1976 ranged from a low of 43,876 in 1971–1972 to a high of 102,090 in 1974–1975. The average price per pelt during this time was $0.58 and the average yearly income from weasel pelts was $43,377. The present demand for weasel pelts is low and there are no fur sales records since the mid-1980s.

The relationship of weasels to humans dates back to antiquity. Native American belief held that if a person captured a least weasel, it was regarded as good fortune and one was destined to have great wealth and power (Seton 1929). Ermine skins decorated the headdress and ceremonial clothing of many Native American tribes in North America, and stuffed skins are found among ceremonial relics (Peterson and Berg 1954; Clark 1976).

The black-footed ferret was never numerous enough to pose a serious threat to domestic stock or to become economically feasible for trade on the fur market. Overall, it benefited the early settlers by helping to keep populations of prairie dogs from expanding on grazing lands. Ironically, human have driven the black-footed ferret to near extinction by reducing the habitat and number of prairie dog towns to levels where an effective breeding population of ferrets does not now exist. Various Native American groups—Sioux, Blackfoot, Crow, Cheyenne, and Pawnee—used the black-footed ferret in a variety of ways. The Blackfoot in Montana used ferret hides as pendants on their chief's headdress. The Crow of Wyoming and Montana used ferrets in their sacred tobacco society. There is evidence that Native peoples also used ferrets for food (Clark 1976). Stuffed ferret skins were included in ceremonial relics of the Crow, suggesting that they were aware of the rarity of the species and attached a mystical significance to it (Peterson and Berg 1954).

As noted, the black-footed ferret is listed as an endangered species. Efforts are being made to protect critical habitat where ferrets have been sighted and evaluate conditions for reintroduction and recovery. The success of the reintroduction and recovery plan is determined by the success of the captive breeding program, the reintroduction protocol, and cooperation between state and federal agencies (Miller et al. 1996). A critical factor in the future of the black-footed ferret is the identification of existing prairie dog towns and preservation of critical habitat to increase the number and size of prairie dog towns that are large enough to maintain a stable population of ferrets (Reading and Matchett 1997; Severson and Plumb 1998).

Following the discovery of a population of black-footed ferrets near Meeteetze, Wyoming, in 1981, interest in conservation of ferrets increased. Subsequently, there was a canine distemper outbreak in the population and all surviving ferrets were placed into a captive breeding program (Williams et al. 1988; Schreiber et al. 1989; Biggins and Godbey 1995). These captive black-footed ferrets were a founder population for captive breeding and reintroduction programs with oversight from state and federal agencies (Biggins and Godbey 1995). No attempted reintroductions into the wild have been completely successful, however (Miller et al. 1994, 1996; Biggins et al. 1998; Vargas and Anderson 1999).

LITERATURE CITED

Aldous, S. E. 1940. Notes on a black-footed ferret raised in captivity. Journal of Mammalogy 21:23–26.

Aldous, S. E., and J. Manweiler. 1942. The winter food habits of the short-tailed weasel in northern Minnesota. Journal of Mammalogy 23:250–55.

Allen, G. M. 1933. The least weasel: A circumboreal species. Journal of Mammalogy 14:316–19.

Anderson, E., S. C. Forrest, T. W. Clark, and L. Richardson. 1986. Paleobiology, biogeography, and systematics of the black-footed ferret, *Mustela nigripes* (Audubon and Backman), 1857. Great Basin Naturalist Memoirs 8:11–62.

Banfield, A. W. F. 1974. The mammals of Canada. University of Toronto Press, Toronto.

Baker, R. H. 1983. Michigan mammals. Michigan State University Press, East Lansing.

Beer, J. R. 1950. Least weasel in Wisconsin. Journal of Mammalogy 31:146–49.

Biggins, D., and J. Godbey. 1995. Black-footed ferrets. Pages 106–8 *in* E. T. LaRoe, G. S. Farris, C. E. Puckett, P. D. Doran, and M. J. Mac, eds. Our living resources: A report to the nation on the distribution, abundance, and health of U.S. plants, animals, and ecosystems. U.S. Department of the Interior, National Biological Survey, Washington, DC.

Biggins, D. E., M. H. Schroeder, S. C. Forrest, and L. Richardson. 1985. Movements and habitat relationships of radio-tagged black-footed ferrets. PP. 11.1–11.17 in Black-Footed Ferret Workshop Proceedings, 18–19 September 1984 (Laramie, Wyo.)

Biggins, D. E., J. L. Godbey, L. R. Hanebury, B. Luce, P. E. Marinari, M R. Matchett, and A. Vargas. 1998. The effect of rearing methods on survival of reintroduced black-footed ferrets. Journal of Wildlife Management 62:643–53.

Bissonnette, T. H., and E. E. Bailey. 1944. Experimental modification and control of molts and changes of coat color in weasels by controlled lighting. Annals of the New York Academy of Science 45:221–50.

Brown, J. H., and R. C. Lasiewski. 1972. Metabolism of weasels: The cost of being long and thin. Ecology 53:939–43.

Brugge, T. 1977. Prooidierkeuze van wezel, Hermelijn en Bunzing in relatie tot geslacht en lichaamsgrootte [Prey selection of weasel, stoat and polecat in relation to sex and size]. Lutra 19:39–49.

Bump, G., R. W. Darrow, F. C. Edminster, and W. F. Crissey. 1947. The ruffed grouse: Life history, propagation, management. New York State Conservation Department, Albany.

Burns, J. J. 1964. Movements of a tagged weasel in Alaska. Murrelet 45:10.

Carpenter, J. W. 1985. Captive breeding and management of black-footed ferrets. Pages 12.1–12.13 *in* S. H. Anderson and D. B. Inkley, eds. Black-footed ferret workshop proceedings. Wyoming Fish and Game Department, Cheyenne.

Clark, T. W. 1976. The black-footed ferret. Oryx 13:275–80.

Day, M. G. 1968. Food habits of British stoats (*Mustela erminea*) and weasels (*Mustela nivalis*). Journal of Zoology (London) 150:485–97.

Dayan, T., D. Simberloff, E. Tchernov, and Y. Yom-tov. 1989. Inter- and intraspecific character displacement in mustelids. Ecology 70:1226–39.

Deanesly, R. 1944. The reproductive cycle of the female weasel (*Mustela nivalis*). Proceedings of the Zoological Society of London 114:339–49.

Dougherty, E. C., and E. R. Hall. 1955. The biological relationships between American weasel (genus *Mustela*) and nematodes of the genus *Skrjabingylus* Petrov, 1927 (Nematoda: Metastrongylidae), the causative organisms of certain lesions in weasel skulls. Revista Iberica de Parasitologie 531–69.

Erlinge, S. 1974. Distribution, territoriality and numbers of the weasel *Mustela nivalis* in relation to prey abundance. Oikos 25:308–14.

Erlinge, S. 1975. Feeding habits of the weasel *Mustela nivalis* in relation to prey abundance. Oikos 26:378–84.

Erlinge, S. 1977a. Spacing strategy in stoat *Mustela erminea*. Oikos 28:32–42.

Erlinge, S. 1977b. Home range utilization and movements of the stoat, *Mustela erminea*. Pages 31–42 *in* Proceedings of the 13th Congress of Game Biologists.

Erlinge, S. 1977c. Agonistic behavior and dominance in stoats (*Mustela erminea* L.) Zeitschrift für Tierpsychologie 44:375–88.

Erlinge, S. 1979. Adaptive significance of sexual dimorphism in weasels. Oikos 33:233–45.

Erlinge, S., and M. Sandell. 1986. Seasonal changes in social organization of male stoats, *Mustela erminea:* An effect of shifts between two decisive resources. Oikos 47:308–14.

Erlinge, S., M. Sandell, and C. Brinck. 1982. Scent-marking and its territorial significance in stoats, *Mustela erminea*. Animal Behaviour 30:811–18.

Errington, P. L. 1936. Food habits of a weasel family. Journal of Mammalogy 17:406–7.

Errington, P. L., F. Hammerstrom, and F. N. Hammerstrom, Jr. 1940. The great horned owl and its prey in north-central United States. Iowa Agricultural Experiment Station Bulletin 277:757–850.

Ewer, R. F. 1973. The carnivores. Cornell University Press, Ithaca, NY.

Fain, A., F. S. Lukoschus, N. J. Kok, and F. V. Clulow. 1974. A key to the genus *Lutrilichus* Fain and description of a new species from the ermine, *Mustela erminea* in Canada (Acarina: Sarcoptiformes). Canadian Journal of Zoology 52:941–44.

Fitzgerald, B. M. 1977. Weasel predation on a cyclic population of the montane vole (*Microtus montanus*) in California. Journal of Animal Ecology 46:367–97.

Forrest, S. C., D. E. Biggins, L. Richardson, T. W. Clark, T. M. Campbell, K. A. Fagerstone, and E. T. Thorne. 1988. Black-footed ferret (*Mustela nigripes*) attributes at Meeteetse, Wyoming, 1981 to 1985. Journal of Mammalogy 69:261–73.

Glover, F. A. 1942. A population study of weasels in Pennsylvania. M.S. Thesis, Pennsylvania State University, University Park.

Goble, F. C. 1942. The guinea-worm in a Bonepart weasel. Journal of Mammalogy 23:221.

Hall, E. R. 1944. Classification of the ermines of eastern Siberia. Proceedings of the California Academy of Science Series 4 23:555–60.

Hall, E. R. 1951. American weasels. University of Kansas Museum of Natural History Publication 4:1–466.

Hall, E. R. 1981. The mammals of North America. John Wiley and Sons, New Jersey.

Hamilton, W. J., Jr. 1933. The weasels of New York. American Midland Naturalist 14:289–337.

Hamlett, G. W. D. 1935. Delayed implantation and discontinuous development in mammals. Quarterly Review of Biology 10:432–47.

Handley, C. O., Jr. 1949. Least weasel, prey of barn owl. Journal of Mammalogy 30:431.

Hartman, L. 1964. The behavior and breeding of captive weasels (*Mustela nivalis* L.) New Zealand Journal of Science 7:147–56.

Hatt, R. T. 1940. The least weasel in Michigan. Journal of Mammalogy 21:412–16.

Heidt, G. A. 1970. The least weasel, *Mustela nivalis* L.: Developmental biology in comparison with other North American *Mustela*. Michigan State University Museum of Natural History Publication 4:227–82.

Heidt, G. A. 1972. Anatomical and behavioral aspects of killing and feeding by the least weasel, *Mustela nivalis* L. Proceedings of the Arkansas Academy of Science 26:53–54.

Heidt, G. A., M. K. Petersen, and G. L. Kirkland, Jr. 1968. Mating and behavior and development of least weasels (*Mustela nivalis*) in captivity. Journal of Mammalogy 49:413–19.

Henderson, F. R., P. F. Springer, and R. Adrian. 1974. The black-footed ferret in South Dakota (Technical Bulletin No. 4). South Dakota Department of Game, Fish, and Parks.

Hillman, C. N. 1968a. Life history and ecology of the black footed ferret in the wild. M.S. Thesis, South Dakota State University, Brookings.

Hillman, C. N. 1968b. Field observations of black-footed ferrets in South Dakota. Transactions of the North American Wildlife and Natural Resources Conference 33:433–43.

Hillman, C. N., and T. W. Clark. 1980. *Mustela nigripes*. Mammalian Species 126:1–3.

Hillman, C. N., and J. W. Carpenter. 1983. Breeding biology and behavior of captive black-footed ferrets. International Zoo Yearbook 23:186–91.

Hillman, C. N., R. L. Linder, and R. B. Dahlgren. 1979. Prairie dog distribution areas inhabited by black-footed ferrets. American Midland Naturalist 102:185–87.

Huff, J. N., and E. O. Price. 1968. Vocalizations of the least weasel, *Mustela nivalis*. Journal of Mammalogy 49:548–50.

Iversen, J. A. 1972. Basal energy metabolism of mustelids. Journal of Comparative Physiology 81:341–44.

Jackson, H. H. T. 1961. Mammals of Wisconsin. University of Wisconsin Press, Madison.

Kavanau, J. L. 1969. Influences of light on activity of small mammals. Ecology 50:548–57.

Kavanau, J. L., and J. Ramos. 1975. Influences of light on activity and phasing of carnivores. American Naturalist 109:391–418.

Keith, L. B., and E. C. Meslow. 1966. Animals using runways in common with snowshoe hares. Journal of Mammalogy 47:541.

King, C. M. 1975. The home range of the weasel (*Mustela nivalis*) in an English woodland. Journal of Animal Ecology 44:639–668.

King, C. M. 1983. Factors regulating mustelid populations. Acta Zoologica Fennica 174:217–20.

King, C. M. 1989. The advantages and disadvantages of small size to weasels, *Mustela* species. Pages 302–34 *in* J. L. Gittleman, ed. Carnivore behavior, ecology, and evolution. Comstock, Ithaca, NY.

King, C. M. 1990. The natural history of weasels and stoats. Comstock, Ithaca, NY.

King, C. M., and C. D. McMillan. 1982. Population structure and dispersal of peak-year cohorts of stoats (*Mustela erminea*) in two New Zealand forests, with special reference to control. New Zealand Journal of Ecology 5:59–66.

Latham, R. 1953. Simple method for identification of least weasel. Journal of Mammalogy 34:385.

Linder, R. L., B. Dahlgren, and C. N. Hillman. 1972. Black-footed ferret–prairie dog interrelationships. Pages 22–23 *in* Proceedings of the symposium on rare and endangered wildlife in the southwestern United States. New Mexico Department of Game and Fish, Santa Fe.

Lockie, J. D. 1966. Territory in small carnivores. Symposia of the Zoology Society of London 18:143–65.

Lyman, C. P. 1942. Control of coat color in the varying hare by daily illumination. Proceedings of the New England Zoology Club 19:75–78.

MacLean, S. F., Jr., B. M. Fitzgerald, and F. A. Pitelka. 1974. Population cycles in arctic lemmings: Winter reproduction and predation by weasels. Arctic and Alpine Research 6:1–12.

Mayr, E. 1966. Animal species and evolution. Belknap Press, Cambridge, MA.

McNab, B. K. 1971. On the ecological significance of Bergmann's rule. Ecology 52:845–54.

McNab, B. K. 1989. Basal rate of metabolism, body size, and food habits in the order Carnivora. Pages 335–54 *in* J. L. Gittleman, ed. Carnivore behavior, ecology, and evolution. Comstock, Ithaca, NY.

Miller, B. J., and S. H. Anderson. 1989. Failure of fertilization following abbreviated copulation in the domestic ferret. Journal of Experimental Zoology 249:85–89.

Miller, B., D. Biggins, L. Hanebury, and A. Vargus. 1994. Reintroduction of the black-footed ferret (*Mustela nigripes*). Pages 455–64 *in* P. J. S. Olney, G. M. Mace, and A. T. C. Feister, eds. Creative conservation: Interactive management of wild and captive animals. Chapman and Hall, London.

Miller, B., R. P. Reading, and S. Forrest. 1996. Prairie night: Black-footed ferrets and the recovery of endangered species. Smithsonian Institution Press, Washington, DC.

Moors, P. L. 1975. The annual energy budget of a weasel (*Mustela nivalis* L.) population in farmland. Ph.D. Dissertation, University of Aberdeen, Aberdeen, Scotland.

Mullen, D. A., and F. A. Pitelka. 1972. Efficiency of winter scavengers in the Arctic. Arctic 25:225–31.

Mumford, R. E. 1969. Distribution of mammals of Indiana. Indiana Academy of Science, Indianapolis.

Mumford, R. E. and J. O. Whitaker, Jr. 1982. Mammals of Indiana. Indiana University Press, Bloomington.

Musgrove, B. F. 1951. Weasel foraging patterns in the Robinson Lake area, Idaho. Murrelet 32:8–11.

Northcott, T. H. 1971. Winter predation of *Mustela erminea* in northern Canada. Arctic 24:142–44.

Nyholm, E. S. 1959. Karpasta ja lumikosta ja niiden Tavisista elinpiireista [Stoats and weasels and their winter habitat]. Suomen Riista 13:106–16.

Oehler, C. 1944. Notes on the temperment of the New York weasel. Journal of Mammalogy 25:198.

Ognev, S. I. 1935/1962. Mammals of U.S.S.R. and adjacent countries. Israel Program for Scientific Translations.

Owen, P. R., C. J. Bell, and E. M. Mead. 2000. Fossils, diet, and conservation of black-footed ferrets (*Mustela nigripes*). Journal of Mammalogy 81:422–33.

Parer, J. T., and J. Metcalfe. 1967. Oxygen transport by blood in relation to body size. Nature 215:653–54.

Parker, R. R. 1934. Recent studies of tick-borne diseases made at the United States Public Health Service Laboratory at Hamilton, Montana. Proceedings 5th Pacific Science Congress B5:3367.

Pearson, O. P. 1971. Additional measurements of the impact of carnivores on California voles (*Microtus californicus*). Journal of Mammalogy 52:41–49.

Peterson, L. A., and E. D. Berg. 1954. Black-footed ferrets used as ceremonial objects by Montana Indians. Journal of Mammalogy 35:593–94.

Petrides, G. A. 1950. The determination of sex and age ratios in fur animals. American Midland Naturalist 43:355–88.

Polder, E. 1968. Spotted skunk and weasel populations den and cover usage in northeast Iowa. Iowa Academy of Science 75:142–46.

Polderboer, E. B. 1948. Predation on the domestic pig by the long-tailed weasel. Journal of Mammalogy 29:295–96.

Polderboer, E. B., L. W. Kuhn, and G. O. Hendrickson. 1941. Winter and spring habitats of weasels in central Iowa. Journal of Wildlife Management 5:115–19.

Powell, R. A. 1973. A model for raptor predation on weasels. Journal of Mammalogy 54:259–63.

Powell, R. A. 1979. Mustelid spacing patterns: Variations on a theme by *Mustela*. Zeitschrift für Tierpsychologie 50:153–65.

Powell, R. A. 1982. Prairie dog coloniality and black-footed ferrets. Ecology 63:1967–68.

Powell, R. A., and C. M. King. 1997. Variation in body size, sexual dimorphism and age-specific survival in stoats, *Mustela erminea* (Mammalia:Carnivora), with fluctuating food supplies. Biological Journal of the Linnean Society 62:165–94.

Progulske, D. R. 1969. Observations of a penned, wild-captured black-footed ferret. Journal of Mammalogy 50:619–21.

Quick, H. F. 1944. Habits and economics of New York weasel in Michigan. Journal of Wildlife Management 8:71–78.

Quick, H. F. 1951. Notes on the ecology of weasels in Gunnison County, Colorado. Journal of Mammalogy 32:281–90.

Ralls, K., and P. H. Harvey. 1985. Geographic variation in size and sexual dimorphism of North American weasels. Biological Journal of the Linnean Society 25:119–67.

Reading, R. P., and R. Matchett. 1997. Attributes of black-tailed prairie dog colonies in northcentral Montana. Journal of Wildlife Management 61:664–73.

Reed, C. A. 1946. The copulatory behavior of small mammals. Journal of Comparative Psychology 39:185–206.

Reig, S. 1997. Biogeographic and evolutionary implications of size variation in North American least weasels (*Mustela nivalis*). Canadian Journal of Zoology 75:2036–49.

Richardson, L., T. W. Clark, S. C. Forrest, and T. M. Campbell. 1987. Winter ecology of the black-footed ferrets at Meeteetse, Wyoming. American Midland Naturalist 117:225–39.

Rosenzweig, M. L. 1966. Community structure in sympatric carnivora. Journal of Mammalogy 47:602–12.

Rust, C. C. 1962. Temperature as a modifying factor in the spring pelage change of short tailed weasels. Journal of Mammalogy 43:323–28.

Rust, C. C. 1965. Hormonal control of pelage cycles in the short-tailed weasel (*Mustela erminea bangsi*). Journal of Comparative Endocrinology 5:222–31.

Rust, C. C., and R. K. Meyer. 1968. Effect of pituitary autograft on hair color in the short-tailed weasel. Journal of Comparative Endocrinology 11:548–51.

Rust, C. C., and R. K. Meyer. 1969. Hair color, molt, and testis size in male, short-tailed weasels treated with melatonin. Science 165:921–22.

Sandell, M. 1984. To have or not to have delayed implantation: The example of the weasel and stoat. Oikos 42:123–26.

Sandell, M. 1986. Movement patterns of male stoats *Mustela erminea* during the mating season: Differences in relation to social status. Oikos 47:63–70.

Sanderson, G. C. 1949. Growth and behavior of a litter of captive long-tailed weasels. Journal of Mammalogy 30:412–15.

Schmidt, G. 1965. *Molineus mustelae* sp. m. (Nematoda: Trichostrongylidae) from the long-tailed weasel in Montana and *Molineus chabaudi* nom. N., with the key to the species of *Molineus*. Journal of Parasitology 51:164–68.

Schmidt-Nielsen, K., and K. Pennycuik. 1961. Capillary density in mammals in relation to body size and oxygen consumption. American Journal of Physiology 200:746–50.

Scholander, P. R. 1955. Evolution of climatic adaptation in homeotherms. Evolution 9:15–26.

Scholander, P. R., V. Walters, R. Hock, and L. Irving. 1950. Body insulation of some arctic and tropical mammals and birds. Biological Bulletin 99:225–36.

Schreiber, A., R. Wirth, M. Riffel, and H. Van Rompaey. 1989. Weasels, civets, mongooses, and their relatives: An action plan for the conservation of mustelids and viverrids. The World Conservation Union Species Survival Commission, Mustelid and Viverrid Specialist Group, Gland, Switzerland.

Seton, E. T. 1929. Lives of game animals. Doubleday, Garden City, NY.

Severson, K. E., and G. E. Plumb. 1998. Comparison of methods to estimate population densities of black-tailed prairie dogs. Wildlife Society Bulletin 26:859–66.

Sheets, R. G., and R. L. Linder. 1969. Food habits of the black-footed ferret (*Mustela nigripes*) in South Dakota. Proceedings of the South Dakota Academy of Science 48:58–61.

Sheets, R. G., R. L. Linder, and R. B. Dahlgren. 1971. Burrow systems of prairie dogs in South Dakota. Journal of Mammalogy 52:451–53.

Simms, D. A. 1979. North American weasels: Resource utilization and distribution. Canadian Journal of Zoology 57:504–20.

Soleman, G. B., and G. S. Warner. 1969. *Trichinella spiralis* in mammals of Mt. Lake, Va. Journal of Parasitology 55:730–32.

Southern, H. N. 1964. The handbook of British mammals. Blackwell, Oxford.

Sperry, C. C. 1941. Food habits of the coyote (Wildlife Research Bulletin No. 4). U.S. Fish and Wildlife Service.

Stoddard, H. L. 1931. The bobwhite quail: Its habits, preservation and increase. Scribner's, New York.

Stroganov, S. U. 1962. Carnivorous mammals of Siberia. Academy of Science of the USSR, Moscow.

Svendsen, G. E. 1976. Vocalizations of the long-tailed weasel (*Mustela frenata*) Journal of Mammalogy 57:398–99.

Teer, J. G. 1964. Predation by long-tailed weasels on eggs of blue-winged teal. Journal of Wildlife Management 28:404–6.

van Soest, R. W. M., J. Van Der Land, and P. J. H. Van Bree. 1972. *Skrjabingylus nasicola* (Nematoda) in skulls of *Mustela erminea* and *Mustela nivalis* (Mammalia) from the Netherlands. Beaufortia 20:85–97.

Vargas, A. 1994. Ontogeny of the endangered black-footed ferret (*Mustela nigripes*) and effects of captive upbringing on predatory behavior and post-release survival for reintroduction. Ph.D. Dissertation, University of Wyoming, Laramie.

Vargas, A., and S. H. Anderson. 1998a. Black-footed ferret (*Mustela nigripes*) behavioral development: Above ground activity and juvenile play. Journal of Ethology 16:29–41.

Vargas, A., and S. H. Anderson. 1998b. Ontogeny of black-footed ferret predatory behavior towards prairie dogs. Canadian Journal of Zoology 76:1696–1704.

Vargas, A., and S. H. Anderson. 1999. Effects of experience and cage enrichment on predatory skills of black-footed ferrets (*Mustela nigripes*). Journal of Mammalogy 80:263–69.

Vaughan, T. A. 1961. Vertebrates inhabiting pocket gopher burrows in Colorado. Journal of Mammalogy 42:171–74.

Wight, H. M. 1932. A weasel attacks a man. Journal of Mammalogy 13:163–64.

Willey, R. 1970. Sound location of insects by the dwarf weasel. American Midland Naturalist 84:563–64.

Williams, E. S., E. T. Thorne, M. J. G. Appel, and D. W. Belitsky. 1988. Canine distemper in black-footed ferrets (*Mustela nigripes*) from Wyoming. Journal of Wildlife Diseases 24:385–98.

Wilson, D. E., and D. M. Reeder. 1993. Mammal species of the world: A taxonomic and geographic reference, 2nd ed. Smithsonian Institution Press, Washington, DC.

Wolf, K. N., D. E. Wildt, A. Vargas, P. E. Marinari, M. A. Ottinger, and J. G. Howard. 2000a. Reproductive inefficiency in male black-footed ferrets (*Mustela nigripes*). Zoo Biology 19:517–28.

Wolf, K. N., D. E. Wildt, A. Vargas, P. E. Marinari, J. S. Dreeger, M. A. Ottinger, and J. G. Howard. 2000b. Age-dependent changes in sperm production, semen quality, and testicular volume in the black-footed ferret (*Mustela nigripes*). Biology of Reproduction 63:179–87.

Wright, P. L. 1942. Delayed implantation in the long-tailed weasel (*Mustela frenata*), the short-tailed weasel (*Mustela cicognaniz*) and the marten (*Martes americana*). Anatomical Record 83:341–49.

Wright, P. L. 1948a. Breeding habits of captive long-tailed weasels (*Mustela frenata*). American Midland Naturalist 39:338–44.

Wright, P. L. 1948b. Preimplantation stages in the long-tailed weasel (*Mustela frenata*). Anatomical Record 100:593–603.

Wright, P. L. 1951. Development of the baculum of the long-tailed weasel. Proceedings of the Society of Experimental Biology and Medicine 75:820–22.

Gerald E. Svendsen, Department of Biological Sciences, Ohio University, Athens, Ohio 45701. Email: svendsen@ohiou.edu.

31

Mink

Mustela vison

Serge Larivière

NOMENCLATURE

COMMON NAMES. American mink, mink
SCIENTIFIC NAME. *Mustela vison*
SUBSPECIES: Hall (1981) recognized 15 subspecies: *M. v. aestuarina, M. v. aniakensis, M. v. energumenos, M. v. evagor, M. v. evergladensis, M. v. ingens, M. v. lacustris, M. v. letifera, M. v. lowii, M. v. lutensis, M. v. melampeplus, M. v. mink, M. v. nesolestes, M. v. vison, and M. v. vulgivaga*

The generic name *Mustela* is Latin for weasel. The specific name *vison* likely originates from the Swedish word *vison,* which means "a kind of weasel" (Lowery 1974). The French name for mink is "vison." Previous reviews of the biology of the mink are provided by Linscombe et al. (1982), Eagle and Whitman (1987), Dunstone (1993), and Larivière (1999). Throughout this account, "mink" refers to the American mink.

DISTRIBUTION

Mink occur throughout the United States except for Arizona and the dry parts of California, Nevada, Utah, New Mexico, and western Texas (Fig. 31.1). In Canada, the mink is present in all provinces and territories. In Newfoundland, mink were brought in 1934 for fur-farming operations (Northcott et al. 1974), and escapees from fur farms have colonized the island since. On Anticosti Island, province of Quebec, mink were purposely released in 1912, but are now believed to be extirpated (Peterson 1966).

Mink have a wide distribution outside of North America. Mink were first brought to Europe in the 1920s for the establishment of fur farms. Subsequently, escapees from fur farms established populations in Denmark, England, France, Germany, Iceland, Ireland, the Netherlands, Norway, Poland, Russia, Scotland, northern Spain, and Sweden (Gerell 1967a; Deane and O'Gorman 1969; Day and Linn 1972; Chanin 1983; Ruprecht et al. 1983; Bevanger and Henriksen 1995; Kauhala 1996; Ruiz-Olmo et al. 1997). The mink is currently colonizing parts of South America, where the species was brought for fur farming in the 1930s (Daciuk 1978).

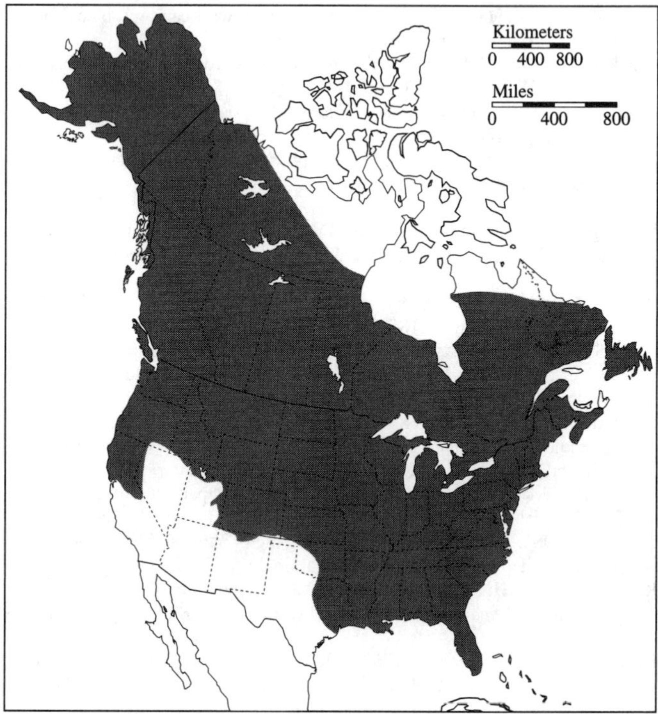

FIGURE 31.1. Distribution of the mink (*Mustela vison*).

DESCRIPTION

The mink has a true weasel shape with a long body, short legs, and short ears. The neck is strong, and as big as or larger in circumference than the head. The tail is one third of the body length. The feet are fully furred and only the pads and toes are naked. All four feet are partially webbed (Jackson 1961).

Size, Weight, and Sexual Dimorphism. The mink is a small mustelid, and typically weighs from 500 to 1500 g. There is great sexual dimorphism in size and body mass, with females about 10% smaller in linear measurements and about 50% lighter in mass (Hall 1981). There is also great geographic variation in average body size; western mink are >50% larger than their eastern counterparts.

In North America, body mass of male and female mink, respectively, averaged 780 and 525 g in Idaho (Whitman 1981), 1523 and 852 g in North Dakota (Eagle et al. 1984), and 1160 and 760 g in Saskatchewan (Larivière et al. 2000). Males and females weighed, respectively, 1091 g (range = 905–1392 g) and 672 g (range = 455–840 g) in Scotland (Hewson 1971) and 1153 g (range = 850–1805 g) and 619 g (range = 450–810 g) in England (Chanin 1983).

Males are also larger than females in linear measurements. In Louisiana, mean external measurements (in mm) of 29 males and 5 females (Lowery 1974), respectively, were as follows: total length, 568 (range = 504–680) and 517 (488–580); length of tail, 184 (167–200) and 172 (152–185); length of hind foot, 68 (60–79) and 52 (50–57); and length of ear, 23 (19–27) and 23 (21–25).

Pelage. Mink are uniformly dark brown. The chin is usually white, and white markings often occur on the throat, chest, or belly. The tail is darker than the body and has longer fur. Coloration does not change with season or age, but old animals may be grizzled with white hairs (Jackson 1961). Albino mink and mink with tan-colored or blond pelts occur infrequently (Lowery 1974). Many pelage colors occur in ranch mink that do not occur in wild mink.

Skull and Dentition. The skull (Fig. 31.2) is somewhat flattened, with a short, broad rostrum and evenly spreading zygomatic arches. The

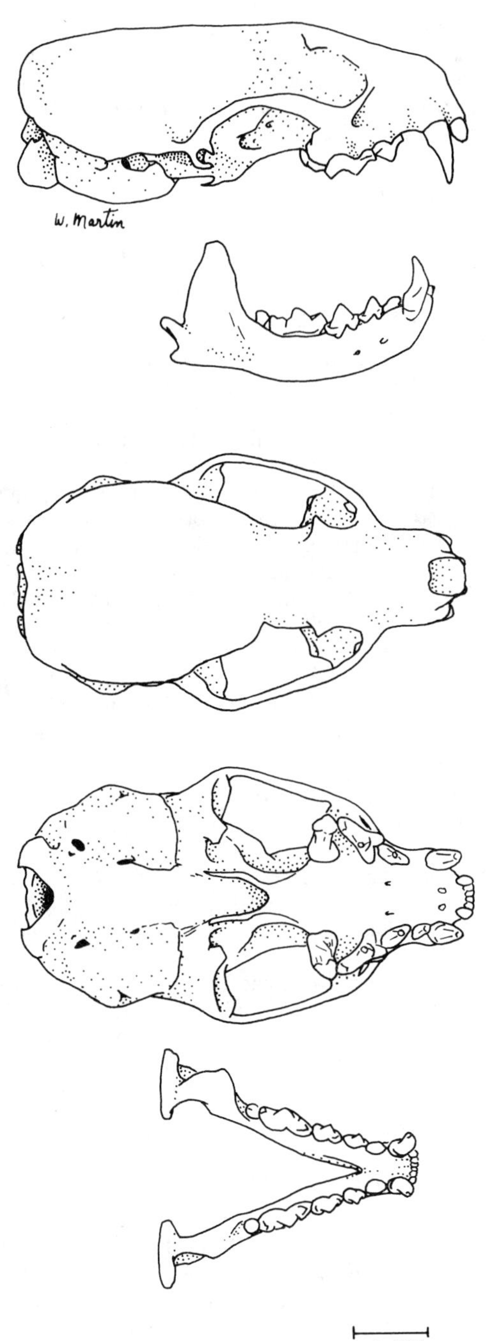

FIGURE 31.2. Skull of the mink (*Mustela vison*). From top to bottom: lateral view of cranium, lateral view of mandible, dorsal view of cranium, ventral view of cranium, dorsal view of mandible.

lambdoidal ridge is well developed in adults, and extends posteriorly as far as the posterior border of the condyle. Auditory bullae are moderately inflated, about 1.5 times longer than wide. The bony palate extends posteriorly to the back molars (Jackson 1961).

Skulls of males and females are sexually dimorphic in size (Wiig 1986). Cranial measurements (in mm) for 54 males and 35 females from Canada (Youngman 1982) averaged, respectively, as follows: condylobasal length, 70.8 (range = 65.3–76.3) and 63.7 (58.7–68.9); mastoid breadth, 36.5 (32.8–40.1) and 31.8 (28.6–36.0); zygomatic breadth, 41.1 (35.9–47.1) and 36.8 (33.6–40.6); palatal length, 32.4 (29.9–35.2) and 28.7 (26.4–31.0); and cranial height, 24.7 (22.4–27.8) and 22.5 (20.6–25.8).

Skulls of ranch mink are larger, have a relatively shorter palate, and have a relatively narrower postorbital constriction than those of wild mink (Lynch and Hayden 1995). Ranch mink also have a slightly smaller brain, heart, and spleen (Kruska and Schreiber 1999). However, comparison of 44 genetic loci between seven wild and seven ranch mink failed to reveal any differences (Kruska and Schreiber 1999).

Dental formula is I 3/3, C 1/1, P 3/3, M 1/2. Deciduous teeth (incisors, canines, premolars) erupt 16–49 days after birth. Molars erupt 44–71 days after birth (Aulerich and Swindler 1968). Teeth are sharp at birth, but canines become more rounded with age.

GENETICS

The mink has $2n = 30$ chromosomes (Lande 1957; Fredga 1961). Both sex chromosomes are submetacentrics, 2 autosomes are acrocentrics, and 26 are either metacentrics, submetacentrics, or subtelocentrics (Hsu and Benirschke 1968). Rarely, diploid–triploid chimerism may produce viable hermaphrodites (Nes 1966). Crossing between *M. vison* and European mink (*M. lutreola*) leads to resorption of embryos (Ternovskii 1977). Microsatellite markers have been developed to facilitate genetic investigations in *M. vison* (Fleming et al. 1999).

ANATOMY

Blood constitutes 7.2% of the body mass. Mean blood pressure averages 153 mm Hg, but pulse pressure is 34.8 mm Hg (Ringer et al. 1974). Males have a lower heart rate than females (250 vs. 300 beats/min, respectively), most likely because of the larger body mass of males. Respiratory rate varies from 18 to 20/min (Ringer et al. 1974).

The penis of adult males is about 5.6 cm long and enclosed in a sheath. Males have a well-developed baculum; the baculum is triangular in cross section and curved at the tip. Bacula of 126 mink from North Dakota averaged 48.0 mm long (range = 41.2–52.5 mm; Burt 1960). The urethra is located in a groove on the ventral side of the baculum. During erection, the baculum protrudes beyond the glans, and the urethral opening is ventral to the tip (Enders 1952). The baculum likely plays a role in sexual selection inside the female tract by helping males dislodge sperm and sperm plugs from other males (S. Larivière and S. H. Ferguson, unpublished data).

Testes remain in the scrotum throughout the year. Testes increase in size from November to March, and there is an associated increase in spermatogenic activity (Onstad 1967; Bostrom et al. 1968). Males have a well-developed prostate gland, which completely surrounds the urethra, but seminal vesicles and bulbo-urethral glands are absent (Linscombe et al. 1982).

The reproductive anatomy of female mink is well known because of the interest in raising mink on fur ranches. Females have six mammae (Peterson 1966). The uterus is bicornate; mean length of vagina is 3.4 cm. Uterine horns connect to the proximal part of the uterus. During anestrus, the uterine horns are threadlike in appearance and about 3 cm long; they become thicker and longer in parous mink (Enders 1952). Placental scars are not considered reliable for determination of breeding history in this species (Elder 1951).

PHYSIOLOGY

The tubular body shape of the mink has several ecological and physiological consequences. First, the long body probably helps mink access the burrows of prey such as muskrats (*Ondatra zibethicus*), rabbits (*Oryctolagus, Sylvilagus*), and hares (*Lepus*). The streamline body shape also helps reduce water resistance while swimming (Williams 1983). However, the great surface area-to-volume ratio caused by this shape makes the species vulnerable to extreme temperatures (Brown and Lasiewski 1972; Segal 1972). The heart rate and basal metabolic rate (BMR = $84.6W^{0.78}$) values for mink are higher than predicted for an animal of this size, likely as a consequence of the fusiform shape (Iversen 1972; Gilbert and Gofton 1982a).

Daily consumption of dry matter (per kilogram of body mass) averages 40 and 53 g for male and female mink, respectively (Bleavins and Aulerich 1981). Stomach capacity is 40–70 cm^3 (Waller 1962; Linscombe et al. 1982). Mean passage time of food averages 187 min for males and females (Bleavins and Aulerich 1981), but undigested material may appear in mink scats only 1 hr after ingestion (Waller 1962; Linscombe et al. 1982). A 1-kg mink requires 152 ± 11 cal of digestible energy/day for maintenance. In comparison, a female nursing five young requires approximately three times that amount for 3 weeks following parturition (Cowan et al. 1957).

Molting occurs twice a year, during spring and autumn (Chanin 1983). The spring molt begins in March–April, and the summer fur, which is shorter and less dense, is acquired by May. Pelage cycles are controlled by photoperiod (Rust et al. 1965; Duby and Travis 1972).

Thick underfur and oily guard hairs render the fur water resistant (Lowery 1974). Mean density of guard hairs from the midback section (780/cm^2) and length of guard hairs (24 mm) are intermediate between those of the more aquatic otters (*Lutra* and *Lontra*) and the strictly terrestrial ferrets (*M. putorius*), which suggests that mink are incompletely adapted to aquatic life (Dunstone 1979).

When diving, *Mustela vison* undergoes rapid bradycardia; heart rate is lower during submersion than during any other behavior (Gilbert and Gofton 1982b). Rapid onset of bradycardia is likely an adaptation to conserve oxygen during the short periods of asphyxia experienced by this unspecialized diver (West and van Vliet 1986; Stephenson et al. 1988).

Mink rely heavily on sight when foraging, and vision of mink is clearer in air than underwater (Sinclair et al. 1974). Auditory acuity is also used to detect prey when foraging; mink are able to hear ultrasonic vocalizations (1–16 kHz) emitted by rodent prey (Powell and Zielinski 1989). The sense of smell is limited; the peripheral olfactory structures of the mink are slightly regressed. Olfactory membranes cover only 14 cm^2; this reduction in the amount of olfactory membrane is likely the result of the semiaquatic lifestyle (Ferron 1973).

Mink have two anal glands, which are used for scent marking when excreting feces or by deliberate rubbing of the anal region on the ground. Anal gland secretions are composed of 2,2-dimethylthietane (main component), 2-ethylthietane, cyclic disulfide, 3,3-dimethyl-1,2-dithiacyclopentane, and indole. The presence of sulfur-containing compounds suggests that the secretions may have a function in defense (Brinck et al. 1983). Mink are able to empty their anal glands when under stress, but the distance reached by the expulsion is <30 cm (Brinck et al. 1978).

Feces have a strong odor, which originates from the proctodeal glands, which open into the rectum. Feces are deposited in prominent places, which is likely to enhance the active range of the scent for marking (Brinck et al. 1978). Urine is greenish-yellow and has a distinct musky odor. The pH of urine is 6.8–7.5. Adult mink excrete 35–60 ml of urine per day (Kubin and Mason 1948).

Mink are extremely sensitive to environmental pollutants. Because of their position at the top of the food chain in aquatic environments, mink accumulate many chemical compounds and heavy metals in their tissue, and for this reason, they are often used as bioindicators of pollution in aquatic environments (Aulerich and Ringer 1979; Halbrook et al. 1996; Smits et al. 1996a, 1996b; Stevens et al. 1997a). They tolerate low levels (<1.0 ppm) of mercury intoxication (Wobeser et al. 1976); however, at higher levels of contamination (>1.8 ppm), severe lesions or death occur (Wobeser and Swift 1976). Clinical signs of heavy mercury intoxication include anorexia, loss of weight, incoordination, tremors, and convulsions (Aulerich et al. 1974). Pollution from heavy metals also produces an increased incidence of morphological abnormalities, parasitism, lower body mass, and lower population densities (Sidorovich and Savcenko 1992). Mink also are sensitive to low quantities (<1 ppm) of dietary polychlorinated biphenyls, and intoxications lead to weight loss, decoloration and necrosis of liver, fibrosis of coronary arteries, and still births (Platonow and Karstad 1973). Exposure to high polychlorinated biphenyl concentrations may also affect growth and development of the baculum, although such effects are less consistent (Harding et al. 1999; Aulerich et al. 2000). Animals exposed to high levels of hexachlorobenzene may experience lower reproductive success (Moore et al. 1997). Similarly, population densities of mink may decrease in areas where intensive acid precipitation affects freshwater fishes (Bevanger and Albu 1986).

REPRODUCTION

Interest in mink as a source of fur has led to numerous physiological studies, especially with regard to reproduction. Thus, several excellent reviews of the physiology of reproduction in mink are available (Hansson 1947; Enders 1952; Venge 1959).

Mating. Mating lasts 3 weeks. It begins from February in the south to April in the north (Hansson 1947; Venge 1959; Chanin 1983; Sidorovich 1993; Ben-David 1997). Ovulation is induced by the presence of males and/or by attempted or successful copulation (Hansson 1947; Venge 1959; Adams 1981). Copulation is vigorous. The male typically bites the female at the nape and holds her with his forefeet. Duration of copulation averages 64 min, but ranges from 10 min to 4 hr (Hansson 1947; Venge 1959). During copulation, the male may ejaculate several times. Ovulation occurs 36–48 hr after copulation (Hansson 1947; Enders 1952). In a study of captive mink, 84% of eggs released implanted, but <50% of eggs released resulted in young (Hansson 1947).

Females are typically receptive at 7- to10-day intervals during the 3-week breeding season. In the wild, multiple matings are believed to occur (Enders 1952). Breeding studies involving multiple mating with several males of different colors indicate that multiple paternity can occur within the same litter (Hansson 1947).

Gestation. Mink exhibit facultative delayed implantation (Hansson 1947). Interestingly, the mink is, along with the striped skunk (*Mephitis mephitis*), one of two species that mate in the spring yet possess a short delay before implantation occurs. Most likely, delayed implantation allows females to track environmental conditions and adjust the timing of parturition to coincide with the optimum time for rearing young (Ferguson et al. 1996).

Gestation averages 51 days, but may vary from 40 to 75 days (Hansson 1947; Enders 1952). Actual embryonic development is 30–32 days (Enders 1952), so implantation delay may vary from 8 to 45 days. Onset of mating and gestation is controlled by photoperiod (Hammond 1951; Duby and Travis 1972).

Parturition and Litter Size. Parturition occurs from April to June (Hansson 1947; Sidorovich 1993). Litter size averages 4 (range = 2–8; Mitchell 1961; Sidorovich 1993), and increases with female age (Sidorovich 1993).

Ontogeny. Neonates are altricial. At birth, the young are blind, possess a fine coat of short, silver-white hairs, and weigh about 6 g (Svihla 1931). Eyes open at 25 days of age, and weaning occurs after 5 weeks. Juveniles begin hunting at 8 weeks of age, but remain with the mother until autumn (Peterson 1966; Poole and Dunstone 1976). Young mink become independent in the fall, spend their first winter alone, and are ready to breed during their first spring, when they are about 10 months old.

ECOLOGY

Habitat Preferences. Mink inhabit a variety of ecoregions including tropical swamps, temperate and boreal forests, prairies, freshwater and saltwater coastal areas, and tundra. Throughout its geographic range, the presence of mink is affected primarily by availability of water and food. In suitable environments, mink activity is concentrated where food abundance and accessibility (ease of capture of prey) is greatest. Along creeks and rivers, mink are more abundant along sections with abundant prey hideouts such as logjams, overhanging banks, and grassy or brushy banks. In prairies, mink are most abundant on large wetlands with irregular shorelines and large areas of open water (Arnold and Fritzell 1990). In Florida, mink abundance is highest in swamp forests, intermediate in saltwater marshes, and lowest in freshwater marshes (Humphrey and Zinn 1982). Use of wetlands and watersheds follows availability of water, and mink frequently move from seasonal to permanent wetlands as the dry season progresses (Humphrey and Zinn 1982). In marine environments, mink select shallow vegetated and tidal slopes and sites protected from waves; beaches with small rocks are avoided because of the low abundance of prey (Ben-David et al. 1996). In Scotland, mink also avoided areas of sand or shingle, and instead foraged mostly in boulder fields and rockpools located in the intertidal zone (Bonesi et al. 2000).

Home Ranges. Few reliable data exist on home ranges of mink in North America, mostly because very few radiotelemetry studies have been conducted on this species. Home range values based on telemetry data are available only for Tennessee and Manitoba. In Tennessee, three males occupied a section of river that averaged 7.5 km long ($SE = 1.8$; Stevens et al. 1997b). In the parklands of Manitoba, mink inhabited numerous small wetlands; mean summer home range of males was 7.7 km^2 (Arnold and Fritzell 1987a).

In Europe, home ranges of mink are better known and it is recognized that adults have larger home ranges than juveniles and males have larger home ranges than females (Gerell 1970). Mean linear home ranges (in km) of adult male and female mink, respectively, were as follows: 2.5 (range = 1.9–2.9, $n = 3$) and 2.2 (1.5–2.9; $n = 2$) in England (Birks and Linn 1982), 5.3 ($n = 1$) and 4.2 ($n = 1$) in Finland (Niemimaa 1995), and 2.6 (1.8–5.0; $n = 4$) and 1.9 (1.0–2.8; $n = 2$) in Sweden (Gerell 1970). In England, home ranges of males and females, respectively, were longest in riverine habitats (2.53 and 2.16 km), intermediate along lakes (1.90 and 1.46 km), and shortest along coastal habitats (1.50 and 1.09 km) (Dunstone and Birks 1985). In archipelagos, home ranges of *M. vison* may include several islands often separated by >500 m of open water (Niemimaa 1995).

Mink are often considered to be territorial, but evidence for territoriality is weak. Few studies of mink in North America have attempted to elucidate this aspect. During spring, males range widely in search of females and many males may visit the same area, again suggesting that mink are not territorial. During autumn, commercial trappers often capture numerous animals at the same trap location when mink are abundant. Thus, the spatial organization of mink may be more a result of limited overlap of home ranges, caused either by low density or possibly by spatial avoidance, than of truly defended territories.

In Europe, mink are reportedly territorial, but again the evidence is unconvincing. For example, home range overlap seems to increase with density, a factor that suggests that low density, instead of territoriality, limits overlap. For example, in riverine and lacustrine habitats of Europe, home ranges of American mink exhibit little intersexual overlap and no intrasexual overlap (Dunstone and Birks 1985). However, overlap increases in marine environments where densities are higher (Dunstone and Birks 1985). Exploitation of variable food sources by mink suggests that territoriality may not be advantageous because distribution of food and mates is spatially and temporally unpredictable. Low densities and spatial avoidance among individuals, instead of territorial defense, may thus provide a better explanation for the low overlap in home ranges of wild mink.

Population Densities. Mink seldom occur at high densities. However, the species occurs on most water bodies within its distribution. Typically, availability of wetlands and water is directly related to the abundance of mink.

Density of mink in England varies from 0.1 to 0.7/km^2 (Halliwell and Macdonald 1996). In England, American mink were most numerous at sites that had high availability of den sites and low emergent vegetation cover (Halliwell and Macdonald 1996).

In North America, few density estimates are available, and most represent the number of mink harvested within a certain area. In Louisiana, density was estimated as high as 25–42 mink/km^2 in cypress–tupelo swamps, with the "average" swamp producing 8–10 mink/km^2 (Linscombe et al. 1982). In comparison, brackish coastal marshes produced 4.1 pelts/km^2. Finally, the peak production in freshwater marshes was 3.7 pelts per km^2 (Linscombe et al. 1982). In Montana, density was 3.3–8.5 mink/km^2 (Mitchell 1961), whereas density was estimated at 1.2 mink/km of river in Michigan (Marshall 1935). In Wisconsin, estimated density varied between 1.6 and 5.4 mink/km^2 (McCabe 1949). Near lakes, mink density decreased with increased cottage development (Racey and Euler 1983).

Harvest data from commercial trappers suggest that mink abundance is not uniform across North America. For example, high mink densities are found in southern states bordering the Mississippi River and also in states adjacent to the Great Lakes (Eagle and Whitman 1987). In the northeast, abundance of mink is much lower and appears variable from year to year. Based on fur harvest records from Canada, it has been suggested that populations of *M. vison* may follow a 10-year cycle (Bulmer 1974; Erb et al. 2000), and that mink populations may peak at the same time as snowshoe hares (*Lepus americanus*) (Keith and Cary 1991). However, causes or reasons for the cyclic nature of mink populations remain poorly understood (Erb et al. 2000).

In prairie states and provinces, abundance of mink appears to be correlated with amount of yearly precipitation. During dry years, abundance of muskrats and mink declines, and remaining animals typically are restricted to permanent and semipermanent wetlands. Following the return of abundant precipitation, mink populations rebound, but only after reestablishment of muskrat populations, so there appears to be a 2- to 3-year delay between the return of water and the reestablishment of abundant mink populations.

Sex Ratio. The sex ratio (M:F) of 32 juveniles captured in Montana was 1.3:1 (Mitchell 1961). Sex ratios favoring males have been reported in numerous locations (Errington 1936; Mitchell 1961), but often result from trapping bias (Buskirk and Lindstedt 1989). During population decreases, the sex ratio of litters favors females (Sidorovich 1993).

Movements. Movements of *M. vison* are either small-scale foraging movements or extensive travel during dispersal, mating, or between dens or foraging areas (Birks and Linn 1982). In the Canadian prairies, nightly movements ranged up to 12 km (Arnold and Fritzell 1987a), whereas in Tennessee, daily movements were <4.3 km (Stevens et al. 1997a). Largest movements are by juveniles during the fall dispersal (≤45 km away from natal areas) and by males during the mating season in spring (Mitchell 1961; Gerell 1970).

Interspecific Competition. In North America, mink and river otter (*Lontra canadensis*) often inhabit the same waterways. It is believed that niche separation occurs through resource partitioning (Ben-David et al. 1996). Mink typically occupy drier sites and consume a lower proportion of fish and invertebrates and a higher proportion of mammals and birds than do river otters (Gilbert and Nancekivell 1982; Humphrey and Zinn 1982). In marine environments, *M. vison* and *L. canadensis* show high dietary overlap (about 80%), but exhibit niche separation through differential habitat preferences (Ben-David et al. 1996). Mink prefer sites with low to medium wave exposure, whereas river otters prefer sites with heavy wave exposure and denser overstory cover (Ben-David et al. 1996).

Mink are extremely adaptable and have colonized numerous habitats outside their original range. As a result, mink have impinged on the distribution of, and now compete with, several other species of carnivores.

In Europe, the best-known and most problematic aspect of the presence of American mink is the resulting competition with European mink. In many areas, the spread of American mink may have contributed to the decline of European mink, especially from marginal habitats (Maran and Henttonen 1995). Nonfertile crossing between male American mink and female European mink may also prevent European mink from successfully reproducing (Maran and Henttonen 1995). There is no indication of competition between polecats (*M. putorius*) and *M. vison* (Gerell 1967a; Lodé 1993; Kauhala 1996). The polecat is strictly terrestrial, and although it is sympatric with the American mink, polecats typically consume more rodents and amphibians, whereas mink consume more fish and birds (Lodé 1993).

Introduced American mink also compete with the European otter (*Lutra lutra*). The diets of both species overlap about 60–70% (Erlinge 1969; Sodorovich et al. 1998), but *M. vison* consumes smaller prey, less fish, and a higher proportion of mammals and arthropods than *L. lutra* (Jenkins and Harper 1980; Chanin 1981; Bueno 1996). Mink also use terrestrial habitats more (Gerell 1967a; Erlinge 1969; Akande 1972; Day and Linn 1972; Chanin and Linn 1980; Wise et al. 1981). Competition between *M. vison* and *L. lutra* is most intense in winter, and high densities of otters may prevent mink from occupying otherwise prime habitats (Erlinge 1972).

In South America, introduced *M. vison* are sympatric with southern river otter (*L. provocax*). However, *M. vison* consumes mostly crustaceans and rodents, whereas *L. provocax* consumes mostly crustaceans and fish (Medina 1997). Finally, habitat overlap is low (5–22%), and there is little evidence for competition between the two species (Medina 1997).

FEEDING HABITS

Mink are strictly carnivorous, and their diet reflects the local prey base. Typically, the diet is fish, amphibians (mostly frogs), crustaceans (crayfish and crabs), muskrats, and small mammals (Errington 1954; Gerell 1967b; Day and Linn 1972; Cuthbert 1979; Chanin and Linn 1980; Birks and Dunstone 1985; Ward et al. 1986; Proulx et al. 1987; Bueno 1994; Ben-David et al. 1997; Sidorovich et al. 1998). Mammals such as muskrats, mice, and voles rank as the most important food of mink during all seasons (Eagle and Whitman 1987). Opportunistically, mink also consume lagomorphs, sciurids, birds and their eggs, reptiles, aquatic insects, earthworms, and snails (Hamilton 1959; Akande 1972; Arnold and Fritzell 1987b). Mink are not very agile in water, and thus capture and consume small and slow-moving fish more than large (>20 cm) or fast-swimming fish such as salmonids (Goodpaster and Hoffmeister 1950; Gerell 1968; Burgess and Bider 1980; Dunstone and Birks 1987). However, salmonids may be more vulnerable during spawning (Melquist et al. 1981).

In North America and Europe, mink are an important predator of waterfowl and their eggs (Eberhardt and Sargeant 1977; Strachan et al. 1998; Ferreras and Macdonald 1999). Adult mink may kill incubating hens on their nests (Arnold and Fritzell 1989). In Manitoba, Canada, it was estimated that a male mink consumed 3–7 adult ducks, 15–25 week-old ducklings, and 18–30 duck eggs during a single waterfowl breeding season (Arnold and Fritzell 1987b). In southern England, American mink obtain 11% of their energy requirements from coots (*Fulica atra*) and moorhens (*Galliinula chloropus*); vulnerability of waterbird species varies according to behavior (Ferreras and Macdonald 1999). Within a season, predation on waterfowl increases during incubation, brood rearing, or molting, when mobility is limited (Arnold and Fritzell 1987b; Sargeant et al. 1973). Mink predation and disturbance also may cause mortality of colonial nesting bird chicks (Burness and Morris 1993; Craik 1997).

In the prairies of North America, the ecology of mink is closely tied to the presence of muskrats. Interestingly, mink also reach their largest mean body size in the prairies, possibly as an adaptation to capture this larger prey. This view is supported by the differential use of muskrats by both sexes, with the larger male mink capturing more muskrats than the female mink (Sealander 1943).

Before weaning (<170 g of body mass), juvenile muskrats are much more vulnerable to mink predation than adult muskrats. Typically, all young muskrats in a litter either survive to become adult or all die of mink predation. As young muskrats get older, their swimming ability increases as well as their body mass, and consequently their vulnerability to mink predation decreases (J. A. Virgl, pers. commun., 2000).

On prairie potholes, mink are most successful at killing muskrats in shallow water, and muskrats inhabiting shallow-water lodges are more vulnerable to predation than muskrats inhabiting deeper water lodges or burrows. When a mink enters a lodge, muskrats rapidly exit the lodge and seek deep water. Deep water may reduce mink activity and increases the odds of muskrats escaping mink predation, and thus the quality of dwelling for muskrats is directly dependent on water depth (Proulx et al. 1987; Messier and Virgl 1992; Virgl and Messier 1996; J. A. Virgl, pers. commun., 2000).

Surplus killing may occur, and *M. vison* may cache food (Gerell 1968; Sargeant et al. 1973; Burness and Morris 1993). For this reason, mink that access chicken coops or other fowl pens can often cause considerable damage. In the wild, frogs, crawfish, and rodents may be stored in old muskrat burrows (Svihla 1931; Yeager 1943). The frequency and importance of food caching are unknown.

BEHAVIOR

Daily and Seasonal Activity. Mink are mostly nocturnal, but daytime activity may occur (Gerell 1969; Birks and Linn 1982; Arnold and Fritzell 1987a; Niemimaa 1995). In captivity, food synchronizes activity (Zielinski 1986). In the wild, nocturnal habits of mink probably are linked to the activity patterns of prey such as small mammals, crayfish, and frogs. Fish may also be easier to capture at night. Mink are active year-round, but activity levels decrease during winter (Birks and Linn 1982) or during periods of cold weather (Marshall 1935; Segal 1972). Based on harvest by commercial trappers, mink apparently increase their activity and movements immediately before weather fronts and storms.

Social Organization. Mink generally are solitary, but pairs may occur during the breeding season. Family groups consisting of one female and her young may occur during late summer and early autumn. Most often, groups comprise young or female–young associations (Mitchell 1961).

Foraging Behavior. Most foraging activity of mink occurs along waterways. When foraging, they travel at a slow gait, stopping frequently to investigate locations that may harbor prey. Along waterways, mink in search of prey will investigate holes in the bank, rock crevices, undercut banks, root systems, logjams, beaver lodges and dams, as well as pools of deep water at the end of shallow riffles. Although most activity occurs within 1–2 m of the shoreline, purely terrestrial activity also occurs. Terrestrial foraging is usually performed by larger males foraging for lagomorphs (Birks and Linn 1982; Dunstone and Birks 1983, 1985; Birks and Dunstone 1985). In the prairies of North America, mink also may forage in upland habitats adjacent to wetlands. On occasion, mink foraging for mice and voles may investigate farmsteads in proximity to wetlands, especially if the farmstead contains grain storage buildings and old hay bales that harbor rodents. When water bodies are small but numerous, overland travel may be common between bodies of water. Along creeks and rivers, overland travel is less common, but may occur as mink travel between drainage systems.

Traveling and foraging mink prefer habitats with overhead cover. For this reason, brushy or grassy shorelines are used more than bare shorelines without vegetation or cover. The preference for cover is because vegetated shorelines harbor more prey and also offer protection against aerial predators.

Mink travel mostly on land, typically following the edge of the shoreline closely. Aquatic prey are located from above the water surface (Poole and Dunstone 1976). When water reflection is a problem, mink may locate prey by immersing their head and scanning for prey (Poole and Dunstone 1976). Aquatic locomotion is frequent while hunting, and mink occasionally search for and capture prey underwater (Sinclair et al. 1974; Dunstone and Clements 1979). Because mink possess few adaptations for underwater foraging, they compensate by focusing on prey refuges (Poole and Dunstone 1976; Dunstone 1978; Dunstone and O'Connor 1979a). They do not stalk or ambush, but instead simply rush on prey (Poole and Dunstone 1976).

Locomotion. Mink walk with the head close to the ground. During bounding, the head is held high and the tail taut and arched upward.

Mean speeds of walking and bounding mink are 48 and 262 cm/sec, respectively (Dunstone 1979).

Mink are agile tree-climbers, capable of ascending large vertical trees, descending head first, and jumping from tree to tree (Larivière 1996). Tree climbing may be used as an escape strategy (Larivière 1996) or to access resting sites such as forks in trees (T. Arnold, pers. commun., 1996) or tree cavities (L. Israelson, unpublished data).

Two forms of swimming occur. When fully submerged, mink alternate use of all four limbs with either diagonally opposite legs or contralateral legs simultaneously. When swimming at the surface, they use only the forelimbs, occasionally aided by a power stroke from the hind limbs for turning or diving. Swimming speeds average 42 and 59 cm/sec for surface and underwater swimming, respectively (Dunstone 1979). Swimming is energetically costly, as both water resistance and oxygen consumption increase curvilinearly with speed. Lack of specialization for swimming contributes to high energetic costs, but enables the mink to effectively forage in aquatic and terrestrial habitats (Williams 1983).

The mink is a good diver, capable of diving 5–6 m deep and swimming underwater for 30–35 m (Peterson 1966). Usually, mink spend 5–20 sec underwater when fishing (Poole and Dunstone 1976), and duration of dives as well as interdive intervals increase with water depth (Dunstone 1983). Although open water is deemed unsuitable for a hunting mink because the species lacks the underwater endurance necessary for effectively pursuing prey (Dunstone and O'Connor 1979b), deepwater pools along creeks and submerged cavities and root systems are often searched for prey. In addition, mink will often visit submerged muskrat burrows either for prey or as resting sites.

Denning Ecology. Mink rarely excavate their own burrows (Birks and Linn 1982); the most commonly used dens in North America are abandoned muskrat burrows (Marshall 1935; Schladweiler and Storm 1969; Sargeant et al. 1973; Arnold and Fritzell 1989). Other den sites include burrows of ground squirrels (*Spermophilus*) or rabbits, cavities under waterside trees, root systems, rockpiles, brush piles, culverts, or bridge foundations (Birks and Linn 1982; Dunstone and Birks 1985). Most dens have two to five entrances (Schladweiler and Storm 1969) and are located <2 m from water (Birks and Linn 1982). Resting mink may also use dense stands of emergent vegetation, forks in trees, and tree cavities up to 10 m above ground (Sargeant et al. 1973; Birks and Linn 1982; Arnold and Fritzell 1989; T. Arnold, pers. commun.; 1996 L. V. Israelson, unpublished data).

Defensive Behavior and Vocalizations. Mink emit defensive screams, warning squeaks, and hissing (Gilbert 1969; Larivière 1996). In addition, chuckling may be audible during the reproductive season and is associated with sexual stimulation (Gilbert 1969). When stressed, *M. vison* will raise its fur, arch its back, and bare its teeth. Defensive behavior is accompanied by high-pitched squeals, hissing, and emptying of anal glands. During the arched-back position, the tail is lifted and moved from side to side, possibly to disperse the strong odor of the anal gland secretions (Brinck et al. 1978).

MORTALITY

In North America, adult mink may be killed by great-horned owls (*Bubo virginianus*), hawks (*Buteo*), fisher (*Martes pennanti*), coyotes (*Canis latrans*), red foxes (*Vulpes vulpes*), bobcats (*Lynx rufus*), lynx (*L. canadensis*), alligators (*Alligator mississippiensis*), and otters (Gerell 1967a; Erlinge 1972; Lowery 1974; Linscombe et al. 1982). However, most mortality occurs through trapping by humans. During years of high fur prices, trapping pressure is high, but even during these years there is no evidence that mink are overharvested. The high reproductive potential of mink as well as their dispersal abilities probably help maintain sustainable populations even in areas of high harvest. Furthermore, many mink trappers trap only at roadside locations (Faler 1988; Spencer 1990), which may help maintain a reservoir of animals away from roads as breeders.

Accidental mortality may occur through roadkills (Eagle and Whitman 1987) or captures in fish cages and gill nets (Gerell 1971). Mink can live up to 8 years in captivity (Dunstone 1993), but in the wild, a complete turnover of a population occurs every 3 years (Mitchell 1961). In Maryland, all but 1 of 169 mink were ≤3 years old (Askins and Chapman 1984).

Diseases. Mink can contract numerous diseases, but many of these diseases have been observed only in captive populations where animals are in closer proximity and more vulnerable. Most common diseases include Aleutian disease, amyloidosis, botulism, distemper, hemorrhagic pneumonia, mink virus enteritis, tularemia, feline panleukopenia, urolithiasis, and canine parvovirus (Tomson 1987; Nieto et al. 1995).

Parasites. The species harbors numerous parasites. Endoparasites include the protozoan *Sarcocystis*; the nematodes *Baylisascaris devosti, Capillaria mucronata, Euparyphium melis, Filaroides martis, Skrjabingylus nasicola,* and *Spirometra erinacei* (Hansson 1967; Sidorovich and Savcenko 1992; Dunstone 1993; Ramos-Vara et al. 1997); and the cestode *Dioctophyma renale* (Wren et al. 1986). Ectoparasites include the ticks (*Ixodes*) and fleas *Ctenophthalmus, Megabothris, Malareus, Nosopsyllus, Paleopsylla,* and *Typhloceras* (Fairley 1980; Chanin 1983; Page and Langton 1996). Linscombe et al. (1982) provided an exhaustive list of mink parasites.

AGE ESTIMATION

Age of mink can be estimated by cementum annuli, sections of mandible, baculum morphology or weight, skull and pelvic girdle measurements, or the weight of various organs (Elder 1951; Lechleitner 1954; Greer 1957; Birney and Fleharty 1968; Franson et al. 1975; Askins and Chapman 1984). Lens weight is a reliable indicator for mink ≤1.5 years old (Pascal and Delattre 1981). The condyle–premaxillae length enables determination of sex by skull alone for animals >10 months old (Birney and Fleharty 1966).

ECONOMIC STATUS, MANAGEMENT, AND CONSERVATION

The fur of mink has always been a symbol of luxury, and the species is a pillar of the North American fur trade. The importance of mink fur also led to the establishment and popularity of mink farms not only in the United States and Canada, but also in many European countries. Mink farming became intensive around 1925. Today, mink remain the most important species raised on fur ranches; 85–90% of the mink fur on the market is produced on farms (Peterson 1966; Thompson 1968; Tomson 1987; Nowak 1999).

Interest in fur farming and raising of mink is still strong in many countries. Consequently, the economic value of the species has generated numerous studies of its behavior (Gilbert and Bailey 1967; 1969; MacLennan and Bailey 1969; Cooper and Mason 2000), metabolism and physiology (Wamberg 1994), reproduction (Hansson 1947; Enders 1952; Sundqvist and Gustafsson 1983; Sundqvist et al. 1988; Lagerkvist et al. 1994; Clausen et al. 1996; Hansen et al. 1996; Blottner et al. 1999), husbandry and veterinary care (Tomson 1987), and its economic value as a fur animal (Lagerkvist 1997).

Mink are also very important in the trade of wild furs, possibly because the market for mink fur has always survived variations in consumer preferences and fashion trends. For this reason, many commercial trappers target mink as the primary species on their traplines. At the same time, they harvest raccoons (*Procyon lotor*) and muskrats, which are captured in many of the same sets. In the commercial trapping season of 1991/92, the harvest in the United States exceeded 129,000 mink, and the pelts sold for an average of $18.47 (Linscombe 1994). Demand for ranch mink affects the price of wild pelts, but about 400,000–700,000 wild mink are taken each year throughout North America, for an annual income exceeding $5 million (Eagle and Whitman 1987).

Although mink are an important furbearer in North America, the situation is much different in Europe. There, the American mink was introduced following escapes and releases from fur farms. It colonized many waterways, where it impinged on the distribution of the European mink. Because European mink are smaller and less aggressive, they have decreased in abundance where American mink occur. In some regions, the European mink has been extirpated. It appears that direct aggression from American mink toward the smaller European mink is responsible for its reduced density and distribution (Sidorovich et al. 1999). The precarious status of European mink is considered a conservation problem of high priority, and survival of this species is considered threatened by continued expansion of the range and increase in abundance of American mink (Maran et al. 1998; Sidorovich et al. 1998).

The American mink is generally abundant throughout its geographic range. Only one subspecies, *M. v. evergladensis* (present only in southern Florida), is rare and may be threatened by human alteration of waterways (Nowak 1999). Elsewhere, mink are generally common to abundant, although their secretive nature and small size make them inconspicuous. For this reason, managers typically infer variations in abundance, even presence or absence, from harvests by commercial fur trappers. Currently, few regulations are imposed on the harvest of mink, and there are probably very few areas where this species is overexploited. Instead, destruction and drainage of wetlands and pollution of waterways probably constitute the most serious threats to the future of mink populations (Linscombe et al. 1982).

It is interesting that a native North American species such as mink has been studied more intensively outside its native range than within. Indeed, most of our current knowledge on this species comes from research done in European countries, with relatively little data for either the United States or Canada. In North America, our knowledge of mink is mostly anecdotal, although there are likely several interesting studies that never were published. The lack of interest in studying mink likely arises from the security of current populations and the low incentive for management agencies to spend money on nonproblematic populations (Eagle and Whitman 1987). Finally, difficulties with live capture and attachment of telemetry transmitters to wild mink probably have contributed to making this species unattractive for theses projects; the few published studies on mink behavior and ecology are typically plagued with small sample sizes (e.g., $n = 3$; Bonesi et al. 2000).

Lack of telemetry studies of mink in North America limits understanding of its ecology in the wild as well as the impact of trapping. Furthermore, there probably is no easy way to determine population densities, except possibly from telemetry data. Thus, harvest management is based more on the stability of harvest from year to year (Eagle and Whitman 1987). In most states and provinces, nonexclusive trapping territories prevent the establishment of quotas on harvest. Instead, harvest of mink is regulated solely by the length of the trapping season in each state, which is typically established based on primeness of pelts. Although this method of regulating the harvest is somewhat crude, stability of the harvest suggests that there are few areas, if any, where mink are overharvested (Linscombe et al. 1982; Eagle and Whitman 1987). With low fur prices, mink are most likely to face conservation challenges from the ongoing destruction of wetland habitats and environmental pollution rather than from direct harvest impacts.

Trapping wild mink in autumn is considered challenging because of their small size and low densities and because traps set near water are often exposed to fluctuations in water level and variations in temperature. For commercial fur harvest, mink can be humanely harvested using foothold traps with drowning sets (Gilbert and Gofton 1982b) or bodygripping traps (Proulx et al. 1990; Proulx and Barrett 1991; Proulx et al. 1993). Most trappers rely on these two trap types, and numerous trapping manuals with precise description of capture methods are available (e.g., Faler 1988; Spencer 1990; Dobbins 1991; Bonecutter 1992).

RESEARCH NEEDS

The ecology of mink is poorly known in North America. Few studies have attempted to radio-track animals in the wild. Furthermore, there are few reliable methods for estimating density or even indexing abundance of mink. It is most probable that in certain areas such as the Northeast, great fluctuations in mink abundance occur but are undetected, except possibly as evidenced by fewer captures by fur trappers. Identifying such fluctuations would help managers understand whether mink populations are truly cyclic. In addition, the lethal and sublethal effects of environmental contaminants on individuals as well as their effects on populations (abundance, sex ratios, proportion of breeders, etc.) in the wild may help identify ecosystems at risk. It is most likely that as more areas increase in urbanization and pollution, what happens to mink may provide a good indication of what is happening to species at lower trophic levels. Monitoring of contaminant levels in mink tissue may continue to increase in importance as a bioindicator of environmental pollution in aquatic systems.

LITERATURE CITED

Adams, C. E. 1981. Observations on the induction of ovulation and expulsion of uterine eggs in the mink, *Mustela vison*. Journal of Reproduction and Fertility 63:241–48.

Akande, M. 1972. The food of feral mink (*Mustela vison*) in Scotland. Journal of Zoology (London) 167:475–79.

Arnold, T. W., and E. K. Fritzell. 1987a. Activity patterns, movements, and home ranges of prairie mink. Prairie Naturalist 19:25–32.

Arnold, T. W., and E. K. Fritzell. 1987b. Food habits of prairie mink during the waterfowl breeding season. Canadian Journal of Zoology 65:2322–24.

Arnold, T. W., and E. K. Fritzell. 1989. Spring and summer prey remains collected from male mink dens in southwestern Manitoba. Prairie Naturalist 21:189–92.

Arnold, T. W., and E. K. Fritzell. 1990. Habitat use by male mink in relation to wetland characteristics and avian prey abundances. Canadian Journal of Zoology 68:2205–8.

Askins, G. R., and J. A. Chapman. 1984. Age determination and morphological characteristics of wild mink from Maryland, USA. Zeitschrift für Säugetierkunde 49:182–89.

Aulerich, R. J., and D. R. Swindler. 1968. The dentition of the mink (*Mustela vison*). Journal of Mammalogy 49:488–94.

Aulerich, R. J., and R. K. Ringer. 1979. Toxic effects of dietary polybrominated biphenyls on mink. Archives of Environmental Contamination and Toxicology 8:487–98.

Aulerich, R. J., R. K. Ringer, and S. Iwamoto. 1974. Effects of dietary mercury on mink. Archives of Environmental Contamination and Toxicology 2:43–51.

Aulerich, R. J., S. J. Bursian, A. C. Napolitano, and T. Oleas. 2000. Feeding growing mink (*Mustela vison*) PCB Aroclor 1254 does not affect baculum (*os-penis*) development. Bulletin of Environmental Contamination and Toxicology 64:443–47.

Ben-David, M. 1997. Timing of reproduction in wild mink: The influence of spawning Pacific salmon. Canadian Journal of Zoology 75:376–82.

Ben-David, M., R. T. Bowyer, and J. B. Faro. 1996. Niche separation by mink and river otters: Coexistence in a marine environment. Oikos 75:41–48.

Ben-David, M., T. A. Hanley, D. R. Klein, and D. M. Schell. 1997. Seasonal changes in diets of coastal and riverine mink: The role of spawning Pacific salmon. Canadian Journal of Zoology 75:803–11.

Bevanger, K., and O. Albu. 1986. Decrease in a Norwegian feral mink *Mustela vison* population—A response to acid precipitation? Biological Conservation 38:75–78.

Bevanger, K., and G. Henriksen. 1995. The distributional history and present status of the American mink (*Mustela vison* Schreber, 1777) in Norway. Annales Zoologici Fennici 32:11–14.

Birks, J. D. S., and N. Dunstone. 1985. Sex-related differences in the diet of the mink *Mustela vison*. Holarctic Ecology 8:245–52.

Birks, J. D. S., and I. J. Linn. 1982. Studies of home range of the feral mink, *Mustela vison*. Symposium of the Zoological Society of London 49:231–57.

Birney, E. C., and E. D. Fleharty. 1966. Age and sex comparisons of wild mink. Transactions of the Kansas Academy of Science 69:139–45.

Birney, E. C., and E. D. Fleharty. 1968. Comparative success in the application of aging techniques to a population of winter-trapped mink. Southwestern Naturalist 13:275–82.

Bleavins, M. R., and R. J. Aulerich. 1981. Feed consumption and food passage time in mink and European ferrets. Laboratory Animal Science 31:268–69.

Blottner, S., H. Roelants, A. Wagener, and U. D. Wenzel. 1999. Testiculat mitosis, meiosis and apoptosis in mink (*Mustela vison*) during breeding and non-breeding seasons. Animal Reproduction Science 57:237–49.

Bonecutter, B. 1992. On target mink trapping. Beaver Pond Publishing, Greenville, PA.

Bonesi, L., N. Dunstone, and M. O'Connell. 2000. Winter selection of habitats within intertidal foraging areas by mink (*Mustela vison*). Journal of Zoology (London) 250:419–24.

Bostrom, R. E., R. J. Aulerich, R. K. Ringer, and P. J. Schaible. 1968. Seasonal changes in the testes and epididymides of the ranch mink (*Mustela vison*). Michigan Agricultural Experiment Station Quarterly Bulletin 50:538–58.

Brinck, C., R. Gerell, and G. Odham. 1978. Anal pouch secretion in mink *Mustela vison*. Oikos 30:68–75.

Brinck, C., S. Erlinge, and M. Sandell. 1983. Anal sac secretion in mustelids: A comparison. Journal of Chemical Ecology 9:727–45.

Brown, J. M., and R. C. Lasiewski. 1972. Metabolism of weasels: The cost of being long and thin. Ecology 53:939–43.

Bueno, F. 1994. Alimentacion del vison Americano (*Mustela vison* Schreber) en el Rio Voltoya (Avila, Cuenca del Duero). Doñana, Acta Vertebrata 21:5–13.

Bueno, F. 1996. Competition between American mink *Mustela vison* and otter *Lutra lutra* during winter. Acta Theriologica 41:149–54.

Bulmer, M. G. 1974. A statistical analysis of the 10-year cycle in Canada. Journal of Animal Ecology 43:701–18.

Burgess, S. A., and J. R. Bider. 1980. Effects of stream habitat improvements on invertebrates, trout populations, and mink activity. Journal of Wildlife Management 44:871–80.

Burness, G. P., and R. D. Morris. 1993. Direct and indirect consequences of mink presence in a common tern colony. Condor 95:708–11.

Burt, W. H. 1960. Bacula of North American mammals. Miscellaneous Publications, Museum of Zoology, University of Michigan 113:1–75.

Buskirk, S. W., and S. L. Lindstedt. 1989. Sex biases in trapped samples of Mustelidae. Journal of Mammalogy 70:88–97.

Chanin, P. 1981. The diet of the otter and its relationships with the feral mink in two areas of southwest England. Acta Theriologica 26:83–95.

Chanin, P. 1983. Observations on two populations of feral mink in Devon, United Kingdom. Mammalia 47:463–76.

Chanin, P., and I. Linn. 1980. The diet of the feral mink (*Mustela vison*) in southwest Britain. Journal of Zoology (London) 192:205–23.

Clausen, T. N., S. Wamberg, and O. Hansen. 1996. Incidence of nursing sickness and biochemical observations in lactating mink with and without dietary salt supplementation. Canadian Journal of Veterinary Research 60:271–76.

Cooper, J. J., and G. J. Mason. 2000. Increasing costs of access to resources cause re-scheduling of behaviour in American mink (*Mustela vison*): Implications for the assessment of behavioural priorities. Applied Animal Behaviour Science 66:135–51.

Cowan, I. McT., A. J. Wood, and W. D. Kitts. 1957. Feed requirements of deer, beaver, bear and mink for growth and maintenance. Transactions of the North American Wildlife Conference 22:179–88.

Craik, C. 1997. Long-term effects of North American mink *Mustela vison* on seabirds in western Scotland. Bird Study 44:303–9.

Cuthbert, J. H. 1979. Food studies of feral mink *Mustela vison* in Scotland. Fisheries Management 10:17–25.

Daciuk, J. 1978. Notas faunísticas y bioecológicas de Península Valdés y Patagonia: IV. Estado actual de las especies de mamíferos introducidos en la Subregión Araucana (Rep. Argentina) y grado de coacción ejercido en algunos ecosistemas surcordilleranos. Anales de Parques Nacionales 14:105–32.

Day, M. G., and I. Linn. 1972. Notes on the food of feral mink *Mustela vison* in England and Wales. Journal of Zoology (London) 167:463–73.

Deane, C. D., and F. O'Gorman. 1969. The spread of feral mink in Ireland. Irish Nature Journal 16:198–202.

Dobbins, C. L. 1991. Mink trapping techniques. Commercial Printing, New Castle, PA.

Duby, R. T., and H. F. Travis. 1972. Photoperiodic control of fur growth and reproduction in the mink (*Mustela vison*). Journal of Experimental Zoology 182:217–26.

Dunstone, N. 1978. Fishing strategy of the mink (*Mustela vison*); time budgeting of hunting effort? Behaviour 67:157–77.

Dunstone, N. 1979. Swimming and diving behavior of the mink. Carnivore 2:56–61.

Dunstone, N. 1983. Underwater hunting behaviour of the mink (*Mustela vison* Schreber): An analysis of constraints on foraging. Acta Zoologica Fennica 174:201–3.

Dunstone, N. 1993. The mink. Poyser Natural History, London.

Dunstone, N., and J. D. S. Birks. 1983. Activity budget and habitat usage by coastal-living mink (*Mustela vison* Schreber). Acta Zoologica Fennica 174:189–91.

Dunstone, N., and J. D. S. Birks. 1985. The comparative ecology of coastal, riverine, and lacustrine mink *Mustela vison* in Britain. Zeitschrift für Angewandte Zoologie 72:52–70.

Dunstone, N., and J. D. S. Birks. 1987. The feeding ecology of mink (*Mustela vison*) in a coastal habitat. Journal of Zoology (London) 212:69–83.

Dunstone, N., and A. Clements. 1979. The threshold for high-speed directional movement detection in the mink, *Mustela vison* Schreber. Animal Behaviour 27:613–20.

Dunstone, N., and R. J. O'Connor. 1979a. Optimal foraging in an amphibious mammal. I. The aqualung effect. Animal Behaviour 27:1182–94.

Dunstone, N., and R. J. O'Connor. 1979b. Optimal foraging in an amphibious mammal. II. A study using principal component analysis. Animal Behaviour 27:1195–1201.

Eagle, T. C., J. Choromanski-Norris, and V. B. Kuechle. 1984. Implanting radio transmitters in mink and Franklin's ground squirrels. Wildlife Society Bulletin 12:180–84.

Eagle, T. C., and J. S. Whitman. 1987. Mink. Pages 615–24 *in* M. Novak, J. A. Baker, M. E. Obbard, and B. Malloch, eds. Wild furbearer management and conservation in North America. Ontario Ministry of Natural Resources, Toronto.

Eberhardt, L. E., and A. B. Sargeant. 1977. Mink predation on prairie marshes during the waterfowl breeding season. Pages 37–43 *in* R. L. Phillips and C. Jonkel, eds. The 1975 predator symposium. Montana Forest Conservation Experiment Station, University of Montana Press, Missoula.

Elder, W. H. 1951. The baculum as an age criterion in mink. Journal of Mammalogy 32:43–50.

Enders, R. K. 1952. Reproduction in the mink (*Mustela vison*). Proceedings of the American Philosophical Society 96:691–755.

Erb, J., N. C. Stenseth, and M. S. Boyce. 2000. Geographic variation in population cycles of Canadian muskrats. Canadian Journal of Zoology 78:1009–16.

Erlinge, S. 1969. Food habits of the otter *Lutra lutra* L. and the mink *Mustela vison* Schreber in a trout water in southern Sweden. Oikos 20:1–7.

Erlinge, S. 1972. Interspecific relations between otter *Lutra lutra* and mink *Mustela vison* in Sweden. Oikos 23:327–35.

Errington, P. L. 1936. Sex ratio and size variation in South Dakota mink. Journal of Mammalogy 17:287.

Errington, P. L. 1954. The special responsiveness of minks to epizootics in muskrat populations. Ecological Monographs 24:377–93.

Fairley, J. S. 1980. Observations on a collection of feral Irish mink (*Mustela vison* Schreber). Proceedings of the Royal Irish Academy 80B:81–92.

Faler, R. E. Jr. 1988. The mink trapper's guide. Beaver Pond Publishing, Greenville, PA.

Ferguson, S. H., J. A. Virgl, and S. Larivière. 1996. Evolution of delayed implantation and associated grade shifts in life history traits of North American carnivores. Ecoscience 3:7–17.

Ferreras, P., and D. W. Macdonald. 1999. The impact of American mink *Mustela vison* on water birds in the upper Thames. Journal of Applied Ecology 36:701–8.

Ferron, J. 1973. Morphologie comparée de l'organe de l'odorat chez quelques mammifères carnivores. Naturaliste Canadien 100:525–41.

Fleming, M. A., E. A. Ostrander, and J. A. Cook. 1999. Microsatellite markers for American mink (*Mustela vison*) and ermine (*Mustela erminea*). Molecular Ecology 8:1351–62.

Franson, J. C., P. A. Dahm, and L. D. Wing. 1975. A method for preparing and sectioning mink (*Mustela vison*) mandibles for age determination. American Midland Naturalist 93:507–8.

Fredga, K. 1961. The chromosomes of the mink. Journal of Heredity 52:91–94.

Gerell, R. 1967a. Dispersal and acclimatization of the mink, *Mustela vison* Schreb. in Sweden. Viltrevy 4:1–38.

Gerell, R. 1967b. Food selection in relation to habitat in mink (*Mustela vison* Schreber) in Sweden. Oikos 18:233–46.

Gerell, R. 1968. Food habits of the mink, *Mustela vison* Schreb. in Sweden. Viltrevy 5:119–94.

Gerell, R. 1969. Activity patterns of the mink *Mustela vison* Schreber in southern Sweden. Oikos 20:451–60.

Gerell, R. 1970. Home ranges and movements of the mink *Mustela vison* Schreber in southern Sweden. Oikos 21:160–73.

Gerell, R. 1971. Population studies on mink, *Mustela vison* Schreber, in southern Sweden. Viltrevy 8:83–114.

Gilbert, F. F. 1969. Analysis of basic vocalizations of the ranch mink. Journal of Mammalogy 50:625–27.

Gilbert, F. F., and E. D. Bailey. 1967. The effect of visual isolation on reproduction in the female ranch mink. Journal of Mammalogy 48:113–18.

Gilbert, F. F., and E. D. Bailey. 1969. Visual isolation and stress in female ranch mink particularly during the reproductive season. Canadian Journal of Zoology 47:209–12.

Gilbert, F. F., and N. Gofton. 1982a. Heart rate values for beaver, mink and muskrat. Comparative Biochemistry and Physiology 73A:249–51.

Gilbert, F. F., and N. Gofton. 1982b. Terminal dives in mink, muskrat and beaver. Physiology and Behavior 28:835–40.

Gilbert, F. F., and E. G. Nancekivell. 1982. Food habits of mink (*Mustela vison*) and otter (*Lutra canadensis*) in northeastern Alberta. Canadian Journal of Zoology 60:1282–88.

Goodpaster, W., and D. F. Hoffmeister. 1950. Bats as prey for mink in Kentucky cave. Journal of Mammalogy 31:457.

Greer, K. R. 1957. Some osteological characters of known-age ranch minks. Journal of Mammalogy 38:319–30.

Halbrook, R. S., A. Woolf, G. F. Hubert, Jr., S. Ross, and W. E. Braselton. 1996. Contaminant concentrations in Illinois mink and otter. Ecotoxicology 5:103–14.

Hall, E. R. 1981. The mammals of North America, 2nd ed. John Wiley, New York.

Halliwell, E. C., and D. W. Macdonald. 1996. American mink *Mustela vison* in the upper Thames catchment: Relationship with selected prey species and den availability. Biological Conservation 76:51–56.

Hamilton, W. J., Jr. 1959. Foods of mink in New York. New York Fish and Game 6:77–85.

Hammond, J., Jr. 1951. Control by light of reproduction in ferrets and mink. Nature 167:150–51.

Hansen, O., S. Wamberg, and T. N. Clausen. 1996. Failure of loop diuretics to induce nursing sickness in mink at weaning. Canadian Journal of Veterinary Research 60:277–80.

Hansson, A. 1947. The physiology of reproduction in mink (*Mustela vison,* Schreb.) with special reference to delayed implantation. Acta Zoologica 28:1–136.

Hansson, I. 1967. Transmission of the parasitic nematode *Skrjabingylus nasicola* (Leuckart 1842) to species of *Mustela* (Mammalia). Oikos 18:247–52.

Harding, L. E., M. L. Harris, C. R. Stephen, and J. E. Elliot. 1999. Reproductive and morphological condition of wild mink (*Mustela vison*) and river otter (*Lutra canadensis*) in relation to chlorinated hydrocarbon contamination. Environmental Health Perspectives 107:141–47.

Hewson, R. 1971. Some aspects of the biology of feral mink *Mustela vison* Schreber in Banffshire. Glascow Naturalist 18:539–46.

Hsu, T. C., and K. Benirschke. 1968. An atlas of mammalian chromosomes, Vol. 2, Folio 81. Springer-Verlag, New York.

Humphrey, S. R., and T. L. Zinn. 1982. Seasonal habitat use by river otters and Everglades mink in Florida. Journal of Wildlife Management 46:375–81.

Iversen, J. A. 1972. Basal energy metabolism of Mustelids. Journal of Comparative Physiology 81:341–44.

Jackson, H. H. T. 1961. Mammals of Wisconsin. University of Wisconsin Press, Madison.

Jenkins, D., and R. J. Harper. 1980. Ecology of otters in northern Scotland II. Analyses of otter (*Lutra lutra*) and mink (*Mustela vison*) faeces from Deeside, N. E. Scotland in 1977–78. Journal of Animal Ecology 49:737–54.

Kauhala, K. 1996. Introduced carnivores in Europe with special reference to central and northern Europe. Wildlife Biology 2:197–204.

Keith, L. B., and J. R. Cary. 1991. Mustelid, squirrel, and porcupine population trends during a snowshoe hare cycle. Journal of Mammalogy 72:373–78.

Kruska, D., and A. Schreiber. 1999. Comparative morphometrical and biochemical–genetic investigations in wild and ranch mink (*Mustela vison*: Carnivora: Mammalia). Acta Theriologica 44:377–92.

Kubin, R., and M. Mason. 1948. Normal blood and urine values for mink. Cornell Veterinarian 38:79–85.

Lagerkvist, G. 1997. Economic profit from increased litter size, body weight and pelt quality in mink (*Mustela vison*). Acta Agriculturae Scandinavica, Section A, Animal Sciences 47:57–63.

Lagerkvist, G., K. Johansson, and N. Lundeheim. 1994. Selection for litter size, body weight, and pelt quality in mink (*Mustela vison*): Correlated responses. Journal of Animal Science 72:1126–37.

Lande, O. 1957. The chromosomes of the mink. Hereditas 43:578–82.

Larivière, S. 1996. The American mink, *Mustela vison* (Carnivora, Mustelidae) can climb trees. Mammalia 60:485–86.

Larivière, S. 1999. *Mustela vison*. Mammalian Species 608:1–9.

Larivière, S., L. R. Walton, and J. A. Virgl. 2000. Field anesthesia of American mink, *Mustela vison,* using halothane. Canadian Field-Naturalist 114:142–44.

Lechleitner, R. R. 1954. Age criteria in mink, *Mustela vison*. Journal of Mammalogy 35:496–503.

Linscombe, G. 1994. U.S. fur harvest (1970–1992) and fur value (1974–1992) statistics by state and region. Louisiana Department of Wildlife and Fisheries, Baton Rouge.

Linscombe, G., N. Kinler, and R. J. Aulerich. 1982. Mink. Pages 629–43 *in* J. A. Chapman and G. A. Feldhamer, eds. Wild mammals of North America: Biology, management, and economics. Johns Hopkins University Press, Baltimore.

Lodé, T. 1993. Diet composition and habitat use of sympatric polecat and American mink in western France. Acta Theriologica 38:161–66.

Lowery, G. H., Jr. 1974. The mammals of Louisiana and its adjacent waters. Louisiana State University Press, Baton Rouge.

Lynch, J. M., and T. J. Hayden. 1995. Genetic influences on cranial form: Variation among ranch and feral American mink *Mustela vison* (Mammalia: Mustelidae). Biological Journal of the Linnean Society 55:293–307.

MacLennan, R. R., and E. D. Bailey. 1969. Seasonal changes in aggression, hunger, and curiosity in ranch mink. Canadian Journal of Zoology 47:1395–1404.

Maran, T., and H. Henttonen. 1995. Why is the European mink (*Mustela lutreola*) disappearing?—A review of the process and hypotheses. Annales Zoologici Fennici 32:47–54.

Maran, T., D. W. Macdonald, H. Kruuk, V. Sidorovich, and V. V. Rozhnov. 1998. The continuing decline of the European mink *Mustela lutreola*: Evidence for the intraguild aggression hypothesis. Symposia of the Zoological Society of London 71:297–323.

Marshall, W. H. 1935. A study of the winter activities of the mink. Journal of Mammalogy 17:382–92.

McCabe, R. A. 1949. Notes on live-trapping mink. Journal of Mammalogy 30:413–23.

Medina, G. 1997. A comparison of the diet and distribution of southern river otter (*Lutra provocax*) and mink (*Mustela vison*) in southern Chile. Journal of Zoology (London) 242:291–97.

Melquist, W. E., J. S. Whitman, and M. G. Hornocker. 1981. Resource partitioning and coexistence of sympatric mink and river otter populations. Pages 187–220 *in* J. A. Chapman and D. Pursley, eds. Proceedings of the worldwide furbearer conference. Frostburg, MD.

Messier, F., and J. A. Virgl. 1992. Differential use of bank burrows and lodges by muskrats, *Ondatra zibethicus,* in a northern marsh environment. Canadian Journal of Zoology 70:1180–84.

Mitchell, J. L. 1961. Mink movements and populations on a Montana river. Journal of Wildlife Management 25:48–54.

Moore, D. R. J., R. L. Breton, and K. Lloyd. 1997. The effects of hexachlorobenzene on mink in the Canadian environment: An ecological risk assessment. Environmental Toxicology and Chemistry 16:1042–50.

Nes, N. 1966. Diploid–triploid chimerism in a true hermaphrodite mink (*Mustela vison*). Hereditas 56:159–70.

Niemimaa, J. 1995. Activity patterns and home ranges of the American mink *Mustela vison* in the Finnish outer archipelago. Annales Zoologici Fennici 32:117–21.

Nieto, J. M., S. Vázquez, M. I. Quiroga, M. López-Peña, F. Guerrero, and E. Gruys. 1995. Spontaneous AA-amyloidosis in mink (*Mustela vison*). Description of eight cases, one of which exhibited intracellular amyloid deposits in lymph node macrophages. European Journal of Veterinary Pathology 1:99–103.

Northcott, T. H., N. F. Payne, and E. Mercer. 1974. Dispersal of mink in insular Newfoundland. Journal of Mammalogy 55:243–49.

Nowak, R. M. 1999. Walker's mammals of the world, 6th ed. Johns Hopkins University Press, Baltimore.

Onstad, O. 1967. Studies on postnatal changes, semen quality, and anomalies of reproductive organs in the mink. Acta Endocrinologia (Supplement) 117:1–117.

Page, R. J. C., and S. D. Langton. 1996. The occurrence of ixodid ticks on wild mink *Mustela vison* in England and Wales. Medical and Veterinary Entomology 10:359–64.

Pascal, M., and P. Delattre. 1981. Comparaison de différentes méthodes de détermination de l'âge individuel chez le vison (*Mustela vison* Schreiber). Canadian Journal of Zoology 59:202–11.

Peterson, R. L. 1966. The mammals of eastern Canada. Oxford University Press, Toronto.

Platonow, N. S., and L. H. Karstad. 1973. Dietary effects of polychlorinated biphenyls on mink. Canadian Journal of Comparative Medicine 37:391–400.

Poole, T. B., and N. Dunstone. 1976. Underwater predatory behaviour of the American mink (*Mustela vison*). Journal of Zoology (London) 178:395–412.

Powell, R. A., and W. J. Zielinski. 1989. Mink response to ultrasound in the range emitted by prey. Journal of Mammalogy 70:637–38.

Proulx, G., and M. W. Barrett. 1991. Evaluation of the Bionic trap to quickly kill mink (*Mustela vison*) in simulated natural environments. Journal of Wildlife Diseases 27:276–80.

Proulx, G., J. A. McDonnell, and F. F. Gilbert. 1987. The effect of water level fluctuations on muskrat, *Ondatra zibethicus,* predation by mink, *Mustela vison*. Canadian Field-Naturalist 101:89–92.

Proulx, G., M. W. Barrett, and S. R. Cook. 1990. The C120 magnum with pan trigger: A humane trap for mink (*Mustela vison*). Journal of Wildlife Diseases 26:511–17.

Proulx, G., I. M. Pawlina, and R. K. Wong. 1993. Re-evaluation of the C120 magnum and Bionic traps to humanely kill mink. Journal of Wildlife Diseases 29:184.

Racey, G. D., and D. L. Euler. 1983. Changes in mink habitat and food selection as influenced by cottage development in central Ontario. Journal of Applied Ecology 20:387–402.

Ramos-Vara, J. A., J. P. Dubey, G. L. Watson, M. Winn-Elliot, J. S. Patterson, and B. Yamini. 1997. Sarcocystosis in mink (*Mustela vison*). Journal of Parasitology 83:1198–1201.

Ringer, R. K., R. J. Aulerich, R. Pittman, and E. A. Cogger. 1974. Cardiac output, blood pressure, blood volume and other cardiovascular parameters in mink. Journal of Animal Science 38:121–23.

Ruiz-Olmo, J., S. Palazon, F. Bueno, C. Bravo, I. Munilla, and R. Romero. 1997. Distribution, status and colonization of the American mink *Mustela vison* in Spain. Journal of Wildlife Research 2:30–36.

Ruprecht, A. L., T. Buchalczyk, and J. M. Wójcik. 1983. The occurence of minks (Mammalia: Mustelidae) in Poland. Przeglad Zoologiczny 27:87–99.

Rust, C. C., R. M. Shackelford, and R. K. Meyer. 1965. Hormonal control of pelage cycles in the mink. Journal of Mammalogy 46:549–65.

Sargeant, A. B., G. A. Swanson, and H. A. Doty. 1973. Selective predation by mink, *Mustela vison,* on waterfowl. American Midland Naturalist 89:208–14.

Schladweiler, J. L., and G. L. Storm. 1969. Den-use by mink. Journal of Wildlife Management 33:1025–26.

Sealander, J. A. 1943. Winter food habits of mink in southern Michigan. Journal of Wildlife Management 7:411–17.

Segal, A. N. 1972. Ecological thermoregulation in the American mink. Soviet Journal of Ecology 3:453–56.

Sidorovich, V. E. 1993. Reproductive plasticity of the American mink *Mustela vison* in Belarus. Acta Theriologica 38:175–83.

Sidorovich, V. E., and V. V. Savcenko. 1992. The effect of pollution on the population of the American mink (*Mustela vison*). Semiaquatische Säugetiere 1992:305–15.

Sidorovich, V. E., H. Kruuk, D. W. Macdonald, and T. Maran. 1998. Diets of semi-aquatic carnivores in northern Belarus, with implications for population changes. Symposia of the Zoological Society of London 71:177–89.

Sidorovich, V. E., H. Kruuk, and D. W. Macdonald. 1999. Body size, and interactions between European and American mink (*Mustela lutreola* and *M. vison*) in eastern Europe. Journal of Zoology (London) 248:521–27.

Sinclair, W., N. Dunstone, and T. B. Poole. 1974. Aerial and underwater visual acuity in the mink *Mustela vison* Schreber. Animal Behaviour 22:965–74.

Smits, J. E. G., B. R. Blakley, and G. A. Wobeser. 1996a. Immunotoxicity studies in mink (*Mustela vison*) chronically exposed to dietary bleached kraft pulp mill effluent. Journal of Wildlife Diseases 32:199–208.

Smits, J. E. G., D. M. Haines, B. R. Blakley, and G. A. Wobeser. 1996b. Enhanced antibody responses in mink (*Mustela vison*) exposed to dietary bleached-kraft pulp mill effluent. Environmental Toxicology and Chemistry 15:1166–70.

Spencer, J. 1990. The mink manual: A common sense approach to mink trapping. Outdoor World Press, Humansville, MO.

Stephenson, R., P. J. Butler, N. Dunstone, and A. J. Woakes. 1988. Heart rate and gas exchange in freely diving American mink (*Mustela vison*). Journal of Experimental Biology 134:435–42.

Stevens, R. T., T. L. Ashwood, and J. M. Sleeman. 1997a. Mercury in hair of muskrats (*Ondatra zibethicus*) and mink (*Mustela vison*) from the U.S. Department of Energy Oak Ridge Reservation. Bulletin of Environmental Contaminants and Toxicology 58:720–25.

Stevens, R. T., T. L. Ashwood, and J. M. Sleeman. 1997b. Fall–early winter home ranges, movements, and den use of male mink, *Mustela vison* in eastern Tennessee. Canadian Field-Naturalist 111:312–14.

Strachan, C., D. J. Jefferies, G. R. Barreto, D. W. Macdonald, and R. Strachan. 1998. The rapid impact of resident American mink on water voles: Case studies in lowland England. Symposia of the Zoological Society of London 71:339–57.

Sundqvist, C., and M. Gustafsson. 1983. Sperm test: A useful tool in breeding work of mink. Journal of the Scientific Agricultural Society of Finland 55:119–31.

Sundqvist, C., L. C. Ellis, and A. Bartke. 1988. Reproductive endocrinology of the mink (*Mustela vison*). Endocrine Reviews 9:247–66.

Svihla, A. 1931. Habits of the Louisiana mink (*Mustela vison vulgivagus*). Journal of Mammalogy 12:366–68.

Ternovskii, D. V. 1977. The biology of the Mustelidae. Akademii Nauk, Novosibirsk, USSR.

Thompson, H. 1968. British wild mink. Annales of Applied Biology 61:345–49.

Tomson, F. N. 1987. Mink. Veterinary Clinics of North America: Small Animal Practice 17:1145–53.

Venge, O. 1959. Reproduction in the fox and mink. Animal Breeding Abstracts 27:129–45.

Virgl, J. A., and F. Messier. 1996. Population structure, distribution, and demography of muskrats during the ice-free period under contrasting water fluctuations. Écoscience 3:54–62.

Waller, D. W. 1962. Feeding behavior of minks at some Iowa marshes. M.S. thesis. Iowa State University, Ames.

Wamberg, S. 1994. Rates of heat and water loss in female mink (*Mustela vison*) measured by direct calorimetry. Comparative Biochemistry and Physiology 107A:451–58.

Ward, D. P., C. M. Smal, and J. S. Fairley. 1986. The food of mink *Mustela vison* in the Irish midlands. Proceedings of the Royal Irish Academy 86B:169–82.

West, N. H., and B. N. Van Vliet. 1986. Factors influencing the onset and maintenance of bradycardia in mink. Physiological Zoology 59:451–63.

Whitman, J. S. 1981. Ecology of the mink (*Mustela vison*) in west-central Idaho. M.S. Thesis, University of Idaho, Moscow.

Wiig, O. 1986. Sexual dimorphism in the skull of minks *Mustela vison,* badgers *Meles meles* and otters *Lutra lutra*. Zoological Journal of the Linnean Society 87:163–79.

Williams, T. M. 1983. Locomotion in the North American mink, a semi-aquatic mammal: I. Swimming energetics and body drag. Journal of Experimental Biology 103:155–68.

Wise, M. H., I. J. Linn, and C. R. Kennedy. 1981. A comparison of the feeding biology of mink *Mustela vison* and otter *Lutra lutra*. Journal of Zoology (London) 195:181–213.

Wobeser, G., and M. Swift. 1976. Mercury poisoning in a wild mink. Journal of Wildlife Diseases 12:335–40.

Wobeser, G., N. O. Nielsen, and B. Schiefer. 1976. Mercury and mink. I. The use of mercury contaminated fish as a food for ranch mink. Canadian Journal of Comparative Medicine 40:30–33.

Wren, C. D., P. M. Stokes, and K. L. Fischer. 1986. Mercury levels in Ontario mink and otter relative to food levels and environmental acidification. Canadian Journal of Zoology 64:2854–59.

Yeager, L. E. 1943. Storing of muskrats and other foods by minks. Journal of Mammalogy 24:100–101.

Youngman, P. M. 1982. Distribution and systematics of the European mink *Mustela lutreola* Linnaeus 1761. Acta Zoologica Fennica 166:1–48.

Zielinski, W. J. 1986. Circadian rhythms of small carnivores and the effect of restricted feeding on daily activity. Physiology and Behavior 38:613–20.

SERGE LARIVIÈRE, Delta Waterfowl Foundation, R.R. #1, Box 1, Site 1, Portage La Prairie, Manitoba, Canada R1N 3A1. Email: slariviere@deltawaterfowl.org.

32

Wolverine

Gulo gulo

Jeffrey P. Copeland
Jackson S. Whitman

NOMENCLATURE

COMMON NAMES. Wolverine, glutton, carcajou, devil bear, skunk bear
SCIENTIFIC NAME. *Gulo gulo*
SUBSPECIES. *G. g. gulo* (Old World); *G. g. luscus* (New World)
Currently, naturalists have widely accepted the taxonomic arrangement of Old World and New World wolverines as being conspecific (Degerbol 1935; Kurten and Rausch 1959; Nowak 1991; Wilson and Reeder 1993). Hall and Kelson (1959) and Hall (1981) considered the New World wolverine a distinct species (*Gulo luscus*) with four recognized subspecies (*katschemakensis, luscus, luteus,* and *vancouverensis*). The name *Gulo* is derived from the Latin *gula* (throat) and *gulosus* (gluttonous).

DISTRIBUTION

Wolverine are a circumboreal species, occurring from Scandinavia eastward across the taiga and forest–tundra zones of Eurasia north of 48° N latitude. In North America, current distribution is limited to the mountainous regions of the western United States, north of 37° N latitude, extending north along the Rocky Mountain corridor into and across the boreal–tundra regions of Canada and Alaska (Fig. 32.1).

In North America, historical records indicate the presence of wolverine broadly across Canada with distribution extending to the northernmost tier of the United States from Maine to Washington State. Southern range extensions were probably limited to montane boreal regions, as distribution gaps are conspicuous across the Great Basin and plains of the midwestern United States and north into the prairie regions of Alberta and Saskatchewan (Wilson 1982). The species was reported south along the Sierra–Cascade axis through Oregon into the southern Sierra Nevada in California, and along the Rocky Mountains into Arizona and New Mexico (Grinnell et al. 1937; Hall 1981; Hash 1987). Hall (1981) referenced historical sightings and collections in California, Colorado, Idaho, Indiana, Maine, Michigan, Minnesota, Nebraska, New Hampshire, New Mexico, New York, North Dakota, Oregon, Utah, Washington, and Wisconsin. Wilson (1982) provided an exhaustive annotated listing of historical records in the contiguous United States and, in addition to the above-mentioned states, added Iowa, Montana, Nevada, Ohio, Pennsylvania, South Dakota, and Vermont. Historical presence of the wolverine in most midwestern, eastern, and southwestern U.S. states is sparse, and probably represents sporadic occurrence during favorable periods of population growth or pioneering movements of dispersing individuals.

Current distribution of the wolverine in the United States appears to constitute several peninsular extensions of Canadian populations (Hash 1987; Banci 1994). Although reports of wolverine sightings persist throughout the Rocky Mountain states (Banci 1994), only Idaho and Montana report populations of known extent. The wolverine is confirmed present in northwestern Wyoming. Its presence in the Wallowa range of eastern Oregon appears likely based on recent sightings (Edelmann and Copeland 1999), although continuing efforts to confirm its presence in the Cascade Range of Oregon have proved unsuccessful. McKay (1991) concluded there was no solid evidence of wolverine presence in Utah, although a Colorado report (Seidel et al. 1998) documented a wolverine shot near Dinosaur, Utah, in 1979. Colorado lists the wolverine's status as "undetermined" with no confirmed presence since the early 1900s. California provides the best evidence for defining the southern extent of the wolverine's historical distribution, with relatively abundant records throughout the southern Sierra Nevada Range (Nowak 1973; Schempf and White 1977). Nowak (1973) believed the wolverine became isolated to the central and southern Sierra Nevada Range by 1933. Sightings and reports of presence have been sporadic and unconfirmed since then. If a significant range decline has indeed occurred, it appears to have done so most notably in California and Colorado. In the northern Rocky Mountain and Cascade ranges, wolverine were believed to be largely extirpated or severely reduced in number prior to the mid-20th century. Reported numbers in Washington (Johnson 1977), IDAHO (Groves 1987), and Montana (Newby and Wright 1955, Newby and McDougal 1964) began to increase

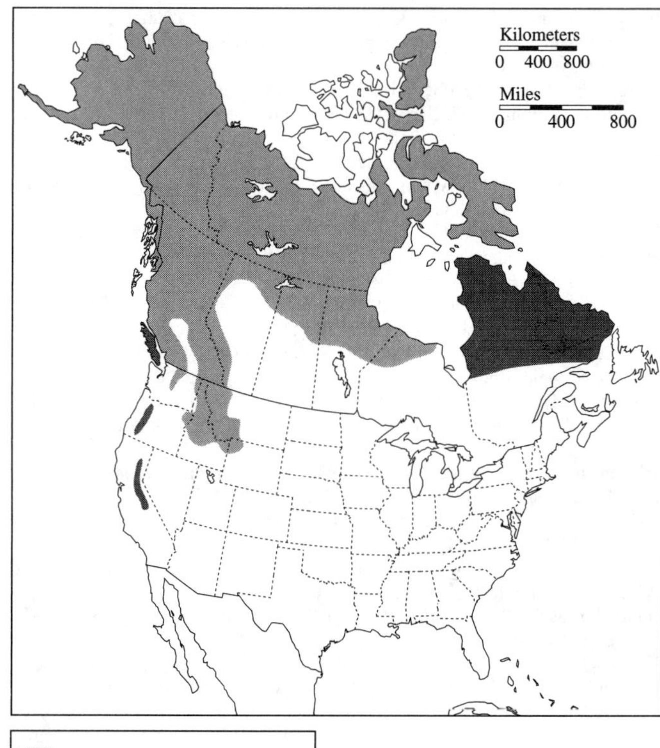

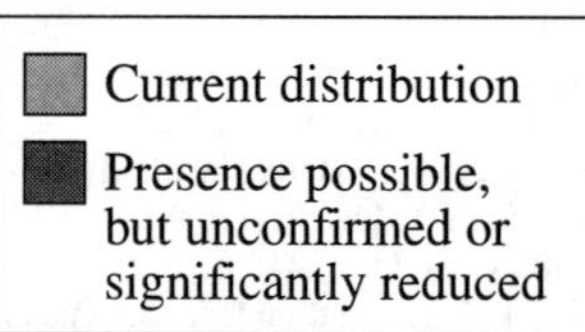

FIGURE 32.1. Distribution of the wolverine (*Gulo gulo*).

post-1950s and appear to be stable or increasing today (Wilson 1982; Hash 1987; Copeland 1996; Edelmann and Copeland 1999). Alaska has viable wolverine populations throughout the state, with the exception of islands in the Bering Sea, the Aleutian chain, Kodiak, Prince William Sound, and outer islands in the Alexander Archipelago. Range reductions in Alaska have not been documented as they have been elsewhere.

In Canada, the wolverine is most notably present in the Yukon, Northwest Territories, and British Columbia (van Zyll de Jong 1975; Kelsall 1981). Wolverine distribution and abundance in these provinces reportedly are relatively unchanged from historical levels (Pasitschniak-Arts and Larivière 1995). Boreal and taiga–tundra communities diminish from west to east along with apparent reduction of the wolverine in eastern Canada (Novak 1975; Prescott 1983). Until about 1900, wolverine were believed to range throughout eastern Canada with the exception of Gaspé Peninsula, Anticosti Island, eastern New Brunswick, Prince Edward Island, and Nova Scotia (Peterson 1966). According to Manning (1943), wolverine were not present on Baffin Island before 1900. Recent records are available only from the Kenora District in Ontario (Peterson 1966), with wolverine status in the province believed to be unchanged since the early 1900s (Novak 1975). In Quebec, the historical range of the wolverine is considered to be north of the 52nd parallel (Moisan 1996). Wolverine are considered very rare in Quebec and the species' status there is uncertain (van Zyll de Jong 1975). Wolverine presence was believed to extend into the aspen parklands of northern Manitoba (van Zyll de Jong 1975), but as with Saskatchewan, populations appear to have receded to the northern portions of the province (Pasitschniak-Arts and Larivière 1995). In Alberta, wolverine distribution was considered very sparse in the east-central portion of the province, gradually increasing toward the northwest and southwest corners (Skinner and Todd 1989).

It is apparent from a variety of sources (Hatler 1989; Banci 1994; Copeland 1996) that wolverine populations are most viable where human activities have done little to alter the landscape. Whether human presence has forced wolverine into remote regions is not well understood. It is likely, however, that large tracts of pristine habitat may be the only assurance of their continued existence.

DESCRIPTION

Size and Weight. The wolverine is the largest terrestrial member of the family Mustelidae. Males are typically 40–60% larger than females; males generally weigh 11 to 18 kg and females 6 to 12 kg (Banci 1994). Based on weights of wolverine from various locations in North America, there possibly is a clinal diminution in wolverine size latitudinally from north to south. From a sample of 8 adult female and 11 adult male wolverine in Idaho, Copeland (1996) found average live weights of 8.2 and 13.1 kg, respectively.

Some sexual dimorphism is apparent, but not to the extent of smaller mustelids. Hall (1981) reported that females average 30% less in weight than males, but it is not clear where his sample was from, nor if animals were from comparable age classes. Copeland (1996) found adult males were almost 60% heavier than adult females in Idaho. Other studies have reported similar results. Hornocker and Hash (1981) reported males in their Montana study area were 53% heavier than females, with mean weights of 12.7 and 8.3 kg, respectively. In northwest Alaska, Magoun (1985) listed mean male and female weights as 14.1 and 9.9 kg, respectively, males being 42% heavier than females. Whitman and Ballard (1983) reported mean weight of seven adult males was 16.1 kg and of five adult females was 11.2 kg (males 44% heavier than females) in south-central Alaska, whereas Golden (1996) reported mean weight of eight adult males was 15.3 kg and of two adult females was 10.5 kg from the same location (males 46% heavier than females).

Standard body measurements, similar to body weight, show considerable variation. Body lengths (not including tail) of wolverine in Siberia range from 650 to 1050 mm (Stroganov 1969). Novikov (1956/1962) reported body lengths from 760 to 860 mm. In the Russian Far East, six males and five females averaged 904 and 760 mm, respectively (Krevoshiev 1984). Hall (1981) reported body lengths of 710–865 mm for males in North America. Copeland (1996) found 10 adult males in Idaho ranged from 780 to 830 mm (mean = 797 mm), whereas six adult females ranged from 640 to 720 mm (mean = 698 mm). The length of the tail in wolverine ranges from 170 to 260 mm (Stroganov 1969; Hall 1981; Copeland 1996). Hind foot measurements of adults range from 145 to 192 mm (Hall 1981; Copeland 1996).

Form and Features. Wolverine appear badger-like rather than long and slender as with the river otter (*Lontra canadensis*), marten (*Martes americana*), or mink and weasels (*Mustela* spp.). They usually present a humped appearance, with head and rump held lower than the arch of the back, perhaps leading to the vernacular name "skunk bear." They generally are active, usually running in a bounding gait rather than walking. The bushy tail is usually held low, except during urination or anal scent marking. They are compact and heavily muscled, with a broad head, short neck, and relatively short legs. The feet are large and well adapted for travel over snow, for digging, and for climbing. Formozov (1946) described the wolverine as a "chioneuphore" or a species that can withstand considerable snow. He felt that the wolverine is not particularly well adapted to deep, soft snow, but is best suited when the snow develops a crust, whether created from wind, freeze–thaw cycles, or the track of another animal. Both forefeet and hind feet have five ivory-colored claws. Stiff hairs occur between the digits and around the foot pads, probably providing additional surface area for travel across snow. Wolverine have relatively small eyes, and according to Jackson (1961), their sight is relatively poor, although their visual capabilities have not been carefully studied. Hearing is probably moderately developed, as suggested by the size of external pinnae. The wolverine is believed to possess an acute sense of smell, which is probably the primary sense used in locating food. Considerable anecdotal accounts describe the wolverine's apparent ability to sense food over long distances through olfaction (Hornocker and Hash 1981). Wolverine also cache food (Magoun 1985), however, which suggests that in some cases, individuals locate food caches through memory rather than olfaction.

Pelage. Wolverine are generally dark. The background color varies from brown to almost black, with a lighter brown or reddish stripe from behind the shoulder, coalescing at the top of the rump, and sometimes extending partially out the dorsal surface of the tail. Light markings on the face are variable, but usually start behind the eyes, fading at or near the ears. Irregular throat and chest patches are common and variable in size and color, from white to pale orange. Generally, smaller irregular light patches occur anterior to the vent. The legs, feet, and most of the tail are usually very dark, although occasionally wolverine are encountered with irregular white markings on front toes or feet.

The 2- to 3-cm long underfur is woolly and coarse. Guard hairs on the wolverine's back and sides are about 10 cm long, except on the top of the rump and bushy tail, where they often exceed 20 cm. Hair on the head is dense and short. A single annual molt occurs in the spring.

Glands. As with other mustelids, wolverine have anal musk glands capable of emitting a yellowish or brownish, odiferous discharge (Pocock 1921). These glands are about the size of a walnut (Coues 1877), and are situated immediately lateral to the anus. A number of authors have described anal musking as a scent-marking method in the wolverine (Haglund 1966; Pulliainen and Ovaskainen 1975; Koehler et al. 1980; Magoun 1985), whereas others have suggested anal glands function primarily as a fear–defense mechanism (Krott 1959; Ewer 1973; Long 1987). In Montana, Koehler et al. (1980) described all scent marks associated with wolverine as "musk," making no mention of urine marks. However, urine was the most common method of scent marking in free-ranging wolverine in Idaho (Copeland 1996) and Alaska (Magoun 1985) and in captive wolverine (Long 1987). Magoun (1985) also suggested that the ventral, or abdominal, gland is probably also used for scent marking in addition to the anal sacs.

The abdominal gland in the wolverine is a general reference to an oval patch of short, dense, yellowish-orange hair, generally 8–10 cm long, on the abdomen immediately prepusal in the male and anterior to the vent in females. Long (1987) described abdominal rubbing as the second most common marking method, following urination, in captive wolverine. Although fisher (*Martes pennanti*) and marten are commonly noted for abdominal rubbing (Hall 1926; Markley and Bassett 1942; Powell 1982; Pulliainen 1982), only Magoun (1985) reported actual observations of this behavior in free-ranging wolverine. de Monte and Roeder (1990) conducted a histological study of abdominal gland tissue in the European pine marten (*Martes martes*) and found it composed exclusively of sebaceous glands.

Buskirk et al. (1986) argued for the presence of plantar glands on the feet of wolverine which may produce a more passive mode of scent-mark transmission.

Deposition of various scents by wolverine may be useful in navigation and orientation or as social advertisements (including reproduction, territoriality, or "ownership" of a resource), but is not well understood. Magoun (1985) suggested that scent marking was important in locating cached food. Koehler et al. (1980) surmised that marking behaviors were important in temporal and spatial separation of individuals. Krott (1959) and Long (1987) suggested that alarm and defense were primary reasons for use of certain scent-producing behaviors.

Tracks. Individual footprints of wolverine appear somewhat doglike, although they are wider, and show the fifth toe and a division of the pads of the sole (Jackson 1961). Depending on the substrate, the stiff, bristle-like hairs often show as well. In snow, tracks are generally parallel and slightly offset, as are those of many mustelids. The belly and tail may leave drag marks in the snow between bounds. Seton (1929) characterized wolverine as digitigrade, although the metatarsal pads of the hind feet rarely contact the ground. Novikov (1956) suggested they are semiplantigrade. Both are plausible characterizations, depending on the substrate.

Skull and Dentition. Wolverine skulls are massive in comparison to those of other mustelids (Fig. 32.2), heavier even than badgers (*Taxidea taxus*). Hall (1981) listed basal lengths of skulls in North America at 127–140 mm, whereas Novikov (1956/1962) listed females at 134–136 mm and males at 144–152 mm from Russia. Greatest width of the skull (zygomatic width) is 89–90 mm for females and 95–107 mm for males in Russia (Novikov 1956/1962). Skulls from Alaska apparently are similar, varying from 90 to 107 mm wide (J. S. Whitman, unpublished data). The rostrum is short, broad, and heavily constructed. The braincase is relatively high anteriorly, broadening evenly behind the postorbital constriction, and is widest at the mastoid processes. The sagittal crest is well developed, and in older males in particular, forms a large posterior protrusion, which overhangs the occipital region of the skull. The occipital and lambdoidal crests also are well developed (Fig. 32.2).

Teeth are robust. The dental formula is I 3/3, C 1/1, P 4/4, M 1/2. Most skulls from animals taken in foothold traps show excessive tooth wear or breakage of incisors, canines, and premolars from chewing on the trap. The upper canines are large, blunt, and canted slightly forward. They are followed by a small, peglike P^1, grading up in size to P^4, the large carnassial, followed by M^1, a typical dumbbell-shaped, somewhat smaller tooth. The lower middle pair of incisors, I_2, have roots which are displaced lingually, allowing heavier, stronger teeth with no increase in space. Lower canines are shorter and more strongly recurved than the uppers. As with the upper toothrow, the premolars increase in size from the small P_1 up to the large M_1 (carnassial), followed by the smaller, almost circular M_2.

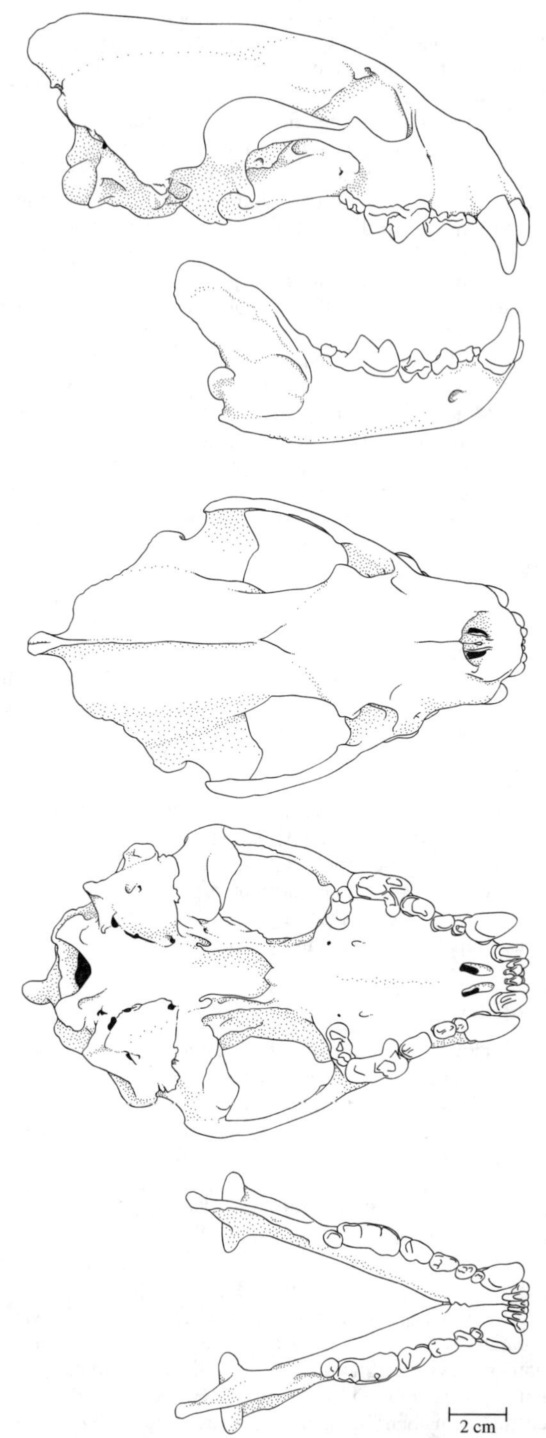

FIGURE 32.2. Skull of the wolverine (*Gulo gulo*). From top to bottom: lateral view of cranium, lateral view of mandible, dorsal view of cranium, ventral view of cranium, dorsal view of mandible.

Baculum. The baculum of adult males is 80–90 mm long, slightly bent midway, and round or oval in cross section. The proximal end is somewhat bulbous, and becomes slightly heavier and ridged with age. The distal end is lightly tuberculated into three prongs (Wright and Rausch 1955).

GENETICS

The diploid chromosome number is 42 (Krevoshiev 1984). Autosomes consist of 24 metacentrics and submetacentrics and 16 acrocentrics. The X chromosome is metacentric and the Y is acrocentric (Hsu and Benirschke 1970). Several studies investigated genetic variability and structure of wolverine across North America. Wilson et al. (2000) examined 46 allozymes and the left domain of the control region (mitochondrial DNA, mtDNA) to assess genetic variability of 43 wolverine from 5 sites across the Northwest Territories, Canada. They reported levels of heterozygosity and polymorphism for allozymes as lower than generally reported for other mammals, but within the range of other mustelids. Mitochondrial DNA analysis revealed 9 haplotypes from 6 polymorphic nucleotide sites; however, the within-site diversity of nucleotides was low, suggesting only slight sequence divergence among the haplotypes. Alternatively, Kyle and Strobeck (2001, 2002) reported varying levels of genetic variation across the North American range of the wolverine. They suggest that genetic variation is constant in all sampled wolverine populations, except Idaho where it is significantly lower.

The allozyme and mtDNA data indicate significant structuring of wolverine in the Northwest Territories. MtDNA data show approximately half of the difference in genetic diversity are attributable to differences among, rather than within, sampling sites. This is consistent with reported side-fidelity of females suggesting that across population gene flow was predominately male-mediated. Microsatellite DNA results appear to conflict with these results, as microsatellite DNA analyses reveal no significant substructure in the north, but increasing levels of substructure at the southern and eastern periphery. These conflicts may, in part, be explained by the fact the mtDNA is female-mediated while microsatellite DNA reflects both male and female movement. Furthermore, substructure, as identified by allozymes in the Northwest Territories was only marginally significant.

PHYSIOLOGY

Unlike most other mammals, which show an increase of the basal metabolic rate in proportion to body weight ($W^{1.0}$), Iversen (1972) found that the rate in the wolverine increased in proportion to the 1.41 power of body weight ($W^{1.41}$) during the first 2.5 months of life. This difference was thought to be associated with rapid development of high energy-producing tissues such as the liver, heart, brain, and kidney during early growth. After 2.5 months of age, consumption of oxygen increased almost linearly with the 0.64 power of body weight ($W^{0.64}$). The timing of the change was assumed to coincide with the period of weaning and a change in diet toward increased intake of carbohydrates. This may also be predicated as due to improved thermoregulatory ability of the young as the mother leaves for longer periods after weaning (Wilson 1982).

REPRODUCTION AND DEVELOPMENT

Reproductive biology of the North American wolverine has been described from in utero examination (Wright and Rausch 1955; Rausch and Pearson 1972; Liskop et al. 1981; Banci 1987; Banci and Harestad 1988; Copeland 1996), captive wolverine (Mead et al. 1991, 1993), and free-ranging wolverine (Hornocker and Hash 1981; Gardner 1985; Magoun 1985; Banci 1987, Copeland 1996).

Breeding Season and Gestation. Wolverine appear to be polygamous. The breeding season extends from about April through August (Rausch and Pearson 1972; Mead et al. 1991). As indicators of estrus, vaginal cornification, vulva size, and serum progesterone all peaked between May and August, as did male plasma testosterone levels in captive wolverine (Mead et al. 1991). Ovulation is induced by coitus (Mead et al. 1993) and implantation is delayed (Wright and Rausch 1955). Nidation generally occurs from November through March, but primarily in January and February (Rausch and Pearson 1972; Banci and Harestad 1988). Two zoo animals had reported total gestations of 215 days (Mohr 1938) and 272 days (Mehrer 1976). Mead et al. (1991) estimated total gestation at 215–272 days. Active gestation (from blastocyst implantation to parturition) is 30–40 days (Rausch and Pearson 1972).

Jackson (1961) reported the presence of eight mammae in the wolverine, whereas Banfield (1974) indicated there are six. Most field biologists agree the latter is the correct number. Mammae enlargement is evident postimplantation (Mead et al. 1993).

Litter Size and Pregnancy Rate. Litter size is normally 2–4 kits born in late winter or early spring (Wright and Rausch 1955; Rausch and Pearson 1972; Magoun 1985). Rausch and Pearson (1972) reported a mean of 3.5 embryos from 54 reproductive females in Alaska and the Yukon Territory. Liskop et al. (1981) found an average litter size of 2.6 in five reproductive tracts from British Columbia wolverines. Hornocker and Hash (1981) found a mean of 2.2 embryos in six Montana wolverines. Data on reproductive potential from in utero studies are substantial, whereas data on reproductive success from free-ranging populations are scarce.

Reported pregnancy rates within age classes and individual females vary. There is general agreement that wolverine do not breed their first summer. For Alaska wolverine, Rausch and Pearson (1972) found 20 of 40 females pregnant and considered them reproductively mature at 16–28 months of age. Banci and Harestad (1988) considered the 12- to 24-month age class of Yukon wolverine as subadult, as only 7.4% (2/27) of females in this age class were pregnant. However, Liskop et al. (1981) found 11 of 13 (85%) British Columbia 12- to 24-month-old wolverine reproductively active as well as a pregnancy rate of 88% (23 of 26) in those females over 24 months. In all cases, pregnancy was based on fetus and corpora lutea counts. Age-specific natality rates measured in the Yukon increased significantly between the second- and third-year age classes, with overall pregnancy rates reaching 73.4% (Banci and Harestad 1988), whereas the Alaska study described an adult pregnancy rate of 92% (Rausch and Pearson 1972). In British Columbia, no 1-year-old females were considered mature, whereas 34% of 2-year-olds were mature (Liskop et al. 1981). In the same study, 88% of females $\geq$2 years old were fecund, lending to the conclusion that adult wolverine in that locale bred every year (Liskop et al. 1981). Measure of female maturity has also been based on thickened uterine horn walls (Wright and Rausch 1955; Liskop et al. 1981). Ewer (1973) and Nowak (1991) reported that wolverines breed every other year because of a relatively long period of parental care. Ingles (1965) stated that females reproduce only every second or third year. Pulliainen (1968) and Rausch and Pearson (1972) found fetal sex ratios did not differ significantly from the expected 1:1.

In her study of free-ranging wolverine, Magoun (1985) used teat length as an indication of age, suggesting a correlation with sexual maturation. She concluded that teat lengths >10 mm suggested sexual maturity. Litter sizes reported from studies of free-ranging wolverine indicate the reproductive potential may be higher than realized reproductive success. Average litter sizes of 1.75 and 2.4 were reported for Alaska (Magoun 1985) and Finland (Pulliainen 1968), respectively. Magoun's (1985) reported reproductive rate was based on three adult females over a 4-year period, with one of these females producing no offspring. In the Idaho study (Copeland 1996), two adult females produced no litters in the two years they were monitored, whereas one female produced three litters in four successive years. Another female contained two unimplanted blastocysts the fall following a successful reproduction. Whereas litter size in Idaho averaged 2.0 kits, the reproductive potential, as measured by the number of postweaning-age kits divided by the number of reproductively mature females was 0.89 kit/female (Copeland 1996). In Lapland, Pulliainen (1968) examined litters in dens, finding an average of 2.5 kits among 161 litters. Landa et al. (1997) reported 61 postweaning-age kits associated with 48 dens, which equates to 1.3 kits/reproductively active female.

Physical condition of the female and food availability during kit rearing undoubtedly influence production and survival, both intrauterine and postpartum. Abundance of small rodents was considered a primary factor in wolverine kit survival in Norway (Landa et al. 1997).

Male reproductive maturation appears to coincide with that of the female, although reports are somewhat variable. Liskop et al. (1981) considered male wolverine mature if the testes exceeded 6.5g and the baculum was >8 cm long. Of 14 one-year-old animals in British Columbia, 4 were considered mature (29%), whereas 7 of 11 (64%) two-year-old males were mature. Banci and Harestad (1988) concluded that age class separation based on baculum measurements was not definitive, and mean testes weights of subadults did not differ from those of young of the year, leading them to conclude that male wolverine reach sexual maturity at 2+ years.

Physiological markers contributing to wolverine breeding behavior indicate an extended breeding season. Signs of early spermatogenesis were noted during January–March (Liskop et al. 1981) as were increases in plasma testosterone levels (Mead et al. 1991). Mead et al. (1991) considered the period of anestrus as September–April and reported maximal testes regression and minimal testosterone levels occurring from mid-August through December.

Growth and Development. Kits are born with soft white natal hair (Seton 1929; Mehrer 1976), which changes to a dull, adult-like pelage in 3–4 weeks. Eyes are closed and teeth unerupted at birth. Mehrer (1976) and Shilo and Tamarovskaya (1981) reported newborn weights of 84–94 g. Crown–rump lengths in these same newborns was 12.1–16 cm. Growth progresses rapidly and kits are weaned at 9–10 weeks (Copeland 1996). Young leave the natal den at 12–14 weeks of age (Magoun 1985) and often attain adult weights by early winter (Rausch and Pearson 1972).

Longevity was reported to be at least 15 years in captive animals by Woods (1944). Jones (1982) reported a 17-year-old captive animal. In wild situations, however, 8–10 years is probably maximum longevity (Jackson 1961).

Denning. Use of reproductive dens begins from early February to late March. Reproductive success may be linked to the availability and quality of reproductive den sites (Banci 1987; Landa et al. 1998a; Magoun and Copeland 1998). Fennoscandian studies provide the earliest data on winter denning habits of the wolverine. Pulliainen (1968) presented the characteristics of 31 reproductive dens in Finland. Eighty-one percent of dens occurred on bare, rocky hillsides of mountain slopes near or above timberline, whereas 6 dens were located in lower elevation spruce (*Picea*) and pine (*Pinus*) peat-bogs. Most of 28 dens in Norway were situated above timberline in deep snow near cliff areas (Myrberget 1968). The general structure of dens in both studies was the same. Den entrances were located in soft snow near trees or rocks, with a vertical tunnel extending 1–5 m to ground level. Lateral tunnels extended for up to 50 m along the ground surface. In most cases, wolverine kits were found at ground level on bare soil.

Data on wolverine denning habits in North America are limited. Rausch and Pearson (1972) described 3 dens in Alaska. Two were above timberline in snow-filled ravines; the third was found in an abandoned beaver house. Magoun and Copeland (1998) provided data on 15 den sites in Alaska and Idaho. They considered reproductive dens as either natal (site of parturition) or maternal (secondary den used subsequent to parturition but before weaning). Dens in arctic Alaska were long, complex snow tunnels without associated trees or boulders, whereas Idaho dens were always associated with boulders or fallen trees (Magoun and Copeland 1998). Thermoregulatory advantages, protection from predators, suitability of the site during spring thaw, and proximity to rearing habitat were factors influencing den site selection in wolverine.

Alaska dens occurred in deep snowdrifts along minor drainages at elevations of 560–625 m (Magoun and Copeland 1998). Idaho wolverines accessed natal dens by tunneling through several meters of snow into the natural chambers and passageways created by the talus and fallen trees (Magoun and Copeland 1998). The site of parturition could not be identified, although it did not appear to have occurred within the snow layer (Copeland 1996). Substantial vegetation caches were present in both natal dens, but evidence of use by wolverine was not present (Copeland 1996). In 10 of 28 dens in Norway, young laid on branches. In the remainder of cases, they were found on the ground or bare snow (Myrberget 1968). In all reported cases in the Finland study (Pulliainen 1968), kits were found lying on bare ground. Fennoscandian studies were based on data collected from wolverine hunters rather than radio-instrumented animals, so it was not always known whether dens were the actual birthing sites.

In some cases, females may use multiple dens before weaning kits. Why dens become unsuitable is not well understood. Fennoscandian studies report den abandonment as a common response to human disturbance. Wolverine hunters in Finland emphasized that pursuit of a pregnant female may result in use of "exceptional" places as birthing sites (Pulliainen 1968). Myrberget (1968) mentions 4 instances of den abandonment due to human disturbance and suggested that secondary dens may be less suitable. Idaho wolverines abandoned natal dens as early as 10 March, moving kits through a series of maternal dens until weaning occurred at 9–10 weeks of age (Copeland 1996). Direct contact occurred with 2 denning females in Idaho in late April and May and resulted in den abandonment in both cases (Magoun and Copeland 1998). Females in arctic Alaska remained at a single natal den until late April or early May and did not appear disturbed by the presence of human observers; however, females associated with maternal dens were more sensitive to human presence (Magoun and Copeland 1998). Ewer (1973) suggests that moves may occur in response to den parasites or attempts by the female to deter predators from locating the den. In both the Idaho and Alaska studies, abandonment of natal dens coincided with a period when maximum daily temperatures rose above freezing for the first time since denning commenced (Magoun and Copeland 1998) suggesting that den abandonment may be a response to increased moisture or humidity.

ECOLOGY

Habitat. Wolverine occur most commonly within boreal forest and taiga communities dominated by black spruce (*Picea mariana*), white spruce (*Picea glauca*), balsam fir (*Abies balsamea*), jack pine (*Pinus banksiana*), tamarack (*Larix laricina*), alpine fir (*Abies lasiocarpa*), lodgepole pine (*Pinus contorta*), white birch (*Betula papyrifera*), and balsam poplar (*Populus balsamifera*) and in tundra ecosystems. Wolverine appear to be most common in regions that receive regular annual snowfall. Across the arctic of northern Alaska and Canada, wolverines occur from sea level to >2000 m elevation. Their presence at southern latitudes appears restricted to high-elevation habitats (Schempf and White 1977).

In south-central Alaska, wolverine preferred spruce (*Picea* spp.) habitats during winter (Gardner 1985; Whitman et al. 1986) and rocky areas during summer. In the Yukon Territory, Banci (1987) found that male wolverines preferred coniferous habitats in summer, but avoided alpine talus slopes. Because of seasonal elevational shifts, Whitman et al. (1986) found spruce forests were avoided in summer. In Montana (Hornocker and Hash 1981) and Idaho (Copeland 1996), 70% of all telemetry relocations occurred in coniferous forest types. Wolverine in Montana were most commonly found in subalpine fir (*Abies lasiocarpa*) communities, with seasonal elevational shifts evident as well (Hornocker and Hash 1981). The most basic habitat relationship noted in Idaho wolverines was the use of mid-elevational coniferous forest types in winter and a preference for high-elevation nonforested cover types during summer (Copeland 1996).

Elevational and habitat shifts in wolverine distribution may be related to prey availability (Gardner 1985; Whitman et al. 1986; Banci 1987; Copeland 1996), human avoidance (Hornocker and Hash 1981; Copeland 1996), or thermoregulatory needs (Hornocker and Hash 1981). Copeland (1996) found a preference for northern aspects by wolverine in central Idaho, further suggesting that cooler habitats may be preferred.

Copeland (1996) noted that habitat use by subadult wolverine varied from that of resident adults, which may reflect subadult interests in pioneering new habitat. This may account for the occasional observation or mortality of wolverine in atypical habitats.

Hatler (1989) suggested that no particular components typify wolverine habitats, and that lack of large-scale refugia may be the limiting factor in their distribution. Banci (1987) thought the presence of large refugia in the form of national parks in southwestern Alberta was largely responsible for the continued population viability of wolverine in that region.

Movement and Activity. Movement and dispersal have not been thoroughly investigated in North American wolverine. It is interesting to note, however, that despite the lack of empirical data, several authors have found wolverine movement noteworthy and somewhat consistent. Novikov (1956/1962) reported that the wolverine may cover dozens of kilometers per day. Munro (1945), Makridin (1964), and Haglund (1966) described wolverine paths as circuitous or rectangular, suggesting the animal repeatedly travels a regular route. Copeland (1996) subjectively described wolverine movement as "direct" in 80% of 127 snowtracking sessions.

Wolverine in the Yukon (Banci 1987) and Montana (Hornocker and Hash 1981) moved greater distances in summer than in winter. Daily movements of wolverine in Idaho ranged from 0 to >35 km, and

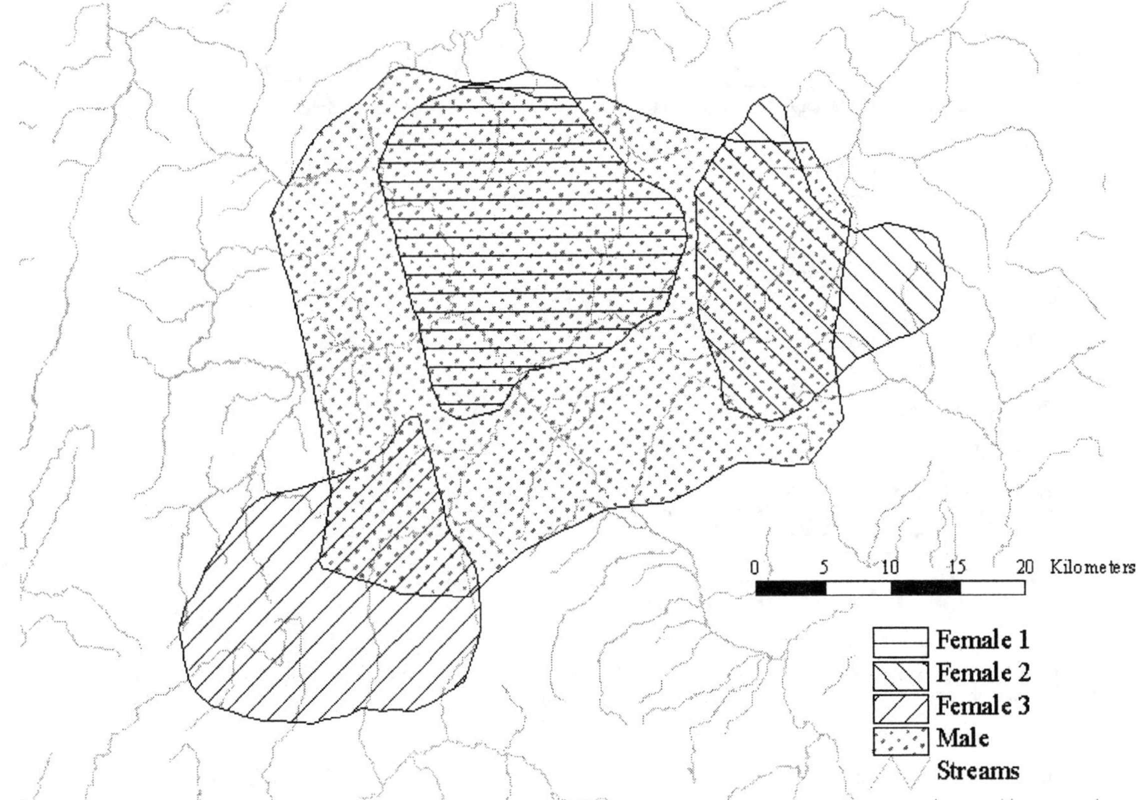

FIGURE 32.3. Core home range distribution and overlap of a resident male and three resident adult female wolverine (*Gulo gulo*) in central Idaho. SOURCE: Copeland (1996).

although seasonal variation in movement distances was not significant, substantial variation was evident among adult males (Copeland 1996).

Extraterritorial movements appear to be common, and movements indicating dispersal in immature animals and young adults seem to be the norm. Gardner et al. (1986) reported a 378-km straight-line movement of a 2-year-old male from south-central Alaska to the Yukon Territory. Magoun (1985), working in northwest Alaska, reported a 300-km movement by an unknown-age female, and dispersals of 100 and 60 km in a 1-year-old male and female, respectively. Copeland (1996) documented three dispersal attempts in 2-year-old males of >200 km. In Scandinavia, wolverine appear to disperse shorter distances, less than 60 km, with emigration occurring in juvenile and subadult age classes (Vangen et al. 2001).

Wolverine were once thought to be primarily nocturnal (Wilson 1982; Nowak 1991). Some investigators, however, have reported alternate periods of sleep and activity of about 3–4 hr (Krott 1960; Copeland 1996). Diel activity may be altered considerably by inclement weather when activities are reduced, or alternately by hunger, when activity periods are probably extended in search of food.

Home Range. Based on previous research, Hatler (1989) summarized several commonalities of wolverine spatial use : males have larger home ranges than females; females without kits have larger home ranges than females with kits; and home range use appears to vary with season. Telemetry investigations have shown that wolverine adult males and females tend to exclude conspecifics of the same sex, allowing cohabitation between sexes. Female home ranges are generally smaller than those of males (especially when they are accompanied by dependent young), allowing multiple females to reside within the home range of single adult male (Fig. 32.3). This pattern of intersexual overlap appears to be one of the few spatial commonalities among previous wolverine studies, whereas degrees of intrasexual overlap varied considerably. Magoun (1985) found most resident female home ranges were maintained exclusive of other females, with spatial separation most prominent during summer. Also in Alaska, Gardner (1985) only found home range overlap within a single adult and subadult male. Hornocker and Hash (1981) found no evidence of exclusive home range use and suggested that trapping harvest may have created behavioral instability in the population, allowing inadequate time for establishment of site tenure.

Mean annual home range sizes for male wolverine in North America vary from 382 km^2 in the Yukon (Banci 1987) to 666 km^2 for arctic Alaska individuals (Magoun 1985) (Fig. 32.4). Female home ranges varied from 104 km^2 in Alaska (Magoun 1985) to 963 km^2 in Montana (Hornocker and Hash 1981). In Scandinavia, Landa et al. (1998b) reported a mean annual male home range of 763 km^2 and a mean female home range of 335 km^2.

Gardner (1985) suggested that wolverine home range size may be related to habitat, topography, and food availability. The availability of den sites may influence home range spacing as well (Krott 1959; Magoun and Copeland 1998). Landa et al. (1998a) argued that reuse of reproductive dens in Norway may indicate denning habitat was limited. Food availability was not measured in any of the North American wolverine study areas, and the relationship between resource dispersion and home range size in wolverine is not well understood.

Sociality has received little attention in wolverine study. Wolverine are reportedly solitary animals, with the obvious exception of females with dependent young and male–female associations during the breeding period (Jackson 1961; Liskop et al. 1981). However, Copeland (1996) found association between an adult male and juvenile and subadults of both sexes, as well as nonbreeding adult male–adult female pairings.

Magoun (1985) found male wolverine near females and offspring on several occasions, although she felt the association was likely related to breeding. Wolverine hunters in Finland commonly found evidence of male wolverine activity near dens in March and April (Pulliainen 1968). Haglund (1966) described the extended presence of a male wolverine near a den in Sweden. Rausch and Pearson (1972), in reference to the Scandinavian observations, entertained the possibility of seasonal monogamy in wolverine, suggesting that a male-biased harvest in such a

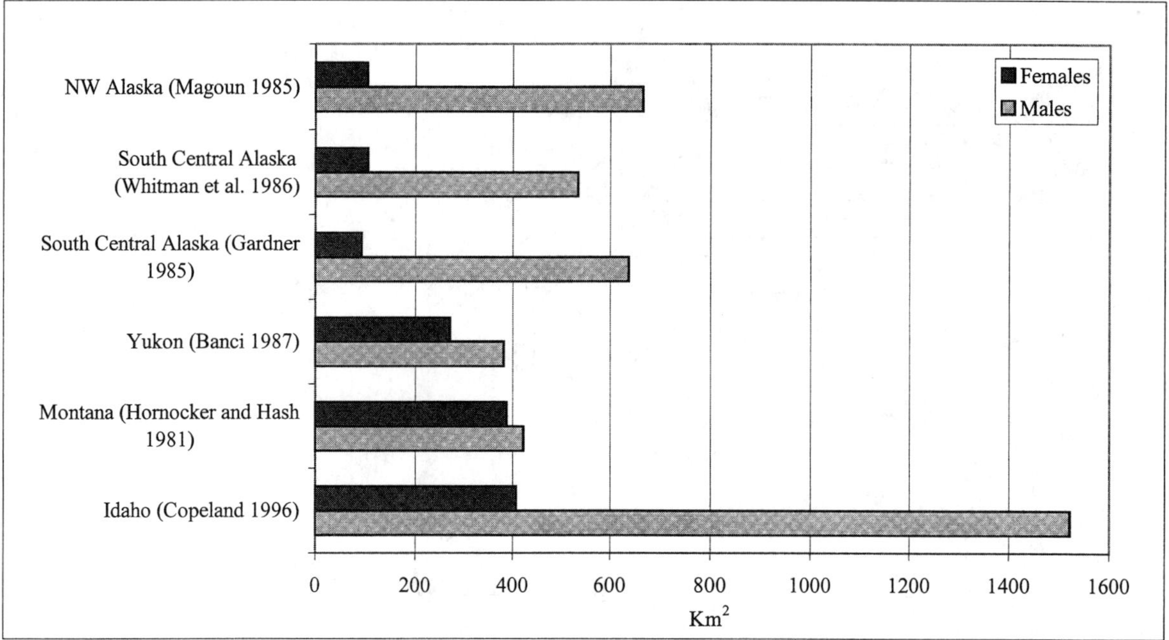

FIGURE 32.4. Home ranges of adult North American wolverines (*Gulo gulo*). Top to bottom: north to south geographic locations of study areas. Home range estimates were all based on minimum convex polygon. Comparisons among studies are difficult because of considerable variation in sample size.

case could seriously impact reproduction. They concluded that wolverine were likely polygamous, adding the possibility that males may be seeking food near den sites.

Population Density. In general, wolverine densities are low relative to those of similar-sized carnivores, with ranges from 1 wolverine/40 km^2 to 800 km^2 (Banci 1994). Magoun (1985) in arctic Alaska and Copeland (1996) in Idaho calculated density based on reproductive potential and home range size. They estimated 1 wolverine/48 to 139 km^2 and 1 wolverine/109 to 304 km^2 respectively. Quick (1953) and Hornocker and Hash (1981) reported population density estimates of 1 wolverine/207 km^2 in British Columbia and 1 wolverine/65 km^2 in northwest Montana, respectively, based on snowtracking and capture data. Banci (1987) estimated Yukon wolverine densities at 1/177 km^2 based on capture data. On the Kenai Peninsula, Alaska, Golden (1996) estimated 1 wolverine/192 km^2, whereas Gardner and Becker (1991) estimated 1/213 km^2 in the Talkeetna Mountains of south-central Alaska.

Home range estimators may bias density estimates due to variability in sample size and home range estimation technique. Comparison among studies should be done with caution. However, these may be the best estimates available until long-term capture–recapture studies can provide precise, empirical estimates.

FEEDING HABITS

Wolverine appear to be largely opportunistic feeders, with a variety of prey and carrion in the diet (Pasitschniak-Arts and Larivière 1995). Based on stomach contents of 121 wolverines, Myhre and Myrberget (1975) found reindeer (*Rangifer tarandus*) were the most important winter food of wolverine in Norway, followed by moose (*Alces alces*), roe deer (*Capreolus capreolus*), red fox (*Vulpes vulpes*), hare (*Lepus* spp.), small rodents, birds, and plant matter. Myrberget (1968) and Myrberget et al. (1969) listed other important dietary components in Norway as voles (*Microtus* spp.), lemmings (*Lemmus* spp.), sheep (*Ovis* spp.), and bird eggs. Krott (1960) added that wolverine in Sweden fed extensively on wasp larvae and various berries during summer.

Rausch (1959) reported moose and caribou were important in the diet in Alaska. Rausch and Pearson (1972) added walrus (*Odobenus rosmarus*), seals, and whales as important carrion in western coastal Alaska, and stated that berries were consumed during summer months. On close inspection of trapped wolverines from interior Alaska, most were found to have imbedded porcupine (*Erethizon dorsatum*) quills in their head, throat, and forelegs (J. S. Whitman, unpublished data), which suggests they prey on porcupines quite often. There are numerous anecdotal accounts of wolverine following established traplines, where they readily consume trapped marten, fox, red squirrels (*Tamiasciurus hudsonicus*), and other furbearers. Copeland (1996) suggested that mule deer (*Odocoileus hemionus*) and elk (*Cervus elaphus*) that were wounded by hunters and not retrieved were used extensively in Idaho. Ognev (1935/1962) reported wolverines in Russia feeding on fish, frogs, and geese in appropriate habitats. In Wisconsin, Jackson (1961) assumed that eastern cottontail rabbits (*Sylvilagus floridanus*), beavers, squirrels (*Sciurus* spp.), eastern chipmunks (*Tamias striatus*), mice, grouse, and waterfowl were important dietary components before 1960. In California, Grinnell et al. (1937) listed yellow-bellied marmots (*Marmota flaviventris*), carrion, pocket gophers (*Thomomys* spp.), and mice as important.

It is likely that a considerable proportion of the larger food items listed above are taken as carrion (Wilson 1982), although most of the literature does not differentiate between items scavenged as carrion and actually killed by wolverine. In Russia, Ognev (1935/1962) suggested that wolverine may feed extensively on kills made by other animals, based on the fact that he often found wolverine tracks accompanying tracks of other carnivores such as bears, foxes, and wolves. In interior Alaska, where wolves are common, wolverine often use moose and caribou carcasses after wolf packs have vacated them.

Depending on the intended prey, wolverine employ differing hunting techniques. Ground squirrels (*Spermophilus* spp.) or marmots are generally taken by digging. Wolverine were apparently able to capture blue grouse (*Dendragapus obscurus*) and ruffed grouse (*Bonasa umbellus*) in Idaho (Copeland 1996). Efforts to catch caribou in tundra habitats were unsuccessful, according to observations made by Magoun (1985), although anecdotal reports of wolverine killing large ungulates are relatively common (Grinnell 1920; Seton 1929).

Caching behavior in wolverine is well developed when a particularly large or easily obtained food source is available. Ognev (1935/1962) reported up to 20 foxes and 100 ptarmigan (*Lagopus*) cached under snow and ice. In northwestern Alaska, caribou and arctic

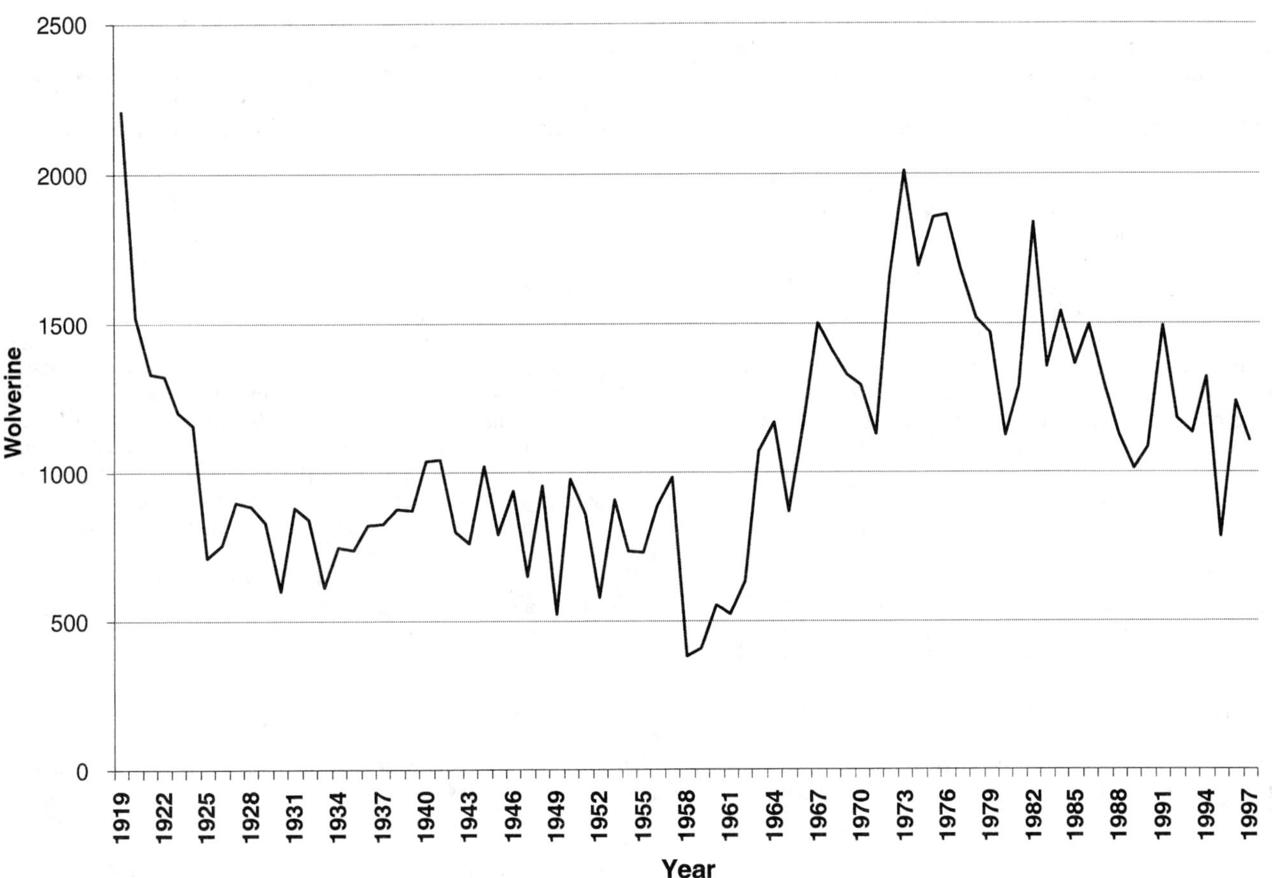

FIGURE 32.5. Number of wolverines (*Gulo gulo*) harvested in North America. SOURCE: Data from Novak et al. (1987).

ground squirrel (*Spermophilus parryii*) carrion were the most important winter food items (Magoun 1985). Caches may be in talus rock piles, beneath snow or ice, or in trees.

MORTALITY

Human-caused mortality, either from trapping and hunting or from habitat alteration (especially urbanization), appears to be the most significant cause of adult wolverine deaths. Recent documented harvests in North America have been between 1200 and 1800 animals annually (Fig. 32.5). Predation by wolves (Boles 1977; Banci 1987; S. D. Whitman unpublished data) and cougars (*Puma concolor;* Copeland 1996) has been documented as well. Scandinavian studies report intraspecific mortality, including infanticide, as a factor that may limit juvenile wolverine survival (Persson et al. 2003). Mortality from starvation also may be a factor in some areas (Hornocker and Hash 1981; Banci 1987; Copeland 1996). Mortality from diseases and parasites does not appear to be a significant factor affecting populations.

Rausch (1959) conducted rather exhaustive helminthological collections from wolverine in Alaska. Sixty-nine of 80 (86%) wolverine examined had helminths. He listed seven species [*Alaria* spp., *Ascaris* (*Baylisascaris*) *devosi, Mesocestoides kirbyi, Physaloptera torquata, Molineus patens, Taenia twitchelli,* and *Trichinella spiralis*]. In addition, Erickson (1946) recorded tapeworms (*Bothriocephalus* spp.), flukes (*Opisthorchis felineus*), and roundworms (*Dioctophyme renale, Soboliphyme baturini*) from wolverine. Williams and Dade (1976) found the heartworm, *Dirofilaria immitis,* in a captive wolverine. Dunagan (1957) reported *Baylisascaris* (*Ascaris*) *columnaris* and *Alaria mustelae*. Shults (1970) reported *Mesocestoides lineatus* from wolverine. In no cases did authors suggest that parasite loads led to, or predisposed wolverine to, increased mortality.

SEX DETERMINATION

Although there are no notable pelage differences between sexes, the sex of live animals or carcasses can be easily determined based on external reproductive organs. The sex of pelts can easily be determined based on examination of the genitalia. During spring, concentrated tracks radiating from a central location may suggest the presence of an adult female with close ties to a natal den, whereas tracks of males generally are wide ranging.

AGE ESTIMATION

Assigning discrete ages or even age classes to wolverine is difficult under field conditions. Field observation of size alone is not a good indicator of age as wolverine reach adult body size by fall of their first year. Relative tooth wear and reproductive organ development may be useful in separating young-of-the-year or sexually immature individuals from older age classes. Teat and testes dimensions appear to provide definitive separation of nulliparous and parous females and a subjective classification of reproductive maturity in males respectively (Magoun 1985; Copeland 1996). Banci (1987) reported that most males and females are not reproductively active until 2+ years of age. As such, presence of descended testicles may be an indicator of sexual maturation in male wolverine. Copeland (1996) aged juvenile wolverine in Idaho based on the pattern of tooth eruption.

A variety of methods have been investigated for age delineation from wolverine carcasses. Eye lens weight; baculum weight; morphology of sagittal, temporal, and lambdoidal crests; condylobasal length of the skull; prominence of the suprasesamoid tubercle; reproductive organ development; and ossification rates of long-bone epiphyseal junctions, have all been investigated with variable usefulness (Rausch and Pearson 1972; Magoun 1985; J. S. Whitman, unpublished data). Banci (1982) felt that the utility of skull measurements was limited to distinguishing young-of-the-year from adults preferring the degree of sagittal crest development and suture closure to separate these age classes. Rausch and Pearson (1972) and Banci (1987) argued that considerable overlap in bacula weight precluded its utility as a measure of age except in differentiating juveniles from older animals. Banci (1987) provided evidence for age-specific natality in wolverine suggesting that reproductive tract examination may provide an indication of age. Testes weight appears to be a good indicator of sexual maturation in male wolverine as well.

Microscopic examination of tooth cementum annuil (Klevezal' and Kleinenberg 1969) for aging has been used widely in both live animals and carcasses (Liskop et al. 1981; Hornocker and Hash 1981; Whitman and Ballard 1984; Gardner 1985; Magoun 1985; Whitman et al. 1986; Copeland 1996) although Banci (1987) cautioned that variability in cementum deposition rates may reduce annuli clarity in wolverine.

ECONOMIC STATUS, MANAGEMENT, AND CONSERVATION

In the United States, wolverine may be harvested only in Alaska and Montana. Outside these areas, a lack of basic information on wolverine distribution and habitat requirements has resulted in little management beyond administrative protection.

Because of its relative scarcity, beauty, long-wearing utility, and frost-shedding ability (Hardy 1948), wolverine fur is highly valued in coastal mainland Alaska and Northwest Territories, Canada, as trim on parkas and mitts. In Alaska, where 600–1000 pelts are harvested annually, most pelts are used domestically as garment trim, although the market for pelts in the taxidermy trade is increasing. Well-handled wolverine pelts in Alaska generally command a price of $250–$350 U.S., whereas average pelt prices at Canadian auction houses seldom exceed $150 U.S. Although attitudes have probably changed somewhat in the ensuing 50 years, Novikov (1956/1962 : 202) reported from the former Soviet Union, "The pelt is of little value, being prepared in small numbers.... Gluttons are destructive animals, and their extermination is permissible throughout the year."

Most wolverine are probably taken as incidental catches during other fur-trapping endeavors.

Although little economic losses from domestic livestock predation are incurred due to wolverine in North America, wolverine predation on semidomestic reindeer is well documented in Scandinavia. Norway spends more than $945,000 U.S. annually to compensate for wolverine depredation on domestic sheep (Landa and Tmmerås 1997). Northern trappers contend with sporadic nuisances because of the wolverine's well-documented propensity to rob trapped furbearers along established traplines, and food caches and cabins of wilderness dwellers are sometimes raided by wolverine.

Life history requirements of the wolverine are tied to the presence and stability of ecosystems lacking broad-scale human influence, which may be most important in providing availability and protection of reproductive denning habitat. Idaho wolverine responded negatively to human disturbance near these den sites (Copeland 1996; Magoun and Copeland 1998). Technological advances in over-snow vehicles and increased interest in winter recreation may displace wolverine from potential denning habitat, and threaten what may be a limited resource.

Vegetative characteristics appear less important to wolverine than physiographic structure of the habitat. Montane coniferous forests, suitable for winter foraging and summer kit rearing, may only be useful if connected with subalpine cirque habitats required for natal denning, security areas, and summer foraging. In addition, these habitats must be available during the proper season. Subalpine cirque areas, important for natal denning, may be made unavailable by winter recreational activities. Conversely, high road densities, timber sales, or housing developments on the fringes of subalpine habitats may reduce potential for winter foraging and kit rearing and increase the probability of human-caused wolverine mortality.

Management practices that reduce carrion may affect wolverine foraging success. A close relationship exists between wolverine and ungulate presence. Ungulate carrion is a primary food item, and activities that decrease large-mammal populations may negatively affect carrion availability. Excessive hunter harvesting and loss of ungulate wintering areas (Banci 1994) as well as displacement of ungulate populations due to excessive timber harvest and urbanization may adversely affect wolverine. Wounding mortality of ungulates from hunting and livestock losses on high-elevation public grazing allotments most likely provides a consistent carrion source.

Trapping of wolverine occurs in British Columbia, Alberta, the Yukon, and the Northwest Territories of Canada and in the U.S. states of Alaska and Montana. Trapping is certainly the largest single influence on wolverine populations, with >1000 harvested annually (Fig. 32.5). Harvest records are not generally a good indicator of the population trend of furbearer populations due to vagaries in fur prices, regulatory changes, or weather anomalies. Variation in long-term trends (Fig. 32.5) appears to be most related to annual fluctuations in harvest in Canadian provinces on the fringe of wolverine distribution (Novak et al. 1987). Sustained harvest in British Columbia, the Yukon, the Northwest Territories, and Alaska is likely due to the presence of adjacent, untrapped refugia.

RESEARCH AND MANAGEMENT NEEDS

Hatler (1989) suggested that appropriately responsive management will require a better knowledge of the nature, extent, and correlates of wolverine occurrence. Zielinski and Kucera (1996) argued that distributional surveys are essential to the generation of habitat-relations models in the evaluation of land-use changes and the effects of human density and disturbance. Surveys should focus on determining occurrence and may include snowtracking surveys or remote camera bait surveys (Zielinski and Kucera 1996). Winter aerial surveys of potential denning habitat may provide an alternative to ground methods.

Refugia may be most important in providing availability and protection of reproductive denning habitat. Based on their conclusion that natural geographic barriers such as mountain ranges and rivers are not confining to movement, Hornocker and Hash (1981) argued for regional rather than local management focus for wolverine. Contiguous undeveloped blocks of habitat in central Idaho most likely contributed to reported movements of >200 km for central Idaho wolverines (Copeland 1996). If such extensive movements are requisite for wolverine persistence, travel corridors may provide critical links for population interspersion. Large-scale habitat alteration may provide barriers to movement, isolating populations and increasing their susceptibility to extinction processes. Developing a better understanding of fine-grained patterns of movement within the home range as well as broad-scale patterns of dispersal for the species should be a priority for research.

Because of their relatively low densities, limited economic desirability, and propensity to inhabit remote country, wolverine ecology is not well understood. Where harvests are still allowed, managers should be encouraged to monitor population fluctuations as well as age and sex specific harvest trends and be prepared to enact regulatory restrictions if necessary. This might best be facilitated through mandatory carcass inspection and collection.

Methodologies should be developed that provide statistically accurate and precise measurements of wolverine density over large tracts of land. The continued viability of existing wolverine populations over broad geographic areas appears to be dependent on large tracts of land where human development is minimal or nonexistent. Maintaining expanses of reasonable size is a significant challenge and will ultimately determine the viability of wolverine populations in North America.

LITERATURE CITED

Banci, V. 1982. The wolverine in British Columbia: Distribution, methods of determining age and status of *Gulo gulo vancouverensis*. Ministries of Environment and Forests, Victoria, B. C. IWIFR-15. 90pp.

Banci, V. A. 1987. Ecology and behavior of wolverine in Yukon. M.S. Thesis, University of British Columbia, Vancouver, Canada.

Banci, V. A. 1994. Wolverine. Pages 99–127 *in* L. F. Ruggiero, K. B. Aubry, S. W. Buskirk, L. J. Lyon, and W. J. Zielinski, eds. The scientific basis for conserving forest carnivores, American marten, fisher, lynx and wolverine, in the western United States (General Technical Report RM-254). USDA Forest Service Rocky Mountain Forest and Range Experiment Station, Fort Collins, CO.

Banci, V. A., and A. Harestad. 1988. Reproduction and natality of wolverine (*Gulo gulo*) in Yukon. Annales Zoologici Fennici 25:265–70.

Banfield, A. W. F. 1974. Mammals of Canada. University of Toronto Press, Toronto.

Boles, B. K. 1977. Predation by wolves on wolverine. Canadian Field-Naturalist 91:68–69.

Buskirk, S., P. Maderson, and R. O'Conner. 1986. Plantar glands in North American Mustelidae.*In* D. Duvall, D. Muller-Schwarze, and R. M. Silverstein, eds. Chemical signals in vertebrates, Vol. 4. Plenum Press, New York.

Copeland, J. P. 1996. Biology of the wolverine in central Idaho. M.S. Thesis, University of Idaho, Moscow.

Coues, E. 1877. Fur-bearing animals: A monograph of North American Mustelidae (Territory Miscellaneous Publication No. 8). U.S. Geological Survey.

Crowe, D. M. 1972. The presence of annuli in bobcat tooth cementum layers. Journal of Wildlife Management 36:1330–32.

Degerbol, M. 1935. Report of mammals collected by the fifth Thule expedition to Arctic North America. Part 1. Systematic notes, Vol. 2, No. 4. Gyldendal, Copenhagen, Denmark.

de Monte, M., and J. J. Roeder. 1990. Histological structure of the abdominal gland and other body regions involved in olfactory communication in pine martens (*Martes martes*). Zeitshrift für Säugetierkunde 55:425–27.

Dunagan, T. T. 1957. Studies on the parasites of the edible animals of Alaska, (Project No. 7955-4). Arctic Aeromed Laboratory.

Edelmann, F., and J. P. Copeland. 1999. Wolverine distribution in the northwestern United States and a survey in the Seven Devils Mountains of Idaho. Northwest Science 73:295–300.

Erickson, A. B. 1946. Incidence of worm parasites in Minnesota Mustelidae and host lists and keys to North American species. American Midland Naturalist 36:494–509.

Ewer, R. F. 1973. The carnivores. Cornell University Press, Ithaca, NY.

Formozov, A. N. 1946. Snow cover as an integral factor of the environment and its importance in the ecology of mammals and birds. Flora and Fauna of the USSR, New Series, Zoology 5:1–152.

Gardner, C. L. 1985. The ecology of wolverine in southcentral Alaska. M.S. Thesis, University of Alaska, Fairbanks.

Gardner, C. L., and E. F. Becker. 1991. Wolf and wolverine density estimation techniques. Progress report (Federal Aid in Wildlife Restoration Project w-23–4). Alaska Department of Fish and Game, Juneau.

Gardner, C. L., W. B. Ballard, and R. H. Jessup. 1986. Long distance movement by an adult wolverine. Journal of Mammalogy 67:603.

Golden, H. N. 1996. Furbearer management technique development (Federal Aid in Wildlife Restoration Research Progress Report. Pittman-Robertson Report. Grants W-24–3 and W-24–4, Study 7.18). Alaska Department of Fish and Game, Juneau.

Grinnell, J., J. S. Dixon, and J. M. Linsdale. 1937. Furbearing mammals of California. University of California Press, Berkeley.

Groves, C. 1987. Distribution of the wolverine (*Gulo gulo*) in Idaho, 1960–1987. Nongame and Endangered Species Program Report, Idaho Dep. Fish and Game, Boise, Idaho. 24 pp.

Haglund, B. 1966. De stora rovjurens vinteervanor [Winter habits of the lynx (*Lynx lynx*) and wolverine (*Gulo gulo*) as revealed by tracking in the snow]. Viltrevy 4:81–299 (In Swedish with English summary).

Hall, E. R. 1926. The abdominal skin gland of *Martes*. Journal of Mammalogy 7:227–29.

Hall, E. R. 1981. The mammals of North America, 2nd ed. John Wiley, New York.

Hall, E. R., and K. R. Kelson. 1959. The mammals of North America. Ronald Press, New York.

Hardy, T. M. P. 1948. Wolverine fur frosting. Journal of Wildlife Management 12:331–32.

Hash, H. S. 1987. Wolverine. Pages 575–85 *in* M. Novak, J. A. Baker, M. E. Obbard, and B. Malloch, eds. Wild furbearer management and conservation in North America. Ontario Ministry of Natural Resources,Toronto.

Hatler, D. F. 1989. A wolverine management strategy for British Columbia (Wildlife Bulletin Number B-60). Wildlife Branch, Ministry of Environment, Victoria, British Columbia, Canada.

Hornocker, M. G., and H. S. Hash. 1981. Ecology of the wolverine in northwestern Montana. Canadian Journal of Zoology 59:1286–1301.

Hsu, T. C., and K. Benirschke. 1970. An atlas of mammalian chromosomes, Vol. 4, Folio 184. Springer-Verlag, New York.

Iversen, J. A. 1972. Basal metabolic rate of wolverine during growth. Norwegian Journal of Zoology 209:317–22.

Ingles, L. G. 1965. Mammals of the Pacific states: California, Oregon, and Washington. Stanford University Press, Stanford, CA.

Jackson, H. H. T. 1961. Mammals of Wisconsin. University of Wisconsin Press, Madison.

Johnson, R. E. 1977. An historical analysis of wolverine abundance and distribution in Washington, U.S.A. The Murrelet 58:13–16.

Johnston, D. H., and I. D. Watt. 1981. A rapid method for sectioning undecalcified carnivore teeth for aging. Pages 407–22 *in* J. A. Chapman and D. Pursley, eds. Proceedings of the worldwide furbearer conference. Frostburg, MD.

Jones, M. L. 1982. Longevity records of captive mammals. Zoologische Garten 52:113–28.

Kelsall, J. P. 1981. Status report on the wolverine, *Gulo gulo* in Canada in 1981. Committee on Status of Endangered Wildlife of Canada, Ottawa, Ontario, Canada.

Klevezal', G. A., and S. E. Kleinenberg. 1969. Age determination of mammals from annual layers in teeth and bones. Israel Program for Scientific Translations, Jerusalem.

Koehler, G. M., M. G. Hornocker, and H. S. Hash. 1980. Wolverine marking behavior. Canadian Field-Naturalist 94:339–41.

Krevoshiev, V. G. 1984. Nazyemnie mlekopitayushchie Dalneygo Vostoka SSSR [Land mammals of the Russian Far East]. Russian Academy of Sciences, Moscow.

Krott, P. 1959. Demon of the north. Alfred A. Knopf, New York.

Krott, P. 1960. Ways of the wolverine. Natural History 69:16–29.

Kurten, B., and R. Rausch. 1959. A comparison between Alaskan and Fennoscandian wolverine (*Gulo gulo* Linnaeus). Acta Arctica 11:5–20.

Landa, A., and B. Å. Tømmerås. 1997. A test of aversive agents on wolverines. Journal of Wildlife Management 61:510–16.

Landa, A., O. Strand, J. D. C. Linnell, and T. Skogland. 1997. Wolverines and their prey in southern Norway. Canadian Journal of Zoology 75:1292–99.

Landa, A., J. Tufto, R. Franzén, T. Bø, M. Lindén, and J. E. Swenson. 1998a. Active wolverine dens as a minimum population estimator in Scandinavia. Wildlife Biology 4:168–78.

Landa, A., O. Strand, J. D. C. Linnell, and T. Skogland. 1998b. Home-range sizes and altitude selection for arctic foxes and wolverines in an alpine environment. Canadian Journal of Zoology 76:448–57.

Liskop, K. S., R. M. S. F. Sadlier, and B. P. Saunders. 1981. Reproduction and harvest of wolverine (*Gulo gulo*) in British Columbia. Pages 469–77 *in* J. A. Chapman and D. Pursley, eds. Proceedings of the worldwide furbearer conference. Frostburg, MD.

Long, C. D. 1987. Intraspecific communication of the wolverine. Page 13 *in* B. Townsend, ed. Abstracts of the fourth northern furbearer conference. Alaska Department of Fish and Game, Juneau.

Magoun, A. J. 1985. Population characteristics, ecology and management of wolverine in northwestern Alaska. Ph.D. Dissertation, University of Alaska, Fairbanks.

Magoun, A. J., and J. P. Copeland. 1998. Characteristics of wolverine reproductive den sites. Journal of Wildlife Management 62:1313–20.

Makridin, B. P. 1964. On the distribution and biology of the wolverine in the Far North. Zoologicheski Zhurnal 43:1688–92.

Manning, T. H. 1943. Notes on the mammals of south and central Baffin Island. Journal of Mammalogy 24:47–59.

Markley, M. H., and C. G. Bassett. 1942. Habits of captive marten. American Midland Naturalist 28:604–16.

McKay, R. 1991. Biological assessment and inventory plan for the wolverine (*Gulo gulo*) in the Uinta Mountains. Unpublished report. Utah Department of Natural Resources, Salt Lake City.

Mead, R. A., M. Rector, G. Starypan, S. Neirinckx, M. Jones, and M. N. DonCarlos. 1991. Reproductive biology of captive wolverine. Journal of Mammalogy 72:807–14.

Mead, R. A., M. Bowles, G. Starypan, and M. Jones. 1993. Evidence for pseudopregnancy and induced ovulation in captive wolverines (*Gulo gulo*). Zoo Biology 12:353–58.

Mehrer, C. F. 1976. Gestation period in the wolverine, *Gulo gulo*. Journal of Mammalogy 57:570.

Mohr, E. 1938. Vom Jarv (*Gulo gulo* L.). Zoologische Garten (Leipzig) 10:14–21.

Moisan, M. 1996. Rapport sur la situation du carcajou (*Gulo gulo*) au Québec. Ministère de l'Environnement et de la Faune, Direction de la faune et des habitats, Quebec City, Quebec, Canada.

Munro, J. A. 1945. Preliminary report on the birds and mammals of Glacier National Park, British Columbia. Canadian Field-Naturalist 59:175–90.

Myhre, R., and S. Myrberget. 1975. Diet of wolverine (*Gulo gulo*) in Norway. Journal of Mammalogy 56:752–57.

Myrberget, S. 1968. The breeding den of the wolverine, *Gulo gulo*. Fauna 21:108–15 (In Norwegian with English summary).

Myrberget, S., B. Groven, and R. Myhre. 1969. Tracking wolverine, *Gulo gulo,* in the Jotunheim Mountains, south Norway, 1965–1968. Fauna 22:237–52. (In Norwegian with English summary.)

Newby, F. E., and J. J. McDougal. 1964. Range extension of the wolverine in Montana. Journal of Mammalogy 45:485–487.

Newby, F. E., and P. L. Wright. 1955. Distribution and status of the wolverine in Montana. Journal of Mammalogy 36:248–253.

Novak, M. 1975. Recent status of the wolverine in Ontario. Ontario Ministry of Natural Resources, Toronto.

Novak, M., M. E. Obbard, J. G. Jones, R. Newmen, A. Booth, A. J. Satterthwaite, and G. Linscombe. 1987. Furbearer harvests in North America, 1600–1984. Supplement to Wild furbearer management and conservation in North America. Ontario Ministry of Natural Resources, Toronto.

Novikov, G. A. 1956/1962. Carnivorous mammals of the fauna of the USSR. Israel Program for Scientific Translations, Jerusalem.

Nowak, R. M. 1973. Return of the wolverine. National Parks and Conservation 47:20–23.

Nowak, R. M. 1991. Walker's mammals of the world, 5th ed. Johns Hopkins University Press, Baltimore.

Ognev, S. I. 1935/1962. Mammals of the USSR and adjacent countries. Vol. 3: Carnivora (Fissipedia and Pinnepedia). Israel Program for Science Translations, Jerusalem, Israel.

Pasitschniak-Arts, M., and S. Larivière. 1995. *Gulo gulo*. Mammalian Species 499:1–10.

Persson, J., T. Willebrand, A. Landa, R. Andersen, and P. Segerström. 2003. The role of intraspecific predation in the survival of juvenile wolverines *Gulo gulo*. Wildlife Biology 9: 21–28.

Peterson, R. L. 1966. The mammals of eastern Canada. Oxford University Press, Toronto.

Pocock, R. I. 1921. On the external characters and classification of the Mustelidae. Proceedings of the Zoological Society of London 1921:803–7.

Powell, R. A. 1982. The fisher. Life history, ecology, and behavior. University of Minnesota Press, Minneapolis.

Prescott, J. 1983. Wolverine, *Gulo gulo,* in Lake St. John area, Quebec. Canadian Field-Naturalist 97:457–58.

Pulliainen, E. 1968. Breeding biology of the wolverine (*Gulo gulo* L.) in Finland. Annales Zoologici Fennici 5:338–44.

Pulliainen, E. 1982. Scent marking in the pine marten (*Martes martes*) in Finnish Forest Lapland in winter. Zeitschrift für Säugetierkunde 47:91–99.

Pulliainen, E., and P. Ovaskainen. 1975. Territory marking by a wolverine (*Gulo gulo*) in northeastern Lapland. Annales Zoologici Fennici 12:268–70.

Quick, H. F. 1952. Some characteristics of wolverine fur. Journal of Mammalogy 33:492–93.

Quick, H. F. 1953. Wolverine, fisher, and marten studies in a wilderness region. Transactions of the North American Wildlife Conference 18:513–32.

Rausch, R. 1959. Studies on the helminth fauna of Alaska. Vol. 36, Parasites of the wolverine, *Gulo gulo* L., with observations on the biology of *Taenia twitchelli* Schwartz, 1924. Journal of Parasitology 45:465–84.

Rausch, R. A., and A. M. Pearson. 1972. Notes on the wolverine in Alaska and the Yukon Territory. Journal of Wildlife Management 36:246–68.

Schempf, P. F., and M. White. 1977. Status of six furbearer populations in the mountains of northern California. U.S. Department of Agriculture, Forest Service, California Region.

Seidel, J., B. Andree, S. Berlinger, K. Buell, G. Byrne, B. Gill, D. Kenvin, and D. Reed. 1998. Draft strategy for the conservation and reestablishment of lynx and wolverine in the southern Rocky Mountains. Unpublished report. Colorado Division of Wildlife, Ft. Collins.

Seton, E. T. 1929. Lives of game animals. Doubleday, Doran, Garden City, NY.

Shilo, R. A., and M. A. Tamarovskaya. 1981. The growth and development of wolverine (*Gulo gulo*) at Novosibirsk zoo. International Zoo Yearbook 21:146–47.

Shults, L. M. 1970. *Mesocestoides kirbyi* and *M. lineatus:* Occurrence in Alaskan carnivores. Transactions of the American Microscopic Society 89:478–86.

Skinner, D. L., and A. W. Todd. 1989. Distribution and status of selected mammals in Alberta as indicated by trapper questionnaires in 1987. (Occasional Paper No. 4). Alberta Fish and Wildlife Division, Edmonton, Canada.

Stroganov, S. U. 1969. Carnivorous mammals of Siberia. Israel Program for Scientific Translation, Jerusalem.

Vangen, K. M., J. Persson, A. Landa, R. Andersen, and P. Segerström. 2001. Characteristics of dispersal in wolverines. Canadian Journal of Zoology 79:1641–49.

van Zyll de Jong, C. G. 1975. The distribution and abundance of the wolverine (*Gulo gulo*) in Canada. Canadian Field-Naturalist 89:431–37.

Walker, □., et al. (1975).

Whitman, J. S., and W. B. Ballard. 1983. Susitna hydroelectric project., 1982 annual report. Vol. VII: Wolverine. Alaska Department of Fish and Game, Anchorage.

Whitman, J. S., and W. B. Ballard. 1984. Susitna hydroelectric project. Phase II final report. Vol. VII. Wolverine. Alaska Department of Fish and Game, Anchorage.

Whitman, J. S., W. B. Ballard, and C. L.Gardner. 1986. Home range and habitat use by wolverines in southcentral Alaska. Journal of Wildlife Management 50:460–63.

Williams, J. F., and A. W. Dade. 1976. *Dirofilaria immitis* infection in a wolverine. Journal of Parasitology 62:174–75.

Wilson, D. E. 1982. Wolverine. Pages 644–52 *in* J. A. Chapman and G. A. Feldhamer, eds. Wild mammals of North America: Biology, management, and economics. Johns Hopkins University Press, Baltimore.

Wilson, D. E., and D. M. Reeder. 1993. Mammal species of the world: A taxonomic and geographic reference, 2nd ed. Smithsonian Institution Press, Washington, DC.

Woods, G. T. 1944. Longevity of captive wolverine. American Midland Naturalist 31:505.

Wright, P. L., and R. Rausch. 1955. Reproduction in the wolverine, *Gulo gulo*. Journal of Mammalogy 36:346–55.

Zielinski, W. J., and T. E. Kucera, eds. 1996. American marten, fisher, lynx, and wolverine: Survey methods for their detection (General Technical Report PSW-GTR-157). USDA Forest Service, Pacific Southwest Research Station.

Jeffrey P. Copeland, Idaho Department of Fish and Game, Rocky Mountain Research Station, Missoula, Montana 59807. Email: jpcopeland@fs.fed.us.

Jackson S. Whitman, Alaska Department of Fish and Game, Sitka, Alaska 99835. Email: jack_5F;whitman@fishgame.state.ak.us.

33

Badger

Taxidea taxus

Frederick G. Lindzey

NOMENCLATURE

Common Names. Badger, North American badger
Scientific Name. *Taxidea taxus*
Subspecies. *T. t. taxus, T. t. jeffersonii, T. t. jacksoni,* and *T. t. berlandieri* (Hall 1981)

DISTRIBUTION

The current range of the badger extends from the northern part of Alberta, Canada, to central Mexico and eastward from the Pacific coast to a line running roughly from East Texas to the central Great Lake states (Fig. 33.1). Badgers are generally associated with treeless regions, prairies, parklands, and cold desert areas. Their range extends from below sea level to greater than 3600 m elevation. The Rocky Mountains and Grand Canyon are geographic barriers associated with the distribution of western subspecies (Long 1973). Recent records suggest, however, an eastward extension of their range (Nugent and Choate 1970). Badgers have extended their range southward since the 1940s in Illinois (Gremillion-Smith 1985; Ver Steeg and Warner 1995) and southerly and westerly from the northern tier of counties since 1900 in Indiana (Whitaker and Gammon 1988).

DESCRIPTION

The badger is readily identifiable by many unique physical characteristics that adapt it for preying on fossorial rodents. The body of the badger is slightly flattened dorsoventrally, with short, stout legs having long, recurved foreclaws and short, shovel-like hind claws. The external auditory meatus is small and rounded; eyes are small with a nictitating membrane present (Long 1973). Badgers have eight mammae: two inguinal, two abdominal, and four pectoral. The skin of the badger is loose, particularly across the front, shoulders, and back.

Size and Weight. Body measurements reported by Long (1973) for sexes combined were total length, 60–73 cm; tail length, 10.5–13.5 cm; and hind foot length, 9.5–12.8 cm. Similar measurements recorded by Messick and Hornocker (1981) in southwestern Idaho for male and female badgers, respectively, were total length, 73.9 and 70.8 cm; body length, 59.9 and 57.8 cm; and hind foot length, 10.7 and 10.3 cm.

Adult males weigh an average of 26% more than adult females (Wright 1966). Average weights for adult male and female badgers, respectively, were 8.4 and 6.4 kg in South Dakota (Wright 1969), 8.7 and 7.1 kg in northern Utah and southern Idaho (Lindzey 1971), and 7.6 and 6.3 kg in southwestern Idaho (Messick and Hornocker 1981). Young badgers grow rapidly until July, and they generally enter their first fall close to adult size. Body weight is least during winter and early spring (Messick and Hornocker 1981). Harlow (1981b) noted maximum levels of fat deposition of 31% of body weight in November, with a 37% reduction in fat stores between November and March in southeastern Wyoming. Large males may exceed 11.5 kg.

Pelage. Color may vary from yellowish brown to silver gray on the dorsal surface and from a light cream to buff ventrally. The feet are black to dark brown. The sides of the face are white with a black triangular patch anterior to the ears (Fig. 33.2). A white medial stripe extends dorsally from the nose. Width and continuity of this stripe are variable, but in *T. t. berlandieri* it continues for the length of the body and terminates at the base of the tail (Long 1973). The amount of guard hairs present is variable and apparently results in the descriptive terms *hair* and *fur* badger.

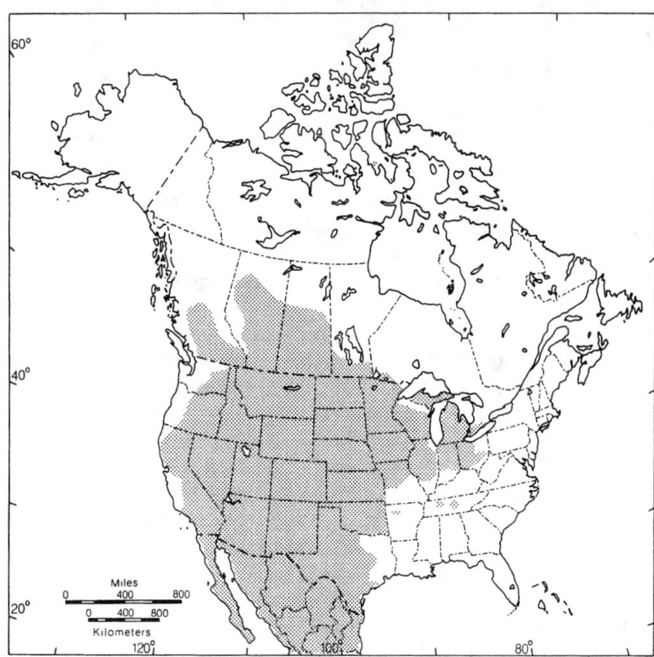

Figure 33.1. Distribution of the badger (*Taxidea taxus*).

Skull and Dentition. The skull is wedge shaped, and is broader posteriorly. Auditory and mastoid bullae are large (Fig. 33.3). Skulls vary from 11.3 to 14.1 cm in length (Long 1973). Skulls of juveniles are shorter than those of older badgers, and skulls of males are generally longer than those of females (Messick and Hornocker 1981). Skulls of old males often do not have a prominent sagittal crest (Wright 1969). The adult dental formula is I 3/3, C 1/1, P 3/3, M 1/2. The formula for deciduous dentition is I 1–2/0, C 1/1, P 3/3 (Long 1973). The triangular upper molar in skulls of adults is an important diagnostic feature for this species. Long (1965) provided a detailed discussion of jaw articulation and wear on canines. Long and Long (1965) discussed dental abnormalities in old badgers.

PHYSIOLOGY

On the basis of energy trials, Jense (1968) estimated that for maintenance, an 8.3-kg badger would require 82.6 g dry weight of cottontail rabbit (*Sylvilagus* spp.) (about 22% of an adult rabbit) or 72.3 g

FIGURE 33.2. Male badger (*Taxidea taxus*), 7 months old. Note the facial characteristics and markings, the claws on the front feet, and the excavated soil.

of ground squirrel (*Spermophilus* spp.)/day. He noted, however, that these estimates were probably below the intake that would be required for maintenance in active badgers. He also found that maintenance requirements of juveniles were as much as 62% greater than those of adult badgers.

Based on maintenance requirements estimated by Jense (45 kcal/kg/day), Messick (1981) determined that an 8-kg badger would need to consume 1.2–1.4 Townsend's ground squirrels (*S. townsendii*)/day. He assumed, however, that a free-ranging badger would have energy requirements 1.8 times greater. Thus, intake would actually need to be 2.3 ground squirrels/day. Harlow et al. (1985) determined the energetic cost of gestation in badgers was low (< 2% increase in a single female). However, lactation would be very costly, about 16 times higher than gestation if the female were nursing two young.

Based on observations and energy requirements determined from monitored heart rate, Lampe (1976) determined the energy costs of various activities. He estimated that it cost a badger actively searching for prey 25.2 kcal/hr and 0.55 kcal/L to displace soil while digging. The cost of resting while underground was, however, only 7.2 kcal/hr. A single pursuit lasting 34 min required expenditure of energy at the rate of 38.4 kcal/hr. On the basis of a diel activity schedule of 10 hr of activity and 14 hrs of rest (east-central Minnesota), Lampe estimated a daily energy requirement, exclusive of energy costs of pursuit activity, of 418 kcal/day. Employing a digestive efficiency of 85.6%, which varies inversely with intake and time spent in each activity and their metabolic costs, a 6.5-kg badger required a minimum of 1.7 pocket gophers/day for maintenance.

Harlow (1981a) determined that a badger under a regime of ad libitum food would spend an average of 15% of its total metabolism on activity, 20% on food processing, and 65% on maintenance. Assimilation efficiencies of badgers fed dry dog food averaged only 66%. This value, however, was low in comparison to efficiencies of captive badgers fed more-natural foods (Jense 1968; Lampe 1976).

To investigate the strategies used by badgers during periods of food shortages, Harlow (1981c) monitored their responses to fast periods of up to 30 days. Total metabolism decreased partially because of reduced energy required for food processing and a decrease in body weight. Reductions of activity metabolism by 37% and maintenance metabolism by 44% together accounted for slightly more than half of the reduction in total metabolism. Harlow concluded that these reductions were adaptive and actually increased a badger's chance of surviving periods of food deprivation. Badgers conserved 17 g of tissue/day after 30 days without food. Harlow calculated that if an individual continued to metabolize at rates observed when food was abundant, it would deplete its body stores and die 26% sooner than a badger that reduced metabolism. Badgers also react to periods of food scarcity by increasing the efficiency with which they use ingested foods (Harlow 1981a). Assimilation efficiency increased 19.9% from pre-fast after a 30-day fast. The amount of food consumed by badgers before and after the trial did not differ, but passage time increased from 520 to 614 min. Caloric content of feces collected during the trial increased after day 10 of the fast, suggesting that the stomach may have acted as a storage organ for 5–10 days.

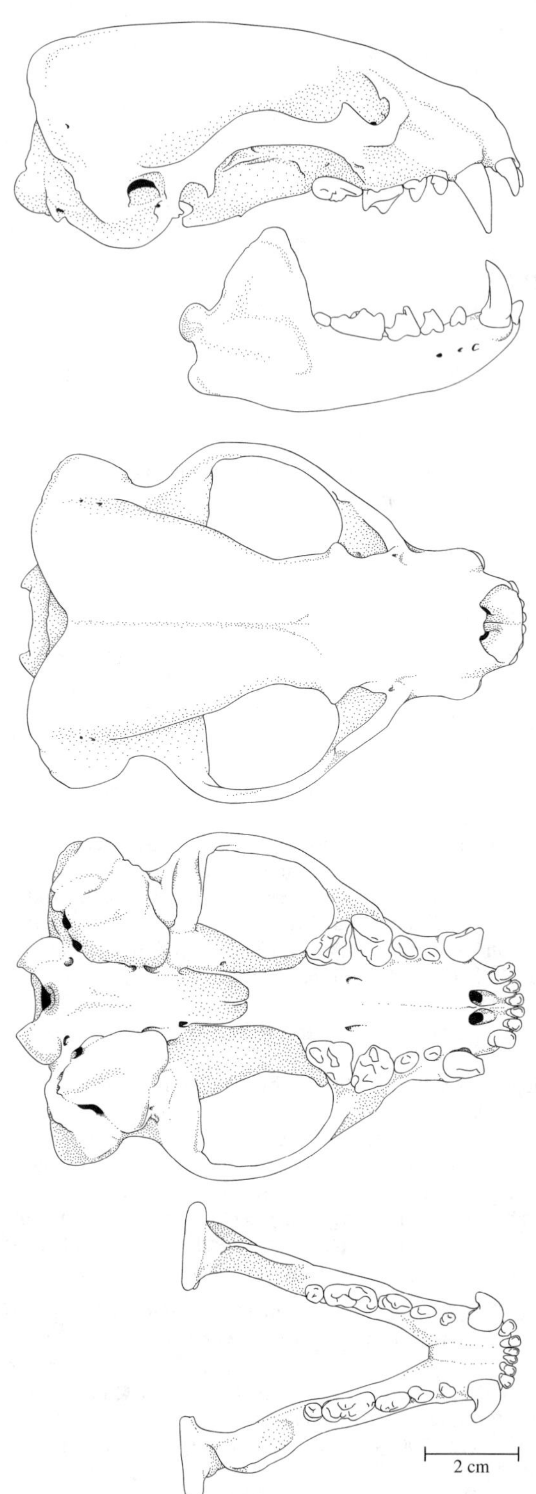

FIGURE 33.3. Skull of the badger (*Taxidea taxus*). From top to bottom: lateral view of cranium, lateral view of mandible, dorsal view of cranium, ventral view of cranium, dorsal view of mandible.

Winter activity of captive badgers indicated they avoided temperatures below −15°C by remaining in the moderated microclimate of their dens (Harlow 1981b). Badgers in outdoor exclosures in southeastern Wyoming reduced aboveground activity by 93% from November through February. While in its den, one badger periodically entered a state of torpor characterized by a 50% reduction in heart rate and a 9°C reduction in body temperature. Each torpor cycle reduced energy costs by 27% (Harlow 1981b).

REPRODUCTION

Badgers reportedly have a promiscuous breeding pattern, with breeding occurring in late July and August (Wright 1966; Messick 1981). A single copulatory sequence observed by Campbell and Clark (1983) lasted 21 min and was characterized by the male holding onto the back of the female's neck with its jaws. Badgers exhibit delayed implantation. Development of the blastocysts is arrested until they are implanted in February (Hamlett 1932; Wright 1966). Parturition occurs in late March or early April, and litter size ranges from one to five. A minority of juvenile females, 4–5 months of age, may breed, but males do not attain full spermatogenesis until 14 months old (Wright 1966, 1969). Messick and Hornocker (1981) noted increased fecundity with age of females. Lindzey (1971) observed that 38% of yearling females did not ovulate; all older females did. Pregnant females or those that had given birth in southwestern and south-central Idaho were 52% and 72%, respectively, of the female population (Messick 1987); percentages varied among years.

Male badgers were aspermatic in late winter, but had fully active testes from late May or June to August (Wright 1969). Although males are capable of breeding for 3 months, the breeding season appears considerably shorter and is apparently determined by the timing of estrus in the females. Swelling of the vulva is noticeable in estrous females (Messick 1981).

Anatomy. The reproductive tract of the female badger and active testes of the male are considerably larger in relation to body size than those of other North American mustelids (Wright 1966, 1969). Ovaries are encapsulated, and the oviduct totally encircles the ovary and forms a part of the capsule. The oviduct hypertrophies after mating. The uterus is bicornate, with horn as long as 60 mm and up to 5 mm in diameter in the early stages of pregnancy. Following implantation of the zygote, corpora lutea increase to nearly twice their inactive size (Wright 1966).

Corpora albacantia persist for a relatively short time and may disappear by early June (Wright 1966; Messick and Hornocker 1981). Placental scars frequently used as indicators of young carried to term may be visible for over 1 year, but their persistence is highly variable (Wright 1966; Messick and Hornocker 1981).

The os clitoris and the baculum are present in badgers, although the os clitoris is not visible externally and requires clearing to be found. The baculum is large, and both its size and ossification increase from the juvenile to the adult age class (Petrides 1950; Wright 1966; Lindzey 1971; Messick and Hornocker 1981). Testes size varies with the stage of spermatogenesis. Wright (1966) found the size of the testes began to increase in late March, developed fully by late May or June, and began to regress by late August. Testes were only one eighth as large during the inactive period as during the breeding season; paired testes weights averaged 43.9 g during the breeding season (Wright 1966).

ECOLOGY

Density. Seton (1929) and Lindzey (1971) reported a minimum estimate of badger density of 1/2.6 km^2. Messick (1981) estimated a density of 5 resident badgers/km^2 in southwestern Idaho and noted densities may be greater during the period when juveniles are dispersing. Estimated density of badgers was 2/km^2 on the National Elk Refuge in northwestern Wyoming (Minta and Mangel 1989; Minta 1993) and 0.8–1.1/km^2 in south-central Wyoming (Goodrich and Buskirk 1998). Locally, densities may vary with prey concentrations.

Sex and Age Structure. Sex ratio of trapped badgers in the Idaho Snake River Birds of Prey Area was essentially 1:1 (Messick and Hornocker 1981). Minta (1993) and Goodrich (1994) reported male-biased adult sex ratios of 1.75 and 1.6–2.8 males/female, respectively. Juveniles constituted 45% of two Idaho populations (Messick 1987) and 35% of a northern Utah population (Lindzey 1971). Minta and Mangel (1989) and Goodrich (1994) urged caution in use of sex and age data because of differential vulnerability of sex and age classes to various capture methods.

The oldest badgers collected by Lindzey (1971), Crowe and Strickland (1975), and Messick (1981) were 8, 13, and 14 years old, respectively. A badger in the National Zoological Park in Washington, D.C., lived 15 years, 5 months (Jackson 1961).

Home Range. Home range size of badgers varies considerably within and among regions of their geographic distribution. Size differences undoubtedly occur because of variable habitat characteristics and prey densities, but also may result from patterns of intraspecific interactions. A radio-collared female badger monitored in Minnesota used a total area of 8.5 km^2 (Sargeant and Warner 1972). Another female monitored in the same area about 5 years later, however, used a 17-km^2 home range (Lampe and Sovada 1981). Two male and three female badgers radio-tracked in northern Utah by Lindzey (1978) occupied home ranges that averaged 5.8 and 2.4 km^2, respectively. In southwestern Idaho, male and female badgers >1 year of age occupied home ranges of 2.4 and 1.6 km^2, respectively (Messick 1981). One-year-old badgers in this same locale had home ranges of 0.6 and 0.8 km^2 for males and females, respectively. Male and female home ranges, respectively, averaged 12.3 and 3.4 km^2 in southeastern Wyoming (Goodrich and Buskirk 1998) and 8.4 and 2.8 km^2 in northwestern Wyoming (Minta 1993). Seasonal ranges of adult males vary more than adult female ranges. Males occupy their largest seasonal range during the breeding season (Minta 1993; Goodrich and Buskirk 1998). Both sexes use the smallest area during winter.

Home ranges of all sex and age classes of badgers may overlap. Larger ranges of adult males typically overlap a number of the smaller female home ranges. Also, transient juveniles may reside temporarily in home ranges of adults. Although Minta (1993) could not demonstrate that intrasexual home range overlap occurred less frequently than intersexual overlap, Goodrich and Buskirk (1998) found male–male overlap less common than female–female overlap. Although female ranges overlapped on their study area, they felt there was a tendency toward female territoriality. This pattern was apparently facilitated by the major prey species of badgers in this area, white-tailed prairie dogs (*Cynomys leucurus*), providing a more stable food resource than prey species relied upon by other badger populations. Hornocker et al. (1983) speculated that the flexible behavior patterns of badgers would allow them to exhibit a traditional territorial system in favorable habitats if human exploitation was minimal.

Males move greater straight-line distances than females, likely a reflection of their larger home ranges. Movement of adult females with dependent young is restricted by the need to return to the natal den. A female may, however, shift her hunting area by moving her young among dens (Lindzey 1971; Lampe and Sovada 1981; Messick and Hornocker 1981). One female used over half of her 210-ha home range and three different dens while caring for her dependent young (Lindzey 1971).

Activity. Badgers are principally nocturnal, foraging at night and remaining underground during daylight hours (Messick and Hornocker 1981; Goodrich 1994). Although all age classes are occasionally active during the day, young of the year tend to be the most diurnal (Messick 1981). Badgers do not hibernate, but rather react to cold weather and reduced prey availability by reducing their aboveground activity. During warmer weather, badgers rarely remain in a den for more than 1 day without emerging, but extended habitations are common during late fall and winter. Extended habitations were significantly more common in Utah between November and May than during the remainder of the year (Lindzey 1978). Messick (1981) observed extended habitations of

2–12 days during winter and reported a badger that apparently did not emerge from its den for 38 days. Harlow (1981b) observed two penned badgers that remained underground for more than 70 consecutive days. Badgers he monitored reduced their aboveground exposure by 93% during winter and were active during the warmest periods of the night.

Den Use. Dens play a central role in the ecology of the badger, functioning as sites for diurnal inactivity, food storage, and parturition and as foci for foraging. Dens are variable in characteristics, with most having only a single, often elliptical entrance. Soil excavated during formation of the den is piled at the entrance. When a den is occupied, particularly during cold weather, the tunnel is often partially plugged with loose soil. Scats frequently occur in the mound of soil at the entrance and in the den itself. Dens are common in areas occupied by badgers and persist for varying lengths of time depending on weather and density of livestock. Lindzey (1978) found an average of 1.6 dens/ha with open entrances in northern Utah. Existing dens were frequently used by badgers as diurnal resting sites, with only 15% of dens used being dug immediately before their use (Lindzey 1978). Some dens were reused often by the same badger, which suggests knowledge of the location of the den. Dens are used only infrequently on successive days following a night's foraging.

The den in which a female gives birth to and raises her young is structurally more complex than other dens and reflects the needs of the family group. In Utah, natal dens had the following characteristics in common: (1) a main tunnel, which branched into two secondary tunnels, which later rejoined; (2) dead-end side tunnels, which projected from the main tunnel, secondary tunnels, and chambers; (3) pockets <15 cm in length in the sides of tunnels and chambers; (4) shallow excavations in the floors of tunnels; and (5) chambers. Branching of the main tunnel presumably allowed badgers to pass each other in the burrow system. Pockets and shallow excavations were filled with feces and covered with soil. Dead-end tunnels contained soil and feces. The soil mound at the entrance to natal dens generally was comparatively larger than those at the entrance to other dens. Badger hair was mixed throughout the soil, presumably because the construction of natal dens coincided with the spring molt in females with young (Lindzey 1976). A natal den excavated by R. P. Lampe (University of Minnesota, pers. commun., 1980) in Minnesota had similar characteristics to those noted above, but it contained no scats. Although Palmer (1954) and Jackson (1961) reported finding chambers lined with grass, no nest material was observed by Lindzey (1976).

Sociality. Badgers are solitary, with the exception of breeding pairs, family groups, and occasional short-term association between siblings after family break-up. Although Messick and Hornocker (1981) cited an incidence of badgers fighting and Minta (1993) found wounds or recent scars apparently inflicted by other badgers on 73% of male and 50% of female badgers, they tend to be solitary and avoid each other. Lindzey (1978) found that 44% of dens reused by badgers were located on the periphery of home ranges, perhaps indicating a respect for boundaries of home ranges of other badgers in the vicinity. Additionally, dens ($n = 3$) used at different times by more than one badger were on the periphery of each badger's home range. Messick and Hornocker (1981) twice noted young badgers sharing a den with an unrelated adult.

It is not known how badgers communicate to establish and maintain observed dispersion patterns. It seems likely, however, that scent plays a major role. Badgers commonly explore each den or old dig site they encounter. In drier areas in particular, the habit of defecating and perhaps urinating in dens or at their entrance, because they are shaded and cooler, provides for extended duration of scent. R. P. Lampe (University of Minnesota, pers. commun., 1982) observed a badger rubbing its abdominal gland on the soil mound at a den entrance. Although Messick and Hornocker (1981) found no evidence of scent marking in wild badgers, they observed a captive female marking and noted that abdominal glands in males and anal glands in both sexes were conspicuous in summer. Because of the central role of dens in badger behavior, they are a likely location for the accumulation and olfactory exchange of information.

The association between a female badger and her young lasts for 10–12 weeks. Females apparently feed their young solid foods during the 5- to 6-week lactation period (Messick and Hornocker 1981). Cubs seldom come aboveground on their own until 4–5 weeks of age, and although they are capable of moving short distances at this age, the female carries them if she changes dens. By late May, female badgers may begin to remain away from their young for 30–36 hr (Lampe and Sovada 1981). Timing of family break-up is variable and is characterized by a progressively looser association among family members. A family monitored by Messick (1981) remained spatially close until 10 June. However, Hetlet (1968) captured four badgers at a den on 10 and 11 July, which he felt were a family group. A radio-collared female badger monitored by Lindzey (1976) remained close to her young until at least 10 June, and he did not observe independent young until 10 July.

Dispersal. Movement of the young after family break-up is erratic and more extensive than that of adults (Messick and Hornocker 1981). Their movements are frequently through areas devoid of other badgers in what would seem unsuitable habitat. Juvenile badgers occasionally remain near their mother's home range, but most disperse. One juvenile female dispersed 52.1 km from the natal range, and a young badger was trapped 5 months later 110 km from its original capture site (Messick and Hornocker 1981).

FEEDING HABITS

The North American badger is uniquely adapted for capturing fossorial prey. However, it is flexible and readily takes advantage of local abundance of more-terrestrial mammals (Table 33.1) as well as avian, reptilian, and insect species. Insects, birds, and reptiles are apparently taken opportunistically (Errington 1937; Snead and Hendrickson 1942). Much of the vegetative material ingested by badgers is probably taken incidentally to the consumption of animals. Jense (1968), however, noted an increase in vegetable matter in the badger's diet in the fall. R. P. Lampe (University of Minnesota, pers. commun., 1980) found corn in 10 of 17 stomachs of badgers collected in the fall in Iowa; 4 of these stomachs were completely filled with corn.

Although ground squirrels are the major food in many locales (Snead and Hendrickson 1942; Jense 1968; Messick and Hornocker 1981; Minta 1990), smaller rodents dominate the diet in other areas (Dearborn 1932; Lindzey 1971). Plains pocket gophers (*Geomys bursarius*) dominated badger diets on two study areas in Minnesota (Lampe 1982). Composition of the diet changes seasonally and yearly in locales as the phenology and abundance of prey species influence their availability to badgers (Snead and Hendrickson 1942; Jense 1968; Messick and Hornocker 1981). Badgers eat carrion and will cache food items, frequently in old dens (Snead and Hendrickson 1942; Lindzey 1971).

Messick and Hornocker (1981) found young badgers consumed more arthropods and birds and fewer mammals than adults did. They felt the difference in diet might be partially accounted for by differences in movement of the relatively sedentary adults and dispersing juveniles and undeveloped predatory skills in young badgers. Foraging patterns undoubtedly differ among regions of the badger's range, as they are adjusted to provide efficiency under varying habitat and prey regimes. Behaviors involved in the search phase of foraging appear to be more general than the often specific behaviors exhibited in the capture of certain species. Badgers tracked in the snow moved from an old den or digging site to another, thoroughly investigating each (Lindzey 1971; Messick 1981). Observations and trapping success during snowless periods suggest that this is a common pattern of movement throughout the year. The value of such a foraging pattern is implied by the frequency with which these sites are used by prey species. Hetlet (1968) noted that badgers often returned to old den sites. Of 112 dens examined, 5.4% were occupied by badgers, 16% by Richardson's ground squirrels (*S. richardsonii*), and 1.8% by least chipmunks (*Eutamias minimus*). Species such as cottontail rabbits, which would generally be unavailable to badgers, occupy old dens and occur in the badger's diet. Snead and

TABLE 33.1. Mammalian prey of the North American badger and lifestyle of prey

Species of Prey	Lifestyle				Reference
	Burrowing	Burrow User	Terrestrial	Arboreal	
Lagomorpha					
Sylvilagus floridanus		X	X		Bailey 1931
S. audubonii		X			Goodrich 1994
S. nuttalli		X	X		Messick and Hornocker 1981
Lepus townsendii		X	X		Snead and Hendrickson 1942
L. californicus		X	X		Lindzey 1971
Rodentia					
Eutamias minimus		X	X		Lindzey 1971
E. amoenus		X	X		Broadbook 1970
Marmota monax	X		X		Snead and Hendrickson 1942
M. flaviventris	X		X		Verbeek 1965
Spermophilus townsendii	X		X		Hall 1946
S. elegans	X		X		Goodrich 1994
S. richardsonii	X		X		Seton 1929
S. armatus	X		X		Balph 1961
S. beldingi	X		X		Grinnell et al. 1937
S. columbianus	X		X		Manville 1959
S. tridecemlineatus	X		X		Seton 1929
S. franklinii	X		X		Errington 1937
S. beecheyi	X		X		Howell 1924
Ammospermophilus leucurus	X		X		Messick and Hornocker 1981
Cynomys ludovicianus	X		X		Bailey 1931
Tamiasciurus hudsonicus				X	Dearborn 1932
Thomomys talpoides	X				Criddle 1930
T. umbrinus	X				Grinnell et al. 1937
Geomys bursarius	X				Bailey 1888
Pappogeomys castanops	X				Baker 1956
Perognathus parvus	X		X		Lindzey 1971
Dipodomys sp.	X		X		Lindzey 1971
Onychomys leucogaster	X		X		Messick and Hornocker 1981
Reithrodontomys megalotis		X	X		Snead and Hendrickson 1942
Peromyscus (*maniculatus or leucopus*)		X	X		Dearborn 1932
Neotoma sp.			X		Lindzey 1971
Microtus (*pennsylvanicus or ochrogaster*)			X		Bailey 1931
M. montanus			X		Lindzey 1971
M. longicaudus			X		Goodrich 1994
Lagurus curtatus			X		Lindzey 1971
Synaptomys cooperi			X		Dearborn 1932
Mus musculus			X		Snead and Hendrickson 1942
Zapus hudsonius			X		Jense 1968
Carnivora					
Canis latrans			X		Young 1951
Mephitis mephitis		X	X		Bailey 1929
Spilogale putorius		X	X		Messick and Hornocker 1981

NOTE: The earliest record of predation for each species is cited from Lampe (1976). More recent records (since 1972) have been added to augment the original table.

Hendrickson (1942) discussed the increased vulnerability to badger predation of species using badger dens in Iowa.

Knopf and Balph (1969) described the selective predation by badgers on family groups of Unitah ground squirrels (*S. armatus*). Over a 30-day period, badgers excavated the burrows of seven family groups while not disturbing adjacent burrows that contained only a single squirrel. Although the disturbance caused by young around the burrow entrance gave it a unique appearance that was detectable by badgers, it seems likely that distinctive olfactory clues were present as well. Badgers captured the squirrels by plugging all but one entrance, then excavating the remaining entrance. Messick (1981) observed a similar pattern of entrance plugging when badgers preyed on Townsend's ground squirrels. Balph (1961) described underground concealment by badgers as a method of predation on Uinta ground squirrels. After plugging a second entrance, the badger apparently dug to a point close to the burrow entrance and waited for squirrels to enter the open entrance. Murie (1992) found that adult female Columbian ground squirrels (*S. columbianus*) generally survived an attack by a badger on the nest burrow, but fewer young eventually emerged from burrows dug by badgers than from undisturbed burrows. He concluded badgers took up to 56% of young produced by this eastern Washington ground squirrel population each year. Lampe (1976) felt that the technique exhibited by badgers to capture plains pocket gophers was used to avoid digging out the entire burrow. The technique entailed penetration of the burrow system at various points. A comparison of the olfactory clues at each point gained by running between points determined the approximate location of the gopher. Extensive digging was then begun to capture the gopher. Based on examination of 30 sites, Lampe determined that badgers were successful at capturing pocket gophers in 73% of attempts. He felt that badgers employed a similar technique to capture ground squirrels.

Despite limited olfactory or visual clues, badgers prey on hibernating mammals. The technique used to capture hibernating prey generally includes only a single excavation (Lampe 1976). Brandt (1994) observed a badger swimming to and investigating dense shoreline vegetation complexes. Although he did not observe the badger killing prey, he concluded it was searching for eggs or young of wetland birds nesting in emergent and shoreline vegetation.

MORTALITY

Predators. Deaths of badgers in many areas often are caused by humans, but deaths caused by other factors may easily go undetected. Larger predators, such as bears (*Ursus* spp.), coyotes (*Canis latrans*), wolves (*C. lupus*), and cougars (*Puma concolor*), occcasionally kill badgers. Messick (1981) listed the major causes of badger deaths in southwestern Idaho as automobiles, farmers, indiscriminate shooting, and fur trappers. Lindzey (1971) reported finding a male badger dead in a den after it had consumed meat from a horse carcass that had been treated with sodium fluoroacetate (compound 1080), a poison used at the time in coyote control programs. Deaths induced by other badgers may be greatest in the juvenile age class. Two of four young badgers found dead in a burrow were partially consumed, apparently by another badger (Lindzey 1971). Messick (1981) reported finding wounds on juvenile badgers that had apparently been inflicted by other badgers. Neither the frequency nor the influence of intraspecific killing in badger populations has been determined.

Parasites. The badger is host to numerous endoparasites including nematodes, trematodes, and cestodes (Table 33.2). Ectoparasites include ticks (*Dermacentor variabilis*: Ellis 1955; *Ixodes cookei*: Whitaker and Goff 1979; and *Amblyomma inornatum*: Gladney et al. 1977; Oliver and Osburn 1985), fleas (*Pulex irritans, Thrassis acamantis*, and *Echidnophaga gailinacea*: (Hubbard 1947; *Oropsylla arctomys*: Whitaker and Goff 1979), and mites (*Androlaelaps fahrenholzi, Hirstionyssus staffordi, Haemogamasus liponyssoides,* and *H. reidi*), and lice (*Neotrichodectes interruptofasciatus*) (Whitaker and Goff 1979). Badgers apparently are susceptible to both tularemia and rabies. Also, Hetlet (1968) found antibodies to *Yersinia pestis,* the infectious agent for plague, in sera of five of six badgers. He postulated that badgers may be instrumental in transmitting plague among rodent

TABLE 33.2. Helminth parasites from the badger in North America

Helminth	Geographic Location and Reference
Trematodes	
Alaria (*Paralaria*) *taxideae*	Minnesota (Swanson and Erickson 1946; Erickson 1946)
Euparyphium melis	North Dakota (Leiby et al. 1971)
Euparyphium sp.	Minnesota (Erickson 1946)
Cestodes	
Atriotaenia (*Ershovia*) *procyonis*	North Dakota (Leiby et al. 1971)
	Wyoming (Keppner 1969b)
	Kansas (Pence and Dowler 1979)
Mesocestoides carnivoricolus	Utah (Grundmann 1956, 1958)
Mesocestoides corti	Kansas and Texas (Pence and Dowler 1979)
Monordotaenia taxidiensis	Colorado (Leiby 1961)
	Montana (Skinker 1935)
	North Dakota (Pederson and Leiby 1969; Leiby et al. 1971)
	Wisconsin (Rausch 1947)
	Wyoming (Honess 1937; Keppner 1967)
Nematodes	
Ancylostoma caninum	Arizona (Hannum 1942)
Ancylostoma taxideae	Kansas (Kalkan and Hansen 1966)
	Texas (Pence and Dowler 1979)
	North Dakota (Leiby et al. 1971)
Angiocaulus gubernaculatus	California (Dougherty 1946)
Ascaris columnaris	Colorado (Leiby 1961)
	Minnesota (Erickson 1946)
	North Dakota (Leiby et al. 1971)
	Kansas and Texas (Pence and Dowler 1979)
Ascaris sp.	Wisconsin (Morgan 1943)
Filaria martis[a]	Kansas (Worley 1961)
	Mexico (Cabellero y C. 1948)
Filaria taxideae	North Dakota (Leiby et al. 1971)
	Wyoming (Keppner 1969a)
Molineus felineus	Utah (Grundmann 1957)
Molineus mustelae	Wyoming (Keppner 1969b)
Molineus patens	Minnesota (Erickson 1946)
	North Dakota (Leiby et al. 1971)
Molineus samueli	Texas (Platt and Pence 1981)
Monopetalonema? eremita	Wyoming (Leidy 1886)
Physaloptera maxillaris	Minnesota (Erickson 1946)
	Unknown (Morgan 1941a)
Physaloptera torquata	Arizona (Hannum 1942)
	California (Morgan 1942)
	Illinois (Morgan 1941b, 1942)
	Kansas and Texas (Pence and Dowler 1979)
	Minnesota (Erickson 1946)
	Montana (Ehlers 1931)
	North Dakota (Leiby et al. 1971)
	Pennsylvania (Leidy 1886; Walton 1927; Canavan 1931)
	Wisconsin (Morgan 1941b, 1942, 1943)
	Wyoming (Leidy 1886)
Trichinella spiralis	Wyoming or New York (Herman and Goss 1940)

SOURCE: Leiby et al. (1971), updated with selected references with citation date after 1972.

[a]Keppner (1969a) stated that nematodes reported as *F. martis* by Worley (1961) should be considered conspecific with *F. taxideae* and also questioned the identity of *F. martis* of Caballero y C. (1948).

colonies. Badgers exhibit only transient infections when exposed to plague, but show a measurable antibody response. Messick et al. (1983) used seralogical testing of badgers to monitor the dynamics of plague in Townsend's ground squirrel populations. Eighty-six percent of 294 badger sera tested in 1975 and 1976 in southwestern Idaho were positive; seropositives declined to 72% in 1977. Badgers also are susceptible to canine distemper virus, but can survive exposure. Badgers in wild populations tested seropositive to canine distemper virus. Also, badgers experimentally vaccinated with modified-live virus survived (Goodrich et al. 1994).

AGE ESTIMATION

Wright (1966) used dried eye-lens weights to separate adult from juvenile badgers, but later found (Wright 1969) that this separation could be made more easily on the basis of cranial suture coalescence. Nearly all cranial sutures are closed at 1 year of age. Juvenile and adult males are easily separated on the basis of testes and bacula weights. Even during the anestrous season, adult testes weighed more than those of juveniles (Wright 1969). Average weights of bacula from juveniles and adults, respectively, reported by Wright (1969) were 1.5 and 4.2 g, by Lindzey (1971) were 1.1 and 4.6 g, and by Messick and Hornocker (1981) were 1.3 and 3.5–5.4 g.

The presence of bands in the cementum layer of canines and in the mandible was originally reported by Wright (1969). Lindzey (1971) and Crowe and Strickland (1975) investigated the relation between these bands and age and concluded they could be used to determine age of badgers. Because the first annulus is not deposited until the badger's second summer or fall, the count of annuli must be increased by 1 to arrive at age in years. Messick and Hornocker (1981) suggested the first dark-staining band was actually formed during the badger's first winter, but became obvious only when the outer, lighter-staining band widens later in the year. Frederickson (1983) used radiographs of teeth to separate juvenile from older badgers based on pulp cavity closure before teeth were processed to determine age in years by cementum layer counts. The amount of wear caused by the upper and lower canines rubbing on each other (Long 1965) and the presence of dental abnormalities most frequent in older-age animals (Long and Long 1965) may discriminate between young and old animals.

RESEARCH AND MANAGEMENT NEEDS

Badgers have never played an important role in the fur trade, but because they are relatively easy to trap and are frequently captured incidentally in traps set for other furbearers, harvests should be monitored as demands for other native furs and trapping pressures change. Numbers of badger pelts sold reached an all-time high in the early 1980s, with about 33,400 sold annually in North America (Obbard et al. 1987), apparently reflecting the demand for other long-haired furs (Messick 1987). Badgers are fully protected in some states and provinces and given no protection in others or managed as furbearers. Most states responding to a questionnaire reported badgers as abundant and their populations stable (Long and Killingley 1983).

Problems caused by badgers primarily result from its digging. Dens pose a threat to livestock, and excavations can lead to the loss of water from earthen irrigation structures and make mowing and plowing of fields more difficult. Badgers only occasionally prey on livestock and poultry. Partially buried fences, habitat modifications, and removal of offending individuals have been used to address problems caused by badgers (Minta and Marsh 1988; Lindzey 1994).

Changing land-use patterns undoubtedly will influence badger populations on a local basis. Clearing of timbered areas in the eastern part of its geographic range may prove beneficial to the badger, whereas clearing of sagebrush (*Artemisia tridentata*) in the cold desert region for dryland farming or cattle grazing may prove detrimental. Both the species composition of small rodents and biomass available to the badger will change as habitats are altered. Badgers may profit from renewed interest in prairie ecosystems. Renewal of interest in these communities, however, has been stimulated by concerns for a number of sensitive species such as black-footed ferret (*Mustela nigripies*), snowy plover (*Charadrius alexandrinus*), burrowing owl (*Speotyto cunicularia*), swift fox (*Vulpes velox*), and prairie dog (*Cynomys* spp.), each of which interacts with the badger. Managing these species and modeling the complex dynamics in prairie ecosystems will require an understanding of the badger's role as predator, competitor, and disease vector and the contribution of its burrows to the life history of many grassland birds and mammals.

LITERATURE CITED

Bailey, V. 1888. Report on some of the results of a trip through parts of Minnesota and Dakota. Pages 426–454 *in* Report to the Commissioner of Agriculture. Washington, DC.

Bailey, V. 1929. Mammals of Sherburne County, Minnesota. Journal of Mammalogy 10:153–64.

Bailey, V. 1931. Mammals of New Mexico. North American Fauna 53:1–412

Baker, R. H. 1956. Mammals of Coahuila, Mexico. University of Kansas Museum of Natural History Publication 9:125–335.

Balph, D. F. 1961. Underground concealment as a method of predation. Journal of Mammalogy 42:423–24.

Brandt, D. A. 1994. Overwater foraging by a badger. Prairie Naturalist 26:171.

Broadbook, H. E. 1970. Populations of the yellow-pine chipmunk, *Eutamias amoenus*. American Midland Naturalist 83:472–88.

Caballero y C., E. 1948. *Filaria martis* Gmelin, 1790 en mamiferos de Neuvo Leon y consideraciones sobre las especies del genero *Filaria* Miller, 1787. Revista de la Sociedad Mexicana de Historia Natural 9:257–61.

Campbell, T. M., and T. W. Clark. 1983. Observation of badger copulatory and agonistic behavior. Southwestern Naturalist 28:107–8.

Canavan, W. P. N. 1931. Nematode parasites of vertebrates in the Philadelphia Zoological Gardens and vicinity. Journal of Parasitology 23:196–229.

Criddle, S. 1930. The prairie pocket gopher, *Thomomys talpoides*. Journal of Mammalogy 11:265–80.

Crowe, D. M., and M. D. Strickland. 1975. Dental annulation in the American badger. Journal of Mammalogy 56:269–72.

Dearborn, N. 1932. Foods of some predatory fur-bearing animals in Michigan (Conservation Bulletin No. 1). University of Michigan School of Forestry.

Dougherty, E. C. 1946. The genus *Aclurostrongylus* Cameron, 1927 (Nematoda: Metastrongylidae), and its relatives; with descriptions of *Parafilaroides*, gen. nov., and *Angiostrongylus gubernaculatus*, sp. nov. Proceedings of the Helminthological Society of Washington 13:16–25.

Ehlers, G. H. 1931. The anthelmintic treatment of infestations of the badger with spirurids *(Physaloptera* sp.). Journal of the American Veterinary Medical Association 31:79–87.

Ellis, L. L., Jr. 1955. A survey of ectoparasites of certain mammals in Oklahoma. Ecology 36:12–18.

Erickson, A. B. 1946. Incidence of worm parasites in Minnesota Mustelidae and host lists and keys to North American species. American Midland Naturalist 36:494–509.

Errington, P. L. 1937. Summer food habits of the badger in northwestern Iowa. Journal of Mammalogy 18:213–16.

Frederickson, L. F. 1983. Use of radiographs to age badger and striped skunk. Wildlife Society Bulletin 11:297–99.

Gladney, W. J., C. C. Dawkins, and M. A. Price. 1977. *Amblyomma inornatum* (Acarina: Ioxidae): Natural hosts and laboratory biology. Journal of Medical Entomology 14:85–88.

Goodrich, J. M. 1994. North American badgers (*Taxidea taxus*) and black-footed ferrets (*Mustela nigripes*): Abundance, rarity and conservation in a white-tailed prairie dog (*Cynomys leucurus*)-based community. Ph.D. Dissertation, University of Wyoming, Laramie.

Goodrich, J. M., and S. W. Buskirk. 1998. Spacing and ecology of North American badgers (*Taxidea taxus*) in a prairie dog (*Cynomys leucurus*) complex. Journal of Mammalogy 79:171–79.

Goodrich, J. M., E. S. Williams, and S. W. Buskirk. 1994. Effects of a modified live virus canine distemper vaccine on captive badgers (*Taxidea taxus*). Journal of Wildlife Diseases 30:492–96.

Gremillion-Smith, C. 1985. Range extension of the badger (*Taxidea taxus*) in southern Illinois. Transactions of the Illinois State Academy of Science 78:111–14.

Grinnell, J., J. S. Dixon, and J. M. Linsdale. 1937. Furbearing mammals of California. University of California Press, Berkeley.

Grundmann, A. W. 1956. A new tapeworm, *Mesocestoides carnivoricolus*, from carnivores of the Great Salt Lake Desert region of Utah. Proceedings of the Helminthological Society of Washington 23:26–28.

Grundmann, A. W. 1957. Nematode parasites of mammals of the Great Salt Lake Desert region of Utah. Journal of Parasitology 43:105–12.

Grundmann, A. W. 1958. Cestodes of mammals from the Great Salt Lake Desert region of Utah. Journal of Parasitology 44:425–29.

Hall, E. R. 1946. Mammals of Nevada. University of California Press, Berkeley.

Hall, E. R. 1981. The mammals of North America. John Wiley, New York.

Hamlett, G. W. 1932. Observations on the embryology of the badger. Anatomy Records 53:283–301.

Hannum, C. A. 1942. Nematode parasites of Arizona vertebrates. Ph.D. Dissertation, University of Washington, Seattle.

Harlow, H. J. 1981a. Effect of fasting on rate of food passage and assimilation efficiencies in badgers. Journal of Mammalogy 62:173–77.

Harlow, H. J. 1981b. Torpor and other physiological adaptations of the badger (*Taxidea taxus*) to cold environments. Physiological Zoology 54:267–75.

Harlow, H. J. 1981c. Metabolic adaptations to prolonged food deprivation by the American badger *Taxidea taxus*. Physiological Zoology 54:276–84.

Harlow, H. J., B. Miller, T. Ryder, and L. Ryder. 1985. Energy requirements for gestation and lactation in a delayed implanter, the American badger. Comparative Biochemical Physiology 82A:885–89.

Herman, C. M., and L. J. Goss. 1940. Trichinosis in an American badger, *Taxidea taxus taxus*. Journal of Parasitology 26:157.

Hetlet, L. A. 1968. Observations on a group of badgers in South Park, Colorado. M.S. Thesis, Colorado State University, Fort Collins.

Honess, R. F. 1937. Un nouveau cestode: *Fossor angertrudae* n. g., n. sp. du bureau d'Amerique *Taxidea taxus taxus* (Schreber 1778). Annals of Parasitology 16:363–66.

Hornocker, M. G., J. P. Messick, and W. E. Melquist. 1983. Spatial strategies in three species of Mustelidae. Acta Zoologica Fennica 174:185–88.

Howell, A. B. 1924. The mammals of Mammoth, Mono County, California. Journal of Mammalogy 5:25–36.

Hubbard, C. A. 1947. Fleas of western North America. Iowa State College Press, Ames.

Jackson, H. H. T. 1961. Mammals of Wisconsin. University of Wisconsin Press, Madison.

Jense, G. K. 1968. Food habits and energy utilization of badgers. M.S. Thesis, South Dakota State University, Brookings.

Kalkan, A., and Hansen, M. F. 1966. *Ancylostoma taxideae* sp. n. from the American badger, *Taxidea taxus taxus*. Journal of Parasitology 52:291–94.

Keppner, E. J. 1967. *Fossor taxidiensis* (Skinker, 1935) n. comb. with a note on the genus *Fossor* Honness, 1937 (Cestoda: Taeniidae). Transactions of the American Microscopical Society 86:157–58.

Keppner, E. J. 1969a. *Filaria taxidea* n. sp. (Filarioidea: Filariidae) from the badger, *Taxidea taxus taxus* from Wyoming. Transactions of the American Microscopical Society 88:581–88.

Keppner, E. J. 1969b. Occurrence of *Atriotaenia procyonis* and *Miloncus mustelae* in the badger, *Taxidea taxus* (Schreber, 1778), in Wyoming. Journal of Parasitology 55:1161.

Knopf, F. L., and D. F. Balph. 1969. Badgers plug burrows to confine prey. Journal of Mammalogy 50:635–36.

Lampe, R. P. 1976. Aspects of the predatory strategy of the North American badger (*Taxidea taxus*). Ph.D. Dissertation, University of Minnesota, St. Paul.

Lampe, R. P. 1982. Food habits of badgers in east central Minnesota. Journal of Wildlife Management 46:790–95.

Lampe, R. P., and M. A. Sovada. 1981. Seasonal variation in home range of a female badger (*Taxidea taxus*). Prairie Naturalist 13:55–58.

Leiby, P. D. 1961. Intestinal helminths of some Colorado mammals. Journal of Parasitology 47:311.

Leiby, P. D., P. J. Sitzmann, and D. C. Kritsky. 1971. Studies on helminths of North Dakota. Parasites of the badger *Taxidea taxus* (Schreber). Proceedings of the Helminthological Society of Washington 38:225–28.

Leidy, J. 1886. Notices of nematoid worms. Proceedings of the Philadelphia Academy of Natural Science 38:308–13.

Lindzey, F. G. 1971. Ecology of badgers in Curlew Valley, Utah and Idaho with emphasis on movement and activity patterns. M.S. Thesis, Utah State University, Logan.

Lindzey, F. G. 1976. Characteristics of the natal den of the badger. Northwest Science 50:178–80.

Lindzey, F. G. 1978. Movement patterns of badgers in northwestern Utah. Journal of Wildlife Management 42:418–22.

Lindzey, F. G. 1994. Badgers. Pages C1–C3 *in* S. E. Hygnstrom, R. M. Timm, and G. E. Larson, eds. Prevention and control of wildlife damage. University of Nebraska, Lincoln.

Long, C. A. 1965. Functional aspects of the jaw-articulation in the North American badger, with comments on adaptiveness of tooth wear. Transactions of the Kansas Academy of Science 68:156–62.

Long, C. A. 1973. *Taxidea taxus*. Mammalian Species 26:1–4.

Long, C. A., and C. A. Killingley. 1983. The badgers of the world. Charles C. Thomas, Springfield, IL.

Long, C. A., and C. F. Long. 1965. Dental abnormalities in North American badgers, genus *Taxidea*. Transactions of the Kansas Academy of Science 68:145–54.

Manville, R. H. 1959. The Columbian ground squirrel in northwestern Montana. Journal of Mammalogy 40:26–45.

Messick, J. P. 1981. Ecology of the badger in southwestern Idaho. Ph.D. Dissertation, University of Idaho, Moscow.

Messick, J. P. 1987. North American badger. Pages 586–97 *in* M. Novak, J. A. Baker, M. E. Obbard, and B. Malloch, eds. Wild furbearer management and conservation in North America. Ontario Trappers Association and Ontario Ministry of Natural Resources, Toronto.

Messick, J. P., and M. G. Hornocker. 1981. Ecology of the badger in southwestern Idaho Wildlife Monograph 76:1–53.

Messick, J. P., G. W. Smith, and A. M. Barnes., 1983. Serological testing of badgers to monitor plague in Southwestern Idaho. Journal of Wildlife Diseases 19:1–6.

Minta, S. C. 1990. The badger, *Taxidea taxus* (Carnivora: Mustelidae): Spatial–temporal analysis, dimorphic territorial polygyny, population characteristics, and human influences on ecology. Ph.D. Dissertation, University of California, Davis.

Minta, S. C. 1993. Sexual differences in spatio-temporal interaction among badgers. Oecologia 96:402–9.

Minta, S., and M. Mangel. 1989. A simple population estimate based on simulation for capture–recapture and capture–resight data. Ecology 70:1738–51.

Minta, S. C., and R. E. Marsh. 1988. Badgers (*Taxidea taxus*) as occasional pests in agriculture. Pages 199–208 *in* A. C. Crabb and R. E. Marsh, eds. Proceedings of thirteenth vertebrate pest conference. Monterey, CA.

Morgan, B. B. 1941a. A summary of the Physalopterinae (Nematoda) of North America. Proceedings of the Helminthological Society of Washington 8:28–30.

Morgan, B. B. 1941b. Additional notes on North American Physalopterinae (Nematoda). Proceedings of the Helminthological Society of Washington 8:63–64.

Morgan, B. B. 1942. The Physalopterinae (Nematoda) of North American vertebrates. Ph.D. Dissertation, University of Wisconsin, Madison.

Morgan, B. B. 1943. New host records of nematodes from Mustelidae (Carnivora). Journal of Parasitology 29:158–59.

Murie, O. J. 1992. Predation by badgers on Columbian ground squirrels. Journal of Mammalogy 73:385–94.

Nugent, R. F., and J. R. Choate. 1970. Eastward dispersal of the badger, *Taxidea taxus*, into the northeastern United States. Journal of Mammalogy 51:626–27.

Obbard, M. E., J. G. Jones, R. Newman, A. Booth, A. J. Satterthwaite, and G. Linscombe. 1987. Furbearer harvest in North America. Pages 1007–34 *in* M. Novak, J. A. Baker, M. E. Obbard, and B. Malloch, eds. Wild furbearer management and conservation in North America. Ontario Trappers Association and Ontario Ministry of Natural Resources, Toronto.

Oliver, J. H., Jr., and R. L. Osburn. 1985. Cytogenics of ticks (Acaria: Ixodoidea): Chromosomes and timing of spermatogenesis in *Amblyomma inornatum*. Journal of Parasitology 71:124–26.

Palmer, R. S. 1954. The mammal guide. Doubleday, Garden City, NY.

Pederson, E. D., and P. D. Leiby. 1969. Studies on the biology of *Monordotaenia taxidiensis*, a taeniid cestode of the badger. Journal of Parasitology 55:759–65.

Pence, D. B., and R. C. Dowler. 1979. Helminth parasitism in the badger, *Taxidea taxus* (Schreber, 1778), from the western Great Plains. Proceedings of the Helminthological Society of Washington 46:245–53.

Petrides, G. A. 1950. The determination of sex and age ratios in fur animals. American Midland Naturalist 43:355–82.

Platt, T. R., and D. B. Pence. 1981. *Molineus samueli* n. sp. (Nematoda: Trichostrongyloidea: Molineidae) from the badger, *Taxidea taxus*. Proceedings of the Helminthological Society of Washinton 48:148–53.

Rausch, R. 1947. A redescription of *Taenia taxidiensis* Skinker, 1935. Proceedings of the Helminthological Society of Washington 14:73–75.

Sargeant, A. B., and D. W. Warner. 1972. Movements and denning habits of a badger. Journal of Mammalogy 53:207–10.

Seton, E. T. 1929. Lives of game animals. Charles Branford, Boston.

Skinker, M. S. 1935. Two new species of tapeworms from carnivores and a

redescription of *Taenia laticollis* Rudolphi, 1819. Proceedings of the U.S. National Museum 83:211–20.
Snead, E., and G. O. Hendrickson. 1942. Food habits of the badger in Iowa. Journal of Mammalogy 23:380–91.
Swanson, G., and A. B. Erickson. 1946. *Alaria taxideae n.* sp., from the badger and other mustelids. Journal of Parasitology 32:17–19.
Verbeek, N. A. 1965. Predation by badger on yellow-bellied marmots in Wyoming. Journal of Mammalogy 46:506.
Ver Steeg, B., and R. E. Warner. 1995. American badgers in Illinois. Pages 108–10 *in* E. T. LaRoe, G. S. Farris, C. E. Puckett, P. S. Doran, and M. J. Mac, eds. Our living resources: A report to the nation on the distribution, abundance, and health of U.S. plants, animals, and ecosystems. U.S. Department of the Interior, National Biological Service, Washington, DC.
Walton, A. 1927. A revision of the nematodes of the Leidy collection. Proceedings of the Philadelphia Academy of Natural Science 79:49–63.
Whitaker, J. O., and R. Goff. 1979. Ectoparasites of wild carnivora of Indiana. Journal of Medical Entomology 15:425–30.
Whitaker, J. O., Jr., and J. R. Gammon. 1988. Endangered and threatened vertebrate animals of Indiana, their distribution and abundance. Indiana Academy of Science, Indianapolis.
Worley, D. E. 1961. The occurrence of *Filaria martis* Gmelin, 1790, in the striped skunk and badger in Kansas. Journal of Parasitology 47:9–11.
Wright, P. L. 1966. Observations on the reproductive cycle of the American badger (*Taxidea taxus*). Pages 37–45 *in* Comparative biology of reproduction in mammals (Symposia of the Zoological Society of London No. 15). Zoological Society, London.
Wright, P. L. 1969. The reproductive cycle of the male American badger (*Taxidea taxus*). Journal of Reproductive Fertility (Supplement) 6:435–45.
Young, S. P. 1951. The clever coyote. Stackpole, Harrisburg, PA.

Frederick G. Lindzey, U.S. Geological Survey, Wyoming Cooperative Fish and Wildlife Research Unit, University of Wyoming, Laramie, Wyoming 82071-3166. Email: flindzey@uwyo.edu.

34

Skunks

Genera *Mephitis, Spilogale,* and *Conepatus*

Rick Rosatte
Serge Larivière

Skunks are members of the order Carnivora and until recently were considered to be members of the family Mustelidae. However, genetic evidence using DNA sequencing techniques suggests that skunks represent a family distinct from other mustelids, and they were placed in a newly created family called Mephitidae (Dragoo and Honeycutt 1997). For the purposes of this chapter, we recognize the family Mephitidae and subfamily Mephitinae.

Skunks are taxonomically subdivided into three genera and 13 species. Their range is restricted to the New World from southern Canada to the Strait of Magellan in South America (Hall and Kelson 1959; Honacki et al. 1982; Rosatte 1987). The genus *Mephitis* is composed of two species, including the striped skunk (*Mephitis mephitis*) and the hooded skunk (*Mephitis macroura*). Spotted skunks (*Spilogale* spp.) and hog-nosed skunks (*Conepatus* spp.) form the other two groups of skunks in North America (Walker 1964; Rosatte 1987; Wilson and Reeder 1993; Dragoo and Honeycutt 1997). The striped skunk will be discussed first, followed by hooded, spotted, and hog-nosed skunks.

NOMENCLATURE

Common Names. Striped skunk, skunk, large striped skunk
Scientific Name. *Mephitis mephitis*

Mephitis means "bad odor," in reference to the pungent odor that is associated with the scent glands of skunks; *skunk* is one of the few Algonquin Indian words to enter the English language,

Subspecies. Thirteen subspecies of striped skunk are recognized in North America. All differ in weight and size according to the geographic location where they occur (Hall and Kelson 1959; Rue 1981); they include *M. m. avia, M. m. elongata, M. m. estor, M. m. holzneri, M. m. hudsonica, M. m. major, M. m. mephitis, M. m. mesomelas, M. m. nigra, M. m. notata, M. m. occidentalis, M. m. spissigrada,* and *M. m. varians*

DISTRIBUTION

The range of striped skunks includes southern Canada from Nova Scotia to British Columbia, including the Hudson Bay area, most of the United States, and portions of northern Mexico (Walker 1964; Godin 1982; Honacki et al. 1982; Rosatte 1987) (Fig. 34.1). They are absent from desert areas of the southwestern United States except in major river corridors. They occur at elevations to 1800 m, but seldom above 4000 m (Rue 1981). Skunks increased their distribution in conjunction with the cutting of forests throughout North America.

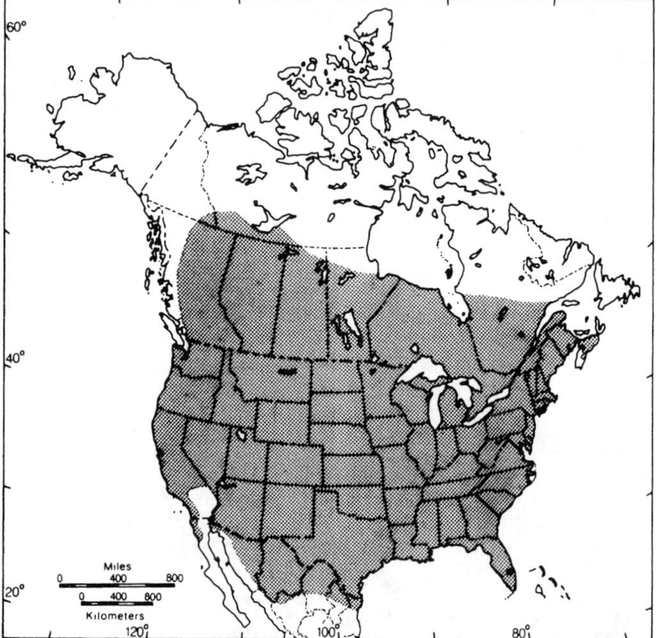

Figure 34.1. Distribution of the striped skunk (*Mephitis mephitis*).

DESCRIPTION

Unique characteristics of the striped skunk include a black-and-white-striped pelage (Fig. 34.2) and an odorous musk. The striped skunk is a medium-sized carnivore with a triangular shaped head, small rounded ears, black eyes, and a somewhat bulbous nose pad. The feet are plantigrade with bare soles, and the forefeet each have five long, curved claws for digging. The hind feet have claws that are shorter and straighter than those on the front feet. Female striped skunks have an average of 12 mammae (range = 10–14). The tail is long and somewhat bushy. They have a keen sense of smell and excellent hearing; however, their poor eyesight is limited to a distance of 6–7 m (Rue 1981; Godin 1982; Rosatte 1987; Nowak 1999).

The striped skunk, as with all skunks, has two well-developed scent glands, one on each side of the anus. Skunks use sphincter muscles to compress the glands and can eject musk several meters through papilla that can be protruded through the anus. The musk is an oily, yellow sulfur-alcohol compound known as butylmercaptan and it contains sulfuric acid. Each musk gland contains about 15 ml of scent (Godin 1982; Rosatte 1987). The musk can cause an intense burning sensation when sprayed into the eyes. However, it does not cause permanent damage (Dice 1921; Rue 1981).

Sexual Dimorphism. Adult male striped skunks are about 10% larger than adult females; the total body length of both sexes is between 520 and 770 mm. Tail lengths vary from 170 to 400 mm and the hind feet are about 55–85 mm (Verts 1967; Godin 1982). However, there is much geographic variation in total body and tail length (Table 34.1). Neck circumferences for striped skunks average 11–17 cm depending on season and sex and age of the skunk (Rosatte et al. 1991). Skunks have long claws for digging (front, 20–30 mm; rear, 10–15 mm). Adult skunks weigh 1.8–4.5 kg, but some may be as heavy as 5.5 kg, and there is much variation on a regional basis (Table 34.1). Mean weights of adult male and female skunks in Illinois were 2.6 and 2.0 kg, respectively (Verts 1967; Rue 1981; Rosatte 1987). In an urban area of Toronto, skunks

FIGURE 34.2. Striped skunk (*Mephitis mephitis*). SOURCE: Photo by R. Rosatte.

averaged 3.0–4.5 kg (Rosatte et al. 1991). In the prairie parklands of Saskatchewan, spring weight of males and females was 2.6 and 2.1 kg, respectively (Larivière and Messier 1996a).

In many northern localities, skunks den during periods of inclement weather and as a result often experience extreme weight loss. In a New York study, males lost 13.8% and females 38% of their summer weight (Hamilton 1937). In Minnesota, skunks lost 55–65% of their fall body weight during the winter (Houseknecht 1969).

Pelage. There is great variation in color patterns among striped skunks; however, the pelt color is generally black with a thin white stripe running from the nose to the back of the forehead. A broad white stripe extends from the crown of the head and usually branches at the shoulders with continuation of the striping along the sides to the rump and onto the tail (Walker 1964). However, some skunks have a white chest patch, whereas others have a white stripe on the outside of the front legs (Godin 1982). The contrasting black and white pelage pattern of both sexes may serve as a warning to potential predators. The pelt of the striped skunk consists of underfur 25–31 mm in length, intermixed with guard hairs, which are usually 38–76 mm long (Verts 1967).

Skunks usually molt once a year. In areas such as Illinois, it begins in April with the shedding of large patches of underfur from the anterior to the posterior of the body. By July, underfur and guard hairs are being replaced, and the molt is completed by early September. Juvenile skunks do not molt until they are about 1 year old (Verts 1967).

Skull and Dentition. Skull measurements of adult striped skunks are variable. Condylobasal lengths range from 70 to 88 mm (Verts 1967). The dental formula is I 3/3, C 1/1, P 3/3, M1/2 (Fig. 34.3).

REPRODUCTION AND DEVELOPMENT

Breeding Season. The majority of breeding for striped skunks occurs during the period from mid-February to mid-April. Breeding occurs later at higher latitudes. Normally, skunks are monestrus, breeding only once a year (Davis 1960; Verts 1967; Banfield 1974; Greenwood and Sargeant 1998). Females that have not bred during their first estrus may have a second cycle about 1 month later, and some skunks may produce two litters/year (Seton 1929). Pregnancy rates of 92–96% have been reported for striped skunks in areas such as Illinois, Ohio, Alberta, and Saskatchewan (Bailey 1931; Verts 1967; Schowalter and Gunson 1982). Confirming these figures, Bjorge et al. (1981) noted that 88% of a sample of adult female skunks in Alberta were lactating or had suckled young.

Anatomy and Physiology. The reproductive organs of females consist of a vagina, a urogenital sinus, a uterus, paired oviducts, ovaries, and uterine horns. The vagina and urethra form the urogenital sinus, which opens to the exterior (Verts 1967; Godin 1982). In late February, ovarian follicles enlarge coincident with estrus. The estrous period usually lasts 9–10 days. Ovulation is coitus induced and occurs 42 hr after insemination (Wade-Smith and Richmond 1978a, 1978b; Rosatte 1987).

Reproductive organs of male striped skunks consist of a penis, a urogenital sinus, a prostate gland, paired *vas deferens,* epididymides, an *os baculum,* and scrotal testes (Verts 1967; Godin 1982). Seasonal reproductive changes are expressed as weight changes in the testes. In late January, the testes begin to enlarge. Males begin searching for females at winter denning sites, sometimes traveling 4 km/night (Rue 1981). Maximum testes size occurs during March; minimum weights occur during August (Verts 1967). Testes weights and the presence of sperm suggested that juvenile skunks were sexually active during their first year of life in a study in Saskatchewan and Alberta (Schowalter and Gunson 1982).

Mating. Male striped skunks are polygamous and may attempt to breed with several females in succession. Normally, males approach an estrous female from the rear, and smell and lick the vulva area. The male will then grasp the female by the nape of the neck with his teeth prior to mounting and copulating (Wight 1931; Verts 1967; Godin 1982). Rutting activities of males last about 36 days during January–March (Ernst 1965). Male skunks will often defend their harem of females against intrusion by other males. Defense is accomplished by hitting another skunk with their shoulders or biting the legs (Rue 1981; Rosatte 1987). Once a female has been successfully bred, she will not allow subsequent breeding to occur and will viciously fight any male attempting to do so. After breeding, male skunks usually remain solitary and attempt to rebuild fat reserves, and females defend their maternity den (Verts 1967; Rosatte 1987; Larivière and Messier 1998a).

Gestation and Parturition. Gestation for striped skunks varies from 59 to 77 days, indicative of a delay in implantation of embryos (Wight 1931; Shadle 1953; Wade-Smith et al. 1980). The young are usually born about mid-May to early June (Shadle 1953; Verts 1967; Godin 1982; Larivière and Messier 1997a).

Litter Sizes. Litter sizes for striped skunks range between 2 and 10 young. Mean litter size is about 5–7 (see Table 34.1), but one litter of 18 was reported in Pennsylvania (Banfield 1974; Lowery 1974; Hall and Kelson 1959; Hall 1981; Schowalter and Gunson 1982). Second litters have been reported for striped skunks (Shadle 1953; Parks 1967).

Growth, Development, and Longevity. Young skunks are born with eyes closed and little body hair, and weigh about 25–40 g. The eyes

TABLE 34.1. Range of measurements, weights, and litter sizes of striped, hooded, spotted, and hog-nosed skunks

Group	Total Body Length (mm)	Weight (kg)	Litter Size	References
Striped skunk	598–719	1.8–4.5	5–9	Verts 1967; Wade-Smith and Richmond 1978a, 1978b; Wade-Smith et al. 1980; Rue 1981; Rosatte 1987
Hooded skunk	558–790	1–2	3–6	Bailey 1931; Davis 1951; Hall and Kelson 1959; Patton 1974; Rosatte 1987
Spotted skunk	250–688	0.5–1.8	1–6	Crabb 1943; Hall and Kelson 1959; Walker 1964; Patton 1974; Rosatte 1987
Hog-nosed skunk	450–900	1.5–4.5	2–5	Davis 1951; Hall and Kelson 1959; Walker 1964; Patton 1974; Rosatte 1987

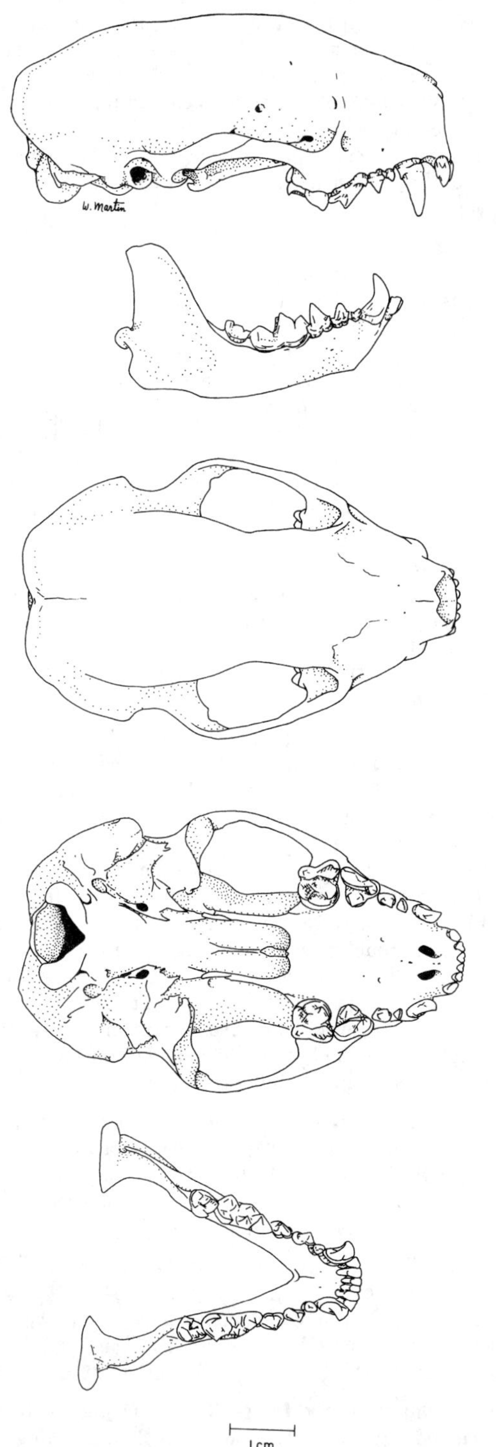

FIGURE 34.3. Skull of the striped skunk (*Mephitis mephitis*). From top to bottom: lateral view of cranium, lateral view of mandible, dorsal view of cranium, ventral view of cranium, dorsal view of mandible.

open and they are capable of hearing at about 2–4 weeks of age, at which time they are also capable of discharging musk from the scent glands (Verts 1967). Young are weaned at about 6–8 weeks of age and are gradually introduced to hunting and foraging by the mother (Hamilton and Whitaker 1979). A family unit consists of an adult female and her litter. Young skunks become independent of the adult female at 2–3 months of age and often disperse from their natal range during July–September (Burns 1953; Bjorge 1977; Andersen 1981; Larivière and Messier 1997a).

Skunks are capable of breeding the spring following their birth and may live as long as 10 years in captivity. However, life span of skunks in the wild is usually 2–3.5 years (Linduska 1947; Schwartz and Schwartz 1959; Verts 1967).

ECOLOGY

Striped skunks occur throughout North America from sea level to elevations of more than 4000 m in a variety of habitats including farmland, grasslands, woodlot edges, ravines, rocky outcrops, fencelines, refuse piles, and urban habitats (Nelson 1918; Grinnell et al. 1937; Godin 1982; Rosatte 1987). They prefer open forest-edge zones and seem to be most abundant on agricultural lands where there is a good supply of food and cover for denning (Walker 1964; Rosatte 1984a, 1984b). In cities, skunks frequent residential areas, fields adjacent to industrial sites, vacant buildings, ground burrows, and dumps (Rosatte 1986). In prairies, skunks use wetlands and woodlands as primary foraging habitats, whereas wetlands and farmsteads are used as primary denning habitats (Larivière and Messier 1998b, 2000).

Types of Dens. Striped skunks use a variety of den types during the year, including maternal dens, summer resting sites, and winter solitary and communal dens (Andersen 1981; Rosatte 1986; Larivière and Messier 1998a). Normally, in northern climates, winter dens are used from November to March and maternal dens and resting sites from April to October (Houseknecht 1971; Storm 1972). Den type varies according to the time of year, habitat, sex and age of the skunks, as well as the social behavior and condition of the animal (Storm 1972). Striped skunks den solitarily or communally in groups of two or more related or unrelated animals. Adult females raise their young in a communal den; however, male skunks may use aboveground resting sites during late spring and summer. Maternal dens may be under buildings such as granaries or in excavated burrows (Allen and Shapton 1942; Verts 1967; Houseknecht 1971; Rosatte 1984a; Lariviere and Messier 1998a, 1999). Aboveground summer resting sites include croplands, fence rows, waterways, and under buildings (Scott and Selko 1939; Storm 1972; Rosatte 1987; Larivière and Messier 1998a, 1998b). However, skunks tend not to den or rest in open areas after mid-September (Verts 1967).

In the late fall and winter both sexes usually use underground communal dens during periods of inclement weather. Winter dens may be used from 75 to 100 days (Houseknecht 1969). These den sites include buildings, burrows, rock piles, and culverts (Selko 1938; Patton 1974; Rosatte 1984a, 1987). In urban areas, skunks will use any protective area including under porches, buildings, commercial sites, and culverts (Rosatte 1986).

Skunks may dig their own burrows as dens or use vacant burrows of other animals including badgers (*Taxidea taxus*), woodchucks (*Marmota monax*), red foxes (*Vulpes vulpes*), and muskrats (*Ondatra zibethicus*) (Selko 1938; Allan and Shapton 1942; Jones 1950). Burrows dug by skunks are usually 2–6 m long and may end in one to three chambers lined with material such as leaves, hay, or grass. Dens are often constructed on ground with slopes of 5–10° or greater, probably for drainage considerations (Verts 1967). Ground burrows may have one to five entrances about 20 cm in diameter (Selko 1938; Allan and Shapton 1942; Verts 1967; Godin 1982).

Home Range and Movements. Home ranges are highly variable for skunks, depending on geographic area, habitat, sex, and season. In rural habitats of Minnesota, Tennessee, Illinois, North Dakota, Ontario, and Alberta, average sizes ranged between 1.2 and 4.9 km^2 (Storm 1972; Bjorge 1977; Rosatte and Gunson 1984a; Greenwood et al. 1997; Bixler and Gittleman 2000). Estimates for the prairie region of Saskatchewan were about 12 km^2 for males and 4 km^2 for females (Larivière and Messier 1998a). Home range estimates for skunks in urban Toronto were exceptionally small, averaging 0.51–0.64 km^2 (Rosatte 1986; Rosatte et al. 1991). Small home ranges were due to an abundance of food and denning sites. In rural habitats of southern Ontario, skunk home range varied between 1 and 3 km^2.

Striped skunks are fairly sedentary animals, but some individuals may be quite mobile (Rosatte and Gunson 1984a; Rosatte et al. 1991, 1992). Andersen (1981) reported extensive juvenile movements of 70 km in Alberta and Saskatchewan. In prairie habitat, Sargeant et al. (1982) reported an adult male moved 119 km. However, those movements are exceptional. Generally skunk movements are 1–3 km, but depend on season and geographic location (Upham 1967; Rosatte and Gunson 1984a; Larivière and Messier 1997a, 1998a). In urban Toronto, the majority of skunk movements were <1 km annually (Rosatte et al. 1991, 1992).

During the breeding period, skunks may move 0.5–2.4 km. (Allan and Shapton 1942). However, in one study, pregnant females remained within 0.4 km of their parturition den (Verts 1967). Movements of adult and juvenile males may exceed female movements during the late summer and autumn in some habitats (Storm 1972). In other areas, such as Alberta, juvenile females moved farthest (Andersen 1981; Bjorge et al. 1981). Conversely, Rosatte and Gunson (1984a) found no differences in movements between males and females in Alberta.

Generally, skunks move very little during periods of inclement weather in northern latitudes, preferring to conserve energy in winter dens. This period is normally from November or December to March, but is dependent on weather (Verts 1967; Gunson and Bjorge 1979). However, male skunks may venture from their dens during late winter in search of females during breeding activities (Rosatte 1987). In southern climates such as Texas, skunks remain active throughout the year (Davis 1951; Verts 1967).

Population Density. Densities for striped skunks are highly variable and depend on habitat, season, geographic location, mortality rate, quantity and quality of forage, and other factors. Skunk densities are generally lowest during the spring because of the impact of winter mortality. Density estimates are also lowest after rabies epidemics. Populations tend to be highest in the early summer as juvenile skunks become active. Therefore, the time of the year in which the population is sampled will directly influence estimated density. Densities in urban habitats of Toronto, estimated during the summer, averaged 2–7 skunks/km^2. However, some habitats, such as fields in close proximity to commercial sites, had densities as high as 38/km^2 (Rosatte et al. 1991, 1992). In the parkland habitat of Alberta, extremely low skunk densities of 0.5–2.4/km^2 were reported by Bjorge (1977). In contrast, Ferris and Andrews (1967) reported skunk densities as high as 13–26/ km^2.

FEEDING HABITS

Striped skunks are omnivorous; their diet depends on the season, but they feed primarily on insects such as beetles (mainly carabids and scarabids), grasshoppers, and crickets associated with grassland areas (Carr 1974; Rosatte 1987; Greenwood et al. 1999). Skunks will also consume earthworms, snails, clams, crayfish, and frogs. Small mammals such as mice, voles, moles, rats, and squirrels often form part of the diet (Hamilton 1936; Selko 1937; Verts 1967). Other items include bird eggs, carrion, and garbage. Striped skunks will also consume a variety of wild fruits including blackberry, black cherry, blueberries, and raspberries as well as grains, corn, and nuts.

Striped skunks usually begin feeding during the early evening and often continue until sunrise, with food located primarily by odor (Godin 1982; Rosatte 1987; Nams 1991, 1997; Larivière 1998). Because skunks are slow moving, they often wait in ambush and pounce on prey such as insects and small mammals. They also dig for insects and earthworms, and usually leave small craters in the ground or on lawns (Rosatte 1987). Skunks will also search scats for insects (Chapman 1946). During such forays, skunks use their keen sense of smell and hearing to capture prey. They will also raid bee hives to consume honey, larvae, and adult bees. They scratch at the hives to attract the bees (Rosatte 1987).

Diets differ seasonally. Skunks prefer insects and mammals as food items during spring and summer. In Michigan, insects constituted 57% of the summer diet (Rue 1981). However, during fall and winter, plant as well as animal matter are consumed (Rosatte 1987). For example, in New York, fruit was the primary food item of skunks during fall and winter (Hamilton 1936). In Canada, the diet during fall comprised 27% fruit, 20% small mammals, 14% carrion, 11% grains and nuts, 7% insects, and 11% vegetation such as grasses. During spring and summer, the diet consisted of 43% insects, 28% fruit, 16% small mammals, and 9% grains (Banfield 1974).

In some areas, striped skunks can have a significant impact on birds through destruction of waterfowl nests (Kalmbach 1938; Crabtree and Wolfe 1988; Larivière and Messier 1997b, 1998a, 1998b, 1998c, 1998d). Interestingly, skunks destroy bird nests incidentally while foraging for their main prey items, usually small mammals and insects (Vickery et al. 1992; Larivière and Messier 1997b, 1998d). At night, skunks discover many nests following the flushing of the incubating hen. Nests may be partially or entirely destroyed, and skunks often rest in the vicinity of destroyed nests (Larivière and Messier 1997b, 1998d; Larivière and Walton 1998). Skunks also destroy eggs of several other game birds, including those of bobwhite quail (*Colinus virginianus*) in the southeastern United States (Stoddard and Komarck 1941).

BEHAVIOR

Striped skunks are normally docile, nonaggressive animals, which do not defend a territory and usually flee if approached by a human (Verts 1967; Rosatte 1987). However, if provoked, skunks may face an adversary and make a variety of noises from low growls, grunts, and snarls to screeches and hisses. If the threat persists, skunks will often assume a defensive posture and discharge their musk; adult males are more prone to spraying than juveniles or adult females (Verts 1967; Godin 1982; Rosatte 1987). Defensive postures include arching the back and stamping front feet on the ground while shuffling backward (Larivière and Messier 1996b).

While discharging musk, a skunk can angle the stream of scent by changing the direction of the nipples of the scent gland, thereby creating a wider zone of coverage both to the side and behind itself. The musk may be expelled as a fine spray or as droplets to a range of 3–5 m (Godin 1982).

Striped skunks are not particularly mobile. However, they can travel at speeds of 10–15 km/hr if necessary. Normally, they are very lethargic, walking at less than 2 km/hr (MacLulich 1936; Verts 1967; Godin 1982; Larivière and Messier 1998c). Skunks are not normally climbers and will avoid water, but are capable of swimming if necessary (Wilber and Weidenbacker 1961; Verts 1967).

Although not social animals, skunks often den together during periods of inclement weather. Males are not tolerant of each other during other seasons, but sometimes will den together during the winter (Seton 1929; Allan and Shapton 1942).

Daily and Seasonal Activities. Striped skunks are usually crepuscular or nocturnal, emerging about sunset. However, they may be active during the daytime (Larivière and Messier 1997a). In areas such as Illinois, skunks commence daily activity about 1800–1900 hr during August–October. Activity may continue through the night without skunks returning to their dens until 0500–0600 hr (Verts 1967; Larivière and Messier 1997a). However, the length of activity may depend on the weather, with forays only lasting a few hours during cold evenings. In rural habitats, during the day, skunks may use hayfields, corn fields, fencerows, pastureland, or watercourses as resting areas (Verts 1967; Storm 1972; Godin 1982; Larivière and Messier 1998a).

During the summer and the fall, skunks spend much of their time accumulating fat reserves for sustenance during the winter denning period. In the northern portions of the range, striped skunks are inactive during the winter, often denning under buildings (Rosatte 1984a). They do not hibernate, but merely go into a deep sleep (Godin 1982). During the summer, they tend to use several dens as resting sites, but usually have only one to two dens during the winter (Storm 1972; Houseknecht and Tester 1978; Larivière et al. 1999).

Skunks usually become active in January and February in preparation for breeding activities. Females may remain in the winter den, but males are active in search of receptive females for breeding. Normally, temperatures do not affect breeding activities during the late winter; however, activity may be dependent on sudden temperature changes, snow cover, crusted snow over dens, sexual drive, and hunger (Smith 1931; Jones 1939; Verts 1967; Godin 1982). Skunk activity and food intake are usually correlated with ambient temperature (Aleksiuk and Stewart 1977).

MORTALITY

Generally, there is a rapid turnover in most striped skunk populations because of high mortality. The primary predators of striped skunks are humans, but other species, including red foxes, coyotes (*Canis latrans*), lynx (*Lynx canadensis*), bobcats (*Lynx rufus*), badgers, mountain lions (*Puma* [*Felis*] *concolor*), fishers (*Martes pennanti*), domestic dogs (*Canis familiaris*), and great horned owls (*Bubo virginianus*), have been known to kill skunks (Errington et al. 1940; Banfield 1974; Godin 1982; Rosatte 1987). Collisions with vehicles and euthanasia by animal control companies also account for a great number of skunk mortalities, especially in urban areas (Rosatte 1987; Rosatte et al. 1991). In addition, skunks often succumb to starvation during prolonged winters (Sunquist 1974; Bjorge 1977; Godin 1982). Mortality also occurs from infectious diseases such as rabies and canine distemper (Rosatte 1988; Davidson and Nettles 1997; Schubert et al. 1998).

Parasites. Striped skunks are often infested with many ectoparasites, including various species of fleas, lice, ticks, and mites (Jackson 1961; Verts 1967; Godin 1982). This is primarily a function of the winter denning activities of skunks, where ectoparasites complete their life cycle and are able to propagate successfully (Stegeman 1939). Striped skunks also harbor a multitude of endoparasites, including protozoans, acanthocephalans, cestodes, nematodes, and trematodes (Babera 1960; Verts 1967; Webster 1967; Davidson and Nettles 1997). Eggs or larvae of the large intestinal roundworm (*Baylisascaris columnaris*) of skunks may be a potential health risk to humans (Davidson and Nettles 1997). Endoparasite infestations in skunks tend to be greater in summer and autumn than in winter and spring. This may be a function of the lack of food intake in skunks during the winter denning period (Stegeman 1939; Godin 1982).

Diseases. Skunks are a major vector of rabies, primarily in the American Midwest and Canadian Prairie provinces as well as Ontario (Rosatte and Gunson 1984b; Pybus 1988; Rosatte 1988; Gremillion-Smith and Wolff 1988; Davidson and Nettles 1997; Greenwood et al. 1997; Krebs et al. 1997). From 1960 through 1995, the number of reported cases of rabid skunks in the United States ranged between about 500 and 4500 cases annually (Krebs et al. 1997). Currently, skunks account for about 20% of the annual rabies cases in the United States (Krebs et al. 1997). They also account for 20–30% of the annual rabies cases in Canada. In Ontario, striped skunks accounted for about 20% of the 56,000 cases of rabid animals reported from 1954 through the mid-1990s (MacInnes 1988; Rosatte 1988). Symptoms of rabies in skunks range from the "dumb form," in which the animal may be docile, to the "furious form," in which extreme aggression may be exhibited.

Striped skunks are very susceptible to canine distemper virus, leptospirosis (*Leptospira* spp.), and infectious canine hepatitis. Leptospirosis is a particularly widespread bacterial disease to which humans are also susceptible (Davidson and Nettles 1997). Other animals, including skunks and humans, become infected with this disease when they come into contact with infected water (contaminated by urine from an infected animal). Prevalence of leptospirosis in skunks is high, ranging from 15% to 60% in Canada and the United States (Roth 1961; Gorman et al. 1962; Verts 1967). Skunks also are infected with a variety of other diseases, including pulmonary aspergillosis and histoplasmosis caused by fungi and listerosis and tularemia caused by bacteria (Durant and Doll 1939; Bohn et al. 1955; Emmons et al. 1955; McKeever et al. 1958; Verts 1967).

AGE ESTIMATION AND SEX DETERMINATION

Probably the most accurate means of estimating the age of skunks is through observation of annual growth lines in the cementum of the teeth using a microscope (Johnston and Watt 1981; Johnston et al. 1982). Casey and Webster (1975) estimated the age of skunks from Ontario using this technique. A cementum line forms in the teeth of skunks during their first fall/winter of life. Succeeding annual lines are formed at the apex of the tooth root each year from October to May. Casey and Webster (1975) also determined the sex of skunks by examining the hippocampal neurons for sex chromatin. They found it appears in the brain of females but not males. Sex of embryonic skunks can be determined by the position of genital papilla after 35 days gestation or by comparing canine teeth measurements in adults (Hamilton 1963; Verts 1967; Frederickson 1983).

Alternative means of determining age class of skunks include radiographs of teeth; length and shape of the *os baculum* to distinguish young males from adults; and color, diameter, and length of teats to distinguish juvenile from parous females (Schwartz and Schwartz 1959; Fredrickson 1983). Petrides (1950) noted that the teats of adult females were at least 2 mm in diameter, 2.5 cm in length, and dark colored when using pelts to determine age class. The teats on the pelts of young female skunks were <1 mm long and flesh colored (Petrides 1950). Age class of skunks can also be determined by using degree of closure of the epiphyseal cartilage of the radius and ulna bones (closed at 8–9 months of age), cranial sutures, and the dry weight of the eye lens (increases with age) (Mead 1967; Verts 1967; Bjorge 1977; Leach et al. 1982; Root and Payne 1984). Generally, however, those techniques are unreliable (Rosatte 1987).

ECONOMIC STATUS

Skunks are valued economically for their pelts by the garment industry, and thousands are trapped each year (Musgrove and Blair 1979; Novak et al. 1987). Skunk harvest for the United States and Canada during 1970–1980 averaged 130,010/year, with a low of 21,485 in 1970–1971 and a high of 372,839 in 1979–1981 (Novak et al. 1987). In the past, the harvest value of skunk pelts in the United States and Canada was about $1 million annually, with an average pelt price in Canada during the 1980s being $3.41 Cdn (Rosatte 1987). Pelt price in Canada ranged from $3.40 to $5.73 Cdn during the 1990s. The pelts are used for the manufacture of coats, jackets, muffs, scarves, and trim (Godin 1982). Skunks are also propagated for use as pets, and the scent is used for musk derivatives.

Skunks are also economically important because of the costs incurred for rabies treatment and control. They are one of the primary vectors of rabies in North America, and resultant costs for rabies diagnosis, human treatment, and vaccination of pets is millions of dollars each year (Rosatte et al. 1986; Rosatte 1987, 1988; MacInnes 1988).

Striped skunks consume considerable volumes of insects, making them an asset to farmers and residential property owners. Insects consumed by skunks include bud worms (*Heliothis virescens*), scarab beetles (Scarabaeidae), June beetles (*Cotinis nitida*), potato beetles (*Leptinotarsa decemlineata*), army worms (*Cirphis unipuncta*), cutworms (*Chorizagrotis* spp.), and moths (Sphingidae) (Chapman 1946; Godin 1982).

Unfortunately, skunks can also be a general nuisance and cause economic damage to property owners by digging in residential lawns for grubs and other invertebrates. In addition, skunks will raid beehives. They often den under residential buildings and associated buildings and the resultant skunk odor, if musk is discharged, can be noxious to humans (Godin 1982). However, the nuisance aspects of skunks should be weighed against the beneficial aspects of consumption of insects that damage agricultural crops (Rosatte 1987).

MANAGEMENT AND CONSERVATION

In the past, skunks have been managed primarily as a nuisance, with little or no protection or regard to the impact of control practices. As a result, skunk management has focused on nuisance control primarily through eradication, census programs, and monitoring of the number of pelts harvested each year (Rosatte 1987). Forms of management included trapping, shooting, poisoning, and other forms of eradication (Rosatte et al. 1986; Rosatte 1987; Pybus 1988). Because of the economic value of skunks due to the great number of insects they consume each year, education programs to inform people of their benefits should be initiated. Proactive efforts should be implemented to prevent skunks from gaining access to areas where they may become a nuisance. This includes installing predator-proof fences to keep skunks out of chicken pens (buried 0.3 m below ground); elevating beehives 1 m above ground, sealing off all entry holes under porches, buildings, and foundations; and removing rubbish piles that may serve as potential denning sites. Insect control in lawns will assist in preventing skunks from digging for grubs and other invertebrates. In addition, to alleviate noxious odor and disease transmission concerns, domestic pets should be confined during the evenings so they do not come in contact with skunks.

If a skunk is denning under a residential building, removal of the animal can be accommodated by sealing off all entry and exit holes except one. When the skunk has left for the evening (determined by examining footprints in flour sprinkled around the entry/exit hole), the hole can be sealed with wire mesh or other materials. However, one must ensure that there are no young skunks left to die of starvation if the adult female has been removed. Alternatively, a livetrap baited with sardines can be placed at the entry/exit hole. When the skunk is captured, the trap can be wrapped in plastic, moved a few meters so the exit hole can be covered with wire mesh, and then the skunk released. Animal control companies can also be hired to remove skunks from residential properties. However, skunks should not be relocated, as this will only promote the transmission and spread of infectious disease to the area where the skunk is released. In Ontario, it is illegal (according to the Fish and Wildlife Conservation Act) to relocate skunks (as well as other wildlife species) greater than 1 km from the point of capture. This legislation is designed to curtail the spread of infectious diseases such as rabies.

RESEARCH NEEDS

Research is needed on the role that skunks play in maintaining rabies at subepizootic levels in areas where the disease is prevalent. Greater understanding also is needed of skunks as rabies vectors. Research is also needed to develop methods to control skunk rabies in the American Midwest and Canadian Prairies. Research also is needed to document range expansion in skunks and reconfirm the classification status of skunks as a separate family. Although genetic aspects of striped skunks were first explored only recently (Dragoo et al. 1993; Dragoo and Honeycutt 1997; Bixler 2000), primers for DNA investigations using microsatellite markers are still unidentified and could greatly enhance our understanding of their spatial organization and social systems.

NOMENCLATURE

COMMON NAMES. Hooded skunk, white-sided skunk, southern skunk
SCIENTIFIC NAME. *Mephitis macroura* (*macroura* is Greek for "large tail")
SUBSPECIES. There are five subspecies of hooded skunk in North America, according to Hall (1981): *Mephitis m. eximus, M. m. macroura, M. m. milleri, M. m. richardsoni,* and *M. m. vittata.*

DISTRIBUTION

The hooded skunk ranges from the southern United States through Mexico to Central America (Fig. 34.4). Specifically, this species is found in southwestern Texas, southeastern Arizona, southwestern New Mexico, Mexico, and Nicaragua (Davis and Russell 1954; Davis and Lukens 1958; Packard 1965; Godin 1982; Rosatte 1987). Range extensions in Texas and Mexico have been reported by Packard (1965, 1974).

FIGURE 34.4. Distribution of the hooded skunk (*Mephitis macroura*).

DESCRIPTION

The hooded skunk (Fig. 34.5) is similar in appearance to the striped skunk. However, the fur is longer and softer and there is a conspicuous tuft of hair on the upper part of the back of the neck and head (hence the vernacular name "hooded"). It is smaller than the striped skunk, but has a long tail, up to 40 cm (Cahalane 1961; Godin 1982; Rosatte 1987). Compared to the striped skunk, there is also a wider separation of the dorsal white stripes in hooded skunks if stripes are present (Wade-Smith and Verts 1982; Nowak 1999).

Sexual Dimorphism. Hooded skunks are normally smaller and more slender than striped skunks (see Table 34.1), with females being about 15% smaller than males. Total body length for male hooded skunks ranges from 56 to 79 cm, tail length from 36 to 40 cm, and hind foot length from 6 to 7 cm (Bailey 1931; Davis 1945; Leopold 1959; Hubbard 1972; Godin 1982; Rosatte 1987). Body weight for males in Mexico was 0.7–0.9 kg and for females was about 0.7 kg (Armstrong et al. 1972).

FIGURE 34.5. Hooded skunk (*Mephitis macroura*). SOURCE: Richard B. Forbes and Jerry W. Dragoo.

Pelage. Hooded skunks have two basic color patterns: white backed or black backed (Godin 1982). However, intermediate patterns may occur. Generally, the pelage of hooded skunks is black with a white back; however, white stripes as well as a mottling of white may be present (Davis 1951). The back and tail are usually black in the black-backed pattern (Godin 1982), with a lateral white stripe on each side. A vertical thin white stripe occurs on the head between the eyes of hooded skunks, but not extending down the nose as in striped skunks (Gharaibeh and Jones 1999).

Skull and Dentition. The skull of hooded skunks is highly arched and can be distinguished from that of the striped skunk by larger tympanic bullae (Hall 1981; Godin 1982). Basilar length of the skull in hooded skunks is 5.6–6.0 cm (Hall and Kelson 1959). The dental formula is the same as for striped skunks: I 3/3, C 1/1, P 3/3, M 1/2.

REPRODUCTION

Very little is known of the life cycle and reproduction of hooded skunks. Normally, breeding occurs during late February to early March and gestation is about 60 days (Patton 1974). There is probably no delayed implantation (Ferguson et al. 1996). Litter size varies from three to six young (Bailey 1931; Reid 1955; Davis and Lukens 1958; Godin 1982; Rosatte 1987) (see Table 34.1). Females have 10 mammae and males have a baculum (Bailey 1931; Cahalane 1961).

ECOLOGY

Hooded skunks are generally uncommon, but occur in a variety of habitats from sea level to about 2500 m (Hubbard 1972; Hwang and Larivière 2001). They are most common in arid lowland areas (Davis and Russell 1954). Preferred habitats include low-elevation desert plains, high-elevation ponderosa pine (*Pinus pondersa*) forests, and riparian vegetation (Findley and Caire 1974; Packard 1974; Schmidly 1974; Godin 1982). They also frequent forested or shrubby uplands and grassland areas (Baker 1956; Davis and Lukens 1958; Findley et al. 1975; Godin 1982). They are more secretive than striped skunks and generally do not occur near human habitation (Rosatte 1987). Denning sites are burrows usually located in rugged, heavily vegetated areas (Patton 1974).

FEEDING HABITS

Because hooded skunks are generally uncommon, there are few published reports on their feeding habits. However, their foraging habits, such as digging, are similar to those of striped skunks and they are considered omnivores (Patton 1974; Rosatte 1987). Generally, hooded skunks are primarily insectivorous, consuming various species of beetles (Bailey 1931; Patton 1974; Rosatte 1987). They also feed on ear-wigs (Forficulidae), grasshoppers (Orthoptera), stink bugs (Pentastomidae), vertebrate material (Dalquest 1953; Patton 1974), and fruits, bird eggs, and garbage (Reid 1955; Hwang and Larivière 2001).

BEHAVIOR

Hooded skunks are very secretive in their habits, and little is known of their behavior. They are considered to be nocturnal and solitary, but may feed communally (Reid 1955). They spend the day denned in rock crevices or burrows (Bailey 1931; Reid 1955; Godin 1982). Cactus plants (*Opuntia* spp.) are sometimes used for hiding places by this species (Reed and Carr 1949). In settled areas, they prefer fields and trails in lower valleys (Godin 1982). The breeding activities and behaviors are apparently similar to those of striped skunks (Patton 1974). Feeding behaviors are also similar to those of striped skunks in that they often pounce on insects such as grasshoppers (Dalquest 1953; Reid 1955). Defensive activities such as spraying scent from anal glands are also similar to those of striped skunks (Larivière and Messier 1996b). It is thought that hooded skunks remain active all year (Bailey 1931).

MORTALITY

Very few mortality data are available for hooded skunks. Great horned owls and coyotes may prey on them (Hwang and Larivière 2001). They have been documented with nematode parasites (*Physaloptera maxillaris*) (Erickson 1946).

ECONOMIC STATUS

The fur of hooded skunks is of low economic value (Bailey 1931). They were hunted for their scent glands, which were used in folk medicine (Reid 1955). Native peoples in Mexico also used the fat of hooded skunks for medicinal purposes (Dalquest 1953).

MANAGEMENT, CONSERVATION, AND RESEARCH NEEDS

Very little is known of the ecology and habits of hooded skunks. Research is needed to determine the dynamics of hooded skunk populations and resultant management implications as well as such specifics as habitat use, mortality factors, and the status of current populations.

NOMENCLATURE

COMMON NAMES. Eastern spotted skunk, western spotted skunk
SCIENTIFIC NAMES. *Spilogale putorius, Spilogale gracilis*

DISTRIBUTION

The eastern spotted skunk (*Spilogale putorius*) occurs from northeastern Mexico through the Great Plains and north to the Canadian border as well as throughout most of the southeastern United States (Dragoo and Honeycutt 1999c) (Fig. 34.6). The western spotted skunk (*S. gracilis*) occurs throughout most of Mexico and through most western states, although only in small parts of Montana and North Dakota (Crooks 1999; Howard and Marsh 1982; Hoffmeister 1986). Both species occur at elevations up to 2500 m (Orr 1943; Baker and Baker 1975). In Washington, *S. gracilis* may occur exclusively near riparian areas at high elevations (Carey and Kershner 1996). Both species of *Spilogale* are sympatric in western Oklahoma, Laramie County in Wyoming, Reeves County in western Texas, and the Hill Country of central Texas (Van Gelder 1959; Patton 1974; Schmidly 1984).

FIGURE 34.6. Distribution of the spotted skunk.

FIGURE 34.7. Spotted skunk. SOURCE: Richard B. Forbes and Jerry W. Dragoo.

DESCRIPTION

Spotted skunks (Fig. 34.7) are the smallest North American skunks. They are slender, have short legs, and seldom exceed 1 kg in weight (see Table 34.1). The ears are large and round and located on the sides of the head. Feet are plantigrade, each with five toes. Forefeet claws are 7 mm long, about twice as long as those of the hind feet. Feet of *Spilogale* are not as specialized for digging as those of *Mephitis* and *Conepatus* (Pocock 1921).

Eastern spotted skunks are larger than their western counterparts. Adult male and female *S. putorius* average, respectively, 720 and 500 g in Iowa (Crabb 1944) and 399 and 283 g in Florida (Kinlaw et al. 1995a). Male and female *S. gracilis* weights average, respectively, 620 and 500 g in California (Crooks 1994a) and 483 and 397 g in Texas (Patton 1974). Spotted skunks are heavier in fall and winter and lightest in spring and summer (Patton 1974).

Sexual Dimorphism. Males are 7–10% larger than females in body and skull measurements and 20–40% heavier (Table 34.2). Males also have wider zygomatic arches and larger sagittal and lambdoidal crests than females (Van Gelder 1959).

Pelage. Spotted skunks are predominantly black with numerous white markings. On the face, a white patch occurs in front of each ear as well as on the forehead. Six distinct white stripes occur on the anterior part of the back. The dorsal stripes parallel the spine and extend from the neck to the tail, joining at the tip. Vertical stripes also occur in some individuals, especially in *S. p. interrupta* (Van Gelder 1959). White markings are more extensive in western spotted skunks. On the back, the black and white stripes are of nearly equal width, whereas black dominates in *S. putorius*. On both species, the fur is fine and dense, and both sexes have similar coloration.

Skull and Dentition. The skull appears flat in dorsal profile. Dental formula is the same as in striped and hooded skunks: I 3/3, C 1/1, P 3/3, M 1/2. The first premolar is small and occasionally absent (Van Gelder 1959).

REPRODUCTION AND DEVELOPMENT

Anatomy and Physiology. Numerous anatomical and physiological differences exist between eastern and western spotted skunks. First, western spotted skunks exhibit delayed implantation, whereas eastern spotted skunks do not (Mead 1968a, 1968b). Second, the *os baculum* is longer and more recurved in eastern than western spotted skunks. The number of chromosomes also differs between the species: *S. putorius* has $2n = 58–64$, whereas *S. gracilis* has $2n = 58–60$ (Hsu and Mead 1969; Owen et al. 1996). For these reasons, the species are reproductively isolated.

Male *S. putorius* have a penis that is 35–40 mm in length. The prostate is well developed, and it is the only accessory sex gland in *Spilogale*. Size of the *os baculum* is about 20 mm for *S. putorius* and about 16 mm in *S. gracilis* (Mead 1967, 1970). Females of both species have 6–10 mammae.

Body temperature of spotted skunks varies from 35.2°C to 37.5°C. Their basal metabolic rate is 0.48 cm^3 O_2/g/hr, which is 30.5% lower than predicted based on their body mass. The nonelongate body morphology and omnivorous diet probably contribute to the lower metabolism of spotted skunks (Knudsen and Kilgore 1990).

Breeding Season and Mating. The mating system of spotted skunks is polygamous, and males and females may breed with several individuals. Pair bonds last until mating is achieved, seldom more than a few days.

Eastern spotted skunks mate in March–April. Females are polyestrous, and estrous cycles occur from September through January. Females are spontaneous ovulators, and it takes 6–7 days for the eggs to reach the uterus. Implantation occurs 14–16 days after mating and is not delayed (Mead 1968a; Greensides and Mead 1973).

Female western spotted skunks come into estrus in September. Testis size and serum testosterone levels peak in September and October, and most breeding occurs during this period (Kaplan and Mead 1993). Adult males have spermatozoa in their testes by June, but young males, like juvenile females, reach sexual maturity in September (Mead 1968b). In southern populations, breeding may occur as early as July, and two litters may be produced yearly.

Gestation and Parturition. The two species of spotted skunk are strikingly different with regard to gestation. In the eastern species, gestation lasts 50–65 days and occurs without delay. In western spotted skunks, implantation is delayed for 200–220 days (Foresman and Mead 1973). After the delay, the uterus is prepared for implantation via the production of leukemia inhibitory factor (Hirzel et al. 1999). Postimplantation gestation lasts 28–31 days for a total gestation of 230–250 days (Foresman and Mead 1973). In western spotted skunks, implantation of the blastocyst typically occurs in April (Mead 1968b). Parturition of both species occurs in late May–early June (Mead 1968a, 1968b).

Litter Sizes. Litter size ranges from 2 to 6 for both species, and averages 5.5 for *S. putorius* and 3.8 for *S. gracilis* (Mead 1968a, 1968b).

Growth, Development, and Longevity. The young are born toothless and with eyes and ears closed. At birth, spotted skunks weigh about 11 g (Constantine 1961). The body is covered with fine hairs and the black and white markings are distinct. Eyes open after 30–32 days. Neonates first display defensive behaviors such as raising the tail 24 days after

TABLE 34.2. Average measurements of eastern spotted skunks and western spotted skunks

	Eastern Spotted Skunk		Western Spotted Skunk	
Measurement	Male	Female	Male	Female
Total length (mm)	519 (463–596)	465 (403–470)	398 (346–446)	370 (331–403)
Tail length (mm)	202 (193–211)	182 (165–193)	138 (101–161)	138 (123–164)
Hindfoot length (mm)	47 (43–51)	44 (39–47)	44 (37–49)	41 (36–43)
Body mass (g)	399 ± (112)	283 ± (39)	336–734	227–482

SOURCE: Eastern spotted skunk, Van Gelder (1959); western spotted skunk, Patton (1974).
NOTE: Range or *SE* is indicated in parentheses.

birth, and musk can be expelled when 46 days old (Davis 1945). Teeth erupt after 35 days and the young are weaned after 54 days (Crabb 1944). Males do not provide parental care.

ECOLOGY

Habitat. Spotted skunks occupy a variety of habitats including wooded areas, tallgrass prairies, and rocky canyons, but seldom occur in low-lying deserts. *S. putorius* avoids wetlands and semiaquatic habitats (Ehrhart 1974; Crooks 1994a, 1994b; Crooks and Van Vuren 1995). In Florida, eastern spotted skunks occur to the coast (Howell 1906; Kinlaw et al. 1995b). Availability of burrows, food, and thick vegetative cover likely is essential for the maintenance of *Spilogale* populations. In Missouri, a radio-collared male *S. putorius* preferred oak–hickory (*Quercus–Carya*) forests more than open fields, and within the forest preferred sites with deeper ground litter (McCullough and Fritzell 1984). In Washington and Oregon, *S. gracilis* was frequently captured in old-growth forests (Carey and Kershner 1996).

Populations of spotted skunks are disjunct and often localized. In an agricultural area of Iowa, density of *S. putorius* was estimated at 2.2/km^2 (Crabb 1948). On the Canaveral National Seashore of Florida, density was estimated at >40 skunks/km^2 (Ehrhart 1974; Kinlaw et al. 1995b). Populations fluctuate widely, and possibly are influenced by the availability of farmland (Polder 1968; Choate et al. 1974) or the incidence of rabies (Johnson 1959).

Movements. Movement rates of eastern spotted skunks vary seasonally. During the spring breeding season, male *S. putorius* traveled 423 m/hr, compared to 252 m/hr in summer. Nightly movements of males were also greater in spring (2807 m) than during summer (1622 m) or autumn (1038 m) (McCullough and Fritzell 1984). Spotted skunks do not defend territories (Crabb 1948). Home range of four male *S. putorius* radio-tracked during various seasons in Missouri averaged 14.2 km^2 and ranged from 0.55 to 43.6 km^2 (McCullough and Fritzell 1984). The extremely large value of 43.6 km^2 was attributed to long-distance movements during the breeding season. Home range of males during seasons other than spring ranged from 0.4 to 1.9 km^2 (McCullough and Fritzell 1984). In Iowa, nightly movements up to 4.8 km were recorded and home range was estimated at 0.65 km^2 (Crabb 1948).

Most information on western spotted skunks is from studies of insular populations off the Californian coast (Crooks 1994a, 1994b, 1999). Western spotted skunks do not defend territories, and home range size is about 30–60 ha (Crooks and Van Vuren 1995).

Denning. In natural areas, spotted skunks may seek shelter in rock crevices, underground burrows, and in hollow trees stumps (Crabb 1948; Frank and Lips 1989; Toland 1991; Crooks 1994a). In Missouri, 10 dens were located in hollow trees with an average diameter at breast height of 26 cm: entrances were 1–7 m aboveground (McCullough and Fritzell 1984). In developed areas, spotted skunks may take refuge in culverts, haystacks, straw piles, grain elevators, woodpiles, or corncribs or under buildings (Crabb 1948). In Texas, 33% of daytime retreats were in human structures, whereas 67% were in natural cavities (Patton 1974). Underground burrows may be selected for thermal reasons: In Missouri, 58% of summer dens faced north, whereas 80% of autumn dens and 100% of winter dens faced south (McCullough and Fritzell 1984). Ground dens typically contain a nest of grass (Crabb 1948). Individuals may use and reuse several dens, and disturbance at dens may cause their abandonment (Crabb 1943, 1948).

FEEDING HABITS

Spotted skunks are primarily insectivorous and consume mostly Orthoptera, Coleoptera, and Hymenoptera (McCullough and Fritzell 1984). Opportunistically, or when insects are unavailable, small mammals are consumed. Birds, carrion, and crops such as corn are consumed whenever available (Crabb 1941a). Spotted skunks may occasionally raid caches of weasels (*Mustela* sp.), but do not cache food (Polder 1968).

Two detailed studies of feeding habits of eastern spotted skunks are available, and both were based on scats collected in Iowa during the 1930s (Selko 1937, Crabb 1941a). The largest sample analyzed was 834 scats (Crabb 1941a). During all seasons except winter, insects such as beetles (adults and larvae), crickets, grasshoppers, and grubs dominated the diet. During winter, small mammals such as cottontails (*Sylvilagus floridanus*), meadow voles (*Microtus pennsylvanicus*), prairie voles (*M. ochrogaster*), and Norway rats (*Rattus norvegicus*) made up 90% of the diet (Crabb 1941a). The importance of small mammals during winter is undoubtedly caused by the scarcity of insects during this season. Nonetheless, insects occurred in about 15% of scats during winter. Corn also was an important item during winter, occurring in about 25% of scats (Crabb 1941a). Other less common food items reported include remains of several birds most likely consumed as carrion, fox squirrels (*Sciurus niger*), wheat and oat kernels, acorns, and apple seeds (Selko 1937).

BEHAVIOR

Spotted skunks are more active and alert than other larger skunks. They also are agile climbers and readily climb trees and even along rough walls (Crabb 1948). Their ability to climb trees is sometimes used to escape predators (Leopold 1965). Vocalizations include grunts and a high-pitched scream similar to that of a blue jay (*Cyanocitta cristata*) (Manaro 1961).

The handstand defensive behavior of spotted skunks is characteristic of the species, and is unique among skunks. When threatened, spotted skunks may stomp their forefeet or run toward the opponent and abruptly stop, simultaneously elevating the hindquarters vertically so they balance directly overhead. By keeping balance with the forefeet, the animals may simultaneously expose the anal region toward the threat while still maintaining visual contact. Animals in this posture may twitch their tail and hiss. Although spotted skunks can spray while in the handstand position, they typically drop on all four legs and curve their body laterally in a horseshoe shape while maintaining the tail raised vertically before spraying (Johnson 1921; Manaro 1961). The defensive spray of spotted skunks is composed primarily of three volatile components: (E)-2-butene-1-thiol, 3-methyl-1-butanethiol, and 2-phenylethanethiol (Wood et al. 1991).

The egg-opening technique of spotted skunks also is peculiar. The animal straddles the egg and attempts to bite through the shell. If unsuccessful, it throws the egg with its forefeet backward between the hind feet. As the egg passes between the hindfeet, the animal may kick it with one back foot. The egg is then chased and inspected for sign of shell fracture or breakage that would facilitate its opening. If the egg is not broken, the procedure may be repeated several times (Van Gelder 1953). Prey such as small mammals and young birds are killed by first restraining with both forepaws and hind paws and then biting at the neck (Ziener 1975).

Daily and Seasonal Activities. Spotted skunks are almost entirely nocturnal, and thus are rarely seen. In northern areas, they retreat and become inactive for 4–8 months. In southern areas, spotted skunks likely remain active all year, although their activity is probably reduced. Males radio-tracked in Missouri moved the least during winter (McCullough and Fritzell 1984).

MORTALITY

Spotted skunks have natural predators. Although they have been reported in the diet of bobcats and great horned owls (Howard and Marsh 1982), most deaths are probably caused by human activities such as automobile collisions, shooting, and trapping of animals as pests (Howard and Marsh 1982; Rosatte 1987). Predation from domestic dogs or cats (*Felis catus*) may also occur (Crabb 1948).

In captivity, spotted skunks may live almost 10 years (Egoscue et al. 1970). In the wild, they probably experience shorter life spans, most likely <5 years (Van Gelder 1959).

Parasites. Ectoparasites include the fleas *Polygenis gwyni, Echidnophaga gallinacea, Pulex simulans, Diamanus montanus, Hoplopsyllus anomalus, Ctenosephalides felis,* and *Foxella ignota;* the louse *Neotrichodectes mephitidis;* and the ticks *Dermacentor variabilis, Ixodes cookei,* and *Amblyomma americanum* (Mead 1963; Layne 1971; Ehrhart 1974; Patton 1974; Kinlaw et al. 1995a). Spotted skunks also harbor numerous endoparasites such as the cestodes *Oochoristica oklahomensis, O. pedunculata, O. wallacei, and Mesocestoides corti* and the nematodes (*Ascaris columaris, Phylasoptera maxillaris, and Skrjabingylus chitwoodorum* (Peery 1939; Tiner 1946; Chandler 1952; Mead 1963).

Diseases. Captive animals may die of pneumonia or coccidiosis (Kinlaw 1995). In the wild, spotted skunks may carry rabies (Johnson 1959), and rabid skunks may attack humans (Aranda and López-de Buen 1999). The first report of rabies in a spotted skunk was in 1826 in Baja, California (Johnson 1959). However, the incidence of rabies in spotted skunks varies temporally and geographically. For example, only 2 of 52 eastern spotted skunks captured in Iowa from 1964 to 1968 were rabid (Hendricks and Seaton 1969). In Texas, the spotted skunk contributes <1% of rabies cases (Kinlaw 1995). The impact of infectious diseases such as rabies on the regulation of populations is unknown.

AGE ESTIMATION AND SEX DETERMINATION

Weight of the eye lens is the best measurement for distinguishing among juveniles, subadults, and adults, with lens weight increasing with age. Individuals <11 months of age may be differentiated from adults by the degree of closure of cranial sutures (Mead 1967). Animals <8 months old may also be identified by examination of the epiphyses of long bones. Less reliably, use of placental scars and weight and shape of the baculum also may help to determine age classes (Mead 1967).

ECONOMIC STATUS

The pelts of spotted skunks are economically insignificant. During 1983–1984, hunting or trapping of spotted skunks occurred in 23 states and British Columbia. During that year, 5588 spotted skunk pelts were sold in the United States (Rosatte 1987). The average pelt price for spotted skunks varied from $2 to $5 in the late 1990s. Because of the low pelt price, few trappers target spotted skunks and most captures are incidental in traps set for red foxes, gray foxes (*Urocyon cinereoargenteus*), or raccoons (*Procyon lotor*).

MANAGEMENT, CONSERVATION, AND RESEARCH NEEDS

Males apparently respond to commercial lures and cat food as trapping bait (Crabb 1948; Ehrhart 1974). The use of more natural bait allows capture of a greater proportion of females and a more even sex ratio (Crabb 1948; Kinlaw 1995). Captured animals may be handled with or without anesthesia (Crabb 1941b; Larivière and Messier 1999). Ketamine hydrochloride at a dose of 15 mg/kg of body mass is suitable for chemical immobilization of spotted skunks (Crooks 1994b).

Many populations of spotted skunks are declining. The eastern spotted skunk is endangered in Missouri, threatened in Kansas and Iowa, and considered rare and in need of special conservation efforts in Montana, North Dakota, and Oklahoma (Choate et al. 1974; Tyler 1994; Kinlaw 1995). Laboratory methods for the conservation of viable sperm through cryopreservation are available (Kaplan and Mead 1992). The status and distribution of the spotted skunk are unclear in several states.

NOMENCLATURE

COMMON NAMES. Common hog-nosed skunk, eastern hog-nosed skunk, rooter skunk
SCIENTIFIC NAMES. *Conepatus mesoleucus, Conepatus leuconotus*

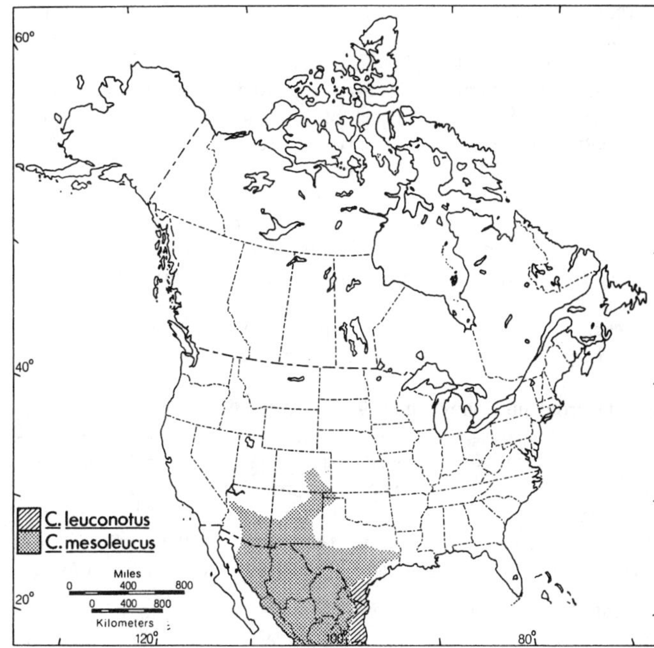

FIGURE 34.8. Distribution of the hog-nosed skunk.

SUBSPECIES. Hall (1981) recognized 10 subspecies of *C. mesoleucus* (*figginsi, filipensis, fremonti, mearnsi, mesoleucus, nelsoni, nicaraguae, sonoriensis, telmalestes,* and *venaticus*) and 2 subspecies of *C. leuconotus* (*leuconotus* and *texensis*). The taxonomic status of the genus is still unclear, and the two North American species (*C. leuconotus* and *C. mesoleucus*) may represent a single species

DISTRIBUTION

The common hog-nosed skunk (*Conepatus mesoleucus*) occurs in parts of western and central Texas, New Mexico, and southern Arizona (Fig. 34.8). The eastern hog-nosed skunk (*Conepatus leuconotus*) occurs in southern Texas and along the eastern coast of Mexico (Merriam 1902; Davis 1966; Manning et al. 1986; Rosatte 1987; Dragoo and Honeycutt 1999a, 1999b).

DESCRIPTION

Hog-nosed skunks (Fig. 34.9) are among the largest of all skunks. Both species possess short, coarse fur. The pelage is black throughout, with

FIGURE 34.9. Hog-nosed skunk. SOURCE: Photo by R. Rosatte.

a broad white stripe on the back. The tail is bushy and predominantly white, and slightly shorter than the body (Ceballos and Miranda 1986).

The nose pad is long, bare, and about three times wider than that of striped skunks. The nostrils are ventral and open downward (Davis 1945, 1951). The size and aspect of the nose led to the "hog-nosed" designation of these species. The nose is used to find prey buried in the ground or under debris. Sense of smell is acute.

The forefeet are strong and adapted for digging. They possess five well-developed claws, which reach 20 mm in length. Claws on the hind feet are about 7 mm (Howard and Marsh 1982). Because of their bone structure and muscle attachment, hog-nosed skunks are apparently the fastest skunks (Patton 1974).

Females have six mammae, one pair inguinal and two pairs pectoral (Bailey 1931). Males possess a baculum, which in adults may reach 13 mm in length.

Dragoo and Honeycutt (1999a) reported that *Conepatus leuconotus* is 20–25% larger than *C. mesoleucus*. The species possesses a narrower white stripe, which is wedge-shaped rather than truncated on the head as in *C. mesoleucus*. The white stripe of *C. leuconotus* is also much narrower or even absent on the rump. The tail of *C. leuconotus* is white on the dorsal side, but black at the base on the ventral surface, whereas the tail of *C. mesoleucus* is white throughout (Davis 1945).

Sexual Dimorphism. Adult weight ranges from 1.5 to 4.5 kg. Like other skunks, male hog-nosed skunks may be up to 10% larger than females in some linear measurements (Hall 1981). In Texas, mean body mass of 12 males (2168 g) was 188% larger than that of females (1152 g, $n = 11$). Four males and two females collected in Sinaloa, Mexico, weighed 1891 g (range = 1445–2300 g) and 2131 g (range = 1812–2450 g), respectively (Armstrong et al. 1972).

Skull and Dentition. The skull of hog-nosed skunks is dimorphic in size: Measurements from males are typically greater in condylobasal length, zygomatic breadth, and mastoidal breadth (Patton 1974). Dental formula is I 3/3, C 1/1, P 2/3, M 1/2. Hog-nosed skunks lack the additional upper premolar that occurs in *Spilogale* and *Mephitis*.

REPRODUCTION AND DEVELOPMENT

Little information is available on the natural history and reproduction of hog-nosed skunks. In North America, mating occurs in February. Hog-nosed skunks are polygynous, and males do not provide parental care.

Gestation. Gestation varies between 42 and 60 days. Delayed implantation has not been reported for hog-nosed skunks and is unlikely to occur, because they live in low-seasonality environments (Ferguson et al. 1996).

Parturition. Parturition occurs in late April–early May (Davis 1945). In Sinaloa, Mexico, a female collected on 27 June was lactating (Armstrong et al. 1972).

Litter Sizes. Although litter size has been reported as three to six (Allen 1906), the maximum number of young observed with a female is four (Davis 1945). Embryonic litter size was reported by Davis (1945) for six females: three females had three embryos, the others had only two embryos. In Texas, a female captured on 4 July had two young (Patton 1974). The small number of mammae in hog-nosed skunks (6) suggests that litter size is consistently lower in *Conepatus* than it is for *Mephitis*, a genus with 12–14 mammae.

Growth, Development, and Longevity. Young are born blind and start crawling in the nest before their eyes open. Musk is present at birth. By June, young weigh about 0.5 kg. They reach adult size by August, at which time they become independent. Sexual maturity occurs at 10–11 months of age (Hayssen et al. 1993). Males presumably attain sexual maturity earlier than females (Dragoo and Honeycutt 1999b).

A captive *Conepatus mesoleucus* lived 8 years, 8 months (Nowak 1999). Longevity in the wild is probably <3–4 years (Patton 1974).

TABLE 34.3. Percentage of various food items based on stomach contents of common hog-nosed skunks (unknown sample size) collected in central Texas

	Autumn	Winter	Spring	Summer
Insects	52	75	82	50
Arachnids	4	12	12	9
Vegetation	38	3	0	31
Reptiles and snails	6	1	6	7
Small mammals	0	9	0	3

SOURCE: Based on data from Davis (1945).

ECOLOGY

The ecology of hog-nosed skunks is poorly known, and much of the current knowledge was obtained from anecdotal observations or during collections of specimens (Goetzee et al. 1995; Goetzee and Nelson 1998). Hog-nosed skunks have yet to be studied with radiotelemetry. For this reason, data on movements, home ranges, and patterns of habitat selection are sparse. The western hog-nosed skunk inhabits the foothills and sparsely timbered or brushy areas throughout its range, but is absent from deserts and heavily timbered areas. The largest population occurs in rocky areas with sparse timber such as the Edwards Plateau (Hill Country) of central Texas; the Chisos, Davis, and Guadalupe Mountains of trans-Pecos Texas; and the isolated mountain ranges in eastern Arizona and New Mexico (Davis 1945). In Texas, *C. mesoleucus* most often occurs along canyons and streambeds (Patton 1974) as well as in oak brush habitat, mesquite brushland, pastures, and areas of semiopen grasslands where thorny brush and cactus dominate (Dragoo and Honeycutt 1999a). In Mexico, *C. leuconotus* occupies mostly the tropical regions of San Luis Potosi, but it also occurs in mountains and coastal plains.

FEEDING HABITS

Hog-nosed skunks are more insectivorous than other skunks. In Texas, insects represent 50–90% of the diet (Table 34.3), regardless of season (Davis 1945; Patton 1974). Preference for beetles (Coleoptera) and their larvae was also observed in New Mexico (Bailey 1931) and Mexico (Leopold 1965). Reptiles, small mammals, or birds seldom account for >10% of diet (Table 34.3). In captivity, hog-nosed skunks did not break chicken eggs, which suggests that eggs may be an uncommon food item in the wild (Patton 1974).

BEHAVIOR

Hog-nosed skunks are typically solitary. Family groups of up to seven animals may occur in August before the dispersal of young (Allen 1906).

Hog-nosed skunks have strong forelimbs, which combined with their long claws, makes them extremely well adapted for digging in rough, rocky terrain (Dragoo and Honeycutt 1999b). They search for food by digging and turning over rocks and debris, a habit that gave them the name "rooter" or "rooter skunk" over much of their distribution (Davis 1945). Because of their forearm strength, hog-nosed skunks are capable climbers, although not as agile as spotted skunks (Dragoo and Honeycutt, 1999b).

Hog-nosed skunks rely on chemical defense when threatened or attacked. Defensive spray is comprises two major components, (E)-2-butene-1-thiol and (E)-S-2-butenyl thioacetate (Wood et al. 1993). The smell of *Conepatus* reportedly is the most offensive odor of all skunks (Patton 1974). However, hog-nosed skunks will typically attempt to escape before facing a threat. Typically, they will seek refuge in brush or under a cactus. If cornered, hog-nosed skunks raise their tail, stomp their forefeet, and scratch the ground in an attempt to warn the assailant. Occasionally, individuals will hiss, bare their teeth, and fling dirt behind them with their front paws. Animals that are further threatened will arch their body in a U to point both their face and their exposed anal region toward the threat. An animal in such a posture is ready to bite or spray, and will do both (Dragoo and Honeycutt 1999b).

Daily and Seasonal Activities. Hog-nosed skunks are mostly nocturnal, but daytime activity is common during winter. During the day, they retreat in underground burrows, brush piles, and rock crevices (Davis 1945; Leopold 1965). Rock crevices also are used as parturition and winter dens. Unlike striped skunks, hog-nosed skunks are not reported to use buildings and human structures as denning sites (Patton 1974). Hog-nosed skunks do not den communally, and individuals seldom share dens (Davis 1945). During fall and early winter, hog-nosed skunks become moderately fat, and it is suspected that they are inactive for short periods in the northern part of their range (Bailey 1931).

MORTALITY

No natural predators are reported (Rosatte 1987). Most mortality is presumably from humans (Howard and Marsh 1982). Possibly, diseases such as rabies may limit populations in remote areas.

Parasites. Two roundworms (*Filaria mastis* and *Physaloptera maxillaris*), a tick (*Ixodes texanus*), and fleas (*Pulex* sp.) are reported as parasites (Tiner 1946; Patton 1974). Presence of the nematode *Skrjabingylus chitwoodorum* was also detected in hog-nosed skunks from Texas by bulging and osteitis over the frontal sinuses (Patton 1974).

Diseases. Hog-nosed skunks may carry rabies, and rabid individuals may attack humans (Aranda and López-de Buen 1999). However, the extent to which these species are involved in maintaining and propagating the rabies virus is unknown. Low densities of hog-nosed skunks probably preclude this species from being a significant rabies reservoir.

AGE ESTIMATION AND SEX DETERMINATION

Radiographs of canine teeth, which are useful for age estimation in striped skunks (Casey and Webster 1975), are reliable for hog-nosed skunks (Patton 1974). Gender is best determined by examination of the external genitalia.

ECONOMIC STATUS

Hog-nosed skunks are not a furbearer of economic importance, as their fur is short and coarse (Davis 1945). In the United States, hog-nosed skunks are captured incidentally by trappers targeting coyotes (*Canis latrans*), bobcats, or foxes. During the 1970s, about 155,000 pelts of *Conepatus* were exported annually at a value of $8.00 each (Broad et al. 1988).

MANAGEMENT, CONSERVATION, AND RESEARCH NEEDS

Hog-nosed skunks are the least known of all skunks. Their low densities, nocturnal habits, and presence in remote areas make them inconspicuous. The sparse information available was obtained mostly from *C. mesoleucus,* and the ecology, behavior, and reproduction of *C. leuconotus* remain largely unknown. Hog-nosed skunks have received little attention because they are neither threatened nor pests. The taxonomic status of the two species is also questioned (Hall 1981). Within the Mephitidae, the two species of *Conepatus* possess the least amount of genetic divergence (Dragoo et al. 1993). Possibly, these taxa warrant only subspecific designation. Further analyses of their genetics should resolve this issue. Wherever they occur, hog-nosed skunks are at low densities (Ceballos and Leal 1984; Ceballos and Miranda 1986). They do not thrive in urban environments and thus are seldom in conflict with humans. The subspecies *C. m. temalestes,* which originally occupied the Big Thicket area of southeastern Texas, may be extinct, possibly because of overharvesting. Surveys to locate surviving populations will help determine their status (Schreiber et al. 1989).

LITERATURE CITED

Aleksiuk, M., and A. Stewart. 1977. Food intake, weight changes and activity of confined skunks (*Mephitis mephitis*) in winter. American Midland Naturalist 98:331–32.

Allen, D. L., and W. W. Shapton. 1942. An ecological study of winter dens, with special reference to the eastern skunk. Ecology 23:59–68.

Allen, J. A. 1906. Mammals of the states of Sinaloa and Jalisco, Mexico, collected by J. H. Batty during 1904 and 1905. Bulletin of the American Museum of Natural History 22:191–262.

Andersen, P. A. 1981. Movements, activity patterns and denning habits of the striped skunk (*Mephitis mephitis*) in the mixed grass prairie. M.S. Thesis, University of Calgary, Calgary, Alberta, Canada.

Anderson, S. 1972. Mammals of Chihuahua, taxonomy and distribution. Bulletin of the American Museum of Natural History 148:149–410.

Aranda, M., and L. López-de Buen. 1999. Rabies in skunks from Mexico. Journal of Wildlife Diseases 35:574–77.

Armstrong, D. M., J. K. Jones, Jr., and E. C Birney. 1972. Mammals from the Mexican state of Sinaloa. III. Carnivora and Artiodactyla. Journal of Mammalogy 53:48–61.

Babera, B. 1960. A survey of parasitism in skunks, *Mephitis mephitis* in Louisiana, with observations on pathological damages due to helminthiasis. Journal of Parasitology 46:26–27.

Bailey, V. 1931. Mammals of New Mexico. North Amerian Fauna 53:1–412.

Baker, R. H. 1956. Mammals of Coahuila, Mexico. University of Kansas Publications, Museum of Natural History 9:125–335.

Baker, R. H., and M. W. Baker. 1975. Montane habitat used by spotted skunk (*Spilogale putorius*) in Mexico. Journal of Mammalogy 56:671–73.

Banfield, A. W. 1974. The mammals of Canada. University of Toronto, Toronto.

Bixler, A. 2000. Genetic variability in striped skunks (*Mephitis mephitis*). American Midland Naturalist 143:370–76.

Bixler, A., and J. Gittleman. 2000. Variation in home range and use of habitat in the striped skunk (*Mephitis mephitis*). Journal of Zoology (London) 251:525–33.

Bjorge, R. R. 1977. Population dynamics, denning and movements of striped skunks in central Alberta. M.S. Thesis, University of Alberta, Edmonton, Canada.

Bjorge, R. R., J. R. Gunson, and W. M. Samuel. 1981. Population characteristics and movements of striped skunks (*Mephitis mephitis*) in central Alberta. Canadian Field-Naturalist 95:149–55.

Bohn, F., J. Turn, S. Richards, and D. Eveleth. 1955. Listerosis of a skunk. Bimonthly Bulletin, North Dakota Agricultural Experimental Station 18:49–50.

Broad, S., R. Luxmoore, and M. Jenkins. 1988. Significant trade in wildlife: A review of selected species in CITES Appendix II. Vol. 1: Mammals. IUCN, Gland, Switzerland.

Burns, E. 1953. The sex life of wild animals. A North American study. Rinehart, New York.

Cahalane, V. E. 1961. Mammals of North America. Macmillan, New York.

Carey, A. B., and J. E. Kershner. 1996. *Spilogale gracilis* in upland forests of western Washington and Oregon. Northwestern Naturalist 77:29–34.

Carr, D. E. 1974. Predatory behavior in the striped skunk (*Mephitis mephitis*). Ph.D. Dissertation, Cornell University, Ithaca, NY.

Casey, G., and W. Webster. 1975. Age and sex determination of striped skunk (*Mephitis mephitis*) from Ontario, Manitoba and Quebec. Canadian Journal of Zoology 53:223–26.

Ceballos, G., and A. Miranda. 1986. Los mamíferos de Chamela, Jalisco. Universidad Nacional Autónoma de México, Mexico.

Ceballos Gonzales, G., and C. Galindo Leal. 1984. Mamiferos silvestres de la Cuenca de Mexico. Edit Limusa, Mexico.

Chandler, A. C. 1952. Two new species of *Oochoristica* from Minnesota skunks. American Midland Naturalist 48:69–73.

Chapman, F. B. 1946. An interesting feeding habit of skunks. Journal of Mammalogy 27:397.

Choate, J. R., E. D. Fleharty, and R. J. Little. 1974. Status of the spotted skunk *Spilogale putorius* in Kansas, USA. Transactions of the Kansas Academy of Science 76:226–33.

Constantine, G. 1961. Gestation period in the spotted skunk. Journal of Mammalogy 42:421–22.

Crabb, W. D. 1941a. Food habits of the prairie spotted skunk in southeastern Iowa. Journal of Mammalogy 22:349–64.

Crabb, W. D. 1941b. A technique for trapping and tagging spotted skunks. Journal of Wildlife Management 5:371–74.
Crabb, W. D. 1943. Ecology and management of the prairie spotted skunk (*Spilogale interupta*) in southeastern Iowa. Ph.D. Thesis, Iowa State University, Ames.
Crabb, W. D. 1944. Growth, development and seasonal weights of spotted skunks. Journal of Mammalogy 25:213–21.
Crabb, W. D. 1948. The ecology and management of the prairie spotted skunk in Iowa. Ecological Monograph 18:201–32.
Crabtree, R. L., and M. Wolfe. 1988. Effects of alternate prey on skunk predation of waterfowl nests. Wildlife Society Bulletin 16:163–69.
Crooks, K. 1994a. Den-site selection in the island spotted skunk of Santa Cruz Island, California. Southwestern Naturalist 39:354–57.
Crooks, K. 1994b. Demography and status of the island fox and the island spotted skunk on Santa Cruz Island, California. Southwestern Naturalist 39:257–62.
Crooks, K. 1999. Western spotted skunk *Spilogale gracilis*. Pages 183–85 *in* D. E. Wilson and S. Ruff, eds. The Smithsonian book of North American mammals. Smithsonian Institution Press, Washington, DC.
Crooks, K., and D. Van Vuren. 1995. Resource utilization by two insular endemic mammalian carnivores, the island fox and island spotted skunk. Oecologia 104:301–7.
Dalquest, W. W. 1953. Mammals of the Mexican State of San Luis Potosi. Louisiana State University Press, Baton Rouge.
Davidson, W., and V. Nettles. 1997. Field manual of wildlife diseases in the Southeastern United States, 2nd ed. Southeastern Cooperative Wildlife Disease Study, University of Georgia, Athens.
Davis, W. B. 1945. Texas skunks. Texas Game and Fish 3(8):8–10, 25–26.
Davis, W. B. 1951. Texas skunks. Texas Game and Fish 9:18–21.
Davis, W. B. 1966. The mammals of Texas (Bulletin No. 41). Texas Parks and Wildlife Department.
Davis, W. B., and P. W. Lukens. 1958. Mammals of the Mexican state of Guerrero, exclusive of Chiroptera and Rodentia. Journal of Mammalogy 39:347–67.
Davis, W. B., and R. J. Russell. 1954. Mammals of the Mexican State of Morelos. Journal of Mammalogy 35:63–80.
Dice, L. R. 1921. Erroneous ideas concerning skunks. Journal of Mammalogy 2:38.
Dragoo, J. W., and R. L. Honeycutt. 1997. Systematics of mustelid-like carnivores. Journal of Mammalogy 78:426–43.
Dragoo, J. W., and R. L. Honeycutt. 1999a. Eastern hog-nosed skunk *Conepatus leuconotus*. Pages 190–91 *in* D. E. Wilson and S. Ruff, eds. The Smithsonian book of North American mammals. Smithsonian Institution Press, Washington, DC.
Dragoo, J. W., and R. L. Honeycutt. 1999b. Western hog-nosed skunk *Conepatus mesoleucus*. Pages 191–92 *in* D. E. Wilson and S. Ruff, eds. The Smithsonian book of North American mammals. Smithsonian Institution Press, Washington, DC.
Dragoo, J. W., and R. L. Honeycutt. 1999c. Eastern spotted skunk *Spilogale putorius*. Pages 185–86 *in* D. E. Wilson and S. Ruff, eds. The Smithsonian book of North American mammals. Smithsonian Institution Press, Washington, DC.
Dragoo, J. W., R. D. Bradley, R. L. Honeycutt, and J. W. Templeton. 1993. Phylogenetic relationships among the skunks: A molecular perspective. Journal of Mammalian Evolution 1:255–67.
Durant, A., and E. Doll. 1939. Pulmonary aspergillosis in a skunk. Journal of the American Veterinary Medical Association 95:645–46.
Egoscue, H. J., J. G. Bittmenn, and J. A. Petrovich. 1970. Some fecundity and longevity records for captive small mammals. Journal of Mammalogy 51:622–23.
Ehrhart, L. M. 1974. Ecological studies of the spotted skunk, *Spilogale putorius* Gray (Carnivora) on the east coast of Florida. Transactions of the First International Theriological Congress 1:154–55.
Emmons, C., D. Rowley, B. Olson, C. Mattern, J. Bell, E. Powell, and E. Marcey. 1955. Histoplasmosis: Proved occurrence of inapparent infection in dogs, cats and other animals. American Journal of Hygiene 61:40–44.
Erickson, A. B. 1946. Incidence of worm parasites in Minnesota Mustelidae and host lists and keys to North American species. American Midland Naturalist 36:494–509.
Ernst, C. H. 1965. Rutting activities in a captive striped skunk. Journal of Mammalogy 46:702–3.
Errington, P. L., F. Hamerstrom, and F. N. Hamerstrom, Jr. 1940. The great horned owl and its prey in northcentral United States. Iowa Agriculture Experimental Station Research Bulletin 227:757–850.
Ferguson, S. H., J. A. Virgl, and S. Larivière. 1996. Evolution of delayed implantation and associated grade shifts in life history traits of North American carnivores. Ecoscience 3:7–17.
Ferris, D. H., and R. D. Andrews. 1967. Parameters of a natural focus of *Leptospira pomona* in skunks and opossums. Bulletin Wildlife Disease Association 3:2–10.
Findley, J. S., and W. Caire. 1974. The status of mammals in the northern region of the Chihuahuan Desert. Pages 127–39 *in* R. H. Wauer and D. H. Riskind, eds. Transactions on the symposium on the biological resources of the Chihuahuan Desert Region (Transactions and Proceedings, Serial No. 3). U.S. National Park Service.
Findley, J. S., A. H. Harris, D. E. Wilson, and C. Jones. 1975. Mammals of New Mexico. University of New Mexico Press, Albuquerque.
Foresman, K. R., and R. A. Mead. 1973. Duration of post-implantation in a western subspecies of the spotted skunk (*Spilogale putorius*). Journal of Mammalogy 54:521–23.
Frank, P. A., and K. R. Lips. 1989. Gopher tortoise burrow use by long-tailed weasels and spotted skunks. Florida Field Naturalist 17:20–22.
Frederickson, L. F. 1983. Use of radiographs to age badger and striped skunk. Wildlife Society Bulletin 11:297–99.
Gharaibeh, B. M., and C. Jones. 1999. Hooded skunk *Mephitis macroura*. Pages 186–87 *in* D. E. Wilson and S. Ruff, eds. The Smithsonian book of North American Mammals. Smithsonian Institution Press, Washington, DC.
Godin, A. J. 1982. Striped and hooded skunks (*Mephitis mephitis* and allies). Pages 674–87 *in* J. Chapman and G. A. Feldhamer, eds. Wild mammals of North America: Biology, management, and economics. Johns Hopkins University Press, Baltimore.
Goetzee, J. R., and A. D. Nelson. 1998. Noteworthy records of mammals from central and south Texas. Texas Journal of Science 50:255–58.
Goetzee, J. R., F. D. Yancey II, C. Jones, and B. M. Gharaibeh. 1995. Noteworthy records of mammals from the Edwards Plateau of central Texas. Texas Journal of Science 47:3–8.
Gorman, G., S. Mckever, and R. Grimes. 1962. Leptospirosis in wild mammals from southwestern Georgia. American Journal of Tropical Medicine and Hygiene 11:518–24.
Greensides, R. D., and R. A. Mead. 1973. Ovulation in the spotted skunk (*Spilogale putorius latifrons*). Biology of Reproduction 8:576–84.
Greenwood, R. J. and A. B. Sargeant. 1998. Age related reproduction in striped skunks (*Mephitis mephitis*) in the upper midwest. Journal of Mammalogy 75:657–62.
Greenwood, R. J., W. E. Newton, G. Pearson, and G. J. Schamber. 1997. Population and movement characteristics of radio collared striped skunks in North Dakota during an epizootic of rabies. Journal of Wildlife Diseases 33:226–41.
Greenwood, R. J., A. B. Sargeant, J. L. Piehl, D. Abuhl, and B. A. Hansen. 1999. Foods and foraging of prairie striped skunks during the avian nesting season. Wildlife Society Bulletin 27:823–32.
Gremillion-Smith, C., and A. Wolff. 1988. Epizootiology of skunk rabies in North America. Journal of Wildlife Diseases 24:620–26.
Grinnell, J., J. S. Dixon, and J. M. Linsdale. 1937. Furbearing animals of California: Their natural history, systematic status, and relations to man, Vol. 1. University of California Press, Berkeley.
Gunson, J. R., and R. R. Bjorge. 1979. Winter denning of the striped skunk in Alberta. Canadian Field-Naturalist 93:252–58.
Hall, E. R. 1981. The mammals of North America, 2nd ed. John Wiley, New York.
Hall, E. R., and K. R. Kelson. 1959. The mammals of North America, Vol. 2. Ronald Press, New York.
Hamilton, W. J., Jr. 1936. Seasonal food of skunks in New York. Journal of Mammalogy 17:240–46.
Hamilton, W. J., Jr. 1937. Winter activity of the skunk. Ecology 18:326–27.
Hamilton, W. J., Jr. 1963. Reproduction of the striped skunk in New York. Journal of Mammalogy 44:123–24.
Hamilton, W. J., and J. O. Whitaker. 1979. Mammals of the eastern United States. Cornell University Press, Ithaca, NY.
Hayssen, V., A. Van Tienhoven, and A. Van Tienhoven. 1993. Asdell's patterns of mammalian reproduction: A compendium of species-specific data. Comstock, Ithaca, NY.
Hendricks, S. L., and V. A. Seaton. 1969. Rabies in wild animals trapped for pelts. Bulletin of the Wildlife Diseases Association 5:231–34.
Hirzel, D. J., J. Wang, S. K. Das, S. K. Dey, and R. A. Mead. 1999. Changes in uterine expression of leukemia inhibitory factor during pregnancy in the western spotted skunk. Biology of Reproduction 60:484–92.
Hoffmeister, D. F. 1986. Mammals of Arizona. University of Arizona Press, Tucson.

Honacki, J. H., K. E. Kinman, and J. W. Koeppl. 1982. Mammal species of the world: A taxonomic and geographic reference. Allen Press, Lawrence, KS.

Houseknecht, C. R. 1969. Denning habits of the striped skunk and the exposure potential for disease. Bulletin of the Wildlife Diseases Association 5:302–6.

Houseknecht, C. R. 1971. Movements, activity patterns and denning habits of striped skunks (*Mephitis mephitis*) and exposure potential for disease. Ph.D. Dissertation, University of Minnesota, Minneapolis.

Houseknecht, C. R., and J. Tester. 1978. Denning habits of striped skunks (*Mephitis mephitis*). American Midland Naturalist 100:424–30.

Howard, W. E., and R. E. Marsh. 1982. Spotted and hog-nosed skunks. Pages 664–73 *in* J. A. Chapman and G. A. Feldhamer, eds. Wild mammals of North America: Biology, management, and economics. Johns Hopkins University Press, Baltimore.

Howell, A. H. 1906. Revision of the skunks of the genus *Spilogale*. North American Fauna 26:1–55.

Hsu, C., and R. A. Mead. 1969. Mechanisms of chromosomal changes in mammalian speciation. Pages 8–17 *in* K. Benirschke, ed. Comparative mammalian cytogenetics. Springer-Verlag, New York.

Hubbard, J. P. 1972. Hooded skunk on the Mongollon Plateau, New Mexico. Southwest Naturalist 16:458.

Hwang, Y., and S. Larivière. 2001. *Mephitis macroura*. American Society of Mammalogists.

Jackson, H. 1961. Mammals of Wisconsin. University of Wisconsin Press, Madison.

Johnson, C. E. 1921. The "hand-stand" habit of the spotted skunk. Journal of Mammalogy 2:87–89.

Johnson, H. N. 1959. The role of the spotted skunk in rabies. Proceedings of the U.S. Livestock Sanitary Association 63:267–74.

Jones, F. H. 1950. Natural history of the striped skunk in Northeastern Kansas. M.A. Thesis, University of Kansas, Lawrence.

Jones, H. W., Jr. 1939. Winter studies of skunks in Pennsylvania. Journal of Mammalogy 20:254–56.

Johnston, D. H., and I. D. Watt. 1981. A rapid method for sectioning undecalcified carnivore teeth for aging. Pages 407–22 *in* J. A. Chapman and D. Pursley, eds. Proceedings of the worldwide furbearer conference, Frostburg, MD.

Johnston, D. H., I. D. Watt, D. Joachim, M. Novak, S. Strathearn, and M. Strickland. 1982. Techniques for aging furbearers using tooth histology. Page 67 *in* G. C. Sanderson, ed. Midwest furbearer management proceedings, 43rd Midwest fish and wildlife conference. Wichita, KS.

Kalmbach, E. R. 1938. A comparative study of nesting waterfowl on Lower Souris Refuge, 1936–1937. Transactions of the North American Wildlife Conference 3:610–23.

Kaplan, J. B., and R. A. Mead. 1992. Evaluation of extenders and cryopreservatives for cooling and cryopreservation of spermatozoa from the western spotted skunk (*Spilogale gracilis*). Zoo Biology 11:397–404.

Kaplan, J. B., and R. A. Mead. 1993. Influence of season on seminal characteristics, testis size and serum testosterone in the western spotted skunk (*Spilogale gracilis*). Journal of Reproduction and Fertility 98:321–26.

Kinlaw, A. 1995. *Spilogale putorius*. Mammalian Species 511:1–7.

Kinlaw, A. E., L. M. Ehrhart, and P. D. Doerr. 1995a. Spotted skunks (*Spilogale putorius ambarvalis*) trapped at Canaveral National Seashore and Merritt Island, Florida. Florida Field Naturalist 23:57–61.

Kinlaw, A., L. M. Ehrhart, P. D. Doerr, K. P. Pollock, and J. E. Hines. 1995b. Population estimate of spotted skunks (*Spilogale putorius*) on a Florida barrier island. Florida Scientist 58:47–54.

Knudsen, K. L., and D. L. Kilgore, Jr. 1990. Temperature regulation and basal metabolic rate in the spotted skunk, *Spilogale putorius*. Comparative Biochemistry and Physiology 97:27–33.

Krebs, J., J. Smith, C. Rupprecht, and J. Childs. 1997. Rabies surveillance in the United States during 1996. Journal of the American Veterinary Medical Association 211:1525–39.

Larivière, S. 1998. The radiating mousing technique of the striped skunk. Blue Jay 56:218–20.

Larivière, S., and F. Messier. 1996a. Immobilization of striped skunks with Telazol. Wildlife Society Bulletin 24:713–16.

Larivière, S., and F. Messier. 1996b. Aposematic behavior in the striped skunk, *Mephitis mephitis*. Ethology 102:986–92.

Larivière, S., and F. Messier. 1997a. Seasonal and daily activity patterns of striped skunks (*Mephitis mephitis*) in the Canadian prairies. Journal of Zoology 423:255–62.

Lariviere, S., and F. Messier. 1997b. Characteristics of waterfowl nest predation by striped skunk (*Mephitis mephitis*): Can predators be identified from nest remains? American Midland Naturalist 137:393–96.

Lariviere, S., and F. Messier. 1998a. Spatial organization of a prairie striped skunk population during waterfowl nesting season. Journal of Wildlife Management 62:199–204.

Lariviere, S., and F. Messier. 1998b. Denning ecology of the striped skunk in the Canadian prairies: Implications for waterfowl nest predation. Journal of Applied Ecology 35:207–13.

Larivière, S., and F. Messier. 1998c. The influence of close-range radio-tracking on the behavior of free-ranging striped skunks, *Mephitis mephitis*. Canadian Field-Naturalist 112:657–60.

Larivière, S., and F. Messier. 1998d. Effect of density and nearest neighbours on simulated waterfowl nests: Can predators recognize high-density nesting patches? Oikos 83:12–20.

Lariviere, S., and F. Messier. 1999. Review and perspective of methods used to capture and handle skunks. Pages 141–54 *in* G. Proulx, ed. Mammal trapping. Alpha Wildlife Research and Management, Sherwood Park, Alberta, Canada.

Larivière, S., and F. Messier. 2000. Habitat selection and use of edges by striped skunks in the Canadian prairies. Canadian Journal of Zoology 78:366–72.

Larivière, S., and L. R. Walton. 1998. Eggshell removal by duck hens following partial nest depredation by striped skunk. Prairie Naturalist 30:183–85.

Larivière, S., L. R. Walton, and F. Messier. 1999. Selection by striped skunks (*Mephitis mephitis*) of farmsteads and buildings as denning sites. American Midland Naturalist 142:96–101.

Layne, J. N. 1971. Fleas (Siphonaptera) of Florida. Florida Entomologist 54:35–51.

Leach, D., B. K. Hall, and A. I. Dagg. 1982. Aging marten and fisher by development of the suprafabellar tubercle. Journal of Wildlife Management 46:246–47.

Leopold, A. S. 1959. Wildlife of Mexico: The game birds and mammals. University of California Press, Los Angeles.

Leopold, A. S. 1965. Fauna silvestre de Mexico. Instituto Mexicano de Recursos Naturales Renovables, Mexico.

Linduska, J. P. 1947. Longevity of some Michigan farm game animals. Journal of Mammalogy 28:126–29.

Lowery, G. H. 1974. The mammals of Louisiana and its adjacent waters. Louisiana State University, Baton Rouge.

MacInnes, C. D. 1988. Control of wildlife rabies: The Americas. Pages 381–405 *in* J. Campbell and K. Charlton, eds. Rabies. Kluwer, Boston.

MacLulich, D. A. 1936. Running speeds of skunks and European hare. Canadian Field-Naturalist 50:92.

Manaro, A. J. 1961. Observations on the behavior of the spotted skunk in Florida. Quarterly Journal of the Florida Academy of Science 24:59–63.

Manning, R. W., J. K. Jones, Jr., and R. R. Hollander. 1986. Northern limits of distribution of the hog-nosed skunk, *Conepatus mesoleucus*, in Texas. Texas Journal of Science 38:289–91.

McCullough, C. R., and E. K. Fritzell. 1984. Ecological observations of eastern spotted skunks on the Ozark Plateau. Transactions of the Missouri Academy of Science 18:25–32.

McKeever, S., G. W. Gorman, J. F. Chapman, M. M. Galton, and D. K. Powers. 1958. Incidence of leptospirosis in wild mammals from southwestern Georgia, with a report of new hosts for six serotypes of leptospires. American Journal of Tropical Medicine and Hygiene 7:646–55.

Mead, R. A. 1963. Some aspects of parasitism in skunks of the Sacramento Valley of California. American Midland Naturalist 70:164–67.

Mead, R. A. 1967. Age determination in the spotted skunk. Journal of Mammalogy 48:606–16.

Mead, R. A. 1968a. Reproduction in eastern forms of the spotted skunk (genus *Spilogale*). Journal of Zoology (London) 156:119–36.

Mead, R. A. 1968b. Reproduction in western forms of the spotted skunk (genus *Spilogale*). Journal of Mammalogy 49:373–90.

Mead, R. A. 1970. The reproductive organs of the male spotted skunk (*Spilogale putorius*). Anatomical Record 167:291–302.

Merriam, C. H. 1902. Six new skunks of the genus *Conepatus*. Proceedings of the Biological Society of Washington 15:161–65.

Musgrove, B., and G. Blair. 1979. Fur trapping. Winchester Press, New York.

Nams, V. O. 1991. Olfactory search images in striped skunks. Behaviour 119:267–84.

Nams, V. O. 1997. Density-dependent predation by skunk using olfactory search images. Oecologia 110:440–48.

Nelson, E. W. 1918. Smaller North American mammals. An intimate study of the smaller wild animals of North America by the foremost authorities. National Geographic Society Magazine 33:477–79.

Novak, M., M. E. Obbard, J. G. Jones, R. Newman, A. Booth, A. J. Satterthwaite, and G. Linscombe. 1987. Furbearer harvests in North America, 1600–1984. Ontario Ministry of Natural Resources and Ontario Trappers Association, Toronto.

Nowak, R. M. 1999. Walker's mammals of the world, 6th ed. Johns Hopkins University Press, Baltimore.

Orr, R. T. 1943. Altitudinal record for the spotted skunk in California. Journal of Mammalogy 24:270.

Owen, J. G., R. J. Baker, and S. L. Williams. 1996. Karyotypic variation in spotted skunks (Carnivora: Mustelidae: *Spilogale*) from Texas, Mexico and El Salvador. Texas Journal of Science 48:119–22.

Packard, R. L. 1965. Range extension of the hooded skunk in Texas and Mexico. Journal of Mammalogy 46:102.

Packard, R. L. 1974. Mammals of the southern Chihuahuan desert: An inventory. Pages 141–53 *in* R. H. Wauer and D. H. Riskind, eds. Transactions of the symposium on the biological resources of the Chihuahuan Desert Region (Serial No. 3). U.S. National Park Service.

Parks, E. 1967. Second litters in the striped skunk. New York Fish and Game Journal 14:208–9.

Patton, R. F. 1974. Ecological and behavioral relationships of the skunks of Trans Pecos, Texas. Ph.D. Dissertation, Texas A&M University, College Station.

Peery, H. J. 1939. A new unarmed tapeworm from the spotted skunk. Journal of Parasitology 25:487–90.

Petrides, G. 1950. The determination of sex and age ratios in fur animals. American Midland Naturalist 43:355–82.

Pocock, R. I. 1921. On the external characters and classification of the Mustelidae. Proceedings of the Zoological Society of London 1921:803–23.

Polder, E. 1968. Spotted skunk and weasel populations den and cover usage in northeast Iowa. Iowa Academy of Science 75:142–46.

Pybus, M. 1988. Rabies and rabies control in striped skunks (*Mephitis mephitis*) in three prairie regions of western North America. Journal of Wildlife Diseases 24:434–49.

Reed, C. A., and W. H. Carr. 1949. Use of cactus as protection by hooded skunk. Journal of Mammalogy 30:79–80.

Reid, F. A. 1955. A field guide to the mammals of Central America and south east Mexico. Oxford University Press, New York.

Root, D. A., and N. E. Payne. 1984. Evaluation of techniques for aging gray fox. Journal of Wildlife Management 48:926–33.

Rosatte, R. C. 1984a. Seasonal occurrence and habitat preference of rabid skunks in southern Alberta. Canadian Veterinary Journal 25:142–44.

Rosatte, R. C. 1984b. The epidemiology, history and control of rabies in Alberta. M.A. Thesis, Norwich University, Montpelier, VT.

Rosatte, R. C. 1986. A strategy for urban rabies control: Social change implications. Ph.D. Dissertation, Walden University, Minneapolis, MN.

Rosatte, R. C. 1987. Striped, spotted, hooded, and hog-nosed skunk. Pages 599–613 *in* M. Novak, J. Baker, M. Obbard, and B. Malloch, eds. Wild furbearer management and conservation in North America. Ontario Trappers Association, Toronto.

Rosatte, R. C. 1988. Rabies in Canada: History, epidemiology and control. Canadian Veterinary Journal 29:362–65.

Rosatte, R. C., and J. R. Gunson. 1984a. Dispersal and home range of striped skunks (*Mephitis mephitis*) in an area of population reduction in southern Alberta. Canadian Field-Naturalist 98:315–19.

Rosatte, R. C., and J. R. Gunson. 1984b. Presence of neutralizing antibodies to rabies virus in striped skunks from areas free of skunk rabies in Alberta. Journal of Wildlife Diseases 20:171–76.

Rosatte, R. C., M. J. Pybus, and J. R. Gunson. 1986. Population reduction as a factor in the control of skunk rabies in Alberta. Journal of Wildlife Diseases 22:459–67.

Rosatte, R. C., M. J. Power, and C. D. MacInnes. 1991. Ecology of urban skunks, raccoons and foxes in Metropolitan Toronto. Pages 31–38 *in* L. W. Adams and D. L. Leedy, eds. Wildlife conservation in metropolitan environments. National Institute for Urban Wildlife, Columbia, MD.

Rosatte, R. C., M. J. Power, and C. D. MacInnes. 1992. Density, dispersion movements and habitat of skunks (*Mephitis mephitis*) and raccoons (*Procyon lotor*) in Metropolitan Toronto. Pages 932–44 *in* D. R. McCullough and R. B. Barrett, eds. Wildlife 2001: Populations. Elsevier, New York.

Roth, E. 1961. Leptospirosis in striped skunks. M.S. Thesis, Texas A&M College, College Station.

Rue, L. L. 1981. Furbearing animals of North America. Crown, New York.

Sargeant, A. B., B. R. Greenwood, J. L. Piehl, and W. B. Bicknell. 1982. Recurrence, mortality and dispersal of prairie striped skunks, *Mephitis mephitis*, and implications to rabies epizootiology. Canadian Field-Naturalist 96:312–16.

Schmidly, D. J. 1974. Factors governing the distribution of mammals in the Chihuahuan Desert Region. Pages 163–92 *in* R. H. Wauer and D. H. Riskind, eds. Transactions of the symposium on the biological resources of the Chihuahuan Desert Region (Serial No. 3). U.S. National Park Service.

Schmidly, D. J. 1984. The furbearers of Texas (Bulletin No. 111). Texas Parks and Wildlife Department.

Schowalter, D., and J. Gunson. 1982. Parameters of population and seasonal activity of striped skunks (*Mephitis mephitis*), in Alberta and Saskatchewan. Canadian Field Naturalist 96:409–20.

Schreiber, A., R. Wirth, M. Riffel, and H. Van Rompaey. 1989. Weasels, civets, mongooses, and their relatives: An action plan for the conservation of mustelids and viverrids. IUCN/SSC Mustelid and Viverrid Specialist Group, Gland, Switzerland.

Schubert, C. A., I. K. Barker, R. C. Rosatte, C. D. MacInnes, and T. D. Nudds. 1998. Effect of canine distemper on an urban raccoon population: An experiment. Ecological Applications 8:379–87.

Schwartz, C. W., and E. R. Schwartz. 1959. The wild mammals of Missouri. University of Missouri Press and Missouri Conservation, Columbia.

Scott, T. G., and L. F. Selko. 1939. A census of red foxes and striped skunks in Clay and Boone Counties, Iowa. Journal of Wildlife Management 3:92–98.

Selko, L. F. 1937. Food habits of Iowa skunks in the fall of 1936. Journal of Wildlife Management 1:70–76.

Selko, L. F. 1938. Notes on the den ecology of the striped skunk in Iowa. American Midland Naturalist 20:455–63.

Seton, E.T. 1929. Lives of game animals. Doubleday, Doran, Garden City, NY.

Shadle, A. R. 1953. Captive striped skunk produces two litters. Journal of Wildlife Management 17:388–89.

Smith, W. P. 1931. Calendar of disappearance and emergence of some hibernating mammals at Wells River, Vermont. Journal of Mammalogy 12:78–79.

Stegeman, L. 1939. Some parasites and pathological conditions of the skunk (*Mephitis mephitis nigra*) in central New York. Journal of Mammalogy 20:493–96.

Stoddard, H. L., and E. V. Komarck. 1941. Predator control in southeastern quail management. Transactions of the North American Wildlife Conference 6:288–93.

Storm, G. L. 1972. Daytime retreats and movements of skunks on farmlands in Illinois. Journal of Wildlife Management 36:31–45.

Sunquist, M. E. 1974. Winter activity of striped skunks (*Mephitis mephitis*) in east-central Minnesota. American Midland Naturalist 92:434–46.

Tiner, J. D. 1946. Some helminth parasites of skunks in Texas. Journal of Mammalogy 27:82–83.

Toland, B. 1991. Spotted skunk use of a gopher tortoise burrow for breeding. Florida Scientist 54:10–12.

Tyler, J. D. 1994. New records for the eastern spotted skunk in Oklahoma. Proceedings of the Oklahoma Academy of Science 74:49–50.

Upham, L. L. 1967. Density, dispersal, and dispersion of the striped skunk (*Mephitis mephitis*) in southeastern Dakota. M.S. Thesis, North Dakota State University, Fargo.

Van Gelder, R. G. 1953. The egg-opening technique of a spotted skunk. Journal of Mammalogy 34:255–56.

Van Gelder, R. G. 1959. A taxonomic revision of the spotted skunks (genus *Spilogale*). Bulletin of the American Museum of Natural History 117:229–392.

Verts, B. J. 1967. The biology of the striped skunk. University of Illinois Press, Urbana.

Vickery, P. D., M. L. Hunter, Jr., and J. V. Wells. 1992. Evidence of incidental nest predation and its effects on nests of threatened grassland birds. Oikos 63:281–88.

Wade-Smith, J., and B. J. Verts. 1982. *Mephitis mephitis*. Mammalian Species 173:1–7.

Wade-Smith, J., R. A. Mead, and H. Taylor. 1980. Hormonal and gestational evidence for delayed implantation in the striped skunk (*Mephitis mephitis*). General and Comparative Endocrinology 42:509–15.

Wade-Smith, R. A., and M. E. Richmond. 1978a. Reproduction in captive striped skunks (*Mephitis mephitis*). American Midland Naturalist 100:452–55.

Wade-Smith, R. A., and M. E. Richmond. 1978b. Induced ovulation, development of the corpus luteum, and tubal transport in the striped skunk (*Mephitis mephitis*). American Journal of Anatomy 153:123–42.

Walker, E. P. 1964. Mammals of the world. Johns Hopkins University Press, Baltimore.

Webster, W. 1967. *Filaroides mephitis* N. sp. (Metastrongyloidea: Filaroididae) from lungs of eastern Canadian skunks. Canadian Journal of Zoology 45:145–47.

Wight, H. M. 1931. Reproduction in the eastern skunk (*Mephitis mephitis nigra*). Journal of Mammalogy 12:42–47.

Wilber, C. G., and G. H. Weidenbacker. 1961. Swimming capacity of some wild mammals. Journal of Mammalogy 42:428–29.

Wilson, D. E., and D. M. Reeder. 1993. Mammal species of the world: A taxonomic and geographic reference, 2nd ed. Smithsonian Institution Press, Washington, DC.

Wood, W. F., C. G. Morgan, and A. Miller. 1991. Volatile components in defensive spray of the spotted skunk, *Spilogale putorius*. Journal of Chemical Ecology 17:1415–20.

Wood, W. F., C. O., Fisher, and G. A. Graham. 1993. Volatile components in defensive spray of the hog-nosed skunk, *Conepatus mesoleucus*. Journal of Chemical Ecology 19:837–41.

Ziener, H. G. 1975. Behavior of striped and spotted skunks. Ph.D. Dissertation, University of California, Davis.

Rick Rosatte, Ontario Ministry of Natural Resources, Wildlife Research and Development Section, Trent University, Peterborough, Ontario, Canada K9J 8N8. Email: rick.rosatte@mnr.gov.on.ca.

Serge Larivière, Delta Waterfowl Foundation, R.R. #1, Box 1, Portage La Prairie, Manitoba, Canada R1N 3A1. Email: slariviere@deltawaterfowl.org.

35

River Otter

Lontra canadensis

Wayne E. Melquist
Paul J. Polechla, Jr.
Dale Toweill

NOMENCLATURE

COMMON NAMES. There are 28 common names in native and exotic languages for *Lontra canadensis* in North America (Table 35.1). The most appropriate name from a biogeographic standpoint (Vaughan et al. 2000) is the Nearctic river otter (Lowery 1974).
SCIENTIFIC NAME. *Lontra canadensis*
SUBSPECIES. *L. c. canadensis, L. c. kodiacensis, L. c. lataxina, L. c. mira, L. c. pacifica, L. c. periclyzomae,* and *L. c. sonora* (van Zyll de Jong 1972)

Otter taxonomy has been problematic since Brisson (1762) first described the genus *Lutra* (Simpson 1945). Schreber (1776) rendered an engraving and then described the "American otter" as *Lutra canadensis.* From that time on, many scientists traversing the North American continent and collecting otters named a new species or subspecies without thorough examination of other scientific specimens. Consequently, many newly described taxa were based on differences regarding sex, age, stage of molt, and condition of the pelt. The confusion grew so large that Pohle (1920) was prompted to conduct a revision of the species. In subsequent years, two species and 20 subspecies of river otters were recognized (Harris 1968). Van Zyll de Jong (1968, 1972) amassed a data set (skull measurements, skeletal meristics, pelt reflectance, baculum shape) of 897 specimens (primarily skulls, skins, and postcranial skeletons) of otters from the New World, 591 of which were considered to be *Lutra canadensis* or *Lutra mira.* He compared these specimens to Old World otters ($n = 158$), including 59 Eurasian otters (*Lutra lutra*), and examined fossil remains. He then statistically analyzed his data and interpreted the results based on the ecological knowledge of the species to that time. The result was a recognition that the New World otters were significantly different from the Old World otters and resurrection of the old available generic name of Gray (1843, cited in Lariviere and Walton 1998). Furthermore, he synonomized *L. canadensis* and *L. mira* into one species and *L. mira* and the 19 previously recognized subspecies of *L. canadensis* (Hall and Kelson 1959) into 7 subspecies. He also concluded that New World otters originated in Eurasia and spread southward from the Bering Land Bridge across North America and the Panamanian Land Bridge. He regarded the Neotropical river otter (*Lontra longicaudis*) as a distinct species. Davis (1978) stated that Van Zyll de Jong's (1972) otter systematics was fraught with difficulties relating to small sample size and natural variability. Davis continued to revise Van Zyll de Jong's otter classification and lumped *Lontra canadensis, Lontra longicaudis,* and *Lontra provocax* into *L. canadensis* based on descriptions of vocalizations, shape of the bacula, and prepucial characteristics from a small sample size. Although he presented an alternative taxonomy of the otters, the evidence Davis (1978) presented was not sufficient by his own standards and that of taxonomists to warrant a valid revision. Although there have been no new data or analyses, several authors still follow Davis' (1978) classification scheme (Kellnhauser 1983; Chanin 1985; Davis 2000; Reed-Smith 2001). Van Zyll de Jong's reanalysis using phylogenetic data (1987) and his statistical reanalysis (1991) caused Wozencraft (1993) to conclude that Van Zyll de Jong's separation of the Nearctic and Eurasian otter genera and the Nearctic and Neotropical otter species should be accepted unless new information becomes available. Based on molecular phylogenetics (mitochondrial DNA nucleotide sequence of the cytochrome *b* gene), Koepfli and Wayne (1998, cited in Reed-Smith 2001) confirmed this relationship. Serfass et al. (1998) questioned the validity of the distinction between *L. c. canadensis* and *L. c. lataxina* in eastern North America (Van Zyll de Jong 1972; Hall 1981) and yet confirmed others. However, they did not offer an alternative designation of subspecies.

DISTRIBUTION

At the time of European settlement, the Nearctic otter (or river otter) inhabited much of the North American continent, excluding extensive permafrost regions and deserts without permanent streams (Hall and Kelson 1959; Hall 1981) (Fig. 35.1). River otters were found in all major waterways of the United States and Canada until at least the nineteenth century. Along with the beaver (*Castor canadensis*) and the timber wolf (*Canis lupus*), the river otter once occupied one of the largest geographic areas of any North American mammal, an area estimated by Anderson (1977) to encompass 20 million km^2. Its historical distribution extended from 25° 08′ N latitude in southern Florida (i.e., Tarpon Springs, Monroe County) to 68° 20′ N latitude in northern Alaska (i.e., Kanayut Lake, 24 km northeast of Tolugak Lake), and from 55° 30′ W longitude in eastern Newfoundland (i.e., Gander River) to 162° 49′ W longitude in western Alaska (i.e., Frosty Peak, Alaskan Peninsula) (Hall 1981; Polechla In press-a). By 1977, their distribution was <75% of the historical range (Fig. 35.2). By 1988, wetland conservation contributed to an otter recovery in >75% of the historical range (Fig. 35.3). Recent reintroduction programs contributed to recovery in nearly 90% of their historical range by 1998 (Fig. 35.4). River otters occur unequivocally in 48 states and 11 Canadian provinces. The status of river otters in New Mexico and Prince Edward Island is uncertain due to the lack of thorough surveys.

Human settlements, habitat destruction, and overharvest were undoubtedly responsible for extirpation from some portions of their range, especially those areas of marginal habitat (Toweill and Tabor 1982; Melquist and Dronkert 1987). Rucker (1983) reported that nearly half of the 87 million ha of wetlands in the United States have been drained for agricultural purposes and developments, with losses continuing at a rate of 185,000 ha/year. Canada is still in the process of inventorying its wetlands (C. Rubec, Canadian Wildlife Service, pers. commun., 2000), which constitute an estimated 25% (127 million km^2) of the world's wetlands. Preliminary data indicate that agricultural drainage accounts for 85% of Canadian wetland loss (Lord 1991). The future of wetlands and riparian areas will be critical to the future status and distribution of river otters in North America.

Otters were extirpated or rare in Arizona, Colorado, Indiana, Iowa, Kansas, Kentucky, Nebraska, New Mexico, North Dakota, Ohio, Oklahoma, South Dakota, Tennessee, Utah, and West Virginia (Park 1971; Endangered Species Scientific Authority 1978). Since the 1970s, reintroduction efforts have helped to restore otters or expand populations in 21 states and Alberta, Canada (Melquist and Dronkert 1987; Raesly

TABLE 35.1. Common names for *Lontra canadensis*

Common Name	Language	Source
American otter	English	Coues 1877
Canada otter	English	Audubon and Bachman 1851; Coues 1877
Canadian otter	English	Harris 1968; Toweill and Tabor 1982
Chah	Navajo	Polechla 2000
Cuilnguq	Yupik Eskimo	M. Kiokun, University of Alaska-Fairbanks, pers. commun., 1994; Kwaraiceus 1994
Fischotter	German	Reuther and Festetics 1980
Fish otter	English	Toweill and Tabor 1982
Kolta	Klamath	Bailey 1936
Ku-tet-tahx	Potawatomi	Reed-Smith 2001
Land Otter	English	Coues 1877; Wilson and Reeder 1993
Loutre	French	Lowery 1974; Polechla 1987a
Loutre de la Carolina	French	Coues 1877
Loutre de Riviere	French	Banfield 1974
Loutre du Canada	French	Coues 1877
Lutra	Latin	Schwartz and Schwartz 1981
Nannocks	Wasco	Bailey 1936
Nearctic river otter	English	Lowery 1974
Neeg-keek	Chippewa	Reed-Smith 2001
Neekeek	Cree, Ojibway, Sauteax	Coues 1877; Seton 1929
Nit-sook	Naskapi	Strong 1930
Nop'-e-ay	Chipewyan	Seton 1929
North American river otter	English	Coues 1877; Lariviere and Walton 1998
Northern river otter	English	Harris 1968; Wilson and Cole 2000
Nutria	Spanish	Gallo and Rojas-B. 1985; Polechla 2000
Odder	Danish	Lowery 1974
Oshan	Choctaw	Arthur 1928
Oter	Middle English	Lowery 1974
Otor	Old English	Lowery 1974
Otter	English, German	Lowery 1974
Pah-hua-pe'na	Keresan	Bailey 1936; Polechla 2000
Pahtsugo	Piute	Bailey 1936
Pat-cukee	Comanche	Polechla 2000
Perro de Agua	Spanish	Gallo and Rojas B. 1985; Polechla 2000
Pe-tang	Yantkton Sioux	Seton 1929
Prince of Wales otter	English	Toweill and Tabor 1982
Ptan	Ogallala Sioux	Seton 1929
River otter	English	Grinnell et al. 1937
Saquenu'ckot	Algonquian	Swanton 1946
See-hah	Zuni	Polechla 2000
Severoamerikanskaya vidra	Russian	Romanenko 1994
Un'-chuch	Montagnais	Seton 1929
Utter	Swedish	Lowery 1974
Xyinixka	Biloxi	Arthur 1928

2001). River otters are still relatively common in suitable habitats along the coasts of the Atlantic Ocean and the Gulf of Mexico, throughout the Pacific Northwest and Great Lake states, and across most of Canada and Alaska.

Magoun and Valkenburg (1977) reported that river otters occurred on the North Slope of Alaska and considered this a range extension. Rausch (1951) found that Inuit people of the area had previously recorded this species in the area, and so river otters may not be new to the region. With global climate change, river otter populations may prosper in these marginal northern areas.

Continent-wide surveys (Deems and Pursley 1978; Polechla 1990, In press-a) of natural resource and wildlife agencies revealed that the status of otter populations has improved during the past two decades (Table 35.2). In the United States, wildlife biologists reported river otter populations were stable in portions or all of 19 states, increasing in 29 states, and unknown in 4 states. In Arizona, populations had been decreasing in the past, but are locally increasing at reintroduction sites. Canadian wildlife biologists reported river otter populations were stable in eight provinces, increasing in two, and unknown in two others. River otters are adapted to existence in freshwater habitats and associated saltwater coastal areas, and because of that adaptation, arid areas, mountain ranges, glaciated areas, and saltwater straits have long been considered barriers to dispersal (Pohle 1920). Thus, the arid southwestern portion of North America may function as a barrier between the Nearctic river otter and Neotropical river otter (*Lontra longicaudis*), and the North Atlantic Ocean and the Bering Sea a barrier between the Nearctic otter and the Eurasian otter.

There are numerous accounts of river otters making extensive overland movements, including over mountain ranges and between drainages (Laughlin 1955; Morejohn 1969; Magoun and Valkenburg 1977; Melquist and Hornocker 1983). Mountains are not absolute barriers for otters, but may act as filters to slow otter migration and gene flow (Serfass et al. 1998). Although otters along the coastal areas of North America make extensive use of estuaries, coastal islands, and the exposed outer coast (Mowbray et al. 1979; Larsen 1983; Woolington 1984; Shannon 1989; Dubuc et al. 1990; Bowyer et al. 1995; Jackson et al. 1998; Blundell 2001), they do not cross the open ocean. Expansive glaciers and sea ice probably limit the northward range of river otters.

DESCRIPTION

In general body conformation, the river otter resembles a long cylinder that reaches its greatest diameter in the thoracic region (Tarasoff et al. 1972). The head is rather blunt, small, and somewhat flattened. It is

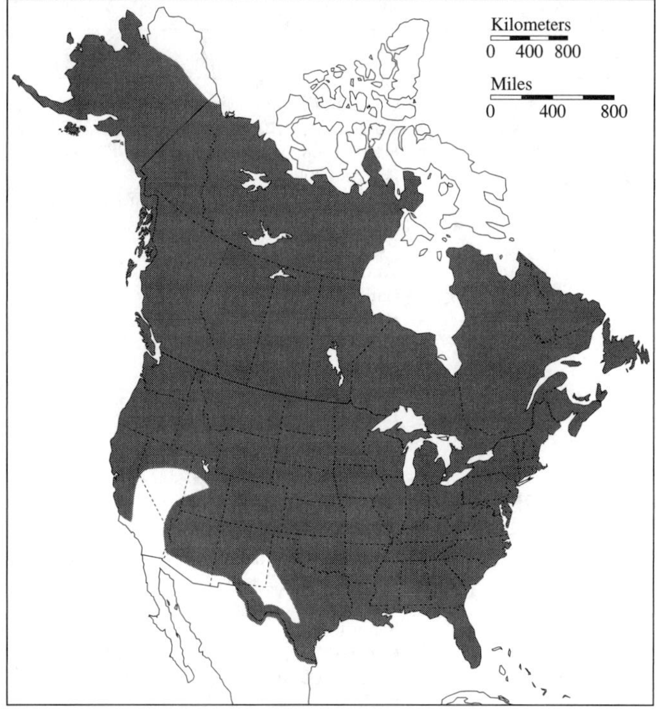

Figure 35.1. Distribution of the river otter (*Lontra canadensis*) during early European settlement. Source: Data from Anderson (1977), Hall (1981), and Polechla (1988, 1990).

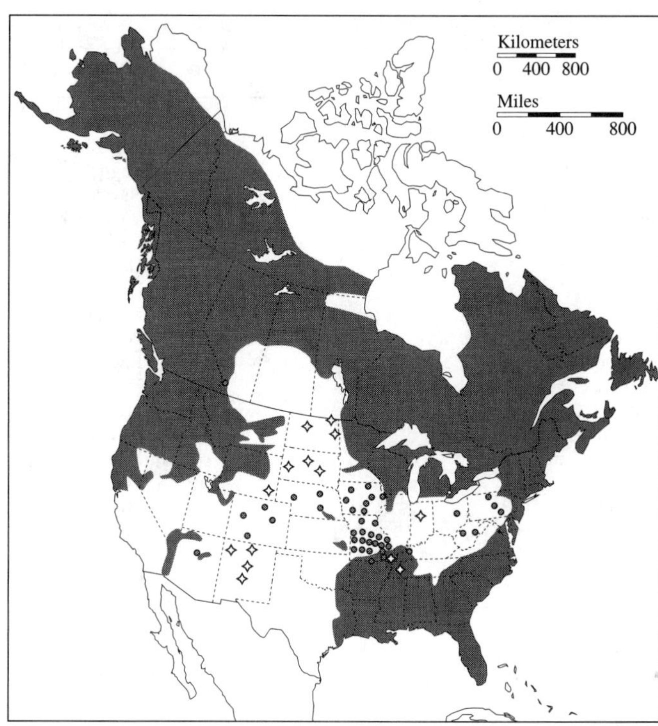

∘ Introductions
✧ Verified reports

Figure 35.3. Distribution of the river otter in 1988. Dots represent reintroduction and restocking sites; stars represent verified reports. Source: Data from Polechla (1990).

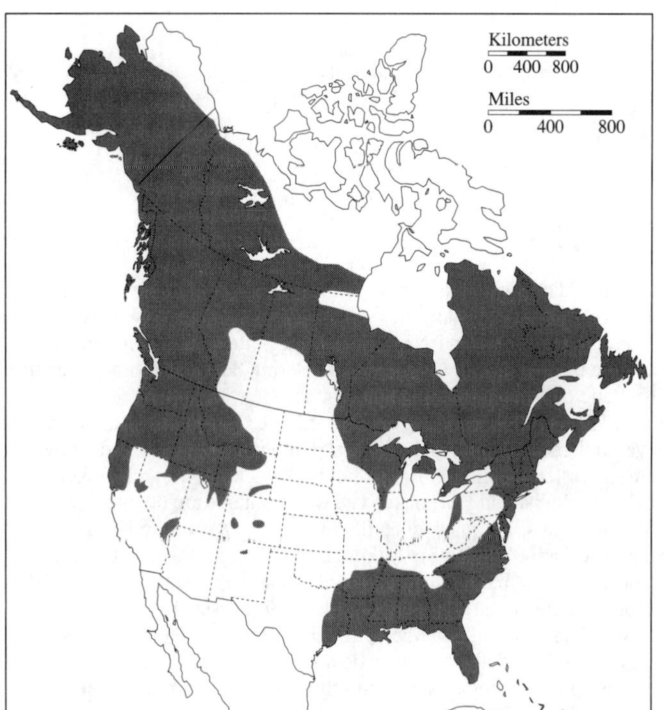

Figure 35.2. Distribution of the river otter in 1977. Source: Data from Deems and Pursley (1978) and Polechla (1990).

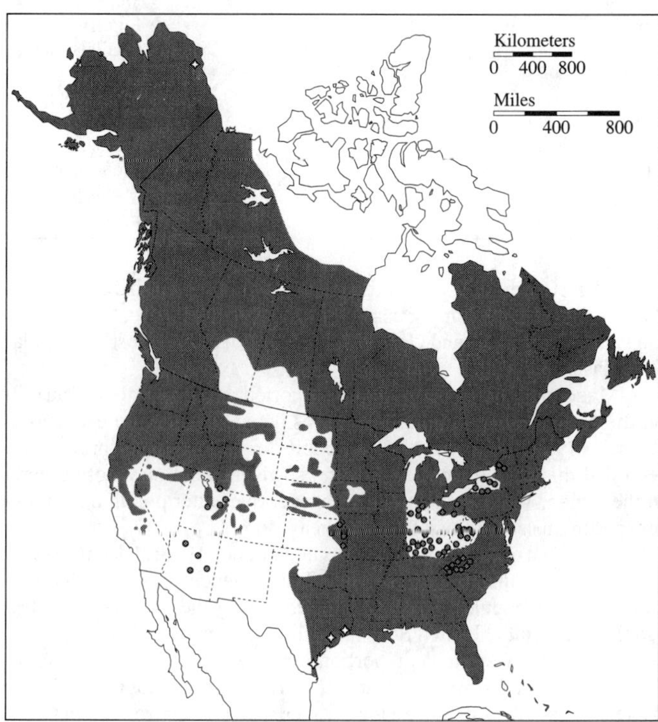

∘ Introduction
✧ Verified reports

Figure 35.4. Distribution of the river otter in 1998–1999. Dots represent reintroduction and restocking sites; stars represent verified reports.

TABLE 35.2. Estimated distribution (km²) of the river otter in North America over three time periods compared to historical distribution during European settlement

Time Period	Figure No.	Source	Estimated Distribution	Percentage of Historical Distribution	Percentage Change from Previous Period
European settlement	1	Anderson 1977; Hall 1981; Polechla 1988, 2000	20,000,000	100	Not applicable
Circa 1977	2	Deems and Pursley 1978	14,220,000	71	−29
Circa 1988	3	Polechla 1990*	15,560,000	78	+7
Circa 1998	4	Polechla In press-a*	17,660,000	88	+10

SOURCE*: Data derived from 100% written returns of wildlife biologists from across North America.

NOTE: Distribution is represented as a portion of the historical distribution at the time of European settlement.

characterized by a bulbous nose on the end of a short muzzle, small rounded ears set well back, and small eyes set high on the head and closer to the nose than to the ears (Fig. 35.5). The muscular neck is thick, cylindrical, and rarely smaller than the head. Legs are short and stocky, with the longer hind legs resulting in a hump-backed gait while traveling on land. Feet are pentadactyl and plantigrade, with interdigital webbing more pronounced on the longer toes of the hind foot. There are four small, rough protuberances on the plantar surface of each hind foot (Fig. 35.6), which appear as clusters of fused columns protruding from a hole in the epidermis. These plantar glands occur on several, but not all North American mustelids, and probably function as chemical transmitters (Buskirk et al. 1986) in addition to providing traction on slippery surfaces (Melquist and Dronkert 1987). The dorsoventrally flattened tail is relatively long and thick and tapers to a point. These features give the river otter a hydrodynamic shape, somewhat distinct from the other mustelids (Davis 1978).

Size and Weight. Adult river otters are 89–137 cm in length, with the tail accounting for 35–40% of the total (Park 1971; Hall 1981; Melquist and Hornocker 1983; Woolington 1984). Maximum length appears to be attained at 3–4 years of age (Stephenson 1977; Melquist and Hornocker 1983). Weight ranges from 3.4 to 15.44 kg for animals in the wild (Hall and Kelson 1959; Wilson 1959; Hall 1981; Woolington 1984; Polechla 1987a). Based on basal length of the skull, the largest of the Nearctic otters, the Prince of Wales Island otter (*L. c. mira*), occurs in southeastern Alaska, whereas the smallest form, *L. c. canadensis,* is found in eastern Canada and the Great Lakes and New England regions. The Southwestern river otter (*L. C. sonora*) of southwestern United States also is a very large subspecies; van Zyll de Jong (1972) identified a specimen with the longest skull (120.2 mm) of any in North America. A clinal decrease in cranial size exists from north to south along the Pacific Coast. However, clinal differences are not evident from west to east (van Zyll de Jong 1972).

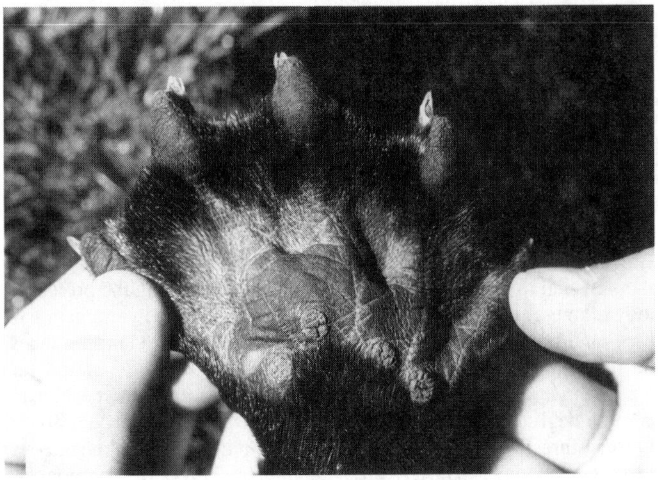

FIGURE 35.5. Plantar glands (four) on the hind foot of a river otter. SOURCE: Photo by Wayne E. Melquist.

Sexual dimorphism in size occurs among most subspecies of the river otter (Harris 1968; van Zyll de Jong 1972), although these differences are not always apparent in southeastern Alaska (Larsen 1983, 1984). Sexual dimorphism is more apparent in the Neotropical than the Nearctic otters (van Zyll de Jong 1972). Weights and measurements taken from otters in different regions indicate that females tend to be from 3% to 21% smaller than males of similar age (Harris 1968; van Zyll de Jong 1972; Stephenson 1977; Hall 1981; Melquist and Hornocker 1983; Polechla 1987a; Duffy et al. 1996; Kollias 1999; Blundell et al. 2002a).

FIGURE 35.6. River otter (*Lontra canadensis*). SOURCE: Photo by Nathan Varley Landis Wildlife Films.

TABLE 35.3. Dorsal coloration of river otter pelts as used by fur graders (Obbard 1987) and matched to naturalist's color guides (Smithe 1975)

Fur Industry Color Category	Naturalist's Color Guide	
	Description	Code
Extra dark	Jet black	89
Dark	Dusky brown	19
Dark brown	Dark grayish brown	20
Brown	Dark grayish brown-fuscous	20–21
Pale	Burnt umber	22
Extra pale	Hair brown	119A
Piebald[a]	"White" with burnt umber–raw umber spots	N.A. 22–23
Albino or lucistic[a]	"White"	N.A.

NOTE: Terms in quotation marks are not true colors.
[a]Biological term (Polechla 1987a).

Pelage. The river otter is characterized by short, dense, soft hair protected by longer, stiff, glossy guard hairs. Its fur is held in high regard by the fur trade, where it is considered a standard (100% durable) by which the durability and quality of all other furs are judged (Peterson 1914; Kaplan 1974; Obbard 1987). Color ranges from a rich, dark chocolate brown to a pale chestnut dorsally and light brown to silver gray ventrally (Obbard 1987; Polechla 1987a) (Table 35.3). Differences in length and density of the fur are related to climate, with northern forms having the longest and most dense pelage (van Zyll de Jong 1972). Similarly, western and southern forms tend to be lighter in color than northern and eastern forms; however, there may be considerable color variation within a region (Obbard 1987; Polechla 1987a). Characteristics of otter fur insulation, density, and patterns of molt are described by Obbard (1987).

Skull, Dentition, and Skeleton. The skull is depressed dorsoventrally and the dorsal edge of the profile presents nearly a straight line (Fig. 35.7). The rostrum is short and blunt. The point of maximum constriction is well forward, just posterior to the broad, flat interorbital area; behind this constriction, the cranium bulges to nearly the posterior margin of the skull. The auditory bullae, housing the inner ears, are flattened.

Dentition, while less massive than that of the sea otter (*Enhydra lutra*), is heavy when compared with that of most other mustelids. Teeth are adapted for crushing and cutting (Grinnell et al. 1937; van Zyll de Jong 1972). The dental formula is I 3/3, C 1/1, P 4/3, M 1/2. Supernumerary premolars have been reported (Dearden 1954; Polechla 1987a), but this condition does not appear to adversely affect populations (Beaver et al. 1981; Polechla 1987a).

River otters have 14 rib-bearing vertebrae and normally 52 vertebrae, including 7 cervical, 14 thoracic, 6 lumbar, 3 sacral, and 22 caudal (van Zyll de Jong 1972). Chevron-shaped bones, found from the fourth and successive caudal vertebrae, are associated with greater vascularization of the tail, and the lack of clavicle bones increases mobility of the forelimbs (Reed-Smith 2001). Fisher (1942) discussed the osteology of river otters in detail and Polechla (1991) and Reed-Smith (2001) contributed additional information on skeletal morphology.

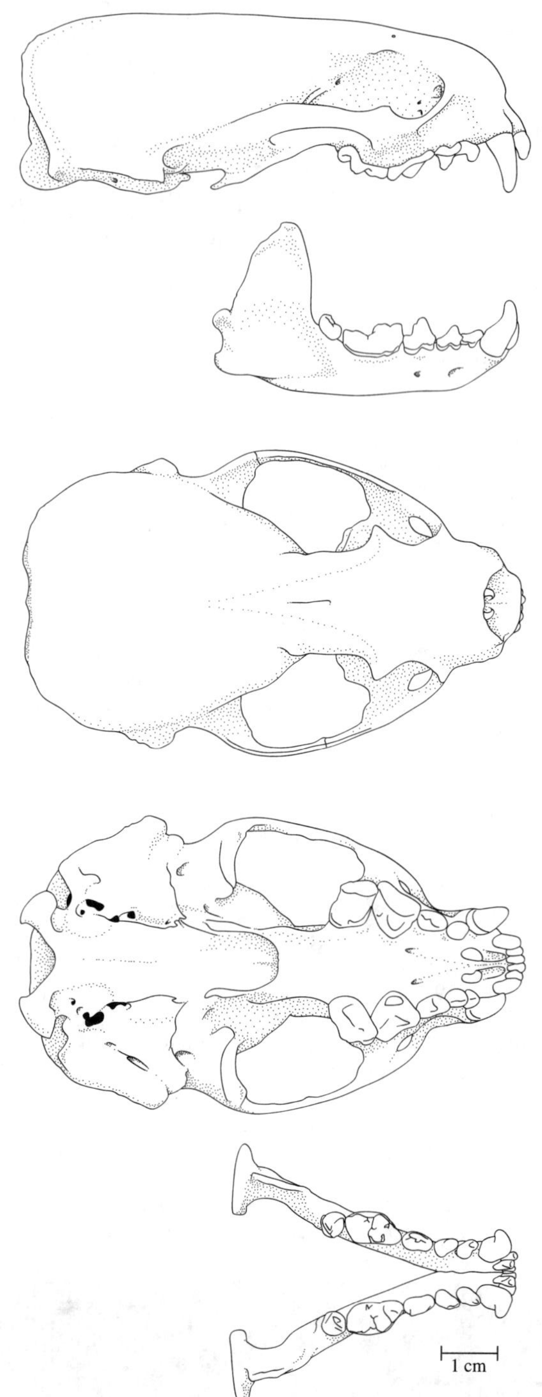

FIGURE 35.7. Skull of the river otter (*Lontra canadensis*). From top to bottom: lateral view of cranium, lateral view of mandible, dorsal view of cranium, ventral view of cranium, dorsal view of mandible.

GENETICS

The diploid (2*n*) number of chromosomes is 38. Thirteen pairs of autosomes are metacentric or submetacentric, while six pairs are acrocentric or subacrocentric (Wurster and Benirschke 1968).

PHYSIOLOGY

Polechla (1991) and Reed-Smith (2001) provided partial reviews of the physiology (and anatomy) of Nearctic otters in comparison to the rest of the members of the subfamily Lutrinae. The integument of Nearctic otters is vital to preventing hypothermia and maintaining the preferred body temperature in these semiaquatic mammals. This integument has three qualities that aid in regulating body temperature: (1) it produces the lipid squalene to enhance the water-repellant quality of the fur (Lindholm et al. 1981); (2) it contains piloerector muscles, which can vary the degree of loft to increase or decrease the insulatory quality of the fur (Tarasoff et al. 1972); and (3) it contains more subcutaneous fat (for insulation) than terrestrial mammals. Through experimentation with Eurasian otters, Green (1977) concluded that the whiskers or moustachial vibrissae aid otters in tactile detection of prey. Nearctic otters with comparable vibrissae are thought to function in a similar manner. In the wild, molt occurs annually (Ling 1970, cited in Worthy et al. 1987) beginning in late winter or early spring (Melquist and Dronkert 1987; Polechla 1987a), apparently proceeds from an anterior to posterior direction (Obbard 1987), and is completed by late fall.

Musculoskeletal System. Fisher (1942) described and drew the muscles of the Nearctic otter with their origins and insertions. A functional/morphological approach has been taken to describe the masticatory (Riley 1985) and locomotory movements of otters (Tarasoff et al. 1972; Fish 1994). Fagen (1986) demonstrated that the bite force of Nearctic otters varies from population to population. Coastal Nearctic otters have a greater bite force than interior Nearctic otters. Propulsion is provided by paddling the feet and thrusting with the tail and vertebral column especially when swimming underwater (Fish 1994). Wilson et al. (1984) tested the amount of torsion that the long bones could withstand for application in lead detection and designing ergonomic live-capture equipment for river otters.

Nervous, Endocrine, and Lymphatic Systems. The tactile senses of river otters are highly developed. In a study of sensory specialization of otter brains, Radinsky (1968) found that the coronal gyrus of the brain was enlarged, which suggests highly developed receptor fields in the head. He concluded that because river otters have expanded coronal gyri in their brains, they must have sensitive vibrissae for excellent somatic sensory reception. Consequently, river otters likely detect prey moving in turbid water through sensations received by the vibrissae (Harris 1968). This contention has been experimentally confirmed for the Eurasian otter (Green 1977).

An indication of the manual dexterity of river otters is suggested by an account of an otter's manipulating a small lead pellet underwater, fondling it in its paws, dropping it, and retrieving it from the bottom (Park 1971). The pellet was <3 mm in diameter.

Visual senses are not acute in the otter. Otters are nearsighted, an adaptation for underwater vision, but apparently can detect movement at considerable distances. River otters are capable of 40 diopters of accommodation due to well-developed intraocular muscles (Ballard et al. 1987). This adaptation allows for better underwater vision, where the light refractive index is relatively higher (Estes 1989). A Eurasian otter tested by Kasprzyk (1984) was able to detect blue, yellow, and green light, but not red. This ability is believed to be an adaptation for underwater vision, where the blue-green wavelength of the spectrum predominates (Estes 1989).

Auditory senses seem to be quite well developed. Harris (1968) reported that the hearing of all otters is exceptionally acute, and a further suggestion of this is found in the variety of noises otters make for communication (Park 1971; Davis 1978).

Little is known about the sense of smell, although indications that it is acute come from the fact that otters regularly communicate by means of scent marking. River otters have also been observed playing "hide and seek" with a newly dead fish in a manner that suggests the fish was located primarily by smell (Park 1971). Likewise, little is known concerning taste discrimination in otters. Only the endocrine system of the Eurasian otter has been studied (Boos 1987). Inflammation of the lymph nodes occurs occasionally (Hoover et al. 1985).

Cardiovascular System. Hematological values are available for the Nearctic otter (Kane 1979; Brandes et al. 1983; Reed-Smith 2001). They have a number of hematological adaptations for aquatic behavior. Fast clotting time and high number of thrombocytes enable otter blood to clot even though they may be in water (Brandes et al. 1983). The oxygen-carrying capacity of the blood is higher than that of terrestrial mammals to ensure adequate oxygen delivery to the tissues during a dive (Brandes et al. 1983).

Resting heart rate varies from 120 to 178 beats/min (Hoover et al. 1985; Spelman 1999; Reed-Smith 2001). Bradycardia, a slowing of the heart rate, is an adaptive dive response in the otter (Harris 1968). Anesthesia (i.e., ketamine hydrochloride and isoflurane) can induce bradycardia (Hoover and Jones 1986). Other adverse reactions to anesthesia are discussed by Kane (1979) and Polechla (1989a).

Respiratory System. The respiratory rate of the Nearctic otter is similar to that of the Cape clawless otter (*Aonyx capensis*) (Kane 1979; Hoover et al. 1985), ranging from 10 to 60 breaths/min (Spelman 1999). Hoover and Jones (1986) observed a decline in respiratory rate from 10 to 83 breaths/min after anesthesia. The long, narrow trachea of the Nearctic otter has overlapping calcified, incomplete rings and bifurcates into a pinnate bronchial tree arrangement of asymmetrical lobes of the lung (Tarasoff and Kooyman 1973a, 1973b). The pulmonary anatomy of the river otter exhibits only a small degree of specialization for aquatic existence compared to more aquatic mammals.

Digestive System. Otters have a highly efficient digestive system. Nearctic otters previously fed bland foods passed exoskeletal remains of crayfish about 1 hr after feeding (Liers 1951b). Davis et al. (1992b) experimentally determined that food transit time was 202 min for a sample size of 12 livetrapped male river otters and that daily food consumption averaged 33.9 g of dry matter/kg body weight. Copious amounts of mucous are found in the gastrointestinal tract, presumably to aid in protecting the inner lining of the stomach and intestines from abrasive food items such as vertebrae, chiton, and scales (Toweill 1974). Anesthetized river otters are particularly susceptible to gastroenteritis (Polechla 1989a).

Excretory System. Compared to the sea otter, the river otter has small kidneys and a relatively low electrolyte flux or urine osmolality (Costa 1982; Hoover and Tyler 1986). Although kidney stones have been noted in Asian small-clawed otters (*Aonyx cinerea;* Calle 1988), they have not been found in Nearctic otters.

Metabolic Rate. Tests conducted on Eurasian otters indicate that otters have rather high metabolic rates for land mammals (Iversen 1972). In fact, mustelids typically have basal metabolic rates about 20% above the standard curve for mammals (Estes 1989). Harris (1968) found that otters in captivity required about 700–900 g of prepared food per day. Serfass et al. (1990) reported that an adult otter consumes 1–1.5 kg fish/day.

Using a bioenergetic approach, Mack (1985) extrapolated from Iversen's (1972) basal metabolic rate adjusted for activity, gender, and sexual condition, and arrived at a standing crop of fish needed to sustain a pair of otters. He found that daily energy costs, based on annual simulations for a reproductively active female otter, varied from 934 to 2385 kcal/day, with basal metabolic rate and cost of reproduction accounting for a major portion of net energy budgets. On a sustained yield basis, an adult female otter would require approximately 2326 kg of available forage within an annual home range. Based on an analysis of the fish available in the Colorado study area, the annual energy availability exceeded energy requirements by a factor of six (Mack 1985).

REPRODUCTION AND DEVELOPMENT

Researchers have delved into many aspects of river otter reproductive biology. The reproductive anatomy and physiology of this species are complex and fascinating. Stenson (1985), Melquist and Dronkert (1987), and Polechla (1987a, In press-b) synthesized earlier knowledge of the reproductive biology of the species.

Reproductive Anatomy. The position of the external genitalia of male and female river otters is typical of most mammals (Thompson 1958). Gender can be properly distinguished among all age groups, even developing embryos (Polechla 1987a).

As in the typical male eutherian mammalian plan, the paired testes are located posterior to the penis and descend into the scrotum on maturity. Long (1969) demonstrated that the anatomy of the penis of the river otter is unique among mustelids in vascularization and musculature. The mature penis of the river otter encloses an ossified structure, the *os penis* or *baculum* (Friley 1949; Polechla 1987a), a characteristic shared by some bats, rodents, most primates, walruses, mustelids, and other carnivores. An ossified baculum may aid in intromission (Long 1969). As the male matures, the baculum changes from a J shape, to a hockey-stick shape, and finally to a structure approaching an S shape (Friley 1949; Polechla 1987a). Age groups can be discerned by baculum length and shape (Friley 1949).

Testes and epididymes with mature sperm are larger than those without (Polechla 1987a). Testes weights are heavier than what one would predict from river otter body weight, an adaptation for polygynous mating systems (Polechla 1987a). The testes are scrotal throughout the year in captive river otters in Kansas and Arkansas (Polechla 1987a), whereas they are scrotal only during the breeding season in more northern latitudes, such as Minnesota (Liers 1951b).

Female river otters possess paired ovaries located along the dorsal wall of the abdominal cavity, caudal to the kidneys, and adjacent to the paired oviducts, which connect with the uteri (Stenson 1985, Polechla 1987a). Their bicornuate uteri have two uterine openings, which join to form the singular cervix and extend to the vagina (Hamilton and Eadie 1964; Vaughan 1972; Sadleir 1973; Lauhachinda 1978; Polechla 1987a). The female river otter has an ossified instead of a cartilaginous clitoris, known as an *os clitoris* or *baubellum,* homologous to the male's baculum (Scheffer 1939; Ewer 1973; Lauhachinda 1978; Polechla 1987a). Mammalian species in which a baculum occurs in the male usually have a baubellum in the female (Feldhamer et al. 1999).

The ovaries increase in size as the female matures (Tabor 1974; Lauhachinda 1978; Polechla 1987a). River otter ovaries do not have fissured surfaces like those of the sea otter (Sinha et al. 1966; Polechla 1987a).

Age at Sexual Maturity. Male river otters exhibit mounting behavior while juveniles (Polechla 1987a), but may not become accomplished breeders until they are 5–7 years old (Liers 1951a). In northern populations, wild male river otters did not produce mature sperm as juveniles (Hamilton and Eadie 1964; Tabor 1974), whereas 10.6% of a southern population in Arkansas did (Polechla 1987a). Wild individuals in Arkansas first produced sperm at an estimated age of 8–20 months.

Estrus lasts from 20 to 56 days (Stenson 1985). Reproductive tract and cementum annuli analysis of dentition of wild river otter carcasses in Arkansas and British Columbia indicate that female river otters bred beginning at 1 year old. Liers (1958) observed a yearling female mating when 451 and 455 days old and giving birth at 766 days of age.

Breeding. Male river otters grasp a receptive female behind the neck during intromission, often leaving a noticeable wound (Liers 1951a; McDaniel 1963; Hediger 1966; Woolington 1984; Stenson 1985; Polechla 1987a; Reed-Smith 2001). Copulation may last from 13 to 73 min and normally occurs in the water, but can take place on land (Liers 1951b; McDaniel 1963; Hediger 1966; Shannon 1991; Tango et al. 1991). Copulation is thought to trigger the release of ova; thus, river otters are induced ovulators (Stenson 1985; Polechla 1987a). The ossified clitoris protrudes from the vulva and may aid in receiving sufficient copulatory stimuli to trigger ovulation (Polechla 1987a, In press-b).

Interbirth Period. The time elapsed between reproductive cycles (from birth to birth) is variable. Annual reproduction has been demonstrated in wild females from Oregon (Tabor and Wight 1977), Idaho (Melquist and Hornocker 1983), and Arkansas (Polechla 1987a), whereas biannual reproduction has been shown for wild populations in Alabama and Georgia (Lauhachinda 1978), Maryland (Mowbray et al. 1979), and Arkansas (Polechla 1987a). Liers (1951a, 1951b) cited examples of 1- and 2-year interbirth periods.

Delayed Implantation. River otters exhibit delayed implantation whereby the developing ball-shaped embryo, called the blastocyst, floats freely in the uterus for an extended period of time. Cocks (1881) observed the average gestation period in captive Eurasian otters to be 63 days. Liers (1951a, 1958) recorded the copulation-to-parturition period in the Nearctic otter to be 336.5 days (±11.2 days, range-289–379 days). Based on the difference of the copulation-to-parturition period between the Eurasian and the Nearctic otters (Liers 1951a) and the predicted gestation period for these similar-sized species (Ferguson et al. 1996), researchers suspected delayed implantation existed. Hamilton and Eadie (1964) confirmed Liers' (1951a) suspicion by finding unimplanted blastocysts in November and December and implanted embryos from February to April. Delayed implantation has been documented in 2 of 13 otter species (subfamily Lutrinae), 18 of 39 species in the weasel family Mustelidae, 7 of 12 families in the order Carnivora, and 6 of 26 orders of mammals (Mead 1989; Wilson and Reeder 1993). Aquatic and semiaquatic mammals such as the river otter have the longest delay period, about 273.5 days (Liers 1951a; Ferguson et al. 1996).

Stenson (1985) demonstrated that the timing of reproductive events is under hormonal control, although the date of implantation at any one locality varies according to latitude (Melquist and Dronkert 1987; Polechla 1987a). Photoperiod is the apparent trigger for the hormonal control of implantation in river otters. The timing of this implantation coincided with a photoperiod of approximately 10.5 hr light:13.5 hr dark throughout their range (Polechla 1987a). From Florida to Alaska, implantation dates vary from 15 November to 19 February (McDaniel 1963; Woolington 1984; Polechla 1987a). The evolutionary advantage of delayed implantation may be to time energy-demanding reproductive events during periods of abundant food availability (Polechla 1987a).

Prenatal Growth. Prenatal development initially shows differentiation in the cranial region (Polechla 1987a). The regions develop in the following order: (1) the primordial eyes, (2) the oral cavity region, (3) the pinnae, and (4) the nasal region. The gender of embryos >37 mm long can be identified by location of external genitalia and presence or absence of nipples. Appendages have well-defined digits beginning when the embryo is 37 mm long, pads (digital, metacarpal/tarsal, and plantar pads) when the embryo is from 40 to 46 mm long, claws when the embryo is >55 mm long, and interdigital webbing when the embryo is 42–153 mm long. Hair for sensory function develops next. The sensory vibrissae appear as nubbins in embryos 37–64 mm long and grow to a length of 1 mm while the embryos are from 75 to 84 mm in length. The vibrissae develop in the following sequence: (1) mystical and superior genial, (2) inferior genial, (3) superciliary, and (4) interramal. The insulatory dorsal guard hair follicles appear as dots when the embryo is from 75 to 95 mm long. When the embryo is from 95 to 153 mm long, the dorsal guard hairs protrude above the follicles. Finally, when the embryo crown–rump length is from 115 to 122 mm, the dorsal guard hairs grow to 2 mm long.

Parturition. Parturition dates for wild populations of river otters occurred from approximately 17 January to 9 May at 10 localities from Florida northward to Alaska (McDaniel 1963; Woolington 1984; Polechla 1987a; Noll 1988). The birth process takes 3–8 hr and is accomplished while the female is standing on all four feet (Liers 1951a). Immediately after birth, the female tightly curls around the newborn (Liers 1951a). Litter size varies from one to six pups or cubs (Polechla 1987a). Female otters are able to successfully bear young even with radiotransmitters implanted in their intraperitoneal cavity (Reid et al. 1986).

Postnatal Growth. The crown–rump lengths of full-term embryos are from 135 to 153 mm. Body weight is from 131.6 to 136.5 g (Hamilton and Eadie 1964). Neonatal crown–rump length, total length, and weight are highly correlated (Polechla 1987a). River otters are born blind, sparsely furred, toothless, and relatively helpless (Hamilton and Eadie 1964; Polechla 1987a). External ears are closely appressed to their head. Claws and webbing are well formed and nail beds of the toes are a pinkish white color. Using Eisenberg's (1981) scale of degree of precocity, Polechla (1987a) assigned river otters a rating of 2.5 on a scale of 1 (highly altricial) to 5 (highly precocial). River otters are similar to most otters, except the sea otter, in development at birth (Polechla 1987a). Sea otters are born with eyes open, fully furred, and capable of limited movement. River otters are more precocial than the mink (*Mustela vison*), but less precocial than the sea otter and beaver.

Polechla (1987a) detailed postnatal growth and development. Their sparse hair at birth develops into a downy growth called lanugo. Captive river otter pups or cubs open their eyes from 22 to 35 days old. On day 12, their deciduous upper and lower canines erupt. They are able to hold their head up between days 33 and 37, first crawl between days 33 and 57, and swim and dive in deep water by 60–165 days of age.

ECOLOGY

Habitat. The vast geographic range in which the river otter is found, from marine environments to high mountain lakes to desert canyons, is testament to their ability to adapt to a variety of aquatic habitats. Otters have been referred to as a "flagship species" for wetlands and aquatic habitats (Foster-Turley 1996) and are an indicator of wetlands with ample and high-quality water (Polechla 2000). Newman and Griffin (1994) suggested that wetland loss and vegetational change in the northeastern United States could affect river otter populations. However, threats to wetlands are not only confined to the Northeast.

Otters are generally most abundant along food-rich coastal areas, estuaries, the lower portions of streams, and coastal marshes, in locations with the least amount of human disturbance (Wilson 1959; Tabor and Wight 1977; Mowbray et al. 1979; Larsen 1983; Foy 1984; Polechla 1990; Testa et al. 1994). The most favorable interior habitats are found in lowland marshes, swamps, and bogs interconnected with meandering streams and lakes (Melquist and Hornocker 1983; Reid et al. 1994). Otters also are relatively common in many of the major nonpolluted river systems and in the lakes and tributaries that feed them. They are less common in heavily settled areas, particularly if the waterways are polluted, and in food-poor mountain streams.

Suitable river otter habitat must also provide sufficient food, as food influences the extent to which different habitats are used (Melquist and Dronkert 1987). Where they occur, logjams are key habitat features used by otters for foraging, feeding, and resting (Melquist and Hornocker 1983). Otters tend to avoid or limit their use of water bodies with gradually sloping shorelines of exposed sand and gravel (Melquist and Hornocker 1983; Reid et al. 1994). These areas, especially water storage reservoirs that are annually drawn down, typically lack suitable escape cover for otters and their prey and suitable den and resting sites. Structural diversity of shorelines tends to increase prey availability and cover (Allen 1987).

Rugged coastal areas with irregular shorelines associated with short intertidal lengths are attractive to otters (Larsen 1983; Woolington 1984). Conversely, coastal shorelines with long intertidal lengths and limited riparian vegetation tend to be avoided. River otters in coastal southeast Alaska showed strong selection for shorelines with old-growth forests (Larsen 1984; Bowyer et al. 1995), where the shoreline is sheltered by the overstory tree canopy and the roots of large conifers provided holes and crevices for otters to seek shelter. Marine populations of river otters in southeastern Alaska avoided clearcut areas, including sites where trees had been harvested 20 years earlier (Larsen 1984; Bowyer et al. 1995). In the arid West, areas with permanent water, high pool-to-riffle rates, high beaver density, good vegetative cover, and low livestock grazing pressure are important habitats for otters (Christensen 1984; Bradley 1986; Allen 1987; Malville 1990).

Winter severity in some temperate zones of North America can also affect habitat use by river otters (Melquist and Hornocker 1983; Route and Peterson 1988; Reid et al. 1994). Lakes and streams that freeze deep and high-elevation areas that receive deep snow accumulations may provide excellent habitat, but are only used by otters on a seasonal basis. In Idaho, otters vacated high mountain lakes and streams during winter, spending the majority of their time in the lower valleys (Melquist and Hornocker 1983). Otters have been known to gain access to beaver ponds during winter by breaching the dam (Reid et al. 1988). In turn, as the water level drops, air space is created and potential prey is concentrated.

Beaver ponds, lodges, and bank dens are important to river otters. A strong correlation between beaver habitat and use by otters has been documented (Tumlison et al. 1982; Melquist and Hornocker 1983; Karnes and Tumlison 1984; Reid 1984; Polechla 1989b; Dubuc et al. 1990; Reid et al. 1994). In Idaho, active and abandoned beaver bank dens and lodges accounted for 38% of the 1283 den and resting sites used by radiotagged otters (Melquist and Hornocker 1983). Tumlison et al. (1982), Reid (1984), and Polechla (1989a) suggested that a facultative commensal relationship exists between beaver and otter, with otters benefiting from past beaver activities that provide den sites and improve habitat for fish by impounding streams. Management plans for river otters need to take beaver management into consideration (Polechla 1987a, 1989b, 2000).

Sex Ratios. Sex ratios (male:female) varied from 0.64:1 to 3.31:1 from river otters trapped in 10 northeastern states (Chilelli et al. 1996) to 1.26:1 in New York (Hamilton and Eadie 1964) and 1.09:1 in Idaho (Melquist and Hornocker 1983). Earlier studies, primarily from harvested otters, generally showed a preponderance of males to females (Knudsen 1956; Wilson 1959; McDaniel 1963; Hamilton and Eadie 1964; Tabor 1974; Stephenson 1977; Lauhachinda 1978; Mowbray et al. 1979; Anderson 1981; Polechla 1987a). Lauhachinda (1978) attributed this difference to males being more vulnerable to trapping because they range more widely than females. A tendency for males to form social groups (Mack et al. 1994; Blundell 2001), whereas females generally remain solitary or with young of the year (Melquist and Dronkert 1987), may also influence the sex ratio in harvested populations.

Density. Establishing density estimates for river otter populations with any degree of accuracy was not possible before the use of radiotelemetry. Density estimates for indigenous river otters range from 1 otter/1.18–3.60 km of coastline in Alaska (Larsen 1983; Woolington 1984; Testa et al. 1994) to 1 otter/71–106 ha in a Texas coastal marsh (Foy 1984) and 1 otter/2.7–5.8 km (mean of 3.85 km) of waterway in Idaho (Melquist and Hornocker 1983). Density estimates from other telemetry-based river otter studies were summarized by Melquist and Dronkert (1987).

Dens and Resting Sites. River otters use a variety of temporary dens and resting sites based largely on availability and convenience, although preference appears to be for sheltered sites that provide protection and seclusion. For example, river otters in the Clearwater River canyon area of northern Idaho used more rock cavities (63%), including railroad (24%) and highway (19%) riprap and natural rock (19%), than any other den type (Mack et al. 1994). Because rock cavities were the most common den types in the study area, otters clearly used them based on their availability. In contrast, in their central Idaho study area of meandering streams and well-developed riparian habitat, Melquist and Hornocker (1983) reported that 67% of the den sites used by otters included beaver bank dens and lodges, logjams, and riparian vegetation.

Melquist and Hornocker (1983) reported that a single otter during a 16-month period used at least 88 different den and resting sites in their central Idaho study area. The authors identified 15 different kinds of resting sites used by instrumented otters. Beaver bank dens and lodges accounted for 38% of the 1283 sites used by otters, indicating the importance of beaver to otter where they coexist. Other den sites used by otters include snow and ice cavities, talus rock, brush and logpiles, natural undercut banks, hollow logs, tree root structures, boat docks, duck blinds, and a variety of others based on availability. River otters do not excavate their own natal dens, but instead use and modify those created by a variety of other animals, or they may use natural shelters and cavities (Lowery 1974; Melquist and Hornocker 1983; Christensen 1984; Griess 1987). Natal dens may be adjacent to and directly accessible from the water or they may be on a high bank above the water, requiring overland travel (Melquist and Hornocker 1983). Otters occasionally build a nestlike structure in aquatic vegetation (Grinnell et al. 1937; Liers 1951a).

Home Range and Movements. A river otter's home range consists of those places where the animal lives, reproduces, and fulfills all life requirements (Melquist and Dronkert 1987). The shape and size of a home range varies because it is a function of the distribution of suitable habitat, available food, weather, topography, reproductive status of individuals, conspecifics, density, season, and perhaps other factors. Activity within the home range is usually concentrated at one or more sites with abundant food and cover (Melquist and Hornocker 1983). The length of time that an otter remains at these important activity centers or core areas is probably a function of available prey. River otters in southeastern Alaska that inhabit marine environments where fish were abundant had smaller home ranges than animals living in freshwater systems with fewer prey (Blundell et al. 2000). Home range size varies among age and sex classes, with lactating females being the most restricted (Melquist and Hornocker 1983).

Males range more widely than females with or without pups, and home range areas overlap, although river otters typically exhibit mutual avoidance (Melquist and Hornocker 1983; Route and Peterson 1988; Shannon 1991; Mack et al. 1994; Reid et al. 1994; Bowyer et al. 1995). In Idaho, home ranges overlapped extensively within and among sex classes (Melquist and Hornocker 1983; Mack et al. 1994). In northern Minnesota, Route and Peterson (1988) reported that adult females maintained exclusive home ranges, except for overlap of two adjacent females during winter water drawdown. Foy (1984) found little evidence of intrasexual home range overlap in Texas coastal marshes. Although Nearctic river otters do not typically defend territories, Hornocker et al. (1983) contended that mustelids, including river otters, are capable of exhibiting flexible spacing strategies.

Home ranges for interior populations reach their greatest extent in northern habitats where resources are more limited and seasons severe. In Alberta, home ranges in excess of 200 km^2 were recorded for male river otters and 70 km^2 for females (Reid et al. 1994). Farther south, in Idaho, Melquist and Hornocker (1983) reported home ranges of 8–78 km of shoreline along streams and lakes; Mack et al. (1994) documented home ranges of 16–148 km of shoreline that did not include lakes.

River otters in Prince William Sound, Alaska, occur at densities of 20–80 animals/100 km of shoreline (Testa et al. 1994), with home ranges encompassing 20–40 km of shoreline (Bowyer et al. 1995). Males had larger home ranges than females in marine and freshwater environments of southeastern Alaska (Blundell et al. 2000). Home ranges of females in coastal Alaska were typically exclusive, used by a single reproductive female, her offspring, and various adult or subadult female "helpers" (Rock et al. 1994), whereas the ranges of coastal males typically broadly overlapped those of females and other males (Blundell 2001).

Sauer et al. (1999) addressed the issue of existing techniques that tend to overestimate linear home range size in river otters, and offered a new application that may be useful for estimating home ranges of otters and other animals approximating a linear distribution of locations. In addition, Blundell et al. (2001) conducted simulations using empirical data from river otters to determine the most accurate method of estimating linear home ranges with kernel analyses and the effects of sample size and autocorrelation on those estimates.

Telemetry studies have advanced our knowledge and understanding of movement and dispersal among river otters. Otters are highly mobile, with movements of up to 42 km reported in 1 day for dispersing animals (Melquist and Hornocker 1983). Daily movements of single animals and family groups tend to be <10 km/day and vary by season. Males and family groups in temperate areas move significantly less during winter than other seasons (Melquist and Hornocker 1983; Reid et al. 1994).

FORAGING AND FEEDING

Foraging. River otters found in coastal areas forage in the intertidal and subtidal zones primarily for marine fish (Larsen 1984; Stenson et al. 1984; Bowyer et al. 1994; Ben-David et al. 1998). Stenson et al. (1984) reported that the most common fish found in their samples were either slow-moving or quick to fatigue, with none capable of sustained swimming to escape predation by otters. The authors concluded that feeding upon intertidal fish is the marine equivalent of the finding by Sheldon and Toll (1964) that otters in a freshwater reservoir feed chiefly in the coves or near the shore.

In most interior streams, especially those that are shallow, fish are forced to seek shelter along the shoreline to avoid potential predators. In small, shallow streams in Idaho, Melquist and Hornocker (1983) observed otters foraging along undercut banks and among logs, overhanging vegetation, or other obstructions that provided some escape cover for fish. When pursued by an otter, fish in exposed areas of a stream would quickly retreat to shelter, where they were often captured. In larger streams with frequent deep pools and shallow riffles, otters typically foraged among logjams located in deep, slow-moving sections where fish tend to concentrate. Otters observed in lakes foraged along the shoreline or among boat docks. In shallow lakes, reservoirs, and ponds with an abundance of slow-swimming fish, otters foraged along the shoreline, where fish were captured by direct pursuit.

During 182 separate observations of foraging and feeding behavior, Melquist and Hornocker (1983) saw no evidence of cooperative foraging among river otters. Similarly, Beckel (1990) observed otters foraging in groups, but did not find evidence of coordinated hunting strategies. Sheldon and Toll (1964) reported observing two otters schooling fish into a shallow cove. Serfass (1995) observed four otters, believed to be a family group (female with offspring), which appeared to herd fish in a stream into the center of a pool where two fish were successfully captured. We concur with the author that such foraging behavior probably developed as the juvenile otters transitioned from reliance on the female parent for securing prey to independent foraging. In Prince William Sound, Alaska, social groups, primarily consisting of males, form to facilitate cooperative foraging, enabling those in the group to obtain a better-quality diet more efficiently (Blundell 2001; Blundell et al. 2002a).

River otters consume a variety of food items reflective of the diversity of prey found in the particular aquatic environment they occupy. Although a diversity of fish species appears in the diet, certain patterns of fish vulnerability to otter predation are evident. The most important is that fish are preyed on in direct proportion to their availability (i.e., occurrence and density) and in inverse proportion to their swimming ability (Ryder 1955; Toweill 1974; Stenson et al. 1984; Serfass et al. 1990). Melquist and Hornocker (1983) considered availability to be synonymous to vulnerability and the degree of availability determined by prey size, abundance, and swimming ability. Although otters may key in on areas where high concentrations of spawning fish occur (Melquist et al. 1981), they capture the first fish they encounter that is not able to escape capture efforts. During controlled feeding studies, Erlinge (1968) found that captive Eurasian otters tended to select larger fish, which were less maneuverable and less able to find effective hiding cover than smaller fish, except when the smaller fish were much more abundant. Although such patterns of selection appear obvious, they contain three general concepts important to the understanding of otter prey selection. First, otters do not select a particular species of fish when foraging (i.e., they do not appear to have a taste preference). Second, slow-swimming species of fish are more vulnerable than fast-swimming species. Third, injured or weakened fish are more vulnerable to otter predation than healthy fish.

Few researchers have documented hunting success rates of otters. While studying otters in Yellowstone National Park, Varley (1998) reported that one female caught a fish during 38% of her dives while foraging in a lake and 62% of her attempts were successful while foraging in an inlet (total $n = 84$). Another female was successful in capturing fish on 40% of 18 dives.

Feeding Habits. River otters are specialists, based on their fish-eating habits (Melquist et al. 1981; Cooley 1983). Feeding habits of river otters, including prey composition, have been studied in many portions of their widespread range. Almost without exception, the bulk of their diet is composed of fish, with crustaceans (primarily crayfish), amphibians, insects, birds, and mammals constituting lesser portions (Table 35.4).

In reviewing the species composition from the food studies identified in Table 35.4 and through an examination of the otter dietary

TABLE 35.4. Frequency of occurrence of major food categories of the river otter in North America

State or Province	Sample Type	Sample Size	Sample Period	Frequency of Occurrence (%) of Food Category							Source
				Fish	Crustacean	Insect	Amphibian	Bird	Mammal	Other	
Alabama	Scat	12	Nov–Feb	92	58			8			Lauhachinda and Hill 1977
Alabama/Georgia	Digestive tract	315	Nov–Feb	83	63	10	5	<1	<1	[a]	Lauhachinda and Hill 1977
Alaska	Scat	272	Four seasons	96[a]	[a]	11		1	<1	[a]	Larsen 1984
	Scat	337	June–Oct	10–37	4–17	[a]		[a]		[a]	Bowyer et al. 1994
Alberta	Scat	1,191	June–May	92	3	54	2	3	1	[a]	Reid et al. 1994
Arizona	Scat	343	Apr–Aug	77–100	2–43	0–24		0–6	0–2	[a]	Christensen 1984
Arkansas	Scat	240	Apr–Mar	14–61	28–83	<1	<1		<1	[a]	Tumlison and Karnes 1987
British Columbia	Scat	528	May–Feb	99[a]	7			4			Stenson et al. 1984
	Stomach	69	Four seasons	87[a]	3			13			Stenson et al. 1984
California	Scat	120	Four seasons	29	98	8[a]	0	38	[a]	15[a]	Grenfell 1974
	Scat	100	June–Aug	70	51	9		6			Modafferi and Yocum 1980
	Scat	302	June–Oct	86–97	0–31	6–59	14–24	2–31		[a]	Reeves 1988
	Scat	94	Apr–Aug	89	16	43	1	5		[a]	Manning 1990
Colorado	Scat	222	Four seasons	100				6	5	[a]	Mack 1985
	Scat	98	Spring–fall	[a]	34					7	Berg 1999
Florida	Stomach	63	Dec–Mar	[a]	9		2				McDaniel 1963
	Digestive tract	187	Dec–Feb	87	75	41	20	2	25		Cooley 1983
Idaho	Scat	1,902	Four seasons	97[a]		8		3	3		Melquist and Hornocker 1983
	Scat	1,367	Four seasons	79	20			<1	<1		Mack et al. 1994
Illinois	Scat	765	Oct–June	98[a]	9	3	9	2			Anderson and Woolf 1984
Kentucky	Scat	333	Four seasons	97	39	[a]	19	3	<1	[a]	Logsdon 1989
Louisiana	Digestive tract	20	Winter	82	[a]		1	2	7	[a]	Holcombe 1980
	Digestive tract	179	Dec–Feb	83	24–42			0–2		[a]	Chabreck et al. 1982
Maine	Scat	200	Winter–Summer	66			[a]	6	4	[a]	Dubuc 1987
Massachusetts	Scat	517	Four seasons	81	31	8	13		5	[a]	Sheldon and Toll 1964
Michigan	Stomach	173	Mar–Apr	[a]	35	13	16	[a]	[a]	4	Lagler and Ostenson 1942
	Intestines	220	Mar–Apr	[a]	59	32	25			1	Lagler and Ostenson 1942
	Stomach	75	Mar–Apr	[a]	22	13	17			[a]	Ryder 1955
	Stomach	18	Feb–Apr	79	29	7	11		0	[a]	Knudsen and Hale 1968
	Intestines	41	Feb–Apr	81	59	22	10		0	[a]	Knudsen and Hale 1968
Minnesota	Stomach	12	Feb–Apr	100	17	33	17		0		Knudsen and Hale 1968
	Intestines	29	Feb–Apr	90	34	38	0		0		Knudsen and Hale 1968
	Scat	796	Summer–Winter	72	71	14	11	<1	<1	[a]	Route and Peterson 1988
Mississippi	Stomach	108	Dec–Feb	65	59						Davis 1981
Montana	Scat	1,374	Four seasons	93	[a]	41	18	5	6	<1	Greer 1955
	Scat	260	Nov–July	99	30				<1		Zackheim 1982
New York	Digestive tract	14	Oct–Apr	70	35	14	25	<1	4		Hamilton 1961
North Carolina	Combination	85	Dec–Feb	91	39	6		3	1	2[a]	Wilson 1954
Ohio	Scat	462	Four seasons	74–99	19–80		34–78	0–2	<1–2		McDonald 1989
Oregon	Stomach	44	Nov–Feb	86	20	[a]	9	9	[a]	[a]	Toweill 1974
	Intestines	75	Nov–Feb	80	27	[a]	12	7	[a]	[a]	Toweill 1974
Pennsylvania	Scat	407	Four seasons	93	44	5	7	1	4	1[a]	Serfass et al. 1990
Tennessee	Scat	75	Apr–Sep	90	95	5				[a]	Griess 1987
	Digestive tract	130	Four seasons	65	65	7	13		6	[a]	Lizotte 1994
Tennessee/North Carolina	Scat	75	Four seasons	64–91	62–100						Miller 1992
Virginia	Scat	209	Four seasons	62	82	20	15	<1		<1[a]	Pierce 1979
Washington	Scat	254	Four seasons	93	46	[a]	[a]	3	[a]	[a]	Toweill and Tabor 1982
Wisconsin	Stomach	131	Four seasons	81	31	8	13		5	[a]	Knudsen and Hale 1968
	Intestines	260	Four seasons	85	40	19	2		3		Knudsen and Hale 1968
Total		15,103									

[a]Consult original publication for details.

literature, one can see some general characteristics of the prey. Slower-swimming species of fish, such as members of the families Amiidae, Catostomidae, Cyprinidae, and Ictaluridae, are prominent in the otter's diet. Also important are fish species that are often abundant and found in large schools, and those bottom-dwelling species that are particularly susceptible to otters because of their habit of remaining immobile until a potential predator is nearly upon them (Stenson et al. 1984; Bowyer et al. 1994). Faster-swimming fish species (e.g., members of the family Salmonidae) appear to be taken by otters in lesser numbers than their abundance might suggest. An exception to this occurs during salmon spawning runs, when large concentrations of spawning and spent salmon become easy prey for otters (Toweill 1974; Melquist and

Hornocker 1983), which could affect overwintering juvenile salmon and trout on coastal streams (Balke 1993a; Dolloff 1993).

Crustaceans, primarily crayfish, make up a major portion of the otter's diet throughout North America (Table 35.4). Crayfish appear to be especially important to otters in some areas during summer (Griess 1987; Manning 1990; Serfass et al. 1990; Mack et al. 1994; Berg 1999), winter (Chabreck et al. 1982; Cooley 1983; Reid et al. 1994), and winter and spring (Tumlison and Karnes 1987). Although crustaceans were abundant in the intertidal water along coastal British Columbia, Stenson et al. (1984) found little consumption by otters and suggested they may constitute a secondary food source when the availability of fish declines. Various crabs and shrimp are also eaten by otters in estuarine areas (Wilson 1954; Toweill 1974; Chabreck et al. 1982; Home 1982; Stenson et al. 1984).

Reptiles and amphibians, particularly frogs (Ranidae), are occasionally eaten by otters (Table 35.4). In Ohio and Virginia, amphibians were most often eaten during warm months, when they were most active (Pierce 1979; McDonald 1989). Otter scats found along a California stream contained the remains of Pacific giant salamanders (*Dicamptodon aterrimus*) (Reeves 1988). River otters ate large two-toed amphiumas (*Amphiumas means*) in the Great Dismal Swamp, Virginia (Pierce 1979). Although turtles may be common in otter habitats and sometimes are consumed (Stophlet 1947; Liers 1951b; Cooley 1983; Manning 1990), many researchers have found no evidence of turtles in the otter diet in areas where turtles were common (Greer 1955; Grenfell 1974; Toweill 1974; Lauhachinda 1978). Griess (1987) found river otters taking turtles in spring when they were laying eggs. Chabreck et al. (1982) found that otters in southern swamps ate water snakes (*Nerodia* spp.). In California, Manning (1990) documented otters preying on garter snakes (*Thamnophis* sp.) and racers (*Masticophis* sp.).

Overall, birds appear to be a relatively unimportant food item for otters (Table 35.4). Only Grenfell (1974) reported otters, in a freshwater habitat, feeding extensively on birds. Birds normally found associated with aquatic ecosystems are the species typically reported in otter food studies. Toweill (1974), Modafferi and Yocum (1980), Melquist et al. (1981), Gilbert and Nancekivell (1982), and Findlay (1992) believe that otters probably eat primarily young, crippled, nesting, breeding, molting, or freshly killed birds. Food studies have often included those periods when waterfowl are being hunted and there is likely a preponderance of hunter-crippled birds and carrion available to otters. However, there are several accounts of otters observed actively hunting and killing healthy birds (Cahn 1937; Meyerriecks 1963; Grenfell 1974). Otters preying on glaucous-winged gull (*Larus glaucescens*) chicks in nesting colonies along the north Pacific coast have been documented (Kennedy 1968; Hayward et al. 1975; Foottit and Butler 1977; Verbeek and Morgan 1978). Marine birds, including several species of petrels, are both preyed upon and scavenged by otters on coastal islands (Quinlan 1983; Speich and Pitman 1984). Predation on ground-nesting colonial birds along the Pacific coast and on coastal islands may be substantial (Verbeek and Morgan 1978; Speich and Pitman 1984) and could be a major cause of nest failure (Quinlan 1983). Balke (1993b) reported the remains of auklet and a large short-legged shorebird in scats collected from Queen Charlotte Islands, British Columbia.

Although various mammals have been reported in the otter's diet, the incidence of mammal remains in otter food studies is consistently low. Melquist and Hornocker (1983) found muskrats (*Ondatra zibethicus*) were the principal mammal preyed on by otters in their Idaho study. Muskrat remains were generally found in scats collected from wetlands where the rodents were most abundant. Findlay (1992) also found muskrats were the dominant mammal eaten by otters in Utah. A higher percentage of muskrats was in the diet of males due to the larger size and greater ability of the males to kill them. Otters infrequently eat beaver (Reid et al. 1994), and the exotic nutria or coypu (*Myocastor coypus*) has been found in otter stomachs from swamps and coastal marshes (Chabreck et al. 1982). Field (1970) presented evidence of otters actively hunting and capturing small mammals in the snow and chasing mammals up to the size of snowshoe hares (*Lepus americanus*).

Various invertebrates, including bivalves, snails, and echinoderms, have been reported in the otter's diet. Because shellfish occur at low frequencies in otter scats and are both primary and secondary prey, they are probably not important food items.

Various aquatic insects recorded in otter food studies are likely ingested as both primary prey and incidental prey when fish are consumed. Insects are staples in most fish diets and normally occur with fish remains in the otter diets, making it difficult to evaluate their true importance as food for otters. Otters have been observed foraging and feeding on large aquatic beetles (Coleoptera) by probing the muddy bottom of shallow backwater sloughs, oxbows, and ponds (Melquist and Hornocker 1983). Reid et al. (1994) found that insects (Odonata and Coleoptera) were a prominent prey of otters during summer in northeastern Alberta. Caddis fly larvae (Trichoptera) were common invertebrate prey of Nearctic otters in coastal California (Reeves 1988).

BEHAVIOR

River otters are intelligent and inquisitive animals, making it easy to train them to perform a variety of activities. Their inclination to make a "game" out of almost any activity (Harris 1968; Park 1971) is likely a product of their intelligence and curiosity. Wild otters appear to spend most of their active time foraging, feeding, and exploring portions of their home range.

Activity Patterns. River otters are active throughout the year, even in northern climates with severe winters. There is no evidence they migrate, although river otters in the interior do make seasonal movements associated with prey availability or environmental conditions (Jackson 1961; Melquist and Hornocker 1983). In spring, summer, and fall in Idaho, activity typically begins near dusk and continues throughout the night and into midmorning (Melquist and Hornocker 1983). However, diurnal activity was common during winter and at other times of the year in areas where otters were rarely disturbed by humans. Others have reported that diurnal activity was common (Larsen 1983; Foy 1984; Woolington 1984).

Daily movements of family groups in Idaho averaged 4.7 km during spring, 4.4 km during summer, and 2.7 km during winter (Melquist and Hornocker 1983), indicating that river otters are quite active. Movements were greatest during spring following snowmelt and least during winter when many small tributaries and ponds were covered in ice and snow.

Locomotion. Terrestrial locomotion patterns of river otters consist in walking, running, and bounding in a pattern typical of terrestrial carnivores, with alternate movements of the opposite forelimbs and hind limbs (Tarasoff et al. 1972). When the animal is walking, limbs are moved parallel to the long axis of the body, which is held rigid with the head and neck outstretched; the distal portion of the tail may touch the ground. The same pattern of limb movement occurs in running, but the tail is held higher above the ground and in a slight arch. Bounding, the most rapid form of movement on land (Liers 1951a), is characterized by the body being held in an arched position and the tail held stiffly. Both forefeet are lifted from the ground simultaneously, but one strikes the ground slightly ahead of the other; both rear feet are moved simultaneously. The tail is used as a balancing organ in all terrestrial movements. In addition, river otters slide or "toboggan" extensively when traveling over snow and ice (Melquist and Hornocker 1983). In sliding, forelimbs are held close to the body and thrust is provided by hind limbs. River otters can move at a top land speed of 24–29 km/hr by running and sliding (Severinghaus and Tanck 1948). Compared to more-aquatic carnivores, the river otter is more efficient in terrestrial locomotion than the sea otter and the harp seal (*Phoca groenlandica*) and less efficient in aquatic locomotion than these two other species (Tarasoff et al. 1972).

The principal mode of aquatic locomotion involves a thrust–recovery movement of one or more of the limbs and flexure of the posterior portion of the body and the tail vertically in the water (Tarasoff et al. 1972). Typically, the movement consists of "paddling" with one or both hind limbs alternated with a period of gliding; forelimbs are used principally in turning movements. Quadrupedal paddling, forelimb paddling, alternate hind limb paddling, simultaneous hind limb paddling, and dorsoventral undulation (wherein the caudal portion of the body and the tail are moved rapidly in a vertical direction) have all been documented (Liers 1951b; Tarasoff et al. 1972; Fish 1994).

Body Care. The most common means of body care in otters is rubbing and rolling, whether in sand, grass, snow, or whatever else is available and relatively dry. This activity is commonly associated with considerable scratching. Both activities apparently serve to clean the animal's fur and thereby maintain its insulative qualities, and to dry the animal quickly after its emergence from the water. Areas used for this activity, called "rolling sites," "scrapes," "haul-outs," or "landings" (Melquist and Hornocker 1979a; Mowbray et al. 1979), are among the most common evidence of otter activity.

Mutual grooming occurs to some extent among otters, especially family groups, but it is probably not common among unrelated individuals. Mutual grooming in captive otters occurred more frequently than the amount required for hygienic purposes (A. Beckel-Kratz, University of Minnesota, pers. commun., 1977), suggesting to that it had other implications, perhaps associated with captivity.

Otters typically have specific "toilets" (Greer 1955) near regular landings for defecation purposes, although single scats may be deposited near rolling areas, scent posts, or logs or on points extending into the water (Melquist and Hornocker 1979a; Mowbray et al. 1979).

Foraging and Feeding. Otters typically feed on fish or crustaceans captured underwater with a quick lunge or as a result of an underwater chase. River otters can swim at a top speed of 10–12 km/hr (Hamilton 1943; Jackson 1961; Park 1971), dive to depths of 18–20 m (Scheffer 1953), and remain underwater for up to 4 min (Harris 1968; Reed-Smith 2001). These features make the river otter well adapted for securing food in the water and maximizes their underwater search effort. Otters also roll or push rocks aside to uncover prey hiding in the substrate of ponds or streams. Other methods documented include "stalking" aquatic birds by approaching them underwater and seizing them from below (Meyerriecks 1963; Harris 1968; Grenfell 1974), stalking birds and mammals on land (Hamilton 1961; Field 1970), and raiding bird nests for eggs and nestlings (Hayward et al. 1975; Verbeek and Morgan 1978). Otters may take waterfowl or fish as carrion (Grinnell et al. 1937; Toweill 1974), but in one instance demonstrated no inclination to feed on white-tailed deer (*Odocoileus virginianus*) carrion (Field 1970).

Social Behavior and Play. River otters have a reputation for being social and playful animals. Sociality beyond the family group is generally not pronounced, however. Unrelated river otters typically show mutual avoidance (Melquist and Hornocker 1983; Shannon 1991), but coastal river otters in Prince William Sound, Alaska, exhibit high variability in social organization (Blundell et al. 2002a).

River otters form social groups where resources are abundant, and there is considerable evidence that cooperative foraging is a key factor influencing social organization of coastal otters (Blundell et al. 2002a). Two types of social groups have been described: "family" and "clan" (Shannon 1998). Families, the most typical group, consist of a reproductively active female and her young of the year, often with one or more helpers, who may be members of previous litters or unrelated adults or yearlings (Melquist and Hornocker 1983; Rock et al. 1994). Clans are composed primarily of males (Blundell 2001); groups consisting of as many as 9–30 otters have been reported (Best 1962; Woolington 1984; Shannon 1989, 1991; Reid et al. 1994; Rock et al. 1994; Testa et al. 1994). In Idaho, Mack et al. (1994) monitored a group of five otters, including three instrumented males and presumably all males, that traveled together and occupied 128 km of the Clearwater River for most of a year.

Associations of males appear to be more frequent in marine environments (Ben-David et al. 1995) and related to cooperative foraging (Blundell et al. 2002a). In coastal Alaska, Blundell et al. (2002b) found that, among social otters, males were social in 46% of their locations and 63% of that time occurred in all-male groups. Females were only social in 26% of locations and were in mixed-sex groups 78% of that time. The authors hypothesized that the time-consuming task of raising offspring prevents females from joining foraging groups. When not raising young, these females may join males to cooperatively forage for better-quality prey (pelagic fishes), which are more difficult to acquire as a solitary forager (Blundell et al. 2002a).

Play behavior, such as repeated sliding, wrestling, retrieving inanimate objects from underwater, and juggling pebbles or sticks while swimming, is commonly observed in captive otters (Reed-Smith 2001), although this may be primarily a displacement behavior in response to confinement. Melquist and Hornocker (1983) documented play behavior during only 17 (6%) of 294 field observations of free-ranging otters in Idaho and reported that this activity was usually associated with immature individuals, primarily juveniles.

Communication. Although the river otter is less vocal than most other members of the subfamily Lutrinae (Harris 1968), various vocalizations have been reported (Reed-Smith 2001). The vocal repertoire typically consists of shrill chirps, soft "chuckles," and low, purring grunts that appear to be contact calls; an explosive nasal snort as an alarm call; defensive snarls and hisses when threatened; and a scream when injured or under extreme duress (Liers 1951b; Jackson 1961; Park 1971). There is also a caterwaul, used by the female during copulation (Liers 1951b; Scheffer 1967; Tango et al. 1991) or when approached by an apparently unwanted male (Polechla 1987a).

Few visual displays have been recorded for the river otter. With their short muzzle and ears and their more or less uniform coat of hair, otters are not well adapted for visual display-based communication. A "threat face," characterized by pulling the ears back and displaying a gaping mouth, is used. The most pronounced visual display is associated with defecating and scent marking (P. J. Polechla, Jr., pers. obs. recorded on tape, 1985; Reed-Smith 2001). Otters exhibiting this behavioral ritual arch their back; undulate their tail; alternately paw, scratch, and tread the substrate with their hind feet; and exude a stream of feces from their anus. If other otters are present, the spectators sniff the scat and then repeat the action, termed the "latrine dance" by Reed-Smith (2001). Beckel (1982) developed a provisional ethogram for captive and free-ranging river otters, but more behavioral studies are needed, especially on wild otters.

Olfaction apparently plays a major role in otter communication. River otters possess anal scent glands, and scent may be released from these glands in times of fear or rage (Liers 1951b; Melquist and Hornocker 1983). Otters also maintain scent posts throughout their range (Grinnell et al. 1937; Mowbray et al. 1976, 1979; Melquist and Hornocker 1979a, 1983). Scent posts apparently do not function as territorial boundaries, but rather as a means of advertising the presence of individuals or groups (Hornocker et al. 1983), allowing otters to minimize intraspecific contact. Mowbray et al. (1979) described scent posts as sites of 1 to 2 m^2 with digging and scratching but no food remains, scats, or beds. In addition to scent posts, they described other areas used by otters as haulouts, bedding sites, rolling sites, scrapes, dens, and diggings. Any of these sites may be used for scent deposition.

MORTALITY

An increase in the number of detailed studies of river otters and other organisms has resulted in a growing list of predators, ectoparasites, endoparasites, bacteria, viruses, and diseases affecting otters. None are known to have severe impacts on otter populations, however. Malnourishment is probably not a limiting factor in most areas, although it could limit some populations in the far North or where fish kills have occurred. By far the most serious causes of mortality among river otters are the direct activities of humans, through harvest of otters for pelts and, far more importantly, indirectly through destruction and modification of habitat.

Predators. Various predators have been reported to kill otters in North America (Table 35.5). No predator has been shown to have a serious

TABLE 35.5. Reported predators of river otters in North America

Predator	Location	Comment	Source
Alligator (*Alligator missippiensis*)	Louisiana	Stomach contents	Vallentine et al. 1972
American crocodile (*Crocodylus acutus*)	—	—	Tabor and Wight 1977
Bald eagle (*Haliaeetus leucocephalus*)	Newfoundland	Both species dead	Rosen 1975
Killer whale or Orca (*Orcinus orca*)	Alaska	Circumstantial	Larsen 1983
Gray wolf (*Canis lupus*)	Minnesota	Scat	Route and Peterson 1988
Gray wolf	Alaska	Scat	Home 1982
Coyote (*Canis latrans*)	California	—	Grinnell et al. 1937
Coyote	Colorado	—	Mack 1985
Coyote	Western USA	—	Sperry 1941
Domestic dog (*Canis familiaris*)	Nebraska	Found dead	Farney and Jones 1978
Domestic dog	Northwest Territories	—	Douglas 1928
Domestic dog	Idaho	—	Melquist and Hornocker 1983
Red fox (*Vulpes vulpes*)	Maryland	Killed juvenile otter	Mowbray et al. 1979
Bobcat (*Lynx rufus*)	—	Stomach contents	Young 1958
Bobcat	Kansas	—	Hall 1955
Mountain lion (*Puma concolor*)	California	Killed in trap	Reeves 1988
Mountain lion	Arizona	Stomach contents	Britt et al. 1984
Black bear (*Ursus americanus*)	Alaska	Unverified report	Home 1982
Brown bear (*Ursus arctos*)	Alaska	Unverified report	Home 1982
Wolverine (*Gulo gulo*)	Alaska	Unverified report	Home 1982

impact on otter populations, and most predation is probably directed toward young animals and/or vulnerable adults.

Parasites and Diseases. Eleven species of ectoparasite, including seven species of tick, one species of sucking louse, one flea, and two species of beetle, have been found on the Nearctic otter (Table 35.6). The two species of beetle are known from beavers, a sympatric semiaquatic mammal. Otters may contract these parasitic beetles while bedding in beaver lodges. Infestation rates of otter ectoparasites are relatively unknown.

Thirty-six species (59 taxa identified to at least the generic level) of helminth endoparasites representing four phyla are known to infect the river otter (Table 35.7). More species and taxa of nematodes (16 species and 30 taxa identified to at least the generic level) infest otters than any other helminth group. Of these, three roundworms, *Strongyloides lutrae* (a small roundworm that lives in the mucosa of the small intestine and completes its life cycle in the lungs), *Skrjabingylus lutrae* (lives in the frontal sinuses), and *Gnathostoma miyazakii* (lives in the collecting tubules of the kidneys), may cause serious but nonlethal pathological damage. The first could cause lesions during its migratory stage, the second could cause damage to the rostral region of the skull, and the third could cause blockage and partial loss of kidney function (Fleming et al. 1977; Lauhachinda 1978; Addison et al. 1988). Instances of lethal infestations of helminths are known only in the nematode *Trichinella spiralis* in the case of acute trichinosis (Rothenbacker 1962; Reed-Smith 2001), in the fluke or trematode *Nanophyetus salmincola* in the case of salmon poisoning diseases (Schlegel et al. 1968), and in the fluke *Paragonimus kellicotti* in the case of paragonimiasis (Stuht 1978). The role river otters play in the parasite–host–predator relationship of the exotic protozoan *Myxobolous cerebralis*, tubifex worms (*Tubifex tubifex*), and the final host (salmonids, including trout and salmon), which results in what is called whirling disease, is unknown (Polechla 2000). Unanswered questions include whether or not otters can be affected in any manner or are able to transmit the parasitic protozoan from infected to uninfected waters.

Although studies of captive individuals have produced a growing body of knowledge about diseases of river otters (Reed-Smith 2001), little is known of diseases of free-ranging river otters. Two taxa of protozoans, four taxa of fungi, at least eight taxa of bacteria, and seven strains of viruses are suspected of infecting otters (Table 35.8). Most are asymptomatic and nonhistopathological, although pneumonia has been implicated in the deaths of 13–29 otters of all species that died in the London Zoo (Harris 1968).

Impact of Humans. The most obvious impact of humans on river otter populations results from trappers harvesting otters for their fur. However, life table analyses of river otter populations in Arkansas (Polechla and Sealander 1985; Polechla 1987a) indicated that under existing harvest levels otter populations were increasing, whereas populations in Oregon (Tabor and Wight 1977) and Maryland (Mowbray et al. 1979) were stable and populations in Alabama and Georgia were decreasing (Lauhachinda 1978). Today, monitoring requirements and a variety of

TABLE 35.6. Ectoparasites found in the river otter in North America

Order	Family	Species	Geographic Location	Source
Acarina	Ixodidae	*Ixodes cookei*	Alabama, Arkansas, Florida, Pennsylvania	Cooney and Hays 1972; Forrester 1992; Polechla 1996
		Ixodes uriae	California	Eley 1977
		Ixodes banksi	Florida, Michigan, Pennsylvania	Forrester 1992; Lawrence et al. 1965
		Amblyoma americanum	Arkansas, Florida	Polechla 1996; Eley 1977
		Dermacentor variabilis	Florida	Eley 1977
	Listrophorideae	*Lutracarus canadensis*	Alaska	Fain and Yunker 1980
		Lynxacarus mustelae	Alaska	Fain and Yunker 1980
Anoplura	Echinophthiriidae	*Latagopthirus rauschi*	Oregon	Kim and Emerson 1974
Siphonaptera	Ceratophyllidae	*Oropsylla arctomys*	Pennsylvania	Serfass et al. 1992
Coleoptera	Leptinidae	*Leptinillus validus*	Minnesota	Route and Peterson 1988
		Platypsyllus castoris	Minnesota	Route and Peterson 1988

SOURCE: Data from Polechla (1996) and Kimber and Kollias (2000)

TABLE 35.7. Species of helminth endoparasites of the river otter and the geographic location and body region parasitized

Phylum	Species	Geographic Location	Location Parasitized
Pentastomida	*Sebekia mississippiensis*	Florida	GI (intestine)
Platyhelminthe			
Cestoidea	*Spirometra mansonoides*	Florida, Georgia	GI (intestine), subcutaneous
	Diphyllobothrium latum	North Carolina	Subcutaneous
	Ligula intestinalis	Montana	GI (feces)
	Protoecephalus perplexus	Alabama	GI (stomach)
	Schistoecephalus solidus	Oregon, Newfoundland	GI (intestine)
	Braunia sp.	New York	GI (feces)
Trematoda	*Euparyphium melis*	Massachusetts, Michican, Minnesota	GI (stomach, small intestine)
	Euparyphium inerme	Oregon, Washington	GI (stomach, small intestine)
	Enhydrodiplostomum alaroides	Alabama, Florida, Georgia, Massachusetts, North Carolina	GI (stomach, small intestine)
	Enhydrodiplostomum fosteri	Alabama, Louisiana	GI (small intestine)
	Enhydrodiplostomum sp.	North Carolina	GI (intestine)
	Bashkirovitrema incrassatum	Alabama, Florida, Georgia, Louisiana, Massachusetts, North Carolina, New York, Tennessee, Ontario	GI (stomach, small intestine)
	Nanophyetus salmincola	Pacific Northwest	Unknown
	Paragonius kellicoti	Michigan	Unknown
	Alaria canis	Ontario	Subcutaneous mesenteric fat
	Crepidostomum cooperi	Alabama	Unknown
	Telorchis gracilis	Alabama	GI (small intestine)
	Telorchis spp.	Alabama	Unknown
	Diplostomium alarioides	Ontario	GI (small intestine)
Acanthocephala	*Pilum* sp.	Unknown	Unknown
	Oncicola spp.	Florida	GI (intestine)
	Acanthocephalus spp.	Alabama, Tennessee	GI (large intestine)
	Pomporhynchus spp.	Alabama	GI (large intestine)
	Corynosoma strumosum	Oregon	GI (stomach, intestine)
	Leptorhynchoides spp.	Alabama	Unknown
	Neoechinorhynchus spp.	Alabama	Unknown
	Paracanthocephalus rauschi	Alaska	GI (intestine)
Nematoda	*Physaloptera* spp.	Alabama, Massachusetts	GI (stomach, intestine)
	Skrjabingylus lutrae	Ontario	Frontal sinuses
	Dracunculus lutrae	Arkansas, Nebraska, New York, Ontario	Subcutaneous tissues, leg
	Dracunculus insignis	Alabama, Florida, Michigan, New York	Facial layers of leg
	Strongyloides lutrae	Alabama, Florida, Louisiana, Oregon, Tennessee, Washington	GI (small intestine), lung
	Crenosoma goblei	Florida, North Carolina	GI (intestine), lung
	Dirofilaria lutrae	Florida, Louisiana, Southeastern USA	Blood, subcutaneous tissue
	Dirofilaria immitis	Louisiana	Heart
	Capillaria plica	Florida, North Carolina	Urinary bladder
	Capillaria aerophilus	Unknown	Lungs
	Capillaria hepatica	Florida	Liver
	Capillaria sp.	Louisiana	Unknown
	Gnathostoma miyazakii	Alabama, Florida, North Carolina, Virginia, Ontario	Kidney, GI (intestine)
	Dioctophyme renale	Unknown	Kidney
	Filaroides canadensis	Ontario	Lungs
	Contracaecum spp.	Oregon	GI (stomach, large intestine)
	Spinitectus gracilis	Alabama, Oregon	GI (small intestine)
	Spinitectus sp.	Oregon	GI (intestine)
	Eustrongylides spp.	Maryland, Oregon	GI (small and large intestine)
	Anisakis simplex	Oregon, Washington	GI (small and large intestine, stomach)
	Anisakis spp.	Oregon	GI (intestine)
	Metabronema sp.	Oregon	GI (intestine)
	Hedruris siredonis	Oregon	GI (small and large intestine)
	Hedruris spp.	Oregon	GI (intestine)
	Cystidicoloides	Washington	GI (intestine)
	Ancylostoma spp.	Louisiana	GI (intestine)
	Uncinaria stenocephala	Unknown	GI (intestine)
	Strongyle sp.	Minnesota	GI (feces)
	Cruzia sp.	Oregon	GI (small and large intestine)

SOURCE: Data from Kimber and Kollias (2000) and U.S. National Museum.
NOTE: GI, Gastrointestinal tract.

TABLE 35.8. Viruses, bacteria, fungi, and protozoans known to inhabit river otters

Disease	Causative Agent	Note	Source
Virus			
Canine distemper	Canine distemper virus	Antibody titers, clinical signs, and distemper serology	Kimber et al. 2000; Reed-Smith 2001
Mink enteritis virus, feline panleukopenia, canine parvovirus	—	Positive antibody titers	Hoover et al. 1985; Kimber et al. 2000
Aleutian disease (plasmacytosis)	Aleutian virus	Challenged, but not clinically ill, no antibody titer	Kenyon et al. 1978; Reed-Smith 2001
Rabies	Rabies virus	24 cases reported	Serfass 1995; Reed-Smith 2001
Infectious canine hepatitis	Adenovirus	Symptoms but no viral isolation confirmation	Kimber et al. 2000; Reed-Smith 2001
Feline rhinotracheitis	Feline herpesvirus-1, feline calicivirus	64 animals tested negative for antibody	Kimber et al. 2000; Reed-Smith 2001
Herpesvirus	Herpesvirus-1	Antibody titers reported	Kimber et al. 2000; Reed-Smith 2001
Bacterium			
Bacterial pneumonia	Unknown	Frequently reported	Hoover 1984; Hoover et al. 1985
Clostridial infection	*Clostidium botulinum*	Susceptible to type C toxin	Reed-Smith 2001
Enteritis	*Clostridium perfringens*	Susceptible during period of stress or dietary conversion	Kollias 1999; Reed-Smith 2001
Tuberculosis	*Mycobacterium bovis*	Reported	Reed-Smith 2001
Purulent pleuritis	*Bacteroides melanigenicus* and others	*Bacteroides melanigenicus* isolated in one case	Griffith et al. 1983; Reed-Smith 2001
Purulent peritonitis	*Klebsiella* pneumonia	Presence of bacteria but no underlying disease	Reed-Smith 2001
Brucellosis	*Brucella* spp.	Presence in lymph node of a road-killed *Lutra lutra*	Foster-Turley 1996; Reed-Smith 2001
Leptospirosis	? bacteria	Believed to be important disease but no histopathology	Chanin 1985; Reed-Smith 2001
Pasteurellosis	*Pasteurella multocida* or *P. pseudotuberculosis*	Clinical signs vary	Wallach and Boever 1983; Reed-Smith 2001
Salmonellosis	*Salmonella* spp.	Isolated from feces of clinically normal otters, no symptoms	Reed-Smith 2001
Fungus			
Dermatomycosis	*Microsporum* spp. and *Trichophyton* spp.	Contagious and potentially zoonotic	Reed-Smith 2001
Coccidiodomycosis	*Coccidioides immitis*	Reported from small-clawed otters	Reed-Smith 2001
Adiaspiromycosis	*Emmonsia crescens*	Reported from Eurasian otters	Simpson and Gavier-Widen 2000; Reed-Smith 2001
Protozoan			
Giardiasis	*Isospora* spp.	Light infestation	Hoover et al. 1985
Coccidiosis	*Giardia* spp.	Reported from other mustelids	Reed-Smith 2001

harvest options help management agencies ensure that otter populations are not adversely impacted (see Management section). As human populations grow, the greatest threat to river otter populations will continue to be loss of habitat.

The primary cause of the decline in otter numbers that led to extirpation from nine states and one Canadian province within recent times was, without exception, habitat destruction (Deems and Pursley 1978). Such destruction may take any of a number of forms. For example, disappearance of otters from West Virginia and parts of Tennessee and Kentucky was attributed to increased acidity of ground water due to mining operations (Lauhachinda 1978). Other causes of habitat destruction include development of waterways for economic or recreational purposes, destruction of riparian habitat for homesites or farmland, and declines in water quality resulting from such things as increased siltation or introduction of pesticide residues into the water as a result of intensive farming operations.

Residues of petroleum products, mercury (Cumbie 1975) and other heavy metals, polychorinated biphenyls, dichlorodiphenyltrichloroethane and its metabolites, Mirex (Hill and Lovett 1976), and aluminum silicates have been reported from river otter tissues (Kimber and Kollias 2000). Ben-David et al. (2001a) assessed the effects of natural sources of mercury contamination, measuring levels of mercury in hair samples from river otters. The authors compared survivorship of otters in two study areas and the influence of age and diet (determined with stable isotope analysis of hair samples) on mercury levels and survivorship. The massive oil spill in Prince William Sound, Alaska, in 1989 appears to have reduced suitable habitat for river otters (Bowyer et al. 1995). The effects of chronic and/or synergistic exposure of these pollutants on otters are unknown.

Because of their susceptibility to pollution, river otters recently have been recognized as an indicator of environmental health in aquatic ecosystems (Melquist and Dronkert 1987; Duffy et al. 1996; Lariviere and Walton 1998). Early studies of the effects of environmental contamination required postmortem evaluation of contaminant levels in organs or tissues. More recently, advances in the use of biomarkers has provided a method of assessing exposure to contaminants in living otters (Taylor et al. 2000; Ben-David et al. 2001b, 2001c). Captive experiments were conducted in which nonlethal doses of weathered crude oil were administered to river otters, simulating chronic exposure to hydrocarbons following an oil spill (Ben-David et al. 2000, 2001b, 2001c, 2002; Ormseth and Ben-David 2000; Taylor et al. 2001). These investigations provided an opportunity to assess the effects of exposure on otter physiology (Ben-David 2000; Ormseth and Ben-David 2000) and develop a better understanding of the use of biomarkers to assess exposure (Ben-David et al. 2001b, 2001c; Taylor et al. 2001).

AGE ESTIMATION

Various physical characteristics have been used to determine the age of river otters. For males, these include growth characteristics of the baculum, development of the testes, and production of spermatozoa. Development of the reproductive tract is useful for determining sexual maturity of females. Age of both sexes may be determined with limited accuracy, using body size, skull characteristics, eye lens weight, and skeletal features. The most reliable technique for estimating age of otters and many other carnivores involves the examination of dental characteristics such as the presence and number of annuli in tooth cementum.

Males. Length, weight, and volume of bacula have been used to determine the age of male otters (Friley 1949; Hooper and Ostenson 1949; Stephenson 1977; Lauhachinda 1978). Bacula of river otters increase rapidly in length until the individual is 3–5 years of age and in weight and volume until an age of 5–6 years. However, there is considerable overlap in older age classes, which limits the usefulness of this technique for otters older than juveniles and yearlings (Stephenson 1977). Size of the testes is useful in distinguishing yearling river otters from older animals (Hamilton and Eadie 1964; Tabor 1974), but is of little use if information on age at first reproduction is desired.

Females. The uterine horns increase in size and degree of vascularization as otters become older or pregnant (Polechla 1987a). Dimensions and weight of the ovaries also may be used to separate juvenile and yearling otters from older animals (Hamilton and Eadie 1964; Tabor 1974; Polechla 1987a).

Both Sexes. Nonreproductive organs and tissues provide the best comparative indications of age in the Nearctic river otter. Crown–rump length may be used as a rough indicator of age for embryos (Hamilton and Eadie 1964). Although no known-age series of river otter embryos is available, Kenyon (1969) presented a hypothetical equation for fetal growth rate for sea otters. Lauhachinda (1978) used the fetal growth rate for mink as a model for river otters. Postnatal growth information was provided by Polechla (1987a).

Because male otters are larger than females of the same age and relative sizes of different subspecies differ, the usefulness of body size as an age criterion is of limited value. Skull measurements show similar patterns of variation, but patterns of suture closure in skull bones are useful in grouping otters into juvenile, yearling, and adult age groups (Hooper and Ostenson 1949; Hamilton and Eadie 1964; Stephenson 1977).

Lauhachinda (1978) found that eye lens weight was highly correlated with age in otters. Differences in eye lens weight between male and female otters existed, but were not significant. Closure of the epiphyses of the long bones is useful in grouping river otters into juvenile, yearling, and adult age classes (Hamilton and Eadie 1964).

A lack of cementum annuli is evident in juvenile river otters and other carnivores (G. Matson, Matson Laboratory, pers. commun., 2001). The ratio of the width of the pulp cavity to the width of the entire canine tooth is indicative of relative age (Kuehn and Berg 1983). Pulp cavities constituting one half or more of the tooth width are considered juveniles, whereas narrower pulp cavities are considered adults. Radiographs showing the pulp cavity require little time in preparation, are easily interpreted, provide an easily stored permanent record, and may be useful in regions where cementum annuli are difficult to interpret. In evaluating river otter harvest and biological data in 10 northeastern states, Chilelli et al. (1996) found that the use of radiographs improved the consistency of age classifications of juveniles or nonjuveniles compared to use of cementum annuli alone. Nonetheless, radiographs are not adequate if the specific age of individuals is needed.

The most reliable and useful age determination technique for river otters results from the annual deposition of dark-staining bands of tooth cementum resulting from an annual period of slowed development (Toweill and Tabor 1982). Tabor (1974) was the first to use this technique with river otters. He found that the initial band was deposited in spring or summer at approximately 1 year of age and that the number of bands equaled the otter's age in years. Stephenson (1977) concluded that the initial band was deposited during summer and also found that the number of bands equaled the otter's age in years; complete agreement was found between cementum annuli counts and age of five known-age otters. Matson's Laboratory (G. Matson, pers. commun., 2001) has sectioned canine teeth from otters throughout North America, and the oldest otters, based on tooth cementum analysis, were 17, 16, and 15 years of age from Nova Scotia, Vermont, and Louisiana, respectively. Lauhachinda (1978) found this technique suitable for age determination of otters from Alabama and Georgia and reported up to 14 cementum annuli in wild-caught otters.

ECONOMIC STATUS

Because of their thick, beautiful, and very durable fur, river otters have been an economically important furbearer since Europeans first arrived in North America (Toweill and Tabor 1982). During 1988–1989, 27 of the 49 continental states and 11 Canadian provinces and territories allowed trappers to harvest otters for their fur (Table 35.9). Although the opportunity to harvest otter pelts in Canada remained consistent throughout the remainder of the twentieth century, some changes occurred in the number of states that allowed a harvest. A survey of Canada and the United States in 1998 by P. J. Polechla (unpublished data) revealed that 28 of the 49 continental states and 10 Canadian provinces and territories allow some form of regulated harvest for river otters in at least a portion of their jurisdiction (Table 35.9).

Obbard et al. (1987) and Polechla (1987a) provided a review of the history of fur harvesting in North America from the 1620s through the 1980s. Native Americans and several companies, including the Hudson's Bay Company, the North West Company, American companies, Dutch companies, French companies, and Russian companies, contributed to the pre-twentieth-century harvest of river otters. Records from these companies indicate that the harvest of river otter pelts fluctuated dramatically from lows of just over 20 pelts/year during the 1670s to an average annual harvest of 43,000 in the 1780s. There was a general decline in furbearer populations in the nineteenth century attributed to overharvest. During that time, the average annual harvest ranged from 16,000 to 24,000 pelts.

Because of past overharvest, states and provinces afforded river otter populations legal protection by implementing management programs during the twentieth century. Populations responded and the annual harvest increased from about 11,000 pelts in the 1920s to more than 41,000 in the 1980s (Obbard et al. 1987). From 1965 to 1980, the number of river otters harvested in North America nearly doubled, with the annual harvest reaching about 50,000 pelts during the late 1970s (Deems and Pursley 1978; Melquist and Dronkert 1987). From 1980 to 1995, otter harvest in the United States ranged from approximately 34,500 in the 1980–1981 trapping season, to nearly 8000 in 1990–1991, to almost 25,000 in 1994–1995 (Fur Resources Committee 1995). Higher harvest numbers during 1980–1995 appeared to correlate with higher pelt values.

Harvest levels are largely influenced by anticipated pelt values and the actual prices paid to trappers for the pelts, which in turn are dictated by fashion and availability (Pursley 1978), economic conditions, and to a lesser extent restrictions on export. In 1977, the river otter was included as an Appendix II species under the Convention on International Trade in Endangered Species of Wild Fauna and Flora (CITES) (Greenwalt 1977). CITES is a treaty that regulates international trade in certain species of animals and plants to ensure that commercial demand does not threaten their survival in the wild. Exports of specimens (live, dead, or parts and products thereof) of animals and plants listed in Appendix II of CITES require an export permit from the country of origin. Consequently, a state could legally trap otters, but without CITES approval from the U.S. Fish and Wildlife Service Management Authority, no pelts could be shipped outside the United States. CITES tagging can affect the value of pelts if the demand for furs is outside the United States, which it typically is, and furbuyers are unable to export the skins.

Total annual raw pelt values for river otters in North America ranged from \$0.6 million to about \$3 million during the 1970s (Pursley 1978; Deems and Pursley 1978). From 1979 to 1995, pelt values in the United States ranged from approximately \$144,000 to \$1.5 million (Fur Resources Committee 1995).

TABLE 35.9. Legal trapping status of the river otter in the United States and Canada from 1919 to 1999

State, Province, or Territory	1919/20	1929/30	1939/40	1949/50	1959/60	1969/70	1979/80	1988/89	1998/99
Alabama	X	?	X	?	X	?	X	X	X
Alaska	X	X	X	X	X	?	X	X	X
Arizona	X						c	c	c
Arkansas	X		X	X	X	X	X	X	X
California	X		X	X	X	o	c	c	c
Colorado	X						c	c	c
Connecticut	X		X	X	X	X	X	X	X
Delaware	X			X	X	X	X	X	X
Florida	X		X	o	X	?	X	X	X
Georgia	X		X	c	?	X	X	X	X
Idaho	X						c	c	c
Illinois	X			c	c	c	c	c	c
Indiana	X		?	?	?	?	c	c	c
Iowa	X			c	c	c		c	c
Kansas	c	X	X	c	c	c	c	c	c
Kentucky	X			c	c	c	c	c	c
Louisiana	X		X	X	X	X	X	X	X
Maine	X		X	X	X	X	X	X	X
Maryland	X*			X	X	X	X	X	X
Massachusetts	X		X	X	X	X	X	X	X
Michigan	X		X	X	X	X	X	X	X
Minnesota	X			X	X	X	X	X	X
Mississippi	X			X	X	?	X	X	X
Missouri	X			c	c	c	c	c	X
Montana	X		o	X	X	X	X	X	X
Nebraska	X						c	c	c
Nevada	X			X	?	X	X	X	X
New Hampshire	X		X	X	X	X	X	X	X
New Jersey	X		X	?	?	?	c	X	X
New Mexico	X		o	c	c	c	c	c	c
New York	?						X	X	X
North Carolina	X		?	X	X	?	X	X	X
North Dakota	c			o			c	c	c
Ohio	X*						c	c	c
Oklahoma	c		o	o	o	o	c	c	c
Oregon	X		?	X	X	X	X	X	X
Pennsylvania	X*				o	?	c	c	c
Rhode Island	X			X	X	X	X	c	c
South Carolina	X		X	X	X	X	X	X	X
South Dakota	c			c	c	c	c	c	c
Tennessee	X*		?	o	c	c	c	c	x/c
Texas	X		?	?	?	?	X	X	X
Utah	c		?	c	c	c	c	c	c
Vermont	X		?	?	?	?	X	X	X
Virginia	X			X	X	o	X	X	X
Washington	X		X	X	X	X	X	X	X
West Virginia	X			c	c	c	c	c	c
Wisconsin	c		?	X	X	X	X	X	X
Wyoming	X	?	?	c	c	o		c	c
USA total	41X, 7c, 1?	2X, 2?	16X, 9?, 2o	22X, 8c, 6?, 5o	23X, 8c, 6?, 2o	18X, 7c, 10?, 4o	18X, 7c, 10?, 4o	27X, 22c	29X, 21c
Alberta	X	X	X	X	X	X	X		X
British Columbia	X	X	X	X	X	X	X		X
Manitoba	X*	X	X	X	X	X	X		x/c
New Brunswick	X	X	X	X	X	X	X		X
Newfoundland	?	?	?	?			?		X
Northwest Territories	X	X	X	X	X	X	X		X
Nova Scotia	X	X	X	X	X	X	X		X
Ontario	X	X	X	X	X	X	X		X
Prince Edward Island	X								c
Quebec	X	X	X	X	X	X	X		X
Saskatchewan	X	X	X	X	X	X	X		X
Yukon Territory	X	X	X	X	X	X	X		X
Canada total	11X, 1?	10X, 1?	10X, 1?	10X, 1?	10x	10x	10X, 1?		11X, 2c

SOURCE: Data from Laut (1921), Nilsson (1980), Novak et al. (1987), Polechla (1990), and P. J. Polechla unpublished data.

NOTE: X, Trapped >1 pelt/year; X*, trapped in some situations or some areas of the state or province; c, closed, protected from trapping; ?, no record reported by sources; o, none trapped during trapping season; blank, no information available.

The twentieth-century harvest of river otters was dominated by Canadian harvests before the 1970s, varying between 51% and 66% of the total North American harvest (Obbard et al. 1987). However, during the 1970s and 1980s, the annual harvest in the United States constituted about 54% of the total. Most river otter pelts are eventually sold to furriers in Central Europe, Russia, and China, where pelts are used in the garment industry.

Because of their large size and specialized requirements, raising of river otters on "fur farms" was never a profitable venture. Training otters for fishing or other activities, as practiced in some parts of Europe and Asia, never became popular in North America (Harris 1968). Although a few animals have been maintained as pets, the largest number of captive otters are maintained in zoos around the world (Reed-Smith 1995, 2001). These captive otters serve a valuable role in education, biological information, and animal husbandry.

River otters continue to be blamed for having serious affects on game fish populations, particularly trout. Numerous studies on the foraging behavior and feeding habits of river otters have revealed that, although otters do take trout, the bulk of their diet consists of nongame fish species. Today, most people view river otters as beneficial to game fish populations based on the assumption that they tend to remove the slower-moving nongame and forage fish that would otherwise compete with game fish for food (Toweill 1974; Lauhachinda 1978; review by Toweill and Tabor 1982; review by Melquist and Dronkert 1987; Manning 1990; Serfass et al. 1990; Mack et al. 1994; Reid et al. 1994; Berg 1999). River otters do occasionally cause severe depredation at fish hatcheries (Serfass et al. 1990) and private ponds where fish have no means of escape.

River otters have also been blamed for damaging the fur resources of certain areas by reportedly preying on beavers (Green 1932) and muskrats (Wilson 1954). In each case where studies of such predation were conducted, otter predation on other furbearers was extremely low.

MANAGEMENT AND CONSERVATION

The harvest of furbearers for food and clothing was a part of early cultures (McGee 1987; Polechla 1987b; Wright 1987). However, it was not until humans began hunting and trapping for commercial purposes that harvest resulted in a significant change in otter populations as market demand replaced individual needs in dictating the extent of the harvest (Obbard et al. 1987). The failure of governments to recognize the potential impacts of a market system had profound effects on populations of many furbearers, including river otters. As demands on natural resources continue to increase, the role of effective management of otters has become increasingly important. At the same time, wildlife management has become increasingly complex, as agencies have attempted to balance their mandates with controversial land and water issues, fiscal constraints, legislative oversight, changing political environments, and a diversity of public opinion. Trapping has come under increasing scrutiny as animal rights groups have challenged wildlife professionals to justify its continued use as an ethical and humane practice (Andelt et al. 1999; Kenyon et al. 1999). Management of river otters becomes further complicated when otters are adopted as a "poster child" by groups that use visual and emotional appeals to further their objectives (Hamilton et al. 2000).

Management options vary according to a combination of biological and sociological needs, and require a considerable investment of time, money, and expertise. Melquist and Dronkert (1987) developed a river otter management program option chart that focused on conservation and population regulation as the key management components.

Conservation. Conservation focuses on establishing populations where they have previously been extirpated or increasing populations in areas where numbers have declined. A translocation is the intentional release of animals to the wild in an attempt to establish, re-establish, or augment a population (IUCN 1987). Reintroduction is a translocation of animals in an attempt to establish a species in an area that was once part of its historical range, but from which it has been extirpated (IUCN 1987). A translocation is a success if it results in a self-sustaining population (Griffith et al. 1989). The reintroduction of river otters to vacant historical range has become increasingly common as a management tool since the mid-1970s, when Colorado initiated the first reintroduction (Tischbein 1976; Mack 1985). Melquist (1985) attributed this increased interest in otter conservation to (1) the inclusion of the river otter in Appendix II of CITES, (2) the development of implant radiotelemetry for otters (Melquist and Hornocker 1979a, 1979b), and (3) the establishment of state nongame programs that provided funding for restoration efforts and research.

Following Colorado's lead, in the early 1980s several states and Alberta initiated river otter restoration efforts (Melquist and Dronkert 1987). By 1990, 17 states had reintroduced or stocked otters, and feasibility studies were underway in others (Polechla 1990; Ralls 1990). River otters have either been reintroduced to areas where they had previously been extirpated or existing populations have been bolstered through translocations of more than 4000 otters in 21 states (Raesly 2001) and Alberta (Melquist and Dronkert 1987). Raesly (2001) conducted a telephone survey of the progress and status of river otter reintroduction projects in the United States and found that, although otters were obtained from a variety of sources, 14 (64%) states used at least some otters from coastal Louisiana. Most biologists felt that reintroductions were successful in restoring extirpated otter populations. We believe that before a restoration effort is deemed successful, management agencies need to document recruitment (reproduction) and population increase, expansion, and stability.

By far the most ambitious and successful restoration effort was undertaken by the Missouri Department of Conservation (Erickson et al. 1984; Erickson and McCullough 1987; Erickson and Hamilton 1988; Hamilton 1998). During an 11-year effort started in 1982, 845 wild-trapped otters, primarily from Louisiana, were released at 43 sites around Missouri (Hamilton 1998). The restoration effort was so successful that a trapping season was initiated in 1996, resulting in an annual harvest of approximately 1000 otters through 1999 (Hamilton et al. 2000). Otters are now widely distributed throughout Missouri's waterways, with the 1999 population estimated to be 11,000–18,000 animals and projected to grow (Hamilton 1998; Gallagher 1999; Hamilton et al. 2000).

Successful wildlife restoration efforts, such as in Missouri, are by no means an accident; they require proper planning, execution, and postrelease monitoring. Griffith et al. (1989) analyzed translocation efforts of native birds and mammals that occurred between 1973 and 1986, which revealed some interesting and valuable information useful to agencies considering some type of translocation. Successful translocations released more animals (mean = 160) than unsuccessful translocations (mean = 54). Areas of high habitat quality were associated with greater success. Translocations into the core of a species historical range were more successful than those on the periphery of, or outside, the historical range (an introduction). Reading and Clark (1996) provide a thorough examination of the complexities of reintroducing different species of carnivores, including the importance of considering human interests, motivations, and local acceptance.

In response to the increasing occurrence of reintroduction projects worldwide, the IUCN developed policy guidelines for reintroduction programs (IUCN 1987, 1995). A reintroduction should only take place where the original causes of extirpation have been removed and the habitat requirements of the species are satisfied (Ralls 1990). There are four basic elements to a river otter reintroduction program, including a feasibility study, a preparation phase, a release or introduction phase, and a follow-up phase. The follow-up phase, which involves monitoring the outcome of the release, is an important aspect of the restoration effort. However, because river otters occur in low densities, are typically secretive and difficult to observe, and tend to occupy poorly accessible habitat, monitoring them is a difficult task (Ralls 1990).

Recovery plans and feasibility studies were completed for many of the river otter restoration projects (Bottorff et al. 1976; Jalkotzy 1980; Choromanski and Fritzell 1982; Goodman 1984; Johnson and Madej 1994; Bluett 1995). Experience resulted in the development and refinement of capture, handling, and release procedures (Melquist and Hornocker 1979a, 1979b; Hoover et al. 1985; Mack 1985; Erickson and McCullough 1987; Polechla 1989b; Beck 1993; Serfass et al. 1993, 1996; Bluett et al. 1999; Blundell et al. 1999), and an examination of genetic considerations for reintroductions (Serfass et al. 1998; K. Koepfli, University of California, Los Angeles, pers. commun. 2000). The handling of hundreds of otters during restoration projects has benefited our understanding of their physiology, medical management, disease, and stress in captured animals (Hoover et al. 1985; Hartup et al. 1999; Kollias 1999). Future river otter restoration projects will benefit from these efforts.

Population Regulation. Population regulation usually involves a regulated harvest or trapping season along with controlling damage at commercial fish hatcheries and private fish ponds. In 1978, no state or Canadian province allowed unregulated taking of otters (Endangered Species Scientific Authority 1978). In 1988–1989, 27 states and 11 Canadian provinces/territories permitted a harvest (Table 35.9). Missouri and Idaho initiated trapping seasons in 1996 and 2000, respectively.

Because river otters were listed as an Appendix II species, the Federal Wildlife Permit Office of the U.S. Fish and Wildlife Service is responsible for setting export quotas on the number of pelts from each state that may enter international commerce. States with CITES approval are required to submit data on river otter populations within their jurisdiction, including the annual harvest and number of pelts tagged. Requiring that each pelt be tagged with CITES export tags provides a mechanism for monitoring characteristics of the harvest and, indirectly, population trends, and may help prevent overharvesting.

Although harvest data are routinely collected by wildlife agencies, there are many limitations associated with these data (Banci and Proulx 1999). Harvest data alone are generally poor short-term indicators of abundance, distribution, and the overall status of furbearers. Annual changes in the harvest may reflect changes in abundance, but such changes are often a function of pelt price, weather conditions that may influence vulnerability, prey cycles, economics, and other factors (Tumlison et al. 1982; Erickson 1982; Hamilton and Fox 1987). Harvest data collected over long periods of time are more reliable in depicting general trends in abundance and distribution, and may be more useful in assessing the impacts of trapping (Obbard et al. 1987). However, Erickson (1982), Polechla (1987a), and Proulx and Barrett (1991) contended that mortality and natality data are needed for reliable interpretation of harvest data. Currently, only carcass analysis of a sample of harvested otters yields statistically reliable natality estimates.

The role of trapping as a furbearer management tool is not easy to assess for most species, yet few other tools to control populations exist. River otters caused losses at 10 of 21 fish hatcheries surveyed in Pennsylvania (Serfass et al. 1990). Properly managed, regulated trapping can help minimize the impact of otters at fish hatcheries, commercial fish ponds, and private ponds. In Missouri, the river otter reintroduction effort was so successful that the Department of Conservation focused the 2000–2001 season on increasing the harvest in areas of the state where otter depredations were adversely affecting commercial fish farming and sport fishing (Hamilton et al. 2000). In other states, trapping restrictions resulting from ballot initiatives have allowed some furbearer populations to grow unchecked, resulting in major depredation and nuisance problems, and costing state agencies considerable money to handle. Conover (2001) contended that if hunting or trapping were to end, some wildlife populations would increase, animals would become more habituated to humans, wildlife damage would increase, and landowner tolerance for wildlife would decrease.

Where trapping occurs, agency biologists have the capability of regulating the harvest to ensure that otter populations are not adversely affected. Regulating trapping season length and establishing area-specific quotas assist wildlife managers in controlling the harvest and preventing otter populations from being adversely affected. Harvest models developed in Minnesota indicate that population stability can be maintained with a harvest of 15–17% (including an estimated 10% poaching factor) of the available autumn otter population (B. Berg and D. Kuehn, Minnesota Department of Natural Resources, pers. commun., 1984). Opening the trapping season for river otters so it coincides with the opening of beaver trapping addresses the issue of otters trapped incidental to beaver trapping. A limited trapping season can benefit trappers and wildlife managers by allowing trappers to keep rather than surrender otters caught while trapping for other aquatic furbearers. Where otters are inadvertently trapped incidental to other trapping, a legal harvest increases the biological data available to state officials and assists them in monitoring the population. A regulated harvest helps wildlife managers deal with depredation problems at fish hatcheries and private fish ponds.

Canada has used a system of registered traplines since the early 1900s as a tool for reducing the intensive competition for furs in an area (Novak 1987), especially when fur prices are high. Registered traplines allow the trapper to manage the fur resources of the area for sustainability. Where there is no control over the number of trappers that can trap a specific area, such as on public lands in the United States, the potential exists for overharvest of those species in high demand.

Survey Methods and Population Monitoring. Population indices of furbearers have long been used by wildlife management agencies to monitor population trends, set harvest regulations, and inform user groups and other constituents about the value of furbearers. CITES reporting requirements and the need to implement a follow-up phase to reintroduction and translocation projects have encouraged wildlife agencies to develop otter monitoring programs. From 1977, when the river otter was included as an Appendix II species under CITES, to 1984, 15 of 55 reported river otter projects in North America focused on the development of census techniques (Melquist 1985). In a 1998–1999 survey, Polechla (In press-a) found that 81% of research efforts in North America involved some form of census technique, ranging from abundance questionnaires, harvest analysis, scat and track surveys, and carcass analysis, to the use of radiotelemetry and radioisotope studies to gather population data. Because states must first show that an otter harvest season will not be detrimental to the resident population before they can receive approval for CITES export tags, the lack of a widely accepted method of monitoring otter relative abundance is cause for concern among state wildlife authorities.

Ideally, the types of data needed to adequately indicate status of furbearer populations include harvest level, catch per unit effort, age-specific pregnancy rates, litter size, and survival (Dixon 1981). However, because of the otter's secretive nature and high degree of mobility, Melquist and Hornocker (1979a) concluded that no simple census method exists for river otters. Arriving at accurate population estimates and trends has proven to be an ongoing challenge because otters are extremely difficult to count and most indices are of questionable accuracy. A well-designed procedure should be adequately tested before implementation and have constant annual bias and reasonable precision in order for managers to anticipate changes in furbearer populations rather than react to such changes (Clark and Andrews 1982). A combination of indices has been recommended (Zackheim 1982; Melquist and Dronkert 1987; Polechla 1987a; Woolf et al. 1997), including carcass collections, sighting reports, sign surveys, population models, radiotelemetry studies, and surveys of trappers or agency biologists. Harvest data, including the number harvested, distribution of the harvest, and sex and age structure, are commonly used in exploited populations (Clark and Andrews 1982; Erickson 1982; Polechla 1987a; Chilleli et al. 1996). Latrine surveys can be used at any time of year as indices of otter occurrence and may be useful where otters are not trapped (Swimley et al. 1998).

Field surveys have been widely accepted as providing the most accurate and objective data on otter distribution and are an important first step in designing appropriate conservation programs (Macdonald 1990). These indirect methods to monitor otter populations typically include winter aerial surveys, bridge-sign surveys, and scent station indices (Kinsey 1981; Humphrey and Zinn 1982; Robson 1982; Foy 1984; Robson and Humphrey 1985; Clark et al. 1987; Reid et al. 1987;

Route and Peterson 1988; Berg and DonCarlos 1998; Gallagher 1999). The use of radiotracer labels (radioactive isotopes) to mark a segment of the population followed by relocation through radioactive feces has been tested (Knaus et al. 1983; Testa et al. 1994). If the number of "labeled" animals is known, a representative sample of labeled and unlabeled scats can be used for mark–recapture analysis (Kruuk et al. 1980). However, because the isotope must be injected in liquid form or implanted as a slow-release tablet (Crabtree et al. 1989) in captured animals, the accuracy of this technique in estimating population densities may not be sufficient to justify its expense (Davison 1980).

Melquist and Dronkert (1987) recommended winter track surveys where snow occurs, but cautioned that otter densities may not correlate with the amount of sign observed. Reid et al. (1987) identified biases in winter aerial surveys. Berg and DonCarlos (1998) concluded that poor snow conditions and inclement weather preventing aerial surveys could compromise their use as a short-term population index. Erb and DePerno (2001) used the winter survey protocol of Berg and DonCarlos (1998) to establish standardized surveys in southern Minnesota. Because otter activity appeared greatest in late winter, surveys were conducted during February and March, 2–4 days following a snowfall of 10 cm or more, under light winds, and by a pilot and an observer in a Bell Jet Ranger helicopter flying at airspeeds of <50 knots and at an altitude of <80 m. The authors concluded that aerial surveys were a reliable method of documenting otter distribution.

Behavioral changes and habituation to scent can affect response rate for scent stations (Robson 1982; Robson and Humphrey 1985). Bridge-sign surveys, while not suitable in roadless areas, appear the most practical and economical of these methods.

The number of otters caught by trappers per trap-night (catch per unit effort, CPUE) and mark–recapture are two of the most effective indices of otter relative abundance (Chilleli et al. 1996; Gallagher 1999). Development of a CPUE index to population level would help state managers assess among-year population trends (Chilleli et al. 1996). CPUE indices increase in effectiveness when trapping intensity and reporting rate are constant (Gallagher 1999).

In an effort to monitor trends in reintroduced otter populations in Missouri, Gallagher (1999) found that neither bridge-sign survey nor CPUE indices showed consistent responses to changes in the otter population. The author recommended that wildlife managers use at least two different methods to monitor otter populations and that one of these methods be a direct measure of relative abundance (CPUE or mark–recapture). An additional benefit of using a direct measure of relative abundance is that wildlife managers are also able to collect direct biological information. Caughley (1974) indicated that population trend and biological data are needed to assess the welfare of a population.

Mark–recapture indices are less effective for otters than for many other species because of the difficulty in recapturing otters (Melquist and Dronkert 1987). In fact, Clark and Andrews (1982) contended that many of the techniques available as population indices for furbearers are not capable of detecting changes of less than 50% of the mean with any degree of confidence. Because increased accuracy normally requires that management agencies expend more money and time, researchers continue to search for cost-effective and less invasive methods to gather population data on otters. Recent developments in molecular genetic techniques have enabled researchers to collect a variety of useful data from feces that enhance our understanding of free-ranging wildlife populations (Kohn and Wayne 1997; Wasser et al. 1997). Analysis of DNA from fecal samples allows researchers to differentiate among species (Reed et al. 1997; Hansen and Jacobsen 1999). DNA can be extracted from river otter intestinal cells shed with feces to generate DNA profiles or "fingerprints" specific to individual animals (Golden 2000). The systematic collection of feces, followed by molecular typing with hypervariable microsatellite markers and sex-specific probes, has provided a rapid estimate of population size and sex ratio in coyotes while avoiding problems of capture and handling (Kohn et al. 1999). M. Cogliano (George Mason University pers. commun., 2001) used microsatellite markers in an effort to estimate river otter population densities and determine dietary differences among individuals. Blundell et al. (2002b) used microsatellite markers in combination with radiotelemetry data to assess the characteristics of sex-biased dispersal and patterns of gene flow among populations of coastal river otters. The authors used these data to evaluate the potential for natural recolonization of otters following a local extirpation.

Fecal analysis also can be used to monitor certain environmental impacts on otter populations. Research following the *Exxon Valdez* oil spill in March 1989 demonstrated that changes in levels of fecal porphryins could serve as a biomarker in detecting different levels of contaminant exposure and contribute to a health assessment of wild otters (Taylor et al. 2000). These techniques show tremendous promise as a noninvasive method of understanding river otter population characteristics and dynamics and will undoubtedly receive further attention by researchers and resource agencies.

Although management policies tend to be conservative, river otter populations remain susceptible to overharvest because individuals travel extensively in restricted avenues provided by watercourses. For this reason, management agencies in those jurisdictions that allow an otter harvest need to continue to monitor population levels and protect riparian habitat (Polechla 1988) to assure the maintenance of healthy and viable populations.

RESEARCH AND MANAGEMENT NEEDS

Considerable research has been conducted on river otter/habitat relationships during the past two decades. Nonetheless, there remains a need for further research and population monitoring. Because river otters are included on CITES, member nations, including Canada and the United States, are required to monitor or restrict trade. The burden of determining whether or not international trade is adversely affecting otter populations rests on the provinces and states that have approved CITES export tags. States requesting CITES authority to export otters must provide evidence that populations can sustain a harvest without being negatively affected. Because it is extremely difficult to census and monitor changes in otter populations, providing adequate and reliable data to CITES authorities remains a challenge for management agencies.

Wildlife management agencies attempting to manage otter populations cannot rely solely on harvest statistics; they must also collect additional information to develop a reliable, cost-effective method of monitoring population status and trends. The best population data are derived from costly and time-intensive telemetry studies, and even though telemetry data have limited use in monitoring year-to-year population trends, when combined with other population indices, they provide the best data set available. Population indices, as described in the Management section, tend to be inexpensive and nonintrusive, but generally fail to provide accurate and reliable population status and trend data. We recommend additional research that will allow testing population indices with known populations to calibrate an index with confidence limits. If inexpensive and reliable population monitoring techniques are not developed, population status and trend data will continue to be suspect by regulatory agencies.

In 1998–1999, Polechla (In press-a) sent a questionnaire to all provincial and state wildlife agencies in Canada and the United States asking for input on research needs. Based on a 100% response, 43% of the states recommended ongoing projects on otter distribution and relative abundance. Demographic studies of river otter populations were recommended by 50% of the Canadian provinces and 24% of the states. Approximately 18% of the states advocated studying otter habitat use and requirements and additional research on the impact of trapping on the population. Biologists from six states (12%) expressed a need to study heavy metal (e.g., mercury) concentrations. Biologists from one inland state and three coastal states felt that feeding habits studies were needed. Other suggestions from Canadian biologists included (1) determining a more detailed account of otter distribution and relative abundance, (2) conducting dietary analyses, (3) delineating habitat use, and (4) investigating environmental toxicants.

Biologists from Arizona, California, and New Mexico, states that encompass the range of the southwestern river otter (*Lontra canadensis sonora*), recommended studying the feasibility of translocating otters into the region. Currently, the states of Utah (D. Mikesic, Navajo Nation Natural Heritage Program, pers. commun., 2003), Colorado (P. Schnurr, Colorado Division of Wildlife, pers. commun., 2003), and possibly New Mexico are considering further actions. The IUCN recommends that, prior to any consideration of translocating otters into a particular area, a thorough survey should be done for otters and their habitat (Ralls 1990). Currently, there is no verified extant population of the southwestern subspecies (Polechla 1993, 2002a; Tolme 2003), and no captive population exists (Reed-Smith 1995). Although reports of otters in the area still surface, few thorough studies have been done in their historical habitat (Polechla 2000, 2002a) and many waters remain unassessed. Prior to any translocation program, it would be necessary to conduct a comprehensive survey of the Colorado River (including the San Juan and Gila tributaries) and nearby watersheds (e.g., the Rio Grande/Pecos River and Canadian River) to locate possible extant populations. Stocking of other subspecies into these waters would likely flood the gene pool of any existing indigenous remnant populations (Polechla 2002b).

As a "flagship species," otter populations should be monitored to provide insight into wetland conditions, water quality, levels of industrial and agricultural pollutants, human disturbance, and other factors that might have an impact on the welfare of populations. There should be a re-evaluation of the potential reproductive contribution of subadult female otters (Chilelli et al. 1996). In the eastern United States, Chilelli et al. (1996) found that at least 35% of the adult females failed to produce litters, which could have important implications for population recruitment and management.

Another area of research that has management implications is the function of otter latrine sites (G. Blundell, University of Alaska, pers. commun., 2000). Although we use the distribution and density of latrine sites to provide information on presence and population characteristics, we do not fully understand what information otters are communicating by scent marking at these sites, the level of participation by otters in marking, and many other aspects of latrine site use. A better understanding of marking behavior is necessary if we plan to use latrine sites to interpret population conditions and trends.

State wildlife agencies, through the Fur Resources Technical Committee of the International Association of Fish and Wildlife Agencies, and trappers are developing "best management practices" for trapping furbearers in the United States as a way to improve animal welfare, sustain regulated trapping, and maintain public acceptance. Of the 23 species identified and prioritized for trap testing needs, the river otter was considered of medium priority. Continued development and refinement of best management practices are important to state wildlife agencies to maintain credibility and public acceptance of trapping as a management tool.

At their meeting in Valdivia, Chile, 20–26 January 2001, members of the Otter Specialist Group of the International Union for the Conservation of Nature and Natural Resources drafted 14 recommendations for river otter research, management, and related activities in North America (C. Reuther pers. commun., 2001). Some of the activities have been identified above, others include the following: (1) conduct long-term research, especially in harvested and nonharvested populations and in areas that are protected and unprotected; (2) assess the effects of completed reintroduction projects on genetic variability as compared with natural populations; (3) initiate and refine methods for monitoring the expansion and persistence of reintroduced populations; (4) develop a predictive model that will estimate the potential for natural recolonization of historical ranges from native populations of river otters versus the need for reintroduction; and (5) investigate the importance of connectivity among all types of aquatic habitats and the influence of land-use patterns within watersheds.

Population models that help wildlife managers set harvest quotas need to include accurate parameters if they are to be useful. Recruitment rates and the role of 1- 2-year-old females in contributing to the annual production are important model parameters. A functional model will also require reliable data on mortality rates and the causes of age- and sex-specific mortality.

LITERATURE CITED

Addison, E. M., M. A. Strickland, A. B. Stephenson, and J. Hoeve. 1988. Cranial lesions possibly associated with *Skrjabingylus* (Nematoda: Metastrongyloidea) infections in martens, fishers, and otters. Canadian Journal of Zoology 66:2155–59.

Allen, A. W. 1987. The relationship between habitat and furbearers. Pages 164–79 *in* M. Novak, J. A. Baker, M. E. Obbard, and B. Malloch, eds. Wild furbearer management and conservation in North America. Ontario Ministry of Natural Resources, Toronto.

Andelt, W. F., R. L. Phillips, R. H. Schmidt, and R. B. Gill. 1999. Trapping furbearers: An overview of the biological and social issues surrounding a public policy controversy. Wildlife Society Bulletin 27:53–64.

Anderson, E. A., and A. Woolf. 1984. River otter food habits in northwestern Illinois. Transactions of the Illinois Academy of Science 80:115–18.

Anderson, K. L. 1981. Population and reproduction characteristics of the river otter in Virginia and tissue concentrations of environmental contaminants. M.S. Thesis, Virginia Polytechnic Institute and State University, Blacksburg.

Anderson, S. 1977. Geographic ranges of North American terrestrial mammals. American Museum Novitates 2629:1–15.

Arthur, S. C. 1928. The fur animals of Louisiana (Bulletin No. 18). Louisiana Department Conservation.

Audubon, J. J., and J. Bachman. 1851. Quadrepeds of North America. V. G. Audubon, New York.

Bailey, V. 1931. Mammals of New Mexico. North American Fauna 53:1–412.

Bailey, V. 1936. The mammals and life zones of Oregon. North American Fauna 55:1–416.

Balke, J. M. E. 1993a. River otter predation on juvenile salmonids in winter: A review (Unpublished report, Project 92.8). Ministry of Forests, Victoria, British Columbia, Canada.

Balke, J. M. E. 1993b. Preliminary report of river otter scat collection and diet analysis in Queen Charlotte Island, November 1992 and February 1993 (Unpublished report, Project 92.8). Ministry of Forests, Victoria, British Columbia, Canada.

Ballard, K., J. G. Sivak, and B. Levy. 1987. Intraocular muscles of mink (*Mustela vison*). Opthamology and Visual Science 28:203.

Banci, V., and G. Proulx. 1999. Resiliency of furbearers to trapping in Canada. Pages 175–203 *in* G. Proulx, ed. Mammal trapping. Alpha Wildlife Research and Management Limited, Sherwood Park, Alberta, Canada.

Banfield, A. W. F. 1974. The mammals of Canada. University of Toronto Press, Toronto.

Beaver, T. D., G. A. Feldhamer, and J. A. Chapman. 1981. Dental and cranial anomalies in the river otter (Carnivora: Mustelidae). Brimleyana 7:101–9.

Beck, T. D. I. 1993. River otter reintroduction procedures. Colorado Division of Wildlife Research Review 2:14–16.

Beckel, A. L. 1982. Behavior of free-ranging and captive river otters in north-central Wisconsin. Ph.D. Dissertation, University of Minnesota, St. Paul.

Beckel, A. L. 1990. Foraging success rates of North American river otters, *Lutra canadensis*, hunting alone and hunting in pairs. Canadian Field-Naturalist 104:586–88.

Ben-David, M., R. T. Bowyer, and J. B. Faro. 1995. Niche separation by mink and river otters: coexistence in a marine environment. Oikos 75:41–48.

Ben-David, M., R. T. Bowyer, L. K. Duffy, D. D. Roby, and D. M. Schell. 1998. Social behavior and ecosystem processes: River otter latrines and nutrient dynamics of terrestrial vegetation. Ecology 79:2567–71.

Ben-David, M., T. M. Williams, and O. A. Ormseth. 2000. Effects of oiling on exercise physiology and diving behavior in river otters: A captive study. Canadian Journal of Zoology 78:1380–90.

Ben-David, M., L. K. Duffy, G. M. Blundell, and R. T. Bowyer. 2001a. Exposure of coastal river otters to mercury: Relation to age, diet and survival. Environmental Toxicology and Chemistry 20:1986–92.

Ben-David, M., L. K. Duffy, and R. T. Bowyer. 2001b. Biomarker responses in river otters experimentally exposed to oil contamination. Journal of Wildlife Diseases 37:489–508.

Ben-David, M., T. Kondratyuk, B. R. Woodin, P. W. Snyder, and J. J. Stegeman. 2001c. Induction of cytochrome P4501A1 expression in captive river otters fed Prudhoe Bay crude oil: Evaluation by immunohistochemistry and quantitative RT-PCR. Biomarkers 6:218–35.

Ben-David, M., G. M. Blundell, and J. E. Blake. 2002. Post-release survival of river otters: Effects of exposure to crude oil and captivity. Journal of Wildlife Management 66:1208–1223.

Berg, J. K. 1999. River otter research project on the upper Colorado River basin in and adjacent to Rocky Mountain National Park, Colorado. Final Report to Rocky Mountain National Park, Grand Lake, CO.

Berg, W. E., and M. DonCarlos. 1998. Experimental river otter population trend survey. Pages 99–105 *in* B. Joselyn, ed. Wildlife populations and research unit reports. Minnesota Department of Natural Resources, St. Paul.

Best, A. 1962. The Canadian otter, *Lutra canadensis,* in captivity. International Zoo Yearbook 4:42–44.

Bluett, R., ed. 1995. Illinois river otter recovery plan. Illinois Department of Natural Resources, Springfield.

Bluett, R. D., E. A. Anderson, G. F. Hubert, Jr., G. W. Kruse, and S. E. Lauzon. 1999. Reintroduction and status of the river otter (*Lutra canadensis*) in Illinois. Transactions of the Illinois State Academy of Science 92:69–78.

Blundell, G. M. 2001. Social organization and spatial relationships in coastal river otters: Assessing form and function of social groups, sex-biased dispersal, and gene flow. Ph.D. Dissertation, University of Alaska, Fairbanks.

Blundell, G. M., J. W. Kern, R. T. Bowyer, and L. K. Duffy. 1999. Capturing river otters: A comparison of Hancock and leg-hold traps. Wildlife Society Bulletin 27:184–92.

Blundell, G. M., R. T. Bowyer, M. Ben-David, T. A. Dean, and S. C. Jewett. 2000. Effects of food resources on spacing behavior of river otters: does forage abundance control home-range size? Pages 325–333 *in* J. H. Eiler, D. Alcorn, and M. Neuman, eds. Biotelemetry 15: Proceedings of the 15th International Symposium on Biotelemetry, Juneau, AK. Wageningen, The Netherlands: International Society on Biotelemetry.

Blundell, G. M., J. A. K. Maier, and E. M Debevec. 2001. Linear home ranges: Effects of smoothing, sample size, and autocorrelation on kernel estimates. Ecological Monographs 71:469–89.

Blundell, G. M., M. Ben-David, and R. T. Bowyer. 2002a. Sociality in river otters: Cooperative foraging or reproductive strategies?. Behavioral Ecology 13:134–41.

Blundell, G. M., M. Ben-David, P. Groves, R. T. Bowyer, and E. Geffen. 2002b. Characteristics of sex-biased dispersal and gene flow in coastal river otters: Implications for natural recolonization of extirpated populations. Molecular Ecology 11:289–303.

Boos, A. 1987. Zur anatomie des fishotters (*Lutra lutra*). Schriftliche Hausarbeit zur Ersten Staatsprüfung für das Lehramtan Gymnasien, Braunschweig, Germany.

Bottorff, J. A., R. A. Wigal, D. Pursley, and J. I. Cromer. 1976. The feasibility of river otter reintroduction in West Virginia (Special Report). West Virginia Department of Natural Resources, Division of Wildlife.

Bowyer, R. T., J. W. Testa, J. B. Faro, C. C. Schwartz, and J. B. Browning. 1994. Changes in diets of river otters in Prince Williams Sound, Alaska: Effects of the *Exxon Valdez* oil spill. Canadian Journal of Zoology 72:970–76.

Bowyer, R. T., J. W. Testa, and J. B. Faro. 1995. Habitat selection and home ranges of river otters in a marine environment: Effects of the *Exxon Valdez* oil spill. Journal of Mammalogy 76:1–11.

Bradley, P. V. 1986. Ecology of river otters in Nevada. M.S. Thesis, University of Nevada, Reno.

Brandes, B., C. Reuther, and J. U. Wieding. 1983. Haematological and biochemical values in comparison to man, dog, and cat. Paper presented at Third international otter symposium. Strasbourg, Germany.

Brisson, M. J. 1762. Regnum animale in classes IX distributum sive synopsis methodica. Editio altera auctior. Leiden, Theodorum Haak.

Britt, T. L., R. Gerhart, and J. S. Phelps. 1984. River otter stocking (Unpublished report, Arizona Federal Aid Project W-53-R-34, WP6, J4:125–34).

Buskirk, S. W., P. F. A. Maderson, and R. M. O'Connor. 1986. Plantar glands in North American Mustelidae. Pages 617–22 *in* D. Duvall, D. Muller-Schwarze, and R. M. Silverstein, eds. Chemical signals in vertebrates IV. Plenum Press, New York.

Cahn, A. R. 1937. The mammals of the Quetico Provincial Park of Ontario. Journal of Mammal Research 18:19–30.

Calle, P. P. 1988. Asian small-clawed otter (*Aonyx cinerea*) urolithiasis prevalence in North America. Zoo Biology 7:233–42.

Caughley, G. 1974. Interpretation of sex ratios. Journal of Wildlife Management 38:557–62.

Chabreck, R. H., J. E. Holcombe, R. G. Lincombe, and N. E. Kinler. 1982. Winter foods of river otters from saline and fresh environments in Louisiana. Proceedings of the Annual Conference of the Southeastern Association of Fish and Wildlife Agencies 36:473–83.

Chanin, P. 1985. The natural history of otters. Facts on File, New York.

Chilelli, M., B. Griffith, and D. J. Harrison. 1996. Interstate comparisons of river otter harvest data. Wildlife Society Bulletin 24:238–46.

Choromanski, J. F., and E. K. Fritzell. 1982. Status of the river otter (*Lutra canadensis*) in Missouri. Transaction of the Missouri Academy of Science 16:43–48.

Christensen, K. M. 1984. Habitat selection, food habits, movements, and activity patterns of reintroduced river otters (*Lutra canadensis*) in central Arizona. M.S. Thesis, Northern Arizona University, Flagstaff.

Clark, J. D., T. Hon, K. D. Ware, and J. H. Jenkins. 1987. Methods for evaluating abundance and distribution of river otters in Georgia. Proceedings of the Annual Conference of the Southeastern Association of Fish and Wildlife Agencies 41:358–64.

Clark, W. R., and R. D. Andrews. 1982. Review of population indices applied in furbearer management. Pages 11–22 *in* G. C. Sanderson, ed. Midwest furbearer management. North Central Section, Central Mountains and Plains Section, and Kansas Chapter, Wildlife Society.

Cocks, A. H. 1881. Notes on the breeding of the otter. Proceedings of the Zoological Society of London 1881:249–50.

Conover, M. R. 2001. Effect of hunting and trapping on wildlife damage. Wildlife Society Bulletin 29:521–32.

Cooley, L. S. 1983. Winter food habits and factors influencing the winter diet of river otter in northern Florida. M.S. Thesis, University of Florida, Gainesville.

Cooney, J. C., and K. L. Hays. 1972. The ticks of Alabama (Ixodidae; Acarina) (Bulletin No. 426). Alabama Experiment Station.

Costa, D. P. 1982. Energy, nitrogen, and electrolyte flux and sea water drinking in the sea otter *Enhydra lutris*. Physiological Zoology 55:35–44.

Coues, E. 1877. The fur-bearing animals: A monograph of North American Mustelidae. Govertnment Printing Office, Washington, D.C.

Crabtree, R. L., F. G. Burton, T. R. Garland, D. A. Cataldo, and W. H. Rickard. 1989. Slow-release radioisotope implants as individual markers for carnivores. Journal of Wildlife Management 53:949–54.

Cumbie, P. M. 1975. Mercury levels in Georgia otter, mink and freshwater fish. Bulletin of Environmental Contamination and Toxicology 14:193–96.

Davis, H. G., R. J. Aulerich, J. G. Sikarskie, and J. N. Stuht. 1992a. Hematologic and blood chemistry values of the northern river otter (*Lutra canadensis*). Scientfur 16:267–71.

Davis, H. G., S. Richard, J. Aulerich, S. J. Bursian, J. G. Sikarskie, and J. N. Stuht. 1992b. Feed consumption and food transit time in northern river otters (*Lutra canadensis*). Journal of Zoo and Wildlife Medicine 23:241–44.

Davis, J. 1978. A classification of otters. Unpaginated *in* Otters. N. Duplaix, ed. Otter Specialist Group, IUCN, Gland, Switzerland.

Davis, J. A. 2000. Mexican otters. River Otter Journal 9:8–9.

Davis, W. S. 1981. Aging, reproduction, and winter food habits of bobcat (*Lynx rufus*) and river otter (*Lutra canadensis*) in Mississippi. M.S. Thesis, Mississippi State University, Mississippi State.

Davison, R. P. 1980. The effects of exploitation on some parameters of coyote populations. Ph.D. Dissertation, Utah State University, Logan.

Dearden, L. C. 1954. Extra premolars in the river otter. Journal of Mammal Research 35:125–26.

Deems, E. F., Jr., and D. Pursely. 1978. North American furbearers: Their management, research and harvest status in 1976. International Association of Fish and Wildlife Agencies and University of Maryland, College Park.

Dixon, K. R. 1981. Data requirements for determining the status of furbearer populations. Pages 1360–73 *in* J. A. Chapman and D. Pursley, eds. Proceedings of the worldwide furbearer conference. Frostburg, MD.

Dolloff, C. A. 1993. Predation by river otters (*Lutra canadensis*) on juvenile Coho salmon (*Oncorhynchus kisutch*) and Dolly Varden (*Salvelinus malma*) in southeast Alaska. Canadian Journal of Fisheries and Aquatic Sciences 50:312–15.

Douglas, W. D. 1928. Natural history notes from Baker Lake, North West Territory. Canadian Field-Naturalist 42:106.

Dubuc, L. J. 1987. Ecology of river otters on Mount Desert Island, Maine. M.S. Thesis, University of Maine, Orono.

Dubuc, L. J., W. B. Krohn, and R. B. Owen, Jr. 1990. Predicting occurrence of river otters by habitat on Mount Desert Island, Maine. Journal of Wildlife Management 54:594–99.

Duffy, L. K., R. T. Bowyer, J. W. Testa, and J. B. Faro. 1996. Acute phase proteins and cytokines in Alaskan mammals as markers of chronic exposure to environmental pollutants. American Fisheries Society Symposia 18:809–13.

Eisenberg, J. F. 1981. The mammalian radiations; an analysis of trends in evolution, adaptation, and behavior. University of Chicago Press, Chicago.

Eley, T. J. 1977. *Ixodes uriae* (Acari; Ixodidae) from a river otter. Journal of Medical Entomology 13:506.

Endangered Species Scientific Authority. 1978. Export of bobcat, lynx, river otters, and American ginseng. Federal Register 43:11082–93.

Erb, J. D., and C. S. DePerno. 2001. Distribution and relative abundance of river otters in southern Minnesota. Pages 19–26 *in* M. W. DonCarlos, R. T. Eberhardt, R. O. Kimmel, and M. S. Lenarz, eds. Wildlife populations and research unit 2000 report. Minnesota Department of Natural Resources, St. Paul.

Erickson, D. W. 1982. Estimating and using furbearer harvest information. Pages 53–65 *in* G. C. Sanderson, ed. Midwest furbearer management. North Central Section, Central Mountains and Plains Section, and Kansas Chapter, Wildlife Society.

Erickson, D. W., and D. A. Hamilton. 1988. Approaches to river otter restoration in Missouri. Transactions of the North American Wildlife and Natural Resources Conference 53:404–13.

Erickson, D. W., and C. R. McCullough. 1987. Fates of translocated river otters in Missouri. Wildlife Society Bulletin 15:511–17.

Erickson, D. W., C. R. McCullough, and W. R. Porath. 1984. River otter investigations in Missouri: Evaluation of experimental river otter reintroductions (Final Report, Missouri Department. Conservation Federal Aid Project Number W-13-R-38 Study Number 63, Job Number 2).

Erlinge, S. 1968. Food studies on captive otters *Lutra lutra* L. Oikos 19:259–70.

Estes, J. E. 1989. Adaptations for aquatic living by carnivores. Pages 242–82 *in* J. L. Gittleman, ed. Carnivore behavior, ecology, and evolution. Cornell University, Ithaca, NY.

Ewer, R. F. 1973. The carnivores. Cornell University Press, Ithaca, NY.

Fagen, J. M. 1986. Ecophenic variation and functional morphology of coastal and interior *Lutra canadensis.* M.S. Thesis, West Chester University, West Chester, PA.

Fain, A., and C. E. Yunker. 1980. *Lutracarus canadensis,* n.g., n. sp. (Acari: Listrophoridae) from the river otter, *Lutra canadensis.* Journal of Medical Entomology 17:424–26.

Farney, J. P., and J. K. Jones. 1978. Recent records of the river otter from Nebraska. Transaction of the Kansas Academy of Science 81:275–76.

Feldhamer, G. A., L. C. Drickamer, S. H. Vessey, and J. F. Merritt. 1999. Mammalogy: Adaptation, diversity, and ecology. W. C. Brown, McGraw-Hill, Boston.

Ferguson, S. H., J. A. Virgl, and S. Lariviere. 1996. Evolution of delayed implantation and associated grade shifts in life history traits of North American carnivores. Ecoscience 3:7–17.

Field, R. J. 1970. Winter habits of the river otter (*Lutra canadensis*) in Michigan. Michigan Academician 3:49–58.

Findlay, W. R. 1992. Ecological aspects and dietary habits of river otter in northeastern Utah. M.S. Thesis, Brigham Young University, Provo, UT.

Fish, F. E. 1994. Association of propulsive swimming mode with behavior in river otters (*Lutra canadensis*). Journal of Mammal Research 75:989–97.

Fisher, E. M. 1942. Osteology and myology of the California river otter. Stanford University Press, Palo Alto, CA.

Fleming, W. J., C. F. Dixon, and J. W. Lovett. 1977. Helminth parasites of river otters (*Lutra canadensis*) from southeastern Alabama. Proceedings of the Helminthological Society of Washington 44:131–35.

Foottit, R. G., and R. W. Butler. 1977. Predation on nesting glaucous-winged gulls by river otter. Canadian Field-Naturalist 91:189–90.

Forrester, D. J. 1992. Parasites and diseases of wild mammals in Florida. University Press of Florida, Gainesville.

Foster-Turley, P. 1996. Making biodiversity conservation happen: The role of environmental education and communication. Environmental Education and Communication Project, U.S. Agency for International Development, Washington, DC.

Foy, M. K. 1984. Seasonal movement, home range, and habitat use of river otter in southeastern Texas. M.S. Thesis, Texas A&M University, College Station.

Friley, C. E. 1949. Age determination, by use of the baculum, in the river otter (*Lutra c. canadensis*) Schreber. Journal of Mammal Research 30:102–10.

Fur Resources Committee. 1995. U.S. Fur Harvest (1970–1995) and fur value (1974–1995) statistics by state and region. International Association of Fish and Wildlife Agencies, Washington, DC.

Gallagher, E. 1999. Monitoring trends in reintroduced river otter populations. M.S. Thesis, University of Missouri, Columbia.

Gallo, J. P., and L. Rojas-B. 1985. Nombres cientificos y comunes de los mamiferos marinos de Mexico. Annals de Instituto Biologia Universidad Nacional Autonoma de Mexico 56:1043–56.

Gilbert, F. F., and E. G. Nancekivell. 1982. Food habits of the mink (*Mustela vison*) and otter (*Lutra canadensis*) in northeastern Alberta. Canadian Journal of Zoology 60:1282–88.

Golden, H. N. 2000. Furbearer management technique development. Job 3: Distribution, trend, habitat use, and harvest potential of coastal river otter populations (Federal Aid in Wildlife Restoration Research Progress Report, 1 July 1999–30 June 2000). Alaska Department of Fish and Game, Juneau.

Goodman, P. 1984. River otter recovery plan. Colorado Division of Wildlife, Denver.

Gray, J. E. 1843. Descriptions of some new genera and species of Mammalia in the British Museum Collection. Annals and Magazine of Natural History, Series I II:117–19.

Green, H. U. 1932. Observations of the occurrence of the otter in Manitoba in relation to beaver life. Canadian Field-Naturalist 46:204–6.

Green, J. 1977. Sensory perception in hunting otters, *Lutra lutra L.* Otters: Journal of the Otter Trust 1977:13–16.

Greenwalt, L. A. 1977. International trade in endangered species of wild fauna and flora. Federal Register 42:10462–88.

Greer, K. R. 1955. Yearly food habits of the river otter in the Thompson Lakes region, northwestern Montana, as indicated by scat analysis. American Midland Naturalist 54:299–313.

Grenfell, W. E., Jr. 1974. Food habits of the river otter in Suisin Marsh, central California. M.S. Thesis, California State University, Sacramento.

Griess, J. M. 1987. River otter reintroduction in Great Smoky Mountains National Park. M.S. Thesis, University of Tennessee, Knoxville.

Griffith, B., J. M. Scott, J. W. Carpenter, and C. Reed. 1989. Translocation as a species conservation tool: Status and strategy. Science 245:477–80.

Griffith, J. W., W. J. White, P. A. Pergrin, and A. A. Darrigrand. 1983. Proliferative pleuritis associated with *Bacteroides melaninogenicus* subsp *asaccharolyticus* infection in a river otter. Journal of the American Veterinary Medical Association 13:1287–88.

Grinnell, J., J. S. Dixon, and J. M. Linsdale. 1937. Furbearing mammals of California, Vol., 1. University of California Press, Berkeley.

Hall, E. R. 1955. Handbook of mammals of Kansas. University of Kansas Museum of Natural History Miscellaneous Publications 7:1–303.

Hall, E. R. 1981. The mammals of North America. John Wiley, New York.

Hall, E. R., and K. R. Kelson. 1959. The mammals of North America. Ronald Press, New York.

Hamilton, D. A. 1998. Missouri river otter population assessment (Unpublished Special Report). Missouri Department of Conservation, Columbia.

Hamilton, D. A., and L. B. Fox. 1987. Wild furbearer management in the midwestern United States. Pages 1100–1116 *in* M. Novak, J. A. Baker, M. E. Obbard, and B. Malloch, eds. Wild furbearer management in North America. Ontario Trappers Association, North Bay, Ontario, Canada.

Hamilton, D. A., D. J. Witter, and T. L. Goedeke. 2000. Balancing public opinion in managing river otters in Missouri. Transactions of the North American Wildlife and Natural Resource Conference 65:292–99.

Hamilton, W. J. 1943. The mammals of eastern United States: An account of recent land mammals occurring east of the Mississippi. Comstock, Ithaca, NY.

Hamilton, W. J., Jr. 1961. Late fall, winter, and early spring foods of 141 otters from New York. New York Fish and Game Journal 8:106–9.

Hamilton, W. J., Jr., and W. R. Eadie. 1964. Reproduction in the otter, *Lutra canadensis.* Journal of Mammal Research 45:242–52.

Hansen, M. M., and L. Jacobsen. 1999. Identification of mustelid species: otter (*Lutra lutra*), American mink (*Mustela vison*) and polecat (*Mustela putorius*), by analysis of DNA from faecal samples. Journal of Zoology (London) 247:177–81.

Harris, C. J. 1968. Otters: A study of the Recent Lutrinae. Weidenfield and Nicolson, London.

Hartup, B. K., G. V. Kollias, M. C. Jacobsen, B. A. Valentine, and K. R. Kimber. 1999. Exertional myopathy in translocated river otters from New York. Journal of Wildlife Diseases 35:542–47.

Hayward, J. L., Jr., C. J. Amlaner, Jr., W. H. Gillett, and J. F. Stout. 1975. Predation on nesting gulls by a river otter in Washington state. Murrelet 56:9–10.

Hediger, H. 1966. Diet of animals in captivity. International Zoo Yearbook 6:37–58.

Hill, E. P., and J. W. Lovett. 1976. Pesticide residues in beaver and river otter from Alabama. Proceedings of the Southeastern Association of Game and Fish Commissioners 29:365–69.

Holcombe, J. E. 1980. Winter food habits of the river otter in southern Louisiana. M.S. Thesis, Louisiana State University, Baton Rouge.

Home, W. S. 1982. Ecology of river otters (*Lutra canadensis*) in marine coastal environments. M.S. Thesis, University of Alaska, Fairbanks.

Hooper, E. T., and B. T. Ostenson. 1949. Age groups in Michigan otter. University of Michigan, Occasional Papers of the Museums of Zoology 518:1–22.

Hoover, J. P. 1984. Surgical implantation of radio telemetry devices in American river otters. Journal of the American Veterinary Medical Association 185:1317–20.

Hoover, J. P., and E. M. Jones. 1986. Physiologic and electrocardiographic responses of American river otters (*Lutra canadensis*) during chemical immobilization and inhalation anesthesia. Journal of Wildlife Diseases 22:557–63.

Hoover, J. P., and R. D. Tyler. 1986. Renal function and fractional clearances of American river otters (*Lutra canadensis*). Journal of Wildlife Diseases 22:547–56.

Hoover, J. P., R. J. Bahr, M. A. Nieves, R. T. Doyle, M. A. Zimmer, and S. E. Lauzon. 1985. Clinical evaluation and prerelease management of American river otters in the second year of a reintroduction study. Journal of the American Veterinary Medicine Association 187:1154–61.

Hornocker, M. G., J. P. Messick, and W. E. Melquist. 1983. Social strategies in three species of Mustelidae. Acta Zoologica Fennica 174:185–88.

Humphrey, S. R., and T. L. Zinn. 1982. Seasonal habitat use by river otters and Everglades mink in Florida. Journal of Wildlife Management 46:375–81.

IUCN. 1987. The IUCN position statement on translocation of living organisms: Introductions, reintroductions, and re-stocking. International Union for the Conservation of Nature and Natural Resources, Gland, Switzerland.

IUCN. 1995. IUCN/SSC guidelines for re-introductions. International Union for the Conservation of Nature and Natural Resources, Gland, Switzerland.

Iversen, J. A. 1972. Basal energy metabolism of Mustelids. Journal of Comparative Physiology 81:341–44.

Jackson, H. H. T. 1961. Mammals of Wisconsin. University of Wisconsin Press, Madison.

Jackson, M. A., D. Fertland, and J. F. Bergan. 1998. Recent records of the river otter (*Lutra canadensis*) along the Texas Gulf Coast. Texas Journal of Science 50:243–47.

Jalkotzy, M. 1980. River otter reintroduction feasibility study, Kananaskis Country, Alberta. Unpublished report. University of Calgary, Calgary, Alberta, Canada.

Johnson, S. A., and K. A. Berkley. 1999. Restoring river otters to Indiana. Wildlife Society Bulletin 27:419–27.

Johnson, S. A., and R. F. Madej. 1994. Reintroduction of the river otter in Indiana: Feasibility study. Indiana Department of Natural Resources, Bloomington.

Kane, K. K. 1979. Medical management of the otter. Pages 100–103 *in* Proceedings of the annual meeting of the American Association of Veterinarians.

Kaplan, D. G. 1974. World of furs. Fairchild, New York.

Karnes, M. R., and R. Tumlison. 1984. The river otter in Arkansas: III. Characteristics of otter latrines and their distribution along beaver-inhabited watercourses in southwest Arkansas. Proceedings of the Arkansas Academy of Sciences 38:56–59.

Kasprzyk, M. 1984. Farbensehen beim fischotter (*Lutra lutra L.*). M.S. Thesis, Zoologisches Institut der Technischen Universität Braunschweig.

Kellnhauser, J. T. 1983. The acceptance of *Lontra* Gray for the New World river otter. Canadian Journal of Zoology 61:278–79.

Kennedy, K. 1968. River otter feeding on glaucous-winged gull. Blue Jay 26:109.

Kenyon, A. J., B. J. Kenyon, and E. C. Hahn. 1978. Protides of the Mustelidae: Immunoresponse of mustelids to Aleutian mink disease virus. American Journal of Veterinarian Research 390:1011.

Kenyon, K. W. 1969. The sea otter in the eastern Pacific Ocean. North American Fauna 68:1–352.

Kenyon, S., R. Southwick, and C. Wynne. 1999. Bears in the backyard, deer in the driveway. Report prepared by Southwick Associates for the International Association of Fish and Wildlife Agencies.

Kim, K. C., and K. C. Emerson. 1974. *Latagophthirus rauschi,* new genus and new species (Anoplura: Echinopthirildae) from the river otter (Carnivora: Mustelidae). Journal of Medical Entomology 11:442–46.

Kimber, K. R., and G. V. Kollias. 2000. Infectious and parasitic diseases and contaminant related problems of North American river otters (*Lontra canadensis*): A review. Journal of Zoo and Wildlife Medicine 31:452–72.

Kimber, K. R., G. V. Kollias, and E. J. Dubovi. 2000. Serologic survey of selected viral agents in recently captured, wild North American river otters (*Lontra canadensis*). Journal of Zoo Wildlife Medicine. 31:168–175.

Kinsey, C. 1981. Experimental otter census. Pages 149–150 *in* B. Joselyn and S. Spoolman, eds. Wildlife populations and research unit project descriptions. Section of Wildlife, Minnesota Department of Natural Resources, St. Paul.

Knaus, R. M., N. Kinler, and R. G. Linscombe. 1983. Estimating river otter populations: The feasibility of 65Zn to label feces. Wildlife Society Bulletin 11:375–77.

Knudsen, G. J. 1956. Forest game and range research project: Preliminary otter investigations (Job Completion Report W-079-R-01). Wisconsin Conservation Department.

Knudsen, K. F., and J. B. Hale. 1968. Food habits of otters in the Great Lakes region. Journal of Wildlife Management 32:89–93.

Kohn, M. H., and R. K. Wayne. 1997. Facts from feces revisited. Trends in Ecology and Evolution 12:223–27.

Kohn, M. H., E. C. York, D. A. Kamradt, G. Haught, R. M. Sauvajot, and R. K. Wayne. 1999. Estimating population size by genotyping faeces. Proceedings of the Royal Society of London B 266:657–63.

Kollias, G. 1999. Health assessment, medical management, and prerelease conditioning of translocated North American river otters. Pages 443–48 *in* M. E. Fowler and R. E. Miller, eds. Zoo and wild animal medicine, current therapy 4. W. B. Saunders, Philadelphia.

Kruuk, H., and R. Hewson. 1978. Spacing and foraging of otters (*Lutra lutra*) in a marine habitat. Journal of Zoology (London) 185:205–12.

Kruuk, H., M. Gorman, and T. Parrish. 1980. The use of 65Zn for estimating populations of carnivores. Oikos 34:206–8.

Kuehn, D. W., and W. E. Berg. 1983. Use of radiographs to age otters. Wildlife Society Bulletin 11:68–70.

Kwaraiceus, J. 1994. Yup'ik terms of the natural world. Alaska Native Language Center, Fairbanks.

Lagler, K. F., and B. T. Ostenson. 1942. Early spring food of the otter in Michigan. Journal of Wildlife Management 6:244–54.

Lariviere, S., and L. R. Walton. 1998. *Lontra canadensis.* Mammalian Species 587:1–8.

Larsen, D. N. 1983. Habitats, movements, and foods of river otters in coastal southeastern Alaska. M.S. Thesis, University of Alaska, Fairbanks.

Larsen, D. N. 1984. Feeding habits of river otters in coastal southeastern Alaska. Journal of Wildlife Management 48:1446–52.

Laughlin, J. 1955. River otter noted east of Sierran crest. California Department of Fish and Game 41:189.

Lauhachinda, V. 1978. Life history of the river otter in Alabama with emphasis on food habits. Ph.D. Dissertation, Auburn University, Auburn, AL.

Lauhachinda, V., and E. P. Hill. 1977. Winter food habits of river otters from Alabama and Georgia. Proceedings of the Annual Conference of the Southeastern Association of Fish and Wildlife Agencies 31:246–53.

Laut, A. C. 1921. The fur trade of America. Macmillan, New York.

Liers, E. E. 1951a. My friends the land otters. Natural History 60:320–26.

Liers, E. E. 1951b. Notes on the river otter (*Lutra canadensis*). Journal of Mammal Research 32:1–9.

Liers, E. E. 1958. Early breeding in the river otter. Journal of Mammal Research 39:438–39.

Lindholm, J. S., J. M. McCormick, S. W. Colton, and D. T. Downing. 1981. Variation of skin surface lipid composition among mammals. Comparative Biochemistry and Physiology 69B:75–78.

Ling, J. K. 1970. Pelage and molting in wild mammals with special reference to aquatic forms. Quarterly Review of Biology 45:16–54.

Lizotte, R. E. 1994. Biology of the river otter (*Lutra canadensis*). M.S. Thesis, Memphis State University, Memphis, TN.

Logsdon, C. W. 1989. Ecology of reintroduced river otters, *Lutra canadensis,* in Land between the Lakes, Kentucky. M.S. Thesis, Murray State University, Murray, KT.

Long, C. A. 1969. Gross morphology of the penis in seven species of the mustelidae. Mammalia 33:145–60.

Lord, J. 1991. The status of wildlife habitat in Canada: Realities and visions. Wildlife Habitat Canada, Ottawa.

Lowery, G. H. 1974. The mammals of Louisiana and its adjacent waters. Louisiana State University, Baton Rouge.

Macdonald, S. 1990. Surveys. Pages 8–10 *in* P. Foster-Turley, S. Macdonald, and C. Mason, eds. Otters: An action plan for their conservation. Kelvyn Press, Broadview, IL.

Mack, C. 1985. River otter restoration in Grand County, Colorado. M.S. Thesis, Colorado State University, Fort Collins.

Mack, C., L. Kronemann, and C. Eneas. 1994. Lower Clearwater aquatic mammal survey (Project Number 90–51). Bonneville Power Administration, Portland, OR.

Magoun, A. J., and P. Valkenburg. 1977. The river otter (*Lutra canadensis*) on the north slope of the Brooks Range, Alaska. Canadian Field-Naturalist 91:303–5.

Malville, L. E. 1990. Movements, distribution, and habitat selection of river otters reintroduced into the Dolores River, southwestern Colorado. M.S. Thesis, University of Colorado, Boulder.

Manning, T. 1990. Summer feeding habits of river otter (*Lutra canadensis*) on the Mendocino National Forest, California. Northwestern Naturalist 71:38–42.

McDaniel, J. C. 1963. Otter population study. Proceedings of the Southeastern Association of Fish and Game Commissioners 17:163–68.

McDonald, K. P. 1989. Survival, home range, movements, habitat use, and feeding habits of reintroduced river otters in Ohio. M.S. Thesis, Ohio State University, Columbus.

McGee, H. F. 1987. The use of furbearers by Native North Americans after 1500. Pages 13–20 *in* M. Novak, J. A. Baker, M. E. Obbard, and B. Malloch, eds. Wild furbearer management and conservation in North America. Ontario Ministry of Natural Resources, Toronto.

Mead, R. A. 1989. The physiology and evolution of delayed implantation in carnivores. Pages 437–64 *in* J. L. Gittleman, ed. Carnivore behavior, ecology, and evolution. Cornell University Press, Ithaca, NY.

Melquist, W. E. 1985. An overview of research on the North American river otter. Idaho Department of Fish and Game, Boise.

Melquist, W. E., and A. E. Dronkert. 1987. River otter. Pages 625–641 *in* M. Novak, J. A. Baker, M. E. Obbard, and B. Malloch, eds. Wild furbearer management and conservation in North America. Ontario Trappers Association, North Bay, Ontario, Canada.

Melquist, W. E., and M. G. Hornocker. 1979a. Methods and techniques for studying and censusing river otter populations (Technical Report 8). Forestry, Wildlife and Range Experiment Station, University of Idaho, Moscow.

Melquist, W. E., and M. G. Hornocker. 1979b. Development and use of a telemetry technique for studying river otter. Pages 104–14 *in* F. M. Long, ed. Proceedings of the second international conference on wildlife biotelemetry. Laramie, WY.

Melquist, W. E., and M. G. Hornocker. 1983. Ecology of river otters in west central Idaho. Wildlife Monographs 83:1–60.

Melquist, W. E., J. S. Whitman, and M. G. Hornocker. 1981. Resource partitioning and coexistence of sympatric mink and river otter populations. Pages 187–220 *in* J. A. Chapman and D. Pursley, eds. Proceedings of the worldwide furbearer conference. Frostburg, MD.

Meyerriecks, A. J. 1963. Florida otter preys on common gallinule. Journal of Mammal Research 44:425–26.

Miller, M. C. 1992. Reintroduction of river otters into Great Smoky Mountains National Park. M.S. Thesis, University of Tennessee, Knoxville.

Modafferi, R., and C. F. Yocum. 1980. Summer food of river otter in north coastal California lakes. Murrelet 61:38–41.

Morejohn, G. V. 1969. Evidence of river otter feeding on freshwater mussels and range extension. California Department of Fish and Game 55:83–85.

Mowbray, E. E., Jr., J. A. Chapman, and J. R. Goldsberry. 1976. Preliminary observations on otter distribution and habitat preferences in Maryland with descriptions of otter field sign. Northeast Fish and Wildlife Conference 33:125–31.

Mowbray, E. E., D. Pursley, and J. A. Chapman. 1979. The status, population characteristics and harvest of the river otter in Maryland (Publications on Wildlife Ecology No. 2). Maryland Wildlife Administration.

Newman, D. G., and C. R. Griffin. 1994. Wetland use by river otters in Massachusetts. Journal of Wildlife Management 58:18–23.

Nilsson, G. 1980. Present status of the river otter in the United States. Pages 126–55 *in* N. Duplaix, ed. Proceedings of the second working meeting of the otter specialist group. Gainesville, FL.

Noll, J. M. 1988. Home range, movement, and natal denning of river otters (*Lutra canadensis*) at Kelp Bay, Baranof Island, Alaska. M.S. Thesis, University of Alaska, Fairbanks.

Novak, M. 1987. Wild furbearer management in Ontario. Pages 1049–61 *in* M. Novak, J. A. Baker, M. E. Obbard, and B. Malloch, eds. Wild furbearer management and conservation in North America. Ontario Ministry of Natural Resources, Toronto.

Novak, M., M. E. Obbard, J. G. Jones, R. Newman, A. Booth, A. J. Satterthwaite, and G. Linscombe. 1987. Furbearer harvests in North America, 1600–1984. Ontario Ministry of Natural Resources, Toronto.

Obbard, M. E. 1987. Fur grading and pelt identification. Pages 717–826 *in* M. Novak, J. A. Baker, M. E. Obbard, and B. Malloch, eds. Wild furbearer management and conservation in North America. Ontario Ministry of Natural Resources, Toronto.

Obbard, M. E., J. G. Jones, R. Newman, A. Booth, A. J. Satterthwaite, and G. Linscombe. 1987. Furbearer harvests in North America. Pages 1007–34 *in* M. Novak, J. A. Baker, M. E. Obbard, and B. Malloch, eds. Wild furbearer management and conservation in North America. Ontario Ministry of Natural Resources, Toronto.

Ormseth, O. A., and M. Ben-David. 2000. Ingestion of oil hydrocarbons: Effects on digesta retention times and nutrient uptake in captive river otters. Journal of Comparative Physiology B 170:419–28.

Park, E. 1971. The world of the otter. J. B. Lippincott, New York.

Peterson, M. 1914. The fur traders and fur bearing animals. Hammond Press, Buffalo, NY.

Pierce, R. M. 1979. The seasonal feeding habits of the river otter (*Lutra canadensis*) in ditches of the Great Dismal Swamp. M.S. Thesis, Old Dominion University, Norfolk, VA.

Pohle, H. 1920. Die Unterfamilie der Lutrinae. Archiv für Naturgeshichte A 9:1–247.

Polechla, P. J., Jr. 1987a. Status of the river otter (*Lutra canadensis*) population in Arkansas with special reference to reproductive biology. Ph.D. Dissertation, University of Arkansas, Fayetteville.

Polechla, P. J. 1987b. Fur trade from Arkansas Factory, Arkansas Territory, 1805–1810. Proceedings of the Arkansas Academy of Science 41:69–72.

Polechla, P. J. 1988. Nearctic river otter (*Lutra canadensis*). Pages 668–682 *in* W. J. Chander and L. Labate, eds. 1988–89 Audubon wildlife report. National Audubon Society, Academic Press, New York.

Polechla, P. J. 1989a. A review of the techniques of radio telemetry of the Nearctic river otter (*Lutra canadensis*). Pages 14–22 *in* C. J. Amlaner, ed. Proceedings of the 10th international symposium on biotelemetry. University of Arkansas Press, Fayetteville.

Polechla, P. J. 1989b. More evidence of a commensal relationship between the river otter and the beaver. Pages 217–36 *in* C. J. Amlaner, ed. Proceedings of the 10th international symposium on biotelemetry. University of Arkansas Press, Fayetteville.

Polechla, P. 1990. Action plan for North American otters. Pages 74–79 *in* P. Foster-Turley, S. Macdonald, and C. Mason, eds. Otters: An action plan for their conservation. Kelvyn Press, Broadview, IL.

Polechla, P. J. 1991. A preliminary review of the anatomy and physiology of otters (Lutrinae; Mustelidae; Carnivora). Pages 85–94 *in* C. Reuther and R. Rochert, eds. Proceedings of the Vth international otter symposium. Hankensbüttel, Germany.

Polechla, P. J. 1993. A review of the Nearctic river otter in southwestern North America: Ethnography, distribution, ecology, and taxonomy. River Otter Journal 3:12–13.

Polechla, P. J. 1996. New host records of ticks (Acarina; Ixodidae) parasitizing the river otter (*Lutra canadensis*). IUCN Otter Specialist Group Bulletin 13:8–13.

Polechla, P. J. 2000. Ecology of the river otter and other wetland furbearers in the upper Rio Grande (Final Report to Bureau of Land Management). Museum of Southwestern Biology, University of New Mexico, Albuquerque.

Polechla, P. J. 2002a. A review of the natural history of the river otter (*Lontra canadensis*) in the southwestern United States with special reference to New Mexico. Final Report to North American Wilderness Recovery, Incorporated. Richmond, VT.

Polechla, P. J. 2002b. River otter (*Lontra canadensis*) and riparian survey of the Los Pinos, Piedra, and San Juan Rivers in Archuleta, Hinsdale, and La Plata Counties, Colorado. Final Report to Colorado Division of Wildlife, Denver, CO.

Polechla, P. J. In press-a. Introduction to the biology of the North American river otter (*Lontra canadensis*). Chapter 1.2.2. Species range. *In* C. Reuther, ed. Otter action plan. IUCN, SSC Otter Specialist Group, Gland, Switzerland.

Polechla, P. J. In press-b. Introduction to the biology of the North American river otter (*Lontra canadensis*). Chapter 1.2.5. Reproduction. *In* C. Reuther, ed. Otter action plan. IUCN, SSC Otter Specialist Group, Gland, Switzerland.

Polechla, P. J., and J. A. Sealander. 1985. An evaluation of the status of the river otter (*Lontra canadensis*) in Arkansas (Final Report, Federal Aid Project No. W-56-23). Arkansas Game and Fish Commission.

Proulx, G., and M. W. Barrett. 1991. Ideological conflict between animal rightists and wildlife professionals over trapping wild furbearers. Transactions of the North American Wildlife and Natural Resources Conference 56:387–99.

Pursley, D. 1978. Economic values of furbearers in North America. Proceedings of the Western Association of Fish and Wildlife Agencies 58:123–40.

Quinlan, S. E. 1983. Avian and river otter predation in a storm-petrel colony. Journal of Wildlife Management 47:1036–43.

Radinsky, L. B. 1968. Evolution of somatic sensory specialization in otter brains. Journal of Comparative Neurology 134:495–506.

Raesly, E. J. 2001. Progress and status of river otter reintroduction projects in the United States. Wildlife Society Bulletin 29:856–62.

Ralls, K. 1990. Reintroductions. Pages 20–21 *in* P. Foster-Turley, S. Macdonald, and C. Mason, eds. Otters: An action plan for their conservation. Kelvyn Press, Broadview, IL.

Rausch, R. 1951. Notes on the Nunamiut Eskimo and mammals of the Anaktuvuk Pass region, Brooks Range, Alaska. Arctic 4:147–95.

Reading, R. P., and T. W. Clark. 1996. Carnivore reintroductions: An interdisciplinary examination. Pages 296–336 *in* J. L. Gittleman, ed. Carnivore behavior, ecology, and evolution, Vol. 2. Cornell University, Ithaca, NY.

Reed, J. Z., D. J. Tollit, P. M. Thompson, and W. Amos. 1997. Molecular scatology: The use of molecular genetic analysis to assign species, sex and individual identity to seal faeces. Molecular Ecology 6:225–34.

Reed-Smith, J. 1995. North American river otter, *Lontra canadensis:* Husbandry notebook. John Ball Zoological Garden, Grand Rapids, MI.

Reed-Smith, J. 2001. North American river otter (*Lontra (Lutra) canadensis*): Husbandry notebook, 2nd edition. John Ball Zoological Garden, Grand Rapids, Michigan.

Reeves, K. A. 1988. Summer diet and status of river otters on Redwood Creek. M.S. Thesis, Humboldt State University, Humboldt, CA.

Reid, D. G. 1984. Ecological interactions of river otters and beavers in a boreal ecosystem. M.S. Thesis, University of Alberta, Calgary, Alberta, Canada.

Reid, D. G., W. E. Melquist, and J. D. Woolington. 1986. Reproductive effects of intraperitoneal transmitter implants in river otters. Journal of Wildlife Management 50:92–94.

Reid, D. G., M. B. Bayer, T. E. Code, and B. McLean. 1987. A possible method for estimating river otter (*Lutra canadensis*) populations using snow tracks. Canadian Field-Naturalist 101:576–80.

Reid, D. G., S. M. Herrero, and T. E. Code. 1988. River otters as agents of water loss from beaver ponds. Journal of Mammalogy 69:100–107.

Reid, D. G., T. E. Code, A. C. H. Reid, and S. M. Herrero. 1994. Spacing, movements, and habitat selection of the river otter in boreal Alberta. Canadian Journal of Zoology 72:1314–24.

Reuther, C., and A. Festetics. 1980. Der fischotter in Europa: Verbreitung, bedrohung, erhaltung. Oderhaus, St. Andreasberg, Germany.

Riley, M. A. 1985. An analysis of masticatory form and function in three mustelids (*Martes americana, Lutra canadensis, Enhydra lutris*). Journal of Mammalogy 65:519.

Robson, M. S. 1982. Monitoring river otter populations: Scent stations vs. sign indices. M.S. Thesis, University of Florida, Gainesville.

Robson, M. S., and S. R. Humphrey. 1985. Inefficacy of scent-stations for monitoring river otter populations. Wildlife Society Bulletin 13:558–61.

Rock, K. P., E. S. Rock, R. T. Bowyer, and J. B. Faro. 1994. Degree of association and use of a helper by coastal river otters, *Lutra canadensis,* in Prince William Sound, Alaska. Canadian Field-Naturalist 108:367–69.

Romanenko, O. 1994. Fishes, birds, and mammals of central Beringia: A taxonomic list in English and Russian. All Russian Institute of Nature Conservation and Reserves, Moscow.

Rosen, M. 1975. Bald eagle and river otter. Canadian Field-Naturalist 89:455.

Rothenbacker, H. 1962. Acute microfilariasis of unknown etiology in an otter. Michigan State University Veterinarian 23:43–45.

Route, W. T., and R. O. Peterson. 1988. Distribution and abundance of river otter in Voyageurs National Park, Minnesota (National Park Service Research/Resources Management Report MWR-10).

Rucker, D. 1983. River otter restoration. IMPACT. TVA-Natural Resources and the Environment (Chattanooga, TN) 6:5–7.

Ryder, R. A. 1955. Fish predation by the otter in Michigan. Journal of Wildlife Management 19:497–98.

Sadleir, R. M. F. S. 1973. The reproduction of vertebrates. Academic Press, New York.

Sauer, T. M., M. Ben-David, and R. T. Bowyer. 1999. A new application of the adaptive-kernel method: Estimating linear home ranges of river otters, *Lutra canadensis.* Canadian Field-Naturalist 113:419–24.

Scheffer, V. B. 1939. The *os clitorides* of the Pacific otter. Murrelet 20:20–21.

Scheffer, V. B. 1953. Otters diving to a depth of sixty feet. Journal of Mammalogy 34:255.

Scheffer, V. B. 1967. Probable mating of otter on Lopez Island, Washington. Murrelet 48:20.

Schlegel, M. W., S. E. Knapp, and R. E. Millemann. 1968. "Salmon poisoning" disease. Part 5: Definitive hosts of the trematode vector, *Nanophyetus salmincola.* Journal of Parasitology 54:770–74.

Schreber, J. C. D. 1776. Die Säugethiere in Abbildungen nach der Natur mit Beschreibungen. Wolfgang Walther, Erlange, Germany.

Schwartz, C. W., and E. R. Schwartz. 1981. The wild mammals of Missouri. University of Missouri Press, Columbia.

Serfass, T. L. 1995. Cooperative foraging by North American river otters, *Lutra canadensis.* Canadian Field-Naturalist 109:458–59.

Serfass, T. L., L. M. Rymon, and R. P. Brooks. 1990. Feeding relationships of river otters in northeastern Pennsylvania. Transactions of the Northeastern Section of the Wildlife Society 47:43–53.

Serfass, T. L., L. M. Rymon, and R. P. Brooks. 1992. Ectoparasites from river otters in Pennsylvania. Journal of Wildlife Diseases 28:138–40.

Serfass, T. L., R. L. Peper, M. T. Whary, and R. P. Brooks. 1993. River otter (*Lutra canadensis*) reintroduction in Pennsylvania: Prerelease care and clinical evaluation. Journal of Zoo and Wildlife Medicine 24:28–40.

Serfass, T. L., R. P. Brooks, T. J. Swimley, L. M. Rymon, and A. H. Hayden. 1996. Considerations for capturing, handling, and translocating river otters. Wildlife Society Bulletin 24:25–31.

Serfass, T. L., R. P. Brooks, J. M. Novak, P. E. Johns, and O. E. Rhodes, Jr. 1998. Genetic variation among populations of river otters in North America: Considerations for reintroduction projects. Journal of Mammalogy 79:736–46.

Seton, E. T. 1929. Lives of game animals, Vol. II, Part 2. Bears, coons, badgers, skunks, and weasels. Doubleday, Doran, New York.

Severinghaus, C. W., and J. E. Tanck. 1948. Speed and gait of an otter. Journal of Mammalogy 29:71.

Shannon, J. S. 1989. Social organization and behavioral ontogeny of otters (*Lutra canadensis*) in a coastal habitat in northern California. IUCN Otter Specialist Group Bulletin 4:8–13.

Shannon, J. S. 1991. Progress on Californian otter research. IUCN Otter Specialist Group Bulletin 6:24–31.

Shannon, J. S. 1998. Behaviour of otters in a marine coastal habitat: Summary of a work in progress. IUCN Otter Specialist Group Bulletin 15:114–17.

Sheldon, W. G., and W. G. Toll. 1964. Feeding habits of the river otter in a reservoir in central Massachusetts. Journal of Mammalogy 45:449–55.

Simpson, G. G. 1945. The principles of classification and a classification of mammals. Bulletin of the American Museum of Natural History 85:1–350.

Simpson, V. R., and D. Gavier-Widen. 2000. Fatal adiaspiromycosis in a wild Eurasian otter (*Lutra lutra*). Veterinary Record 147:239–41.

Sinha, A. A., C. H. Conaway, and K. W. Kenyon. 1966. Reproduction in the female sea otter. Journal of Wildlife Management 30:121–30.

Smithe, F. B. 1975. Naturalist's color guide. American Museum of Natural History, New York.

Speich, S. M., and R. L. Pitman. 1984. River otter occurrence and predation on nesting marine birds in the Washington Islands wilderness. Murrelet 65:25–27.

Spelman, L. 1999. Otter anesthesia. Pages 436–43 *in* M. Fowler and R. E. Miller, eds. Zoo and wild animal medicine: Current therapy 4. W. B. Saunders, Philadelphia.

Sperry, C. C. 1941. Food habits of the coyote (Wildlife Research Bulletin No. 4). U.S. Department of Interior Fish and Wildlife Service.

Stenson, G. B. 1985. The reproductive cycle of the river otter, *Lutra canadensis,* in the marine environment of southwestern British Columbia. Ph.D. Dissertation, University of British Columbia, Vancouver, Canada.

Stenson, G. B., G. A. Badgero, and H. D. Fisher. 1984. Food habits of the river otter *Lutra canadensis* in the marine environment of British Columbia. Canadian Journal of Zoology 62:88–91.

Stephenson, A. B. 1977. Age determination and morphological variation of Ontario otters. Canadian Journal of Zoology 55:1577–83.

Stophlet, J. J. 1947. Florida otters eat large terrapin. Journal of Mammal Research 28:183.

Strong, W. D. 1930. Notes on mammals of the Labrador Interior. Journal of Mammalogy 11:1–10.

Stuht, J. N. 1978. Paragonimiasis in a river otter. Michigan Department of Natural Resources, Wildlife Division Report 2824:1–2.

Swanton, J. R. 1946. The Indians of the southeastern United States. U.S. Government Printing Office, Washington, DC.

Swimley, T. J., T. L. Serfass, R. P. Brooks, and W. M. Tzilkowski. 1998. Predicting river otter latrine sites in Pennsylvania. Wildlife Society Bulletin 26:836–45.

Tabor, J. E. 1974. Productivity, survival, and population status of river otter in western Oregon. M.S. Thesis, Oregon State University, Corvallis.

Tabor, J. E., and H. M. Wight. 1977. Population status of river otter in western Oregon. Journal of Wildlife Management 41:692–99.

Tango, P. J., E. D. Michael, and J. I. Cromer. 1991. Mating and first-season births in interstate transplanted river otters, *Lutra canadensis* (Carnivora: Mustelidae). Brimleyana 17:53–55.

Tarasoff, F. J., and G. L. Kooyman. 1973a. Observations on the anatomy of the respiratory system of the river otter, sea otter, and harp seal. Part 1, The topography, weight, and measurements of the lungs. Canadian Journal of Zoology 51:163–70.

Tarasoff, F. J., and G. L. Kooyman. 1973b. Observations on the anatomy of the respiratory system of the river otter, sea otter, and harp seal. Part 2, Trachea and bronchial tree. Canadian Journal of Zoology 51:171–77.

Tarasoff, F. J., A. Basaillon, J. Pierard, and A. P. Whitt. 1972. Locomotory patterns and external morphology of the river otter, sea otter, and harp seal (Mammalia). Canadian Journal of Zoology 50:915–29.

Taylor, C., L. K. Duffy, R. T. Bowyer, and G. M. Blundell. 2000. Profiles of fecal porphyrins in river otters following the *Exxon Valdez* oil spill. Marine Pollution Bulletin 40:1132–38.

Taylor, C., M. Ben-David, R. T. Bowyer, and L. K. Duffy. 2001. Response of river otters to experimental exposure of weathered crude oil: Fecal porphyrin profiles. Environmental Science and Technology 35:747–52.

Testa, J. W., D. F. Holleman, R. T. Bowyer, and J. B. Faro. 1994. Estimating populations of marine river otters in Prince William Sound, Alaska, using radio-tracer implants. Journal of Mammalogy 75:1021–32.

Thompson, D. R. 1958. Field techniques for sexing and aging game animals. Ohio Department of Natural Resources, Division of Wildlife, Columbus.

Tischbein, G. 1976. More river otter transplanted to Colorado's west slope. Wildlife News-Colorado Division of Wildlife 1:2.

Tolme, P. 2003. Looking for the lost river otters of the southwest. National Wildlife World Edition. 41 (4):48–55.

Toweill, D. E. 1974. Winter food habits of river otters in western Oregon. Journal of Wildlife Management 38:107–11.

Toweill, D. E., and J. E. Tabor. 1982. River otter: *Lutra canadensis.* Pages 688–703 *in* J. A. Chapman and G. A. Feldhamer, eds. Wild mammals of North America: Biology, management, and economics. Johns Hopkins University Press, Baltimore.

Tumlison, R., and M. Karnes. 1987. Seasonal changes in food habits of river otters in southwestern Arkansas beaver swamps. Mammalia 51:225–31.

Tumlison, R., M. Karnes, and A. W. King. 1982. The river otter in Arkansas: II. Indications of a beaver-facilitated commensal relationship. Proceedings of the Arkansas Academy of Science 36:73–75.

Vallentine, J. M., Jr., J. R. Walther, K. M. McCartney, and L. M. Ivy. 1972. Alligator diets on the Sabine National Wildlife Refuge, Louisiana. Journal of Wildlife Management 36:809–15.

Van Zyll de Jong, C. G. 1968. A systematic study of the Nearctic and Neotropical river otters (Order: Carnivora; Family: Mustelidae). Ph.D. Dissertation, University of Toronto, Toronto.

Van Zyll de Jong, C. G. 1972. A systematic review of the Nearctic and Neotropical river otters (Genus *Lutra,* Mustelidae, Carnivora). Royal Ontario Museum, Life Sciences Contributions 80:1–104.

Van Zyll de Jong, C. G. 1987. A phylogenetic study of the Lutrinae (Carnivora; Mustelidae) using morphological data. Canadian Journal of Zoology 65:2536–44.

Van Zyll de Jong, C. G. 1991. A brief review of the systematics and a classification of the Lutrinae. Pages 79–83 *in* C. Reuther and R. Rochert, eds. Proceedings of the Vth international otter symposium. Hankensbüttel, Germany.

Varley, N. 1998. Yellowstone's river otters: Enigmatic water weasels. Yellowstone Science 6:15–18.

Vaughan, T. A. 1972. Mammalogy. W. B. Saunders, Philadelphia.

Vaughan, T. A., J. M. Ryan, and N. J. Czaplewski. 2000. Mammalogy. W. B. Saunders, Fort Worth, TX.

Verbeek, N. A. M., and J. L. Morgan. 1978. River otter predation on glaucous-winged gulls on Mandarta Island, British Columbia. Murrelet 59:92–95.

Wallach, J. D., and W. J. Boever. 1983. Diseases of exotic animals. W.B. Saunders, Philadelphia.

Wasser, S. K., C. S. Houston, G. M. Koehler, G. G. Cadd, and S. R. Fain. 1997. Techniques for application of faecal DNA methods to field studies of ursids. Molecular Ecology 6:1091–97.

Wilson, D. E., and F. R. Cole. 2000. Common names of the mammals of the world. Smithsonian Institution Press, Washington, DC.

Wilson, D. E., and D. M. Reeder. 1993. Mammal species of the world: A taxonomic and geographic reference. Smithsonian Institution Press, Washington DC.

Wilson, J. H., K. L. Anderson-Bledsoe, J. L. Baker, and P. F. Scanlon. 1984. Mechanical properties of river otter limb bones. Zoo Biology 3:27–34.

Wilson, K. A. 1954. The role of mink and otter as muskrat predators in northeastern North Carolina. Journal of Wildlife Management 18:199–207.

Wilson, K. A. 1959. The otter in North Carolina. Proceedings of the Southeastern Association of Fish and Game Commissioners 13:267–77.

Woolf, A., R. S. Halbrook, D. T. Farrand, C. Schieler, and W. Weber. 1997. Survey of habitat and otter population status (Final Report, Federal Aid in Wildlife Restoration Project, W-122-R-3). Illinois Department of Natural Resources, Springfield.

Woolington, J. D. 1984. Habitat use and movements of river otters at Kelp Bay, Baranof Island, Alaska. M.S. Thesis, University of Alaska, Fairbanks.

Worthy, G. A. J., J. Rose, and F. Stormshak. 1987. Anatomy and physiology of fur growth: The pelage priming process. Pages 827–41 *in* M. Novak, J. A. Baker, M. E. Obbard, and B. Malloch, eds. Wild furbearer management and conservation in North America. Ontario Ministry of Natural Resources, Toronto.

Wozencraft, W. C. 1993. Order Carnivora in mammal species of the world: A taxonomic and geographic reference. Pages 279–348 *in* D. E. Wilson and D. M. Reeder, eds. Mammal species of the world: A taxonomic and geographic reference. Smithsonian Institution Press, Washington, DC.

Wright, J. V. 1987. Archaeological evidence for the use of furbearers in North America. Pages 3–12 *in* M. Novak, J. A. Baker, M. E. Obbard, and B. Malloch, eds. Wild furbearer management and conservation in North America. Ontario Ministry of Natural Resources, Toronto.

Wurster, D. H., and K. Benirschke. 1968. Comparative cytogenetic studies in the order Carnivora. Chromosoma (Berlin) 24:336–82.

Young, S. P. 1958. The bobcat of North America. Stackpole, Harrisburg, PA, and Wildlife Management Institute, Washington, DC.

Zackheim, H. 1982. Ecology and population status of the river otter in southwestern Manitoba. M.S. Thesis, University of Montana, Missoula.

Wayne E. Melquist, Department of Fish and Wildlife Resources, University of Idaho, Moscow, Idaho 83844. Email: melquist@uidaho.edu.

Paul J. Polechla, Jr. Department of Biology, Museum of Southwestern Biology, University of New Mexico, Albuquerque, New Mexico 87131-1091. Email: ppolechl@sevilleta.unm.edu.

Dale Toweill, Bureau of Wildlife, Idaho Department of Fish and Game, P.O. Box 25, 600 S. Walnut Street, Boise, Idaho 83707-0025. Email: dtoweill@idfg.state.id.us.

36

Sea Otter

Enhydra lutris

James L. Bodkin

NOMENCLATURE

COMMON NAME. Sea otter
ORDER. Carnivora
FAMILY. Mustelidae
SCIENTIFIC NAME. *Enhydra lutris*
SUBSPECIES. *E. l. nereis, E. l. kenyoni,* and *E. l. lutris* (Wilson et al. 1991)

PHYLOGENY AND CLASSIFICATION

Sea otters are the only completely marine species of the aquatic lutrinae, or otter subfamily of the family Mustelidae (skunks, weasels, minks, badgers, and honey badgers) (Wozencraft 1993). Based largely on skull morphology and characterized by dietary differences, the lutrinae were purported to be a monophyletic group represented by three clades: (1) the genera *Lutra, Lontra,* and *Pteroneura* (fish eaters); (2) the genus *Aonyx* (crab eaters); and (3) the genus *Enhydra* (the sea otter) (Berta and Morgan 1985). Based on nucleotide sequence of the mitochondrial cytochrome *b* gene, Koepfli and Wayne (1998) placed *Enhydra* in one of three reorganized lutrine clades. The sea otter was the earliest lineage to diverge within the clade that includes *Lutra* (Eurasian and Cape clawless otters) and *Aonyx* (small-clawed and spotted-neck otters), about 13 million years ago. Two lineages of sea otter are recognized. One led to the extinct *Enhydriodon,* the other to *Enhydritherium* and subsequently to *Enhydra*. Fossil evidence of ancestral sea otters has been found in Africa, Eurasia, and North America. The *Enhydritherium* lineage apparently arose in Eurasia, although clear evidence of routes of immigration is lacking. Early specimens of *Enhydra,* dating to the early Pleistocene, 1–3 million years ago, have been found from the Pacific Rim, and the genus has apparently remained confined to that basin (Riedman and Estes 1990).

Advances in multivariate analyses and molecular genetics in recent decades have helped resolve previous controversy about the subspecific taxonomy of sea otters. Early taxonomy was based primarily on comparison of skull morphology between sea otters from Alaska and California, with two divergent interpretations of the data. One interpretation considered small but significant regional differences to represent a simple clinal gradient corresponding to latitudinal influences, and concluded that the California population (*E. l. nereis*) did not warrant subspecies designation (Scheffer and Wilke 1950; Roest 1973, 1976; Kenyon 1982). Davis and Lidicker (1975), recognizing differences in skull morphology and considering reproductive isolation and possible behavioral differences, recommended the California population retain the *nereis* subspecies classification. An exhaustive systematic review and analysis of sea otter skull morphology concluded there are three subspecies, *E. lutris lutris* from Asia to the Commander Islands, *E. l. nereis* from California, and the new subspecies *E. l. kenyoni* from Alaska (Wilson et al. 1991).

The subspecific taxonomy suggested by morphological analyses is largely supported by subsequent molecular genetic data. Analysis of mitochondrial DNA (mtDNA) variation among eight geographically isolated populations identified four major groups generally corresponding with the three recognized subspecies (Fig. 36.1) (Cronin et al. 1996; Scribner et al. 1997). The haplotype frequency in the Commander Islands population of *E. l. lutris* is more similar to that observed in the Aleutian–Kodiak grouping, *E. l. kenyoni,* than to the Asian subspecies, *E. l. lutris,* with which it was aligned by skull morphology. In addition, the Prince William Sound population differs from the other Alaska populations in haplotype frequency. The distribution of mtDNA haplotypes suggests little or no recent female-mediated gene flow among populations sampled. However, populations separated by large geographic distances shared some haplotypes (e.g., the Kuril and Kodiak Islands), which is suggestive of common ancestry and some level of historical gene flow. In a review of sea otter studies using genetic markers, Scribner et al. (1997) concluded that populations are highly differentiated genetically, although limited sequence divergence and lack of phylogeographic concordance suggest an evolutionarily recent common ancestor and some degree of gene flow throughout their range. The differences we see in genetic markers among contemporary sea otter populations likely reflect periods of habitat fragmentation and consolidation during Pleistocene glacial advance and retreat, some effect of reproductive isolation over large spatial scales, and the recent history of harvest-related reductions and subsequent recolonization.

DISTRIBUTION AND ABUNDANCE

The sea otter occurs only in the North Pacific Ocean (Fig. 36.2). The historical range includes coastal habitats around the Pacific Rim between central Baja California and northern Japan. The northward limits in distribution appear related to the southern limits of sea ice, which can preclude access to foraging habitat. Seasonal and interannual variation in the southern extent of sea ice results in concordant constriction and expansion of the sea otter's northern range. During periods of advancing winter sea ice along their northern range, sea otters occasionally become trapped. Ice-related mortality has been reported from Russia and Alaska (Nikolaev 1966; Schneider and Faro 1975). Sea otters attempting to

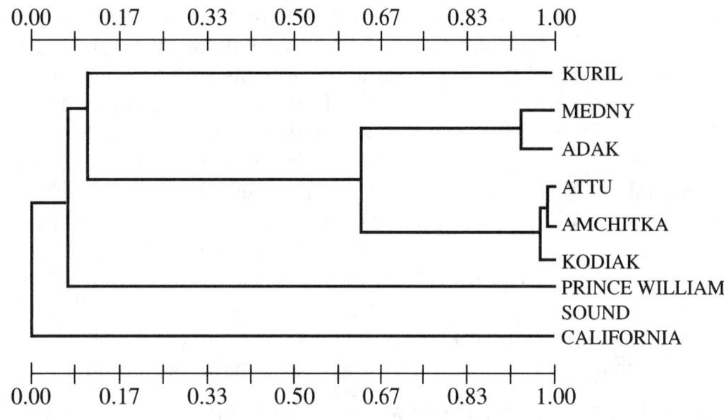

FIGURE 36.1. Dendrogram of mtDNA haplotype relations among eight sea otter populations. SOURCE: Data from Cronin et al. (1996).

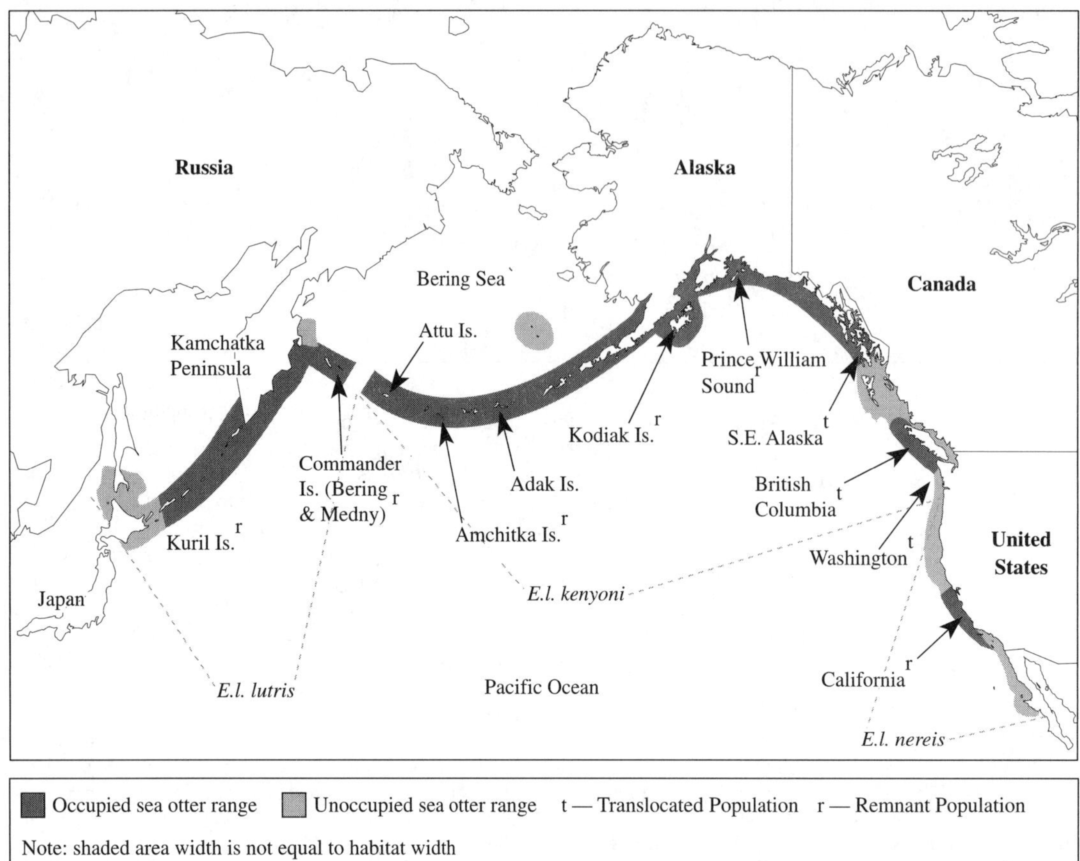

FIGURE 36.2. Historical and current distribution of sea otter (*Enhydra lutris*) in the Pacific Ocean.

travel tens of kilometers over the Alaska Peninsula to access the ice-free Pacific were observed in 1971 and 1972 (Schneider and Faro 1975) and again in 1982, 1999, and 2000. Although some otters may succeed, most apparently die in the journey. Mortality appears to result largely from starvation, although terrestrial predators, including wolves (*Canis lupus*), red foxes (*Vulpes vulpes*), and wolverines (*Gulo gulo*), will take sea otters that are traveling overland and likely in a weakened condition. Southern range limits are less well understood, but appear to coincide with the southern limits of coastal upwelling, associated canopy-forming kelp forests, and the 20–22°C isotherm (Kenyon 1969; Estes 1980). It seems likely that the adaptations that allow sea otters to reside near the ice edge in the high latitudes may also limit the extent of their range into the lower latitudes.

DESCRIPTION

Morphology. The sea otter is the largest mustelid and the sexes are moderately dimorphic. Adult males attain weights of 46 kg and total lengths of 148 cm. Adult females attain weights of 36 kg and total lengths of 140 cm. Weights reported from populations below equilibrium density exceed those from populations at or near equilibrium density by 28% for males and 16% for females (Kenyon 1969). At Bering Island, Russia, mean weights of adult male sea otters declined from 32.1 kg in 1980 to 25.1 in 1990, coinciding with the population exceeding carrying capacity and a 41% reduction in population size (Bodkin et al. 2000). Although sex of individuals is difficult to determine in the field, males can be identified by the presence of penile and testicle bulges and females by a pair of abdominal mammae. At birth, pups weigh about 1.7–2.3 kg and are about 60 cm in total length.

Pelage. Fur and the air trapped within it provide the primary source of insulation and buoyancy for the sea otter, as compared with the fat or blubber layer used by other marine mammals. The pelage consists of relatively sparse outer guard hairs and a shorter, very dense underfur at a ratio of about 1:70. Hair densities range from nearly 26,000/cm^2 on the hind flipper to 165,000/cm^2 on the foreleg (Williams et al. 1992). Sebaceous glands secrete oil, which aids in water repulsion. The absence of *arrector pili* muscles in the epidermis permits the guard hair to lie nearly parallel to the skin, and although submerged in water, this allows the underfur to remain dry. The ability of the sea otter to thermoregulate is dependent on its maintaining the integrity of the pelage. This requires a nearly constant, yet gradual molt as well as frequent and vigorous grooming. The color of the pelage ranges from light brown to nearly black. As animals age, they can attain a grizzled appearance, with whitening occurring in the head, neck, and upper torso regions (Fig. 36.3). Newborn pups possess a pale brown, woolly natal pelage.

FIGURE 36.3. Adult male sea otter (*Enhydra lutris*).

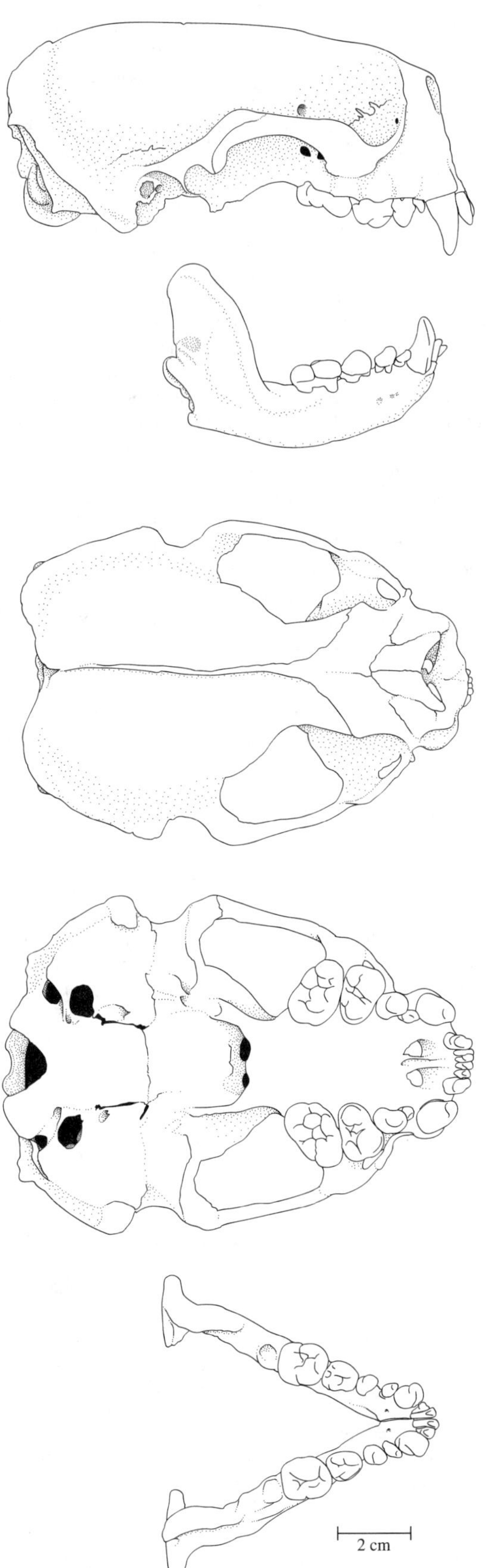

FIGURE 36.4. Skull of sea otter (*Enhydra lutris*). From top to bottom: lateral view of cranium, lateral view of mandible, dorsal view of cranium, ventral view of cranium, dorsal view of mandible.

Skull and Dentition. The skull is broad and blunt (Fig. 36.4). Adult dental formula is I 3/2, C 1/1, P 3/3, M 1/2. The dentition differs from that of most other carnivores in being adapted to crush prey, as opposed to shearing. The canines are long, rounded, and blunt and are used to puncture and pry open prey such as bivalve mollusks. The molars are bunodont, broad, and flat with rounded crowns effective in grinding hard-bodied prey such as crustaceans and gastropods. The incisors and canines are used to scrape tissues out of shelled prey. A vestigial premolar is present and can be used to estimate the age of sea otters based on annual deposits of cementum (Bodkin et al. 1997).

Appendages. The forelegs of the sea otter are short and powerful with sensitive paws used to locate and acquire prey. Forelegs are not used in propulsion. Claws in the forelimbs can be extruded. A fold of skin at the axilla (armpit) of each forelimb is used to store and transport prey gathered while foraging. Prey are held and manipulated by the forepaws while being consumed on the sea surface. Sea otters are one of the few species known to use tools, such as rocks, as anvils or hammers to break open hard-shelled prey. The hind limbs are flattened and flipper-like. While swimming, the posterior margins of the hind flippers approximate the lunate pattern and undulating movement of the caudal fin of cetaceans. The tail is long, horizontally flattened, and used in swimming, particularly during slow movements while on the surface. The ears are short and usually held erect while on the surface. While diving, the ears are held downward, presumably to exclude water.

PHYSIOLOGY

Occupying an aquatic environment requires adaptations to overcome the high thermal conductivity of water, which results in rapid heat loss. The general mammalian problem of maintaining a constant and elevated body temperature is exacerbated in the sea otter because of its small body size (relative to other marine mammals) and the resulting high surface-to-volume ratio as well as the generally cold water temperatures found in high-latitude marine environments. In addition to using air in the pelage as an external insulator to reduce heat loss, metabolic heat production in the sea otter is about 2.4–3.2 times that predicted in a terrestrial mammal of similar size (Costa and Kooyman 1982, 1984). To maintain an average body temperature of about 38°C, a standard metabolic rate of about 0.72 cm^3 O_2/g body weight/hr has been measured in the sea otter (Morrison et al. 1974). To maintain the elevated metabolic rate, energy intake must also be elevated, requiring consumption of prey equal to about 20–33% of their body weight/day (Kenyon 1969; Costa 1982). Although the air layer in the fur is an efficient insulator, it is also inflexible, requiring a mechanism to dissipate heat during periods of exercise. This appears to be accomplished through the broad, highly vascularized, and sparsely furred hind flippers.

Some of the physiological adaptations evident in sea otters result from their residing solely in a saltwater environment and foraging under hyperbaric conditions. Sea otters have little access to fresh water and feed primarily on benthic marine invertebrates that are isotonic with seawater. They are able to consume seawater and possess a large and efficient kidney for producing concentrated urine. Modifications of the respiratory system for diving are evident. The lungs are nearly 2.5 times larger than in similar-sized terrestrial mammals, and serve to store oxygen and provide buoyancy (Costa and Kooyman 1982). Oxygen–hemoglobin affinity is relatively high, thus increasing their blood-oxygen storage capacity. Hemoglobin, red blood cell, and hematocrit values in sea otters are similar to values in pinnipeds and cetaceans, reflecting adaptations to aquatic existence (Bossart and Dierauf 1990). The tracheal length–width ratio is reduced compared to other otters, allowing rapid and complete air replacement.

LIFE HISTORY

Male sea otters gain access to estrous females by establishing and maintaining territories from which other males are excluded (Kenyon 1969; Garshelis et al. 1984; Jameson 1989). Male territories vary in size from about 20 to 80 ha. Territories may be located in or adjacent to female resting or feeding areas or along travel corridors between those areas. Territories are occupied continuously or intermittently over time (Loughlin 1981; Garshelis et al. 1984; Jameson 1989). Male occupancy in a territory may extend for 6–9 years (Riedman and Estes 1990). Male territoriality results in segregation of the sexes, and males that do not occupy territories reside in dense aggregations (Kenyon 1969; Bodkin et al. 2000). Males that do not defend territories (transients) may gain access to receptive females by traveling through or adjacent to male territories and female areas. Male aggregation areas identified by Kenyon in 1962 at Amchitka persisted through at least 1995. Female choice in mate selection is facilitated through females traveling among male territories, although males may try to sequester estrous females in their territory.

REPRODUCTION

Sea otters display polygynous reproductive behavior. Male sea otters can attain sexual maturity by age three but likely do not attain the social maturity required for successful reproduction until >5 years of age (Garshelis 1983). Variation in reproductive success among males or between territorial and transient reproductive strategies is unknown. Female sea otters attain sexual maturity as early as age 2, and by 3 years of age most females are sexually mature. Where food resources may be limiting population growth, sexual maturation may be delayed to 4–5 years of age. Annual adult female reproductive rates range from 0.80 to 0.94 (Siniff and Ralls 1991; Bodkin et al. 1993; Jameson and Johnson 1993; Riedman et al. 1994; Monson and DeGange 1995; Monson et al. 2000b). Among areas where sea otter reproduction has been studied, reproductive rates appear to be largely invariant, despite differences in resource availability. Gestation, including a period of delayed implantation, requires about 6 months. Although copulation and subsequent pupping can take place at any time of year, there appears to be a positive relation between increasing latitude and reproductive synchrony. In California, pupping is weakly synchronous to nearly uniform across months, whereas in Prince William Sound, a distinct peak in pupping occurs in late spring.

Females give birth to a single pup, although rare instances of twinning have been observed (Jameson and Bodkin 1986). The average length of time of pup dependency and weaning is about 6 months, resulting in a reproductive interval of approximately 1 year from copulation to weaning. Occasional instances of pup dependency may exceed 9 months. Female experience apparently is important to the survival of her offspring; primiparous females are generally less successful in weaning a pup (Riedman et al. 1994; Monson and DeGange 1995). If a female loses her pup before weaning, she will soon enter estrus and breed again. Copulation occurs in the water, during which time a pair bond is formed that can last several days (Riedman and Estes 1990), although a female may breed with more than one male. Distinctive and sometimes severe wounds can result from males biting the nose of the female during copulation.

MORTALITY

Whereas reproductive output remains relatively constant over a broad range of ecological conditions, pup survival appears to be more closely influenced by resource availability. At Amchitka Island, a population at or near equilibrium density, dependent pup survival ranged from 22% to 40%, compared to nearly 85% at Kodiak Island, where food was not limiting and the population was increasing (Monson et al. 2000b). Postweaning annual survival is variable among populations and years, ranging from 18% to nearly 60% (Monson et al. 2000b). Factors affecting survival of young sea otters, as opposed to reproductive rates, may be important in ultimately regulating sea otter population size. Once a sea otter survives its first year of life, there is a good probability of its surviving to old age. Survival of sea otters more than 2 years of age is generally high, approaching 90%, but gradually declines over time (Bodkin and Jameson 1991; Monson et al. 2000b). Most mortality, other than human related, occurs during late winter and spring (Kenyon 1969; Bodkin and Jameson 1991; Bodkin et al. 2000). Maximum ages, based on tooth annuli, are about 22 years for females and 15 years for males.

Fetal sex ratio does not differ from parity (Kenyon 1982; Bodkin et al. 1993), yet sea otter populations generally consist of more females than males. Age-specific survival of sea otters is generally lower among males (Kenyon 1969, 1982; Siniff and Ralls 1991; Monson and DeGange 1995; Bodkin et al. 2000), resulting in the female-biased adult population. Survival of juvenile (postweaning) males in California exceeds that of juvenile females (Siniff and Ralls 1991), although at Amchitka the opposite was found (Kenyon 1969).

Causes of mortality in sea otter populations are inherently difficult to determine, and the probability of detecting and assigning cause of death depends on the cause. For example, the carcass of a sea otter that dies of starvation is more likely to be recovered than one killed by a predator. Sources of mortality include predation, starvation, disease, contaminants, incidental take by humans, intentional human harvest, and intraspecific aggression. As noted previously, sea ice is a mortality factor as well. Recognized sea otter predators include the white shark (*Carcharadon carcharias*), brown bear (*Ursus arctos*), wolf, red fox, wolverine, killer whale (*Orca orcinus*), and bald eagle (*Haliaeetus leucocephalus*) (Kenyon 1969; Ames and Morejohn 1980; Riedman and Estes 1990; Monson and DeGange 1995; Hatfield et al. 1998; Bodkin et al. 2000). Bald eagles prey primarily on young pups, and Gelatt et al. (2002) found mothers with young pups fed less often overall, and when they did feed it was often nocturnally, apparently to avoid eagle predation. Declining sea otter populations across the Aleutian Archipelago during the 1990s were attributed to increased predation by killer whales (Estes et al. 1998).

Sea otter population densities generally are thought to be limited by prey availability, with mortality being density dependent and increasing during periods of food shortage and inclement seas. This was observed at Amchitka Island (Kenyon 1969). In the Commander Islands, the sea otter population declined by 41% in a single year, following 10 years of increasing density, declining prey populations, and declining weights of adult male otters (Bodkin et al. 2000). In California, infectious disease was implicated in nearly 40% of the 195 mortalities studied between 1992 and 1995 (Thomas and Cole 1996). Factors contributing to mortality included peritonitis induced by acanthocephalan parasites, protozoan encephalitis, coccidioidomycosis, and bacterial infections. Between 1997 and 2001, 62% of 107 dead, and 42% of 116 live sea otters tested positive for the protozoan parasite, *Toxoplasma gondii*, implicating pathogen pollution as a factor limiting recovery of the California population (Miller et al. 2002). It is possible that some of the disease-related mortality in California is ultimately linked to prey availability.

FEEDING HABITS

The sea otter is a generalist predator, known to consume more than 150 different prey species (Kenyon 1969; Riedman and Estes 1990; Estes and Bodkin 2002). With few exceptions, their prey generally consist of sessile or slow-moving benthic invertebrates such as mollusks, crustaceans, and echinoderms. Foraging occurs in marine habitats between the high intertidal zone to depths slightly in excess of 100 m. Preferred foraging habitat is generally in depths less than 25 or 40 m (Riedman and Estes 1990), although studies in southeast Alaska have found that some animals forage mostly at depths from 40 to 80 m. Average dive times range from about 60 to 120 sec depending on habitat and depth. Maximum reported dive durations are 246 sec in California and 260 sec in Alaska, and swimming speed during foraging dives averages about 1 m/sec. A sea otter may forage several times daily, with feeding bouts averaging about 3 hr, separated by periods of rest that also average about 3 hr. Generally, the amount of time a sea otter allocates toward foraging is positively related to sea otter density and inversely related to prey availability (Estes et al. 1982; Garshelis et al. 1986; Gelatt et al. 2002).

Although the sea otter is known to prey on a large number of species, only a few tend to predominate in the diet. Prey selection depends on location, habitat type, season, length of occupation, and can vary among individuals within populations (Estes et al. 2003). In California, otters foraging over rocky substrates and in kelp forests mainly consume decapod crustaceans, gastropod and bivalve mollusks, and echinoderms (Ebert 1968; Estes et al. 1981). In protected bays with soft sediments, otters mainly consume infaunal bivalves (*Saxidomus nuttallii* and *Tresus nuttallii*) (Kvitek et al. 1988). Along exposed coasts of soft sediments, the Pismo clam (*Tivela stultorum*) is a common prey (Stephenson 1977). Important prey in Washington include crabs (*Cancer* spp., *Pugettia* spp.), octopus (*Octopus* spp.), intertidal clams (*Protothaca* spp.), sea cucumbers (*Cucumaria miniata*), and the red sea urchin (*Strongylocentrotus franciscanus*) (Kvitek et al. 1989). The predominately soft-sediment habitats of southeast Alaska, Prince William Sound, and Kodiak Island support populations of clams that are the primary prey of sea otters. Throughout most of southeast Alaska, burrowing bivalve clams (species of *Saxidomus, Protothaca, Macoma,* and *Mya*) predominate in the sea otter's diet (Kvitek et al. 1993). They account for more than 50% of the identified prey, although urchins (*S. droebachiensis*) and mussels (*Modiolis modiolis, Musculus* spp.) can also be important. In Prince William Sound and Kodiak Island, clams account for 34–100% of the otter's prey (Calkins 1978; Doroff and Bodkin 1994; Doroff and DeGange 1994). Mussels (*Mytilus trossulus*) apparently become more important as the length of occupation by sea otters increases, ranging from 0% in the diet at newly occupied sites at Kodiak to 22% in long-occupied areas (Doroff and DeGange 1994). Crabs (*C. magister*) were once important sea otter prey in eastern Prince William Sound, but apparently have been depleted by otter foraging and are no longer eaten in large numbers (Garshelis et al. 1986). Sea urchins are minor components of the sea otter's diet in Prince William Sound and the Kodiak Archipelago. In contrast, the sea otter's diet in the Aleutian, Commander, and Kuril Islands is dominated by sea urchins and a variety of fin fish (including hexagrammids, gaddids, cottids, perciformes, cyclopterids, and scorpaenids) (Kenyon 1969; Estes et al. 1982). Sea urchins tend to dominate the diet of low-density sea otter populations, whereas fish are consumed in populations near equilibrium density (Estes et al. 1982). For unknown reasons, fish are rarely consumed by sea otters in regions east of the Aleutian Islands.

Sea otters also exploit episodically abundant prey such as squid (*Loligo* spp.) and pelagic red crabs (*Pleuroncodes planipes*) in California and smooth lumpsuckers (*Aptocyclus ventricosus*) in the Aleutian Islands. On occasion, sea otters attack and consume sea birds, including teal (*Anas crecca*), scoters (*Melanita perspicillata*), loons (*Gavia immer*), gulls (*Larus* spp.), grebes (*Aechmophorus occidentalis*), and cormorants (*Phalacrocorax* spp.) (Kenyon 1969; Riedman and Estes 1990).

ECOLOGY

Habitat. Sea otters occupy nearly all coastal marine habitats, from fine sediment bays and estuaries to rocky shores exposed to oceanic swells. The width of habitat they occupy is defined by the intertidal zone and extends offshore to about the 100 m contour. Habitat area depends on the slope of the sea floor, and, where depth contour intervals are widely spaced, may extend far offshore to include shallow areas. Highest densities of sea otters occur in water less than 40 m deep, although they can be found in water up to 200 m deep (Bodkin and Udevitz 1999).

Community Ecology. Sea otters are known for the effects their foraging has on the structure and function of coastal marine communities. They provide an important example of a "keystone species" concept (Power et al. 1996). In the absence of sea otter foraging during the twentieth century, populations of several species of urchins (*Strongylocentrotus* spp.) became extremely abundant. Grazing activities of urchins effectively limited kelp populations, resulting in deforested areas known as "urchin barrens" (Lawrence 1975; Estes and Harrold 1988). Because sea urchins are a preferred prey item, as otters recovered they dramatically reduced the sizes and densities of urchins as well as other prey

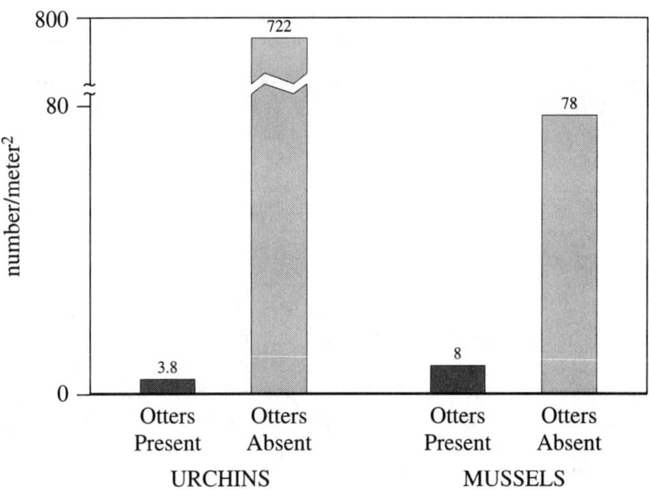

FIGURE 36.5. Densities of mussels and urchins in the presence and absence of sea otters in the Aleutian Islands, Alaska. SOURCE: Data from Estes and Palmisano (1974).

such as mussels (*Mytilus* spp.) (Fig. 36.5). Released from the effects of urchin-related herbivory, populations of macroalgae responded, resulting in diverse and abundant populations of understory and canopy-forming kelp forests. Although other factors, both abiotic and biotic, can also limit sea urchin populations (Foster and Schiel 1988; Foster 1990), the generality of the sea otter effect in reducing urchins and increasing kelp forests is widely recognized (reviewed in Estes and Duggins 1995). Further cascading effects of sea otters in coastal rocky subtidal communities may stem from the proliferation of kelp forests. Following sea otter recovery, kelp forests provide food and habitat for other species, including fin-fish (Simenstead et al. 1978; Ebeling and Laur 1988), that can provide forage for birds and mammals. Furthermore, kelps provide an additional source of detrital organic carbon to the coastal marine community (Duggins et al. 1989).

Effects of sea otter foraging also are documented in rocky intertidal and soft-sediment marine communities. In Prince William Sound, Alaska, the size-class distribution of mussels was strongly skewed toward animals <40 mm in shell length where otters were present, whereas mussels >40 mm in length made up a large component of the population where sea otters were absent (VanBlaricom 1988). In soft-sediment coastal communities, sea otters forage on epifauna (crustaceans, echinoderms, and mollusks) and infauna (primarily clams). They generally select the largest individuals. This causes declines in prey abundance and reductions in size-class distributions, although the deepest burrowing clams (e.g., *Tresus nuttallii* and *Panopea generosa*) may attain refuge from some sea otter predation (Kvitek and Oliver 1988; Kvitek et al. 1992). Community-level responses to reoccupation by sea otters are much less well studied in soft-sediment habitats, which dominate much of the northern Pacific and need additional research.

MANAGEMENT, CONSERVATION, AND RESEARCH NEEDS

Early Exploitation. We have little direct knowledge of the relations that existed between the indigenous people of the northern Pacific and the sea otter. However, two lines of evidence support the premise that early coastal dwellers used sea otters. First, skeletal remains of sea otters are common in coastal middens, and accumulated refuse suggests periods of overexploitation at sites of human occupation (Simenstad et al. 1978). Second, coastal people, particularly in the Aleutian Archipelago, had developed an elaborate and efficient method for hunting sea otters long before the use of modern firearms (Ogden 1941). Development of a sophisticated means of cooperative hunting provides good evidence of the importance of sea otters long before the arrival of Europeans into the northern Pacific.

TABLE 36.1. Recent counts or estimates of sea otter (*Enhydra lutris*) abundance in the northern Pacific

Subspecies	Area	Year(s)	Number	Status
E. l. lutris	Russia	1995–2002	17,000	Decreasing in Kuril and stable in Commander Islands, and Kamchatka
E. l. kenyoni	Alaska	1994–2002	70,700	Declining in Aleutians, uncertain in Gulf of Alaska, and increasing in southeast Alaska
	British Columbia	1998	2,500	Increasing
	Washington	1997	500	Increasing
E. l. nereis	California	2003	2,500	Uncertain
Total			93,200	

The discovery of abundant sea otter populations in Alaska by Bering in 1741 initiated a commercial sea otter harvest that lasted 170 years. Overexploitation reduced a population of about 300,000 to perhaps several hundreds by 1911 (Lensink 1962; Kenyon 1969; Bodkin et al. 1999). The harvest was initiated by Russians, began in the eastern Pacific, and eventually included Spanish, British, and American ships. The harvest progressed to the east as sea otters were depleted from successive island groups in the Aleutians. Only by acquiring the hunting skills of indigenous coastal hunters, sometimes through slavery, was a commercial harvest possible, at least before the use of modern firearms. An estimated 500,000–900,000 sea otters were harvested between 1741 and 1911 (Lensink 1962; Kenyon 1969).

Recovery. The end of the commercial harvest of sea otters probably occurred late in the nineteenth century, as their abundance declined to a point where the harvest was no longer economically feasible. In the early nineteenth century, annual maximum harvests exceeded 15,000/year, whereas by 1910 the average annual harvest had declined to less than 50, despite extensive effort (Lensink 1962). When afforded legal protection by the Fur Seal Treaty of 1911, sea otters existed in 13 remnant populations, each probably numbering a few dozen to possibly a few hundred individuals (Fig. 36.2) (Kenyon 1969; Estes 1980; Bodkin et al. 1999). All remnant populations occurred between Prince William Sound and the Kuril Islands, with the exception of a group in California. Long-term annual growth rates among remnant sea otter populations averaged about 1.09 and ranged from 1.06 in California to 1.13 at Amchitka Island in the Aleutians (Bodkin et al. 1999). The first population considered to have recovered was at Amchitka Island in about 1940 (Kenyon 1969). More recently, the Commander Islands sea otter population apparently reached carrying capacity about 1990 (Bodkin et al. 2000). Other populations, notably those in California, have experienced periods of increase and decline, and continue to struggle to recover their prior range and numbers (Estes and Bodkin 2002).

In the central Aleutian Archipelago of Alaska, where sea otter populations had largely recovered, a decline of nearly 40,000 animals (80% of the population) was observed during the 1990s (Estes et al. 1998). The cause of the decline has been linked to large-scale changes in the northern Pacific that ultimately resulted in changing forage fish populations. Concurrent with altered fish populations, pinniped populations (Steller sea lions, *Eumatopias jubatus;* and harbor seals, *Phoca vitulina*) underwent dramatic declines beginning about 1970. In response to fewer pinnipeds, killer whale (*Orca orcinus*) predation may have shifted to sea otters, contributing to their decline (Estes et al. 1998).

To aid recovery of the species, sea otters were translocated to unoccupied habitat between 1956 and 1987 (Jameson et al. 1982; Rathbun et al. 1990). Sea otters were translocated from Amchitka and Prince William Sound to Oregon, Washington, British Columbia, southeast Alaska, and the Pribilof Islands, and from central California to San Nicolas Island off southern California. Post-translocation mortality and emigration resulted in founding population sizes from 15% to 36% of the number originally released (Estes et al. 1989). Growth in translocated sea otter populations has been significantly greater than in remnant populations (Estes et al. 1989), averaging 21% annually and ranging from 18% to 24% (Bodkin et al. 1999). The high growth rates observed in successful translocations likely result from unlimited food and space resources available following long periods of sea otter absence.

The reasons for different growth rates between remnant and translocated populations are not well understood. It appears likely that continued illegal harvest following protection in 1911 may have contributed to delayed recovery in some remnant populations and that incidental mortality related to fisheries reduced growth rates later in the twentieth century. Translocations to British Columbia, southeast Alaska, and British Columbia have been very successful, whereas those to Oregon and the Pribilof Islands failed. The fate of the San Nicolas translocation remains uncertain because of small population size. Although effective conservation tools, the need for future translocations will diminish as existing populations continue to expand their ranges.

Current Status. Sea otters currently number about 95,000 (Table 36.1) and occupy much of their historical range in the Pacific (Fig. 36.2). Remaining large areas of unoccupied habitat exist south of Santa Barbara, California, between central California and Washington, between Vancouver Island and southeast Alaska, and along the Kamchatka Peninsula of Russia. It seems likely that sea otters will continue to recolonize this unoccupied habitat in the future through natural dispersal. It is possible the total population size exceeds 95,000, because most surveys are counts, rather than estimates, and do not account for animals not detected. Alternatively, some populations may have experienced declines since the most recent surveys.

Contaminants are of increasing concern in the conservation and management of sea otter populations throughout the north Pacific. Concentrations of organochlorines, similar to levels causing reproductive failure in captive mink (*Mustela vison*), occurred in the Aleutian Islands and California, whereas otters from southeast Alaska were relatively uncontaminated (Estes et al. 1997; Bacon et al. 1998). Elevated levels of butyltin residues and organochlorine compounds have been associated with sea otter mortality caused by infectious disease in California (Kannan et al. 1998; Nakata et al. 1998). Changes in stable lead isotope compositions from preindustrial and modern sea otters in the Aleutians reflect changes in the sources of lead in coastal marine food webs. In preindustrial samples, lead was from natural deposits, whereas in contemporary sea otters, lead is primarily from Asian and North American industrial sources (Smith et al. 1990).

Susceptibility of sea otters to oil spills, because of the reliance on their fur for thermoregulation, has long been recognized (Kenyon 1969; Siniff et al. 1982; Williams et al. 1988). In 1989, the T/V *Exxon Valdez* spilled about 41 million liters of crude oil in Prince William Sound (Morris and Loughlin 1994). Accurate estimates of acute mortality resulting from that spill are not available, but nearly 1000 sea otter carcasses were recovered in the months following the spill (Ballachey et al. 1994). Estimates of carcass recovery rates ranged from 20% to 59% (DeGange et al. 1994; Garshelis 1997), indicating mortality of up to several thousand animals (Ballachey et al. 1994). Through at least 1997, sea otter populations had not completely recovered in areas where oil deposition was heaviest and persistent, and sea otter mortality following the spill was nearly complete (Bodkin and Udevitz 1994; Dean et al. 2000; Bodkin et al. 2002). Long-term effects of the spill include reduced sea otter survival for at least a decade following the spill (Monson et al. 2000a), likely a result of sublethal oiling in 1989, chronic exposure to residual oil in the years following the spill, and spill-related effects on invertebrate prey populations (Ballachey et al. 1994; Fukuyama et al. 2002; Peterson 2001). As human populations increase, exposure to acute and chronic environmental contaminants will likely increase. Improved understanding of the effects of contaminants on populations and ecosystems may be valuable in understanding how and why ecosystems change.

Human activities contribute to sea otter mortality throughout the Pacific Rim. Incidental mortality occurs in the course of several commercial fisheries. In California, an estimated annual take of 80 sea otters in gill and trammel nets likely contributed to a lack of population growth during the 1980s (Wendell et al. 1986a). Developing fisheries and changing fishing techniques continue to present potential problems to recovering sea otter populations. In Alaska, sea otters are taken incidental to gillnet, seine, and crab trap fisheries throughout the state, but estimates of total mortality have not been made (Rotterman and Simon-Jackson 1988). In Alaska, indigenous people are permitted to harvest sea otters for subsistence purposes; however, the harvest is largely unregulated. Moreover, an illegal harvest of unknown magnitude and consequence continues throughout much of the geographic range of sea otters.

Interactions with Fisheries. Based on reduced abundance and size of many prey species following the return of sea otters, it seems reasonable to conclude that, during the sustained absence of sea otters over more than a century, some of their prey would have increased in abundance and size. This is likely the case for several invertebrate species for which humans have developed economically or socially valuable fisheries in recent decades. These fisheries include several species of crustaceans (*Cancer* spp., *Chionecetes* sp., *Paralithoides* spp., and *Panulirus interruptus*), mollusks (*Haliotis* spp., *Saxidomus* spp., *Tresus* spp., *Siliqua patula,* and *Tivela sultorum*), and echinoderms (*Strongylocentrotus* spp.). In some cases there is evidence that human fisheries for *Haliotis, Strongylocentrotus,* and *Tivela* are incompatible with populations of sea otters (Lowery and Pearse 1973; Wendell et al. 1986b). As sea otters continue to recolonize previous habitat, it is likely that conflicts with humans will arise as they compete for these invertebrate resources. Whereas some prey may attain refuge from sea otter predation either through depth in the water (e.g., *Paralithoides*) or in the sediment (e.g., *Panopea*), it remains unlikely that most of these fisheries can coexist with sea otters. In many cases, these conflicts can be predicted and management alternatives to mediate conflict considered. Potential actions may include increasing human harvests before a conflict and supporting transitions to alternative fisheries.

The California Population. Because of its small population size, limited range, and potential for a catastrophic oil spill, the California population of sea otters remains listed as "threatened" under the Endangered Species Act. Reasons for the prolonged slow growth rate and recent declines of this population late in the twentieth century are unknown, but may be related to human activities. These include accidental mortality associated with fisheries, contaminants affecting health, disease, and perhaps increased predation. It remains important to understand and correct causes of delayed recovery in California.

Harvesting by Humans. In Alaska, where sea otters are legally harvested, the reported take since 1993 has been highly variable. Harvest exceeded 1200 in 1993, with most of that from a few, relatively small areas. There are no limits on the harvest, and an unreported illegal harvest undoubtedly exists in Alaska and elsewhere. It is important to recognize that sea otters are relatively sedentary, with low reproductive potential. Overexploitation probably can occur at spatial scales on the order of tens to hundreds of kilometers of coastline. Successive overharvest of contiguous segments of shoreline could result in large-scale depletion. Cronin et al. (1996) and Scribner et al. (1997) suggested that differences in genetic markers could be used, in conjunction with other data, to identify populations of sea otters that could be managed as independent units. Gorbics and Bodkin (2001) used distribution, population response (i.e., life history rates), and phenotype and genotype data to evaluate the potential discrimination of sea otter stocks, or management units, within the state of Alaska. They concluded that three distinct populations of sea otters could be recognized, and populations should be conserved at geographic scales relative to their natural history. Given the good understanding of the life history of sea otters, a sustainable management plan on appropriate geographic scales should be developed, reviewed, and implemented in Alaska. One of the challenges in sea otter conservation and management is accurately estimating population sizes, either to manage a harvest or establish conservation objectives. Accurate and precise survey methods are available and should be employed, particularly where harvests are occurring.

Sea otters may be one of the best studied of the marine mammals because of their ecological role in nearshore marine communities, the historic value of their fur to humans, and the competition between otters and humans over some marine invertebrate populations. Although the recovery of sea otters in the northern Pacific constitutes a remarkable success story in the conservation of a species that neared extinction, large areas of habitat remain unoccupied, and some populations (e.g., California) remain in need of protective status. Our view of sea otters during the twentieth century was largely influenced by rapidly recovering populations. As populations reach carrying capacity and attain equilibrium densities with food and space resources, they will likely be subjected to new and different factors that will regulate their numbers. It may not be reasonable to assume that sea otter populations will maintain equilibrium densities at some constant value; instead, populations may fluctuate on different temporal and spatial patterns. Because of the broad life history and ecological knowledge concerning sea otters, they may represent a good case study of mechanisms that regulate long-lived mammalian populations as well as the effects of changing predator populations on their prey. One of our long-term challenges in understanding sea otters and their coastal marine communities will lie in maintaining the level of study undertaken in recent decades to continue our education of how coastal marine ecosystems function.

ACKNOWLEDGMENT

This chapter built on that of Karl Kenyon in the 1982 edition of this book. Karl's career contributed toward our better understanding of sea otters, and I gratefully acknowledge his contribution toward this revised chapter.

LITERATURE CITED

Ames, J. A., and G. V. Morejohn. 1980. Evidence of white shark, *Carcharodon carcharias,* attacks on sea otters, *Enhydra lutris*. California Fish and Game 66:196–209

Bacon, C. E., W. M. Jarman, J. A. Estes, M. Simon, and R. J. Norstrom. 1998. Comparison of organochlorine contaminants among sea otter (*Enhydra lutris*) populations in California and Alaska. Environmental Toxicology and Chemistry 18:452–58.

Ballachey, B. E., J. L. Bodkin, and A. R. DeGange. 1994. An overview of sea otter studies. Pages 47–59 *in* T. R. Loughlin, ed. Marine mammals and the *Exxon Valdez*. Academic Press, San Diego, CA.

Berta, A., and G. S. Morgan. 1985. A new sea otter (Carnivora: Mustelidae), from the late Miocene and early Pliocene (Hemphillian) of North America. Journal of Paleontology 59:808–19.

Bodkin, J. L., and R. Jameson. 1991. Patterns of seabird and marine mammal carcass deposition along the central California coast, 1980–1986. Canadian Journal of Zoology 69:1149–55.

Bodkin, J. L., and M. S. Udevitz. 1994. An intersection model for estimating sea otter mortality along the Kenai Peninsula. Pages 81–95 *in* T. R. Loughlin, ed. Marine mammals and the *Exxon Valdez*. Academic Press, San Diego, CA.

Bodkin, J. L., and M. S. Udevitz. 1999. An aerial survey method to estimate sea otter abundance. Pages 13–26 *in* G. W. Garner, S. C. Amstrup, J. L. Laake, B. F. J. Manly, L. L. McDonald, and D. G. Robertson, eds. Marine mammal survey and assessment methods. Balkema Press, Rotterdam, the Netherlands.

Bodkin, J. L., D. Mulcahy, and C. J. Lensink. 1993. Age specific reproduction in the sea otter (*Enhydra lutris*): an analysis of reproductive tracts. Canadian Journal of Zoology 71:1811–15.

Bodkin, J. L., J. A. Ames, R. J. Jameson, A. M. Johnson, and G. M. Matson. 1997. Estimating age of sea otters with cementum layers in the first premolar. Journal of Wildlife Management 61:967–73.

Bodkin, J. L., B. E. Ballachey, M. A. Cronin, and K. T. Scribner. 1999. Population demographics and genetic diversity in remnant and re-established populations of sea otters. Conservation Biology 13:1278–1385.

Bodkin, J. L., A. M. Burdin, and D. A. Ryzanov. 2000. Age and sex specific mortality and population structure in sea otters. Marine Mammal Science 16:201–19.

Bodkin, J. L., B. E. Ballachey, T. A. Dean, A. K. Fukuyama, S. C. Jewett, L. McDonald, D. H. Monson, C. E. O'Clair, and G. R. VanBlaricom. 2002. Sea outter population status and the process of recovery from the 1989 "Exxon Valdez" oil spill. Marine Ecology Progress Series 241:237–253.

Bossart, G. D., and L. A. Dierauf. 1990. Marine mammal clinical laboratory medicine. Pages 1–52 *in* L. Dierauf, ed. Handbook of marine mammal medicine: Health disease and rehabilitation. CRC Press, Boca Raton, FL.

Calkins, D. G. 1978. Feeding behavior and major prey species of the sea otter, *Enhydra lutris,* in Montague Strait, Prince William Sound, Alaska. Fishery Bulletin 76:125–31.

Costa, D. P. 1982. Energy, nitrogen and electrolyte flux and sea-water drinking in the sea otter *Enhydra lutris*. Physiological Zoology 55:34–44.

Costa, D. P., and G. L. Kooyman. 1982. Oxygen consumption, thermoregulation, and the effect of fur oiling and washing on the sea otter *Enhydra lutris*. Canadian Journal of Zoology 60:2761–67.

Costa, D. P., and G. L. Kooyman. 1984. Contribution of specific dynamic action to heat balance and thermoregulation in the sea otter *Enhydra lutris*. Physiological Zoology 57:199–203.

Cronin, M. A., J. Bodkin, B. Ballachey, J. Estes, and J. C. Patton. 1996. Mitochondrial-DNA variation among subspecies and populations of sea otters (*Enhydra lutris*). Journal of Mammalogy 72:546–57.

Davis, J., and W. Z. Lidicker, Jr. 1975. The taxonomic status of the southern sea otter. Proceedings of the California Academy of Sciences 40:429–37.

Dean, T. A., J. L. Bodkin, S. C. Jewett, D. H. Monson, and D. Jung. 2000. Changes in sea urchins and kelp following a reduction in sea otter density as a result of the *Exxon Valdez* oil spill. Marine Ecology Progress Series 199:281–91.

DeGange, A. R., A. M. Doroff, and D. H. Monson. 1994. Experimental recovery of sea otter carcasses at Kodiak Island, Alaska, following the *Exxon Valdez* oil spill. Marine Mammal Science 10:492–96.

Doroff, A. M., and J. L. Bodkin. 1994. Sea otter foraging behavior and hydrocarbon levels in prey. Pages 193–208 *in* T. R. Loughlin, ed. Marine mammals and the *Exxon Valdez*. Academic Press. San Diego, CA.

Doroff, A. M., and A. R. DeGange. 1994. Sea otter, *Enhydra lutris,* prey composition and foraging success in the northern Kodiak Archipelago. Fishery Bulletin 92:704–10.

Duggins, D. O., C. A. Simenstead, and J. A. Estes. 1989. Magnification of secondary production by kelp detritus in coastal marine necosystems. Science 245:170–73.

Ebeling, A. W., and D. R. Laur. 1988. Fish populations in kelp forests without sea otters: Effects of severe storm damage and destructive sea urchin grazing. Pages 169–91 *in* G. R. VanBlaricom and J. A. Estes, eds. The community ecology of sea otters. Springe-Verlag, Berlin.

Ebert, E. E. 1968. A food habits study of the southern sea otter, *Enhydra lutris nereis*. California Fish and Game 54:33–42.

Estes, J. A. 1980. *Enhydra lutris*. Mammalian Species 133:1–8.

Estes, J. A., and J. F. Palmisano. 1974. Sea otters: Their role in structuring nearshore communities. Science 185:1058–60.

Estes, J. A., and C. Harrold. 1988. Sea otters, sea urchins, and kelp beds: Some questions of scale. Pages 116–42 *in* G. R. VanBlaricom and J. A. Estes, eds. The community ecology of sea otters. Springer-Verlag, Berlin.

Estes, J. A., and D. O. Duggins. 1995. Sea otters and kelp forests in Alaska: Generality and variation in a community ecology paradigm. Ecological Monographs 65:75–100.

Estes, J. A., and J. L. Bodkin. 2002. Marine otters. Pages 842–858 *in* W. F. Perrin, B. Wursig, and H. G. M. Thewissen, eds. Encyclopedia of Marine Mammals. Academic Press, San Diego, CA.

Estes, J. A., R. J. Jameson, and A. M. Johnson. 1981. Food selection and some foraging tactics of sea otters. Pages 606–41 *in* J. A. Chapman and D. Pursley, eds. Worldwide furbearer conference proceedings. Frostburg, MD.

Estes, J. A., R. J. Jameson, and E. B. Rhode. 1982. Activity and prey selection in the sea otter: Influence of population status on community structure. American Naturalist 120:242–58.

Estes, J. A., D. O. Duggins, and G. B. Rathbun. 1989. The ecology of extinctions in kelp forests. Conservation Biology 3:252–64.

Estes, J. A., C. E. Bacon, W. M. Jarman, R. J. Norstrom, R. G. Anthony, and A. K. Miles. 1997 Organochlorines in sea otters and bald eagles from the Aleutian Archipelago. Marine Pollution Bulletin 34:486–90.

Estes, J. A., M. T. Tinker, T. M. Williams, and D. F. Doak., 1998. Killer whale predation on sea otters linking oceanic and nearshore ecosystems. Science 282:473–76.

Estes, J. A., M. L. Riedman, M. M. Staedler, M. T. Tinker, and B. E. Lyon. 2003. Individual variation in prey selection by sea otters: Patterns, causes and implications. J. Animal Ecology. 72:144–155.

Foster, M. S. 1990. Organization of macroalgal assemblages in the northeast Pacific: The assumption of homogeneity and the illusion of generality. Hydrobiologia 192:21–33.

Foster, M. S., and D. R. Schiel. 1988. Kelp communities and sea otters: Keystone species or just another brick in the wall? Pages 92–108 *in* G. R. VanBlaricom and J. A. Estes, eds. The community ecology of sea otters. Springer-Verlag, Berlin.

Fukuyama, A. K., G. Shigenaka, and R. F. Hoff. 2000. Effects of residual *Exxon Valdez* oil on intertidal *Protothaca staminea:* Mortality, growth and bioaccumulation of hydrocarbons in transplanted clams. Marine Pollution Bulletin. 40:1042–1050.

Garshelis, D. L. 1983. Ecology of sea otters in Prince William Sound, Alaska. Ph.D. Dissertation, University of Minnesota, Minneapolis.

Garshelis, D. L. 1997. Sea otter mortality estimated from carcasses collected after the *Exxon Valdez* oil spill. Conservation Biology 11:905–16.

Garshelis, D. L., A. M. Johnson, and J. A. Garshelis. 1984. Social organization of sea otters in Prince William Sound, Alaska. Canadian Journal of Zoology 62:2648–58.

Garshelis, D. L., J. A. Garshelis, and A. T. Kimker. 1986. Sea otter time budgets and prey relationships in Alaska. Journal of Wildlife Management 50:637–47.

Gelatt, T. S., D. B. Siniff, and J. A. Estes. 2002. Activity patterns and time budgets of the declining sea otter population at Amchitka Island, Alaska. Journal of Wildlife Management. 66:29–39.

Gorbics, C. S., and J. L. Bodkin. 2001. Stock identity of sea otters in Alaska. Marine Mammal Science. 17:632–647.

Hatfield, B. B., D. Marks, M. T. Tinker, K. Nolan, and J. Peirce. 1998. Attacks on sea otters by killer whales. Marine Mammal Science 14:888–94.

Jameson, R. J. 1989. Movements, home ranges, and territories of male sea otters off central California. Marine Mammal Science 5:159–72.

Jameson, R. J., and J. L. Bodkin. 1986. An incidence of twinning in the sea otter (*Enhydra lutris*). Marine Mammal Science 2:305–9.

Jameson, R. J., and A. M. Johnson. 1993. Reproductive characteristics of female sea otters. Marine Mammal Science 9:156–67.

Jameson, R. J., K. W. Kenyon, A. M. Johnson, and H. M. Wight. 1982. History and status of translocated sea otter populations in North America. Wildlife Society Bulletin 10:100–107.

Kannan, K., K. S. Guruge, N. J. Thomas, S. Tanabe, and J. P. Giesy. 1998. Butyltin residues in southern sea otters (*Enhydra lutris nereis*) found dead along California coastal waters. Environmental Science and Technology 32:1169–75.

Kenyon, K. W. 1969. The sea otter in the eastern Pacific Ocean. North American Fauna 68:1–352.

Kenyon, K. W. 1982. Sea otter, *Enhydra lutris*. Pages 704–10 *in* J. A. Chapman and G. A. Feldhamer, eds. Wild mammals of North America. Johns Hopkins University Press, Baltimore.

Koepfli, K.-P., and R. K. Wayne. 1998. Phylogenetic relationships of otters (Carnivora: Mustelidae) based on mitochondrial cytochrome *b* sequence. Journal of Zoology (London) 246:401–16.

Kvitek, R. G., and J. S. Oliver. 1988. Sea otter foraging habits and effects on prey populations and communities in soft-bottom environments. Pages 22–47 *in* G. R. VanBlaricom and J. A. Estes, eds. The community ecology of sea otters. Springer-Verlag, Berlin.

Kvitek, R. G., A. K. Fukuyama, B. S. Anderson, and B. K. Grimm. 1988. Sea otter foraging on deep-burrowing bivalves in a California coastal lagoon. Marine Biology 98:157–67.

Kvitek, R. G., D. Shull, D. Canestro, E. C. Bowlby, and B. L. Troutman. 1989. Sea otters and benthic prey communities in Washington State. Marine Mammal Science 5:266–80.

Kvitek, R. G., J. S. Oliver, A. R. DeGange, and B. S. Anderson. 1992. Changes in Alaskan soft-bottom prey communities along a gradient in sea otter predation. Ecology 73:413–28.

Kvitek, R. G., C. E. Bowlby, and M. Staedler. 1993. Diet and foraging behavior of sea otters in southeast Alaska. Marine Mammal Science 9:168–81.

Lawrence, J. M. 1975. On the relationship between marine plants and sea urchins. Oceanography and Marine Biology Annual Review 13:213–86.

Lensink, C. J. 1962. The history and status of sea otters in Alaska. Ph.D. Dissertation, Purdue University, Lafayette, IN.

Loughlin, T. R. 1981. Home range and territoriality of sea otters near Monterey, California. Journal of Wildlife Management 44:576–82.

Lowery, L. F., and J. S. Pearse. 1973. Abalones and sea urchins in an area inhabited by sea otters. Marine Biology 23:213–19.

Miller, M. A., I. A. Gardner, C. Kreuder, D. M. Paradies, K. R. Worcester, D. A. Jessup, E. Dodd, M. D. Harris, J. A. Ames, A. E. Packham, and P. A. Conrad. 2002. Coastal freshwater runoff is a risk factor for *Toxoplasma gondii* infection of southern sea otters (*Enhydra lutris nereis*). International Journal for Parasitology 32:997–1006.

Monson, D. H., and A. R. DeGange. 1995. Reproduction, preweaning survival, and survival of adult sea otters at Kodiak Island, Alaska. Canadian Journal of Zoology 73:1161–69.

Monson, D. H., D. F. Doak, B. E. Ballachey, A. M. Johnson, and J. L. Bodkin. 2000a. Long-term impacts of the *Exxon Valdez* oil spill on sea otters, assessed through age-dependent mortality patterns. Proceedings of the National Academy of Sciences of the USA 97:6562–67.

Monson, D. H., J. A. Estes, J. L. Bodkin, and D. B. Siniff. 2000b. Life history plasticity and population regulation in sea otters. Oikos 90:457–68.

Morris, B. F., and T. R. Loughlin. 1994. Overview of the *Exxon Valdez* oil spill, 1989–1992. Pages 1–22 *in* T. R. Loughlin, ed. Marine mammals and the *Exxon Valdez*. Academic Press, San Diego, CA.

Morrison, P., M. Rosenmann, and J. A. Estes. 1974. Metabolism and thermoregulation in the sea otter. Physiological Zoology 47:218–29.

Nakata, H., K. Kannan, L. Jing, N. J. Thomas, S. Tanabe, and J. P Giesey. 1998. Accumulation pattern of organochlorine pesticides and polychlorinated biphenyls in southern sea otters (*Enhydra lutris nereis*) found stranded along coastal California, USA. Environmental Pollution 103:45–53.

Nikolaev, A. M. 1966. On the feeding of the Kurile sea otter and some aspects of their behavior during the period of ice. Pages 231–36 *in* E. N. Palovskii and B. A. Zenkovich, eds. Marine mammals. National Marine Mammal Laboratory, Seattle, WA.

Ogden, A. 1941. The California sea otter trade, 1784–1848. University of California Press, Berkeley.

Peterson, C. H. 2001. The web of ecosystem connections to shoreline habitats as revealed by the *Exxon Valdez* oil spill perturbation: A synthesis of acute direct vs indirect and chronic effects. Advances in Marine Biology. 39:1–103.

Power, M. E., D. Tilman, J. A. Estes, B. A. Menge, W. J. Bond, L. S. Mills, G. Daily, J. C. Castilla, J. Lubchenco, and R. T. Paine. 1996. Challenges in the quest for keystones. Bioscience 46:609–20.

Rathbun, G. B., R. J. Jameson, G. R. VanBlaricom, and R. L. Brownell. 1990. Reintroduction of sea otters to San Nicolas Island, California: Preliminary results for the first year. Pages 99–114 *in*P. J. Bryant and J. Remmington, eds. Endangered wildlife and habitats in Southern California (Memoir No. 3) Natural History Foundation of Orange County.

Riedman, M. L., and J. A. Estes. 1990. The sea otter (*Enhydra lutris*): Behavior, ecology and natural history [Biological Report 90(14)]. U.S. Fish and Wildlife Service.

Riedman, M. L., J. A. Estes, M. M. Staedler, A. A. Giles, and D. R. Carlson. 1994. Breeding patterns and reproductive success of California sea otters. Journal of Wildlife Management 58:391–99.

Roest, A. I. 1973. Subspecies of the sea otter, *Enhydra lutris*. Los Angeles City Museum, Contributions in Science 252:1–17.

Roest, A. I. 1976. Systematics and the status of sea otters, *Enhydra lutris*. Bulletin of the California Academy of Sciences 75:267–70.

Rotterman, L. M., and T. Simon-Jackson. 1988. Sea otter. Pages 237–75 *in* J. W. Lentfer, ed. Selected marine mammals of Alaska. Marine Mammal Commission, Washington, DC.

Scheffer, V. B., and F. Wilke. 1950. Validity of the subspecies, *Enhydra lutris nereis*, the southern sea otter. Journal of the Washington Academy of Science 40:269–72.

Schneider, K. B., and J. B. Faro. 1975. Effects of sea ice on sea otters (*Enhydra lutris*). Journal of Mammalogy 56:91–101.

Scribner, K. T., J. Bodkin, B. Ballachey, S. R. Fain, M. A. Cronin, and M. Sanchez. 1997. Population genetic studies of the sea otter (*Enhydra lutris*): A review and interpretation of available data. Pages 197–208 *in* A. E. Dizon, S. J. Chivers, and W. F. Perrin, eds. Molecular genetics of marine mammals (Special Publication 3). Society for Marine Mammalogy, Lawrence, KS.

Simenstad, C. A., J. A. Estes, and K. W. Kenyon. 1978. Aleuts, sea otters, and alternate stable state communities. Science 200:403–11.

Siniff, D. B., and K. Ralls. 1991. Reproduction, survival and tag loss in California sea otters. Marine Mammal Science 7:211–29.

Siniff, D. B., T. D. Williams, A. M. Johnson, and D. L. Garshelis. 1982. Experiments on the response of sea otters, *Enhydra lutris,* to oil. Biological Conservation 23:261–72.

Smith, D. R., S. Niemeyer, J. A. Estes, and A. R. Flegal. 1990. Stable lead isotopes evidence of anthropogenic contamination in Alaskan sea otters. Environmental Science and Technology 24:1517–21.

Stephenson, M. D. 1977. Sea otter predation on Pismo clams in Monterey Bay, California Department of Fish and Game 54:100–107.

Thomas, N. J., and R. A. Cole. 1996. The risk of disease and threats to the wild population. Endangered Species Update, University of Michigan 13 (12):23–27.

VanBlaricom, G. R. 1988. Effects of foraging by sea otters on mussel-dominated intertidal communities. Pages 48–91 *in* G. R. VanBlaricom and J. A. Estes, eds. The community ecology of sea otters. Springer-Verlag, Berlin.

Wendell, F. E., and R. A. Hardy, and J. A. Ames. 1986a. An assessment of the accidental take of sea otters, *Enhydra lutris,* in gill and trammel nets (Marine Resources Technical Report 54). California Department of Fish and Game, Sacramento.

Wendell, F. E., R. A. Hardy, J. A. Ames, and R. T. Burge. 1986b. Temporal and spatial patterns in sea otter (*Enhydra lutris*) range expansion and in the loss of the Pismo clam fisheries. California Fish and Game 72:197–212.

Williams, T. D., D. D. Allen, J. M. Groff, and R. L. Glass. 1992. An analysis of California sea otter (*Enhydra lutris*) pelage and integument. Marine Mammal Science 8:1–18.

Williams, T. M., R. A. Kastelein, R. W. Davis, and J. A. Thomas. 1988. The effects of oil contamination and cleaning on sea otters (*Enhydra lutris*). I. Thermoregulatory implications based on pelt studies. Canadian Journal of Zoology 66:2776–81.

Wilson, D. E., M. A. Bogan, R. L. Brownell, Jr., and A. M. Burdin and M. K. Maminov. 1991. Geographic variation in sea otters, *Enhydra lutris*. Journal of Mammalogy 72:22–36.

Wozencraft, W. C. 1993. Carnivora. Pages 279–348 *in* D. E. Wilson and D. M. Reeder, eds. Mammal species of the world: A taxonomic and geographic reference. Smithsonian Institution Press, Washington, DC.

James L. Bodkin, U.S. Geological Survey, Alaska Science Center, 1011 E. Tudor Road, Anchorage, Alaska 99503. Email: james_bodkin@usgs.gov.

37

Mountain Lion

Puma concolor

Becky M. Pierce
Vernon C. Bleich

NOMENCLATURE

COMMON NAMES. Mountain lion, puma, panther, catamount, and cougar
SCIENTIFIC NAME. *Puma concolor*

TAXONOMY

Mountain lions have a complicated taxonomic history. Numerous subspecies have been described, largely based on morphometric differences. For example, Young and Goldman (1946) listed a total of 30 subspecies, and Culver et al. (2000) noted that 32 named subspecies existed. Of those listed by Young and Goldman (1946), at least 13 occurred in Canada or the United States (and adjacent Mexico); the remainder had geographic distributions restricted to Mexico, Central America, or South America.

The mountain lion was first described from a specimen collected in Brazil (Marcgrave 1648); Linnaeus reclassified the mountain lion as *Felis concolor* in 1771. Jardine (1834) placed the species in the genus *Puma,* where it remains (Wozencraft 1993). The historical taxonomy of the mountain lion was reviewed in detail by Young and Goldman (1946) and Currier (1983).

Based on molecular and morphological investigations, the mountain lion is thought to have evolved from an ancestor in common with the cheetah (*Acinonyx jubatus*) and jaguarundi (*Herpailurus yaguaroundi*) (Van Valkenburgh et al. 1990; Janczewski et al. 1995; Johnson and O'Brien 1997; Pecon-Slattery and O'Brien 1998). Mountain lions are represented in the North American fossil record dating back 300,000 years (Turner 1997). Based on an extensive analysis of mitochondrial DNA haplotypes and microsatellite alleles from throughout the geographic range of *Puma concolor,* Culver et al. (2000) speculated that a mass extinction during the Pleistocene (Martin 1989), which eliminated the majority of large mammals in North America, was followed by a colonization of mountain lions of South American stock. Furthermore, Culver et al. (2000) concluded that mountain lions (north of Nicaragua) today represent a single subspecies, *Puma concolor couguar,* in lieu of the 15 that formerly were recognized. Five other subspecies of *Puma concolor* are recognized throughout their southern range (Culver et al. 2000).

DESCRIPTION

Although the sexes are similar in appearance, male and female mountain lions are dimorphic in size. Gay and Best (1995) examined specimens from throughout the range of mountain lions and reported that males were significantly larger than females in 14 cranial and 5 mandibular measurements. In California, mean body weight of adult ($\geq$2 years of age) males was $\bar{X} = 53.4$ kg ($\pm$8.5 kg [*SD*]). Adult females weighed 30–40% less ($\bar{X} = 35.8 \pm 7.7$ kg) (Charlton et al. 1998). Based on a sample of 1076 specimens from Oregon, Kohlmann and Green (1999) reported that males averaged about 50% heavier than females of equivalent age. Gay and Best (1995) found no evidence that sexual dimorphism among mountain lions varied geographically. Data compiled by Anderson (1983) show that similar degrees of sexual size dimorphism occur among mountain lions throughout their range. Although the degree of sexual dimorphism did not vary geographically, the size of mountain lions generally increased with an increase in latitude (Kurten 1973; Iriarte et al. 1990).

The dorsal pelage of a mountain lion is tawny, and they generally are white on the ventral surface (Fig. 37.1). Slight variation in color is common, and their coats can have reddish, yellowish, or grayish tinges. There are no obvious contrasting markings on the coats of adult lions, other than black markings at the base of the vibrissae on the muzzle, on the dorsal surface of the ears, and on the tip of the tail. In Latin, the specific epithet *concolor* means "single color."

In contrast to adults, young mountain lions have darker facial markings and are heavily spotted at birth (Fig. 37.2), but that pattern fades as kittens mature. The spotted pelage of kittens becomes less obvious at about 9 months of age. By the time young are approximately 2 years old, the spotting has largely disappeared (Russell 1978). The eyes are closed at birth, but open by 2 weeks of age (Young and Goldman 1946). Logan and Sweanor (1999) reported that the light blue eye color of kittens changes to the amber color typical of adults when kittens are as young as 5 months.

Despite the large difference in body weight typical of adult males and females, distinguishing the sexes from a distance can be difficult. Unless males are in an older age category, with well-developed shoulder, neck, and facial musculature, body conformation of young adult males can be confused with that of older adult females. Genders can be distinguished by observing the genitalia, but male lions frequently are misclassified as females, even by professional wildlife biologists and law enforcement personnel, because the scrotum and penis are not obvious to the untrained observer. Dark coloration is associated with hair surrounding the penis sheath of males, and is an important clue to the gender of lions (Logan and Sweanor 1999).

FIGURE 37.1. Adult mountain lion (*Puma concolor*). SOURCE: Photo by Becky M. Pierce.

FIGURE 37.2. Three-week-old mountain lion (*Puma concolor*) kittens in the birth nursery. SOURCE: Photo by Becky M. Pierce.

As with all felids, except cheetahs, mountain lions have retractable claws (Fig. 37.3) at the terminus of each digit. These claws function primarily to grasp prey, rather than as an aid in locomotion (Dixon 1982). The rostrum of felids is short and the orbits are large (Fig. 37.4) compared to those of canids. These skull adaptations allow for a more powerful bite, and felids are more dependent on sight than are canids to detect prey (Vaughan et al. 2000). The adult dental formula is I 3/3, C 1/1, P 3/2, M 1/1.

DISTRIBUTION

Although they occur at low densities, mountain lions are the most abundant large felid occupying North America and are second in size only to the jaguar (*Felis onca*) (Russell 1978). Historically, mountain lion distribution encompassed most of the western hemisphere (Young and Goldman 1946). They occurred from the Atlantic to the Pacific oceans and from approximately 50° N to 50° S latitude, and from near sea level to about 4000 m elevation (Young and Goldman 1946). In essence, mountain lions ranged from approximately the Cassiar Range of northern British Columbia to southern Chile and Argentina, and from ocean to ocean on the continents of North and South America.

In North America, the distribution of mountain lions has been reduced by as much as two thirds of its historical range (Fig. 37.5), largely because of conflicts with humans as settlers migrated westward (Ross et al. 1997). Currently, mountain lions occur in suitable habitat throughout Mexico, in the majority of the western United States, and in western Canada; a small population also occurs in southern Florida

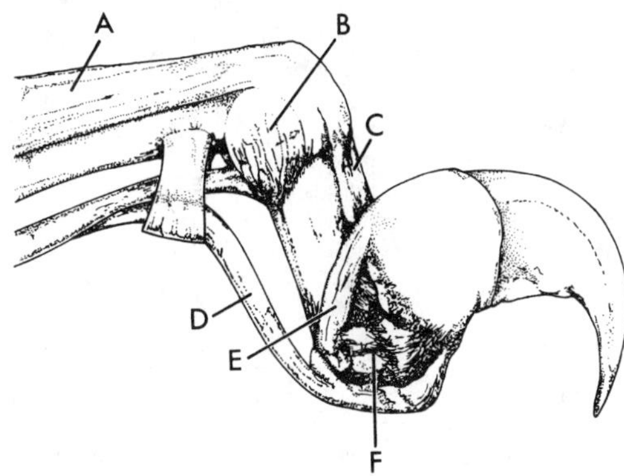

FIGURE 37.3. Retractile claw mechanism of the mountain lion (*Puma concolor*). A, Extensor expansion; B, middle interphalangeal joint; C, extensor tendon; D, flexor digitorum profundus tendon; E, lateral dorsal elastic ligament; F, distal interphalangeal joint. SOURCE: Adapted from Gonyea and Ashworth (1975).

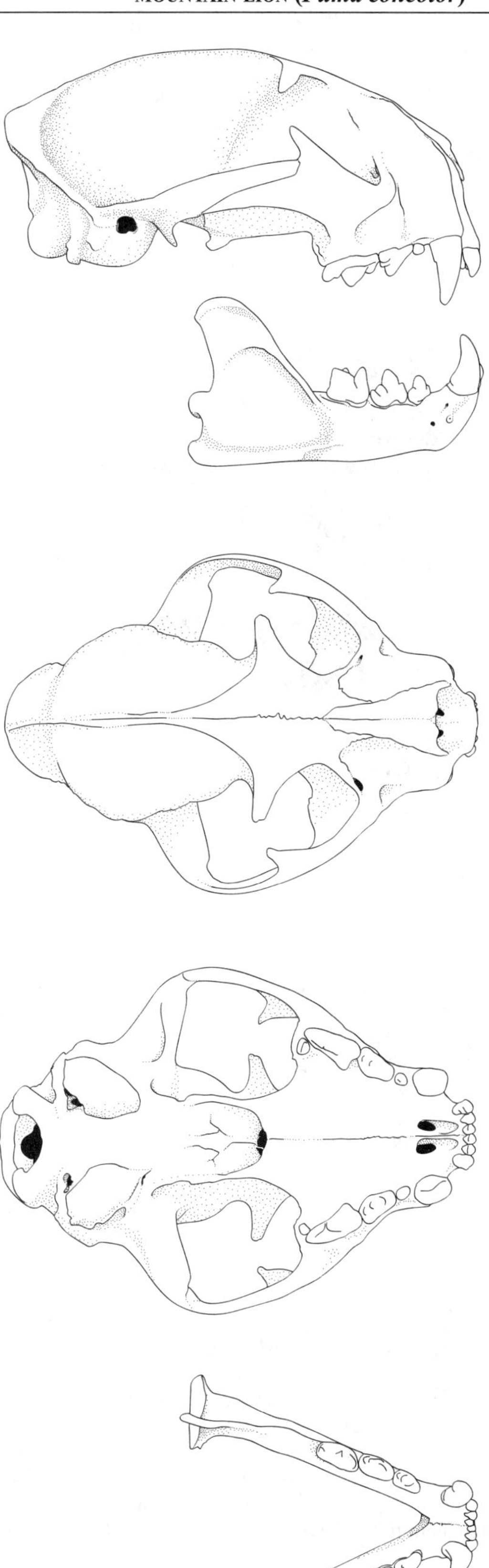

FIGURE 37.4. Skull of the mountain lion (*Puma concolor*). From top to bottom: lateral view of cranium, lateral view of mandible, dorsal view of cranium, ventral view of cranium, dorsal view of mandible.

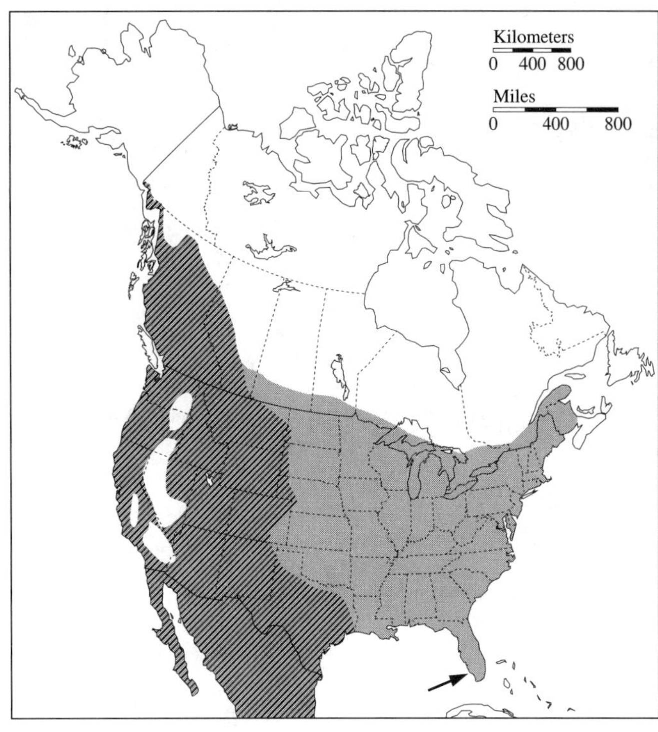

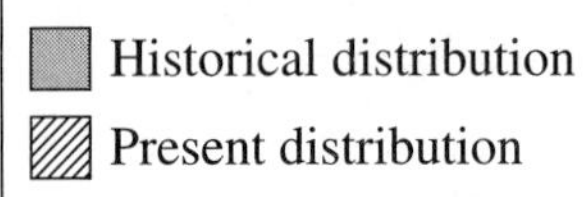

FIGURE 37.5. Past and present distribution of the mountain lion (*Puma concolor*).

(Fig. 37.5). Recently, mountain lions have been observed in southeastern Alaska and southwestern Minnesota (Nowell and Jackson 1996), and they inhabit western North Dakota (Jensen 2001) and Nebraska (Genoways and Freeman 1996) and are thought to be colonizing previously occupied habitat in Oklahoma (Pike et al. 1997). Recent evidence of a mountain lion in New Brunswick, Canada, was indisputable, but it could not be ascertained whether the animal had escaped from captivity (Cumberland and Dempsey 1994). The persistence of mountain lions in the maritime regions of Canada and the United States remains speculative (Hansen 1992). Anderson (1983) provided a detailed compilation of distributional records for mountain lions in the United States and Canada.

PHYSIOLOGY

There has been a paucity of information on blood parameters for free-ranging mountain lions. Anderson et al. (1992) sought to rectify this by providing hematological and biochemical reference values for ≤ 7 adult mountain lions from Colorado. Prior to that, information was limited to reports by Currier and Russell (1982) or Hawkey and Hart (1986). Paul-Murphy et al. (1994) provided serum biochemical reference ranges for 19 free-ranging mountain lions from throughout California, and concluded that the values were similar to those in domestic cats and captive exotic felids. Dunbar et al. (1997) described hematological and serum biochemical values for mountain lions from Florida. Most values were consistent with those established for mountain lions in Colorado (Currier and Russell 1982) and California (Paul-Murphy et al. 1994), but some differences were apparent (Table 37.1).

Dunbar et al. (1997) attributed higher packed cell volume, hemoglobin, and red blood cell values reported by Currier and Russell (1982) to the higher elevations in Colorado than in Florida. They also reported higher values for these three parameters within one "population" of Florida panthers compared to a nearby "population," and speculated that differences could be a result of generalized differences in health and nutritional status between the two groups. It may prove useful for mountain lion researchers to examine hematological and serological data from their study animals, given differences reported within localized areas as well as among widely separated populations (Dunbar et al. 1997). For example, mountain lions that use high-elevation passes to follow migratory prey (Pierce et al. 1999b) may exhibit differences that are not evident in mountain lions from the same population(s) that do not use high-elevation habitats. Further, data on hematological and serological parameters may reflect the general health status of mountain lions existing under different nutritional regimes (e.g., abundant vs. uncommon primary prey) that may occur as lion populations fluctuate with populations of mule deer (*Odocoileus hemionus*).

Mountain lions, like most cats, are well adapted to approaching prey closely and then rushing forward quickly to subdue their intended target (Guggisberg 1975). As such, they are not adapted to running long distances. Harlow et al. (1992) investigated the effects of pursuit on mountain lions. Animals responded physiologically to stresses experienced during pursuit by a depression in adrenal responsiveness. They suggested that frequent pursuit could result in deleterious physiological changes and urged caution by wildlife agencies in setting pursuit seasons until more information becomes available.

BEHAVIOR

Activity Patterns. Most felids are nocturnal predators (Kitchener 1991), and mountain lions are most active during crepuscular periods (Ackerman 1982; Hopkins 1989; Sweanor 1990). That pattern appears to shift to a more nocturnal pattern for mountain lions in the proximity of human disturbance (Van Dyke et al. 1986). Mountain lions will

TABLE 37.1. Blood serum biochemical values for free-ranging mountain lions from California ($n = 19$) and Florida ($88 \leq n \leq 94$)

Parameter	Units	Location ($\bar{X} \pm SD$)	
		California	Florida
Albumin	g/dl	3.13[a] ± 0.32[b]	3.70 ± 0.36
Alanine aminotransferase	IU/L	58.8 ± 16.7	60.2 ± 35.0
Alkaline phosphatase	IU/L	22.6 ± 11.3	35.4 ± 38.6
Aspartate aminotransferase	IU/L	—	73.4 ± 77.8
Calcium	mg/dl	9.53 ± 0.66	9.92 ± 0.66
Carbon dioxide	mEq/L	12.53 ± 1.75	14.33 ± 4.00
Cholesterol	mg/dl	155.1 ± 29.9	147.9 ± 26.7
Chloride	mEq/L	—	115.5 ± 4.3
Creatine phosphokinase	IU/L	—	515.6 ± 415.1
Creatinine	mg/dl	2.05 ± 0.45	1.84 ± 0.54
Gamma glutamine transferase	IU/L	—	1.6 ± 1.4
Globulin	g/dl	3.45 ± 0.41	—
Glucose	mg/dl	110.6 ± 37.3	154.4 ± 51.0
Inorganic phosphorus	mg/dl	5.66 ± 1.15	5.77 ± 1.51
Iron	μg/dl	—	65.1 ± 33.5
Lactate dehydrogenase	IU/L	—	269.7 ± 173.2
Potassium	mEq/L	—	4.60 ± 0.48
Sodium	mEq/L	—	152.6 ± 3.4
Total bilirubin	mg/dl	0.30 ± 0.25	0.26 ± 0.61
Total protein	g/dl	6.58 ± 0.67	7.35 ± 0.67
Triglycerides	mg/dl	—	54.9 ± 103.4
Urea nitrogen	mg/dl	32.9 ± 6.4	37.7 ± 14.1
Uric acid	mg/dl	—	0.55±0.59

SOURCE: Data have been adapted from Paul-Murphy et al. (1994) and are reported directly from Dunbar et al. (1997).

[a]The number of digits to the right of the decimal point has been reduced, by rounding, for consistency with the data reported by Dunbar et al. (1997) for mountain lions from Florida.

[b]Paul-Murphy et al. (1994) reported means and two standard deviations for serum biochemistry values. Standard deviations reported here were calculated by halving, and rounding as appropriate, data from Paul-Murphy et al. (1994) for mountain lions from California.

repeatedly move and wait in ambush during the night when hunting (Beier et al. 1995), but that behavior is suspended when cached prey is available (Beier et al. 1995; Pierce et al. 1998). Movement to prey caches at night occurs earlier for females with small young than for other social categories of mountain lions (Pierce et al. 1998). That behavior could be a strategy to avoid adult males, a response to greater energetic needs of the mother, or a tendency to kill prey closer to the daytime resting location of the young. Female mountain lions without young cover more area and make longer movements than females with young (Sweanor 1990). Males make the longest movements (Sweanor 1990; Beier et al. 1995), cover the greatest area (Seidensticker et al. 1973; Logan et al. 1986a; Pierce et al. 1998), and are more likely to leave a prey cache before it is completely consumed (Ross 1994). The more extensive movements of males likely are necessary to monitor as many females as possible and result in large home ranges.

Adult mountain lions spend a majority of time resting during a diel cycle (Beier et al. 1995). The long periods of rest exhibited by most felids may reduce energy expenditure for a species that often kills prey as large as or larger than itself and frequently has prey caches available. Mountain lions will gorge when a large amount of food is available (Danvir and Lindzey 1981; Ross 1994) and move more frequently when they are unsuccessful at making a kill (Beier et al. 1995). Mountain lions also inhabit regions of extreme temperatures: resting may be the most efficient way to prevent overheating in desert environments or conserving fat stores in extremely cold climates. The negative relationship between mass-specific metabolic rate and body size likely influences the behavior patterns of large carnivores (Robbins 1993) and accounts for long periods of inactivity by mountain lions.

Communication. Mountain lions are the largest felid that make the low rumbling noise called a "purr." Purring, chirping, and whistling vocalizations have been described for adult females when they return to their offspring (Logan et al. 1996). Defensiveness and aggression are expressed by a suite of similar behaviors shared among felids (Kitchener 1991): ears layed down in a flattened position, growling, opening the mouth wide while making a hissing sound, tail twitching, and piloerection. Female mountain lions in estrus make a loud "caterwauling" vocalization to advertise their condition to males, and this behavior continues throughout any resulting mating association (Rabb 1959; Padley 1991; Beier et al. 1995; Logan et al. 1996).

Home range areas of mountain lions are marked by scrapes made by pawing the ground in a backward motion, drawing dirt and ground litter into a small pile behind two parallel grooves (Seidensticker et al. 1973). Males make scrapes more frequently than females (Cunningham et al. 1995), and scrapes often have feces or urine deposits associated with them (Smith 1981; Logan et al. 1996). Scrapes occur most frequently in locations where topography funnels a number of mountain lions into the same area (Smith 1981) and may facilitate mutual avoidance among individuals with overlapping home ranges (Hornocker 1969). Mountain lions display a Flehman response when smelling urine deposits in scrapes (Seidensticker et al. 1973). Females may scrape only when in estrus (Maehr 1997) and, if so, such scrapes may be used by males to monitor the reproductive cycles of females as well as home range occupancy. Mountain lions also scratch logs and trees repeatedly with their front claws. Scratch marks may serve the purpose of marking home range areas (Schaller 1972), but may simply be for sharpening claws (Seidensticker et al. 1973). In addition, felids have glands in the cheek, which often are rubbed against objects and family members as another means of communication (Macdonald 1985).

Hunting and Feeding. Mountain lions can jump vertically ≥3 m (Anderson 1983) and are capable tree climbers (Hornocker 1970). When pursued by canids, mountain lions may seek refuge in trees or rocks (Hornocker 1970; Seidensticker et al. 1973; Davis et al. 1996). Most reports of mountain lions ambushing prey have described them stalking from the ground (Koford 1946; Hornocker 1970; Wilson 1984; Beier et al. 1995) with very few instances of attacking prey from an elevated perch (Connolly 1949). The effect of hunting by humans on the behavior of mountain lions is unknown. Pursuit of mountain lions with hounds for research does not appear to cause changes in home range use or abandonment of prey caches in most instances (Seidensticker et al. 1973; Maehr 1997; Pierce et al. 1998, 1999b). Mountain lions remain close to human activity directly after being pursued, captured, and then released for research purposes (Seidensticker et al. 1973; Beier et al. 1995). However, they also may avoid areas where they are repeatedly harassed (Hebert and Lay 1997; Janis and Clark 2002).

Mountain lions move quietly and low to the ground when hunting and stalking prey (Koford 1946; Wilson 1984; Bank and Franklin 1998). Ears also may be lowered to reduce detection while they move slowly and deliberately, frequently freezing in position while waiting to approach closely enough to ambush the intended prey (Leyhausen 1979). When prey is within a suitable distance, they rush at the target. Pursuit of prey by mountain lions is relatively short compared to the long chases of coursing predators, such as wolves (*Canis lupus*). If prey is caught, it usually is attacked at or near the neck. When attacking ungulates, mountain lions most frequently grip the anterior ventral portion of the neck, delivering a crushing bite to the trachea and causing suffocation (Kitchener 1991). The dorsal portion of the neck also may be bitten, and the spinal cord can be severed. After killing prey, it is usual for mountain lions to open the body cavity of the carcass and remove the digestive tract (Hornocker 1970). Other internal organs including the heart, liver, and lungs often are consumed first (Danvir and Lindzey 1981). Gorging on the most nutritious tissues during the first day may be an important strategy for mountain lions that can lose food to scavengers (Danvir and Lindzey 1981) or need to move to locate mates (Ross 1994).

In an examination of hunter-killed mountain lions, Robinette et al. (1959) reported that 30% had empty stomachs. Mountain lions may consume >10 kg of meat during a single feeding (Hornocker 1970; Danvir and Lindzey 1981; Ackerman et al. 1986), with more being consumed during the first day than on consecutive days following the kill (Danvir and Lindzey 1981). Results of field sampling and trials with captive animals suggest that mountain lions consume an average of 4.4 kg of meat/day (Danvir and Lindzey 1981; Ackerman et al. 1986). In most instances, prey carcasses are dragged, sometimes >200 m, to locations with vegetative or topographic cover (Beier et al. 1995). When lions finish feeding, they usually conceal the prey by covering it with sticks, grass, and other material until the carcass is consumed or abandoned. Mountain lions normally consume 73–79% of a deer carcass (Danvir and Lindzey 1981; Ackerman 1982). Covering caches may slow decomposition of the meat and hide it from scavengers (Beier et al. 1995). Prey caches may be guarded to protect them from scavengers, but mountain lions can use day-beds several kilometers from kill sites, returning only at night to feed (Beier et al. 1995; Pierce et al. 1998, 2000b).

REPRODUCTION

Mountain lions are polygynous and males do not contribute to rearing young. Like all felids, except African lions (*Panthera leo*), adult mountain lions are solitary except when raising young, dispersing with siblings, or mating. Males tend to have relatively large home ranges, which overlap with those of more than one female (Seidensticker et al. 1973; Logan et al. 1986a; Pierce et al. 1999b, 2000b). Male and female mountain lions likely rely on auditory and olfactory signals to locate each other for mating (Currier 1983) (see Behavior).

Estrous cycles in females with overlapping home ranges may be synchronous (Padley 1990). Associations for mating generally last 2–5 days (Beier et al. 1995), during which time the pair may copulate up to 70 times a day (Eaton 1976). Female estrous cycles last approximately 8 days (Rabb 1959; Eaton and Velander 1977) and gestation is 82–96 days (Young and Goldman 1946; Anderson 1983; Currier 1983). Mountain lions may reproduce during any season (Robinette et al. 1961; Ashman et al. 1983), but many studies have identified seasonal birth pulses (Robinette et al. 1961; Ashman et al. 1983; Barnhurst and Lindzey 1984; Logan et al. 1986a; Ross 1994). Timing of reproduction in mountain lions may be affected by climate or prey abundance (Logan et al. 1996; Pierce et al. 2000a).

Females have eight teats, but only six produce milk (Lechleitner 1969). The number of kittens born per litter typically ranges from two to four (Robinette et al. 1961; Ashman et al. 1983; Ross and Jalkotzy 1992; Logan et al. 1996), but this may vary with prey availability, demography, and population density. Females may average more offspring per litter during their first year of reproduction and fewer toward the end of their lives (Logan et al. 1996). Most studies of mountain lions have reported an equal sex ratio among young (Donaldson 1975; Anderson et al. 1992; Logan et al. 1996). Padley (1991), however, reported that for a population of mountain lions in California, with 12 adult females and seven litters, six of seven kittens whose sex he was able to determine were male. In that population, adult males were scarce, and estrous females were unable to locate an adult male with which to mate for almost a year.

Kittens are brought to kill sites by their mother as early as 8 weeks of age, but are not weaned until they are 2–3 months old. Canine teeth first appear between 20 and 30 days and molars at about 40 days of age. Some permanent teeth erupt at 5–6 months of age and permanent canines appear during the eighth month, when primary and permanent canines are present. At weaning, kittens may weigh 3–4 kg (Currier 1983) with little variation in weight between the sexes until about 30 weeks of age (Robinette et al. 1961).

Mountain lion dens generally are located in rocky terrain (Logan et al. 1996) or thick vegetation (Beier et al. 1995; Bleich et al. 1996). Dens are not elaborate, but do provide hiding cover (Ross 1994; Beier et al. 1995) and thermal benefits (Bleich et al. 1996). Females with kittens have restricted movement patterns until the young are old enough to travel away from the den (Beier et al. 1995). A mother may move her young to several different dens before they are weaned (Shaw 1989).

ECOLOGY

Habitat. The broad distribution of mountain lions attests to their ability to adapt to a wide variety of habitats. Because of their wide distribution (see Fig. 37.5), defining preferred habitat for mountain lions is limited to general characteristics specific to distinct regions, and tremendous variation occurs among populations. Despite the large range of habitats used by mountain lions, relatively few quantitative studies of habitat selection exist; many studies present descriptive information without reference to habitat availability. For those studies that quantified habitat selection, prey availability, vegetation, and topography are the general characteristics that determine suitability for mountain lions. Within regions where a variety of habitat types occurs, structure of the vegetation and topography appear to be the most important criteria for determining habitat use (Lindzey 1987).

Most studies of habitat selection for mountain lions in western North America suggest that vegetative and topographic cover, in addition to steep slopes and higher elevations, are preferred resting, hunting, and denning sites, whereas open agricultural lands, sagebrush (*Artemesia tridentata*) grasslands, and open meadows and pastures are avoided (Murphy 1983; Logan and Irwin 1985; Laing 1988; Lindzey et al. 1989). Ashman et al. (1983) suggested that thermal characteristics cause mountain lions to select north-facing slopes at high elevations, with more vegetation and cooler temperatures, in the summer. South-facing slopes with little snow cover were selected in winter. Those habitats were strongly correlated with the density of deer. Rugged or steep terrain, with adequate vegetative cover, may be critical for stalking and catching prey (Ashman et al. 1983; Logan and Irwin 1985; Laing 1988) or for providing hiding cover and protection from thermal extremes for young at denning sites (Belden et al. 1988; Ross 1994; Bleich et al. 1996).

Dense hiding cover is preferred by mountain lions for hunting (Russell 1978; Beier et al. 1995). Indeed, mountain lions were more successful hunting pronghorn (*Antilocapra americana*) that inhabited rugged terrain with more vegetation than pronghorn that occurred in open prairie habitat (Ockenfels 1994). Mountain lions, however, can inhabit open or sparsely vegetated habitats, such as the plateaus of Patagonia (Wilson 1984; Bank and Franklin 1998) and the deserts of the southwestern United States (Cunningham et al. 1995; Logan and Sweanor 2001), by successfully using the limited available cover to catch prey. In addition, Pierce (1999) noted that mountain lions were more successful killing mule deer in relatively open habitat compared to that in which deer foraged. The pursuit of deer by mountain lions likely began in areas with vegetative cover, but deer that fled into more open areas without obstacles during the chase were more likely to be caught than those that remained in heavier vegetation.

Mountain lions can thrive in extremely dry climates with limited rainfall (Shaw et al. 1987; Cunningham et al. 1995; Logan et al. 1996; Pierce et al. 2000a); however, the severely cold temperatures of northern Canada and Alaska may limit their distribution despite the availability of ungulate prey. Mountain lions may avoid human disturbance when possible (Van Dyke et al. 1986), but can persist near human development (Beier et al. 1995; Torres et al. 1996).

Laing (1988) suggested that habitat use by mountain lions did not differ significantly during a 24-hr cycle, but that pattern may depend on the availability of habitats. When habitat used by mountain lions for denning or for caching prey is different than that used for resting, then they will use different habitats during a diel cycle. Beier et al. (1995) reported that mountain lions with young hunted throughout the night, but returned to the den during the day. Den sites were located in nearly impenetrable vegetation. Pierce et al. (1998) found that mountain lions fed on cached prey primarily after sunset and often rested long distances from the cache site during the day. In addition, Beier et al. (1995) reported that, in some instances, mountain lions rested >4 km from their cached prey. This behavior also was consistent with the reported use of high-elevation cliffs during the daytime and use of lower elevations with flatter terrain for hunting deer at night (Pierce et al. 1998, 2000b).

Several studies have described seasonal changes in elevation and habitat use associated with changes in home range for mountain lions preying on migrating populations of deer and elk (*Cervus elaphus*) (Rasmussen 1941; Seidensticker et al. 1973; Ackerman 1982; Ashman et al. 1983; Murphy 1983; Anderson et al. 1992; Pierce et al. 1999b). Indeed, differences in forage availability associated with seasonal ranges of prey leads to potential differences in the tradeoff between foraging benefit and predation risk for deer and elk. Consequently, there may exist differences in the effects of mountain lion predation on the seasonal foraging behavior and habitat selection of those large ungulates (Pierce 1999).

Dispersal. Young remain with their mother for 12–18 months. Dispersal may result from the mother abandoning the young (Seidensticker et al. 1973), but it is likely that there is some aggression by the mother directed toward the young to prevent them from following her (Hansen 1992). Dispersal of young from the mother is coincident with the female coming into estrous; adult males will kill young (Young and Goldman 1946; Ackerman et al. 1984; Spreadbury 1989; Logan and Sweanor 2001). During the period of dispersal, mountain lions are often referred to as "transients" (Hornocker 1970). Transients, as opposed to "residents," are individuals that do not have a defined home range area within a population.

Dispersal is a relatively risky period during the life cycle for many species (Baker 1978), and survivorship likely declines for mountain lions during the dispersal period (Logan and Sweanor 2001). Transients are more likely than resident mountain lions to be involved in depredation incidents or conflicts with humans, as they attempt to locate food without the advantage of an established home range (Torres et al. 1996). Males are more likely to disperse from their natal ranges than are females, and males tend to disperse further (Anderson et al. 1992; Sweanor et al. 2000).

Mountain lions can disperse nearly 500 km from their natal home range (Logan and Sweanor 1999). Consequently, mountain lion habitat throughout North America is considered almost contiguous on a large scale, with the exception of the isolated populations in southeastern Florida. Nevertheless, mountain lion populations can be isolated by residential development or geographic features such as desert basins. Reduction in the size of habitat "islands" for populations of any species increases the chance of extirpation for that population (MacArthur and Wilson 1967).

Sweanor et al. (2000) described a metapopulation structure (Levins 1970) for mountain lions living in basin and range habitats of the Southwest. Mountain ranges provide islands of suitable habitat that support semi-isolated populations of mountain lions. Those populations are maintained through immigration by dispersers (primarily male) from other mountain ranges and recruits (primarily female) from within the population. The basins between the mountains are not considered suitable mountain lion habitat (Germaine et al. 2000) and act as barriers that may limit gene flow. Protecting source populations and habitat stepping stones may be exceptionally important in conserving subpopulations of mountain lions that exist in a metapopulation structure (Sweanor et al. 2000). Although concerns about habitat fragmentation and interruption of gene flow are legitimate, Ernest et al. (2003) have cautioned that it has not been demonstrated that mountain lions conform to the assumptions inherent in basic metapopulation models.

Survivorship of adults likely varies within most populations on an annual basis (Lindzey et al. 1988), but can be affected significantly by hunting (Anderson 1983). Destruction of habitat also can have large impacts on the demography of mountain lion populations. For example, fragmentation of habitat can inhibit dispersal and increase competition for resources (Beier et al. 1995; Maehr 1997). Furthermore, real estate development may limit prey populations, increase depredation events (Shaw 1980; Cunningham et al. 1995; Torres et al. 1996), and increase collisions with automobiles (Maehr 1997; Beier et al. 1995).

Disease and Parasites. Anderson (1983) suggested that the widely held opinion that mountain lions are relatively free of parasites and diseases reflected a lack of specific research rather than reality. For example, based on his literature review, Anderson (1983) reported the occurrence of feline panleukopenia virus was low, but Paul-Murphy et al. (1994) detected titers to the virus in nearly 100% of the animals that they tested. Furthermore, Adaska (1999) reported the first case of coccidioidomycosis detected in a mountain lion, and Jessup et al. (1993) documented the first occurrence of feline leukemia virus infection in a free-ranging mountain lion. Evermann et al. (1997) reported on the occurrence of puma lentivirus specifically in mountain lions in the state of Washington. Yamamoto et al. (1998) found overall seroprevalance of bartonellosis in 26 of 74 mountain lions they examined from California, among the first evidence of this condition detected in wild felids. Hence, it appears that Anderson's (1983) admonition was insightful.

Mountain lions potentially are susceptible to many infectious agents that affect domestic cats (Paul-Murphy et al. 1994) as well as to other diseases (summarized by Anderson 1983). Some of these, including feline panleukopenia virus, have the potential to affect morbidity and mortality (Paul-Murphy et al. 1994), may limit growth of wild populations (Anderson 1983), and have implications for the conservation and persistence of a viable population of the endangered Florida panther (Roelke et al. 1993). Foley (1997) concluded that epizootic diseases likely are not a primary threat to populations of mountain lions in the western United States. From a human health perspective, rabies has been detected among mountain lions (Storer 1923) and should be of concern in all cases involving attacks on humans (Kadesky et al. 1998).

Parasitic infections among mountain lions are not uncommon; Anderson (1983) listed $\geq$40 species of parasites that have been collected. Forrester et al. (1985) and Waid (1990) listed additional internal parasites not noted by Anderson (1983). The majority of parasites recorded have been internal (protozoans, trematodes, cestodes, or nematodes), but mites, ticks, and insects are known to parasitize mountain lions. Pence et al. (1987) noted a high degree of overdispersion (i.e., a few mountain lions harbored the majority of parasites collected and most had few or no parasites) in helminth species collected from the viscera of mountain lions in Texas. In general, mountain lions have been described as remarkably free of external parasites (Currier 1983), but our experience has been that ticks are commonly encountered during handling.

FEEDING HABITS

Mountain lions kill and eat vertebrate prey almost exclusively (Lindzey 1987). Vegetation often is ingested because it adhered to a carcass or to help with passage of parasites and hair from the gut (Robinette et al. 1959; Anderson 1983). The historical distribution of mountain lions throughout North America coincided with the distribution of their primary prey—deer. Most studies in North America identified deer as the most frequent prey in diets of mountain lions (Hornocker 1970; Shaw 1980; Ackerman 1982; Logan et al. 1996; Ross et al. 1997; Pierce et al. 2000a).

Despite the congruence of deer and mountain lion distributions in North America, lions are generalist predators (Anderson et al. 1992; Logan and Sweanor 1999). They do prey on other large ungulates including moose (*Alces alces;* Ross and Jalkotzy 1996), elk (Hornocker 1970), feral horses (*Equus caballus;* Turner et al. 1992), feral pigs (*Sus scrofa;* Harveson 1997; Sweitzer 1998), and wild sheep (*Ovis canadensis;* Cronemiller 1948; Harrison 1990; Wehausen 1996; Ross et al. 1997; Schaefer et al. 2000). Mountain lions also feed on smaller prey such as lagomorphs (Shaw 1980; Ackerman et al. 1984); ground squirrels (*Spermophilus* spp.; Seidensticker et al. 1973); beavers (*Castor canadensis;* Padley 1991); porcupines (*Erethizon dorsatum;* Sweitzer et al. 1997); small carnivores including raccoons (*Procyon lotor*), gray foxes (*Urocyon cinereoargenteus*), and skunks (Cashman et al. 1992; Beier et al. 1995); and birds such as wild turkeys (*Meleagris gallopavo*), grouse, and quail (Ashman et al. 1983; Ackerman et al. 1986; Maehr et al. 1990). Currier (1983) listed 41 nondomestic species of vertebrates identified in the feces of mountain lions from North and South America. Domestic sheep, goats, cattle, horses, dogs, and cats also are consumed when populations of mountain lions are in proximity to livestock operations or areas of human habitation (Shaw 1980; Torres et al. 1996). Attacks on humans by mountain lions are rare (Beier 1991); however, in most instances, victims of fatal attacks were treated as prey and either dragged to cover or partially consumed (Beier 1991; J. Banks, California Department of Fish and Game, pers. commun., 1996). Remains of bobcats (*Lynx rufus*), coyotes (*Canis latrans*), and other competing carnivores occasionally occur in feces of mountain lions. Also, species that scavenge on lion kills, including eagles and turkey vultures (*Cathartes aura*), can be killed by mountain lions guarding a prey cache. Although intraspecific aggression leading to death in mountain lions may occur as a result of competition for resources or mating opportunities (Robinette et al. 1961; Hornocker 1970; Sweanor 1990; Pierce et al. 1998), instances of cannibalism by mountain lions, in which the carcass has been largely consumed, have been reported frequently (Lesowski 1963; Donaldson 1975; Ackerman et al. 1984; Spreadbury 1989).

Selection of Prey. Numerous authors have described the sex, age class, and condition of prey killed by mountain lions. Hornocker (1970), Spalding and Lesowski (1971), Shaw (1977), Ackerman (1982), Ackerman et al. (1984), and Murphy (1998) all suggested that vulnerability of individual prey may be the most important factor in their selection by mountain lions. Those investigations identified individuals in younger or older age classes or in poorer condition as being selected by mountain lions. Among adult deer and bighorn sheep, mountain lions also may prey on males selectively (Hornocker 1970; Ackerman 1982; Harrison 1990), especially when males are in a weakened condition following the rut (Robinette et al. 1959; Shaw 1977; Harrison 1990) or during drought conditions (Logan and Sweanor 2001). Studies of prey selection, however, require comparisons with availability of prey. Most studies of prey selection by mountain lions have been limited by lack of information on availability of prey or by biases in detecting prey.

Anderson (1983) attempted to compare sex and age ratios of deer killed by mountain lions to the estimated ratios in the populations of deer for six study areas, and concluded that the sampling methods used in those studies prevented clear testing of any hypothesis regarding prey selection. Since then, Ross et al. (1997) reported selection for young bighorn sheep by mountain lions. Pierce et al. (2000a) reported that mountain lions selected deer in young and old age classes, and that females were selected among adult deer. Preference for females among adult mule deer also has been reported for other populations of mountain lions (Bleich and Taylor 1998). More recent investigations suggested that mountain lions do not select prey in poor condition (Kunkel et al. 1999; Pierce et al. 2000a); however, bone marrow fat has been the primary index used to determine prey condition by most investigators. Although percentage bone marrow fat was not related to prey selection by mountain lions (Kunkel et al. 1999; Pierce et al. 2000a), they selected for older animals. Thus, percentage marrow fat may not adequately reflect other forms of weakness detected by mountain lions (Pierce et al. 2000a).

Kunkel et al. (1999) determined that mountain lions killed deer with shorter diastemae than those killed by wolves or hunters, but values of marrow fat did not differ between deer killed by mountain lions and wolves. Size of prey also appears to affect selection by mountain lions. They selected deer over elk, and elk over moose in Montana (Kunkel et al. 1999). Ross and Jalkotzy (1996) reported that male mountain lions were more likely to kill moose than were females. Body size, however, did not appear to be important for selection of mule deer preyed on by mountain lions when compared with those killed by coyotes during winter, when young deer were unavailable (Pierce et al. 2000a).

Sex and age of mountain lions likely affect their diet. Solitary mountain lions may be more likely to eat smaller prey than do females with kittens (Ackerman 1982). Ackerman suggested that killing large prey to provide for their offspring would be a necessary strategy for mothers, and that populations of mountain lions could not exist in areas devoid of large ungulate prey. Pierce et al. (2000a), however, reported that female mountain lions with kittens (≤6 months old) were significantly more likely to kill young deer (<1 year of age) than were single adult females or males. Birth pulses of mountain lion populations often coincide with the birth pulse of their primary prey (Logan and Sweanor 2001; Pierce et al. 2000a), suggesting that timing of reproduction in mountain lions may be dependent on the availability of vulnerable, young prey. Adult females that are lactating may not be able to fast for extended periods between unsuccessful attempts to catch larger prey (Pierce et al. 2000a).

Latitude correlates strongly with size of prey selected by mountain lions (Iriarte et al. 1990; Maehr et al. 1990). In temperate zones, they tend to kill larger ungulates, whereas mountain lions in tropical environments have a higher frequency of small prey in their diet. The complicating effect, however, of competition with the jaguar on this pattern of latitudinal prey variation is unknown. Additionally, variation in prey selection is notable for some populations of mountain lions as vulnerability or availability of prey changes. During the wet season in coastal California, mountain lions increased predation rates on feral pigs (Craig 1986) and, in Utah, they killed twice as many black-tailed jackrabbits (*Lepus californicus*) during winter (Ackerman et al. 1984) than during summer. Adult male deer may be taken at a higher rate during winter when they are in weakened condition from the rut (Robinette et al. 1959; Shaw 1977). The vulnerability of newborn calves results in high depredation rates on cattle in Arizona (Shaw 1977; Cunningham et al. 1995).

POPULATION DENSITY AND DYNAMICS

Mountain lions coexist in a system of individual home ranges with varying amounts of overlap (Seidensticker et al. 1973; Hemker et al. 1984). Mean home range size and distribution of mountain lions can be affected by sex (Seidensticker et al. 1973) or availability of resources such as prey (Pierce et al. 1999b, 2000a). Anderson et al. (1992), however, concluded there was no relationship between prey density and frequency of use by mountain lions of areas inhabited by deer and elk. They felt that other factors, such as suitable habitat for hunting, may affect prey availability independently of prey density (Kruuk 1986) and may ultimately affect the home range size necessary for mountain lions to be successful.

Home ranges of mountain lions are delineated by visual and olfactory marking behaviors, and adult males tend to have larger home ranges than females (Seidensticker et al. 1973; Logan et al. 1986a; Pierce et al. 1998) (Table 37.2) (see Behavior). Male home ranges often overlap those of several females but have limited overlap with those of other males (Seidensticker et al. 1973; Murphy 1983; Logan and Sweanor 2001), whereas the home ranges of adult females often overlap those of other females extensively (Seidensticker et al. 1973; Sweanor et al. 1996; Pierce et al. 1999b, 2000b; Logan and Sweanor 2001). Extensive overlap among male home ranges (Anderson et al. 1992; Sweanor 1990) and limited overlap of female home ranges (Harrison 1990) also have been reported. Additionally, mountain lions may avoid each other temporarily, a behavior termed "mutual avoidance" (Hornocker 1969), but no investigations have tested for the dominance reversal necessary to demonstrate true territoriality in mountain lions (Kitchen 1974).

Hornocker (1969, 1970) and Seidensticker et al. (1973) suggested that mountain lion populations were self regulating (see Predator–Prey Dynamics). They described social behaviors such as mutual avoidance, territorial marking, and cannibalism as mechanisms that limited densities of mountain lions. Transient mountain lions were unable to secure permanent home range areas unless the home range area of a resident adult became vacant. This pattern occurred independently of prey densities. Lindzey et al. (1994) also concluded that the density of a mountain

TABLE 37.2. Estimated home range size and average density of mountain lions from North American studies using the minimum convex polygon method (Mohr 1947)

Study	Location	Males		Females		Lions/100 km^2
		n	Range (km^2)	n	Range (km^2)	
Ross and Jalkotzy 1992	Alberta	6	334	21	140	1.9 (RA, H), 4.2 (T)
Cunningham et al. 1995	Arizona	5	196	2	109	— (H)
Beier and Barrett 1993	California	2	767	12	218	1.1 (RA)
Hopkins 1989	California	4	199	7	84	3.6 (T)[a]
Anderson et al. 1992	Colorado	6	256	7	309	1.1 (RA, H)
Maehr et al. 1991	Florida	8	558	10	191	—
Seidensticker et al. 1973	Idaho	1	453	3	233	0.6 (RA[b], H)
Ashman et al. 1983	Nevada	7	574	6	178	— (H)
Logan et al. 1996	New Mexico	23	187	29	74	1.7 (RA)
Murphy 1983	Montana	2	462	5	202	7.1 (RA, H)
Pittman et al. 2000	Texas	6	349	5	206	0.4 (RA, H)
Pence et al. 1987	Texas	1	628	5	143	—
Hemker et al. 1984	Utah	1	826	4	685	0.4 (T)
Logan 1983	Wyoming	2	320	2	73	1.5 (RA, H), 3.9 (T)

NOTE: RA, Density of resident adults; T, density of total population; H, hunted population.
[a]From Anderson et al. (1992), Table 23.
[b]Estimated from Seidensticker et al. (1973), Table 6, Fig. 16.

lion population in Utah was independent of the density of their primary prey, mule deer, because the population of mountain lions did not increase at the rate predicted by the increase in prey. Nonetheless, the population of mountain lions did increase by one third after an increase in the deer population. Long time lags between fluctuations in prey populations and responses in mountain lion populations make determining the true relationship between predator and prey difficult (Schaller 1972), and emphasize the need for long-term research when working with large carnivores (Pierce 1999).

To invoke social regulation for a population (Hornocker 1969, 1970; Seidensticker et al. 1973), it must be demonstrated that the population is not limited by competition for resources (Watson and Moss 1970). Mutual avoidance does not prevent passive competition for resources. Pierce et al. (2000b) demonstrated that competition for prey by mountain lions occurred in a pattern that would be expected for nonterritorial species that do not exclude each other from prey resources. Furthermore, if mountain lions engaged in a territorial system that excluded individuals from resources beyond what would be expected for a population limited by competition, such costly behavior could have evolved only if it maximized reproduction for individuals that were territorial and, ultimately, would not lead to population limitation (McCullough 1979). Hemker et al. (1984) concluded that population size of mountain lions was primarily limited by the population of prey, a conclusion supported by Pierce et al. (2000b) and Logan and Sweanor (2001).

The density of males may be limited by intraspecific aggression to maximize access to females, but mountain lion populations on a whole likely are not limited by social behavior. In addition, dispersal by young mountain lions appears to be independent of population density (Seidensticker et al. 1973; Hemker et al. 1984; Ross and Jalkotzy 1992; Logan et al. 1996), and mountain lion population numbers parallel their prey populations (Lindzey et al. 1994; Logan et al. 1996; Cox and Stiver 1997) (Fig. 37.6). Furthermore, Logan and Sweanor (2001) reported a pattern of density-dependent growth for a mountain lion population, providing additional evidence that competition is an important factor affecting their population dynamics.

PREDATOR–PREY DYNAMICS

Few long-term studies of predator–prey dynamics have been conducted that include mountain lions. Unfortunately, many studies have

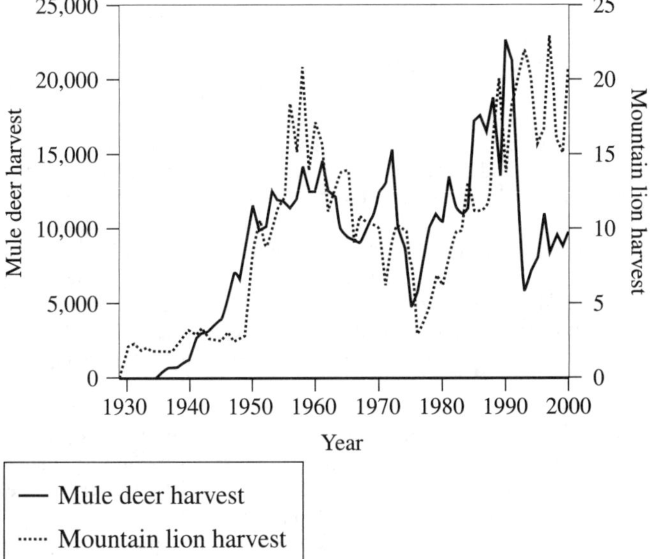

FIGURE 37.6. Harvest of mule deer (*Odocoileus hemionus*) and mountain lions (*Puma concolor*) in Nevada. Mule deer harvest shown includes males only. Mountain lion harvest shown includes depredation removals and sport hunting (sport hunting estimated to 1967). SOURCE: Data from S. Stiver, Nevada Department of Wildlife, pers. commun., 2001.

independently focused on either the predator or the prey, thereby limiting conclusions that can be drawn about interspecific effects (Pierce 1999). In most instances, viable populations of mountain lions are dependent on populations of large ungulate prey (Ackerman 1982; Anderson 1983). Most ungulates that are preyed on by mountain lions, including deer, elk, and moose, are affected by density-dependent processes (McCullough 1979; Boyce 1989; Bowyer et al. 1999; Keech et al. 2000). Therefore, determining the effect of predation by mountain lions on the population dynamics of their prey requires an understanding of those density-dependent processes and all factors potentially limiting the population including weather, resources, and other predators.

In the ecological systems that mountain lions inhabit, both exploitative and interference competition with other predators likely occur. Niches of wolves (Kunkel et al. 1999), coyotes (Pierce et al. 2000a; Harrison 1990), bears (Murphy 1998), bobcats (Koehler and Hornocker 1991), and jaguars (Iriarte et al. 1990) overlap extensively with that of mountain lions. Niche partitioning via habitat use and prey selection may allow for coexistence of mountain lions with those other carnivores. In some instances, scavenging of their food caches requires mountain lions to increase predation rates (Harrison 1990). In addition, mountain lions often kill smaller species of carnivores, such as bobcats, without feeding on them (Koehler and Hornocker 1991). Conversely, mountain lions have been killed by packs of wolves (White and Boyd 1989; Boyd and Neale 1992). Those observations also suggest that competition is an important factor affecting the predator–prey dynamics of multipredator systems that include mountain lions.

Mountain lions are an important limiting factor for some ungulate species (Hornocker 1970; Shaw 1980). Shaw (1980) estimated that lions annually removed 15–20% of the mule deer population on the Kaibab Plateau in Arizona. Anderson et al. (1992) estimated that mountain lions annually killed 8–12% of the mule deer population on the Uncompahgre Plateau, Colorado. In Alberta, a single mountain lion killed 8.7% ($n = 11$) of the early-winter bighorn sheep population and 26.1% ($n = 6$) of the young (Ross et al. 1997).

Mountain lion predation may reduce the severity of fluctuations in prey populations (Hornocker 1970) by slowing the growth rate and reducing overshoots of resource-based carrying capacity (K). Limitation and regulation, however, are distinctly different mechanisms. Limitation implies that the number of prey is reduced through decreased production or increased loss, and therefore any source of predation is limiting (Sinclair 1989; Boutin 1992). Limitation slows the increase of a population from the growth rate it would have achieved in the absence of the predator. However, if the prey population continues to increase, the proportion of prey removed by the predator is reduced. Consequently, prey populations limited by predators can continue to grow, although at a slower rate than in the absence of predation, until they reach K.

Regulation implies that rate of removal by the predator changes with the population of prey; as prey numbers increase, so does the rate at which predators kill prey. This phenomenon results from an increase in the number of predators (numerical response), the ease with which they find and kill prey (functional response), or a combination of both (Holling 1959). Regulation by a predator eventually stops the increase of the prey population and causes a decline to lower densities, which ultimately results in lower predation rates. Such feedback mechanisms can cause a prey population to fluctuate between low- and high-density equilibria, both of which are below K. This multiple-equilibria scenario is frequently termed a "predator pit" (Haber 1977; Bergerud et al. 1983; Messier and Crete 1985). Functional responses of the predator may ultimately determine whether a species has the capacity to regulate a prey population (Messier 1994). Nonetheless, determining the functional response curve for a species requires an unbiased measure of predation rate throughout the range of population densities of the prey, and controlled removal studies are necessary to determine the conditions that may cause multiple equilibria (i.e., regulation) to occur in predator–prey systems (Van Ballenberghe and Ballard 1994).

Attempts to measure predation rates of mountain lions have been made for static populations of prey. Using an energetics model, Ackerman et al. (1986) estimated that adult male mountain lions would kill a deer every 8–11 days, adult females would kill every 14–17 days, and a female with three juveniles would kill every 3.3 days. Connolly (1949) reported that an adult mountain lion killed one deer every 9.7 days. Predation rates on bighorn sheep and mule deer by females with kittens ranged from 0.7 to 2.1 ungulates/week (Harrison 1990). Beier et al. (1995) determined that mountain lions killed 48 large and 58 small mammals/year. Rate of predation by mountain lions can vary extensively (Ross and Jalkotzy 1996), and is dependent on a number of factors including the number and age of young (Ackerman et al. 1986), season (Pierce 1999), and extent of competition from other predators (Kunkel et al. 1999) or scavengers (Harrison 1990). Those studies give insight into the behavior of mountain lions and their possible effects on prey populations, but the question of whether mountain lions can independently regulate a prey population remains unanswered.

Mountain lions have been implicated as regulating a wild horse population (Turner et al. 1992), and are thought to have caused populations of porcupines to decline through predation (Sweitzer et al. 1997). In addition, some bighorn sheep populations may have declined due to avoidance of high-quality forage in response to predation risk by mountain lions (Wehausen 1996). In those instances, mule deer were the primary prey for mountain lions. Systems with multiple predators and multiple prey are more likely to be regulated by predation and to produce predator pits (Gasaway et al. 1992). Mule deer populations can keep mountain lion numbers high. If populations of alternative prey increase, mountain lions encounter those species more frequently and predation rates may increase, causing alternative prey to decline. Bleich et al. (1997) noted, however, that unless female mountain sheep were killed in significant proportions, predation by mountain lions probably is not an important factor in the population dynamics of those ungulates.

Several studies examining the population dynamics of deer and mountain lions suggest that mountain lions do not determine the ultimate size of deer populations. Instead, other factors (especially forage resources) appear to be more important determinants of deer populations. For example, Logan and Sweanor (2001) determined that mountain lions were not responsible for the decline of mule deer in New Mexico, but that a lack of forage resources because of drought was responsible. Pierce et al. (1999a) came to a similar conclusion for a population of mule deer in the eastern Sierra Nevada of California. In that study, mountain lion numbers declined precipitously as did the deer numbers, but with a lag of 8 years. Long time lags between changes in prey populations and responses in mountain lion populations increase the difficulty of determining the relationships between mountain lion populations and prey populations (Schaller 1972), and emphasize the need for long-term studies (Pierce 1999).

MANAGEMENT AND CONSERVATION

Management Techniques. Demographic information is difficult to obtain for populations of large, cryptic predators like mountain lions. Track surveys have been a traditional method of determining presence or absence of mountain lions as well as establishing estimates of relative abundance. Various investigators have attempted to distinguish among lions using track characteristics unique to individual animals (Currier et al. 1977; Fitzhugh and Gorenzel 1985; Van Dyke et al. 1986) or track measurements (Currier et al. 1977; Shaw 1983; Fitzhugh and Gorenzel 1985) to recognize individuals. None of these investigators believed it was reliable to distinguish among individual mountain lions using only tracks.

Smallwood and Fitzhugh (1993) used a sophisticated statistical technique (multiple group discriminant analysis) in an attempt to develop a method to discriminate among individual animals based on measurements and characteristics of tracks. However, Grigione et al. (1999) determined that Smallwood and Fitzhugh (1993) had used an inappropriate method. Based on additional analyses, they concluded it was unlikely that individual mountain lions could be detected using discriminant analysis. Furthermore, Grigione et al. (1999) concluded that the most commonly used measurement (width of heel pad) for distinguishing among mountain lions based on their tracks was also the least variable, calling into question the utility of that measurement. Despite the problematic nature of discriminating among individual lions, Beier and Cunningham (1996) conducted several detailed analyses. They felt it is possible to detect changes in relative abundance of mountain lions using tracking techniques at the local level. Nonetheless, the migratory nature of some mountain lions makes it problematic to define populations (Pierce et al. 1999b), with serious implications for the use of track surveys for estimating changes in population size. Track surveys could be useful for estimating relative use of a specific geographic area by mountain lions, but inferences about population size over time are confounded because it is difficult to define an open population. The use of track surveys over large geographic scales to index relative abundance of mountain lions (Smallwood 1994; Smallwood and Fitzhugh 1995) has been attempted, but likely does not provide the power necessary to detect such changes.

Mountain lions frequently are captured to allow attachment of radiotransmitters. Capture techniques often involve the pursuit of animals with hounds, followed by chemical immobilization (Logan et al. 1986b; Jessup et al. 1986). Physical retrieval of immobilized animals (Hornocker 1970) is necessary if they are sedated in a location high above the ground. McCown et al. (1990) described a portable cushion that lessened the probability of injury to mountain lions falling from trees. Davis et al. (1996) developed a method to safely lower immobilized mountain lions from trees and cliffs. Foot snares have been used safely and successfully by some investigators (Logan et al. 1999; Pierce et al. 2000a), and are advantageous in some situations.

Radiotelemetry has been used extensively to determine the distribution of mountain lions in numerous studies, from which density estimates have been inferred. Capture and recapture methods also have been used to estimate population densities. Reliability of estimates obtained with those techniques is variable (Logan and Sweanor 1999), but densities have been estimated at 0.3–2.2 adult mountain lions/100 km^2. Pierce et al. (2000b) estimated the abundance of mountain lions by maintaining an intensive and constant effort to capture and radio-collar all lions detected on a mule deer winter range. They then used aerial telemetry to estimate the number of mountain lions present per aerial-telemetry day, and were able to determine their absolute abundance from year to year. Smallwood (1997) noted that most variation in densities of mountain lions can be attributed to the spatial extent of individual study areas, and suggested that field studies would be more meaningful if they spanned larger areas, a variety of land uses and habitats, and greater periods of time than most conventional studies.

Molecular techniques have been developed that allow investigators to identify individual mountain lions using DNA obtained from blood, tissue, or hair samples (Ernest 2000a, 2000b) and from feces (Ernest et al. 2000). These techniques depend on obtaining good-quality DNA, which is more difficult when using feces (Ernest 2000a). Fecal DNA has been used to identify individual lions occurring within certain geographic areas (Ernest et al. 2000) or associated with particular kill sites for large mammals (Hayes et al. 2000). Use of this technique to determine the absolute number of mountain lions occurring in a particular geographic area is subject to biases associated with their distribution and frequency in the area. Although it is possible to estimate minimum numbers of individuals present, the probability of detecting all individuals using molecular techniques is a function of sampling effort.

The literature is replete with estimates for home range sizes of mountain lions that have been determined using land-based or aerial telemetry. However, global positioning system (GPS) technology raises questions regarding previous studies. Bleich et al. (2000) described the results from GPS collars on two adult male mountain lions. Patterns of movement were vastly different and home ranges were much larger than were estimates from simultaneous aerial telemetry flights conducted on a weekly basis. Some GPS collars incorporate switches (Bleich et al. 2000) that have the potential to be useful in predicting mountain lion activity; Janis et al. (1999) described and quantified the use of traditional very high frequency transmitters equipped with tip switches to estimate activity of mountain lions.

Translocation of mountain lions as a technique for managing "problem" lions (i.e., animals that have preyed on livestock or that occupy areas adjacent to human habitation) is controversial and is inconsistent with management policies in some states because of potential liabilities. Generally, survival of translocated lions is very low. For example, Ruth et al. (1998) described an experiment in which 14 animals were moved and released different distances ($\bar{X} = 477$ km) from their locations of capture. Nine of those individuals died and, overall, survival was lower than that of reference animals that remained within their original home ranges (Ruth et al. 1998). They concluded that translocation was most successful when mountain lions were 12–27 months old. Translocation as a management technique for mountain lions is not widely implemented, but may become increasingly important as a result of changing attitudes toward predator control (Hancock 1980) and loss or fragmentation of habitat (Nowell and Jackson 1996).

Conservation. Conservation of a large carnivore that preys on other large mammals, frequently is involved in the killing of domestic livestock, and occasionally is a threat to human safety presents challenges that seem overwhelming in contemporary times. The primary prey of mountain lions require large tracts of open and undeveloped land, both of which are becoming increasingly uncommon. Mountain lions themselves similarly depend on large tracts of land to meet their needs. Indeed, the conservation and management challenge for the future will be to assure the presence of mountain lions and their prey, despite the certain loss of habitat and increases in human numbers and activity (Torres 1997).

In western North America, mountain lion populations generally remain healthy (Hornocker 1992), but conservationists will be faced with the task of maintaining large, contiguous tracts of suitable habitat that are linked to other such areas. Populations of mountain lions may occur in a metapopulation structure (Sweanor et al. 2000; but, see Ernest et al. 2003). Islands of habitat suitable for supporting permanent populations sometimes are separated by vast areas of habitat that are not permanently occupied, but instead provide opportunities for movement between habitat islands. Thus, habitat fragmentation is a major concern. Beier (1993) and Beier et al. (1995) described a situation where mountain lion habitat in the Santa Ana Mountains of southern California had become isolated from other suitable areas and immigration likely was reduced as a result of human development. Immigration and emigration of individuals are necessary for the maintenance of genetic diversity. Further, opportunities for recolonization must be provided in the event of extirpations from islands of suitable habitat. Maintaining adequate space for mountain lions and their prey as well as linkages between such areas clearly is a pressing issue that must be addressed if mountain lions are to retain their current status.

Depredations on livestock and other domestic animals likely will increase in frequency, especially as the human population expands (Torres 1997). Depredation may be coincident with increases in lion populations (Torres et al. 1996), but may also be a function of declines in populations of primary prey (Pierce et al. 1999a, Kamler et al. 2002). Whatever the cause, management of depredating mountain lions will remain an important issue. Current management of depredating individuals usually involves killing of the offending animal, a strategy unlikely to change in the immediate future. Even in densely populated areas, such events are relatively uncommon (Torres 2000), and such removals likely do not present a threat to existing populations of mountain lions.

Mountain lion populations apparently have expanded during recent years, and there has been increasing concern about the potential for an increase in human encounters. Indeed, such concerns have resulted in publication of pamphlets and books (Torres 1997) designed to minimize the probability of such incidents. Young mountain lions, dispersing between areas of suitable habitat, may be most apt to encounter humans; indeed, the majority of attacks on humans have involved subadult or apparently underweight lions (Aune 1991; Beier 1991). As human populations increase and expand into currently occupied lion habitat, the rate at which such encounters occur may be expected to increase. Meeting human safety objectives while simultaneously providing habitat for viable populations of mountain lions will become increasingly important (Torres 1997).

Sport hunting of mountain lions occurs in the majority of states that have viable populations of mountain lions, and harvest is a method by which wildlife managers attempt to control lion numbers. Increasingly, however, the concept of sport hunting has come under criticism, and hunting of mountain lions has been eliminated in California, and made extremely difficult in other states such as Oregon and Washington by restrictions on the use of hounds to bring animals to bay. Nonetheless, carefully regulated sport hunting remains an important recreational pursuit and is employed as a method of sustaining mountain lion populations at viable levels in a majority of western states (Logan and Sweanor 1999).

Conservation of mountain lions has a long and varied history, and has been couched largely in terms of political expediency (Torres 1997). Efforts to control or eliminate these obligate carnivores by unrestricted harvest and bounty systems were unsuccessful throughout much of the United States, but their efficacy has been attested to by the absence of mountain lions from much of their historical range. In those areas, such as western North America, where large tracts of essential habitat remained in relatively pristine conditions, intensive methods of controlling mountain lions gave way to regulated sport hunting, with an emphasis on ensuring viable populations with the capability of maintaining a sustainable harvest. Selective removal of individual lions that have preyed on livestock or domestic animals has been compatible throughout much of the range of mountain lions during the recent past. Elimination of lions involved with human safety incidents largely has achieved public acceptance.

RESEARCH NEEDS

Mountain lions occur at low densities and are secretive and cryptic; thus, they are among the most difficult large mammals to study. Furthermore, costs of research associated with the study of large carnivores are great, confounding efforts to obtain meaningful answers to difficult questions. Although many aspects of the ecology of mountain lions have been investigated, few authors have conducted the long-term investigations necessary to begin to understand the relationships among habitat, prey densities, and the dynamics of the populations of these secretive predators and their prey.

Mountain lions have been implicated as factors important in the dynamics of some populations of mountain sheep (Kamler et al. 2002), including two population segments (Wehausen 1996; Hayes et al. 2000) listed as endangered by the federal government. Impacts of lion predation on mountain sheep populations are most probable where mule deer and mountain sheep occur sympatrically (Schaefer et al. 2000; Kamler et al. 2002). Nonetheless, clear linkages among the dynamics of populations of primary prey (mule deer), mountain lions, and secondary prey (mountain sheep) are yet to be established, and warrant serious investigation. Such investigations must, however, be conducted over a temporal scale adequate to elucidate changes in habitat quality and resultant responses among prey and predators in such systems (Pierce 1999). As a result, population modeling may become an even more useful predictive tool, (Beier 1993) with more potential for application than currently recognized.

The genetic structure of populations of mountain lions has been examined only recently (Walker et al. 2000; Ernest et al. 2003), and additional opportunities to define populations from a genetic perspective are needed. Furthermore, the spatial structure of lion populations has been reported only on a local scale (Germaine et al. 2000; Sweanor et al. 2000), and landscape-level efforts to more clearly define populations of mountain lions clearly are warranted. Moreover, the potential metapopulation structure (Sweanor et al. 2000) of mountain lions warrants further investigation, and has important implications for the conservation of these large felids (Ernest et al. 2003). GPS technology can be incorporated into telemetry collars for investigations of mountain lion movements and habitat selection (Bleich et al. 2000; Anderson and Lindzey 2003) and can be combined with sophisticated genetic analyses (Ernest 2000a; Ernest et al. 2000) to enhance the probability of obtaining meaningful results in landscape-level investigations of lion ecology.

Efforts to improve methods of estimating relative numbers of mountain lions in specific geographic areas should continue. Although investigators have developed a number of indirect methods of assessing abundance (Germaine et al. 2000; Pierce et al. 2000b) and some of these are robust (Beier and Cunningham 1996), they are not without problems associated with the definition of populations (Pierce et al. 1999b). Moreover, methods that rely on indirect evidence, such as tracks to identify individuals, are problematic and subject to statistical vagaries (Grigione et al. 1999). Efforts to determine relationships between indirect evidence of mountain lion abundance and genetic evidence from individuals known to be present in specific geographic areas (Ernest et al. 2000) may be a productive field of endeavor, and should be initiated.

As the human population increases, encounters between humans and mountain lions will occur more frequently (Torres 1997). The potential development of nonlethal methods to decrease such encounter rates has implications for the conservation of mountain lions in urbanized areas as well as for some endangered taxa on which they prey. Indeed, the question of aversive conditioning of mountain lions has not been adequately explored, and warrants investigation.

Preliminary findings (Foley 1997) have suggested that epizootic processes are not likely to affect populations of mountain lions in the western United States. Anderson (1983) felt that the relative absence of pathogens among wild mountain lions has largely reflected insufficient efforts to detect diseases, and the results of recent investigations have been consistent with that notion. Nonetheless, some diseases detected among mountain lions are similar to those affecting domestic cats, and further research is needed to determine whether they are the same as those carried by domestic cats or are endemic in wild populations.

Research on mountain lions has often been considered only in a management context, with publications limited to agency reports. Investigators have frequently failed to incorporate concurrent research on prey populations (Pierce 1999) or approach questions from the evolutionary perspective necessary (Pierce 1999; Bleich and Oehler 2000) to ensure that their findings have widespread applicability. The most meaningful knowledge provided by future investigations of mountain lion ecology will incorporate large-scale spatial and temporal components. Investigators will continue, however, to face challenges posed by the high costs and logistical constraints associated with working on a large, cryptic predator (Pierce 1999). Nonetheless, as mountain lions become a more significant and controversial management challenge, funding for the long-term, landscape-level research necessary to meet both proximate and ultimate conservation needs could be forthcoming. Investigators must be ready and willing to proceed along the most meaningful lines of endeavor.

LITERATURE CITED

Ackerman, B. B. 1982. Cougar predation and ecological energetics in southern Utah. M.S. Thesis, Utah State University, Logan.

Ackerman, B. B., F. G. Lindzey, and T. P. Hemker. 1984. Cougar food habits in southern Utah. Journal of Wildlife Management 48:147–55.

Ackerman, B. B., F. G. Lindzey, and T. P. Hemker. 1986. Predictive energetics model for cougars. Pages 333–52 *in* S. D. Miller and D. Everett, eds. Cats of the world: Biology, conservation, and management. National Wildlife Federation, Washington, DC.

Adaska, J. M. 1999. Peritoneal coccidioidomycosis in a mountain lion in California. Journal of Wildlife Diseases 35:75–77.

Anderson, A. E. 1983. A critical review of literature on puma (*Felis concolor*) (Special Report No. 54). Colorado Division of Wildlife, Fort Collins.

Anderson, A. E., D. C. Bowden, and D. M. Kattner. 1992. The puma on Uncompahgre Plateau, Colorado (Technical Publication 40). Colorado Division of Wildlife, Fort Collins.

Anderson, C. R., Jr. and F. G. Lindzey. 2003. Estimating cougar predation rates from GPS location clusters. Journal of Wildlife Management 67:307–316.

Ashman, D. L., G. C. Christensen, M. L. Hess, G. K. Tsukamoto, and M. S. Wickersham. 1983. The mountain lion in Nevada (Federal Aid in Wildlife Restoration Final Report W-48-15, Study S&I 1, Job 5 and Study R-V, Job 1). Nevada Department of Wildlife, Reno.

Aune, K. E. 1991. Increasing mountain lion populations and human–mountain lion interactions in Montana. Page 86–94 *in* C. E. Braun, ed. Mountain lion–human interaction: Symposium and workshop. Colorado Division of Wildlife, Denver.

Baker, R. R. 1978. The evolutionary ecology of animal movement. Hodder and Stoughton, London.

Bank, M. S., and W. L. Franklin. 1998. Puma (*Puma concolor patagonica*) feeding observations and attacks on guanacos (*Lama guanicoe*). Mammalia 62:599–605.

Barnhurst, D., and F. G. Lindzey. 1984. Utah—Cougar research report. Pages 185–88 *in* J. Roberson and F. Lindzey, eds. Proceedings of the second mountain lion workshop. Utah Division of Wildlife Research Unit and Utah Cooperative Wildlife Research Unit, Zion National Park.

Beier, P. 1991. Cougar attacks on humans in the United States and Canada. Wildlife Society Bulletin 19:403–12.

Beier, P. 1993. Determining minimum habitat areas and habitat corridors for cougars. Conservation Biology 7:94–108.

Beier, P., and R. H. Barrett. 1993. The cougar in the Santa Ana Mountain Range, California. (Final Report, Orange City Mountain Lion Study). University of California, Berkeley.

Beier, P., and S. C. Cunningham. 1996. Power of track surveys to detect changes in cougar populations. Wildlife Society Bulletin 24:540–46.

Beier, P., D. Choate, and R. H. Barrett. 1995. Movement patterns of mountain lions during different behaviors. Journal of Mammalogy 76:1056–70.

Belden, R. C., W. B. Frankenberger, R. T. McBride, and S. T. Schwikert. 1988. Panther habitat use in Southern Florida. Journal of Wildlife Management 52:660–63.

Bergerud, A. T., W. Wyett, and J. B. Snider. 1983. The role of wolf predation in limiting a moose population. Journal of Wildlife Management 47:977–88.

Bleich, V. C., and M. W. Oehler. 2000. Wildlife education in the United States: Thoughts from agency biologists. Wildlife Society Bulletin 28:542–45.

Bleich, V. C., and T. J. Taylor. 1998. Survivorship and cause-specific mortality in five populations of mule deer. Great Basin Naturalist 58:265–72.

Bleich, V. C., B. M. Pierce, J. L. Davis, and V. L. Davis. 1996. Thermal characteristics of mountain lion dens. Great Basin Naturalist 56:276.

Bleich, V. C., R. T. Bowyer, and J. D. Wehausen. 1997. Sexual segregation in mountain sheep: Resources or predation? Wildlife Monographs 134:1–50.

Bleich, V. C., B. M. Pierce, S. G. Torres, and T. Lupo. 2000. Using space age technology to study mountain lion ecology. Outdoor California 61(3):24–25.

Boutin, S. 1992. Predation and moose population dynamics: A critique. Journal of Wildlife Management 56:116–27.

Bowyer, R. T., M. C. Nicholson, E. M. Molvar, and J. B. Faro. 1999. Moose on Kalgin Island: Are density-dependent processes related to harvest? Alces 35:73–89.

Boyce, M. S. 1989. The Jackson elk herd: Intensive wildlife management in North America. Cambridge University Press, New York.

Boyd, D. K., and G. K. Neale. 1992. An adult cougar, *Felis concolor,* killed by gray wolves, *Canis lupus,* in Glacier National Park, Montana. Canadian Field-Naturalist 106:524–25.

Cashman, J. L., M. Peirce, and P. R. Krausman. 1992. Diets of mountain lions in southwestern Arizona. Southwestern Naturalist 37:324–26.

Charlton, K. G., D. W. Hird, and E. L. Fitzhugh. 1998. Physical condition, morphometrics, and growth characteristics of mountain lions. California Fish and Game 84:104–11.

Connolly, E. J., Jr. 1949. Food habits and life history of the mountain lion (*Felis concolor hippolestes*). M.S. Thesis, University of Utah, Salt Lake City.

Cox, M. K., and S. Stiver. 1997. Status and management of mountain lions in Nevada. Pages 17–18 *in* W. D. Padley, ed. Proceedings of the fifth mountain lion workshop. San Diego, CA.

Craig, D. L. 1986. The seasonal food habits in sympatric populations of puma (*Puma concolor*), coyote (*Canis latrans*), and bobcat (*Lynx rufus*) in the Diablo Range of California. M.A. Thesis, San Jose State University, San Jose, CA.

Cronemiller, F. P. 1948. Mountain lion preys on bighorn. Journal of Mammalogy 29:68.

Culver, M., W. E. Johnson, J. Pecon-lattery, and S. J. O'Brien. 2000. Genomic ancestry of the American puma (*Puma concolor*). Journal of Heredity 91:186–97.

Cumberland, R. E., and J. A. Dempsey. 1994. Recent confirmation of a cougar, *Felis concolor,* in New Brunswick. Canadian Field-Naturalist 108:224–26.

Cunningham, S. C., L. A. Haynes, C. Gustavson and D. D. Haywood. 1995. Evaluation of the interaction between mountain lions and cattle in the Aravaipa–Klondyke area of southeast Arizona (Technical Report 17). Arizona Game and Fish Department.

Currier, M. J. P. 1983. *Felis concolor*. Mammalian Species 200:1–7.

Currier, M. J. P., and K. R. Russell. 1982. Hematology and blood chemistry of the mountain lion (*Felis concolor*). Journal of Wildlife Diseases 18:99–104.

Currier, M. J. P., S. L. Sheriff, and K. R. Russell. 1977. Mountain lion population and harvest near Canon City, Colorado, 1974–77. (Colorado Division of Wildlife Special Report 42). Colorado Cooperative Wildlife Research Unit, Colorado State University, Fort Collins.

Danvir, R. E., and F. G. Lindzey. 1981. Feeding behavior of a captive cougar on mule deer. Encyclia 58:50–56.

Davis, J. L., C. B. Chetkiewicz, V. C. Bleich, G. Raygorodetsky, B. M. Pierce, J. W. Ostergard, and J. D. Wehausen. 1996. A device to safely remove immobilized mountain lions from trees and cliffs. Wildlife Society Bulletin 24:537–39.

Dixon, K. R. 1982. Mountain lion. Pages 711–27 *in* J. A. Chapman and G. A. Feldhamer, eds. Wild mammals of North America: Biology, management and economics. Johns Hopkins University Press, Baltimore.

Donaldson, B. R. 1975. Mountain lion research (1971–75) (Federal Aid in Wildlife Restoration Final Report Project W-93-R-17). New Mexico Department of Game and Fish, Santa Fe.

Dunbar, M. R., P. Nol, and S. B. Linda. 1997. Hematologic and serum biochemical reference intervals for Florida panthers. Journal of Wildlife Diseases 33:783–89.

Eaton, R. L. 1976. Why some felids copulate so much. Pages 74–94 *in* R. L. Eaton, ed. The world's cats, Vol. 3. Carnivore Research Institute, University of Washington, Seattle.

Eaton, R. L., and K. A. Velander. 1977. Reproduction in the puma: Biology, behavior and ontogeny. Pages 45–70 *in* R. L. Eaton, ed. The world's cats, Vol. 3. Carnivore Research Institute, University of Washington, Seattle.

Ernest, H. B. 2000a. DNA analysis for mountain lion conservation. Outdoor California 61(3):16–19.

Ernest, H. B. 2000b. DNA sampling and research techniques. Outdoor California 61(3):20–21.

Ernest, H. B., W. M. Boyce, V. C. Bleich, B. May, S. J. Stiver, and S. G. Torres. 2003. Genetic structure of mountain lion (*Puma concolor*) populations in California. Conservation Genetics 4:353–366.

Ernest, H. B., M. C. T. Penedo, B. P. May, M. Syvanen, and W. M. Boyce. 2000. Molecular tracking of mountain lions in the Yosemite Valley region in California: Genetic analysis using microsatellite and faecal DNA. Molecular Ecology 9:433–41.

Evermann, J. F., W. J. Foreyt, B. Hall, and A. J. McKeirnan. 1997. Occurrence of puma lentivirus infection in cougars from Washington. Journal of Wildlife Diseases 33:316–20.

Fitzhugh, E. L., and W. P. Gorenzel. 1985. Design and analysis of mountain lion track surveys. Cal-Neva Wildlife Transactions 1985:78–87.

Foley, J. E. 1997. The potential for catastrophic infectious disease outbreaks in populations of mountain lions in the western Unitd States. Pages 29–36 *in* W. D. Padley, ed. Proceedings of the fifth mountain lion workshop. San Diego, CA.

Forrester, D. J., J. A. Conti, and R. C. Belden. 1985. Parasites of the Florida panther (*Felis concolor coryi*). Proceedings of the Helminthological Society of Washington 52:95–97.

Gasaway, W. C., R. D. Boertje, D. V. Grangaard, D. G. Kelleyhouse, R. O. Stephenson, and D. G. Larsen. 1992. The role of predation in limiting moose at low densities in Alaska and Yukon and implications for conservation. Wildlife Monographs 120:1–59.

Gay, S. W., and T. L. Best. 1995. Geographic variation in sexual dimorphism of the puma (*Puma concolor*) in North and South America. Southwestern Naturalist 40:148–59.

Genoways, H. H., and P. W. Freeman. 1996. A recent record of a mountain lion in Nebraska. Prairie Naturalist 28:143–45.

Germaine, S. S., K. D. Bristow, and L. A. Haynes. 2000. Distribution and population status of 59 mountain lions in southwestern Arizona. Southwestern Naturalist 45:333–38.

Gonyea, □., and □. Ashworth. 1975.

Grigione, M. M., P. Burman, V. C. Bleich, and B. M. Pierce. 1999. Identifying individual mountain lions (*Felis concolor*) by their tracks: Refinement of an innovative technique. Biological Conservation 88:25–32.

Guggisberg, C. A. W. 1975. Wild cats of the world. Taplinger, New York.

Haber, G. C. 1977. Socio-ecological dynamics of wolves and prey in a subarctic ecosystem. Ph.D. Dissertation, University of British Columbia, Vancouver, Canada.

Hancock, L. 1980. A history of changing attitude toward *Felis concolor*. M.S. Thesis, Simon Fraser University, Vancouver, British Columbia, Canada.

Hansen, K. 1992. Cougar: The American lion. Northland, Flagstaff, AZ.

Harlow, H. J., F. G. Lindzey, W. D. Van Sickle, and W. A. Gern. 1992. Stress response of cougars to nonlethal pursuit by hunters. Canadian Journal of Zoology 70:136–39.

Harrison, S. 1990. Cougar predation on bighorn sheep in the Junction Wildlife Management Area, British Columbia. M.S. Thesis, University of British Columbia, Vancouver, Canada.

Harveson, L. 1997. Ecology of a mountain lion population in southern Texas. Ph.D. Dissertation, Texas A&M University, Kingsville.

Hawkey, C. M., and M. G. Hart. 1986. Haematological reference values for adult pumas, lions, tigers, leopards, jaguars and cheetahs. Research in Veterinary Science 41:268–69.

Hayes, C. J., E. S. Rubin, M. C. Jorgensen, R. A. Botta, and W. M. Boyce. 2000. Mountain lion predation of bighorn sheep in the peninsular ranges, California. Journal of Wildlife Management 64:954–59.

Hebert, D., and D. Lay. 1997. Cougar–human interactions in British Columbia. Pages 44–45 *in* W. D. Padley, ed. Proceedings of the fifth mountain lion workshop. San Diego, CA.

Hemker, T. P., F. G. Lindzey, and B. B. Ackerman. 1984. Population characteristics and movement patterns of cougars in southern Utah. Journal of Wildlife Management 48:1275–84.

Holling, C. S. 1959. The components of predation as revealed by a study of small mammal predation of the European pine sawfly. Canadian Entomologist 91:293–320.

Hopkins, R. A. 1989. Ecology of the puma in the Diablo Range, California. Ph.D. Dissertation, University of California, Berkeley.

Hornocker, M. G. 1969. Winter territoriality in mountain lions. Journal of Wildlife Management 33:457–64.

Hornocker, M. G. 1970. An analysis of mountain lion predation upon mule deer and elk in the Idaho Primitive Area. Wildlife Monographs 21:1–39.

Hornocker, M. G. 1992. Learning to live with mountain lions. National Geographic 182:38–65.

Iriarte, J. A., W. L. Franklin, W. E. Johnson, and K. H. Redford. 1990. Biogeographic variation of food habits and body size of the America puma. Oecologia 85:185–90

Janis, M. W., J. D. Clark, and C. S. Johnson. 1999. Predicting mountain lion activity using radiocollars equipped with mercury tip-sensors. Wildlife Society Bulletin 27:19–24.

Janis, M. W., and J. D. Clark. 2002. Responses of Florida panthers to recreational deer and hog hunting. Journal of Wildlife Management 66:839–848.

Janczewski, D. N., W. S. Modi, J. C. Stephens, and S. J. O'Brien. 1995. Molecular evolution of mitochondrial 12S RNA and cytochrome *b* sequences in the Pantherine lineage of Felidae. Molecular Biology and Evolution 12:690–707.

Jardine, W. 1834. Naturalist library. Volume 2: Mammals, Felidae.

Jensen, B. 2001. A brief natural history of North Dakota: 1804 to present. North Dakota Outdoors 63(9):10–19.

Jessup, D. A., W. E. Clark, and M. A. Fowler. 1986. Wildlife restraint handbook, 3rd ed. California Department of Fish and Game, Rancho Cordova.

Jessup, D. A., K. C. Pettan, L. J. Lowenstine, and N. C. Pedersen. 1993. Feline leukemia virus infection and renal spirochetosis in a free-ranging cougar (*Felis concolor*). Journal of Zoo and Wildlife Medicine 24:73–79.

Johnson, W., and S. J. O'Brien. 1997. Phylogenetic reconstruction of the Felidae using 16S RNA and NADH-5 mitochondrial genes. Journal of Molecular Evolution 44 (Supplement):S98–116.

Kadesky, K. M., C. Manarey, G. K. Blair, J. J. Murphy III, C. Verchere, and K. Atkinson. 1998. Cougar attacks on children: Injury pattern and treatment. Journal of Pediatric Surgery 33:863–65.

Kamler, J. F., R. M. Lee, J. C. deVos, Jr., W. B. Ballard, and H. A. Whitlaw. 2002. Survival and cougar predation of translocated bighorn sheep in Arizona. Journal of Wildlife Management 66:1267–1272.

Keech, M. A., R. T. Bowyer, J. M. Ver Hoef, R. D. Boertje, B. W. Dale, and T. R. Stephenson. 2000. Life-history consequences of maternal condition in Alaskan moose. Journal of Wildlife Management 64:450–62.

Kitchen, D. W. 1974. Social behavior and ecology of the pronghorn. Wildlife Monographs 38:1–96.

Kitchener, A. 1991. The natural history of the wild cats. Cornell University Press, Ithaca, NY.

Koehler G. M., and M. G. Hornocker. 1991. Seasonal resource use among mountain lions, bobcats, and coyotes. Journal of Mammalogy 72:391–96.

Koford, C. B. 1946. A California mountain lion observed stalking. Journal of Mammalogy 27:274–75.

Kohlmann, S. G., and R. L. Green. 1999. Body size dynamics of cougars (*Felis concolor*) in Oregon. Great Basin Naturalist 59:193–94.

Kruuk, H. 1986. Interactions between Felidae and their prey species: A review. Pages 353–74 *in* S. D. Miller and D. D. Everett, eds. Cats of the world: Biology, conservation and management. National Wildlife Federation, Washington, DC.

Kunkel, K. E., T. K. Ruth, D. H. Pletscher, and M. G. Hornocker. 1999. Winter prey selection by wolves and cougars in and near Glacier National Park. Journal of Wildlife Management 63:901–10.

Kurten, B. 1973. Geographic variation in size in the puma (*Felis concolor*). Commentationes Biologicae 63:3–8.

Laing, S. P. 1988. Cougar habitat selection and spatial use patterns in southern Utah. M.S. Thesis, University of Wyoming, Laramie.

Lechleitner, R. R. 1969. Wild mammals of Colorado. Pruett, Boulder, CO.

Leopold, A. 1949. A sand county almanac. Oxford University Press, New York.

Lesowski, J. 1963. Two observations of cougar cannibalism. Journal of Mammalogy 44:586.

Levins, R. 1970. Extinction. Pages 77–107 *in* M. Gesternhaber, ed. Some mathematical questions in biology. American Mathematical Society, Providence, RI.

Leyhausen, P. 1979. Cat behaviour. Garland STPM Press, New York.

Lindzey, F. G. 1987. Mountain lion. Pages 657–68 *in* M. Novak, J. A. Baker, M. E. Obbard, and B. Malloch, eds. Wild furbearer management and conservation in North America. Ontario Trappers Association, Ontario Ministry of Natural Resources, Toronto.

Lindzey, F. G., B. B. Ackerman, D. Barnhurst, and T. P. Hemker. 1988. Survival rates of mountain lions in southern Utah. Journal of Wildlife Management 52:664–67.

Lindzey, F. G., B. B. Ackerman, D. Barnhurst, T. Becker, T. P. Hemker, S. P. Laing, C. Mecham, and W. D. VanSickle. 1989. Boulder–Escalante Cougar Project. Final report. Utah Division of Wildlife Resources.

Lindzey, F. G., W. D. Van Sickle, B. B. Ackerman, D. Barnhurst, T. P. Hemker, and S. P. Laing. 1994. Cougar population dynamics in southern Utah. Journal of Wildlife Management 58:619–24.

Logan, K. A. 1983. Mountain lion population and habitat characteristics in the Big Horn Mountains of Wyoming. M.S. Thesis, University of Wyoming, Laramie.

Logan, K. A., and L. L. Irwin. 1985. Mountain lion habitats in the Big Horn Mountains, Wyoming. Wildlife Society Bulletin 13:257–62.

Logan, K. A., and L. Sweanor. 1999. Puma. Pages 347–77 *in* S. Demarais and P. R. Krausman, eds. Ecology and management of large mammals in North America. Prentice-Hall, Englewood Cliffs, NJ.

Logan, K. A., L. L. Irwin, and R. Skinner. 1986a. Characteristics of a hunted mountain lion population in Wyoming. Journal of Wildlife Management 50:648–54.

Logan, K. A., E. T. Thorne, L. L. Irwin, and R. Skinner. 1986b. Immobilizing wild mountain lions (*Felis concolor*) with ketamine hydrochloride and xylazine hydrochloride. Journal of Wildlife Diseases 22:97–103.

Logan, K. A., L. Sweanor, and M. Hornocker. 1996. Cougar population dynamics. Chapter 3 *in* Cougars in the San Andres Mountains, New Mexico (Project No. W-128-R, Final Report). New Mexico Department of Game and Fish, Santa Fe.

Logan, K. A., L. Sweanor, J. F. Smith, and M. G. Hornocker. 1999. Capturing pumas with foot-hold snares. Wildlife Society Bulletin 27:201–8.

Logan, K., and L. Sweanor. 2001. Desert Puma: Evolutionary ecology and conservation of an enduring carnivore. Hornocker Wildlife Institute, Island Press, Washington.

MacArthur, R. H., and E. O. Wilson. 1967. The theory of island biogeography. Princeton University Press, Princeton, NJ.

Macdonald, D. W. 1985. The carnivores: Order Carnivora. Pages 619–722 *in* R. E. Brown and D. W. Macdonald, eds. Social odours in mammals. Clarendon Press, Oxford.

Maehr, D. S. 1997. The Florida panther: Life and death of a vanishing carnivore. Island Press, Washington, DC.

Maehr, D. S., R. C. Belden, E. D. Land, and L. Wilkins. 1990. Food habits of panthers in southwest Florida. Journal of Wildlife Management 54:420–23.

Maehr, D. S., E. D. Land, and J. C. Roof. 1991. Social ecology of Florida panthers. National Geographic Research and Exploration 7:414–31.

Marcgrave, G. 1648. *Historiae rerum naturalium Brasiliae* (cited by Young and Goldman 1946).

Martin, L. D. 1989. Fossil history of terrestrial Carnivora. Pages 536–68 *in* J. L. Gittleman, ed. Carnivore behavior, ecology, and evolution. Cornell University Press, Ithaca, NY.

McCown, J. W., D. S. Maehr, and J. Roboski. 1990. A portable cushion as a wildlife capture aid. Wildlife Society Bulletin 18:34–36.

McCullough, D. R. 1979. The George Reserve deer herd: Population ecology of a *K*-selected species. University of Michigan Press, Ann Arbor.

Messier, F. 1994. Ungulate population models with predation: A case study with the North American moose. Ecology 75:478–88.

Messier, F., and M. Crete. 1985. Moose–wolf dynamics and the natural regulation of moose populations. Oecologia 65:503–12.

Mohr, C. O. 1947. Table equivalent populations of North American mammals. American Midland Naturalist 37:223–49.

Murphy, K. M. 1983. Relationships between a mountain lion population and hunting pressure in western Montana. M.S. Thesis, University of Montana, Missoula.

Murphy, K. M. 1998. The ecology of the cougar (*Puma concolor*) in the northern Yellowstone ecosystem: Interactions with prey, bears, and humans. Ph.D. Dissertation, University of Idaho, Moscow.

Nowell, K., and P. Jackson, eds. 1996. Wild cats: Status survey and conservation action plan. IUCN/SSC Cat Specialist Group, Gland, Switzerland.

Ockenfels, R. A. 1994. Mountain lion predation on pronghorn in central Arizona. Southwestern Naturalist 39:305–6.

Padley, W. D. 1990. Home range and social interactions of mountain lions (*Felis concolor*) in the Santa Ana Mountains, California. M.S. Thesis, California State Polytechnic University, Pomona.

Padley, W. D. 1991. Mountain lion ecology in the southern Santa Ana mountains, California (Final Contract Report). California State Polytechnic University, Pomona.

Paul-Murphy, J., T. Work, D. Hunter, E. McFie, and D. Fjelline. 1994. Serologic survey and serum biochemical reference ranges of the free-ranging mountain lion (*Felis concolor*) in California. Journal of Wildlife Diseases 30:205–15.

Pecon-Slattery, J., and S. J. O'Brien. 1998. Patterns of Y and X chromosome DNA sequence divergence during the Felidae radiation. Genetics 148:1245–55.

Pence, D. B., R. J. Warren, D. Waid, and M. J. Davin. 1987. Aspects of the ecology of mountain lions (*Felis concolor*) in Big Bend National Park. Final report. National Park Service, Santa Fe, NM.

Pierce, B. M. 1999. Predator–prey dynamics between mountain lions and mule deer: Effects on distribution, population regulation, habitat selection, and prey selection. Ph.D. Dissertation, University of Alaska, Fairbanks.

Pierce, B. M., V. C. Bleich, C.-L. B. Chetkiewicz, and J. D. Wehausen. 1998. Timing of feeding bouts of mountain lions. Journal of Mammalogy 79:222–26.

Pierce, B. M., V. C. Bleich, and R. T. Bowyer. 1999a. Population dynamics of mountain lions and mule deer: Top-down or bottom-up regulation? Final report. Deer Herd Management Plan Implementation Program, California Department of Fish and Game, Sacramento.

Pierce, B. M., V. C. Bleich, J. D. Wehausen, and R. T. Bowyer. 1999b. Migratory patterns of mountain lions: Implications for social regulation and conservation. Journal of Mammalogy 80:986–92.

Pierce, B. M., V. C. Bleich, and R. T. Bowyer. 2000a. Selection of mule deer by mountain lions and coyotes: Effects of hunting style, body size, and reproductive status. Journal of Mammalogy 81:462–72.

Pierce, B. M., V. C. Bleich, and R. T. Bowyer. 2000b. Social organization of mountain lions: Does a land-tenure system regulate population size? Ecology 81:1533–43.

Pierce, B. M., R. T. Bowyer, and V. C. Bleich. 1999. Habitat selection by mule deer: Forage benefits or risk of predation by mountain lions? Journal of Wildlife Research.

Pike, J. R., J. H. Shaw, and D. M. Leslie Jr. 1997. The mountain lion in Oklahoma and surrounding states. Proceedings of the Oklahoma Academy of Science 77:39–42.

Pittman, M. T., G. J. Guzman, and B. P. Mckinney. 2000. Ecology of the mountain lion on Big Bend Ranch State Park in the Trans-Pecos region of Texas. Final report (Project No. 86). Texas Parks and Wildlife, Wildlife Division.

Rabb, G. B. 1959. Reproductive and vocal behavior in captive pumas. Journal of Mammalogy 40:616–17.

Rasmussen, D. I. 1941. Biotic communities of the Kaibab Plateau, Arizona. Ecological Monographs 11:229–75.

Robbins, C. T. 1993. Wildlife feeding and nutrition. Academic Press, San Diego, CA.

Robinette, W. L., J. S. Gashwiler, and O. W. Morris. 1959. Food habits of the cougar in Utah and Nevada. Journal Wildlife Management 23:261–73.

Robinette, W. L., J. S. Gashwiler, and O. W. Morris. 1961. Notes on cougar productivity and life history. Journal of Mammalogy 42:204–17.

Roelke, M. E., D. J. Forrester, E. R. Jacobson, G. V. Kollias, F. W. Scott, M. C. Barr, J. F. Evermann, and E. C. Pirtie. 1993. Seroprevalence of infectious disease agents in free-ranging Florida panthers (*Felis concolor coryi*). Journal of Wildlife Diseases 29:36–49.

Ross, P. I. 1994. Lions in winter. Natural History 103:52–59.

Ross, P. I. and M.G. Jalkotzy. 1992. Characteristics of a hunted population of cougars in southwestern Alberta. Journal of Wildlife Management 56:417–26.

Ross, P. I., and M. G. Jalkotzy. 1996. Cougar predation on moose in southwestern Alberta. Alces 32:1–8.

Ross, P. I., M. G. Jalkotzy, and M. Festa-Bianchet. 1997. Cougar predation on bighorn sheep in southwestern Alberta during winter. Canadian Journal of Zoology 74:771–75.

Russell, K. R. 1978. Mountain lion. Pages 207–25 *in* J. L. Schmidt and D. L. Gilbert, eds. Big game of north America: Ecology and management. Stackpole, Harrisburg, PA.

Ruth, T. K., K. A. Logan, L. L. Sweanor, M. G. Hornocker, and L. J. Temple. 1998. Evaluating cougar translocation in New Mexico. Journal of Wildlife Management 62:1264–75.

Schaefer, R. J., S. G. Torres, and V. C. Bleich. 2000. Survivorship and cause-specific mortality in sympatric populations of mountain sheep and mule deer. California Fish and Game 86:127–35.

Schaller, G. B. 1972. The Serengeti lion. University of Chicago Press, Chicago.

Seidensticker, J. C., IV., M. G. Hornocker, W. V. Wiles, and J. P. Messick. 1973. Mountain lion social organization in the Idaho Primitive Area. Wildlife Monographs 35:1–60

Shaw, H. G. 1977. Impact of mountain lion on mule deer and cattle in northwestern Arizona. Pages 17–32 *in* R. L. Phillips and C. J. Jonkel, eds. Proceedings of the 1975 predator symposium. Montana Forest and Conservation Experiment Station, University of Montana, Missoula.

Shaw, H. G. 1980. Ecology of the mountain lion in Arizona. Final report (P-R Project W-78-R, Work Plan 2, Job 13). Arizona Game and Fish Department, Phoenix.

Shaw, H. G. 1983. Mountain lion field guide (Special Publication 9). Arizona Game and Fish Department, Phoenix.

Shaw, H. G. 1989. Soul among lions: The cougar as peaceful adversary. Johnson, Boulder, CO.

Shaw, H. G., N. G. Woolsey, J. R. Wegge, and R. L. Day, Jr. 1987. Factors affecting mountain lion densities and cattle depredation in Arizona. Final report (Project W-78-R, Wk. Pl. 2, Job 29). Arizona Game and Fish Department, Phoenix.

Sinclair, A. R. E. 1989. Population regulation in animals. Pages 197–241 *in* J. M. Cherrett, ed. Ecological concepts: The contribution of ecology to an understanding of the natural world. Blackwell, Oxford.

Smallwood, K. S. 1994. Trends in California mountain lion populations. Southwestern Naturalist 39:67–72.

Smallwood, K. S. 1997. Interpreting puma (*Puma concolor*) population estimates for theory and management. Environmental Conservation 24:283–89.

Smallwood, K. S., and E. L. Fitzhugh. 1993. A rigorous technique for identifying individual mountain lions *Felis concolor* by their tracks. Biological Conservation 65:51–59.

Smallwood, K. S., and E. L. Fitzhugh. 1995. A track count for estimating mountain lion *Felis concolor californica* population trend. Biological Conservation 71:251–59.

Smith, T. E. 1981. Food habits and scrape site characteristics of mountain lions in the Diablo Range of California. M.A. Thesis, San Jose State University, San Jose, CA.

Spalding, D. J., and J. Lesowski. 1971. Winter food of the cougar in south-central British 169 Columbia. Journal of Wildlife Management. 35:378–81.

Spreadbury, B. 1989. Cougar ecology and related management implications and strategies in southeastern British Columbia. Masters Thesis, University of Calgary, Calgary, Alberta, Canada.

Storer, T. I. 1923. Rabies in a mountain lion. California Fish and Game 9:45–48.

Sweanor, L. L. 1990. Mountain lion social organization in a desert environment. M.S. Thesis, University of Idaho, Moscow.

Sweanor, L. L., K. A. Logan, and M. Hornocker. 1996. Cougar social organization. Chapter 4 *in* Cougars in the San Andres Mountains, New Mexico (Project No. W-128-R, Final Report). New Mexico Department of Game and Fish, Santa Fe.

Sweanor, L. L., K. A. Logan, and M. Hornocker. 2000. Cougar dispersal patterns, metapopulation dynamics, and conservation. Conservation Biology 14:798–808.

Sweitzer, R. A. 1998. Conservation implications of feral pigs in island and mainland ecosystems, and a case study of feral pig expansion in California. Pages 29–34 *in* R. O. Baker and A. C. Crabb, eds. Proceedings of the 18th vertebrate pest conference. University of California, Davis.

Sweitzer, R. A., S. H. Jenkins, and J. Berger. 1997. Near-extinction of porcupines by mountain lions and consequences of ecosystem change in the Great Basin Desert. Conservation Biology 11:1407–17.

Torres, S. 1997. Mountain lion alert. Falcon Books, Helena, MT.

Torres, S. 2000. Counting cougars in California. Outdoor California 61(3):7–9.

Torres, S., T. M. Mansfield, J. E. Foley, T. Lupo, and A. Brinkhaus. 1996. Mountain lion and human activity in California: Testing speculations. Wildlife Society Bulletin 24:451–60.

Turner, A. 1997. The big cats and their fossil relatives. Columbia University Press, New York.

Turner, J. W., Jr., M. L. Wolfe, and J. F. Kirkpatrick. 1992. Seasonal mountain lion predation on a feral horse population. Canadian Journal of Zoology 70:929–34.

Van Ballenberghe, V., and W. B. Ballard. 1994. Limitation and regulation of moose populations: The role of predation. Canadian Journal of Zoology 72:2071–77.

Van Dyke, F. G., R. H. Brocke, H. G. Shaw, B. B. Ackerman, T. P. Hemker, and F. G. Lindzey. 1986. Reactions of mountain lions to logging and human activity. Journal of Wildlife Management 50:95–102.

Van Valkenburgh, B., F. Grady, and B. Kurten. 1990. The Plio-pleistocene cheetah-like cat *Miracinonyx inexpectatus* of North America. Journal of Vertebrate Paleontology 10:434–26.

Vaughan, T. A., J. A. Ryan, and N. J. Czaplewski. 2000. Mammalogy, 4th ed. Harcourt, Fort Worth, TX.

Waid, D. D. 1990. Movements, food habits, and helminth parasites of mountain lions in southwestern Texas. Ph.D. Dissertation, Texas Tech University, Lubbock.

Walker, C. W., L. A. Harveson, M. T. Pittman, M. E. Tewes, and R. L. Honeycutt. 2000. Microsatellite variation in two populations of mountain lions (*Puma concolor*) in Texas. Southwestern Naturalist 45:196–203.

Watson, A., and R. Moss. 1970. Dominance, spacing behavior and aggression in relation to population limitation in vertebrates. Pages 167–220 *in* A. Watson, ed. Animal populations in relation to their food resources. Blackwell, Oxford.

Wehausen, J. D. 1996. Effects of mountain lion predation on bighorn sheep in the Sierra Nevada and Granite Mountains of California. Wildlife Society Bulletin 24:471–79.

White, P. A., and D. K. Boyd. 1989. A cougar, *Felis concolor,* kitten killed and eaten by gray wolves, *Canis lupus,* in Glacier National Park, Montana. Canadian Field-Naturalist 103:408–9.

Wilson, P. 1984. Puma predation on Guanacos in Torres del Pain National Park, Chile. Mammalia 48:515–22.

Wozencraft, W. C. 1993. Order Carnivora. Pages 286–346 *in* D. E. Wilson and D. M. Reeder, eds. Mammal species of the world: A taxonomic and geographic reference, 2nd ed. Smithsonian Institution Press, Washington, DC.

Yamamoto, K., B. B. Chomel, L. J. Lowenstine, Y. Kikuchi, L. G. Phillips, B. C. Barr, P. K. Swift, K. R. Jones, S. P. D. Riley, R. W. Kasten, J. E. Foley, and N. C. Pedersen. 1998. *Bartonella henselae* antibody prevalence in free-ranging and captive wild felids from California. Journal of Wildlife Diseases 34:56–63.

Young, S. P., and E. A. Goldman. 1946. The puma, mysterious American cat. American Wildlife Institute, Washington, DC.

Becky M. Pierce, Sierra Nevada Bighorn Sheep Recovery Program, California Department of Fish and Game, Bishop, California 93514. Email: bmpierce@dfg.ca.gov.

Vernon C. Bleich, Sierra Nevada Bighorn Sheep Recovery Program, California Department of Fish and Game, Bishop, California 93514. Email: vbleich@dfg.ca.gov.

38

Bobcat and Lynx

Lynx rufus and *Lynx canadensis*

Eric M. Anderson
Matthew J. Lovallo

NOMENCLATURE

COMMON NAMES.

Bobcat: barred bobcat, bay lynx, bob-tailed cat, cat o' the mountain, cat lynx, catamount, *chat sauvage* (French Canadian), *chat sauvage de la nouvelle cosae* (French Canadian), *gato monte* (Mexican), *loupcervier* (French Canadian), lynx cat, *lynx roux* (French), pallid bobcat, *pichou* or *pichu* (French Canadian), red lynx, and wildcat (Jackson 1961; McCord and Cardoza 1982; Banfield 1987)

Lynx: *be-jew* or *pe-zu* (Chippewa), Canada lynx, Canadian lynx, catamount, gray wildcat, *le chat* (French Canadian), *loupcervier* (French Canadian), lynx cat, and *pichu* (French Canadian) (Jackson 1961; McCord and Cardoza 1982; Banfield 1987)

SCIENTIFIC NAME.

Bobcat: *Lynx rufus* (also *Felis rufus*)

Lynx: *Lynx canadensis* (also *Felis lynx*)

SUBSPECIES.

Bobcat: *L. r. baileyi, L. r. californicus, L. r. escuinapae, L. r. fasciatus, L. r. floridanus, L. r. gigas, L. r. oaxacensis, L. r. pallescens, L. r. peninsularis, L. r. rufus, L. r. superiorensis, L. r. texensis*

Lynx: *L. c. canadensis, L. c. subsolanus*

TAXONOMY

There is little consensus on the generic names of the bobcat and the lynx (*Lynx* vs. *Felis*). Originally placed in the genus *Lynx,* they have a distinctly shorter tail and lack an upper premolar (P^2) found in other members of *Felis*. Based on morphologic, phylogenetic, and genetic evidence, a number of authors have retained them in *Lynx* (Hall 1981; Werdelin 1981; Wozencraft 1993; Johnson and O'Brien 1997), whereas others have found the differences insufficient to warrant a separate genus (Jones et al. 1975; Tumlison 1987; Nowak 1999). Regardless of the status of the genus of the two species, there is abundant evidence that the lynxes of Asia and Europe are congeners with the bobcat and the lynx of North America (Johnson and O'Brien 1997).

The bobcat and the lynx probably developed from a common ancestor, which emerged from Africa into Eurasia (Werdelin 1981). Phylogenetic research using mitochondrial DNA sequences suggests that the lynx differentiated from its Eurasian equivalent much more recently than did the bobcat, which may have differentiated up to 5.3 million years ago (Johnson and O'Brien 1997). This supports an earlier theory that the two species emerged as a result of two separate invasions of North America by the same ancestral cat from Asia across the land bridge of Berengia. The original invasion that produced the bobcat may have occurred during a postulated mammalian dispersal event of the late Pliocene (2.6 million years ago); the oldest fossil remains in North America have been dated to 2.4–2.5 million years ago. During the ensuing time, the bobcat evolved as a carnivorous generalist and expanded to occupy a variety of habitats. A secondary invasion of North America occurred sometime within the last 200,000 years and resulted in the lynx (Werdelin 1981). Competitive exclusion by the well-established bobcat may have prevented spread of the lynx farther south into North America.

FIGURE 38.1. Distribution of the bobcat (*Lynx rufus*).

DISTRIBUTION

The bobcat is the most widely distributed native felid in North America (Fig. 38.1). It ranges as far north as central British Columbia (55°N) and south to Oaxaca, Mexico (17°N), although the distribution in Mexico is not well documented (Rolley 1987). With the exception of Delaware, the bobcat occurs in all the contiguous United States, although its distribution is greatly restricted in Illinois, Indiana, Iowa, Michigan, Missouri, and Ohio (Woolf and Hubert 1998). Historically, the bobcat occurred in all 48 contiguous states (Young 1958). During the last century, it expanded into northern Minnesota, southern Ontario, and Manitoba as lumbering, fire, and farming opened the dense, unbroken coniferous forests of these areas (Rollings 1945). However, concurrent with the bobcat expansion, the range of the lynx retreated northward, making the causal factor of the expansion unclear. In the midwest, populations declined during the twentieth century probably in response to habitat changes brought by intensive agriculture and persecution by humans (Erickson 1981).

There are 12 recognized subspecies of the bobcat in North and Central America (Hall 1981). Samson (1979) used multivariate statistical analyses of cranial characters to substantiate the number of recognized subspecies, although he suggested that *L. r. rufus* was different enough to divide into an eastern and a western subspecies. In contrast, a detailed morphological study of bobcat skulls from the south-central United States led Read (1981) to suggest that there were far fewer valid intraspecific taxa than Hall (1981) had recognized, because the bobcat

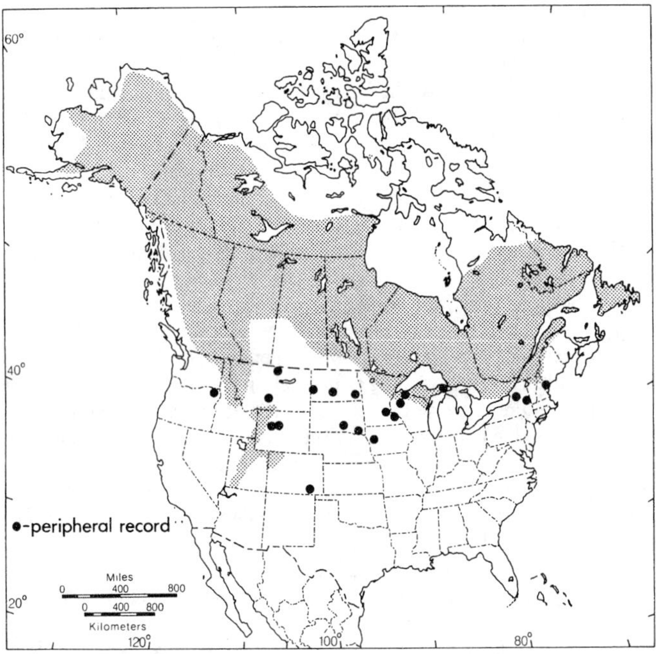

FIGURE 38.2. Distribution of the lynx (*Lynx canadensis*).

is fairly continuous in its distribution with no clear geographic breaks. McCord and Cardoza (1982) suggested that the differences between subspecies are so minor that they have no biological or managerial significance. The notable exception is *L. r. escuinapae* of central Mexico, which was designated as endangered by the U.S. Fish and Wildlife Service in 1976 and is on Appendix I of the Convention on International Trade in Endangered Species (CITES).

The lynx occurs throughout the boreal forests of North America from treeline south to approximately the United States/Canada border (Fig. 38.2). Historically, the species appeared to have ranged into the southern extensions of the boreal forest in the contiguous United States. Lynx were reported in 24 states as far south as Nevada, Utah, and Colorado (McKelvey et al. 2000a). Like the bobcat, habitat changes and human persecution probably extirpated the lynx from much of the contiguous United States and southern regions of Manitoba to Alberta. It also appears to be extirpated from Prince Edward Island and the mainland of Nova Scotia (Koehler and Aubry 1994). The largest populations of lynx in the lower 48 states reside in Washington and Montana. The species also is found intermittently in several other northern U.S. states. Its occurrence in those areas appears to be associated with patterns of lagged synchrony of snowshoe hare (*Lepus americanus*) declines that occur across Alaska and Canada. However, it is unclear whether their appearance is due to the increase in immigrant dispersers from farther north or increased production by local populations (Thiel 1987; McKelvey et al. 2000a).

There are only two recognized subspecies of the lynx in North America: *L. c. canadensis,* found throughout the mainland of North America, and the isolated *L. c. subsolanus* of Newfoundland (Hall 1981). It has been postulated that the long dispersal distances of lynx (see Movement, Activity Patterns, and Dispersal) have allowed the species to remain relatively undifferentiated throughout its range (Koehler and Aubry 1994).

DESCRIPTION

Morphology. Morphologically, the bobcat and the lynx are distinguished from the other felids by a short tail, tufted ears, a flared facial ruff, long legs relative to body length, relatively small head, and absence of the second upper premolars, giving them 28 teeth instead of the normal 30 found in most members of *Felis* (Anderson 1987).

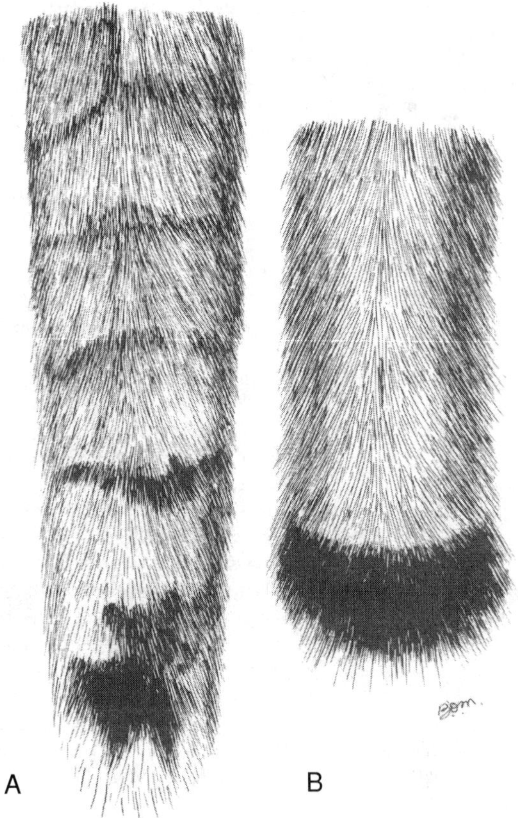

FIGURE 38.3. Tails of (A) the bobcat (*Lynx rufus*) and (B) the lynx (*L. canadensis*). SOURCE: Jackson (1961).

Lynx can be distinguished visually from bobcats by their large furry foot pads, slightly shorter tail, longer black ear tufts (>2.5 cm), and less well defined spotting on the coat. The pelage of the lynx is generally grayer than the reddish-brown color of the bobcat. The tail of the bobcat is banded on the upper surface only (Fig. 38.3A) (Lariviere and Walton 1997). The tail of the lynx is brownish or pale buff white and ends in a black tip that completely encircles the tail (Fig 38.3B).

Both species are digitigrade with sharp, retractile claws. They have four functional toes on the front and hind feet (McCord and Cardoza 1982). The lynx paw can spread in excess of 10 cm and is well furred (Murie 1975). Buskirk et al. (2000a) calculated the average foot loading of bobcat versus lynx and concluded that bobcats have nearly four times the weight per unit area on their foot than does the lynx. This crucial adaptation allows lynx to access their primary prey, snowshoe hares, in deep snows. Snow depth appears to limit the northern distribution of the bobcat and creates a nearly allopatric distribution of the two species (Quinn and Parker 1987).

Size. Adult bobcat weights vary considerably throughout their range. Adult males average 9.6 (6.4–18.3) kg and adult females weigh 6.8 (4.1–15.3) kg (Banfield 1987). The largest verifiable record for an adult male bobcat was from Minnesota, which weighed 17.6 kg (Berg 1979). Total length of males and females, respectively, is 869 (475–1252) and 786 (610–1092) mm (McCord and Cardoza 1982). Bobcat body size appears to generally follow Bergmann's rule, with size increasing with latitude and elevation (Sikes and Kennedy 1992). Analysis of 950 bobcat skulls led Wiggington and Dobson (1999) to postulate that bobcat body size is probably related to seasonality in food abundance and energy demands rather than the thermal advantages of a lower surface-to-volume ratio classically put forward as an explanation (Boyce 1979).

On average, adult lynx are slightly larger than bobcats. Mean body weight is about 10 (6.3–17.2) kg for males and about 8.5 (4.9–11.7) kg for females (Banfield 1987; Quinn and Parker 1987). However, because of the range of body size for bobcats, they often exceed the size of lynx in areas of sympatry (Buskirk et al. 2000a). Among adult males on Cape Breton Island, Nova Scotia, Parker et al. (1983) found mean weight of bobcats exceeded that of lynx by 40%. Total body length of lynx averages 850 mm for males and 820 mm for females (Quinn and Parker 1987).

Sex-related dimorphism is pronounced in the bobcat and to a lesser extent in the lynx. Among adult bobcats, males may weigh from 25% to 80% more and are up to 10% longer than females (Anderson 1987). Sikes and Kennedy (1993) used 26 skull measurements of 1056 adult bobcat museum specimens from the eastern United States to explore the geographic variation in dimorphism. They found the greatest dimorphism occurred in mountainous areas and the least in areas of little topographic relief.

A number of theories have been offered to explain male-favored sexual size dimorphism. Because bobcats have a polygynous mating system where home ranges of several males may overlap a single female range, larger males can more effectively compete for breeding opportunities and therefore are favored by natural selection. It has also been suggested that because male and female home ranges generally overlap, dimorphism may reduce intraspecific competition by permitting optimal exploitation of different-sized prey items. Evidence from several regions supports this niche-partitioning theory, indicating that females take a larger proportion of smaller prey items than do males (Fritts and Sealander 1978a; Knick et al. 1984; Litvaitis et al. 1984). Unfortunately, it is difficult to determine cause and effect with regard to feeding habits because when one sex is larger than the other there will inevitably be differences in feeding habits. In addition, Sikes and Kennedy (1993) found no change in cranial characteristics associated with feeding activities between males and females, casting further doubt on the hypothesis. They did suggest that dimorphism might result from a combination of selection for large size in males to maximize reproductive success and selection for smaller size in females to minimize the energy expense associated with locomotion in the mountains.

Male lynx and bobcats possess rudimentary bacula (Tumlison and McDaniel 1984b; Tumlison 1987). Based on 16 samples from adult bobcats in Arkansas, Tumlison and McDaniel (1984b) concluded that the structure of the baculum was similar to that of the European lynx, but sufficiently different from other felids to support the validity of the genus *Lynx*. The glans penis of the bobcat is short and generally barbed with a backwardly directed spiny papilla (Lariviere and Walton 1997). Females of both species have four mammae.

McKinney and Dunbar (1976) reported an intriguing asymmetry in the size of bobcat adrenal glands. The right glands were significantly smaller than the left. In addition, the glands of females, which are generally larger than those of males in most mammals, were smaller than that of males in the bobcat. The size of the adrenal gland in males closely followed their reproductive status as indicated by testicular weight.

Pelage. The winter fur of the lynx is long and dense. The upper body is generally grizzled grayish brown mixed with buff or pale brown. The summer pelage is darker and more reddish to gray brown. The ventral surface, legs, and feet are grayish white to buffy white throughout the year and are often speckled with brownish-black spots, particularly on the inside of the legs (McCord and Cardoza 1982). Lynx pelts reach their greatest primeness during December and January (Stains 1979).

Bobcat fur is shorter, but equally dense. The fur comes in a variety of colors and patterns, probably reflecting the breadth of habitat types bobcats occupy. The upper body is generally yellowish or reddish brown, while the venter is white with black spots. During the late 1970s, trapping pressure increased dramatically on the bobcat as the fur industry substituted the underbelly fur of the bobcat for the newly banned fur of endangered spotted cats (see Fur Harvest and Value). Their summer coats are generally reddish, whereas winter coats are grayer (Peterson and Downing 1952), suggesting that bobcats undergo two annual molts (McCord and Cardoza 1982). Bobcat pelts reach maximum primeness during January and February (Stains 1979).

Winter fur provides considerable thermal insulation for both species. In an energetics study in New Hampshire, the bobcat's winter pelage allowed it to survive temperatures that were 20°C colder than summer without increasing its standard metabolic rate (Mautz and Pekins 1989).

Melanistic and albinistic color phases have been reported for the bobcat. Several reports of melanistic bobcats have come from Florida (Ulmer 1941; Regan and Maehr 1990). Young (1958) described an albino bobcat in a Texas zoo, which had survived in the wild for 4 years before being captured. Partial albinism, restricted to the forefeet of a bobcat from Washington, also has been reported (Schantz 1939).

Skull and Dentition. Bobcat skulls can be identified by the presence of both a narrow presphenoid bone (<6 mm) and a confluence of the anterior condyloid foramen (hypoglossal canal) and the posterior lacerate (jugular) foramen. Lynx skulls have an inflated presphenoid bone and the anterior condyloid and posterior lacerate foramina are separated (Figs. 38.4 and 38.5) (Jackson 1961; Tumlison 1987). In addition, Ommundsen (1991) identified three other morphometric measures for distinguishing skulls, of which the angle of the infraorbital foramen was most diagnostic. In the bobcat, the long axis is nearly horizontal and intersects the nasal bone, whereas it is closer to vertical in the lynx. Sex-related dimorphism is also evident in cranial structure (Sikes and Kennedy 1993). The dental formula for adults of both species is I 3/3,

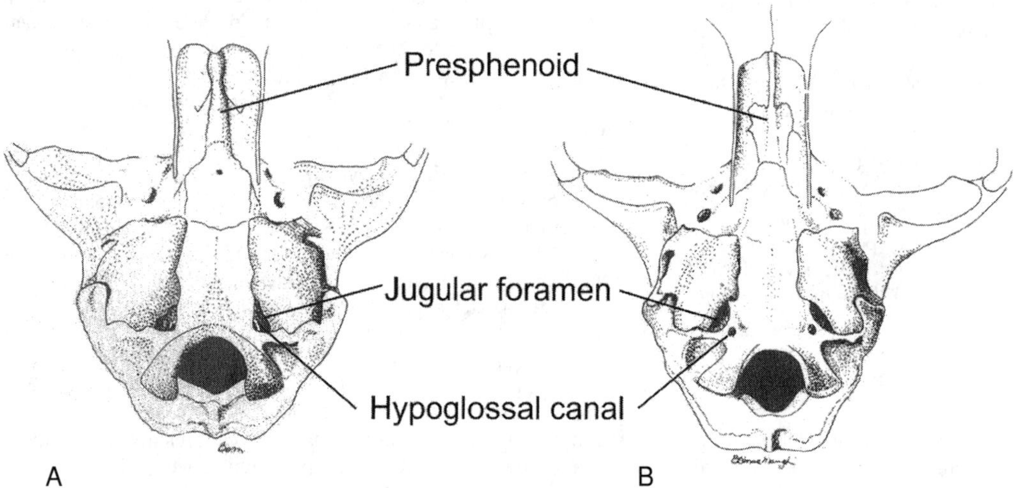

FIGURE 38.4. Basiventral portion of the skulls of (A) the bobcat (*Lynx rufus*) and (B) the lynx (*L. canadensis*).
SOURCE: Jackson (1961).

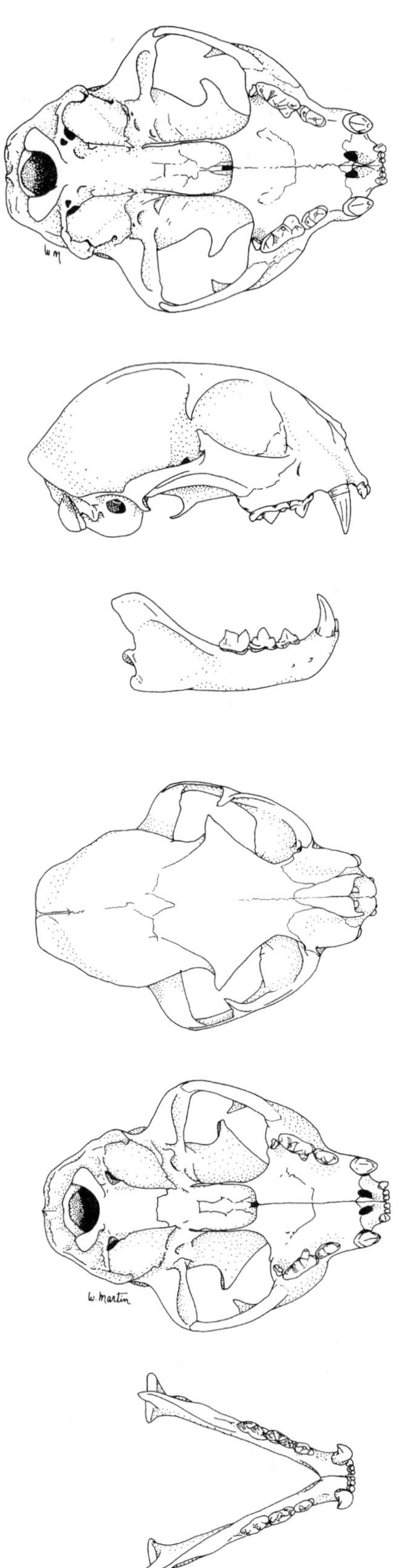

FIGURE 38.5. Skulls of the lynx (*Lynx canadensis*) and the bobcat (*L. rufus*). From top to bottom: ventral view of lynx cranium, lateral view of bobcat cranium and mandible, dorsal view of bobcat cranium, ventral view of bobcat cranium, dorsal view of bobcat mandible.

C 1/1, P 2/2, M 1/1, whereas deciduous dentition is I 3/3, C 1/1, P 2/2, M 0/0 (Saunders 1964; Jackson et al. 1988). Patterns and timing of dental replacement for the bobcat are described by Tumlison and McDaniel (1984a).

Bregmatic bones are a common cranial anomaly in bobcats, occurring in >15% of skulls. However, they are rarely found in the lynx (Manville 1959; Anderson 1987). A variety of other cranial suture and dental anomalies occur in bobcats (Tumlison and McDaniel 1981).

Blood Parameters. Blood parameters have been described for wild bobcats (Fuller et al. 1985c; Knick et al. 1993) and captive animals of both species (Weaver and Johnson 1995; Miller et al. 1999). Based on samples from 56 healthy captive bobcats, Miller et al. (1999) concluded that bobcat blood parameters were similar to reported values for domestic cats (*Felis catus*). Knick et al. (1993) described hematological, biochemical, and endocrine characteristics of bobcats during a prey decline in southeastern Idaho. As lagomorph prey declined, phosphorus and insulin levels dropped, whereas hemoglobin, erythrocyte counts, and packed cell volume increased.

GENETICS

Bobcat and lynx have a diploid chromosome number of $2N = 38$. The autosomal chromosomes are composed of 16 metacentrics, 16 submetacentrics, and 4 acrocentrics. Both sex choromosomes are submetacentric, with a large X and smaller Y (Hsu and Benirschke 1970, 1974). Hybridization between lynx and bobcat has not been reported (Quinn and Parker 1987), although instances of crosses of male bobcats and female domestic cats are known (Young 1958; Gashwiler et al. 1961).

In the bobcat, minisatellite DNA profiles using the multilocus human probe 33.6 indicated that males have significantly more fingerprinting bands than females, and that 30% (6) of the bands were found exclusively in males (Domingo-Roura et al. 1997). This suggests that not only species, but also gender can be clearly identified from DNA samples. However, DNA fingerprinting does not appear to provide an accurate way of assessing relatedness and can only differentiate closer relatives from unrelated individuals (Domingo-Roura et al. 1997).

FEEDING HABITS AND PREDATOR-PREY RELATIONSHIPS

Bobcat Diet. Bobcats are almost exclusively carnivorous (Table 38.1). They most frequently kill prey that weighs between 700 g and 5.5 kg, although their diet is not restricted to that size class (Rosenzweig 1966). Throughout most of their range, rabbits and hares constitute the largest portion of their diet, sometimes exceeding 90% (Dearborn 1932; Bailey 1979; Parker and Smith 1983). However, there are regional variations. In the northern areas, snowshoe hare and white-tailed deer (*Odocoileus virginianus*) predominate (Nussbaum and Maser 1975; Berg 1979; Parker and Smith 1983), whereas in the southeast and southern Central Plains, cotton rats (*Sigmodon* sp.) may be the major prey item (Kight 1962; Beasom and Moore 1977; Miller and Speake 1979). In western Washington, probably because of their abundance, the mountain beaver (*Aplodontia rufa*) constitutes the majority of the bobcat diet (Knick et al. 1984).

Deer represent an important food source for bobcats, particularly in the northern portion of their range, where winter snow depth may make them more vulnerable to predation. As a consequence, deer is the only bobcat food item that shows consistent seasonal shifts in use in many areas, with consumption highest in the winter (Matson 1948; Erickson 1955; Fritts and Sealander 1978a; Miller and Speake 1979; May 1981; Dibello et al 1990). A number of authors (Dearborn 1932; Rollings 1945; Pollack 1951a; Erickson 1955) suggest that the majority of deer eaten by bobcats represent carrion that becomes available following hunting season or due to winter starvation. However, numerous accounts of bobcats killing deer and other ungulates have been given (Marston 1942; Dill 1947; Matson 1948; Erickson 1955; Young 1958; Cook et al. 1971; Beale and Smith 1973; McCord 1974). The majority

Table 38.1. Bobcat feeding habits studies, reported as percentage occurrence in sample

Location	Season	Technique[a] (Sample Size)	Hare	Rabbit	Squirrel	Cotton Rats	Woodrats	Mice, Voles, Shrews	Ungulates	Birds	Other spp.	Reference
Northeast												
Maine	W–Sp	SCAT (248)	75					11	22	6		Litvaitis et al 1986
Maine	Su–F	SCAT (98)	75					25	9	9		Litvaitis et al 1986
New Hampshire	W	STOM (388)	22	27	19			12	22			Litvaitis et al. 1984
Nova Scotia	W	STOM (377)	85						11	7		Parker and Smith 1983
Minnesota	W	STOM (73)	40		16			9	24		12[b]	Berg 1979
Vermont	F–W	STOM (140)	16		9				32			Hamilton and Hunter 1939
New England	W	STOM (208)	38	22					32		18[b]	Pollack 1951a
New England	Annual	SCAT (250)	20	32				14	28			Pollack 1951a
Southeast												
Arkansas	W	STOM (150)		39	22	7	5	9	7			Fritts and Sealander 1978a
Georgia	Annual	SCAT (317)		27		39				11		Kight 1962
Central												
Nebraska	W	STOM (45)		68	10			10		16		Epperson 1978
Oklahoma	W	STOM (81)		41		29	19	23				Whittle 1979
Southwest												
Texas (lower prey)	W–Sp	STOM (51)		25		73	10	16	6			Beasom and Moore 1977
Texas (high prey)	W–Sp	STOM (74)		16		81						Beasom and Moore 1977
Arizona	Annual	SCAT (176)		38		67[c]		3	4			Jones and Smith 1979
Northwest												
Washington (eastern)	F–W	STOM (123)	26								42[d]	Knick et al. 1984
Washington (western)	F–W	STOM (324)	20	15				11	11			Knick et al. 1984
Oregon (Coast Range)	Annual	SCAT (143)		44			11	13	2			Nussbaum and Maser 1975
Oregon (Cascades)	Annual	SCAT (34)	71		24			12	3			Nussbaum and Maser 1975
Idaho (mountains)	W	STOM (197)	62					20[c]	12	16		Bailey 1979
Idaho (plains)	W	STOM (36)	67					17		25		Bailey 1979

[a] STOM, Stomach contents; SCAT, feces.
[b] Porcupine or includes porcupine.
[c] All rodents combined.
[d] Mountain beaver.

of deer taken is fawns or does that are generally in poor physical condition, although healthy adult bucks have also been killed (Matson 1948; Young 1958). Deer predation is not limited to winter. Bobcats were an important cause of mortality among deer fawns on Steens Mountain, Oregon (Trainer 1975), accounting for 10% of fawn mortality. Epstein et al. (1983), working on two islands off the South Carolina coast, found 54% ($n = 26$) of their radio-collared white-tailed deer fawns succumbed to predation. Where the predator could be identified, bobcats were responsible for 67% of the mortalities. Nelms et al. (2001) suggested that predation by reintroduced bobcats on Cumberland Island resulted in a decline in deer density. Beale and Smith (1973) reported bobcats took 23% of the pronghorn (*Antilocapra americana*) fawn crop during a 5-year period in a dense scrub area of the southwest United States.

The remaining mammalian portion of the bobcat diet includes an assortment of rodents, which vary with habitat and availability. Several species of squirrels are used in small, but consistent amounts, with the exception of the southwest, where bobcats seem to depend heavily on woodrats (*Neotoma* sp.). Voles (*Microtus* sp.) and mice also are taken regularly (Anderson 1987).

Bobcats also consume a variety of game and nongame birds. Bailey (1979) reported birds in 25% of the scats examined from the sagebrush steppe of Idaho, although most studies have found remains of birds in <5% of their samples (Nussbaum and Maser 1975; Jones and Smith 1979; Parker and Smith 1983).

Bobcats also eat reptiles, fish, amphibians, insects, and eggs. Delibes et al. (1997) surveyed 38 bobcat feeding habits studies and found that the occurrence of reptiles increased as latitude decreased. Only 5% of the studies north of 40° N latitude reported reptiles as a food item, whereas 78% of the studies south of 40°N latitude did. Reptiles constituted up to 15% of the bobcat diet in some areas.

Although the heavy dependence on lagomorphs throughout their range suggests that bobcats are not purely opportunistic in their prey selection, several studies have shown that the type and diversity of prey consumed is influenced by availability. In southeastern Idaho, during two dramatic declines in black-tailed jackrabbit (*Lepus californicus*) populations, the proportion of small mammals and birds in bobcat diets increased while jackrabbits decreased substantially (Bailey 1981; Knick 1990). A similar switching to alternate prey was observed in Florida when cotton rat populations declined (Maehr and Brady 1986). In southern Texas, a sudden increase in the abundance of the main prey items of the bobcat (cotton rat and cottontail rabbits) decreased the number of species consumed from 21 to 6 (Beasom and Moore 1977). In contrast, Jones and Smith (1979) in central Arizona found that the monthly occurrence of lagomorphs and rodents in bobcat scats collected throughout a year did not vary significantly, even though lagomorph and rodent populations varied considerably.

Lynx Diet. Lynx are heavily dependent on snowshoe hare (Table 38.2). Hares constitute 35–100% of their diet and significantly influence their distribution and abundance (Koehler and Aubry 1994). Other prey items are taken opportunistically and include red squirrels (*Tamiasciurus* sp.), grouse, mice and voles, ungulates, flying squirrels (*Glaucomys* sp.), ground squirrels (*Spermophilus* sp.), beavers (*Castor canadensis*), muskrats (*Ondatra zibethicus*), fish, waterfowl, ptarmigans (*Lagopus* sp.), and other birds (Quinn and Parker 1987; Koehler and Aubry 1994). Lynx appear to exhibit a Type 2 functional response to increasing hare numbers (Mowat et al. 2000). As the density of hare increases, the number consumed per lynx also increases, but at some point levels off even with further increase in hare density (Holling 1959).

Switching to alternate prey during snowshoe hare declines has been extensively documented. Brand and Keith (1979) observed the biomass of hare in the lynx diet decline from 97% when hares were abundant to 65% when they were scarce. There was a corresponding increase in consumption of ruffed grouse (*Bonasa umbellus*), red squirrels, ungulate carrion, mice, and voles. In southwestern Yukon, red squirrels made up only 0–4% of the biomass of the lynx diet during high snowshoe hare densities, but rose to 20–44% during years of the lowest hare density (O'Donoghue et al. 1998b). Mowat et al. (2000) felt that red squirrels in particular were an important alternate food source for lynx in the North during snowshoe hare declines.

The summer and fall diets of lynx also generally contain less hare and more alternate prey, although very few feeding habits studies have been conducted during snow-free periods (Aubry et al. 2000a). In Newfoundland, Saunders (1963a) found a greater percentage of voles (14–30%) and birds (28–47%) in the summer diet than the winter diet (voles 5–9%, birds 3–22%). On the Kenai Peninsula, Alaska, Staples (1995) compared percentage occurrence of prey items found in winter versus summer scats. Hare declined from 64% to 34% and red squirrel increased from 11% to 28%. Similarly in the Gaspé Peninsula of Quebec, researchers found the percentage of hare declined from 85% during winter to 58% during snow-free periods, with small mammals and red squirrels constituting the majority of the alternate prey (Mowat et al. 2000).

Lynx occasionally prey on ungulates, particularly fawns and calves, but they are generally a relatively minor part of their diet (Bergerud 1971; Parker et al. 1983; Koehler 1990; Stephenson et al. 1991). Deer, caribou (*Rangifer tarandus*), Dall's sheep (*Ovis dalli*), moose (*Alces alces*), and bison (*Bison bison*) have all been reported as prey items, although they may be scavenged primarily as carrion. Locally and at certain time periods, consumption of ungulates may be substantial. Saunders (1963a) found moose made up 71% of the stomach volume of nine lynx taken in the fall on Newfoundland. In general, ungulate consumption appears to increase in fall and winter and during periods of low hare abundance. The only notable exception to the minor impact lynx have on ungulate populations comes from Newfoundland. Following the crash of a newly introduced snowshoe hare population, predation on caribou calves by lynx was so heavy that it became the primary factor limiting growth of the caribou population (Bergerud 1971).

Bobcats depredate domestic livestock, although their impact is generally minor and localized. Sheep, goats, and chickens are particularly susceptible. Young (1958) reported that 12% of 3990 bobcat stomachs collected in the western United States from 1918 to 1922 contained sheep or goat tissues. He also reported that a single bobcat was reputed to have killed 38 lambs in one night. In contrast, Neale et al. (1998) found that none of 64 predator kills of sheep in north-coastal California were attributable to bobcats, although the species was common in the area. In the western United States, bobcat depredation accounts for <10% of all sheep and goat losses (Virchow and Hogeland 1994). Lynx predation on livestock is rare (McCord and Cardoza 1982).

Grass is a small but consistent component in bobcat diets, occurring in up to 66% of scats examined (Miller and Speake 1979). It has also been reported in lynx feeding habits studies, although at a lower level of incidence (Saunders 1963a; Stewart 1973). Some grass and other vegetative materials are probably ingested incidentally with the digestive system of prey items, or intentionally while the bobcat is caught in a trap. However, the relatively large, unaltered boluses of grass found in some bobcats suggest that grass is often ingested intentionally and may act as a purgative as with domestic cats (Rollings 1945; Story et al. 1982).

Impact on Prey. The impact of lynx predation on hares is significant. Lynx respond to higher hare numbers by increasing their numbers (numerical response) and by increasing their individual kill rates (functional response). Keith et al. (1977) and O'Donoghue et al. (1998b) estimated that individual lynx killed three to five times the number of hares during the cyclic peak than during the lows. Combining the numerical and functional responses by lynx gave an estimated percentage of hares consumed by lynx that varied from 2% to 32%. The greatest predation rates occurred 1–4 years after the cyclic peak. A study in southwest Yukon experimentally excluded lynx and coyotes (*Canis latrans*) while providing extra food for the hares. In the absence of the predators, hare densities were 11 times higher than on the control areas (Krebs et al. 1995). Lynx can also have an impact on ungulate populations under rare circumstances. In Newfoundland, lynx were experimentally removed in an area, with a nearly 100% increase of caribou calf survival (Bergerud 1971).

TABLE 38.2. Lynx feeding habits studies, reported as percentage occurrence in sample

Location	Season	Technique[a] (Sample Size)	Hare	Red Squirrel	Small Mammals	Grouse/ Ptarmigan	Other Birds	Ungulates	Other spp.	Reference
Southern boreal forests										
Washington	Annual	SCAT (29)	79	24	3			7		Koehler 1990
Nova Scotia	W	STOM (75)	97	1	3	3		5		Parker et al. 1983
Nova Scotia	W	SCAT (55)	93		7	4	2	5		Parker et al. 1983
Nova Scotia	Su	SCAT (441)	70	4	4–5	1	6–7	9	<1	Parker et al. 1983
Northern boreal forests (low hare density)										
Central Alberta	W	STOM	35	12	33	6	6	3	7	Brand and Keith 1979
Southwest Yukon	W	SCAT (101)	79	50	21				7	O'Donoghue et al. 1998b
Alaska	W	SCAT (161)	91	15	13	10	1	7	6	Staples 1995
Alaska	Su	SCAT (42)	67	50	29	12	10		10	Staples 1995
Northern boreal forests (high hare density)										
Central Alberta	W	STOM	90		5		3		1	Brand and Keith 1979
Southwest Yukon	W	SCAT (239)	92	5	16				20	O'Donoghue et al. 1998b
Alaska	W	SCAT (41)	100	5	7		10			Kesterson 1988
Alberta and NWT	W	STOM (52)	79	2	10	10	13	12	2	van Zyll de Jong 1966a
Alberta and NWT	Su	STOM (23)	52	9	39	8	26			van Zyll de Jong 1966a

SOURCE: Adapted from Aubry et al. (2000a).
[a] STOM, Stomach contents; SCAT, feces.

Although not as well studied as in the lynx/hare interaction, bobcat predation may also significantly affect prey populations temporally and spatially. Schnell (1968) reported that hawks, fox, and bobcats held a cotton rat population at a "predator-limited carrying capacity" in South Carolina. Pronghorn numbers in some areas in Texas were considered limited by a high incidence of bobcat predation on fawns (Beale and Smith 1973).

Influence of Sex and Age. Several investigators have reported differences in the diet of male and female bobcats. Fritts and Sealander (1978a) analyzed bobcat stomachs from carcasses of known age and sex from Arkansas and found that females consumed a significantly higher percentage of rats and mice than did males. Litvaitis et al. (1984) in New Hampshire found that males consumed significantly more deer and fewer cottontails than did females. Similarly, Rolley and Warde (1985) reported that male bobcats in Oklahoma consumed more cotton rats, tree squirrels (*Sciurus* sp.), and large mammals, whereas females ate more cottontails and deer mice (*Peromyscus maniculatus*). Differences in diet may be the result of sexual size dimorphism (see Morphology) or different foraging strategies promoted by social organization and energetic demands of maternity (Sikes and Kennedy 1993).

The age of individuals also influences the type of prey consumed. Litvaitis et al. (1984) reported that yearlings and adults consumed white-tailed deer significantly more often than did juveniles (<1 year old). Young bobcats were probably not as skillful in capturing larger prey and therefore used smaller, more easily captured prey. Whittle (1979) in Oklahoma and Toweill (1982) in Oregon also found that rodents occurred more frequently and rabbits and hares less frequently in the diet of juveniles than in that of adults and yearlings. Sweeney (1978) found that stomachs of male bobcats from Washington contained approximately twice the weight of prey items than those of females, suggesting that females may feed more frequently.

Similar studies of sex- and age-specific differences in the lynx diet have not been done (Mowat et al. 2000). However, a study of the similar European lynx (*Lynx lynx*) found males consumed more deer and fewer hare than females in southwest Finland, although there were no differences in diet in east Finland (Pulliainen et al. 1995).

Energetics. The energetic and nutritive value of bobcat prey items varies. In a comparison of the digestibility of white-tailed deer (meat and viscera only), snowshoe hare, gray squirrels (*Sciurus carolinensis*), and small mammals by bobcats, Powers et al. (1989) found that 77.3% of the gross energy contained in snowshoe hares was digested, 82.6% of that in the small mammals, 86.8% of that in the squirrels, and 95.4% of that in the deer. They also found considerable difference in the nutritional value of the selected prey items. Most of the items contained similar levels of crude protein (60–66%) and fat (15–21%), although snowshoe hare had significantly higher crude protein (73%) and lower fat (2.9%). The metabolizable energy/g dry weight was highest for the deer diet (5.58 kcal/g) and lowest for the hare (2.93 kcal/g).

The standard metabolic rate for bobcats in the thermoneutral zone (79 kcal/kg$^{0.75}$/day) is constant among seasons based on research in New Hampshire (Mautz and Pekins 1989). However, bobcats have a relatively high lower critical temperature in winter (−2.2°C), which means that below this temperature, energetic costs of thermoregulation rise steeply. In the northern portion of their range, this may substantially increase their winter food requirements. It also may promote behaviors that moderate the ambient temperature, such as seeking more favorable microclimates or actively sunning (Gustafson 1984; Mautz and Pekins 1989).

Using Mautz and Perkins's (1989) estimates of standard metabolic rates, Powers et al. (1989) combined them with digestibility trials to estimate the minimum prey requirements for bobcats. A resting adult male (15 kg) bobcat during a 120-day winter season would need to consume 1.8 deer, 63 snowshoe hares, 3633 small mammals, or 102 squirrels. Considering that free-ranging large mammals generally have daily energy expenditures two to three times their fasting metabolic rate (Robbins 1993) and that winter temperatures may add additional thermoregulatory costs, prey requirements may actually be more than three times higher. In addition, lactating females with sole responsibility for rearing kittens require considerably more prey. Kitchener (1991) suggested that smaller lactating felids like bobcats might require two to three times the prey resources of nonlactating individuals.

Energetic requirements of lynx have not been well studied. Based on snow tracking and captive animals, Saunders (1963a), Nellis et al. (1972), and Brand et al. (1976) estimated the consumption rate of prey at 593, 600, and 960 g/day, respectively. Saunders (1963a) estimated that lynx would need to consume 170–200 adult hare/year, which is consistent with the estimated prey requirements for bobcats.

All felids appear unable to convert beta-carotene into fat-soluble vitamin A (retinol). Therefore, all vitamin A must be obtained from the liver, lungs, adrenals, or kidneys of their prey (Scott 1968). A lack of vitamin A appears to, among other things, adversely affect egg implantation, which may explain the mechanism that reduces conception rates in bobcats and lynx during prey declines.

REPRODUCTION AND DEVELOPMENT

Estrus. Bobcat and lynx are polygamous, which partly explains the observed sexual size dimorphism in both species (Sikes and Kennedy 1993). Bobcat females are considered seasonally polyestrus, as are lynx (Pollack 1950; Duke 1954; Gashwiler et al. 1961; Fritts and Sealander 1978b; Crowe 1975a; Mehrer 1975; Stys and Leopold 1993). A study of carcasses in Wyoming suggested that bobcats probably experienced up to three estrous cycles from March through June if they were not impregnated during one of the ovulations (Crowe 1975a). Stys and Leopold (1993) observed a similar trend in captive bobcats. Females cycled a second or third time as a result of unsuccessful breeding, prevention of breeding, aborted or resorbed litters, or the death or removal of kittens. Using estrogen levels in urine and the cytological appearance of the vagina, Mehrer (1975) also tracked each of three bobcats through five estrous periods during 2 years. Length of estrous cycles averaged 7.6 days. The entire estrous cycle lasts approximately 44 days, with females in estrus 5–10 days (Crowe 1975a; Mehrer 1975).

Bobcats probably are spontaneous ovulators (Woshner 1988). Initial researchers assumed that, like domestic cats, bobcats were induced ovulators (Asdell 1946; Colby 1974), and there is some limited evidence to support their contention. Male bobcats generally possess a barbed penis and engage in coitus repeatedly (Mehrer 1975), both characteristic of induced ovulators. However, as noted, if a female in estrus is not bred, she may cycle up to three times a year (Crowe 1975a; Fritts and Sealander 1978b; Stys and Leopold 1993). Clearly, bobcats can ovulate without the stimulation of a male, but coitus may induce or hasten ovulation. Less is known about the reproductive biology of the lynx, although ovulation is assumed to occur in much the same way (Quinn and Parker 1987).

The histology of the ovaries of the lynx and bobcat differ substantially from those of most other mammals. In most mammals, the ruptured follicle becomes a corpus luteum, which then degenerates into a corpus albicans and disappears within a year. Bobcat and lynx, however, retain the corpora lutea indefinitely, never producing true corpora albicantia. The corpora lutea from the most recent ovulation (called the corpora lutea of previous cycles) are lighter and have a yellowish composition. They have been used to estimate the most recent reproductive effort (Saunders 1961; Crowe 1975a).

Breeding Season. The majority of lynx breeding activity occurs in March and April (Saunders 1961; Nava 1970), although it has been reported as late as May in Alberta (Nellis et al. 1972). Breeding apparently can occur in bobcats any time, because litters have been reported in every month (Duke 1954; Young 1958; Gashwiler et al. 1961; Fritts 1973; Crowe 1975a). The majority of bobcat breeding, however, occurs during February and March, although in more southern latitudes, it commences earlier and continues longer (Blankenship and Swank 1979; Parker and Smith 1983). The breeding season probably varies in both species with latitude, longitude, altitude, climate, photoperiod, and prey availability (McCord and Cardoza 1982).

Sexual Maturity. Bobcat and lynx females are capable of reproducing during their first year (9–12 months of age), although rarely do (Crowe 1975a; Rolley 1985). The onset of sexual maturity may be influenced by prey availability. In Oklahoma, Rolley (1985) observed the rate of yearling pregnancies in bobcats fluctuated with prey availability. A similar relationship has been demonstrated in the lynx, where the percentage of juveniles conceiving depends on the abundance of snowshoe hare (Nava 1970; Brand and Keith 1979; Parker et al. 1983). Evidence also suggests that bobcat and lynx yearlings cycle later and generally have lower pregnancy rates than adults (Fritts 1973; Crowe 1975a; Knick et al. 1985; Anderson 1987; Stys and Leopold 1993; Koehler and Aubry 1994). Stys and Leopold (1993) speculated that because they are not yet full grown, yearling female bobcats might not be physically or energetically capable of maintaining a pregnancy. They also do not have corpora lutea of previous cycles yet, so they may lack the necessary hormonal secretions to maintain pregnancy.

Juvenile males of both species are not sexually active their first year, although nearly all males appear capable of breeding during their second winter (Saunders 1961; Crowe 1975a). Crowe (1975a) found evidence that spermatogenesis in bobcat and lynx may be reduced or arrested during July and August, but recommences in September or October. A similar reduction in gametogenesis during summer and fall was reported for lynx (Saunders 1961; van Zyll de Jong 1963). Males and females in natural populations are sexually active until death (Crowe 1975a).

Gestation, Litter Size, and Pregnancy Rate. Gestation in the bobcat and lynx is comparable to that in domestic cats. Estimates range from 63 to 70 days for both species (Saunders 1961; Ewer 1973; Mehrer 1975). Using 8 years of observations on captive bobcats, Stys and Leopold (1993) found the average length of gestation was 65.8 days from the first observed copulation and 61.7 days from the last observed copulation.

Average litter sizes are estimated through counts of either corpora lutea, placental scars, embryos, or live litters. However, there is a discrepancy between average litter size based on placental scars and the ovulation rate as indicated by the number of recent corpora lutea. This suggests either follicles degenerate without ovulating or all the ova shed do not implant. Placental scar counts also probably overestimate actual fecundity due to intrauterine and postnatal mortality. Beeler (1985) found that average litter size in wild Mississippi bobcats was 3.2 based on all placental scars. However, if only dark black scars were counted, the litter sizes averaged only 2.5 kittens/litter. This compared with observed average litter size for captive bobcats of 2.0 kittens/litter. The discrepancy between placental scars and actual field productivity seems particularly marked in yearlings (Knick 1990).

Despite the problems with estimating average litter size, a number of authors have reported means. Anderson (1987) surveyed 21 bobcat studies and found that average litter sizes ranged from 1.7 to 3.6 kittens/litter, with a mean of 2.7. There were no apparent regional trends. Lynx also have from 1 to 5 kittens, with an average litter size of 3.7 (McCord and Cardoza 1982). In both species, yearling females consistently produce smaller litter sizes than do older adults (Brand and Keith 1979; Anderson 1987; Koehler and Aubry 1994).

Bobcats and lynx generally have only one litter/year, although if a bobcat litter is lost shortly after parturition, the female is capable of cycling again and producing a second litter (Winegarner and Winegarner 1982; Beeler 1985; Stys and Leopold 1993). In the lynx, there is some evidence to suggest that during low hare abundance some individuals may only have litters in alternate years (Saunders 1961).

The rate of pregnancy and average litter size, in addition to being influenced by age, may be related to the availability of prey or other density-dependent factors. In Idaho during a decline in jackrabbits, Knick (1990) observed the pregnancy rate of adult bobcats decrease from 100% to 12.5%. A similar pattern has been observed in lynx. Summarizing three studies, Koehler and Aubry (1994) reported that during periods of hare abundance, pregnancy rates ranged from 33% to 79% for yearlings and 73% to 92% for adults, whereas during lows in the hare cycle only 0–10% of yearlings and 33–64% of adults became pregnant. Similarly, average litter size (as measured by placental scars) was 3.5–3.9 for yearlings and 4.4–4.8 for adults during peaks in hare numbers, but decreased to 0.2 for yearlings and 1.4–3.4 for adults when hares were scare. Density of bobcats and the concomitant change in social interactions also may affect pregnancy rates. Lembeck and Gould (1979) observed that only half of the females on their study area became pregnant when population density was highest, compared to 100% when density was low.

Neonatal Size and Development. Kittens are born blind and helpless. Bobcat kittens commonly weigh 150–340 g and fully open their eyes in 9–18 days (Pollack 1950; Young 1958; Stys and Leopold 1993). Lynx kittens are of a similar size and take 10–17 days to open their eyes (Tumlison 1987). The first deciduous teeth (incisors and canines) begin appearing in 11–14 days and are fully erupted by 9 weeks. Permanent dentition appears at 16–19 weeks and is completed by 34 weeks (Jackson et al. 1988). Crowe (1975a) developed growth curves for male and female bobcats in Wyoming based on age in days (X): male weight (g) $= 11{,}652/(1 + 22.13^{e-0.0135X})$ and female weight (g) $= 8976/(1 + 2.79^{e-0.6056X})$. The curves indicated that males grew faster than females and achieved their maximum weight in approximately 500 days, whereas it took 700 days for females. This conflicts with other literature, which suggests females may achieve full size in a shorter period of time than males (Colby 1974; Read 1981).

Bobcat kittens emerge from the den in 33–42 days and begin to eat solid food shortly afterward (Stys and Leopold 1993). Young bobcats generally start accompanying the female when they are 3 months old (Bailey 1979). Kittens of both species generally stay with the female until the next breeding season. The young often remain in their natal range for several more months until they disperse or settle locally (see Movement, Activity Patterns, and Dispersal).

ECOLOGY

It has become increasing clear that lynx in the southern montane portion of their range are ecologically distinct from those in the northern boreal area. The classic 10-year cycle of rise and fall in hare and lynx populations is conspicuously absent from populations in the southern portion of their range (Koehler and Aubry 1994). The southern populations tend to consistently display feeding habits, home range sizes, densities, and reproductive characteristics similar to those that occur in the northern boreal lynx populations during periods of low hare numbers. The differences appear to be related to the noncycling nature of snowshoe hare populations, their chronic low abundance, a wider diversity of alternate prey, habitat patchiness, and a wider array of competitors and predators (Aubry et al. 2000a). Because of its impact on management, whenever the data exist, the differences in lynx ecology between regions have been noted.

Age and Sex Structure. Sex ratios of lynx and bobcat kittens are normally 1:1 (Anderson 1987; Stys and Leopold 1993; Poole 1994; Slough and Mowat 1996). Much of the information on yearling and adult sex ratios comes from harvest data and may reflect relative trapping vulnerability rather than actual sex ratios in the wild. In both species, male-dominated harvests suggest that males are more vulnerable to trapping mortality than females (Mech 1980; Parker at al. 1983; Quinn and Thompson 1987). Harvest records from numerous studies indicate that in exploited populations males are taken more frequently in the younger age cohorts, whereas females make up a larger percentage of the older cohorts (Crowe and Strickland 1975; Fritts and Sealander 1978b; Brand and Keith 1979; Parker and Smith 1983). Gilbert (1979) suggested that sex ratios in bobcats might reflect the intensity of harvest. He expected lightly and moderately exploited populations to show a preponderance of males in the harvest, but greater harvest pressure would result in a more even sex ratio. Knick et al. (1985) found that the proportion of male bobcats in a sample from western Washington increased as the harvest season progressed. They suggested that as the breeding season approached, males moved more frequently between female ranges to assess their breeding status and therefore made themselves more vulnerable to trapping. McCord and Cardoza (1982) warned, however, that much of the sex ratio data from harvest is suspect because misidentification of sex is common among field personnel.

A number of studies that examined the age distribution of harvested bobcat populations have found that kittens are underrepresented in the sample and that the proportion of yearlings usually exceeds that of kittens (Bailey 1979; Brittell et al. 1979; Parker and Smith 1983; Parker et al. 1983). It is unknown whether this is due to differences in capture vulnerability or differences in reporting rates by hunters and trappers. Pelts from juvenile bobcats are commercially valuable, and pelts of juvenile lynx are more valuable than pelts of adults because their fur is finer and silkier (Obbard 1987). Blankenship and Swank (1979) found that the proportion of kittens represented in the harvest increased from 6.7% in November to 31% by February. Similarly, the proportion of kittens harvested from Cape Breton Island, Nova Scotia, increased from a little over 10% in November to more than half the sample by March (Parker and Smith 1983). The increasing vulnerability of kittens to harvest probably results from a combination of their independence from the protection of the adult female, increased movements as they begin to disperse, and increased vulnerability after the adult female is removed by trapping.

The proportion of young animals (<2 years old) in a population is closely related to the intensity of harvest. Unexploited populations are largely composed of older individuals, whereas younger animals dominate exploited populations. This may result from increased reproduction, higher adult mortality, or both. Lembeck and Gould (1979) found 16% of the bobcats in an unexploited population in California were <2 years old, compared to 43% for an exploited population in similar habitat. In some areas of intense harvest and low densities, 1- and 2-year-olds may make up as much as 76% of the population (Fredrickson and Rice 1979).

Prey density is directly related to the proportion of kittens in the lynx harvest. Because reproduction and kitten survival are so low during declines in the hare population, the number of kittens in the harvest dramatically decreases. O'Conner (1984) saw the proportion of kittens and yearlings in the harvest ($n = 745$) decrease from 40% for both age groups during a period of high hare abundance to 0% and 8%, respectively, at low hare abundance. Under similar prey declines, Brand and Keith (1979) reported the proportions of kittens decreased from 31% to 7%, and Parker et al. (1983) noted a decrease from 29% to 2%. As a result of decreased recruitment, the average age of trapped lynx increases as hare numbers decline.

Density of bobcat populations may also affect sex ratios. Lembeck and Gould (1979) noted that in an unharvested population in California at the highest density, the sex ratio was 2.1 males/female. The ratio decreased to 0.86 male/female when density was at its lowest. Zezulak and Schwab (1979) observed an extremely skewed sex ratio of 7 males/female in adult bobcats in the Mojave Desert. They hypothesized that males were selected for at high densities when competition was intense. The unbalanced sex ratio would limit reproduction until mortality, emigration, or environmental shifts reduced the population density and ameliorated conspecific interactions and competition.

The maximum age attained by a bobcat in captivity is 32.2 years (Jones 1982) and for a lynx is 26.9 years (Rich 1983). Longevity in the wild is significantly less, with neither species surviving more than 16 years (Knick et al. 1985; Quinn and Parker 1987).

Mortality. Rates and causes of mortality can be most meaningfully considered in three separate periods: kitten (<1 year old), yearling (1–2 years old), and adult (>2 years old). Kitten survival is poorly understood and often inferred indirectly because they are difficult to radio-instrument or monitor frequently without potentially influencing their survival rates. They are also generally underrepresented in the harvest, which makes survival rates calculated from life tables unreliable at best. Placental scars also appear to overestimate actual birth rates, and hence inflate kitten mortality rates. Yearlings are often in a "transient period." During this time, bobcats and lynx leave maternal care, seek their own home range, and are subjected to a wide array of mortality factors. Adults are usually "resident" animals with their own home range and generally lower mortality rates. Unfortunately, although the mortality rates may be considerably different, the adult and yearling classes have been combined and are reported simply as an adult survival rate.

Survival rates of bobcat and lynx kittens, like those of most juvenile mammals, are generally lower than that of adults and are highly variable. Crowe (1975b) used life tables to estimate that kitten survival rates in Wyoming fluctuated from 18% to 71% and averaged 26% from 1948 to 1973. Blankenship and Swank (1979) reported a 29% survival rate for kittens in Texas, and Hoppe (1979) estimated 33% in Michigan.

Kitten survival rates are strongly influenced by prey abundance. During declines in rabbit numbers in Idaho, no bobcat kittens survived to the fall, even though survival had been high during previous years of greater prey abundance (Bailey 1974; Knick 1990). Bailey (1974) speculated that adults fed themselves first, leaving the young of the year to succumb to starvation-related deaths. Zezulak (1981) also observed that two of three radio-collared juveniles died of malnutrition and parasitism. Similar findings have been reported for lynx kittens during periods of snowshoe hare scarcity (Nellis et al. 1972). Brand and Keith (1979) observed a 65–95% mortality rate for kittens in Alberta during a 3-year period of low hare numbers, presumably due to starvation. Kitten mortality or failure to reproduce appears to be the primary mechanism responsible for declining lynx numbers during periods of snowshoe hare declines (see Lynx–Hare Cycle).

Several techniques have been used to estimate adult and yearling survival rates, but daily survival rates calculated from radiotelemetry data (Heisey and Fuller 1985) probably give the best estimates. Excluding heavily exploited populations or during periods of dramatic prey declines, adult bobcat survival rates range from 56% to 67% (Table 38.3). Similar results have been obtained using life tables. In Wyoming, Crowe (1975b) estimated a 67% adult survival rate, and in South Dakota, Fredrickson and Rice (1979) estimated 60% for adult bobcats. Rolley (1985) estimated adult survival in Oklahoma at 53%, which was similar to his radiotelemetry estimate of 56%. Using simulation techniques with his population model, Knick (1990) determined that his Idaho bobcat population could not sustain itself when adult female survival was <52%.

Few data are available on adult survival rates in unexploited populations because even in closed areas, populations are affected by both illegal harvest and legal harvest in adjacent areas. However, adult survival rates in unexploited populations appear to be much higher than in exploited populations. Bailey (1974) observed only three natural mortalities among 35 resident adults in his 3-year study on the closed Idaho National Engineering Laboratory (INEL), resulting in an apparent annual adult survival rate of 97% (Crowe 1975b). Knick (1990) also felt that an adult survival rate of 78% that he observed in the same area of the INEL was near the maximum for wild bobcats. Chamberlain et al. (1999) observed a similar survival rate of 80% in an unexploited population in Mississippi.

Human exploitation, both legal and illegal, appears to be the most prevalent cause of mortality in bobcats and in some populations of lynx subject to annual harvest. In Minnesota, 82% of mortalities in two bobcat studies were attributable to harvest (Berg 1979; Fuller et al. 1985a). At one of those study sites, 100% of mortality was due to legal and illegal harvest and adult male annual survival was only 8% (Fuller et al. 1985a). Rolley (1985) also observed that all mortality in his bobcat study in Oklahoma was due to exploitation. Hamilton (1982) found in Missouri that 50% of juvenile and 80% of adult bobcat mortality was human caused. In an analysis of eight radiotelemetry bobcat survival studies, Fuller et al. (1995) determined that 47% of all deaths were due to legal harvest.

Harvest of bobcats and lynx appears to be primarily additive to other forms of mortality. The highest adult survival rates come from populations that are essentially unexploited (see Table 38.3). However, the effect on population growth is not as evident because overall population size may remain relatively constant, although the proportion of yearlings and the percentage breeding increase with lower adult

TABLE 38.3. Annual bobcat survival rates and mortality factors of exploited and unexploited populations based on radiotelemetry data

	Annual Survival Rate				Cause-Specific Mortality (%)						
Location[a]	Males	Females	Total	n[b]	Hunt/trap	Poach	Starve	Disease	Predation	Other[c]	Reference
Maine	0.70	0.55	0.67	11	55					45	Litvaitis et al. 1987
Massachusetts	0.50	0.70	0.62	6		21	16	37	26		Fuller et al. 1995
Minnesota (BV)	0.35	0.08	0.19	5	42	58					Fuller et al. 1985a
Minnesota (HC)	0.73	0.57	0.61	9	82					18	Fuller et al. 1985a
Missouri			0.58	11	30	42	20			7	Hamilton 1982 (as calculated by Fuller et al. 1995)
Mississippi[d]	0.75	0.84	0.80	31	45			4	6[e]	45	Chamberlain et al. 1999
Oklahoma			0.56–0.66	5–7	100						Rolley 1985
Idaho (IN)[d]	0.70	0.63	0.67	10	12	21	24		33	24	Knick 1990
Idaho (BC)	0.14	0.62	0.49	7	55	12				29	Knick 1990
California				29	14		10	17	31	28	Lembeck 1986

SOURCE: Adapted from Fuller et al. (1995).
[a] BV, Bearville; HC, Hill City; IN, Idaho National Engineering Laboratory; BC, Box Canyon.
[b] Total of dead radio-marked bobcats.
[c] Includes vehicle collisions, accidents, study-related deaths, unknown causes.
[d] Area closed to harvest.
[e] Intraspecific strife.

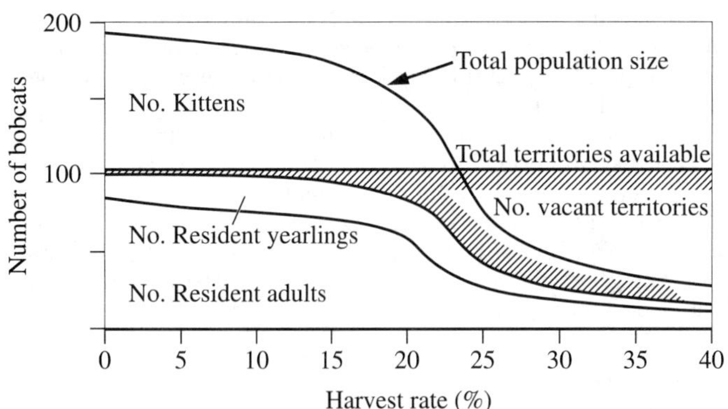

FIGURE 38.6. Relationship between harvest intensity and population size, age categories, and number of occupied territories. The shaded area indicates the number of vacant territories at varying harvest intensities. SOURCE: Data from Knick (1990).

survival rates. Knick (1990) specifically created a population model to describe bobcat population responses to different harvest levels. His model, based on 7 years of intensive research on bobcats in southeastern Idaho, indicated little impact on population size until harvest exceeded 20% of the population. Beyond that, even small increases in harvest led to large population declines (Fig. 38.6).

Brand and Keith (1979) similarly concluded that human-caused mortality was largely additive to natural mortality in lynx populations. Intense exploitation, particularly when hare densities were low, could significantly depress lynx populations. In a study of a heavily exploited lynx population on the Kenai Peninsula, Alaska, Bailey et al. (1986) reported that 80% of their marked individuals were trapped in a single year. Seasonal mortality rates were as high as 86% and were higher for adult males (88%) than for adult females (45%). Parker et al. (1983) estimated that 65% of their study population was removed by trapping. Mowat et al. (2000) felt that trapping mortality was compensatory during the first 2 years of lynx decline, when natural mortality was high.

Harvest appears to be somewhat independent of density, making bobcat and lynx populations vulnerable to overexploitation. Levels of exploitation are driven by pelt price and accessibility. In addition, weather plays a role in the vulnerability of bobcats to trapping. In Minnesota (Petraborg and Gunvalson 1962) and Idaho (Koehler and Hornocker 1989), bobcats were more vulnerable to trapping during severe winters.

Much of the impact of harvest on kittens is indirect. Kittens are generally underrepresented in the harvest. However, because kittens are dependent on females until they become 8–10 months old, premature removal of maternal support by harvest can lead to high kitten mortality rates. In Idaho, bobcat kittens orphaned before independence did not survive (Knick 1990). Similarly in Alaska, lynx kittens that lost their mothers to trapping died of apparent starvation (Bailey et al. 1986).

Adult mortality rates are not constant over all age cohorts. In exploited populations, mortality rates decrease after the first year and either continue to decrease or remain fairly constant at low levels until age 4 or 5 years, when they increase again (Fritts and Sealander 1978b; Blankenship and Swank 1979; Litvaitis et al. 1987). Much of the decrease in mortality rates during the first several years of life can be explained by juveniles improving their hunting efficiency and eventually settling into permanent home ranges (Bailey 1974).

Sex-related differences in mortality rates also are apparent in exploited populations of both species. Male mortality is generally higher than that of females, particularly during the first several years as adults (Mech 1980; McCord and Cardoza 1982; Parker at al. 1983; Quinn and Thompson 1987). This trend may be the result of the higher susceptibility of males to trapping due to their larger home ranges and more extensive daily movements.

Because in many bobcat populations more than half of the mortality can be attributed to human harvest, it is not surprising that survival rates are lowest during the winter months when hunting and trapping seasons are generally open. All non-study-related deaths reported by Rolley (1985) occurred during the furbearer season, and were directly attributable to harvest. Likewise, most of the 14 mortalities observed in north-central Minnesota occurred during the December–January bobcat trapping and hunting season, whereas no deaths were reported during June–September (Fuller et al. 1985a). In addition, winter and early spring are the most likely periods of starvation because lagomorph and rodent populations are lowest and environmental stresses are the greatest (Petraborg and Gunvalson 1962). This period is particularly crucial to kittens/yearlings because maternal support is being withdrawn and their hunting skills are still developing (Bailey 1974).

In addition to trapping and hunting, a number of natural causes of mortality affect bobcat and lynx survival. Starvation, predation, disease, and vehicle collisions constitute the majority of other forms, with starvation being the major cause of mortality of lynx kittens during periods of hare decline.

Predation by other carnivores on bobcats and lynx has been reported. Coyotes have been repeatedly identified as direct predators, although their impact on population dynamics is unclear (Young 1958; Knick 1990; Fedriani et al. 2000). Particularly in human-dominated areas, domestic dogs can be a significant form of mortality (Lembeck 1986; Knick 1990). Working just outside San Diego, Lembeck (1986) reported that dogs were responsible for 20% ($n = 6$) of bobcat mortalities. Young (1958) also described several instances of mountain lion (*Puma concolor*) predation on bobcats (see Competitive Interactions).

In some populations, diseases and parasites may be an important form of direct mortality. Pollack (1951b) thought disease, particularly rabies and feline panleukopenia, might be an important decimator of bobcats. In a dense bobcat population in California, feline panleukopenia was responsible for 17% of the observed mortalities. In Massachusetts, 37% of bobcat mortalities were attributable to gastric enteritis (Fuller et al. 1995). Mitchell and Beasom (1974) found severe infestations of hookworms (*Ancylostoma caninum*) in bobcats from Texas and suspected they accounted for some mortality in wild populations. However, it is likely that disease and parasites play a much larger, but difficult to identify, role in predisposing individuals to other sources of mortality such as starvation, predation, and accidents.

A few instances of cannibalism among bobcats and lynx have been reported. Remains of bobcat flesh and fur were found in several stomachs of bobcats collected in Utah and eastern Nevada (Gashwiler et al. 1960), and Zezulak (1981) reported an adult male feeding on a bobcat it apparently killed. Litvaitis et al. (1982) found an adult female bobcat feeding on a juvenile bobcat during late January in Maine. Although the female was underweight and may have consumed the other bobcat due to hunger and opportunity, the capture and kill occurred on the edge of the adult's home range and may have been influenced by territorial behavior. Elsey (1954) also reported an incidence of cannibalism in the lynx. Nonetheless, cannibalism is extremely rare in either species, and does not appear to substantially influence their population dynamics.

Several unusual forms of mortality for bobcats have been described. Six kittens, representing 30% of total observed mortalities, were electrocuted while climbing power-line poles in Idaho (Bailey 1974). Several studies from the Northeast have reported bobcat mortalities due to injuries from porcupine (*Erethizon dorsatum*) quills (Berg 1979; Fuller et al. 1985a).

Diseases and Parasites. Bobcat and lynx are infected with a wide array of diseases and parasites. Their impact on the population dynamics and life history of both species is poorly understood; however, in some populations they clearly play a significant role (see Mortality). Whereas some diseases and parasites may be fatal, in many cases they are contributing factors to other forms of mortality.

Viral diseases reported for lynx and bobcat most commonly include rabies, feline panleukopenia (also known as feline distemper and feline infectious enteritis), rhinotracheitis, feline leukemia, feline calicivirus, and feline infectious peritonitis (McCord and Cardoza 1982; Roelke 1990). Rabies is relatively uncommon in lynx, but may occur more commonly in bobcats (Quinn and Parker 1987). In a survey of the carnivores infected with rabies in the United States from 1960 to 1997, bobcats were the seventh most common species reported, accounting for 402 cases (Krebs et al. 1999). Feline panleukopenia, reported in both species, is highly infectious and often fatal (Povey and Davis 1977).

Bacterial diseases in bobcats are primarily represented by sylvatic plague (Poland et al. 1973), tularemia (Bell and Reilly 1981), salmonellosis, leptospirosis (Labelle et al. 2000), brucellosis (Witter 1981), and possibly tuberculosis (Bruning-Fann et al. 2001). Protozoan parasites include toxoplasma (Riemann et al. 1975) and feline infectious anemia (Glenn et al. 1982).

Bobcats are infected by an array of endoparasites. The most common helminth parasites are *Toxocara cati* and *Toxascaris leonina* (Rollings 1945; Pollack 1951b), *Physaloptera* spp. (Hamilton and Hunter 1939; Stone and Pence 1978; Whittle 1979), *Ancylostoma caninum* (Little et al. 1971; Mitchell and Beasom 1974), and a variety of tapeworms (*Taenia* spp.) (Bursey and Burt 1970; Whittle 1979). Similar helminth parasites have been reported for the lynx (van Zyll de Jong 1966b; Smith et al. 1986).

Bobcats and lynx also host a much smaller array of ectoparasites, perhaps as a result of regularly bedding in different sites (McCord and Cardoza 1982). Van Zyll de Jong (1966b) reported six species of fleas on North American lynx. A variety of fleas (Pollack 1951b; Stone and Pence 1977) and ticks (Stone and Pence 1977; Wehinger et al. 1995) have been found on the bobcat, including lice (Lovallo et al. 1993) and sarcoptic mites (Pollack 1951b; Pence et al. 1982).

Because bobcats and lynx are harvested and occur in human-dominated landscapes, they may represent a potential route for infection of humans with zoonoses. Rabies, sylvatic plague, and toxoplasmosis are several of the diseases of concern that have been transmitted to humans from wild bobcats and lynx.

Lynx–Hare Cycles. The dramatic 8- to 11-year cycle of snowshoe hare and lynx populations in the northern portion of their range is well known and is a crucial aspect of the population biology of lynx (Elton and Nicholson 1942). Hare density between highs and lows may vary by 10–25 times, and lynx numbers may change by 3–17 times. As the number of snowshoe hare varies, lynx respond both numerically—by changes in their rates of survival, recruitment, and movement—and functionally—by changes in their individual kill rate of hares. The response of the lynx generally lags a year behind changes in the snowshoe hare numbers. This delayed density dependence gives rise to the cyclic nature of the lynx populations. The cycle is relatively synchronous across large geographic regions and appears to radiate out from central Canada. This means that not all of North America experiences high lynx numbers at the same time, but that when numbers peak in Manitoba and Saskatchewan, the same peak generally occurs 2–4 years later in western and eastern populations. The mechanism for synchrony is poorly understood, but may result from long-distance dispersal of lynx and other predators during the hare declines (Mowat et al. 2000).

Keith (1983) proposed that food shortages at peak densities initiated the hare decline. However, recent experimental manipulations of food and predators in the Yukon suggest that it is a more complex interaction of food supply, hares, and predators throughout the cycle that causes the decline (Krebs et al. 1995; Hodges 2000). In response to the diminished hare population, lynx populations decline due to a collapse in recruitment, higher adult mortality, and increased dispersal. Recruitment plummets due to little or no survival of kittens, reduced pregnancy rates in adults and particularly yearlings, and decreased litter sizes (Brand and Keith 1979; Parker et al. 1983). In the absence of predation and with the recovery of vegetation, hares increase, followed within several years by a subsequent rise in lynx numbers. Growth in the lynx population is a result of high fecundity, high kitten survival, and low adult mortality (Mowat et al. 2000).

As discussed earlier, hare populations and consequently lynx populations do not cycle in all parts of their range (see Ecology). Lynx in the southern portion of their range fluctuate little and generally exhibit characteristics of populations in the North during years of low hare abundance. The result is lynx with generally larger home ranges, lower reproduction, lower densities, and more consistent use of alternate prey (Aubry et al. 2000a).

Bobcats, particularly in areas where hare populations fluctuate widely, may also exhibit some of the characteristics of a cycling lynx population. Bailey (1974) and Knick (1990) were able to study the impact on bobcats of two periods of cyclic decline in jackrabbits in southeastern Idaho. Bobcat populations declined ninefold within 3 years. Home ranges enlarged, extraterritorial forays increased, adult mortality rates increased, and recruitment declined to zero.

BEHAVIOR

Home Range and Social Organization. Bobcats and lynx are essentially solitary, with direct social interactions being brief and infrequent. The exceptions are females with kittens and adult males and females during the breeding season. Three social classes appear to exist in all populations: residents, transients, and kittens. Most adults are considered residents and generally reside in a single home range. Transients are frequently yearling individuals dispersing from their natal home ranges and are generally distinguished from adults by their lower weight and shorter total body length. Kittens include all individuals still under maternal care (Bailey 1974; Rolley 1983).

Sizes of bobcat and lynx home ranges vary widely within and among studies (Tables 38.4 and 38.5). Caution should be exercised in comparing home range sizes among studies. Variations in the number of locations used, the season of sampling, number of individuals tracked and their social status, and the kind of home range estimators used can confound comparisons. However, there are several consistent generalizations that can be made about bobcat and lynx home range sizes.

Home ranges of bobcats in the northern latitudes are considerably larger than those from the South, probably due to lower prey populations, increased thermal demands, and larger body size in the North. Litvaitis et al. (1987) found male ranges averaged 112.0 km^2 in Maine, whereas male ranges in Alabama averaged only 2.6 km^2 (Miller and Speake 1979).

Average male bobcat home ranges are generally two to three times larger than those of females, although some studies have reported size differences as large as four to five times (Hall and Newsom 1976; Major 1983; Witmer and DeCalesta 1986). Increased metabolic demands of larger body size of the male cannot explain the magnitude of difference in range sizes between the sexes, particularly because female energetic demands of lactation and subsequent feeding of kittens probably far exceed male metabolic needs during spring and summer. Female home range size may be more closely tied to prey availability, whereas male range size is more influenced by the number of mating opportunities (female home ranges) within the range. The need for females to more intensively exploit the food resources in their ranges to meet energetic demands compared to males may also partially explain the observed sex-related differences in feeding habits (see Feeding Habits).

Home range size appears to be most strongly related to variations in habitat quality and associated prey abundance. Bobcats studied on the Savannah River Plant, South Carolina, in 1966 (Marshall and Jenkins 1966) and again in 1978–1979 (Buie et al. 1979) showed changes in home range size as the vegetation in the area matured from abandoned pastures and open fields to pine forests and managed tree plantations. As a result, biomass of the small- mammal community declined. Home range size for bobcats in the area increased from an average of 2.5–4.6 km^2 for both sexes in 1966 to 20.8 km^2 for males and 10.3 km^2 for females by 1979, which strongly suggests that additional area was needed as prey populations decreased.

Time-in-residence of a home range may also affect estimates of home range size for bobcats as it relates to resource acquisition, foraging efficiency, and breeding opportunities. Conner et al. (1999) found that female bobcat home range size decreased over time in Mississippi and suggested that this was a function of increased hunting skills, familiarity

TABLE 38.4. Average home range size and density of bobcat populations[a]

Location	Sample Size		Average Home Range Size (km^2)		Density	Reference
	M	F	M (*SE*)	F (*SE*)	(*n*/100 km^2)	
Minnesota	16	6	62	38	4–6	Berg 1979
Minnesota (BV)	4	3	46	49		Fuller et al 1985b
Minnesota (HC)	18	8	61	32		Fuller et al 1985b
Wisconsin	6	6	60.4	28.5		Lovallo and Anderson 1996a
Maine	9	1	138.6	27.5		Major 1983
Maine (CF)	4	6	71.1 (27.2)	32.5 (3.8)		Litvaitis et al. 1986
Maine (PP)	6	2	112.2 (18.5)	33.2 (5.7)		Litvaitis et al. 1986
Kansas	1	3	20.0	7.5		Kamler and Gipson 2000
Missouri	9	8	60.4	16.1	6–10	Hamilton 1982
Pennsylvania[b]	17	17	42.2[c]	17.2[c]		Lovallo et al. 2001
Tennessee	2	3	76.8[d]	25.9[d]		Kitchings and Story 1984
Alabama	6	6	2.6	1.1		Miller and Speake 1979
Louisiana			5.2	1.0		Hall and Newson 1978
California[b]			2–6	1.4	115–153	Lembeck and Gould 1979
Arizona	2	5	9.1	4.8	25.6	Lawhead 1984
Arizona					24–28	Jones and Smith 1979
Nevada (HH)			22.8	7.7	20	Golden 1982
Nevada (LH)			21.6	26.6		Golden 1982
Utah	3	3	22.5 (11.5)	16.4 (8.7)	6.2	Karpowitz 1981
Idaho[b]	4	8	42.1	19.3	5.4	Bailey 1974
Idaho (HPD)	3	5	20.4	11.6		Knick 1990
Idaho (LPD)	2	2	123	69.7		Knick 1990
Oregon	1	5	11	2	77	Witmer and DeCalesta 1986

NOTE: Home ranges are estimated using 100% minimum convex polygon (Mohr 1947) unless otherwise noted.
[a] M, Males; F, females; BV, Bearville; HC, Hill City; CF, Cherryfield; PP, Pierce Pond; HH, heavy harvest; LH, light harvest; HPD, high prey density; LPD, low prey density.
[b] Area closed to harvest.
[c] Median value.
[d] Jennrich and Turner (1969) home range estimator.

of resource distribution, and changes in social pressures and habitat quality.

Lynx home range size also appears to be linked to prey availability, although not in a linear fashion. Mowat et al. (2000), using data from two multiple-year studies in the Yukon and the Northwest Territories, found no evidence of a linear relationship between lynx home range size and hare density the previous year. They did observe, however, a threshold hare density (0.5–1.0 hare/ha) below which lynx home range sizes increased significantly.

Studies on bobcats and lynx at low prey densities suggest that at some threshold, prey density becomes too low to support any kind of home range, and nomadic movements are necessary (Ward and Krebs 1985; Knick 1990; Mowat et al. 2000). Knick (1990) studied bobcats during a prey decline in Idaho. He found that residents did not leave their home ranges until the second year of a jackrabbit decline when the prey base was substantially reduced. In Alberta, lynx remained on their home ranges even when hare densities decreased to 0.7 hare/ha (Brand et al. 1976). However, in the northern Yukon, when hare densities decreased to <0.5 hare/ha, lynx became nomadic (Ward and Krebs 1985).

Although prey abundance and distribution may be the most general factors influencing bobcat home range size, a myriad of other factors including bobcat density, energetic demands, and the availability of

TABLE 38.5. Average home range size and density for lynx populations relative to hare density[a]

Location	Sample Size		Average Home Range Size (km^2)		Density	Reference
	M	F	M (*SD*)	F (*SD*)	(*n*/100 km^2)	
Southern boreal forests						
Washington[b]	5	2	69 (28)	39 (2)	3	Koehler 1990
Montana	4	2	238 (99)	115 (50)		Squires and Laurion 2000
Southern Canadian Rocky Mountains	3	3	277 (71)	135 (124)		Apps 2000
Northern boreal forests (low hare density)						
South-central NWT	2	2	44 (14)	63 (28)	3	Poole 1994
South-central Yukon	2	4	266 (106)	506 (297)	3	Slough and Mowat 1996
Southwestern Manitoba[b]	1	2	221	158		Carbyn and Patriquin 1983
Northern boreal forests (high hare density)						
South-central NWT	3	10	17(2)	18 (3)	30	Poole 1994
South-central Yukon	12	13	44 (23)	13 (4)	45	Slough and Mowat 1996
Southwestern Yukon[c]	2	2	14 (1)	13 (7)		Ward and Krebs 1985

SOURCE: Adapted from Aubry et al. (2000a).
NOTE: Home ranges are estimated using 95% minimum convex polygon (Mohr 1947) unless otherwise noted.
[a] M, Males; F, females; NWT, Northwest Territories.
[b] 100% minimum convex polygon.
[c] 90% minimum convex polygon.

escape cover, hunting cover, den sites, and mating opportunities also influence home range size and distribution.

Seasonal differences in bobcat home range size have been reported. Anderson (1987) speculated that male ranges should be largest during the breeding season to procure as many breeding opportunities as possible; female ranges should be smallest during kitten rearing. Small female ranges during the kitten-rearing periods have been reported by a number of investigators (Bailey 1974; Kitchings and Story 1984; Litvaitis et al. 1987), and Witmer and DeCalesta (1986) saw a slight increase in a male's range during the breeding season. However, Litvaitis et al. (1987) in Maine found that although female home ranges were smallest during nursing (16 May–15 June) and largest during gestation (16 March–15 May), males also followed a similar pattern, suggesting that factors other than those associated with reproduction were influencing home range size. They speculated that bobcats were responding to the lack of prey in late winter and the subsequent abundance during spring.

The relationship of home range use among individuals has illuminated some of the details about social organization of bobcat and lynx populations. With bobcats, most studies have reported significant intersexual overlap of home ranges with varying degrees of intrasexual overlap. Generally, adult female home ranges exclude other adult females, whereas adult male home ranges may extensively overlap each other and encompass the ranges of two or three females. However, there are a number of cases where female ranges overlap or male ranges are exclusive (Zezulak and Schwab 1979; Kitchings and Story 1984). Occasionally, adult males and adult females will maintain intrasexually exclusive ranges (Miller and Speake 1979; Rolley 1983). Transients appear to be tolerated in adult home ranges regardless of the exclusivity of the ranges to other adults (Provost et al. 1973; Miller and Speake 1979).

Bobcat social organization appears to be influenced by climate, habitat, food resources, and population density. Buie et al. (1979), repeating a study of bobcats on the Savannah River Plant conducted 12 years earlier (Marshall and Jenkins 1966), found larger home ranges and less intrasexual overlap, suggesting that bobcat density, prey availability, and the degree of home range overlap were related. Studies found the highest bobcat densities and smallest home ranges were also associated with exclusive female home ranges (Lembeck and Gould 1979) as well as exclusive male home ranges (Miller and Speake 1979). Bailey (1981) suggested that in warm regions where prey and cover are abundant and evenly spaced, female home ranges should be small and exclusive, with male ranges being of similar size and only slightly overlapping those of other males. In environments where the climate is more severe and food and cover are seasonally limiting and unevenly distributed, home ranges of females should be less exclusive of those of other females, and male home ranges should be considerably larger, with a high degree of intrasexual overlap. Unfortunately, the true relationship of bobcat density to degree of overlap may be obscured by sampling problems. If all resident bobcats in an area are not radio-collared or transients are misclassified as adult residents, observations on overlap are meaningless. It has also been suggested that relatedness among individuals could influence the degree of overlap and should be evaluated or quantified before examining spatial distribution (Chamberlain and Leopold 2001).

In the lynx, male and female home ranges generally completely overlap, while intrasexual overlap of both sexes is limited. Related females seem to tolerate greater overlap.

Both species show strong range fidelity over years (Bailey 1974; Litvaitis et al. 1987; Brietenmoser et al. 1993; Poole 1994; Slough and Mowat 1996). Generally, individuals move ranges only when either adjacent individuals die or lack of food resources cause the social organization to dissolve.

Land tenure in bobcats appears to be based on prior rights with little displacement apart from changes created by mortality. Vacancies created by the death of resident individuals, whether from harvest or natural mortality, are filled either by transient bobcats or adjacent residents. This has been observed for males (Bailey 1974; Miller and Speake 1979; Hamilton 1982; Anderson 1988) and less frequently in females (Bailey 1974; Hamilton 1982; Lovallo and Anderson 1995). The shifting of adult home ranges when vacancies occur suggests that bobcats are cognizant of a hierarchy of "quality" in home ranges. It also indicates that bobcat habitat use is influenced by, among other things, the location of adjacent conspecifics.

Overt agonistic behavior between bobcats or between lynx is generally avoided by use of visual contact and scent marking, which is characteristic of territory maintenance by other solitary felids (Hornocker and Bailey 1986; Kitchener 1991; Mellen 1993). Bailey (1974) found that adult bobcats scent marked with feces, urine, scrapes, and anal glands. He reported finding "fecal marking locations" where bobcats regularly defecate. He suspected that females with kittens were responsible for the majority of sites, although some sites were found in areas of heavy bobcat use away from den sites. Similar concentrations of scats were reported by Kight (1962) in South Carolina, who found 254 scats in a single marking site. The location of fecal marking sites led Bailey (1974) to conclude that they were used primarily by females to demarcate special places within their home range, contrary to Provost et al. (1973), who felt that feces were used by a female to mark the edge of her home range. Guenther (1980) observed a seasonal shift in marking behavior with significantly more scats deposited in scrapes at marking sites during February than in July and August.

Despite mechanisms for avoiding physical contacts, several intraspecific encounters have been reported for both species, which suggests that males may engage in aggressive interactions far more frequently than females. While snow tracking bobcats in February, Erickson (1955) observed an area in the snow where one large and two smaller bobcats had a severe fight, which left the two smaller bobcats bleeding profusely. Provost et al. (1973) reported an adult male pursuing a young male up a tree, where the two growled and spat at each other until the adult male was disturbed by the approaching researchers and fled. Hamilton (1982), during January in Missouri, watched an adult and juvenile male hissing and screaming at each other for >2 min. The encounter ended without physical contact when the adult male turned and walked away. He also observed a "vicious battle" between two adult males during March. Kalmer et al. (2000) observed an aggressive encounter in mid-December between an adult female and a male offspring that had become independent 11 months earlier. The adult female growled, hissed, screamed, and then lunged at the male, who was submissive. He dispersed shortly afterward and was harvested off-site. Similar encounters, although rare, have been reported for lynx (Poole 1995; Mowat and Slough 1998). Most reported encounters occurred around the breeding season when spermatogenesis peaks and testosterone levels are the highest.

Because bobcats are polygynous and several male ranges may overlap that of a single female, it is not surprising that competition for matings occurs. However, the general lack of aggressive female–female encounters combined with evidence that females maintain home ranges exclusive of each other suggests that strong, sex-specific territorial marking occurs among female bobcats.

Movement, Activity Patterns, and Dispersal. Daily and weekly distances moved by bobcats differ considerably according to region, sex, weather conditions, and individuals. Bailey (1974) noted that male bobcats averaged 1.8 km between consecutive daily radiotelemetry locations, whereas females averaged 1.2 km. Daily movements of bobcats in eastern Tennessee averaged 4.5 km for males and 1.2 km for females (Kitchings and Story 1979). Similarly, in Montana, males moved 4.9 km and females moved 1.1 km in a day (Knowles 1985). Much of the difference in distances moved has been attributed to the difference in home range size between males and females, suggesting that both sexes traverse their home ranges in a similar amount of time. The slower movement rates by females may also indicate that they are hunting more intensively for smaller prey items than are males, as suggested by sex-related differences in feeding habits (see Feeding Habits).

Lynx movement is also highly variable and tied closely to snow characteristics and prey density (Nellis and Keith 1968; Ward and Krebs 1985). When hare numbers were high, Ward and Krebs (1985) found the average distance between consecutive 24-hr relocations of lynx was

2.7 km, but increased to 5.4 km when hares became scarce. Poole (1994) observed similar results in the Northwest Territories, although in the Yukon, O'Donoghue et al. (1998b) did not.

Both species are generally considered nocturnal, although the majority of activity is centered around the crepuscular periods at sunrise and sunset (Saunders 1963b; Hall and Newsom 1976; Buie et al. 1979). Not surprisingly, these activity periods generally coincide with the peaks of activity of lagomorphs, their primary prey item.

Buie et al. (1979) noted that bobcats in South Carolina increased their daylight movements during the winter. Males used their home range more extensively in the winter by traveling greater distances in 24-hr periods while maintaining the same home range size. In contrast, however, Rolley (1983) in Oklahoma found that daylight movements of males and females and 24-hr movements of males were the least during fall and winter.

Timing of dispersal by juveniles of both species is highly variable, but is often initiated by separating from the mother before the litters of the following year are born (Crowe 1975b; Bailey 1981; Griffith et al. 1981, Kitchings and Story 1984). Griffith et al. (1981) observed two juvenile male bobcats leave their natal home ranges in early spring in the South Carolina Coastal Plain region and begin a pattern of nomadic movement characterized by small, temporary activity areas occupied for 30–60 days and then abandoned. A study in Tennessee showed a similar gradual movement of juvenile bobcats away from their natal range during the spring following their birth year (Kitchings and Story 1984). Klamer et al. (2000) chronicled the dispersal of six juvenile bobcats (two males, four females) from northeastern Kansas. They became independent of their mothers by February of their first winter, but stayed in the area until they dispersed. Males did not disperse until July (1.5 years of age), and females waited until the following fall and winter (1.7–2 years old). Half of the dispersers adopted a straight-line dispersal, whereas the other half were transients for 4 months to >2 years. In heavily exploited populations, widespread dispersal probably does not occur because abundant unused areas are readily available (Crowe 1975b). Females have also been known to settle in portions of their mother's home range. Generally, males disperse before females (Bailey 1981) and usually move farther than females (Robinson and Grand 1958; Griffith et al. 1981; Hamilton 1982; Kitchings and Story 1984; Knick and Bailey 1986).

In the lynx, there are two types of dispersal: "innate" or juvenile dispersal and "environmental" dispersal, when adults emigrate in response to significant prey declines. Timing of juvenile dispersal is similar to that of the bobcat. Juveniles often disperse shortly after they become independent from their mothers, although, as with bobcats, timing is variable. Juveniles constitute the bulk of emigrants during the increase, peak, and initial decline in hare abundance. Adults, however, make up the majority of the migrant population during periods of low hare density. Most adult dispersal is initiated from March to June in years following the initiation of hare declines and in midwinter for up to 2 years following that (Slough and Mowat 1996; Poole 1997). Poole (1997) found no differences in dispersal distances between sexes or between adults versus kittens–yearlings.

Various distances traveled by dispersing bobcats have been reported. Robinson and Grand (1958) found the mean recovery distance of 48 of 81 tagged adult and juvenile bobcats was 6.6 km with a maximum movement of 37 km. In Missouri, Hamilton (1982) followed eight dispersing juveniles and found an average movement of 33.4 km before individuals were trapped or established a new home range. An adult female abandoned her home range in northern Minnesota and resettled 136 km away (Berg 1979). The longest dispersal distances are associated with bobcats in the northern portion of their range and during periods of low prey availability. Knick and Bailey (1986) documented two young male bobcats dispersing 182 and 158 km during a cyclic crash of the jackrabbit population in Idaho.

Lynx dispersal is characterized by much longer movements, particularly during hare declines. The longest straight-line dispersal documented was 1100 km, with 15 documented cases of dispersal >500 km (Mowat et al. 2000).

The ability of bobcats and lynx to disperse, sometimes very long distances, has profound implications for the population dynamics and management of both species. The long-distance dispersal of lynx has been offered as an explanation for the regional synchrony of the hare cycle (Lack 1954; O'Donoghue 1997). For both species, the long dispersal distances imply that when areas, particularly those that are isolated, have been overharvested or suffered dramatic losses, new immigrants can easily recolonize them.

Breeding Behavior. Most of what is known of bobcat and lynx breeding behaviors comes from observations of captive animals. Behaviorally, estrus in bobcats can be recognized by increased cheek rubbing on objects, increased scent marking, loud and frequent vocalizations, tail flicking, holding the tail erect to indicate receptiveness, and increased interest in males (Jackson et al. 1988). During anestrus, the female physically rejects approaches by a male, often biting or clawing him around the head. As proestrus ensues, the female frequently urinates and rubs against objects with her head and neck, makes loud and frequent vocalizations, flicks her tail and keeps it erect, and generally appears more interested in the male (Mehrer 1975; Stys and Leopold 1993). During estrus, the female assumes the coital position by crouching on the ground with her perineal region raised and tail deflected. The male grasps the female by the nape of the neck, straddles her, swaying continuously on his hind legs, then brings his perineal region to that of the female by a series of thrusts that end in intromission. The female then rises, displacing the male, and begins urinating, rolling, and rubbing on objects. Although coitus seldom exceeds 5 min, it is repeated up to 16 times/day (Mehrer 1975). Eaton (1976) speculated that high rates of copulation might have evolved to allow females to evaluate male vigor in polygynous mating systems. Following the estrous period, females move into metestrus, where they may allow mounting, but refuse intromission, often freeing themselves from the males with sudden and violent movements (Mehrer 1975).

Hunting Behavior. Bobcat and lynx, like many felids, hunt either by stalking or ambush, and often use a combination of the two. Brand et al. (1976) found that lynx typically either followed well-used hare runways or lay in wait in short-term "hunting beds" for prey to come by. Saunders (1963a) and Brand et al. (1976) found lynx used the "hunting bed" ambush technique more commonly, although choice of technique is probably related to prey density, availability of cover, snow conditions, and individual preference. The same techniques have been described for the bobcat (Rollings 1945; Marshall and Jenkins 1966; McCord 1974). Lookouts and "hunting beds" were commonly found in areas that had high cottontail and hare activity. Marshall and Jenkins (1966) detailed the killing of a cotton rat by a bobcat on a roadside in South Carolina. The bobcat took 13 min to move 1 m before pouncing on its prey.

Snow tracking has revealed that hunting success is variable for lynx. Success appears to be heavily influenced by bearing strength of the snow as well as experience of the lynx and its familiarity with an area (Nellis and Keith 1968). In Newfoundland (Saunders 1963a), Nova Scotia (Parker et al. 1983), and Washington (Koehler 1990), success rates of lynx in winter were 42%, 24% and 33%, respectively. Murray and Boutin (1991), working in the Yukon, found that lynx caught hares on 32 of their 52 attempts, for a success rate of 61%. Surprisingly, the rate of capture of hares does not appear to be influenced by their abundance. Capture rates were 24% when hares were numerous, whereas during a population low rates were 24–35% (Nellis and Keith 1968; Brand et al. 1976).

Both species occasionally cache prey items (McCord and Cardoza 1982). Nellis and Keith (1968) surveyed 418 km of lynx trails in snow and found that 54% of the captured hares were cached. This contrasts with the findings of O'Donoghue et al. (1998a), who found that lynx only cached 1.6% of radio-collared hares they killed. Typically, lynx form caches on the surface of the snow by pulling snow over them and retrieve them within 2 days (Mowat et al. 2000).

Bobcats and lynx are primarily solitary, however, there are several recorded incidences of cooperative hunting. McCord (1974) snow tracked two bobcats that were presumably a breeding pair as they hunted

10–15 m apart through an area with snowshoe hare. In Newfoundland, two lynx were tracked as they made an unsuccessful attack on a small group of caribou (Saunders 1963a). In Montana, Barash (1971) also found two adults and a nonparticipating juvenile lynx hunting Columbia ground squirrels (*Spermophilus columbianus*) together. Group size also seems to influence success rates. Parker et al. (1983) observed that capture success increased from 14% to 55% as the size of the group of lynx (usually a female with kittens) increased from one to four.

HABITAT

Bobcats occur in a variety of habitats, from bottomland forests of Alabama to arid deserts of New Mexico, and from northern boreal forests of Minnesota to the humid tropical regions of Florida. They generally prefer rough, rocky country interspersed with dense cover (Pollack 1951b; Erickson 1955; Young 1958; Zezulak and Schwab 1979; Karpowitz 1981; Golden 1982)—any habitat that produces abundant prey and allows hunting by either ambush or stalking. McCord (1974) tracked bobcats through snow in Massachusetts and found that roads, cliffs, spruce plantations, and hemlock–hardwoods were used most in relation to their abundance. He attributed the use of hemlock–hardwoods to high deer densities and the use of spruce plantations to abundant snowshoe hare and protection from the wind. Similarly, Fuller et al. (1985a), in Minnesota, found a disproportionate use of coniferous areas, which also supported the highest density of snowshoe hares and white-tailed deer, the bobcat's main prey in that region. Bobcats in Missouri preferred bluffs, brushy fields, and second-growth oak habitats (Hamilton 1982). Bluffs were apparently selected for social reasons as well as the physiological advantages of cover; brushy fields and oak regeneration offered high densities of prey. In Wisconsin, lowland coniferous forests were consistently selected by both sexes during all seasons, although there were sex-related and seasonal differences in selection of other habitats (Lovallo and Anderson 1996a).

Lynx are much more restricted in their habitat selection and generally occur in association with boreal forests. On the Kenai Peninsula, Alaska, the dominant tree types where they occur are white (*Picea glauca*) and black spruce (*P. mariana*), paper birch (*Betula papyrifera*), willow (*Salix* spp.), and quaking aspen (*Populus tremuloides*) (Bailey et al. 1986). In southwestern Yukon, they occur in white spruce-dominated forests (Ward and Krebs 1985). In central Alberta, they are found in quaking aspen, balsam poplar (*Populus balsamifera*), and spruce stands. In Manitoba, lynx occupy quaking aspen forests (Carbyn and Patriquin 1983). On Cape Breton Island, Nova Scotia, they are associated with balsam fir (*Abies balsamea*), white and black spruce, and paper birch forests (Parker et al. 1983). In northern Minnesota, lynx occur in jack pine (*Pinus banksiana*), balsam fir, black spruce, quaking aspen, and paper birch forests (Mech 1980), and in Montana and Washington, they can be found in forests of Engelmann spruce (*Picea engelmannii*), subalpine fir (*Abies lasiocarpa*), lodgepole pine (*Pinus contorta*), and quaking aspen (Koehler 1990). In the southern extension of the lynx range in the western mountains, Koehler and Aubry (1994) felt lynx required two structurally different forest types to survive. They needed early-successional stages to provide good prey (primarily hare) habitat for foraging and late-successional forests to provide the blowdowns for den and kitten-rearing sites. Bobcats and lynx appear to avoid large open areas, even though they may have abundant potential prey (Rolley 1987; Koehler and Aubry 1994).

Habitat characteristics directly influence the diversity, abundance, and stability of prey populations, and consequently partially regulate bobcat and lynx density and home range size. The highest bobcat densities and smallest home ranges are in the thick chaparral vegetation of southern California (Lembeck and Gould 1979), the rough, dissected desert scrub/desert grassland regions of Arizona (Jones and Smith 1979), and openings in the bottomland hardwood forests of southern Alabama (Miller and Speake 1979). In contrast, some of the lowest densities have been reported from areas with low productivity: the coniferous forests of Minnesota (Berg 1979; Fuller et al. 1985b), the sagebrush–grasslands of southeastern Idaho (Bailey 1974), and the oak–pine forests of the Ozark Mountains (Hamilton 1982) (see Table 38.4).

To a large extent, snowshoe hare habitat is also lynx habitat. High stem density is directly related to the presence of hares, and hence lynx (Litvaitis et al. 1985). There are two exceptions, however. Hares select some of the densest stands, which are impenetrable by lynx. This may explain why hares manage to survive in pockets of dense habitat during their cyclic lows (Wolff 1980). Hares also are found in pockets of dense shrub that also apparently exclude lynx (Mowat et al. 2000).

Although prey abundance is probably the most important factor in habitat selection for bobcats and lynx, protection from severe weather, availability of resting and denning sites, dense cover for hunting and escape, and freedom from disturbance are also important factors in determining habitat use (Pollack 1951b; Erickson 1955; Bailey 1974). Knowles (1985) found that bobcats in Montana generally selected habitat types with $\geq$52% visual obscurity. Although prey densities were highest in those types, she felt that cover was crucial for the bobcat's effective use of ambushing and stalking hunting methods. Similarly, Lovallo (1999) in Pennsylvania found that bobcats were strongly associated with eastern to southeastern exposures on 7–8° slopes. McCord (1974) felt that behavioral factors, such as hunting habits or social interactions, also dictate the temporal and spatial use of habitat types.

Deep snow directly influences patterns of habitat use by bobcats and lynx. Marston (1942) observed that movements of bobcats became restricted when snow accumulated to depths >13 cm, and Hamilton (1982) reported increased use of protected rock ledges and small caves during and after winter storms. McCord (1974) found that bobcats in Massachusetts walked normally in snow <15 cm deep, but consistently avoided deeper snow by traveling in trails of other animals, on logs, or in plowed roads or snowmobile trails.

Bailey (1981) suggested that female bobcats use better quality habitat than males because they require more prey from a smaller area, particularly during the physiologically demanding period of kitten rearing. In Pennsylvania, Lovallo (1999) found that males used a wider range of habitat conditions than females, which resulted in more than twice the suitable habitat for males in the state than for females. Hamilton (1982) reported similar findings from bobcats in the Ozark Mountains, where breeding females were located in areas where preferred habitats occurred nearby in relatively large amounts. Rolley and Warde (1985) in Oklahoma also showed sex-related differences in habitat use; females preferred deciduous or mixed pine–deciduous forests and males preferred grass fields and brush. Sex-related differences were also shown by Lovallo and Anderson (1996a) in Wisconsin.

There is growing interest in habitat modeling and landscape-scale analysis of bobcat and lynx habitat selection. Lovallo (1999) used radiotelemetry-determined locations, geographic information systems, multivariate modeling techniques, and remotely sensed landcover and physiographic data to model bobcat habitat selection in Pennsylvania and to predict the state-wide distribution of suitable habitat conditions. He found that home range size of females was inversely related to amounts and pattern of the most suitable habitat patches. Similarly, Conner et al. (2001) developed multivariate models of habitat selection for bobcats in Mississippi. The application of these habitat selection models provides an information source for habitat-based management decisions and conservation strategies, and serves as a basis for developing further hypotheses concerning local- and landscape-level habitat associations.

Den and Resting Sites. Bobcat kittens are often born in caves, rock shelters, or dense piles of brush (Bailey 1974; Hamilton 1982;, Kitchings and Story 1984), but have been found in abandoned buildings (Bailey 1974) and abandoned beaver lodges (Lovallo et al. 1993). In Bailey's (1974) study area, denning sites were limited and not uniformly distributed, and therefore influenced the size and configuration of female home ranges.

Lynx dens are usually placed in dense tangles of wind-fallen trees. Describing 39 maternal dens in south-central Yukon, Slough (1999) found 34 under blowdowns; 3 in young, bushy subalpine firs; and 2 in dense willow thickets. Use of sites with dense woody debris is common to most descriptions of dens (Berrie 1973; Kesterson 1988; Koehler 1990). Selection for stand type is more variable. Koehler (1990) in Washington found 4 dens in mature conifer forests. Of the 39 dens Slough (1999) found, 37 were in 30-year-old regenerating burned stands and 2 were in mature stands. Koehler and Brittell (1990) recommended that stands protected for den sites should be mature forest at least 1 ha in size in close proximity to early-successional forests (for foraging habitat) and have minimal human disturbance.

Den sites of bobcats and lynx are moved several times while females are rearing kittens. Females move kittens from their natal den to auxiliary dens up to five times (Bailey 1979). Generally, dens are not used in consecutive years (Slough 1999), although in areas where sites are scarce, both species may regularly reuse sites (Bailey 1981; Koehler and Aubry 1994).

Bobcats generally move from one rest site to another every day. In Colorado, rest sites occurred on steep-sloped, rocky areas with dense vertical cover and sparse herbaceous ground cover (Anderson 1990). Other sites include rock piles (Rolling 1945; Bailey 1974), brush piles (Kitchings and Story 1984), blowdowns, hollow snags and trees, overhanging roots, and rocky cliffs (Rollings 1945).

Impact of Humans. Bobcats are frequently observed in areas of human habitation, and established populations have been observed in rural, suburban, and urban habitats. In a survey of more than 2500 households in three separate residential areas of New Mexico, 29% reported seeing bobcats, most often <25 m from houses (Harrison 1998). However, Neilsen and Woolf (2001) found that bobcats in southern Illinois avoided habitats around human dwellings and suggested that bobcats selected core areas within home ranges that minimized human activity. Similarly, Riley (2001) in California found that bobcats, unlike gray fox (*Urocyon cinereoargenteus*), did not use available habitats in adjacent developed areas. Lynx also seem tolerant of human presence and disturbance, although they are much less likely to occur in areas with year-round human habitation (Staples 1995; Aubry et al. 2000a; Mowat et al. 2000).

Roads and trails affect both species in several ways. They can be a source of direct mortality, may influence spatial use, may allow increased access to areas by hunters or trappers, or may provide access during winter for competing carnivores. In some areas, collisions with vehicles can be a significant form of mortality for bobcats (see Mortality). For lynx, translocated individuals appear more vulnerable to traffic deaths. Brocke et al. (1991) found vehicles were the primary cause of death in lynx reintroduced to the Adirondack Mountains of New York. Likewise, 14% of the deaths of transplanted lynx in Colorado involved vehicle collisions (Shenk 2001). As for the influence on spatial use, Lovallo and Anderson (1996b) found that bobcats selected home ranges with higher densities of trails and lower densities of secondary highways and crossed paved roads at a lower than expected rate. McKelvey et al. (2000b) concluded that road density did not affect lynx habitat selection in north-central Washington and that lynx did not avoid crossing roads. However, Apps (2000) found that all six radio-collared resident lynx in southeastern British Columbia crossed highways less frequently than expected. Like the bobcat, lynx use some roads for hunting and travel (Koehler and Aubry 1994). However, roads increase human access to areas, which increases the vulnerability of both species to intentional or unintentional shooting and trapping.

COMPETITIVE INTERACTIONS

Competition between species occurs when an organism negatively affects another by consuming or controlling access to a required resource that is in limited supply. Interspecific competition can take two forms. *Exploitative competition* occurs when different species use a common resource that is in short supply. *Interference competition* occurs when a species acts aggressively toward another and denies it access to a resource (Krebs 2001). Both forms of competition may influence the distribution and population dynamics of bobcats and lynx.

Ecologically, the bobcat and the lynx are very similar, particularly relative to their feeding habits. Consistent with Gause's competitive exclusion principle, which suggests ecologically similar species cannot exist in the same area at the same time, bobcat and lynx are rarely sympatric. It appears that lynx have the advantage in areas of deep snow, whereas bobcats dominate in all other areas. In the mountains of western Montana during winter, bobcat home ranges were at significantly lower elevations than lynx ranges, although there was no altitudinal segregation during spring and fall (Smith 1984). A suggestive example of bobcat dominance over lynx comes from a fortuitous "experiment" that happened on Cape Breton Island, Nova Scotia, during the last century. Up through the mid-1900s, lynx were common throughout the island. However, in the early 1960s, bobcats began to appear, perhaps as the result of access provided by a causeway built to the island in 1955, and rapidly colonized the island's lowlands. Concurrently, lynx densities declined and their range constricted until they only occurred in the highlands, where snow depth apparently prevented bobcats from successfully establishing populations (Parker et al. 1983).

Body size in areas of sympatry also may explain the dominance of bobcats over lynx. In regions where both species are present, the bobcat often is larger than the lynx. For example, Parker et al. (1983) found mean body weight of male bobcats was 40% heavier than that of adult male lynx on Cape Breton Island, Nova Scotia. Similarly in Montana, Buskirk et al. (2000a) reported that the largest male bobcat was 2–4 kg larger than the largest male lynx. The discrepancy in body size suggests that interference competition favoring the bobcat is possible, although exploitative competition for their common food resource, snowshoe hare, also is probable.

In some ecosystems, coyotes are a formidable competitor of bobcats and, to a lesser extent, lynx. Coyotes often use the same habitats as bobcats or lynx and prey extensively on the same food items—lagomorphs. Ecologically, coyotes have a distinct advantage due to their larger size, broader diet, wider habitat tolerance, higher reproductive rates, wider behavioral plasticity, and higher human tolerance (Buskirk et al. 2000a).

There is abundant circumstantial evidence that coyotes sometimes influence bobcat population size either through exploitative or interference competition. Robinson (1961), using records from predator control operations in New Mexico, Colorado, and Wyoming, was the first to notice an inverse relationship between the number of coyotes versus those of bobcats and other sympatric carnivores. Nunley (1978) searched annual U.S. Fish and Wildlife Service Animal Damage Control trapping records from 1916 through 1976 to examine relative bobcat and coyote abundance in the western United States through time. He found that not only bobcat, but also skunk, fox, and badger harvest declined as coyote harvest increased. Litvaitis and Harrison (1989) identified a similar pattern in Maine, where a sharp decline of bobcat harvest between 1974 and 1985 was highly correlated with a dramatic increase in coyote harvest. More recently, in a 3-year controlled experiment in western Texas, coyote populations were artificially reduced in an area with bobcats. This led to an increase in rodent density and biomass as well as an increase in the relative density of badgers, bobcats, and gray foxes, whereas no population changes were observed in the control area (Henke and Bryant 1999).

Several studies have attempted to quantify the nature and extent of competition between bobcats and coyotes by measuring overlap of resource use. Small (1971) compared the feeding habits and habitat use of bobcats, coyotes, and gray foxes in Arizona. Although some food items were similar (cottontails and rodents, primarily), the percentage of these items in the diets of the three carnivores varied significantly, suggesting that little competition was occurring. Witmer and DeCalesta (1986) in Oregon and Major and Sherburne (1987) in Maine used scat analysis and radiotelemetry to show there was extensive dietary overlap (>92% in Oregon, 68% in Maine), similar habitat use, and simultaneously overlapping home ranges between bobcats and coyotes. Despite the opportunity for interference competition, neither study found any evidence of aggression or avoidance. Witmer and DeCalesta (1986) concluded that abundant prey minimized any competition. Major and

Sherburne (1987) felt that abundant prey, as well as evolutionary and taxonomic separation, kept the two species from being in direct competition. Subsequently, a feeding habits study in the same area concluded that there was high overlap of hare in the diets of bobcats, coyotes, and red foxes and that some level of exploitative competition may be occurring (Dibello et al. 1990).

Litvaitis and Harrison (1989), also working in Maine, more definitively illustrated exploitative competition between coyotes and bobcats. They coupled their data on the high levels of dietary, temporal, and spatial overlap of coyotes and bobcats with harvest data of both species for the previous 10 years. There was a strong negative correlation between bobcat and coyote harvest, suggesting a significant bobcat decline in response to exploitative competition by the increasing coyote population, particularly during winter and spring. Fedriani et al. (2000), in the Santa Monica Mountains of California, also demonstrated high seasonal food overlap between bobcats, gray foxes, and coyotes. They concluded that coyotes were dominant in the system because they used more foods and more habitats and were more abundant than either of the other species. Also, behaviorally, coyotes illustrated interference competition and dominance of bobcats by preying on two adults. Because there have been several documented examples of predation by coyotes on bobcats, interference competition undoubtedly occurs in some ecosystems (Young 1958; Anderson 1987; Litvaitis and Harrison 1989; Fedriani et al. 2000).

There is substantially less known about the competitive interactions of coyotes and lynx. Clearly there is an opportunity for exploitative competition because in some areas of overlap, and during highs in snowshoe hare abundance, coyotes kill hares at a higher rate than do lynx (O'Donoghue et al. 1998b). Competition may be mitigated somewhat in winter months by deep snow, which can spatially segregate coyotes and lynx (Murray and Boutin 1991). Nonetheless, circumstantial evidence suggests that coyote competition may be substantial. O'Donoghue (1997) compared densities of lynx, coyotes, and hares in his Yukon study area with those in central Alberta (Keith et al. 1977), and found that lynx were more abundant where coyotes were less common rather than where hare density was the highest. There is also the potential for interference competition because direct predation by coyotes on young and malnourished lynx has been documented (O'Donoghue et al. 1998a).

Summarizing what was known about lynx–coyote interactions, Buskirk et al. (2000a) felt exploitative competition by the coyote, particularly in the southern part of lynx range, could be substantial and might reduce lynx numbers. They speculated that human-prompted ecosystem changes, such as expansion of coyote range into the northeastern and northwestern United States, as well as access to deeper snow areas provided by snowmobile trails and roads may increase the threat of coyotes to lynx.

Interactions of bobcats and lynx with other sympatric mesocarnivores have been less studied. Gilbert and Keith (2001) evaluated spatial and dietary overlap among sympatric fishers (*Martes pennanti*) and bobcat populations in northern Wisconsin. They found that competition was relatively weak, but suggested that encounter competition in areas where fishers are most abundant might result in increased bobcat kitten mortality and subsequent reduction in bobcat population growth.

A number of other predators also kill bobcats and lynx. Cougars take lynx (Squires and Laurion 2000) as well as bobcats (Koehler and Hornocker 1991), although the extent of predation and its impact on the population dynamics of either species is unknown (Buskirk et al. 2000a). Wolves (*Canis lupus*) also have been documented as lynx predators in North America (Banfield 1987) and in Europe (Pulliainen 1965). Buskirk et al. (2000a) pointed out, however, that lynx may actually benefit from the presence of wolves. Wolves are fierce competitors with coyotes, reducing coyote numbers in areas where both species occur. Because wolves do not prey heavily on rabbits and hares, they may actually reduce competition with the lynx by lowering coyote numbers. A similar argument could be made for the bobcat. The expanding gray wolf populations in the United States should provide an opportunity to test the theory.

ECONOMIC STATUS, MANAGEMENT, AND CONSERVATION

Age Estimation. Dental characteristics provide the most accurate technique for estimating age for bobcats and lynx. Age of kittens (<240 days) can be fairly accurately determined using a deciduous tooth replacement schedule published by Jackson et al. (1988) for bobcats and Saunders (1964) for lynx. Root canals of permanent canines remain open until 13–18 months of age, providing an effective criterion for distinguishing between juveniles and adults collected during the harvest season (Saunders 1961; Crowe 1975a). Johnson et al. (1981) warned that assigning bobcats that had teeth with open root canals to young-of-the-year class, at least in Kansas, would bias this age class upward. They found bobcats in their harvest sample that had open root canals but had already completed one reproductive cycle. They suggested instead that a ratio of tooth length to root canal width was a better predictor of juvenile age.

Counting the layers of cementum annuli on cross sections of canine (or other) teeth has provided the best means for estimating adult ages of both species (Conley and Jenkins 1969; Nellis et al. 1972; Crowe 1975a). Dark-staining bands begin to appear during an animal's second winter and presumably are only added at the rate of one per year. Although the technique is the most widely used, it may not present clear or accurate results. Mahan (1979) cautioned that it might be difficult to distinguish the first or last layers on a tooth due to their proximity to other tissue types. Also, interpretation can be confused by the presence of false annuli. Most tooth samples are taken from carcasses; however, an incisor, the first bottom premolar (P_2), or the first upper molar (M^1) may be extracted from a live animal to estimate age.

Other age-estimation techniques for the bobcat and lynx have been developed that allow classification of animals into general age categories. Skull measurements and morphology permitted Conley and Jenkins (1969) to distinguish among juveniles, yearlings, and adult bobcats. Mahan (1979) was able to distinguish between the same age categories based on the ratio of the diameter of the pulp cavity to the tooth width, provided the sex of the bobcat was known. In both species, epiphyseal and cranial suture closings also provide some age discrimination. Adults exhibit complete fusion of epiphyseal sutures in long bones and the basioccipital–basisphenoid suture, juveniles show no fusion, and yearlings exhibit partial closure (van Zyll de Jong 1963; Conley and Jenkins 1969; Nava 1970; Mahan 1979). Eye lens weight has not been fully explored as a criterion for estimating age, although Conley and Jenkins (1969) felt that it could be useful. Body measurements are generally too variable and overlap age categories so much that they are not good indicators of age (Conley and Jenkins 1969).

Juvenile and adult lynx also can be distinguished by pelt size. Quinn and Gardner (1984) found that pelts <81 cm from tip of the nose to the base of the tail (categorized as small) were exclusively kittens, and very few kittens had pelts large enough to be placed in the next size category. The technique offers an inexpensive way to monitor the growth and early decline phases of lynx populations.

Friedrich et al. (1984) developed an index to distinguish between male and female bobcats by multiplying the maximum canine root width by the maximum root thickness. They were able to correctly predict the sex of 88.7% of juveniles and 96.9% of adults by the method.

Population Estimation. Perhaps the most significant hindrance to better management of lynx and bobcats is lack of an effective technique for estimating population size. Both species exist at low densities, are widely dispersed, and are secretive. Actual censusing of their populations is essentially impossible and many of the techniques are imprecise and inaccurate (Quinn and Parker 1987; Rolley 1987; Diefenbach et al. 1994). However, effective management and monitoring of either species requires reasonable estimates of numbers or relative changes in population size. Techniques for estimating the absolute and relative abundance of bobcats can be considered in five categories: (1) total enumeration, (2) mark–recapture, (3) life table analyses and computer population models, (4) questionnaires and surveys, and (5) track indices.

TABLE 38.6. Methods used to monitor abundance of bobcats in the United States as of 1996

Method	Number of States
Hunter/trapper surveys	31
Harvest data (e.g., catch/hunter-trapper, pelt sales/tagging)	26
Employee opinion	20
Sighting reports	19
Life table analysis	13
Computer population model	13
Archer's index	8
Sign/track survey	8
Scent-station survey	6
Prey survey	4
Other techniques	10

SOURCE: Adapted from Bluett et al. (2001).

Telemetry provides the best estimate of absolute numbers of bobcat or lynx in a limited study area. If all individuals in the area are captured, density is easily calculated. Generally not all individuals are caught and marked, so by using average adult home range size, percentage home range overlap, local sex ratios, and percentage of population made up of juveniles, an estimate of density is possible. Unfortunately, conventional very high frequency telemetry is costly and labor-intensive, and the results cannot be safely extrapolated beyond the study area to different habitat types or areas under different harvest regimes.

Estimating population size using mark–recapture techniques has been attempted on bobcats with radioactively labeled feces (Pelton 1979; Conner et al. 1983). Bobcats are injected with a radioisotope (usually ^{65}Zn or ^{54}Mn), which is slowly excreted in their feces. Feces are then collected from throughout the study area, and the ratio of labeled to nonlabeled feces provides an estimate of abundance using a modified Lincoln/Petersen estimator (Pelton and Marcum 1977). The technique is relatively inexpensive, but requires that all animals are labeled within a relatively short time and that feces are relatively easy to locate. Also, use of radioisotopes may be severely restricted in areas open to the public.

Life table analyses and computer population models are also used to estimate bobcat and lynx population sizes. However, they generally require assumptions that are difficult to meet. These techniques are discussed more fully in the Harvest Management section.

There are a number of techniques currently used that provide a relative index of bobcat and lynx numbers. Unfortunately, in most cases, it is unclear how sensitive these techniques are to real changes in the population size. Because detection rates are so low and variability so high, the methods generally only detect short-term changes that are >50% with any confidence (Rolley 1987). Because of the imprecision and inaccuracy of these indices, most states (94%) use two or more methods for assessing abundance of bobcats (Bluett et al. 2001) (Table 38.6).

Questionnaires or surveys are used by many states to corroborate the results of other bobcat population indices. Questionnaires requesting information on bobcat sightings are either sent to landowners, biologists, wildlife conservation officers, hunters, or trappers or distributed to hunters at check stations. Results compiled over several years may be useful for indicating regional population trends. In Oklahoma, Hatcher and Shaw (1981) compared mail questionnaires with two types of scent-station surveys for estimating furbearer abundance. They found the questionnaire was better than either of the scent-station surveys when populations were low.

An archer's index is a special modification of the questionnaire for indexing bobcats. Archery deer hunters are provided with logbooks for their observations and activities. The index is computed as the number of bobcat sightings/1000 hr of archery hunting (Hamilton et al. 1990). The technique generates large confidence limits, but is useful for monitoring long-term trends (Bluett et al. 2001).

In some northern states and Canadian provinces where snow is present for extended periods of time, track transects have provided information on the relative changes in bobcat and lynx numbers. Within a specified time period following a fresh snowfall (usually one night of track registry), preselected road routes are driven and the number of furbearer tracks intercepted is recorded (Thompson et al. 1989).

Scent post stations also have been used to track population changes of bobcats, although their validity has not been well established (Roughton and Sweeny 1982; Sargeant et al. 1998). Scent stations usually consist of a 0.91-m-diameter circle of sifted sand with an attractant (often bobcat urine) at the center. Several stations are equally spaced along a transect and left for varying periods of time. Visitation rates by bobcats are usually extremely low. Rust (1980) considered the rates to be so low in Tennessee that the technique was of little practical value. Diefenbach et al. (1994) used the reintroduction of bobcats to Cumberland Island, Georgia, as an opportunity to evaluate various index techniques, including scent stations. They concluded that increases in the bobcat population on the island were not reflected in the scent post survey. Sargeant et al. (1998) suggested that long-term trends in visitation rates probably reflected real changes in population size, although the usefulness of the technique was limited by poor spatial and temporal resolution, susceptibility to confounding error, and low statistical power. Also, the results of scent-station transects generally are not considered valid for making among-area comparisons (Roughton and Sweeney 1982).

Conner et al. (1983) compared different population indexing methods for bobcats in a small study area in northeastern Florida. They evaluated the results from scent stations, trapping, radioisotope feces tagging, and radiotelemetry collected over a 24-month period. All the estimators moved in synchrony with the number of bobcats on the area as indicated by radiotelemetry.

New techniques hold potential promise for indexing lynx and bobcat populations. Hair snares or traps combined with the use of lures have been used to index lynx populations even at low densities (McDaniel et al. 2000). The technique is equally applicable to bobcats, although in areas where both species overlap, DNA analysis is required to distinguish the hairs (Foran et al. 1997). The hair traps also provide a possible technique for estimating population size of either species by the application of mark–recapture theory to hair samples collected at different time intervals. DNA analysis genetically identifies each individual that leaves hair on the trap. Subsequent sessions of hair trapping constitute the resampled population. The ratio of previously identified individuals in the resample to the total number of individuals resampled can be used to estimate population size (Foran et al. 1997).

Reintroductions. Several attempts have been made to translocate bobcats and lynx into formerly occupied areas. A successful reintroduction of bobcats to Cumberland Island, Georgia, occurred in 1988 and 1989, when 31 bobcats trapped from the coastal plains were "hard released" to the island (Diefenbach et al. 1993). Because the original population of bobcats had been extirpated since the early 1900s, the "known" number of bobcats on the island provided an ideal opportunity to study bobcat census techniques (Diefenbach et al. 1994) (see Population Estimation).

Bobcats also were reintroduced into New Jersey from March 1978 through November 1980. Captured in Maine, 15 bobcats (7 males, 8 females) were released in the northern part of the state (McConnel 1981). Subsequent to the releases, it was discovered that a remnant population of bobcats already existed in that portion of the state (M. Valent, New Jersey Division of Fish, Game, and Wildlife, pers. commun., 2001). As of 2001, bobcats were well established in the northern portions of the state.

Reintroduction of lynx into North America has been attempted only twice. The first attempt was in the Adirondack Mountains of New York State from 1989 to 1991, when 83 lynx were translocated from the Yukon (Brocke et al. 1990, 1993). Wide dispersal by the lynx resulted in high mortality and movement away from the study site. Within 2 years, nearly half of the individuals had died, with 16 killed by collisions with vehicles (Brocke et al. 1991; Aubry et al. 2000a). Despite considerable search effort since the releases, no lynx were found (Ruggiero et al. 2000a).

Reintroduction efforts in Colorado were initiated in 1998. Spearheaded by the Colorado Division of Wildlife, 96 lynx (57 females, 39 males) were captured in British Columbia, the Yukon, and Alaska and released in southwestern Colorado in 1999 and 2000. As of the summer of 2001, there were 37 known mortalities; 24% were due to starvation, 14% were hit by vehicles, and 14% were shot. Reproduction has not yet been documented (Shenk 2001).

Fur Harvest and Value. Bobcat research and management are relatively recent developments. Due to their nocturnal habits and generally secretive nature, bobcats are not a highly visible species. Economically, they were of little historical importance. The average value of a bobcat pelt between 1950 and 1970 was only $5.00. With the exception of occasional instances of bobcat depredation on sheep or chickens, a general lack of interest in the species provided little incentive for state or federal wildlife agencies to research or actively manage the species. As of 1970, 40 of the contiguous states did not offer protection to the bobcat, including 10 states that still offered bounties (Faulkner 1971).

The bobcat became the center of intense political and ecological debate during the 1970s and early 1980s. With the passage of the Endangered Species Conservation Act in the late 1960s, the United States prohibited the import of fur of endangered felids. Consequently, an increasing amount of commercial attention was diverted to the nonthreatened lynx and bobcat. Between 1970 and 1976, the annual harvest of bobcats in the United States rose from 10,882 to 35,990, while the average price per pelt escalated from $10 to $125. Bobcat pelts are used for coats, trim, and accessories, with the spotted belly fur being most valuable. Bobcat pelts from northern or mountainous areas are sometimes referred to as "lynx cats" by fur graders or buyers, whereas pelts from southern or lowland areas are simply referred to as "bob cats" [*sic*] (Obbard 1987).

In 1975, the United States, along with 79 other countries, signed the Convention on International Trade in Endangered Species (CITES) to provide international protection for endangered species. The bobcat was listed in Appendix II, which required member countries to prove that exporting of bobcat pelts would "not be detrimental to the survival of that species." In 1977, the Endangered Species Scientific Authority in the United States placed a temporary ban on the export of pelts until more evidence of "no detriment" could be found. Although the Authority moved to lift the export ban in 1978, Defenders of Wildlife filed suit in U.S. District Court in 1979 to stop exportation of the pelts and won a temporary export ban in five states and portions of two others. The legal battles continued until the spring of 1981, when the court issued an order that prohibited the export of all bobcat pelts until reliable population estimates and harvest limits could be established. In the reauthorization of the Endangered Species Act in late 1982, Congress negated the CITES requirement that reliable bobcat population estimates were prerequisite to the "no detriment" finding and the Defenders case was dismissed. The following summer, CITES convened in Botswana and permitted the United States to move the bobcat to another subsection of Appendix II. The new listing permitted management of the bobcat because it appears similar to other endangered felids, and not as an endangered species itself. During this entire period of turmoil, interest in the biology, ecology, and status of bobcat populations by state wildlife agencies became intense and numerous research projects were initiated. Also amid the controversy and debate, harvest of the species increased dramatically. From 1977 to 1984, 94,000 bobcats were harvested annually from the United States and Canada at an average pelt price of $125 (Funderburk 1986).

However, with inclusion of the bobcat in Appendix II of CITES and the subsequent improvement in management and monitoring of the species in the United States, the threat of overexploitation subsided. As of 1996, populations were considered stable in 22 states and increasing in 20, with no states reporting overall declines (Woolf and Hubert 1998). Demand for fur also decreased. By 1991–1992, the total number of pelts harvested was only 22,077 (approximately 25% of 1977–1984 average harvest) and sold for an average of only $63 (Linscombe 1994). Currently, 38 states allow harvest of bobcats, and they are protected by a continuously closed season in 9 others. In Indiana, Iowa, New Jersey, and Ohio, they are listed as endangered (Woolf and Hubert 1998).

The value of lynx pelts in Canada has paralleled that of bobcat pelts, although harvest has been much more influenced by the cyclic populations of hare than by changes in pelt price. From 1953/54 through 1978/79, the average price of a lynx pelt rose from $3.62 to $336.36 and peaked at nearly CDN$600/pelt in 1984 (Nowak 1999). During the 1990s, prices were generally <$100 and as of May 2001, mean pelt price was US $53.44 (North American Fur Auctions 2001).

In the United States, the recent history of the lynx has been highly contentious. Beginning in 1991, a number of organizations sequentially petitioned the U.S. Fish and Wildlife Service to list the lynx in the contiguous United States as endangered. After several decisions by the Fish and Wildlife Service that the designation was not warranted and subsequent appeals, the Service announced in May 1997 that listing was warranted, but precluded by actions on other species of higher taxonomic importance. A lawsuit brought by Defenders of Wildlife and others resulted in the U.S. Fish and Wildlife Service agreeing to propose the listing of the contiguous U.S. population at the latest by June 1998. The final listing decision, slated for July 1999, was extended to allow a Lynx Science Team time to produce an interim conservation strategy. The contiguous U.S. population of the Canada lynx was finally listed as a threatened species in March 2000 (U.S. Fish and Wildlife Service 2000). Before the listing, Montana was the last state to allow legal harvest of lynx.

Harvest Management. Annual harvest estimates for states and provinces that permit harvesting of bobcats or lynx generally are derived through mandatory registration of carcasses. Bobcats and lynx are listed in Appendix II of CITES. The Appendix was established to offer better protection of look-alike species that were or could be threatened with extinction without strict regulation of trade. Under CITES, all pelts to be exported are required to be tagged with a permanently attached numbered tag, which identifies the species, state of origin, and year of harvest. However, because some harvested animals are taken as trophies and do not need to be tagged for export, many states and provinces require the mandatory registration of harvested animals regardless of their ultimate disposition. Carcasses are often collected and used to gather annual information on fecundity rates and age structure of the population.

In the absence of reliable information on changes in bobcat and lynx population size from year to year, age distributions from carcass collections have been used to infer changes in the population size. However, without additional data, changes in the age structure can imply an increasing, a decreasing, or a stable population (Caughley 1977). Knick (1990) suggested for bobcats that combining the yearling-to-adult ratio with information about prey base production could be a reliable index of exploitation intensity. If the prey base is constant, indicating a constant reproductive effort, then an increased yearling-to-adult ratio in a series of harvests would signal increased exploitation intensities. Unfortunately, a multitude of other factors independent of prey could also potentially affect the reproduction or survival of bobcats. With lynx, however, the ratio of kittens to adults can indicate the approximate location of the population in the 10-year cycle.

The most reliable way of assessing changes in the size of bobcat and lynx populations is to measure reproduction and survival. As discussed previously, darker placental scars indicate in utero litter sizes. However, actual litter sizes will generally be smaller than those estimated by placental scars due to abortions, resorbed fetuses, and stillbirths.

Survival rates have commonly been estimated from age structure of the harvest. However, there are a number of assumptions of life table analyses that are difficult to meet. If different age cohorts are differentially vulnerable to harvest (e.g., kittens are less vulnerable to trapping than adults), they are either under- or overrepresented. Also, both time-specific and composite life tables require either knowledge of the rate of change or that the population is numerically stable, neither of which is easily determined.

Radiotelemetry is probably the most accurate way to assess survival rates. Unfortunately, it is expensive and time consuming, and generally applies to only a relatively small area of study.

Ideally, population models should include information on all the variables that significantly influence growth of the population. Fuller et al. (1995) noted that bobcat population models could be much more sensitive if they included area-specific data on bobcat harvest, prey abundance and change, habitat composition and change, and bobcat demography.

Managers have generally assumed that bobcat populations can sustain a 20% harvest rate without declining (Knick 1990). However, depending on environmental conditions, levels of poaching, prevalence of disease, density of competitors, and prey availability, 20% may be excessive. Survival rates in moderately exploited populations are often close to 60%. If legal harvest is set at 20%, the remaining 20% of mortality will be due to other causes including poaching, disease, predation, or accidental death. If poaching or other forms of mortality increase and exceed the level of legal harvest, the population may decline. Obviously, calculating 20% of the fall population requires a reasonable estimate of total population size, a figure that is rarely known accurately.

Harvest management strategies for lynx in Canada involve three major elements: (1) tracking lynx population changes and curtailing or eliminating trapping 3–4 years following initiation of the lynx decline (Brand and Keith 1979), (2) maintaining permanent untrapped areas or refuges (Slough and Mowat 1996), and (3) using season and/or quota restrictions (Parker et al. 1983; Poole 1994; Slough and Mowat 1996).

Lynx are vulnerable to overexploitation when densities are lowest because recruitment essentially stops and no new individuals are added to the population (Brand and Keith 1979). The tracking harvest strategy allows managers to track the population as it enters the lowest portion of the cycle and suspend harvest. Mowat et al. (2000) recommend that managers consider curtailing harvest beginning 1–2 years after lynx numbers begin to decline. Following complete closure, seasons should not open for 3–4 years or until there is significant increase in the juvenile portion of the population. Pelt length (juvenile-to-adult ratio) in the harvest can track the decline (Quinn and Gardner 1984; Slough 1996), and track counts can reveal when juvenile recruitment increases (Thompson et al. 1989).

The concept of maintaining refugia is based on protecting viable populations from overexploitation during population lows. Unfortunately, given the large home ranges and extremely low densities of lynx during the cyclic lows, huge areas of thousands of square kilometers would be needed to protect a core population. The strategy is advisable only where large reserves currently exist (Mowat et al. 2000).

The size and impact of harvest on the lynx population can be controlled by the traditional techniques of adjusting the length and timing of the trapping season. Seasons should be set to correspond with periods of pelt primeness and should be early enough not to interfere with breeding activities, but late enough so harvesting of adult females will not orphan still-dependent kittens. Quotas may be the most effective way to limit the take in heavily harvested areas or during periods of low lynx density.

With the exception of the tracking harvest strategy, all the same tools can be applied to good bobcat harvest management. Key to any successful management strategy, however, is monitoring of changes in population size. Carcass collection, track counts, scent stations, and questionnaires can be used in collaboration to detect significant changes in population size, and harvest can be adjusted accordingly.

Habitat Management. Habitat management for the lynx generally involves providing a suitable forest environment for snowshoe hares while providing adequate blowdown for den and kitten-rearing sites (Quinn and Parker 1987). Because hare density is directly related to woody stem density, habitat prescriptions generally involve opening the forest canopy to stimulate understory growth. Quinn and Parker (1987) described ideal lynx habitat as uneven-age forests with a relatively open canopy as well as "patchy" areas of disturbed forest. Although strip or blockcutting within dense forest could provide that ideal habitat mix, they warned that extensive clearcuts would not meet lynx habitat requirements.

More recently, Buskirk et al. (2000b) advised that the creation of early-successional forest might not be the best way to provide habitat for lynx prey. Although on mesic sites the highest hare densities are associated with sapling-stage forests following disturbance, those sites lack habitat for the red squirrel, an important alternate prey for lynx during hare declines. They suggested that old gap-phase forests might be a preferred habitat type because they provide a moderate density of hares and high densities of squirrels while providing an ample supply of blowdown for den sites. On drier sites in the West, early-successional forest may not provide adequate shrub and sapling densities to support high densities of hares. Because postdisturbance regeneration is erratic, gap-phase forests may provide a higher density of smaller stems compared to earlier successional stages on the same site. McKelvey et al. (2000c) outlined a management model that would provide the same components, but based on the restoration of historical patterns and processes.

A specific prescription for bobcat habitat management is more difficult because they are relatively ubiquitous across a wide range of habitat types, and specific recommendations will vary from site to site. In general, any treatments that improve the habitat for hares or rabbits, supply ample cover for stalking and ambush, and provide or protect den sites should improve the quality of the habitat for bobcats.

RESEARCH AND MANAGEMENT NEEDS

Research projects and publications on bobcats rose dramatically during the 1980s in response to a number of factors. Listing of the bobcat in Appendix II of CITES and subsequent political, legal, and biological debate highlighted our lack of basic knowledge about the species. At the same time, radiotelemetry was just coming of age and provided the tools to allow detailed studies of nocturnal and secretive species like the bobcat. Despite several well-organized and extensive efforts, the pressure to generate results produced very few well-coordinated and systematic research efforts (Anderson 1987). Recent efforts have been much more focused and executed at larger geographic scales.

Lynx populations, particularly at the core of their range, have seen a more consistent research effort. The cyclic nature of the lynx/hare relationship has motivated a fair amount of research by ecologists since the cycle was described by Elton and Nicholson (1942). However, the U.S. populations of lynx received very little attention and were the subject of only five studies. In the 1990s, the intensity of lynx research in both Canada and the United States began to escalate. Pressure in the United States to consider listing the lynx as a threatened species led to the publication of an excellent review of lynx ecology by Koehler and Aubry (1994). They identified very specific research needs in six broad areas. It was followed by publication of a very thorough analysis of the scientific basis for lynx conservation compiled by a team of government and university scientists (Ruggiero et al. 2000b). Within that publication, in an attempt to avoid the haphazard approach that characterized early bobcat research, Aubry et al. (2000b) provided a systematic framework and rationale for conducting future research to "maximize the applicability and utility of new information." Their suggestions covered nine areas of research needed on the U.S. population of lynx:

1. Distribution and relative abundance. What is the distribution and abundance of lynx in the southern portion of the range and what factors control it?
2. Habitat relationships. Little is known about lynx habitat relationships at any spatial scale. To manage them effectively will require knowledge of the impact of timber harvest strategies and effects of other disturbances.
3. Movements and dispersal. The southern lynx population may be a "metapopulation" and require dispersal to maintain subpopulations. However, little is known about the necessity of dispersal, average rates, use of corridors, anthropogenic barriers, and associated factors.

4. Demography and population dynamics.
5. Relationships with prey. What are the patterns of prey abundance in the southern boreal forest and what impact do they have on lynx ecology?
6. Community interactions. Large questions remain about the impact of predation and competition on lynx. How do changes in the landscape pattern affect those processes?
7. Human impacts. Direct and indirect impacts of humans on lynx numbers and distribution are poorly understood. What are the impacts of trapping? How does outdoor recreation affect the species?
8. Snowshoe hare habitat relationships. How patterns of hare abundance change among forest types, moisture regimes, and successional stages is poorly understood.
9. Snowshoe hare population dynamics. Although suspected not to cycle in the southern portion of their range, how variable are populations and what drives the changes?

A similar analysis of key research questions has been done for the bobcat (Dyer 1979; Bluett et al. 2001; Woolf and Nielsen 2001). A survey of state agencies in 1996 indicated that the top research needs reported by bobcat managers paralleled those for the lynx; (1) reliable survey methods, (2) demographics (e.g., mortality, recruitment), (3) distribution and abundance, (4) habitat availability and use, and (5) interactions with coyotes and other carnivores. They also identified the top five management needs: (1) control harvest to better match geographic/temporal difference in abundance, (2) monitor abundance, (3) protect or improve habitat, (4) improve public knowledge of and support for management techniques, and (5) evaluate effectiveness of and need for federal oversight (Bluett et al. 2001).

Clearly, one of the most pressing needs for both species is to develop a more accurate and precise index of abundance that is comparable across a variety of habitats and climates. In a survey of wildlife biologists in the contiguous 48 states, 63% of respondents indicated a need for more reliable survey techniques and better information about the distribution, abundance, and demographics of bobcats (Woolf and Hubert 1998).

As with most scientific inquiry, bobcat and lynx research could greatly benefit from long-term studies, spanning decades instead of months. Much of the conflicting results among studies might be reconciled by continuous views of a population through time. Similarly, many of the generalizations about lynx and bobcat biology have been extrapolated from research on very few individuals in very few environments. Increasing sample sizes should help us understand the variability and consistency of observations across a range of temporal and spatial scales.

The role of disease in lynx and bobcat population dynamics and behavior is relatively unknown. Understanding the impact and being able to monitor the movement of disease and parasite loads through populations is likely to be an important aspect of future management.

The taxonomy of the species also needs careful reconsideration. If populations of southern lynx are isolated in subpopulations, then they may represent evolutionarily significant units, which should be given special consideration. Alternatively, DNA analysis may reduce the number of bobcat subspecies into more biologically significant groupings.

LITERATURE CITED

Anderson, E. M. 1987. Critical review and annotated bibliography of the literature on the bobcat (Special Report No. 62). Colorado Division of Wildlife.

Anderson, E. M. 1988. Effects of male removal on bobcat spatial distribution. Journal of Mammalogy 69:637–41.

Anderson, E. M. 1990. Characteristics of bobcat diurnal loafing sites in southeastern Colorado. Journal of Wildlife Management 54:600–602.

Apps, C. D. 2000. Space-use, diet, demographics, and topographic associations of lynx in the southern Canadian Rocky Mountains: A study. Pages 351–71 *in* L. F. Ruggiero, K. B. Aubry, S. W. Buskirk, G. M. Koehler, C. J. Krebs, K. S. McKelvey and J. R. Squires, eds. Ecology and conservation of lynx in the United States. University Press of Colorado, Boulder.

Asdell, S. A. 1946. Patterns of mammalian reproduction. Comstock Press, Ithaca, NY.

Aubry, K. B., G. M. Koehler, and J. R. Squires. 2000a. Ecology of Canadian lynx in southern boreal forests. Pages 373–96 *in* L. F. Ruggiero, K. B. Aubry, S. W. Buskirk, G. M. Koehler, C. J. Krebs, K. S. McKelvey and J. R. Squires, eds. Ecology and conservation of lynx in the United States. University Press of Colorado, Boulder.

Aubry, K. B., L. F. Ruggiero, J. R. Squires, K. S. McKelvey, G. M. Koehler, S. W. Buskirk, and C. J. Krebs. 2000b. Conservation of lynx in the United States: A systematic approach to closing critical knowledge gaps. Pages 455–70 *in* L. F. Ruggiero, K. B. Aubry, S. W. Buskirk, G. M. Koehler, C. J. Krebs, K. S. McKelvey and J. R. Squires, eds. Ecology and conservation of lynx in the United States. University Press of Colorado, Boulder.

Bailey, T. N. 1974. Social organization in a bobcat population. Journal of Wildlife Management 38:435–46.

Bailey, T. N. 1979. Den ecology, population parameters and diet of eastern Idaho bobcats. Pages 62–69 *in* P. C. Escherich and L. Blum, eds. Proceedings of the 1979 bobcat research conference (Science and Technology Series 6). National Wildlife Federation, Washington, DC.

Bailey, T. N. 1981. Factors of bobcat social organization and some management implications. Pages 984–1000 *in* J. A. Chapman and D. Pursley, eds. Proceedings of the worldwide furbearer conference. Frostburg, MD.

Bailey, T. N., E. E. Bangs, M. F. Portner, J. C. Malloy, and R. J. McAvinchey. 1986. An apparent overexploited lynx population in the Kenai Peninsula, Alaska. Journal of Wildlife Management 50:279–90.

Banfield, A. W. F. 1987. The mammals of Canada, 2nd ed. University of Toronto Press, Toronto.

Barash, D. P. 1971. Cooperative hunting in the lynx. Journal of Mammalogy 52:480.

Beale, D. M., and A. D. Smith. 1973. Mortality of pronghorn antelope fawns in western Utah. Journal of Wildlife Management 37:343–52.

Beasom, S. L., and R. A. Moore. 1977. Bobcat food habit response to a change in prey abundance. Southwestern Naturalist 21:451–57.

Beeler, I. E. 1985. Reproductive characteristics of captive and wild bobcats (*Felis rufus*) in Mississippi. M.S.Thesis, Mississippi State University, Mississippi State.

Bell, J. F., and J. R. Reilly. 1981. Tularemia. Pages 213–31 *in* J. W. Davis, L. H. Karstad and D. O. Trainer, eds. Infectious diseases of wild mammals, 2nd ed. Iowa State University Press, Ames.

Berg, W. E. 1979. Ecology of bobcats in northern Minnesota. Pages 55–61 *in* P. C. Escherich and L. Blum, eds. Proceedings of the 1979 bobcat research conference (Science and Technology Series 6). National Wildlife Federation, Washington, DC.

Bergerud, A. T. 1971. The population dynamics of the Newfoundland caribou. Wildlife Monographs 25:1–55.

Berrie, P. M. 1973. Ecology and status of the lynx in interior Alaska. World's Cats 1:4–41.

Blankenship, T. L., and W. G. Swank. 1979. Population dynamic aspects of the bobcat in Texas. Pages 116–22 *in* P. C. Escherich and L. Blum, eds. Proceedings of the 1979 bobcat research conference (Science and Technology Series 6). National Wildlife Federation, Washington, DC.

Bluett, R. D., G. F. Hubert, Jr., and A. Woolf. 2001. Perspectives on bobcat management in Illinois. Pages 67–73 *in* A. Woolf, C. K. Neilsen and R. D. Bluett, eds. Proceedings of a symposium on current bobcat research and implications for management. Wildlife Society, Nashville, TN.

Boyce, M. S. 1979. Seasonality and patterns of natural selection for life histories. American Naturalist 114:569–83.

Brand, C. J., and L. B. Keith. 1979. Lynx demography during a snowshoe hare decline in Alberta. Journal of Wildlife Management 43:827–49.

Brand, C. J., L. B. Keith, and C. A. Fischer. 1976. Lynx responses to changing snowshoe hare densities in Alberta. Journal of Wildlife Management 40:416–28.

Breitenmoser, U., P. Kavczensky, M. Dotterer, C. Breitenmoser-Wiirsten, S. Capt, F. Bernhart, and M. Liberek. 1993. Spatial organization and recruitment of lynx in a re-introduced population in the Swiss Jura Mountains. Journal of Zoology 231:449–64.

Brittell, J. D., S. J. Sweeney, and S. T. Knick. 1979. Washington bobcats: Diet, population dynamics, and movement. Pages 107–10 *in* P. C. Escherich and L. Blum, eds. Proceedings of the 1979 bobcat research conference (Science and Technology Series 6). National Wildlife Federation, Washington, DC.

Brocke, R. J., K. A. Gustafson, and A. R. Major. 1990. Restoration of lynx in New York: Biopolitical lessons. North American Wildlife and Natural Resource Conference 55:590–98.

Brocke, R. J., K. A. Gustafson, and L. B. Fox. 1991. Restoration of large predators: Potentials and problems. Pages 303–15 *in* D. J. Decker, M. E. Krasny, G. R. Goff, C. R. Smith and D. W. Gross, eds. Challenges in the conservation of biological resources: A practitioner's guide. Westview Press, Boulder.

Brocke, R. J., J. Belant, and K. Gustafson. 1993. Lynx population and habitat survey in the White Mountain National Forest, New Hampshire. State University of New York, College of Environmental Sciences and Forestry, Syracuse.

Bruning-Fann, C. S., S. M. Schmitt, S. D. Fitzgerald, J. S. Fierke, P. D. Friedrich, J. B. Kaneene, K. A. Clark, K. L. Butler, J. B. Payeur, and D. L. Whipple. 2001. Bovine tuberculosis in free-ranging carnivores from Michigan. Journal of Wildlife Diseases 37:58–64.

Buie, D. E., T. T. Fendley, and H. McNab. 1979. Pages 42–46 *in* P. C. Escherich and L. Blum, eds. Proceedings of the 1979 bobcat research conference (Science and Technology Series 6). National Wildlife Federation, Washington, DC.

Bursey, C. C., and M. D. B. Burt. 1970. *Taenia macrocystis* (Diesing 1850), its occurrence in eastern Canada and Maine, and its life cycle in wild felines (*Lynx rufus* and *Lynx canadensis*) and hares (*Lepus americanus*). Canadian Journal of Zoology 48:1287–93.

Buskirk, S. W., L. F. Ruggiero, and C. J. Krebs. 2000a. Habitat fragmentation and interspecific competition: Implications for lynx conservation. Pages 83–100 *in* L. F. Ruggiero, K. B. Aubry, S. W. Buskirk, G. M. Koehler, C. J. Krebs, K. S. McKelvey and J. R. Squires, eds. Ecology and conservation of lynx in the United States. University Press of Colorado, Boulder.

Buskirk, S. W., L. F. Ruggiero, K. B. Aubry, D. E. Pearson, J. R. Squires, and K. S. McKelvey. 2000b. Comparative ecology of lynx in North America. Pages 397–417 *in* L. F. Ruggiero, K. B. Aubry, S. W. Buskirk, G. M. Koehler, C. J. Krebs, K. S. McKelvey and J. R. Squires, eds. Ecology and conservation of lynx in the United States. University Press of Colorado, Boulder.

Carbyn, L. N., and D. Patriquin. 1983. Observations on home range sizes, movements and social organization of lynx, *Lynx canadensis,* in Riding Mountain National Park, Manitoba. Canadian Field-Naturalist 97:262–67.

Caughley, G. 1977. Analysis of vertebrate populations. John Wiley, New York.

Chamberlain, M. J., and B. D. Leopold. 2001. Spatio-temporal relationships among adult bobcats in central Mississippi. Pages 45–55 *in* A. Woolf, C. K. Neilsen and R. D. Bluett, eds. Proceedings of a symposium on current bobcat research and implications for management. Wildlife Society, Nashville, TN.

Chamberlain, M. J., B. D. Leopold, L. W. Burger, Jr., B. W. Plowman, and M. L. Conner. 1999. Survival and cause-specific mortality of adult bobcats in central Mississippi. Journal of Wildlife Management 63:613–20.

Colby, E. D. 1974. Artificially induced estrus in wild and domestic felids. World's Cats 2:126–47.

Conley, R. H., and J. H. Jenkins. 1969. An evaluation of several techniques for determining the age of bobcat (*Lynx rufus*) in the Southeast. Proceedings of the Annual Conference of the Southeastern Association of Fish and Wildlife Agencies 23:104–9.

Conner, L. M., B. Plowman, B. D. Leopold, and C. Lovell. 1999. Influence of time-in-residence on home range and habitat use of bobcats. Journal of Wildlife Management 63:261–69.

Conner, L. M., B. D. Leopold, and M. J. Chamberlain. 2001. Multivariate habitat models for bobcats in southern forested landscapes. Pages 51–55 *in* A. Woolf, C. K. Neilsen and R. D. Bluett, eds. Proceedings of a symposium on current bobcat research and implications for management. Wildlife Society, Nashville, TN.

Conner, M. C., R. F. Labisky, and D. R. Progulske, Jr. 1983. Scent-station indices as measures of population abundance for bobcats, raccoons, gray foxes, and opossums. Wildlife Society Bulletin 11:146–52.

Cook, R. S., M. White, D. O. Trainer, and W. C. Glazener. 1971. Mortality of young white-tailed deer fawns in south Texas. Journal of Wildlife Management 35:47–56.

Crowe, D. M. 1975a. Aspects of aging, growth, and reproduction of bobcats from Wyoming. Journal of Mammalogy 56:177–98.

Crowe, D. M. 1975b. A model for exploited bobcat populations in Wyoming. Journal of Wildlife Management 39:408–15.

Crowe, D. M., and D. Strickland. 1975. Population structure of some mammalian predators in southeastern Wyoming. Journal of Wildlife Management 39:449–50.

Dearborn, N. 1932. Food of some predatory furbearing animals of Michigan (Conservation Bulletin No. 1). University of Michigan School of Forestry.

Delibes, M., M. C. Blazquez, R. Rodriguez-Estrella, and S. C. Zapata. 1997. Seasonal food habits of bobcats (*Lynx rufus*) in subtropical Baja California Sur, Mexico. Canadian Journal of Zoology 75:478–83.

Dibello, F. J., S. M. Arthur, and W. B. Krohn. 1990. Food habits of sympatric coyotes, *Canis latrans,* red foxes, *Vulpes vulpes,* and bobcats, *Lynx rufus,* in Maine. Canadian Field-Naturalist 104:403–8.

Diefenbach, D. R., L. A. Baker, W. E. James, R. J. Warren, and M. J. Conroy. 1993. Reintroducing bobcats to Cumberland Island, Georgia. Restoration Ecology 12:241–47.

Diefenbach, D. R., M. J. Conroy, R. J. Warren, W. E. James, L. A. Baker, and T. Hon. 1994. A test of the scent-station survey technique for bobcats. Journal of Wildlife Management 58:10–17.

Dill, H. H. 1947. Bobcat preying on deer. Journal of Mammalogy 28:63.

Domingo-Roura, X., H. A. Jacobson, and R. F. Weaver. 1997. Sex linkage of minisatellite bands in bobcats (*Felis rufus*). Journal of Heredity 88:527–31.

Duke, K. L. 1954. Reproduction in the bobcat *Lynx rufus*. Anatomical Record 120:816–17.

Dyer, M. I. 1979. Conference summary: Current status of North American bobcat programs. Pages 134–37 *in* P. C. Escherich and L. Blum, eds. Proceedings of the 1979 bobcat research conference (Science and Technology Series 6). National Wildlife Federation, Washington, DC.

Eaton, R. L. 1976. Why some felids copulate so much. World's Cats 3:73–94.

Elsy, C. A. 1954. A case of cannibalism in Canada lynx (*Lynx canadensis*). Journal of Mammalogy 35:129.

Elton, C., and M. Nicholson. 1942. The ten-year cycle in numbers of lynx in Canada. Journal of Animal Ecology 11:215–44.

Epperson, C. J. 1978. The biology of the bobcat (*Lynx rufus*) in Nebraska. M.S. Thesis, University of Nebraska, Lincoln.

Epstein, M. B., G. A. Feldhamer, and R. L. Joyner. 1983. Predation by white-tailed deer fawns by bobcats, foxes, and alligators: Predator assessment. Proceedings of the Annual Conference of the Southeastern Association of Fish and Wildlife Agencies 37:161–72.

Erickson, A. W. 1955. An ecological study of the bobcat in Michigan. M.S. Thesis, Michigan State University, East Lansing.

Erickson, D. W. 1981. Furbearing harvest mechanics: An examination of variables influencing fur harvests in Missouri. Pages 1469–91 *in* J. A. Chapman and D. Pursley, eds. Proceedings of the worldwide furbearer conference. Frostburg, MD.

Ewer, R. E. 1973. The carnivores. Cornell University Press, Ithaca, NY.

Faulkner, C. E. 1971. The legal status of the wildcats in the United States. Pages 124–25 *in* S. E. Jorgensen and L. D. Mech, eds. Symposium on the native cats of North America (36th North American Wildlife Resource Conference). Bureau of Sport Fisheries and Wildlife, Minneapolis, MN.

Fedriani, J. M., T. K. Fuller, R. M. Sauvajot, and E. C. York. 2000. Competition and intraguild predation among three sympatric carnivores. Oecologia 125:258–70.

Foran, D. R., S. C. Minta, and K. S. Heinemeyer. 1997. DNA-based analysis of hair to identify species and individuals for population research and monitoring. Wildlife Society Bulletin 25:840–47.

Fredrickson, L. F., and L. A. Rice. 1979. Bobcat management survey study in South Dakota, 1977–79. Pages 32–36 *in* P. C. Escherich and L. Blum, eds. Proceedings of the 1979 bobcat research conference (Science and Technology Series 6). National Wildlife Federation, Washington, DC.

Friedrich, P. D., G. E. Burgoyne, T. M. Cooley, and S. M. Schmitt. 1984. Use of lower canine teeth for determining the sex of bobcats in Michigan. Page 15 *in* J. A. Litvaitis and J. A. Bissonette, eds. Eastern bobcat workshop. Maine Cooperative Wildlife Research Unit, Orono.

Fritts, S. H. 1973. Age, food habits, and reproduction in the bobcat (*Lynx rufus*) *in* Arkansas. M.S. Thesis, University of Arkansas, Fayetteville.

Fritts, S. H., and J. A. Sealander. 1978a. Diets of bobcats in Arkansas with special reference to age and sex differences. Journal of Wildlife Management 42:533–39.

Fritts, S. H., and J. A. Sealander. 1978b. Reproductive biology and population characteristics of bobcats in Arkansas. Journal of Mammalogy 59:347–53.

Fuller, T. K., W. E. Berg, and D. W. Kuehn. 1985a. Survival rates and mortality factors of adult bobcats in north-central Minnesota. Journal of Wildlife Management 49:292–96.

Fuller, T. K., W. E. Berg, and D. W. Kuehn. 1985b. Bobcat home range size and daytime cover-type use in north-central Minnesota. Journal of Mammalogy 66:568–71.

Fuller, T. K., K. D. Kerr, and P. D. Karns. 1985c. Hematology and serum chemistry of bobcats in north central Minnesota. Journal of Wildlife Diseases 21:29–32.

Fuller: T. K., S. L. Berendzen, T. A. Decker, and J. E. Cardoza. 1995. Survival and cause-specific mortality rates of adult bobcats (*Lynx rufus*). American Midland Naturalist 134:404–8.

Funderburk, S. 1986. International trade in U.S. and Canadian bobcats, 1977–81. Pages 489–501 *in* S. D. Miller and D. D. Everett, eds. Cats of the world: Biology, conservation, and management. National Wildlife Federation, Washington, DC.

Gashwiler, J. S., W. L. Robinette, and O. W. Morris. 1960. Foods of bobcats in Utah and western Nevada. Journal of Wildlife Management 24:226–29.

Gashwiler, J. S., W. L. Robinette, and O. W. Morris. 1961. Breeding habits of bobcats in Utah. Journal of Mammalogy 42:76–84.

Gilbert, J. H., and L. B. Keith. 2001. Impacts of reintroduced fishers on Wisconsin's bobcat populations. Pages 18–31 *in* A. Woolf, C. K. Neilsen and R.D. Bluett, eds. Proceedings of a symposium on current bobcat research and implications for management. Wildlife Society, Nashville, TN.

Gilbert, J. R. 1979. Techniques and problems of population modeling and analysis of age distribution. Pages 130–33 *in* P. C. Escherich and L. Blum, eds. Proceedings of the 1979 bobcat research conference (Science and Technology Series 6). National Wildlife Federation, Washington, DC.

Glenn, B. L., R. E. Rolley, and A. A. Kocan. 1982. Cytauxzoon-like piroplasms in erythrocytes of wild-trapped bobcats in Oklahoma. American Veterinary Medical Association Journal 181:1251–53.

Golden, H. 1982. Bobcat populations and environmental relationships in northwestern Nevada. M.S. Thesis, University of Nevada, Reno.

Griffith, M. A., D. E. Buie, T. T. Fendley, and D. A. Shipes. 1981. Preliminary observations of subadult bobcat movement behavior. Proceedings of the Annual Conference of the Southeastern Association of Fish and Wildlife Agencies 34:563–71.

Guenther, D. D. 1980. Home range, social organization and movement patterns of the bobcat, *Lynx rufus,* from spring to fall in south-central Florida. M.S. Thesis, University of Southern Florida, Tampa.

Gustafson, K. A. 1984. The winter metabolism and bioenergetics of the bobcat in New York. M.S. Thesis, State University of New York, Syracuse.

Hall, E. R. 1981. The mammals of North America. John Wiley, New York.

Hall, H. T., and J. D. Newsom. 1978. Summer home ranges and movement of bobcats in bottomland hardwoods of southern Louisiana. Proceedings of the Annual Conference of the Southeastern Association of Fish and Wildlife Agencies 30:427–36.

Hamilton, D. A. 1982. Ecology of the bobcat in Missouri. M.S. Thesis, University of Missouri, Columbia.

Hamilton, D. A., T. G. Kulowiec, and D. Erickson. 1990. Archer's index to upland furbearer populations and sign station indices—A comparison. Page 39 *in* T. G. Kulowiec, ed. Proceedings of the 7th Midwest and 3rd Southeast furbearer workshop. Missouri Department of Conservation, Jefferson City, MO.

Hamilton, W. J., and R. P. Hunter. 1939. Fall and winter food habits of Vermont bobcats. Journal of Wildlife Management 3:99–103.

Harrison, R. L. 1998. Bobcats in residential areas: Distribution and homeowner attitudes. Southwestern Naturalist 43:469–75.

Hatcher, R. T., and J. H. Shaw. 1981. A comparison of 3 indices to furbearer populations. Wildlife Society Bulletin 9:153–56.

Heisey, D. M., and T. K. Fuller. 1985. Evaluation of survival and cause-specific mortality rates using telemetry data. Journal of Wildlife Management 49:668–74.

Henke, S. E., and F. C. Bryant. 1999. Effects of coyote removal on the faunal community in western Texas. Journal of Wildlife Management 63:1066–81.

Hodges, K. E. 2000. The ecology of snowshoe hares oin northern boreal forests. Pages 117–61 *in* L. F. Ruggiero, K. B. Aubry, S. W. Buskirk, G. M. Koehler, C. J. Krebs, K. S. McKelvey and J. R. Squires, eds. Ecology and conservation of lynx in the United States. University Press of Colorado, Boulder.

Holling, C. S. 1959. The components of predation as revealed by a study of small mammal predation of the European pine sawfly. Canadian Entomologist 91:293–320.

Hoppe, R. T. 1979. Population dynamics of the Michigan bobcat (*Lynx rufus*) with reference to age structure and reproduction. Pages 111–15 *in* P. C. Escherich and L. Blum, eds. Proceedings of the 1979 bobcat research conference (Science and Technology Series 6). National Wildlife Federation, Washington, DC.

Hornocker, M., and T. Bailey. 1986. Natural regulation in three species of felids. Pages 211–20 *in* S. D. Miller and D. D. Everett, eds. Cats of the world: Biology, conservation, and management. National Wildlife Federation, Washington, DC.

Hsu, T. C., and K. Benirschke. 1970. *Lynx rufus*. An atlas of mammalian chromosomes, Vol. 4, Folio 187. Springer-Verlag, New York.

Hsu, T. C., and K. Benirschke. 1974. *Felis lynx*. An atlas of mammalian chromosomes, Vol. 8, Folio 385. Springer-Verlag, New York.

Jackson, D. L., E. A. Gluesing, and H. A. Jacobson. 1988. Dental eruption in bobcats. Journal of Wildlife Management 52:515–17.

Jackson, H. H. T. 1961. Mammals of Wisconsin. University of Wisconsin Press, Madison.

Jennrich, R. I., and F. B. Turner. 1969. Measurement of noncircular home range (terrestrial vertebrates). Journal of Theoretical Biology 22:227–37.

Johnson, N. F., B. A. Brown, and J. C. Bosomworth. 1981. Age and sex characteristics of bobcat canines and their use in population assessment. Wildlife Society Bulletin 9:203–6.

Johnson, W. E., and S. J. O'Brien. 1997. Phylogenetic reconstruction of the Felidae using 16S rRNA and NADH-5 mitochondrial genes. Journal of Molecular Evolution 44:S98–116.

Jones, J. H., and N. S. Smith. 1979. Bobcat density and prey selection in central Arizona. Journal of Wildlife Management 43:666–72.

Jones, J. K., Jr., D. C. Carter, and H. H. Genoways. 1975. Revised checklist of North American mammals north of Mexico. Occasional Papers, the Museum, Texas Tech University 28:1–14.

Jones, M. L. 1982. Longevity of captive mammals. Zoologische Garten 52:113–28.

Kamler, J. F., and P. S. Gipson. 2000. Home range, habitat selection, and survival of bobcats, *Lynx rufus,* in a prairie ecosystem in Kansas. Canadian Field-Naturalist 114:388–94.

Kamler, J. F., P. S. Gipson, and T. R. Snyder. 2000. Dispersal characteristics of young bobcats from northeastern Kansas. Southwestern Naturalist 45:543–546.

Karpowitz, J. F. 1981. Home range and movements of Utah bobcats with reference to habitat selection and prey base. M.S. Thesis, Brigham Young University, Provo, UT.

Keith, L. B. 1983. Role of food in hare population cycles. Oikos 40:385–95.

Keith, L. B., A. W. Todd, C. J. Brand, R. S. Adamcik, and D. H. Rusch. 1977. An analysis of predation during a cyclic fluctuation of snowshoe hares. Proceedings of the International Congress of Game Biologists 13:151–75.

Kesterson, M. B. 1988. Lynx home range and spatial organization in relation to population density and prey abundance. M.S. Thesis, University of Alaska, Fairbanks.

Kight, J. 1962. An ecological study of the bobcat *Lynx rufus* (Schreber), in west-central South Carolina. M.S. Thesis, University of Georgia, Athens.

Kitchener, A. 1991. The natural history of the wild cats. Cornell University Press, Ithaca, NY.

Kitchings, J. T., and J. D. Story. 1979. Home range and diet of bobcats in eastern Tennessee. Pages 47–54 *in* P. C. Escherich and L. Blum, eds. Proceedings of the 1979 bobcat research conference (Science and Technology Series 6). National Wildlife Federation, Washington, DC.

Kitchings, J. T., and J. D. Story. 1984. Movement and dispersal of bobcats in east Tennessee. Journal of Wildlife Management 48:957–61.

Knick, S. T. 1990. Ecology of bobcats relative to exploitation and a prey decline in southeastern Idaho. Wildlife Monographs 108:1–42.

Knick, S. T., and T. N. Bailey. 1986. Long-distance movements by two bobcats from southeastern Idaho. American Midland Naturalist 116:222–23.

Knick, S. T., S. J. Sweeney, J. R. Alldredge, and J. D. Brittell. 1984. Autumn and winter foods habits on bobcats in Washington State. Great Basin Naturalist 44:70–74.

Knick, S. T., J. D. Brittell, and S. J. Sweeney. 1985. Population characteristics of bobcats in Washington State. Journal of Wildlife Management 49:721–28.

Knick, S. T., E. C. Hellgren, and U. S. Seal. 1993. Hematologic, biochemical, and endocrine characteristics of bobcats during a prey decline in southeastern Idaho. Canadian Journal of Zoology 71:1448–53.

Knowles, P. R. 1985. Home range size and habitat selection of bobcats, *Lynx rufus,* in north-central Montana. Canadian Field-Naturalist 99:6–12.

Koehler, G. M. 1990. Population and habitat characteristics of lynx and snowshoe hares in north central Washington. Canadian Journal of Zoology 68:845–51.

Koehler, G. M., and K. B. Aubry. 1994. Lynx. Pages 74–98 *in* L. F. Ruggiero, K. B. Aubry, S. W. Buskirk, L. J. Lyon and W. J. Zielinski, eds. The scientific basis for conserving forest carnivores: American marten, fisher, lynx, and wolverine in the Western United States (General Technical Report RM-254). U.S. Forest Service.

Koehler, G. M., and J. D. Brittell. 1990. Managing spruce–fir habitat for lynx and snowshoe hares. Journal of Forestry 88 (10):10–14.

Koehler, G. M., and M. G. Hornocker. 1989. Influences of seasons on bobcats in Idaho. Journal of Wildlife Management 53:197–202.

Koehler, G. M., and M. G. Hornocker. 1991. Seasonal resource use among mountain lions, bobcats, and coyotes. Journal of Mammalogy 72:391–96.

Krebs, C. J. 2001. Ecology: The experimental analysis of distribution and abundance, 5th ed. Benjamin Cummings, San Francisco.

Krebs, C. J., S. Boutin, R. Boonstra, A. R. E. Sinclair, J. N. M. Smith, M. R. T. Dale, K. Martin, and R. Turkington. 1995. Impact of food and predation on the snowshoe hare cycle. Science 269:1112–15.

Krebs, J. W., M. Smith, C. E. Rupprecht, and J. E. Childs. 1999. Rabies among non-reservoir, carnivorous mammals in the United States, 1960–1997. American Journal of Tropical Medicine and Hygiene 61:172–73.

Labelle, P., M. Igor, D. Martineau, S. Beaudin, N. Blanchette, R. Lafond, and S. St-Onge. 2000. Seroprevalence of leptospirosis in lynx and bobcats from Quebec. Canadian Veterinary Journal 41:319.

Lack, D. 1954. Cyclic mortality. Journal of Wildlife Management 18:25–37.

Lariviere, S., and L. R. Walton. 1997. *Lynx rufus*. Mammalian Species 563:1–8.

Lawhead, D. N. 1984. Bobcat *Lynx rufus* home range, density and habitat preferences in south-central Arizona. Southwestern Naturalist 29:105–13.

Lembeck, M. 1986. Long term behavior and population dynamics of an unharvested bobcat population in San Diego County. Pages 305–10 *in* S. D. Miller and D. D. Everett, eds. Cats of the world: Biology, conservation, and management. National Wildlife Federation, Washington, DC.

Lembeck, M., and G. I. Gould, Jr. 1979. Dynamics of harvested and unharvested bobcat populations in California. Pages 53–54 *in* P. C. Escherich and L. Blum, eds. Proceedings of the 1979 bobcat research conference (Science and Technology Series 6). National Wildlife Federation, Washington, DC.

Linscombe, G. 1994. U.S. fur harvest (1970–1992) and fur value (1974–1992) statistics by state and region. Louisiana Department of Wildlife and Fisheries, Baton Rouge.

Little, J. W., J. P. Smith, F. F. Knowlton, and R. R. Bell. 1971. Incidence and geographic distribution of some nematodes in Texas bobcats. Texas Journal of Science 22:403–7.

Litvaitis, J. A., and D. J. Harrison. 1989. Bobcat–coyote niche relationships during a period of coyote population increase. Canadian Journal of Zoology 67:1180–88.

Litvaitis, J. A., J. A. Sherburne, M. O'Donoghue, and D. May. 1982. Cannibalism by a free-ranging bobcat, *Felis rufus*. Canadian Field-Naturalist 96:476–77.

Litvaitis, J. A., C. L. Stevens, and W. W. Mautz. 1984. Age, sex, and weight of bobcats in relation to winter diet. Journal of Wildlife Management 48:632–35.

Litvaitis, J. A., J. A. Sherburne, and J. A. Bissonette. 1985. Influence of understory characteristics on snowshoe hare habitat use and density. Journal of Wildlife Management 49:866–73.

Litvaitis, J. A., J. A. Sherburne, and J. A. Bissonette. 1986. Bobcat habitat use and home range size in relation to prey density. Journal of Wildlife Management 50:110–17.

Litvaitis, J. A., J. T. Major, and J. A. Sherburne. 1987. Influence of season and human-induced mortality on spatial organization of bobcats (*Felis rufus*) in Maine. Journal of Mammalogy 68:100–106.

Lovallo, M. J. 1999. Multivariate models of bobcat habitat selection for Pennsylvania landscapes. Ph.D. Dissertation, Pennsylvania State University, University Park.

Lovallo, M. J., and E. M. Anderson. 1995. Range shift by a female bobcat (*Lynx rufus*) after removal of neighboring female. American Midwest Naturalist 134:409–12.

Lovallo, M. J., and E. M. Anderson. 1996a. Bobcat (*Lynx rufus*) home range size and habitat use in northwest Wisconsin. American Midland Naturalist 135:241–52.

Lovallo, M. J., and E. M. Anderson. 1996b. Bobcat movements and home ranges relative to roads in Wisconsin. Wildlife Society Bulletin 24:71–76.

Lovallo, M. J., J. H. Gilbert, and T. M. Gehring. 1993. Bobcat, *Felis rufus*, dens in an abandoned beaver lodge. Canadian Field-Naturalist 107:108–9.

Lovallo, M. J., G. L. Storm, D.S. Klute, and W. M. Tzilkowski. 2001. Multivariate models of bobcat habitat selection for Pennsylvania landscapes. Pages 4–17 *in* A. Woolf, C. K. Neilsen and R. D. Bluett, eds. Proceedings of a symposium on current bobcat research and implications for management. Wildlife Society, Nashville, TN.

Maehr, D. S., and J. R. Brady. 1986. Food habits of bobcats in Florida. Journal of Mammalogy 67:133–38.

Mahan, C. J. 1979. Age determination of bobcats by means of canine pulp cavity ratios. Pages 126–29 *in* P. C. Escherich and L. Blum, eds. Proceedings of the 1979 bobcat research conference (Science and Technology Series 6). National Wildlife Federation, Washington, DC.

Major, J. T. 1983. Ecology and interspecific relationships of coyotes, bobcats, and red foxes in western Maine. Ph.D. Dissertation, University of Maine, Orono.

Major, J. T., and J. A. Sherburne. 1987. Interspecific relationships of coyotes, bobcats, and red foxes in western Maine. Journal of Wildlife Management 51:606–16.

Manville, R. H. 1959. Bregmatic bones in North American lynx. Science 130:1254–55.

Marshall, A. D., and J. H. Jenkins. 1966. Movements and home ranges of bobcats as determined by radio-tracking in the upper coastal plain of west-central South Carolina. Proceedings of the Annual Conference of the Southeast Game and Fish Commission 20:206–14.

Marston, M. A. 1942. Winter relations of bobcat to white-tailed deer in Maine. Journal of Wildlife Management 6:328–37.

Matson, J. R. 1948. Cat kills deer. Journal of Mammalogy 29:69–70.

Mautz, W. W., and P. J. Pekins. 1989. Metabolic rate of bobcats as influenced by seasonal temperatures. Journal of Wildlife Management 53:202–5.

May, D. W. 1981. Habitat utilization by bobcats in eastern Maine. M.S. Thesis, University of Maine, Orono.

McConnel, P. A. 1981. Bobcat restoration project: March 1, 1978–November 31, 1980 (Final Report W-59-R-3). New Jersey Division of Fish and Wildlife, Trenton.

McCord, C. M. 1974. Selection of winter habitat by bobcats (*Lynx rufus*) on the Quabbin Reservation, Massachusetts. Journal of Mammalogy 55:428–37.

McCord, C. M., and J. E. Cardoza. 1982. Bobcat and lynx (*Felis rufus* and *F. lynx*). Pages 728–66 *in* J. A. Chapman and G. A. Feldhamer, eds. Wild mammals of North America: Biology, management, and economics. Johns Hopkins University Press, Baltimore.

McDaniel, G. W., K. S. McKelvey, J. R. Squires, and L. F. Ruggerio. 2000. Efficacy of lures and hair snares to detect lynx. Wildlife Society Bulletin 28:119–23.

McKelvey, K. S., K. B. Aubry, and Y. K. Ortega. 2000a. History and distribution of lynx in the contiguous United States. Pages 207–64 *in* L. F. Ruggiero, K. B. Aubry, S. W. Buskirk, G. M. Koehler, C. J. Krebs, K. S. McKelvey and J. R. Squires, eds. Ecology and conservation of lynx in the United States. University Press of Colorado, Boulder.

McKelvey, K. S., Y. K. Ortega, G. M. Koehler, K. B. Aubry, and J. D. Brittell. 2000b. Canada lynx habitat and topographic use patterns in north central Washington: A reanalysis. Pages 307–36 *in* L. F. Ruggiero, K. B. Aubry, S. W. Buskirk, G. M. Koehler, C. J. Krebs, K. S. McKelvey and J. R. Squires, eds. Ecology and conservation of lynx in the United States. University Press of Colorado, Boulder.

McKelvey, K. S., K. B. Aubry, J. K. Agee, S. W. Buskirk, L. F. Ruggiero, and G. M. Koehler. 2000c. Lynx conservation in an ecosystem management context. Pages 419–41 *in* L. F. Ruggiero, K. B. Aubry, S. W. Buskirk, G. M. Koehler, C. J. Krebs, K. S. McKelvey and J. R. Squires, eds. Ecology and conservation of lynx in the United States. University Press of Colorado, Boulder.

McKinney, T. D., and M. R. Dunbar. 1976. Weight of adrenal glands in the bobcat (*Lynx rufus*). Journal of Mammalogy 57:378–80.

Mech, L. D. 1980. Age, sex, reproduction, and spatial organization of lynxes colonizing northeastern Minnesota. Journal of Mammalogy 61:261–67.

Mehrer, C. F. 1975. Some aspects of reproduction in captive mountain lions (*Felis concolor*), bobcats (*Lynx rufus*), and lynx (*Lynx canadensis*). Ph.D. Dissertation, University of North Dakota, Grand Forks.

Mellen, J. D. 1993. A comparative analysis of scent-marking, social and reproductive behavior in 20 species of small cats (*Felis*). American Zoologist 33:151–66.

Miller, D. L., B. D. Leopold, M. J.Gray, and B. J. Woody. 1999. Blood parameters of clinically normal captive bobcats (*Felis rufus*). Journal of Zoo and Wildlife Medicine 30:242–47.

Miller, S. D., and D. W. Speake. 1979. Demography and home range of bobcat in south Alabama. Pages 123–24 *in* P. C. Escherich and L. Blum, eds. Proceedings of the 1979 bobcat research conference (Science and Technology Series 6). National Wildlife Federation, Washington, DC.

Mitchell, R. L., and S. L. Beasom. 1974. Hookworms in south Texas coyotes and bobcats. Journal of Wildlife Management 38:455–58.

Mohr, C. O. 1947. Table of equivalent populations of North American small mammals. American Midland Naturalist 37:223–49.

Mowat, G., and B. G. Slough. 1998. Some observations on the natural history and behavior of the Canada lynx, *Lynx canadensis*. Canadian Field-Naturalist 112:32–36.

Mowat, G., K. G. Poole, and M. O'Donoghue. 2000. Ecology of lynx in northern Canada and Alaska. Pages 265–306 *in* L. F. Ruggiero, K. B. Aubry, S. W. Buskirk, G. M. Koehler, C. J. Krebs, K. S. McKelvey and J. R. Squires, eds. Ecology and conservation of lynx in the United States. University Press of Colorado, Boulder.

Murie, O. J. 1975. A field guide to animal tracks, 2nd. ed. Houghton Mifflin, Boston.

Murray, D. L., and S. Boutin. 1991. The influence of snow on lynx and coyote movements: Does morphology affect behavior? Oecologia 88:463–69.

Nava, J. A., Jr. 1970. The reproductive biology of Alaska lynx (*Lynx canadensis*). M.S. Thesis, University of Alaska, Fairbanks.

Neale, J. C. C., B. N. Sacks, M. M. Jaeger, and D. R. McCullough. 1998. A comparison of bobcat and coyote predation on lambs in north-coastal California. Journal of Wildlife Management 62:700–706.

Neilsen, C. K., and A. Woolf. 2001. Bobcat habitat use relative to human dwellings in southern Illinois. Pages 40–44 *in* A. Woolf, C. K. Neilsen and R. D. Bluett, eds. Proceedings of a symposium on current bobcat research and implications for management. Wildlife Society, Nashville, TN.

Nellis, C. H., and L. B. Keith. 1968. Hunting activities and success of lynxes in Alberta. Journal of Wildlife Management 32:718–22.

Nellis, C. H., S. P. Wetmore, and L. B. Keith. 1972. Lynx–prey interactions in central Alberta. Journal of Wildlife Management 36:320–29.

Nelms, M. G., L. A. Hansen, R. J. Warren, J. J. Brooks, and D. R. Diefenbach. 2001. Deer herd trends, bobcat food habits, and vegetation change over 18 years on Cumberland Island, Georgia, before and after bobcat restoration. Page 80 *in* A. Woolf, C. K. Neilsen and R. D. Bluett, eds. Proceedings of a symposium on current bobcat research and implications for management. Wildlife Society, Nashville, TN.

North American Fur Auctions. 2001. Wild fur sales results: May 13 and 14, 2001. Available at http://www.nafa.ca/sales/results_maywild2001.asp. Accessed 25 August 2001.

Nowak, R. M. 1999. Walker's mammals of the world, 6th ed. Johns Hopkins University Press, Baltimore.

Nunley, G. L. 1978. Present and historical bobcat population trends in New Mexico and the West. Pages 77–84 *in* Proceedings of the 8th vertebrate pest conference.

Nussbaum, R. A., and C. Maser. 1975. Food habits of the bobcat (*Lynx rufus*) in the Coast and Cascade Ranges of western Oregon in relation to present management policies. Northwest Science 49:261–66.

Obbard, M. E. 1987. Fur grading and pelt identification. Pages 717–826 *in* M. Novak, J. A. Baker, M. E. Obbard and B. Malloch, eds. Wild furbearer management and conservation in North America. Ontario Trappers Association, North Bay, Canada.

O'Conner, R. M. 1984. Population trends, age structures, and reproductive characteristics of female lynx in Alaska, 1961 through 1973. M.S. Thesis, University of Alaska, Fairbanks.

O'Donoghue, M. 1997. Responses of the coyote and lynx to the snowshoe hare cycle. Ph.D. Dissertation, University of British Columbia, Vancouver, Canada.

O'Donoghue, M., S. Boutin, C. J. Krebs, D. L. Murray, and E. J. Hofer. 1998a. Behavioral responses of coyotes and lynx to the snowshoe hare cycle. Oikos 82:169–83.

O'Donoghue, M., S. Boutin, C. J. Krebs, G. Zuleta, D. L. Murray, and E. J. Hofer. 1998b. Functional responses of coyotes and lynx to the snowshoe hare cycle. Ecology 79:1193–1208.

Ommundsen, P. D. 1991. Morphological differences between lynx and bobcat skulls. Northwest Science 65:248–50.

Parker, G. R., and G. E. J. Smith. 1983. Sex- and age-specific reproductive and physical parameters of the bobcat (*Lynx rufus*) on Cape Breton Island, Nova Scotia. Canadian Journal of Zoology 61:1771–82.

Parker, G. R., J. W. Maxwell, L. D. Morton, and G. E. J. Smith. 1983. The ecology of the lynx (*Lynx canadensis*) on Cape Breton Island. Canadian Journal of Zoology 61:770–86.

Pelton, M. R. 1979. Potential use of radio-isotopes for determining densities of bobcats. Pages 97–100 *in* P. C. Escherich and L. Blum, eds. Proceedings of the 1979 bobcat research conference (Science and Technology Series 6). National Wildlife Federation, Washington, DC.

Pelton, M. R., and L. C. Marcum. 1977. The potential use of radioisotopes for determining densities of black bears and other carnivores. Pages 221–37 *in* R. L. Phillips and C. Jonkel, eds. Proceedings of the 1975 predator symposium. Montana Forest Conservation Experimental Station, University of Montana, Missoula.

Pence, D. B., F. D. Matthews, and L. A. Windberg. 1982. Notoedric mange in the bobcat, *Felis rufus,* from south Texas. Journal of Wildlife Diseases 18:47–50.

Peterson, R. L., and S. C. Downing. 1952. Notes on the bobcat (*Lynx rufus*) of eastern North America with the description of a new race. Contributions of the Royal Ontario Museum 33:1–23.

Petraborg, W. H., and V. E. Gunvalson. 1962. Observations on bobcat mortality and bobcat predation on deer. Journal of Mammalogy 43:430–31.

Poland, J. D., A. M. Barnes, and J. J. Herman. 1973. Human bubonic plague from exposure to a naturally infected wild carnivore. American Journal of Epidemiology 97:332–37.

Pollack, E. M. 1950. Breeding habits of the bobcat in northeastern United States. Journal of Mammalogy 31:327–30.

Pollack, E. M. 1951a. Food habits of bobcats in New England states. Journal of Wildlife Management 15:209–13.

Pollack, E. M. 1951b. Observations on New England bobcats. Journal of Mammalogy 32:356–58.

Poole, K. G. 1994. Characteristics of an unharvested lynx population during a snowshoe hare decline. Journal of Wildlife Management 58:608–18.

Poole, K. G. 1995. Spatial organization of a lynx population. Canadian Journal of Zoology 73:632–41.

Poole, K. G. 1997. Dispersal patterns of lynx in the Northwest Territories. Journal of Wildlife Management 61:497–505.

Povey, R. C., and E. W. Davis. 1977. Panleukopenia and respiratory virus infection in wild felids. Pages 120–28 *in* R. L. Eaton, ed. The world's cats, Vol. 3. Carnivore Research Institute, University of Washington, Seattle.

Powers, J. G., W. M. Mautz, and P. J. Pekins. 1989. Nutrient and energy assimilation of prey by bobcats. Journal of Wildlife Management 53:1004–8.

Provost, E. E., C. A. Nelson, and A. D. Marshall. 1973. Population dynamics and behavior in the bobcat. Pages 42–67 *in* R. L. Eaton, ed. The World's cats, Vol. 1. World Wildlife Safari, Winston, OR.

Pulliainen, E. 1965. Studies of the wolf in Finland. Acta Zoologica Fennica 2:215–59.

Pulliainen, E., E. Lindgren, and P. S. Tunkkari. 1995. Influence of food availability and reproductive status on the diet and body condition of the European lynx in Finland. Acta Theriologica 40:181–96.

Quinn, N. W. S., and J. F. Gardner. 1984. Relationship of age and sex to lynx pelt characteristics. Journal of Wildlife Management 48:953–56.

Quinn, N. W. S., and G. Parker. 1987. Lynx. Pages 683–94 *in* M. Novak, J. A. Baker, M. E. Obbard and B. Malloch, eds. Wild furbearer management and conservation in North America. Ontario Trappers Association, North Bay, Canada.

Quinn, N. W. S., and J. E. Thompson. 1987. Dynamics of an exploited Canada lynx population in Ontario. Journal of Wildlife Management 51:297–305.

Read, J. A. 1981. Geographic variation in the bobcat (*Felis rufus*) in the south-central United States. M.S. Thesis, Texas A & M University, College Station.

Regan, T. W., and D. S. Maehr. 1990. Melanistic bobcats in Florida. Florida Field Naturalist 18:84–87.

Rich, M. S. 1983. The longevity record for *Lynx canadensis* Kerr, 1792. Zoologische Garten 53:365.

Riemann, H. P., J. A. Howarth, R. Ruppanner, C. E. Franti, and D. E. Behymer. 1975. Toxoplasma antibodies among bobcats and other carnivores of northern California. Journal of Wildlife Diseases 11:272–76.

Riley, S. D. 2001. Spatial and resource overlap of bobcats and gray foxes in urban and rural zones of a national park. Pages 32–39 *in* A. Woolf, C. K. Neilsen and R.D. Bluett, eds. Proceedings of a symposium on current bobcat research and implications for management. Wildlife Society, Nashville, TN.

Robbins, C. T. 1993. Wildlife feeding and nutrition, 2nd ed. Academic Press, San Diego, CA.

Robinson, W. B. 1961. Population changes of carnivores in some coyote control areas. Journal of Mammalogy 42:510–15.

Robinson, W. B., and E. F. Grand. 1958. Comparative movement of bobcats and coyotes as disclosed by tagging. Journal of Wildlife Management 22:117–22.

Roelke, M. E. 1990. Florida panther biomedical investigation (July 1, 1986– June 30, 1990). (Final Performance Report, Study No. 7506). Florida Game and Fresh Water Fish Commission, Tallahassee.

Rolley, R. E. 1983. Behavior and population dynamics of bobcats in Oklahoma. Ph.D. Dissertation, Oklahoma State University, Stillwater.

Rolley, R. E. 1985. Dynamics of a harvested bobcat population in Oklahoma. Journal of Wildlife Management 49:283–92.

Rolley, R. E. 1987. Bobcat. Pages 671–81 *in* M. Nowak, J. A. Baker, M. E. Obbard and B. Malloch, eds. Wild furbearer management and conservation in North America. Ontario Ministry of Natural Resources, Toronto.

Rolley, R. E., and W. D. Warde. 1985. Bobcat habitat use in southeastern Oklahoma. Journal of Wildlife Management 49:913–20.

Rollings, C. T. 1945. Habits, foods and parasites of the bobcat in Minnesota. Journal of Wildlife Management 9:131–45.

Rosenzweig, M. L. 1966. Community structure in sympatric Carnivora. Journal of Mammalogy 47:602–12.

Roughton, R. D., and M. D. Sweeny. 1982. Refinements in scent-station methodology for assessing trends of carnivore populations. Journal of Wildlife Management 46:217–29.

Ruggiero, L. K., M. K. Schwartz, K. B. Aubry, C. J. Krebs, A. Stanley, and S. W. Buskirk. 2000a. Species conservation and natural variation among populations. Pages 101–16 *in* L. F. Ruggiero, K. B. Aubry, S. W. Buskirk, G. M. Koehler, C. J. Krebs, K. S. McKelvey and J. R. Squires, eds. Ecology and conservation of lynx in the United States. University Press of Colorado, Boulder.

Ruggiero, L. K., K. B. Aubry, S. W. Buskirk, G. M. Koehler, C. J. Krebs, K. S. McKelvey, and J. R. Squires, eds. 2000b. Ecology and conservation of lynx in the United States. University Press of Colorado, Boulder.

Rust, W. D. 1980. Scent-station transects as a means of indexing bobcat population fluctuations. M.S. Thesis, Tennessee Technological University, Cookeville.

Samson, F. B. 1979. Multivariate analysis of cranial characters among bobcats, with a preliminary discussion of the number of subspecies. Pages 80–86 *in* P. C. Escherich and L. Blum, eds. Proceedings of the 1979 bobcat research conference (Science and Technology Series 6). National Wildlife Federation, Washington, DC.

Sargeant, G. A., D. H. Johnson, and W. E. Berg. 1998. Interpreting carnivore scent-station surveys. Journal of Wildlife Management 62:1235–45.

Saunders, J. K. 1961. The biology of the Newfoundland lynx. Ph.D. Dissertation, Cornell University, Ithaca, NY.

Saunders, J. K. 1963a. Food habits of lynx in Newfoundland. Journal of Wildlife Management 27:384–90.

Saunders, J. K. 1963b. Movements and activities of lynx in Newfoundland. Journal of Wildlife Management 27:390–400.

Saunders, J. K. 1964. Physical characteristics of the Newfoundland lynx. Journal of Mammalogy 45:36–47.

Schantz, V. S. 1939. A white-footed bobcat. Journal of Mammalogy 20:106.

Schnell, J. H. 1968. The limiting effect of natural predation on experimental cotton rat populations. Journal of Wildlife Management 32:698–711.

Scott, P. P. 1968. The special features of nutrition in cats, with observation on wild Felidae nutrition in the London zoo. Symposia of the London Zoological Society 21:21–36.

Shenk, T. 2001. Post-release monitoring of lynx: July 1, 2000–June 30, 2001 (Job Progress Report W-153-R-13). Colorado Division of Wildlife, Fort Collins.

Sikes, R. S., and M. L. Kennedy. 1992. Morphologic variation of the bobcat (*Felis rufus*) in the eastern United States and its association with selected environmental variables. American Midland Naturalist 128:313–24.

Sikes, R. S., and M. L. Kennedy. 1993. Geographic variation in sexual dimorphism of the bobcat (*Felis rufus*) in the United States. Southwestern Naturalist 38:336–44.

Slough, B. G. 1996. Estimating lynx population age ratio with pelt length data. Wildlife Society Bulletin 24:495–99.

Slough, B. G. 1999. Characteristics of Canada lynx, *Lynx canadensis,* maternal dens and denning habitat. Canadian Field-Naturalist 113:605–8.

Slough, B. G., and G. Mowat. 1996. Population dynamics of lynx in a refuge and interactions between harvested and unharvested populations. Journal of Wildlife Management 60:946–61.

Small, R. L. 1971. Interspecific competition among three species of Carnivora on the Spider Ranch, Yavapai County, Arizona. M.S. Thesis, University of Arizona, Tuscon.

Smith, D. S. 1984. Habitat use, home range, and movements of bobcats in western Montana. M.S. Thesis, University of Montana, Missoula.

Smith, J. D., E. M. Addison, D. G. Joachim, L. M. Smith, and N. W. S. Quinn. 1986. Helminth parasites of Canada lynx (*Felis canadensis*) from northern Ontario. Canadian Journal of Zoology 64:358–64.

Squires, J. R., and T. Laurion. 2000. Lynx home range and movements in Montana and Wyoming: Preliminary results. Pages 337–49 *in* L. F. Ruggiero, K. B. Aubry, S. W. Buskirk, G. M. Koehler, C. J. Krebs, K. S. McKelvey and J. R. Squires, eds. Ecology and conservation of lynx in the United States. University Press of Colorado, Boulder.

Stains, H. J. 1979. Primeness in North American furbearers. Wildlife Society Bulletin 7:120–24.

Staples, W. R. 1995. Lynx and coyote diet and habitat relationships during a low hare population on the Kenai Peninsula, Alaska. M.S. Thesis, University of Alaska, Fairbanks.

Stephenson, R. O., D. V. Grangaard, and J. Burch. 1991. Lynx, *Felis lynx,* predation on red foxes, *Vulpes vulpes,* caribou, *Rangifer tarandus,* and Dall sheep, *Ovis dalli,* in Alaska. Canadian Field-Naturalist 105:255–62.

Stewart, R. R. 1973. Age distributions, reproductive biology and food habits of Canada lynx, *Lynx canadensis* Kerr, in Ontario. M.S. Thesis, University of Guelph, Guelph, Ontario, Canada.

Stone, J. E., and D. B. Pence. 1977. Ectoparasites of the bobcat from west Texas. Journal of Parasitology 63:463.

Stone, J. E., and D. B. Pence. 1978. Ecology of helminth parasitism in the bobcat from west Texas. Journal of Parasitology 64:295–302.

Story, J. D., W. I. Galbraith, and J. T. Kitchings. 1982. Food habits of bobcats in eastern Tennessee. Journal of the Tennessee Academy of Science 57:25–28.

Stys, E. D., and B. D. Leopold. 1993. Reproductive biology and kitten growth of captive bobcats in Mississippi. Proceedings of the Southeastern Association of Fish and Wildlife Agencies 47:80–89.

Sweeney, S. J. 1978. Diet, reproduction, and population structure of the bobcat in western Washington. M.S. Thesis, University of Washington, Seattle.

Thiel, R. P. 1987. The status of Canada lynx in Wisconsin, 1865–1980. Wisconsin Academy of Sciences, Arts and Letters 75:90–96.

Thompson, I. D., I. J. Davidson, S. O'Donnell, and F. Brazeau. 1989. Use of track transects to measure the relative occurrence of some boreal mammals in uncut forest and regeneration stands. Canadian Journal of Zoology 67:1816–23.

Toweill, D. E. 1982. Winter foods of eastern Oregon bobcats. Northwest Science 56:310–15.

Trainer, C. E. 1975. Direct causes of mortality in mule deer fawns during summer and winter periods on Steens Mountain, Oregon. Proceedings of the Western Association of State Game and Fish Commissions 55:163–70.

Tumlison, R. 1987. *Felis lynx*. Mammalian Species 269:1–8.

Tumlison, R., and V. R. McDaniel. 1981. Anomalies of bobcat skulls (*Felis rufus*) in Arkansas. Proceedings of the Arkansas Academy of Science 35:94–96.

Tumlison, R., and V. R. McDaniel. 1984a. Morphology, replacement, and functional conservation in dental replacement patterns of the bobcat (*Felis rufus*). Journal of Mammalogy 65:111–17.

Tumlison, R., and V. R. McDaniel. 1984b. A description of the baculum of the bobcat (*Felis rufus*), with comments on its development and taxonomic implications. Canadian Journal of Zoology 62:1172–76.

U.S. Fish and Wildlife Service. 2000. Determination of threatened status for the contiguous U.S. distinct population segment of the Canada lynx: Final rule. Federal Register 65 (58):16051–86.

Ulmer, F. A. Jr. 1941. Melanism in the Felidae, with special reference to the genus *Lynx*. Journal of Mammalogy 22:285–88.

van Zyll de Jong, C. G. 1963. The biology of the lynx, *Felis canadensis* (Kerr) in Alberta and the Mackenzie District, N.W.T. M.S. Thesis, University of Alberta, Calgary, Canada.

van Zyll de Jong, C. G. 1966a. Food habits of the lynx in Alberta and the Mackenzie District, N. W. T. Canadian Field-Naturalist 80:18–23.

van Zyll de Jong, C. G. 1966b. Parasites of the Canada lynx, *Felis* (*Lynx*) *canadensis* (Kerr). Canadian Journal of Zoology 44:499–509.

Virchow, D., and D. Hogeland. 1994. Bobcat. Pages 35–43 *in* S. E. Hygnstrom, R. M. Timm and G. E. Larson, eds. Prevention and control of wildlife damage. University of Nebraska Cooperative Extension, Lincoln.

Ward, R. M. P., and C. J. Krebs. 1985. Behavioral responses of lynx to declining snowshoe hare abundance. Canadian Journal of Zoology 63:2817–24.

Weaver, J. L., and M. R. Johnson. 1995. Hematologic and serum chemistry values of captive Canadian lynx. Journal of Wildlife Diseases 31:212–15.

Wehinger, K. A., M. E. Roelke, and E. C. Greiner. 1995. Ixodid ticks from panthers and bobcats in Florida. Journal of Wildlife Diseases 31:480–85.

Werdelin, L. 1981. Evolution of the lynxes (*Lynx* spp.). Annales Zoologici Fennici 18:37–71.

Whittle, R. K. 1979. Age in relation to the winter food habits and helminth parasites of the bobcat in Oklahoma. M.S. Thesis, Oklahoma State University, Stillwater.

Wigginton, J. D., and F. S. Dobson. 1999. Environmental influences on geographic variation in body size of western bobcats. Canadian Journal of Zoology 77:802–13.

Winegarner, C. E., and M. S. Winegarner. 1982. Reproductive history of a bobcat. Journal of Mammalogy 63:680–82.

Witmer, G. W., and D. S. DeCalesta. 1986. Resource use by unexploited sympatric bobcats and coyotes in Oregon. Canadian Journal of Zoology 64:2333–38.

Witter, J. F. 1981. Brucellosis. Pages 280–87 *in* J. W. Davis, L. H. Karstad and D. O. Trainer, eds. Infectious diseases of wild mammals, 2nd ed. Iowa State University Press, Ames.

Wolff, J. O. 1980. The role of habitat patchiness in the population dynamics of snowshoe hares. Ecological Monographs 50:111–30.

Woolf, A., and G. F. Hubert, Jr. 1998. Status and management of bobcats in the United States over three decades: 1970's–1990's. Wildlife Society Bulletin 26:287–94.

Woolf, A., and C. K. Nielsen. 2001. Bobcat research and management: Have we met the challenge? Pages 1–3 *in* A. Woolf, C. K. Neilsen and R. D. Bluett, eds. Proceedings of a symposium on current bobcat research and implications for management. Wildlife Society, Nashville, TN.

Woshner, V. M. 1988. Aspects of reproductive physiology and luteal function in the female bobcat (*Felis rufus*). M.S. Thesis, Mississippi State University, Mississippi State.

Wozencraft, W. C. 1993. Order Carnivora: Felidae. Pages 288–99 *in* D. E. Wilson and D. M. Reeder, eds. Mammal species of the world. Smithsonian Institution Press, Washington, DC.

Young, S. P. 1958. The bobcat of North America. Wildlife Management Institute, Washington, DC.

Zezulak, D. S. 1981. Northeastern California bobcat study (Federal Aid Wildlife Restoration Project W-54-R-R, Job IV-3). California Department of Fish and Game, Sacramento.

Zezulak, D. S., and R. G. Schwab. 1979. A comparison of density, home range, and habitat utilization of bobcat populations at Lava Beds and Joshua Tree National Monuments, California. Pages 74–79 *in* P. C. Escherich and L. Blum, eds. Proceedings of the 1979 bobcat research conference (Science and Technology Series 6). National Wildlife Federation, Washington, DC.

Eric M. Anderson, College of Natural Resources, University of Wisconsin–Stevens Point, Stevens Point, Wisconsin 54481. Email: eanderson@uwsp.edu.

Matthew J. Lovallo, Pennsylvania Game Commission, Bureau of Wildlife Management, 2001 Elmerton Avenue, Harrisburg, Pennsylvania 17110-9797. Email: mjlovallo@earthlink.net.

VI

Seals and Manatee

39

Seals

Phocidae, Otariidae, and Odobenidae

Keith Ronald
Barra L. Gots

NOMENCLATURE (as set out in Rice 1998)

ORDER. Carnivora
SUBORDER. Caniformia
TAXON. Pinnipedia
FAMILY. Phocidae (earless seals, true seals)
SUBFAMILY. Phocinae (northern phocids)
TRIBE. Phocini
GENERA. *Phoca* Linnaeus, 1758; *Pagophilus* Gray, 1844; *Pusa* Scopoli, 1771; *Histriophoca* Gill, 1873; *Halichoerus* Nilsson, 1820
TRIBE. Cystophorini
GENUS. *Cystophora* Nilsson, 1820
TRIBE. Erignathini
GENUS. *Erignathus* Gill, 1866
SUBFAMILY. Monachinae (southern phocids)
TRIBE. Monachini
GENERA. *Monachus* Fleming, 1822; *Mirounga* Gray, 1827
FAMILY. Otariidae (eared seals)
SUBFAMILY. Otariinae (sea lions)
GENERA. *Eumetopias* Gill, 1866; *Zalophus* Gill, 1866
SUBFAMILY. Arctocephalinae (fur seals)
GENERA. *Callorhinus* Gray, 1859; *Arctocephalus* E. Geoffroy Saint-Hilaire and F. Cuvier, 1826
FAMILY. Odobenidae (walrus)
GENUS. *Odobenus* Brisson, 1762

GENERAL CHARACTERISTICS

In the immediate vicinity of continental North America, there are 16 different species of pinnipeds (true seals, sea lions, fur seals, and the walrus) whose distributions are within the political jurisdiction of the United States and Canada. Because of the number of species, certain biological and physiological topics that are pertinent to these mammals are limited to discussion in general terms. Individual characteristics of the species and distribution ranges are discussed separately.

Because management of these animals is complex, the topic is divided into those systems through which management operates, the policies that exist in the two countries, and the current status of each species. These species, with few exceptions, have been the center of considerable public interest and at times even marked and bitter controversy. It is hoped that in the future the emotional aspects of management can be minimized and the basic need to obtain enough scientific knowledge is not hindered, so as to ensure an adequate conservation strategy for seals. The scope of this topic is so great and the quantity of available literature so vast that reference has been made to only some key publications and ideas. Some of the material in this chapter is based on original research; hence, it is unreferenced and should be considered as work carried out by the University of Guelph. Further bibliographic details are in Ronald et al. (1976a, 1983, 1991).

DESCRIPTION

Pinnipeds have not made the complete transition to a full aquatic environment. Most usually haul-out on a solid surface like land or ice to reproduce, and adaptive streamlining qualities ideal for marine life make them appear clumsy when out of water. In general, the pinnipeds are adapted to the aquatic and usually cold environment in which most of them exist. Their large size has evolved mainly in response to this cold environment (Scheffer 1958).

The entire body is spindle shaped and streamlined. The rounded head tapers smoothly into the trunk without a constriction at the neck. The head is sometimes flattened—an advantage for diving—and external ears are reduced or absent; large eyes are situated well forward and sometimes rather close together. Although the neck is thick and muscular, it is very flexible. There are no sharp protuberances on the trunk and the small tail is tucked neatly between the hind flippers. External projections have been reduced. In male pinnipeds, the penis lies in an internal sheath with no external projection. The nipples can also be retracted to achieve less resistance while swimming. A layer of fatty tissue or blubber under the skin smoothes the general contours of the body (Harrison and King 1965; Scheffer 1958).

Forelimbs and hind limbs are modified and used for propulsion. The arm and leg bones are relatively short and contained within the body, whereas the bones in the hands and feet are greatly elongated. The shortened humerus and femur are swimming modifications and allow the limbs to be almost completely withdrawn alongside the body. This configuration streamlines the body and brings the source of power through muscle insertion closer to the body. A web of skin joins the digits to provide a larger surface for propulsion (Harrison and King 1965).

The epidermis is thick and tough. The coat consists of bundles of hairs, each having a long, deep-rooted guard hair and a variable number of finer, shorter underhairs or fur fibers. These bundles are associated with sebaceous glands, which help make the fur somewhat water repellent. In fur seals, there can be as many as 50–100 underhairs per hair canal and such dense fur makes it impossible for water to reach the skin surface. The walrus and adult phocids have less dense fur, and elephant and monk seals lack underhairs altogether (Pabst et al. 1999).

Although seals are found in water with various saline concentrations, there has been little internal change to ensure ionic regulation. The basic mammalian system, therefore, apparently can accommodate moderate changes in the aquatic medium of seals.

Each of the three pinniped families, Phocidae, Otariidae, and Odobenidae, has its own distinguishing characteristics. North American phocids have no ear pinnae, their testes are internal, and their mammae have two teats, with the exception of the bearded seal (*Erignathus barbatus*), which has four. Phocids have a dense single layer of hair plus a thick layer of blubber under the skin to prevent heat loss (Maxwell 1967). The flippers are furred on top and bottom, and the large hind flippers face backward. These provide the means for aquatic propulsion by rhythmic lateral movements (Howell 1970; Fish et al. 1988). Locomotion is performed on land by a caterpillar-like flexion of the body. The hind flippers are elevated and the foreflippers

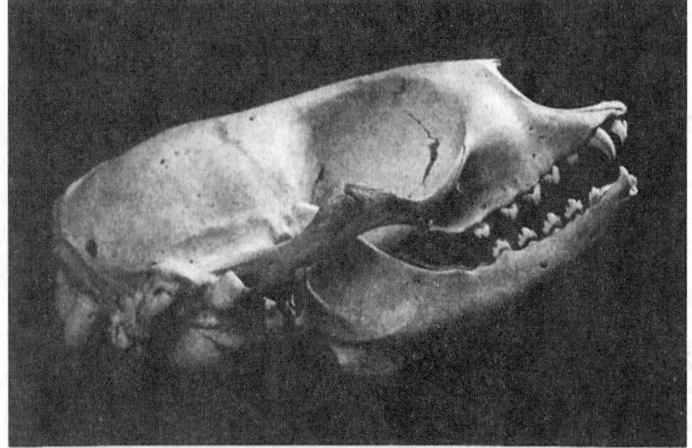

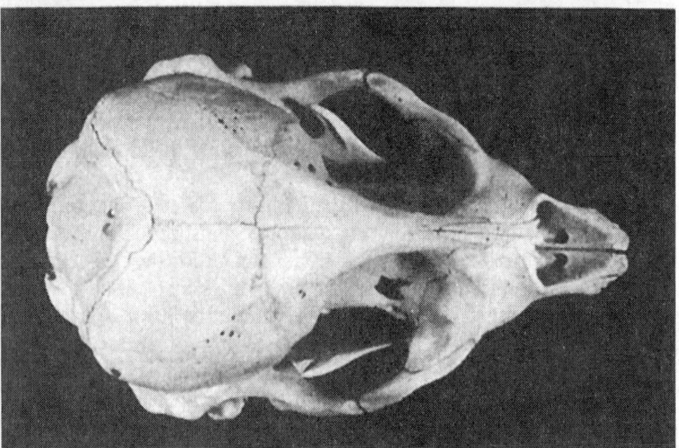

FIGURE 39.1. Skull of a young phocid. Top: lateral view of cranium and mandible. Bottom: dorsal view of cranium. Note that the sutures are open.

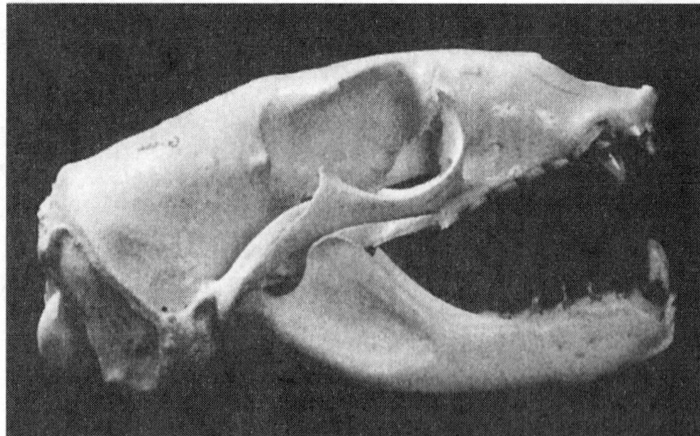

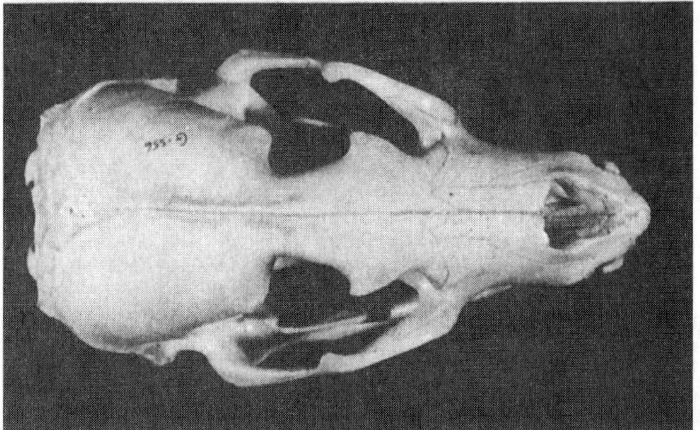

FIGURE 39.2. Skull of an adult otariid. Top: lateral view of cranium and mandible. Bottom: dorsal view of cranium.

are used for traction and negotiating rough surfaces or for swimming. The nails are the same size on all five digits (King 1964). Postorbital processes of the skull are absent (Fig. 39.1), as is the alisphenoid canal. The interorbital region of phocids is narrowly constricted relative to that of otariids and the walrus. The canines are elongated, and postcanine teeth usually have three or more cusps. The dental formula is I 2–3/1–2, C 1/1, PM 4–6/4–5.

The Otariidae, or eared seals, have prominent vestigial ear pinnae. This family comprises fur seals, which are distinguished by their dense, two-layered fur made up of long, coarse guard hairs and a fine, thick undercoat; and sea lions, which have a sparser and coarser coat. Like phocids, sea lions lack underfur (Maxwell 1967; Food and Agriculture Organization [FAO] of the United Nations, Advisory Committee on Marine Resources Research 1976). The testes of otariids are scrotal and their mammae have four teats. Foreflippers and hind flippers are used in limited terrestrial locomotion (Howell 1970). The hind limbs can be rotated forward on land, enabling the animal not only to stand on four legs, but also to manage a kind of gallop. Most of the power for swimming comes from the forelimbs; the hind limbs are rarely active (Maxwell 1967).

Otariids have incompletely furred flippers, with only a sparse growth of short hair on the dorsal surface. Their foreflippers have five rudimentary nails. There are three claws on each hind flipper on the three central digits. The tail is distinct and free from the body. The tip of the tongue is notched and the dentition is unspecialized (King 1964; Jefferson et al. 1993).

In contrast to that of phocids, the otariid skull has a postorbital process (Fig. 39.2) and an alisphenoid canal. The interorbital region is relatively wide. The first and second incisors are small; the third incisor is canine-like. Canines are large, conical, and recurved; the cheek teeth have one main cusp. The dental formula is I 3/2, C 1/1, P 4/4, M 1–3/1.

The Odobenidae, or walrus (*Odobenus rosmarus*), has no ear pinnae, its testes are internal, and its mammae have four teats. Walruses can also rotate their hind limbs forward, and their terrestrial locomotion is similar to that of otariids. When the walrus is swimming, its hind flippers move from side to side while its foreflippers are used alternately (Howell 1970). The foreflippers have five small distinct nails. The tail of the walrus is enclosed in a web of skin. The tip of its tongue is rounded. Its integument is nearly bare. For warmth, the walrus depends on its large quantities of blubber (Maxwell 1967). Nearly all vibrissae are on the end (as opposed to the sides) of the muzzle (Jefferson et al. 1993). The upper canines are extremely enlarged and form ever-growing tusks (Fig. 39.3). The remainder of the dentition is smaller in size, similar, and forms a continuous row. The dental formula of adults is variable, but is usually I 1/0, C 1/1, P 3/3, M 0/0.

PHYSIOLOGY

Growth. In an ideal environment, every animal has the potential to reach its maximum size. Growth is a feedback mechanism that reflects the variables in the environment. These include weather, food availability, herd density, and disease. Growth rates reflect changes in the environment, the herd, and the individual. In seals, mass and chest girths change seasonally because of reproductive stress, availability of food, molting, and other factors. Therefore, they are not a reliable measure of size. Also, these measurements are difficult to obtain in the larger species. A pinniped's size is usually expressed in terms of various length measurements (Laws 1959) (Table 39.1). Girth measurements are valuable as an expression of blubber thickness.

Compared to other mammals, most seals receive a relatively short period of maternal care because they are born well developed. Neonates of the different species are all different weights at birth in relation to their mother's mass, but some seals accelerate prenatal growth to

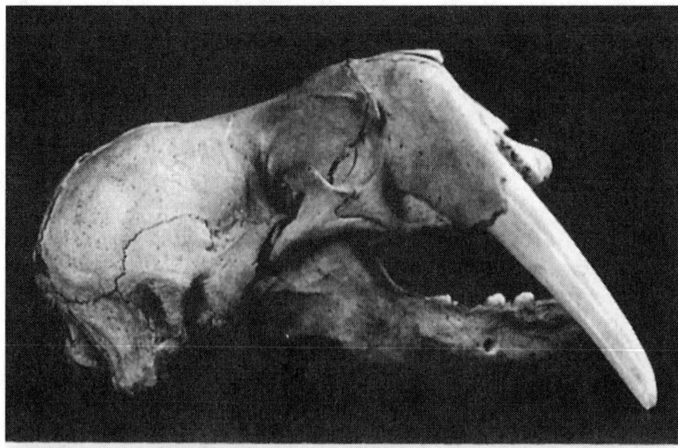

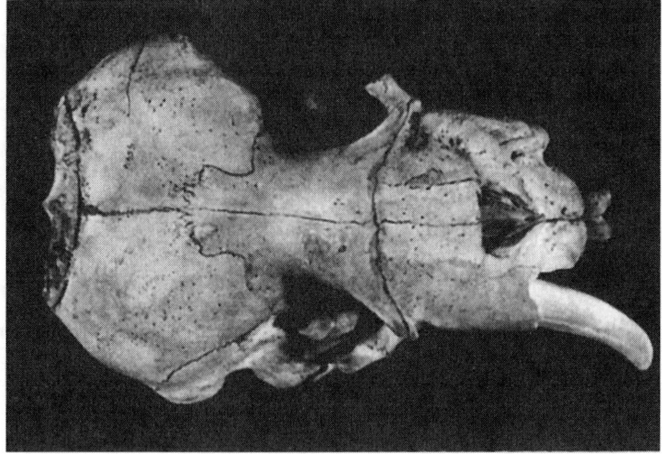

FIGURE 39.3. Skull of a young odobenid. Top: lateral view of cranium and mandible. Bottom: dorsal view of cranium.

accommodate for difference in size. They do not prolong gestation as land mammals do (Laws 1959). In fact, the period of active growth during gestation is basically the same for all pinnipeds except the walrus (Table 39.2).

Neonatal growth in the harp seal (*Pagophilus groenlandicus*) illustrates some of the growth patterns of newborn seals. There are two critical periods in the pagophilic ("ice loving") harp seal's early life: the first few hours after birth and the time period after weaning. Immediately after birth, the pup must cope with a harsh environment, which taxes its thermoregulatory capacities because it lacks a well-developed insulating blubber layer (Blix et al. 1975). Nonshivering thermogenesis (see Thermoregulation) helps it to survive initially. Within minutes or hours of birth, the mother's fat-rich milk provides the energy necessary for the enormous weight gain experienced during lactation. All seals experience this rapid postnatal weight gain, but the rate is species specific. Harp seal pups gain approximately 2.5 kg/day for 9 days. Newborn otariids, large at birth, nurse for several months. The majority of the weight gain in the harp seal is deposited as a subcutaneous layer of fat; the rest is gained in the core body mass. The sculp (skin and blubber weight) is 60% of the total body weight at the gray coat stage.

A weaned pup no longer gains energy from its mother's milk. This loss is partially compensated for by accumulation of external energy by the fur in the form of solar radiation. Delayed weaning corresponds to delay in dentition development. This is aptly demonstrated in the walrus, whose tusks, which are needed to assume adult dietary habits, are not developed until after the 1- or 2-year period of nursing (Laws 1959). Phocids, however, shed their milk teeth in utero and are born with adult dentition.

As the newly weaned pup learns to forage in the aquatic environment, further stress is placed on its energy requirements and it loses body weight. In the harp seal, it is lost from the body core, not from the subcutaneous fat reserves as is often suggested. These reserves are critical for the prevention of heat loss. Juvenile mortality at this time might be significant and should be taken into account in management strategies (Lavigne and Kovacs 1988).

Once this postweaning weight loss is overcome, the nutritionally independent seal grows steadily. Final body length is attained at approximately 5 years of age in the harp seal (Lavigne and Kovacs 1988). Overall, a pinniped generally reaches sexual maturity, but not necessarily reproductive activity, at about 86% of its final length, but there are variations among species (Table 39.1).

Some species demonstrate sexual dimorphism, the male being larger than the female. This is mainly characteristic of otariids; however, the northern elephant seal (*Mirounga angustirostris*) is also sexually dimorphic. Attainment of this larger size by the male occurs during a postpubertal growth spurt (Laws 1959; Bryden 1972), although some species show sexual size differentiation as early as 1 month of age (Payne 1979; Lavigne and Kovacs 1988). This later growth period accounts for the time lapse between sexual maturity and reproductive activity (Bryden 1972). Additional data on the growth of individual species are given in Table 39.1.

Changes in per capita food consumption can result in significant differences in growth rates. This can have important ecological consequences regarding population parameters and the impact of the population on its ecosystem. An increase in stock size may result in increased intraspecific and interspecific competition for food, which in turn causes a decrease in per capita consumption. A large population will consume less per capita. This decreased food consumption results in slower growth and delayed attainment of sexual maturity. Per capita food consumption may change as a result of either increased population size or interspecific competition. Growth rates, age of maturation, and age of first whelping significantly affect the energy requirements of the entire population over time and correspondingly affect the ecosystem (Lavigne and Kovacs 1988).

Molting. Periodic growth and replacement of the pelage of mammals is necessary if individuals are to survive and function efficiently. Pinnipeds renew their coats annually by molting (Ling 1970).

The pelage of seals performs three functions: waterproofing, streamlining, and insulating. Whereas the adult pelage is hydrodynamic in function (Ling and Button 1975), the coat of the newborn seal is more important as an insulator (Ling 1974). In all pinnipeds, the primary or guard hairs are flattened. Phocids have few, if any, underfur or secondary hairs. Otariids possess a thick, water-repellent underfur, which insulates against heat loss in both air and water (Ling 1970).

Molting cannot be considered as an independent process. It merely marks the end of a complex pelage cycle and is the outward sign of more basic subcutaneous phenomena (Ling 1970). This pelage cycle is a morphogenetic process occurring within the annual cycle and is influenced not only by environmental factors, but also by the physiology and the behavior of the individual (Ling 1974). The direct proximate stimulus of light acting through neuroendocrine pathways determines the pelage cycle and regulates it with respect to season. Such factors as temperature, behavior, and nutritional and reproductive status modify the influence of photoperiod (Ling 1970; Ronald et al. 1970). The pelage cycle is independent of the reproductive cycle (Ling 1974), and molting usually occurs between parturition/lactation and implantation of the new blastocyst (Ling 1970). Molting is also affected by the ability of the seal to store energy sources in a subcutaneous layer of fat (Ling 1974).

In some pinnipeds, the first molt occurs in utero shortly before birth, as in the harbor seal (Ling 1970), probably Steller's sea lion (Ling and Button 1975), and the hooded seal. It is possible that prenatal molting of this first pelage allows the newborn to enter the water and swim sooner than other species can (Mohr 1950). In pinnipeds that breed on ice, a rapid postnatal molt is typical, and allows them to live in the aquatic environment (Ling and Button 1975). Scheffer (1958) correlated color patterns of newborn seals with ancestral habitats. He suggested that the prenatal molt is an adaptation by species that now

TABLE 39.1. Growth parameters of pinnipeds

Pinniped	Birth Weight (kg)	Birth Length (m)	Nursing Period (weeks)	Weaned Weight (kg)	Weaned Length (m)	Average Weight Increase over Nursing Period (times)	Length at Sexual Maturity (m)	Length at Physical Maturity (m)	Length at Sexual Maturity as % of Final Length	Female Adult Weight (kg)	Weight of Pup as % of Mother's Weight: Birth	Weight of Pup as % of Mother's Weight: Weaned
Harbor seal (*Phoca vitulina*)	10[a]	0.82[a]	5–6[a]	24[g]	0.97[a]	2.6[b]	1.55,[a] 1.40 F[a]	1.61 M,[a] 1.48 F[a]	87.5[c]	89[b]	11.2[b]	28.6[b]
Larga seal (*Phoca largha*)	7–12[d]	0.77–0.92[d]	3–4[e]	—	—	2.0+[e]	—	1.42–1.70[e]	—	81–109[e]	—	—
Harp seal (*Pagophilus groenlandicus*)	11[a]	0.90[a]	1.5[a]	33[a]	1.04[a]	3–4[f]	1.70 M[a]	1.76 M[a]	92.3[c]	120–130[d,g]	8.0–9.1[d,g]	27.0[g]
Ringed seal (*Pusa hispida*)	4[a]	0.65[a]	8–10[a]	16[a]	0.80[a]	2.7[b]	1.23 M,[a] 1.17 F[a]	1.38 M,[a] 1.35 F[a]	86.0[c]	68[b]	6.7[b]	18.0[b]
Ribbon seal (*Histriophoca fasciata*)	10.5[e]	0.86[e]	4–6[b]	27–30[e]	—	2.4[b]	—	1.7 M[h] (mean length)	92.4[i]	81[b]	13.6[b]	32.4[b]
Gray seal (*Halichoerus grypus*)	14[a]	0.89[a]	2–3[a]	40[a]	1.07[a]	—	2.00 M,[a] 1.85 F[a]	2.40 M,[a] 2.02 F[a]	83[c]	—	—	—
Hooded seal (*Cystophora cristata*)	12[a]	0.91[a]	>1[a]	—	1.20[a]	2.0[f]	2.36 M,[a] 2.02 F[a]	—	—	—	—	—
Bearded seal (*Erignathus barbatus*)	45[a]	1.2[a]	1.7–2.2[a]	90[a]	1.52[a]	2.5[b]	2.20 M,[a] 2.10 F[a]	2.30 M,[a] 2.30 F[a]	80.6[c]	270[b]	12.3–12.9[b]	31.2[b]
Hawaiian monk seal (*Monachus schauinslandi*)	16[a]	1.0[a]	5[a]	50[a]	1.38[a]	—	—	2.1 M,[d] 2.4 F[d]	—	Up to 272[d]	—	—
Caribbean monk seal (*Monachus tropicalis*)	16–18[d]	1.0[d]	—	—	—	—	—	At least 2.4[d]	—	170–270[d]	—	—
Northern elephant seal (*Mirounga angustirostris*)	36 M,[j] 31.5 F[j]	1.53 M,[j] 1.47 F[j]	4[k]	—	—	—	—	4.5 M,[j] 3.6 F[j]	—	400–800[d]	—	—
Steller's sea lion (*Eumetopias jubatus*)	12[a]	0.98[a]	12,[a] variable[e]	20[a]	—	—	—	3.20 M,[a] 2.70 F[a]	—	180–230[l]	—	—
California sea lion (*Zalophus californianus*)	5–6[j]	0.75[j]	20–48[j]	—	—	—	—	2.2 M,[j] 1.8 F[j]	—	70–110[l]	—	—
Northern fur seal (*Callorhinus ursinus*)	5[a]	0.64[a]	12[a]	13[a]	—	—	1.4 M,[a] 1.10 F[a]	2.00 M,[a] 1.30 F[a]	88.2[c]	800[b]	4.3–5.3[b]	—
Guadalupe fur seal (*Arctocephalus townsendi*)	—	—	Extended maternal care[m]	—	—	—	—	1.9 M,[m] 1.4 F[m]	—	45–55[d]	—	—
Walrus (*Odobenus rosmarus*)	55[a]	1.2[a]	100,[a] Up to 2.5 years[e]	36[a]	2.00[a]	8.8[b]	2.6 M,[a] 2.4 F[a]	3.00 M,[a] 2.60 F[a]	89.8[c]	560[a]	—	41.9[b]

[a]Bryden 1972.
[b]Irving 1972.
[c]Laws 1956.
[d]Jefferson et al. 1993.
[e]Hull 1994.
[f]Seal Conservation Society 1999.
[g]Lavigne and Kovacs 1988.
[h]Popov 1976.
[i]Burns 1971.
[j]Burns 1978; DeLong 1978b.
[k]FAO 1976.
[l]Olesiuk and Bigg 1988.
[m]Fleischer 1978a.

Table 39.2. Reproductive parameters of pinnipeds

Species	Age to Sexual Maturity (years)		Pregnancy Rate[a]	Pupping Season	Lactation	Mating	Gestation[b]	Delayed Implantation	Comments
	Female	Male							
Harbor seal	3–6 (Seal Conservation Society 1999)	3–7 (Seal Conservation Society 1999)	—	Duration 1.5–5.2 mo Alaska: May–mid-July BC & Washington: June–Sep Mexico; Early Feb–Mar. (Seal Conservation Society 1999) Atlantic Canada: May–June Eastern US: mid-Apr–mid-June (Hannah 2000) Ungava: mid-Apr–mid-May (Environment Canada 1999a)	May last 4–6 wk, after which pups weaned (Hannah 2000) Alaska: 1 mo (Hull 1994)	30 d (Boulva and McLaren 1979), most shortly after pup weaned (Hull 1994)	10.5–11 mo (FAO 1976)	2 mo (FAO 1976)	—
Larga seal	3–4 (Hull 1994)	4–5 (Hull 1994)	—	Feb–May depending on location (Seal Conservation Society 1999) Alaska: early Apr–first part of May; peak during first 1 wk of Apr (Hull 1994)	3–4 wk (Hull 1994)	Late Apr–early May (pairs are formed as early as late Feb; break up after breeding season; Hull 1994)	—	—	—
Harp seal	~4 (Hannah 2000), 4–6 (Seal Conservation Society 1999)	7–8 (Seal Conservation Society 1999)	~94% in 1979, compared with 85% in 1952	Front: early Mar (Sergeant 1975) Gulf of St. Lawrence: 15 Feb–15 Mar (King 1964)	2–4 wk (Terhune et al. 1979), av. 12 d (Seal Conservation Society 1999)	At end of 2nd wk after parturition (Sivertsen 1941)	11.5 mo (FAO 1976)	4.5 mo (FAO 1976)	The maturity of the harp seal appears to have responded to density pressures within the herd, so the female probably reached a mean whelping age in 1979 of 4.5 yr as compared to a mean age of 6.2 yr in 1952. Fertility rate and mean age at maturity appear to be density dependent
Ringed seal	Canada: 4+ (Hannah 2000) Alaska; most at 5 or 6 (Hull 1994)	7–9 (Burns 1978; Hull 1994; Hannah 2000)	85% ovulation rate; >90% (FAO 1976)	Mid-Mar–late Apr, whenever stable land-fast or pack ice is available (Hull 1994; Hannah 2000)	4–6 wk (depends on stability of ice where birth lair built; Burns 1978) Alaska: ~2 mo (Hull 1994)	4–6 wk postparturition (FAO 1976) Alaska: most within 1 mo after birth of pup (Hull 1994)	11 mo (Burns 1978)	3.5 mo (Burns 1978)	Newborn ringed seals stay in the birth lair until they are weaned; if early ice break-up destroys the lairs, they are abandoned and usually starve (Burns 1978).
Ribbon seal	2–4 (Hull 1994)	5 (Burns 1971), 3–6 (Hull 1994)	85% (FAO 1976)	Late Mar–mid-Apr (FAO 1976) Alaska: early Apr to about mid-May; peak between Apr 5 and 15 (Hull 1994)	Arctic: 4–6 wk (Irving 1972) Alaska: 3–4 wk (Hull 1994)	Late Apr–early May (FAO 1976) About time pups are weaned (Hull 1994)	10.5–11 mo (FAO 1976)	2.5–3 mo (Hull 1994)	Annual reproductive rate 26% (Propov 1976)

(*Continues*)

TABLE 39.2—*Continued*

Species	Age to Sexual Maturity (years): Female	Male	Pregnancy Rate[a]	Pupping Season	Lactation	Mating	Gestation[b]	Delayed Implantation	Comments
Gray seal	4–7 (Hewer 1964), 3–5 (Seal Conservation Society 1999)	Likely by 3, but because of competition, most do not breed until about 8 (Hannah 2000)	85% for >6-yr-old females (Mansfield and Beck 1977; Mansfield 1978)	Late Dec–Feb; peak: mid-Jan (Mansfield 1978; Bruemmer 1979) North America; mid-Dec–Early Feb (Hannah 2000)	Weaned at 2 wk (Mansfield and Beck 1977); 16- to 17-d nursing period (Hannah 2000)	~7 wk after lactation ends (FAO 1976)	11.5 mo (FAO 1976)	3 mo (FAO 1976)	In some crowded colonies females mature as late as 9 yr; males mature physically, but not socially, at the same time as females (Platt et al. 1974; Mansfield 1978). Ovulation appears to be spontaneous, as in the southern elephant seal (Mansfield 1978).
Hooded seal	3 (first pup at 4) (FAO 1976), 3–6 (Seal Conservation Society 1999)	~5 (Hannah 2000), 5–7 (Seal Conservation Society 1999)	95% (FAO 1976)	Second half of Mar–early Apr (FAO 1976; Seal Conservation Society 1999)	4 d (Hannah 2000), av 3.8 d (Seal Conservation Society 1999)	Toward end of lactation (Sergeant 1976b)	11.7 mo (FAO 1976)	>4 mo (FAO 1976)	
Bearded seal	3–6 (Hannah 2000), reproductive maturity at 5 or 6 (Hull 1994)	6–7 (Burns 1978)	83% (Burns 1978) Alaska: ~85% (Hull 1994)	Bering Strait: mid-Mar–1st wk May; later further north (Burns 1978) Canada: mid-Mar–early May (Hannah 2000)	Pups born on ice nursed 1.7–2.2 wk (Burns 1978) Canada: 18–24 d (Hannah 2000) Alaska: 12–18 d (Hull 1994)	Within 2 wk after pup weaned (Hull 1994)	11 mo (Burns 1978)	2–2.5 mo (Burns 1978; Hull 1994)	Age to maturity higher in Pacific seal, perhaps because of higher rate of exploitation (Burns 1978) Banfield (1974) stated there is no postpartum estrus in this species, so some females do not ovulate until early June when males are not potent; these same females would mate the following year and thus pup once every second year; if this is true, it is unique in phocids According to Burns (1967), the reproduction rate or annual growth of the population as a whole is 22–25%.
Hawaiian monk seal	3 possibly; first pup usually at 5–10 (FAO 1976; Seal Conservation Society 1999)	3 (FAO 1976)	See Comments	Late Dec–mid-Aug, with a peak late Mar–June (Jefferson et al. 1993)	6 wk (FAO 1976)	At sea and rarely seen Courting behavior observed early Mar–July (King 1964)	Unknown	—	Pregnancy rate is 56% of females over 2 yr with 34% of these breeding in both seasons, 32% in the first year only, and 34% in the second year only (FAO 1976)
Caribbean monk seal	—	—	—	About beginning of Dc (Seal Conservation Society 1999)	—	—	—	—	—

Northern elephant seal	Give birth at 3–5 (FAO 1976)	4–5, socially mature at 9 or 10 (FAO 1976)	95% cows reproductively active until death (FAO 1976)	Late Dec–Mar (Jafferson et al. 1993)	~4 wk (DeLong 1978b)	Near time of weaning when female comes into estrus (DeLong 1978b); about 3–5 d (Seal Conservation Society 1999)	11.3 mo (FAO 1976)	3 mo (FAO 1976)	—
Steller's sea lion	3–6 (Pitcher and Calkins 1981)	3–7 (Perlov 1971), socially and physically competitive at 7–9 (FAO 1976)	See Comments 60%, Gulf of Alaska during Apr–May 1985 (Calkins and Goodwin 1988) Can be as high as 85% for 7- to 19-yr-old females in a low-density population (FAO 1976).	Mid-May–late June (FAO 1976) Gulf of Alaska: late May–early July (Pitcher and Calkins 1981)	32–44 wk (occasionally a female will nurse both a newborn and a yearling; (FAO 1976) Variable: some pups weaned in first year, others may suckle until 3 yr of age (Hull 1994)	8–14 days postpartum (FAO 1976), most breed annually (Pitcher and Calkins 1981)	12 mo (FAO 1976)	3–4 mo (Pitcher and Calkins1981)	Pregnancy rate is variable according to crowding and age distribution
California sea lion	4–5 (Seal Conservation Society 1999)	4–5; usually mate at 8–9, when they can hold a breeding territory (Seal Conservation Society 1999)	—	Mexico and California: late May–late June (FAO 1976)	20–48 wk (FAO 1976) 4–8 mo, although some are nursed for over 1 yr (Seal Conservation Society 1999)	14 d after parturition (FAO 1976) California: peak mating is mid-July	11.9 mo (FAO 1976)	—	—
Northern fur seal	3–5, peak breeding between 7 and 15 (FAO 1976; Hull 1994)	5–6, do not breed until 9–10 when socially and physically capable (Hull 1994)	60% (FAO 1976)	Late June–early Aug, with a peak in early July (Fiscus 1978)	16 wk (Fiscus 1978) Alaska: 3–4 mo (Hull 1994)	5–10 d after parturition (Fiscus 1978)	—	3.25–4 mo (FAO 1976)	—
Guadalupe fur seal	—	—	In 1977, 56% of females observed breeding (Fleischer 1978b)	Early June–early Aug (Seal Conservation Society 1999)	8–9 mo (Seal Conservation Society 1999)	7–8 d after parturition (Seal Conservation Society 1999)	—	See Comments	Little is known about the reproduction of this species; there is almost certainly some period of delayed implantation (Fleischer 1978b)
Walrus	av 7 (Reeves 1978), 4–10 (Seal Conservation Society 1999)	av 6 (Reeves 1978); 6–10, usually cannot compete successfully until about 15 (Seal Conservation Society 1999)	See Comments	Mid-April–mid-June (FAO 1976) Alaska: late Apr, early May (Hall 1994)	Minimum of 78 wk (Reeves 1978), most nurse 2 or 3 yr after birth	Feb, Mar (FAO 1976) Alaska: Jan, Feb (Hull 1994)	380 d (Reeves 1978), about 15 mo (Hull 1994)	3 mo (Reeves 1978)	This species has a low level of productivity; 80% of females give birth every 2 yr and 15% every 3 yr (FAO 1976)

[a]Incidence of pregnancy in sexually mature adult females.
[b]Includes delayed implantation stage.

inhabit climatic zones other than those in which they originated. In each seal family, the adult pelage is different from that of newborns. Among phocids, only the harbor and hooded seals molt in utero, whereas the others shed their first pelage within a few weeks of birth. Otariid pups have dark, wettable body fur. This is shed a few months after birth and replaced by a more adult-like pelage. The light coat of a newborn odobenid is molted after a few weeks, and after a delay of some months, the growth of the adult pelage is slowly accomplished (Ling 1974).

Although pinnipeds molt annually, only phocids replace their entire coat each time. In northern elephant, hooded, and monk seals, molting is rapid. The hairs are shed along with large sheets of cornified epidermis (Ling 1970). The northern fur seal replaces only 75% of its guard hair each year, whereas 25% of the underfur fibers of mature males and 35% of that of mature females is renewed annually (Scheffer and Johnson 1963).

To molt, pinnipeds must modify their behavior to maintain the necessary skin temperatures. Accordingly, additional rest is essential for heightened epidermal mitotic activity and for high tissue temperatures of around 37°C (Feltz and Fay 1966). Ling (1970) gave a detailed description of cellular activity during the hair cycle.

Thermoregulation. Many factors affect heat balance in animals. From a purely physical aspect, it is possible to produce a heat balance model for a homeothermic animal using simple heat transfer principles (Øritsland 1978). However, there are physiological, behavioral, and environmental complexities, many of which are only now being studied in humans and marine mammals.

Different species of seals share some of the same variables affecting thermoregulation. All seals must be able to maintain a uniform body temperature in two very different media, air and water. Because water has a greater thermal capacity than air, it is a more effective cooling agent. However, the different species of seal live throughout a range of habitats and climates from the Arctic to the subtropics, which imposes different stresses. For example, seals such as the northern fur seal migrate annually from temperate waters to more northerly latitudes. Other pinnipeds, like the walrus, are gregarious, and crowd together by the thousands in rookeries where they are able to either benefit or suffer from their neighbor's thermoregulatory capacities.

Diving behavior places some demand on the animal's metabolism and the need to conserve heat becomes critical, a factor that is compensated for by certain circulatory adaptations (see Diving Physiology). Disposing of heat can also be a problem. Because sweating is not an appropriate method for marine mammals (Hemplemen and Lockwood 1978), the seal has made specific accommodations in its flippers.

Phocids are generally the smallest pinnipeds, whereas otariids are medium to large in size, as is the walrus. Adult phocids have thick blubber, their newborn little or none. The walrus has a thicker layer than phocids to accommodate for its lack of fur, but it also has a very thick, rugose hide. Otariids have little or no blubber, as they do not need the excessive insulation, but rely on their fur for its insulative properties.

Pelage is important in establishing the heat balance of pinnipeds, except in the walrus. Insulative properties of pelage (otariids have air trapped between the layers, whereas phocids normally have wettable hair), its color (often white in newborn phocids unless the initial layer has been shed in utero, gray to black in otariid young), and its structure are complexities that interact with solar and other environmental conditions (Øritsland and Ronald 1978a). Apparently, arctic fur allows more solar heating at skin level (producing a "greenhouse effect") than nonarctic fur (Øritsland 1978). In phocids, only pigments hinder transmittance of radiation through the hairs, because the hair is solid with no medulla and has a very transparent cuticle and cortex, in contrast to that of otarids. In mammals, hair transmittance is very important in solar heating (Øritsland et al. 1978). At skin level, heating is a function primarily of fur depth, coupled with hair transmittance and coat reflectance (Øritsland and Ronald 1978b).

Environmental conditions such as wind, radiation, and precipitation affect metabolism, but to an unknown extent. Øritsland and Ronald (1977) produced a computer simulation model that attempted to include all the variables involved in heat balance in harp seals (Øritsland 1978). Some of the factors studied were ambient temperature, solar radiation (absorbance, reflectance, and transmittance), wind chill, and precipitation (rain, hail, and snow), which can cause severe cold stress in neonates.

Because of their diving habit and often extreme environmental conditions, seals have certain physiological adaptations for thermoregulation. Heat loss in harp seals is fairly localized, so expired heat loss and convective heat loss from the body are <25% of the metabolic heat production (Gallivan 1977). During a dive, the seal's general metabolic rate declines (Blix 1976). The seal, however, can still maintain heat balance without increased muscle activity or metabolic rate at low ambient water temperatures, for example, 13°C for the harbor seal and 5°C for the gray seal. The harp seal can even maintain its body temperature while remaining motionless in the water, possibly because of a higher metabolic rate; whereas at 0°C, otariids, such as the California sea lion, which do not have the insulative properties of phocids, must remain active to balance heat loss (Hempleman and Lockwood 1978). Water temperature normally does not exert a significant influence on respiration or diving patterns in harp seals. Measurements of metabolism and respiration in relation to water temperature showed no difference in metabolic rate, core temperature, ventilation, gas exchange, or diving pattern over a range of water temperature from 1.8°C to 28.2°C. The cooling power of the aquatic environment always greatly exceeded the heat production of the seal, and internal conduction rather than external convection governed heat loss from body surfaces (Gallivan and Ronald 1979). The insulative properties are such that the core temperature of a seal in ice water conditions extends only to a depth of 5 cm from the skin surface (Hempleman and Lockwood 1978).

Primary sources of heat loss are the head and the flippers (Gallivan 1977). The flippers are not insulated, but heat loss is controlled by the parallel structure of the veins and arteries, which allows for a countercurrent blood flow (Hempleman and Lockwood 1978). Phocids have a system of arteriovenous anastomoses in the superficial layer of the dermis (occurring in greater densities in harp seals than in hooded seals), which also seems to act as a temperature regulatory device (Bryden 1978; Blix et al. 1979).

During a dive, the seal shunts its blood supply away from the surface areas and internal organs to supply the heart and brain, which are critically in need of oxygen. Some heat loss is prevented in this way and the large venous plexus in the neck is embedded in large deposits of brown adipose tissue, a condition also found on the pericardium and around the kidneys. It is speculated that this "venous plexus–brown fat complex might function as a high-efficiency tubular heat exchanger" (Blix et al. 1975; Blix 1976:176).

Phocids have parabdominal testes, which are located outside of the abdomen but wedged between muscle and a thick layer of blubber. Otariids possess scrotal testes, so they are more like terrestrial mammals. It is thought that the close relation between the bones of the hind limb and the pelvis of a seal provides the opportunity for venous blood vessels to form shunts that are not present in terrestrial mammals. These allow cool blood from the hind flipper to be carried directly into the pelvic and abdominal cavities and establish a thermal gradient, allowing direct heat transfer from the testes and the adjacent muscles. In turn, pelvic and abdominal venous plexuses, which receive cool blood from the hind flipper, surround the seal uterus, and these veins provide a large thermal mass, which can directly cool the fetus (Rommel et al. 1998). The blood volume of some phocids is double that of a human of similar weight.

As harp seal pups are born, they leave the 37°C warmth of the uterus (Grav et al. 1974) for a climate where ambient temperatures can be as low as −15°C and winds can be up to 30 m/sec. Their white coat is wet and they have no blubber, so they shiver vigorously for over 1 hr until their coat is dry (Blix et al. 1979). There are also physiological adaptations that produce heat and help them to survive this critical period. They appear to have a high tolerance to low body core temperatures under cold conditions (Øritsland and Ronald 1978b), and the presence of loosely coupled mitochondria in brown fat and in dark muscle fibers allows for a process called nonshivering thermogenesis (George and Ronald 1973, 1975; Øritsland and Ronald 1978a; Blix et al. 1979). The "brown fat" actually appears to have an ultrastructure between brown and white fat, which might indicate the transformation to blubber, which takes place during lactation. Glycogen stores are also believed to be depleted during this process (Blix et al. 1979). For the newborn exposed to prolonged direct sunlight, hyperthermia can be as severe a problem as cold conditions. Pups alleviate the condition up to ambient temperatures of 15±1°C by heat dissipation through the flippers, because they are unable to lower their metabolism sufficiently (Øritsland and Ronald 1978b).

Vision. The large, well-developed eyes of pinnipeds are an adaptation to aquatic and terrestrial environments and are indicative of the importance of visually guided behavior (Walls 1942; Hobson 1966; Schusterman 1972). The ability to see well above and below water under a variety of light intensities is essential to the survival of these animals. They use their vision for finding landmarks when migrating, establishing territories, or hauling out to give birth; avoiding obstacles and predators; exploring the ocean floor; locating prey; and recognizing conspecific individuals (Schusterman 1975). Behavior is visually guided, but because each species's feeding and social habits differ, so do their visual capabilities (Schusterman 1972).

Anatomically, the seal eye is adapted in numerous ways to function in the two different optical media. Lavigne et al. (1977) believed that the seal retina is duplex, or has two types of photoreceptors (rods and cones), which allow individuals to function on land and underwater during day and night. This theory is supported by the examination of various aspects of seal vision, such as visual acuity, spectral sensitivity, pupilomotor response, and critical flicker frequency. The retina's cone receptors permit detailed vision under high light intensities, whereas a rod-dominated retina with its increased sensitivity gives effective perception underwater, especially in low luminance. It is possible that rod and cone receptors function in surface and underwater vision and that the transition from cone to rod vision occurs when background light is decreased. The presence of a Purkinje shift (Purkinje 1825) in the seal eye supports the idea of a duplex retina and at least two types of photopigment with different absorption spectra, one functioning under photopic and the other under scotopic conditions. The pupil responses, which give evidence of the different physiological responses of both rods and cones, occur at lower light intensities in seals than in humans, and point to a more sensitive duplex retina. The critical flicker frequency response contour of the seal shows a shift in function from one type of photoreceptor to another, again indicating a duplex retina (Lavigne et al. 1977).

Other anatomical adaptations include a highly developed tapetum (retinal reflecting layer), which provides increased visual sensitivity, especially under water (Nagy and Ronald 1975; Lavigne et al. 1977), and a large, spherical lens similar to that in fish (Walls 1942; Munz 1971; Jamieson and Fisher 1972). This allows for sufficient accommodation underwater, in the absence of the corneal refracting surface (the refractive indices of the cornea and the water are the same), to focus an image on the retina (Schusterman 1972; Lavigne et al. 1977).

Seal visual pigments and the shape of the pupil are also important adaptations. Some pinnipeds are nocturnal, opportunistic feeders and will approach their prey from below as it is silhouetted against the surface light above. Under relatively clear coastal water conditions, some seals can recognize small food items such as herring or sardines even under a cloudy sky at a depth of 200 m. Under ideal conditions in the open ocean, the same species can be seen from depths slightly greater than 1 km (Schusterman 1975). The absorption spectra of rod visual pigments tend to correlate with the spectral distribution of radiant energy underwater, thus maximizing contrast for silhouetting prey species against the ambient surface light (Lavigne et al. 1977). Its photic environment determines the type of visual pigment each seal has. For example, the spectral region of greatest intensity in temperate and polar seas is shifted toward greener wavelengths when compared to the blue of tropical oceans. Thus, the harp seal would have a greener visual pigment adaptation, whereas a tropical species such as the southern elephant seal (*Mirounga leonina*) would have a bluer visual pigment (Lavigne et al. 1977). Pinniped visual pigments are adapted to the underwater environment that facilitates feeding.

A recent study suggests the possibility of color-blindness in seals. Peichl et al. (2001), using visual pigment-specific antibodies, reported the absence of short-wave-sensitive (commonly blue) S-cones in the retinas of all species studied [the Australian fur seal (*Arctocephalus pusillus*), northern fur seal, southern sea lion (*Otaria byronia*), grey seal, hooded seal, harbor seal, and ringed seal]. They have only long-to-middle-wave-sensitive (commonly green) L-cones (cone monochromacy) and hence are essentially color-blind. It was presumed by Peichl et al. (2001) that the loss of S-cones is universal for all pinnipeds. Some capacity for color discrimination, however, has been reported for several pinniped species (e.g., harp seal, Lavigne and Ronald 1972; California sea lion, Griebel and Schmid 1992), but the underlying mechanism is obscure.

The pupil, which is narrow and "teardrop in shape" (Jamieson and Fisher 1972; Lavigne and Ronald 1972), performs an important function. Its size varies with the ambient light intensity, independent of whether the seal is on the surface or underwater (Lavigne et al. 1977). In air, the corneal astigmatism inherent in seals is effectively reduced when the pupil closes to a narrow vertical slit, which acts as a pinhole providing the eye with a huge depth of focus in that meridian. Under dim light, this astigmatism causes a loss of visual acuity, making the eye strongly myopic (Schusterman 1972; Lavigne et al. 1977).

Finally, dark adaptation in the seal is important because of the animal's transition from high light intensities on land or ice during sunny days to much lower levels of light underwater, at night, or during the dark polar winter. Its narrow pupil reduces the amount of light energy reaching the retina, thus minimizing the effects of prolonged exposure to bright light, such as the sun on ice and snow (Lavigne et al. 1977). In humans, exposure to sunlight can cause temporary and cumulative effects on night vision (Hecht et al. 1948). For pinnipeds, the more sensitive rod photoreceptors, which are operative at low light levels, greatly outnumber the cones in the retina. The ringed seal has some 400,000–520,000 rods/mm^2 and 1.5–1.8% cones (Peichl and Moutairou 1998). Although the harp seal initially adapts quickly to dark, it takes 30–40 min for maximum sensitivity to be reached. This rate of adaptation could affect feeding habits, limiting the diving depth during daylight hours and accounting for the common occurrence of night feeding among seals when the low illumination in air allows for sufficient time to dark-adapt before the dive (Lavigne et al. 1977). In contrast, the northern elephant seal, a deep-diving seal, which regularly dives to 300–700 m and reaches these foraging depths in about 6 min (Le Boeuf and Laws 1994), is capable of adapting from daylight conditions to maximum sensitivity within this short time period (Levenson and Schusterman 1999). Thus, as also confirmed by pupillometric findings, the visual capabilities of the shallow- and deep-diving pinniped species are directly related to the needs incurred by their respective diving behaviors, and even the deepest-diving seals can function visually at depth (Levenson and Schusterman 1997, 1999).

Hearing. Pinnipeds use their sense of hearing in a variety of ways: to communicate with conspecific individuals, find prey, avoid predators, and navigate by means of ice and shallow water wave noises (Terhune and Ronald 1974).

Seals are capable of hearing in air and underwater, although hearing is more sensitive underwater. For phocids, the range of hearing extends from 0.7 to 32 kHz in air and from 1 to 100 kHz underwater, with the best frequencies being 17–25 kHz (Terhune and Ronald 1974). Only the elephant seal has good to moderate hearing below 1 kHz; underwater, low-frequency hearing thresholds are significantly better than for other pinnipeds tested (Kastak and Schusterman 1995). For otariids tested, functional underwater high-frequency hearing limits are 35–40 kHz, with peak sensitivities from 15 to 30 kHz (Fay 1988; Richardson et al. 1995).

Although more may be known about behavioral responses of pinnipeds to airborne sounds, these data cannot be applied directly to determine the underwater response. Reeves et al. (1996) noted that pinnipeds tend to accommodate reasonably quickly to loud noise, which might be explained either by threshold shifts in hearing or by "habituation," perhaps both. Sound does not appear to be particularly aversive to pinnipeds, except at very high intensities.

Hearing capabilities of seals and humans have similar features, including pitch discrimination abilities (Møhl 1968b), critical ratios (influence of background noises) (Terhune and Ronald 1971), and ability to localize sounds (Møhl 1964; Terhune 1973).

The seal ear is fully adapted to the aquatic environment (Møhl 1964, 1968a; Terhune and Ronald 1971, 1972). Many of its anatomical adaptations are described here, but for a more detailed description, see Ramprashad (1975). When a phocid submerges, water is prevented from entering the ear by the loss of pinna in the outer ear and by changes in the auricular muscles that open and close the outer ear. There is an internal constriction of the membranous part of the outer ear at the internal pinna and a subsequent closing of the external orifice. Adaptations to accommodate increased pressure during a dive are an elastic, cartilaginous outer ear canal and presence of large blood sinuses within the wall of the outer ear canal. When engorged with blood, these sinuses could reduce the volume of the outer ear canal by acting as a pressure-regulating device, and they might also form a fluid wall around the outer ear canal, thus preventing its complete collapse during diving (Ramprashad 1975).

Adaptations within the middle ear are associated with pressure regulation. The presence of cavernous tissue within the middle ear mucosa (Tandler 1899) and the thick medial wall of the auditory tube are the most prominent features (Møhl 1968a; Kooyman et al. 1970). During a dive, the distension of this cavernous tissue reduces the pressure difference between the middle ear and the nasopharynx, thus facilitating muscular opening of the lateral wall of the auditory tube by its associated muscles. Air at ambient pressure within the nasopharynx would accomplish pressure regulation in a manner similar to that of other mammals (Møhl 1968a; Ramprashad et al. 1973).

No specific aquatic adaptations with the seal's vestibular system are known. The size of the macula utriculi and the presence of the crista neglecta (not found in terrestrial mammals) are, however, unique features. The large macula is probably important for body orientation underwater due to the stimulation of the otolithic receptors in response to gravity. The crista neglecta acts as an extrasensory area, which may be important in rotational acceleration of the seal's body (Ramprashad et al. 1972). Morphological features of the phocid cochlea are similar to those of other high-frequency-hearing mammals. A major difference within the phocid cochlea is the position and size of the round window, which may be significant for hearing at high frequencies (Ramprashad 1975).

In air, hearing results from sound conducted through the meatal orifice or, if this is blocked, via the superficial tissues ventral to the orifice; these latter are important in underwater hearing. Thus, there is a possibility of two parallel inputs that merge in front of the middle ear complex, the underwater input being best adapted (Møhl and Ronald 1975). Because the external ear is closed underwater, hearing cannot be accomplished by conventional air conduction mechanisms. Blood sinuses of the outer ear may aid in transmission of sound energy to the tympanic membrane and hence to the cochlea via the ossicular chain. Sound energy may also be transmitted through the cavernous tissue of the middle mucosa (Ramprashad 1975). This feature may explain the seal's acute directional hearing capabilities underwater.

Humans are continually introducing new sounds into the seal's underwater world that are a great deal louder than natural sounds. This "noise pollution" could adversely affect the seal's hearing and subsequent behavior as these animals make efforts to avoid such disturbances.

Echolocation and Phonation. Echolocation is an animal's use of the echo from its phonation to determine the location, distance, or characteristics of the echoing object (Norris 1969). Although clicks are produced, there is no clear evidence from experiments that indicate pinnipeds can echolocate (Renouf et al. 1980; Schusterman 1981; Wartzok et al. 1984). Norris (1969) supported the probability of an echolocation skill in seals, although the degree of development and extent of use likely vary with the habitat and the individual animal.

Strongly supporting the probability of an echolocation skill is the fact that pinnipeds, particularly those in the icy polar seas, tend to be very vocal and as yet few alternative functions for these sound emissions have been found. However, Ray et al. (1969) stated that the frequency-modulated calls of bearded seal males in breeding condition are perhaps used in establishing territories. Also, the underwater bell-like sound of the walrus is believed to be associated with sexual activity. This "song," produced by adult males, probably has territorial and courtship functions (Ray and Watkins 1975).

Although little is known of the functions of pinniped phonations, many species's sounds have been recorded. Calls have been described as grunts, barks, rasps, rattles, growls, creaky doors, and warbles, in addition to the conventional whistles, clicks, and pulses (Beier and Wartzok 1979; Ralls et al. 1985; Watkins and Wartzok 1985; Miller and Job 1992). Phonations characteristic of a particular species include the growls and barks of *Zalophus* (Schusterman et al. 1967) and the bell-like sound of the walrus (Schevill et al. 1966), made with the help of its pharyngeal pouches (Fay 1960).

Phocid calls are commonly between 100 Hz and 15 kHz, with peak spectra <5 kHz, but can range as high as 40 kHz (Ketten 1998). Typical source levels in water are estimated to be near 130 dB re 1 μPa (decibel levels normalized to 1 micropascal, a standard used in underwater sound measurement) (Richardson et al. 1995), but levels as high as 178 have been reported for the bearded seal (Ray et al. 1969), 164 for the harp seal (Møhl et al. 1975), and 160 for the ribbon-seal (Watkins and Ray 1977). Shipley et al. (1992) indicate that infrasonic to seismic-level vibrations are produced by northern elephant seals while vocalizing in air.

Most otariid calls are in the 1- to 4-kHz range. In-air harmonics that may be important in communication range up to 6 kHz (Ketten 1998). Walrus phonation is generally in the low sonic range (fundamentals near 500 Hz; peak <2 kHz). Although the signal type is commonly described as bell-like, whistles have also been reported (Schevill et al. 1966; Ray and Watkins 1975; Verboom and Kastelein 1995).

For further details on phonation, see the behavior sections under individual species.

Circulatory System. Diving animals have a larger blood volume than nondiving animals, to allow for an increased oxygen capacity during a dive (Elsner 1969; Ronald et al. 1969; Vallyathan et al. 1969), and blood volume correlates somewhat with the species's diving capabilities (Elsner 1999). For example, for a relatively brief diver, the northern fur seal, it corresponds to 11% of body weight (Lenfant et al. 1970); for a longer diver, the southern elephant seal, it corresponds to 22% of body weight (Bryden and Lim 1969).

A seal's heart, however, is not proportionately larger than those of terrestrial animals (de Kleer 1972). In fact, Bryden (1972) calculated that, in phocids, average heart weights were 0.68% of body weight. In form, the bifid heart has a very deep cleft, a fetal characteristic that persists until nearly the adult stage (de Kleer 1972). It is shorter and broader than that of terrestrial carnivores, such as the dog. Drabek (1975) considered the broad form as a specific adaptation to diving because of the extreme hydrostatic pressure encountered in deep dives. Because the flexible ribs allow the thorax to be highly collapsible and because the diaphragm is at an oblique angle, the bifid heart is more advantageous because it is less subject to deformation. Placement of the heart in the thoracic cavity is symmetrical, a condition that, along with the even distribution of the lungs, favors stability (de Kleer 1972). A large aortic bulb, a stretched enlargement of the root of the aorta roughly equal to the volume of the left ventricle, is the most important external feature of the heart. Its size is directly correlated to the individual species's habitat and needs. It is functionally significant during a dive because its extreme elasticity allows and ensures blood perfusion of the coronary system and brain during diastole. It may also be important for storage during diastole, ensuring that blood pressure does not overly increase (Drabek 1975; Rhode et al. 1986; Shadwick and Gosline 1995).

A seal's vagus nerve is more prominent than that in domestic animals, and unlike other animals, it has no escape mechanism for rest during prolonged vagal stimulation (de Kleer 1972). Such stimulation occurs in a dive and an escape mechanism could kill a seal, which might never come out of the dive. Resting time between heartbeats is highly irregular in the seal.

In general, seals have long, ventricular chambers, which are narrower than those of any fissiped family yet examined. The Weddell seal (*Leptonychotes weddellii*), a deep diver, has the longest and narrowest right ventricle of those seals so far examined; the leopard seal (*Hydrurga leptonyx*), a shallow diver, has a broad structure. The ventricular wall is thicker on the left than on the right, and the mass of the right is a lesser percentage of the total heart weight. It is possible that a thinner walled and less massive right ventricle, with a long, narrow chamber, would be better designed for dilation in response to pulmonary resistance during diving. The harp seal's right ventricle is extremely dilated during a dive when the pulmonary vascular resistance is greatly increased. This fact would explain the compensatory mechanism of the alteration of the ventricular structure (Drabek 1975).

Most adaptations in the venous system are for the habit of diving. In pinnipeds, the jugular system is not highly developed (Harrison and Tomlinson 1956). Blood drains from the cranium by two hypocondylar veins, which join to form a large sinus (King 1964). Vessels leaving this sinus on either side of the spinal cord join together to form the major venous exit from the brain, the extradural vein (Ronald et al. 1977a). The extradural vein also communicates with almost every other branch of the venous system (King 1964). Anteriorly there are branches from the dorsal muscular and intercostal veins. More of the dorsal intercostal veins are connected with plexuses to the abdominal walls and in the renal plexus. Posteriorly the extradural vein receives branches from the renal and pelvic plexuses, the dorsal muscles, and the veins of the abdominal walls (King 1964).

There are vast systems of these venous plexuses in a pinniped's neck, pericardium, and abdominal wall adjacent to the kidney, and one exists as a stellate renal plexus. In the harp seal, these are imbedded in functional brown adipose tissue containing mitochondria that behave like loosely coupled mitochondria, sustaining a high rate of heat production. It is speculated that this brown adipose tissue–venous plexus complex functions as a high-efficiency tubular heat exchange system that warms venous blood returning to the already cooled core after a dive. This is probably affected by an activation of the sympathetic innervation (Blix et al. 1975).

The vascular system of the harp seal has been fully described in detail (de Kleer 1972; Ronald et al. 1977a).

Diving Physiology. Because of its habitat, a seal is forced to submerge without breathing for protracted periods for protection, food, escape from predators, and locating new breathing holes. Whereas the lung oxygen capacity of seals is comparable to that of humans (5 L/kg of body weight), seals display a number of physiological adaptations for homeostatic compensation during dives of which humans are certainly incapable (Ronald et al. 1977a; Hempleman and Lockwood 1978). For example, many pinniped species are capable of surviving regular dives lasting 20–30 min, and longer than 60 min in the Weddell (Kooyman 1966, 1981, 1989) and Baikal (*Pusa sibirica*) seals. Hooded seals can remain submerged longer than 52 min (Folkow and Blix 1995), and maximum diving times to 2 hr have been noted for the southern elephant seal (Hindell et al. 1991a, 1991b). The surfactants of the harp seal's lungs inner surfaces have greater gas exchange capability than for any other species so far studied.

As a seal's face is submersed, the trigeminal nerve receptors around the nose are stimulated, causing a reflex apnea that is reinforced by the collapsed lungs from a predive expiration. Apnea then causes reduced activity in the pulmonary vagal system, which in turn results in bradycardia (the process of decreasing the heart's rate of beating) and vasoconstriction. As artenalhypoxia and hypercapnia develop, arterial chemoreceptors are stimulated, which provide a response reinforcement for cardioinhibition, vasoconstriction, and humoral responses (Angell James and de Burgh Daly 1972; de Burgh Daly 1984). All these mechanisms turn the circulatory system of a seal into a heart–brain system, so that oxygen may be conserved and therefore supplied to those tissues that are relatively sensitive to even short periods of hypoxia and rationed to the others.

Hol et al. (1975) found that the redistribution of blood during a dive was a result of the vena caval sphincter, a function independent of the dive response. This sphincter acted as a mediastinal bypass, protecting the anterior caval vein and the thermoregulatory flipper veins against venous stasis as well as a possible creation of large venous reservoirs. Reservoir formation in the hepatic sinuses and the abdominal section of the posterior vena cava occur both before and up to 40 sec after constriction of the posterior vena caval sphincter. The tissues not receiving blood are almost completely ischemic. This condition eliminates normal *vis a tergo,* the force that moves extensive volumes of oxygenated blood held in the posterior vena cava (PVC), splanchnic circulation, and many tributaries of the venae cavae into the heart–brain circulation. At the same time, the oxygen-depleted blood is removed, allowing passage of oxygenated blood from the hepatic sinus to the heart–brain circulation (Ronald et al. 1977a). Meanwhile, the cervical vertebral venous system, the major connection between the extradural intervertebral vein (EIV) and the anterior vena cava (AVC), is closed during a dive, thus increasing the resistance to the flow of blood from the EIV into the AVC.

In the splanchnic system, the blood is apparently transferred to the PVC by a profound peristaltic venoconstriction. Muscle pumping aids the movement of blood into the diving circulation from the tributary veins of the vena cava. The hepatic sinus expands, on diving, to accommodate the large volumes of blood being moved out of other veins. The caval sphincter appears to contract, retaining the blood engorging the hepatic sinus (Ronald et al. 1977a).

During a dive, the heart is supplied with blood from both venae cavae. On postsystolic expansion of the heart, a stream of blood enters the heart from the AVC. Synchronously, the caval sphincter appears to open briefly, allowing a bolus of blood to pass from the hepatic sinus to the heart. The ratio of blood from the AVC to the PVC appears to be regulated by the occasional prevention of the opening of the caval sphincter on expansion of the heart and the occlusion of the cervical vertebral venous system, with a resultant increase in resistance to blood flow from the extradural intravertebral vein to the AVC (Ronald et al. 1977a).

Apparently circulation through muscles is shut off early in the dive, and lactic acid remains in the muscles until the end of the dive (Scholander 1940). Blood also contains high levels of lactic dehydrogenase, which Vallyathan et al. (1969) suggested could be oxidized to pyruvate. This then transfers to the liver for conversion to glycogen, or proceeds to the muscles and becomes part of the Krebs cycle via aerobic muscular activity, or forms lactate anaerobically (George and Ronald 1975).

Diving imposes stress on a seal different from that of the terrestrial environment. The animal must be able to withstand greater pressure, must conserve oxygen, and yet must have energy for muscular action and thermoregulation. Adaptations that accommodate this diving take the form of morphological, physiological, and biochemical changes. They are also part of an integrated system, so they function synchronously.

A physiological compensation that all diving mammals make during water submergence is variability of cardiac rhythms (Casson and Ronald 1975). A normal circulatory response to the asphyxia encountered during a dive is bradycardia. This can reduce the heart rate during a dive to about 8–12 beats/min (bpm). A following tachycardia (increase in heartbeats) may raise it to as much as 200–250 bpm. When the face of a seal is submerged, the heart rate decreases immediately, and as apnea occurs, a further slowing results from a response to blood gas changes. On resurfacing, the heart rate and cardiac output are greater than predive rates, though they gradually return to normal (Casson and Ronald 1975; Hempleman and Lockwood 1978). There also are extended apneic pauses in the normal respiratory pattern, a condition that is reflected in the irregular occurrence of bradycardia even while on land (Dykes 1974). Dykes (1974) speculated that exhalation in the first 15–30 sec might establish a maximum diving bradycardia after immersion. He also suggested that the anticipatory onset of bradycardia seconds before a dive may be not part of the dive response, but rather an autonomic reaction to threatening stimuli.

Contrast between cardiac function and breathing in surface swimming and diving has been measured in unrestrained immature captive harp seals. Heart rate averaged 122 bpm while swimming on the surface, but only 50 bpm during a dive. In seals trained to dive on command, the diving heart rates were 34.9–43.0 bpm, as opposed to resting rates of 125.5 bpm when the seal was floating on the surface (Casson and Ronald 1975). Bradycardia is usually quite profound in forced laboratory dives ("psychological" diving, 10 bpm) and less intense during free dives ("biological" diving, 40 bpm). Extremely low heart rates, however, have been recorded during free dives of other phocids, such as harbor seals (Murdaugh et al. 1961; Jones et al. 1973), ringed seals (Elsner et al. 1989), and northern elephant seals (Andrews et al. 1995). For quietly resting, submerged gray seals, Thompson and Fedak (1993) recorded heart rates as low as ever determined for that species. Elsner et al. (1989) noted that ringed seals, whether swimming under the ice at about 1 m/sec or just resting underwater at a breathing hole, had similarly extreme bradycardia. The heart rate response to diving can vary among different pinniped species, and even individuals within the same species depending on the circumstances, suggesting some voluntary control over the cardiovascular system, possibly to establish a level appropriate for a particular diving situation (Elsner 1999).

Diving imposes certain metabolic problems, which seals have compensated for in three ways: they store large amounts of oxygen, they are able to use these reserves economically, and they have biochemical adaptations that improve anaerobic metabolism (Blix 1976). Not only do some seals possess twice the blood volume of many terrestrial mammals (Ronald et al. 1977a), but they also have high hematocrit values (percentage of blood volume that consists of red blood cells), which increase total oxygen-carrying capacity (Hempleman and Lockwood 1978).

In air, seals ventilate their lungs more fully with each breath than do land mammals, to remove excess O_2 quickly, so that blood and body fluids can more quickly be completely reoxygenated after a dive (Hempleman and Lockwood 1978). Not only is the oxygen used during a dive stored in the blood, but it is also bound to high concentrations of myoglobin in muscle. In addition, when oxygen supplies in the lungs are insufficient for the requirements of a dive, the metabolic rate of the seal falls, conserving the O_2 stores in the blood (Ronald et al. 1977a; Hempleman and Lockwood 1978). During long dives, an oxygen debt occurs, which is repaid when the seal surfaces and resumes regular breathing. The urge to surface and breathe can be resisted longer by seals, mainly because they are less sensitive to CO_2 and have a larger capacity (hemoglobin level) to absorb excess CO_2 than terrestrial mammals (Hempleman and Lockwood 1978). Seals can tolerate high lactic acid and carbon dioxide concentration in their blood, and have intracellular zones of glycogen concentration (Pfeiffer and Viers 1995). Two- to threefold-enhanced glycogen levels have been observed in cardiac tissue (Kerem et al. 1973), and seal myocardium releases profuse amounts of lactate in the diving seal (Kjekshus et al. 1982).

Oxygen conservation results not only from the metabolic response already noted, but also from the restriction of circulation to tissues that are capable of sustaining short periods of hypoxia (Ronald et al. 1977a). Filtration in the kidneys ceases during a dive, and waste from the tissues not receiving blood is held until circulation returns to normal (Hempleman and Lockwood 1978). The heart and brain, which are incapable of prolonged anaerobic respiration, continue to receive an adequate supply (Ronald et al. 1977a; Hempleman and Lockwood 1978). This redistribution of blood has the advantage of saving the O_2 reserves for those tissues easily damaged by hypoxia, but also the disadvantage of causing a dramatic decrease in heat production due to the change from aerobic to anaerobic metabolism in the deprived tissues. This cooling is further reinforced by the increased circulation to the poorly insulated brain (Irving 1938). As a result, deep body temperature decreases during a dive (Scholander et al. 1942) and the venous return must be warmed to minimize cooling of the central body core (Hempleman and Lockwood 1978). To accomplish this, the venous return is shunted through the venous plexus, which is imbedded in brown adipose tissue with its thermogenic potential (Hol et al. 1975).

When a seal dives deeply, the increased hydrostatic pressure could increase the risk of rupture in the gas-filled spaces as well as the rate that nitrogen enters the blood, a condition that on resurfacing could cause the "bends." Anatomical adaptations that accommodate increased hydrostatic pressure are flexible lungs and chest wall (flexible rib attachments); absence of connections between the thoracic wall and the lungs; strong, cartilaginous main airways; presence of *retia mirabilia* lining the sinuses and the thoracic wall region; and a reduction of sinuses in the skull (excluding those of the middle ear) (Hempleman and Lockwood 1978). The retia expand with the blood during a dive and partially compensate for the change in volume that occurs as the lungs collapse. In general, seals exhale just before diving, which limits the amount of available atmospheric gases (Scholander 1940; Kooyman et al. 1970; Hempleman and Lockwood 1978), although shallow divers, such as sea lions, dive with lungs full of air (Dormer et al. 1977). Deep-diving seals exhale most, or all, of the air in their lungs before they dive to avoid excess nitrogen absorption into the blood (Hempleman and Lockwood 1978).

The lungs are designed to collapse under the pressures experienced on deep dives. The terminal airways are reinforced with cartilage or thickened muscle. When alveoli collapse, air is forced back into the reinforced windpipe, where the blood cannot absorb the nitrogen. Nitrogen is prevented from penetrating into the blood not only by the exclusion of gas from the lungs, but also by the large amounts of fat in which it is soluble. Thus, nitrogen is absorbed into lipid bodies rather than into more critical areas. Excess nitrogen is also accommodated in the fatty foam in sinuses and other air spaces (Hempleman and Lockwood 1978).

REPRODUCTION

Ancestral pinnipeds slowly evolved a dependence on marine resources. There was, therefore, less motivation to return either inland or upriver. Social interaction and migrations were oriented along coastal regions. Present reproductive physiology and behavior were developed in response to these changes and to seasonal climatic variations and food availability. Individual communication was difficult during the periods of dispersal throughout the sea, so there was strong selection pressure to return to the same sites to reproduce. Access to land was obstructed by ice, which forced breeding on ice. This resulted in selection pressures for short, synchronized suckling and mating periods, a less rigid social organization, and monogamy in some instances (Stirling 1975). Delayed implantation, an evolutionary response to this problem, resulted in synchronized periods in one location each year (Maxwell 1967). An impregnated seal experiences both a period of delayed implantation and a "true gestation," or actual fetal growth period (Maxwell 1967). Most biologists, however, include this period of delay in their estimates of gestation. Because habitats are different, reproductive behavior is variable among species, as are individual maturity, gestation length, and delayed implantation periods (Table 39.2).

ECOLOGY AND FEEDING HABITS

Pinnipeds are carnivores and feed on fish and invertebrates, which are supported by a multilevel trophic food chain. Seals live in areas of the ocean where upwelling of currents occurs, bringing nutrients to the surface that nourish their food source. Because pinnipeds require large quantities of food near their breeding and hauling-out sites, these sites are located near those regions in the ocean of high food productivity. These areas include the eastern sides of ocean basins and areas adjacent to high-latitude landmasses (Lipps and Mitchell 1976).

For proper management of commercial fisheries, it is important to understand and predict ecological interactions between seals, as predators, and their prey (Boyle 1997). Feeding behavior and food will be discussed in detail for each species. Table 39.3 gives the dietary range of the pinnipeds in North America.

MORTALITY

The primary cause of death in several species of pinniped is human related. The other common causes of mortality are parasitism, predation, trauma or infectious agents, premature weaning, early abandonment,

Table 39.3. Dietary range of North American pinnipeds

	Harbor Seal	Larga Seal	Harp Seal	Ringed Seal	Ribbon Seal	Gray Seal	Hooded Seal	Bearded Seal	Hawaiian Monk Seal	Caribbean Monk Seal	Northern Elephant Seal	Steller's Sea Lion	California Sea Lion	Northern Fur Seal	Guadalupe Fur Seal	Walrus
Invertebrates																
Phylum Coelenterata or Cnideria (jellyfishes, anemones, corals, hydras)												X				X
Phylum Annelida (segmented worms)												X				X
Phylum Mollusca (molluscs)												X				
Class Gastropoda (gastropods—abalones, limpets, snails, slugs)													X			X
Class Pelecypoda or Bivalvia (bivalves: clams, mussels)								X								X
Class Cephalopoda (cuttlefish, squid, octopuses)	X	X	X		X	X	X	X	X		X	X	X	X	X	X
Phylum Arthropoda (arthropods)																
Subphylum Crustacea (crustaceans)			X				X									X
Class Malacostraca (malacostracans)	X	X	X	X	X		X									X
Order Decapoda (crabs, lobsters, shrimps)	X	X	X	X	X	X		X	X			X				X
Phylum Echinodermata (echinoderms)																X
Class Asteroidea (starfish)	X															
Class Holothuroidea (sea cucumbers)																
Class Echinoidea (sea urchins, sand dollars)												X				X
Chordates																
Subphylum Urochordata (tunicates)								X								X
Subphylum Vertebrata																
Class Cephalaspidomorphi																
Family Petromyzontidae (lampreys)												X		X		
Class Chondrichthyes (cartilaginous fishes)											X					
Family Squalidae (dogfish sharks)											X					
Family Rajidae (skates)	X					X					X					
Family Chimaeridae (chimaeras)											X		X			
Class Actinopterygii (ray-finned fishes)										X						X
Family Anguillidae (freshwater eels)									X		X	X		X		
Family Clupeidae (herrings)	X	X	X			X						X		X	X	
Family Engraulidae (anchovies)													X	X		
Family Salmonidae (trouts)	X					X						X	X	X		
Family Osmeridae (smelts)	X	X	X	X	X	X	X		X					X		
Family Myctophidae (lanternfishes)	X													X	X	
Family Gadidae (cods)	X	X	X	X	X	X	X	X				X	X	X		
Family Merlucciidae (merluccid hakes)											X		X			
Family Zoarcidae (eelpouts)					X											
Family Scomberesocidae (sauries)														X		
Family Embiotocidae (surfperches)												X				X
Family Stichaeidae (pricklebacks)					X											
Family Ammodytidae (sand lances)	X	X	X	X		X	X					X		X		X
Family Scombridae (mackerels)	X					X							X	X	X	
Family Scorpaenidae (scorpionfishes)	X						X				X	X	X	X		
Family Anoplopomatidae (sablefishes)												X				
Family Hexagrammidae (greenlings)	X											X				
Family Cottidae (sculpins)	X	X			X			X								
Family Cyclopteridae (snailfishes)					X							X				
Family Pleuronectidae (righteye flounders)	X	X	X			X	X	X				X	X	X		
Family Bothidae (lefteye flounders)	X	X				X	X	X				X	X	X		

or an unsuccessful struggle against the stresses imposed by the marine environment (Sweeney and Geraci 1979), such as heavy precipitation on unprotected neonates. Stress, a physiological and biochemical response to changing environmental demands, weakens the animal and causes it to be more susceptible to disease and parasitic infections (Sweeney and Geraci 1979).

AGE ESTIMATION

Laws (1962) outlined the various methods used in determining the age of a pinniped. The distinctive coat of a newborn indicates a very young seal, and in some species, pelage distinguishes a subadult from an adult. Other characteristics from which to determine age include the degree of scarring in polygamous species; tooth size in the walrus; color, growth, or wear of the vibrissae; and general body proportions. The degree of suture closure demonstrates skull development; however, a study of the epiphyseal fusion of other parts of the skeleton, such as the manus or pelvic bones, will result in only a rough approximation of the age of a seal (Sumner-Smith et al. 1972). The numbers of ovarian scars and laminations in certain bones have limited value.

The most accurate method of age determination of seals is examination of the annual increments of dentine, which form rings in the teeth. When these dentine layers are not distinguishable, similar growth layers in the cementum can be used. These visible annular interruptions in the deposition of cementum and dentine are caused by fasting periods (McLaren 1958, Kenyon and Fiscus 1963). They are accurate indicators of age in most pinnipeds, with the exception of the bearded seal, whose teeth begin to deteriorate at an earlier age (Banfield 1974). Use of growth layer groups (GLGs) in teeth has been validated, with reference to individuals of known age, in the northern fur seal, harp seal, and gray seal (Bowen and Siniff 1999). The use of GLGs has been extended to incisors, which can be removed from living individuals, enabling age determination for animals used in long-term studies (Arnbom et al. 1992; Bernt et al. 1996).

Baculum length is a good indicator of age in the bearded seal. Development of the *os penis* is stimulated by the increased production of androgens during the breeding season throughout the animal's life (Burns 1967). Age can also be determined in *E. barbatus* by examination of its foreclaws, which show annual rings. During the breeding season, a light band forms, and during the subsequent molt, a ridge appears (Banfield 1974). Age of the ringed seal can also be determined by this method. Alternating light and dark bands mark its claws in spring and early summer, respectively. Age determination by claws is useful only up to 10 years of age in the ringed seal, however. Beyond this age, claws become too worn (McLaren 1958).

NOMENCLATURE, DISTRIBUTION, AND GENERAL CHARACTERISTICS OF SPECIES OF PHOCIDS IN NORTH AMERICA

The following species of phocid seals have been reported as resident in the waters surrounding continental North America: harbor seal (*Phoca vitulina*), larga seal (*Phoca largha*), harp seal (*Pagophilus groenlandicus*), ringed seal (*Pusa hispida*), ribbon seal (*Histriophoca fasciata*), gray seal (*Halichoerus grypus*), hooded seal (*Cystophora cristata*), bearded seal (*Erignathus barbatus*), Hawaiian monk seal (*Monachus schauinslandi*), Caribbean monk seal (*Monachus tropicalis*), and northern elephant seal (*Mirounga angustirostris*).

HARBOR SEAL

Phoca vitulina

NOMENCLATURE

COMMON NAMES. Harbor (or harbour) seal, spotted seal, common seal, hair seal
SUBSPECIES. *P. v. vitulina, P. v. concolor, P. v. richardii, P. v. stejnegeri,* and *P. v. mellonae* (Rice 1998).

FIGURE 39.4. Distribution of the harbor seal (*Phoca vitulina*).

DISTRIBUTION

Only three subspecies of *Phoca vitulina* are found in North American waters (Fig. 39.4). *P. v. concolor* occurs on coasts of the western North Atlantic from James Bay, Hudson Bay, Hudson Strait, and Admiralty Inlet on northern Baffin Island south to southern New England and New York and occasionally the Carolinas, as well as on the coasts of Greenland. There was a population in Lake Ontario, but it was extirpated by the early 1800s. Vagrants are found occasionally as far south as Georgia and Florida (Mansfield 1967a; Gilbert and Guldager 1998; Rice 1998; Hannah 2000). Breeding and pupping normally occur in waters north of the New Hampshire/Maine border, although in the early part of the 1900s breeding occurred as far south as Cape Cod (Temte et al. 1991; Katona et al. 1993).

P. v. richardii, the coastal seal, occurs in more temperate, ice-free waters extending along the coasts of the eastern North Pacific, south from the eastern Aleutian Islands, the Pribilof Islands, and Kuskokwim Bay in the Alaska Sea to Isla Asuncion in northern Baja California Sur. This subspecies is present year-round in freshwater Iliamna Lake in Alaska (reached from Bristol Bay), and is vagrant to Isla Guadalupe, Laguna San Ignacio, and the southern Golfo de California (Los Frailes and Los Islotes) (Rice 1998). In April 1998, two harbor seal pups were observed near Puerto San Carlos, Magdalena Bay (24°47' N, 112°06' W), Mexico, suggesting that a small breeding colony exists near the Bay (Chávez-Rosales and Gardner 1999). In California, there are approximately 400–500 known harbor seal haul-out sites, widely distributed along the mainland and on offshore islands, including intertidal sandbars, rocky shores, and beaches (Hanan 1996).

P. v. mellonae, the Ungava seal, is a landlocked group of seals found only in certain freshwater rivers and connecting lakes that flow into southeastern Hudson Bay and James Bay: Rivière Nastapoca, Petite Rivière de la Baleine, Grande Rivière de la Baleine, and La Grande Rivière (Rice 1998). It is prominent in Lacs des Loups Marins (Upper and Lower Seals Lakes), an area 160 km east of Hudson Bay on the Ungava peninsula of northern Quebec (Environment Canada 1999a).

A sedentary species, the harbor seal spends most of its time offshore in winter. They may make some short, seasonal migrations away from the occurrence of fast ice to areas of open water, because they do not maintain breathing holes (Boulva 1976). Harbor seals have occasionally been observed up to 81 km offshore (Hull 1994).

The total number of harbor seals is difficult to estimate because they can be accurately counted only when they are hauled out, and the proportion of the total population hauled out at any given time is largely unknown. Population size can be estimated by counting the number of seals ashore during the peak haul-out period (e.g., the May/June molt) and multiplying this figure by the inverse of the estimated fraction of seals on land. In the 1970s, the world population was thought to be approximately 600,000–900,000 and in equilibrium, although decreasing around human habitation (Scheffer 1977). A more recent estimate placed the worldwide harbor seal abundance at around 400,000 animals (Northridge 1984). Of this, about 285,000 make up the total eastern Pacific population (Seal Conservation Society 1999).

In the eastern North Pacific, the number of harbor seals in Mexico is considered to be >1000 individuals (Aurioles-Gamboa 1993). Based on the most recent surveys, the population in California is estimated to be slightly over 30,000, with a suggested minimum size of almost 28,000 (Hanan 1996; Forney et al. 2000). Since the U.S. Marine Mammal Protection Act of 1972 and termination of the harbor seal bounty program, the count number for harbor seals along the Washington and Oregon coasts has increased from 6389 in 1977 to 17,111 in 1997. Based on a correction factor of 1.53 calculated by Huber (1995), the minimum population was estimated at 24,705 for the Oregon and Washington coast stock of harbor seals in 1997 (Forney et al. 2000). Net production appears to be slowing in California, Oregon, and Washington. Within the inland waters of Washington State, population size was estimated at 16,056 (minimum 15,174) for 1997 (Forney et al. 2000).

Along the coast of British Columbia, there were approximately 35,000 harbor seals in the 1970s (Newby 1978). In the transboundary waters of British Columbia and Washington, harbor seals are the only pinniped species that breeds in the area. Recently, approximately 27,000 harbor seals occurred in this area, with a population growth of 5–15%/year (Calambokidis and Baird 1994).

In Alaska, from the British Columbia border to Cape Suckling (southeast Alaska, 144° W), trend analyses strongly indicate that the number of harbor seals has been increasing since at least 1983 (Small et al. 1997). Hill and DeMaster (1999) estimated about 37,500 seals in this area. Conversely, since the mid-1970s, the number of harbor seals has declined in several areas within the Gulf of Alaska and Prince William Sound. At Tugidak Island, near Kodiak, for example, numbers declined from approximately 11,000 seals to 1000 by the mid-1990s (Hull 1994). In 1996, the count at Tugidak increased slightly to 1420, but this is still only a fraction of its historical size (Small 1996; Withrow and Loughlin 1997). At present, the population estimate for the Gulf of Alaska, including Columbia Bay, is about 30,000 harbor seals. Also, the number of seals in the Bering Sea (waters north of Unimak Pass) is thought to be declining, but data are inconclusive; recent estimated abundance is 13,312 animals (Hill and DeMaster 1999).

In the Atlantic, although *Phoca vitulina* once penetrated the rivers and lakes of eastern Canada, exploitation and displacement by humans had reduced the population to about 12,700 seals in semi-isolated groups in the waters south of Labrador by 1973. Overall, a decline of 4%/year occurred between 1950 and 1970 in eastern Canada (Boulva and McLaren 1979). From 1981 to 1997, an annual increase in numbers and in pup counts was noted along the Maine coast (Gilbert and Guldager 1998). For 1997, the minimum population was estimated at 30,990 seals, based on an uncorrected total count along the Maine coast. Furthermore, the best estimate of harbor seals in southern New England was 6083 in spring of 1999, which is 23% higher than the peak count (4915) reported by Payne and Selzer (1989).

Thus, in the western Atlantic, the total U.S. and Canadian populations are thought to be about 60,000–70,000 animals. There is no reliable estimate for the small Greenland population. The number of Ungava seals is unknown, but could be around 100–600, probably at the lower end of the range (Seal Conservation Society 1999).

Habitat. The harbor seal is an adaptable species with a wide tolerance of temperatures and water salinities (Boulva 1976), and can therefore occupy a variety of environments (Scheffer 1977). In eastern North America, its habitat consists of inlets, islets, reefs, and sandbars, which may be important for breeding (Boulva and McLaren 1979). In the Pacific, it occurs in areas of tidal mud flats, sand bars, shoals, river deltas, estuaries, bays, coastal rocks, and offshore islets (Johnson and Jeffries 1977), and commonly is seen some distance inland. These seals tend to select a protected location with unobstructed access to water and at low tide are seen on sand spits, coastal rocks, and small reefs (Newby 1978).

P. v. mellonae lives exclusively in freshwater. This subspecies depends on strong currents to free some areas of ice, on ice fissures, and on air pockets partly created by the complex geometry of the shoreline (Environment Canada 1999a).

DESCRIPTION

The harbor seal has a short, heavy body and a rounded, smooth head. Valvular nostrils and large, convex eyes are dorsally situated (Fig. 39.5). Its face, which is slightly canine in appearance, has flattened, beaded mystacial vibrissae on each side of the muzzle (Banfield 1974). There is some sexual size dimorphism, with adult males averaging 1.6 m in length and 87.6 kg in body weight, and adult females about 1.5 m in length and 64.8 kg in body weight (FAO 1976). Newborn pups are at least 0.75 m long and weigh 9 kg (Fay et al. 1979). Coat color and pattern vary considerably, and pups are born with an adult-type pelage (FAO 1976; Fay et al. 1979), their silvery gray lanugo being shed in utero. The coat pattern is basically a mottle of dark spots on a lighter background, but in some animals the spots coalesce, particularly on the back (FAO 1976). Thus, color ranges from nearly black with scattered, whitish rings to pale and spotted with some whitish rings on the back (Fay et al. 1979). There are usually a moderate number of light rings around spots; the animal is more heavily spotted above than below. The face is generally light, unlike the back (Jefferson et al. 1993). The pelage consists of an outer coat of stiff hairs 11 mm long and an undercoat of sparse, curly hairs 5 mm long (Banfield 1974). Male harbor seals live up to 20–25 years and females up to 30–35 years (Seal Conservation Society 1999).

The Ungava harbor seal is small and dark in color. The curvature of the lower jaw is enlarged. It is mainly identified, however, by its freshwater distribution, which is geographically isolated from that of the oceanic harbor seal (Environment Canada 1999a).

FEEDING HABITS

Harbor seals are generally close inshore, shallow-water feeders with catholic tastes (FAO 1976). Their food base consists of pelagic, demersal, anadromic, and catadromic fishes, cephalopods, and crustaceans (Table 39.3). Dietary items of economic importance to humans are gadoids, clupeids, pleuronectids, and salmonids (FAO 1976). In the Bering Sea and northwestern Pacific Ocean, walleye pollock (*Theragra*

FIGURE 39.5. Harbor seal (*Phoca vitulina*).

chalcogramma) is a most important prey species, and in the eastern Bering Sea, a wide range of sizes are taken (Lowry et al. 1996). They swallow small fish whole underwater, but take the larger ones to the surface, where they eat them in pieces, usually leaving the heads. Because these seals are opportunistic feeders, there is a great variation in meal size, although they generally have one large meal/day (Boulva and McLaren, 1979). In captivity, they bite the center of a large fish and break off the tail and head so as to disable the prey.

BEHAVIOR

Harbor seals tend to occur in small groups of about 30–80 animals, although larger groups are found in areas where food is plentiful (FAO 1976). Although gregarious on land (although not lying in close contact with each other), these seals have no developed social structure and in the water they tend to disperse and forage for food alone (Banfield 1974; FAO 1976). In Atlantic Canada, haul-out behavior is related to the tides and the weather; seals rest on land at low tide, reentering the water as the tide returns (Johnson and Jeffries 1977). Few seals haul out if high winds are causing rough seas and almost none stay ashore except for females with pups and those molting (Boulva and McLaren 1979). In winter when wind chill temperatures are below −15°C, these pinnipeds stay in the water. When the inlets become frozen, they stay away from the mainland, as they do not maintain breathing holes (Boulva and McLaren 1979).

Haul-out sites must have immediate access to deep water and some protection from human activities and other disturbances. They most commonly consist of offshore rocks and islets, high-banked grassy areas, log booms, and anchored beach floats (Johnson and Jeffries 1977). Harbor seals are very wary and will retreat to the water immediately if alarmed. They have an odd behavior trait of clapping their foreflippers on their chests, causing a large splash. Phonations include a variety of grunts and growls and a high-pitched yapping bark usually heard on land; while in the water, they emit snorts and snuffles (Banfield 1974). Harbor seals do not form harems and are probably promiscuous (Bigg 1969; Bonner 1972). In the North Pacific, *P. v. richardii* is polygamous and females form nursery areas for the 2-week period of lactation (Newby 1978).

MORTALITY

Newborn harbor seals are preyed on by sea eagles (*Haliaeëtus*), golden eagles (*Aquila chrysaëtos*), and red foxes (*Vulpes vulpes*); adults are attacked by sharks, killer whales (*Orcinus orca*), bears, and walruses (FAO 1976). Predation on pups by polar bears (*Ursus maritimus*) and arctic foxes (*Alopex lagopus*) is heavy; that by killer whales and walruses is incidental.

In eastern Canada, there is a high natural mortality rate for *P. vitulina* of 17.5%/year (postweaning), possibly as a result of shark predation and overexposed breeding sites, which make these seals particularly vulnerable (Boulva and McLaren 1979). The three major causes of pup mortality are stillbirth, desertion by the mother, and shark kills (Boulva and McLaren 1979). Human disturbance can cause desertion of the pup by its mother. Other causes of death are shooting, underwater blasting, propeller wounds, internal hemorrhage, and infection (Johnson and Jeffries 1977). An outbreak of a distemper virus claimed an estimated 18,000 seals in the European population (Jefferson et al. 1993). Overall, the major causes of mortality can be attributed to human-related incidents, either intentional or by disturbance.

LARGA SEAL

Phoca largha

NOMENCLATURE

Common Names. Larga seal, spotted seal; "larga" is the name of this seal in the Tungus language of eastern Siberia (Rice 1998); native names are *issuriq* in central Yupik, *gazigyaq* in St. Lawrence Island Yupik, and *gazigyaq* in northern Inupiaq (Hull 1994)

Larga seals were formerly considered to be conspecific with the harbor seal.

DISTRIBUTION

P. largha, an ice-breeding seal, is found in the pack-ice zone of the North Pacific Ocean and adjacent seas, where it occurs along the continental shelf of the Beaufort, Chukchi, Bering, and Okhotsk Seas, and south to the northern Huanghai Sea and western Sea of Japan (Newby 1978; Seal Conservation Society 1999). In North American waters, in spring, the larga seal ranges south as far as the eastern Aleutian Islands, and in summer, north into the Chukchi Sea and Herschel Island, Yukon Territory (Rice 1998) (Fig. 39.6). By late summer–early autumn, they are found along the entire northwestern coast of Alaska, although not abundant in the Beaufort Sea, and are not limited to the coastal zone. In general, in autumn–early winter, the seals move southward and away from the coast just before and during freeze-up, so by midwinter they are again concentrated in the southern part of the Bering Sea ice pack (Hull 1994).

In the late 1970s, the Bering Sea population of larga seals was estimated at 200,000–250,000 individuals. The Russian estimate of the number at Karaginski Bay is about 80,000 (Hull 1994). Based on relative abundance data by location presented by Braham et al. (1984), Perez (1990) considered that 70% of the Bering Sea population occurred in the eastern basin from November to April, and that 10% remained in this region from May to October. Burns (1973) suggested a world population size of 335,000–450,000, although a reliable estimate of overall abundance is not available (Rugh et al. 1995).

Habitat. The larga seal is strongly associated with sea ice from autumn to late spring–early summer, and coastal areas, including river mouths, in late summer and autumn. The seals breed exclusively and haul-out

Figure 39.6. Distribution of the larga seal (*Phoca largha*).

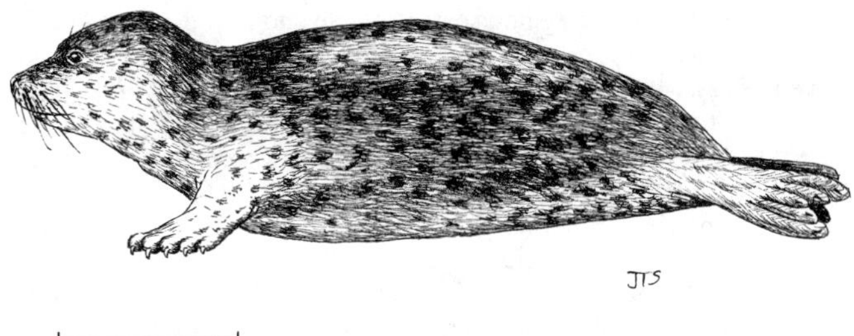

FIGURE 39.7. Larga seal (*Phoca largha*).

regularly on ice, but do come ashore on beaches and sandbars (Jefferson et al. 1993).

DESCRIPTION

The larga seal is intermediate size, similar in appearance to the harbor seal (Fig. 39.7). They can reach 123 kg, but most physically mature adults of both sexes weigh 81–109 kg. Length of grown seals is 1.4–1.7 m. The coat pattern consists of numerous dark, irregularly shaped spots (sometimes encircled by a faint ring) on a lighter background, which shows considerable variation, ranging from gray-white to gray. Spots are most numerous on the back and upper flanks. The face is generally dark, like the back. Pups are born with a dense coat of whitish hair, which is shed by 4–5 weeks of age. The new hair is similar in appearance to that of adults. The maximum life span of the larga seal is about 35 years, although few apparently live beyond the age of 25 years (Hull 1994).

FEEDING HABITS

In general, principal foods are schooling fishes, although there is quite a variation in prey items eaten (Table 39.3). In Karaginski Bay and the Gulf of Anadyr, major prey items during the spring–early autumn period include Arctic cod (*Arctogadus glacialis*), sand lance (*Ammodytes* spp.), sculpins, flatfishes, cephalopods (mainly octopus), and a variety of shrimps. In the northern Bering Sea, Arctic cod is a major food item, whereas in the central and southeastern parts, pollock and capelin (*Mallotus villosus*) are dominant items. Along the Alaskan coast, these seals mainly feed on herring (*Clupea harengus pallasi*), capelin, saffron cod (*Eleginus gracilis*), some salmon (*Oncorhynchus* spp.), and smelts. They dive to near the bottom to feed (Hull 1994).

Composition of the diet also varies with the age of the seal. Newly weaned pups feed on small crustaceans, then advance to schooling fishes, larger crustaceans, and octopuses, and finally to bottom-dwelling fish and cephalopods (Jefferson et al. 1993).

BEHAVIOR

Larga seals form large aggregations on the ice and certain preferred locations on land. In Alaska, several thousand seals may be hauled out together in Kasegaluk Lagoon in the Chukchi Sea, near Cape Espenburg in Kotzebue Sound, and on bars and shoals in Kuskokwim Bay (Hull 1994). Phonations consist of a variety of sounds described as growls, barks, moans, and roars. *P. largna* is monogamous and forms family groups during the breeding season (Newby 1978). Pairs separate after the breeding season.

MORTALITY

Normal natural mortality occurs as a result of starvation, predation, trauma, and disease. Walruses occasionally take young larga seals (Lowry and Fay 1984).

HARP SEAL

Pagophilus groenlandicus

NOMENCLATURE

COMMON NAMES. Greenland seal, harp seal, saddleback seal (King 1964); the Inuktitut name for the harp seal is *kaigulik;* the Innu-Aimun name is *pitshuatshuk*
SUBSPECIES. *P. g. groenlandicus* breeds around Newfoundland and Jan Mayen; *P. g. oceanicus* breeds in the White Sea.

DISTRIBUTION

Harp seals occur in the pack-ice zone of the North Atlantic (Fig. 39.8). Whelping and mating take place on three separate breeding grounds: the White Sea, the Greenland Sea between Jan Mayen and Svalbard, and the Northwest Atlantic. This latter population is divided during the breeding season into two major groups, one near the Magdalen Islands and the other off the coast of Newfoundland–Labrador (Lavigne and Kovacs 1988). There are only slight differences between the Newfoundland and Jan Mayen populations, but the White Sea population is considered sufficiently distinct to be treated as a separate subspecies, *P. g. oceanicus* (Rice 1998).

As the ice recedes each spring, *P. g. groenlandicus* adults migrate north along the east coast of Canada from the breeding grounds to the cooler waters surrounding the Canadian Arctic Archipelago, and occur in Foxe Basin, Lancaster Sound, Jones Sound, Baffin Bay, the Greenland Sea, and Svalbard (Lavigne 1978; Rice 1998). During the summer, immature harp seals (bedlamers), in particular, are found on the west coast of Greenland (especially near Disko Island) and north to Upimavik at 73° N (FAO 1976). In the fall, harp seals return ahead of the arctic pack ice, reaching Labrador and the Gulf of St. Lawrence in late December or early January (Lavigne 1978). During winter, they range to Nova Scotia, Newfoundland, southern Greenland, Iceland, Jan Mayen, and northern Norway (Rice 1998). Stragglers occur south to Virginia, Scotland, Germany, and France. In the 1990s, juvenile harp seals apparently became more regular visitors off the northeastern coast of the United States, especially during the winter. Recently, adults and

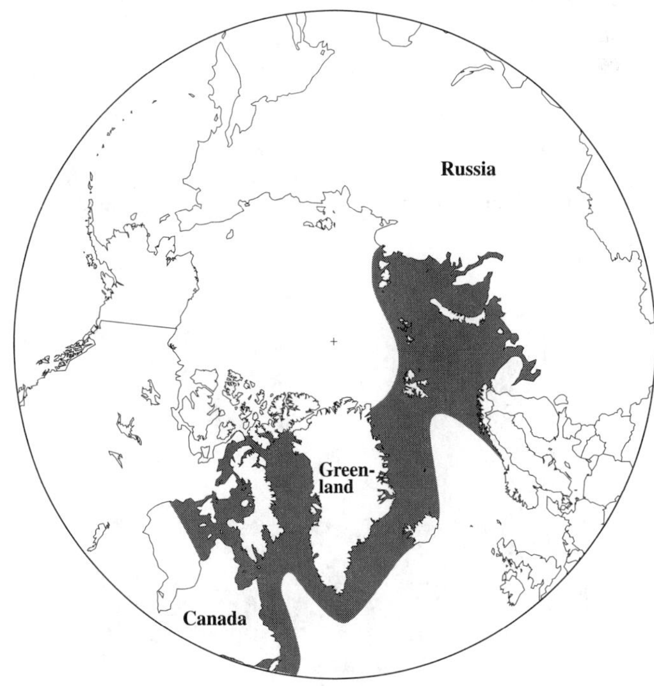

FIGURE 39.8. Distribution of the harp seal (*Pagophilus groenlandicus*).

juveniles have been found in the Gulf of Maine, in Massachusetts Bay, and off the coast of New Jersey (Hannah 2000).

By the mid-1970s, the numbers of harp seals in the three populations were thought to be as follows: White Sea, 0.5–0.7 million; Jan Mayen, 0.1 million (Lavigne 1978); and Newfoundland, 1.3 million (Ronald et al. 1970). By the 1980s, the world harp seal population was conservatively estimated at 2.25–3.0 million (Ronald and Dougan 1982).

In the early 1990s, the population size was estimated at about 500,000 or less in the White/Barents Sea, 100,000–400,000 around Jan Mayen Island, and 2.4–4.2 million in the northwestern Atlantic (Reijnders et al. 1993). Estimated number of harp seals at Newfoundland and in the Gulf of St. Lawrence, based on photographic surveys in March 1994 and computer modeling, was 4.8 million (range = 3.4–5.0 million) seals. The replacement yield that could be taken without changing the total population was estimated at 286,700 (range = 170,000–300,000) animals (Shelton et al. 1995, 1996). More recently, based on mean pup production, the total Barents Sea harp seal stock was estimated to number 2.22 million individuals in 1998 and 2.18 million (range = 1.79–2.58 million) in 1999 (Nilssen et al. 2000).

For the Northwest Atlantic population, various estimates of annual pup production, as given by Reijnders et al. (1993), are 500,000–550,000 (mark–recapture, up to early 1980s), 562,000 ± 78,000 (1990 aerial survey), 536,400 ± 115,300 (*SE*) (aerial photographic surveys, 1990), and 577,900 ± 38,800 (*SE*) (combination photographic and visual estimates, 1990). Different methodologies make it difficult to analyze trends. A March 1994 aerial photographic/visual survey gave an estimate of 702,900 (*SE* = 63,600) pups produced in the Northwest Atlantic (Stenson et al. 1996b).

Habitat. The harp seal's habitat is the drifting pack ice of the North Atlantic. In summer, they use open water and loose ice. During spring, they occur on more solid ice in the Gulf of St. Lawrence or on the front off Newfoundland–Labrador on the thick winter ice at the edge of the advancing ice sheet. Harp seals use channels or leads that penetrate deep into the ice cover to gain access to breeding sites. Because these leads are also important places from which seals can mount the ice, their location often determines the haul-out sites. Harp seals inhabit rough, hummocky ice at least 0.25 m thick (Ronald et al. 1976b).

DESCRIPTION

The harp seal has a small head and short neck with limited mobility (Fig. 39.9), a condition that is related to its mode of propulsion as well as to its dietary habits (Bisaillon et al. 1976). Large eyes dominate the skull (King 1972), and the face has a few beaded whiskers (King 1964). The paddle-like forefeet and hind feet are covered on each surface with fur (Green 1972). On each of the five digits are nails of equal size, and the third digit of the forelimbs is longer than the first or second (King 1964). There is no cartilaginous extension beyond the bony framework of the digits in the forelimbs as there is in those of the hind flippers (Green 1972).

FIGURE 39.9. Harp seal (*Pagophilus groenlandicus*).

The average adult female harp seal is 1.7–1.8 m in length, and the male is 1.7–1.9 m in length (Smirnov 1924). Newborn harp seal pups are 0.97–1.08 m long and the molting pups are from 0.90 to 1.2 m in length. The newborn pups weigh on the average about 11.8 kg at birth, but increase their weight rapidly to about 22.8 kg within 4–5 days. Their weight continues to increase at the same rapid rate to a maximum of 40 kg at weaning. Adult male harp seals average 135 kg and females about 119.7 kg (Sivertsen 1941). There are circannial weight changes in all age groups.

The newborn "whitecoat," so named from its first, preweaning pelage color, molts through several "color" changes before reaching adulthood. The clarity of the "color" of the background fur, spots, and the distinguishing harp on the back is dependent on the medium (air or water) in which the animal is observed and whether the coat is wet or dry. A full-grown adult with a defined harp on its dorsal surface, if observed in water, appears as a silver animal with a deep jet-black marking. The same animal seen wet in air would appear as generally dark gray in color with markings that are indistinct and dull. An adult that has a dry coat and is observed in air appears to have a dark chocolate-brown or black harp on a gray background tinged with brown. Maturity of the pelage differs between males and females, so males may develop the harp pattern earlier, but it is usually distinct by age 7 years. There is such individual variation in markings among adult females, however, that only generalizations may be made. The harp begins to form in the females by age 5–8 years and usually closes by age 15 years. Spots may persist until age 19 years. Usually a lighter-faced seal with a dark harp is a female, but a definitely dark-faced seal, although probably male, could also be female.

FEEDING HABITS

A review by Wallace and Lawson (1997) indicated that more than 100 taxa of crustaceans, cephalopods, and teleosts have been identified from stomach content samples of harp seals collected in the Northwest Atlantic. Pelagic fish, especially capelin, and some benthic fish, however, form the major part of the diet (Sergeant 1973) (Table 39.3). Small animals are sucked in, tail first, but larger ones are bitten. Feeding habits vary by age, season, and region. For example, harp seals feed mainly on capelin and a few euphausiids off western Greenland, whereas in the colder waters of northwestern Greenland they consume various crustaceans and polar cod (*Boreogadus saida*) (Kapel 1973). Weaned pups, after an initial period of apparently voluntary fasting, begin to feed in late April. They feed chiefly on pelagic crustaceans, such as the euphausid *Thysanoessa,* the amphipod *Anonyx,* and a small amount of fish for their first year of their life, changing to include pelagic schooling fish at 1 year of age. Juveniles (>1 year old) feed at intermediate depths on capelin. Adult harp seals dive deeper to feed on bottom fish such as herring and cod, and can dive to a depth of 200–400 m for food. Renouf et al. (1990) noted that harp seals consume large quantities of fresh water daily in the form of snow and ice.

During the winter and summer, *P. groenlandicus* feeds intensively, but adults eat little or nothing during spring and fall migrations and during the whelping and molting periods (Mansfield 1967b; Sergeant 1973). Pregnant females are found in midwinter in the best feeding grounds and remain separated from other harp seals during lactation, feeding on decapods.

In general, individual captive seals consume about 2% of their body weight per day; intake is about 800–900 kg annually, allowing for 65 days of fasting. The estimated ratio of production to ingestion (or gross efficiency of yield to ingestion) is 3.2–5.2%; they are as efficient converters of fish flesh into usable energy as are other seals and terrestrial mammals.

The number of Atlantic cod (*Gadus morhua*) taken by harp seals in 1994 is estimated at 88,000 tonnes at Newfoundland and 20,000 tonnes in the Gulf of St. Lawrence (with wide confidence intervals), and is considered a small component of the diet, about 3% at Newfoundland and 5.6% in the Gulf. Overall, 40 years of studies show that Atlantic cod is a minor constituent of the harp seal diet (Wallace and Lawson 1997). In Newfoundland waters, harp seals select capelin over other prey species, regardless of abundance, when given the choice. Arctic cod were also preferred in nearshore areas, but not in the offshore. They are neutrally selective toward Atlantic cod, American plaice (*Hippoglossoides platessoides*), and Greenland halibut (*Reinhardtius hippoglossoides*) (Lawson et al. 1998). Similarly, in West Greenland waters, harp seal diet consists mainly of pelagic crustaceans and small fish species like capelin, sand eel (*Ammodytes* spp.), polar cod, and Arctic cod. Important commercial species, such as northern prawn (*Pandalus borealis*), Atlantic cod, and Greenland halibut, are only a minor item in the diet (Kapel 2000).

Food energy requirements are needed for growth, maintenance, and temperature regulation. With a mean energy density of prey of 1500 kcal/kg, the corresponding food consumption is 9 kg/day for females and 7.4 kg/day for males (Markussen and Øritsland 1991). Food requirement also depends on the energy content of the prey species. Important prey items of Northeast Atlantic seals, such as capelin, herring, and krill, apparently undergo a large variation in energy density during the year. Assessments of harp seal yearly food intake based on energy requirement models would be affected if the changes in caloric value in the prey (seasonally and interyearly) were not accounted for (Markussen and Øritsland 1991; Mårtensson et al. 1996).

BEHAVIOR

Harp seals are very gregarious, except that old males tend to separate themselves from the herd. Although their terrestrial locomotion appears limited, they are agile and powerful swimmers, capable of speeds up to 5 m/sec. Because of their social nature, the many sounds that these seals emit probably have a communicative function (Møhl et al. 1975). Fifteen underwater phonations have been recorded, but the only known air phonations are a hoarse hissing made by lactating females, distressed adults, and pups (Maxwell 1967; Møhl et al. 1975) and the wailing of the pups, which is similar to the cry of a human infant (Terhune and Ronald 1970). Repetitive calling increases during the breeding season. The structurally complex phonations used are easily distinguished at close range in the underwater environment and probably are adapted to contrast with ambient noise (Møhl et al. 1975; Watkins and Schevill 1979; Terhune and Ronald 1986). Breeding sites occur near water leads and natural or seal-made holes in the ice, which are used for exit and breathing and which the seals attempt to maintain. These holes, 60–90 cm at the top and widening toward the base, are often communal, with up to 40 seals occurring in a pool around the hole (Ronald et al. 1976b). Females form "whelping patches," the size and shape of which change as the ice floes drift. Before mating, males will often fight, using their teeth and flippers (Popov 1966).

During parturition, harp seals tend to stay close to a lead and near females with pups of the same age (Dorofeev 1939). Males usually keep together in groups of about 12 (Maxwell 1967). Pups nurse five or six short times a day soon after birth, and change to fewer but longer periods later in the 9 days of lactation. There is not necessarily any correlation, however, between length of time and quantity of milk obtained. Females leave the pups occasionally during the first week. After this, they return to the water, coming out periodically to feed their pups and thermoregulate. If threatened while in the water, a few females will defend their pups, but most will swim away. If threatened on the ice, however, about 10% will either defend their pups vigorously or "play possum," as do the pups (Ronald et al. 1970). On clear windless days, males and females bask in the sun, but they tend to spend more time in the water in snowy and foggy weather (Ronald et al. 1976b). In April and May, adults and immature 1-year-olds haul out to molt, forming dense aggregations. These molting groups are even larger than the breeding colonies and may form groups of >10,000 seals.

MORTALITY

Predation by polar bears, Greenland sharks (*Somniosus microcephalus*), and killer whales on harp seals is low. Mortality rates from parasitism and disease are also incidental. Harp seals have a life span of 30 years or more, although the average is about 15–20 years (FAO 1976).

RINGED SEAL

Pusa hispida

NOMENCLATURE

COMMON NAMES. Ringed seal, floe rat, fjord seal, jar seal; ringed seals are referred to as *natchek* in Inupiat and *niknik* in Bering Sea Yupik (Hull 1994); the scientific name *hispida* is derived from the Latin meaning rough or bristly, which refers to the stiff hairs of the coat (Burns 1978)
SUBSPECIES. *P. h. hispida,* of the Arctic Ocean and the confluent Bering Sea; *P. h. ochotensis,* of the Okhotsk Sea; *P. h. botnica,* of the northern Baltic Sea, including the gulfs of Bothnia and Finland, south to Stockholm, Sweden, and Riga, Latvia; *P. h. saimensis,* landlocked in a series of interconnected lakes in Finland: Saimaa, Haukivesi, Orivesi, Puruvesi, and Pyhäselkä; and *P. h. ladogensis,* of Lake Ladoga in Russia (Rice 1998)

Taxonomy of ringed seal populations in the Arctic basin and Bering Sea (now all considered as *P. h. hispida*) is likely complex, and the relationship between offshore and coastal animals remains unresolved (Rice 1998). Fedoseev (1975) discovered slight morphological differences between ringed seals that live in drifting pack ice and those that live in adjacent shorefast ice in the Chukchi, Bering, and Okhotsk Seas. Also, a population of ringed seals inhabits the pack ice of Baffin Bay that is morphologically distinguishable from adjacent coastal animals and appears to be reproductively isolated from them (Finley et al. 1983).

DISTRIBUTION

The range of the ringed seal is circumpolar, from the southern edge of the pack ice to at least within 2 km of the North Pole. Along the North American coast, they range south with the formation of ice to James Bay, the Strait of Belle Isle (to Harrington Harbour, Quebec), Kap Farvel in Greenland, and northern Bristol Bay in Alaska (Rice 1998). In the Bering Sea, *P. hispida* occurs in the fast ice gulfs and bays. When the ice disappears, they remain sedentary along the coastal inlets of the Chukotsk and Kamchatka Peninsula and the Commander Islands (Popov 1976). The ringed seal also inhabits freshwater Nettilling Lake (31 m above sea level) and the 85 km-long Koukdjuak River, its outlet into Foxe Basin, on the west side of Baffin Island (Fig. 39.10). Stragglers have been found in New Jersey and southern California (Rice 1998).

The ringed seal is the most abundant and widely distributed arctic seal. Because of the vast area of its occurrence, however, it is impossible to obtain an accurate population estimate, but total population size has been estimated in excess of 5 million (FAO 1976), or tentatively at 6–7 million animals (Stirling and Calvert 1979). Ringed seals present in Alaska waters have ranged from 1 to 1.5 million (Kelly 1988). At least 250,000 occur on the shorefast ice in the Bering Sea (Frost 1985). Status of the worldwide population is variable, depending on location, with numbers in some areas increasing and in other areas decreasing (Jefferson et al. 1993).

Habitat. The ringed seal maintains a year-round association with ice, occupying the land-fast or shore ice in winter and migrating with the annual advance and retreat of the ice pack in other seasons (Burns 1978). Adults tend to remain on the stable, inshore ice, whereas subadults are found further offshore in areas of shifting but relatively stable ice (FAO 1976).

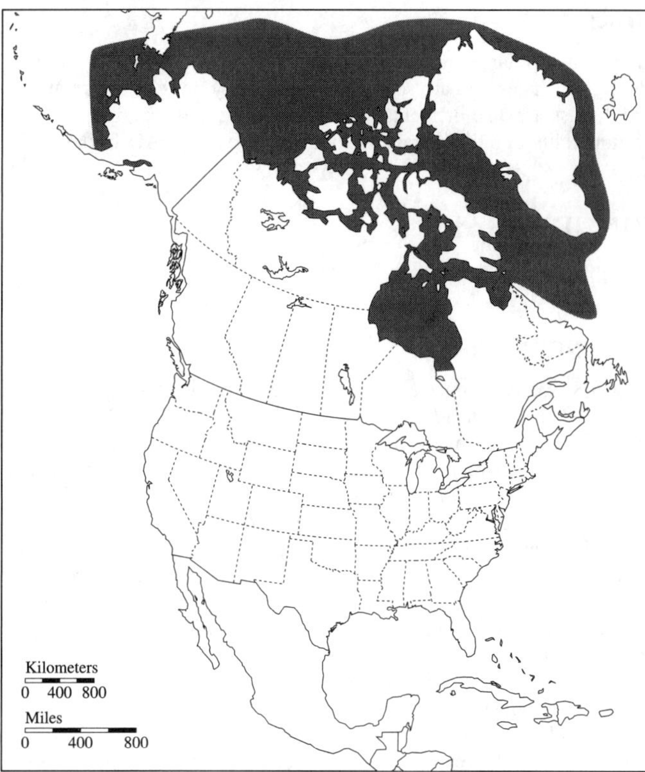

FIGURE 39.10. Distribution of the ringed seal (*Pusa hispida*).

DESCRIPTION

The ringed seal, *P. hispida,* is the smallest North American pinniped. It is similar in appearance to the harbor seal, although its head is rounder and its snout more pointed, which gives its face a feline character (Banfield 1974) (Fig. 39.11). There is some sexual size dimorphism, adult males being slightly larger than females (FAO 1976). There are also some size differences and color distinctions between ringed seals from the North Pacific and those from the Northeast Atlantic and the Arctic. The former are about 1.4 m long, with a mass of about 63–66 kg, whereas the latter are 1.2–1.5 m long and up to 80 kg. Pups are similar in size, being about 0.62 m long with weight of 5 kg. Mass varies seasonally owing to substantial changes in the proportion of blubber (FAO 1976). The layer of blubber reaches a maximum of 40% of the body mass in late autumn and a minimum of 23% during the spring fast (Banfield 1974).

The basic pelage pattern consists mostly of small round to oval spots, which, especially on the back and sides, are circled with lighter rings; generally there are no, or very few, spots on the undersides (Jefferson et al. 1993). The North Pacific ringed seal is gray brown, sometimes with a light greenish yellow tint and light strips along the back and sides forming irregular rings. Northeast Atlantic and Arctic adults vary from brown to gray with an olive shading dorsally to dark gray and almost black. The ventral surface is light gray with silver shading. Light-colored veins are interlaced on the main background, forming the net or lace design that produces oval rings. The newborn coat is fluffy and white, but changes to the short, adult-colored pelage after the first molt by the age of 8 weeks (Popov 1976).

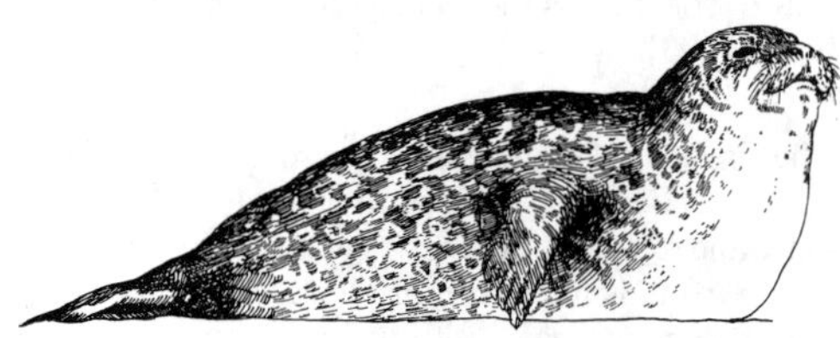

FIGURE 39.11. Ringed seal (*Pusa hispida*).

FEEDING HABITS

Depending on the season and the area, the ringed seal consumes at least 72 different food items (Table 39.3) (McLaren 1958). There is significant seasonal and regional variation in the diet (Lowry et al. 1980). Major prey species include amphids, mysids, euphausiids, and schooling fish such as saffron and polar cods, capelin, and sand lance (Burns 1978), indicating that prey species that occur in concentrations are of particular importance in the overall nutrition of ringed seals (Lowry et al. 1980). Food could be limited in some areas and at certain times if these kinds of prey are not available (Lowry et al. 1980), although patterns of feeding suggest that generally food is not a limiting factor in the ringed seal's distribution and abundance (McLaren 1958). In deeper, offshore waters, polar cod and zooplankton are most commonly eaten, whereas in other areas, small fish and shrimp form the diet (Burns 1978). In the eastern Bering Sea, crustaceans make up a major portion of the diet, especially in young seals (Lowry and Frost 1981). Intense fasting in June and early July, when most of the seals are basking on fast ice, follows a reduction of feeding in the early spring. Feeding is resumed and blubber restored after departure from the ice (McLaren 1958).

BEHAVIOR

The ringed seal is solitary, although it is occasionally observed in loosely dispersed groups (Banfield 1974). Phonations include whining and moaning sounds when on land and a threatening growl when trapped (Banfield 1974). The first author has observed a ringed seal slapping the water with its flipper as he approached. Its two most distinctive behavioral traits are using the strong foreflipper claws to maintain breathing holes in thick, stable ice and to excavate lairs in the snow for resting and parturition (Burns 1978). The haul-out lair is single chambered and round, whereas the birth lair contains an extensive network of tunnels made by the pup (Smith and Stirling 1975). Use of such lairs results from a need for protection from predators and cold. Newborn pups accumulate blubber slowly and thus depend on the lair for thermal protection (Smith and Stirling 1975). Ringed seals do not normally haul out on land in Alaskan waters, but remain on the fast ice (Burns 1978).

MORTALITY

The main enemies of ringed seals, apart from humans, are the polar bear (chiefly) and arctic fox, both of which can locate seal lairs by their smell, and enter the lairs in an attempt to kill and eat the pups. Other predators include killer whales, red foxes, dogs, wolves (*Canis lupus*), wolverines (*Gulo gulo*), and ravens (Hull 1994). Also, walruses sometimes prey on young pups (King 1964; Lowry and Fay 1984). The life span of the ringed seal is believed to be about 40 years (FAO 1976).

RIBBON SEAL

Histriophoca fasciata

NOMENCLATURE

COMMON NAMES. Ribbon seal, banded seal; Alaska Native names are *qasruliq* in central Yupik, *kukupak* in St. Lawrence Island Yupik, and *qaigullik* in northern Inupiaq (Hull 1994); the scientific name of this seal describes its appearance; *Histriophoca* is derived from the Latin word *Histro,* meaning a stage player, and the Greek word *Phoca,* meaning seal; *fasciata* comes from the Latin *fascia,* meaning a band or ribbon (Burns 1978)

DISTRIBUTION

The range of *H. fasciata* is Pacific–Arctic, where it inhabits the pack-ice zone except during the summer, when the species becomes pelagic (Rice

1998). Ribbon seals occur from the East Siberian Sea and the Chukchi Sea, southeast to Bristol Bay and Unaslaska Island, and southwest along the coast of Kamchatka and the Ostrova Kuril'skiye as far as northern Hokkaido, including the Sea of Okhotsk south to Tatarskiy Proliv. Vagrants have been found at Cordova, Alaska, and Morro Bay, California (Rice 1998). In the Bering Sea, the primary hauling-out grounds are in the Anadyr Gulf and adjacent southeastern St. Lawrence Gulf, and on the ice massifs (central ridges) near St. Mathew Island as well as in the Bering Strait (Popov 1976) (Fig. 39.12). During an unusual winter freeze-up, ribbon seals were seen moving overland at Cape Prince of Wales, Alaska (King 1964). They are generally associated with the spring and winter ice front in the Bering Sea and range to only 150 km north of its southern periphery (Fay 1974). Ribbon seal migration consists of passive movement on the ice during the breeding season and a shift to solid ice areas preceding spring molting period (Popov 1976).

In the late 1950s, before intensive commercial hunting by (then) Soviet sealers, the Bering Sea stock of ribbon seals was estimated to be about 120,000. From 1961 to 1967, about 13,000 seals were taken annually from Bering Sea waters, resulting in a population decline. Population estimates in the Bering Sea were 80,000–90,000 in 1964 and 60,000 in 1969. The harvest quota was set at 3000 in 1969. Since then, the population has increased and may be approaching preexploitation levels (Popov 1976; Hull 1994). Burns (1981a) estimated the Bering Sea population at 90,000–100,000. In May 1969, there were also about 133,000 ribbon seals in the Sea of Okhotsk (Popov 1976). The worldwide population has been estimated at 200,000–250,000 animals (Stirling 1979). Current reliable estimates of abundance for ribbon seals are not available.

Habitat. Ribbon seals tend to remain in areas with open water or very thin ice, possibly because of their feeding habits or because they are incapable of maintaining breathing holes (Fay 1974). They usually haul out on firm pack ice, far from shore, where cracks and leads are found. Choice of habitat is obviously influenced by the availability of food and the ice conditions (Popov 1976). In the summer, they stay in ice-free waters near, but usually not in, the permanent ice pack (Fay 1974), which they use for parturition, lactation, breeding, and molting (Popov 1976). The large, heavy claws of the ribbon seal, typical of northern phocids, are an adaptation to its icy habitat (Fay 1974).

FIGURE 39.12. Distribution of the ribbon seal (*Histriophoca fasciata*).

FIGURE 39.13. Ribbon seal (*Histriophoca fasciata*).

DESCRIPTION

Compared to other Bering Sea phocids, the ribbon seal is slender and medium in size (Fig. 39.13). Its eyes are wide (Burns 1971). There is little sexual dimorphism in size, with adults attaining a maximum length of 1.9 m and a mean length of 1.7 m. Adults may weigh up to 100 kg, with the average about 70–80 kg (Popov 1976). Newborns are 0.80–0.90 m in length, with a weight of about 9 kg (Popov 1976). Blubber and skin account for 27–35% of the total body mass (Burns 1978). The pelage shows a striking degree of sexual dimorphism. Two distinctive pelage stages occur depending on age (Burns 1971, 1978). Adult males are dark brown to black with four white or yellowish ribbonlike bands, 10–12 cm in width, circling the neck and the hind end of the body. They form a large ring around each foreflipper, which extends from the shoulder to the midtrunk area. Females are lighter in color, with less distinctive markings (King 1964; Popov 1976). Newborns have a shaggy, wooly, whitish lanugo, and they molt after 4–5 weeks to a pale yellow pelage that lacks ribbons. Subadults are dark gray dorsally and lighter ventrally. The face is pale and the end of the muzzle, the lower jaw, and the chin are dark (Popov 1976; Jefferson et al. 1993). Full adult color is distinct by age 3 years (Burns 1978).

FEEDING HABITS

Ribbon seals can dive 200 m for food. They eat a variety of foods, but demersal fish appear to be most important in the diet (Table 39.3). During breeding and molting periods, this species subsists mainly on crustaceans plus some fish and cephalopods (primarily squids). Young seals feed mainly on crustaceans. Ribbon seals of the Bering Sea have a more diverse diet than those of the Sea of Okhotsk, and consume 14 species of crustaceans, mainly *Pandalus gonionus, Temisto* sp., *Pandalopsis* sp., *Enalus gaimardi,* and amphipods, as well as 10 species of fish, such as polar cod, lumpfish (*Cyclopterus* sp.), navaga (=saffron cod), and capelin (Popov 1976). From March to June, the major prey items are pollock, capelin, and eelpouts, with saffron cod, pricklebacks, snailfishes (*Liparis* sp.), sculpins, shrimps, octopus, and squid also being taken (Burns 1978). The centers of ribbon seal abundance usually coincide with regions where pollock are also abundant, and studies have shown that in regions where pollock are present, they usually constitute the major single prey item (Hull 1994). In winter and spring when ribbon seals inhabit the drifting ice of the Bering Sea, they feed on pollock, and in the Sea of Okhotsk, pollock is a significant food item for ribbon seals (Lowry et al. 1996).

BEHAVIOR

Ribbon seals are solitary and form groups only during the breeding season. However, they are often observed in widely dispersed aggregations during the spring molt (Fay 1974). They frequently rest on the ice far from water, and for long periods of time do not lift their heads to look around. Also, mothers tend to leave their pups unattended for lengthy periods. These traits suggest that ribbon seals mainly occupy regions relatively free of natural predators and humans (Hull 1994).

Ribbon seals are able to move rapidly on ice, using slashing side-to-side motions (Jefferson et al. 1993). When moving on land, they slide or slither across the ice with their head and neck held low. The body is pulled forward by alternate extension of the powerful foreflippers, and the pelvis and hind flippers are moved from side to side (Hull 1994).

MORTALITY

Ribbon seal pups are not subject to high predation by polar bears and arctic foxes because they are born outside their normal range. During their time at sea, however, sharks and killer whales prey on them (Burns 1978). Ribbon seals probably live for 22–25 years (FAO 1976).

GRAY SEAL

Halichoerus grypus

NOMENCLATURE

COMMON NAMES. Gray (or grey) seal, horsehead seal, Atlantic seal; Inuit names include *hodge* and *apa* (Mansfield and Beck 1977)
SUBSPECIES. *H. g. grypus,* found in the eastern and western Atlantic Ocean; *H. g. macrohynchus,* in the Baltic Sea from the gulfs of Bothnia and Finland south to Denmark (Rice 1998)

DISTRIBUTION

The subspecies *H. g. grypus* occurs in the western Atlantic from Cape Chidley in Labrador south to Nantucket Island in Massachusetts, including Newfoundland and the Gulf of St. Lawrence. In the eastern Atlantic, it is found from the Murman coast of Russia southwest along the coast to Stavanger in Norway, in Iceland, the Faeroes, the Shetland Islands, the Orkney Islands, western Great Britain, Ireland, the Netherlands, and the coast of Bretagne, France. It is vagrant north to Disko Bugt in Greenland, and south as far as New Jersey and Portugal (Rice 1998). The two Atlantic stocks of gray seals (western and eastern) are distinguished by geographic isolation and differences in breeding season (FAO 1976).

In North America, there is a well-established gray seal population in eastern Canada (Fig. 39.14), with the largest colony occupying Sable Island and small breeding colonies located at the Magdalen Islands (Deadman Island), on Amet Island in Northumberland Strait, on Point Michaud on the east coast of Cape Breton Island, and on the fast ice along the western Cape Breton shore from the Strait of Canso to Inverness. During the summer, gray seals disperse around the Gulf of St. Lawrence and coasts of Newfoundland and the Maritime Provinces, ranging from Hebron in Labrador to Nantucket Island. Small numbers of gray seals and pups have been observed on several isolated islands along the Maine coast and in Nantucket–Vineyard Sound, Massachusetts, in particular on the islands of Muskeget and Tuckernuck off Nantucket (Mansfield and Beck 1977; Rough 1995; Hannah 2000).

There is limited evidence of migration in adult gray seals, although their breeding and feeding grounds are usually distinct. However, pups tend to disperse from their birthplace to all parts of eastern North America from New Jersey to northern Labrador. Tagging studies indicate that gray seals migrate up to 50 km/day. Most seals probably return to their birthplace as breeding adults, though some may move to other colonies (Mansfield and Beck 1977).

In the 1970s, the total world population of gray seals was estimated at about 88,000–99,000 (FAO 1976). This rose to 120,000–124,000 and was increasing by the late 1980s (Reijnders et al. 1993). Currently, the overall world population is thought to be around 290,000–300,000 individuals (Seal Conservation Society 1999).

In eastern Canadian waters of the North Atlantic, there was a marked increase in the number of gray seals after 1962. By 1979, the numbers had likely climbed to over 43,000 seals, with about 10,000 of these being young of the year (Gray and Beck 1979). Stobo and Zwanenburg (1990) estimated that between 100,000 and 130,000 individuals $\geq$1 year old were on Sable Island, Nova Scotia, in 1986. In 1993, the estimate for Sable Island and Gulf of St. Lawrence stocks was 143,000 animals. Pup production on Sable Island has been about 13%/year since 1962 (Mohn and Bowen 1994), and about 57% of the total western North Atlantic gray seal population is considered to be from the Sable Island stock (Waring et al. 1999).

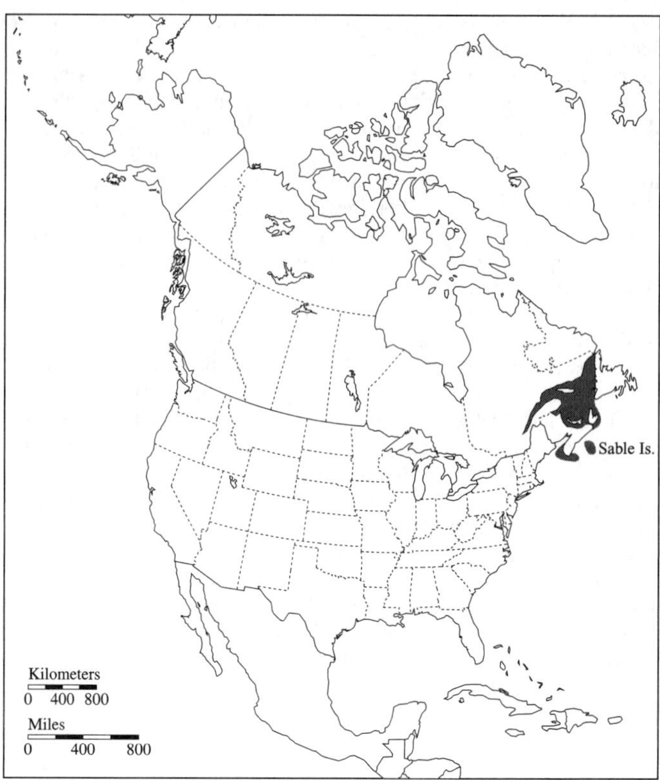

FIGURE 39.14. Distribution of the gray seal (*Halichoerus grypus*).

In the United States, gray seals in waters off Maine have increased from about 30 in the early 1980s to between 500 and 1000 in 1993, and recently 29–49 pups/year have been recorded in Penobscot Bay. For a winter breeding colony on Muskeget Island, west of Nantucket Island, counts taken during the spring molt in the 1970s never exceeded 13 any year. These rose during the 1980s, however, and reached 1549 in 1993. Aerial counts in April and May 1994 recorded a peak count of 2010 gray seals for Muskeget Island and Monomoy (Cape Cod) combined. This count is thought to include an unknown fraction of gray seals from Sable Island that were overwintering in the Nantucket Sound region. As such, percentage increase in abundance for U.S. waters is unknown, but the population is certainly thought to be increasing. Five pups were born at Muskeget in 1988, and this increased to 59 in 1994 (Rough 1995; Waring et al. 1999).

Habitat. The gray seal is essentially a coastal species, moving out to sea only to feed. It is generally associated with heavily indented, rocky coasts where there are small islands and reefs, but has also been observed in some estuaries and lagoons where sandbanks are found (Mansfield and Beck 1977). Within local populations of gray seals, there has been some divergence in certain life history and behavioral traits. For example, those in the Gulf of St. Lawrence whelp on shorefast sea ice, whereas gray seals east of Nova Scotia whelp on ice-free islands (Davies 1957; Mansfield and Beck 1977). This suggests that the species can assume pagophilic habits under heavy colonization pressure.

DESCRIPTION

The gray seal is a moderately large pinniped whose long, broad, straight snout gives it an equine appearance (Fig. 39.15). Its flexible forelimbs

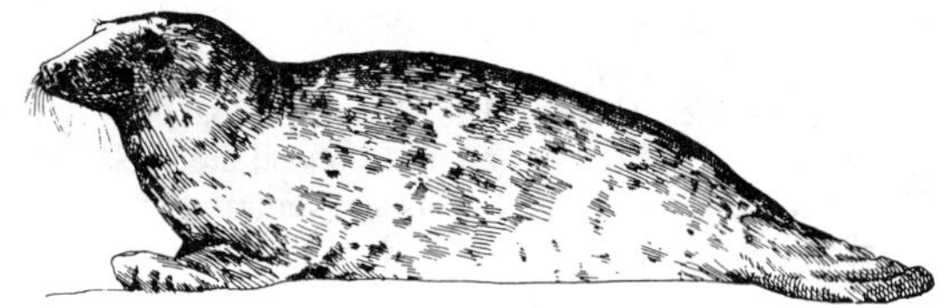

FIGURE 39.15. Gray seal (*Halichoerus grypus*).

and long, slender claws make it the most adept on land of all phocids (Banfield 1974). There is marked sexual dimorphism in coat pattern, size, and appearance (FAO 1976). More massive than the female, the adult male has a swollen, wrinkled neck and a broad, slightly "roman" snout. Older males acquire a prominent sagittal crest (Banfield 1974). Males average 2 m in length and 233 kg in weight; females are about 1.8 m in length and weigh about half as much as the males (FAO 1976; Mansfield and Beck 1977). Newborn pups are 0.90–1.0 m long. Body weight ranges from 11 to 20 kg, and averages 14.5 kg (FAO 1976; Jefferson et al. 1993). After 2–4 weeks, pups molt their silky, creamy white lanugo and become blue-gray dorsally and paler ventrally. With successive annual molts, they gradually achieve the adult color and pattern. Adult coat color varies greatly, with many shades of dark and light gray, brown, and silver (King 1964). Adults are generally darker dorsally, shading lighter ventrally, and males tend to be darker than females (FAO 1976).

FEEDING HABITS

Essentially a coastal species, *H. grypus* spends part of its time at sea feeding on shallow, offshore fishing banks. It feeds primarily on bottom fish, but takes several kinds of demersal, anadromous, and pelagic species during their inshore migrations (Table 39.3). In fact, more than 40 taxa of fish and invertebrates were identified in gray seal stomachs examined from eastern Canada (Benoit and Bowen 1990). In general, however, a high percentage of the biomass consumed by gray seals is accounted for by only a few species. In Canada, near Sable Island, on the eastern Scotian Shelf, for example, in stomach samples collected, two to four species made up more than 80% of total prey biomass eaten (Bowen and Harrison 1994).

Occasionally the gray seal will eat invertebrates, particularly squid and epibenthic crustaceans such as crabs and shrimp, but rarely lobsters. The most important food species available all year are benthic skates and flounders (flatfishes), whereas herring, Atlantic cod, squid, and mackerel (*Scomber* sp.) become important when they begin their inshore migrations in the spring (Mansfield and Beck 1977). They occasionally take seabirds (Jefferson et al. 1993). Gray seals fast during the breeding season—females for at least 2 weeks and adult males even longer. Feeding is minimal during February, March, and April when important food species such as herring, cod, and mackerel are not abundant inshore. Gray seals eat about 3% of their body weight daily. The Canadian gray seal stock has been reported as eating approximately 47,083 tonnes/year. These seals "compete" for food with harbor seals, some porpoises and larger cetaceans, and migrating harp and hooded seals early in the year. The quantities of food they consume are not considered to present significant competition to commercial fisheries (Mansfield and Beck 1977), although they are a nuisance factor to fishermen on a local basis. Gray seals are the preferred host of the codworm, *Terranova decipiens,* which causes economic loss in the inshore fishery.

BEHAVIOR

The gray seal is very gregarious, feeding in groups and hauling out to breed and molt in dense colonies. It dominates the smaller harbor seal, taking over the best hauling-out spots at low tide. It drives off intruders with a variety of phonation calls, from hisses and snarls to short barks and mournful hoots (Banfield 1974). Gray seal males tend to be polygamous and fiercely territorial. A harem is usually made up of two or three females and one male, but most females lie widely dispersed, each with a male in close attendance. Older females often return to specific spots where they bore pups in previous years (Bruemmer 1979), a behavior that allows for selected population control using a chemosterilent. This is one of the few pinnipeds that is susceptible to birth control techniques. The gray seal, like other seals, has been observed sleeping on the bottom in shallow water and apparently surfaces to breathe without awakening.

MORTALITY

Aside from predation by humans, there is only the insignificant loss of gray seals by killer whales (FAO 1976), and possibly sharks. Records exist of a 46-year-old female (Bonner 1971), and a male gray seal 26 years old has been reported (Platt et al. 1974).

HOODED SEAL

Cystophora cristata

NOMENCLATURE

COMMON NAMES. Hooded seal, crested seal, bladdernose seal (King 1964)

DISTRIBUTION

The hooded seal inhabits the pack-ice zone of the North Atlantic from the Gulf of St. Lawrence, insular Newfoundland, and Labrador in the west (Fig. 39.16) to the eastern limit near Novaya Zemlya and Kanin Peninsula in the Barents Sea (Popov 1976). Hooded seals in the winter occur along the Atlantic coast of the United States to Long Island, New York, and into New Jersey (Hannah 2000).

The population is concentrated in three discrete areas during the breeding season: off the coast of Newfoundland–Labrador and in the

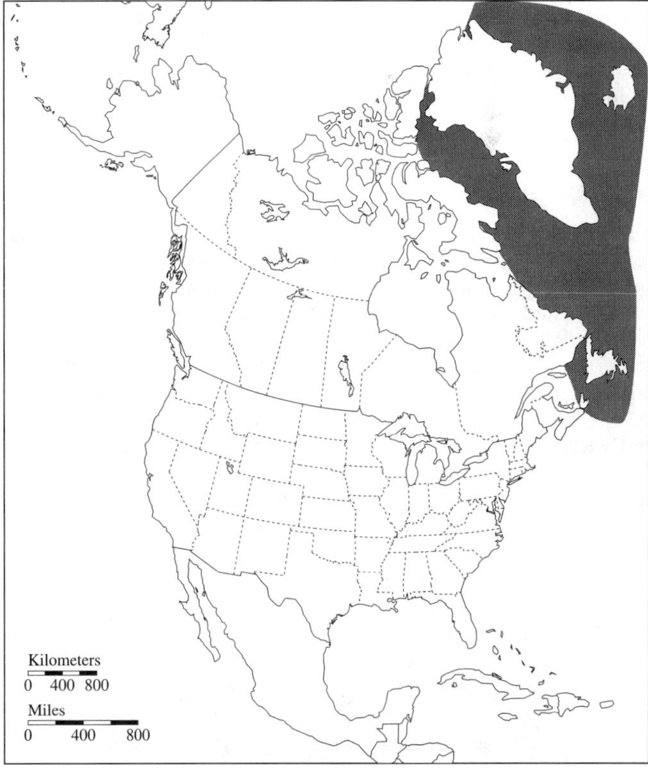

FIGURE 39.16. Distribution of the hooded seal (*Cystophora cristata*).

Gulf of Saint Lawrence (these can be split into two breeding herds, Front and Gulf, respectively); in the Davis Strait; and on the West Ice (around Jan Mayen Island) off eastern Greenland (Sergeant 1976a; Rice 1998). They assemble in February, and parturition and subsequent copulation occur at about the same time in March for each area. By July and August, the seals have migrated to their molting grounds east of Greenland in Denmark Strait and also from 72° to 74° N (Sergeant 1976a). Hooded seals move from Newfoundland to southern, southwestern, and southeastern Greenland for the summer (Sergeant 1976a). For management purposes, two stocks are recognized, one in the Northwest Atlantic and the other in the Greenland Sea.

The hooded seal shows a great tendency toward wandering. A number of stragglers have moved west through the Canadian Arctic Ocean to the Beaufort Sea and thence south through the Bering Strait to southeastern Alaska and even to southern California. To the south, they have traveled along the Atlantic seaboard of the United States as far as Florida, Puerto Rico, and the Virgin Islands and down the European side as far as Denmark, the British Isles, the Bay of Biscay in Portugal, and southwestern Spain (Rice 1998).

The total hooded seal world population is thought to number at least 500,000–600,000 (Popov 1976, 1982). The actual number in the western North Atlantic is unknown, although seasonal abundance estimates are available. There have been several pup production estimates for hooded seals off Newfoundland–Labrador (Front herd), ranging from 25,000 to 32,000 annually between 1966 and 1977 (Benjaminsen and Øritsland 1975; Sergeant 1976b; Lett 1977; Winters and Bergflodt 1978; Stenson et al. 1996b), with a slight decrease to 26,000 in 1978 (Winters and Bergflodt 1978), followed by increases after 1978 to reach 62,000 by 1984 (Bowen et al. 1987). Hammill et al. (1992) estimated pup production was about 82,100 in 1990. Considering recent pup production estimates, the total hooded seal population in the Gulf and Front herds could be 400,000–450,000 individuals (Stenson 1994; Stenson et al. 1996a). For the Davis Strait area, Bowen et al. (1987) estimated pup production at 18,600 in 1984. In general, though, there are insufficient data to permit adequate evaluation of current stock size.

Habitat. A pelagic animal, *C. cristata* inhabits areas of open sea and drifting shore ice (Popov 1976). Its habitat is limited by the fact that it dives deeply for its food, which consists of larger organisms than that of the harp seal, whose range is similar (Sergeant 1976a). The only time this species remains sedentary is during its breeding and molting periods, when it is concentrated on two areas of drift ice (Popov 1976).

DESCRIPTION

The most distinguishing feature of the hooded seal is the male proboscis (Fig. 39.17). This nasal sac is actually the greatly enlarged skin of the snout, which is inflated in times of anger or excitement (Dunbar 1949). Mature males, intolerant of each another, may in anger also extrude a fiery red bladder through the nostril, which is formed by the inflation of a very elastic portion of the internasal septum (Dunbar 1949). Sexual dimorphism is evident: adult males are up to 2.8 m long and weigh >300 kg, whereas females are up to 2.3 m (average about 2 m) long and weigh 145–300 kg. Newborn pups are between 1.0 and 1.5 m long and weigh 20–30 kg (Popov 1976; Jefferson et al. 1993). The adult pelage is silvery gray with scattered black blotches and spots (Fay et al. 1979) and tends to be lighter ventrally (Dunbar 1949). The head is dark in both sexes from merged blotches, creating a hooded appearance (Jefferson et al. 1993). The embryonic coat is shed in utero, so pups are dark gray with silver and blue tints on the back, sides, and dorsal sides of the back flippers; the chest and ventral surface are white. Following postpartum molt at 1 year, adult markings appear (Popov 1976).

FIGURE 39.17. Hooded seal (*Cystophora cristata*).

FEEDING HABITS

Because of their deep-diving capabilities, their food is mainly squid (notably *Gonatus fabricii*), redfish (*Sebastes* spp.), capelin, and polar cod (Table 39.3) (FAO 1976). In the Barents Sea, bottom fishes like halibut, cods, redfish, and flounders are taken (Popov 1976). In West Greenland waters, in addition to small fish species like capelin, sand eel, and polar and Arctic cods, larger demersal fishes such as Greenland halibut, Atlantic cod, redfish, and wolffish (*Anarhichas minor*) are apparently important prey items (Kapel 2000). Pups initially feed independently near the ice edge on crustaceans and squid (Popov 1976), with the pelagic crustacean *Parathemisto* spp., especially *P. libellula,* being dominant. Diet also includes *Gammarus* spp., with krill (*Thysanoessa* sp.) of variable importance in yearly samples (Haug et al. 2000). Overall, nutritional habits of hooded seals have no marked effect on any commercial fishing interests (FAO 1976).

BEHAVIOR

Hooded seals spend most of the time swimming in deep water and are very active (Maxwell 1967); when hauled out on land to whelp or molt, they form "families" consisting of a female, her pup, and one or several competing males (FAO 1976). Females are protective and will actively defend their pups. The male hooded seal is often maligned for supposed aggressiveness. Such aggression only occurs toward humans, when they stand upright near the female and neonate. The seal's vocal repertoire is limited to a few sounds (Terhune and Ronald 1973). Proboscis inflation and nasal bladder extrusion by the male are thought to be behavioral displays (Maxwell 1967).

MORTALITY

Major causes of death in hooded seals include predation by polar bears at whelping sites, predation by Greenland sharks in the molting areas, heartworm infection (FAO 1976), and humans. A hooded seal lives for approximately 20 years (Sergeant 1976a).

BEARDED SEAL

Erignathus barbatus

NOMENCLATURE

COMMON NAMES. Bearded seal, square flipper (because of the unusual shape of its foreflippers) (Banfield 1974); Native names are *mukluk* in Yupik and *oogruk* in Inupik (Hull 1994), and *ugjuk* (or *udjuk*) in Inuktitut; *Erignathus,* derived from Greek, refers to this animal's rather deep jaw, and *barbatus* is from the Latin word *barba,* meaning beard, and refers to its numerous vibrissae

SUBSPECIES. Two intergrading subspecies are recognized, *E. b. barbatus,* in the eastern Arctic and subarctic; and *E. b. nauticus,* in the western Arctic and subarctic (Burns 1967, Rice 1998)

DISTRIBUTION

The bearded seal occurs in the circumpolar region of the moving pack ice and along all coasts of northern Eurasia and northern North America (Fig. 39.18). The range of the subspecies *E. b. barbatus*

FIGURE 39.18. Distribution of the bearded seal (*Erignathus barbatus*).

includes suitable habitat from the central Canadian Arctic, east to the central Arctic coast of Eurasia. It ranges north to Jones Sound in the Canadian Arctic Archipelago, Kap York in western Greenland, Nordostrundingen in eastern Greenland, and Svalbard, Zemlya Frantsa, Iosifa, and Novaya Zemlya; and south to James Bay, northern Newfoundland, Kap Farvel in Greenland, Iceland, Jan Mayen, Bjørnøya, and Vesterålen in northern Norway. This subspecies has also occasionally been found in the St. Lawrence estuary, Cape Cod, British Isles, France, Spain, and Portugal (Rice 1998). In general, bearded seals rarely travel south of northeastern Newfoundland (Hannah 2000), and are considered relatively rare on the Labrador coast (Stenson 1994).

The subspecies *E. b. nauticus* occurs from the central Canadian Arctic Archipelago westward to the Laptev Sea (Burns 1967). It ranges north to Paluostrov Taymyr, Severnaya Zemlya, Novosibirskiye Ostrova, Ostrov Vrangelvy, Banks Island, and Victoria Island; and ranges south to Karaginskiy Zaliv in Kamchatka and Bristol Bay in Alaska. A disjunct population occurs in the northern and western Sea of Okhotsk, south to Tatarskiy Proliv and northern Hokkaido. This subspecies is vagrant to Zhejiang, China, and Honshu, Japan (Rice 1998). In North America, the largest concentrations of bearded seals are in the Bering Sea on the inshore ice of St. Lawrence, St. Mathew, and Hall Islands (Popov 1976). In the Bering and Chukchi Seas area, they stay in the region of the Bering–Chukchi platform, where they can reach the bottom to feed. Their seasonal movements and distribution patterns are primarily the result of ice conditions overlying this platform.

From January to April, *E. barbatus* is sparsely distributed throughout the Chukchi and northern Bering Seas. A northward migration, which follows the ice edge to the Bering Strait area, begins in April. The southward fall migration is concurrent with the southward movement of sea ice (Burns 1967). In Canada, in many areas, bearded seals are sedentary with only local movements in response to ice conditions. Otherwise, they follow the seasonal advance and retreat of ice cover (Cleator 1996). In central Labrador, for example, they migrate north in the summer and return in the fall (Schwartz 1977).

The Bering Sea population has been estimated at approximately 250,000 (Popov 1976). A large group in the neighboring Sea of Okhotsk could comprise about 200,000 animals. It is also thought that the current population of bearded seals in Canadian waters is stable, and is estimated at about 190,000 (Cleator 1996). Overall, the total world population probably exceeds 500,000 animals (FAO 1976; Stirling and Archibald 1979).

Habitat. A seal's habitat is based on food resources and breeding and hauling-out sites. The bearded seal is a benthic or bottom feeder and therefore is restricted to relatively shallow water. Such regions have heavy offshore ice that is in motion because it is influenced by winds, currents, and coastal features. Such a region is often known as a "flaw zone" (Burns 1978). Animals remain in these open water situations until their breathing holes are destroyed by shifting ice (Burns 1967). Adult bearded seals are almost always associated with ice. Young seals, however, may sometimes remain in ice-free areas where they frequent bays and estuaries (Hull 1994).

DESCRIPTION

The bearded seal is characterized by a conspicuous "moustache" composed of long, regular vibrissae, which tend to curl at the tips when dry (Banfield 1974) (Fig. 39.19). Its foreflippers are an unusual shape, with the digits almost equal in length. Other distinguishing features are its prominent ear orifices and thickened neck (Banfield 1974). The bearded seal is the only northern phocid with four retractable mammae instead of two (Jefferson et al. 1993). Males and females are similar in size and weigh ≥300 kg. The mean length is 2.2 m and the maximum is 2.6 m. Newborn pups are 1.2–1.4 m long and weigh 27–35 kg (Popov 1976). The pelage of the pup is dark gray-brown, with whitish dorsal blotches and forelimbs. Adults are pale grayish to buff with a slightly darker saddle and are sometimes a rusty color around the head and neck (Fay et al. 1979). This brown facial and upper body color is thought to be a result of mud stains caused by their benthic feeding habits (FAO 1976).

FEEDING HABITS

The bearded seal is a benthic feeder, consuming mainly epibenthic organisms, that is, those that live on the surface of the sea floor. It also feeds on organisms that live in the bottom sediments. Its major prey items are a variety of crabs, shrimp, clams, benthic fish, and schooling demersal (near-bottom-dwelling) fish (Table 39.3). The feeding habits of this pinniped restrict it to waters <200 m deep, such as the Bering–Chukchi continental shelf (Burns 1978). In the eastern Bering Sea, clams, crabs, and shrimp were the bulk of the diet of bearded seals (Lowry and Frost 1981). They do eat some pollock, although this prey species is of little importance in the diet. More pollock is consumed in the southern part of the Sea of Okhotsk than in the north (Lowry et al. 1980). The average volume of food found in a single bearded seal stomach was 854 ml, including sand, pebbles, parasites, and food items. Certain food items may aid in the removal of stomach parasites (Burns 1967).

BEHAVIOR

Erignathus barbatus is not gregarious and is rarely observed in large numbers (Banfield 1974). When basking, it remains widely distributed and faced away from others. Old males are often seen fighting on the ice floes, using their foreflippers as weapons; frequent observations of

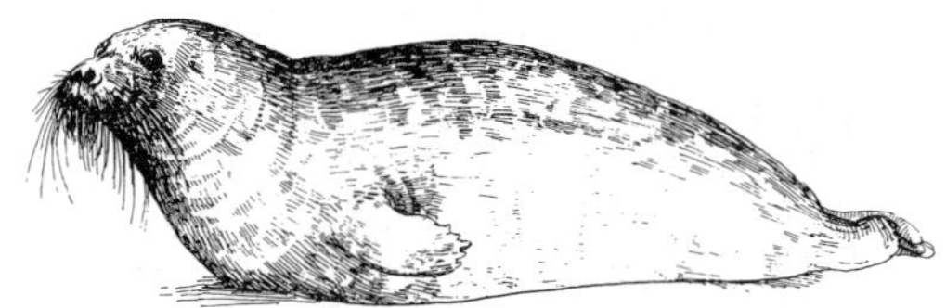

FIGURE 39.19. Bearded seal (*Erignathus barbatus*).

scarred seals of both sexes further confirm their asocial traits (Burns 1967). The bearded seal is a curious animal, but when surprised it becomes immobilized with fear and is easy prey for a polar bear or Inuit (Maxwell 1967). During the spring breeding season, adult males begin a loud, distinct, and highly characteristic "singing" underwater, which is a complex, frequency-modulated whistle, parts of which are audible to humans. This "song" is thought to be connected with courtship behavior (Burns 1967; Hull 1994).

MORTALITY

The major predators of the bearded seal are polar bears and humans (Burns 1978); occasional deaths are attributed to killer whales (FAO 1976). The oldest seal found was 31 years old.

HAWAIIAN MONK SEAL

Monachus schauinslandi

NOMENCLATURE

COMMON NAMES. Hawaiian monk seal, Laysan seal (King 1964)

DISTRIBUTION

The breeding range of the Hawaiian monk seal is essentially limited to the atolls and islands of the Northwestern (Leeward) Hawaiian Islands from Nihoa Island to Kure Atoll in the Hawaiian Archipelago (Fig. 39.20). The six main breeding sites are Kure Atoll, Midway Atoll, Pearl and Hermes Reef, Lisianski Island, Laysan Island, and French Frigate Shoals. About 40–50% of the total population and total number of births occur at French Frigate Shoals. A few births also occur regularly at Nihoa and Necker Islands. The monk seal also wanders to Maro Reef and Gardner Pinnacles (Rice 1998; Marine Mammal Commission 2000).

There is no evidence that monk seals were ever abundant throughout the main Hawaiian Islands, and there are no records of monk seal presence in Polynesian history. It seems likely, however, that before the arrival of the Polynesians, all of the main Hawaiian Islands were important breeding areas. A number of sightings have occurred on Kauai and, in 1991, two births were recorded in the main Hawaiian Islands, on Oahu and Kauai. In 1997, about five births occurred in the main Hawaiian Islands east of Niihau (Marine Mammal Commission 2000). Overall, as many as 70 monk seals may be scattered along the coasts of Hawaii, Oahu, Maui, and Kauai, according to sources at NOAA Fisheries (Anonymous 2000). The monk seal is also occasionally found on Wake Island, Johnston Island, and Palmyra Island (Rice 1998).

The population was considered nearly extinct by the 1900s; none were seen on Laysan Island during a 1911 expedition (Dill and Bryan 1912). From then, the remnant population on the Northwestern Hawaiian Islands remained essentially undisturbed until the 1940s and World War II. A naval station was built at Midway Atoll, and while the monk seal colonies on other atolls were apparently thriving, the colony at Midway dwindled to about 70 animals by 1957 (Kenyon 1972). In the late 1950s, approximately 1200 individuals were found during the first systematic beach counts of monk seals throughout their range (Kenyon and Rice 1959; Rice 1960). By 1968, the regularly breeding colony at Midway had disappeared (Kenyon 1972). Population counts in 1976 and 1977 were 695 and 625 seals, respectively, with total population estimates considered to be about 1000 (Kenyon 1978b). In 1983, when the total population (including pups) was estimated at nearly 1500 animals, beach counts were roughly half those recorded in 1958. A new estimate of about 1750 seals was reported in 1988, derived from beach counts. At present, it appears the monk seal population is perhaps 1300–1400 animals. In summary, the status of the Hawaiian monk seal population is variable. Some colonies are stable, some are declining, and some are increasing (Marine Mammal Commission 2000).

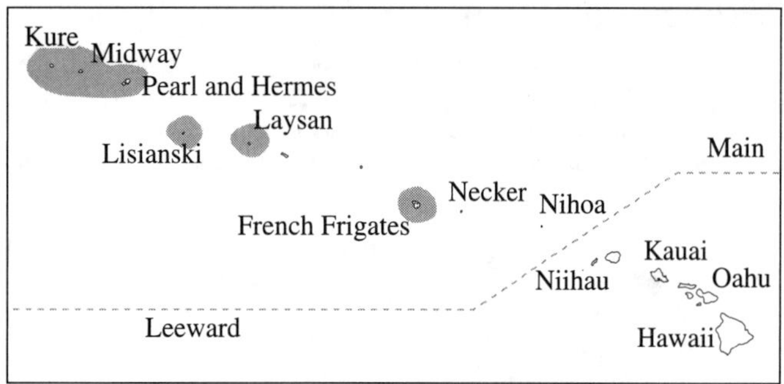

FIGURE 39.20. Distribution of the Hawaiian monk seal (*Monachus schauinslandi*).

Habitat. Critical habitat of the Hawaiian monk seal consists of dry coral sand beaches bordered by sheltering vegetation and shallow, calm water. Here the young can learn to swim and the seal can thermoregulate during the heat of the day in wet sand or shallow water. On cold, windy nights it can seek shelter and warmth on the dry sand and under the vegetation. It feeds in the surrounding atoll lagoons and shallows. This type of habitat is essentially undisturbed by humans and their activities because of its isolation. This factor has enabled the seal to survive after an intense period of overexploitation (Kenyon 1973).

DESCRIPTION

Monk seals, the most primitive living seals (Kenyon 1978b), are the only phocids that live permanently in tropical waters. The Hawaiian monk seal has evolved in remote oceanic islands, as a trusting and tame animal. This is a detriment to its survival, because human disturbance now threatens its existence. Physiologically, the seal has not made great adaptations to the warm climate. The blubber layer is the same thickness as that of other phocids. The pelage of the pup is less woolly and silkier, however, than that of its northern relatives (King 1964). The head is relatively small, and the body is robust, with short flippers (Fig. 39.21). The muzzle is wide and compressed from top to bottom. The nostrils are situated on top of the muzzle, unlike any other North Pacific phocid. There are four retractable mammary teats (Jefferson et al. 1993).

Monachus schauinslandi exhibits sexual dimorphism in size, with females growing slightly larger than males as adults and also varying more in weight (King 1964). Typical adult males have an average weight of 200 kg and reach lengths of about 2.1 m. Females reach 272 kg in weight and 2.4 m in length. Newborn pups are about 1 m long and weigh 16–18 kg (Jefferson et al. 1993). The adult pelage is light silvery gray ventrally and slate gray dorsally, with males tending to be darker than females. At 3–5 weeks of age, certainly completely by the sixth week, the pups lose their soft, black birth coat; their new pelage is silvery blue dorsally, shading to silvery white ventrally (King 1964). Adults and juveniles can have a greenish or reddish cast from algal growth. The first molt is a shedding of individual hairs, but each successive annual molt is a more dramatic epidermal molt of hair and skin, which detaches in patches (Jefferson et al. 1993).

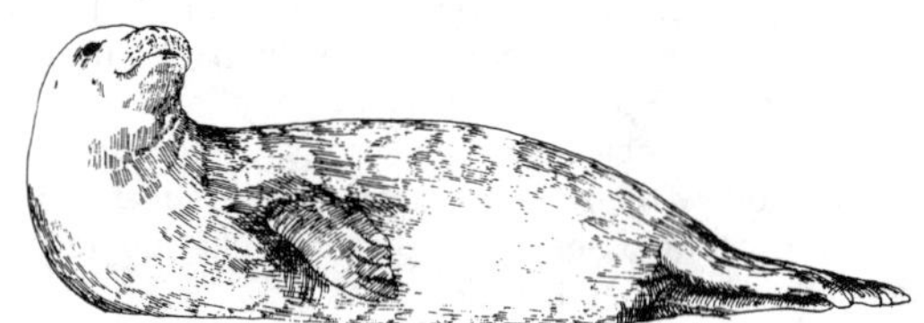

FIGURE 39.21. Hawaiian monk seal (*Monachus schauinslandi*).

FEEDING HABITS

Hawaiian monk seals feed on bottom-inhabiting forms such as octopuses and crayfish, as well as a variety of reef fish (Table 39.3). Eels, which are numerous in the atolls, are an important food item (Kenyon and Rice 1959). These seals also feed on lobster (*Panulirus* and *Scylarides* spp.), although to what degree is not known (MacDonald 1982; DeLong et al. 1984). While at sea, monk seals may travel many kilometers among islands, so they probably forage in the open sea as well as in the shallow lagoons, where they have been observed eating at all hours of the day and night. The diverse diet and wide-ranging movements suggest that the monk seal is an opportunistic forager or foraging generalist, eating prey as it is encountered (Goodman-Lowe 1998). In general, most foraging dives are of short duration and relatively shallow. Dives commonly range from 10 to 15 min (Kenyon 1978b). At Lisianski Island, more than 60% of recorded dives were <40 m and the rest were <150 m (DeLong et al. 1984; Schlexer 1984). At French Frigate Shoals, nearly all foraging occurred on the relatively level 50- to 100-m terraces on the deep slopes. Some seals forage below the photic layer (>300 m), where they may be visiting the few precious-coral beds (300–500 m deep) found in the area. These long-duration dives last around 17 min (Parrish et al. 2000).

BEHAVIOR

Hawaiian monk seals spend most of their time in the relatively shallow water around the reefs they inhabit or basking in the sun on the sandy beaches. They are generally solitary; even when gathering together on land, they are not gregarious. Only mothers and pups regularly make physical contact. Nocturnal in their feeding activities, monk seals are most likely on land in the afternoon, though some are in the water at all times of the day (Kenyon and Rice 1959). Because of its isolated environment, the Hawaiian monk seal evolved free of terrestrial enemies and is therefore innately docile and easily approached by humans, a trait that could be important in its survival (Kenyon 1978b). A characteristic habit of the monk seal is its tendency to roll and root on coral sand beaches until a 15-cm depression is formed. These "sand wallows" are always evident on beaches frequented by this species, and the seals resting in them have faces and eyes caked with sand from using their heads to burrow (Kenyon and Rice 1959).

MORTALITY

Human disturbance of Hawaiian monk seals during the nursing period may affect survival of the neonates. Greater pup mortality has been observed on islands occupied by humans. Seals are often seen with healed scars from shark bites. The rates of shark predation vary among islands, and the main predators are Galapagos sharks (*Carcharhinus galapagensis*), tiger sharks (*Galeocerdo cuvier*), grey reef sharks (*Carcharhinus amblyrhynchos*), and whitetip reef sharks (*Triaenodon obesus*). Pups are particularly vulnerable; for example, it is possible that 30% of pup mortality at French Frigate Shoals in 1998 was due to sharks (Marine Mammal Commission 2000). Galapagos shark and tiger shark predation, in particular, appears to cause significant injury and mortality in some areas.

Naturally occurring biotoxins cause deaths. Between December 1977 and July 1978, 22 seals died from ciguatera poisoning on Laysan Island, and an additional 7 seals were found dead on other islands (DeLong 1978a). Ciguatoxin is a toxic substance produced by a dinoflagellate, concentrated through the food chain so that the flesh of affected fish is highly toxic to mammals. Outbreaks often occur after coral reefs have been disturbed.

Availability of food resources may be limiting the recovery of monk seals at French Frigate Shoals, and prey availability may be affected by overfishing and natural factors. Information on prey species and feeding areas is scant; studies are ongoing to identify prey eaten by monk seals and investigate distribution, abundance, and productivity of prey species.

Adult females and immature seals of both sexes are subject to injury or death from "mobbing" by adult males attempting to mate. Mobbing incidents have been most apparent at Laysan Island, but have also been seen on Lisianski Island and French Frigate Shoals, with incidents apparently increasing in recent years (Hiruki et al. 1993; Atkinson et al. 1994). Such imbalances in the adult sex ratio are more likely to occur when populations are reduced (Starfield et al. 1995). The life span of this species is unknown, although a female aged 11 years and a male aged 20 years have been recorded (FAO 1976).

CARIBBEAN MONK SEAL

Monachus tropicalis

NOMENCLATURE

COMMON NAMES. Caribbean monk seal and West Indian monk seal; early explorers called it "sea-wolf"

DISTRIBUTION

Probably extinct, *M. tropicalis* formerly occurred on small islets and cays in the Caribbean region. The last reliable report for this species was from Serranilla Bank in 1952. Known localities of occurrence were an islet off northern Veracruz, Arrecifé Triangulos off Campeche, Arrecifé Alacran off Yucatan, the Dry Tortugas, Key West, Cay Sal Bank in the Strait of Florida, Cape Canaveral in Florida, the Bahamas, Isla de Providencia, Isla de Juventud (=Isla de Pinos), Rosalind Bank, Serranilla Bank, Pedro Cays south of Jamaica, Isla Alto Velo (near Isla Beata) south of Hispaniola, and Guadeloupe (Rice 1998). As indicated by bones recovered in Indian middens, the species also ranged to Pinellas, Lee, Dade, and Brevard Counties in Florida; Cumberland Island in Georgia; Puerto Rico; St. Eustatius; and Nevis (Wing 1992).

Historically, the Caribbean monk seal was quite abundant. Early reports mention the capture of 100 seals in one night on the Bahama Islands in the early eighteenth century. In January 1911, about 200 seals were killed on Arrecife Triangulos (Mexico), but only 4 were seen there in 1948 and none in 1950. They were easily taken by hunters, such that even by the late 1880s, they were so scarce as to be called an "almost mythical species" (Allen 1887). The last recorded Caribbean monk seal in the United States was killed in 1922 off the coast of Key West in Florida. In 1952, there was a small breeding colony on the Serranilla bank off the Yucatan Peninsula of Mexico (halfway between Jamaica and Honduras), which was thought to be the last (Rice 1973).

The species is now considered extinct, but there is always a remote possibility that it continues to exist. A few searches have been undertaken. No seals were seen during aerial surveys flown in 1950 over reefs north of Yucatan or in 1969 off the Chincorro Reef, eastern Yucatan. Circulars in English and Spanish were distributed about 1973 offering a $500 reward for information on recent sightings of the species, but none were reported. Aerial surveys covering 6377 km, including islands and atolls off Campeche, Yucatan, Quintana Roo (Mexico), Belize, Honduras, Nicaragua, and the central Caribbean to Jamaica, produced no monk seals or evidence of monk seals (Kenyon 1977).

In 1974, at least two seals were sighted by fishermen in international waters near Cay Verde and Cay Burro in the southern Bahamas (22° N, 75° W), an area 300 km northwest of Haiti and 50 km east of Ragged Island on the edge of the Great Bahamian Bank. In 1979, a seal was sighted at Tamarindo Beach in Aguadilla (Puerto Rico). However, a survey conducted from 13 to 25 April 1980 of several sandy islands in the southeastern Bahamas between the Dominican Republic and Nassau revealed no seals. It was concluded that a seal was present at the mouth of the Baie de l'Acul in 1981 in the region of Ile Rat (north coast of Haiti), but it was not possible to determine whether it was a Caribbean monk seal. In September 1984, a search of island groups off the north coast of the Yucatan Peninsula (Islas Triangulos, Cayo

Arenas, Arrecife Alacran, and Cayo Arcas) showed signs of frequent human visits, but no seals (Le Boeuf et al. 1986).

During 1997, an assessment was made of the likelihood that monk seals survive in the region of northern Haiti and Jamaica. Of 93 fishermen asked to select marine mammals known to them from randomly arranged pictures, 22.6% choose the monk seal. When these respondents were questioned further, in particular about the size and color of the seal, results suggested that 16 of the 21 had seen at least one monk seal in the past 1–2 years (Boyd and Stanfield 1998). It is more plausible that sightings of extralimital arctic seals account for reports of phocids in this area (Mignucci-Giannoni and Odell 2001). There is also the possibility that some could actually be sightings of escaped California sea lions (*Zalophus californianus*), but nothing is confirmed. In 1999, a visit to likely sites for monk seals on the eastern coast of Cuba and interviews with local fishermen revealed no reports or recollection of seals ever being sighted. It seems probable that monk seals have not been present in eastern Cuba for several decades at least (Reijnders 2000).

Habitat. The Caribbean monk seal required some shallow lagoons and reefs in which to feed, beaches on which to haul out, and sheltered beach areas on which to pup. Preferred habitat was islands abounding in coral rocks and sand. The monk seal habitat was, and is still, increasingly invaded by tourists and yachtsmen, making secluded undisturbed beaches difficult to find.

DESCRIPTION

Gray (1850) made the first scientific description of the species, nearly 350 years after its discovery by a European: Columbus in 1494. Females are slightly smaller than males. Nose–tail lengths are available for four adult females (220, 216, 211, and 198 cm), and one male (226 cm). They are generally described as 2–2.5 m in length from nose to tail.

Coloration of adults is given as grayish brown or grizzled on the back, ochreous yellow to yellowish white ventrally. Females seem to have much less yellow or white on the ventral surface. Caribbean monk seals were sometimes spotted with a green-flecked back, probably caused by an algal growth. The flippers are umber-brown, lips are bordered with white and whiskers are described as yellowish white. Pups are intense, ebony black in color. In adults, the hairs of the coat lie close to the body and are extremely short; the longest, those of the sides, are about 1 cm long. Pups are born with a woolly coat. The pelvis is short, and shoulder blades short and broad (Fig. 39.22).

FEEDING HABITS

The mainstay of the diet was likely slow-moving reef fish.

BEHAVIOR

Nothing is known about the behavior of the Caribbean monk seal, except to speculate that it would have been similar to that of *Monachus schauinslandi* and/or *Monachus monachus*.

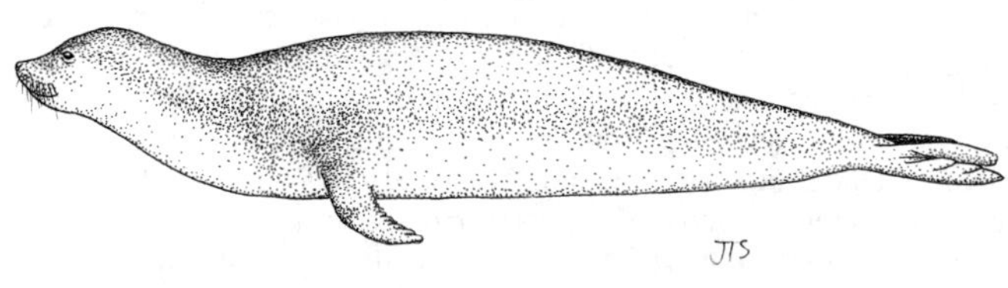

FIGURE 39.22. Caribbean monk seal (*Monachus tropicalis*).

MORTALITY

Sharks were considered likely natural predators. The Caribbean monk seal had a lack of enemies on shore—it was a "genetically tame species"—so was particularly vulnerable to the approach of humans. A large, rapidly growing and mostly indigent human population within the seal's range is considered to be the main factor for the reduction or extinction of the Caribbean monk seal. Many human inhabitants make a living from the sea and would probably kill any seal encountered.

NORTHERN ELEPHANT SEAL

Mirounga angustirostris

NOMENCLATURE

COMMON NAMES. Northern elephant seal, northern sea-elephant

DISTRIBUTION

M. angustirostris is a pelagic species, which ranges throughout the northeastern Pacific from 40° N, north to the Aleutian Islands and Gulf of Alaska, and west to 173° W. The breeding range of the northern elephant seal is restricted to the offshore islands from central Baja California north to central California, or from Isla Cedros, Mexico, to Point Reyes, California (Fig. 39.23). There is no mass migration of these seals, but a widespread northward dispersion occurs as soon as the pups leave their birthplace at about 3 months of age (FAO 1976). Vagrants have been found in Nii-jima in the Izu-shotō of Japan, Midway Atoll in the northwestern Hawaiian Islands, and the Golfo de California as far north as Isla Angel de la Guarda (29°30′ N) (Rice 1998).

Populations of northern elephant seals now in the United States and Mexico are all derived from a small number (tens or hundreds) of individuals surviving in Mexico after being nearly hunted to extinction (Stewart et al. 1994). The growth rate measured for the whole

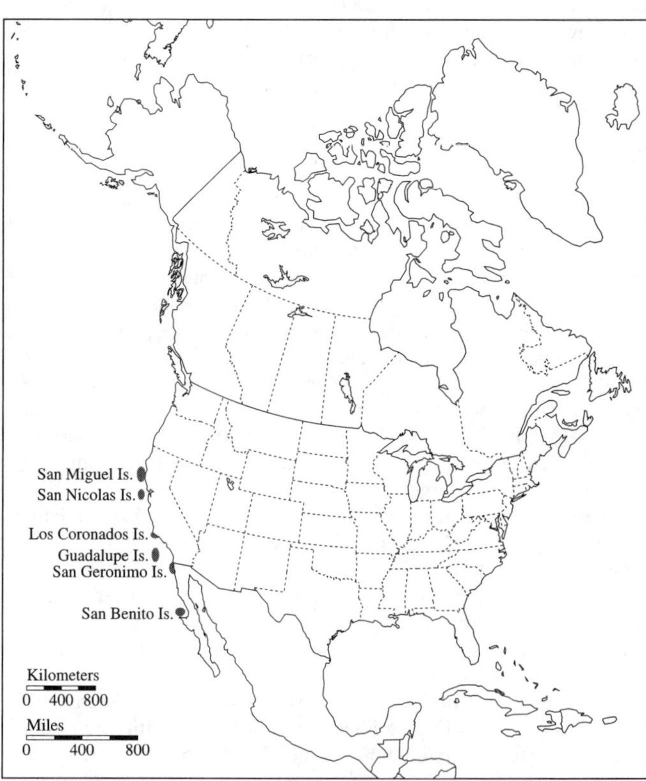

FIGURE 39.23. Distribution of the northern elephant seal (*Mirounga angustirostris*).

U.S./Mexico population was 8.3% between 1965 and 1977 (Cooper and Stewart 1983). By the late 1970s, the population was about 45,000, and the largest groupings were at Isla Guadalupe (15,000–20,000), Islas San Benito (5000–10,000), and San Miguel Island. A colony was established in the southeast Farallon Islands in 1972, and 60 pups were born there in 1976. The Año Nuevo Island population was reestablished in 1961 and numbered 2000 (DeLong 1978b); a new rookery began on the mainland at Año Nuevo Point, where 16 females gave birth in 1977. This movement to Año Nuevo Point likely reflected the continued growth and recovery of the elephant seal population at this time (LeBoeuf and Panken 1977). Other rookeries were found at Isla Cedros, Los Coronados, San Nicolas Island, and Point Reyes Peninsula, and individuals were seen as far north as Alaska and up to 240 km from shore (FAO 1976).

Because all age classes are not ashore at the same time, a complete population count of elephant seals is not possible. The population size, however, is typically estimated by counting the number of pups produced and multiplying by the inverse of the expected ratio of pups to total animals (McCann 1985). Thus, Stewart et al. (1994) arrived at a population estimate of 127,000 elephant seals in the United States and Mexico for 1991 using a multiplier of 4.5. Based on the estimated 24,000 pups born in California annually over 1994–1996, and a more appropriate multipler of 3.5 for a rapidly growing population (Boveng 1988; Barlow et al. 1993), the California stock was approximately 84,000 in 1996. The conservative minimum population is given as 51,625 (Forney et al. 2000). Calambokidis and Baird (1994) reported that in the transboundary waters of British Columbia and Washington, elephant seals occurred in small numbers, but have increased in recentred years.

Habitat. The northern elephant seal spends most of its time at sea, but hauls out on sandy beaches to breed. *M. angustirostris* has been seen to cohabit peacefully in areas with the California sea lion, Steller's sea lion (*Eumetopias jubatus*), Pacific harbor seal, and northern fur seal (*Callorhinus ursinus*). However, this species breeds at a time of year when the others are at sea (FAO 1976). Gulls and cormorants are usually found in close association with these seals, walking among them or alighting on their backs. The birds are more alert to the approach of danger and act as sentinels for the elephant seals (Bartholomew 1952).

DESCRIPTION

Elephant seals are the largest of all pinnipeds. They are characterized by a very long proboscis, which, when relaxed, hangs over the mouth (Fig. 39.24). This is fully developed in the male only and when erected is used as an instrument of phonation to threaten other seals. Another male trait is the rugose, corrugated neck shield, which also becomes more pronounced with age. There is considerable sexual dimorphism in size, but none in coat pattern. Adult males are 4.5 m long and weigh 1.8–2.7 tonnes, whereas females are 3.6 m long and weigh about 0.7 t. Newborn male pups are about 1.53 m long and weigh 36.0 kg; females are approximately 1.47 m long and weigh 31.5 kg (FAO 1976). Pups are born with a long, woolly black lanugo, which is shed at about 3 weeks of age to reveal a silver-gray coat, similar to that of adults. Adults and subadults undergo an epidermal molt, which usually starts in the axillary region and progresses around the body (Jefferson et al. 1993).

FIGURE 39.24. Northern elephant seal (*Mirounga angustirostris*).

FEEDING HABITS

The northern elephant seal is a deep-diving pinniped. Time–depth recording devices have recorded dives of 1580 m and 80 min. After leaving the rookeries, most of these seals spend 80–90% of their time underwater (Jefferson et al. 1993). Also, they tend to feed about 50 km offshore (DeLong 1978b). Bottom-dwelling marine life such as skates (Rajidae), rays (Dasyatidae, Myliobatidae), ratfish (*Hydrolagus colliei*), frilled sharks (Chlamydoselachidae), squids, and Pacific hake (*Merluccius productus*) are part of their diet (Huey 1930; Morejohn and Baltz 1970) (Table 39.3).

BEHAVIOR

A solitary animal, this species spends most of the year at sea. When hauled out for the annual breeding and molting periods, they are gregarious and maintain an orderly social structure. Retention of social position is very important and results in fierce competition between males, who deal crushing blows to each other with their head and neck (Bartholomew 1952). They are polygamous in their mating behavior, the males gathering harems. Phonations include snorts, sneezes, grunts, and yawns. The most characteristic sound of the males is a prolonged loud snort, which is produced with the help of the arched proboscis, the end pointed down the animal's throat (Maxwell 1967). Females and subadults make a vomiting cough when threatened (Bartholomew 1952). The animal is indifferent to the approach of humans, but will rapidly retreat to the water if frightened (FAO 1976). The placid nature of these seals, apart from male aggression during the breeding season, is remarkable and must have evolved from hundreds of thousands of years of existence in remote areas where they were the supreme land mammals. On land, they remain either asleep or inert and will periodically suspend breathing for about 5 min. To gain relief from dry skin and surface parasites, they often engage in sand-flipping behavior, throwing sand over their backs with their foreflippers (Maxwell 1967). This also may provide thermoregulatory relief.

MORTALITY

Predators of the northern elephant seal include killer whales and sharks (FAO 1976). Great white sharks (*Carcharodon carcharias*) prey on juvenile and subadult elephant seals hauled out on the South Farallon Islands, California, during fall (Pyle et al. 1996). Elephant seals live about 14 years (FAO 1976).

OTARIIDS IN NORTH AMERICA

The following species of otariid inhabit the waters adjacent to continental North America: Steller's sea lion (*Eumetopias jubatus*), California sea lion (*Zalophus californianus*), northern fur seal (*Callorhinus ursinus*), and Guadalupe fur seal (*Arctocephalus townsendi*).

STELLER'S SEA LION

Eumetopias jubatus

NOMENCLATURE

COMMON NAMES. Steller's (or Steller) sea lion, northern sea lion

DISTRIBUTION

Steller's sea lions occur in coastal and immediate offshore waters of the cool–temperate North Pacific. Historically, they bred along the western coast of North America from San Miguel Island in the California Channel Islands, northwest to the Gulf of Alaska, along the Alaska Peninsula, and throughout the Aleutian and Pribilof Islands (Rice 1998). No adults have been seen at San Miguel Island, however, since 1983, and no births have been recorded since 1981, making Año Nuevo Island the southernmost breeding site at present (National Marine Fisheries Service 1992). Off Asia, they are found in the Kuril Islands, Kamchatka, and the islands in the Sea of Okhotsk. Beyond this range, there are rare occurrences of Steller's sea lions entering the lower reaches of coastal rivers in Washington and Oregon. Stragglers have gone to Herschel Island (69°35′ N, 139°05′ W) in the Beaufort Sea and to Jiangsu, China (Rice 1998). The worldwide population is considered to be composed of two stocks, with an east–west division near Cape Suckling, Alaska (144° W), based on genetic information (Bickham et al. 1996), population dynamics, and morphological comparisons (Loughlin 1997).

In the eastern Pacific, to the southern extent of their range (Fig. 39.25), adult males undergo a distinct postbreeding seasonal northward migration. Some females may also participate in this northward movement during the winter (FAO 1976). Some animals move as far north as St. Lawrence Island in the Bering Sea in the summer (Gentry and Withrow 1978). On the other hand, when the far northern breeding aggregations disperse in August, the males of these rookeries leave immediately in a general southward migration from Alaska. The newborn pups disperse among the islands of the North Pacific.

In the 1970s, the total worldwide Steller's sea lion population was about 281,800 (Hull 1994), and was also estimated at 245,000–290,000 animals between 1975 and 1980 (Loughlin et al. 1984). There were 5000–7000 in California (includes fewer than 100 in the Channel Islands, 200 on the Farallon Islands, and 1600–2000 on Año Nuevo Island), 2000 in Oregon, 600 in Washington, and 5000 in British Columbia (2500 around Vancouver Island, 2000 in the area of Cape St. James and the Queen Charlottes, and 500 along the northern British Columbia coast) (FAO 1976). The California population had been declining since the 1920s, with a sharp decrease since the late 1960s (Gentry and Withrow 1978); it remained at about 2000–2500 adults between 1980 and 1990. Counts of Steller's sea lions in Oregon have been relatively stable since 1981 at about 2000–3000 animals (National Marine Fisheries Service 1992). The number of Steller's sea lions in the Strait of Georgia, Puget Sound, and the Juan de Fuca Strait area appears to be stable, although well below historical levels (Calambokidis and Baird 1994).

Kilometers
0 400 800
Miles
0 400 800
Herschel Is.
Santa Barbara Is.

FIGURE 39.25. Distribution of the Steller's sea lion (*Eumetopias jubatus*).

In Alaska, Steller's sea lions, adults and pups, numbered around 242,000 in the 1970s (Hull 1994). The Gulf of Alaska populations were considered stable, but the eastern Aleutian Island population (10,000 in 1967) (Maxwell 1967) had experienced a drastic decline (Gentry and Withrow 1978). From the mid-1970s to the mid-1980s, populations in the Aleutian Islands and the Gulf of Alaska declined at rates of about 5–9% annually (Loughlin and Nelson 1986; Merrick et al. 1987; Loughlin et al. 1990). Adult and juvenile counts showed a mean reduction of 63% from 1985 to 1989, or about 16%/year (Loughlin 1989; Loughlin et al. 1990). In contrast, counts of sea lions in southeast Alaska showed a stable or possibly increasing trend. For example, pup counts reportedly increased from 2220 in 1979 to 4164 in 1991, notably at Hazy Islands and Forrester Island (National Marine Fisheries Service 1992).

Overall, for the western U.S. Pacific stock, 1996 surveys resulted in an estimated 39,500 Steller's sea lions for the entire Gulf of Alaska, Aleutian Island, and Bering Sea region. A comparable estimate for the eastern Pacific stock is not available; however, counts from southeast Alaska, British Columbia, and the Oregon region indicate a population of a least 30,400 Steller's sea lions (National Marine Fisheries Service 1999).

Habitat. Steller's sea lions have well-defined, traditionally used rookeries. During the winter, they prefer shallow waters free of fast ice near the coast and gravel beaches and ice flows that are not too far out to sea (King 1964). This is the only otariid that habitually hauls out on sea ice (Rice 1998). Throughout their range, they are primarily found from the coast to the outer continental shelf, but do frequent deep oceanic waters in some parts (Jefferson et al. 1993).

DESCRIPTION

Typical of other otariids, *Eumetopias jubatus* has large, naked flippers with small nails (Fig. 39.26). All ages and sexes have contrasting black flippers. There are four mammary teats (Fay et al. 1979). The female is slim, but the male has massive forequarters (Banfield 1974) and a thick, muscular neck with a substantial mane of long, coarse hairs (Maxwell 1967). Its head is bearlike, with a straight muzzle and small, stiff, pointed ear pinnae. Long (50 cm), stiff, pale mystacial vibrissae are present (Banfield 1974). Their eyes are very unusual because there is a circle of white around the outer edges of the iris (Maxwell 1967). As already mentioned, there is sexual dimorphism in size and appearance, adult males reaching lengths of 3 m and weights of 900 kg, whereas females grow up to 2 m and weigh 300 kg. Newborn pups are 1 m long

FIGURE 39.26. Steller's sea lion (*Eumetopias jubatus*).

and weigh 16–23 kg, and have a thick blackish brown lanugo, which is molted by about 4–6 months of age. Subadults are silver to light brown when wet (FAO 1976). Adults are pale yellow to light tan above, darkening to brown and shading to rust below. The pelage is short and coarse and lacks an undercoat (Fay et al. 1979).

FEEDING HABITS

Steller's sea lions feed all year long, except during the breeding season, when the males in the harem fast for 2 or 3 months. At this time, females with pups feed principally at night (Higgins et al. 1988). Foraging dives tend to be relatively shallow, with few dives recorded to depths >250 m. For individual adult females, maximum depths recorded are 100–250 m in summer and >250 m in winter (Merrick and Loughlin 1997).

Their food consists of a variety of invertebrate marine life such as coelenterates, sand dollars, worms, and molluscs, and fish such as pollock, Atka mackerel (*Pleurogrammus monopterygius*), Pacific cod (*Gadus macrocephalus*), rockfish (*Sebastes* spp.), herring, halibut, and salmon (Banfield 1974; Jones 1981; Treacy 1985) (Table 39.3). Most of the top-ranked prey is off-bottom, schooling species (Hull 1994). In the southern Bering Sea, Aleutian Islands, and the Gulf of Alaska, Pacific cod is a top prey item, particularly during the winter months. Both male and female Steller's sea lions prey on other pinnipeds, specifically ringed seal, larga seal, harbor seal, bearded seal, and northern fur seal pups (Bowen and Siniff 1999). In fact, on St. George Island (Pribilof Islands), they may take 3–6% of the northern fur seal pups born each year (Gentry and Johnson 1981). Overall, the food and energy needs of Steller's sea lions are not well known. These requirements can vary greatly depending on the energy content of the food and the physiological status of the individual (Innes et al. 1987). Higgins et al. (1988) determined that pups at Año Nuevo Island (California) consumed 1.5–2.4 L of milk (23–25% fat content) per day while nursing.

BEHAVIOR

Steller's sea lions are highly gregarious, and crowd together on breeding rookeries, while swimming, and when hauling out in groups the rest of the year. They haul out in sunny, calm weather, but stay at sea when it is rough and stormy. They have been observed at play in the water, but on land appear to be very quarrelsome. Although this is a polygamous species, only the larger and older sea lion males have their own harem (Banfield 1974). They arrive at the breeding grounds first, and they establish and defend their territories with the use of many threat displays (Gentry and Withrow 1978). Rarely, if ever, do they leave their territories during the breeding period and as a result may fast for up to 60 days. Females are more gregarious and show no attachment to a specific male or territory. They usually wean their pups in the first year, but occasionally suckle a yearling (Gentry and Withrow 1978). There is a definite social structure in breeding grounds, with barren females, bachelor males, and yearlings remaining on the periphery (Banfield 1974). This species is normally very wary of humans and will hurry to the water as soon as an intruder is seen, except during the breeding season, when the females lose their shyness and will defend their pups (Maxwell 1967).

MORTALITY

Killer whales and sharks probably eat Steller's sea lions, but any possible impact on populations is unknown. Pups are lost due to drowning, starvation caused by separation from their mother, crushing by larger animals, disease, predation, and biting by females other than the mother (Orr and Poulter 1967; Edie 1977). There is the likelihood that certain diseases in Steller's sea lions can result in reproductive failure and neonate, juvenile, and adult mortality, but the prevalence is difficult to evaluate. Antibodies to two types of bacteria (*Leptospira* and *Chlamydia*), one marine calicivirus (San Miguel sea lion virus), and seal herpesvirus were found in blood taken from Steller's sea lions in Alaska (Barlough et al. 1987; Vedder et al. 1987; Calkins and Goodwin 1988). Females may live to 30 years and males to about 20 years (Calkins and Pitcher 1982).

CALIFORNIA SEA LION

Zalophus californianus

NOMENCLATURE

This species is probably the most widely known sea lion in the world because it performs in shows, circuses, and zoos around the world, and frequently escapes or is released indiscriminately.

There are two geographic divisions recognized within the population of the California sea lion, one on the Pacific coast and one in the Golfo de California. There has been long genetic isolation between the two groups, as determined from an analysis of the control region of the mitochondrial DNA (Maldonado et al. 1995). No cranial differences, however, are evident between animals from the two regions (Orr et al. 1970). Conversely, the morphological differences between the Galapagos (*Z. wollebaeki*) and California (*Z. californianus*) populations, along with their geographic isolation and certain dissimilarities in behavior, resulted in the conclusion that they should be considered specifically distinct (Rice 1998).

DISTRIBUTION

On the Pacific coast, California sea lions range mainly in near-shore waters, with hauling grounds located on coastal islands from Solander Island (49°57′ N) on the west coast of Vancouver Island and Denman Island (49°50′ N) in the Strait of Georgia south to Cabo San Lucas, Baja California Sur. There are also two hauling grounds far offshore on oceanic islands—Islote Zapato (28°50′ N, 118°20′ W) off the southern tip of Isla Guadalupe, and Rocas Alijos (24°57′ N, 115°45′ W). Only males occupy the hauling grounds north of southern7 California.

Some subadult and adult male California sea lions undergo a northward postbreeding migration and are often seen as far north as Bull Harbour, Vancouver Island (51° N), during the winter. This is 1000 km north of the northern rookery at San Miguel Island, California (34° N). Vagrants have been reported up to Prince William Sound, Alaska. During the winter season, some individual males regularly enter the lower reaches of coastal rivers in northern California, Oregon, and Washington, including Lake Washington (Mate 1976; Rice 1998). Females have been reported as far north as Año Nuevo Island (37° N). Regular rookeries, however, occur only from Point Piedras Blancas (35°39′ N), California, south to Punta Lobos (23°25′ N), Baja California Sur (Rice 1998) (Fig. 39.27).

In the Gulf of California, males, females, and young are found throughout the entire breeding range all year. Rookeries are located from Roca Consag (31°03′ N, 114°28′ W) south to Los Islotes (24°33′ N, 110°26′ W). Following the breeding season, some adult males and subadults move south along mainland Mexico to Manzanillo (19° N), and some have been sighted at 23° N (Los Frailes) and Chiapas (14°42′ N) (Mate 1976; Rice 1998).

In the North Atlantic, free-ranging California sea lions have been seen once in Newfoundland, and more often along the eastern seaboard from Virginia to Louisiana (Gunter 1968; Schmidly 1981). These sightings are all likely former captives that escaped or were freed, and there is no evidence of any breeding (Rice 1998).

The North American California sea lion population is probably increasing, possibly approaching its historical level. Populations in southern California and along the Pacific coast of Baja California declined during the nineteenth and early twentieth centuries due to hunting, but there is no evidence that the Gulf of California population was reduced (Townsend 1918; Stewart et al. 1993). Sea lion numbers have

FIGURE 39.27. Distribution of the California sea lion (*Zalophus californianus*).

increased rapidly since the cessation of commercial exploitation in the mid-1900s, particularly in southern California, and they were estimated to number over 174,000 in U.S. and Mexican waters in the early 1980s (Aurioles-G. et al. 1983; Le Boeuf et al. 1983). As of the late 1990s, from 85,000 to 180,000 California sea lions were reported to breed on the Channel Islands (Wong 1997), with up to 80,000 on San Miguel Island alone. Washington State's population grew from occasional sightings in the 1970s to 400–500 by 1995. The wintering population in British Columbia has been estimated to be about 3000 (Olesiuk and Bigg 1988). The number of California sea lions occupying the transboundary waters between Washington and British Columbia increased in the 1980s and appears to have stabilized (Calambokidis and Baird 1994).

Habitat. Pupping and breeding occurs on island-based rookeries consisting of sandy beach areas (Mate 1978), although some rookeries in the Gulf of California are semiaquatic or fully aquatic but are always close to the mainland or an island (Mate 1976). They frequent bays, harbors, and river mouths, and regularly haul out on buoys and jetties. Also, they can occasionally be found up to several hundred kilometers offshore (Jefferson et al. 1993).

Competition with the Steller's sea lion for food, habitat, and other resources may significantly affect the distribution of both species. There are indications of mutual shifts in breeding range, short periods of cohabitation, use of similar hauling-out grounds, and similar prey species (FAO 1976).

DESCRIPTION

The California sea lion is more slender than Steller's sea lion and does not have the overly thickened neck of that species. The ear pinnae are stiff but inconspicuous (Fig. 39.28), and the mystacial vibrissae are long and stiff (Banfield 1974). At the age of 5 years, the males develop a distinctive sagittal crest. This crest becomes increasingly prominent during the animal's adult life and is often more striking in older sea lions

FIGURE 39.28. California sea lion (*Zalophus californianus*).

because the normally dark hair turns to light brown or tan in this area (Mate 1978). Males are noticeably larger than females, being 2.2 m long and weighing 275 kg, whereas the females are 1.8 m long and weigh 91 kg. Newborn pups are 0.75 m long and weigh 5–6 kg (FAO 1976). The pelage is short, dense, and chocolate brown (Fay et al. 1979).

FEEDING HABITS

California sea lions feed day and night in shallow water. Observations suggest that they are opportunistic feeders, shifting their diets with local variations in the abundance of diverse prey species (Antonelis et al. 1984; Lowry et al. 1991). However, there is evidence that wild sea lions at times largely ignore an abundant food item in preference for another (King 1983). A preliminary study by Cox et al. (1996) suggested that they may have the ability to evaluate food nutrient content and thus select accordingly.

They compete with the Steller's sea lion and other near-shore pinnipeds for the same food items (Mate 1976), such as squid, octopus, abalone, and a variety of fish such as herring, sardines, rockfishes, hakes, and ratfishes (Table 39.3). Studies in Oregon, California, and Baja California indicate that the more frequent food types consumed are anchovy (*Engraulis mordax*), Pacific hake (*Merluccius productus*), jack mackerel (*Trachurus symmetrycus*), rockfish, and market squid (*Loligo opalescens*) (Antonelis et al. 1984; De Anda 1985; Lowry et al. 1990, 1991). In addition, adult and subadult male California sea lions tend to aggregate near the mouths of freshwater rivers and fish ladders along the Pacific Northwest coast, where they prey on salmon returning to spawn (Fraker and Mate 1999).

In the southern Gulf of California, main prey are deep-sea fishes such as flagfins (Aulopidae: *Aulopus* sp.), cusk-eels (Ophidiidae: *Neobythites* sp.), and sea basses (Serranidae: *Pronotogrammus* sp.) (Aurioles-G. et al. 1984); in the northern part of the Gulf, main prey are hake, mackerel (*Scomber* sp.), rockfish, and anchovy (Orta 1988). California sea lions bring their prey to the surface, bite through the neck, snap off the head with a powerful shake, and quickly swallow the body "head first." They have learned to steal fish from commercial fishing nets and on lines at sea, damaging the fishing gear in the process (Banfield 1974). California sea lions have also been observed feeding on chicks of the common murre (*Uria aalge*).

BEHAVIOR

California sea lions are very gregarious, though wary of intruders (Banfield 1974). In the water, they are very playful, chasing each other and leaping out of the water. They are also active on land, partly because of parasitic infections that irritate their skin (Maxwell 1967). At sea, they often "raft" at the surface alone or in groups, frequently raising their flippers out of the water. Phonations include a honking bark by the males, a quavering howl by the females, and a bleating sound by the pups. A distinctive behavioral trait of this species is their ability to catch objects in the air with their teeth, and this has been exploited in animals in captivity (Banfield 1974). California sea lions are polygamous and form loosely organized harems. Males patrol the waters off their beach territories and make threatening gestures toward intruders. In the water, males appear to be very protective toward the pups (Maxwell 1967).

MORTALITY

Predators of California sea lions include sharks and killer whales. Drowning in rough seas and injuries received on crowded rookeries (Mate 1976) cause pup mortality.

El Niño is a phenomenon that occurs at irregular intervals in the eastern tropical Pacific, and in characterized by the warming of surface waters. It alters the distribution or abundance of prey species, and pinnipeds are unable to find sufficient food. Females must dive deeper and travel farther for food, thus expending more energy and requiring more time away from their pups. In turn, the mother's milk is undernourished and not as plentiful. There have been five strong El Niño events since 1970: 1972, 1983, 1987, 1992, and 1997. Other than 1987, all have had dramatic deleterious effects on rates of pup births, pup growth, and pup survival (DeLong et al. 1998; Forney et al. 2000). However, with population abundance close to historical levels, the population has quickly rebounded from any El Niño effects.

The bacterium *Leptospira pomona,* which is known to cause abortion in many domestic mammalian species, was first found in California sea lions in 1972. Leptospirosis, caused by the bacteria *Leptospira,* was initially identified in 1970. In 1984, a mass mortality attributed to leptospirosis occurred in California (Dierauf et al. 1985). More recently, from July to December 1991, a total of 144 animals stranded live or washed up dead along the north-central coast of California. During October 1997, more than 100 California sea lions with symptoms of leptospirosis stranded along the north-central California coast (Marine Mammal Commission 1998). Leptospirosis outbreaks occur periodically, so are not judged to be alarming, and seemingly tend to occur during the fall, following El Niño events (Lander and Herskovitz 1997). The San Miguel sea lion virus has been associated with reproductive failures or neonatal deaths in California sea lions (Gilmartin et al. 1976). *Z. californianus* has a life span of around 15–24 years. In the wild, it averages 18–20 years; in captivity, up to 25 years or more.

NORTHERN FUR SEAL

Callorhinus ursinus

NOMENCLATURE

Common Names. Northern fur seal (Scheffer 1958), Pribilof fur seal, Alaska fur seal (King 1964)

DISTRIBUTION

C. ursinus is a pelagic species, which ranges in the subarctic waters of the North Pacific Ocean (Fig. 39.29) from the southeastern Sea of Okhotsk, the southern Bering Sea, and the northern Gulf of Alaska south to about 35° N in the Sea of Japan (Fiscus 1978; Rice 1998). They occur as far south as San Miguel Island, a common home for this seal, the northern and California sea lions, and the Guadalupe fur seal. There is competition for space, and aggressive behavior has been observed between *Callorhinus* and *Zalophus*. Occasionally there is interaction between *Callorhinus* and *Arctocephalus* males (DeLong 1975). The colony of northern fur seals on San Miguel Island was discovered in 1968. This population originated from the Pribilof Islands, likely in the late 1950s or early 1960s (DeLong 1982).

The main herd leaves the North Pacific and Bering Sea in October and migrates south down the North American coast as far as San Francisco (Fiscus 1978). Females and young of both sexes make the

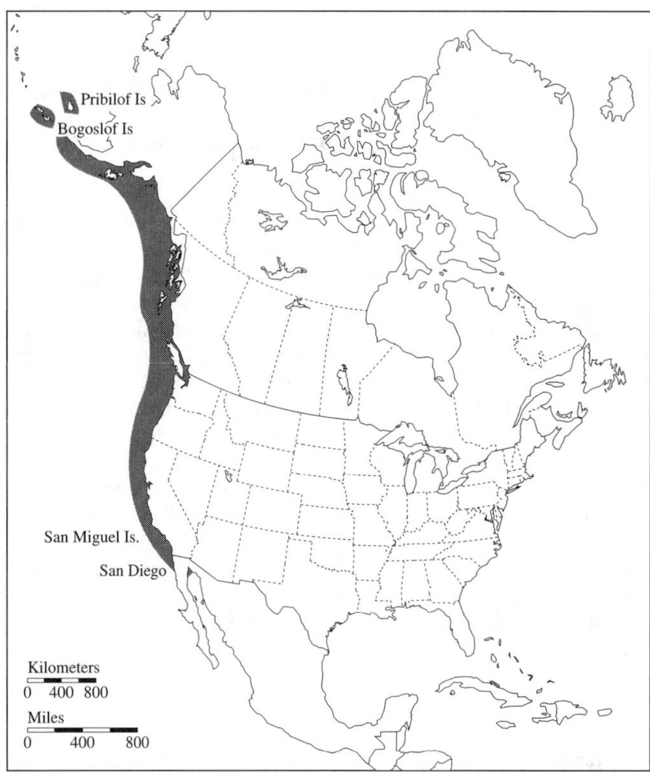

Figure 39.29. Distribution of the northern fur seal (*Callorhinus ursinus*).

longest migration, whereas adult males remain north near the Aleutian chain and the Gulf of Alaska (FAO 1976). The northward migration from California begins in March. The migration patterns of fur seals follow the movement of schools of prey species, whose abundance influences the length of time they may stay in an area (Fiscus 1978).

During the breeding season, approximately 74% of the worldwide population is on the Pribilof Islands in the southern Bering Sea, with the remaining animals spread throughout the North Pacific Ocean (Lander and Kajimura 1982). Outside of the Pribilofs, in U.S. waters, about 1% of the population occurs on Bogoslof Island in the southern Bering Sea and San Miguel Island off southern California (National Marine Fisheries Service 1993). Northern fur seals haul out at other sites in Alaska, British Columbia, and islets along the coast of the continental United States, but this is generally temporary and outside of the breeding season (Fiscus 1983).

The world population of northern fur seals could be around 2 million individuals. The total population in United States rookeries was estimated at 1,004,000 in 1998, with 973,000 on the Pribilof Islands (Robson 2000). In Russian waters, estimates give a population size of 210,000 in the Commander Islands, 75,000 on Robben Island, and 47,000 in the Kuril Islands (Anonymous 1985a, 1985b). Based on 1999 counts and an expansion factor of 4.0 (DeLong 1982), the size of the San Miguel Island (California) stock was estimated at 4336 northern fur seals. The minimum population estimate is 2336 animals (Forney et al. 2000).

Habitat. Apart from the breeding period when northern fur seals are hauled out on rocky island beaches, these animals spend most of their time at sea. They are usually seen offshore along the continental slope and in areas where the topography causes upwelling of nutrient-rich water and an abundance of prey species (Fiscus 1978). They range through waters with a surface temperature of 1–15°C and are most abundant in waters of 8–12°C, probably because of food availability (Baker et al. 1970).

DESCRIPTION

The northern fur seal has a small head with large eyes; a short, pointed snout; a high forehead; and moderately long, stiff, slender ear pinnae (Banfield 1974) (Fig. 39.30). The ears are tightly rolled cylinders, each having a wax-coated orifice to prevent the entrance of water. Its most distinctive features are its thick, waterproof underfur and its unusually large hind flippers, about one fourth of the total body length (Fiscus 1978; Jefferson et al. 1993). The vibrissae are long, often extending to beyond the ears, and are white in adults. Newborns have black vibrissae, which change gradually to become white in subadults (Jefferson et al. 1993).

There is great sexual dimorphism, with males having a massive body and neck and females being more gracefully proportioned (Banfield 1974). Adult males average 2.13 m in length and weigh 181.3 kg; females are 1.4 m long and weigh 43–50 kg. Newborn males are 0.66 m long and weigh 5.4 kg, and newborn females are 0.63 m long and weigh 4.5 kg (FAO 1976). Pelage color varies between the sexes. Males are dark brown, except for the mane, which has a grayish tinge, whereas females are slate gray dorsally and a lighter reddish gray ventrally. The throat of both sexes is lighter whereas the rest of the coat and the underfur are a chestnut color (Maxwell 1967). Pups are blackish at birth, with variable oval areas of buff. After 3–4 months, pups molt to the color of adult females and subadults (Jefferson et al. 1993). The thick waterproof underfur coat has 56,900 hairs/cm^2 (Fiscus 1978).

FEEDING HABITS

Northern fur seals obtain some of their food in moderately deep water and have been recorded diving to a depth of 180 m (Hull 1994). They tend to feed during the evening, night, and early morning, and sleep during the day. Northern fur seals eat a wide variety of prey species (Table 39.3), although prey selected varies with the season, the seal's age, and the area (FAO 1976). Over 60 taxa, mainly fish, have been identified from stomach contents (Kajimura 1985). The main prey is anchovy, herring, capelin, and sand lance (Kajimura 1984; Perez and Bigg 1986), although squid, pollock and lanternfishes are also considered common food items (FAO 1976). In the eastern Bering Sea, pollock, squid, and capelin account for 70% of the diet (Perez and Bigg 1986).

Fur seals will swallow small fish (25 cm) while underwater, but will bring larger prey to the surface to be torn into chunks by shaking before it is eaten (Fiscus 1978). This species is believed to consume about 10% of its body weight in food per day. Perez and Mooney (1986) determined that the average daily feeding rate for lactating northern fur seals was 1.6 times higher than for nonlactating females. Major changes in the Bering Sea ecosystem as a result of intensive commercial fishing may have a profound effect on this species (FAO 1976).

BEHAVIOR

Usually this species swims alone, although groups of two or three are common (Fiscus 1978). On land, they are very gregarious, and a complex social structure exists on the breeding rookery, where territorial harems are the rule and subadults remain on the outskirts. They spend the day resting, preening, or swimming, and feed at night. They are able to sleep either on land or in water (Banfield 1974). A characteristic pose when sleeping at sea is the "jug handle position," where the seal lies on its back, with the hind flippers folded forward and held by a foreflipper (Fiscus 1978). Hind-flipper waving is probably a thermoregulatory device (Banfield 1974). Phonations consist of loud coughs, roars, barks, and bleating sounds (Maxwell 1967).

FIGURE 39.30. Northern fur seal (*Callorhinus ursinus*).

MORTALITY

Mortality rates of northern fur seal pups since 1963 have varied from 5% to 12%. Major causes of death are hookworm infection and malnutrition; injuries, congenital defects, and bacterial infections are additional factors. Hookworms cause severe anemia and subsequent death. Other parasites probably kill a few, and some pups starve when their mother is killed at sea. Violent weather and the inability to obtain adequate food cause mortality during the seal's first year at sea. Up to 85% die before the age of 3 years in some cohorts. Predators of northern fur seals include killer whales and great white sharks. Because this seal will put its head through looplike objects floating at sea, scraps of synthetic netting can be dangerous. If caught around the animal's head or neck, they can impede feeding or cut deeply, causing crippling infection or death (Baker et al. 1970).

The northern fur seals on San Miguel Island are negatively affected by El Niño events, which occur periodically along the California coast. The 1982–83 El Niño event, for example, resulted in a 60.3% decline in the population (DeLong and Antonelis 1991), which required 7 years to recover because adult female mortality occurred in addition to pup mortality (Melin and DeLong 1994). Because of a severe El Niño event from July 1997 through May 1998, it appears that 87% of the pup production in 1997 on San Miguel Island (over 3000 pups) died before weaning, and total production in 1998 was only 627 pups. Total production increased to 1084 in 1999, but rate of recovery could be slow if adult female mortality occurred in addition to the high pup mortality in 1997 and 1998 (Forney et al. 2000; Melin and DeLong 2000).

The northern fur seal lives for 20 years or more (FAO 1976). In Alaska waters, tagged females as old as 26 years have been recaptured; one was reproductively active at age 25 years. No males have been found older than 17 years (Hull 1994).

GUADALUPE FUR SEAL

Arctocephalus townsendi

NOMENCLATURE

The Guadalupe fur seal was first identified as a new species in 1897, at which time it was already thought to be extinct. Later, King (1954) believed that *A. townsendi* was conspecific with *A. philippii,* the Juan Fernandez fur seal, and Scheffer (1958) classified it as a subspecies of the latter. Currently, *A. townsendi* is recognized as a full species (Repenning et al. 1971; Rice 1998). Of the eight species of *Arctocephalus,* only *A. townsendi* occurs in North America.

DISTRIBUTION

The main breeding colony of *A. townsendi* is on the east shore of Isla Guadalupe off Baja California, Mexico. In 1997, a second rookery was discovered at Isla Benito del Este, Baja California (Maravilla-Chavez and Lowry 1999), and a pup was born at San Miguel Island, California (Melin and DeLong 1999).

Guadalupe fur seals range along the coast of California at least 500 km north and 270 km east of the Isla Guadalupe rookery as far as the Farallon Islands and Sonoma County (38°26′ N); they haul out regularly on San Miguel Island (Hanni et al. 1997; Rice 1998). They also move south around Cabo San Lucas into the Golfo de California as far north

FIGURE 39.31. Distribution of the Guadalupe fur seal (*Arctocephalus townsendi*).

as Bahía de Bacochibampo (27°55′ N. Rice 1998) (Fig. 39.31). Individuals have been sighted as far south as Zihuatanejo, Mexico (17°39′ N; 101°34′ W) (Aurioles-G. and Hernadez-Camacho 1999).

The former breeding range is difficult to know because early records failed to distinguish among fur seal species. Rice (1998) considers that the populous rookeries reported on the Islas San Benito and Isla Cedros in the early 1800s were most likely made up of *A. townsendi,* as were all of those on Isla Guadalupe, because these islands lie well to the south of the usual migratory range of *Callorhinus ursinus*. Bones recovered from aboriginal middens (from late 1700s and 1800s) indicate that rookeries or hauling-out grounds probably occurred on Catalina Island (Bickford and Martz 1980), San Miguel Island (Walker and Craig 1979), and the southern California mainland at Newport Bay (Lagenwalter 1981) and Point Mugu (Lyon 1937; Repenning et al. 1971).

Commercial sealing during the late eighteenth and early nineteenth centuries drastically reduced the population of the Guadalupe fur seal. By 1825, none was found in California waters, and by 1894, they were considered extinct throughout their range. In 1928, a few dozen fur seals were discovered on Isla Guadalupe, of which two adult males were captured and brought to the San Diego Zoo (Townsend 1931). In 1949, there was a report of one male Guadalupe fur seal living on San Nicolas Island (McClung 1978). The species was not seen again until 1954, when the breeding colony at Isla Guadalupe was reestablished (Hubbs 1956). Estimates of their historical population levels range from 20,000 to 100,000 animals (Wedgeforth 1928; Hubbs 1956; Fleischer 1987). A 1977 census of these seals on Isla Guadalupe indicated 1073 individuals (Fleischer 1978a). Gallo (1994) estimated the population was about 7408 in 1993, and through analysis of counts made during the breeding season from 1954, indicated that the population was increasing exponentially at an average annual growth rate of 13.7%.

Habitat. The Guadalupe fur seal prefers rocky habitat. The volcanic caves along the east side of the Guadalupe Islands provide shelter from prevailing winds, launching spots from which to swim in hot weather, and suitable places to breed (Fleischer 1978b).

DESCRIPTION

The Guadalupe fur seal is distinguished by its large head and extremely long, fleshy snout and pointed muzzle (Repenning et al. 1971; Fleischer 1978b) (Fig. 39.32). Adults have moderate-length, whitish cream vibrissae. The long prominent ear pinnae are scroll shaped and capable of limited movement, and the muscles allow the pinna to be closed tightly when the animal is submerged. This species is characterized by its large flippers, which are very dark in color and hairless to the area of the metacarpals (Fleischer 1978a). Sexual dimorphism occurs, with males about 1.9 m long and weighing 159 kg, whereas females are approximately 1.4 m long and weigh 45 kg (Fleischer 1978b). Very little is known about the young of this species. Adult male Guadalupe fur seals are dark brown to black, with lighter hair on the neck, shoulders, and breast, and possess a heavy mane. Females vary more in color, with shades of gray, brown, and cream on various areas of their body. Pups are born with a black coat, similar to the adult coat; coloration is dark dorsally and gray-brown ventrally with lighter faces (Fleischer 1978a).

FEEDING HABITS

Little has been studied on the feeding habits of the Guadalupe fur seal. They are known to feed on squid and fish such as lanternfishes (myctophids) and mackerel. From examination of the feces and gastrointestinal tracts of stranded Guadalupe fur seals, Hanni et al. (1997) concluded that the diet consisted of different squid and teleost species. Gallo (1994) noted that they fed on species of mackerel (Scombridae) and herring (Clupeidae) and vertically migrating squid in shallow depths. Diving activity occurs predominately at night.

About 7–8 days after the birth of the pup, the mother mates and then leaves to feed at sea. For a period of about 8–9 months, she spends an average of 9–13 days at sea, then returns to land to nurse her pup for about 5–6 days. Observations of a group of females indicated that they fed in the California Current south of Isla Guadalupe, making round trips of 704–4092 km (average =2375 km). Most foraging dives are shallow and of relatively short duration. One female dove to an average depth of 17 m, for an average duration of about 2.5 min, to feed (Seal Conservation Society 1999). A similar result was found for a rehabilitated female satellite-tracked off California, where the majority of dives were <20.2 m and ranged from 2 to 4 min (Lander et al. 2000).

BEHAVIOR

A. townsendi is polygamous, the males forming territories during the breeding season. A typical territory consists of one adult territorial male, perhaps several nonterritorial fringe males (Fleischer 1978a), and two or three, but occasionally as many as six or eight, females with pups (Fleischer 1978b). A great deal of aggression has been observed between territorial males during the breeding season (Fleischer 1978a). Although the nursery areas are very noisy, seal phonations are limited and probably have some important social functions. Adult males growl,

FIGURE 39.32. Guadalupe fur seal (*Arctocephalus townsendi*).

bark, whimper, and cough, whereas pregnant females are very quiet. Mothers are heard calling to their pups as a means of locating them. A distinctive behavioral trait of the Guadalupe fur seal is floating and grooming itself in the water, much like the sea otter (*Enhydra lutris*). This grooming allows the water to penetrate the seal's thick fur, giving a cooling effect. Also related to this seal's thermoregulation is its habit of floating. It floats with one flipper extended above the water, on its back with its snout above the surface, or upside down with its hind flippers in the air (Fleischer 1978a).

MORTALITY

Sharks are known to prey on Guadalupe fur seals. Oceanographic conditions also have an impact. There was 33% pup mortality in 1992 because of El Niño and Hurricane Darby (Seal Conservation Society 1999).

ODOBENIDS IN NORTH AMERICA

WALRUS

Odobenus rosmarus

NOMENCLATURE

SCIENTIFIC. NAME. *Odobenus* means tooth-walker, which refers to the tusks

SUBSPECIES. Two subspecies have long been recognized by most scientists: *Odobenus rosmarus rosmarus* (Atlantic Ocean) and *Odobenus rosmarus divergens* (Pacific Ocean), both occurring in North American waters; these two subspecies are physically and reproductively isolated; based on specimens from the Laptev Sea in the Pacific Ocean, a third subspecies of walrus, *Odobenus rosmarus laptevi,* has been admitted as valid (Rice 1998), although it is weakly differentiated: Whereas *O. r. laptevi* has skull characteristics similar to *O. r. divergens,* its size is intermediate to the Atlantic and Pacific subspecies (Fay 1982)

DISTRIBUTION

Atlantic walruses occur from eastern Canada to the Kara Sea (Fig. 39.33). There are four populations: eastern Canadian Arctic from Lancaster Sound, Jones Sound, and the Kane Basin, south to the Belcher Islands in Hudson Bay, Ungava Bay, and Godthåb on the west coast of Greenland; the east coast of Greenland from Kronprins Christian Land south to Angmagssalik; Svalbard archipelago and Zemlya Frantsa Iosifa; and eastern Barents Sea and western Kara Sea bordering Novaya Zemlya (Rice 1998). There has been a great diminution in the range of the Atlantic walrus in recent history. It used to occur as far south as New York City and southwest Britain, but has deserted much of its former habitat as a result of human activities rather than natural factors (Reeves 1978). In Canada, it has been extirpated from the Northwest Atlantic, the Mackenzie Delta, and the St. Lawrence River. It is vagrant to Iceland and along the coasts of Europe south to the Netherlands, Belgium, the British Isles, and the Bay of Biscay (Rice 1998).

The range of the Pacific walrus includes the Chukchi Sea from Mys Shelagskiy in Siberia east to Point Barrow in Alaska and from the Bering Sea south to Karaginskiy Zaliv in Kamchatka and Bristol Bay in Alaska (see Fig. 39.33). It is vagrant east to the Beaufort Sea and Bathurst Inlet; southwest to southern Kamchatka, the northern Sea of Okhotsk, and Honshu; and southeast to Unalaska Island, the south side of the Alaska Peninsula, Kodiak Island, Cook Inlet, and Yakutat Bay (Rice 1998).

Although migration occurs, not all individuals move every year. Atlantic walruses move north as the ice edge retreats in the summer and move south as the ice advances in October. Pacific walruses also move north in the spring, migrating on floating ice (King 1964). A segment of the population, mostly adult males, remains year-round in the

FIGURE 39.33. Distribution of the walrus (*Odobenus rosmarus*).

southeastern Bering Sea and Bristol Bay, where four land-based haul-out sites (Round Island, Cape Pierce, Cape Newenham, and Cape Seniavin) are used during the summer (Marine Mammal Commission 2000).

Most walrus populations were decimated in the nineteenth and early twentieth centuries. Although the Pacific population has recovered, the Atlantic population is still at low levels (Jefferson et al. 1993). There are no estimates on the number of Atlantic walruses in Canada (Environment Canada 1999b). Based on a range-wide survey conducted jointly by United States and Russian scientists in 1990, the population of the Pacific walrus was estimated to be at least 188,316 animals, and probably more than 200,000 animals (Marine Mammal Commission 2000).

Habitat. Major requirements for walrus survival include adequate hauling-out platforms of land or ice and suitable adjacent feeding banks. The range and distribution are determined by the conformation of the shore and temperature (Reeves 1978). In the summer, walruses haul out on all available kinds of habitat, such as cobble or rock beaches and large boulders (Miller 1976), but on hot days spend most of their time in the water. In air temperatures from about −20°C to +15°C, adult walruses are thermoneutral (Fay and Ray 1968).

Atlantic walruses inhabit areas of the Arctic where there are ice floes drifting on the shallow waters above the continental shelf. These waters are less than 80–100 m deep. During the winter, walruses live in areas where the water does not freeze, such as polynyas, and haul out on the ice (Environment Canada 1999b).

DESCRIPTION

The walrus is instantly recognizable from other pinnipeds by its long tusks (Fig. 39.34). The Pacific walrus is larger than the Atlantic walrus and has a wider skull. The main visible difference between the Atlantic and the Pacific odobenid subspecies is in their tusk size and shape. Tusks of the Atlantic walrus are shorter, slimmer, and straighter (Reeves 1978). Among pinnipeds, walruses are second in size to the elephant seal. They have a small, square head with a broad muzzle, which bears a heavy, bristly moustache (Banfield 1974). Their eyes are small and

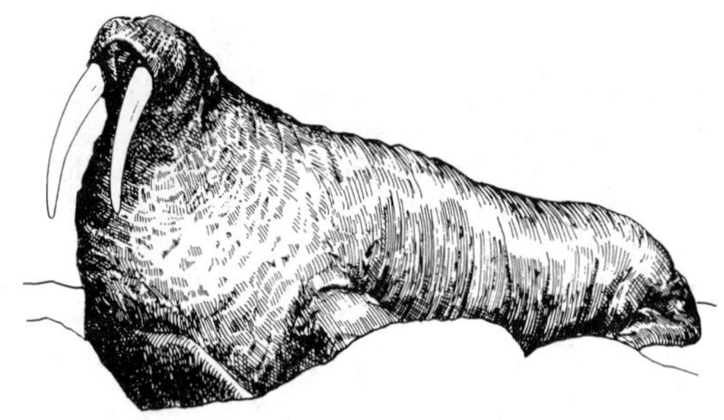

FIGURE 39.34. Walrus (*Odobenus rosmarus*).

frequently bloodshot. They possess no external ears, but a fold of skin protects the meatus (King 1964). The large, massive neck of a mature male is often covered with coarse tubercles, which is a secondary sexual characteristic (Banfield 1974). Located in the neck are a pair of air sacs or pharyngeal pouches, which can be inflated and used as buoys in the water when the walrus is sleeping or wounded (King 1964; Kenyon 1978a). Muscular whisker pads, a highly vaulted mouth, and tusks are all adaptations to their feeding habitat. The tusks, which are large and conspicuous in both sexes, but are usually more slender in females, are used as an aid when hauling out onto the pack ice and also as a means of propulsion as the animal feeds on the bottom (Kenyon 1978a). A single tusk may reach a length of 1 m and a weight of 5 kg (King 1964). Walrus calves are born without tusks, but they erupt at an early age.

Adult males are larger than the females. They are 3–3.6 m long and weigh 1.2–1.6 tonnes. Females are 2.5–2.6 m long and weigh 0.75–1.0 t (FAO 1976). As animals increase in age, so does the thickness of their skin, reaching 2.5–5.08 cm in adults and as much as 6.35 cm on the neck of a mature male (King 1964). Blubber accounts for one third of the total body weight in the walrus (Maxwell 1967). Walruses have a scanty coat of reddish brown hair, although old animals may be nearly naked (King 1964). When they are in the water, the skin appears whitish in color, but when resting in the sunlight, blood circulation increases and the skin turns pink (Kenyon 1978a). Pups are a slate gray color, and change to the reddish brown after the first molt (King 1964).

FEEDING HABITS

Tusk development delays weaning in this species until the animal is between 1 and 2 years old. Adult walruses use their snouts to root on the bottom, with their vibrissae helping to locate food. A powerful sucking action, resulting from a highly vaulted mouth working together with the tongue (Kenyon 1978a), extracts worms and tunicates or pulls clams and snails from their shells (Reeves 1978). This feeding process is an important means of recycling nutrients by stirring up the sediment on the ocean floor and makes walruses the "earthworms" of the subpolar seas (Reeves 1978). Furrows, pits, and discarded bivalve shells leave a distinct record of feeding activities and can be used to indicate the number of prey consumed in a single dive. In the Bering Sea, observations from continuous furrow-pit systems suggest that a walrus can locate, excavate, and consume more than six clams/min (Oliver et al. 1983). The aerobic dive limit for walruses is around 10 min (Wiig et al. 1993; Nowicki et al. 1997), although foraging dives usually last 6–9 min, with a 1-min surface interval (Kastelein et al. 2000).

Walruses feed on a broad array of prey (Sheffield et al. 2001), primarily benthic invertebrates usually found at water depths of <100 m (Fay 1982; Fay and Burns 1988) (Table 39.3). Occasionally, bony fishes are consumed. In general, molluscs, such as clams and mussels, are considered the primary prey. In the eastern Bering Sea, walruses feed almost exclusively on clams (Lowry and Frost 1981), each consuming on the average about 27 kg/day (Kenyon 1978a). They will sometimes eat seal flesh or blubber (ringed, larga, harp, and bearded), narwhals (*Monodon monoceros*), and belugas (*Delphinapterus leucas*) (Reeves 1978; Lowry and Fay 1984; Timoshenko and Popov 1990). Most consumption of seals is predation, rather than scavenging of carrion, and is common behavior. In the Chukchi Sea in summer, seal remains occurred in about 10% of walrus stomachs examined (Lowry and Fay 1984). Seabirds are also taken, such as thickbilled murre, *Uria lomvia* (Donaldson et al. 1995), and black guillemot, *Cepphus grille* (Fay et al. 1990).

A walrus stomach can process in one day at least twice as much as the average daily food intake (taken over a year), which would allow it to eat more during autumn and early winter. Stomach volumes (1-year-olds, about 10 L; 10-year-olds, about 40 L) suggest that daily energy requirements are obtainable in one meal, if necessary (Kastelein et al. 2000). There is a linear relationship between lean body mass (total body mass − blubber mass) and stomach volume in Atlantic walruses (Knutsen and Born 1991). Walruses feed in the early morning and haul out for the rest of the day, occasionally fasting for up to 1 week during pleasant weather after extensive feeding. In the autumn, they spend much of their time feeding in shallow water until the bays freeze over, at which point they try to keep breathing holes open by bunting the new, soft ice. Eventually they retreat to the shore leads (1.6 km offshore in winter) (Banfield 1974). During the breeding season, food consumption falls for both male and female walruses. Males fast for long periods to maintain mobile territories near females in estrus.

Pregnant and lactating female walruses consume $\geq 90\%$ more food than nonpregnant, nonlactating females of a similar weight (Fay 1982; Gehnrich 1984; Kastelein et al. 2000). Females with newborn calves, however, consume little food (Fay 1982). The mother's feeding activity likely is restricted because of the need to stay near and tend to the calf, especially until it is capable of deep, prolonged dives. Food intake then increases to meet the combined energetic needs of the mother and her calf. Walruses nurse their young for a relatively long period. Fisher and Stewart (1997) noted, through examination of stomach contents, that 3-year-old Atlantic walruses consumed mostly milk, although some benthic invertebrates were taken.

BEHAVIOR

Walruses are very gregarious, feeding, hauling out, and migrating in densely packed groups (Banfield 1974). In cool weather, they lie in passive body contact on land, the dominant walruses (those with long, unbroken tusks) taking the best positions while the most subordinate ones lie on the periphery of the herd. This body contact provides warmth and aids the molting process (Miller 1976). Individual bickering occurs (Banfield 1974), but if a member is attacked, the rest of the herd will defend it fiercely (Maxwell 1967). On land, they are usually placid, but males, especially if they are wounded, have been known to attack boats (Banfield 1974). A sentinel warns the entire herd of possible danger with a low whistling bellow, which sends the walruses into the sea (Maxwell 1967). In addition to helping walruses climb out of the water onto the ice, the tusks may also have a role in the social structure of the herd by establishing dominance (Reeves 1978). The distinctive male gong or bell-like phonation (Schevill et al. 1966) is thought to be associated with courtship display, as are songlike responses (Ray and Watkins 1975). Walruses do not have territories or harems. Mating occurs during the spring migration. Mothers will defend their young and will often clasp their young calves to themselves with their foreflippers to protect them. If the calf becomes tired in the water, the mother will sometimes carry it on her back (McClung 1978).

MORTALITY

Nonhuman predation is not a major cause of death for walruses, as the adults make a formidable adversary for both killer whales and polar bears. *Odobenus rosmarus* appears to be relatively free of parasitism and disease. Death of young by crushing during a stampede to the sea and as a result of interspecific aggression among males does occur (Reeves 1978). Examination of teeth has shown that walruses can reach 40 years of age (FAO 1976; Hull 1994).

PINNIPED MANAGEMENT IN NORTH AMERICA

Pinniped management in North America is often complicated. It may involve two or more countries, active legislation, and people with political, economic, scientific, and other interests. This section cannot deal with the topic in its entirety, but will serve as an informative introduction.

WHO IS INVOLVED

Marine mammals do not recognize international boundaries, so any discussion of the management and conservation of seals, sea lions, and walruses in North America must encompass the management policies of the United States and Canada. To do this, it is important to understand the laws and agencies from which these policies arise. It is also important to be aware of some of the international agreements that have bearing on the policies regarding these animals.

Like all furbearing animals, some seals have long been used as a source of wearing apparel for either practical or aesthetic reasons. Recent interest groups, intent on preserving animals from this type of exploitation, have become part of seal management. Further complications arise because seals and humans compete for the same food source, so the rights of commercial fishermen become important considerations.

Unable to obtain enough fuel on land, humans are now exploring the seas for oil, particularly in the northern climes. The transport of oil products poses certain direct problems. The livelihood of the North American Inuit as well as a group of other Native peoples, the Aleuts, has been intimately tied to marine life for many centuries. Preserving the right to maintain this tie has been an important consideration founded in the laws of each country. In Canada, certain people are also permitted to hunt seals as a means of income. These people have taken worldwide criticism and have faced on-the-job problems that most industries have never had to face.

Curious, and intent on seeking answers, scientists have been studying pinnipeds, puzzling over their existence and making recommendations that often oppose opinions of other groups involved. Finally, government civil servants have been charged with the responsibility of applying the laws of the land and sea, weighing the interests of all, and then attempting to produce a satisfactory plan of resource allocation and program administration.

HISTORICAL BACKGROUND

A complicated cast of characters figures in the management of pinnipeds. Therefore, it is helpful to define where they fit in the system and how the system works. Enough has been written on the historical exploitation of some pinnipeds in North America for most people to realize that between 1800 and 1960 many seal populations were threatened with extinction. Ample documentation outlines and recounts the history of the great slaughters. Suffice it to say here, management of seals began after World War II in a very minor way. By the late 1960s and early 1970s, legislation began to appear to improve the life expectancy of the seals. Most of this legislation, which is examined in the next section, was part of a larger development of people's understanding of the environment.

UNITED STATES

Legislation. Four pieces of legislation are the foundations of the U.S. government pinniped management program. The first is the National Environmental Policy Act of 1969 (NEPA; P.L. 91–190, 42 U.S.C. 4321–4347, Jan. 1, 1970, 83 Stat. 852). The impetus for this act came from the Committee on Merchant Marine and Fisheries, which saw many conflicts and overlapping research projects regarding the environment and little cooperation among the departments involved. The purpose of the act was to declare a national policy and to promote preventative measures to protect the environment and the biosphere for the health and welfare of humankind. To carry out the creation of an understanding of the ecological system and natural resources of the United States, the act established the Council on Environmental Quality to review government policies and programs for conformity with the NEPA (Musgrave et al. 1998).

The second piece of U.S. legislation important to pinnipeds is the Marine Mammal Protection Act of 1972 (MMPA; 16 U.S.C. 1361–1407, P.L. 92-522, Oct. 21, 1972, 86 Stat. 1027). Formerly, states were responsible for the marine mammals on lands and waters under their jurisdiction. The MMPA investes marine mammal management authority in the federal government. This act recognizes that certain species of marine mammal have become depleted because of human activities, and seeks to ensure that they remain a significant part of the ecosystem. This might require measures to replenish depleted stocks and the protection of "areas of significance" for such stocks. The MMPA recognizes the aesthetic and economic value of marine mammals, and thus directs that the protection and conservation afforded to them must be conducted with the most sound resource management policies and that the primary objectives of such management should be "to maintain the health and stability of the marine ecosystem" (Musgrave et al. 1998).

Major provisions are as follows:

1. A moratorium is declared on the taking and importation of marine mammals and marine mammal products; permits are available for such purposes as scientific research, public display, photography, or enhancing the survival or recovery of a species.
2. During commercial fishing operations, incidental kill of marine mammals may be allowed under permit, provided that the method and equipment used meet certain requirements. The "immediate goal," however, in this instance would be to reduce such takings to "insignificant levels approaching zero mortality." The secretary of the Treasury could ban the import of any commercial fish from countries that did not meet the U.S. standards of incidental kill, and could demand reasonable proof from foreign governments that such standards were met.
3. The secretary may waive taking and importing if such a recommendation is made by the best scientific advice with sound principles of resource protection and conservation. However, products from nations whose criteria do not meet the aims of this act may not be imported, even for processing.
4. Endangered species may not be taken except with written permission for research purposes.
5. Indians, Aleuts, or Eskimos (Natives) who reside in Alaska and dwell on the coast of the North Pacific Ocean or Arctic Ocean are exempt from the provisions of this act if the taking of marine mammals is for subsistence purposes or for creating or selling authentic native handicrafts. Such taking must not be wasteful, and if the species is endangered, certain regulations may be made about their take by Native peoples.
6. Pelagic hunting is banned.
7. International treaties, conventions, and agreements made before this act was effective are still viable. (Certain recommendations and exceptions are discussed below in the section International Agreements.)
8. Persons and vessels are banned from taking marine mammals or using ports for importing, and they may not offer any such animals or their products for sale.
9. It is unlawful to import any marine mammal that at the time of taking is pregnant, nursing, <8 months old, taken from a depleted species, or taken in an inhumane manner.
10. The secretaries of the departments involved are charged with making regulations about the taking and importing to ensure that such taking will not disadvantage the species or stocks. These regulations are to be made with the best scientific advice and evidence and in consultation with the Marine Mammal Commission. Consideration must be given to what effects the regulations will have on: (a) existing and future levels of marine mammal species

and population stocks; (b) existing international treaty and agreement obligations; (c) the marine ecosystem and related environmental considerations; (d) the conservation, development, and use of fishery resources; and (e) the economic and technological feasibility of implementation.
11. The secretaries of the Department of Commerce and the Department of the Interior must report annually through the *Federal Register* on the current status of all marine mammal species and stocks and on any actions taken on their behalf.
12. Permits for the taking or import of marine mammals may be granted by the secretary, subject to several stipulations.

There are severe monetary fines and prison terms for infractions against the MMPA. The act defines the term "take" to mean "to harass, hunt, capture, or kill, or attempt to harass, hunt, capture, or kill any marine mammal."

Three agencies act as administrative bodies for this act. The Department of Commerce, in which the National Oceanic and Atmosphere Administration (NOAA) operates, has jurisdiction over all pinnipeds except the walrus. The Department of the Interior has jurisdiction over the walrus. Title II of the act established an independent body, the Marine Mammal Commission, whose responsibility is to develop and review information, actions, and policies to achieve the objectives set forth in the MMPA.

A statute added in 1992 requires the secretary to establish a Marine Mammal Health and Stranding Response Program. This program deals with (1) collection and dissemination of reference data on marine mammal health and health trends; (2) correlation of health data with physical, chemical, and biological parameters; and (3) effective responses to unusual mortality events. The secretary must monitor strandings, collect information on rescue and rehabilitation procedures, collect tissues for analysis, and develop objective criteria for deciding when a rehabilitated marine mammal can be released to the wild. Agreements can be entered into with persons to take marine mammals in response to a stranding (Musgrave et al. 1998).

In 1994, the MMPA was amended (P.L. 103-238), establishing a new vehicle to govern the taking of marine mammals incidental to commercial fishing, which included "the preparation of stock assessments for all marine mammal stocks in waters under U.S. jurisdiction, development and implementation of take reduction plans for stocks that may be reduced or are being maintained below their optimum sustainable population levels due to interactions with commercial fisheries, and studies of pinniped–fishery interactions" (Buck 1995). Also, Section 120 was added to address interactions between pinnipeds and fishery resources. States may obtain lethal-take authority to protect certain depleted salmonid stocks, and the National Marine Fisheries Service (NMFS) must assess and submit reports on pinniped–fishery interactions along the west coast and in the Gulf of Maine aquaculture industry (Marine Mammal Commission 2000). Since the 1994 amendments became law, NMFS has published several regulations in the *Federal Register* to implement requirements under the act. An amendment in 1997, the "good samaritan exemption," allows the taking of a marine mammal in cases where it is entangled in fishing gear or debris and can be freed and released with reasonable care (Musgrave et al. 1998).

Another important piece of legislation is the Endangered Species Act of 1973 (ESA; 16 U.S.C. 1531–1544, 87 Stat. 884). Although the United States was previously involved in international agreements about endangered flora and fauna, this act provides national recognition that such species exist in a state of extinction or near extinction because of economic growth and development and the lack of adequate conservation measures to protect such animals. The act recognizes that such endangered species are part of an ecosystem, and that ecosystems are in need of protection and preservation as well. The secretary of the interior is charged with the responsibility of using any viable method to bring threatened populations to the point where they are no longer considered endangered. The ESA was amended in 1982 to include habitat protection plans (Green-Hammond 1980; Musgrave et al. 1998).

"Species" is defined by the act to mean either a species, a subspecies, or, for vertebrates only, a distinct population. Once a species is listed, the ESA requires the development and implementations of recovery conservation plans, which identify mitigation measures to be initiated to improve the species's status (Buck 1995).

The fourth piece of legislation is the Fishery Conservation and Management Act of 1976 (FCMA; P.L. 94-265, approved Apr. 13, 1976; 16 U.S.C. 1801–1882; 90 Stat. 331). Later renamed the Magnuson Fishery Conservation and Management Act, this act was a response to the need for a program of conservation and management to protect the fish stocks in U.S. waters from overfishing and to develop those fisheries that were not being used. Exclusive federal management was vested in NMFS, within NOAA of the Department of Commerce, and the secretary had jurisdiction over the Fishery Conservation Zone established by the act. This zone coincides with the U.S. territorial waters of 200 nautical miles (approximately 350 km) from the coast. The fisheries conservation zone was superseded by an Exclusive Economic Zone (EEZ), by presidential proclamation in 1983. The EEZ excludes near-shore waters under state jurisdiction.

Conservation and management were defined in the act as measures designed to avoid irreversible or long-term effects on the marine environment and measures useful in rebuilding, restoring, or maintaining any fishery resource and the marine environment. Fishery resources include stocks and their habitats. The act specifies that these fishery resources are to be used to get the optimum yield but prevent overfishing. To implement this policy of management and conservation, the act required the development of fishery management plans (FMPs), which were to be the responsibility of the new agencies established by the FCMA, namely, eight regional fishery management councils (Green-Hammond 1980; Buck 1995).

With the passage of the Sustainable Fisheries Act in 1996, the act was extensively amended and renamed by Congress the Magnuson–Stevens Fishery Conservation and Management Act. The act now contains, in particular, provisions addressing a transition to sustainable fisheries. All fishery management plans must include a description and identification of essential fish habitat, adverse impacts, and actions to conserve and enhance habitat (Musgrave et al. 1998).

The Marine Protected Areas Executive Order (#13158), signed in May 2000, directed federal agencies to take a number of specific actions to strengthen existing marine protected areas (MPAs) and build a national system of MPAs. The order defines MPAs as "any area of the marine environment that has been reserved by Federal, State, territorial, tribal or local laws or regulations to provide lasting protection for part of all the natural and cultural resources therein" (*Federal Register* 2000). The goal is to use MPAs, in combination with other management tools, to preserve, protect, and sustainably manage ocean and coastal resources (Salm et al. 2000).

Administrative Bodies. In the United States, four separate government bodies have jurisdiction, input, or advisory capabilities for management of seal and walrus populations and their environment. The Departments of Commerce and Interior and the Marine Mammal Commission are, however, the primary management agencies. Their responsibilities for marine mammals are set out in the MMPA, which may be regarded as the cornerstone of conservation and management of marine mammals in the United States. The other agency that has some impact on marine mammal populations is the Council on Environmental Quality.

It is important to note that whereas of these departments or agencies may work together on certain aspects of seal management, they are legislatively separate. The Marine Mammal Commission and the Council on Environmental Quality act as watchdogs, reporting to the Congress on the two departments charged with administering the management policies. As objective observers, they offer another perspective to many areas, and may point out areas for research or management.

The Department of Commerce contains the NOAA and its subdivision the NMFS, or "NOAA Fisheries." NOAA and NMFS manage all seal populations in the United States. NOAA's responsibilities with regard to marine mammals are best defined by the Coastal Zone Management Act, 1972; the Marine Mammal Protection Act,

1972; the Marine Protection, Research and Sanctuaries Act, 1972; the Endangered Species Act, 1973; and the Magnuson–Stevens Fishery Conservation and Management Act (MFCMA), 1976; and subsequent amendments.

As stated, under provisions of the MFCMA, eight regional fishery management councils were established under the auspices of the Department of Commerce; seals are specifically under NMFS management. Their districts are New England, Mid-Atlantic, South Atlantic, Gulf of Mexico, Caribbean, Pacific, North Pacific, and Western Pacific. Their jurisdiction extends outside of state waters, but within the fishery zone created by MFCMA. Each council consists of voting members—those appointed by the secretary of commerce, the governors of pertinent states, and the regional director of NMFS—and nonvoting members—representatives of the U.S. Fish and Wildlife Service, the Coast Guard, the Marine Fishery Commission, and the State Department. Each council has its own scientific and statistics committee (Green-Hammond 1980; Buck 1995).

Certain species of seals occur within the jurisdiction region of each council, and certain fishery management plans could have broadly based effects on seal populations. The breakdown of the relevant groups is as follows.

North Pacific Council. Northern fur seal, bearded seal, Steller's sea lion, walrus, ribbon seal, ringed seal, larga seal, harbor seal, northern elephant seal.

North Pacific FMPs: The council has been developing an ecosystem approach for management of North Pacific groundfish fisheries, and measures have been implemented to reduce potential impacts of localized depletion of prey for various seal species. Certain areas around rookeries and haulouts have been closed to avoid disrupting marine mammals and to reduce competition from fisheries. Regulations have been established to reduce direct mortality based on the criterion that incidental mortality should have negligible impacts on the affected species or stock (Witherell et al. 2000).

Pacific Council. Northern fur seal, Steller's sea lion, northern elephant seal, harbor seal, California sea lion, Guadalupe fur seal.

Western Pacific Council. Hawaiian monk seal.

Western Pacific FMPs (S. Spalding, Media and Education Specialist, Western Pacific Management Council, NMFS, NOAA, pers. commun., 2000): Crustaceans and Bottomfish and Seamount Groundfish Fisheries FMP: Provisions exist to ensure that these fisheries do not affect the Hawaiian monk seal, such as by establishing observer programs, 0- to 10-fathom area closures for the lobster fishery, and protected species workshops.

Pelagics FMP: The Hawaiian longline fishery is not allowed within 50 miles (80 km) of the Northwestern Hawaiian Islands.

Coral Reef Ecosystem FMP: This puts strict limits on any potential future fishery for coral reef resources and includes no-take zones that constitute 14% of the coral reef habitat in federal waters of the Northwestern Hawaiian Islands (foreign fishing was removed from the area after passage of the Magnuson Act in 1976).

New England Council. Harbor seal, hooded seal, gray seal, harp seal.

Mid-Atlantic Council. Harbor seal (transients: hooded seal, harp seal, gray seal).

Mid-Atlantic FMPs: Summer Flounder, Scup and Black Sea Bass FMP; Atlantic Mackerel, Squid and Butterfish FMP; and Bluefish FMP: Each of these could potentially involve prey for seals and could affect seal prey resources. Interactions between seals and mid-Atlantic fisheries, however (either directly through bycatch or indirectly through prey competition), are considered insignificant (R. Merrick, Chief, Protected Species Branch, NMFS, NOAA, pers. commun., 2000).

South Atlantic Council. Transients: harbor seal, hooded seal.

The department of the interior contains the U.S. Fish and Wildlife Service (USFWS). The USFWS has held its present structure and sphere of influence since 1974, when Congress transferred the old Bureau of Commercial Fisheries to the Department of Commerce. The Bureau of Sport Fisheries was incorporated into the new USFWS. Accordingly, it maintains a responsibility for wild birds, mammals (except certain marine mammals), inland sport fisheries, and specific fisheries research activities. Specifically, it has jurisdiction over the walrus (U.S. Office of Federal Register 1979).

By virtue of Title II of the MMPA, the Marine Mammal Commission was established as an independent body. It reviews and makes recommendations on domestic and international actions and policies of all federal agencies with respect to marine mammal protection and conservation and with carrying out a research program. The commission consists of three presidential-appointed commissioners, who in turn appoint a nine-member committee of scientific advisers knowledgeable in marine ecology and mammalian affairs (Marine Mammal Commission 2000).

The Council on Environmental Quality may conduct studies and research related to ecology. The president directly appoints the three council members. They make recommendations to the president on national environmental policies, monitor changing trends in the environment, and appraise any government policies affecting the environment (U.S. Office of Federal Register 1979).

In August 2000, the Oceans Act of 2000 was approved (P.L. 106-256) to establish a Commission on Ocean Policy, with 16 members. The purpose of this legislation is to "comprehensively evaluate concerns that cannot be viewed effectively through current processes or through privately commissioned studies." The intention of the commission is to assess U.S. oceans and coastal policies and tackle the complex issues of marine conservation.

CANADA

In Canada, commercial sealing is part of a national fishery industry. In the United States, only subsistence sealing is allowed. Harp and hooded seals are legally harvested in the eastern Canadian provinces by commercial sealers who depend on sealing for a living. The vast difference between the legislation of the two countries is that, in the United States, lawmakers have responded to various groups who sought to protect all marine mammals from commercial harvest. In Canada, considering the needs of people involved, there is no impetus by government to call a moratorium on sealing. The seal herds continue to be a renewable resource; management is based on the best scientific knowledge available.

Legislation. For Canada, the major piece of legislation affecting seals is the Fisheries Act (R.S., c. F-14, s.1.), first enacted in 1868, which defines "fish" to include marine mammals, and any parts and juvenile stages of marine mammals. Under the act, a number of marine mammal protection regulations were produced between 1966 and 1982. The Seal Protection Regulations, 1966 (which were amended annually), governed seal management in Canada. Specific Walrus Protection Regulations were made by Order in Council in 1980. The Fisheries Act delegates to the minister of fisheries and oceans various provisions regarding fisheries in Canada. With regard to seal fisheries, the act clearly provides protection for the seals and the sealers: "No one shall with boat or vessel or in any other way during the time of fishing for seals knowingly or willfully disturb, impede or interfere with any seal fishery or prevent or impede the shoals of seals from coming into such fishery or knowingly or willfully frighten such shoals" (Fish Act. Amended List 1979).

The Seal Protection Regulations (as with those for the walrus) had more to do with regulating the exploitation of seals than with providing increased protection for species and their habitats. These regulations detailed every aspect of seal hunting and protection of seal herds, from season opening and closing dates, to types of instruments approved, to animals permitted to be taken. In 1993, individual protection regulations were consolidated into a single set of regulations called the Marine Mammal Regulations (SOR/93-56, under the Fisheries

Act): "These Regulations apply in respect of the management and control of (a) fishing for marine mammals and related activities in Canada or in Canadian fisheries waters; and (b) fishing for marine mammals from Canadian fishing vessels in the Antarctic."

Provisions include the following:

1. A clear definition is given of those permitted to hunt (fish) for seal or walrus, and the fee for licenses is set out. There are a great many requirements for permits of sealing licenses, each one clearly stated. Generally speaking, residents, native Inuit or Indians, and persons holding a valid sport-sealing license are permitted to hunt. Provision is also made for scientific research. (The exclusion of Native peoples from usual restrictions has become a cause leading to high-seas conflicts).
2. An annual amendment sets out which pinniped species may or may not be hunted, announces a quota (based on a total allowable catch, or TAC) on each species per person or group, and specifies in which areas the seals or walrus may be hunted and in what manner.
3. For seals, the methods of transportation to and from the hunt areas are indicated, including types and lengths of vessels and aircraft, safety factors, and equipment specifications. Permits for such vessels are granted entirely at the discretion of the minister of fisheries and oceans. Specifications are made about the distance at which aircraft may land from the seal herds.
4. Observers and nonparticipants of the hunt are carefully regulated by permit only, and application of such permits must give definite information on purpose, equipment, and length of stay.
5. Season opening and closing dates may be changed by ministerial decree annually.
6. Hunting techniques are very clearly spelled out, with strict definitions of instruments, methods of humane slaughter, and regional particularities.
7. The regulations detail the means to judge when an animal is dead, and specifies methods and timing of skinning the dead animal.
8. Certain areas in the eastern coastal waters are specified where seals may not be taken.
9. Tagging, marking, and moving of live seals are forbidden without written permit.
10. Time periods in the day when seals may be hunted in the Gulf of St. Lawrence and the Front off Newfoundland–Labrador are set out.
11. The vessel master is specified as being in charge of everyone on his boat with regard to the provisions of these regulations.

In 1996–1997, the federal government tried unsuccessfully to pass national endangered species legislation. In April 2000, the government again introduced legislation, the Species at Risk Act (Bill C-33). Both failed to pass before an election call. At present, the list of species at national risk in Canada is determined by the Committee on the Status of Endangered Wildlife in Canada (COSEWIC), of which the Canadian Nature Federation is a founding member. COSEWIC is a national scientific body, which considers scientific status reports of species and then evaluates their status. A species can be at risk within one area of Canada, but have healthy populations in another. It is felt that any new federal endangered species legislation should include a requirement that the status of species be assessed and the legal list of species at risk be determined by COSEWIC scientists, not politicians. Also, legislative proposals should provide assurance that the habitats of nationally endangered species will be protected.

The 1997 Oceans Act [C.S.C. Ch.O-2.4, 1996, c. 31, in force Jan. 1997 (SI/97–21)] formally recognizes Canada's jurisdiction over its ocean areas through the declaration of an Exclusive Economic Zone and a Contiguous Zone. The act required the Department of Fisheries and Oceans to lead the development of a National Oceans Management Strategy, based on the principles of sustainable development, integrated management of activities in estuaries and coastal and marine waters, and the precautionary approach (a commitment to err on the side of caution) (Fisheries and Oceans Canada 1997). The act also provides for the consolidation of most federal ocean responsibilities under one organization. In fisheries, this builds on an already strong base, including extensive at-sea observer programs and a strong scientific program as described by Parsons (1993). Under the authority of the Oceans Act, Race Rocks off British Columbia, home to a unique diversity of Pacific marine life including sea lions and seals, was designated as Canada's first Marine Protected Area in September 2000.

Administrative Bodies. In Canada, one department or ministry has a legislative mandate to manage pinnipeds, Fisheries and Oceans. There is an advisory committee as well, but, unlike the U.S. watchdog agency, the Canadian contingent reports to the ministry that created it. The duties of the Department of Fisheries and Oceans include all matters in the area of renewable resources that are designated through the Parliament of Canada and are not by law assigned to another department. Pinnipeds are part of the fisheries resources. In April 1995, the Department of Fisheries and Oceans merged with the Canadian Coast Guard with a vision "to be a world leader in ocean and aquatic resources management" (Fisheries and Oceans Canada 1997).

In 1977, with the extension of the offshore fisheries jurisdiction to 200 nautical miles, the Canadian Atlantic Fisheries Scientific Advisory Committee (CAFSAC) was established as a forum for the management of the Atlantic fisheries resources. In 1993, CAFSAC, along with the Atlantic Groundfish Advisory Committee, combined to form the Fisheries Resource Conservation Council (FRCC), a new partnership of scientists, academics, and the various sectors of the Atlantic fishing industry. The council holds public hearings on resource assessments and conservation measures, provides the federal minister of fisheries and oceans with public written advice on recommended harvest levels and conservation measures, and advises the minister on matters dealing with straddling and transboundary stocks that are managed by international organizations, such as the Northwest Atlantic Fisheries Organization (NAFO). The FRCC generally advocates a "precautionary" and "ecosystems" approach to decision making.

In 1971, observers at the annual seal hunt in the Gulf of St. Lawrence, Canada, felt that there was a need for a direct and effective channel to the then minister of fisheries regarding sealing recommendations. These observers were from such diverse fields as the Canadian Federation of Humane Societies, university scientists, and representatives of the International Society for the Protection of Animals. As a result, the Committee on Seals and Sealing (COSS) was formed. This committee of six persons included three scientists and three nonscientists. It functioned in an advisory capacity and reported directly to the minister of fisheries and oceans. Recommendations were based on investigations into the socioeconomic, biological, ecological, and humane aspects of the seal hunts in Canada, as well as any international aspects of sealing that affect Canada.

The objectives of the committee were (1) to advise on quota changes before the harp and hooded seal hunt, and over a long time period to observe the different phases of the hunt; (2) to evaluate the size and composition of the herd with any suitable techniques, and to carry out and evaluate the methods of statistical analysis and make recommendations on changes if any; and (3) to investigate hunting methods, and carry out research and necessary changes. Although COSS mainly concerned itself with harbor, gray, harp, and hooded seals, it made recommendations on several species, including those in Antarctica. The committee operated for 12 years and, on completion of its main work, was disbanded. Since COSS, there have been many government-created advisory committees and commissions formed to direct the government on policy concerning seal management and the seal hunt.

INTERNATIONAL AGREEMENTS

Both Canada and the United States have unilateral and multilateral agreements regarding marine mammals as well as membership in many international conventions whose ratification entails conservation and protection of, and research into, either the animals or their habitat. Some examples follow.

Interim Convention on Conservation of North Pacific Fur Seals Convention, 1957; and As Amended. By 1911, pelagic sealing had seriously depleted the North Pacific fur seal herds, and a meeting in Washington involving the United States, Great Britain (acting for Canada), Japan, and the (then) Soviet Union resulted in a convention that protected the species and banned pelagic sealing. The United States and the Soviet Union would conduct their own hunts, but would each provide Japan and Great Britain with 15% of the take. The United States banned commercial sealing for 3 years. The herds increased so remarkably that by 1941 Japan broke the agreement, claiming the seals could interfere with its commercial fisheries concerns. From 1942 until 1957, the seals were protected by a provisional agreement between Canada and the United States. In 1957, Canada, Japan, the Soviet Union, and the United States convened a new North Pacific Fur Seal Convention. Research and management of the North Pacific fur seal is coordinated by the Fur Seal Commission, a body provided by the convention, consisting of representatives of the four governments. Research programs and harvesting rates are recommended by a standing scientific committee on an annual basis. From 1957, the commercial harvest on the Pribilof Islands was conducted by the United States under the authority of the Convention.

To effect the Convention on the domestic level, the Fur Seal Act of 1966 (16 U.S.C. 1151–1187) was enacted; it renewed the authority of the secretary of the interior to administer management of the Pribilof Islands as federal territory. In 1970, that authority was transferred to the Department of Commerce. The Fur Seal Act generally prohibits the taking of fur seals except for Native peoples dwelling on the coasts of the North Pacific Ocean. In addition, the act codifies the obligations of the United States as agreed to at the Interim Convention, and as subsequently amended, and required the establishment of the Pribilof Islands Trust for the benefit of the Natives of the Islands. The trust's purpose, at present, is to promote a "stable, self-sufficient and diversified economy not dependent on sealing" (Musgrave et al. 1998).

Under the MMPA, the secretary of commerce, through the secretary of state, is charged with initiating agreements with nations to protect or conserve marine mammals, and where the United States is specifically involved in multilateral agreements such as the North Pacific Fur Seal Convention, to initiate amendments if necessary that comply with the policies and purposes of the act. At the time the act was enacted, the secretary of commerce was to ensure, in conference with the Marine Mammal Commission, that the North Pacific fur seals were at their optimum sustainable populations and to review the North Pacific Fur Seal convention for consistency with the MMPA. If necessary, the secretary of state could then initiate negotiations with other parties of the Convention to protect the herds. In 1973, largely because of depressed population levels on the Pribilof Islands, a moratorium on commercial harvesting of fur seals had been established at St. George Island.

In August 1979, a bill was introduced into the U.S. Congress calling for the termination of the Interim Convention on the Conservation of the North Pacific Fur Seal and an end to the harvest. It also recommended the necessary changes with member nations to the treaty, creation of the Pribilof Islands as a wildlife refuge and marine sanctuary, and a buffer around the 200-mile limit in this area (National Marine Fisheries Service 1999).

A 1984 protocol amending the Interim Convention was agreed on, but the convention expired before the protocol came into force. NOAA determined that a commercial harvest could not take place under domestic law. Aleuts, however, were still allowed to take subadult males for subsistence purposes. The governments of Canada, Japan, USSR (Russia), and the United States did want the North Pacific Fur Seal Commission to continue its contribution to the conservation and management of North Pacific fur seals and the ecosystem that supports those resources. In 1984, a moratorium on commercial harvesting was imposed on St. Paul Island. Since 1984, studies have been conducted independently by cooperating former convention member nations.

Northwest Atlantic Fisheries Organization. The International Commission for the Northwest Atlantic Fisheries (ICNAF) was established under the International Convention for the Northwest Atlantic Fisheries, which came into force in 1950. The first meeting of ICNAF was held in 1951 (Anderson 1998). The organization's members were from those countries that had coastal states in the area and whose interests were in the species of fish, seal, and other marine mammals within the waters of the Northwest Atlantic Ocean. This formation effectively brought the management of the northwest Atlantic population of harp and hooded seals in Canada within ICNAF's area of concern. These seals migrate within the boundaries of Canada and Denmark (Greenland). Norway and the (then) USSR were also interested in the Northern Atlantic populations. A Standing Committee on Research and Statistics (STACRES) was created, which provided personnel who were charged with making annual recommendations to the commission about seal management. Harp and hooded seals were added as a species group to the convention by a protocol in 1966 (Parsons and Beckett 1998).

In 1976, Canada (and other nations) extended the zone of fisheries jurisdiction to 200 nautical miles from the coast. This meant that there were changes in the management authority for seals, especially because harp seal populations were residing within the districts of Canada and Denmark (Greenland). This required a restructuring of ICNAF. The Convention on Future Multilateral Cooperation in the Northwest Atlantic Fisheries came into effect in 1979, and provided for the establishment of the Northwest Atlantic Fisheries Organization (NAFO). ICNAF was officially dissolved at the end of 1979. STACRES was replaced with its counterpart, the Scientific Council of NAFO, which provided scientific advice and assessments on harp and hooded seal management. In the case of stocks within Canadian waters, all final management decisions are with the Department of Fisheries and Oceans (Anderson 1998; Parsons and Beckett 1998). The Fisheries Resource Conservation Council (FRCC) is responsible for advising the minister of fisheries and oceans on Canada's position with respect to straddling and transboundary stocks under the jurisdiction of international bodies, such as NAFO. Total allowable catches for seals are now negotiated annually between Canada and the European Economic Community (representing Denmark). In northern Canada and Greenland, this effectively amounts to a very small, locally important, but relatively insignificant harvest by Native peoples.

INTERNATIONAL CONVENTION PROTOCOLS

Canada and the United States also subscribe to several international conventions concerned with the environment or with marine life and its habitats, which can affect seals. These are briefly listed below, with the exception of the International Union for the Conservation of Nature and Natural Resources (IUCN–The World Conservation Union), which is discussed more fully. More detailed information can be found in the United Nations Environment Programme (1977), Register of International Conventions and Protocols in the Field of Environment, and its supplements, and through the Center for International Earth Science Information Network (1996–2001).

LUCN–The World Conservation Union. IUCN–The World Conservation Union was founded in 1948 and is stationed in Gland, Switzerland. It is an independent international body whose membership is composed of member states, government bodies, private institutions, international organizations, scientists, and experts irrespective of political or social systems. Its objective is to influence, encourage, and assist societies throughout the world to preserve the natural environment and its resources, conserve the integrity and diversity of nature, and ensure that any use of natural resources is equitable and ecologically sustainable. The IUCN maintains active conservation programs, in cooperation with such agencies as UNESCO and FAO. The World Wildlife Fund (WWF), also located in Gland, is the basic fundraiser for IUCN, whose scientific and technical advice it receives. The WWF allocates its funds to projects either directly or through IUCN publicity programs, particularly focusing on the education of young people. There are national branches of the WWF in Canada and the United States.

Of the six IUCN volunteer commissions, the Species Survival Commission (SSC) is the world's largest species conservation network (Global Conservation Network for Species Survival). The SSC seeks to mobilize action by the world conservation community for species conservation, particularly those species threatened with extinction and those of importance for human welfare. To help to identify conservation priorities, the commission produces assessments like the "IUCN Red List of Threatened Species."

Within the framework of global conventions, the IUCN has helped more than 75 countries to prepare and implement national conservation and biodiversity strategies. Canada (CAN) and/or the United States (USA) have signed the following international conventions that are directly or indirectly relevant to pinnipeds. Country and date of entry are indicated.

1. Convention on Nature Protection and Wild Life Preservation in the Western Hemisphere (USA, 1942).
2. International Convention for the High Seas Fisheries of the North Pacific Ocean (CAN and USA, 1953); and Amending Protocol (CAN and USA, 1979).
3. International Convention for the Prevention of Pollution of the Sea by Oil, 1954, as Amended in 1962 and 1969 (CAN, 1958; USA, 1961).
4. Convention on the Continental Shelf (CAN, 1970; USA, 1964).
5. Convention on Fishing and Conservation of the Living Resources of the High Seas (USA, 1966).
6. Convention on the High Seas (USA, 1962).
7. The Antarctic Treaty (CAN, 1988; USA, 1961).
8. Convention on the Conservation of Antarctic Marine Living Resources (CAN, 1988; USA, 1982).
9. International Convention Relating to Intervention on the High Seas in Cases of Oil Pollution Casualties (USA, 1975).
10. Convention on the Prevention of Marine Pollution by Dumping of Wastes and Other Matter (CAN and USA, 1975).
11. Convention on International Trade in Endangered Species of Wild Fauna and Flora (CITES); (adopted 1 January 1974) (addresses the need by governments to complement conservation efforts with stronger support for the sustainable use of wildlife; objective is to protect certain endangered species from overexploitation by means of a system of import/export permits) (CAN and USA, 1975).
12. Convention for the Conservation of Antarctic Seals (CAN, 1990; USA, 1978).
13. Convention for the International Council for the Exploration of the Sea (amending protocol 1975) (CAN, 1968; USA, 1973).
14. Treaty on the Prohibition of the Emplacement of Nuclear Weapons and Other Weapons of Mass Destruction on the Sea-Bed and the Ocean Floor and in the Subsoil Thereof (CAN and USA, 1972).
15. Convention Concerning the Protection of the World Cultural and Natural Heritage (The World Heritage Convention) (CAN, 1976; USA, 1975).
16. Convention on the Law of the Sea (opened for signature 10 December 1982; entered into force 16 November 1994) [this treaty governs all aspects of the law of the sea, with substantive provisions regarding protection and preservation of the marine environment; provisions require prevention, reduction, and control of pollution of the marine environment; the conservation of living resources is entrusted to coastal states; in addition, special protection may be provided for marine mammals (Schiffman 1998)] (CAN, 1982).
17. Convention Establishing a Marine scientific Organization for the North Pacific Region (PICES) (CAN and USA, 1992).
18. Convention on Biological Diversity (opened for signature June 1992; entered into force 29 December 1993) (this convention requires legislation protecting endangered species and their habitat; National strategies are to be developed to conserve and use biological diversity in a sustainable manner, and these strategies made a part of overall national development strategies). (CAN, 1993; USA, 1994).

Management Policy and Application. How any group of animals is managed, wild or domestic, is dependent on the idea of what use those animals will assume. Whether it is decided to use the animal for food or clothing, to preserve and protect it for aesthetic or biological reasons, or to keep it as a companion, that idea of use must be formulated into a policy either formally stated or clearly understood. Ideas of "conservation [are complex issues] not solely related to policies of federal and provincial government for the encouragement of restraint and curtailment of demand, but more importantly to choices freely made by the public at large on life styles and on social goals, as well as to economic considerations" (National Energy Board 1977:1–67 to 1-68). The policies governing seal management in the United States and Canada differ in a very important way. The United States has made a policy statement that is definitely philosophical in nature, firmly stated and defined in legislation, and part of a much larger policy that affords protection to the environment. Canadian policy on pinnipeds is not philosophical in nature. It is somewhat more difficult to ascertain what Canadian policy is, because, unlike the Marine Mammal Protection Act, the Canadian Marine Mammal Regulations do not state moral values about marine mammals. Nevertheless, the policy is there, but it is formulated in a more abstract form. The Fisheries Act and its seal regulations are rules and regulations that apply to a policy that is understood as a basic premise. In other words, if there are rules about hunting seals or protecting seals, the people as a whole believe in hunting or protecting seals. Canadian policy is usually made known by statements by the government in power, or the philosophy of policy often exists in such forms as commission reports. For example, when the Mackenzie Valley Pipeline Commission (Berger 1977) made its recommendations to halt a decision for at least 10 years to settle Native claims, it was adopted as policy by the Canadian government (National Energy Board 1977). The philosophy of the policy is clearly stated.

Essentially, the United States and Canada share the same philosophy about pinnipeds that the herds must be protected. Canadians have chosen a middle-of-the-road position, which allows a small segment of their population to supplement their income from regulated hunting of harp and hooded seals. The United States has chosen a more extreme position, which does not allow any hunting, except by Native peoples.

If a policy is society's ideal, achievement of that ideal, or application of it, depends on the people's understanding of the concept, the correct methods for applying it, and the basic perception that the policy is correct or workable. It is in this area that breakdowns sometimes occur. This is an interesting time in the history of seal management because both the United States and Canada are discarding old methods, and new ones are being formulated to deal with the myriad problems facing marine populations as a whole. The policies are defined, but not yet the means to carry them out.

Many of the difficulties are common to all periods of transition. "Changes normally proceed more slowly than some elements in society wish" (National Energy Board 1977:1-68), and it is a matter of fact that the rate of change in all areas of our lives over the past few decades has been unequaled in human history (Toffler 1970). Computers have caused major changes whose effects are still being felt and are still to be realized. It was predicted in the late 1960s that management decisions would be made from a wide computer database by 1980. However, George (1979) found in a survey that major management decisions in all spheres of society were still being made by individuals at one or two top levels, with no indication that such data were widely used. A common problem, then and now, is that the data may be there, but interpretation is often difficult, ignored, or not relevant to those people in top decision-making positions.

Upper management decisions, particularly in government areas, are often a function of territoriality, with the unwritten law being "stay out of my area and I' ll stay out of yours." Many government decisions are politically motivated. Compounded with these realities are the complexities of the data, new opinions, and theories. In the United States, administrators are faced with stringent laws and a profusion of powerful and knowledgeable interest groups. Marine mammal management for them demands a tightrope act of balance between interpretation of the laws and the watchdogs, especially since the major shift in policy in the 1970s. In Canada, other factors, such as the failure of methods that have resulted in the collapse of certain fish stocks, have acted as an impetus in this regard. Whereas the legislation in the United States is based on a multispecies-oriented philosophy, single-species strategies, often disguised as multidimensional, have been used in pinniped and fisheries management.

The fate of seals is in fact intimately tied to fisheries management. Although, on paper, seals are protected from excessive hunting by humans, they are at the top of the food chain. Consequently, they compete with humans for food resources as well as being in danger from exploitation, fisheries interactions, resource exploration, transportation factors, and pollution. It is apparent that management of pinnipeds, or any other marine animal, can no longer be spoken of in isolation. The ecosystem in which seals, fur seals, sea lions, and the walrus live is being used consistently and intensively by humans, and it is an environment that is interdependent on all levels. It is also a world that is largely unfamiliar to most people.

Current Trends in Management Methods. Because the policies of pinniped management in Canada and the United States and the concepts that are the basis for these policies were formulated in the late 1960s and early 1970s, it is critical to understand some of the old concepts. Pinniped and fisheries management have traditionally been based on a theory of maximum sustainable yield (MSY). This strategy has undergone marked change. MSY is defined as "the greatest harvest that can be taken from a self-regenerating stock of animals year after year, while maintaining constant average size of the stock" (Holt and Talbot 1978). This procedure has been an integral part of the methods used to harvest and preserve seal and fish stocks, but it has been overused; it was never intended to be the "sole conceptual basis for management" (Green-Hammond 1980:G-5). There were many faults with the MSY strategy. The concept has tunnel vision, ignoring trophic-level relationships, symbiotic or commensal relationships, environmental changes, and human influences (Holt and Talbot 1978). Most of the methods used for the MSY calculations include such factors as recruitment, growth, natural mortality, fish mortality, and fishing effort (Ricker 1975), and most of this information comes from fishing. If there is no exploitation, information stops. In general, collection of data has been difficult and calculations have been erroneous, resulting in overexploitation.

Perhaps the most challenging conclusion that could be drawn from this problem is that there was no alternative that dealt with the environment as a whole (Gulland 1976). Fisheries management plans around the world were largely based on a preventive and trial-and-error approach, yet some fisheries collapsed, indicating the weakness of this approach. In addition, fishery management had considerable impact on seal populations. The time was ripe for innovative methods.

In the United States, there are some very powerful groups and organizations (e.g., Greenpeace Foundation, Center for Marine Conservation, Earthjustice, and Natural Resources Defense Council) with considerable knowledge, wealth, and lobbying abilities, which monitor all management decisions about marine animals. Many of these organizations were instrumental in making some of the very early fundamental policy changes about marine mammals, and continue to campaign through science-based advocacy, research, and public education. The Hawaiian Environmental Alliance (KAHEA), for example, was instrumental in the movement to secure the creation of the Northwestern Hawaiian Islands Coral Reef Ecosystem Reserve, primary habitat for the Hawaiian monk seal.

In Canada, groups such as the International Marine Mammal Association (IMMA), the Canadian Marine Environmental Protection Society (CMEPS), and the Lifeforce Foundation, are only now having far-reaching impact. Various organizations, such as the International Fund for Animal Welfare (IFAW Canada) and the Animal Protection Institute (API), are specifically dedicated to bringing an end to the commercial East Coast seal hunt, largely through education, debates, and demonstrations. Seal hunting is a right afforded by law to certain groups of people in Canada; dissenters and observers have been ineffective in changing that existing law, but have forced humane slaughtering techniques to be adopted (Lavigne 1978).

In the Marine Mammal Regulations, the Canadian minister of fisheries and oceans is granted the right to implement any measure seen fit to protect the sealers or the seals. The opinions of dissenters have always been welcomed in open forums, but management decisions have been changed only through the structured channels of government. Although world attention has focused on the seal hunt for many years, the emotionalism of the issue often exceeds reason or fact.

Watchdog agencies in Canada have only advisory capabilities. Although they are invited to make recommendations, their abilities to change policy are limited. The U.S. watchdog agencies have an impact that is somewhat more effective, if less direct, as they do not report to the body they are reviewing.

To illustrate the effectiveness of new management methods in certain situations, two examples of current management situations are detailed. Fisheries management plans encompass a wide scale of variables that are addressed in an overall management plan.

FISHERIES MANAGEMENT PLANS

The terms of the Magnuson–Stevens Fishery Conservation and Management Act (MSFCMA) require that fisheries management plans be formulated and approved before any new fisheries industry is implemented. Thanks to new legislation and international agreements, fishery management is moving toward a new level where MSY is treated as a limit to be avoided rather than a target that can be exceeded (Witherell et al. 2000). Ecosystem-based management strategies are starting to be adopted for marine ecosystems (National Research Council 1999). The basic ecosystem consideration is a precautionary principle "developed over the past ten years as a policy measure to address sustainability of natural resources in the face of uncertainty" (Witherell et al. 2000). Because the impact of human activity cannot be known precisely, a conservative approach should be applied (Dovers and Handmer 1995), especially if there is much uncertainty and potential for irreversible high-cost mistakes (Garcia 1996).

Alternatively, a "precautionary approach" can be used, which is a relaxation of the precautionary principle developed to deal with systems like fisheries that are slowly reversible but often difficult to control, not well understood, and may be subject to changing environment and human values. The approach is defined as "applying judicious and responsible fisheries management practices, based on sound scientific research and analysis, proactively (to avoid or reverse overexploitation) rather than reactively (once all doubt has been removed and the resource is severely overexploited), to ensure the sustainability of fishery resources and associated ecosystems for the benefit of future as well as current generations" (Restrepo et al. 1999).

The goals of current fishery management plans as stated by Witherell et al. (2000) include conserving fishery resources for optimum yield, maintaining productive fish habitats, and minimizing interactions with other elements of the ecosystem. In the North Pacific, for example, specification of harvest limits for groundfish is done in a precautionary manner (Thompson 1996). Total removals of groundfish are controlled by annual catch limits established for each stock. For each target stock, three harvest levels are set, based on calculated definitions, corresponding to the overfishing level (OFL, unacceptable harvest), the acceptable biological catch (ABC, acceptable from a biological perspective), and total allowable catch (TAC, annual catch limits for the fishery) (Witherell et al. 2000). All fish caught in any fishery, including by-catch, whether landed or discarded, are counted toward TAC for that stock. Fisheries can be closed before the TAC is reached, based on comprehensive onboard observer data and reports provided by the fleet (Volstad et al. 1997). In addition, management measures have been taken in the groundfish fisheries to minimize potential impacts on other ecosystem components such as marine mammals (Witherell et al. 2000).

POPULATION AND STOCK ASSESSMENT

A need to develop a reliable estimate of the total population size and trends is likely the greatest problem facing any pinniped conservation program. The lack of accurate population data makes it difficult to determine the effects of natural events and human-caused activities on a species.

In the United States, under amendments to the MMPA (Section 117, Stock Assessments), the establishment of three regional scientific review groups was required, to advise and report on the status of marine mammal stocks within Alaska waters, along the Pacific Coast (including Hawaii), and along the Atlantic Coast (including the Gulf of Mexico). Basically, each stock assessment report includes (1) a description of the stock's geographic range, (2) a minimum population estimate, (3) current population trends, (4) current and maximum net productivity rates, (5) optimum sustainable population levels and allowable removal levels, and (6) estimates of annual human-caused mortality and serious injury through interactions with commercial fisheries and subsistence hunters. These data help managers evaluate the progress of each fishery toward the goal of zero-fishery-related mortality and serious injury of marine mammals. Stock assessment reports are reviewed annually for stocks designated as strategic (e.g., Steller's sea lion, northern fur seal), annually for stocks where significant new information is available, and at least once every 3 years for all others. The initial criterion of reliable estimates of abundance is essential, however, before any accurate predictions on stock trends can be made.

New concepts and new methods of determining the population of the northwestern Atlantic harp seal population have been developed in the last 3 decades. The Canadian government revises the TAC of harp seals annually based on the best population estimates of the size and condition of the herd. Accurate estimates, therefore, are important. Methods of obtaining such figures have ranged from eyeball "guestimates" to survivorship indices (Sergeant 1969; Benjaminsen and Øritsland 1975), sequential population analyses (Lett and Benjaminsen 1977), and other visual methods relying on a systematic fixed method of estimation (Lett and Benjaminsen 1977). The development of ultraviolet photography, which uses absorbance of ultraviolet radiation by the hair of a polar species, ultimately improved population estimates. The method also encompasses an aerial survey and ground truthing, which results in a high level of accuracy (Lavigne 1975; Lavigne et al. 1982). White-coated pups, which were not seen in previous methods, can therefore be photographed, even against the white snow and ice conditions.

One of the issues that has arisen from the collection of survey data has been the development of computer models that attempt to accommodate all variables. The resulting disputes over which models are better and more reliable have been active and at times prolonged. For the harp seal, the first model (Allen 1975) still stands undisputed, however. One thorough discussion of modeling and other management methods for seal populations was the IUCN-sponsored workshop held at the University of Guelph in 1979 (Beddington and Lavigne 1979).

Modeling also serves to illustrate the changing role of the biologist in management situations. Computer simulation experiments are highly sophisticated, complex, and always in innovative stages of development. As a result, there are very few persons conversant with the technicalities of the entire situation. The new biologist must be part statistician, part mathematician, part political scientist, and part traditional biologist. The realities of the situation demand liaison people who can brief other managerial persons and the public in simpler terms about the problems or theories behind such complex models.

For management purposes, it is also vital to recognize whether there are separate stocks within each population of a given species, and, if so, that each is treated accordingly. Stock structure can be determined by various means, including geography, morphology, and life history. One means of classification is based on the phylogeographic approach of Dizon et al. (1992), where data on distribution, population response, phenotype, and genotype are combined to help separate stocks.

Present-day management decisions are no longer made on how well a worldwide or even oceanic population is doing. The health of a population is geared to critical habitats. Each stock is considered unique, the loss of which is irreversible in terms of specific adaptions to a specific place. Reintroduction from another region, no matter how plentiful, is considered a last resort for conservation of a species and its habitats. Population estimates and stock determinations are the first stepping stone toward learning the status of a species. Standardized surveys must continue on a regular basis so trends can be followed, and population models must be adaptable and reliable.

INTERNATIONAL MANAGEMENT

A response to worldwide pollution and international industrialization cannot be made using only methods that are local in scale. Certain issues and concerns are important nationally and can therefore be handled at that level. Where the problem is international, however, local or national remedies are often inadequate.

In 1980, the IUCN, in cooperation with the United Nations Environmental Programme (UNEP), the World Wildlife Fund (WWF), FAO, and UNESCO, released a document entitled "World Conservation Strategy," which stressed the need for a global conservation strategy for management of shared resources (International Union for the Conservation of Nature and Natural Resources, United Nations Environment Programme, and the World Wildlife Fund [IUCN, UNEP, WWF] 1980). It explains the contribution of living resource conservation to human survival and to sustainable development, identifies the priority conservation issues and the main requirements for dealing with them, and proposes ways for effectively achieving the Strategy's aim. "Human beings in their quest for economic development and enjoyment of the riches of nature, must come to terms with the reality of resource limitation and the carrying capacities of the ecosystems, and must take account of the needs of future generations. This is the message of conservation" (IUCN, UNEP, WWF 1980: Preamble and Guide, Foreword). Many countries have adopted conservation strategies formulated within the guidelines suggested in this document.

Most coastal nations have extended their territory to include a 200-mile limit. The time is ripe to renegotiate international agreements regarding regional fisheries commissions to ensure that such areas will be managed as ecological entities. The United Nations Convention on the Law of the Sea of 1982 provided several mechanisms for promoting responsible management of marine fisheries, but it was not until the 1990s that work began on developing a precautionary approach to fisheries management (FAO 1995a, 1995b).

The World Conservation Strategy endorsed special recommendations for some regions that are critical habitats, such as the Arctic. The Arctic would take a very long time to recover from any damage, and should be considered a priority by member nations. IUCN suggested mapping of critical ecological areas, marine and terrestrial; long-term management guidelines; and protected areas. Measures should be taken, including joint research, to improve protection of migratory species breeding within the Arctic and wintering inside or outside the region. Studies are needed on the impact of fisheries and other economic activities in the northern seas on ecosystems and nontarget species. The successful Agreement on Conservation of Polar Bears serves as a model for developing agreement among the arctic nations on the conservation of the region's vital biological resources.

Pinnipeds cross open ocean and, as the World Conservation Strategy states, the open ocean and its living resources may be exploited by anyone. It is a critical habitat to some species, and other species that live in coastal areas travel through it. Such species should be regarded as common resources and special provision for both groups is needed in a formal international agreement. The strategy also cautions against deep-sea mining, including oil exploitation, until the effects of such actions are better understood.

Summary of Seal Management. The complexities of seal management have only been introduced here. The rate of change is so great that even as this material is being written, much of it is becoming outdated.

One fact becomes apparent in reviewing management philosophies, plans, and methods: Often people do not want to accept responsibility for their actions. Management of marine resources must be shared. People using the resources must also accept responsibility. The IUCN suggests that industry be encouraged to analyze the resource base and work in cooperation with governments and other commercial sectors of society to ensure sustainable use and ecosystem preservation. Informed use may breed responsibility.

Scientists are often caught in the middle of management conflicts. Faced with a database that is deficient in many aspects, they are called on by bureaucrats who like to make management decisions based on the certainty of numbers. Many assumptions, however, are not empirically testable and the data in fact are far from ideal. The danger, of course, is that selection can be made of data that support only one view. Lavigne (1978) cautioned that future management decisions must be mindful of uncertainty, such as climatic change and human-made interferences that could have chain reactions in the ecosystem. Feeling secure with data collections by the most sophisticated methods from any stock is an infirm foundation (Lavigne 1978).

Recalling the time lag between policy and application, it is interesting to note a statement by Toffler (1970:5), which is applicable to seal management: "The rate of change has implications quite apart from, and sometimes more important than, the directions of change." He suggests " that there must be balance not merely between rates of change in different sectors, but between the pace of environmental change and the limited pace of human response." North American seals are in no extreme danger at present, but given the transitions, it is a good time to reassess our priorities and be mindful not only of where we are going and how we are getting there, but also how fast we are going to travel before we know the risks involved. Sometimes common sense is overlooked in the haste to develop.

RESEARCH AND MANAGEMENT NEEDS

An overview of the current status of seals in North America clearly points out that environmental conditions are of prime concern. The most common human-caused problems facing seals are heavy metal contamination and pollution and oil exploration. It is therefore important to review this situation and be aware that, as with all marine life, the seal's habitats need protection and continued monitoring from human effects (Johnson and Jeffries 1977).

Heavy Metal Contamination. Heavy metal contamination in the environment has been a concern since the 1960s. Such contaminants as mercury, DDT, dieldrin, selenium, and polychlorinated biphenyls (PCBs) are taken up by various organisms within the food web. They occur, however, in their heaviest concentrations in such terminal predators as seals and large pelagic fish (Mansfield and Sergeant 1973), and ultimately humans. Because DDT and PCBs affect steroid reproductive hormones, thereby often prolonging estrus and decreasing successful implantation, the implications are that heavy metals could be the cause of reproductive failure in northern seals and sea lions (Holden 1978).

The highest concentrations of mercury are found in the liver of wild animals (Mansfield and Sergeant 1973; Holden 1978). Deaths at high exposure levels, however, are normally associated with renal rather than hepatic failure (Ronald et al. 1977b). There is a wide range of contamination levels among species, and there is an age and sex correlation as well. For example, Addison (1973) found that lactating females showed lower organochlorine levels than other adult females, which indicated that the contaminants are probably passed on to the pup through the milk. Variations of mercury levels also correspond to diet. Harp seals, which eat small pelagic fish and crustaceans, had lower levels than gray or harbor seals, which feed on large pelagic and benthic fish and cephalopods (Sergeant and Armstrong 1973).

Seals may not suffer ill effects from some of these contaminants, depending, of course, on concentrations, because of certain protective mechanisms (Holden 1978). Selenium occurs at higher levels in seals than in other animals, and shows a positive correlation with levels of mercury (Koeman et al. 1972, 1975). This may be similar to the situation in humans, where dietary selenium has a protective effect against heavy metal toxicities.

It appears that seals are able to demethylate mercury (Freeman and Horne 1973; Ronald et al. 1977b; Holden 1978), perhaps with an enzyme system (Freeman and Horne 1973). Evidence of demethylation has been found in kidney and liver and in low amounts in the small intestine of the harp seal (Ronald et al. 1977b).

Although high concentrations of several organochlorine compounds occur in the lipid of seals and other marine mammals (Holden 1978), there does not seem to be any evidence of adverse effects. However, Newby (1978) indicated that PCB concentrations between 0.6 and 3 ppm cause reproductive dysfunction. Such a link may be the cause of the abnormally high rate of abortion in California sea lions (Holden 1978).

Urbanization, Oil Exploration, Drilling, and Spills. Much of the habitat of pinnipeds, especially that of the northern hemisphere species, is being threatened by human activities. Incidental human disturbance, offshore oil drilling, the concentration of industry in the Northern Hemisphere, and the increased human population may ultimately result in direct overexploitation of seals, environmental contamination, or intense competition for fish (FAO 1976).

Offshore oil drillings and transportation of fossil fuels and their by-products pose the most serious ecological threats to seals. These bring not only human disturbance, but also the possibility of oil spills. Until recently, oil had not caused any serious effects on populations of marine mammals, possibly because seals are highly mobile most of the year and are capable of avoiding heavily polluted areas. Any breeding, and therefore sedentary, population could be affected, however, in a harmful way (Hyland et al. 1977). Such a situation could have disastrous consequences on the neonates.

Adverse effects of an oil spill in the Arctic Ocean, the habitat of many pinnipeds, would tend to be heightened and prolonged compared to those of spills in temperate waters (Percy 1977). Biodegradation in this climate is slow and complex, and clean-up operations in inclement weather are impossible. Harmful side effects of an oil spill on seals could include the following situations. The rich algae on and under the ice surface in sea-ice habitat would be subject to severe contamination, and pinnipeds are part of the food chain based on this vegetation. Large quantities of oil could collect as slicks in the breathing holes, leads, and breeding areas, and oil tends to settle along the shoreline, which would affect animals hauling out (Percy 1977). This could result in ingestion of toxic oil droplets during grooming, loss of thermal insulation from coating, and irritation of the eyes and exposed mucous membranes (Hyland et al. 1977), areas that show extreme sensitivity (Percy 1977). Solar heating of the skin increases in the presence of oil, which may be significant during haulout. Under conditions of natural stress, such as molting, seals are particularly sensitive to an oil spill. Physiological repercussions seem to be kidney damage and behavioral changes (Percy 1977).

Underwater Acoustic Effects. There is increasing concern about the possible effect of extraneous underwater noise on marine mammal welfare. What matters most is whether a certain underwater sound pressure level "hurts" a marine mammal (Chapman and Ellis 1998). Some scientific research has already been done to determine the "community" noise standards for marine animals (Richardson et al. 1995). There is concern, also, of biological impact if the natural underwater sounds made by wild marine mammals were masked by noise (i.e., elicit a change in perception of intraspecific signaling).

Most pinniped species have relatively good sensitivity in the 1- to 15-kHz range. In fact, there is the potential for direct acoustic impact from low- to mid-sonic devices on the majority of pinnipeds (Ketten 1998). Although more may be known about behavioral responses of pinnipeds to airborne sounds, these data cannot be applied directly to determine the underwater response. Reeves et al. (1996) noted that pinnipeds tend to accommodate reasonably quickly to loud noise, which might be explained either by threshold shifts in hearing or by habituation, perhaps both. Sound does not appear to be particularly aversive to pinnipeds except at very high intensities. In this respect, transportation of liquid natural gas from the High Arctic to terminals produces a peculiar problem. There is reason to believe that in an area such as the Davis Strait in the North Atlantic, the power requirements of the vessels would produce an acoustic barrage of such intensity as possibly to modify seal behavior and movements (B. Møhl, University of Århus, Denmark, pers. commun., 1998).

Initial responses of marine mammals to sounds may not be indicative of long-term effects. They may habituate to the sounds if there is no perceived harmful effect and re-enter regions having dangerously high sound levels. Hearing loss induced by exposure to intense sound is painless. Alternatively, they may become sensitized to the signals, and show an increasing level of disturbance with time. Short exposures to intense sounds could produce a temporary hearing loss that recovers to normal, the time depending on the magnitude of the exposure and the individual. After repeated exposure, hearing loss could become permanent. The combination of sound-pressure level and duration of exposure that can produce temporary hearing loss (temporary threshold shift) or permanent hearing loss (permanent threshold shift) is unknown for marine mammal species. Some knowledge of the process of temporary and permanent threshold shifts is needed to predict the consequences of exposure to intense, low-frequency sounds. Ketten (1998) stated that data suggested that a received level of 80–140 dB over species-specific threshold for a narrowband source will induce temporary to permanent loss for hearing in and near that band in pinnipeds. Certainly a much better enhanced audiometric database is required for all species.

Harbor Seal. The harbor seal is protected from commercial harvest and harassment by legislation throughout its Pacific and Atlantic ranges. It is hunted to a small extent only by Native peoples, which results in minimal impact on arctic populations. The flesh is occasionally used as food for humans and dogs; the liver is especially high in vitamin A (Banfield 1974). The species also is hunted for skins, which produce a high-quality fur. The manufacture of trinkets and handicrafts is of minor importance. *P. vitulina* represents a popular tourist attraction in aquaria and zoos, and is also the most commonly used seal in experimental research (FAO 1976).

In Canada, the Atlantic and Arctic coastal populations were designated in 1999 as "data deficient" species at risk. The Ungava harbor seal was designated in 1996 as a population of "special concern," and, with the unusual situation of a marine mammal in freshwater, it is not clear whether this population falls within provincial or federal jurisdiction for legal protection (Environment Canada 1999a). Since passage of the MMPA in 1972, the number of harbor seals along the U.S. Atlantic coast (New England) has increased nearly fivefold. The species is not listed as threatened or endangered under the ESA or considered as a strategic stock for management purposes (National Marine Fisheries Service 1999).

In the United States, before federal and state protection through the Marine Mammal Protection Act, the Pacific harbor seal population was greatly reduced. Commercial hunting, especially during the nineteenth century, reduced the number of harbor seals along the California coast to only a few hundred individuals in a few isolated areas (Bonnot 1928, 1951; Bartholomew and Boolootian 1960). Harbor seals in Oregon and Washington apparently decreased during the 1940s and 1950s due to bounty hunting. For example, >17,000 seals were killed in Washington between 1943 and 1960 (Newby 1973) and >3800 were removed in Oregon between 1925 and 1972 by state and bounty hunters (Pearson 1968).

Within Pacific waters, there are geographically identifiable stocks from California to Alaska, based on studies of mitochondrial DNA (Huber et al. 1994; Burg 1996; Lamont et al. 1996) and movement patterns (Jeffries 1985; Brown 1988). Management stocks need to be defined to avoid depletion of local populations. For assessment purposes, six stocks are recognized along the west coast of the continental United States: California, Oregon/Washington outer coastal waters, Washington inland waters, and three stocks in Alaskan coastal and inland waters. A small number of harbor seals also occurs along the west coast of Baja California. They are not considered a part of the California stock, however, because there are no international agreements in place for the joint management of this species by the United States and Mexico (Forney et al. 2000).

Harbor seals are not popular with commercial or sport fishermen because they take fish from nets or bite and tear caught fish. They also damage fishing gear and gill nets, often becoming entangled in them. They also feed on salmon during upriver spawning periods (Banfield 1974) and on herring and trout (Johnson and Jeffries 1977). The incidence of bite marks (scars and open wounds) observed on spawning salmon gives an indication of losses from direct seal predation. Further losses occur if the wounded fish die as they ascend the river. Park (1993) reported that in 1991 and 1992, 20.9% and 17.4%, respectively, of the Snake River spring chinook (*O. tshawytscha*) were scarred; bites were identified as caused by harbor seals.

The Canadian populations are also protected, but for many years the Canadian government allowed a bounty system on this species for commercial fishermen on both coasts (Banfield 1974). Similarly, harbor seals were bounty hunted in New England waters until the late 1960s (Katona et al. 1993). Such bounties are always ineffectual. In Alaska, for example, harbor seals were culled at the Copper River Delta in the 1960s to reduce predation on salmon. The immediate result, however, was a failure of the local razor clam (*Siliqua patula*) fishery, rather than an increase in numbers of salmon caught. After the fact, it was discovered that the seals fed primarily on starry flounder (*Platichthys stellatus*), which in turn fed on razor clams; without the seals, the flounder population grew unchecked (Trites 1997).

Recently, mariculture operators, under a pilot program implemented by the Canadian government, are permitted to use acoustic deterrents or shoot problem seals. In the United States, interactions between seals and salmon aquaculture operations in Maine appear to be increasing, although the magnitude has not been quantified (Anonymous 1996).

Humans interfere with the harbor seal's habitat. Some traditional hauling-out grounds, such as the Nisqually Delta, Washington State, where 200 seals used to breed until the 1950s, were abandoned because of hunting pressures. Many inland waterways that are frequented by harbor seals are being taken over by human activity (Newby 1978). In San Francisco Bay, for example, haul-out sites are used by only a small number of animals. A site at Strawberry Spit was abandoned during the 1980s because of increased disturbance levels and a decrease in local food resources (Allen 1991; Kopec and Harvey 1995). Human-related activities may also have a negative effect on behavior.

Because *P. vitulina* is a shallow-water dweller, it is more affected by pollution than other species (FAO 1976). High concentrations of contaminants, especially chlorinated hydrocarbons and some metals, have been identified in harbor seals from southern Puget Sound (Calambokidis and Baird 1994). On Gertrude Island, Washington, PCB levels in harbor seal blubber and liver tissues were as high as 400 ppm. A high incidence of birth defects also was noted in this population (Newby 1978). Organochloride pollutants cause reproductive failure in harbor seals in the Dutch Wadden Sea (Reijnders 1987).

Oil spills are also a concern, but the impact is sometimes difficult to assess. The 1989 *Exxon Valdez* oil spill in Prince William Sound, Alaska, caused some harbor seal mortality. Frost et al. (1994), in comparing postspill findings with prespill baseline data from "oiled" and "unoiled" sites, estimated that at least 302 seals had died as a result of the spill, although only 14 were recovered, 11 of which were newborn pups (Spraker et al. 1994). It is known that the mothers came under a great deal of stress from the oil and from human activity associated with the clean-up. In this situation, however, it is difficult to determine how many harbor seals were actually lost directly because of the spill and what loss was part of an ongoing unexplained decline already occurring in the population. Between the 1984 and 1988 "trend-count" aerial surveys, there had been an overall 40% decrease in seal numbers in Prince William Sound. On Tugidak Island, south of Kodiak, for example, the population decreased from 10,000 to 2000 (Hodgson 1990). The fact that the number of harbor seals in Prince William Sound was declining before the *Exxon Valdez* oil spill therefore complicates both the assessment of impact because of the spill and the definition of recovery (Frost et al. 1999). In addition, when comparing sites, it cannot be assumed that harbor seals will have 100% site fidelity under adverse conditions. They often respond to disturbance by short- and long-turn movement away from impact areas and by abandonment of pups, and are not actually "lost" (Hoover-Miller 1994; Hoover-Miller et al. 2001).

Between November 1979 and May 1980, 437 dead harbor seals were recovered from an area between Cape Cod and southern Maine. The cause of the fatalities was diagnosed as an influenza virus, possibly complicated by a species of mycoplasm that manifested itself as pneumonia. The outbreak originally occurred around Cape Cod, but the virus migrated with the seals 500–600 km up the coast (Geraci et al. 1981). More recently, a massive die-off in the Wadden Sea has been associated with immuneohypea, more likely associated with high pollution levels in the northwestern European rivers. The associated viruses also occur in North American pinnipeds. Ham-Lammé et al. (1999), in testing west coast harbor seals for morbilliviruses, found them to be seronegative, which indicated that this disease in not endemic in the population. The Pacific harbor seal therefore is considered to be quite susceptible to a morbillivirus epidemic.

Current management needs exist in the area of taxonomic information based on morphometry, development of better stock assessment methods, and the study of haul-out behavior to assist with on-land counts (FAO 1976). In Washington State, the population is stable and, to keep the status quo, an active research program should be encouraged. All locations that are currently used as haul-out sites must be evaluated and possibly designated as critical habitats. Because there is conflict between harbor seals and commercial fishermen, a need exists for recognition and accurate documentation of this problem.

Larga Seal. Reliable estimates of the minimum population, the potential biological removal, and human-caused mortality and serious injury are not available. The Alaska stock is not classified as a strategic stock (Hill and DeMaster 1999). Lack of accurate population data makes it difficult to determine the effects of natural events and human-caused activities on the species.

Since about 1985, there has been evidence of a shift in regional weather patterns in the arctic region (Tynan and DeMaster 1996), and with it comes a concern about the potential for arctic climate change. Ice-associated seals, like the larga seal, likely would be particularly sensitive to any change in weather and sea-surface temperatures that affects their ice habitats. However, no reliable predictions can be made.

Oil and gas exploration and extraction in many parts of the larga seal range, particularly in the Chukchi and Beaufort Seas, has the potential for disturbing the seals as well as possible pollution of the habitat and food supply. Commercial fishing in the Bering Sea, particularly for pollock and herring, may already be causing problems with food availability for larga seals (Seal Conservation Society 1999).

Larga seals are an important species for subsistence hunters. In Alaska, they are harvested by coastal-based hunters, primarily in the Bering Strait and Yukon–Kuskokwim regions, who may take a few thousand annually. Hunting in Russian waters includes vessel- and shore-based efforts, with an annual take of several thousand each year. Past and present levels of harvest have had little influence on population size (Hull 1994).

Harp Seal. Harp seals are the income base for local residents of the Gulf of St. Lawrence and Newfoundland–Labrador as well as West Greenland and aboriginals of the Canadian High Arctic. In fact, they are hunted every day of the year along their 6000- to 8000-km cyclic migration return route.

The "whitecoat" pups were hunted for their white lanugo fur, which must be taken when they are between 2 and 10 days old. In Canada and elsewhere, a public outcry over clubbing of whitecoats began to discourage commercial sealing as early as 1984. Catch of whitecoats is forbidden in the Newfoundland and Jan Mayen area, but it is still legal to harvest weaned pups and adults. When molt is complete (about 6 days after weaning), the pup has a spotted coat, which is of more commercial value than the whitecoat.

In 1992, the northern cod fishery collapsed and the Canadian government declared a moratorium on the cod harvest. Faced with a formidable economic disaster, government officials were accused of following the "most prudent" course of increasing the harp seal quota (TACs) with the intention of reducing the size of the seal herds in order to increase fish stocks, despite objections from many fishery scientists that the depletion of fish stocks had nothing to do with seals, but was due to overfishing (Lavigne 1996). The government countered with the statement that, although the growth in the seal population had become a concern for the fishing industry, this is not the basis for Canada's seal harvest. It was felt that residents of coastal communities devastated by groundfish moratoria could sustainably harvest the abundant seal population (Anonymous 1997).

Although there is an assumption that the feeding behavior of harp seals adversely affects fishery yields, often there is no supportive evidence for this or for the supposition that a reduction in harp seal numbers would result in increased yields. Overfishing of the prey species that harp seals feed on, however, could adversely affect harp seal populations. In addition, thousands of harp seals are killed as incidental catches in commercial fishing gear (Woodley and Lavigne 1991).

The Atlantic Canada quota for harp seals was set at 186,000 animals in 1992. Since 1988, kills by "landsmen" had been 50,000–95,000/year, excluding those unreported. The quota had not been reached largely because of limited markets for products. In 1995, despite subsidies and other incentives, sealers in Canadian waters harvested about 65,000 animals, again reflecting the limited market demand for seal products worldwide. In 1996, the TAC was raised to 250,000, and new emerging markets allowed a harvest of 242,000 seals in that year. All age classes were taken, but young of the year made up 40–50% of the take. The TAC for 1997 was 275,000 harp seals and the actual harvest was about 262,000, well within the estimated replacement yield of 287,000 (number of animals that can be harvested each year without reducing the existing population).

Some scientists think that the actual kill is 38–89% higher than official figures. Lavigne (1999a, 1999b) maintained that more harp seals have been killed in the Northwest Atlantic than would be considered prudent under a "precautionary approach to management." In 1998, for example, the TAC was 275,000 for Canada and Greenland, but the total reported landed catch was 327,240–357,662 harp seals as shown from catch statistics. However, not all seals taken are reported (e.g., number struck and lost, incidental catches, etc.), making the actual exploitation rate between 400,000 and >500,000 each year. Despite everything, the harp seal hunt quota in 1999 and 2000 remained at 275,000.

The harp seal has received a great deal of management attention, and it is probably one of the best-studied pinnipeds with respect to its biology. This does not mean, however, that it is completely understood; even though scientists are aware of such activities as those relating to its phonation, audiogram, visual acuity, and dive responses, there is a great deal left to understand about its biology. Ongoing Canadian research includes satellite tracking of tagged seals for movement and behavior patterns, study of their diet and consumption behavior, partitioning of cod mortality, contraceptive research, and fatty acids and stable isotope studies.

Estimations by the government of Canada of harp seal population size and rate of increase are disputed. The latest definitive count was carried out in 1994. There is no COSEWIC designation, but the species is under review (Gaskin 1997). One of the recommendations arising from observations made by Canadian Veterinary Medical Association (CVMA) observers at the hunt in 1998 and 1999 was that population studies be carried out to ensure the survival of the species (Seal Conservation Society 2000).

Ringed Seal. The ringed seal is the most important seal species for Alaskan coastal residents; it is relied on most heavily for fur, meat, and other commercial products (Alaska Department of Fish and Game 1976; FAO 1976). Its stocks are presently underexploited, the average annual take in Alaska being about 9000–13,000. Russia also harvests ringed seals in the White Sea, with a limit of 3500/year. These totals are well within the limits for the productivity of ringed seals (Alaska Department of Fish and Game 1976; Popov 1976). Hunting is actually controlled by regional climatic conditions and the availability of *P. hispida* (Alaska Department of Fish and Game 1976). In Canada, seal hunting is important to the continuation of the Inuit's resource-based way of life. Eighty percent of the annual subsistence seal harvest is ringed seals, mainly from the Baffin region of the Canadian Arctic. In Labrador, subsistence landings ranged from 670 to 1639 (average =1157) ringed seals annually between 1993 and 1999. Because the ringed seal has had such little contact with humans, it is very abundant and its numbers are probably at original levels.

The only potential for a significant threat to this species is gas and oil exploration on the outer continental shelf (Burns 1978). Investigations on effects of this industry in the Arctic point to a definite threat to the Beaufort Sea population of ringed seals. Surface contact with oil would cause eye and kidney damage, stress, and interference in thermoregulation, especially for pups (Geraci and Smith 1976). In Canada, a population of ringed seals near Norman Wells, Northwest Territories, has been tested for tolerance to crude oil. It was found that hydrocarbons are absorbed rapidly, from both ingestion and immersion. Although the liver appears to be the main detoxification and excretion organ, no serious lesions developed from the amounts tested. Results indicate that for limited exposure, accumulations occur in body tissues and fluids that can be excreted, but more research is needed to establish the limits of safety (Englehardt et al. 1977). A high level of PCBs causes uterine occlusion, leading to infertility (Helle et al. 1976).

The mere presence, however, of a large offshore oil field may affect the seasonal pattern of movement of ringed seals and reduce their ability to survive adverse natural conditions (Geraci and Smith 1976). Furthermore, chronic low-volume releases of petrochemicals and the occasional major oil spill may have detrimental effects on the food of ringed seals, including the eggs and larvae of the saffron and polar cods and various kinds of zooplankton, which are highly susceptible to petrochemical pollution (Burns 1978).

Pollutant levels in ringed seals have risen even in the more northerly locations, which is indicative of a general rise in pollution of the marine environment on a global scale. Also, any arctic climate change could affect the seal's ice habitat, and the species is very sensitive to changes in ice conditions (Seal Conservation Society 1999).

Current research on ringed seals encompasses population dynamics, censusing methods, interspecific relationships on the pack ice, caloric requirements, heavy metal and toxic chemical uptake and pollution, and physiology. Areas of need include improved censusing techniques and the study of female reproductive cycles, migrations and dispersal, the influence of environmental factors on reproduction and ecological relationships with other phocids (FAO 1976), and polar bears, for which the seal is a prime food source.

Ribbon Seal. The ribbon seal is important economically to the residents of northern Alaska and represents part of the subsistence base of all the Native settlements along the Alaskan coast from Kuskokuim Bay to Demarkation Point. It provides food and fiber for everyday use and is a source of income through pelt sales. In comparison with ringed, bearded, and harbor seals, however, it constitutes an insignificant part of the Alaskan native harvest. In contrast, Russian hunters overexploited this species in recent years (Alaska Department Fish and Game 1976). From the 1950s to the 1980s, Russian commercial sealing vessels killed 6500–23,000 ribbon seals each year for their pelts, oil, and meat in the Sea of Okhotsk and Bering Sea. Since the break-up of the Soviet Union, however, commercial hunting on the high seas has largely stopped (Seal Conservation Society 1999).

The only threat to the ribbon seal population exists in the continental shelf development near their wintering and breeding areas in Bristol Bay and Saint George Basin. Although this will not result in direct seal mortality, it may reduce their food sources (Burns 1978). Certainly, ribbon seals could be affected by disturbance from any oil and gas exploration and extraction.

Ribbon seals remain the least known of the arctic seals. There is a need for information on numbers, reproductive cycle, general biology, seasonal distribution, and migration routes (FAO 1976).

Gray Seal. In New England, human occupation and erosion by storms have depleted the haul-out areas (Gilbert 1977) of the gray seal, reducing their population. In most of its North American range, *H. grypus* is said to be undisturbed by humans, as its island and pack-ice colonies are rarely visited (Mansfield and Beck 1977). In eastern Canada, only a small number are hunted each year. Gray seals pelts are less valuable than harp seal pelts. Sealing is limited to a traditional commercial hunt in an area off the Magdalen Islands and small commercial hunts in other areas. No commercial hunting is permitted on Sable Island, Nova Scotia. In 1998, sealers from the Magdalen Islands took 206 gray seals and those from Cape Breton, Nova Scotia, harvested 69. Only 98 were landed in 1999 (Fisheries and Oceans Canada 2000).

The main concern with management of this species is competition for food resources. Although it competes with the harbor seal for food (FAO 1976), its real competition is with humans. The gray seal has a considerable effect on inshore fisheries, especially those of mackerel, herring, cod, and salmon (*Salmo salar*), because it damages nets and gear and eats or mutilates the trapped fish. It is also the primary host of the codworm, which infests the flesh of groundfish, especially cod, thereby decreasing their commercial value.

In 1967, the Conservation and Protection Branch of the Canadian Department of Fisheries began culling gray seals (mostly molted pups) in an attempt to reduce the population and thereby limit competition with commercial fisheries. Despite a take of about 800 animals annually (maximum 2300 taken in 1975), the population continued to expand. In April 1976, the Canadian government also placed a bounty on the gray seal. By 1977, over 600 seals had been taken in this manner, and there were probably an equal number lost through sinking (Mansfield and Beck 1977). Both the culling and the bounty programs ceased after 1983. In 1986, the Commission on Seals and Sealing recommended a cull of gray seals as an economic advantage (Malouf 1986). With the reduction in cod quotas in the Gulf of St. Lawrence and Scotia–Fundy regions to conserve overexploited stocks, the fishing industry renewed requests for a cull (Anonymous 1992); however, the government announced in 1994 that it would not consider a return to seal culls (Anonymous 1994). It is generally a mistake to assume that a reduction in the abundance of a seal population will actually result in a simultaneous increase in fishery yield. All major relevant factors must be taken into account in evaluating cull proposals. In addition to scientific evaluation, there are social and economic implications that must be addressed.

Concern has been expressed about the effects of marine pollution, in particular contaminants such as organochlorines, on the heath of gray seals. The species can also be affected by oil spills. Biologists observed an oil spill in Wales. It occurred at the beginning of the gray seal breeding season and resulted in oiling of some pups and an obvious interference with the mother–pup relationship. These pups did not gain weight as rapidly as unoiled pups (Davis and Anderson 1976). No such oil spill has occurred in North American waters to threaten gray seals. PCBs, methyl mercury, arsenic, cadmium, and selenium, however, have been found in *H. grypus* in levels that may have a significant effect on their physiology and that already appear to have altered their steroid hormone metabolism (Freeman et al. 1975).

The gray seal is on a collision course with human needs. The next confrontation will probably arise regarding colonies on Sable Island. Attempts are being made to control the population using antifertility drugs. Long-term studies are needed on range, distribution, migration patterns and corridors, whelping sites, competition, and predation.

Hooded Seal. In the Newfoundland–Labrador area, the hooded seal has been hunted annually since the early eighteenth century. Stock assessment was not attempted until the early 1960s. In 1961, the Canadian government introduced opening and closing dates for the hunt. Hunting in the Gulf of St. Lawrence (below 50° N) was prohibited in 1964. Adult females have been protected on whelping areas since 1965. The ICNAF recommended quotas on the species in 1971, which were then put into effect. The stocks are thought to be increasing, because they have been exploited at below their sustainable yield levels (estimated at

24,000–34,000 in 1990, depending on the age of the animals hunted) since the early 1960s. For example, in 1974, the total allowable catch was 15,000 seals, but only about 12,000 were actually taken. The maximum was harvested only during the 1975 season. During the next 3 years, the proportion of adult seals allowed in the catch was decreased progressively to 5%. In the 1980s, the TAC was reduced (Stenson 1994), and then increased again in the 1990s. A TAC of 8000 was set for 1993 for Canadian waters and held at that level through 1997. In 1996, catches (25,754) were more than three times the allowable quota because of good ice conditions and a strong market demand (Anonymous 1998a). Of this number, 22,800 juvenile (blueback) hooded seals were taken, and >100 charges were laid. In contrast, the harvest in 1997 (7058) was slightly below the TAC. In 1998, the TAC was increased to 10,000. Official reports indicate that 10,148 hooded seals were taken in the 1998 east coast hunt and an estimated 9896 in Greenland. In 1999, however, because of a high inventory of hooded seal pelts, only 201 hooded seals were harvested along the Newfoundland–Labrador coast. The combined hunts in Canada and Greenland are considered to be below the replacement yield (Fisheries and Oceans Canada 2000). As with the harp seal, however, the actual kill in any year could be 38–89% higher.

In the Jan Mayan breeding area, the Norwegians exploited the hooded seal until 1961, when the (then) USSR and Norway agreed that since the population was decreasing, management measures were in order. Now, they do not allow the taking of adult and immature animals at the molting grounds in Denmark Strait or a spring harvest season.

There is speculation that *C. cristata* became increasingly available from the 1950s to the 1980s because of the existence of a reserve in the Davis Strait. No commercial hunting of hooded seals is permitted there. Hunting pressure increased on hooded seals as the availability of harp seals decreased (FAO 1976). This hunt is associated with the same vessels carrying out the harp seal hunt, with the hooded seal being the species of choice. The hooded seal is prized for its pelts, which are of particularly high quality, especially those of the young bluebacks. The only other commodity that enters the commercial market is oil.

Current and maximum net productivity are unknown. Pup production in Canada may be increasing slowly (5% annually), but due to wide confidence intervals and lack of understanding regarding stock dynamics, it is possible that pup production is stable or declining (Stenson 1994). The hooded seal has minimal effect on any commercial fish species (FAO 1976). The total fishery-related mortality and serious injury is believed to be very low relative to the population size.

Bearded Seal. The bearded seal is vitally important in the economy of coastal subsistence hunters because of its large size (Burns 1967; Banfield 1974), the high quality of its meat, its blubber, and its strong, durable skin (Burns 1967). The meat often poses a health risk because it may harbor the nematode *Trichinella,* a parasite causing trichinosis if meat is eaten uncooked. The liver of this species contains high levels of vitamin A, like that of the polar bear, so overindulgence could result in poisoning (Banfield 1974). Estimated annual harvests for the subsistence hunt in Alaska were 1784 from 1966 to 1977 (Burns 1981b); however, the actual number of bearded seals currently taken for subsistence is unknown. In Canada, they are taken each year in the subsistence hunt in the Canadian Arctic and the northern Atlantic. Only small numbers of bearded seals are taken by Labrador subsistence hunters, usually <100/year; for example, in 1998, 56 were landed, and in 1999, 61 (Fisheries and Oceans Canada 2000).

Reliable data on trends in population abundance in U.S. waters are unavailable, although there is no evidence that abundance is declining. Any changes in weather and sea-surface temperatures, a shift suggested by Tynan and DeMaster (1996), could have adverse affects on the ice habitats of the bearded seal.

When hunted, *E. barbatus* is often shot, resulting in many animals being wasted because they sink when killed and cannot be retrieved. In the northern Pacific, two types of hunting are practiced: subsistence hunting along the Siberian and Alaskan coasts, and commercial sealing by Russia and Japan. It is estimated that half the annual harvest is lost to sinkage. Annual take is dependent on the seasonal availability of the migrating seal. Up to 2000 seals/year were killed in the Bering and Okhotsk Seas; however, the current extent of the hunt is likely much lower, although accurate figures are not available (Seal Conservation Society 1999).

Future management problems may occur because of some of the prey species of this seal, especially pandalid and crangonid shrimps and eithode crabs, are also a focus of fisheries (FAO 1976). Although no serious competition for these food resources has arisen, development of a clam fishery in the bearded seal's range may present future problems (Burns 1978). Habitat of this seal is also potentially threatened by extensive offshore oil exploration and production, interisland pipeline construction, and the shipment of oil by tanker in arctic waters (FAO 1976). These may affect either the seal itself or its food web, which comprises mainly invertebrates with highly susceptible larval stages (Burns 1978).

Hawaiian Monk Seal. In 1976, the Hawaiian monk seal was designated "depleted" under the Marine Mammal Protection Act and as "endangered" under the Endangered Species Act. The species was listed as endangered after a 50% decline in beach counts occurred from the late 1950s to the 1970s. Under the ESA, the allowable take of monk seals is zero. In addition, Hawaiian monk seals are listed as endangered under Hawaiian state law. Internationally, they are protected through provisions in CITES. This treaty protects them against trade in their parts or products. Although management activities and research focus on single island and atoll populations, this species is managed as, and considered to be, a single stock (National Marine Fisheries Service 1999).

Because the monk seal is a particularly sensitive pinniped, it is remarkable that it survived the period of intensive marine mammal exploitation during the eighteenth and nineteenth centuries, when it might have been eliminated. The first recorded observations of seals in the Northwestern Hawaiian Islands date from the early 1800s. By the mid- to late 1800s, their numbers had been significantly depleted due to scavenging by shipwrecked sailors and bird hunters as well as commercial sealers. It was thought that sealers took the last monk seals in 1824, but some seals escaped and another vessel took 1500 skins in 1859 (Kenyon 1981). After this, the monk seal probably survived because its habitat was so isolated and its low population levels and solitary habitats were not appealing commercially.

A shy animal, the monk seal has disappeared from its former habitats, which are now inhabited by humans. A definite correlation exists between the presence of humans and the absence of monk seals. An increase in pup mortality has been observed when nursing mothers are disturbed (Kenyon 1978b). When humans intrude on their habitat, these seals desert favored beaches and haul out on isolated shifting sandpits. Here the young are exposed to extremes of wind and tidal conditions and are also nearer to deep water, which is frequented by sharks. When human activity became extensive on Midway Atoll during World War II, the monk seal population began to decline. The breeding population disappeared between 1958 and 1968, primarily because the personnel at a large naval base disturbed the nursing females (Kenyon 1978b). If protected and undisturbed, however, a seal population can begin repopulating ancestral breeding grounds. The Laysan Islands have been repopulated this way (Kenyon 1973).

As a result of the U.S. government's declaring the monk seal an endangered species in 1976, the Hawaiian Islands National Wildlife Refuge was created to encompass all of the Leeward Hawaiian Islands except Midway and Kure atolls, which constitute a separately protected area. In the same year, the U.S. National Marine Fisheries Service and the U.S. Fish and Wildlife Service began an intensive study. The results of that survey indicated that many populations were declining because of human disturbance of nursing females in all areas and that minor contributions were shark attack and ciguatera poisoning (Kenyon 1978b).

More recently, the migration of monk seals from other islands in the Hawaiian Archipelago, in particular from Pearl and Hermes Reef, led to the reestablishment of a breeding population on Midway Atoll.

The main reason for increased births has been the restriction of human access to beaches coupled with an overall reduction in human activity (Meisenheimer 1998). In turn, the population at French Frigate Shoals has declined because of poor juvenile survival rates and the resulting low numbers of young females reaching reproductive age.

Fishing interests in the Leeward Hawaiian Islands, developed and expanded during the 1970s, have led to direct and serious threats to monk seals. Interactions include operations and gear conflict, potential entanglement in fisheries debris, seal consumption of potentially toxic discard, and competition for prey. Establishment of a Protected Species Zone around the Northwestern Hawaiian Islands appears to have helped reduce the potential for direct fisheries interactions. There could be possible indirect interactions with fisheries, however, such as competition for lobster or the degradation of foraging habitat associated with precious coral harvesting (National Marine Fisheries Service 1999).

The monk seal decline at French Frigate Shoals, attributed to a decrease in juvenile survival rates, started shortly after commercial lobster fishing began in the Northwestern Hawaiian Islands early in the 1980s, concentrated at Necker Island, Maro Reef, and Gardner Pinnacles. The most likely explanation was decreased prey availability, but the importance of lobsters in monk seal diets was difficult to assess. From 1994 to 1998, the Marine Mammal Commission recommended each year to the NMFS that lobster fishing should be suspended at French Frigate Shoals, but no steps were taken. In 1998, a new bank-specific guideline for lobster fishing was adopted, whereby separate catch limits were set, rather than single area-wide quotas. For the first time, lobster fishing was reported at French Frigate Shoals as well as at other banks supporting major monk seal colonies. In 1999, results of studies based on analyses of prey fatty acid signatures in monk seal blubber samples revealed that lobsters probably constitute a significant percentage of the diet of most juvenile and adult female monk seals at French Frigate Shoals, but only a small proportion of the diet of adult male monk seals. These data supported recommendations to close all breeding atolls to lobster fishing (Marine Mammal Commission 2000).

The IUCN/SSC Seal Specialist Group recommends the following conservation measures: continuation of recovery activities, reduction of mobbing behavior of male monk seals, research to determine monk seal foraging and movement patterns, and monitoring and mitigating the impacts of fishing (Reijnders et al. 1993). It is listed as endangered in the IUCN Red List of Threatened Animals.

Caribbean Monk Seal. In the Caribbean, *M. tropicalis* is almost certainly extinct, primarily because of hunters and fishermen (Kenyon 1978b). Columbus first came upon the Caribbean monk seal in 1494. From then on it was increasing heavily exploited by European explorers and hunters. For example, in 1494, 8 on the island of Alta Vela, to the south of Haiti; on 21 June 1513, 14 on the Dry Tortugas, Florida; early in the eighteenth century, sometimes 100/night on the Bahamas; and in January 1911, about 200 on Arrecife Triangulos (Rice 1973). They were easily taken, such that by the late 1880s, they were so scarce as to be called an "almost mythical species" (Allen 1887). The last recorded Caribbean monk seal in the United States was killed in 1922 off the coast of Key West, Florida.

No effective conservation methods have ever been applied. It is legally protected in Jamaica. The Caribbean monk seal was listed as endangered throughout its range on 10 April 1979. No recovery effort is being made for this species (Office of Protected Resources 1996).

The Caribbean monk seal is included in the IUCN Red List as being "extinct" and as an Appendix 1 species under CITES. The species is listed as "endangered" under the U.S. Endangered Species Act and "depleted" under the U.S. Marine Mammal Protection Act (Marine Mammal Commission 2000).

Northern Elephant Seal. During the early nineteenth century, the northern elephant seal was abundant along the western U.S. coast from Point Reyes to Cabo San Lazaro. For at least 40 years at the beginning of the nineteenth century, this species was exploited continually, primarily for its blubber, which entered the market as oil. However, by 1860, after intensive slaughter, this oil was no longer a valuable commodity. By 1869, the species was considered to be extinct, but an occasional individual was seen on San Benito and Isla Guadalupe until 1880. It was not until 1892 that more were seen, and from that time until 1930 the species bred on Guadalupe in very small numbers. The Mexican government granted protection to the 264 seals living on the island in 1922. By 1932, they began a slow dispersal along the Baja California coast. The United States also protected the elephant seal, and it increased to 13,000 by 1957, its population tripling in 20 years. In the 1970s, four new colonies were established, but only two colonies, those on Guadalupe and San Benito, had reached equilibrium by 1977 (LeBoeuf and Panken 1977). In addition, range expansion and colonization of mainland beaches occurred in central and northern California and Oregon. The abundance of elephant seals on South Farallon Islands, California, may now be limited by available and accessible high-quality breeding habitat, but populations at newer locations continue to grow (Sydeman and Allen 1999). No international agreements exist for the joint management of this species by the United States and Mexico. The California breeding stock is considered to be separate from the Mexican breeding stock for management purposes (Forney et al. 2000).

Despite the rapid renewal in growth, the population of elephant seals, once so abundant, has suffered irreparable damage because the current generation has reduced genetic variability. Depleted populations contain only a small fraction of the total genetic variability of the parent population (LeBoeuf and Panken 1977).

Tourists do not have a chance to disturb the rookeries, but the elephant seals do have a great appeal for the public. Many people come to see them every year off the California mainland at Año Nuevo Point (FAO 1976). Human presence does not appear to bother male elephant seals, but if the females are disturbed before parturition, they may give birth elsewhere. Fortunately, females do not abandon their pups even if disturbed (LeBoeuf and Panken 1977).

Offshore oil exploration could disturb rookeries (FAO 1976) and any oil spills could also be harmful, although elephant seal pups exposed to an oil spill on San Miguel Island survived and later dispersed normally (Hyland et al. 1977).

Steller's Sea Lion. In the United States, both the western and eastern Pacific stocks are currently listed under the Endangered Species Act, the former as "endangered," the latter as "threatened" (U.S Department of Commerce 1997). Both are classified as strategic stocks.

Steller's sea lions were (and likely still are) considered serious competitors by commercial fishermen. They may destroy nets, traps, and other gear (Banfield 1974), although some scientists have pointed out that no commercially valuable fish is a major item in their diet (Gentry and Withrow 1978). However, walleye pollock, a prime target for commercial fisheries, is now a major prey item consumed by the western population of Steller's sea lions (Trites et al. 1998). The Canadian government allowed a control program in British Columbia in 1959–1960, which consisted of an intensive program to reduce the population because of commercial fishing conflicts. As a result, numbers decreased from a high in 1956 of 12,000 sea lions to a low of 4000 by 1969 (Banfield 1974) in this area. Unintentional harassment by sport and commercial fishermen, divers, photographers, and tourists is ongoing in various degrees, although the consequences of this are unknown (FAO 1976).

In the mid- to late 1970s, a climatic regime shift occurred in the North Pacific, and as a result there was an alteration in the abundance of key fish species. The composition of the Bering Sea and Gulf of Alaska ecosystem changed from an abundance of pelagic fishes in the 1950s–1960s to one dominated by pollock and large flatfish by the 1980s (Trites 1998). Within this same period, the western population of Steller's sea lions declined by at least 80%. This decline is partly attributed to increased commercial fishery effort concentrated in sea lion critical habitat. Lavigne (1999c) pointed out that the distribution of the pollock fishery overlapped extensively with that of foraging sea lions (Anonymous 1998b), and there were correlations between the decline in Steller's sea lion numbers and the increase in size of the pollock fisheries (e.g., Loughlin and Merrick 1989; Trites and Larkin 1992; Trites et al. 1998).

Alternatively, it is considered that the ecosystem shift affected the quantity and quality of prey species available for Steller's sea lions in the area, and malnutrition (nutritional stress) is a major factor in the decline of the western population. Small fatty fish like capelin, sandlances, and herring formed the main part of the diet before the 1970s. With the declining abundance of small forage species, Steller's sea lions became increasing dependent on pollock, a species with low fat content (Alverson 1992; Merrick 1995; Trites 1998). Merrick et al. (1997) noted a significant negative correlation between the diversity of diet and the rate of the Steller's sea lion population decline across the Aleutian Islands and Gulf of Alaska, with the sharpest declines in areas with the lowest diversity of diet. Bioenergetic results indicate that sea lions would have to eat ~35–65% more pollock than herring to achieve the same gross energy intake. Heavy reliance on pollock, therefore, may have detrimental affects on the health of Steller's sea lions, either owing to its low energy, its low fat, or some other intrinsic factor. Furthermore, the specific composition of pollock (e.g., amino acids, vitamins, minerals) could lead to health problems independent of energy density (Rosen and Trites 2000). Low-fat diets are known to result in vitamin E deficiency in pinnipeds (Geraci 1975). Competition for food with large flatfish and adult pollock may also be a factor impeding the recovery of Steller's sea lions, given the large overlaps in their diets (Trites 1998). Merrick (1995) concluded, "what is clear is that the Steller sea lion probably will not recover unless a fundamental change occurs in the prey availability in the North Pacific Ocean."

In the United States, management actions recently implemented to reduce Steller's sea lion interactions with human activities include no-entry buffer zones around rookeries, prohibition of groundfish trawling within 10–20 nautical miles of certain rookeries, and spatial and temporal allocation of Gulf of Alaska pollock catches (National Marine Fisheries Service 1999). In the North Pacific, TACs for pollock and Atka mackerel, both important prey species for Steller's sea lions, were adjusted to prevent prey removals from occurring all at once or in localized areas. In the Aleutian Islands region, all pollock fishing has been prohibited to eliminate any potential competition with sea lions. To further protect them, no trawling is allowed year-round within 18.5 km of numerous rookeries. Trawling for pollock is prohibited near haulouts, a number of which extend 37 km from shore on a seasonal basis. Incidental catch limits have been established for Steller's sea lions based on the criterion that incidental mortality should have a negligible impact (Witherell et al. 2000).

There is evidence, however, that commercial fisheries and Steller's sea lions are targeting different portions of the pollock population, and consequently any changes in the timing of pollock harvest should be approached with caution (Boyd 1995). Trites (1998) stated that "changing the amount and timing of commercial catches of pollock can have unexpected and undesirable consequences for Steller sea lions. The most precautionary action for Steller sea lions may be to protect the fattier fishes and maintain a status quo on catching adult pollock."

Significant levels of heavy metals and chlorinated hydrocarbons have been found in Steller's sea lions in the southern extent of their eastern Pacific range, but the effects are unknown (FAO 1976). The stock has declined from historical numbers in California, especially in the southern and central regions. Counts between 1927 and 1947 ranged between 5000 and 7000 nonpups, but have subsequently declined by over 50%, remaining at 1500–2000 nonpups during 1980–1998 (Hill and DeMaster 1999). At Año Nuevo (central California), for example, there was an 85% decline in the breeding population between 1970 and 1987 (Le Boeuf et al. 1991). Causes of the decline include reduced prey availability, contaminants, and disease (Sydeman and Allen 1999). There has been a decrease in organochlorine and trace metal contaminants in central California Steller's sea lion pups, but levels are still elevated (Jarman et al. 1996).

Alaskan Natives have hunted Steller's sea lions for subsistence for centuries. Recent data gathered on the subsistence harvest for the western stock show a decline in total take (struck and lost and harvested) from about 549 sea lions in 1992 to about 178 in 1998. Consideration has been given to the development of a co-management agreement between the NMFS and the tribal government of St. Paul Island. Under this, annual management plans could be developed, monitoring and research programs recommended and considered, and local decision making provided on the harvest, including which rookery to harvest, numbers that are allowed to be taken, and the timing of the harvest. A tribal ecosystem officer would be, among other things, responsible for overseeing the harvest and involved in a Steller's sea lion biosampling program (Marine Mammal Commission 2000).

Subsistence hunters in Canada harvest an unknown number of Steller's sea lions from the eastern Pacific stock; the magnitude is believed to be small. Only a very small percentage (<1%) of the total subsistence take in Alaska was ever from the eastern U.S. stock. Alaskan Native subsistence hunters have initiated talks with Canadian hunters to quantify their respective Steller's sea lion subsistence harvests and identify any effect these harvests may have on the cooperative management process (Hill and DeMaster 1999).

California Sea Lion. The California sea lion is protected under the Marine Mammal Protection Act against any killing or harassment in the United States. Killing of California sea lions was banned in Mexico in 1969 and Canada in 1970. Mexico offers the species protection, although small numbers are taken. In Canada, on the coast of British Columbia, there is no management program, but fishermen are allowed to take a sea lion if it continually interferes in their fishing area. No international agreements for joint management of California sea lions exist between the United States, Mexico, and Canada.

Although exploitation of California sea lions occurred in the nineteenth century, it was not as intense as that for the Guadalupe fur seal or the northern elephant seal. From 1860 to 1888, there was a trade in seal oil. Later a trade developed in the manufacture of glue from their hides (Banfield 1974). The dried "trimmings" were sold in China as medicinal cure-alls. A minor reason for taking these animals was to make dog food. In the 1920s, a small but lucrative trade developed in capturing them for zoos, aquaria, and traveling shows (Banfield 1974; Mate 1978).

Present populations inhabit their entire former range. This species has increased in numbers much more than the Steller's sea lion, which in the 1930s was the more abundant. Whether such circumstances as food or space have changed to favor California sea lions is not fully understood (Mate 1978).

For management purposes, the California sea lion population is separated into three stocks corresponding to the regional breeding areas: U.S. stock (United States/Mexico border northward into Canada); Western Baja California stock (United States/Mexico border south to the tip of the Baja California peninsula); and Gulf of California stock (tip of Baja California peninsula, through Gulf of California, along mainland to southern Mexico) (Lowry et al. 1992). Because of the significant phylogenetic division between California sea lions in the Gulf of California and those along the Pacific coast of North America (Maldonado et al. 1995), these two regional stocks should certainly be treated as separate genetic units for conservation management.

For the U.S. stock (only), counting only sea lions ashore or hauled out during the breeding season, the minimum population estimate for 1999 was about 109,850 individuals. Alternatively, pups are counted in July after all are born, and the number of live births is estimated from the pup counts (as adjusted for a calculated pre-census mortality rate). The size of the population is then estimated from the number of births and the proportion of pups in the population, adjusted for growth rate. From this method, the 1999 population estimates for the United States stock range from 204,000 to 214,000 sea lions. Between 1975 and 1999, the counts of pups increased at an overall annual rate of 5.0%. Since 1983, the rate has shown an increase of 6.2% annually. The PBR level for this stock is calculated at about 6590 sea lions/year. The total human-caused mortality is less than the PBR, although not considered insignificant (Forney et al. 2000).

Because of reduced prey resources from overfishing and the destruction of fish spawning habitat due to poor forest practices, hydroelectric power dams, and pollution, California sea lions are now using upriver areas to feed, resulting in competition with humans, especially sport fishing (Mate 1978). California sea lion predation on endangered anadromous fish is a direct result of human development, which has resulted in the loss of spawning streams. Historically, these fish entered the freshwater systems more quickly because of an abundance of natural streams and inlets and were less vulnerable to sea lion predation than when they are concentrated at fishways. Their survival, despite the historical numbers of sea lions, attests to this fact. The anadromous salmonid stocks found in California, Oregon, and Washington are in decline because of habitat degradation and overfishing, not pinniped predation. Although pinniped predation may slow or prevent recovery, reducing predation will not result in recovery if factors responsible for the salmonid decline are not adequately addressed (Marine Mammal Commission 1998, 2000).

Although human competition for food is somewhat less intense with California sea lions than some other marine mammal species (Mate 1982), human commercial fishing activities reduce the carrying capacity of the ecosystem for the sea lion. As sea lions returned to their historical numbers, they competed with commercial fishing interests that had moved into the fisheries that had traditionally supported sea lions. In addition, California sea lions eat commercial fish in nets and on lines, damaging the fishing gear and caught fish in the process. Based on 1994–98 data, the minimum total annual mortality in U.S. commercial fisheries was set at 1208 California sea lions, largely from the California driftnet and set gillnet fisheries (Forney et al. 2000). They also become entangled in fishing gear (nets and hooks) and plastic debris after it is lost or discarded. Near La Paz, Mexico, 3.9–7.9% of sea lions examined were entangled or showed evidence of previous entanglement (Seal Conservation Society 1999). In the United States, the proportion of entangled California sea lions observed at rookeries and haulouts ranged from 0.08% to 0.35% of those present on land, with the majority (52%) entangled in monofilament gillnet material (Stewart and Yochem 1987; Oliver 1991). The overall mortality of these injured animals is unknown. In addition, the number of each pinniped species taken in the Mexican fisheries is not available.

Aquaculture operations also attract California sea lions to captive salmon being reared in net-cages in the ocean as well as discharge points around hatcheries and pond culture systems. In British Columbia, sea lions do not pose a major problem, but there have been some significant interactions. Acoustic deterrent devices, producing signals (12–17 kHz) in the range of maximum hearing sensitivity of the target pinnipeds, are used to deter seals and sea lions from the vicinity of fish farms, but effectiveness diminishes with time (Iwama et al. 1997).

The potential for sea lion conflicts with human economic interests will need to be addressed. Coastal communities often built facilities in areas that had in times of marine mammal abundance served as rookeries or other components of the habitat. Thus, the resurgence of the California sea lion population has resulted in coastal conflicts. Dolan (1990), in addressing the problem, mentioned that "California sea lions, once hunted to the brink of extinction, have made such a comeback that they are becoming a nuisance here [Monterey] and in other coastal communities." Regulations alone will not solve these conflicts. Conversely, intentional and unintentional human disturbance of haulouts and rookeries may reduce the suitability of certain sites as rookeries and eliminate some entirely. Such interference is suspected to have caused some colonies to move their breeding rookeries to more-aquatic sites (FAO 1976; Mate 1982).

Proximity to industrialization also exposes them to toxic wastes through their prey species. California sea lions spend much of the year in the waters off southern California, where the marine organisms are highly contaminated with PCBs and DDT and its metabolites from the Los Angeles sewage discharge (Buhler et al. 1975). A correlation was observed between premature pupping and high tissue organochlorine levels in the early 1970s. Metabolites of PCBs and DDT were two to eight times higher in the blubber and liver tissues of aborting females and their pups than in the tissues of full-term animals (DeLong et al. 1973; Gilmartin et al. 1975). There is, however, no evidence this affects populations (Reijnders et al. 1993).

Oil exploration and drilling have serious implications for marine mammal species, but exploration is afforded special status (incidential "take" of pinnipeds is authorized), and challenges have failed to inhibit oil activities. The increase in environmental toxins may have negative effects on the population. The most plausible cause for the deaths of 51 California sea lions (including 43 large pups), along with various cetaceans and sea birds, in the Upper Gulf of California, Mexico, in 1995 was the discharge of an unknown toxic substance in the water in or near where sea lion prey were concentrated (Vidal and Gallo-Reynoso 1996).

Northern Fur Seal. On the Pribilof Islands, full-scale exploitation of northern fur seals began in 1786, and by 1787 an overwhelming 2.5 million skins had been taken. It is not difficult to imagine that the population was reaching dangerously low proportions by 1834. From 1835 until 1867, only males on land could be taken, and the killing of females was forbidden. The result was an increase in the population, although sealers were turning their interests to other species of seal as well. In 1867, the United States acquired Alaska and the Pribilofs, and until 1869, sealing in this area was very unorganized. In 1870, the first of two 20-year leases of sealing privileges was awarded. In the next 19 years, a total of 1,854,029 skins was officially shipped from these islands. From 1890 to 1909, during the second lease period, 342,651 animals were taken. Basically, then, from 1869 to 1911 there were few restrictions on the taking of fur seals. Historically, it stands as the zenith of the fur sealing trade, when pelagic sealing was also practiced regularly. Most of the seals taken were females (U.S. Department of Commerce 1977).

The 1867 population estimate of 4.7 million fur seals on the Pribilof Islands had decreased by 1910 to 125,000 survivors. This swift decline encouraged Britain, Japan, Russia, and the United States to finally realize that management was an integral part of resource use. The 1911 North Pacific Fur Seal Convention (see International Agreements) was held and formalized (Marx 1967). In 1912, 3764 fur seals were harvested; in 1941, 95,000 were harvested. The killing of females was terminated in 1968, and the population continued to recover (National Marine Fisheries Service 1999).

In 1973, the United States declared St. George Island a research study area where no seals could be harvested. Growth and behavior of this unharvested population could be compared with the harvested population on St. Paul's Island to ensure practical management of the herd. In 1977, there was unusually high mortality among young fur seals. There was speculation that a reduced availability of food caused by the intense trawl fishing of pollock in the Bering Sea near the Pribilof rookeries may have been responsible (Green-Hammond 1980). With reduced food supply, nursing females have to travel further to find food and young seals may be forced to search out their own food supply sooner than normal.

In 1984, a moratorium on commercial harvesting of fur seals was also imposed on St. Paul's Island because of depressed population levels. Subsequently, in 1988, the eastern Pacific stock of northern fur seals within United States waters was designated as depleted under the MMPA due to declining numbers on St. George Island and no population growth on St. Paul Island (Sinclair 1994). A conservation plan was written to outline reasonable actions to protect the species (National Marine Fisheries Service 1993), and in 1994, fisheries regulations were implemented [50 CFR 679.22(a)(6)] to create a Pribilof Islands Area Habitat Conservation Zone, in part to protect the northern fur seal. Under the MMPA, this stock will remain listed as depleted until levels reach at least the lower limit of its optimum sustainable population (National Marine Fisheries Service 1999).

Alaskan Natives residing on the Pribilof Islands are allowed an annual subsistence harvest of only juvenile males, with the number determined from annual household surveys of requirements. The mean annual subsistence take of northern fur seals during the 3-year period from 1994 to 1996 was 1708 animals from St. Paul (87.2% of take) and St. George Islands. Subsistence take in other areas occurred, but was considered minimal (Hill and DeMaster 1999).

The northern fur seal also faces other problems. Debris, such as scraps of fishnets, twine, and plastic wrapping bands, is having an increasingly serious effect on seals, which ingest these items or become entangled in them. Mortality arising from entanglement has been implicated in the decline observed in the Pribilof Islands population during the 1970s and early 1980s (Fowler 1987; Swartzman et al. 1990). Approximately 3500 fur seals were accidentally caught each year in high-seas gill nets (FAO 1976) or in net fragments. This injury often means that the animals die of starvation (Fiscus 1978). Surveys indicated that the rate of entanglement among subadult males on St. Paul Island was 0.4% during 1976–1985 and 0.2% from 1988 to 1992 (Fowler and Ragen 1990; Fowler et al. 1994) and from 1995 to 1997 (Hill and DeMaster 1999).

An oil spill would have a particularly detrimental effect on this species. A small amount of crude oil on the fur increases thermal conductance and heat loss. The fur seals must consequently increase their metabolic rate to maintain body temperature. Oil encountered at sea in any amount will probably harm fur seals by rendering their dense underfur ineffective as an insulator (Fiscus 1978). Recent rapid human development on the Pribilof Islands has increased the potential for impact on habitat used by northern fur seals. On St. Paul Island, pup production has declined in rookeries in closest proximity to human habitation and outfalls from sewer and seafood processing plants (Hill and DeMaster 1999).

Guadalupe Fur Seal. A massive and unrelenting slaughter in the late 1700s and early 1800s almost exterminated this species by 1834, and reduced it to near extinction in 1894 (Townsend 1931). Commercial sealing continued in Mexican waters through 1894. In 1928, a very small herd was rediscovered on Isla Guadalupe; however, no others were seen for 26 years (McClung 1978; Fleischer 1978b). In 1954, a small colony was found on Isla Guadalupe (Hubbs 1956). Since 1922, the island has been a wildlife sanctuary under Mexican protection, and was designated a pinniped sanctuary in 1975. The Guadalupe fur seal is now fully protected by Mexican national legislation (Seal Conservation Society 1999).

In the U.S. portion of its range, the state of California lists the Guadalupe fur seal as a fully protected mammal in the Fish and Game Code. It is listed as a threatened species in the Fish and Game Commission California Code of Regulations. It was designated as "threatened" throughout its range under the Endangered Species Act in 1985, which automatically qualifies it as a "depleted" and "strategic" stock under the Marine Mammal Protection Act (Forney et al. 2000). The species is also listed as "vulnerable" on the IUCN Red List and as an Appendix 1 species under CITES (Seal Conservation Society 1999).

Drift and gillnet fisheries may cause incidental mortality of Guadalupe fur seals in Mexico and the United States. There have been no reports of mortalities or injuries in U.S. commercial fisheries, however, and species-specific information is not available for the Mexican fisheries (Forney et al. 2000). However, juvenile female Guadalupe fur seals have stranded in central and northern California with net abrasions around the neck and entangled with fish hooks, monofilament line, and polyfilament string (Hanni et al. 1997).

Walrus. Because the walrus has tusks, its valuable ivory has been marketed in a similar manner to seal fur. Use of the walrus as a natural resource by Alaskan native peoples has posed some of the most difficult management problems under the MMPA.

Whalers and sealers hunted the Pacific walrus in the nineteenth century and often abandoned the slaughtered animals after taking only their tusks. Klondike gold seekers used the walrus as an object of sport killing, similarly wasting the carcass. By 1900, the walrus was virtually extirpated south of Nunivak Island. The last, and one of the most intense, commercial harvesting episodes occurred between the 1930s and early 1950s. Overall, from 1650 to 1960, over 3.5 million animals were killed on the Pacific coast. In contrast to the early days of proliferation, by the mid-1950s, the west coast walrus population was thought to number only 40,000–50,000, perhaps as high as 100,000, individuals. In 1962, Alaskan Natives took another 12,000 animals, and many of these sank when killed (McClung 1978; Marine Mammal Commission 2000).

In the 1960s, under conservation measures initiated independently by the former Soviet Union and the state of Alaska, the Pacific walrus population rebounded. The Walrus Islands, a group of seven islands in northeast Bristol Bay, Alaska, which are the only regular summer hauling-out grounds of the Pacific walrus, became the Walrus Islands State Game Sanctuary in 1960. By the mid-1970s, the animals started to return to their ancestral hauling-out grounds on the Pribilof Islands, a place they had inhabited before the slaughters of the previous century. The Marine Mammal Protection Act of 1972 banned trophy hunting of the walrus, and permitted the taking of this species by native peoples provided that the take is for subsistence reasons or for the creation of handicrafts and is not wasteful (Kenyon 1978a). The state of Alaska placed restrictions on the site of the taking, however. Walrus Island State Game sanctuary and another state game management unit in Bristol Bay were prohibited areas. Alaska fought for control of the management of marine mammals under the provisions of the MMPA, and Congress granted this right in April 1976.

In April 1977, the Alaskan native people sued the U.S. Department of the Interior for allowing the state of Alaska to forbid them the right to hunt on these traditional hunting grounds. Their argument was that it was necessary for them to hunt in these areas to fulfill their essential dietary and economic needs. Government officials working in the area stated that they believed that the overriding motivation was not subsistence, but ivory. A Native organization, the Eskimo Walrus Commission, was established in 1978 to help conserve the walrus population. In April 1979, a U.S. district court judge ruled that Alaskan Natives could hunt walruses despite the state regulations against it, which in effect negated Alaska's walrus management program. After this judgment, the state of Alaska, in June 1979, returned the responsibility for all marine mammal management to the federal government. The U.S. Fish and Wildlife Service, under the auspices of the Department of the Interior, accordingly issued an emergency statement effective 8 July 1979, which ruled that the federal restraints laid down under the MMPA be reinstated and that henceforth all recreational hunting, importation, and other activities that the state had permitted now were prohibited. Federal permits were required to take walruses. Natives were still allowed to take walruses for subsistence reasons or to create authentic handicraft articles or clothing.

The Fish and Wildlife Service is the lead federal agency responsible for conserving Pacific walrus under the MMPA. Pacific walrus abundance, estimated in 1990 at about 200,000 individuals, is considered to be within the optimum sustainable population range. The current status, however, is uncertain. In 1994, the Service adopted a Pacific Walrus Conservation Plan, which was prepared following recommendations by the Marine Mammal Commission and the Alaska Native Community, to help guide walrus research and management actions. The Biological Resources Division of the U.S. Geological Survey also conducts research on walruses in support of the walrus conservation program. A cooperative agreement was signed with the Eskimo Walrus Commission in 1997, and pursuant to this, the Fish and Wildlife Service works with Native communities and hunters to manage and monitor the subsistence hunt of walruses and collect biological samples for walrus research (Marine Mammal Commission 2000).

To estimate the number of Pacific walrus taken in Alaska, an annual walrus harvest monitoring program has been undertaken cooperatively by the Fish and Wildlife Service, the Eskimo Walrus Commission, and Alaskan Native hunters since 1980 (not funded in 1991 and 1992). Harvests at Round Island in northern Bristol Bay have been monitored since 1995, when a limited walrus take resumed there after 35 years. In 1988, the Service also began a marine mammal marking, tagging, and reporting program to help improve data on subsistence harvest and trade (Marine Mammal Commission 2000). With a method that used data from both programs (Garlich-Miller and Burns 1999), annual catch levels are estimated to have ranged from 1003 to 2501 walruses during

1992–1998. The 1999 total catch was likely about 2500 animals (Marine Mammal Commission 2000).

In addition to walruses landed, some are shot but escape or sink before they are retrieved. It is thought most walruses shot and lost soon die of their wounds. Data collected between 1952 and 1972 suggest that 42% of the walruses shot during that period were not recovered. Applying that ratio to recent U.S. and Russian catch data, the total number killed in Native subsistence hunts in both Russia and Alaska between 1992 and 1998 would range from about 3800 to 6100 animals/year (Fay and Bowlby 1994; Marine Mammal Commission 2000).

A need to develop a reliable estimate of the total population size and trends is likely the greatest problem facing the Pacific walrus conservation program. Between 1975 and 1990, a series of range-wide aerial and shipboard surveys were conducted at 5-year intervals in remote areas of the Bering and Chukchi Seas in summer jointly by U.S. and Russian researchers. No surveys have been conducted since 1990, however, in part because of their expense. Sampling problems associated with the patchy distribution of walruses over large areas caused population estimates from past surveys to be imprecise and of limited value in determining trends. It is hoped that potentially viable survey approaches for estimating the size of the Pacific walrus population and the critical information needs and assumptions associated with each approach can be identified and successfully applied in the future (Marine Mammal Commission 2000).

Harassment by humans presents many problems resulting from extensive exploration for oil and minerals throughout the walrus's range. A major oil spill would be very serious (FAO 1976). Under North Pacific fishery management plans, to prevent disrupting the walrus at rookeries and haulouts, fishery vessels are prohibited in that part of the Bering Sea within 22 km of Round Island, the Twins, and Cape Pierce in northern Bristol Bay during summer (Witherell et al. 2000).

The Atlantic walrus is mainly a Canadian management challenge. Exploitation of the Atlantic walrus during the seventeenth to nineteenth centuries eradicated the species from some of its important centers of abundance, such as Bear Island, where it went from the height of its abundance in 1604 to near extinction by 1613 (McClung 1978). Other areas of depletion were most of Svalbard, Sable Island, and the Gulf of St. Lawrence (Reeves 1978). The Baffin Island herds were left untouched until the 1920s, when 175,000 animals were taken between 1925 and 1931. In 1928, under the Fisheries Act, regulations were passed prohibiting walrus hunting in the area of Baffin Island and Hudson Bay, except for food for Inuit. This was amended in 1931 by setting quotas of seven/year for Inuit with dependents and four/year for others.

Other countries, such as Norway and Russia, that had traditionally taken part in the walrus hunt have passed restrictive legislation to protect these animals. There are two main populations, the Kara Sea to East Greenland region and from West Greenland to Canadian waters and land. Since 1956, Greenland (Denmark) has allowed hunting by permanent residents only with strict regulations (McClung 1978). In 1952, Norway passed the Norwegian Walrus Decree, which forbade hunting for any purpose. Since 1973, it has created many marine sanctuaries encompassing habitats of the walrus. In May 1974, the Canadian government established a national park in some of the critical habitats of the walrus (Reeves 1978).

In Canada, in 1993, the Walrus Protection Regulations were revoked and the walrus was placed under the new Marine Mammal Regulations. Native people in the eastern Arctic have the right to hunt walruses without permits for subsistence use, a condition that has been in effect since 1928, but annual quotas are established for each community (Environment Canada 1999b). Hides may not be exported, nor may unworked ivory. No edible part can be wasted.

The most threatening condition for the walrus is still human encroachment, accidental or otherwise, on the hauling-out grounds. Behavioral research is needed to determine the effects of such trespassing. Investigation is also needed to determine whether there might be any conflict in harvesting of the invertebrate food supply of the walrus in the Canadian eastern Arctic (Reeves 1978), as is happening in the Pacific.

The economic aspects of walrus consumption by humans are varied. Use varies by region, but on the average only 35% is consumed as human food, the rest being used as dog food (FAO 1976). The skin was used for the tips of billiard cues in Greenland (Reeves 1978). Although there is still trade with the Hudson's Bay Company for ivory, much is sold privately. In the 1970s, quotes for raw ivory in Canada ranged from $44 to $55/kg (Reeves 1978). Traditional uses were much less wasteful; almost every part of the animal being is used for subsistence purposes from food to wearing apparel, rope, tools, and fuel (Alaska Deptartment of Fish and Game 1976; Reeves 1978). Life has changed for the Native peoples, and new items have replaced traditional ones. One thing that cannot be taken away or replaced is the part that the walrus hunt takes in their culture. The hunt still provides them with pride, self-sufficiency, and a connection with the past.

Finally, certain principles must be considered in the management of pinniped species.

1. The management approach for species should meet international standards of taking a precautionary approach to the exploitation of wild living resources (Restrepo et al. 1999).
2. Population models must be reevaluated, updated, and perfected continuously, as they are used as a basis for management decisions. Programs to monitor long-term changes in population sizes should account for factors that cause short-term variations in indices of abundance. The accuracy of monitoring programs can be improved by the inclusion of such factors as covariates in models (Frost et al. 1999).
3. Studies on food habits should be conducted throughout the year, using bioenergetics to properly assess the impact on commercial species of fish. Contents of stomachs should be analyzed and locations, energy requirements, and diet by age, population, and season taken into account. Development of simulation models to estimate energy requirements of populations is critical.
4. A systems approach to management should be adopted; trophic interactions are complex. It is naive to manage exploited populations on the basis of single-species population models. A change in one trophic level will have effects on other levels within the system. For example, the harp seal, as the most abundant seal species and an important predator in the Barents Sea, is included in a multispecies model (MULTSPEC), which may provide the basis for future management of marine resources in this sea (Bogstad et al. 1995).
5. It is necessary to apply and integrate biological, ecological, economic, sociological, political, and institutional knowledge, suggested by guiding principles for marine mammal conservation offered by Meffe et al. (1999).
6. Impact studies require sampling of impacted and reference areas over time. Sampled sites, however, as stated by Hoover-Miller et al. (2001), need to cover a geographic region sufficiently large enough to include most of the population being assessed. Sample timing and effort need to be as consistent as possible among years. There are meticulous demands on site selection, sampling protocols, data collection, and analyses, Furthermore, associated ecological and statistical assumptions should be evaluated.

In conclusion, most pinniped species found in North America receive considerable public support, although to a slightly lesser extent than that displayed for whales. A study undertaken by the Humane Society of the United States (Kellert 1999) revealed strong support for marine mammal protection. Findings showed willingness on the part of people to make significant sacrifices to sustain and enhance marine mammal populations and species. There was concern for various commercially important ocean activities, including fishing and oil and gas extraction, but these interests did not supercede the public's inclination to protect marine mammals. Instead, there was a desire to modify or alter these and other human activities in the marine environment to protect marine mammals, even if it necessitated sacrifice, as in higher consumer prices for fish or gasoline. It was considered important to maintain the Marine Mammal Protection Act and the Marine Mammal Commission to ensure domestic and international well-being of marine mammals. These findings clearly indicate that marine mammals possess considerable aesthetic, scientific, and moral support among the great majority of residents of the United States.

LITERATURE CITED

Addison, R. F. 1973. Organochlorine residues in Canadian arctic marine mammals (Council Meeting, Marine Mammals Commission, N:4). International Council for the Exploration of the Sea, Copenhagen.

Alaska Department of Fish and Game. 1976. Ice inhabiting phocid seals Scientific consultation. *in* Symposium: Scientific consultation on marine mammals. Food and Agriculture Organization of the United Nations, Advisory Committee on Marine Resources Research, Marine Mammals. Bergen, Norway.

Allen, J. A. 1887. The West Indian seal (*Monachus tropicalis,* Gray). Bulletin of the American Museum of Natural History 2:1–34.

Allen, R. L. 1975. A life table for harp seals in the Northwest Atlantic. Pages 303–11 *in* K. Ronald and A. W. Mansfield, eds. Biology of the seal. International Council for the Exploration of the Sea, Copenhagen.

Allen, S. G. 1991. Harbor seal habitat restoration at Strawberry Spit, San Francisco Bay (Report PB91–212332/GAR). Point Reyes Bird Observatory.

Alverson, D. L. 1992. A review of commercial fisheries and the Steller sea lion (*Eumetopias jubatus*): The conflict arena. Reviews in Aquatic Science 6:203–56.

Anderson, E. D. 1998. The history of fisheries management and scientific advice: The ICNAF/NAFO history from the end of World War II to the present. Journal of Northwest Atlantic Fishery Science 23:75–94.

Andrews, R. D., D. R. Jones, J. D. Williams, D. E. Crocker, D. P. Costa, and B. J. Le Boeuf. 1995. Metabolic and cardiovascular adjustments to diving in northern elephant seals (*Mirounga angustirostris*) [Abstract]. Physiological Zoology 68:105.

Angell James, J. E., and M. de Burgh Daly. 1972. Some mechanisms involved in the cardiovascular adaptations to diving. Pages 313–41 *in* M. A. Seligh and A. G. MacDonald, eds. Symposium: Effects of pressure on organisms. Proceedings of the 26th symposium of the Society of Experimental Biology, Wales. Academic Press, New York.

Anonymous. 1985a. Final environmental impact statement on the Interim Convention on Conservation of North Pacific Fur Seals. Environmental Impact Statement, U.S. Department of Commerce, NOAA, Washington, DC.

Anonymous. 1985b. North Pacific fur seals: Current problems and opportunities concerning conservation and management. A special report to the President and the Congress. National Advisory Committee on Oceans and Atmosphere, Washington, DC.

Anonymous. 1992. Marine mammal/fisheries interactions: Analysis of cull proposals. Report of the second meeting of the Scientific Advisory Committee of the Marine Mammal Action Plan, Liège, Belgium, 27 November–1 December 1992. Marine Mammal Action Plan Secretariat, United Nations Environment Programme.

Anonymous. 1994. Marine mammal/fisheries interactions: Analysis of cull proposals. Report of the third meeting of the Scientific Advisory Committee of the Marine Mammal Action Plan, Crowborough, UK 24–27 August 1994. Marine Mammal Action Plan Secretariat, United Nations Environment Program.

Anonymous. 1996. Report of the Gulf of Maine Aquaculture–Pinniped Interaction Task Force. Office of Protected Resources, National Marine Fisheries Service, Silver Spring, MD.

Anonymous. 1997. The seal story at a glance. Understanding the seal fishery (No. 1). Canada Department of Fisheries and Oceans, Ottawa.

Anonymous. 1998a. Report of the Joint ICES/NAFO Working Group on Harp and Hooded Seals, 28 August–3 September 1997 (ICES CM 1998/Assess: 3). Copenhagen.

Anonymous. 1998b. Endangered Species Act. Section 7: Consultation—Biological opinion. Authorization of an Atka mackerel fishery under the BSAI groundfish Fishery Management Plan between 1999 and 2002; Authorization of a walleye pollock fishery under the Bering Sea—Aleutian Island groundfish Fishery Management Plan between 1999 and 2002, and Authorization of a walleye pollock fishery under the Gulf of Alaska groundfish Fishery Management Plan between 1999 and 2002. National Marine Fisheries Service—Alaska Region.

Anonymous. 2000. Rare birth on Kauai. Hawaiian news. Monachus Guardian 3(November): 7. Now available at www.monachus.guardian.org.

Antonelis, G. A., C. H. Fiscus, and R. L. DeLong. 1984. Spring and summer prey of California sea lions, *Zalophus californianus,* at San Miguel Island, California, 1978–79. Fisheries Bulletin 82:67–76.

Arnbom, T. A., N. J. Lunn, I. L. Boyd, and T. Barton. 1992. Aging live Antarctic fur seals and southern elephant seals. Marine Mammal Science 8:37–43.

Atkinson, S., B. L. Becker, J. R. Johanos, J. R. Pietraszek, and B.C.S. Kuhn. 1994. Reproductive morphology and status of female Hawaiian monk seals (*Monachus schauinslandi*) fatally injured by adult male seals. Journal of Reproductive Fertility 100:225–30.

Aurioles-Gamboa, D. 1993. Bioversidad y estado actual de los mamiferos marinos en México. Revista de le Sociedad Mexicana de la Naturaleza Especial XLIV:397–412.

Aurioles-G., D., F. Sinsel, C. Fox, E. Alvarado, and O. Maravilla. 1983. Winter migration of subadult male California sea lions (*Zalophus californianus*) in the southern part of Baja California. Journal of Mammalogy 64:513–18.

Aurioles-G., D., C. Fox, F. Sinsel, and G. Tanos. 1984. Prey of Californian sea lions (*Zalophus californianus*) in the Bay of La Paz, Baja California Sur, Mexico. Journal of Mammalogy 65:519–21.

Aurioles-G., D., and C. J. Hernadez-Camacho. 1999. Notes on the southernmost records of Guadalupe fur seal, *Arctocephalus townsendi,* in Mexico. Marine Mammal Science 15:581–83.

Baker, R. C., F. Wilke, and C. H. Baltzo. 1970. The northern fur seal, reprint ed. (Circular 336). Bureau of Commercial Fisheries, U.S. Fish and Wildlife Service, Washington, DC.

Banfield, A. W. F. 1974. The mammals of Canada. University of Toronto Press, Toronto.

Barlough, J. E., E. S. Berry, E. A. Goodwin, R. F. Brown, R. L. DeLong, and A. W. Smith. 1987. Antibodies to marine caliciviruses in the Steller sea lion (*Eumetopias jubatus,* Schreber). Journal of Wildlife Disease 23:34–44.

Barlow, J., P. Boveng, M. S. Lowry, B. S. Stewart, B. J. Le Boeuf, W. J. Sydeman, R. J. Jameson, S. G. Allen, and C. W. Oliver. 1993. Status of the northern elephant seal population along the U.S. West Coast in 1992 (Administrative Report LJ-93–01). Southwest Fisheries Science Center, National Marine Fisheries Service, La Jolla, CA.

Bartholomew, G. A. 1952. Reproductive and social behavior of the northern elephant seal. University of California Publications in Zoology 47:369–472.

Bartholomew, G. A., and R. A. Boolootian. 1960. Numbers and population structure of the pinnipeds on the California Channel Islands. Journal of Mammalogy 41:366–75.

Beddington, J. R., and D. M. Lavigne, eds. 1979. International workshop on biology and management of Northwest Atlantic harp seals. Proceedings of the symposium at the University of Guelph, Guelph, Ontario, 3–6 December. World Wildlife Fund, Morges, Switzerland.

Beier, J. C., and D. Wartzok. 1979. Mating behavior of captive spotted seals (*Phoca largha*). Animal Behavior 27:772–81.

Benjaminsen, T., and T. Øritsland. 1975. The survival of year-classes and estimates of production and sustainable yield of Northwest Atlantic harp seals (Research Document 75/121, Series No. 3625). International Commission for the Northwest Atlantic Fisheries.

Benoit, D., and W. D. Bowen. 1990. Seasonal and geographic variation in the diet of gray seals (*Halicheorus grypus*) in eastern Canada. Pages 215–42 *in* W. D. Bowen, ed. Population biology of sealworm (*Pseudoterranova decipiens*) in relation to its intermediate and seal hosts (Canadian Bulletin of Fisheries and Aquatic Sciences No. 222). Ottawa.

Berger, T. R. 1977. Northern frontier, northern homeland; the report of the Mackenzie Valley pipeline inquiry. Vol. 1. Supply and Services Canada, Ottawa.

Bernt, K. E., M. O. Hammill, and K. M. Kovacs. 1996. Age determination of gray seals (*Halichoerus grypus*) using incisors. Marine Mammal Science 12:476–82.

Bickford, B., and P. Martz. 1980. Test excavations at Cottonwood Creek Catalina Island, California. Pacific Coast Archaeological Society Quarterly 16:106–24.

Bickham, J. W., J. C. Patton, and T. R. Loughlin. 1996. High variability for control-region sequences in a marine mammal; implications for conservation and biogeography of Steller sea lions (*Eumetopias jubatus*). Journal of Mammalogy 77:95–108.

Bigg, M. A. 1969. The harbour seal in British Columbia (Bulletin 172). Fisheries Research Board of Canada.

Bisaillon, A., I. Picard, and N. La Rivière re. 1976. Le segment cervical des carnivores (Mammalia: Carnivora) adaptés à la vie aquatique. Canadian Journal of Zoology 54:431–36.

Blix, A. S. 1976. Metabolic consequences of submersion asphyxia in mammals and birds. Biochemistry Society Symposium 41:169–78.

Blix, A. S., H. J. Grav, and K. Ronald. 1975. Brown adipose tissue and the significance of the venous plexuses in pinnipeds. Acta Physiologica Scandinavica 94:133–35.

Blix, A. S., H. J. Grav, and K. Ronald. 1979. Some aspects of temperature regulation in newborn harp seal pups. American Journal of Physiology 236:188–97.

Bogstad, B., K. H. Hauge, and Ø. Ulltang. 1995. Results of simulation studies using a multispecies model for the Barents Sea (MULTISPEC) (SC/47/NA1). International Whaling Commission Cambridge, UK.

Bonner, W. N. 1971. An aged grey seal (*Halichoerus grypus*). Journal of Zoology 164:261–62.

Bonner, W. N. 1972. The grey seal and common seal in European waters. Oceanography and Marine Biology 10:461–507.

Bonnot, P. 1928. Report on the seals and sea lions of California (Fisheries Bulletin No. 14). California Division of Fish and Game, Sacramento.

Bonnot, P. 1951. The sea lions, seals and sea otter of the California coast. California Fish and Game 37:371–89.

Boulva, J. 1976. *Phoca vitulina concolor*. Scientic consultation 47 *in* Symposium: Scientific consultation on marine mammals. Food and Agriculture Organization of the United Nations, Advisory Committee on Marine Resources Research, Marine Mammals. Bergen, Norway.

Boulva, J., and I. A. McLaren. 1979. Biology of the harbour seal, *Phoca vitulina,* in eastern Canada (Bulletin 200). Fisheries Research Board of Canada.

Boveng, P. 1988. Status of the northern elephant seal population on the U.S. West Coast (Administrative Report LJ-88-05). Southwest Fisheries Science Center, National Marine Fisheries Service, La Jolla, CA.

Bowen, W. D., and G. Harrison. 1994. Offshore diet of gray seals *Halichoerus grypus* near Sable Island, Canada. Marine Ecology Progress Series 112:1–11.

Bowen, W. D., and D. B. Siniff. 1999. Distribution, population biology, and feeding ecology of marine mammals. Pages 423–84 *in* J. E. Reynolds, III and S. A. Rommel, eds. Biology of marine mammals, Smithsonian Institution Press, Washington, DC.

Bowen, W. D., R. A. Myers, and K. Hay. 1987. Abundance estimation of a dispersed, dynamic population: Hooded seals (*Cystophora cristata*) in the Northwest Atlantic (Research Document 75/121). International Commission for the Northwest Atlantic Fisheries.

Boyd, I. L. 1995. Steller Sea lion research. A report prepared for the U.S. National Marine Fisheries Service, National Marine Mammal Laboratory, Seattle, WA, 89p. Available from Alaska Fisheries Science Center, 7600 Sand Point Way NE, Seattle, WA 98115.

Boyd, I. L., and M. P. Stanfield. 1998. Circumstantial evidence for the presence of monk seals in the West Indies. Oryx 32:310–16.

Boyle, G. J. 1997. Studying prey selection by seals: The utility of prey preference experiments. Journal of Northwest Atlantic Fishery Science 22:115–17.

Braham, H. W., J. J. Burns, G. A. Fedoseev, and B. D. Krogman. 1984. Habitat partitioning by ice-associated pinnipeds: Distribution and density of seals and walruses in the Bering Sea, April 1976. Pages 25–47 *in* F. H. Fay and G. A. Fedoseev, eds. Soviet–American cooperative research on marine mammals, Vol. 1: Pinnipeds (NOAA Technical Report NMFS-12). U.S. Department of Commerce, Washigton, DC.

Brown, R. F. 1988. Assessment of pinniped populations in Oregon (Process Report 88–05). National Marine Fisheries Service, Northwest Alaska Fisheries Center, Seattle, WA.

Bruemmer, F. 1979. The homecoming: The grey seals of Sable. Nature Canada 8:48–53.

Bryden, M. M. 1972. Growth and development of marine mammals. Pages 2–79 *in* L. J. Harrison, ed. Functional anatomy of marine mammals, Vol. 1. Academic Press, New York.

Bryden, M. M. 1978. Arteriovenus anastomoses in the skin of seals. Part 3: The harp seal *Pagophilus groenlandicus* and the hood seal *Cystophora cristata* (Pinnipedia: Phocidae). Aquatic Mammals 6:67–75.

Bryden, M. M., and G. H. K. Lim. 1969. Blood parameters of the southern elephant seal (*Mirounga leonina*) in relation to diving. Comparative Biochemistry and Physiology 28:139–48.

Buck, E. H. 1995. Summaries of major laws implemented by the National Marine Fisheries Service (Congressional Research Service Issue Brief for Congress). National Council for Science and the Environment, Washington, DC.

Buhler, D. H., R. R. Clacys, and B. R. Mate. 1975. Heavy metal and chlorinated hydrocarbon residues in California sea lions *(Zalophus californianus californianus*). Journal of the Fisheries Research Board of Canada 32:2391–97.

Burg, T. M. 1996. Genetic analysis of eastern Pacific harbor seals (*Phoca vitulina richardsi*) from British Columbia and parts of Alaska using mitochondrial DNA and microsatellites. Master's Thesis, University of British Columbia, Vancouver, Canada.

Burns, J. J. 1967. The Pacific bearded seal (Annual Project Segment Report 8, Federal Aid in Wildlife Restoration, Project Report W-6-R, W-14-R). Alaska Department of Fish and Game, Juneau.

Burns, J. J. 1971. Biology of the ribbon seal, *Histriophoca fasciata,* in the Bering Sea [Abstract]. Page 135 *in* Symposium: Adaptation for northern life. Proceedings of the 22nd Alaskan science conference. Fairbanks, AK.

Burns, J. J. 1973. Marine mammal report (Pittman-Robertson Project Report W-17-3, W-17-4, and W-17-5). Alaska Department of Fish and Game, Anchorage.

Burns, J. J. 1978. Ice seals. Pages 192–205 *in*D. Haley, ed. Marine mammals of eastern North Pacific and arctic waters. Pacific Search Press, Seattle, WA.

Burns, J. J. 1981a. Ribbon seal *Phoca fasciata*. Pages 89–109 *in* S. H. Ridgway and R. J. Harrison, eds. Handbook of marine mammals, Vol. 2: Seals. Academic Press, New York.

Burns, J. J. 1981b. Bearded seal *Erignathus barbatus* Erxleben, 1977. Pages 145–70 *in* S. H. Ridgway and R. J. Harrison, eds. Handbook of marine mammals. Vol. 2: Seals. Academic Press, New York.

Calambokidis, J., and R. W. Baird. 1994. Status of marine mammals in the Strait of Georgia, Puget Sound, and the Juan de Fuca Strait, and potential human impacts. Canadian Technical Report of Fisheries and Aquatic Sciences 1948:282–300.

Calkins, D. G., and E. Goodwin. 1988. Investigation of the declining sea lion population in the Gulf of Alaska. Alaska Department of Fish and Game, Anchorage.

Calkins, D. G., and K. W. Pitcher. 1982. Population assessment, ecology and trophic relationships of Steller sea lions in the Gulf of Alaska. Pages 447–546 *in* Environmental assessment of the Alaskan continental shelf. Final report. U.S. Department of Commerce and U.S. Department of the Interior Washington, DC.

Casson, D. M., and K. Ronald. 1975. The harp seal *(Pagophilus groenlandicus* Erxleben, 1777). XIV. Cardiac arrythmias. Comparative Biochemistry and Physiology 50A:307–14.

Center for International Earth Science Information Network. 1996–2001. Environmental treaties and resource indicators (ENTRI). Available at http://sedac.ciesin.org/entri/.

Chapman, D.M.F., and D. D. Ellis. 1998. The elusive decibel: Thoughts on sonars and marine mammals. Canadian Acoustics 26:29–31.

Chávez-Rosales, S., and S. C. Gardner. 1999. Recent harbour seal (*Phoca vitulina richardsi*) pup sightings in Magdalena Bay, Baja California, Mexico. Aquatic Mammals 25:169–71.

Cleator, H. J. 1996. The status of the bearded seal, *Erignathus barbatus,* in Canada. Canadian Field-Naturalist 110:501–10.

Cooper, C. F., and B. S. Stewart. 1983. Demography of northern elephant seals, 1911–1982. Science 219:969–71.

Cox, M., E. Gaglione, P. Prowten, and M. Noonan. 1996. Food preferences communicated via symbol discrimination by a California sea lion (*Zalophus californianus*). Aquatic Mammals 22:3–10.

Davies, J. L. 1957. The geography of the gray seal. Journal of Mammalogy 38:297–310.

Davis, J. E., and S. S. Anderson. 1976. Effects of oil pollution on breeding grey seals. Marine Biology Bulletin 7:115–18.

De Anda, H. 1985. Habitos alimenticios del lobo marino de California (*Zalophus californianus*) en las islas Los Coronados B.C., de Noviembre 1983 de Octubre 1984. Tesis de Licenciatura en Ciencias Marinas, Universidad Autónoma de Baja California, Mexico.

de Burgh Daly, M. 1984. Breath-hold diving: Mechanisms of cardiovascular adjustments in the mammal. Pages 201–45 *in* P. F. Baker, ed. Recent advances in physiology. Churchill Livingstone, London.

de Kleer, V. S. 1972. The anatomy of the heart and the electrocardiogram of *Pagophilus groenlandicus*. M.S. Thesis, University of Guelph, Guelph, Ontario, Canada.

DeLong, R. L. 1975. Interspectific reproductive behavior among four otariids at San Miguel Island, California [Abstract]. Page 13 *in* Symposium: Conference on the biology and conservation of marine mammals, 4–7 December. University of California, Santa Cruz.

DeLong, R. L. 1978a. Investigations of Hawaiian monk seal mortality at Laysan, Lisianski, French Frigate Shoals, and Necker Island, May 1978. Interim report. NOAA, National Marine Fisheries Service, Northwest Alaska Fisheries Center, Seattle, WA.

DeLong, R. L. 1978b. Northern elephant seal. Pages 206–11 *in* D. Haley, ed. Marine mammals of eastern North Pacific and arctic waters. Pacific Search Press, Seattle, WA.

DeLong, R. L. 1982. Population biology of northern fur seals at San Miguel Island, California. Ph.D. Dissertation, University of California, Berkeley.

DeLong, R. L., and G. A. Antonelis. 1991. Impacts of the 1982–1983 El Niño on the northern fur seal population at San Miguel Island, California. Pages 75–83 *in* F. Trillmich and K. Ono, eds. Pinnipeds and El Niño: Responses to environmental stress. Springer-Verlag, New York.

DeLong, R. L., W. G. Gilmartin, and J. G. Simpson. 1973. Premature births in California sea lions: Association with high organochloride pollutant residue levels. Science 181:1168–70.

DeLong, R. L., G. L. Kooyman, W. G. Gilmartin, and T. R. Loughlin. 1984. Hawaiian monk seal diving behavior. Acta Zoologica Fennica 172:129–31.

DeLong, R., J. Laake, and S. Melin. 1998. The impacts of El Niño on pinnipeds in the California Channel Islands. National Marine Mammal Laboratory, U.S. Department of Commerce/NOAA/NMFS/AFSC. Available at nmml01.afsc.noaa.gov/El_Nino/.

Dierauf, L. A., D. J. Vandenbroek, J. Roletto, M. Koski, L. Amaya, and L. J. Gage. 1985. An epizootic of leptospirosis in California sea lions. Journal of the American Veterinary Medical Association 187:1145–48.

Dill, H. R., and W. A. Bryan. 1912. Report of an expedition to Laysan Island in 1911. U.S. Department of Agriculture Bureau of Biology Survey Bulletin 42:9.

Dizon, A. E., C. Lockyer, W. F. Perrin, D. P. DeMaster, and J. Sisson. 1992. Rethinking the stock concept: A phylogeographic approach. Conservation Biology 6:24–36.

Dolan, M., 1990. Sea lions come roaring back. Los Angeles Times, 18 November, 1990, Part A, p. 3.

Donaldson, G. M., G. Chapdelaine, and J. D. Andrews. 1995. Predation of thick-billed murres, *Uria lomvia,* at two breeding colonies by polar bears, *Ursus maritimus,* and walruses, *Odobenus rosmarus*. Canadian Field-Naturalist 109:112–14.

Dormer, K. J., M. J. Denn, and H. L. Stone. 1977. Cerebral blood flow in the sea lion (*Zalophus californianus*) during voluntary dives. Comparative Biochemistry and Physiology 58A:11–18.

Dorofeev, S. V. 1939. The influence of ice conditions on the behavior of harp seals. Zoologicheskii Zhurnal 18:748–61 (In Russian).

Dovers, S. R., and J. W. Handmer. 1995. Ignorance, the precautionary principle, and sustainability. Ambio 24:92–97.

Drabek, C. M. 1975. Some anatomical aspects of the cardiovascular system of antarctic seals and their possible functional significance in diving. Journal of Morphology 145:85–106.

Dunbar, M. J. 1949. The Pinnipedia of the arctic and subarctic. Fisheries Research Board of Canada Bulletin 85:1–22.

Dykes, R. W. 1974. Factors related to the dive reflex in harbor seals: Sensory contributions from the trigeminal region. Canadian Journal of Physiology and Pharmacology 52:259–65.

Edie, A. G. 1977. Distribution and movements of Steller sea lion cows (*Eumetopias jubata*) on a pupping colony. M.S. Thesis, University of British Columbia, Vancouver, Canada.

Elsner, R. 1969. Cardiovascular adjustments to diving. Pages 117–45 *in* H. T. Andersen, ed. The biology of marine mammals. Academic Press, New York.

Elsner, R. 1999. Living in water: Solutions to physiological problems. Pages 73–116 *in* J. E. Reynolds, III and S. A. Rommel, eds. Biology of marine mammals. Smithsonian Institution Press, Washington, DC.

Elsner, R., D. Wartzok, N. B. Sonafrank, and B. P. Kelly. 1989. Behavioral and physiological reactions of arctic seals during under-ice pilotage. Canadian Journal of Zoology 67:2506–13.

Engelhardt, F. R., J. R. Geraci, and T. G. Smith. 1977. Uptake and clearance of petroleum hydrocarbons in the ringed seal, *Phoca hispida*. Journal of the Fisheries Research Board of Canada 34:1143–47.

Environment Canada. 1999a. Harbour seal, Lac des Loups Marins population (Species at Risk Fact Sheet). Environment Canada, Canadian Wildlife Service, Ottawa.

Environment Canada. 1999b. Atlantic walrus, Atlantic coastal waters population (Species at Risk Fact Sheet). Environment Canada, Canadian Wildlife Service, Ottawa.

Fay, F. H. 1960. Structure and function of the pharyngeal pouches of the walrus (*Odobenus rosmarus* L.) Mammalia 24:362–71.

Fay, F. H. 1974. The role of ice in the ecology of marine mammals of the Bering Sea. University of Alaska, Institute of Marine Science, Occasional Papers 2:383–97.

Fay, F. H. 1982. Ecology and biology of the Pacific walrus, *Odobenus rosmarus divergens* Illiger. North American Fauna 74:1–279.

Fay, F. H., and E. E. Bowlby. 1994. The harvest of Pacific walrus, 1931–1989 (Technical Report MMM 94.2). Fish and Wildlife Service, Anchorage, AK.

Fay, F. H., and J. J. Burns. 1988. Maximal feeding depth of walruses. Arctic 41:239–40.

Fay, F. H., and C. Ray. 1968. Influence of climate on the distribution of walruses, *Odobenus rosmarus* (Linnaeus). I. Evidence from thermoregulatory behavior. Zoologica 53:1–14.

Fay, F. H., L. M. Shults, and R. A. Dieterich. 1979. A field manual of procedures for postmortem examination of Alaskan marine mammals (Occasional Publications on Northern Life No. 3). University of Alaska, Institute of Arctic Biology.

Fay, F. H., J. L. Sease, and R. L. Merrick. 1990. Predation on a ringed seal, *Phoca hispida,* and a black guillemot, *Cepphus grille,* by a Pacific walrus, *Odobenus rosmarus divergens*. Marine Mammal Science 6:349–50.

Fay, R. R. 1988. Hearing vertebrates: A psychophysics databook. Hill-Fay Associates, Winnetka, IL.

Federal Register. 2000, May 31. Presidential documents. Executive Order 13158 of May 26, 2000. 65(105):34909–11.

Fedoseev, G. A. 1975. Ecotypes of the ringed seal (*Pusa hispida* Schreber, 1777) and their reproductive capabilities. Pages 156–60 *in* K. Ronald and A.W. Mansfield, eds. Biology of the seal. International Council for the Exploration of the Sea, Copenhagen.

Feltz, E. T., and F. H. Fay. 1966. Thermal requirements in vitro of epidermal cells from seals. Cryobiology 3:261–64.

Finley, K. J., G. A. Miller, R. A. Davis, and W. R. Koski. 1983. A distinctive large breeding population of ringed seals (*Phoca hispida*) inhabiting the Baffin Bay pack ice. Arctic 36:162–73.

Fiscus, C. H. 1978. Northern fur seal. Pages 152–59 *in* D. Haley, ed. Marine mammals of eastern North Pacific and arctic waters. Pacific Search Press, Seattle, WA.

Fiscus, C. H. 1983. Fur seals. Background paper submitted by the United States to the 26th annual meeting of the Standing Scientific Committee of the North Pacific Fur Seal Commission, Washington, DC, March 28–April 5, 1983. Available from National Marine Mammal Laboratory, Alaska Fisheries Science Center, Seattle, WA.

Fish, F. E., S. Innes, and K. Ronald. 1988. Kinematics and estimated thrust production of swimming harp and ringed seals. Journal of Experimental Biology 137:157–73.

Fisher, K. I., and R. E. A. Stewart. 1997. Summer foods of Atlantic walrus, *Odobenus rosmarus rosmarus,* in northern Fox Basin, Northwest Territories. Canadian Journal of Zoology 75:1166–75.

Fisheries and Oceans Canada. 1997. The role of the federal government in the oceans sector (DFO/5265). Communications Directorate, Department of Fisheries and Oceans, Ottawa.

Fisheries and Oceans Canada. 2000. II. Overview of the Atlantic seal hunt. Atlantic seal hunt, 2000 Management Plan. Fisheries Management, Department of Fisheries and Oceans, Ottawa.

Fleischer, L. A. 1978a. The distribution, abundance, and population characteristics of the Guadalupe fur seal, *Arctocephalus townsendi* (Mcriam, 1897). M.S. Thesis, University of Washington, Seattle.

Fleischer, L. A. 1978b. Guadalupe fur seal. Pages 160–65 *in* D. Haley, ed. Marine mammals of eastern North Pacific and arctic waters. Pacific Search Press, Seattl, WA.

Fleischer, L. A. 1987. Guadalupe fur seal, *Arctocephalus townsendi*. Pages 43–48 *in* J. P. Croxall and R. L. Gentry, eds. Status, biology, and ecology of fur seals. Proceedings of an international symposium and workshop, Cambridge, England (NOAA Technical Report NMFS 51). U.S. Department of Commerce, Washington, DC.

Folkow, L. P., and A. S. Blix. 1995. Distribution and diving behaviour of hooded seals. Pages 193–202 *in* A. S. Blix, L. Walløe, and Ø. Ullyang, eds. Whales, seals, fish and man, Elsevier, Amsterdam.

Food and Agriculture Organization of the United Nations. 1976. Mammals in the seas. Ad hoc group III on seals and marine otters, report. Advisory Committee on Marine Resources Research, Scientific Consultation on Marine Mammals, Bergen, Norway. (FAO-FI–ACMRR/MM/SC/4).

Food and Agriculture Organization of the United Nations. 1995a. Precautionary approach to fisheries. Part I: Guidelines on the precautionary approach to capture fisheries and species introductions. Elaborated by the technical consultation on the precautionary approach to capture fisheries (including species introductions). Lysekil, Sweden (Technical Paper 350, Part 1). Food and Agriculture Organization, Rome.

Food and Agriculture Organization of the United Nations. 1995b. Code of conduct for responsible fisheries. Food and Agriculture Organization, Rome.

Forney, K. A., J. Barlow, M. M. Muto, M. Lowry, J. Baker, G. Cameron, J. Mobley, C. Stinchcomb, and J. V. Carretta. 2000. U.S. Pacific marine mammal stock assessments, 2000 (Technical Memorandum NOAA-TM-NMFS-SWFSC-300). U.S. Department of Commerce, Southwest Fisheries Science Center, La Jolla, CA.

Fowler, C. W. 1987. Marine debris and northern fur seals: A case study. Marine Pollution Bulletin 18:326–35.

Fowler, C. W., and T. J. Ragen. 1990. Entanglement studies, St. Paul Island, 1989: Juvenile male roundups (NWAFC Process Report 90–06). U.S. Department of Commerce, Washington, DC.

Fowler, C. W., J. D. Baker, R. Ream, B. W. Robson, and M. Kiyota. 1994. Entanglement studies on juvenile male northern fur seals, St. Paul Island, 1992. Pages 100–136 *in* E. H. Sinclair, ed. Fur seal investigations, 1992 (NOAA Technical Memorandum NMFS-AFSC-45). U.S. Department of Commerce, Washington, DC.

Fraker, M. A., and B. R. Mate. 1999. Seals, sea lions, and salmon in the Pacific Northwest. Pages 156–78 *in* J. R. Twiss and R. R. Reeves, eds. Conservation and management of marine mammals. Smithsonian Institution Press, Washington, DC.

Freeman, H. C., and D. A. Horne. 1973. Mercury in Canadian seals. Bulletin of Environmental Contamination and Toxicology 10:172–80.

Freeman, H. C., O. Sangalang, and J. F. Uthe. 1975. A study of the effects of contaminants on steroidogenesis in Canadian grey and harp seals (Council Meeting, Marine Mammals Committee, N:7). International Council for the Exploration of the Sea, Copenhagen.

Frost, K. J. 1985. The ringed seal (*Phoca hispida*). Alaska Department of Fish and Game, Technical Bulletin 7:79–87.

Frost, K. J., L. F. Lowry, E. H. Sinclair, J. Ver Hoef, and D. C. McAllister. 1994. Impacts on distribution, abundance, and productivity of harbor seals. Pages 97–118 *in* T. R. Loughlin, ed. Marine mammals and the *Exxon Valdez*. Academic Press, San Diego, CA.

Frost, K. J., L. F. Lowry, and J. M. Ver Hoef. 1999. Monitoring the trend of harbor seals in Prince William Sound, Alaska, after the *Exxon Valdez* oil spill. Marine Mammal Science 15:494–506.

Gallivan, G. J. 1977. Temperature regulation and respiration in the freely diving harp seal (*Phoca groenlandica*). M.S. Thesis, University of Guelph, Guelph, Ontario, Canada.

Gallivan, G. J., and K. Ronald. 1979. Temperature regulation in the freely diving harp seal (*Phoca groenlandica*). Canadian Journal of Zoology 57:2256–63.

Gallo, R. J. P. 1994. Factors affecting the population status of Guadalupe fur seal, *Arctocephalus townsendi* (Merriam, 1897), at Isla de Guadalupe, Baja California, Mexico. Ph.D. Dissertation, University of California, Santa Cruz.

Garcia, S. M. 1996. The precautionary approach to fisheries and its implications for fishery research, technology and management: An updated review. Pages 1–76 *in* Precautionary approach to fisheries. Part 2: Scientific papers (Fisheries Technical Paper No. 350, Part 2). Food and Agriculture Organization, Rome.

Garlich-Miller, J. L., and D. M. Burns. 1999. Estimating the harvest of Pacific walrus, *Odobenus rosmarus divergens,* in Alaska. Fisheries Bulletin 97:1043–46.

Gaskin, D. E. 1997. Marine biodiversity monitoring. Monitoring protocol for marine mammals in Canadian waters. Report by the Marine Biodiversity Monitoring Committee (Atlantic Maritime Ecological Science Cooperative, Huntsman Marine Science Centre) to the Environmental Monitoring and Assessment Network of Environment Canada.

Gehnrich, P. H. 1984. Nutritional and behavioral aspects of reproduction in walruses. Master's Thesis, University of Alaska, Fairbanks.

Gentry, R. L., and J. H. Johnson. 1981. Predation by sea lions on fur seal neonates in Alaska (Report 28). U.S. Fish and Wildlife Service.

Gentry, R. L., and D. E. Withrow. 1978. Steller sea lion. Pages 166–71 *in* D. Haley, ed. Marine mammals of eastern North Pacific and arctic waters. Pacific Search Press, Seattle, WA.

George, J. C., and K. Ronald. 1973. The harp seal, *Pagophilus groenlandicus* (Erxleben, 1777). XXV. Ultrastructure and metabolic adaptation of skeletal muscle. Canadian Journal of Zoology 51:833–39.

George, J. C., and K. Ronald. 1975. The harp seal, *Pagophilus groenlandicus* (Erxleben, 1777). XXVII. The structure and metabolic adaptation of the caval sphincter muscle with some observations on the diaphragm. Acta Anatomica 93:88–99.

George, R. E. 1979. The senior manager's self-reported use of information sources for selected decision types. Ph.D. Dissertation, University of Waterloo, Waterloo, Ontario, Canada.

Geraci, J. R. 1975. Pinniped nutrition. Pages 312–23 *in* K. Ronald and A. W. Mansfield, eds. Biology of the seal. International Council for the Exploration of the Sea, Copenhagen.

Geraci, J. R., and T. G. Smith. 1976. Direct and indirect effects of oil on ringed seals (*Phoca hispida*) of the Beaufort Sea. Journal of the Fisheries Research Board of Canada 33:1976–84.

Geraci, J. R., D. J. St. Aubin, and I. K. Barker. 1981. Mass mortality of harbor seals: Pneumonia associated with influenza A virus. Science 215:1129–31.

Gilbert, J. R. 1977. Past and present status of grey seals in New England (Council Meeting, Marine Mammals Committee, N:14). International Council for Exploration of the Sea, Copenhagen.

Gilbert, J. R., and N. Guldager. 1998. Status of harbor and gray seal populations in northern New England. Final report (NMFS/NER Cooperative Agreement 14-16-009–1557). National Marine Fisheries Service, Northeast Fisheries Science Center, Woods Hole, MA.

Gilmartin, W. G., J. C. Sweeney, and R. D. Gunnels. 1975. Effects of certain environmental pollutants on the California sea lion. (Contract MM5ACO14, Final Report). U.S. Marine Mammal Commission, Washington, DC.

Gilmartin, W. G., R. L. DeLong, A. W. Smith, J. C. Sweeney, B. W. DeLappe, R. W. Risebrough, L. A. Griner, M. D. Dailey, and D. B. Peakall. 1976. Premature parturition in the California sea lion. Journal of Wildlife Diseases 12:104–15.

Goodman-Lowe, G. 1998. Diet of Hawaiian monk seal (*Monachus schauinslandi*) from the northwestern Hawaiian Islands during 1991–1994. Marine Biology 132:535–46.

Grav, H. J., A. S. Blix, and A. Pasche. 1974. How do seal pups survive birth in arctic winter? Acta Physiologica Scandinavica 92:427–29.

Gray, D. F., and B. Beck. 1979. Eastern Canadian grey seal: 1978 research report and stock assessment (Research Document 79/I). Canadian Atlantic Fisheries Science Advisory Committee.

Gray, J. E. 1850. Catalogue of the specimens of Mammalia in the collection of the British Museum. Part 2: Seals. British Museum (Natural History), London.

Green, R. F. 1972. Observations on the anatomy of some cetaceans and pinnipeds. Pages 247–97 *in* S. H. Ridgway, ed. Mammals of the sea: Biology and medicine. C. C. Thomas, Springfield, IL.

Green-Hammond, K. A. 1980. Fisheries management under the Fishery Conservation and Management Act, the Marine Mammal Protection Act, and the Endangered Species Act. Report to the U.S. Marine Commission, no. MMC-78/12. NTIS Publication PB80-180599.

Griebel, U., and A. Schmid. 1992. Color vision in the California sea lion (*Zalophus californianus*). Vision Research 32:477–82.

Gulland, J. A. 1976. A note on the strategy of the management of marine mammals. Scientific consultation 82 *in* Symposium: Scientific consultation on marine mammals. Food and Agriculture Organization of the United Nations, Advisory Committee on Marine Resources Research, Marine Mammals, Bergen, Norway.

Gunter, G. 1968. The status of seals in the Gulf of Mexico with a record of feral otariid seals off the United States Gulf Coast. Gulf Research Reports 2:301–7.

Ham-Lammé, K. D., D. P. King, B. C. Taylor, C. House, D. A. Jessup, S. Jeffries, P. K. Yochem, F. M. D. Gulland, D. A. Ferrick, and J. L. Stott. 1999. The application of immuno-assays for serological detection of morbillivirus exposure in free ranging harbor seals (*Phoca vitulina*) and sea otters (*Enhydra lutris*) from the western coast of the United States. Marine Mammal Science 15:601–8.

Hammill, M., G. B. Stenson, and R. A. Myers. 1992. Hooded seal (*Cystophora cristata*) pup production in the Gulf of St. Lawrence. Canadian Journal of Fisheries and Aquatic Science 49:2546–50.

Hanan, D. A. 1996. Dynamics of abundance and distribution for Pacific harbor seal, *Phoca vitulina richardsi,* on the coast of California. Ph.D. Dissertation, University of California, Los Angeles.

Hannah, J. 2000. Seals of Atlantic Canada and the northeastern United States, 2nd ed. (rev.). International Marine Mammal Association, Guelph, Ontario, Canada.

Hanni, K. D., D. J. Long, R. E. Jones, P. Pyle, and L. E. Morgan. 1997. Sightings and strandings of Guadalupe fur seals in central and northern California, 1988–1995. Journal of Mammalogy 78:684–90.

Harrison, R. J., and J. E. King. 1965. Marine mammals. Hutchinson, London.

Harrison, R. J., and J. D. W. Tomlinson. 1956. Observations on the venous system in certain pinnipedia and cetacea. Proceedings of the Zoological of Society of London 126:205–31.

Haug, T., K. T. Nilssen, and L. Lindblom. 2000. First independent feeding of harp seal (*Phoca groenlandica*) and hooded seal (*Cystophora cristata*) pups in the Greenland Sea. North Atlantic Marine Mammal Commission (NAMMCO) Scientific Publications 2:29–39.

Hecht, S., C. D. Hendley, S. Ross, and P. N. Richmond. 1948. The effect of exposure to sunlight on night vision. American Journal of Ophthalmology 31:1573–80.

Helle, E., M. Olsson, and S. Jensen. 1976. PCB levels correlated with pathological changes in seal uteri. Ambio 5:261–63.

Hempleman, H. V., and A. P. M. Lockwood. 1978. The physiology of diving in man and other animals (Institute of Biology, Studies in Biology 99). Edward Arnold, London.

Hewer, H. R. 1964. The determination of age, sexual maturity, longevity, and a life table in the grey seal (*Halichoerus grypus*). Proceedings of the Zoological Society of London 142:593–624.

Higgins, L. V., D. P. Costa, A. C. Huntley, and B. J. Le Boeuf. 1988. Behavioral and physiological measurements of maternal investment in the Steller sea lion, *Eumetopias jubatus*. Marine Mammal Science 4:44–58.

Hill, P. S., and D. P. DeMaster. 1999. Alaska marine mammal stock assessments, 1999. National Marine Mammal Laboratory, Alaska Fisheries Science Center, Seattle, WA.

Hindell, M. A., D. J. Slip, and H. R. Burton. 1991a. The diving behaviour of adult male and female southern elephant seals, *Mirounga leonina*. Australian Journal of Zoology 39:595–619.

Hindell, M. A., D. J. Slip, H. R. Burton, and M. M. Burton. 1991b. Physiological implications of continuous, prolonged, and deep dives of the southern elephant seal, *Mirounga leonina*. Canadian Journal of Zoology 70:370–79.

Hiruki, L. M., W. G. Gilmartin, B. L. Becker, and I. Stirling. 1993. Wounding in Hawaiian monk seals (*Monachus schauinslandi*). Canadian Journal of Zoology 71:464–68.

Hobson, E. S. 1966. Visual orientation and feeding in seals and sea lions. Nature 210:326–27.

Hodgson, B. 1990. Alaska's big spill: Can the wilderness heal? National Geographic 177(1):5–43.

Hol, R., A. S. Blix, and H. O. Myhre. 1975. Selective redistribution of the blood volume in the diving seal (*Pagophilus groenlandicus*). Pages 423–32 *in* K. Ronald and A. W. Mansfield, eds. Biology of the seal. International Council for the Exploration of the Sea, Copenhagen.

Holden, A. V. 1978. Pollutants and seals. Mammalogy Review 8:53–66.

Holt, S. J., and L. M. Talbot. 1978. New principles for the conservation of wild living resources. Wildlife Monographs 59:1–33.

Hoover-Miller, A. 1994. Harbor seal (*Phoca vitulina*) biology and management in Alaska (Contract No. T75134749). U.S. Marine Mammal Commission, Washington, DC.

Hoover-Miller, A., K. R. Parker, and J. J. Burns. 2001. A reassessment of the impact of the *Exxon Valdez* oil spill on harbor seals (*Phoca vitulina richardsi*) in Prince William Sound, Alaska. Marine Mammal Science 17:111–35.

Howell, A. B. 1970. Aquatic mammals, their adaptions to life in the water, reprint ed. Dover, New York.

Hubbs, C. L. 1956. The Guadalupe fur seal still lives! Zoonooz (South Dakota Zoological Society) 29(12):6–9.

Huber, H. 1995. The abundance of harbor seals (*Phoca vitulina richardsi*) in Washington, 1991–1993. M.S. Thesis, University of Washington, Seattle.

Huber, H., S. Jeffries, R. Brown, and R. DeLong. 1994. Harbor seal stock assessment in Washington and Oregon 1993 (Annual Report to the MMPA Assessment Program). Office of Protected Resources, National Marine Fisheries Service, Silver Spring, MD.

Huey, L. M. 1930. Capture of an elephant seal off San Diego, California, with notes on stomach contents. Journal of Mammalogy 11:229–31.

Hull, C., ed. 1994. Alaska Department of Fish and Game's Wildlife Notebook Series. Alaska department of Fish and Game, Public Communications Section, Juneau, AK. Available at www.state.ak.us/adfg/notebook/notehome.htm.

Hyland, J. L., E. D. Schneider, and E.R.L. Nananyunsett. 1977. Petroleum hydrocarbons and their effects on marine organisms, populations, communities and ecosystems (Council Meeting, Marine Mammals Committee, E:64) International Council for the Exploration of the Sea, Copenhagen.

Innes, S., D. M. Lavigne, W. M. Earle, and K. M. Kovacs. 1987. Feeding rates of seals and whales. Journal of Animal Ecology 56:115–30.

International Union for the Conservation of Nature and Natural Resources, United Nations Environment Programme, and the World Wildlife Fund. 1980. World conservation strategy: Living resource conservation for sustainable development. IUCN, Gland, Switzerland.

Irving, L. 1938. Vascular adjustments of diving animals during apneoa. Science 88:502.

Irving, L. 1972. Arctic life of birds and mammals including man. Springer-Verlag, New York.

Iwama, G., L. Nichol, and J. Ford. 1997. Aquatic mammals and other species (British Columbia Salmon Aquaculture Review, Discussion Paper). Environmental Assessment Office, Victoria, British Columbia, Canada.

Jamieson, G. S., and H. D. Fisher. 1972. The pinniped eye: A review. Pages 245–61 *in* R. J. Harrison, ed. Functional anatomy of marine mammals, Vol. 1. Academic Press, New York.

Jarman, W. M., K. A. Hobson, W. J. Sydeman, C. E. Bacon, and E. B. McLaren. 1996. Influence of trophic position and feeding location on contaminant levels in the Gulf of the Farallones food web as revealed by stable isotope analysis. Environmental Science and Technology 30:654–60.

Jefferson, T. A., S. Leatherwood, and M. A. Webber. 1993. Marine mammals of the world. FAO species identification guide. United Nations Environment Programme, Food and Agriculture Organization, Rome.

Jeffries, S. J. 1985. Occurrence and distribution patterns of marine mammals in the Columbia River and adjacent coastal waters of northern Oregon and Washington. Marine mammals and adjacent waters, 1980–1982 (Process Report 85–04). National Marine Fisheries Service, Northwest Alaska Fisheries Center, Seattle, WA.

Johnson, M. L., and S. J. Jeffries. 1977. Population evaluation of the harbour seal (*Phoca vitulina richardii*) in the waters of the state of Washington (Report No. MMC-75/05). U.S. Marine Mammal Commission, Washington, DC.

Jones, D. R., H. D. Fisher, S. McTaggart, and N. H. West. 1973. Heart rate during breath-holding and diving in the unrestrained harbor seal, *Phoca vitulina richardsi*. Canadian Journal of Zoology 51:671–80.

Jones, R. E. 1981. Food habits of smaller marine mammals from northern California. Proceedings of the California Academy of Science 42:409–33.

Kajimura, H. 1984. Opportunistic feeding of the northern fur seal, *Callorhinus ursinus,* in the eastern North Pacific Ocean and eastern Bering Sea (Technical Report NMFS SSRF-779). U.S. Department of Commerce, Washington, DC.

Kajimura, H. 1985. Opportunistic feeding by the northern fur seal (*Callorhinus ursinus*). Pages 301–18 *in* J. R. Beddington, R. J. H. Beverton, and D. M. Lavigne, eds. Marine mammals and fisheries. Allen & Unwin, Cambridge, MA.

Kapel, F. O. 1973. Some second-hand reports on the food of harp seals in west Greenland waters (Council Meeting, Marine Mammals Committee, N:8). International Council for Exploration of the Sea, Compenhagon.

Kapel, F. O. 2000. Feeding habits of harp and hooded seals in Greenland waters. North Atlantic Marine Mammal Commission (NAMMCO) Scientific Publications 2:50–64.

Kastak, D., and R. J. Schusterman. 1995. Aerial and underwater hearing thresholds for 100 Hz pure tones in two pinniped species. Pages 71–81 *in* R. A. Kastelein, J. A. Thomas, and P. E. Nachtigall, eds. Sensory systems of aquatic mammals, De Spil, Woerden, the Netherlands.

Kastelein, R. A., N. M. Schooneman, and P. R. Wiepkema. 2000. Food consumption and body weight of captive Pacific walruses (*Odobenus rosmarus divergens*). Aquatic Mammals 26:175–90.

Katona, S. K., V. Rough, and D. T. Richardson. 1993. A field guide to whales, porpoises, and seals from Cape Cod to Newfoundland. Smithsonian Institution Press, Washington, DC.

Kellert, S. R. 1999. American perceptions of marine mammals and their management. Humane Society of the United States, Washington, DC.

Kelly, B. P. 1988. Ringed seal, *Phoca hispida*. Pages 57–75 *in* J. W. Lentfer, ed. Selected marine mammals of Alaska. Species accounts with research and management recommendations. U.S. Marine Mammal Commission, Washington, DC.

Kenyon, K. W. 1972. Man versus the monk seal. Journal of Mammalogy 53:687–96.

Kenyon, K. W. 1973. The Hawaiian monk seal (*Monachus schauinslandi*). Pages 88–97 *in* Seals: Proceedings of a working meeting of seal specialists on threatened and depleted seals of the world, Guelph, Ontario (Supplementary Paper No. 39). International Union for Conservation of Nature and Natural Resources, Morges, Switzerland.

Kenyon, K. W. 1977. Caribbean monk seal extinct. Journal of Mammalogy 58:97–98.

Kenyon, K. W. 1978a. Walrus. Pages 178–83 *in* D. Haley, ed. Marine mammals of eastern North Pacific and arctic waters. Pacific Search Press, Seattle, WA.

Kenyon, K. W. 1978b. Hawaiian monk seal. Pages 212–16 *in* D. Haley, ed. Marine mammals of eastern North Pacific and arctic waters. Pacific Search Press, Seattle, WA.

Kenyon, K. W. 1981. Monk seals *Monachus* Fleming, 1822. Pages 195–220 *in* S. H. Ridgway and R. J. Harrison, eds. Handbook of marine mammals. Vol. 2: Seals. Academic Press, New York.

Kenyon, K. W., and C. H. Fiscus. 1963. Age determination in the Hawaiian monk seal. Journal of Mammalogy 44:280–82.

Kenyon, K. W., and D. W. Rice. 1959. Life history of the Hawaiian monk seal. Pacific Science 13:215–52.

Kerem, D., D. D. Hammond, and R. Elsner. 1973. Tissue glycogen levels in the Weddell seal, *Leptonychotes weddelli:* A possible adaptation to asphyxial hypoxia. Comparative Biochemistry and Physiology 45:731–37.

Ketten, D. 1998. Marine mammal auditory systems: A summary of audiometric and anatomical data and its implications for underwater acoustic impacts. (Technical Memorandum NOAA-TM-NMFS-SWFSC-256). U.S. Department of Commerce, Washington, DC.

King, J. E. 1954. The otariid seals of the Pacific coast of North America. Bulletin of the British Museum (Natural History), Zoology 2:309–37.

King, J. E. 1964. Seals of the world. British Museum (Natural History), London.

King, J. E. 1972. Observations on phocid skulls. Pages 81–115 *in* R. J. Harrison, ed. Functional anatomy of marine mammals, Vol. 1. Academic Press, London.

King, J. E. 1983. Seals of the world, 2nd ed. British Museum (Natural History), Oxford University Press, Oxford.

Kjekshus, J., A. S. Blix, R. Elsner, R. Hol, and E. Amundsen. 1982. Myocardial blood flow and metabolism in the diving seal. American Journal of Physiology 242:R97–104.

Koeman, J. H., W. H. M. Peters, C. J. Smit, P. S. Tjioc, and J. J. M. de Goeij. 1972. Persistent chemicals in marine mammals. TNO-Nieuws 27:570–78.

Koeman, J. H., W. S. M. Van de Ven, J. J. M. de Goeij, P. S. Tioie, and J. L. Van Haaften. 1975. Mercury and selenium in marine mammals and birds. Science of the Total Environment 3:279–87.

Kooyman, G. L. 1966. Maximum diving capacities of the Weddell seal. Science 151:1553–54.

Kooyman, G. L. 1981. Weddell seal: Consummate diver. Cambridge University Press, Cambridge.

Kooyman, G. L. 1989. Diverse divers. Springer-Verlag, Berlin.

Kooyman, G. L., D. D. Hammond, and J. P. Schroeder. 1970. Bronchograms and tracheograms of seals under pressure. Science 168:82–84.

Kopec, D. and J. Harvey. 1995. Toxic pollutants, health indices, and population dynamics of harbor seals in San Francisco Bay, 1989–91: A final report. Moss Landing Marine Labs, Moss Landing, CA.

Knutsen, L. O., and E. W. Born. 1991. Growth, body composition and insulative characteristics of Atlantic walruses. Greenland Fisheries Research Institute, Copenhagen.

Lagenwalter, P. E. 1981. Excavations at ORA-193, Newport Bay, California. Appendix 3: The reptiles and mammals from ORA-193. Pacific Coast Archaeological Society Quarterly 17:100–118.

Lamont, M. M., J. T. Vida, J. T. Harvey, S. Jeffries, R. Brown, H. H. Huber, R. DeLong, and W. K. Thomas. 1996. Genetic substructure of the Pacific harbor seal (*Phoca vitulina richardsi*) off Washington, Oregon, and California. Marine Mammal Science 12:402–13.

Lander, M. E., and B. Herskovitz. 1997. El Niño. Information report. Marine Mammal Center, Marin Headlands, Sausalito, CA. Available at www.tmmc.org/elnino.htm.

Lander, R. H., and H. Kajimura. 1982. Status of northern fur seals. FAO Fisheries Series 5:319–45.

Lander, M. E., F. M. D. Gulland, and R. L. DeLong. 2000. Satellite tracking a rehabilitated Guadalupe fur seal (*Arctocephalus townsendi*). Aquatic Mammals 26:137–42.

Lavigne, D. M. 1975. Harp seal, *Pagophilus groenlandicus,* production in the western Atlantic during March 1975 (Research Document 75/XII/150, Series 3728). International Commission for the Northwest Atlantic Fisheries.

Lavigne, D. M. 1978. The harp seal controversy reconsidered. Queen's Quarterly 85:377–88.

Lavigne, D. M. 1996. The sudden ecologists. BBC Wildlife 14:57.

Lavigne, D. M. 1999a. Estimating total kill of Northwest Atlantic harp seals, 1994–1998. Marine Mammal Science 15:871–78.

Lavigne, D. M. 1999b. Harp seals and Atlantic cod: Notes for a presentation to the Standing Committee on Fisheries and Oceans, House of Commons, Ottawa. International Marine Mammal Association (IMMA), Guelph, Ontario, Canada.

Lavigne, D. M. 1999c. Comments on the 3 December 1998 biological opinion regarding interactions between Steller sea lions and pollock fisheries in the North Pacific. North Pacific Fishery Management Council, independent review of the scientific basis for the 3 December 1998 biological opinion regarding interactions between Steller Sea lions and pollock fisheries in the North Pacific. Seattle, WA.

Lavigne, D. M., and K. M. Kovacs. 1988. Harps and hoods: Ice breeding seals of the northwest Atlantic. University of Waterloo Press, Waterloo, Ontario, Canada.

Lavigne, D. M., and K. Ronald. 1972. The harp seal, *Pagophilus groenlandicus* (Erxleben, 1777). XXIII. Spectral sensitivity. Canadian Journal of Zoology 50:1197–1206.

Lavigne, D. M., and K. Ronald. 1975. Pinniped visual pigments. Comparative Biochemistry and Physiology 52B:325–29.

Lavigne, D. M., C. S. Bernholtz, and K. Ronald. 1977. Functional aspects of pinniped vision. Pages 135–73 *in* R. J. Harrison, ed. Functional anatomy of marine mammals, Vol. 3. Academic Press, New York.

Lavigne, D. M., S. Innes, K. Kalpakis, and K. Ronald. 1982. An aerial census of western Atlantic harp seals (*Pagophilus groenlandicus*) using ultraviolet photography. Pages 295–302 *in* FAO Advisory Committee on Marine Resources Research, Working Party on Marine Mammals, Mammals in the Seas. Vol. IV: Small cetaceans, seals, sirenians and otters (FAO Fisheries Series No. 5). Food and Agriculture Organization, Rome.

Laws, R. M. 1956. Growth and sexual maturity of aquatic mammals. Nature 178:193–94.

Laws, R. M. 1959. Accelerated growth in seals, with special reference to the phocidae. Norsk Hvalfangst-Tidsskrift 48:425–52.

Laws, R. M. 1962. Age determination of pinnipeds with special reference to growth layers in the teeth. Zeitschrift für Säugetierkunde 27:129–46.

Lawson, J. W., J. T. Anderson, E. L. Dalley, and G. B.Stenson. 1998. Selective foraging by harp seals *Phoca groenlandica* in nearshore and offshore waters of Newfoundland, 1993 and 1994. Marine Ecology Progress Series 163:1–10.

Le Boeuf, B. J., and R. M. Laws, eds. 1994. Elephant seals. University of California Press, Berkeley.

Le Boeuf, B. J., and K. J. Panken. 1977. Elephant seals breeding on the mainland in California. Proceedings of the California Academy of Science 41:267–80.

Le Boeuf, B. J., D. Aurioles-G., R. Condit, C. Fox, R. Gisner, R. Romero, and F. Sinsel. 1983. Size and distribution of the California sea lion population in Mexico. Proceedings of the California Academy of Science 43:77–85.

Le Boeuf, B. J., K. W. Kenyon, and B. Villa-Ramirez. 1986. The Caribbean monk seal is extinct. Marine Mammal Science 2:70–72.

Le Boeuf, B. J., K. A. Ono, and J. Reiter. 1991. History of the Steller sea lion population at Año Nuevo Island, 1961–1991 (Administration Report LJ-91-45C). Southwest Fisheries Science Center, National Marine Fisheries Service, La Jolla, CA.

Lenfant, C., K. Johansen, and J. D. Torrance. 1970. Gas transport and oxygen storage capacity in some pinnipeds and the sea otter. Respiration Physiology 9:277–86.

Lett, P. F. 1977. A model to determine stock size and management options for the Newfoundland hooded seal stock (Research Document 77/25). Canadian Atlantic Fisheries Scientific Advisory Committee.

Lett, P. F., and T. Benjaminsen. 1977. A stochastic model for the management of the northwestern Atlantic harp seal (*Pagophilus groenlandicus*) population. Journal of the Fisheries Research Board of Canada 34:1155–87.

Levenson, D. H., and R. J. Schusterman. 1997. Pupillometry in seals and sea lions: Ecological implications. Canadian Journal of Zoology 75:2050–57.

Levenson, D. H., and R. J. Schusterman. 1999. Dark adaptation and visual sensitivity in shallow and deep-diving pinnipeds. Marine Mammal Science 15:1303–13.

Ling, J. K. 1970. Pelage and molting in wild animals with special reference to aquatic forms. Quarterly Review of Biology 45:16–54.

Ling, J. K. 1974. The integument of marine mammals. Pages 1–44 *in* R. J. Harrison, ed. Functional anatomy of marine mammals, Vol. 2. Academic Press, New York.

Ling, J. K., and C. E. Button. 1975. The skin and pelage of grey seal pups *(Halichoerus grypus* Fabricus), with a comparative study of foetal and neonatal molting in the pinnipedia. Pages 112–32 *in* K. Ronald and A. W. Mansfield, eds. Biology of the seal. International Council for the Exploration of the Sea, Copenhagen.

Lipps, J. H., and E. Mitchell. 1976. Trophic model for the adaptive radiations and extinctions of pelagic marine mammals. Paleobiology 2:147–55.

Loughlin, T. R. 1989. Northern sea lions: Status update 1989. National Marine Mammal Laboratory, Alaska Fisheries Science Center, Seattle, WA.

Loughlin, T. R. 1997. Using the phylogeographic method to identify Steller sea lion stocks. Pages 329–41 *in* A. Dizon, S. J. Chivers, and W. Perrin, eds. Molecular genetics of marine mammals (Special Publication No. 3). Society for Marine Mammalogy, Lawrence, KS.

Loughlin, T. R., and R. L. Merrick. 1989. Comparison of commercial harvest of walleye pollock and northern sea lion abundance in the Bering Sea and Gulf of Alaska. Pages 679–700 *in* Proceedings of the international symposium on the biology and management of walleye pollock (Alaska Sea Grant Program, AK-SG-89-01). Anchorage, AK.

Loughlin, T. R., and R. Nelson, Jr. 1986. Incidental mortality of northern sea lions in Shelifof Strait, Alaska. Marine Mammal Science 2:14–33.

Loughlin, T. R., D. J. Rugh, and C. H. Fiscus. 1984. Northern sea lion distribution and abundance: 1956–80. Journal of Wildlife Management 48:729–40.

Loughlin, T. R., A. S. Perlov, and V. A. Viadimirov. 1990. Survey of northern sea lions (*Eumetopias jubatus*) in the Gulf of Alaska and Aleutian Islands during June 1989 (NOAA Technical Memorandum, Document F/NWC-176). National Marine Fisheries Service. Washington, DC.

Lowry, L. F., and F. H. Fay. 1984. Seal eating by walruses in the Bering and Chukchi Seas. Polar Biology 3:11–18.

Lowry, L. F., and K. J. Frost. 1981. Feeding and trophic relationships of phocid seals and walruses in the eastern Bering Sea. Pages 813–24 *in* D. W. Hood and J. A. Calder, eds. The Eastern Bering Sea shelf: Oceanography and resources, Vol. 2. Office of Marine Pollution Assessment, NOAA, Rockville, MD.

Lowry, L. F., K. J. Frost, and J. J. Burns. 1980. Variability in the diet of ringed seals (*Phoca hispida*) in Alaska. Canadian Journal of Fisheries and Aquatic Sciences 37:2254–61.

Lowry, L. F., V. N. Burkanov, and K. J. Frost. 1996. Importance of walleye pollock (*Theragra chalcogramma*) in the diet of phocid seals in the Bering Sea and northwestern Pacific Ocean. U.S. Department of Commerce, NMFS, NOAA Technical Report 126:141–51.

Lowry, M. S., C. W. Oliver, C. Macky, and J. B. Wexler. 1990. Food habits of California sea lions *Zalophus californianus* at San Clemente Island, California, 1981–1986. Fisheries Bulletin 88:509–21.

Lowry, M. S., C. W. Oliver, C. Macky, and J. B. Wexler. 1991. Seasonal and annual variability in the diet California sea lions *Zalophus californianus* at San Nicolas Island, California, 1981–1986. Fisheries Bulletin 89:331–36.

Lowry, M. S., P. Boveng, R. J. DeLong, C. W. Oliver, B. S. Stewart, H. DeAnda, and J. Barlow. 1992. Status of the California sea lion (*Zalophus californianus californianus*) population in 1992 (Administrative Report LJ-92-32). Southwest Fisheries Science Center, National Marine Fisheries Service, La Jolla, CA.

Lyon, G. M. 1937. Pinnipeds and a sea otter from the Point Mugu shell mound of California. Publications of the University of California (Los Angeles) Biological Science 1(8):133–68.

MacDonald, C. D. 1982. Predation by Hawaiian monk seals on spiny lobsters. Journal of Mammalogy 63:700.

Maldonado, J. E., F. O. Davila, B. S. Stewart, E. Geffen, and R. K. Wayne. 1995. Intraspecific genetic differentiation in California sea lions (*Zalophus californianus*) from Southern California and the Gulf of California. Marine Mammal Science 11:46–58.

Malouf, A., ed. 1986. Seals and sealing in Canada. Report of the Royal Commission. Supply and Services Canada, Ottawa.

Mansfield, A. W. 1967a. Distribution of the harbour seal, *Phoca vitulina,* Linnaeus, in Canadian waters. Journal of Mammalogy 48:249 57.

Mansfield, A. W. 1967b. Seals of arctic and eastern Canada (Bulletin 137). Fisheries Research Board of Canada.

Mansfield, A. W. 1978. Reproduction of the grey seal, *Halichoerus grypus,* in eastern Canada (Council Meeting, Marine Mammals Committee, N:13). International Council for Exploration of the Sea, Copenhagen.

Mansfield, A. W., and B. Beck. 1977. The grey seal in eastern Canada (Technical Report No. 704). Environment Canada, Fisheries and Marine Service.

Mansfield, A. W., and D. E. Sergeant. 1973. Seals as indicators of pollution (Council Meeting, Marine Mammals Committee, E:25). International Council for the Exploration of the Sea, Copenhagen.

Maravilla-Chavez, M. O., and M. S. Lowry. 1999. Incipient breeding colony of Guadalupe fur seals at Isla Benito del Este, Baja California, Mexico. Marine Mammal Science 15:239–41.

Marine Mammal Commission. 1998. Annual report of the Marine Mammal Commission, calendar year 1997. Report to Congress, 31 January 1998. Bethesda, MD.

Marine Mammal Commission. 2000. Annual report of the Marine Mammal Commission, calendar year 1999. Report to Congress, 31 January 2000. Bethesda, MD.

Markussen, N. H., and N. A. Øritsland. 1991. Food energy requirements of the harp seal (*Phoca groenlandica*) population in the Barents and White Seas. Pages 603–8 *in* E. Sakshaug, C. C. E. Hopkins, and N. A. Øritsland, eds. Proceedings of the pro mare symposium on polar marine ecology, Trondheim, 12–16 May 1990. Polar Research 10(2).

Mårtensson, P. E., A. R. Lager Gotaas, E. S. Nordøy, and A. S. Blix. 1996. Seasonal changes in energy density of prey of Northeast Atlantic seals and whales. Marine Mammal Science 12:635–40.

Marx, W. 1967. The frail ocean. Ballantine, New York.

Mate, B. R. 1976. *Zalophus californianus californianus*. Ad Hoc Group III/19 *in* Symposium: Scientific consultation on marine mammals. Food and Agriculture Organization of the United Nations. Advisory Committee on Marine Resources Research, Marine Mammals. Bergen, Norway.

Mate, B. R. 1978. California sea lion. Pages 172–77 *in* D. Haley, ed. Marine mammals of eastern North Pacific and arctic waters. Pacific Search Press, Seattle, WA.

Mate, B. R. 1982. History and present status of the California sea lion, *Zalophus californianus*. Pages 303–9 *in* AO Advisory Committee on Marine Resources Research, Working Part on Marine Mammals, Mammals in the Seas. Vol. IV: Small cetaceans, seals, sirenians and otters (FAO Fisheries Series No. 5). Food and Agriculture Organization, Rome.

Maxwell, G. 1967. Seals of the world. Constable, London.

McCann, T. S. 1985. Size, status and demography of southern elephant seal (*Mirounga leonina*) populations. Pages 1–17 *in* J. K. Ling and M. M. Bryden, eds. Studies of sea mammals in south latitudes. South Australian Museum, Adelaide, Australia.

McClung, R. M. 1978. Hunted mammals of the sea. Morrow, New York.

McLaren, I. A. 1958. The biology of the ringed seal *(Phoca hispida* Schreber) in the eastern Canadian arctic (Bulletin 118). Fisheries Research Board of Canada.

Meffe, G. K., W. F. Perrin, and P. K. Dayton. 1999. Marine mammal conservation: Guiding principles and their implementation. Pages 437–54 *in* J. R. Twiss Jr., and R. R. Reeves, eds. Conservation and management of marine mammals. Smithsonian Institution Press, Washington, DC.

Meisenheimer, P. 1998. Midway's monk seals. Monachus Guardian 1(2):23–25.

Melin, S. R., and R. L. DeLong. 1994. Population monitoring of northern fur seals on San Miguel Island, California. Pages 137–41 *in* E. H. Sinclair, ed. Fur seal investigations, 1992 (NOAA Technical Memorandum NMFS-AFSC-45). U.S. Department of Commerce, Washington, DC.

Melin, S. R., and R. L. DeLong. 1999. Observations of a Guadalupe fur seal (*Arctocephalus townsendi*) female and pup at San Miguel Island, California. Marine Mammal Science 15:885–88.

Melin, S. R., and R. L. DeLong. 2000. Population monitoring studies of northern fur seals at San Miguel Island, California. Pages 41–51 *in* B. W. Robson, ed. Fur seal investigations, 1998. (NOAA Technical Memorandum NMFS-AFSC-113). U.S. Department of Commerce, Washington, DC.

Merrick, R. L. 1995. The relationship of foraging ecology of Steller sea lions (*Eumetopias jubatus*) to their population decline in Alaska. Ph.D. Dissertation, University of Washington, Seattle.

Merrick, R. L., and T. R. Loughlin. 1997. Foraging behavior of adult female and young-of-the-year Steller sea lions (*Eumetopias jubatus*) in Alaskan waters. Canadian Journal of Zoology 75:776–86.

Merrick, R. L., T. R. Loughlin, and D. G. Calkins. 1987. Decline in the abundance of the northern sea lion, *Eumetopias jubatus,* in Alaska, 1956–86. Fisheries Bulletin (U.S.) 85:351–65.

Merrick, R. L., M. K. Chumblely, and G. V. Byrd. 1997. Diet diversity of Steller sea lions (*Eumetopias jubatus*) and their population decline in Alaska: A potential relationship. Canadian Journal of Fisheries and Aquatic Sciences 54:1342–48.

Mignucci-Giannoni, A. A., and D. K. Odell. 2001. Tropical and subtropical records of hooded seals (*Cytophora cristata*) dispel the myth of extant Caribbean monk seals (*Monachus tropicalis*). Bulletin of Marine Science 68:47–58.

Miller, E. H. 1976. Walrus ethology. Part 2: Herd structure and activity budgets of summering males. Canadian Journal of Zoology 54:704–15.

Miller, E. H., and D. A. Job. 1992. Airborne acoustic communication in the Hawaiian monk seal, *Monachus schauinslandi*. Pages 485–531 *in* J. A. Thomas, R. A. Kastelein, and A. Y. Supin, eds. Marine mammal sensory systems, Plenum Press, New York.

Møhl, B. 1964. Preliminary studies on hearing in seals. Dansk Naturhistorisk Forening Videnskabelige Meddelelser 127:283–94.

Møhl, B. 1968a. Hearing in seals. Pages 172–95 *in* R. J. Harrison, R. C. Hubbard, R. S. Peterson, C. E. Rice, and R. S. Schusterman, eds. The behaviour and physiology of pinnipeds. Appleton-Century-Crofts, New York.

Møhl, B. 1968b. Auditory sensitivity of the common seal in air and water. Journal of Auditory Research 8:27–38.

Møhl, B., and K. Ronald. 1975. The peripheral auditory system of the harp seal, *Pagophilus groenlandicus* (Erxleben, 1777). Pages 514–23 *in* K. Ronald and A. W. Mansfield, eds. Biology of the seal. International Council for the Exploration of the Sea, Copenhagen.

Møhl, B., K. Ronald, and J. M. Terhune. 1975. Underwater calls of the harp seal, *Pagophilus groenlandicus*. Pages 533–43 *in* K. Ronald and A. W. Mansfield, eds. Biology of the seal. International Council for the Exploration of the Sea, Copenhagen.

Mohn, R., and W. D. Bowen. 1994. A model of grey seal predation on 4VsW cod and its effects on the dynamics and potential yield of cod (Atlantic Fisheries Research Document 94/64). Department of Fisheries and Oceans, Canada, Ottawa.
Mohr, E. 1950. Behasrung und haarwechsel der robben. Neue Ergeb. Probl. Zool. 1950:602–14. (Klatt.-Festchr.)
Morejohn, G. V., and D. M. Baltz. 1970. Contents of the stomach of an elephant seal. Journal of Mammalogy 51:173–74.
Munz, F. W. 1971. Vision: Visual pigments. Pages 1–32 *in* W. S. Hoar and D. J. Randall, eds. Fish physiology. Part 5: Sensory systems and electric organs. Academic Press, New York.
Murdaugh, H. V., J. C. Seabury, and W. L. Mitchell. 1961. Electrocardiogram of the diving seal. Circulation Research 9:358–61.
Musgrave, R. S., J. A. Flynn-O'Brien, P. A. Lambert, A. A. Smith, and Y. D. Marinakis. 1998. Federal wildlife and related laws handbook. Center for Wildlife Law, Institute of Public Law, School of Law, University of New Mexico, Santa Fe.
Nagy, A. R., and K. Ronald. 1975. A light and electromicroscopic study of the structure of the retina of the harp seal, *Pagophilus groenlandicus* (Erxleben, 1777). Pages 92–96 *in* K. Ronald and A. W. Mansfield, eds. Biology of the seal. International Council for the Exploration of the Sea, Copenhagen.
National Energy Board. 1977. Reasons for decision on northern pipeline Vol. 1. Supply and Services Canada, Ottawa.
National Marine Fisheries Service. 1992. Recovery plan for the Steller sea lion (*Eumetopias jubatus*). Prepared by the Steller Sea Lion Recovery Team for the National Marine Fisheries Service, Silver Spring, MD.
National Marine Fisheries Service. 1993. Final conservation plan for the northern fur seal (*Callorhinus ursinus*). Prepared by the National Marine Mammal Laboratory, Alaska Fisheries Science Center, Seattle, WA, and the Office of Protected Resources, National Marine Fisheries Service, Silver Spring, MD.
National Marine Fisheries Service. 1999. Our living oceans. Report on the status of U.S. living marine resources, 1999 (NOAA Technical Memorandum NMFS-F/SPO-41). U.S. Department of Commerce, Washington, DC. Available at http://spo.nwr.noaa.gov/olo99.htm.
National Research Council. 1999. Sustaining marine fisheries. National Academy Press, Washington, DC.
Newby, T. C. 1973. Changes in Washington State harbor seal population, 1942–1972. Murrelet 54:5–6.
Newby, T. C. 1978. Pacific harbor seal. Pages 184–91 *in* D. Haley, ed. Marine mammals of eastern North Pacific and arctic waters. Pacific Search Press, Seattle, WA.
Nilssen, K. T., O. P. Pedersen, L. P. Folkow, and T. Haug. 2000. Food consumption estimates of Barents Sea harp seals. NAMMCO Scientific Publications 2:9–28.
Norris, K. S. 1969. The echolocation of marine mammals. Pages 391–423 *in* H. T. Anderson, ed. The biology of marine mammals. Academic Press, New York.
Northridge, S. P. 1984. World reviews of interactions between marine mammals and fisheries (FAO Fisheries Technical Paper 251). Food and Agriculture Organization, Rome.
Nowicki, S. N., I. Stirling, and B. Sjare. 1997. Duration of stereotyped underwater vocal displays by male Atlantic walruses in relation to aerobic dive limit. Marine Mammal Science 13:566–75.
Office of Protected Resources. 1996. Pinnipeds: Seals and sea lions. U.S. Office of Protected Resources, National Marine Fisheries Service, Silver Spring, MD.
Olesiuk, P. F., and M. A. Bigg. 1988. Seals and sea lions of the British Columbia coast. Canada Department of Fisheries and Oceans, Ottawa.
Oliver, C. W. 1991. 1988–1991 field studies on pinnipeds at San Clemente Island (Administrative Report LJ-91-27). Southwest Fisheries Science Center, National Marine Fisheries Service, La Jolla, CA.
Oliver, J. S., P. N. Slattery, E. F. O' Conner, and L. F. Lowry. 1983. Walrus, *Odobenus rosmarus,* feeding in the Bering Sea: A benthic perspective. Fisheries Bulletin 81:501–12.
Øritsland, N. 1978. Some applications of thermal values of fur samples in expressions for in vivo heat balance. University of Oslo, Institute of Zoophysiology, Oslo, Norway.
Øritsland, N. A., and K. Ronald. 1977. A simulation programme for heat balance in marine mammals (Council Meeting, Marine Mammals Commission N:8). International Council for Exploration of the Sea, Copenhagen.
Øritsland, N. A., and K. Ronald. 1978a. Aspects of temperature regulation in harp seal pups evaluated by in vivo experiments and computer simulations. Acta Physiologica Scandinavica 103:263–69.
Øritsland, N. A., and K. Ronald. 1978b. Solar heating of mammals: Observations of hair transmittance. International Journal of Biometeorology 22:197–201.
Øritsland, N. A., D. M. Lavigne, and K. Ronald. 1978. Radiative surface temperatures of harp seals. Comparative Biochemistry and Physiology 61A:9–12.
Orr, R. T., and T. C. Poulter. 1967. Some observations on reproduction, growth, and social behavior in the Steller sea lion. Proceedings of the California Academy of Sciences 35:193–226.
Orr, R. T., J. Schonewald, and K. W. Kenyon. 1970. The California sea lion: Skull growth and a comparison of two populations. Proceedings of the California Academy of Sciences, Fourth Series 37:381–94.
Orta, D. F. 1988. Habitos alimenticios del lobo marino de California en El Racito, Golfo de California, Septiembre 1986–Octubre 1987. Tesis de Licenciatura en Ciencias Marinas, Universidad Autónoma de Baja California, Mexico.
Pabst, D. A., S. A. Rommel, and W. A. McLellan. 1999. The functional morphology of marine mammals. Pages 15–72 *in* J. E. Reynolds, III., and S. A. Rommel, eds. Biology of marine mammals, Smithsonian Institution Press, Washington, DC.
Park, D. L. 1993. Effects of marine mammals on Columbia River salmon listed under the Endangered Species Act. Recovery issues for threatened and endangered Snake River salmon (Technical Report 3 of 11: DOE/BP-99654-3). U.S. Department of Energy, Bonneville Power Administration, Portland, OR.
Parrish, F. A., M. P. Craig, T. J. Ragen, G. J. Marshall, and B. M. Buhleier. 2000. Identifying diurnal foraging habitat of endangered Hawaiian monk seals using a seal-mounted video camera. Marine Mammal Science 16:392–412.
Parsons, L. S. 1993. Management of marine fisheries in Canada. Canadian Bulletin of Fisheries and Aquatic Sciences 255:1–763.
Parsons, L. S., and J. S. Beckett. 1998. The NAFO model of international collaborative research, management and cooperation. Journal of Northwest Atlantic Fishery Science 23:1–18.
Payne, M. R. 1979. Growth in the Antarctic fur seal, *Arctocephalus gazella*. Journal of Zoology 187:1–20.
Payne, P. M., and L. A. Selzer. 1989. The distribution, abundance and selected prey of the harbor seal, *Phoca vitulina concolor,* in southern New England. Marine Mammal Science 5:173–92.
Pearson, J. P. 1968. The abundance and distribution of harbor seals and Steller sea lions in Oregon. M.S. Thesis, Oregon State University, Corvallis.
Peichl, L., and K. Moutairou. 1998. Absence of short-wavelength sensitive cones in the retinae of seals (Carnivora) and African giant rats (Rodentia). European Journal of Neuroscience 10:2586–94.
Peichl, L., G. Behrmann, and R. H. H. Kroeger. 2001. For whales and seals the ocean is not blue: A visual pigment loss in marine mammals. European Journal of Neuroscience 13:1520–28.
Percy, J. A. 1977. Effects of oil on arctic marine organisms: A review of studies conducted by the Arctic Biological Station (Council Meeting, Fisheries Improvement Committee, E:26). International Council for Exploration of the Sea, Copenhagen.
Perez, M. A. 1990. Review of marine mammal and prey population information for Bering Sea ecosystem studies (NOAA Technical Memorandum NMFS F/NWC-186). U.S. Department of Commerce, Washington, DC.
Perez, M. A., and M. A. Bigg. 1986. Diet of northern fur seals (*Callorhinus ursinus*) off western North America. Fishery Bulletin 84:957–71.
Perez, M. A., and E. E. Mooney. 1986. Increased food and energy consumption of lactating northern fur seals, *Callorhinus ursinus*. Fisheries Bulletin 84:371–81.
Perlov, A. S. 1971. The onset of sexual maturity in sea lions. Proceedings of the All-Union Institute of Marine Fisheries and Oceans 80:174–87.
Pfeiffer, C. J., and V. Viers. 1995. Cardiac ultrastructure of the ringed seal, *Phoca hispida* and harp seal, *Phoca groenlandica*. Aquatic Mammals 21:109–19.
Pitcher, K. W., and D. G. Calkins. 1981. Reproductive biology of Steller sea lions in the Gulf of Alaska. Journal of Mammalogy 63:599–605.
Platt, N. E., J. H. Prime, and T. R. Witthames. 1974. The age of the grey seal at the Farne Islands (Council Meeting, Marine Mammals Committee, N:3). International Council for Exploration of the Sea, Copenhagen.
Popov, L. A. 1966. On an ice floe with the harp seals: Ice drift of biologists in the White Sea. Priroda 9:93–101 (In Russian) [Fisheries Research Board of Canada Translation Series 814, 1907].
Popov, L. A. 1976. Status of main ice forms of seals inhabiting waters of the U.S.S.R. and adjacent to the country marine areas. Scientific consultation 51, *in* Symposium: Scientific consultation on marine mammals. Food and Agriculture Organization of the United Nations, Advisory Committee on Marine Resources Research, Marine Mammals. Bergen, Norway.

Popov, L. A. 1982. Status of the main ice-living seals inhabiting inland waters and coastal marine areas of the USSR. Pages 361–81 *in* FAO Advisory Committee on Marine Resources Research, Working Party on Marine Mammals, Mammals in the Seas. Vol. IV: Small cetaceans, seals, sirenians and otters (FAO Fisheries Series No. 5). Food and Agriculture Organization, Rome.

Purkinje, J. E. 1825. Neue Beitrage zur Kenntnis des Sehen. Berlin.

Pyle, P., S. Anderson, and D. G. Ainley. 1996. Trends in white shark predation at the South Farallon Islands, 1968–1993. Pages 375–79 *in* A. P. Klimley and D. G. Ainley, eds. Great white sharks: The biology of *Carcharodon carcharias*. Academic Press, San Diego, CA.

Ralls, K. P., P. Fiorelli, and S. Gish. 1985. Vocalizations and vocal mimicry in captive harbor seals, *Phoca vitulina*. Canadian Journal of Zoology 63:1050–56.

Ramprashad, F. 1975. Aquatic adaptations in the ear of the harp seal *(Pagophilus groenlandicus)* (Erxleben, 1777). Pages 102–11 *in* K. Ronald and A. W. Mansfield, eds. Biology of the seal. International Council for the Exploration of the Sea, Copenhagen.

Ramprashad, F., K. E. Money, and K. Ronald. 1972. The harp seal, *Pagophilus groenlandicus* (Erxleben, 1777). XXI. The structure of the vestibular apparatus. Canadian Journal of Zoology 50:1357–61.

Ramprashad, F., S. Corey, and K. Ronald. 1973. Anatomy of the seal's ear (*Pagophilus groenlandicus* Erxleben, 1777). Pages 264–305 *in* R. J. Harrison, ed. Functional anatomy of marine mammals, Vol. 1. Academic Press, New York.

Ray, G. C., and W. A. Watkins. 1975. Social function of underwater sounds in the walrus *Odobenus rosmaris*. Pages 524–26 *in* K. Ronald and A. W. Mansfield, eds. Biology of the seal. International Council for the Exploration of the Sea, Copenhagen.

Ray, G. C., W. A. Watkins, and J. J. Burns. 1969. The underwater song of *Erignathus* (bearded seal). Zoologica 54:79–83.

Reeves, R. R. 1978. Atlantic walrus (*Odobenus rosmarus rosmarus*): A literature survey and status report (Wildlife Research Report 10). U.S. Department of the Interior, Fish and Wildlife Service, Washington, DC.

Reeves, R. R., R. J. Hofman, G. K Silber, and D. Wilkinson, eds. 1996. Acoustic deterrence of harmful marine mammal–fishery interactions: Proceedings of a workshop held in Seattle, Washington (NOAA Technical Memorandum NMFS-OPR-10). U.S. Department of Commerce, Washington, DC.

Reijnders, P. J. H. 1987. Reproductive failure in common seals feeding on fish from polluted coastal waters. Nature 324:456–57.

Reijnders, P. J., and H., chair. 2000. Seal Specialist Group. Species (Newsletter of the Species Survival Commission IUCN-The World Conservation Union) 34(Fall):89–92.

Reijnders, P., S. Brasseur, J. van der Toorn, P. van der Wolf, I. Boyd, J. Harwood, D. Lavigne, and L. Lowry. 1993. Seals, fur seals, sea lions, and walrus. Status survey and conservation action plan. IUCN/SSC Seal Specialist Group, IUCN, Gland, Switzerland.

Renouf, D., G. Galway, and L. Gaborko. 1980. Evidence for echolocation in harbour seals. Journal of Marine Biology Association 60:1039–42.

Renouf, D., E. Noseworthy, and M. C. Scott. 1990. Daily fresh water consumption by captive harp seals (*Phoca groenlandica*). Marine Mammal Science 6:253–57.

Repenning, C. A., R. S. Peterson, and C. L. Hubbs. 1971. Contributions to the systematics of the southern fur seals, with particular reference to the Juan Fernandez and Guadalupe species. Pages 1–34 *in* W. H. Burt, ed. Antarctic Pinnipedia (Antarctic Research Series 18). American Geophysical Union.

Restrepo, V. R., P. M. Mace, and F. M. Serchuk. 1999. The precautionary approach: A new paradigm, or business as usual. Feature article 1. National Marine Fisheries Service (NMFS), Our living oceans, Report on the status of U.S. living marine resources, 1999 (NOAA Technical Memorandum NMFS-F/SPO-41). U.S. Department of Commerce, Washington, DC. Available at http://spo.nwr.noaa.gov/olo99.htm.

Rhode, E. A., R. Elsner, T. M. Peterson, K. B. Campbell, and W. Spangler. 1986. Pressure–volume characteristics of aortas of harbor and Weddell seals. American Journal of Physiology 251:R174–R180.

Rice, D. W., 1960. Population dynamics of the Hawaiian monk seal. Journal of Mammalogy 41:376–85.

Rice, D. W. 1973. Caribbean monk seal (*Monachus tropicalis*). IUCN, Survival Service Commission, IUCN Publications New Series Supplementary Paper 39:98–112.

Rice, D. W. 1998. Marine mammals of the world: Systematics and distribution (Special Publication No. 4). Society for Marine Mammalogy, Lawrence, KS.

Richardson, W. J., C. R. Greene Jr., C. I. Malme, and D. H. Thomson. 1995. Marine mammals and noise. Academic Press, New York.

Ricker, W. E. 1975. Mortality and production of harp seals with reference to a paper of Benjaminsen and Øritsland (1975) (Series 3716). International Commission for the Northwest Atlantic Fisheries.

Robson, B. W., ed. 2000. Fur seal investigations, 1998 (NOAA Technical Memorandum NMFS-AFSC-113). U.S. Department Commerce, Wasington. DC.

Rommel, S. A., D. A. Pabst, and W. A. McLellan. 1998. Reproductive thermoregulation in marine mammals. American Scientific 86:440–48.

Ronald, K., and J. L. Dougan. 1982. The ice lover: Biology of the harp seal (*Phoca groenlandica*). Science 214:928–33.

Ronald, K., M. E. Foster, and E. Johnson. 1969. The harp seal, *Pagophilus groenlandicus* (Erxleben, 1777). II. Physical blood properties. Canadian Journal of Zoology 47:461–68.

Ronald, K., E. Johnson, M. E. Foster, and D. VanderPol. 1970. The harp seal, *Pagophilus groenlandicus* (Erxleben 1777). 1. Methods of handling, moult, and diseases in captivity. Canadian Journal of Zoology 48:1035–40.

Ronald, K., L. M. Hanly, P. J. Healey, and L. J. Selley. 1976a. An annotated bibliography of the Pinnipedia. International Council for the Exploration of the Sea, Charlottenlund, Denmark.

Ronald, K., P. J. Healey, and H. D. Fisher. 1976b. The harp seal *Pagophilus groenlandicus* Scientific consultation 36, *in* Symposium: Scientific consultation on marine mammals. Food and Agriculture Organization of the United Nations, Advisory Committee on Marine Resources Research, Marine Mammals. Bergen, Norway.

Ronald, K., R. and McCarter, L. J. Selley. 1977a. Venous circulation in the seal, *Pagophilus groenlandicus*. Pages 1–38 *in* R. J. Harrison, ed. Functional anatomy of marine mammals, Vol. 3. Academic Press, London.

Ronald, K., S. V. Tessaro, J. F. Uthe, H. C. Freeman, and R. Frank. 1977b. Methylmercury poisoning in the harp seal (*Pagophilus groenlandicus*). Science of the Total Environment 8:1–11.

Ronald, K., P. J. Healey, J. L. Dougan, L. J. Selley, and L. Dunn, L. 1983. An annotated bibliography of the Pinnipedia, Suppl. 1. International Council for the Exploration of the Sea, Copenhagen.

Ronald, K., B. L. Gots, J. D. Lupson, C. J. Willings, and J. L. Dougan. 1991. An annotated bibliography on seals, sea lions, and walrus, Suppl. 2. International Council for the Exploration of the Sea, Copenhagen.

Rosen, A. S., and A. W. Trites. 2000. Pollock and the decline of Steller sea lions: Testing the junk-food hypothesis. Canadian Journal of Zoology 78:1243–50.

Rough, V. 1995. Gray seals in Nantucket Sound, Massachusetts, winter and spring, 1994. Final report (NTIS Publication PB95-191391). U.S. Marine Mammal Commission, Washington, DC.

Rugh, D. J., E. W. Shelden, and D. E. Withrow. 1995. Spotted seals sightings in Alaska 1992–93. (Annual Report to the MMPA Assessment Program). Office of Protected Resources, National Marine Fisheries Service Silver Spring, MD.

Salm, R. V., J. Clark, and E. Siirila. 2000. Marine and coastal protected areas: A guide for planners and managers. IUCN–The World Conservation Union, Washington, DC.

Scheffer, V. B. 1958. Seals, sea lions, and walruses: A review of the Pinnipedia. Stanford University Press, Stanford, CA.

Scheffer, V. B. 1977. A lesson in survival. Animal Kingdom 80:6–12.

Scheffer, V. B., and A. M. Johnson. 1963. Molt in the northern fur seal (Special Science Report, Fisheries Series 450). U.S. Fish and Wildlife Service, Washington, DC.

Schevill, W. E., W. A. Watkins, and G. C. Ray. 1966. Analysis of underwater *Odobenus* calls with remarks on the development and function of the pharyngeal pouches. Zoologica 51:103–6.

Schiffman, H. 1998. The dispute settlement mechanism of UNCLOS: A potentially important apparatus for marine wildlife management. Journal of International Wildlife Law Policy 1:293–306.

Schlexer, F. V. 1984. Diving patterns of the Hawaiian monk seal, Lisianski Island, 1982 (NOAA Technical Memorandum NMFS-SWFSC-41). U.S. Department of Commerce, Washington, DC.

Schmidly, D. J. 1981. Marine mammals of the southeastern United States coast and the Gulf of Mexico. (FWS/OBS-80/41). U.S. Fish and Wildlife Service, Office of Biological Services, Washington, DC.

Scholander, P. F. 1940. Experimental investigations on the respiratory function in diving mammals and birds. Hvalrådets Skrifter 22:1–131.

Scholander, P. F., L. Irving, and S. W. Grinnell. 1942. On the temperature and metabolism of the seal during diving. Journal of Cellular and Comparative Physiology 21:53–63.

Schusterman, R. J. 1972. Visual acuity in pinnipeds. Pages 469–92 *in* H. E. Winn and B. L. Olls, eds. Behaviour of marine animals: Current perspectives in research, Vol. 2. Plenum Press, New York.

Schusterman, R. J. 1975. Pinniped sensory perception. Pages 165–69 *in* K. Ronald and A. W. Mansfield, eds. Biology of the seal. International Council for the Exploration of the Sea, Copenhagen.

Schusterman, R. J. 1981. Behavioral capabilities of seals and sea lions: A review of their hearing, visual, learning and diving skills. Psychological Record 31:125–43.

Schusterman, R. J., R. Gentry, and J. Schmook. 1967. Underwater sound production by captive California sea lions, *Zalophus californianus*. Zoologica 52:21–24.

Schwartz, F. 1977. Land use in the Makkovik Region. Pages 239–78 *in* C. Brice-Bennett, ed. Our footprints are everywhere. Labrador Inuit Association, Nain, Labrador, Canada.

Seal Conservation Society. 1999. Pinniped species information pages. Seal Conservation Society, Aberdeen, UK. Available at www.pinnipeds.org/species/species.htm.

Seal Conservation Society. 2000. Canadian seal hunt news. 2000 news digest. Seal Conservation Society, Aberdeen, UK.

Sergeant, D. E. 1969. On the population dynamics and size of stocks of harp seals in the northwestern Atlantic (Research Document 69/31, Series 2171). International Commission for the Northwest Atlantic Fisheries.

Sergeant, D. E. 1973. Feeding, growth, and productivity of Northwest Atlantic harp seals *(Pagophilus groenlandicus)*. Journal of the Fisheries Research Board of Canada 30:17–29.

Sergeant, D. E. 1975. Estimating numbers of harp seals. Pages 274–80 *in* K. Ronald and A. W. Mansfield, eds. Biology of the seal. International Council for the Exploration of the Sea, Copenhagen.

Sergeant, D. E. 1976a. History and present status of populations of harp and hooded seals. Scientific consultation 1 *in* Symposium: Scientific consultation marine mammals. Food and Agriculture Organization of the United Nations, Advisory Committee on Marine Resources Research, Marine Mammals. Bergen, Norway.

Sergeant, D. E. 1976b. Research on hooded seals *Cystophora cristata* Erxleben in 1976 (Research Document 76/x/126). International Commission of Northwest Atlantic Fisheries.

Sergeant, D. E., and F. A. J. Armstrong. 1973. Mercury in seals from eastern Canada. Journal of the Fisheries Research Board of Canada 30:843–46.

Shadwick, R. E., and J. M. Gosline. 1995. Arterial windkessels in marine mammals. Pages 243–52 *in* C. P. Ellington and T. J. Pedley, eds. Biological fluid dynamics (Symposia of the Society for Experimental Biology, Vol. 49). Cambridge University Press, Cambridge.

Sheffield, G., F. H. Fay, H. Feder, and B. P. Kelly. 2001. Laboratory digestion of prey and interpretation of walrus stomach contents. Marine Mammal Science 17:310–30.

Shelton, P. A., G. B. Stenson, B. Sjare, and W. G. Warren. 1995. Model estimates of harp seal numbers at age for the northwest Atlantic (Atlantic Fisheries Research Document 95/21). Canada Department of Fisheries and Oceans, Ottawa.

Shelton, P. A., G. B. Stenson, B. Sjare, and W. G. Warren. 1996. Model estimates of harp seal numbers at age for the Northwest Atlantic. Northwest Atlantic Fisheries Organization, Scientific Council Studies 26:1–14.

Shipley, C., B. S. Stewart, and J. Bass. 1992. Seismic communication in northern elephant seals. Pages 553–62 *in* J. A. Thomas, R. A. Kastelein, and A. Y. Supin, eds. Marine mammal sensory systems. Plenum Press, New York.

Sinclair, E. H. 1994. Introduction. Pages 1–8 *in* 994 Fur seal investigations (NOAA Technical Memorandum NMFS-AFSC-46). U.S. Department of Commerce, Washington, DC.

Sivertsen, E. 1941. On the biology of the harp seal, *Phoca groenlandica* Erxl., investigations carried out in the White Sea. Hvalrådets Skrifter 26.

Small, R. J. 1996. Population assessment of harbor seals in Alaska: A report of a workshop held in Fairbanks, Alaska. National Marine Mammal Laboratory, Alaska Fisheries Science Center, National Marine Fisheries Service, Seattle, WA.

Small, R. J., G. W. Pendelton, and K. M. Wynne. 1997. Harbor seal population trends in the Ketchikan, Sitka, and Kodiak Island areas of Alaska. Pages 7–12 *in* Annual report: Harbor seal investigations in Alaska. Alaska Department of Fish and Game, Division of Wildlife Conservation, Douglas, AK.

Smirnov, N. A. 1924. On the eastern harp seal *Phoca* (*Pagophoca*) *groenlandica*. Tromsø Museum Arsberetnins Naturhistorisk avd 47:3–11.

Smith, T. G., and I. Stirling. 1975. The breeding habits of the ringed seal (*Phoca hispida*): The birth, air, and associated structures. Canadian Journal of Zoology 53:1297–1305.

Spraker, T. R., L. F. Lowry, and K. J. Frost. 1994. Gross necropsy and histopathologic lesions found in harbor seals. Pages 281–311 *in* T. R. Loughlin, ed. Marine mammals and the *Exxon Valdez*. Academic Press, San Diego, CA.

Starfield, A. M., J. D. Roth, and K. Ralls. 1995. "Mobbing" in Hawaiian monk seals (*Monachus schauinslandi*): The value of simulation modeling in the absence of apparently crucial data. Conservation Biology 9:166–74.

Stenson, G. B. 1994. The status of pinnipeds in the Newfoundland region. North Atlantic Fisheries Organization, Science Council Studies 21:115–19.

Stenson, G. B., R. A. Myers, I.-H. Ni, and W. G. Warren. 1996a. Pup production of hooded seals (*Cystophora cristata*) in the Northwest Atlantic. North Atlantic Fisheries Organization, Science Council Studies 26:105–14.

Stenson, G. B., B. Sjare, W. G. Warren, R. A. Myers, M. O. Hammill, and M.C.S. Kingsley. 1996b. 1994 Pup production of the Northwest Atlantic harp seal, *Phoca groenlandica*. North Atlantic Fisheries Organization, Science Council Studies 26:47–62.

Stewart, B. S., and P. K. Yochem. 1987. Entanglement of pinnipeds in synthetic debris and fishing net and line fragments at San Nicolas and San Miguel Islands, California, 1978–1986. Marine Pollution Bulletin 18:336–39.

Stewart, B. S., P. K. Yochem, R. L. DeLong, and G. A. Antonelis. 1993. Status and trends in abundance of pinnipeds on the Southern California Channel Islands. Pages 501–16 *in* F. G. Hochberg, ed. Third California islands symposium: Recent advances in research on the California islands. Santa Barbara Museum of Natural History, Santa Barbara, CA.

Stewart, B. S., P. K. Yochem, H. R. Huber, R. L. DeLong, R. J. Jameson, W. J. Sydeman, S. G. Allen, and B. J. Le Boeuf. 1994. History and present status of the northern elephant seal population. Pages 29–48 *in* B. J. LeBeouf and R. M. Laws, eds. Elephant seals: Population ecology, behavior, and physiology. University of California Press, Los Angeles.

Stirling, I. 1975. Factors affecting the evolution of social behaviour in the pinnipedia. Pages 205–12 *in* K. Ronald and A. W. Mansfield, eds. Biology of the seal. International Council for the Exploration of the Sea, Copenhagen.

Stirling, I. 1979. Ribbon seal. FAO Fisheries Series 5 [Mammals in the seas] 2:81–82.

Stirling, I., and R. Archibald. 1979. Bearded seal. FAO Fisheries Series 5 [Mammals in the seas] 2:83–85.

Stirling, I., and W. Calvert. 1979. Ringed seal. FAO Fisheries Series 5 [Mammals in the seas] 2:66–69.

Stobo, W. T., and C. T. Zwanenburg. 1990. Grey seal (*Halichoerus grypus*) pup production on Sable Island and estimates of recent production in the Northwest Atlantic. Pages 171–84 *in* W. D. Bowen, ed. Population biology of sealworm (*Pseudoterranova decipiens*) in relation to its intermediate and seal hosts (Canadian Bulletin of Fisheries and Aquatic Sciences No. 222). Ottawa.

Sumner-Smith, G., P. W. Pennock, and K. Ronald. 1972. The harp seal, *Pagophilus groenlandicus* (Erxleben, 1777). XVI. Epiphyseal fusion. Journal of Wildlife Diseases 8:29–32.

Swartzman, G. L., C. A. Ribic, and C. P. Haung. 1990. Simulating the role of entanglement in northern fur seal, *Callorhinus ursinus*, population dynamics. Pages 513–30 *in* R. S. Shomura and M. L. Godfrey, eds. Proceedings of the second international conference on marine debris, Honolulu, Hawaii. (NOAA Technical Memorandum NMFS-SWFSC-154). U.S. Department of Commerce, Washington, DC.

Sweeney, J. C., and J. R. Geraci. 1979. Medical care and strandings. Pages *in* J. R. Geraci and D. J. St. Aubin, eds. Biology of marine mammals: insights through strandings (U.S. Marine Mammal Commission Report No. MMC/771 13; NIST PB-293 890). U.S. Department of Commerce, Washington, DC.

Sydeman, W. J., and S. G. Allen. 1999. Pinniped population dynamics in central California: Correlations with sea surface temperature and upwelling indices. Marine Mammal Science 15:446–61.

Tandler, J. 1899. Ucher em corpus cavernosum tympanicum beim seehund. Monattschrift für Ohrenheilkunde, Kehlkopfheilkunde, Nasen- und Rachenkrankhait, Organ Oesterreich Otologie Gesellschaft 33:437–40.

Temte, J. L., M. A. Bigg, and O. Wiig. 1991. Clines revisited: The timing of pupping in the harbour seal (*Phoca vitulina*). Journal of Zoology (London) 224:617–32.

Terhune, J. M. 1973. Aspects of hearing and acoustical communication of seal. M.S. Thesis. Åarhus University, Åarhus, Denmark.

Terhune, J. M., and K. Ronald. 1970. The audiogram and calls of the harp seal *(Pagophilus groenlandicus)* in air. Pages 133–43 *in* Symposium: Biological sonar and diving mammals. Stanford Research Institute, Menlo Park, CA.

Terhune, J. M., and K. Ronald. 1971. The harp seal, *Pagophilus groenlandicus* (Erxleben, 1777). X. The air audiogram. Canadian Journal of Zoology 49:385–90.

Terhune, J. M., and K. Ronald. 1972. The harp seal, *Pagophilus groenlandicus*. III. The underwater audiogram. Canadian Journal of Zoology 50:565–69.

Terhune, J. M., and K. Ronald. 1973. Some hooded seal *(Cystophora cristata)* sounds in March. Canadian Journal of Zoology 51:319–21.

Terhune, J. M., and K. Ronald. 1974. Underwater hearing of phocid seals (Council Meeting, Marine Mammals Committee, N:5). International Council for the Exploration of the Sea, Copenhagen.

Terhune, J. M., and K. Ronald. 1986. Distant and near-range functions of harp seal underwater calls. Canadian Journal of Zoology 64:1065–70.

Terhune, J. M., M. E. Terhune, and K. Ronald. 1979. Location and recognition of pups by adult female harp teals. Applied Animal Ethnology 5:375–80.

Thompson, D., and M. A. Fedak. 1993. Cardiac responses of grey seals during diving at sea. Journal of Experimental Biology 174:139–64.

Thompson, G. 1996. The precautionary principle in North Pacific groundfish management (AFSC Quarterly Report, July 1996). National Marine Fisheries Service, Washington, DC.

Toffler, A. 1970. Future shock. Random House, New York.

Timoshenko, Y., and L. A. Popov. 1990. On predatory habits of the Atlantic walrus. Pages 177–78 *in* F. H. Fay, B. P. Kelly, and B. A. Fay, eds. The ecology and management of walrus populations (Report PB91-100479). Marine Mammal Commission, Washington, DC.

Townsend, C. 1918. Sea lions and fishery industry. NY Zoological Society Bulletin 12:1679–82.

Townsend, C. H. 1931. The fur seal of the California islands with new descriptive and historical matters. Zoologica 9:443–57.

Treacy, S. D. 1985. Feeding habits of marine mammals from Grays Harbor, Washington to Netarts Bay, Oregon. Pages 149–98 *in* R. J. Beach, A. C. Geiger, S. J. Jeffries, and B. L. Troutman, eds. Marine mammals and their interactions with fisheries of the Columbia River and adjacent waters (NWAFC Process Report 85-04). U.S. Department of Commerce, Washington, DC.

Trites, A. W. 1997. The role of pinnipeds in the ecosystem. Pages 31–39 *in* G. Stone, J. Goebel, and S. Webster, eds. Pinniped populations, eastern north Pacific: Status, trends and issues. A symposium of the 127th annual meeting of the American Fisheries Society. New England Aquarium, Conservation Department, Boston.

Trites, A. W. 1998. Steller sea lions (*Eumetopias jubatus*): Causes for their decline and factors limiting their restoration. Marine Mammal Research Unit, Fisheries Centre, University of British Columbia, Vancouver, Canada.

Trites, A. W., and P. A. Larkin. 1992. The status of Steller sea lion populations and the development of fisheries in the Gulf of Alaska and Aleutian Islands. Report of the Pacific States Marine Fisheries Commission, pursuant to NOAA Award No. NA 17FD0177.

Trites, A. W., J. Money, and P. A. Larkin. 1998. The decline of Steller sea lions (*Eumetopias jubatus*) and the development of commercial fisheries in the Gulf of Alaska and Aleutian Islands from 1950 to 1990. Marine Mammal Research Unit, Fisheries Centre, University of British Columbia, Vancouver, Canada.

Tynan, C., and D. P. DeMaster. 1996. Observations and predictions of Arctic climate change (Document SC/48/O 21). International Whaling Commission, Cambridge, UK.

U.N. Environment Programme. 1977. Register of international conventions and protocols in the field of the environment (UNEP/GC/Information No. 76-3464). UNEP Governing Council, 5th Session, Nairobi.

U.S. Department of Commerce. 1997. Threatened fish and wildlife; change in listing status of Steller sea lions under the Endangered Species Act. Federal Register 62:24345–55.

U.S. Department of Commerce. 1977. The story of the Pribilof fur seals. U.S. Government Printing Office, Washington, DC.

U.S. Office of Federal Register. 1979. Government manual, 1979–80. National Archives and Records Service, General Services Administration, Washington, DC.

Vallyathan, N. V., J. C. George, and K. Ronald. 1969. The harp seal, *Pagophilus groenlandicus* (Erxleben, 1777). V. Levels of haemoglobin, iron and certain metabolites and enzymes in blood. Canadian Journal of Zoology 47:1193–97.

Vedder, L., R. Zarnke, I. Spijkers, and A. Osterhaus. 1987. Prevalence of a virus neutralizing antibodies to seal herpesvirus (phocid herpesvirus) in different pinniped species [Abstract]. *In* Abstracts of the seventh biennial conference on the biology of marine mammals. Miami, FL.

Verboom, W. C., and R. A. Kastelein. 1995. Rutting whistles of a male Pacific walrus (*Odobenus rosmarus divergens*). Pages 287–99 *in* R. A. Kastelein, J. A. Thomas, and P. E. Nachtigall, eds. Sensory systems of aquatic mammals. DeSpil, Woerden, the Netherlands.

Vidal, O., and J.-P. Gallo-Reynoso. 1996. Die-offs of marine mammals and sea birds in the Gulf of California, México. Marine Mammal Science 12:627–35.

Volstad, J. H., W. Richkus, S. Gaurin, and R. Easton. 1997. Analytical and statistical review of procedures for collection and analysis of commercial fishery data used for management and assessment of groundfish stocks in the U.S. exclusive economic zone off Alaska. Versar, Columbia, MD.

Wallace, S. D., and J. W. Lawson. 1992. A review of stomach contents of harp seals (*Phoca groenlandica*) from the Northwest Atlantic (Technical Report 92-03). International Marine Mammal Association, Guelph, Ontario, Canada.

Wallace, S. D., and J. W. Lawson. 1997. A review of stomach contents of harp seals (*Phoca groenlandica*) from the Northwest Atlantic: An update (IMMA Technical Report 97-01). International Marine Mammal Association, Guelph, Ontario, Canada.

Walker, P. L., and S. Craig. 1979. Archaeological evidence concerning the prehistoric occurrence of sea mammals at Point Bennett, San Miguel Island. California Fish and Game 65:50–54.

Walls, G. L. 1942. The vertebrate eye. Cranbrook Press, Bloomfield Hills, MI.

Waring, G. T., D. L. Palka, P. J. Clapham, S. Swartz, M. C. Rossman, T. V. N. Cole, L. J. Hansen, K. D. Bisack, K. D. Mullin, R. S. Wells, D. K. Odell, and N. B. Barros. 1999. U.S. Atlantic and Gulf of Mexico marine mammal stock assessments, 1999 (NOAA Technical Memorandum NMFS-NE-153). U.S. Department of Commerce, Washington, DC.

Wartzok, D., R. J. Schusterman, and J. Gailey-Phipps. 1984. Seal echolocation? Nature 308:753.

Watkins, W. A., and G. C. Ray. 1977. Underwater sounds from ribbon seal, *Phoca (Histriophoca) fasciata*. Fisheries Bulletin 75:450–53.

Watkins, W. A., and W. E. Schevill. 1979. Distinctive characteristics of underwater calls of the harp seal, *Phoca groenlandica,* during the breeding season. Journal of the Acoustical Society of America 66:983–88.

Watkins, W. A., and D. Wartzok. 1985. Sensory biophysics of marine mammals. Marine Mammal Science 1:219–60.

Wedgeforth, H. M. 1928. The Guadalupe fur seal (*Arctocephalus townsendi*). Zoonooz [South Dakota Zoological Society] 3(3):4–9.

Wiig, Ø., I. Gjertz, D. Griffiths, and C. Lydersen. 1993. Diving patterns of an Atlantic walrus (*Odobenus rosmarus rosmarus*) near Svalbard. Polar Biology 13:71–72.

Wing, E. S. 1992. West Indian monk seal (*Monachus tropicalis*). Pages 35–40 *in* S. R. Humphrey, ed. Rare and endangered biota of Florida. Vol. I: Mammals. University Press of Florida, Gainesville.

Winters, G. H., and B. Bergflodt. 1978. Mortality and productivity of the Newfoundland hooded seal stock (Research Document 78/xi/91). International Commission for the Northwest Atlantic Fisheries.

Witherell, D., C. Pautzke, and D. Fluharty. 2000. An ecosystem-based approach for Alaska groundfish fisheries. International Council for the Exploration of the Sea, Journal of Marine Science 57:771–77.

Withrow, D. E., and T. R. Loughlin. 1997. Abundance and distribution of harbor seals (*Phoca vitulina richardsi*) along the southside of the Alaska Peninsula, Shumagin Islands, Coon Inlet, Kenai Peninsula, and the Kodiak Archipelago in 1996. Annual report. MMPA Assessment Program, Office of Protected Resource, National Marine Fisheries Service, Silver Spring, MD.

Woodley, T., and D. M. Lavigne. 1991. Incidential capture of pinnipeds in commercial fishing gear (Technical Report 91-01). International Marine Mammal Association, Guelph, Ontario, Canada.

Wong, J. 1997. Thousands of seals, sea lions starving. The Seattle Times, Nation & World, Thursday, 11 December 1997.

Keith Ronald, Oceanographic Center, Nova Southeastern University, Dania Beach, Florida 33004-3078. Email: kronald@nova.edu.

Barra L. Gots, Sol Circ Incorporated, 15 Howitt Street, Guelph, Ontario, Canada NIE 3C6. Email: johnbarra@sympatico.ca.

40

West Indian Manatee

Trichechus manatus

Daniel K. Odell

NOMENCLATURE

COMMON NAMES. West Indian manatee, manatee, sea cow, Caribbean manatee, Florida manatee, Antillean manatee
SCIENTIFIC NAME. Family: Trichechidae; species: *Trichechus manatus*
SUBSPECIES. *T. m. manatus, T. m. latirostris*

Harlan (1824) named the two subspecies; however, for many years biologists thought that there were insufficient data to warrant subspecific designation of *T. manatus* (Moore 1951a; Gunter 1954). Domning and Hayek (1986) conducted an extensive morphological examination of *T. manatus* and concluded that the subspecific designation was valid. Garcia-Rodriguez et al. (1998) discussed the phylogeography of *T. manatus*. Other extant trichechids are the Amazonian manatee (*T. inunguis*) and the West African manatee (*T. senegalensis*); the fourth extant sirenian is the dugong (Dugongidae: *Dugong dugon*) (Reynolds and Odell 1991; Reynolds et al. 1999).

DISTRIBUTION

The West Indian manatee is currently distributed in the southern United States (south of latitude 30° N), throughout the Caribbean region, along the east coast of Central America, and on the northeast coast of South America as far south as Marque Seca, Brazil (latitude 12° S) (Reynolds and Odell 1991; Lefebvre et al. 2001) (Fig. 40.1). This species is coastal in distribution and can be found in fresh, brackish, and saltwater habitats. It moves freely between the salinity extremes. Florida is essentially the northern end of the year-round range, although individuals have been reported as far north as Rhode Island on the Atlantic coast (latitude 41.5° N) and as far west as Texas on the Gulf coast (Reid 1996; U.S. Fish and Wildlife Service 2001; Lefebvre et al. 2001). At the southern end of the range, manatees have been reported as far south as Espirito Santo, Brazil (latitude 20° S) (U.S. Fish and Wildlife Service 1978). The northern and southern extremes of distribution apparently are limited by air and water temperatures. The exact distribution of the manatee is not known because many areas within the overall range have not been surveyed.

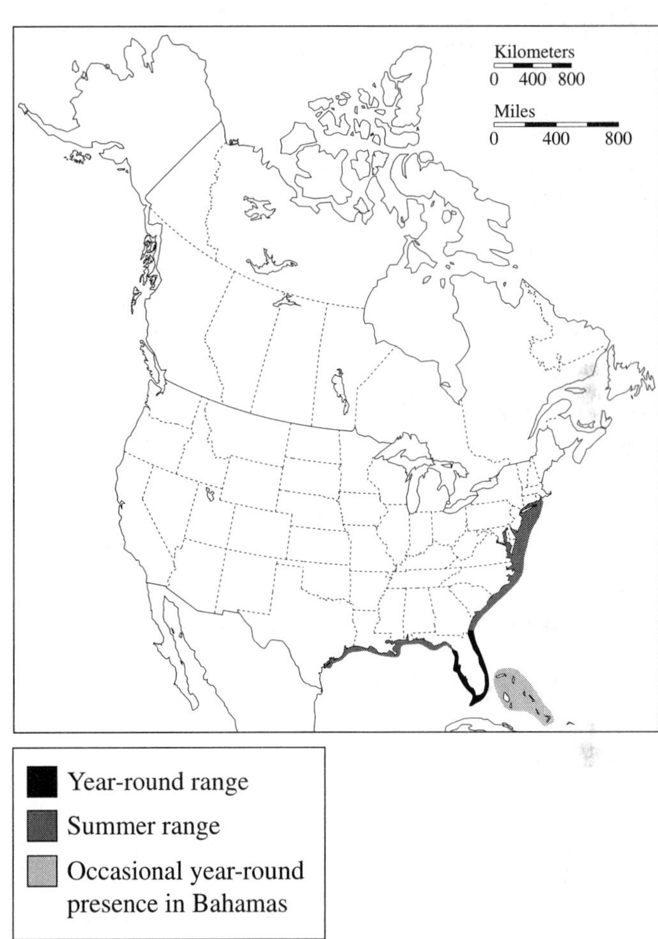

FIGURE 40.1. Year-round and summer distribution of the West Indian in the US & Bahamian water.

DESCRIPTION

Published records indicate that adult manatees can reach a length of 450 cm (Gunter 1941). The longest manatee collected in recent years was a 411-cm individual (Ackerman et al. 1995). The heaviest individual recorded was a 375-cm female, which weighed 1620 kg (Rathbun et al. 1990). The overall body shape is fusiform and flattened dorsoventrally (Fig. 40.2). A large, rounded, spatulate tail is the major organ of propulsion. In accordance with streamlining, external protuberances that would lead to turbulent water flow are reduced or absent. Manatees lack auditory pinnae and pelvic appendages. The long bones of the pectoral appendages are shortened, but the flippers remain flexible. Nails are present. Hairs are sparsely scattered over the body surface, but the facial vibrissae are prominent. Skin color of the adult is best described as grayish, whereas newborn calves are nearly black. The epidermis is continually sloughing off, perhaps to prevent "excessive" accumulation of algae or epizoites. The only externally apparent, sexually dimorphic characteristic is the placement of the genital openings. They are slightly posterior to the umbilicus in males and slightly anterior to the anus in females. Females appear to attain larger weights and lengths than males (Odell 1977; Odell et al. 1981; Ackerman et al. 1995).

Skeleton. The skeleton is massive, composed of dense pachyostotic bone (Fawcett 1942a), which is an important factor in buoyancy (Domning and Buffrénil 1991). There are no marrow cavities in the ribs or the long bones of the pectoral appendage. The vertebrae formula is unusual: there are 6 cervical, 17–19 thoracic, and 27–29 lumbocaudal vertebrae (Hatt 1934; Jones and Johnson 1967; Kaiser 1974). The bones of the skull are as dense as those of the postcranial skeleton (Fig. 40.3). Functional incisors and canines are absent, although

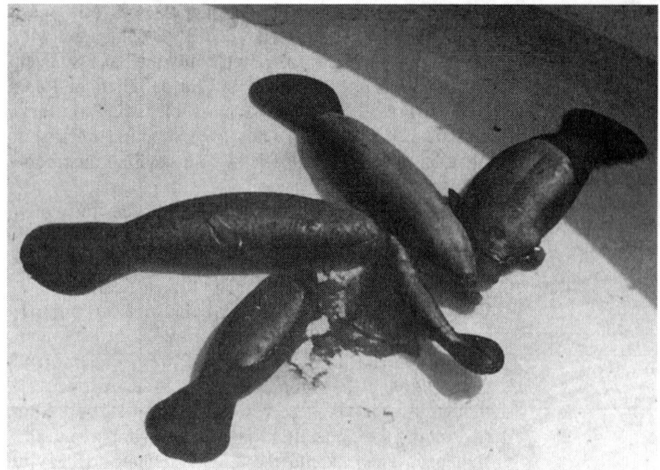

FIGURE 40.2. The West Indian manatee (*Trichechus manatus*) colony at the Miami Seaquarium. The animal on the left has a healed propeller scar. The photograph was taken when the water level was lowered to facilitate cleaning the tank. Photo by D. K. Odell

an unerupted pair of vestigial upper incisors occurs in young animals. Adults usually have six or seven erupted (functional) molariform teeth in each quadrant. The crowns are bilophodont with two major and two minor cusps. Maxillary teeth have tripartite roots, and mandibular teeth have bipartite roots. Tooth replacement is in the horizontal plane, posterior to anterior (Domning and Magor 1977). It is not clear whether tooth movement is a discontinuous or a continuous function. Bone mineral is apparently resorbed at the anterior edge of alveoli. Miller et al. (1980) suggested that the forces of occlusion are the impetus behind tooth movement. As a particular tooth moves forward, the crown is worn down and the root is resorbed. In adults, the anterior tooth in each row is often a thinly worn crown with no root, loosely held in place by the gums. This extreme degree of wear seems to be found only in the Florida population and may be due to the nature of the substrate on which the food plants are growing. The tooth pattern in the ramus of a young manatee, in which there has been no tooth replacement, may be contrasted with the dental pattern in an adult (Fig. 40.4).

As in cetaceans, the otic bullae are not fused to the skull. They are massive and dense, and lie in intimate contact with the squamosal bones. It is noteworthy that the zygomatic process of the squamosal is the only part of the skull containing oil. The auditory sensitivity of the Amazonian manatee is greatest in those areas of the head close to the zygomatic process (Bullock et al. 1980; also see Ketten et al. 1992 and Pabst et al. 1999).

GENETICS

The diploid chromosome number of the West Indian manatee is 48 (White et al. 1976) and, before the widespread use of modern molecular biology techniques (Duffield and Amos 2001), that was the limit of our knowledge of manatee genetics. Garcia-Rodriguez et al. (1998) used molecular genetics to investigate the phylogeography of *T. manatus*. Cashman et al. (1996) characterized the cDNA coding of interleukin 2 from *T. m. latirostris*. McClenaghan and O'Shea (1988) examined, via gel electrophoresis, the genetic variability in *T. manatus* using the allozyme phenotypes of 24 structural loci.

PHYSIOLOGY AND ANATOMY

The physiology of *T. manatus* has not been examined in any detail. Speculation is extensive and based on limited data. New studies are highly restricted because of the manatee's endangered status. Scholander and Irving (1941) documented the response of restrained individuals to forced submersion. The animals exhibited the typical "diving response" (bradycardia and apparent peripheral vasoconstriction) found in marine mammals subjected to similar conditions (Elsner 1969, 1999). Under field conditions, adult manatees may remain submerged for up to 20 min (Reynolds 1977; Reynolds 1981a; Wells et al. 1999), but when active they breathe every 2–3 min. Manatee body temperature is about 35.6–36.4°C (Costa and Williams 1999). Basal metabolic rate (300 cm^3 O_2/kg/min) apparently is lower than in other marine mammals and is about 25–30% of predicted values (Scholander and Irving 1941; Irving 1969; Worthy 2001). The low metabolic rate may be linked with hypothyroidism (Fawcett 1942a; Harrison and King 1965), but Ortiz et al. (2000) reported that thyroid hormone concentrations in captive manatees were comparable to those for other terrestrial and marine mammals.

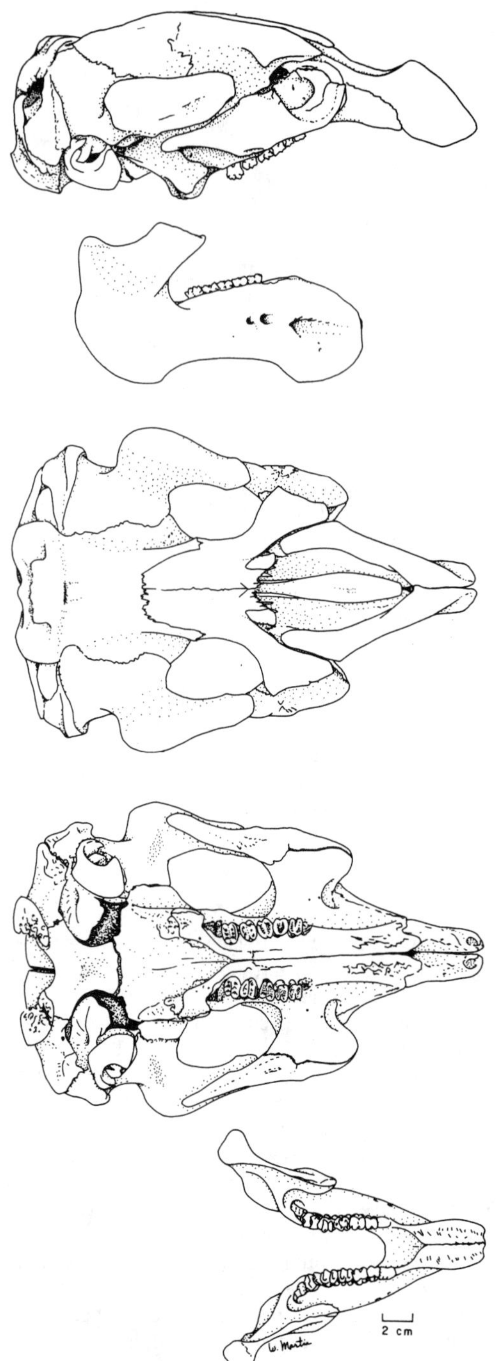

FIGURE 40.3. Skull of the West Indian manatee (*Trichechus manatus*). From top to bottom: lateral view of cranium, lateral view of mandible, dorsal view of cranium, ventral view of cranium, dorsal view of mandible.

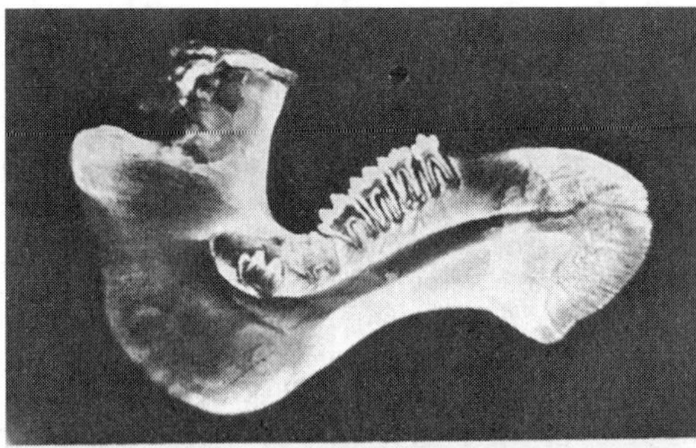

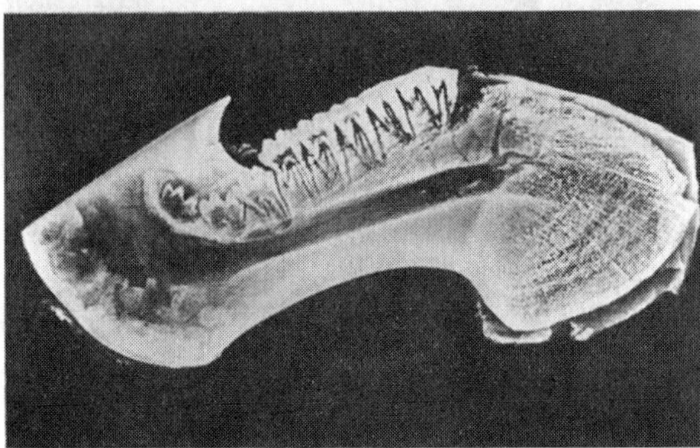

FIGURE 40.4. Xeroradiographs of the mandibles of (top) juvenile (M-76-30; male, 181 cm) and (bottom) adult (M-76-7; female, 316 cm) Florida manatees, showing the pattern of tooth development, crown wear, and root resorbtion. The juvenile had not lost any teeth.

Harrison (1969) reported thyroid glands ranged in weight from 0.11 to 0.13 g/kg body mass. Odell et al. (1981) reported values from 0.06 to 0.32 (mean = 0.22 ± 0.10) g/kg. There also is limited evidence that manatees in Florida may undergo seasonal weight changes. They are heaviest in late summer, fall, and early winter (Odell et al. 1981). If this is the case, organ weights should be compared with fat-free body weight. Low metabolic rate has also been linked with winter mortality in Florida (Campbell and Irvine 1981). It may prove most interesting to examine manatee thyroid structure and function on a seasonal basis and in comparison with detailed metabolic studies.

Odell et al. (1981) derived the following weight–length relationship: $W_{kg} = 4.28 \times 10^{-6} L_{cm}^{3.26}$ ($r^2 = 0.98$). They also gave the following organ weights (g organ/kg body weight; mean ± *SD*): heart 3.25 ± 0.87, liver 15.4 ± 5.68, kidneys 4.08 ± 1.52, adrenal glands 0.06 ± 0.04, spleen 0.06 ± 0.08, and pancreas 0.31 ±. 0.11. The spleen is extremely small and the vertebrae contain some hematopoietic tissue. The lungs are long (exceeding 1 m in adults), unlobed, and dorsoventrally flattened. They extend the length of the body cavity (Wislocki 1935a). The major bronchi extend nearly the full length of the lungs, as in other marine mammals (Bergey 1986; Bergey and Baier 1987). Similarly, the diaphragm extends the length of the body cavity. The diaphragm is actually composed of two hemidiaphragms, which are parallel to the body axis, in contrast to the oblique situation in other mammals (Reynolds 1980). This unique condition unquestionably relates to buoyancy regulation and the horizontal attitude of manatees in the water (Pabst et al. 1999).

The digestive tract occupies a major portion of the body cavity, as one would expect in a herbivore. On first inspection, the stomach appears to be multicompartmental (see plates in Murie 1872). In fact, the stomach proper is a single, highly muscular compartment with a single-lobed "digestive gland" located along the greater curvature near the esophageal opening. The second "stomach compartment" is the duodenal ampulla with two small diverticulae at the proximal end. The small and large intestines are of nearly equal length, joined by a cecum with two hornlike diverticulae. The combined length of the intestines can exceed 40 m (Reynolds 1979) and the weight of the hindgut, including contents, may represent 14% of the total body weight (Reynolds 1980; Reynolds and Rommel 1996). Most, if not all, of the cellulose digestion probably occurs in the hindgut (Burn 1986; Burn and Odell 1987), most likely with the aid of a bacterial and/or protozoan microbiota. This has been suggested for the West African manatee (Lemire 1968) and has been documented for the dugong (Murray et al. 1977).

The structure of the pancreas has not been fully described, but it appears similar to that of other mammals. At least three morphologically distinct types of endocrine cells have been found in the islets of Langerhans (Hinkley et al. 1979).

Lomolino (1977) studied the digestive efficiency of manatees (*T. manatus*) fed water hyacinth (*Eichornia* sp.) and lettuce. Mean digestive efficiencies for hyacinth in terms of dry weight, energy, and nitrogen were 82.6%, 80.0%, and 78.1%, respectively. Similar values for lettuce were 91.4%, 88.8%, and 93.8%, respectively. However, Best (1979) reported digestibility (percentage dry matter) of *Panicum, Cabomba,* and *Pistia* to be 44%, 55%, and 68%, respectively, in *T. inunguis*.

Walsh and Bossart (1999) gave the composition of manatee milk as about 74% water, 16% fat, 8% protein, 1% carbohydrate, and 1% ash, which provide about 189 kcal/100 g milk. Walsh et al. (1996) found manatee milk was high in 12:0, 16:0, and 18:1 fatty acids.

The manatee kidney is a multilobulated, flattened organ located between the peritoneum and the diaphragm (Hill and Reynolds 1989; Maluf 1989; Pabst et al. 1999). It is not reniculate, as in cetaceans and pinnipeds, but grossly similar to that of domestic cattle. The manatee's ability to move between salinity extremes and apparently function perfectly well in freshwater or saltwater suggests that the kidneys play a role. Ortiz et al. (1998, 1999) studied wild and captive manatees in fresh, brackish, and saltwater. They suggested that manatees are good osmoregulators.

The myology of the manatee has not been detailed (Murie 1872; Ronald et al. 1978; Pabst et al. 1999; Rommel and Lowenstine 2001), but is probably grossly similar to that of *T. inunguis* (Domning 1978). Cardiovascular anatomy and physiology also are poorly documented (Pabst et al. 1999). The heart is unique in that the ventricular apices are separate (Quiring and Harlan 1953). Murie (1872) and Fawcett (1942b) described the general vascular network. Vascular bundles, rather than a major artery and vein, extend into the flippers. Similar bundles (one on each side) extend posterior from the thoracic region, in the ventrolateral body wall, to coalesce in the pelvic region and enter the hemal canal of the posterior vertebrae. White et al. (1976) described the basic hematological parameters and Farmer et al. (1979) compared some of these parameters with those of *T. inunguis*. Bossart et al. (2001) reviewed West Indian manatee hematology.

The manatee nervous system has partially been described (Murie 1872; Quiring and Harlan 1953; Harrison and King 1965; Verhaart 1972). The brain has relatively unconvoluted cerebral hemispheres (see plates in Murie 1872). Schneyer et al. (1979) examined the anterior pituitary at the light microscopic and ultrastructural levels. It is similar to that of other mammals, and at least six distinct cell types have been found. Reep et al. (1989) and Reep and O'Shea (1990) gave detailed descriptions of the manatee brain based on extremely fresh specimens obtained from the manatee stranding and carcass salvage program in Florida.

Manatees have extremely small eyeballs (about 15 mm diameter) for such a large animal. Cohen et al. (1982) found that the manatee retina contains rods and cones. Mass et al. (1997) described ganglion layer topography and retinal resolution in the Florida manatee. Griebel and Schmid (1996) found that manatees most likely possess dichromic (blue and green) color vision. Griebel and Schmid (1997) documented brightness discrimination in the manatee.

REPRODUCTION

The female has a bicornuate uterus, and the ovaries are located inguinally within a bursa (Hill 1945). The adult ovaries are sheetlike (Mossman and Duke 1973:381), often with numerous, large follicles up to 5 mm in diameter (Marmontel 1988). The corpus luteum is not grossly apparent in a pregnant animal. The smallest female known to have given birth was 260 cm long. This may correspond with an age of about 8 years (Odell et al. 1981), although Hartman (1979) suggested that sexual maturity may occur at 4 years. Studies (Marmontel et al. 1992; Marmontel 1995; Rathbun et al. 1995) have shown that some female manatees reach sexual maturity at 3–4 years of age. The estrous cycle has not been examined (Larkin 2000). Gestation is thought to be about 1 year (Dekker 1977; Rathbun et al. 1995; Reid et al. 1995). The placenta is hemichorial and deciduate (Wislocki 1935b; O'Shea and Hartley 1995; Reid et al. 1995; Boyd et al. 1999).

The testes are large and located abdominally, posterolateral to the kidneys. In mature males, they may weigh up to 1 kg combined (Odell et al. 1981). Testis weight increases rapidly at a body length of about 275 cm. Sperm occur in animals as small as 250 cm long, and Hernandez et al. (1995) demonstrated that some males reach sexual maturity at 3–4 years of age. Evidence that testicular activity may be seasonal (Odell et al. 1981) was verified by Hernandez et al. (1995).

Neonates occur throughout the year, but, at least in Florida, more calves seem to be born in spring and summer (Irvine et al. 1981; Rathbun et al. 1995). Newborn are black and rely on their flippers more than on their tails for locomotion. Average birth length is about 125 cm (Rathbun et al. 1995) and weight about 30 kg. Calves suckle from the axillary teats of the mother. Although the newborn may ingest small amounts of vegetation within a few days of birth (Odell 1981), the age at nutritional independence is not completely known. They appear to remain with the mother for up to 2 years. Twinning has been documented (Rathbun et al. 1995). The calving interval is about 3–5 years. Rathbun et al. (1995) described the average interval between births as 2.5 winter seasons, which equates to about 2.5 years. See Odell (2002) for a review of sirenian life history.

ECOLOGY

The West Indian manatee generally occurs in marine habitats also occupied by humans. This is particularly true in Florida, where inevitable conflicts arise. Manatees are restricted to near-shore tropical and subtropical areas where seagrasses or freshwater vegetation occurs. Environmental temperatures delineate the northern and southern ends of the range. Most studies concerning manatee ecology have been done in Florida and may not be an accurate representation of the situation in other parts of the range. The size of the population of *T. manatus* is not known. The minimum size of the Florida population, based on aerial survey counts in January 2001, was 3276 (Florida Marine Research Institute, cited in U.S. Fish and Wildlife Service 2001). We do not know what proportion of the population this number represents. Some geographic components of the Florida population are stable or growing, whereas other components are stable at best and possibly declining.

Movements. In Florida, some manatees move northward and westward in the summer. They have been found as far north as Rhode Island on the Atlantic coast and as far west as Texas on the Gulf coast. The extent of seasonal movements is becoming much better known since the development of reliable radio and satellite telemetry (Deutsch et al. 2000; Lander et al. 2001).

Manatees move into natural and artificial warm water areas in Florida during winter (Moore 1951b; U.S. Fish and Wildlife Service 2001). These movements probably are cued by decreasing air and/or water temperatures. The best-known natural warm spring is Crystal River on the Gulf coast (Hartman 1971, 1979). Power plants are the major artificial source of warm water refugia. As power plants are retired and as the power generation industry enters a period of deregulation, the future and stability of these refugia become uncertain. Since generations of manatees have come to rely on these artificial refugia, a long-term strategy is necessary if the power plants are to be decommissioned and the warm water source not replaced.

FEEDING HABITS

West Indian manatees feed on bottom, midwater, and floating vegetation and on overhanging and bank vegetation. They have split, "prehensile" upper lips covered with stiff vibrissae (Marshall et al. 2000). The highly flexible forelimbs are often used to hold vegetation (Hartman 1971, 1979). Manatees spend a large portion of their time feeding (6–8 hr/day) and may consume 10–15% of their body weight/day. Food selection and the nutritional content of the various food items have not been examined in detail.

Manatees eat a variety of submerged, emergent, floating, and overhanging vegetation. Marine vegetation consumed includes *Thalassia* (turtle grass), *Syringodium* (manatee grass), *Diplanthera, Halophila, Halodule* (shoalgrass), and *Ruppia* (widgeongrass). Freshwater vegetation includes *Hydrilla, Eichornia* (water hyacinth), *Cabomba* (fanwort), *Alternathera, Vallisneria* (wild celery), *Elodea, Ceratophyllum* (coontail), *Myriophyllum* (watermilfoil), *Potamogeton* (pond weed), and *Najas* (naiad) (Hartman 1974, 1979). Manatees also eat the leaves and seeds of mangroves (*Rhizophora*) (Maynard 1872; Ledder 1986), as well as a great diversity of bank vegetation that is within their reach. In captivity, they will eat lettuce, cabbage, apples, bananas, carrots, and other fruits and vegetables. Manatees essentially are herbivores, but they undoubtedly ingest a variety of organisms that live on the vegetation they consume. However, there is evidence that they occasionally will eat fish (Powell 1978).

BEHAVIOR

Until recently, few field studies have been done on the behavior of manatees, because they are difficult to observe (Husar 1978). Hartman (1971, 1979) made the most extensive set of field observations on manatees that congregate during the winter at the clear, warm springs in Crystal River, Florida. He described many of their routine behavior patterns. These studies are being continued by the U.S. Geological Survey. Hartman's work was limited because the manatees only congregated during the winter and apparently were widely dispersed during the rest of the year. Reynolds (1977, 1979, 1981b) conducted a similar study on a semi-isolated colony of manatees in a freshwater lake system. These animals could be observed throughout the year. The latter situation is probably close to what can be expected in other parts of the range. However, these studies were done in Florida, at the northern end of the range, and may not be representative of the species. The advent of workable radio and satellite telemetry systems has added another dimension to the study of manatee behavior (Reid et al. 1995; Deutsch et al. 1998, 2000). Manatees can be tracked throughout 24-hr periods, revealing the details of movement patterns and habitat preferences. Newer global positioning system telemetry will provide even greater detail.

Manatees are perhaps best described as being "weakly social" and do not occur in large herds, with the exception of winter congregations in Florida. Odell (1979) found a mean herd size of 2.55 (range =1–15) based on 302 herds observed from the air in the Everglades National Park over a 2.5-year period. Reynolds (1977, 1981b) found a mean herd size of 2.54. Hartman (1971, 1979) and Reynolds (1977, 1981b) agreed that the most stable social group is the female–calf pair. In captivity, a calf may suckle for >3 years (D. K. Odell unpublished data, 1979). From the air, I have often seen groups of three animals: a small calf, a large adult (presumably the calf's mother), and an animal of intermediate size, which might be an older calf of the same female. The period of social dependence could be quite long. See Wells et al. (1999) for an overview of sirenian behavior.

A commonly observed grouping is the "estrous herd" (Hartman 1971, 1979; Wells et al. 1999). This consists of one female in estrus and a variable number of males trying to mate with her. Presumably only the most persistent male or males mate.

Manatees produce a simple, squeaklike vocalization, which is audible to the human ear. Most of the sound energy is below 5 kHz and there are apparently no ultrasonic components (Schevill and Watkins 1965). Until recently, there was only circumstantial evidence that manatees had excellent hearing capabilities (Hartman 1979). From anatomical observations, Ketten et al. (1992) concluded preliminarily that *T. manatus* has a low-frequency ear with a relatively narrow range, poor sensitivity, and poor localization ability (also see Wartzok and Ketten

1999). Gerstein et al. (1999) conducted underwater audiograms on two captive male Florida manatees. They concluded that the best range of hearing was in the 6- to 20- kHz range. Because manatees often inhabit turbid waters, hearing probably plays a large role in herd cohesion, particularly of the female–calf group. Reynolds (1977) recorded a lengthy vocal interchange between a female and her calf.

Chemoreception probably plays an important role in manatee behavior (Hartman 1971, 1979). Animals often "nibble" each other's skin and anal and genital regions. Feces ingestion is common. Chemoreception may be important for males to locate estrous females. There is confusion in the literature as to whether manatees have a vomeronasal organ, but Mackay-Sim et al. (1985) found no evidence of one. The turbinate bones are reduced. The histological structure of the olfactory epithelium has not been examined. Taste buds have been found on the tongue (Barrett 1979).

MORTALITY

Manatee mortality is of great concern because of the endangered status of the species. Aside from hunting throughout the range, mortality factors have been studied only in Florida (Irvine et al. 1981). Large predators such as crocodiles, alligators, and sharks may take an occasional small manatee, but this is not documented. Human activities, particularly boating, are the major identifiable cause of manatee mortality in Florida (Hartman 1971; O'Shea et al. 1985; Ackerman et al. 1995; Wright et al. 1995) (Fig. 40.5). Although many manatees have scars from cuts caused by boat propellers, the incidence of mortality due to boats was not documented until recently. In 1974 a carcass salvage program was started in Florida. For 4082 manatee carcasses recovered, 31% of the deaths were attributed to human activities, as follows: watercraft, 24.3%; navigation locks and flood control gates (Fig. 40.6) (Odell and Reynolds 1979), 4.1%; and other human causes, including drowning in nets, ingestion of fish hooks (Forrester et al. 1975) and other debris, hunting/shooting, and entanglement in ropes, 2.1%. Causes of the remainder of the deaths were undetermined, but included natural mortality. Many carcasses are too decomposed to determine cause of death if factors like broken bones or entanglements are not present.

Some manatee mortality is related to cold weather. This was suggested by earlier researchers (Cahn 1940; Krumholz 1943; Layne 1965), but documentation was not available until a comprehensive carcass salvage program was undertaken (O'Shea et al. 1985; Ackerman et al. 1995). The nature of the relationship between manatee mortality and cold weather is unclear. Speculation includes increased susceptibility to disease, including pneumonia, and inability to produce enough heat metabolically to compensate for increased heat loss, resulting in death from hypothermia. In any event, the trend of increased mortality with cold weather is clear.

FIGURE 40.5. A Florida manatee (M-75-1; male, 305 cm) killed as a result of collision with a watercraft. Propeller cuts are clearly visible on the animal's back.

FIGURE 40.6. A Florida manatee (M-77-20; female, 229 cm) killed in an automatic flood control dam in Dade County, Florida. The impression left by the bottom of the dam gate is down the center of the animal's back.

Parasites. Beck and Forrester (1988) reviewed the helminths of the Florida manatee and found six species including *Heterocheilus tunicatus* (Nematoda: Ascaridoidea), *Anoplocephala* sp. (Cestoda: Cyclophyllidea), *Cochleotrema cochleotrema* (Digenea: Opisthotrematidae), *Chiorchis fabaceus* (Digenea: Paramphistomatidae), *Nudacotyle undicola* (Digenea: Nudacotylidae), and *Moniligerum blairi* (Digenea: Opisthotrematidae). Seventy-three percent of the animals examined were infected with at least one species of helminth. Mignucci-Giannoni et al. (1999) reviewed the parasites and commensals from the West Indian manatee in Puerto Rico. Upton et al. (1989) reported two previously undescribed species of coccidian protozoans in the feces of Florida manatees: *Eimeria nodulosa* and *Eimeria manatus* (Apicomplexa: Eimeriidae). Also see Dailey (2001) and Bossart (2001) for reviews. Epizoites include a copepod (*Harpacticus pulex*) (Humes 1964), the diatoms *Zygnema* and *Navicula*, the remora (*Remora*), and numerous other commensals living in the cracks and crevices of the skin (Husar 1978). Species of at least two genera of algal epiphyte (*Lyngbya* and *Compsapogon*) (Hartman 1971) and a barnacle (*Chelonibia manati*) (Ross and Newman 1967; W. Newman, Scripps Institution of Oceanography, pers. commun., 1979) are associated with the manatee. Darwin (1854) reported the barnacle *Platylepas bissexlobata* (de Blainville 1824) (= *P. hexastylos* [Fabricius 1798]) on the manatee from "Honduras" (H. R. Spivey, Florida State University, pers. commun., 1979). Stubbings (1965) reported *P. hexastylos, C. manati,* and *Balanus trigonus* from *T. senegalensis* from Senegal.

Diseases. Historically, naturally occurring diseases in the manatee were difficult to study, primarily because fresh carcasses were difficult to obtain. However, the extensive carcass salvage and rescue/rehabilitation efforts in Florida have provided an abundance of specimens for study. As one would expect, manatees may contract a variety of bacterial and viral diseases (Buergelt et al. 1984; Bossart 2001). In 1982 and 1996, red tide toxin (from *Karenia breve*) was responsible for a significant number of manatee deaths in Florida (O'Shea et al. 1991; Bossart et al. 1998). Duignan et al. (1995) reported morbillivirus exposure in Florida manatees. Ewing et al. (1997) and Bossart et al. (2002) reported the first cases of a cutaneous papilloma virus. See Miller et al. (2001), Bossart (2001), and Bossart et al. (2001) for overviews of manatee diseases.

Anthropogenic Factors. Manatees are herbivores and the levels of various pesticides and other anthropogenic compounds in their tissues are low (O'Shea et al. 1984; Ames and VanVleet 1996; O'Shea 1999; O'Hara and O'Shea 2001; also see O'Shea et al. 1999). However, the possible effects of even low levels of these chemicals on manatee immune competency and reproduction cannot be ignored. Beck and Barros (1991) reported on the impact of debris ingested by manatees.

AGE ESTIMATION

Manatees lack tusks used for age estimation of dugongs (Mitchell 1973, 1976, 1978), and the molar teeth are not permanent. Marmontel (1995) and Marmontel et al. (1996) found that growth layer groups in the periotic earbones could be used reliably to estimate manatee age and that maximum longevity was about 60 years. A manatee born in the Miami Aquarium in 1948 was still alive at the South Florida Museum in Bradenton, Florida, as of July 2003.

ECONOMIC STATUS, MANAGEMENT, AND CONSERVATION

The manatee cannot be considered a game or pest animal at the present time because of its endangered status, but hunting for meat, bone, hides, and fat previously caused severe population reduction (Bertram and Bertram 1973; Peterson 1974). Although the manatee is protected by law throughout most of its range (Husar 1978), it is impossible to prevent all poaching. Even in Florida, where the manatee has been protected since 1893, poaching may still occur (Irvine et al. 1981). Florida legislation enacted in 1978 created regulations (e.g., slow-speed zones, no- and restricted-access zones) for boating activities in areas of manatee congregations during the winter months. In the United States, the manatee is also protected by the Marine Mammal Protection Act of 1972 and the Endangered Species Act of 1973 (U.S. Fish and Wildlife Service 1978, 2001). The Marine Mammal Protection Act also requires that the ecosystems in which the manatees live remain healthy and stable.

Perhaps the greatest potential economic use of manatees is as a biological agent for the clearance of aquatic weeds (Vietmeyer 1976). The effectiveness of their voracious appetites has been demonstrated in Guyana (Allsopp 1960) and Florida (Sguros 1966). The Florida studies were terminated because some of the manatees were killed by vandals and others apparently died from pneumonia during cold weather. Before manatees can be seriously considered for extensive aquatic weed clearance, their biology must be better understood and the species must be recovered enough to be removed from the endangered list. In time, it may be possible to match the food intake of a group of manatees with the vegetation growth rate in a particular lake, river, or canal. If the group reproduces, the offspring could be removed for use in other areas. Assuming that manatee physiology has been taken into consideration, human activities (primarily vandalism) would be the most serious threat to the weed clearance proposals. Some have suggested that manatees could be produced commercially as a source of meat (Guyana National Science Research Council 1974). Their endangered status attests to the fact that they are good to eat (Bertram and Bertram 1973). There are no current plans to use manatees for either aquatic weed control or as a commercial source of meat.

CURRENT RESEARCH AND MANAGEMENT NEEDS

Great progress has been made in our understanding of manatee biology and in identifying major mortality factors (O'Shea et al. 1985; Ackerman et al. 1995; U.S. Fish and Wildlife Service 2001). However, much remains to be learned, especially in the areas of environmental physiology and reproductive biology. The most important management area is that of direct and indirect human impact on manatees and their habitat. Power plants and other industrial activities have created artificial warm springs where manatees congregate in the winter (Moore 1951b; Hartman 1974; U.S. Fish and Wildlife Service 2001). These artificial springs may keep manatees north of their historical winter range, thus exposing them to weather conditions more severe than they would normally encounter and resulting in an increased mortality rate (Campbell and Irvine 1981). What would happen to manatees that were using a heated effluent if the discharge suddenly stopped during a period of severe cold weather? This question cannot be answered until the manatee's metabolic physiology is better understood. Certainly manatees are attracted to warm water, both natural and artificial, and increased mortality may be associated with these areas because marginal or sick animals, as well as normal animals, seek the warm water. These congregations would also seem to increase the chances of manatees being killed by watercraft or exposed to disease. Projects are under way to eliminate mortality caused by flood control dams and navigation locks (Odell and Reynolds 1979; Ackerman et al. 1995; U.S. Fish and Wildlife Service 2001). Boat speeds are regulated by Florida law in areas used for manatee feeding, travel, and winter congregation. However, enforcement has been less than adequate due to a chronic shortage of law enforcement personnel. Refuges cannot be fully established until migratory patterns and feeding habits are known and thermoregulatory and osmoregulatory physiology have been studied. Finding survey techniques for monitoring changes in the population is complicated by the fact that manatees can remain submerged for >20 min (Reynolds 1981a) and often inhabit turbid waters. Winter surveys in Florida give only minimum population sizes. That manatees inhabit waters under the jurisdiction of many countries further complicates survey efforts.

The ultimate goal of the Florida Manatee Recovery Plan is to remove the species from the Endangered Species List by improving the status of the population. O'Shea et al. (1995) reviewed the data available (roughly through 1991) on aspects of Florida manatee population biology and reported significant progress (also see Langtimm et al. 1998 and Marmontel et al. 1997). However, at that time we still lacked sufficient information to determine whether or not the population could be downlisted to "threatened." O'Shea et al. (2001) reviewed the work that has been accomplished since 1992 and discussed some of the uncertainties about manatee population status. Although further progress clearly had been made, we still lack adequate information on Florida manatee population biology.

The Florida Manatee Recovery Plan–Third Revision (U.S. Fish and Wildlife Service 2001) elaborates in great detail a wide range of biological research and management tasks that need to be accomplished to ultimately remove the manatee from the endangered species list. Downlisting (from endangered to threatened) criteria include the *identification* (but not protection) of minimum spring flows, foraging habitats, and other important manatee areas; protection of selected warm-water refuges; and reduction of unauthorized human-caused "take." This begs the question of what authorized "take" may be. Downlisting criteria also include statistical confidence, over the most recent 10-year period, in a 90% average annual adult survival rate; that 40% of adult females seen in winter are accompanied by first- or second-year calves; and that the average annual population growth rate is not less than zero. These criteria are less than the recommendation of the Manatee Population Status Working Group of 94% average annual adult mortality and a 4% average annual population growth rate (U.S. Fish and Wildlife Service 2001:Appendix A). Delisting criteria in the recovery plan include actual protection of the areas identified in the downlisting criteria and the same population benchmarks as in the downlisting criteria. The U.S. Fish and Wildlife Service estimated that it would be at least 14 years before the standards for assessing downlisting could be met and an additional 10 years before the criteria for delisting could be met. A population assessment workshop was held in the spring 2002 to review the most recent data sets.

However, even within Florida there is a highly controversial move (as of May 2003) to downlist the manatee from endangered to threatened on the state's endangered species list, which has criteria greatly different from those of the federal recovery plan.

In the meantime, the human population of Florida continues to grow and, with it, the demand for more water, more coastal development, and more use of manatee habitat for recreation. Reams of biological studies will be of little value if human population growth and land development cannot be managed to protect the manatee and its habitat.

LITERATURE CITED

Ackerman, B. B., S. D. Wright, R. K. Bonde, D. K. Odell, and D. J. Banowetz. 1995. Trends and patterns in mortality of manatees in Florida, 1974–1992. Pages 223–58 *in* T. J. O'Shea, B. B. Ackerman, and H. F. Percival, eds. Population biology of the Florida manatee (Information and Technology Report 1). U.S. Department of the Interior, National Biological Service, Washington, DC.

Allsopp, W. H. L. 1960. The manatee: Ecology and use for weed control. Nature 188:762.

Ames, A. L., and E. S. VanVleet. 1996. Organochlorine residues in the Florida manatee (*Trichechus manatus latirostris*). Marine Pollution Bulletin 32:374–77.

Barrett, S. 1979. Taste receptors in the West Indian manatee, *Trichechus manatus*. [Abstract]. *In* 3rd biennial conference on the biology of marine mammals. Seattle, WA.

Beck, C. A., and N. B. Barros. 1991. The impact of debris on the Florida manatee. Marine Pollution Bulletin 22:508–10.

Beck, C., and D. J. Forrester. 1988. Helminths of the Florida manatee, *Trichechus manatus latirostris,* with a discussion and summary of the parasites of sirenians. Journal of Parasitology 74:628–37.

Bergey, M. R. 1986. Lung structure and mechanics of the West Indian manatee (*Trichechus manatus*). M.S. thesis, University of Miami, Coral Gables, FL.

Bergey, M., and H. Baier. 1987. Lung mechanical properties in the West Indian manatee (*Trichechus manatus*). Respiration Physiology 68:63–76.

Bertram, G. C. L., and C. K. R. Bertram. 1973. The modern Sirenia: Their distribution and status. Biological Journal of the Linnaean Society 5:297–338.

Best, R. 1979. Food and feeding habits of wild and captive Sirenia (Projeto Peixe-Boi). INPA, Manaus, Brazil.

Bossart, G. D. 2001. Specific medicine and husbandry of marine mammals: Manatees. Pages 939–60 *in* L. A. Dierauf and F. M. D. Gulland, eds. CRC handbook of marine mammal medicine, 2nd ed. CRC Press, Boca Raton, FL.

Bossart, G. D., D. G. Baden, R. Y. Ewing, B. Roberts, and S. D. Wright. 1998. Brevitoxicosis in manatees (*Trichechus manatus latirostris*) from the 1996 epizootic: Gross, histological and immunohistochemical features. Toxicologic Pathology 26:276–82.

Bossart, G. D., T. Reidarson, L. A. Dierauf, and D. Duffield. 2001. Pathology of marine mammals: Clinical pathology. Pages 383–436, *in* L. A. Dierauf and F. M. D. Gulland, eds. CRC handbook of marine mammal medicine, 2nd ed. CRC Press, Boca Raton, FL.

Bossart, G. D., R. Y. Ewing, M. Lowe, M. Sweat, S. J. Decker, C. J. Walsh, S. Ghim, and A. B. Johnson. 2002. Viral papillomatosis in Florida manatees (*Trichechus manatus latirostris*). Experimental and Molecular Pathology 72:37–48.

Boyd, I. L., C. Lockyer, and H. D. Marsh. 1999. Reproduction in marine mammals. Pages 218–86 *in* J. E. Reynolds III and S. A. Rommel, eds. Biology of marine mammals. Smithsonian Institution Press, Washington, DC.

Buergelt, C. D., R. K. Bonde, C. A. Beck, and T. J. O'Shea. 1984. Pathologic findings in manatees in Florida. Journal of the American Veterinary Medical Association 185:1331–34.

Bullock, T. H., D. Domning, and R. C. Best. 1980. Evoked potentials demonstrate hearing in a manatee (*Trichechus inunguis*). Journal of Mammalogy 61:130–33.

Burn, D. M. 1986. The digestive strategy and efficiency of the West Indian manatee, *Trichechus manatus*. Comparative Biochemistry and Physiology 85A:139–42.

Burn, D. M., and D. K. Odell 1987. Volatile fatty acid concentrations in the digestive tract of the West Indian manatee, *Trichechus manatus*. Comparative Biochemistry and Physiology 88B:47–49.

Cahn, A. R. 1940. Manatees and the Florida freeze. Journal of Mammalogy 21:222–23.

Campbell, H. W., and A. B. Irvine 1981. Manatee mortality during the unusually cold winter of 1977–1978. Pages 86–91 *in* R. L. Brownell, Jr. and K. Ralls, eds. Proceedings of the West Indian manatee workshop.

Cashman, M. E., T. L. Ness, W. B. Roess, W. G. Bradley, and J. E. Reynolds III. 1996. Isolation and characterization of cDNA encoding interleukin 2 from the Florida manatee, *Trichechus manatus latirostris*. Marine Mammal Science 12:89–98.

Cohen, J. L., G. S. Tucker, and D. K. Odell. 1982. The photoreceptors of the West Indian manatee. Journal of Morphology 173:197–202.

Costa, D. P., and T. M. Williams. 1999. Marine mammal energetics. Pages 176–217 *in* J. E. Reynolds III and S. A. Rommel, eds. Biology of marine mammals. Smithsonian Institution Press, Washington, DC.

Dailey, M. D. 2001. Infectious diseases of marine mammals: Parasitic diseases. Pages 357–79 *in* L. A. Dierauf and F. M. D. Gulland, eds. CRC handbook of marine mammal medicine, 2nd ed. CRC Press, Boca Raton, FL.

Darwin, C. 1854. Monograph on the subclass Cirripedia., Vol. 2. Royal Society of London.

Dekker, D. 1977. Zeekoegeborte. Artis 23:111–19.

Deutsch, C. J., R. K. Bonde, and J. P. Reid. 1998. Radio-tracking manatees from land and space: Tag design, implementation, and lessons learned from long-term study. Marine Technology Society Journal 32:18–29.

Deutsch, C. J., J. P. Reid, R. K. Bonde, D. E. Easton, H. I. Kochman, and T. J. O'Shea. 2000. Seasonal movements, migratory behavior, and site fidelity of West Indian manatees along the Atlantic coast of the United States as determined by radio-telemetry (Work Order No. 163). Unpublished report. Florida Cooperative Fish and Wildlife Research Unit, U.S. Geological Survey and University of Florida.

Domning, D. P. 1978. The myology of the Amazonian manatee, *Trichechus inunguis* (Natterer) (Mammalia: Sirenia). Acta Amazonica 8 (Supplement 1): 1–81.

Domning, D. P., and D. de Buffrénil. 1991. Hydrostasis in the Sirenia: Quantitative data and functional comparisons. Marine Mammal Science 7:331–68.

Domning, D. P., and L.-A. C. Hayek. 1986. Interspecific and intraspecific morphological variation in manatees (Sirenia: *Trichechus*). Marine Mammal Science 2:87–144.

Domning, D. P., and D. Magor. 1977. Taxa de substituicão horizontal de dentes no peixe-boi. Acta Amazonica 7:435–38.

Duffield, D. A., and W. Amos. 2001. Genetic analyses. Pages 271–81 *in* L. A. Dierauf and F. M. D. Gulland, eds. CRC handbook of marine mammal medicine, 2nd ed. CRC Press, Boca Raton, FL.

Duignan, P. J., C. House, M. T. Walsh, T. Campbell, G. D. Bossart, N. Duffy, P. J. Fernandes, B. K. Rima, S. Wright, and J. R. Geraci. 1995. Morbillivirus infection in manatees. Marine Mammal Science 11:441–51.

Elsner, R. 1969. Cardiovascular adjustments to diving. Pages 117–45 *in* H. T. Andersen, ed. The biology of marine mammals. Academic Press, New York.

Elsner, R. 1999. Living in the water: Solutions to physiological problems. Pages 73–116 *in* J. E. Reynolds III and S. A. Rommel, eds. Biology of marine mammals. Smithsonian Institution Press, Washington, DC.

Ewing, R., G. D. Bossart, and M. Lowe. 1997. Cutaneous viral papillomatosis in a West Indian manatee (*Trichechus manatus latirostris*) [Abstract]. *In* 46th annual Wildlife Disease Association Conference. St. Petersburg, FL.

Farmer, M., R. E. Weber, J. Bonaventura, R. C. Best, and D. P. Domning. 1979. Functional properties of hemoglobin and whole blood in an aquatic mammal, the Amazonian manatee (*Trichechus inunguis*). Comparative Biochemistry and Physiology 62A:231–38.

Fawcett, D. W. 1942a. The amedullary bones of the Florida manatee (*Trichechus latirostris*). American Journal of Anatomy 71:271–309.

Fawcett, D. W. 1942b. A comparative study of blood-vascular bundles in the Florida manatee (*Trichechus latirostris*) and in certain cetaceans and edentates. Journal of Morphology 71:105–33.

Forrester, D. J., F. H. White, J. C. Woodard, and N. P. Thompson. 1975. Intussception in a Florida manatee. Journal of Wildlife Disease 11:566–68.

Garcia-Rodriguez, A. I., B. W. Bowen, D. P. Domning, A. A. Mignucci-Giannoni, M. Marmontel, R. A. Montoya-Ospina, B. Morales-Vela, M. Rudin, R. K. Bonde, and P. M. McGuire. 1998. Phylogeography of the West Indian manatee (*Trichechus manatus*): How many populations and how many taxa? Molecular Ecology 7:1137–49.

Gerstein, E. R.,L. Gerstein, S. E. Forsythe, and J. E. Blue. 1999. The underwater audiogram of the West Indian manatee (*Trichechus manatus*). Journal of the Acoustical Society of America 105:3575–83.

Griebel, U., and A. Schmid. 1996. Color vision in the manatee (*Trichechus manatus*). Vision Research 36:2747–3757.

Griebel, U., and A. Schmid. 1997. Brightness discrimination ability in the West Indian manatee (*Trichechus manatus*). Journal of Experimental Biology 200:1587–92.

Gunter, G. 1941. Occurrence of the manatee in the United States, with records from Texas. Journal of Mammalogy 22:60–64.

Gunter, G. 1954. Mammals of the Gulf of Mexico. Pages 543–551 *in* P. Galtsoff, ed. Gulf of Mexico, its origin, waters and marine life. Fishery Bulletin 55.

Guyana National Science Research Council. 1974. An international centre for manatee research. Report of a workshop held 7–13 February 1974, Georgetown, Guyana. National Science Research Council, Guyana (Georgetown).

Harlan, R. 1824. On a species of lamantin (*Manatus latirostris* n. s.) resembling the *Manatus senegalensis* (Cuvier) inhabiting the east coast of Florida. Journal of the Philadelphia Academy of Natural Sciences 3:390–94.

Harrison, R. J. 1969. Endocrine organs: Hypophysis, thyroid, and adrenal. Pages 349–90 *in* H. T. Andersen, ed. The biology of marine mammals. Academic Press, New York.

Harrison, R. J., and J. E. King. 1965. Marine mammals. Hutchinson, London.

Hartman, D. S. 1971. Behavior and ecology of the Florida manatee, *Trichechus manatus latirostris* (Harlan), at Crystal River, Citrus County. Ph.D. Dissertation, Cornell University, Ithaca, NY.

Hartman, D. S. 1974. Distribution, status and conservation of the manatee in the United States (Final Report, Contract 14-16-0008-748). U.S. Fish and Wildlife Service, National Fish and Wildlife Laboratory.

Hartman, D. S. 1979. Ecology and behavior of the manatee (*Trichechus manatus*) in Florida (Special Publication 5). American Society of Mammalogists, Lawrence, KS.

Hatt, R. T. 1934. A manatee collected by the American Museum Congo expedition, with observations on the recent manatees. Bulletin of the American Museum of Natural History 66:533–66.

Hernandez, P., J. E. Reynolds III, H. Marsh, and M. Marmontel. 1995. Age and seasonality in spermatogenesis of Florida manatees. Pages 84–97 *in* T. J. O'Shea, B. B. Ackerman, and H. F. Percival, eds. Population biology of the Florida manatee. (Information and Technology Report 1). U.S. Department of the Interior, National Biological Service, Washington, DC.

Hill, D. A., and J. E. Reynolds, III. 1989. Gross and microscopic anatomy of the kidney of the West Indian manatee, *Trichechus manatus* (Mammalia: Sirenia). Acta Anatomica 135:53–56.

Hill, W. C. O. 1945. Notes on the dissection of two dugongs. Journal of Mammalogy 26:153–75.

Hinkley, R. E., A. Schneyer, J. Reynolds, and D. K. Odell. 1979. Structural organization of the pancreas of the West Indian manatee [Abstract]. *In* 3rd biennial conference on the biology and conservation of marine mammals. Seattle, WA.

Humes, A. G. 1964. *Harpacticus pulex,* a new species of copepod from the skin of a porpoise and a manatee in Florida. Bulletin of Marine Science of the Gulf and Caribbean 14:517–28.

Husar, S. L. 1978. *Trichechus manatus*. Mammalian Species 93:1–5.

Irvine, A. B., and H. W. Campbell. 1978. Aerial census of the West Indian manatee, *Trichechus manatus,* in the southeastern United States. Journal of Mammalogy 59:613–17.

Irving, L. 1969. Temperature regulation in marine mammals. Pages 147–74 *in* H. T. Andersen, ed. The biology of marine mammals. Academic Press, New York.

Jones, J. K. Jr., and R. R. Johnson. 1967. Sirenians. Pages 366–373 *in* S. Anderson and J. K. Jones, eds. Recent mammals of the World: A Synopsis of Families. Ronald Press Co., New York.

Kaiser, H. E. 1974. Morphology of the Sirenia: A macroscopic and x-ray atlas of the osteology of recent species. S. Karger, Basel, Switzerland.

Ketten, D. R., D. K. Odell, and D. P. Domning. 1992. Structure, function, and adaptation of the manatee ear. Pages 77–95 *in* J. Thomas, ed. marine mammal sensory systems. Plenum Press, New York.

Krumholz, L. A. 1943. Notes on manatees in Florida waters. Journal of Mammalogy 24:272–73.

Lander, M. E., A. J. Westgate, R. K. Bonde, and M. J. Murray. 2001. Tagging and tracking. Pages 851–80 *in* L. A. Dierauf and F. M. D. Gulland, eds. CRC handbook of marine mammal medicine, 2nd ed. CRC Press, Boca Raton, FL.

Langtimm, C. A., T. J. O'Shea, R. Pradel, and C. A. Beck. 1998. Estimates of annual survival probabilities for adult Florida manatees (*Trichechus manatus latirostris*). Ecology 79:981–97.

Larkin, I. L. V. 2000. Reproductive endocrinology of the Florida manatee (*Trichechus manatus latirostris*): Estrous cycles, seasonal patterns and behavior. Ph.D. Dissertation, University of Florida, Gainesville.

Layne, J. N. 1965. Observations on marine mammals in Florida waters. Bulletin of the Florida State Museum, Biological Sciences 9:131–81.

Ledder, D. A. 1986. Food habits of the West Indian manatee (*Trichechus manatus latirostris*) in south Florida. M.S. Thesis, University of Miami, Coral Gables, FL.

Lefebvre, L. W., M. Marmontel, J. P. Reid, G. B. Rathbun, and D. P Domning. 2001. Status and biography of the West Indian 2001. Pages 425–74 *in* C. A. Woods and F. F. Sergile, eds. Biogeography of the West Indies: Patterns and perspectives. CRC Press, Boca Raton, FL.

Lemire, M. 1968. Particularités de l'estomac du lamantin *Trichechus manatus* Link (Sireniens, Trichechides). Mammalia 32:475–524.

Lomolino, M. V. 1977. The ecological role of the Florida manatee (*Trichechus manatus latirostris*) in water hyacinth-dominated ecosystems. M.S. Thesis. University of Florida, Gainesville.

Mackay-Sim, A., D. Duvall, and B. M. Graves. 1985. The West Indian manatee, *Trichechus manatus,* lacks a vomeronasal organ. Brain, Behavior and Evolution, 27:186–94.

Maluf, N. S. R. 1989. Renal anatomy of the manatee, *Trichechus manatus* Linnaeus. American Journal of Anatomy 184:269–86.

Marmontel, M. 1988. The reproductive anatomy of the female manatee, *Trichechus manatus latirostris* (Linnaeus 1758) based on gross and histologic observations. M.S. thesis, University of Miami, Coral Gables, FL.

Marmontel, M. 1995. Age and reproduction in female Florida manatees. Pages 98–119 *in* T. J. O'Shea, B. B. Ackerman, and H. F. Percival, eds. Population biology of the Florida manatee (Information and Technology Report 1). U.S. Department of the Interior, National Biological Service, Washington, DC.

Marmontel, M., D. K. Odell, and J. E. Reynolds, III. 1992. Reproductive biology of South American manatees. Pages 295–312 *in* W. C. Hamlett, ed. Reproductive biology of South American vertebrates. Springer-verlag, New York.

Marmontel, M., T. J. O'Shea, H. I. Kochman, and S. R. Humphrey. 1996. Age determination in manatees using growth-layer-group counts in bone. Marine Mammal Science 12:54–88.

Marmontel, M., S. R. Humphrey, and T. J. O'Shea. 1997. Population viability analysis of the Florida manatee (*Trichechus manatus latirostris*). Conservation Biology 11:467–81.

Marshall, C. D., P. S. Kublis, G. D. Huth, V. M. Edmonds, D. L. Halin, and R. L. Reep. 2000. Food-handling ability and feeding-cycle length of manatees feeding on several species of aquatic plants. Journal of Mammalogy 81:649–58.

Mass, A. M., D. K. Odell, D. R. Ketten, and A. Ya. Supin. 1997. Ganglion layer topography and retinal resolution of the Caribbean manatee *Trichechus manatus latirostris*. Doklady Biological Sciences 355:392–94.

Maynard, C. J. 1872. Catalogue of the mammals of Florida, with notes on their habits, distribution, etc. Bulletin of the Essex Institute 4:135–50.

McClenaghan, Jr., L. R., and T. J. O'Shea. 1988. Genetic variability in the Florida manatee (*Trichechus manatus*). Journal of Mammalogy 69:481–88.

Mignucci-Giannoni, A. A., C. A. Beck, R. A. Montoya-Ospina, and E. H. Williams, Jr. 1999. Parasites and commensals of the West Indian manatee from Puerto Rico. Journal of the Helminthological Society of Washington 66:67–69.

Miller, D. L., R. Y. Ewing, and G. D. Bossart. 2001. Emerging and resurging diseases. Pages 15–30 *in* L. A. Dierauf and F. M. D. Gulland, eds. CRC handbook of marine mammal medicine, 2nd ed. CRC Press, Boca Raton, FL.

Miller, W. A., G. D. Sanson, and D. K. Odell. 1980. Molar progression in the manatee (*Trichechus manatus*). Anatomical Record 196:128A.

Mitchell, J. 1973. Determination of relative age in the dugong *Dugong dugon* (Müller) from a study of skulls and teeth. Zoological Journal of the Linnaean Society 53:1–23.

Mitchell, J. 1976. Age determination in the dugong, *Dugong dugon* (Müller). Biological Conservation 9:25–28.

Mitchell, J. 1978. Age growth layers in the dentine of dugong incisors (*Dugong dugon* [Müller]) and their application to age determination. Zoological Journal of the Linnaean Society 62:317–48.

Moore, J. C. 1951a. The status of the manatee in the Everglades National Park, with notes on its natural history. Journal of Mammalogy 32:22–36

Moore, J. C. 1951b. The range of the Florida manatee. Quarterly Journal of the Florida Academy of Sciences 14:1–19.

Mossman, H. W., and K. L. Duke. 1973. Comparative morphology of the mammalian ovary. University of Wisconsin Press, Madison.

Murie, J. 1872. On the form and structure of the manatee (*Manatus americanus*). Transactions of the Zoological Society of London 8:127–202.

Murray, R. M., H. Marsh, G. E. Heinsohn, and A. V. Spain. 1977. The role of the mid-gut caecum and the large intestine in the digestion of sea grasses by the dugong (Mammalia: Sirenia). Comparative Biochemistry and Physiology 56A:7–10.

Odell, D. K. 1977. Age determination and biology of the manatee (Final Report, Contract 14-16-0008-930.). U.S. Fish and Wildlife Service, National Fish and Wildlife Laboratory.

Odell, D. K. 1979. Distribution and abundance of marine mammals in the waters of the Everglades National Park. Pages 673–78 *in* R. M. Linn, ed. Proceedings of the first conference on scientific research in the national parks (Transactions and Proceedings Series 5). U.S. Department of the Interior, National Park Service.

Odell, D. K. 1981. Growth of a West Indian manatee (*Trichechus manatus*) born in captivity. Pages 131–140 *in* R. L. Brownell, Jr. and K. Ralls, eds. Proceedings of the West Indian manatee workshop.

Odell, D. K. 2002. Sirenian life history. Pages 1086–88 *in* W. F. Perrin, B. Würsig, and J. G. M. Thewissen, eds. Encyclopedia of marine mammals. Academic Press, San Diego, CA.

Odell, D. K., and J. E. Reynolds. 1979. Observations on manatee mortality in south Florida. Journal of Wildlife Management 43:572–77.

Odell, D. K., D. Forrester, and E. D. Asper. 1981. Preliminary analysis of organ weights and sexual maturity in the West Indian manatee, *Trichechus manatus*. Pages 52–65 *in* R. L. Brownell, Jr. and K. Ralls, eds. Proceedings of the West Indian manatee workshop.

O'Hara, T. M., and T. J. O'Shea. 2001. Toxicology. Pages 471–520 *in* L. A. Dierauf and F. M. D. Gulland, eds. CRC handbook of marine mammal medicine, 2nd ed. CRC Press, Boca Raton, FL.

Ortiz, R. M., G. A. J. Worthy, and D. S. MacKenzie. 1998. Osmoregulation in wild and captive West Indian manatees (*Trichechus manatus*). Physiological Zoology 71:449–57.

Ortiz, R. M., G. A. J. Worthy, and F. M. Byers. 1999. Estimation of water turnover rates of captive West Indian manatees (*Trichechus manatus*) held in fresh and saltwater. Journal of Experimental Biology 202:33–38.

Ortiz, R. M., D. S. MacKenzie, and G. A. J. Worthy. 2000. Thyroid hormone concentrations in captive and free-ranging West Indian manatees (*Trichechus manatus*). Journal of Experimental Biology 203:3631–37.

O'Shea, T. J. 1999. Environmental contaminants and marine mammals. Pages 485–564 *in* J. E. Reynolds III and S. A. Rommel, eds. Biology of marine mammals. Smithsonian Institution Press, Washington, DC.

O'Shea, T. J., and W. C. Hartley. 1995. Reproduction and early-age survival of manatees at Blue Spring, upper St. Johns River, Florida. Pages 157–70 *in* T. J. O'Shea, B. B. Ackerman, and H. F. Percival, eds. Population biology of the Florida manatee (Information and Technology Report 1). U.S. Department of the Interior, National Biological Service, Washington, DC.

O'Shea, T. J., J. F. Moore, and H. I. Kochman. 1984. Contaminant concentrations in manatees (*Trichechus manatus*) in Florida. Journal of Wildlife Management 48:741–48.

O'Shea, T. J., C. A. Beck, R. K. Bonde, H. I. Kochman, and D. K. Odell. 1985. An analysis of manatee mortality patterns in Florida 1976–1981. Journal of Wildlife Management 49:1–11.

O'Shea, T. J., G. B. Rathbun, R. K. Bonde, C. D. Buergelt, and D. K. Odell. 1991. An epizootic of Florida manatees associated with a dinoflagellate bloom. Marine Mammal Science 7:165–779.

O'Shea, T. J., B. B. Ackerman, and H. F. Percival, eds. 1995. Population biology of the Florida manatee (Information and Technology Report 1). U.S. Department of the Interior, National Biological Service, Washington, DC.

O'Shea, T. J., R. R. Reeves, and A. K. Long, eds. 1999. Marine mammals and persistent ocean contaminants: Proceedings of the Marine Mammal Commission workshop. Keystone, CO.

O'Shea, T. J., L. W. Lefebvre, and C. A. Beck. 2001. Florida manatees: Perspectives on populations, pain and protection. Pages 31–43 *in* L. A. Dierauf and F. M. D. Gulland, eds. CRC handbook of marine mammal medicine, 2nd ed. CRC Press, Boca Raton, FL.

Pabst, D. A., S. A. Rommel, and W. A. McLellan. 1999. The functional morphology of marine mammals. Pages 15–72 *in* J. E. Reynolds III and S. A. Rommel, eds. Biology of marine mammals. Smithsonian Institution Press, Washington, DC.

Peterson, S. L. 1974. Man's relationship with the Florida manatee, *Trichechus manatus latirostris* (Harlan): An historical perspective. M.A. Thesis, University of Michigan, Ann Arbor.

Powell, J. A., Jr. 1978. Evidence of carnivory in manatees (*Trichechus manatus*). Journal of Mammalogy 59:442.

Quiring, D. P., and C. F. Harlan. 1953. On the anatomy of the manatee. Journal of Mammalogy 4:193–203.

Rathbun, G. B., J. P. Reid, and G. Carowan. 1990. Distribution and movement patterns of manatees (*Trichechus manatus*) in northwestern peninsular Florida (Florida Marine Research Publications No. 48). Florida Marine Research Institute, Florida Department of Natural Resources, St. Petersburg.

Rathbun, G. B., J. P. Reid, R. K. Bonde, and J. A. Powell. 1995. Reproduction in free-ranging manatees. Pages 135–56 *in* T. J. O'Shea, B. B. Ackerman, and H. F. Percival, eds. Population biology of the Florida manatee (Information and Technology Report 1). U.S. Department of the Interior, National Biological Service, Washington, DC.

Reep, R. L., and T. J. O'Shea. 1990. Regional brain morphometry and lissencephaly in the Sirenia. Brain, Behavior and Evolution 35:185–94.

Reep, R. L., J. I. Johnson, R. C. Switzer, and W. I. Welker. 1989. Manatee cerebral cortex: Cytoarchitecture of the frontal region in *Trichechus manatus latirostris*. Brain, Behavior and Evolution 34:365–86.

Reid, J. P. 1996. Chessie the manatee: From Florida to Rhode Island. Argos Newsletter 51:13.

Reid, J. P., R. K. Bonde, and T. J. O'Shea. 1995. Reproduction and mortality of radio-tagged and recognizable manatees on the Atlantic coast of Florida. Pages 171–91 *in* T. J. O'Shea, B. B. Ackerman, and H. F. Percival, eds. Population biology of the Florida manatee (Information and Technology Report 1). U.S. Department of the Interior, National Biological Service, Washington, DC.

Reynolds, J. E., III. 1977. Aspects of the social behavior and herd structure of a semiisolated colony of Florida manatees (*Trichechus manatus*). M.S. Thesis, University of Miami, Coral Gables, FL.

Reynolds, J. E., III. 1979. Internal and external morphology of' the manatee (sea cow). Anatomical Record 193:663.

Reynolds, J. E., III. 1980. Aspects of the structural and functional anatomy of the gastrointestinal tract of the West Indian manatee, *Trichechus manatus*. Ph.D. Dissertation, University of Miami, Coral Gables, FL.

Reynolds, J. E., III. 1981a. Behavior patterns of the West Indian manatee, with emphasis on feeding and diving. Florida Scientist 44:233–242.

Reynolds, J. E., III. 1981b. Aspects of the social behaviour and herd structure of a semi-isolated colony of West Indian manatees, *Trichechus manatus*. Mammalia 45:431–51.

Reynolds, J. E., III. and D. K. Odell. 1991. Manatees and dugongs. Facts on File, New York.

Reynolds, J. E., III. and S. A. Rommel. 1996. Structure and function of the gastrointestinal tract of the Florida manatee, *Trichechus manatus latirostris*. Anatomical Record 245:539–58.

Reynolds, J. E., III., D. K. Odell, and S. A. Rommel. 1999. Marine mammals of the world. Pages 1–14 *in* J. E. Reynolds III. and S. A. Rommel, eds. Biology of marine mammals. Smithsonian Institution Press, Washington, DC.

Rommel, S. A., and J. Lowenstine. 2001. Anatomy and physiology of marine mammals: Gross and microscopic anatomy. Pages 129–64 *in* L. A. Dierauf and F. M. D. Gulland, eds. CRC handbook of marine mammal medicine, 2nd ed. CRC Press, Boca Raton, FL.

Ronald, K. L., J. Selley, and E. C. Amoroso. 1978. Biological synopsis of the manatee (IDRC-TSI3e). International Development Research Centre, Ottawa.

Ross, A., and W A. Newman. 1967. Eocene Balanidae of Florida, including a new genus and species with a unique plan of "turtle-barnacle" organization. American Museum Novitates 2288:1–21.

Schevill, W. E., and W A. Watkins. 1965. Underwater calls of *Trichechus* (manatee). Nature 205:373–74.

Schneyer, A., R. Hinkley, and D. Odell. 1979. Structural organization of the anterior pituitary gland of the West Indian manatees [Abstract]. *In* 3rd biennial conference on the biology and conservation of marine mammals. Seattle, WA.

Scholander, P. F., and L. Irving. 1941. Experimental investigations on the respiration and diving of the Florida manatee. Journal of Cellular and Comparative Physiology 17:169–91.

Sguros, P. L. 1966. Research report and extension proposal submitted to the Central and Southern Florida Flood Control Board on the use of the Florida manatee as an agent for suppression of aquatic and bankweed growth in essential inland waterways. Department of Biological Science, Florida Atlantic University, Boca Raton.

Stubbings, H. G. 1965. West African Cirripedia in the collections of the Institute Francais d'Afrique Noire, Dakar, Senegal. Bulletin IFAN 27:876–907.

Upton, S. J., D. K. Odell, G. D. Bossart, and M. T. Walsh. 1989. Description of the oocysts of two new species of *Eimeria* (Apicomplexa: Eimeriidae) from the Florida manatee, *Trichechus manatus* (Sirenia: Trichechidae). Journal of Protozoology 36:87–90.

U.S. Fish and Wildlife Service. 1978. Administration of the Marine Mammal Protection Act of 1972, June 22, 1977 to March 31, 1978. U.S. Fish and Wildlife Service, Washington, DC.

U.S. Fish and Wildlife Service. 2001. Florida Manatee Recovery Plan. 3rd Revision. 30 October 2001. U.S. Fish and Wildlife Service, Southeast Region, Atlanta, GA.

Verhaart, W. J. C. 1972. The brain of the seacow *Trichechus*. Psychiatria, Neurologia Neurochirgeria (Amsterdam) 75:271–92.

Vietmeyer, N., ed. 1976. Making aquatic weeds useful: Some perspectives for developing countries. National Academy of Sciences, Washington, DC.

Walsh, M. T., and G. D. Bossart. 1999. Manatee medicine. Pages 507–16 *in* M. E. Fowler and R. E. Miller, eds. Zoo and wildlife medicine, W. B. Saunders, Philadelphia.

Walsh, M. T., O. T. Oftedal, G. A. J. Worthy, Q. R. Rodgers, S. M. Innis, and T. W. Campbell. 1996. Manatee milk analyses: Changes throughout nursing. Page 67 *in* Proceedings of the 27th annual conference of the International Association for Aquatic Animal Medicine. Chattanooga, TN.

Wartzok, D., and D. R. Ketten. 1999. Marine mammal sensory systems. Pages 117–75 *in* J. E. Reynolds III and S. A. Rommel, eds. Biology of marine mammals. Smithsonian Institution Press, Washington, DC.

Wells, R. S., D. J. Boness, and G. B. Rathbun. 1999. Behavior. Pages 324–422 *in* J. E. Reynolds III and S. A. Rommel, eds. Biology of marine mammals. Smithsonian Institution Press, Washington, DC.

White, J. R., D. R. Harkness, R. E. Isaaks, and D. A. Duffield. 1976. Some studies on blood of the Florida manatee, *Trichechus manatus latirostris*. Comparative Biochemistry and Physiology 55A:413–17.

Wislocki, G. B. 1935a. The lungs of the manatee (*Trichechus manatus*) compared with those of older aquatic mammals. Biological Bulletin 68:385–96.

Wislocki, G. B. 1935b. The placentation of the manatee (*Trichechus latirostris*). Memoirs of the Museum of Comparative Zoology, Harvard University 54:159–78.

Worthy, G. A. J. 2001. Nutrition and energetics. Pages 791–827 *in* L. A. Dierauf and F. M. D. Gulland, eds. CRC handbook of marine mammal medicine, 2nd ed. CRC Press, Boca Raton, FL.

Wright, S. D., B. B. Ackerman, R. K. Bonde, C. A. Beck, and D. J. Banowetz. 1995. Analysis of watercraft-related mortality of manatees in Florida, 1979–1991. Pages 259–68 *in* T. J. O'Shea, B. B. Ackerman, and H. F. Percival, eds. Population biology of the Florida manatee (Information and Technology Report 1). U.S. Department of the Interior, National Biological Service, Washington, DC.

DANIEL K. ODELL, Hubbs-SeaWorld Research Institute, Orlando, Florida 32821-8043. Email: dodell@hswri.org.

VII

Hoofed Mammals

41

Collared Peccary

Tayassu tajacu

Eric C. Hellgren
John A. Bissonette

NOMENCLATURE

COMMON NAMES. Collared peccary, javelina
SCIENTIFIC NAME. *Tayassu tajacu* (Linnaeus 1758)
SUBSPECIES. *T. t. angulatus, T. t. bangsi, T. t. crassus, T. t. crusnigrum, T. t. humeralis, T. t. nanus, T. t., nelsoni, T. t. nigrescens, T. t. sonoriensis,* and *T. t. yucatanensis.*

Subspecies designations are from Hall (1981) for North and Central American forms only, with a change from *Dicotyles tajacu* in Bissonette (1982b) to *Tayassu tajacu* based on a review by Donkin (1985). The subspecies *T. t. angulatus* occurs in Texas and possibly southeastern New Mexico, and *T. t. sonoriensis* in Arizona and southwestern New Mexico (Mearns 1907; Miller and Kellogg 1955). The collared, white-lipped (*T. pecari*), and Chacoan (*Catagonus wagneri*) peccaries compose the three living species in the family Tayassuidae (or Dicotylidae, which takes nomenclatural precedence; Donkin 1985). Only *T. tajacu* is found in North America. All three species are sympatric in the Chaco region of Paraguay (Wetzel and Lovett 1974; Wetzel et al. 1975; Mayer and Brandt 1982). See Theimer and Keim (1998) for a phylogenetic evaluation of the extant tayassuids as two clades of three genera.

DISTRIBUTION

Collared peccaries occur in a range of habitats from Argentina northward to Texas, New Mexico, and Arizona (Sowls 1978). These habitats include thornscrub (or chaparral) in southern Texas, oak woodland and desert scrub in western Texas and Arizona, and tropical wet and dry forests in Central and South America. Peccaries inhabit the southern part of Texas from the Gulf Coast county of Refugio, northwest to southern Nolan County, southwest to northern Ward County, northwest to the New Mexico border in central Loving County, west along the border to western Culberson County, and south to the Mexican border in southeastern Hudspeth County (Davis and Schmidly 1994) (Fig. 41.1). Highest densities are in the Rio Grande Plains of southern Texas. Recent introductions of peccaries have been made in several northern Texas counties bordering the Red River (Davis and Schmidly 1994). Peccary distributions are disjunct around the larger metropolitan areas.

Arizona populations are primarily in the southeastern quarter of the state, and are generally contiguous with those in southwestern New Mexico. From the southeastern corner of the state in Cochise County, the distribution extends northwesterly into Graham, Pinal, Gila, Santa Cruz, Pima, and Yavapai Counties (Arizona Game and Fish Department 2000). The northern edge of the distribution lies at the Coconino–Yavapai County line at the southern tip of the Kaibab National Forest. Peccaries are distributed in northern and southeastern Maricopa County around the Phoenix metropolitan area. They also have become abundant in southeastern Mohave County. The range is slowly expanding north and west.

In New Mexico, peccaries occur mainly in the southern counties. Based on harvest, they are likely most abundant in the southwestern counties of Hidalgo, Grant, and Luna (the former two counties are adjacent to Arizona). They also occur in parts of Socorro, Catron, and Sierra Counties, and are common in the Gila National Forest of west-central New Mexico. There are isolated populations of peccaries in the Cedar, Tres Hermanas, Florida, and San Andres Mountains in south-central New Mexico; they occur in the southeastern counties of Eddy and Lea near the Texas border (Findley et al. 1975; B. Thompson, USGS/New Mexico State University, pers. commun., 2000).

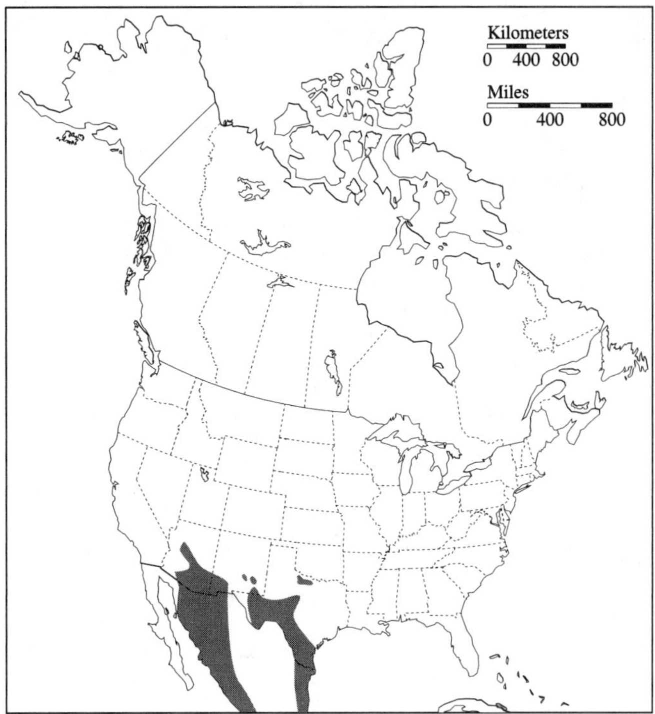

FIGURE 41.1. Distribution of the collared peccary (*Tayassu tajacu*). SOURCE: Data from Findley et al. (1975), Davis and Schmidly (1994), Arizona Game and Fish Department (2000).

Peccaries in the northernmost parts of their range in all three states, but especially in the mountains of western Texas, New Mexico, and Arizona, are subject to occasional periods of snow and very cold weather and probably undergo repeated local extirpation and repopulation. No work has been conducted to document the dynamics of this phenomenon.

The distribution of the collared peccary in Mexico is very extensive, ranging along both sets of coastal states in northern Mexico and throughout southern Mexico, including Yucatan (Donkin 1985). The high plateaus of north-central Mexico are not populated by peccaries, except in deep canyons such as in Durango. Populations in the thornscrub communities of Coahuila, Tamalipas, and San Luis Potosi are continuous with those of southern Texas (Donkin 1985).

FIGURE 41.2. Collared peccary (*Tayassu tajacu*). SOURCE: Photo by Eric Hellgren.

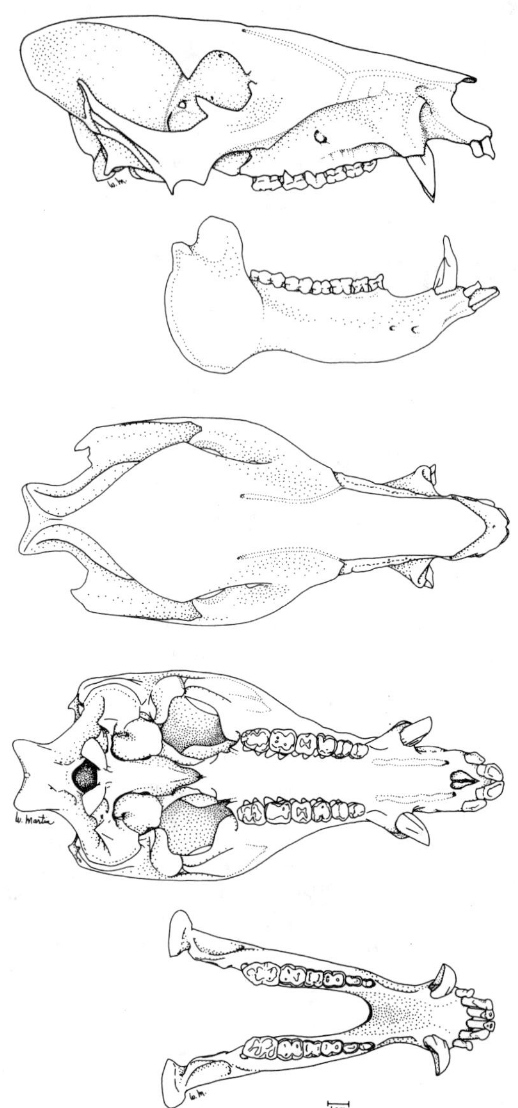

FIGURE 41.3. Skull of the collared peccary (*Tayassu tajacu*). From top to bottom: lateral view of cranium, lateral view of mandible, dorsal view of cranium, ventral view of cranium, dorsal view of mandible.

DESCRIPTION

Peccaries are small ungulates with short legs and a piglike snout (Fig. 41.2). The tail is about 30 mm long (Sowls 1966) and barely noticeable. Total length of adult animals is 86–96 cm. Height at the shoulder ranges from 50 to 60 cm. The front feet have four digits and the hind feet have three digits, unlike suids, which have four digits on each foot. Most Chacoan peccaries have only two digits on the hind foot (Wetzel 1977b). Weights of adult animals average 20 kg, with a range from 13 to 31 kg. From the side view, the head and forequarters appear overly large for the delicate legs and hindquarters. Collared peccaries are sexually monomorphic in general appearance and body mass, but subtle differences occur for tail length (longer in females), dorsal scent gland (larger in males), and skeletal mass of forequarters (heavier in males) (Lochmiller et al. 1986a, 1987b). No sexual dimorphism in coloration has been reported.

Peccaries have enlarged upper and lower canines approximately 30–35 mm in length. Only the tips of the upper tusks protrude beyond the lips. Unlike those of wild suids, the tusks are not sharply curved and flared, but are straight or slightly curved. Four pairs of mammae are apparent, but only the two posterior pairs appear to be functional (Neal 1959; Sowls 1966). Sowls (1965) found that the second anterior pair of mammae secreted milk through galactophores (milk ducts). Peccaries have a subcutaneous scent gland, described by Epling (1956), located on the dorsal midline 20–25 cm anterior to the tail. The scent gland is used in several behaviors and is one of the few sexually dimorphic characters in collared peccaries; as noted, it is larger in males (Lochmiller et al. 1986a).

In adults, the pelage is light gray or brown to almost black with a salt-and-pepper effect caused by distinctive banding of individual bristles. Occasionally, reddish-colored adults are seen. When the dark-colored dorsal bristles are erect, a white stripe formed by the coordination of the white bands of individual bristles is apparent (Schweinsburg 1969). A light-colored collar, for which the species is named, is variable in shade and pattern.

Individual bristles are bicolored with white or beige, and black or dark brown bands (Wetzel 1977a). There usually are four to six distinct bands for each hair shaft. Band width is variable and appears to be related to the location of the bristle either on the collar, dorsal midline, or body. Bristles surrounding the dorsal scent gland are often stained yellow. Peccaries <2–3 months old are reddish brown with a dark dorsal midline stripe and distinctive collar.

Skull and Dentition. Adult peccary skulls are roughly triangular shaped in lateral view (Fig. 41.3). The rostrum slopes upward to a prominent lambdoidal crest. The zygomatic arches are thick and flare posteriorly. The mandible is sturdy. Herring (1974) discovered that the sequence of suture closure in peccaries differs from that of most other mammals in the early fusion of the palatal and facial sutures. She postulated that the early fusion strengthens the snout and is related to rooting and feeding activities.

The peccary dental formula is I 2/3, C 1/1, P 3/3, M 3/3. The upper and lower canines are well developed in adults of both sexes. The posterior portion of the lower canines occludes with the anterior portion of the uppers, resulting in flat contact surfaces with sharp edges. In old animals, canines are well worn, often <15 mm long. Molars are bunodont and have changed little from ancestral times, in contrast to suid molars, which have changed substantially from their original pattern (Herring 1972). Diastemas are present between the second upper incisors and the canines, and between the upper and lower canines and anterior premolars. Herring (1972) compared the functional anatomy of tayassuid and suid skulls and dentition and discussed the role of canine morphology in suid–tayassuid evolution. She postulated that peccary canines evolved for use as weapons. However, Woodburne (1968) found canines were slightly longer and wider in males, leading Bissonette (1982b) to posit that canines functioned in competitive interactions between males. A third postulated function of canines is to stabilize the jaw for breaking tough tropical nuts (Kiltie 1981).

PHYSIOLOGY

Peccaries are not a truly xeric-adapted species (Zervanos and Hadley 1973), having evolved in mesic environments (Byers and Bekoff 1981), but they can survive for periods of up to 6 days without water. Turnover rates for peccaries with access to water averaged 1.58 L of water/day, whereas dehydrated peccaries averaged 0.55 L/day, indicating adaptive physiological mechanisms for conserving fluids under stress. Free-ranging peccaries lost 1.35 L of water/day during the summer and 1.17 L in winter (Zervanos and Day 1977). Zervanos and Hadley (1973) demonstrated that respiratory evaporation was the main avenue of water loss. During dehydration, peccaries reduced evaporative water loss by 68% and urinary water loss by 93%. Peccaries also have relatively larger kidneys, thicker kidney medullae, and a greater capacity to concentrate urine than sympatric feral pigs. This difference may provide a competitive advantage in water-limited semiarid environments (Gabor et al. 1997).

Peccaries seem capable of maintaining full hydration in desert environments without free water. Minnamon (1962) reported that absence of water did not alter home range movements or activities. Prickly pear (*Opuntia* sp.) cladophylls (pads) appear to supply most water requirements. Zervanos and Day (1977) reported that an 18.2-kg peccary required approximately 1.5 kg wet weight (0.3 kg dry weight) of *Opuntia* (78% water by weight) per day to maintain water balance. In addition, Zervanos and Hadley (1973) reported skin temperatures of free-ranging peccaries were labile and ranged from 37.5°C to 40.9°C during all seasons. Skin temperatures were always above ambient temperatures (Zervanos and Day 1977).

Daily energy requirements are higher in winter (435 kJ/kg$^{0.75}$/day) than in summer (376 kJ/kg$^{0.75}$/day) because of added thermoregulatory costs in winter and a possible change in basal metabolism (Zervanos and Hadley 1973). Cardiac output of five captive animals was estimated to be 120 ml/min/kg with mean arterial blood pressure of 139 mm Hg and heart rate of 121 beats/min (Schilling and Stone 1969). Nitrogen requirements are 0.81–0.84 g N/kg$^{0.75}$/day, which corresponds to 6.8% crude protein at dry matter intake rates (Gallagher et al. 1984; Carl and Brown 1985).

REPRODUCTION

Male Biology. The dominant, or alpha, male copulates with most of the estrous females in the herd (Bissonette 1982b; but see Byers and Bekoff 1981 for an alternative view). Bissonette (1982a) reported that the alpha male was involved in five of six successful intromissions observed in free-ranging peccaries. In captivity, dominant males readily confront or attack subordinate males attempting to breed (Lochmiller et al. 1984). Dominant males also mounted females during fertile periods more often than expected (Packard et al. 1991). Subordinates may be successful in copulating only when two or more females are in estrus synchronously and perhaps only when ovulation is synchronized (Packard et al. 1991). During the breeding season, an adult male peccary forms a short tending bond with the estrous female that lasts from a few hours to several days (Low 1970; Bissonette 1976). During this time, the male remains within a few meters and does not allow other males near the female; however, other females may approach. Subordinate males do not leave the group and do not form separate male groups. McCulloch (1955) and Sowls (1966) reported that conflict among males was not noticeably greater when receptive females were present than at other times, perhaps because of prior establishment of a dominance hierarchy (Sowls 1984). Although no clear yearly pattern of interaction rates is apparent, there is a tendency for more interactions to occur just before and during parturition in early summer and during the breeding season in late fall and winter. Highly aggressive interactions such as fights were typically observed in western Texas peccaries only when a female was in estrus (Bissonette 1982a).

Captive male peccaries in Texas exhibited a low-amplitude rhythm in serum testosterone concentrations, testicular size, and scrotal circumference, with lowest levels of all variables in summer (Hellgren et al. 1989). A similar rhythm was not seen in wild animals (Low 1970). Summer is also the period of least reproductive activity in wild peccaries in Texas (Ellisor and Harwell 1979). Nevertheless, males remain fertile year-round, based on sperm counts and activity (Hellgren et al. 1989). Puberty in males has not been established in the field, but occurs at 10–15 months of age in captive animals (Sowls 1966; Hellgren et al. 1989).

Female Biology. Collared peccaries evolved in South America (Woodburne 1968; Byers and Bekoff 1981) in what were probably mesic and relatively predictable environments. Breeding and parturition may have occurred without significant peaks throughout the year in these climates. In the unpredictable and arid climate of the American Southwest, peccaries may breed throughout the year but strong seasonal peaks in farrowing have been reported (Neal 1959; Sowls 1965, 1974; Schweinsburg 1969; Low 1970; Bissonette 1976; Hellgren et al. 1995). In southern Texas, 77% of harvested female peccaries (n = 197) were pregnant in February, with a peak parturition period predicted to be April and May (Hellgren et al. 1995) based on patterns of fetal development (Smith and Sowls 1975). Low (1970) termed February–June the obligatory period of parturition in southern Texas, and concluded that precipitation-regulated forage conditions affected the timing of winter breeding and probability of rebreeding in late spring. Only 12 of 197 female peccaries sampled in February were both lactating and pregnant in a Texas population (Hellgren et al. 1995). Most peccaries in Arizona are born in July and August, when vegetation is most abundant following rains (Knipe 1957; Neal 1959; Sowls 1966, 1978). Pregnancy rate is related not only to yearly forage conditions, but also to the success of previous parturitions that year. Females whose young are lost soon after parturition may breed again.

The estrous cycle for the collared peccary ranges from 22 to 26 days, and females can repeat the estrous cycle if they fail to become pregnant (Sowls 1966). Ovulation occurs 12–24 hr after the beginning of estrus (Coscarelli 1985) and is sensitive to nutritional condition (Lochmiller et al. 1986b).

The gestation period is about 145 days (Low 1970; Sowls 1961a, 1965), and is not affected by moderate nutritional restriction (Lochmiller et al. 1987a). Usually two young are produced (Jennings and Harris 1953; Knipe 1957; Sowls 1966, 1978; Low 1970; Smith and Sowls 1975), but litters may include three or four young. Halloran (1945) reported a female with five fetuses. Litter size tends to be smaller in yearlings than adults (Hellgren et al. 1995) and can be reduced by severe nutritional stress (Lochmiller et al. 1986b).

Sowls (1965) reported that captive female peccaries in Arizona bred as early as 33 weeks of age (9 months), and the earliest parturition observed was in a female 54 weeks old (12.5 months). Low (1970) stated the earliest observed female copulation for Texas animals occurred at 44 weeks of age (10 months); however, the earliest conception occurred at 48–49 weeks (11 months). In Texas, ovarian analysis allowed back-dating of the first corpora lutea of nonpregnancy scars to 32–33 weeks (Low 1970). In wild Texas peccaries, primiparity did not occur before 20 months of age and the youngest reported pregnant female was 17 months old (Hellgren et al. 1995). Yearling pregnancy rates range from 7% to 35% (Low 1970; Hellgren et al. 1995).

Lactation lasts 6–8 weeks (Lochmiller 1984). Sowls et al. (1961) and Brown et al. (1963) studied the chemical composition and physical properties of mature and colostrum milk and milk fat of collared peccaries. Comparisons with domestic sows indicated that peccary milk was lower in fat (4.7% vs. 6.9%) but higher in protein (5.6% vs. 4.5%). Lochmiller et al. (1985) provided additional details on milk composition, including mineral concentrations. Concentrations of calcium, phosphorus, and magnesium were lower in peccary milk than porcine milk, but sodium and potassium were higher (Lochmiller et al. 1985).

ECOLOGY

Activity. Activity patterns of peccary are strongly influenced by temperature (Bigler 1974; Bissonette 1978). In western Texas, the phenological periods of November–February (winter), March–April and September–October (spring and fall), and May–August (summer)

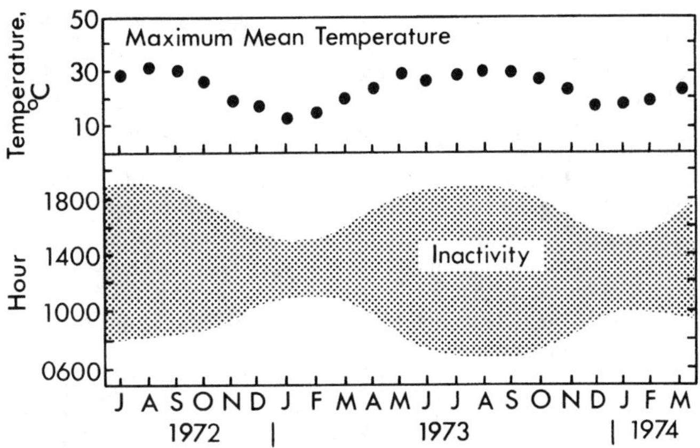

FIGURE 41.4. Relationship of ambient temperature to diurnal and crepuscular activity patterns in the peccary. SOURCE: Bissonette (1978).

were characterized by increasingly warmer temperatures; mean maximum temperatures equaled 18.8°C, 26.2°C, and 32.0°C, respectively (Bissonette 1976). During summer, peccaries were primarily crepuscular and nocturnal (Elder 1956; Schweinsburg 1969). With cooler fall and winter temperatures, herds were active for longer periods during midday (Fig. 41.4). In midwinter, when night temperatures approached freezing, they often huddled for warmth during at least part of the night (Zervanos 1972; Zervanos and Hadley 1973). In western Texas, most activity during winter was concentrated on the open bajada (open, gently sloping areas) or flat desert country. Less activity occurred on the bajada during daylight hours in summer. Spring and fall appeared to be transitional, with periods of activity and resting during day and night. Activity patterns of collared peccaries in southern Texas were very similar to those detailed above. Individuals were mainly crepuscular with slight seasonal shifts in peak activity periods (Ilse and Hellgren 1995a).

Social Organization and Herd Dynamics. Peccaries are territorial (Ellisor and Harwell 1969; Schweinsburg 1969; Bissonette 1982a) and are organized into mixed-sex herds, which appear highly stable (Fig. 41.5). Herds are composed of roughly equal numbers of males and females (Bissonette 1976; Hellgren et al. 1984), although Sowls (1974, 1978) reported some variation in sex ratio. Herd sizes appear to be larger in unhunted areas (mean = 13.6), based on data from Byers and Bekoff (1981), Bissonette (1982a), Day (1985), and Ilse and Hellgren (1995b), than in hunted areas (mean = 9.4), based on data from Knipe (1957), Sowls (1984), Day (1985), and Gabor (1997). Herds generally are larger in Texas than Arizona.

FIGURE 41.5. Herd of collared peccaries (*Tayassu tajacu*) in southern Texas. SOURCE: Photo courtesy of Timothy Gabor.

Territory size and home range size in collared peccaries are essentially equivalent. Most estimates of territory size are 1–3 km^2, although estimates as large as 6.7 km^2 in Texas (Ilse and Hellgren 1995b) and 8.1 km^2 in Arizona (Bigler 1974) have been reported. Territory size may be affected by resource availability or presence of potential competitors (Gabor and Hellgren 2000). Interestingly, in western Texas, herd sizes, but not territory size, varied with resource availability (Bissonette 1982a), whereas in southern Texas, both variables were flexible (Gabor and Hellgren 2000).

Little overlap occurs between herds and territories are defended by more than one herd member (Bissonette 1976). Day (1985) reported observing territorial defense; aggressive behaviors occur when two herds are simultaneously in the overlap zone between territories. Bissonette (1982a) also reported reversals in dominance between peccary herds as boundary lines were crossed. However, Day (1985) did not observe aggression at water sources that were shared by two or more herds. Herds were not seen more than 185 m inside the boundary of another herd (Schweinsburg 1971). There is evidence of territorial marking by peccaries using scent and fecal piles (Schweinsburg 1971; Bissonette 1982a).

Subgrouping is common among peccary herds. Intensive study of subgrouping behavior, however, has confirmed characterization of herds as stable and cohesive (Oldenburg et al. 1985). Feeding subgroups may remain apart for as long as 2 weeks (Day 1977; Bissonette 1982a), although most subgroups and single individuals are separated for <12 hr (Oldenburg et al. 1985). In southern Texas, all radio-collared members of a study herd (37–55% of adult herd members) were together 41% of the time (Oldenburg et al. 1985). Subgroups often exchange members with other subgroups or the main group (Oldenburg et al. 1985). Interchange between territorial groups was infrequent, although Schweinsburg (1969, 1971) documented its occurrence.

Subgroups change in size and composition throughout the year. These changes may be influenced by reproductive behavior, habitat, season, and temperature. In western Texas, peccaries breed primarily in late November to mid-January. Males do not herd females, yet subgroups tended to coalesce into larger territorial groups during this time (Bissonette 1982a). Small subgroups were seen less frequently at this time, a phenomenon also observed in southern Texas (Oldenburg et al. 1985). During the cool winter months in western Texas, peccaries also tended to use bajadas. Group size on the bajada in Texas was larger ($x = 13$) than that observed in dense vegetation ($x = 10$). In contrast, mean territorial herd size for western Texas peccaries over a 2-year period was 17.5 animals. Peccary subgroup and herd sizes generally were smaller when sympatric with feral pigs (Ilse and Hellgren 1995b; Gabor and Hellgren 2000) and in areas of more open habitat with fewer patches of dense cover.

Formation of new territorial groups in peccaries appears to occur by herd fission and is influenced by the existence of subgroups. In Arizona, Supplee (1981) and Day (1985) reported formation of new herds and recolonization of available habitats by subgroups splintering from larger herds. Bissonette (1982a) hypothesized that subgroups form the nucleus for new territorial groups. As subgroups spend more time apart, each group restricts its movements to a portion of the territory, and range extensions into adjacent territories become evident. Subgroups in southern Texas appeared to use parts of the territory preferentially (Oldenburg et al. 1985). Interactions between subgroups can become characterized by increasing aggression. The decisive step for territorial group formation occurs when subgroups do not rejoin during the breeding season, in effect creating an additional breeding alpha male (Bissonette 1982a).

BEHAVIOR

Peccaries are social animals. Partially as a result of year-round group living and perhaps high genetic relatedness, they have evolved complex vocal and behavioral repertoires. Schweinsburg (1969) classed peccary behavior into broad categories including intraspecific aggressive behavior, sexual behavior, and maternal behavior, among others, and further subdivided aggressive behavior into distinct categories. Sowls (1974)

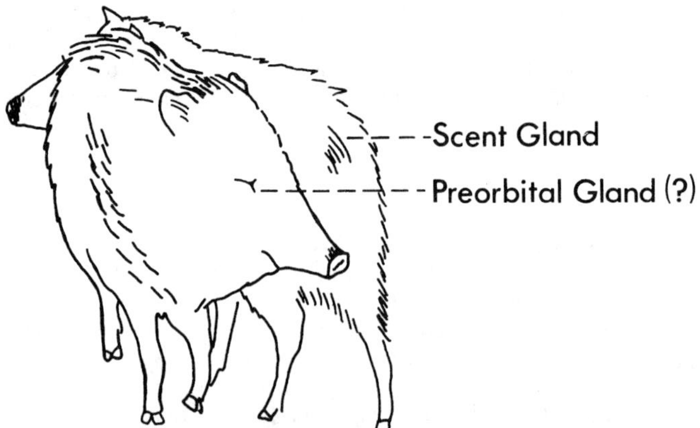

FIGURE 41.6. Reciprocated grooming by two collared peccaries (*Tayassu tajacu*) SOURCE: Bissonette (1982a).

described behavioral patterns for Arizona peccaries. Byers and Bekoff (1981) classified behavioral acts into 147 detailed classifications within 18 major categories. Bissonette (1982a) described three broad classes of vocalization, including aggressive, submissive, and alarm calls, which included 15 distinct, graded, or linked vocal patterns. In addition, six behavioral categories were described, which included 31 distinct, graded, or linked behavioral patterns. Additional literature with detailed behavioral examinations of peccaries includes Schweinsburg and Sowls (1972), Byers (1983), and Babbitt and Packard (1990a, 1990b).

The most common behavioral pattern exhibited by peccaries is reciprocal grooming, also known as the mutual rub (Fig. 41.6). The animals approach and stand alongside each other, head to tail. Each peccary rubs its head along the hind legs, rump, and scent gland of the other. In western Texas, 43% of all interactions involved reciprocal grooming. In Arizona, it was the most common social act after play (Byers and Bekoff 1981). This interaction probably is most important in maintenance of the social bond, including intragroup recognition of territory members and social rank. The mutual rub is probably important in allowing individuals to determine their spatial position with respect to other herd members (Byers and Bekoff 1981). Peccaries used this behavior more often when moving, when leaving a bedground, and when alarmed. The dorsal scent gland, originally described by Tyson (1683), produces secretions that likely convey information relative to herd recognition and reproduction (Hannon et al. 1991).

Sowls (1974) described peccaries as contact animals. Few large mammals, especially ungulates, show the high percentage of contact interaction exhibited by peccaries. Living in closely knit social groups year round may directly affect this phenomena, as does the hypothesized high genetic relationship between herd members. Bissonette (1982a) reported that of 31 different behavior patterns, 55% were characterized by contact. Of 709 recorded interactions involving these patterns, 67% involved contact, and of these, 82% were not aggressive in nature. Grooming, including reciprocal and nonreciprocal rubbing, and head rubbing were involved in 49% (350 of 709) of all encounters and composed 90% of all nonaggressive contact interactions.

Peccaries in western Texas exhibited linear dominance hierarchies that included males and females (Bissonette 1982a). Separate male or female hierarchies do not appear to exist. Sowls (1974) suggested that among penned peccaries, females were usually dominant over males in most situations. Dominance hierarchies in free-ranging populations were stable and closely related to the size of the animal (Bissonette 1982a). Exceptions occurred during squabbles over food. Possession of food seemed to confer an advantage unless there was a large disparity in size between interacting animals. Subordinates won approximately half of the encounters with more dominant animals when they were in possession of a disputed food item.

Byers and Bekoff (1981), in contrast, reported that exclusive courtship and mating rights for males did not exist in the seven herds that they studied in Arizona. They also stated that no field studies of collared peccaries quantitatively documented a dominance hierarchy, although Bissonette (1982a) reported such a hierarchy based on his dissertation work (Bissonette 1976). In addition, Packard et al. (1991) observed sexual competition among captive male peccaries. Byers and Bekoff (1981) proposed that the lack of sexual dimorphism in body and canine size coupled with the lack of other characters that might influence fighting ability in peccaries reflects an evolutionary history of reduced male–male competition. They concluded that group living in peccaries initially was favored selectively as an antipredator adaptation. Subsequently, reduced intraherd competition and increased cooperative behavior, such as communal nursing (Byers and Bekoff 1981), resulted from kin selection. This hypothesis rests on high relatedness among herd members and low rates of mixing among herds. Genetic relationships among herd members have not been described, and dispersal remains poorly documented in the species (Day 1985). Recent evidence from Texas indicated that movement by individuals between adjacent herds is not uncommon among subadult males (Gabor and Hellgren 2000).

An alternative explanation for the lack of sexual dimorphism in collared peccaries that incorporates observations of subtle male–male competition and exclusive male breeding rights (Bissonette 1982a; Packard et al. 1991) is that strong natural selection for other processes overrides the pressure of sexual selection. For example, Wright (1993) suggested that the reduction in the degree of sexual dimorphism in canines and zygomatic processes of extant peccary species relative to fossil forms may be because of strong natural selection against these processes in females, producing a similar response in males. The argument of Kiltie (1981), that canines evolved to stabilize the jaw in breaking hard tropical nuts, describes a natural selective pressure acting against sexual dimorphism. Both sexes would require similar canine structure to help crack nuts. Packard et al. (1991) added that the peccary mating system may be flexible, switching between polygyny and promiscuity based on ecological factors affecting estrous synchrony and group size. For example, males in small groups may be able to exclusively mate females, but this control is lessened in larger groups. Such flexibility also would reduce sexual selection and male–female dimorphism.

FEEDING HABITS

Collared peccaries are generalist herbivores. They are intermediate between ruminant and monogastric ungulates in gastrointestinal anatomy, with a complex stomach containing two blind sacs, a forestomach, and a glandular stomach (Langer 1978, 1979). Peccaries can digest fiber as efficiently as ruminants (Carl and Brown 1986; Commizzoli et al. 1997), and have populations of protozoa (Carl and Brown 1983) and bacteria (Lochmiller et al. 1989) in their stomachs.

In North America, where peccaries generally inhabit semiarid regions, they consume primarily fruits and vegetative parts of succulent species such as prickly pear cactus (*Opuntia* spp.), agaves (*Agave* spp.), and yuccas, as well as forbs and browse (Everitt et al. 1981; Bissonette 1982a; Corn and Warren 1985). In tropical regions, peccaries are more likely to consume hard nuts. Their crania are adapted to crack such foods (Kiltie 1981, 1985).

Regionally, the diet of peccaries varies with available foods. Contrary to popular belief, little if any animal matter is taken regularly. In southern Texas, prickly pear pads compose 30–100% of the diet seasonally (Gallagher et al. 1984; Corn and Warren 1985; Ilse and Hellgren 1995a). Other major food items that are seasonally important include prickly pear fruit, mesquite (*Prosopis glandulosa*) beans, and a variety of forbs (*Portulaca mundula, Lepidium* spp., *Sida* spp, *Zexmania* spp, *Salsola iberica, Ambrosia* spp., *Malvastrum* spp., *Helianthus* spp.). Browse, roots, tubers, and grass compose <12% of the diet. Peccary diets in western Texas are similarly dominated by prickly pear pads, but large amounts of roots and leaves of *Agave lechuguilla,* a succulent, also are eaten (Bissonette 1982a). Diets of collared peccaries in Arizona consistently contain succulent plants and beans of leguminous species such as mesquite and acacias (*Acacia* spp.). The presence of roots, tubers, and oak (*Quercus* spp.) acorns, as well as the predominance of prickly pear, varies with habitat type (Eddy 1961; Day 1985).

The importance of prickly pear cactus in peccary diets has led to much interest in their use of this food. Prickly pear pads are low in protein, digestible energy, vitamins, and minerals. They also contain oxalic acid, a compound that binds calcium to form calcium oxalates, which can cause kidney damage. Peccaries fed ad libitum only on prickly pear pads consumed a daily portion amounting to one third of their body weight and existed on this diet for about 5 months (Sowls 1966). However, they also developed renal disease (Sowls 1984). Theimer and Bateman (1992) reported that peccaries selected prickly pear pads based on water content, but avoided them based on spinescence, calcium oxalate crystal content, and neutral detergent fiber content. It is likely that peccaries eat succulent plants, such as *Opuntia,* to meet water requirements, but consume other foods of higher nutritional value to make up deficits in energy and protein (Sowls 1984).

MORTALITY

Parasites. Sowls (1997:105–10) reviewed the literature on parasites from collared peccaries. Numerous species of ectoparasites from the order Acarina (ticks, mites, chiggers), order Mallophaga (biting lice), order Anoplura (sucking lice), and order Siphonaptera (fleas) occur on collared peccaries. Endoparasites reported from peccaries include nematodes, flukes, tapeworms, protozoans, and blood parasites (*Eperythrozoon* spp.). A key reference on ecto- and endoparasites of peccaries is Corn et al. (1985). Samuel and Low (1970) discussed the prevalence and abundance of ectoparasites in Texas. Wilber et al. (1996) described three new species of *Eimeria* in southern Texas. They suggested that because the collared peccary has such a wide geographic range, it may serve as a useful model for examining parasite community change over large spatial scales. The role of parasites in peccary population dynamics is unknown, as is the case for most mammalian species.

Diseases. Sowls (1997:110–18) reviewed the literature on diseases and infections reported from collared peccaries. Interest in peccary disease processes and susceptibility centers on possible transmission from peccaries to domestic livestock, especially cattle and pigs. Dardiri et al. (1969) directly inoculated collared peccaries to a variety of disease agents found in swine. They found that clinical signs and disease courses of foot-and-mouth disease, vesicular stomatitis, vesicular exanthema of swine, and hog cholera in peccaries were milder and of shorter duration than in pigs. Loan and Storm (1968) found that although peccaries inoculated with hog cholera virus developed antibodies, the disease was not transmitted to pen mates. Thus, peccaries are not likely to act as a disease reservoir. Corn et al. (1987) reported that 8% (18 of 218) of wild peccaries had antibody titers to vesicular stomatitis virus. Collared peccaries were resistant to three strains of African swine fever and were unlikely to become carriers of this disease (Dardiri et al. 1969). Rinderpest virus was lethal to collared peccaries. There is a single report of rabies for the collared peccary (Corn et al. 1987).

Corn et al. (1987) also conducted serological surveys for pseudorabies virus (1 of 218 peccaries had positive antibody responses), brucellosis (0 of 194), and leptospirosis (48 of 213) in free-ranging peccary populations in Arizona. Peccaries are apparently more resistant than domestic swine to pseudorabies (Crandell et al. 1986). Corn et al. (1987) emphasized that serological surveys only provide estimates of the prevalence of exposure of wildlife species to selected pathogens. Exposure to certain viral and bacterial agents does not provide information on the role of peccary populations as reservoirs or transmitters of these diseases.

As with parasites, the role of diseases in peccary population dynamics is unknown. Sowls (1997), however, described evidence for a form of encephalitis in collared peccaries in Arizona and Texas that has caused local and widespread die-offs since about 1980. This disease, which may be similar to canine distemper virus (Sowls 1997 and references therein), causes symptoms such as paralysis (especially in the hind limbs), lack of fear of humans, lethargy, and lesions in the central nervous system. An encephalitis-like disease outbreak that killed a large number of peccaries in southern Texas in 1983–1984 was transmitted to captive animals at Texas A&M University from wild-captured peccaries that had been placed in the facility (E. C. Hellgren, unpublished obs.).

POPULATION DYNAMICS

Life-History Characteristics. Reproductive and survival rates of collared peccaries are similar to those expected for an ungulate of its size. Based on a 10-year data set from a harvested population, Hellgren and Lochmiller (1999) reported that annual reproductive rate (female young/female) for adult peccaries ranged between 0.7 and 1.4, with mean annual fecundity near 1.0 female/female. The range in values was because of variation in precipitation-induced forage availability (Low 1970; Hellgren et al. 1995). Work with captive peccaries under controlled conditions indicates that poor nutrition can interfere with normal reproduction (Lochmiller et al. 1986a). Although peccaries can produce two litters/year, one litter of two young is the mode. Yearling productivity was 30–40% of adult productivity (Hellgren et al. 1995).

Annual survival rates of adults have been estimated from life table analyses (Low 1970; Day 1985) and from survival of telemetered animals (Hellgren et al. 1995; Gabor and Hellgren 2000). Estimates were 0.65–0.73 in heavily harvested sites, 0.73–0.85 in moderately harvested sites, and 0.87–0.90 in unharvested sites. Hunting appears to be at least partially additive to natural mortality, but juvenile survival may be compensatory to hunting mortality within a range of harvest intensity (Hellgren et al. 1995). Survival rates of juveniles have not been determined, but they are likely highly dynamic. Population modeling suggested that juvenile survival rates must average 0.3–0.4 annually to maintain stationary populations (Hellgren et al. 1995).

Sex ratios at birth have ranged from female biased (44% male, Arizona, $n = 186$ young: Sowls 1984) to male biased (56% male, Texas, $n = 57$: Lochmiller et al. 1984; 53% male, Texas, $n = 213$: Hellgren et al. 1995). Sex ratio of the harvest is typically male biased (53% male: Sowls 1984, Hellgren et al. 1995). Because of the difficulty of determining the sex of peccaries in the field at a distance and their size monomorphism, a male bias by hunters is unlikely.

Dispersal by collared peccaries is poorly described. Natal dispersal rarely has been reported. Sparse data suggest that it is uncommon and normally involves males (Gabor and Hellgren 2000). Interherd movements of adults also occur, with reported rates of herd interchange ranging from 10–12% in Arizona (Day 1985) to 22–41% in southern Texas (Ellisor and Harwell 1979). Formation of new herds by fission is discussed in Social Organization and Herd Dynamics, and is important in modeling population dynamics.

Density. Density estimates for peccaries have been based on a number of methods, including strip censuses, hunter reports, long-term herd observations, sightings, and mark–recapture (Sowls 1984; Gabor and Hellgren 2000). Estimates generally range from 2 to 12 animals/km^2 in North America (Table 41.1). Estimates in tropical parts of the geographic range are similar (Sowls 1997).

The chief paradigm of density dependence, namely, that animal density is positively correlated with food resources or other life requisites, logically leads to predictive statements about the factors controlling animal numbers. If close relationships exist, assessments of resource abundance should give a close estimate of the density of the population. Certain species, because of their life history characteristics, are better suited to investigations of this kind. Territorial herbivores with clearly defined food preferences are easily studied. For collared peccaries, group size in western Texas was positively related to percentage vegetation cover and percentage cover of available forage (prickly pear, *Agave lechuguilla,* and forbs) on their territories. Group size was negatively correlated with percentage woody cover—mostly plants of low nutritional and cover value (Bissonette 1982a). Similarly, Gabor and Hellgren (2000) found that peccary densities were six times higher on areas with greater landscape patchiness, canopy cover, and prickly

TABLE 41.1. Summary of representative herd or population densities for collared peccary (*Tayassu tajacu*) populations in the southwestern United States

Density (animals/km^2)	Type	Region	Source
2.6–3.9	Population	Gulf Coastal Prairie, Texas	Low 1970
2.0	Population	Gulf Coastal Prairie, Texas	Ilse and Hellgren 1995b
3.9–10.9	Population	Eastern South Texas Plains	Low 1970
2.6–3.1[a]	Population	Western South Texas Plains	Gabor and Hellgren 2000
13.0–15.5[b]	Population	Western South Texas Plains	Gabor and Hellgren 2000
1.2–2.5	Population	Trans-Pecos, western Texas	Low 1970
3.4–10.9	Herd	Trans-Pecos, western Texas	Bissonette 1982a
12.6	Herd	Southern Arizona	Schweinsburg 1971
11.9	Herd	Central Arizona (3-Bar)	Day 1985

NOTE: Herd density estimate refers to densities calculated from territory size and number of herd members; population density estimate refers to density calculated for an entire population.

[a]Feral hogs (*Sus scrofa*) present.

[b]Feral hogs (*Sus scrofa*) absent.

pear abundance. However, these latter findings were confounded by the presence of feral pigs, a potential competitior, on the low-density areas.

According to Low (1970), drought, with its concomitant effects on plant quality and availability, is the ultimate factor controlling peccary populations in Texas. Availability of forbs and highly nutritious seeds, fruits, and nuts probably are most important in maintaining a positive energy balance for successful reproduction. Lochmiller et al. (1986b, 1987a) demonstrated the role of nutrition in peccary reproduction. Low (1970) suggested that predation and parasitism are probably of minor importance in regulating populations. These effects are density dependent, having proportionately greater influence at high densities.

AGE ESTIMATION

Kirkpatrick and Sowls (1962) described the pattern of tooth replacement in collared peccaries. They are born with all four deciduous canine teeth and the posterior pair of lower incisors already erupted. In sequence, eruption of the third pairs of upper (p^3) and lower (p_3) premolars, the first pair of lower incisors (i_1), the second pairs of upper (p^2) and lower (p_2) premolars, the first pair of upper incisors (i^1), the fourth upper (p^4) and lower (p_4) pairs of premolars, and the second upper (i^2) and lower (i_2) pairs of incisors complete the temporary dentition between 2 and 3 months of age. The first upper (m^1) and lower (m_1) permanent molars erupt at about 5 months of age, followed by replacement of the canines at 8–9 months. The third pair of lower incisors (i_3) then appears, and the second upper (m^2) and lower (m_2) pairs of molars are replaced. The remaining incisors are replaced and the permanent premolars erupt. Permanent dentition is complete at about 20–21 months of age with eruption of the third upper (m^3) and lower (m_3) pairs of molars (Kirkpatrick and Sowls 1962).

Adults (>21 months old) have traditionally been assigned to five age classes based on tooth wear, ranging from age class 1, showing "slight wear but no particular teeth showing more wear than the others," to age class 5, with "very heavy wear with some teeth or nearly all teeth missing" (Sowls 1961b). Estimation of age by cementum annuli was conducted by Low (1970), who found that age class 1 corresponded to 2- to 3-year olds, age class 2 to 3- to 5-year olds, age class 3 to 5- to 7-year olds, age class 4 to >6-year olds, and age class 5 to >7-year olds. Overlap in age classes is a consequence of differential tooth wear among individuals, and is probably related to silicaceous content of consumed vegetation.

MANAGEMENT AND CONSERVATION

Conservation of the collared peccary in the southwestern United States is considered secure, based on regulated and monitored hunting seasons (Sowls 1997). Indeed, populations in Arizona are expanding and causing nuisance problems in suburban areas. Surveys are conducted in Texas and Arizona (often in conjunction with other wildlife surveys) by foot, vehicle, horseback, or aircraft to assess population status and trends, and distribution of peccaries. Conservation and management of peccaries in Central and South America is a matter of concern, however. Unregulated hunting, timber harvest, land cleared for agriculture, and other factors are reducing peccary habitat and populations. See Oliver (1993) and Sowls (1997) for discussions of these issues.

Peccaries are legal game in all three states in which they occur. During the 1999/2000 season, an estimated 21,377 peccaries were harvested legally in the United States. Texas accounted for the largest number of animals, with 17,889 peccaries killed during the 1999/2000 season. In total, >44,000 hunters went afield for 281,000 days hunting for collared peccaries. Hunter success is lower in Arizona than Texas and New Mexico because of a focus on archery and handgun hunts versus rifle hunting (Table 41.2). Throughout the 1990s, hunter effort and harvest have been fairly stable in Arizona, but have steadily declined in Texas (Table 41.3).

The herd structure of peccary populations is a key feature relative to management. In Arizona, number of herds, average herd size, and age ratios are key indices used by managers to make harvest recommendations (Day 1985). Accessible herds may be eliminated by local harvest. The Arizona Game and Fish Department instituted a harvest quota system in 1972 to prevent local extirpations in management units. If herd sizes increase in an area, the number of hunting permits can be increased (J. Phelps, Arizona Game and Fish Department, pers. commun., 2000). Variation in observed herd and group sizes (Green et al. 1984) may have important management implications because it may lead to underestimates of population size, especially when subgroups are smaller (Hellgren and Lochmiller 1999).

The effects of herd reduction on herd dynamics is an important management concern. Herd sizes in hunted areas stabilize at lower levels than in unhunted areas (Day 1985), and herds recover slowly from

TABLE 41.2. Summary of annual collared peccary (*Tayassu tajacu*) harvest statistics for the continental United States, 1997–2000

State	Average Number Harvested/Year	Average Number of Hunters Afield	Percentage Hunter Success
Arizona	2,842	10,970	25.9
New Mexico	427	817	52.3
Texas	16,151	35,847	45.5

SOURCE: Data from J. Phelps, Arizona Game and Fish Department, unpublished data; D. Dolcino, New Mexico Department of Game and Fish, unpublished data; J. Moore, Federal Aid Report, Javelina Harvest Regulations, Texas Parks and Wildlife Department, unpublished data. Arizona and Texas data represent years 1997, 1998, and 1999. New Mexico data represent 1997/98, 1998/99, and 1999/2000 seasons.

TABLE 41.3. Collared peccary harvest and hunter effort in Arizona and Texas, 1991–1999

	Harvest		Hunter-days	
Year	Arizona	Texas	Arizona	Texas
1991	2,856	20,640	31,618	282,534
1992	3,178	18,552	32,408	281,303
1993	3,157	21,003	29,318	283,621
1994	3,585	19,482	31,961	311,230
1995	2,797	19,582	32,126	253,453
1996	3,509	17,470	31,996	252,380
1997	3,036	14,276	30,717	198,611
1998	2,594	16,288	28,220	253,213
1999	2,895	17,889	28,636	249,650

SOURCE: Data from J. Phelps, Arizona Game and Fish Department, unpublished data; J. Moore, Federal Aid Report, Javelina Harvest Regulations, Texas Parks and Wildlife Department, unpublished data.

heavy harvest. Hellgren et al. (1995) proposed that, within limits, peccaries may exhibit density-dependent recruitment in response to harvest mortality. In southern Texas, peccaries appear to withstand 20–30% harvests in dense mesquite habitat, but only 15% harvest in open mesquite savannas (Ellisor and Harwell 1979; Hellgren et al. 1995). Green et al. (1985) reported that harvest rates of 8.5/km^2 in dense chaparral habitat and 2.7/ km^2 in open mesquite habitat had no detrimental effects on herds; however, percentage harvest rates were unknown. In Arizona, 12% harvest annually over 7 years did not affect overall population size, but 16% harvest over 5 years led to a population decline (Day 1985). Day (1977) suggested that it is unlikely that severely depleted herds can quickly regain their former numbers through recruitment, herd interchange, or invasion of unoccupied range by a neighbor herd. If this is correct, closing heavily hunted areas may be the preferred management strategy for sustained harvest and maintenance of adequate population numbers.

Habitat quality is likely important in determining the rate of recovery in hunted herds (Ellisor and Harwell 1979). In southern Texas, peccaries select vegetation communities of dense cover, including chaparral mixed-grass, live oak–chaparral, and mesquite woodlands. They avoid communities of open canopy cover such as open mesquite–grass savannas (Ilse and Hellgren 1995a; Gabor and Hellgren 2000; Gabor et al. 2001). Gabor and Hellgren (2000) found that peccary populations with the highest densities and the largest herds were found in association with contiguous patches of dense cover and abundant prickly pear cactus. Recruitment varied from 1.08 young/adult female in areas of dense cover to 0.36 young/female in open habitat (Ellisor and Harwell 1979).

Habitat selection by peccaries in arid western Texas and Arizona showed similar patterns. Herds preferred mesquite-dominated associations for foraging and avoided more open associations in Arizona (Bellantoni and Krausman 1993). In addition, Ockenfels and Day (1990) found that the number of adults in a herd was positively related to prickly pear cactus density within the herd territory. In western Texas, Bissonette (1982a) reported that peccary group size was related to cover composition of forbs and succulent plants, but inversely related to woody cover. However, woody cover in that area was dominated by nonbrowse species.

Management decisions involving habitat manipulations on landscapes containing peccaries can have dramatic influences on populations. However, the effects of these manipulations are not clearly described. Managers at the Peccary Workshop in Tucson (Ockenfels et al. 1985) identified important questions relative to habitat management and its effects on peccary populations. For instance, "at what level of livestock grazing is peccary population dynamics impacted?" (Ockenfels et al. 1985:74). In southern Texas, large areas of thornscrub have been converted to pasture for livestock, with subsequent loss of habitat elements favorable to peccaries, such as key forages (forbs, cactus), thermal cover, and escape cover. Decreased canopy cover and increased grass production in such areas favors ecological replacement of peccaries with feral pigs, a keystone invading species (Gabor and Hellgren 2000; Gabor et al. 2001). Conversely, grazing-induced brush invasion of desert grasslands and mesquite savannas may increase the amount of peccary habitat. In addition, increased economic value of other wildlife, such as white-tailed deer (*Odocoileus virginianus*), in southern Texas has slowed the pressure for brush clearing.

The complex direct and indirect effects of brush removal by burning, chemical, or mechanical means on peccaries are also unknown. Ellisor and Harwell (1979) suggested that up to 50% of available shrub cover could be cleared without negatively influencing peccary populations, as long as remaining patches of dense brush were interconnected. Burning, which increases forb production, may increase food availability for peccaries. However, Gabor and Hellgren (2000) reported that peccary populations are more dense in homogeneous thornscrub cover than where cover is patchy. Also, patchy, open habitats are more suited to feral pigs.

Brush control in more arid areas of the peccary distribution in western Texas and Arizona is less of a habitat management concern. Timber harvest, though a major consideration for peccary range in the Tropics, is a minor issue in peccary management in North America (Hellgren and Lochmiller 1999). Finally, the effects of water management strategies on population dynamics of peccaries never have been studied (Hellgren and Lochmiller 1999). Although free water in the form of stock tanks, guzzlers, and ponds are used by peccaries, the availability of water in cactus and other succulents in the diet of peccaries suggests that free water is not required (Zervanos and Day 1977). In addition, the efficacy of water developments for other wildlife species in the American Southwest has recently been questioned (Broyles 1995; Broyles and Cutler 1999).

RESEARCH AND MANAGEMENT NEEDS

A developing problem concerning peccary management in Arizona is an increase in negative interactions between humans and peccaries. Nuisance complaints in semiurban and semirural areas are becoming a serious problem in the greater Tucson and Phoenix areas as well as other Arizona cities. Urban herds are increasing in size and new techniques to address the problem are necessary (J. Phelps, Arizona Game and Fish Department, pers. commun., 2000). Possible solutions include trapping and translocation, harvest, and contraception. Translocation is an expensive strategy with high mortality, and suitable sites for placement of herds are scarce. Harvest in the urban environment is not a preferred alternative because of liability and negative public sentiment. Contraception is a possible long-term solution, but does not address acute nuisance problems. Educating the public to coexist with peccaries may be the best solution.

Research and management needs in Texas differ from those in Arizona. The best peccary habitat is dense thornscrub, and population surveying is difficult. In addition, data suggest the occurrence of a statewide peccary decline. Potential contributing factors to a decline include competition with feral pigs (which show a complementary, increasing long-term trend; R. B. Taylor, Texas Parks and Wildlife Department, unpublished data, 1995) and illegal harvest by hunters and landowners. The increasing economic value of wildlife in Texas, especially leasing rights to hunt white-tailed deer and northern bobwhite (*Colinus virginianus*), may be indirectly increasing illegal peccary harvest. Many Texan landowners provide pelleted feed and corn to deer to increase body mass and antler size, and consider peccaries and other species competitors for this supplemental food. Illegal killing may occur because of the perception of peccaries as nuisance animals.

Development of an efficient, precise, and accurate censusing technique is needed. Managers in Texas have proposed conducting research to develop correction factors for estimating peccary population densities using available deer census techniques (J. Moore and D. R. Synatzske, Texas Parks and Wildlife Department, pers. commun., 2000). Another management need involves increasing demand to hunt peccaries. The relative novelty of peccaries, especially to northern hunters, suggests that they represent an unrealized economic opportunity to private landowners in Texas.

Several other research questions are of interest to either basic or applied ecologists. For example, what conditions favor formation of new herds by fission of subgroups and what is the importance of this process in colonizing unoccupied habitat? Given the limited amount of natal dispersal documented among peccaries, what is the genetic structure of peccary herds and how is inbreeding avoidance maintained? What is the competitive effect of invasive feral pigs on native peccaries? Several years of work in Texas have shown that peccary and pig densities are inversely related, yet direct negative effects of pigs on peccaries have not been documented (Ilse and Hellgren 1995b; Gabor and Hellgren 2000). This latter question has important implications for the conservation of peccaries in the Neotropics, where the bulk of their distribution lies.

LITERATURE CITED

Arizona Game and Fish Department. 2000. Javelina distribution [Map.] Arizona Game and Fish Department, Phoenix.

Babbitt, K. J., and J. M. Packard. 1990a. Suckling behavior of the collared peccary (*Tayassu tajacu*). Ethology 86:102–15.

Babbitt, K. J., and J. M. Packard. 1990b. Parent–offspring conflict relative to phase of gestation. Animal Behaviour 40:765–73.

Bellantoni, E. S., and P. R. Krausman. 1993. Habitat use by collared peccaries in an urban environment. Southwestern Naturalist 38:345–51.

Bigler, W. J. 1974. Seasonal movements and activity patterns of the collared peccary. Journal of Mammalogy 55:851–55.

Bissonette, J. A. 1976. The relationship of resource quality and availability to social behavior and organization in the collared peccary. Ph.D. Dissertation, University of Michigan, Ann Arbor.

Bissonette, J. A. 1978. The influence of extremes of temperature on activity patterns of peccaries. Southwestern Naturalist 23:339–46.

Bissonette, J. A. 1982a. Social behavior and ecology of the collared peccary in Big Bend National Park (Science Monograph No. 16). National Park Service.

Bissonette, J. A. 1982b. Collared peccary. Pages 841–50 *in* J. A. Chapman and G. A. Feldhamer, eds. Wild mammals of North America: Biology, management, and economics. Johns Hopkins University Press, Baltimore.

Brown, W. H., J. W. Stull, and L. K. Sowls. 1963. Chemical composition of the milk fat of the collared peccary. Journal of Mammalogy 44:112–13.

Broyles, B. 1995. Desert wildlife water developments: Questioning use in the Southwest. Wildlife Society Bulletin 23:663–75.

Broyles, B., and T. L. Cutler. 1999. Effect of surface water on desert bighorn sheep in the Cabeza Prieta National Wildlife Refuge, southwestern Arizona. Wildlife Society Bulletin 27:1082–88.

Byers, J. A. 1983. Social interactions of juvenile collared peccaries, *Tayassu tajacu* (Mammalia: Artiodactyla). Journal of Zoology (London) 201:83–96.

Byers, J. A., and M. Bekoff. 1981. Social, spacing, and cooperative behavior of the collared peccary. Journal of Mammalogy 62:767–85.

Carl, G. R., and R. D. Brown. 1983. Protozoa in the forestomach of the collared peccary (*Tayassu tajacu*). Journal of Mammalogy 64:709.

Carl, G. R., and R. D. Brown. 1985. Protein requirements of adult collared peccaries. Journal of Wildlife Management 49:351–55.

Carl, G. R., and R. D. Brown. 1986. Comparative digestive efficiency and feed intake of the collared peccary. Southwestern Naturalist 31:79–85.

Commizzoli, P., J. Peiniau, C. Dutertre, P. Planquette, and A. Aumaitre. 1997. Digestive utilization of concentrated and fibrous diets (*Tayassu peccari* [*sic*], *Tayassu tajacu*) raised in French Guyana. Animal Feed Science Technology 64:215–26.

Corn, J. L., and R. J. Warren. 1985. Seasonal food habits of the collared peccary in south Texas. Journal of Mammalogy 66:155–59.

Corn, J. L., D. B. Pence, and R. J. Warren. 1985. Factors affecting the helminth community structure of adult collared peccaries in southern Texas. Journal of Wildlife Diseases 21:254–63.

Corn, J. L., R. M. Lee, G. A. Erickson, and C. D. Murphy. 1987. Serologic survey for evidence of exposure to vesicular stomatitis virus, pseudorabies virus, brucellosis, and leptospirosis in collared peccaries from Arizona. Journal of Wildlife Diseases 23:551–57.

Coscarelli, K. P. 1985. Ovulation in the collared peccary as documented by laparoscopy. M.S. Thesis, Texas A&M University, College Station.

Crandell, R. A., R. M. Robinson, and P. G. Hannon. 1986. Pseudorabies infection in collared peccaries (*Tayassu tajacu*). Southwest Veterinarian 37:193–95.

Dardiri, A. H., R. J. Yedloutschnig, and W. D. Taylor. 1969. Clinical and serologic response of American white-collared peccaries to African swine fever, foot-and-mouth disease, vesicular stomatitis, vesicular exanthema of swine, hog cholera, and rinderpest viruses. Proceedings of the United States Animal Health Association 73:437–52.

Davis, W. B., and D. J. Schmidly. 1994. The mammals of Texas. Texas Parks and Wildlife, Austin.

Day, G. I. 1977. Javelina activity patterns (Pittman-Robertson Report W-78R-29). Arizona Game and Fish Department, Phoenix.

Day, G. I. 1985. Javelina research and management in Arizona. Arizona Game and Fish Department, Phoenix.

Donkin, R. A. 1985. The peccary—With observations on the introduction of pigs to the New World. Transactions of the American Philosophical Society 75(5):1–152.

Eddy, T. A. 1961. Foods and feeding patterns of the collared peccary in southern Arizona. Journal of Wildlife Management 25:248–57.

Elder, J. B. 1956. Watering patterns of some desert game animals. Journal of Wildlife Management 20:368–78.

Ellisor, J. E., and W. F. Harwell. 1969. Mobility and home range of collared peccary in southern Texas. Journal of Wildlife Management 33:425–27.

Ellisor, J. E., and W. F. Harwell. 1979. Ecology and management of javelina in south Texas (Federal Aid Report Series No. 16). Texas Parks and Wildlife Department, Austin.

Epling, G. P. 1956. Morphology of the scent gland of the javelina. Journal of Mammalogy 37:246–48.

Everitt, J. H., C. L. Gonzalez, M. A. Alaniz, and G. V. Latigo. 1981. Food habits of the collared peccary on south Texas rangeland. Journal of Range Management 34:141–44.

Findley, J. S., A. H. Harris, D. E. Wilson, and C. Jones. 1975. Mammals of New Mexico. University of New Mexico Press, Albuquerque.

Gabor, T. M. 1997. Ecology and interactions of sympatric collared peccaries and feral pigs. Ph.D. Dissertation, Texas A&M University, Kingsville and College Station.

Gabor, T. M., and E. C. Hellgren. 2000. Variation in peccary populations: Landscape composition or competition by an invader? Ecology 81:2509–25.

Gabor, T. M, E. C. Hellgren, and N. J. Silvy. 1997. Renal morphology of sympatric suiforms: Implications for competition. Journal of Mammalogy 78:1089–95.

Gabor, T. M., E. C. Hellgren, and N. J. Silvy. 2001. Multi-scale habitat partitioning in sympatric suiforms. Journal of Wildlife Management 65:99–110.

Gallagher, J. F., L. W. Varner, and W. E. Grant. 1984. Nutrition of the collared peccary in south Texas. Journal of Wildlife Management 48:749–61.

Green, G. E., W. E. Grant, and E. Davis. 1984. Variability of observed group sizes within collared peccary herds. Journal of Wildlife Management 48:244–48.

Green, G. E., W. E. Grant, and E. Davis. 1985. Effects of hunting on javelina in south Texas. Wildlife Society Bulletin 13:149–53.

Hall, E. R. 1981. The mammals of North America, Vol. 2, 2nd ed. John Wiley, Sons, New York.

Halloran, A. 1945. Five fetuses of *Pecari angulatus* from Arizona. Journal of Mammalogy 26:434.

Hannon, P. G., D. M. Dowdell, R. L. Lochmiller, and W. E. Grant. 1991. Dorsal-gland activity in peccaries at several physiological states. Journal of Mammalogy 72:825–27.

Hellgren, E. C., and R. L. Lochmiller. 1999. Collared peccary. Pages 429–46 *in* S. Demarais and P. R. Krausman, eds. Ecology and management of large mammals in North America. Prentice-Hall, Upper Saddle River, NJ.

Hellgren, E. C., R. L. Lochmiller, and W. E. Grant. 1984. Demographic, morphologic, and reproductive status of a herd of collared peccaries (*Tayassu tajacu*) in south Texas. American Midland Naturalist 112:402–7.

Hellgren, E. C., R. L. Lochmiller, M. S. Amoss, Jr., S. W. J. Seager, S. J. Magyar, K. P. Coscarelli, and W. E. Grant. 1989. Seasonal variation in serum testosterone, testicular measurements, and semen characteristics in the collared peccary (*Tayassu tajacu*). Journal of Reproduction and Fertility 85:677–86.

Hellgren, E. C., D. R. Synatszke, P. W. Oldenburg, and F. S. Guthery. 1995. Demography of a collared peccary population in south Texas. Journal of Wildlife Management 59:153–63.

Herring, S. W. 1972. The role of canine morphology in the evolutionary divergence of pigs and peccaries. Journal of Mammalogy 53:500–512.

Herring, S. W. 1974. A biometric study of suture fusion and skull growth in peccaries. Anatomy and Embryology 146:167–80.

Ilse, L. M., and E. C. Hellgren. 1995a. Resource partitioning in sympatric populations of collared peccaries and feral hogs in southern Texas. Journal of Mammalogy 76:784–99.

Ilse, L. M., and E. C. Hellgren. 1995b. Spatial use and group dynamics in sympatric populations of collared peccaries and feral hogs in southern Texas. Journal of Mammalogy 76:993–1002.

Jennings, W. S., and J. T. Harris. 1953. The collared peccary in Texas (Federal Aid Report Series No. 12). Texas Game and Fish Commission, Austin.

Kiltie, R. A. 1981. The function of interlocking canines in rain forest peccaries (Tayassuidae). Journal of Mammalogy 62:459–69.

Kiltie, R. A. 1985. Craniomandibular differences between rain forest and desert collared peccaries. American Midland Naturalist 113:384–87.

Kirkpatrick, R. D., and L. K. Sowls. 1962. Age determination of the collared peccary by the tooth replacement pattern. Journal of Wildlife Management 26:214–17.

Knipe, T. 1957. The javelina in Arizona (Arizona Game and Fish Wildlife Bulletin No. 2). Arizona Game and Fish Department, Phoenix.

Langer, P. 1978. Anatomy of the stomach of the collared peccary, *Dicotyles tajacu* (L. 1798) (Mammalia, Artiodactyla). Zeitschrift für Saügetierkunde 43:42–59.

Langer, P. 1979. Adaptational significance of the forestomach of the collared peccary, *Dicotyles tajacu* (L. 1758) (Mammalia: Artiodactyla). Mammalia 43:235–45.

Loan, R. W., and M. M. Storm. 1968. Propagation and transmission of hog cholera virus in nonporcine hosts. American Journal of Veterinary Research 29:807–11.

Lochmiller, R. L. 1984. Nutritional influences on growth and reproduction and physiological assessment of nutritional status in the collared peccary. Ph.D. Dissertation, Texas A&M University, College Station.

Lochmiller, R. L., E. C. Hellgren, and W. E. Grant. 1984. Selected aspects of collared peccary (*Dicotyles tajacu*) reproductive biology in a captive Texas herd. Zoo Biology 3:145–49.

Lochmiller, R. L., E. C. Hellgren, W. E. Grant, L. W. Greene, and C. W. Dill. 1985. Description of collared peccary (*Tayassu tajacu*) milk composition. Zoo Biology 4:375–79.

Lochmiller, R. L., E. C. Hellgren, and W. E. Grant., 1986a. Absolute and allometric relationships between internal morphology and body mass in the adult collared peccary, *Tayassu tajacu* (Tayassuidae). Growth 50:296–316.

Lochmiller, R. L., E. C. Hellgren, and W. E. Grant. 1986b. Reproductive responses to nutritional stress in adult female collared peccaries. Journal of Wildlife Management 50:295–300.

Lochmiller, R. L., E. C. Hellgren, and W. E. Grant. 1987a. Influences of moderate nutritional stress during gestation on reproduction of collared peccaries. Journal of Zoology (London) 211:321–16.

Lochmiller, R. L., E. C. Hellgren, and W. E. Grant. 1987b. Physical characteristics of neonate, juvenile, and adult collared peccaries (*Tayassu tajacu angulatus*) from south Texas. Journal of Mammalogy 68:188–94.

Lochmiller, R. L., E. C. Hellgren, J. F. Gallagher, L. W. Varner, and W. E. Grant. 1989. Volatile fatty acids in the gastrointestinal tract of the collared peccary (*Tayassu tajacu*). Journal of Mammalogy 70:189–91.

Low, W. A. 1970. The influence of aridity on reproduction of the collared peccary (*Dicotyles tajacu*) (Linn.) in Texas. Ph.D. Dissertation, University of British Columbia, Vancouver, Canada.

Mayer, J. J., and P. N. Brandt. 1982. Identity, distribution, and natural history of the peccaries, Tayassuidae. Pages 433–55 *in* M. A. Mares and H. H. Genoways, eds. Mammalian biology of South America (Special Publication Series, Vol. 6). Pymatuning Laboratory in Ecology, University of Pittsburgh.

McCulloch, C. Y. 1955. Breeding record of javelina, *Tayassu angulatus* in southern Arizona. Journal of Mammalogy 36:146.

Mearns, E. A. 1907. Mammals of the Mexican boundary of the United States. Part 1. U.S. National Museum Bulletin 56:159–69.

Miller, G. S., Jr., and R. Kellogg. 1955. List of North American Recent mammals (Bulletin 205). U.S. National Museum, Smithsonian Institution, Washington, DC.

Minnamon, P. S. 1962. The home range of the collared peccary *Pecari tajacu* (Mearns) in the Tucson Mountains. M.S. Thesis, University of Arizona, Tucson.

Neal, B. J. 1959. A contribution on the life history of the collared peccary in Arizona. American Midland Naturalist 61:177–90.

Ockenfels, R. A., G. I. Day, and V. C. Supplee. 1985. Proceedings of the peccary workshop. Arizona Game and Fish Department, Phoenix.

Ockenfels, R. A., and G. I. Day. 1990. Determinants of collared peccary home range size in central Arizona. Pages 76–81 *in* P. R. Krausman and N. S. Smith, eds. Managing wildlife in the Southwest. Arizona Chapter of the Wildlife Society, Phoenix.

Oldenburg, P. W., P. J. Ettestad, W. E. Grant, and E. Davis. 1985. Structure of collared peccary herds in south Texas: Spatial and temporal dispersion of herd members. Journal of Mammalogy 66:764–70.

Oliver, W. L. R., ed. 1993. Pigs, peccaries, and hippos. International Union for the Conservation of Nature, Gland, Switzerland.

Packard, J. M., K. J. Babbitt, K. M. Franchek, and P. M. Pierce. 1991. Sexual competition in captive collared peccaries (*Tayassu tajacu*). Applied Animal Behaviour Science 29:319–26.

Samuel, W. M., and W. A. Low. 1970. Parasites of the collared peccary from Texas. Journal of Wildlife Diseases 6:16–23.

Schilling, P. W., and R. L. Stone. 1969. Comparative cardiovascular measurements in the javelina (*Tayassu tajacu*). Laboratory Animal Care 19:331–35.

Schweinsburg, R. 1969. Social behavior of the collared peccary (*Pecari tajacu*) in the Tucson Mountains. Ph.D. Dissertation, University of Arizona, Tucson.

Schweinsburg, R. 1971. Home range, movements, and herd integrity of the collared peccary. Journal of Wildlife Management 35:455–60.

Schweinsburg, R., and L. K. Sowls. 1972. Aggressive behavior and related phenomena in the collared peccary. Zeitschrift für Tierpsychologie 30:132–45.

Smith, N. S., and L. K. Sowls. 1975. Fetal development of the collared peccary. Journal of Mammalogy 56:619–25.

Sowls, L. K. 1961a. Gestation period of the collared peccary. Journal of Mammalogy 42:425–26.

Sowls, L. K. 1961b. Hunter-checking stations for collecting data on the collared peccary (*Pecari tajacu*). Transactions of the North American Wildlife Conference 26:496–505.

Sowls, L. K. 1963. Chemical composition of the milk fat of the collared peccary. Journal of Mammalogy 44:112–13.

Sowls, L. K. 1965. Reproduction in the collared peccary, *Tayassu tajacu*. Journal of Reproduction and Fertility 9:371–72.

Sowls, L. K. 1966. Reproduction in the collared peccary (*Tayassu tajacu*). Pages 155–72 *in* I. W. Rowlands, ed. Comparative biology of reproduction in mammals. Zoological Society of London, London.

Sowls, L. K. 1974. Social behavior of the collared peccary *Dicotyles tajacu* (L.). Pages 144–65 *in* V. Geist and F. Walther, eds. The behavior of ungulates and its relation to management. International Union for the Conservation of Nature, Morges, Switzerland.

Sowls, L. K. 1978. Collared peccary. Pages 191–205 *in* J. L. Schmidt and D. L. Gilbert, eds. Big game of North America. Stackpole, Harrisburg, PA.

Sowls, L. K. 1984. The peccaries. University of Arizona Press, Tucson.

Sowls, L. K. 1997. Javelinas and other peccaries: Their biology, management, and use. Texas A&M University Press, College Station.

Sowls, L. K., V. R. Smith, R. Jenness, R. E. Sloan, and E. Regehr. 1961. Chemical composition and physical properties of the milk of the collared peccary. Journal of Mammalogy 42:245–51.

Supplee, V. C. 1981. The dynamics of collared peccary dispersion into available range. M.S. Thesis, University of Arizona, Tucson.

Theimer, T. C., and G. C. Bateman. 1992. Patterns of prickly pear herbivory by collared peccaries. Journal of Wildlife Management 36:234–40.

Theimer, T. C., and P. Keim. 1998. Phylogenetic relationships of peccaries based on mitochondrial cytochrome *b* DNA sequences. Journal of Mammalogy 79:566–72.

Tyson, E. 1683. Anatomy of the Mexico musk hog. Philosophical Transactions of the Royal Society of London 13:359–85.

Wetzel, R. M. 1977a. The extinction of peccaries and a new case of survival. Annals of the New York Academy of Science 288:538–44.

Wetzel, R. M. 1977b. The chacoan peccary (Bulletin No. 3). Carnegie Museum of Natural History.

Wetzel, R. M., and J. W. Lovett. 1974. A collection of mammals from the Chaco of Paraguay. University of Connecticut Occasional Papers 2 (13):203–16.

Wetzel, R. M., R. E. Dubos, R. L. Martin, and P. Myers. 1975. Catagonus, an 'extinct' peccary, alive in Paraguay. Science 189:379–91.

Wilber, P. G., E. C. Hellgren, and T. M. Gabor. 1996. Coccidia of the collared peccary (*Tayassu tajacu*) in southern Texas with descriptions of three new species of Eimeria (Apicomplexa: Eimeriidae). Journal of Parasitology 82:624–29.

Woodburne, M. O. 1968. The cranial myology and osteology of *Dicotyles tajacu*, and its bearing on classification. Memoirs of the California Academy of Science 7:1–48.

Wright, D. B. 1993. Evolution of sexually dimorphic characters in peccaries (Mammalia, Tayassuidae). Paleobiology 19:52–70.

Zervanos, S. M. 1972. Thermoregulation and water relations of the collared peccary (*Tayassu tajacu*). Ph.D. Dissertation, Arizona State University, Tempe.

Zervanos, S. M., and G. I. Day. 1977. Water and energy requirements of captive and free-living collared peccaries. Journal of Wildlife Management 41:527–32.

Zervanos, S. M., and N. F. Hadley. 1973. Adaptational biology and energy relationships of the collared peccary (*Tayassu tajacu*). Ecology 54:759–74.

Eric C. Hellgren, Department of Zoology, Oklahoma State University, Stillwater, Oklahoma 74078. Email: ehellgr@okstate.edu.

John A. Bissonette, Cooperative Fish and Wildlife Research Unit, College of Natural Resources, Utah State University, Logan, Utah 84322-5290. Email: j.bissonette@cnr.usu.edu.

42

Wapiti

Cervus elaphus

James M. Peek

NOMENCLATURE

Common Names. Wapiti, elk
Scientific Name. *Cervus elaphus*
Subspecies. *C. e. nelsoni,* Rocky Mountain elk; *C. e. canadensis,* Eastern elk (extinct); *C. e. manitobensis,* Manitoba elk; *C. e. merriami,* Merriam elk (extinct); *C. e. roosevelti,* Roosevelt elk; *C. e. nannodes,* Tule elk.

The North American wapiti or elk is considered part of the red deer complex, with four extant and two extinct subspecies (Nowak 1999). However, there are significant behavioral, ecological, morphological, and genetic differences between red deer and wapiti (Geist 1971, 1998; Polziehn and Strobeck 1998). Evaluations of mitochondrial DNA sequences of Siberian, European, and North American red deer and wapiti by Polziehn and Strobeck (1998) suggested that North American wapiti and the closely related *Cervus elaphus sibiricus* of eastern Siberia are sufficiently divergent from the Eurasian red deer complex to warrant specific status as *Cervus canadensis.* Polziehn et al. (1998) concluded that Roosevelt and Tule wapiti warrant subspecific status, but that the Rocky Mountain and Manitoban subspecies should be combined. If these phylogenetic investigations are accepted, there will likely be four subspecies of wapiti, including one in eastern Asia, that will be separated from the red deer complex. The eastern Asia subspecies is thought to have crossed from North America back into Siberia approximately 52,000 years ago, during the last connection of the Bering land bridge. Wapiti are thought to have differentiated from red deer after crossing into North America across the land bridge during the Illinoian glaciation, perhaps 246,000 years ago (Guthrie 1966; Polziehn and Strobeck 1998).

DISTRIBUTION

Wapiti are increasing their range, largely through introductions (Fig. 42.1). For instance, the single population of approximately 600 Tule wapiti that existed in the Owens Valley, California, for years has increased to more than 2500 in 22 different populations (McCullough 1969; McCullough et al. 1996). Manitoba wapiti have been reintroduced to the Pembina Gorge of North Dakota, and Rocky Mountain wapiti have been introduced to Arkansas, Kentucky, Michigan, Nevada, North Carolina, Oklahoma, Pennsylvania, Tennessee, and Wisconsin (Rocky Mountain Elk Foundation web site).

Wapiti now occupy more habitat than at any time in the past century, and populations are estimated at 782,000 individuals (Peek 1999). More range is currently occupied by wapiti in British Columbia, Colorado, Oregon, and Washington than in 1800. All states and provinces that have hunted wapiti populations report more occupied range now than in 1930. Wapiti have become naturally established in shrub–steppe in south-central Washington (Rickard et al. 1977) and southern Idaho (Strohmeyer and Peek 1996), reoccupied the Mount St. Helens blast zone within a year after the eruption (Merrill et al. 1987), and were introduced into previously unoccupied habitat in central Nevada (Bryant and Maser 1982).

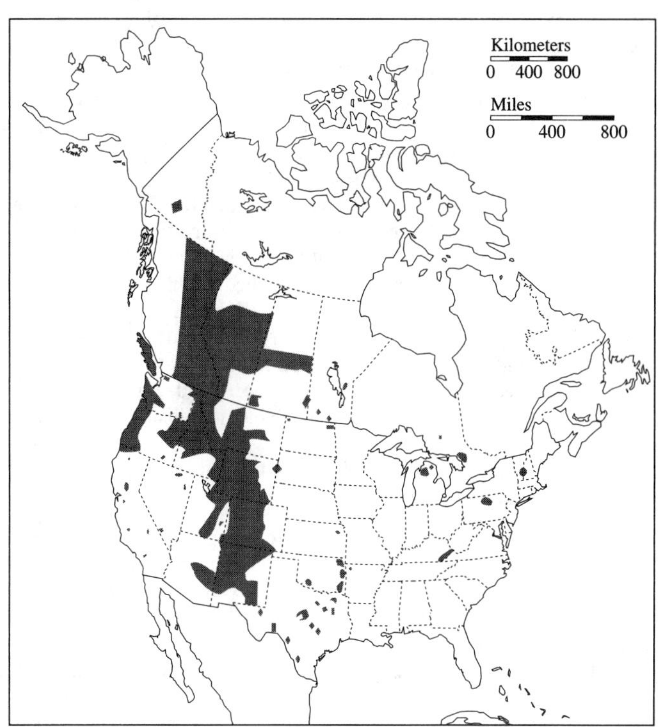

Figure 42.1. Distribution of the wapiti (*Cervus elaphus*).

Wapiti originally occupied most of the United States (Murie 1951; Wisdom and Cook 2000). The misconception that wapiti were originally a "plains" animal, forced to the mountains by the onslaughts of advancing civilization, still prevails. However, wapiti were present in the Yellowstone region before significant intrusion by humans (Lovaas 1970; Boyce 1989). Evidence that wapiti were used by aboriginal people as far back as 10,000 years ago occurs in archaeological excavations in the central Idaho mountains (Sappington 1994; Sappington and Carley 1989). Lewis and Clark took wapiti in 1805–1806 in the upper Missouri River, Columbia River, and Yellowstone River (Martin and Szuter 1999). The species was hunted to extirpation in much of its range east of the Rocky Mountains, but occurred in the mountains as well. It was extirpated from plains and deciduous forests, but is not a recent colonizer of mountainous and forested terrain. Lewis and Clark took Roosevelt wapiti at Fort Clatsop at the mouth of the Columbia River in 1805. McCorquodale (1985) reported wapiti remains from 10 archaeological sites scattered across the Columbia Basin east of the Cascade Range, with dates ranging from 10,000 years ago to recent times.

DESCRIPTION

Morphological comparisons of *C. e. nannodes, C. e. nelsoni,* and *C. e. roosevelti* are shown in Table 42.1 (McCullough 1969).

TABLE 42.1. Some morphological comparisons of Tule (*C. e. nannodes*), Rocky Mountain (*C. e. nelsoni*), and Roosevelt wapiti (*C. e. roosevelti*)

Character	Tule Wapiti	Rocky Mountain Wapiti	Roosevelt Wapiti
Antler shape	Light and spreading, branches curved	Light and spreading, branches straight	Heavy and crowning, branches short
Pelage color	Light	Dark	Dark
Skull form	Short and broad	Intermediate	Long and slender
Length of toothrow	Longest	Shorter	Shortest

SOURCE: Adapted from McCullough (1969).

Size and Weight. Substantial variation in size and body weight attributable to environmental factors exists among individuals and populations. Newborn calves weigh about 15 kg (Johnson 1951), and virtually double their weight within 2 weeks after birth (Table 42.2). Yearlings in their second fall are about 65–70% of adult weight. Adult cows weigh about 80% of what adult bulls weigh. Authenticated records indicate that bulls may weigh over 450 kg; a bull on Afognak Island was estimated to weigh 590 kg (Troyer 1960).

The range in body weight of the subspecies overlaps considerably, but Tule wapiti are significantly smaller. However, specimens from recently introduced populations are larger. Average live weight of seven Tule wapiti bulls was 194 kg, but some large individuals may exceed 318 kg. Weights of wapiti introduced to new habitats may exceed the average weights of those from long-established populations.

Weights of adult bulls of other subspecies average over 300 kg. Flook (1970) reported that at Banff and Jasper National Parks in Alberta, body weights of ≤4-year-old and older bulls were 275–375 kg (mean = 333 kg) and those of females >3 years old were 230–260 kg (mean = 259 kg). Bulls are heaviest in August or September just before or early in the rut, and lightest in March (Flook 1970; Mitchell et al. 1976). Males tend to reach maximum weights at 5–9 years and females at 3–7 years old (Greer and Howe 1964; Flook 1970).

Pelage. Calves are born with cream-colored spots on a russet coat. Spots become progressively less apparent through the summer, and disappear in August. The rump patch is deep yellow to almost orange, fading to cream color in winter. Adults are dark brown to brownish red in summer; however, winter pelage acquired in late summer is characteristically dimorphic. Adult bulls are easily recognized by the light cream-colored coat contrasted with a darker mane. Yearling bulls also exhibit this characteristic, but not as pronounced. Cows remain darker colored, although coat color fades during the winter. Roosevelt wapiti tend to be darker than the Rocky Mountain subspecies.

Antlers. The wapiti is a "six-point" deer (Geist 1971, 1998), reflecting the most common number of tines per antler on a mature bull. Male calves commonly grow "buttons" that are 2–3 cm long and recognizable only on close inspection. Yearling bulls may grow "spikes" or exhibit up to five tines, but only rarely have a brow tine. Colorado yearling bulls ($N = 1317$) examined by Boyd (1970) were 72% spikes, whereas 28% had two or more tines per antler. Nutritional status affects antler growth. Yearling males that experience difficult winter or other conditions as calves grow fewest tines and small spikes. Spike antlers may vary widely within the range of 15–90 cm in length.

Two-year-old bulls, often called "raghorns," produce antlers with up to six tines, including a brow tine. Blood and Lovaas (1966) reported three Manitoba wapiti that had antlers with three to four tines at 2 years of age. The antlers weighed 1.8–2.8 kg and were 48–85 cm long. Two-year-olds on the Arid Lands Ecology Reserve in south-central Washington had three to six points per antler; 46% had five points and 42% had six points (McCorquodale et al. 1989). This population was growing rapidly at that time, and it appears that 2-year-old bulls typically have fewer than six points (Boyd 1970; Flook 1970; Murie 1951).

Mature bull antlers weigh 7.1–15.5 kg (Geist 1998). Largest antlers are produced by bulls 7–10 years old (Flook 1970; McCorquodale et al. 1989; Wolfe 1983), and decline thereafter. The mean weight of antlers from nine bulls 10.5 years old from New Mexico was 10.1 kg (Wolfe 1983). McCorquodale et al. (1989) reported single antler weights of 4–6 kg from the south-central Washington population. More variation within populations occurs than among the subspecies, except for the smaller Tule wapiti. The record Roosevelt wapiti scored by the Boone and Crockett Club had 11 points on its right antler; a bull taken in Manitoba had 14 points on one antler (Byers and Bettas 1999).

Antlers are shed primarily in late March and early April. They may be shed earlier after mild winters than after severe winters. Bulls in good condition shed earlier than those in poor condition, and younger bulls shed earlier than older bulls. Antler regrowth becomes apparent in late May. They are fully formed by early August, at which time rubbing of velvet begins.

Records of the Boone and Crockett Club indicate that antlers of the Rocky Mountain and the Manitoban wapiti are longer than those of the Roosevelt wapiti (Byers and Bettas 1999). Antler beam lengths up to 162 cm are recorded for Rocky Mountain bulls, 144 cm for Roosevelt bulls, and 93.5 cm for Tule wapiti (Geist 1998).

Dentition. The dental formula for a wapiti with a complete set of adult teeth is I 0/3, C 1/1, P 3/3, M 3/3 (Fig. 42.2). Neonates have deciduous or "milk" incisors and the canine, milk premolars erupted, and the first molar either covered by membrane or just protruding (Johnson 1951). From 8 days to 4 months, the incisors, canines, premolars, and first molar are evident. At age 4 months, the calf dentition is I 0/3, C 1/1, PM 3/3, M 1/1, with incisors, canines, and premolars deciduous. At 16 months, yearlings retain deciduous third incisors, canines, and premolars, and the last molar has not yet erupted, with variation in timing of replacement occurring in the first two incisors and whether or not the second molar is erupted completely (Quimby and Gaab 1957). At 28 months, dental formulas range from the complete set of adult dentition to retention of the deciduous canines and the last molar only partially erupted. Permanent dentition occurs at 3 years of age. Although wear patterns of premolars and molars may be used to estimate the age of adults, examination of annulations in dental cementum is more reliable (Keiss 1969).

REPRODUCTION

The female reproductive organs of members of the genus *Cervus* are generally similar to those of the Bovidae. The uterus is bipartite, similar to that of a cow (*Bos taurus*) (Eckstein and Zuckerman 1956). The average conception period for Montana wapiti was the first week in October (Morrison et al. 1959). McCullough (1969) reported variations in the rut of Tule wapiti from August to September. Flook (1970) found mean conception dates of 11, 19, and 28 September in Banff and Jasper National Parks. One 5-year old cow apparently conceived in November. Wisdom and Cook (2000) reported breeding dates from 29 September to 7 October for captive cow wapiti on adequate nutrition, whereas a cow on a low-nutrition diet did not breed until 27 October. Wishart (1981) reported a near-term fetus in a cow shot in Alberta in late September, suggesting a January breeding date.

Wapiti are spontaneous ovulators, and apparently exhibit a reproductive cycle of about 20 days (Morrison 1960; Wisdom and Cook 2000). A "silent" heat or short estrous cycle of about 11 days may occur before the first complete estrus (Morrison 1960; Wisdom and Cook 2000). Initiation of ovulation is apparently related to declining daylight.

TABLE 42.2. Weights and measurements of North American wapiti

Subspecies	Location	No.	Age	Sex	Whole Weight (kg)	Length (cm) Total	Tail	Hind Foot	Ear	Reference
nelsoni	Montana	23	1 d	Both	14 (9–20)	97 (96–112)	5 (4–7)	39 (35–43)	11 (10–13)	Johnson 1951
nelsoni	Montana	47	5–7 d	Both	20 (15–27)	108 (97–117)	6 (5–7)	41 (39–44)	12 (11–13)	Johnson 1951
nelsoni	Wyoming	1	2 mo	F	59	—	—	—	—	Murie 1951
nelsoni	Colorado	2	8–9 mo	F	114 (107–123)	—	—	—	—	Boyd 1970
nelsoni	Colorado	4	8–9 mo	M	111 (91–120)	176 (169–184)	8 (6–10)	43 (29–57)	13 (9–17)	Boyd 1970
manitobensis	Manitoba	1	5 mo	F	134	—	—	—	—	Blood and Lovaas 1966
manitobensis	Manitoba	1	8 mo	M	133	—	—	—	—	Blood and Lovaas 1966
nelsoni	Missouri	3	5–6 mo	F	123 (118–132)	—	—	—	—	Murphy 1963
nelsoni	Missouri	2	5–6 mo	M	104 (91–118)	—	—	—	—	Murphy 1963
roosevelti	Alaska	2	15–16 mo	F	272 (250–293)	—	—	—	—	Troyer 1960
nelsoni	Missouri	4	16–17 mo	F	157 (143–195)	—	—	—	—	Murphy 1963
nelsoni	Missouri	3	16–17 mo	M	197 (182–222)	—	—	—	—	Murphy 1963
nelsoni	Colorado	2	18–20 mo	M	146 (137–163)	210 (205–215)	12 (11–12)	64 (63–65)	19 (19–20)	Boyd 1970
nelsoni	Colorado	3	18–20 mo	F	178 (143–220)	—	—	—	—	Boyd 1970
nelsoni	Montana	2	19 mo	F	—	201 (190–211)	13	62 (61–62)	21 (20–21)	Quimby and Johnson 1951
nelsoni	Montana	4	19 mo	M	—	200 (196–207)	12 (9–13)	60 (57–62)	19 (18–20)	Quimby and Johnson 1951
manitobensis	Manitoba	1	18 mo	M	220	—	—	—	—	Blood and Lovaas 1966
nelsoni	Colorado	16	3 yr + winter	M	237 (193–285)	221 (207–239)	11 (9–19)	64 (62–67)	20 (18–22)	Boyd 1970
nelsoni	Montana	11	3 yr + winter	M	255 (245–292)	227 (208–248)	14 (10–18)	63 (60–67)	20 (18–22)	Quimby and Johnson 1951
nelsoni	Montana	10	3 yr + winter	F	331 (298–373)	242 (231–251)	14 (13–16)	67 (64–69)	21 (19–23)	Quimby and Johnson 1951
roosevelti	California	9	3 yr + winter	M	215 (171–292)	221 (206–234)	13 (10–17)	66 (64–69)	22 (20–22)	Harper et al. 1967
roosevelti	California	9	3 yr + winter	F	254 (178–326)	234 (206–246)	12 (11–13)	69 (66–74)	21 (19–22)	Harper et al. 1967
roosevelti	Alaska	10	4 yr + winter	F	381 (336–497)	—	—	—	—	Troyer 1960
manitobensis	Manitoba	8	3 yr + winter	F	353 (288–478)	241 (234–262)	12 (10–17)	69 (68–72)	22 (21–22)	Blood and Lovaas 1966
manitobensis	Manitoba	4	3 yr + winter	M	275 (258–289)	224 (198–239)	11 (8–14)	67 (62–69)	20 (20–21)	Blood and Lovaas 1966

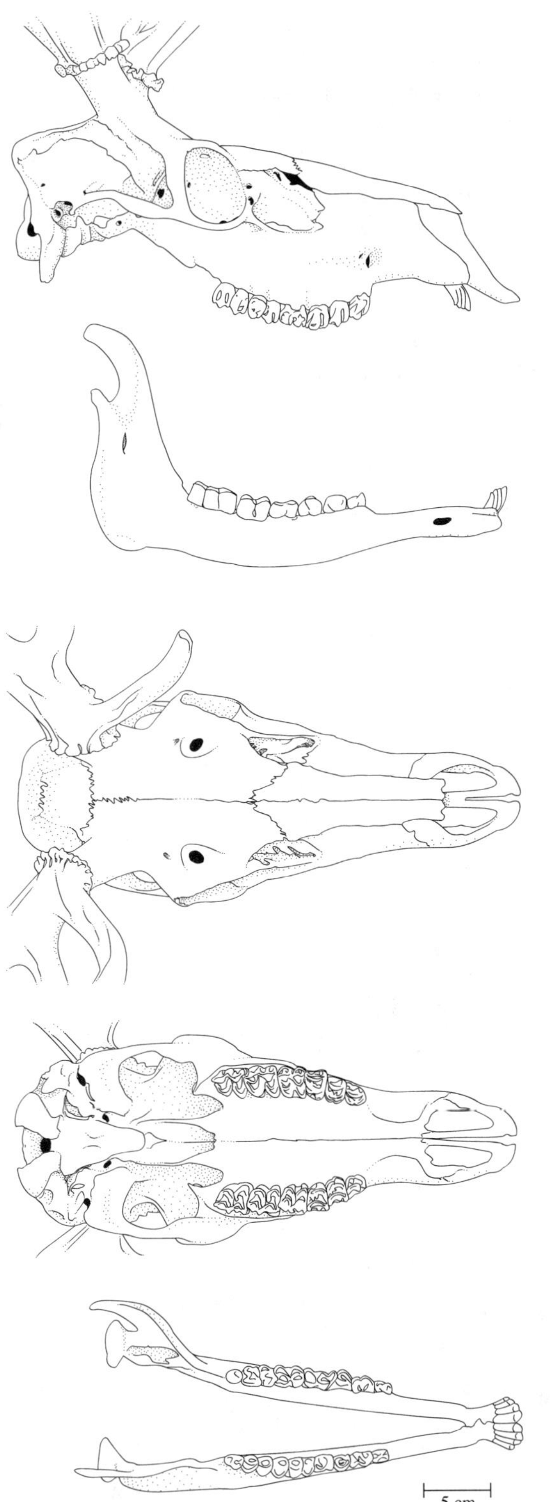

FIGURE 42.2. Skull of the wapiti (*Cervus elaphus*). From top to bottom: lateral view of cranium, lateral view of mandible, dorsal view of cranium, ventral view of cranium, dorsal view of mandible.

Calving occurs around 1 June (Smith et al. 1996), so if most breeding occurs near 1 October, gestation is approximately 244 days. Noyes et al. (1996) reported conception dates ranging from 9 September to 24 November among cow wapiti in northeastern Oregon between 1989 and 1993. Yearling males can be fertile (Conaway 1952) and contribute substantially to breeding, especially when adult bulls are absent or greatly reduced in number because of hunting. When bulls 5 years old and presumably older are the primary breeders, the rut may occur earlier, and may be reduced by as much as 1 month (Follis 1972; Squibb et al. 1986; Noyes et al. 1996). When yearling bulls are the primary breeders, the rut may extend to 71 days.

Wapiti may exhibit a postconception ovulation (Halazon and Buechner 1956; Douglas 1966). Ovaries of anestrous wapiti are characterized by the absence of corpora lutea or follicles with >2 mm diameter (Halazon and Beuchner 1956). Proestrous ovaries collected in August and September had several follicles 7–11 mm in diameter. Estrous ovaries had one freshly ruptured follicle. After ovulation and during pregnancy, a second period of follicular development occurs. One follicle ovulates to form a secondary corpus luteum, which is smaller than the primary corpus luteum formed by the initial ovulation. Primary corpora lutea range from 10 to 18 mm in diameter, whereas secondary corpora lutea average 5–10 mm. About 55% of the secondary corpora lutea appear in the same ovary as the primary one. About 60% of 288 female wapiti Halazon and Buechner (1956) examined had a secondary corpus luteum. The postconception ovulation may be caused by insufficient placental hormone production before implantation of the embryo to suppress secretion of gonadotropins.

Yearling females may be fertile (Coffin and Remington 1953; Buechner and Swanson 1955; Gogan and Barrett 1987). The percentage of yearlings that are capable of breeding depends on the nutritional status of the population. Animals that are born late and develop slowly through late summer drought or a severe winter are not as likely to reach sexual maturity as those that are born early and have better nutrition.

Pregnancy rates for yearling females averaged 24% and ranged from 0% to 81% for 24 investigations reviewed by Follis (1972). Pregnancy rates of adults >1 year old averaged 86% and ranged from 43% to 100%. In areas where pregnancy rates are low, nutritional status of the population likely is low (Greer 1966; Thorne et al. 1976; Gogan and Barrett 1987). Cows >8 years old may be less fertile, especially during less favorable nutritional conditions (Greer 1966; Lowe 1969). Thorne et al. (1976) reported that cows that lost more than 3% of their body weight between January and late May produced small calves, which were less likely to survive. Mitchell and Brown (1973) reported that lactating red deer cows in Scotland were less likely to breed the following year, again depending on nutritional status of the individual. Wisdom and Cook (2000) showed that cows on a low nutritional plane did not undergo a reproductive cycle and did not ovulate.

One calf is produced annually, usually in late May or early June; twinning is rare (Greer 1966). Sex ratios of neonates tend to be biased toward males (Peek et al. 1967; Flook 1970; Clutton-Brock et al. 1982; Smith et al. 1996). Smith et al. (1996) concluded that winter stress on gravid females reduced survival of male fetuses, and survival of males after birth is lower than for females in all age classes.

FEEDING HABITS

Wapiti diets are highly variable and depend on local availability of forage. Feeding habits overlap those of other ungulates, and wapiti are able to vary their diets on the same area according to season and availability of different items. Kufeld (1973) reported 159 forbs, 59 grasses, and 95 shrubs have been recorded in the wapiti diet. Wapiti may be classified as intermediate, opportunistic, mixed feeders (Hofmann 1989). Species that fit within this category choose a mixed diet, practice a high degree of forage selectivity, avoid fibrous plants, can increase food intake when food is plentiful, and go through lactation, juvenile growth, and accumulation of energy reserves during periods when forage is most nutritious and available. They reduce metabolism and food intake during winter when fibrous plants of lower quality are the only foods available (Hofmann 1989). Mould and Robbins (1982) reported that secondary plant compounds found in many forbs and shrubs that depress their digestibility are used by wapiti in a mixed diet. Mould and Robbins (1981) reported elk consume 63–896 g of protein/day, which is indicative of how much variation in intake can occur. Thus, maintaining diversity in plant communities is an important consideration in management of wapiti habitat.

Grasses or shrubs constitute the major winter diet. Roosevelt elk in Oregon subsisted on 56% browse with trailing blackberry (*Rubus* spp.), grasses, sedges, and salal (*Gaultheria* spp.) as preferred species (Harper 1971). The White River wapiti in Colorado also used 56% browse in winter, including oak brush (*Quercus gambelli*), aspen (*Populus tremuloides*), serviceberry (*Amelanchier alnifolia*), big sagebrush (*Artemisia tridentata*), and snowberry (*Symphoricarpos albus*) (Boyd 1970). Hash (1974) reported that browse constituted 92% of the wapiti diet in the Lochsa River area of Idaho. Redstem ceanothus (*Ceanothus sanguineus*), mountain maple (*Acer glabrum*), scouler willow (*Salix scouleriana*), shiny leaf ceanothus (*Ceanothus velutinus*), and serviceberry were preferred items. Montana studies summarized by Rognrud and Janson (1971) are representative of wapiti wintering on grasslands. Climax grasses such as bluebunch wheatgrass (*Pseudoroegneria spicata*), Idaho fescue (*Festuca idahoensis*), western wheatgrass (*Agropyron smithii*), and rough fescue (*Festuca scabrella*) were preferred forage species. Grasses constituted 65–100% of the winter diet on these grassland ranges.

The spring diet reflects a transition from winter to summer foods, with grasses often being important. As summer nears, forbs become important, although leaves of browse species and grasses may continue to be important. Pale agoseris (*Agoseris glauca*), wild onion (*Allium* spp.), arnicas (*Arnica cordifolia*), fireweed (*Epilobium angustifolium*), balsamroot (*Balsamorhiza sagitata*), dandelions (*Taraxacum* spp.), and lupines (*Lupinus* spp.) are among the important species. However, Baker and Hobbs (1982) found wapiti on alpine and subalpine summer range in central Colorado to take graminoids most frequently. Fall diets often revert to predominately grass or browse.

Merrill et al. (1987) reported that wapiti using the Mount St. Helens blast zone in southwestern Washington made primary use of forbs in summer, including horsetail (*Equisetum* spp.), catsear (*Hypochaeris radicata*), pearly everlasting (*Anaphalis margaritacea*), and fireweed (*Epilobium angustifolium*). They switched to grasses (*Festuca arundinacea* and *Elymus* spp.) in the fall. Shrubs, including willow (*Salix* spp.), elderberry (*Sambucus racemosa*), salmonberry (*Rubus spectabilis*), and maples, were common in the diet during all months.

Baker and Hobbs (1982) reported that summer diets provide abundant energy and protein, but during drought years quality may decline rapidly. Cook et al. (1996) noted that diet quality in late summer and fall was a major influence on growth of wapiti calves and was linked to overwinter survival. Severe, prolonged winters that reduce forage availability and quality also reduce growth and survival, especially when coupled with low forage quality in late summer.

Roosevelt elk inhabiting the Hoh Valley inside Olympic National Park made extensive use of spruce–hemlock forests (Jenkins and Starkey 1982). In these coastal forests, clearcuts provide extensive forage. However, the forage in the openings has higher levels of tannins, which decrease the digestibility of the plants (Happe et al. 1990). The lower levels of understory forage biomass in old-growth forest are offset by the higher digestibility than in the clearcuts and openings.

POPULATION CHARACTERISTICS

Elk can live more than 20 years (Quimby and Gaab 1957), but mean life expectancies are much lower, and are significantly different between the sexes (Peek et al. 1967; Flook 1970; Kimball and Wolfe 1974). Populations where exploitation is low, characterized by older animals, may have greater disparity in sex ratios because bulls do not live as long as cows. Higher mortality rate of bulls is related to rutting activities, which cause great energy expenditures and predispose bulls to malnutrition and death, especially when winter energy intake is inadequate (Mitchell et al. 1976). Populations where exploitation is high for bulls and low for cows may show very imbalanced sex ratios as well. Sex ratios may vary from as low as 4:100 in heavily hunted populations to >40:100 in relatively unexploited populations with quality forage (Freddy 1987). Berger and Gompper (1999) found no strong relationship between presence of predators and adult sex ratio in ungulates. They concluded that selection processes operating differently on the two sexes were responsible for differences in survival of the sexes.

Calf survival is also variable among years and populations. The typical criterion used to estimate production and survival is the cow:calf ratio. This ratio is not indicative of population trend (Caughley 1974), but represents some combination of survival and production of calves and survival of adults. Although adult female survival has the greatest influence on rate of population increase (Nelson and Peek 1982), calf production and survival may be more variable in most populations (e.g., Sauer and Boyce 1983; Coughenour and Singer 1996). Given this assumption, the cow:calf ratio reflects changes in calf production and survival more than changes in the adult population. Winter cow:calf ratios have ranged from over 100:70 (Boyd 1970) to less than 100:10 (Dekker et al. 1995). Variation is attributed to fluctuating survival patterns of neonates rather than to prepartum mortality. Pregnancy rates of the Northern Yellowstone population fluctuated little over a 6-year period (Greer 1966) in comparison with cow:calf ratios (Houston 1982).

Thorne et al. (1976), Smith et al. (1996), and Singer et al. (1997) reported that survival of elk calves was related to their weight at birth. Calves weighing less than 11.4 kg at birth had only a 50% chance of survival (Thorne et al. 1976). Even if postpartum losses are more critical than prepartum mortality, the nutritional status of the pregnant cow is involved in calf survival.

Predation by wolves (*Canis lupus*), coyotes (*Canis latrans*), and bears (*Ursus* spp.) can be a major mortality factor on wapiti calves. Dekker et al. (1995) reported substantial predation by wolves on wapiti in Jasper National Park, and Hamer and Herrero (1991) reported grizzly bear predation on calves in Banff National Park. Singer et al. (1997) noted that calves dying from predation weighed less at birth than those surviving in summer in Yellowstone National Park. Calves born late were less likely to survive than those born early. Predation was primarily by grizzly bears (*Ursus arctos*) and coyotes. Schlegel (1976) found that black bears (*Ursus americanus*) and cougars (*Puma concolor*) were the major mortality factors on calves in central Idaho. However, Hornocker (1970) did not detect that mountain lion predation was sufficient to keep a wapiti population in central Idaho from increasing. Wapiti constitute a primary food source for wolves in some areas, and individuals that are preyed on are not necessarily "culls," diseased, young, or old (Cowan 1947). Thus, predation can affect population dynamics of wapiti, especially through its impact on calf survival. However, wapiti have evolved patterns of behavior and survival that allow populations to coexist with predators. The influence of predation is an integral and essential component of the species's ecology.

Brucellosis is a major factor in reducing calf production and survival in wapiti in Jackson Hole, Wyoming (Thorne et al. 1976). In 29 cows that were pregnant and infected, 12 (41%) lost a calf. First-time breeders are most likely to lose calves, and subsequent pregnancies tend to be successful. Over 30% of the wapiti on feed grounds in the Jackson Hole region may test positive for brucellosis.

For most wapiti populations, hunter harvest and associated extralegal losses such as crippling and poaching are the major mortality factors. Losses to malnutrition can be significant during severe winters in some populations, whereas predation may be important in others. Some mortality occurs because of combat between bulls during the rut (Flook 1970; Geist 1986), and accidents, such as drowning (Martinka 1969a, 1969b), have been reported.

The maximum rates of increase for wapiti populations are documented for colonizing populations (Eberhardt et al. 1996). A population colonizing the south-central Washington shrub–steppe increased at a rate of 24%. A maximum sustainable rate of increase for elk could be as high as 28%. Merrill et al. (1987) reported a rate of increase of 0.35 during 1983–1986 for wapiti colonizing the Mount St. Helens area after the eruption of the volcano. Pregnancy rates were 31% for yearling cows, 33–50% for 2-year-old cows, and 87% for older cows in this population.

MOVEMENT

Some wapiti undertake long seasonal migrations between summer and winter ranges, whereas other populations are nonmigratory. In migratory herds, there may be segments that do not migrate. Populations in and around Yellowstone National Park provide examples of both. Movements between the upper limits of summer range and lower limits of winter range for the Gallatin population extend over 80 km (Brazda 1953). The Jackson Hole population wintering on the National Elk Refuge may migrate as far as 100 km to summer range (Boyce 1989). The Northern Yellowstone population wintering near the north-central boundary also migrates long distances (Craighead et al. 1972). Other migratory populations include the Sun River, Montana (Picton 1960), and the White River, Colorado (Boyd 1970), herds.

In contrast, a population in the Madison River drainage inside Yellowstone National Park is essentially nonmigratory and exhibits local shifts in habitat use rather than pronounced seasonal movements (Craighead et al. 1973). The Roosevelt wapiti occupying Boyes Prairie, California, are nonmigratory, but populations in the Siskiyou Mountains were migratory before they were extirpated (Harper et al. 1967). The Manitoba wapiti occupying Riding Mountain National Park are apparently nonmigratory, but do exhibit movements of at least 8 km to agricultural areas during more severe winters (Blood and Lovaas 1966). The Wind Cave National Park population is confined by fence and survives well without migration (Varland et al. 1978). Wapiti in the Missouri River Breaks of northeastern Montana are nonmigratory, but use different portions of their range at different times of the year (Mackie 1970). Some populations, such as those in the Selway River, Idaho, have segments that exhibit small seasonal shifts in habitat use, whereas other segments exhibit long movements between summer and winter ranges (Dalke et al. 1965a). Almost all populations that are migratory contain segments that are associated with winter ranges all year long, as with the Jackson Hole population (Martinka 1969b) and the White River, Colorado, population (Boyd 1970).

Calving areas in the Gallatin River, Montana (Johnson 1951), the Selway River, Idaho (Young and Robinette 1939), and the Sun River, Montana (Picton 1960), are on the upper elevational limits of winter ranges. Johnson (1951) reported newborn calves on the Gallatin River area were located predominantly in interspersed sagebrush–timber. They were within 70 m of timber if in sagebrush, and within 9 m of sagebrush if in timber. The importance of the ecotone between open and dense cover for calving was apparent.

Mackie (1970) reported cows and calves used Douglas fir (*Pseudotsuga menziesii*)–juniper (*Juniperus horizontalis*) stands in late May and early June in the Missouri River Breaks. Calves were hidden in sagebrush–grass areas in central Idaho (Davis 1970). Harper et al. (1967) found 10 calves born in meadows and grasslands, 6 in riverside hardwoods or brush, and 2 along a spruce–salmonberry meadow edge in the Boyes Prairie.

Interspersion of cover with open areas appears to be critical for young calves that are hidden by their dams. Newborn calves hide under shrubs and taller herbaceous vegetation. The actual calving site may be less important than subsequent sites used by calves during the "hiding" period.

Movement to the upper elevation limits of summer range in mountainous terrain appears to be related to the presence of tabanid flies and vegetation development (Brazda 1953). Cows with calves appear to lag behind other members of a population in their timing of spring movements. Natural and artificial salt licks on winter range do not appear to affect the timing of movements to summer range in the Selway River area (Dalke et al. 1965b).

Movements onto winter range often are initiated with the first snowstorms. Mature bulls frequently appear first, but most of a population will migrate as snow progressively eliminates higher elevations as foraging areas. Snow accumulations elsewhere may force elk onto south-facing slopes and associated river bottoms at the lowest elevations of the annual range. However, some population segments in the Yellowstone and in other areas remain at very high elevations on grassy, windblown slopes in winter.

Home ranges of wapiti are highly variable relative to vegetation and region (Table 42.3). Areas as small as 3 km^2 were used by wapiti during March–September in California redwoods (Franklin et al. 1975). Irwin and Peek (1983) found 8-km^2 home ranges for radio-collared wapiti using cedar (*Thuja plicata*)–hemlock (*Tsuga heterophylla*) forests in northern Idaho. Martinka (1969a) reported mean home ranges of 12 km^2 for wapiti in Jackson Hole. In xeric sagebrush steppe in southeastern Idaho, Strohmeyer and Peek (1996) reported 245-km^2 mean home ranges for July–September. McCorquodale et al. (1989) reported home ranges of 95–118 km^2 during June–August in south-central Washington sagebrush steppe. They concluded that annual precipitation was inversely correlated with size of home range. Strohmeyer and Peek (1996) suggested that forage availability, juxtaposition of resources, cover quality, ambient temperature, difficulty of travel, population density, plant phenology, and human disturbance also influenced size of home range.

BEHAVIOR

Wapiti are very gregarious, but great variation in degree of sociality exists between seasons, sexes, and populations. Murie (1951) and Altmann (1952) observed cows seeking seclusion from other wapiti before and

TABLE 42.3. Summary of mean summer home range areas for adult wapiti, standardized across the investigations

Area (km^2)	Period	Elk	Habitat	Source
3[a]	June–Aug	Herd	Beech–redwood	Franklin et al. 1975
6	Apr–Aug	Females	Prairie–oak	Waldrip and Shaw 1979
8	16 June–8 Sep	Females	Cedar–hemlock	Irwin and Peek 1983
10	21 June–23 Sep	Females	Lodgepole pine	Craighead et al. 1973
11[b]	5 June–15 Sep	Females	Spruce–hemlock	Jenkins and Starkey 1982
12	June–Sep	Females	Spruce–fir–alpine	Boer et al. 1989
12	6 May–15 Sep	Males	Conifer	Martinka 1969
12	6 May–15 Sep	Females	Conifer	Martinka 1969
16	June–Sep	Males	Spruce–fir–alpine	Boer et al. 1989
20	June–Aug	Males	Conifer–chaparral	Wallace 1991
31	June–Aug	Females	Conifer–chaparral	Wallace 1991
95	June–Aug	Females	Shrub–steppe	McCorquodale et al. 1989
118	June–Aug	Males	Shrub–steppe	McCorquodale et al. 1989
165	July–Sep	Herd	Sagebrush	Strohmeyer and Peek 1996
245	July–Sep	Herd	Juniper–lava	Strohmeyer and Peek 1996

NOTE: Minimum convex polygon procedure used unless indicated otherwise.
[a]Boundaries based on vegetation types.
[b]95% probability ellipse.

during parturition. The calf hides for the first 18–20 days of its life, associating with the cow for short periods to nurse (Lent 1974). Calves join "nursery herds" a few weeks after parturition, and then remain in cow–calf groups of various sizes through the summer (Altmann 1952). Martinka (1969a) indicated that by mid-August, a decrease in group size and an attachment of adult bulls to cow–calf–yearling bull groups occurred at Jackson Hole.

Aggregations appear to be related to vegetation density, with the largest groups occurring in the most open habitats (Knight 1970). In summer, cow–calf bands of up to 400 individuals including young bulls have been observed in White River, Colorado, and in Yellowstone National Park. Conversely, mean group sizes in the Sun River, Montana, population, which occupies more densely forested terrain, were 7.2 in open areas and 2.5 in timbered areas (Picton 1960). Varland et al. (1978) reported that aggregations were lowest in summer and highest in winter in Wind Cave National Park, a typical observation.

A possible reason for variation in aggregation size relative to vegetative cover is that the large group in open terrain provides security for individuals that substitutes for cover (Crook 1970; Kie 1999). Protection against predators, availability of forage, population density, breeding activities, and weather and snow conditions are other influences on aggregation patterns.

Wapiti are polygamous, the bull collecting a harem of cows and calves. Altmann (1952) reported that bulls actively searched for and joined existing cow–calf groups. Adult bulls may tolerate yearling bulls in harems but not older bulls, and in some cases will chase male yearlings out as well (Clutton-Brock et al. 1982). Knight (1970) and Clutton-Brock et al. (1982) found that bulls with and without harems present defended areas during the rut. Scent may be used to identify territories or the presence of the bull (Graf 1956).

Struhsaker (1967) reported rutting groups in fall consisted of 2–26 animals with one adult bull associated with a harem of cows, calves, and occasionally a yearling bull. Large groups of yearling and young adult bulls, apparently nonbreeding, were observed during the rut. Solitary adult bulls also occur, often those searching for harems or displaced from rutting activities by other adult bulls. In the Northern Yellowstone herd, the harem may be replaced by a large mixed-sex-and-age rutting group. This occurs at high densities, in open terrain, where there are many males relative to females.

Mature bulls are hostile toward each other during the rut. Clutton-Brock et al. (1982) reported that mature red deer stags on Rhum, Scotland, fought at least once every five days during the rut, and fought at least 5 times during a breeding season. Fights between mature bulls are distinguished from sparring matches that usually take place between unequally matched individuals. Fights may be preceded by a vocal contest and a parallel walk (Clutton-Brock et al. 1982). Contestants may fight for a considerable time. Eventually one is pushed backward, breaks contact, and runs off. Extensive wounding, in some cases leading to death, occurs during these fights.

Sexes tend to occupy separate areas (Peek and Lovaas 1968). Adult bulls on winter range in the Gallatin, Montana, area tended to concentrate on the fringes of the range, whereas cows, calves, and younger bulls most frequently occupied the central portions. This segregation also is evident at other times of the year, except during the breeding season. Main et al. (1996) and Kie and Bowyer (1999) concluded that sexual segregation was related to predator avoidance as a means to increase reproduction and survival, rather than to competition between sexes for food.

Rutting behavior of bulls consists in bugling, thrashing, digging, rubbing of antlers, wallowing, sparring, and a series of other aggressive displays reflecting various levels of physiological intensity. Struhsaker (1967) identified the following displays: head extension (a means of herding the harem), sexual approach (approach to a cow with head held upright and muzzle tilted downward), muzzling (the perineum of a cow or calf), flehmen (lip curling or holding the mouth open), jawing (by cows, opening and closing the mouth rapidly), head lowering (head lowered to ground, ears pressed back against the head), throat over back (bull over a cow), and mounting for purposes of copulation. Agonistic behavior not necessarily associated with one sex or with breeding activity included one individual displacing another from a site, kicking forward with stiff foreleg (primarily by cows and calves), elevated muzzle, and raising of hind legs. Alarm calls are a bark. A squeal, emitted by all sex and age classes, serves to retain group cohesion.

Although wapiti are highly gregarious, aggregations are not consistently composed of the same individuals (Knight 1970). Considerable interchange among groups occurs, and associations appear to be short termed. A cow with a calf up to 3 months old may be the most consistent, stable social group in this species.

MANAGEMENT, CONSERVATION, AND RESEARCH NEEDS

Management of wapiti populations is intensive because demand by hunters is very high. Changes in habitat and population management are related to the affects that hunters have on populations. Reintroduction of the gray wolf into the northwestern United States has implications for management. Farming of wapiti and resulting potential for spread of diseases also affect management.

Population management should be predicated on the assumption that recruitment and juvenile survival are density dependent (Dennis and Taper 1994). This has been demonstrated for wapiti in the Jackson Hole (Boyce 1989) and the Northern Yellowstone populations (Coughenour and Singer 1996). Populations that are at high levels relative to forage resources will not be as productive as populations that are at lower levels. When the Northern Yellowstone population numbered about 5000, summer calf recruitment rate was 0.5–0.6 calf/cow. When the population increased to 15,000, summer calf recruitment was 0.3. Conversely, winter calf mortality rate was negligible when the population was at 5000 and increased to over 0.60 when the population was 15,000. Cow:calf ratios in Grand Teton National Park averaged approximately 100:45 when the population of cows, calves, and yearling males was approximately 500. The ratio declined to approximately 100:25 when the population was 1300 (Boyce 1989). When the White River, Colorado, population was composed of young bulls and cows in the 1960s, high cow:calf ratios of 100:60 were typical. As the population aged and became larger, cow:calf ratios declined to lower levels (Freddy 1987).

The problem of low survival of bull elk in hunted populations is now a major concern (Bender and Miller 1999). A number of populations have exhibited declining calf production and survival as sex ratios have declined, including the White River, Colorado, population (Freddy 1987) and populations in Oregon (Noyes et al. 1996) and Washington (Bender and Miller 1999). Hines and Lemos (1979) and Prothero et al. (1979) reported that yearling bulls did not reach sexual maturity until the breeding season was in progress. In situations where yearlings do most of the breeding, calves were born in July instead of early June. Calf production and survival also tended to be lower when yearling bulls were the primary breeders. Noyes et al. (1996) concluded that conception dates occurred earlier as bull survival increased to 5 years from 1 year in an experimental, semiconfined population where individuals were known. The rut was more synchronous and shorter when older bulls were principal sires; pregnancy rates increased from 89% to 97% as age increased. Bull survival has to be skewed downward appreciably before calf production and survival is affected. Kimball and Wolfe (1974) observed no change in pregnancy rates of a Utah population when harvest of yearling bulls, an indication of proportion in the population, increased from 36% to 75% of males taken.

The type of hunting season has an effect on bull survival (Thelen 1991; Bender and Miller 1999). Limited-entry or permitted hunts appear to be the most successful means of increasing bull survival. In southwestern Washington, mortality of bulls declined over 50% compared to unlimited hunting seasons that included taking any bull and those with three or more antler points (Bender and Miller 1999). Much of the variability depends on how much access to wapiti habitat is available and how many hunters are present. Hunting seasons based on restriction of harvest to elk with a minimum number of points appear to be least successful in retaining genetic diversity of bulls (Thelen 1991).

Peek (1986) projected harvest from different model wapiti populations that represent commonly encountered situations in hunted areas. Populations were set at 2000 with different sex and age ratios, and age structures were partitioned according to the ratios with the assumption that life expectancy would be 10 years, a low estimate. For a highly productive elk population with adult survival rates of 95% and calf survival rates at 80%, a harvest of 75% of the bulls for a period of 10 years would reduce the sex ratio from 25:100 to 7:100. The population could be stabilized by harvesting 10% of the calves and 15% of the cows along with the 75% bull harvest.

An old-age population with a cow:calf ratio of 100:34 and a sex ratio of 74:100, approximating a red deer population studied by Lowe (1969), showed a reduction in sex ratio to 5:100 when 75% of the bulls were harvested for a 5-year period. Harvesting 1% of the cow–calf population resulted in a population decline, resembling a bull-only season where some extralegal loss from poaching and wounding takes place. In this case, the 75% bull harvest resulted in just under 9% of the population being harvested.

These results suggest that harvest of bulls must be <75% of the population to maintain 6-year-old bulls in a population. It is evident that unlimited-entry hunting seasons for bull wapiti are not adequate to maintain them in populations. Efforts to maintain more bulls by restricting seasons to yearling bulls only may be temporarily satisfactory, but eventually quotas on bulls must be established for each population to adequately control harvest.

In areas where wolves and other predators cause extensive mortality to wapiti populations, hunter harvest may be affected. Vales and Peek (1995) concluded that a reduction in hunter harvest of cows might be a consequence of the presence of wolves in an area where mule deer (*Odocoileus hemionus*) and white-tailed deer (*O. virginianus*) are also prey species. Although bull harvest might not be affected by the presence of wolves, the probabilities that cows and calves will have to be left unhunted will increase, especially when severe winters or other adverse conditions affect populations.

Elk census techniques vary extensively with the nature of the terrain involved and the extent of forest cover. Early surveys in the Blue Mountains of Washington by Beuchner et al. (1951) showed the utility of using fixed-wing aircraft in conjunction with ground counts. Lovaas et al. (1966) used repeated fixed-wing surveys in Montana, considering them adequate to reflect trends if the weather conditions that affect visibility of the elk were considered. Weckerly and Kovacs (1998) did not consider helicopter surveys of elk in densely forested habitats in coastal California to be as effective as systematic surveys from the ground. Staines and Ratcliffe (1987) concluded that direct observation of red deer in open habitat was preferred, but that fecal counts were more useful for forested habitats.

Total counts of individuals always provide minimum estimates of population size and are not subject to statistical analysis. The failure to count all individuals, the visibility bias, can be corrected for if the factors that lead to not counting animals can be assessed. Cook and Martin (1974) demonstrated that the number of animals in a group affects their visibility, with the larger groups having a greater probability of being observed than the smaller groups. In addition, LeResche and Rausch (1974) demonstrated that observer variation contributes to variability in seeing animals. Samuel et al. (1987) developed a "sightability" model using group size and percentage vegetative cover to estimate the probability of an animal being seen. This model was incorporated into an aerial survey procedure by Unsworth et al. (1991). Unsworth et al. (1990) demonstrated the efficacy of the procedure by comparing the results of the survey with a known number of elk on the National Bison Range in Montana. Otten et al. (1993) concluded that a stratified random sampling based on relative elk density coupled with a visibility bias correction using group size and forest cover provided an estimate comparable to a complete ground count that was less expensive and quantified the uncertainty of the estimate for a Michigan population. Eberhardt et al. (1998) dispensed with the development of a visibility bias curve by using a Peterson index involving use of marked groups of elk, and concluded that group sizes of elk were highly variable. Weckerly (1996) also used a mark–recapture procedure effectively on coastal Roosevelt elk in California. Anderson et al. (1998) found application of the Samuel et al. (1987) model was most effective in summer surveys in northwestern Wyoming when elk were less gregarious and a variant was more effective when large groups of elk were present.

Elk census must be tailored to the specific conditions involved. Methods that provide estimates with an indication of accuracy and precision are preferable to methods that provide total enumeration, especially for populations that occupy forested or highly dissected terrain where counting is most difficult. Population size, aggregation patterns, forest cover, snow cover, observer experience, and observer ability all affect estimates of elk population size.

Habitat management for wapiti is highly complex and complicated by the interactions involving vulnerability to hunters and human access. Wapiti are habitat generalists, occurring in dry shrub–steppe rangeland as well as in coastal rainforest. Fundamental population regulation theory regarding native ungulates centers on the roles of forage and predation. Keith (1974) concluded that native ungulates lack self-regulating mechanisms, and noted that large predators can account for an appreciable amount of total mortality. Removal of predators could result in population increases. In contrast, Cole (1971) concluded that wapiti in the Yellowstone system were naturally regulated by a weather–forage interaction, and predators were superfluous to the system. This was before the introduction of the gray wolf in 1995 into that system (Bangs et al. 1998). Intensive browsing pressure by wapiti populations at high density may alter vegetation composition in the absence of wolves (e.g., Singer et al. 1998; Wambolt and Sherwood 1999; Riggs et al. 2000). Wolves may cause changes in distribution and numbers of wapiti and subsequent changes in vegetation conditions (Murre 1944; Hornocker 1970; Peek 1980).

Forest cover is obviously not required for wapiti, because productive populations persist in areas devoid of trees (Peek et al. 1982; McCorquodale et al. 1989; Strohmeyer et al. 1999). However, where forest cover is lacking or sparse, wapiti require minimal disturbance to persist (Basile and Lonner 1976; Lyon 1983; Strohmeyer et al. 1999). The hypothesis that a weather-sheltering effect of dense forest overstory cover reduces energy expenditure and enhances survival and reproduction was not supported by extensive studies with captive wapiti in controlled conditions (Cook et al. 1998). Much of the work that was purported to support the hypothesis involved field investigations of free-ranging, radio-collared wapiti. A utilization-availability analysis showed seasonal preference for mature closed canopy forests. However, Fretwell and Lucas (1970) and Hobbs and Hanley (1990) cautioned that population density had an effect on habitat use patterns. Field observations coupled with a utilization-availability analysis need to be considered in the perspective of population density, available habitat, and the level at which selection occurs (e.g., Johnson 1980). Peek (1986) concluded that the actual habitat requirement cannot be determined from observational field study and would have to be studied in experimental conditions. The field observations essentially produced descriptions of habitat use that imply the most-used habitats were required for the population to persist, but were not sufficient to assess cause-and-effect relationships (Cook et al. 1998). The role of overstory cover is thus related to its effects on vulnerability of wapiti to hunting and to the presence of understories used as forage. In areas where mature forests occur, wapiti will find security from disturbance and will demonstrate a preference for their use.

Experience in Idaho with changing access conditions illustrates the relationships of roads and vulnerability of wapiti to hunter harvest. Theissen (1976) reported that road densities of 5 linear miles of road/mi^2 (3.1 km/km^2) commonly have been constructed to facilitate timber harvest. Bull harvest was reduced and increasingly restricted to the least accessible areas in a southern Idaho drainage, where no restrictions on hunter participation and any-bull seasons prevailed. Unsworth et al. (1993) reported that mortality rates of bulls were lower in areas with fewer hunters and less roads and in highly fragmented or dissected terrain. Cause-specific mortality of 121 bulls that were radio-collared included 43 rifle kills, 8 rifle-wounding losses, 4 archery-wounding losses, 2 recovered archery kills, 3 illegal kills, and 9 undetermined mortalities, with 86% of all wapiti deaths associated with hunting.

A comparison of hunter success relative to density of open roads was reported by Gratson and Whitman (2000). Daily hunter density in an open-road area with 1.54 km/km^2 of roads was 0.57 hunter/km^2, whereas in a managed access area of 0.56 km/km^2 roads, daily hunter density was 0.14 hunter/km^2, and a least-roaded area of 0.23 km of roads/km^2, had a daily hunter density of 0.18 hunter/km^2. Hunter success ranged from 14.8% in the area of highest access to 24.8% in the least-roaded area. Access management in hunted areas is an important means of managing hunter harvest, but limited-entry hunting seasons will ultimately provide the best means of ensuring adequate survival of bulls.

Wapiti may be conditioned to the presence of humans (Thompson and Henderson 1998), particularly in national parks (Schultz and Bailey 1978). However, even in the national parks, direct disturbance by recreationists may cause them to abandon important habitat (Cassirer et al. 1992). In areas where wapiti are hunted, distribution away from roads during the nonhunting seasons is still evident (Rowland et al. 2000), which shows the need to minimize disturbance at any time of the year for areas that are known to be important wapiti habitat. Phillips and Alldredge (2000) reported that repeated disturbance of female wapiti for 3–4 weeks during calving caused significant declines in survival of calves.

The major disturbance to much forested habitat used by wapiti has been timber harvest activities. Although these activities have contributed to better forage conditions, associated access and human disturbance have precluded use of otherwise suitable habitat. Lyon (1979) reported that elk abandoned active logging areas and were slow to reestablish their presence if the activity continued for >4 years. Current trends in forest management that provide more retention of forest cover when logging occurs and take into account the needs of wapiti should alleviate these problems. However, there is still a need for population managers to coordinate methods of harvest with forest managers, so when additional habitat is modified, hunter distribution and numbers can be regulated to minimize dislocation of wapiti from suitable habitat.

In the late 1950s and early 1960s, concern over competition for forage between wapiti and livestock increased. Investigations revealed that the potential for competition for forage among wapiti, cattle, and domestic sheep was indeed great (Blood 1966; Skovlin et al. 1968; Stevens 1966), especially when rangeland conditions were poor. Efforts to coordinate grazing programs and wapiti habitat needs and to improve rangeland conditions have contributed to the increases in wapiti populations since then.

Concerns over livestock grazing on rangelands have been redirected toward the influences on riparian zones and woody vegetation. However, as wapiti populations have increased, concerns are prevalent over their influences on rangelands where livestock, primarily cattle, are being reduced (Hobbs et al. 1996). One investigation in eastern Idaho by Kelly (1995), where wapiti populations had increased five-fold since 1980, demonstrated that cattle removed approximately 90% of the forage that was grazed; other herbivores, including wapiti, removed just over 10%. Wapiti diets were primarily forbs during the growing season, whereas cattle selected grasses. Total forage removal was below objectives and the conclusion that rangeland conditions would improve seemed tenable.

Interactions between wapiti and cattle are complex (Hobbs et al. 1996; Wisdom and Thomas 1996). The potential for competition is highest on winter and spring–fall ranges that include low-elevation bottomlands and adjacent foothills, where forage and habitats used are limited when compared with habitats available at other times of the year. However, competition for forage during late summer and fall may increase following periods of prolonged seasonal drought. In some cases, wapiti and cattle distribute themselves spatially and temporally so competition is minimized; differences in diets also may reduce competition.

Rangeland conditions in some areas have been affected by chronically high wapiti populations as well as by past heavy grazing by livestock (Irwin et al. 1994). In the Blue Mountains of Oregon and Washington, long-term herbivory by large mammals, primarily cattle and wapiti, caused reductions of shrubs and forage productivity, with resulting reductions in calf production and survival of wapiti and weight gains by cattle. Current conditions must be viewed in the context of the century-old history of land use, and attempts to improve rangeland conditions need to continue. Obviously, reductions in livestock use in some areas need to be complemented by reductions in wapiti populations if the goal of restoring rangelands to productive status includes restoration of shrubs as well as grasses.

Experiments in Colorado by Hobbs et al. (1996) illustrate some of the relationships between wapiti and livestock. They demonstrated that winter and early-spring grazing by elk can reduce weight gains by cattle in spring. The effects of wapiti population density were not proportional to weight losses of cattle, suggesting threshold levels in removal of forage. When removal was about 45 g/m^2 and wapiti densities were 9/km^2, interactions were most intense. Wapiti density, variable plant production related to rainfall, and grazing intensity all interacted. Interactions between wapiti and livestock need to be evaluated in the context of range condition, weather patterns, and population density on each area where concerns are expressed.

As mule deer have declined and wapiti have increased, concerns over interactions of the two species also have increased. An exhaustive review by Lindzey et al. (1997) revealed no consistent trends in populations of these species in cases where they were sympatric. Research provided highly variable results as well, and most studies reviewed failed to demonstrate competition between the species. However, Lindzey et al. (1997) concluded that because of the high adaptability of wapiti and their generalized feeding habits, habitat use patterns, and larger size, it must be assumed that wapiti at high densities may reduce habitat quality for mule deer. Johnson et al. (2000) reported that mule deer avoided areas used by wapiti. Manipulation of wapiti population densities and monitoring of mule deer responses were needed to further examine the cause–effect relationships. As with the interactions with livestock, wapiti populations and habitats need to be managed with other values in mind.

Enough is known about wapiti population and habitat management to initiate an adaptive management approach that involves prediction of population response to environmental change and population manipulation. This may entail major changes in management philosophy in many instances (Levin 1993; Holling and Meffe 1996). Major limiting factors can be incorporated into predictive models, and populations, habitats, or both can be manipulated and monitored to determine response. The predictive models can be modified as new information is obtained. This approach (Walters 1986; Lancia et al. 1996) acknowledges that uncertainty exists, and that we must continue to learn by monitoring. A variety of models based on stock-recruitment investigations (Ricker 1954) may be expanded to incorporate appropriate environmental parameters, such as rainfall (Dennis and Otten 2000).

Elk populations should be expected to respond to different environmental variables in each region, and local knowledge obtained can help determine which variables will be most useful to assess. For instance, late-summer precipitation may be most useful in eastern Oregon and adjacent areas (Cook et al. 1996), whereas winter temperatures and snow depth and persistence may be useful in the Yellowstone area (Coughenour and Singer 1996). Models that are initially developed should not be expected to provide as accurate or precise predictions of population performance as subsequent modifications. The goal is to understand population performance under the range of conditions that exist in the wapiti's environment and then use that understanding to improve the prediction. Alternatives in habitat manipulation, whether intended specifically for elk or for other purposes, may be incorporated into such models. Alternative hunter harvest strategies could be examined. A complete adaptive management strategy would involve public participation to help ensure broad understanding of procedures and goals and what uncertainties exist.

Wapiti are a very popular and sought-after wildlife species in western North America. They are highly adaptable, but sensitive to human activities. Opportunities to coordinate and manage habitats, human activities, and wapiti abound, and must be considered on a case-by-case basis. Wapiti populations will be managed more and more as a component of the ecosystem in which they occur. Understanding the effects of high wapiti populations on vegetation production and composition is increasingly important if proper coordination with other uses and maintenance of biodiversity are to be achieved. The high value that society places on this species is a major factor in ensuring its perpetuation across its range.

LITERATURE CITED

Altmann, M. 1952. Social behavior of elk, *Cervus canadensis nelsoni,* in the Jackson Hole area of Wyoming. Behaviour 4:116–43.

Anderson, C. R., Jr., D. S. Moody, B. L. Smith, F. G. Lindzey, and R. P. Lanka. 1998. Development and evaluation of sightability models for summer elk surveys. Journal of Wildlife Management 62:1055–65.

Baker, D. L., and N. T. Hobbs. 1982. Composition and quality of elk summer diets in Colorado. Journal of Wildlife Management 46:694–703.

Bangs, E. E., S. H. Fritts, J. A. Fontaine, D. W. Smith, K. M. Murphy, C. M. Mack, and C. C. Niemeyer. 1998. Status of gray wolf restoration in Montana, Idaho, and Wyoming. Wildlife Society Bulletin 26:785–98.

Basile, J. V., and T. N. Lonner. 1979. Vehicle restrictions influence elk and hunter distribution in Montana. Journal of Forestry 77:155–59.

Bender, L. C., and P. J. Miller. 1999. Effects of elk harvest strategy on bull demographics and herd composition. Wildlife Society Bulletin 27:1032–37.

Berger, J., and M. E. Gompper. 1999. Sex ratios in extant ungulates: Products of contemporary predation or past life histories? Journal of Mammalogy 80:1084–1113.

Beuchner, H. K., and C. V. Swanson. 1955. Increased natality resulting from lowered population densities among elk in southeastern Wyoming. Transactions of the North American Wildlife Conference 20:561–67.

Beuchner, H. K., I. O. Buss, and H. F. Bryan. 1951. Censusing elk by airplane in the Blue Mountains of Washington. Journal of Wildlife Management 15:81–87.

Blood, D. A. 1966. Range relationships of elk and cattle in Riding Mountain National Park, Manitoba (Wildlife Management Bulletin Series 1, No. 19). Canadian Wildlife Service.

Blood, D. A., and A. L. Lovaas. 1966. Measurements and weight relationships in Manitoba elk. Journal of Wildlife Management 30:135–40.

Boer, A. H., G. Redmond, and T. J. Pettigrew. 1989. Loran-C: A navigation aid for aerial surveys. Journal of Wildlife Management 53:228–30.

Boyce, M. S. 1989. The Jackson elk herd. Cambridge University Press, New York.

Boyd, R. J. 1970. Elk of the White River Plateau, Colorado. Colorado Division of Game, Fish and Parks.

Brazda, A. R. 1953. Elk migration patterns and some of the factors affecting movements in the Gallatin River drainage, Montana. Journal of Wildlife Management 17:9–23.

Bryant, L. D., and C. M. Maser. 1982. Classification and distribution. Pages 1–59 *in* J. W. Thomas and D. E. Toweill, eds. Elk of North America: Ecology and management. Stackpole, Harrisburg, PA.

Byers, C. R., and G. A. Bettas eds. 1999. Records of North American big game, 11th ed. Boone and Crockett Club, Missoula, MT.

Cassirer, E. F., D. J. Freddy, and E. D. Ables. 1992. Elk responses to disturbance by cross-country skiers in Yellowstone National Park. Wildlife Society Bulletin 20:375–81.

Caughley, G. 1974. Interpretation of age ratios. Journal of Wildlife Management 38:557–62.

Clutton-Brock, T. H., F. E. Guinness, and S. D. Albon. 1982. Red deer: Behavior and ecology of two sexes. University of Chicago Press, Chicago.

Coffin, A. L., and J. D. Remington. 1953. Pregnant yearling cow elk. Journal of Wildlife Management 17:223.

Cole, G. F. 1971. An ecological rationale for the natural or artificial regulation of native ungulates in parks. Transactions of the North American Wildlife Conference 36:417–25.

Conaway, C. P. 1952. The age at sexual maturity of male elk. Journal of Wildlife Management 16:313–15.

Cook, R. D., and F. B. Martin. 1974. A model for quadrat sampling with "visibility bias". Journal of the American Statistical Association 69:345–49.

Cook, J. G., L. J. Quinlan, L. L. Irwin, L. D. Bryant, R. A. Riggs, and J. W. Thomas. 1996. Nutrition-growth relations of elk calves during late summer and fall. Journal of Wildlife Management 60:528–41.

Cook, J. G., L. L. Irwin, L. D. Bryant, R. A. Riggs, and J. W. Thomas. 1998. Relations of forest cover and condition of elk: A test of the thermal cover hypothesis in summer and winter. Wildlife Monographs 141:1–61.

Coughenour, M. B., and F. J. Singer. 1996. Elk population processes in Yellowstone National Park under the policy of natural regulation. Ecological Applications 6:577–93.

Cowan, I. McT. 1947. The timber wolf in the Rocky Mountain National Parks of Canada. Canadian Journal of Research 25(Section D):139–74.

Craighead, J. J., G. Atwell, and B. W. O' Gara. 1972. Elk migration in and near Yellowstone National Park. Wildlife Monographs 29:1–48.

Craighead, J. J., G. Atwell, and B. W. O' Gara. 1973. Home ranges and activity patterns of nonmigratory elk of the Madison drainage herd as determined by biotelemetry. Wildlife Monographs 33:1–50.

Crook, J. H. 1970. The socio-ecology of primates. Pages 103–66 in J. H. Crook, ed. Social behavior in birds and mammals. Academic Press, New York.

Dalke, P. D., R. D. Beeman, F. J. Kindel, R. J. Robel, and T. R. Williams. 1965a. Seasonal movement of elk in the Selway River drainage, Idaho. Journal of Wildlife Management 29:333–38.

Dalke, P. D., R. D. Beeman, F. J. Kindel, R. J. Robel, and T. R. Williams. 1965b. Use of salt by elk in Idaho. Journal of Wildlife Management 29:319–22.

Davis, J. L. 1970. Elk use of spring and calving range during and after controlled logging. M.S. Thesis, University of Idaho, Moscow.

Dekker, D., W. Bradford, and J. R. Gunson. 1995. Elk and wolves in Jasper National Park, Alberta, from historical times to 1992. Pages 85–94 *in* L. N. Carbyn, S. H. Fritts, and D. R. Seip, eds. Ecology and conservation of wolves in a changing world (Occasional Publication No. 35). Canadian Circumpolar Institute, Edmonton, Alberta, Canada.

Dennis, B., and M. R. M. Otten. 2000. Joint effects of density dependence and rainfall on abundance of San Joaquin kit fox. Journal of Wildlife Management 64:388–400.

Dennis, B., and M. L. Taper. 1994. Density dependence in time series observations of natural populations: Estimation and testing. Ecological Monographs 64:205–24.

Douglas, M. J. W. 1966. Occurrence of accessory corpora lutea in red deer, *Cervus elaphus*. Journal of Mammalogy 47:152–53.

Eberhardt, L. E., L. L. Eberhardt, B. L. Tiller, and L. L. Cadwell. 1996. Growth of an isolated elk population. Journal of Wildlife Management 60:369–73.

Eberhardt, L. L., R. G. Garrott, P. J. White, and P. J. Gogan. 1998. Alternative approaches to aerial censusing of elk. Journal of Wildlife Management 62:1046–54.

Eckstein, P., and S. Zuckerman. 1956. Morphology of the reproductive tract. Pages 43–155 in A. S. Parkes, ed. Marshall's physiology of reproduction, Vol. I. Longman's Green, London.

Flook, D. R. 1970. A study of sex differential in the survival of wapiti (Report Series 11). Canadian Wildlife Service.

Follis, T. B. 1972. Reproduction and hematology of the Cache elk herd (Research Publication No. 72–8). Utah Division of Wildlife.

Franklin, W. L., J. W. Lieb, A. S. Mossman, and M. Dole. 1975. Social organization and home range of Roosevelt elk. Journal of Mammalogy 56:102–18.

Freddy, D. J. 1987. The White River elk herd: A perspective, 1960–1985 (Technical Publication 37). Colorado Division of Wildlife.

Fretwell, S. D., and H. L. Lucas. 1970. On territorial behaviour and other factors influencing habitat distribution in birds. Acta Biotheoretica 19:16–36.

Geist, V. 1971. The relation of social evolution and dispersal in ungulates during the Pleistocene, with emphasis on the old world deer and the genus *Bison*. Quarternary Research 1:285–315.

Geist, V. 1986. New evidence of high frequency of antler wounding in cervids. Canadian Journal of Zoology 64:380–84.

Geist, V. 1998. Deer of the world: Their evolution, behavior and ecology. Stackpole, Mechanicsburg, PA.

Gogan, P. J. P., and R. H. Barrett. 1987. Comparative dynamics of introduced Tule elk populations. Journal of Wildlife Management 51:20–27.

Graf, W. 1956. Territorialism in deer. Journal of Mammalogy 37:165–70.

Gratson, M. W., and C. L. Whitman. 2000. Effects of road closures on density and success of elk hunters in Idaho. Wildlife Society Bulletin 28:302–11.

Greer, K. R. 1966. Fertility rates of the northern Yellowstone elk populations. Proceedings of the Western Association State Fish and Game Commissioners 46:123–28.

Greer, K. R., and R. E. Howe. 1964. Winter weights of northern Yellowstone elk, 1961–1962. Transactions of the North American Wildlife Conference 29:238–48.

Guthrie, R. D. 1966. The extinct wapiti of Alaska and Yukon territory. Canadian Journal of Zoology 44:47–57.

Halazon, G. C., and H. K. Beuchner. 1956. Postconception ovulation in elk. Transactions of the North American Wildlife Conference 21:545–54.

Hamer, D., and S. Herrero. 1991. Elk calves as grizzly bear food in Banff National Park, Alberta. Canadian Field-Naturalist 105:101–3.

Happe, P. J., K. J. Jenkins, E. E. Starkey, and S. H. Sharrow. 1990 Nutritional quality and tannin astringency of browse in clear-cuts and old-growth forests. Journal of Wildlife Management 54:557–66.

Harper, J. A. 1971. Ecology of Roosevelt elk. Oregon Game Commission, Portland.

Harper, J. A., J. H. Harn, W. W. Bentley, and C. F. Yocom. 1967. The status and ecology of the Roosevelt elk in California. Wildlife Monographs 16:1–49.

Hash, H. S. 1974. Movements and food habits of the Lochsa elk. M.S. Thesis, University of Idaho, Moscow.

Hines, W. W., and J. C. Lemos. 1979. Reproduction performance by two age classes of male Roosevelt elk in southwestern Oregon (Research Report No. 8). Oregon Department of Fish and Wildlife.

Hobbs, N. T. 1996. Modification of ecosystems by ungulates. Journal of Wildlife Management 60:695–713.

Hobbs, N. T., and T. A. Hanley. 1990. Habitat evaluation: Do use/availability data reflect carrying capacity? Journal of Wildlife Management 54:515–21.

Hobbs, N. T., D. L. Baker, G. D. Bear, and D. C. Bowden. 1996. Ungulate grazing in sagebrush grassland: Mechanisms of resource competition. Ecological Applications 5:207–24.

Hofmann, R. R. 1989. Evolutionary steps of ecophysiological adaptation and diversification of ruminants: A comparative view of their digestive system. Oecologia 78:443–57.

Holling, C. S., and G. K. Meffe. 1996. Command and control and the pathology of natural resource management. Conservation Biology 10:328–37.

Hornocker, M. G. 1970. An analysis of mountain lion predation upon mule deer and elk in the Idaho Primitive Area. Wildlife Monographs 21:1–39.

Houston, D. B. 1982. The northern Yellowstone elk: Ecology and management. Macmillan, New York.

Irwin, L. L., and J. M. Peek. 1983. Elk habitat use relative to forest succession in Idaho. Journal of Wildlife Management 47:664–72.

Irwin, L. L., J. G. Cook, R. A. Riggs, and J. M. Skovlin. 1994. Effects of long-term grazing by big game and livestock in the Blue Mountains Forest Ecosystems, (General Technical Report PNW-GTR-325). U.S. Department of Agriculture, Forest Service.

Jenkins, K. J., and E. E. Starkey. 1982. Social organization of Roosevelt elk in an old-growth forest. Journal of Mammalogy 63:331–34.

Johnson, B. K., J. W. Kern, M. J. Wisdom, S. L. Findholt, and J. G. Kie. 2000. Resource selection and spatial separation of mule deer and elk during spring. Journal of Wildlife Management 64:685–97.

Johnson, D. E. 1951. Biology of the elk calf, *Cervus canadensis nelsoni*. Journal of Wildlife Management 15:396–410.

Johnson, D. H. 1980. The comparison of usage and availability measurements for evaluating resource preference. Ecology 61:65–71.

Keiss, R. E. 1969. Comparison of eruption-wear patterns and cementum annuli as age criteria in elk. Journal of Wildlife Management 33:175–80.

Keith, L. B. 1974. Population dynamics of mammals. International Congress of Game Biologists VI:17–58.

Kelly, S. M. 1995. Elk and cattle range relations on the Lemhi Mountains, Idaho. M.S. Thesis, University of Wyoming, Laramie.

Kie, J. 1999. Optimal foraging and risk of predation: Effects on behavior and social structure in ungulates. Journal of Mammalogy 80:1114–29.

Kie, J. G., and R. T. Bowyer. 1999. Sexual segregation in white-tailed deer: Density-dependent changes in use of space, habitat selection and dietary niche. Journal of Mammalogy 80:1004–20.

Kimball, J. F., Jr., and M. L. Wolfe. 1974. Population analysis of a northern Utah elk herd. Journal of Wildlife Management 38:161–74.

Knight, R. R. 1970. The Sun River elk herd. Wildlife Monographs 23:1–66.

Kufeld, R. C. 1973. Foods eaten by the Rocky Mountain elk. Journal of Range Management 26:106–13.

Lancia, R. A., C. E. Braun, M. W. Colopy, R. D. Dueser, J. G. Kie, C. J. Martinka, J. D. Nichols, T. D. Nudds, W. R. Porath, and N. G. Tilghman. 1996. ARM! For the future: Adaptive resource management in the wildlife profession. Wildlife Society Bulletin 24:436–42.

Lent, P. C. 1974. Mother–infant relationships in ungulates. Pages 14–55 *in* V. Geist and F. Walther, eds. The behaviour of ungulates and its relation to management. International Union for the Conservation of Nature, Morges, Switzerland.

LeResche, R. E., and R. A. Rausch. 1974. Accuracy and precision of aerial moose censusing. Journal of Wildlife Management 38:175–82.

Levin, S. A., ed. 1993. Forum: Science and sustainability. Ecological Applications 3:546–89.

Lindzey, F. G., W. G. Hepworth, T. A. Mattson, and A. F. Reese. 1997. Potential for competitive interactions between mule deer and elk in the western United States and Canada: A review. Wyoming Cooperative Fisheries and Wildlife Research Unit, Laramie.

Lovaas, A. L. 1970. People and the Gallatin elk herd. Montana Fish and Game Department, Helena.

Lovaas, A. L., J. L. Egan, and R. R. Knight. 1966. Aerial counting of two Montana elk herds. Journal of Wildlife Management 30:364–69.

Lowe, V. P. W. 1969. Population dynamics of the red deer (*Cervus elaphus* L.) on Rhum. Journal of Animal Ecology 38:425–57.

Lyon, L. J. 1979. Habitat effectiveness for elk as influenced by roads and cover. Journal of Forestry 77:658–60.

Lyon, L. J. 1983. Road density models describing habitat effectiveness for elk. Journal of Forestry 81:592–95.

Mackie, R. J. 1970. Range ecology and relations of mule deer, elk and cattle in the Missouri River Breaks, Montana. Wildlife Monographs 20.

Main, M. B., F. W. Weckerly, and V. C. Bleich. 1996. Sexual segregation in ungulates: New directions for research. Journal of Mammalogy 77:449–61.

Martin, P. S., and C. R. Szuter. 1999. War zones and game sinks in Lewis and Clark's west. Conservation Biology 13:36–45.

Martinka, C. J. 1969a. An incident of mass elk drowning. Journal of Mammalogy 50:640–41.

Martinka, C. J. 1969b. Population ecology of summer resident elk in Jackson Hole, Wyoming. Journal of Wildlife Management 33:465–81.

McCorquodale, S. M. 1985. Archaeological evidence of elk in the Columbia Basin. Northwest Science 59:192–97.

McCorquodale, S. M., L. E. Eberhardt, and G. A. Sargeant. 1989. Antler characteristics in a colonizing elk population. Journal of Wildlife Management 53:618–20.

McCullough, D. R. 1969. The Tule elk: Its history, behavior, and ecology (University of California Publications in Zoology, Vol. 88). University of California Press, Berkeley.

McCullough, D. R., J. K. Fischer, and J. D. Ballou. 1996. From bottleneck to metapopulation: Recovery of the Tule elk in California. Pages 375–404 *in* D. R. McCullough, ed. Metapopulations and wildlife conservation. Island Press, Washington, DC.

Merrill, E. H., K. D. Raedeke, and R. D. Taber. 1987. The population dynamics and habitat ecology of elk in the Mount St. Helens blast zone. College of Forest Resources, University of Washington, Seattle.

Mitchell, B., and D. Brown. 1973. The effects of age and body size on fertility of female red deer (*Cervus elaphus* L.). International Congress of Game Biologists 11:89–98.

Mitchell, B., D. McGowan, and I. A. Nicholson. 1976. Annual cycles of body weight and condition in Scottish red deer, *Cervus elaphus*. Journal of Zoology (London) 180:107–27.

Morrison, J. A. 1960. Ovarian characteristics in elk of known breeding history. Journal of Wildlife Management 24:197–207.

Morrison, J. A., C. E. Trainer, and P. L. Wright. 1959. Breeding season in elk as determined from known-age embryos. Journal of Wildlife Management 23:411–14.

Mould, E. D., and C. T. Robbins. 1981. Nitrogen metabolism in elk. Journal of Wildlife Management 45:323–34.

Mould, E. D., and C. T. Robbins. 1982. Digestive capabilities of elk compared to white-tailed deer. Journal of Wildlife Management 46:22–29.

Murie, A. 1944. The Wolves of Mount McKinley. US Dep. Interior, National Park Service Fauna, PA.

Murie, O. J. 1951. The elk of North America. Stackpole, Harrisburg. Series 4 200 pp.

Murphy, D. A. 1963. A captive elk herd in Missouri. Journal of Wildlife Management 27:27–34.

Nelson, L. J., and J. M. Peek. 1982. Effect of survival and fecundity on rate of increase of elk. Journal of Wildlife Management 46:535–40.

Nowak, R. M. 1999. Walker's mammals of the world, Vol. II, 6th ed. Johns Hopkins University Press, Baltimore.

Noyes, J. H., B. K. Johnson, L. D. Bryant, S. L. Findholt, and J. W. Thomas. 1996. Effects of bull age on conception dates and pregnancy rates of cow elk. Journal of Wildlife Management 60:508–17.

Otten, M. R. M., Hauffler, J. B., S. R. Winterstein, and L. C. Bender. 1993. An aerial censusing procedure for elk in Michigan. Wildlife Society Bulletin 21:73–79.

Peek, J. M. 1986. A review of wildlife management. Prentice-Hall, Englewood Cliffs, NJ.

Peek, J. M. 1999. Elk, *Cervus elaphus*. Pages 327–28 *in* D. E. Wilson and S. Ruff, eds. The Smithsonian book of North American mammals. Smithsonian Institution Press, Washington, DC.

Peek, J. M., and A. L. Lovaas. 1968. Differential distribution of elk by sex and age on the Gallatin winter range, Montana. Journal of Wildlife Management 32:553–57.

Peek, J. M., A. L. Lovaas, and R. A. Rouse. 1967. Population changes within the Gallatin elk herd, 1932–1965. Journal of Wildlife Management 32:304–16.

Peek, J. M. 1980. Natural regulation of regulates (what constitutes a real wilderness? Wildlife Society Bulletin 8:217–227.

Peek, J. M., M. D. Scott, L. J. Nelson, D. J. Pierce, and L. L. Irwin. 1982. Role of cover in habitat management for big game in northwestern United States. North American Wildlife and Natural Resources Conference 47:363–73.

Phillips, G. E., and A. W. Alldredge. 2000. Reproductive success of elk following disturbance by humans during calving season. Journal of Wildlife Management 64:521–30.

Picton, H. D. 1960. Migration patterns of the Sun River elk herd, Montana. Journal of Wildlife Management 24:279–90.

Polziehn, R. O., and C. Strobeck. 1998. Phylogeny of wapiti, red deer, sika deer, and other North American cervids as determined from mitochondrial DNA. Molecular Phylogenetics and Evolution 10:249–58.

Polziehn, R. O., J. Hamr, F. F. Mallory, and C. Strobeck. 1998. Phylogenetic status of North American wapiti (*Cervus elaphus* subspecies). Canadian Journal of Zoology 76:998–1010.

Prothero, W. L., J. J. Spillett, and D. F. Balph. 1979. Rutting behavior of yearling and mature bull elk: Some implications for open bull hunting. Pages 160–65 *in* M. S. Boyce and L. D. Hayden- Wing, eds. North American elk: Ecology, behavior and management. University of Wyoming, Laramie.

Quimby, D. C., and J. E. Gaab. 1957. Mandibular dentition as an age indicator in Rocky Mountain elk. Journal of Wildlife Management 21:435–51.

Quimby, D. C., and D. E. Johnson. 1951. Weights and measurements of Rocky Mountain elk. Journal of Wildlife Management 15:57–62.

Rickard, W. H., J. D. Hedlund, and R. E. Fitzner. 1977. Elk in the shrub-steppe region of Washington: An authentic record. Science 196:1009–10.

Ricker, W. E. 1954. Stock and recruitment. Journal of the Fisheries Research Board of Canada 11:559–623.

Riggs, R. A., A. R. Tiedemann, J. G. Cook, T. M. Ballard, P. J. Edgerton, M. Vavra, W. C. Krueger, F. C. Hall, L. D. Bryant, L. L. Irwin, and T. Delcurto. 2000. Modification of mixed-conifer forests by ruminant herbivores in the Blue Mountains Ecological Province (Research Paper PNW-RP-527). U.S. Department of Agriculture, Forest Service.

Rognrud, M., and R. Janson. 1971. Elk. Pages 39–51 *in* T. W. Mussehl and F. W. Howell, eds. Game management in Montana. Montana Department of Fish, Wildlife and Parks, Helena.

Rowland, M. M., M. J. Wisdom, B. K. Johnson, and J. G. Kie. 2000. Elk distribution and modeling in relation to roads. Journal of Wildlife Management 64:672–84.

Samuel, M. D., E. O. Garton, M. W. Sclegel, and R. G. Carson. 1987. Visibility bias during aerial surveys of elk in northcentral Idaho. Journal of Wildlife Management 51:622–30.

Sappington, R. L. 1994. The prehistory of the Clearwater River region, north central Idaho (Anthropological Reports 95). University of Idaho, Moscow.

Sappington, R. L., and C. D. Carley. 1989. Archaeological investigations at the Beaver Flat and Pete King Creek sites, Lochsa River, north central Idaho (Anthropological Reports 89). University of Idaho, Moscow.

Sauer, J. R., and M. S. Boyce. 1983. Density dependence and survival of elk in northwestern Wyoming. Journal of Wildlife Management 47:31–37.

Schlegel, M. 1976. Factors affecting calf elk survival in northcentral Idaho. Annual Conference of the Western Association of State Game and Fish Commissioners 56:342–55.

Schultz, R. D., and J. A. Bailey. 1978. Responses of national park elk to human activity. Journal of Wildlife Management 42:91–100.

Singer, F. J., A. Harting, K. K. Symonds, and M. B. Coughenour. 1997. Density dependence, compensation, and environmental effects on elk calf mortality in Yellowstone National Park. Journal of Wildlife Management 61:12–25.

Singer, F. J., L. C. Zeigenfuss, R. G. Cates, and D. T. Barnett. 1998. Elk multiple factors, and persistence of willows in national parks. Wildlife Society Bulletin 26:419–29.

Skovlin, J. M., P. J. Edgerton, and R. W. Harris. 1968. The influence of cattle management on deer and elk. Transactions of the North American Wildlife and Natural Resources Conference 33:169–81.

Smith, B. L., R. L. Robbins, and S. H. Anderson. 1996. Adaptive sex ratios: Another example? Journal of Mammalogy 77:818–25.

Squibb, R. C., J. F. Kimball, and D. R. Anderson. 1986. Bimodal distribution of estimated conception dates in Rocky Mountain elk. Journal of Wildlife Management 50:118–22.

Staines, B. W., and P. R. Ratcliffe. 1987. Estimating the abundance of red deer (*Cervus elaphus* L.) and roe deer (*Capreolus capreolus* L.) and their current status in Great Britain. Symposia of the Zoological Society of London 58:131–52.

Stevens, D. R. 1966. Range relationships of elk and livestock, Crow Creek drainage, Montana. Journal of Wildlife Management 32:349–63.

Strohmeyer, D. C., and J. M. Peek. 1996. Wapiti home range and movement patterns in a sagebrush desert. Northwest Science 70:79–87.

Strohmeyer, D. C., J. M. Peek, and T. R. Bowlin. 1999. Wapiti bed sites in Idaho sagebrush steppe. Wildlife Society Bulletin 27:547–51.

Struhsaker, T. T. 1967. Behavior of elk (*Cervus canadensis*) during the rut. Zeitschrift für Tierpsycholgie 24:80–114.

Thiessen, J. 1976. Some elk–logging relationships in southern Idaho. Pages 3–5 *in* Proceedings of the elk–logging-roads symposium. University of Idaho, Moscow.

Thelen, T. H. 1991. Effects of harvest on antlers of simulated populations of elk. Journal of Wildlife Management 55:243–49.

Thompson, M. J., and R. E. Henderson. 1998. Elk habituation as a credibility challenge for wildlife professionals. Wildlife Society Bulletin 26:477–83.

Thorne, E. T., R. E. Dean, and W. G. Hepworth. 1976. Nutrition during gestation in relation to successful reproduction in elk. Journal of Wildlife Management 40:330–335.

Troyer, W. A. 1960. The Roosevelt elk on Afognak Island, Alaska. Journal of Wildlife Management 24:15–21.

Unsworth, J. W., L. Kuck, and E. G. Garton. 1990. Elk sightability model validation at the National Bison Range, Montana. Wildlife Society Bulletin 18:113–15.

Unsworth, J. W., F. A. Leban, G. A. Sargeant, E. O. Garton, M. A. Hurley, and J. R. Pope. 1991. Aerial survey: User's manual with practical tips for designing and conducting aerial big game surveys. Idaho Department of Fish and Game, Boise.

Unsworth, J. W., L. Kuck, M. D. Scott, and E. O. Garton. 1993. Elk mortality in the Clearwater drainage of northcentral Idaho. Journal of Wildlife Management 57:495–502.

Vales, D. J., and J. M. Peek. 1995. Projecting the potential effects of wolf predation on elk and mule deer in the east front portion of the northwest Montana wolf recovery area. Pages 211–22 *in* L. N Carbyn, S. H. Fritts, and D. R. Seip, eds. Ecology and conservation of wolves in a changing world, (Occasional Publication No. 35). Canadian Circumpolar Institute, Edmonton, Alberta, Canada.

Varland, K. L., A. L. Lovaas, and R. B. Dahlgren. 1978. Herd organization and movements of elk in Wind Cave National Park, South Dakota (Natural Resources Report 13). U.S. Department of Interior, National Park Service.

Waldrip, G. P., and J. H. Shaw. 1979. Movements and habitat use by cow and calf elk at the Wichita Mountains National Wildlife Refuge. Pages 177–84 *in* M. S. Boyce and L. D. Hayden-Wing, eds. North American elk: Ecology, behavior and management. University of Wyoming, Laramie.

Wallace, M. C. 1991. Elk habitat use in the White Mountains, Arizona. Ph.D. Dissertation, University of Arizona, Tucson.

Walters, C. J. 1986. Adaptive management of renewable resources. Macmillan, New York.

Wambolt, C. L., and H. W. Sherwood. 1999. Sagebrush response to ungulate browsing in Yellowstone. Journal of Range Management 52:363–69.

Weckerly, F. W. 1996. Roosevelt elk along the Prairie Creek drainage: An evaluation of estimating abundance and herd composition. California Fish and Game 82:175–81.

Weckerly, F. W., and K. E. Kovacs. 1998. Use of military helicopters to survey an elk population in north coastal California. California Fish and Game 84:44–47.

Wisdom, M. J., and J. G. Cook. 2000. North American elk. Pages 694–735 *in* S. Demarais and P. R. Krausman, eds. Ecology and management of large mammals in North America. Prentice-Hall, Upper Saddle River, NJ.

Wisdom, M. J., and J. W. Thomas. 1996. Elk. Pages 157–81 *in* P. R. Krausman, ed. Rangeland wildlife. Society for Range Management, Denver, CO.

Wishart, W. D. 1981. January conception in an elk in Alberta. Journal of Wildlife Management 45:544.

Wolfe, G. J. 1983. The relationship between age and antler development in wapiti. Pages 29–36 *in* R. D. Brown, ed. Antler development in Cervidae. Caesar Kleberg Wildlife Research Institute, Kingsville, TX.

Young, V. A., and W. L. Robinette. 1939. A study of the range habits of elk in the Selway River game preserve. University of Idaho Bulletin 34:1–48.

James M. Peek, College of Natural Resources, University of Idaho, Moscow, Idaho 83844-1136. Email: peek@uidaho.edu.

43

Mule Deer

Odocoileus hemionus

Richard J. Mackie
John G. Kie
David F. Pac
Kenneth L. Hamlin

NOMENCLATURE

COMMON NAMES. Mule deer, black-tailed deer
SCIENTIFIC NAME. *Odocoileus hemionus*
SUBSPECIES. *Odocoileus hemionus hemionus,* Rocky Mountain mule deer; *O. h. californicus,* California mule deer; *O. h. fuliginatus,* southern mule deer; *O. h. peninsulae,* peninsula mule deer; *O. h. crooki* (possibly *O. h. eremicus),* desert mule deer; *O. h. columbianus,* Columbian black-tailed deer; and *O. h. sitkensis,* Sitka black-tailed deer

Mule and black-tailed deer comprise two groups of subspecies or races of mule deer, characterized broadly by the extremes in external appearance and behavior within the species. Wallmo (1978:31) noted "mule deer and black-tailed deer are sufficiently different to justify distinct common names, yet similar enough to be included in one species." Indeed, mule and black-tailed deer were characterized by mitochondrial DNA exhibiting high divergence among and low divergence within subspecies (Cronin 1992). This suggests limited gene flow between populations of mule and black-tailed deer since the Pleistocene and supports the validity of the subspecies concept for *Odocoileus hemionus*.

Within the two groups, Cowan (1956) recognized 11 subspecies [the seven subspecies noted above plus *Odocoileus hemionus inyoensis,* Inyo mule deer; *O. h. eremicus,* burro deer (*sensu* Cowan 1956); *O. h. sheldoni,* Tiburón Island mule deer; and *O. h. cerrosensis,* Cedros Island mule deer]. Inyo mule deer and burro mule deer are no longer generally recognized. The taxonomic status of Tiburón Island mule deer and Cedros Island mule deer is not clear. These two insular subspecies were thought to be peninsula mule deer by Wallmo (1981) and Anderson and Wallmo (1984). However, because of the proximity of Tiburón Island to mainland Mexico, Tiburón mule deer are more likely to be desert mule deer. The status of desert mule deer as *Odocoileus hemionus crooki* was questioned by Heffelfinger (2000). He argued that the description of this subspecies was based on a specimen that was a hybrid between desert mule deer and Coues white-tailed deer (*Odocoileus virginianus couesi*). Hence, the oldest available name for desert mule deer is *Odocoileus hemionus eremicus*.

Mule deer are members of order Artiodactyla, suborder Ruminantia, family Cervidae. This family dates from the Miocene in the Old World and probably reached North America in the latter part of that epoch. The genus *Odocoileus,* however, has its phylogenetic foundation based strictly in the New World. The biology of *Odocoileus* is so different from that of Old World cervids that any similarity is likely to be analogous rather than homologous (Geist 1981). In North America, white-tailed deer (*Odocoileus virginianus*) evolved in the mesic, deciduous forests of the east, whereas mule deer evolved in the dry, rugged badlands and mountains of the west. Today, although the two species overlap broadly in geographic range and to some extent in local distribution, where occasional hybridization occurs, they remain remarkably distinct in their biological, ecological, and behavioral attributes. Mule deer are distinguished from white-tailed deer by a number of characteristics. Externally, these include the form and color of the tail, the size of the metatarsal glands, the form of the antlers in males, and overall appearance. Cranial and dental characteristics, such as the shape and the size of lachrymal pits and the size and the form of the lower incisors, also differ.

DISTRIBUTION

Mule deer and black-tailed deer occur over much of western North America from 23° N northward to 60° N, and from the Pacific coast eastward to the 100th meridian (Fig. 43.1). The most southerly records are from northern San Luis Potosi, Mexico, and the most northerly from Yukon Territory, Canada (Cowan 1956; Wallmo 1981; Kie and Czech 2000). In the east, the range extends into central North Dakota,

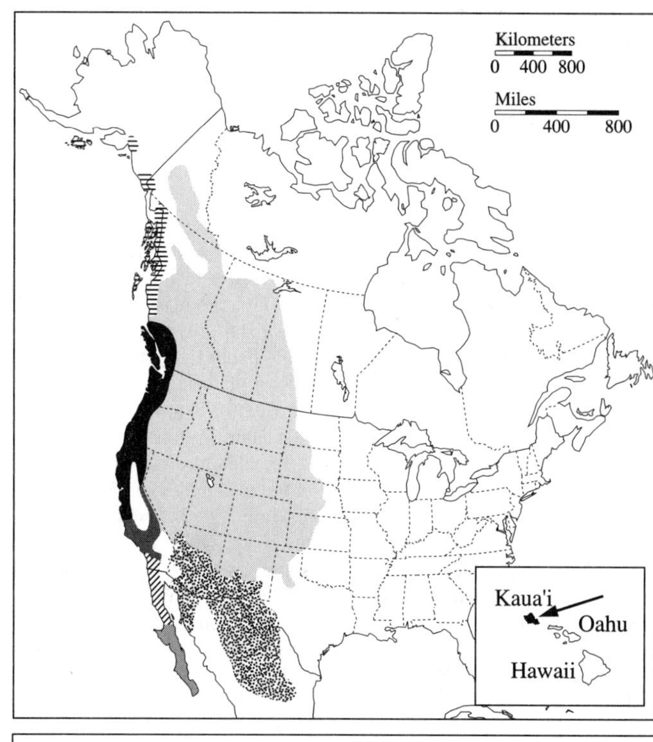

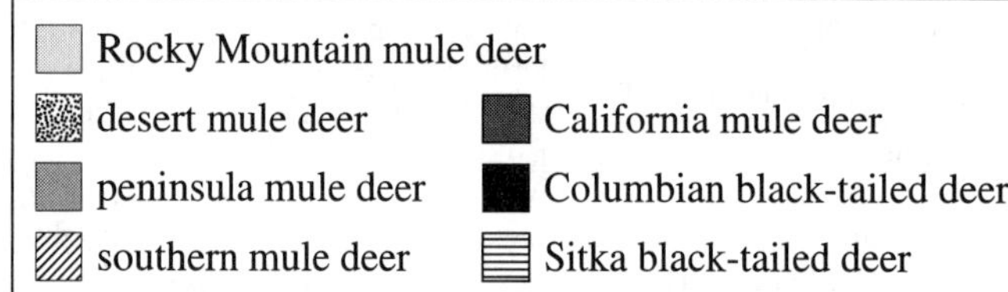

FIGURE 43.1. Distribution of subspecies of mule and black-tailed deer: (1) Rocky Mountain mule deer, (2) desert mule deer, (3) peninsula mule deer, (4) southern mule deer, (5) California mule deer, (6) Columbian black-tailed deer, and (7) Sitka black-tailed deer.

east-central South Dakota and Nebraska, west-central Kansas, and portions of western Oklahoma and the panhandle of Texas. Scattered populations have been reported as far east, however, as western Minnesota and Iowa.

Rocky Mountain mule deer range from the Yukon to Arizona and New Mexico. Desert mule deer extend from Arizona and New Mexico south into Mexico as far as San Luis Potosi. Peninsula mule deer occupy Baja California del Sur. Southern mule deer extend from Baja California del Norte northward into southern California. California mule deer are found throughout central and southern California. Columbian black-tailed deer range from central California to central British Columbia. A small population of Columbian black-tailed deer has become established on the island of Kaua'i, Hawaii, following several introductions from western Oregon in the 1960s (Tomich 1986; Telfer 1988). Sitka black-tailed deer occur in northern British Columbia and southeastern Alaska, and in disjunct populations elsewhere in Alaska such as on Afognak and Kodiak Islands and on islands in Prince William Sound, also as the result of introductions (Wallmo 1978).

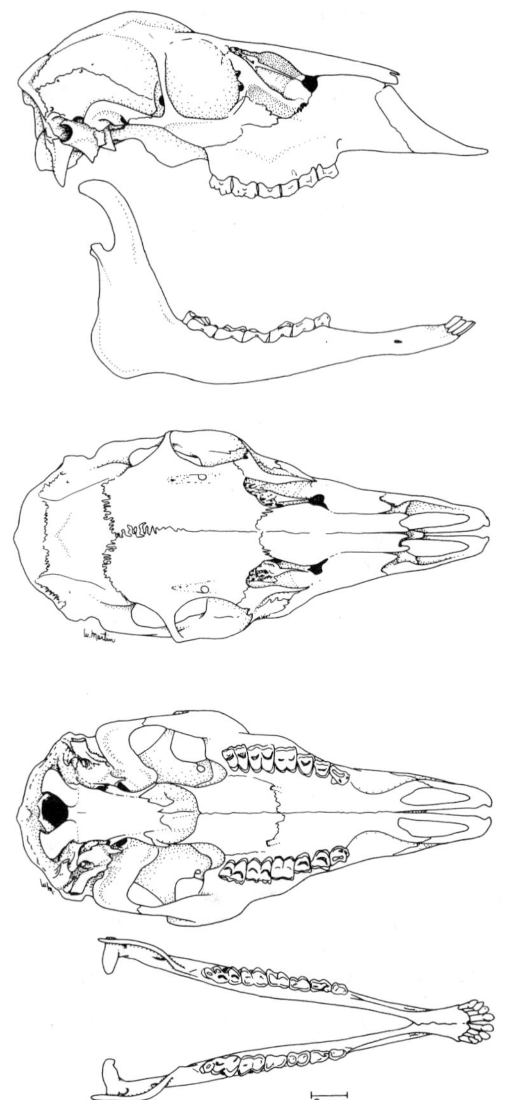

FIGURE 43.2. Skull of the mule deer (*Odocoileus hemionus*). From top to bottom: lateral view of cranium, lateral view of mandible, dorsal view of cranium, ventral view of cranium, dorsal view of mandible.

DESCRIPTION

Size and Growth. Mule deer are quite variable in size. The largest individuals occur in the Rocky Mountains among *O. h. hemionus*. Anderson et al. (1974a) listed average field-dressed carcass weights of 51 males ≥1.5 years old as 74.0 kg. For 91 females, the average was 59.0 kg. Hunter (1924) reported maximum eviscerated carcass weight of wild bucks was 172 kg, whereas Mackie (1964) reported eviscerated carcass weights for does averaged 59 kg. Body lengths of 51 males and 90 females ≥1.5 years old averaged 152.3 and 142.4 cm, respectively, and shoulder heights averaged 96.6 cm for males and 90.8 cm for females (Anderson et al. 1974a). The smallest individuals occur among black-tailed deer. Field-dressed carcass weights of adult males and females as low as 50 and 32 kg, respectively, have been recorded (Cowan and Guiguet 1965). Body size and weight may vary greatly among individual mule and black-tailed deer, depending on age, time of year, reproductive status, and environmental factors that affect the nutritional status and growth of animals.

Newborn mule deer fawns range in weight from about 2.3 to 5.0 kg. Mean birth weight for Rocky Mountain mule deer fawns, weighed during nine different studies, ranged from 2.7 kg to about 4.0 kg. Prenatal growth, reflected in birth weights, and postnatal growth are influenced by environmental, ecological, and physiological factors that affect the nourishment received by fetuses and growing fawns. Although the effect of the dam's nutritional status on birth weights of mule deer fawns is not clear, studies by Robinette et al. (1973), Hamlin and Mackie (1989), and others indicate that well-nourished females have a shorter gestation period and/or give birth earlier than females on a lower nutritional plane during the year before parturition. Mule deer fawns grow rapidly and by 5–6 months of age average about 30 kg, with males tending to be slightly heavier than females. Yearlings generally weigh 50–60 kg; females average about 10% less than males. Anderson et al. (1974a) indicated that males continued to gain weight and presumably grow throughout most of their life, whereas females achieved their maximum weight at about 8 years of age. Mackie (1964) concluded that males gained weight at least to the age of 7.5 years, whereas weights of females changed very little after 2.5 years.

Body weights fluctuate seasonally among adult mule deer of all races and both sexes, increasing during summer and early autumn and decreasing during late fall and winter. Carcass weights of mature males peaked during October and reached a low in March (Anderson et al. 1974a). Female weights also peaked in October, but were lowest in April. Mean seasonal weight losses were 19% and 22% of peak weights for males and females, respectively. Wood et al. (1962) reported overwinter weight losses of 20–22% for a young adult male mule deer on a high-quality diet.

Dentition. The dentition of mule deer is heterodont. The dental formula for adults with complete permanent dentition is I 0/3, C 0/1, P 3/3, M 3/3 (Fig. 43.2). Upper incisors, upper canines in most deer, and upper and lower first premolars are missing. In mule deer, as in other cervids, the lower canines are incisiform.

In fawns, deciduous or milk incisors and canines are fully erupted 10 days after birth, whereas mineralization and eruption of premolars requires 2.5–3 months (Rees et al. 1966). Cusps of M_1 first become evident at approximately 2.5–3 months and eruption is complete by 12–14 months of age (Robinette et al. 1957). Cusps of M_2 first appear at about 8 months, with eruption complete at 20–23 months, whereas M_3 begins to emerge at 15 months and is complete at 28 months of age, when the full permanent adult dentition is achieved. These tooth eruption and replacement rates are somewhat slower than those reported for white-tailed deer by Severinghaus (1949).

Tooth replacement patterns provide a reliable means of determining age in mule deer through about 2.5 years of age. For older deer, age may be estimated approximately by evaluating wear on mandibular cheek teeth (Robinette et al. 1957). However, recent studies (Hamlin et al. 2000) showed that eruption–wear criteria correctly determined only 62% of known-age mandibles. Accuracy decreased rapidly with increased age of the animal. Age may also be determined in the laboratory by sectioning the root of an incisor and enumerating annual rings in the cementum (Low and Cowan 1963). This method accurately determined age of 92.6% of known-age Rocky Mountain mule deer to 14 years (Hamlin et al. 2000). However, McCullough (1996) found that annuli may become obscure in teeth of deer in populations where deer density was greatly reduced and quality of seasonal diet increased.

Antlers and Antler Growth. Male mule deer have deciduous antlers. When mature and completely ossified, antlers consist of a thick sheath and base of hard, compact bone around a core of spongy bone. Antlers apparently evolved as a social adaptation associated with individual reproductive success (Geist 1968) and may provide the major stimulus for social interaction (Bubenik 1968).

In mule deer, the antlers of adults branch equally, or dichotomously, above the base typically to form four major tines, or points, per side. A relatively short brow tine may also occur. In contrast, the antlers of adult white-tailed deer consist of a series of tines arising from a continuous, forecurving main beam.

Antler development in Rocky Mountain mule deer begins with the formation of antler primordia, or pedicels, in the fetus 73–83 days after conception (Hudson and Browman 1959). Male fawns typically have buttons, which first appear at about 3 months of age (Davis 1962), but are recognizable only by close examination. In subsequent years, particularly among yearlings, growth and development appear to be sensitive to the nutritional quality of the diet, with body growth taking precedence over antler growth (French et al. 1955). Yearling males on adequate diets will normally have forked antlers with two points on a side, whereas deficient diets result in thin spike antlers. Fawns born as singletons may also develop larger antlers as yearlings than those born as one of a set of twins (Robinette et al. 1977). Yearling males having three or more points on a side are uncommon. The typical four-point antler of adults frequently is first obtained at 2 years, although three points may be more common and larger two-point antlers may yet occur at that age. Subsequent growth is most prominent in the diameter and length of the antler, and brow tines and additional, nontypical tines may also appear. Nontypical antlers having 20 or more points on a side have been recorded. Age of deer cannot be determined by the number of antler points.

The annual antler cycle begins in April or May with the appearance of velvety antler bulbs. Initially, growth is slow, but by summer it becomes extremely rapid and may exceed 1 cm/day (Goss 1963). Growth is completed by late summer. The velvet dries and is usually shed in late August and September. Antler shedding dates are highly variable. Northern Rocky Mountain mule deer usually drop their antlers in late January and early February, although individual bucks may shed as early as mid-December or as late as the end of March. Mature bucks may shed antlers earlier than yearlings, and bucks that are undernourished or in poor physical condition often shed earlier than those on high-quality diets.

Pelage. Adult mule deer molt twice annually, in late spring and in early fall. The characteristic dark gray winter coat with a woolly undercoat grows from September through January. It is shed during May to early June and is replaced by the reddish coat of summer. The autumnal molt begins on the outer ear pinna and generally spreads downward and backward from the head to neck, shoulders, flanks, abdomen, legs, and the rump patch, which remains white throughout the year. The spring molt occurs as approximately the reverse of the autumnal molt. The phenology of molting may vary locally and among individual deer. Barren does molt earlier than pregnant individuals, and deer in poor condition may experience an extended molt (Cowan and Raddi 1972).

Fawns are born spotted, with the pelage replaced between July and October. In summer, mule deer fawns typically have a medium–dark brown coat in contrast to the lighter, orange coat of white-tailed deer fawns. Robinette et al. (1977) noted that Rocky Mountain mule deer fawns lost their spots at an average age of 86 days.

Vocalizations. Mule deer emit a variety of sounds involving communication among conspecifics. Studies suggest that communications between adults of either sex include low-pitched vocalizations of undisturbed animals, whistling snorts of alarm, and coughing grunts of bucks challenging rivals. When deer are distressed, they may emit very loud sounds ranging from bleats to sharp barks to deep guttural roars depending on age, sex, and degree of disturbance (Cowan 1956).

Gait. Mule deer have a characteristic type of stiff-legged gait, referred to as stotting. The stot is a four-footed bound, which may carry the animal 60 cm or more above the ground and span a distance of 3–5 m. Speeds up to 40 km/hr can be attained for short durations. More important, perhaps, this gait enables mule deer to move sure-footedly with extremely high maneuverability in the steep rocky terrain in which they typically occur. If necessary, they can turn in any direction or completely reverse the direction of movement in the course of a single bound.

Another characteristic gait of mule deer is a deliberate, stilted, stiff-legged walk, which seems to be used mostly when an animal is alarmed but not aware of the cause or is curiously investigating an unknown element in its environment. Undisturbed mule deer walk normally, trot, lope, and even gallop occasionally in the course of normal movements associated with feeding, travel, play, and courtship.

PHYSIOLOGY

Thermal Requirements and Climatic Relations. Across their range, mule deer are subject to vast extremes in climate and weather. Average temperatures in mule deer habitats range from −15°C or less in January in the northern Rocky Mountains to greater than 30°C in July on hot deserts. Extreme temperatures may be below −60°C and above 50°C (Wallmo 1981). Mule deer employ a number of physical, physiological, and behavioral mechanisms and adaptations to cope with these extremes, which occur in all environments.

Mule deer are homoiothermal with an average body core temperature ranging from 37.1°C to 40.6°C (Leopold et al. 1951; Cowan and Wood 1955; Thorne 1975). They lack an extensive system of sweat glands, which essentially prevents heat loss due to evaporation (Loveless 1967). Thermoregulation is achieved by altering their basic metabolic rate, panting, shivering, changing posture, erecting or fluffing hairs, undergoing vasoconstriction, or adjusting blood flow through the ears. Deer also alter feeding habits, and use plant cover and physiographic sites that afford shade in summer and protection from wind and radiant heat loss in winter.

Daily and seasonal activity patterns and movements may be strongly influenced or induced by air temperature regimes (Loveless 1967); however, wind, relative humidity, and snow depths, individually or in combination with temperature and each other, may also be important. In Colorado, mule deer tended to move about in their principal winter habitat, seeking out the most "comfortable" temperature zones, which appeared to be from −9°C to 7°C. Studies have shown that deer increase their metabolic rate when air temperature decreases below about 7°C, although mule deer may not significantly increase this rate until temperature decreases to −23°C (Mautz et al. 1985).

The reactions of mule deer to climate or weather factors (except for snow depth) seem to vary, especially in relation to ambient temperatures. Wind, for example, has little apparent influence on movements and activities except during periods of cold weather, when high winds force deer to seek shelter and reduce their activities (Wood 1988). Humidity may influence the comfort of deer and thus the extent to which they are active (Linsdale and Tomich 1953), but again it probably does not act independently of other weather factors. Loveless (1967) observed that deer appeared slightly more nervous than usual during periods of low atmospheric pressure. This behavior also can be influenced by temperature gradients and, to some extent, by precipitation and wind currents, which affect deer movements and activity. Precipitation, either rain or snow, apparently has little direct effect on mule deer except when severe storms force animals to seek shelter.

Snow depth probably has more influence on the behavior of mule deer than any other single weather element. In autumn, depths of 15–30 cm may be sufficient to initiate major migratory movements or shifts in habitat use. Depths of 25–30 cm may impede movements, especially among young animals, whereas >50 cm will essentially preclude the use of an area by deer (Loveless 1967). The time, rate, and amount of snow accumulation thus become major factors in determining both the time and the extent of mule deer concentrations on winter ranges throughout the mountainous area. In the Bridger Mountains of Montana, "normal" snow depths and patterns restrict mule deer to <20% of their total yearlong range in winter, and under severe snow conditions only 20–50% of the winter range may be usable (Pac et al. 1991). Generally in these situations, mule deer move down from higher elevations and up from areas of low relief onto steep slopes of southern and western exposure. They also may move to windswept ridges or into timber cover, where snow accumulations are less and/or clearing occurs more rapidly after storms. In addition to these influences, snow depth greatly affects the kinds and amounts of food plants available to mule deer.

Water Requirements and Relations. Mule deer evolved in a generally arid region and appear well adapted to cope with limited amounts of free water in their environments. Enough free water usually occurs to have little or no effect on mule deer distribution and abundance across most of the species range. In the semiarid Missouri Breaks region of Montana, seasonal changes in the availability of free water did not influence the distribution of mule deer (Mackie 1970; Hamlin and Mackie 1989). Furthermore, where ungulates live in more arid environments they display both behavioral and physiological adaptations to the scarcity of free water. For example, in Arizona, mule deer averaged one visit to water/day and consumed 5–6 L of water/visit during the hot summer months. Visitation rates and amount of water consumed/visit declined during cooler seasons of the year (Hervert and Krausman 1986; Hazam and Krausman 1988). Mule deer also obtain water from succulent plant material, dew on the surface of plants, and from metabolic processes (Anderson 1981). Feeding at night in hot, arid environments may provide relief from thermal stress. It also may be timed to take advantage of diel cycles in plant water content (Taylor 1969).

Whether mule deer actually require free surface water has been debated (Severson and Medina 1983). Nonetheless, when access to free water is severely restricted in penned white-tailed deer, they reduce their consumption of forage (Lautier et al. 1988). Therefore, although deer in the wild may exist for some periods of time without access to standing water, this likely represents a marginal, survival situation (Severson and Medina 1983).

The abundance and spacing of water sources does influence the distribution of mule deer in arid environments. In Arizona and New Mexico, mule deer are usually found within 2.4 km of free water (Hanson and McCulloch 1955; Swank 1958; Wood et al. 1970). Mule deer in northern California averaged 1.1–1.5 km away from water sources, with a mean greatest distance of 2.4 km (Boroski and Mossman 1996). Female mule deer drink more water than males during late summer (Hazam and Krausman 1988). Females are often found closer to sources of water than males, presumably because of the demands of lactation (Bowyer 1984; Boroski and Mossman 1996; Main and Coblentz 1996), although in one study, females remained close to water sources throughout the year and not just during lactation (Fox and Krausman 1994). Hamlin and Mackie (1989) concluded that distributional differences between adult males and adult females and fawns probably were not related to a requirement for lactating females and fawns to range closer to water. Other factors influencing distribution and habitat use apparently were overriding.

REPRODUCTION

Breeding and Gestation. Most mule deer attain sexual maturity and become capable of breeding at approximately 1.5 years of age. Pregnancy among female fawns is uncommon (McCullough 1997). The minimum breeding age of females is influenced by their physical condition, and in some severe or nutritionally impoverished environments, successful conception may not occur until the third fall or even later. Although yearling males are physically capable of breeding, the presence of many older, more mature males in a population may limit their participation in the rut.

Breeding occurs in autumn and early winter; the exact dates vary widely by subspecies and location (Anderson 1981). Rocky Mountain mule deer breed primarily during November and December; however, some breeding may occur as late as February and early March. Southern races tend to breed somewhat later, although southern mule deer at Camp Pendleton, California, apparently breed from mid-September through early November with a peak in mid-October (Bischoff 1957). Breeding among black-tailed deer begins in October, up to 1 month earlier than for most mule deer, and extends into mid-December. However, Thomas and Cowan (1975) found ovulation occurred primarily during November on Vancouver Island; the mean date for first ovulations was during 14–17 November.

It is generally believed that the timing of reproductive activity in deer is controlled at least broadly by day length. However, variation in reproductive chronology occurs across regions too small to be affected by photoperiodism. This variation may reflect local adaptation to environmental conditions such that fawning occurs at the time most favorable to early growth and survival. Local influences of nutrition, weather, population structure, and other factors also affect reproductive activity and breeding success.

The estrous cycle in female mule deer is 22–28 days. The period of estrus, when the doe is receptive to the buck, lasts 24–36 hr. More than one ovum may be shed during ovulation. If conception does not occur, the cycle may be repeated several times. In female black-tailed deer, estrous cycles can reoccur at intervals of 12–23 days. Conception rarely if ever occurs at first ovulation; most females conceive at a second ovulation 6–9 days later (Thomas and Cowan 1975).

The annual reproductive cycle in males begins with the regrowth of antlers in spring and summer. Antler growth and other reproductive activity is initiated and controlled by levels of testosterone in the blood. With the onset of the rut in autumn, the neck swells and aggressive behavior becomes apparent. Also at this time, the males become hyperactive, movements increase greatly, and food intake declines. In Rocky Mountain mule deer, the most intensive rutting behavior and breeding occur from mid-November through mid-December. Potency spans the entire period of female fertility, however, and viable, mature sperm may be present in the seminiferous tubules of some males each month of the year (Anderson and Medin 1964). In the Missouri River Breaks, Montana, conception occurred from 15 November to 24 December, with 75% of all litters conceived between 21 November and 1 December (Hamlin and Mackie 1989).

Gestation for mule deer varies from 183 to 218 days with a mean of about 203 days (Robinette et al. 1977). In northern environments, parturition occurs primarily from late May to mid-July, although fawns born early or late are occasionally observed. In the Missouri River Breaks, parturition occurred between 27 May and 15 July with a median of 10–12 June (Hamlin and Mackie 1989). Within this framework, mean birth date varied up to 10 days among years, apparently in accord with the nutritional plane of females during the year before parturition. Earliest fawning occurred in association with above-average range and nutritional conditions; late fawning with below-average to poor conditions (Hamlin and Mackie 1989). In southern environments, parturition occurs later, in July and August.

Litter Sizes. Adult female mule and black-tailed deer commonly conceive twins, whereas yearlings usually carry a single fetus. Robinette et al. (1955) reported litter sizes for 492 adult females as no fetuses (2%), one fetus (37%), twins (60%), and triplets (1%). Reproductive potential, however, may vary quite widely among populations on different ranges, as well as yearly on the same range, according to local environmental conditions and the nutritional status of does. Thus, Hamlin and Mackie (1989) observed litters averaging 44% singles, 55% twins, and 1% triplets during a 12-year period. However, for individual years, the respective percentages varied from 75%, 25%, and 0% under very poor range conditions to 27%, 70%, and 3% when excellent conditions prevailed.

Pregnancy rates range from 70% to >90% among females ≥1 year old. Fetal rates range from about 1.3 to 1.8 fetuses/doe. Pregnancy and fetal rates are generally lower for yearling females than for older adults. Fetal rates for black-tails may be slightly lower than those reported for mule deer. Mule deer fawns (6 months to 1 year of age) rarely breed, although under captive conditions, about 50% of female black-tailed deer conceived in their first year of life (Mueller and Sadleir 1979). McCullough (1997) reported breeding by fawns occurred in wild black-tailed females when population density was greatly reduced.

The full reproductive potential of mule deer is only rarely reflected in the recruitment of fawns and, especially, yearlings into a population. The reproductive performance of does may be strongly influenced by their physical condition and nutritional status at the time of breeding and during gestation, especially the last 2–3 months of pregnancy. Total fawn production in a population also will be determined by the age distribution of females, including the proportions of fawns, yearlings, and older adults. Hamlin and Mackie (1991) reported age-specific effects related to reproductive history and environmental conditions. In that study, adult females experienced decreased reproduction after their fifth reproductive year, possibly as a result of physiological exhaustion associated with reproduction in a semiarid environment. Reproduction by females ≥8 years stabilized at a level similar to that of 3-year-olds, but 12–21% lower than levels for 4-, 5-, and 6-year olds, which constitute the most efficient segment of the population in terms of reproduction and recruitment.

Postpartum survival of fawns may be influenced by many factors. The most important of these is their physical condition at birth as determined by the quality of the diet of pregnant does on winter and spring ranges, nutritional plane of the mothers during lactation on summer range, and quality of late-summer, fall, and winter diets after weaning. Other factors regulating fawn survival include predation, disease, weather extremes, hunting, and accidents. Overall, it is not uncommon that 25–30% of the fawns produced by a mule deer population are lost by autumn, 50% or more by early winter, and up to 75% or more by spring (Mackie et al. 1998).

Longevity. Records for tame and captive mule deer indicate that does may live as long as 22 years and bucks may live to 16 years (Cowan 1956). Maximum life expectancy of wild deer is poorly known, but does and bucks in natural populations often live less than half as long as their captive counterparts. Extensive mortality accrues annually to both sexes such that most individuals die from various causes before reaching "old" age. Records for individually tagged mule deer in the Bridger Mountains of Montana show that females rarely live beyond 12–14 years of age, whereas males seldom live beyond 8 years (Pac et al. 1991; Mackie et al. 1998). Only 1% of 1790 adult mule deer bucks harvested on a Utah range were estimated to be ≥8 years old (Robinette et al. 1977).

Tooth wear, which influences feeding efficiency, is an important factor determining the maximum potential life span of deer and may be responsible for differences between captive and wild mule deer, and between wild deer on different ranges. Differences in soil texture and food types affect the rate of tooth wear. Reasons for the apparently greater maximum and average longevity of mule deer females as compared with males have not been studied extensively. However, males in all age classes have a higher natural mortality rate than females of their respective cohorts, even in the absence of hunting. In Montana studies (Mackie et al. 1998), adult males experienced mortality rates ranging from 41% to 61%, more than twice the rates experienced by adult females. Hunters selected heavily for males over females and for older, larger antlered males over young, smaller antlered individuals. Yearling males and males >6 years old are particularly vulnerable to mortality during severe winters.

FEEDING HABITS

Mule deer are herbivores, possessing a four-chambered ruminating stomach in which vegetation is reduced to usable form by microbial fermentation. This process, though efficient in extracting energy and nutrients from plant tissue, is slow. Both the amount of vegetation that can be ingested, which is determined by the size of the rumen and reticulum (the first two chambers), and the rate at which it can be passed through the gut, which is determined by the digestibility of the plant tissues consumed, are extremely important in nutrition. Compared with other ruminants, mule deer have small rumens and gut length relative to body size; thus, they must eat small volumes of high-quality, easily digested food. Hoffman (1985) classified deer as concentrate selectors in feeding, as compared with larger ruminants, or bulk feeders, which can eat larger volumes of forage of lower nutritional quality and digestibility. Even when browsing on woody plant material, deer select mainly the terminal portions of new twig growth, buds, and other tissues that are highest in quality and most digestible among plant materials available.

In addition to energy and protein, ingested plant tissues provide dietary vitamins and minerals as well as water. Deer apparently can store or synthesize all vitamins needed for metabolism such that nutritional deficiencies can seldom, if ever, be traced to this source. Minerals important in growth, development, and metabolism include calcium, phosphorus, sodium, potassium, chlorine, magnesium, sulfur, iron, iodine, copper, selenium, zinc, and manganese. Of these, phosphorus, iodine, and selenium could be limiting for mule deer on some western rangelands. Thus, mule deer commonly visit salt and other mineral licks, especially during spring, where some minerals are obtained by direct ingestion.

The specific foods eaten by mule deer are extremely varied. Kufeld et al. (1973) listed 788 species of plant, among which were 202 shrubs and trees, 484 forbs, and 84 grasses, sedges, and rushes eaten by Rocky Mountain mule deer alone. Even this list is incomplete because of studies not cited. In addition, plant species used only in small amounts often are not listed in reports. Not all plants are deer food, but many different plants are used at some time, in some places, by mule deer as individuals and populations adapt to the vegetationally diverse environments they occupy. Some plants may be eaten in one area and not in another, only in certain seasons or stages of growth, or under certain environmental conditions. Some species may be eaten only in association with, or in the absence of, other forage species, or only when deer may periodically use the sites on which those plants occur. Deer seem to have the ability to select plant parts and plants from certain soil types or sites that are highest in nutritional content (Swift 1948; Bissell 1959). The kinds and amounts of different plants eaten may also vary between individual mule deer, reflecting individual preferences and/or differences in kinds of plants available within individual home ranges (Willms and McLean 1978). Because of this, it usually is not feasible to generalize about the kinds and quantities of forage used, even by broad forage classes or habitat types.

Mule and black-tailed deer traditionally have been thought of as browsers. This concept of browsers by choice is among the oldest and most persistent myths in deer ecology and management (Gill 1976). Given access to seasonally abundant, nutritious, herbaceous plants of high digestibility, deer will tend to select those plants rather than browse species of lower digestibility. For example, diets of Sitka black-tailed deer on Admiralty Island, Alaska, ranged from 57% to 79% nonbrowse items such as forbs, ferns, grasses, sedges, lichens, algae, and mosses depending on season, although during a winter with deep snow, browse consumption peaked at 87% (Hanley et al. 1989). Diets of Columbian black-tailed deer on oak (*Quercus* spp.) woodland–annual grass ranges in northern California included as much as 62% newly germinated annual grasses during winter when those species were green and highly digestible (Taber and Dasmann 1958). Finally, black-tailed deer introduced on the island of Kaua'i, Hawaii, are primarily frugivorous in autumn, with diets in September and October consisting of 58% passionfruit (*Passiflora* spp.) and guava (*Psidium* spp.), both fruits of nonnative species (Telfer 1988). Other seasonally important forages for mule and black-tailed deer include acorns (Beale and Darby 1991), mistletoe (Urness 1969), lichens (Hanley et al. 1989), mushrooms (Beale and Darby 1991), and succulents (Krausman et al. 1997). Many of those forages are often underrepresented in diet studies because of their high digestibility or deficiencies in sampling and analytical methodology (Beale and Darby 1991).

The nutritional requirements of mule deer differ seasonally and according to the sex, age, activity, condition, and reproductive status of individuals as well as with environmental conditions. The natural environments in which mule deer occur are extremely variable, and they are never static. Foods eaten vary in kind, quantity, and nutritional quality as well as in digestibility among seasons, among years, and among locations. Thus, the costs of obtaining energy and ultimately the net energy gained also vary. Sometimes only small and seemingly insignificant differences in the kinds, amounts, growth, and chemical composition of plant tissues available may make the difference between a negative and a positive energy balance for deer feeding on them.

Generally, in spring and early summer, when herbaceous vegetation is immature and succulent, plant tissues are easily fermented, the total concentration of volatile fatty acids is greatest, and usable energy is high. Plant protein contents also are high. In late summer and autumn, herbaceous plants become dry and woody vegetation ceases to grow, resulting in diets progressively lower in protein, higher in carbohydrates, and much higher in lignin and cellulose content. Digestibility also decreases, rumen turnover time and energy costs for fermentation increase, and the quantity of usable energy and nutrients derived from forage is reduced.

Deer can adapt physiologically and behaviorally to these changes. In winter, at least in severe northern environments, metabolic rate is reduced, food intake is voluntarily limited, habitats favorable to energy conservation are selected, and activity becomes restricted. During this period, mule deer may predominantly use woody browse and other low-quality forages. They typically experience an energy deficit, which must be met by drawing on body fat deposits accumulated in late summer and fall. Their survival is governed not by the potential of winter forage supplies to meet energy needs, but by body condition entering winter in relation to temperature, snow conditions, and quality and quantity of forage available to slow the drain on fat reserves. Temperature and snow conditions determine forage availability, energy expenditures, and the costs for energy gained in feeding. The amount of fat that an animal carries and the severity and duration of the winter determine whether finite fat deposits will carry it through winter (Wallmo et al. 1977). When fat reserves are expended before a positive energy balance is again achieved in spring, the animal will begin to assimilate structural body proteins. In general, if this source of maintenance is required for more than a short time, death will occur.

Although winter is usually considered the most critical period in the annual nutritional cycle of mule deer, body size, condition, and fat reserves are related directly to environmental conditions influencing the kinds, quantity, and quality of forage available during prior seasons. Therefore, deer physical condition is a product of all habitats used throughout the year. Torbit et al. (1985:84) concluded that "seasonal ranges can no longer be considered to have only isolated effects on herd performance." Recruitment of mule deer fawns may be related directly to weather conditions during the previous year (Hamlin and Mackie 1989; Wood et al. 1989), apparently through effects on nutrition and body size (Unsworth et al. 1999).

BEHAVIOR

Behavior is an integral part of the process by which deer adapt to living with each other and within their diverse and variable environments. Distribution, movements, feeding habits, activity patterns, habitat selection, reproduction, and population dynamics all reflect behavior or are closely linked to behavioral adaptation (Geist 1981; Mackie et al. 1998).

Socializations. The degree of sociality in mule deer varies according to season, sex, population, and subspecies. In general, these deer are neither socially gregarious like elk (*Cervus elaphus*) nor strictly solitary like moose (*Alces alces*). Instead, social structure is functionally organized around family groups consisting of two or more generations of related females and their offspring (Hamlin and Mackie 1989; Pac et al. 1991; Mackie et al. 1998). The dominant member or matriarch is a mature female with a history of successful reproduction. In late spring and early summer, adult females drive off the previous year's young and isolate themselves on fawning or "parturition" territories (Ozoga et al. 1982) in reproductive habitat. There, through late summer or fall, maternal females restrict their movement and direct all of their energy to successfully rearing young. Other social associations during summer include small, dispersed groups of adult males, mixed yearling–adult male groups, yearling males and/or females wandering together, and nonproductive adult females that may have lost young postpartum. Sometimes apparent larger groups may occur where home ranges overlap and individuals or small family or other groups come together on common feeding sites.

Mixed family groups of maternally related females and offspring begin to reform in late summer and fall. Yearlings of either sex are accepted back into family groups, related does and fawns from adjacent home ranges at least temporarily aggregate in sharing resources, and related females that were barren or lost fawns come together within commonly used habitat. Later in fall, winter, and spring, when forced to concentrate on winter range or use common feeding areas or cover types, mule deer may aggregate in large groups, sometimes numbering several hundred animals. Such large groups rarely endure, however. The largest groups typically occur in the north during midwinter to late winter, when snow depths greatly restrict available range and forage supplies. Large groups also occur during early spring, when the animals frequently concentrate their feeding on certain open range sites where new green forage first appears and is most available.

Miller (1974) reported that black-tailed deer also showed a strong tendency to form cohesive groups led by females within which individuals were socially bonded and each held distinct social ranking. The groups were distinct social units from winter into the prefawning period, whereas male groups were distinct from the time of antler casting in winter to the late prerut period.

Sexual Segregation. Adults of both sexes establish and traditionally use relatively small individual home ranges within suitable habitat seasonally or year-long. Males older than yearlings, however, tend to occur somewhat grouped, with closely overlapping if not common home ranges, within certain "buck habitats" (Mackie 1970). In some areas at least, buck habitats may be located peripheral to areas used by does. Does tend to be more widely and uniformly distributed across available habitat. Because of this, individual mature bucks may be seen in the company of other males more frequently or consistently from winter through early autumn than is the case with does. The apparent socialization among males breaks down during the rut, when individuals become highly aggressive toward each other and widely dispersed across all available habitats. Miller (1974) believed that black-tailed deer exhibited individual and group territoriality in their distribution and use of range resources. This phenomenon has not been evident or documented in studies of the behavior and habitat use of mule deer.

Why adult males and females use different ranges has been a controversial topic. Main et al. (1996) summarized previous literature and classified concepts about sexual segregation into three groups: body-size hypotheses, reproductive-strategy hypotheses, and social-factor hypotheses. One body-size hypothesis posits that smaller bodied females can more efficiently use habitats with closely cropped forages, thereby excluding larger males through intersexual competition. Main and Coblentz (1996) provided some evidence for this hypothesis for mule deer by noting that the biomass of nutritious forage was lower on range occupied by females than on areas inhabited by males. Another gastrocentric model proposes that male deer may perform better on areas with lower quality forage than would be ideal for females because of their larger body and rumen–body size ratio (Barboza and Bowyer 2000). This body-size hypothesis may help explain instances of sexual segregation where male mule deer occupy areas characterized by forage that appears to be less than optimum for females. A reproductive-strategy hypothesis suggests the need to bear and raise young pose constraints on habitat use by adult females, thereby limiting their use of space (Kie and Bowyer 1999).

Breeding Behavior. Mule and black-tailed deer are polygamous, the males wandering about extensively, seeking and pursuing individual does in estrus. Mature bucks become particularly mobile and are highly aggressive and antagonistic toward other males as they actively seek to breed as many receptive females as possible. Miller (1974) suggested that superior, dominant black-tailed males established rutting territories within which other bucks were not tolerated. In most populations, older, dominant males are likely to accomplish most of the breeding. However, bucks as young as yearlings are fully capable of breeding, and occasionally, especially in areas where few mature males occur, participate extensively in the rut.

Typical rutting behavior of bucks includes snorting, urine marking, thrashing and rubbing antlers in shrubs and trees, sparring and antler fighting with other males, and, occasionally, herding individual does. Does may also be more active during the rut. Females may use various behaviors such as urine marking to signal their location and the onset of estrus to the male. Actual courtship apparently involves a number of different behavioral patterns on the part of the buck and the doe, culminating in breeding. A single doe may be bred several times during the estrous period, and a single buck may breed many, if not all, receptive does over a large area during the period of rut. In black-tailed deer, mature males may court and breed primarily mature does, whereas yearling does are courted and bred mainly by yearling and younger adult males (Miller 1974).

Other Behavior. Mule deer tend to be creatures of habit in many aspects of their activities. The size and shape of home ranges and timing and extent of seasonal and year-long movements are behavior patterns that "adapt" each individual to the particular environment in which it settles. Daily activity patterns also tend to be adaptive and habitual, as are the kinds and amounts of forage selected and the particular habitat or cover types used.

Mule deer apparently are quite capable of learning and adjusting their activity patterns and habits to accommodate at least some changes in their environment, including the presence and activities of humans. Thus, in the Missouri River Breaks, mule deer became extremely alert and relatively few were observed feeding following a period of intense hunting (Mackie 1970). Similarly, alertness of mule deer and their response to disturbance may vary between populations or habitats in relation to the occurrence of natural predators, previous disturbing experiences, or the security of an area.

Unlike white-tailed deer, mule deer do not necessarily attempt to hide from predators. Instead, they attempt to detect danger at long range with their large ears and excellent vision, and use generally open habitats to outmaneuver their enemies. In doing so, they effectively use characteristics of the terrain, including steep slopes, boulders, ledges, trees, brush, and deadfalls, together with their stotting gait, to place obstacles between themselves and the predator. This strategy requires precise timing and very calm, unexcitable individual behavior, and probably explains the greater use of open habitats by mule deer as compared with white-tailed deer (Geist 1981).

HABITAT

Mule deer are broadly adapted and occur in all major climatic and vegetational zones of western North America except arctic and tropical areas and most extreme desert. Thus, their habitats are many and extremely diverse, and complete description is a complex if not elusive task.

Generally, mule deer frequent semiarid, open forest, brush, and shrub lands associated with steep, broken, or otherwise rough terrain. Their stronghold may be the mountain–foothill habitats that extend from northern New Mexico and Arizona north into British Columbia and Alberta along the Rocky Mountains and other mountain ranges. However, extensive populations also occur in prairie breaks and badlands habitats, especially in the Great Plains, and in semidesert shrub habitats of the southwest.

The mountain–foothill habitats occupied by mule deer span a broad range of latitudes and elevations. Climates and topography are diverse and, correspondingly, vegetational components vary considerably. Mule deer associated with these habitats may be year-long residents in foothills, occupy seasonal home ranges that include footslopes in winter and adjacent mountain slopes in summer and autumn, or exhibit migrations between distinct, widely separated winter and summer–fall home ranges (Pac et al. 1991). In summer, they usually occur widely distributed over all suitable or available habitat within a mountain–foothill complex. In winter, however, extreme snow depths usually preclude use of higher elevations, and mule deer are forced to concentrate, often at very high densities, on lower south-facing slopes where snow depths are less and stands of shrubby vegetation provide winter forage. Generally, montane and subalpine forest communities dominate summer ranges; open, shrub-dominated slopes and ridges characterize the primary wintering areas.

In the prairies of the northern United States and southern Canada, level and rolling plains dominated by grasslands provide little habitat for mule deer. Instead, the rough, timbered or nontimbered breaks along river drainages, heavily dissected badlands, and brushy streamcourses and draws, especially where interspersed by agricultural land, provide deer habitat (Severson and Carter 1978; Hamlin and Mackie 1989; Wood et al. 1989). Mule deer do not occur at high densities over broad areas of prairie habitat. Instead, their occurrence is patchy, with locally high densities found mostly in rough breaks or badlands, along streamcourses, and on certain local prairie–agricultural complexes. In these environments, deer often tend to be nonmigratory, but various migratory and other seasonal movement patterns may be employed to fully use available habitats (Hamlin and Mackie 1989; Wood et al. 1989). Ponderosa pine (*Pinus ponderosa*) and Rocky Mountain juniper (*Juniperus scopulorum*) provide cover in timbered breaks and badlands, whereas cottonwoods (*Populus* spp.), green ash (*Fraxinus pennsylvanica*), and box elder (*Acer negundo*) are important along streamcourses and in draws. Common and important shrubs for mule deer food and cover include sagebrush (*Artemisia* spp.), rabbitbrush (*Chrysothamnus* sp.), skunkbush sumac (*Rhus trilobata*), snowberry (*Symphoricarpos* spp.), rose (*Rosa* spp.), chokecherry (*Prunus virginiana*), buffaloberry (*Shepherdia argentea*), and willow (*Salix* spp.).

In the Southwest, mule deer are associated with two types of semidesert range; both are arid, sparsely vegetated environments dominated by shrubs. One, occurring in southern Arizona and New Mexico, western Texas, and parts of Mexico, is characterized by creosote bush (*Larrea divaricata*), mesquite (*Prosopis* spp.), greasewood (*Sarcobatus* sp.), and several species of cactus. In some areas, various species of oak (*Quercus* spp.) and chaparral occur. The other, the northern or Great Basin type, occurs in parts of Nevada, western Utah, and southeastern Oregon. Common plants are sagebrush, saltbush (*Atriplex* spp.), cliffrose (*Cowania stansburiana*), and winterfat (*Ceratoides* sp.). Juniper–pinyon woodlands and pine forests may occur at higher elevations.

Black-tailed deer inhabit temperate, coniferous forests occurring along the northern Pacific Coast from northern California north to southeastern Alaska, though important populations also occur in woodland–chaparral habitats of central coast ranges in California. Coastal rain forest habitats are characterized by dense coniferous forests and a marine climate with cool temperatures, many cloudy days, and high precipitation. In California and southern Oregon, redwoods (*Sequoia sempervirens*) and Douglas fir (*Pseudotsuga menziesii*) typify these habitats. To the north, Sitka spruce (*Picea sitchensis*), western red cedar (*Thuja plicata*), western hemlock (*Tsuga heterophylla*), and Douglas fir predominate. Forest succession is very rapid, and the kinds and amounts of understory present vary widely depending on successional stage. Habitat values and use by black-tailed deer also vary with successional stage. For example, in western Washington, deer use generally was highest between 10 and 30 years after the forest was opened (Brown 1961). On Vancouver Island, British Columbia, and in coastal forests of southeastern Alaska, however, deer use of mature forests generally is much higher than use of logged habitats (Schoen and Wallmo 1978).

California woodland–chaparral habitats consist of woodlands characterized by oak and pine, a variety of shrubs, and numerous grassy openings. Chaparral is dominated by numerous shrubs, including chamise (*Adenostoma fasciulatum*) and manzanita (*Arctostaphylos* spp.). In the absence of fire, chaparral becomes extremely dense and is little used by deer. Use of such areas can only be restored through fire or mechanical disruptions of the vegetation.

Mule deer use of any habitat is determined by resource needs of the animals (food, cover, water, space) and where these resources are distributed in the environment. Mule deer require a diversity of plant species as food during the course of the year, such that several individual vegetation types must be available. A diversity of vegetation may also be required to meet their various needs for hiding, escape, and thermal cover. Because of this, the juxtaposition and interspersion of different habitat types may be more important than the occurrence of individual types.

MOVEMENTS AND HOME RANGE

Movements by mule deer can include dispersal, migration, and daily or other local movements. Dispersal is the movement of an individual out of an area, often that in which it is born, with little likelihood of return. Unidirectional movements >5 km were considered dispersal among black-tailed deer, and most occurred among individuals 1–2 years of age (Bunnell and Harestad 1983). Dispersal among young males averaged 15.2 km; that for females averaged 12.2 km. Dispersal movements >12 km were rare, although maximum dispersal distances were 32 km for males and 30 km for females (Harestad and Bunnell 1983). In the Missouri River Breaks, Montana, 70% of yearling males dispersed from natal home ranges and 51% left their natal population; among females, only 16% of the yearlings dispersed from natal home ranges and 14% left their natal populations (Hamlin and Mackie 1989). Dispersal distances averaged 25.7 km (range = 7–83 km) for males and 36.9 km (range = 5–69 km) for females. One young male mule deer in Arizona dispersed >44 km (Scarbrough and Krausman 1988). Dispersal among deer >2 years old apparently is rare.

Migration occurs as regular, predictable, seasonal movements between disjunct ranges. Where resources are abundant and appropriately interspersed and climate is mild, mule and black-tailed deer can be residents throughout the year (Taber and Dasmann 1958). However, deer often exhibit migratory behavior, especially in mountain–foothill habitats. There, deer commonly move to high-elevation montane ranges during summer to take advantage of seasonally abundant herbaceous forages and retreat to lower elevation ranges in fall and winter. Distances covered in migration may vary from a few kilometers to >160 km. The timing of migratory movements also may vary. Fall migration usually is influenced by snowfall and the depth of snow on summer and intermediate ranges, whereas timing of spring migrations typically is associated with snow melt and the reappearance of succulent forage on transition and summer ranges. In the Bridger Mountains, Montana, the traditional nature of home range and movement patterns was established and maintained by interactions among related members of family groups (Pac et al. 1991).

Occasionally, mule and black-tailed deer cross open bodies of water during migration (Cowan 1956; Einarsen 1956; Geist 1981). In southeast Alaska, blacktails swim across Frederick Sound at distances >22 km (Einarsen 1956); in northern California, groups of black-tailed deer regularly cross a reservoir, swimming for periods as long as 23 min across distances >1 km (Boroski 1998).

Mule deer populations may comprise migratory and nonmigratory individuals (Nicholson et al. 1997) as well as individuals following different migratory strategies (Pac et al. 1991; Mackie et al. 1998). Nicholson et al. (1997) suggested that climatic variability likely allows the maintenance of different migratory and other movement patterns in a single population of deer. Alternatively, Mackie et al. (1998) theorized that different movement strategies arise in the process of adaptation of individuals and family groups to the total environment and allow deer to efficiently use all available habitat in an area. The general type of movement patterns employed by individual deer depended on the spatial arrangement of important habitat components they attempted to exploit.

Seasonally, movements of mule deer are largely within individual home ranges where minimum travel is necessary to meet their resource needs. Daily and other short-term movements within a single season and area typically define an individual's home range. Because home range size is part of the habitat use strategy employed by individual deer to exploit the diverse environments they occupy, extreme variability seems characteristic of individual home ranges in all environments (Mackie et al. 1998).

Overall, reported home range sizes for mule deer vary from a low of 39 ha for a yearling female black-tailed deer in western Oregon (Miller 1970) to 13,850 ha for an adult female mule deer in open plains of eastern Colorado (Kufeld and Bowden 1995). In general, (1) larger home ranges have been reported in open, simple, and more variable habitats as compared with closed, diverse, and stable environments; (2) males have larger home ranges than females; and (3) mule deer have larger home ranges than black-tailed deer. The latter, however, may reflect differences in the structure of typical habitats rather than inherent subspecies differences.

In most migratory situations, distinct summer and winter home ranges are established and used. In the Bridger Mountains of Montana, winter home ranges averaged 248 ± 48 ha for bucks and 244 ± 84 ha for does (Pac et al. 1991). Summer home ranges of does averaged 177 ± 55 ha, whereas adult males ranged over 240 ± 123 ha. Other studies in mountain environments indicated seasonal home ranges 40–100 ha in size (Leopold et al. 1951; White 1960).

Mule deer are not considered migratory in most prairie habitats, but they may migrate locally or use some parts of their year-long range more intensively than others. In addition, movement out of normal home ranges may occur during periods of unusually severe environmental conditions. In a nontimbered northern prairie environment, Wood et al. (1989) reported mule deer females ranging over areas 1830–6599 ha in size yearly; male home ranges varied from 83 to 8410 ha. In comparison, in timbered, prairie breaks habitat, Hamlin and Mackie (1989) found the average home range of does throughout the year was 660 (150–3020) ha, whereas mature bucks averaged 2710 (540–2920) ha. In Colorado, one nonmigratory mule deer population in open grassland and pinyon–juniper habitat occupied average home ranges of 1220 ha (Gerlach 1987). Another sedentary population on river bottom habitat along the South Platte River had a median home range size of only 390 ha, although one individual, which lived primarily on adjacent plains habitat, had an annual home range of 13,850 ha (Kufeld and Bowden 1995). Yearly home ranges averaging 1060 and 1240 ha were reported for females and males, respectively, on semidesert range (Rogers et al. 1978).

Reasons for variations in the sizes of home ranges in deer are not well understood, but have been related to individual behavior, sex, age, body mass, season, race, habitat occupied, environmental characteristics and conditions, population density, computational method, and other factors (Mackie et al. 1982; Anderson and Wallmo 1984). Hamlin and Mackie (1989) found that home range sizes of adult females and fawns were not related to population density, forage condition, or age of the females. The presence of fawns-at-side overrode all other factors in determining home range size of adult females. Females with fawns-at-side had significantly smaller home ranges than barren females under all conditions. Hervert and Krausman (1986) observed that, when denied access to water, desert mule deer females enlarged their home ranges to search for alternative sources outside their normal home ranges. In California, cattle grazing led to increases in the size of home ranges among mule deer on summer range (Loft et al. 1993) and black-tailed deer on winter range (Kie and Boroski 1995).

Measures of landscape heterogeneity, such as number of different types of habitat patches, habitat patch shape and spatial arrangement, and contrast between adjacent patches, may also play important roles in determining home range size in deer. Landscape metrics accounted for over half the variation in home range size in female mule and black-tailed deer on five sites in California (Kie et al. 2002).

POPULATION ECOLOGY

Historical Trends. Seton (1929) estimated that as many as 10 million mule deer and 3 million black-tailed deer may have existed in North America in pre-settlement times. Most authorities today view these estimates as too high. Deer thrive on disturbed ranges in intermediate stages of plant succession and may have become more abundant and widely distributed in North America during the past half-century than ever before. Early settlers found deer scarce in many parts of the West. Archaeological evidence (Jennings 1957) suggested that mule deer historically may not have occurred in much of the Great Basin area of Nevada and Utah where they have occurred abundantly in recent years. Because of this, Wagner (1978) estimated that no more than 5 million, and possibly even fewer, mule and black-tailed deer occurred in the western United States during pre-Columbian times.

Populations of both forms, but especially mule deer, declined drastically and became quite localized over most of the West following settlement. This was the result of unrestricted hunting and use of wild animals for food and disturbance and preemption of deer habitats by agriculture. By the beginning of the twentieth century, mule deer were generally scarce, although lows in some areas in the north may not have occurred until the 1920s and 1930s (Hamlin and Mackie 1989). The disturbances of settlement, especially widespread and often abusive livestock grazing, logging, and burning, ultimately proved beneficial, however, as range and forest vegetation was opened up and became more diversified. Many plants more palatable to mule deer than those that dominated the original vegetation (Longhurst et al. 1968, 1976) either invaded or increased greatly in abundance. At the same time, competitors and predators were removed or greatly reduced in distribution and abundance.

Mule deer responded rapidly to these changes and, under restrictive hunting regulations, increasingly effective law enforcement, widespread predator control, and perhaps generally favorable weather conditions, increased in numbers and distribution throughout the West. By the early 1920s, mule deer were extremely abundant in some parts of the Southwest, for example, the Kaibab Plateau, Arizona (Rasmussen 1941). From the 1920s through the 1950s, increases spread north and west to the point where mule deer were of unprecedented abundance (Wagner 1978). Although reliable population estimates have never been made, mule deer numbers probably reached all-time highs between 1935 and 1955 (Gill 1999), and it seems likely that there were at least 7.5 million mule and black-tailed deer throughout the West in the early 1960s (Mackie 1987).

The great abundance of the 1940s and 1950s could not be sustained, and within a few years after peak populations occurred in most areas, numbers began to decline. Deer filled all available habitats and overpopulation in many areas reduced the abundant and nutritious forage plants on which they had originally thrived. Fawn production and/or survival declined and malnutrition and starvation losses during winter or other critical periods became common (Leopold et al. 1947). Hunting regulations were generally liberalized in most areas in management efforts to balance deer numbers with food supplies, prevent depredations to agricultural crops and products, and/or harvest the large surplus of deer that prevailed.

Beginning during the mid-1960s, and especially during the early 1970s, mule deer appeared to decline sharply in numbers and distribution over most of the West. This decline, like the increases, occurred somewhat later in the north than in the south. Efforts were made to tie this decline to many factors, including hunting; range deterioration or habitat changes due to overbrowsing; succession, and livestock grazing; predation; competition with livestock and other game animals; destruction or loss of habitat due to human development; diseases; and climate/weather changes. None of these satisfactorily explained population declines in all areas where they apparently occurred and no consensus on causes was reached (Workman and Low 1976).

By the late 1970s, mule deer populations generally had stabilized and, in many areas, began to increase through the 1980s to the early 1990s. Trends varied widely between populations and habitats. In an analysis of mule deer population trends in Colorado, Gill (1999) concluded that mule deer populations in general have been declining since the late 1950s and 1960s and numbers of mule deer in 1999 were <50% of those during the peak in the 1940s. Similar trends may have characterized other mule deer ranges. In mountain–foothill and montane forest environments of Montana, some populations have declined over the past 50 years, whereas others have exhibited increase and decrease phases since the mid-1970s (Mackie et al. 1998). However, Unsworth et al. (1999) reported that mule deer populations generally rebounded over a 15-year period from the mid- to late 1970s into the 1990s. In portions of the northern Great Plains, numbers recovered rapidly from lows in the mid-1970s to relative highs in the mid-1980s and fluctuated or continued to increase to perhaps unprecedented highs in the early 1990s (Mackie et al. 1998).

Another widespread decline in mule deer populations occurred across much of the West beginning in about 1992. These declines sparked renewed concern about causes and the future of the species in many areas. General analyses of trends and possible causes (Gill 1999) were conducted, as were detailed studies of survival rates for mule deer fawns and adults and their effect on population trends and dynamics in Colorado, Idaho, and Montana (Unsworth et al. 1999). Some evidence suggested that important mule deer habitats have deteriorated through time and survival of fawns influenced by annual environmental conditions is a primary factor influencing population trends. Nonetheless, "the answer to the question 'What caused mule deer numbers to decline ?' remains both speculative and controversial" (Gill 1999).

Population Characteristics and Dynamics. The abundance of mule deer is determined by the number of deer that can be supported per unit area of habitat and by the total amount of habitat available. Factors that influence either local densities or the amount of area available to mule deer are important in determining population size and trends.

Habitats occupied by mule and black-tailed deer vary greatly in their ability to support deer, depending on how well each habitat satisfies the environmental requirements, adaptations, and tolerances of the animals (Mackie 1978a). Prairie habitats are characterized by low relief and relatively few vegetation types. They also tend to be quite variable with respect to meeting seasonal resource needs of deer. Mule deer usually occur either widely dispersed on large individual home ranges or concentrated in local areas where inclusions of badlands, riparian, or agricultural habitats satisfy their diverse needs more locally. As a result, deer densities are typically low and vary temporally and spatially. Wood et al. (1989) found annual average mule deer densities ranging from 0.6 to 2.0/km^2 during a 12-year study on an eastern Montana prairie, compared with average densities ranging from 0.1 to 4.4/km^2 reported for various other northern prairie or plains habitats. Higher densities, averaging 7.5/km^2, have been reported only for a prairie mule deer population inhabiting a surface mine reclamation–suburban habitat complex (Fritzen 1995).

Prairie breaks and badlands habitats, typified by greater diversity of topography and vegetation, meet the diverse resource needs of mule deer more consistently than open plains, and higher average densities usually occur. Densities in the timbered Missouri Breaks, Montana, varied from 1.4 to 6.2/km^2 during a 28-year period (Hamlin and Mackie 1989). Irby et al. (2000) estimated a mean density of 3.0/km^2 for mule deer in badlands habitat in Theodore Roosevelt National Park, North Dakota, during 1985–1996. In Arizona, Swank (1958) estimated densities of about 4 mule deer/km^2 in chaparral habitats just before the hunting season, whereas Smith et al. (1969) estimated 5.5–10.3 mule deer/km^2 in desert shrub habitat on the Three Bar Wildlife Management Area during autumn 1961–1976.

Mountain–foothill-type habitats tend to be highly complex and diverse environments in which the seasonal resource needs of mule deer can be consistently met within very small areas and many such areas are available to deer (Mackie 1978a). Thus, these habitats often support relatively high densities. Winter population estimates for the Bridger Mountains, Montana, indicated 4–7 deer/km^2 of year-long habitat at a time when mule deer numbers were relatively low (Pac et al. 1991).

Robinette et al. (1977) estimated that the average posthunting season density of mule deer on a mountain–foothill-type range in Utah was about 16/km^2 during the years 1947–1956, a period of extreme mule deer abundance.

Mule deer typically do not occur in extremely high densities based on the broad areas they inhabit year-long. However, in mountainous environments, winter concentrations up to 22–70 deer/km^2 were documented on a study area in Colorado (Unsworth et al. 1999). Occasionally, mule deer densities up to 130/km^2 occur for extended periods during severe winters when migratory populations aggregate on $\leq$20% of their total year-long range (Pac et al. 1991).

Black-tails may occur in higher density populations than mule deer. Dasmann (1956) reported summer densities in California chaparral ranging from a high of about 55 deer/km^2 immediately following a fire to about 30/km^2 several years later, which he considered more normal for this habitat. Anderson et al. (1974) estimated black-tail densities averaging 21/km^2 in Mendocino County, California, between 1958 and 1970 and 29/km^2 on the Hopland Field Station during 1964–1966. Winter densities of two populations inhabiting unmanaged and managed (burned and seeded) chaparral were about 10 and 23 deer/km^2, respectively (Taber 1956). Black-tailed deer densities in coastal forest habitats are dependent on the time and extent of logging or burning (Brown 1961). For one area, Brown (1961) estimated summer–fall densities from 13 to 22 deer/km^2, whereas on another area they ranged from about 6 to 10 deer/km^2, the higher values representing more favorable successional stages.

Most mule deer populations are almost constantly fluctuating in relation to fawn production and recruitment and adult mortality. At least four different patterns of change can be recognized: seasonal, annual, periodic, and long term. Seasonal changes result from the annual dynamics of populations. Populations are highest in early summer as fawns are added. Mortality accrues through summer, fall, winter, and spring to reduce populations to their lowest level immediately before fawning the following year. The magnitude of seasonal change depends on the numbers of fawns born and seasonal mortality patterns and rates among both fawns and adults.

Fawn mortality often is high at or immediately following parturition. Generally, lower mortality prevails among fawns and adults through summer and early fall. Exceptions occur on some southern, semidesert mule deer ranges and black-tail ranges in California woodland–chaparral, where drought conditions during fall may sharply reduce numbers of fawns and adults during that season. Most mule and black-tailed deer populations are hunted during fall; harvests reduce numbers through that period. In northern environments, winter is often a critical time in the survival of mule deer and some mortality, especially among fawns and old adults, is typical. Occasionally, it may be extensive, resulting in the loss of as many as 33–50% of all animals present in early winter. Early spring also may be an important mortality period in some areas or during some years. Average annual survival rates for telemetry-collared fawns and adult females in Colorado, Idaho, and Montana were 0.44 and 0.85, respectively (Unsworth et al. 1999). Annually, survival of fawns varies greatly in relation to environmental conditions, whereas adult female survival is typically high and stable among years. Annual survival of adult males varies by environment and in relation to hunting, which typically selects for adult males. In Montana, average annual survival rates of males ranged from 0.39 to 0.59 (Mackie et al. 1998).

Annual fluctuations in population size occur as a result of differences in fawn production and/or seasonal mortality rates between years. Because of this, density during a particular season may be different from year to year.

Periodic fluctuations are population increases and decreases spanning several years. They generally reflect short-term, accrued effects of high or low fawn production and recruitment and mortality rates. That is, several years of relatively high recruitment and/or below-average adult mortality tend to result in population increases, whereas low recruitment and/or above-average adult mortality result in a population decline. This type of fluctuation is illustrated in Figure 43.3, which shows the general trend in a mule deer population in the Missouri River Breaks, Montana, from 1960 through 1995. Data for this population (Hamlin and Mackie 1989; Mackie et al. 1998) show that the average annual mortality rate, from winter to winter, has been about 30–33%. Above average mortality was associated with downward population trends, especially those occurring during years or periods of severe environmental conditions such as extreme drought during 1961 and 1962 and severe winters in 1964/65 and 1971/72. Catastrophic mortality also occurred during winter 1994/95. Fawn production and survival, together with catastrophic mortality, were the major factors influencing trends. Low fawn recruitment, such that <30–33% of the early winter population was made up of this age class, was invariably associated with a decreasing trend. Higher recruitment led to increases, and where it occurred over several years, as from 1967 to 1970, the population grew

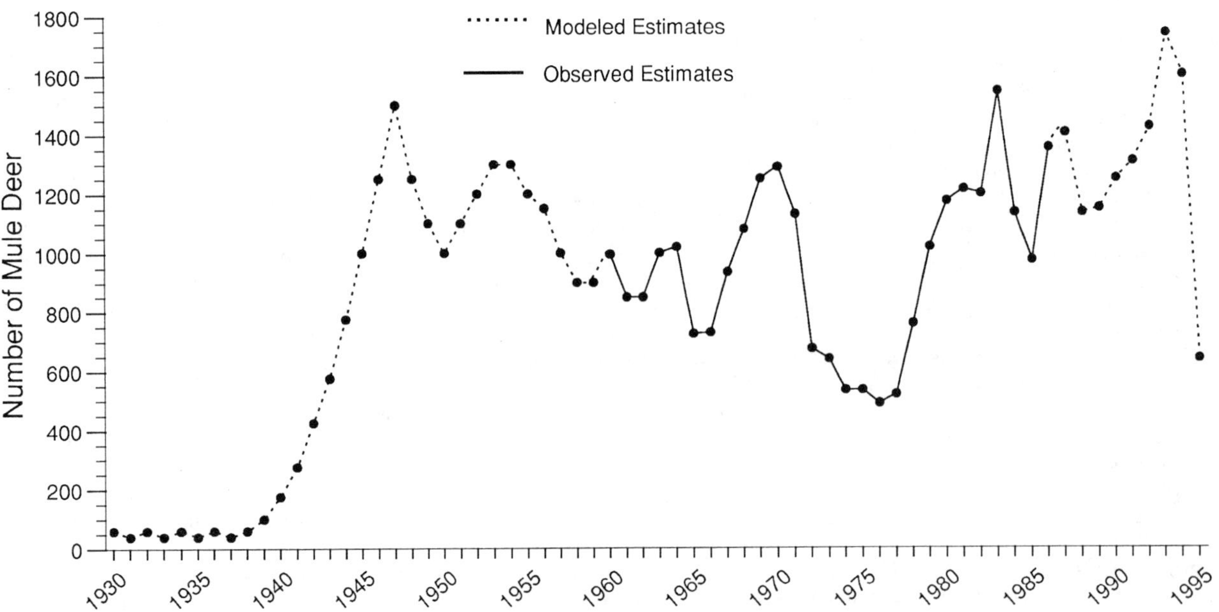

FIGURE 43.3. Estimated numbers of mule deer on the Missouri River Breaks, Montana, study area during early winter, 1930–1995. SOURCE: Data from Mackie et al. (1998).

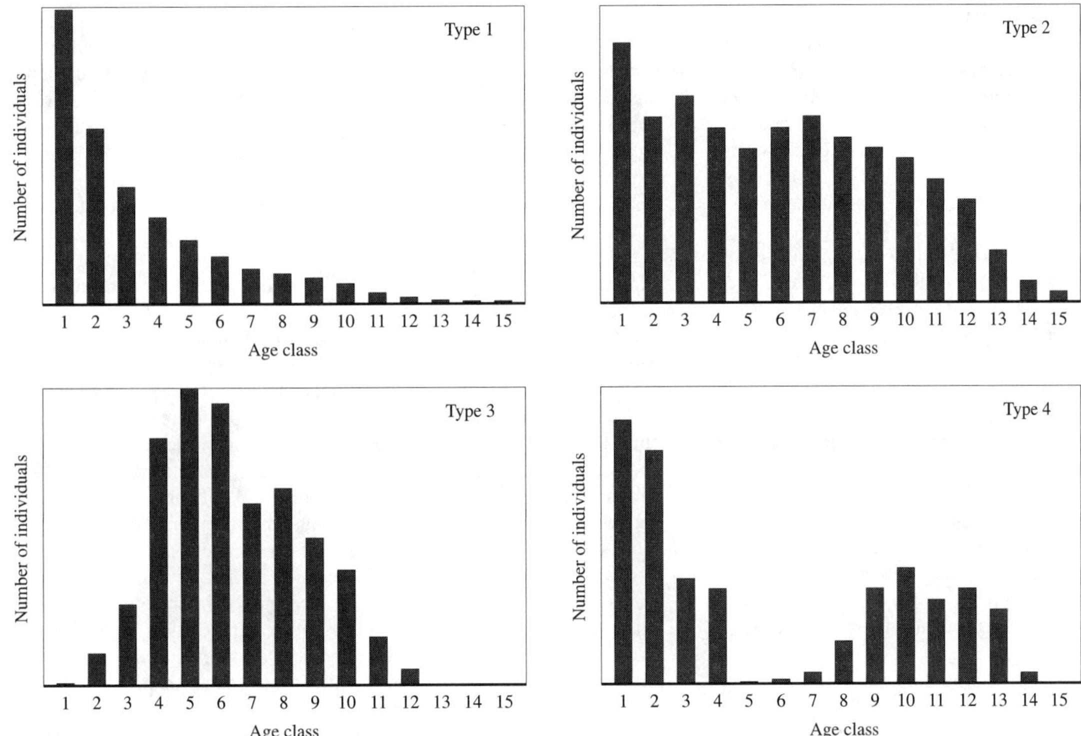

FIGURE 43.4. Conceptualized examples of four types of age structures observed in Montana mule deer: Type 1, pyramidal; type 2, flat or low pyramid; type 3, convex; type 4, concave or U shaped. SOURCE: Data from Mackie et al. (1998).

rapidly. It should also be noted, however, that periodic trends can be influenced directly by fluctuations in the amount of habitat available to deer. Habitat availability is affected by the presence and activities of other animals and humans or by environmental factors and conditions.

Long-term trends in mule deer populations are those that accrue as broad changes in numbers and distribution over time. Thus, a population or several populations in an area or region may fluctuate annually or periodically within either an expanding or a shrinking trend. Such trends reflect long-term, often very slow and subtle changes in the kinds, quality, and amounts of habitat available to deer as suggested by Gill (1999) for long-term declines of 50% or more in Colorado mule deer populations since the 1940s.

Sex and age structure provides an ongoing record of the dynamics of a deer population in the particular environment it occupies. Any variation in environmental conditions that affects natality and/or mortality patterns or rates can also affect population structure. For this reason, there is no typical sex or age structure that will generally characterize all populations or any one population over time. Mackie et al. (1998) conceptualized four types of age structure observed in Montana mule deer (Fig. 43.4). Type 1 is a pyramidal age structure heavily skewed to young age classes and considered characteristic of populations experiencing rapid increase as a result of high recruitment of young and/or relatively low annual adult survival (i.e., high annual turnover). Type 2 is a relatively flat or low pyramid age structure resulting from periods of low and stable recruitment and high longevity among adults (i.e., low annual turnover). Type 3 is a convex age structure dominated by middle or "prime" age classes and associated with sharply reduced or declining populations. Type 4 is a concave or U-shaped age structure containing more individuals in young and old age classes than in middle age classes and associated with population recovery following a decline.

In natural populations, adult females generally outnumber adult males by more than 2:1, even where no hunting occurs. Heavily hunted populations may show winter sex ratios of <5–20 adult males:100 females, whereas light to moderately hunted populations may have ratios of 20–40 males:100 females. However, natural changes in age structure can also result in ratios of only 10–20 males:100 females in winter populations.

Extrinsic Factors Influencing Populations. Many factors play a role in the population ecology of mule and black-tailed deer and may limit populations over time. Although habitat and nutritional limitations often are of foremost importance, weather, diseases, parasites, predation, competition with other wild and domestic ungulates, and hunting and other anthropogenic activities also may play important roles.

Weather exerts an important yet highly variable influence on deer populations. It can act as both a stress factor during severe winter weather and as a nutritional–forage factor through precipitation and temperature effects on the kinds, quantity, and quality of forage available. Because of its variability, weather often will influence population changes seasonally, annually, and over periods of years. Through its pervasive influence on fawn survival, weather conditions can trigger widespread declines and/or foster general increases in mule deer populations (Unsworth et al. 1999).

The role of diseases and parasites in limiting deer numbers is not well understood and is difficult to measure. Hibler (1981) discussed most of the diseases and parasites known to occur in mule and black-tailed deer. Although all may cause, directly or indirectly, the death of individual animals, there has been no documentation of massive population impacts of these factors in wild populations. Mule deer and black-tailed deer have coevolved with most of the diseases and parasites infecting them, and populations today are likely to contain a high proportion of resistant animals. This, together with the fact that mule deer live in dry environments, which are less conducive to chronic disease and parasitic problems than humid areas, decreases the chance for important population consequences.

Some diseases and parasites are always present in deer populations, with their importance varying in relation to other predisposing factors. For example, a deer on a poor nutritional regime is more likely to die as a result of infection by disease or parasites than an animal living on a good nutritional regime. Hemorrhagic diseases, including epizootic hemorrhagic disease and blue tongue, are perhaps the most important viral diseases in mule deer, especially in areas where mule deer and white-tailed deer are sympatric. Mortality attributable to outbreaks of hemorrhagic disease has been estimated as ± 50% over a period of several weeks in late summer and fall (Gill 1999). Of growing concern are diseases or parasites not previously detected in mule deer or introduced from some other area, from captive animals on game farms, or from species with which the deer did not coevolve. Recent examples that could possibly have dramatic population consequences include the appearance of chronic wasting disease in Colorado and Wyoming (Williams and Young 1980, 1992), systematic adenovirus infection (Woods et al. 1996) in California, and bovine tuberculosis (Rhyan et al. 1995) introduced from captive cervids on game farms in several states and Canadian provinces.

Although losses of deer to predators are well documented, the population effects of predation remain uncertain (Ballard et al. 2001). There has been much research indicating that, in at least some situations and at some times, predation can have population impacts. Coyotes (*Canis latrans*) occur, often in abundance, throughout the range of mule deer. They primarily kill young deer, but may kill adults in some situations. Mountain lions (*Puma concolor*) are capable of killing any deer, given the right situation, and have increased greatly in distribution and abundance since the 1970s. Bobcats (*Lynx rufus*) and golden eagles (*Aquila chrysaetos*) occasionally kill young deer. Reintroduction and management to encourage expansion of gray wolves (*Canis lupus*) in Montana, Idaho, Wyoming, and elsewhere have expanded the amount of contact between wolves and mule deer and heightened concern about population consequences. Whether losses to predation are compensatory with other mortality or directly affect ultimate population levels of mule deer has yet to be determined in most cases.

The influences of sport hunting usually depend on how well the harvest is regulated in relation to population size and dynamics and on the vulnerability of deer to hunting. Flexible and opportunistic regulations allow for the harvest of surpluses when they are available. When segments of the population with high natural mortality rates are not available, liberal regulations and more intense hunting may shift the harvest to population segments with low natural mortality rates. Persistent, intense hunting in low-security habitats may eliminate deer from some areas and also directly limit the total population size. Security of deer has traditionally been related only to habitat characteristics. More recently, however, vulnerability of deer to hunters has increased with changes in leisure time, mobility, financial status, and equipment used by hunters.

Mule deer may compete for food, cover, and space with all other wild and domestic ungulates on western rangelands (Mackie 1976, 1978b). The extent to which competition actually occurs or is important in the population ecology of mule deer depends on many factors, so generalization is difficult, if not impossible. In recent years, resource managers have increasingly questioned the effect of elk on mule deer as elk populations have increased and spread across much of their historical range in the West. Although, as Lindzey et al. (1997) concluded, it seems prudent to assume that elk can reduce the quality of ranges for deer, there remains a paucity of data documenting competition between the species.

Humans and their related activities have been, and continue to be, influential in mule deer population changes. As described earlier, widespread population declines can be traced to human activities during the middle to late 1800s. Conversely, extensive population increases in more recent times can be partially traced to human management activities combined with the beneficial consequences of natural systems working on habitats formerly altered by humans. The significance of past anthropogenic influences on mule deer lies in the fact that the alterations were all within the limits of tolerance of the species. In modern times, humans have the ability to alter environments beyond these limits by completely usurping space needed by deer, which can effectively eliminate their basis for existence.

MANAGEMENT AND CONSERVATION

Management actions and practices applied to mule and black-tailed deer have varied greatly over the past century as well as among different parts of their range. The earliest efforts, in the late 1800s and early 1900s, were to protect dwindling populations. These included close regulation or prohibition of hunting, establishment of refuges or protective game ranges, and predator control programs. Later, in some places, efforts to restore populations in suitable habitat included trapping and transplanting deer. The importance of these approaches in influencing historical population trends in any one area may be debated. Collectively, they are credited with at least helping to speed the recovery and spread of mule deer throughout the West.

As populations increased, often to the point where damage to range forage plants and agricultural crops was apparent, management emphasis shifted from rigid protection to population control. Almost everywhere, efforts were directed toward liberalizing hunting regulations to balance deer populations with their forage supplies on important or key habitats. Liberal harvests were also implemented to prevent depredations to agricultural crops and provide recreational hunting and the use of the large annual surplus of deer that might otherwise be lost to starvation and other natural mortality. As a basis for management, emphasis was also placed on research studies and surveys to learn more about the biology and ecology of mule deer and their habitats. This new science gradually provided a framework of methods, concepts, and criteria for deer management. Thus, management efforts came to focus on the systematic collection of data on deer population characteristics, harvests, and use and condition trends of important forage plants, especially on winter ranges. These data, interpreted in light of existing knowledge of the habitat relationships of deer, provided the basis for ascertaining management needs and developing and selling harvest recommendations. At the same time, education programs to inform sportsmen and the general public about management efforts and develop public acceptance of these efforts became an integral part of mule deer management.

Generally, these same approaches are followed in management of mule and black-tailed deer today. However, individual programs have become more refined with accumulating knowledge of the biology, behavior, and ecology of deer, their habitats, and habitat relationships. New technology, including radiotelemetry, geographic information and global positioning systems, powerful computers, new and more powerful modeling systems, and new and improved methods for surveying deer population trends and characteristics, annual harvests, and habitat conditions also have led to many refinements in management concepts and practice.

Management typically includes many different programs and activities depending on the species, time, place, and agency involved. The essential components are (1) a sound research program to provide and expand knowledge about the biology, behavior, and ecology of deer and the application for management; (2) management surveys to determine and assess habitat and population characteristics and trends; (3) information and education to further public understanding and support of management programs; (4) law enforcement to assure that laws and regulations developed to manage deer populations are followed; and (5) habitat management, including acquisition, maintenance, and physical alteration.

Within this broad framework, there is no generally accepted philosophy for management of mule and black-tailed deer. In addition, specific management can vary widely between subspecies, geographic areas, and broad habitat types, and even between local populations and habitats. Discussions of the merits of the many approaches and specific techniques that have been applied fill a voluminous literature and are similar in many respects to those discussed for white-tailed deer (Chapter 44). It should be noted, however, that what is useful in one environment may be inappropriate in another. Each technique must be carefully selected and evaluated with respect to the particular management situation to which it is to be applied.

Management can be viewed in the context of two temporal perspectives that must necessarily be integrated. Population and harvest management is conducted within a short-term perspective that must recognize and respond to the dynamic nature of deer populations and the diverse environments they occupy as well as to diverse recreational, social, and economic considerations. Habitat management is usually conducted from a longer term perspective in which habitats are protected and basic components are maintained and enhanced to sustain both quantity and quality of this basic resource over time.

Population management through regulation of hunting dominates the management efforts of state wildlife agencies. In part, this reflects their basic legal responsibility; however, it is equally important that hunting is the primary factor influencing deer populations that most state wildlife agencies can directly control.

Declining mule deer populations throughout the West during the 1960s and 1970s, together with a growing public concern for management of all natural resources on public lands in particular, paved the way for increased emphasis on habitat management. Following those declines, concern was not with too many deer on most ranges, but with too few, and with the reasons for this. As a result, much effort has been directed to assessing the relationships between deer populations and habitat factors and to protecting important habitats threatened by conflicting land uses. Public concerns fostering increased involvement of federal agencies in wildlife management have led to greater consideration of the habitat needs of mule and black-tailed deer and to expanded opportunities for deer habitat management on public lands. Federal agencies have hired more field-level biologists to evaluate habitat conditions and problems and develop land management policies. As a result, cooperative efforts to actively manage, maintain, and enhance deer habitats have been established and expanded in most areas.

Logging and fire have been considered responsible for producing high-quality habitat for mule and black-tailed deer (Wallmo 1978). It is unlikely, however, that large forest clear-cuts will be a silvicultural option on public lands in the future. In addition, use of prescribed fire at a large enough spatial scale to have a significant effect on deer populations is expensive and may not see widespread use, while let-burn policy for wildfire is hampered by a formidable set of social and political factors (Czech 1996). The extent of early and midsuccessional vegetation seen in the early twentieth century as a result of wildfire, logging, and livestock grazing practices is unlikely to reoccur in the foreseeable future. If resource managers want to improve conditions for deer, they will need to develop tools and techniques for maximizing benefits for deer in mid and late successional habitats.

Seeding, fertilizing, and prescribed burning have all been used as tools to improve habitat for mule and black-tailed deer (Kie and Czech 2000). Such activities are expensive, difficult to accomplish on a scale large enough to substantially increase deer numbers, and may conflict with other land management goals. Direct habitat improvement projects may provide an illusion of actively benefitting deer. However, efforts directed at coordinating deer habitat management with other land uses such as livestock grazing, timber management, roads and recreation, and home site development have far greater potential to provide benefits to mule and black-tailed deer (Kie and Czech 2000).

Livestock management practices and factors that can affect deer include livestock numbers, timing and duration of grazing, animal distribution, livestock types, and specialized grazing systems (Kie et al. 1994). The timing and duration of grazing, along with livestock distribution, are the bases for various specialized grazing systems, including continuous grazing, deferred-rotation, and rest-rotation grazing. Timing of livestock grazing is critical to mule and black-tailed deer, and they are particularly susceptible to adverse effects during fawning season (Loft et al. 1987, 1991, 1993; Kie et al. 1991). Therefore, carefully designed deferred-rotation and rest-rotation grazing systems have great potential benefit to deer (Kie et al. 1994), at least in some areas.

With the exception of black-tailed deer in British Columbia and southeast Alaska, which are dependent on canopy cover in mature forests, mule deer generally are considered to benefit from the early-successional vegetation that often becomes established following timber harvest. Timber management programs treat large acreage of forested lands each year and provide a potentially important source for indirect management of deer habitats. The array of activities associated with timber management changes the structural characteristics of forests and strongly influences the amount, diversity, and distribution of forage and cover on deer ranges. Well-planned timber management offers an opportunity for restoring or improving forest habitats in many areas. Conversely, poorly planned projects may seriously degrade important habitats.

Silvicultural systems influence vegetation structure and species composition within managed stands. Although site responses are highly variable, the kind and amount of understory response tend to vary inversely with the amount of overstory remaining after harvest (Ffolliot and Carey 1972). The length of time that successional vegetation is of value to deer and other early-successional animal species varies with specific forest type, soil quality, climate, and other factors. In general, however, truncating the successional sequence by use of herbicides, planting of seedlings, and other artificial regeneration practices will shorten the time period over which benefits to deer will occur (Leopold 1978).

In some desert and other arid southwestern environments, water development may be an appropriate habitat management practice for mule deer. Recommendations for the spacing of free water sources include a maximum of 4.0–4.8 km in New Mexico (Wood et al. 1970) and preferred spacing in northern California of <3.2 km with a maximum of 4.6 km (Boroski and Mossman 1996). However, water developments in arid regions may have some negative ecological consequences by concentrating deer, and their use should be considered carefully (Broyles 1995; Krausman and Czech 1998).

Information regarding the response of mule and black-tailed deer to roads and vehicular traffic is scarce and imprecise. Perry and Overly (1977) found that main roads had the greatest impact on mule deer and elk, and primitive roads the least impact. Roads through meadow habitats reduced deer and elk use, whereas roads through open forest habitat reduced habitat use by elk but not by mule deer.

Where large numbers of mule deer are killed in collisions with motor vehicles along paved highways, a variety of techniques is available for reducing deer mortality and damage to vehicles. Clearing screening cover from beside highways and reducing deer forage availability in the right-of-way can minimize deer–vehicle collisions. Simply providing a fenced, cross-walk system that forces deer to cross highways at well-marked points where motorists can anticipate them can reduce deer mortality 37–43% (Lehnert and Bissonette 1998).

In many places throughout the western United States, people are building houses in areas occupied by deer seasonally or throughout the year. Mule deer and black-tailed deer can become habituated to the presence of humans and may even thrive and become nuisances when they feed in ornamental plantings and gardens. However, the general effect of the development of home sites in most areas is to reduce deer use immediately adjacent to the structures. Smith et al. (1989) found reductions in mule deer use close to rural houses in northern California. Within a 90-m radius of the homes, deer use averaged only 61% of overall mean deer use (Smith et al. 1989). Vogel (1989) also noted that the number of mule deer seen per study plot in Montana decreased as a curvilinear function of the number of houses within 900 m. The greatest adverse effect on mule deer occurred when houses were evenly distributed. Therefore, concentrating new home sites on areas already developed might lessen the impact on deer (Vogel 1989).

RESEARCH NEEDS

Since the earliest days of mule deer management on the Kaibab Plateau of Arizona, conservation-minded people have recognized that knowledge about these animals, their habitats, and habitat relationships is essential to sound conservation and management. One of the first tasks of new deer management programs in the 1920s and 1930s was to establish research and other information-gathering studies. From early studies of food and habitat requirements, mule deer research gradually expanded to include the spectrum of basic biological, behavioral, and ecological studies necessary to understand the animals and how populations operate and should be managed in the diverse environments where they occur. Over the years, studies have evolved from natural history investigations to more rigorously designed and controlled experiments.

Mule deer populations and the environments influencing their occurrence and abundance have changed through time. Much of the theory and practice of mule deer management evolved from research and management studies conducted during the mid-1900s, a period when mule deer populations were expanding and widely abundant. At that time, other wild ungulates were scarce, major predators were controlled, hunting was restricted, and human populations were low in mule deer habitat. Since the mid-1970s, mule deer populations have stabilized and declined, whereas those of other wild ungulates, especially elk and white-tailed deer, have increased greatly. Predator control was relaxed and major predators, especially coyotes and mountain lions, increased. Wolves were protected and reintroduced to some historical ranges. Hunter numbers increased and they became more efficient with new and better equipment. Human populations and developments spread into mule deer habitat throughout the rural West; land uses such as agriculture, grazing, and logging were intensified; and habitats in many areas became fragmented and reduced. The importance of most, if not all, of these factors to mule deer ecology and management is unclear and being debated in scientific, management, sporting, and other conservation circles. Answers can only come from continuing and new research.

Examples of needed research include further studies of interspecific interactions and the possible impacts of vastly expanding elk populations on mule deer throughout the West. Similar studies relative to white-tailed deer populations and mule deer also are needed. The re-emergence of mountain lions and wolves, complementing coyotes as major predators of mule deer, may require new research to evaluate and understand the influence of increased predator populations and multiple-predator systems on mule deer management, including opportunity for harvest by humans. New studies also are needed to evaluate the impacts of hunter harvest on mule deer populations in their changing environment and to better adapt harvests to population size and characteristics that determine harvest opportunities across the spectrum of environments they occupy. The increasing overlap with elk, white-tailed deer, and other wild ungulate populations may also prescribe research into how harvest strategies for one species affect the harvest of others. In many areas, housing and other human developments have either spread into mule deer habitat or attracted increasing numbers of mule deer and their predators. There is no effective measure of the impacts of subdevelopment on deer populations and habitats. New research may be necessary to understand and design strategies for management of the human–wildlife interface as well as to effectively control deer and their predators in these environments.

Habitat is the base on which mule and black-tailed deer exist, and habitat relationships have been studied across their range. However, additional research is needed to understand how landscape patterns and characteristics affect behavior, life history, and other biological and ecological traits of the species. Recent changes in fire regimes across the West, and the apparent increasing incidence of wildfires, may provide opportunity for further research into fire relationships, including the influences of large-scale fires on habitat quality and long-term trends in deer populations. Studies also will be needed to determine how postfire forest management and rehabilitation practices affect mule deer.

The recent emergence of widespread concern about diseases and the potential long-term impacts of newly introduced diseases such as chronic wasting disease on mule deer populations has signaled a need for research to understand diseases and their role in the ecology and management of the species. The potential transmission of disease organisms from game farms to wild mule deer populations and from deer to domestic animals or humans ultimately could be the single most important factor determining the future of mule deer as one of the premier wildlife species in the West.

LITERATURE CITED

Anderson, A. E. 1981. Morphological and physiological characteristics. Pages 27–97 *in* O. C. Wallmo, ed. Mule and black-tailed deer of North America. University of Nebraska Press, Lincoln.

Anderson, A. E., and D. E. Medin. 1964. Reproductive studies. Pages 239–69 *in* Colorado Game and Fish Department Game Research Report (Federal Aid Project W105-R-3).

Anderson, A. E., and O. C. Wallmo. 1984. *Odocoileus hemionus*. Mammalian Species 219:1–9.

Anderson, A. E., D. E. Medin, and D. C. Bowden. 1974a. Growth and morphometry of the carcass, selected bones, organs, and glands of mule deer. Wildlife Monographs 39:1–122.

Anderson, F. M., G. E. Connolly, A. N. Halter, and W. M. Longhurst. 1974. A computer simulation study of deer in Mendocino County, California (Technical Bulletin 130). Agricultural Experiment Station, Oregon State University, Corvallis.

Ballard, W. B., D. Lutz, T. W. Keegan, L. H. Carpenter, and J. C. deVos, Jr. 2001. Deer–predator relationships: A review of recent North American studies with emphasis on mule and black-tailed deer. Wildlife Society Bulletin 29:99–115.

Barboza, P. S., and R. T. Bowyer. 2000. Sexual segregation in dimorphic deer: A new gastrocentric hypothesis. Journal of Mammalogy 81:473–89.

Beale, D. M., and N. W. Darby. 1991. Diet composition of mule deer in mountain brush habitat of southwestern Utah (Publication 91–14). Utah Division of Wildlife Resources.

Bischoff, A. I. 1957. The breeding season of some California deer herds. California Fish and Game 43:91–96.

Bissell, H. 1959. Interpreting chemical analyses of browse. California Fish and Game 45:57–58.

Boroski, B. B. 1998. Development and testing of a wildlife–habitat relationships model for Columbian black-tailed deer, Trinity County, California. Ph.D. Dissertation, University of California, Berkeley.

Boroski, B. B., and A. S. Mossman. 1996. Distribution of mule deer in relation to water sources in northern California. Journal of Wildlife Management 60:770–76.

Bowyer, R. T. 1984. Sexual segregation in southern mule deer. Journal of Mammalogy 65:410–17.

Brown, E. R. 1961. The black-tailed deer of western Washington (Biological Bulletin 13). Washington Department of Game.

Broyles, B. 1995. Desert wildlife water developments: questioning the use in the Southwest. Wildlife Society Bulletin 23:663–75.

Bubenik, A. G. 1968. The significance of the antlers in the social life of the Cervidae. Deer 1:208–14.

Bunnell, F. L., and A. S. Harestad. 1983. Dispersal and dispersion of black-tailed deer: Models and observations. Journal of Mammalogy 64:201–9.

Cowan, I. McT. 1956. Life and times of the coast black-tailed deer. Pages 523–617 *in* W. P. Taylor, ed. The deer of North America. Stackpole, Harrisburg, PA.

Cowan, I. McT., and C. J. Guiguet. 1965. The mammals of British Columbia. Handbook II. British Columbia Provincial Museum, Victoria, Canada.

Cowan, I. McT., and A. G. Raddi. 1972. Pelage and molt in the black-tailed deer (*Odocoileus hemionus,* Rafinesque). Canadian Journal of Zoology 50:639–47.

Cowan, I. McT., and A. J. Wood. 1955. The growth rate of the black-tailed deer (*Odocoileus hemionus columbianus*). Journal of Wildlife Management 19:331–36.

Cronin, M. A. 1992. Intraspecific variation in mitochondrial DNA of North American cervids. Journal of Mammalogy 73:70–82.

Czech, B. 1996. Challenges to establishing and implementing sound natural fire policy. Renewable Resources Journal 14:14–19.

Dasmann, R. F. 1956. Fluctuations in a deer population in California chaparral. Transactions of the North American Wildlife Conference 21:487–99.

Davis, R. W. 1962. Studies on antler growth in mule deer (*Odocoileus hemionus hemionus*, Rafinesque). Pages 61–64 *in* Proceedings of the first national white-tailed deer disease symposium. University of Georgia, Athens.

Einarsen, A. S. 1956. Life of the mule deer. Pages 363–429 *in* W. P. Taylor, ed. The deer of North America. Stackpole, Harrisburg, PA.

Ffolliott, P. F., and W. P. Clary. 1972. A selected and annotated bibliography of understory–overstory relationships (Technical Bulletin 198). University of Arizona Agricultural Experiment Station, Tucson.

Fox, K. B., and P. R. Krausman. 1994. Fawning habitat of desert mule deer. Southwestern Naturalist 39:269–75.

French, C. E., L. C. McEwen, N. D. Magruder, R. H. Ingram, and R. W. Swift. 1955. Nutritional requirements of white-tailed deer for growth and antler development (Bulletin 600). Pennsylvania Agricultural Experiment Station.

Fritzen, D. E. 1995. Ecology and behavior of mule deer on the Rosebud Coal Mine, Montana. Ph.D. Dissertation, Montana State University, Bozeman.

Geist, V. 1968. Horn-like structures as rank symbols, guards and weapons. Nature 220:813–14.
Geist, V. 1981. Behavior: Adaptive strategies in mule deer. Pages 157–223 *in* O. C. Wallmo, ed. Mule and black-tailed deer of North America. University of Nebraska Press, Lincoln.
Gerlach, T. P. 1987. Population ecology of mule deer on Pinon Canyon maneuver site, Colorado. M.S. Thesis, Virginia Polytechnic Institute and State University, Blacksburg.
Gill, R. B. 1976. Mule deer management myths and the mule deer population decline. Pages 99–106 *in* G. W. Workman and J. B. Low, eds. Mule deer decline in the west. College of Natural Resources, Utah State University, Logan.
Gill, R. B. 1999. Declining mule deer populations in Colorado: Reasons and responses. A Report to the Colorado Legislature, November 1999. Colorado Division of Wildlife, Denver.
Goss, R. J. 1963. The deciduous nature of deer antlers. Pages 339–69 *in* P. Sognnaes, ed. Mechanisms of hard tissue destruction. American Association for Advancement of Science, Washington, DC.
Hamlin, K. L., and R. J. Mackie. 1989. Mule deer in the Missouri River Breaks, Montana: A study of population dynamics in a fluctuating environment (Final Report, Federal Aid in Wildlife Restoration Project W-120-R). Montana Department of Fish, Wildlife and Parks, Helena.
Hamlin, K. L., and R. J. Mackie. 1991. Age-specific reproduction and mortality in female mule deer: Implications to population dynamics. Page 569–73 *in* B. Bobek, K. Perzanowski, and W. Reglin, eds. Global trends in wildlife management. Proceedings XVIII congress of the International Union of Game Biology, Vol. 1. Swiat Press, Krakow, Poland.
Hamlin, K. L., D. F. Pac, C. A. Sime, R. M. DeSimone, and G. L. Dusek. 2000. Evaluating the accuracy of ages obtained by two methods for Montana ungulates. Journal of Wildlife Management 64:441–49.
Hanley, T. A., C. T. Robbins, and D. E. Spalinger. 1989. Forest habitats and the nutritional ecology of Sitka black-tailed deer: A research synthesis with implications for forest management (General Technical Report PNW-GTR-230). U.S. Forest Service, Pacific Northwest Research Station, Portland, OR.
Hanson, W. R., and C. Y. McCulloch. 1955. Factors influencing the distribution of mule deer on Arizona rangelands. Transactions of the North American Wildlife Conference 20:568–88.
Harestad, A. S., and F. L. Bunnell. 1983. Dispersal of a yearling male black-tailed deer. Northwest Science 57:45–48.
Hazam, J. E., and P. R. Krausman. 1988. Measuring water consumption of desert mule deer. Journal of Wildlife Management 52:528–34.
Heffelfinger, J. R. 2000. Status of the name *Odocoileus hemionus crooki* (Mammalia: Cervidae). Proceedings of the Biological Society of Washington 113:319–33.
Hervert, J. J., and P. R. Krausman. 1986. Desert mule deer use of water developments in Arizona. Journal of Wildlife Management 50:670–76.
Hibler, C. P. 1981. Diseases. Pages 129–55 *in* O. C. Wallmo, ed. Mule and black-tailed deer of North America. University of Nebraska Press, Lincoln.
Hoffman, R. R. 1985. Digestive physiology of the deer: Their morphophysiological specialization and adaptation. Pages 393–407 *in* P. F. Fennessy and K. R. Drew, eds. Biology of deer production (Bulletin 22). Royal Society of New Zealand, Wellington.
Hudson, P., and L. Browman. 1959. Embryonic and fetal development of the mule deer. Journal of Wildlife Management 23:295–304.
Hunter, J. W. 1924. Deer hunting in California. California Fish and Game 10:18–24.
Irby, L. R., J. E. Norland, M. G. Sullivan, J. A. Westfall, Jr., and P. Anderson. 2000. Dynamics of green ash woodlands in Theodore Roosevelt National Park. Prairie Naturalist 32:77–102.
Jennings, J. D. 1957. Danger cave. University of Utah Anthropological Papers 27:1–328.
Kie, J. G., and B. B. Boroski. 1995. The effects of cattle grazing on black-tailed deer during winter on the Tehama Wildlife Management Area (Final Report PSW-89-CL-030). U.S. Forest Service, Pacific Southwest Research Station, Fresno, CA.
Kie, J. G., and R. T. Bowyer. 1999. Sexual segregation in white-tailed deer: Density dependent changes in use of space, habitat selection, and dietary niche. Journal of Mammalogy 80:1004–20.
Kie, J. G., and B. Czech. 2000. Mule and black-tailed deer. Pages 629–57 *in* S. Demarais and P. R. Krausman, eds. Ecology and management of large mammals in North America. Prentice-Hall, Upper Saddle River, NJ.
Kie, J. G., C. J. Evans, E. R. Loft, and J. W. Menke. 1991. Foraging behavior by mule deer: The influence of cattle grazing. Journal of Wildlife Management 55:665–74.
Kie, J. G., V. C. Bleich, A. L. Medina, J. D. Yoakum, and J. W. Thomas. 1994. Managing rangelands for wildlife. Pages 663–88 *in* T. A. Bookhout, ed. Wildlife management techniques, 5th ed. Wildlife Society, Bethesda, MD.
Kie, J. G., R. T. Bowyer, B. B. Boroski, M. C. Nicholson, and E. R. Loft. 2002. Landscape heterogeneity at differing scales: Effects on spatial distribution of mule deer. Ecology 83:530–44.
Krausman, P. R., and B. Czech. 1998. Water developments and desert ungulates. Pages 138–54 *in* R. Pearlman, ed. Proceedings of a symposium on environmental, economic, and legal issues related to rangeland water development. Center for the Study of Law, Arizona State University, Tempe.
Krausman, P. R., A. J. Kuenzi, R. C. Etchberger, K. R. Rautenstrauch, L. L. Ordway, and J. J. Hervert. 1997. Diets of desert mule deer. Journal of Range Management 50:513–22.
Kufeld, R. C., and D. C. Bowden. 1995. Mule deer and white-tailed deer inhabiting eastern Colorado Plains River Bottoms (Technical Publication No. 41). Colorado Division of Wildlife.
Kufeld, R. C., O. C. Wallmo, and C. Feddema. 1973. Foods of the Rocky Mountain mule deer (Research Paper RM-111). U.S. Department of Agriculture, Forest Service.
Lautier, J. K., T. V. Dailey, and R. B. Brown. 1988. Effect of water restriction on feed intake of white-tailed deer. Journal of Wildlife Management 52:602–6.
Lehnert, M. E., and J. A. Bissonette. 1998. Effectiveness of highway crossing structures at reducing deer–vehicle collisions. Wildlife Society Bulletin 25:809–19.
Leopold, A., L. K. Sowls, and D. L. Spencer. 1947. A survey of overpopulated deer ranges in the United States. Journal of Wildlife Management 11:162–77.
Leopold, A., T. Riney, R. McCain, and L. Tevis, Jr. 1951. The Jawbone deer herd (Bulletin 4). California Department of Fish and Game, Sacramento.
Leopold, A. S. 1978. Wildlife and forest practice. Pages 108–20 *in* H. P. Brokaw, ed. Wildlife and America. Council on Environmental Quality, Washington, DC.
Lindzey, F. G., W. G. Hepworth, T. A. Mattson, and A. F. Reeves. 1997. Potential for competitive interactions between mule deer and elk in the western United States and Canada: A review. Wyoming Cooperative Fish and Wildlife Research Unit, Laramie.
Linsdale, J. M., and P. Q. Tomich. 1953. A herd of mule deer: A record of observations made on the Hastings Natural History Reservation. University of California Press, Berkeley.
Loft, E. R., J. W. Menke, J. G. Kie, and R. C. Bertram. 1987. Influence of cattle stocking rate on the structural profile of deer hiding cover. Journal of Wildlife Management 51:655–64.
Loft, E. R., J. W. Menke, and J. G. Kie. 1991. Habitat shifts by mule deer: The influence of cattle grazing. Journal of Wildlife Management 55:16–26.
Loft, E. R., J. G. Kie, and J. W. Menke. 1993. Grazing in the Sierra Nevada: Home range and space use patterns of mule deer as influenced by cattle. California Fish and Game 79:145–66.
Longhurst, W. M., H. K. Oh, M. B. Jones, and R. E. Kepner. 1968. A basis for the palatability of deer forage plants. Transactions of the North American Wildlife Conference 33:181–89.
Longhurst, W. M., E. O. Garton, H. F. Heady, and G. E. Connolly. 1976. The California deer decline and possibilities for restoration. Transactions of the Western Section of the Wildlife Society 12:1–41.
Loveless, C. M. 1967. Ecological characteristics of a mule deer winter range (Technical Bulletin 20). Colorado Game, Fish and Parks Department, Denver.
Low, W. O., and I. McT. Cowan. 1963. Age determination of deer by annular structure of dental cementum. Journal of Wildlife Management 27:466–71.
Mackie, R. J. 1964. Montana deer weights. Montana Wildlife 1964 (Winter):9–14.
Mackie, R. J. 1970. Range ecology and relations of mule deer, elk and cattle in the Missouri River Breaks, Montana. Wildlife Monographs 20:1–79.
Mackie, R. J. 1976. Interspecific competition between mule deer, other game animals and livestock. Pages 49–54 *in* G. W. Workman and J. B. Low, eds. Mule deer decline in the West: A symposium. Utah State University, College of Natural Resources, Logan.
Mackie, R. J. 1978a. Natural regulation of mule deer populations. Pages 112–25 *in* F. L. Bunnell, D. S. Eastman, and J. M. Peek, eds. Symposium on natural regulation of wildlife populations. Forest, Wildlife, and Range Experiment Station, University of Idaho, Moscow.
Mackie, R. J. 1978b. Impacts of livestock grazing on wild ungulates. Transactions of the North American Wildlife and Natural Resources Conference 43:462–76.

Mackie, R. J. 1987. Mule deer. Pages 265–71 *in* H. Kallman, ed. Restoring America's wildlife, 1937–1987. U.S. Department of Interior, Fish and Wildlife Service, Washington, DC.

Mackie, R. J., K. L. Hamlin, and D. F. Pac. 1982. Mule deer. Pages 862–77 *in* J. A. Chapman, and G. A. Feldhamer, eds. Wild mammals of North America: Biology, management, economics. Johns Hopkins University Press, Baltimore.

Mackie, R. J., D. F. Pac, K. L. Hamlin, and G. L. Dusek. 1998. Ecology and management of mule deer and white-tailed deer in Montana. Montana Department of Fish, Wildlife and Parks, Helena.

Main, M. B., and B. R. Coblentz. 1996. Sexual segregation in Rocky Mountain mule deer. Journal of Wildlife Management 60:497–507.

Main, M. B., F. W. Weckerly, and V. C. Bleich. 1996. Sexual segregation in ungulates: New directions for research. Journal of Mammalogy 77:449–61.

Mautz, W. W., P. J. Pekins, and J. A. Warren. 1985. Cold temperature effect on metabolic rate of white-tailed, mule, and black-tailed deer in winter coat. Pages 453–57 *in* P. K. Fennessy and K. R. Drew, eds. Biology of deer production (Bulletin 22). Royal Society of New Zealand, Wellington.

McCullough, D. R. 1996. Failure of the tooth cementum aging technique with reduced population density of deer. Wildlife Society Bulletin 24:722–24.

McCullough, D. R. 1997. Breeding by female fawns in black-tailed deer. Wildlife Society Bulletin 25:296–97.

Miller, F. L. 1970. Distribution patterns of black-tailed deer (*Odocoileus hemionus columbianus*) in relation to environment. Journal of Mammalogy 51:248–60.

Miller, F. L. 1974. Four types of territoriality observed in a herd of black-tailed deer. Pages 644–60 *in* V. Geist and F. R. Walther, eds. The behavior of ungulates and its relation to management. International Union for Conservation of Nature and Natural Resources, Morges, Switzerland.

Mueller, C. C., and R. M. F. S. Sadleir. 1979. Age at first conception in black-tailed deer. Biology of Reproduction 21:1099–1104.

Nicholson, M. C., R. T. Bowyer, and J. G. Kie. 1997. Habitat selection and survival of mule deer: Tradeoffs associated with migration. Journal of Mammalogy 78:483–504.

Ozoga, J. J., L. J. Verme, and C. S. Bienz. 1982. Parturition behavior and territoriality in white-tailed deer: Impact on neonatal mortality. Journal of Wildlife Management 46:1–11.

Pac, D. F., R. J. Mackie, and H. E. Jorgensen. 1991. Mule deer population organization, behavior, and dynamics in a northern Rocky Mountain environment (Final Report, Federal Aid in Wildlife Restoration Project W-120-R). Montana Department of Fish, Wildlife and Parks, Helena.

Perry, C., and R. Overly. 1977. Impact of roads on big game distributions in portions of the Blue Mountains of Washington, 1972–1973 (Applied Research Bulletin No. 11). Washington Department of Game, Olympia.

Rasmussen, D. I. 1941. Biotic communities of Kaibab Plateau, Arizona. Ecological Monographs 3:229–75.

Rees, J. W., R. A. Kainer, and R. W. Davis. 1966. Chronology of mineralization and eruption of mandibular teeth in mule deer. Journal of Wildlife Management 30:629–31.

Rhyan, J., K. Aune, B. Hood, R. Clarke, J. Payeur, J. Jarnagin, and L. Stackhouse. 1995. Bovine tuberculosis in a free-ranging mule deer (*Odocoileus hemionus*) from Montana. Journal of Wildlife Disease 31:432–35.

Robinette, W. L., J. S. Gashwiler, D. A. Jones, and H. S. Crane. 1955. Fertility of mule deer in Utah. Journal of Wildlife Management 19:115–36.

Robinette, W. L., D. A. Jones, G. Rogers, and J. S. Gashwiler. 1957. Notes on tooth development and wear for Rocky Mountain mule deer. Journal of Wildlife Management 21:134–53.

Robinette, W. L., C. H. Baer, R. E. Pillmore, and C. E. Knittle. 1973. Effects of nutritional change on captive mule deer. Journal of Wildlife Management 37:312–26.

Robinette, W. L., N. V. Hancock, and D. A. Jones. 1977. The Oak Creek mule deer herd in Utah. Utah Division of Wildlife Resources, Salt Lake City.

Rogers, K. J., P. F. Ffolliott, and D. R. Patton. 1978. Home range and movement of five mule deer in a semidesert grass–shrub community (Research Paper RM-355). U.S. Department of Agriculture, Forest Service.

Scarbrough, D. L., and P. R. Krausman. 1988. Sexual segregation by desert mule deer. Southwestern Naturalist 33:157–65.

Schoen, J. W., and O. C. Wallmo. 1978. Timber management and deer in southeast Alaska: Current problems and research direction. Pages 69–85 *in* O. C. Wallmo and J. W. Schoen, eds. Sitka black-tailed deer: Proceedings of a conference in Juneau, Alaska. U.S. Department of Agriculture Forest Service and Alaska Department of Fish and Game, Juneau.

Seton, E. T. 1929. Lives of game animals, Vol. 3, Part I. Doubleday, Doran, Garden City, NY.

Severinghaus, C. W. 1949. Tooth development and wear as criteria of age in white-tailed deer. Journal of Wildlife Management 13:195–216.

Severson, K. E., and A. V. Carter. 1978. Movements and habitat use by mule deer in the northern Great Plains, South Dakota. Pages 466–68 *in* Proceedings of the 1st international rangelands congress. Society for Range Management, Denver, CO.

Severson, K. E., and A. L. Medina. 1983. Deer and elk habitat management in the southwest. Journal of Range Management Monograph No. 2.

Smith, D. O., M. Connor, and E. R. Loft. 1989. The distribution of winter mule deer use around homesites. Transactions of the Western Section of the Wildlife Society 25:77–80.

Smith, R. H., T. J. McMichael, and H. G. Shaw. 1969. Decline of a desert mule deer population. Arizona Game and Fish Department Wildlife Digest Abstracts 3:1–8.

Swank, W. G. 1958. The mule deer in Arizona chaparral. Arizona Game and Fish Department Wildlife Bulletin 3:1–109.

Swift, R. W. 1948. Deer select most nutritious forages. Journal of Wildlife Management 12:109–10.

Taber, R. D. 1956. Deer nutrition and population dynamics in the North Coast Range of California. Transactions of the North American Wildlife Conference 21:159–72.

Taber, R. D., and R. F. Dasmann. 1958. The black-tailed deer of the chaparral; its life history and management in the North Coast Range of California (Game Bulletin 8). California Fish and Game Department, Sacramento.

Taylor, C. R. 1969. The eland and the oryx. Scientific American 220:88–95.

Telfer, T. C. 1988. Status of black-tailed deer on Kauai. Transactions of the Western Section of The Wildlife Society 24:53–60.

Thomas, D. C., and I. McT. Cowan. 1975. The pattern of reproduction in female Columbian black-tailed deer, *Odocoileus hemionus columbianus*. Journal of Reproductive Fertility 44:261–72.

Thorne, E. T. 1975. Normal body temperature of pronghorn antelope and mule deer. Journal of Mammalogy 56:697–98.

Tomich, P. Q. 1986. Mammals in Hawaii, 2nd ed. Bishop Museum Press, Honolulu, Hawaii.

Torbit, S. C., L. H. Carpenter, D. M. Swift, and A. W. Alldredge. 1985. Differential loss of fat and protein by mule deer during winter. Journal of Wildlife Management 49:80–85.

Unsworth, J. W., D. F. Pac, G. C. White, and R. M. Bartman. 1999. Mule deer survival in Colorado, Idaho, and Montana. Journal of Wildlife Management 63:315–26.

Urness, P. J. 1969. Nutritional analyses and in vitro digestibility of mistletoes browsed by deer in Arizona. Journal of Wildlife Management 33:499–505.

Vogel, W. O. 1989. Response of deer to density and distribution of housing in Montana. Wildlife Society Bulletin 17:406–13.

Wagner, F. H. 1978. Effects of livestock grazing and the livestock industry on wildlife. Pages 121–45 *in* H. Brokaw, ed. Wildlife and America. Council on Environmental Quality, Washington, DC.

Wallmo, O. C. 1978. Mule and black-tailed deer (*Odocoileus hemionus*). Pages 31–42 *in* D. L. Gilbert and J. L. Schmidt, eds. Big game of North America: Ecology and management. Stackpole, Harrisburg, PA.

Wallmo, O. C. 1981. Mule and black-tailed deer distribution and habitats. Pages 1–25 *in* O. C. Wallmo, ed. Mule and black-tailed deer of North America. University of Nebraska Press, Lincoln.

Wallmo, O. C., L. H. Carpenter, W. L. Regelin, R. B. Gill, and D. L. Baker. 1977. Evaluation of deer habitat on a nutritional basis. Journal of Range Management 30:122–27.

White, K. L. 1960. Differential range use by mule deer in the spruce–fir zone. Northwest Science 34:118–26.

Williams, E. S., and S. Young. 1980. Chronic wasting disease of captive mule deer: A spongiform encephalopathy. Journal of Wildlife Disease 16:89–98.

William, E. S., and S. Young. 1992. Spongiform encephalopathy in Cervidae. Revue scientifique et Technique, Office International des Epizootics 11:551–67.

Willms, W., and A. McLean. 1978. Spring forage selection by tame mule deer on big sagebrush range, British Columbia. Journal of Range Management 31:192–99.

Wood, A. J., I. McT. Cowan, and H. C. Nordan. 1962. Periodicity of growth in ungulates as shown by deer of the genus *Odocoileus*. Canadian Journal of Zoology 40:593–603.

Wood, A. K. 1988. Use of shelter by mule deer during winter. Prairie Naturalist 20:15–22.

Wood, A. K., R. J. Mackie, and K. L. Hamlin. 1989. Ecology of sympatric populations of mule deer and white-tailed deer in a prairie environment. Montana Department of Fish, Wildlife and Parks, Helena.

Wood, J. E., T. S. Bickle, W. Evans, J. C. Germany, and V. W. Howard, Jr. 1970. The Fort Stanton mule deer herd (Bulletin No. 567). New Mexico State University, Agricultural Experiment Station.

Woods, L. W., P. K. Swift, B. C. Barr, M. C. Horzinek, R. W. Nordhausen, M. H. Stillian, J. F. Patton, M. N. Oliver, K. R. Jones, and N. J. MacLachlan. 1996. Systematic adenovirus infection associated with high mortality in mule deer (*Odocoileus hemionus*) in California. Veterinary Pathology 33:125–32.

Workman, G. W., and J. B. Low, eds. 1976. Mule deer decline in the West, a symposium. College of Natural Resources, Utah State University, Logan.

Richard J. Mackie, Montana State University, Bozeman, Montana 59717. Email: mackie@montana.com.

John G. Kie, U.S. Forest Service, Forestry and Range and Science Lab, 1401 Gekeler Lane, LaGrande, Oregon 97850. Email: jkie@fs.fed.us.

David F. Pac, Fish, Wildlife, and Parks Research, Montana State University, Bozeman, Montana 59718. Email: dpac@montana.edu

Kenneth L. Hamlin, Fish, Wildlife and Parks Research, Montana State University, Bozeman, Montana 59718. Email: khamlin@montana.edu.

44

White-tailed Deer

Odocoileus virginianus

Karl V. Miller
Lisa I. Muller
Stephen Demarais

NOMENCLATURE

COMMON NAMES. Virginia deer, whitetail, white-tailed deer
SCIENTIFIC NAME. *Odocoileus virginianus* (Zimmermann, 1780)
SUBSPECIES NORTH OF MEXICO. *Odocoileus virginianus virginianus, O. v. borealis, O. v. carminis, O. v. clavium, O. v. couesi, O. v. dacotensis, O. v. hiltonensis, O. v. leucurus, O. v. mcilhennyi, O. v. macrourus, O. v. nigribarbis, O. v. ochrourus, O. v. osceola, O. v. seminolus, O. v. taurinsulae, O. v. texanus, O. v. venatorius* (Hall 1981; Baker 1984)

The genus *Odocoileus* first appeared on the North American landscape 3.5 million years ago (Kurten and Anderson 1980), with the white-tailed deer or a very similar form appearing more than 3 million years ago (Kurten, in Geist 1998). The center of origin likely was Middle America. Although range expansion into South America occurred in the early Pleistocene (Geist 1998), expansion into northern latitudes occurred relatively recently (Hershkovitz 1972).

Taxonomic authorities generally recognize 17 subspecies of white-tailed deer in North America north of Mexico, 13 additional subspecies in Mexico and Central America (Hall 1981; Baker 1984) and 8 subspecies from South America (Halls 1978; Smith 1991). Hesselton and Hesselton (1982) did not include *O. v. carminis* from near the Rio Grande of Texas (Goldman and Kellogg 1940) as occurring north of Mexico.

Current subspecies designations of white-tailed deer are based on morphometric characters such as pelage color, cranial dimensions, and antler size and shape (Barbour and Allen 1922). However, these morphometric characteristics can be influenced by habitat features such as soil region (Strickland and Demarais 2000). In addition, deer were extirpated throughout much of their North American range by the early 1900s. Extensive restoration efforts in the United States in the 1930s–1950s often introduced subspecies from disparate areas into areas that were void of deer or had only remnant populations. Several states restocked deer from sources that may have represented as many as seven subspecies (McDonald and Miller 1993). As a result, current subspecific status of white-tailed deer in North America is confounded and uncertain.

Development and widespread use of molecular genetic approaches to study deer populations (Honeycutt 2000) will allow a much-needed reexamination of white-tailed deer subspecific designations. However, initial genetic analyses of deer populations have shown conflicting results. Leberg et al. (1994) indicated that deer reintroductions could affect allozyme variation for 3–40 generations after the stocking event. In contrast, Ellsworth et al. (1994a, 1994b) found that restocking programs had little influence on local genetics. Ellsworth et al. (1994b) examined mitochondrial DNA and allozyme variation to characterize genetic patterns of white-tailed deer across the southeastern United States. They concluded that restocking programs had little effect on local genetics and that expansion of native herds resulted in the restoration of deer in the Southeast. However, further analyses suggested that translocations have had substantial and persistent effects on the genetic composition of localized deer populations (Leberg and Ellsworth 1999).

In a study of the taxonomic status and genetics of the geographically isolated Columbian white-tailed deer (*O. v. leucurus*) based on allelic frequencies, Gavin and May (1988) concluded that the genetic distance of this subspecies from other subspecies may not be unique enough to warrant subspecific designation. Clearly, white-tailed deer subspecies designation may need to be reexamined throughout the range as genetic techniques are refined.

DISTRIBUTION

White-tailed deer are widely distributed throughout North America (Fig. 44.1), extending from just inside the Yukon and Northwest Territories across the southern provinces of Canada and south into Central and South America. The species is present in 45 of the contiguous states and probably rare or absent in Utah, Nevada, and California (Hesselton and Hesselton 1982). They are found in all Canadian provinces except Newfoundland, Labrador, and Prince Edward Island.

Changing land use patterns and successful wildlife management programs have affected the distribution and population size of the broadly adaptable whitetail. For example, whitetails probably did not inhabit the upland prairies of western Montana until agricultural crops were established. Distribution in the western United States is limited

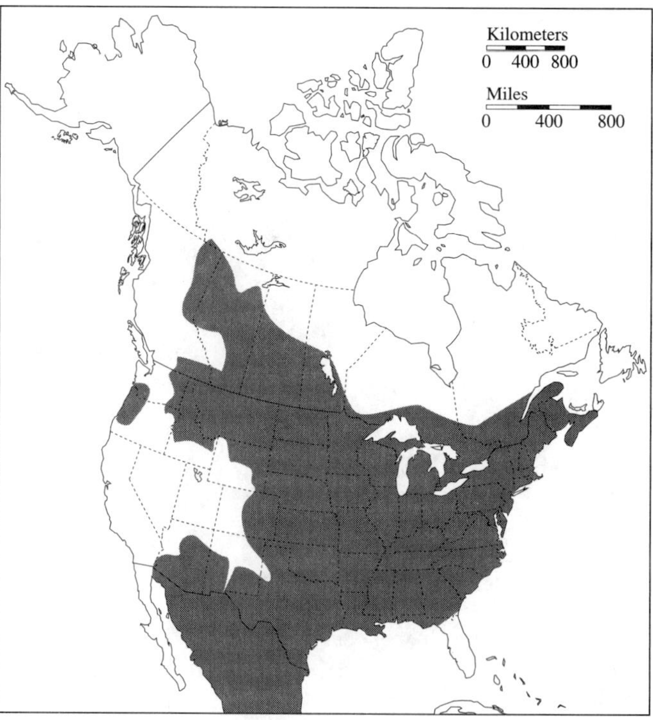

FIGURE 44.1. Distribution of the white-tailed deer (*Odocoileus viginianus*).

to upland prairie croplands adjacent to well-developed hardwood cover (Swenson et al. 1983). Similarly, the northern Great Lakes Region of the United States and southern Canada originally was marginal whitetail habitat. Removal of the virgin forests coupled with wildfires improved habitat for whitetails, and by the 1930s and 1940s, they were abundant in these regions (Blouch 1984).

In contrast, some local populations have decreased as woodland habitats were lost, such as with the limited distribution of Columbian white-tailed deer within the Columbia River drainage of the Pacific Northwest (Smith 1985). Population status along the white-tailed deer's western range in Arizona is affected by severity of drought during early summer and fall (Brown and Henry 1981). However, it can be generalized that there are viable populations of white-tailed deer throughout most of its range wherever there are adequate woodlands interspersed with early-successional habitats. In fact, expanded distributions and population sizes have had negative effects on other resources and human use of these resources in some areas. Overabundance of white-tailed deer is one of the most challenging problems facing wildlife managers (Warren 1997).

DESCRIPTION

Size and Growth. A wide range of body sizes would be expected within a species that ranges from the northern temperate forests of Canada to tropical forests in Central and South America. Within North America, the extremes are found in the northern woodland (*O. v. borealis*) and the Florida Key (*O. v. clavium*) subspecies. Eviscerated body mass ranges from <23 kg for Florida Key deer to ≥136 kg for the northern subspecies, with exceptional weights >170 kg (eviscerated) having been reported (Rue 1978).

Within regions and subspecies, there is great variation in body size related to inherent habitat productivity, anthropogenic features, and effectiveness of population and habitat management. In Mississippi, eviscerated body mass in the fertile Delta region was 30–40% higher than that of equivalent-aged deer in the less fertile Coastal Flatwood region (Strickland and Demarais 2000). Mature, nonpregnant females weigh roughly 60–75% of what adult bucks weigh. Eviscerated body mass of females was consistently 65% of that of equivalent-aged males in Mississippi.

Neonates weigh about 2–4 kg. Singletons generally are larger than twins or triplets, males are larger than females, and northern specimens are larger than those from the south. Growth from birth to maturity occurs in a predictable pattern, generally described by a sigmoid curve in which most of the growth occurs during a relatively linear intermediate phase (Robbins 1993). Nutrition of pregnant females affects early development of their fawns (Verme 1963). After parturition, fawns grow rapidly. In Oklahoma, they gained 0.27 kg/day during their first 3 weeks of life (Bartush and Garner 1979). Similarly, Rawson et al. (1992) estimated fawns gained 0.2 kg daily during their first 3 months. Females at roughly 6 months of age have reached about 50% of their ultimate eviscerated body mass, whereas males have obtained only about 35% of their ultimate mass.

Growth rate to maturity is influenced by many variables, including sex of the animal, population density, and habitat conditions. Females achieve mature body mass sooner because of earlier epiphyseal closure (Purdue 1983) and the physiological demands of fawn production. Males contribute little to offspring production, so they can devote more energy to body mass accumulation (Leberg et al. 1992). Females reach maximum body mass about 1 year earlier than males; females stabilize at 3–4 years of age, whereas males stabilize at 4–5 years of age (Roseberry and Klimstra 1975; Knowlton et al. 1979; Sauer 1984). In Minnesota, deer body mass continued to increase until 7.4 and 3.4 years of age in male and female deer, respectively (Fuller et al. 1989). Differences in the growth period of male and female deer also were observed in South Carolina (about 5.0 and 2.2 years of age for males and females, respectively; Leberg et al. 1992), and Mississippi (4.5 and 3.5 years of age for males and females, respectively; Strickland and Demarais 2000).

In addition to the effects of sex and age, time required to reach peak body weight is affected by soil fertility (Strickland and Demarais 2000). Similar to the impact of lower nutrition, high deer density causes males to grow more slowly and reach lower maximum weight (Leberg and Smith 1993). Management strategies for optimizing deer growth and antler size should consider the effects of sex, age, soil fertility, and density for each management unit. In South Carolina, males were more sensitive than females to population density effects on growth rate, which could decrease sexual dimorphism in size at greater population densities (Leberg and Smith 1993).

Pelage. Two complete seasonal molts produce distinct seasonal variation in pelage characteristics of white-tailed deer. The summer coat of short, thin hairs of reddish-brown color is replaced during August and September. The winter coat consists of longer, thicker, hollow gray or brown hairs. These hollow hairs, combined with a wooly undercoat, provide significant insulation. Winter hairs are molted in April and May. According to Rue (1978), the winter coat (at the top of the neck) contains in excess of 2500 hairs/in.^{2}. Individual hairs are 15–17 mm long and 0.18 mm in diameter. Summer hairs are more numerous (>5000/in.^{2}), but shorter (25 mm) and thinner (0.086 mm).

Neonates are reddish brown with several hundred small white spots, which provide cryptic coloration within forested habitats. Rows of spots occur along either side of the backbone and randomly along the sides and flanks. Spot size averages 0.6–1.3 cm in diameter (Sauer 1984). Fawns molt to adult coloration at 3–4 months of age. Although brown is the predominant color on the deer's body, white is common around the fringe and from under the raised tail, which gives rise to the species's common name. White also distinguishes the belly, under the chin, inside the ears, and around the muzzle and eyes. A single, or less commonly double, white throat patch occurs on the ventral surface of the neck.

Normal variations of adult hair color from reddish to brown to gray occur throughout the range of white-tailed deer. However, two unique coloration conditions are restricted in frequency and distribution. The piebald or white deer with brown spots has been described locally in New York (Sauer 1984) and occurs somewhat rarely throughout the southeastern United States. The piebald condition varies from only slight deviations in normal hair color to extreme deviations in color. It is often combined with skeletal deformation such as a "Roman" nose, arched back, and short legs. True albino deer and melanistic deer are rare. However, 8.5% of the animals in some localized populations in central Texas have been reported with melanism (Baccus and Posey 1999).

Skull and Dentition. The white-tailed deer skull may be distinguished from other cervid skulls by the vomer dividing the nares into two separate chambers posteriorly. In addition, the width of the slightly inflated auditory bullae is equal to the length of the bony tube leading to the meatus (Godin 1977). The lacrimal vacuity is very large, and the lacrimal fossae are small and shallow (Fig. 44.2). The dental formula of adult deer with permanent dentition is I 0/3, C 0/1, P 3/3, M 3/3. The upper incisors and canine are absent, and lower canines are incisiform. However, small upper incisors are reported infrequently.

Antlers. Antler size and conformation are highly variable and depend on age, nutrition, and genetics. They range in size from smaller antlers of Coues and tropical subspecies to the more massive antlers of the northern subspecies. Large antlers, however, are not restricted to northern areas. Typical antler configuration includes a main beam with single points arising directly from the beam, in contrast to the dichotomous branching found in mule deer (*Odocoileus hemionus*).

In most regions, annual antler growth begins in mid-March or April. Because the growing bone tissue is highly vascularized and innervated, it is sensitive to touch and injuries can result in profuse bleeding. The growing antlers are covered with a soft, hairy skin called velvet, which is shed in late-August or September in preparation for the rut. A complex array of hormones is involved in regulating antler growth; however, changes in testosterone concentrations as a result of changing

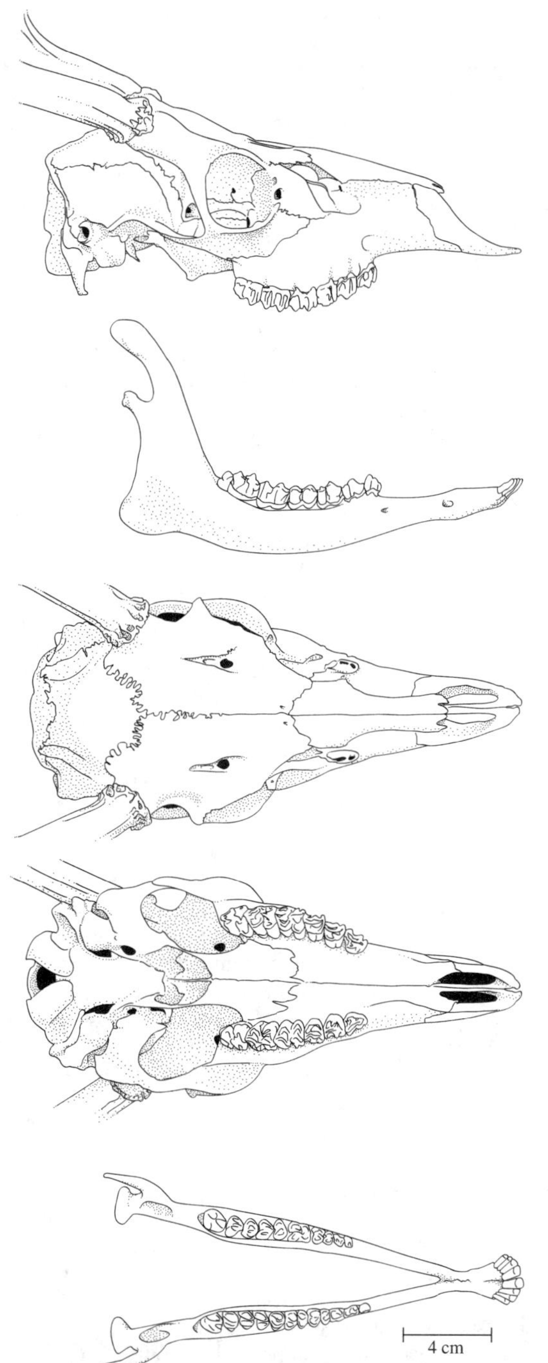

FIGURE 44.2. Skull of the white-tailed deer (*Odocoileus viginianus*). From top to bottom: lateral view of cranium, lateral view of mandible, dorsal view of cranium, ventral view of cranium, dorsal view of mandible.

photoperiods are the primary cue dictating the antler cycle. Hardening of the antler and shedding of velvet are a direct result of rising testosterone levels cued by shortening daylength. Following the rut, decreasing testosterone levels trigger antler casting. Typically, antlers are cast in late-December–early February, although occasionally bucks will retain their antlers until spring (Zagata and Moen 1974).

Overall size, spread, number of points, and presence of nontypical points increase as males mature, with a maximum antler size at 5–7 years of age. The first antlers typically occur on yearling bucks. However, hardened antlers developed in fawns in a supplementally fed population in Michigan (Ozoga 1988), and 20–30% of buck fawns in southern states develop small hardened antler nubs during their first January or February (Jacobson 1995). A yearling (1.5 years of age) buck's antlers are only 15–25% of the ultimate size potential.

Antler development is affected greatly by nutritional intake before and during antler growth. Early work on nutrition–antler relationships showed that buck fawns fed 4.5% or 9.5% protein through 1.5 years grew smaller antlers than bucks fed 16% protein (French et al. 1956). Because of the sensitivity of yearling antler growth to dietary quality, antler measurements such as antler beam diameter have proven to be sensitive measures of deer population status relative to carrying capacity.

Older bucks also show antler responses to changes in diet quality; antlers of 2-year-old bucks fed 16% protein had almost twice the mass of those of bucks fed 8% protein (Harmel et al. 1989). Diets containing as little as 10% protein may fulfill the requirements for antler development in adult bucks (Asleson et al. 1996), perhaps because they have completed their body growth and because of their ability to recycle nitrogen on low-protein diets.

Potential for antler development is genetically regulated, although debate continues over the relative heritability of specific antler features and the age at which ultimate antler size can be predicted (Williams et al. 1994; Lukefahr and Jacobson 1998). The expression of genetic potential for antler development by yearling bucks can be masked by other sources of variation, such as poor nutrition and late birth date. However, selective-harvest strategies based on antler characteristics of young bucks can affect antler size at maturity for surviving bucks (Strickland et al. 2001).

Glands. Arguably, olfaction is the most important mechanism for communication among deer. Skin glands of the white-tailed deer that have been identified include the interdigital, tarsal, metatarsal, forehead, preorbital, nasal, and preputial glands. However, communicative significance has only been demonstrated or suggested for the forehead, tarsal, and interdigital glands.

Glandular activity in the forehead region of all deer increases during the rut. Not surprisingly, greatest activity is found in the forehead glands of dominant males. A simultaneous rise in antler rubbing activity among mature males suggests that this gland produces socially significant odors used by males to communicate presence and/or social status to conspecifics. Forand and Marchinton (1989) speculated that social grooming of the forehead region of dominants by subordinate animals allows them to remember the smell of the adversary in the future and allows identification of the animal's dominance areas that have been marked by antler rubs.

The interdigital glands are pockets between the primary digits of both fore- and hind feet. Presumably, compounds arising from the interdigital gland mark a trail while a deer walks and conveys home range familiarity. Chemical analyses of the interdigital secretions have identified a number of compounds that occur in significantly greater concentrations in dominant males than in subordinates. Thus, chemical signals associated with the interdigital glands may signal presence of a dominant male while also acting as a generalized trail marker (Gassett et al. 1996).

Individual identity, dominance status, physical condition, and reproductive status likely are conveyed via the tarsal gland. This gland consists of elongated hairs associated with greatly enlarged sebaceous glands. Deer frequently urinate over these hairs in a behavior termed rub-urination. During the nonbreeding season, all deer (bucks, does, and fawns) urinate onto the tarsal glands about once per day, but more frequently when encountering unknown individuals (Sawyer et al. 1993). During the breeding season, older and more aggressive males rub-urinate more frequently, thereby acquiring the rank, musky odor of a rutting buck. During the rut, urine stains the tarsals of most bucks a dark rust color. Odors produced on the gland likely arise from interactions among the glandular secretions, urine deposited on the gland, and bacteria residing on the hairs of the tarsal gland (Gassett et al. 2000; Osborn et al. 2000). The skin underneath the tarsal tuft also has well-developed erector pili muscles, which allow an excited or stressed deer to flare the tarsal hairs to release a burst of scent.

Little is known about the communicative significance of the other glands. The preorbital gland has been suggested as a source of scent left on overhanging branches at scrapes, although definitive studies have not been done. There is evidence that the metatarsal glands of black-tailed deer (*O. hemionus columbianus*) are a source of an alarm pheromone (Müller-Schwarze et al. 1984), but efforts to demonstrate a similar function in whitetails have been unsuccessful.

Vocalizations. Whitetails produce a variety of vocal and nonvocal sounds associated with interspecific and intraspecific communication. Studies by Atkeson et al. (1988) identified 12 behaviorally distinct vocalizations, which include alarm and distress calls, agonistic calls, maternal–neonatal calls, mating calls, and a contact call. Unquestionably, the most distinct call of the white-tailed deer is the snort produced by an alerted or fleeing animal. It is not actually a call, but a sound produced by air forcibly passed through the nasal passages. Members of doe groups are much more likely to snort when potential danger arises than members of buck groups (Hirth and McCullough 1977). The other alarm or distress call is the bawl, which can be given by deer of any age, but only in times of extreme distress.

Agonistic calls consist of a low, guttural grunt with additional elements as the intensity of the encounter increases. The low grunt is given by both bucks and does throughout the year, and often is used with other displays to displace lower ranking individuals. The grunt–snort and the grunt–snort–wheeze typically are produced by males during the rut and often are accompanied with an intense dominance display indicating a willingness to fight.

A female approaching a fawn's bedding site often gives the maternal grunt. In response, fawns leave their beds and may produce a care-soliciting mew or a more demanding bleat. Intensity of the bleat is related to the degree of need of the fawn and may be heard up to 100 m away.

Hooves and Locomotion. The two-pronged track of the whitetail identifies the animal as a member of the Artiodactyla, the even-toed ungulates. Hooves consist of an outer wall of highly keratinized material and a sole having a softer, spongy surface. Dew claws on the front legs are closer to the hooves than are those on the rear legs. Also, the front feet tend to be larger than the rear feet, particularly in adult bucks. Average length of the front hoof of an adult buck is about 75 mm and width typically is 35–50 mm.

White-tailed deer have two basic gaits, the walk and the gallop. When relaxed, deer move at a leisurely walk, or technically a cross-walk (Geist and Lingle 1994). The right front foot is lifted followed by the left rear. As the right front moves to a new position, power is provided by the right rear, weight is supported by the left front, and the left rear is finishing its stride and is ready to be lifted. Movement of the left front foot is accompanied by mirrored movements of the other legs. This same movement pattern can be accelerated to a faster trot.

The gallop, or technically the rotary gallop, consists of a sequence of steps from the left rear to the right rear, followed by the right front and finally the left front. The same sequence is repeated in the next stride, although the lead foot may change midstride depending on terrain. The typical whitetail gallop attains speeds of 6.3–10.9 m/sec (14.1–24.5 mph), although speeds approaching 18 m/sec (40 mph) have been recorded (Geist and Lingle 1994).

PHYSIOLOGY

Metabolism. White-tailed deer have evolved an annual cycle of fat deposition and use that is adapted to food availability, reproduction, and potentially stressful environmental conditions. Mautz (1978) used the analogy of a deer pulling a sled up a bushy hillside to describe the period of fat accumulation that occurs before the winter stress period. Rate of fat metabolism varies with sex, age, and physiological state. Lactating does have higher energy demands (i.e., a heavier sled) than barren does, but both will put effort into fat accumulation before the winter. Young animals on restricted diets will allocate resources into fat production at the expense of body growth (Verme and Ozoga 1980a). Males must accumulate surplus energy stores before the breeding season. Bucks exhibit elevated energy expenditures during the rut because of increased activity and reduced food consumption associated with breeding behavior.

Fat deposition is related to available nutrition (Verme and Ozoga 1980a) and photoperiod (Abbott et al. 1984). Budde (1983) observed increased growth rates of fawns 2 weeks after the animals were switched from long days (16 hr light and 8 hr dark) to short days (8 hr light and 16 hr dark). Using similar photoperiod treatments, Abbott et al. (1984) found short-day fawns accumulated more fat and weighed more than fawns under long-day photoperiod. Heavier fawns ate more, but also were more efficient with the calories consumed. Heavier fawns on short daylength also had lower serum prolactin concentrations, which may be an important endocrine mediator of fat accumulation and use.

During winter, previously accumulated fat reserves are used. Fat reserves may be conserved by addition of winter browse. However, these foods are typically poor in nutritional value and may be unevenly distributed within habitats. Therefore, it may be more energetically expensive to forage on low-quality browse than to reduce food consumption and limit movements. Deer decrease their food consumption in winter even if highly nutritious food is available ad libitum. Worden and Pekins (1995) found metabolizable energy intake peaked in September and October and decreased 54% by February in New Hampshire. Ozoga and Verme (1982) observed that supplementally fed adult does and prime-age bucks lost an average of 8% and 18% of their prewinter body mass, respectively, during the winter.

Food consumption changes by season and may be accompanied by changes in metabolic rate to conserve energy during winter stress periods (Silver et al. 1969). However, Mautz et al. (1992) reported no change in fasting metabolic rate and felt previous results were an artifact of experimental design.

Digestion. Rumen microfauna allow deer to digest complex carbohydrates found in fibrous plant material. Deer chew their food to allow mixing with saliva for ease in swallowing. Saliva also acts to buffer rumen content and counteract plant defensive compounds such as soluble phenolics. Food passes down the esophagus to the rumen, the first compartment of their four-part stomach. Rumen microbes are responsible for breaking the chemical bonds of the compound plant material into usable nutrients through anaerobic fermentation. Large boluses of food enter the reticulum, the second compartment, and are regurgitated via muscular contractions for further mastication and returned to the rumen for additional fermentation. This process occurs until forage particle size is reduced. In addition to cellulose digestion, microfauna also are involved in nitrogen metabolism and vitamin synthesis.

The third compartment is the muscular omasum, where much of the water is absorbed from the fluid contents of the rumen. The omasum acts as a sieve, permitting passage of food items by particle size, retaining larger, less digested particles in the rumen–reticulum. The final compartment, the abomasum, is considered the true stomach and contains acids and enzymes to aid digestion of proteins. Digesta leave the abomasum and pass into the small intestine. Whereas some nutrients such as volatile fatty acids are absorbed in the rumen, digestion of simple sugars, peptides, and lipids and absorption of nutrients occurs in the small intestine. Digesta then travel through the large intestine, where some additional hindgut fermentation occurs in the cecum, and more water and small water-soluble nutrients are absorbed.

Neonates have a reticular groove to pass milk directly from the esophagus to the abomasum because fermentation of milk is not required (Robbins 1993). As the fawns begin eating plant matter, the rumen develops structurally and functionally. Inoculation of the rumen with bacteria and protozoa probably occurs through interaction of the young and mother or from environmental sources.

Rumen microbial concentration and species composition may change depending on plant material eaten. Starvation causes a reduction of protozoa and bacteria, diminishing the animal's ability to digest cellulose. Feeding of highly nutritious food after extensive fasting may be detrimental or even deadly. The rumen microbes are not adapted to the food, and animals may die with full stomachs (Cowan and Clark 1981).

Nutrient Requirements. White-tailed deer nutritional requirements include water, minerals, vitamins, essential fatty acids, protein, and a source of energy. The most critical nutrients for wild white-tailed deer appear to be related to energy and protein. Protein (or a source of non-protein nitrogen) is important for growth and metabolism. However, Verme and Ozoga (1980b) observed that fawn growth was unaffected by low dietary protein if energy levels were adequate. Deer are able to recycle urea to the rumen, where it is converted to microbial protein, thus conserving nitrogen when consuming low-protein forage. Urea recycling increased from 40.6% to 92.3% when dietary protein decreased from 26% to 5% in captive fawns (Robbins et al. 1974).

Although deer may offset low dietary protein through urea recycling, a review by Thompson et al. (1973) reported that dietary protein levels below 9% caused poor fawn growth. Intermediate growth occurred with 10–13% protein and optimum growth occurred with 13–20% dietary protein. Male fawns require higher dietary protein for maximum growth than do female fawns (Ullrey et al. 1967). Protein requirements also are reduced with age and vary by sex. Holter et al. (1979) found maintenance protein requirement for yearling deer was only 5.8% crude protein. Asleson et al. (1996) found protein requirements differed between maintenance and antler growth in males (5.8% and 9.9% crude protein, respectively).

Energy requirements also vary by age, sex, and season. Energy affects body growth and fat storage in fawns. Thompson et al. (1973) found fawns required 154 kcal/kg body weight$^{0.75}$ of metabolizable energy. Energy requirements were highest in August and lowest in January. Low winter energy requirements are related to physiological changes associated with winter survival mechanisms. Energy requirements also increase with gestation and lactation in females (Pekins et al. 1998).

The macroelements calcium and phosphorus are important for growth and development, as these minerals are closely associated with skeletal formation. Ullrey et al. (1973) concluded that a diet containing 0.40% calcium and 0.25–0.27% phosphorus was adequate for growth and antler development of young animals. Grasman and Hellgren (1993) also evaluated dietary phosphorus requirements. They reported that although requirements changed seasonally, phosphorus was not considered limiting for antler growth in adult males.

Little information exists on specific requirements for other macroelements or vitamins. Jones and Weeks (1985) described calcium, magnesium, and phosphorus concentrations in forage of deer in south-central Indiana.

REPRODUCTION

Breeding Season. The breeding season (rut) of deer is timed to maximize successful production and survival of offspring. Short-day photoperiod is the proximate factor used to time reproduction and subsequent survival of offspring (Goss 1983). Shortening daylength triggers a chain of hormonal events initiated by the pineal gland, which typically ensures conception occurs in the fall. In temperate white-tailed deer populations, optimal fawning conditions generally occur during the spring when there is adequate forage for the dams to meet their nutritional needs and produce sufficient milk. The fawns must be able to grow, escape predators, and put on reserves before the harsh winter months.

Breeding tends to occur in a more discrete, synchronous period in northern populations, becoming more protracted the farther south until there is continuous breeding near the equator (Table 44.1). However,

TABLE 44.1. Timing of the breeding season for adult (≥1.5 years of age) white-tailed deer in selected areas

Area[a]	Latitude (°N)	Peak Breeding Season (Range or Mean ± *SD* Where Available)	Reference
Honduras	15	July–Nov	Klein 1982
St. Croix, Virgin Islands	17	Apr–May Every month	Webb and Nellis 1981
Rotenberger WMA, FL	26.5	10 Aug ± 16	Richter and Labisky 1985
Tosohatchee State Preserve, FL	28.5	7 Oct ± 17	
Camp Blanding WMA, FL	30.0	2 Nov ± 17	
Eglin Air Force Base, FL	30.5	22 Feb ± 16	
Delta Refuge, LA	~29	14 Dec–29 Dec	Roberson and Dennett 1967
Avery Island, LA	~30	1 Oct–31 Oct	
West Bay GMA, LA	~31	16 Oct–31 Oct	
Evangeline GMA, LA	~31	18 Oct–31 Oct	
Red Dirt GMA, LA	~31.5	1 Nov–15 Nov	
Jackson-Bienville GMA, LA	~32	1 Dec–15 Dec	
Tensas Parish, LA	~32	1 Jan–15 Jan	
Blackbeard Island, GA	~31.5	16 Sep–15 Nov	Osborne et al. 1992
Mississippi	~30–35	21 Dec–21 Jan (>80%) 20 Nov–15 Mar	Jacobson et al. 1979
Savannah River Plant, Aiken, SC	~33.5	13 Nov ± 15	Rhodes et al. 1991a
Radford Army Ammunition Plant, Dublin, VA	~37	11 Nov ± 3[b]	McGinnes and Downing 1977
Missouri	~36–40	18 Nov ± 1.0[c] (yearling) 16 Nov ± 0.6 (adults) 10 Dec ± 2.2 (fawns)	Hansen et al. 1996
Ohio	~40	Early Nov Begins late Oct	Nixon 1971
New York	42–44	10–23 Nov	Cheatum and Morton 1946
South of Houghton Lake, MI	~44	18 Nov ± 0.8	Verme et al. 1987
Manitoba, Canada	49–51	Late Oct–Dec	Ransom 1967

[a]GMA, Game Management Area; WMA, Wildlife Management Area (WMA).
[b]McGinnes and Downing (1977) reported fawning dates. Conception dates were backdated using an estimated 200 days for gestation.
[c]*SE*.

there are many exceptions to this trend because timing of reproduction is potentially affected by environmental factors, nutrition, age, exposure to mates, and genetics.

Although rutting activity is mediated by seasonal shifts in photoperiod, there is considerable geographic variation in the timing of breeding, particularly in southerly regions (McDowell 1970; Richter and Labisky 1985). Photoperiod changes more abruptly in northern regions than in southern latitudes. In northern climates, a short, precisely timed breeding season is critical for fawn survival. In areas with harsh winters, fawning too early in spring may expose the young to freezing temperatures. Furthermore, the dam may not be capable of producing sufficient milk before the spring growth of vegetation. Fawning late may not allow ample time for sufficient body growth and development of fawns before severe winter conditions.

In more southerly regions, seasonal climatic fluctuations are less severe and reproductive timing is less critical. In these regions, changing photoperiod apparently opens a reproductive window during which breeding can occur, but the precise timing of reproduction can vary dramatically among populations within the same latitude. In these southern ranges, timing of breeding can vary according to a variety of factors such as herd demographics (Miller and Ozoga 1997), genetics (Jacobson 1994), and likely physiological cues passed from mother to daughter. In southern states, peak breeding may occur in September in portions of South Carolina, November in Georgia, and December–January in portions of Alabama and Mississippi. In peninsular Florida, breeding occurs during midsummer, and in Central and South America, deer reproduce throughout the year.

Harsh winters are not the only limiting environmental factor deer must face. Fawning in some areas of Florida and Louisiana may be timed around seasonal flooding (Roberson and Dennett 1967; Richter and Labisky 1985). In the Everglades, the peak of rutting activity is probably tied to the hydrologic cycle (Richter and Labisky 1985). These fawns are born during the driest season in February–March.

Breeding may occur throughout the year in some tropical deer populations (Webb and Nellis 1981). Animals may be in all stages of reproductive development through the year, even though there are still peaks in breeding activity. It is not known what controls the reproductive cycle of each individual deer in these areas.

The physiological status of the animal may affect photoperiodic input into the timing of breeding. Nutrition plays an important role in fine tuning the exact time of reproduction. Animals with poor nutrition and inadequate body reserves may delay reproduction. Wentworth et al. (1990) found reproductive activity of deer in the Southern Appalachians was closely tied to acorn crops. When acorns were rare, kidney fat index was lower and conception dates were later than in years with high mast production. Cothran et al. (1987) found late breeders had lower total body fat than early breeders (late versus early defined as does breeding one standard deviation after versus before the mean conception date, respectively). If there are insufficient resources for body maintenance, then survival of both parent and offspring may be jeopardized. It may be a better investment of resources to delay reproduction until the female is in better physiological condition. This may mean a delay or skipping reproduction for that year.

Age of the doe also affects reproductive timing. If fawn does breed their first fall, they generally do so later than the adult does (Rhodes et al. 1991a). Where yearling females have nutritional constraints, they may also breed later. Alternatively, yearlings with excellent nutrition that did not breed earlier in life may enter breeding condition sooner than older females. The older females may be delayed because of the associated energetic demands of reproduction and lactation from the previous year.

Reproductive status of females also may be affected by male pheromones. Miller et al. (1991) theorized that dominant bucks may leave pheromone priming scents at signposts, which stimulate does to enter reproductive condition earlier or promote synchronous cycling of all does in an area. Verme et al. (1987) found that does constantly exposed to a rutting buck in a small pen bred earlier than does in a large, forested exclosure. They hypothesized that the biostimulation from the bucks caused the does to enter estrus at the onset of the rut. Male pheromones may provide fine tuning of the timing of estrus.

Adult sex ratios may affect the timing and duration of the rut. Many hunted deer populations have skewed adult sex ratios greatly in favor of females. The male segment of the population has a young age structure and few older males exist. The lack of mature bucks may prevent pheromone priming and provide inadequate numbers of males for breeding as the does come into estrus. If a doe fails to conceive, she will cycle again about 26 days later (21–30 days; Knox et al. 1988). Knox et al. (1988) found does were capable of recycling two to seven times if not bred.

Does that have to recycle before being inseminated extend the breeding season. Gruver et al. (1984) felt that the extensive bucks-only harvest early in the season contributed to the prolonged breeding season in Mississippi. However, physiological and behavioral mechanisms may occur to ensure all females are bred in areas with low numbers of males. Prolonged estrus and increased female movements during estrus (Labisky and Fritzen 1998) may facilitate contact between bucks and does during the rut and thereby limit the number of does not bred on their first reproductive cycle.

Genetics may affect timing of reproduction. This makes evolutionary sense because deer genetically programmed to fawn at suboptimal times would be selected against. It is difficult to isolate this effect from other factors influencing the timing of breeding. Furthermore, the influence of genetics on timing of the rut may be difficult to interpret because of previous restocking efforts.

Age at Sexual Maturity. Young animals must obtain sufficient body growth and development before investing energy in reproduction. Doe fawns may breed their first fall in areas with excellent nutrition. Rhodes et al. (1991b) found 16–69% of doe fawns bred at the Savannah River Site near Aiken, South Carolina. Proportions of fawns breeding were higher in upland sites, where food resources were greater, than in lowland swamp habitats. However, deer populations in both areas were characterized by a young age structure, and reproduction by fawns added significantly to the annual recruitment. These young animals bred about 45 days later than the adult does (Rhodes et al. 1991a), probably because they did not attain large enough body size for reproduction to be physically possible until late in the breeding season. The mild winter climate in this region probably enhanced survival of their late-born fawns.

Doe fawn breeding has been reported at more northern latitudes in areas where diet quality is high. Significant fawn breeding occurred in Ohio (Nixon 1971), Iowa (Haugen 1975), Michigan (Verme 1991), and New York (Morton and Cheatum 1946). Morton and Cheatum (1946) reported 36.3% of fawns from the southern region of New York were pregnant. However, only 4.2% of fawns bred in the Adirondack region farther north where food resources were limited. Nixon (1971) found 76.7% of the female fawns examined in Ohio had ovulated, with an average of 1.29 fetuses/doe fawn. The peak breeding in these doe fawns occurred about 1 month later than in adult does.

There are probably many factors that influence fawn breeding in wild populations. Low nutrition and social stress may limit attainment of puberty. Breeding may occur as long as sufficient body reserves are obtained during normal photoperiod constraints. Budde (1983) found ovulation of doe fawns was dependent on photoperiod. Verme and Ozoga (1987) found similar reproductive output in doe fawns subjected to different photoperiod treatments in the fall, but the fawns on an extra 9 weeks of long-day photoperiod exhibited delayed breeding. In areas of poor nutrition, yearling females may function like doe fawns and not reproduce. Where yearling females have nutritional constraints, they may also have lower productivity (Verme 1969).

Other cryptic factors may also influence fawn breeding. Lack of fawn breeding has been reported in many parts of the United States including island habitats (Osborne et al. 1992), the Southern Appalachian mountains in years of high or low acorn abundance (Wentworth et al. 1990), and supplementally fed deer in Upper Michigan (Ozoga and Verme 1982).

Sexual maturity of males may occur as early as their first reproductive season when they are 6 months old. Schultz and Johnson (1992) demonstrated successful breeding by male fawns. However, this study was done in captivity with no older, mature bucks in the pens. In a wild setting in New York, Cheatum and Morton (1946) found that males reached sexual maturity at about 18 months of age. They found no sperm in buck fawns collected throughout the breeding season. Even if buck fawns attain breeding condition, it will probably occur later in the season than for prime-aged animals (Miller et al. 1987). Socially subordinate bucks may have little opportunity for breeding in natural populations even if they are physiologically capable because mature males may monopolize access to estrous females.

Gestation. In temperate deer, demands of gestation are closely tied to energy balance and timing of spring green-up. Pekins et al. (1998) observed a 16.4% increase in fasting metabolic rate over the entire gestation period. However, most (92.2%) of the increase occurred during the last trimester, coinciding with cessation of winter (77% of the increase occurred after spring green-up). Thus, fetal development proceeds slowly during winter stress periods and increases dramatically when quality nutrition becomes available.

Duration of gestation is probably influenced by litter size, nutrition, and mass of fawns. Haugen (1959) observed a mean gestation period in 12 captive white-tailed deer in Alabama of 193.8 days (range = 187–198 days). The average weight of newborn female and male fawns was 2.5 and 2.8 kg, respectively. Verme (1969) observed a mean gestation period in Upper Michigan of 202.1 ± 4.1 days (range = 196–213 days, $n = 55$), which was probably influenced by the larger fawns produced (3.4 ± 0.6 kg).

Pregnancy Rate and Number of Young. Available nutrition and energy resources greatly affect the number of fawns produced. Cothran et al. (1987) found that does with a greater percentage of body fat at conception have more twin fetuses. As expected, the highest conception rates among adult does are recorded from the agricultural areas of the midwestern United States. Fetal rates (fetuses/doe) for does (≥2.5 years old) in these areas are 2.04 in Ohio (Nixon 1971), 2.10 in Iowa (Haugen 1975), and 1.82–1.97 in Missouri (Hansen et al. 1996).

Nutritional constraints may influence young breeders more than mature animals. For example, Wentworth et al. (1990) reported that changes in acorn abundance among years greatly affected reproductive output of yearling does. Ovulation incidence (corpora lutea/doe) was 1.07 and 1.31/yearling doe in years of poor and good acorn production, respectively. Ovulation incidence was greater in adult does (≥2.5 years) regardless of acorn abundance (1.59 vs. 1.68 for years of poor and good acorn production, respectively).

Deer from islands with abundant but low-quality forage typically have low reproductive rates. Deer examined from the barrier islands of Georgia had a low number of fetuses/doe for yearlings and adults (1.0 and 1.14, respectively; Osborne et al. 1992). Miller (1988) found similar low reproductive rates on a Georgia barrier island with poor habitat conditions, but found no difference between yearlings and adults (1.08 corpora lutea/doe). Key deer live in isolated, poor-quality habitat with limited nutrition and also exhibit very low reproductive output (0.76 fetus/doe ≥1 year of age; Folk and Klimstra 1991).

Ransom (1967) reported high fertility rates in Manitoba deer populations. However, weather may significantly affect the number of young that are born and survive. Winter conditions caused malnutrition, which resulted in many stillbirths and postnatal losses of fawns. Therefore, weather was more important than habitat quality in determining reproductive output.

Primary Sex Ratio. Sex ratio of white-tailed deer may vary at conception and/or birth according to population demographics and nutrition (Verme 1983; Degayner and Jordan 1987). The adaptive significance of sex ratios different from 50:50 (males:females) is debated. Trivers and Willard (1973) proposed that the sex ratio of offspring was related to the parent's ability to invest resources into the young. Females in good condition with excellent resources should produce more male offspring in polygynous mating systems. Therefore, if a female has extra resources, investment in sons should provide an advantage for the sons to ultimately attain dominance and be reproductively successful. Daughters may have lower total reproductive potential, but are likely to breed regardless of initial parental investment.

Local resource competition also may select for an adaptive change in fetal sex ratio. Female white-tailed deer are typically philopatric and young females establish home ranges overlapping those of their mothers (Ozoga et al. 1982; Mathews and Porter 1993). Closely related, philopatric female groups have familiarity with food and cover resources and better ability to defend prime habitat, which is important for neonatal survival (Ozoga et al. 1982). Therefore, an older, dominant doe should have female offspring to inherit her access to the best habitat (Degayner and Jordan 1987). However, females in poorer condition without access to good habitat should produce sons who would disperse and have the potential for reproductive success in new areas

There may be a continuum of maternal-investment ability and the optimal sex and number of fawns to be produced. Kucera (1991) examined sex ratio variation in a population of nutritionally stressed mule deer. A greater kidney fat index (better opportunity for maternal investment) was associated with more male offspring. However, there was a continuum, with does in the poorest condition producing single females; does in better condition producing single males, followed by twin females, and mixed male and female twins; and finally mothers in the best condition producing twin males.

We have little understanding of possible physiological mechanisms mediating primary sex ratio variation. However, Verme and Ozoga (1981) reported that captive does bred early in the estrous cycle had fewer male fawns (14.3%), whereas those bred later in the cycle produced predominantly male offspring (80.8%). Perhaps the timing of insemination relative to ovulation affects the fetal sex ratio.

FEEDING HABITS

Optimal types and amounts of forage eaten by white-tailed deer can be predicted based on the characteristics of the gastrointestinal tract (Hoffman 1985). Their anatomical, behavioral, and physiological adaptations allow these "concentrate selectors" to select the most nutritious forage available and then obtain the easily digestible nutrients quickly. Compared to larger ungulates, they have a small rumen relative to body size and as a result they tend to require a higher quality diet. Forbs and various types of mast typically are highly digestible sources of nutrients, which readily fulfill deer forage requirements. Their large salivary glands produce enzymes that deactivate secondary plant compounds that might inhibit digestion of browse and hard mast. High-fiber, low-quality forages, such as mature grasses, are not digested effectively because they require more extended fermentation within the rumen.

Forages eaten by white-tailed deer are as varied as the habitats and latitudes occupied by this adaptable species. Extensive research throughout their range has documented the feeding habits of whitetails. A sampling of feeding habits studies includes those in Maine (Crawford 1982), Michigan (Stormer and Bauer 1980), Wisconsin (Dahlberg and Guettinger 1956), Illinois (Strole and Anderson 1992), Florida (Harlow 1965), the Southern Appalachians (Harlow and Hooper 1972; Johnson et al. 1995), and South Carolina (Castleberry et al. 1999). Long lists of plants are required to fully describe the feeding habits in any particular area (Nudds 1980). However, the majority of the whitetail's diet comprises only a portion of the complete list of forage species. For example, 93% of their diet consisted of only 47 of the 154 species or species groups consumed in Oklahoma (Gee et al. 1991), and forbs and other browse made up 85% of the annual diet (Fig. 44.3). Similarly, Warren and Hurst (1981) evaluated deer use of grasses, forbs, legumes, vines, woody plants, sedges, and ferns in Mississippi. Of the 521 species of plants examined, 267 (>51%) received moderate to high use. More than 60% of the forbs and legumes had moderate to high use. Few sedges, rushes, and ferns were eaten.

Whitetails seem to have an uncanny ability to select the most nutritious forages. Deer are opportunistic, concentrate selectors, and

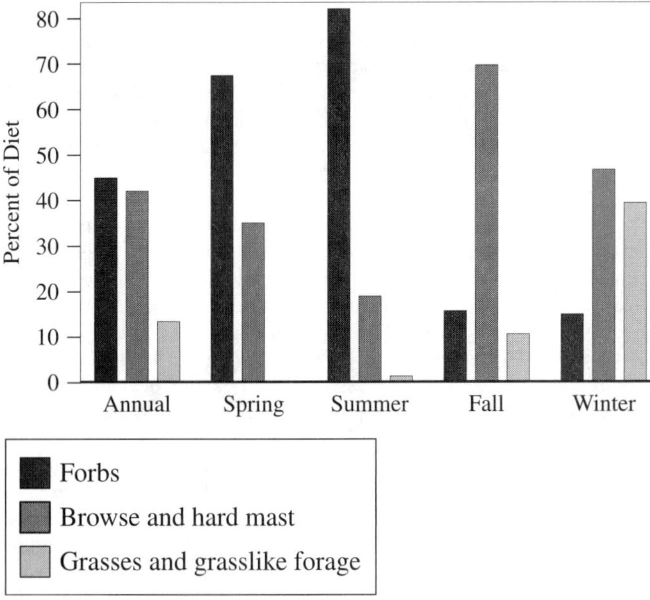

FIGURE 44.3. Seasonal use of forage classes by white-tailed deer in the Cross Timbers Region of Oklahoma. SOURCE: Data from Gee et al. (1991).

regular sampling from a wide range of species allows them to continually evaluate new sources of nutrition across seasons and years. Many researchers have observed captive deer select forages by apparently tasting them first. Some leaves or other plant parts may be accepted, whereas others are rejected. There also have been numerous anecdotal observations from biologists and agronomists who describe deer selecting fertilized versus unfertilized plantings and particular varieties of crops. Preference for acorns from a particular species of oak or even individual trees also has been observed. Females may select a higher quality diet than males (McCullough et al. 1989; Weckerly and Nelson 1990) due to selection of higher quality forages within the same habitat or to differential use of available foraging habitats.

Browse, mast, and forbs in varying amounts make up the majority of the diet throughout the range of white-tailed deer. However, diet selection varies in response to seasonal changes in forage abundance and quality and the metabolic demands of the animal. When resource levels are low, deer tend to be more generalist feeders (Nudds 1980). When quality and quantity of forage resources are not limiting, deer are more selective in their feeding (Murden and Risenhoover 1993). Varying intake of forbs, browse, mast, and grasses during spring and summer allows deer to satisfy their changing nutritional needs. Highly digestible forbs and grasses address protein and phosphorus requirements, browse provides rapidly fermented cell solubles, and mast provides a concentrated energy source (Vangilder et al. 1982).

Mast is important when available, whether it is soft mast such as fruits and berries during the summer or hard mast such as acorns during fall. Acorns are a major component of the whitetail's diet in oak (*Quercus* spp.) woodlands, and, when abundant, they can constitute >70% of the November and December diet. They are a highly palatable source of energy, but protein content is low. In many habitats, oak mast is heavily used, but is not a critical component of the whitetail's diet. However, in some southern ranges such as the Southern Appalachians or the southern Coastal Plain, acorns are a crucial component of the diet and whitetail population dynamics can be driven by the vagaries of mast production. Other common sources of mast during fall include beechnuts (*Fagus grandifolia*), persimmons (*Diospyros virginiana*), grapes (*Vitis* spp.), and apples (*Malus* spp.). Summer mast producers include blueberries (*Vaccinium* spp.), blackberries and raspberry (*Rubus* spp.), pokeberry (*Phytolacca americana*), and cherry (*Prunus* spp.), among others.

Mushrooms are used extensively throughout the year. When available, they may constitute >15% of the diet. They are high in phosphorus and protein and can be a significant dietary source of these nutrients in some areas such as the southeastern Coastal Plain. In a review of mushroom consumption by wildlife, Miller and Halls (1969) reported that across studies, mushrooms were found in 10–100% of rumen samples examined and volumes ranged from a trace to 83%. Across these studies, crude protein content ranged from 23% to 30% and phosphorus ranged from 0.20% to 0.68%

Agricultural crops can be a significant component of the diet for populations across the range, but most notably in the agricultural regions of the midwestern United States. It is no surprise that the largest deer and the most productive populations are frequently found in agricultural areas. However, this preference for agricultural crops by deer can cause significant economic damage.

ECOLOGY AND BEHAVIOR

Few mammalian species occupy such a wide range of latitudes or inhabit such a diverse array of habitats as the white-tailed deer. As such, one would expect a great deal of behavioral plasticity among populations in disparate regions or habitats, and there are numerous examples of these differing behavioral responses to environmental variables. In northern latitudes, many populations migrate between fawning and wintering ranges, whereas similar seasonal movements are absent in southerly populations. Group size is influenced by habitat structure. Whereas aggregations of deer may be common in open habitats, group size decreases in forested habitats (Hirth 1977). Even social behavior can vary with herd demographic characteristics. Marchinton and Atkeson (1985) suggested this plasticity is a key adaptive characteristic that allows this species to thrive in a variety of habitats at a wide range of densities.

Social Organization. White-tailed deer social structure during the nonreproductive period centers on two basic social groups. Family groups, or matriarchal groups, consist of a maternal doe, her young of the year, and female offspring from previous years. Adult males (>1.0 year old) form loose-knit fraternal, or bachelor, groups of varying size. Mixed-sex groups occur occasionally, particularly as temporary feeding assemblages at a concentrated food source. However, sex-specific groupings typically are isolated socially, geographically, or temporally (McCullough et al. 1989). Maternal groups often fuse into larger groups during fall and winter, particularly in northern regions where deer aggregate in "yards" during winter. Smaller winter aggregations often consist of related individuals (Mathews and Porter 1993). Larger winter aggregations may function as demes, or subpopulations, with historic wintering areas and migration routes (Nelson and Mech 1987).

Parturient females isolate themselves preceding fawn birth and actively defend fawning territories from intrusion by other deer. During the first few days after birth, females rarely venture more than 100 m from their fawns. At 3–4 weeks of age, fawns begin to accompany their mothers, and by 8–10 weeks they and their mothers form new family groups. As fawns mature during summer, does become increasingly tolerant of related females and their offspring. In forested landscapes, matriarchal groups tend to remain small, except during winter. However, in some of the more open habitats in the western portion of the whitetail's range, larger groups of females with their fawns are common.

Young parturient females in some intensively farmed areas of the midwestern United States where forested cover is limited often disperse long distances to locate suitable fawning habitat. For example, in an Illinois study, 50% of female fawns and 21% of yearling females dispersed to new home ranges outside their natal range (Nixon et al. 1991a). In Nebraska, 16% of a sample of all-aged does emigrated (VerCauteren 1998). In northeastern Minnesota, 20% of yearling females dispersed (Nelson 1993). In other regions, female dispersal rates generally are very low (<5%) regardless of deer density.

In contrast to the sedentary nature of young females, yearling males typically disperse from their natal areas and seek association with bachelor groups. Dispersal distances typically range from 3 to 10 km, but distances >150 km have been reported. Although antagonism by mature males and competition for breeding privileges may be partially responsible for yearling dispersal (Marchinton and Hirth 1984), aggression from the yearling's dam and other related females might be the primary impetus for dispersal (Holzenbein and Marchinton 1992a, 1992b).

During the spring and summer seasons, whitetail bucks typically are more social than does. Bachelor groups typically consist of two to five individuals, but group membership is fluid (Hirth 1977), particularly among younger members. Behavior of members in bachelor groups is congenial, and social grooming is common.

Social organization may vary considerably in relation to habitat and predation. For example, in more-open habitats such as in agricultural areas or savannah habitats, larger aggregations are common and may represent a distinct social structure unknown in denser habitats. Maternal–neonatal bonds may not be as strong in these open habitats (Marchinton and Hirth 1984). Similarly, social structure, deer dispersion, and behavior may vary as an adaptation to reduce predation. For example, Nelson and Mech (1981) suggested that the use of traditional summer and winter ranges allows deer to maximize use of buffer zones along wolf pack territories.

Social Hierarchy. Social relationships among deer are based on a social hierarchy that is established and maintained through a variety of postures, threats, and aggressive interactions. Dominance hierarchies minimize aggression and conflict within groups, thereby minimizing energy expenditure, injury risk, and exposure to predation. Social rank typically is related to age among females, although social rank may be transmitted from mother to daughter. In contrast, rank among males is primarily dependent on body size. Physically superior and behaviorally mature bucks ≥3.5 years old typically dominate younger bucks.

Except during the rut, males and females use similar visual cues to communicate information such as dominance status, aggressive intent, and alarm (Hirth 1977). Most cues are quite subtle and often go unnoticed by human observers. Aggressive potential is conveyed through threats that signal aggressive intent or dominance displays designed to intimidate others. Simple threats are communicated with a pinning back of the ears accompanied by a direct stare. A head-high or a head-low posture signals increased aggression and an intent to rear or chase. Subordinates typically avoid direct eye contact with higher ranking animals and either move when approached directly or display submissive behaviors.

During the breeding season, dominance displays are used primarily by males and always are accompanied by threatening postures (Hirth 1977). Intense displays include pinning back the ears, piloerection, flared nostrils and preorbitals, and a sidling approach. Beads of moisture may appear on the buck's muzzle and there often is noticeable drooling. These dominance displays often are accompanied by vocal cues such as the grunt–snort–wheeze and olfactory signals such as rub-urination. Most aggressive encounters end when one of the rivals turns away indicating submissiveness. Infrequently, neither buck will accept the subordinate role, which can result in vicious, but typically brief, dominance fights.

Social aggression among males declines sharply at the end of the breeding season coincident with reduced testosterone levels and casting of the antlers. Throughout the remainder of the annual cycle, males are very tolerant of each other. Very few overt interactions can be observed among members of buck groups, although smaller bucks sometimes can be seen stepping out of the way of a larger buck (Marchinton and Hirth 1984).

Activity, Movements, and Home Range. White-tailed deer typically are crepuscular, although their activity patterns are flexible and are influenced by a variety of environmental and anthropogenic factors. In northern climates, deer tend to be more diurnal during winter, with a single late afternoon activity peak. In other seasons, deer are crepuscular with activity peaks at dusk and just after dawn. The evening peak starts before sunset in spring, summer, and fall, but ends at sunset in winter. Secondary periods of deer activity have been reported during the middle of the night and during midday. In addition to the morning and evening peaks, Michael (1970) reported two lesser peaks in activity at noon and at midnight among Texas deer. By using activity counters to monitor the activity of penned deer in Michigan, Ozoga and Verme (1970) suggested a five-peaked activity pattern of whitetails. From their studies, they concluded that deer fed most around sunrise, midday, and twice during the night, particularly during winter.

The total distance that deer move during a 24-hr period has been reported as 1.5 km in Iowa (Zagata and Haugen 1973), 4.5 km in Florida (Bridges 1968), and 5.7 km (males) and 5.6 km (females) in Illinois (Nixon et al. 1991a). Movements by mature bucks increase dramatically during the breeding season, and some may travel widely outside of their normal range. Deer activity is depressed by extreme environmental conditions such as heavy rain (Hawkins and Klimstra 1970), strong or gusty winds (Michael 1970), above-average summer temperatures (Michael 1970), and winter temperatures and snow depth (Beier and McCullough 1990). Although associations between deer movements and lunar phase have been proposed, no scientific studies have been published that demonstrate any clear relationships.

Deer movements vary greatly according to the sex and age of the individual as well as season, habitat, and weather. Mature males have the greatest daily movements during the breeding season, whereas females have the least movements during parturition (Pledger 1975). Mature males may travel greater distances than other deer, particularly during the breeding season. In an Alabama study, does increased activity during the rut, but total distance moved decreased (Ivey and Causey 1988). However, estrous does may undertake a brief "breeding excursion" outside of their normal range, apparently in search of a suitable breeding partner (Sawyer 1981).

On a yearly basis, deer are least active during midwinter (January and February), with another moderate lull in activity during the summer months. Seasonal deer activity consistently peaks in spring and fall. The rise in activity levels likely is associated with an increase in thyroid function and rising metabolism.

Whitetail home range patterns usually are elliptical, although circular and irregularly shaped ranges have been reported. Home range size also varies by sex and age of the individual as well as habitat and season. Reported home range size varies greatly over various regions (Table 44.2), but typically a deer's home range size approximates 2.6 km^2. Home ranges tend to be smaller in diverse, productive habitats. Small home ranges also have been reported in high-density populations and among urban/suburban deer. Typically, the annual home range size of adult females is approximately 50% that of adult males.

Migrations and Yarding. In extreme northern ranges, whitetails often have pronounced seasonal migrations in response to snow depth and cold weather. Migrations to wintering habitats usually are short. Mean migration distances range from 6 to 23 km, although migrations of much greater length have been reported from some areas such as Minnesota and the Upper Peninsula of Michigan (Marchinton and Hirth 1984).

In northern ranges, deer frequently congregate in sheltered areas during winter. These deer "yards" typically consist of closed-canopy coniferous forests that provide thermal cover, reduced wind velocities, and decreased snow depths compared to adjacent areas. Mature northern white cedar (*Thuja occidentalis*) is the preferred forest type in deer yards because it provides excellent thermal protection, a firm snow pack, and significantly reduced winds, and is a preferred and nutritious winter browse (Ozoga 1995). Other coniferous types that are used as winter yards include spruce (*Picea* spp.), hemlock (*Tsuga canadensis*), and balsam fir (*Abies balsamea*).

Although most winter yards tend to be relatively small (20–40 ha) in areas of relatively light snowfall, much larger yards are necessary in areas receiving heavy winter snowfall. For example, the 940-km^2 Mead Deer Yard in Upper Michigan serves as winter cover for deer occupying 3600 km^2 of summer range. This single deer yard supported an estimated 43,000 deer during the winter of 1987 (Ozoga 1995). Unfortunately, the thick coniferous cover combined with heavy deer use can limit available browse in deer yards. Poor deer population and habitat management coupled with prolonged winter conditions can make winter yards death traps. During the severe winter of

TABLE 44.2. Reported home range size for white-tailed deer by sex and age class in different locations

Location	Animal (Sex/Age Class)	No. Deer	Home Range (ha)	Reference
Florida Everglades	Male	10	700 ± 140 *SE*	Sargent 1992
		23	290 ± 40 *SE*	
Florida	Female (yearling)	7	2458	Labisky et al. 1991
	Female (adult)	6	344	
	Male	5	701	
Mississippi	Female	4	737 ± 219 *SD*	Mott et al. 1985
	Male	5	1511 ± 571 *SD*	
Texas	Female	27	84 ± 8 *SE*	Inglis et al. 1979
		14	139 ± 37 *SE*	
Georgia	Female (summer)	21	140.9 ± 16.2 *SE*	Rogers 1996
	Female (winter)	20	167.4 ± 18.5 *SE*	
	Male (summer)	15	234.7 ± 44.1 *SE*	
	Male (winter)	13	206.1 ± 27.8 *SE*	
Kentucky	Female	6	642 ± 132 *SE*	Pais et al. 1991
Illinois	Female	7	50.8 ± 23 *SE*	Cornicelli et al. 1996
Nebraska	Female	14	170 (CI = 38)	VerCauteren and Hygnstrom 1998
Washington	Female	18	103.6 ± 16.5 *SE*	Gavin et al. 1984
	Male	7	208.6 ± 24.6 *SE*	
Montana	Female (adult)	11	71 ± 18.4 *SE*	Leach and Edge 1994
	Female (yearling)	5	91 ± 30.4 *SE*	
Michigan	Female	—	45	Beier and McCullough 1990
	Male	—	142	
New York	Female (summer)	64	221 ± 19.0 *SE*	Tierson et al. 1985
	Female (winter)	45	132 ± 18.3 *SE*	
	Male (summer)	34	233 ± 23.4 *SE*	
	Male (winter)	12	150 ± 31.6 *SE*	
Connecticut	Female (suburban)	9	158	Swihart et al. 1995
	Female (urban)	12	67	
Connecticut	Female	25	43.2 ± 2.7 *SE*	Kilpatrick and Spohr 2000
Wisconsin	Female	15	178 ± 102 *SD*	Larson et al. 1978

1985–1986, an estimated 11,000 deer died in the Mead Deer Yard (Ozoga 1995).

Travel within yards often is confined to frequently used trails, which minimizes energy expenditure. Daily movements covering <0.4 ha occur in winter yards during severe weather.

As soon as snowmelt allows ease of travel, deer leave yarding areas for summer ranges, although some individuals may linger on the winter range before migrating to summer habitats (Nelson and Mech 1986). As with the fall migration, fawns must be led back to summer ranges by their mother or other adult females. Fawns that have been orphaned over winter may wander during spring and summer (Nelson and Mech 1986).

Although migrations in response to winter weather are pronounced on northern ranges, seasonal shifts in activity also have been reported in other regions. In Illinois, 19.6% of adult does on a study area during winter migrated an average of 13 km away in spring and returned in fall (Nixon et al. 1991a). Similarly, in a Nebraska study, 13% of a marked population of does migrated seasonally (VerCauteren 1998). Seasonal shifts also can occur in response to habitat conditions and hunting pressure (Kammermeyer and Marchinton 1976) or seasonal flooding (Joanen et al. 1985; Morrison 1985).

Predator Avoidance. As a prey species, whitetails use various behaviors to avoid predation, ranging from remaining motionless to group cohesion and flight behavior (Smith 1991). In response to perceived danger, very young fawns may have dramatically reduced heart rates (alarm bradycardia; Jacobson 1979) and reduced movements and sounds. As fawns mature and begin accompanying their mothers, flight becomes the primary escape mechanism.

When whitetails detect danger or potential danger, they use a series of visual and auditory displays. The most conspicuous of these is the erection of the tail exposing the white underside, from which the species gets its name. When an unknown object is encountered, deer will assume an alert posture with the head held erect and the ears directed toward the focal object. The tail may be held partially or fully erect to alert other members of the social group. An alerted whitetail often will stamp one or alternating front feet. This foot stamp may serve to further warn other group members, urge the potential predator to move and reveal itself, or release scent from the interdigital gland. Accompanying the foot stamp is the snort, an explosive sound resulting from air being forcibly passed through the nasal passages.

If the unknown object is not identified, deer may either retreat with a bounding, tail-wagging gait or further investigate the potential danger. Whitetails apparently have poor depth perception, and typically will move parallel or toward an object, often moving their head from side to side to get a glimpse from another angle.

Unless greatly alarmed, whitetails expose their tail underside when fleeing, and bucks and does are equally likely to flag when alarmed. Hirth and McCullough (1977) reported that buck groups flagged on 91% of encounters with humans, whereas doe groups flagged on 95% of encounters. Tail flagging would seem to be a maladaptive trait. However, Hirth and McCullough (1977) suggested a logical explanation of the causes and consequences of tail flagging. Among ungulates, many social species in open habitats have evolved conspicuous white rump patches as a means of maintaining group cohesiveness. In contrast, solitary species that live in dense vegetation lack a conspicuous rump patch. Whitetails can be viewed as intermediate between these two extremes; they are a social animal that typically lives in dense vegetation. Therefore, they have evolved an "on-again, off-again" rump patch. At normal times, the patch is concealed and deer can rely on their cryptic coloration to help avoid detection by predators. However, when danger is encountered, the tail patch can be "turned on" to help maintain group cohesiveness while fleeing.

Social groups of bucks or does may react differently to potential predators, depending on the circumstances of the encounter. If danger is detected early and escape is assured, does typically will snort an alarm to warn other members of the social group about the possible danger. Because does typically are members of matriarchal groups, warning others would help ensure the survival of related individuals, thereby maximizing their inclusive fitness. In contrast, bachelor groups of bucks are composed of unrelated individuals. In low-risk encounters, bucks may retreat without warning others within the group because there is no advantage to such a warning.

If the danger is more imminent and escape certainly would be detected, both buck and doe groups may flee with tail-wagging and snorting. Tail wagging helps maintain group cohesion, thereby reducing an individual's chance of falling prey. In addition, the combined senses of several animals make a group more difficult to approach than a solitary animal.

Reproductive Behavior: The Rut. Because social groups are small and often transitory, deer use olfactory signposts to communicate throughout the year. During spring and summer, bucks communicate by depositing scent at communal licking branches located near trails or at field borders. Following velvet shedding and throughout the breeding season, bucks make "antler rubs" by rubbing small trees and shrubs with their antlers and forehead region. The forehead skin contains glands that become more active during the breeding season and presumably are the source of an odor of communicative significance (Atkeson and Marchinton 1982). Other bucks and does often investigate antler rubs (<3 days old) by smelling, licking, or rubbing them with their foreheads (Sawyer et al. 1989).

Scrapes are complex signposts composed of three consecutive behaviors: marking an overhanging branch, pawing a shallow depression, and rub-urinating over the pawed depression. Scraping activity typically peaks 2 weeks before peak breeding occurs (Miller and Marchinton 1999). Scraping and visits to scrapes decline dramatically during the peak of rutting, although activity may increase again following the rut.

Scrape sites may be defended in the presence of a dominant male, although they may be visited and re-marked by several males during the breeding season. Females also visit and investigate scrape sites throughout the breeding season. Most visits to scrapes by bucks and does appear to be nocturnal.

Bucks find females in estrus via visual and olfactory cues. Other than an increased tolerance of a buck's approaches, females are not known to display any overt visual signals to the male indicating their reproductive status and willingness to breed. Volatile compounds produced in the female's reproductive tract apparently contain a sexually attractive odor (Whitney et al. 1992). Urinary compounds of low volatility likely are analyzed by the vomeronasal system during a behavior called flehmen, although vomerofaction does not appear to signal reproductive attractiveness. Anosmic males can detect estrous females by behavioral cues, whereas males with occluded vomeronasal organs are attracted to estrous females even in the absence of behavioral cues (Gassett et al. 1998). Vomerofaction may mediate physiological changes that synchronize male and female reproductive condition, although this concept is speculative.

White-tailed deer exhibit a tending-bond mating system where males pursue and court individual females. Does apparently produce sexually attractive odors well before they are willing to stand for copulation. A tending buck may test a female's receptiveness with a courtship approach. At close range, a buck follows a doe with his neck extended and lowered and his chin slightly elevated. Decreased avoidance of a male's approaches indicates a willingness to stand for copulation. Intromission is brief and terminates with a single hard thrust by the buck, which may knock the female from underneath him. Captive white-tailed does bred to fertile bucks exhibited estrous behavior for around 24 hr (Verme 1965; Warren et al. 1978). However, White et al. (1995) found estrous duration was extended in does not allowed to copulate with males.

As noted (Gestation), fawns are born following approximately 200 days gestation. Before parturition, does establish small, exclusive fawning territories (Ozoga and Verme 1982) with the dominant females occupying the choice areas. Fawns are born front feet and head first while the doe is either standing or prostrate. Siblings follow the birth of the first fawn by about 15–20 min. Does immediately clean the amniotic membranes from the fawns and lick the fawns dry. The first nursing bout occurs almost immediately.

Newborn fawns spend most of their time bedded. The doe returns every few hours to nurse and groom the fawn, and then move it to a new location. By 2–3 weeks of age, fawns begin experimenting with forages and by 4 months they are totally weaned.

MORTALITY

Longevity. Survival of white-tailed deer is influenced by various mortality factors. In hunted populations, turnover rates can be exceptional, particularly among males. In many populations, a male older than $3\frac{1}{2}$ years is uncommon.

The normal life span of females usually is much longer than that of males. The rut is particularly stressful for bucks because of the large amounts of energy expended and the subsequent loss of body reserves (Warren et al. 1981). Moen (1994) suggested males were much more susceptible to winter stress after the rut than females. These stressors may lead to higher mortality rates in males.

Extreme records of longevity include accounts of captive does living to ages of 17–19 years (Rue 1978). However, it is unlikely that does would live to these ages in wild populations, because of excessive tooth wear. Masters and Mathews (1990) recorded white-tailed does surviving beyond 9 years and remaining reproductively active in an unhunted population. Ozoga (1969) reported age records of 9–15 years for 14 wild female deer in an area with low antlerless deer harvest. Does on a refuge surrounded by hunted lands in Illinois survived a mean of 6.6 years (Nixon et al. 1991b).

The Florida Key deer (*O. v. clavium*) was listed as endangered in 1967 and survival and longevity has been extensively monitored. Lopez et al. (2000) reported a maximum age for female Key deer of 19 years (mean life expectancy 6.2 years; $n = 34$). Male Key deer have a lower life expectancy (11 years maximum; mean = 3.0 years; $n = 43$).

Mortality. Understanding rates and causes of mortality can provide important input into the selection of harvest management goals for deer populations and can improve the management of related natural resources. Mortality rates due to hunting or other factors can vary greatly among populations (Table 44.3), and the success of management activities may be affected by these factors. For example, the high rate of natural mortality in South Texas makes management for mature-aged males an inefficient harvest goal (DeYoung 1990). In contrast, Coggin (1998) reported that natural mortality rates for adult males (>2.5 years old) in a Mississippi study area ranged from 5% to 15%. Clearly, management strategies designed to produce mature males must be cognizant of nonharvest mortality rates so that realistic expectations can be established.

Harvest. Population size will change if mortality and recruitment are not balanced over time. Legal and illegal harvest along with wounding loss can be significant sources of mortality. Because mortality rate can be manipulated by varying hunting pressure, regulation of harvest is the principal tool for manipulating deer population size, sex ratio, and age structure.

Legal harvest is a major factor affecting population size and remains the only practical option available to limit population size in most management scenarios. In New Brunswick, annual survival rates of adult white-tailed deer in exploited populations were dependent on legal harvest rates (Whitlaw et al. 1998). In Minnesota, the legal harvest mortality rate was 34% and 28% for yearling and adult males, respectively, and accounted for more than 50% of all mortality (Nelson and Mech 1986). Harvest accounted for 26 of 33 deaths of yearling males in Maryland (Rosenberry 1999). In eastern Montana, harvest accounted for 81% of mortality among deer >6 months old. In this case, harvest was particularly important to population regulation; winter forage supplies did not affect demographic rates because this intensively farmed alluvial plain with an undisturbed riparian forest–shrub complex provided a stable habitat (Dusek et al. 1989).

Wounding loss is difficult to estimate due to problems locating dead animals, and there is great variation in wounding rates among

TABLE 44.3. Reported mortality factors for adult (≥2.5 year old) white-tailed deer in selected areas

Area	Mortality Factor (% of total mortality)					Total Annual Mortality Rate
	Legal Hunting	Unrecovered Kills	Poaching	Predation	Other	
Central Minnesota						
Bucks	61	10	9	12	9	0.54
Does	49	5	18	24	5	0.31
Lower Yellowstone River, Montana						
Bucks	91	—	1	9	—	0.43
Does	77	—	—	22	—	0.27
Central Illinois						
Bucks	48	17	13	—	22	0.61
Does	34	25	16	—	25	0.32

SOURCE: Adapted from Jacobson and Guynn (1995) and using data from Dusek et al. (1989), Fuller (1990), and Nixon et al. (1991).

harvest methods. Archers lost 7% of the deer they killed in New York (Severinghaus 1963) and 50% in Georgia (Downing 1971). Crippling loss by hunters using shotguns varied from 7% for hunters using slugs to 36% for hunters using buckshot (Downing 1971). In Mississippi, 8% of radio-collared males shot legally by hunters were not recovered (Coggin 1998).

Illegal harvest is difficult to estimate accurately due to the cryptic nature of the participants. Many state agencies consider poaching a major mortality factor, which may occasionally approach the magnitude of legal harvest (Gladfelter 1984). Annual mortality due to poaching was 4–5% for yearling deer in Minnesota (Nelson and Mech 1986). In Mississippi, 13% of radio-collared males were shot by poachers (Coggin 1998).

Predation. Body size and the extensive distribution of white-tailed deer make them available and susceptible to the full North American complement of large predators. Susceptibility to predation is affected by physiological and environmental factors. In the arid South Texas plains, coyote (*Canis latrans*) predation rates are influenced by range condition and drought as they affect hiding cover for fawns and abundance of alternative prey populations (Carroll and Brown 1977). Coyote predation of adult males is significant during the months immediately following the breeding season (DeYoung 1989). Deep snow and cold temperatures predispose animals to wolf (*Canis lupus*) predation. Adult male deer may be more susceptible than younger males, probably because they depleted fat reserves during rut (Nelson and Mech 1986). Deep snow and cold temperatures also may increase deer vulnerability to coyote predation (Whitlaw et al. 1998).

White-tailed deer are a significant component of the prey base of wolves wherever they coexist. Wolf numbers are related to ungulate biomass, and where deer are the primary prey, wolf territory size is related to deer density. Winter consumption in Minnesota averaged 2.0 kg deer/wolf/day (Fuller 1989). A wolf population in Ontario responded primarily to deer availability, even though moose (*Alces alces*) were availabile and were a major food item (Forbes and Theberge 1996). A wolf population in Minnesota changed behavior in response to snow-induced changes in deer distribution and mobility (Fuller 1991b). Wolves appeared to negatively affect legal harvest of male white-tailed deer in some of Minnesota's poor deer habitats, but not in better habitats (Mech and Nelson 2000). On this study area, wolves killed about 20% and hunters about 30% of the legal male deer.

The omnivorous feeding habits of coyotes make them less dependent on white-tailed deer as a primary prey item, although they clearly consume deer during some seasons. Coyotes affected deer numbers in Quebec; they preyed primarily on fawns and older deer, but not necessarily deer in poorer physical condition (Messier et al. 1986). Coyotes and domestic dogs accounted for 69% of natural fawn mortality in Illinois (Nelson and Woolf 1987). Predators, primarily coyotes, were responsible for 88% of fawn losses in Oklahoma (Bartush and Lewis 1981). Population manipulation studies to reduce or exclude coyotes significantly increase fawn production and winter survival (Kie et al. 1979; Messier et al. 1987). Coyotes may have replaced gray wolves as predators of white-tailed deer in northeastern North America (Ballard et al. 1999).

Cougar (*Puma concolor*), bobcat (*Lynx rufus*), and black bear (*Ursus americanus*) predation on white-tailed deer can be a significant source of mortality under some circumstances. Whitetails were the most important prey species for Florida panther based on kill and scat analyses (Dalrymple and Bass 1996). In Montana, 87% of cougar kills consisted of white-tailed deer (Kunkel et al. 1999). Bobcat predation appeared to be the principal factor limiting the abundance of deer in Everglades National Park (Labisky et al. 1995). Black bear accounted for 49% of fawn mortality in Minnesota (Kunkel and Mech 1994).

Diseases and Parasites. White-tailed deer host >100 species of internal and external parasites (Samuel 1994). Greatest parasite diversity tends to occur on southern ranges, although high parasite burdens commonly are associated with nutritionally stressed herds. The number of species of protozoa, flukes, tapeworms, roundworms, and arthropods found on whitetails has ranged from 23 in Alberta to 70 in Florida (Samuel 1994). Parasite burdens sufficient to cause direct or indirect pathology can occur at high deer densities, such as with the lungworm (*Dictyocaulus viviparous*)–pneumonia complex in the southern United States (Davidson and Doster 1997). The large stomach worm (*Haemonchus cortortus*) also can have high prevalence in the southern United States, with fawns more susceptible than adults. Other important helminth parasites include the liver flukes (*Fascioloides magna* and *F. hepatica*), the arterial worm (*Elaeophora schneideri*), the muscle worm (*Parelaphostrongylus andersoni*), and stomach worms (*Ostertagia* spp. and *Trichostrongylus* spp.) The meningeal worm (*Parelaphostrongylus tenuis*) does not cause significant pathology in its normal host, white-tailed deer, but causes a fatal neurological disease in other ungulates that share range with infected whitetails (Lankester 2001). The abdominal worm (*Setaria yehi*) is a large nematode commonly found in the abdominal cavity of deer from the southeastern United States.

Approximately 20 species of ticks have been documented on whitetails and likely are the most important external parasites of deer. Tick infestations are rarely a cause of mortality, although significant fawn mortality has been attributed to lone star tick (*Amblyomma americanus*) infestations. Ticks are important, however, because they are the vectors of several pathological agents of human health concern. The lone star tick is the suspected carrier of human granulocytic ehrlichiosis, whereas the deer tick (*Ixodes scapularis*) is the vector of the Lyme disease spirochete.

Nasal bot (*Cephenemyia* spp.) larvae are commonly found in the retropharyngeal pouches and nasal passages of deer. Infestation rates vary across the country, but typically 10–30% of deer will host bot fly larvae, with slightly higher incidence during winter. The short-lived adult flies mate and the female deposits small larvae in the nostrils of deer, which then migrate to the nasal passages. Mature larvae are expelled from the nares by sneezing and soon pupate. Deer apparently tolerate nasal bot infestations without serious deleterious effects. However, deer appear cognizant of egg-laying attacks by adult flies and may panic in attempts to escape these persistent parasites.

Four species of deer keds (*Lipoptena* spp. and *Neolipoptena* spp.) have been reported from white-tailed deer. Hunters often confuse these flattened, wingless flies with ticks. Deer keds do not attach to the hosts as do ticks, and are very mobile on the deer's body. Highest concentrations occur in the groin and belly region.

Several species of lice, including sucking lice (*Solenopotes ferrisi* and *S. binipilosus*) and chewing lice (*Tricholiperus lipeuroides* and *T. parallelus*), occur on whitetails. The blood-feeding sucking lice are most prevalent during the winter, and greatest infestations are observed on unthrifty or debilitated animals. Chewing lice feed on epidermal scales, skin exudation, and other matter on the host skin. They also are most common during winter. Although it is unlikely that high infestations by either group of lice are debilitating, they likely are a nuisance by causing skin irritation and hair loss.

A variety of biting flies (Tabanidae), mosquitoes (Culicidae), black flies (Simuliidae), and midges (Ceratopogonidae) torment deer throughout the summer months. Although most commonly a nuisance, mass attacks certainly can affect deer physically and behaviorally. These biting flies also are vectors or intermediate hosts of a variety of diseases and parasites such as epizootic hemorrhagic disease and the arterial worm.

Whitetails are susceptible to a variety of diseases, although population-level impacts in most cases are limited. However, when deer are stressed by other problems, such as severe winter weather or malnutrition, normally benign diseases may become pathogenic (Samuel 1994). The expansion of deer farming and ranching during the 1980s and 1990s fueled concern that disease outbreaks would result from commercial translocation and intensive production of cervids (Samuel and Demarais 1993). Several infectious diseases found in white-tailed deer could have significant management implications.

Epizootic hemorrhagic disease (EHD) and blue tongue (BT) are closely related viral diseases that have been recognized for decades to cause significant mortality in southern populations annually. The disease (EHD/BT) was first reported during a 1955 epizootic in New Jersey. Five serotypes of BT and two serotypes of EHD occur in the United States. Mortality rates typically are <15%, but mortality rates >50% have occurred (Davidson and Doster 1997). Although prevalence is greater at southern latitudes, disease severity increases with latitude. Die-offs due to EHD have occurred as far north as Alberta and British Columbia (Trainer and Karstad 1970).

Most cases of EHD/BT occur during late summer or early fall coincident with peak populations of *Culicoides* spp., which serve as the vector. Genetics may influence susceptibility to hemorrhagic disease, as captive deer of northern lineage were less resistant than deer with southern origins (Jacobson and Lukefahr 1998).

Outbreaks of bovine tuberculosis identified within commercial, captive cervid herds in eight states and four provinces during 1990–1992 (Miller and Thorne 1993) and within free-ranging white-tailed deer in Michigan during 1994 (Schmitt et al. 1997) are of significant concern to wildlife agencies and agricultural entities. In Michigan, it is believed that winter supplemental feeding and the increased focal densities promote disease transfer (Schmitt et al. 1997). Bovine tuberculosis does not appear to be a limiting disease in deer populations, but wildlife could serve as a reservoir for infection of domestic ruminants, with significant implications (Davidson and Nettles 1997).

Chronic wasting disease or transmissible spongiform encephalopathy has been documented in captive and free-ranging elk (*Cervus elaphus*), mule deer, and white-tailed deer in Colorado and Wyoming and several commercial, captive cervid herds in the United States and Canada (Miller et al. 2000; Williams et al. 2001). Relatively little is known about the epidemiology and the impact of this disease in free-ranging deer populations. Concerns and perceptions of management agencies and the public about human health risks associated with the transmissible encephalopathy diseases may ultimately influence population management decisions in endemic areas (Williams et al. 2001).

A commonly reported, but usually benign disease of white-tailed deer is infectious cutaneous fibroma. Also called warts or papilloma, the widespread occurrence of these cutaneous fibromas suggests that the virus is enzootic over much of the range of the whitetail. Fibromas are either warty or smooth, raised, pigmented, hairless growths, which may occur singly or in groups. They typically are external; deeper tissues are rarely invaded by the tumors. On most deer, fibromas naturally regress through dessication, contraction, and finally sloughing of the lesion.

Accidents. Collisions with vehicles are a significant mortality factor for white-tailed deer in many regions, and a major human safety concern. The number of deer road-kills in the United States has been estimated to range from 500,000 (Romin and Bissonette 1996) to 1.5 million annually (Conover et al. 1995). Most accidents occur at dawn, dusk, or after dark, and seasonal peaks coincide with rutting activity and hunter disturbance (Allen and McCullough 1976). Property damage to vehicles along with human injuries and fatalities resulting from these collisions can be significant. With average damage estimates ranging from $1200 to $1881/accident, total annual economic loss may range from $600 million to >$2 billion annually. In addition, more than 100 people are killed annually in animal–vehicle collisions, with the majority involving white-tailed deer. Hansen (1983) in Michigan and Stoll et al. (1985) in Ohio reported that 4–5% of deer–vehicle collisions resulted in human injury. If these statistics are representative nationwide, deer–vehicle accidents annually may result in 20,000 or more human injuries.

Weather and Malnutrition. Starvation and chronic malnutrition can cause mortality in many regions where population density exceeds habitat carrying capacity. Malnutrition has greater impacts on population demographics when associated with high density, diseases and parasites, and extreme winter weather. Starvation caused significant natural mortality in Texas, ranging from 4% to 52%, during periods of drought and excessive deer densities (Teer 1984). Mortality due to undernutrition in northern regions is influenced by population density relative to carrying capacity, snow depth, and duration of snow cover (Severinghaus 1972; Boyer et al. 1986). During a severe winter in southeastern Canada, the mortality rate reached 40% for the population and a loss of 50% of the fawns (Huot et al. 1984). A winter severity index that incorporates wind chill, snow depth, and the ability of the snow to support the body weight of a deer is highly correlated with winter mortality (Verme 1968). Severity of weather during December affects fawn survival over winter, and extension of winter conditions into April negatively affects neonate fawn survival in Michigan (Verme 1977).

POPULATION ASSESSMENT

Age Estimation. The tooth eruption and wear pattern described by Severinghaus (1949) and in standard wildlife management techniques manuals (e.g., Dimmick and Pelton 1994) is used extensively to estimate age of harvested whitetails throughout their range (Fig. 44.4). At birth, incisors typically are erupted, and by 4 weeks of age all incisiform teeth (three incisors, one canine) as well as two premolars are fully erupted. The last premolar erupts at approximately 10 weeks of age. Fawns during the hunting season (6–7 months) are distinguished by eruption of permanent incisors and presence of one permanent molar. The second permanent molar erupts at 13 months of age. Yearlings (1.5 years old) are characterized by a three-cusped deciduous premolar, which is replaced by the permanent premolar at roughly 1 year and 7 months; a permanent third molar is completing eruption.

Relative wear patterns of the permanent teeth are used to estimate age for animals older than yearlings. Eruption and wear criteria have been developed for differentiating year classes beyond 2.5 years old; however the accuracy of these techniques has been questioned (Hamlin et al. 2000). Nevertheless, tooth wear and replacement remains the most common age determination method because it is rapid and because highly accurate age determination for animals >2.5 years old is not necessary for most management decisions.

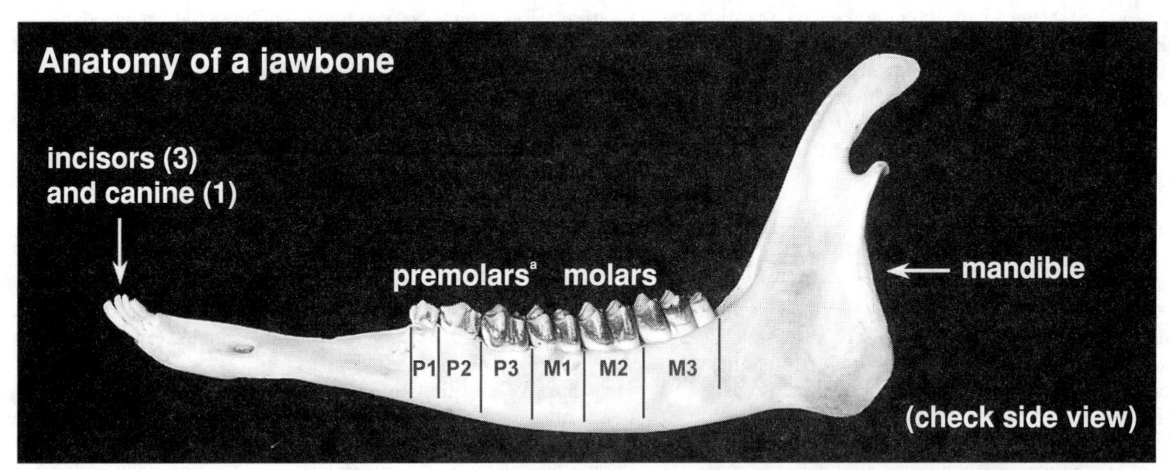

1) P3 is a 3-cusped temporary tooth[b]
2) M1 and/or M2 are at some stage of eruption
3) M3 has not begun to erupt

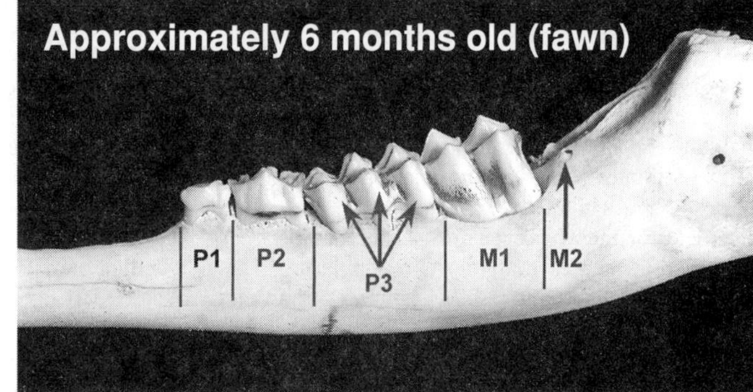

1) P3 is usually a 3-cusped temporary tooth (can be a 2-cusped permanent tooth in older yearlings).
2) M3 is usually not fully erupted (gum line is usually high on the rear cusp).

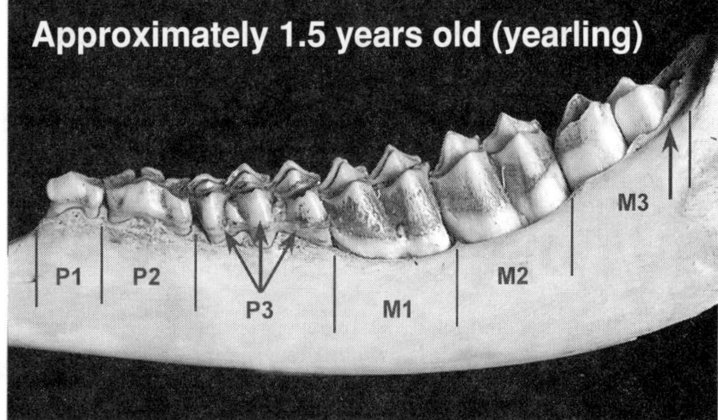

1) All permanent premolars and molars are fully erupted and showing varying degrees of wear[c]
2) P3 is 2-cusped

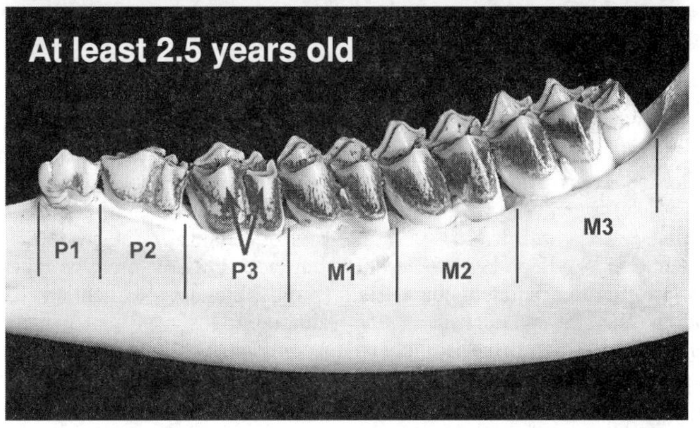

[a] Most scientific sources refer to the premolars as P2, P3, and P4 due to the evolutionary loss of the first premolar

[b] Young white-tailed deer have a set of temporary premolars which are replaced with permanent premolars. Molars are always permanent.

[c] Specific age class estimations for 2.5 years old and beyond are usually based on molar and premolar wear

FIGURE 44.4. Age estimation of white-tailed deer by tooth eruption and replacement. SOURCE: Adapted from Gee et al. (1991).

Counts of cementum annuli for year-class assignment are more accurate than tooth wear and replacement, although accuracy may differ with latitude (Hackett et al. 1979; Hamlin et al. 2000). Cementum annuli counts are used less frequently for management purposes because they require laboratory preparation and are much more costly than tooth wear evaluation. Clearly, the selection of an age determination method should include consideration of desired accuracy level above 2.5 years, comparative cost, and geographic location.

Population Estimation. Reliable estimates of population parameters are needed to understand the status of the population relative to management objectives and evaluate the effectiveness of management actions. Population parameters of interest include adult sex ratio, age structure, age-specific fecundity, and an estimate of density or at least relative abundance. Evaluations of methods to estimate or index population parameters have been attempted with varying levels of success because the actual population parameters were unknown, there were no benchmarks, and the researchers were restricted to comparing techniques with each other. For a more complete review of population parameters and their estimation, see Lancia et al. (2000).

Selection of the most appropriate deer survey method must consider the level of information needed (e.g., density vs. relative abundance vs. index to relative abundance), availability of resources to collect and interpret the data, harvest-based considerations that affect sampling bias, and habitat- and animal-based considerations that affect observation probability. Estimates of parameter variation should be incorporated whenever possible. As with any estimate based on sampling protocol, primary consideration must be given to ensuring that the areas or animals that are sampled are representative of the population. This consideration applies equally to direct and indirect estimates of animal abundance within their habitat and estimates based on harvest statistics. There is not a single technique to estimate population parameters accurately and precisely. A suite of techniques should be selected that produce the information needed for management decisions regarding the deer population and the habitat. More complete descriptions of factors to consider when selecting a technique are provided in Caughley (1977) and Seber (1982).

Efforts to develop methods to monitor deer populations began simultaneous with restoration efforts. Early deer survey techniques included direct estimates using drive counts and indirect estimates using counts of pellets and tracks (Dice 1941). Line-transect sampling based on King's strip method was used for deer (Haync 1949) and later improved significantly to incorporate sighting probability functions (Burnham et al. 1980). The pellet-group count technique has been used widely for estimating abundance and relative abundance, but is sensitive to variation in defecation rates, deterioration of pellet groups, and observer bias (Neff 1968; Fuller 1991a). Deer track counts are limited as an estimator of abundance, but may accurately reflect population changes (Harlow and Downing 1967). In a Texas study, use of track-count estimates failed to detect a population increase on one area, but did reflect a large difference in density between two areas (DeYoung et al. 1988). Night counts with spotlights are attractive to managers because they allow visual contact with individual deer. However, they are sensitive to weather, habitat-based biases, and animal behavior-based variation (Progulske and Duerre 1964; Gunson 1979). Line-transect sampling methodology has been used with fixed-wing and helicopter aerial counts. Accuracy varies with habitat types and environmental conditions (Beasom et al. 1986; Leon 1986). Thermal infrared sensing systems have the resolution to accurately count and identify age and sex of deer populations (Wiggers and Beckerman 1993), although costs are much higher than methods such as the spotlight count (Naugle et al. 1996). Infrared-triggered cameras set on a sampling grid over bait can be used to estimate density, sex ratio, and age structure (Jacobson et al. 1997).

Population parameters can be estimated using results of removal or mark–recapture of deer. Removal methods of population estimation are appealing because animals removed by harvest are manipulated in conjunction with other data to produce the population estimate (Lancia et al. 2000). Harvest estimates alone are not an accurate estimator of density because of the influence of hunter effort and success. Removal methods are characterized by whether the removals are selective (i.e., change-in-ratio techniques) or nonselective (i.e., catch-per-unit effort). Mark–recapture methods are based on Lincoln–Petersen methods. A review of removal and mark–recapture methodology is provided in Lancia et al. (2000).

White-tailed deer populations can affect the prevalence and abundance of plant species within their habitat. As such, examination of the relative browse pressure by deer on the more preferred species within the plant community provides an index to deer population density. An intensive browse use survey can be used to estimate biomass removed by deer, although for management purposes a less intensive estimate of percentage of twigs browsed can provide valuable trend information.

Health and Condition Indicators. Successful deer management depends on accurately assessing herd health and body conditions, which depend on habitat quality and deer density. Blood chemistry parameters often are unreliable indicators of health and body condition because variations occur with handling procedures, season, nutritional state, and reproductive status (Wesson et al. 1979a, 1979b; Warren et al. 1981; Waid and Warren 1984). Seal et al. (1981) and Franzmann (1985) reviewed reference values for blood chemistry and hematology of white-tails under a variety of conditions. Sams et al. (1998) found that ratios of serum albumin concentration or blood urea nitrogen to creatinine provided a sensitive indicator of nutritional stress. Blood urea nitrogen concentration is the most widely used blood parameter because it is relatively unaffected by handling or immobilization drugs and is a good indicator of protein nutrition provided energy intake is constant (Wesson et al. 1979a, 1979b). Blood indices for evaluating energy intake have not proven consistent enough to be reliable across a range of energy intake levels (Harder and Kirkpatrick 1994).

Urinary urea nitrogen, when expressed as a ratio to creatinine levels, can provide an indicator of nutritional status. This technique has the added advantage that samples may be obtained from urine deposited on snow without animal handling (DelGuidice 1995). However, utility of this technique may be limited because identification of the individual or even the sex or age of the individual sampled may be unknown.

Eve and Kellogg (1977) reported that the number of abomasal parasites (stomach worms) is directly related to deer density. Population status relative to carrying capacity can be inferred by collecting a sample of deer from an area and counting stomach worms in a sample from the abomasums. These abomasal parasites counts are used widely in the southeastern United States to evaluate herd condition.

Carcass and organ mass, and fat indices also are used extensively to monitor herd health (Kie et al. 1983; Sams et al. 1998). Although various fat indices have been proposed to assess body condition, the most commonly used are the femur (or mandibular) fat index and the kidney fat index. Marrow fat is among the last sites of fat depletion in nutritionally stressed animals, and thus reduced marrow fat in the femur or the mandible is indicative of severe malnutrition.

Various methods of assessing fat deposition surrounding the kidneys have been proposed, but all can provide a useful indicator of lower levels of nutritional stress. An index of kidney weight to attached perirenal fat may assess early stages of malnutrition (Warren and Kirkpatrick 1982). Kidney fat reserves appear more sensitive to initial fat metabolism than femur or mandibular marrow fat. However, care must be taken to consider sex and age of the animal and season of collection (Johns et al. 1984).

Changes in weight of various glands such as the thymus, adrenal, and thyroid also are correlated with changes in animal condition. All management programs should measure whole or dressed carcass weights (particularly of fawns and yearlings) as an indicator of changes in herd condition. Weights should be recorded by season, age, and sex of the animal. Examination of year-to-year changes in body mass records can reveal population trends as a result of habitat or herd population management. However, as with other condition indicators, mass data can be biased depending on when the animal is harvested. In areas with short hunting seasons, interpretation of mass data presents little problem. However, in some areas of the southern United States where seasons may extend over several months, mass data must be grouped according to when the animals were harvested.

Because antler size within a particular age class is an excellent reflection of the nutritional status of the animal, measurements of antler development can provide an excellent indicator of herd health conditions. Yearling antler beam diameter is a good predictor of range conditions and can be used to estimate conception rates of females (Moen 1994). Similarly, antler measurements such as number of antler points, main beam length, and greatest antler spread sometimes can be predictors of buck age, at least until 3.5 years of age. However, age–antler size relationships vary according to range conditions and must be established for each location.

POPULATION ECOLOGY AND MANAGEMENT

Deer populations are components of larger dynamic systems, which consist of interacting biotic communities, humans, and the physical world (Fryxell and Sinclair 2000). Although researchers should not overly emphasize the importance of individual regulatory factors within the long-term dynamics of deer populations, practical management efforts usually involve emphasis on certain parameters. The effectiveness of management cannot be assessed without reliable data on population parameters. Data on birth, death, immigration, and emigration rates are valuable because these parameters directly affect population size, and changes in population size typically are the "currency" by which management decisions are evaluated (Lancia et al. 2000).

Historic Trends. Since human colonization of North America, whitetails have been inexorably linked with human exploitation. Native Americans relied heavily on the white-tailed deer, as seen by their abundant remains in archaeological sites. McCabe and McCabe (1997) argued that 3.5 adult whitetail hides/person/year were required for clothing, and that consumption rates were 9.5 whitetails/person/year. In addition, whitetails supplied antlers for tools, arrow points, and decorations, fat for soap and lamplight, and hides for blankets and currency.

Based on the historical range of the species, estimates of human consumption rates, and assumptions about nonhuman predation rates, McCabe and McCabe (1984) estimated that before European colonization of North America, whitetail populations ranged from 23.6 to 32.8 million animals, or roughly 3.1–4.2 deer/km^2 across the whitetail's range. These estimates fall well within the range earlier suggested by Seton (1929) of 20–40 million animals. Clearly, however, deer densities varied considerably across North America and were dependent on habitat conditions, climate, and Native American population distribution.

European colonization sparked a shift from subsistence hunting to market hunting by Native Americans and European colonists. By 1800, market hunting and habitat alteration had reduced white-tailed deer populations to approximately half of precolonial estimates. McCabe and McCabe (1984) suggested that populations rebounded briefly during the early 1800s as a result of increased colonization of the continent's interior and subsequent reduced Native American influence. This respite was short-lived, however, and the period from 1850 to around 1900 saw deer populations reduced to record low levels by market hunting, deforestation, fire, and other factors. By 1900, there were only approximately 350,000 animals remaining in isolated areas.

Deer restoration programs began in the late 1800s, but few states had organized efforts or the requisite funding for these efforts to be effectual. Pennsylvania was an exception, and by 1925 Pennsylvania's deer population had grown to the point that liberal antlerless harvests were required to control massive starvation. Major restoration efforts in most states did not begin until passage of the Pittman–Robertson Act in 1937 provided the funding mechanisms to allow larger, organized restoration efforts. Deer populations increased to an estimated 6 million animals by 1948 (Bartlett 1949) and to 15 million by 1978 (Hesselton and Hesselton 1982). Jacobson and Kroll (1994) provided a population estimate of 26 million animals in the United States in 1993 and voiced concerns that deer populations in many areas had grown to levels that were dangerously close to, or exceeding, the carrying capacity of the habitat. Ten states reported populations in excess of 1 million animals, and certainly additional states have been added to these rolls since 1993. Nationwide, >5 million animals, or roughly 20% of the estimated population, are harvested annually. Nevertheless, populations in many states continue to increase.

Harvest Management. Deer population dynamics are driven by reproduction, mortality, dispersal, and movement patterns. Therefore, deer management typically involves the manipulation of mortality and natality to maintain the population within the capacity of its habitat. Mortality is manipulated via harvest regimes that are dependant on management goals, population size, and habitat productivity.

Important reproductive parameters including timing of the breeding season, fecundity rates (number of young/female), fertility rates (percentage of females breeding), age at first breeding, and sex ratio of offspring. These, in turn, are affected by population density, habitat conditions, and perhaps genetics. On good range, does ≥2.5 years of age have pregnancy rates of >90% and reproductive rates of 1.5–2.0 fetuses/doe. Examples of reported reproductive rates (fetuses/doe) include 1.61 in Mississippi (Jacobson et al. 1979), 1.72 in South Carolina (Rhodes et al. 1985), 1.83–2.01 in Missouri (Hansen et al. 1996), and 1.88 in Illinois (Roseberry and Klimstra 1970). On poorer ranges, reproductive rates are lower and include 1.42–1.5 in the Southern Appalachian Mountains (Wentworth 1989), 1.14 on Blackbeard Island, Georgia (Osborne et al. 1992), and 1.26–1.38 in various areas of Florida (Richter and Labisky 1985). These habitats typically have abundant low-quality forage, and reproductive parameters are little affected by changes in herd density.

Similar to reproductive rates, breeding by fawns (6 months of age) varies regionally. Reported fawn breeding rates include 38% in North Carolina (Chiavetta 1958), 41% in Illinois (Roseberry and Klimstra 1970), 41% in South Carolina (Rhodes et al. 1986), 2.9% in Mississippi (Jacobson et al. 1979), and 27.3% in Missouri (Hansen et al. 1996). In contrast, on some inherently poorer ranges such as the Coastal Plain of the southeastern United States, or in overpopulated herds, some females may not breed until 2.5 years of age.

As would be expected with a species that inhabits a wide range of habitats, population management varies. In most regions, white-tailed deer population management is based on density-dependent population growth models such as that developed for the George Reserve by McCullough (1979). In these habitats, competition for forages increases as population density increases toward carrying capacity, reproduction decreases, and mortality increases. Also, as a population approaches carrying capacity, the amount and quality of available forage decrease and an equilibrium between the plant community and the deer population is reached. Density-dependent effects often lag behind relative density, allowing populations to exceed the carrying capacity, which results in substantial impact on the vegetation. The degree to which a herbivore population overshoots carrying capacity depends on the reproductive rate of the species. Thus, white-tailed deer are more likely to degrade habitat quality than are many other large herbivores.

Carrying capacity often is related to the successional stage of the plant community. For example, winter carrying capacity in forested areas of Pennsylvania is estimated at 23 deer/km^2 in sapling stands, 7.7 deer/km^2 in mature hardwood forests, and only 1.9 deer/km^2 in pole timber stands (Drake and Palmer 1991). Similarly, carrying capacity in areas such as the Southern Appalachian mountains can be highly variable and depends on yearly variations in acorn mast production (Wentworth 1989).

The predominant paradigm of density-dependent growth models consists of an initial population irruption and crash, followed by a recovery to a reduced carrying capacity through a series of dampened oscillations. However, not all deer populations respond in accord with this theory. McCullough (1997) provided examples of populations that exhibited a series of repeated irruptions. These repeated irruptions appear to be the norm where a resilient plant community dominated by herbaceous vegetation constitutes the forage resources as a result of either natural or anthropogenic factors.

Managers typically evaluate carrying capacity of a habitat by either direct measurement of forage quantity and quality or indirect evaluations of herd condition. Neither method, however, provides a singular tool for assessing population status relative to carrying capacity and allows precise management prescriptions. Management of deer populations requires an integration of all tools available to a knowledgeable biologist, along with adaptive management based on outcomes of previous management prescriptions.

Habitat Management. Habitat management techniques are as varied as the habitats that white-tailed deer inhabit and the objectives of the responsible individual, organization, or governmental agency. Nevertheless, habitat management typically involves maintaining or enhancing both the quantity and the quality of forage resources.

In most forested landscapes, enhancement of deer habitat is best accomplished through a variety of silvicultural techniques. Contrary to popular belief, the best habitat conditions for whitetails do not occur in mature forests. Instead, this species prefers earlier successional or edge-type habitats where forage is abundant. Thus, it is of little surprise that habitat enhancement in forests typically encompasses techniques such as forest harvesting, thinning, prescribed burning, and creation of openings. The response of deer habitat to each of these techniques varies among forest types and conditions, and extensive research has examined forage responses in various parts of the whitetail's range. The following discussion should be viewed as general guidelines. Because responses are habitat specific, detailed recommendations should be sought from a wildlife biologist with experience in the region and vegetation types being managed.

Clearcutting, or even-aged forest management, can produce abundant, high-quality deer forage. It is best suited to regenerate shade-intolerant tree species (those that require full sunlight to grow) such as the southern pines (*Pinus* spp.), hemlock (*Tsuga canadensis*), white pine (*Pinus strobus*), northern white cedar (*Thuga occidentalis*), black cherry (*Prunus serotina*), and tulip poplar (*Liriodendron tulipifera*) among others. Clearcutting maximizes the response of herbaceous forages (forbs and grasses) as well as the production of soft mast (berries and other fruits), which are important components of the whitetail's summer diet. In northern hardwood stands, browse production peaks about 8 years after clearcutting (Cooperrider and Behrend 1982). However, mature forests provide seasonally abundant and important mast production in many regions as well as winter thermal cover on northern ranges. As the result of a long-term study of deer habitats in the Southern Appalachian forests, Johnson et al. (1995) concluded that clearcuts and older forests provide complementary benefits to deer.

In most commercial forests that are managed by clearcutting, deer habitat responses are influenced by the size and shape of the regenerating stands. Smaller, irregularly shaped clearcuts will optimize deer use of clearcuts by maximizing the amount of forest edge and minimizing large open areas. Within a particular regeneration unit, forage responses are affected by silvicultural techniques used to establish the stands. Mechanical site preparation, herbicide application, fire, and spacing of planting can be positive or negative influences depending on the type of application, application rate, or intensity. For example, in some southern pine systems, a single herbicide application for site preparation can enhance deer forage responses when compared to untreated control sites. However, some herbicide combinations can reduce abundance and diversity of woody and herbaceous species (Miller et al. 1999) that could be important deer forages. Repeated applications also can limit the development of the early-successional forbs and vines that constitute the bulk of deer forages in these habitats.

Uneven-aged forest management includes a variety of techniques such as single-tree selection, group selection (patch) harvesting, and shelterwood. As with the even-aged systems, deer habitat responses to these types of cutting systems depend on the forest type and on deer range conditions. As a general rule, forage responses are directly related to the percentage of the canopy removed. The removal of small (<0.50 ha) patches of trees is particularly suited to management of many bottomland hardwood forests as well as northern wintering areas.

Habitat management in many forested regions requires knowledge of specialized habitat features. In some areas of the southern United States such as the Southern Appalachians and the Ouachita and the Ozark Mountains, and some coastal islands, deer herd productivity is highly dependent on acorn mast production. Retention of forested stands with high mast production capabilities is requisite in these areas because clearcutting does not improve winter deer range. Similarly, extensive timber harvesting in deer wintering areas can reduce winter survival and negatively affect deer populations over fairly broad areas. In yarding areas, cutting units should be relatively small (2–4 ha) and well distributed (Pruitt and Pruitt 1986).

Forest thinnings can stimulate forage production in managed forests. In southern pine forests, thinnings can result in a >15-fold increase in summer forage production. Maximal forage response occurs within 2 years of a thinning and declines thereafter (Hurst and Warren 1982). However, on some infertile soils such as the southern Coastal Plain, thinnings may only increase an already abundant supply of low-quality forages and would thus be of little benefit to deer.

Prescribed burning can be an integral part of forest management in fire-dependent ecosystems. In the southeastern United States, more than 1 million ha is burned each year to reduce wildfire hazard, manage competition, and stimulate forage production. When used in open-canopied forests or combined with silvicultural thinnings, prescribed fire can increase the abundance of legumes and other important forages, increase soft mast production, and temporarily improve the nutritional quality of forage plants. Prescribed fire generally is recommended on a 3- to 5-year basis to provide maximal benefit to deer.

Prescribed fire also is used to manage habitats in other regions. For example, winter burning is used in the Texas Post Oak Belt on a 6-year cycle to improve range conditions for deer and cattle (Yantis et al. 1983).

Since the early 1980s, there has been a tremendous increase in the use of agricultural plantings to enhance deer habitat in areas where management objectives are to improve herd quality or raise carrying capacity. Although their effectiveness was initially questioned, recent research overwhelmingly supports food plots as a valuable tool for deer management. Supplemental plantings are not a cure for poor population management; however, intensively managed agricultural plantings on as little as 5% of an area can stimulate harvest and improve deer condition (Kammermeyer and Thackston 1995). Even planting 1% of an area can improve deer diets and enhance reproduction, growth, and antler development (Johnson et al. 1987).

A variety of agricultural crops are commonly planted for deer; the most common are the small grains (wheat, oats, rye, and ryegrass), corn, grain sorghum, and legumes such as soybeans, cowpeas, alfalfa, and clovers. Many biologists recommend a clover–grass mixture such as ladino (*Trifolium repens*) and red clover (*T. pratense*) plus wheat, rye, and ryegrass. Managed properly, these plots may last up to 10 years before reseeding is required (Kammermeyer and Thackston 1995).

Where legal, supplemental feeding with corn, commercial deer feeds, or hay can provide nutritional benefits. However, supplemental feeding can be expensive and concentration of deer at feeding sites can raise significant herd health concerns. Extensive winter feeding of deer in Michigan was implicated in the spread of bovine tuberculosis among free-ranging deer (Schmitt et al. 1997).

ECONOMIC STATUS

Deer provide abundant recreational opportunities for hunters and wildlife watchers. A 1996 survey indicated that more than $9.7 billion were spent on big game hunting excluding the cost of hunting licenses (U.S. Department of the Interior and U.S. Department of Commerce 1996). Estimates indicate >10.7 million hunters spent an average of 12 days each hunting for deer. Each hunter spent an average of $860. More than 62.9 million people in the United States also spent more than $29.2 billion for wildlife watching. Many of these people just enjoy seeing and photographing deer.

Several studies have attempted to determine the economic value of deer herds on various areas or in specific states. However, only two studies have conducted a thorough analysis of the value of white-tailed deer on a national level. Williamson and Doster (1981) presented an economic analysis of the positive values associated with populations of white-tailed deer, mule deer, black-tailed deer, and axis deer (*Axis axis*). Specific to white-tailed deer, they estimated total hunting expenditure value at $1.01 billion, values received by hunters (recreation, consumable meat) at $1.8 billion, and values received by nonhunters at $5.4 billion, resulting in a total annualized value of $8.2 billion. If capitalized at 30%, or a typical rate of increase of a deer herd, the value of the white-tailed population in the United States in 1982 was an estimated $27.3 billion, or roughly three fourths of the farm value of cattle and calves.

More recently, Conover (1997) provided a benefit/cost analysis to more accurately depict the net value of deer to society. His results suggest that deer have a net annual monetary value of >$12 billion. This results from subtracting >$2 billion in negative values (i.e., $1 billion in car damage + >$100 million in crop damage + >$750 million in forest damage + >$250 million in damage to households) from the >$14 billion in recreational value (expenses by recreationists + consumer surplus).

Although deer are a highly valuable natural resource, the negative impacts associated with overabundant deer populations are creating many new management challenges. The white-tailed deer is the leading wildlife species associated with agricultural damage (Conover 1998). For example, deer were responsible for losses of field crops for more than 41% of producers in the northeastern and north-central United States (Wywialowski 1994).

White-tailed deer are keystone herbivores, which when at high densities affect (1) the distribution and abundance of other species, (2) community structure by altering patterns of relative abundance among competitors, or (3) community structure by modifying the abundance of species at multiple trophic levels (Waller and Alverson 1997). Browsing by white-tailed deer can significantly alter forest vegetation structure and composition (Alverson et al. 1988). Deer foraging on sprouts and seedlings can negatively affect forest regeneration (Alverson et al. 1988; Buckley et al. 1998), with the degree of impact directly correlated to deer density (Tilghman 1989). This concern about deer herbivory is not strictly limited to regeneration of commercial forest species. For example, Anderson and Katz (1993) suggested that deer herbivory encourages the development of forests dominated by browse-tolerant species, such as sugar maple (*Acer saccharum*), at the expense of those sensitive to browsing, such as eastern hemlock. Computer simulation models estimate that in <150 years, deer browsing will reduce hemlock to only a minor portion of forested ecosystems in the northeastern United States (Frelich and Lorimer 1985). Thus, the negative impacts of herbivory by high-density deer herds include economic losses associated with failed regeneration of commercially valuable tree species in addition to losses that are more difficult to quantify economically, but perhaps more important ecologically.

The effects of these changes on vegetation composition and structure on other wildlife species can be profound (Casey and Hein 1983; DeGraaf et al. 1991; McShea and Rappole 1997). DeCalesta (1994) found that deer browsing reduced woody vegetation that provided important habitat for intermediate-canopy-nesting songbirds. Species richness and abundance of these songbirds were reduced when white-tailed deer densities exceeded 7.9/km^2.

Arguably, white-tailed deer are the most economically important wildlife species in North America. However, this economic valuation is positive as well as negative. As argued by Conover (1997), the goal of any deer management program should be to keep populations at levels where the net positive benefit is highest. In many unhunted populations, such as in parks and urban/suburban areas, and perhaps in some hunted populations as well, populations may greatly exceed the point of maximal net value. At some point, populations can rise to levels where negative values outweigh positive values and a traditionally valued wildlife resource becomes viewed as a pest. Deer herd management should seek to maintain populations not only within their biological carrying capacity, but also well within their cultural carrying capacity (Minnis and Peyton 1995).

RESEARCH NEEDS

The previous century witnessed white-tailed deer populations brought back from near extirpation to abundant levels. Whitetail populations are now at an all time high across the species's geographic range and have expanded into some previously unoccupied range. However, populations of some genetically distinct subspecies, such as the Florida Key deer, remain at risk, and research efforts focused on management of these populations should be continued and perhaps expanded. Concurrently, the taxonomic validity of the subspecific designations of other discrete populations should be examined with modern genetic techniques, as has been done for the Columbian white-tailed deer (Gavin and May 1988).

Successful management programs, combined with the extreme adaptability of the species, have resulted in new challenges in deer management. Although most states and provinces have stabilized or even reduced deer populations in efforts to maintain ecological balance, negative impacts of burgeoning deer populations are increasing. Problems with deer overabundance likely will increase rapidly as the human population expands and hunting opportunities are challenged in urban/suburban areas. Vehicle collisions, crop depredation, urban/suburban deer damage, and the negative impacts of overabundant deer populations on forest communities are important issues that must be addressed. In fact, deer overabundance has become such a common problem in many parts of North America that it probably will represent one of the greatest challenges facing wildlife professionals during the coming decades (Warren 1997). In addition, emerging disease issues, such as bovine tuberculosis and chronic wasting disease, certainly will demand the attention of research biologists, resource managers, and wildlife agencies in the future, particularly if public fear reduces hunting pressure below what is necessary to control deer populations.

Most states use a variety of techniques to minimize deer–vehicle collisions, including signs, modified speed limits, roadside fencing, over/underpasses, reflective devices, habitat alteration, and public awareness programs. However, few techniques have been thoroughly evaluated. Of those tested, neither lighted nor unlighted deer crossing signs, mirrors, or Swareflex reflectors have proven successful, although fencing along rights-of-way can reduce collision rates (Feldhamer et al. 1986). Insufficient research has been conducted on deer–vehicle collisions, and many previous studies have suffered from poorly replicated designs or lack of sufficient funding. The magnitude of the deer–vehicle collision problem emphasizes the importance of developing a coordinated and collaborative research initiative to provide credible and defensible management recommendations.

As North America changes from a rural to an urban society, public support for hunting is waning. This change increasingly estranges the public from the life and death requirements for regulating natural systems. Programs designed to educate the public, particularly youth, about deer management are an urgent, although traditionally neglected, need among wildlife agencies. Implicit in the development of these programs is the need for research to identify communication channels between resource professionals and the public, develop methods to motivate youths to participate in outdoor sports, and understand factors affecting hunter retention rates.

We must develop methods to facilitate hunter harvest in situations where deer populations are overabundant, while concurrently developing nontraditional techniques for management of these populations. Sharpshooting or capture and euthanasia are valid, cost-effective techniques in some circumstances. Contraception, although theoretically plausible, is still in the early stages of development and evaluation as a management tool. In most cases, it certainly will be cost-prohibitive and ineffective.

Management strategies that integrate white-tailed deer social behavior, as proposed by McNulty et al. (1997), have potential applicability in localized areas. However, the behavioral plasticity of this species in response to changing habitat and demographic conditions, as well as a paucity of research studies on its behavioral responses to these techniques, may limit their applicability. Studies incorporating deer behavioral biology and management techniques should be encouraged and ultimately may increase the success of deer management programs.

With the backdrop of deer overabundance problems in many regions, research on methods to enhance deer herd productivity and condition in other regions nevertheless are still valid. Similarly, research on the impact of other land management activities on whitetail habitat should be considered. For example, increased intensity of forest management activities may reduce the duration of optimal habitat conditions in many industrial forests.

Little is known about the relationship between deer population density and the amount of deer damage in various landscapes. However, deer movements and use of surrounding habitat can greatly affect the amount of damage to agriculture. We need to correlate levels of deer damage with deer population densities, movements, and use of surrounding habitat.

Unquestionably, the percentage of hunters in the population is declining and the absolute number of hunters appears stable or slightly declining. This trend is not unique to North America, but appears global. At the Annual Congress of the International Union of Game Biologists in Godollo, Hungary, Bobek (1991) warned that "the isolation of hunters as a social group will deepen . . . and in the not-far future, the silhouette of the hunter will disappear from our forests and fields." Research on the motivations of hunters, and in particular reasons for relinquishing their hunting heritage, should strive to identify methods to encourage hunter retention and recruitment rates. Without this major tool of deer population management, the ability to maintain deer numbers within a biological or a cultural carrying capacity appears bleak.

If there is one answer to the management of deer in the future, it lies in the education of all affected parties. The emergence of philosophies such as quality deer management (Miller and Marchinton 1995) that foster stewardship through education is an encouraging development that must be amplified if the white-tailed deer is to maintain its status as the premier game species in North America.

ACKNOWLEDGMENT

The authors thank R. DeYoung for editorial comments.

LITERATURE CITED

Abbott, M. J., D. E. Ullrey, P. K. Ku, S. M. Schmitt, D. R. Romsos, and H. A. Tucker. 1984. Effect of photoperiod on growth and fat accretion in white-tailed doe fawns. Journal of Wildlife Management 48:776–87.

Allen, R. E., and D. R. McCullough. 1976. Deer–car accidents in southern Michigan. Journal of Wildlife Management 40:317–25.

Alverson, W. S., D. M. Waller, and S. L. Solheim. 1988. Forests too deer: Edge effects in northern Wisconsin. Conservation Biology 2:348–58.

Anderson, R. C., and A. J. Katz. 1993. Recovery of browse sensitive tree species following release from white-tailed deer *Odocoileus virginianus* browsing pressure. Biological Conservation 63:203–8.

Asleson, M. A., E. C. Hellgren, and L. W. Varner. 1996. Nitrogen requirements for antler growth and maintenance in white-tailed deer. Journal of Wildlife Management 60:744–52.

Atkeson, T. D., and R. L. Marchinton. 1982. Forehead glands in white-tailed deer. Journal of Mammalogy 63:613–17.

Atkeson, T. D., R. L. Marchinton, and K. V. Miller. 1988. Vocalizations of white-tailed deer. American Midland Naturalist 120:194–200.

Baccus, J. T., and J. C. Posey. 1999. Melanism in white-tailed deer in central Texas. Southwestern Naturalist 44:184–92.

Baker, R. H., 1984. Origin, classification and distribution of the white-tailed deer. Pages 1–18 *in* L. K. Halls, ed. White-tailed deer: Ecology and management. Stackpole, Harrisburg, PA.

Ballard, W. B., H. A. Whitlaw, S. J. Young, R. A. Jenkins, and G. J. Forbes. 1999. Predation and survival of white-tailed deer fawns in northcentral New Brunswick. Journal of Wildlife Management 63:574–79.

Barbour, T., and G. M. Allen. 1922. The white-tailed deer of eastern United States. Journal of Mammalogy 3:65–78.

Bartlett, I. H. 1949. White-tailed deer resources: United States and Canada. Transactions of the North American Wildlife Conference 14:543–52.

Bartush, W. S., and G. W. Garner. 1979. Physical characteristics of white-tailed deer fawns in southwestern Oklahoma. Proceedings of the Annual Conference of the Southeastern Association of Fish and Wildlife Agencies 33:250–58.

Bartush, W. S., and J. C. Lewis. 1981. Mortality of white-tailed deer fawns in the Wichita Mountains. Proceedings of the Oklahoma Academy of Science 61:23–27.

Beasom, S. L., F. G. Leon III, and D. R. Synatzke. 1986. Accuracy and precision of counting white-tailed deer with helicopters at different sampling intensities. Wildlife Society Bulletin 14:364–68.

Beier, P., and D. R. McCullough. 1990. Factors influencing white-tailed deer activity patterns and habitat use. Wildlife Monographs 109:1–51.

Blouch, R. I. 1984. Northern Great Lakes States and Ontario forests. Pages 391–410 *in* L. K. Halls, ed. White-tailed deer ecology and management. Stackpole, Harrisburg, PA.

Bobek, B. 1991. Wildlife management concepts—Present trends and perspectives for 21st century. Transactions of the Annual Congress of the International Union of Game Biologists 20:17–24.

Boyer, R. T., M. E. Shea, and S. A. McKenna. 1986. The role of winter severity and population density in regulating northern populations of deer (Miscellaneous Publication No. 689). Maine Agricultural Experiment Station, Orono.

Bridges, R. J. 1968. Individual white-tailed deer movement and related behavior during winter and spring in northwestern Florida. M.S. Thesis, University of Georgia, Athens.

Brown, D. E., and R. S. Henry. 1981. On relict occurrences of white-tailed deer within the Sonoran Desert in Arizona. Southwestern Naturalist 26:147–52.

Buckley, D. S., T. L. Sharik, and J. G. Isebrands. 1998. Regeneration of northern red oak: Positive and negative effects of competitor removal. Ecology 79:65–78.

Budde, W. S. 1983. Effects of photoperiod on puberty attainment of female white-tailed deer. Journal of Wildlife Management 47:595–604.

Burnham, K. P., D. R. Anderson, and J. L. Laake. 1980. Estimation of density from line transect sampling of biological populations. Wildlife Monographs 72:1–202.

Carroll, B. K., and D. L. Brown. 1977. Factors affecting neonatal fawn survival in southern-central Texas. Journal of Wildlife Management 41:63–69.

Casey, D., and D. Hein. 1983. Effects of heavy browsing on a bird community in deciduous forest. Journal of Wildlife Management 47:829–36.

Castleberry, S. B., W. M. Ford, K. V. Miller, and W. P. Smith. 1999. White-tailed deer browse preferences in a southern bottomland hardwood forest. Southern Journal of Applied Forestry 23:78–82.

Caughley, G. 1977. Analysis of vertebrate populations. John Wiley, New York.

Cheatum, E. L., and G. H. Morton. 1946. Breeding season of white-tailed deer in New York. Journal of Wildlife Management 10:249–63.

Chiavetta, K. J. 1958. Harvest antlerless deer! North Carolina Wildlife 22:16–19.

Coggin, D. S. 1998. Survival and mortality of adult male white-tailed deer in Mississippi. M.S. Thesis, Mississippi State University. Mississippi State.

Conover, M. R. 1997. Monetary and intangible valuation of deer in the United States. Wildlife Society Bulletin 23:298–305.

Conover, M. R. 1998. Perceptions of American agricultural producers about wildlife on their farms and ranches. Wildlife Society Bulletin 26:597–604.

Conover, M. R., W. C. Pitt, K. K. Kessler, T. J. DuBow, and W. A. Sanborn. 1995. Review of human injuries, illnesses, and economic losses caused by wildlife in the United States. Wildlife Society Bulletin 23:407–14.

Cooperrider, A. Y., and D. F. Behrend. 1982. Simulation of forest dynamics and deer browse production. Journal of Forestry 78:85–88.

Cornicelli, L., A. Woolf, and J. L. Roseberry. 1996. White-tailed deer use of a suburban environment in southern Illinois. Transactions of the Illinois State Academy of Science 89:93–103.

Cothran, E. G., R. K. Chesser, M. H. Smith, and P. E. Johns. 1987. Fat levels in female white-tailed deer during the breeding season and pregnancy. Journal of Mammalogy 68:111–18.

Cowan, R. L., and A. C. Clark. 1981. Nutritional requirements. Pages 72–86 *in* W. R. Davidson, F. A. Hayes, V. F. Nettles, and F. E. Kellog, eds. Diseases and parasites of white-tailed deer. Tall Timbers Research Station, Tallahassee, FL.

Crawford, H. S. 1982. Seasonal food selection and digestibility by tame white-tailed deer in central Maine. Journal of Wildlife Management 46:974–82.

Dahlberg, B. L., and R. C. Guettinger. 1956. The white-tailed deer in Wisconsin (Technical Wildlife Bulletin 14). Wisconsin Conservation Department, Game Management Division, Madison.

Dalrymple, G. H., and O. L. Bass, Jr. 1996. The diet of the Florida panther in Everglades National Park, Florida. Bulletin of the Florida Museum of Natural History 39:173–93.

Davidson, W. R., and G. L. Doster. 1997. Health characteristics and white-tailed deer population density in the southeastern United States. Pages 164–84 *in* W. J. McShea, H. B. Underwood, and J. H. Rappole, eds. The science of overabundance: Deer ecology and population management. Smithsonian Institution Press, Washington, DC.

Davidson, W. R., and V. F. Nettles. 1997. Field manual of wildlife diseases in the southeastern United States, 2nd ed. Southeastern Cooperative Wildlife Disease Study, Athens, GA.

DeCalesta, D. S. 1994. Effect of white-tailed deer on songbirds within managed forests in Pennsylvania. Journal of Wildlife Management 58:711–18.

Degayner, E. J., and P. A Jordan. 1987. Skewed fetal sex ratios in white-tailed deer: Evidence and evolutionary speculations. Pages 178–88 *in* C. M. Wemmer, ed. Biology and management of the Cervidae. Smithsonian Institution Press, Washington, DC.

DeGraaf, R. M., W. M. Healy, and R. T. Brooks. 1991. Effects of thinning and deer browsing on breeding birds in New England oak woodlands. Forest Ecology and Management 41:179–91.

DelGiudice, G. D. 1995. Assessing winter nutritional restriction of northern deer with urine in snow: Considerations, potentials, and limitations. Wildlife Society Bulletin 23:687–93.

DeYoung, C. A. 1989. Mortality of adult male white-tailed deer in South Texas. Journal of Wildlife Management 53:513–18.

DeYoung, C. A. 1990. Inefficiency in trophy white-tailed deer harvest. Wildlife Society Bulletin 18:7–12.

DeYoung, C. A., J. R. Heffelfinger, S. P. Coughlin, and S. L. Beasom. 1988. Accuracy of track counts to estimate white-tailed deer abundance. Proceedings of the Annual Conference of the Southeastern Association of Fish and Wildlife Agencies 42:464–69.

Dice, L. R. 1941. Methods for estimating populations of mammals. Journal of Wildlife Management 5:398–407.

Dimmick, R. W., and M. R. Pelton. 1994. Criteria of sex and age. Pages 169–214 *in* T. A. Bookhout, ed. Research and management techniques for wildlife and habitats. Wildlife Society, Bethesda, MD.

Downing, R. L. 1971. Comparison of crippling losses of white-tailed deer caused by archery, buckshot, and shotgun slugs. Proceedings of the Annual Conference of the Southeastern Association of Game and Fish Commissioners 25:77–82.

Drake, W. E., and W. L. Palmer. 1991. Overwintering feeding capacities of mixed oak forests in central Pennsylvania (Final Report, Job No. 06218). Federal Aid in Wildlife Restoration.

Dusek, G. L., R. J. Mackie, J. D. Herriges, Jr., and B. B. Compton. 1989. Population ecology of white-tailed deer along the lower Yellowstone River, USA. Wildlife Monographs 104:1–68.

Ellsworth, D. L., R. L. Honeycutt, N. J. Silvy, J. W. Bickham, and W. D. Klimstra. 1994a. Historical biogeography and contemporary patterns of mitochondrial DNA variation in white-tailed deer from the southeastern United States. Evolution 48:122–36.

Ellsworth, D. L., R. L. Honeycutt, N. J. Silvy, M. H. Smith, J. W. Bickham, and W. D. Klimstra. 1994b. White-tailed deer restoration to the southeastern United States: Evaluating genetic variation. Journal of Wildlife Management 58:686–97.

Eve, J. H., and R. E. Kellogg. 1977. Management implications of abomasal parasites in southeastern white-tailed deer. Journal of Wildlife Management 41:169–77.

Feldhamer, G. A., J. E. Gates, D. M. Harman, A. J. Loranger, and K. R. Dixon. 1986. Effects of interstate highway fencing on white-tailed deer activity. Journal of Wildlife Management 50:497–503.

Folk, M. J., and W. D. Klimstra. 1991. Reproductive performance of female Key deer. Journal of Wildlife Management 55:386–90.

Forand, K. J., and R. L. Marchinton. 1989. Patterns of social grooming in adult white-tailed deer. American Midland Naturalist 122:357–64.

Forbes, G. J., and J. B. Theberge. 1996. Response of wolves to prey variation in central Ontario. Canadian Journal of Zoology 74:1511–20.

Franzmann, A. W. 1985. Assessment of nutritional status. Pages 239–60 *in* R. J. Hudson and R. G. White, eds. Bioenergetics of wild herbivores. CRC Press, Boca Raton, FL.

Frelich, L. E., and C. G. Lorimer. 1985. Current and predicted long-term effects of deer browsing in hemlock forests in Michigan, USA. Biological Conservation 34:99–120.

French, C. E., L. C. McEwen, N. D. Magruder, R. H. Ingram, and R. W. Swift. 1956. Nutrient requirements for growth and antler development in white-tailed deer. Journal of Wildlife Management 20:221–32.

Fryxell, J. M., and A. R. E. Sinclair. 2000. A dynamic view of population regulation. Pages 156–74 *in* S. Demarais and P. R. Krausman, eds. Ecology and management of large mammals in North America. Prentice-Hall, Upper Saddle River, NJ.

Fuller, T. K. 1989. Population dynamics of wolves in north-central Minnesota, USA. Wildlife Monographs 105:1–41.

Fuller, T. K. 1990. Dynamics of a declining white-tailed deer population in north-central Minnesota. Wildlife Monographs 110:1–37.

Fuller, T. K. 1991a. Do pellet counts index white-tailed deer numbers and population change? Journal of Wildlife Management 55:393–96.

Fuller, T. K. 1991b. Effect of snow depth on wolf activity and prey selection in north central Minnesota, USA. Canadian Journal of Zoology 69:283–87.

Fuller, T. K., R. M. Pace, III., J. A. Markl, and P. L. Coy. 1989. Morphometrics of white-tailed deer in north-central Minnesota. Journal of Mammalogy 70:184–88.

Gassett, J. W., D. P. Wiesler, A. G. Baker, D. A. Osborn, K. V. Miller, R. L. Marchinton, and M. Novotny. 1996. Volatile compounds from the forehead region of male white-tailed deer (*Odocoileus virginianus*). Journal of Chemical Ecology 23:569–78.

Gassett, J. W., K. A. Dasher, D. A. Osborn, and K. V. Miller. 1998. What the nose knows: Detection of oestrus by male white-tailed deer. Page 52 *in* Z. Zomborszky, ed. Proceedings of the fourth international deer biology congress. Kaposvar, Hungary.

Gassett, J. W., K. A. Dasher, K. V. Miller, D. A. Osborn, and S. M. Russell. 2000. White-tailed deer tarsal glands: Sex and age-related variation in microbial flora. Mammalia 64:371–77.

Gavin, T. A., and B. May. 1988. Taxonomic status and genetic purity of Columbian white-tailed deer. Journal of Wildlife Management 52:1–10.

Gavin, T. A., L. H. Suring, P. A. Vohs, Jr., and E. C. Meslow. 1984. Population characteristics, spatial organization, and natural mortality in the Columbian white-tailed deer. Wildlife Monograph 91:1–41.

Gee, K. L., M. D. Porter, S. Demarais, F. C. Bryant, and G. Van Vreede. 1991. White-tailed deer: Their foods and management in the Cross Timbers, 2nd ed. Samuel Roberts Noble Foundation, Ardmore, OK.

Geist, V. 1998. Deer of the world: Their evolution, behaviour, and ecology. Stackpole, Mechanicsburg, PA.

Geist, V., and S. Lingle. 1994. Taking their world in stride. Pages 76–81 *in* D. Gerlach, S. Atwater, and J. Schnell, eds. Deer. Stackpole, Harrisburg, PA.

Gladfelter, H. L. 1984. Midwest agricultural region. Pages 427–40 *in* L. K. Halls, ed. White-tailed deer: Ecology and management. Stackpole, PA.

Godin, A. J. 1977. Wild mammals of New England. Johns Hopkins University Press, Baltimore.

Goldman, E. A., and R. Kellogg. 1940. Ten new white-tailed deer from North and Middle America. Proceedings of the Biological Society of Washington 53:81–89.

Goss, R. J. 1983. Deer antlers: Regeneration, function, and evolution. Academic Press, New York.

Grasman, B. T., and E. C. Hellgren. 1993. Phosphorus nutrition in white-tailed deer: Nutrient balance, physiological responses, and antler growth. Ecology 74:2279–96.

Gruver, B. J., D. C. Guynn, Jr., and H. A. Jacobson. 1984. Simulated effects of harvest strategy on reproduction in white-tailed deer. Journal of Wildlife Management 48:535–41.

Gunson, J. R. 1979. Use of night-lighted census in management of deer in Alberta and Saskatchewan. Wildlife Society Bulletin 7:259–67.

Hackett, E. J., D. C. Guynn, and H. A. Jacobson. 1979. Differences in age structure produced by two aging techniques. Proceedings of the Annual Conference of the Southeastern Association of Fish and Wildlife Agencies 33:25–29.

Hall, E. R. 1981. The mammals of North America, 2nd ed. John Wiley, New York.

Halls, L. K. 1978. White-tailed deer. Pages 43–65 *in* J. L. Schmidt and D. L. Gilbert, eds. Big game of North America: Ecology and management. Stackpole, Harrisburg, PA.

Hamlin, K. L., D. F. Pac, C. A. Sime, R. M. DeSimone, and G. L. Dusek. 2000. Evaluating the accuracy of ages obtained by two methods for Montana ungulates. Journal of Wildlife Management 64:441–49.

Hansen, C. S. 1983. Costs of deer–vehicle accidents in Michigan. Wildlife Society Bulletin 11:161–64.

Hansen, L. P., J. Beringer, and J. H. Schulz. 1996. Reproductive characteristics of female white-tailed deer in Missouri. Proceedings of the Annual Conference of the Southeastern Association of Fish and Wildlife Agencies 50:357–66.

Harder, J. D., and R. L. Kirkpatrick. 1994. Physiological methods in wildlife research. Pages 275–306 *in* T. A. Bookhout, ed. Research and management techniques for wildlife and habitats, 5th ed. Wildlife Society, Bethesda, MD.

Harlow, R. F. 1965. Food habits. Pages 74–107 *in* R. F. Harlow and F. K. Jones, Jr., eds. The white-tailed deer in Florida (Technical Bulletin 9). Florida Game and Freshwater Fish Commission, Tallahassee.

Harlow, R. F., and R. L. Downing. 1967. Evaluating the deer track census method used in the Southeast. Proceedings of the Annual Conference of the Southeastern Association of Fish and Wildlife Agencies 21:39–41.

Harlow, R. F., and R. G. Hooper. 1972. Forages eaten by deer in the Southeast. Proceedings of the Annual Conference of the Southeastern Association of Game and Fish Commissioners 25:18–46.

Harmel, D. E., J. D. Williams, and W. E. Armstrong. 1989. Effects of genetics and nutrition on antler development and body size of white-tailed deer (FA Report Series 26, Projects W-56-D, W-76-R, W-109-R, and W-14-C). Texas Parks and Wildlife Department, Austin.

Haugen, A. O. 1959. Breeding records of captive white-tailed deer in Alabama. Journal of Mammalogy 40:108–13.

Haugen, A. O. 1975. Reproductive performance of white-tailed deer in Iowa. Journal of Mammalogy 56:151–59.

Hawkins, R. E., and W. D. Klimstra. 1970. A preliminary study of the social organization of white-tailed deer. Journal of Wildlife Management 34:407–19.

Hayne, D. W. 1949. An examination of the strip census method for estimating animal populations. Journal of Wildlife Management 13:145–57.

Hershkovitz, P. 1972. The Recent mammals of the Neotropical region: A zoological and ecological review. Pages 311–421 *in* A. Keast, F. C. Erk, and B. Glass, eds. Evolution, mammals and southern continents. State University of New York Press, Albany.

Hesselton, W. T., and R. M. Hesselton. 1982. White-tailed deer (*Odocoileus virginianus*). Pages 878–901 *in* J. A. Chapman and G. A. Feldhamer, eds. Wild mammals of North America. Johns Hopkins University Press, Baltimore.

Hirth, D. H. 1977. Social behavior of white-tailed deer in relation to habitat. Wildlife Monographs 57:1–55.

Hirth, D. H., and D. R. McCullough. 1977. Evolution of alarm signals in ungulates with special reference to white-tailed deer. American Midland Naturalist 111:31–42.

Hoffman, R. R. 1985. Digestive physiology of the deer: Their morphophysiological specialization and adaptation. Pages 393–407 *in* P. F. Fennessy and K. R. Drew, eds. Biology of deer production (Bulletin 22). Royal Society of New Zealand, Wellington.

Holter, J. B., H. H. Hayes, and S. H. Smith. 1979. Protein requirement of yearling white-tailed deer. Journal of Wildlife Management 43:872–79.

Holzenbein, S., and R. L. Marchinton. 1992a. Emigration and mortality of orphaned male white-tailed deer. Journal of Wildlife Management 56:147–53.

Holzenbein, S., and R. L. Marchinton. 1992b. Spatial integration of maturing-male white-tailed deer into the adult population. Journal of Mammalogy 73:326–34.

Honeycutt, R. L. 2000. Genetic applications for large mammals. Pages 233–59 *in* S. Demarais and P. R. Krausman, eds. Ecology and management of large mammals in North America. Prentice-Hall, Upper Saddle River, NJ.

Hurst, G. A., and R. C. Warren. 1982. Deer forage in 13-year-old commercially thinned and burned loblolly pine plantations. Proceedings of the Annual Conference of the Southeastern Association of Fish and Wildlife Agencies 36:420–26.

Huot, J., F. Potvin, and M. Belanger. 1984. Southeastern Canada. Pages 293–304 *in* L. K. Halls, ed. White-tailed deer ecology and management. Stackpole, Harrisburg, PA.

Inglis, J. M., R. E. Hood, B. A. Brown, and C. A. DeYoung. 1979. Home range of white-tailed deer in Texas coastal prairie brushland. Journal of Mammalogy 60:377–89.

Ivey, T. L., and M. K. Causey. 1988. Social organization among white-tailed deer during the rut. Proceedings of the Annual Conference of the Southeastern Association of Fish and Wildlife Agencies 42:266–71.

Jacobson, H. A. 1994. Reproduction. Pages 98–108 *in* D. Gerlach, S. Atwater, and J. Schnell, eds. Deer. Stackpole, Harrisburg, PA.

Jacobson, H. A. 1995. Age and quality relationships. Pages 103–11 *in* K. V. Miller and R. L. Marchinton, eds. Quality whitetails: The why and how of quality deer management. Stackpole, Mechanicsburg, PA.

Jacobson, H. A., and D. C. Guynn. 1995. A primer. Pages 81–102 *in* K. V. Miller and R. L. Marchinton, eds. Quality whitetails: The why and how of quality deer management. Stackpole, Mechanicsburg, PA.

Jacobson, H. A., and J. C. Kroll. 1994. The white-tailed deer (*Odocoileus virginianus*): "The most managed and mismanaged species." Pages 25–35 *in* J. A. Milne, ed. Recent developments in deer biology. Macaulay Land Use Research Institute, Craigiebuckler, Aberdeen, UK.

Jacobson, H. A., and S.D. Lukefahr. 1998. Genetics research on captive deer at Mississippi State University. Pages 46–50 *in* D. Rollins, ed. The role of genetics in white-tailed deer management. Texas A&M University, College Station.

Jacobson, H. A., D. C. Guynn, Jr., R. N. Griffin, and D. Lewis. 1979. Fecundity of white-tailed deer in Mississippi and periodicity of corpora lutea and lactation. Proceedings of the Annual Conference of the Southeastern Association of Fish and Wildlife Agencies 33:30–35.

Jacobson, H. A., J. C. Kroll, R. W. Browning, B. H. Koerth, and M. H. Conway. 1997. Infrared-triggered cameras for censusing white-tailed deer. Wildlife Society Bulletin 25:547–56.

Jacobson, N. K. 1979. Alarm bradycardia in white-tailed deer fawns (*Odocoileus virginianus*). Journal of Mammalogy 60:343–49.

Joanen, T., L. McNease, and D. Richard. 1985. The effects of winter flooding on white-tailed deer in southwestern Louisiana. Proceedings of the Louisiana Academy of Sciences 48:109–15.

Johns, P. E., M. E. Smith, and R. K. Chesser. 1984. Annual cycles of the kidney fat index in a southeastern white-tailed deer herd. Journal of Wildlife Management 48:969–73.

Johnson, A. S., P. E. Hale, W. M. Ford, J. M. Wentworth, J. R. French, O. F. Anderson, and G. B. Pullen. 1995. White-tailed deer foraging in relation to successional stage, overstory type and management of Southern Appalachian forests. American Midland Naturalist 133:18–35.

Johnson, M. K., B. W. Delany, S. P. Lynch, J. A Zeno, S. R. Schultz, T. W. Keegan, and B. D. Nelson. 1987. Effects of cool-season agronomic forages on white-tailed deer. Wildlife Society Bulletin 15:330–39.

Jones, R. L., and H. P. Weeks, Jr. 1985. Ca, Mg, and P in the annual diet of deer in south-central Indiana. Journal of Wildlife Management 49:129–33.

Kammermeyer, K. E., and R. L. Marchinton. 1976. Notes on dispersal of male white-tailed deer. Journal of Mammalogy 57:776–78.

Kammermeyer, K. E., and R. Thackston. 1995. Habitat management and supplemental feeding. Pages 129–54 *in* K. V. Miller and R. L. Marchinton, eds. Quality whitetails: The why and how of quality deer management. Stackpole, Mechanicsburg, PA.

Kie, J. G., M. White, and F. F. Knowlton. 1979. Effects of coyote predation on population dynamics of white-tailed deer. Pages 65–82 *in* D. L. Drawe, ed. Proceedings of the first Welder Wildlife Foundation Symposium. Rob and Bessie Welder Wildlife Foundation, Sinton, TX.

Kie, J. G., M. White, and D. L. Drawe. 1983. Condition parameters of white-tailed deer in Texas. Journal of Wildlife Management 47:583–94.

Kilpatrick, H. J., and S. M. Spohr. 2000. Spatial and temporal use of a surburban landscape by female white-tailed deer. Wildlife Society Bulletin 28:1023–29.

Klein, E. H. 1982. Phenology of breeding and antler growth in white-tailed deer in Honduras. Journal of Wildlife Management 46:826–29.

Knowlton, F. F., M. White, and J. G. Kie. 1979. Weight patterns of wild white-tailed deer in southern Texas. Pages 55–64 *in* D. L. Drawe, ed. Proceedings of the first Welder Wildlife Foundation Symposium. Rob and Bessie Welder Wildlife Foundation, Sinton, TX.

Knox, W. M., K. V. Miller, and R. L. Marchinton. 1988. Recurrent estrous cycles in white-tailed deer. Journal of Mammalogy 69:384–86.

Kucera, T. E. 1991. Adaptive variation in sex ratios of offspring in nutritionally stressed mule deer. Journal of Mammalogy 72:745–49.

Kunkel, K. E., and L. D. Mech. 1994. Wolf and bear predation on white-tailed deer fawns in northeastern Minnesota. Canadian Journal of Zoology 72:1557–65.

Kunkel, K. E., T. K. Ruth, D. H. Pletscher, and M. G. Hornocker. 1999. Winter prey selection by wolves and cougars in and near Glacier National Park, Montana. Journal of Wildlife Management 63:901–10.

Kurten, B., and E. Anderson. 1980. Pleistocene mammals of North America. Columbia University Press, New York.

Labisky, R. F., and D. E. Fritzen. 1998. Spatial mobility of breeding female white-tailed deer in a low-density population. Journal of Wildlife Management 62:1329–34.

Labisky, R. F., D. E. Fritzen, and J. C. Kilgo. 1991. Population ecology and management of white-tailed deer in the Osceola National Forest, Florida. Final report to the Florida Game and Fresh Water Fish Commission. Department of Wildlife Ecology and Conservation, University of Florida, Gainesville.

Labisky, R. F., M. C. Boulay, K. E. Miller, R. A. Sargent, Jr., and J. M. Zultowsky. 1995. Population ecology of white-tailed deer in Big Cypress National Preserve and Everglades National Park. Final report to National Park Service. Department of Wildlife Ecology and Conservation, University of Florida, Gainesville.

Lancia, R. A., C. S. Rosenberry, and M. C. Conner. 2000. Population parameters and their estimation. Pages 64–83 *in* S. Demarais and P. R. Krausman, eds. Ecology and management of large mammals in North America. Prentice-Hall, Upper Saddle River, NJ.

Lankester, M. W. 2001. Extrapulmonary lungworms of cervids. Pages 228–78 *in* W. M. Samuel, M. J. Pybus, and A. A. Kocan, eds. Parasitic diseases of wild mammals. Iowa State University Press, Ames.

Larson, T. J., O. J. Rongstad, and F. W. Terbilcox. 1978. Movement and habitat use of white-tailed deer in southcentral Wisconsin. Journal of Wildlife Management 42:113–17.

Leach, R. H., and W. D. Edge. 1994. Summer home range and habitat selection by white-tailed deer in the Swan Valley, Montana. Northwest Science 68:31–36.

Leberg, P. L., and D. L. Ellsworth. 1999. Further evaluation of the genetic consequences of translocations on southeastern white-tailed deer populations. Journal of Wildlife Management 63:327–34.

Leberg, P. L., and M. H. Smith. 1993. Influence of density on growth of white-tailed deer. Journal of Mammalogy 74:723–31.

Leberg, P. L., M. H. Smith, and I. L. Brisbin, Jr. 1992. Influence of sex, habitat, and genotype on the growth patterns of white-tailed deer. Pages 343–50 *in* R. D. Brown, ed. The biology of deer. Springer-Verlag, New York.

Leberg, P. L., P. W. Stange, H. O. Hillestad, R. L. Marchinton, and M. H. Smith. 1994. Genetic structure of reintroduced wild turkey and white-tailed deer populations. Journal of Wildlife Management 58:698–711.

Leon, F. G., III. 1986. Evaluation of white-tailed deer survey methods. M.S. Thesis, Texas A&I University, Kingsville.

Lopez, R. R., N. J. Silvy, and P. A. Frank. 2000. Survival and longevity of Florida Key deer. Abstracts of the Wildlife Society Annual Conference 7:137.

Lukefahr, S. D., and H. A. Jacobson. 1998. Variance component analysis and heritability of antler traits in white-tailed deer. Journal of Wildlife Management 62:262–68.

Marchinton, R. L., and T. D. Atkeson. 1985. Plasticity of socio-spatial behavior of white-tailed deer and the concept of facultative territoriality. Pages 375–77 *in* P. F. Fennessy and K. R. Drew, eds. Biology of deer production (Bulletin 22). Royal Society of New Zealand, Wellington.

Marchinton, R. L., and D. H. Hirth. 1984. Behavior. Pages 129–68 *in* L. K. Halls, ed. White-tailed deer ecology and management. Stackpole, Harrisburg, PA.

Masters, R. D., and N. E. Mathews. 1990. Notes on reproduction of old (≥9 years) free-ranging white-tailed deer, *Odocoileus virginianus*, in the Adirondacks, New York. Canadian Field-Naturalist 105:286–87.

Mathews, N. E., and W. F. Porter. 1993. Effect of social structure on genetic structure of free-ranging white-tailed deer in the Adirondack Mountains. Journal of Mammalogy 74:33–43.

Mautz, W. W. 1978. Sledding on a bushy hillside: The fat cycle in deer. Wildlife Society Bulletin 6:88–90.

Mautz, W. W., J. Kanter, and P. J. Pekins. 1992. Seasonal metabolic rhythms of captive female white-tailed deer: A reexamination. Journal of Wildlife Management 56:656–61.

McCabe, T. R., and R. E. McCabe. 1984. Of slings and arrows: An historical retrospection. Pages 19–72 *in* L. K. Halls, ed. White-tailed deer ecology and management. Stackpole, Harrisburg, PA.

McCabe, T. R., and R. E. McCabe. 1997. Recounting whitetails past. Pages 11–26 *in* W. J. McShea, H. B. Underwood, and J. H. Rappole, eds. The science of overabundance: Deer ecology and population management. Smithsonian Institution Press, Washington, DC.

McCullough, D. R. 1979. The George Reserve deer herd: Population ecology of a *K*-selected species. University of Michigan Press, Ann Arbor.

McCullough, D. R. 1997. Irruptive behavior in ungulates. Pages 69–98 *in* W. J. McShea, H. B. Underwood, and J. H. Rappole, eds. The science of overabundance: Deer ecology and population management. Smithsonian Institution Press, Washington, DC.

McCullough, D. R., D. H. Hirth, and S. J. Newhouse. 1989. Resource partitioning between sexes in white-tailed deer. Journal of Wildlife Management 53:277–83.

McDonald, J. S., and K. V. Miller. 1993. A history of white-tailed deer restocking in the United States, 1878 to 1992 (Research Publication 93-1). Quality Deer Management Association, Watkinsville, GA.

McDowell, R. D. 1970. Photoperiodism among breeding white-tailed deer (*Odocoileus virginianus*). Transactions of the Northeast Section of the Wildlife Society 1970:19–38.

McGinnes, B. S., and R. L. Downing. 1977. Factors affecting the peak of white-tailed deer fawning in Virginia. Journal of Wildlife Management 41:715–19.

McNulty, S. A., W. F. Porter, N. E. Mathews, and J. A. Hill. 1997. Localized management for reducing white-tailed deer populations. Wildlife Society Bulletin 25:265–71.

McShea, W. J., and J. H. Rappole. 1997. Herbivores and the ecology of forest understory birds. Pages 298–309 *in* W. J. McShea, H. B. Underwood, and J. H. Rappole, eds. The science of overabundance: Deer ecology and population management. Smithsonian Institution Press, Washington, DC.

Mech, L. D., and M. E. Nelson. 2000. Do wolves affect white-tailed buck harvest in northeastern Minnesota? Journal of Wildlife Management 64:129–36.

Messier, F., C. Barrette, and J. Hout. 1986. Coyote *Canis latrans* predation on a white-tailed deer *Odocoileus virginianus* population in southern Quebec Canada. Canadian Journal of Zoology 64:1134–36.

Messier, F., F. Potvin, and F. Duchesneau. 1987. Feasibility of an experimental reduction in coyote population in order to increase a white-tailed deer population. Naturaliste Canadien (Quebec) 114:477–86.

Michael, E. D. 1970. Activity patterns of white-tailed deer in South Texas. Texas Journal of Science 21:417–28.

Miller, H. A., and L. K. Halls. 1969. Fleshy fungi commonly eaten by southern wildlife (Research Paper SO-49). U.S. Department of Agriculture, Forest Service.

Miller, J. H., R. S. Boyd, and M. B. Edwards. 1999. Floristic diversity, stand structure, and composition 11 years after herbicide preparation. Canadian Journal of Forest Research 29:1073–83.

Miller, K. V., R. L. Marchinton, and W. M. Knox. 1991. White-tailed deer signposts and their role as a source of priming pheromones: A hypothesis. Transactions of the Congress of International Union of Game Biologists 18:455–58.

Miller, K. V., and R. L. Marchinton, eds. 1995. Quality whitetails: The why and how of quality deer management. Stackpole, Mechanicsburg, PA.

Miller, K. V., and R. L. Marchinton. 1999. Temporal distribution of rubbing and scraping by a high-density white-tailed deer, *Odocoileus virginianus*, population in Georgia. Canadian Field-Naturalist 113:519–21.

Miller, K. V., and J. J. Ozoga. 1997. Density effects on deer sociobiology. Pages 36–150 *in* W. J. McShea, H. B. Underwood, and J. H. Rappole, eds. The science of overabundance: Deer ecology and population management. Smithsonian Institution Press, Washington, DC.

Miller, K. V., O. E. Rhodes, Jr., T. R. Litchfield, M. H. Smith, and R. L. Marchinton. 1987. Reproductive characteristics of yearling and adult male white-tailed deer. Proceedings of the Annual Conference of the Southeast Association of Fish and Wildlife Agencies 41:378–84.

Miller, M. W., and E. T. Thorne. 1993. Captive cervids as potential sources of disease for North America's wild cervid populations: Avenues, implication, and preventive management. Transactions of the North American Wildlife and Natural Resources Conference 58:460–67.

Miller, M. W., E. S. Williams, C. W. McCarty, T. R. Spraker, T. J. Kreeger, C. T. Larsen, and E. T. Thorne. 2000. Epidemiology of chronic wasting disease in free-ranging cervids. Journal of Wildlife Diseases 36:676–90.

Miller, S. K. 1988. Reproductive biology of white-tailed deer on Cumberland Island, Georgia. M.S. Thesis, University of Georgia, Athens.

Minnis, D. L. and R. B. Peyton. 1995. Cultural carrying capacity: Modeling a notion. Pages 19–34 *in* J. B. McAninch, ed. Urban deer: A manageable resource? Proceedings of a symposium, 55th Midwest Fish and Wildlife Conference.

Moen, A. N. 1994. Whitetail population dynamics. Pages 208–17 *in* D. Gerlach, S. Atwater, and J. Schnell, eds. Deer. Stackpole, Harrisburg, PA.

Morrison, P. A. 1985. Habitat utilization by white-tailed deer on Davis Island. M.S. Thesis, Mississippi State University, Mississippi State.

Morton, G. H., and E. L. Cheatum. 1946. Regional differences in breeding potential of white-tailed deer in New York. Journal of Wildlife Management 10:242–48.

Mott, S. E., R. L. Tucker, D. C. Guynn, and H. A. Jacobson. 1985. Use of Mississippi bottomland hardwoods by white-tailed deer. Proceedings of the Annual Conference of the Southeastern Association of Fish and Wildlife Agencies 39:403–11.

Müller-Schwarze, D., R. Altieri, and N. Porter. 1984. Alert odor from skin gland in deer. Journal of Chemical Ecology 10:1707–29.

Murden, S. B., and K. L. Risenhoover. 1993. Effects of habitat enrichment on patterns of diet selection. Ecological Applications 3:497–505.

Naugle, D. E., J. A. Jenks, and B. J. Kernohan. 1996. Use of thermal infrared sensing to estimate density of white-tailed deer. Wildlife Society Bulletin 24:37–43.

Neff, D. J. 1968. The pellet-group count technique for big game trend, census, and distribution: A review. Journal of Wildlife Management 32:597–614.

Nelson, M. E. 1993. Natal dispersal and gene flow in white-tailed deer in northeastern Minnesota. Journal of Mammalogy 74:316–22.

Nelson, M. E., and L. D. Mech. 1981. Deer social organization and wolf predation in northeastern Minnesota. Wildlife Monographs 77:1–53.

Nelson, T. A., and L. D. Mech. 1986. Mortality of white-tailed deer in northeastern Minnesota. Journal of Wildlife Management 50:691–98.

Nelson, M. E., and L. D. Mech. 1987. Demes within a northeastern Minnesota deer population. Pages 27–40 *in* B. D. Chepko-sade and Z. T. Halpin, eds. Mammalian dispersal patterns. University of Chicago Press, Chicago.

Nelson, T. E., and A. Woolf. 1987. Mortality of white-tailed deer fawns in southern Illinois USA. Journal of Wildlife Management 51:326–29.

Nixon, C. M. 1971. Productivity of white-tailed deer in Ohio. Ohio Journal of Science 71:217–25.

Nixon, C. M., L. P. Hansen, P. A. Brewer, and J. E. Chelsvig. 1991a. Ecology of white-tailed deer in an intensively farmed region of Illinois. Wildlife Monograph 118:1–77.

Nixon, C. M., L. P. Hansen, P. A. Brewer, and J. E. Chelsvig. 1991b. Longevity of female white-tailed deer on a refuge in Illinois. Transactions of the Illinois State Academy of Science 84:84–91.

Nudds, T. D. 1980. Forage "preference": Theoretical considerations of diet selection by deer. Journal of Wildlife Management 44:735–40.

Osborn, D. A., K. V. Miller, D. M. Hoffman, W. H. Dickerson, J. W. Gassett, and C. F. Quist. 2000. Morphology of the white-tailed deer tarsal gland. Acta Theriologica 45:117–22.

Osborne, J. S., A. S. Johnson, P. E. Hale, R. L. Marchinton, C. V. Vansant, and J. M. Wentworth. 1992. Population ecology of Blackbeard Island white-tailed deer (Bulletin 26). Tall Timbers Research Station, Tallahassee, FL.

Ozoga, J. J. 1969. Some longevity records for female white-tailed deer in northern Michigan. Journal of Wildlife Management 33:1027–28.

Ozoga, J. J. 1988. Incidence of "infant" antlers among supplementally-fed white-tailed deer. Journal of Mammalogy 69:393–95.

Ozoga, J. J. 1995. Whitetail winter. Willow Creek Press, Minocqua, WI.

Ozoga, J. J., and L. J. Verme. 1970. Winter feeding patterns of penned white-tailed deer. Journal of Wildlife Management 34:431–39.

Ozoga, J. J., and L. J. Verme. 1982. Physical and reproductive characteristics of a supplementally-fed white-tailed deer herd. Journal of Wildlife Management 46:281–301.

Ozoga, J. J., L. J. Verme, and C. S. Bienz. 1982. Parturition behavior and territoriality in white-tailed deer: Impact on neonatal mortality. Journal of Wildlife Management 46:1–11.

Pais, R. C., W. C. McComb, and J. Phillips. 1991. Habitat associated with home ranges of female *Odocoileus virginianus* (Mammalia: Cervidae) in eastern Kentucky. Brimleyana 17:57–66.

Pekins, P. J., K. S. Smith, and W. W. Mautz. 1998. The energy cost of gestation in white-tailed deer. Canadian Journal of Zoology 76:1091–97.

Pledger, J. M. 1975. Activity, home range, and habitat utilization of white-tailed deer (*Odocoileus virginianus*) in southeastern Arkansas. M.S. Thesis, University of Arkansas, Fayetteville.

Progulske, D. R., and D. C. Duerre. 1964. Factors influencing spotlighting counts of deer. Journal of Wildlife Management 28:27–34.

Pruitt, L. R., and S. E. Pruitt. 1986. Effects of northern hardwood management on deer and other large mammal populations. Pages 363–79 *in* R. D. Nyland, ed. Managing northern hardwoods (Miscellaneous Publication No. 13). State University of New York, Faculty of Forestry.

Purdue, J. R. 1983. Epiphyseal closure in white-tailed deer. Journal of Wildlife Management 47:1207–13.

Ransom, A. B. 1967. Reproductive biology of white-tailed deer in Manitoba. Journal of Wildlife Management 31:114–23.

Rawson, R. E., G. D. DelGiudice, H. E. Dziuk, and L. D. Mech. 1992. Energy metabolism and hematology of white-tailed deer fawns. Journal of Wildlife Diseases 28:91–94.

Rhodes, O. E., Jr., K. T. Scribner, M. H. Smith, and P. E. Johns. 1985. Factors affecting the number of fetuses in a white-tailed deer herd. Proceedings of the Annual Conference of the Southeastern Association of Fish and Wildlife Agencies 39:380–88.

Rhodes, O. E., Jr., J. M. Novak, M. H. Smith, and P. E. Johns. 1986. Assessment of fawn breeding in a South Carolina deer herd. Proceedings of the Annual Conference of the Southeastern Association of Fish and Wildlife Agencies 40:430–37.

Rhodes, O. E., Jr., J. M. Novak, M. H. Smith, and P. E. Johns. 1991a. Frequency distribution of conception dates in a white-tailed deer herd. Acta Theriologica 36:131–40.

Rhodes, O. E., Jr., M. H. Smith, K. T. Scribner, and P. E. Johns. 1991b. Factors affecting productivity of a white-tailed deer herd. Transactions of the 18th International Union of Game Biologists congress. Swiat Press, Krakow-Warszawa, Poland.

Richter, A. R., and R. F. Labisky. 1985. Reproductive dynamics among disjunct white-tailed deer herds in Florida. Journal of Wildlife Management 49:964–71.

Robbins, C. T. 1993. Wildlife feeding and nutrition. Academic Press, San Diego, CA.

Robbins, C. T., R. L. Prior, A. N. Moen, and W. J. Visek. 1974. Nitrogen metabolism of white-tailed deer. Journal of Animal Science 38:186–91.

Roberson, J. H, Jr., and D. Dennett, Jr. 1967. Breeding season of white-tailed deer in Louisiana. Proceedings of the Southeastern Association of Game and Fish Commissioners 20:123–30.

Rogers, C. L. 1996. Utilization of cedar glades by white-tailed deer at Chickamauga Battlefield Park. M.S. Thesis, University of Georgia, Athens.

Romin, L. A., and J. A. Bissonette. 1996. Deer–vehicle collisions: Status of state monitoring activities and mitigation efforts. Wildlife Society Bulletin 24:276–83.

Roseberry, J. L., and W. D. Klimstra. 1970. Productivity of white-tailed deer on Crab Orchard National Wildlife Refuge. Journal of Wildlife Management 34:23–28.

Roseberry, J. L., and W. D. Klimstra. 1975. Some morphological characteristics of the Crab Orchard deer herd. Journal of Wildlife Management 39:48–58.

Rosenberry, C. S. 1999. Dispersal ecology and behavior of yearling male white-tailed deer. Ph.D. Dissertation, North Carolina State University, Raleigh.

Rue, L. L., III. 1978. The deer of North America. Outdoor Life Books/Crown, New York.

Sams, M. G., R. L. Lochmiller, C. W. Qualls, Jr., and D. M. Leslie, Jr. 1998. Sensitivity of condition indices to changing density in a white-tailed deer population. Journal of Wildlife Diseases 34:110–25.

Samuel, W. M. 1994. The parasites and diseases of whitetails. Pages 233–35 *in* D. Gerlach, S. Atwater, and J. Schnell, eds. Deer. Stackpole, Mechanicsburg, PA.

Samuel, W. M., and S. Demarais. 1993. Conservation challenges concerning wildlife farming and ranching in North America. Transactions of the North American Wildlife and Natural Resources Conference 58:445–47.

Sargent, R. A., Jr. 1992. Movement ecology of adult male white-tailed deer in hunted and non-hunted populations in the wet prairie of the Everglades. Ph.D. Dissertation, University of Florida, Gainesville.

Sauer, P. R. 1984. Physical characteristics. Pages 73–90 *in* L. K. Halls, ed. White-tailed deer: Ecology and management. Stackpole, Harrisburg, PA.

Sawyer, T. G. 1981. Behavior of female white-tailed deer with emphasis on pheromonal communication. M.S. Thesis, University of Georgia, Athens.

Sawyer, T. G., R. L. Marchinton, and K. V. Miller. 1989. Response of female white-tailed deer to scrapes and antler rubs. Journal of Mammalogy 70:431–33.

Sawyer, T. G., K. V. Miller, and R. L. Marchinton. 1993. Patterns of urination and rub-urination in female white-tailed deer. Journal of Mammalogy 74:477–79.

Schmitt, S. M., S. D. Fitzgerald, T. M. Cooley, C. S. Bruning-Fann, L. Sullivan, D. Berry, T. Carlson, R. B. Minnis, J. B. Payeur, and J. Sikarskie. 1997. Bovine tuberculosis in free-ranging white-tailed deer from Michigan. Journal of Wildlife Diseases 33:749–58.

Schultz, S. R., and M. K. Johnson. 1992. Breeding by male white-tailed deer fawns. Journal of Mammalogy 73:148–50.

Seal, U. S., L. J. Verme, and J. J. Ozoga. 1981. Physiologic values. Pages 17–34 *in* W. R. Davidson, F. A. Hayes, V. F. Nettles, and F. E. Kellog, eds. Diseases and parasites of white-tailed deer. Tall Timbers Research Station, Tallahassee, FL.

Seber, G. A. F. 1982. The estimation of animal abundance and related parameters, 2nd ed. Macmillan, New York.

Seton, E. T. 1929. Lives of game animals, Vol III. Doubleday, Doran, Garden City, NY.

Severinghaus, C. W. 1949. Tooth development and wear as criteria of age in white-tailed deer. Journal of Wildlife Management 13:195–216.

Severinghaus, C. W. 1963. Effectiveness of archery in controlling deer abundance on the Howland Island Game Management Area. New York Fish and Game Journal 10:186–93.

Severinghaus, C. W. 1972. Weather and the deer population. The Conservationist 28:28–31.

Silver, H., N. F. Colovos, J. B. Holter, and H. H. Hayes. 1969. Fasting metabolism of white-tailed deer. Journal of Wildlife Management 33:490–98.

Smith, W. P. 1985. Current geographic distribution and abundance of Columbian white-tailed deer *Odocoileus virginianus leucurus*. Northwest Science 59:243–51.

Smith, W. P. 1991. *Odocoileus virginianus*. Mammalian Species 388:1–13.

Stoll, R. J., Jr., W. L. Culbertson, and M. W. McClain. 1985. Deer–vehicle accidents in Ohio 1977–1983. Ohio Department of Natural Resources, Division of Wildlife Resources.

Stormer, F. A., and W. A. Bauer. 1980. Summer forage use by tame deer in northern Michigan. Journal of Wildlife Management 44:98–106.

Strickland, B. K., and S. Demarais. 2000. Age and regional differences in antlers and mass of white-tailed deer. Journal of Wildlife Management 64:903–11.

Strickland, B. K., S. Demarais, L. E. Castle, J. W. Lipe, W. H. Lunceford, H. A. Jacobson, D. Frels, and K. V. Miller. 2001. Effects of selective-harvest strategies on white-tailed deer antler size. Wildlife Society Bulletin 29:509–20.

Strole, T. A., and R. C. Anderson. 1992. White-tailed deer browsing: Species preferences and implications for central Illinois forests. Natural Areas Journal 12:139–44.

Swenson, J. E., S. J. Knapp, and H. J. Wentland. 1983. Winter distribution and habitat use by mule deer and white-tailed deer in southeastern Montana. Prairie Naturalist 15:97–112.

Swihart, R. K., P. M. Picone, A. J. DeNicola, and L. Cornicelli. 1995. Ecology of urban and suburban white-tailed deer. Pages 35–44 *in* J. B. McAninch, ed. Urban deer: A manageable resource? Proceedings of a Symposium, 55th Midwest fish and wildlife conference.

Teer, J. G. 1984. Lessons from the Llano Basin, Texas. Pages 261–90 *in* L. K. Halls, ed. White-tailed deer: Ecology and management. Stackpole, Harrisburg, PA.

Thompson, C. B., J. B. Holter, H. H. Hayes, H. Silver, and W. E. Urban, Jr. 1973. Nutrition of white-tailed deer. I. Energy requirements of fawns. Journal of Wildlife Management 37:301–11.

Tierson, W. C., G. F. Mattfeld, R. W. Sage, Jr., and D. F. Behrend. 1985. Seasonal movements and home range of white-tailed deer in the Adirondacks. Journal of Wildlife Management 49:760–69.

Tilghman, N. G. 1989. Impacts of white-tailed deer on forest regeneration in northwestern Pennsylvania. Journal of Wildlife Management 53:524–32.

Trainer, D. O., and L. H. Karstad. 1970. Epizootic hemorrhagic disease. Pages 50–54 *in* J. W. Davis, L. H. Karstad, and D. O. Trainer, eds. Infectious diseases of wild mammals. Iowa State University Press, Ames.

Trivers, R. L., and D. E. Willard. 1973. Natural selection of parental ability to vary the sex ratio of offspring. Science 179:90–92.

Ullrey, D. E., W. G. Youatt, H. E. Johnson, L. D. Fay, and B. L. Bradley. 1967. Protein requirement of white-tailed deer fawns. Journal of Wildlife Management 31:679–85.

Ullrey, D. E., W. G. Youatt, H. E. Johnson, L. D. Fay, B. L. Schoepke, W. T. Magee, and K. K. Keahey. 1973. Calcium requirements of weaned white-tailed deer fawns. Journal of Wildlife Management 37:187–94.

U.S. Department of the Interior and U.S. Department of Commerce. 1996. 1996 National survey of fishing, hunting, and wildlife-associated recreation. U.S. Government Printing Office, Washington, DC.

Vangilder, L. D., O. Torgerson, and W. R. Porath. 1982. Factors influencing diet selection by white-tailed deer. Journal of Wildlife Management 46:711–18.

VerCauteren, K. C. 1998. Dispersal, home range fidelity, and vulnerability of white-tailed deer in the Missouri River Valley. Ph.D. Dissertation, University of Nebraska, Lincoln.

VerCauteren, K. C., and S. E. Hygnstrom. 1998. Effects of agricultural activities and hunting on home ranges of female white-tailed deer. Journal of Wildlife Management 62:280–85.

Verme, L. J. 1963. Effect of nutrition on growth of white-tailed deer fawns. Transactions of the North American Wildlife and Natural Resources Conference 28:431–43.

Verme, L. J. 1965. Reproduction studies on penned white-tailed deer. Journal of Wildlife Management 29:74–79.

Verme, L. J. 1968. An index of winter weather severity for northern deer. Journal of Wildlife Management 32:566–74.

Verme, L. J. 1969. Reproductive patterns of white-tailed deer related to nutritional plane. Journal of Wildlife Management 33:881–87.

Verme, L. J. 1977. Assessment of natal mortality in Upper Michigan deer. Journal of Wildlife Management 41:700–708.

Verme, L. J. 1983. Sex ratio variation in *Odocoileus*: A critical review. Journal of Wildlife Management 47:573–82.

Verme, L. J. 1991. Decline in doe fawn fertility in southern Michigan deer. Canadian Journal of Zoology 69:25–28.

Verme, L. J., and J. J. Ozoga. 1980a. Effects of diet on growth and lipogenesis in deer fawns. Journal of Wildlife Management 44:315–24.

Verme, L. J., and J. J. Ozoga. 1980b. Influence of protein-energy intake on deer fawns in autumn. Journal of Wildlife Management 44:305–14.

Verme, L. J., and J. J. Ozoga. 1981. Sex ratio of white-tailed deer and the estrus cycle. Journal of Wildlife Management 45:710–15.

Verme, L. J., and J. J. Ozoga. 1987. Relationship of photoperiod to puberty in doe fawn white-tailed deer. Journal of Mammalogy 68:107–10.

Verme, L. J., J. J. Ozoga, and J. T. Nellist. 1987. Induced early estrus in penned white-tailed deer does. Journal of Wildlife Management 51:54–56.

Waid, D. D., and R. J. Warren. 1984. Seasonal variations in physiological indices of adult female white-tailed deer in Texas. Journal of Wildlife Diseases 20:212–19.

Waller, D. M., and W. S. Alverson. 1997. The white-tailed deer: A keystone herbivore. Wildlife Society Bulletin 25:217–26.

Warren, R. C., and G. A. Hurst. 1981. Ratings of plants in pine plantations as white-tailed deer food (Information Bulletin 18). Mississippi Agricultural and Forest Experiment Station, Mississippi State.

Warren, R. J. 1997. The challenge of deer overabundance in the 21st century. Wildlife Society Bulletin 25:213–14.

Warren, R. J., and R. L. Kirkpatrick. 1982. Evaluating nutritional status of white-tailed deer using fat indices. Proceedings of the Annual Conference of the Southeastern Association of Fish and Wildlife Agencies 36:463–72.

Warren, R. J., R. W. Vogelsang, R. L. Kirkpatrick, and P. F. Scanlon. 1978. Reproductive behavior of captive white-tailed deer. Animal Behaviour 26:179–83.

Warren, R. J., R. L. Kirkpatrick, A. Oelschlaeger, P. F. Scanlon, and F. C. Gwazdauskas. 1981. Dietary and seasonal influences on nutritional indices of adult male white-tailed deer. Journal of Wildlife Management 45:926–36.

Webb, J. W., and D. W. Nellis. 1981. Reproductive cycle of white-tailed deer of St. Croix, Virgin Islands. Journal of Wildlife Management 45:253–58.

Weckerly, F. W., and J. P. Nelson, Jr. 1990. Age and sex differences of white-tailed deer diet composition, quality, and calcium. Journal of Wildlife Management 54:532–38.

Wentworth, J. M. 1989. Deer-habitat relationships in the Southern Appalachians. Ph.D. Dissertation, University of Georgia, Athens.

Wentworth, J. M., A. S. Johnson, and P. E. Hale. 1990. Influence of acorn use on nutritional status and reproduction in the Southern Appalachians. Proceedings of the Annual Conference of the Southeastern Association of Fish and Wildlife Agencies 44:142–54.

Wesson, J. A., III., P. F. Scanlon, and H. S. Mosby. 1979a. Influence of chemical immobilization and physical restraint on packed cell volume, total protein, glucose, and blood urea nitrogen in blood of white-tailed deer. Canadian Journal of Zoology 57:756–67.

Wesson, J. A., III., P. F. Scanlon, and H. S. Mosby. 1979b. Influence of time of sampling after death on blood measurements of white-tailed deer. Canadian Journal of Zoology 57:777–80.

White, L. M., D. A. Hosack, R. J. Warren, and R. A. Fayrer-Hosken. 1995. Influence of mating on duration of estrus in captive white-tailed deer. Journal of Mammalogy 76:1159–63.

Whitney, M. D., D. L. Forster, K. V. Miller, and R. L. Marchinton. 1992. Sexual attraction in white-tailed deer. Pages 327–33 *in* R. D. Brown, ed. Biology of deer. Springer-Verlag, New York.

Whitlaw, H. A., W. B. Ballard, D. L. Sabine, S. J. Young, R. A. Jenkins, and G. J. Forbes. 1998. Survival and cause-specific mortality rates of adult white-tailed deer in New Brunswick. Journal of Wildlife Management 62:1335–41.

Wiggers, E. P., and S. F. Beckerman. 1993. Use of thermal infrared sensing to survey white-tailed deer populations. Wildlife Society Bulletin 21:263–68.

Williams, E. S., J. K. Kirkwood, and M. W. Miller. 2001. Transmissible spongiform encephalopathies. Pages 292–301 *in* E. S. Williams and I. K. Barker, eds. Infectious diseases of wild mammals, 3rd ed. Iowa State University Press, Ames.

Williams, J. D., W. F. Krueger, and D. H. Harmel. 1994. Heritabilities for antler characteristics and body weight in yearling white-tailed deer. Heredity 73:78–83.

Williamson, L. L., and G. L. Doster. 1981. Socio-economic aspects of white-tailed deer diseases. Pages 434–39 *in* W. R. Davidson, ed. Diseases and parasites of white-tailed deer (Miscellaneous Publication 7), Tall Timbers Research Station, Tallahassee, FL.

Worden, K. A., and P. J. Pekins. 1995. Seasonal change in feed intake, body composition, and metabolic rate of white-tailed deer. Canadian Journal of Zoology 73:452–57.

Wywialowski, A. P. 1994. Agricultural producers' perceptions of wildlife-caused losses. Wildlife Society Bulletin 22:370–81.

Yantis, J. H., C. D. Frentress, W. S. Daniel, and G. H. Veteto. 1983. Deer management in the Post Oak Belt. Texas Parks and Wildlife, Austin.

Zagata, M. D., and A. O. Haugen. 1973. Influence of light and weather on observability of Iowa deer. Journal of Wildlife Management 38:220–28.

Zagata, M. D., and A. N. Moen. 1974. Antler shedding by white-tailed deer in the Midwest. Journal of Mammalogy 55:656–59.

Karl V. Miller, Wildlife Ecology and Management, D. B. Warnell School of Forest Resources, University of Georgia, Athens, Georgia 30602. Email: kmiller@smokey.forestry.uga.edu.

Lisa I. Muller, Department of Forestry, Wildlife and Fisheries, University of Tennessee, Knoxville, Tennessee 37996. Email: limuller@agmail.ag.utk.edu.

Stephen Demarais, Department of Wildlife and Fisheries, Mississippi State University, Mississippi State, Mississippi 39762-9690. Email: sdemarais@cfr.msstate.edu.

45

Moose

Alces alces

R. Terry Bowyer
Victor Van Ballenberghe
John G. Kie

NOMENCLATURE

Common Name. Moose in North America, elk in Eurasia
Scientific Name. *Alces alces*
Subspecies. Moose are in the suborder Ruminatia, infraorder Pecora, family Cervidae, subfamily Odocoileinae, and tribe Alcini. *Alces* (Gray 1821) is a monotypic genus with four subspecies recognized in North America (Peterson 1952, 1955; Hundertmark et al. 2003). The Alaskan moose (*A. a. gigas*) occupies Alaska, United States, and western Yukon Territory, Canada. Moose in southeastern Alaska appear distinct from other subspecies based on their mitochondrial DNA, but traditionally have been classified with *A. a. gigas* (Hundertmark et al. 2003). The northwestern moose (*A. a. andersoni*) occurs from British Columbia to Ontario in Canada and into the northern tier of the lower United States. Shiras moose (*A. a. shirasi*) inhabits mostly mountainous areas of southern British Columbia, northeastern Washington, Idaho, Montana, and Wyoming, and now also occurs in Colorado and Utah. The eastern moose (*A. a. americana*) occurs from Ontario eastward to the Atlantic seaboard (Fig. 45.1). Franzmann (1981) provided a complete taxonomic account of *Alces,* including synonyms.

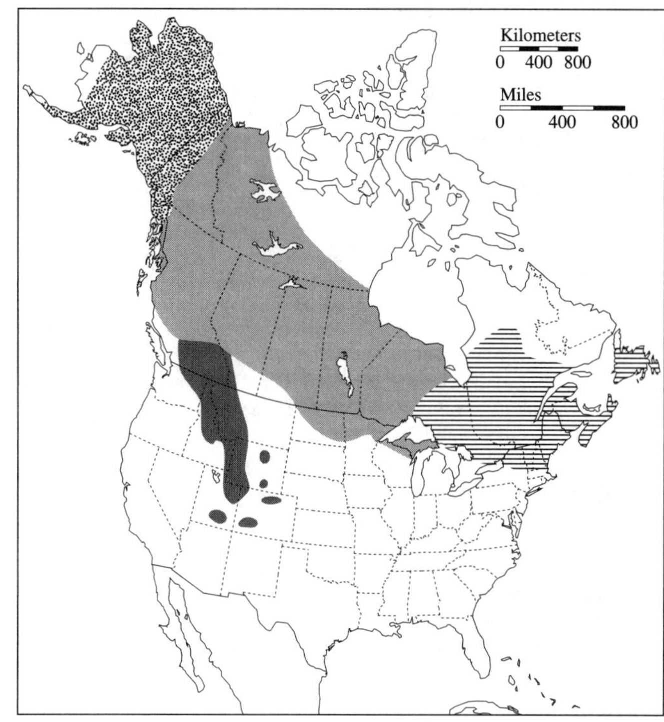

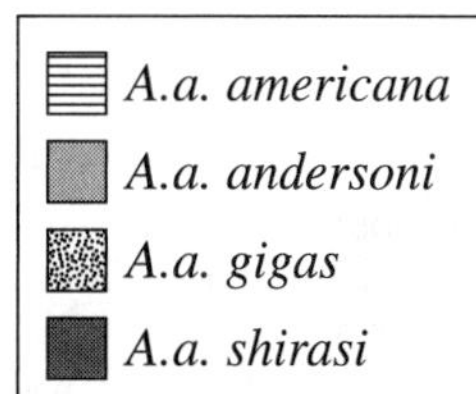

Figure 45.1. Distributions of subspecies of moose (*Alces alces*). Intergradation is common along zones of contact between subspecies. Source: Data from Hall and Kelson (1959), Hall (1981), Peek and Morris (1998), and Franzmann (2000).

EVOLUTION AND DISTRIBUTION

Alces likely evolved in Eurasia during the early Pleistocene (about 1.5 million years ago) with a direct lineage from *A. gallicus* to *A. latifrons* to *A. alces* (Lister 1993). Although *A. alces* may have existed in Europe as early as 100,000–150,000 years ago (Lister 1993), remains of *A. latifrons* from Beringia may be as recent as 35,000 years old (Guthrie 1990, 1995). Thus, *A. latifrons* may have been isolated from *A. alces* and persisted for much longer in Beringia than in Eurasia.

Cervalces (= *Alces*) *scotti* occurred in North America from as early as 100,000–200,000 to 11,000 years ago. Those mooselike cervids (stag moose) are not thought to be progenitors of modern moose (Lister 1993), and likely diverged from *A. latifrons* during the Wisconsinan glaciation. The earliest remains of modern moose (*A. alces*) in Alaska date from only about 9700 years ago (Guthrie 1990). Thus, *A. alces* colonized the New World via the Bering land bridge during the late Pleistocene. Moose then likely dispersed down an ice-free corridor between the Cordilleran and Laurentide ice shields into suitable habitat south of the ice, which still covered much of the north (Bowyer et al. 1991; Hundertmark et al. 2003). Glaciation clearly played an important role in the evolution of moose in North America, but not in the manner previously proposed. Several hypotheses forwarded to explain differences in subspecies of moose in North America require that *A. a. gigas* be separated from the other subspecies by the Wisconsinan glaciation (Peterson 1955; Kelsall and Telfer 1974; Bubenik 1998; Geist 1998). Those hypotheses, however, are not consistent with the late dates for *A. alces* in Alaska and their absence south of the ice sheet, and, consequently, do not provide an adequate explanation for subspeciation of moose in the New World (Bowyer et al. 1991; Hundertmark et al. 2002, 2003).

Systematics of moose are complicated further by European moose possessing a karyotype of $2n = 68$ (Gustavsson and Sundt 1968), whereas moose from North America are typified by $2n = 70$ (Hsu and Benirschke 1973). Moreover, North American moose have a 75-base pair indel in their mitochondrial control region thought to be lacking in European moose (Mikko and Andersson 1995). Those differences have led to speculation that a back-colonization occurred in moose from North America to the Russian Far East (Coady 1982), or that moose on opposite sides of the Bering Strait were separate species (Boeskorov 1997). Recently, however, the North American indel was documented for moose inhabiting eastern Asia (Boeskorov 1993, 1997; Udina et al. 2002; Hundertmark et al. 2002), indicating that moose colonizing the New World were a subset of the races occurring in Asia. Thus, moose occupying opposite sides of the Bering Strait are best characterized as distinct subspecies, which are not closely related; morphological similarities between those subspecies are the result of convergent evolution (Hundertmark et al. 2002, 2003). Moreover, the hypothesis that a double migration of the New World occurred from Asia (Bubenik 1998) is not supported by genetic relationships among subspecies of moose (Hundertmark et al. 2002, 2003). Evidence indicates, however, that moose colonizing the New World across the Bering land bridge underwent extremely rapid evolution, with differences in morphology and behavior evident among subspecies developing in about 10,000 years (Bowyer et al. 1991). Moose in the Alexander Archipelago in southeastern Alaska probably colonized that area via movements down major river corridors from interior British Columbia, Canada (Hundertmark et al. 2003), rather than from the south following retreat of the ice sheet, as did many other mammals (Klein 1965). Those moose are genetically distinct from *A. a. gigas* in Alaska and *A. a. shirasi* and *A. a. andersoni* in British Columbia; indeed, mitochondrial DNA indicates differences among all subspecies of moose (Hundertmark et al. 2002, 2003).

Moose are a prominent component of boreal landscapes throughout the Holarctic. In North America, these large herbivores are associated with forests dominated by spruce (*Picea*), fir (*Abies*), and pine (*Pinus*) (Telfer 1984; Karns 1998). Likewise, moose may occupy tundra where suitable forage is plentiful or occur in mountain zones with a more open overstory than in dense taiga. Fire plays a crucial role in the ecology of moose, with populations tracking forest succession in many areas (Peek 1974a; Bangs and Bailey 1980; Loranger et al. 1991; Weixelman et al. 1998). Forage and snow cover likely influence the northern distribution of moose, whereas a hot climate may limit their distribution to the south (Kelsall and Telfer 1974; Renecker and Hudson 1986; Karns 1998).

Moose populations are stable or increasing throughout North America; estimates compiled by Karns (1998) indicate about 939,000 moose in 1960 and 975,000 moose in 1990. Timmermann (In press) estimated about 973,000 moose in 2001. Their geographic range recently has expanded southward down the Rocky Mountains (Smith 1985) as well as down the east coast of the United States (Peek and Morris 1998). Historical records on the distribution of moose are summarized in Seton (1927) and Peterson (1955). Translocations have aided the expansion of moose in some areas (Pimlott 1961; Burris and McKnight 1973; Dodds 1974; Duvall and Schoonveld 1988; Aho and Hendrickson 1989; Franzmann 1998; Olterman and Kenvin 1998).

DESCRIPTION

Numerous accounts describing the biology of moose are available (Murie 1934; Peterson 1955; Houston 1968; Krefting 1974a; Franzmann 1978, 1981; Coady 1982; Bowyer et al. 1997; Franzmann and Schwartz, 1998; Franzmann 2000). Moreover, the journal *Alces* is dedicated to understanding the biology and management of moose. Those sources, as well as other publications referenced in this chapter, provide additional information and relevant citations for this unique mammal.

The moose is the largest extant member of the Cervidae. Adult males have massive palmate antlers, which typically are absent in females. Indeed, like many polygynous ruminants (Weckerly 1998; Loison et al. 1999a), moose exhibit extreme sexual dimorphism, with males being >40% heavier than females (Schwartz et al. 1987). *A. a. gigas* is the largest subspecies of moose (including subspecies in Eurasia). In North America, *A. a. andersoni* is next largest, followed by *A. a. americana,* and *A. a. shirasi.*

Adult moose have long legs supporting massive bodies with a pronounced shoulder hump. Their nose is disproportionately long and slightly pendulous compared with that of other cervids, and they have an overhanging upper lip with a small, inverted triangular area of bare skin. Moose have long ears (250 mm) and short tails (80–120 mm). Moose also have a long dewlap of hair-covered skin, termed the bell, which hangs beneath their neck (Bubenik 1983; Miquelle and Van Ballenberghe 1985). Among adults, bells become shorter with age. Especially long bells may freeze off during winter (Franzmann 1981; Timmermann et al. 1985, 1988). Unlike those of most other cervids, the ears of male moose are positioned above the main beam of the antlers. The neck of male moose swells markedly in preparation for the mating season (rut). Moose may appear awkward and ungainly until they are observed negotiating tussock tundra or deep snow; their gait and smooth movement attest to their unique adaptations to a northern environment.

Moose have a suborbital (lachrymal) gland below each eye and, compared with other cervids, have small tarsal glands on the inside of their hind legs. They lack metatarsal glands on the outside of their legs (Franzmann 2000). Moose have four mammae. Contrary to most published accounts, however, moose have interdigital glands (Chapman 1985). Unlike *Odocoileus* and *Rangifer,* moose lack nasal glands (Atkeson et al. 1988). Sokolov and Chernova (1987) described the morphology of the skin.

Pelage. Young are born with a reddish-brown pelage, which lacks spots or other markings more typical of adults. Coloration of pelage for adult moose varies among subspecies, with *A. a. gigas* possessing a shiny black rump patch, which extends forward toward the withers. Legs are black with white "stockings"; pelage across the back and down the sides can be dark gray to grizzled brown (Bowyer et al. 1991). Adult males have darker facial hair than do females (Timmermann 1993). Adult females possess a line of white hair around the vulva, which can be used to identify them during aerial surveys (Mitchell 1970; Roussel 1975). A white color morph of moose that is not an albino has been described (Franzmann 1981). White moose may give birth to normal-colored young (Armstrong and Brown 1986). Moose molt in spring and acquire a coat of short, shiny hair. They grow their winter pelage by late September. Winter pelage is characterized by long (20 cm), hollow guard hairs and a substantial undercoat, which provides excellent insulation (Franzmann 1981). The pelage becomes gradually lighter during winter as hairs become bleached and broken. *A. a. shirasi* and *A. a. americana* are much darker and less distinctly marked than *A. a. gigas,* with *A. a. andersoni* possessing intermediate pelage characteristics. Franzmann et al. (1977) discussed mineral composition of moose hair.

Weight and Measurements. Male *A. a. gigas* may attain >770 kg and females >570 kg in body weight when mature. Males achieve asymptotic body weight by 8 years old, whereas females typically do so by 4 years of age (Schwartz et al. 1987). Other subspecies are smaller (Breckenridge 1946; Blood et al. 1967; Doutt 1970; Schladwieler and Stevens 1973; Peterson 1974; Sæther 1985; Quinn and Aho 1989; Lynch et al. 1995). Substantial annual variation occurs in body weight, especially for males, which may loose 12–18% of their weight during rut (Schwartz et al. 1987), when they become hypophagic for about 18 days (Miquelle 1990). Moose may loose 20–55% of their maximal body weight attained in autumn by late winter (Franzmann et al. 1978); a maintenance diet cannot be obtained by feeding on browse in winter (Schwartz and Renecker 1998; Spalinger 2000). Karns (1976) and Haigh et al. (1980) correlated body measurements with weights of moose. Hundertmark and Schwartz (1998) provided equations for predicting body weight from body measurements and indices of condition, and Hundertmark and Schwartz (2002) predicted body fat from bioelectrical impedance. Adult female *A. a. gigas* averaged 302 cm in body length and 201 cm in chest girth, and were 182 cm high at the shoulder (Franzmann et al. 1978). Sexual dimorphism also occurs in the morphology of phalanges and the metapodials (Iregren 1985). Regional variation can occur in skeletal measurements of moose even within the same subspecies (Crête 1988).

Mean birth weight for male *A. a. gigas* was 18 kg, with females weighing about 2 kg less. Average weight of twins was 3–4 kg less than singletons (Keech et al. 2000). By 10 months of age, female Alaskan moose had attained a mean body weight of 148.9 kg (Keech et al. 1999). Mean total body length was 205.7 cm and mean metatarsus length reached 50.3 cm. Neonates with low body weight at birth remained among the smallest individuals in their cohort 10 months later (Keech et al. 1999).

Skull and Dentition. Facial bones of moose skulls are elongated in comparison with those of other cervids (Fig. 45.2). Frontal bones, which support antler pedicles in males, are anchored to the parietal with an unusually strong suture, which likely helps support massive antlers and strenuous forces placed on the skull during rutting activities (Bubenik 1998). The orbits protrude from the skull and possess a pronounced supraorbital rim; moose undoubtedly have a wide-ranging view of their surroundings. Nasals are relatively short with a space between those bones and the anterior edge of the premaxillae, which likely was associated with the development of a muscular and somewhat prehensile nose (Bubenik 1998). The maxilla is long and slender; the vomer is V-shaped and separates the nasal cavity into two chambers. Turbinate bones are well developed, and probably indicate enhanced olfactory capabilities (Bubenik 1998). The lower mandible narrows distally toward the front teeth and possesses a distinct diastema, which separates the front teeth from premolars and molars. Diastema length is correlated with the total length of the skull (Peterson 1955). Moose exhibit sexual dimorphism in their cranial development (Bartowsiewvicz 1987).

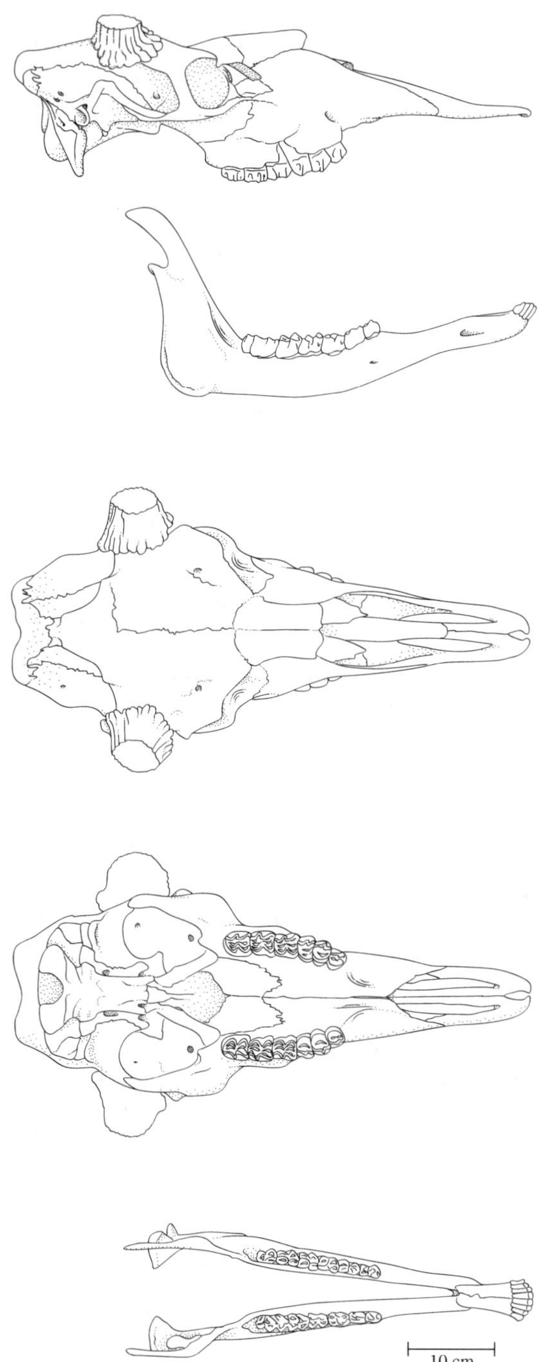

FIGURE 45.2. Skull of a male moose (*Alces alces andersoni*). From top to bottom: lateral view of cranium, lateral view of mandible, dorsal view of cranium, ventral view of cranium, dorsal view of mandible.

Moose possess hypsodont cheek teeth, with enamel and dentin interfolded to help reduce wear from browsing on woody vegetation; premolars and molars are broad. The typical dentition of moose is I 0/3, C 0/1, P 3/3, M 3/3. Moose normally lack upper incisors and canines. Supernumerary incisiform teeth have been reported (Peterson 1955; Steele and Parama 1979), as have rudimentary upper canines (Bubenik 1998), though both are rare. The lower incisors and incisiform canines bite against a callous pad in the upper jaw. Breadth of incisiform teeth reaches an asymptote at about 4 years of age in males and females. Moose have a much narrower muzzle relative to their body weight than do other ruminants, which likely relates to their feeding selectively on browse (Spaeth et al. 2001). Tooth eruption and replacement of deciduous teeth (incisors, canines, and premolars) have been used to determine age in moose (Peterson 1955; Peterson et al. 1983; Hindelang and Peterson 1993). The first permanent incisors may erupt by 6 months of age, the first molars emerge at 4–6 months, second molars at 10–13 months, and the third molars by 16–19 months of age. By 16–19 months, all premolars have been replaced and full permanent dentition has been attained (Peterson 1955). Thereafter, tooth wear may be used to estimate age of moose. Nonetheless, tooth wear and breakage of incisiform teeth (Smith 1992) can be exacerbated in populations at high density feeding on coarse foods, leading to early senescence (Young and Marty 1986; Hindelang and Peterson 1994; Bowyer et al. 1999a). Counts of cementum annuli are thought to provide a more accurate estimate of age than patterns of tooth wear and eruption (Sergeant and Pimlott 1959; Wolfe 1969; Gasaway et al. 1978; Haagenrud 1978), but also may be influenced by effects of population density on physical condition in cervids (McCullough 1996).

Antlers. Male moose possess the largest antlers of any living cervid, attaining weights of up to 35 kg (Bubenik 1998); antlered females are rare (Wishart 1990). Antler growth, size, and conformation are dependent on a combination of genetics, age, and nutrition (Goss 1983). Antler abnormalities can be caused by injury to growing antlers or to other parts of the body during antlerogenesis, as well as by abnormal hormone levels (Bubenik 1998). Accessory antlers occasionally occur in moose (Bubenik and Hundertmark 2002). Antlers of moose in North America are strongly palmate (palmicorn), with *A. a. gigas* possessing a "butterfly" configuration involving the main palm and brow palm, which is absent or less evident in other subspecies (Gasaway et al. 1987). Antler size and body weight are correlated among cervids (McCullough 1982; Sæther and Haagenrud 1985; Bowyer 1986; Solberg and Sæther 1994; Stewart et al. 2000). Accordingly, size of antlers at maturity generally follows patterns of body size noted for subspecies of moose in North America, with *A. a. gigas* possessing the largest antlers and *A. a. shirasi* the smallest. Although nutrition affects the size and confirmation of moose antlers, clear differences in the size of antlers exist among subspecies of moose (Gasaway et al. 1987; Bowyer et al. In press).

Antlers are regrown and cast each year. During antler growth, which may begin as early as March in large males, antlers are covered with skin and short, fine hair referred to as "velvet," and are highly vascularized and innervated—paths of blood vessels can be seen in the hardened antler following shedding of velvet in late August. Antlers are among the fastest growing of all vertebrate tissues and during June and July, when most growth occurs, may increase in length by nearly 2.5 cm each day (Van Ballenberghe 1983a). Antler thrashing of trees and shrubs coincides with velvet shedding, and strips of velvet may be removed during that dominance display (Fig. 45.3); sometimes males consume the shed velvet (Van Ballenberghe and Miquelle 1996). Hardened antlers are composed of an outer layer of compact bone and an inner core of cancellous bone. Growth and ultimately mineralization occur as osteoblasts build on a cartilaginous matrix. Demands for calcium

FIGURE 45.3. Antler thrashing by a male moose (*Alces alces gigas*). This behavior is a dominance display used to intimidate rivals. The primary function of antler thrashing is not shedding of velvet; although velvet may be stripped away during this aggressive activity, antler thrashing continues long after velvet shedding is completed. SOURCE: Photo by V. Van Ballenberghe.

and phosphorus (deposited in a ratio of about 2:1) are high during mineralization of antlers (Moen and Pastor 1998), and may be provided by temporary decalcification of the skeleton (Hillman et al. 1973). Such high demands for those minerals may result in osteoporosis in moose (Hindelang and Peterson 1996, 2000; Hindelang et al. 1998).

Following the mating season (rut), large males may begin antler casting as early as late November, with most large males having completed that process by early January. The phenology of the antler cycle in moose differs by age class, with larger, older males typically growing, shedding velvet, and casting antlers earlier than their smaller, younger counterparts (Table 45.1).

Casting of antlers likely is achieved by osteoclasts demineralizing bone at the distal ends of the pedicles, thereby allowing the antlers to drop off. Blood sometimes is observed on the pedicle immediately following casting. Antler pedicles usually are obvious in young moose by their first autumn, and some males may produce a small, mineralized area at the tip of the antler (Bubenik 1998). Thereafter, sets of antlers grow rapidly in size each year until maximal size is attained at 7–11 years old, with a subsequent senescence in antler size through about 18 years of age (Bubenik et al. 1978; Gasaway et al. 1978; Bowyer et al. 2001a) (Fig. 45.4). Moose invest differentially in antler grow relative to their body weight, with maximal investment occurring after body growth is complete at about 8 years of age Fig. 45.5), but declining slightly in senescent individuals (Stewart et al. 2000).

Moose possess slightly more tines on left than right antlers (directional asymmetry), but the functional significance of that difference is unknown (Bowyer et al. 2001a). Likewise, moose exhibit fluctuating asymmetry (small random departures from perfect bilateral symmetry) in palm length and width, but not in circumference of antler beams. That finding is contrary to research on fluctuating asymmetry based solely on counts of antler tines (Solberg and Sæther 1993; Nygrén 2000). Moreover, an inverse relation exists between overall size of antlers and relative fluctuating asymmetry, an outcome expected for a secondary sexual characteristic such as antlers. Consequently, antler size may serve as an index to male quality in moose both among and within age cohorts (Bowyer et al. 2001a).

TABLE 45.1. Generalized phenology of the moose antler cycle in Alaska

Time Period	Phenological Event
Late March	Initiation of antler growth by mature males
Mid-April	Last casting of antlers by small-antlered males
Late April	Mature males with antlers in stalk plus bulb stage; up to 25 cm of new growth evident
Mid-May	Some nonyearling males with no obvious new growth; mature males with small palms forming
Early June	Midpoint of growth period for mature males; antlers about 25% grown
Late June	Some yearling males with about 15 cm of new growth
Early July	Antlers of mature males about 75% grown
Late July	Some yearling males still in stalk plus bulb stage
Early August	Antlers of mature males 90–95% grown
Mid-August	Termination of antler growth by mature males
Late August	Velvet shedding begins
Mid-September	Velvet shedding complete
Late November	Antler casting begins
Early January	Most mature males antlerless
Late January	Antler castings virtually complete; a few small-antlered males retain antlers for 60–80 additional days

SOURCE: Van Ballenberghe (1983a).

PHYSIOLOGY

Moose are uniquely adapted to northern environments (Kelsall and Telfer 1974; Schwartz 1992; Van Ballenberghe 1992; Bowyer et al. 1997). Their massive body size and the insulating properties of their pelage buffer them against cold temperatures. The lower limit of the thermal-neutral zone of a moose (the temperature below which energy must be expended beyond basal metabolism to maintain body heat) has never been measured, but lies below −30°C (Renecker and Hudson 1986). Likewise, their long legs help them negotiate deep snow; only depths > 70 cm hinder movements of moose (Coady 1974, 1982). Travel through deep snow is energetically costly (Telfer and Kelsall 1979), and snow of sufficient depth or with a crusty surface that will not support moose may increase energetic expenditures markedly or curtail movements during winter. Moose are susceptible to heat stress (Renecker and Hudson 1990). In winter pelage, heat stress may begin when temperatures reach −5°C. In summer coats, heat stress begins at 14°C, and panting occurs at 20°C (Renecker and Hudson 1990). Thus, high temperatures may limit the use of some habitats by moose in summer or even influence the extent of their southern distribution.

Moose inhabit boreal environments with extreme seasonal variation in climate as well as forage abundance and quality. The growing season is short and winter typically is long and severe; snow may persist for up to 9 months at the northern limit of their distribution. Consequently, moose accumulate large stores of fat during the brief growing season, which they rely on to meet their energetic needs during the long winter (Fong 1981; Ballard and Whitman 1987; Cederlund et al. 1989; Schwartz 1992). Digestibility and crude-protein content of the diet vary seasonally, with the highest values occurring from May to August (Regelin et al. 1987). Moose may consume aquatic plants that are high in sodium during spring and summer to replenish mineral reserves depleted in winter and meet requirements for lactation (Belovsky and Jordan 1981; Jordan 1987). Moose do occur in areas without substantial amounts of aquatic plants, where they often consume a diet high in willows (*Salix*) (Van Ballenberghe et al. 1989), which also are higher in sodium content than many other terrestrial plants (Staaland and White 2001). Moose also may obtain sodium from natural mineral licks, especially in spring and early summer (Jordan et al. 1973; Fraser and Hristienko 1981; Fraser et al. 1982; Tankersley and Gasaway 1983; Risenhoover and Peterson 1986). Moose also will use roadside salt licks where sodium has been applied to hasten melting of snow and ice from highways (Miller and Litvaitis 1992a); placement of salt blocks can affect patterns of browsing by moose (Risto and Harkonen 1998).

Beeler et al. (1959), Flynn and Franzmann (1987), and Treble and Thompson (1998) discussed minerals in moose; copper (Flynn et al. 1977; Frank et al. 1994; Galgan and Petersson 1994; Barboza and Blake 2001; O'Hara et al. 2001) and chromium (Frank et al. 1994; Galgan and Petersson 1994) deficiencies have been reported. Cadmium also has been reported in kidneys, liver, and muscles of moose, with the highest concentrations occurring in kidneys (Scanlon et al. 1986; Glooschenko et al. 1988; Brazil and Ferguson 1989; Paré et al. 1999; Crichton and

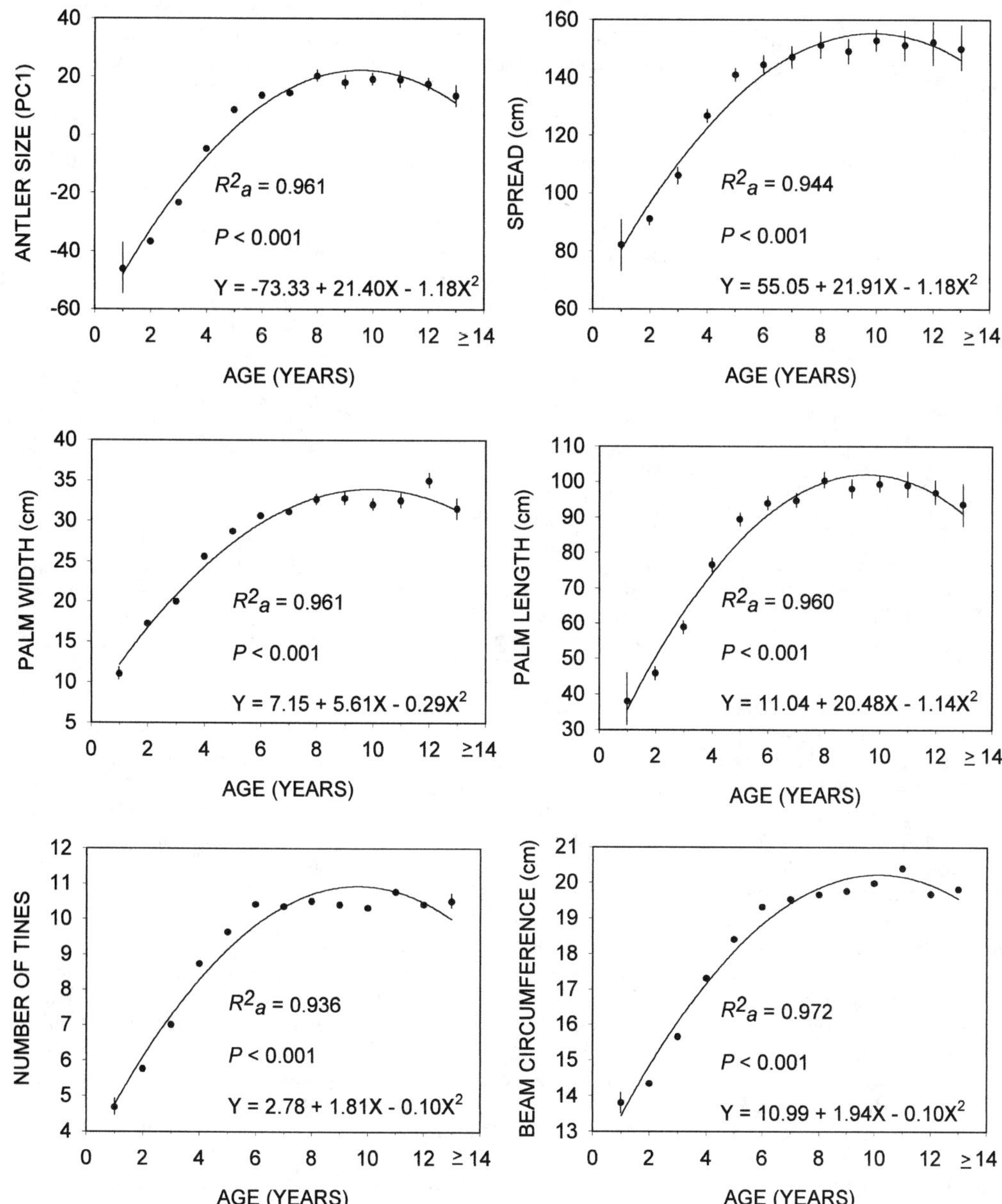

FIGURE 45.4. Changes in size of antler characteristics of male moose (*Alces alces gigas*) with increasing age. Note that senescence occurs in antlers of older moose, producing a curvilinear relation; PC1 (principal component 1) provides an overall index to antler size. SOURCE: Bowyer et al. (2001a). Reproduced with permission from Alliance Communications Group.

Paquet 2000; Gustafson et al. 2000). Contamination may be from either anthropogenic or natural sources. In some areas, concentrations of cadmium in kidneys and liver are sufficiently high that hunters are warned against eating those organs too frequently (Crichton 1998b). Lead, mercury, and nickel have been reported in other northern ungulates (Duffy et al. 2001).

Females must meet the high costs of lactation and recover and replenish body reserves necessary to ovulate and to survive harsh winter conditions. Sufficient fat reserves in females near the end of winter and in early spring are a necessity for successfully bearing and provisioning young (Keech et al. 2000). Males must accumulate resources in spring and summer necessary to support intense rutting activities in autumn and likewise survive winter. The diet of moose in winter is mostly woody browse with a low content of crude protein (5–7%), which will not meet maintenance requirements (Schwartz 1992). In addition, moose reduce their metabolic rate and food intake during winter as an adaptation to conserve energy (Schwartz et al. 1988c). Energy metabolism and expenditure in moose has been studied extensively (Belovsky and Jordan 1978; Regelin et al. 1985, 1986; Renecker and Hudson 1986; Schwartz et al. 1988c, 1991; Hjeljord et al. 1994, 1996). Studies of fat reserves in moose have benefited from recent developments in ultrasonography (Stephenson et al. 1998a). DelGiudice et al. (1991) described urine chemistry and its relation to physical condition of moose during winter.

Milk of moose contains about 1.5 kcal/g (Franzmann et al. 1976a; Renecker 1987). Peak production of milk occurs from 21 to 31 days following birth, with a maximal output of 5.5 kg/day; twins receive 67% more milk than a singleton (Schwartz and Renecker 1998). Energetic costs of lactation far outweigh other components of maternal investment (White and Luick 1984). Cook et al. (1970) described the composition, fatty acids, and mineral constitution of milk from moose. Reese and Robbins (1994) further characterized lactation and neonatal growth in moose.

Moose are ruminants. Unlike mammals with simple digestive systems, moose have a specialized four-chambered structure, which allows them to feed on vegetation high in structural carbohydrates that

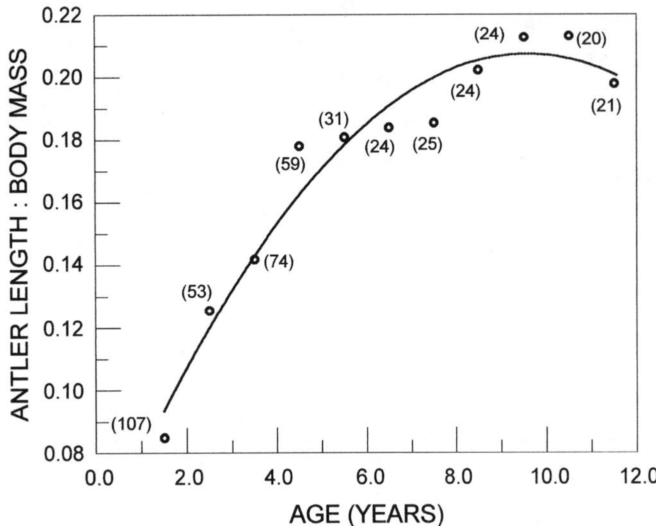

FIGURE 45.5. Antler length:body mass in relation to age in male moose (*Alces alces*). Sample size in parentheses. Note that maximal investment in antlers does not occur until about 8 years of age, when males complete body growth. SOURCE: Stewart et al. (2000).

monogastric mammals cannot digest in their stomach (Schwartz and Renecker 1998). The rumen and reticulum (the first two chambers) form an interconnected compartment where anaerobic conditions coupled with nitrogen and other nutrients recycled in saliva promote microbial fermentation of cellulose and hemicellulose. Moose further aid the digestive process by accumulating coarse vegetation into a bolus (cud) and eructing it from the rumenoreticular complex into the mouth for additional mastication, which reduces particle size. Moose rechew coarser foods with high cell-wall content longer than more succulent vegetation (Schwartz and Renecker 1998). Microbial breakdown of structural carbohydrates results in production of volatile fatty acids, which are absorbed mostly by papillae on the rumen surface (Hofmann and Nygren 1992; Pehrson et al. 1997), and constitute the primary energy source for moose and other ruminants (Russell and Rychlik 2001). Gasaway and Coady (1974) discussed composition of volatile fatty acids in moose, and Dehority (1974) listed the ciliate fauna of the rumen. Moose also benefit by digesting rumen microbes (bacteria, protozoans, and fungi). The omasum (the third chamber) removes water and prevents passage of large particles of vegetation into the abomasum (the fourth chamber and true stomach), where gastric digestion occurs.

Generally, shorter rumination times are associated with higher-quality diets, and longer retention times with foods of lower quality (Schwartz et al. 1988b). Soluble sugars and other contents within the plant cell (cell-solubles) are fermented rapidly by rumen microbes, whereas cell walls require much longer fermentation times (Spalinger 2000). Lignin content also reduces digestibility of forages, as can tannins and other plant secondary compounds (Danell et al. 1990; Bryant et al. 1991, 1992). Moose, however, possess tannin-binding proteins in their saliva (Hagerman and Robbins 1993; Juntheikki 1996).

The longer forages remain in the rumen, the more completely they are digested, but increasing rumination time reduces the amount of forage that can be processed. Hence, moose face a tradeoff between turnover (and hence digestion) of forage and its rate of passage; those parameters must be optimized for maximal extraction of energy from the diet. Consequently, moose and other ruminants do not technically die from starvation even in winters with extremely deep snow and limited access to forage. Instead, they die from malnutrition with their rumens filled with large and mostly indigestible stems of winter browse (Coady 1982).

Sexual dimorphism (Weckerly 1998; Loison et al. 1999a) coupled with the needs of the sexes to follow different life-history strategies (Bowyer 1984; Bleich et al. 1997; Kie and Bowyer 1999) have led

TABLE 45.2. Summary statistics for variables used to examine the relationship between rump fat and blood-serum components from 38 pregnant, adult moose from the Tanana Flats, Alaska, March 1996

Variable	$\overline{X}$	*SD*	CV (%)	Range
Maximum depth of rump fat (cm)	1.54	1.01	65.6	0–3.8
Sodium (meq/L)	134.53	4.78	3.6	123–147
Potassium (meq/L)	8.29	1.4	16.9	5.7–10.9
Chlorine (meq/L)	95.42	3.91	4.1	87–103
Glucose (mg/dl)	97.87	19.53	20.0	64–153
Blood urea nitrogen (mg/dl)	3.38	1.15	34.0	2–5
Creatinine (mg/dl)	2.08	0.26	12.5	1.4–2.6
Blood urea nitrogen:creatinine (ratio)	1.55	0.40	25.8	0.8–3.3
Calcium (mg/dl)	11.07	1.09	9.9	8.2–13.4
Phosphorus (mg/dl)	4.77	1.13	23.7	2.5–8.5
Cholesterol (mg/dl)	72.27	11.39	15.8	54–117
Total bilirubin (mg/dl)	0.29	0.07	24.1	0.2–0.6
Protein (gm/dl)	6.87	0.47	6.8	5.4–7.6
Albumin (gm/dl)	3.78	0.36	9.5	2.8–4.5
Globulin (gm/dl)	3.09	0.24	7.8	2.5–3.8
Albumin:globulin (ratio)	1.23	0.13	10.6	0.9–1.5
Aspartate aminotransferase (U/L)	79.11	13.61	17.2	52–107
Alanine aminotransferase (U/L)	71.55	17.62	24.6	43–124
Total lactate dehydrogenase (U/L)	626.29	118.92	19.0	343–907
Creatine kinase (U/L)	50.82	28.87	56.8	20–122
Glutamyl transferase (U/L)	21.18	4.30	19.7	16–39
Alkaline phosphate (U/L)	43.87	19.07	43.5	25–135

CV, Coefficient of variation.

to allometric differences in the digestive systems of moose and other cervids, which promote spatial segregation of the sexes (Barboza and Bowyer 2000, 2001). Thus, the sexes of moose exhibit differences in diet selection or habitat requirements outside the mating season to satisfy their respective nutrient demands during periods of segregation (Miller and Litvaitis 1992b; Miquelle et al. 1992; Bowyer et al. 2001b), a point we return to later.

Mean body temperature of moose exhibiting no or slight excitability was 38.7°C; heart rate was 76.5 beats/min and respiratory rate was 19 respirations/min (Franzmann et al. 1984). Excitability affected those physiological values, as did season and the drug used to immobilize moose. Values that were considered sufficiently high to require intervention and corrective actions for immobilized moose were body temperature ≥40.2°C, heart rate ≥102 beats/min, and respiratory rate ≥40 respirations/min (Franzmann et al. 1984). Numerous authors have provided data on blood values of moose (Houston 1969; Franzmann and LeResche 1978; Crête et al. 1982; Franzmann et al. 1987; Ballard et al. 1996; Addison et al. 1998b). A suite of blood-serum values for pregnant females reported by Keech et al. (1998) is provided in Table 45.2. Ballard et al. (1996) noted that some blood variables differed between severe winters and mild ones, and Franzmann et al. (1987) reported differences in blood characteristics between moose populations on differing nutritional planes. Keech et al. (1998) related blood parameters to depth of rump fat and, hence, physical condition. They noted, however, that those results should be used with caution because of potential variability caused by differences in sex, age, season, reproductive status, and the effects of handling on individual moose.

REPRODUCTION

Estrous Cycle and Gestation. Although differences in length of the estrous cycle and gestation for moose have been reported (reviewed by Schwartz 1998), the most reliable data are from Schwartz and Hundertmark (1993). Those authors reported that estrus in captive female moose ranged from 28 September to 12 October. Moose are polyestrus and females will recycle if not bred in their first estrus (Edwards and Ritcey 1958; Markgren 1969). Indeed, Schwartz and Hundertmark (1993) observed a second period of estrus from 19 October to 5 November. The estrous cycle ranged from 22 to 28 days, with females typically being receptive for 15–26 hr (Schwartz and Hundertmark 1993). Females usually mate only once; observations of copulations with more than one male are uncommon (Van Ballenberghe and Miquelle 1996). Gestation length was 231 days and ranged from 216 to 240 days, with most fetal growth occurring during the last one third of gestation (Schwartz and Hundertmark 1993). Second-estrus females shortened length of gestation markedly (Schwartz and Hundertmark 1993), likely to have sufficient time during the short summer at high latitudes to provision young (Bowyer et al. 1998; Keech et al. 2000). Female ungulates altering length of gestation to cope with an unpredictable environment or short growing season may be more common than previously thought (Rachlow and Bowyer 1991; Berger 1992).

Productivity and Age at First Reproduction. Female moose on a high nutritional plane typically mate as yearlings and give birth near their second birthday (Schwartz 1998). Although ovulation by young (<1 year old) moose has been reported (Simkin 1965; Addison 1975), mating and conception does not occur (Schwartz 1998). Most moose have ovulated by either 16 or 28 months of age (Sand and Cederlund 1996; Schwartz 1998), but females in poor physical condition may delay ovulation until they are 40 months old (Albright and Keith 1987). Boer (1992) noted that pregnancy rates for yearlings were affected by the relationship of the population to carrying capacity of the habitat (K), with female yearlings from populations well below K reaching rates as high as 64%. Female yearlings from populations beyond K had low rates of pregnancy. Pregnant yearlings typically give birth to a single young (near their second birthday), although twins occur rarely (Pimlott 1959; Blood 1974). The proportion of yearlings pregnant in a population is positively correlated with the proportion of adult females giving birth to twins (Boer 1992).

Pregnancy rates among adult females are high (>70%), and adult moose typically give birth to either a single young or twins (for review Schwartz 1998). Some females in poor physical condition, however, may not reproduce in consecutive years (Albright and Keith 1987). Indeed, the energetic costs of successfully provisioning young may affect reproduction in the following year (Testa 1998). Twinning rates for moose are variable and depend on good physical condition of females as affected by the relation of the population to K (Gasaway et al. 1992; Sand 1996; Heard et al. 1997; Nygrén and Kojola 1997; Testa and Adams 1998; Keech et al. 2000). Twinning rate for a particular population of moose provides an index to their physical condition (Franzmann and Schwartz 1985; Keech et al. 2000). Triplets occur occasionally (Hosley and Glaser 1952; Peterson 1955; Franzmann and Schwartz 1985; Bowyer et al. 1998), and the possibility of quadruplets exists (Martin 1989). Physical condition of maternal females affects litter size, weight of neonates, and date of birth, all of which can influence survivorship of young (Keech et al. 2000). Reproductive senescence occurs from about 12 years of age onward in female moose (Ericsson et al. 2001), ostensibly as a result of tooth wear (Ericsson and Wallin 2001); senescence is typical in other ungulates (Loison et al. 1999b).

The sex ratio at birth tends toward parity in moose, but substantial variation may occur (39–62% male; Boer 1992). Several authors have reported more males (Reuterwall 1981; Albright and Keith 1987; Ballard et al. 1991), others more females (Boer 1987; Larsen et al. 1989), and some no difference in sex ratios of neonates (Schwartz and Hundertmark 1993; Keech et al. 2000). Crichton (1992) noted that the sex ratio of fetuses conceived before the peak of rut were predominantly (66%) females, whereas males were more common (55%) in adult females bred later in the mating season.

Male moose become sexually mature as yearlings (Schwartz et al. 1982). Nonetheless, yearlings seldom gain an opportunity to mate because of their small body and antler size (Van Ballenberghe and Miquelle 1993, 1996; Stewart et al. 2000; Bowyer et al. 2001a). Asymptotic body weight is not reached in males until about 8 years of age, and maximal antler size is attained between 7 and 11 years old. Hence, age at first reproduction tends to be delayed for male moose because of their polygynous mating system, and the inability of small males to compete successfully for mates.

Timing and Synchrony of Parturition. Moose give birth in an extremely synchronous manner in late May (Gasaway et al. 1983; Schwartz and Hundertmark 1993; Bowyer et al. 1998; Keech et al. 2000). The median date of birth in interior Alaska was 25 May (Fig. 45.6); 85% of births occurred during an 11-day period and 95% of births during a 16-day period (Bowyer et al. 1998). Remarkably, there is little variation in timing of births by moose across a latitudinal gradient in North America (Sigouin et al. 1997; Bowyer et al. 1998; Schwartz 1998), a pattern that is much different in Fennoscandia, where parturition at far northern latitudes is delayed (Sæther et al. 1996). Predation is not the primary factor promoting synchrony in parturition among moose (Bowyer et al. 1998; Keech et al. 2000). The need to give birth early so there is sufficient time during the short growing season for young to acquire body reserves necessary to survive a harsh and long winter determines their timing and synchrony of parturition (Bowyer et al. 1998; Keech et al. 2000). Young moose born too late in spring have very high rates of mortality compared with others in their cohort (Keech et al. 2000). Moose did not alter timing of births in response to marked variability in winter or spring weather, and may be susceptible to climate change (Bowyer et al. 1998), a topic of increasing interest to those studying northern ungulates (Post and Stenseth 1998, 1999; Lenart et al. 2002).

BEHAVIOR

Rutting Dynamics and Mating Systems. The mating season (rut) in moose peaks in late September and early October and is highly synchronized (Lent 1974; Schwartz and Hundertmark 1993; Van Ballenberghe and Miquelle 1993, 1996). Group size increases as rut approaches (Dodds 1958; Molvar and Bowyer 1994), with aggregations being

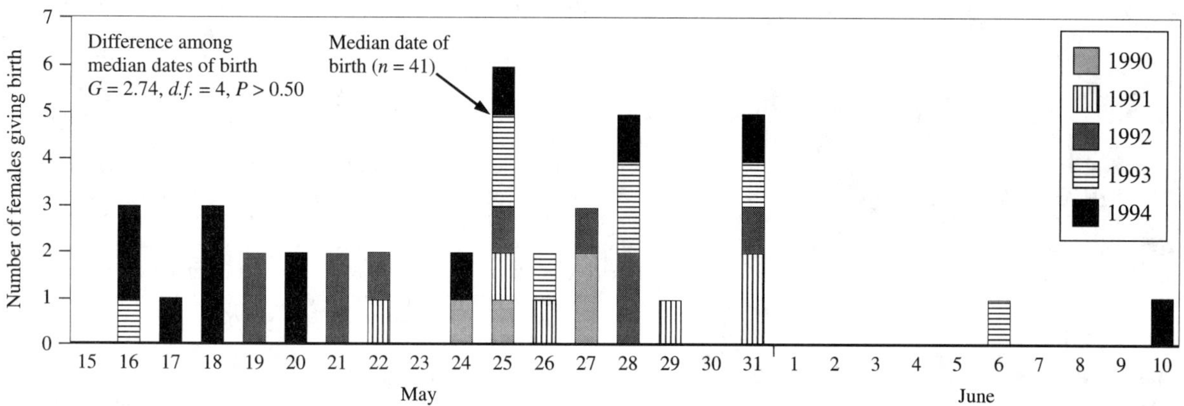

FIGURE 45.6. Timing of parturition in 41 moose (*Alces alces gigas*) from interior Alaska. Note the extremely synchronous nature of parturition for an ungulate that does not congregate to give birth. No differences occurred in timing or synchrony of births among years. SOURCE: Data from Bowyer et al. (1998).

largest at peak of rut, and then declining again late in the mating season (Van Ballenberghe and Miquelle 1996). Among polygynous, sexually dimorphic cervids, the largest males with the largest antlers typically mate most often (Hirth 1977; McCullough 1982; Bowyer 1986; Weckerly 1998), a generalization that also holds for moose (Peek et al. 1986; Van Ballenberghe and Miquelle 1993; Stewart et al. 2000). Aggregations of rutting moose occupy more open habitat in Alaska than elsewhere (Molvar and Bowyer 1994), and *A. a. gigas* tends to occur in larger groups than other subspecies (Peek et al. 1974).

Moose throughout most of North America exhibit a tending-bond system of mating similar to that of white-tailed deer (*Odocoileus virginianus*; Hirth 1977), wherein a dominant male defends an estrous female until he mates with her, and then moves on in search of additional mates (Altmann 1959; Geist 1963). Moose are not territorial (Peterson 1955). Moose in Alaska, however, exhibit a harem mating system more typical of North American elk (*Cervus elaphus*; Bowyer 1981). A dominant male herds and defends a group of females without regard to their state of estrus and does not permit sexually mature males in his rutting group. Harem masters court females as they come into estrus (Van Ballenberghe and Miquelle 1993; Molvar and Bowyer 1994; Van Ballengerghe and Miquelle 1996). In Alaska, a second bout of rutting behavior occurs in late October and early November, ostensibly for mating with females that were not bred in their first estrus or came into estrus late. During that second rut, the mating system reverts to a tending bond, probably because there are too few estrous females to form sufficiently large groups to make harem mating a worthwhile strategy. Only about 11% of females are bred in their second estrus (Schwartz and Hundertmark 1993).

Aggressive Behavior by Males. General accounts of the social behavior of moose during rut are provided by Altmann (1959), Geist (1963), de Vos et al. (1967), Lent (1974), Peek et al. (1986), Van Ballenberghe and Miquelle (1993, 1996), and Bubenik (1998). In addition, other authors have considered particular behaviors or focused on time periods surrounding rut (Dodds 1958; Houston 1974; Miquelle and Van Ballenberghe 1985; Miquelle 1990, 1991; Bowyer et al. 1994; Molvar and Bowyer 1994; Whittle et al. 2000). Our descriptions of rut-related behaviors rely on all those publications, but draw most heavily from Van Ballenberghe and Miquelle (1993, 1996).

Sparring (ritualized fighting) occurs mostly before and after the peak of rut and often involves opponents of different size and dominance status, and sometimes more than two individuals (Fig. 45.7). Antlers are presented to solicit sparring, and participants carefully position the antlers together; the ears are not held back and downward as in more serious encounters. Such interactions are of low intensity, and antler positioning and pushing are done gently. Subordinate individuals may give a submissive whine if the level of aggression escalates. That vocalization often lowers the level of aggression or terminates the interaction.

FIGURE 45.7. Three bachelor male moose (*Alces alces gigas*) about to engage in sparring. Sparring is a low-key behavior, and differs markedly in intensity from serious fights among older males during the peak of rut. SOURCE: Photo by V. Van Ballenberghe.

Antler thrashing, which is a dominance display between males, does not function principally to remove velvet; thrashing behavior continues long after all velvet has been removed in moose as well as other cervids (Bowyer 1986; Van Ballenberghe and Miquelle 1993). Once velvet shedding begins, antlers no longer are innervated or supplied with blood—they are dead bone that function primarily as weapons for male–male combat (Geist 1966). Consequently, hardened antlers lack feeling; cervids do not thrash their antlers to masturbate (Bowyer and Kitchen 1987). Large males may antler thrash to intimidate small opponents, or antler thrashing may precede more serious encounters between opponents of more equal size and dominance status. In contrast to North American elk (Bowyer and Kitchen 1987), moose do not release urine when antler thrashing.

Large males often exhibit a swaying gait in which the antlers are rocked from side to side as they approach or challenge each other during rut. That behavior may be followed by broadside displays of the body or lateral displays of the antlers. If one opponent does not retreat, such dominance displays may lead to a serious fight. Fighting typically occurs over access to estrous females. Male moose may antler thrash, direct antler threats at opponents, or threaten them by approaching with their head below the level of the withers and their ears held downward; hair along the dorsal midline also may be piloerected. Males also may rush or charge a rival with the antlers being lowered and tines tipped downward as the opponent is approached. Males seldom kick toward each other with a stiff foreleg during rut.

Serious fighting usually occurs between large males of near-equal dominance around the peak of rut. Fighting is dangerous and smaller individuals typically flee dominant males, thereby avoiding violent and risky encounters. Fighting often is initiated by two large males presenting their antlers with their ears held back and downward. Lateral displays of antlers, accomplished by turning the head sideways rather than by repositioning the body, and pawing the ground often accompany this behavior. Males may engage in jousting, whereby they place their antlers together without pushing, ostensibly so an individual can gain a more favorable positioning of his tines relative to the shield provided by the antler palms of an opponent. Violent clashes occur after males take several quick steps together, or thrust forward following jousting; antler tines may be broken during these fierce encounters. Males attempt to gore each other during combat, either by twisting the neck of an opponent sideways, thereby breaking antler contact, and then thrusting tines into the side of an opponent's body, or by driving antlers into the rump of an adversary as he turns to flee. Males that are being gored in the side may whirl in a tight circle to break contact with opponents, and then slam their antlers back into those of rivals to regain a more favorable position and continue the fight. Losers of fights typically flee, often with the victor in close pursuit attempting to gore the retreating opponent. If opponents are equally matched, fights can last for hours with bouts of displaying and jousting intermixed with vigorous pushing.

Fighting is hazardous and males may be killed outright or die later from wounds sustained during combat. Rarely, moose will perish with antlers locked (Peterson 1955). In addition, rut-related injuries may make males more vulnerable to predators or impair their ability to survive harsh winter conditions. Large male moose cease feeding (i.e., become hypophagic) for about 18 days during the peak of rut (Miquelle 1990), which may further compromise their energy reserves and their ability to endure winter. Sex ratios among adult polygynous ungulates that strongly favor females may result, in part, from increased mortality among males associated with strenuous and hazardous rutting activities (Bowyer 1981, 1991; Berger and Gompper 1999).

Scent Marking. Scent marking by moose during rut is typified by two behaviors: digging of rutting pits (Miquelle 1991; Van Ballenberghe and Miquelle 1996; Whittle et al. 2000) and rubbing of trees (Bowyer et al. 1994). Scent urination is widespread among ungulates (McCullough 1969), and in moose, digging and wallowing in rutting pits characterizes

FIGURE 45.8. Pit-digging behavior in moose (*Alces alces gigas*). The dominant male has just dug a rutting pit by pawing a depression in the ground with his forefeet, and is squatting and urinating in the pit. The urine of dominant males is odiferous and highly attractive to females, which are approaching the pit. This strong-smelling urine likely contains a pheromone that primes estrus in females. SOURCE: Photo by V. Van Ballenberghe.

this behavior. A large number of old pits may be observed on traditional rutting grounds of moose (Woodin 1956). Large males paw a shallow depression (pit) in the ground with their forefoot and then squat and urinate into the pit (Fig. 45.8). The urine of large males has a pungent odor during rut and can be smelled by humans at great distances. Males then stomp with a forefoot, splashing the mixture of mud and urine onto the underside of their face, neck, and antlers. Males often bed (wallow) in the pit, and sometimes rock from side to side to impregnate the pelage on their underside with the aromatic urine; they do not roll onto their backs when wallowing. Males also may slap their antlers into the pit, further spreading urine over their forequarters and darkening and discoloring the underside of their antlers. Scent urination in ungulates may serve either a male–male or male–female function (Bowyer and Kitchen 1987); in moose, this behavior clearly is directed toward females (Miquelle 1991). Indeed, females respond immediately to pit-digging behavior by males and rapidly approach the pit. Females then attempt to gain access to the urine, sometime interfering with the male, which can lead to an aggressive response by the male. Females then wallow, often engaging in aggressive interactions to gain access to the pit. Urine of large males undoubtedly contains a pheromone that is attractive to females, which likely helps prime estrus (Whittle et al. 2000). Smaller but sexually mature males do not become hypophagic during rut and do not possess strong-smelling urine or dig rutting pits. Moreover, large males begin feeding again before the second rutting period in late October and early November, and do not engage in pit-digging behavior during that period. Catabolism of body reserves during hypophagia and subsequent elimination of metabolites in urine of males offer a suite of substances that may contain the pheromone involved, but more research is needed (Whittle et al. 2000).

Moose also scent mark small trees and shrubs by stripping away the bark, which is not eaten, with their lower incisors or antlers and rubbing their forehead and preorbital glands into the mark (i.e., sign posting), a behavior common to many ungulates (Bowyer et al. 1994). This behavior is unrelated to territoriality in moose (Weckerly 1992; Bowyer et al. 1994); many mammals that engage in scent marking are not territorial (Ralls 1971). Both sexes of moose rub trees. Females rub during the peak of rut and males rub mostly during the second rutting period in late October and early November. The rubbed tree may be a visual and olfactory cue that indicates the presence of an estrous female at the peak of rut (Bowyer et al. 1994). Trees rubbed by males are attractive to females during the brief second rutting period, and may serve a function similar to rutting pits with odiferous urine—priming of estrus (Bowyer et al. 1994). Rubbing of trees (scent marking) should not be confused with bark stripping of trees and shrubs for feeding during winter (Bowyer et al. 1994).

Courtship. Males repeat a soft croak or grunt when courting females. This same vocalization may be given more forcefully when males are traveling alone or sometimes in response to other males (Van Ballenberghe and Miquelle 1996). Courtship is characterized by tongue flicking as the male approaches the female, sometimes in a low-stretch posture. Large males salivate copiously during rut, and their saliva is high in androstenones (Schwartz et al. 1990). Females often give plaintive moans in response to male courtship, especially if the male is small. Males typically smell and may lick the perianal region of females they are courting and often exhibit flehmen. This behavior involves the vomeronasal organ, which likely allows the male to determine the state of estrus in the female (Estes 1973). Males also may flehmen after smelling or licking urine of a female from the ground (Fig. 45.9). If the female does not move away, the male places his chin over her rump and moves his head up and down or from side to side. Females may respond by rubbing their head along the flanks of the male. Males mount females by rising on their rear legs and clasping the female with their forelegs; females can terminate such encounters by walking forward. Successful copulation can be accomplished in one pelvic thrust; ejaculation coincides with penile intromission. Females rarely have been observed to accept copulations from more than one male. In Alaskan moose, large

FIGURE 45.9. Flehmen behavior by a male moose (*Alces alces gigas*). This display involves the vomeronasal organ, which likely allows a male to assess the state of estrus in a female from metabolic by-products in her urine. SOURCE: Photo by V. Van Ballenberghe.

males herd females to block their departure from the harem. Males may give a herding vocalization, an explosive, short "oh-wah," which induces females to form a tight group. Under extreme stress, moose may emit a loud growling roar, which is not limited to the mating season (Franzmann 1981).

Other Behaviors. Other aggressive behaviors performed by females, or sometimes by males outside the mating season, include a direct stare. This behavior often is accompanied by the dominant individual dropping its ears and aversion of the head by the subordinate. Likewise, females will intimidate opponents with a head-high threat in which the nose is lifted upward and the ears are laid down and backward against the neck. If the opponent does not withdraw, the dominant female may rush toward an opponent and deliver a kick with a stiff foreleg, or rear onto her hind legs and flail with her forelegs. Play in moose often involves exaggerated aggressive behaviors, chases, and running through and splashing in water. Indeed, most play activities involved running (Geist 1963). Comfort behaviors of moose are similar to those of other cervids and were described by Geist (1963).

Parturition and Maternal Behavior. Behavior associated with parturition and care of young has been described by several authors (Altmann 1958, 1963; Stringham 1974; Cederlund 1987; Addison et al. 1993; Bubenik 1998). Maternal females become asocial and drive their yearlings away several weeks before giving birth (Peterson 1955). Indeed, as the birthing season approaches, group size declines (Molvar and Bowyer 1994), and the sexes remain spatially segregated (Miquelle et al. 1992). Immediately before parturition, some females may make long, erratic movements, which presumably are an adaptation to hinder predators in locating their birth sites (Bowyer et al. 1999b). Nonetheless, moose use traditional areas for giving birth (Addison et al. 1990; Bowyer et al. 1999b). Females seek secluded sites for parturition. Observations of births are rare, but in several instances young were dropped within 15–20 min of the initiation of labor; twins were born about 30 min apart (Bubenik 1998). Parturition is accomplished either standing or more commonly by laying down. Females lick the neonate clean and eat the afterbirth. Young may begin nursing 1.5 hr following birth. Length of nursing bouts averaged 6.6 min for neonates, but about 42 sec for older infants (Stringham 1974). Nursing is initiated by the neonate bunting the udder of the female, which causes milk letdown. Females remain close to the birth site for several weeks or longer (Addison et al. 1990; Bowyer et al. 1999b). Females and their young communicate with cohesion calls, but young attempt to avoid predators by behaving cryptically at the birth site. Weaning typically occurs between September and early December (Denniston 1956; Altmann 1958), but Molvar (1993) reported nursing by a yearling.

Activity Patterns. Moose are crepuscular for much of the year, with peaks of activity occurring around sunrise and sunset (McMillian 1954; deVos 1958; Best et al. 1978; Belovsky 1981a; Cederlund 1989; Renecker and Hudson 1989; Bevins et al. 1990). This pattern tends to be less pronounced during winter, especially at high latitude, where hours of daylight are limited (Risenhoover 1986; Gillingham and Klein 1992). Furthermore, summer activity patterns in the far north may exhibit little synchrony and become free running (Van Ballenberghe and Miquelle 1990). Marked differences in activity occur between the sexes, especially during the mating season, when males engage in strenuous rutting activities (Miquelle 1990; Van Ballenberghe and Miquelle 1996).

FEEDING HABITS

Moose are browsers, eating mostly the stems and twigs of woody plants in winter and the leaves and succulent shoots of those shrubs and trees during the remainder of the year. Peek (1974b) and Renecker and Schwartz (1998) reviewed plant species consumed by moose. More than 220 different plant genera or species were eaten, with willows (*Salix*), birch (*Betula*), and alder (*Alnus*) predominating in diets of moose across North America where those species were available. Willows are preferred forage for moose (Peek 1974b; Pierce 1984; Renecker and Schwartz 1998). During winter, however, moose in eastern North America consume large amounts of conifers. Ludewig and Bowyer (1985) reported 73% of winter diet of moose in Maine was composed of conifers, mostly balsam fir (*Abies balsamea*), but also including white spruce (*Picea glauca*). Moose living in the midcontinent select hardwoods, but also consume balsam fir during winter (Peek 1974b; Crête and Bedard 1975; Thompson and Vukelich 1981; Risenhoover and Maass 1987). Moose also will eat fallen leaves of deciduous trees in winter where the boreal forest has a strong component of hardwoods (Renecker and Schwartz 1998). In Alaska, white spruce is common within the distribution of moose, but spruce or other conifers seldom are consumed (Oldemeyer 1983; Risenhoover 1989; Van Ballenberghe et al. 1989; Weixelman et al. 1998). This east-west gradient in the willingness of moose to eat conifers likely relates to potent secondary metabolites in trees and shrubs (Palo 1991; Sunnerheim-Sjoberg and Hamalainen 1992) in some western species that help deter feeding by moose and other herbivores (Bryant et al. 1994; Bowyer et al. 1997). Population density of moose, however, affects availability of forage and can alter diets (Risenhoover and Maass 1987; Brandner et al. 1990). At sufficiently high density relative to *K*, resource depletion occurs (Edenius 1991), and under such circumstances even Alaskan moose may consume white spruce (Bowyer et al. 1999a). Low-quality or limited food resources can affect body condition and consequently population status of moose (Messier and Crête 1984; Ferguson et al. 2000). Moose at lower density relative to *K*, however, may have less effect on their food supply (Crête and Jordan 1982a).

During winter, moose may break main stems of shrubs or smaller branches on trees by biting them with their premolars and molars or by using their body to bend and break stems. They then feed on the more nutritious twigs of current and second-year growth that are then within their reach (Telfer and Cairns 1978). Moose also strip bark from trees and shrubs for feeding during winter (Miquelle and Van Ballenberghe 1989; Risenhoover 1989; Faber and Edenius 1998; Scharf and Hirth 2000) and also strip bark of woody species around birth sites in early spring (Miquelle and Van Ballenberghe 1989; Bowyer et al. 1999b).

In spring and summer, moose often consume aquatic vegetation where it is available (Murie 1934; Peterson 1955; Peek et al. 1976; Fraser et al. 1980, 1982, 1984; Timmermann and Racey 1989; MacCracken et al. 1993). Aquatic vegetation provides sodium (Jordan 1987), but moose also may forage on aquatic plants because of the substantial biomass available in some areas and the high digestibility of aquatic vegetation compared with many terrestrial species (MacCracken et al. 1993). Aquatic plants, however, are not necessary for productive moose populations; moose endure where they feed principally on browse in summer with little access to aquatic vegetation (Van Ballenberghe et al. 1989).

Belovsky (1978, 1981b) employed an optimal-foraging approach to predict selection of forage classes by moose. Other mechanistic approaches also have been valuable in understanding use of vegetation patches, intake rates, and diet optimization by moose (Danell et al. 1991; Shipley and Spalinger 1992, 1995; Shipley et al. 1998, 1999). Moreover, substantial information exists on the nutritional composition and quality of plants eaten by moose (Cowan et al. 1950; Kubota et al. 1970; Oldemeyer 1974; Oldemeyer et al. 1977; Hjeljord et al. 1982; Crête and Jordan 1982b; Eastman 1983; Schwartz et al. 1988; Spaeth et al. In press) as well as fecal indices of forage quality (Leslie et al. 1989). Quality of forage during the growing season is much higher than during winter. Moreover, plants growing in the shade tend to have larger leaves and be more nutritious than those growing in direct sunlight (Hjeljord et al. 1990; Bø and Hjeljord 1991; Hjeljord 1992; Molvar et al. 1993). Forage species selected by moose, however, have been difficult to predict based solely on their nutritional composition, especially during winter (Weixelman et al. 1998).

Quality of most browse declines from current annual growth to woody tissue laid down 2–3 years previously (Cowan et al. 1950; Spaeth et al. In press)—the older and larger the stem diameter, the lower is its forage quality. Accordingly, selection of particular stem diameters for a browse species by moose is thought to represent a tradeoff between biomass and quality obtained in a particular bite (Vivas and Sæther 1987; Molvar et al. 1993), resulting in moose eating an optimal twig size (Vivas et al. 1991). Several factors, however, complicate interpreting bite-size selection in moose. Species of browse available and the size of their stems or twigs undoubtedly affect the size of a bite taken. Moreover, browsing in winter positively affects the size of stems regrown the following spring (Bowyer and Bowyer 1997), provided that browsing is not too severe wherein regrowth takes on a hedged appearance with smaller twigs (Molvar et al. 1993). Moose preferentially feed on stems regrown from those browsed in previous years (Bergstrom and Danell 1987; Bowyer and Bowyer 1997). Conversely, defoliation of shrubs (often by leaf stripping) by moose and subsequent regrowth of foliage in summer results in low-biomass foliage, which moose avoid (Miquelle 1983). These processes alter the size of twigs available for moose to forage on and, undoubtedly, their twig-size selection. Snow also can affect the availability of browse (Coady 1974; Weixelman et al. 1998), thereby affecting size of twigs moose are willing to forage on and whether they will eat leaders browsed previously that year. The sexes of moose also have incisor arcades of differing size (Spaeth et al. 2001); adult males take larger bites than do adult females (Miquelle et al. 1992; Bowyer et al. 2001b). Thus, knowledge of which sex predominates in a particular area may be necessary to interpret bite-size selection in moose.

Risk of predation also influences feeding behavior of moose (Edwards 1983; Berger 1999; Kie 1999; Berger et al. 2001a, 2001b; White et al. 2001). Molvar and Bowyer (1994) reported that stem diameter at the point of browsing increased with the distance moose moved from the edge of the forest (Fig. 45.10). Moose presumably took larger, less-nutritious bites as they ventured farther from concealment cover and, in consequence, foraged less selectively because of predation risk (Molvar and Bowyer 1994). Moreover, larger groups of moose formed at greater distances from cover, but foraged less efficiently than moose in smaller groups or those nearer cover (Molvar and Bowyer 1994). Females with young were especially sensitive to risk of predation and altered their feeding behavior more than other sex and age classes (Molvar and Bowyer 1994). Weixelman et al. (1998) also reported that moose altered patterns of diet selection and foraged on less-preferred species at greater distances from concealment cover in winter.

ECOLOGY

Home Range, Migration, and Dispersal. Young moose accompany, and consequently have the same home range as, their mother during their first year of life (Altmann 1958). Females with young may have smaller home ranges, however, than those without them (Ballard et al. 1991). Separation of young from the mother typically occurs at 12–16 months of age (Ballard et al. 1991). Yearlings may wander and not occupy traditional home ranges, but moose usually have established relatively permanent home ranges by 2–3 years of age (Houston 1968; Addison et al. 1980). The ultimate sizes of home ranges occupied by young are related linearly to the home-range size of their mother (Ballard et al. 1991). Female moose tend to be philopatric and are more likely to establish home ranges overlapping with those of their mothers than are young males (Ballard et al. 1991).

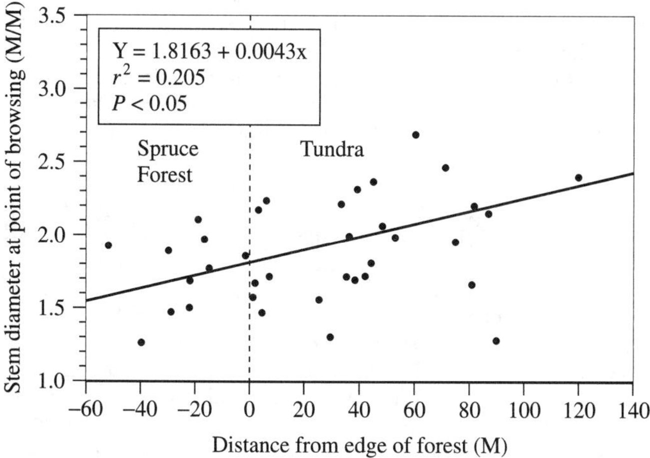

FIGURE 45.10. Diameter at point of browsing on twigs of feltleaf willow (*Salix pulchra*) eaten by moose (*Alces alces gigas*) in interior Alaska. Note that quality of forage eaten declines (larger diameters) as moose venture farther from concealment cover provided by the forest into open tundra; this outcome likely results from predation risk. SOURCE: Data from Molvar and Bowyer (1994).

Hundertmark (1998) reviewed home-range size in moose and reported variation (excluding migratory movements) from 3.6 to 92 km^2. Moose may inhabit the same range throughout the year or migrate to separate summer and winter ranges; considerable variation occurs among populations as to which seasonal range is larger (Hundertmark 1998). Moose also may use traditional rutting and calving areas in autumn and spring, respectively (Houston 1968; Van Ballenberghe and Miquelle 1996; Bowyer et al. 1999b). The sexes of moose are spatially segregated during winter and spring (Miller and Litvaitis 1992a; Miquelle et al. 1992; Bowyer et al. 2001b).

A trend exists for increasing size of summer and winter home ranges with increasing northern latitude (Hundertmark 1998); home ranges at $>60°$ N latitude are exceptionally large (Stenhouse et al. 1995). Population density also plays a role—home ranges tend to be smaller at higher densities than at lower ones (Sweanor and Sandegren 1989; Mytton and Keith 1981). Furthermore, deep snow can limit movements and thereby home-range size during winter (Sweanor and Sandegren 1989).

Size of home range also differs based on gender. Ballard et al. (1991) reported that males had larger home ranges than females, but Phillips et al. (1973), Hauge and Keith (1981), and Sweanor and Sandegren (1989) did not observe that difference. Cederlund and Sand (1994), however, noted that male, but not female moose increased the size of their home range with age. Consequently, comparison of home-range size between the sexes needs to account for age.

Individual moose may migrate seasonally. Such movements are common among cervids and often involve seasonal home ranges separated by long distances. Sinclair (1983) suggested that migratory movements place an animal in an area with abundant food resources before mating, thereby enhancing individual reproductive fitness. Unlike elk and caribou, which migrate in herds, moose typically move as individuals, although timing of migration may be synchronized by weather events, including snowfall. Migratory movements are directional and often follow traditional routes (Andersen 1991), but the pattern of migration may vary greatly from year to year depending on factors such as extent and duration of snowfall. Individuals may leave their summer ranges in response to accumulation of deep snow when it occurs, or remain on summer ranges if snow depths are shallow, or may even undergo partial migration under conditions of moderate snowfall (Van Ballenberghe 1977). Variability in snowfall has been implicated in maintaining migratory and nonmigratory elements in the same population of mule deer (*Odocoileus hemionus*; Nicholson et al. 1997). Moose populations also may contain segments that display long-distance migrations and other segments that do not migrate. Sinclair (1983) asserted that in stable populations, migratory and nonmigratory segments reach equilibrium whereby the benefits of leaving versus staying were equal. Migration in moose is apparently a learned behavior, because young moose follow the movement patterns of their mothers and acquire not only seasonal home ranges, but also migration routes. Migratory females in a moose population in Alberta, Canada, had higher fecundity than nonmigrant females (Mytton and Keith 1981), indicating that eventually migratory moose would outnumber those failing to move.

Distances between summer and winter home ranges in migratory populations vary greatly and may depend on terrain features and habitat dispersion (Sandegren and Sweanor 1988). Often, moose move along elevational gradients seeking lower elevations in winter where snow depth and hardness may be less, and returning to high-elevation summer ranges where forage quality and quantity is greater. Phillips et al. (1973) observed migration distances of 14–34 km in northwestern Minnesota, longer than those (2–13 km) of moose in northwestern Ontario (Addison et al. 1980) or Quebec (7 km; Roussel et al. 1975). Mytton and Keith (1981) observed males in Alberta migrating a mean of 13.4 km, whereas females moved a mean of 6.8 km. Moose in Alaska generally migrated much longer distances. Van Ballenberghe (1977) documented movements of 8–94 km, whereas Ballard et al. (1991) reported movements of 16–93 km for populations of moose in a mountainous area of south-central Alaska. Mauer (1998) reported maximum distances between summer and winter range were 18–196 km in northern Alaska; movements to summer range were underway in late March, and migration to winter range was completed by early October. Atypical migratory patterns also may occur in which moose move to lowland habitats in late winter and remain there during summer, returning to higher elevations in time for the rut (Gasaway et al. 1983). Such patterns are poorly explained, but are apparently related to exploitation of spatial distribution of favorable habitats and seasonal changes in quality of forage.

Unlike migration, which involves a repeated pattern of movements between traditionally used areas, dispersal results in animals leaving previously used areas to settle in new areas where they may remain and reproduce (or attempt to do so) for the remainder of their lives. Migration and dispersal may affect the reproductive fitness of an individual moose, potentially through providing access to improved food quality and environmental conditions, as well as by increasing the probability of finding suitable mates compared with animals that do not undertake such movements. Dispersal of young males also reduces inbreeding (Hjeljord 2001). As with migration, individuals in the same moose population may display different tendencies to disperse. Young may undertake extensive dispersal movements or not disperse at all.

Gasaway et al. (1985a) and Cederland and Sand (1992) used a conceptual model of dispersal postulated for small mammals by Lidicker (1975), who contrasted dispersal from growing populations below *K* to that from populations at or near *K*. These two types of dispersal were termed presaturation and saturation dispersal, respectively. Stenseth (1983), however, recognized that some individuals dispersed regardless of population density ("ambient" dispersal) and suggested that pattern and presaturation dispersal were adaptive. Nonadaptive dispersal was defined as that occurring as individuals were forced from existing home ranges by competition and social pressure in high-density populations. The latter would include younger, immature animals and nonreproducing adults. Both classcs would suffcr poor survival when forced to disperse.

Studies of moose do not indicate clear patterns of dispersal or definitively explain factors that drive animals to disperse. Gasaway et al. (1985a) noted that only 1 of 36 collared moose established a home range that did not overlap that of its mother in a moderately dense population in interior Alaska. In south-central Alaska, Ballard et al. (1991) reported 5 of 15 young displayed a similar pattern in a high-density population. Dispersers moved to areas of lower population density. Dispersal rates between males and females did not differ in Sweden (Cederland et al. 1987). Lynch (1976) reported higher rates of dispersal by subadults versus adults in Alberta. Many (33% of 147) yearlings that dispersed from reserves to areas where hunting was permitted in Quebec were harvested the following year as adults (Labonté et al. 1998).

Dispersal distances vary greatly. A mean dispersal distance of 3 km was noted by Gasaway et al. (1985a) in Alaska, compared with a female that moved 177 km in another area of Alaska (Ballard et al. 1991). Four young moose in Alberta moved at least 50 km and one male dispersed 250 km (Mytton and Keith 1981). Kufeld and Bowden (1996) reported that female dispersal from natal areas ranged from 13 to 120 km in Colorado. Studies have not been conducted to examine survival and reproductive success of long-distance dispersers in moose. Global positioning systems should aid in future studies of moose movements (Rodgers et al. 1996).

Dispersal may play an important role in population dynamics as animals immigrate to and emigrate from different areas. Rolley and Keith (1980) studied a moose population in Alberta that increased about 40-fold in a 14-year period. During the first part of that period, the observed finite rate of increase exceeded that possible by survival and reproduction. The reverse occurred in the later years, leading to the conclusion that immigration (by dispersers) accounted for the high rate of increase initially, in contrast to dispersal out of the population (emigration) in the later years.

Dispersal also may result in population expansion of moose, either on a small scale, whereby moose occupy areas of newly created habitat, or on a large scale, where populations may expand greatly or pioneer into previously unused areas. An example of the former was documented in Minnesota by Peek (1974a), who observed greatly increased use of a burned area. Coady (1980) and Mercer and Kitchen (1968) provided examples of the latter, where distribution of moose populations expanded at rates up to 18 km/year.

Effects on Ecosystem Structure and Function. Moose can markedly modify structure and function of ecosystems they inhabit and thereby govern the well-being of other species. Those abilities to alter ecosystem processes make moose a keystone species (Simberloff 1998). The nature of such change, however, is associated with the population density of moose relative to *K* of the environment (Bowyer et al. 1997). Correspondingly, moose are a critical component of northern environments, and knowledge of their ecology is prerequisite to understanding and managing productivity and species composition of boreal regions (Bowyer et al. 1997). The trend is to ignore megafauna in managing species diversity (Crichton 1998a). Such a view is shortsighted where large herbivores drive ecosystem dynamics (McNaughton 1984; Ruess and McNaughton 1987; Frank and McNaughton 1993). Understanding the role of moose and other large herbivores in such systems offers unique opportunities to unite the best features of single-species and ecosystem management (Simberloff 1998; Snaith and Beazley 2002).

Foraging behavior of moose may alter environments they inhabit, especially by adjusting patterns of plant structure and succession (Risenhoover and Maass 1987; Pastor and Naiman 1992; Pastor et al. 1993). On Isle Royale, Michigan, browsing by moderate to high densities of moose accelerated rates of succession from hardwoods to conifers (Risenhoover and Maass 1987; Brandner et al. 1990; Pastor and Naiman 1992; McLaren and Peterson 1994), thereby lowering rates of nutrient cycling in conifer-dominated habitats (Pastor et al. 1993). High density of moose can lower species diversity and alter composition of plant communities (Connor et al. 2000). A low-density population of moose inhabiting floodplain habitat in interior Alaska also influenced patterns of early succession and lowered rates of nutrient cycling (Kielland et al. 1997; Kielland and Bryant 1998), including negative effects on root dynamics (Ruess et al. 1998). Cascading effects of moose browsing have been reported on invertebrates (Suominen et al. 1999a, 1999b) and birds (Berger et al. 2001a).

Conversely, moderate densities of moose in interior Alaska living in treeline habitat, which was not undergoing rapid succession, had a positive affect on rates of nitrogen mineralization, ostensibly because of deposition of urine and feces (Molvar et al. 1993). Likewise, regrowth of stems and leaves following browsing by moose may be larger, more nutritious, and decompose more rapidly in terrestrial (Bryant et al. 1983; Molvar et al. 1993) and aquatic (Irons et al. 1991) systems. Indeed, moose may positively affect their food supply in three ways. First, browsing in autumn and winter release stems from apical dominance, resulting in regrowth of larger stems with larger leaves the following spring (Bergstrom and Danell 1987; Molvar et al. 1993). Likewise, browsing may enhance adventitious growth. Willows responded by increasing biomass per growing point with overall levels of browsing on the entire plant (Molvar et al. 1993) or from browsing on individual leaders of current annual growth (Bowyer and Bowyer 1997). Saliva of moose also may promote branching of saplings (Bergman 2002). Moose selected current annual growth from stems that had been browsed in previous years over those that had not been foraged on (Bowyer and Bowyer 1997). Second, moderate levels of browsing affect the carbon–nitrogen balance of shrubs and trees. Regrowth is poorly defended by secondary compounds, and has a lower lignin-to-nitrogen ratio, allowing more rapid decomposition of litter from browsed versus unbrowsed plants (Molvar et al. 1993; Bowyer et al. 1997). Third, moose fertilize the plants they feed on via deposition of urine and feces (Molvar et al. 1993). Nitrogen content of feces increases with increasing forage quality (Leslie and Starkey 1985). These processes combine to enhance rate of nitrogen cycling on areas frequented by moose and other large herbivores (Ruess and McNaughton 1987; Molvar et al. 1993).

Such alterations of ecosystems by moose are best viewed in terms of the "herbivore optimization" curve of Hik and Jefferies (1990), where herbivores have negative effects on productivity at low and high population densities, but promote positive ones at moderate densities. Moreover, presence or absence of large mammalian carnivores has a substantial effect on feeding behavior (Berger 1999; Berger et al. 2001b) and population dynamics (Gasaway et al. 1992) of moose, outcomes that have far-reaching implications for managing ecosystems. Comprehending the nature of these complex processes is critical for understanding the ecology of northern systems inhabited by moose, and is a necessity for the wise management these unique ungulates (Bowyer et al. 1997).

Habitat Use and Selection. Telfer (1984) and Peek (1998) provide overviews of habitats used by moose in North America. Telfer (1984) categorized the primary habitats of moose into five broad vegetation communities: (1) boreal forest in which fire plays a dominant role, (2) mixed forest representing the coniferous–deciduous ecotone, (3) large delta floodplains, (4) tundra and subalpine zones, and (5) stream–valley shrub including riparian zones. Delta floodplains provide a high biomass of forage for moose and can be exceptionally productive habitats (Telfer 1984; MacCracken et al. 1997; Peek 1998; Bowyer et al. 2002a In press). Floodplains offer relatively permanent habitat for moose compared with more successional types of vegetation (Peek 1998). Stream–valley shrub likewise provides riparian habitat for moose because rivers and streams drive patterns of primary succession via ice action and flooding, which favor plants eaten by moose. Aspen (*Populus*) parklands are used by moose (Peek 1998), as are mature coniferous forest in some areas (Stevens 1970; Pierce and Peek 1984; Van Dyke et al. 1995; Balsom et al. 1996). The value of boreal forest to moose is strongly influenced by fire and subsequent plant succession, which provides abundant forage following burns. Fire, however, is less important in treeline habitats in the boreal forest (Van Ballenberghe 1992) and in boreal regions with higher precipitation and, consequently, longer fire cycles (Jandt 1992). Summer habitat also may include aquatic foraging areas associated with riparian zones around streams, rivers, oxbows, and ponds (Peek 1998).

In many areas, fire plays an important role in the ecology of moose year-round. Quality and abundance of forage, and ultimately moose density, often increase following fire (Aldous and Krefting 1946; Cowan et al. 1950; Spencer and Chatelain 1953; Dewitt and Derby 1955; Spencer and Hakala 1964; Peek et al. 1976; Oldemeyer et al. 1977; Bangs and Bailey 1980; Bangs et al. 1985). In the Kenai Mountains of Alaska, some production of browse occurred by 7–10 years after fire, peaked following 20–30 years of succession, and then declined markedly by 70–80 years after burning (Weixelman et al. 1998) (Fig. 45.11). Fire cycles and the time necessary for production of abundant and succulent vegetation in warmer areas of Alaska (Wolff 1978; MacCracken and Viereck 1990), however, may be more rapid than on the Kenai Peninsula; specifically how moose respond to that more-rapid pattern of succession requires additional study. Peterson (1978) noted that moose on Isle Royale, Michigan, concentrated in areas with more forage as vegetation in burns grew out of their reach (Krefting 1974a). Indeed, Loranger et al. (1991) reported an inverse relation between moose density and years following fire in Alaska. Weixelman et al. (1998) and Peek (1998) cautioned that moose populations held at low density by predation would be less likely to respond to fire than those nearer *K*. Such low-density populations already are reproducing at near maxima and would be unlikely to benefit from enhanced forage abundance (Kie et al. 2003). Kie et al. (2002) noted the importance of habitat heterogeneity in affecting the distribution of mule deer; similar investigations of moose have yet to be undertaken.

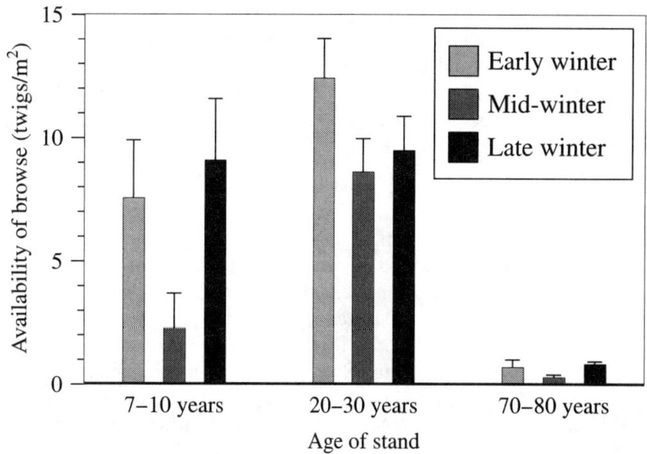

FIGURE 45.11. Relationship between age of burns and available forage for moose (*Alces alces gigas*) on the Kenai Peninsula, Alaska. Note that maximal availability of browse occurs in the 20- to 30-year-old burns, and that snow covers forage during midwinter 7–10 years after fire. SOURCE: Data from Weixelman et al. (1998).

Substantial overstory cover, which intercepts snow, characterizes winter habitat for moose throughout much of their distribution (Peek 1998); yarding is prevalent in the East (Telfer 1967a; Proulx 1983; Crête and Jordan 1982a, 1982b). Yards may occur in level, low-elevation sites (Proulx 1983) or on steep terrain at higher elevations (Kelsall and Prescott 1971). Moose also concentrate in areas without substantial overstories where abundant food is available during winter (Telfer 1978), including river deltas (MacCracken et al. 1997) and riverine areas or riparian zones with willows (Risenhoover 1989; Miquelle et al. 1992; Bowyer et al. 2001b). Nonetheless, snow characteristics play a critical role in winter selection of habitat by moose (Peek 1998), which may cover browse in early successional stages of burns (Weixelman et al. 1998). Depth of snow in late winter also influences selection of bedding sites (Minzey and Robinson 1991). Depths of snow $\geq$70 cm may cause moose to seek winter shelter or other microhabitats with lower snow depths (Kelsall and Prescott 1971; Coady 1974; Hundertmark et al. 1990). Crusted snow also can impede movements and injure legs. Moose will break through snow not capable of supporting 1 kg/cm^2 (Kelsall and Prescott 1971).

Habitat used for parturition by moose is critical for the survival of neonates (Bowyer et al. 1999b). Risk of predation is thought to play a major role in selection of habitats for giving birth (Bailey and Bangs 1980; Stephens and Peterson 1984; Leptich and Gilbert 1986; Addison et al. 1990; Wilton and Garner 1991; Langely and Pletscher 1994; Chekchak et al. 1998; Welch et al. 2000; Testa et al. 2000a). In areas with islands, females may seek those secluded sites for parturition as an antipredator strategy (Addison et al. 1990). Moose also may select high-elevation areas to give birth (Wilton and Garner 1991), and such sites likely provide a better view of approaching predators (Bowyer et al. 1999b). Bowyer et al. (1999b), however, rejected the hypothesis that moose moved to high elevations to "space away" from predators. Langley and Pletscher (1994) and Bowyer et al. (1999b) also noted that moose did not give birth near human developments to avoid predation. Birth sites were farther from human developments than random sites, and proximity of such developments did little to dissuade predators from killing neonates (Bowyer et al. 1998, 1999b). If undisturbed, females remain at birth sites with young for several weeks, likely to help thwart predators (Addison et al. 1990; Bowyer et al. 1999b). Females remaining near the birth site require substantial food to support the high costs of lactation; forage, especially willows, was far more abundant at birth than random sites (Bowyer et al. 1999b).

Bowyer et al. (1999b) hypothesized that maternal females behaved unpredictably to thwart the hunting tactics of predators. Female moose made erratic movements before giving birth and did not select any of the broadly distributed vegetation communities for parturition; birth sites were neither clumped nor overly dispersed spatially (Bowyer et al. 1999b). Instead, females selected microsites with more food, a view of approaching predators, and a southeasterly exposure to give birth, a strategy that balanced the need for forage against the risk of predation (Bowyer et al. 1999b), a pattern common to northern ungulates (Barten et al. 2001).

Sexual Segregation. Sexual segregation is the differential use of space by the sexes outside the mating season (Barboza and Bowyer 2000), and often includes differential use of habitats or forages. Spatial segregation of the sexes is prevalent among sexually dimorphic and polygynous ruminants (Bleich et al. 1997; Mysterud 2000), including all North American cervids (McCullough 1979; Bowyer 1984; Bowyer et al. 1996; Kie and Bowyer 1999; Barboza and Bowyer 2001). This phenomenon is especially pronounced in moose and plays a crucial role in their ecology (Miller and Litvaitis 1992a; Miquelle et al. 1992; Bowyer et al. 2001b). Sexual dimorphism, which is positively related to the degree of polygyny (Weckerly1998; Loison et al. 1999a), affects digestive morphology and physiology (Barboza and Bowyer 2000, 2001) as well as susceptibility to predation (Bleich et al. 1997) differentially for the genders. Many other hypotheses forwarded to explain sexual segregation in ungulates have been rejected (Miquelle et al. 1992; Bleich et al. 1997; Barboza and Bowyer 2001). Risk of predation and differences in digestive capability between the sexes may be the only hypotheses necessary to explain spatial segregation of the sexes.

In moose, spatial separation of the sexes occurs because adult males select habitats with greater forage abundance and females select areas with more concealment cover during winter (Miquelle et al. 1992; Bowyer et al. 2001b). Such differences in the use of space have implications for sampling of moose populations (Peek 1998; Bowyer et al. 2002). Moreover, Kie and Bowyer (1999) suggested that the habitat requirements of the sexes of white-tailed deer were sufficiently different that they should be managed as if they were different species. Indeed, Bowyer et al. (2001b) reported that mechanical crushing of willows markedly enhanced forage abundance. Males, however, moved into the opening to forage, whereas females and young remained in areas with less forage but more cover. Thus, that habitat modification for moose likely benefited one sex at the expense of the other because the manipulation resulted in a net loss of winter habitat for females (Bowyer et al. 2001b). Similarly, spatial segregation of the sexes has implications for population dynamics of ruminants because females compete more intensely with each other and young than do males, which are spatially separated from females for most of the year. Consequently, density of females relative to K has a greater affect on recruitment of young than does density of males (McCullough 1979; Bowyer et al. 1999a).

Density Dependence and Population Regulation. Most life-history characteristics of large mammals are influenced by density-dependent mechanisms (McCullough 1979, 1999; Fowler 1987; Kie et al. 2003). Convincing empirical evidence of density dependence and its role in population regulation exists for cervids (Klein 1968, 1981; McCullough 1979; Clutton-Brock et al. 1982; Skogland 1985; Boyce 1989), including moose (Edwards and Ritchey 1958; Pimlott 1959, 1961; Coady 1982; Bowyer et al. 1999a; Ferguson et al. 2000). These studies provide strong support for conceptual models forwarded by Bowyer et al. (1997), Van Ballenberghe and Ballard (1998), and Kie et al. (2003) that density dependence as modified by predation, hunting, other mortality factors, and severe weather offer the best explanation for population dynamics among large herbivores. Evidence that intrinsic mechanisms regulate populations of large mammals is nil; likewise, documentation that altruistic behavior mediated via group selection plays a role in population processes of ungulates or their predators is lacking (McCullough 1979; Van Ballenberghe and Ballard 1998; Pierce et al. 2000). Some populations of northern cervids were thought to lack strong density dependence (Bergerud 1983a, 1983b; Gasaway et al. 1983; Boertje et al. 1996). That view resulted, in part, from difficulties in understanding complex interactions with weather, not examining populations over sufficient time or densities relative to K, and mistaking an overharvest for a lack of density dependence (McCullough 1990; Bowyer et al. 1999a).

Effects of population density on dynamics of moose populations are mediated principally through the influence of nutrition on reproduction and successful recruitment of young into the population (Simkin 1974; Sæther and Haagenrud 1983; Schwartz and Hundertmark 1993; Keech et al. 2000). Quantity and quality of forage typically regulate moose populations at high density (Messier 1991). Nutritional effects on reproduction are the result of lax intraspecific competition at low population densities and intense competition at high densities relative to K (McCullough 1979; Kie et al. 2003). The outcome of that relationship is a parabolic function between recruitment of young and population size (Bowyer et al. 1999a) (Fig. 45.12). Maximum sustained yield (MSY), at the tip of the recruitment parabola (Fig. 45.12), is the maximal harvest from any source (or combination of causes) that a population can sustain on an annual basis without being driven to low density or extirpation. Kie et al. (2003) offered criteria for determining where the population is with respect to K.

Density-dependent mechanisms also may affect survivorship of adults, but reported survival rates generally are high (92%, Bangs et al. 1989; 91%, Larsen et al. 1989; 94.8%, Ballard et al. 1991; 88%, Bertram and Vivion 2002). Van Ballenberghe and Ballard (1998) noted that survival rates for adults generally ranged between 75% and 94%, depending on degree of human harvest. Moose populations exhibit a classic U-shaped curve with respect to mortality rate and age (Peterson 1978). For example, Gasaway et al. (1983) noted that survivorship decreased from 93% among 6- to 10-year olds to 79% among moose >10 years old.

How rapidly a moose population can increase and its potential yield are a result of the maximal intrinsic rate of increase (r_{max}). Van Ballenberghe (1983b) believed that $r_{max} = 0.35$ might be the limit moose could achieve under natural conditions. Cederlund and Sand (1991) observed $r = 0.40$ for a moose population not experiencing heavy predation and suggested that $r_{max} = 0.47$ might be possible. Bowyer et al. (1999b) reported values of r_{max} between 0.35 and 0.44 for moose on an island without predators or a severe winter climate. Such high values of r_{max} and the associated high potential yields indicate losses of moose to predation and severe weather might be higher than previously suspected (Bowyer et al. 1999b). Van Ballenberghe and Ballard (1998) provide additional population metrics for moose and equations for calculating those parameters.

Ballard et al. (1991) reported that crude densities of moose in Alaska ranged from 0.05 to 1.2 animals/km^2. Gasaway et al. (1992) noted that moose densities were much lower (0.05–0.4 moose/km^2)

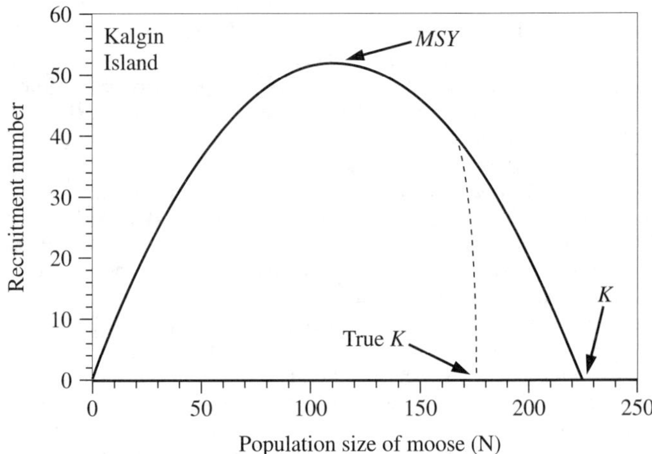

FIGURE 45.12. A recruitment parabola, illustrating yearlings successfully added to the population at differing population sizes for adult moose (*Alces alces gigas*), Kalgin Island, Alaska. Positions of maximum sustained yield (MSY) and ecological carrying capacity (K) are depicted; true K is a correction for an overestimate of K. SOURCE: Data from Bowyer et al. (1999b).

across broad regions of Alaska where populations were subjected to heavy predation, but substantially higher densities (0.2–1.4 moose/km^2) occurred where predators had been controlled. Peek et al. (1976) reported densities of 0.6–0.8 moose/km^2 in Minnesota, and Crête (1987) noted densities of 0.3–2.0 moose/km^2 in eastern Canada. Bowyer et al. (1999a) noted a peak of 3.5 moose/km^2 on Kalgin Island in Cook Inlet, Alaska. Comparisons of crude densities across broad areas, however, should be interpreted with caution because localized densities, such as moose congregating on rutting grounds, can be much higher (Molvar et al. 1993). Moreover, moose concentrating in riparian habitats, including surrounding areas that received less use, can confound assessments of crude density. Comparisons of crude density across populations also can be misleading because *K* may vary among populations, and some populations may be well below *K* (Weixelman et al. 1998; Bowyer et al. 1999b; Kie et al. 2003). Van Ballenberghe and Ballard (1998) emphasized that some populations might temporarily reach 5 moose/km^2, but such high densities probably could not be sustained (Gasaway et al. 1992; Peterson 1999). Overshoots of *K* can be caused by populations irrupting after release from an important source of mortality and rapidly approaching *K*, an outcome that has implications for stability at *K*. McCullough (1979) noted that an overshoot of *K* resulted in a disproportionate decline in the ability of the habitat to support large herbivores.

Severe weather, especially deep snow, holds the potential to affect moose populations adversely (Peterson and Allen 1974; Peterson 1978; Mech et al. 1987). Young, old, and male moose may be most susceptible to mortality, including predation, during winters with deep snow (Coady 1982). Effects of severe weather, however, are seldom independent of population density (Bowyer et al. 1999b; Kie et al. 2003). Moose populations at low density relative to *K* may exhibit high twining rates even following winters with unusually deep snow (Bowyer et al. 1998). Conversely, malnourished populations near *K* may suffer substantial mortality during winter, especially among young moose (Bishop and Rausch 1974). Understanding complex interactions between stochastic environmental conditions and population dynamics of large herbivores is a topic requiring more study (Sæther 1997).

Interspecific Competition. The geographic distribution of moose overlaps that of eight native ungulates as well as range cattle and, less commonly, domestic sheep. Competitive interactions with hares (*Lepus* spp.) and beaver (*Castor canadensis*) also are possible because of their similar range distributions and use of habitats (Boer 1998). Little interspecific competition is likely for moose and bighorn sheep (*Ovis canadensis*), Dall's sheep (*O. dalli*), or mountain goat (*Oreamnos americanus*) because those mountain ungulates occupy rugged, precipitous habitats at higher elevations than those generally used by moose (Bowyer et al. 2000; Krausman and Shackelton 2000; Peek 2000). Similarly, limited competition is expected between moose and muskoxen (*Ovibos moschatus*) because muskoxen inhabit tundra regions at the northern extent of the distribution of moose, and those bovids feed principally on graminoids (Klein 2000). Bison (*Bison bison*) similarly are primarily grazers and exhibit disparate food habits from moose where the two species are sympatric (Cairns and Telfer 1980). Nonetheless, willows compose a substantial proportion of the diet of bison translocated to interior Alaska (Campbell and Hinkes 1983), which might lead to dietary competition with moose.

Competition between caribou (*Rangifer tarandus*) and moose is thought to be slight (Davis and Franzmann 1979; Boonstra and Sinclair 1984), in part because of the more specialized diet of caribou and their more extensive use of lichens (Peterson 1955; Boer 1998). Indeed, analysis of stable isotope ratios (δ^{13}C and δ^{15}N) from moose and caribou indicate dietary differences between those large herbivores (Ben-David et al. 2001). Moreover, introduction of moose into Newfoundland did not reduce caribou numbers (Peterson 1955). Seip (1992) postulated that in multiple-prey ecosystems, high densities of moose could maintain wolf populations at levels that would intensify predation on caribou. Dale et al. (1994), however, noted that wolves preferred caribou to moose, and Coady (1980) ascribed an increase in moose numbers in northern Alaska to wolves preferring caribou. Multiple-prey systems are complex (Gasaway et al. 1992; Dale et al. 1994) and variation in abundance of one ungulate is likely to have effects on that of another.

The distributional overlap between moose and mule deer is not extensive, although these cervids are sympatric in portions of the northern Rocky Mountains and along the southern border of the boreal forest (Boer 1998). Mule deer consume more forbs than do moose during summer (Kie and Czech 2000). Although both species eat browse in winter and some competition may occur (Prescott 1974), mule deer often overwinter at lower elevations than moose, thereby limiting competitive interactions (Telfer 1978; Boer 1998). Elk are primarily grazers, but have broad, flexible feeding habit, which include browse, especially in winter (Houston 1982; Boyce 1989; Boer 1998; Stewart et al. 2002). Cowan (1950) and Flook (1964) believed that elk possessed the capability to outcompete moose for food. More research is needed to elucidate interspecific relationships between these two cervid species.

White-tailed deer and moose are sympatric near the northern distribution of white-tailed deer and the southerly extent of moose (Boer 1998). The northern (and altitudinal) distribution of white-tailed deer is limited by winter conditions, especially snow depth, whereas moose are more affected by availability of browse than winter severity (Telfer 1970, 1978; Prescott 1974; Cairns and Telfer 1980; Telfer and Cairns 1986). In the Northeast, white-tailed deer yard in areas with coniferous overstories, which intercept snow and provide for easier travel (Telfer and Kelsall 1979); moose habitat is more dependent on stands of balsam fir. Competition for food between these cervids is only likely when moose move to areas with substantial overstory cover (Telfer 1970; Peek et al. 1976). Even where the spatial distribution of moose and white-tailed deer coincide and both consume largely conifers, only a 41.2% overlap occurred in their winter diets (Ludewig and Bowyer 1985). Habitat and dietary differences indicate strong niche partitioning between these cervids. Nonetheless, white-tailed deer in the Northeast may affect moose adversely via parasite-mediated competition by harboring meningeal worms (*Parelaphostrongylus tenuis*), which are benign in white-tailed deer, but can cause a neurological disorder and death in moose (Lankester and Peterson 1996).

Hares also may compete with moose for browse (Dodds 1960; Wood 1974). A high degree of overlap in winter diets has been reported (Telfer 1972; Oldemeyer 1975, 1983). Some partitioning of niche by hares and moose may occur based on foraging height (Telfer 1972; Oldemeyer 1974), but Belovsky (1984) noted that accumulating snow may allow hares access to higher branches because they can walk on top of snow. Competition between snowshoe hares (*Lepus americanus*) and moose likely occurs at peak hare densities (Wolff 1980). Beavers also may affect moose by creating aquatic habitat favored by moose, but browsing by moose may suppress growth of large trees preferred by beavers (Wolfe 1974). Because beavers are alternative prey for wolves, high densities of beavers may lessen predation on young moose in summer, but intensify predation on moose in winter when beavers are less susceptible to wolves (Shelton and Peterson 1983).

MORTALITY

Predation. Considerable controversy exists over whether predation can regulate populations of moose, as well as the intensity of predation and species of predators that might be necessary to bring about regulation (Van Ballenberghe and Ballard 1998). Van Ballenberghe (1987), Skogland (1991), Boutin (1992), Messier (1994), Van Ballenberghe and Ballard (1994), and Ballard and Van Ballenberghe (1998) discussed hypotheses, study designs, and types of data critical to answering this ecologically important question. Whatever the exact mechanism and most appropriate model (McLaren and Peterson 1994; Peterson et al. 1998; Eberhardt and Peterson 1999; Eberhardt 2000; Messier and Joly 2000; Post et al. 2002; Vucetich et al. 2002), there is little doubt that large mammalian carnivores have the capability of holding moose populations at densities well beneath *K* (Crête and Jordan 1982a, 1982b; Bergerud et al. 1983a; Gasaway et al. 1983; Messier 1991; Gasaway et al. 1992; Bergerud and Elliott 1998). Predator control has brought about increases in moose populations in some instances (Gasaway et al. 1992; Boertje et al. 1996), but has been less effective in others (Van Ballenberghe and Ballard 1998). Demonstrating population regulation (which requires density dependence) is difficult, however, and Crête and Courtois (1997) cautioned that limiting factors could obscure evidence of regulation in relatively unproductive habitats. Moreover, most studies considering moose–wolf relationships follow the method of Fuller (1989) by combining ungulate species into a single value for biomass available to wolves. That procedure does not consider differences in r_{max} among ungulate species. Thus, effects of population dynamics of prey on their abundance and, hence, availability to predators are not incorporated in such approaches (Person et al. 2001).

Debate also exists over the type of functional response (Holling 1959) that wolves exhibit with regard to changing densities of prey, including moose. Empirical evidence indicates a type II functional response, which has implication for potential equilibria between wolves and their ungulate prey (Dale et al. 1994; Messier 1994; Hayes and Harestad, 2000). Marshal and Boutin (1998) noted, however, that gathering data necessary to determine the shape of the functional response at low densities of prey (where type II and III curves differ most) was problematic. Person et al. (2001) reported that the shape of the functional-response curve affected equilibria only at extremely low densities of ungulate prey; relation of the ungulate population to K was more influential at higher densities.

Predation is the primary source of mortality for young moose in many populations, with most neonates killed by predators perishing within 6 weeks of birth (Franzmann et al. 1980; Ballard et al. 1981, 1990, 1991; Franzmann and Schwartz 1986; Larsen et al. 1989; Osborne et al. 1991; Testa et al. 200b). In those studies, brown and grizzly bears (*Ursus arctos*) or black bears (*U. americanus*), rather than gray wolves (*Canis lupus*), were largely responsible for killing neonatal moose. A review by Ballard and Van Ballenberghe (1998) reported that grizzly bears killed 3–52% and black bears 2–50% of young moose. Bertram and Vivion (2002) reported that black bears (45%) and grizzly bears (39%) killed most young moose in interior Alaska. Bowyer et al. (1998) noted that survivorship of young moose in a low-density population in Denali National Park and Preserve, Alaska, was reduced to about 0.2 after only 30 days of life, largely the result of predation by grizzly bears. Keech et al. (2000), however, observed nearly equal mortality of young caused by grizzly bears, black bears, and wolves in a growing population of moose in interior Alaska, where survivorship was about 0.5 after 1 year of life. Kill rates for grizzly bears on moose were 0.14–0.85 young/bear/day, whereas black bears killed moose at a rate of 0.02–0.09 young/bear/day in spring and summer (Ballard 1992).

Wolf predation on young moose is most pronounced in winter when bears are denning (Ballard et al. 1987; Gasaway et al. 1992). Wolves kill young moose disproportional to their occurrence in the population during that season (Ballard et al. 1987; Hayes et al. 2000). Mortality from wolves may reach 18% of young killed in some populations (Gasaway et al. 1983); Hayes et al. (2000) reported 12–55% of young moose killed by wolves, with a kill rate of 0.024 young/wolf/day during winter (Hayes and Harestad 2000). Mountain lions (*Puma concolor*) also may kill young and sometimes adult moose (Kunkel et al. 1999); principally male mountain lions kill moose (Ross and Jalkotzy 1996).

Losses of young moose to predators may be either additive or compensatory, depending on the population density of moose in relation to K (Ballard 1992; Ballard and Van Ballenberghe 1998; Person et al. 2001; Kie et al. 2003). Mortality in low-density populations tends to be additive, whereas populations near K exhibit higher levels of compensatory mortality, wherein one source of mortality is substituted for another (McCullough 1979; Ballard and Van Ballenberghe 1998; Person et al. 2001; Kie et al. 2003). Grizzly bears also may prey on adult moose at rates of about 0.02 moose/bear/day (Ballard 1992); black bears occasionally kill adult moose, but are not considered important predators on adults (Ballard and Van Ballenberghe 1998). Wolves kill older adults at rates of one adult /5–16 days (Peterson et al. 1984). Hayes et al. (2000) reported an overall kill rate of 0.045 moose/wolf/day during winter. Physical condition of moose killed by wolves may be either poor or good (Peterson et al. 1984; Ballard et al. 1987), likely depending on the relationship of the population to K and winter conditions. Peterson et al. (1984) reported higher kill rates of moose by wolves in winter than during summer, but Ballard et al. (1987) observed no seasonal differences.

Ratios of predators to prey have been used widely to interpret effects of wolf predation on their ungulate prey (Person et al. 2001). If densities of wolves were greater than predicted, or if ungulate–wolf ratios were less than predicted, then wolves were presumed to be causing a decline in the ungulate population. Person et al. (2001) cautioned that these models assumed that wolf and ungulate populations were at or near equilibrium, and that the density of prey with respect to K had no effect on the number of wolves that could be supported. Moreover, Ballard et al. (1995) noted that predator–prey ratios were too variable to be predictive. Theberge (1990) argued that variation in the functional response of wolves to changing density of prey, prey-switching behavior by wolves, and the relation of the prey population to K would complicate interpretation of such ratios. Person et al. (2001) reported that equilibrium ratios between prey and wolves increased rapidly as the ungulate population grew beyond MSY toward K, and that wide confidence intervals occurred around wolf–ungulate ratios, necessitating caution in even a general interpretation of such values.

Incidental Mortality. Child (1998) provided an overview of incidental causes of mortality in moose. Collisions with automobiles (Grenier 1973; Child et al. 1991; Oosenbrug et al. 1991; DelFrate and Spraker 1991; McDonald 1991; Belant 1995) and trains (Child 1983; Child et al. 1991; Modaferi 1991; Belant 1995; Gundersen and Andreassen 1998) by far cause the most losses in this category of mortality. For instance, in the 1990s, automobiles killed 713, 600, 460, and 265 moose/year in British Columbia, Minnesota, Newfoundland, and Quebec, respectively. Similarly, annual mortality from collisions with trains killed >700 moose in one deep-snow winter in Alaska (Child 1998). Characteristics of highways and railways, including creating successional habitat and foods favored by moose and providing travel routes through deep snow, often lure moose to these dangerous thoroughfares. Moose naturally fluoresce under ultraviolet light, and car manufactures in Fennoscandia and Europe have equipped automobiles with ultraviolet headlights to make moose more visible at night (Child 1998).

Moose occasionally drown, especially in association with hydroelectric projects and reservoirs (Child 1998). Like most mammals, moose also suffer from accidental falls (Child 1998). Near Fairbanks, Alaska, a moose perished when it fell through the top of a septic tank and became tightly wedged in the tank and subsequently frozen into contents therein. Moose also are killed by avalanches and by becoming entangled in wire, cables, fencing, and even a garden hose (Child 1998). Attempts to mitigate some of this mortality have included creating pathways leading away from dangerous roads and railways, creating barriers to prevent moose from reaching such areas, and using light reflectors (Schwartz and Bartley 1991; Child 1998). Reducing speed of trains or having a pilot car precede the train has reduced accidental mortality in moose (Child 1998). Partial drainage of roadside salt pools, however, did not reduce moose–vehicle collisions in Quebec (Jolicoeur and Crête 1994).

Diseases, Parasites, and Pests. Anderson and Lankester (1974), Samuel et al. (1976), Lankester (1987), and Lankester and Samuel (1998) provided extensive reviews of those organisms capable of infecting or infesting moose. Population density of moose and their resultant physical condition, severe weather and its effects on moose as well as on intermediate hosts of parasites or directly on parasites and pests themselves, and the abundance of additional definitive hosts that serve as vectors all may exacerbate or limit diseases in moose populations. Indeed, moose are beset with a plethora of pests and parasites, some of which can cause serious diseases (Table 45.3). Only some of those disease-causing agents, however, are suspected of having the potential to adversely affect populations of moose. Although chronic wasting

TABLE 45.3. Selected parasites, pests, and diseases of moose (*Alces alces*)

Disease-Causing Organisms	Comment	Reference
Viruses		
Arboviruses		
Epizootic hemorrhagic disease (EHDV)	Transmitted by sand flies	Trainer and Jochim 1969
Bluetongue (BTV)	—	Hoff and Trainer 1978; Kocan et al. 1986b
Bovine respiratory viruses		
Bovine rhinotracheitis (IBRV)	Moose are seropositive, but no disease obvious	Kocan et al. 1986a
Parainfluenza type 3 (PI-3)	—	Lankester 1987
Bovine viral diarrhea (BVDV)	—	Kocan et al. 1986b
Other viruses		
Malignant catarrhal fever (MCF)	Domestic sheep likely vector; also present in free-ranging ungulates in Alaska	Thorne and Honess 1982; Williams et al. 1984; Zarnke et al. 2002
Contagious ecthyma	Experimental infection only	Zarnke et al. 1983
Rabies	Rare	Lankester and Samuel 1998
Adenovirus	In captive moose only	Shilton et al. 2002
Spirochaetes		
Leptospirosis (*Leptospira interrogans*)	May cause abortion in severe cases	Thorne and Honess 1982; Kocan et al. 1986b
Fibromas and other tumors		
Infectious cutaneous fibromas	Infection by direct contact or maybe biting flies; self-limiting	Lankester and Samuel 1998
Lymphosarcoma	In abdominal cavity	Garner and Schwartz 1969
Myxoma	Cauliflower-like growth	Lankester and Bellhouse 1982
Bacteria		
Brucellosis (*Brucella abortus, B. suis, B. ovis*)	Can cause abortion, stillbirths, and sterility; often infects joints	McCorquodale and DiGiacomo 1985; Dieterich et al. 1991; Honour and Hickling 1993; Forbes et al. 1996; O'Hara et al. 1998
Listeriosis (*Listeria monocytogenes*)	Spread by direct contact or biting arthropods; rare in moose	Archibald 1960
Johne disease (*Mycobacterium paratuberculosis*)	Lower digestive track; rare in moose	Soltys et al. 1967
Anthrax (*Bacillus anthracis*)	Rare in moose	Choquette 1970
Tularemia (*Francisella tularensis*)	More common in lagomorphs and rodents; rare in moose	Bourque and Higgins 1984
Erysipelas (*Erysipelothrix rhusiopathiae*)	Cutaneous infection associated with hair loss caused by ticks	Campbell et al. 1994
Pinkeye (*Moraxella* sp.)	From Wyoming	Thorne and Honess 1982
Actinomycosis (*Actinomyces* sp.)	Abscesses and bone erosion of lower jaw (lumpy jaw)	Ritcey and Edwards 1958
Foot rot (*Fusobacterium necrophorum*)	Necrotizing lesions in hooves; abscess formation in oral cavity and internal organs	Lankester 1987
Puss pockets or abscesses (*Staphylococcus aureus, Acinetobacter* sp., *Actinomyces* sp.)	Often located on flank	Thorne and Honess 1982
Protozoans		
Toxoplamosis (*Toxoplasma gondii*)	Often spread by felids, can infect humans	Dubey 1981; Kocan et al. 1986a; Siepierski et al. 1990; Zarnke et al. 2000
Other coccidia		
Sarcocystis sp.	Spread by canids; causes unsightly cysts in muscles of moose	Dubey 1980; Dubey and Williams 1980; Lankester 1987
S. alceslatrans		
Other protozoans		
Trypanosoma cervi	Transmitted by horse flies; blood flagellate	Kingston 1981b
Amoeba resembling *Entamoeba bovis*	In cecum	Kingston 1981a
Trematodes		
Fascioloidiasis (*Fascioloides magna*)	Does not reach maturity in liver of moose; can cause substantial pathology; aquatic snail intermediate host	Lankester 1974, 1987
Paramphistomiasis (*Paramphistomum cervi, P. liorochis*)	No clinical disease, but large numbers of rumen flukes reported; snail is intermediate host	Snider and Lankester 1986
Other flukes		
Zygocotyle lunata	Occurs in intestine of waterfowl and occasionally moose; snail is intermediate host	Samuel et al. 1976; Stock and Barrett 1983; Hoeve et al. 1988
Cestodes		
Hydatid cysts (*Echinococcus granulosus*)	Adult tapeworm in wolves; cysts occur mostly in lungs and may debilitate moose; eggs in wolf feces infective to humans, but cysts in moose are not	Anderson and Lankester 1974; Lankester 1987; McNeill and Rau 1987; Messier et al. 1989
Cysticerci (*Taenia hydatigena, T. ovis krabbei*)	Wild canids definitive host; cysts occur in liver and muscles of moose, including heart; not substantial pathogens to moose; cysts in moose not infective to humans	Samuel 1972; Addison et al. 1979; Pybus 1990

(*Continues*)

Table 45.3—*Continued*

Disease-Causing Organisms	Comment	Reference
Fringed tapeworm (*Thysanosoma actinoides*)	Bark lice intermediate host; occurs in small intestine and bile ducts; no obvious pathology	Stock and Barrett 1983
Moniezia benedeni, M. expansa	Free-living lice intermediate host; tapeworms can be long, but no obvious pathology to intestine	Thorne and Honess 1982
Nematodes		
Parelaphostrongylosis (*Parelaphostrongylus tenuis*)	A parasite of white-tailed deer; gastropods are intermediate hosts; can cause severe neurological disorder in moose	Anderson 1972; Lankester and Anderson 1968; Anderson and Lankester 1974; Anderson and Prestwood 1981; Lankester 1987; Lankester and Samuel 1998
Elaphostrongylosis		
Tissue worm (*Elaphostrongylus rangiferi*)	Restricted to Newfoundland; experimental infection produced neurological disease, but not yet observed in free-ranging moose	Lankester 1977; Lankester and Fong 1989; Olsson et al. 1995; Steen et al. 1997
Lung nematodes		
Dictyocaulus viviparous	Direct life cycle; adults occur in smaller bronchi; no obvious disease; most common in young	Gupta and Gibbs 1971
Orthostrongylus macrotis	Intermediate host slug or snail; parasite of mule deer reported in moose only from Alberta	Samuel et al. 1976; Stock and Barrett 1983
Gastrointestinal nematodes		
Nematodirella alcidis	Direct life cycle; occur in abomasum and duodenum; can cause enteritis and scouring	Stock and Barrett 1983; Fruetel and Lankester 1988; Hoeve et al. 1988
Trichostrongylus spp.	In abomasum; no obvious disease observed in moose	Stock and Barrett 1983
Ostertagia spp.		
Whipworm (*Trichuris discolor*)	In cecum and colon; parasite of cattle; rarely associated with death in moose	Lankester and Samuel 1998
Filarioid nematodes		
Arterial worm (*Elaeophora schneideri*)	Transmitted by horseflies; first-stage microfilariae in skin or blood; parasite of mule deer infects elk and occasionally moose; causes blindness	Worley et al. 1972; Madden et al. 1991
Legworm (*Onchocerca cervipedis*)	Likely transmitted by black flies; adults occur in subcutaneous tissue around legs and feet; only reported from western Canada and Alaska, but may be more widespread	Samuel et al. 1976; Pledger et al. 1980
Abdominal worm (*Setaria yehi*)	Perhaps transmitted by mosquito or stable fly; adults free in body cavity; may cause peritonitis	Anderson and Lankester 1974; Samuel et al. 1976
Rumen filarioids (*Rumenfilaria andersoni*)	Vector unknown; in folds of rumen; only from northwestern Ontario; no obvious pathology	Lankester and Snider 1982
Ticks		
Winter tick (*Dermacentor albipictus*)	All three parasitic stages occur on moose; spread by seed ticks on vegetation; grooming associated with ticks causes hair loss, which can kill moose; absent from Alaska and Newfoundland	Welch et al. 1991; Samuel and Welch 1991; Lankester and Samuel 1998
Flies		
Throat and nasal bot flies (*Cephenemyia jellisoni, C. phobifera*)	Larvae deposited on muzzle, migrate to pharyngeal pouches at back of throat; not significant pathogen	Bennett and Sabrosky 1962
Moose fly (*Haematobosca alcis*)	Closely associated with moose; both males and females feed on moose, often on rump or hindquarters; females deposit eggs in moose feces; may cause skin lesions on back legs of moose	Anderson and Lankester 1974; Burger and Anderson 1974; Pledger et al. 1980; Lankester and Sein 1986
Other biting flies	Most information from captive moose	Smith et al. 1970; Pledger et al. 1980
Stable fly (*Stomoxys calcitrans*)		
Deer fly (*Chrysops* spp.)		
Horse fly (*Hybomitra* spp., *Tabanus* spp.)		
Black fly (*Simulium* spp., *Prosimulium* spp.)		
Mosquitoes (Culicidae)		

Source: Adapted from Lankaster and Samuel (1998).

disease occurs in other free-ranging cervids in North America, moose have not been reported to harbor that disease (Williams et al. 2002). Lankester and Samuel (1998) proposed that pathogens capable of affecting populations of moose included the meningeal worm in eastern North America, the winter tick (*Dermacentor albipictus*), and perhaps the arterial worm (*Elaephora schneideri*) in localized areas of western North America. In addition, the tapeworm that causes hydatid disease (*Echinococcus granulosus*) may debilitate moose, thereby making them more susceptible to wolf predation (Messier et al. 1989). We restrict our discussion to those important diseases of moose, although a broader treatment based primarily on Lankester and Samuel (1998) is provided in Table 45.3.

The nematode *Parelaphostrongylus tenuis* (meningeal worm), which can cause a severe neurological disorder in moose, normally infects and is spread by white-tailed deer (Anderson 1972; Anderson and Lankester 1974; Anderson and Prestwood 1981; Lankester 1987, 2001; Dumont and Crête 1996; Nankervis et al. 2000). Lankester and Samuel (1998) provided an excellent overview of the life cycle of *P. tenuis*. First-stage larvae are passed in the feces of infected white-tailed deer. Those larvae are present only in the mucous layer on the surface of fecal pellets (Lankester and Anderson 1968). First-stage larvae are extremely resistant to freezing and can withstand some drying (Shostak and Samuel 1984). Melting snow or rain release larvae into the soil, where some penetrate the foot of woodland snails or slugs; *P. tenuis* larvae then molt to the infective third stage. Those third-stage larvae are long lived and can overwinter in gastropods (Lankester and Anderson 1968). Numerous species of snails or slugs can serve as intermediate hosts. The small snails *Discus cronkhitei*, *Zonitoides* spp., *Succinea* spp., and *Cochlicopa* spp. and the slugs *Deroceras leave* and *D. reticulatum,* which are inadvertently eaten by ungulates, serve as primary sources for infection.

Once consumed by the herbivore, *P. tenuis* larvae penetrate the wall of the abomasum, move into the abdominal cavity, and migrate along nerves in the body wall to the vertebral canal. Larvae enter the tissue of the spinal cord and dorsal horns of the gray matter, where they molt to fourth and fifth stages. Adults ultimately migrate into the cranium and position themselves along large veins and sinuses in the dura. Following mating, female worms lay eggs, which are carried by the blood via the jugular to the right side of the heart, through the pulmonary artery, and into the lungs. First-stage larvae develop in the lungs and are moved by cilia on bronchi into the oral cavity, where they are swallowed and ultimately occur in deer feces. The life cycle requires 90–137 days from the time of infection (Lankester and Samuel 1998).

In white-tailed deer, *P. tenuis* causes no obvious disease, and deer may develop immunity to this parasite on initial infection (Slomke et al. 1995). In moose, however, considerable tissue damage to the spinal cord as well as a strong cellular reaction occur (Lankester and Samuel 1998); adult worms also may enter the brain tissue causing further neurological damage (Anderson and Prestwood 1981). The result for infected moose can include an abnormal tilting of the head, weakness in the hindquarters, aimless circling, and often death.

Although *P. tenuis* may complete its life cycle in moose with first-stage larvae released in feces, maintenance of this disease complex is thought to require white-tailed deer. For instance, several authors related frequency of disease in moose to the density of white-tailed deer (Karns 1967; Telfer 1967b; Gilbert 1974; Whitlaw and Lankester 1994a, 1994b). Strong circumstantial evidence exists that *P. tenuis* can limit moose numbers—infection with meningeal worms was long thought to be fatal for moose. Indeed, experimental infections with large numbers of larvae had catastrophic effects on moose, especially young animals (Lankester and Samuel 1998).

Expanding populations of moose in eastern North America in areas with substantial numbers of white-tailed deer (Lautenschlager and Bowyer 1985; Upshall et al. 1987; Thomas and Dodds 1988), however, caused a rethinking of traditional ideas concerning *P. tenuis* (Clark and Bowyer 1986; Nudds 1990; Lankester and Samuel 1998). Indeed, moose have increased elsewhere in the presence of high population densities of white-tailed deer (Aho and Hendrickson 1989). One explanation for that increase is that low-dose infections with *P. tenuis* provide a degree of immunity to moose (Lankester and Samuel 1998; Lankester 2001). That explanation alone, however, will not account for recent expansions of moose when they failed to do so in the past, even though adequate habitat was available. Consequently, some alteration in pathogenicity of *P. tenuis,* or an improved ability of moose to accommodate infection with this nematode more synergistically than in the past, is necessary to explain population expansion by moose; both are viable hypotheses (Clark and Bowyer 1986).

Winter ticks infest moose throughout much of their distribution, especially south of 60° N latitude. These parasites are absent from Newfoundland and Alaska (Lankester and Samuel 1998), although experimental evidence indicates winter ticks could survive in Alaska if introduced (Zarnke et al. 1990). Winter ticks have been recovered from moose in northern British Columbia (Samuel and Welch 1991). Winter ticks spend their three parasitic stages (larva, nymph, and adult) on a single host, which most often includes moose (Welch et al. 1991). Adult females drop off their host in late March or early April and lay eggs (Addison et al. 1998a), which do not hatch until summer. Infestation of moose occurs in late summer and autumn when moose brush against clumps of small (1 mm) seed ticks (larvae with three rather than four pairs of legs) that have aggregated in large clumps at the tips of vegetation (Lankester and Samuel 1998).

Once larval winter ticks are on a moose, those parasites begin feeding on blood, which fuels the molt to the nymphal stage in November; nymphs (1.7 mm in length) likewise feed on blood from their host (Lankester and Samuel 1998). By mid-February, nymphs reach peak numbers and molt into adults. Adult males are 6.5 mm in length and are dark brown with a white cross-hatching on their backs, whereas females (7 mm) are dark reddish brown. Large (≤9.8 mm) grayish ticks observed in late winter and spring are females engorged with blood. Adults reach peak numbers in late March–April; all ticks have dropped off moose by mid-May, and moose are free of ticks during summer (Lankester and Samuel 1998).

Samuel and Welch (1991) estimated the mean number of individual winter ticks on a moose from various areas in western North America was 32,500, with 18% of 212 moose with >50,000 ticks. This massive infestation with ticks results in excessive grooming by moose with concomitant lose of hair. Grooming begins in late February and results in a distinctive pattern of hair loss: damage to the coat starts at the neck and shoulders and progresses toward the withers and tail region (Samuel et al. 1986; Samuel 1991). Samuel and Welch (1991) reported that 89% of 724 moose from across North American exhibited evidence of hair loss from infestations with winter ticks. In addition to suffering hair loss, moose infested with winter ticks exhibit restlessness and chronic anemia, and have reduced fat reserves (Glines and Samuel 1989; Samuel 1991; Addison and McLaughlin 1993; DelGiudice et al. 1997; Addison et al. 1998b). Also, growth is reduced in young moose infested with ticks (Addison et al. 1994). Moose may help reduce infestation with ticks by evading seed ticks on vegetation, tolerating tick-foraging birds, and via grooming, which can reduce tick numbers (Samuel et al. 2000). Direct evidence that winter ticks affect population dynamics of moose is difficult to obtain. Nonetheless, losses of moose have coincided with major epizootics of winter tick (Lankester and Samuel 1998). Moreover, tick-induced loss of hair and other associated maladies likely interact with other factors, such as nutrition and climate, to affect mortality in moose (Lancaster and Samuel 1998).

Mule deer are thought to be the usual host for the arterial worm, which is transmitted to moose by horse flies (*Hybomytra* and *Tabanus*), in localized areas of northwestern North America (Worley et al. 1972; Worley 1975; Madden et al. 1991). This nematode, which also infects domestic sheep, can complete its life cycle in moose. Adult arterial worms reside in small branches of the common carotid and internal maxillary arteries, where they can persist for up to 4 years (Hibler and Adcock 1971). First-stage larvae (microfilariae) are released and carried to capillaries of the skin on the face and forehead, where they are transmitted to horse flies seeking a blood meal. Few instances of this disease have been reported in moose, but arterial worms can be serious parasites of elk, where they cause blindness, necrosis of ears and muzzle, antler deformities, and damage to the central nervous system (Hibler and Adcock 1971). Infection of moose by arterial worms is restricted to where their distribution overlaps mule deer and black-tailed deer; whether this parasite can affect moose populations is uncertain (Lankester and Samuel 1998).

Echinococcus granulosus are diminutive, three-segmented tapeworms that reside in the small intestine of wolves, where these cestodes cause little harm. Proglottids of adult worms are passed in wolf feces, and eggs (onchospheres) are released into the environment, where they are consumed inadvertently by foraging moose (Lankester and Samuel 1998). Ultimately, eggs develop into hydatid cysts, which are bladder-like structures filled with fluid and numerous larvae (protoscoleces) resembling grains of sand. The most common location for hydatid cysts is the lungs, but cysts also occur in the liver of moose (Lankester 1987). The life cycle is completed when a wolf eats an infected moose; this parasite is uncommon outside the distribution of wolves (Anderson and Lankester 1974; Lankester 1987). Hydatid disease occurs in numerous ungulates and can infect domestic dogs. Human also can acquire this parasite, but only from eggs released in feces of canids, not from hydatid cysts in moose (Lankester and Samuel 1998).

The number and size of hydatid cysts, which range from pea sized to several centimeters in diameter, increase with the age of moose (McNeill and Rau 1987; Messier et al. 1989). Severity of infection with hydatid cysts also intensifies with increasing population density of moose (Messier et al. 1989). Cysts in the lungs make moose more prone to predation by reducing their stamina during pursuits by wolves (McNeill and Rau 1987), and wolves may recognize the vulnerability of infected individuals (Messier et al. 1989). Because predation has the potential to hold moose populations at low density (Gasaway et al. 1992), this parasite may affect dynamics of moose.

MANAGEMENT

Habitat. Modification of habitat to benefit moose has included logging, mechanical crushing and disturbance, prescribed fire, and application of fertilizer and herbicides (for reviews Eastman and Ritcey 1987; Joyal 1987; Peek et al. 1987; Thompson and Euler 1987; Thompson and Stewart 1998). Outcomes from many management techniques, however, were not evaluated and their value to moose was uncertain (Oldemeyer and Regelin 1980). More recent studies have rectified that shortcoming (Lautenschlager et al. 1997; Rempel et al. 1997; Collins and Schwartz 1998; Stephenson et al. 1998b; Weixelman et al. 1998; Bowyer et al. 2001b), but much remains to be learned about manipulating moose habitat, especially from opportunities offered through adaptive management of boreal forests (Thompson and Stewart 1998).

Throughout much of the distribution of moose in North America, logging and subsequent production of forest products are the dominant land use; moose management often is a byproduct of industrial processes (Thompson and Stewart 1998). Use of herbicides following logging also can affect composition and abundance of browse in regenerating stands (Lautenschlager et al. 1989; Eschholz et al. 1996; Raymond et al. 1996). Clearly, logging can create successional habitats of value to moose. Peek et al. (1976) suggested that logging was more important than fire in producing high-quality habitat for moose in Minnesota. Indeed, clearcutting of forests has resulted in an increase in use of those areas by moose (Parker and Morton 1978; Monthey 1984).

Moose respond to regenerating browse in understories resulting from clearcutting; clearcuts with insufficient browse receive little use by moose (Matchett 1985; Kufeld and Bowden 1996). Moreover, even clearcuts with adequate browse may receive limited use by moose if those logged areas are too large and consequently have regrowing shrubs and trees far from concealment cover. Numerous authors have noted negative effects of distance from cover on habitat use and foraging behavior by moose (Hamilton et al. 1980; Mastenbrook and Cumming 1989; Molvar and Bowyer 1994; Tyers and Irby 1995; Weixelman et al. 1998). Kufeld and Bowden (1996) recommended that no point in a clearcut be >100 m from concealment cover. Bowyer et al. (2001b) further reported that male moose used an open area (200–800 m wide) of crushed willows with substantial forage, but females and young did so less often, ostensibly because of predation risk. Moose were more likely to be killed by wolves away from edges of small forest patches, indicating that sites farther from cover were more dangerous (Kunkel and Pletscher 2000). Alternatively, logged areas situated too near forest cover may lessen their value for moose. Todesco et al. (1985) reported that strip cuts 40–50 m wide accumulated more snow than nearby clearcuts, thereby making forage less available to moose in winter on those narrow cuts. Bowyer et al. (2001c) observed that foraging efficiency of mule deer declined both near and far from concealment cover, presumably because they were more vulnerable to ambush predators near cover and coursing predators farther away from cover. Moose also, may be susceptible to ambush predators such as mountain lions on small clearcuts and alter their behavior accordingly, but that outcome remains to be demonstrated.

Not all types of habitat manipulations will produce equivalent amounts of browse for moose. For instance, Lautenschlager et al. (1997) noted that forage production and digestibility of browse following cutting and controlled burning were similar. Those treatments were greater than on control plots or those defoliated by spruce budworm (*Choristoneura fumiferana*) in Maine. Stephenson et al. (1998) reported that mechanical manipulation of plant communities without willow understories in Alaska were prone to colonization with unpalatable bluejoint reedgrass (*Calmagrostis canadensis*) unless mineral soil was exposed to allow seeds of willows to germinate. Collins and Schwartz (1998) reached similar conclusions for logged areas in that state. Temporal variation in the availability of browse also occurs on different types of habitat manipulations. Clearcutting and fire may take 15–20 years for maximal browse production for moose (Timmermann and Buss 1998; Weixelman et al. 1998), whereas Bowyer et al. (2001b) reported a threefold increase in dry weight of current annual growth for willows 3 years following mechanical crushing compared with an uncrushed site.

Habitat suitability models also have been developed and evaluated for moose (Allen et al. 1991). Timmermann and Buss (1998) emphasized that forest practices, especially clearcutting, needed to consider proximal relationships among cover, water, mineral licks, aquatic vegetation, and human disturbances if they are to be of value to moose. Indeed, Colescott and Gillingham (1998) reported that moose within 300 m of snowmobile activity altered their behavior and were displaced to less favorable habitats. Creating access for hunters also can affect moose populations. Several studies have reported disproportionally high moose harvests associated with increased road access (Lynch 1973; Eason et al. 1981; Bangs et al. 1984). Schneider and Wasel (2000) also noted effects of human settlement on density of moose. Moreover, manipulation of habitat to produce browse will have little benefit for populations held well below K by predation or overharvest (Bowyer et al. 2001b; Kie et al. 2003). Consequently, long-term planning for manipulations and subsequent production of suitable habitat is essential for productive populations of moose (Timmermann and Buss 1998).

Harvest. In the 1990s, moose were managed in 2 territories and 9 provinces in Canada and in 11 states of the United States; moose had stable or expanding populations in 3 additional states; in 2001, moose were harvested in Vermont (Timmermann and Buss 1995, 1998; Timmermann 2003). Goals for moose management include subsistence and sport harvests as well as nonconsumptive uses such as viewing (Timmermann and Buss 1998). Satisfying multiple objectives of landowners, hunters, and wildlife agencies is not simple (Messmer et al. 1998). Co-management of moose involving indigenous people is common in Canada (Marshal 1999). Wildlife species, including moose, have become an increasingly important component of the tourism industry (Snepenger and Bowyer 1990).

Population assessment underpins moose management throughout North America, and includes total counts, estimates, and indices of abundances as well as information on sex and age composition and harvests (Crichton 1987; Timmermann and Buss 1998). Groups of moose fecal pellets have been used widely as an index to population density (Bergerud and Manuel 1968; Krefting 1974b; Franzmann et al. 1976a). Defecation rates for moose in winter range from 13 to 21 groups/moose/day (Timmermann and Buss 1998); females defecate less often (14.6 groups/day) than males (19.6 groups/day) (Franzmann et al. 1976b, 1976c). Tracks in snow also have been used as an index to moose abundance (Weixelman et al. 1998; Bowyer et al. 2001); Lautenschlager and Jordan (1993) combined track and pellet group counts to improve the methodology. Controversy exists over use of those indirect methods to estimate moose numbers, especially counts of pellet groups (Timmermann and Buss 1998). Both techniques are best used as indices to population size rather than estimates. The volume of fecal pellets in winter also has been used to determine sex and age classes of moose (MacCracken and Van Ballenberghe 1987).

Ground-based surveys of moose (Peterson 1955; Pimlott 1959) as well as sightings by hunters or others have been used to index moose abundance, although dense vegetation may reduce detection of moose in some habitats (Timmermann and Buss 1998). Crichton (1993) noted the need for consistency in recording such data for valid comparisons among years. Road kills also have been employed as an index to population size (Alexander 1993; Hicks 1993). Hunter effort and success (catch per unit effort) have been used to detect changes in population size of moose (Mercer and Manuel 1974; Crête et al. 1981; Crête and Dussault 1987; Hatter 1998, 2001). Bowyer et al. (1999b) cautioned, however, that catch per unit effort should not be used without an independent measure of population size because potential variation in skill of hunters over time led to anomalies in this measure. Where data on age structure are available, population reconstructions, including cohort analysis, also have been employed to estimate population size of moose from harvest (Fryxell et al. 1988; Ferguson 1993; Bowyer et al. 1999b). Such reconstructions assume equal catchability, as well as fixed fecundity and survivorship, which result in a stable age distribution; those assumptions are difficult to meet (Bowyer et al. 1999b).

Aerial surveys, including sampling transects and quadrats, are the most widespread method of estimating the size of moose populations. Timmermann and Buss (1998) provided a historical account detailing the development of methods that culminated in modern survey techniques for moose. Those techniques involve helicopters and fixed-wing aircraft (Crête et al. 1986; Smits et al. 1994). LeResche and Rausch (1974) and Bisset and Rempel (1991) noted that numerous variables could affect accuracy and precision of aerial surveys of moose, and that survey conditions needed to be similar among years to yield comparable results. Snow conditions and openness of habitat, in particular, affect census results (Gasaway et al. 1986). Modern techniques that stratify moose density before intensive sampling, thereby reducing variability in counts and increasing precision (Ward et al. 2000), stem from methods developed by Gasaway et al. (1986). These surveys recently have been improved by geostatistical analyses of spatial data, including use of the cokriging method to make predictions (Ver Hoef and Barry 1998; Ver Hoef 2000). Some aerial surveys have been augmented with infrared thermal imagery (Bontaites et al. 2000). Aerial surveys often incorporate a sightability correction factor (Gasaway et al. 1985b; Timmermann 1993; Anderson and Lindzey 1996; Lenarz 1998) to account for moose missed in the initial survey. This correction factor usually is obtained by intensively sampling some portion of a quadrat with an airplane or more efficient helicopter (Gasaway et al. 1986) or using radio-collared or obviously marked moose to account for missed individuals (Oosenbrug and Ferguson 1992; Peterson and Page 1993). Data on sex and age classes of moose often are obtained during aerial surveys; Timmermann (1993) and Timmermann and Buss (1998) provided criteria for determining sex and age classes of moose. Such data may supply information valuable for evaluating harvest (Timmermann and Rempel 1998), although composition estimates may be biased by the sexes spatially segregating during winter and some sex or age classes of moose being more difficult to observe than others (Peek et al. 1974; Miquelle et al. 1992). Data on age ratios, however, should not be used to estimate population size without some independent measure of rate of increase (Caughley 1974).

Timmermann and Buss (1998) discussed seasons, bag limits, firearm and other restrictions, including hunter eligibility and education, for legally taking moose in North America. Traditionally, harvests of moose have emphasized adult males. Alaska has restrictive regulations designed to prevent low male-to-female ratios (Schwartz et al. 1992). Those regulations allow the harvest of small males with a spike or fork on either antler (usually yearlings) or of large males with ≥50-in. spreads (127 cm) or with at least three tines on the brow palm. The largest antlered males mate most often in moose (Van Ballenberghe and Miquelle 1993, 1996); hence, some hunting regulations based on antler size have the potential to eliminate high-quality males from the population. Simulations involving the genetics of antler growth coupled with density-dependent population responses indicated that particular hunting regulations could adversely affect the size of moose antlers, but that such changes would be small, slow, and easily altered by changes in management (Hundertmark et al. 1998).

Male-only harvests work well for populations held well below K by predation (Gasaway et al. 1992); the goal is to obtain the maximum number of recruits from females, and that outcome cannot be achieved if females are harvested from populations at low density (Fig. 45.12). Populations near K, however, benefit little from a harvest of all males if the management objective is to maximize harvest. Males in populations of ungulates have a limited effect on recruitment rates of young, and high densities of females depress recruitment (McCullough 1979; Bowyer et al. 1999b; Kie et al. 2003). Consideration should be given to total sources of mortality in managing populations because harvest, crippling of animals in the hunting season that ultimately perish (Gasaway et al. 1983), predation (Gasaway et al. 1992), and other sources of mortality are likely to be additive at lower population densities with respect to K (Bowyer et al. 1999b; Kie et al. 2003). Obtaining data necessary to assess density-dependent recruitment (Fig. 45.12) or evaluate sources of mortality, however, can be a daunting task (Kie et al. 2003).

Value and Hunting. Moose have long played a role in myth, human subsistence, and recreation as well as served important roles in ecosystem structure and function. Wherever they have occurred in North America, moose have been intertwined inextricably with human culture since the late Pleistocene (Guthrie 1995; Reeves and McCabe 1998). These large mammals have provided spiritual, cultural, and subsistence values to indigenous people for centuries, and those values continue among many North American tribes throughout the range of moose (Feit 1987; Wolfe 1987).

Crichton (1998b) provided an excellent review of the means and methods of hunting moose as well as field dressing and the proper care and processing of hide and meat. In 1991, hunters across the northern United States and Canada killed an estimated 88,630 moose (Table 45.4). This harvest represents an approximate increase of 22% over the period from 1972 to 1982, when the annual harvest remained relatively stable. Timmermann (In press) estimated 82,619 moose harvested in North America during 2001, a decline of about 7% since 1991. Today, moose provide recreational hunting opportunities and are important components in local economies (Condon and Adamowicz 1995). Moose hunting in Alaska generates about $31 million/year for the economy (Regelin and Franzmann 1998). In Ontario, hunters contribute >$50 million annually (Ontario Ministry of Natural Resources 1980). Continent-wide, moose hunters spend about CDN$464 million annually in pursuit of their sport (Bisset 1987). On a per-hunter basis, those expenditures amount to about $600 per resident hunter, and

TABLE 45.4. Moose harvest in North America, 1972–2001

Jurisdiction	1972	1982	1991	2001
Alaska	5,700	5,900	6,100	5,509
Washington	—	—	8	64
Idaho	90	150	490	774
Utah	70	90	290	175
Wyoming	1,300	1,300	1,400	1,365
Montana	400	360	114	596
North Dakota	—	20	109	117
Colorado	—	—	7	64
Minnesota	370	760	410	125
Maine	—	880	960	2,550
Vermont	—	—	—	137
New Hampshire	—	—	89	378
Yukon[a]	—	1,000	640	743
Northwest Territories	—	130	140	1,400
British Columbia	14,300	12,800	13,500	9,200
Alberta	9,400	12,400	12,200	7,971
Saskatchewan	4,100	2,600	4,100	3,412
Manitoba	2,100	1,700	1,100	1,000
Ontario	13,800	10,700	11,000	11,000
Quebec	6,800	11,800	11,900	14,000
New Brunswick	1,000	1,300	1,700	2,537
Nova Scotia	400	160[b]	113	180
Newfoundland	11,000	7,000	21,000	19,322
Total	70,830	71,070	88,370	82,619

SOURCE: Data from Timmermann and Buss (1995, 1998) and Timmermann (In press).

[a]Not counting harvest by Natives.

[b]Data from 1981.

more than $1000 per nonresident hunter (Timmermann and Buss 1998). Moose hunting, however, occasionally has engendered controversy (Lautenschlager and Bowyer 1985; Crichton 1998b). Moose also have aesthetic values as well. Being able to see moose when afield or simply knowing they are present in boreal ecosystems is an important, if intangible, value attributable to moose (Franzmann 2000).

Moose can have negative economic effects as well. Each year, moose are involved in collisions with automobiles, trains, and even airplanes. Conservative estimates indicate 3500 moose/year are killed in vehicle collisions across North America (Child 1998). Negative aspects of such collisions, economic and otherwise, can include reductions in moose numbers, loss of recreational opportunities, increased risk to human life and property, rising insurance costs, and increased public dissatisfaction in seeing moose on roadways (Child 1998).

Perhaps most important, moose play significant roles in healthy ecosystems throughout their range. These herbivores can influence species composition dramatically among plants, alter forest structure, influence rates of nutrient cycling, and modify biological diversity (Molvar et al. 1993; Kie et al. 2003). Moose justifiably can be considered keystone species and as such, used as indicators of forest-management practices (Wallis de Vries 1995; Hobbs 1996).

MANAGEMENT, CONSERVATION, AND RESEARCH NEEDS

Determining where a moose population is with respect to *K* (Kie et al. 2003) remains a major challenge. That knowledge is essential for the wise management of populations and for determining when predator control may be effective. Indeed, understanding that relationship is essential to managing predator–prey systems. Losses of young to predators in moose populations near *K* have little affect on population dynamics because of their compensatory nature (i.e., young likely would have perished from other causes anyway), but fatalities of young in low-density populations are added to other sources of mortality and can limit or regulate populations (Ballard and Van Ballenberghe 1998; Person et al. 2001; Kie et al. 2003). Interactions with harvest and severe weather may further complicate matters. Sorting among those factors and how they relate to the size of moose populations in relation to *K* is a fundamental management need. Moreover, there is a need to develop spatially explicit models for moose populations (McKenney et al. 1998).

Populations of moose should no longer be managed as if the sexes have similar requirements; they do not. The failure to consider sexual segregation in the management of moose has implications for estimates and indices of population size, understanding habitat requirements, habitat manipulation, and allowable harvest. Kie and Bowyer (1999) suggested the sexes of polygynous ungulates should be managed as if they were separate species. We concur.

Understanding how moose interact with their environment and their role in trophic cascades is an essential research need. Evidence exists that moose browsing and their deposition of urine and feces may have either positive or negative effects on nutrient cycling and other ecosystem processes. Knowledge of how moose population density influences that process is essential. We cannot manage all components of diverse ecosystems, and defining the role of keystone species, such as moose, is critical to the future management of boreal regions.

LITERATURE CITED

Addison, E. M., and R. F. McLaughlin. 1993. Seasonal variation and effects of winter ticks (*Dermacentor albipictus*) and consumption of food by captive moose (*Alces alces*). Alces 29:219–24.

Addison, E. M., A. Fyvie, and F. J. Johnson. 1979. Metacestodes of moose, *Alces alces,* of the Chapleau Crown Preserve, Ontario. Canadian Journal of Zoology 57:1619–23.

Addison, E. M., J. D. Smith, R. F. McLaughlin, D. J. H. Fraser, and D. G. Joachim. 1990. Calving sites of moose in Central Ontario. Alces 26:142–53.

Addison, E. M., R. F. McLaughlin, D. J. H. Fraser, and M. E. Buss. 1993. Observations of pre- and post-partum behaviour of moose in central Ontario. Alces 29:27–33.

Addison, E. M., R. F. McLaughlin, and J. D. Broadfoot. 1994. Growth of moose calves (*Alces alces americana*) infested and uninfested with winter ticks (*Dermacentor albipictus*). Canadian Journal of Zoology 72:1469–76.

Addison, E. M., D. G. Joachim, R. F. McLaughlin, and D. J. H. Fraser. 1998a. Ovipositional development and fecundity of *Dermacentor albipictus* (Acari: Ixodidae) from moose. Alces 34:165–72.

Addison, E. M., R. F. McLaughlin, and J. D. Broadfoot. 1998b. Effects of winter tick (*Dermacentor albipictus*) on blood characteristics of captive moose (*Alces alces*). Alces 34:189–99.

Addison, R. B. 1975. Reproduction of moose measured by pigmented ovarian scars. Proceedings of the North American Moose Conference and Workshop 11:369–90.

Addison, R. B., J. C. Williamson, B. P. Sanders, and D. Fraser. 1980. Radiotracking of moose in the boreal forest of northwestern Ontario. Canadian Field-Naturalist 94:269–76.

Aho, R. W., and J. Hendrickson. 1989. Reproduction and mortality of moose translocated from Ontario to Michigan. Alces 25:75–80.

Aldous, S. E., and L. W. Krefting. 1946. The present status of moose on Isle Royale. Transactions of the North American Wildlife Conference 11:296–308.

Alexander, C. E. 1993. The status and management of moose in Vermont. Alces 29:187–95.

Allen, A. W., J. W. Terrell, W. L. Mangus, and E. L. Lindquist. 1991. Application and partial validation of a habitat model for moose in the Lake Superior region. Alces 27:50–64.

Altmann, M. 1958. Social integration of the moose calf. Animal Behaviour 6:155–59.

Altmann, M. 1959. Group dynamics in Wyoming moose during the rutting season. Journal of Mammalogy 40:420–24.

Altmann, M. 1963. Naturalistic studies of maternal care in moose and elk. Pages 233–53 *in* H. L. Rheingold, ed. Maternal behavior in mammals. John Wiley, New York.

Albright, C. A., and L. B. Keith. 1987. Population dynamics of moose, *Alces alces,* on the south-coast barrens of Newfoundland. Canadian Field-Naturalist 101:373–87.

Andersen, R. 1991. Habitat deterioration and the migratory behaviour of moose (*Alces alces* L.) in Norway. Journal of Applied Ecology 28:102–8.

Anderson, C. R., Jr., and F. G. Lindzey. 1996. Moose sightability model developed from helicopter surveys. Wildlife Society Bulletin 24:247–59.

Anderson, R. C. 1972. The ecological relationships of meningeal worm and native cervids in North America. Journal of Wildlife Diseases 8:304–9.

Anderson, R. C., and M. W. Lankester. 1974. Infectious and parasitic diseases and arthropod pests of moose in North America. Naturaliste Canadien 101:23–50.

Anderson, R. C., and A. K. Prestwood. 1981. Lungworms. Pages 266–317 *in* W. R. Davidson, F. A. Hayes, V. F. Nettles, and F. E. Kellogg, eds. Diseases and parasites of white-tailed deer (Miscellaneous Publication No. 7). Tall Timbers Research Station, Tallahassee, FL.

Archibald, R. M. 1960. *Listeria monocytogenes* from a Nova Scotia moose. Canadian Veterinary Journal 1:225–26.

Armstrong, E. R., and G. Brown. 1986. White moose, *Alces alces,* sightings in northern Ontario. Canadian Field-Naturalist 100:262–63.

Atkeson, T. D., V. F. Nettles, R. L. Marchinton, and W. V. Branan. 1988. Nasal glands in the Cervidae. Journal of Mammalogy 69:153–56.

Bailey, T. N., and E. E. Bangs. 1980. Moose calving areas and use on the Kenai National Moose Range, Alaska. Proceedings of the North American Moose Conference and Workshop 16:289–313.

Ballard, W. B. 1992. Bear predation on moose: A review of recent North American studies and their management implications. Alces Supplement 1:162–76.

Ballard, W. B., and V. Van Ballenberghe. 1998. Moose–predator relationships: Research and management needs. Alces 34:91–105.

Ballard, W. B, and J. S. Whitman. 1987. Marrow fat dynamics in moose calves. Journal of Wildlife Management 51:66–69.

Ballard, W. B., T. H. Spraker, and K. P. Taylor. 1981. Causes of neonatal moose calf mortality in south central Alaska. Journal of Wildlife Management 45:335–42.

Ballard, W. B., J. S. Whitman, and C. L. Gardner. 1987. Ecology of an exploited wolf population in south-central Alaska. Wildlife Monographs 98:1–54.

Ballard, W. B., S. D. Miller, and J. S. Whitman. 1990. Brown and black bear predation on moose in southcentral Alaska. Alces 26:1–8.

Ballard, W. B., J. S. Whitman, and D. J. Reed. 1991. Population dynamics of moose in south-central Alaska. Wildlife Monographs 114:1–49.

Ballard, W. B., M. E. McNay, C. L. Gardner, and D. J. Reed. 1995. Use of line-intercept track sampling for estimating wolf densities. Pages 469–80 *in* L. N. Carbyn, S. H. Fritts, and D. R. Seip, eds. Ecology and conservation of wolves in a changing world (Occasional Publication No. 35). Canadian Circumpolar Institute, Edmonton, Alberta, Canada.

Ballard, W. B., P. J. MacQuarrie, A. W. Franzmann, and P. R. Krausman. 1996. Effects of winters on physical condition of moose in south-central Alaska. Alces 32:51–59.

Balsom, S., W. B. Ballard, and H. A. Whitlaw. 1996. Mature coniferous forest as critical moose habitat. Alces 32:131–40.

Bangs, E. E., and T. N. Bailey. 1980. Interrelationships of weather, fire, and moose on the Kenai National Moose Range, Alaska. Proceedings of the North American Moose Conference and Workshop 16:255–74.

Bangs, E. E., T. N. Bailey, and M. F. Portner. 1984. Bull moose behavior and movements in relation to harvest on the Kenai National Wildlife Reffuge, Alaska. Alces 20:187–208.

Bangs, E. E., S. A. Duff, and T. N. Bailey. 1985. Habitat differences and moose use of two large burns on the Kenai Peninsula, Alaska. Alces 21:17–35.

Bangs, E. E., T. N. Bailey, and M. F. Portner. 1989. Survival rates of adult female moose on the Kenai Peninsula, Alaska. Journal of Wildlife Management 53:557–63.

Barboza, P. S., and J. E. Blake. 2001. Ceruloplasmin as an indicator of copper reserves in wild ruminants at high latitudes. Journal of Wildlife Diseases 37:324–31.

Barboza, P. S., and R. T. Bowyer. 2000. Sexual segregation in dimorphic deer: A new gastrocentric hypothesis. Journal of Mammalogy 81:473–89.

Barboza, P. S., and R. T. Bowyer. 2001. Seasonality of sexual segregation in dimorphic deer: Extending the gastrocentric model. Alces 37:275–92.

Barten, N. L., R. T. Bowyer, and K. J. Jenkins. 2001. Habitat use by female caribou: Tradeoffs associated with parturition. Journal of Wildlife Management 65:77–92.

Bartowsiewvicz, L. 1987. Sexual dimorphism in the cranial development of Scandinavian moose (*Alces alces* (L.) *alces*). Canadian Journal of Zoology 65:747–50.

Beeler, D. A., D. A. Benson, and W. M. Langille. 1959. Mineral analyses of livers and kidneys of moose (*Alces americana*). Journal of Wildlife Management 23:356–58.

Belant, J. L. 1995. Moose collisions with vehicles and trains in northeastern Minnesota. Alces 31:45–52.

Belovsky, G. E. 1978. Diet optimization in a generalist herbivore: The moose. Theoretical Population Biology 14:105–34.

Belovsky, G. E. 1981a. Optimal activity times and habitat choice of moose. Oecologia 48:22–30.

Belovsky, G. E. 1981b. Food plant selection by a generalist herbivore: The moose. Ecology 62:1020–30.

Belovsky, G. E. 1984. Moose and snowshoe hare competition and a mechanistic explanation from foraging theory. Oecologia 61:150–59.

Belovsky, G. E., and P. A. Jordan. 1978. The time–energy budget of a moose. Theoretical Population Biology 14:76–104.

Belovsky, G. E., and P. A. Jordan. 1981. Sodium dynamics and adaptations of a moose population. Journal of Mammalogy 62:613–21.

Ben-David, M., E. Shochat, and L. G. Adams. 2001. Utility of stable isotope analysis in studying foraging ecology of herbivores: Examples from moose and caribou. Alces 37:412–34.

Bennett, G. F., and C. W. Sabrosky. 1962. The Nearctic species of the genus *Cephenemyia* (Diptera, Oestridae). Canadian Journal of Zoology 40:431–48.

Berger, J. 1992. Facilitation of reproductive synchrony by gestation adjustment in gregarious mammals: A new hypothesis. Ecology 73:323–29.

Berger, J. 1999. Anthropogenic extinction of top carnivores and interspecific animal behaviour: Implications of the rapid decoupling of a web involving wolves, bears, moose and ravens. Proceedings of the Royal Society of London B 266:2261–67.

Berger, J., and M. E. Gompper. 1999. Sex ratios in extant ungulates: Product of contemporary predation or past life histories? Journal of Mammalogy 80:1084–1113.

Berger, J., P. B. Stacey, L. Bellis, and M. P. Johnson. 2001a. A mammalian predator–prey imbalance: Grizzly bear and wolf extinction affect avian Neotropical migrants. Ecological Applications 11:947–60.

Berger, J., J. E. Swenson, and I.-L. Persson. 2001b. Recolonizing carnivores and naive prey: Conservation lessons for Pleistocene extinctions. Science 291:1036–39.

Bergerud, A. T., and J. P. Elliott. 1998. Wolf predation in a multiple-ungulate system in northern British Columbia. Canadian Journal of Zoology 76:1551–69.

Bergerud, A. T., and F. Manuel. 1968. Moose damage to balsam fir–white birch forests in central Newfoundland. Journal of Wildlife Management 32:729–46.

Bergerud, A. T., M. J. Nolan, K. Curnew, and W. E. Mercer. 1983a. Growth of the Avalon Peninsula, Newfoundland caribou herd. Journal of Wildlife Management 47:989–98.

Bergerud, A. T., W. Wyett, and B. Snider. 1983b. The role of wolf predation in limiting a moose population. Journal of Wildlife Management 47:977–88.

Bergman, M. 2002. Can saliva from moose, *Alces alces,* affect growth responses in the sallow, *Salix caprea?* Oikos 96:164–68.

Bergstrom, R., and K. Danell. 1987. Effects of simulated winter browsing by moose on morphology and biomass of two birch species. Journal of Ecology. 75:533–44.

Bertram, M. R., and M. T. Vivion. 2002. Moose mortality in eastern interior Alaska. Journal of Wildlife Management 66:747–56.

Best, D. A., G. M. Lynch, and O. Rongstad. 1978. Seasonal activity patterns of moose in the Swan Hills, Alberta. Proceedings of the North American Moose Conference and Workshop 14:109–25.

Bevins, J. S., C. C. Schwartz, and A. W. Franzmann. 1990. Seasonal activity patterns of moose on the Kenai Peninsula, Alaska. Alces 26:14–23.

Bishop, R. H, and R. A. Rausch. 1974. Moose population fluctuations in Alaska, 1950–1972. Naturaliste Canadien 101:559–93.

Bisset, A. R. 1987. The economic importance of moose (*Alces alces*) in North America. Swedish Wildlife Research Supplement 1:677–98.

Bisset, A. R., and R. S. Rempel. 1991. Linear analysis of factors affecting the accuracy of moose aerial inventories. Alces 27:127–39.

Bleich, V. C., R. T. Bowyer, and J. D. Wehausen. 1997. Sexual segregation in mountain sheep: Resources or predation? Wildlife Monographs 134:1–50.

Blood, D. A. 1974. Variation in reproduction and productivity of an enclosed herd of moose (*Alces alces*). Transactions of the International Congress of Game Biologists 11:59–66.

Blood, D. A., J. R. McGillis, and A. L. Lovaas. 1967. Weights and measurements of moose in Elk Island National Park, Alberta. Canadian Field-Naturalist 81:263–69.

Bø, S., and O. Hjeljord. 1991. Do continental moose ranges improve during cloudy summers? Canadian Journal of Zoology 69:1875–79.

Boer, A. H. 1987. Reproductive productivity of moose in New Brunswick. Alces 23:49–60.

Boer, A. H. 1992. Fecundity of North American moose (*Alces alces*): A review. Alces Supplement 1:1–10.

Boer, A. H. 1998. Interspecific relationships. Pages 337–49 *in* A. W. Franzmann and C. C. Schwartz, eds. Ecology and management of the North American moose. Smithsonian Institution Press, Washington, DC.

Boertje, R. D., P. Valkenburg, and M. E. McNay. 1996. Increases in moose, caribou, and wolves following wolf control in Alaska. Journal of Wildlife Management 60:474–89.

Boeskorov, G. G. 1993. Karyotype of moose (*Alces alces* L.) from northeastern Asia. Proceedings of the Russian Academy of Sciences 329:506–8 (in Russian).

Boeskorov, G. G. 1997. Chromosomal differences in moose (*Alces alces* L., Artiodactyla, Mammalia). Genetika 33:974–78 (in Russian).

Boonstra, R., and A. R. E. Sinclair. 1984. Distribution and habitat use of caribou, *Rangifer tarandus caribou,* and moose, *Alces alces andersoni,* in the Spatsizi Plateau Wilderness Area, British Columbia. Canadian Field-Naturalist 98:12–21.

Bontaites, K. M., K. A. Gustafson, and R. Makin. 2000. A Gasaway-type moose survey in New Hampshire using infrared thermal imagery: Preliminary results. Alces 36:69–76.

Bourque, M., and R. Higgins. 1984. Serologic studies on brucellosis, leptospirosis and tularemia in moose (*Alces alces*) in Quebec. Journal of Wildlife Diseases 20:95–99.

Boutin, S. 1992. Predation and moose population dynamics: A critique. Journal of Wildlife Management 56:116–27.

Bowyer, J. W., and R. T. Bowyer. 1997. Effects of previous browsing on the selection of willow stems by Alaskan moose. Alces 33:11–18.

Bowyer, R. T. 1981. Activity, movement, and distribution of Roosevelt elk during rut. Journal of Mammalogy 62:574–82.

Bowyer, R. T. 1984. Sexual segregation in southern mule deer. Journal of Mammalogy 65:410–17.

Bowyer, R. T. 1986. Antler characteristics as related to social status of male southern mule deer. Southwestern Naturalist 31:289–98.

Bowyer, R. T. 1991. Timing of parturition and lactation in southern mule deer. Journal of Mammalogy 72:138–45.

Bowyer, R. T., and D. W. Kitchen. 1987. Significance of scent-marking by Roosevelt elk. Journal of Mammalogy 68:418–23.

Bowyer, R. T., J. L. Rachlow, V. Van Ballenberghe, and R. D. Guthrie. 1991. Evolution of a rump patch in Alaskan moose: An hypothesis. Alces 27:12–23.

Bowyer, R. T., V. Van Ballenberghe, and K. R. Rock. 1994. Scent marking by Alaskan moose: Characteristics and spatial distribution of rubbed trees. Canadian Journal of Zoology 72:2186–92.

Bowyer, R. T., J. G. Kie, and V. Van Ballenberghe. 1996. Sexual segregation in black-tailed deer: Effects of scale. Journal of Wildlife Management 60:10–17.

Bowyer, R. T., V. Van Ballenberghe, and J. G. Kie. 1997. The role of moose in landscape processes: Effects of biogeography, population dynamics, and predation. Pages 265–87 *in* J. A. Bissonette, ed. Wildlife and landscape ecology: Effects of pattern and scale. Springer-Verlag, New York.

Bowyer, R. T., V. Van Ballenberghe, and J. G. Kie. 1998. Timing and synchrony of parturition in Alaskan moose: Long-term versus proximal effects of climate. Journal of Mammalogy 79:1332–44.

Bowyer, R. T., M. C. Nicholson, E. M. Molvar, and J. B. Faro. 1999a. Moose on Kalgin Island: Are density-dependent processes related to harvest? Alces 35:73–89.

Bowyer, R. T., V. Van Ballenberghe, J. G. Kie, and J. A. K. Maier. 1999b. Birth-site selection by Alaskan moose: Maternal strategies for coping with a risky environment. Journal of Mammalogy 80:1070–83.

Bowyer, R. T., D. M. Leslie, Jr., and J. L. Rachlow. 2000. Dall's and Stone's sheep. Pages 491–544 *in* S. Demarais and P. R. Krausman, eds. Ecology and management of large mammals in North America. Prentice-Hall, Upper Saddle River, NJ.

Bowyer, R. T., K. M. Stewart, J. G. Kie, and W. C. Gasaway. 2001a. Fluctuating asymmetry in antlers of Alaskan moose: Size matters. Journal of Mammalogy 82:814–24.

Bowyer, R. T., B. M. Pierce, L. K. Duffy, and D. A. Haggstrom. 2001b. Sexual segregation in moose: Effects of habitat manipulation. Alces 37:109–22.

Bowyer, R. T., D. R. McCullough, and G. E. Belovsky. 2001c. Causes and consequences of sociality in mule deer. Alces 37:371–402.

Bowyer, R. T., K. M. Stewart, B. M. Pierce, K. J. Hundertmark, and W. C. Gasaway. In press. Geographical variation in antler morphology of Alaska moose: Putative effects of habitat and genetics. Alces.

Bowyer, R. T., K. M. Stewart, S. A. Wolfe, G. M. Blundell, K. L. Lehmkkuhl, P. J. Joy, T. J. McDonough, and J. G. Kie. 2002. Assessing sexual segregation in deer. Journal of Wildlife Management 66:536–44.

Boyce, M. S. 1989. The Jackson elk herd: Intensive wildlife management in North America. Cambridge University Press, New York.

Brandner, T. A., R. O. Peterson, and K. L. Risenhoover. 1990. Balsam fir on Isle Royale: Effects of moose herbivory and population density. Ecology 71:155–64.

Brazil, J., and S. Ferguson. 1989. Cadmium concentrations in Newfoundland moose. Alces 25:52–57.

Breckenridge, W. J. 1946. Weights of a Minnesota moose. Journal of Mammalogy 27:90–91.

Bryant, J. P., F. S. Chapin III, and D. R. Klein. 1983. Carbon/nutrient balance of boreal plants in relation to vertebrate herbivory. Oikos 40:357–68.

Bryant, J. P., F. D. Provenza, J. Pastor, P. B. Reichardt, T. P. Clausen, and J. T. du Toit. 1991. Interactions between woody plants and browsing mammals. Annual Review of Ecology and Systematics 22:431–46.

Bryant, J. P., P. B. Reichardt, and T. P. Clausen. 1992. Chemically mediated interactions between woody plants and browsing mammals mediated by secondary metabolites. Journal of Range Management 45:18–24.

Bryant, J. P., R. K Swihart, P. B. Reichardt, and T. P. Clausen. 1994. Biogeography of woody plant chemical defense against snowshoe hare browsing: Comparison of Alaska and eastern North America. Oikos 70:385–94.

Bubenik, A. B. 1983. Behavioral significance of the moose bell. Alces 19:238–45.

Bubenik, A. B. 1998. Evolution, taxonomy, and morphophysiology. Pages 77–123 *in* A. W. Franzmann and C. C. Schwartz, eds. Ecology and management of the North American moose. Smithsonian Institution Press, Washington, DC.

Bubenik, A. B., O. Williams, and H. T. Timmermann. 1978. Some characteristics of antlerogensis in moose. A preliminary report. Proceedings of the North American Moose Conference and Workshop 14:157–77.

Bubenik, G. A., and K. J. Hundertmark. 2002. Accessory antlers in male Cervidae. Zeitschrift für Jagdwissenschaft 48:10–21.

Burger, J. F., and J. R. Anderson. 1974. Taxonomy and life history of the moose fly, *Haematobosca alcis,* and its association with the moose, *Alces alces shirasi* in Yellowstone National Park. Annals of the Entomological Society of America 67:204–14.

Burris, O. E., and D. E. McKnight. 1973. Game transplants in Alaska. Alaska, (Wildlife Technical Bulletin 4). Department of Fish and Game, Juneau.

Cairns, A. L., and E. S. Telfer. 1980. Habitat use by four sympatric ungulates in boreal mixedwood forest. Journal of Wildlife Management 44:849–57.

Campbell, B. H., and M. Hinkes. 1983. Winter diets and habitat use of Alaska bison after wildfire. Wildlife Society Bulletin 11:16–21.

Campbell, G. D., E. M. Addison, I. K. Barker, and S. Rosendal. 1994. *Erysipelothrix rhusiopathiae,* serotype 17, septicemia in moose (*Alces alces*) from Algonquin Park, Ontario. Journal of Wildlife Diseases 30:436–38.

Caughley, G. 1974. Interpretation of age ratios. Journal of Wildlife Management 38:557–62.

Cederlund, B.-M. 1987. Parturition and early development of moose (*Alces alces* L.) calves. Swedish Wildlife Research Supplement 1:399–422.

Cederlund, G. 1989. Activity patterns in moose and roe deer in a north boreal forest. Holarctic Ecology 12:39–45.

Cederlund, G., and H. Sand. 1991. Population dynamics and yield of a moose population without predators. Alces 27:31–40.

Cederlund, G., and H. Sand. 1992. Dispersal of subadult moose (*Alces alces*) in a nonmigratory population. Canadian Journal of Zoology 70:1309–14.

Cederlund, G., and H. Sand. 1994. Home-range size in relation to age and sex in moose. Journal of Mammalogy 75:1005–12.

Cederlund, G., F. Sandgren, and K. Larsson. 1987. Summer movements of female moose and dispersal of their offspring. Journal of Wildlife Management 51:342–52.

Cederlund, G. N., R. L. Bergstrom, and K. Danell. 1989. Seasonal variation in mandible marrow fat in moose. Journal of Wildlife Management 53:587–92.

Chapman, D. M. 1985. Histology of the moose (*Alces alces andersoni*) interdigital glands and associated green hairs. Canadian Journal of Zoology 63:899–911.

Chekchak, T., R. Courtois, J.-P. Ouellet, L. Breton, and S. St.-Onge. 1998. Calving site characteristics for moose (*Alces alces*). Canadian Journal of Zoology 76:1663–70 (in French).

Child, K. N. 1983. Railways and moose in the central interior in British Columbia: A recurrent management problem. Alces 19:118–35.

Child, K. N. 1998. Incidental mortality. Pages 275–301 *in* A. W. Franzmann and C. C. Schwartz, eds. Ecology and management of North American moose. Smithsonian Institution Press, Washington, DC.

Child, K. N., S. P. Barry, and D. A. Aitken. 1991. Moose mortality on highways and railways of British Columbia. Alces 27:41–49.

Choquette, L. P. E. 1970. Anthrax. Pages 256–60 *in* J. W. Davis, L. H. Karstad, and D. O. Trainer, eds. Infectious diseases of wild mammals. Iowa State University Press, Ames.

Clark, R. A., and R. T. Bowyer. 1986. Occurrence of protostrongylid nematodes in sympatric populations of moose and white-tailed deer in Maine. Alces 22:313–22.

Clutton-Brock, T. H., F. E. Guninness, and S. D. Albon. 1982. Red deer: Behavior and ecology of two sexes. University of Chicago Press, Chicago.

Coady, J. W. 1974. Influence of snow on behavior of moose. Naturaliste Canadien 101:417–36.

Coady, J. W. 1980. History of moose in northern Alaska and adjacent regions. Canadian Field-Naturalist 94:61–68.

Coady, J. W. 1982. Moose. Pages 902–22 *in* J. A. Chapman and G. A. Feldhamer, eds. Wild mammals of North America: Biology, management, and economics. Johns Hopkins University Press, Baltimore.

Colescott, J. H., and M. P. Gillingham. 1998. Reaction of moose (*Alces alces*) to snowmobile traffic in the Greys River Valley, Wyoming. Alces 34:329–38.

Collins, W. B., and C. C. Schwartz. 1998. Logging in Alaska's boreal forest: Creation of grasslands or enhancement of moose habitat. Alces 34:355–74.

Condon, B., and W. Adamowicz. 1995. The economic value of moose hunting in Newfoundland. Canadian Journal of Forest Research 25:319–28.

Connor, K. J., W. B. Ballard, T. Dilworth, S. Mahoney, and D. Anions. 2000. Changes in structure of a boreal forest community following intense herbivory by moose. Alces 36:111–32.

Cook, H. W., R. A. Rausch, and B. E. Baker. 1970. Moose (*Alces alces*) milk. Gross composition, fatty acid, and mineral constitution. Canadian Journal of Zoology 48:213–15.

Cowan, I. McT. 1950. Some vital statistics of big game on overstocked mountain range. Transactions of the North American Wildlife Conference 15:581–88.

Cowan, I. McT., W. S. Hoar, and J. Hatter. 1950. The effect of forest succession upon the quantity and upon the nutritive values of woody plants used as food by moose. Canadian Journal of Research D 28:249–71.

Crête, M. 1987. The impact of sport hunting on North American Moose. Swedish Wildlife Research Supplement 1:553–63.

Crête, M. 1988. Regional variation in the skeletal measurements of *Alces alces* in three game reserves in Quebec. Alces 24:102–11.

Crête, M., and J. Bedard. 1975. Daily browse consumption of moose in the Gaspé Peninsula, Quebec. Journal of Wildlife Management 39:368–73.

Crête, M., and R. Courtois. 1997. Limiting factors might obscure population regulation of moose (Cervidae: *Alces alces*) in unproductive boreal forests. Journal of Zoology (London) 242:765–81.

Crête, M., and C. Dussault. 1987. Using hunting statistics to estimate density, cow–calf ratio and harvest rate of moose in Quebec. Alces 23:227–42.

Crête, M., and P. A. Jordan. 1982a. Population consequences of winter forage resources for moose, *Alces alces,* in southwestern Quebec. Canadian Field-Naturalist 96:467–75.

Crête, M., and P. A. Jordan. 1982b. Production and quality of forage available to moose in southwestern Quebec. Canadian Journal of Forest Research 12:151–59.

Crête, M., R. J. Taylor, and P. A. Jordan. 1981. Optimization of moose harvest in southwestern Quebec. Journal of Wildlife Management 45:598–611.

Crête, M., A. Tremblay, and P. A. Jordan. 1982. Hunter collected blood samples for comparing the physical condition of two Quebec moose populations. Alces 18:25–44.

Crête, M., L. Rivest, J. Jolicoeur, J. Brassard, and F. Messier. 1986. Predicting and correcting helicopter counts of moose with observations made from fixed-wing aircraft in southern Quebec. Journal of Applied Ecology 23:751–61.

Crichton, V. 1987. Moose management in North America. Swedish Wildlife Research Supplement 1:541–51.

Crichton, V. 1992. Six year (1986/87–1991/92) summary of in utero productivity of moose in Manitoba, Canada. Alces 28:203–14.

Crichton, V. 1993. Hunter effort and observations—The potential for monitoring trends of moose populations: A review. Alces 29:181–86.

Crichton, V. 1998a. Moose and ecosystem management in the 21st century—Does the king have a place? A Canadian perspective. Alces 34:467–77.

Crichton, V. 1998b. Hunting. Pages 617–53 *in* A. W. Franzmann and C. C. Schwartz, eds. Ecology and management of the North American moose. Smithsonian Institution Press, Washington, DC.

Crichton, V., and P. C. Paquet. 2000. Cadmium in Manitoba's wildlife. Alces 36:205–16.

Dale, B. W., L. G. Adams, and R. T. Bowyer. 1994. Functional response of wolves preying on barren-ground caribou in a multiple-prey ecosystem. Journal of Animal Ecology 63:63:644–52.

Danell, K., R. Gref, and R. Yazdani. 1990. Effects of mono- and diterpenes in Scots pine needles on moose browsing. Scandinavian Journal of Forest Research 5:535–39.

Danell, K., L. Edenius, and P. Lundberg. 1991. Herbivory and tree stand composition: Moose patch use in winter. Ecology 72:1350–57.

Davis, J. L., and A. W. Franzmann. 1979. Fire–moose–caribou interrelationships: A review and assessment. Proceedings of the North American Moose Conference and Workshop 15:80–118.

Dehority, B. A. 1974. Rumen ciliate fauna of Alaskan moose *Alces americana,* musk ox *Ovibos moschatus* and Dall mountain sheep *Ovis dalli*. Journal of Protozoology 21:26–32.

DelFrate, G. G., and T. H. Spraker. 1991. Moose–vehicle interactions and an associated public awareness program on the Kenai Peninsula, Alaska. Alces 27:1–7.

DelGiudice, G. D., R. O. Peterson, and U. S. Seal. 1991. Differences in urinary chemistry profiles of moose on Isle Royale during winter. Journal of Wildlife Diseases 27:407–16.

DelGiudice, G. D., R. O. Peterson, and W. M. Samuel. 1997. Trends of winter nutritional restriction, ticks, and numbers of moose on Isle Royale. Journal of Wildlife Management 61:895–903.

Denniston, R. H. 1956. Ecology, behaviour and population dynamics of the Wyoming or Rocky Mountain moose (*Alces alces shirasi*). Zoologica 41:105–18.

de Vos, A. 1958. Summer observations on moose behaviour in Ontario. Journal of Mammalogy 39:128–39.

de Vos, A., P. Brokx, and V. Geist. 1967. A review of social behavior of the North American cervids during the reproductive period. American Midland Naturalist 77:390–417.

Dewitt, J. B., and J. V. Derby, Jr. 1955. Changes in nutritive value of browse plants following forest fires. Journal of Wildlife Management 19:65–70.

Dieterich, R. A., J. K. Morton, and R. L. Zarnke. 1991. Experimental *Brucella suis* biovar 4 infection in moose. Journal of Wildlife Diseases 27:470–72.

Dodds, D. G. 1958. Observations of pre-rutting behavior in Newfoundland moose. Journal of Mammalogy 39:412–16.

Dodds, D. G. 1960. Food competition and range relationships of moose and snowshoe hare in Newfoundland. Journal of Wildlife Management 24:52–60.

Dodds, D. G. 1974. Distribution, habitat and status of moose in the Atlantic provinces of Canada and northeastern United States. Naturaliste Canadien 101:51–65.

Doutt, J. K. 1970. Weights and measurements of moose, *Alces alces shirasi*. Journal of Mammalogy 51:808.

Dubey, J. P. 1980. *Sarcocystis* species in moose (*Alces alces*), bison (*Bison bison*), and pronghorn (*Antilocapra americana*) in Montana, USA. American Journal of Veterinary Research 41:2063–65.

Dubey, J. P. 1981. Isolation of encysted *Toxoplasma gondii* from musculature of moose and pronghorn in Montana. American Journal of Veterinary Research 42:126–27.

Dubey, J. P., and C. S. F. Williams. 1980. *Hammondia heydorni* infection in sheep, goats, moose, dogs, and coyotes. Parasitology 81:123–27.

Duffy, L. K., C. Kaiser, C. Ackley, and K. S. Richter. 2001. Mercury in hair of large Alaskan herbivores: Routes of exposure. Alces 37:293–301.

Dumont, A., and M. Crête. 1996. The meningeal worm, *Parelaphostrongylus tenuis,* a marginal limiting factor for moose, *Alces alces,* in southern Quebec. Canadian Field-Naturalist 110:413–18.

Duvall, A. C., and G. S. Schoonveld. 1988. Colorado moose: Reintroduction and management. Alces 24:188–94.

Eason, G., E. R. Thomas, R. Jerrard, and K. Oswald. 1981. Moose hunting closure in a recently logged area. Alces 17:111–25.

Eastman, D. S. 1983. Seasonal changes in crude protein and lignin of ten common forage species of moose in north-central British Columbia. Alces 19:36–70.

Eastman, D. S., and R. Ritcey. 1987. Moose habitat relationships and management in British Columbia. Swedish Wildlife Research Supplement 1:101–18.

Eberhardt, L. L. 2000. Reply: Predator–prey ratio dependence and regulation of moose populations. Canadian Journal of Zoology 78:511–13.

Eberhardt, L. L., and R. O. Peterson. 1999. Predicting the wolf–prey equilibrium point. Canadian Journal of Zoology 77:494–98.

Edenius, L. 1991. The effect of resource depletion on the feeding behaviour of a browser; winter foraging by moose on Scots pine. Journal of Applied Ecology 28:318–28.

Edwards, J. 1983. Diet shifts in moose due to predator avoidance. Oecologia 60:185–89.

Edwards, R. Y., and R. W. Ritcey. 1958. Reproduction in a moose population. Journal of Wildlife Management 22:261–68.

Ericsson, G., and K. Wallin. 2001. Age-specific moose (*Alces alces*) mortality in a predator-free environment: Evidence for senescence in females. Ecoscience 8:157–63.

Ericsson, G., K. Wallin, J. P. Ball, and M. Broberg. 2001. Age-related reproductive effort and senescence in free-ranging moose, *Alces alces*. Ecology 82:1613–20.

Eschholz, W. E., F. A. Servello, B. Griffith, K. S. Raymond, and W. B. Krohn. 1996. Winter use of glyphosate-treated clearcuts by moose in Maine. Journal of Wildlife Management 60:764–69.

Estes, R. D. 1973. The role of the vomeronasal organ in mammalian reproduction. Mammalia 36:315–40.

Faber, W. E., and L. Edenius. 1998. Bark stripping by moose in commercial forests of Fennoscandia: A review. Alces 34:261–68.

Feit, H. A. 1987. North American native hunting and management of moose populations. Swedish Wildlife Research Supplement 1:25–42.

Ferguson, S. H. 1993. Use of cohort analysis to estimate abundance, recruitment and survivorship for Newfoundland moose. Alces 29:99–113.

Ferguson, S. H., A. R. Bisset, and F. Messier. 2000. The influences of density on growth and reproduction in moose *Alces alces*. Wildlife Biology 6:31–39.

Flook, D. R. 1964. Range relationships of some ungulates native to Banff and Jasper national parks, Alberta. Pages 119–28 *in* D. J. Crisp, ed. Grazing in terrestrial and marine environments. Blackwell Scientific, Oxford.

Flynn, A., and A. W. Franzmann. 1987. Mineral element studies in North American moose. Swedish Wildlife Research Supplement 1:289–300.

Flynn, A., A. W. Franzmann, P. D. Arneson, and J. L. Oldemeyer. 1977. Indications of copper deficiency in a subpopulation of Alaskan moose. Journal of Nutrition 107:1182–89.

Fong, D. W. 1981. Seasonal variation of marrow fat content from Newfoundland moose. Journal of Wildlife Management 45:545–48.

Forbes, L. B., S. V. Tessaro, and W. Lees. 1996. Experimental studies on *Brucella abortus* in moose (*Alces alces*). Journal of Wildlife Diseases 32:94–104.

Fowler, C. W. 1987. A review of density dependence in populations of large mammals. Pages 401–41 *in* H. Genoways, ed. Current mammalogy 1. Plenum Press, New York.

Frank, A., V. Galgan, and L. R. Petersson. 1994. Secondary copper deficiency, chromium deficiency and trace element imbalance in the moose (*Alces alces* L.): Effect of anthropogenic activity. Ambio 23:315–17.

Frank, D. A., and S. J. McNaughton. 1993. Evidence for the promotion of aboveground grassland production by native large herbivores in Yellowstone National Park. Oecologia 96:157–61.

Franzmann, A. W. 1978. Moose. Pages 67–81 *in* J. L. Schmidt and D. L. Gilbert, eds. Big game of North America: Ecology and management. Stackpole, Harrisburg, PA.

Franzmann, A. W. 1981. Alces alces. Mammalian Species 154:1–7.

Franzmann, A. W. 1998. Restraint, translocation and husbandry. Pages 519–57 *in* A. W. Franzmann and C. C. Schwartz, eds. Ecology and management of the North American moose. Smithsonian Institution Press, Washington, DC.

Franzmann, A. W. 2000. Moose. Pages 578–600 *in* S. Demarais and P. R. Krausman, eds. Ecology and management of large mammals in North America. Prentice-Hall, Upper Saddle River, NJ.

Franzmann, A. W., and R. E. LeResche. 1978. Alaskan moose blood studies with emphasis on condition evaluation. Journal of Wildlife Management 42:334–51.

Franzmann, A. W., and C. C. Schwartz. 1985. Moose twinning rates: A possible population condition assessment. Journal of Wildlife Management 49:394–96.

Franzmann, A. W., and C. C. Schwartz. 1986. Black bear predation on moose calves in highly productive versus marginal moose habitats on the Kenai Peninsula, Alaska. Alces 22:139–54.

Franzmann, A. W., and C. C. Schwartz, eds. 1998. Ecology and management of the North American moose. Smithsonian Institution Press, Washington, DC.

Franzmann, A. W., A. Flynn, and P. D. Arneson. 1976a. Moose milk and hair element levels and relationships. Journal of Wildlife Diseases 12:202–7.

Franzmann, A. W., P. D. Arneson, and J. L. Oldemeyer. 1976b. Daily winter pellet groups and beds of Alaskan moose. Journal of Wildlife Management 40:374–75.

Franzmann, A. W., J. L. Oldemeyer, P. D. Arneson, and R. K. Seemel. 1976c. Pellet-group count evaluation for census and habitat use of Alaskan moose. Proceedings of the North American Moose Conference and Workshop 12:127–42.

Franzmann, A. W., A. Flynn, and P. D. Arneson. 1977. Alaskan moose hair element values and variability. Comparative Biochemistry and Physiology 57A:299–306.

Franzmann, A. W., R. E. LeResche, R. A. Rausch, and J. L. Oldemeyer. 1978. Alaskan moose measurements and weights and measurement–weight relationships. Canadian Journal of Zoology 56:298–306.

Franzmann, A. W., C. C. Schwartz, and R. O. Peterson. 1980. Moose calf mortality in summer on the Kenai Peninsula, Alaska. Journal of Wildlife Management 44:764–68.

Franzmann, A. W., C. C. Schwartz, and D. C. Johnson. 1984. Baseline body temperatures, heart rates, and respiratory rates of moose in Alaska. Journal of Wildlife Diseases 20:333–37.

Franzmann, A. W., C. C. Schwartz, and D. C. Johnson. 1987. Monitoring status (condition, nutrition, health) of moose via blood. Swedish Wildlife Research Supplement 1:281–88.

Fraser, D., and H. Hristienko. 1981. Activity of moose and white-tailed deer at mineral springs. Canadian Journal of Zoology 59:1991–2000.

Fraser, D., D. Arthur, and B. K. Thompson. 1980. Aquatic feeding by moose *Alces alces* in a Canadian lake. Holarctic Ecology 3:218–23.

Fraser, D., B. K. Thompson, and D. Arthur. 1982. Aquatic feeding by moose: Seasonal variation in relation to plant chemical composition and use of mineral licks. Canadian Journal of Zoology 60:3121–26.

Fraser, D., E. R. Chavez, and J. E. Paloheimo. 1984. Aquatic feeding by moose: Selection of plant species and feeding areas in relation to plant chemical composition and characteristics of lakes. Canadian Journal of Zoology 62:80–87.

Fruetel, M., and M. W. Lankester. 1988. *Nematodirella alcidis* (Nematoda: Trichostronglyoidea) in moose in northwestern Ontario. Alces 24:159–63.

Fryxell, J. M., W. E. Mercer, and R. B. Gellately. 1988. Population dynamics of Newfoundland moose using cohort analysis. Journal of Wildlife Management 52:14–21.

Fuller, T. K. 1989. Population dynamics of wolves in north-central Minnesota. Wildlife Monographs 105:1–41.

Galgan, V., and L. R. Petersson. 1994. Secondary copper deficiency, chromium deficiency and trace element imbalance in the moose: Effect of anthropogenic activity. Ambio 23:315–17.

Garner, F. M., and L. W. Schwartz. 1969. Spontaneous hematopoietic neoplasms of free living and captive wild mammals. National Cancer Institute Monographs 32:153–56.

Gasaway, W. C., and J. W. Coady. 1974. Review of energy requirements and rumen fermentation in moose and other ruminants. Naturaliste Canadien 101:227–62.

Gasaway, W. C., D. B. Harkness, and R. A. Rausch. 1978. Accuracy of moose age determination from incisor cementum layers. Journal of Wildlife Management 42:558–63.

Gasaway, W. C., R. O. Stephenson, J. L. Davis, P. E. K. Shepherd, and O. E. Burris. 1983. Interrelationships of wolves, prey, and man in interior Alaska. Wildlife Monographs 84:1–50.

Gasaway, W. C., S. D. Dubois, D. J. Preston, and D. G. Reed. 1985a. Home range information and dispersal of subadult moose in interior Alaska (Federal Aid to Wildlife Restoration Project W-22-2). Alaska Department of Fish and Game, Juneau.

Gasaway, W. C., S. D. Dubois, and S. J. Harbo. 1985b. Biases in aerial transect surveys for moose during May and June. Journal of Wildlife Management 49:777–84.

Gasaway, W. C., S. D. Dubois, D. J. Reed, and S. Harbo. 1986. Estimating moose population parameters from aerial surveys. Biological Papers of the University of Alaska 22:1–108.

Gasaway, W. C., D. J. Preston, D. J. Reed, and D. D. Roby. 1987. Comparative antler morphology and size of North American moose. Swedish Wildlife Research Supplement 1:311–25.

Gasaway, W. C., R. D. Boertje, D. V. Grangaard, D. G. Kelleyhouse, R. O. Stephenson, and D. G. Larsen. 1992. The role of predation in limiting moose at low densities in Alaska and Yukon and implications for conservation. Wildlife Monographs 120:1–59.

Geist, V. 1963. On the behaviour of the North American moose (*Alces alces andersoni* Peterson, 1950) in British Columbia. Behaviour 20:377–416.

Geist, V. 1966. The evolution of horn-like organs. Behaviour 27:175–214.

Geist, V. 1998. Deer of the world: Their evolution, behaviour, and ecology. Stackpole, Mechanicsburg, PA.

Gilbert, F. F. 1974. *Parelaphostrongylus tenuis* in Maine: II—Prevalence in moose. Journal of Wildlife Management 38:42–46.

Gillingham, M. P., and D. R. Klein. 1992. Late-winter activity patterns of moose (*Alces alces gigas*) in western Alaska. Canadian Journal of Zoology 70:293–99.

Glines, M. V., and W. M. Samuel. 1989. The effect of *Dermacentor albipictus* (Acarina: Ixodidae) on blood composition, weight gain and hair coat of moose, *Alces alces*. Experimental and Applied Acarology 6:197–213.

Glooschenko, V., C. Downes, R. Frank, H. E. Braun, E. M. Addison, and J. Hickie. 1988. Cadmium levels in Ontario moose and deer in relation to soil sensitivity to acid precipitation. Science of the Total Environment 71:173–86.

Goss, R. J. 1983. Deer antlers: Regeneration, function, and evolution. Academic Press, New York.

Gray, J. E. 1821. On the natural arrangement of vertebrose animals. London Medical Repository 15:296–310.

Grenier, P. A. 1973. Moose killed on the highway in the Laurentides Park, Quebec, 1962–1972. Proceedings of the North American Moose Conference and Workshop 9:155–94.

Gundersen, H., and H. P. Andreassen. 1998. The risk of moose *Alces alces* collision: A predictive logistic model for moose–train accidents. Wildlife Biology 4:103–10.

Gupta, R. P., and H. C. Gibbs. 1971. Infectivity of *Dictyocaulus viviparous* (moose strain) to calves. Canadian Veterinary Journal 12:56.

Gustafson, K. A., K. M. Bontaites, and A. Major. 2000. Analysis of tissue cadmium concentrations in New England moose. Alces 36:35–40.

Gustavsson, I., and C. O. Sundt. 1968. Karyotypes in five species of deer (*Alces alces* L., *Capreolus capreolus* L., *Cervus elaphus* L., *Cervus nippon nippon* Temm., and *Dama dama* L.). Hereditas 60:233–48.

Guthrie, R. D. 1990. New dates in Alaskan Quaternary moose, *Cervalces–Alces*: Archaeological, evolutionary and ecological implications. Current Pleistocene Research 7:111–12.

Guthrie, R. D. 1995. Mammalian evolution in response to the Pleistocene–Holocene transition and the break-up of the mammoth steppe: Two case studies. Acta Zoologica Cracoviensia 38:139–54.

Haagenrud, H. 1978. Layers in secondary dentine of incisors as age criteria in moose (*Alces alces*). Journal of Mammalogy 59:857–58.

Hagerman, A. E., and C. T. Robbins. 1993. Specificity of tannin-binding salivary proteins relative to diet selection by mammals. Canadian Journal of Zoology 71:628–33.

Haigh, J. C., R. R. Stewart, and W. Mytton. 1980. Relations among linear measurements and weights for moose (*Alces alces*). Proceedings of the North American Moose Conference and Workshop 16:1–10.

Hall, E. R. 1981. The mammals of North America, Vol. II, 2nd ed. John Wiley, New York.

Hall, E. R., and K. R. Kelson. 1959. The mammals of North America, Vol. II. Ronald Press, New York.

Hamilton, G. D., P. D. Drysdale, and D. L. Euler. 1980. Moose winter browsing patterns on clear-cuttings in northern Ontario. Canadian Journal of Zoology 58:1412–16.

Hatter, I. W. 1998. A Bayesian approach to moose population assessment and harvest decisions. Alces 34:47–58.

Hatter, I. W. 2001. An assessment of catch per unit effort to estimate rate of change in deer and moose populations. Alces 37:71–77.

Hauge, T. M., and L. B. Keith. 1981. Dynamics of moose populations in northeastern Alberta. Journal of Wildlife Management 45:573–97.

Hayes, R. D., and A. S. Harestad. 2000. Wolf functional response and regulation of moose in the Yukon. Canadian Journal of Zoology 78:60–66.

Hayes, R. D., A. M. Baer, U. Wotschikowsky, and A. S. Harestad. 2000. Kill rate by wolves on moose in the Yukon. Canadian Journal of Zoology 78:49–59.

Heard, D., S. Barry, G. Watts, and K. Child. 1997. Fertility of female moose (*Alces alces*) in relation to age and body composition. Alces 33:165–76.

Hibler, C. P., and J. L. Adcock. 1971. Elaeophorosis. Pages 263–78 *in* J. W. Davis and R. C. Anderson, eds. Parasitic diseases of wild mammals. Iowa State University Press, Ames.

Hicks, A. C. 1993. Using road-kills as an index to moose population change. Alces 29:243–47.

Hik, D. S., and R. L. Jefferies. 1990. Increases in the net above-ground primary production of a salt-marsh forage grass: A test of the predictions of the herbivore-optimization model. Journal of Ecology 78:180–95.

Hillman, J. R., R. W. Davis, and Y. Z. Abdelbaki. 1973. Cyclic bone remodeling in deer. Calcified Tissue Research 12:323–30.

Hindelang, M., and R. O. Peterson. 1993. Relationship of mandibular tooth wear to gender, age and periodontal disease of Isle Royale moose. Alces 29:63–73.

Hindelang, M., and R. O. Peterson. 1994. Moose senescence related to tooth wear. Alces 30:9–12.

Hindelang, M., and R. O. Peterson. 1996. Osteoporotic skull lesions in moose at Isle Royale National Park. Journal of Wildlife Diseases 32:105–8.

Hindelang, M., and R. O. Peterson. 2000. Skeletal integrity in moose at Isle Royale National Park: Bone mineral density and osteopathology related to senescence. Alces 36:61–68.

Hindelang, M., G. Jayaraman, and R. O. Peterson. 1998. Biomechanical properties, osteoporosis, and fracture potential of metatarsal trabecular bone from moose in Isle Royale National Park. Alces 34:311–18.

Hirth, D. H. 1977. Social behavior of white-tailed deer in relation to habitat. Wildlife Monographs 53:1–55.

Hjeljord, O. 1992. Sunny and shaded growth sites: Influence on moose forage quality. Alces Supplement 1:112–14.

Hjeljord, O. 2001. Dispersal and migration in northern forest deer: Are there unifying concepts? Alces 37:353–70.

Hjeljord, O., F. Sundstol, and H. Haagenrud. 1982. The nutritional value of browse to moose. Journal of Wildlife Management 46:333–43.

Hjeljord, O., N. Høvik, and H. B. Pedersen. 1990. Choice of feeding sites by moose during summer, the influence of forest structure and plant phenology. Holarctic Ecology 13:281–92.

Hjeljord, O., B.-E. Sæther, and R. Andersen. 1994. Estimating energy intake of free-ranging moose cows and calves through collection of feces. Canadian Journal of Zoology 72:1409–15.

Hjeljord, O., B.-E. Sæther, and R. Andersen. 1996. Erratum: Estimating energy intake of free-ranging moose cows and calves through collection of feces. Canadian Journal of Zoology 74:585.

Hobbs, N. T. 1996. Modification of ecosystems by ungulates. Journal of Wildlife Management 60:695–713.

Hoeve, J., D. G. Joachim, and E. M. Addison. 1988. Parasites of moose (*Alces alces*) from an agricultural area of eastern Ontario. Journal of Wildlife Diseases 24:371–74.

Hoff, G. L., and D. O. Trainer. 1978. Blue tongue and epizootic hemorrhagic disease viruses: Their relationship to wildlife species. Advances in Veterinary Science and Comparative Medicine 22:111–32.

Hofmann, R. R., and K. Nygren. 1992. Morphophysiological specialization and adaptation of the moose digestive system. Alces Supplement 1:91–100.

Holling, C. S. 1959. The components of predation as revealed by a study of small-mammal predation of the European pine sawfly. Canadian Entomologist 91:293–320.

Honour, S., and K. M. H. Hickling. 1993. Naturally occurring *Brucella suis* biovar 4 infection in a moose (*Alces alces*). Journal of Wildlife Diseases 29:596–98.

Hosley, N. W., and F. S. Glaser. 1952. Triplet Alaskan moose calves. Journal of Mammalogy 33:247.

Houston, D. B. 1968. The Shiras moose in Jackson Hole, Wyoming. Grand Teton Natural History Association Technical Bulletin 1:1–110.

Houston, D. B. 1969. A note on the blood chemistry of the Shiras moose. Journal of Mammalogy 50:826.

Houston, D. B. 1974. Aspects of the social organization of moose. Pages 690–96 *in* V. Geist and F. Walther, eds. The behaviour of ungulates and its relation to management. International Union for the Conservation of Nature and Natural Resources, Morges, Switzerland.

Houston, D. B. 1982. The northern Yellowstone elk: Ecology and management. Macmillan, New York.

Hsu, T. C., and K. Benirschke. 1973. An atlas of mammalian chromosomes. Springer-Verlag, New York.

Hundertmark, K. J. 1998. Home range, dispersal and migration. Pages 303–35 *in* A. W. Franzmann and C. C. Schwartz, eds. Ecology and management of the North American moose. Smithsonian Institution Press, Washington, DC.

Hundertmark, K. J., and C. C. Schwartz. 1998. Predicting body mass of Alaskan moose (*Alces alces gigas*) using body measurements and condition assessment. Alces 34:83–90.

Hundertmark, K. J., and C. C. Schwartz. 2002. Evaluation of bioelectrical impedance analysis as an estimator of moose body composition. Wildlife Society Bulletin 30:915–21.

Hundertmark, K. J., W. L. Eberhardt, and R. E. Ball. 1990. Winter habitat use by moose in southeastern Alaska: Implications for forest management. Alces 26:108–14.

Hundertmark, K. J., T. H. Thelen, and R. T. Bowyer. 1998. Effects of population density and selective harvest on antler phenotype in simulated moose populations. Alces 34:375–83.

Hundertmark, K. J., G. F. Shields, I. G. Udina, R. T. Bowyer, A. A. Danilkin, and C. C. Schwartz. 2002. Mitochondrial phylogeography of moose (*Alces alces*): Late Pleistocene divergence and population expansion. Molecular Phylogenetics and Evolution 22:375–87.

Hundertmark, K. J., R. T. Bowyer, G. F. Shields, and C. C. Schwartz. 2003. Mitochondrial phylogeography of moose (*Alces alces*) in North America. Journal of Mammalogy 84:718–28.

Hundertmark, K. J., F. G. Shields, R. T. Bowyer, and C. C. Schwartz. In press-b. Genetic relationships deduced from cytochrome-*b* sequences among moose. Alces.

Iregren, E. 1985. Sexual dimorphism in the extremities and weight of the elk (*Alces alces*) in central Sweden. Acta Zoologica Fennica 170:105–8.

Irons, J. G., III., J. P. Bryant, and M. W. Oswood. 1991. Effects of moose browsing on decomposition rates of birch leaf litter in a subarctic stream. Canadian Journal of Fisheries and Aquatic Sciences 48:442–44.

Jandt, R. R. 1992. Modeling moose density using remotely sensed habitat variables. Alces 28:41–58.

Jolicoeur, H., and M. Crête. 1994. Failure to reduce moose–vehicle accidents after a partial drainage of roadside salt pools in Quebec. Alces 30:81–89.

Jordan, P. A. 1987. Aquatic foraging and the sodium ecology of moose: A review. Swedish Wildlife Research Supplement 1:119–37.

Jordan, P. A., D. B. Botkin, A. S. Dominski, H. W. Lowendorf, and G. E. Belovsky. 1973. Sodium as a critical nutrient of the moose of Isle Royale. Proceedings of the North American Moose Conference and Workshop 9:13–42.

Joyal, R. 1987. Moose habitat investigations in Quebec and management implications. Swedish Wildlife Research Supplement 1:139–52.

Juntheikki, M.-R. 1996. Comparison of tannin-binding proteins in saliva of Scandinavian and North American moose (*Alces alces*). Biochemical Systematics and Ecology 24:595–601.

Karns, P. D. 1967. *Pneumostrongylus tenuis* in deer in Minnesota and implications for moose. Journal of Wildlife Management 31:299–303.

Karns, P. D. 1976. Relationships of age and body measurements in moose weight in Minnesota. Proceedings of the North American Moose Conference and Workshop 12:274–84.

Karns, P. D. 1998. Population distribution, density and trends. Pages 125–39 *in* A. W. Franzmann and C. C. Schwartz, eds. Ecology and management of the North American moose. Smithsonian Institution Press, Washington, DC.

Keech, M. A., T. R. Stephenson, R. T. Bowyer, V. Van Ballenberghe, and J. M. Ver Hoef. 1998. Relationships between blood-serum variables and depth of rump fat in Alaskan moose. Alces 34:173–79.

Keech, M. A., R. D. Boertje, R. T. Bowyer, and B. W. Dale. 1999. Effects of birth weight on growth of young moose: Do low-weight neonates compensate? Alces 35:51–57.

Keech, M. A., R. T. Bowyer, J. M. Ver Hoef, R. D. Boertje, B. W. Dale, and T. R. Stephenson., 2000. Life-history consequences of maternal condition in Alaskan moose. Journal of Wildlife Management 64:450–62.

Kelsall, J. P., and W. Prescott. 1971. Moose and deer behavior in snow in Fundy National Park, New Brunswick. Canadian Wildlife Service Report Series 15:1–27.

Kelsall, J. P., and E. S. Telfer. 1974. Biogeography of moose with particular reference to western North America. Naturaliste Canadien 101:117–30.

Kie, J. G. 1999. Optimal foraging and risk of predation: Effects on behavior and social structure in ungulates. Journal of Mammalogy 80:1114–29.

Kie, J. G., and R. T. Bowyer. 1999. Sexual segregation in white-tailed deer: Density-dependent changes in use of space, habitat selection, and dietary niche. Journal of Mammalogy 80:1004–20.

Kie, J. G., and B. Czech. 2000. Mule and black-tailed deer. Pages 629–57 *in* S. Demarais and P. R. Krausman, eds. Ecology and management of large mammals in North America. Prentice-Hall, Upper Saddle River, NJ.

Kie, J. G., R. T. Bowyer, M. C. Nicholson, B. B. Boroski, and E. R. Loft. 2002. Landscape heterogeneity at differing scales: Effects on spatial distribution of mule deer. Ecology 83:530–44.

Kie, J. G., R. T. Bowyer, and K. M. Stewart. 2003. Ungulates in western forests: Habitat requirements, population dynamics, and ecosystem processes. Pages 296–340 *in* C. Zablel and R. Anthony, eds. Mammal community dynamics in western coniferous forests: Management and conservation. Johns Hopkins University Press, Baltimore.

Kielland, K., and J. P. Bryant. 1998. Moose herbivory in taiga: Effects on biogeochemistry and vegetation dynamics in primary succession. Oikos 82:377–83.

Kielland, K., J. P. Bryant, and R. W. Ruess. 1997. Moose herbivory and carbon turnover of early successional stands in interior Alaska. Oikos 80:25–30.

Kingston, N. 1981a. Protozoan parasites. Pages 193–236 *in* W. R. Davidson, F. A. Hayes, V. F. Nettles, and F. E. Kellogg, eds. Diseases and parasites of white-tailed deer (Miscellaneous Publication 7). Tall Timbers Research Station, Tallahassee, FL.

Kingston, N. 1981b. Trypanosoma. Pages 166–69 *in* R. A. Dieterich, ed. Alaskan wildlife diseases. University of Alaska, Fairbanks.

Klein, D. R. 1965. Postglacial distribution patterns of mammals in the southern coastal regions of Alaska. Arctic 18:7–20.

Klein, D. R. 1968. The introduction, increase, and crash of reindeer on St. Matthew Island. Journal of Wildlife Management 32:350–67.

Klein, D. R. 1981. The problem of overpopulation of deer in North America. Pages 119–27 *in* P. A. Jewell and S. Holt, eds. Problems in management of locally abundant wild mammals. Academic Press, New York.

Klein, D. R. 2000. The muskox. Pages 545–58 *in* S. Demarais and P. R. Krausman, eds. Ecology and management of large mammals in North America. Prentice-Hall, Upper Saddle River, NJ.

Kocan, A. A., S. J. Barron, J. C. Fox, and A. W. Franzmann. 1986a. Antibodies to *Toxoplasma gondii* in moose (*Alces alces* L.) from Alaska. Journal of Wildlife Diseases 22:432.

Kocan, A. A., A. W. Franzmann, K. A. Waldrup, and G. J. Kubat. 1986b. Serological studies of select infectious diseases of moose (*Alces alces* L.) from Alaska. Journal of Wildlife Diseases 22:418–20.

Krausman, P. R., and D. M. Shackleton. 2000. Bighorn sheep. Pages 517–44 *in* S. Demarais and P. R. Krausman, eds. Ecology and management of large mammals in North America. Prentice-Hall, Upper Saddle River, NJ.

Krefting, L. 1974a. The ecology of the Isle Royale moose (Technical Bulletin 297). University of Minnesota Agricultural Experiment Station.

Krefting, L. 1974b. Moose distribution and habitat selection in north central North America. Naturaliste Canadien 101:81–100.

Kubota, J., S. Reiger, and V. A. Lazar. 1970. Mineral composition of herbage browsed by moose in Alaska. Journal of Wildlife Management 34:565–69.

Kufeld, R. C., and D. C. Bowden. 1996. Movements and habitat selection of Shiras moose (*Alces alces shirasi*) in Colorado. Alces 32:85–99.

Kunkel, K. E., and D. H. Pletscher. 2000. Habitat factors affecting vulnerability of moose to predation by wolves in southeastern British Columbia. Canadian Journal of Zoology 78:150–57.

Kunkel, K. E., T. K. Ruth, D. H. Pletscher, and M. G. Hornocker. 1999. Winter prey selection by wolves and cougars in and near Glacier National Park, Montana. Journal of Wildlife Management 63:901–10.

Labonté, J., J.-P. Quellet, R. Courtois, and F. Belislé. 1998. Moose dispersal and its role in the maintenance of harvested populations. Journal of Wildlife Management 62:225–35.

Langley, M. A., and D. H. Pletscher. 1994. Calving areas of moose in northwestern Montana and southeastern British Columbia. Alces 30:127–35.

Lankester, M. W. 1974. *Parelophostrongylus tenuis* (Nematoda) and *Fascioloides magna* (Trematoda) in moose in southeastern Manitoba. Canadian Journal of Zoology 52:235–39.

Lankester, M. W. 1977. Neurologic disease in moose caused by *Elaphostrongylus cervi* Cameron 1931 from caribou. Proceedings of the North American Moose Conference and Workshop 13:177–90.

Lankester, M. W. 1987. Pests, parasites, and diseases of moose (*Alces alces*) in North America. Swedish Wildlife Research Supplement 1:461–90.

Lankester, M. W. 2001. Extrapulmonary lungworms of cervids. Pages 228–78 *in* W. M. Samuel, M. J. Pybus, and A. A. Kocan, eds. Parasitic diseases of wild mammals, 2nd ed. Iowa State University Press, Ames.

Lankester, M. W., and R. C. Anderson. 1968. Gastropods as intermediate hosts of *Pneumostrongylus tenuis* Dougherty of white-tailed deer. Canadian Journal of Zoology 46:373–83.

Lankester, M. W., and T. J. Bellhouse. 1982. Pathological anomalies in moose of northwestern Ontario. Alces 18:17–24.

Lankester, M. W., and D. Fong. 1989. Distribution of elaphostrongyline nematodes (Metastrongyloidae: Protostrongylidae) in Cervidae and possible effects of moving *Rangifer* spp. into and within North America. Alces 25:133–45.

Lankester, M. W., and W. J. Peterson. 1996. The possible importance of wintering yards in the transmission of *Parelaphstrongylus tenuis* to white-tailed deer and moose. Journal of Wildlife Diseases 32:31–38.

Lankester, M. W., and W. M. Samuel. 1998. Pests, parasites and diseases. Pages 479–517 *in* A. W. Franzmann and C. C. Schwartz, eds. Ecology and management of North American moose. Smithsonian Institution Press, Washington, DC.

Lankester, M. W., and R. D. Sein. 1986. The moose fly, *Haematobosca alcis,* and skin lesions on *Alces alces*. Alces 22:361–75.

Lankester, M. W., and J. B. Snider. 1982. *Rumenfilaria andersoni* (Nematoda: Filaricoedea) in moose, *Alces alces* (L), from northwestern Ontario, Canada. Canadian Journal of Zoology 60:2455–58.

Larsen, D. G., D. A. Gauthier, and R. L. Markel. 1989. Causes and rate of moose mortality in the southwest Yukon. Journal of Wildlife Management 53:548–57.

Lautenschlager, R. A., and R. T. Bowyer. 1985. Wildlife management by referendum: When professionals fail to communicate. Wildlife Society Bulletin 13:564–70.

Lautenschlager, R. A., and P. A. Jordan. 1993. Potential use of track-pellet group counts for moose censusing. Alces 29:175–80.

Lautenschlager, R. A., D. E. White, and M. L. McCormack, Jr. 1989. Browse availability after conifer release in Maine's spruce-fir forests. Journal of Wildlife Management 53:643–49.

Lautenschlager, R. A., H. S. Crawford, M. R. Stokes, and T. L. Stone. 1997. Forest disturbance type differentially affects seasonal moose forage. Alces 33:49–73.

Lenart, E. A., R. T. Bowyer, J. Ver Hoef, and R. W. Ruess. 2002. Climate change and caribou: Effects of summer weather on forage. Canadian Journal of Zoology 80:664–78.

Lenarz, M. S. 1998. Precision and bias of aerial moose surveys in northeastern Minnesota. Alces 34:117–24.

Lent, P. C. 1974. A review of rutting behavior in moose. Naturaliste Canadien 101:307–23.

Leptich, D. J., and J. R. Gilbert. 1986. Characteristics of moose calving sites in northern Maine: A preliminary investigation. Alces 22:69–81.

LeResche, R. E., and R. A. Rausch. 1974. Accuracy and precision of aerial moose censusing. Journal of Wildlife Management 38:175–82.

Leslie, D. M., Jr., and E. S. Starkey. 1985. Fecal indices to dietary quality of cervids in old-growth forests. Journal of Wildlife Management 49:142–46.

Leslie, D. M., Jr., J. A. Jenks, M. Chilelli, and G. R. Lavigne. 1989. Nitrogen and diaminopimelic acid in deer and moose feces. Journal of Wildlife Management 53:216–18.

Lidicker, W. Z., Jr. 1975. The role of dispersal in the demography of small mammals. Pages 103–28 *in* F. B. Golley, K. Petrusewicz, and L. Ryszkowski, eds. Small mammals: Their productivity and population dynamics. Cambridge University Press, Cambridge.

Lister, A. M. 1993. Evolution of mammoths and moose: The Holarctic perspective. Pages 178–204 *in* A. D. Barnosky, ed. Morphological change in Quaternary mammals of North America. Cambridge University Press, Cambridge.

Loison, A. M., J.-M. Gaillard, C. Pélabon, and N. G. Yoccoz. 1999a. What factors shape sexual size dimorphism in ungulates? Evolutionary Ecology Research 1:611–33.

Loison, A., M. Festa-Bianchet, J.-M. Gaillard, J. T. Jorgenson, and J.-M. Jullien. 1999b. Age-specific survival in five populations of ungulates: Evidence of senescence. Ecology 80:2539–54.

Loranger, A. J., T. N. Bailey, and W. W. Larned. 1991. Effects of forest succession after fire in moose wintering habitats on the Kenai Peninsula, Alaska. Alces 27:100–109.

Ludewig, H. A., and R. T. Bowyer. 1985. Overlap in winter diets of sympatric moose and white-tailed deer in Maine. Journal of Mammalogy 66:390–92.

Lynch, G. M. 1973. Influence of hunting on an Alberta moose herd. Proceedings of the North American Moose Conference and Workshop 9:123–35.

Lynch, G. M. 1976. Some long-range movements of radio-tagged moose in Alberta. Proceedings of the North American Moose Conference and Workshop 12:220–35.

Lynch, G. M., B. Lajeunesse, J. Willman, and E. S. Telfer. 1995. Moose weights and measurements from Elk Island National Park, Canada. Alces 31:199–207.

MacCracken, J. G., and V. Van Ballenberghe. 1987. Age- and sex-related differences in fecal pellet dimensions of moose. Journal of Wildlife Management 51:360–64.

MacCracken, J. G., and L. A. Viereck. 1990. Browse regrowth and use by moose after fire in interior Alaska. Northwest Science 64:11–18.

MacCracken, J. G., V. Van Ballenberghe, and J. M. Peek. 1993. Use of aquatic plants by moose: Sodium hunger or foraging efficiency? Canadian Journal of Zoology 71:2345–51.

MacCracken, J. G., V. Van Ballenberghe, and J. M. Peek. 1997. Habitat relationships of moose on the Copper River Delta in coastal south-central Alaska. Wildlife Monographs 136:1–52.

Madden, D. J., R. T. Spraker, and W. J. Adrian. 1991. *Elaeophora schneideri* in moose (*Alces alces*) from Colorado. Journal of Wildlife Diseases 27:340–41.

Markgren, G. 1969. Reproduction of moose in Sweden. Viltrevy 6:127–299.

Marshal, J. P. 1999. Co-management of moose in the Gwich'in settlement area, Northwest Territories. Alces 35:151–58.

Marshal, J. P., and S. Boutin. 1998. Power analysis of wolf–moose functional responses. Journal of Wildlife Management 63:396–402.

Martin, C. J. 1989. Observation of a female moose, *Alces alces,* accompanied by possible quadruplet calves at Isle Royale National Park, Michigan. Canadian Field-Naturalist 103:418–19.

Mastenbrook, B., and H. Cumming. 1989. Use of residual strips of timber by moose within cutovers in northwestern Ontario. Alces 25:146–55.

Matchett, M. R. 1985. Habitat selection by moose in the Yaak River drainage, northwestern Montana. Alces 21:161–89.

Mauer, F. J. 1998. Moose migration: Northeastern Alaska to northwestern Yukon Territory, Canada. Alces 34:75–81.

McCorquodale, S. M., and R. F. DiGiacomo. 1985. The role of wild North American ungulates in the epidemiology of bovine brucellosis: A review. Journal of Wildlife Diseases 21:351–57.

McCullough, D. R. 1969. The tule elk: Its history, behavior, and ecology. University of California Publications in Zoology 88:1–209.

McCullough, D. R. 1979. The George Reserve deer herd: Population ecology of a *K*-selected species. University of Michigan Press, Ann Arbor.

McCullough, D. R. 1982. Antler characteristics of George Reserve white-tailed deer. Journal of Wildlife Management 46:821–26.

McCullough, D. R. 1990. Detecting density dependence: Filtering the baby from the bathwater. Transactions of the North American Wildlife and Natural Resources Conference 55:534–43.

McCullough, D. R. 1996. Failure of the tooth cementum aging technique with reduced population density of deer. Wildlife Society Bulletin 24:722–24.

McCullough, D. R. 1999. Density dependence and life-history strategies of ungulates. Journal of Mammalogy 80:1130–46.

McDonald, M. G. 1991. Moose movement and mortality associated with the Glenn Highway expansion, Anchorage, Alaska. Alces 27:208–19.

McKenney, D. W., R. S. Rempel, L. A. Venier, Y. Wang, and A. R. Bisset. 1998. Development and application of a spatially explicit moose population model. Canadian Journal of Zoology 67:1922–31.

McLaren, B. E., and R. O. Peterson. 1994. Wolves, moose, and tree rings on Isle Royale. Science 266:1555–58.

McMillan, J. F. 1954. Some observations on moose in Yellowstone Park. American Midland Naturalist 52:392–99.

McNaughton, S. J. 1984. Grazing lawns: Animals in herds, plant form, and coevolution. American Naturalist 124:863–86.

McNeill, M. A., and M. E. Rau. 1987. *Echinococcus granulosus* (Cestoda: Taeniidae) infections in moose (*Alces alces*) from southwestern Quebec. Journal of Wildlife Diseases 23:418–21.

Mech, L. D., R. E. McRoberts, R. O. Peterson, and R. E. Page. 1987. Relationships of deer and moose populations to previous winters' snow. Journal of Animal Ecology 56:615–27.

Mercer, W. E., and D. A. Kitchen. 1968. A preliminary report on the extension of moose range in the Labrador Peninsula. Proceedings of the North American Moose Conference and Workshop 5:62–81.

Mercer, W. E., and F. Manuel. 1974. Some aspects of moose management in Newfoundland. Naturaliste Canadien 101:657–71.

Messier, F. 1991. The significance of limiting and regulating factors on the demography of moose and white-tailed deer. Journal of Animal Ecology 60:377–93.

Messier, F. 1994. Ungulate population models with predation: A case study with the North American moose. Ecology 75:478–88.

Messier, F., and M. Crête. 1984. Body condition and population regulation by food resources in moose. Oecologia 65:44–50.

Messier, F., and D. O. Joly. 2000. Comment: Regulation of moose populations by wolf predation. Canadian Journal of Zoology 78:506–10.

Messier, F., M. E. Rau, and M. A. McNeill. 1989. *Echinococcus granulosus* (Cestoda:Taeniidae) infection and moose–wolf population dynamics in southwestern Quebec. Canadian Journal of Zoology 67:216–19.

Messmer, T. A., C. E. Dixon, W. Shields, S. C. Barras, and S. A. Schroeder. 1998. Cooperative wildlife management units: Achieving hunter, landowner, and wildlife management agency objectives. Wildlife Society Bulletin 26:325–32.

Minzey, R. T., and W. L. Robinson. 1991. Characteristics of winter bed sites of moose in Michigan. Alces 27:150–60.

Mikko, S., and L. Andersson. 1995. Low major histocompatibility complex class II diversity in European and North American moose. Proceedings of the National Academy of Sciences 92:4259–63.

Miller, B. K., and J. A. Litvaitis. 1992a. Use of roadside salt licks by moose, *Alces alces,* in northern New Hampshire. Canadian Field-Naturalist 106:112–17.

Miller, B. K., and J. A. Litvaitis. 1992b. Habitat segregation by moose in a boreal forest ecotone. Acta Theriologica 37:41–50.

Miquelle, D. G. 1983. Browse regrowth and consumption following summer defoliation by moose. Journal of Wildlife Management 47:17–24.

Miquelle, D. G. 1990. Why don't bull moose eat during the rut? Behavioral Ecology and Sociobiology 27:145–51.

Miquelle, D. G. 1991. Are moose mice? The function of scent urination in moose. American Naturalist 138:460–77.

Miquelle, D. G., and V. Van Ballenberghe. 1985. The moose bell: A visual or olfactory communicator? Alces 21:191–213.

Miquelle, D. G., and V. Van Ballenberghe. 1989. Impact of bark stripping by moose on aspen-spruce communities. Journal of Wildlife Management 53:577–86.

Miquelle, D. G., J. M. Peek, and V. Van Ballenberghe. 1992. Sexual segregation in Alaskan moose. Wildlife Monographs 122:1–57.

Mitchell, H. 1970. Rapid aerial sexing of antlerless moose in British Columbia. Journal of Wildlife Management 34:645–46.

Modafferi, R. D. 1991. Train moose-kill in Alaska: Characteristics and relationship with snowpack depth and moose distribution in lower Susitna Valley. Alces 27:193–207.

Moen, R., and J. Pastor. 1998. A model to predict nutritional requirements for antler growth in moose. Alces 34:59–74.

Molvar, E. M. 1993. Nursing by a yearling moose, *Alces alces gigas,* in Alaska. Canadian Field-Naturalist 107:233–35.

Molvar, E. M., and R. T. Bowyer. 1994. Costs and benefits of group living in a recently social ungulate: The Alaskan moose. Journal of Mammalogy 75:621–30.

Molvar, E. M., R. T. Bowyer, and V. Van Ballenberghe. 1993. Moose herbivory, browse quality, and nutrient cycling in an Alaskan treeline community. Oecologia 94:472–79.

Monthey, R. W. 1984. Effects of timber harvesting on ungulates in northern Maine. Journal of Wildlife Management 48:279–85.

Murie, A. 1934. The moose of Isle Royale. Miscellaneous Publications of the Museum of Zoology of the University of Michigan 25:1–44.

Mysterud, A. 2000. The relationship between ecological segregation and sexual body size dimorphism in large herbivores. Oecologia 124:621–30.

Mytton, W. R., and L. B. Keith. 1981. Dynamics of moose populations near Rochester, Alberta 1975–1978. Canadian Field-Naturalist 95:39–49.

Nankervis, P. J., W. M. Samuel, S. M. Schmitt, and J. G. Sikarskie. 2000. Ecology of meningeal worm, *Parelaphostrongylus tenuis* (Nematoda), in white-tailed deer and terrestrial gastropods of Michigan's upper peninsula with implications for moose. Alces 36:163–81.

Nicholson, M. C., R. T. Bowyer, and J. G. Kie. 1997. Habitat selection and survival of mule deer: Tradeoffs associated with migration. Journal of Mammalogy 78:483–504.

Nudds, T. D. 1990. Retroductive logic in retrospect: The ecological effects of meningeal worms. Journal of Wildlfie Management 54:396–402.

Nygrén, K. 2000. Directional asymmetry in moose. Alces 36:147–54.

Nygrén, T., and I. Kojola. 1997. Twinning and fetal sex ratio in moose: Effects of maternal age and mass. Canadian Journal of Zoology 75:1945–48.

O'Hara, T. M., J. Dau, G. Carroll, J. Bevins, and R. L. Zarnke. 1998. Evidence of exposure to *Brucella suis* biovar 4 in northern Alaska moose. Alces 34:31–40.

O'Hara, T. M., G. Carroll, P. Barboza, K. Mueller, J. Blake, V. Woshner, and C. Willetto. 2001. Mineral and heavy metal status as related to a mortality event and poor recruitment in moose population in Alaska. Journal of Wildlife Diseases 37:509–22.

Oldemeyer, J. L. 1974. Nutritive value of moose forage. Naturaliste Canadien 101:217–26.

Oldemeyer, J. L. 1975. Characteristics of paper birch saplings browsed by moose and snowshoe hares. Proceedings of the North American Moose Conference and Workshop 11:53–62.

Oldemeyer, J. L. 1983. Browse production and its use by moose and snowshoe hares at the Kenai Moose Research Center, Alaska. Journal of Wildlife Management 47:486–96.

Oldemeyer, J. L., and W. L. Regelin. 1980. Response of vegetation to tree crushing in Alaska. Alces 16:429–43.

Oldemeyer, J. L., A. W. Franzmann, A. L. Brundage, P. D. Arneson, and A. Flynn. 1977. Browse quality and the Kenai moose population. Journal of Wildlife Management 41:533–42.

Olterman, J. H., and D. W. Kenvin. 1998. Reproduction, survival, and occupied ranges of Shiras moose transplanted to southwestern Colorado. Alces 34:41–46.

Olsson, I.-M., R. Bergstrom, M. Steen, and F. Sandegren. 1995. A study of *Elaphostrongylus alces* in an island moose population with low calf body weights. Alces 31:61–75.

Ontario Ministry of Natural Resources. 1980. Moose management in Ontario, a report of open house public meetings. Wildlife Branch, Ontario Ministry of Natural Resources, Toronto.

Oosenbrug, S. M., and S. H. Ferguson. 1992. Moose mark–recapture survey in Newfoundland. Alces 28:21–29.

Oosenbrug, S. M., E. W. Mercer, and S. H. Ferguson. 1991. Moose-vehicle collisions in Newfoundland—Management considerations for the 1990's. Alces 27:220–25.

Osborne, T. O., T. F. Paragi, J. L. Bodkin, A. J. Loranger, and W. N. Johnson. 1991. Extent, causes, and timing of moose calf mortality in western interior Alaska. Alces 27:24–30.

Palo, R. T. 1991. Effects of phenols in birch *Betula* spp. on digestion and metabolism in browsing mammals: A review. Transactions of the Congress of International Union of Game Biologists 18:151–54.

Paré, M., R. Prairie, and M. Speyer. 1999. Variations of cadmium levels in moose tissues from the Abitibi-Témiscamingue region. Alces 35:177–90.

Parker, G. R., and L. D. Morton. 1978. The estimation of winter forage and its use by moose on clearcuts in northcentral Newfoundland. Journal of Range Management 31:300–304.

Pastor, J., and R. J. Naiman. 1992. Selective foraging and ecosystem processes in the boreal forests. American Naturalist 139:690–705.

Pastor, J., B. Dewey, R. J. Naiman, P. F. McInnes, and Y. Cohen. 1993. Moose browsing and soil fertility in the boreal forests of Isle Royale National Park. Ecology 74:467–80.

Peek, J. M. 1974a. Initial response of moose to a forest fire in northeastern Minnesota. American Midland Naturalist 91:435–38.

Peek, J. M. 1974b. A review of moose food habits studies in North America. Naturaliste Canadien 101:195–215.

Peek, J. M. 1998. Habitat relationships. Pages 351–75 *in* A. W. Franzmann and C. C. Schwartz, eds. Ecology and management of the North American moose. Smithsonian Institution Press, Washington, DC.

Peek, J. M. 2000. Mountain goat. Pages 467–90 *in* S. Demarais and P. R. Krausman, eds. Ecology and management of large mammals in North America. Prentice-Hall, Upper Saddle River, NJ.

Peek, J. M., and K. I. Morris. 1998. Status of moose in the contiguous United States. Alces 34:423–34.

Peek, J. M., R. E. LeResche, and D. R. Stevens. 1974. Dynamics of moose aggregations in Alaska, Minnesota, and Montana. Journal of Mammalogy 55:126–37.

Peek, J. M., D. L. Urich, and R. J. Mackie. 1976. Moose habitat selection and relationships to forest management in northeastern Minnesota. Wildlife Monographs 48:1–65.

Peek, J. M., V. Van Ballenberghe, and D. G. Miquelle. 1986. Intensity of interactions between rutting bull moose in central Alaska. Journal of Mammalogy 67:423–26.

Peek, J. M., D. J. Pierce, D. C. Graham, and D. L. Davis. 1987. Moose habitat use and implications for forest management in northcentral Idaho. Swedish Wildlife Research Supplement 1:195–99.

Pehrson, A., R. T. Palo, H. Staaland, and P. A. Jordan. 1997. Seasonal variation in weight of functional segments of the gastrointestinal tract and its contents in young moose (*Alces alces*). Alces 33:1–10.

Person, D. K., R. T. Bowyer, and V. Van Ballenberghe. 2001. Density dependence of ungulates and functional responses of wolves: Effects on predator–prey ratios. Alces 37:253–73.

Peterson, R. L. 1952. A review of the living representatives of the genus *Alces*. Contributions to the Royal Ontario Museum of Zoology and Paleontology 34:1–30.

Peterson, R. L. 1955. North American moose. University of Toronto Press, Toronto.

Peterson, R. L. 1974. A review of the general life history of moose. Naturaliste Canadien 101:9–21.

Peterson, R. O. 1978. Wolf ecology and prey relationships on Isle Royale. (Scientific Monograph Series No. 11). U.S. National Park Service.

Peterson, R. O. 1999. Wolf–moose interaction on Isle Royale: The end of natural regulation? Ecological Applications 9:10–16.

Peterson, R. O., and D. L. Allen. 1974. Snow conditions as a parameter in moose–wolf relationships. Naturaliste Canadian 101:481–92.

Peterson, R. O., and R. E. Page. 1993. Detection of moose in midwinter from fixed-wing aircraft over dense forest cover. Wildlife Society Bulletin 21:80–86.

Peterson, R. O., C. C. Schwartz, and W. B. Ballard. 1983. Eruption patterns of selected teeth in three North American moose populations. Journal of Wildlife Management 47:884–88.

Peterson, R. O., T. N. Bailey, and J. D. Woolington. 1984. Wolves of the Kenai Peninsula, Alaska. Wildlife Monographs 88:1–52.

Peterson, R. O., N. J. Thomas, J. M. Thurber, J. A. Vucetich, and T. A. Waite. 1998. Population limitation and the wolves of Isle Royale. Journal of Mammalogy 79:828–41.

Phillips, R. L., W. E. Berg, and D. B. Siniff. 1973. Moose movement patterns and range use in northwestern Minnesota. Journal of Wildlife Management 37:266–78.

Pierce, B. M., V. C. Bleich, and R. T. Bowyer. 2000. Social organization of mountain lions: Does a land-tenure system regulate population size? Ecology 81:1533–43.

Pierce, D. J. 1984. Shiras moose forage selection in relation to browse availability in north central Idaho. Canadian Journal of Zoology 62:2404–9.

Pierce, D. J., and J. M. Peek. 1984. Moose habitat use and selection patterns in north central Idaho. Journal of Wildlife Management 48:1335–43.

Pimlott, D. H. 1959. Reproduction and productivity of Newfoundland moose. Journal of Wildlife Management 23:381–401.

Pimlott, D. H. 1961. The ecology and management of moose in North America. Terre Vie 2:246–65.

Pledger, D. J., W. M. Samuel, and D. A. Craig. 1980. Black flies (Diptera: Simuliidae) as possible vectors of legworm (*Onchocerca cervipedis*) in moose of central Alberta. Proceedings of the North American Moose Conference and Workshop 16:171–202.

Post, E., and N. C. Stenseth. 1998. Large-scale climatic fluctuation and population dynamics of moose and white-tailed deer. Journal of Animal Ecology 67:537–43.

Post, E., and N. C. Stenseth. 1999. Climatic variability, plant phenology, and northern ungulates. Ecology 80:1322–39.

Post, E., N. C. Stenseth, R. O. Peterson, J. A. Vucetich, and L. M. Ellis. 2002. Phase dependence and population cycles in a large-mammal predator–prey system. Ecology 83:2997–3002.

Prescott, W. H. 1974. Interrelationships of moose and deer of the genus *Odocoileus*. Naturaliste Canadien 101:493–504.

Proulx, G. 1983. Characteristics of moose (*Alces alces*) winter yards on different exposures and slopes in southern Quebec. Canadian Journal of Zoology 61:112–18.

Pybus, M. J. 1990. Survey of hepatic and pulmonary helminths of wild cervids in Alberta, Canada. Journal of Wildlife Diseases 26:453–59.

Quinn, N. W. S., and R. W. Aho. 1989. Whole weights of moose from Algonquin Park, Ontario, Canada. Alces 25:48–51.

Rachlow, J. L., and R. T. Bowyer. 1991. Interannual variation in timing and synchrony of parturition in Dall's sheep. Journal of Mammalogy 72:487–92.

Ralls, K. 1971. Mammalian scent marking. Science 171:443–49.

Raymond, K. S., F. A. Servello, B. Griffith, and W. E. Eschholz. 1996. Winter foraging ecology of moose on glyphosate-treated clearcuts in Maine. Journal of Wildlife Management 60:753–63.

Reese, E. O., and C. T. Robbins. 1994. Characteristics of moose lactation and neonatal growth. Canadian Journal of Zoology 72:953–57.

Reeves, H. M., and R. E. McCabe. 1998. Of moose and man. Pages 1–75 *in* A. W. Franzmann and C. C. Schwartz, eds. Ecology and management of North American moose. Smithsonian Institution Press, Washington, DC.

Regelin, W. L., and A. W. Franzmann. 1998. Past, present, and future moose management and research in Alaska. Alces 34:279–86.

Regelin, W. L., C. C. Schwartz, and A. W. Franzmann. 1985. Seasonal energy metabolism of adult moose. Journal of Wildlife Management 49:388–93.

Regelin, W. L., C. C. Schwartz, and A. W. Franzmann. 1986. Energy cost of standing in adult moose. Alces 22:83–90.

Regelin, W. L., C. C. Schwartz, and A. W. Franzmann. 1987. Effects of forest succession on nutritional dynamics of moose forage. Swedish Wildlife Research Supplement 1:247–64.

Rempel, R. S., P. C. Elkie, A. R. Rodgers, and M. J. Gluck. 1997. Timber-management and natural-disturbance effects on moose habitat: Landscape evaluation. Journal of Wildlife Management 61:517–24.

Renecker, L. A. 1987. The composition of moose milk following a late parturition. Acta Theriologica 32:129–33.

Renecker, L. A., and R. J. Hudson. 1986. Seasonal energy expenditures and thermoregulatory responses of moose. Canadian Journal of Zoology 64:322–27.

Renecker, L. A., and R. J. Hudson. 1989. Seasonal activity budgets of moose in aspen-dominated boreal forests. Journal of Wildlife Management 53:296–302.

Renecker, L. A., and R. J. Hudson. 1990. Behavioral and thermoregulatory responses of moose to high ambient temperatures and insect harassment in aspen-dominated forests. Alces 26:66–72.

Renecker, L. A., and C. C. Schwartz. 1998. Food habits and feeding behavior. Pages 403–39 *in* A. W. Franzmann and C. C. Schwartz, eds. Ecology and management of the North American moose. Smithsonian Institution Press, Washington, DC.

Reuterwall, C. 1981. Temporal and spatial variability of the calf sex ratio in Scandinavian moose *Alces alces*. Oikos 37:39–45.

Risenhoover, K. L. 1986. Winter activity patterns of moose in interior Alaska. Journal of Wildlife Management 50:727–34.

Risenhoover, K. L. 1989. Composition and quality of moose winter diets in interior Alaska. Journal of Wildlife Management 53:568–77.

Risenhoover, K. L., and S. A. Maass. 1987. The influence of moose on the composition and structure of Isle Royale forests. Canadian Journal of Forest Research 17:357–64.

Risenhoover, K. L., and R. O. Peterson. 1986. Mineral licks as a sodium source for Isle Royale moose. Oecologia 71:121–26.

Risto, H., and S. Harkonen. 1998. The effects of salt stones on moose browsing in managed forests in Finland. Alces 34:435–44.

Ritcey, R. W., and R. Y. Edwards. 1958. Parasites and diseases of the Wells Gray moose herd. Journal of Mammalogy 39:139–45.

Rodgers, A. R., R. S. Rempel, and K. F. Abraham. 1996. A GPS-based telemetry system. Wildlife Society Bulletin 24:559–66.

Rolley, R. E., and L. B. Keith. 1980. Moose population dynamics and winter habitat use at Rochester, Alberta, 1965–1979. Canadian Field-Naturalist 94:9–18.

Ross, P. I., and M. G. Jalkotzy. 1996. Cougar predation on moose in southwestern Alberta. Alces 32:1–8.

Roussel, Y. E. 1975. Aerial sexing of antlerless moose by white vulval patch. Journal of Wildlife Management 39:450–51.

Roussel, Y. E., E. A. Audy, and F. Potvin. 1975. Preliminary study of seasonal moose movements in Laurentides Provincial Park, Quebec. Canadian Field-Naturalist 89:47–52.

Ruess, R. W., and S. J. McNaughton. 1987. Grazing and the dynamics of nutrient and energy regulated microbial processes in the Serengeti grasslands. Oikos 49:101–10.

Ruess, R. W., R. L. Hendrick, and J. P. Bryant. 1998. Regulation of fine root dynamics by mammalian browsers in early successional Alaskan taiga forests. Ecology 79:2706–20.

Russell, J. B., and J. L. Rychlik. 2001. Factors that alter rumen microbial ecology. Science 292:1119–22.

Sæther, B.-E. 1985. Annual variation in carcass weight of Norwegian moose in relation to climate along a latitudinal gradient. Journal of Wildlife Management 49:977–83.

Sæther, B.-E. 1997. Environmental stochasticity and population dynamics of large herbivores: A search for mechanisms. Trends in Ecology and Evolution 12:143–49.

Sæther, B.-E., and H. Haagenrud. 1983. Life history of the moose (*Alces alces*): Fecundity rates in relation to age and carcass weight. Journal of Mammalogy 64:226–32.

Sæther, B.-E., and H. Haagenrud. 1985. Geographical variation in the antlers of Norwegian moose in relation to age and size. Journal of Wildlife Management 49:983–86.

Sæther, B.-E., R. Anderson, O. Hjeljord, and M. Heim. 1996. Ecological correlates of regional variation in life history of the moose *Alces alces*. Ecology 77:1493–1500.

Samuel, W. M. 1972. *Taenia krabbei* in the musculature of moose: A review. Proceedings of the North American Moose Conference and Workshop 8:18–41.

Samuel, W. M. 1991. Grooming by moose (*Alces alces*) infected with the winter tick, *Dermacentor albipictus* (Acari): A mechanism for premature loss of winter hair. Canadian Journal of Zoology 69:1255–60.

Samuel, W. M., and D. A. Welch. 1991. Winter ticks on moose and other ungulates: Factors influencing their population size. Alces 27:169–82.

Samuel, W. M., M. W. Barrett, and G. M. Lynch. 1976. Helminths in moose of Alberta. Canadian Journal of Zoology 54:307–12.

Samuel, W. M., D. A. Welch, and M. L. Drew. 1986. Shedding of the juvenile and winter hair coats of moose (*Alces alces*) with emphasis on the influence of the winter tick, *Dermacentor albipictus*. Alces 22:345–60.

Samuel, W. M., M. S. Mooring, and O. I. Aaslangdong. 2000. Adaptations of winter ticks (*Dermacentor albipictus*) to invade moose and moose to evade ticks. Alces 36:183–95.

Sand, H. 1996. Life history patterns in female moose (*Alces alces*): The relationship between age, body size, and fecundity and environmental conditions. Oecologia 106:212–20.

Sand, H., and G. Cederlund. 1996. Individual and geographical variation in age at maturity in female moose (*Alces alces*). Canadian Journal of Zoology 74:954–64.

Sandegren, F., and P. Y. Sweanor. 1988. Migration distances of moose populations in relation to river drainage length. Alces 24:112–17.

Scanlon, P. F., K. I. Morris, A. G. Clark, N. Fimreite, and S. Lierhagen. 1986. Cadmium in moose tissue: Comparison of data from Maine, U. .A., and from Telemark, Norway. Alces 22:303–12.

Scharf, C. M., and D. H. Hirth. 2000. Impact of moose bark stripping on mountain ash in Vermont. Alces 36:41–52.

Schladweiler, P., and D. R. Stevens. 1973. Reproduction of Shiras moose in Montana. Journal of Wildlife Management 37:535–44.

Schneider, R. R., and S. Wasel. 2000. The effects of human settlement on the density of moose in northern Alberta. Journal of Wildlife Management 64:513–20.

Schwartz, C. C. 1992. Physiological and nutritional adaptations of moose to northern environments. Alces Supplement 1:139–55.

Schwartz, C. C. 1998. Reproduction, natality, and growth. Pages 141–71 *in* A. W. Franzmann and C. C. Schwartz, eds. Ecology and management of the North American moose. Smithsonian Institution Press, Washington, DC.

Schwartz, C. C., and B. Bartley. 1991. Reducing incidental moose mortality: Considerations for management. Alces 27:227–31.

Schwartz, C. C., and K. J. Hundertmark. 1993. Reproductive characteristics of Alaskan moose. Journal of Wildlife Management 57:454–68.

Schwartz, C. C., and L. A. Renecker. 1998. Nutrition and energetics. Pages 441–78 *in* A. W. Franzmann and C. C. Schwartz, eds. Ecology and management of the North American moose. Smithsonian Institution Press, Washington, DC.

Schwartz, C. C., W. L. Regelin, and A. W. Franzmann. 1982. Male moose successfully breed as yearlings. Journal of Mammalogy 63:334–35.

Schwartz, C. C., W. L. Regelin, and A. W. Franzmann. 1987. Seasonal weight dynamics of moose. Swedish Wildlife Research Supplement 1:301–10.

Schwartz, C. C., W. L. Regelin, and A. W. Franzmann. 1988a. Estimates of digestibility of birch, willow, and aspen mixtures in moose. Journal of Wildlife Management 52:33–37.

Schwartz, C. C., W. L. Regelin, A. W. Franzmann, R. G. White, and D. F. Holleman. 1988b. Food passage rate in moose. Alces 24:97–101.

Schwartz, C. C., M. E. Hubbert, and A. W. Franzmann. 1988c. Energy requirements of adult moose for winter maintenance. Journal of Wildlife Management 52:26–33.

Schwartz, C. C., A. B. Bubenik, and R. Claus. 1990. Are sex-pheromones involved in moose breeding behavior? Alces 26:104–7.

Schwartz, C. C., M. E. Hubbard, and A. W. Franzmann. 1991. Energy expenditure in moose calves. Journal of Wildlife Management 55:391–93.

Schwartz, C. C., K. J. Hundertmark, and T. H. Spraker. 1992. An evaluation of selective bull harvest on the Kenai Peninsula, Alaska. Alces 28:1–13.

Seip, D. R. 1992. Factors limiting woodland caribou populations and their interrelationships with wolves and moose in southeastern British Columbia. Canadian Journal of Zoology 70:1494–1503.

Sergeant, D. E., and D. H. Pimlott. 1959. Age determination in moose from sectioned incisor teeth. Journal of Wildlife Management 23:315–21.

Seton, E. T. 1927. The lives of game animals, Vol. 4. Doubleday, Doran, New York.

Shelton, P. C., and R. O. Peterson. 1983. Beaver, wolf and moose interactions in Isle Royale National Park, USA. Acta Zoologica Fennica 174:265–66.

Shilton, C. M., D. A. Smith, L. W. Woods, G. J. Crawshaw, and H. D. Lehmkuhl. 2002. Adenoviral infection in captive moose (*Alces alces*) in Canada. Journal of Zoo and Wildlife Medicine 33:73–79.

Shipley, L. A., and D. E. Spalinger. 1992. Mechanics of browsing in dense food patches: Effects of plant and animal morphology on intake rate. Canadian Journal of Zoology 70:1743–52.

Shipley, L. A., and D. E. Spalinger. 1995. Influence of size and density of browse patches on intake rates and foraging decisions of young moose and white-tailed deer. Oecologia 104:112–21.

Shipley, L. A., S. Blomquist, and K. Danell. 1998. Diet choices made by free-ranging moose in northern Sweden in relation to plant distribution, chemistry, and morphology. Canadian Journal of Zoology 76:1722–33.

Shipley, L. A., A. W. Illius, K. Danell, N. T. Hobbs, and D. E. Spalinger. 1999. Predicting bite size selection of mammalian herbivores; a test of a general model of diet optimization. Oikos 84:55–68.

Shostak, A. W., and W. M. Samuel. 1984. Moisture and temperature effects on survival and infectivity of first-stage larvae of *Parelaphostrongylus odocoilei* and *P. tenuis* (Nematoda: Metastrongyloidae). Journal of Parasitology 70:261–69.

Siepierski, S. J., C. E. Tanner, and J. A. Embil. 1990. Prevalence of antibody to *Toxoplasma gondii* in moose (*Alces alces americana* Clinton) of Nova Scotia, Canada. Journal of Parasitology 76:136–38.

Sigouin, D., J.-P. Ouellet, and R. Courtois. 1997. Geographical variation in the mating and calving periods of moose. Alces 33:85–95.

Simberloff, D. 1998. Flagships, umbrellas, and keystones: Is single-species management passé in the landscape era? Biological Conservation 83:247–57.

Simkin, D. W. 1965. Reproduction and productivity of moose in northwestern Ontario. Journal of Wildlife Management 29:740–50.

Simkin, D. W. 1974. Reproduction and productivity of moose. Naturaliste Canadien 101:517–26.

Sinclair, A. R. E. 1983. The function of distance movements in vertebrates. Pages 240–58 *in* I. R. Swingland and P. J. Greenwood, eds. The ecology of animal movement. Clarendon Press, Oxford.

Skogland, T. 1985. The effects of density-dependent resource limitations on the demography of wild reindeer. Journal of Animal Ecology 54:359–74.

Skogland, T. 1991. What are the effects of predators on large ungulate populations? Oikos 61:401–11.

Slomke, A. M., M. W. Lankester, and W. J. Peterson. 1995. Infrapopulation dynamics of *Parelaphostrongylus tenuis* in white-tailed deer. Journal of Wildlife Diseases 31:125–35.

Smith, B. L. 1985. Moose and their management on the Wind River Indian Reservation, Wyoming. Alces 21:359–91.

Smith, S. M., D. M. Davies, and V. I. Golini. 1970. A contribution to the bionomics of the Tabanidae (Diptera) of Algonquin Park, Ontario: Seasonal distribution, habitat preferences, and biting records. Canadian Entomologist 102:1461–73.

Smith, T. E. 1992. Incidence of incisiform tooth breakage among moose from the Seward Peninsula, Alaska, USA. Alces Supplement 1:207–12.

Smits, C. M. M., R. M. P. Ward, and D. G. Larsen. 1994. Helicopter or fixed-winged aircraft: A cost–benefit analysis for moose surveys in Yukon Territory. Alces 30:45–50.

Snaith, T. V., and K. F. Beazley. 2002. Moose (*Alces alces americana* [Gray Linneaus Clinton] Peterson) as focal species for reserve design in Nova Scotia, Canada. Natural Areas Journal 22:235–40.

Snepenger, D. J., and R. T. Bowyer. 1990. Differences among nonresident tourists making consumptive and nonconsumptive uses of Alaskan wildlife. Arctic 43:262–66.

Snider, J. B., and M. W. Lankester. 1986. Rumen flukes (*Paramphistomum* spp.) in moose of northwestern Ontario. Alces 22:323–44.

Sokolov, V. E., and O. F. Chernova. 1987. Morphology of the skin of moose (*Alces alces* L.) . Swedish Wildlife Research Supplement 1:367–75.

Solberg, E. J., and B.-E. Sæther. 1993. Fluctuating asymmetry in the antlers of moose (*Alces alces*): Does it signal male quality? Proceeding of the Royal Society of London B 254:251–55.

Solberg, E. J., and B.-E. Sæther. 1994. Male traits as life-history variables: Annual variation in body mass and antler size in moose (*Alces alces*). Journal of Mammalogy 75:1069–79.

Soltys, M. A., C. E. Andress, and A. L. Fletch. 1967. Johne's disease in a moose (*Alces alces*). Wildlife Diseases Association Bulletin 3:183–84.

Spaeth, D. F., K. J. Hundertmark, R. T. Bowyer, P. S. Barboza, T. R. Stephenson, and R. O. Peterson. 2001. Incisor arcades of Alaskan moose: Is dimorphism related to sexual segregation? Alces 37:217–26.

Spaeth, D. F., R. T. Bowyer, T. R. Stephenson, P. S. Barboza, and V. Van Ballenberghe. In press. Nutritional quality of willows for moose: Effects of twig age and diameter. Alces.

Spalinger, D. E. 2000. Nutritional ecology. Pages 108–39 *in* S. Demarais and P. R. Krausman, eds. Ecology and management of large mammals in North America. Prentice-Hall, Upper Saddle River, NJ.

Spencer, D. L., and E. F. Chatelain. 1953. Progress in the management of the moose of south-central Alaska. Transactions of the North American Wildlife Conference 18:539–52.

Spencer, D. L., and J. Hakala. 1964. Moose and fire on the Kenai. Proceedings of the Tall Timbers Fire Conference 3:10–33.

Staaland, H., and R. G. White. 2001. Regional variation in mineral contents of plants and its significance for migration by arctic reindeer and caribou. Alces 37:497–509.

Steele, D. G, and W. D. Parama. 1979. Supernumerary teeth in moose and variations in tooth number in North American Cervidae. Journal of Mammalogy 60:852–54.

Steen, M., C. G. M. Blackmore, and A. Skorping. 1997. Cross-infection of moose (*Alces alces*) and reindeer (*Rangifer tarandus*) with *Elaphostrongylus alces* and *Elaphostrongylus rangiferi* (Nematoda, Protostrongylidae): Effects on parasite morphology and prepatent period. Veterinarian Parasitology 71:27–38.

Stenhouse, G. B., P. B. Latour, L. Kutny, N. MacLean, and G. Glover. 1995. Productivity, survival, and movements of female moose in a low-density population, Northwest Territories, Canada. Arctic 48:57–62.

Stenseth, N. C. 1983. Causes and consequences of dispersal in small mammals. Pages 63–101 *in* I. R. Swingland and P. J. Greenwood, eds. The ecology of animal movement. Clarendon Press, Oxford.

Stephens, P. W., and R. O. Peterson. 1984. Wolf-avoidance strategies of moose. Holarctic Ecology 7:239–44.

Stephenson, T. R., K. J. Hundertmark, C. C. Schwartz, and V. Van Ballenberghe. 1998a. Predicting body fat and body mass in moose with ultrasonography. Canadian Journal of Zoology 76:717–22.

Stephenson, T. R., V. Van Ballenberghe, and J. M. Peek. 1998b. Response of moose forages to mechanical cutting on the Copper River Delta, Alaska. Alces 34:479–94.

Stevens, D. R. 1970. Winter ecology of moose in the Gallatin Mountains, Montana. Journal of Wildlife Management 34:37–46.

Stewart, K. M., R. T. Bowyer, J. G. Kie, and W. C. Gasaway. 2000. Antler size relative to body mass in moose: Tradeoffs associated with reproduction. Alces 36:77–83.

Stewart, K. M., R. T. Bowyer, J. G. Kie, N. J. Cimon, and B. K. Johnson. 2002. Temporospatial distributions of elk, mule deer, and cattle: Resource partitioning and competitive displacement. Journal of Mammalogy 83:229–44.

Stock, T. M., and M. W. Barrett. 1983. Helminth parasites of the gastrointestinal tracts and lungs of moose (*Alces alces*) and wapiti (*Cervus elaphus*) from Cypress Hills, Alberta, Canada. Proceedings of the Helminthological Society of Washington 50:246–51.

Stringham, S. F. 1974. Mother–infant relations in moose. Naturaliste Canadien 101:325–69.

Sunnerheim-Sjoberg, K., and M. Hamalainen. 1992. Multivariate study of moose browsing in relation to phenol pattern in pine needles. Journal of Chemical Ecology 18:659–72.

Suominen, O., K. Danell, and R. Bergstrom. 1999a. Moose, trees, and ground-living invertebrates: Indirect interactions in Swedish pine forests. Oikos 84:215–26.

Suominen, O., K. Danell, and J. P. Bryant. 1999b. Indirect effects of mammalian browsers on vegetation and ground-dwelling insects in an Alaskan floodplain. Ecoscience 6:505–10.

Sweanor, P. Y., and F. Sandegren. 1989. Winter range philopatry of seasonally migratory moose. Journal of Applied Ecology 26:25–33.

Tankersley, N. G., and W. C. Gasaway. 1983. Mineral lick use by moose in Alaska. Canadian Journal of Zoology 61:2242–49.

Telfer, E. S. 1967a. Comparison of moose and deer winter range in Nova Scotia. Journal of Wildlife Management 31:418–25.

Telfer, E. S. 1967b. Comparison of a deer yard and a moose yard in Nova Scotia. Canadian Journal of Zoology 45:485–90.

Telfer, E. S. 1970. Winter habitat selection by moose and white-tailed deer. Journal of Wildlife Management 34:553–59.

Telfer, E. S. 1972. Forage yield and browse utilization on logged areas in New Brunswick. Canadian Journal of Forest Research 2:346–50.

Telfer, E. S. 1978. Cervid distribution, browse and snow cover in Alberta. Journal of Wildlife Management 42:352–61.

Telfer, E. S. 1984. Circumpolar distribution and habitat requirements of moose (*Alces alces*). Pages 145–82 *in* R. Olson, R. Hastings, and F. Geddes, eds.

Northern ecology and resource management. University of Alberta Press, Edmonton, Canada.
Telfer, E. S., and A. Cairns. 1978. Stem breakage by moose. Journal of Wildlife Management 42:639–42.
Telfer, E. S., and A. Cairns. 1986. Resource use by moose versus sympatric deer, wapati and bison. Alces 22:113–37.
Telfer, E. S., and J. P. Kelsall. 1979. Studies of morphological parameters affecting ungulate locomotion in snow. Canadian Journal of Zoology 57:2153–59.
Telfer, E. S., and J. P. Kelsall. 1984. Adaptations of some large North American mammals for survival in snow. Ecology 65:1828–34.
Testa, J. W. 1998. Compensatory response to changes in calf survivorship: Management consequences of a reproductive cost in moose. Alces 34:107–15.
Testa, J. W., and G. P. Adams. 1998. Body condition and adjustments to reproductive effort in female moose (*Alces alces*). Journal of Mammalogy 79:1345–54.
Testa, J. W., E. F. Becker, and G. R. Lee. 2000a. Movements of female moose in relation to birth and death of calves. Alces 36:155–62.
Testa, J. W., E. F. Becker, and G. R. Lee. 2000b. Temporal patterns in the survival of twin and single moose (*Alces alces*) calves in southcentral Alaska. Journal of Mammalogy 81:162–68.
Theberge, J. B. 1990. Potentials for misinterpreting impacts of wolf predation through prey:predator ratios. Wildlife Society Bulletin 18:188–92.
Thomas, J. E., and D. G. Dodds. 1988. Brainworm, *Parelaphostrongylus tenuis,* in moose, *Alces alces,* and white-tailed deer, *Odocoileus virginianus,* of Nova Scotia. Canadian Field-Naturalist 102:639–42.
Thompson, I. D., and D. L. Euler. 1987. Moose habitat in Ontario: A decade of change in perception. Swedish Wildlife Research Supplement 1:181–93.
Thompson, I. D., and R. W. Stewart. 1998. Management of moose habitat. Pages 377–401 *in* A. W. Franzmann and C. C. Schwartz, eds. Ecology and management of North American moose. Smithsonian Institution Press, Washington, DC.
Thompson, I. D., and M. F. Vukelich. 1981. Use of logged habitats in winter by moose cows with calves in northeastern Ontario. Canadian Journal of Zoology 59:2103–14.
Thorne, E. T., and R. F. Honess. 1982. Diseases of wildlife in Wyoming, 2nd ed. Wyoming Game and Fish Department, Cheyenne.
Timmermann, H. R. 1993. Use of aerial surveys for estimating and monitoring moose populations: A review. Alces 29:35–46.
Timmermann, H. R. In press. The status of moose management in North America, circa 2001. Alces.
Timmermann, H. R., and M. E. Buss. 1995. The status and management of moose in North America, early 1990's. Alces 31:1–14.
Timmermann, H. R., and M. E. Buss. 1998. Population and harvest management. Pages 559–615 *in* A. W. Franzmann and C. C. Schwartz, eds. Ecology and management of North American moose. Smithsonian Institution Press, Washington, DC.
Timmermann, H. R., and G. D. Racey. 1989. Moose access routes to aquatic feeding site. Alces 25:104–11.
Timmermann, H. R., and R. S. Rempel. 1998. Age and sex structure of hunter harvested moose under two harvest strategies in northcentral Ontario. Alces 34:21–30.
Timmermann, H. R., M. W. Lankester, and A. B. Bubenik. 1985. Morphology of the bell in relation to sex and age of moose (*Alces alces*). Alces 21:419–46.
Timmermann, H. R., M. W. Lankester, and A. B. Bubenik. 1988. Vascularization of the moose bell. Alces 24:90–96.
Todesco, C. J., H. G. Cumming, and J. G. McNicol. 1985. Winter moose utilization of alternate strip cuts and clearcuts in northwestern Ontario: Preliminary results. Alces 21:447–74.
Trainer, D. O., and M. M. Jochim. 1969. Serologic evidence of bluetongue in wild ruminants of North America. American Journal of Veterinary Research 30:2007–11.
Treble, R. G., and T. S. Thompson. 1998. Trace metals in moose (*Alces alces*) liver. Bulletin of Environmental Contamination and Toxicology 60:531–37.
Tyers, D. B., and L. R. Irby. 1995. Shiras moose winter habitat use in the upper Yellowstone River Valley prior to and after the 1988 fires. Alces 31:35–43.
Udina, I. G., A. A. Danilkin, and G. G. Boeskorov. 2002. Genetic diversity of moose (*Alces alces* L.) in Eurasia. Russian Journal of Genetics 38:951–57.
Upshall, S. M., M. D. B. Burt, and T. G. Dilworth. 1987. *Parelaphostrongylus tenuis* in New Brunswick: The parasite in white-tailed deer (*Odocoileus virginianus*) and moose (*Alces alces*). Journal of Wildlife Diseases 23:682–85.
Van Ballenberghe, V. 1977. Migratory behavior of moose in southcentral Alaska. International Congress of Game Biologists 13:103–9.
Van Ballenberghe, V. 1983a. Growth and development of moose antlers in Alaska. Pages 37–48 *in* R. D. Brown, ed. Antler development in Cervidae. Caesar Kleberg Wildlife Research Institute, Kingsville, TX.
Van Ballenberghe, V. 1983b. Rate of increase in moose populations. Alces 19:98–117.
Van Ballenberghe, V. 1987. Effects of predation on moose numbers: A review of recent North American studies. Swedish Wildlife Research Supplement 1:431–60.
Van Ballenberghe, V. 1992. Behavioral adaptations of moose to treeline habitats in subarctic Alaska. Alces Supplement 1:193–206.
Van Ballenberghe, V., and W. B. Ballard. 1994. Limitation and regulation of moose populations: The role of predation. Canadian Journal of Zoology 72:2071–77.
Van Ballenberghe, V., and W. B. Ballard. 1998. Population dynamics. Pages 223–45 *in* A. W. Franzmann and C. C. Schwartz, eds. Ecology and management of North American moose. Smithsonian Institution Press, Washington, DC.
Van Ballenberghe, V., and D. G. Miquelle. 1990. Activity of moose during spring and summer in interior Alaska. Journal of Wildlife Management 54:391–96.
Van Ballenberghe, V., and D. G. Miquelle. 1993. Mating in moose: Timing, behavior and male access patterns. Canadian Journal of Zoology 71:1687–90.
Van Ballenberghe, V., and D. G. Miquelle. 1996. Rutting behavior of moose in central Alaska. Alces 32:109–30.
Van Ballenberghe, V., D. G. Miquelle, and J. G. MacCracken. 1989. Heavy utilization of woody plants by moose during summer in Denali National Park, Alaska. Alces 25:31–35.
Van Dyke, F., B. L. Probert, and G. M. Van Beek. 1995. Seasonal habitat use characteristics of moose in southcentral Montana. Alces 31:15–26.
Ver Hoef, J. 2002. Sampling and geostatistics for spatial data. Ecoscience 9:152–61.
Ver Hoef, J. M., and R. P. Barry. 1998. Constructing and fitting models for cokriging and multivariable spatial prediction. Journal of Statistical Planning and Inference 69:275–94.
Vivas, H. J., and B.-E. Sæther. 1987. Interactions between a generalist herbivore, the moose *Alces alces,* and its food resources: An experimental study of winter foraging behaviour in relation to browse availability. Journal of Animal Ecology 56:509–20.
Vivas, H. J., B.-E. Sæther, and R. Andersen. 1991. Optimal twig-size selection of a generalist herbivore, the moose *Alces alces*: Implications for plant–herbivore interactions. Journal of Animal Ecology 60:395–408.
Vucetich, J. A., R. O. Peterson, and C. L. Schaefer. 2002. The effect of prey and predator densities on wolf predation. Ecology 83:3003–13.
Wallis de Vries, M. F. 1995. Large herbivores and the design of large-scale nature reserves in western Europe. Conservation Biology 9:25–33.
Ward, R. M. P., W. C. Gasaway, and M. M. Dehn. 2000. Precision of moose density estimates derived from stratification survey data. Alces 36:197–204.
Weckerly, F. W. 1992. Territoriality in North American deer: A call for a common definition. Wildlife Society Bulletin 20:228–31.
Weckerly, F. W. 1998. Sexual size dimorphism: Influence of mass and mating systems in the most dimorphic mammals. Journal of Mammalogy 79:33–52.
Weixelman, D. A., R. T. Bowyer, and V. Van Ballenberghe. 1998. Diet selection by Alaskan moose during winter: Effects of fire and forest succession. Alces 34:213–38.
Welch, D. A., W. M. Samuel, and C. J. Wilke. 1991. Suitability of moose, elk, mule deer, and white-tailed deer as hosts for winter ticks (*Dermacentor albipictus*). Canadian Journal of Zoology 69:2300–2305.
Welch, I. D., A. R. Rodgers, and R. S. McKinley. 2000. Timber harvest and calving site fidelity of moose in northwestern Ontario. Alces 36:93–103.
White, K. S., J. W. Testa, and J. Berger. 2001. Behavioral and ecologic effects of differential predation pressure on moose in Alaska. Journal of Mammalogy 82:422–29.
White, R. G., and J. R. Luick. 1984. Plasticity and constraints in the lactational strategy of reindeer and caribou. Symposia of the Zoological Society of London 51:215–32.
Whitlaw, H. A., and M. W. Lankester. 1994a. A retrospective evaluation of the effects of parelaphostrongylosis on moose populations. Canadian Journal of Zoology 72:1–7.
Whitlaw, H. A., and M. W. Lankester. 1994b. The co-occurrence of moose, white-tailed deer and *Parelaphostrongylus tenuis* in Ontario. Canadian Journal of Zoology 72:819–25.
Whittle, C. L., R. T. Bowyer, T. P. Clausen, and L. K. Duffy. 2000. Putative pheromones in urine of rutting male moose (*Alces alces*): Evolution of honest advertisement? Journal of Chemical Ecology 26:2747–62.

Williams, E. S., E. T. Thorne, and H. A. Dawson. 1984. Malignant catarrhal fever in a Shiras moose (*Alces alces shirasi* Nelson). Journal of Wildlife Diseases 20:230–32.
Williams, E. S., M. W. Miller, T. J. Kreeger, R. H. Kahn, and E. T. Thorne. 2002. Chronic wasting disease of deer and elk: A review with recommendations for management. Journal of Wildlife Management 66:551–63.
Wilton, M. L., and D. L. Garner. 1991. Preliminary findings regarding elevation as a major factor in moose calving site selection in south central Ontario, Canada. Alces 27:111–17.
Wishart, W. D. 1990. Velvet antlered female moose (*Alces alces*). Alces 26:64–65.
Wolfe, M. L. 1969. Age determination in moose from cemental layers of molar teeth. Journal of Wildlife Management 33:428–31.
Wolfe, M. L. 1974. An overview of moose coactions with other animals. Naturaliste Canadien 101:437–56.
Wolfe, M. L. 1987. An overview of the socioeconomics of moose in North America. Swedish Wildlife Research Supplement 1:659–75.
Wolff, J. O. 1978. Burning and browsing effects on willow growth in interior Alaska. Journal of Wildlife Management 42:135–40.
Wolff, J. O. 1980. Moose snowshoe hare competition during peak hare densities. Proceedings of the North American Moose Conference and Workshop 16:238–54.
Wood, T. J. 1974. Competition between arctic hares and moose in Gros Morne National Park, Newfoundland. Proceedings of the North American Moose Conference and Workshop 10:215–37.
Woodin, H. E. 1956. The appearance of a moose rutting ground. Journal of Mammalogy 37:458–59.
Worley, D. E. 1975. Observations on epizootiology and distribution of *Elaeophora schneideri* in Montana ruminants. Journal of Wildlife Diseases 11:486–88.
Worley, D. E., C. K. Anderson, and K. R. Greer. 1972. Elaeophorosis in moose from Montana. Journal of Wildlife Diseases 8:242–44.
Young, W. G., and T. M. Marty. 1986. Wear and microwear on the teeth of a moose (*Alces alces*) population in Manitoba, Canada. Canadian Journal of Zoology 64:2467–79.
Zarnke, R. L., R. A. Dieterich, K. A. Neiland, and G. Ranglack. 1983. Serologic and experimental investigations of contagious ecthyma in Alaska. Journal of Wildlife Diseases 19:170–74.
Zarnke, R. L., W. M. Samuel, A. W. Franzmann, and R. Barrett. 1990. Factors influencing the potential establishment of the winter tick (*Dermacentor albipictus*) in Alaska. Journal of Wildlife Diseases 26:412–15.
Zarnke, R. L., J. P. Dubey, O. C. H. Kwok, and J. M. Ver Hoef. 2000. Serologic survey for *Toxoplasma gondii* in selected wildlife species from Alaska. Journal of Wildlife Diseases 36:219–24.
Zarnke, R. L., H. Li, and T. B. Crawford. 2002. Serum antibody prevalence of malignant catarrhal fever viruses in seven wildlife species from Alaska. Journal of Wildlife Diseases 38:500–504.

R. Terry Bowyer, Institute of Arctic Biology, University of Alaska, Fairbanks, Fairbanks, Alaska 99775-7000. Email: ffrtb@uaf.edu.

Victor Van Ballenberghe, 8941 Winchester Street, Anchorage, Alaska, 99570. Email: vicvanb@alaska.com.

John G. Kie, U.S. Forest Service, Forestry and Range Science Laboratory, 1401 Gekeler Lane, LaGrande, Oregon 97850. Email: jkie@fs.fed.us.

46

Caribou

Rangifer tarandus

Frank L. Miller

NOMENCLATURE

COMMON NAMES. Caribou, reindeer, "deer"
SCIENTIFIC NAME. *Rangifer tarandus*

Although caribou and reindeer are classified taxonomically as a single Holarctic species, I believe that there are important behavioral and range-use differences between them, especially between domesticated reindeer and free-ranging caribou, which would meaningfully influence their respective conservation and management. Therefore, I have made a distinction between caribou and reindeer throughout this chapter, with a few exceptions.

SUBSPECIES. *R. t. groenlandicus,* barren-ground caribou or American tundra reindeer; *R. t. granti,* Alaskan barren-ground caribou or Grant's caribou; *R. t. caribou,* American woodland caribou or woodland caribou; *R. t. pearyi,* Peary caribou or Peary reindeer (Banfield 1961).

The Queen Charlotte Island's caribou (*R. t. dawsoni*) probably became extinct shortly after 1910 (Banfield 1963). The causes for its extinction are unknown, but supposedly habitat deterioration through climate moderation and loss of genetic plasticity through isolation likely were more important than hunting or other human interference (Banfield 1963).

Caribou originated from South American deer of the Tertiary age (65 to 2 million years ago), which gave rise to two North American groups, *Navahoceros* (a "mountain deer" with short, stocky limbs and simple three-tined antlers) and *Rangifer* (Harington 1999). The oldest known caribou are from the early Pleistocene of Eastern Beringia (Harington 1999). Subsequently, *Rangifer* became the only member of the deer family (Cervidae) to adapt to life in the harsh Arctic and Subarctic of North America, and, like reindeer in Eurasia (*R. t. tarandus* and *R. t. fennicus*), to have year-round populations north of the treeline. The terms "population" and "herd" are used synonymously throughout this chapter, more or less in accordance with their respective use by previous authors, but clear distinctions are needed (see Current Research and Management Needs). Caribou remains from Ice Age Beringia are common. Beringian caribou gave rise to the barren-ground race of caribou, which apparently includes the Peary caribou and other arctic-island caribou offshoots. Woodland caribou arose from caribou that took refuge in a broad region paralleling the front of the Laurentide ice sheet in the continental United States (Banfield 1961; Harington 1999).

The caribou is a primitive cervid (Banfield 1961, 1974). Features that suggest the primitive nature of *Rangifer,* if not secondarily acquired, include antlers on both sexes, long metapodial bones, well-marked tarsal and interdigital glands, and relatively simple crests on the cheek teeth (Banfield 1974). Banfield's (1961) revision of *Rangifer* is an excellent source for vernacular names and previous revisions; supplementary information is in Kelsall (1968).

Genetic relationships among many North American caribou populations were first investigated by examining frequencies of transferrin alleles in blood serum (Røed and Whitten 1986; Røed et al. 1986, 1991; Røed and Thomas 1990; Røed 1991). Findings from those studies generally agreed with Banfield's (1961) subspecific classifications. Barren-ground caribou and woodland caribou were well separated: the large genetic distance in transferrin loci among subspecies supported survival during the Wisconsin glaciation in different refugia. The belief that Peary caribou had survived in a northern refugium other than Beringia and recolonized their former High Arctic range after the last Great Ice Age remained intact (Røed et al. 1986; Røed 1991), but was also questioned (Røed et al. 1991). A south-to-north cline among caribou from the mainland to the High Arctic Islands continued support for the belief that those caribou on southern Arctic Islands resulted from mainland barren-ground caribou interbreeding with Peary caribou (Manning 1960; Banfield 1961; Thomas and Everson 1982; Røed et al. 1986).

Investigations of caribou genetics using mitochondrial DNA (mtDNA) and nuclear DNA (microsatellites) challenge the conventional caribou taxonomy (and at least some of the findings from transferrin alleles). Recent limited mtDNA analyses reveal genetic relationships different from established taxa. Investigation of the genetic relationship between barren-ground and woodland caribou yielded two distinct mtDNA clades: a monophyletic or paraphyletic northern (barren-ground) clade and a monophyletic southern (woodland) clade, which only correspond in part with existing subspecific classifications (Dueck 1998). Nearly all of the individual barren-ground caribou sampled from populations classified taxonomically as barren-ground caribou belonged to the northern mtDNA clade, but some were from the southern (woodland) clade. Contrary to existing classifications for woodland caribou, all individuals from Yukon herds; most from British Columbia and Alberta; some from northern Labrador, Quebec's Ungava Peninsula, and northeastern Ontario; and a few from Saskatchewan belonged to the northern (barren-ground) mtDNA clade (Dueck 1998). All caribou sampled from Newfoundland and southern Ontario fell in the southern (woodland) clade (Dueck 1998). Nearly all of the caribou sampled from the South Purcell woodland caribou herd, just above the United states–Canada border in British Columbia, actually belonged to the northern (barren-ground) mtDNA clade (Dueck 1998), an extreme example of misidentification of the origin of those caribou. Contrary to Banfield (1961), the findings for the South Purcell herd, along with findings for the other woodland herds in British Columbia and the Yukon, suggest that it was more a matter of barren-ground caribou penetrating southward than woodland caribou moving north after the last glaciation in the western part of the continent.

Other studies using mtDNA indicate that subspecific recognition of Peary caribou and arctic-island caribou solely on the basis of genetics is not warranted, as those caribou have a polyphyletic origin (Gravlund et al. 1998; J. Eger, Royal Ontario Museum, Toronto, pers. commun., 1996; A. Gunn, Department of Resources, Wildlife and Economic Development, Government of the Northwest Territories, Yellowknife, Canada, pers. commun., 2000). Microsatellite analyses of nuclear DNA to investigate genetic relatedness and levels of genetic diversity will continue to resolve these questions (C. Strobeck and K. Zittlau, Department of Biological Sciences, University of Alberta, Edmonton, Alberta, Canada, pers. commun. 2000).

Current taxonomy alone does not allow us to clearly understand what form of caribou we are attempting to manage, conserve, or preserve. Geist (1991) argued that the taxonomic criteria that would reflect genetic differences in caribou are the "social insignia" exhibited during the rut: hair, color, gland patterns, and to a lesser extent antlers. He believed that a subspecies can be defined by the same body characteristics. Such an approach requires extremely strict standards and categories for making visual judgments.

It has been suggested that ecotyping would lead to a functional classification system, and some initial efforts have been made (Bergerud 1978b; Edmonds 1988, 1991; Davis and Valkenburg 1991; Thomas 1995; Mallory and Hillis 1998; Seip 1998). Although a consideration of caribou populations by ecotype is useful, it does not completely resolve the matter of fully identifying the population(s) under consideration. A truly functional biological and ecological system requires a composite approach. This composite should start with recognition of the genetic stock as a necessary first step toward biologically meaningful separation. There are important distinctions, with application to herds across the North American continent. We have no way of knowing the probable importance of the genetic contribution in terms of population dynamics. That is, are there any meaningful differences between barren-ground caribou living like woodland caribou versus woodland caribou living like woodland caribou in terms of their respective collective ability to cope with the stresses of their environment? Essentially, it is the age-old question: What counts the most, genetics or environment? And can the two ever be isolated for evaluation of their relative importance in different environments?

Ideally, the second stage of the composite approach would involve ecotyping each geographic population (grouping populations if each cannot be separated) within each jurisdiction, using an array of ecologically meaningful criteria. Such a system should clearly identify special exceptions, whether they involve major or minor populations, as any fully functional classification system must be able to recognize such special exceptions.

It is inherently weak to rely on a single taxonomic system, the as-yet-limited DNA information, and the levels of ecotyping proposed to date. A functional system most likely is a composite approach involving genetics; morphological, ecological, and behavioral adaptations of the individuals; the environmental setting; predator–prey relationships; and other circumstances of each population within each region. Unfortunately, we may not yet have the complete criteria necessary for such a workable composite approach to classifying caribou populations within each jurisdiction. Most jurisdictions do not have the necessary biological and ecological information for each herd, or necessary funding. Thus, our current state of knowledge best supports the simplistic conclusion that there are different gene pools exposed to different environmental settings. This results in several ecotypes of North American *Rangifer,* which may or may not originate from different genetic stocks (clades). Accordingly, each of the major caribou populations within each geographic region warrants management, conservation, and preservation for the maintenance of maximum biodiversity among North American *Rangifer* and each geographic population should serve as the basic management unit.

DISTRIBUTION AND ABUNDANCE

Until the beginning of the twentieth century, essentially all of the northern half of North America, except prairie areas, was potential caribou range. Then, human settlement, logging, fires, northward advancement of the white-tailed deer (*Odocoileus virginianus*), and introduction of domestic livestock pushed the southern boundary northward in the boreal forest across the continent. Caribou were all but lost from the contiguous 48 United States, except for a few small, isolated pockets near the Canada–United States border and a montane projection into northern Washington, Idaho, and extreme northwestern Montana.

In the early 1990s, there were about 2.2 million North American caribou (e.g., Ferguson and Gauthier 1992). Updated estimates by jurisdiction (Table 46.1) suggest that number has increased to about 4.0 million. These caribou occur in nearly 200 recognized herds (populations or subpopulations) that range in size from <100 to >800,000 animals. Details of individual herd size, within- and among-year variation in numbers, and other details are given in the references listed in Table 46.1 and their respective literature citations.

TABLE 46.1. Approximate overall estimates of the recent numbers of North American caribou (*Rangifer tarandus*), by jurisdiction

Jurisdiction[a]	Approximate Estimate[b] (×1000)	Compilation Dates[c]	Source[d]
Northwest Territories/Nunavut[e]	1827	1985, 1991, 1994, 1998	1, 2, 3
Quebec/Labrador[f,g]	1107	1991, 1993	2, 4, 5
Alaska[f,h]	904	1995, 2000	6, 7
Insular Newfoundland[f]	90	2000	8
Yukon[f,i]	32	1997	9
Ontario[f]	21	1996	10
British Columbia[f]	18	1996	11
Manitoba[f]	13	1991, 1994	2, 12
Alberta[f]	5	1991, 1996	13, 14
Saskatchewan[f]	3	1984, 1996	15, 16

[a]Some major herds are shared between or among jurisdictions (see source references in note d below for specific details by individual herd).

[b]Based on mean point values when an estimated range was given or a composite of multiple sources when necessary to update overall estimate; estimates rounded up to the nearest 1000 animals.

[c]When not identified, publication date is used.

[d]SOURCES: 1. Williams and Heard 1986; 2. Ferguson and Gauthier 1992; 3. A. Gunn, Department of Resources, Wildlife and Economic Development, Fisheries and Wildlife Division, Government of Northwest Territories, Yellowknife, Northwest Territories, pers. commun., 2000; 4. Couturier et al. 1996; 5. S. Couturier, Quebec Wildlife and Parks, pers. commun., 2000; 6. Valkenburg 1998; 7. P. Valkenburg, Alaska Department of Fish and Game, Fairbanks, Alaska, pers. commun., 2000; 8. C. M. Doucet, Newfoundland and Labrador Department of Forest Resources and Agrifoods, Wildlife Branch, pers. commun., 2001; 9. Farnell et al. 1998; 10. Cumming 1998; 11. Heard and Vagt 1998; 12. Scholten 1994 in Abraham and Thompson 1998; 13. Edmonds 1991; 14. Edmonds 1998; 15. Kelsall 1984; 16. Rettie et al. 1998.

[e]Includes only Canadian form of barren-ground caribou (*R. t. groenlandicus*).

[f]Woodland caribou (*R. t. caribou*) only.

[g] It is probable that the number of caribou in Quebec was markedly reduced by the year 2000, with the George River caribou herd likely at only about half its 1993 maximum mean estimate of 823,000 animals (S. Couturier, Quebec Wildlife and Parks, pers. commun., 2000).

[h]Includes only Alaskan form of barren-ground caribou (*R. t. granti*), which includes the four herds shared with Canada.

[i]Includes herds shared with Northwest Territories, British Columbia, and Alaska.

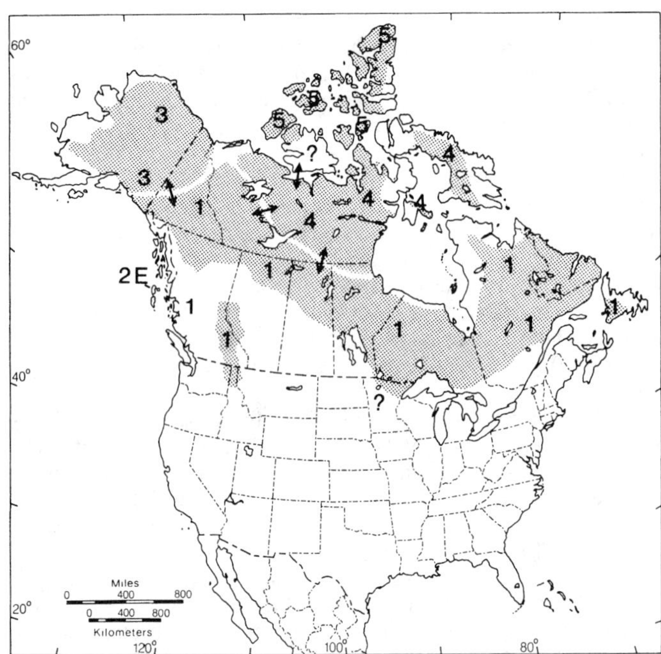

FIGURE 46.1. Distribution of caribou (*Rangifer tarandus*): 1, *R. t. caribou;* 2, *R. t. dawsoni* (E = extinct); 3, *R. t. granti;* 4, *R. t. groenlandicus;* 5, *R. t. pearyi* (includes the range of arctic-island caribou). The actual boundaries between the subspecies are not well defined; seasonal overlap did and probably still does occur in many areas. Arrows indicate areas of probable subspecific overlap. SOURCE: After Banfield (1974).

Woodland Caribou. Woodland caribou occupy the boreal forest and alpine tundra extensions of suitable mountainous habitats. They once occurred in Maine, New Hampshire, Vermont, Michigan, Minnesota, New Brunswick, Nova Scotia, and Prince Edward Island. The range of the woodland caribou has decreased considerably since the 1800s, probably because of destruction of the climax forests and overhunting. Northward extension of white-tailed deer following loss of the climax forests likely also contributed to the elimination of woodland caribou from the southern portion of their range. These deer carry "moose sickness," the parasitic meningeal worm that is fatal to woodland caribou: *Elaphostrongylus tenuis* of Smith et al. (1964) or *Pneumostrongylus tenuis* of Anderson (1971).

Now their range has been pushed northward to Newfoundland; Quebec north of the St. Lawrence River, except in the Shickshock Mountains of the Gaspe Peninsula; Ontario; Manitoba; Saskatchewan; and Alberta; they are also present as remnant populations in the Salmo River–Selkirk area of Idaho, and possibly Montana, Washington State, and British Columbia (Fig. 46.1) (Freddy and Erickson 1975; Johnson 1976; Freddy 1979), although currently, most likely only the Selkirk herd is extant. To the east, woodland caribou range extends north to the Ungava Labrador Peninsula and Newfoundland; to the west, to the District of Mackenzie, Northwest Territories, Yukon Territory, and the Copper River area of Alaska. Woodland caribou supposedly no longer exist on the Kenai Peninsula (Banfield 1961). Although the Chisana caribou herd, which is shared between Alaska and the Yukon, is listed as woodland caribou in the Yukon, they are still considered barren-ground caribou in Alaska. This creates an interesting dilemma, as all caribou in the Yukon supposedly belong to the northern (barren-ground) clade (Dueck 1998). R. Farnell (Yukon Department of Renewable Resources, Fish and Wildlife Branch, pers. commun., 2001) believed that both the genetics and the ecological behavior of caribou in the Chisana herd places them in the woodland caribou group. The range of woodland caribou is sympatric with those of white-tailed deer, black-tailed deer (*O. hemionus columbianus*), mule deer (O. *h. hemionus*), moose (*Alces alces*), and elk (*Cervus elaphus*).

Woodland caribou constitute about 32% of the estimated caribou in Table 46.1. Nearly 65% of them occur in one herd: the George River herd of Quebec/Labrador. This is or was the largest caribou herd in North America and likely the largest in the world (maximum mean estimate of 823,000 in 1993: Couturier et al., 1996). As a result, this herd has been studied extensively (Parker 1981; Messier et al. 1988; Huot 1989; Couturier et al. 1990, 1996; Crête et. al. 1990, 1991, 1993, 1996; Crête and Huot 1991, 1993). However, it is probable that the George River herd experienced a marked reduction of 50% or more by the year 2000 (S. Couturier, Quebec Wildlife and Parks, Quebec, pers. commun., 2000).

The Committee on the Status of Endangered Wildlife in Canada recognizes only six "populations" of woodland caribou grouped by different-sized geographic areas rather than individually identified herds within each jurisdiction. As of the year 2000, three of those groupings are listed as being at risk: (1) the Gaspésie population in Quebec is considered "endangered," whereas (2) the southern mountain populations in British Columbia and Alberta and (3) the boreal populations from British Columbia across Canada to Quebec and southern Labrador are all considered "threatened."

Barren-Ground Caribou. Barren-ground caribou of Alaska and Canada range over thousands of square kilometers of Arctic tundra. Some herds remain on the tundra year-round, whereas others migrate south into the boreal forest (or taiga) for the winter. Some barren-ground caribou that are in the boreal forests in autumn may intergrade with northern populations of woodland. As noted, some so-called woodland caribou are, in fact, barren-ground caribou living like woodland caribou (Dueck 1998). Barren-ground caribou share portions of their range with moose and muskoxen (*Ovibos moschatus*), and thinhorn sheep (*Ovis dalli*) and mountain goats (*Oreamnos americanus*) on some western ranges.

All of the caribou, about 2.7 million, listed for the Northwest Territories/Nunavut and Alaska in Table 46.1, are considered barren-ground caribou. They represent about 68% of all North American caribou: 67% by the Canadian form (*groenlandicus*) and 33% by the Alaskan form (*granti*). The largest herd of barren-ground caribou in Canada (496,000 in 1994) is currently the Qamanirjuaq herd (A. Gunn, Department of Resources, Wildlife and Economic Development, Government of the Northwest Territories, Yellowknife, pers. commun., 2000), formerly the Kaminuriak population. The largest herd (430,000 in 1999) in Alaska is the Western Arctic herd (Valkenburg 1998).

In Canada, barren-ground caribou occur on the Arctic Islands of the Baffin Island region, the islands in Foxe Basin and Hudson Bay, Victoria Island, and King William Island (e.g., Banfield 1961). On the mainland, they occur from west of Hudson Bay in the District of Keewatin, Nunavut (previously Northwest Territories), and northern Manitoba westward across the District of Mackenzie, Northwest Territories, and Nunavut, northern Saskatchewan, and occasionally northeastern Alberta to the Mackenzie River (Fig. 46.1).

In the 1970s, only nine herds or populations of barren-ground caribou were recognized in the Northwest Territories (including what is now Nunavut) (Calef 1978). Those herds include three that seasonally range into the provinces to the south: Bathurst, 349,000 in 1996; Beverly, 277,000 in 1994; and Qamanirjuaq (A. Gunn, Department of Resources, Wildlife and Economic Development, Government of the Northwest Territories, Yellowknife, pers. commun., 2000). At that time, based on aerial surveys, Calef (1978) suggested that there were then about 624,000 barren-ground caribou west of Hudson Bay and east of the Mackenzie River. Subsequently, the overall estimated number of Canadian barren-ground caribou rose to over 1 million in the late 1980s. Densities appeared to be increasing into the 1990s (Williams and Heard 1986; Ferguson and Gauthier 1992), reaching about 1.8 million in 21 herds in the mid-1990s (Table 46.1) (A. Gunn, Department of Resources, Wildlife and Economic Development, Government of the Northwest Territories, Yellowknife, pers. commun., 2000). Continuing increase in the number of recognized herds in this jurisdiction and throughout North America results from more investigation and advances in radio and satellite telemetry and apparently not from actual changes in the number of herds present.

Closeness of the timing between surveys of the same herds and repeated use of survey results for the estimates of Canadian barren-ground caribou (Williams and Heard 1986; Ferguson and Gauthier 1992) do not allow appreciation for how these herds appear to change in size over short time intervals (a few years, and sometimes just from one year to the next). Subsequent mean estimates from aerial surveys of some of these herds do, however, suggest probable major increases, if we ignore imprecision of the estimates. Thomas (1998d) offered a good review of the problems of imprecision and questionable accuracy in aerial survey results for the Beverly and the Qamanirjuaq herds. Supposed major changes in herd size based on mean population estimates for these two herds (Thomas 1998d) are suspect and occasionally biologically impossible based on reproduction. In addition, confidence intervals overlap within either set of estimates. It is no surprise that caribou hunters, especially Native users, question the value of these survey estimates. For example, the estimate for the Qamanirjuaq herd in 1994 was between 310,000 and 682,000 caribou. A population estimate range of 372,000 inspires little confidence. Most important, the herd would have to decrease or increase by about 200,000 animals before we could say with statistical confidence that the change was real.

At the Mackenzie River, the Canadian form of barren-ground caribou *(groenlandicus)* becomes and apparently sometimes intergrades with the Alaskan form (*granti*). The Alaskan barren-ground caribou ranges west and north across the Yukon Territory, onto the North Slope of Alaska, and throughout Arctic Alaska. Its survival as a pure stock on the Alaskan Peninsula is questionable, however, because of extensive interbreeding with introduced domestic reindeer (Banfield 1961). Earlier estimates placed the number of Alaskan barren-ground caribou at about 600,000 (Skoog 1968; Hemming 1971). The estimated number then decreased to about 240,000 (Davis 1978). Overhunting and predation by wolves (*Canis lupus*) have been suggested as predominent causes for that decline in the 1970s (Davis et al. 1978; Valkenburg et al. 1994). Most recently, during the last two decades, the overall estimated number of Alaskan barren-ground caribou has increased by nearly 200% and is approaching 1 million (Table 46.1) (Davis 1980; Williams and Heard 1986; Valkenburg 1998; P. Valkenburg, Alaska Department of Fish and Game, Fairbanks, pers. commun., 2000). Densities vary markedly among and within herds over time. Two herds among the 33 currently recognized in Alaska—the Western Arctic herd (430,000 in 1999) and the Porcupine herd (130,000 in 1998)—have consistently contributed 63–70% of all the individuals estimated in Alaska (Davis 1980; Williams and Heard 1986; Valkenburg 1998; P. Valkenburg, Alaska Department of Fish and Game, Fairbanks, pers. commun., 2000). The Mulchatna herd (200,000 in 1994) has exhibited the most rapid rate of growth (17% annually since the mid-1970s), however, and has become the second largest caribou herd in Alaska (Van Daele, pers. commun., in Valkenburg 1998). Four of the Alaskan herds are shared with the Yukon—the Fortymile (34,600 in 2000), Nelchina (33,800 in 2000), Mentasta (350 in 1999), and Chisana (325 in 1999) herds—and one—the Porcupine herd—is shared with the Yukon and the Northwest Territories (Valkenburg 1998; P. Valkenburg, Alaska Department of Fish and Game, Fairbanks, pers. commun., 2000).

Peary Caribou and Arctic-Island Caribou. Peary caribou and arctic-island caribou occupy the Canadian Arctic Archipelago, except the Baffin Island region and the islands in Foxe Basin and Hudson Bay, where barren-ground caribou occur (Gates et al. 1986; Miller 1990b; Ferguson and Messier 2000). Of the 4.0 million estimated caribou, only about 0.1% are Peary caribou and 0.1% are arctic-island caribou. Peary caribou on the Queen Elizabeth Islands and arctic-island caribou on Banks Island are listed as endangered in Canada. The remaining arctic-island caribou on the other islands (and some on the Boothia Peninsula on the mainland) are listed as threatened.

The Peary caribou, or more likely a similar form of arctic-island caribou from Banks Island, has ranged as far south as the mainland coast of the Arctic Ocean in the Northwest Territories and Yukon (Youngman 1975) and supposedly has been found in Old Crow, Yukon. They were believed to have intergraded with the Canadian mainland form of barren-ground caribou (Banfield 1961; Manning 1960). However, DNA studies do not support intergradation (A. Gunn, Department of Resources, Wildlife and Economic Development, Government of the Northwest Territories, Yellowknife, pers. commun., 2000). In turn, some mainland barren-ground caribou have migrated north to the more southerly islands of the Arctic Archipelago and shared portions of summer ranges of arctic-island caribou. Most of the supposed overlap in range use had apparently ceased by the end of the 1920s, a period during which Native hunters supposedly took excessive harvests from migrating herds (Hoare 1927; Manning 1960). Caribou movements between southern islands and the coastal mainland were limited for about the next 40 years, then began increasing in the 1970s (Gunn et al. 2000).

The over estimate of Peary caribou (first range-wide aerial survey) on the Queen Elizabeth Islands (Fig. 46.1) in the Canadian Arctic Archipelago was about 26,000 in 1961 (Tener 1963). By the mid-1970s, the over estimate was about 4000 (Miller et al. 1977a). Total Peary caribou in the Canadian High Arctic probably does not now exceed 2000–3000 (Miller 1998). Melville, Prince Patrick, and Bathurst and their respective satellite islands of the western Queen Elizabeth Islands remain the heartland of Peary caribou. However, overall mean density is reduced by 20-fold (Miller 1998; Gunn et al. 2000; Gunn and Dragon 2002). Peary caribou on the western Queen Elizabeth Islands were estimated at about 24,000 in 1961 (Tener 1963), declined by about 89% in 1974, then recovered somewhat over the next 20 years before crashing between 1994 and 1997 to only 5% of their 1961 estimated over number (Miller et al. 1977a; Miller 1995b, 1998; Gunn et al. 2000; Gunn and Dragon 2002).

Arctic-island caribou (commonly referred to by some as Peary caribou or island caribou) on the southern tier of Arctic Islands (Banks, northwest Victoria, Prince of Wales, and Somerset) occurred at relatively high numbers in the 1970s and 1980s. In the 1990s, those populations were drastically reduced or almost disappeared (Urquhart 1973; Kevan 1974; Fischer and Duncan 1976; Gunn and Miller 1983; Jakimchuk and Carruthers 1983; Nagy et al. 1996; Miller 1997a; Gunn and Dragon 1998; J. Nagy, department of Resources, Wildlife and Economic Development, Government of the Northwest Territories, Inuvik, pers. commun., 2000). At their current numbers, none of the arctic-island caribou populations on the southern tier of the Arctic Islands can support the Native hunter's desired levels of annual harvest.

DESCRIPTION

Caribou are medium-sized deer with relatively long legs, large hooves, and broad muzzles. Their heads are elongated, with forehead-nose profiles that vary from almost straight to the "Roman" noses of some mature males. The muzzle is blunt, except in Peary caribou, and well haired except for the small oval rhinarium (Pocock 1923). Physical appearances of caribou differ among subspecies; many of the differences are given in Banfield (1961). Detailed descriptions of caribou are given by Dugmore (1913) for woodland caribou of Newfoundland, Murie (1935) for Alaskan–Yukon caribou, and Harper (1955) and Kelsall (1968) for barren-ground caribou of the Northwest Territories and Nunavut. Banfield (1961) described all of the North American subspecies (as well as the Eurasian subspecies of reindeer).

Size. Males weigh from about 110 kg for Peary caribou (season unknown; Banfield 1961), to 169 kg for barren-ground caribou in the Northwest Territories and the Arctic Slope of Alaska (Bergerud 1978a), to 299 kg for Alaskan prerut barren-ground caribou (Skoog 1968). Autumn weights for females usually vary between 91 and 136 kg (Bergerud 1978a). Seasonal differences in body weights, especially for adult males, are so great that much of the range of differences between males and females is lost. Generally, however, mature female caribou are about 10–15% smaller and weigh 10–50% less than adult males. Variation in whole body weights by sex, age, and season of the year is detailed in Dauphine (1976) and over winter by Thomas and Kiliaan (1998a) for Canadian barren-ground caribou. The heaviest female weighed 113 kg and the heaviest male 172 kg. Late-winter (March–April) mean weights for >3-year-old female and male Peary caribou varied were about 51–54 kg and 66–69 kg, respectively, and for arctic-island caribou were about 59–68 kg and 78–92 kg, respectively (Thomas et al. 1977). However, summer weights for some adult males of a large arctic-island form on Prince of Wales Island averaged 203 kg (Thomas and Everson 1982).

Pelage. The forest-dwelling woodland caribou have the darkest pelage. Peary caribou have the lightest, with more gray than brown in their coats. Barren-ground caribou have many variations in intermediate shades of color relative to the other two subspecies. The underparts of caribou are lighter, leg coloring and socks are pronounced, and flank stripes vary from prominent to lacking. Descriptions of pelages using standard color guides are given in Manning (1960) and Banfield (1961).

Each spring, winter pelage falls away in patches, revealing darker hair beneath and leaving the caribou with a ragged appearance. Prime bulls are usually the first to develop sleek summer coats, followed by juveniles and yearlings, by calfless cows, and last by maternal cows. By August, all but cows in poor condition are in new coats, which are the darkest. In autumn, the hair lengthens and longer, white-tipped guard hairs grow to form the winter coat. By prerut (September–early October), bulls have developed handsome white manes and the pelage shows the maximum contrast of an individual's coloration. As winter progresses, guard hairs are bleached and many of the hair tips are broken off, resulting in a general lightening and loss of contrast in the pelage. Finally, the winter coat falls away and the cycle begins again.

Calves are born with a light brown or red-brown body and a dark brown to black dorsal stripe from the neck to the tail. Their undersides are white or light gray (Kelsall 1968).

Antlers. Males and females have antlers, although the male's are larger and can be impressive. The main beams are long and curved toward the animal's posterior with a cylindrical to flattened cross section. Usually one and sometimes two brow tines are widely palmated (the "shovels") and the dominant tine extends vertically over the face. The brow and second (bez) tines extend anteriorly and the terminal tines extend posteriorly, and can be either pointed or palmate.

Antlers of females and young caribou are smaller and simpler. Females do, however, sometimes grow miniature replicas of the male's large antlers. The number of females in a herd with only one antler or no antlers ("genetically bald") varies considerably and sometimes can be relatively high (e.g., Bergerud 1976). Calves grow "spike" antlers, which remain in velvet for their first winter and spring.

Banfield (1961) followed Jacobi (1931) and separated barren-ground forms from woodland forms based on the cross-sectional shape of their antlers: cylindricornus versus compressicornus. However, only the color of the antler velvet consistently separates all woodland and barren-ground caribou from Peary and arctic-island caribou. The former have chocolate-brown velvet, whereas the latter have slate-gray velvet antler coverings.

Shapes of caribou antlers vary greatly, and no two sets are the same. One antler of a pair is not the mirror image of the other (Banfield 1954a). Variation among populations usually negates use of antlers in detailed taxonomic work (Banfield 1961), although Bubenik (1975b) argued for their use as a taxonomic characteristic. The annual cycle of antler development (Banfield 1954a; Harper 1955; Moisan 1959; Bubenik 1975b) varies with the sex, reproductive status, and age of the caribou.

Males that are active breeders shed their antlers at the end of the autumn–early-winter rut or shortly after. The role of testosterone in the antler cycle of caribou is minor, and under deep physical stress some older breeding bulls can drop their antlers before the end of the rut (Murie 1935; Bubenik 1975b). After casting their antlers, bulls immediately lose their sex drive and dominance and do not participate further in rutting activities (Espmark 1964a). Antlerless bulls begin foraging and recover their physical condition more rapidly than other prime bulls that participated in the entire rut.

Female caribou retain their antlers through the winter and most often shed them shortly before, after, or about the time of calving in this species (Lent 1965a; Skoog 1968; Bergerud 1976; Whitten 1995). Some cows, supposedly on good range, can shed their antlers and start new growth before they calve. Gagnon and Barrette (1992) found 13.5% shed their antlers ≥2 weeks before giving birth. Yearly and herd variation in the retention of antlers makes its use as an indicator of natality only an approximation. The presence of osteolytic resorption in antlers in December suggests that the retained antlers of females and juveniles could serve as a calcium bank during the winter (Belanger et al. 1967).

The annual cycle of antler velvet and antler shedding varies with sex and age. Bergerud (1976) illustrated the antler cycle for woodland caribou in Newfoundland. Like Skoog (1968), I have observed slight subspecific variations from woodland caribou in the annual antler cycle of barren-ground caribou: (1) mature bulls begin to grow antlers in early March, although a few carry hard antlers until late February; (2) yearlings begin to grow new antlers in late April or early May, although some carry old antlers until April or May; (3) many non-pregnant barren-ground females cast their antlers earlier than pregnant females and begin growing new antlers as early as March–April, but some carry last year's antlers until May or June; and (4) many pregnant barren-ground females start antler growth about 1 week after calving. Whereas many pregnant females carry old antlers to calving, some few cast their antlers as early as March–April (based on shot samples from the Qamanirjuaq herd; F. L. Miller, unpublished data, 1966–1968). Among-year and among-herd variation in timing of antler casting could reflect changes in range condition. On poor range in West Greenland, females with antlers were less likely to have calves than were females without antlers (Thing et al. 1986).

Hooves. Caribou hooves are well adapted to spongy muskegs and frozen snow-covered ground. Width of the hoof may be greater than length; dew claws are large and add greatly to the bearing surface of the foot. The hoof is curved inward and abrupt at the tip. In winter, the soft pads of the hoof deteriorate and the pad area becomes concave. The hoof becomes sharp edged, and hair between the hooves elongates to cover the pad. Those changes result in a spreading, "nonskid" support, which is effective on snow and ice (Kelsall 1968). A polydactylous hoof was found on a Canadian barren-ground caribou (Miller and Broughton 1971).

Skull and Dentition. Banfield (1961:26–27) described the skull characteristics (Fig. 46.2) of *Rangifer* as follows: "Cranium moderately expanded, extreme postorbital position of the pedicels which encroach on the parietal bone, nasals prominent and expanded proximally, preorbital pit in the lachrymal bone moderate in size, lachrymal vacuity generally moderate in size. Premaxillae prevented from reaching nasals by a small lobe on the maxillae."

Caribou have heterodont dentition (Loomis 1925; Frick 1937). The permanent dentition is subhypsodont, selenodont, and deerlike in general, but seemingly more adapted for grazing than browsing (Kelsall 1968; Skoog 1968). This dental adaptation has been questioned by Miller (1974a, 1974c) because of the varied feeding habits of caribou. The dental formula is I 0/3, C 1/1, P 3/3, M 3/3. The upper canine is not exposed.

Neonatal caribou have a set of functional deciduous milk teeth that are smaller than their permanent counterparts. The first permanent tooth to erupt is the first molar at 3–5 months, followed by the incisors and the second molar at 10–15 months, then the premolars and third molar at 22–29 months (Miller 1972, 1974a, 1974c). Wear on the teeth is slight until after the third year of life. Further attrition is not marked until the animal is 5–7 years old; by 10 years, the wear is appreciable on most teeth and the molars appear to be losing their usefulness. Empirically, I suggest that Peary caribou exhibit accelerated attrition of their teeth relative to other forms of caribou on more southerly ranges.

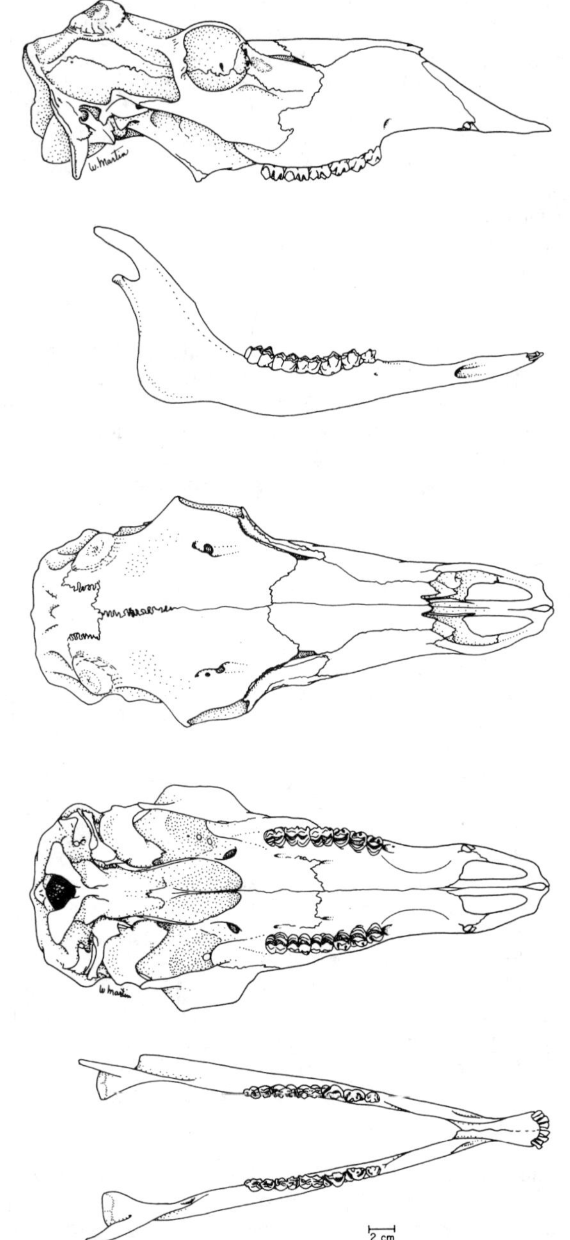

FIGURE 46.2. Skull of the caribou (*Rangifer tarandus*). From top to bottom: lateral view of cranium, lateral view of mandible, dorsal view of cranium, ventral view of cranium, dorsal view of mandible.

Dental anomalies include morphological variations, supernumeraries, and agenesis (Banfield 1954a, 1961; Miller and Tessier 1971). Many dental anomalies are congenital, whereas others may be caused by trauma and subsequent repair.

Skin Glands. Caribou have preorbital, tarsal, interdigital, and caudal (tail) glands, which all secrete odorous material (Quay 1955; Lewin and Stelfox 1967; Kelsall 1968; Anderson et al. 1975; Muller-Schwarze et al. 1978). Preorbital glands are located medial to the eye sockets, over the lachrymal pit in the lachrymal bone. They are almost hidden by a dense covering of hair. A tarsal gland is located on the inside of the hock at the tarsal joint on each hind leg. They are fringed by tufts of stiff, whitish hair, usually stained yellowish by the secretion. Deep-socketed interdigital glands occur on each hind foot, and a smaller one occurs on each front foot, medial and posterior to the primary digits. The caudal gland is located underneath the distal portion of the tail.

These scent glands most likely play an important role in social organization. Caribou can track each other by scent, which enables separated members to relocate the group if they lose sight of each other. It is likely that scent glands function in many ways to provide olfactory signals to caribou under various conditions.

Longevity. Caribou are moderately long lived ungulates. Some extremes of longevity for North American caribou are 18–20 years (McEwan 1963), 15+ years (Skoog 1968), 17 years (Bergerud 1971c), and 17+ years (Miller 1974c).

PHYSIOLOGY

Nutrition. Physiological studies of caribou nutrition provide some insight into the adaptation of caribou to their Arctic and Subarctic environments (Chapin 1980; Person et al. 1980; Thomas and Kroeger 1980; Whitten and Cameron 1980; Kuropat and Bryant 1983; Thomas et al. 1984; Adamczewski et al. 1987, 1988, 1993; Klein 1990; Rominger and Robbins 1996). Lichens (Ascomycetes), usually the main winter forage, are low in protein and high in carbohydrates (Scotter 1965, 1972; McEwan and Whitehead 1970). They are fermented by specifically adapted rumen microorganisms (Dehority 1975a, 1975b); this fermentation produces heat. A low concentration of rumen ammonia also results, which assists recycling of nitrogen by increasing the rate of urea transfer to the rumen from the plasma (Wales et al. 1975). The increase of urea recycling in winter (Wales et al. 1975) compensates for the low protein content of lichens and also has a marked effect on the water flux (White 1975). The total amount of body water increases principally because water replaces decreasing body fat (Cameron and Luick 1972; Cameron et al. 1975). The increased volume of body water could function as a thermal buffer, especially to the temperature changes caused by food and snow intake (Cameron et al. 1975). The decrease in nitrogen intake and the increase in urea recycling are also accompanied by increases in glucose resynthesis (White 1975). A high rate of glucose resynthesis is particularly important in pregnant or lactating females (Luick and White 1975).

Reduced forage intake during winter (McEwan 1968), probably in response to lower energy requirements (White 1975), in itself reduces energy expenditure. Thing (1977) calculated that the energy costs of foraging in snow increase markedly in late winter. By eating less, caribou save the energy required both to find food and to warm it once eaten—an additional cost estimated to be 20–25% of the fasting metabolic rate (White and Yousef 1974; White 1975; Young and McEwan 1975). Daily time required for particle size breakdown of injested forage by rumination likely places a restriction on the portion of the day remaining for forage intake (Trudell-Moore and White 1983).

Caribou conserve energy in winter through the insulating qualities of their pelage and by heating the blood returning from the distal joints of their legs with outgoing blood in the proximal arteries (Jacobi 1931; Scholander et al. 1950a, 1950b, 1950c; Irving and Krog 1955; Moote 1955; Lentz and Hart 1960; Hart et al. 1961; Øritsland 1974). Caribou feet remain flexible in subzero temperatures because the fatty tissues formed there remain soft, whereas marrow fats further up the leg are solid even at room temperature.

In summer, caribou lose heat by panting, from patches of "bare" skin on their bodies (areas of new hair growth), and from the sparsely haired extremities (McEwan et al. 1965; Irving 1966; Krog and Wika 1975; Yousef and Luick 1975). Stonehouse (1968) suggested that growing antlers in velvet serve to dissipate body heat. However, Krog and Wika (1975) believed that heat loss from the antlers was unavoidable because of the blood flow necessary for rapid antler growth.

Growth. Caribou calves grow rapidly in the first 5 months after birth; body weights increase from about 6 kg to 48 kg (Dauphine 1976). Parker (1989) found that the winter plateau in calf body weight occurred at about 170 days of age in all calves, bottle raised or free ranging, in two herds (Porcupine and Delta). Variations in body weight may reflect the condition of the cow in late gestation; nutritionally stressed cows gave birth to small calves, which had low survival rates (Skoog 1968;

Bergerud 1975; Adamczewski et al. 1987; Cameron et al. 1993; Thomas and Kiliaan 1998b).

Caribou normally grow between April–June and October–December of each year (Dauphine 1976). Overwinter skeletal growth apparently occurs in captive calves (McEwan and Wood 1966) and in some free-ranging calves (Larter and Nagy 1995). Growth rate depends on their nutritional plane in relation to other demands such as hair and antler growth, reproduction, weather, and movements, including responses to insect harassment (Skoog 1968). Growth rates are likely influenced by genetics and environmental conditions. Thus, it seems that there would be no basis for comparing body weights of calves of the same age between or among herds to assess the relative condition of calves by herd.

The age at which skeletal growth stops depends on which bone is measured (McEwan and Wood 1966; Dauphine 1976). Mandibles continue to lengthen significantly until 5.5 years of age in cows and 6 years in bulls (Miller and McClure 1973), but body weights reach maximums at 4.5 and 6 years for cows and bulls, respectively (Dauphine 1976). There are marked seasonal fluctuations in body weight because of changes in fat deposits (Dauphine 1976), organs, muscles, and bones (Adamczewski et al. 1987). Female caribou gain weight annually during late summer through early winter. Then their weights stabilize or decline in winter and are lowest during early summer due to pregnancy and lactation. Adult males gain the most weight from summer to the beginning of the rut. Mature prime males typically lose up to 25% of their weight during the rut (Bergerud 1978a); in winter, their weight usually stabilizes or declines (Dauphine 1976; Thomas and Kiliaan 1998b). In some years, however, some of them regain weight after the rut. Young animals reach maximum weights between early summer and early winter. Although their weights may stabilize, they often decline during winter, as they usually lack sufficient fat reserves to maintain body weights until spring.

REPRODUCTION

Caribou rely on high-quality forage on their summer range for their reproduction, growth, and winter survival. Pregnant and lactating females compensate in their feeding because of their additional metabolic demands (Cameron et al. 1991). The success of such compensatory feeding is determined by prevailing environmental conditions that influence forage intake (Cameron et al. 1991). Female caribou with low fat reserves in the autumn forego breeding to build up their fat reserves for the following breeding season. Body size and fat reserves of females are also related to the age of first conception; although most cows conceive at 2.5 years of age, some will conceive at 1.5 years of age if they are in good condition (Dauphine 1976; Parker 1981; Thomas and Kiliaan 1998b). Conception varies among years and populations. Apparently, more Alaskan than Canadian caribou conceive at early ages (Skoog 1968). Alaskan yearling female caribou are, on average, considerably larger (heavier) than their Canadian counterparts, which may influence the apparent respective difference in reproductive effort by young females. Caribou are relatively slow to mature and rarely bear twins (McEwan 1971; Shoesmith 1976; McDonald and Martell 1981). Females over 2 years of age often are fertile, however, commonly with pregnancy rates of about 80% or greater (Bergerud 1978a). The chief cause of reproductive failure is a failure to conceive; in utero mortality is rare (Dauphine 1976). However, annual initial production of calves and/or early survival of calves varies from essentially zero under extremely poor conditions to 100% in highly favorable environmental conditions (Miller et al. 1977a; Gates et al. 1986; Miller 1995b, 1998; Thomas and Kiliaan 1998b).

Insufficient build-up of fat reserves in summer and subsequent failure to conceive may at least partially explain fluctuations in pregnancy rates of some caribou (Cameron et al. 1991, 1993; Thomas and Broughton 1978; Thomas 1982). Although females can be above the threshold of physical condition necessary to conceive, subsequent deterioration of the cow's condition can contribute to calf mortality at birth or shortly thereafter. Death of the calf soon after birth relieves the cow of the metabolic cost of lactation. This relief allows her to build up fat reserves sufficiently to breed again the following autumn, but sometimes not sufficiently to produce and rear a calf successfully. Because of this, Dauphine (1976) suggested a direct relationship between high calf mortality and high pregnancy rates.

Breeding and Gestation. The rut and calving are highly synchronized in caribou (Dauphine and McClure 1974). They have two or three full estrous cycles, and one or more of those result in ovulation without overt estrous behavior. Such "silent heats" could generate and synchronize the endocrine system and thus mating (McEwan and Whitehead 1972; Bergerud 1975). Although dates of rutting vary among different caribou populations, the rut is usually in October and early November, and local breeding synchrony is apparent.

Average gestation is 225–235 days according to Skoog (1968), or 227–229 days according to Bergerud (1978a), and synchronized breeding results in a strongly peaked distribution of births. Usually 80–90% of caribou calves are born in a 10-day period in late May or early June (Dauphine and McClure 1974; Bergerud 1975). In general, calving is earlier in Alaska than in the Northwest Territories or Nunavut west of Hudson Bay. This timing appears reversed on the east side of Hudson Bay. The trend toward later calving progresses northward in Canada to the Arctic Islands. Peary caribou appear to have the latest and possibly the most variable calving period among North American *Rangifer*. Peak calving for Peary caribou on southwestern Queen Elizabeth Islands often is during the third to the fourth week of June. In the most favorable years, the peak for calving among those Peary caribou can come in the second to the third week of June. However, it can be as late as the last days of June and the first days of July in years with extreme snow/ice conditions (Miller 1998).

Parturition. Parturition has been observed and reported by several workers (Kelsall 1957; de Vos 1960; Lent 1966a; Bergerud 1974d; McDonald and Martell 1981). The newborn calf usually stands and walks within the first hour of life and nurses shortly thereafter. The mother vigorously licks and cleans the neonate. She nibbles and pulls the afterbirth from around the calf and often eats it (Pruitt 1960a; Miller and Parker 1968).

Miller and Parker (1968) suggested that maternal caribou frequently consume the afterbirth. There are two possible reasons: (1) removal of the tissue reduces odors associated with the birth site and thus the chances of predation and (2) maternal females derive nutritional benefit from eating the afterbirth. The latter supposition seems most tenable for caribou, but not exclusively so. Neonates follow their mothers and the birthsite is usually abandoned within hours. Also, the attending cow on tundra calving grounds would be sighted by a predator long before the predator was close enough to detect the odors of the birth site. Reducing the chances of predation by eating the afterbirth would be more beneficial for forest-dwelling woodland caribou calving in cover and would be much more beneficial for any ungulate that hides its neonate(s) and then move away from them to feed. There is some evidence that woodland caribou calves may hide on occasion rather than always follow the dam, even at 3–4 weeks of age (Chubbs 1993).

ECOLOGY

Habitat. The geographic distribution of caribou encompasses two of the largest biomes in North America: tundra and boreal forest (including taiga) and their southern extensions in mountainous areas. Variations in climate, geology, and topography from Alaska to Newfoundland result in differences in the plant communities of these biomes; thus, caribou occur in diverse habitats (Edwards and Ritcey 1960; Bergerud 1974a; Klein 1982; Miller et al. 1982; Gates et al. 1986; Adamczewski et al. 1988; Thomas et al. 1998, 1999; Rettie and Messier 2000). Habitat-specific knowledge of foraging habits and foods eaten is required for biologically sound management (Rominger and Oldemeyer 1990). This consideration is particularly true for the low importance of lichens on the Arctic Islands, where lichens are lacking in general and inadequate for providing the bulk of the caribou's winter diet, regardless of accessibility (Thomas et al. 1999).

The typical range of eastern woodland caribou is climax stands of boreal forest with mixed-age stands of black and white spruces (*Picea mariana, P. glauca*), balsam fir (*Abies balsamea*), and white birch (*Betula papyrifera*) with tree and ground lichens (Cringan 1957; Bergerud 1972). Bogs and jack pine (*Pinus banksiana*) stands are important, particularly in postburn areas (Schaefer and Pruitt 1991).

The habitat of many western woodland caribou often includes mountain summits above the timber line or elevated table lands with alpine meadows and open subalpine forest (Edwards 1958; Edmonds and Bloomfield 1984; Simpson et al. 1985; Edmonds 1988; Rominger and Oldemyer 1989, 1990; Seip 1992). In high-snowpack mountain areas, caribou use Engelmann spruce (*P. engelmanni*)–subalpine fir (*A. lastocarpa*) and western red cedar (*Thuja plicata*)–western hemlock (*Tsuga heterophylla*) communities. Southern aspects commonly include mixed-conifer stands of Douglas fir (*Pseudotsuga menziesii*), western larch (*Larix occidentalis*), lodgepole pine (*P. contorta*), and western white pine (*P. montocola*). During deep-snow periods, alectoroid arboreal lichens (*Alectoria sarmentosa, Bryoria* spp.) from windthrown trees and myrtle boxwood (*Pachistima myrsinites*) are the important, if not the critical, part of the mountain caribou's diet. At lower elevations, boreal mixed-wood and peatland vegetation and fen complexes were important in northeastern Alberta (Stuart-Smith et al. 1997), and in Saskatchewan, peatlands and black spruce dominated stands (Rettie et al. 1996; Rettie and Messier 2000).

In the taiga toward the treeline, the trees become more widely spaced and the ground cover of lichens, ericaceous shrubs, willows (*Salix* spp.), and dwarf birch (*B. glandulosa*) increases. North of the treeline, the black spruce–moss muskeg and sedge (*Carex* spp.) bogs of poorly drained areas are replaced by willow and alder (*Alnus* spp.) thickets along watercourses, and by sedge in grassland communities on wet areas. Dwarf shrub–heath communities occupy large areas of the tundra except on drier sites, where fruticose lichens and mosses dominate with dwarf willow, birch, and rhododendron (*Rhododendron lapponicum*) (Rowe 1959; Miller 1976). On the Arctic Islands, Peary caribou and arctic-island caribou show a preference for polar "desert" and similar dry to mesic range types with sparse vegetation of willow, sedges, grasses, and forbes (Fischer and Duncan 1976; Parker and Ross 1976; Wilkinson et al. 1976; Thomas et al. 1976, 1977, 1999; Parker 1978; Shank et al. 1978; Thomas and Broughton 1978; Russell et al. 1978; Miller et al. 1982; Thomas and Edmonds 1983, 1984; Larter and Nagy 1997).

The climate of caribou ranges is characterized by long, cold winters; short, cool summers; and low precipitation compared to more temperate regions. Much of the precipitation falls as snow, which can cover the ground for 7–9 months. Snow cover data are especially important to understanding caribou ecology, but show great local variation. Caribou contend with snow depths usually >50 cm, except in the High Arctic and on the open tundra. Usually, 60–70 cm is considered a threshold critical snow depth for foraging caribou (e.g., Pruitt 1959; Henshaw 1968b; Bergerud 1974b; LaPerriere and Lent 1977). Many woodland caribou experience much deeper snow cover during mid- to late winter. Brown and Theberge (1990) reported that during December–April, woodland caribou in Labrador dug through snow ("crater depth") with mean depths up to 123 cm. Vandal and Barrette (1985) reported a mean crater depth of 99.9 cm during February in Quebec. The combined effect of deep snow cover, hard snowpack resulting from wind action, and thaws and winter rains challenges caribou in insular Newfoundland (Tucker et al. 1991).

Autumn and winter thaws and rainstorms can occur in almost any northern area. These sometimes cause serious restrictions on forage availability (Miller et al. 1977a; Miller 1998). Spring melting and refreezing of snow leading to crust formation often restricts caribou feeding and movements (Pruitt 1959; Henshaw 1968b; Bergerud 1974b; Stardom 1975; Miller 1976; Miller and Gunn 1978; Miller et al. 1982). In extreme cases, however, relative forage unavailability could be so restrictive and widespread as to cause environmentally forced movements (Miller 1990a, 1998).

Fancy and White (1985:992) concluded that caribou have evolved an energetically efficient mechanism for obtaining food from under snow cover. They concluded, "it is evident the 'chopping' movement used by caribou to break through crusted snow requires a much higher expenditure of energy than a more sweeping movement used in loose snow of the same density." If this is so, I suggest that caribou on the Arctic Islands, particularly the High Arctic Islands, experience much higher cratering costs. Although snow cover on feeding sites is generally shallower on islands than on the mainland, snow hardness and density (plus icing on, in, and under the snowpack) commonly reach much higher values on islands than on mainland ranges (Thomas et al. 1976, 1977; Thomas and Broughton 1978; Miller et al. 1982; Thomas and Edmonds 1983; Adamczewski et al. 1988). Thus, Peary caribou and arctic-island caribou must often employ the chopping movement when cratering (Miller et al. 1982; F. L. Miller, Canadian Wildlife Service, Edmonton, Alberta, pers. obs., 1974, 1990–1992). This greater energy expenditure could explain why Peary caribou stop cratering in shallow but extremely hard-packed snow in late May and June and favor snow-free but poorly vegetated foraging sites at that time. Peary caribou also extend their foraging into better vegetated sites at the periphery of the snow-free patches at that time by breaking large chunks of hard packed snow away with chisel-like blows from their front hoof. They then push the blocks of snow back into the snow-free area underneath them to expose the forage that had been covered by the dense but shallow snow pack.

The short growing season on caribou ranges is partially offset by increased solar radiation and increased day length characteristic of high latitudes (Klein 1964). In addition, plants have higher nitrogen, phosphorus, and carbohydrate levels than plants at lower latitudes (Chapin et al. 1975). Some Arctic plants are adapted to the short growing season by having wintergreen and evergreen leaves (Bell and Bliss 1977), which are preferred winter forage (Kelsall 1968). Caribou sometimes expose, trample, and eat feeding push-ups constructed on ice in winter by muskrats (*Ondatra zibethicus*) (Skoog 1968; Kelsall 1970). Caribou use mineral licks and ice licks high in mineral content (Skoog 1968; Calef and Lortie 1975; Heard and Williams 1990; F. L. Miller, Canadian Wildlife Service, Edmonton, Alberta, pers. obs., 1981–1983; A. Gunn, Department of Resources, Wildlife and Economic Development, Government of the Northwest Territories, Yellowknife, pers. obs., 1981–1983).

Summer and Winter Ranges. Most caribou populations have distinct summer and winter ranges; the latter are characterized by tree and/or shrub cover and lichens. Caribou are not solely dependent on lichens in winter (Murie 1935; Skoog 1968; Bergerud 1972, 1996). Availability of food on winter range is usually regarded as the limiting factor for caribou populations. Factors modifying or reducing the availability of winter range have been of particular interest to wildlife managers. The relative importance of summer range has become more apparent recently (Couturier et al. 1990; Crête et al. 1990; Crête and Huot 1993; Bergerud 1996; Valkenburg 1997). A population's annual range is only as good as its lowest quality seasonal link, which can vary by year and place.

All Peary caribou and arctic-island caribou remain on tundra ranges throughout the year. Some barren-ground caribou herds remain all year on the tundra, whereas other herds move to forested ranges in winter. Woodland caribou, even the migratory herds of the Ungava Peninsula, usually move to forested winter ranges. However, even forest-dwelling herds will move in winter to tundra or tundra-like (alpine) areas if excessively deep snow cover or snow pack restrict forest forage supplies on a widespread and prolonged basis.

Movements between tundra and forested ranges are learned traditions of particular herds that have resulted in favorable long-term survival. Yearly differences in movements reflect annual variations in snow cover and icing conditions, which influence annual availability of forage supplies en route.

The effect of forest fires on winter ranges of caribou has long been a controversial subject. There are two theories concerning the importance of forest fires in limiting or reducing numbers and changing distributions of caribou. One is that forest fires cause extensive detrimental impact on forage supply and subsequent regrowth (Leopold and Darling 1953; Edwards 1954; Cringan 1957; Banfield and Tener 1958; Scotter 1964, 1967a; Kelsall 1968). The other suggests that only certain fires under certain conditions are important to caribou. Some researchers feel that fires not only are beneficial to caribou, but also are necessary to promote the heterogeneity of vegetation needed to perpetuate mixed forage supplies (Skoog 1968; Bergerud 1971a, 1971b, 1974a; Rowe and Scotter 1973; Bunnell et al. 1975; Johnson and Rowe 1975; Miller 1976, 1980; Schaefer and Pruitt 1991; Thomas 1998b). Probably, a composite view best reflects the realities of range–caribou relations. Production and availability of forage are at least theoretically the ultimate factors in governing numbers and distributions of caribou.

Movements and Migrations. Survival through adaptive movements and migrations characterizes the ecology of caribou. Bergerud (1974d) believed that wolf predation led to gregarious behavior. Movements and migrations followed as a result, so that caribou could maintain themselves in relation to their varying forage supplies. The caribou's movements and migrations are further governed temporally and spatially by weather (especially snow cover and icing conditions), blood-sucking and biting insects, and various physiological and psychological drives. Snow cover most often is the dominant influence on movements by caribou during most of the year. Evaluations of the snow (nival) environment of caribou are numerous (Banfield 1949; Pruitt 1959; Edwards and Ritcey 1960; Henshaw 1968b; Bergerud 1971a; Stardom 1975; LaPerriere and Lent 1977; Thing 1977; Thomas 1991; Russell et al. 1993; Thomas et al. 1998).

On the mainland, woodland and barren-ground caribou use frozen lakes and rivers during seasonal migrations and environmentally forced movements (Banfield 1954b). Seasonal migration and sporadic movement between the mainland and southern Arctic Islands by barren-ground and arctic-island caribou are accomplished by travel over frozen straits (Banfield 1954b). The ultimate use of ice is exhibited by caribou on the Canadian Arctic Archipelago, where the sea remains frozen for most of each year and essentially year-round on the northern fringe of those islands. There, the frozen sea plays the major role in allowing the distribution and resultant use of range by Peary and arctic-island caribou by annually permitting seasonal interisland migrations and sporadic movements (Miller et al. 1977a, 1977b, 1982; Miller and Gunn 1978, 1980; Miller 1990a). Interisland movements are made even during the brief open-water period of the year (Miller 1995a, 1997b).

The annual cycle of movements of migratory barren-ground caribou begins in spring. As the amount of daylight increases and snow begins to recede, wintering bands begin to coalesce, and these large aggregations move northward. In the early stages of the spring migration, caribou move sporadically until the apparent urge to return to the calving ground grips the parturient cows and they move steadily north. The number of juveniles, and especially yearlings, accompanying the females to the calving ground varies annually depending on the difficulties of migration. In some years, deep snow and slush prevent most young animals from migrating with the cows. They drop behind and move northward with the bulls as traveling conditions improve. On reaching the traditional calving grounds, the cows and their followers continue to move around the area in aggregations. When parturition is imminent, some parturient cows disperse. They may calve in relative isolation or close to groups, apparently depending on timing of their arrival and the peak of calving.

Maternal cows and their newborn calves form nursery bands after the peak of calving (Pruitt 1960a). Nursery bands then merge into large postcalving aggregations, which move off the calving ground to summer ranges and begin to mix with the rest of the herd. During summer months, they move extensively, often 500 km or more. Some herds return south into the open taiga or even the closed boreal forest, whereas others move northward to tundra and coastal areas. In autumn, there are prerut movements and the prime bulls establish their dominance hierarchies. With the beginning of the rut, bulls join cow–juvenile groups and remain in their company until the cows become receptive. Breeding often takes place while the aggregations are on the move, sometimes during autumn migration to wintering grounds (Henshaw 1970). After the rut, caribou move to their winter ranges. Adult bulls often separate from the cow–juvenile groups, and some groups of bulls move farther into the boreal forest.

Seasonal movements of more sedentary forest-dwelling woodland caribou and those of Peary and other arctic-island caribou are small scale compared to those of migratory barren-ground and migratory woodland caribou. Some Peary and other arctic-island caribou do, however, make seasonal migrations of 200–500 km (Miller et al. 1977b). Local alterations in movement patterns of caribou probably reflect seasonal variations in distribution and availability of forage, and habitat differences.

On western ranges, where elevation can vary by over 2000 m, habitat influences seasonal movements of woodland caribou (Edwards 1954, 1958; Edwards and Ritcey 1959, 1960; Edmonds and Bloomfield 1984; Edmonds 1988; Simpson et al. 1985; Servheen and Lyon 1989; Rominger and Oldemeyer 1989, 1990). The 30 or so remaining woodland caribou in the Selkirk Mountains of northern Idaho, northeastern Washington, and southern British Columbia stopped making distinct seasonal shifts in elevation in the 1970s (Freddy 1979). Woodland caribou on mountainous western ranges are found on high meadows and in adjacent, open subalpine forests above 2000 m during summer and fall. The first deep snow of winter forces them down to mature forests in valleys, where they prefer poorly drained sites interspersed with open bogs, meadows, and ponds. As winter progresses and the snow settles, hardens, and becomes crusted, the caribou return to their summer haunts on the high, wind-swept ridges and remain there until spring. Softening snow drives them into lowland forest again, where they remain until they can return to the snow-free uplands in May and June.

Most, if not all, Peary caribou and arctic-island caribou make migration-like trips at about calving time (Banfield 1954b; Miller et al. 1977a, 1977b; Miller and Gunn 1978, 1980; Miller 1990a). They may travel among islands or around one island. Movements are not necessarily days or weeks before calving, as displayed by barren-ground and migratory woodland caribou. Peary caribou and arctic-island caribou prefer certain areas of different islands for calving. The more favorable snow and ice conditions on those areas immediately before, during, and after calving time can be critical to early survival of calves in some years.

Peary caribou and arctic-island caribou constantly move and feed on their summer ranges until early fall. In some instances, at least, Peary caribou include interisland swims of at least 2–4 km in their July–August movements (Miller 1995a, 1997b). Perhaps the most interesting aspect of these open-water interisland crossings is that maternal cows sometimes make these swims in frigid, often ice-clogged, sea water when their calves are only 2–3 weeks old.

There appears to be a prerut movement, or "autumn shuffle," in late August to September, but directional movements to rutting areas can be delayed until late September–late October. Nothing is known concerning the rut in Peary caribou or arctic-island caribou. The finding of cast male antlers along coastal areas suggests that there is some preference for the coast during the rut or shortly thereafter. This would facilitate male-female contact, particularly in low-density populations (Miller and Barry 1992). Peary caribou on Melville Island (42,000 km^2) can reduce their search effort for receptive females during the rut by 14 times or more if they search only within 1 km of the perimeter of the island rather than range back and forth across the island.

Herds that use more than one island annually probably return to winter range shortly after freeze-up in late autumn or early winter following the rut. The rut often occurs on the island where the caribou summered. Peary caribou and arctic-island caribou that remain on the same island all year usually shift from summer range to winter ranges on different parts of the island (Freeman 1975; Miller et al. 1977a;

Miller 1997b, 1998). Satellite telemetry location data indicate that some Peary caribou, at least in favorable years, remain year-round on the same ranges (Miller 1997b). Even relatively sedentary individuals must sometimes move to adjacent islands to survive when deep hard-packed snow and/or icing reduces forage availability on large areas of their home ranges. Satellite telemetry location data indicate that some individual arctic-island caribou annually return to the same seasonal ranges (Gunn and Fournier 2000).

Biologists have described caribou movements and migrations as nomadic (Skoog 1968; Bergerud 1974d). That description should not connote herds wandering aimlessly about tundra and taiga. Their movements are structured and completed in an orderly sequence (Heape 1931). Caribou herds, or segments of herds, prefer specific sections of ranges at certain times for calving grounds, wintering areas, seasonal staging areas, rutting areas, and migrational paths. Fidelity to calving grounds is particularly strong for migratory barren-ground caribou (Gunn and Miller 1986). These areas most likely represent more favorable environmental settings over the long run, but any of them could fail the animals in extremely severe years (Miller 1998).

To maintain such traditional range use requires a refined state of *Ortstreue,* or fidelity of the offspring to the land of the parents. The apparent wanderings of individuals and groups of caribou are strongly orientated and directed.

Herd Composition. Most caribou are highly gregarious, with weaker tendencies for large aggregations among relatively sedentary forest-dwelling woodland caribou. The zenith of their gregarious behavior is reached in the postcalving herds composed of thousands to tens of thousands of barren-ground caribou and tundra-dwelling woodland caribou. Sights and sounds of such a mass of moving animals are unforgettable. Thousands of grunting cows are answered by their bleating calves. There is a continuous clicking of hooves, and a myriad of coughs, sneezes, and belches coming from the moving mass. Such aggregations sometimes bunch so tightly as they mill about that from the air they appear like a swarm of bees.

Large herds of migrating caribou are temporary gatherings of many social units (groups or bands). Some social order may be maintained within aggregations by an interacting hierarchy of dominant animals from the bands that form the aggregations. Reasons for different social groupings were discussed by Miller (1982).

Interspecific Relations. Most interspecific interactions between caribou and sympatric species are predator–prey, parasite–host, or competitive. Animals that do not interact directly with caribou benefit by scavenging on them.

The wolf is the principal predator of caribou. On many western ranges, grizzly bears (*Ursus arctos*), mountain lions (*Puma concolor*), and golden eagles (*Aquila chrysaetos*) can be the primary predator. On eastern woodland caribou ranges, especially where wolves have been reduced or are absent, black bears (*U. americanus*), lynx (*Lynx canadensis*), and coyotes (*C. latrans*) become important predators. In addition to the major predators, wolverines (*Gulo gulo*), red foxes (*Vulpes vulpes*), bobcats (*L. rufus*), and ravens (*Corvus corax*) sometimes prey on newborn calves. The possible predation on crippled or infirm individuals is of no consequence, as those prey animals would be predisposed to death from other causes.

Most potential competitors include muskoxen, lemmings (*Dicrostonyx* spp. and *Lemmus* spp.), arctic hares (*Lepus arcticus*), and snowshoe hares (*L. americanus*). Minor or theoretical competitors could include other ungulates, other rodents, and grazing waterfowl (Kelsall 1968; Skoog 1968).

Scavengers other than the previously named predators include polar bears (*U. maritimus*), arctic foxes (*Alopex lagopus*), gulls (*Larus* spp.), and jaegers (*Sterocorarius* spp.). Other potential scavengers include mustelids and carrion-eating birds (Kelsall 1968; Skoog 1968).

Insect Season. Perhaps no other aspect of caribou ecology is so stressful as encounters with biting and blood-sucking insects during summer. A year when such insects are numerous is indeed a time of madness for the caribou. There are no large areas free of harassing insects on the inland tundra. Caribou must constantly dash about wildly to escape the ever-present hordes of insects during diurnal peaks of insect activity. Fortunately, in all but the worst insect years, there is some temporary relief during the cooler early and late hours of the day and during rainy periods. Caribou occupying coastal summering areas often seek relief from mosquitoes by moving out onto the mud flats and simply standing motionless with their heads down, muzzles nearly to the ground (or water). Caribou harassed by biting flies will gallop short distances, buck, and spin about in vigorous attempts to avoid contact.

Windswept ridges, glaciers, and lingering snow drifts on tundra ranges of some barren-ground caribou and land areas above the timber line on western woodland caribou range are relatively free of insects. On those sites, caribou usually remain huddled closely together throughout much of the day, dispersing to feed in the cooler late and early hours. In Newfoundland, Bergerud (1974d) observed that woodland caribou dispersed into the forest because biting flies were less active in the shade. This behavior appears to be the only option available to most woodland caribou of eastern North America unless they are on coastal areas or the shores of large lakes.

Helle and Aspi (1983), using reindeer dummies that emitted CO_2, concluded that herd formation (clumping) reduced the serious effects of insects and that herding at those times had high selection value. Russell et al. (1993) found that it was advantageous for caribou to be in the center of the group rather than the periphery at low wind speeds, but that the opposite was true at higher wind speeds. Moderate to high insect harassment results in formation and maintenance of large aggregations, a significant decrease in time spent feeding, and increases in times spent standing and moving, walking, trotting, and running collectively (Baskin 1970; White et al. 1975; Fancy 1983; Thing 1984; Murphy 1988; Russell et al. 1993).

Arctic-island caribou on the more southern islands of the Canadian Arctic Archipelago experience insect seasons of short duration that seemingly would rarely have much impact on their physical condition. Insects in certain years could stress caribou, but this would be of minimal impact compared to mainland situations.

Only the Peary caribou on the High Arctic Islands are essentially continuously free of harassment by insects (with the exception of minimal insect activity in the "heat-sink area" on northern Ellesmere Island). Peary caribou are at the northern edge of the species's range and are subjected to a short period of the year in which body condition and fat reserves can be accumulated. Therefore, they might not survive the rigors of the High Arctic, at least in some extremely unfavorable years, if they had to withstand the additional stress of severe harassment by biting and blood-sucking insects.

FEEDING HABITS

Caribou, like most North American cervids, feed on a broad range of plants, including lichens, fungi, sedges, grasses, forbs, and twigs and leaves of woody plants. Their preference for lichens is unique among North American ungulates, and it appears to be the key to caribou survival in many areas when populations are at high densities. For example, lichens represented 87–90% relative density based on rumen and fecal samples of the winter diet of caribou in the Beverly herd in north-central Canada (Thomas 1998a). However, lichens are generally lacking on the Queen Elizabeth Islands and at least on Banks and Victoria islands to the south (Thomas and Edmonds 1983; Larter and Nagy 1997; Thomas et al. 1999; A. Gunn, Department of Resources, Wildlife and Economic Development, Yellowknife, pers. commun., 2000). Thus, lichens are relatively unimportant in the diets of caribou on those islands (references given below).

Descriptions of feeding habits are based on direct observations of free-ranging and captive animals, and also on examination of fecal and rumen samples, fistula experiments, and examinations of craters and other feeding areas. Feeding habits and preferences vary according to range type. Much information has been compiled, for example, for eastern woodland caribou (Cringan 1957; Simkin 1965; Des Meules and Heyland 1969a, 1969b; Bergerud and Nolan 1970; Bergerud 1971a, 1972, 1974b, 1977; Stardom 1975; Darby and Pruitt 1984; Crête et al. 1990), western woodland caribou (Edwards and Ritcey 1960; Edwards et al. 1960; Fuller and Keith 1981; Edmonds and Bloomfield 1984; Simpson et al. 1985; Stevenson and Hatler 1985; Hatler 1986; Rominger and Oldemyer 1989, 1990, 1991; Servheen and Lyon 1989; Bradshaw et al. 1995; Thomas et al. 1996; Stuart-Smith et al. 1997), Canadian barren-ground caribou (Banfield 1954a; Scotter 1967b; Kelsall 1968; Miller 1974, 1976, 1980; Thomas et al. 1984; Thomas and Hervieux 1986; Adamczewski et al. 1988; Ouellet et al. 1996; Terry et al. 1996; Thomas 1998a), Alaskan barren-ground caribou (Skoog 1968; Klein 1970a, 1970b; White et al. 1975; Holleman et al. 1979; Chapin 1980; Kuropat and Bryant 1980, 1983; White and Trudell 1980; Whitten and Cameron 1980; Boertje 1984; Martell et al. 1986; Fleischman 1991; Russell et al. 1993), and Peary caribou and arctic-island caribou in Canada (Fischer and Duncan 1976; Parker and Ross 1976; Thomas et al. 1976, 1977, 1999; Wilkinson et al. 1976; Parker 1978; Russell et al. 1978; Shank et al. 1978; Thomas and Kroeger 1980; Miller et al. 1982; Thomas and Edmonds 1983, 1984; Larter and Nagy 1997).

Most caribou ranges are considered fragile, but that actually relates only to the lichen mat component, which is slow to recover after overuse or fire. Productivity can be high seasonally, but the growing season is short, so annual production is low. Also, lichens are especially susceptible to trampling (Pegau 1970, 1975). It is unlikely, however, that caribou destroy their own forage supplies (e.g., Skoog 1968) because (1) their cursory feeding behavior prevents excessive use of individual plants; (2) they select mostly the newer growth parts of plants, which are readily replaced; (3) they have diverse feeding habits and use a wide range of plants; (4) deep snow often covers large areas of ranges, causing caribou to move to other areas and thus protecting the plants; and (5) snow disturbed from cratering and tracking becomes too compacted and prevents or reduces later use in the same winter.

The wide range of habitats and snow conditions that determines caribou feeding habits limits the value of generalizations. Nevertheless, some comments are applicable to many caribou, at least on mainland ranges. Fruticose and foliose lichens dominate their diet in the fall and winter. Even crustose lichens are used by caribou on the Arctic Islands, especially in late winter. In treed areas with deep snow, terrestrial lichens of early winter are replaced by arboreal lichens as snow depth increases. Many woodland caribou in the western United States and Canada rely on arboreal lichens for winter survival (Edwards and Ritcey 1960; Edwards et al. 1960; Simpson et al. 1985; Rominger and Oldemeyer 1989, 1990, 1991; Servheen and Lyon 1989). Stardom (1975) reported that woodland caribou in Manitoba fed mainly on arboreal lichens in open tamarack (*L. laricina*) bogs during early winter until snow cover appeared to hinder travel. They then moved to mature jackpine and fed mainly on the ground lichens (*Cladonia* spp.). Woodland caribou fed along shorelines on sedges and on ground lichens on southeast-facing slopes of rocky lake shores during spring. Barren-ground caribou select fungi in fall (Kelsall 1968; Miller 1976). The winter diet includes woody twigs of shrubs such as *Vaccinium* spp. and some trees (willows), evergreen leaves, and graminoids that have retained some green leaves (Kelsall 1968; Russell et al. 1993; Thomas 1998a).

As snow melts, caribou seek exposed sites to feed on leaves and graminoids, which tend to dominate the diet, with little or no lichens as summer progresses. Caribou especially select plant species according to phenology of greening leaf buds and flower buds during summer. Selectivity is closely tied to the nutritive status and chemical defense of the part of the plant (e.g., Klein 1970c; Kuropat and Bryant 1983). Caribou feed while on the move, and tend to concentrate on one or two species at a particular time (e.g., Kuropat and Bryant 1980; Miller et al. 1982). They use sight and smell to detect preferred plant parts (Wright, in Klein and White 1978). Little is known about how selectively caribou feed during fall and winter, but limited availability may reduce selectivity. Captive caribou and reindeer prefer different lichen species (Des Meules and Heyland 1969a, 1969b; Holleman and Luick 1977; Rominger and Robbins 1996).

Caribou probably use their keen sense of smell to detect the lichens under snow. In deeper snow, they may smell lichens through air vents caused by shrubs reaching the snow's surface (Bergerud 1974b). Caribou will move soft, shallow snow with their noses, but as snow depths increase, they dig craters of varying sizes with their front feet to expose vegetation. If the snow is crusted, caribou may break the crust with their front feet. Thresholds of depth and crust hardness—caused by wind, surface melting and refreezing, or freezing rain—that hinder or prevent cratering vary according to habitat. In the boreal forest or taiga, where snow tends to be deep and soft, caribou stop cratering when snow hardness exceeds 50 g/cm^2 (Pruitt 1959). On the tundra, however, caribou continued to crater until snow hardness reached 6500–9000 g/cm^2 in Alaska (Henshaw 1968b; Thing 1977) and 4000–10,000 g/cm^2 on Arctic Islands (Thomas et al. 1977; Miller et al. 1982; Thomas and Edmonds 1983; Adamczewski et al. 1988). Critical snow depths for cratering by caribou supposedly range from 60 cm (Pruitt 1959) to 75 cm (Thing 1977), but much greater cratering depths have been reported for eastern woodland caribou (Brown et al. 1986; Brown and Theberge 1990).

BEHAVIOR

It is easy to agree with the suggestion of Davis et al. (1986) that individuals with different behavioral tendencies may exist within a caribou herd. Whether these tendencies are inherent in individuals or facultative responses to the environment (Valkenburg et al. 1988) is debatable. I suggest they are both. See Davis et al. (1986) for a learned discussion of empirical and theoretical considerations toward a model for caribou socioecology.

Caribou communicate by a combination of visual, auditory, and olfactory modes. These cues are manifested as acts of aggression, displays of dominance, vocalizations, skeletal sounds, and secretions of pheromones. Supposedly, all these messages have the meaning of either "come closer" (affin type) or "move away" (difrug type) (Lent 1974).

Leadership and group cohesion have been demonstrated in caribou (Miller et al. 1972) and reindeer (Baskin 1970). Play has been described in caribou (Lent 1966a; Miller and Gunn 1981; Muller-Schwarze and Muller-Schwarze 1983) and reindeer (Espmark 1971).

Caribou and reindeer antlers deserve special consideration in descriptions of behavior, as they have high social significance. Behavioral posturing or gestures in which antlers have a role have been described by Dugmore (1913), Pruitt (1960a), Espmark (1964b), Lent (1965a, 1965b), Henshaw (1968a), Bergerud (1973, 1974c), and Bubenik (1975b).

Caribou and reindeer antlers are known for their asymmetry and particularly for development by adult males of at least one, shovel-like, enlarged brow tine (e.g., Murie 1935; Banfield 1954a; Pruitt 1966; Skoog 1968; Davis 1973, 1974; Bubenik 1975a; Goss 1980, 1983; Miller 1986). Morphological dominance of the brow tines in each antler pair and the possible function of the enlarged shovel (brow tine) remain open to debate. After comparing 1168 pairs of antlers from *Rangifer,* Miller (1986) concluded that the apparent predominance of enlarged (dominant) left brow tines over right ones for the entire species cannot yet be accepted with confidence. Any future studies of brow tine dominance could be improved by clearly identifying the sex and at least approximate age of the donor animals to allow evaluations of antler pairs by sex and age class (Miller 1986).

Pruitt (1966) believed that the well-developed brow tine (shovel) protects the eyes during bouts of "bush-thrashing" (stereotyped movements of the head, swinging antlers back and forth through bushy vegetation). However, all other North American male cervids are equally vigorous bush or tree thrashers without the benefit of any such elaborate protection, and they apparently are not plagued by eye injuries. Bubenik (1975b) believed that the brow tine is used as an offensive weapon and that the second tine protects the eyes and facial region. Whatever the function of the enlarged brow tine, there are three major points to consider: (1) development is delayed to sexual maturity, (2) most mature males possess at least one enlarged brow tine, and (3) the antlers of mature males are shed in autumn/early winter and not fully regrown until prerut during the following autumn (i.e., fully developed antlers are carried for only about 3 months of the year). These conditions lead to the likelihood that the probable function of the enlarged brow tine in mature males is linked to their courtship behavior (Miller 1986).

Henshaw (1968a) believed that retention of antlers by pregnant females enhances their social ranking and therefore their ability to compete for restricted forage supplies, thus contributing to their survival and that of the fetus. Although the occurrence of antlers on females is acceptably explained by Henshaw, one might ask (1) why are 1- to 3-year-old males that often retain their antlers throughout most or all of the winter permitted socially to remain in close association with wintering groups of pregnant females and their young, and (2) why do many female caribou, especially in Newfoundland, not possess antlers (Bergerud 1971a, 1976)? Henshaw (1968a) also believed that possession of antlers by calves allowed them to exhibit a marked degree of precocity and independence from maternal care during winter. Although such behavior by calves is likely necessary, as pregnant maternal cows offer little or no nutritional care, their small, usually velvet-covered, spike antlers are probably of little significance in encounters, except probably with other calves. Cows are often even antagonistic toward their previous year's offspring during gestation, especially in winter if forage becomes restricted. Female caribou without antlers often attack antlered animals quite effectively by kicking with their forelegs.

Vocalization and Other Sounds. Caribou of both sexes and all ages produce many sounds, especially when in large postcalving aggregations and during the rut (Murie 1935; Banfield 1954a; de Vos 1960; Pruitt 1960a; Espmark 1964a; Lent 1965b, 1966b, 1975; Kelsall 1968; Bergerud 1973, 1974c; Erickson 1975). Cows give coughlike grunts and calves have bleating cries; rutting males grunt, snort, hoot, slurp, cough, sneeze, and pant. When caribou move, their hooves make a unique clicking sound supposedly caused by the sesamoid bones slipping over each other. This noise is particularly noticeable when a large number of caribou are on the move and, in calm weather, at a distance over 100 m, even when mixed with assorted vocalizations.

Locomotion. Caribou have several gaits, including a walk, fast walk, trot, pace, and gallop. They walk while feeding and moving between feeding areas. Although walking during maintenance activities seems leisurely, it has been timed for barren-ground caribou at 7 km/hr over long distances and on rough terrain (Pruitt 1960b). Peary caribou traveled at rates of about 3–4 km/hr while essentially continually foraging on broken, rolling terrain (Miller et al. 1982). Fast walks often occur during migration and periods of initial alertness. Trotting often follows an alarm or stresses such as predators or insect harassment. If the alarm or stress situation continues and/or increases, caribou may pace or, more commonly, gallop.

Caribou are adept climbers and will ascend cliffs and traverse glacial snowfields. They climb sheer walls of hard-packed snow by digging steps with their front hooves as they make their way up and over the barriers. Surprisingly, they often make those steep climbs when there are pathways available nearby that offer little or no resistance.

Caribou are strong swimmers and readily cross swift rivers or lakes during their travels. Their broad hooves and dew claws act as efficient paddles. They swim with their heads held high out of the water at speeds ranging from 3 to 11 km/hr (Seton 1927; Banfield 1954a; Kelsall 1968). On many occasions, I have watched individual caribou and groups travel along river banks, bypassing easy crossing sites (with shallow or deep, relatively slow-moving water) only to jump into fast-flowing water to cross at relatively dangerous, treacherous sites.

Reproductive Behavior. In September and early October, the bulls strip the velvet from their antlers, become irritable, and engage in sparring bouts with other similar-sized males to establish social rank. Bulls that appear dominant seldom actually fight subordinates, but effectively use their antlers and posturing to reinforce their status (Bubenik 1975a). Other common breeding displays seen during this period are bush gazing (standing motionless, with a fixed distant gaze) (Lent 1965b), bush thrashing, hock rubbing or tramping (taking up a hunched position and in most cases urinating on the hocks) (Espmark 1964a; Lent 1965b), mock battles, rearing, and flailing.

FIGURE 46.3. (Top) Threat pose, exhibited by a female caribou (*Rangifer tarandus*); (bottom) modification of threat pose by male caribou in courtship display. SOURCE: L. J. Miller.

The intensities and kinds of behavior exhibited during the rut vary according to subspecies, habitat, and possibly the nutritional state of the breeders. Open habitat appears to be preferred, if available. Barren-ground bulls do not gather harems (Pruitt 1960a; Lent 1965b; Skoog 1968). Dugmore (1913) suggested that male woodland caribou gather harems; however, Bergerud (1973, 1974c) stated that such groupings are not true harems because there was "no fixed social attachment between individuals," that is, individuals were commonly coming into and going from the breeding groups.

The preliminary phase of the rut involves testing the estrous state of females; males drive females, who flee before them. A modified form of threat (Figs. 46.3 and 46.4) is often exhibited by the male as he drives a female (Pruitt 1960a; Lent 1965b). Pursuits of females by males are often interrupted by fights with other males. In contrast to sparring during prerut, fights are often vigorous encounters (Lent 1965b). Banfield (1954a) reported mortality from rutting fights, and Bergerud (1971c) attributed high mortality among prime bulls to fighting during the rut. Males also stop pursuing the females to bush thrash and bush gaze.

As the rut peaks, prime bulls concentrate on tending estrous females and markedly reduce their foraging. A bull tends only one cow at a time and follows it wherever it goes (Lent 1965b). Tending bulls usually show agonistic behavior only when another animal approaches the tended cow. Tending bulls tramp and bush gaze.

Copulation is rapid and has rarely been observed and reported (Kelsall 1968). Bergerud (1974c) suggested that the much greater weight of the male requires that copulation be brief. Espmark (1964a) believed that for Swedish reindeer, copulation occurred only once with each cow and mostly at dawn or dusk. However, I observed copulation in the Porcupine herd in Yukon during late morning and early afternoon.

Most female barren-ground caribou exhibit fidelity to specific calving grounds (Skoog 1968; Hemming 1971; Fleck and Gunn 1982;

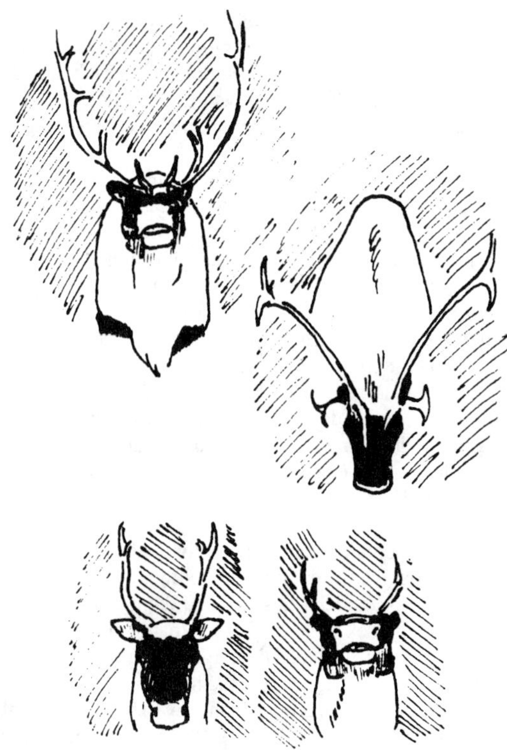

FIGURE 46.4. (Top) Outline and tonal pattern of threat (left) and attack (right) poses as presented by a bull to another caribou (*Rangifer tarandus*); (bottom) outline and tonal pattern of normal (left) and threat (right) poses by a cow to another caribou. SOURCE: L. J. Miller.

FIGURE 46.5. Head-bobbing pose, exhibited by a maternal cow to its newborn calf caribou (*Rangifer tarandus*). SOURCE: L. J. Miller.

Valkenburg et al. 1983; Whitten and Cameron 1984; Davis et al. 1985a, 1986; Cameron et al. 1986; Gunn and Miller 1986; Fancy and Whitten 1991; Gunn and Fournier 2000). On calving grounds, pregnant cows space themselves in smaller groups (maternity bands) (Pruitt 1960a). In Alaska, calving groups were open and transient, but showed both intragroup and intergroup social facilitation (Lent 1966a). Females that drop their calves early or late tend to separate from other caribou, probably because the groups remain on the move. Lent (1966a), Kelsall (1968), Skoog (1968), and Bergerud (1975) all suggested that, during peak calving time, cows about to give birth do not actively seek isolation. Pruitt (1960a) stated, however, that females giving birth during the peak of calving move off a short distance to calve and then rejoin their groups. My limited observations tend to favor Pruitt's, if the cow's group remains in the area, but there is likely considerable individual variation (possibly related to the cow's previous social rank in her group and the strength of her social ties).

Valkenburg and Davis (1986) documented calving distributions of the Fortymile caribou herd from the mid-1950s to the mid-1980s. Although they thought that the Fortymile herd might be a case of infidelity to calving grounds, they acknowledged that such long-term calving ground infidelity is not common. They believed that there must have been good reasons for the shifts, at least initially, although they could not pinpoint any causes. Perhaps the shifts represented a long-term rotation in the herd's use of its calving grounds, which could not be detected because of the relatively short temporal frame of reference. Valkenburg and Davis (1986) concluded that the unpredictability of recent calving locations warrants reconsidering the merit of protecting only calving grounds previously known to be important. I would agree that we should extend range protection well beyond known calving grounds, but not at the expense of not protecting the known ones.

A caribou calf moves off with its mother within hours after birth. The maternal cow uses head bobbing (moving its head up and down on a vertical plane from the ground to about the height of the calf; Fig. 46.5) and vocalizations to strengthen the calf's following response (Pruitt 1960a; Lent 1966a). After calves are mobile, "nursery bands" composed initially almost entirely of the calves and their maternal cows are formed (Pruitt 1960a). Last year's calves (yearlings), nonpregnant females, and some subadult males are first to rejoin cow–calf groups as the maternal cow's intolerance toward them wanes. By midsummer, some bulls and more subadult males join the mixed sex/age groups, but many bulls and some subadult males remain in male-only groups until at least prerut.

Herds of forest-dwelling caribou are likely to exhibit different patterns of dispersal and behavior at calving time than are tundra-dwelling caribou. Different environmental settings demand that caribou use an array of temporal, spatial, and social strategies to minimize predation on their newborn calves (e.g., Adams et al. 1995a). Females of four woodland caribou herds in Quebec and Labrador did not use specific calving grounds (Brown and Theberge 1985; Paré and Huot 1985; Brown et al. 1986). Instead, the parturient cows space-out and become solitary to calve in forest–wetland habitat throughout each herd's range, which may reduce the relative vulnerability of an individual to predation (e.g., Cumming 1975; Bergerud et al. 1984; Bergerud and Elliot 1986; Bergerud and Page 1987; Bergerud 1992). Brown and Theberge (1985) found that the radio-collared cows did, however, individually show a high degree of calving-area fidelity, which apparently involved learning and traditions (e.g., Bergerud 1974d). Similar distributions and patterns of behavior at calving time were found in Alberta and British Columbia (Edmonds and Bloomfield 1984; Hatler 1986; Seip 1990).

The mother licks and feeds its calf within the first minutes of the calf's life, initiating the mother–young bond. As the calf grows, it solicits care from the mother and further develops that bond. Bonding and individual recognition involve visual, auditory, and olfactory stimuli. A strong mother–young bond is necessary for the survival of offspring during the first 6 months of life. Without a strong mother–young bond, most newborn calves would die during the rapid and extensive post-calving movements of the large herds. The mother–young bond assures that a mother provides its young with (1) passive immunity through colostrum and milk, (2) nutrition, (3) thermoregulation by direct contact and indirectly by licking and drying, (4) assistance in traversing difficult terrain, (5) "defense" against predators, (6) optimum environment for rapid development and learning, and (7) behavior patterns that shield the infant from extremes of social and nonsocial stimuli (de Vos 1960; Pruitt 1960a; Lent 1966a; Kelsall 1968; Skoog 1968).

After the death of a mother or its young, the surviving member will retain the bond, sometimes for several days (Banfield 1954a; Lent 1966a; Miller and Broughton 1973). Cows with calves do not readily accept other calves during the postpartum period (de Vos 1960; Pruitt 1960a; Lent 1966a). Pruitt (1960a) observed, however, that some cows deprived of their calves will accept strange calves.

During the first 2 days after a calf is born, it suckles about every 18 min, usually for periods of <1 min. Calves >1 week old nurse about one third as frequently as younger calves (Lent 1966a). In arctic-island caribou, frequency and duration of suckling bouts were significantly reduced within 30 days of birth and further by mid-August compared to late June and mid-July (Miller and Gunn 1982). Lavigueur and Barrette (1992) found that suckling rate and total suckling time for woodland caribou rapidly decreased during the first 20 days and slowly thereafter, whereas suckle duration remained constant and decreased gradually thereafter. They divided the weaning process into metabolic weaning and behavioral weaning, and suggested that metabolic weaning starts at 15–20 days and ends at 40–45 days, but behavioral weaning was not yet completed at 160 days. See Lavigueur and Barrette (1992) for a detailed account of suckling, weaning, and growth in captive woodland caribou.

The most common position for nursing is reverse parallel, which allows the mother to lick and nibble the calf's anogenital and back regions (Lent 1966a; Miller and Gunn 1982). Such caregiving behavior from the mother prolongs the nursing period (Lent 1966a). Calves also nurse from the rear, sometimes while the mother is walking.

Kelsall (1968) stated that suckling was greatly reduced by early July. He thought that biting insects would greatly disrupt nursing after July, so weaning must occur at about that time. Although his conclusion is based on considerable observation, it is difficult to accept. Insect activity should usually be at its worst in July, not after July, and the energy demands are high for the rapid growth of the calf during its first 4–6 months of life. Skoog (1968) suggested that calves are weaned during September–December, most likely before November. Subsequent investigations have shown that weaning does not occur before the end of autumn, at least in woodland caribou in Newfoundland (Butler 1983) and Quebec (Lavigueur and Barrette 1992). The calf probably associates with its mother during most of the first year of life, not only because she might be a source of high nutritional energy, but also because the mother–young bond fosters psychological well-being. Formation of "peer groups" by short-yearlings (calves of the previous year) contributes to deterioration of the mother–young bond and facilitation of the calf's assimilation into its mother's social group.

Agonistic Behavior. Pruitt (1960a) described and illustrated threat and attack poses by caribou (also see Fig. 46.6). He suggested that threat posturing (muzzle extended, antlers back) is used by females to ward off strange calves or adults, by calfless females showing antagonism toward males during the calving period, and by males to challenge other males. Threat posturing also is part of courtship displays during the rut, and is used by dominant animals to defend or scare subordinates away from feeding craters in winter. Attacks (head lowered, antlers presented) occur when the threat pose fails to intimidate or when the individuals are more highly motivated. Attacks commonly occur when two females compete over a single calf, when juvenile animals are testing their strength or playing, and when two or more males are sparring before or during the rut.

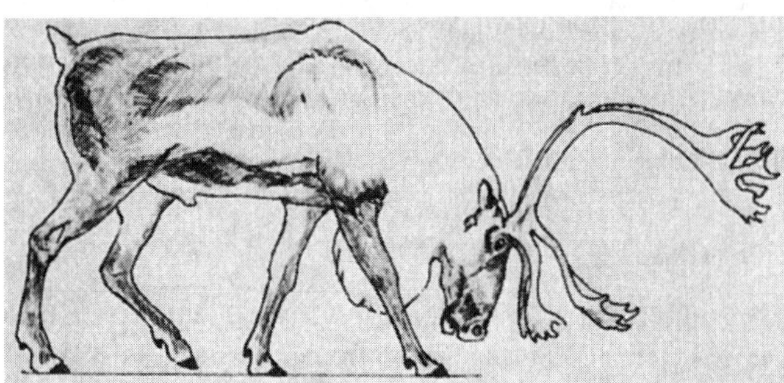

FIGURE 46.6. Attack pose, exhibited by a male caribou (*Rangifer tarandus*). SOURCE: L. J. Miller.

A system of rank order probably exists involving all individuals within a social unit. Age, sex, size, and possession of antlers are all important criteria in determining rank order. Males in prime condition are usually dominant over all other animals during the rut: older, larger bulls with bigger antlers are dominant over younger, smaller bulls with small antlers.

In general, subordinate animals are not allowed into the feeding craters of dominant animals in winter (Henshaw 1968a). Also, dominant caribou will displace subordinate animals from feeding craters. Maternal cows are usually dominant over other caribou in winter. Adult animals in general dominate all juveniles and yearlings. As winter progresses, the social status of short-yearlings (6+ months old) changes as the mother–young bond weakens.

Like most generalizations, there are notable exceptions in caribou behavior in small populations where individual recognition is most probable (and probably also in large populations). Vandal and Barrette (1985) found in the Grand Jardins woodland caribou herd in Quebec that regardless of the sex or age of the initiator and the target of an interaction, the initiator displaced the target and used the resource over 80% of the time. They concluded that the high success rate of initiators resulted from those caribou having a tendency to select targets that they were likely to beat (Vandal and Barrette 1985). If so, this suggests that individual recognition played a major role in their actions and subsequent responses, thereby minimizing the likelihood of injury. I believe this can also apply to individuals in large populations.

Lent (1966a) argued that caribou recognize sex and age classes and that dominance hierarchies are based on classes rather than on individuals. Parker (1972b) and Miller et al. (1975a), however, obtained evidence for long-term associations between caribou, suggesting that caribou do indeed recognize individuals. Bergerud (1973) suggested that establishment of a hierarchy depends on frequent reinforcement through interactions, but it may be necessary to have intensive interactions only for relatively brief periods of the year.

Pruitt (1960a) described and illustrated the unique alarm stance (one hind leg spread, head erect) and excitation jump (rearing, pivoting, and discharging of interdigital glands). Both acts are highly noticeable to other members of the group and thus warn animals that do not actually see the danger. The alarm stance occurs when a caribou is alerted to some unidentified stimulus. The excitation jump usually occurs after the individual has investigated the foreign stimulus by sight or smell and has begun to flee. Low-intensity alarms include head low alerts (brief poses during feeding activities) and head high alerts (more intensive and prolonged poses made by moving, bedded, or feeding caribou).

Predator–Prey Behavior. The caribou has evolved with the wolf, and undoubtedly much of the caribou's behavior has developed as a result of that relationship (e.g., Messier 1995; Seip 1991, 1992, 1995; Thomas 1995). The gregariousness of caribou, especially after calving, probably resulted from wolf–caribou encounters (Bergerud 1971b, 1974d; Miller 1974b). The caribou's use of openings, frozen bogs, and lakes for resting areas is also a learned behavior that seemingly resulted from wolf–caribou interactions. The caribou's apparent reluctance to enter riparian willows and other heavy brush, and its state of alertness when passing through, suggest that it associates such cover with attack by wolves and bears. Caribou have learned to distinguish between hunting and nonhunting wolves. The development of specific flight releasers (Pruitt 1965) and threshold distances allows caribou to conserve energy by not taking unnecessary escape during wolf–caribou encounters. However, female caribou living with wolves in Alaska exhibited vigilance about six times more often per hour than did females with calves on wolf-free ranges in West Greenland (Bøving and Post 1997). Vigilance against predators is at the expense of feeding and resting.

Sexual segregation apparently occurs as a result of differing energetic and reproductive causes (Main and Coblentz 1990). The resultant differences in preferences for habitat types exhibited by female versus male caribou influences their relative exposure to predators, especially at calving time and during winter (e.g., Geist 1982; Bowyer 1984; Clutton-Brock et al. 1987; Jakimchuck et al. 1987; Fancy and Whitten 1991). It appears that because males are intent on optimizing forage resources, they select habitats that are favored by predators. Parturient and maternal females, however, select habitat more removed from most predators and thus more suitable for rearing their calves. Selection by females appears to be an antipredator strategy, whereas selection by males appears to reflect their greater need for an early start on new plant growth and their ability to cope with deeper snow cover. More recently, Young and McCabe (1998) found no avoidance of river corridors by calving barren-ground caribou of the Porcupine herd or preference for river corridors at that time of the year by the grizzly bears. They therefore proposed that those "caribou reduce the risks of predation to neonates by migrating to a common calving grounds, where swamping is the operational antipredator strategy." I think this proposition is only half true. Swamping helps, but it is the selection of calving grounds where relatively few wolves occur during calving (predator avoidance) that results in major reductions in the deaths of neonates to wolves by greatly improving the ratio of neonates to wolves and thus making swamping much more effective. They hypothesized that nutritional demands, not predator avoidance, ultimately regulate habitat use patterns. This is necessarily true only at a threshold level below which the caribou cannot produce and rear viable young. Calving caribou of the major migratory herds in at least the Northwest Territories and Nunavut could obtain more forage with less effort (earlier snowmelt) and higher quality (earlier new growth) forage if they stayed south of their calving grounds—where the bulls and many subadult males are found at that time. Thus, it seems that avoidance of wolves is the major operative strategy for calving caribou as well as calving woodland caribou (Bergerud and Page 1987).

POPULATION DYNAMICS

Reproduction, mortality, ingress, and egress limit the number of caribou in a population. Mass emigrations are not predictable within a specific time frame. If caribou in a population are suffering advanced nutritional stress and reduced reproduction and subsequent survival of their young over several years or more, they will eventually change their range or die. Also, the reliability of any evaluation of change in range use depends on the amount of information that is available for the population. We have not yet studied any caribou population long enough or in enough detail to know its long-term (50–100+ years) range-use pattern(s).

Dispersal may be related to density, that is, a threshold density triggers social intolerance and leads to dispersal of some segment of the population. This has not been demonstrated for caribou. Also, it is difficult to perceive because of the gregariousness of caribou in general, and especially because of the large aggregations during seasonal migrations and the formation of huge (±10,000) postcalving aggregations, which suggest that caribou are tolerant to extremely high densities (e.g., Whitten 1996). In addition, and most important, since "density" is generally defined as the number of animals divided by the area used, density becomes a confusing and usually poorly measured parameter. However, advanced nutritional stress may be all the catalyst that is needed to trigger dispersal. In some cases, even exceptionally severe predator harassment could cause dispersal.

Mass movements into or out of a caribou herd are so rare that it is reasonable to think of a herd as a closed population. Therefore, a population increases when recruitment exceeds mortality and decreases when the opposite occurs. This basic assumption is the key to all caribou monitoring programs.

Natality. Fertility and birth rates in a caribou population usually have been considered similar among years (Skoog 1968; Parker 1972a; Bergerud 1974a; Dauphine 1976). However, among-year variation can be and often is appreciable (about 50–100%) throughout many years (Miller et al. 1977a; Parker 1981; Whitten and Cameron 1983; Cameron et al. 1988; Davis et al. 1988, 1991; Farnell and McDonald 1988; Couturier et al. 1990; Crête et al. 1996; Whitten 1996; Valkenburg 1997; Miller 1998; Thomas and Kiliaan 1998b). Although generally considered rare in caribou, in some years disease or poor nutrition can cause relatively high intrauterine mortality in some herds (Skoog 1968; McGowen 1966; Neiland et al. 1968). When that happens, the birth rate declines relative to the fertility rate.

Extreme within- or among-year variation in annual birth and survival rates can and does occur in caribou populations on the Arctic Islands (Parker et al. 1975; Thomas et al. 1976, 1977; Miller et al. 1977a; Thomas and Broughton 1978; Thomas 1982; Miller 1995b, 1998). Pregnancy rates for caribou sampled on the southern tier of Arctic Islands were 40% in 1974, 92% in 1975, 73% in 1976, and 100% in 1977. Pregnancy rates for Peary caribou sampled on the High Arctic Islands remained fixed at only 6–7% in 1974–1976 after the cataclysmic winter and spring of 1973–1974, but rebounded to 88% in 1977. Newborn calves represented <1% of all caribou seen during aerial surveys of Peary caribou on the western Queen Elizabeth Islands in summers 1972, 1974, 1996, and 1997 (Miller et al. 1977a; Miller 1998; Gunn et al. 2000; Gunn and Dragon 2002). In contrast, in the favorable year of 1993–94, Peary caribou calves on south-central Queen Elizabeth Islands represented 28% of 2400 caribou segregated in August 1993 and there still were 97 calves:100 breeding cows (Miller 1995b). Rates of early survival of Peary caribou and arctic-island caribou calves and recruitment to 1 year of age exhibit the same degree of extreme yearly fluctuation (Miller 1990b).

Skoog (1968), Kelsall (1968), Dauphine (1976), and Parker (1981) found an age-dependent pregnancy rate in caribou. In addition, Dauphine (1976), Thomas (1982), Reimers (1983), Allaye-Chan and White (1991), Cameron et al. (1993), Cameron (1994), and Thomas and Kiliaan (1998b) found relationships between body weight and fat reserves and pregnancy in caribou and reindeer. Poor condition of females on Coats Island was associated with light fetuses in May 1984 (Adamczewski et al. 1987). Davis et al. (1991) suggested that in addition to body weight (and I would think fat reserves), age plays an important role in determining pregnancy probabilities. Body weight, seemingly, must be a relative consideration, as caribou of different subspecies, forms, or herds obviously exhibit different weight ranges from birth onward. Adamczewski et al. (1987) concluded that substantial changes in rumen fill made total body weight a much poorer predictor of condition than carcass weight. Cameron et al. (1991) emphasized that changes in body weight did not necessarily clarify the role of condition in the reproductive process and that more emphasis should be placed on reliable *in vivo* techniques for estimating body composition.

Female caribou periodically or sporadically fail to reproduce every year: on average, once in 4 years in the Alaskan Central Arctic herd (Cameron 1994); once in 5 years in the Kaminuriak herd of north-central Canada (Dauphine 1976); and once in 9 years in the Alaskan Delta herd (Davis et al. 1991). Davis et al. (1991) investigated "pregnancy resting" in females that first reproduced at 2 years of age and in older females. They did not find clear patterns and concluded that "it is unlikely that this phenomenon is important if lactating costs over the summer, rather than merely producing a calf, is the mechanism resulting in insufficient body condition to breed during the fall rut." They also suggested, "Even if pregnancy resting is important, individual variation may be so great as to mask its effects on productivity." However, Cameron (1994) thought that a shift in mean frequency of reproductive pauses by adult females was an important but generally unrecognized mechanism that could cause the parturition rate of a herd to change markedly.

Free-ranging caribou calves are seldom bred, and it is unlikely that many females bred as calves could produce and rear young successfully, because they are physiologically unsuited for the task. Even the yearling females in many populations are not ready to be mothers. They usually lack the necessary fat reserves and they are still growing during the second winter of life. Also, young, primiparous mothers may lack the psychological adjustment for calving, and the subsequent mother–young bond either does not develop or is weak (Skoog 1968; Miller and Broughton 1974). Most females have to live at least 3 years before they successfully produce and rear their young. Thomas and Kiliaan (1998b) found that over an 8-year period, about 55% of the calves in the Beverly herd were produced by females 3–6 years of age. Fecundity varied markedly during those 8 years: 0–33% in yearlings and 64–100% in older females (Thomas and Kiliaan 1998b).

Forage. The ultimate limiting (regulating) factor on caribou populations is theoretically their maximum seasonal forage supplies. Forage production and availability give a true approximation of the range's maximum ability to support caribou. Although the density dependence of this consideration is seemingly obvious, threshold values would vary according to relative quality and quantity of range and have not yet been documented satisfactorily.

Bergerud (1974a) stated that a winter shortage of forage for free-ranging caribou has not changed birth or death rates in any populations; he has held that position throughout the years (Bergerud 1996). However, data on those rates are difficult to obtain and the lack of evidence is not conclusive proof that such changes have not occurred. Relative forage shortages occur in winter due to adverse snow cover and icing conditions. Absolute forage production as a regulating factor would be density dependent. Other mortality factors probably do not allow caribou numbers to reach such critically high population levels on most ranges that absolute forage availability becomes the regulating factor (e.g., Caughley and Gunn 1993; Valkenburg et al. 1996; Whitten 1996; Miller 1998; Gunn et al. 2000; Gunn and Dragon 2002). Major to cataclysmic winter/spring die-offs among Peary caribou have resulted in as much as 50–85% declines in the estimated population sizes in 1 year and a cumulative 95% decline in 3 consecutive years (Miller et al. 1977a; Miller 1998; Gunn and Dragon 2002). Gates et al. (1986) documented a 71% winter die-off in barren-ground caribou on Coats Island along with a near total loss of calves in the following spring. Major losses to near total calf crop failures also are associated with those years and can occur when no appreciable increase in 1+-year-old mortality is detected. These losses can best be explained by extremely unfavorable and prolonged snow and ice conditions leading to extreme undernutrition and starvation due to widespread relative unavailability of forage (Miller et al. 1977a; Gates et al. 1986; Miller 1998; Gunn et al. 2000; Gunn and Dragon 2002).

Weather. Weather, alone or combined with mortality factors, could hold a caribou population at a low level where density-dependent factors are not effective (Skoog 1968; Miller 1974c, 1998; Miller et al. 1977a; Caughley and Gunn 1993; Valkenburg et al. 1996; Whitten 1996; Gunn et al. 2000; Gunn and Dragon 2002). Ultimately, caribou likely are governed more by precipitation in the form of snow cover, icing, freezing rains, and cold rains than by all other environmental factors. To some degree, snow and ice determine where, how, and if caribou can live for most of the year. Precipitation that greatly restricts forage availability, combined with low temperatures and high winds, likely results in high mortality. Prolonged and severe restrictions of forage supplies lead to malnutrition and subsequent death by starvation or predation. Although there is little direct evidence for starvation caused by adverse weather conditions among all sex and age classes in free-ranging caribou on the mainland, there is evidence for it among Peary caribou on the Canadian High Arctic Islands (Parker et al. 1975; Miller et al. 1977a; Miller 1998; Gunn et al. 2000; Gunn and Dragon 2002) and for barren-ground caribou on Coats Island (Gates et al. 1986). In addition, timing and amount of precipitation will govern the absolute abundance and quality of the annual forage supply. Even if mortality is attributed to starvation caused indirectly by weather conditions, the limiting factor (regulating factor of Skoog 1968) is the existing weather, not the absolute food supply.

Bergerud (1996:105) stated, "Weather is not a sufficient mortality factor, either nutrition or predation are necessary interactions to bring death and influence population dynamics." That view seems to ignore reality when prolonged, extreme undernutrition leading to death or markedly greater susceptibility to predation is strongly influenced or essentially caused by weather. When the caribou would not die in the absence of extremely unfavorable weather, then the paramount factor is the weather!

Behnke (2000 : 142) suggested that caribou are in a nonequilibrium grazing system where "population crashes represent one of the mechanisms that promote the persistence of such systems despite the instability and harshness of the environments in which they operate." The caribou population in such a system appears to be mainly weather driven, that is, to be regulated mainly by sporadic (unpredictable) abiotic variables (e.g., Caughley and Gunn 1993; Behnke 2000). Whitten (1996 : 45), working on the Porcupine caribou herd in the Yukon, declared that "There is no true density dependent regulation and no equilibrium in this system." Caughley and Gunn (1993) suggested that in such weather-driven systems, we should compare rate of increase in the animal population and density of forage biomass rather than animals per unit area. Whitten (1996) concurred by suggesting that we should be thinking of available resources per caribou rather than caribou/km^2. I agree; we should recognize the shortcomings of thinking solely in terms of density of caribou per unit area without having any measure of the available forage in that unit area. Instead, we should turn our attention to the feasibility of obtaining accurate measures of the available forage per caribou over huge areas of caribou range (ideally, both quality and quantity)—the density of forage biomass of Caughley and Gunn (1993). In addition, the overall mean density of a caribou population based on its supposedly known annual home range size is subject to too many probable errors both in terms of the accuracy of the measurement of annual home range size and lacking or limited year-round knowledge of how the animals actually use that range on daily, seasonal, and annual bases. Unfortunately, knowing what to measure is only the first step. The paramount problem remains: How do we make such measurement over meaningfully large areas of range at all seasons at satisfactorily frequent intervals on an ongoing basis among years to obtain usable data for the biologically sound management and conservation of caribou populations?

If limiting factors drive the system most of the time and regulating factors seldom do, should we be more concerned about limiting or regulating factors? It could be argued that regulating factors never would come into play in a properly managed caribou population because densities would be kept below thresholds where regulation would or could be meaningful (e.g., Skoog 1968; Miller et al. 1977a; Caughly and Gunn 1993; Valkenburg et al. 1996; Whitten 1996; Miller 1998; Gunn et al. 2000; Gunn and Dragon 2002).

MORTALITY

Calf Mortality. High calf mortality on calving grounds is common among all forms of caribou (Banfield 1954a; Pruitt 1961; Kelsall 1968; Skoog 1968; Bergerud 1971c; Miller and Broughton 1974; Bergerud and Elliot 1986; Edmonds 1988; Miller et al. 1988a; Valkenburg et al. 1991, 1996, 1999; Whitten et al. 1992; Adams et al. 1995a, 1995b; Boertje and Gardner 1998, 1999; Stuart-Smith et al. 1997; Valkenburg 1997). Most neonatal mortality occurs mainly within the first hours or days of life.

It appears that most mortality among calves on the calving ground, and probably throughout the first year of life, results from predation. Predation by wolves on many calving grounds can be severe and often exceeds the wolves' needs (Miller and Broughton 1974; Miller et al. 1985, 1988a, 1988b; Adams et al. 1995a; Wildlife biologist L. Camps in Bergerud 1996). On some western calving grounds, grizzly bears and/or golden eagles can be the dominant predators (Whitten et al. 1992; Adams et al. 1995b). In the mountains of British Columbia, Alberta, northern Idaho, and northeastern Washington, the mountain lion is likely an important predator of calves and other caribou (e.g., Compton et al. 1995). As noted previously, black bears, lynx, and coyotes can be important predators on some caribou, particularly on eastern ranges (Bergerud 1971c; Mahoney et al. 1990; Crête et al. 1991).

Predation on newborn calves varies spatially and temporally. It has been suggested, but not proven as an absolute, that wolves kill mostly those young that would not survive anyway. Although wolves readily take sick and lame calves, they must take healthy calves when there are few sick or lame animals available. Therefore, evolution of antipredator strategies should have long-term benefits at or about the time of calving by reducing annual losses of neonates when they are at their most vulnerable (Bergerud et al. 1984; Bergerud 1985, 1988, 1992; Bergerud and Elliot 1986; Bergerud and Page 1987; Jakimchuk et al. 1987; Seip 1990).

An interesting supposition proposed by K. A. Neiland (in Klein and White 1978) is that wolves perpetuate the occurrence of parasites and disease in caribou populations by their interactions with caribou, assuring themselves of an easily obtainable food source. Considering the transmission of brucellosis in reindeer in Alaska, arctic foxes also may perpetuate parasites and disease (Morton 1986). This condition may then explain the usual rapid departure of cows with their calves at heel from the calving grounds shortly after calving, thus considerably reducing the chance of exposure to parasites and disease than if they lingered at relatively high densities in these proportionally small areas of their range.

Other causes of caribou calf mortality are varied and include fetal and neonatal atelectasis, pathophysiological disorders, pneumonia, shock, stress (adverse weather and/or entrapment in frigid waters or slushy mires along the shores of lakes and streams), drownings (while crossing streams and lakes), malnutrition/starvation, broken leg(s), hydrocephalus and other anomalies, and separation from or abandonment by the mother.

Precipitation, cold, and wind are a deadly combination for newborn calves, and often result in hypothermia or respiratory problems such as pneumonia and subsequent death. Soft, deep snow traps calves and they subsequently die. Weather also severely limits the availability of forage to parturient cows during gestation and indirectly causes calf mortality. This is critically important to Peary caribou in the Canadian High Arctic, where near-total calf crop failures may occur, sometimes due to the failure of many or most cows to get pregnant (Thomas et al. 1976, 1977; Thomas and Broughton 1978; Miller et al. 1977a; Miller 1998). The relative importance to barren-ground caribou populations of adverse weather during and shortly after calving is unclear. Even though the importance of adverse weather is often speculated on, it remains debatable based on the limited existing samples (Miller and Gunn 1986b).

Most accidental deaths and some predation of calves result from their investigative behavior, lack of maternal care, or the agonistic or escape behaviors of other animals (intraspecific factors; Skoog 1968). A clear distinction should be made between mortality of newborn calves due to interspecific strife and deaths caused by accidental injury or predation (e.g., Miller and Gunn 1986a). Other deaths are due to drownings, trampling, injuries caused by hostile adults (foreleg kicks and/or antler punctures), falls from cliffs, and sprained or broken limbs that prevent the calves from keeping up with their group. Maternal cows usually remain with their injured calves, but such young are usually easy prey to predators.

Poor maternal nutrition can result in death of the mother before parturition, malpresentation, stillbirth, physiological disorders of various organ systems, physical abnormalities of the calf, or neonatal death through malnutrition of the fetus. Mothers can transmit fatal diseases to their young. The frequency of disease in calves is not known, but epidemic brucellosis has occurred in barren-ground caribou of the Arctic herds in Alaska (Neiland et al. 1968). Disease was the single most important killer recorded for reindeer calves in Russia during a 20-year period (Baskin 1983). Baskin (1983) concluded, however, that the major cause of reindeer calf mortality was in fact ultimately the lack of a stable mother–young bond, which often led to calves becoming diseased. Predation ranked only 5th among 11 factors and accounted for only about 10% of the deaths in Russian reindeer.

Adult Mortality. Factors that predispose caribou to accidental death or predation are their investigative behavior; their nutritional state; advanced (old) age; adverse weather (precipitation, cold, and wind); predation by wolves, bears, mountain lions, and other predators; and limited forage. Forage is limited possibly by production, but usually by weather-related availability. Much of adult mortality due to accidents and predation is the direct or indirect result of intraspecific behavioral and physiological characteristics. Investigative behavior can lead to drowning by falling through weak ice, falling from cliffs, and being caught in snow slides. Sprained or broken limbs from travel on dangerous terrain leads to subsequent death by predation. Agonistic and reproductive behaviors can lead to death by kicking or antler punctures, or cause injuries that lead to subsequent death by predation. Examples of mortality proximally caused by wolf predation but ultimately the result of intraspecific physiological characteristics that increase vulnerability include the following: (1) bulls that fatten before the rut can run fast for any distance only with great difficulty and tire quickly, (2) prime bulls are often very lean after the rut and do not build up sufficient reserves in all years to carry them through to spring, and (3) near-term cows have a heavy fetus and high energy requirements in late winter and spring (Skoog 1968).

Environmental (interspecific; Skoog 1968) mortality factors should take the greatest toll on caribou over the long term, but there has been extended uncertainty about the principal mortality factors among herds. Many early workers (Leopold and Darling 1953; Edwards 1954; Scotter 1964; 1967a) believed that range deterioration, caused mostly by fires and land clearing and logging, on southern ranges was the principal cause for the decline of caribou throughout much of North America. Cringan (1957) and Banfield and Tener (1958) blamed much of the decline on overhunting or range deterioration. Banfield (1954a), Banfield and Tener (1958), Sonnenfield (1960), and Bergerud (1974a) all believed that predation by wolves or humans was the principal cause. All of these mortality factors, along with weather and likely parasites and disease, have had marked impacts on caribou numbers at some time or another.

Predation. Caribou evolved with the wolf. Only in relatively recent times have wolves been extirpated or seriously reduced in numbers on the southern and eastern ranges of caribou. Wolves are not currently on insular Newfoundland. Wolf densities remain high among most barren-ground caribou populations. They are possibly high relative to the low numbers of Peary caribou and arctic-island caribou, and muskoxen are alternative prey in High Arctic regions. Moose, thinhorn sheep, and mountain goats are alternative prey for wolves on most Alaskan and Yukon ranges. Deer and moose are common on most of the more southerly ranges of woodland caribou, as are elk, sheep, and goats on southwestern ranges. The presence of alternative ungulate prey species can serve at times as a buffer to the caribou, but at other times it can increase predation on the associated caribou, especially when the caribou represent only a minor proportion of the total ungulate prey base.

Wolves are a natural part of the caribou's environment. However, wolves are large carnivores, which require considerable amounts of protein to reproduce and survive. Caribou are important prey for wolves (Murie 1944; Banfield 1954a; Kelsall 1968; Kuyt 1972; Bergerud 1978a, 1983, 1988; Davis et al. 1983; Gauthier and Theberge 1986; Seip 1992; Dale et al. 1995; Mech et al. 1995; Farnell et al. 1996; Rettie and Messier 1998). A wolf requires 11–14 caribou annually (Clarke 1940; Kelsall 1968; Parker 1972b) if caribou are its major prey. Empirical and theoretical ratios for wolves being in balance with their ungulate prey base have been reported as 1 wolf:50 caribou (Bergerud and Elliot 1986), 1:100 (Bergerud 1980), and 1:137 (Walters et al. 1978). Thomas (1995) discussed these wolf:caribou ratios, wolf densities versus caribou dynamics, and other wolf–caribou relationships in Canada. This issue is further complicated by the fact that on occasion wolves kill in excess of their needs (Eide and Ballard 1982; Miller et al. 1985, 1988a, 1988b).

Large numbers of wolves have a potentially large impact on a caribou population, based on estimates of a wolf's required annual kill of caribou. Supposedly, <1 wolf/100 caribou could control the growth of a barren-ground caribou population (Parker 1972a). Skoog (1968) stated that predation was the greatest single mortality factor in most wild populations of Rangifer.

Although Crisler (1956) and Kelsall (1968) stated that caribou, except the incapacitated and calves, could normally outrun single wolves, Murie (1944), Banfield (1954a), Skoog (1968), and other field workers observed that wolves are capable of overtaking healthy caribou. Such feats are not always necessary, as wolves often take prey by surprise or ambush (Miller 1974, 1975). When in pursuit, wolves also take advantage of caribou that leave hard snow and attempt to cross soft snow areas. Wolves also run relays and involve several pack members in herding and ambushing prey. Skoog (1968) suggested that wolves can obtain whatever caribou are available, and that the final selection is made more by chance than by design. Mech et al. (1995) found, however, that vulnerability of prey was related to youth, old age, poor condition, and hindrance by snow and that it varied by sex, time of year, and snow depth. Sick and lame animals may be selected, if available, but if the caribou population is healthy and there are few incapacitated individuals, the wolves will take healthy animals. Wolves can and sometimes do control and even depress caribou populations. Also, it is possible that when a caribou population has been reduced to a remnant ($\leq$30 animals), wolves could extirpate such an extremely small population.

Diseases and Parasites. Although much information has been collected in the past 20 years, the importance of disease and parasitism as a control of caribou numbers is not yet satisfactorily known. Much literature exists on this subject, especially for domestic reindeer, because of economic considerations and close-herding practices. A detailed bibliography of parasites, diseases, and disorders of *Rangifer* and other ungulates was compiled by Neiland and Dukeminier (1972). There is more recent information from Alaska by Dieterich (1981) and Zarnke (1996) and by Gunn et al. (1991) for the Kitikmeot Region of the Northwest Territories/Nunavut. Disease and parasite burdens in caribou are endemic and normal in most populations. Epidemic disease in a free-ranging population of caribou occurred in Alaska (Neiland et al. 1968; Skoog 1968). *Brucella* organisms were associated with orchitis–epididymitis, bursitis–synovitis, and metritis, singly or in combination in Alaskan caribou (Neiland et al. 1968). Examination of barren-ground caribou in Canada found that brucellosis was not a serious threat (Broughton et al. 1970).

In West Greenland, Clausen et al. (1980) found that colibacillosis with polyarthritis from *Escherichia coli* (*E. coli* 055) caused high summer mortality among caribou calves. They assumed it was caused by overstocking and intense feeding by concentrations of caribou on blue grass (*Poa pratensus*) short-grass swards ("golf greens"). Therefore, they concluded that the problem would be self-correcting when the caribou density decreased on the preferred habitat of the blue grass.

Another important disease in Alaskan caribou is necrobacillosis (Skoog 1968). The causative agent (*Spherophorus necrophorus*) usually enters through lesions in the feet or mouth, and results in hoof rot or necrotic stomatitis. Banfield (1954a) considered that actinomycosis (lumpy jaw) was widespread in Canadian caribou. Miller et al. (1975b) and Doerr and Dieterich (1979) suggested, however, that most, if not all, mandibular deformities are not of actinomycotic origin. Fibropapillomas occur in Alaskan (Skoog 1968) and Canadian caribou (Broughton et al. 1972).

The caribou is host to relatively few species of external parasite, but the species that do parasitize them are common and cause considerable harassment during summer. Harassment by blood-sucking and parasitic flies can lead to injuries when the caribou flee, and subsequently to death by predation. Severe attacks by mosquitoes (*Culicidae*) and black flies (*Simuliidae*) have reportedly directly resulted in caribou deaths (Skoog 1968).

The warble fly (*Oedamagena tarandi*) and the nose bot (*Cephenomyia trompe*) parasitize caribou (Banfield 1954a; Skoog 1968; Kelsall 1975; Dieterich and Haas 1981; Parker 1981; Thomas and Kiliaan 1990), but they probably are often more important in influencing the ecology of caribou than as pathologic agents. Warble larvae could affect caribou by causing allergenic responses, nutritional imbalance, and secondary infections (Dieterich and Haas 1981). Thomas and Kiliaan (1998a) suggested that it was not the presence of warble larvae that caused lower fat reserves in female caribou of the Beverly herd, but the summer fly harassment that results in lower fat reserves during the next winter. Heavy infestations of nose bot larvae severely distress caribou and may weaken infected animals.

Internal parasites include various cestodes, some nematodes, and at least one trematode (Banfield 1954a; Skoog 1968; Parker 1981; Huot and Beaulieu 1985; Thomas 1996). The common forms are *Taenia hydatigena,* found mosty in the liver; *T. krabbei,* mostly in the muscle; and *Echinococcus granulosus,* mostly in lungs and occasionally in the liver. Again, the importance of these internal parasites is not known, but they have general debilitating effects on the host and could predispose them to predation. The thread lungworm (*Dictyocaulus hadweni*) is associated with debilitated caribou and supposedly causes their death. Skoog (1968) suggested that this parasite might cause mortality to nutritionally stressed caribou, but there is no satisfactory evidence for it. Lungworm is probably linked to some recent herd reductions in Alaska and it should still be considered a serious potential problem (P. Valkenburg, Alaska Department of Fish and Game, Fairbanks, Alaska, pers. commun., 2000). Besnoitiosis occurs in barren-ground caribou in Canada (Choquette et al. 1967) and Q fever in Alaskan caribou (Hopla 1975). The giant liver fluke (*Fascioloides magna*) was common in caribou of the George River herd (Lankester and Luttich 1988). Fruetal and Lankester (1989) identified 21 species of parasitic helminths in wild and captive caribou. Lankester and Hauta (1989) concluded that identification of *Paraelaphostrongylus andersoni* from three widely separated regions of Canada supports the belief that the parasite might have originated in caribou rather than in white-tailed deer. Two protostrongylid nematodes (*E. rangiferi* and *P. andersoni*) are present in Newfoundland woodland caribou and can be important in the deaths of individuals <1 year old, especially males (Lankester and Fong 1998).

Dauphine (1975) concluded that 51 woodland caribou captured in Quebec and released in the Cape Breton Highlands National Park of Nova Scotia most likely succumbed to neurological disease caused by the meningeal worm *P. tenuis.* Lankester and Fong (1989) concluded that cerebrospinal elaphostrongylosis is a significant disease in Newfoundland caribou caused by an introducted parasite, *E. rangiferi.*

Accidents. Although accidents usually contribute little to overall mortality in a caribou population, on some occasions they cause a substantial number of deaths. Animals may drown while crossing treacherous stretches of river, by falling through weak ice, or by becoming trapped in the water by high ice shelves (e.g., Kelsall 1968; Doucet et al. 1988; Miller and Gunn 1986c). Many accidents occur when caribou are in large migrating aggregations and traveling over unfamiliar terrain. They often result in injuries that increase vulnerability to predation. This is more important when aggregations are harassed by predators and especially when they are exposed to intense novel stimuli such as harassment by low-flying aircraft and all-terrain vehicles.

Physiological Disorders. Physiological disorders cause very few deaths among adult caribou. Most such deaths occur among females experiencing complications in fetal delivery (Miller and Broughton 1974; Miller et al. 1988a). Additional mortality can result from organ failures brought on by stress in caribou of either sex or any age.

ECONOMIC STATUS

In the past, caribou were all-important to Arctic and Subarctic Native cultures. Caribou meat was the staple diet for humans and dogs; caribou hide and sinew provided materials for clothing, tents, and sleeping bags; and hides, sinews, antlers, and bone were used for many tools and weapons. Caribou, especially the migratory herds, were the key to the very existence of Inuit and Indian (Dene) cultures on the central tundra and adjacent taiga (Kelsall 1968). Indians (Dene) in the boreal forests might have survived without the caribou, but to do so they would have had to alter their cultures markedly. Abundance of caribou meant good times and shortage meant hardships and sometimes slow death from malnutrition and starvation.

Caribou remain to this day a staple in the diets of many northern people and are still valued as an integral part of their chosen (traditional) ways of life. Until those Native people choose an alternative lifestyle that does not have a great cultural dependency on caribou, it will be difficult to measure the replacement value of caribou.

To estimate the aesthetic value of caribou is a nebulous exercise. The price would be appreciable. But is this market-price approach to the value of caribou a legitimate consideration beyond the point of sustained annual yields and/or recreational pursuits? The caribou is a natural entity in Arctic and Subarctic environs. Would those regions not lose their very character with the passing of caribou?

MANAGEMENT AND CONSERVATION

This chapter is concerned with the management and conservation of caribou as a renewable resource, not merely with caribou as an entity in Arctic and Subarctic natural systems. Biologists are interested in the sustained use of the resource by humans as well as the continued perpetuation of the species (which includes wolf predation on caribou). Therefore, the concept of dynamic equilibrium is not a functional basis on which to accept the wolf–caribou relationship.

Caribou literature before 1980 was relatively sparse compared to the body of literature that has accumulated in the last 20 years. By 1980, only two international reindeer/caribou meetings had taken place, but six more have been held since then. The first North American Caribou Workshop was held in 1983, and eight have taken place since then. There also is a wealth of information in Alaska Department of Fish and Game, Federal Aid in Wildlife Restoration reports produced before 1980, and especially since then.

Students and all caribou biologists should familiarize themselves with the proceedings from these symposia, conferences, and workshops. These proceedings, together with literature from various scientific journals cited in the articles in the proceedings, give a solid foundation to work from. It will become apparent that the caribou (and reindeer) literature is relatively long on personal opinions, general suppositions, and poorly thought out and formulated hypotheses, and relatively short on strictly objective facts and validations. This is meant more as an objective observation than a criticism per se, as this shortcoming seems inevitable when people work with limited, fragmentary data sets and often no data. This is particularly true when they try to make detailed interpretations and overall evaluations between or among subspecies, phenotypes, or populations. This shortcoming is most easily seen in papers dealing with comparisons among various forms of caribou, their ranges, numbers, and reasons for population increases or declines. For example, use of extremely small samples of animals (e.g., ≤10–20) from a population of 50,000–100,000 or more to interpret and evaluate significant changes in various population parameters such as pregnancy rates should be strongly discouraged.

Miller (1982) extensively discussed determining sex and age structure of a population, evaluating physical condition, and livecapture and marking of caribou. Those techniques and procedures still apply, but our capture and marking efforts have advanced considerably in the last 20 years. New immobilizing drugs have been used with success. Perhaps most important, highly successful aerial net-gunning has been developed (e.g., Barrett et al. 1982) and used by many caribou biologists in preference to drugs. Also, those agencies that can afford it have contracted the services of talented professional capture teams. Especially meaningful advances have been in satellite telemetry, which has allowed us to gain previously unattainable information on daily, seasonal, and annual range use patterns and movements/migrations (e.g., Gunn and Tomkiewicz 1991). Recently, the addition of global positioning system (GPS) software in combination with satellite telemetry packages has refined old and created new applications. We should, however, remember to temper our zeal for investigation with the reality of limitations imposed by even this advanced technology. Unrealistic fine-grained assessments that look good in the short term may contain erroneous perspectives that may go undetected and confound future endeavors, interpretations, and evaluations.

Many attempts at managing big game species in North America on a biological basis have been plagued by overriding sociopolitical considerations. Perhaps no other ungulate species has suffered as much from sociopolitical pressures (including Native rights and provincial, territorial, state, federal, and international jurisdictional questions) in the face of opposing biological concerns as the migratory caribou of North America. Co-management between Native user groups and wildlife agencies is now being heralded as a significant step forward in caribou conservation and management, especially of the large migratory herds. As yet, however, co-management efforts are essentially untested, and it will require Natives going beyond cultural value systems and accepting the value and usefulness of wildlife techniques. In turn, wildlife biologists must give due consideration to the possible value of Native positions on the subject. Although there is no doubt that the need for cooperation between users and biologists is now mandatory for successful conservation and management of caribou, the potential for meaningful, large-scale efforts in the near future remains problematic.

Traditional ecological knowledge (TEK, ideally from Native people who lived full-time on the land) has moved to the forefront in caribou issues as a pertinent part of management exercises (e.g., Ferguson and Messier 1997; Ferguson et al. 1998). We must be cautious about TEK perspectives and subject them to rigorous evaluation, much in the manner of Western science (excluding statistical testing). Some caribou biologists are active disciples of TEK, but the main literature has been prepared by social scientists. Many social scientists have not understood the biology or ecology of caribou or what has actually occurred. Social scientists accuse wildlife biologists of not listening to Native hunters. Yet most caribou biologists easily identify with Native hunters because of their own personal strong interests in hunting, fishing, and being on the land. Usher (2000:183) broke stride with other social scientists when he concluded, "It is important that TEK be comprehensible and testable as a knowledge claim in public reviews, and usable for ongoing public monitoring and co-management processes." The acid test is having TEK provide us with something that wildlife biologists do not know that will meaningfully improve management efforts.

If we cannot regulate the harvests of caribou on a biological basis, we cannot hope to manage the resource. Fortunately, during periods of population highs, our lack of control is moot, as annual demands for use of caribou do not exceed levels that the various populations can sustain. Unfortunately, at some point, annual demands for use of caribou will exceed sustainable levels. It appears that caribou are at best to be managed only during periods of crisis, and then only if the users can be convinced that there is a crisis. The matter is even more complicated when organized sport hunting run by Native or non-Native 'outfitters' and/or commercial harvests by Native groups represents a large portion of an annual harvest. This extension of the exploitation of the caribou beyond traditional uses can create serious problems when the caribou population is in a state of decline. Ironically, the desire to obtain ongoing profits from this nontraditional utilization of the caribou resource can lead to greater emphasis being placed on sociopolitical concerns rather than due attention to biologically sound management for the long-term use of the caribou population at sustainable levels.

There is an overall lack of effectiveness in management and conservation of caribou at the level of the geographic population. The estimated number of caribou in North America has essentially doubled since the 1970s. The significant increases in caribou populations over the past 20–30 years have occurred independent of any endeavor by caribou biologists, their agencies, or any other groups or individuals involved; some increases occurred in spite of human actions. One possible exception is the Western Arctic caribou herd in Alaska, where at a critical time the annual harvest was reduced about 10-fold apparently through enforced regulation by the Alaska Department of Fish and Game (Davis et al. 1978, 1980, 1985b). Hunting of some woodland caribou populations was stopped or severely reduced (D. C. Thomas, Canadian Wildlife Service, Edmonton, Alberta, pers. commun., 2000).

Management of caribou in North America is unique in that, in addition to being seasonally sought after by meat, trophy, and non-Native subsistence hunters, caribou remain a staple in the diets of many, perhaps most, northern Natives who have built their cultures around the species. Native harvest of caribou is almost unrestricted, or not enforced at a satisfactory level. In Alaska, there was some regulation (Davis et al. 1978), but that appears to have weakened considerably with time (Valkenburg 1998). Freedom to harvest an unrestricted number of caribou in Canada has proven detrimental to caribou populations since the introduction of the high-powered rifle (Hoare 1927; Banfield 1957; Banfield and Tener 1958; Manning 1960; Kelsall 1968). Most aboriginal populations are growing and are now concentrated in settlements. The shift in most Native communities to a partially wage-based economy has increased the Native ownership of rifles, snowmobiles or similar machines, and even aircraft to be used in hunting. In particular, hunting from aircraft could lead to excessive harvests of caribou during population lows because of the range that can be covered and the ease and speed with which carcasses can be returned to the settlement.

Native people often state their desire to continue their traditional ways of life, and few people would deny them that wish. Unfortunately, however, these traditions now include the use of power boats to replace kayaks, snowmobiles and aircraft instead of dog teams, and high-powered rifles with telescopic scopes instead of bows and spears. The common use of two-way radios now allows Native hunters to track caribou along hundreds of kilometers as the caribou move from tundra to forested winter ranges or back out to the tundra. Each day, different snowmobile-mounted hunters can intercept the caribou and kill whatever number they want. Such kills can be high because the hunters can call in aircraft to haul the carcasses back to the settlements. Caribou populations cannot indefinitely withstand unregulated harvests employing such modern equipment and logistics. This will be especially true in the future with rapidly increasing human populations of aboriginal users that retain the right to use caribou, essentially without enforced restrictions, unless they practice self-restraint.

A second, unique consideration in management of migratory caribou is that many caribou migrate across international and national (state, provincial, and/or territorial) boundaries. For example, (1) barren-ground caribou of the Porcupine herd move among Alaska, Yukon, and the western Northwest Territories; (2) barren-ground caribou of the Bathurst, Beverly, and Qamanirjuak herds move between or among the Northwest Territories, Nunavut, Manitoba, and Saskatchewan and in some years have penetrated northeastern Alberta; and (3) woodland caribou of the George River herd move between Quebec and Labrador (governed by Newfoundland). Remnants of the Selkirk Mountains herd (±34 animals, includes introduced caribou from British Columbia) sometimes move seasonally between or among southern British Columbia, northern Idaho, and northwestern Washington (J. Almack, Washington State Department of Fish and Wildlife, Metaline Falls, pers. commun., 2000). Those transboundary movements necessitate agreement and cooperation at all levels of government and between or among the various resource and land agencies and the people concerned with the welfare of caribou, their habitat, and the people who use those resources. Well-balanced programs are difficult, if not impossible during population lows, to achieve when confronted with a multitude of social, economic, and political concerns among the different jurisdictions.

Management Regions. One of the first problems faced by managers of caribou is to determine what unit of caribou on what range represents a manageable area. Fortunately, because of their gregarious nature and general affinities for seasonal ranges, migration paths, and calving grounds, most groupings of woodland, barren-ground, Peary, and arctic-island forms of caribou are discrete enough to be recognized by managers as herds or populations. Unfortunately, caribou biologists do not always agree on what constitutes a herd or a population (see Current Research and Management Needs).

In Alaska, biologists have believed that movements among their caribou units are frequent and great enough to classify them as herds, and they think of all caribou in Alaska as a population (e.g., Skoog 1968; Valkenburg 1998). In Canada, biologists working on barren-ground caribou first thought of them as herds; then, after learning more, they thought of them as populations (infrequent and negligible movements among units). Now some biologists in Canada have reverted to the herd category as a matter of preference. This convention includes the debatable extension of a definition for a population to apply to a herd.

Distributions of caribou can be considered on a geographic rather than a subspecific or ecotypic basis, although distributions of subspecies do, for the most part, parallel geographic distributions based on current taxa (but less clear based on mtDNA investigations). I suggest three management regions for caribou (Miller 1982): (1) the Arctic Islands of Canada, (2) mainland tundra and taiga of Canada and Alaska, and (3) boreal forests. Subdivisions of these regions on an ecologically meaningful basis could serve as management units.

Most management problems within and among each of those regions would be similar enough to warrant exchange of information and ideas. Biologists should not lose sight of important ecological differences within or among each distinct area or region. Such differences include (1) seasonal quality of forage, (2) seasonal availability of forage, (3) levels of harassing insects, (4) importance of climate, (5) variation in factors causing or potentially causing range deterioration, (6) species and densities of animal predators, (7) levels of hunting pressures, (8) amount of industrial and resource development, and (9) land ownership.

Management Practices. Two important conditions are necessary for the proper management of caribou: the authority (ability) to regulate the harvests of all users, and a biological basis on which to set the regulations. For the most part, neither has been obtained at a level that would maintain populations at the desired sizes. The need has been discussed for controlling population growth in some major herds that appear to be at densities exceeding the apparent carrying capacity of their respective ranges (e.g., Crête and Huot 1991), but nothing has been done.

Until the annual harvest of caribou by all users is regulated, there can be no true management of the species on a sustained yield basis. This has been strongly stated on several occasions for the management of barren-ground caribou in Canada (Banfield 1957; Kelsall 1968; Thomas 1981; Miller 1982, 1983, 1987). However, no concerted meaningful actions have been taken during the last 40 years. The problem is real, however, and it is only a matter of time before the ever increasing human-user population, much greater human accessibility to more caribou range, and incompatible nonrenewable resource development activities on those ranges militate against continued survival of some caribou populations at sizes that can sustain desired levels of future annual harvests.

In Canada, education of Natives to the realities of current caribou use has been and will continue to be a long-term process. It is unlikely, if the educational process is successful, that it will be accomplished without some caribou populations first being extirpated or nearly so. Even a catastrophic loss of caribou would not necessarily convince the Natives of the need to regulate harvests. Because of their beliefs, they most likely would not recognize or accept that they were responsible for the loss. Therefore, educational programs and co-management efforts will or should become the major concern of caribou managers.

Management of caribou, like that of other game species, has generally followed a five-step sequence (Kelsall 1968): (1) regulating hunting, (2) instituting predator control, (3) giving special status to land areas for wildlife, (4) transplanting and reintroductions, and (5) placing corrective and preventive controls on the environment.

Regulation of hunting has been only partially implemented, as Native peoples are not subject to such regulations, at least in Canada. Meaningful enforcement of legal and illegal hunting is questionable on essentially the entire range. Although the situation appeared better in Alaska in the 1970s, it was still far from being totally satisfactory (Miller 1982). Enforcement of existing regulations on caribou hunting has been and continues to be, with few exceptions, inadequate or nonexistent.

I summarized the regulation of hunting of caribou in Alaska and Canada up until the 1980s (Miller 1982). Since that time, many regulations have been made, mostly regarding non-Native hunting of caribou, but relatively few have been effectively enforced. Few regulations have been imposed on Native users and essentially none have been enforced.

In Alaska, Valkenburg (1998:126–28) summarized the primary caribou management problem as conflicting management authority between state and federal agencies. The U. S. Congress passed in 1980 the Alaska National Interest Lands Conservation Act, Title 8, which permits federal control of what was previously a state mandate, which in turn has led to state and federal conflicts. As a result, management decisions have become inefficient and costly, and planning is impossible amid the chaos. There is no single law to point to in Canada, but generically a similar situation exists. This condition results mainly from the new-found power of Native groups, arising primarily through recent large land claim settlements on caribou range and initiation of the devolution of political powers. Territorial governments are mainly controlled by Native representatives, and the federal government and various departments within federal, provincial, and territorial governments see caribou (and other wildlife) as a "plum" to be given away in negotiations to win other concessions. Native land claim settlements provide for direct involvement in wildlife management by Native agencies. Thus, any attempt at biologically and ecologically sound caribou management is now more subjected to sociopolitical considerations than it ever has been in the past. Sociopolitical positions are now backed by official law and regulation as well as unofficial higher government political posturing. Management practices based on the biology and the ecology of the caribou will not be applied without first being accepted by Native groups that now control much of the caribou range, and for political reasons, even on caribou range that they do not control.

One very important factor in the setting of harvest constraints is control of the land over which the caribou range. In the past, U.S. Park Service lands were the only areas over which the Alaska Department of Fish and Game did not have wildlife management jurisdiction. Now, however, with the National Park Service Organic Act and the Endangered Species Act, the priority for "rural residents" enacted by the U.S. Congress, and other legal precedents, jurisdiction over wildlife management is in question in Alaska. In Canada, most ranges of barren-ground caribou were controlled not by wildlife agencies, but by the Department of Indian and Northern Affairs. Most recently, much of the land in the Northwest Territories and Nunavut has been transferred to private ownership of Native groups as a result of huge land claim settlements with the Canadian federal government. As a consequence, neither the Northwest Territories and Nunavut governments nor the Department of the Environment can act effectively, even if they so desired, to curtail Native hunting without full support from the Native owners and the Department of Indian and Northern Affairs. Whether the Northwest Territories or Nunavut governments could act unilaterally is debatable, but there is no reason to believe that they would go against the wishes of their people.

Somehow, the Natives must be convinced that game management agencies are acting on their behalf, and that suggested courses of action to manage caribou are necessary to conserve them. Perhaps co-management efforts or even total Native control will be the answer. Biologically and ecologically sound principles must be identified, accepted, and established in programs, depending on whether populations are declining, stable, or increasing beyond carrying capacity.

Predator control on caribou ranges in the past has essentially meant lethal control of wolves (Miller 1982). Predation is a principal limiting factor in the dynamics of caribou populations. Wolves are the major predator; therefore, I believe there are times when wolf control is a valid management tool, which has been demonstrated in the past (Farnell and McDonald 1988; Boertje et al. 1996). However, public sentiment and thus the lack of political support will most likely prevent lethal control, and other means must be employed. These could include diversionary feeding of wolves at their dens just before and at calving time (Valkenburg et al. 1999), sterilization of dominant pairs, and translocation of remaining wolves in some packs (Boertje and Gardner 1999).

Public opinion currently demands that nonlethal control of wolf numbers be developed if we want to effectively use this management tool (Boertje et al. 1995; Boertje and Gardner 1998). Our greatest challenge is to convince the general public that in the long run both wolf and caribou populations will benefit from control of wolf numbers in accordance with the size of the available ungulate prey base. Naturally, this consideration is greatly complicated in a multiungulate species prey base situation.

A well-planned and properly executed wolf maintenance program is a justifiable management tool when annual net losses of a caribou population constantly exceed annual recruitment (see Fig. 6 in Miller et al. 1988a). From a wildlife management standpoint, it is unfortunate that wolf control as a management tool has become misunderstood and distasteful to usually well-meaning but often misinformed and misguided individuals.

The main consideration should be that wolf control programs not have irreversible impacts on the populations of any of the species involved. We should develop our thinking on the subject and continue gathering evidence to support its judicial use in special situations. The basic question remains: Who should receive priority in using caribou—predators or humans? In the last century, essentially every major barren-ground caribou population and woodland caribou population experienced periods of decline and increase. This will continue. However, when the caribou populations are low, and if we still have ever-increasing, unregulated hunting, something must give! Ideally, both the predators and the hunters must be regulated or the caribou populations will suffer more pronounced declines. This will markedly reduce the sustained level of use by predators and humans. In some cases, it could jeopardize the persistence of some caribou populations at usable sizes.

Wolves and caribou can benefit from an objective program for maintenance of wolf numbers (e.g., Boertje et al. 1996). Constraints on the harvests by humans in such situations are also a valid management procedure. In reality, there are only two factors in the ecology of a declining caribou population that we can actually control when given the proper authority—wolves and humans (hunters, nonconsumptive recreationalists, developers, and exploiters).

Caribou have benefited, at least indirectly, from the establishment of special wildlife management areas, sanctuaries, preserves, and parks. One of the best examples is the Arctic National Wildlife Range (ANWR; 3.6 million ha), set aside in 1960 in northeastern Alaska. Calving grounds and summer ranges of the Porcupine caribou herd occur within the ANWR. However, Prudhoe Bay lies only about 100 km west of ANWR, and interest is strong for petroleum exploration. Thus, the future of the ANWR as a wildlife refuge remains tenuous in the face of strong pressures by the petroleum industry to explore and develop any reserves within the refuge.

In 1926, the Arctic Islands Game Reserve was created. It covered all of the Peary and arctic-island caribou ranges on the Arctic Islands of Canada and some barren-ground caribou ranges in the former Northwest Territories. Natives could hunt within it, but no others. However, the reserve was rescinded in 1966.

The Thelon Game Sanctuary was established in 1927 in the central mainland Barren-Grounds of the Northwest Territories. Its primary purpose was to protect muskoxen, but barren-ground caribou of the Beverly population summer and sometimes calve within the sanctuary (Hoare 1930; Clarke 1940). The mineral industry has lobbied to rescind the Thelon Game Sanctuary and, at other times, to change its boundaries so that they can explore it and develop any worthwhile finds. Continued challenges to open up the Sanctuary will undoubtedly be made in the future.

Unfortunately, such areas actually offer little year-round benefit to caribou populations: (1) the caribou use them only seasonally, then range over unprotected areas; (2) Natives are not restricted from hunting caribou on those areas (in Canada, Natives can hunt within national parks north of 60° N latitude and in Alaska, all "qualified rural residents" can hunt in all parks); and (3) most of the established refuges and sanctuaries are in areas that were not heavily hunted by Natives. Migratory caribou cannot be protected effectively by reserves, except seasonally and locally, because of the huge areas over which they range (Kelsall 1968). However, protection of calving grounds, postcalving areas, and migration routes by special reserve land status or by special land use regulations could be beneficial in giving maternal cows and newborn calves an added degree of protection during those time periods (Miller 1974a). Such areas were traditionally afforded protection by their remoteness and ruggedness, but the airplane, all-terrain vehicle, and snowmobile now make them readily accessible. Encroachment by resource exploiters and the added pressures of the periodic looming of North American energy demands make the special status of wildlife areas tenuous at best.

As Kelsall (1968) mentioned, to help assure success when transplanting barren-ground caribou, restocking should be done in an area where (1) the caribou cannot stray, (2) there is little or no harvesting, and (3) there is a nonconflicting need for caribou. Relatively few such places exist for barren-ground caribou in Alaska or Canada.

Bergerud and Mercer (1989) reviewed 33 introductions of caribou in eastern North America between 1924 and 1985. They found that 20 introductions resulted in sustained populations and 13 failed. The major cause of failures outside Newfoundland was the presence of white-tailed deer infected with meningeal worm and relatively high densities of wolves on the release areas. In Newfoundland, all of the failures were attributed to the introduced animals joining other herds.

Caribou calves were introduced to Adak Island in the Aleutians during 1958 and 1959. Bergerud (1978a) reported that harvests were as high as 30% on the Adak population in 1973 and that the 1973 harvest was a record for sustained yield. I believe that the estimates for the Adak population in 1970 and 1977 (Davis 1978) suggest that the population stagnated for several years and possibly even temporarily declined slightly after the 1973 record kill, then rebounded and reached an estimated 1500 in 1993 (Valkenburg 1998). Valkenburg et al. (2000) reviewed the success of five transplants of Alaskan caribou (including Adak), with emphasis on mean rates of population increase, maximum summer density achieved, and mean weights of 10-month-old females and "adult" males. All of the transplants were successful. Three herds ($\lambda = 1.38$, 1.37, and 1.30) increased at or slightly below the theoretical maximum rate of annual increase for the species ($\lambda = 1.375$, based on a ratio of 60:40 1+ year-old females to 1+ year-old males), the fourth herd increased at about 69% ($\lambda = 1.26$) of the theoretical maximum annual rate, and the fifth herd increased slowly at about 13% ($\lambda = 1.05$) of the theoretical maximum annual rate.

Bergerud and Mercer (1989) reported 384 woodland caribou were released on 22 sites in Newfoundland between 1961 and 1982. Their greatest problem was preventing adult caribou from straying; however, hand-reared calves did not attempt to migrate or stray from the transplant sites. They concluded that success varied among the transplants. Caribou were harvested from seven transplant sites and illegal hunting was widespread at several other sites. Bergerud (1978a) suggested that releases, particularly of bottle-fed calves (short-yearlings), should succeed on islands free of disease and predators. Thus, the primary consideration should be that we are ever mindful of the possibility of introducing diseased or parasitized animals to new areas throughout the caribou's entire range (Bergerud and Mercer 1989; Lankester and Fong 1989, 1998).

Kelsall (1968) pointed out that Southampton Island in northwestern Hudson Bay was an excellent site for reintroduction of caribou. They were abundant on the island until 1924, but were almost totally gone by 1930, and believed extirpated shortly thereafter. The Canadian Wildlife Service at the request of the government of the Northwest Territories captured 52 caribou on Coats Island (about 80 km south of Southampton Island) and successfully relocated 48 of them to Southampton Island in summer 1967. The Southampton Island introduction has been an outstanding success. The mean population estimate of 1+-year-old caribou was about 1200 in 1978, 4000 by June 1987, 13,700 in June 1991, and 29,000 in summer 1997 (Heard and Ouellet 1994; R. Mulders, Department of Resources, Wildlife and Economic Development, Government of Northwest Territories, Yellowknife, pers. commun., 2000). Unfortunately, there are relatively few large, highly suitable, isolated, predator-free areas like Southampton Island (43,000 km^2) for restocking with caribou. Interestingly, Heard (1990) found that the lower primary productivity and longer duration of snow cover on Southampton Island did not affect the intrinsic rate of increase among caribou introduced to the island when compared to caribou and reindeer introduced to islands with better range and more favorable (milder) environmental settings. Heard (1990:171) concluded that "caribou do well as long as the energetic costs of obtaining food (e.g., cratering) are compensated for by the energy derived from their forage." The paramount consideration is actually that the energy derived from their forage exceeds their energy expenditures. Ouellet et al. (1996) suggested that in the absence of predation or high human harvest, competition for food would regulate the caribou abundance on Southampton Island.

Sixty woodland caribou were translocated from southern British Columbia to northern Idaho between 1987 and 1992 to assist in recovery of the endangered Selkirk caribou population (Compton et al. 1995). Annual survival rates for the introduced caribou between 1987–88 and 1991–92 ranged from 0.65 to 0.94: 27 animals died and 6 emigrated. It appears that the introduced caribou are doing no better than the resident Selkirk caribou and likely represent a declining collective population. Only 34 resident and introduced caribou were estimated within the range of the Selkirk population in 2000 (J. Almack, Washington State Department of Fish and Wildlife, Metaline Falls, pers. commun., 2000).

Restocking of islands and possibly small areas of the mainland with caribou has some potential as a management practice. However, the main aim should be to prevent local loss of caribou so that such practices are not necessary. If restocking or creating new herds is justifiable, we should remember that domestic (or wild) reindeer are not a legitimate substitute for caribou. Introduction by any wildlife agency of reindeer on former caribou range and especially to range where caribou are still present is unjustified. If the landowner(s) want reindeer, they should do it as an agricultural pursuit, but not under the guise of wildlife management (and they should acknowledge the potential for conflict with caribou). This position extends to indiscriminate moving of one indigenous form of caribou onto another form's range (e.g., barren-ground caribou to the High Arctic Islands or Peary caribou to the mainland).

A basic prerequisite to all translocations should be that each animal be held and tested for disease and parasites. Ideally, other ungulates in the area where an introduction will be made should also be tested. This requirement to test at least donor stock should be legally mandatory, particularly when introductions involve exotic forms, but also when moving caribou within North America. If caribou still occur where the transplant is being made, their genetic makeup should be determined and compared to the DNA profile of animals being introduced to the area. This is the only way to know what genetic stock we are moving around. This is particularly crucial when involved in an endangered species program (e.g., introductions to the Selkirk caribou range). It is critical to introduce animals of the same genetic makeup as the endangered resident animals. Introducing animals of different genetic makeup will actually genetically contaminate the resident endangered stock rather than bolster survival of the endangered animals.

Forest fire suppression on winter ranges of caribou will be necessary if the rates of wildfires increase and large areas are burned. Kelsall (1968) believed that the need to suppress fires on caribou winter ranges was well recognized, but opposition to that belief has been raised by Bergerud (1971a, 1972, 1974a, 1978a) and others. Kelsall's (1968) plan called for mapping unburned winter ranges and ranking them in relation to (1) quantity and quality of lichens present, (2) their location relative to caribou use, (3) potential effects of fires on each mature forest area, and (4) relative potential of fire-damaged immature forest. Although much information has been collected in the past 30 years (e.g., Kelsall et al. 1977; Klein 1982; Schaefer and Pruitt 1991; Thomas 1998b, 1998c; Thomas and Armbruster 1998; Thomas and Kiliaan 1998c), the impact of forest fires on caribou winter range remains open to debate. Forest stands can become productive for caribou forage as early as 40–50 years after fire, but caribou most use150- to 200 -year-old stands in the taiga (Thomas and Kiliaan 1998c). The burn rate and the stocking density of caribou are important considerations in deciding whether fire control is needed (Thomas 1998c). Managers must remain aware that wildfires can be important in the ecology of caribou. If extensive, fire is detrimental in the short term, but perhaps necessary or at least desirable in the long term (e.g., Miller 1976, 1980; Klein 1982; Schaefer and Pruitt 1991; Thomas 1998a). If rates of burning approach the time required for rotation of successional ranges of vegetation to provide sufficient forage supplies for existing and future numbers of caribou, protective actions should be taken to minimize destruction of ranges.

Kelsall (1968) believed that the long-term average of caribou lost annually by drowning did not exceed the number taken by only a few Native hunters. Although drownings are commonly associated with fast-water sites, caribou also drown on quiet-water crossings when struck by sudden gales or by breaking through ice (Kelsall 1968; Miller and Gunn 1986c). Diversion of caribou from one area may simply cause their deaths elsewhere. The drowning of about 10,000 caribou of the George River herd on the Caniapiscau River in late September 1984 at Calcaire Falls stands out as a massive one-time loss (Doucet et al. 1988; Messier et al. 1988) linked to natural events and human-induced activities on caribou range. Because of the large size of the George River herd, the loss of even 10,000 animals would not have had any significant biological influence on the herd's dynamics. An important question is, How often do large-scale drownings occur without our detecting them? Fencing of hazardous water crossings has been suggested by many interested groups. Such ventures would be very costly and probably would create as many hazards as they removed.

Forested winter ranges could be fertilized, but the value of such action and the possible undesirable side effects are unknown. Such extreme management practices demand study on small areas before extensive programs are considered. We must consider the possible toxicity to caribou and techniques for application to caribou range (Eriksson 1980; Rajala and Westerling 1980; Nordkvist and Erne 1983).

Klein and Vlasova (1992:26) concluded that "increased exploitation of mineral resources in the circumpolar north and associated industrial development, expansion of human populations, pollution from point sources, and global increases in atmospheric pollution pose a major threat to lichens and their use as food for the world's population of reindeer and caribou." A concern that goes well beyond wildlife management, although it has been catastrophic to wildlife, is widespread contamination by radioactivity induced by humans. The accident at the Chernobyl nuclear power plant on 26 April 1986 affected a large area of northern Europe; however, no decrease was found in wildlife during the first 3 years after Chernobyl (Johanson 1990).

A major concern that becomes more critical as more human activities take place on caribou range is the importance of human–caribou interactions. In addition to naturally occurring limiting factors, some caribou populations are subjected to potential detrimental influences from timber harvesting; peat moss extraction; agricultural expansion; road building and much greater access; vehicle collisions; high levels of legal (Native) and illegal (non-Native) hunting; and mining, oil, and gas exploration and development. Potential effects of harassment from human-induced novel stimuli in the caribou's environment could vary from minor increased energy expenditures to reduced reproduction and/or survival of neonates, death of 1+-year-old individuals, and finally the decline of the population (see Table 85 in Miller and Gunn 1979).

Many disturbance studies were conducted along the trans-Alaska oil pipeline, which was much more accelerated and of larger scale than any developments in Canada. There were associated disturbance studies in Canada during the 1970s and into the 1980s when exploration activities were at their highest level in the Canadian Arctic. Since then, studies have been less frequent. The following references and their respective literature sections are pertinent, mainly for caribou in Alaska and Canada, but also for some Eurasian reindeer (Klein 1971, 1980; Urquhart 1973; Child 1974; Johnson and Todd 1977; Cameron et al. 1979, 1992, 1995; Cameron and Whitten 1979, 1980; Miller and Gunn 1979; Horejsi 1981; Cameron 1983; Fancy 1983; Curatolo 1985; Gunn et al. 1985; Curatolo and Murphy 1986; Dau and Cameron 1986; Shideler 1986; Shideler et al. 1986; Harrington and Veitch 1992; Cronin et al. 1994, 1998; Bradshaw et al. 1997, 1998; Cumming and Hyer 1998; Nellemann and Cameron 1998; Farnell 2000; Griffith and Cameron 2000; James and Stuart-Smith 2000; Smith et al. 2000). A great deal of additional information is available in the 18 papers on caribou and human activity published in the proceedings of the first North American Caribou Workshop held at Whitehorse, Yukon, Canada, in September 1983 (Martell and Russell 1983). Recent pertinent general information for reindeer and caribou is available in the proceedings of the Human Role in Reindeer/Caribou Systems Workshop held in Rovaniemi, Finland, in February 1999 (Forbes and Kofinas 2000).

An important point is that whatever level of influence human activities have on caribou and caribou range, it will be in addition to the natural environmental stress. In at least some cases, human interference could be "the straw that breaks the camel's back." Many people with vested interests and even some caribou biologists argue that the caribou's ability to adapt will allow them to cope and live with humans in a highly altered environment. That exploiters and developers of nonrenewable resources ignore the need for a precautionary approach is understandable, but not biologists. In all likelihood, the concept of "compatible sustainable development" is a myth; unfortunately, extensive development on caribou range is a reality. The future will bring large-scale developments of nonrenewable resources well beyond any seen to date in the Canadian Arctic and will likely surpass those currently in Alaska. There is no panacea in Native ownership of the land. Native groups and individuals have already demonstrated their collective willingness to subject the caribou and their range to human activities and alterations for appreciable financial rewards.

At current and foreseeable future levels of caribou management, it is hard to conceive of extensive use of environmental controls in caribou management unless rapid industrial development brings networks of roads and pipelines onto caribou ranges. Alterations (mitigation) will probably be made then only if the funds come from industry and governments. In such an event, it may be necessary to build artificial aids such as crossing devices to allow free flow of caribou past physical or psychological barriers such as roads, pipelines, airstrips, or drilling pads. In the event of large-scale developments, we will find out if large herds of migratory caribou (and possibly even relatively sedentary herds of woodland caribou) can live with associated human activities.

Perhaps the single most important phenomenon that will reshape the fate of many caribou populations is global warming, especially populations on the Arctic Islands, but to some degree throughout the caribou's entire range. Changes brought on by global warming will supposedly be the strongest in the western Arctic (Maxwell 1997) at the edges of caribou distribution (Gunn et al. 2000). Increased frequency of extremely unfavorable deep, hard-packed, windblown snow and/or heavy icing conditions likely will initiate a cascade of changes as plants, caribou, and other herbivores adjust (Miller 1998; Gunn et al. 2000; Gunn and Dragon 2002). As noted by Gunn et al. (2000), it appears likely that environmental changes are underway and by the time we have determined how plant and animal populations will respond, they will already be affected. Although we can probably do little or nothing to mitigate detrimental effects of global warming on caribou, it will be irresponsible to ignore the probably unfavorable outcomes of global warming on caribou populations, wildlife in general, and especially entire ecosystems.

RESEARCH AND MANAGEMENT NEEDS

The dire need for defining a population precisely for use as a meaningful research or management unit has been discussed by Wells and Richmond (1995). They concluded, "Indeed, in many circumstances it will be difficult to make meaningful progress in understanding biological systems without precise terminology" (Wells and Richmond 1995:462). Caribou biologists have vacillated for >40 years about whether we are studying and managing herds or populations of caribou. Is there a real difference between a herd and a population or are they the same? Should we define each as separate units or should we use the two terms interchangeably? The need for appropriate terminology should be extended to all aspects of research and management of caribou. Future caribou research and management would benefit greatly by us defining and standardizing and then collectively, rigorously, cooperatively, and consistently using appropriate terminology. Unfortunately, we cannot always detect discrepancies that might seriously detract from interpreting technical literature. Our failure to do so has been responsible for much of the confusion and resultant misinterpretation that has crept into the caribou literature. For example, it is not always possible to determine how an author(s) is identifying an "adult female" (cow) caribou. To some, any female over 1 year of age is an adult female, to others its a 2+-year-old female, and still others a 3+-year-old female. Of course, a more important problem arises from our inability to consistently visually determine the age of any female beyond 1 year old. Is any female that is bred, regardless of age, an adult female? Then, what is a

5- or 6-year-old nonpregnant female? Is a "cow" any female or should the term "cow" be restricted to only a female within a certain age range? Does it make sense to call a female calf or yearling a cow calf or cow yearling and, if so, why?

There are dozens, perhaps hundreds, of terms that need stringent definition and subsequent faithful use to foster our collective understanding and advancement. For example, caribou biologists write about "emigration" with no evidence of whether the animals returned to origin or not. By definition, emigration connotes moving away and not returning, and, thus, in its correct and exacting use, should demand such knowledge. We use "interchange, exchange, or mixing" (which connotes two-way movements in the same time period) when what has actually happened is merely that some animals moved from one place or herd to another (one-way movement) without evidence of two-way movements during that time period. Most Alaskan caribou biologists collectively use "cohort" as an age class, whereas most Canadian caribou biologists use cohort for the animals born in a specific year, two very different considerations, and sometimes surprisingly confusing.

Collective paramount efforts should be made within, between, and among jurisdictions to develop and employ relatively accurate and cost-effective standardized methods for inventorying caribou populations (Edmonds 1998). Alaskan caribou biologists believe that their aerial survey results (postcalving photo-census) for major herds (>5000) can be treated as minimum herd estimates and are precise and accurate enough to allow them to manage those herds (P. Valkenburg, Alaska Department of Fish and Game, Fairbanks, pers. commun., 2000). The greater difficulty of counting forest-dwelling caribou rather than those living on tundra range is obvious. Past results from aerial surveys of caribou on tundra in Canada indicate, however, that there still are problems with obtaining precise estimates.

The Qamanirjuaq caribou herd that is aerial surveyed on the Barren Grounds of north-central Canada stands out as an example of the imperfection and/or variation in quality of aerial surveys for obtaining estimations of total population size. Most important, this condition apparently applies equally to all the major barren-ground caribou herds in the Northwest Territories and Nunavut. Formerly known as the Kaminuriak caribou population, the Natives, for reasons known only to them, did not believe the rapidly declining numbers obtained from the survey efforts in the 1960s and 1970s. The early survey estimates looked like "textbook" results to me (Miller 1987): survey biologists were predicting future population sizes, and subsequent estimates fell right on the projected line. According to population estimates from aerial survey (Gates 1985; Heard and Calef 1986; Heard and Jackson 1990 in Thomas 1998d; Williams 1995 in Thomas 1998d), the herd went from 63,000 in 1968 down to 39,000 in 1980. Then the trend reversed unexplainably and the estimate went up to 180,000 in 1982, to 230,000 in 1983, and to 272,000 in 1985; then it went down to 221,000 in 1988, and then supposedly rebounded to 496,000 in 1994. It turned out the Natives were right to question survey results. The huge, and truly unexplainable, variation in population size led to speculation that emigration and/or infidelity to calving grounds was the cause (Gates 1985; Heard and Calef 1986; Heard and Jackson 1990 in Thomas 1998d; Williams 1995 in Thomas 1998d). Shortcomings (flaws) in procedure and unsatisfied assumptions in the survey techniques indicate why these endeavors misled us (Thomas 1998d). Improved and consistent survey procedures combined with broader knowledge of caribou on the landscape are necessary for us to obtain reliable estimates.

Thomas (1998d) argued that less counting is needed, but that more ecological studies are required. I agree with his concern about the imprecision and unknown accuracy of many of the aerial surveys to date. I disagree, however, with his conclusion that it would be better to carry out more ecological studies in place of monitoring (census) efforts. I believe the best approach to the proper management of a renewable resource is through periodically obtaining reasonably good approximations of the number of individuals.

Many populations of other species of ungulate are managed on the basis of annual harvest statistics. Caribou do not lend themselves to this approach. Although millions of dollars have been spent to collect kill statistics, I know of no data set of annual harvest statistics that would allow biologically sound management of any large caribou herd in North America. Also, the "quirks" of migratory caribou behavior and choices by Native hunters would likely distort the kill statistics in terms of representation of the population. Efforts to collect annual kill statistics have been relatively short term and fragmentary, and have often failed to get reliable total numbers. Furthermore, they have failed to get sex/age composition of the kill and in some settlements, have not determined from which population(s) caribou were harvested.

Wise management of caribou herds also demands that movements and migrations of caribou be studied in detail to determine whether individual caribou show strong seasonal affinities for specific sections of their ranges. If different caribou from a herd return to the same areas every year, continued heavy annual harvest of them would not be detrimental (assuming that the population could sustain it) because different individuals would be filling the voids each year. If, however, individual caribou showed strong affinities for specific sections of the population's ranges, continued heavy annual harvest would soon markedly reduce or destroy that segment of the population.

Previously (Miller 1982), I said that the proceedings edited by Klein and White (1978) served as the most complete and up-to-date thinking on research needs for caribou. If a large group of caribou biologists gathered in the near future, they would list the same set of research questions as necessary for an understanding of the population dynamics of caribou: including (1) population size, (2) population structure, (3) age-specific birth and death rates, (4) mortality to predators, (5) other natural mortality, (6) dispersal, (7) human harvest, (8) distribution (seasonal movements), and (9) status of other interacting herbivores (competitors and alternate prey of predators). Even considering the great body of literature that has been produced about caribou in the last 20 years, I must still draw the same conclusion. No jurisdiction has, to the best of my knowledge, realized these objectives in any meaningful way at the population level on an ongoing basis for any major or even minor population of caribou in North America. We have conducted many studies and made considerable strides forward in many of these areas of investigation. We have failed, however, in most, if not all, cases to apply our findings at even a level of basic management, let alone a more refined level.

In hindsight, although the above nine areas of concern are necessary for a refined level of management, they are not mandatory for the biologically sound basic management of a caribou population. Unfortunately, what is necessary is still extremely difficult to obtain because of funding and technical limitations and, at least in most cases, currently impossible to achieve because of sociopolitical constraints.

We could carry out satisfactory basic management of a caribou population if we could accomplish the following on an ongoing basis:

1. Periodic, reasonably precise and accurate measures of population size, ideally including an adequate general knowledge of variations in the population's annual range size and locations.
2. An accurate measure of annual kill from each population by all hunters, and, ideally, the harvest statistics should include sex/age composition of the annual kill.
3. The ability and the will to regulate and enforce the size of annual harvest. We must limit it during periods when the population cannot sustain the annual level of kill and increase it when there is range deterioration and/or markedly reduced reproductive performance that cannot be explained by weather events, disease, or parasites.

Our management efforts for a caribou population would be more satisfactory when we also could accomplish the following:

1. Regulate predator numbers when the population is declining and the current number of caribou would not support the desired or anticipated annual harvest.
2. Limit or control human activities on the range of the caribou population, especially the calving grounds. Satisfactory range at all seasons of the year is crucial to the continued survival and growth of the population. Annual range is only as good as its weakest seasonal link.

3. Have an ongoing, long-term program of ecological studies on a designated "research herd" in each jurisdiction, carried out with special external funding or with agency funds that are available after the three management steps are satisfactorily met in the basic management program.

ACKNOWLEDGMENTS

Working with caribou has given me the opportunity to witness some of the most marvelous sights in nature, and has reinforced in me the belief that the Arctic would indeed be an empty land without all caribou and particularly the migratory caribou. Naturally, I have a special empathy for the Peary caribou—the tough little guy on the block!

I am particularly grateful to other caribou biologists who have taken the time over the years to debate matters pertaining to ecological and behavioral relations of caribou to their environment. I offer special thanks to those colleagues who argued the "gray areas" of those considerations; much of what I think I know about caribou has been developed out of those conversations.

I thank J. Almack, Washington State Department of Fish and Wildlife; C. M. Doucet, Newfoundland and Labrador Department of Forest Resources and Agrifoods, Wildlife Branch; R. Farnell, Yukon Department of Renewable Resources, Wildlife Branch; E. J. (Edmonds) Ficht, Alberta Environment, Natural Resources Service, Wildlife Branch; A. Gunn, Department of Resources, Wildlife and Economic Development, Fisheries and Wildlife Division, Government of Northwest Territories; S. Couturier, Quebec Wildlife and Parks; M. W. Lankester, Department of Biology, Lakehead University; R. Mulders, Department of Resources, Wildlife and Economic Development, Fisheries and Wildlife Division, Government of Northwest Territories; C. Strobeck and K. Zittlau, Biological Sciences, University of Alberta; E. S. Telfer and D. C. Thomas, Canadian Wildlife Service; and P. Valkenburg, Alaska Department of Fish and Game, for providing information and/or comments on earlier draft text for this second edition of my caribou chapter.

Much of this text is liberally awash with my personal opinions and beliefs; the material does not necessarily reflect the opinions, beliefs, or policies of the Canadian Wildlife Service, the Department of the Environment, or the Government of Canada. Virtually all of my generalizations have exceptions. The number of exceptions about a specific point is usually directly proportional to the amount of investigation that has been conducted on it.

LITERATURE CITED

Abraham, K. F., and J. E. Thompson. 1998. Defining the Pen Islands caribou herd of southern Hudson Bay. Rangifer (Special Issue) 10:33–40.

Adamczewski, J. Z., C. C. Gates, R. J. Hudson, and M. A. Price. 1987. Seasonal changes in body composition of mature female caribou and calves (*Rangifer tarandus groenlandicus*) on an Arctic island with limited winter resources. Canadian Journal of Zoology 65:1149–57.

Adamczewski, J. Z., C. C. Gates, B. M. Soutar, and R. J. Hudson. 1988. Limiting effects of snow on seasonal habitat use and diets of caribou (*Rangifer tarandus groenlandicus*) on Coats Island, Northwest Territories, Canada. Canadian Journal of Zoology 66:1986–96.

Adamczewski, J. Z., R. J. Hudson, and C. C. Gates. 1993. Winter energy balance and activity of female caribou on Coats Island, Northwest Territories: The relative importance of foraging and body reserves. Canadian Journal of Zoology 71:1221–29.

Adams, L. G., B. W. Dale, and L. D. Mech. 1995a. Wolf predation on caribou calves in Denali National Park, Alaska. Pages 245–60 *in* L. N. Carbyn, S. H. Fritts, and D. R. Seip, eds. Ecology and conservation of wolves in a changing world (Occasional Paper 35). Canadian Circumpolar Institute, Edmonton, Alberta, Canada.

Adams, L. G., F. J. Singer, and B. W. Dale. 1995b. Caribou and calf mortality in Denali National Park, Alaska. Journal of Wildlife Management 59:584–94.

Allaye-Chan, A. C., and R. G. White. 1991. Body condition variations among adult females of the Porcupine caribou herd. Proceedings of North American Caribou Workshop 4:103–8.

Anderson, G., K. Andersson, A. Brundin, and C. Rappe. 1975. Volatile compounds from the tarsal scent gland of reindeer (*Rangifer tarandus*). Journal of Chemical Ecology 1:275–81.

Anderson, R. C. 1971. Neurological disease in reindeer (*Rangifer tarandus*) introduced into Ontario. Canadian Journal of Zoology 49:159–66.

Banfield, A. W. F. 1949. The present status of North American caribou. Transactions of the North American Wildlife Conference 14:477–91.

Banfield, A. W. F. 1954a. Preliminary investigation of the barren-ground caribou (Wildlife Management Bulletin, Series 1, 10B). Canadian Wildlife Service.

Banfield, A. W. F. 1954b. The role of ice in the distribution of mammals. Journal of Mammalogy 35:104–7.

Banfield, A. W. F. 1957. The plight of the barren-ground caribou. Oryx 4:5–20.

Banfield, A. W. F. 1961. A revision of the reindeer and caribou genus *Rangifer.* (Bulletin 177, Biological Series 66). National Museum of Canada, Ottawa.

Banfield, A. W. F. 1963. The disappearance of the Queen Charlotte Island's caribou. National Museum of Canada Bulletin 185:40–49.

Banfield, A. W. F. 1974. The mammals of Canada. University of Toronto Press, Toronto.

Banfield, A. W. F., and J. S. Tener. 1958. A preliminary study of the Ungava caribou. Journal of Mammalogy 39:560–73.

Barrett, M. W., J. W. Nolan, and L. D. Roy. 1982. Evaluation of a hand-held net-gun to capture large mammals. Wildlife Society Bulletin 10:108–14.

Baskin, L. M. 1970. Reindeer: their ecology and behaviour. Nauka, Moscow (English transl., Canadian Wildlife Service, Ottawa).

Baskin, L. M. 1983. The causes of calf reindeer mortality. Acta Zoologica Fennica 175:133–34.

Belanger, L. F., L. P. E. Choquette, and J. G. Cousineau. 1967. Osteolysis in reindeer antlers: Sexual and seasonal variations. Calcified Tissue Research 1:37–43.

Bell, K. L., and L. C. Bliss. 1977. Overwinter phenology of plants in a polar semi-desert. Arctic 30:118–21.

Behnke, R. H. 2000. Equilibrium and non-equilibrium models of livestock population dynamics in pastoral Africa: Their relevance to Arctic grazing systems. Proceedings of Arctic Ungulate Conference 10:141–52.

Bergerud, A. T. 1971a. Abundance of forage on the winter range of Newfoundland caribou. Canadian Field-Naturalist 85:39–52.

Bergerud, A. T. 1971b. Hunting of stag caribou in Newfoundland. Journal of Wildlife Management 35:71–75.

Bergerud, A. T. 1971c. The population dynamics of Newfoundland caribou. Wildlife Monographs 25:1–55.

Bergerud, A. T. 1972. Food habits of Newfoundland caribou. Journal of Wildlife Management 36:913–23.

Bergerud, A. T. 1973. Movement and rutting behavior of caribou (*Rangifer tarandus*) at Mount Albert, Quebec. Canadian Field-Naturalist 87:357–69.

Bergerud, A. T. 1974a. Decline of caribou in North America following settlement. Journal of Wildlife Management 38:757–70.

Bergerud, A. T. 1974b. Relative abundance of food in winter for Newfoundland caribou. Oikos 25:379–87.

Bergerud, A. T. 1974c. Rutting behavior of Newfoundland caribou. Pages 394–435 *in* V. Geist and F. Walters, eds. The behavior of ungulates and its relation to management, Vol. 1. International Union for the Conservation of Nature and Natural Resources, Morges, Switzerland.

Bergerud, A. T. 1974d. The role of the environment in the aggregation, movement, and disturbance behavior of caribou. Pages 552–84 *in* V. Geist and F. Walters, eds. The behavior of ungulates and its relation to management, Vol. 2. International Union for the Conservation of Nature and Natural Resources, Morges, Switzerland.

Bergerud, A. T. 1975. The reproductive season in Newfoundland caribou. Canadian Journal of Zoology 53:1213–21.

Bergerud, A. T. 1976. The annual antler cycle in Newfoundland caribou. Canadian Field-Naturalist 90:449–63.

Bergerud, A. T. 1977. Diets of caribou. Pages 243–66 *in* M. Recheigl, Jr., ed. Diets for mammals, Vol. 1. CRC Press, Cleveland, OH.

Bergerud, A. T. 1978a. Caribou. Pages 83–101 *in* J. L. Schmidt and D. L. Gilbert, eds. Big game of North America: Ecology and management. Stackpole, Harrisburg, PA.

Bergerud, A. T. 1978b. The status and management of caribou in British Columbia. Report to the Ministry of Recreation and Conservation, Victoria, British Columbia, Canada.

Bergerud, A. T. 1980. A review of the population dynamics of caribou and wild reindeer in North America. Proceedings of International Reindeer/Caribou Symposium 2:556–81.

Bergerud, A. T. 1983. The natural population control of caribou. Pages 14–61 *in* F. L. Bunnell, D. S. Eastman, and J. M. Peek, eds. Symposium on natural regulation of wildlife populations. Forestry, Wildlife and Range Experimental Station, University of Idaho, Moscow.

Bergerud, A. T. 1985. Antipredator strategies of caribou: Dispersion along shorelines. Canadian Journal of Zoology 63:1324–29.

Bergerud, A. T. 1988. Caribou, wolves and man. Trends in Ecological Evolution 3:68–72.

Bergerud, A. T. 1992. Rareness as an antipredator strategy to reduce risk for moose and caribou. Pages 1008–21 *in* D. R. McCullough and R. H. Barrett, eds. Wildlife 2001: Populations. Elsevier, New York.

Bergerud, A. T. 1996. Evolving perspectives on caribou population dynamics, have we got it right yet? Rangifer (Special Issue) 9:95–115.

Bergerud, A. T., and J. P. Elliot., 1986. Dynamics of caribou and wolves in northern British Columbia. Canadian Journal of Zoology 64:1515–29.

Bergerud, A. T., and W. E. Mercer. 1989. Caribou introductions in eastern North America. Wildlife Society Bulletin 17:111–20.

Bergerud, A. T., and M. J. Nolan. 1970. Food habits of hand-reared caribou *Rangifer tarandus* L. in Newfoundland. Oikos 21:348–50.

Bergerud, A. T., and R. E. Page. 1987. Displacement and dispersion of parturient caribou at calving as an antipredator tactic. Canadian Journal of Zoology 65:1597–1606.

Bergerud, A. T., H. E. Butler, and D. R. Miller. 1984. Antipredator tactics of calving caribou: Dispersion in mountains. Canadian Journal of Zoology 62:1566–75.

Boertje, R. D. 1984. Seasonal diets of the Denali caribou herd, Alaska. Arctic 37:161–65.

Boertje, R. D., and C. L. Gardner. 1998. Factors limiting the Fortymile caribou herd (Federal Aid in Wildlife Restoration Project W-24-4, Study 3.38, Final Report). Alaska Department of Fish and Game.

Boertje, R. D., and C. L. Gardner. 1999. Reducing mortality on the Fortymile caribou herd. (Federal Aid in Wildlife Restoration Project W-27–1, Study 3.43, Progress Report). Alaska Department of Fish and Game.

Boertje, R. D., D. G. Kellyhouse, and R. D. Hayes. 1995. Methods for reducing natural predation on moose in Alaska and Yukon: An evaluation. Pages 505–13 *in* L. N. Carbyn, S. H. Fritts, and D. R. Seip, eds. Ecology and conservation of wolves in a changing world. (Occasional Paper No. 35). Canadian Circumpolar Institute, Edmonton, Alberta, Canada.

Boertje, R. D., P. Valkenburg, and M. E. McNay. 1996. Increases in moose, caribou, and wolves following wolf control in Alaska. Journal of Wildlife Management 60:474–89.

Bøving, P. S., and E. Post. 1997. Vigilance and foraging behaviour of female caribou in relation to predation risk. Rangifer 17:55–63.

Bowyer, R. T. 1984. Sexual segregation in southern mule deer. Journal of Mammalogy 65:410–17.

Bradshaw, C. J. A., D. M. Hebert, A. B. Rippin, and S. Boutin. 1995. Winter peatland habitat selection by woodland caribou in northeastern Alberta. Canadian Journal of Zoology 76:1319–24.

Bradshaw, C. J. A., S. Boutin, and D. M. Hebert. 1997. Effects of petroleum exploration on woodland caribou in northeastern Alberta. Journal of Wildlife Management 61:1127–33.

Bradshaw, C. J. A., S. Boutin, and D. M. Hebert. 1998. Energetic implications of disturbance caused by petroleum exploration to woodland caribou. Canadian Journal of Zoology 76:1319–24.

Broughton, E., L. P. E. Choquette, J. G. Cousineau, and F. L. Miller. 1970. Brucellosis in reindeer, *Rangifer tarandus* L., and the migratory barren-ground caribou, *Rangifer tarandus groenlandicus* (L.), in Canada. Canadian Journal of Zoology 48:1023–27.

Broughton, E., F. L. Miller, and L. P. E. Choquette. 1972. Cutaneous fibropapillomas in migratory barren-ground caribou. Journal of Wildlife Diseases 8:138–40.

Brown, W. K., and J. B. Theberge. 1985. The calving distribution and calving-area fidelity of a woodland caribou herd in Central Labrador. Proceedings of North American Caribou Workshop 2:57–67.

Brown, W. K., and J. B. Theberge. 1990. The effect of extreme snow cover on feeding-site selection by woodland caribou. Journal of Wildlife Management 54:161–68.

Brown, W. K., J. Huot, P. Lamothe, S. Luttich, M. Parè, G. St. Martin, and J. B. Theberge. 1986. The distribution and movement patterns of four woodland caribou herds in Quebec and Labrador. Rangifer (Special Issue) 1:43–49.

Bubenik, A. B. 1975a. Significance of antlers in the social life of barren-ground caribou. Proceedings of International Reindeer/Caribou Symposium 1:436–61.

Bubenik, A. B. 1975b. Taxonomic value of antlers in genus *Rangifer,* H. Smith. Proceedings of International Reindeer/Caribou Symposium 1:41–63.

Bunnell, F., T. C. Dauphine, R. Hilborn, D. R. Miller, F. L. Miller, E. H. McEwan, G. R. Parker, R. Peterman, G. W. Scotter, and C. J. Walters. 1975. Preliminary report on computer simulation of barren-ground caribou management. Proceedings of International Reindeer/Caribou Symposium 1:189–93

Butler, H. E. 1983. Fall suckling behavior in woodland caribou (*Rangifer tarandus caribou*). Acta Zoologica Fennica 175:109–11.

Calef, G. W. 1978. Population status of caribou in the Northwest Territories. Pages 9–16 *in* D. R. Klein and R. G. White, eds. Parameters of caribou population ecology in Alaska (Biological Papers, University of Alaska, Special Report No. 3). University of Alaska, Fairbanks.

Calef, G. W., and G. M. Lortie. 1975. A mineral lick of the barren-ground caribou. Journal of Mammalogy 56:240–42.

Cameron, R. D. 1983. Issue: Caribou and petroleum development in Arctic Alaska. Arctic 36:227–31.

Cameron, R. D. 1994. Reproductive pauses by female caribou. Journal of Mammalogy 75:10–13.

Cameron, R. D., and J. R. Luick. 1972. Seasonal changes in total body water, extra-cellular fluid, and blood volume in grazing reindeer. Canadian Journal of Zoology 50:107–16.

Cameron, D. R., and K. R. Whitten. 1979. Caribou distribution and group composition associated with construction of the trans-Alaska pipeline. Canadian Field-Naturalist 93:155–62.

Cameron, R. D., and K. R. Whitten. 1980. Influence of the trans-Alaska pipeline corridor on the local distribution of caribou. Proceedings of International Reindeer/Caribou Symposium 2:475–84.

Cameron, R. D., R. G. White, and J. R. Luick. 1975. The accumulation of water in reindeer during winter. Proceedings of International Reindeer/Caribou Symposium 1:374–78.

Cameron, R. D., K. R. Whitten, W. T. Smith, and D. D. Roby. 1979. Caribou distribution and group composition associated with construction of the trans-Alaskan pipeline. Canadian Field-Naturalist 93:155–62.

Cameron, R. D., K. R. Whitten, and W. T. Smith. 1986. Summer range fidelity of radio-collared caribou in Alaska's Central Arctic herd. 1986. Rangifer (Special Issue) 1:51–55.

Cameron, R. D., W. T. Smith, and R. T. Shideler. 1988. Variations in initial calf production of the Central Arctic caribou herd. Proceedings of North American Caribou Workshop 3:1–7.

Cameron, R. D., W. T. Smith, and S. G. Fancy. 1991. Comparative body weights of pregnant/lactating and non-pregnant female caribou. Proceedings of North American Caribou Workshop 4:109–14.

Cameron, R. D., D. J. Reed, J. R. Dau, and W. T. Smith. 1992. Redistribution of calving caribou in response to oil field development on the Arctic slope of Alaska. Arctic 45:338–42.

Cameron, R. D., W. T. Smith, S. G. Fancy, K. L. Gerhart, and R. G. White. 1993. Calving success of female caribou in relation to body weight. Canadian Journal of Zoology 71:480–86.

Cameron, R. D., E. A. Lenart, D. J. Reed, K. R. Whitten, and W. T. Smith. 1995. Abundance and movements of caribou in the oilfield complex near Prudhoe Bay, Alaska. Rangifer 15:3–7.

Caughley, G., and A. Gunn. 1993. Dynamics of large herbivores in deserts: Kangaroos and caribou. Oikos 67:47–55.

Chapin, S. F., III. 1980. Effect of clipping upon nutrient status and forage value of tundra plants in arctic Alaska. Proceedings of International Reindeer/Caribou Symposium 2:19–25.

Chapin, S. F., III, K. Van Cleve, and L. L. Tieszen. 1975. Seasonal nutrient dynamics of tundra vegetation at Barrow, Alaska. Arctic Alpine Research 7:209–26.

Child, K. N. 1974. Reaction of caribou to various types of simulated pipelines at Prudhoe Bay, Alaska. Pages 805–12 *in* V. Geist and F. Walters, eds. The behaviour of ungulates and its relation to management. International Union for the Conservation of Nature and Natural Resources, Morges, Switzerland.

Choquette, L. P. E., E. Broughton, F. L. Miller, H. C. Gibbs, and J. G. Cousineau. 1967. Besnoitiosis in barren-ground caribou in northern Canada. Canadian Veterinary Journal 8:282–87.

Chubbs, T. E. 1993. Observations of calf-hiding behavior by female woodland caribou, *Rangifer tarandus caribou,* in east-central Newfoundland. Canadian Field-Naturalist 107:368–69.

Clarke, C. H. D. 1940. A biological investigation of the Thelon Game Sanctuary (Bulletin 96, Biological Series 25). National Museum of Canada, Ottawa.

Clausen, B., A. Dam, E. Elvestad, H. V. Krogh, and H. Thing. 1980. Summer mortality among caribou calves in West Greenland. Nordisk Veterinaermedicin 32:291–300.

Clutton-Brock, T. H., G. R. Iason, and F. E. Guinness. 1987. Sexual segregation and density-related changes in habitat use in male and female red deer (*Cervus elaphus*). Journal of Zoology (London) 211:275–89.

Compton, B. B., P. Zager, and G. Servheen. 1995. Survival and mortality of translocated woodland caribou. Wildlife Society Bulletin 23:490–96.

Couturier, S., J. Brunelle, D. Vandal, and G. St-Martin. 1990. Changes in the population dynamics of the George River caribou herd, 1976–87. Arctic 43:9–20.

Couturier, S., R. Courtois, H. Crépeau, L.-P. Rivest, and S. Luttich. 1996. Calving photo-census of the Riviére George caribou herd and comparison with an independent census. Rangifer (Special Issue) 9:283–96.

Crête, M., and J. Huot. 1991. Recent changes in the population dynamics of the Rivière George caribou herd: What is the best management strategy? Proceedings of North American Caribou Workshop 4:180–83.

Crête, M., and J. Huot. 1993. Regulation of a large herd of caribou: Summer nutrition affects calf growth and body reserves of dams. Canadian Journal of Zoology 71:2291–96.

Crête, M., J. Huot, and L. Gauthier. 1990. Food selection during early lactation by caribou Calving on the tundra in Quebec. Arctic 43:60–65.

Crête, M., C. Banville, D. Le Henaff, J. Levesque, and H. Ross. 1991. High calf mortality endangers the Gaspesie Park caribou herd. Proceedings of North American Caribou Workshop 4:178–79.

Crête, M., J. Huot, R. Nault, and R. Patenaude. 1993. Reproduction, growth and body composition of Rivière George caribou in captivity. Arctic 46:185–96.

Crête, M., S. Couturier, B. J. Hearn, and T. E. Chubbs. 1996. Relative contribution of decreased productivity and survival to recent changes in the demographic trend of the Rivière George caribou herd. Rangifer (Special Issue) 9:27–36.

Cringan, A. T. 1957. History, food habits and range requirements of the woodland caribou of continental North America. Transactions North American Wildlife Conference 22:485–501.

Crisler, L. 1956. Observations of wolves hunting caribou. Journal of Mammalogy 37:337–346.

Cronin, M. A., W. B. Ballard, J. Truett, and R. Pollard. 1994. Mitigation of the effects of oil field development and transportation corridors on caribou. Final Report of the Alaska Steering Committee. LGL Alaska Research Associates, Anchorage, AK.

Cronin, M. A., W. B. Ballard, J. D. Bryan, B. J. Pierson, and J. D. McKendrick. 1998. Northern Alaska oil fields and caribou: A commentary. Biological Conservation 83:195–208.

Curatolo, J. A. 1985. Sexual segregation and habitat use by the Central Arctic caribou herd during summer. Proceedings of North American Caribou Workshop 2:193–98.

Curatolo, J. A., and S. M. Murphy. 1986. The effects of pipelines, roads, and traffic on the movements of caribou, *Rangifer tarandus*. Canadian Field-Naturalist 100:218–24.

Cumming, H. G. 1975. Clumping behavior and predation with special reference to caribou. Proceedings of International Reindeer/Caribou Symposium 1:474–97.

Cumming, H. G. 1998. Status of woodland caribou in Ontario: 1996. Rangifer (Special Issue) 10:99–104.

Cumming, H. G., and B. T. Hyer. 1998. Experimental log hauling through a traditional caribou wintering area. Rangifer (Special Issue) 10:241–58.

Dale, B. W., L. G. Adams, and R. T. Bowyer. 1995. Winter wolf predation in a multiple ungulate prey system, Gates of the Arctic National Park, Alaska. Pages 223–30 *in* L. N. Carbyn, S. H. Fritts, and D. R. Seip, eds. Ecology and conservation of wolves in a changing world (Occasional Paper 35). Canadian Circumpolar Institute, Edmonton, Alberta, Canada.

Darby, W. R., and W. O. Pruitt, Jr. 1984. Habitat use, movements, and grouping behaviour of woodland caribou, *Rangifer tarandus caribou,* in southeastern Manitoba. Canadian Field-Naturalist 98:184–90.

Dau, J. R., and R. D. Cameron. 1986. Effects of a road system on caribou distribution during calving. Rangifer (Special Issue) 1:95–101.

Dauphine, T. C., Jr. 1975. The disappearance of caribou reintroduced to Cape Breton Highlands National Park. Canadian Field-Naturalist 89:299–310.

Dauphine, T. C., Jr. 1976. Biology of the Kaminuriak population of barren-ground caribou. Part 4. Growth, reproduction, and energy reserves (Report Series 38). Canadian Wildlife Service.

Dauphine, T. C., Jr., and R. L. McClure. 1974. Synchronous mating in Canadian barren-ground caribou. Journal of Wildlife Management 38:54–66.

Davis, J. L. 1978. History and current status of Alaska caribou herds. Pages 1–8 *in* D. R. Klein and R. G. White, eds. Parameters of caribou population ecology in Alaska (Biological Papers, University of Alaska, Special Report No. 3). University of Alaska, Fairbanks.

Davis, J. L. 1980. Status of *Rangifer* in the USA. Proceedings of International Reindeer/Caribou Symposium 2:793–97.

Davis, J. L., and P. Valkenburg. 1991. A review of caribou population dynamics in Alaska emphasizing limiting factors, theory and management implications. Proceedings of North American Caribou Workshop 4:184–209.

Davis, J. L., C. Grauvogal, H. Reynolds, and P. Valkenburg. 1978. Human utilization of the Western Arctic caribou herd (Federal Aid in Wildlife Restoration Projects W-17-8 and W-17-9, Final Report). Alaska Department of Fish and Game.

Davis, J. L., P. Valkenburg, and H. V. Reynolds. 1980. Population dynamics of Alaska's Western Arctic caribou herd. Proceedings of International Reindeer/Caribou Symposium 2:595–604.

Davis, J. L., P. Valkenburg, and R. D. Boertje. 1983. Demography and limiting factors of Alaska's Delta caribou herd, 1954–1981. Acta Zoologica Fennica 175:135–37.

Davis, J. L., P. Valkenburg, and R. D. Boertje. 1985a. Disturbance and the Delta caribou herd. Proceedings of North American Caribou Workshop 1:2–6.

Davis, J. L., C. A. Grauvogel, and P. Valkenburg. 1985b. Changes in subsistence harvest of Alaska's Western Arctic caribou herd, 1940–1984. Proceedings of North American Caribou Workshop 2:105–18.

Davis, J. L., P. Valkenburg, and R. D. Boertje. 1986. Empirical and theoretical considerations toward a model for caribou socioecology. Rangifer (Special Issue) 1:103–9.

Davis, J. L., P. Valkenburg, and D. J. Reed. 1988. Mortality of Delta herd caribou to 24 months of age. Proceedings of North American Caribou Workshop 3:38–51.

Davis, J. L., L. G. Adams, P. Valkenburg, and D. J. Reed. 1991. Relationships between body weight, early puberty, and reproductive histories in central Alaskan caribou. Proceedings of North American Caribou Workshop 4:115–42.

Davis, T. A. 1973. Asymmetry of reindeer antlers. Forma et Functio 6:373–82.

Davis, T. A. 1974. Further notes on asymmetry of reindeer antlers. Forma et Functio 7:55–58.

Dehority, B. A. 1975a. Characterization studies of rumen bacteria isolated from Alaskan reindeer (*Rangifer tarandus*). Proceedings of International Reindeer/Caribou Symposium 1:228–40.

Dehority, B. A. 1975b. Rumen ciliate protozoa of Alaskan reindeer and caribou (*Rangifer tarandus* L.). Proceedings of International Reindeer/Caribou Symposium 1:241–50.

Des Meules, P., and J. Heyland. 1969a. Contribution to the study of the food habits of caribou. Part 2: Daily consumption of lichens. Nature Canada 96:333–36.

Des Meules, P., and J. Heyland. 1969b. Contribution to the study of the food habits of caribou. Part 1: Lichen preferences. Nature Canada 96:317–31.

de Vos, A. 1960. Behavior of barren-ground caribou on their calving grounds. Journal of Wildlife Management 24:250–58.

Dieterich, R. A., ed. 1981. Alaskan wildlife diseases. Institute of Arctic Biology, University of Alaska, Fairbanks.

Dieterich, R. A., and G. E. Haas. 1981. Warbles. Pages 179–82 *in* R. A. Dieterich, ed. Alaskan wildlife diseases. Institute of Arctic Biology, University of Alaska, Fairbanks.

Doerr, J. G., and R. A. Dieterich. 1979. Mandibular lesions in the Western Arctic caribou herd of Alaska. Journal of Wildlife Diseases 15:309–18.

Doucet, G. J., M. Julien, D. Messier, and G. Hayeur. 1988. Compatibility between reservoir downstream flow regime and caribou ecology in northern Quebec. Proceedings of North American Caribou Workshop 3:173–84.

Dueck, G. S. 1998. Genetic relations and phylogeography of woodland and barrenground caribou. Thesis, University of Alberta, Edmonton, Alberta, Canada.

Dugmore, A. A. R. 1913. The romance of the Newfoundland caribou. J. P. Lippincott, Philadelphia.

Edmonds, E. J. 1988. Population status, distribution, and movements of woodland caribou in west central Alberta. Canadian Journal of Zoology 66:817–26.

Edmonds, E. J. 1991. Status of woodland caribou in western North America. Rangifer (Special Issue) 7:91–107.

Edmonds, E. J. 1998. Status of woodland caribou in Alberta: 1996. Rangifer (Special Issue) 10:111–15.

Edmonds, E. J., and M. I. Bloomfield. 1984. A study of woodland caribou (*Rangifer tarandus caribou*) in west central Alberta, 1979 to 1983. Alberta Energy and Natural Resources, Fish and Wildlife Division, Edmonton, Alberta, Canada.

Edwards, R. Y. 1954. Fire and the decline of a mountain caribou herd. Journal of Wildlife Management 18:521–26.

Edwards, R. Y. 1958. Land form and caribou distribution in British Columbia. Journal of Mammalogy 39:408–12.

Edwards, R. Y., and R. W. Ritcey. 1959. Migration of caribou in a mountainous area in Wells Gray Park, British Columbia. Canadian Field-Naturalist 73:21–25.

Edwards, R. Y., and R. W. Ritcey. 1960. Foods of caribou in Wells Gray Park, British Columbia. Canadian Field-Naturalist 74:3–7.

Edwards, R. Y., J. Soos, and R. W. Ritcey. 1960. Quantitative observations on epidendric lichens used as foods by caribou. Ecology 41:425–31.

Eide, S., and Ballard, W. B. 1982. Apparent case of surplus killing of caribou by gray wolves. Canadian Field-Naturalist 96:87–88.

Erickson, C. A. 1975. Some preliminary observations on interspecific acoustic communication of semi-domestic reindeer, with emphasis on the mother–calf relationship. Proceedings of International Reindeer/Caribou Symposium 1:387–97.

Eriksson, O. 1980. Effects of forest fertilization on the cratering intensity of reindeer. Proceedings of International Reindeer/Caribou Symposium 2:26–40.

Espmark, Y. 1964a. Rutting behavior in reindeer (*Rangifer tarandus* L.). Animal Behaviour 12:159–63.

Espmark, Y. 1964b. Studies in dominance–subordination relationship in a group of semi-domestic reindeer (*Rangifer tarandus* L.). Animal Behaviour 12:420–26.

Espmark, Y. 1971. Mother–young relationship and ontogeny of behaviour in reindeer (*Rangifer tarandus*). Zeitschrift für Tierpsychologie 29:42–81.

Fancy, S. G. 1983. Movements and activity budgets of caribou near oil drilling sites in the Sagavanirktok River plain, Alaska. Arctic 36:193–97.

Fancy, S. G., and R. G. White. 1985. Energy expenditure by caribou while cratering in snow. Journal of Wildlife Management 49:987–93.

Fancy, S. G., and K. R. Whitten. 1991. Selection of calving sites by Porcupine herd caribou. Canadian Journal of Zoology 69:1736–43.

Farnell, R. 2000. Panel Discussion: Human developments and their effects on caribou. Rangifer (Special Issue) 12:115–22.

Farnell, R., and J. McDonald. 1988. The influence of wolf predation on caribou mortality in Yukon's Finlayson caribou herd. Proceedings of North American caribou workshop 2:52–70.

Farnell, R., N. Barichello, K. Egli, and G. Kuzyk. 1996. Population ecology of two woodland caribou herds in the southern Yukon. Rangifer (Special Issue) 9:63–72.

Farnell, R., R. Florkiewicz, G. Kuzyk, and K. Egli. 1998. The status of *Rangifer tarandus caribou* in Yukon, Canada. Rangifer (Special Issue) 10:131–37.

Ferguson, M. A. D., and L. Gauthier. 1992. Status and trends of *Rangifer tarandus* and *Ovibos moschatus* populations in Canada. Rangifer 12:127–41.

Ferguson, M. A. D., and F. Messier. 1997. Collection and analysis of traditional ecological knowledge about a population of Arctic tundra caribou. Arctic 50:17–28.

Ferguson, M. A. D., and F. Messier. 2000. Mass emigration of Arctic tundra caribou from a traditional winter range: Population dynamics and physical condition. Journal of Wildlife Management 64:168–78.

Ferguson, M. A. D., R. G. Williamson, and F. Messier. 1998. Inuit knowledge of long-term changes in a population of Arctic tundra caribou. Arctic 51:201–19.

Fischer, C. A., and E. A. Duncan. 1976. Ecological studies of caribou and muskoxen in the Arctic Archipelago and northern Keewatin. Renewable Resources Consulting Service, Edmonton, Alberta, Canada.

Fleck, E. S., and A. Gunn. 1982. Characteristics of three barren-ground caribou calving grounds in the Northwest Territories (Progress Report 7). Northwest Territories Wildlife Service.

Fleischman, S. J. 1991. Lichen availability on the range of an expanding caribou population in Alaska. Proceedings of North American Caribou Workshop 4:423–28.

Forbes, B. C., and G. Kofinas, eds. 2000. Proceedings of the human role in reindeer/caribou systems workshop. Polar Research 19 (1):1–142.

Freddy, D. J. 1979. Distribution and movements of Selkirk caribou, 1972–1974. Canadian Field-Naturalist 93:71–74.

Freddy, D. J., and A. W. Erickson. 1975. Status of the Selkirk Mountain caribou. Proceedings of International Reindeer/Caribou Symposium 1:221–27.

Freeman, M. M. R. 1975. Assessing movement in an Arctic caribou population. Journal of Environmental Management 3:251–57.

Frick, C. 1937. Horned ruminants of North America (Bulletin 69). American Museum of Natural History.

Fruetal, M., and M. W. Lankester. 1989. Gastrointestinal helminths of woodland and barren ground caribou (*Rangifer tarandus*) in Canada, with keys to species. Canadian Journal of Zoology 67:2253–69.

Fuller, T. K., and L. B. Keith. 1981. Woodland caribou population dynamics in northeastern Alberta. Journal of Wildlife Management 45:197–213.

Gagnon, L., and C. Barrette. 1992. Antler casting and parturition in wild female caribou. Journal of Mammalogy 73:440–42.

Gates, C. C.. 1985. The fall and rise of the Karminuriak caribou population. Proceedings of North American Caribou Workshop 2:215–28.

Gates, C. C., J. Adamczewski, and R. Mulders. 1986. Population dynamics, winter ecology, and social organization of Coats Island caribou. Arctic 39:216–22.

Gauthier, D. A., and J. B. Theberge. 1986. Wolf predation in the Burwash caribou herd, southwest Yukon, Canada. Rangifer (Special Issue) 1:137–44.

Geist, V. 1982. Adaptive behavioral strategies. Pages 219–77 *in* J. W. Thomas and D. Toweill, eds. Elk of North America: Ecology and management. Stackpole, Harrisburg, PA.

Geist, V. 1991. Taxonomy: On an objective definition of subspecies, taxa as legal entities, and its application to *Rangifer tarandus* Lin. 1758. Proceedings of North American Caribou Workshop 4:1–36.

Goss, R. J. 1980. Is antler asymmetry in reindeer and caribou genetically determined? Proceedings of International Reindeer/Caribou Symposium 2:364–72.

Goss, R. J. 1983. Deer antlers: Regeneration, function and evolution. Academic Press, New York.

Gravlund, P., M. Meldgaard, S. Paabo, and P. Arctander. 1998. Polyphyletic origin of the small-bodied subspecies of tundra reindeer (*Rangifer tarandus*). Molecular Phylogenetics and Evolution 10:155–59.

Griffith, B., and R. D. Cameron. 2000. Shifts in the distribution of calving caribou: Developing a model for assessing the impacts of development. Rangifer (Special Issue) 12:103.

Gunn, A. 1990. The decline and recovery of caribou and muskoxen on Victoria Island. Volume 2. Pages 590–607 *in* C. R. Harington, ed. Canada's missing dimension: Science and history in the Canadian Arctic Islands. Canadian Museum of Nature, Ottawa.

Gunn, A., and J. Dragon. 1998. Status of caribou and muskox populations within the Prince of Wales–Somerset Island–Boothia Peninsula complex, July–August 1995 (File Report 122). Northwest Territories Department of Resources, Wildlife and Economic Development, Yellowknife, Canada.

Gunn, A., and J. Dragon. 2002. Peary cariou and muskox abundance and distribution on the western Queen Elizabeth Islands, Northwest Territories and Nunavut June-July 1997. (File Report 130) Northwest Territories Department of Resources, Wildlife and Economic Development, Yellowknife, Canada.

Gunn, A., and B. Fournier. 2000. Caribou herd delimitation and seasonal movements based on satellite telemetry on Victoria Island 1987–89. (File Report 125). Northwest Territories Department of Resources, Wildlife and Economic Development, Yellowknife, Canada.

Gunn, A., and F. L. Miller. 1983. Size and status of an inter-island population of Peary caribou. Acta Zoologica Fennica 175:153–54.

Gunn, A., and F. L. Miller. 1986. Traditional behavior and fidelity to caribou calving grounds by barren-ground caribou. Rangifer (Special Issue) 1:151–58.

Gunn, A., and S. M. Tomkiewicz, Jr. 1991. Potential applications for satellite telemetry: Where do we go from here? Proceedings of North American Caribou Workshop 4:362–81.

Gunn, A., F. L. Miller, R. Glaholt, and K. Jingfors. 1985. Behavioural responses of barren- ground caribou cows and calves to helicopters on the Beverly Herd calving ground, Northwest Territories. Proceedings of North American Caribou Workshop 1:10–14.

Gunn, A., T. Leighton, and G. Wobeser. 1991. Wildlife diseases and parasites in the Kitikmeot Region, 1984–1990. (File Report 104). Northwest Territories Department of Renewable Resources, Yellowknife, Canada.

Gunn, A., A. Buchan, B. Fournier, and J. Nishi. 1997. Victoria Island caribou migrations across Dolphin and Union Strait and Coronation Gulf from the Mainland Coast, 1976–94 (Manuscript Report 94). Department of Renewable Resources, Government of Northwest Territories, Yellowknife, Canada.

Gunn, A., F. L. Miller, and J. Nishi. 2000. Status of endangered and threatened caribou on Canada's Arctic Islands. Rangifer (Special Issue) 12:39–50.

Harington, C. R. 1999. Ancient caribou. Yukon Tourism, Heritage Branch, Beringian Research Notes 12:1–4.

Harper, F. 1955. The barren-ground caribou of Keewatin (Miscellaneous Publication 6). University of Kansas Museum of Natural History, Lawrence.

Harrington, F. H., and A. M. Veitch. 1992. Calving success of woodland caribou exposed to low-level jet fighter overflights. Arctic 45:213–18.

Hart, J. S., O. Heroux, W. H. Cottle, and C. A. Mills. 1961. The influence of climate on metabolic and thermal responses of infant caribou. Canadian Journal of Zoology 39:845–56.

Hatler, D. F. 1986. Studies of radio-collared caribou in the Spatsizi Wilderness Park area, British Columbia, 1980–1984 (Report 3). Spatsizi Association for Biological Research.

Heape, W. 1931. Migration, emigration and nomadism. W. Heffer, Cambridge.

Heard, D. C. 1990. The intrinsic rate of increase of reindeer and caribou populations in arctic environments. Rangifer (Special Issue) 3:169–73.

Heard, D. C., and G. W. Calef. 1986. Population dynamics of the Kaminuriak caribou herd, 1968–1985. Rangifer (Special Issue) 1:159–66.

Heard, D. C., and J.-P. Ouellet. 1994. Dynamics of an introduced caribou population. Arctic 47:88–95.

Heard, D. C., and K. L. Vagt. 1998. Caribou in British Columbia: A 1996 status report. Rangifer (Special Issue) 10:117–23.

Heard, D. C., and T. M. Williams. 1990. Ice and mineral licks used by caribou in winter. Rangifer (Special Issue) 3:203–6.

Helle, T., and J. Aspi. 1983. Does herd formation reduce insect harassment among reindeer; a field experiment with animal traps. Acta Zoologica Fennica 175:129–31.

Hemming, J. E. 1971. The distribution and movement patterns of caribou in Alaska (Wildlife Technical Bulletin 1). Alaska Department of Fish and Game.

Henshaw, J. 1968a. A theory for the occurrence of antlers in females of the genus *Rangifer.* Journal of British Deer Society 1:222–26.

Henshaw, J. 1968b. The activities of wintering caribou in north-western Alaska in relation to weather and snow conditions. International Journal of Biometeorology 12:21–27.

Henshaw, J. 1970. Consequences of travel in the rutting of reindeer and caribou (*Rangifer tarandus*). Animal Behaviour 18:256–58.

Hoare, W. H. B. 1927. Report on investigations affecting Eskimo and wild life, District of Mackenzie, 1925–1926, together with general recommendations. Department of Interior, Northwest Territories and Yukon Branch, Ottawa.

Hoare, W. H. B. 1930. Conserving Canada's musk-oxen (being an account of an investigation of Thelon Game Sanctuary, 1928–29, with a brief history of the area and an outline of known facts regarding the musk-ox). Department of the Interior, Northwest Territories and Yukon Branch, Ottawa.

Holleman, D. F., and J. R. Luick. 1977. Lichen species preference by reindeer. Canadian Journal of Zoology 55:1368–69.

Holleman, D. F., J. R. Luick, and R. G. White. 1979. Lichen intake estimates for reindeer and caribou during winter. Journal of Wildlife Management 43:192–201.

Hopla, C. E. 1975. Q fever and Alaskan caribou. Proceedings of International Reindeer/Caribou Symposium 1:498–506.

Horejsi, B. L. 1981. Behavioural response of barren-ground caribou to a moving vehicle. Arctic 34:180–85.

Huot, J. 1989. Body composition of the George River caribou (*Rangifer tarandus caribou*) in fall and late winter. Canadian Journal of Zoology 67:103–7.

Huot, J., and M. Beaulieu. 1985. Relationship between parasite infection levels and body fat reserves in George River caribou in spring and fall. Proceedings of North American Caribou Workshop 2:317–27.

Irving, L. 1966. Adaptation to cold. Scientific American 214:94–101.

Irving, L., and J. Krog. 1955. Temperature of skin in the Arctic as a regulator of heat. Journal of Applied Physiology 7:355–64.

Jacobi, A. 1931. Das Rentier: Eine zoologische Monographie der Gattung *Rangifer.* Akademie Verlag, Leipzig, Germany.

Jakimchuk, R. D., and D. R. Carruthers. 1983. Caribou on Victoria Island, Northwest Territories, Canada. Acta Zoologica Fennica 175:149–51.

Jakimchuk, R. D., S. H. Ferguson, and L. G. Sopuck. 1987. Differential habitat use and sexual segregation in the Central Arctic caribou herd. Canadian Journal of Zoology 65:534–41.

James, A. R. C., and A. K. Stuart-Smith. 2000. Distribution of caribou and wolves in relation to linear corridors. Journal of Wildlife Management 64:154–59.

Johanson, K. J. 1990. The consequences in Sweden of the Chernobyl accident. Rangifer (Special Issue) 3:9–10.

Johnson, D. R. 1976. Mountain caribou: Threats to survival in the Kootenay Pass region, British Columbia. Northwest Science 50:97–101.

Johnson, D. R., and M. C. Todd. 1977. Summer use of a highway crossing by mountain caribou. Canadian Field-Naturalist 91:312–14.

Johnson, E. A., and J. S. Rowe. 1975. Fire in the sub-arctic wintering ground of the Beverly caribou herd. American Midland Naturalist 94:1–14.

Kelsall, J. P. 1957. Continued barren-ground caribou studies (Management Bulletin Series 1, No. 12). Canadian Wildlife Service.

Kelsall, J. P. 1968. The migratory barren-ground caribou of Canada (Monograph 3). Canadian Wildlife Service.

Kelsall, J. P. 1970. Interaction between barren-ground caribou and muskrats. Canadian Journal of Zoology 48:605.

Kelsall, J. P. 1975. Warble fly distribution among some Canadian caribou. Proceedings of International Reindeer/Caribou Symposium 1:509–17.

Kelsall, J. P. 1984. Status report on woodland caribou *Rangifer tarandus dawsoni* and *Rangifer tarandus caribou,* in Canada, in 1982. Committee on the Status of Endangered Wildlife in Canada, Canadian Wildlife Service, Ottawa.

Kelsall, J. P., E. S. Telfer, and T. D. Wright. 1977. The effects of fire on the ecology of the Boreal Forest, with particular reference to the Canadian North: A review and selected bibliography (Occasional Paper 32). Canadian Wildlife Service.

Kevan, P. G. 1974. Peary caribou and muskoxen on Banks Island. Arctic 27:256–64.

Klein, D. R. 1964. Range-related differences in growth of deer reflected in skeletal ratios. Journal of Mammalogy 45:226–35.

Klein, D. R. 1970a. Food selection by North American deer and their response to overutilization of preferred plant species. Pages 25–46 *in* A. Watson, ed. Animal populations in relation to their food resources (British Ecological Society Symposium No. 10). Blackwell, Oxford.

Klein, D. R. 1970b. Interactions of *Rangifer tarandus* (reindeer and caribou) with its habitat in Alaska. Transaction of International Congress of Game Biologists 8:289–93.

Klein, D. R. 1970c. Tundra ranges north of the boreal forests. Journal of Range Management 23:8–14.

Klein, D. R. 1971. Reaction of reindeer to obstructions and disturbances. Science 173:393–98.

Klein, D. R. 1980. Reaction of caribou and reindeer to obstructions: A reassessment. Proceedings of International Reindeer/Caribou Symposium 2:519–27.

Klein, D. R. 1982. Fire, lichens, and caribou. Journal of Range Management 35:390–95.

Klein, D. R. 1990. Variation in quality of caribou and reindeer forage plants associated with season, plant part, and phenology. Rangifer (Special Issue) 3:123–30.

Klein, D. R., and T. J. Vlasova. 1992. Lichens, a unique forage resource threatened by air pollution. Rangifer 12:21–27.

Klein, D. R., and R. G. White, eds. 1978. Parameters of caribou population ecology in Alaska. (Biological Papers of the University of Alaska, Special Report 3). University of Alaska, Fairbanks.

Krog, J. O., and M. Wika. 1975. The circulation in the growing reindeer antlers. Proceedings of International Reindeer/Caribou Symposium 1:368–73.

Kuropat, P., and J. P. Bryant. 1980. Foraging behavior of cow caribou on the Utukok calving grounds in northwestern Alaska. Proceedings of International Reindeer/Caribou Symposium 2:64–70.

Kuropat, P., and J. P. Bryant. 1983. Digestibility of caribou forage in Arctic Alaska in relation to nutrient, fiber, and phenolic constituents. Acta Zoologica Fennica 175:51–52.

Kuyt, E. 1972. Food habits of wolves on barren-ground caribou range, (Report Series 21). Canadian Wildlife Service.

Lankester, M. W., and D. Fong. 1989. Distribution of elaphostrongyline nematodes (Metastrongyloidea: Protostrongylidae) in Cervidae and possible effects of moving *Rangifer* spp. into and within North America. Alces 25:133–45.

Lankester, M. W., and D. Fong. 1998. Protostrongylid nematodes in caribou (*Rangifer tarandus caribou*) and moose (*Alces alces*) of Newfoundland. Rangifer (Special Issue) 10:73–83.

Lankester, M. W., and P. L. Hauta. 1989. *Paraelaphostrongylus andersoni* (Nematoda: Protostrongylidae) in caribou (*Rangifer tarandus*) of northern and central Canada. Canadian Journal of Zoology 67:1966–75.

Lankester, M. W., and S. Luttich. 1988. *Fascioloides magna* (Trematoda) in woodland caribou (*Rangifer tarandus caribou*) of the George River herd, Labrador. Canadian Journal of Zoology 66:475–79.

LaPerriere, A. J., and P. C. Lent. 1977. Caribou feeding sites in relation to snow characteristics in northeastern Alaska. Arctic 30:101–8.

Larter, N. C., and J. A. Nagy. 1995. Evidence of overwinter growth in Peary caribou, *Rangifer tarandus pearyi,* calves. Canadian Field-Naturalist 109:446–48.

Larter, N. C., and J. A. Nagy. 1997. Peary caribou, muskoxen and Banks Island forage: Assessing seasonal diet similarities. Rangifer 17:9–16.

Lavigueur, L., and C. Barrette. 1992. Suckling, weaning, and growth in captive woodland caribou. Canadian Journal of Zoology 70:1753–66.

Lent, P. C. 1965a. Observations on antler shedding by female barren-ground caribou. Canadian Journal of Zoology 43:553–58.

Lent, P. C. 1965b. Rutting behavior in a barren-ground caribou population. Animal Behaviour 13:259–64.

Lent, P. C. 1966a Calving and related social behavior in the barren-ground caribou. Zeitschrift für Tierpsychologie 23:701–56.

Lent, P. C. 1966b. The caribou of northwestern Alaska. Pages 481–517 *in* N. J. Wilimovsky and J. N. Wolfe, eds. Environment of the Cape Thompson region, Alaska. U.S. Atomic Energy Commission, Washington, DC.

Lent, P. C. 1974. Mother–infant relationships in ungulates. Pages 14–55 *in* V. Geist and F. Walthers, eds. The behavior of ungulates and its relation to management, Vol. 1. International Union for the Conservation of Nature and Natural Resources, Morges, Switzerland.
Lent, P. C. 1975. A review of acoustic communication in *Rangifer tarandus*. Proceedings of International Reindeer/Caribou Symposium 1:398–408.
Lentz, C. P., and J. S. Hart. 1960. The effect of wind and moisture on heat loss through the fur of newborn caribou. Canadian Journal of Zoology 38:679–88.
Leopold, A. S., and F. F. Darling. 1953. Wildlife in Alaska. Ronald Press, New York.
Lewin, V., and J. G. Stelfox. 1967. Functional anatomy of the tail and associated behavior in woodland caribou. Canadian Field-Naturalist 1:63–66.
Loomis, F. G. 1925. Dentition of artiodactyls. Bulletin of the Geological Society of America 36:583–604.
Luick, J. R., and R. G. White. 1975. Glucose metabolism in female reindeer. Proceedings of International Reindeer/Caribou Symposium 1:379–86.
Mahoney, S. P., H. Abbott, L. H. Russell, and B. R. Porter. 1990. Woodland caribou calf mortality in insular Newfoundland, Vol. 2. Pages 592–99 *in* Transactions of 19th International Union of Game Biologists congress. Trondheim, Norway.
Main, M. R., and B. E. Coblentz. 1990. Sexual segregation among ungulates: A critique. Wildlife Society Bulletin 18:204–10.
Mallory, F. F., and T. L. Hillis. 1998. Demographic characteristics of circumpolar caribou populations: Ecotypes, ecological constraints, releases, and population dynamics. Rangifer (Special Issue) 10:49–60.
Manning, T. H. 1960. The relationship of the Peary and barren-ground caribou (Technical Paper 4). Arctic Institute of North America.
Martell, A. M., and D. E. Russell, eds. 1983. Proceedings of the first North American Caribou workshop. Whitehorse, Yukon, Canada.
Martell, A. M., W. Nixon, and D. E. Russell. 1986. Distribution, activity and range use of male caribou in early summer in Northern Yukon, Canada. Rangifer (Special Issue) 1:181–89.
Maxwell, B. 1997. Responding to global climate change in Canada's Arctic. Environment Canada, Downsview, Ontario, Canada.
McDonald, E. J., and A. M. Martell. 1981. Twinning and postpartum activity in barren-ground caribou (*Rangifer tarandus*). Canadian Field-Naturalist 95:354–55.
McEwan, E. H. 1963. Seasonal annuli in the cementum of the teeth of barren-ground caribou. Canadian Journal of Zoology 41:111–13.
McEwan, E. H. 1968. Hematological studies of barren-ground caribou. Canadian Journal of Zoology 46:1031–36.
McEwan, E. H. 1971. Twinning in caribou. Journal of Mammalogy 52:479.
McEwan, E. H., and P. E. Whitehead. 1970. Seasonal changes in the energy and nitrogen intake in reindeer and caribou. Canadian Journal of Zoology 48:905–13.
McEwan, E. H., and P. E. Whitehead. 1972. Reproduction in female reindeer and caribou. Canadian Journal of Zoology 50:43–46.
McEwan, E. H., and A. J. Wood. 1966. Growth and development of the barren-ground caribou. Part 1: Heart girth, hind foot length, and body weight relationships. Canadian Journal of Zoology 44:401–11.
McEwan, E. H., A. J. Wood, and H. C. Nordan. 1965. Body temperature of barren-ground caribou. Canadian Journal of Zoology 43:683–87.
McGowen, T. A. 1966. Caribou studies in northwestern Alaska. Proceedings of the Western Association of State Game and Fish Commissioners 46:57–66.
Mech, L. D., T. J. Meier, J. W. Burch, and L. G. Adams. 1995. Pages 231–43 *in* L. N. Carbyn, S. H. Fritts, and D. R. Seip, eds. Ecology and conservation of wolves in a changing world (Occasional Paper 35). Canadian Circumpolar Institute, Edmonton, Alberta, Canada.
Messier, F. 1995. On the functional and numerical responses of wolves to changing prey density. Pages 187–97 *in* L. N. Carbyn, S. H. Fritts, and D. R. Seip, eds. Ecology and conservation of wolves in a changing world (Occasional Paper 35). Canadian Circumpolar Institute, Edmonton, Alberta, Canada.
Messier, F., J. Huot, D. Le Henaff, and S. Luttich. 1988. Demography of the George River caribou herd: Evidence of population regulation by forage exploitation and range expansion. Arctic 41:279–87.
Miller, D. R. 1974. Seasonal changes in the feeding behaviour of barren-ground caribou on the Taiga winter range. Pages 744–55 *in* V. Geist and F. Walters, eds. The behaviour of ungulates and its relation to management, Vol. 2. International Union for the Conservation of Nature and Natural Resources, Morges, Switzerland.
Miller, D. R. 1975. Observations of wolf predation on barren ground caribou in winter. Proceedings of International Reindeer/Caribou Symposium 1:209–20.
Miller, D. R. 1976. Biology of the Kaminuriak population of barren-ground caribou. Part 3. Taiga winter range relationships and diet (Report Series 36). Canadian Wildlife Service.
Miller, D. R. 1980. Wildfire effects on barren-ground caribou wintering on the taiga of north-central Canada: A reassessment. Proceedings of International Reindeer/Caribou Symposium 2:84–98.
Miller, F. L. 1972. Eruption and attrition of mandibular teeth in barren-ground caribou. Journal of Wildlife Management 36:606–12.
Miller, F. L. 1974a. Age determination of caribou by annulations in dental cementum. Journal of Wildlife Management 38:47–53.
Miller, F. L. 1974b. A new era: Are migratory barren-ground caribou and petroleum exploitation compatible? Transactions of Northeastern Section of the Wildlife Society 31:45–55.
Miller, F. L. 1974c. Biology of the Kaminuriak population of barren-ground caribou. Part 2. Dentition as an indicator of sex and age; composition and socialization of the population (Report Series 31). Canadian Wildlife Service.
Miller, F. L. 1982. Caribou *Rangifer tarandus*. Pages 923–59 *in* J. A. Chapman and G. A. Feldhamer, eds. Wild mammals of North America: Biology, management, and economics. Johns Hopkins University Press, Baltimore.
Miller, F. L. 1983. Restricted caribou harvest or welfare: Northern native's dilemma. Acta Zoologica Fennica 175:171–75.
Miller, F. L. 1986. Asymmetry in antlers of barren-ground caribou, Northwest Territories, Canada. Rangifer (Special Issue) 1:195–202.
Miller, F. L. 1987. Management of barren-ground caribou (*Rangifer tarandus groenlandicus*) in Canada. Pages 523–34 *in* C. M. Wemmer, ed. Biology and management of the Cervidae. Research symposium. National Zoological Park, Smithsonian Institute, Front Royal, VA.
Miller, F. L. 1990a. Inter-island movements of Peary caribou: A review and appraisement of their ecological importance. Volume 2. Pages 608–32 *in* C. R. Harington, ed. Canada's missing dimension: Science and history in the Canadian Arctic Islands. Canadian Museum of Nature, Ottawa.
Miller, F. L. 1990b. Peary caribou status report. Environment Canada, Canadian Wildlife Service, Edmonton, Alberta.
Miller, F. L. 1995a. Inter-island water crossings by Peary caribou, south-central Queen Elizabeth Islands. Arctic 48:8–12.
Miller, F. L. 1995b. Peary caribou studies, Bathurst Island complex, Northwest Territories, Canada, 1993 (Technical Report Series 230). Canadian Wildlife Service.
Miller, F. L. 1997a. Late winter absence of caribou on Prince of Wales, Russell, and Somerset islands, Northwest Territories, April–May 1996 (Technical Report Series 291). Canadian Wildlife Service.
Miller, F. L. 1997b. Peary caribou conservation studies, Bathurst Island complex, Northwest Territories, April–August 1994 and June–July 1995. (Technical Report Series 295). Canadian Wildlife Service.
Miller, F. L. 1998. Status of Peary caribou and muskox populations within the Bathurst Island complex, south-central Queen Elizabeth Islands, Northwest Territories, July 1996 (Technical Report Series 317). Canadian Wildlife Service.
Miller, F. L., and S. J. Barry. 1992. Nonrandom distribution of antlers cast by Peary caribou bulls, Melville Island, Northwest Territories. Arctic 45:252–57.
Miller, F. L., and E. Broughton. 1971. Polydactylism in a barren-ground caribou from northwestern Manitoba. Journal of Wildlife Diseases 7:307–9.
Miller, F. L., and E. Broughton. 1973. Behaviour associated with mortality and stress in maternal–filial pairs of barren-ground caribou. Canadian Field-Naturalist 87:21–25.
Miller, F. L., and E. Broughton. 1974. Calf mortality on the calving ground of Kaminuriak caribou (Report Series 26). Canadian Wildlife Service.
Miller, F. L., and A. Gunn. 1978. Interisland movements of Peary caribou south of Viscount Melville Sound, Northwest Territories. Canadian Field-Naturalist 92:327–33.
Miller, F. L., and A. Gunn. 1979. Responses of Peary caribou and muskoxen to helicopter harassment (Occasional Paper 40). Canadian Wildlife Service, Ottawa.
Miller, F. L., and A. Gunn. 1980. Inter-island movements of Peary caribou (*Rangifer tarandus pearyi*) south of Viscount Melville Sound and Barrow Strait, Northwest Territories, Canada. Proceedings of International Reindeer/Caribou Symposium 2:89–114.
Miller, F. L., and A. Gunn. 1981. Play by Peary caribou calves before, during, and after helicopter harassment. Canadian Journal of Zoology 59:823–27.
Miller, F. L., and A. Gunn. 1982. Nursing and associated behavior of Peary caribou, *Rangifer tarandus pearyi*. Canadian Field-Naturalist 96:200–202.
Miller, F. L., and A. Gunn. 1986a. Caribou calf deaths from intraspecific strife: A debatable diagnosis. Rangifer (Special Issue) 1:203–9.

Miller, F. L., and A. Gunn. 1986b. Effect of adverse weather on neonatal caribou survival: A review. Rangifer (Special Issue) 1:211–17.

Miller, F. L., and A. Gunn. 1986c. Observations of barren-ground caribou travelling on thin ice during autumn migration. Arctic 39:85–88.

Miller, F. L., and R. L. McClure. 1973. Determining age and sex of barren-ground caribou from dental variables. Transactions of Northeastern Section of the Wildlife Society 30:79–100.

Miller, F. L., and G. R. Parker. 1968. Placental remnants in the rumens of maternal caribou. Journal of Mammalogy 49:778.

Miller, F. L., and G. D. Tessier. 1971. Dental anomalies in barren-ground caribou. Journal of Mammalogy 52:164–74.

Miller, F. L., C. J. Jonkel, and G. D. Tessier. 1972. Group cohesion and leadership response by barren-ground caribou to man-made barriers. Arctic 25:193–202.

Miller, F. L., F. W. Anderka, C. Vithayasai, and R. L. McClure. 1975a. Distribution, movements and socialization of barren-ground caribou radio-tracked on their calving and post-calving areas. Proceedings of International Reindeer/Caribou Symposium 1:423–35.

Miller, F. L., A. J. Cawley, L. P. E. Choquette, and E. Broughton. 1975b. Radiographic examination of mandibular lesions in barren-ground caribou. Journal Wildlife Diseases 11:465–70.

Miller, F. L., R. H. Russell, and A. Gunn. 1977a. Distributions, movements, and numbers of Peary caribou and muskoxen on western Queen Elizabeth Islands, Northwest Territories, 1972–74 (Report Series 40). Canadian Wildlife Service.

Miller, F. L., R. H. Russell, and A. Gunn. 1977b. Inter-island movements of Peary caribou (*Rangifer tarandus pearyi*) on western Queen Elizabeth Islands, Arctic Canada. Canadian Journal of Zoology 55:1029–37.

Miller, F. L., E. J. Edmonds, and A. Gunn. 1982. Foraging behaviour of Peary caribou in response to springtime snow and ice conditions (Occasional Paper 48). Canadian Wildlife Service.

Miller, F. L., A. Gunn, and E. Broughton. 1985. Surplus killing as exemplified by wolf predation on newborn caribou. Canadian Journal of Zoology 63:295–300.

Miller, F. L., E. Broughton, and A. Gunn. 1988a. Mortality of migratory barren-ground caribou on the calving grounds of the Beverly herd, Northwest Territories, 1981–1983. (Occasional Paper 66). Canadian Wildlife Service.

Miller, F. L., A. Gunn, and E. Broughton. 1988b. Utilization of carcasses of newborn caribou killed by wolves. Proceedings of North American Caribou Workshop 3:73–87.

Moisan, G. 1959. The caribou of Gaspe. Northeastern Section of the Wildlife Society Conference 10:201–7.

Moote, I. 1955. The thermal insulation of caribou pelts. Textile Research Journal 25:837.

Morton, J. K. 1986. Role of predators in reindeer brucellosis in Alaska. Rangifer (Special Issue) 1:368.

Muller-Schwarze, D., and C. Muller-Schwarze. 1983. Play behaviour in free-ranging caribou, *Rangifer tarandus*. Acta Zoologica Fennica 175:121–24.

Muller-Schwarze, D., L. Kallquist, T. Mossing, A. Brundin, and G. Andersson. 1978. Responses of reindeer to interdigital secretions of conspecifics. Journal of Chemical Ecology 4:325–36.

Murie, A. 1944. The wolves of Mount McKinley (Fauna Series 5). U.S. National Park Service.

Murie, O. J. 1935. Alaska–Yukon caribou. North American Fauna 54:1–93.

Murphy, S. M. 1988. Caribou behavior and movements in the Kuparuk oilfield: Implications for energetic and impact analyses. Proceedings of North American Caribou Workshop 3:196–210.

Nagy, J. A., N. C. Larter, and V. P. Fraser. 1996. Population demography of Peary caribou and muskox on Banks Island, N.W.T., 1982–1992. Rangifer (Special Issue) 9:213–22.

Neiland, K. A., and C. Dukeminier. 1972. A bibliography of the parasites, diseases, and disorders of several important wild ruminants of the northern hemisphere (Wildlife Technical Bulletin 3). Alaska Department of Fish and Game.

Neiland, K. A., J. A. King, B. E. Huntley, and R. O. Skoog. 1968. The diseases and parasites of Alaskan wildlife populations, Part 1: Some observations on brucellosis in caribou. Bulletin of the Wildlife Disease Association 4:27–36.

Nellemann, C., and R. D. Cameron. 1998. Cumulative impacts of an evolving oilfield complex on the distribution of caribou. Canadian Journal of Zoology 78:1425–30.

Nordkvist, M., and K. Erne. 1983. The toxicity of forest fertilizers (ammonium nitrate) to reindeer. Acta Zoologica Fennica 175:101–5.

Øritsland, N. A. 1974. A windchill and solar radiation index for homeotherms. Journal of Theoretical Biology 47:413–20.

Ouellet, J.-P., D. C. Heard, and R. Mulders. 1996. Population ecology of caribou populations without predators: Southampton and Coats Island herds. Rangifer (Special Issue) 9:17–25.

Paré, M., and J. Huot. 1985. Seasonal movements of female caribou of the Caniapiscau Region, Quebec. Proceedings of North American Caribou Workshop 2:47–56.

Parker, G. R. 1972a. Biology of the Kaminuriak population of barren-ground caribou. Part 1. Total numbers, mortality, recruitment, and seasonal distribution (Report Series 20). Canadian Wildlife Service.

Parker, G. R. 1972b. Distribution of barren-ground caribou harvest in northcentral Canada (Occasional Paper 15). Canadian Wildlife Service.

Parker, G. R. 1978. The diets of muskoxen and Peary caribou on some islands in the Canadian High Arctic (Occasional Paper 35). Canadian Wildlife Service.

Parker, G. R. 1981. Physical and reproductive characteristics of an expanding woodland caribou population (*Rangifer tarandus caribou*) in northern Labrador. Canadian Journal of Zoology 59:1929–40.

Parker, G. R., and R. K. Ross. 1976. Summer habitat used by muskoxen (*Ovibos moschatus*) and Peary caribou (*Rangifer tarandus pearyi*) in the Canadian High Arctic. Polarforschung 46:12–25.

Parker, G. R., D. C. Thomas, E. Broughton, and D. R. Gray. 1975. Crashes of muskox and Peary caribou populations in 1973–74 in the Parry Islands, Arctic Canada (Progress Notes 56). Canadian Wildlife Service.

Parker, K. 1989. Growth rates and morphological measurements of Porcupine caribou calves. Rangifer 9:9–13.

Pegau, R. E. 1970. Effect of reindeer trampling and grazing on lichens. Journal of Range Management 23:95–97.

Pegau, R. E. 1975. Analysis of the Nelchina caribou range. Proceedings of International Reindeer/Caribou Symposium 1:316–23.

Person, S. J., R. G. White, and J. R. Luick. 1980. Determination of nutritive value of reindeer–caribou range. Proceedings of International Reindeer/Caribou Symposium 2:224–39.

Pocock, R. I. 1923. On the external characters of *Elaphurus, Hydropotes, Pudu,* and other Cervidae. Proceedings of the Zoological Society of London 1923:181–207.

Pruitt, W. O., Jr. 1959. Snow as a factor in the winter ecology of barren-ground caribou (*Rangifer arcticus*). Arctic 12:159–79.

Pruitt, W. O., Jr. 1960a. Behavior of the barren-ground caribou (Biological Papers 3). University of Alaska, Fairbanks.

Pruitt, W. O., Jr. 1960b. Locomotor speeds of some large northern mammals. Journal of Mammalogy 41:112.

Pruitt, W. O., Jr. 1961. On postnatal mortality in barren-ground caribou. Journal of Mammalogy 42:550–51.

Pruitt, W. O., Jr. 1965. A flight releaser in wolf–caribou relations. Journal of Mammalogy 46:350–51.

Pruitt, W. O., Jr. 1966. The function of the brow tine in caribou antlers. Arctic 19:111–13.

Quay, W. B. 1955. Histology and cytochemistry of skin gland areas in the caribou, Rangifer. Journal of Mammalogy 36:187–201.

Rajala, P., and B. Westerling. 1980. Response of corral-fed reindeer to some commonly-used wood fertilizers in Finland. Proceedings of International Reindeer/Caribou Symposium 2:240–43.

Reimers, E. 1983. Reproduction in wild reindeer in Norway. Canadian Journal of Zoology 61:211–17.

Rettie, W. J., and F. Messier. 1998. Dynamics of woodland caribou populations at the southern limit of their range in Saskatchewan. Canadian Journal of Zoology 76:251–59.

Rettie, W. J., and F. Messier. 2000. Hierarchical habitat selection by woodland caribou: Its relationship to limiting factors. Ecography 23:466–78.

Rettie, W. J., J. W. Sheard, and F. Messier. 1996. Identification and description of forested vegetation communities available to woodland caribou: Relating wildlife habitat to forest cover data. Forest Ecology and Management 93:245–60.

Rettie, J., T. Rock, and F. Messier. 1998. Status of woodland caribou in Saskatchewan. Rangifer (Special Issue) 10:105–9.

Røed, K. H. 1991. Genetic differentiation and evolution of tundra reindeer and caribou. Proceedings of North American Caribou Workshop 4:64–76

Røed, K. H., and D. C. Thomas. 1990. Transferrin variation and evolution of Canadian barren-ground caribou. Rangifer (Special Issue) 3:385–89.

Røed, K. H., and K. R. Whitten. 1986. Transferrin variation and evolution of Alaska reindeer and caribou, *Rangifer tarandus* L. Rangifer (Special Issue) 1:247–251.

Røed, K. H., H. Staaland, E. Broughton, and D. C. Thomas. 1986. Transferrin variation in caribou (*Rangifer tarandus* L.) on the Canadian Arctic Islands. Canadian Journal of Zoology 64:94–98.

Røed, K. H., M. A. D. Ferguson, M. Crête, and A. T. Bergerud. 1991. Genetic variation in transferrin as a predictor for differentiation and evolution of caribou from eastern Canada. Rangifer 11:65–74.

Rominger, E. M., and J. L. Oldemeyer. 1989. Early-winter habitat of woodland caribou, Selkirk Mountains, British Columbia. Journal of Wildlife Management 53:238–43.

Rominger, E. M., and J. L. Oldemeyer. 1990. Early-winter diet of woodland caribou in relation to snow accumulation, Selkirk Mountains, British Columbia. Canadian Journal of Zoology 68:2691–94.

Rominger, E. M., and J. L. Oldemeyer. 1991. Arboreal lichen on windthrown trees: A seasonal forage resource for woodland caribou, Selkirk Mountains, British Columbia. Proceedings of North American Caribou Workshop 4:475–80.

Rominger, E. M., and C. T. Robbins. 1996. Generic preference and in-vivo digestibility of alectorioid arboreal lichens by woodland caribou. Rangifer (Special Issue) 9:379–80.

Rowe, J. S. 1959. Forest regions of Canada (Forestry Branch Bulletin 123). Department of Northern Affairs and Natural Resources, Ottawa.

Rowe, J. S., and G. W. Scotter. 1973. Fire in the boreal forest. Quaternary Research 3:444–64.

Russell, D. E., A. M. Martell, and W. A. C. Nixon. 1993. Range ecology of the Porcupine caribou herd in Canada. Rangifer (Special Issue) 8:1–168.

Russell, R. H., E. J. Edmonds, and J. Roland. 1978. Caribou and muskoxen habitat studies (ESCOM A1-26). Environmental-Social Program, Northern Pipelines, Minister of Indian and Northern Affairs and Minister of State, Ottawa.

Schaefer, J. A., and W. O. Pruitt, Jr. 1991. Fire and woodland caribou in southeastern Manitoba. Wildlife Monographs 116:1–36.

Scholander, P. F., R. Hock, V. Walters, and L. Irving. 1950a. Adaptations to cold in Arctic and tropical mammals and birds in relation to body temperature, insulation and basal metabolic rate. Biological Bulletin 99:259–71.

Scholander, P. F., R. Hock, V. Walters, F. Johnson, and L. Irving. 1950b. Heat regulation in some Arctic and tropical mammals and birds. Biological Bulletin 99:237–58.

Scholander, P. F., V. Walters, R. Hock, and L. Irving. 1950c. Body insulation of some Arctic and tropical mammals and birds. Biological Bulletin 99:225–36.

Scotter, G. W. 1964. Effects of forest fires on the winter range of barren-ground caribou in northern Saskatchewan (Wildlife Management Bulletin Series 1, No. 18). Canadian Wildlife Service.

Scotter, G. W. 1965. Chemical composition of forage lichens from northern Saskatchewan as related to use by barren-ground caribou. Canadian Journal of Plant Science 45:246–50.

Scotter, G. W. 1967a. Effects of fire on barren-ground caribou and their forest habitat in northern Canada. Transactions of North American Wildlife and Natural Resources Conference 32:246–59.

Scotter, G. W. 1967b. The winter diet of barren-ground caribou in northern Canada. Canadian Field-Naturalist 81:33–39.

Scotter, G. W. 1972. Chemical composition of forage plants from the Reindeer Reserve, Northwest Territories. Arctic 25:21–27.

Seip, D. R. 1990. Ecology of woodland caribou in Wells Gray Provincial Park (Wildlife Bulletin B-68) British Columbia Ministry of Environment, Victoria, Canada.

Seip, D. R. 1991. Predation and caribou populations. Rangifer (Special Issue) 7:46–52.

Seip, D. R. 1992. Factors limiting woodland caribou populations and their interrelationships with wolves and moose in southeastern British Columbia. Canadian Journal of Zoology 70:1494–1503.

Seip, D. R. 1995. Introduction to wolf–prey interactions. Pages 179–86 *in* L. N. Carbyn, S. H. Fritts, and D. R. Seip, eds. Ecology and conservation of wolves in a changing world (Occasional Paper 35). Canadian Circumpolar Institute, Edmonton, Alberta, Canada.

Seip, D. R. 1998. Ecosystem management and the conservation of caribou habitat in British Columbia. Rangifer (Special Issue) 10:203–11.

Servheen, G., and L. J. Lyon. 1989. Habitat use by woodland caribou in the Selkirk Mountains. Journal of Wildlife Management 53:230–37.

Seton, E. T. 1927. Hoofed animals. Pages 53–150 *in* Lives of game animals, Vol. 3. Doubleday Page, New York.

Shank, C. C., P. F. Wilkinson, and D. F. Penner. 1978. Diet of Peary caribou, Banks Island, NWT. Arctic 31:125–32.

Shideler, R. T. 1986. Impacts of human developments and land use on caribou: A literature review. Volume 2: Impacts of oil and gas development on the Central Arctic herd (Technical Report 86-3). Alaska Department of Fish and Game, Habitat Division.

Shideler, R. T., J. F. Robus, J. F. Winters, and M. Kuwada. 1986. Impacts of human developments and land use on caribou: A literature review. Volume 1: A worldwide perspective (Technical Report 86-2). Alaska Department of Fish and Game, Habitat Division.

Shoesmith, M. W. 1976. Twin fetuses in woodland caribou. Canadian Field-Naturalist 90:498–99.

Simkin, D. W. 1965. A preliminary report of woodland caribou study in Ontario (Section Report [Wildlife] 59). Ontario Department of Lands and Forests, Toronto.

Simpson, K., G. P. Woods, and K. B. Hebert. 1985. Critical habitats of caribou (*Rangifer tarandus caribou*) in the mountains of Southern British Columbia. Proceedings of North American Caribou Workshop 2:177–91.

Skoog, R. O. 1968. Ecology of the caribou (*Rangifer tarandus granti*) in Alaska. Dissertation, University of California at Berkeley, Berkeley.

Smith, H. J., R. M. Archibald, and A. H. Corner. 1964. Elaphostrongylosis in Maritime moose and deer. Canadian Veterinary Journal 5:287–96.

Smith, K. G., E. J. Ficht, D. Hobson, and D. Hervieux. 2000. Woodland caribou distribution on winter range in relation to clear-cut logging in west central Alberta—preliminary analysis. Rangifer (Special Issue) 12:111.

Sonnenfeld, J. 1960. Changes in an Eskimo hunting technology: An introduction to implement geography. Annals of the Association of American Geographers 50:172–86.

Stardom, R. R. P. 1975. Woodland caribou and snow conditions in southeast Manitoba. Proceedings of International Reindeer/Caribou Symposium 1:324–34.

Stonehouse, B. 1968. Thermoregulatory function of growing antlers. Nature 218:870–72.

Stevenson, S. K., and D. F. Hatler. 1985. Woodland caribou and their habitat in southern and central British Columbia (Report 23[1]. British Columbia Ministry of Forests and Land Management, Victoria, Canada.

Stuart-Smith, A. K., C. J. A. Bradshaw, S. Boutin, D. M. Hebert, and A. B. Rippin. 1997. Woodland caribou relative to landscape patterns in northeastern Alberta. Journal of Wildlife Management 61:622–33.

Tener, J. S. 1963. Queen Elizabeth Islands game survey, 1961 (Occasional Paper 4). Canadian Wildlife Service.

Terry, E., B. McLellan, G. Watts, and J. Flaa. 1996. Early winter habitat use by Mountain caribou in the North Cariboo and Columbia Mountains, British Columbia. Rangifer (Special Issue) 9:133–40.

Thing, H. 1977. Behavior, mechanics and energetics associated with winter cratering by caribou in northwestern Alaska (Biological Papers, University of Alaska, No. 18). University of Alaska, Fairbanks.

Thing, H. 1984. Feeding ecology of the West Greenland caribou (*Rangifer tarandus groenlandicus*) in the Sisimiut–Kangerlussuaq region. Danish Review of Game Biology 12 (3):1–54.

Thing, H., C. R. Olesen, and P. Aastrup. 1986. Antler possession by West Greenland female caribou in relation to population characteristics. Rangifer (Special Issue) 1:297–304.

Thomas, D. C. 1981. At the crossroads of caribou management in northern Canada (Special Publication 10). Canadian Nature Federation.

Thomas, D. C. 1982. The relationship between fertility and fat reserves of Peary caribou. Canadian Journal of Zoology 60:597–602.

Thomas, D. C. 1991. Adaptations of barren-ground caribou to snow and burns. Proceedings of North American Caribou Workshop 4:482–500.

Thomas, D. C. 1995. A review of wolf–caribou relationships and conservation implications in Canada. Pages 261–73 *in* L. N. Carbyn, S. H. Fritts, and D. R. Seip, eds. Ecology and conservation of wolves in a changing world (Occasional Paper 35). Canadian Circumpolar Institute, Edmonton, Alberta, Canada.

Thomas, D. C. 1996. Prevalence of *Echinococcus granulosus* and *Taenia hydatigena* in caribou in north-central Canada. Rangifer (Special Issue) 9:331–35.

Thomas, D. C. 1998a. Fire–caribou relationships: (V) Winter diet of the Beverly herd in northern Canada, 1980–87 (Technical Report Series 313). Canadian Wildlife Service.

Thomas, D. C. 1998b. Fire–caribou relationships: (VII) Fire management on winter range of the Beverly herd: Final conclusions and recommendations (Technical Report Series 315). Canadian Wildlife Service.

Thomas, D. C. 1998c. Fire–caribou relationships: (VIII) Background information (Technical Report Series 316). Canadian Wildlife Service.

Thomas, D. C. 1998d. Needed: Less counting of caribou and more ecology. Rangifer (Special Issue) 10:15–23.

Thomas, D. C., and H. J. Armbruster. 1998. Fire–caribou relationships: (VI) Fire history of winter range of the Beverly herd (Technical Report Series 314). Canadian Wildlife Service.

Thomas, D. C., and E. Broughton. 1978. Status of three Canadian caribou populations north of 70 in winter 1977 (Progress Notes 85). Canadian Wildlife Service.

Thomas, D. C., and E. J. Edmonds. 1983. Rumen contents and habitat selection of Peary caribou in winter, Canadian Arctic Archipelago. Arctic and Alpine Research 15:97–105.

Thomas, D. C., and E. J. Edmonds. 1984. Competition between caribou and muskoxen, Melville Island, N.W.T., Canada. Biological Papers, University of Alaska, Special Reports 4:93–100.

Thomas, D. C., and P. Everson. 1982. Geographic variation in caribou on the Canadian Arctic islands. Canadian Journal of Zoology 60:2442–54.

Thomas, D. C., and D. P. Hervieux. 1986. The late winter diet of barren-ground caribou in north-central Canada. Rangifer (Special Issue) 1:305–10.

Thomas, D. C., and H. P. L. Kiliaan. 1990. Warble infestations in some Canadian caribou and their significance. Rangifer (Special Issue) 3:409–17.

Thomas, D. C., and H. P. L. Kiliaan. 1998a. Fire–caribou relationships: (I) Physical characteristics of the Beverly herd, 1980–87 (Technical Report Series 309). Canadian Wildlife Service.

Thomas, D. C., and H. P. L. Kiliaan. 1998b. Fire–caribou relationships: (II) Fecundity and physical condition of the Beverly herd (Technical Report Series 310). Canadian Wildlife Service.

Thomas, D. C., and H. P. L. Kiliaan. 1998c. Fire–caribou relationships: (IV) Recovery of habitat after fire on winter range of the Beverly herd (Technical Report Series 312). Canadian Wildlife Service.

Thomas, D. C., and P. Kroeger. 1980. In vitro digestibilities of plants in rumen fluids of Peary caribou. Arctic 33:757–67.

Thomas, D. C., R. H. Russell, E. Broughton, and P. L. Madore. 1976. Investigations of Peary caribou populations on Canadian Arctic Islands (Progress Notes 64). Canadian Wildlife Service.

Thomas, D. C., R. H. Russell, E. Broughton, E. J. Edmonds, and A. Gunn. 1977. Further studies of two populations of Peary caribou in the Canadian Arctic (Progress Notes 80). Canadian Wildlife Service.

Thomas, D. C., P. Kroeger, and D. P. Hervieux. 1984. In vitro digestibilities of plants utilized by barren-ground caribou. Arctic 37:31–36.

Thomas, D. C., E. J. Edmonds, and W. K. Brown. 1996. The diet of woodland caribou populations in west-central Alberta. Rangifer (Special Issue) 9:337–42.

Thomas, D. C., H. P. L. Kiliaan, and T. W. P. Trottier. 1998. Fire–caribou relationships: (III) Movement patterns of the Beverly herd in relation to burns and snow (Technical Report Series 311). Canadian Wildlife Service.

Thomas, D. C., E. J. Edmonds, and H. J. Armbruster. 1999. Range types and their relative use by Peary caribou and muskoxen on Melville Island, NWT (Technical Report Series 343). Canadian Wildlife Service.

Trudell-Moore, J., and R. G. White. 1983. Physical breakdown of food during eating and rumination in reindeer. Acta Zoologica Fennica 175:47–49.

Tucker, B., S. Mahoney, B. Green, E. Menchenton, and L. Russell. 1991. The influence of snow depth and hardness on winter habitat selection by caribou on the southwest coast of Newfoundland. Rangifer (Special Issue) 7:160–63.

Urquhart, D. R. 1973. Oil exploration and Banks Island wildlife—A guideline for the preservation of caribou, muskox, and arctic fox on Banks Island, NWT. Northwest Territories Wildlife Service, Yellowknife, Canada.

Usher, P. J. 2000. Traditional ecological knowledge in environmental assessment and management. Arctic 53:183–93.

Valkenburg, P. 1997. Investigation of regulating and limiting factors in the Delta caribou herd (Federal Aid in Wildlife Restoration Projects W-23-5, W-24-1 through W-24-4, Final Report). Alaska Department of Fish and Game.

Valkenburg, P. 1998. Herd size, distribution, harvest, management issues, and research priorities relevant to caribou herds in Alaska. Rangifer (Special Issue) 10:125–29.

Valkenburg, P., and J. L. Davis. 1986. Calving distribution of Alaska's Steese–Fortymile caribou herd: A case of infidelity. Rangifer (Special Issue) 1:315–23.

Valkenburg, P., J. L. Davis, and R. D. Boertje. 1983. Social organization and seasonal range fidelity of Alaska's Western Arctic caribou—Preliminary findings. Acta Zoologica Fennica 175:125–26.

Valkenburg, P., J. L. Davis, and D. J. Reed. 1988. Distribution of radio-collared caribou from the Delta and Yanert herds during calving. Proceedings of North American Caribou Workshop 3:14–32.

Valkenburg, P., J. L. Davis, and D. V. Grangaard. 1991. Wolf predation and growth of the Fortymile caribou herd. Proceedings of North American Caribou Workshop 4:282–87.

Valkenburg, P., D. G. Kelleyhouse, J. L. Davis, and J. M. Ver Hoef. 1994. Case history of the Fortymile caribou herd, 1920–1990. Rangifer 14:11–22.

Valkenburg, P., J. L. Davis, J. M. Ver Hoef, R. D. Boertje, M. E. McNay, R. M. Eagan, D. J. Reed, C. L. Gardner, and R. W. Tobey. 1996. Population decline in the Delta caribou herd with reference to other Alaskan herds. Rangifer (Special Issue) 9:53–62.

Valkenburg, P., B. Dale, R. W. Tobey, and R. A. Sellers. 1999. Investigation of regulating and limiting factors in the Delta caribou herd (Federal Aid in Wildlife Restoration Project W-27-1, Progress Report). Alaska Department of Fish and Game.

Valkenburg, P., T. H. Spraker, M. T. Hinkes, L. H. Van Daele, R. W. Tobey, and R. A. Sellers. 2000. Increases in body weight and nutritional status of transplanted Alaskan caribou. Rangifer (Special Issue) 12:133–38.

Vandal, D., and C. Barrette. 1985. Snow depth and feeding interactions at snow craters in woodland caribou. Proceedings of North American Caribou Workshop 2:199–212.

Wales, R. A., L. P. Milligan, and E. H. McEwan. 1975. Urea recycling in caribou, cattle and sheep. Proceedings of International Reindeer/Caribou Symposium 1:297–307.

Walters, C. J., R. Hilborn, R. Peterman, M. Jones, and B. Everitt. 1978. Porcupine caribou workshop. Draft report on submodels and scenarios. Unpublished report. University of British Columbia, Institute of Animal Research and Ecology, Vancouver, Canada.

Wells, J. V., and M. E. Richmond. 1995. Populations, metapopulations, and species populations: What are they and who should care? Wildlife Society Bulletin 23:458–62.

White, R. G. 1975. Some aspects of nutritional adaptations of Arctic herbivorous mammals. Pages 239–68 *in* F. J. Vernberg, ed. Adaptions to the environment. Educational Publishers, New York.

White, R. G., and J. Trudell. 1980. Habitat preference and forage consumption by reindeer and caribou near Atkasook, Alaska. Arctic and Alpine Research 4:511–29.

White, R. G., and M. K. Yousef. 1974. Energy cost of locomotion in reindeer. Pages 38–42 *in* Studies on the nutrition and metabolism of reindeer–caribou in Alaska with special interest in nutritional and environmental adaptation (Progress Report, July 1973–December 1974). Institute of Arctic Biology, University of Alaska Fairbanks, Fairbanks.

White, R. G., B. R. Thomson, T. Skogland, S. J. Person, D. E. Russell, D. F. Holleman, and J. R. Luick. 1975. Ecology of caribou at Prudhoe Bay, Alaska. Pages 150–201 *in* J. Brown, ed. Ecological investigations of the tundra biome in the Prudhoe Bay region, Alaska (Biological Papers, University of Alaska, Special Report 2). University of Alaska, Fairbanks.

Whitten, K. R. 1995. Antler loss and udder distention in relation to parturition in caribou. Journal of Wildlife Management 59:273–77.

Whitten, K. R. 1996. Ecology of the Porcupine caribou herd. Rangifer (Special Issue) 9:45–51.

Whitten, K. R., and R. D. Cameron. 1980. Nutrient dynamics of caribou forage on Alaska's Arctic Slope. Proceedings of International Reindeer/Caribou Symposium 2:159–66.

Whitten, K. R., and R. D. Cameron. 1983. Population dynamics of the Central Arctic herd, 1957–1981. Acta Zoologica Fennica 175:159–61.

Whitten, K. R., and R. D. Cameron. 1984. Calving distribution and initial productivity in the Porcupine caribou herd, 1982. Pages 600–608 *in* G. W. Garner and P. E. Reynolds, eds. 1983 update report, baseline study of the fish, wildlife, and their habitats. U.S. Fish and Wildlife Service, Anchorage, AK.

Whitten, K. R., G. W. Garner, F. J. Mauer, and R. B. Harris. 1992. Productivity and early calf survival in the Porcupine caribou herd. Journal of Wildlife Management 56:201–12.

Wilkinson, P. F., C. C. Shank, and D. F. Penner. 1976. Muskox–caribou summer range relations on Banks Island, N.W.T. Journal of Wildlife Management 40:151–62.

Williams, T. M., and D. C. Heard. 1986. World status of wild *Rangifer tarandus* populations. Rangifer (Special Issue) 1:19–28.

Young, B. A., and E. H. McEwan. 1975. A method for measurement of energy expenditure in unrestrained reindeer and caribou. Proceedings of International Reindeer/Caribou Symposium 1:355–59.

Young, D. D., Jr., and T. R. McCabe. 1998. Grizzly bears and calving caribou: What is the relation with river corridors? Journal of Wildlife Management 62:255–61.

Youngman, P. M. 1975. Mammals of the Yukon Territory (Publication 10). National Museum of Canada, Ottawa.

Yousef, M. K., and J. R. Luick. 1975. Responses of reindeer, *Rangifer tarandus*, to heat stress. Proceedings of International Reindeer/Caribou Symposium 1:360–67.

Zarnke, R. L. 1996. Serologic survey of Alaska wildlife for microbial pathogens (Federal Aid in Wildlife Restoration Projects W-23-5, W-24-1 through W-24-4, Final Report). Alaska Department of Fish and Game.

FRANK L. MILLER, Canadian Wildlife Service, Prairie and Northern Region, Edmonton, Alberta, Canada T6B 2X3. Email: frank.miller@ec.gc.ca.

47

Pronghorn

Antilocapra americana

John A. Byers

NOMENCLATURE

COMMON NAMES. Pronghorn, antelope, pronghorn antelope, prongbuck
SCIENTIFIC NAME. *Antilocapra americana*
SUBSPECIES. *A. a. americana, A. a. mexicana, A. a. oregona, A. a. peninsularis, A. a. sonoriensis* (Hall 1981)

This species is the sole survivor of the family Antilocapridae. The common name suggests an affinity to the antelopes of Africa, and some (e.g., O'Gara and Matson 1975) suggested that pronghorn should be assigned to a subfamily within the Bovidae. However, all recent systematic studies have retained the family Antilocapridae, and most have concluded that this family is most closely related to the Cervidae (Leinders 1984; Groves and Grubb 1987; Janis 1988; Solounias 1988; Irwin et al. 1991; Blake et al. 1997). The Antilocapridae, known only from North America, comprises two subfamilies, the small-bodied Merycodontinae (middle to late Miocene) and the larger Antilocaprinae (end of Miocene to present). Antilocaprinae apparently evolved from and then quickly replaced the Merycodontinae (Frick 1937; Janis 1982). The genus is known from the middle Pliocene (Webb 1973); the date of origin of the species is uncertain.

Although one common and four local subspecies are named (Bailey 1932; Goldman 1945; Paradiso and Nowak 1971), there is little evidence that the degree of difference among subspecies is greater than that among geographically isolated populations (Lee et al. 1994). In addition, transplantations likely blurred some of the original genetic differences (Halloran 1957). Pronghorn may show slight local variation in anatomy, but the roles of genetic versus environmental effects in producing these differences are not known. Nevertheless, two named subspecies, *A. a. peninsularis* and *A. a. sonoriensis,* are listed as endangered (Cancino et al. 1996; Sanabria et al. 1996; Woodley 1997) under the Endangered Species Act.

DISTRIBUTION

Historically, pronghorn occurred throughout the Great Plains, from southern Canada to Texas, throughout the Great Basin, and in high-elevation deserts of the Southwest into Mexico and Baja California (Quinn 1930; Blair 1982). That range was reduced drastically when European migrants settled in and plowed the grasslands. Today, pronghorn occur principally in high deserts and grasslands, where low annual precipitation precludes crops. About 1 million pronghorn are scattered across 15 U.S. states, 2 Canadian provinces, and 4 Mexican states (Fig. 47.1). However, about one half of the North American pronghorn occur in Wyoming.

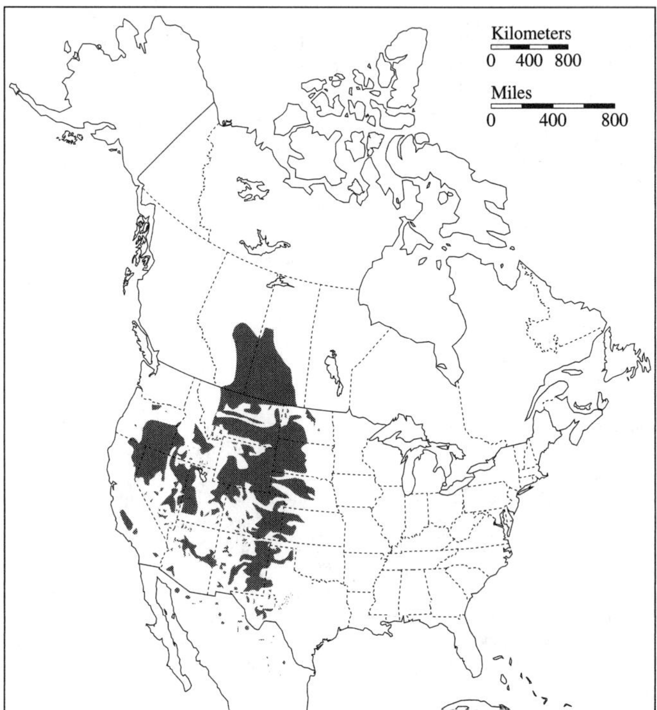

FIGURE 47.1. Distribution of the pronghorn (*Antilocapra americana*).

DESCRIPTION

Pronghorn are medium-sized ungulates, adapted for running in open country. Thus, the general appearance is of an almost cylindrical trunk from which long slender limbs project. Pronghorn are reddish-brown dorsally to about halfway down the side, and white ventrally. Eyes are large in proportion to body size, and are set far back on the head. Males bear conspicuous black curved horns. Female horns are short spikes.

Integument and Pelage. Pronghorn begin to molt in late winter. In healthy, nonlactating animals, molt is completed in August and individuals have a sleek appearance, with clear color patterning. The hairs are hard and brittle, with large internal air cells. Hair length increases well into the winter, and the reddish-brown coat becomes lighter in color.

The rump is white, with long hairs, which are erected when the animal is alarmed. The small tail is partly or entirely brown. The neck is reddish brown above, with a darker erectile mane. The ventral side of the neck is reddish brown with two white crescents, which vary in size and shape to the extent that the underside of the neck may be almost entirely brown or almost entirely white. Just posterior to the junction of the lower crescent and the white underparts, another white stripe projects along the lateral base of the neck for varying distances toward the dorsal midline. The specific pattern of these three ventral stripes is constant in individuals throughout life.

Coloration of the head is complex, but, like that of the trunk, is generally dark dorsally and white ventrally. The muzzle is black. The roughly triangular space from the muzzle to the median sides of the eyes is tan, brown, or black, with differing patterns of these colors. Hair around the lips and along the side of the head to below the eyes generally is white. Pinnae, nearly always held erect, are covered with hair inside and out and vary in coloration from almost white to dark brown. Typically, there is a fringe of black hair along the top anterior margin of each pinna. A small rectangle of white hair extends directly beneath each pinna. Hair between the horns varies in color from white to grizzled gray, to dark brown, to almost black. The top of the head posterior to the horns usually is white, interrupted by two narrow, dark brown stripes, which form a T.

Skin Glands. Pronghorn have an interdigital gland on each foot (Moy 1971) and two rump glands (Moy 1970). Rump glands appear to release odor when the hairs of the white rump patch are erected (Müller-Schwarze and Müller-Schwarze 1972). Males possess conspicuous black subauricular glands, located at the angle of the jaw below the ear, and a median gland, located on the dorsal midline just anterior to the white rump patch. During courtship, males display the subauricular glands on both sides to females and erect a U-shaped strip of hairs around the median gland. Males also scent mark by rubbing the subauricular glands on vegetation.

Horns. Like the horns of bovids, pronghorn horns consist of a keratinous integumentary sheath surrounding a horn core of bone. Solounias (1988) showed that the bladelike core of pronghorn develops as a direct outgrowth of the frontal bone, as do the antler pedicels of deer. In contrast, horn cores of bovids arise by the fusion of outgrowths of the frontal bones with dermal centers of ossification (the ossa cornua). This is compelling evidence that pronghorn belong in the superfamily Cervoidea.

Both sexes have horns. Female horns are short, simple spikes that are rarely >7 cm long. Male horns are about 30 cm long. The central shaft of male horns curves markedly near the upper end of each horn and narrows to form a hooked, sharply pointed horn tip. About halfway up each shaft, a sort shelf, or prong, projects forward. The horn core extends only to the base of the prong.

Males shed the horn sheath annually, in early fall, following the rut (O'Gara et al. 1971). The old horn sheath is pushed off by a new sheath growing underneath. The new horn is complete by late winter. Females tend to shed the sheath in midsummer, but on a much more variable schedule. Sometimes, females accumulate two or three old sheaths, like stacked paper cups, on the horn core.

Body Mass and Dimensions. Pronghorn have a gracile build, which evolved under selection imposed by the diverse and dangerous predators of the North American savanna (Byers 1997a). Consequently, pronghorn are relatively lightweight for their size, and sexual dimorphism in body mass is slight. In Alberta, mean body masses of males and females in late fall were 52 and 47 kg, respectively (Mitchell 1980). At the National Bison Range in northwestern Montana, late fall masses were 54 and 51 kg (J. A. Byers, unpublished data). In Yellowstone National Park, mean body mass of females in midwinter was 47 kg (J. A. Byers, unpublished data). At the Sevilleta National Wildlife Refuge in the Rio Grande Valley of New Mexico, mean body mass of females in late fall was 42 kg (J. A. Byers, unpublished data). In West Texas (time of year not reported), mean body masses of males and females were 43 and 40 kg, respectively (Buechner 1950). These and some other data (Mitchell 1980) suggest that pronghorn may exemplify Bergmann's rule (Bergmann 1847), but data are insufficient to show true clinal variation.

Total length of adult pronghorn is 1.3–1.4 m, and the sexes do not differ (Mitchell 1980). Hind foot length in males ranges from 394 to 451 mm; in females, the range is 394–409 mm (Mitchell 1980; J. A. Byers, unpublished data). Shoulder height is 873–948 mm in males and 830–914 mm in females (Mitchell 1980). In Alberta pronghorn, mean circumference of the chest immediately posterior to the front legs was 970 mm in males and 954 in females (Mitchell 1980). The neck girth of males is noticeably greater than that of females, especially in late summer.

Appendicular Skeleton. In general, the limb bones of pronghorn are delicate and lightweight, as first noted by Murie (1870:353): "no ruminant skeleton of equal size possesses such delicacy of osseus texture.... If one were to speculate on this fact it might be given as one reason for the extraordinary fleetness of the creature." The ulna is a thin, but complete splint. The fibula is a thin, incomplete splint. The cannon bone of the carpus is as long as the radius, and in the tarsus is only about 3 cm shorter than the tibia. Neither cannon bone shows any trace of lateral digits ("dew claws"). In the tarsus, the cuboid and navicular are fused, as are the ectocuneiform and middle cuneiform.

Skull and Dentition. In a sample of 22 adult male and 7 adult female skulls measured in Alberta, Mitchell (1980) reported mean total lengths of 283 and 277 mm, respectively. In most measurements, male skulls also were slightly wider than female skulls. These sex differences emerge during the first year of life, when skulls of males grow in length and width faster than those of females (Mitchell 1980). The pronghorn skull is widest just posterior to the orbits. The tubular preorbital rostrum makes up about two thirds of the total length of the skull. The dorsal aspect of the skull thus is roughly triangular. The orbits are large, with a strong postorbital bar (Fig. 47.2). The orbits resemble turrets and the top of each is dorsal to the skull roof. The turreted orbits, placed high and far back in the skull, allow pronghorn to see almost directly to the rear when the head is up and far to the side when the head is lowered. The horn cores are directly superior to the orbits. At the medial sides of each horn core, a large supraorbital foramen provides a passage for veins from the ears, superficial muzzle, and deep muzzle, which pass to the cavernous sinus of the orbit (Carlton and McKean 1977). The scrolled turbinate bones occupy the upper two thirds of the nasal cavity. Length of the upper jaw anterior to the teeth is greater than the length of the toothrow.

The angle of the dentary is wide and gently curved. The coronoid process is slender and shallowly concave medially. The mandibular condyle is roughly rectangular and dished dorsally; its articulation with the cranium is at a shallow saddle. There is no angular process.

The dental formula is I 0/3, C 0/1, P 3/3, M 3/3. The six incisors and two incisiform canines in the lower jaw are spatulate and form an arcade, which works in conjunction with the upper lips and the dental pad to clip vegetation. The upper and lower molariform teeth are hypsodont and selenodont, a condition found in antilocaprids from the earliest known fossils to the present. Pronghorn chew only briefly when feeding; the cheekteeth are used most in rumination.

Dow and Wright (1962) described dentition and its development in pronghorn. At birth, only the first deciduous incisor is visible. Deciduous incisors, canines, and premolars are present by 44 days of age. The first molar also is partly erupted at this time. The first permanent incisor and the three molars are erupted when individuals are about 18–20 months of age. Permanent incisors 2 and 3 and permanent premolars 1–3 erupt in the third year. The last deciduous tooth to be replaced is the canine; this occurs at 3.5–4 years of age. Infundibula on permanent premolar 1 are not lost until the 4th year. This pattern of tooth replacement and wear allows age estimation of pronghorn, through age 4 years, by a quick visual inspection of the teeth.

PHYSIOLOGY

Metabolic Rate and Body Temperature. Wesley et al. (1970) reported a mean fasting metabolic rate in 108- to 180-day-old fawns of 385 kJ/kg$^{0.75}$/day. Wesley et al. (1973) found the rate was 318 kJ/kg$^{0.75}$/day for mature animals. Wesley et al. (1970) measured the incremental cost of standing at 191 kJ/kg$^{0.75}$/day.

Lonsdale et al. (1971) implanted transmitters in the deep hip musculature of two pregnant females held in small pens and recorded body temperature. The mean value was 38.5°C. Thorne (1975) used the same methods to record temperatures of one adult and one yearling female. The adult's temperatures ranged from 36.2°C to 42.2°C and averaged 36.9. The yearling's temperature ranged from 36.8°C to 42.2°C and averaged 38.3°C. Barrett and Chalmers (1979) recorded rectal temperatures of fawns captured by hand and of adults that had been herded by aircraft into a capture corral and then restrained by hand. The mean fawn temperature was 39.6°C; the mean adult temperature was 40.8°C. Wesley et al. (1973) measured a lower critical temperature at −12 to

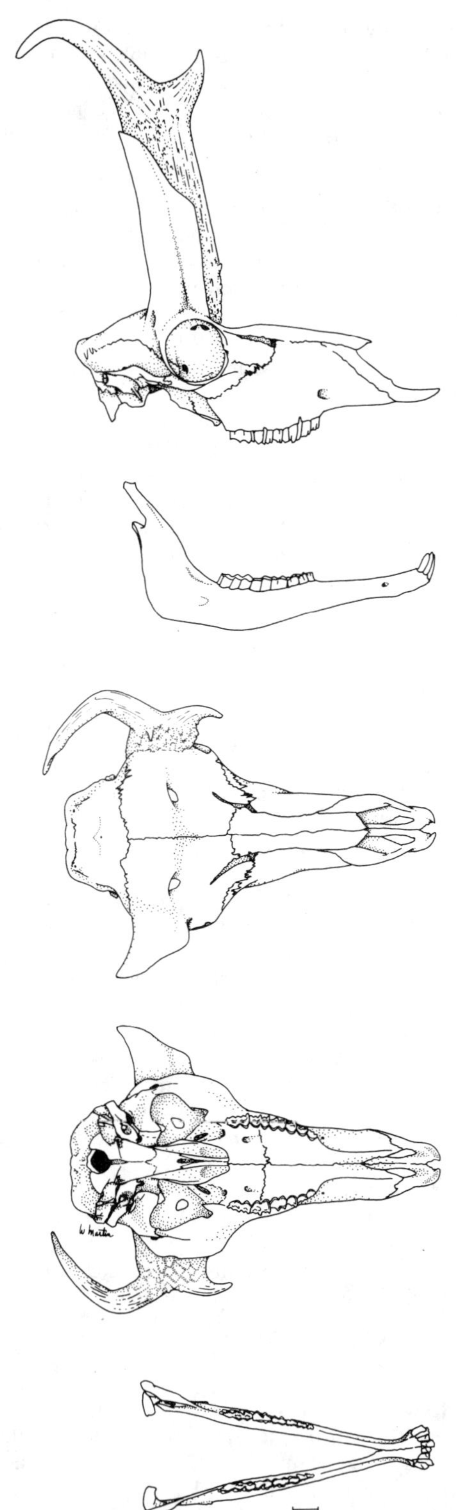

FIGURE 47.2. Skull of the pronghorn (*Antilocapra americana*). From top to bottom: lateral view of cranium, lateral view of mandible, dorsal view of cranium, ventral view of cranium, dorsal view of mandible. The right horn sheath has been removed.

−23°C for a captive animal on feed and an upper critical temperature of about 30°C.

Pronghorn behaviorally thermoregulate by erecting hairs in the cold, by opening the mouth and slightly protruding the fleshy black tongue when hot, and by adopting postures that use the upper dark and ventral white pelage to advantage. In cold conditions, pronghorn recline with the legs tucked underneath and the brown back exposed to the sun. Often, they orient when reclined to face away from the wind (Bruns 1977). On hot days, pronghorn recline with the white rump patch presented toward the sun, or recline more onto the side, presenting the white side patch to the sun. Pronghorn likely are able to keep the temperature of the brain cooler than the temperature of the body by virtue of the carotid and orbital retia (Carlton and McKean 1977). Other ungulates have such heat exchangers, in which the retia of ascending arteries are arranged in countercurrent to plexuses of veins that drain the ears and muzzle. The surface area for heat exchange in the carotid and orbital retia of pronghorn is greater than that in mule deer (*Odocoileus hemionus*) or elk (*Cervus elaphus*) (Carlton and McKean 1977). An efficient heat exchanger likely is important for pronghorn because they are prodigious runners, able to use the large running muscles for extended periods of time.

Aerobic Capacity. Pronghorn are legendary not only for running speed but also for endurance. McKean and Walker (1974) first studied the physiology that supports extended high-speed running. They compared heart size, hemoglobin concentration, blood volume, hematocrit, airway resistance, and blood buffering of pronghorn to those of an unspecialized runner, the domestic goat. All comparisons (Table 47.1) suggested that pronghorn have an unusually high aerobic capacity. Lindstedt et al. (1991) measured the $\dot{V}O_2$ max of pronghorn that ran on a treadmill. The measured value was five times that of goats and deviated from the body mass allometric expectation more than that of any other mammal. Lindstedt et al. (1991) also showed that pronghorn have unusually high lung volume and diffusing capacity, cardiac output, blood hemoglobin concentration, and muscle mitochondrion density and volume. The extreme values of these features, and not unusual locomotor efficiency, seem to account for the sustained running that pronghorn can accomplish.

Blood. Barrett and Chalmers (1977a, 1977b, 1979) presented descriptive data on blood ions, enzyme activity levels, cholesterol, glucose, creatinine, urea nitrogen, cell counts, hematocrit, and hemoglobin in pronghorn adults and fawns from Alberta. For many variables, values differed significantly with sex, year of capture, and duration of handling time. Dunbar et al. (1999) presented a similar data set for pronghorn at the Hart Mountain National Antelope Refuge in Oregon. Clemens et al. (1987) presented similar data on captive pronghorn.

Digestion. Pronghorn are ruminants with relatively small rumens and a digestive system anatomy and physiology that Hofmann (1989) reported as indicative of his "intermediate opportunistic mixed feeders" category. My own observations (Byers 1997a) and those of many other workers at numerous sites across pronghorn range indicate that pronghorn are very selective foragers (Rouse 1941; Hoover 1971; Schwartz et al. 1977; Kessler et al. 1981; Courtney 1989; Wood 1989). Nagy and Williams (1969) reported that volatile fatty acid profiles of the rumen contents of pronghorn generally were similar to those of domestic ungulates on equivalent diets. Profiles of pronghorn fawns at

TABLE 47.1. Comparison of components of O_2 transport in pronghorn and domestic goat

Component	Pronghorn	Goat
Heart weight/body weight (g/100 g)	0.948	0.495
Hemoglobin (g/100 ml)	18.8	11.5
Blood volume as a percentage of body weight	9.29	6.59
Hematocrit	0.44	0.31
Airway resistance (cm H_2O/L/sec)	1.34	2.85
Blood buffering[a]	0.43	0.55
Relative lung volume	100	44
Relative cardiac output	100	34
Relative $\dot{V}O_2$ max	100	20

SOURCE: Data from McKean and Walker (1974) and Lindstedt et al. (1991).
[a]pH drop of 15 ml of blood to which 2 ml of 0.10 *M* HCl is added.

16 weeks of age were indistinguishable from those of pronghorn adults. The annual reproductive effort of pronghorn females, measured by the ratio of litter metabolic mass to maternal metabolic mass, is higher than that of most other ungulates (Byers and Moodie 1990; Byers and Hogg 1995). Nevertheless, pronghorn subsist in many areas on poor range; this suggests that digestive efficiency is higher than that of many ungulates. A study of digestive efficiency in nature would be interesting. However, such a study is not possible at present because there is no known noninvasive method to accurately quantify what pronghorn eat in nature. Diet composition measured by microscopic examination of feces does not always agree with diet composition measured by microscopic examination of rumen contents (Kessler et al. 1981). Also, because pronghorn evolved in North America (Frick 1937; Webb 1977), their ability to avoid or to neutralize the antidigestive and toxic effects of the secondary compounds of North American plants may be more advanced than the abilities of domestic ruminants or more recent immigrants, such as mountain sheep (*Ovis canadensis*).

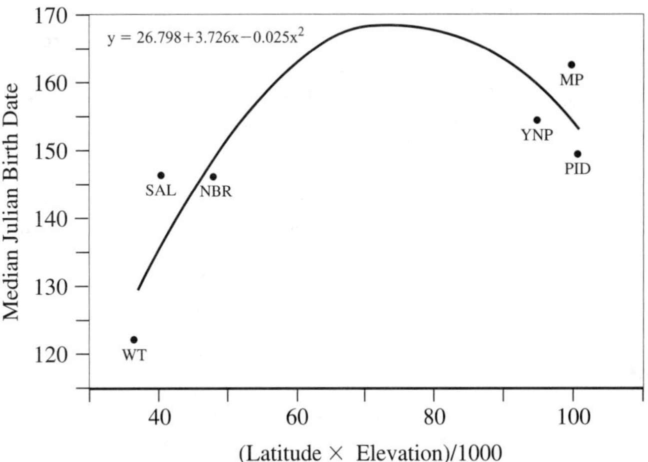

FIGURE 47.3. Pronghorn show clinal variation in the timing of birth dates. Latitude multiplied by elevation indexes the severity of winter and the relative lateness of spring. WT, West Texas (Buechner 1950); SAL, Southern Alberta (Mitchell 1980); NBR, National Bison Range (Byers 1997a); YNP, Yellowstone National Park (J. A. Byers, unpublished data); MP, Middle Park, Colorado (Fairbanks 1993); PID, Pahsimeroi Valley, Idaho (Autenrieth and Fichter 1975).

REPRODUCTION

Females generally have the first estrus when about 16 months of age. Rarely, females enter estrus in the first year, when 5 months of age (Mitchell 1967; Byers 1997a), and breed in mid-October, about 1 month after the main rut. Males begin spermatogenesis when 12 weeks of age (Mitchell 1980), but because of male–male competition and female mate choice, they rarely have the opportunity to mate until 3 years old.

The pronghorn rut lasts about 3 weeks, from mid-September to early October, and female estrous dates are clumped within this period. At the National Bison Range, 90% of estrous dates occur during 12–25 September (Byers 1997a). Females practice active mate choice, visiting several males and usually making repeat visits to particular males before mating. Females control mating and almost always mate once per estrus. Males cannot force copulation, although they are sometimes able to block or temporarily thwart a female's direction of travel. At the National Bison Range (Byers 1997a) and in Alberta (Mitchell 1980), females enter estrus and conceive every year until death. A similar pattern appears to exist in female pronghorn in Yellowstone National Park (J. A. Byers, unpublished data). Females that fail to conceive in September have a second estrus 28 days later, in October (Pojar and Miller 1984; J. A. Byers, pers. obs.). On the National Bison Range, this occurs in about 1 in 300 estruses.

Males make serious attempts to breed starting at age 3 years and continue annually until death. However, roughly one half of the males in any cohort never breed. Male variance in lifetime reproductive success is influenced in part by longevity, but most heavily by variation in number of matings and variation in offspring survival (Byers 1997a). Male mating success is not correlated with body size or horn size (Byers 1997a). The characteristics that make some males particularly successful breeders are unknown, although factors that contribute to running speed and endurance likely are important.

When they are yearlings, females ovulate two to seven eggs at estrus. Older females ovulate three to nine eggs. All or most eggs are fertilized, but intrauterine competition and mortality reduces litter size to two, with one embryo in each uterine horn (O'Gara 1969; Mitchell 1980). Where studied, the typical litter size is two, but the litter size in most pronghorn populations is unknown. Hepworth and Blunt (1966) reported a mean gestation of 252 days for captive pronghorn. A 95% confidence interval for gestation, based on observed estrous and birth dates of known individuals on the National Bison Range, was 250 ± 9 days (Byers 1997a). Because of the large variation in gestation length, the date distribution of births is wider than the distribution of estrous dates. Estrous dates and corresponding birth dates show slight clinal variation (Fig. 47.3). Among populations, estrous dates seem to be timed so that fawns are born as early as possible in the spring green-up, but the mechanism that results in such timing is unknown.

Fawns gain 80% of their birth mass in the final one third of gestation. Fawn prenatal growth is slowed and hence the length of gestation is increased when late-summer rains fail in the year of conception (Byers and Hogg 1995). The sexes do not differ in mass at birth or in mass gain to 12 weeks of age (Mitchell 1980; Fairbanks 1993; Byers 1997a). Mass gain to 12 weeks is linear and is described by the equation (Byers 1997a) $\text{Mass}_{(\text{kg})} = 3.53 + 0.245\text{Age}_{(\text{days})}$. Although mass gain of the sexes is equivalent, male linear dimensions grow slightly faster than those of females (Mitchell 1980).

Birth masses of pronghorn at the National Bison Range (Byers and Hogg 1995; Byers 1997a), Yellowstone National Park (J. A. Byers, unpublished data), and Middle Park, Colorado (Fairbanks 1993), are greater than the birth masses reported for captive, well-fed pronghorn (Wild et al. 1994; Martin and Parker 1997). Postnatal growth rates at the National Bison Range (Byers 1997a) also are greater than those reported for captive, bottle-fed, or dam-reared pronghorn (Wild et al. 1994; Martin and Parker 1997). Reasons for these wild versus captive differences in pre- and postnatal growth rates are not known.

Pronghorn are classical hiders (Lent 1974). Females become solitary to give birth. They eat the placentae and other extraembryonic membranes and lick neonates extensively. They rest with young for 1–2 hr after birth and then lead the young away from the birth site. Apparently at a signal from the mother, the neonates walk away and choose a place to recline. Canon and Bryant (1997) described some of the bed site characteristics of fawns. The mother then moves away from the young and remains away for 2–3 hr. For the next 2–3 weeks, mothers visit fawns at these intervals. Between visits, mothers are unusually alert, and maintain a distance from the reclined fawns that renders systematic search, based on the mother's location, energetically unprofitable for coyotes or larger predators (Byers and Byers 1983). The time that fawns spend up and active when mothers visit them increases with age, so that there is a gradual transition out of hiding. At about 3 weeks of age, mothers bring fawns into social groups of other mothers and fawns. These groups are fairly stable for the next 2 months.

Weaning defined as the steeply descending part of the suckling rate curve occurs between postnatal weeks 4 and 6. Weaning defined as the age at which the fawn takes its final suckle occurs at about 12 weeks of age (Byers 1997a). Milk collected from a single captive female pronghorn from postpartum day 4 to day 74 averaged 21% dry matter, 7% fat, and 6.3% protein, but showed a significant decrease in dry matter, fat, and protein levels at dates corresponding to the ages of the first weaning event (Martin and Parker 1997). These values then rebounded somewhat after postpartum day 60.

At the National Bison Range, a cost of reproduction, measurable by delay of estrous date, lengthened gestation, lighter birth mass, or decreased probability of survival of the subsequent litter, is not detectable. This is somewhat paradoxical, because pronghorn reproductive expenditure, measured as the ratio of metabolic litter mass or growth to maternal mass, is at the upper end of these indices in ungulates (Robbins and Robbins 1979; Byers and Moodie 1990), whereas a cost of reproduction is detectable in some ungulates with lower levels of expenditure (Clutton-Brock et al. 1983; Lee and Moss 1986; Berger 1989). Whether pronghorn are capable of such impressive annual reproductive effort in all habitats is not known. However, preliminary data indicate that expenditure is comparable in Yellowstone National Park, where winter temperatures are much colder and winter range poorer than at the Bison Range (J. A. Byers, unpublished data).

ECOLOGY

Pronghorn evolved in the later Tertiary savannah of North America (Frick 1937; Webb 1977), and the species shows many adaptations for living in open grassland. Pronghorn prefer open country with unrestricted directions of travel and lines of sight. They are selective browsers, adapted to exploit the diverse and patchy mosaic of grasses and forbs created by bison (*Bison bison*) (Krueger 1986) and other large, and now extinct, grassland grazers. Settlers who plowed the grasslands destroyed the prime habitat of pronghorn. Today, pronghorn persist at the fringes of their former range, but seem remarkably efficient at exploiting dryer, less diverse habitats. Management guidelines typically describe acceptable pronghorn habitat as open terrain with at least 50% plant cover (mixture of native grasses, forbs, and shrubs), and availability of winter range where snow depth does not prevent foraging or free movement (Hailey et al. 1966; Sundstrom 1968; Autenrieth 1978; Yoakum 1980; Cook and Irwin 1985; Irwin and Cook 1985; Ryder and Irwin 1987). Pronghorn do not always require a source of drinking water. When forage is succulent, individuals may not drink for several weeks (Beale and Smith 1970). During hot weather, when plant moisture content declines, pronghorn tend to visit water sources once per day (J. A. Byers, pers. obs.), and the spatial distribution of herds begins to match the distribution of water sources (Sundstrom 1968; Deblinger and Alldredge 1991).

Pronghorn in many populations make seasonal movements between summer and winter range (Martinka 1967; Hoskinson and Tester 1980; Bryant and Huber 1998; J. A. Byers, pers. obs.). In general, these migrations move animals in winter to areas where the climate is milder and snow depth less, and in summer to moister areas where forb diversity and abundance is greater. On summer range, individuals tend to be site faithful, within and among seasons (Autenrieth and Fichter 1975; Mitchell 1980; Fichter 1987; Byers 1997a). In fall 1988, the National Bison Range staff captured 16 male and 24 female pronghorn and gave them to local Salish and Kootenai tribal wildlife officials, who released them in the Hot Springs Valley, 50 km to the west. By the following spring, almost all of these animals had returned to the Bison Range; the return trip required them to swim across the Flathead River.

Seasonal habitat shifts are associated with shifts in group size, sex composition, and individual activity budgets (Byers 1997a). In summer, group sizes are smallest. Males >3 years old tend to be solitary or with a group of females and young. Males that are 1–3 years old tend to be in all-male herds (often referred to as bachelor herds). Much larger, mixed-sex herds form in fall and winter. Activity budgets of males and females vary with season (Fig. 47.4). In winter, when individual energy budgets tend to be negative (Bear 1971), about 80% of a pronghorn's day is spent either feeding or reclining.

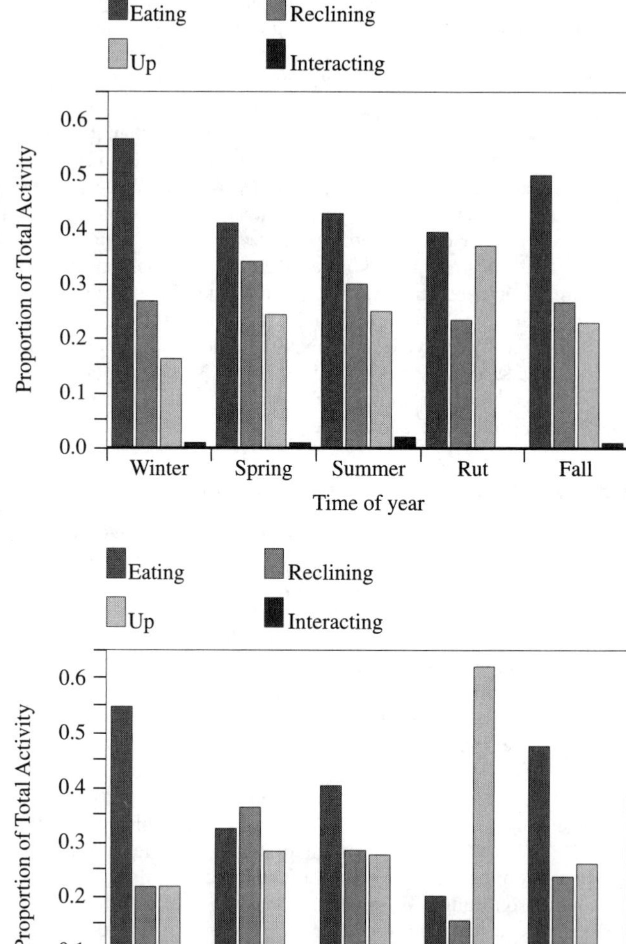

FIGURE 47.4. Yearly variation in activity budgets of adult pronghorn (A) females and (B) males. SOURCE: Data from Byers (1997a).

FEEDING HABITS

All who have studied pronghorn food habits agree on two things. First, pronghorn are very selective feeders. The percentage plant species composition in the habitat does not predict the plant species composition of the diet (Schwartz and Nagy 1976; Schwartz et al. 1977; Kessler et al. 1981; Schwartz and Ellis 1981), and individuals obviously change the plant species composition of the diet in response to seasonal succession or other alterations of local plant productivity (Rouse 1941; Dirschl 1963; Bayless 1969; Hoover 1971; Mitchell and Smoliak 1971; Stephenson et al. 1985; Courtney 1989; Wood 1989). Second, it is exceedingly difficult to quantify what pronghorn eat. One cannot directly observe what pronghorn eat in nature, even at close range. One cannot follow an individual's foraging path and then quantify diet by plant damage, because plant damage is almost impossible to detect (Rouse 1941). Microhistological analysis of feces gives only a rough approximation of species composition in the diet (Kessler et al. 1981). Robinson et al. (2002) conducted a diet restriction experiment on captive pronghorn and showed that fecal nitrogen and fecal diaminopimelic acid were fairly accurate indicators of dietary nitrogen and energy.

BEHAVIOR

Gait and Locomotion. Adult pronghorn exhibit at least nine distinct gaits (Bullock 1974); five symmetrical gaits are used at slow speeds, and four asymmetrical gaits are used at greater speeds. At the slowest speeds, typical of movement during foraging, pronghorn use a gait that Bullock (1974) called the very slow diagonal walk. At the highest speeds, pronghorn use a rotary gallop, and flex the vertebral column in the sagittal plane to increase stride length (Bullock 1982). Pronghorn show an apparently unique ability to smoothly shift gaits by shifting the front and rear lead feet independently (Bullock 1974). This ability in part supports the synchronization of gait and lead foot in running groups, which results in one of the most beautiful sights in nature. A group of running pronghorn packs into a tight ellipse that flows smoothly over the ground. Bullock (1974) reported that, within such groups, there was 97% synchronization of the gait used and 92% synchronization of the lead foot used. Anecdotal accounts of top speed in the range of 90–100 kph exist (e.g., Einarsen 1948), but accurate data are not available. Bullock (1974) noted that a speed of 100 kph would require a stride length of 8.9 m for pronghorn (assuming that the measured stride rate at slightly slower speeds represents the maximum stride rate), and that such a stride length seemed mechanically impossible.

Pronghorn are more likely to crawl under a barrier, if possible, than to jump over it. However, pronghorn occasionally jump across small streams or other low obstacles (J. A. Byers, pers. obs.) and they may leap over fences (Bullock 1974; M. Robinson, University of Idaho, pers. commun., 1998).

Grouping. Pronghorn are almost always found in social groups. Exceptions occur when females become solitary just before and for about 2 weeks after birth, and when >3-year-old males become solitary and site faithful in summer. Group size varies substantially and predictably throughout the year. In summer, group sizes are smallest. Groups consist of bands of females and young (often accompanied by a single mature male) or of bands of 1- to 3-year-old males. After the rut, pronghorn coalesce into much larger, mixed-sex groups, which persist until early spring. Group composition is fluid (Fichter 1987) and individuals move among groups independently. Social bonds that influence the patterns of association between individuals do not exist (Byers 1997a).

Within female groups, all social interactions are aggressive, and the rate of these interactions, (>1/female/hr), is higher than that recorded for any ungulate except mountain goats (*Oreamnos americanus*) (Fournier and Festa-Bianchet 1995). Paradoxically, although group size varies throughout the year, the aggressive interaction rate remains constant. Byers (1997a) presented evidence that pronghorn leave and join groups to maintain an upper limit on the rate of aggressive interaction. In winter, large groups may form because there are few prime food resource areas that would promote competition. Among females, aggressive interactions are feeding displacement, in which one individual forces another to relinquish a feeding site; bedding displacement, in which one individual forces another reclined individual to stand and move away; and simple displacement, in which one individual causes another to move or to turn away, with no obvious resource contested. Most of these interactions involve either no physical contact or perhaps brief nudges. Females experience many such interactions daily. The interactions reinforce dominance relations. A female's dominance rank is stable throughout her life, and is largely a consequence of her birth date rank in the fawn social group where she spent her first summer (Byers 1997a). Early-born fawns become dominant and late-born fawns become subordinate.

The three forms of aggressive interaction that occur in female groups also occur in bachelor herds. In these all-male groups, there is also another salient form of social interaction called sparring. Sparring resembles fighting, but at very low intensity, with no attempt to injure the opponent. Miller and Byers (1998) presented evidence that sparring is a largely cooperative activity in which individuals practice the motor skills involved in fighting.

Communication. The skin glands of pronghorn communicate a variety of social messages. The rump glands appear to release a substance that combines with the visual signal of the erected rump patch hairs to indicate alarm or danger. The median gland of males is used in courtship, as are the paired subauricular glands. Males also use the subauricular glands to scent mark objects within the home range or territory. Males create combined visual and olfactory marks with the scrape–urinate–defecate sequence (Kitchen 1974): A male first scrapes a spot bare with a forefoot, leans forward to urinate on the spot, then steps forward to defecate in the middle of the urine. Females appear to signal estrus with compounds in urine. Males show flehmen in response to female urine, especially during rut. Females also seem to signal estrus with volatile compounds produced by glands of the vagina or external genitalia (Byers 1997a; Murphy et al. 1994). During rut, males herd groups of females into concealed locations and attempt to keep them there. Males scrape–urinate–defecate over each female urine spot, apparently as part of the harem concealment strategy (Moodie and Byers 1989). It is unknown whether the interdigital glands have any function in social communication.

Mothers call to fawns with a quiet, high-pitched grunt. Fawns signal alarm with a loud bleat. When manually restrained by humans, adults produce a lower pitched, louder, quite startling version of this bleat (J. A. Byers, pers. obs.). Adults of both sexes produce an explosive, breathy snort (Waring 1969) in response to an alarming stimulus, and they often erect the rump patch simultaneously. In males, a slightly modified form of the alarm snort serves as the introductory note of the snort–wheeze. The introductory note generally is higher pitched than the alarm snort, and its duration is about twice as long. It is followed by a pause, then a series of 10–20 regularly spaced chugs, which fall off in intensity toward the end of the sequence. Many pronghorn males have snort–wheezes that are individually identifiable to humans, and Chao (1985) presented evidence that pronghorn males recognize the snort–wheezes of neighbors. Territorial males and harem-holding males in rut direct the snort–wheeze at distant pronghorn (males or females) and sometimes toward females at the beginning of a courtship sequence. In rut, males produce a growl or roar, which they direct toward other pronghorn (males or females) that they are chasing. During a courtship approach, males emit a groaning whine and may lip smack.

Pronghorn have a number of postures and movements that signal alarm, aggressive intent, submission, instruction to move in a particular direction, and sexual intent. Most were described by Kitchen (1974). Alarm is signaled by snorting, erecting the white hairs of the rump patch, staring fixedly at the source of the alarm, and prancing back and forth. Aggressive intent is signaled, with increasing intensity, by a direct stare, an approach with the head held upright, a lowering of the head in a butting intention movement, and by butting. Submission is signaled by turning away and walking or running away. Males direct other males and females to stop and to change their direction of travel by walking broadside to and in front of the other animal, holding the neck at about a 30° angle above horizontal. When the recipient stops, the male then turns toward it, lowers his head, and walks toward the recipient. Territorial males display at boundaries and males display before fighting with the same broadside posture, the two males walking parallel to each other. Males signal sexual intent by walking toward a female with the head tipped back slightly as it is slowly turned from side to side displaying the subauricular glands to the female.

Reproductive Behavior. In rut, when males are 1 and 2 years old, they approach female groups and sometimes run into them and court hurriedly before they are ejected by the harem male. They patrol widely and sometime intercept females that are in transit between harems. In rare instances, these young males copulate by using one of these two tactics, but in general they are easily displaced by older males and are decisively rejected by females. When they are 3–4 years old, males become solitary and site faithful in summer and they attempt, with increasing vigor as summer progresses, to retain groups of females. Each male home range tends to have a hiding place where a group of females can be concealed from distant view. As the rut begins, males with female groups push the group into the hiding place, scent mark over female urine, and scan frequently for distant males, which they challenge with a snort–wheeze and then usually chase. When females in the harem enter estrus, the male's attention is divided. If he chases rivals too much, the estrous female and possibly the entire harem will leave. If he does not keep rivals at a sufficient distance, the harem is likely to leave. The most successful males are those that maintain a large harem for many days during rut (Byers et al. 1994; Byers 1997a). To accomplish this, a male must be very active, repel rivals, keep the harem compact and out of sight, and court receptive females. In some populations, males are territorial—beginning in early summer, they chase all males from a large area, and they defend that area through the rut. In other populations, such defense of space does not occur (Byers and Kitchen 1988; Maher 1991, 1994, 2000). On the National Bison Range, a system of male territoriality changed to harem defense in the early 1980s (Bromley 1969; Byers and Kitchen 1988), and the population still practices harem defense (Byers 1997a). The event that started the shift apparently was a large winter kill of mature males, which reduced the proportion of males that defended territories. Suddenly, most males were young and nonterritorial, and in rut they put unrelenting pressure on the boundaries of the few remaining territories. It became physically impossible for territorial males to defend, and over the course of three ruts these males eventually gave up the territorial strategy. The shift showed that the success of the territorial strategy is frequency dependent, or at least density dependent. Maher (1994) suggested that the significant difference between territorial and nonterritorial populations was density. The cues that affect this aspect of male behavior remain unknown. Kitchen (1974) argued that pronghorn show resource-defense territoriality and presented some data to support the hypothesis that territories contained better food resources. Byers (1997a) critically reexamined Kitchen's data and concluded that the evidence for resource-defense territoriality in pronghorn does not hold up under scrutiny.

Males fight only when in the presence of a female that is ready to copulate. In fights, males attempt to push the opponent backward and sideways to gain an opening where the horns are used to injure the opponent. Fights are brutal and dangerous. On the National Bison Range, 12% of fights result in death of one of the contestants (Byers 1997a). The most common injury in fights is partial or complete dislocation of a limb joint.

Males cannot force copulation. Mating in pronghorn is under female control, and females actively select mates using one of two distinct behavioral strategies. Females that use the "quiet" strategy move to join a male in an out-of-the-way, isolated location before the rut begins, remain with that male throughout the rut, and copulate with him when they enter estrus. On the National Bison Range, this is the minority strategy. In some populations where density is low and the distances between harems are large, female groups in rut are stable, and essentially all females practice the "quiet" strategy (Maher 1994). On the National Bison Range, most females use a "sampling" strategy. In the 10 days before estrus, sampling females increase the rate at which they move among harem males (Byers et al. 1994). Females typically return to a specific male or a few males several times, but never remain with a male that fails to keep other males well away from the harem. At estrus, some sampling females run from the controlling male toward other males and thus incite fights. They watch the fights and aggressive competition closely, and then they mate with the winning male. Females, with rare exceptions, mate once per estrus. The successful male guards his mate and continues to court her for about 1 hr after copulation. Then he permits the female to leave his harem. She often wanders away, now unmolested by other males that would have pursued her relentlessly 2 hr earlier. It is unknown how females are able to turn off the olfactory signal of estrus so rapidly.

MORTALITY

Survivorship curves for the sexes, based on 1680 individuals followed on the National Bison Range from birth to death, are shown in Figure 47.5. Most mortality occurs in the first year, and mostly in the first 45 days after birth. Schedules of fawn mortality are similar in Yellowstone National Park (J. A. Byers, unpublished data). Low fawn recruitment at these sites and elsewhere in the West is attributable to predation by coyotes (*Canis latrans*), golden eagles (*Aquila chrysaetos*), and sometimes bobcats (*Lynx rufus*) (Byers 1997b; Von Gunten 1978). On the National Bison Range from 1981 to 1996, fawn mortality varied from 56% to 99%, and there was a significant positive correlation between percentage survival and the number of coyotes removed from pronghorn habitat in May and June (Fig. 47.6) (Byers 1997b). In Arizona, Smith et al. (1986) also found a sharp increase

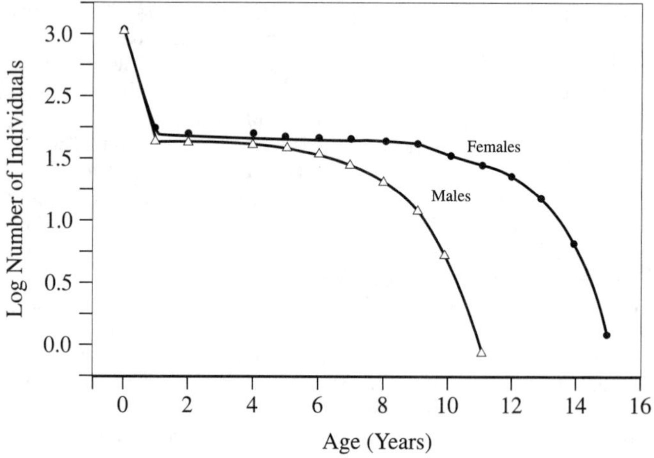

FIGURE 47.5. Survival of marked males and females on the National Bison Range, 1981–1994. SOURCE: Data from Byers (1997a).

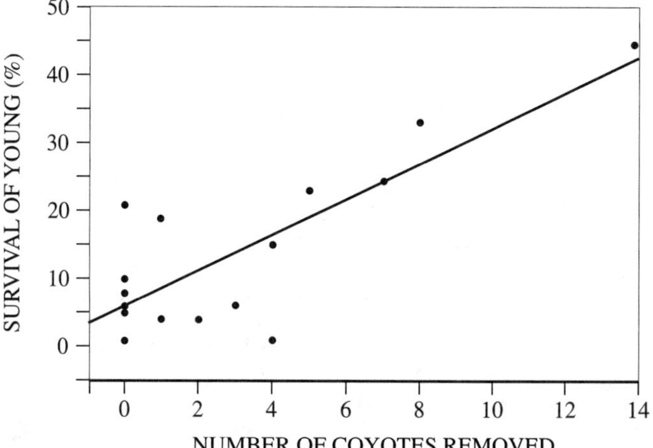

FIGURE 47.6. Annual survival to weaning of pronghorn fawns on the National Bison Range as a function of the number of coyotes removed from pronghorn habitat in May and June.

in pronghorn fawn recruitment in response to removal of coyotes by helicopter gunning in spring, just before fawn births.

As Figure 47.5 shows, mortality of females to age 10 years and of males to age 7 years is insignificant. Male survivorship is lower than that of females due to risks of injury in rut, energy expenditure in rut and the consequent lower fat reserves at the start of winter (Bear 1971), higher parasite loads in males than in females (Byers 1997a), and the fact that males senesce faster than females (Byers 1997a).

In some populations, massive mortality occurs during severe winters with deep snow and very cold temperatures (Martinka 1967; Mitchell 1980). In these die-offs, mortality is male biased (Martinka 1967; Barrett 1982; Byers and Kitchen 1988). In the Southwest, similar levels of adult mortality may follow severe drought (Buechner 1950).

The effect of predators on adult mortality seems slight. Mountain lions (*Puma* [*Felis*] *concolor*) occasionally kill pronghorn when terrain or cover permits a close stalk (Ockenfels 1994; J. A. Byers, unpublished data). Wolves rarely pursue pronghorn when slower prey are available (D. Smith, pers. commun.). Large coyotes can kill pronghorn adults (J. A. Byers, pers. obs.), but such events are rare. Healthy pronghorn in their preferred habitat experience very little risk from any modern predator.

Bluetongue virus was first described in Africa in 1905, and was first described in the United States in 1952 in domestic sheep (Lance and Pojar 1984). In North America, the virus is transmitted between hosts by a biting gnat. Thus, bluetongue is to pronghorn as measles was to Native Americans. Bluetongue virus causes significant mortality in some populations (Thorne et al. 1988). In Wyoming, 3200 pronghorn deaths in 1976 and 300 deaths in 1984 were attributed to bluetongue. Pronghorn are hosts to a variety of protozoans, such as *Eimeria* and *Isospora,* and to several flukes, tapeworms, and nematodes (Lance and Pojar 1984). None of these parasites is reported to be a significant source of mortality.

AGE ESTIMATION

The most accurate method for estimating the age of a pronghorn is to make histological sections of an incisor tooth to count the number of cementum rings (McCutchen 1969; Kerwin and Mitchell 1971). Sectioning of teeth for cementum annuli is thus the appropriate technique for estimating population age structure from individuals obtained from hunter check stations or from catastrophic mortalities. If pronghorn are to be handled and released, a reasonably accurate estimate of age is obtained using criteria from Dow and Wright (1962). Kerwin and Mitchell (1971) showed that the Dow and Wright method tends to under-age animals, but is satisfactory for grouping animals into 3-year age classes (Table 47.2).

In late summer, some age classes of pronghorn can be detected visually. Fawns of the year are noticeably smaller than adults. Yearling females are more slender than adults, and have slightly smaller heads, which gives the appearance of proportionately large ears. Yearling males have spikelike horns with only a small prong, if any, and little curvature of the horn tip. In 2-year-old males, the horn tip curves more and the prong is better developed, but the horns are noticeably smaller that those of males 3 years old. At 3 years of age and beyond, body and horn sizes of males do not indicate age.

ECONOMIC STATUS AND MANAGEMENT

Photographs of pronghorn often appear on the cover of big game hunting regulations in the western states. Pronghorn are a popular game animal, and activity associated with pronghorn hunting is a significant source of revenue to some states. In 1998, about 177,000 hunter-days resulted in a harvest of about 70,500 pronghorn; if each hunter-day represented an expenditure of $200, then pronghorn hunting brought a premultiplier revenue of $35.4 million. Also, tourists increasingly come to western states to see pronghorn and other high-profile wildlife species. Unfortunately, the hunting season in most areas coincides with the rut. Because pronghorn females prefer mates that successfully defend space and/or harems, disruption by hunters may alter the natural distribution of male mating success. Whether such an effect diminishes population health and vigor is unknown.

In some areas, pronghorn eat crops (Cole 1956; Cole and Wilkins 1958; Buechner 1960), but the economic impact is rarely measured. Cole (1956) reported that pronghorn damage to alfalfa may be substantial. Torbit et al. (1993) showed that pronghorn in Colorado fed on winter wheat in fall and winter, but then shifted to other forage in spring. Consequently, pronghorn had no detectable effect on yield.

Pronghorn are affected by domestic livestock when stocking rates are high enough to deplete forbs. Overstocked sheep have the most severe effect. A more pernicious effect of domestic livestock is through the requirement for fences. Pronghorn are disturbed by fences and they do not cross any kind readily. Fences disrupt daily and seasonal movement patterns, and may separate mothers and fawns during the period when fawns are most vulnerable to coyotes. Fences sometimes create densities of pronghorn mothers that are high enough to give coyotes a clear advantage in searching for fawns. A fence with the bottom rung at least 46 cm above ground allows pronghorn to cross by crawling underneath.

TABLE 47.2. Criteria to for classifying age of pronghorn based on inspection of teeth

Criterion	Age Class (years)
Deciduous canine present	0–3
Permanent canine present, posterior infundibulum on molar 2 present	3–6
Posterior infundibulum on molar 2 worn away	>6

SOURCE: Based on Dow and Wright (1962).

Management decisions rest primarily on estimates of population size and recruitment. Counts of the numbers of adult males, adult females, and fawns tend to vary with the counting method used (Guenzel 1994; Kraft et al. 1995; Pojar et al. 1995; Woolley and Lindzey 1997). Fawn:doe ratios in late summer are the standard measure of recruitment. It is important to recognize that this ratio reflects the summed effects of female fecundity and fawn survival. Although annual production of twins by each female is described where pronghorn have been intensively studied (Mitchell 1980; Byers 1997a), it is not known whether pronghorn show this level of productivity in all habitats.

CURRENT RESEARCH AND CONSERVATION NEEDS

A critical reevaluation of the systematic rank of the named subspecies is needed. It would be inappropriate and ultimately against conservation interests to use the Endangered Species Act to defend these populations if the real biological differences among them and *A. a. americana* are slight or if the differences are largely due to environmental effects. I am not suggesting that these populations have no value or that they do not warrant protection. I am asserting that their protection should be based on objective biological understanding. Similarly, the conservation significance of the fragmentation of pronghorn distribution needs attention. Small populations may be susceptible to extinction due to random environmental effects such as a severe winter that follows an unusually dry summer. The relationship between population size and probability of stochastic extinction is unknown in pronghorn. Small populations also may show inbreeding depression, hastening a slide toward extinction. The magnitude of such a threat is difficult to evaluate because the genetic load of pronghorn is unknown. The National Bison Range pronghorn population is closed and under random breeding it should have lost one half of its initial heterozygosity since it was established in the 1950s. However, this population shows no evidence of inbreeding depression (Byers 1997a). Estimates of the degree of inbreeding depression in other populations are needed. Smaller populations at the edge of the species's range, especially those in southern Arizona and Mexico, currently seem most vulnerable. In the longer term, one can envision many of the smaller peripheral populations disappearing as the species range contracts toward its center in Wyoming and Montana. In this process, local adaptations may be lost. For many reasons, therefore, including the potential to evaluate the validity of subspecies, a comprehensive study of the biology and genetics of many pronghorn populations is needed.

A noninvasive method, more accurate than microhistological analysis of feces, is needed to describe diet composition. This would allow a more accurate stipulation of habitat requirements and would allow for correlation between diet and population condition. Use of fecal nitrogen as an index of dietary nitrogen and dietary energy needs further validation. When methods to describe diet composition and quality are worked out, it should then be possible to assign a value scale to the array of habitats where pronghorn currently occur. Subsequently, it will be possible to examine how habitat quality affects annual female fecundity and reproductive effort.

Investigation of lower prenatal and postnatal growth rates in captivity than in nature might yield interesting results. Perhaps the hay and pellet rations fed in captivity do not match the diet quality that pronghorn can attain by selective feeding in nature. Alternatively, perhaps pronghorn in captivity have chronic levels of stress that result in lower growth rates.

Studies are needed to examine the significance of female mate choice and of the consequent variation in male mating success. Do offspring of the most successful males have higher fitness than offspring of infrequently chosen males? If so, then policies that set the pronghorn hunting season coincident with the pronghorn rut may need to be reevaluated.

ACKNOWLEDGMENTS

Research and manuscript preparation were supported by NSF IBN 9808377 and 0097115, National Geographic Society Grant 6396-98, and National Park Service Cooperative Agreement No. 1443CA157099002. I thank the National Park Service (Yellowstone National Park) and the U.S. Fish and Wildlife Service (National Bison Range) for cooperation and other assistance.

LITERATURE CITED

Autenrieth, R. E. 1978. Guidelines for the management of pronghorn antelope. Pronghorn Antelope Workshop Proceedings 8:473–526.

Autenrieth, R. E., and E. Fichter. 1975. On the behavior and socialization of pronghorn fawns. Wildlife Monographs 42:1–111.

Bailey, V. 1932. The Oregon antelope. Proceedings of the Biological Society of Washington 45:45–46.

Barrett, M. W. 1982. Distribution, behavior, and mortality of pronghorns during a severe winter in Alberta. Journal of Wildlife Management 46:991–1002.

Barrett, M. W., and G. A. Chalmers. 1977a. Clinicochemical values for adult free-ranging pronghorns. Canadian Journal of Zoology 55:1252–60.

Barrett, M. W., and G. A. Chalmers. 1977b. Hematological values for adult free-ranging pronghorns. Canadian Journal of Zoology 55:448–55.

Barrett, M. W., and G. A. Chalmers. 1979. Hematological and clinicochemical values for free-ranging pronghorn fawns. Canadian Journal of Zoology 57:1757–66.

Bayless, S. R. 1969. Winter food habits, range use, and home range of antelope in Montana. Journal of Wildlife Management 33:538–51.

Beale, D. M., and A. D. Smith. 1970. Forage use, water consumption and productivity of pronghorn antelope in western Utah. Journal of Wildlife Management 34:570–82.

Bear, G. D. 1971. Seasonal trends in fat levels of pronghorns, *Antilocapra americana,* in Colorado. Journal of Mammalogy 52:583–89.

Berger, J. 1989. Female reproductive potential and its apparent evaluation by male mammals. Journal of Mammalogy 70:347–58.

Bergmann, C. 1847. Uber die Verhaltnisse der Warmeokonomie der Thiere zu ihrer Grosse. Göttinger Studien 1:595–708.

Blair, N. 1982. An historical account of the pronghorn antelope. Wyoming Wildlife 46:3–7.

Blake, R. D., J. Z. Wang, and L. Beauregard. 1997. Repetitive sequence families in *Alces alces americana*. Journal of Molecular Evolution 44:509–20.

Bromley, P. T. 1969. Territoriality in pronghorn bucks on the National Bison Range, Moiese, Montana. Journal of Mammalogy 50:81–89.

Bruns, E. H. 1977. Winter behavior of pronghorns in relation to habitat. Journal of Wildlife Management 41:560–71.

Bryant, F. C., and G. E. Huber. 1998. Classification of pronghorn fawning habitat using Landsat thematic mapper data. Texas Journal of Science 50:3–16.

Buechner, H. K. 1950. Life history, ecology, and range use of the pronghorned antelope in Trans-Pecos, Texas. American Midland Naturalist 43:257–354.

Buechner, H. K. 1960. Regulation of numbers of pronghorn antelope in relation to land use. Pages 266–285 *in* Proceedings of the technical meetings in Warsaw–Cracow. International Union for Conservation of Nature and Natural Resources, Morges, Switzerland.

Bullock, R. E. 1974. Functional analysis of locomotion in pronghorn antelope. Pages 274–305 *in* V. Geist and F. Walther, eds. The behavior of ungulates and its relation to management. International Union for Conservation of Nature and Natural Resources, Morges, Switzerland.

Bullock, R. E. 1982. An analysis of locomotor body movements in pronghorn antelope. Canadian Journal of Zoology 60:1871–80.

Byers, J. A. 1997a. American pronghorn. Social adaptations and the ghosts of predators past. University of Chicago Press, Chicago.

Byers, J. A. 1997b. Mortality risk to pronghorns from handling. Journal of Mammalogy 78:894–99.

Byers, J. A., and K. Z. Byers. 1983. Do pronghorn mothers reveal the locations of their hidden fawns? Behavioral Ecology and Sociobiology 13:147–56.

Byers, J. A., and J. T. Hogg. 1995. Environmental effects on prenatal growth rate in pronghorn and bighorn: Further evidence for energy constraint on sex-biased maternal expenditure. Behavioral Ecology 6:451–57.

Byers, J. A., and D. W. Kitchen. 1988. Mating system shift in a pronghorn population. Behavioral Ecology and Sociobiology 22:355–60.

Byers, J. A., and J. D. Moodie. 1990. Sex-specific maternal investment in pronghorn, and the question of a limit on differential provisioning in ungulates. Behavioral Ecology and Sociobiology 26:157–64.

Byers, J. A., J. D. Moodie, and N. Hall. 1994. Pronghorn females choose vigorous mates. Animal Behaviour 47:33–43.

Cancino, J., A. Ortega-Rubio, and J. A. Sanchez-Pacheco. 1996. Status of an endangered subspecies: The peninsular pronghorn at Baja California. Journal of Arid Environments 32:463–67.

Canon, S. K., and F. C. Bryant. 1997. Bed-site characteristics of pronghorn fawns. Journal of Wildlife Management 61:1134–41.

Carlton, C., and T. McKean. 1977. The carotid and orbital retia of the pronghorn, deer and elk. Anatomical Record 189:91–107.

Chao, J. 1985. Responses of territorial pronghorn to snort–wheeze vocalizations. M.S. Thesis, University of Idaho, Moscow.

Clemens, E. T., K. L. Meyer, M. P. Carlson, and N. R. Schneider. 1987. Hematology, blood chemistry and selenium values of captive pronghorn antelope, white-tailed deer and American bison. Comparative Biochemistry and Physiology C 87:167–70.

Clutton-Brock, T. H., F. C. Guinness, and S. D. Albon. 1983. The costs of reproduction to red deer hinds. Journal of Animal Ecology 52:367–83.

Cole, G. F. 1956. The pronghorn antelope: Its range use and food habits in central Montana with special reference to alfalfa. Montana Fish and Game Department Technical Bulletin 516:1–63.

Cole, G. F., and E. T. Wilkins. 1958. The pronghorn antelope: its range use and food habits in central Montana with special reference to wheat. Montana Fish and Game Department Technical Bulletin 2:1–39.

Cook, J. G., and L. L. Irwin. 1985. Validation and modification of a habitat suitability model for pronghorns. Wildlife Society Bulletin 13:440–48.

Courtney, R. F. 1989. Pronghorn use of recently burned mixed prairie in Alberta. Journal of Wildlife Management 53:302–5.

Deblinger, R. D., and A. W. Alldredge. 1991. Influence of free water on pronghorn distribution in a sagebrush/steppe grassland. Wildlife Society Bulletin 19:321–26.

Dirschl, H. J. 1963. Food habits of the pronghorn in Saskatchewan. Journal of Wildlife Management 27:81–93.

Dow, S. A., Jr., and P. L. Wright. 1962. Changes in mandibular dentition associated with age in pronghorn antelope. Journal of Wildlife Management 26:1–18.

Dunbar, M. R., R. Velarde, M. A. Gregg, and M. Bray. 1999. Health evaluation of a pronghorn antelope population in Oregon. Journal of Wildlife Diseases 35:496–510.

Einarsen, A. S. 1948. The pronghorn antelope and its management. Wildlife Management Institute, Washington, DC.

Fairbanks, W. S. 1993. Birthdate, birthweight, and survival in pronghorn fawns. Journal of Mammalogy 74:129–35.

Fichter, E. 1987. Pronghorn groups: On social organization. Tebiwa 23:11–22.

Fournier, F., and M. Festa-Bianchet. 1995. Social dominance in adult female mountain goats. Animal Behaviour 49:1449–59.

Frick, C. 1937. Horned ruminants of North America. Bulletin of the American Museum of Natural History 69:1–669.

Goldman, E. A. 1945. A new pronghorn antelope from Sonora. Proceedings of the Biological Society of Washington 58:3–4.

Groves, C. P., and P. Grubb. 1987. Relationships of living deer. Pages 21–59 *in* C. M. Wemmer, ed. Biology and management of the Cervidae. Smithsonian Institution Press, Washington, DC.

Guenzel, R. J. 1994. Adapting new techniques to population management: Wyoming's pronghorn experience. Transactions of the North American Wildlife Natural Resources Conference 59:189–200.

Hailey, T. L., J. W. Thomas, and R. M. Robinson. 1966. Pronghorn die-off in trans-Pecos Texas. Journal of Wildlife Management 30:488–96.

Hall, E. R. 1981. The mammals of North America. John Wiley, New York.

Halloran, A. F. 1957. A note on the Sonoran pronghorn. Journal of Mammalogy 38:423.

Hepworth, W. G., and F. Blunt. 1966. Research findings on Wyoming antelope. Wyoming Wildlife 30:24–29.

Hofmann, R. R. 1989. Evolutionary steps of ecophysiological adaptation and diversification of ruminants: A comparative view of their digestive system. Oecologia 78:443–57.

Hoover, J. P. 1971. Food habits of pronghorn antelope on Pawnee National Grasslands. M.S. Thesis, Colorado State University, Fort Collins.

Hoskinson, R. L., and J. R. Tester. 1980. Migration behavior of pronghorn in southeastern Idaho. Journal of Wildlife Management 44:132–44.

Irwin, D. M., T. D. Kocher, and A. C. Wilson. 1991. Evolution of the cytochrome-B gene of mammals. Journal of Molecular Evolution 32:128–44.

Irwin, L. L., and J. G. Cook. 1985. Determining appropriate variables for a habitat suitability model for pronghorns. Wildlife Society Bulletin 13:434–40.

Janis, C. 1982. Evolution of horns in ungulates: ecology and paleoecology. Biological Reviews 57:261–318.

Janis, C. M. 1988. New ideas in ungulate phylogeny and evolution. Trends in Ecology and Evolution 3:291–97.

Kerwin, M. L., and G. J. Mitchell. 1971. The validity of the wear–age technique for Alberta pronghorns. Journal of Wildlife Management 35:743–47.

Kessler, W. B., W. F. Kasworm, and W. L. Bodie. 1981. Three methods compared for analysis of pronghorn diets. Journal of Wildlife Management 45:612–19.

Kitchen, D. W. 1974. Social behavior and ecology of the pronghorn. Wildlife Monographs 38:1–96.

Kraft, K. M., D. H. Johnson, J. M. Samuelson, and S. H. Allen. 1995. Using known populations of pronghorn to evaluate sampling plans and estimators. Journal of Wildlife Management 59:129–37.

Krueger, K. 1986. Feeding relationships among bison, pronghorn, and prairie dogs: An experimental analysis. Ecology 67:760–70.

Lance, W. R., and T. M. Pojar. 1984. Diseases and parasites of pronghorn: A review (State Publication Code R-S-57-'84, Special Report Number 57). Colorado Division of Wildlife.

Lee, P. C., and C. J. Moss. 1986. Early maternal investment in male and female African elephant calves. Behavioral Ecology and Sociobiology 18:353–61.

Lee, T. E. Jr., J. W. Bickham, and M. D. Scott. 1994. Mitochondrial DNA and allozyme analysis of North American pronghorn populations. Journal of Wildlife Management 58:307–18.

Leinders, J. 1984. Hoplitomerycidae fam. nov. (Ruminantia, Mammalia) from Neogene fissure fillings in Gargano (Italy). Part 1: The cranial osteology of *Hoplitomeryx* gen. nov. and a discussion on the classification of pecoran families. Scripta Geologica 70:1–68.

Lent, P. C. 1974. Mother–infant relationships in ungulates. Pages 14–55 *in* V. Geist and F. Walther, eds. The behavior of ungulates and its relation to management. International Union for Conservation of Nature and Natural Resources, Morges, Switzerland.

Lindstedt, S. L., J. F. Hokanson, D. J. Wells, S. D. Swain, H. Hoppeler, and V. Navarro. 1991. Running energetics in the pronghorn antelope. Nature 353:748–50.

Lonsdale, E. M., B. Bradach, and E. T. Thorne. 1971. A telemetry system to determine body temperature in pronghorn antelope. Journal of Wildlife Management 35:747–51.

Maher, C. R. 1991. Activity budgets and mating system of male pronghorn antelope at Sheldon National Wildlife Refuge, Nevada. Journal of Mammalogy 72:739–44.

Maher, C. R. 1994. Pronghorn male spatial organization: Population differences in degree of nonterritoriality. Canadian Journal of Zoology 72:455–64.

Maher, C. R. 2000. Quantitative variation in ecological and hormonal variables correlated with spatial organization of pronghorn (*Antilocapra americana*) males. Behavioral Ecology and Sociobiology 47:327–38.

Martin, S. K., and K. L. Parker. 1997. Rates of growth and morphological dimensions of bottle-raised pronghorns. Journal of Mammalogy 78:23–30.

Martinka, C. J. 1967. Mortality of northern Montana pronghorns in a severe winter. Journal of Wildlife Management 31:159–64.

McCutchen, H. E. 1969. Age determination of pronghorns by the incisor cementum. Journal of Wildlife Management 33:172–75.

McKean, T., and B. Walker. 1974. Comparison of selected cardiopulmonary parameters between the pronghorn and the goat. Respiratory Physiology 21:365–70.

Miller, M. N., and J. A. Byers. 1998. Sparring as play in young pronghorn males. Pages 141–60 *in* M. Bekoff and J. A. Byers, eds. Animal play: Evolutionary, comparative, and ecological perspectives. Cambridge University Press, Cambridge.

Mitchell, G. J. 1967. Minimum breeding age of female pronghorn antelope. Journal of Mammalogy 48:489–90.

Mitchell, G. J. 1980. The pronghorn antelope in Alberta. G. J. Mitchell, Regina, Saskatchewan, Canada.

Mitchell, G. J., and S. Smoliak. 1971. Pronghorn antelope range characteristics and food habits in Alberta. Journal of Wildlife Management 35:238–50.

Moodie, J. D., and J. A. Byers. 1989. The function of scent marking by males on female urine in pronghorns. Journal of Mammalogy 70:812–14.

Moy, R. F. 1970. Histology of the subauricular and rump glands of the pronghorn (*Antilocapra americana* Ord.) . American Journal of Anatomy 129:65–88.

Moy, R. F. 1971. Histology of the forefoot and hindfoot interdigital and median glands of the pronghorn. Journal of Mammalogy 52:441–46.

Müller-Schwarze, D., and C. Müller-Schwarze. 1972. Social scents in hand-reared pronghorns (*Antilocapra americana*). Zoologica Africana 7:257–71.

Murie, J. 1870. Notes on the anatomy of the prongbuck, *Antilocapra americana*. Proceedings of the Zoological Society of London 25:334–68.

Murphy, B. P., K. V. Miller, and R. L. Marchinton. 1994. Sources of reproductive chemosignals in female white-tailed deer. Journal of Mammalogy 75:781–86.

Nagy, J. G., and G. L. Williams. 1969. Rumino reticular VFA content of pronghorn antelope. Journal of Wildlife Management 33:437–39.

Ockenfels, R. A. 1994. Mountain lion predation on pronghorn in central Arizona. Southwestern Naturalist 39:305–6.

O'Gara, B. W. 1969. Unique aspects of reproduction in female pronghorn. American Journal of Anatomy 125:217–31.

O'Gara, B. W., and G. Matson. 1975. Growth and casting of horns by pronghorns and exfoliation of horns by bovids. Journal of Mammalogy 56:829–46.

O'Gara, B. W., R. F. Moy, and G. D. Bear. 1971. The annual testicular cycle and horn casting in the pronghorn (*Antilocapra americana*). Journal of Mammalogy 52:537–44.

Paradiso, J. L., and R. M. Nowak. 1971. Taxonomic status of the Sonoran pronghorn. Journal of Mammalogy 52:855–58.

Pojar, T. M., and L. L. W. Miller. 1984. Recurrent estrus and cycle length in pronghorn. Journal of Wildlife Management 48:973–79.

Pojar, T. M., D. C. Bowden, and R. B. Gill. 1995. Aerial counting experiments to estimate pronghorn density and herd structure. Journal of Wildlife Management 59:117–28.

Quinn, D. 1930. The antelope's S.O.S. Emergency Conservation Committee, New York.

Robbins, C. T., and B. L. Robbins. 1979. Fetal and neonatal growth patterns and maternal reproductive effort in ungulates and subungulates. American Naturalist 114:101–16.

Robinson, M., M. Wild, and J. Byers. 2002. Relationships between diet quality and fecal nitrogen, fecal diaminopime acid and behavior in a captive group of pronghorn. Proceedings of the Pronghorn Antelope Workshop 19:28–44.

Rouse, C. H. 1941. Notes on winter foraging habits of antelopes in Oklahoma. Journal of Mammalogy 22:57–60.

Ryder, T. J., and L. L. Irwin. 1987. Winter habitat relationships of pronghorns in southcentral Wyoming. Journal of Wildlife Management 51:79–85.

Sanabria, B., C. Arguelles-Mandez, and A. Ortega-Rubio. 1996. Occurrence of the endangered pronghorn *Antilocapra americana penisularis* [*sic*] in coyote diets from northwestern Mexico. Texas Journal of Science 48:159–62.

Schwartz, C. C., and J. E. Ellis. 1981. Feeding ecology and niche separation in some native and domestic ungulates on the shortgrass prairie. Journal of Applied Ecology 18:343–53.

Schwartz, C. C., and J. G. Nagy. 1976. Pronghorn diets relative to forage availability in northern Colorado. Journal of Wildlife Management 40:469–78.

Schwartz, C. C., J. G. Nagy, and R. W. Rice. 1977. Pronghorn dietary quality relative to forage availability and other ruminants in Colorado. Journal of Wildlife Management 41:161–68.

Smith, R. N., D. J. Neff, and N. G. Woolsey. 1986. Pronghorn response to coyote control: A benefit:cost analysis. Wildlife Society Bulletin 14:226–31.

Solounias, N. 1988. Evidence from horn morphology on the phylogenetic relationships of the pronghorn (*Antilocapra americana*). Journal of Mammalogy 69:140–43.

Stephenson, T. E., J. L. Holechek, and C. B. Kuykendall. 1985. Drought effect on pronghorn and other ungulate diets. Journal of Wildlife Management 49:146–51.

Sundstrom, C. 1968. Water consumption by pronghorn antelope and distribution related to water in Wyoming's Red Desert. Antelope States Workshop Proceedings 3:39–46.

Thorne, E. T. 1975. Normal body temperature of pronghorn antelope and mule deer. Journal of Mammalogy 56:697–98.

Thorne, E. T., E. S. Williams, T. R. Praker, W. Helms, and T. Segarstrom. 1988. Bluetongue in free-ranging pronghorn antelope (*Antilocapra americana*) in Wyoming, 1976 and 1984. Journal of Wildlife Diseases 24:113–19.

Torbit, S. C., R. B. Gill, A. W. Alldredge, and J. C. Liewer. 1993. Impacts of pronghorn grazing on winter wheat in Colorado. Journal of Wildlife Management 57:173–81.

Von Gunten, B. L. 1978. Pronghorn fawn mortality on the National Bison Range. Proceedings of the Biennial Pronghorn Antelope Workshop 8:394–416.

Waring, G. H. 1969. The blow sound of pronghorns (*Antilocapra americana*). Journal of Mammalogy 50:647–48.

Webb, S. D. 1973. Pliocene pronghorns of Florida. Journal of Mammalogy 54:203–21.

Webb, S. D. 1977. A history of savanna vertebrates in the new world. Part I: North America. Annual Review of Ecology and Systematics 8:355–80.

Wesley, D. E. 1973. Energy metabolism of pronghorn antelopes. Journal of Wildlife Management 37:563–73.

Wesley, D. E., K. L. Knox, and J. G. Nagy. 1970. Energy flux and water kinetics in young pronghorn antelope. Journal of Wildlife Management 34:908–12.

Wild, M. A., M. W. Miller, D. L. Baker, N. T. Hobbs, R. B. Gill, and B. J. Maynard. 1994. Comparing growth rates of dam- and hand-raised bighorn sheep, pronghorn, and elk neonates. Journal of Wildlife Management 58:340–47.

Wood, A. K. 1989. Comparative distribution and habitat use by antelope and mule deer. Journal of Mammalogy 70:335–40.

Woodley, C. 1997. The Sonoran pronghorn: The Air Force's strongest adversary. Dickinson Journal of Environmental Law and Policy 6:299–319.

Woolley, T. P., and F. G. Lindzey. 1997. Relative precision and sources of bias in pronghorn sex and age composition surveys. Journal of Wildlife Management 61:57–63.

Yoakum, J. D. 1980. Habitat management guides for the American pronghorn antelope (Technical Note 347). U.S. Department of the Interior, Bureau of Land Management.

John A. Byers, Department of Biological Science, University of Idaho, Moscow, Idaho 83844-3051. Email: jbyers@uidaho.edu.

48

Bison

Bison bison

Hal W. Reynolds
C. Cormack Gates
Randal D. Glaholt

NOMENCLATURE

Common Names. American bison, plains bison, prairie bison, bison, buffalo, wood bison, or woodland bison
Scientific Name. *Bison bison*
Subspecies. *B. b. bison, B. b. athabascae*

The bison is a member of the family Bovidae, to which domestic cattle, muskox (*Ovibos moschatus*), sheep, and goats belong. Both sexes of bovids possess true horns, which are never shed. These horns are composed of a bony core and a hard, outer sheath of epidermis. Fossil bovids date to the Lower Miocene (Feldhamer et al. 1999). Bovids are grazers primarily and browsers secondarily, and they possess a four-chambered, ruminating stomach. They inhabit major grassland, shrubland, forest, and tundra ecosystems. They feed by biting off forage with forward-projecting incisors, which are present only in the lower jaw.

The genus *Bison* is characterized by a short, broad forehead with a narrowed muzzle and pointed nasal bones (Hall 1981). Bison are particularly noted for a massive head, short neck, a high hump at the shoulders, and short, curved, rounded horns, which exhibit annual growth patterns (Soper 1964). The moderate-length tail is haired, with a terminal tassel of long hair (Banfield 1981). There is a distinct beard formed by long, woolly hair on the chin. The hair on the head, neck, and shoulders is brownish black. Body pelage is generally brown, varying moderately with season to light brown. Although technically a misnomer, the popular name *buffalo* has been used interchangeably for North American bison since early European explorers first encountered the species on the continent. True buffalo, while they are bovids, belong to genera distinctly different from bison and do not possess the shoulder hump characteristic of bison. The African cape buffalo (*Syncerus caffer*) is native only to Africa, whereas the water buffalo (*Bubalus bubalis*) is native to Asia.

TAXONOMY

Bison taxonomy has been controversial for many years, and classification to the subspecies level continues to be a subject of debate. Mayr (1963:348) defined a subspecies as "an aggregate of local populations of a species, inhabiting a geographic subdivision of the range of the species, and differing taxonomically from other populations of the species," noting that differing taxonomically means "by diagnostic morphological characters." Avise and Ball (1990:59–60) established the following subspecies definitional guidelines: "Subspecies are groups of actually or potentially interbreeding populations phylogenetically distinguishable from, but reproductively compatible with, other such groups. Importantly, the evidence for phylogenetic distinction must normally come from the concordant distributions of multiple, independent, genetically based traits." In 1991, a subcommittee of the Committee on the Status of Endangered Wildlife in Canada (COSEWIC) prepared guidelines for listing populations below the species level. It recommended the following criteria for including infraspecies groups, such as subspecies, on the COSEWIC list: "Geographical distinctiveness as indicated by barriers, distribution gaps, behavioural isolating mechanisms, different modes of relating to the environment, or other compelling evidence." Furthermore, the Canadian Species at Risk Act (SARA), (Proclamation into law June 5, 2003), provides the following definition of a wildlife species: "Wildlife species means a species, subspecies, variety or geographically or genetically distinct population of animal, plant or other organism, other than a bacterium or virus, that is wild by nature."

Bison Subspecies. Currently, two subspecies of North American bison have been scientifically described, the plains bison (*Bison bison bison* Linnaeus, 1758) and the wood bison (*Bison bison athabascae* Rhoads, 1897). Allen (1876), Seton (1886), and Ogilvie (1893) were among the first to describe wood bison as distinct from plains bison by their larger size and darker color. Hornaday (1889) failed to recognize a subspecific status for wood bison. Rhoads (1897) was the first to scientifically describe wood bison as a separate subspecies. Many other authors (Raup 1933; Soper 1941; Skinner and Kaisen 1947; Banfield and Novakowski 1960; Flerov 1965; Karsten 1975; Geist and Karsten 1977; Cook and Muir 1984; van Zyll de Jong 1986, 1993; Wood Bison Recovery Team 1987; Gates et al. 1992a, 2001b; van Zyll de Jong et al. 1995) have since acknowledged subspecific status for wood bison in agreement with Rhoads. However, Graham (1923), Seibert (1925), Garretson (1927), and Geist (1991) recognized the larger size and darker color of wood bison, but attributed this to environmental influences and not to genetics. In addition, conflicting evidence about the size and color of wood bison has been reported (Peden and Kraay 1979).

Skinner and Kaisen (1947) made a preliminary revision of the genus *Bison,* based on skull characteristics, measurements, and identifiable patterns of horn core growth. They classified mountain bison as a southern extension of the woodland race into mountainous habitat along the Rockies. However, Skinner and Kaisen (1947:165) apparently recognized an important shortcoming in their work when they wrote, "The population sample of true *athabascae* skulls is too small to present a comprehensive understanding of the amount of variation possible within this mountain or woodland race." Further work with cranial characteristics of plains bison suggested that a revision of the genus may be necessary (Shackleton et al. 1975). Furthermore, Peden and Kraay (1979) questioned the taxonomy of bison on the basis of their research on blood characteristics. An osteological study to clarify systematics of extant forms of *Bison,* evolutionary trends in late Pleistocene and postglacial fossil *Bison,* and possible origins of extant forms, in particular the wood bison, was completed by van Zyll de Jong (1986).

Morphological studies of plains bison and wood bison showed significant differences in cranial and skeletal characteristics (van Zyll de Jong 1986) as well as in the anterior slope of the hump, location of the highest point on the hump, angle of the hump, cape variegation and demarcation, upper front leg hair, frontal display hair, ventral neck mane, and beard. However, Geist (1991) suggested that the subspecific status is not warranted and that those observed differences are environmentally induced. A study by van Zyll de Jong et al. (1995) showed that these traits are not affected by geographic location, suggesting that phenotypic differences are genetically controlled. In an osteological study of the genus *Bison,* van Zyll de Jong (1986) concluded that subspecific status for wood bison was warranted, based on the differing cranial and morphological characteristics. Of 19 cranial variables measured, wood bison differed significantly from plains bison in seventeen of those measurements (van Zyll de Jong 1986). Similarly, of 23 postcranial variables measured, wood bison differed significantly from plains bison in twenty-two of those measurements (van Zyll de Jong 1986). In a DNA microsatellite study, Wilson and Strobeck (1999) reported that the distinctness of wood bison is supported because between-subspecies distances were, in most cases, larger than within-subspecies distances. This suggests that genetic differences exist between wood and plains bison, which can be maintained by continuing to manage them as separate entities.

Wood bison appear to meet accepted criteria for their classification as a valid subspecies, and they have been a key component in the diversity of natural life forms in northern Canada and Alaska. Free-ranging wood bison populations in northern Canada are particularly important because they are the only herds existing in relatively unaltered and intact ecosystems. The Wood Bison Recovery Team recognizes wood bison as a distinct subspecies based on morphological characteristics and molecular genetics and strongly supports the position that publicly owned wood bison continue to be maintained and managed separately from plains bison (Gates et al. 2001b). Furthermore, the cooperating management jurisdictions within the national Wood Bison Recovery Program have accepted wood bison as a separate subspecies, in spite of some still controversial aspects of bison phylogeny and taxonomy. Currently, there is an option to manage the most representative wood bison available as a separate subspecies and/or as distinct subpopulations of a common species. If a decision to withdraw subspecific status is made and genetic mixing is permitted, then that decision becomes irreversible. Considerable thought, research, and confirmatory scientific data are prerequisite to a decision of that significance.

***Bos* Versus *Bison*.** The wood bison has been documented as a valid subspecies of North American bison (Cook and Muir 1984; van Zyll de Jong 1986). However, some taxonomists believe that, because of the close morphological and genetic similarity of *Bos* and *Bison,* they should be united in a single genus (Simpson 1961; Van Gelden 1977). Morphological (Groves 1981) and genetic evidence (Miyamoto et al. 1989; Wall et al. 1992; Janecek et al. 1996) support this view, in addition to the incomplete fertility of cattle × bison hybrids (Van Gelden 1977). Furthermore, Jones et al. (1997:5) included bison and cattle in a single genus in their checklist of North American mammals, as they followed up on the "somewhat unpopular decision to use the name combination of *Bos bison* for the American bison." However, under the auspices of the American Society of Mammalogists, a checklist of the mammal species of the world (available online at http://www.nmnh.si.edu/msw/) was compiled and lists the American bison (*Bison bison*) and the European bison or wisent (*Bison bonasus*) as a separate genus from *Bos,* following the taxonomy of Wilson and Reeder (1993). Because of the lack of consensus on the taxonomy of *Bos* and *Bison,* the Canadian Wood Bison Recovery Team has favored recognition of separate genera and continues to refer to wood bison as *Bison bison athabascae* and to American or plains bison as *Bison bison bison* (Gates et al. 2001b).

External Morphology. Geist (1991) argued that extant North American subspecies are "ecotypes" with morphological differences reflecting local environmental influences rather than heritable traits. Mayr (1963:354) defined an ecotype as "the product arising as a result of the genotypical response of an ecospecies to a particular habitat." A recent study of external morphological characteristics in 11 bison populations in North America revealed three significantly different groups: plains bison (6 populations), wood bison (4 populations), and an intermediate form, the Pine Lake subpopulation in central Wood Buffalo National Park, Alberta (van Zyll de Jong et al. 1995). The Peace–Athabasca Delta in Wood Buffalo National Park, the Slave River Lowlands, the Mackenzie, and the southern Elk Island National Park bison populations were the four herds classified as wood bison. Furthermore, plains bison transferred from the National Bison Range, Montana, to Delta Juction, Alaska, in the 1940s maintained their plains bison pelage and body confirmation characteristics for more than 60 years within the known range distribution and habitat of wood bison. This phenotypic study indicated that the shape of the hump and the pelage characters are independent and, in addition, four of the five differentiating pelage characteristics showed significant associations, which suggests that they may be genetically linked (van Zyll de Jong et al. 1995). The persistence of wood and plains bison external morphological characteristics in varying environments demonstrates that phenotypic characteristics of wood and plains bison are genetically controlled and not induced by environmental factors (van Zyll de Jong et al. 1995).

Skeletal Morphology. Skeletal material from a number of wood and plains bison populations has been analyzed in considerable detail (McDonald 1981; van Zyll de Jong 1986). These studies indicated that geographic variation in historical populations of North American plains bison was mainly continuous (clinal) through a series of overlapping populations, manifested along a north–south axis. In general, bison in the eastern and southern United States tended to be smaller than those found in the north-central United States or on the Canadian prairies. However, a continuous gradation of intermediate types between the extremes makes meaningful delineation of northern and southern plains bison types impossible. In contrast to the clinal variation observed in plains bison, phenotypic discontinuity was evident between wood bison and plains bison populations. This discontinuity was reflected not only in size, but also in morphological differences (van Zyll de Jong 1986).

Morphometric analysis of bison from the northwest area of Wood Buffalo National Park (Nyarling River) demonstrated that they are similar to the original *B. b. athabascae,* although some evidence of interbreeding with *B. b. bison* is apparent (van Zyll de Jong 1986). Because Nyarling River bison are much closer to *athabascae* morphometrically than they are to *bison,* they should be assigned to the former taxon as a valid subspecies. van Zyll de Jong (1986) concluded that wood bison differed demonstrably from plains bison in cranial and morphological characters.

Blood Characteristics. Several researchers have used serological characteristics (blood groups and other proteins) of bison to elucidate phylogenetic relationships (Stormont et al. 1960; Braend 1963; Naik and Anderson 1970; Ying and Peden 1977; Buckland and Evans 1978a, 1978b; Fulton et al. 1978; Peden and Kraay 1979). Peden and Kraay (1979) examined 10 blood characteristics from each of five herds of North American bison by using 13 cattle bloodtyping reagents and carbonic anhydrase electrophoresis. They failed to differentiate between the two subspecies and concluded that further tests for red cell antigens and blood enzymes were required. However, Peden and Kraay (1979) reported a significant variation in blood characteristics of six bison herds, including four plains bison and two wood bison herds. They were able to distinguish plains bison in Canada from those in the United States. In another study, Zamora (1983) was unable to distinguish subspecies of bison based on analysis of erythrocyte antigens and blood proteins. Based on 15 blood protein and enzyme systems, the genetic distance between *B. b. bison* and *B. bonasus* was comparable to that between local populations of a species (Hartl and Reimoser 1988).

Conservation Status. Like the plains bison, wood bison were nearly eliminated during the late 1800s. The history of their near extinction and subsequent conservation effort is thoroughly documented in the 1987 status report on wood bison prepared for COSEWIC (Wood Bison Recovery Team 1987). Wood bison were first designated as "endangered" in 1978, but progress toward their recovery during the subsequent decade resulted in them being downlisted to "threatened" status in 1988, based on the 1987 status report. Wood bison also are recognized as a threatened subspecies in Appendix II in the Convention on International Trade in Endangered Species of Wild Flora and Fauna (CITES). By definition, Appendix II species are not necessarily threatened with extinction, but do require some control of trade to avoid commercial uses that are incompatible with their survival. Further recovery of wood bison is required before they can be downlisted from threatened status and removed from the list of species at risk.

The American bison is listed in the Lower Risk, Conservation Dependent category in the Red List of Threatened Species (2000) maintained by the World Conservation Union (IUCN), formally known as the International Union for the Conservation of Nature and Natural Resources (Hilton-Taylor 2000). The status of plains bison in Canada has not been officially determined by COSEWIC, but a Conservation Status Survey project to review the conservation status of bison in North America and to develop recommendations for enhancing bison conservation has commenced (Boyd and Gates 2001). This will be done in conjunction with the development of an action plan for North American bison by the Bison Specialist Group, North America, which operates under the Species Survival Commission (SSC) of the IUCN. The Bison Specialist Group will evaluate the status of North American bison using the IUCN Red List Criteria and will recommend a listing designation for each subspecies of bison to the IUCN/SSC (Boyd and Gates 2001).

Application of subspecies definitional guidelines established by Avise and Ball (1990) and O'Brien and Mayr (1991) leads to the following conclusions regarding the status of North American bison subspecies. Historically, wood bison differed from other bison populations with regard to multiple morphological and genetic characteristics; intrusion of plains bison into the range of wood bison in 1925–1928 was entirely human caused; the two North American bison subspecies continue to be morphologically and genetically distinct, despite some hybridization in the 1920s, thus modern-day wood bison and their descendants continue to constitute populations of a valid subspecies of bison.

DISTRIBUTION

Historical. Various forms of the genus *Bison* have been important elements in the fauna of northwestern North America and Siberia for a period exceeding 400,000 years (Gates et al. 2001d). Bison were one of the most common large herbivores during much of the last 100,000 years and, until recently, the Bering Land Bridge linked these regions. Therefore, it is not surprising that the last types of bison to exist in these regions were similar, small-horned forms represented by the surviving wood bison. Various human societies interacted with bison and other large herbivores in Eurasia for tens of thousands of years, and in North America before and after the glacial recession some 10,000–15,000 years ago (Gates et al. 2001d). It is clear that humans were responsible for the near extinction of plains bison on the Great Plains of North America (Roe 1970; Isenberg 2000) and wood bison from northern Alberta, British Columbia, and the southwestern Northwest Territories (Gates et al. 1992a). Humans likely played a role in the extirpation of bison from Alaska and Yukon between 200 and 400 years ago (Stephenson et al. 2001).

The earliest known fossil records of the genus *Bison* appeared in the Villafranchian deposits of India and China (Geist 1971). These late Pliocene forms eventually gave rise to the present-day bison. It is believed that bison first immigrated to North America over the Bering Land Bridge (Beringia) during the early and middle Pleistocene near the time of the Illinoian glacial and Sangamon interglacial (mid-Pleistocene) periods and then again with the recession of the Wisconsin glaciation during the late Pleistocene. This latter glaciation is believed to have mediated much of bison evolution and zoogeography. Fossil bison have been found throughout northeastern Asia, Alaska, Yukon, and most of western and central North America, and show evidence of considerable variation in body size, conformation, and horn growth. Controversial theories have been presented to elucidate the evolutionary history of bison. Definitive work on the evolution and zoogeography of the genus has been done by Flerov and Zablotski (1961), Guthrie (1970, 1980, 1990), Geist (1971), Geist and Karsten (1977), Harington (1980), Hillerud (1980), and Wilson (1980). Others have reported findings on fossil bison, theories on evolution of bison, and relationships of bison and humans in North America (Fuller and Bayrock 1965; Schultz and Hillerud 1977; Frison 1980; Stephenson et al. 2001).

Fossil evidence indicates there was a single species of bison in Eurasia and North America during the middle and late Pleistocene (Guthrie 1990). During the last (Wisconsin) glaciation, there were two separate populations of bison in North America (Guthrie 1990; van Zyll de Jong 1993). The steppe bison (*B. priscus:* McDonald 1981; Guthrie 1990; or *B. b. priscus:* van Zyll de Jong 1986), a relative newcomer from Eurasia, occupied Beringia and was adapted to the rigors of the cold steppe. *Bison antiquus,* a descendant of an invasion of bison from Eurasia during the preceding Illinoian glaciation, was adapted to a temperate climate and open woodlands, and persisted south of the continental ice sheet (Gates et al. 2001d). By about 14,000 years ago, Eurasia and Alaska were separated by rising sea levels resulting from climatic warming. Subsequently, bison on both sides of the Bering Isthmus appear to have undergone parallel evolution from large-horned to small-horned forms (van Zyll de Jong 1993). Three bison taxa (wood bison, plains bison, and European bison) continue to exist. The two North American subspecies, wood and plains bison, represent the most recent variants on this continent, and the European bison, which survived only in the forests of eastern Europe, represents the only extant bison on the Eurasian continent (Gates et al. 2001d). A northern form of small-horned bison, similar to wood bison, became extinct in eastern Siberia by the late Holocene (van Zyll de Jong 1993); however, the taxonomy of this form is not well defined.

A corridor in the ice sheet separating Beringia from central North America began to form about 13,000 years ago toward the late Pleistocene, after which northern and southern forms of bison apparently dispersed and intermingled (McDonald 1981). Existing North American bison are descendants of these two Pleistocene lines. However, the contribution of the southern *antiquus* and the northern *priscus* in the evolution of modern North American bison is not well understood (Gates et al. 2001d). The southern *antiquus* was more widely distributed and abundant than *priscus,* and may have played a larger role in the evolution of modern North American bison in the southern part of their range (van Zyll de Jong 1993).

At the end of the last glaciation and the beginning of the Holocene about 10,000 years ago, the barriers separating the two populations disappeared and a shorter horned bison (*B. occidentalis*) appeared (van Zyll de Jong 1993). Thus, the fusion of northern and southern forms produced the early Holocene form, *B. occidentalis* (van Zyll de Jong 1986), which underwent rapid evolutionary change and gave rise to the two modern North American subspecies by about 5000 years ago (Gates et al. 2001d). The living North American bison occupy an intermediate position relative to *priscus* and *antiquus,* indicating that they descended from a mixed *priscus–antiquus* line evidenced by skull and vertebrae data (van Zyll de Jong 1993). Mixing of northern and southern stocks coupled with the profound environmental changes occurring then created selection pressures that led to modern-day North American bison (van Zyll de Jong 1993). Adapted to northern woodlands and meadows, the nonmigratory wood bison evolved in the northwestern section of the species's range, whereas the migratory plains bison evolved in the extensive grasslands of central and southern North America. Radiometric data spanning much of the past 10,000 years, together with archaeological data, demonstrate that the mid- to late Holocene distribution of wood bison included much of eastern Alaska, southern Yukon, the western Northwest Territories, and possibly western Alaska (Stephenson et al. 2001).

The taxonomic status of Holocene bison in eastern Siberia is less clear, but evidence suggests they were morphologically similar, and ecologically equivalent, to wood bison (Gates et al. 2001d). A well-preserved skull from the Kolyma River lowlands, believed to be of late Pleistocene or early Holocene age, is taxonomically intermediate between *B. b. occidentalis* and *B. b. athabascae,* demonstrating a close affinity with these forms (van Zyll de Jong 1986, 1993). Bison remains also have been recorded at archaeological sites dating to as late as 900 A.D. in southern Yakutia (Lazarev et al. 1998).

Disappearance of bison and some other large herbivores from northern areas on either side of the Bering Strait during the mid- to late Holocene likely involved the combined effects of changes in habitat distribution and hunting by humans (Ward 1997; Stephenson et al. 2001). Disappearance of bison and other large herbivores may have had important consequences for the structure and function of northern ecosystems (Zimov et al. 1995).

Over millennia, the "Great Bison Belt" that extended across Eurasia into Beringia and southward across North America diminished in size (Guthrie 1980, 1990:51). Evidence suggests that hunting by humans played a central role in the history of northern bison, first in Siberia, where bison were extirpated during the past 2000 years, then in North America, where only remnant populations remained by the end of the nineteenth century (Gates et al. 2001d). Although bison likely disappeared from Siberia by the late Holocene, they were still widely distributed in North America, including the Great Plains, in eastern woodlands, and in northwestern Canada and Alaska (Soper 1941; Dary 1989; Stephenson et al. 2001). Bison in northern habitats were not as numerous as bison on the Great Plains. The phenomenon of a large migratory population was unique to the Holocene Great Plains, with its huge expanse of contiguous grasslands (Guthrie 1980, 1982, 1990). Beginning in the 1700s, the availability of horses altered the mobility and hunting strategies of Native groups (Isenberg 2000). Horses eventually became sufficiently abundant and widespread to compete with bison for forage (Flores 1996; Fisher and Roll 1998). A number of studies describe the importance of hunting by humans in the dynamics of late Holocene plains bison populations (Roe 1970; Guthrie 1980; Speth 1983; Flores 1991, 1996; Belue 1996; Dobak 1996; Fisher and Roll 1998; Haynes 1998; Morgan 1998; Martin and Szuter 1999; Isenberg 2000).

Written records document the occurrence of bison in the late eighteenth and early nineteenth centuries in the southern Yukon, western Northwest Territories, Alberta, and British Columbia (Gates et al. 1992a, 2001d; Lotenberg 1996). Bison were apparently scarce or had disappeared from Alaska and western Yukon before their presence was recorded in written records. However, oral narratives provide insight into the early historical distribution, human use, and disappearance of wood bison in Alaska and adjacent Yukon (Gates et al. 2001d). Historical accounts from Native elders in interior Alaska describe how bison were a source of food and raw materials, which suggests that substantial populations of wood bison declined or disappeared from Alaska by the early to mid-1800s, with small numbers ocurring in the eastern interior as late as the early 1900s (Stephenson et al. 2001). These accounts are in general agreement with oral narratives obtained from First Nation elders in the Yukon (Lotenberg 1996), which suggested a decline in bison numbers during the last 400 years and the eventual extirpation of wood bison by the early twentieth century.

Historical accounts and paleontological and archaeological data indicate that humans hunted wood bison until their disappearance from Alaska and Yukon during the last few hundred years (Gates et al. 2001d). Wood bison apparently had become scarce shortly before early Euroamerican explorers, naturalists, and entrepreneurs entered the region from the east and before firearms became widely available (Jennings 1968; Holmes and Bacon 1982; Guthrie 1990). Wood bison were extirpated from most of their original range in North America by the late nineteenth or early twentieth century, persisting only in the area south of Great Slave Lake, where the Canadian Government eventually afforded some protection. This decline coincided with the historical period during which plains bison were extirpated from the woodlands east of the Mississippi River (Dary 1989; Belue 1996).

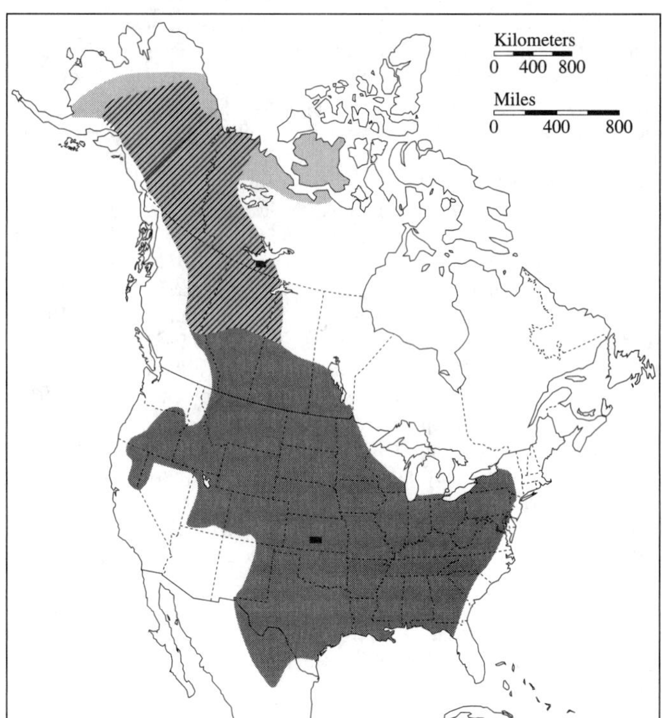

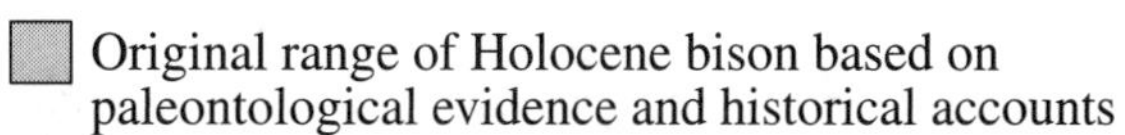

Original range of Holocene bison based on paleontological evidence and historical accounts

Original distribution of wood bison based on zooarchaeological and paleontological evidence and existing oral and written accounts

Original distribution of plains bison based on historical accounts and fossil evidence of horn core occurrence

Type specimen locality

FIGURE 48.1. Original distribution of plains bison (*Bison bison bison*) and wood bison (*Bison bison athabascae*) in North America, based on available zooarchaeological, paleontological, oral, and written historical accounts. SOURCE: Data from van Zyll de Jong (1986) and Stephenson et al. (2001).

Plains Bison. The American bison ranged throughout much of North America (Fig. 48.1). They were formerly widespread from western and northern Canada across the United States and into northern Mexico (Meagher 1986). Ernest Thompson Seton estimated that 75 million were in North America before White settlers arrived (Dary 1989). However, McHugh (1972) estimated that 30 million bison was the maximum number that available range could support. Although millions of bison once roamed this region, few free-ranging herds of North American bison remain. Descriptions of historical distribution patterns for plains bison are provided by Allen (1876), Hornaday (1889), Skinner and Kaisen (1947), Roe (1970), and Meagher (1986). The original distribution of the American bison included most of central North America, from northern Mexico to Great Slave Lake and from Washington to the Rocky Mountain states (Banfield 1981).

In Canada, the northeastern boundary for the original range of plains bison is roughly outlined by a line extending from north-central Saskatchewan in a southeastward direction to the southern shore of the Great Lakes. The northern boundary for bison in central Canada is also approximated by this line (Fig. 48.1).

In the United States, the eastern boundary was that of the Allegheny Mountains extending south through the states of Maryland, Virginia, North Carolina, and South Carolina. The southern boundary for plains bison extended from Alabama across southern Mississippi and Louisiana and continued westward along the southeastern coast of Texas and into Mexico. It was in southeastern Texas near present-day Houston that the American bison was first seen by a European, Cabeza de Vaca, in 1530 (Hornaday 1889).

The western boundary of North American plains bison distribution generally extended northward from north-central Mexico, merging with the original range for wood bison in northeastern British Columbia near the central western border of Alberta, Canada (Fig. 48.1). Here, the western and northern limits of plains bison range approximated the boundary of the ecotone between grassland, aspen parkland, and boreal forest habitat. To the north of this interface is boreal forest habitat, which is the beginning of the original range of wood bison.

Wood Bison. Based on additional information since 1987, the original range of wood bison in North America has been expanded to include a large area northwest of what was previously designated "historic range" (Gates et al. 2001b), including much of Alaska, the Yukon, and the western portion of the Northwest Territories (Fig. 48.1). Oral narratives obtained from aboriginal people and radiometric data of bison bones from archaeological and paleontological locations (van Zyll de Jong 1986; Harington 1990) indicate that bison were present in the Yukon and Alaska during the last few hundred years and persisted in small numbers into the early twentieth century (Lotenberg 1996; Stephenson et al. 2001), similar to areas in northeastern British Columbia and the southwestern Northwest Territories (Gates et al. 1992a). In view of the historical documentation and physical evidence demonstrating that wood bison inhabited this region for several thousand years, the area properly constitutes "historic" range. Therefore, in the absence of objective and biologically meaningful criteria, the dichotomy between "historic" and "prehistoric" range is not useful for the purposes of conservation and recovery, and the continuum of history should be recognized to avoid this artificial distinction (Stephenson et al. 2001). The descriptive term "original range" better represents this concept (Gates et al. 2001b).

The original range of wood bison included most of the boreal regions of northern Alberta, northeastern British Columbia, a small portion of northwestern Saskatchewan, the western Northwest Territories, most of the Yukon, and much of Alaska (Fig. 48.1) (Harington 1977; van Zyll de Jong 1986; Wood Bison Recovery Team 1987; Guthrie 1990; Gates et al. 2001b; Stephenson et al. 2001). Populations persisted in the area south of Great Slave Lake in Canada, but wood bison were extirpated in other parts of their range. The coniferous forests and aspen (*Populus*) parkland with interspersed grass and sedge (*Carex* sp.) meadows and prairies, typical of this area, constituted the main habitat for the wood bison throughout this expansive region.

DESCRIPTION

Plains Bison. The unmistakable appearance of plains bison is characterized by a massive, heavy head with a short, broad nasal area. The large head appears to be carried low because of the high shoulder hump and massive forequarters (Pattie and Fisher 1999). A short, thick neck and a high shoulder hump leave the impression that the forequarters are out of proportion to the much smaller appearing hindquarters. The hindquarters of plains bison are lighter than the forequarters, a disproportion that is further accented by differing pelages between front and rear. Pelage is long over the forehead, neck, hump, and forequarters, but short over the rear and tail (Meagher 1986). There is a tufted tail of moderate length (Banfield 1981). The short, round, black horns rise laterally from the side of the head and curve inward over the head. Horns of the female are more slender and tend to curve inward to a greater degree than those of the male. The eyes are located anterolaterally on the head, and the ears are nearly invisible as they are buried under the long thick pelage of the head (Meagher 1986). Plains bison have rather short legs and large, rounded hooves, which leave tracks similar to those of domestic cattle. Sexual dimorphism is evident among adults, with females being smaller and slighter. In general, however, females resemble males in color, body configuration, and presence of permanent horns. Males have larger, more evenly curving horns, with bases that are buried in head hair; a larger hump and thicker neck; longer pelage on the forehead, chin (beard), ventral mane, and chaps of the forelegs; and a better defined, demarcated cape of longer hair on the forequarters ending abruptly with shorter hair on the flanks and rear (Meagher 1986).

Wood Bison. Wood bison possess the same general characteristics as plains bison except for some minor differences in body morphology, general conformation, pelage, and skeletal measurements. For example, wood bison have larger horn cores and exhibit differences in other cranial elements. Karsten (1975) reported that wood bison possess denser fur than plains bison, they are larger and heavier (within similar age and sex classes as verified by greater mass), more elongated in the forequarters, and darker in color, and they have a squarish hump with a more gently sloping back contour than the plains bison. Geist and Karsten (1977) described how the wood bison bull and cow differ significantly in external morphological characteristics from their prairie counterparts. There is less sexual dimorphism in wood bison compared with plains bison in respect to body size, horn structure, pelage characteristics, and body proportions. A summary of these differing characteristics follows.

1. Hair on top of the head, around the horns, in the beard, and in the midventral neck area is significantly shorter and less dense in wood bison bulls than in their plains counterparts of the same age. Plains bison generally exhibit a dense, woolly bonnet of hair between the horns, whereas in wood bison, the dark forelock of hair tends to hang in strands over the forehead. Thus, the head of the wood bison appears smaller, the horns longer, and the ears more noticeable. The beard of the wood bison is thinner and more pointed (Fig. 48.2), and the long, full throat mane, which extends from the beard to the brisket on the plains bison, is rudimentary or absent in the wood bison. The head and neck of the wood bison generally are darker in color than those of the plains bison.
2. Long hair in the area of the "chaps" on the front legs is well developed, forming a skirt on the plains bison, but is reduced or absent on the wood bison (Fig. 48.2). This most striking difference in pelage between wood and plains bison partly accounts for the more massive appearance of the plains bison in the front quarters.
3. The "robe" or cape of the shoulders, hump, and neck region of the plains bison is more distinct and lighter (golden) colored than that of the wood bison. The well-demarcated cape of the plains bison is composed of longer hair, which forms an obvious boundary with the rest of the body fur just posterior to the shoulders and is generally lighter in color than that of the wood bison. There is no clear cape demarcation in the wood bison (Fig. 48.2) and the hair is usually darker than that of the plains bison.
4. The tail of the wood bison is usually longer and more heavily haired than that of the plains bison.

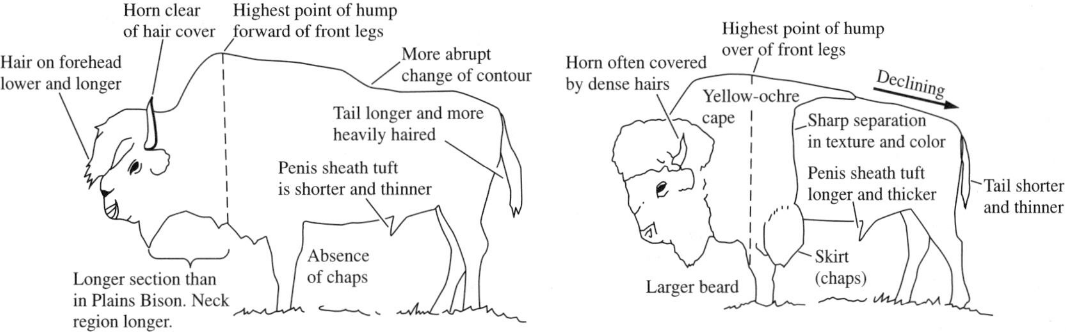

FIGURE 48.2. Basic pelage and morphological difference between (left) a male wood bison (*Bison bison athabascae*) and (right) a male plains bison (*Bison bison bison*). SOURCE: Based on character states after van Zyll de Jong et al. (1995).

5. The penis sheath tuft of the wood bison is usually shorter and thinner than that of the plains bison.
6. Wood bison tend to be taller at the hump, which is squarer than the hump of the plains bison. The highest point of the hump is more forward in the wood bison, just ahead of the front legs, whereas in the plains bison, it is farther back and centered over the front legs (Fig. 48.2). The back contour of the wood bison has a more abrupt change at the hump, but is more gently sloping or flat overall (Fig. 48.2).

Roe (1970) cited observations of two early frontiersmen that contradicted numerous claims that wood bison were larger than plains bison. Nonetheless, most biologists who have had the opportunity to view wood and plains bison at Elk Island National Park, Alberta, agree with the descriptions presented by Karsten (1975) and Geist and Karsten (1977). Historical physical description accounts of wood bison, most of which related to size and color, were usually compiled by explorers. Therefore, one can expect some degree of controversy in reports from these untrained (biological) observers, especially when describing characteristics that are quite variable. Still, the contradictory reports are in the minority, whereas the majority of historical accounts agree with more recent physical descriptions of living wood bison.

Morphology. Overall, the body of the bison is massive, tall, but proportionately narrow in transverse profile, where its massiveness and height are accentuated by the shoulder hump (Fig. 48.2). The hump is supported by tall, elongated spines on the thoracic vertebrae and musculature of the shoulder and pectoral girdle (Banfield 1981; Meagher 1986). The head is massive and appears to be carried low because of the high shoulder hump, massive forequarters, and the short, thick neck.

TABLE 48.1. Body measurements and weights of plains bison from several locations in North America and wood bison weights from Elk Island National Park, Alberta

		Body Measurement (cm)					
Sex	Age Class (years)	Total Length	Tail Length	Hind Foot	Height at Shoulder	Weight (kg)	Reference
M and F	Calf-birth	—	—	—	—	15–25	Rutberg 1984
M and F	Calf-birth	—	—	—	—	18	Pattie and Fisher 1999
M; F	Calves	—	—	—	—	140; 138[a]	Rothstein and Griswold 1991
M and F	Yearlings	—	—	—	—	227–318	Meagher 1973
M; F	Yearlings	—	—	—	—	233; 235[a]	Rothstein and Griswold 1991
F	>2	—	—	—	—	413[a]	Rothstein and Griswold 1991
M; F	2	—	—	—	—	407; 376[a]	Rothstein and Griswold 1991
M	Adult (12.5)	318	—	—	186	814[b]	Halloran 1961
F	Adult (6.5)	—	—	—	157[c]	488[b]	Halloran 1961
M	Adult	340	43	61	178	816–998	Soper 1964
F	Adult	d	d	d	d	363–544	Soper 1964
M and F	Adult	210–350	50–60	—	260–280	450–1350	Walker et al. 1975
M	Adult	304–380	33–91	58–68	167–186	544–907	Meagher 1986
F	Adult	213–318	30–51	50–53	152–157	318–545	Meagher 1986
M	Adult	304–380	43–48	56–68	167–182	460–720	Banfield 1981
F	Adult	213	45	53	152	360–460	Banfield 1981
Mixed	Mixed	—	—	—	—	450[e]	Telfer and Scotter 1975
M	Adult	—	—	—	—	600–860	Rutberg 1984
F	Adult	—	—	—	—	350–550	Rutberg 1984
F	Adult (5+)	—	—	—	—	255–410[f]	Lott and Galland 1987
M	Adult (4+)	—	—	—	—	615–682[g]	Berger and Peacock 1988
F	Adult (4+)	—	—	—	—	410–470[g]	Berger and Peacock 1988
F	Adult (6+)	—	—	—	—	352–605[h]	Green and Rothstein 1991
F	Adult (6+)	—	—	—	—	424[i]	Green and Rothstein 1991
F	Adult (6+)	—	—	—	—	518[j]	Green and Rothstein 1991
M	Adult (≥7)	—	—	—	—	750–945[k]	Wolff 1998
M and F	Adult	240–390	28–39	—	130–180	360–1090	Pattie and Fisher 1999
M	Adult (7.5)	—	—	—	—	727[l]	Towne 1999
F	Adult (7.5)	—	—	—	—	455[l]	Towne 1999
M	Adult (9.5)	—	—	—	—	858[m]	Towne 1999
F	Adult (8.5)	—	—	—	—	542[n]	Towne 1999
M, plains	Adult	—	—	—	—	591–769[o]	Olson 2002
F, plains	Adult	—	—	—	—	417–454[o]	Olson 2002
M, **wood**	Adult	—	—	—	—	642–910[o]	Olson 2002
F, **wood**	Adult	—	—	—	—	493–567[o]	Olson 2002

[a]Mean weight by sex and age class in November 1983, Wind Cave National Park, South Dakota.
[b]Heaviest in sample of 510, Wichita Mountains Wildlife Refuge, Oklahoma.
[c]5.5-year-old female.
[d]Females were 25–30% smaller than males.
[e]Majority of animals in sample were adult females, Elk Island National Park, Alberta.
[f]Weight range of 11 adult (5+ years) cows, Santa Catalina Island, California.
[g]Mean seasonal plains bison (4+ years) weights for winter and fall, Badlands National Park, South Dakota.
[h]Body weight range in females aged 6 years and older (over 8 years), Wind Cave National Park, South Dakota.
[i]Mean weight of paturient females aged 6 years and older (over 8 years), Wind Cave National Park, South Dakota.
[j]Mean weight of nonreproductive females aged 6 years and older (over 8 years), Wind Cave National Park, South Dakota.
[k]Weight range of ≥7-year-old bulls in September at the Fort Niobrara National Wildlife Refuge, Valentine, Nebraska.
[l]Average weight for age class, Konza Prairie Research Natural Area, northeastern Kansas.
[m]Heaviest bull in herd, Konza Prairie Research Natural Area, northeastern Kansas.
[n]Heaviest cow in herd, Konza Prairie Research Natural Area, northeastern Kansas.
[o]Mean weight range from age class 4+ years to a maximum mean weight by age class, Elk Island National Park, Alberta.

The forequarters appear out of proportion to the slim hindquarters, a disproportion that is also accented by pelage differences between the front and the rear (Fig. 48.2). The legs are short, stout, clothed in shaggy hair, and end with large, rounded hooves (Banfield 1981; Meagher 1986; Pattie and Fisher 1999). Morphological studies of plains bison and wood bison show differences in cranial and skeletal characteristics (van Zyll de Jong 1986). Furthermore, phenotypic differences exist between plains bison and wood bison in the anterior slope of the hump, location of the highest point on the hump, angle of the hump, cape variegation and demarcation, chap hair on the front legs, frontal display hair on the head, and ventral neck mane and beard hair. These differences occur regardless of geographic location and environmental conditions, indicating that they are genetically controlled and not environmentally induced (van Zyll de Jong et al. 1995).

Size and Weight. Bison are the largest native terrestrial mammal in North America. Plains bison are smaller than wood bison, based on body mass. However, bison weights and measurements differ considerably by age and sex among different localities (Table 48.1). In studying the relationship of weight to chest girth, Kelsall et al. (1978) discovered that males were 9.1% heavier than females of equal chest girth in an approximately linear relationship. However, the maximum amount of variation for weight explained by chest girth in a study of live male and female bison in Badlands National Park, South Dakota, was only 64% and 33%, respectively (Berger and Peacock 1988). Although chest girth may be a good measure for approximating relative weight within a population, such information is difficult to collect, therefore, an alternative method for estimating weights is to rely on differences in head length or width or body length (Berger and Peacock 1988).

McHugh (1958) reported that plains bison bulls approach maximum size by 5–6 years of age, with small yearly increments for a few years thereafter. Banfield (1981) stated that male plains bison reached adult size at 6 years of age, whereas females attained maximum size at about 4 years. Weight data for plains and wood bison herds at Elk Island National Park, Alberta, have been sporadically collected over a 30-year period since 1962 and annually since the early 1980s providing data by which growth could be compared between the two subspecies and between males and females (Fig. 48.3). Male bison continued to grow until they were 8–10 years of age (Fig. 48.3A), whereas females reached mature weight at between 5 and 6 years (Fig. 48.3B). Wood bison females became markedly heavier than plains bison females after 3 years of age; male weights diverged after about 6 years, when wood bison males became markedly heavier than their plains counterparts, based on mean weights by age category.

During the routine, midwinter weighing of bison at Elk Island National Park, mean weights of wood bison were greater than those for plains bison (Table 48.1) for each sex and age category (Olson 2002). The asymptotic mean weight for mature plains bison males was 739 ± 10.0 kg, which was attained at age 8–9 years (Fig. 48.3A). The greatest overall mean weight for plains bison males of 769 kg was reached at 13 years of age. The asymptotic mature mean weight for wood bison males was 880 ± 15.1 kg and this was attained at age 8 years, similar to the plains bison (Fig. 48.3A). The greatest overall mean weight for wood bison males of 910 kg was reached at age 13 years. The asymptotic mean weight for mature plains bison females was 440 ± 2.1 kg and this was attained at 6 years (Fig. 48.3B). The greatest overall mean weight for plains bison females of 454 kg was reached at 10 years of age. The asymptotic mean weight for mature wood bison females was 540 ± 5.7 kg, exactly 100 kg greater than that for plains bison, and this was attained at 7 years of age (Fig. 48.3B). The maximum overall mean weight for wood bison females of 567 kg was reached at 12 years of age, 2 years older than when plains bison females reached their maximum weight.

A wide range of intrapopulation and interpopulation variability in body weight and growth occurs in bison (Berger and Peacock 1988). Although some studies have implicated age, sex, and season as variables (Halloran 1961; Lott 1979; Rutberg 1983; Olson 2002), other factors such as population density, nutrition, weather, reproductive effort, and inbreeding should be considered as possible influences at the individual and the population levels (Berger and Peacock 1988). Food as a limiting factor may lead to smaller animals in poorer condition. On Santa Catalina Island, California, where bison forage is limited, the largest plains bison mature cow weighed only 410 kg and was lighter than the smallest (427 kg) of 18 mature cows studied by Rutberg (1983) in the National Bison Range, Montana (Lott and Galland 1987). The average cow in the National Bison Range weighed 482 kg, compared to only 362 kg for an average Santa Catalina cow, a difference attributed to poorer nutrition on Santa Catalina Island (Lott and Galland 1987).

Pelage. The pelage of bison is composed of long, coarse guard hairs with a thick, woolly undercoat (Banfield 1981). The hair on the head, shoulders, and forequarters is long, shaggy, manelike, and dark brown to black, and abruptly becomes shorter and lighter brown behind the shoulders and on the hindquarters (Pattie and Fisher 1999). Chin hair usually resembles the shape of a goatee-type beard. The head is very dark, almost black, with little or no color contrast. There are usually two seasonal molts: one in early spring and one in late summer (Banfield 1981). Albino and gray hair colors are rarely observed in bison, with a speculated proportion varying from 1 in 100,000 to 1 in millions (McHugh 1972). Historically, the former type was held in great reverence by the Plains Indians. Newborn calves are reddish to orange brown, but this changes to the typical dark brown at about 2.5 months of age (Meagher 1978, 1986).

Skull. Skulls of male bison are larger and more massive and have longer, more pronounced horn cores and burrs than those of females (Skinner and Kaisen 1947). The muzzle is narrow with long, pointed nasal bones, which do not reach the premaxillae (Fig. 48.4, top right).

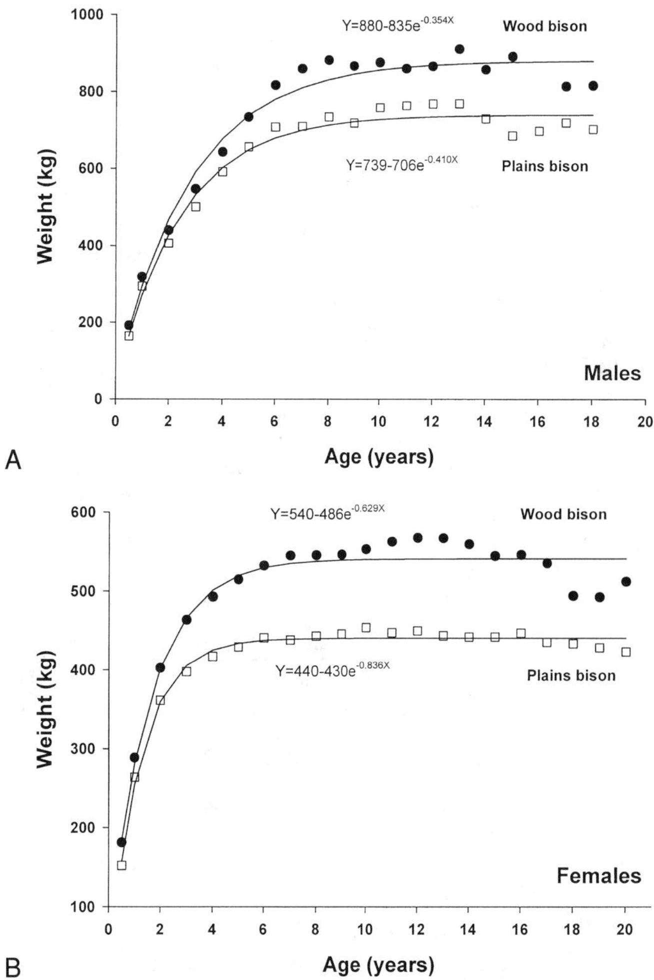

FIGURE 48.3. Growth of wood and plains bison (A) males and (B) females at Elk Island National Park. SOURCE: Average weight for age-specific data from Olson (2002).

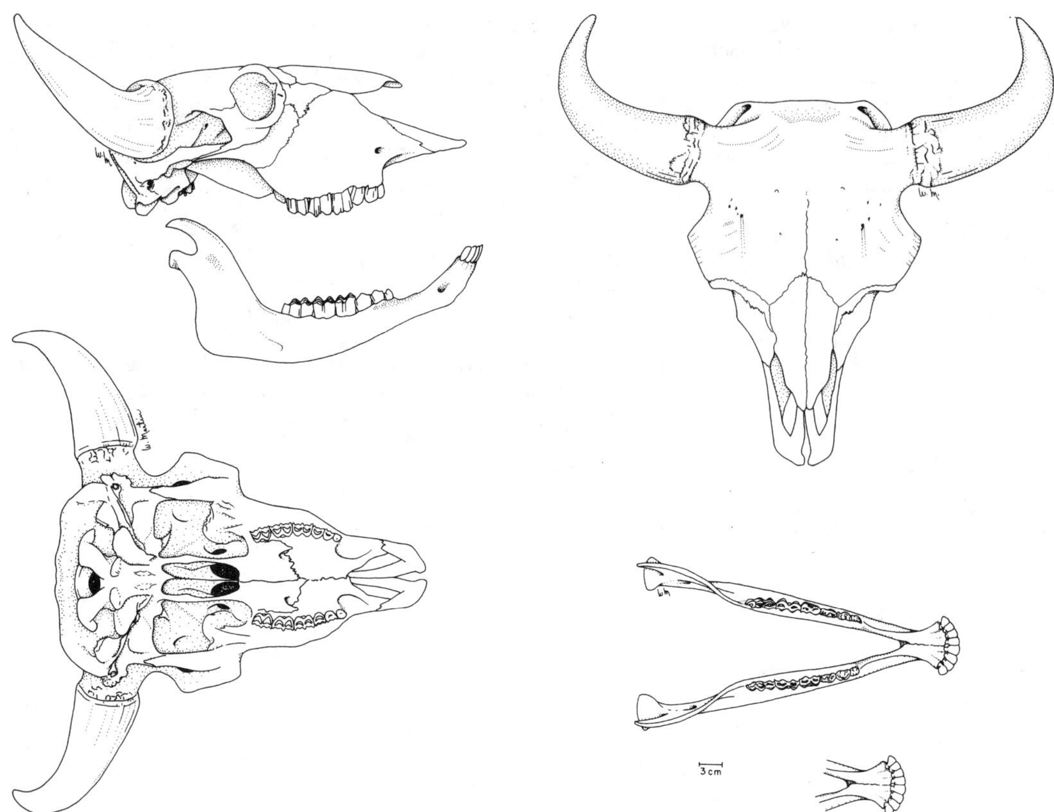

FIGURE 48.4. Skull of the bison (*Bison bison*). Top: (left) lateral view of cranium and mandible and (right) dorsal view of cranium; Bottom: (left) ventral view of cranium and (right) dorsal view of mandible and incisor bar.

Unlike cervids and antilocaprids, bison do not have preorbital vacuities in the skull. The length of 12 male bison skulls ranged from 500 to 600 mm and the width varied from 240 to 280 mm (Allen 1876). Overall length of 29 plains bison skulls varied from 491 to 570 mm, and the greatest postorbital width ranged from 271 to 343 mm (Skinner and Kaisen 1947). Hall (1981) reported that the greatest skull length of male plains bison ranged between 491 and 595 mm. Length for 81 specimens of adult male plains bison from Elk Island National Park, Alberta, ranged from 476 to 570 mm, and the greatest postorbital width ranged between 282 and 352 mm (Shackleton et al. 1975). The overall length for 33 adult female skulls from the park population varied from 425 to 500 mm, and the greatest postorbital width for 35 specimens ranged from 237 to 275 mm (Shackleton et al. 1975). Wood bison and plains bison skulls resemble each other; however, wood bison, generally, are larger (Meagher 1986). A few sets of measurements of male skulls for wood and plains bison, respectively, are as follows: maximum length, 562–604 and 500–583 mm; spread of horn cores, 542–848 and 510–778 mm; and least width at frontals between horn cores and orbits, 273–313 and 237–318 mm (McDonald 1981). Some mean measurements of male skulls for original wood bison (pre-1925), modern-day wood bison, and plains bison, respectively are as follows: basilar length, 531, 518, and 486 mm; spread of horns tip to tip, 701, 653, and 594 mm; and width of cranium between horn cores and orbits, 295, 285, and 265 mm (van Zyll de Jong 1986). Of 19 cranial variables measured by van Zyll de Jong, the single measurements that best discriminate between wood and plains bison skulls are basilar length, width of the cranium between horn cores and orbits, vertical diameter of the horn core at right angles to the longitudinal axis, and circumference of horn-core at right angle to longitudinal axis (van Zyll de Jong 1986).

Bayrock and Hillerud (1964) used the standard skull measurements of Skinner and Kaisen (1947) to describe three *Bison bison athabascae* skulls that had been collected in Wood Buffalo National Park, Alberta, in 1925. Two of these skulls exhibited measurements exceeding the maximum and the minimum for the subspecies, thereby appreciably extending the range of variation for cranial measurements of wood bison. Skinner and Kaisen (1947) concluded that horn cores of fossil bison appeared to provide the best distinguishing criterion once growth, age, and individual variation were taken into consideration. In contrast, horn core dimensions were the most variable of the craniometric measurements taken from 157 known-age skulls of plains bison (Shackleton et al. 1975). This suggested that further studies were required to determine intraspecific and interspecific variability of these and other skull characteristics (Shackleton et al. 1975).

Selected mean measurements for 6 of 23 postcranial variables for male bison for original wood bison (pre-1925), modern-day wood bison, and plains bison are, respectively: metacarpal length, 228, 226, and 208 mm; radius length, 361, 370, and 337 mm; proximal width of radius, 109, 111, and 98 mm; tibia length, 452, 447, and 411 mm; humerus length, 404, 404, and 368 mm; and femur length, 483, 476, and 430 mm (van Zyll de Jong 1986). These six measurements were the best univariate discriminators of the 23 postcranial variables to distinguish between wood and plains bison. The mean size of different postcranial elements between male and female wood and plains bison both averaged about 12.5%, indicating that sexual dimorphism of size of postcranial elements is quite similar in both subspecies (van Zyll de Jong 1986).

Dentition. Bovids have hypsodont premolars and no upper incisors (Fig. 48.4). The upper canines are reduced or absent (Fig. 48.4, top left). The selenodont molars of bison have a median style or enamel fold between the anterior and posterior lobes (Fig. 48.4, bottom right), which tends to disappear with age and wear (Skinner and Kaisen 1947). These molars are used to section and finely grind vegetation (Feldhamer et al. 1999). The dentition of bovids is phylogenetically advanced and probably developed in association with grazing habits. The dental formula for bison and all other bovids is I 0/3, C 0/1, P 3/3, M 3/3 (Soper 1964, Hall 1981).

Abnormal dentition, such as supernumerary teeth or missing teeth, is generally infrequent in bison. In five reported cases of dental anomalies, four from the Henry Mountains, Utah, and one from Delta Junction, Alaska, these herds were founded with few individuals, suggesting the influence of in-breeding and genetic drift (Van Vuren 1984a). The rate of tooth wear in northern wood bison was much lower than that from northern plains archaeological sites. This was attributed to a grittier diet and possible mineral deficiencies that may cause increased tooth wear in bison from southern, more arid habitats (Haynes 1984).

Horns. In bison, both sexes possess short, round, curved black horns, which rise laterally on the side of the head (Fig. 48.4, top left and right; bottom left), and with inward-curving tips, which taper to circular, relatively sharp tips (Banfield 1981; Meagher 1986; Pattie and Fisher 1999). The female's horns are more slender, whereas males have proportionately stouter, more evenly curving horns, which have cores that are burred at the base (Meagher 1986). Based on horn growth, Fuller (1959) recognized four age classes for female bison and five age classes for male bison: (1) calf, <1 year; (2) yearling, <2 years; (3) spikehorn, 2 to <4 years; (4) young adult, 4 to 7 or 8 years; and (5) adult and aged, greater than 7 or 8 years. The spikehorn stage was not recognized in classification of females. Counting annual growth rings on horns was not a useful criterion for age estimation. Only gross age classification is possible for free-roaming bison as determined by horn development and body conformation.

GENETICS

Cronin (1986) examined mitochondrial DNA (mtDNA) restriction fragments in Nyarling River wood bison and plains bison and concluded they could not be distinguished. With the notable exception of mule deer (*Odocoileus hemionus hemionus*) and black-tailed deer (*O. h. columbianus*), Cronin (1991) also was unable to distinguish between other subspecies within several cervid and bovid taxa. Strobeck et al. (1993) compared sequence divergence in a section of the D-loop in mtDNA of a small number of Nyarling River wood bison and plains bison. They found that differences between wood bison and plains bison are approximately the same as or less than those within the plains bison group. Their gene tree did not reveal monophyletic separation between wood and plains bison alleles. However, separate groups are not expected to have monophyletic mtDNA trees until $4N$ generations have passed (where N represents population size), assuming constant and equal population sizes. Thus, there is no reason to expect a monophyletic tree for bison mtDNA unless N is less than 250, assuming a generation time of 5 years and that wood and plains bison existed separately for the last 5000 years.

MtDNA is maternally inherited and therefore reflects the contribution of the maternal population that gave rise to Nyarling River bison and other bison populations in the vicinity of Wood Buffalo National Park. Because plains bison cows greatly outnumbered wood bison cows following the introduction of plains bison from 1925 to 1928 (Lothian 1981), it is not surprising that mtDNA in extant bison populations in this area is similar to that of plains bison. Furthermore, the rate of sequence divergence in mtDNA is 1–2%/million years (Wilson et al. 1985). Because divergence of wood and plains bison occurred only during the last 5000–10,000 years, there has been little evolutionary time for development of significant differences in bison mtDNA.

Molecular studies that have examined restriction fragment length polymorphisms (RFLP) of various nuclear genes have provided additional insights regarding the phylogeny of bison subspecies (Bork et al. 1991; Cronin and Crocket 1993; Morris et al. 1994). Bork et al. (1991) found significant differences in RFLP frequencies in 2 of 28 fragments in wood and plains bison. The low number of net nucleotide substitutions in these two populations suggests recent divergence, a finding that supports the idea that wood and plains bison existed as reproductively isolated populations during the last 5000–10,000 years, which is a relatively short time in evolutionary terms. However, protein-level methods detect little or no variation and usually only few differences when determining taxonomic differentiation of bison (Bork et al. 1991; Cronin and Crockett 1993; Stormont 1993).

A system more able to detect genetic variation uses DNA microsatellites, which are highly polymorphic nuclear markers. These have been employed to analyze genetic relationships among bison populations to address conservation genetics (Wilson 2001). Wilson and Strobeck (1999) investigated genetic variability, diversity, and heterogeneity in 11 microsatellite loci of genomic DNA in 11 public herds of North American bison. Wood bison herds originating in Wood Buffalo National Park formed a single group, although some subpopulations showed significant differences in allele frequencies. In particular, allele frequencies in the Pine Lake subpopulation were significantly different from those in the other subpopulations. This subpopulation also was identified by van Zyll de Jong et al. (1995) as being intermediate between wood bison and plains bison based on external morphology. Genetic distances between wood bison and plains bison populations were larger than those within either of the two subspecies populations. Wilson and Strobeck (1999) and Wilson (2001) concluded that genetic clustering of the three wood bison populations implies that they are functioning as a genetic entity, distinct from plains bison, and that wood and plains bison populations should continue to be managed separately.

Low levels of genetic variability have been reported in many bison herds (McClenaghan et al. 1990; Wilson and Strobeck 1999). The near extinction of both subspecies in the late 1800s may be the initial factor causing reduced variability. However, subsequent establishment of numerous populations from few individuals and prevention of gene flow among populations has resulted in further reduced genetic variation and diversity from founder effects and genetic drift. There is a positive correlation between the number of founders for a population and the genetic diversity within that population (Wilson 2001). Reduced variability resulting from the effects of small number of founders and genetic drift are present in many public herds. For example, the Elk Island National Park and the Mackenzie wood bison populations showed less genetic variation than their parent population in Wood Buffalo National Park, reflecting the small number of founders used to establish the former two populations. The most genetically diverse bison populations in North America are the Greater Wood Buffalo National Park Ecosystem and the Greater Yellowstone Ecosystem herds (Wilson and Strobeck 1999).

The karyotype of wood bison and plains bison is characterized by the same number of chromosomes ($2n = 60$) which resembles that of domestic cattle in that the diploid complement comprises 60 chromosomes (Bhambhani and Kuspira 1969; Ying and Peden 1977). Bison and cattle differ only in that the Y chromosome is telocentric in bison and metacentric in domestic cattle (Bhambhani and Kuspira 1969). This karyotype is nearly identical to that of the European bison or wisent, which is indigenous to Europe (Orlov and Chudinovokaya 1979). When G-banding patterns were compared for wood bison and plains bison, 20 pairs plus the sex chromosomes were found to be homologous and patterns for the 9 remaining pairs of chromosomes could not be distinguished (Ying and Peden 1977).

All living forms of bison are completely interfertile and lack intrinsic isolating mechanisms, which suggests that they are members of one species. Similarity of the karyotypes of bison and cattle justifies the grouping of bison and domestic cattle within a single genus (Bhambhani and Kuspira 1969). However, because bison and cattle are not completely interfertile, they should be considered as separate species (Gates et al. 2001b). For a more detailed explanation, see the section *Bos* Versus *Bison*. In conclusion, karyotype comparison suggests that these two members of the family *Bovidae,* domestic cattle and plains bison, are closely related species (Bhambhani and Kuspira 1969).

PHYSIOLOGY

Metabolism. Like other northern wild ruminants, bison exhibit seasonal variation in energy metabolism. Rutley and Hudson (2000) found that metabolizable energy intake was lower in pen and field trials with yearling bison during the winter than in the summer. In field trials, estimated metabolizable energy requirements varied from 146 ± 22 kJ $W^{-0.75}$/day in December to 478 ± 45 kJ $W^{-0.75}$/day in June, where W is weight in kilograms.

Bison and cattle respond differently to cold. Christopherson et al. (1979) observed that in winter at −30°C, the metabolic rate of bison (718 kJ kg^{-1} metabolic body weight/day) was less than that of cattle (830 kJ kg^{-1} metabolic body weight/day). At 10°C, the metabolic rate of bison (934 kJ kg^{-1} metabolic body weight/day) was greater than that of cattle (659 kJ kg^{-1} of metabolic body weight/day), indicating that in winter a temperature of 10°C exceeds the upper critical temperature of bison. An increase in metabolic rate is considered a normal response to cold exposure, although this response may be reduced by cold acclimation (Slee 1972). However, Christopherson et al. (1979) found that yearling bison decreased their metabolic rate from 748 kJ kg^{-1} metabolic body weight/day at 10°C to 584 kJ kg^{-1} metabolic body weight/day at −30°C. They attributed the reduced metabolic rate largely to a reduction in activity.

Bison calves are as cold tolerant at 6 months of age as Hereford yearlings are between 13 and 17 months of age (Christopherson et al. 1979). Yearling Herefords that were acclimatized to ambient spring temperatures increased their metabolic rate from 760 kJ kg^{-1} metabolic body weight/day at 10°C to 938 kJ kg^{-1} metabolic body weight/day at −30°C. The greater cold tolerance of bison compared to cattle can be attributed primarily to greater insulating pelage. Peters and Slen (1964) observed that in winter, bison had a greater weight of hair per unit area, a greater density of hair, and a greater fineness of hair than Hereford, Angus, or Shorthorn cattle. Dziurdzik (1978) observed a similar contrast between the hair of European bison and cattle. Cattle × bison hybrids (cattalo) were intermediate in these pelage characteristics and were more cold tolerant than cattle, resulting in a greater capacity for foraging on open range in the winter (Smoliak and Peters 1955).

Growth and Performance. Plains bison weigh 14–18 kg at birth (McHugh 1972). By 6 months of age, male bison calves are heavier than female calves (Towne 1999). At Konza Prairie (tallgrass prairie in Kansas), by 7.5 years of age, female plains bison weighed an average of 455 kg and males averaged 727 kg (Towne 1999). Male and female calves gained an average of 2.6% of their body mass over their first winter, but older bison lost an average of 11.3% of their body mass over winter. Winters were relatively mild during the study.

Bison have a lower rate of weight gain than cattle under conditions favorable for growth of cattle. When fed a complete finishing ration under feedlot conditions, male and female bison calves gained 0.64 and 0.50 kg/day, respectively, whereas Hereford calves gained 0.91 and 0.82 kg/day, respectively (Peters 1958). Cattalo calves were intermediate in performance, gaining 0.86 and 0.68 kg/day for males and females, respectively. Feed conversion efficiencies are higher for cattle (Young et al. 1977). The average gain from November to October of Hereford calves was 215 kg, compared to 160 kg in a female bison calf and 163 kg in a male bison calf. The ration was an alfalfa–bromegrass mixture. Christopherson et al. (1979) fed one male and one female bison calf and two male and two female Hereford calves a ration comprised of 40% alfalfa–brome hay and 60% concentrate at a rate of 100 g of feed/unit metabolic weight/day. Average daily gain of the bison calves over 1 year, approximated from graphically presented data, was 0.5 kg/day, whereas that of cattle calves was 0.9 kg/day.

Peters and Slen (1966) reported birth weights of calves surviving to weaning as 32.7 kg for Herefords, 31.5 kg for cattalo from dams less than one-fourth bison, 30.5 kg for cattalo from bison dams, and 26.4 kg for cattalo from F_1 hybrid dams. Thus, birth weight decreased with an increased proportion of bison in the dam. The rate of gain of calves was inversely correlated with the birth weight. Average daily gain of calves was 0.69 kg from Hereford dams, 0.74 kg from bison dams, and 0.77 kg for calves from F_1 hybrid dams. Keller (1980) calculated that the effect of percentage bison parentage of the dam on calf daily gain was negligible, despite the fact that it had a negative effect on total milk yield. Total milk yield decreased by over 10 kg for each 1% increase in bison parentage.

Reynolds et al. (1982) reported that the rate of gain of Hereford steers exceeded that of bison steers from approximately 4 months to 31 months under feedlot conditions. The average growth rate of six bison steer calves from September 1975 to November 1977 was 0.4 kg/day. The greatest rate of gain in the bison was 0.5 kg/day from March to June. Feed intake and feed conversion were 5.5 kg/day and 9.8%, respectively. The ration, fed ad libitum, contained 12.4% crude protein (CP), 68% total digestible nutrients (TDN), 0.7% calcium, and 0.4% phosphorus. The rate of gain, feed intake, and feed conversion of eight Hereford steer calves during this period that received the same ration were 1.1 kg/day, 9.9 kg/day, and 10.9%, respectively. When receiving a finishing ration of 13.4% CP and 72% TDN at approximately 20 months of age, cattle gained 1.1 kg/day and bison 0.4 kg/day. These data may not have been representative of maximum animal performance, because of the intensive handling of animals and the interspersion of experiments involving different rations. Hawley et al. (1981b) observed that bison steers gained 0.4 kg/day on sedge hay in the summer, whereas Hereford steers did not show any appreciable growth. When ration composition and intake only marginally met the nutrient requirements for growth of cattle, the rate of gain of bison exceeded that in cattle. Dry matter (DM) intake rates on a body weight basis were similar. Intake of digestible energy on a metabolic body weight basis was slightly greater for bison because of their greater ability to digest sedge hay.

Hawley et al. (1981b) observed a significantly ($p < .05$) lower DM intake of sedge hay and a numerically lower gain in bison steers than in cattle steers during winter. Inappetence, reduced metabolism, and reduced growth in the winter are common in wild ungulates (Wood et al. 1962; Silver et al. 1969; Ozoga and Verme 1970; Kirkpatrick et al. 1975; Westra and Hudson 1979), and are viewed as adaptive strategies developed to reduce nutritional requirements. Seasonal inappetence and reduced growth rate may depend on age and ration quality. Bison calves that were fed high-quality rations grew considerably in their first winter (Christopherson et al. 1979). Richmond et al. (1977) suggested that seasonal changes in growth and the stress of handling and confinement could have contributed to the relatively poor performance they observed in bison fed sedge and grass hays and the good performance of bison relative to cattle when fed alfalfa.

Carcass Characteristics. Bison have similar hot carcass weight as a proportion of liveweight (dressing percentage), slightly higher cooler shrink, and a higher proportion of saleable yield compared to beef (Aalhus et al. 1992). Bison dress from 55.7% to 61.9% of liveweight (Aalhus and Janz 2001). The range reported for beef carcasses is 58.6–61.8% for cattle raised on diverse feeding regimes (Aalhus et al. 1992). Shrinkage in bison carcasses during cooling ranges from 0.98% to 2.25% (Aalhus and Janz 2001), whereas shrinkage in beef carcasses ranges from 1.14% to 1.62% (Aalhus et al. 1992). The distribution of fat cover (finish) on bison carcasses tends to be uneven and is located mainly over the shoulders and loins (Hawley 1986; Koch et al. 1995). This provides less protection from evaporation for lean tissue than in beef carcasses, which have a more evenly distributed cover of fat. Aalhus and Janz (2001) reported that total bison carcass saleable yield was 78%. Similarly, Hawley (1986) reported a mean saleable yield of 77% from six bison 2.5 years of age at slaughter with an average liveweight of 444 kg. Aalhus and Janz (2001) and Koch et al. (1995) reported that total saleable yield from bison was greater than that from beef. Koch et al. (1995) found that bison had less fat trim in all cuts except in the rib section, where subcutaneous fat accumulates in bison.

The massive front shoulder of bison gives the impression that a large proportion of the body weight is in the forequarter. Peters (1958) compared carcass proportions between yearling bison and Hereford bulls, and reported a similar weight distribution in the forequarters (54% in bison and 52.5% in Hereford bulls). In contrast, Berg and Butterfield (1976) observed that the proportion of body muscle in the proximal pelvic limb area of a bison bull was greater than that in domestic cattle bulls, and attributed the difference to muscles connecting the neck to the forelimbs and in the hump. Koch et al. (1995) explained that the large dorsal spinous processes in the bison hump result in more meat in the shoulder region in bison than in beef. The exaggerated size of the forequarter in bison creates the appearance of a disproportionately small hindquarter. However, there is little difference between bison and beef hindquarter cuts (Berg and Butterfield 1976; Koch et al. 1995).

The flavor and texture of meat from identically finished bison and cattle is similar. Meat from carcasses of 2.5-year-old bison was compared to meat from A2, C1, and D1 grade cattle carcasses (Cox 1978); panelists could not tell the difference between bison and cattle meat. Bison meat has a similar shear force (tenderness), darker meat color, similar moisture and protein content, and lower intramuscular fat compared to beef (Aalhus and Janz 2001). In a controlled comparison of bison and beef, Koch et al. (1995) reported bison meat had a lower mean shear force and a greater acceptability rating for tenderness.

Digestion and Nutrition. Much of what is known about bison digestion and nutrition has been extrapolated from beef cattle requirements and digestive physiology. Applied information is generally available in books and government reports published for the commercial bison production industry (Feist 2000). Specific energy, protein, mineral, and vitamin requirements have not been developed for bison.

Characteristic of grazing ruminants, bison have a four-compartment stomach designed for efficient processing and digesting of a roughage diet made up mostly of grass and grasslike plants. The first two stomach chambers are the rumen and reticulum, which together form the forestomachs or reticulorumen. A bison's rumen is structured into broadly connected compartments or sacs, which retain forage for long periods of time for more complete microbial digestion. Rumination, or rechewing of food returned to the mouth as a bolus by facultative muscular contractions of the esophagus, reduces the size of food particles. As microbial fermentation proceeds in the rumen, forage is further reduced to smaller and smaller particles. Contraction of the muscular rumen wall constantly flushes suspended solids back into the rumen. Small, dense material is flushed into the cranial sac of the rumen and then into the reticulum.

The highly liquefied, microbe-rich fluid in the reticulum flows through the reticulo-omasal orifice into the omasum, or third stomach. The omasum absorbs residual volatile fatty acids and bicarbonate. Fluids pass rapidly through the omasal canal, but particulate matter is trapped. Material passes from the omasum into the fourth stomach—the abomasum. The abomasum is the glandular stomach, which secretes acid and digestive enzymes similar to the stomach of a monogastric. One important functional specialization of this organ that differentiates it from a monogastric stomach is the requirement to digest large masses of microbes produced in the reticulorumen. In a ruminant, the abomasum secretes lysozyme, an enzyme that efficiently digests bacterial cell walls.

The rate of passage of forage through the digestive system of a bison is constrained by digestion in the reticulorumen. Poor-quality feeds are broken down into small particles by microbes more slowly than high-quality forage, reducing the rate of passage from the forestomach. As a result, bison only feed four to nine times a day, consuming large quantities of forage per feeding, then resting and ruminating until the mass of ingesta in the reticulorumen is reduced.

Bison retain low-quality, high-fiber, low-protein forage in their digestive system longer and are more efficient in digesting it than are cattle (Hawley et al. 1981a; Hawley 1987). However, differences between bison and cattle are not as great for high-quality, high-protein feeds like alfalfa or alfalfa brome hay (Hawley et al. 1981b). The fiber level in alfalfa-based forages is typically lower than in grasses and sedges. The poorest quality ration used in comparative trials by Hawley et al. (1981a, 1981b) was native winter forages from northern Canada containing 6% CP content. The highest quality ration was alfalfa hay (18.7% CP). The weighted average DM digestibility values for native forages reported by Hawley et al. (1981a) were 52% and 39% for bison and cattle, respectively.

Digestion of slough sedge (*Carex atherodes*) also was compared in bison and cattle through total fecal collection (Hawley et al. 1981b). The CP and acid-detergent fiber (ADF) contents of the sedge were about 8% and 46%, respectively. All nutrients tested were digested to a greater extent by bison than by cattle; organic matter digestibilities averaged 55% and 49% for bison and cattle, respectively.

Hawley et al. (1981a) compared the digestion of five native forages in bison using a nylon bag technique. The forage samples ranged from 36.4% crude fiber (CF) to 70.1% CF. The mean DM digestibilities of forages for bison were as follows: willow (*Salix* spp.), 58%; slough sedge, 59%; baltic rush (*Juncus balticus*), 55%; aleppo avens (*Geum aleppicum*), 45%; and northern reed grass (*Calamagrostis inexpansa*), 46%.

Peden et al. (1974) used a nylon bag technique to measure the digestion of native forages grazed by bison and cattle on shortgrass prairie. Digestibilities were greater in bison than in cattle for fall and winter forages in which the CF content was high and the CP content was <7%. For spring and winter forage, in which the CP content exceeded 7% and the CF content was low, differences in digestibilities between bison and cattle were not evident.

Richmond et al. (1977) used acid-insoluble ash as an indicator to compare the digestion of alfalfa, grass, and sedge hays in bison, Hereford, and yak (*Bos grunniens*) yearlings. The CP and ADF contents of the alfalfa, sedge, and grass hays were, respectively, 18.7% and 30.5%, 8.3% and 39.2%, and 6.6% and 40.3%. Grass and sedge hays were digested more efficiently by bison than by cattle, but there was little difference in digestion of alfalfa.

Young et al. (1977) used chromic oxide as an indicator to compare the digestibility of a pelleted alfalfa–bromegrass mixture (15.1% CP, 25.1% ADF) in pairs of bison, yak, and three breeds of cattle. The greatest digestibility was observed in Holstein cattle, whereas Hereford cattle, Highland cattle, and bison displayed similar digestibilities.

Differences in nitrogen recycling or passage rate could explain differences in forage digestion between bison and cattle (Reynolds et al. 1982). When nitrogen in the rumen is limiting, greater recycling of nitrogen to the rumen may enhance microbial fermentation and hence forage digestion (Peden et al. 1974; Church 1988). Crude protein levels below about 6% appear to limit rumen metabolism in domestic ruminants (Gilchrist and Clark 1957; Glover and Dougall 1960). Peden et al. (1974) observed greater forage digestibilities in bison than in cattle only when CP levels were less than about 7% and suggested that these differences might be attributable to animal species differences in nitrogen recycling. This hypothesis has not been tested. The research of el Shazly et al. (1961) suggested that the availability of nitrogen markedly affects digestion of fibrous substances within the rumen.

Differences in nitrogen recycling presumably would produce different rumen environments and therefore different rumen microbial populations. Pearson (1967) observed that the rumen bacteria and ciliated protozoa from range-fed bison killed in fall were similar in kind and number to those found in domestic livestock. Reynolds et al. (1982) reported the results of an unpublished study in which only small differences existed in the number of rumen protozoa in bison and cattle receiving a high-quality finishing ration under feedlot conditions.

A lower rumen passage rate generally increases the digestion of fibrous feedstuffs (Church 1988) because feed is maintained in the presence of gastrointestinal microflora for a longer period of time. Young et al. (1977) observed forestomach retention times of 38.7 hr in bison and an average of 30.6 hr in three breeds of cattle. However, the longer retention time in bison was not manifested by greater forage digestion on good-quality ration. Dziuk (1965) concluded, based on sampling periods of 2–3 hr, that reticulorumen motility in bison was similar to that in domestic cattle. During in situ nylon bag experiments (Peden et al. 1974; Hawley et al. 1981a), samples remained within the rumen for equal lengths of time, thereby obviating the direct effect of the rate at which material leaves the rumen on the digestion of that material, and all forages tested were better digested by bison than by cattle.

In general, reducing DM intake increases the digestibility of fibrous foods (Schneider and Flatt 1975). In some instances, intake has been observed to be less in bison than in cattle (Peters 1958; Christopherson et al. 1976, 1978; Hawley et al. 1981b). Thus, greater forage digestion in bison might be associated with lower rates of intake.

However, greater digestibility coefficients were observed in bison than in cattle when intake rates did not differ significantly (Richmond et al. 1977; Hawley et al. 1981b).

Rice et al. (1974) estimated greater intake rates for bison than for cattle or sheep on native range. Forage grazed by bison was less digestible than that grazed by cattle, which, in turn, was less digestible than forage grazed by sheep. They suggested that intake was greater to compensate for a poorer quality ration. Because bison were less selective grazers (Peden et al. 1974; Rice et al. 1974), availability of acceptable forage may have been greater for bison than for the other, more selective grazers.

Galbraith et al. (1998) measured the digestibility of alfalfa pellets and methane and heat production in bison, wapiti (*Cervus elaphus*), and white-tailed deer (*Odocoileus virginianus*) in February–March and April–May. No difference in heat production or estimated energy requirements for maintenance could be detected between species, although animals numerically produced 40% more heat in April–May when feed intakes were higher than in February–March. Voluntary dry matter intake of native ungulates was higher in spring than winter. Methane emissions per unit feed consumed were highest with bison and least with white-tailed deer.

Bergman et al. (2001) developed a forage intake model for bison in which the two constraints were forage quality and availability. They tested the predictions of the model on wood bison grazing on naturally occurring slough sedge. In penned trials, digestibility and intake rate decreased with sedge biomass, which varied with phenology. However, short-term intake increased with biomass. The model predicted that daily energy gain should be maximized by grazing patches with a biomass of 10 g/m^2, whereas the minimum daily foraging time needed to fulfill energy requirements could be achieved by cropping patches with a biomass of 279 g/m^2. Observations of bison grazing mosaics of sedge patches varying in quality and quantity indicated that patches with a biomass $<$120 g/m^2 were avoided, whereas patches with a biomass of 156 and 219 g/m^2 were highly preferred, with the greatest preference for the latter. The results of the study indicated that bison behaved as time minimizers rather than energy maximizers when grazing sedge.

Post et al. (2001) observed differences in content and quality of the diet between age and sex classes of plains bison. They found that bulls had a significantly higher proportion of C_4 (warm season) grasses in their diet than cows, juveniles, or calves. Diets of calves were of higher quality than diets of bulls, cows, or juveniles. These results support the hypothesis that sexual dimorphism in body size allows for nutritional and habitat segregation between the sexes in large ungulates.

As with metabolism, digestion varies seasonally in bison. Rutley and Hudson (2000) reported that total digesta turnover time was longer in December (46.4 $\pm$ 1.4 hr) than in June (24.9 $\pm$ 2.7 hr). The difference was related to seasonal variation in transit time (18.2 $\pm$ 1.2 and 4.0 $\pm$ 0.8 hr, respectively) rather than differences in reticulorumen and lower gastrointestinal tract turnover time.

Blood and Urine Chemistry and Hematology. Marler (1975) analyzed jugular and tail vein blood samples from 77 bison in two herds in Kansas. No differences in chemistry or hematology were noted between sexes, but there were differences between bison $<$2 years old and those $\geq$2 years old. Most blood components measured were at levels comparable to those of domestic cattle, but hematologic values were generally greater in bison. The mean packed cell volume for all bison was 50% and the mean hemoglobin (Hgb) concentration was 17 g/100 ml, compared with values of approximately 35% and 11 g/100 ml for domestic cattle (Schalm et al. 1975). Mehrer (1976) observed a mean hematocrit of 47% and a Hgb concentration of 17 g/100 ml for 163 bison ranging in age from 1 to 5 years and sampled from five different herds in five states. The density of red blood cells, packed cell volume, and Hgb can increase with excitement (Searcy 1969; Swenson 1970). Because wild animals are likely to be more excited than domestic animals during sampling, the higher erythrocytic values of bison might be attributable in part to excitation. Haines et al. (1977) recorded an oxygen-carrying capacity for adult bison of 22.2 ml/100 ml blood and a Hgb concentration of 17.1 g/100 ml blood. These values were equivalent to values reported for some cervids and exceeded values reported for several domestic ungulates.

Blood urea nitrogen (BUN) was higher in bison than in cattle under a variety of ration and season conditions (Hawley 1978). Nitrogen recycling may be related to the level of BUN over a wide range of concentrations (Houpt 1970). A greater level of BUN could thereby contribute to greater recycling of nitrogen, and this has been suggested as one reason for the greater digestive capacity of bison on low-protein rations (Peden et al. 1974). However, BUN is not the only parameter influencing nitrogen recycling. Keith (1977) reported that rumen ammonia levels and salivary urea concentrations were more strongly influenced by factors other than BUN.

As in other ungulates (DelGiudice et al. 1989, 1991; Saltz and White 1991), urine chemistry has been used as an index of nutritional status in free-roaming bison, particularly for monitoring increases in endogenous protein catabolism during the winter (DelGiudice et al. 1994). Chemical analysis of urine suspended in snow provides the opportunity to collect large sample sizes necessary to study free-ranging populations in natural environments. In bison, urinary urea: creatinine ratios of 4 mg: mg or higher indicate either an unnaturally high-protein, high-energy diet or accelerated body protein catabolism in response to severe dietary energy restriction (Keith et al. 1981; DelGiudice et al. 1989, 1994). In Yellowstone National Park, urea:creatinine ratios increased over the winter on most bison ranges (DelGiudice et al. 1994). Urinary potassium:creatinine and phosphorus:creatinine ratios declined under nutritional deprivation. Low dietary sodium in winter forage was reflected by low sodium:creatinine values.

Means and standard deviations of serum glutamate oxaloacetate transaminase (GOT) levels reported by Marler (1975) for bison were 99 $\pm$ 8 mU/ml in adults and 128 $\pm$ 31 mU/ml in animals $<$2 years old. Keith et al. (1978) observed levels of 57 $\pm$ 25 to 121 $\pm$ 63 IU/L in adults. These values are somewhat higher than the normal range for cattle (Kaneko and Cornelius 1970). Serum GOT levels can increase markedly with rough handling of untractable animals, which could explain higher GOT levels in bison. Alkaline phosphatase was higher in young bison than in adults (Marler 1975). Higher levels in young animals also have been observed in pronghorn antelope (*Antilocapra americana*) by Barrett and Chalmers (1977) and in cattle and sheep by Kaneko and Cornelius (1970).

BEHAVIOR

Reproductive Behavior. During the rut (breeding season), there is a marked increase in herd size and activity. Activities of rutting bulls may include sexual investigation, exhibiting flehmen, tending cows, incomplete and fertile mountings, threat posturing, fighting, horning, wallowing, and loud vocalizations.

Mature bulls tend to form their own groups apart from the cow–calf herd. However, during the rut, they often enter the cow–calf herd to investigate cows sexually, and temporarily stay within the herd to tend a cow approaching or in estrus (McHugh 1958; Fuller 1960; Egerton 1962; Shackleton 1968; Shult 1972; Meagher 1973). Bulls methodically check herd cows by sniffing their vulvas and often prod resting females to stand for a more thorough examination (McHugh 1958). It has been suggested that stimulation by bulls may induce estrus in cows (Petropavlovskii and Rykova 1958).

Flehmen refers to a reflexive facial expression manifest by bison and many other ungulates (Egerton 1962; Geist 1963; Alexander et al. 1974; Mahan et al. 1978). It often occurs during the rut when a bull sniffs and/or licks the vulva or urine of a cow; however, bulls, females, and immature animals will initiate flehmen over other odors. During flehmen, the upper lip is curled upward and the neck is extended, an expression that may last several seconds. It is believed that flehmen makes the vomeronasal organ more effective (Estes 1972). A cow's urine on the ground, a bloody wound, amniotic fluid, rotted skeletons, bison hair, new calves, and human urine may also stimulate flehmen (Egerton 1962; Herrig and Haugen 1969; Shult 1972; Lott 1974).

Tending is defined as a temporary bond between a cow and a bull (McHugh 1958), which can last from a few minutes to several days. In consideration of the short-term nature of the tending bond, Seton (1929) and Soper (1941) described the mating system in bison as polygamous, whereas McHugh (1958:24) referred to it as "temporary monogamous mateship." Atypical tending bonds, where bulls tended cows for short time periods or tended calves or young bulls, also have been observed. In the latter case, McHugh (1958) observed a bull tend a yearling for approximately 4 hr and attempt mounting with penis unsheathed.

While tending, the bull usually tries to keep the tended cow peripheral to the main herd by keeping himself between the herd and the cow. A bull will occasionally use considerable force to keep the cow sexually isolated (Lott 1974). Close proximity of other males is not tolerated. Female bison appear to actively participate in mate selection. Wolff (1998) reported that cows often approach high-ranking bulls, but attempt to run away from low-ranking bulls. Tending by adult males appears to be a more reliable indicator of estrus in females than when subadult males are observed tending females (Komers et al. 1994). Wolff (1998) noted that bulls attempted to mate with any female in estrus.

Immediately before copulation, bison pairs may engage in amatory behavior such as mutual licking and butting. The cow may also attempt to mount the bull. A cow mounting another cow or mounting the tending bull is often a sign of estrus in domestic cattle (Dukes 1937; Schein and Fohrman 1955). The bull indicates his intention to mount by swinging his head up onto the rump of the cow. He next rears up to embrace the lower ribcage of the cow with his forelimbs and follows with penetration. The cow and bull may start to run while copulating. Insemination is usually achieved on the first few thrusts, after which the cow often displays "servicing symptoms" (Jaczewski 1958), which may last up to several hours (McHugh 1958; Shult 1972). Typically, the cow arches her back and holds her tail at some angle to the body. Urine and/or semen are often voided. Most breeding is done by "prime bulls," generally those animals between 6 and 9 years old (Egerton 1962; Lott 1974). Copulation is usually a crepuscular or nocturnal activity, a behavioral phenomenon that may enhance physiological performance, such as increased semen viability, or coincide with a time when the animals are less conspicuous to predators.

Wood bison herds tend to form smaller groups than plains bison and herd size declines during the rut as opposed to large herd sizes observed for plains bison (Soper 1941; Melton et al. 1989). In this manner, smaller herd sizes are more controllable by fewer dominant males which may result in a different mating system between wood and plains bison (Calef and Van Camp 1987). It is not evident whether there is a genetic basis for these differences in group size between wood and plains bison or whether it is a response to environmental influences of differing habitats (Melton et al. 1989). Small habitat patches in northern boreal forest areas are less likely to be grazed by large herds of bison. Furthermore, wood bison appear to be more solitary during the rut than plains bison and the greatest interactions occur when a new lone male wanders into a cow herd and clashes with the existing dominant male (Melton et al. 1989). Wood bison may use a harem formation system, whereby a small harem is formed which is defended against intruders, versus the dominance hierarchy system employed by plains bison (larger groups which are more difficult to defend) indicating that a smaller proportion of wood bison males should be reproductively successful each year (Wilson et al. 2002). However, dominant males within harems do change throughout the breeding season (Komers et al. 1992b).

Calving Season. Immediately before parturition, the behavior of the cow and nearby members of the herd often changes. McHugh (1958) observed that cows close to parturition became restless and excitable as well as exhibited marked physical changes such as viscous, mucous discharge from the vagina, swelling of the vulva into a heart-shaped flaccid mass, and filling of the udder. Before calving, cows often wander away from the herd for 1 or more days, although calving may occur within the herd (Audubon and Bachman 1849; McHugh 1958; Egerton 1962). Lott (1991) reported that cows in open country choose to calve within close proximity to other herd members, whereas in habitat with more tree and shrub growth, cows generally give birth away from the herd and take advantage of the available cover.

Bison usually give birth while lying down. The amniotic membranes, portions of the umbilical cord, and the placenta often are eaten by the cow following birth of the calf (McHugh 1958; Egerton 1962; Mahan 1978). Egerton (1962) noted that cows licked their calves frequently for several hours postpartum and that such licking seemed to stimulate activity in the calf. Licking also may serve to dry and warm the calf and thus lessen the stress imposed by harsh climatic conditions. Calves have been observed standing within a few minutes (Egerton 1962) to 85 min (Mahan 1978) following birth.

Suckling behavior appears to be the first directed action of the newborn calf (Egerton 1962; Mahan 1978), and may motivate the calf to stand and gain mobility. First suckling may take anywhere from 12 to 95 min to be initiated (Mahan 1978).

Calves usually nurse by standing parallel to the cow and facing her posterior. Suckling periods last an average of 6.3 min (Mahan 1978) and range up to 10 min (McHugh 1958). Suckling periods are usually short and erratic in newborn calves and are similar for those older than 3 months (McHugh 1958; Egerton 1962; Mahan 1978). Disturbance near a calf often induces suckling behavior (Egerton 1962). Yearlings may suckle occasionally (Hornaday 1889; McHugh 1958; Egerton 1962; Mahan 1978), although most bison are weaned before this stage. McHugh (1958) reported weaning within 7–8 months, whereas Mahan (1978) believed it occurred within 9–12 months. Green et al. (1993) reported that nursing may last between 12 and 24 months. Mature bison cows do not force wean their calves. Older cows tend to nurse their young longer and are less likely to terminate nursing bouts than are young cows despite the probability that young cows likely produce less milk during their first few lactations (Green 1986).

Cow–calf pairs maintain their closest contact during the first week after parturition, a pattern that tends to be reduced in subsequent weeks. Calf independence and distance separation from the maternal cow increase with age and are accelerated by the presence of other conspecifics and herd members (Green 1992). This independence may potentially increase the feeding efficiency of the cow and the resting time for the calf, the former being particularly important where range productivity may be limited (Green 1992). Lott (2002) noted that cohesion between bull calves and their mothers was less than that for heifer calves and that the mother—daughter bond or close association appeared to last longer on higher quality range. McHugh (1958) believed that, during the first year of life, cohesion between the calf and the cow was sufficiently evident to identify them during most periods of the day. Recognition between the cow and the calf may include the use of scent, sight, and/or sound, although calves occasionally may follow the wrong cow.

Herd members may focus considerable interest on new cow–calf pairs shortly after parturition and occasionally will come to sniff and lick the newborn calf (Egerton 1962; Engelhard 1970). However, cows will not hesitate to defend their calves against intruding bison or other animals. Egerton (1962) and Mahan (1978) observed that cows usually try to keep themselves between their calf and other herd members. The ability of a cow to defend its calf from investigation by other bison may depend on the position of the cow in the dominance hierarchy (Egerton 1962).

McHugh (1958) observed cows defending their calves by quick charges or slow advances when confronted by other species. Similar defensive behavior was reported by Hornaday (1889) and Garretson (1938). Carbyn and Trottier (1987) observed that calves tended to gather around cows when under attack. They also noted that when groups of bison with calves were being pursued by wolves, the calves tended to be positioned in a group more toward the front half of the herd than the rear. This behavior was suggested as reducing the likelihood of calf mortality. A cow occasionally will abandon its calf when the calf drops behind after a long chase. When the precipitating disturbance has stopped, the cow usually returns in search of its calf (Seton 1929; Soper 1941; McHugh 1958).

Protection of calves may be shared by other herd members. At Wood Buffalo National Park, Alberta, a small, mixed herd of bison defended a calf against a wolf attack for 36 hr; however, defensive response to wolf predation overall was observed to be quite variable

(Carbyn 1998). Although he never observed a bull conspicuously defend a calf, McHugh (1958) did observe one incident where a mixed herd, including two older bulls, had clustered around a corral housing a lone calf and could not be chased away. Several instances have been reported where calves as young as 2 days old tried to defend themselves (Hornaday 1889; Inman 1899; McHugh 1958). Bison calves also occasionally hide in foliage as a defensive behavior (Allen 1876; Grinnell 1904; McHugh 1958).

Play in bison, as in other mammals, appears to occur with a frequency inversely proportional to age. It is manifest by seemingly purposeless frolicking, including chasing, battling, butting, mounting, kicking, and racing. To the casual observer, the motivation to engage in play appears to be "for the sake of the activity itself"; however, its inverse relationship with age would, for example, also hasten muscle development and coordination essential in later life.

Agonistic and Dominance Relations. Both sexes and all age groups of bison may engage in threat displays and fighting. The most frequent and dramatic participants are usually bulls 4 years of age (McHugh 1958; Fuller 1960; Lott 1974). Although agonistic behavior may occur at any time of year, it is much more common during the rut when herds are larger.

Threat postures may be the prelude to fighting, although they are usually sufficient to terminate an encounter before serious physical contact. Such postures include elevation of the tail, broadside threats, pawing and wallowing, aggressive advances and lunges, and the nod threat (McHugh 1958; Lott 1974; Komers et al. 1992a). During the broadside threat, one or both animals stand broadside to each other, presumably to display their size, disposition, and intent. Aggressive advances involve one or both animals approaching the other using a slow foot-by-foot walk, which may, if they do not result in displacement, lead to physical exchange. Following such advances and between fights, the pair of combatants may bob their heads up and down in what Lott (1974) described as the nod threat.

Fighting may involve butting, horn locking, shoving, and hooking. The thick cushion of hair on the head helps to reduce the impact from butting (McHugh 1958). Hooking can result in serious injury. In this respect, bison differ somewhat from other North American bovids and cervids. The frequency with which serious injury occurs suggests that bison have better perfected offensive strategies than defensive ones. Bison occasionally take advantage of exposure of the flanks of an opponent and, rather than resume head-to-head combat, follow through to gore the animal in the side and belly. Such goring may result in broken bones, lacerations, punctured organs, general trauma, and death (McHugh 1958; Lott 1974). Approximately 50% of the bison carcasses examined at a herd-reduction slaughter at Elk Island National Park during the winter of 1971 showed evidence of healed previously-fractured ribs. The majority of these were assumed to be a direct result of fighting.

Fighting between sexes or involving more than two animals may also occur (McHugh 1958; Lott 1974). Bulls may temporarily stop tending a cow to cross the herd and fight. Disturbance by an external source such as an approaching vehicle or placing animals in confinement may also provide the stimulus for fighting.

Submissive display can include turning and/or running away, backing up with head swinging side to side, and sudden resumption of grazing and tail wagging. The victor also may attempt to mount the loser (Lott 1974; Meagher 1978; Komers et al. 1992a).

Agonistic behavior was observed by Coppedge et al. (1997) following the release of 43 bison calves without their mothers into a 288-member resident bison population. They reported significantly more aggression against these calves than between any other age class in the resident population, with most aggression attributable to the resident yearling bison.

Rothstein and Griswold (1991) observed that yearlings, but not calves, overtly differentiated between sexes. Yearlings showed more aggression toward male conspecifics and initiated more olfactory investigation of females than did calves. They suggested that the experiences gained during this period provide important social training for bison as they mature.

Expression of intraspecific dominance in bison appears after the first few weeks of life (McHugh 1958; Mahan 1978). McHugh (1958) found no correlation between the position of dominance of the calf and its seniority in the calf group, morphological differences, or the mother's position in the dominance hierarchy (derived dominance). Male calves tend to dominate female calves, as is the norm for older animals.

It appears that dominance among calves is primarily a function of inherent disposition. Early disposition may predispose an individual bison to develop specific physical and behavioral attributes significant in maintaining or advancing dominance position later in life. In a 9-year study at Wind Cave National Park, Green and Rothstein (1993a) observed that an individual bison's subsequent dominance, growth, and reproductive success typically was enhanced for earlier-born individuals. At the National Bison Range in Montana, neither age nor weight was correlated with dominance in mature bison bulls; however, among females, age was positively correlated with dominance (Lott 1979; Rutberg 1983). Similar findings were reported by Wolff (1998) at Fort Niobrara Wildlife Refuge in Nebraska, where dominance was not correlated with age for bulls in the 7- to 13-year-old age classes nor was it correlated with size for bulls >750 kg.

McHugh (1958) observed that dominance expressions by bison at Jackson Hole Wildlife Park were either "passive" or "aggressive." Passive dominance did not involve the use of force, whereas aggressiveness involved the use of force or threat. Of the 1027 dominance interactions observed, 73% were passive and 27% were aggressive.

After 3 years of studying a herd of 14 bison, McHugh (1958) did not observe any permanent reversals in hierarchial position between the herd's two mature bulls and among the seven most dominant cows. At the National Bison Range, Lott (1974) observed that 12% of aggressive interactions resulted in hierarchial reversals in a breeding herd containing 35 mature bulls. A number of dominance triangles also were observed. Instability in dominance relationships may be a function of habitat attributes, size of age class, and fatigue (Lott 1974). Expression of dominance tends to be more intense between animals whose positions are close together in dominance hierarchy.

Bulls, and cows with calves, are the most prone to be involved in intraspecific interaction (Egerton 1962). Almost any desired resource can elicit expression of dominance. The expression of dominance may occur as a result of competition for some obvious resource or may be precipitated by sudden disturbance. Bull groups at Elk Island National Park, Alberta, have been observed to initiate fighting, mounting, and horning when vehicles stopped near them. Similar occurrences have been observed in Wyoming (McHugh 1958) and Wood Buffalo National Park (Fuller 1960).

Horning and Wallowing. During the rut, bulls frequently horn trees including pine (*Pinus* spp.) or spruce (*Picea* spp.) (McHugh 1958). Rutting bulls at Elk Island National Park thrash and horn shrubs and saplings of several species. Following horning, the trees may be used for rubbing and/or simply uprooted. Rubbing posts are frequently sought by both sexes throughout the year. Horning also has been observed in cows just before parturition (McHugh 1958). Although the activity is not restricted to calving and rutting seasons, the increased frequency at those times suggests that it could be important in physically conditioning the animal for particularly stressful periods.

Wallowing is practiced by both sexes and all age classes of bison. Wallows are usually in dry sites, although wet, muddy wallows may be used. Wallowing may have a role in grooming, sensory stimulation, alleviating skin irritations, and reproductive behavior. As with horning, wallowing may help precondition the animal for periods of physical stress. Dust, which packs into the hair as a result of wallowing, appears to minimize the effect of biting insects (Lott 1974). Bulls wallow more frequently during the rut and occasionally urinate in the wallow before engaging in the activity (McHugh 1958). Studies in Alaska have indicated that such behavior is linked to priming of estrus in females (Bowyer et al. 1998). Urine odor may advertise the physical condition of the rutting bull. Lott (1974) suggested that the odor of urine from a bull may permit it to use preestablished dominance relations more effectively in the dark, a tenable hypothesis, considering the crepuscular nature of rutting activity and the highly developed olfactory sense of bison.

Vocalization. The repertoire of sounds made by bison includes soft to loud grunts, bleats, roars, snorts, sneezes, foot stamping, and tooth grinding (McHugh 1958; Fuller 1960; Lott 1974; Gunderson and Mahan 1980). Calves bleat and issue piglike grunts in response to grunting by or separation from the dam, when playing, and in response to other stimuli. When searching for a calf, cows often snort or give a loud grunt similar to their "threat grunt." Bulls are prone to giving loud, lion-like roars or bellows, particularly during the rut (Shult 1972; Lott 1974; Meagher 1978; Gunderson and Mahan 1980). Such roars may be audible from 5 km (McHugh 1958) to 16 km (Audubon and Bachman 1849). Bison bulls at Badlands National Park in South Dakota bellowed less frequently following copulation and more frequently on days when females were in estrus (Berger and Cunningham 1991). In this same study, bulls did not bellow unless other bulls were present, and smaller bulls tended to bellow more frequently than large bulls. Berger and Cunningham (1991) concluded that bellowing is more likely an intrasexual display than a display to attract females.

The roar produced by bison is the result of a single forceful exhalation over the vocal cords, in contrast to the two-way system in domestic cattle (Gunderson and Mahan 1980). Bulls often roar while tending cows, before fighting, while moving through or approaching bull subgroups or mixed herds, in answer to a roar from another bull, and less commonly at other times, for example, when loafing, when disturbed by vehicles, and in response to imitated roars or distant thunder. Occasionally, cows with newborn calves will roar when approached. Bulls also use snorts and foot stamping as part of their agonistic behavior. McHugh (1958) observed bison producing a squeaking noise by grinding their teeth. Bison are usually more vocal during herd movements, as are domestic cattle.

Tail Posture. Bison tail posture appears to reflect their social state and likely forms part of bison "body language" used to convey information to other herd members. Tail postures of free-ranging wood bison vary from tail wagging during grazing to holding the tail horizontally or vertically during periods of sexual behavior, aggression, or danger (Komers et al. 1992a). During fights between bulls, "tail up" was associated with dominance, whereas tail wagging was associated with submission. Cows more often held their tail up when disturbed by predators; bulls more often held their tail up during sexual encounters. As previously discussed, cows hold their tail out at some angle to the body for up to several hours following copulation.

Disposition. The disposition and approachability of individual bison and of herds are a function of the many environmental, genetic, and sociological conditions impinging on the individual. Bison have been described as having personalities (Shinn 1978). Although bison have been trained to do tricks and pull carts, they become more aggressive and intolerant with maturity and should be treated with considerable respect. "Pet" bison, once mature, have been known to kill their owners (Garretson 1927). At Elk Island National Park during the late 1970s, a park warden, who was experienced in working with bison, was gored severely by a bull bison after releasing it from a squeeze chute. The attack came without any warning and minimal threat posturing. Many similar attacks have been reported (McHugh 1958, 1972). During handling, bison can become enraged and can inflict serious damage on other animals, on themselves, and on property. At other times, these animals can be docile and shy.

Movements and Migration. Bison often undertake annual migrations that may be elevational or directional. In the American Northwest, Garretson (1927) noted that there was a definite movement of herds from the plains region in the east to the foothill areas of the Rockies in the winter. The reverse occurred in spring. Regarding these migrations, Hornaday (1889:423–24) stated that "the buffalo had settled migratory habits.... At the approach of winter the whole great system of herds which ranged from the Peace River to the Indian Territory moved south a few hundred miles, and wintered under more favorable circumstances than each band would have experienced at its farthest north." In mountainous areas of Wyoming (Aune et al. 1998) and in northeastern British Columbia, seasonal movements from higher elevation habitats in summer and fall to lower elevation habitats in winter and spring are common. Snipe flies (*Symphoromyia* sp.) may be responsible for some elevational movements in Yellowstone bison herds during the summer (Meagher 1973). Large, wind-swept prairies also may be chosen in summer for similar relief. Directional movements occur annually at Wood Buffalo National Park (Reynolds 1976) and at Yellowstone National Park (Meagher 1978). Bison, particularly cows, show strong affinity to return to traditional winter range (Meagher 1973).

In studies of bison home range size in northern Canada, the median home range size varied with age, sex, and forage availability (Larter and Gates 1994). Median home range size for young of the year (712 km^2), immature males (706 km^2), and adult females (1240 km^2) on lower productivity range was significantly larger than median home range size for two age classes of mature bulls (younger, 434 km^2; and older, 170 km^2) and as well as for adult females on more productive range (398 km^2). This difference may relate to differing energy requirements or other behavioral adaptations. In Yellowstone National Park, adult female home range size averaged 541 km^2 (Aune et al. 1998). In an unconfined setting, the home range of wild bison varies with habitat productivity. Temporal and spatial variations in range use are likely related to such factors as tradition, forage availability and nutritional quality, macroclimatic and microclimatic variations, open water, shelter, and insect harassment. An unusually severe winter in 1975/76 in Yellowstone National Park appeared to provide the impetus that led to the major westward movements or stress dispersal of bison on the northern winter range (Meagher 1989b).

Bison usually travel via the most practical direct route and will rapidly establish trails to do so (Garretson 1927; McHugh 1958). Forest and shrub areas often are used as daily and seasonal travel corridors in northern Canada (Reynolds 1976). When crossing rivers, bison usually take shallow fords with gradual approaches. However, they will not hesitate to cross large, swift-flowing rivers in northern Canada, such as the Peace, Liard, and Nahanni. Distance traveled on a daily basis can vary from <1 km to considerably more. Carbyn (1997) documented a herd of bison in Wood Buffalo National Park moving 81.5 km in a 24-hr period following wolf predation on a calf. In Yellowstone National Park, use of a plowed road in winter for easier and energy-efficient travel as well as acquired knowledge of areas having less snow appeared to be a stimulus that strongly influenced bison movements (Meagher 1989b). The attraction of bison to road corridors in northern Canada has lead to animal-vehicle collisions in some instances and likely has affected dispersal and movement patterns.

During inclement weather, bison often head into the wind, unlike domestic cattle, a behavior perhaps related to the greater amount of hair insulating their head and forequarters. If unable to avoid traveling in deep snow, bison will form a line, with the lead animals plunging to create deep trenches (Meagher 1973).

Older cows tend to be the most wary and often take the lead in sudden herd movements (McHugh 1958; Fuller 1960). In the family group, the cow is usually the leader (Walker et al. 1975). At Elk Island National Park and Wood Buffalo National Park, cows generally are the leaders during herd movements, although some observers have ascribed this role to bulls (Seibert 1925; Soper 1941).

Disturbance of a wary group of bison within or near a herd may precipitate sudden movements or stampedes. Bison are capable of fleeing at speeds up to 60 km/hr (McHugh 1958; Fuller 1960). The initiating stimulus can be seemingly insignificant, making the stampede appear spontaneous. Stimulus to stampede may be caused by the sudden running of one animal toward the herd after being alarmed or by any other sudden external stimulus.

Encounters with fences can act as a deterrent to the movement of bison. Buck-and-pole style fencing virtually stops bison and elk movement while being much less of a barrier to pronghorn antelope and mule deer (Scott 1992). In general, efforts, such as hazing or herding, physical barriers (short fences and cattleguards), and scare devices, made over a 12-year period to contain bison within the boundaries of Yellowstone National Park proved to be ineffective (Meagher 1989a).

TABLE 48.2. Size and composition of matriarchal and breeding groups at Lamar and Hayden valleys in Yellowstone National Park, Wyoming, and at Wind Cave National Park, South Dakota

Item	Group Location and Season			
	Lamar (Jan–Mar)	Lamar (May; Calving)	Wind Cave (Calving)	Hayden (Rut)
Group size				
Number of groups	18	15	17	36
Mean number/group	23	23.6	21.9	175.3
Standard deviation	11.4	18.4	21.2	108
Range	10–50	4–63	3–76	19–480
Group composition				
Number of groups censused	12	10	14	3
Mean number/group	20.3	17.3	16.8	115.2
Mean grouping tendency				
Cows ≥2 years old	12.1[a]	8.6[b]	7.4	39.3
Yearlings	4.4	4.8	4.0	20.0
Calves	0	3.9	3.9	25.3
Bulls 2–3 years old	3.8	3.7	1.1	16.3
Bulls ≥4 years old	0	0.2	0.4	14.3
Percentage of bulls 4 years old	0	1.0	2.1	9.2[c]

SOURCE: After McHugh (1958).
[a] 19.3% 2-year-olds.
[b] 19.5% 2-year-olds.
[c] Computed from census of eight groups.

The conclusion was that hazing and herding techniques served to move bison only where they wanted to go and may treat immediate problems at specific locations, but did not change the overall direction of bison movement through the river corridor (Meagher 1989a). When primary routes were blocked by human activities, the bison moved around barriers by crossing steeper terrain or traveling along tributary drainages (Meagher 1989b). Page wire game fencing is used routinely in the bison farming industry to contain captive bison herds. Commercial bison ranches have also had success with five-strand barbed wire fences and electric high-tensile fences (Deshano 2002). Animals that have learned to break fences, often bulls, can subsequently become much more difficult to restrain (Grandin 1999).

Aggregation. Bison are gregarious animals, but group size varies. Over the course of a year, three types of groups can be observed: matriarchal groups (cows, calves, yearlings, and sometimes a few older bulls), bull groups (including solitary bulls), and breeding groups (a combination of the first two groups).

Individual matriarchal groups vary little in size for most of the year. During the rut, they are joined by breeding bulls and other matriarchal groups to form breeding groups (Table 48.2). In Wood Buffalo National Park, Fuller (1960) suggested that the average group size for matriarchal groups ranged from 11 to 20 individuals among years and that there was considerable flexibility in the size of groups. In a study of bison group dynamics on a predator-free, forested summer range in southern Utah, Van Vuren (1983) noted that group size tended to be small, similar to what was seen for bison in other forested locations, but found no evidence of a basic social unit of the size reported by Fuller (1960). Van Vuren (1983) observed that group size varied with habitat openness, with larger groups associated with more-open habitats. He noted that bison usually grazed 3 m or more apart and that this behavioral phenomenon could be associated with smaller group size in small open habitats. Social groups include nursery groups composed of cow–calf pairs, calf groups, bull groups, barren cows, and yearlings, accompanied by some 2-year-olds (McHugh 1958). Nursery groups with 100–200 individuals have been reported in Yellowstone National Park (Aune et al. 1998). There is some controversy over whether matriarchal groups are consanguineous; Seton (1929) and Soper (1941) suggested that they are, whereas Garretson (1938), McHugh (1958), and Fuller (1960) believed otherwise. Green et al. (1989) showed a more prolonged association of cows with older daughters, which lends some support to the suggestion of Soper (1941) and Seton (1929). Aune et al. (1998) noted that, in general, group fidelity was temporal and that associations between marked bison suggested a very fluid and dynamic group structure.

Mature bulls seldom form groups of more than a few animals and seem to become less gregarious with increasing age. Solitary bulls are common even during the rut. Occasionally, one or two cows are found with bull groups.

McHugh (1958) reported that breeding herds in Yellowstone ranged in size from 19 to 480 and averaged 175 animals (Table 48.2). Egerton (1962) reported that the size of herds in Wood Buffalo National Park varied from 11 to 30 animals. At the National Bison Range, Lott (1974) observed an average group size during the rut of 57 with a maximum of 174, whereas on Catalina Island, California, he found an average breeding group size of 17. Shackleton (1968) observed smaller breeding groups at Elk Island National Park than on the National Bison Range.

Bison herds often show a remarkable degree of herd fidelity even after temporarily mixing with other herds (Fuller 1960). Fences separating groups of animals will not discourage their attempts to form a single herd. McHugh (1958) and Meagher (1973) observed that disturbance by aircraft and certain physical phenomena such as shade and rain would increase herd cohesiveness.

Foraging Behavior. McHugh (1958) observed that feeding activity was mostly a diurnal occurrence, with the night spent loafing or occasionally feeding and traveling. Similar behavior has been reported in cattle (Hancock 1953). The amount of time spent foraging appears to be seasonally influenced. Hudson and Frank (1987) observed that during summer, a greater proportion of grazing occurred at night. Study of bison foraging ecology and behavior in the aspen boreal habitat of the Beaver Hills in central Alberta documented an increase in the amount of time bison spent grazing from 9 hr/day to 11 hr/day from summer through fall (Hudson and Frank 1987). On the Slave River Lowlands, Reynolds (1976) observed that foraging was done in open meadows, whereas loafing and ruminating occurred in forest habitat. During winter, bison in that area foraged in small, sheltered meadows and along river and creek beds. Such areas offered less-severe snow conditions and supported preferred forages.

Melton et al. (1990) concluded that wood bison foraging behavior was affected by sex, season/rut, biting flies, social status, and forage biomass/habitat type. Presence of moderate to heavy concentrations of biting flies on northern bison ranges results in a shift from foraging to simply standing (Melton et al. 1989).

When lakeshores are free of ice and emergent vegetation is available, bison often forage in chest-deep water, similar to feeding behavior in moose (*Alces alces*). Bison usually prefer to seek open waters or break through thin ice to get water rather than eat snow (McHugh 1958). Unlike all other North American bovids and cervids, bison prefer to use their massive heads to sweep snow away from forage rather than to paw. For additional information on feeding patterns, refer to the Feeding Habits and Ecology, Forage and Habitat Requirements sections.

REPRODUCTION

Breeding Season. In most regions, the breeding season, or rut, for bison generally occurs from July to October with the peak occurring between late July and mid-August (Garretson 1927; Soper 1941; Fuller 1966; Halloran 1968; Lott 1972; Meagher 1973; Banfield 1981; Haugen 1974). Temporal variability in onset and duration of the rut may be related to variation in climate, photoperiod, habitat, population density, and genetic expression. Female bison tend to be seasonally polyestrus, with a cycle of approximately 3 weeks duration (Fuller 1966; Banfield 1981; Kirkpatrick et al. 1993); however, unseasonal matings sometimes occur (Soper 1941; McHugh 1958; Banfield 1981). If conception does not occur with the first mating, the female may breed again 19–21 days later (Wolff 1998).

The rutting season varies in length depending on herd location. Rut has been observed to last from 15 June to 30 September at Hayden Valley, Yellowstone National Park, Wyoming (McHugh 1958), but generally it occurs in Yellowstone between mid-July and early September (Meagher 1978; Kirkpatrick et al. 1993); from 23 June to 14 September at Wind Cave National Park, South Dakota (McHugh 1958); from 21 July to 15 August for peak breeding activity and from 25 August to 3 September for a second, smaller peak of activity in Wind Cave National Park and Custer State Park, South Dakota (Haugen 1974); is confined to a 4- to 6-week period centering around early August at the Fort Niobrara National Wildlife Refuge, Nebraska (Mahan 1978; Wolff 1998); from 1 June to 30 July at the Wichita Mountains Wildlife Refuge, Oklahoma (Halloran and Glass 1959); and from 1 July to 30 September in Wood Buffalo National Park, Alberta (Fuller 1960; Banfield 1981).

Calving and Pregnancy Rates. Calving rate is determined by dividing the total number of calves produced by the total number of mature females ≥3 years old in the herd (Rutberg 1986; Shaw and Carter 1989; Towne 1999). Pregnancy rate can be determined by necropsy of culled animals, rectal palpation in live, restrained animals (Wolfe et al. 1999), serum assay for pregnancy-specific protein B of blood samples drawn from immobilized live animals (Gates and Larter 1990), and urinary estrone conjugates measured by enzyme immunoassay of urine samples and fecal total estrogens measured by radioimmunoassay of fecal samples (Kirkpatrick et al. 1993). Urinary and fecal steroids or their metabolites can be used to detect pregnancy and ovulation in free-ranging bison with a reasonable level of accuracy (Kirkpatrick et al. 1992). Similarly, wood bison estrous cycles and synchronization can be monitored using either urine or feces (Matsuda et al. 1996). Calving and pregnancy rates for mature female bison vary considerably in free-ranging and semiwild herds throughout North America, ranging from a low of 35% to a high of 88% (Table 48.3). Calving and pregnancy rates for bison vary according to age, with reproductive vigor highest in animals between age 3 years and the onset of old age, presumed to commence at about 12–13 years of age, after which calving rates decline (Fuller 1961; Shaw and Carter 1989). In a reproductive success study of wood bison at Elk Island National Park, Alberta, Wilson et al. (2002) reported that the peak reproductive age of wood bison females was from 5 to 14 years indicating a decline in calving rates at age 15–16 years. Within this age span of 9–10 years, calving rates for young and older adults did not differ appreciably (Shaw and Carter 1989). Wild bison normally produce two calves every 3 years (Soper 1941; Fuller 1961, 1966; Halloran 1968; Banfield 1981).

The low reproductive rates observed in the Santa Catalina Island and Antelope Island bison herds, 35% and 46%, respectively (Table 48.3), were attributed to nutritional deficiency and poor-quality forage (Lott and Galland 1987; Wolfe et al. 1999). Nutrition likely affected productivity in the Henry Mountains bison population (Van Vuren and Bray 1986). Higher population densities may impose nutritional stresses that could explain an alternate-year pattern in calving (Shaw and Carter

TABLE 48.3. Calving and pregnancy rates of several free-ranging herds of North American bison

Herd Location	Calving Rate[a] (%)	Pregnancy Rate (%)	Reference
National Bison Range, Montana	—	78–100	McHugh 1958
National Bison Range, Montana	88.2	—	Rutberg 1986
Fort Niobrara National Wildlife Refuge, Nebraska	83.0	—	Wolff 1998
Fort Niobrara National Wildlife Refuge, Nebraska and Wind Cave National Park, South Dakota	78.4	—	Haugen 1974
Konza Prairie Research Natural Area, Kansas	74.4	—	Towne 1999
Wichita Mountains Wildlife Refuge, Oklahoma	71.8	—	Shaw and Carter 1989
Mackenzie Bison Sanctuary, Northwest Territories	—	70.0[b]	Gates and Larter 1990
Wood Buffalo National Park, Northwest Territories	67.0	—	Fuller 1966
Wichita Mountains Wildlife Refuge, Oklahoma	66.9	—	Halloran 1968
Badlands National Park, South Dakota	64.3	—	Berger and Cunningham 1994
Henry Mountains, Utah	62.0	—	Van Vuren and Bray 1986
Yellowstone National Park, Wyoming	—	52.0[c]	Meagher 1973
Yellowstone National Park, Wyoming	—	48.2[d]	Kirkpatrick et al. 1993
Antelope Island State Park, Utah	—	46.2[e]	Wolfe et al. 1999
Slave River Lowlands, Northwest Territories	<50.0[f]	—	Van Camp and Calef 1987
Santa Catalina Island, California	35.0	—	Lott and Galland 1987

[a]Number of calves produced based on females ≥3 years old.

[b]Based on necropsies of 28 adult females and serum pregnancy-specific protein B testing of 16 adult females in March 1987 and 1988 from an increasing herd of wood bison.

[c]Necropsies of animals from herd reductions.

[d] Based on measurements of urinary estrone conjugates (E_1C) and fecal total estrogens from 255 random urine and fecal samples collected from two subpopulations of lactating and nonlactating bison in 1989 and 1990.

[e]Mean annual pregnancy rate based on rectal palpation examination of live animals for the period 1987–1997 (range = 32.5–66.6%).

[f]Declining herd of diseased wood bison based on herd segregation counts in 1978.

1989). Environmental factors can have direct effects on reproductive success in bison and other large ungulates (Kirkpatrick et al. 1993) and may contribute to prolonged lactation and delayed ovulation the following season, especially in younger cows (Kirkpatrick et al. 1996). Population age structure differences may also account for some of the variation in pregnancy rates among herds. In Yellowstone National Park, overall reproductive performance was influenced by three variables: environmental conditions, which appear to have the most impact; age; and lactational status (Kirkpatrick et al. 1996). In Wind Cave National Park, reproductive strategies appeared to change with age, whereby older females trade offspring quantity for quality and are more likely to reproduce in alternate years (Green and Rothstein 1991). Furthermore, earlier-born females at Wind Cave had higher fecundity during their first 9 years and birth date not only was important to survival, but likely had positive fitness consequences for the survivors (Green and Rothstein 1993a). In areas where there are severe winters, calving and pregnancy rates often may be <50% (Van Camp and Calef 1987; Kirkpatrick et al. 1993). At Elk Island National Park, female wood bison reproductive success depended on age, mass, and prior success and was also affected by environmental differences between years (Wilson et al. 2002).

At the Wichita Mountains Wildlife Refuge, Oklahoma, Halloran (1968) attributed the low overall reproductive rate (52%) to high calf survival. In this herd, calves were observed nursing into their second year of life. This prolonged the physiological stress on the cows, which could adversely affect pregnancy and reproductive rates. Lactating cows at Wood Buffalo National Park carried smaller midwinter fetuses than did nonlactating cows, which suggested that bison cows with calves bred later in the year than did dry cows (Fuller 1961). Loss of a calf before the end of the breeding season probably increases the chances of breeding for the cow (Fuller 1966). Loss of a calf at any time before weaning may enhance the vigor of a calf in utero by improving the energy balance of the cow during the critical winter period. In Yellowstone National Park, during the 1960s, pregnancy rates averaged 52% (Meagher 1973). Visual observations in 1989–1990 indicated a calving rate that varied between 35% and 55%, implying an every-other-year or every 3-year pattern of calving in two subpopulations of bison at Yellowstone National Park (Kirkpatrick et al. 1993). The 2-year combined pregnancy rate, based on endocrine evidence of ovulation and pregnancy, was 48.2%, which was consistent with the visual observations and suggested that lactating cows with calves have significantly reduced fertility. Lactational suppression of ovarian activity likely is the primary mechanism by which fecundity is reduced (Kirkpatrick et al. 1993).

Diseases such as brucellosis and tuberculosis can influence pregnancy rate. Brucellosis causes abortion of calves and temporary sterility in cattle (Choquette et al. 1978). This disease in bison, in most aspects, is similar to that in cattle, where it also causes abortion in adults, and it can be associated with death in bison calves at least 2 weeks of age (Rhyan et al. 2001). From 1959 to 1974 in Wood Buffalo National Park, brucellosis infection rates averaged 30.2% (Broughton 1987), and from 1997 to 1999, 30.9% of bison tested were seropositive for brucellosis (Joly and Messier 2001a). From 1950 to 1967, the rate of tuberculosis infection in bison in Wood Buffalo National Park, based on postmortem examination, averaged 38.8% (Broughton 1987), and from 1997 to 1999, 49% of bison tested were positive for tuberculosis (Joly and Messier 2001a). The presence of bovine tuberculosis and brucellosis affected late-winter pregnancy rates in Wood Buffalo National Park bison. Animals that tested positive for tuberculosis and also had a high titer for brucellosis had a reduced pregnancy rate compared to bison with only one or neither disease (Joly and Messier 2001a). Furthermore, bison that tested positive for tuberculosis were significantly less likely to be pregnant than bison that tested negative (Joly and Messier 2001a). Based on population simulations using field data, Joly and Messier (2001b) described a "disease–predation" hypothesis where the presence of tuberculosis and brucellosis reduces bison survival and reproduction, thereby shifting bison abundance from a high-density equilibrium where food competition is regulatory to a low-density equilibrium where predation by wolves is regulatory. These data suggest that tuberculosis and brucellosis are responsible for the sustained decline of bison abundance in Wood Buffalo National Park during the last 30 years (Joly and Messier 2001b). In Yellowstone National Park, abortions have occurred in bison since 1917 (Meagher 1973). Because the population continues to increase, the impact of this disease on fecundity appears to be minimal (Dobson and Meagher 1996). Although interpopulation variation in birth synchrony occurs in Yellowstone, evidence suggests that bison exposed to brucellosis failed to deviate significantly from two nondiseased populations (Badlands National Park and Wind Cave National Park), suggesting that food plays a greater role in the timing of reproduction in bison than does brucellosis exposure alone (Berger and Cain 1999). However, prudent conservation planning demands that disease transmission risks be minimized and ideally eliminated by rendering bison disease-free (Berger and Cain 1999).

Sexual Maturity. The age at which bison cows first conceive varies considerably among locations and often within herds from the same region (Fuller 1961, 1966; Halloran 1968; Meagher 1973; Haugen 1974). In Fort Niobrara National Wildlife Refuge, Nebraska, 79% of 2-year-olds breed (Wolff 1998). Few bison cows conceive as yearlings, giving birth to a calf at 2 years of age, varying from 5% in Wood Buffalo National Park (Fuller 1966), to 6% in Nebraska and South Dakota (Haugen 1974), to 12–13% in the Wichita Mountains Wildlife Refuge (Halloran 1968; Shaw and Carter 1989). Sexual maturity was attained earlier in bison cows at the Wichita Mountains Wildlife Refuge and in herds from Nebraska and South Dakota than in herds from Wood Buffalo National Park. In the Wichita Mountains Wildlife Refuge and in Nebraska and South Dakota bison herds, 73% and 87%, respectively, conceived as 2-year olds, producing first calves at age 3 years (Halloran 1968; Haugen 1974), whereas in Wood Buffalo National Park, only 52% of the cows conceived for the first time as 3-year-olds (Fuller 1966). During a 5-year study at Wind Cave National Park in the 1980s, 5% of the females began calving at age 2 years, 79% at age 3 years, and 17% at age 4 years (Green and Rothstein 1991). In general, bison cows usually breed when they are 2 years old and give birth to their first calf when they are 3 years of age (McHugh 1958; Fuller 1961; Shaw and Carter 1989; Green and Rothstein 1991). However, at Elk Island National Park over a 4-year reproductive study on wood bison, only two 2-year-olds successfully produced offspring and the mean age at first reproduction for females was 3.7 years suggesting that younger female wood bison in Elk Island National Park were not as fecund as plains bison in other populations (Wilson et al. 2002). Reduced range nutrition resulting from environmental influences and competition with an increasing elk population may have been a factor.

Attainment of sexual maturity in male bison was similar to that for female bison in Wood Buffalo National Park, varying from a low percentage of yearlings, to approximately 33% of 2-year-olds, to the majority of bison ≥3 years old (Fuller 1961). In the Wichita Mountains Wildlife Refuge, two experimental bulls were not effective herd sires as yearlings, but were effective at 2 years of age (Halloran 1968). In Nebraska and South Dakota, three of six yearling males and >75% in subsequent age classes were in breeding condition (Haugen 1974). In a 1985 study at the Fort Niobrara National Wildlife Refuge, the curve depicting male reproductive effort is flat from age 0 to 6 years, increases to age 8 years, flattens out again to about age 10 years, and then gradually decreases (Maher and Byers 1987). Because reproductive effort begins about age 6 years, contrasted with sexual maturation, which occurs at age 3 years (Maher and Byers 1987), this confirms that male bison attain sexual maturity well in advance of becoming part of the active breeding population (Fuller 1960; Meagher 1973; Rothstein and Griswold 1991). In Elk Island National Park over a 4-year study, only two 5-year-old male wood bison were reproductively successful as most of the successful males were in the 7- to 14-years age classes; however, the range of reproductively successful males was from 5 to 14 years (Wilson et al. 2002). In Fort Niobrara, reproductive success of 6+-year-old bulls ranged from 0 to 16 calves sired per bull over 3 years and was positively correlated with dominance rank (Wolff 1998). At Elk Island National Park, mature wood bison males produced a mean of 3.8 offspring over the 4-year reproductive study period with a range of 0–24 (Wilson et al. 2002). In Badlands National Park, South Dakota, Berger (1989) demonstrated that bison males were capable of discrimination among females based on the reproductive potential of the females, whereby males ≥6 years outnumber younger males early in the rut, and older males actively select and copulate with females that have higher probabilities of bearing calves in the next season. At Elk Island National Park, only mass and prior success were useful in predicting male wood bison reproductive success (Wilson et al. 2002).

Number of Young. One calf is usual, as twinning in bison is rare (Garretson 1927; McHugh 1958; Fuller 1961, 1966; Halloran 1968; Banfield 1981; Van Vuren and Bray 1986; Rutberg 1986; Dary 1989; Green and Rothstein 1991). One instance of twin bison calves was reported by McHugh (1958) in the Lamar herd in Yellowstone and one set of twins was recorded in the plains bison herd at Wichita Mountains Wildlife Refuge in 1965 (Halloran 1968). In Wood Buffalo National Park from 1952 to 1956, no twins were observed in 481 gravid uteri examined (Fuller 1966), nor were twins observed in the 1964–1968 productivity study at Yellowstone National Park (Meagher 1973). In Elk Island National Park, Alberta, during the period 1945–1960, three sets of plains bison twins were observed. Between 1947 and 1971, examination of herd-reduction slaughter records of 4500 plains bison at Elk Island yielded only one case of twin fetuses (Wes Olson, Elk Island National Park, pers. commun., 2002). During round-ups of plains and wood bison at the park from 1984 to 2002, there was only one case of suspected twins where a cow with two calves was observed; however, the calves were never seen nursing (Wes Olson, Elk Island National Park, pers. commun., 2002). There was an extremely rare case of triplets reported from a farmed bison herd in Alberta in 2000, but two of the three calves were abandoned by the cow and had to be bottle-raised (Wes Olson, Elk Island National Park, pers. commun., 2002).

Calving Season. Most calves are born in a 3- to 4-week period from late April to early June, with an occasional calf being born as late as July (Rutberg 1984). However, conception and therefore parturition can occur at any time of year. From 1937 to 1950, calving at Yellowstone National Park had commenced by mid-April, whereas more recent studies indicated a later calving season, with most calves being born in the first half of May (Meagher 1973). McHugh (1958) noted that a few late calves were born from June through October in herds at Yellowstone National Park, the Crow Reservation, Wind Cave National Park, and the National Bison Range. Calving season became prolonged as the size of the population increased (Meagher 1973).

In the Wichita Mountains Wildlife Refuge herd, the first bison calf was always recorded within the period from 10 March to 7 April (Halloran 1968). In Wood Buffalo National Park, calving was observed from mid-April until the beginning of June (Egerton 1962; Banfield 1981). Soper (1941) noted that most calving in this park occurred around mid-May. Egerton (1962) observed that the calving period in a captive herd at Waterton Lakes National Park, Alberta, occurred from April through July. In the National Bison Range, the mean length of the calving season, the minimum period during which 80% of the young are born, was 23 days (Rutberg 1984), compared with 54 days, from early April to late May, in Wind Cave National Park (Green and Rothstein 1993b). In the National Bison Range, births were concentrated into the last week of April and the first 2 weeks of May, where 97% of the calves were born in the 6 weeks between 20 April and 2 June (Rutberg 1984). In Fort Niobrara National Wildlife Refuge, Nebraska, the first calves were born about 20 April each year and the last calves born by 24 August, with about 90% born between 20 April and 1 June each year (Wolff 1998). The high degree of synchrony of breeding, exhibited by the short calving season, at the National Bison Range appears to be the result of climatic and energetic factors rather than antipredator adaptations and is not limited by nutrition (Rutberg 1984). Bison breed less synchronously than previously reported, as indicated by the longer birth season (54 days) at Wind Cave National Park, a longer than expected time frame for an ungulate species with follower young (Green and Rothstein 1993b).

In Elk Island National Park, the calving period for wood bison commences during the latter part of April and lasts through mid-August, whereas for plains bison it usually begins by the first week in May and lasts until late August, with occasional late calves born in September and sometimes October. The calving season for plains bison became prolonged as herd size increased. The peak calving season for wood and plains bison is from 1 May to 15 June at Elk Island National Park. Calving season in northern bison occurs about 2 weeks later than in more southern herds, a phenomenon likely related to variations in climate and photoperiod between regions (Egerton 1962).

Gestation. Gestation for bison is usually 9–9.5 months and is similar to that for domestic cattle (Garretson 1927; Soper 1941; Walker et al. 1975; Halloran 1968; Rutberg 1986). Haugen (1974) reported that gestation was about 285 days for bison herds in South Dakota and Nebraska, and Banfield (1981) indicated that, in general, gestation is between 270 and 300 days. The length of gestation in bison approximates 262–272 days (Towne 1999).

POPULATION DYNAMICS

Longevity. Longevity of bison is not well documented. However, reports exist of bison living beyond 20 years of age (Halloran 1968; Meagher 1973; Berger and Peacock 1988) and even up to age 41 years (Dary 1989). In wild populations, by the time a bison has reached age 15 years it can be considered to have entered old age (Fuller 1966); in captivity, life span increases.

Calf and Adult Survival. In the Henry Mountains, Utah, calf survival was high at 93% and when hunting mortality was excluded, average survival rates for bulls (95%) and cows (96%) were similarly high (Van Vuren and Bray 1986). In Wood Buffalo National Park, calf (through 6 months of age) mortality was estimated at 50% and survival to the yearling category ranged from 5% to 8% of the herd per year (Fuller 1966). Wolf predation in the northern herd was considered to be the primary cause for the differential mortality rate in calves. During the peak growth period for the increasing Mackenzie wood bison population near Fort Providence, Northwest Territories, between 1970 and 1980, survival must have been greater than 95% among both calves and adults to achieve the observed population increase (Calef 1984). The adult survival rate in the Mackenzie wood bison herd during 1986–1987 was 97%, whereas the calf survival rate was 55% (Gates and Larter 1990).

Calf Percentage. In Yellowstone National Park, spring calf percentages (expressed as percentages of mixed herds) are normally 18–20% (Meagher 1973). In the spring of 1965, after herd reductions had removed large numbers of cows, the lowest calf percentages of the study (7–14%) were recorded for three herds. Calf percentages for these populations increased after 1965. Pooled percentages for the three herds in 1967 (20%) and 1968 (19%) suggested that the calf-producing segment of the population was leveling off and if so, then the proportion of newborn calves in mixed bison herds in Yellowstone approximated 20% (Meagher 1973). Nonselective herd reductions may alter calf percentages by taking an imbalanced harvest from the more easily slaughtered mixed herd groups rather than over the entire population. In Wood Buffalo National Park, a potential calf crop of 20–25% of the herd is expected during the latter part of June and early July. The calf crop declines by approximately 2%/month until December, when calves make up <10% of the herd (Fuller 1966). In the Mackenzie wood bison population during its peak growth years between 1970 and 1980, calves accounted for approximately 20% of the total number of animals, yearlings were 16%, 2- to 3-year olds were 13%, and ≥4-year olds were about 50% of the population (Calef 1984).

The reproductive rate (calf crop expressed as a percentage of all herd females less calves and yearlings) for the bison herd in the Wichita Mountains Wildlife Refuge varied from 47% to 60% between 1960 and 1966, with an average of 52% (Halloran 1968). In that study, the rate for 198 experimental animals from six age classes was 67%. The higher value for the experimental animals compared with the entire herd was likely attributable to the higher percentages of younger animals (prime breeding age) in the former group (Halloran 1968).

Based on observations at Lake One in Wood Buffalo National Park in 1980, the recorded calf/cow ratio was extremely low at 0.12 (12 calves/100 cows) and was deemed not to be representative of the rest of the park population (Carbyn and Trottier 1987). The mean spring calf/cow ratio for bison herds south of the Peace River in Wood Buffalo National Park from 1989 to 1996 was 0.30, indicating an overall low rate of reproduction in the Park bison population (Carbyn et al. 1998). The low calf production and extremely low yearling survival may have been caused by a combination of factors such as disease, an aging population because of low recruitment, wolf predation, and variable habitat conditions (Carbyn et al. 1998). In contrast, over a 35-year period from 1960 to 2001, the annual calf/cow ratio for plains bison at Elk Island National Park varied from a high of 1.04 to a low of 0.28 and showed a significant ($p < .001$) inverse relationship with the number of breeding-age females ($\geq$3 years) in the population (Fig. 48.5A). Likewise, over a 35-year period from 1967 to 2001 the annual calf/cow ratio for wood bison at Elk Island National Park varied from a high of 1.12 to a low of 0.47 and also showed a significant inverse relationship with the number of breeding-age females in the population (Fig. 48.5B).

The yearling-to-cow ratio for the increasing Mackenzie wood bison population near Fort Providence, Northwest Territories, during 1984 to 1988 averaged 0.30 (Gates and Larter 1990). The average ratio of males to females in the adult population ($\geq$2 years old) was 0.76 (Gates and Larter 1990).

The mean exponential growth rate of the Mackenzie wood bison herd was 0.21, a rate that exceeded that of most other North American bison populations and is considered to be close to the maximum rate of increase for bison under natural conditions in the Northwest Territories (Calef 1984; Gates and Larter 1990). For example, the Henry Mountains herd grew at the exponential rate of 0.09 between 1977 and 1983 (Van Vuren and Bray 1986).

Sex Ratio. At Yellowstone National Park during the 1930s, Rush (1932) reported that primary (*in utero*) sex ratios for bison ranged from 108 to 163 males:100 females. The average in a sample of 294 fetuses was 56% males. In bison herds in Nebraska and South Dakota, of 101 embryos examined, 54.5% were males (Haugen 1974). Between 1952 and 1956 in Wood Buffalo National Park, the primary sex ratio for 472 fetuses examined was 112 males:100 females (Fuller 1966). Palmer (1916) reported 119 males:100 females in a sample of 460 plains bison fetuses. The fetal sex ratio in 82 culled pregnant females from the National Bison Range herd from 1964 to 1967 was 51:31 (Rutberg 1986). The reported primary sex ratios from four of these five herds are similar. With the exception of the National Bison Range sample, the different herd fetal sex ratios vary only within a 6% range. A slight excess in favor of males in the primary sex ratio is common among mammals (Fuller 1961, 1966; Halloran 1968; Haugen 1974; Feldhamer et al. 1999).

At the Wichita Mountains Wildlife Refuge between 1908 and 1966, 51% of the 5633 bison born at the Refuge were males (Halloran 1968). In the National Bison Range during fall round-ups from 1965 to 1968, the number of calves born to the herd showed a sex ratio of 162 males:184 females or 47% males, which differed substantially from the fetal proportion of 62% males from the culled sample (Rutberg 1986). Selective culling of nonlactating females may distort estimates of natural sex ratios *in utero* and at birth (Rutberg 1986). On the Konza Prairie in northeastern Kansas between 1991 and 1997, out of 317 calves born, 47% were male and 53% were female, and the number of male and female calves born each year did not deviate from an expected 50:50 ratio, with only one exception in 1997 (Towne 1999). In the Fort Niobrara National Wildlife Refuge, Nebraska, the sex ratio of calves did not deviate from 50:50 in any year, but the 3-year overall sex ratio slightly favored females (Wolff 1998). The sex ratio of bison calves at birth is not dependent on maternal condition and generally does not deviate from parity (Shaw and Carter 1989; Towne 1999). Sex ratio adjustment is more closely correlated with previous reproductive effort than it is to food source competition (Wolff 1998).

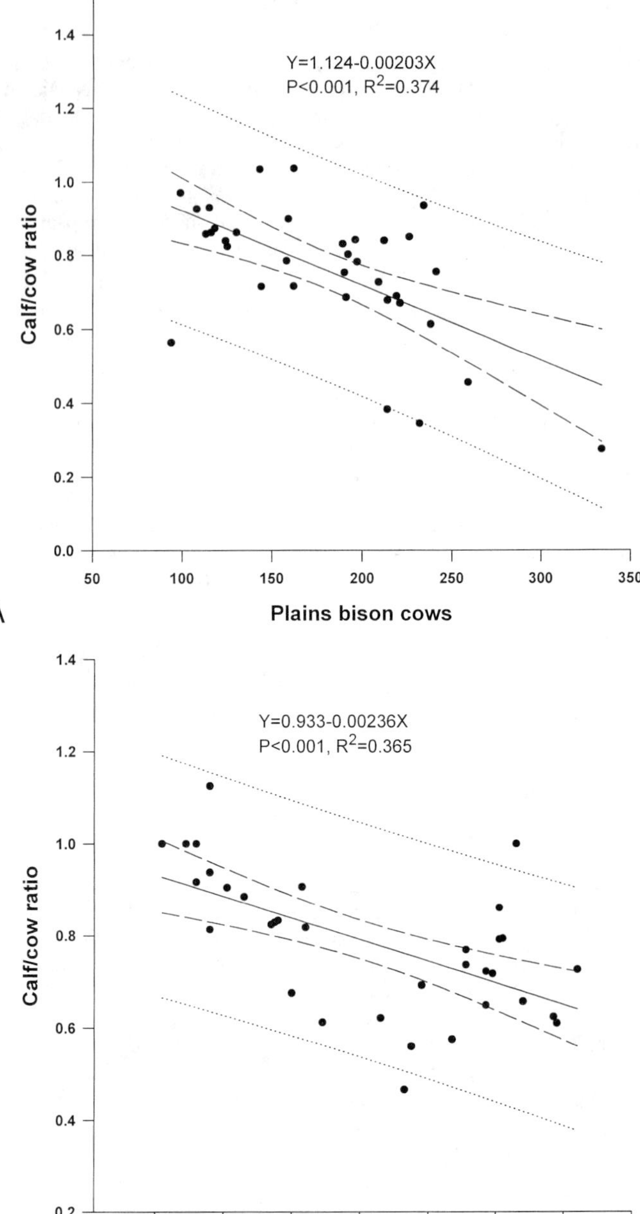

FIGURE 48.5. Relationship between the calf/cow ratio and the number of (A) plains bison (*Bison bison bison*) cows and (B) wood bison (*Bison bison athabascae*) cows aged 3 years and older at Elk Island National Park. SOURCE: Data from Olson (2002).

MORTALITY

Predation. A few circumstances have suggested occasional predation by the grizzly bear (*Ursus arctos*) on bison calves and adults in Yellowstone National Park, Wyoming (McHugh 1958; Meagher 1973). In the past, predation on bison by wolves in Yellowstone has not been a problem, as evidenced by the long survival time of injured and solitary animals and by the fact that wolves were rare and never observed in packs (Meagher 1973). However, with the reintroduction of 31 Canadian wolves into Yellowstone National Park in 1995–1996, this situation is expected to change with an increasing number of wolf–bison interactions and kills made during spring (Smith et al. 2000). Between April 1995 and March 1999, 14 bison kills by wolves were documented; however, the first wolf-killed bison was not observed until 25 months after the release (Smith et al. 2000). Furthermore, all kills were made during late winter when bison were more vulnerable because of poor condition, injuries, or young age (Smith et al. 2000).

Wolf predation on bison is an important mortality factor in Wood Buffalo National Park, Alberta, (Fuller 1961, 1966; Oosenbrug and Carbyn 1982; Carbyn and Trottier 1987, 1988; Carbyn et al. 1993, 1998; Carbyn 1997). Fuller (1961) concluded that bison form the staple diet of Wood Buffalo National Park wolves during summer and winter. He found that 80% of the summer wolf scats examined contained bison hair, and in an early-winter sample of 59 wolf stomachs, 65% of the contents was bison (Fuller 1966). During winter 1978/79 in Wood Buffalo National Park, a study of radio-collared wolves indicated that bison were their major prey item. The Hornaday River wolf pack, composed of 10 members, killed an average of one bison every 7.8 days from 12 February to 31 March 1979. This equaled an estimated consumption rate of 5.3 kg/wolf/day and an estimated winter predation of 19 bison or 9.9% of the study area population (Oosenbrug and Carbyn 1982). Similar consumption rates were observed in the Slave River Lowlands, Northwest Territories, where wolves killed 13 adult female bison, 2 adult males, and 4 calves during 33 pack-weeks of effort (Van Camp 1987).

In Wood Buffalo National Park during the 1950s, wolves selectively preyed on calves or old animals, but predation was not considered to be detrimental to the bison population (Fuller 1961, 1966). However, during the 1978–1979 wolf study in the park, a higher proportion of bison killed by wolves were adult males (Oosenbrug and Carbyn 1982), whereas in the Slave River Lowlands, cows and calves were killed more often (Van Camp 1987). However, packs of wolves in other areas of Wood Buffalo National Park preferentially attacked bison herds with calves (Carbyn and Trottier 1987, 1988). Similarly, wolves associated with the Mackenzie wood bison population exhibited a preference for bison calves (Gates and Larter 1990). Recently reintroduced wolves in Yellowstone National Park showed a decided preference for bison that were more vulnerable because of poor condition, injury, or young age (Smith et al. 2000). Prey species, such as bison, are more vulnerable in winter and represent a greater amount of resource per unit effort at that time than do smaller prey species.

In the Slave River Lowlands, free-ranging bison herds numbered at least 2000 animals in 1971, but had declined to an estimated 750 by March 1977. Wolf numbers in the region were estimated to be between 64 and 76 during winter 1976/77 (Van Camp 1987). During this time, six packs of wolves were observed operating in the Lowlands, of which four were radio-marked. During 1975–1977, bison was the most important of six major prey types of wolves in this region and represented 88% of prey weight of wolf diets in winter (Van Camp 1987). Based on estimates for 22 weeks of winter in 1976/77, wolf predation accounted for approximately 31% of the adult and subadult mortality and about 27% of the calf mortality. When combined with hunting mortality, it accounted for at least 70% of the adult and subadult bison lost that year (Van Camp 1987). Wolf predation was exerting a major role in the continued decline of the Slave River Lowlands bison population. A wolf control program selectively removed 72 wolves from the region during the winters of 1977–1979, but the bison population continued to decline (Van Camp 1987).

In conclusion, wolf predation is a significant limiting factor for some bison populations (Van Camp 1987; Carbyn and Trottier 1988; Carbyn et al. 1993, 1998) and may regulate diseased herds at low density (Gates 1993). The "disease–predation" hypothesis described by Gates (1993) and Joly and Messier (2001b) suggests that tuberculosis and brucellosis reduce bison survival and reproduction, thereby increasing vulnerability of some animals to predation and causing populations to decline to chronically low densities where predation by wolves can be regulatory. In contrast, disease-free bison populations may be regulated at a high-density equilibrium where interspecific food competition can be regulatory, despite the presence of wolves.

Hunting. During regulated seasons from 1968 to 1977, sport hunting of bison was permitted in the Slave River Lowlands north of Fort Smith, Northwest Territories. The reported number of bison kills from license returns averaged 179/year from 1969 to 1974, which approximated 8.8% of the population/year (Van Camp 1987). During 1974–1976, the average annual recruitment to the bison population declined to an average of only 3%/year (Van Camp and Calef 1987), but hunters killed at least 8% of the bison population annually (Van Camp 1987). Hunting alone could have caused the bison population to enter into a period of negative annual increments. In fall 1977, the recreational hunting season was closed and general hunting license holders voluntarily agreed to reduce their harvest. However, based on kill return data, the general hunting license holders accounted for approximately 5% of the total bison herd annually since the 1970s. Therefore, hunting harvest rates exceeded the average annual recruitment rate, at least during the period of major decline (1973–1976), and a decreasing population was inevitable (Van Camp 1987). Wolf predation and human hunting were the two major mortality factors contributing to the continued decline of the Slave River Lowlands bison population and appeared to be additive rather than compensatory mortality (Van Camp 1987). In a 7-year study (1977–1983) in the Henry Mountains, Utah, hunting mortality averaged 9% of the population, but was heavily biased toward males (Van Vuren and Bray 1986).

Accidents. Accidental drowning often occurs as a result of animals falling through thin ice in spring and fall. Whole herds of bison have succumbed to such fatalities (Raup 1933; Meagher 1973). Drowning was considered an important mortality factor of the plains bison (Roe 1970). Spring flooding in the Peace–Athabasca River Delta in Wood Buffalo National Park in 1958 caused the death of about 500 bison, autumn flooding in 1959 resulted in the death of an estimated 3000, and spring floods in 1961 resulted in the death of >1100 bison (Fuller 1966). Several thousand bison also were drowned in the delta region of Wood Buffalo National Park during a nontypical spring flood in 1974. In Yellowstone National Park, a few bison from all age classes have drowned in bogs or by falling into hot pools (Meagher 1973). In Elk Island National Park, Alberta, occasionally a few bison fall through the ice and drown as a result of traveling too close to beaver (*Castor canadensis*) houses when crossing frozen lakes.

Forest fires commonly occur in northern bison ranges without causing appreciable mortality (Soper 1941; Fuller 1966). The main effect on bison is loss of cover; however, in many situations, feeding habitat is improved and sometimes created by fire. Prescribed burning has been employed as a management tool to enhance habitat in the Hook Lake and Mackenzie bison ranges in the Northwest Territories (Chowns et al. 1998). Wildfires can be catastrophic to herds of bison, although such events are rare. Roe (1970) describes two historical accounts of bison being destroyed by fire where many animals were killed, blinded, or badly burned.

A few bison fatalities from motor vehicles have occurred along access roads within Wood Buffalo National Park (Fuller 1966) and along highways through the Mackenzie Bison Sanctuary and the Liard River Valley, Northwest Territories. In the Konza Prairie Research Natural Area, northeastern Kansas, annual mortality from lightning strikes, injury, or advanced age averages approximately 4% of the plains bison herd (Towne 1999).

Climate. Above-average snowfall, long periods of subzero temperatures, and midwinter thaws followed by severe freezing conditions result in winter-caused mortality in bison. Such severe winter conditions reduce forage availability, which often leads to poor animal condition and subsequent death. Extreme snowfall, in excess of 4 m, on bison ranges in northern Canada during the early 1800s resulted in the loss of thousands of animals (Soper 1941). In 1928, an early spring thaw followed by freezing temperatures in Wood Buffalo National Park caused crusting snow conditions that forced bison to remain in forests and sheltered areas to feed (Raup 1933). In Yellowstone National Park, Meagher (1973) defined winterkill as the combined effects of climatic stress, reduced forage availability, and physiological condition of individual animals and, considered it to be the main cause of bison mortality in the park. Weather, independent of other factors, is not usually a significant cause of mortality, but is an additional physiological stress, which, when

TABLE 48.4. Parasites reported from *Bison bison*

Parasite	Site of Infection[a]	Location[b]	Reference[c]
Arthropoda			
Cochliomyia macellaria	S	?	37, 38
Damalinia sedecimdecembrii	S	WBNP	14, 37
Demodex spp.	S	—	32
Dermacentor albipictus	S	EINP	26, 36
D. andersoni	S, NV (neurotoxin)	M	17
D. nigrolineatus (= *D. albipictus*)	S	WMWR	24
Hypoderma bovis	S, SC	MI	33
Hypoderma lineatum	S, E	WMWR, NY, OK, YNP	24, 27
Hypoderma sp.[d]	S, E, D	WBP, NBR	5, 6, 18
Speleognathus australis	NS	WMWR	11, 38
Cestoda			
Echinococcus granulosus (suspected)	LV	WBP	5
Moniezia benedeni	SI	SD, M, YNP, NBR, WMWR, EINP	3, 9, 10, 12, 15, 18, 24, 26
M. planissima (probably)	SI	WBP	5
Protozoa and rickettsia			
Anaplasma marginale	B	NBR	28
Babesia bigemina	B	Experimental infection	29, 30
B. major	B	Experimental infection	29, 30
Eimeria sp.	I	LZ	22
E. auburnensis	I	EINP, W	20, 25, 31
E. bovis	I	EINP, M	20, 31
E. brasiliensis	I	W	25, 31
E. bukidnonensis	I	—	21
E. canadensis	I	W	25, 31
E. ellipsoidalis	I	W	25,31
E. zurnii	I	W, M	31
Sarcocystis sp.	HM	WBP, NY	5, 6, 27
Toxoplasma gondii	Serology	M	42
Nematoda			
Chabertia ovina	C/CO	SD, ZGP	3, 16
Cooperia bisonis[e]	SI	WBP, EINP	5, 10, 15
C. oncophora	AB, SI	SD, EINP, NY	3, 4, 12, 26, 27
C. surnabada (= *C. mcmasteri*)	AB, SI	SD, EINP	3, 4, 12, 26
Cooperia sp.	SI	WBP	6
Dictyocaulus filaria	LG	WBP	5
D. viviparus (*hadweni* syn.)	LG	SD, WBP, K, NBR, EINP, NY, WMWR	3, 7, 8, 10, 12, 13, 15, 18, 24, 26, 27
Dictyocaulus sp.	LG	WBNP, YNP	14, 19

(Continues)

combined with predation and disease, can increase the rate of mortality (Fuller 1961).

Parasites and Diseases. Bison carry many parasites (Table 48.4) and diseases that also occur in their close relative, domestic cattle (Tessaro 1989). Controversies have arisen over the risk of transmission of diseases between bison and domestic livestock in Utah (Popov and Low 1950; Nelson 1965), the Greater Yellowstone Ecosystem (Thorne et al. 1991), and northern Canada (Gates et al. 1992a). Despite the similarity and close genetic relatedness of bison and cattle, knowledge of disease organisms in cattle cannot necessarily be used to infer their epidemiology or pathobiology in bison (Tessaro 1989; Meagher and Meyer 1994; Meyer and Meagher 1995). For example, Van Vuren and Scott (1995) found that endoparasite prevalence differed between sympatric bison and cattle in the Henry Mountains, Utah. They suggested that host–parasite relationships may differ between bison and cattle because of genetic differences (Cronin and Crockett 1993), metabolism (Christopherson et al. 1978), digestive physiology (Schaefer et al. 1978), or different rumen microbial populations (Towne 1988). Ecological and behavioral differences between bison and cattle may affect transmission. When bison and cattle occur in the same area, they may forage in different localities because of differing responses to forage availability, slope, and distance to water (Van Vuren 1982).

ENDOPARASITES. Anaplasmosis is caused by *Anaplasma marginale,* a rickettsia that parasitizes red blood cells of host animals. Insects, like the tick *Dermacentor andersoni,* transmit the organism between hosts (Radostits et al. 2000). Bison have been experimentally infected with *A. marginale* (Zaugg and Kuttler 1985; Zaugg 1986). In the National Bison Range, Montana, 15.7% of bison tested positive for anaplasmosis (Zaugg and Kuttler 1985). Clinical signs of anaplasmosis in cattle include anemia, jaundice, emaciation, and debility (Radostits et al. 2000). Bison may be more resistant than cattle, because experimentally infected bison calves demonstrated only mild clinical signs (Zaugg and Kuttler 1985).

Babesiosis is caused by the protozoans *Babesia bigemina* and *B. major*. These two species of *Babesia* are the only ones reported to cause disease in bison (Locker 1953; Zaugg and Kuttler 1987). *Babesia* spp. are transmitted by ticks. Babesiosis has been induced in bison by experimental infection with *B. bigemina* (Zaugg and Kuttler 1987). Natural infections of bison have not been reported. The organism causes intravascular hemolysis where clinical signs include fever, jaundice, hemoglobinuria, heavy breathing, and anorexia (Locker 1953; Zaugg and Kuttler 1987).

Coccidiosis is caused by six species of protozoans in the genus *Eimeria,* including *E. auburnesis, E. bovis, E. brasiliensis, E. canadensis, E. ellipsoidalis,* and *E. zurnii* (Ryff and Bergstrom 1975; Penzhorn et al. 1994). This disease has been reported infrequently in bison calves. Short-yearling bison in a captive herd of wood bison in the Slave River Lowlands in northern Canada experienced severe coccidiosis in March 1994 (C. Gates, pers. obs.); two calves died after experiencing bloody

TABLE 48.4—*Continued*

Parasite	Site of Infection[a]	Location[b]	Reference[c]
Haemonchus contortus	AB	SD, K, NY, WMWR	3, 10, 12, 13, 23, 24, 27
Nematodirella longispiculata	SI	EINP	15
Nematodirus helvetianus	SI	EINP, NY	15, 27
Oesophagostomum radiatum	CO	K, EINP, WMWR	10, 12, 13, 24, 26
Oesophagostomum sp.	CO	WBP	5
Oesophagostomum sp.	SI, LI	Y	27
Ostertagia bisonis	AB	SD, WBP, K, EINP	3, 7, 10, 12, 13, 26, 41
O. lyrata (=*Grosspiculagia lyrata*)	AB, SI	EINP	12, 26
O. ostertagi	AB	EINP, NY	10, 12, 26, 27
O. trifurcata	AB	SD	3
Setaria labiatopapillosa	CE	ND, M, NZP, WBP, WBNP, EINP, NY	1, 5, 12, 14, 15, 26, 27
S. yehi	CE	ND	1
Setaria sp.	CE	WBP	6, 10
Strongylus sp.	I	H, YNP	34, 35
Trichostrongylus axei	CE, SI	EINP, NY	10, 12, 26, 27
T. lerouxi	SI	NY	27
Trichuris discolor	C	NY	27
T. ovis[f]	C	NBR	12, 18
Trichuris sp. (eggs)	F	EINP	26
Trematoda			
Fasciola hepatica	BD/LV	NBR, W, H	12, 18, 2, 34, 35
Fascioloides magna	LV	WBP	5, 6, 10, 40
Paramphistomum microbothrioides	RU	—	39

[a]AB, Abomasum; B, blood; BD, bile duct; C, cecum; CE, coelom; CO, colon; D, diaphragm; E, esophagus; F, feces; HM, heart muscle; I, intestine; LI, large intestine; LV, liver; LG, lung; M, muscle; RU, rumen; NS, nasal sinus; NV, nervous system; S, skin; SC, spinal cord; SI, small intestine.

[b]EINP, Elk Island National Park; K, Kansas; H, Henry Mountains, Utah; LZ, Leningrad Zoo; M, Montana; MI, Michigan; NBR, National Bison Range; ND, North Dakota; NY, New York State; NZP, National Zoological Park (District of Columbia); OK, Oklahoma; SD, South Dakota; WBP, Wainwright Buffalo Park; WMWR, Wichita Mountains Wildlife Refuge; WBNP, Wood Buffalo National Park; W, Wyoming; YNP, Yellowstone National Park; ZGP, Zoological Garden of Prague.

[c]1, Becklund and Walker 1969; 2, Bergstrom 1967; 3, Boddicker and Hugghins 1969; 4, Burtner and Becklund 1971; 5, Cameron 1923; 6, Cameron 1924; 7, Chapin 1925; 8, Corner and Connell 1958; 9, Dikmans 1934; 10, Dikmans 1939; 11, Drummond and Medley 1964; 12, Dunn 1968 (literature review only); 13, Frick 1951; 14, Fuller 1966; 15, J. Holmes, 1981, pers. commun.; 16, Jaros et al. 1966; 17, Kohls and Kramis 1952; 18, Locker 1953; 19, Meagher 1973; 20, L. Morgantini, unpublished data; 21, Pellerdy 1963; 22, Pellerdy 1974; 23, Ransom 1911; 24, Roudabush 1936; 25, Ryff and Bergstrom 1975; 26, W. Samuel, University of Alberta, Edmonton, 1981, pers. commun.; 27, Wade et al. 1979; 28, Zaugg and Kuttler 1985; 29, Findlay and Begg 1977; 30, Zaugg and Kuttler 1987; 31, Penzhorn et al. 1994; 32, Vestweber et al. 1999; 33, Schillhorn van Veen et al. 1991; 34, Zaugg et al. 1993; 35, Van Vuren and Scott 1995; 36, Mooring and Samuel 1998; 37, McHugh 1972; 38, Tessaro 1989; 39, Herd and Hull 1981; 40, Swales 1936; 41, Lichtenfels and Pilitt 1991; 42, Dubey 1985.

[d]Other species of parasitic Diptera (e.g. those within the Culicidae, Muscidae, Rhagionidae, and Tabanidae) have been observed in association with bison (Meagher 1973).

[e]*Cooperia bisonis* (Cram 1925) was reported as *Haemonchus ostertagi* by Cameron (1923).

[f]Identification suspect, since it was based on examination of female worms only (Locker 1953).

diarrhea and emaciation. Others were treated and survived. The disease is associated with overcrowding, fecal build-up in pens, fecal contamination of water sources, a build-up of coccidia oocysts in the environment, and consumption of large numbers of oocysts by calves (Radostits et al. 2000). Stress associated with weaning, transport and mixing of calves, and cold weather conditions can induce coccidiosis. Coccidiosis was diagnosed in a captive wood bison herd established in northeast Alberta during 2000, resulting from stress and overcrowding. It was believed to be a contributing factor in the death of several animals. Treatment administered in drinking water along with reduced levels of stress alleviated the problem. Although there are no reports of this disease in wild bison, *Eimeria bovis* and *E. zurnii* are common among bison in Montana (Penzhorn et al. 1994). In spite of coccidia oocysts being found in cattle grazed sympatrically with bison in Utah, no oocysts were found in bison feces (Van Vuren and Scott 1995).

Sarcocysticosis is caused by protozoans in the genus *Sarcocystis,* for which bison and other ungulates serve as intermediate hosts. Oocysts or sporocytes are released in the feces of an infected wolf, coyote, or dog, which serves as the definitive host. Intermediate hosts become infected when they consume contaminated food or water. This protozoan has been reported in the muscle tissue of 94% of bison in Alberta and 13% in Montana (Pond and Speer 1979; Dubey 1980a, 1980b; Mahrt and Colwell 1980; Fayer et al. 1982). Clinical signs have not been described for bison. However, bison calves experimentally infected with large doses of sporocytes became anemic, anorexic, and lethargic, and had elevated body temperatures (Dubey 1982).

The term *helminth* refers to a number of phyla with superficial similarities, usually grouped into three categories: tapeworms (Cestoda), flukes (Trematoda), and roundworms (Nematoda).

The tapeworms *Taenia hydatigenia* and *Echinococcus granulosus* cause hydatid disease in mammals, which is expressed as fluid-filled cysts in the liver or lungs. The adults of these tapeworms live in the intestinal tracts of their primary hosts, wild and domestic carnivores. Eggs are passed in the feces. If the eggs are consumed by a secondary host, including cattle, sheep, bison, and other herbivores, they hatch in the intestinal tract and migrate to the liver, lungs, and peritoneal cavity. Within these organs, cysts, called cysticerci or hydatids, are formed. The resistance of bovids to *E. granulosus* may account for its absence in bison at Wood Buffalo National Park despite its occurrence in dogs, wolves, caribou, and moose in the region (Fuller 1966). Hydatid cysts commonly occur in cervids at Elk Island National Park, but have not been seen in the resident bison. Choudary et al. (1987) reported *E. granulosus,* in an American bison in a zoo in India; pathological changes associated with infection included hepatomegaly and hydatid cysts in the liver and the lungs.

The tapeworm *Monezia benedeni* lives as an adult in the gastrointestinal tract of its herbivore host. It is a common parasite in bison and has been reported widely in North America (Cameron 1923; Dikmans 1934; Boddicker and Hugghins 1969; Wade et al. 1979). Neither clinical symptoms nor pathology associated with this parasite have been reported in bison.

Both the common bile duct liver fluke *Fasciola hepatica* and the deer fluke *Fascioloides magna* have been reported from bison (Table 48.4). *F. hepatica* has been isolated from bison fecal samples in Utah (Locker 1953; Van Vuren and Scott 1995) and Yellowstone National Park (Zaugg et al. 1993). The life cycle of *F. hepatica* requires the development of immature stages in snails, which are the intermediate host. Larval flukes leave the snail and encyst on vegetation. Bison, cattle, and sheep become infected when they consume infected snails or cysts. Clinical signs of liver fluke infection have not been described for bison. Bison are not considered to be a preferred host for the deer fluke (Swales 1936).

The rumen fluke (*Paramphistomum microbothrioides*) has been reported in bison (Herd and Hull 1981). In domestic cattle, severe diarrhea and weight loss have been associated with intestinal paramphistomiosis.

A variety of parasitic roundworms occurs in the gastrointestinal tracts or peritoneal cavity of bison (Table 48.4). Although reduced growth rate, loss of productivity, lack of vitality, and unthriftiness have been reported in cattle that were heavily infected with gastrointestinal nematodes, the only species reported to have caused clinical disease in bison is *Ostertagia* spp. (Wade et al. 1979). *Ostertagia* larvae can penetrate the glands of the abomasum, where they enter into a dormant, hypobiotic state. If large numbers of larvae emerge from hypobiosis, they can cause serious damage to the abomasal mucosa, which can then result in clinical disease and death. This form of ostertagiosis (Type II) was reported in three bison herds in New York State (Wade et al. 1979).

In ranched bison in northern Alberta, *Cooperia* spp. was the predominant parasite identified in feces, accounting for 96% of all parasites found in calves and 92% of all parasites found in cows (Dies 1998). Van Vuren and Scott (1995) reported a high prevalence of strongylida in bison feces collected from the Henry Mountains herd in Utah. This is similar to results from Yellowstone National Park, where 80% of bison examined carried *Strongylus* spp. (Zaugg et al. 1993). A low prevalence of *Fasciola hepatica* and *Trichuris* spp. was reported in several studies (Locker 1953; Zaugg et al. 1993; Van Vuren and Scott 1995).

The nematodes *Dictyocaulus hadweni, D. filaria,* and *D. viviparus* live and reproduce as adults in the bronchi of lungs. Eggs produced in the lungs develop into larvae, which are coughed up, swallowed, and passed in the feces. *Dictyocaulus* spp. is a direct life cycle parasite. Larvae are consumed with vegetation by grazing bison. They then migrate through the wall of the intestines, enter the venous drainage, and are transported through the circulatory system to the lungs. *Dictyocaulus* spp. have been isolated from bison in Kansas (Frick 1951), Wyoming (Bergstrom 1982), Montana (Locker 1953), Oklahoma (Roudabush 1936), Yellowstone National Park (Meagher 1973), and Wood Buffalo National Park (Fuller 1966). Clinical signs associated with lungworm in bison include increased respiratory rate, coughing, slight nasal discharge, increased heart rate, and mild fever (Berezowski n.d.). Severe infections may cause death. Postmortem findings include pulmonary edema, emphysema, and large quantities of bloody froth in the trachea and bronchi containing adult lungworms (Berezowski n.d.).

Ectoparasites. Demodex mange, or demodecosis, is caused by mites (*Demodex* spp.) infesting hair follicles and sebaceous glands in the skin. The disease is characterized by small, 7- to 9-mm pus-filled nodules around the eyes, perineum, and ventral surface of the tail, or as small, palpable lesions on the neck, flank, and shoulders (Vestweber et al. 1999).

The winter tick (*Dermacentor albipictus*) and the Rocky Mountain tick (*D. andersoni*) occur in bison (Table 48.4). Ticks cling to the hair and suck blood by attaching to and piercing the skin of the host. In large numbers, they can cause severe anemia and debility and negatively affect host populations (Samuel and Welch 1991). Bison are less susceptible to tick infestation than are cervids. Bison at Elk Island National Park exhibited only light infestations in the presence of moose and elk that were infested with high numbers of *D. albipictus*. This probably reflected the tick-reducing effect of the grooming behavior of bison (Mooring and Samuel 1998). The only reported clinical effect resulting from tick infestation in bison is tick paralysis (Kohls and Kramis 1952). This condition is caused by a neurotoxin produced by some ticks. Initially, bison experiencing tick paralysis exhibit an unsteady gate and jerky movements followed by an inability to get up. Tick paralysis has only been reported in bison calves and yearlings (Kohls and Kramis 1952).

Warbles, the late larval stage of the parasitic fly *Hypoderma bovis* and *H. lineatum,* have been reported in bison from several locations (Roudabush 1936; Locker 1953; Schillhorn van Veen et al. 1991). Adult warble flies attach their eggs to the hair on the legs and lower body of the host. After hatching, the larvae penetrate the skin and migrate to the esophagus (*H. lineatum*) or the spine (*H. bovis*). The larvae grow throughout the winter in these locations, then in the spring they migrate to just under the skin on the back. They eventually make a hole in the skin, emerge, and drop to the ground, where they pupate and metamorphose into adult flies (Radostits et al. 2000). In bison from Montana and Yellowstone National Park, dead, discolored larvae were found under the skin, which suggests that they may not have been able to penetrate the thick hide (Locker 1953). Clinical signs of warble infestation in bison have not been reported.

Other ectoparasites reported from bison include the chewing louse (*Damalinia sedecimdecembrii*) recorded by Fuller (1966) in Wood Buffalo National Park, the nasal mite (*Speleognathus australis*) (Drummond and Medley 1964; McHugh 1972), and the screw worm (McHugh 1972; Tessaro 1989). Biting insects cause serious harrassment of bison, including mosquitoes (*Aedes* spp.), horse and deer flies (*Tabanus* spp.), black flies (*Simulium* spp.), the face fly (*Musca autumnalis*), the stable fly (*Stomoxys calcitrans*), and snipe flies (*Symphoromyia* spp.) (Bay et al. 1968; Burger and Anderson 1970; Meagher 1973; Tessaro 1989).

Bacterial Diseases. Actinobacillosis is caused by *Actinobacillus lignieresii,* which is found in the mouth of clinically normal ruminants (Radostits et al. 2000). Abrasions of the oral mucosa associated with foraging on course feed may allow entry into the oropharyngeal. Eruption of teeth also may create an opportunity for infection with the bacterium. Actinobacillosis has been reported in bison, but the clinical signs associated with infection have not been described (Choquette et al. 1961). In cattle, actinobacillosis is known as wooden tongue disease (Radostits et al. 2000). It is characterized by a swollen, hard tongue and excessive salivation.

Anthrax is an infectious, often fatal disease of mammals, including humans. It is caused by the endospore-forming bacterium *Bacillus anthracis* (Dragon and Rennie 1995; Gates et al. 2001a). The life history strategy of *B. anthracis* is different from that in many other co-evolved host–parasite relationships in that its persistence depends on extreme virulence, the death of its host, and the long-term survival of highly resistant endospores in the environment. Anthrax epidemics occur regularly around the world. The bacillus is endemic to many areas in Europe, Asia, Africa, Australia, and North, Central, and South America (Gates et al. 2001a). In North America, outbreaks of anthrax occur regularly in bison herds in the Slave River Lowlands, Northwest Territories, and in Wood Buffalo National Park (Tessaro 1989; Broughton 1992; Dragon and Elkin 2001). Over the period from 1962 to 1978, at least 1086 bison deaths in these populations were attributable to anthrax (Reynolds et al. 1982). An outbreak occurred in a free-ranging wood bison herd near Fort Providence west of Great Slave Lake in 1993, killing at least 172 bison (Gates et al. 1995). The disease disproportionately affects male bison and is not considered to be a significant factor in population dynamics (Gates et al. 1995). Anthrax is not treatable in free-ranging wildlife, but captive bison can be vaccinated (Gates et al. 2001a). Anthrax Emergency Response Plans, preplanned logistical frameworks for responding effectively to outbreaks to minimize spread of the disease, reduce environmental contamination, and minimize risk to public health, are now being developed by responsible jurisdictions (Nishi et al. 2002c).

Clostridium chauvei causes the disease blackleg in bison. Other bacteria may cause a similar disease in cattle (Tessaro 1989). *C. chauvei* is a soil-borne bacterium. Similar to *B. anthracis,* it can survive in the soil for many years. The bacterium enters the body when consumed with contaminated forage. In clinically normal animals, bacteria occur in the spleen, the liver, and the intestinal tract. Disease may occur when bacterial spores that are lodged in normal tissue, typically muscle, proliferate following bruising of the tissue. The bacteria rapidly increase, producing toxins that kill tissue and cause toxemia. In cattle, blackleg leads rapidly to death of the host. The clinical signs of blackleg in bison have not been described (Tessaro 1989).

Brucella abortus is a coccobacillus that lives as a facultative intracellular parasite (Radostits et al. 2000) and causes a disease known as bovine brucellosis. It is transmitted in mammals primarily by oral contact with aborted fetuses, contaminated placentas, and uterine discharges. The organism also may be excreted in milk. *B. abortus* is thought to have been introduced to bison from cattle (Meagher and Meyer 1994). Lesions associated with brucellosis are similar in bison and cattle, including abortion (Williams et al. 1993; Rhyan et al. 1994), retained placenta, and endometritis (Tessaro 1987; Rhyan et al. 2001), although there may be some interspecies differences in tissue colonization (Rhyan et al. 2001). Brucella-induced abortion occurs in >90% of female bison during the first pregnancy following infection, depending on inoculum size (Davis et al. 1990, 1991). The abortion rate subsequently declines to near zero after the third pregnancy due to naturally acquired immunity. Some females may be lost as a result of acute metritis associated with retained placentas (Broughton 1987). Infected calves surviving to birth may be weak, with low survival rates. Pathology described in neonatal bison includes broncointerstitial pneumonia, focal splenic infarction, and purulent nephritis (Rhyan et al. 2001). In male bison, *B. abortus* causes orchitis, seminal vesiculitis, and epididymitis and sterility in advanced cases (Tessaro 1989). In Wood Buffalo National Park, localization of brucellae in joints causes chronic suppurative arthritis and lameness (Tessaro 1989). Fuller (1966) reported that arthritis occurred in 2% of adult bison slaughtered in Wood Buffalo National Park during the 1950s. Debility caused by chronic septic arthritis may increase susceptibility to predation. The disease was formerly found in public bison herds in Elk Island National Park (Corner and Connell 1958), and the National Bison Range, Montana (Creech 1930). *B. abortus*-infected public bison herds continue to exist in and near Yellowstone National Park, Grand Teton National Park, and Wood Buffalo National Park, where the disease has been the center of management controversies for many years (Gates et al. 1992a, 2001b, 2001c; Inserro 1997; Keiter 1997).

Pasteurellosis refers to a number of localized and systemic infections caused by bacteria in the genera *Pasturella* and *Mannheimia*. These organisms are nonmotile, small, pleiotropic, Gram-negative rods or coccobacilli. Three species are potentially important for bison, *P. multocida, P. trehalosi,* and *P. haemolytica* (reclassified as *Mannheimia haemolytica;* Angen et al. 1999). All of them are common fauna of the respiratory tract in bison (Jaworski et al. 1998) in captivity (Ward et al. 1999a) and in the wild (Taylor et al. 1996). Diseases caused by these organisms include localized abscesses arising from scratch or bite wounds and respiratory disease or septicemia. *Pasteurella* spp. can be considered as typically commensal organisms, which function as endemic, opportunistic pathogens. The organisms become pathogenic, causing respiratory or septicemic disease following exposure to predisposing factors such as severe weather, crowding, or other environmental stresses (Carter and de Alwis 1989). Three epidemics of septicemic pasturellosis occurred in bison between 1911 and 1965 (Mohler and Eichorn 1912–1913; Gochenour 1924; Heddleston et al. 1967). The disease occurs in explosive outbreaks with a high mortality rate (Gochenour 1924; Heddleston and Gallagher 1969). Young animals are most susceptible to pneumonic pasteurellosis (Dyer and Ward 1998). Pathology associated with *P. hemolytica* pneumonia includes fibrinopurulent bronchopneumonia, with fibrinous pleuritis and pericarditis (Dyer and Ward 1998). Pasteurellosis has a sudden onset. Clinical signs can include an extremely high fever, profuse salivation, hemorrhages in mucous membranes, severe depression, and swelling of the throat, brisket, and perineum. Death is commonly the first clinical sign reported in a herd with peracute septicemic and pneumonic pasteurellosis.

Hemophilosis is a disease caused by the bacterium *Hemophilus somnus*. The organism is commonly present in the respiratory system of cattle; in a Canadian study, 25% of cattle were serologically positive (Radostits et al. 2000). Similar to *Mannheimia haemolytica* and *Pasteurella multocida, H. somnus* is associated with disease in livestock and should be regarded as a potential pathogen for bison, particularly in animals stressed by management practices such as herding, crowding, and shipping (Ward et al. 1999b). Tissue analysis from 21 bison with bronchopneumonia suggested that *H. somnus* can be a respiratory pathogen in bison (Dyer 2001). Berezowski (n.d.) reported meningitis in bison caused by *H. somnus* and referred to the same disease in cattle as infectious thromboembolic menigioencephalitis or ITEME. Outbreaks in captive bison occur most commonly in the fall and winter and are associated with extremely cold weather and stressful situations such as handling.

Paratuberculosis (Johne's disease) is caused by *Mycobacterium avium paratuberculosis*. It occurs around the world in a wide range of domestic and wild mammals (Williams 2001) including bison (Buergelt et al. 2000). *M. a. paratuberculosis* primarily infects the digestive tract, which reflects its primary mode of transmission through contaminated feed and water. It has a long incubation period and exhibits prolonged clinical duration. Gross lesions reported in bison included enlarged mesenteric lymph nodes and intestinal mucosal thickening. Histology reveals noncaseating granulomatous inflammatory infiltrates and acid-fast bacilli characteristic of *M. a. paratuberculosis* (Buergelt et al. 2000). Morphological changes of subclinical Johne's disease in bison are characterized by microgranulomas composed of epithelioid macrophages and individual multinucleate giant cells of Langhans type occasionally containing individual cytoplasmic acid-fast bacilli compatible with *M. a. paratuberculosis* (Buergelt and Ginn 2000). The most common clinical signs of Johne's disease in cattle are chronic diarrhea and loss of body weight leading to emaciation and death (Radostits et al. 2000).

Bovine tuberculosis is a chronic infectious disease, which causes death in advanced cases. The etiological agent is the bacterium *Mycobacterium bovis*. Depending on the location of the infection in the host, the organism can be excreted in exhaled air, sputum, feces, milk, urine, and vaginal and uterine discharges (Radostits et al. 2000). Inhalation and consumption of contaminated feed or water are the principal modes of transmission and determine the primary locations of lesions. Consuming infected milk may infect young animals. Intrauterine infection of the fetus occurs via the umbilical vessels. Inhalation is the primary route of infection in bison (Tessaro et al. 1990). The organism may spread to any part of the body after invading the blood or lymph vessels. The pathogenesis of mammalian tuberculosis is well described (Thoen and Bloom 1995). Infection with *M. bovis* generally results in development of chronic granulomatous lesions, which sometimes become necrotic, caseous, and calcified (Radostits et al. 2000). The pathology, pathogenesis, and epidemiology of bovine tuberculosis are the same in bison and cattle (Thoen et al. 1988; Tessaro et al. 1990).

Tuberculosis was first reported in bison in 1923 at Wainwright Buffalo Park in east-central Alberta (Cameron 1924). The source of infection was probably local domestic cattle rather than bison herds in Montana from which the Wainwright bison originated (Hadwen 1942; Tessaro et al. 1990). In 1922, >6000 bison were translocated from Wainwright Buffalo Park to Wood Buffalo National Park in northern Alberta, thereby introducing *M. bovis* to a wood bison population existing in the region. Tuberculosis was first documented in Wood Buffalo National Park bison in 1952 (Fuller 1966) and was verified in subsequent postmortem examinations (Choquette et al. 1961). Bison in and around Wood Buffalo National Park remain the only public herds infected with *M. bovis* in North America. The Mackenzie wood bison herd located 100 km northeast of Wood Buffalo National Park is considered to be at a high risk of infection. Tessaro et al. (1993) found it to be free of tuberculosis when it was tested in the late 1980s. Another free-ranging population in the region, the Hay-Zama herd in northwestern Alberta, has not been tested (Gates et al. 2001c), and its tuberculosis status remains unknown. There have been occasional outbreaks among fenced herds in New Brunswick, Manitoba, Ontario, Alberta, South Dakota (Stumpff et al. 1985), and Pennsylvania (Tuckerman 1955).

The effects of tuberculosis on bison population dynamics were studied in Wood Buffalo National Park (Joly and Messier 2001a, 2001b). Considered individually, neither brucellosis nor tuberculosis had significant effects on reproduction or survival rates. However, bison that were both positive for tuberculosis and had a high titer for antibodies against *B. abortus* had lower reproductive and survival rates in the presence of wolves. A model of the tuberculosis–brucellosis/wolf–bison system demonstrated that the combined effects of these diseases and wolf predation could have caused the decline of the Wood Buffalo National Park bison population observed since the cessation of intensive management, including wolf control, during the late 1960s (Joly and Messier 2001b).

Viral Diseases. Bovine virus diarrhea/mucosal disease (BVD) is caused by an RNA virus in the pestivirus group. Positive serological evidence has been found in Yellowstone National Park bison in Wyoming (Taylor et al. 1997) and in bison at Elk Island National Park in Alberta (Cool 1999; Gates et al. 2001b). In Yellowstone National Park, positive antibody titers were detected in 31% of bison tested (Taylor et al. 1997). BVD was detected in the Elk Island National Park plains bison herd in 1996, prompting a serological survey of the plains bison and wood bison herds (Cool 1999; Gates et al. 2001b). Forty-seven percent of 561 plains bison tested seropositive for BVD, with only 1 testing positive for the virus antigen. At least six plains bison deaths in the Park have been attributed to the BVD virus (Cool 1999). The type 1 virus was isolated from tissues of plains bison from Elk Island National Park that were submitted to the Animal Disease Research Institute, Lethbridge, Alberta (Tessaro and Deregt 1999). None of 352 wood bison in Elk Island National Park tested seropositive for BVD. Both populations are vaccinated for BVD during annual round-ups. However, calves to be used in translocations are not vaccinated to allow future screening for BVD in the recipient herd.

The BVD virus is highly contagious and is transmitted by direct contact between animals, in contaminated feed and water, and will cross the placenta to the fetus (Van Campen et al. 2001). In cattle, BVD virus is shed in nasal discharge, saliva, semen, feces, urine, tears, milk, and discharges following abortion of a fetus (Radostits et al. 2000). Some individuals remain persistently infected and shed large quantities of the virus for the rest of their lives. Chronically infected individuals are the main source of infection in cattle herds (Radostits et al. 2000). Little is known about the pathology of BVD in bison. Acute BVD is seen in bison and can cause mortality at any age (Berezowski n.d.). Infection of the fetus can cause abortion, stillbirths, weak calves, and fetal abnormalities. Mucosal disease in cattle may occur in animals that are persistently infected. Persistently infected calves may be unthrifty and may develop lameness associated with foot lesions. Commercial BVD vaccines are widely available for use in cattle. Modified live BVD vaccines can cause diarrhea in recently weaned bison calves (Berezowski n.d.).

Malignant catarrhal fever (MCF) is caused in North American bison by ovine herpes virus type 2. The natural host for this virus is domestic sheep, in which it does not cause disease. Li et al. (1996) found that 61% of goats and 53% of sheep in a sample in the United States were seropositive to MCF virus. Recently, MCF has become an important disease in the commercial bison industry. Infected herds have experienced up to 100% mortality (Schultheiss et al. 2001). Infected bison exhibit rapid clinical disease. With the acute fatal form of disease, they usually die within 7–10 days of infection or within 48 hr of the onset of illness. With the chronic fatal form, they may die up to 156 days after infection. Some animals recover and remain persistently infected for an unknown length of time (Schultheiss et al. 1998).

Clinical signs in bison include hemorrhagic cystitis, colitis, conjunctivitis, ocular discharge, nasal discharge, excess salivation, anorexia, diarrhea, melaena, hematuria, multifocal ulceration of the oral mucosa, fever, circling, ataxia, behaviors suggestive of blindness, lameness, and difficulty urinating (Ruth et al. 1977; Liggitt et al. 1980; Schultheiss et al. 1998). Lymphadenomegaly and corneal opacity occur in fewer than half the cases (Schultheiss et al. 2001). Direct contact with sheep is considered to be the most likely source of infection, although it has not been demonstrated in all bison herds experiencing MCF. The common means of transmission of this disease is unknown.

Parainfluenza 3 (PI3) virus has been detected by serology in bison in Yellowstone National Park (Taylor et al. 1997), the National Bison Range in Montana (Heddleston and Wessman 1973), and the free-ranging Delta Junction herd in Alaska (Zarnke and Erickson 1990). In the latter case, a recent expansion of the cattle population in the area was considered the likely source of PI3 infection in bison. Clinical signs of PI3 infection have not been reported in bison. In cattle, infection has been associated with pneumonia (Radostits et al. 2000).

FEEDING HABITS

In the majority of situations, North American bison are grazers. Because bison are located in widely varied habitats throughout their range, it is most useful to identify their diets by association with geographic area. Seasonal diets from eight different populations of bison have been compared by forage class (Table 48.5). Grasses and sedges were the most important foods of free-roaming bison in most of these populations. Peden (1976) confirmed that bison on the shortgrass plains in northeastern Colorado consume mainly grasses during all seasons. Similarly, on semidesert range in southwestern Colorado at the Colorado National Monument, grass was the dominant forage used by bison during summer and exceeded 27% during all seasons. In a shrub–steppe plant community in southern Utah, Van Vuren (1982) found that bison diets were composed of 93% grasses and sedges. Sedges also have been noted as common dietary items of several bison populations (Fuller 1966; Meagher 1973; Reynolds et al. 1978). In Yellowstone National Park, Wyoming, and in northern Canada, sedges constituted the highest proportion of bison diets in all seasons, with grasses second in importance. In northeastern Colorado, sedges were important to bison only during spring. In other bison herds located at Wood Buffalo National Park, Alberta (Soper 1941) and Elk Island National Park, Alberta (Holsworth 1960), bison fed on grasses in summer and sedges in winter. In some areas, forbs are seasonally important to foraging bison (Soper 1941; Nelson 1965; Wasser 1977; Banfield 1981). In semidesert range in southwestern Colorado, forbs were common food items during all seasons, but never exceeded 17% in any season (Wasser 1977). In Yellowstone National Park and in northern Canada, forbs were important to bison only during summer, whereas in northeastern Colorado, forbs were important dietary items during fall and winter. Bison in the Colorado National Monument, southwestern Colorado, primarily forage on grass in summer and browse in winter, and eat both in spring and fall (Wasser 1977). Unlike other populations, these bison used browse as the major source of food (67%) during winter. Browse was not eaten by bison on the shortgrass plains in northeastern Colorado and was of minor importance to bison in Yellowstone National Park and in northern Canada.

On the shortgrass plains of the Pawnee National Grasslands in Colorado, on which blue grama (*Bouteloua gracilis*) is the dominant species, Peden (1976) observed 36 plant species in the diets of bison, but only 11 contributed significantly to the total. Blue grama and buffalo grass (*Buchloe dactyloides*) were the most abundant plants in the habitat and also in the diet. A preference for western wheatgrass (*Agropyron smithii*) over blue grama was noted. Other commonly consumed species were red threeawn grass (*Aristida longiseta*), sun sedge (*Carex heliophila*), scarlet mallow (*Sphaeralcea coccinea*), sand dropseed (*Sporobolus cryprandrus*), and needle and thread (*Stipa comata*). Grasses were the main component of bison diets on the shortgrass plains, making up in excess of 79% of the diet in all seasons (Table 48.5). Sedges were important only in the spring, whereas forbs contributed from 3% to 9% of the diet during summer, fall, and winter.

On semidesert range in the Colorado National Monument, southwestern Colorado, Wasser (1977) reported strong selectivity by bison for preferred forages. The major plant communities in the study area were sagebrush (*Artemisia* spp.), Utah juniper (*Juniperus osteosperma*), mixed sagebrush and juniper, and saltbush (*Atriplex* sp.).

TABLE 48.5. Percentage composition of the diet of bison by forage class and season as averaged and summarized from eight populations in the United States and Canada

Herd Location[a]	Season and Forage Class[b]																							
	Winter						Spring						Summer						Fall					
	G	S+R	F	S	L	O	G	S+R	F	S	L	O	G	S+R	F	S	L	O	G	S+R	F	S	L	O
NE Col	87	T	9	—	—	—	79	16	T	—	—	—	88	1	3	—	—	—	89	—	9	—	—	—
SW Col	27	T	3	67	—	—	57	1	17	25	—	—	72	4	7	17	—	—	57	2	11	29	—	—
S Utah	—	—	—	—	—	—	—	—	—	—	—	—	96	3	1	—	—	—	—	—	—	—	—	—
YNP	34	65	T	1	—	—	46	50	3	T	—	—	32	59	6	2	—	—	30	69	T	T	—	—
N Can-PANP	34	59	—	—	—	7	35	65	—	—	—	—	26	73	—	—	—	1	17	63	—	—	—	20
N Can-SRL	36	63	—	1	—	—	16	81	1	2	—	—	24	59	8	8	—	—	21	71	4	2	—	—
N Can-MBS	2	96	—	2	—	—	6	68	—	26	—	—	11	53	2	28	6	—	32	15	4	12	37	—
Alaska	29	70	—	1	—	—	—	—	—	—	—	—	3	2	—	95	—	—	38	36	—	26	—	—

[a]NE Col, Northeastern Colorado, Pawnee National Grasslands, shortgrass plains, after Peden (1976); SW Col, southwestern Colorado, Colorado National Monument, semidesert range, after Wasser (1977); S Utah, southern Utah, Henry Mountains, shrub–steppe plant community, after Van Vuren (1984b); YNP, Yellowstone National Park, forest with interspersed grass–sedge meadows, after Meagher (1973); N Can-PANP, northern Canada, Prince Albert National Park, Saskatchewan, boreal aspen forest interspersed with sedge–grass meadows, after Fortin et al. (2002); N Can-SRL, northern Canada, Slave River Lowlands, Northwest Territories, boreal forest interspersed with sedge–grass meadows, after Reynolds (1976), Reynolds et al. (1978); N Can-MBS, northern Canada, Mackenzie Bison Sanctuary, Northwest Territories, boreal forest and old glacial lake beds interspersed with wet sedge meadows and willow savannas, after Larter and Gates (1991); Alaska, Farewell herd, boreal forest river flood plains and dry glacial lake beds interspersed with grass–sedge meadows—winter diets, after Campbell and Hinkes (1983), summer and fall diets, after Waggoner and Hinkes (1986). T, Trace amount or <1%.

[b]Forage class: G, grass; S+R, sedge and rush; F, forbs; S, shrubs; L, lichen; O, other.

The most common plant species in the bison diet during most seasons was fourwing saltbush (*Atriplex canescens*), followed by needle and thread, which was important during cooler months. Sand dropseed and Galleta grass (*Hilaria jamesii*) were prominent in the diet in warmer seasons. Prickly pear cacti (*Opuntia* spp.) were among the 10 top forages during all seasons except summer (Wasser 1977). The only forbs used significantly during all seasons except winter were mallows (*Malva* spp.). Some of the most common plants in the habitat, cheatgrass (*Bromus tectorum*), Utah juniper, and big sagebrush (*Artemisia tridentata*), were the least preferred forages. Grasses contributed from a low of 27% of the diet in winter to a high of 72% of the diet in summer (Table 48.5). Shrubs were the major dietary component during winter at 67% and were the second most important dietary item in fall, spring, and summer at 29%, 25%, and 17%, respectively.

On a shrub–steppe plant community range in the Henry Mountains in southern Utah, Van Vuren (1982) reported summer diets of bison comprised 93% grasses and sedges, 5% forbs, and 1% browse. In another area on the west side of the mountains, where dominant shrubs in the habitat were big sagebrush, black sagebrush (*A. nova*), and snowberry (*Symphoricarpos* spp.) and the grasses were mainly native perennials, including bluegrass (*Poa* spp.), needlegrass (*Stipa* spp.), wheatgrass (*Agropyron* spp.), fescue (*Festuca* spp.), fringed brome (*Bromus ciliatus*), squirreltail (*Sitanion hystrix*), and june-grass (*Koeleria cristata*), bison diets comprised 99% grasses and sedges and 1% forbs (Table 48.5; Van Vuren 1984b). Grasses contributed 96% of the bison diet, with bluegrass (66%) the most important, followed by june-grass (13%), fescue (10%), fringed brome (3%), and wheatgrass (2%), whereas sedges contributed only 3% of the bison diet (Van Vuren 1984b). Total forbs contributed just 1% and shrubs were present only in a trace (<0.5%) amount. Grasses predominated in the summer diet of bison in the Henry Mountains.

In Yellowstone National Park, much of the habitat consists of lodgepole pine (*Pinus contorta*) forests. Interspersed throughout the area are meadows dominated by sedges and grasses. Sedges were the main bison forage in all seasons and grasses were the next most common forage (Meagher 1973). Sedge content in the diet varied from 50% in spring to a high of 69% in fall, whereas the grass content varied from 30% to 46% in fall and spring, respectively (Table 48.5). Minor quantities of forbs (6%) and browse (2%) were consumed, but this occurred mainly during summer.

In a Boreal Aspen Forest habitat in Prince Albert National Park in northern Saskatchewan, Fortin et al. (2002) observed that the bison diet was almost entirely composed of sedges and grasses, varying from the lowest of 80% of the fall diet to 100% of the spring diet (Table 48.5). The seven most abundant plant groups in the habitat were two sedge species, slough sedge and water sedge (*Carex aquatilis*); four species of grass, wheatgrass, northern reed grass, foxtail (*Hordeum jubatum*), and whitetop (*Scolochloa festucacea*); and one rush species, wire rush (*Juncus balticus*). These species represented 50–72% of the total biomass available throughout the year and contributed 81–99% of the bison diet (Fortin et al. 2002). Sedges were the most important dietary component to bison in all seasons, varying from the low of 59% of the winter diet to the high of 73% in summer (Table 48.5). Slough sedge was the major food item throughout the year, as it was highly selected for and used disproportionately to its availability (Fortin et al. 2002). Grasses were the second most important forage item, contributing from a low of 17% in the fall diet to a high of 35% in spring (Table 48.5). Fortin et al. (2002) concluded that bison dietary decisions were made in preference for rapid energy acquisition over long-term gains.

In the Northwest Territories, bison habitat along the Slave River Lowlands is within the Boreal Forest region of Canada, where white spruce (*Picea glauca*) forests separate vast open meadows supporting sedge and grass communities. Bison diets from this area contained 29 different plants, of which 12 were present in quantities exceeding 1% in any one season. Slough sedge was by far the most abundant plant in the diet, varying from 42% in winter to 77% in spring (Reynolds 1976). The second most common food was reed grass (*Calamagrostis* spp.), which varied from 15% of the diet in spring to 35% in winter. Slough sedge and reed grasses contributed >70% of the bison diet at all seasons (Reynolds et al. 1978). Although slough sedge was the most abundant plant in the bison diet, it was second to reed grass in abundance in the habitat. Together, these two plants were the most abundant meadow plants in the Slave River Lowlands. Sedges and grasses collectively contributed in excess of 80% of bison diets in all seasons, varying from the low of 83% in summer to the high of 99% in winter (Table 48.5). Forbs and browse appeared to be of minor importance and were consumed in small amounts, mainly in summer and fall.

In the Boreal Forest habitat of the Mackenzie Bison Sanctuary in the Northwest Territories, Larter and Gates (1991) observed that seasonal bison diets varied. In winter, it was dominated by sedge (96–99%); in summer, it became a more diverse mixture of sedges, grasses, and shrubs. In fall, it became the most diverse, with lichen being a major component (34–41%). In wet meadow habitats, high-biomass stands of slough sedge and water sedge were the dominant plant species. In grass–sedge associations, other plant species were reed grass, wheatgrass, foxtail, reed canary grass (*Phalaris arundinacea*), rough hair grass (*Agrostis scabra*), and willows (Larter and Gates 1991). Sedges and grasses collectively contributed the major component of the bison diet in all seasons except fall, varying from the low of 47% in fall to the high of 98% of the winter diet (Table 48.5). Sedges were the most important dietary item throughout the year except during fall, when lichens and grass dominated the diet at 37% and 32%, respectively. Grasses were important to bison in fall (32%), less so in summer (11%), relatively unimportant during spring (6%), and almost nonexistent in winter (2%) (Table 48.5). The diet of the wood bison in the Mackenzie Bison Sanctuary was more diverse than the grass-dominant diets of southern plains bison (Peden 1976; Wasser 1977; Van Vuren 1984b) or of other northern Canadian herds of wood and plains bison (Reynolds et al. 1978; Fortin et al. 2002). These dietary differences basically reflect relative differences in forage availability in the various areas (Larter and Gates 1991). Shrubs contributed 26% and 28% of spring and summer diets, respectively (Table 48.5). This is similar to what is found for the Farewell herd in Alaska (Waggoner and Hinkes 1986), but differs from findings for Yellowstone National Park bison and other northern bison herds in Canada in the Slave River Lowlands (Reynolds et al. 1978) and in Prince Albert National Park (Fortin et al. 2002). However, the winter diet was similar to that of other populations of bison (Meagher 1973; Reynolds et al. 1978; Fortin et al. 2002).

In the northern Boreal Forest river flood plains habitat of Alaska, Campbell and Hinkes (1983) reported the winter diet of the Farewell plains bison herd as primarily sedges and rushes, and grasses (fescue), collectively constituting 99% (Table 48.5). Grass and sedge-dominated communities made up approximately 38% of the study area before a major fire, and the habitats being used by bison were dominated by bluejoint reed grass (*Calamagrostis canadensis*) and sedges, as well as other grasses and low-shrub-dominated communities. Range expansion was enhanced because of the increase in new sedge–grass areas created by the fire, which linked summer with prefire winter ranges. The summer diet of the Farewell bison herd was composed of 95% shrubs (Table 48.5). The fall diet from the river flood plains was 68% shrubs, whereas the fall diets from the dry lake beds and the burn area were dominated by sedges (74%) and grasses (75%), respectively (Waggoner and Hinkes 1986). Grasses and sedges were used heavily when available during migration movements from summering to wintering areas. In general, Alaskan plains bison diets were comprised mainly of three forage classes, grasses (38%), sedges and rushes (36%), and shrubs (26%) (Table 48.5).

ECOLOGY

Forage and Habitat Requirements. Bison, over most of their North American range, typically show strong selection for open grassland or meadow habitat types. Use of forested areas for escape cover, thermal cover, or other purposes such as calving also occurs to varying degrees. However, habitat selection by bison appears to be driven more by the interplay between their nutritional requirements and plant phenology, forage biomass, and snow depth (Larter and Gates 1991) than by predator avoidance or climatic condition. When these latter elements intensify, temporary shifts in habitat selection often occur. In the Mackenzie Bison Sanctuary in northern Canada, wood bison consistently selected wet and mesic meadows that were dominated by key forage species, slough sedge, in association with other grasses and sedges (Larter and Gates 1991). During summer and winter, they selected habitats that yielded the greatest amount of protein, and habitat selection did not change between years (Larter and Gates 1991).

In Aspen Parklands habitat, bison have shown a preference in summer and fall for grazing in upland meadows despite relatively low pasture biomass and potential dry matter intake rates (Hudson and Frank 1987). On the Slave River Lowlands, Reynolds (1976) observed that foraging was done in open meadows, whereas loafing and ruminating occurred in forest habitat. During winter, bison in that northern habitat foraged in small, sheltered meadows and along river and creek beds. Such areas offered less severe snow conditions and tended to support preferred forage. Although forested habitats, especially coniferous types, were avoided during summer and winter in the Mackenzie Bison Sanctuary (Larter and Gates 1991), shifts to greater use of coniferous habitat during the fall are believed related more to food availability (greater biomass of lichen) than to factors such as rut, climate, or biting flies (Melton et al. 1989; Larter and Gates 1991). During fall, bison dispersed among all habitats at a time when forage quantity and quality became more homogeneous throughout these habitat types (Larter and Gates 1991).

The recent burn history of a landscape has a strong influence on bison habitat selection, although there is some variability in their response. In the Wichita Mountains Wildlife Refuge in southwestern Oklahoma, bison increased use of a portion of their annual range following a prescribed burn (Shaw and Carter 1990). In a tallgrass prairie in Oklahoma that was subjected to seasonally and spatially variable burning regimes, mixed groups of bison showed significantly more use of burned areas than unburned areas (Coppedge and Shaw 1998). However, bull groups exhibited significantly more use of unburned areas than burned areas. Positive response to burns on tallgrass prairie appear particularly marked during the summer season, when bison concentrate on warm-season, perennial C_4 grasses, but this response is less apparent during fall and winter, when bison concentrate on cool-season, C_3 grasses (Vinton et al. 1993). In Nebraska, bison significantly increased their use of burned areas during the growing season 1–3 years post-fire, but did not select for these areas during the nongrowing season (Biondini and Steuter 1998).

In Alaska, bison in the Farewell study area showed a marked shift in range use within 4–5 years following wildfire, which included areas that had been previously forested with open black spruce (*Picea mariana*), but which became dominated with grass and sedge communities after the fire (Campbell and Hinkes 1983). Interestingly, studies of bison diet within the Farewell herd showed relatively high use of shrubs during summer at higher elevation ranges where only browse was readily available (Waggoner and Hinkes 1986). Periodic range burning has been effectively used to enhance bison habitat in the Northwest Territories of Canada (Chowns et al. 1998).

In a study on the Blue Mountain Bison Ranch in Colorado from August through September, bison selected different sites for nocturnal grazing than for diurnal grazing (Hein and Preston 1998). In this study, nocturnal grazing accounted for 15% of the daily grazing pattern of the herd. Also, Hudson and Frank (1987) noted that a greater percentage of summer foraging occurred at night in Aspen Parklands habitat in central Alberta.

Bison generally are less selective in what they eat than are other ungulates under similar environmental conditions. In the Henry Mountains of southern Utah, bison selected similar forage to cattle, but tended to move more; did not overgraze preferred feeding areas; and made greater use of steep slopes (Nelson 1965; Van Vuren 1982). Diets of bison on the shortgrass plains of Colorado resembled cattle diets more than sheep diets in areas of light grazing (Peden 1972). Bison are less selective than cattle and are therefore better adapted to use herbage of the shortgrass prairie more fully (Peden et al. 1974). A study on the Konza Prairie Research Natural Area in Kansas showed that the ratio of cool-season to warm-season grasses (C_3:C_4) in the diet of all bison groups changed as the seasonal availability changed (Post et al. 2001). Consumption of cool-season (C_3) grasses showed two peaks, the first in spring and the second in fall. In Yellowstone National Park, Wyoming, forage availability regulated bison feeding patterns within the preferred sedge and upland feeding sites (Meagher 1973). Bison were usually the least selective and cattle were the most selective during feeding trial experiments involving bison, yak, and cattle when provided sedge, grass, and alfalfa hays (Richmond et al. 1977). Bison have shown less grazing selectivity than cattle (Rice et al. 1974). In northern Canada, forage consumption by bison was not directly proportional to plant availability, indicating light to moderate feeding selection (Reynolds et al. 1978). Bison were superior to cattle in the digestion of all northern forages, including a greater ability to digest sedge hay, which resulted in superior performance (Hawley 1987). Therefore, bison are better adapted than cattle to use poorer quality rangeland forages and to survive harsh winter conditions and severe northern environments (Reynolds and Peden 1987; Hawley 1987).

In Alaska and northern Canada, although grasses and sedges were usually heavily selected for, dietary shifts to a high content of shrubs and lichens during summer and fall indicated the capability of bison to switch from a diet high in sedge when sedge biomass is scarce (Waggoner and Hinkes 1986; Larter and Gates 1991). It is apparent that where grasses and sedges are available in the habitat, they are selectively grazed by bison, and where they are sparse, browse may be substituted. Dietary shifts from grasses to sedges and back again within a habitat type usually are directly related to plant phenology. Bison generally may be considered as specialized grazers.

Effects of Bison on Habitat. It is likely that the vast historical bison herds had a significant effect on vegetation and soil nutrient regime within their traditional ranges through grazing, nutrient cycling, and sheer physical disturbance. Recent exclosure studies on migratory ungulates in Yellowstone National Park have shown the importance of ungulates in controlling soil nitrogen cycling (Frank and Groffman 1998; Tracy and Frank 1998). For specific details on the ability of bison to recycle nitrogen, see the section Physiology, Digestion and Nutrition.

Bison directly affect vegetation communities through grazing, urinating, defecating, trampling, rubbing, horning, and wallowing, and indirectly through incidental seed dispersal. These activities, in addition to climate and fire, help maintain grassland ecosystems on which bison and other species depend (Weaver and Clements 1938; Larson 1940; Axelrod 1985; Campbell et al. 1994; Coppedge and Shaw 1997). Larson (1940) observed that shortgrass vegetation thrived under moderately heavy grazing, unlike the taller bunch grasses such as sand dropseed and needle and thread.

Ungulate grazing on bluebunch wheatgrass (*Agropyron spicatum*) and Idaho fescue (*Festuca idahoensis*) in Yellowstone National Park results in compensatory growth following grazing, such that plants grazed in winter and spring have similar biomass as do the ungrazed plants (Merril et al. 1994). Response to grazing also included lower root biomass, increased foliar nitrogen levels, and alterations in soil nematode fauna. In a similar study of bison grazing effects on big bluestem (*Andropogon gerardii*) and switch grass (*Panicum virgatum*), compensatory increase in relative growth rate was observed in grazed tillers in the first year of grazing such that lost tissue was replaced by the end of the season (Vinton and Hartnett 1992). This response was enhanced on burned prairie for big bluestem, but unaffected for switch grass. At the same time, relative growth rates, mass, and survivorship of grazed big bluestem was lower than that of ungrazed individuals in the year following grazing.

Bison grazing patterns and selection for little bluestem (*Andropogon scoparius*) is affected by burn history and has potential to contribute to the decline of little bluestem under persistant grazing (Pfeiffer and Hartnett 1995). Following fire in tallgrass prairie dominated by little bluestem and big bluestem, bison increased their use of little bluestem threefold. In part, this related to removal of standing dead tillers associated with mature bunchgrass. Frequency of grazing on neighboring big bluestem was unaffected by fire (Pfeiffer and Hartnett 1995).

In studies of bison in relation to burn patterns and tallgrass prairie, Knapp et al. (1999) noted that bison grazing increased the spatial variation in tallgrass prairie in which fire tended to burn ungrazed patches. Preferential grazing of dominant species altered plant competition among grasses and forbs.

Recent field experiments with bison urine in South Dakota have demonstrated that aboveground biomass, root mass, and foliar nitrogen were higher on urine patches than off patches for Kentucky bluegrass (*Poa pratensis*) and little bluestem (Day and Detling 1990). Aboveground use by herbivores in this study was greater on urine patches than off urine patches.

Relative to cattle, bison exhibit habitat selection patterns that reinforce or otherwise help maintain habitat in a productive state. Studies by Wuerthner (1998) indicated that bison are less apt to regraze a site during a single growing season, wander more, eat drier, rougher forage, and spend less time in riparian areas and wetlands. Such behavior likely reduces impact on native grasslands and riparian ecosystems.

Localized stands of trees, particularly those not tightly clumped, may be significantly affected by the horning and thrashing of bison during the rut and at other times of the year. McHugh (1958) estimated that 51% of the lodgepole pine in some areas of Yellowstone National Park had been horned by bison. Bison horning of vegetation in a tallgrass prairie landscape had its greatest effect on saplings and shrubs, as opposed to trees, and killed or severely damaged 4% of woody plants, caused moderate injury to 13%, and caused light injury to 12% (Coppedge and Shaw 1997). Such activity by bison and/or elk may inhibit succession of prairie to forest and thus serve to maintain grasslands (Moss 1932; Patten 1963; Meagher 1973).

Bison wallows support a characteristic vegetation distinct from that found on adjacent prairie and appear to be related to site morphology, soil texture, soil moisture, available phosphorous, and pH (Polley and Collins 1984). In central and southwestern Oklahoma, wallows were dominated by spike rush (*Eleocharis* spp.) (Polley and Collins 1984).

The thick fur on the head and forequarters of a bison is ideally suited for dispersal of awned, barbed, or sticky seed-bearing structures. The seed of buffalo grass, cockle burs (*Xanthium italicum*), and St. John's wort (*Hypericum perforatum*) readily adhere to bison fur. The dissemination of St. John's wort throughout the National Bison Range is believed to have been caused by bison (McHugh 1958).

Where bison trails or wallows are cut into steep hillsides, considerable water and wind erosion can occur (McHugh 1958; Capp 1964; Meagher 1973). Hillside trails can serve as drainage channels, effectively lowering the water table in upland areas and subsequently causing a change in the vegetation. Where trails cut near the top of steep sandy hills, erosion and slippage may produce barren areas.

Interspecific Relationships. For the most part, habitat niches of plains bison and moose do not overlap. However, in northwestern Canada, moose coexist with the majority of free-ranging wood bison herds, and sometimes at high densities. Whereas bison diets are largely dominated by grasses and/or sedges, they will forage on willows and other woody browse, particularly when more-preferred forage is of poor quality (Larter et al. 1994). Potential for forage competition with moose could occur under such circumstances. In Farewell, Alaska, bison spent more time browsing in shrublands when grasslands were unavailable (Waggoner and Hinkes 1986). Because of the difference in height, moose would be able to take advantage of taller browse than bison. In general, moose are primarily browsers and bison are primarily grazers and therefore are considered to be more complementary than competitive in feeding habits.

Bison and moose generally appear to tolerate each other. However, at Jackson Hole Wildlife Park, Wyoming, bison killed a 7-month-old moose calf soon after it had been introduced into the park (McHugh 1958). The intense aggression of bison toward other wildlife species observed at the Jackson Hole Wildlife Park may be more frequent than that normally occurring in the wild because of the semiconfined nature of the animals at the park.

In a study of ungulate habitat relationships at Wind Cave National Park, South Dakota, there was relatively little overlap in range use between bison and elk (Wydeven and Dahlgren 1985). At Elk Island National Park, Alberta, bison winter diet was strongly different than that of sympatric elk, white-tailed deer, and moose (Telfer and Cairns 1986). Compared with elk, winter bison diets included a much higher percentage of sedge (82% vs. 22%) and much lower percentage of browse (1% vs. 64%). Within aspen parkland habitat where bison and elk have overlapping range, bison forage cropping rates are higher than those of elk at an equivalent forage biomass, although both species spend similar amounts of time daily foraging (Hudson and Frank 1987).

In Jackson Hole Wildlife Park, McHugh (1958) considered bison at the top of the interspecific dominance hierarchy, followed by elk, mule deer, pronghorn antelope, moose, and white-tailed deer. Bison are usually dominant over elk, and McHugh (1958) noted that bison calves could displace six-point bull elk. However, at the edge of bison herds, five- and six-point bull elk could displace bison cows and yearling bulls. Aggression by bison reversed any elk dominance. Bison occasionally forced elk into deep snow in winter and chased them from feed in summer. The usually wary elk sometimes were caught and butted. Bison harass and kill elk calves in Jackson Hole Wildlife Park, Fort Niobrara National Wildlife Refuge, Wind Cave National Park, and Yellowstone National Park (Rush 1942; McHugh 1958; Mahan 1977). At the National Bison Range, Montana, a bottle-raised bull elk calf formed an attachment to two cow bison and, when mature, rounded up a herd of cow bison and bugled. On another occasion, a harem bull elk killed a yearling bison (McHugh 1958). At Yellowstone National Park, elk and bison have been reported within 10 m of each other despite their seeming intolerance. At Elk Island National Park and Yellowstone National Park, bison interfere with elk livetrapping programs, in part because bison are attracted to the high-quality hay bait lines used for trapping elk; elk will not enter traps when bison are nearby.

Bison and white-tailed deer at Colorado National Monument did not appear to compete for food (Capp 1964). Wydeven and Dahlgren (1985) observed relatively little range overlap between bison and mule deer. Winter diet of bison at Elk Island National Park included substantially less browse than did that of white-tailed deer (1% vs. 57%) and more herbage (99% vs. 43%) (Telfer and Cairns 1986). Bison charge and strike mule deer at Jackson Hole Wildlife Park and Yellowstone National Park (McHugh 1958).

There is potential for bison to compete with woodland caribou (*Rangifer tarandus*) on northern ranges. Dietary overlap has been documented within the Mackenzie Bison Sanctuary in northern Canada, where bison selectively feed on terrestrial lichen from August to November (Larter and Gates 1991). In southern Yukon, Fischer (2002) assessed the potential for exploitative competition between bison and woodland caribou in winter by determining resource selection and overlap in resource use across three spatial scales: the landscape, habitat, and feeding site. Additionally, bison and caribou diets were analyzed and overlap in forage species use was assessed. Bison and caribou used resources in a differential non-random and selective manner at all levels. At the level of the landscape, caribou used higher elevations and areas that were more rugged and further from water than bison. Similar trends were also evident at the habitat level. Bison selected for graminoid-dominated habitats and caribou selected for coniferous/shrub-dominated habitats. Overlap in resource use at feeding sites was also minimal, since bison foraged in areas with high graminoid abundance while caribou foraged in areas with high lichen abundance. The dietary overlap was less than 10%; lichens made up nearly 80% of caribou diets while over 80% of bison diets consisted of graminoids. Fischer (2002) concluded that the potential for exploitative competition between bison and caribou in the late winter was low.

In a study of niche overlap in Yellowstone National Park that examined a group of ungulates including bison and bighorn sheep (*Ovis canadensis*), bison and sheep diets were not significantly associated with each other (Singer and Norland 1994). Traditional bighorn sheep range in much of North America typically is located in terrain not associated with bison use. However, in northern British Columbia, bison in summer will graze and browse in high-elevation, alpine and subalpine meadows as well as shrublands also used occasionally by Stone's sheep (*Ovis dalli stonei*). The two species occasionally comingle. Three instances of an older bighorn ram being associated with a bison bull were observed at the National Bison Range (McHugh 1958).

Pronghorn antelope are highly selective feeders (Schwartz et al. 1977), whereas bison are more flexible in choice of diet. Whereas pronghorn diets are 50–80% forbs, bison diets are usually <20% forbs (Peden 1972). In studies at Yellowstone National Park, bison and pronghorn antelope diets were not significantly associated with each other (Singer and Norland 1994). The theory that large and small ruminants will not compete with each other for food resources (Bell 1971) is further affirmed by similarity in sheep and pronghorn diets and their dissimilarity to bison diets (Peden 1972). Wydeven and Dahlgren (1985) suggested there was greater potential for competition between pronghorn and bison than between bison and either mule deer or elk. In a study of bison, pronghorn antelope, and prairie dog (*Cynomys* sp.) feeding relationships at Wind Cave National Park, bison preferentially grazed on graminoids located around the margins of prairie dog colonies; pronghorn focused on the forb-dominated central portions of prairie dog colonies (Krueger 1986).

In Wyoming, Bryant (1885) noted that pronghorns and deer would seek protection from wolves in bison herds. At Wind Cave National Park, pronghorn at times pass near or through bison herds unhindered. However, at the Jackson Hole Wildlife Park, bison disturbed a group of resting pronghorns when passing within 50 m (McHugh 1958). Bison occasionally charge pronghorns. McHugh (1958) observed a bison kill an 8-month-old pronghorn buck.

Bison diets more closely resemble those of cattle than those of sheep under light grazing conditions; however, the differences become less distinct under heavy grazing (Peden 1972). Diets of feral horses (*Equus caballus*) near Sundre, Alberta, had a higher sedge component than did diets of free-ranging cattle (Salter 1978). In this respect, feral horse diets show greater similarity to diets of northern bison than do cattle diets. Herbivores, however, appear to select for higher proportions of plant groups that are best digested (Peden et al. 1974). This may provide a mechanism whereby bison and other herbivores can avoid competition, particularly at low to moderate stocking rates. In the past, there seems to have been a reasonably good balance between bison and their forage supply, and competition with other species for food has not been a major factor limiting population growth (Longhurst 1961).

Draft horses at the Jackson Hole Wildlife Park were dominant over bison cows and yearlings at a salt lick during October; however, by March, they had lost some dominance (McHugh 1958). All bison exhibited dominance over a saddle horse. In Yellowstone National Park, the head animal keeper observed only one horse killed by bison in 23 years. Numerous instances of horses being killed or charged when used to pursue bison have been reported from other areas (McHugh 1958, 1972). Cattle ranged with bison during summer on a ranch in South Dakota and appeared to be compatible; the two species essentially ignored each other (Colman 1978).

In some regions, bison may form the prey base on which wolf populations depend, although this appears dependent in part on the abundance and diversity of alternate prey sources. In a study of wolf predation in northeastern British Columbia, Weaver and Haas (1998) found that bison constituted 26.7% of the prey biomass and 10.3% of the prey items in wolf diets. This compared with 32.9% and 34.7% for caribou, respectively, and 17.2% and 13.4% for moose. Within this study area, bison are among the more abundant prey species. Bison are, for the most part, indifferent to the presence of wolves until attacked (Fuller 1960; McHugh 1972). When a bison population decreases below a certain critical level, wolf predation—in the absence of alternate prey—may take more than the annual increment and effectively reduce the population. Bison may actively chase wolves (Smith et al. 1996) and defend their young from wolves, although the occurrence of this latter phenomenon appears difficult to predict (Carbyn 1998).

At Yellowstone National Park, coyotes stayed closer to bison in winter than in summer (McHugh 1958). These coyotes may have caught small mammals that were trapped in snow craters made by feeding bison or they may have been waiting for scavenging opportunities. On two occasions, coyotes wandering into a herd of bison at the National Bison Range were horned and trampled to death (McHugh 1958).

The importance of scavenged or killed bison in the diets of grizzly bears is considerably reduced from what it likely was when the plains grizzly and large bison herds coexisted on the Great Plains, although bison are significant contributors to grizzly bear diets where these species still coexist. A radiotelemetry study of grizzly bear in Yellowstone revealed that 95% of the energy requirements of grizzly bears were being derived from the largest bodied ungulate species; scavenged adult bison made up 16% of that total (Mattson 1997). Winter-killed bison and other ungulates may be important food sources to bears in early spring after they emerge from dens (Meagher 1978). In Yellowstone National Park, three grizzly bears captured an elk calf within 213 m of a bison herd without causing alarm to the bison (McHugh 1958). On occasion, bison prove too large an adversary for grizzly bears and may kill their would-be predators (McHugh 1972).

The role of bison in black bear ecology can be expected to parallel that of grizzly bear where black bear and bison ranges overlap. During a wolf study from 1975 to 1978 in the Slave River Lowlands, Northwest Territories, observations of black bears following bison groups and feeding on bison carcasses during the calving and postcalving seasons were made, but no observations were made of black bears actually chasing, attacking, or killing bison (Van Camp 1987). Black bears are opportunistic scavengers and probably scavenge on bison carcasses when available.

Humans have hunted North American bison for more than 12,000 years (McHugh 1972). The ecological relationship between the two species has remained that of a classic predator and its prey for approximately 11,900 of those years. In this regard, an early pioneer stated that for the Indian, bison were "meat, drink, shoes, houses, fire, vessels, and their Master's whole substance" (McHugh 1972:4). Bison influenced the settlement of North America more than any other endemic species (Roe 1970) with the possible exception of beaver. Whereas aboriginal cultures spiritually embraced the bison, European pioneers inhabiting buffalo country often viewed them as "an insufferable nuisance" (McHugh 1972:xxii). It was this dichotomy that, in part, led to the disruption of what had been an ecological balance between humans and bison. In less than 100 years, an estimated 40–60 million North American bison had been reduced to just over 1000 animals (Hornaday 1889). Only the dedication of a few conservationists in Canada and the United States and the physical isolation of bison in the Mackenzie basin (Raup 1933) saved them from extinction.

Bison have a special attraction to and relationship with prairie dog (*Cynomys ludovicianus*) colonies. Colonies are preferentially grazed by bison and are used for grooming and wallowing. Part of the attraction for bison relates to the lack of extensive dead standing graminoids such that foraging is more efficient. Heavily grazed prairie dog colonies at Wind Cave National Park had lower plant biomass and were dominated by plants that had higher leaf nitrogen than plants outside the colony (Detling 1998). Bison at Wind Cave National Park showed particularly strong attraction to prairie dog colonies by midsummer. They tended to graze in portions of colonies that were <8 years old and rested in the oldest parts (>26 years old) of the colony (Coppock et al. 1983). These observations were consistent with those of Krueger (1986), who observed bison foraging efforts concentrated around the margins of prairie dog colonies. Following experimental 2-year exclusion of prairie dogs and bison from prairie dog colonies, Cid et al. (1991) observed a 32–36% increase in aboveground biomass (mostly graminoid), whereas species diversity, equitability, and dominance remained similar for all treatments.

In Theodore Roosevelt National Park, North Dakota, extensive use of prairie dog colonies by bison during the breeding season occurs (Radtke et al. 1999). In one instance, 273 bison aggregated on a prairie dog town. Krueger (1986) observed that bison activity appeared to reduce foraging time available for prairie dogs. Prairie dogs benefit as a result of bison keeping the vegetation at a more favorable structural level (Shaw 1998). To what other extent prairie dogs may benefit from

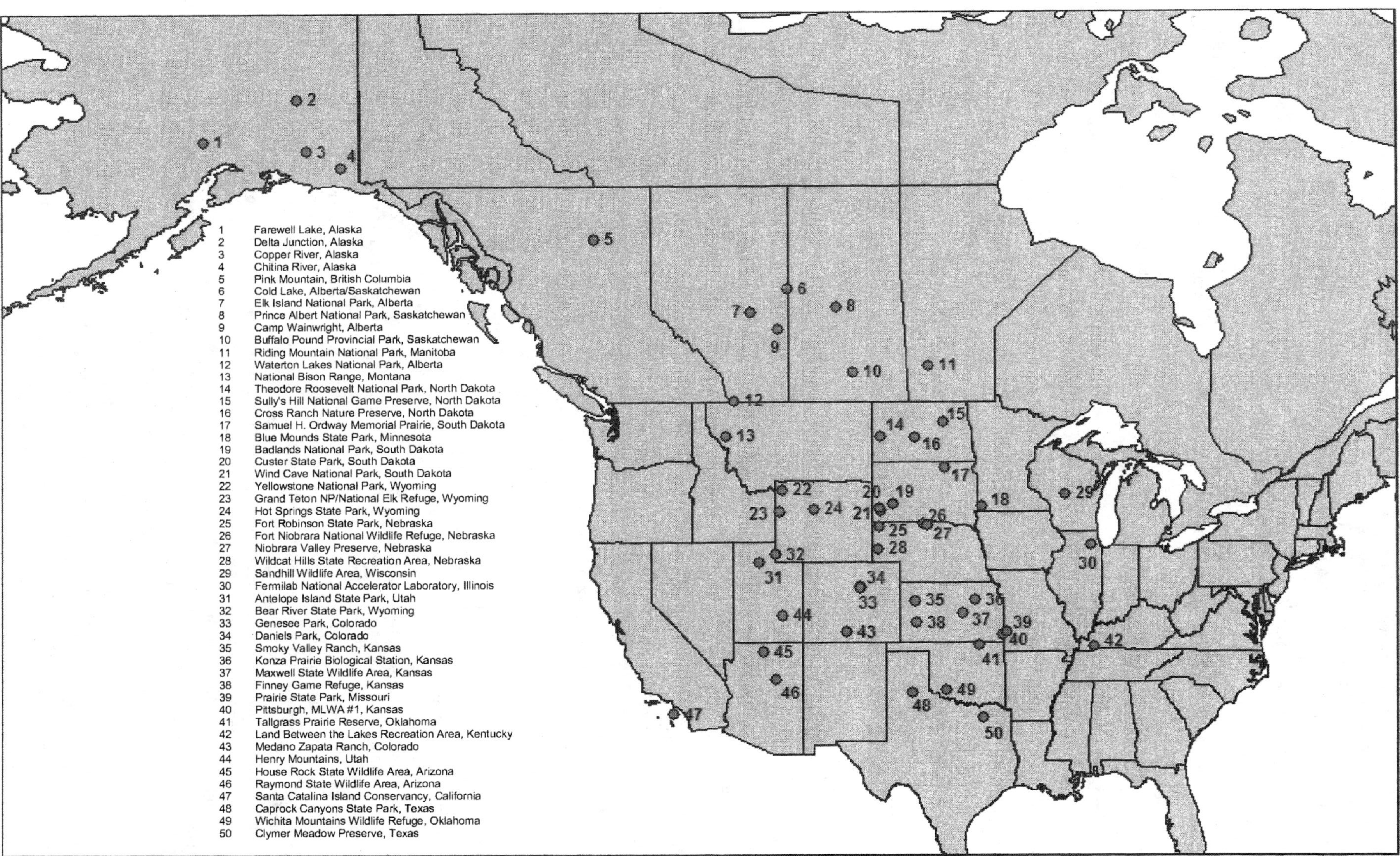

FIGURE 48.6. Distribution of public herds of plains bison (*Bison bison bison*) in North America. The current status of these herds is under investigation by the IUCN Bison Specialist Group. SOURCE: Delaney Boyd, Officer of the Bison Specialist Group, University of Calgary.

bison foraging is unclear, although potential mechanisms could include forage fertilizing and perhaps interference with predators.

Bison substantively shift their use from prairie dog colonies to recently burned areas where such an option exists (Coppock and Detling 1986). This phenomenon could be used as a management tool to mitigate negative impacts of bison on prairie dog colonies.

In a comparison of diet and distribution between bison and plains pocket gophers (*Geomys bursarius*) in north-central Nebraska, diets did not overlap. Highest pocket gopher activity was in forb patches, which appeared to be a direct response to bison activity (Steuter et al. 1995).

In a review of the close relationship between the brown-headed cowbird (*Molothrus ater*) and bison, Chace and Cruz (1998) suggested that cowbirds likely had a larger historical range and occupied a broader elevational range than at present. In turn, this effect could have influenced the intensity and extent of nest parasitism on high-elevation songbirds (Chace and Cruz 1998). Creation of small ephemeral wetlands through wallowing and attraction of insects to bison has likely had an attendant historical effect on avian species diversity and distribution within the prairie ecosystem where bison were common.

AGE ESTIMATION

Skinner and Kaisen (1947) developed a system of wear classification for bison teeth by recognizing six general age categories based on tooth wear and the sequence of molar eruption: immaturity, early adolescence, late adolescence, early maturity, full maturity, and old age. Fuller (1959) was able to identify five yearly age classes based on sequence of tooth eruption and replacement, and he could identify three general age classes based on tooth wear (young adult, adult, and aged). However, the precise age of animals older than 4 years could not be determined based on tooth eruption.

A more precise technique for determining the age of bison older than 4 years was developed by counting annual cementum layers in the roots of the fourth premolar (Novakowski 1965). In bison, cementum deposition on PM4 begins at age 4 years, making yearly age estimates possible after this. Cementum band counts using optical and scanning electron microscope examinations on the first mandibular molar were used to successfully determine the age of 20 wood bison (Haynes 1984). The first mandibular molar was selected for age determination because it is a large tooth, which is usually present even in weathered jaws; it is fully erupted and functional at 12 months of age, and thin layers of cementum are laid down annually on its roots (Haynes 1984). Although assessment of three classes of teeth—incisiforms, premolars, and molars—determined that all classes contained annuli in their cementum that accurately reflected age of the animal, premolar teeth provided the lowest standard deviation of cementum ages (± 1.04 years) and were considered to be best for age determination (Moffitt 1998). Premolars are more likely to remain attached to the aveolar bone after death because of their slightly more complex root system, which results in greater recovery of these teeth after an extended period post-death (Moffitt 1998).

Frison and Reher (1970) compared patterns of mandibular tooth eruption and wear with known-age samples to establish seven broad age categories for bison specimens collected from the Glenrock Jump in Wyoming: 0.5, 1.5, 2.5, 3.5, and 4.5 years, mature (5.5–9.5 years), and old age (10.5–13.5 years). Based on tooth eruption, the first five age categories for the Glenrock specimens could be distinctly classified with no overlap. Based on tooth wear, mature and old-age categories could be distinguished, but because of the subjective nature of wear classifications, definite ages could not be determined.

Age estimation from the postcranial elements of bison has always been difficult. Because of the need to estimate age and sex for such elements from archaeological sites, Duffield (1973) developed an age determination technique using a table showing ages associated with degree of epiphysial closure. It was possible to assign skeletal remains into yearly age categories up to 11 years using this table. However, precision depended on the degree of articulation of the elements examined (Duffield 1973). Fuller (1959) distinguished calf and yearling categories based on differences in body size and conformation.

ECONOMIC STATUS, MANAGEMENT, AND CONSERVATION

Current Status. Jennings (1978) estimated that there were approximately 65,000 plains bison in North America during the mid-1970s, with only a small proportion of those animals being free-ranging and wild. Most of these bison were confined in fenced areas of various sizes in parks, in nature preserves, and on private lands. By 1990, there were an estimated 90,000 plains bison in the United States, of which about 70,000 (78%) were in privately owned herds (National Buffalo Association 1990). In 2002, there were approximately 400,000–480,000 plains bison in the United States, of which about 385,000–465,000 (96%) were in privately owned herds. The total number of plains bison in Alberta, in Canada, and in North America, respectively, is estimated to be approximately 100,000–120,000, 200,000–240,000, and 600,000–720,000 animals (Gerald Hauer, Bison Specialist, Bison Centre of Excellence, pers. commun., 2002). Within North America, about 580,000–700,000, or 97% of the total number of plains bison, are held in privately owned herds. Only about 3% of the total number of continental plains bison are maintained in publicly owned and conservation-oriented herds. The distribution of public plains bison herds in North America is shown in Fig. 48.6. The origin, current status, conservation objectives, and management regime for each of these herds are presently under investigation by the World Conservation Union Bison Specialist Group.

Wood bison have achieved modest recovery and are no longer threatened by imminent extinction. However, the scope for further recovery in the wild in Canada is constrained by a variety of factors, of which the presence of diseased herds, habitat loss, and the increasing number of commercial bison ranches are the main limitations (Gates et al. 2001b). Reestablishment of additional herds within the original range of the subspecies in Canada and in Alaska will continue to play a key role in the conservation and recovery of wood bison. In addition, these initiatives will continue to contribute to ecological restoration and create opportunities for bison to play a major role in the culture and economies of northern peoples, as they have in the past.

Status of Wood Bison Recovery Herds in Canada. As of 2002, there were approximately 3154 wood bison in six free-ranging, disease-free herds in northwestern Canada (Table 48.6 and Fig. 48.7). There were 1029 in five captive-breeding herds, three of which are public herds and two of which are privately owned herds with conservation objectives; and about 4495 in infected, or presumed infected, free-ranging herds in and around Wood Buffalo National Park (Tables 48.6 and 48.7). In addition, there are at least 500–700 wood bison on 45–60 private ranches in Canada, mainly in the Prairie provinces. However, privately owned herds are managed for commercial production and are not regarded as contributing to the Wood Bison Recovery Program unless they are directly linked to conservation projects. The number and size of privately owned herds is increasing rapidly, but this is driven mainly by commercial interests. Two reestablished wild herds, the Mackenzie and the Yukon herds, have exceeded the minimum individual population objective of 400 animals (as set by the Wood Bison Recovery Team) and are the only disease-free, free-ranging wood bison populations for which hunting is permitted.

Wood bison conservation and recovery herds are classified into four categories: public herds that are free-ranging and not exposed to or infected with bovine tuberculosis and brucellosis; captive-breeding herds that are under public ownership or are being co-managed; captive-breeding herds that are under private ownership, but are directly linked to the recovery program as conservation projects (Table 48.6); and public herds that are free-ranging and infected or presumed to be infected with bovine tuberculosis and/or brucellosis (Table 48.7). The status of the six public wood bison herds that are free-ranging and disease-free follows.

The Mackenzie population (Fig. 48.7), the first disease-free herd to be reestablished in the wild, was founded in 1963 with the transfer of 18 wood bison from Wood Buffalo National Park to the Fort Providence area west of Great Slave Lake, Northwest Territories (Gates and Larter 1990). The herd currently is about 2000 animals and is the largest of six disease-free wild populations (Table 48.6). Regulated hunting has been

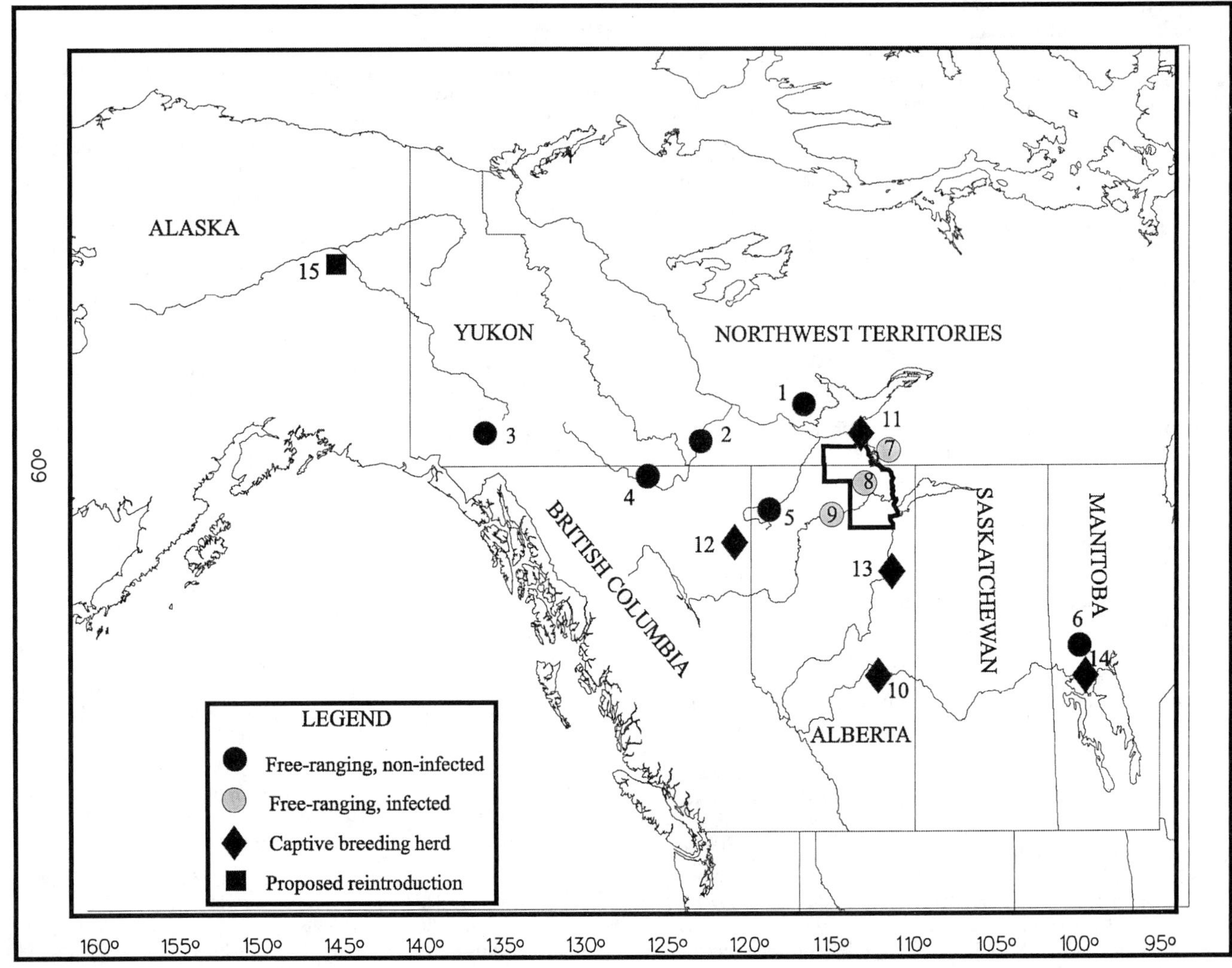

FIGURE 48.7. Distribution of free-ranging and captive-breeding wood bison (*Bison bison athabascae*) herds in Canada and the location of a proposed reintroduction in Alaska. Numbers provide a cross-reference to the following list of herds and locations: 1, Mackenzie (2000); 2, Nahanni (200); 3, Yukon (530); 4, Nordquist (62); 5, Hay-Zama (262); 6, Chitek Lake (100); 7, Slave River Lowlands (518); 8, Wood Buffalo National Park (3870); 9, Caribou–Lower Peace (107) (Wentzel–Wabasca); 10, Elk Island National Park (350); 11, Hook Lake Recovery Project (132); 12, Etthithun Lake (49) (Note: now free-ranging); 13, Syncrude Canada Ltd./Beaver Creek Wood Bison Ranch (200); 14, Waterhen Wood Bison Ranches Ltd. (298); 15, Yukon Flats, Alaska (proposed).

permitted since 1988. All other reestablished, free-ranging, disease-free herds were founded with stock originating or derived from Elk Island National Park.

The Nahanni population (Fig. 48.7) was established in 1980, when 28 wood bison were directly released to the wild. It was supplemented with 12 bison in 1989 and 60 in 1998. The Nahanni herd is estimated to be approximately 200 (Table 48.6). It ranges in the Liard River drainage in the southwestern Northwest Territories and northeastern British Columbia. Hunting of this herd is not permitted in either jurisdiction.

The Yukon herd was established by seven independent releases for a total of 170 wood bison from 1986 through 1992. Most of the released bison were transferred from Elk Island National Park, with small numbers being transferred from the Moose Jaw Wild Animal Park in 1989 and the Toronto Zoo in 1992. The Aishihik herd free-ranging in the Yukon is about 530 (Fig. 48.7 and Table 48.6). Based on winter herd counts, annual calf recruitment from 1998 to 2003 has maintained a rate of 18–20%/year. Hunting (annual harvest of 70–90 animals) has been instituted as a management tool to limit the growth of the herd and maintain a population of approximately 500 (Government of Yukon 1998).

Wood bison were reintroduced into northeastern British Columbia in March 1995 to establish the Nordquist herd. Forty-nine were transferred from Elk Island National Park to the Nordquist Flats area in the Upper Liard River Valley at a site approximately 80 km from the southern part of the Nahanni herd's range (Fig. 48.7). The present population is estimated at 62 animals (Table 48.6). The Nahanni and Nordquist herds eventually are expected to coalesce to establish a population of at least 400 bison. The management goals for wood bison in British Columbia are to reestablish additional herds through translocation, maintain separation from plains bison, maintain disease-free status, and permit populations to increase to a level sufficient to support nonconsumptive and consumptive use (Harper et al. 2000).

A project to reestablish the Hay-Zama wood bison herd in northwestern Alberta (Fig. 48.7) was initiated in 1981 in cooperation with the Dene Tha First Nation. A free-ranging herd was established in 1993 when 48 bison escaped prematurely from the fenced compound. An aerial survey (Morton 2003) estimated the population at 262 in winter 2002/2003 (Table 48.6). The Government of Alberta established a 36,000-km^2 management area in the northwest part of the province to protect the Hay-Zama wood bison herd because all other bison in the province are designated as "livestock" in the Alberta Wildlife Act. This management area has the potential to support at least 400 wood bison, but plans are to maintain the herd well below 400 animals until the northern bison disease issue is resolved.

TABLE 48.6. Status of public, free-ranging, disease-free wood bison herds in Canada, public and co-managed captive-breeding herds, and privately owned herds that have conservation objectives

Location	Herd Type	Status	Year of Estimate	Source of Stock[a]	Year Established	Number Released[b]
Mackenzie	Free-ranging/public	2000	2000	NR	1963	18
Yukon (Aishihik herd)	Free-ranging/public	530	2002	EINP	1986	170
Nahanni	Free-ranging/public	200	2002	EINP	1980	100
Chitek Lake	Free-ranging/public	100	2002	EINP	1991	22
Hay-Zama	Free-ranging/public	262	2003	EINP	1993	48
Nordquist	Free-ranging/public	62	2003	EINP	1995	49
Subtotal		3154				
Elk Island National Park	Captive/public	350	2002	NR	1965	21
Hook Lake Wood Bison Recovery Project	Captive/co-managed	132	2003	SRL	1996	62
Etthithun Lake	Captive/public[c]	49	2003	EINP	1999	43
Subtotal		531				
Waterhen Wood Bison Ranches Ltd.	Private/conservation	298	2003	EINP	1984	34
Syncrude Canada Ltd./Beaver Creek Wood Bison Ranch	Private/conservation	200	2003	EINP	1993	30
Subtotal		498				
Total		4183				

[a]NR, Nyarling River; EINP, Elk Island National Park; SRL, Slave River Lowlands.
[b]Number released to establish free-ranging herd or initial number to establish captive-breeding herd.
[c]Note: now free-ranging herd.

A free-ranging herd of wood bison was established in the northern Interlake Region of Manitoba (Fig. 48.7) from a total of 22 wood bison released near Chitek Lake, north of the Waterhen Wood Bison Ranch, from 1991 to 1993. The population was estimated to be approximately 100 during winter 2002–2003 (Table 48.6). The Chitek Lake region has the potential to support 400–500 wood bison. The province of Manitoba has established a protected area for those wood bison and their habitat.

The status of the three captive-breeding herds of wood bison that either are under public ownership or are co-managed for conservation and recovery follows.

The Elk Island National Park wood bison herd (Fig. 48.7) has played a key role in the recovery of wood bison in Canada, providing stock directly or indirectly for the establishment of five wild populations. The 2002 fall population was approximately 350 (Table 48.6). This semiwild herd is fenced in a 65-km^2 area, relies on natural forage, and interacts with its environment under natural conditions. However, this herd is not subjected to predation by wolves or any other large carnivores.

In 1991, the Deninu Kue' First Nation initiated a program to restore a disease-free herd of wood bison in the Hook Lake area of the eastern Slave River Lowlands, Northwest Territories. Specific objectives of the Hook Lake Wood Bison Recovery Project which was implemented in 1996 are to raise a disease-free herd of captive wood bison from salvaged Hook Lake calves, conserve genetic integrity, salvage disease-free wood bison from the wild Hook Lake herd, and use the captive herd as a source stock to establish a disease-free wild population (Nishi et al. 2001; Nishi et al. 2002a). The project involves three phases including habitat renewal, propagation of a disease-free captive-breeding herd in Fort Resolution (Fig. 48.7), and eventual reestablishment of a disease-free herd in the wild (Gates et al. 1998; Nishi et al. 2001). By July 2001, 62 calves had been removed from the wild herd yielding a captive-breeding herd of 58 founders plus another 50 captive-born (Nishi et al. 2002a). This has resulted in a captive-breeding herd of at least 132 disease-free wood bison (Table 48.6). The number of founders used to establish the Hook Lake captive-breeding herd represents more genetic diversity than two previous attempts to salvage wood bison genetics from Wood Buffalo National Park (Nishi et al. 2001).

TABLE 48.7. Status of public, free-ranging wood bison herds in Canada that are infected or presumed infected with bovine tuberculosis and brucellosis

Location	Herd Type	Status	Year of Estimate	Source of Stock
Slave River Lowlands	Free-ranging/public	518	2000	Indigenous
Wood Buffalo National Park	Free-ranging/public	3870	2002	Indigenous
Caribou-Lower Peace, Wentzel	Free-ranging/public	44	2002	Indigenous
Caribou-Lower Peace, Wabasca	Free-ranging/public	63	2002	Indigenous
Total		4495		

In March 1999, 19 wood bison were transferred from Elk Island National Park to a fenced 850-ha enclosure in the Etthithun Lake area in northeastern British Columbia (Fig. 48.7). In February 2000, 24 additional wood bison were moved to the captive herd from Elk Island National Park. In January 2003, the total herd, including 7 calves, numbered 49 (Table 48.6). However, at press time, it was learned that the Etthithun Lake wood bison are no longer contained by fencing and constitute a new free-ranging herd.

The status of the two captive-breeding herds that are privately owned, but have conservation objectives with direct linkages to the Wood Bison Recovery Program follows.

In 1984, Elk Island National Park provided founding stock for a wood bison ranch in the northern Interlake region of Manitoba (Fig. 48.7). In 1991 and 1993, 22 wood bison from this herd were used to establish a free-ranging herd in the Chitek Lake area. The Skownan First Nation manages the Waterhen Wood Bison Ranches Ltd. captive herd of approximately 298 animals (Table 48.6) as a commercial wood bison ranch.

The Syncrude Canada Ltd. herd was established in 1993 with 30 wood bison from Elk Island National Park to determine whether restored soil on reclaimed oil sands could support forage crops and a productive bison herd. Syncrude Canada Ltd. and the Fort McKay First Nation cooperatively manage the Beaver Creek Wood Bison Ranch herd of wood bison in fenced pastures encompassing 2.6 km^2 of oil sands reclamation property north of Fort McMurray (Fig. 48.7). This captive-breeding herd presently numbers about 200 (Table 48.6). In the future, there is potential to convert 1800 ha of reclaimed oil sands lands to a mosaic of grassland, wetland, and shrubland habitats for maintenance of a commercial and/or conservation herd exceeding 400 animals.

Disease testing has confirmed that bison herds in the vicinity of Wood Buffalo National Park and the Slave River Lowlands have been exposed to or are infected with bovine tuberculosis and brucellosis, which were transferred with plains bison that were introduced from

Wainwright, Alberta, in the 1920s (Tessaro 1987). The status of the four public, free-ranging populations of bison that are infected or presumed to be infected with bovine tuberculosis and brucellosis follows.

In 1970, there were about 2500 bison on the Slave River Lowlands, Northwest Territories, adjacent to Wood Buffalo National Park. The Hook Lake herd on the east side of the Slave River (Fig. 48.7) declined from 1700 to about 200 by the late 1980s. Poor calf production, infection with bovine tuberculosis and brucellosis, wolf predation, and hunting contributed to this decline (Reynolds and Hawley 1987). In 2000, the Slave River Lowlands population, including both the east and west side of the Slave River, was estimated at 518 bison (Table 48.7).

The Wood Buffalo National Park wood bison population (Fig. 48.7) has continued to decline since the 1970s from at least 10,000 to a low of about 2200 in 2000. However, in winter 2002, a more intensive coverage aerial survey resulted in a minimum count of 3870 bison (Table 48.7) in Wood Buffalo National Park (Mark Bradley, Wood Buffalo National Park, pers. commun., 2003). The metapopulation of bison associated with Wood Buffalo National Park, which includes the Slave River Lowlands herds, is approximately 4300–4400 bison. It is infected with bovine tuberculosis and brucellosis (Joly and Messier 2001a, 2001b). The long-term viability of Wood Buffalo National Park subpopulations and the impact on wood bison recovery needs to be addressed. There is a requirement for a collaborative, multistakeholder planning process to develop a management plan that will lead to eradication of these bovine diseases and recovery of healthy increasing herds of wood bison in the Greater Wood Buffalo Region.

There is relatively little information concerning bison populations in the Caribou-Lower Peace region of Alberta adjacent to the southwest corner of Wood Buffalo National Park (Fig. 48.7). A Wentzel Lake bison herd ranges in the boundary area between Wood Buffalo National Park and Wentzel Lake, north of the Peace River. An estimate of the number of animals residing in the Wentzel area in winter 2002 was 44 (Table 48.7). During an aerial survey in March 2002, a minimum of 63 bison was counted in two herds located between the Mikkwa and Wabasca Rivers (Kim Morton, Alberta Sustainable Resource Development, pers. commun., 2003). These animals constitute a second bison population known as the Wabasca herd (Table 48.7). There is a need to assess the size and structure, the disease and genetic status, habitat use, and movement patterns of these two wild herds in the Caribou-Lower Peace region of Alberta.

Habitat Management. Expansion of resource extraction, forestry, and associated human activities in the boreal forests of northern Alberta and British Columbia present challenges and opportunities for wood bison recovery in these areas. Although habitat loss and degradation and increased hunting are likely to result from development activities, there is some potential for positive effects. For example, the Syncrude Canada Ltd. captive-breeding herd was established on reclaimed grasslands in northeastern Alberta, and planting of forage species in oil and gas exploration right-of-ways in northeastern British Columbia was commissioned to enhance habitat for wood bison.

Disease Management. Two exotic diseases are problematic for wild bison conservation in North America. Bovine brucellosis and bovine tuberculosis are bacterial diseases originating in cattle. Bison are susceptible and the pathobiology of the infectious organisms is similar in cattle and bison. Free-ranging bison populations in and adjacent to Wood Buffalo National Park are infected with *Mycobacterium bovis,* the causative agent for tuberculosis. It was introduced in the region with the transfer of plains bison in the 1920s. Bison populations in northern Canada and those in Yellowstone and Grand Teton National Parks are infected with *Brucella abortus,* which causes the disease brucellosis. These pathogens are zoonoses, as they can cause disease in humans. In addition, they have been the subject of largely successful eradication programs in domestic livestock in Canada and the United States. Brucellosis in Yellowstone bison is of immediate concern to the cattle industry in Montana, Wyoming, and Idaho. The presence of tuberculosis and brucellosis in bison in the Greater Wood Buffalo Region is of concern for wood bison conservation (Tessaro 1986; Gates et al. 1992a, 1997, 2001c), the commercial bison industry, and the cattle industry in northern Canada. Proposals to eradicate these pathogens from bison have been the subject of significant controversy since the mid-1980s.

In December 2000, following several years of collaborative planning by the Greater Yellowstone Interagency Brucellosis Committee, the U.S. Department of the Interior, National Park Service and the U.S. Department of Agriculture, Animal and Plant Health Inspection Service and Forest Service signed a Record of Decision on a Joint Management Plan for bison in Yellowstone National Park and Montana (*Federal Register* 2001). This plan was designed to preserve the Yellowstone bison population and minimize the risk of transmission of brucellosis between bison and cattle, but was not designed to achieve disease eradication. The plan allows for limited use of public lands outside the park by bison during winter when cattle are not present. The potential for mixing of bison with cattle is reduced by hazing bison back into the park when possible. Bison that remain outside the park are subject to capture or removal and cattle are not allowed to return to public lands until a sufficient amount of time passes to ensure that bacteria are no longer viable on the range. In an area where there is a large degree of concern about bison leaving the park, bison may be captured and tested for antibodies to *B. abortus,* and seropositive bison may be slaughtered. The Joint Management Plan prescribes a spring bison population of up to 3000 animals be maintained. The signatory agencies have agreed to increase the use of nonlethal management measures should severe winter conditions result in a large removal for management purposes or a natural winter die-off. Cattle in areas north and west of Yellowstone National Park must be vaccinated against brucellosis. Vaccination of bison inside the park with a safe and effective vaccine delivered by remote injection is also prescribed. In addition, research initiatives are being undertaken in support of the management plan. Initiatives include development and testing of the efficacy and environmental safety of vaccines for use in bison and elk, pathobiology, histopathology, and epidemiology and population effects of brucellosis in bison, environmental persistence of *B. abortus* in tissues and contaminated ground, detection of *B. abortus* in tissues, the potential for transmission of the disease between ungulate species, and factors affecting seasonal movements and distribution of bison.

In Canada, tuberculosis and brucellosis are widespread among bison herds in the vicinity of Wood Buffalo National Park (Tessaro et al. 1990). All subpopulations within Wood Buffalo National Park are infected. Two herds in the Slave River Lowlands, adjacent to Wood Buffalo National Park, are infected with these two diseases (Broughton 1987). Evidence of disease was also detected in bison outside the southwest corner of Wood Buffalo National Park in the Caribou–Lower Peace region (Tessaro et al. 1990). The disease status in other small remnant herds reported to exist south of Wood Buffalo National Park has not been assessed (Gates et al. 2001c). Also, there is concern that livestock in the region are at risk of becoming infected. In 1985, Canada's national cattle herd was declared free of bovine brucellosis, and bovine tuberculosis had nearly been eliminated (Gates et al. 1997). Consequently, wildlife reservoirs of these two diseases became of increasing interest to the livestock industry and agencies responsible for regulating infectious animal diseases. These infected bison herds are seen as the last unregulated source for potential reinfection of cattle in Canada. Agricultural activity near the southwest corner of Wood Buffalo National Park has substantially expanded over the past 30 years. Not only are local farmers grazing cattle and bison in the area, but livestock is transported under subsidy to and from community grazing pastures in the region. In recent years, the commercial bison industry has rapidly expanded in northern Alberta, with an increase in the risk of infection through contact with wild bison from infected herds (Gates et al. 2001c). Free-ranging bison have been seen in a large area west and south of Wood Buffalo National Park and in the agricultural zone west of the park (Gainer 1985; Gates et al. 1992b, 2001c). Clearly, there is a need to better integrate conservation biology with agricultural livestock policy to develop management options and better address the unique conservation challenges that are presented by diseased, free-ranging bison populations in Northern Canada (Nishi et al. 2002b).

Unlike the Yellowstone Ecosystem, there are several healthy wild bison populations in northern Canada for which there is concern about the risk of infection (Gates et al. 2001b, 2001c). Two captive-breeding herds of wood bison also have been established in the region. Among the free-ranging herds, the Mackenzie population is in closest proximity to the infected herds; about 100 km separates it from Wood Buffalo National Park. The Government of the Northwest Territories, in cooperation with Wood Buffalo National Park, established a buffer zone and ongoing surveillance program in 1987 between the disease-free Mackenzie herd and Wood Buffalo National Park to reduce the risk of disease transmission. The program attempts to exclude bison from the buffer zone through active surveillance and culling (Gates et al. 1992b). In northern Alberta, bison around Wood Buffalo National Park are not protected from hunting to encourage depopulation of the area and to reduce the risk of contact between healthy and infected bovine populations.

Although a collaborative research program has been completed in Wood Buffalo National Park to define the role of the diseases in bison population dynamics, there is presently no mechanism for developing a long-term approach to bison conservation and disease risk management. Research has been completed on the role of the two diseases in limiting population growth of northern bison (Joly and Messier 2001a, 2001b) and a disease risk assessment was completed by the Animal, Plant and Food Health Risk Assessment Network (1998). Gates et al. (2001c) compiled and mapped local ecological knowledge of bison in the Northwest Territories and Alberta, described community-based initiatives and interests, and calculated spatially explicit movement corridors through northern Alberta. They suggested that the information gained from this research provides a sufficient basis for initiating a planning process to define actions to deal with disease risk management and bison conservation. Gates et al. (2001c) recommended that a collaborative planning process be developed in which the interests of northern communities, the livestock industry, conservation groups, and government agencies are represented. They recommended implementing a process similar to the collaborative resource management process described by Wondolleck and Yaffee (2000), a stakeholder consensus decision-making process that involves the people who use the resources (affected interests) in developing decisions about those resources. Collaborative resource management is the most open and accessible process available for planning and decision making in natural resource management and has the potential to produce the most widely endorsed plan.

It is clear that any effective solution to the diseased bison problems in the Greater Yellowstone Ecosystem and northern Canada must satisfy interested parties who hold differing and often opposing points of view. There is no apparent easy solution to either situation. In our view, continued dialogue and collaborative planning should be accompanied by research aimed at addressing key information gaps while safeguarding values at risk such as the genetic and ecological value of infected bison populations as well as the disease-free status of free-ranging bison herds and livestock. Essential to this process will be the need for increased integration and collaboration among conservation biologists, veterinary practitioners, wildlife management authorities, the general public, and all affected stokeholders in a truly collaborative management planning process (Nishi et al. 2002b).

Emerging Issues. Chronic wasting disease (CWD), a form of spongiform encephalopathy, was first detected in North American wildlife in association with captive mule deer at a research facility run by the Colorado Division of Wildlife in 1967 (Madson 1998). The infective agent of CWD is a class of proteins now referred to as "prions," which can be transmitted where animals consume contaminated ruminant proteins in feed and possibly by other means (Prusiner 1995). This disease and its relatives such as scrapie, bovine spongiform encephalopathy, and the human variant, Creutzfeldt–Jakob disease, typically result in an incurable neuropathological condition resulting in death. Among North American wildlife, CWD has not been reported in bison, pronghorn, or bighorn sheep, but has been diagnosed in elk, mule deer, white-tailed deer, and black-tailed deer (Williams and Young 1982; Spraker et al. 1997).

Conservation and Recovery Management. The bison was a keystone component of the North American plains and is considered to be a keystone species in tallgrass prairie (Knapp et al. 1999). Bison provided sustenance and the very "life blood" for aboriginal residents in North America and were the staple food for early explorers, fur traders, and many European settlers. The plains bison was nearly exterminated during the late 1800s because of indiscriminate killing resulting from market hunting for meat to satisfy the increasing population growth of European settlement and hide hunters for the fur trade. Threats to conservation and recovery of wild bison continue to exist even though bison numbers have significantly increased because of conservation and recovery efforts since the early 1900s as well as the rapid expansion of commercial bison production. Further recovery of bison as a wildlife species on public lands is limited by a variety of factors. These include habitat loss from agricultural development, urbanization, and other competitive intensive land uses; commercial bison production; loss of genetic diversity; bovine diseases; reintroduction difficulties; problems with maintaining genetic distinctness between subspecies; introgression of cattle DNA; and, in some instances, legislation, regulations, and policies that are inadequate for the long-term protection and conservation of bison (Boyd and Gates 2001). In general, there is a misconception that the North American bison as a wildlife species is secure and will survive in perpetuity.

Conservation Methods. In North America, bison management practices vary considerably, depending on the objectives of the individuals and agencies controlling the animals. Bison have been managed to preserve the species and subspecies, for commercial meat and commodities production in farming and ranching operations, as a game species of wildlife, as tourist attractions for nonconsumptive users, and for their historical significance. A variety of international, national, and local organizations and initiatives are involved with conservation and management of bison. A listing and brief description of the main organizations and initiatives follow.

THE WORLD CONSERVATION UNION/SPECIES SURVIVAL COMMISSION. The World Conservation Union (IUCN) was founded in 1948 and brings together more than 10,000 scientists and experts from 181 countries in a unique worldwide partnership. The IUCN mission is to influence, encourage, and assist societies throughout the world to conserve the integrity and diversity of nature. The Species Survival Commission (SSC), the largest of six commissions within the IUCN, cooperatively functions to provide information and recommendations on the status and conservation of species to managers, agencies, educational institutions, and others capable of implementing conservation actions. The Bison Specialist Group (BSG), which operates under the SSC, focuses specifically on bison conservation. There are two sections of the BSG, North America and Europe, each focusing on bison conservation in their respective jurisdictions and chaired by a regional co-chair. In 2002, the BSG, North America, commissioned a bison conservation status survey. This project is to review the conservation status of North American bison subspecies, the plains bison and the wood bison, and develop recommendations for enhancing bison conservation.

CONVENTION ON INTERNATIONAL TRADE IN ENDANGERED SPECIES OF WILD FAUNA AND FLORA. The Convention on International Trade in Endangered Species of Wild Fauna and Flora (CITES) is an international agreement between governments to ensure that international trade in specimens of wild animals and plants does not threaten their survival. No species protected by CITES has become extinct as a result of trade since the Convention was ratified, and its member parties now exceed 150. The species protected under CITES are listed in three appendices according to the degree of protection they need. In 1977, wood bison were placed on Appendix I, the highest level of protection. In June 1997, the subspecies was downlisted to Appendix II because of the absence of a threat from international trade, progress toward recovery in the wild, and the rapidly expanding ranching industry. In addition, there would be inherent problems with regulating exports and imports of wood bison and their products produced on commercial ranches if wood bison were to remain at the Appendix I level (Gates et al. 2001b). Plains bison are not on the CITES list.

Committee on the Status of Endangered Wildlife in Canada. The Committee on the Status of Endangered Wildlife in Canada (COSEWIC) is a federal–provincial committee created in 1977 at the Conference of Federal, Provincial, and Territorial Government Wildlife Directors. This Canadian national committee, composed of scientific representatives from government as well as nongovernment scientific experts, operates at arm's length from governments and is responsible for assessing and classifying Canadian wildlife species that are suspected of being at risk based on the best available scientific, community and aboriginal traditional knowledge. With the proclamation into law of the Species at Risk Act in June 2003, COSEWIC has now been established as a legal entity. Wood bison were recognized by COSEWIC as an endangered subspecies of Canadian wildlife in 1978 (Gates et al. 2001b). This status was changed to threatened in June 1988, based on recovery progress as reported in the COSEWIC status report (Wood Bison Recovery Team 1987). Plains bison are currently being assessed and therefore do not yet have a status designation by COSEWIC.

Recovery of Nationally Endangered Wildlife. Canada launched a national recovery program in 1988 to rescue wildlife species at risk of extinction and to prevent other species from becoming at risk, named RENEW (the acronym for REcovery of Nationally Endangered Wildlife). This program involves federal, provincial, and territorial government agencies, wildlife management boards authorized by a land claims agreement, aboriginal organizations, other nongovernment organizations, and interested individuals working together for the recovery of endangered, threatened, or, where possible, extirpated species that have been designated by COSEWIC. The recovery process includes formation of national recovery teams, development of recovery strategies and action plans, cooperative recovery actions, and program evaluation. Wood bison is on the priority list for RENEW, and the National Wood Bison Recovery Plan was produced under its authority and direction (Gates et al. 2001b).

Wood Bison Recovery Program. Wood bison probably were never as numerous as plains bison, although they inhabited a vast region in the boreal forest biome during the late Holocene. Most early observations of northern bison were recorded by explorers during expeditions throughout northern Canada in the late 1700s (Wood Bison Recovery Team 1987). The total population of wood bison in 1800 was about 168,000 animals, based on estimated carrying-capacity potential (Soper 1941). Wood bison were nearly eliminated from their remaining range in Canada during the late 1800s, coinciding with the rapid decline of plains bison between 1840 and 1900 (Raup 1933). By 1891, a population of only 300 wood bison remained in the wilderness region between Great Slave Lake and the Peace–Athabasca Delta (Ogilvie 1893). This population further declined to an estimated low of approximately 250 animals during 1896–1900 (Soper 1941). After 1900, wood bison occasionally were observed in parts of their original range, but numbers were insignificant except in the Slave River Lowlands and the area designated later as Wood Buffalo National Park (Gates et al. 1992a, 2001d). Heavy exploitation following the advent of the fur trade played a major role in the decline of wood bison in Canada (Gates et al. 1992a, 2001b).

Bison conservation efforts began in Canada in 1877 with passage of the Buffalo Protection Act (Hewitt 1921). However, this measure was largely ineffective because of lack of enforcement. The first police outpost was established along the Slave River in 1907, and six Buffalo Rangers were appointed in 1911 to patrol the wood bison range. The wood bison population began to increase slowly to approximately 500 by 1914 (Banfield and Novakowski 1960). By 1922, when Wood Buffalo National Park was established to save the wood bison from extinction and to protect their habitat, the total number was estimated at between 1500 and 2000 (Seibert 1925; Raup 1933; Soper 1941; Lothian 1979).

In 1905, the largest privately owned herd of plains bison in North America was threatened by the loss of grazing rights. Following negotiations between the Canadian government and the owner, 410 plains bison were purchased from the Pablo-Allard herd and shipped from northern Montana to Elk Island National Park, Alberta, in 1907 in a major conservation effort (Lothian 1981). On completion of the newly enclosed Buffalo Park at Wainwright, Alberta, in 1909, 325 plains bison from Elk Island, 218 plains bison from the original herd in Montana, and 77 plains bison from the exhibition herd in Banff were brought to Wainwright (Lothian 1981). Additional shipments over the next 5 years brought the total number of plains bison introduced at Wainwright to 748. In 1913, the Wainwright Buffalo Park herd had grown to 1188 plains bison, and by 1923 it had increased to 6780, where overcrowding caused range depletion. This led to a planned phased slaughter to control the herd, but the idea received such intense public criticism it was quickly abandoned, and a more publicly acceptable solution proposed to ship surplus plains bison north to the newly established Wood Buffalo National Park (Graham 1924). From 1925 to 1928, 6673 young plains bison were transported by rail from Wainwright to the Waterways railway terminus at Fort McMurray, Alberta, where they were then taken by barge down the Athabasca and Slave Rivers to Wood Buffalo National Park near Hay Camp (Lothian 1981). Plains bison were released at several sites along the west bank of the Slave River, south and north of Hay Camp (Fig. 48.8), into range already occupied by wood bison (Soper 1941). The total number of animals actually released into the Park was substantially lower than the number shipped, because of injuries and mortalities.

The proposed introduction of plains bison into wood bison range was challenged by the American Society of Mammalogists (Howell

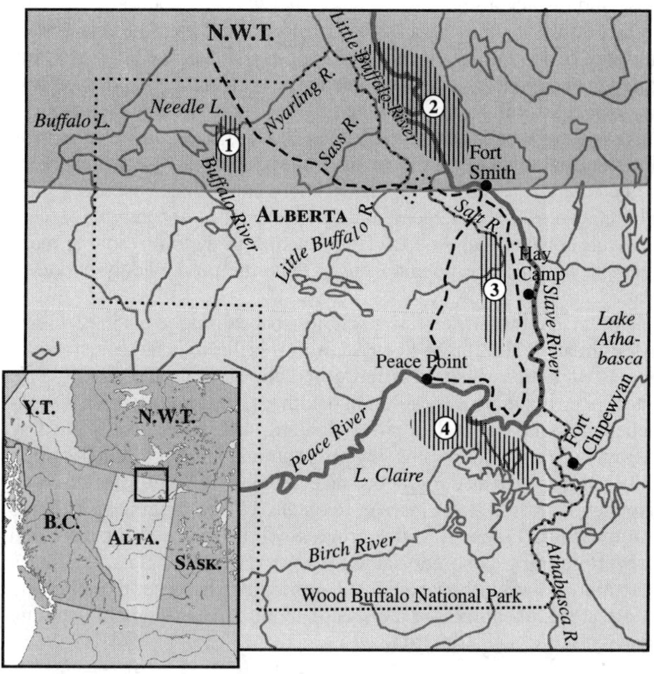

Figure 48.8. Distribution of bison (*Bison bison*) herds in the Greater Wood Buffalo National Park region in 1957 and the location of the Nyarling River wood bison (*Bison bison athabascae*) herd.

1925) and by individual biologists (Harper 1925; Saunders 1925). They believed interbreeding would result in the loss of both subspecies of bison, and wood bison would become infected with tuberculosis, which was known to be present in the Wainwright herd. The plains bison carried *Mycobacterium bovis* and *Brucella abortus,* the causative agents of bovine tuberculosis and brucellosis, respectively. The scientific challenges went unheeded and the Canadian government gave in to public pressure some 78 years ago and proceeded with what turned out to be a serious conservation and biological mistake (Fuller 2002). Bison herds in the Greater Wood Buffalo Region remain infected to this day (Tessaro et al. 1990; Fuller 1991). The number of bison in Wood Buffalo National Park increased to an estimated 12,000 by 1934 (Soper 1941), and wolf control was used to maintain the population at this level until the late 1960s. In 1959, 5 bison collected from a herd of about 200 animals near the Nyarling River in the northwest corner of the Park (Fig. 48.8) were determined to be morphologically representative of wood bison (Banfield and Novakowski 1960). Despite some hybridization, Wood Buffalo National Park bison and herds founded with stock from the Nyarling River area remain genetically and morphologically distinguishable from plains bison (van Zyll de Jong et al. 1995; Wilson and Strobeck 1999) and warrant conservation as separate forms (Gates et al. 2001b; Wilson 2001; see the section Taxonomy).

Since 1960, there have been three major efforts to salvage wood bison from the Greater Wood Buffalo Region to establish disease-free captive-breeding herds or wild populations. During the winter of 1963, 77 wood bison were captured in the Nyarling River and Buffalo Lake area in northwestern Wood Buffalo National Park (Fig. 48.8) to establish a captive-breeding herd near Fort Smith, Northwest Territories. After being tested for diseases, 19 of the captured wood bison were transported to a holding corral near Fort Smith, and in June 1963, 18 of these were transferred to establish the Mackenzie herd. The second salvage of wood bison was conducted again in northwestern Wood Buffalo National Park in 1965 and 21 animals were transferred to Elk Island National Park, Alberta. Unfortunately, the relocated wood bison carried bovine tuberculosis and brucellosis, but a rigorous management protocol involving isolation, quarantine of neonates, and elimination of all original stock managed to eradicate these diseases by 1971. The third salvage and recovery effort was initiated in 1996. Over a 3-year period, 62 newborn calves were captured from the Hook Lake herd in the Slave River Lowlands, Northwest Territories, maintained in enclosures near Fort Resolution, Northwest Territories, and treated prophylactically for exposure to tuberculosis and brucellosis (Gates et al. 1998; Nishi et al. 2001, 2002a). These bison are managed under quarantine as an experimental captive-breeding herd, and the goal is to restore a disease-free population of wood bison in the Slave River Lowlands (Nishi et al. 2001).

The Wood Bison Recovery Team officially was established in 1988 under the Canadian initiative RENEW. In 1987, a recovery goal of four herds of 200 or more wood bison was established (Wood Bison Recovery Team 1987), which later was revised to four free-ranging populations of at least 400 each, presently the number believed to represent a minimum viable population (Gates et al. 2001b).

The Recovery Team completed a National Recovery Plan for wood bison in October 2001 with the ultimate goal of removing wood bison from the list of species at risk (Gates et al. 2001b). Prescribed activities in the Recovery Plan are consistent with principles embraced in the World Conservation Strategy (International Union for Conservation of Nature and Natural Resources [IUCN] 1980), the World Conservation Strategy for Canada (Pollard and McKechnie 1986), guidelines established by the IUCN for reintroducing indigenous species to original range (IUCN 1987), and the Wildlife Policy for Canada (Environment Canada 1990). Three conservation principles endorsed by the Wood Bison Recovery Team that were embodied in the Plan are to preserve intraspecific diversity within the context of biodiversity conservation by continuing to manage wood and plains bison separately, to restore interactions between wood bison and their environment to allow for continued natural selection and evolution, and to promote recovery of either subspecies of bison in suitable habitat within its original geographic range and secure from factors that might threaten its long-term survival.

Issues and Challenges for Wood Bison Recovery. The presence of cattle diseases, tuberculosis and brucellosis, in bison herds in the Greater Wood Buffalo Region continues to be the major obstacle to further recovery of healthy populations in northern Canada (Fuller 1991; Gates et al. 1992a, 2001b). Total depopulation of Wood Buffalo National Park and replacement with disease-free wood bison was recommended in 1990 by the Environmental Assessment Panel (1990). However, as in 1923–25 this recommendation to slaughter and restock has recieved so much public opposition that definitive action has not yet taken place (Fuller 2002). We question, as does Fuller (2002), why the serious error made in 1923 must be repeated because there is no apparent logical reason for further delay. The risk of infection of the Mackenzie and Hay-Zama wood bison herds is significant and ways to mitigate disease risk for wild and domestic herds of bison in this region need to be implemented (Gates et al. 2001c). Wood bison are protected in northwestern Alberta in a wildlife management area that was designated in 1993. However, recovery in adjacent areas is constrained by the presence of disease-infected herds and by legislation that exempts bison from being managed as wildlife throughout the rest of the province (Gates et al. 2001d).

The bison ranching industry has grown rapidly in northwestern Canada, in terms of both the number of bison and the number of new ranches. Presence of captive and free-roaming plains bison within the historical range of wood bison creates an obstacle to implementation of the recovery plan. The existence of free-ranging wood bison adjacent to commercial bison operations is a potential conflict, creating a need for wildlife managers and the ranching industry to jointly address this issue.

Establishing additional free-ranging herds of wood bison within original range in Canada is a high conservation priority. The number of wood bison currently in herds that are subject to evolutionary selection in the wild is small relative to historical levels, and the degree to which existing herds can expand is limited by a variety of factors (Gates et al. 2001d). Furthermore, the risk of infection of several existing herds with cattle diseases remains high. The long-term conservation of wood bison and its full recovery require an increase in the number of wild herds. Availability of release sites where conflict with agriculture and urban activities are minimized is a major limiting factor to the establishment of new wild populations, although some habitat exists in northeastern British Columbia, central and northern Yukon, and the lower Mackenzie Valley in the Northwest Territories. In addition, extensive high-quality habitat for wood bison exists in Alaska (Berger et al. 1995).

The conservation of genetic resources is an important issue for wood bison conservation and recovery. The national captive-breeding herd at Elk Island National Park is genetically less diverse than other wood bison populations, including the parent population in Wood Buffalo National Park (Wilson and Strobeck 1999; Wilson 2001). Studies of the Hook Lake Wood Bison Recovery Project indicate that a significant portion of the genetic variability from the parent population in the Slave River Lowlands has been successfully salvaged and is represented in the captive herd, one of the most genetically variable herds that has been salvaged from the Greater Wood Buffalo Region (Nishi et al. 2001). However, the Hook Lake wood bison recovery herd, the Mackenzie herd, and the Elk Island National Park source herd of wood bison are all genetically less variable than the parent population in Wood Buffalo National Park (Nishi et al. 2001). Incorporating and maintaining as much genetic diversity as possible within existing and newly established herds is of paramount importance for successful recovery.

Hunting. Authorized hunting of bison occurs in several locations throughout North America. Harvest of bison by hunting combines meat production with recreation, which can generate considerable revenue through trophy fees. The demand for bison hunting opportunities far exceeds supply (National Buffalo Association 1990). Bison are hunted on public lands in three areas in the contiguous United States, four areas in Alaska, and in five areas in Canada.

The Henry Mountains, Utah, is home to the only free-roaming herd of plains bison in the 48 contiguous United States that is subject to regulated hunting. Over 400 buffalo roam the lower areas of the Henry Mountains in the winter and range in the higher regions during the summer. Permits are issued to hunt bison annually by the Utah Division of Wildlife Resources, with hunts usually occurring in October–December. The first sanctioned hunt was held in 1950, and the second occurred in 1960. Hunts have occurred annually with the exception of 4 years in the mid-1960s and early 1970s (National Buffalo Association 1990). Forty-two bull bison and hunter's choice bison permits were issued for the Henry Mountains in 2002.

A herd of approximately 450 plains bison is maintained on Antelope Island, where annual round-ups identify surplus animals for culling by live sale. However, a number of mature bulls that are designated as surplus are removed through a hunting program. An estimated 10 bulls/year may be available for hunting based on a population size of 450. In 1987, 15 mature bull permits were issued. In 1988, due to an increasing population, 22 cow permits were issued for an October hunt and 22 hunter-choice permits were issued for a November/December hunt (National Buffalo Association 1990).

Management hunts for bison were held during the 1980s and 1990s in areas of Montana adjacent to Yellowstone National Park. Montana's first bison hunt was held in 1986 for a total of 57 animals, most of which were taken from the Gardiner area (National Buffalo Association 1990). Hunts were staged to manage the risk of infection to domestic livestock with brucellosis carried by Yellowstone bison moving out of the park, primarily during the winter. The hunts were highly controversial and were stopped in response to protests. Under a recent management plan (*Federal Register* 2001), the potential for mixing of bison with cattle outside the park is managed first by hazing bison back into the park. Bison remaining outside the park are then subject to capture and return to the park or lethal removal by management authorities. Public hunting is no longer permitted.

The Custer State Park plains bison herd originated from the transfer of 36 animals from the Scotty Philip herd (Pete Dupree herd) in 1914 with additional animals from the Pine Ridge Indian Reservation during the late 1940s, and later from Wind Cave National Park. The managed herd size at fall round-up is approximately 1400. This is reduced to about 950 after surplus animal removals in October for the annual live auction sale in November. In mid-December to mid-January, the removal of 8- to 10-year old bulls takes place through a 3-day guided trophy bull hunt with hunters selected through an application and draw process (National Buffalo Association 1990).

Plains bison are hunted in four different herds in Alaska. Permits are allocated through draws and hunters are assigned specific hunting periods as determined by the order drawn. The largest population is located along the Delta River near Delta Junction. This herd was established in 1928 when 23 plains bison were translocated from the National Bison Range in Montana. By 1947, the herd had increased to 400. Hunting was initiated in 1951 and is used to limit the herd to a precalving population of 275–300. The Delta Junction herd is maintained at this level to manage the risk of damage to agricultural crops and to keep the herd within the carrying capacity of its summer range. Between 6000 and 11,000 people apply annually for permits to hunt the Delta herd. In the 2002/2003 hunting season, 70 bull permits and 65 cow permits were allocated.

Three other Alaskan populations were established with translocations from the Delta Junction herd. The Copper River herd was established in 1950 with 17 bison from the Delta herd. In 1999, it numbered 87; 12 hunting permits were let in 2002/2003. In 1962, 35 bison were transplanted from the Delta herd to the Chitina River drainage. Poor habitat and winter severity have limited the growth of this population and it remains at between 30 and 50. A small quota of two tags is allocated for this herd. The fourth Alaskan population was established near the western end of the Alaska Range along the south fork of the Koyukuk River in the Farewell Lake area. Introduced with releases in 1965 and 1968, the Farewell herd has grown to about 300. Forty either-sex permits were allocated in the 2002/2003 hunting season.

Bison hunting was first disallowed in Canada in 1877 with the passing of the federal Buffalo Protection Act. Hunting was reinstated in the mid-1950s in the Slave River Lowlands in the Northwest Territories. Regulated, outfitted hunts were permitted annually between 1956 and 1962, when an outbreak of anthrax forced a temporary closure of hunting. Between 1968 and 1977, sport hunters took an average of 123 bison/year during regulated seasons in the Slave River Lowlands. At the same time, hunting by General Hunting Licence (GHL) holders, which include eligible aboriginal residents and a few long-term nonaboriginal residents, had no season or quota limit. Between 1973 and 1976, the wood bison population in the Slave River Lowlands dramatically declined and sport hunting was closed in 1977, but hunting by GHL holders has remained unregulated, although voluntary restrictions were applied.

Regulated hunting for male bison was first permitted in the Mackenzie wood bison population beginning in 1988. In 2001, a quota of 47 permits was allocated among the residents of Fort Providence (20) and Rae-Edzo (3), a limited-entry draw for other eligible hunters in the Northwest Territories (15), and 9 permits were allocated to Fort Providence to provide outfitting and guiding services for nonresident bison hunters.

The Yukon Department of Renewable Resources first allowed wood bison hunting in 1997, when five permits were provided for Yukon residents. The 1998 management plan stipulates that the herd will be maintained at a size of 500 through special-permit hunting (Government of Yukon 1998). Ninety permits were issued in 2001–2002.

Regulated hunting of plains bison is allowed in northeastern British Columbia. The Pink Mountain herd was established in 1971 with the escape of captive plains bison from a ranch. The province assumed management authority for the herd in 1988. In February 2003, a total count survey resulted in 900 in the herd, which ranges in the Sikani Chief and Halfway River drainages. Five bull and 115 either-sex hunting permits were available for the 2002/2003 hunting season.

Free-ranging bison are found across northern Alberta. Bison hunting is not permitted in a special management zone that was established in northwestern Alberta to protect the reintroduced Hay-Zama wood bison population. Small herds and individual bison are found outside of this zone in areas near and adjacent to Wood Buffalo National Park. These bison are not recognized as wildlife under the Alberta Wildlife Act and are subject to unregulated hunting.

Nonconsumptive Use. Although difficult to evaluate, the direct and indirect human benefits derived from free-roaming bison herds should not be underestimated. The aesthetic value of bison also makes them a popular and valuable tourist attraction, especially in private and public zoos and parks. Grazing bison on public lands should make these areas more productive, and potentially would make them more attractive to public users.

Cultural and Ecological Revitalization. The Intertribal Bison Cooperative (ITBC) was initiated in the United States in 1990 to coordinate and assist tribes in returning bison to Indian country. The ITBC is a nonprofit tribal organization committed to reestablishing bison herds on Indian lands in a manner that promotes cultural enhancement, spiritual revitalization, ecological restoration, and economic development. The current membership of the ITBC is 51 tribes in 16 state regions, with an estimated collective herd of more than 8000 bison. Bison always have held great meaning for North American Indian people. Therefore, they feel reintroduction of healthy bison populations to tribal lands will help heal the spirit of both the Indian people and the bison.

The shortgrass expanse of the Great Plains, situated in the central United States between the Rocky Mountains on the West and the tallgrass prairies on the Midwest and South, is an already sparsely populated region, which is experiencing rapid depopulation as a result of a warming climate and cooling economy (Popper and Popper 1991). This situation has provided an opportunity for restoration of the Great Plains frontier by deprivatizing and reverting the most rural, environmentally fragile parts of the region to its native grasslands thereby creating a great "Buffalo Commons" as the world's largest historical preservation project (Popper and Popper 1991). Development of an operational strategy for a Buffalo Commons will require at least several decades to initiate and implement, whether it is governmentally assisted or privately induced on recently abandoned prairie lands. The Buffalo Commons land base could be greatly expanded from its present status of national parks, grasslands, public grazing lands, wildlife refuges, and tribal lands as well as their state counterparts and many private and land conservancy holdings (Popper and Popper 1991).

Bison Farming. Bison farming is one of the fastest growing industries in agriculture. It is based mainly on consumer demand for healthy, lean, nutritious bison meat, which in North America is primarily destined for domestic markets. Bison producers exist in all 50 states, every Canadian province, and many countries around the world. Farms range in size from hobby herds of a few animals to major operations of 5000 or more head. Plains bison are unique among North American native ungulates in that they are generally not considered to be wildlife and, because of this status, commercial production of bison has been facilitated (Hawley 1989). The current legal classification of bison is important to the industry because regulations controlling the sale, slaughter, and transport of animals and the processing and sale of meat significantly differ between wildlife and domestic species (Hawley 1989). Bison can be processed in domestic animal abattoirs rather than requiring special facilities and regulations, a distinct advantage over wildlife species. Increasing interest in bison ranching has resulted in development of specialized handling facilities and the publication of specialty magazines. In 1978, the industry designed and published its first book on bison history and husbandry which was intended for use as a bison rancher's handbook (Jennings 1978). In 1990, this handbook was revised and updated into the Buffalo Producer's Guide to Management and Marketing (National Buffalo Association 1990), which was intended for use as a resource tool for bison farmers. Several organizations have been established throughout North America to support the industry. In the United States, the National Bison Association is the main organization whereas in Canada, the Canadian Bison Association is the national organization.

The National Bison Association is a nonprofit association of more than 2400 members from all 50 states and 16 foreign countries, which promotes the preservation, production, and marketing of bison. It is dedicated to the future of the industry and strives to better inform and educate its members and the general public about bison through its activities and services. The Canadian Bison Association provides services to more than 1400 bison producers across Canada as well as to any others involved in the industry.

Productive Advantages of Bison. Because of their intractable nature, cold-hardiness, and ability to digest poor-quality forages, bison appear to be most efficiently used as commercial meat animals under natural range conditions. The characteristics of bison that make them desirable as a source of meat primarily involve their ability to be productive under range conditions that are not optimal for cattle. One of the greatest assets of bison are their ability to graze native range throughout the year because of their cold weather resistance and ability to forage through deep snow (Hawley and Reynolds 1987; Hawley 1989). Furthermore, the superior digestion of low-quality forages allows bison to use range forage that would be considered marginal for cattle. Bison tend to be less selective grazers than cattle (Peden et al. 1974). This digestive difference can provide a production advantage to bison when feed quality is low (Hawley et al. 1981b), but this advantage may be reversed when feed quality is high (Hawley 1989). Although bison on range consume poorer quality forage than do cattle under similar conditions, they will readily consume grain and high-quality feeds (McHugh 1958, Peters 1958).

Feed requirements for bison are generally similar to those for cattle with the exception of a few important differences in relation to feed quality and intake (Hawley 1989). For example, bison usually eat less than cattle, especially during winter. Lower feed consumption may be an adaptive strategy to reduce growth and limit activity and intake at a time of year when forage is difficult to obtain and quality is low (Hawley et al. 1981a; Hawley 1987). In a minimum-input production system, this seasonal adjustment in intake by bison would serve to reduce feed costs during winter (Hawley 1989).

Bison can be raised under feedlot conditions, but this obviates some of the physiological productive advantages of bison over cattle. During the 1980s, there was a trend toward "finishing" bison on grain for 60–90 days to increase carcass fat and render bison meat more comparable to that of cattle (Hawley 1989). This may have enhanced product attractiveness in the domestic market in the past, but more recently, this is considered counterproductive because the low-fat content of bison meat is desirable in marketing it as a health food. The primary marketing feature of bison meat is that it is a healthy and delicious alternative source of red meat that is lower in fat, calories, and cholesterol than beef, pork, or skinless chicken.

Crossbreeding. Crossbreeding bison with domestic cattle to take advantage of the hardiness of bison under adverse climatic and range conditions and the meat characteristics of domestic cattle dates back to the mid-eighteenth century (McHugh 1972). The government of Canada probably contributed one of the greatest sustained efforts of crossbreeding research experiments from 1916 to 1964, in which bison were crossed with Hereford, Angus, Shorthorn, and Holstein cattle. This work demonstrated that outstanding winter hardiness, as measured by hair coat density and performance of cattle–bison hybrids (cattalo) on winter range could be obtained by combining the two species (Peters 1978). However, major problems of hydromacy in domestic cows bred to bison bulls, mating indifference, and infertility of hybrid bulls developed (Hawley 1989). Hydromacy problems were alleviated by breeding bison cows to domestic bulls, but sterility problems of F_1 and F_2 bulls remained. Other disadvantages of cattalo to domestic cattle included lower calving frequencies, lower birth rates, and lower rates of post-weaning growth of hybrid calves under feedlot conditions (Peters 1958; Peters and Slen 1966; Hawley 1989). The major production advantage of hybrids was their winter hardiness (Smoliak and Peters 1955), but the disadvantages far outweighed the advantages and cattalo did not develop as an independent breed (Hawley 1989). More recently, beefalo, a mix of bison, Charolais, and Hereford, has renewed some interest in crossbreeding, but the genetic makeup and production advantages of beefalo have yet to be proven. Although the full potential of crossbreeding may not yet be realized, efforts so far have failed to produce a hybrid with outstanding production characteristics. It would appear that crossbreeding is probably not a viable production alternative (Hawley 1989).

Economics. Bison farming has been an active agricultural industry for several decades; however, it is still developing, which makes a meaningful economic evaluation difficult. Prices of stock and meat vary regionally and temporally (Hawley 1989). During the mid-1990s, animal prices were subject to a significant upward swing, especially for breeding stock, and the 1998 selling prices reflected that trend. Bull calves sold in the US$600–800 range, yearling bulls sold in the US$700–900 range, and 2-year old bulls averaged around US$1100. For females, the prices increased to higher levels, where heifer calves sold in the price range of US$2000, yearling heifers sold in the US$2500–3000 range, and bred cows sold in the US$3000–4000 range. The market is subject to a series of fluctuations, which can be quite severe, as experienced during 2002, when prices for breeding stock declined by up to 50%. At the Elk Island National Park annual live-sale in December 2001, male plains bison calves (33) averaged US$155, yearling males (5) averaged US$296, female calves (33) averaged US$163, and yearling females (16) averaged US$211 (Olson 2002). In February 2002, wood bison prices were somewhat higher than those of plains bison, but these prices were down nearly 50% from previous years. Male wood bison calves (27) averaged US$591, 2-year-old males (7) averaged US$1268, wood bison female calves (18) averaged US$1940, yearling females (2) averaged US$1640, and 2-year-old females (2) averaged US$1800 (Olson 2002).

The commercial value of bison by-products (heads, robes, hides, wool, skulls, and horns) contributes significantly to the economic prospects of the industry and can provide considerable additional revenue compared with traditional livestock operations. There are markets in some areas of the United States for raw bison hides (US$50–75), prime tanned hides (US$400–800), and skulls of 2- or 3-year old bulls (US$50–75) and 4+-year-old bulls (US$75–125) (National Buffalo Association 1990). A limited market also exists for hooves, teeth, bones, tails, and bladders. Bison heads for display mounts have a limited market. Depending on size and quality, they are worth a few hundred dollars in an unprocessed state to thousands of dollars in a completed taxidermy mount.

Operating costs may be similar for bison and cattle under intensive farming production systems, but lower for bison under extensive ranching or less intensive farming systems (Hawley 1989). Capital investment for proper enclosures and handling facilities (exclusive of land) requires a greater initial expenditure for raising bison than for cattle in both types of production systems. Prime bison meat animals are bulls between 18 and 30 months of age, and the price of bison meat is generally 1.5–2 times that of the equivalent cuts of beef. Bison meat is a specialty product and will likely remain as such until supply is no longer limited. Bison meat is becoming more consistently available in the retail market. As production continues to increase, which will provide more stability in supply, the market for bison meat will likely increase. Demand is expected to exceed supply for many years to come (Hawley 1989).

The greatest potential for bison farming may be to complement cattle production rather than trying to replace it, because the greatest complement is derived by exploiting the environmental hardiness of bison (Hawley 1989). Bison are a suitable alternative on large areas of range that would be only marginal for cattle production because of severe winters, insect harassment, predation, and poor-quality forage (Hawley and Reynolds 1987). The primary constraints to the growth of the bison farming industry are the higher initial start-up costs and availability of breeding stock (Hawley 1989). The potential for growth and expansion of the bison farming industry is high, but will continue to be somewhat dependent on effective promotion.

RESEARCH NEEDS

Taxonomy and Genetics. The debate over the validity of bison subspecies in North America (Geist 1991; van Zyll de Jong et al. 1995) has the potential to result in genetic homogenization of wild bison if the distinction between these forms is denied as proposed by some authors (Bork et al. 1991; Geist 1991; Polziehn et al. 1996). The rapid development of molecular genetic techniques has led many investigators to regard these data as essential to conservation management, often to the exclusion of other data. Depending on the number and type of markers, there is a risk that molecular methods alone may not adequately define populations that merit separate management. Molecular methods that detect apparent reciprocal monophyly can be used to argue erroneously that morphologically or otherwise ecologically distinguishable populations should be managed as a unit. Natural or anthropogenic hybridization between closely related forms may further confound this problem, for example, the red wolf and coyotes (Nemecek 1996), wood bison × plains bison (Wilson and Strobeck 1999), and bison × cattle (Ward et al. 1999c).

Crandall et al. (2000) offered a method for diagnosing distinct populations, which emphasizes variation in phenotypes, allowing preservation of important adaptive characteristics and associated underlying genetic variation. The method incorporates both ecological and genetic exchangeability and considers whether each is recent or historical. There likely are sufficient available data to support this analysis for North American bison. In recognition of the pressure from taxonomic "lumpers," this pragmatic and conservative analysis is urgently needed for the species. A principle that should govern genetic management actions for the two forms of North American bison is preservation of the adaptive diversity of the species and the evolutionary processes that remain across its geographic range. A lack of evidence of nonexchangeability, either genetic or ecological (*sensu* Crandall et al. 2000), should not be used to promote homogenization or tampering with the gene pool of populations or to eliminate populations to reduce the geographic range of the species to make way for other uses of the land such as commercial bison production and other agricultural or industrial enterprises.

Diseases. Bovine tuberculosis and brucellosis remain the most important diseases of immediate concern for bison conservation because of their importance to human and domestic animal health and regulatory requirements for their control in both Canada (Gates et al. 1998) and the United States (Morrisette 2000). In addition, the effects of the diseases on wild bison populations are of concern. The presence of both diseases, in synergism with wolf predation, may cause bison populations to exist at a low density relative to food carrying capacity (Joly and Messier 2001a). Management approaches that could either eliminate the diseases from wild bison or contain them within designated ranges without risk to disease-free bison or cattle populations remain problematic and further research is required. Approaches being studied include the development of vaccines (Roffe et al. 1999), improvements in the sensitivity and specificity of tests for disease (Lin et al. 1996), and salvage and veterinary treatment to establish disease-free herds (Gates et al. 1997) and mitigate the risk of spread of diseases from infected populations (Gates et al. 2001c; Bjornlie and Garrott 2001). Research on effective vaccines against brucellosis have been undertaken for the Yellowstone National Park herd (Roffe et al. 1999). This work is focused on efficacy in preventing infection with pathogenic field strains of *B. abortus* (Olsen et al. 1998) and biosafety of a new vaccine in the environment (Roffe et al. 1999; Cook et al. 2001). There was no published evidence of research on vaccine development for tuberculosis in bison. However, research is being undertaken to develop improved tests for detecting the presence of antibodies in bison (Joly and Messier 2001a).

Predator–Prey Dynamics. The recent reintroduction of wolves into the Yellowstone ecosystem has provided an opportunity to study in great detail the dynamics of a multiple-prey system and to add to knowledge gained from previous research on bison, other prey, and wolves in northern Canada (Carbyn and Trottier 1987; Van Camp 1987; Larter et al. 1994; Joly and Messier 2000). Research in Yellowstone National Park has shown that elk and bison female groups become more vigilant when sympatric with wolves (Laundre et al. 2001), wolves learn rapidly how to kill bison, and wolves prefer elk over bison (Smith et al. 2000). Wolves select bison calves over other classes and show a preference for weak individuals, including bulls (Smith et al. 2000). Carbyn (1983) found a similar pattern of prey selection within a guild of ungulates present in Riding Mountain National Park, Manitoba.

Although behavioral interactions among wolves and prey species are becoming better understood, less is known about the regulation by wolves of species within multiple-prey assemblages in which bison are represented. Messier (1996) demonstrated that for predation to be regulatory, there must be density dependence in both the functional (number of prey killed per wolf) and numerical responses of wolves to prey abundance, except when learning or prey switching is expressed as a sigmoidal functional response. In systems where wolves do not exhibit a numerical response to a particular species of prey, the relationship between prey density and predation rate can be inversely proportional and nonlinear (Dale et al. 1994; Messier 1996). This can cause the extirpation of a more highly preferred prey species in a guild, for example, woodland caribou in a moose–wolf system (Seip 1992). Indeed, Larter et al. (1994) found that although bison formed a larger proportion of the wolf diet because of their greater biomass, moose were the preferred prey species in the Mackenzie Bison range in the Northwest Territories. Furthermore, they suggested that woodland caribou may have declined substantially since bison were reintroduced in the study area.

Interactions between selectivity by wolves, relative and absolute abundance of each species, predation rate, and prey population dynamics is an important area of research. With its diversity of prey species, the Yellowstone ecosystem represents a unique opportunity to undertake research on predator–prey dynamics. Several interesting questions can be addressed in this system. What is the relationship between wolf density and ungulate biomass? Is the relationship between elk biomass and predation rate density dependent or will it be uncoupled if elk decline and bison become an increasing proportion of the wolves' diet? If the predation rate on bison increases as other species decline, will the relatively higher availability of bison lead to near extirpation of more preferred prey species? Will the system shift to a predominantly bison–wolf system? Will the numerical response of wolves to bison be linear if other prey species decline? The Yellowstone ecosystem offers a unique "laboratory" in which to answer these and many other questions regarding predators and prey.

ACKNOWLEDGMENTS

We gratefully acknowledge Environment Canada, Canadian Wildlife Service; the University of Calgary, Faculty of Environmental Design; and Tera Environmental Consultants Ltd. for supporting and encouraging us throughout the preparation of this chapter. We thank the biological and warden staff at Elk Island National Park for their assistance and cooperation; in particular, we are extremely grateful to Norm Cool and Wes Olson. We especially thank Greg Wilson for his helpful review comments on the manuscript, in particular, with the taxonomy and genetics sections. We are deeply indebted to Roberta McCarthy, Environment Canada Library, Downsview, Ontario, and Terri Fraser and Susan Blackman, Canadian Wildlife Service Librarians, Edmonton, Alberta, for their excellent and continuous library assistance and support. We thank Vi Jespersen, Canadian Wildlife Service, for electronic input of the initial draft manuscript. We further express our great appreciation of those many individuals, government agencies, and nongovernment organizations that have been involved with the conservation of both subspecies of the North American bison.

LITERATURE CITED

Aalhus, J., S. D. Jones, A. K. Tong, L. E. Jeremiah, W. M. Robertson, and L. L. Gibson. 1992. The combined effects of time on feed, electrical stimulation and aging on beef quality. Canadian Journal of Animal Science 72:525–36.

Aalhus, J. L., and J. A. M. Janz. 2001. Bison: Meating the beef challenges. Pages 273–84 *in* B. D. Rutley, ed. Bison are back, 2000. Proceedings of the second international bison conference. Edmonton, Alberta, Canada.

Alexander, G., J. P. Signoret, and E. S. E. Hafez. 1974. Sexual and maternal behavior. Pages 222–54 *in* E. S. E. Hafez, ed. Reproduction in farm animals, 3rd ed. Lea & Febiger, Philadelphia.

Allen, J. A. 1876. The American bisons, living and extinct. Memoirs of the Museum of Comparative Zoology, Harvard College [Reprinted 1974, Arno Press, New York].

Angen, O., R. Mutters, D. A. Caugant, J. E. Olson, and M. Bisgaard. 1999. Taxonomic relationships of the *[Pasteurella] haemolytica* complex as evaluated by DNA hybridizations and 16S rRNA sequencing with proposal of *Mannheimia haemolytica* gen.nov., comb. nov., *Mannheimia granulomatis* comb. nov., *Mannheimia glucosida* sp. nov., *Mannheimia ruminalis* sp. nov. and *Mannheimia varigena* sp. nov. International Journal of Systematic Bacteriology 49:67–86.

Animal, Plant and Food Health Risk Assessment Network. 1998. Risk assessment on bovine brucellosis and tuberculosis in Wood Buffalo National Park and area. Canadian Food Inspection Agency, Ottawa, Ontario, Canada.

Audubon, J. J., and J. Bachman. 1849. *Bos americanus.* Pages 292–95 *in* The quadrupeds of North America. V. G. Audubon, New York.

Aune, K., T. Roffe, J. Rhyan, J. Mack, and W. Clark. 1998. Preliminary results on home range, movements, reproduction and behavior of female bison in northern Yellowstone National Park. Pages 61–70 *in* L. Irby and J. Knight, eds. International symposium on bison ecology and management in North America. Montana State University, Bozeman.

Avise, J. C., and R. M. Ball, Jr. 1990. Principles of genealogical concordance in species concepts and biological taxonomy. Pages 45–67 *in* D. Futuyma and J. Autonovics, eds. Oxford surveys in evolutionary biology. Oxford University Press, New York.

Axelrod, D. I. 1985. Rise of the grassland biome, Central North America. Botanical Review 51:163–201.

Banfield, A. W. F. 1981. The mammals of Canada. University of Toronto Press, Toronto, Ontario, Canada.

Banfield, A. W. F., and N. S. Novakowski. 1960. The survival of the wood bison (*Bison bison athabascae* Rhoads) in the Northwest Territories (Natural History Papers, No. 8). National Museum of Canada, Ottawa, Ontario, Canada.

Barrett, M. W., and G. A. Chalmers. 1977. Clinicochemical values for adult free-ranging pronghorns. Canadian Journal of Zoology 55:1252–60.

Bay, D. E., C. W. Pitts, and G. Ward. 1968. Oviposition and development of the face fly in feces of six species of animals. Journal of Economic Entomology 61:1733–35.

Bayrock, L. A., and J. M. Hillerud. 1964. New data on *Bison bison athabascae* Rhoads. Journal of Mammalogy 45:630–32.

Becklund, W. W., and M. L. Walker. 1969. Taxonomy, hosts, and geographic distributions of the *Setaria* (Nematoda: Filarioidea) in the United States and Canada. Journal of Parasitology 55:359–68.

Bell, R. H. V. 1971. A grazing ecosystem in the Serengeti. Scientific American 225:86–93.

Belue, T. F. 1996. The long hunt: Death of the buffalo east of the Mississippi. Stackpole, Mechanicsburg, PA.

Berezowski, J. (n.d.). Diseases of bison. Department of Large Animal Clinical Studies, Western College of Veterinary Medicine, Saskatoon, Saskatchewan, Canada.

Berg, R. T., and R. M. Butterfield. 1976. New concepts of cattle growth. Sydney University Press, Sydney, Australia.

Berger, J. 1989. Female reproductive potential and its apparent evaluation by male mammals. Journal of Mammalogy 70:347–58.

Berger, J., and S. L. Cain. 1999. Reproductive synchrony in brucellosis-exposed bison in the southern Greater Yellowstone Ecosystem and in noninfected populations. Conservation Biology 13:357–66.

Berger, J., and C. Cunningham. 1991. Bellows, copulations and sexual selection in bison (*Bison bison*). Behavioural Ecology 2:1–6.

Berger, J., and C. Cunningham. 1994. Bison: Mating and conservation in small populations. Columbia University Press, NY.

Berger, J., and M. Peacock. 1988. Variability in size–weight relationships of *Bison bison*. Journal of Mammalogy 69:618–24.

Berger, M., R. O. Stephenson, P. Karczmarczyk, and C. C. Gates. 1995. Habitat inventory of the Yukon Flats as potential wood bison range. Unpublished report. Alaska Department of Fish and Game, Fairbanks, AK.

Bergman, C. M., J. M. Fryxell, C. C. Gates, and D. Fortin. 2001. Ungulate foraging strategies: Energy maximizing or time minimizing? Journal of Animal Ecology 70:289–300.

Bergstrom, R. C. 1967. Sheep liver fluke, *Fasciola hepatica* L. 1758, from buffalo, *Bison bison* (L. 1758) in western Wyoming. Journal of Parasitology 53:724.

Bergstrom, R. C. 1982. Nematodes: Lungworms of ruminants. Pages 206–8 *in* E. T. Thorne, N. Kingston, W. R. Jolley, and C. Bergstrom, eds. Diseases of wildlife in Wyoming, 2nd ed. Wyoming Game and Fish Department Special Publications Section, Cheyenne.

Bhambhani, R., and J. Kuspira. 1969. The somatic karyotypes of American bison and domestic cattle. Canadian Journal of Genetics and Cytology 11:243–49.

Biondini, M. E., and A. A. Steuter. 1998. Spatial distribution of bison grazing as a function of fire and range site. Pages 71–80 *in* L. Irby and J. Knight, eds. International symposium on bison ecology and management in North America. Montana State University, Bozeman.

Bjornlie, D. D., and R. A. Garrott. 2001. Effects of winter road grooming on bison in Yellowstone National Park. Journal of Wildlife Management 65:560–72.

Boddicker, M. L., and E. J. Hugghins. 1969. Helminths of big game mammals in South Dakota. Journal of Parasitology 55:1067–74.

Bork, A. M., C. M. Strobeck, F. C. Yeh, R. J. Hudson, and R. K. Salmon. 1991. Genetic relationship of wood and plains bison based on restriction fragment length polymorphisms. Canadian Journal of Zoology 69:43–48.

Bowyer, R. T., X. Manteca, and A. Hoymork. 1998. Scent marking in American bison: Morphological and spatial characteristics of wallows and rubbed trees. Pages 81–91 *in* L. Irby and J. Knight, eds. International symposium on bison ecology and management in North America. Montana State University, Bozeman.

Boyd, D., and C. Gates. 2001. Bison Specialist Group—North America. Species 36:18.

Braend, M. 1963. Haemoglobin and transferrin types in the American buffalo. Nature 197:910–11.

Broughton, E. 1987. Diseases affecting bison. Pages 34–38 *in* H. Reynolds and A. Hawley, eds. Bison ecology in relation to agricultural development in the Slave River lowlands, N.W.T. (Occasional Paper No. 63; Catalogue No. CW69-1/63E). Minister of Supply and Services Canada, Ottawa, Ontario, Canada.

Broughton, E. 1992. Anthrax in bison in Wood Buffalo National Park. Canadian Veterinary Journal 33:134–35.

Bryant, E. 1885. Rocky Mountain adventures. Worthington, NY.

Buckland, R. A., and H. J. Evans. 1978a. Cytogenetic aspects of phylogeny in the Bovidae. Part 1: G-banding. Cytogenetics and Cell Genetics 21:42–63.

Buckland, R. A., and H. J. Evans. 1978b. Cytogenetic aspects of phylogeny in the Bovidae. Part 2: C-banding. Cytogenetics and Cell Genetics 21:64–71.

Buergelt, C. D., and P. E. Ginn. 2000. The histopathologic diagnosis of subclinical Johne's disease in North American bison (*Bison bison*). Veterinary Microbiology 77:325–31.

Buergelt, C. D., A. W. Layton, P. E. Ginn, M. Taylor, J. M. King, P. L. Habecker, E. Mauldin, R. Whitlock, C. Rossiter, and M. T. Collins. 2000. The pathology of spontaneous paratuberculosis in the North American bison (*Bison bison*). Veterinary Pathology 37:428–38.

Burger, J. F., and J. R. Anderson. 1970. Association of the face fly, *Musca autumnalis,* with bison in western North America. Annals of the Entomological Society of America 63:635–39.

Burtner, R. H., and W. W. Becklund. 1971. Prevalence, geographic distribution, and hosts of *Cooperia surnabada* Antipin, 1931, and *C. oncophora* (Raillet, 1898) Ransom, 1907, in the United States. Journal of Parasitology 57:191–92.

Calef, G. W. 1984. Population growth in an introduced herd of wood bison (*Bison bison athabascae*). Pages 183–200 *in* R. Olson, R. Hastings, and F. Geddes, eds. Northern ecology and resource management. University of Alberta Press, Edmonton, Alberta, Canada.

Calef, G. W., and J. Van Camp. 1987. Seasonal distribution, group size and structure, and movements of bison herds. Pages 15–20 *in* H. Reynolds and A. Hawley, eds. Bison ecology in relation to agricultural development in the Slave River Lowlands, N.W.T (Occasional Paper No. 63; Catalogue No. CW 69-1/63E). Minister of Supply and Services Canada, Ottawa, Ontario, Canada.

Cameron, A. E. 1923. Notes on buffalo: Anatomy, pathological conditions, and parasites. British Veterinary Journal 79:331–36.

Cameron, A. E. 1924. Some further notes on buffalo. British Veterinary Journal 80:413–17.

Campbell, B. H., and M. Hinkes. 1983. Winter diets and habitat use of Alaska bison after wildfire. Wildlife Society Bulletin 11:16–21.

Campbell, C., I. D. Campbell, C. B. Blyth, and J. H. McAndrews. 1994. Bison extirpation may have caused aspen expansion in western Canada. Ecography 17:360–62.

Capp, J. C. 1964. Ecology of the bison of Colorado National Monument. U.S. Government Printing Office, Washington, DC.

Carbyn, L. 1983. Wolf predation on elk in Riding Mountain National Park, Manitoba. Journal of Wildlife Management 47:963–75.

Carbyn, L. N. 1997. Unusual movement by bison, *Bison bison*, in response to wolf, *Canis lupus*, predation. Canadian Field-Naturalist 111:461–62

Carbyn, L. N. 1998. Some aspects regarding wolf predation on bison in Wood Buffalo National Park. Pages 92–95 *in* L. Irby and J. Knight, eds. International symposium on bison ecology and management in North America. Montana State University, Bozeman.

Carbyn, L. N., and T. Trottier. 1987. Responses of bison on their calving grounds to predation by wolves in Wood Buffalo National Park. Canadian Journal of Zoology 65:2072–78.

Carbyn, L. N., and T. Trottier. 1988. Descriptions of wolf attacks on bison calves in Wood Buffalo National Park. Arctic 41:297–302.

Carbyn, L. N., S. Oosenbrug, and D. Anions. 1993. Wolves, bison and the dynamics related to the Peace–Athabasca Delta in Canada's Wood Buffalo National Park (Circumpolar Research Series No. 4). Canadian Circumpolar Institute, Edmonton, Alberta, Canada.

Carbyn, L. N., N. J. Lunn, and K. Timoney. 1998. Trends in the distribution and abundance of bison in Wood Buffalo National Park. Wildlife Society Bulletin 26:463–70.

Carter, G. R., and M. C. L. de Alwis. 1989. Hemorrhagic septicemia. Pages 131–60 *in* C. Adlam and J. M. Rutter, eds. *Pasteurella* and pasteurellosis. Academic Press, San Diego, CA.

Chace, J. F., and A. Cruz. 1998. Range of the brown-headed cowbird in Colorado: Past and present. Great Basin Naturalist 58:245–49.

Chapin, E. A. 1925. New nematodes from North American mammals. Journal of Agriculture Research 30:677–81.

Choquette, L. P. E., J. F. Gallivan, J. L. Byrne, and J. Pilipavicius. 1961. Parasites and diseases of bison in Canada. Part 1: Tuberculosis and some other pathological conditions in bison at Wood Buffalo and Elk Island National Parks in the fall and winter of 1959–60. Canadian Veterinary Journal 2:168–74.

Choquette, L. P. E., E. Broughton, J. G. Cousineau, and N. S. Novakowski. 1978. Parasites and diseases of bison in Canada. Part 4: Serologic survey for brucellosis in bison in Canada. Journal of Wildlife Diseases 14:329–32.

Choudary, C., B. Narasimhaswamy, V. Shivasankar, M. R. Krishnamohan Rao, and J. Hararam Das. 1987. Hydatidosis in lungs and liver of an American bison (*Bison bison*). Indian Veterinary Journal 64:713–14.

Chowns, T., C. Gates, and F. Lepine. 1998. Large scale free burning to improve wood bison habitat in northern Canada. Pages 205–10 *in* L. Irby and J. Knight, eds. International symposium on bison ecology and management in North America. Montana State University, Bozeman.

Christopherson, R. J., R. J. Hudson, and R. J. Richmond. 1976. Feed intake, metabolism, and thermal insulation of bison, yak, Scottish Highland, and Hereford calves during winter. Pages 51–52 *in* 55th annual Feeder's Day report. University of Alberta, Edmonton, Alberta, Canada.

Christopherson, R. J., R. J. Hudson, and R. J. Richmond. 1978. Comparative winter bioenergetics of American bison, yak, Scottish Highland, and Hereford calves. Acta Theriologica 23:49–54.

Christopherson, R. J., R. J. Hudson, and M. K. Christophersen. 1979. Seasonal energy expenditures and thermoregulatory responses of bison and cattle. Canadian Journal of Animal Science 59:611–17.

Church, D. C. 1988. The ruminant animal. Digestive physiology and nutrition. Waveland Press, Prospect Heights, IL.

Cid, M. S., J. K. Detling, A. D. Whicker, and M. A. Brizuela. 1991. Vegetational responses of a mixed-grass prairie site following exclusion of prairie dogs and bison. Journal of Range Management 44:100–105.

Colman, D. 1978. Roy Phillips has a "home" for buffalo. Buffalo 6:14.

Cook, F. R., and D. Muir. 1984. The Committee on the Status of Endangered Wildlife in Canada (COSEWIC): History and progress. Canadian Field-Naturalist 98:63–70.

Cook, W. E., E. S. Williams, E. T. Thorne, S. K. Taylor, and S. Anderson. 2001. Safety of *Brucella abortus* strain RB51 in deer mice. Journal of Wildlife Diseases 37:621–25.

Cool, N. L. 1999. Infectious diseases in wildlife populations, a case study of BVD in plains bison and elk in Elk Island National Park (Parks Canada Agency Report). Elk Island National Park, Site 4, RR 1, Fort Saskatchewan, Alberta, Canada.

Coppedge, B. R., and J. H. Shaw. 1997. Effects of horning and rubbing behavior by bison (*Bison bison*) on woody vegetation in a tallgrass prairie landscape. American Midland Naturalist 138:189–96.

Coppedge, B. R., and J. H. Shaw. 1998. Bison grazing patterns on seasonally burned tallgrass prairie. Journal of Range Management 51:258–64.

Coppedge, B. R., T. S. Carter, J. H. Shaw, and R. G. Hamilton. 1997. Agonistic behavior associated with orphan bison (*Bison bison* L.) calves released into a mixed resident population. Applied Animal Behaviour Science 55: 1–10.

Coppock, D. L., and J. K. Detling. 1986. Alteration of bison and black-tailed prairie dog grazing interaction by prescribed burning. Journal of Wildlife Management 50:452–55.

Coppock, D. L., J. E. Ellis, J. K. Detling, and M. I. Dyer. 1983. Plant–herbivore interactions in a North American mixed-grass prairie. II. Responses of bison to modification of vegetation by prairie dogs. Oecologia 56:10–15.

Corner, A. H., and R. Connell. 1958. Brucellosis in bison, elk, and moose in Elk Island National Park, Alberta, Canada. Canadian Journal of Comparative Medical Veterinary Science 22:9–21.

Cox, B. L. 1978. Comparison of meat quality from bison and beef cattle. Undergraduate Thesis, Department of Home Economics, University of Saskatchewan, Saskatoon, Saskatchewan, Canada.

Cram, E. B. 1925. *Cooperia bisonis,* a new nematode from the buffalo. Journal of Agriculture Research 30:571–73.

Crandall, K. A., O. R. Bininda-Edmonds, G. M. Mace, and R. K. Wayne. 2000. Considering evolutionary processes in conservation biology. Tree 15:290–95.

Creech, B. T. 1930. *Brucella abortus* infection in a male bison. North American Veterinarian 11:35–36.

Cronin, M. A. 1986. Genetic relationships between white-tailed deer, mule deer, and other large mammals inferred from mitochondrial DNA analysis. M.S. Thesis, Montana State University, Bozeman.

Cronin, M. A. 1991. Mitochondrial-DNA phylogeny of deer (Cervidae). Journal of Mammalogy 72:533–66.

Cronin, M., and N. Crockett. 1993. Kappa-casein polymorphisms among cattle breeds and bison herds. Animal Genetics 24:135–38.

Dale, B. W., L. G. Adams, and R. T. Bowyer. 1994. Functional response of wolves preying on barren-ground caribou in a multiple-prey system. Journal of Animal Ecology 63:644–52.

Dary, D. A. 1989. The buffalo book: The full saga of the American animal. Swallow Press/Ohio University Press, Chicago.

Davis, D. S., J. W. Templeton, T. A. Ficht, J. D. Williams, J. D. Kopec, and L. G. Adams. 1990. *Brucella abortus* in captive bison. I. Serology, bacteriology, pathogenesis, and transmission to cattle. Journal of Wildlife Diseases 26:360–71.

Davis, D. S., J. W. Templeton, T. A. Ficht, J. D. Huber, R. D. Angus, and L. G. Adams. 1991. *Brucella abortus* in bison. II. Evaluation of strain 19 vaccination of pregnant cows. Journal of Wildlife Diseases 27:258–64.

Day, T. A., and J. K. Detling. 1990. Grassland patch dynamics and herbivore grazing preference following urine deposition. Ecology 71:180–88.

DelGiudice, G. D., L. D. Mech, and U. S. Seal. 1989. Physiological assessment of deer populations by analysis of urine in snow. Journal of Wildlife Management 53:284–91.

DelGiudice, G. D., U. S. Seal, and L. D. Mech. 1991. Indicators of severe undernutrition in urine of free-ranging elk during winter. Wildlife Society Bulletin 19:106–10.

DelGiudice, G. D., F. J. Singer, U. S. Seal, and G. Bowser. 1994. Physiological responses of Yellowstone bison to winter nutritional deprivation. Journal of Wildlife Management 58:24–34.

Deshano, S. 2002. Raising bison. Available at Showdown Bison Ranch web site, http://www.showdownbison.com/raising-bison.html.

Detling, J. K. 1998. Mammalian herbivores: Ecosystem-level effects in two grassland national parks. Wildlife Society Bulletin 26:438–48.

Dies, K. H. 1998. Are bison deworming practices based on sound scientific principals? Smoke Signals 9:83–95.

Dikmans, G. 1934. New records of helminth parasites. Proceedings of the Helminthological Society of Washington 1:63–64.

Dikmans, G. 1939. Helminth parasites of North American semidomesticated and wild ruminants. Proceedings of the Helminthological Society of Washington 6:97–101.

Dobak, W. 1996. Killing the Canadian buffalo. Western History Quarterly 27:33–52.

Dobson, A., and M. Meagher. 1996. The population dynamics of brucellosis in the Yellowstone National Park. Ecology 77:1026–36.

Dragon, D. C., and B. T. Elkin. 2001. An overview of early anthrax outbreaks in northern Canada: Field reports of the Health of Animals Branch, Agriculture Canada, 1962–71. Arctic 54:32–40.

Dragon, D. C., and R. P. Rennie. 1995. The ecology of anthrax spores: Tough but not invincible. Canadian Veterinary Journal 36:295–301.

Drummond, R. O., and J. G. Medley. 1964. Occurrence of *Speleognathus australis* Womersley (Acarina: Speleognathidae) in the nasal passages of bison. Journal of Parasitology 50:655.

Dubey, J. P. 1980a. *Sarcocystis* species in moose (*Alces alces*), bison (*Bison bison*), and pronghorn (*Antilocapra americana*) in Montana. American Journal of Veterinary Research 41:2063–65.

Dubey, J. P. 1980b. Coyote as a final host for *Sarcocystis* species of goats, sheep, cattle, bison, and moose in Montana. American Journal of Veterinary Research 41:1227–29.

Dubey, J. P. 1982. Sarcocystosis in neonatal bison fed *Sarcocystis cruzi* sporocysts derived from cattle. Journal of the American Veterinary Medical Association 181:1272–74.

Dubey, J. P. 1985. Serologic prevalence of toxoplasmosis in cattle, sheep, goats, pigs, bison, and elk in Montana. Journal of the American Veterinary Medical Association 186:969–70.

Duffield, L. F. 1973. Aging and sexing the postcranial skeleton of bison. Plains Anthropologist 18:132–39.

Dukes, H. H. 1937. The physiology of domestic animals. Comstock, Ithaca, NY.

Dunn, A. M. 1968. The wild ruminant as reservoir host of helminth infection. Symposia of the Zoological Society of London 24:221–48.

Dyer, N. W. 2001. *Haemophilus somnus* bronchopneumonia in American bison (*Bison bison*). Journal of Veterinary Diagnostic Investigation 13:419–21.

Dyer, N. W., and A. C. S. Ward. 1998. Pneumonic pasteurellosis associated with *Pasteurella hemolytica* serotype A6 in American bison (*Bison bison*). Journal of Veterinary Diagnostic Investigation 10:360–62.

Dziuk, H. E. 1965. Eructation, regurgitation, and reticuloruminal contraction in the American bison. American Journal of Physiology 208:343–46.

Dziurdzik, B. 1978. Histological structure of the hair in hybrids of European bison and domestic cattle. Acta Theriologica 23:277–84.

Egerton, P. J. M. 1962. The cow–calf relationship and rutting behavior in the American bison. M.S. Thesis, University of Alberta, Edmonton, Alberta, Canada.

el Shazly, K., B. A. Dehority, and R. R. Johnson. 1961. Effect of starch on the digestion of cellulose in vitro and in vivo by rumen microorganisms. Journal of Animal Science 20:268–73.

Engelhard, J. G. 1970. Behavior patterns of American bison calves of the National Bison Range, Moiese, Montana. M.S. Thesis, Central Michigan University, Mt. Pleasant.

Environment Canada. 1990. A wildlife policy for Canada. Canadian Wildlife Service for Wildlife Minister's Council of Canada, Ottawa, Ontario, Canada.

Environmental Assessment Panel. 1990. Northern diseased bison. Federal Environmental Assessment Review Office. Ottawa, Ontario, Canada.

Estes, R. D. 1972. The role of the vomeronasal organ in mammalian reproduction. Mammalia 36:315–41.

Fayer, R., J. P. Dubey, and R. G. Leek. 1982. Infectivity of *Sarcocystis* spp. from bison, elk, moose, and cattle for cattle via sporocysts from coyotes. Journal of Parasitology 68:681–85.

Federal Register. 2001, January 22. Notices. Federal Register 66(14):6665–66. Available at Federal Register Online, wais.access.gpo.gov.

Feist, M. 2000. Bison nutrition. Saskatchewan Agriculture and Food, Extension Research Unit, Saskatoon, Saskatchewan, Canada.

Feldhamer, G. A., L. C. Drickamer, S. H. Vessey, and J. F. Merritt. 1999. Mammalogy: Adaptation, diversity and ecology. WCB McGraw-Hill, Boston.

Findlay, C. R., and T. B. Begg. 1977. Redwater of American bison caused by *Babesia major*. Veterinary Record 100:406.

Fischer, L. A. 2002. Late winter resource selection and the potential for competition between wood bison and woodland caribou in the Yukon. Master of Environmental Design, University of Calgary, Calgary, Alberta, Canada.

Fisher, J. W., Jr., and T. E. Roll. 1998. Ecological relationships between bison and Native Americans during late prehistory and the early historic period. Pages 283–302 *in* L. Irby and J. Knight, eds. International symposium on bison ecology and management in North America. Montana State University, Bozeman.

Flerov, C. C. 1965. Comparative craniology of recent representatives of the genus *Bison*. Bulletin of the Moscow Nature Society LXX (1):1–17.

Flerov, C. C., and Zablotski, M. A. 1961. On the causative factors responsible for the change in the bison range. Bulletin of the Moscow Nature Society LXVI (6):99–109.

Flores, D. 1991. Bison ecology and bison diplomacy. Journal of American History 78:465–85.

Flores, D. 1996. The great contraction: Bison and Indians in northern plains environmental history. Pages 3–22 *in* C. E. Rankin, ed. Legacy: New perspectives on the Battle of the Little Bighorn. Montana Historical Society Press, Helena.

Fortin, D., J. M. Fryxell, and R. Pilote. 2002. The temporal scale of foraging decisions in bison. Ecology 83:970–82.

Frank, D. A., and P. M. Groffman. 1998. Ungulate vs landscape control of soil C and N processes in grasslands of Yellowstone National Park. Ecology 79:2229–41.

Frick, E. J. 1951. Parasitism in bison. Journal of the American Veterinary Medical Association 119:386–87.

Frison, G. C. 1980. Man and bison relationships in North America. Canadian Journal of Anthropology 1:75–76.

Frison, G. C., and C. A. Reher. 1970. Age determination of buffalo by teeth eruption and wear. Pages 46–50 *in* G. Frison, ed. The Glenrock buffalo jump: Late Prehistoric period buffalo procurement and butchering (Plains Anthropologist Memoir 7).

Fuller, W. A. 1959. The horns and teeth as indicators of age in bison. Journal of Wildlife Management 23:342–44.

Fuller, W. A. 1960. Behavior and social organization of the wild bison of Wood Buffalo National Park, Canada. Arctic 13:3–19.

Fuller, W. A. 1961. The ecology and management of the American bison. Terre et la Vie 2:286–304.

Fuller, W. A. 1966. The biology and management of the bison of Wood Buffalo National Park. Canadian Wildlife Service Wildlife Management Bulletin Series 1 (16):1–52.

Fuller, W. A. 1991. Disease management in Wood Buffalo National Park, Canada: Public attitudes and management implications. Transactions of the 56th North American Wildlife and Natural Resources Conference.

Fuller, W. A. 2002. Canada and the "buffalo," *Bison bison:* A tale of two herds. Canadian Field-Naturalist 116:141–59.

Fuller, W. A., and L. A. Bayrock. 1965. Late Pleistocene mammals from central Alberta, Canada. Pages 53–63 *in* Vertebrate paleontology in Alberta. University of Alberta, Edmonton, Alberta, Canada.

Fulton, R. D., J. Caldwell, and D. F. Weseli. 1978. Methemoglobin reductase in three species of Bovidae. Biochemical Genetics 16:635–40.

Gainer, B. 1985. Free-roaming bison in northern Alberta. Alberta Naturalist 15:86–87.

Galbraith, J. K., G. Mathison, R. J. Hudson, T. A. McAllister, and K. J. Cheng. 1998. Intake, digestibility, methane and heat production in bison, wapiti, and white-tailed deer. Canadian Journal of Animal Science 78:681–91.

Garretson, M. S. 1927. A short history of the American bison. American Bison Society, New York.

Garretson, M. S. 1938. The American bison: The story of its extermination as a wild species and its restoration under federal protection. New York Zoology Society, NY.

Gates, C. C. 1993. Biopolitics and pathobiology: Diseased bison in northern Canada. Pages 271–88 *in* R. E. Walker, ed. and comp. Proceedings of the North American public bison herds symposium. Custer State Park, Custer, SD.

Gates, C. C., and N. C. Larter. 1990. Growth and dispersal of an erupting large herbivore population in northern Canada: The Mackenzie wood bison (*Bison bison athabascae*). Arctic 43:231–38.

Gates, C. C., T. Chowns, and H. Reynolds. 1992a. Wood buffalo at the crossroads. Pages 139–65 *in* J. Foster, D. Harrison, and I. S. MacLaren, eds. Alberta: Studies in the Arts and Sciences. 3: Special issue on the buffalo. University of Alberta Press, Edmonton, Alberta, Canada.

Gates, C. C., B. Elkin, L. Kearey, and T. Chowns. 1992b. Surveillance of the bison-free management area, January–June 1992 (Manuscript Report, No. 65). Northwest Territories Department of Renewable Resources, Yellowknife, Northwest Territories, Canada.

Gates, C. C., B. T. Elkin, and D. C. Dragon. 1995. Investigation, control and epizootiology of anthrax in a geographically isolated, free-roaming bison population in northern Canada. Canadian Journal of Veterinary Research 59:256–64.

Gates, C. C., B. T. Elkin, and L. Carbyn. 1997. The diseased bison issue in northern Canada. Pages 120–32 *in* E. T. Thorne, M. S. Boyce, P. Nicoleti, and T. J. Kreeger, eds. Brucellosis, bison, elk and cattle in the Greater Yellowstone area: Defining the problem, exploring solutions. Wyoming Game and Fish Department, Cheyenne.

Gates, C. C., B. T. Elkin, and D. C. Beaulieu. 1998. Initial results of an attempt to eradicate bovine tuberculosis and brucellosis from a wood bison herd in northern Canada. Pages 221–28 *in* L. Irby and J. Knight, eds. International symposium on bison ecology and management in North America. Montana State University, Bozeman.

Gates, C. C., B. T. Elkin, and D. C. Dragon. 2001a. Anthrax. Pages 396–412 *in* E. S. Williams, I. K. Barker, and E. T. Thorne, eds. Infectious diseases of wild mammals, 3rd ed. Iowa State University Press, Ames.

Gates, C. C., R. O. Stephenson, H. W. Reynolds, C. G. van Zyll de Jong, H. Schwantje, M. Hoefs, J. Nishi, N. Cool, J. Chisholm, A. James, and B. Koonz. 2001b. National recovery plan for the wood bison (*Bison bison athabascae*) (National Recovery Plan No. 21). Recovery of Nationally Endangered Wildlife, Ottawa, Ontario, Canada.

Gates, C. C., J. Mitchell, J. Wierzchowski, and L. Giles. 2001c. A landscape evaluation of bison movements and distribution in northern Canada. Axys Environmental Consulting, Calgary, Alberta, Canada.

Gates, C. C., R. O. Stephenson, S. Zimov, and M. C. Chapin. 2001d. Wood bison recovery: Restoring grazing systems in Canada, Alaska, and eastern Siberia. Pages 82–102 *in* B. D. Rutley, ed. Bison are back, 2000. Proceedings of the second international bison conference. Edmonton, Alberta, Canada.

Geist, V. 1963. On the behaviour of the North American moose (*Alces alces andersoni* Peterson 1950) in British Columbia. Behaviour 20:377–416.

Geist, V. 1971. The relation of social evolution and dispersal in ungulates during the Pleistocene, with emphasis on the Old World deer and the genus *Bison*. Quaternary Research 1:285–315.

Geist, V. 1991. Phantom subspecies: The wood bison *Bison bison athabascae* Rhoads 1897 is not a valid taxon, but an ecotype. Arctic 44:283–300.

Geist, V., and P. Karsten. 1977. The wood bison (*Bison bison athabascae* Rhoads) in relation to hypotheses on the origin of the American bison (*Bison bison* Linnaeus). Zeitschrift für Säugetierkunde 42:119–27.

Gilchrist, F. M. C., and R. Clark. 1957. Refresher courses in physiology. Part 3: The microbiology of the rumen. Journal of the South African Veterinary Medical Association 28:295–309.

Glover, J., and H. W. Dougall. 1960. The apparent digestibility of the non-nitrogenous components of ruminant feeds. Journal of Agriculture Science 55:391–94.

Gochenour, W. S. 1924. Hemorrhagic septicemia studies: The development of a potent immunizing agent (natural aggressin) by the use of highly virulent strains of hemorrhagic septicemia organisms. Journal of the American Veterinary Medical Association 65:433–41.

Government of Yukon. 1998. Yukon bison management plan, 1998–2003. Unpublished report. Department of Renewable Resources, Government of Yukon, Whitehorse, Yukon, Canada.

Graham, M. 1923. Canada's wild buffalo. Canada Department of the Interior, Ottawa, Ontario, Canada.

Graham, M. 1924. Finding range for Canada's buffalo. Canadian Field-Naturalist 38:189.

Grandin, T. 1999. Safe handling of large animals (cattle and horses) (Occupational Medicine: State of the Art Reviews 14, No. 2). Colorado State University, Fort Collins. Available at http://ansci.colostate.edu/res/environ/safe.html.

Green, W. C. H. 1986. Age-related differences in nursing behavior among American bison cows (*Bison bison*). Journal of Mammalogy 67:739–41.

Green, W. C. H. 1992. The development of independence in bison: Pre-weaning spatial relations between mothers and calves. Animal Behaviour 43:759–73.

Green, W. C. H., and A. Rothstein. 1991. Trade-offs between growth and reproduction in female bison. Oecologia 86:521–27.

Green, W. C. H., and A. Rothstein. 1993a. Persistent influences of birth date on dominance, growth and reproductive success in bison. Journal of Zoology 230:177–86.

Green, W. C. H., and A. Rothstein. 1993b. Asynchronous parturition in bison: Implications for the hider–follower dichotomy. Journal of Mammalogy 74:920–25.

Green, W. C. H., J. G. Griswold, and A. Rothstein. 1989. Post-weaning associations among bison mothers and daughters. Animal Behaviour 38:847–58.

Green, W. C. H., A. Rothstein, and J. G. Griswold. 1993. Weaning and parent–offspring conflict: Variation relative to interbirth interval in bison. Ethology 95:105–25.

Grinnell, G. B. 1904. The bison. Pages 111–66 *in* C. Whitney, G. B. Grinnell, and O. Wister, eds. Muskox, bison, sheep, and goat. Macmillan, NY.

Groves, C. P. 1981. Systematic relationships in the Bovini (Artiodactyla, Bovidae). Zeitschrift für zoologischen Systematik und Evolutionsforschung 19:264–78.

Gunderson, H. L., and B. R. Mahan. 1980. Analysis of sonograms of American bison (*Bison bison*). Journal of Mammalogy 61:379–81.

Guthrie, R. D. 1970. Bison evolution and zoogeography in North America during the Pleistocene. Quarterly Review of Biology 45:1–15.

Guthrie, R. D. 1980. Bison and man in North America. Canadian Journal of Anthropology 1:55–73.

Guthrie, R. D. 1982. Mammals of the mammoth steppe as paleoenvironmental indicators. Pages 307–29 *in* D. M. Hopkins, J. V. Matthews, Jr., C. E. Schweger, and S. B. Young, eds. Paleoecology of Beringia. Academic Press, New York.

Guthrie, R. D. 1990. Frozen fauna of the mammoth steppe: The story of Blue Babe. University of Chicago Press, Chicago.

Hadwen, S. 1942. Tuberculosis in the buffalo. Journal of the American Veterinary Medical Association 100:19–22.

Haines, H., H. G. Chichester, and H. I. Landreth. Jr. 1977. Blood respiratory properties of *Bison bison*. Respiratory Physiology 30:305–10.

Hall, E. R. 1981. The mammals of North America, 2nd ed. John Wiley, New York.

Halloran, A. F. 1961. American bison weights and measurements from the Wichita Mountains Wildlife Refuge. Proceedings of the Oklahoma Academy of Science 41:212–18.

Halloran, A. F. 1968. Bison (Bovidae) productivity on the Wichita Mountains Wildlife Refuge, Oklahoma. Southwestern Naturalist 13:23–26.

Halloran, A. F., and B. P. Glass. 1959. The carnivores and ungulates of the Wichita Mountains Wildlife Refuge, Oklahoma. Journal of Mammalogy 40:360–70.

Hancock, J. 1953. Grazing behavior of cattle. Commonwealth Bureau of Animal Breeding and Genetics 21:1–13.

Harington, C. R. 1977. Pleistocene mammals of the Yukon Territory. Ph.D. Dissertation, University of Alberta, Edmonton, Alberta, Canada.

Harington, C. R. 1980. Faunal exchanges between Siberia and North America: Evidence from Quaternary land mammal remains in Siberia, Alaska, and the Yukon Territory. Canadian Journal of Anthropology 1:45–49.

Harington, C. R. 1990. Arctic bison. Biome 10:4.

Harper, F. 1925. Letter to the editor of the Canadian Field-Naturalist. Canadian Field-Naturalist 39:45.

Harper, W. L., J. P. Elliott, I. Hatter, and H. Schwantje. 2000. Management plan for wood bison in British Columbia. British Columbia Ministry of Environment, Lands and Parks, Victoria, British Columbia, Canada.

Hartl, G. B., and F. Reimoser. 1988. Biochemical variation in Roe deer (*Capreolus capreolus* L.): Are *R*-strategists among deer genetically less variable than *K*-strategists? Heredity 6:221–27.

Haugen, A. O. 1974. Reproduction in the plains bison. Iowa State Journal of Research 49:1–8.

Hawley, A. W. L. 1978. Comparison of forage utilization and blood composition of bison and Hereford cattle. Ph.D. Dissertation, University of Saskatchewan, Saskatoon, Saskatchewan, Canada.

Hawley, A. W. L. 1986. Carcass characteristics of bison (*Bison bison*) steers. Canadian Journal of Animal Science 66:293–95.

Hawley, A. W. L. 1987. Bison and cattle use of forages. Pages 49–52 *in* H. Reynolds and A. Hawley, eds. Bison ecology in relation to agricultural development in the Slave River Lowlands, N.W.T. (Occasional Paper No. 63; Catalogue No. CW69-1/63E). Minister of Supply and Services Canada, Ottawa, Ontario, Canada.

Hawley, A. W. L. 1989. Bison farming in North America. Pages 346–62 *in* R. J. Hudson, K. R. Drew, and L. M. Baskin, eds. Wildlife production systems: Economic utilisation of wild ungulates. Cambridge University Press, Cambridge.

Hawley, A. W. L., and H. W. Reynolds. 1987. Management alternatives for ungulate production in the Slave River Lowlands. Pages 63–66 *in* H. Reynolds and A. Hawley, eds. Bison ecology in relation to agricultural development in the Slave River Lowlands N.W.T. (Occasional Paper No. 63; Catalogue No. CW69-1/63E). Minister of Supply and Services Canada, Ottawa, Ontario, Canada.

Hawley, A. W. L., D. G. Peden, H. W. Reynolds, and W. R. Stricklin. 1981a. Bison and cattle digestion of forages from the Slave River Lowlands, Northwest Territories, Canada. Journal of Range Management 34:126–30.

Hawley, A. W. L., D. G. Peden, and W. R. Stricklin. 1981b. Bison and Hereford steer digestion of sedge hay. Canadian Journal of Animal Science 61:165–74.

Haynes, G. 1984. Tooth wear rate in northern bison. Journal of Mammalogy 65:487–91.

Haynes, T. 1998. Bison hunting in the Yellowstone River drainage, 1800–1884. Pages 303–11 *in* L. Irby and J. Knight, eds. International symposium on bison ecology and management in North America. Montana State University, Bozeman.

Heddleston, K. L., and J. E. Gallagher. 1969. Septicemic pasturellosis (hemmorhagic septicemia) in the American bison: A serologic survey. Bulletin of the Wildlife Disease Association 5:206–7.

Heddleston, K. L., and G. Wessman. 1973. Vaccination of American bison against *Pasteurella multocida* serotype 2 infection (hemorrhagic septicemia). Journal of Wildlife Diseases 9:306–10.

Heddleston, K. L., K. R. Rhoades, and P. A. Rebers. 1967. Experimental pasteurellosis: Comparative studies on *Pasteurella multocida* from Asia, Africa, and North America. American Journal of Veterinary Research 28:1003–12.

Hein, F. J., and C. R. Preston. 1998. Summer nocturnal movements and habitat selection by *Bison bison* in Colorado. Pages 96–102 *in* L. Irby and J. Knight, eds. International symposium on bison ecology and management in North America. Montana State University, Bozeman.

Herd, R. P., and B. L. Hull. 1981. *Paramphistomum microbothrioides* in American bison and domestic beef cattle. Journal of the American Veterinary Medical Association 179:1019–20.

Herrig, D. M., and A. O. Haugen. 1969. Bull bison behavior traits. Iowa Academy of Science 76:245–62.

Hewitt, C. G. 1921. The buffalo or bison: Its present, past, and future. Pages 113–42 *in* C. G. Hewitt, ed. The conservation of the wildlife of Canada. Charles Scribner's Sons, New York.

Hillerud, J. M. 1980. Bison as indicators of geologic age. Canadian Journal of Anthropology 1:77–80.

Hilton-Taylor, C., comp. 2000. 2000 IUCN red list of threatened species. IUCN, Gland, Switzerland.

Holmes, C. E., and G. Bacon. 1982. Holocene bison in central Alaska: A possible explanation for technological conservatism. Paper presented at the 9th annual meeting of the Alaska Anthropological Association, April 2–3, 1982. Fairbanks, AK.

Holsworth, W. N. 1960. Interactions between moose, elk, and buffalo in Elk Island National Park, Alberta. M.S. Thesis, University of British Columbia, Vancouver, Canada.

Hornaday, W. T. 1889. The extermination of the American bison, with a sketch of its discovery and life history. Pages 367–548 *in* Report of the United States National Museum, 1886–87, Part 2 [1889].

Houpt, T. R. 1970. Transfer of urea and ammonia to the rumen. Pages 119–31 *in* A. T. Phillipson, ed. Physiology of digestion and metabolism in the ruminant. Oriel Press, Cambridge, UK.

Howell, A. B. 1925. Letter to the editor of the Canadian Field-Naturalist from the Corresponding Secretary of the American Society of Mammalogists, April 13, 1925. Canadian Field-Naturalist 39:118.

Hudson, R. J., and S. Frank. 1987. Foraging ecology of bison in aspen boreal habitats. Journal of Range Management 40:71–75.

Inman, H. 1899. Buffalo Jones' forty years of adventure. Crane, Topeka, KS.

Inserro, J. C. 1997. States, agencies discuss solutions for handling Yellowstone bison, brucellosis. Journal of the American Veterinary Medical Association 210:593–95.

International Union for Conservation of Nature and Natural Resources. 1980. World conservation strategy: Living resource conservation for sustainable development. IUCN, Gland, Switzerland.

International Union for Conservation of Nature and Natural Resources. 1987. Translocation of living organisms: IUCN position statement, 4 September 1987. IUCN, Gland, Switzerland.

Isenberg, A. C. 2000. The destruction of the bison; an environmental history, 1750–1920. Cambridge University Press, Cambridge, MA.

Jaczewski, Z. 1958. Reproduction of the European bison, *Bison bonasus* (L.), in reserves. Acta Theriologica 1:333–76.

Janecek, L. L., R. L. Honeycutt, R. M. Adkins, and S. K. Davis. 1996. Mitochondrial gene sequences and the molecular systematics of the Artiodactyl subfamily Bovinae. Molecular and Phylogenetics and Evolution 6:107–19.

Jaros, Z., Z. Valenta, and D. Zajicek. 1966. A list of helminths from the section material of the Zoological Garden of Prague in the years 1954–1964. Helminthologia 7:281–90.

Jaworski, M. D., D. L. Hunter, and A. C. S. Ward. 1998. Biovariants of isolates of *Pasteurella* from domestic and wild ruminants. Journal of Veterinary Diagnostic Investigation 10:49–55.

Jennings, D. C. 1978. Buffalo history and husbandry. Pine Hill Press, Freeman, ND.

Jennings, J. D. 1968. Prehistory of North America. McGraw-Hill, New York.

Joly, D. O., and F. Messier. 2000. A numerical response of wolves to bison abundance in Wood Buffalo National Park, Canada. Canadian Journal of Zoology 78:1101–4.

Joly, D., and F. Messier. 2001a. Limiting effects of bovine brucellosis and tuberculosis on bison within Wood Buffalo National Park. Final report dated March 2001. Department of Biology, University of Saskatchewan, Saskatoon, Saskatchewan, Canada.

Joly, D., and F. Messier. 2001b. Limiting effects of bovine brucellosis and tuberculosis on bison within Wood Buffalo National Park. Testing hypotheses of bison population decline in Wood Buffalo National Park. Addendum to the final report dated April 2001. Department of Biology, University of Saskatchewan, Saskatoon, Saskatchewan, Canada.

Jones, C., R. S. Hoffmann, D. W. Rice, R. J. Baker, M. D. Engstrom, R. D. Bradley, D. J. Schmidly, and C. A. Jones. 1997. Revised checklist of North American mammals north of Mexico, 1997 (Occasional Paper No. 173). Museum of Texas Tech University, Lubbock.

Kaneko, J. J., and C. E. Cornelius. 1970. Clinical biochemistry of domestic animals, 2nd ed. Academic Press, New York.

Karsten, P. 1975. Don't be buffaloed... by a bison: History of Alberta herds. Dinny's Digest, Calgary Zoological Society 2:313.

Keiter, R. B. 1997. Greater Yellowstone's bison: Unraveling of an early American wildlife conservation achievement. Journal of Wildlife Management 61:1–11.

Keith, E. O. 1977. Urea metabolism of North American bison. M.S. Thesis, Colorado State University, Fort Collins.

Keith, E. O., J. E. Ellis, R. W. Phillips, and M. M. Benjamin. 1978. Serologic and hematologic values of bison in Colorado. Journal of Wildlife Diseases 14:493–500.

Keith, E. O., J. E. Ellis, R. W. Phillips, M. I. Dyer, and G. M. Ward. 1981. Some aspects of urea metabolism in North American bison. Acta Theriologica 26:257–68.

Keller, D. G. 1980. Milk production in cattalo cows and its influence on calf gains. Canadian Journal of Animal Science 60:19.

Kelsall, J. P., E. S. Telfer, and M. C. S. Kingsley. 1978. Relationship of bison weight to chest girth. Journal of Wildlife Management 42:659–61.

Kirkpatrick, R. L., D. E. Buckland, W. A. Abler, P. E. Scanlon, J. B. Whelan, and H. E. Burkhan. 1975. Energy and protein influences on blood urea nitrogen of white-tailed deer fawns. Journal of Wildlife Management 39:692–98.

Kirkpatrick, J. F., K. Bancroft, and V. Kincy. 1992. Pregnancy and ovulation detection in bison (*Bison bison*) assessed by means of urinary and fecal steroids. Journal of Wildlife Diseases 28:590–97.

Kirkpatrick, J. F., D. F. Gudermuth, R. L. Flagan, J. C. McCarthy, and B. L. Lasley. 1993. Remote monitoring of ovulation and pregnancy of Yellowstone bison. Journal of Wildlife Management 57:407–12.

Kirkpatrick, J. F., J. C. McCarthy, D. F. Gudermuth, S. E. Shideler, and B. L. Lasley. 1996. An assessment of the reproductive biology of Yellowstone bison (*Bison bison*) subpopulations using noncapture methods. Canadian Journal of Zoology 74:8–14.

Knapp, A. K., J. M. Blair, J. M. Briggs, S. L. Collins, D. C. Hartnett, L. C. Johnson, and E. G. Towne. 1999. The keystone role of bison in the North American tallgrass prairie. BioScience 49:39–50.

Koch, R. M., H. G. Jung, J. D. Crouse, V. H. Varal, and L. V. Cundiff. 1995. Growth, digestive capability, carcass, and meat characteristics of *Bison bison, Bos taurus,* and *Bos × Bison.* Journal of Animal Science 73:1271–81.

Kohls, G. M., and N. J. Kramis. 1952. Tick paralysis in the American buffalo *Bison bison* (Linn.). Northwest Science 26:61–64.

Komers, P. E., K. Roth, and R. Zimmerli. 1992a. Interpreting social behaviour of wood bison using tail postures. Zeitschrift für Saügetierkunde 57:343–50.

Komers, P. E., F. Messier, and C. C. Gates. 1992b. Search or relax: The case of bachelor wood bison. Behavioral Ecological Sociobiology 31:195–203.

Komers, P. E., F. Messier, P. F. Flood, and C. C. Gates. 1994. Reproductive behavior of male wood bison in relation to progesterone level in females. Journal of Mammalogy 75:757–65.

Krueger, K. 1986. Feeding relationships among bison, pronghorn and prairie dogs: An experimental analysis. Ecology 67:760–70.

Larson, F. 1940. The role of the bison in maintaining the shortgrass plains. Ecology 21:113–21.

Larter, N. C., and C. C. Gates. 1991. Diet and habitat selection of wood bison in relation to seasonal change in forage quantity and quality. Canadian Journal of Zoology 69:2677–85.

Larter, N. C., and C. C. Gates. 1994. Home-range size of wood bison: Effects of age, sex, and forage availability. Journal of Mammalogy 75:142–49.

Larter, N. C., A. R. E. Sinclair, and C. C. Gates. 1994. The response of predators to an erupting bison, *Bison bison athabascae,* population. Canadian Field-Naturalist 108:318–27.

Laundre, J., W. L. Hernandez, and K. B. Altendorf. 2001. Wolves, elk, and bison: Reestablishing the "landscape of fear" in Yellowstone National Park, U.S.A. Canadian Journal of Zoology 79:1401–9.

Lazarev, P. A., G. G. Boeskorov, and A. I. Tomskaya. 1998. Mlekopitauschie antropogena Yakutii. *In* Yu. V. Labutin, ed. Yakutsk. Yakutskii nauchnyi tsentr, Sibirskoe otdelenie Rossiiskoi Akademii Nauk.

Li, H., D. T. Shen, D. A. Jessup, D. P. Knowles, J. R. Gorham, T. Thorne, D. O'Toole, and T. B. Crawford. 1996. Prevalence of antibody to malignant catarrhal fever virus in wild and domesticated ruminants by competitive-inhibition elisa. Journal of Wildlife Diseases 32:437–43.

Lichtenfels, J. R., and P. A. Pilitt. 1991. A redescription of *Ostertagia bisonis* (Nematoda: Trichostrongyloidea) and a key to species of Ostertaginae with a tapering lateral synlophe from domestic ruminants in North America. Journal of the Helminthological Society of Washington 58:231–44.

Liggitt, H. D., A. E. McChesney, and J. C. DeMartini. 1980. Experimental transmission of bovine malignant catarrhal fever to a bison (*Bison bison*). Journal of Wildlife Diseases 16:299–304.

Lin, M., E. A. Sugden, M. E. Jolley, and K. Stilwell. 1996. Modification of the *Mycobacterium bovis* extracellular protein MPB70 with fluorescin for rapid detection of specific serum antibodies by fluorescence polarization. Clinical and Diagnostic Laboratory Immunology 3:438–43.

Linnaeus, C. 1758. Systema naturae per regna tria naturae, secundum classes, ordines, genera, species cum characteribus, differentiis, synonymis, locis. 10th ed. Stockholm: Laurentii Salvii.

Locker, B. 1953. Parasites of bison in northwestern U.S.A. Journal of Parasitology 39:58–59.

Longhurst, W. M. 1961. Big game and rodent relationships for forests and grasslands in North America. Terre et la Vie 2:305–26.

Lotenberg, G. 1996. History of wood bison in the Yukon: A reevaluation based on traditional knowledge and written records. Unpublished report submitted to the Yukon Renewable Resources Department. Boreal Research Associates, Whitehorse, Yukon, Canada.

Lothian, W. F. 1979. A history of Canada's national parks, Vol. III. Parks Canada, Supply and Services Canada, Ottawa, Ontario, Canada.

Lothian, W. F. 1981. A history of Canada's national parks, Vol. IV. Parks Canada, Supply and Services Canada, Ottawa, Ontario, Canada.

Lott, D. F. 1972. Bison would rather breed than fight. Natural History 81:40–45.

Lott, D. F. 1974. Sexual and aggressive behavior of adult male American bison (*Bison bison*). Pages 382–93 *in* V. Geist and F. Walther, eds. The behavior of ungulates and its relation to management. Morges, Switzerland.

Lott, D. F. 1979. Dominance relations and breeding rate in mature male American bison. Zeitschrift für Tierpsychologie 49:418–32.

Lott, D. F. 1991. American bison socioecology. Applied Animal Behaviour Science 29:135–45.

Lott, D. F. 2002. American bison. A natural history. University of California Press, Berkeley.

Lott, D. F., and J. C. Galland. 1987. Body mass as a factor influencing dominance status in American bison cows. Journal of Mammalogy 68:683–85.

Madson, C. 1998. Chronic wasting disease. Wyoming Wildlife Magazine 5:7.

Mahan, B. R. 1977. Harassment of an elk calf by bison. Canadian Field-Naturalist 91:418–19.

Mahan, B. R. 1978. Aspects of American bison (*Bison bison*) social behavior at Fort Niobrara National Wildlife Refuge, Valentine, Nebraska, with special reference to calves. M.S. Thesis, University of Nebraska, Lincoln.

Mahan, B. R., M. P. Munger, and H. L. Gunderson. 1978. Analysis of the flehmen display in American bison (*Bison bison*). Prairie Naturalist 10:33–42.

Maher, C. R., and J. A. Byers. 1987. Age-related changes in reproductive effort of male bison. Behavioral Ecology and Sociobiology 21:91–96.

Mahrt, J. L., and D. D. Colwell. 1980. Sarcocystis in wild ungulates in Alberta. Journal of Wildlife Diseases 16:571–76.

Marler, R. J. 1975. Some hematologic and blood chemistry values in two herds of American bison in Kansas. Journal of Wildlife Diseases 11:97–100.

Martin, P. S., and C. R. Szuter. 1999. War zones and game sinks in Lewis and Clark's West. Conservation Biology 13:36–45.

Matsuda, D. M., A. C. Bellem, C. J. Gartley, V. Madison, W. A. King, R. M. Liptrap, and K. L. Goodrowe. 1996. Endocrine and behavioral events of estrous cyclicity and synchronization in wood bison (*Bison bison athabascae*). Theriogenology 45:1429–41.

Mattson, D. J. 1997. Use of ungulates by Yellowstone grizzly bears. Biological Conservation 81:161–77.

Mayr, E. 1963. Animal species and evolution. Harvard University Press, Cambridge, MA.

McClenaghan, L. R., J. Berger, and H. D. Truesdale. 1990. Founding lineages and genetic variability in plains bison (*Bison bison*) from Badlands National Park, South Dakota. Conservation Biology 4:285–89.

McDonald, J. N. 1981. North American bison, their classification and evolution. University of California Press, Berkeley.

McHugh, T. 1958. Social behavior of the American buffalo (*Bison bison bison*). Zoologica 43(1):1–40.

McHugh, T. 1972. The time of the buffalo. Alfred A. Knopf, New York.

Meagher, M. M. 1973. The bison of Yellowstone National Park. National Park Service Scientific Monograph Series 1:1–161.

Meagher, M. M. 1978. Bison. Pages 123–33 *in* J. L. Schmidt and D. L. Gilbert, eds. Big game of North America: Ecology and management. Stackpole, Harrisburg, PA.

Meagher, M. M. 1986. *Bison bison.* Mammalian Species 266:1–8.

Meagher, M. M. 1989a. Evaluation of boundary control for bison of Yellowstone National Park. Wildlife Society Bulletin 17:15–19.

Meagher, M. M. 1989b. Range expansion by bison of Yellowstone National Park. Journal of Mammalogy 70:670–75.

Meagher, M. M., and M. E. Meyer. 1994. On the origin of brucellosis in bison of Yellowstone National Park: A review. Conservation Biology 8:645–53.

Mehrer, C. F. 1976. Some hematologic values of bison from five areas of the United States. Journal of Wildlife Diseases 12:713.

Melton, D. A., N. C. Larter, C. C. Gates, and J. A. Virgl. 1989. The influence of rut and environmental factors on the behavior of wood bison. Acta Theriologica 34:179–93.

Melton, D. A., C. C. Gates, N. C. Larter, and J. A. Virgl. 1990. Foraging behavior of wood bison in an expanding population. Transactions of the Congress of the International Union of Game Biologists 19:44–47.

Merril, E. H., N. L. Stanton, and J. C. Hak. 1994. Responses of bluebunch wheatgrass, Idaho fescue, and nematodes to ungulate grazing in Yellowstone National Park. Oikos 69:231–40.

Messier, F. 1996. On the functional and numerical responses of wolves to changing prey density. *in* L. N. Carbyn, S. H. Fritts, and D. R. Seip, eds. Ecology and conservation of wolves in a changing world. University of Alberta Press, Edmonton, Alberta, Canada.

Meyer, M. E., and M. M. Meagher. 1995. Brucellosis in free-ranging bison (*Bison bison*) in Yellowstone, Grand Teton, and Wood Buffalo National Parks: A review. Journal of Wildlife Diseases 31:579–98.

Miyamoto, M. M., S. M. Tanhauser, and P. J. Laipis. 1989. Systematic relationships in the artiodactyl tribe Bovini (family Bovidae), as determined from mitochondrial DNA sequences. Systematic Zoology 38:342–49.

Moffitt, S. A. 1998. Aging bison by the incremental cementum growth layers in teeth. Journal of Wildlife Management 62:1276–80.

Mohler, J. R., and A. Eichhorn. 1912–1913. Immunization against hemorrhagic septicemia. American Veterinary Review 42:409–18.

Mooring, M. S., and W. M. Samuel. 1998. Tick defense strategies in bison: The role of grooming and hair coat. Behaviour 135:693–718.

Morgan, R. G. 1998. The destruction of the northern bison herds. Pages 312–25 *in* L. Irby and J. Knight, eds. International symposium on bison ecology and management in North America. Montana State University, Bozeman.

Morris, B. G., M. C. Spencer, S. Stabile, and J. N. Dodd. 1994. Restriction fragment length polymorphism (RFLP) of exon 2 of the MhcBibi-DRB3 gene in American bison (*Bison bison*). Animal Genetics (Supplement 1) 25:91–93.

Morrisette, P. 2000. Is there room for free-roaming bison in greater Yellowstone? Ecology Law Quarterly 27:467–519.

Morton, K. 2003. Population surveys in the Hay-Zama Lowlands, wood bison (*Bison bison athabascae*), February 24, 2003. Unpublished report, Alberta, Sustainable Resource Development, Fish and Wildlife Division. March 31, 2003. High Level, Alberta, Canada.

Moss, E. H. 1932. The vegetation of Alberta. Journal of Ecology 20:380–415.

Naik, S. N., and D. E. Anderson. 1970. Study of glucose-6-phosphate dehydrogenase and 6-phosphoglucanate dehydrogenase in the American buffalo (*Bison bison*). Biochemical Genetics 4:651–54.

National Buffalo Association. 1990. Buffalo producer's guide to management and marketing. R. R. Donnelley, Chicago.

Nelson, K. L. 1965. Status and habits of the American buffalo (*Bison bison*) in the Henry Mountain area of Utah (Publication 652). Utah State Department of Fish and Game.

Nemecek, S. 1996. Return of the red wolf: Controversy over taxonomy endangers protection efforts. Scientific American 274 (January):31–32.

Nishi, J. S., B. T. Elkin, T. R. Ellsworth, G. A. Wilson, D. W. Balsillie, and J. van Kessel. 2001. An overview of the Hook Lake Wood Bison Recovery Project: Where have we come from, where are we now, and where we would like to go? Pages 215–33 *in* B. D. Rutley, ed. Bison are back, 2000. Proceedings of the second international bison conference. Edmonton, Alberta, Canada.

Nishi, J. S., B. T. Elkin, and T. R. Ellsworth. 2002a. The Hook Lake Wood Bison Recovery Project. Can a disease-free captive wood bison herd be recovered from a wild population infected with bovine tuberculosis and brucellosis? Annals of the New York Academy of Sciences 969:229–35.

Nishi, J. S., C. Stephen, and B. T. Elkin. 2002b. Implications of agricultural and wildlife policy on management and eradication of bovine tuberculosis and brucellosis in free-ranging wood bison of northern Canada. Annals of the New York Academy of Sciences 969:236–44.

Nishi, J. S., D. C. Dragon, B. T. Elkin, J. Mitchell, T. R. Ellsworth, and M. E. Hugh-Jones. 2002c. Emergency response planning for anthrax outbreaks in bison herds of northern Canada. A balance between policy and science. Annals of the New York Academy of Sciences 969:245–50.

Novakowski, N. S. 1965. Cemental deposition as an age criterion in bison, and the relation of incisor wear, eye-lens weight, and dressed bison carcass weight to age. Canadian Journal of Zoology 43:173–78.

O'Brien, S. J., and E. Mayr. 1991. Bureaucratic mischief: Recognizing endangered species and subspecies. Science 231:1187–88.

Ogilvie, W. 1893. Report on the Peace River and tributaries in 1891. Pages 1–44 *in* Annual report, Department of the Interior of Canada for 1892, Part 7.

Olsen, S. C., M. V. Palmer, J. C. Rhyan, and T. Gidlewski. 1998. Biosafety of *Brucella abortus* strain RB51 in adult bison bulls and efficacy as a calfhood vaccine in bison. Proceedings of the Annual Meeting of the United States Animal Health Association 102:142–44.

Olson, W. E. 2002. Plains and wood bison weight and population dynamics in Elk Island National Park for 2001–2002. Annual report. Elk Island National Park, Fort Saskatchewan, Alberta, Canada.

Oosenbrug, S. M., and L. N. Carbyn. 1982. Winter predation on bison and activity patterns of a wolf pack in Wood Buffalo National Park. Pages 43–53 *in* F. H. Harrington and P. C. Paquet, eds. Wolves of the world. Perspectives of behavior, ecology, and conservation. Noyes, Park Ridge, NJ.

Orlov, V. N., and G. A. Chudinovokaya. 1979. Chromosome set of European bison and other bovini. Pages 435–41 *in* V. E. Sokolov, ed. European bison. Morphology, systematics, evolution, ecology. Nauka, Moscow.

Ozoga, J. J., and L. J. Verme. 1970. Winter feeding patterns of penned white-tailed deer. Journal of Wildlife Management 34:431–39.

Palmer, T. S. 1916. Our national herds of buffalo. Annual Report of the American Bison Society 10:40–62 [Cited in Fuller. 1966].

Patten, D. T. 1963. Vegetational pattern in relation to environments in the Madison Range, Montana. Ecological Monographs 33:375–406.

Pattie, D., and C. Fisher. 1999. Mammals of Alberta. Lone Pine, Edmonton, Alberta, Canada.

Pearson, H. A. 1967. Rumen microorganisms in buffalo from southern Utah. Applied Microbiology 15:1450–51.

Peden, D. G. 1972. The trophic relations of *Bison bison* to the shortgrass plains. Ph.D. Dissertation, Colorado State University, Fort Collins.

Peden, D. G. 1976. Botanical composition of bison diets on shortgrass plains. American Midland Naturalist 96:225–29.

Peden, D. G., and G. J. Kraay. 1979. Comparison of blood characteristics in plains bison, wood bison, and their hybrids. Canadian Journal of Zoology 57:1778–84.

Peden, D. G., G. M. Van Dyne, R. W. Rice, and R. M. Hansen. 1974. The trophic ecology of *Bison bison* L. on shortgrass plains. Journal of Applied Ecology 11:489–98.

Pellerdy, L. P. 1963. Catalogue of Eimeriidea (Protozoa; Sporozoa). Akademia Kiado, Budapest.

Pellerdy, L. P. 1974. Coccidia and coccidiosis, 2nd ed. Paul Parey, Berlin.

Penzhorn, B. L., S. E. Knapp, and C. A. Speer. 1994. Enteric coccidia in free-ranging American bison (*Bison bison*) in Montana. Journal of Wildlife Diseases 30:267–69.

Peters, H. F. 1958. A feedlot study of bison, cattalo, and Hereford calves. Canadian Journal of Animal Science 38:87–90.

Peters, H. F. 1978. Utilization of bison as a genetic resource for meat production. *In* 13th international symposium on zootechny. Accademia Nazionale di Agricoltura, Milan, Italy.

Peters, H. F., and S. B. Slen. 1964. Hair coat characteristics of bison, domestic × bison hybrids, cattalo, and certain domestic breeds of beef cattle. Canadian Journal of Animal Science 44:48–57.

Peters, H. F., and S. B. Slen. 1966. Range calf production of cattle, bison, cattalo, and Hereford cows. Canadian Journal of Animal Science 46:157–64.

Petropavlovskii, V. V., and A. I. Rykova. 1958. The stimulation of sexual functions in cows. Tr. Vljjarov. Sel. Hoz. Institute 5:193–99 (In Russian) [English abstract: Animal Breeders Abstracts 29:168 (1961)].

Pfeiffer, K. E., and D. C. Hartnett. 1995. Bison selectivity and grazing response of little bluestem in tallgrass prairie. Journal of Range Management 48:26–31.

Pollard, D. F., and M. R. McKechnie. 1986. World conservation strategy, Canada. A report on achievements in conservation. Conservation and Protection, Environment Canada, Ottawa. Ontario, Canada.

Polley, H. W., and S. L. Collins. 1984. Relationship of vegetation and environment in buffalo wallows. American Midland Naturalist 112:178–86.

Polziehn, R. O., R. Beech, J. Sheraton, and C. Strobeck. 1996. Genetic relationships among North American bison populations. Canadian Journal of Zoology 74:738–49.

Pond, D. B., and C. A. Speer. 1979. Sarcocystis in free-ranging herbivores on the National Bison Range. Journal of Wildlife Diseases 15:51–53.

Popov, B. H., and J. B. Low. 1950. Game, fur, and fish introductions into Utah (Miscellaneous Publication 4). Utah State Department of Fish and Game.

Popper, F. J., and D. E. Popper. 1991. The restoration of the Great Plains frontier. Earth Island Journal 6:34–36.

Post, D. M., T. S. Armbrust, E. A. Horne, and J. R. Goheen. 2001. Sexual segregation results in differences in content and quality of bison (*Bos bison*) diets. Journal of Mammalogy 82:407–13.

Prusiner, S. B. 1995. The prion diseases. Scientific American 272:48–57.

Radostits, O. M., C. C. Gay, D. C. Blood, and K. W. Hinchcliff. 2000. Veterinary medicine: A textbook of the diseases of cattle, sheep, pigs, goats and horses, 9th ed. W. B. Saunders, Philadelphia.

Radtke, K. M., H. R. Taylor, D. M. Stockrahm, and K. J. Scoville. 1999. Bison habitat use and herd composition in Theodore Roosevelt National Park in the North Dakota Badlands. Proceedings of the North Dakota Academy of Science 53:191.

Ransom, B. H. 1911. The nematodes parasitic in the alimentary tract of cattle, sheep, and other ruminants (Bulletin 127). U.S. Department of Agriculture, Bureau of Animal Industry.

Raup, H. M. 1933. Range conditions in the Wood Buffalo Park of western Canada with notes on the history of the wood bison (Special Publication 1[2]). American Committee for International Wildlife Protection.

Reynolds, H. W. 1976. Bison diets of Slave River Lowlands, Canada. M.S. Thesis, Colorado State University, Fort Collins.

Reynolds, H. W., and A. W. L. Hawley, eds. 1987. Bison ecology in relation to agricultural development in the Slave River Lowlands, N.W.T (Occasional Paper No. 63; Catalogue No. CW69-1/93E). Canadian Wildlife Service, Ottawa, Ontario, Canada.

Reynolds, H. W. and D. G. Peden. 1987. Vegetation, bison diets, and snow cover. Pages 39–44 *in* H. W. Reynolds and A. W. L. Hawley, eds. Bison ecology in relation to agricultural development in the Slave River Lowlands, N.W.T (Occasional Paper No. 63; Catalogue No. CW69-1/93E). Canadian Wildlife Service, Ottawa, Ontario, Canada.

Reynolds, H. W., R. M. Hansen, and D. G. Peden. 1978. Diets of the Slave River Lowland bison herd, Northwest Territories, Canada. Journal of Wildlife Management 42:581–90.

Reynolds, H. W., R. D. Glaholt, and A. W. L. Hawley. 1982. Bison: *B. bison bison*. Pages 972–1007 *in* J. A. Chapman and G. A. Feldhamer, eds. Wild mammals of North America: Biology, management, and economics. Johns Hopkins University Press, Baltimore.

Rhoads, S. N. 1897. Notes on living and extinct species of North American Bovidae. Proceedings of the Academy of Natural Sciences 49:483–502.

Rhyan, J. C., J. W. Quinn, S. L. Stackhouse, J. J. Henderson, D. R. Ewalt, J. B. Payeur, M. Johnson, and M. Meagher. 1994. Abortion caused by *Brucella abortus* biovar 1 in a free-ranging bison (*Bison bison*) from Yellowstone National Park. Journal of Wildlife Diseases 30:445–46.

Rhyan, J. C., T. Gidlewski, T. J. Roffe, K. Aune, L. M. Philo, and D. R. Ewalt. 2001. Pathology of brucellosis in bison from Yellowstone National Park. Journal of Wildlife Diseases 37:101–9.

Rice, R. W., R. E. Dean, and J. E. Ellis. 1974. Bison, cattle, and sheep dietary quality and food intake. Journal of Animal Science 38:1332.

Richmond, R. J., R. J. Hudson, and R. J. Christopherson. 1977. Comparison of forage intake and digestibility by American bison, yak, and cattle. Acta Theriologica 22:225–30.

Roe, F. G. 1970. The North American buffalo: A critical study of the species in its wild state, 2nd ed. University of Toronto Press, Toronto, Ontario, Canada.

Roffe, T., J. Olsen, C. Steven, T. Gidlewski, A. E. Jensen, M. V. Palmer, and R. Huber. 1999. Biosafety of parenteral *Brucella abortus* RB51 vaccine in bison calves. Journal of Wildlife Management 63:950–55.

Rothstein, A., and J. G. Griswold. 1991. Age and sex preferences for social partners by juvenile bison bulls, *Bison bison*. Animal Behaviour 41:227–37.

Roudabush, R. L. 1936. Arthropod and helminth parasites of the American bison (*Bison bison*). Journal of Parasitology 22:517–18.

Rush, W. M. 1932. Bang's disease in the Yellowstone National Park buffalo and elk herds. Journal of Mammalogy 13:371–72.

Rush, W. M. 1942. Wild animals of the Rockies. Harper, New York.

Rutberg, A. T. 1983. Factors influencing dominance status in American bison (*Bison bison*). Zeitschrift für Tierpsychologie 63:206–12.

Rutberg, A. T. 1984. Birth synchrony in American bison (*Bison bison*): Response to predation or season? Journal of Mammalogy 65:418–23.

Rutberg, A. T. 1986. Lactation and fetal sex ratios in American bison. American Naturalist 127:89–94.

Ruth, G. R., D. E. Reid, C. A. Daley, M. W. Vorhies, K. Wohlgemuth, and H. Shave. 1977. Malignant catarrhal fever in bison. Journal of the American Veterinary Medical Association 171:913–17.

Rutley, B. D., and R. J. Hudson. 2000. Seasonal energetic parameters of free-grazing bison (*Bison bison*). Canadian Journal of Animal Science 80:663–71.

Ryff, K. L., and R. C. Bergstrom. 1975. Bovine coccidia in American Bison. Journal of Wildlife Diseases 11:412–14.

Salter, R. E. 1978. Ecology of feral horses in western Alberta. M.S. Thesis, University of Alberta, Edmonton, Alberta, Canada.

Saltz, D., and G. C. White. 1991. Urinary cortisol and urinary nitrogen responses in irreversibly undernourished mule deer fawns. Journal of Wildlife Diseases 27:41–46.

Samuel, W. M., and D. A. Welch. 1991. Winter ticks on moose and other ungulates: Factors influencing their population size. Alces 27:169–82.

Saunders, W. E. 1925. Letter to the editor of the Canadian Field-Naturalist. Canadian Field-Naturalist 39:118.

Schaefer, A. L., B. A. Young, and A. M. Chimwano. 1978. Ration digestion and retention times of digesta in domestic cattle (*Bos taurus*), American bison (*Bison bison*), and Tibetan yak (*Bos grunniens*). Canadian Journal of Zoology 56:2355–58.

Schalm, O. W., N. C. Jain, and E. J. Carroll. 1975. Veterinary hematology, 3rd ed. Lea and Febiger, Philadelphia.

Schein, M. W., and M. H. Fohrman. 1955. Social dominance relationships in a herd of dairy cattle. British Journal of Animal Behaviour 3:45–55.

Schillhorn van Veen, T. W., T. P. Mullaney, A. L. Trapp, and R. F. Taylor. 1991. Fatal reactions in bison following systemic organophosphate treatment for the control of *Hypoderma bovis*. Journal of Veterinary Diagnostic Investigation 3:355–56.

Schneider, B. H., and W. P. Flatt. 1975. The evaluation of feeds through digestibility experiments. University of Georgia Press, Athens.

Schultheiss, P. C., J. K. Collins, L. E. Austgen, and J. C. DeMartini. 1998. Malignant catarrhal fever in bison, acute and chronic cases. Journal of Veterinary Diagnostic Investigation 10:255–62.

Schultheiss, P. C., J. K. Collins, M. W. Miller, G. Brundige, T. K. Bragg, J. G. Patterson, and M. T. Jessen. 2001. Malignant catarrhal fever in bison. Veterinary Pathology 38:577.

Schultz, C. B., and J. M. Hillerud. 1977. The antiquity of *Bison latifrons* (Harlan) in the Great Plains of North America. Transactions of the Nebraska Academy of Science 4:103–16.

Schwartz, C. C., J. G. Nagy, and R. W. Rice. 1977. Pronghorn dietary quality relative to forage availability and other ruminants in Colorado. Journal of Wildlife Management 41:161–68.

Scott, M. D. 1992. Buck-and-pole fence crossings by 4 ungulate species. Wildlife Society Bulletin 20:204–10.

Searcy, R. L. 1969. Diagnostic biochemistry. McGraw-Hill, New York.

Seibert, F. V. 1925. Some notes on Canada's so-called wood buffalo. Canadian Field-Naturalist 39:204–6.

Seip, D. R. 1992. Factors limiting woodland caribou populations and their interrelationships with wolves and moose in southeastern British Columbia. Canadian Journal of Zoology 70:1494–1503.

Seton, E. T. 1886. The wood buffalo. Proceedings of the Royal Canadian Institute, Series 3(3):114–17.

Seton, E. T. 1929. The buffalo. Pages 639–703 *in* Lives of game animals, Vol. 3, Part 2. Doubleday, Doran, Garden City, NY.

Shackleton, D. M. 1968. Comparative aspects of social organization of American bison. M.S. Thesis, University of Western Ontario, London, Ontario, Canada.

Shackleton, D. M., L. V. Hills, and D. A. Hutton. 1975. Aspects of variation in cranial characters of plains bison (*Bison bison bison* Linnaeus) from Elk Island National Park, Alberta. Journal of Mammalogy 56:871–87.

Shaw, J. H. 1998. Bison ecology: What we do and do not know. Pages 113–20 *in* L. Irby and J. Knight, eds. International symposium on bison ecology and management in North America. Montana State University, Bozeman.

Shaw, J. H., and T. S. Carter. 1989. Calving patterns among American bison. Journal of Wildlife Management 53:896–98.

Shaw, J. H., and T. S. Carter. 1990. Bison movements in relationship to fire and seasonality. Wildlife Society Bulletin 18:426–30.

Shinn, R. 1978. Buffalo in the news. Buffalo 6:26.

Shult, M. J. 1972. American bison behavior patterns at Wind Cave National Park. Ph.D. Dissertation, Iowa State University, Ames.

Silver, H., N. F. Colovos, J. B. Holter, and H. H. Hayes. 1969. Fasting metabolism of white-tailed deer. Journal of Wildlife Management 33:490–98.

Simpson, G. G. 1961. Principles of animal taxonomy. Columbia University Press, New York.

Singer, F. J., and J. E. Norland. 1994. Niche relationships within a guild of ungulate species in Yellowstone National Park, Wyoming, following release from artificial controls. Canadian Journal of Zoology 72:1383–94.

Skinner, M. F., and O. C. Kaisen. 1947. The fossil bison of Alaska and preliminary revision of the genus. Bulletin of the American Museum of Natural History 89:123–256.

Slee, J. 1972. Habituation and acclimatization of sheep to cold following exposures of varying length and severity. Journal of Physiology 227:51–70.

Smith, D. W., M. K. Phillips, and B. Crabtree. 1996. Interactions of wolves and other wildlife in Yellowstone National Park. Pages 140–45 *in* Proceedings of the Defenders of Wildlife's Wolves of America conference. Albany, NY.

Smith, D. W., L. D. Mech, M. Meagher, W. E. Clark, R. Jaffe, M. K. Phillips, and J. A. Mack. 2000. Wolf–bison interactions in Yellowstone National Park. Journal of Mammalogy 81:1128–35.

Smoliak, S., and H. F. Peters. 1955. Climatic effects on foraging performance of beef cows on winter range. Canadian Journal of Agricultural Science 35:213–16.

Soper, J. D. 1941. History, range, and home life of the northern bison. Ecological Monographs 11:349–412.

Soper, J. D. 1964. The mammals of Alberta. Hamly Press, Edmonton, Alberta, Canada.

Speth, J. D. 1983. Bison kills and bone counts. University of Chicago Press, Chicago.

Spraker, T. R., M. W. Miller, E. S. Williams, D. M. Getzy, W. J. Adrian, G. G. Schoonveld, R. A. Spowart, K. I. O'Rourke, J. M. Miller, and P. A. Merz. 1997. Spongiform encephalopathy in free ranging mule deer (*Odocoileus hemionus*), white-tailed deer (*Odocoileus virginianus*), and rocky mountain elk (*Cervus elaphus nelsoni*) in northcentral Colorado. Journal of Wildlife Diseases 33:1–6.

Stephenson, R. O., S. C. Gerlach, R. D. Guthrie, C. R. Harington, R. O. Mills, and G. Hare. 2001. Wood bison in late Holocene Alaska and adjacent Canada: Paleontological, archaeological and historical records. Pages 125–59 *in* S. C. Gerlach and M. S. Murray, eds. People and wildlife in northern North America: Essays in honor of R. Dale Guthrie (British Archaeological Reports, International Series 994). Hadrian, Oxford, UK.

Steuter, A. A., E. M. Steinauer, G. L. Hill, P. A. Bowers, and L. L. Tieszen. 1995. Distribution and diet of bison and pocket gophers in a sandhills prairie. Ecological Applications 5:756–66.

Stormont, C. J. 1993. An update on bison genetics. Pages 15–37 *in* R. E. Walker, ed. and comp. Proceedings of the North American public bison herds symposium. Custer State Park, Custer, SD.

Stormont, C., W. J. Miller, and Y. Suzuki. 1960. Blood groups and taxonomic status of American buffalo and domestic cattle. Evolution 15:196–208.

Strobeck, C., R. O. Polziehn, and R. Beech. 1993. Genetic relationship between wood and plains bison assayed using mitochondrial DNA sequence. Pages 209–27 *in* R. E. Walker, ed. and comp. Proceedings of the North American public bison herds symposium. Custer State Park, Custer, SD.

Stumpff, C. D., M. A. Essey, D. H. Person, and D. Thorpe. 1985. Epidemiologic study of *M. bovis* in American bison. Proceedings of the Annual Meeting of the U.S. Animal Health Association 89:564–70.

Swales, W. E. 1936. Further studies on *Fascioloides magna* (Bassi, 1875) Ward, 1917, as a parasite of ruminants. Canadian Journal of Research 14(D8):83–95.

Swenson, M. J. 1970. Physiologic properties, cellular and chemical constituents of blood. Pages 21–61 *in* M. J. Swenson, ed. Dukes' physiology of domestic animals, 8th ed. Cornell University Press, Ithaca, NY.

Taylor, S. K., A. C. S. Ward, D. L. Hunter, K. Gunther, and L. Kortge. 1996. Isolation of *Pasteurella* spp. from free-ranging American bison (*Bison bison*). Journal of Wildlife Diseases 32:322–25.

Taylor, S. K., V. M. Lane, D. L. Hunter, K. G. Eyre, S. Kaufman, S. Fyre, and M. R. Johnson. 1997. Serologic survey for infectious pathogens in free-ranging American bison. Journal of Wildlife Diseases 33:308–11.

Telfer, E. S., and A. L. Cairns. 1986. Resource use by moose versus sympatric deer, wapiti and bison. Alces 22:113–38.

Telfer, E. S., and G. W. Scotter. 1975. Potential for game ranching in boreal aspen forests of western Canada. Journal of Range Management 28:172–80.

Tessaro, S. V. 1986. The existing and potential importance of brucellosis and tuberculosis in Canadian wildlife: A review. Canadian Veterinary Journal 27:119–24.

Tessaro, S. V. 1987. A descriptive and epizootiologic study of brucellosis and tuberculosis in bison in northern Canada. Ph.D. Dissertation, University of Saskatchewan, Saskatoon, Saskatchewan, Canada.

Tessaro, S. V. 1989. Review of the diseases, parasites, and miscellaneous pathological conditions of North American bison. Canadian Veterinary Journal 30:416–22.

Tessaro, S. V., and D. Deregt. 1999. Pathogenesis of bovine viral diarrhea (BVD) viruses in elk and characterization of BVD viruses isolated from elk and bison in Elk Island National Park. Unpublished final report. Animal Diseases Research Institute, Canadian Food Inspection Agency. Lethbridge, Alberta, Canada.

Tessaro, S. V., L. B. Forbes, and C. Turcotte. 1990. A survey of brucellosis and tuberculosis in bison in and around Wood Buffalo National Park, Canada. Canadian Veterinary Journal 31:174–80.

Tessaro, S. V., C. C. Gates, and L. B. Forbes. 1993. The brucellosis and tuberculosis status of wood bison in the Mackenzie Bison Sanctuary, Northwest Territories, Canada. Canadian Journal of Veterinary Research 57: 231–35.

Thoen, C. O., and B. R. Bloom. 1995. Pathogenesis of *Mycobacterium bovis*. Pages 3–14 *in* C. O. Thoen and J. H. Steele, eds. *Mycobacterium bovis* infection in animals and humans. Iowa State University Press, Ames.

Thoen, C. O., K. J. Throlson, L. D. Miller, E. M. Himes, and R. L. Morgan. 1988. Pathogenesis of *Mycobacterium bovis* infection in American bison. American Journal of Veterinary Research 49:1861–65.

Thorne, E. T., M. Meagher, and R. Hillman. 1991. Brucellosis in free-ranging bison: Three perspectives. Pages 275–87 *in* R. B. Keiter and M. S. Boyce, eds. The Greater Yellowstone Ecosystem: Redefining America's wilderness heritage. Yale University Press, New Haven, CT.

Towne, E. 1988. Ruminal fermentation characteristics and microbial comparisons between bison and cattle. Ph.D. Dissertation, Kansas State University, Manhattan.

Towne, E. G. 1999. Bison performance and productivity on tallgrass prairie. Southwestern Naturalist 44:361–66.

Tracy, B. F., and D. A. Frank. 1998. Herbivore influence on soil microbial biomass and nitrogen mineralization in a northern grassland ecosystem: Yellowstone National Park. Oecologia 114:556–62.

Tuckerman, E. D. 1955. Report of bovine tuberculosis infection in the buffalo herd of the Lehigh County Game Preserve. University of Pennsylvania, Veterinary Extension Quarterly 130:108–10.

Van Camp, J. 1987. Predation on bison. Pages 25–33 *in* H. Reynolds and A. Hawley, eds. Bison ecology in relation to agricultural development in the Slave River Lowlands, N.W.T. (Occasional Paper No. 63; Catalogue No. CW69-1/63E). Minister of Supply and Services Canada, Ottawa, Ontario, Canada.

Van Camp, J., and G. W. Calef. 1987. Population dynamics of bison. Pages 21–23 *in* H. Reynolds and A. Hawley, eds. Bison ecology in relation to agricultural development in the Slave River Lowlands, N.W.T. (Occasional Paper No. 63; Catalogue No. CW69-1/63E). Minister of Supply and Services Canada, Ottawa, Ontario, Canada.

Van Campen, H., K. Frölich, and M. Hofmann. 2001. Pestivirus infections. Pages 230–42 *in* E. S. Williams, I. K. Barker, and E. T. Thorne, eds. Infectious diseases of wild mammals, 3rd ed. Iowa State University Press, Ames.

Van Gelden, R. G. 1977. Mammalian hybrids and generic limits. American Museum Novitates 2635:1–25.

Van Vuren, D. 1982. Comparative ecology of bison and cattle in the Henry Mountains, Utah. Pages 449–57 *in* J. M. Peek and P. D. Dalke, eds. Proceedings of the wildlife–livestock relationships symposium. Forest, Wildlife and Range Experiment Station, University of Idaho, Moscow.

Van Vuren, D. 1983. Group dynamics and summer range of bison in southern Utah. Journal of Mammalogy 64:329–32.

Van Vuren, D. 1984a. Abnormal dentition in the American bison, *Bison bison*. Canadian Field-Naturalist 98:366–67.

Van Vuren, D. 1984b. Summer diets of bison and cattle in southern Utah. Journal of Range Management 37:260–61.

Van Vuren, D., and M. P. Bray. 1986. Population dynamics of bison in the Henry Mountains, Utah. Journal of Mammalogy 67:503–11.

Van Vuren, D., and C. A. Scott. 1995. Internal parasites of sympatric bison, *Bison bison,* and cattle, *Bos taurus*. Canadian Field-Naturalist 109:467–69.

van Zyll de Jong, C. G. 1986. A systematic study of recent bison, with particular consideration of the wood bison (Publications in Natural Sciences No. 6). National Museum of Natural Sciences, Ottawa, Ontario, Canada.

van Zyll de Jong, C. G. 1993. Origin and geographic variation of recent North American bison. Pages 21–35 *in* Alberta: Studies in the Arts and Sciences, Vol. 3, No. 2. University of Alberta Press, Edmonton, Alberta, Canada.

van Zyll de Jong, C. G., C. Gates, H. Reynolds, and W. Olson. 1995. Phenotypic variation in remnant populations of North American bison. Journal of Mammalogy 76:391–405.

Vestweber, J. G., R. K. Ridley, J. C. Nietfeld, and M. J. Wilkerson. 1999. Demodicosis in an American bison. Canadian Veterinary Journal 40:417–18.

Vinton, M. A., and D. C. Hartnett. 1992. Effects of bison grazing on *Andropogon gerardii* and *Panicum virgatum* in burned and unburned tallgrass prairie. Oecologia 90:374–82.

Vinton, M. A., D. C. Hartnett, E. J. Finck, and J. M. Briggs. 1993. Interactive effects of fire, bison (*Bison bison*) grazing and plant community composition in tallgrass prairie. American Midland Naturalist 129:10–18.

Wade, S. E., W. M. Haschek, and J. R. Georgi. 1979. Ostertagiosis in captive bison in New York State: Report of nine cases. Cornell Veterinarian 69:198–205.

Waggoner, V., and M. Hinkes. 1986. Summer and fall browse utilization by an Alaskan bison herd. Journal of Wildlife Management 50:322–24.

Walker, E. P., F. Warnick, S. E. Hamlet, K. I. Lange, M. A. Davis, H. E. Uible, and P. F. Wright. 1975. Genus: *Bison*. Page 1433 *in* E. P. Walker, ed. Mammals of the world, Vol. 2, 3rd ed. Johns Hopkins University Press, Baltimore.

Wall, A. D., S. K. Davis, and B. M. Reed. 1992. Phylogenetic relationships in the subfamily bovinae (Mammalia—Artiodactyla) based on ribosomal DNA. Journal of Mammalogy 73:262–75.

Ward, P. 1997. The call of distant mammoths: Why the ice age mammals disappeared. Springer-Verlag, New York.

Ward, A. C. S., N. W. Dyer, and B.W. Fenwick. 1999a. Pasteurellaceae isolated from tonsillar samples of commercially-reared American bison (*Bison bison*). Canadian Journal of Veterinary Research 63:161–65.

Ward, A. C. S., N. W. Dyer, and L. B. Corbeil. 1999b. Characterization of putative *Haemophilus somnus* isolates from tonsils of American bison (*Bison bison*). Canadian Journal of Veterinary Research 63:166–69.

Ward, T. J., J. P. Bielawski, S. K. Davis, J. W. Templeton, and J. N. Derr. 1999c. Identification of domestic cattle hybrids in wild cattle and bison species: A general approach using mtDNA markers and the parametric bootstrap. Animal Conservation 2:51–57.

Wasser, C. H. 1977. Bison induced stresses in Colorado National Monument. Pages 28–36 *in* Final report. National Park Service Contract PX 120060617.

Weaver, J. W., and F. E. Clements. 1938. Plant ecology. McGraw-Hill, NY.

Weaver, J. L., and G. T. Haas. 1998. Bison in the diet of wolves denning amidst high diversity of ungulates. Pages 141–44 *in* L. Irby and J. Knight, eds. International symposium on bison ecology and management in North America. Montana State University, Bozeman.

Westra, R., and R. J. Hudson. 1979. Urea recycling in wapiti. Pages 236–39 *in* M. S. Boyce and L. D. HaydenWing, eds. North American elk: Ecology, behavior, and management. University of Wyoming, Laramie.

Williams, E. S. 2001. Paratuberculosis. Pages 361–71 *in* E. S. Williams, I. K. Barker, and E. T. Thorne, eds. Infectious diseases of wild mammals, 3rd ed. Iowa State University Press, Ames.

Williams, E. S., and S. Young. 1982. Spongiform encephalopathy of Rocky Mountain elk. Journal of Wildlife Diseases 18:465–71.

Williams, E. S., E. T. Thorne, S. L. Anderson, and J. D. Herriges. 1993. Brucellosis in free-ranging bison (*Bison bison*) from Teton County, Wyoming. Journal of Wildlife Diseases 29:118–222.

Wilson, M. 1980. Morphological dating of late Quaternary bison on the northern plains. Canadian Journal of Anthropology 1:81–85.

Wilson, G. A. 2001. Population genetic studies of wood and plains bison populations. Ph.D. Dissertation, University of Alberta, Edmonton, Canada.

Wilson, A. C., R. L. Cann, S. M. Carr, M. George, U. B. Gyllensten, K. M. Helm-Gychowski, R. G. Higuchi, S. R. Palumbi, E. M. Prager, R. D. Sage, and M. Stoneking. 1985. Mitochondrial DNA and two perspectives on evolutionary genetics. Biological Journal of the Linnean Society 26:375–400.

Wilson, D. E., and D. M. Reeder. 1993. eds. Mammal species of the world. Smithsonian Institution Press, Washington, DC.

Wilson, G., and C. Strobeck. 1999. Genetic variation within and relatedness among wood and plains bison populations. Genome 42:483–96.

Wilson, G., W. Olson, and C. Strobeck. 2002. Reproductive success in wood bison (*Bison bison athabascae*) established using molecular techniques. Canadian Journal of Zoology 80:1537–48.

Wolfe, M. L., M. P. Shipka, and J. F. Kimball. 1999. Reproductive ecology of bison on Antelope Island, Utah. Great Basin Naturalist 59:105–11.

Wolff, J. O. 1998. Breeding strategies, mate choice, and reproductive success in American bison. Oikos 83:529–44.

Wondolleck, J., and S. Yaffee. 2000. Making collaboration work: Lessons from innovation in natural resource management. Island Press, Washington, DC.

Wood, A. J., I. McT. Cowan, and H. C. Nordan. 1962. Periodicity of growth in ungulates as shown by deer of the genus *Odocoileus*. Canadian Journal of Zoology 40:593–603.

Wood Bison Recovery Team. 1987. Status report on endangered wildlife in Canada 1987: Wood Bison. Canadian Wildlife Service, Ottawa, Ontario, Canada.

Wuerthner, G. 1998. Are cows just domestic bison? Behavioral and habitat use differences between cattle and bison. Pages 374–83 *in* L. Irby and J. Knight, eds. International symposium on bison ecology and management in North America. Montana State University, Bozeman.

Wydeven, A. P., and R. B. Dahlgren. 1985. Ungulate habitat relationships in Wind Cave National Park. Journal of Wildlife Management 49:805–13.

Ying, K. L., and D. G. Peden. 1977. Chromosomal homology of wood bison and plains bison. Canadian Journal of Zoology 55:1759–62.

Young, B. A., A. Schaefer, and A. Chimwano. 1977. Digestive capacities of cattle, bison, yak. Pages 31–34 *in* 56th annual Feeders' Day report. University of Alberta, Edmonton, Alberta, Canada.

Zamora, L. E. 1983. An analysis of bison erythrocyte antigens and blood proteins. M.S. Thesis, Texas A&M University, College Station.

Zarnke, R. L., and G. A. Erickson. 1990. Serum antibody prevalence of parainfluenza 3 virus in a free-ranging bison (*Bison bison*) herd from Alaska. Journal of Wildlife Diseases 26:416–19.

Zaugg, J. L. 1986. Experimental anaplasmosis in American bison: Persistence of infections of *Anaplasma marginale* and non-susceptibility to *A. ovis*. Journal of Wildlife Diseases 22:169–72.

Zaugg, J. L., and K. L. Kuttler. 1985. *Anaplasma marginale* infections in American bison: Experimental infection and serologic study. American Journal of Veterinary Research 46:438–41.

Zaugg, J. L., and K. L. Kuttler. 1987. Experimental infections of *Babesia bigemina* in American bison. Journal of Wildlife Diseases 23:99–102.

Zaugg, J. L., S. K. Taylor, B. C. Anderson, D. L. Hunter, J. Ryder, and M. Divine. 1993. Hematologic, serologic values, histopathologic and fecal evaluations of bison from Yellowstone Park. Journal of Wildlife Diseases 29:453–57.

Zimov, S. A., V. I. Chuprynin, A. P. Oreshko, F. S. Chapin III, J. F. Reynolds, and M. C. Chapin. 1995. Steppe–tundra transition: A herbivore-driven biome shift at the end of the Pleistocene. American Naturalist 146:765–94.

Hal W. Reynolds, Canadian Wildlife Service, Environment Canada, Prairie and Northern Region, Edmonton, Alberta, Canada T6B 2X3. Email: hal.reynolds@ec.gc.ca.

C. Cormack Gates, University of Calgary, Calgary, Alberta, Canada T2N 1N4. Email: ccgates@nucleus.com.

Randal D. Glaholt, Biotechnics International, Ltd., 6031 Bowwater Crescent N.W., Calgary, Alberta, Canada T3B 2E5. Email: rglaholt@teraenv.com.

49

Mountain Goat

Oreamnos americanus

Steeve D. Côté
Marco Festa-Bianchet

NOMENCLATURE

COMMON NAMES. Mountain goat, Rocky Mountain goat, snow goat, white goat
SCIENTIFIC NAME. *Oreamnos americanus*

Mountain goats belong to the family Bovidae, subfamily Caprinae, and tribe Rupicaprini (*Rupes* = rock, *capra* = goat). Despite their name, mountain goats are not a true goat and do not belong to the genus *Capra*. There is considerable disagreement about the phylogenetic position of *Oreamnos* (Hassanin et al. 1998). Some studies based on molecular and morphological characteristics place the mountain goat close to wild sheep (*Ovis* spp; Gatesy et al. 1997). Most taxonomists, however, classify the mountain goat in a clade grouping *Capricornis, Nemorhaedus,* and *Ovibos* (Cronin et al. 1996; Groves and Shields 1996; Hassanin et al. 1998). Other rupicaprins include the goral (*Nemorhaedus goral*), serow (*Capricornis sumatraensis*), and Japanese serow (*Capricornis crispus*) from Asia and two species of chamois (*Rupicapra rupicapra* and *R. pyrenaica*) from Europe, Turkey, and the Caucasus. *O. americanus* is the only rupicaprin in North America.

Mountain goats were classified into four subspecies (*americanus, kennedyi, missoulae,* and *columbiae*), but currently no subspecies of *O. americanus* are recognized (Cowan and McCrory 1970). A smaller species, *Oreamnos harringtoni,* went extinct about 11,000 years ago, at the end of the Rancholabrean (Mead and Lawler 1994).

DISTRIBUTION

The ancestors of mountain goats are thought to have come from Asia, crossing the Bering Land Bridge between Alaska and Siberia during the Pleistocene (Cowan and McCrory 1970; Rideout and Hoffmann 1975). They colonized the mountains during the glaciations and remained there when the glaciers withdrew. Ancestors of mountain goats apparently evolved specialized adaptations to mountainous environments at least 100,000 years ago. Fossil distribution is poorly documented, but was likely wider than historical distribution; for example, fossils have been recovered from Vancouver Island (Nagorsen and Keddie 2000).

Mountain goats use alpine and subalpine areas throughout northwestern North America. They occur primarily on the Rocky Mountains and foothills, as well as along the main coastal mountain ranges in British Columbia and southern Alaska (Fig. 49.1). Most native mountain goat populations are in British Columbia and Alaska (Table 49.1). Although their historical range has been reduced, native populations also exist in Washington, Montana, Idaho, Alberta, Yukon, and the Northwest Territories (Fig. 49.1 and Table 49.1). Goat populations have been introduced in some states outside their known historical range, including Colorado, Oregon, Nevada, South Dakota, Utah, and Wyoming, mostly during 1940–1970 (Johnson 1977). They have also been reintroduced in parts of Alaska, Idaho, Montana, Washington, and Alberta. Including native and nonnative herds, there are an estimated 75,000–110,000 mountain goats in North America (Table 49.1).

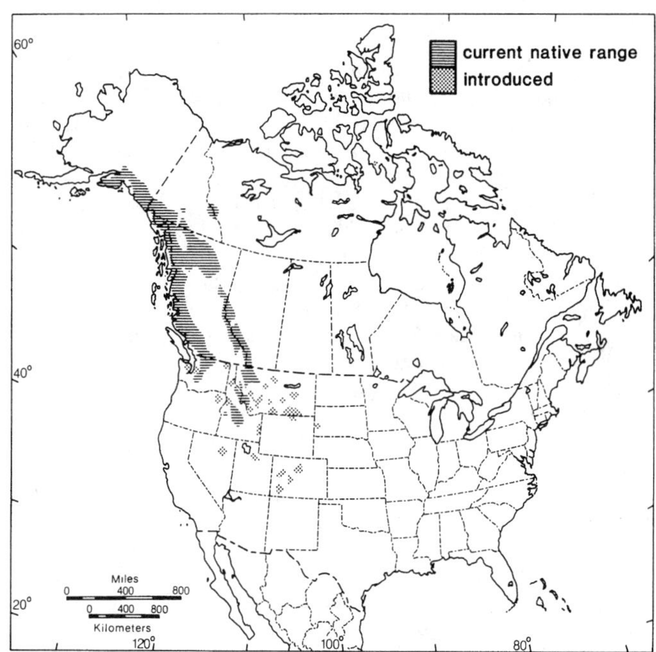

FIGURE 49.1. Distribution of mountain goats.

DESCRIPTION

Male and female mountain goats are alike in appearance and characterized by a stocky body with short legs, a coat of long and coarse white hairs, conspicuous sharp black horns, and a short tail (Fig. 49.2). Their forequarters are large, with massive muscles to help pull the animal up rocky ledges. Their paired hooves are separated by a large interdigital cleft and their dew claws are well developed. The hooves have a soft pad protruding beyond the outer cornified shell, which gives good traction in precipitous terrain (Brandborg 1955). The horns are conical in shape and slightly curved posteriorly. Black supraoccipital glands, swollen during the rut and more developed in males than in females, are located just behind the horns. The glands are thought to be used for scent marking during the rutting season, although no precise information exists on their role (Geist 1964).

Body Size and Growth. Although little information is available, mountain goat kids probably weigh 2.5–3.5 kg at birth (Brandborg 1955; Lentfer 1955; S. D. Côté, unpublished data.). Carl and Robbins (1988) reported an average mass of 4.1 kg for seven captive-born kids at 1 week of age. In the Caw Ridge mountain goat population in west-central Alberta, male and female kids weigh 13.9±2.9 and 13.2±2.7 kg, respectively, at 1.5 months of age (Table 49.2). Growth rate during the first summer is similar in both sexes, averaging 195 g/day (Fig. 49.3). On Caw Ridge, kid mass at weaning (approximately 4 months of age) was not affected by birthdate or maternal characteristics such as age and

TABLE 49.1. Estimates of mountain goat abundance for jurisdictions with >100 individuals

Jurisdiction	Year 2000 Estimate	Source[a]
British Columbia	36,000–63,000	I. Hatter, pers. commun., 2000
Alaska	24,193–29,820	Alaska Department of Fish and Game 2000
Washington	4,000	Washington Department of Fish and Wildlife 2000
Montana	2,295–3,045	J. McCarthy, pers. commun., 2000
Idaho	2,700	D. Toweill, pers. commun., surveys conducted in 2000
Alberta Provincial Lands	1,650	Alberta Fisheries and Wildlife Management Division 2000
Alberta National Parks	800	G. Mercer and W. Glasgow, pers. commun., 2000
Colorado	1,620	J. Ellenberger, pers. commun., 2000 estimates
Yukon	1,400	Hoefs et al. 1977; J. Carey, pers. commun., 2000
Northwest Territories	400+	Johnson 1977; A. Veitch, A. Gunn, and R. Case, pers. commun., 2000
South Dakota	150–170	T. A. Benzon, pers. commun., 2000

[a]Named sources (excluding literature citations) are individuals on staff with the respective jurisdictional wildlife agencies.

social rank (Côté and Festa-Bianchet 2001a). Forage quality measured by fecal crude protein in late spring, however, affected kid mass: kids born in years when forage quality was high during early lactation were heavier in summer than kids born in years of poor-quality forage (Côté and Festa-Bianchet 2001a).

As yearlings, males are about 10% heavier than females and their skeletal size, as measured by total length and hind foot length, is about 4% larger (Table 49.2). Sexual dimorphism in body mass develops after weaning, increasing gradually up to at least 6 years of age (Table 49.2 and Fig. 49.4) (Houston et al. 1989). Females complete their mass gain at 6 years of age (Côté 1999), whereas males may continue to increase in mass with age. At ≥5 years of age, males are about 40–60% heavier than females (Houston et al. 1989; Côté 1999); midsummer body mass is about 95–115 and 60–75 kg for males and females, respectively (Table 49.2). Total lengths are about 155–180 cm for adult males and 140–170 cm for adult females (Table 49.2). At Caw Ridge, Alberta, shoulder height is about 5% and chest girth 7% larger in adult males than in adult females: Shoulder height for males is 96.3±8.0 (*SD*) cm ($n = 6$) and for females is 91.9±4.0 cm ($n = 43$); chest girth for males is 111.2±9.7 cm ($n = 32$) and for females is 103.9±6.5 cm ($n = 73$). Comparative data indicate differences among populations (Table 49.2). For example, mountain goats on Caw Ridge are larger than individuals from several populations in the United States. Comparisons, however, are difficult to make because only the Caw Ridge study accounted for the substantial effects of season on body mass. For goats of all sex–age classes, mass increases during the forage growing season and then decreases through the winter (Houston et al. 1989). In midsummer at Caw Ridge, adult males gain >400 g/day and adult females >200 g/day (S. D. Côté and M. Festa-Bianchet, unpublished data).

FIGURE 49.2. Adult female mountain goat in summer coat in August. SOURCE: Photo by S. D. Côté.

Pelage. Mountain goats have a long and shaggy white coat. The winter pelage consists of long and coarse guard hairs (often >20 cm) and an underlayer of wool about 3–5 cm thick. There is a manelike dorsal ridge of hair on the center of the back, and their pointed beard grows with age (B. L. Smith 1988). The winter coat is shed in May–August and the new hairs start growing before the molt is completed (Brandborg 1955). The summer coat is short (guard hairs 2–5 cm long), and hairs grow from June to autumn. Growth of the winter coat is completed by November or early December (Holroyd 1967; B. L. Smith 1988). Adult males finish shedding their coat before females. At Caw Ridge, males ≥3 years old completed their molt on 16 July ± 7 days ($n = 60$) on average, whereas females of the same age finished shedding on 9 August ± 12 days ($n = 270$). Adult females with kids completed their molt slightly later than nonlactating females (12 August ± 9 days, $n = 145$ compared to 4 August ± 14 days, $n = 114$). This later molt may reflect an energy or protein constraint imposed by a greater expenditure in lactating females (Robbins 1993; Byers 1997). In the same population, juvenile males and females (1- and 2-year-olds) completed their molt about a week before adult females with kids (males: 5 August ± 14 days, $n = 50$; females: 5 August ± 13 days, $n = 84$).

Mountain goats have thick dermal shields protecting their rump. Geist (1967) suggested that dermal shields were an adaptation to the high risk of injury caused by the goat's typical antiparallel circle fighting technique, during which the sharp horns of each goat aim for the opponent's flanks.

Skull and Dentition. The skull of mountain goats (Fig. 49.5) is light and bones are thin (Rideout and Hoffmann 1975). The horn cores are approximately one third the length of the horn sheath and are conical in shape and straight, with a round base and a sharp tip (Fig. 49.5). Rostral length is about 5% longer in adult males (193 mm) than in females (183 mm) (B. L. Smith 1988). The dental formula is I 0/3, C 0/1, P 3/3, M 3/3 (Rideout and Hoffmann 1975). Kids are born with 18 milk teeth, lacking deciduous incisiform canines (Rideout and Hoffmann 1975). The I1 incisors are replaced at 15–16 months of age, the I2 incisors at 2 years of age, the I3 at 3 years of age, and the incisiform canines at 4 or 5 years of age (Brandborg 1955; Rideout and Hoffmann 1975; Wigal and Coggins 1982). Premolars and molars are hypsodont and selenodont (Wigal and Coggins 1982). Goats >8 years of age often show severe tooth wear and can lose some or all of their incisors and canines (Brandborg 1955; B. L. Smith 1986; S. D. Côté, pers. obs.).

Horns. Both sexes have sharp black horns (Fig. 49.6) and use them prominently in social interactions (Geist 1967). Mountain goats display an unusually high level of female–female aggression (Fournier and Festa-Bianchet 1995; Côté 2000a). However, they do not fight through horn contact (Geist 1964), and the high level of female–female aggression may make horns just as useful for females as for males (Côté et al. 1998a). Horn growth starts at birth and continues throughout life, decreasing with advancing age and stopping during winter (Fig. 49.7)

TABLE 49.2. Body mass, total length, and hind foot length in mountain goats of different age–sex classes

Location	Sex	Age (years)	Mass (kg)	Total Length (cm)	Hind Foot Length (cm)	Reference and Comments
Alberta[a] (Caw Ridge)	M	Kid	13.9 ± 2.9 (41)	92.4 ± 6.3 (13)	23.6 ± 1.4 (36)	S. D. Côté and M. Festa-Bianchet, unpublished data
	F	Kid	13.2 ± 2.7 (53)	90.2 ± 6.6 (33)	22.4 ± 1.4 (50)	
	M	1	35.7 ± 4.5 (51)	126.7 ± 6.4 (13)	30.4 ± 1.7 (45)	
	F	1	32.5 ± 3.7 (45)	124.4 ± 3.6 (13)	29.2 ± 1.3 (39)	
	M	2	53.0 ± 4.4 (37)	144.8 ± 6.0 (6)	33.8 ± 1.3 (29)	
	F	2	46.2 ± 3.6 (32)	140.0 ± 4.8 (10)	32.6 ± 1.5 (29)	
	M	3	65.3 ± 5.5 (12)	—	35.4 ± 1.6 (12)	
	F	3	55.9 ± 3.9 (14)	144.5 ± 5.3 (6)	32.9 ± 1.1 (14)	
	M	4	83.1 ± 6.7 (5)	—	36.7 ± 1.3 (6)	
	F	4	62.2 ± 8.4 (13)	146.1 ± 9.2 (9)	34.3 ± 1.6 (11)	
	M	≥5	102.3 ± 10.1 (16)	—	37.3 ± 1.4 (16)	
	F	≥5	70.6 ± 5.9 (47)	160.2 ± 11.9 (31)	34.5 ± 1.4 (49)	
Washington[b] (Olympic National Park)	M	Kid	12.6 ± 2.6 (11)	—	—	Calculated from Fig. 2 in Houston et al. (1989)
	F	Kid	13.6 ± 3.2 (17)	—	—	
	M	1	30.7 ± 6.7 (20)	—	—	
	F	1	33.1 ± 5.2 (16)	—	—	
	M	2	46.5 ± 9.3 (10)	—	—	
	F	2	40.3 ± 6.2 (15)	—	—	
	M	3	60.0 ± 8.9 (11)	—	—	
	F	3	52.3 ± 10.8 (8)	—	—	
	M	4	72.6 ± 14.4 (6)	—	—	
	F	4	57.6 ± 6.2 (17)	—	—	
	M	≥5	~111.6 ± 4.8 (9)	—	—	
	F	≥5	~61.9 ± 5.2 (40)	—	—	
Montana[c] (Sapphire Mountains)	M	Kid	16	90.8	23.9	Rideout (1978)
	F	Kid	15	85.1	24.8	
	M	1	33.6	119.4	29.2	
	F	1	32.4	118.2	28.0	
	M	2	41.5	134.3	29.7	
	F	2	38.6	128.6	29.5	
	F	3	49.6	136.3	31.6	
	M	≥4	68.6	154.0	32.8	
	F	≥4	56.5	141.0	30.8	
Montana (Crazy Mountains)	M	1	35 (5)	120.1 (5)	28.4 (5)	Lentfer (1955)
	F	1	30.5 (3)	114.3 (3)	28.7 (3)	
	M	2	46.8 (2)	143.5 (2)	32.8 (2)	
	F	2	44.5 (2)	142.2 (2)	31.8 (2)	
	F	3	62.3 (2)	148.8 (2)	32.0 (2)	
	M	≥4	82.0 (2)	178.6 (4)	36.6 (4)	
	F	≥4	71.6 (2)	154.4 (5)	32.0 (5)	
Idaho	M	≥4	69.9 (5)	153.7 (5)	—	Brandborg (1955)
	F	≥4	53.1 (6)	141.0 (6)	—	

NOTE: Means ± *SD* are shown with sample sizes in parentheses.

[a]The Caw Ridge data were collected from 1988 to 2000. Body mass was adjusted to midsummer (15 July) for each individual using age- and sex-specific growth rates (Côté 1999). Total length and hind foot length of kids and yearlings also were adjusted to 15 July using age- and sex-specific growth rates. Only means based on more than five individuals are presented.

[b]Goats were captured and weighed between 25 June and 16 July 1981–1984.

[c]Total sample size is 28, including all age–sex classes.

(Côté et al. 1998a). Goats complete about 93% of horn growth by 3 years of age (Côté et al. 1998a). B. L. Smith (1988) reported average horn lengths of 232 mm in adult males and 222 mm in adult females from Idaho and Montana. Males have longer first increments (grown from birth to 18 months of age) than females, whereas females grow more horn than males in their third year (Fig. 49.7C) (Cowan and McCrory 1970; Hoefs et al. 1977; Côté et al. 1998a). After 3 years of age, the annual horn increments of males and females are similar and very short (Fig. 49.7C). Males have larger horn circumference than females at all ages (Fig. 49.7B) (Hoefs et al. 1977; B. L. Smith 1988; Côté et al. 1998a).

Horns do not show directional asymmetry or antisymmetry in total length or base circumference (Côté et al. 1998a; Côté and Festa-Bianchet 2001b). Horn length, however, exhibits fluctuating asymmetry, which refers to small random deviations from perfect bilateral symmetry. Because symmetry reflects the ability of individuals to undergo stable development, fluctuating asymmetry may be a potential measure of individual quality (Møller and Swaddle 1997; Gangestad and Thornhill 1999). Absolute asymmetry increases with horn length in both sexes, but relative asymmetry in horn length does not vary with either sex or age (Côté and Festa-Bianchet 2001b). A study of fluctuating asymmetry in mountain goat horns indicated that asymmetry in horn length revealed individual quality in females, because those with highly symmetrical horns had a higher social rank, were heavier, and produced more young than asymmetrical females (Côté and Festa-Bianchet 2001b). In adult males, however, there was no strong

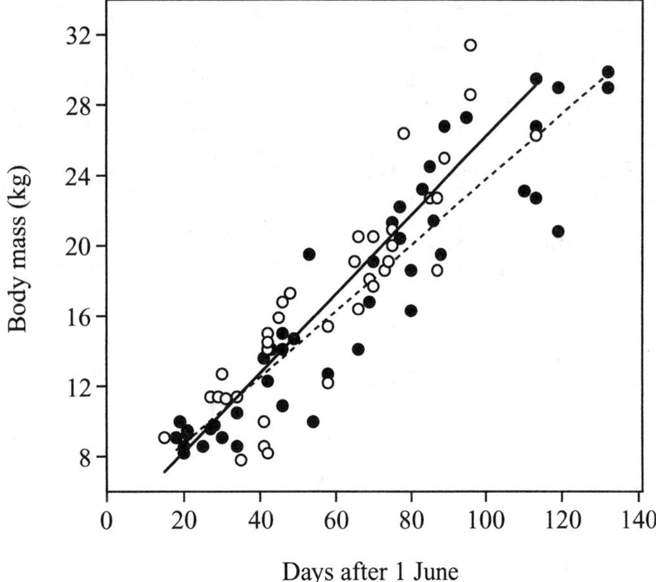

FIGURE 49.3. Summer mass gain by mountain goat kids at Caw Ridge, Alberta, 1988–1997. Males are indicated by open circles and solid line, females by closed circles and dashed line. Each point refers to an individual kid. SOURCE: Côté and Festa-Bianchet (2001a).

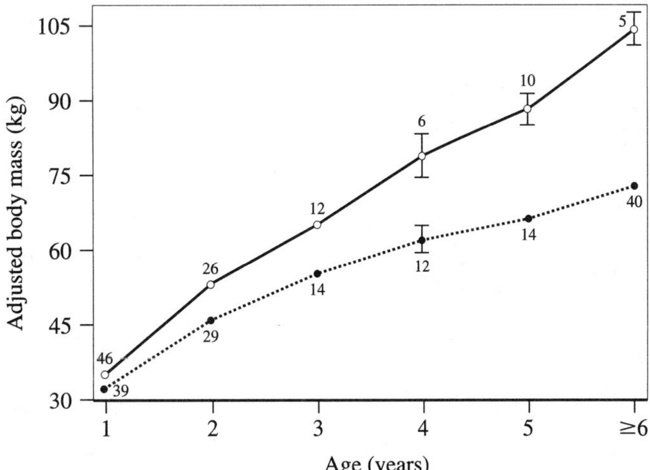

FIGURE 49.4. Development of sexual dimorphism in body mass of mountain goats from Caw Ridge, Alberta, 1988–1999. Body mass was adjusted to midsummer (15 July) using the sex-specific growth rate of each age class. Mean body mass (±*SE*) for males (open circles) and females (closed circles) is accompanied by sample size. SOURCE: Data from Côté (1999).

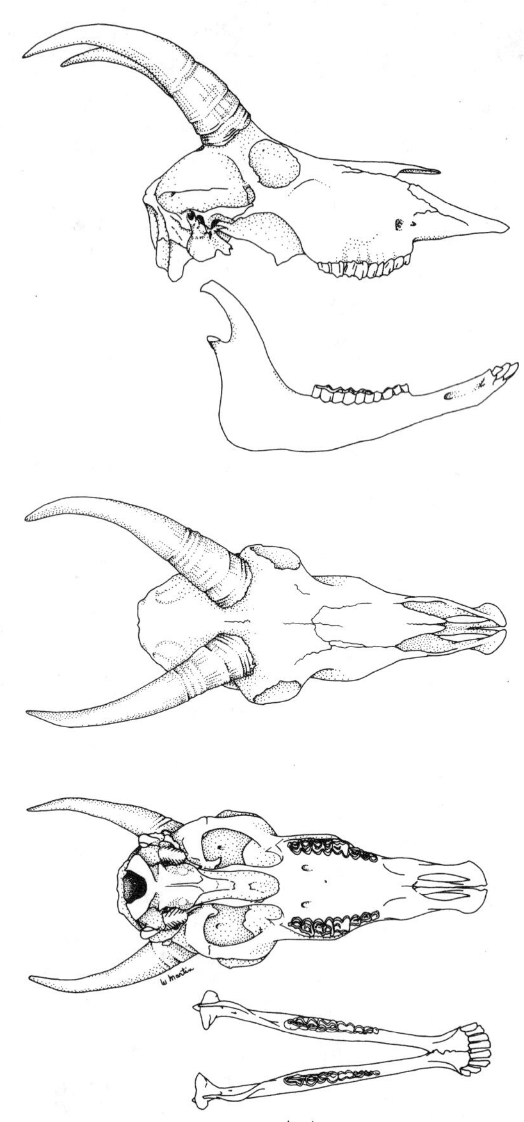

FIGURE 49.5. Skull of the mountain goat (*Oreamnos americanus*). From top to bottom: lateral view of cranium, lateral view of mandible, dorsal view of cranium, ventral view of cranium, dorsal view of mandible.

relationship between horn asymmetry and life-history traits (Côté and Festa-Bianchet 2001b).

SEX DETERMINATION AND AGE ESTIMATION

Compared to other ungulates, it is difficult to determine the sex and age of mountain goats. The most useful characteristic for determining the sex of mountain goats is horn shape. As stated in the previous section, males have larger horn base circumferences than females (also see Fig. 49.6). A useful rule is that horn base circumference is about the size of the eye in females, but is much bigger than the eye in males. In addition, horn curvature in the male is distributed regularly along the horn's entire length (see profile in Fig. 49.6) (Brandborg 1955; B. L. Smith 1988), whereas the female's horns are comparatively straight, but the tips often display a distinctive backward "crook" (Fig. 49.6) (B. L. Smith 1988). Two-year-olds have intermediate rostral length between yearlings and adults, and this characteristic can be used to identify them (B. L. Smith 1988). Males are larger than females, but differences in size are useful for field identification only if animals of different age–sex classes are present for comparison. Determining sex of kids from a distance is very difficult. Two criteria could be used: observation of the vulvar patch in females and urination posture. When urinating, females squat, whereas males stretch. Determining sex of yearlings is difficult, especially at 12–15 months of age, but is possible for an experienced observer familiar with the subtle differences in horn shape. Accuracy of classified counts obtained from aerial surveys is poor, and even experienced observers may confuse kids and yearlings (Gonzalez-Voyer et al. 2001). Most of the time, sex cannot be determined during aerial surveys.

Age of mountain goats can be estimated by counting the horn annuli (Brandborg 1955). The first distinct annual growth ring is formed at the beginning of the second winter, when the goat is ca. 1.5 years old; thereafter, each subsequent ring is formed in early winter (Brandborg 1955; B. L. Smith 1988). Age can be estimated by adding 1 year to the number of distinct rings observed at capture (Stevens and Houston 1989). Only the first seven to eight annuli can be measured because

FIGURE 49.6. Variation in horn shape of (top) female and (bottom) male mountain goats (*Oreamnos americanus*). SOURCE: Adapted from B. L. Smith (1988).

later growth rings are often indistinct (Brandborg 1955; Stevens and Houston 1989).

Horn measurements also can be used for estimating body mass and size. Côté et al. (1998a) provided equations for estimating body mass, chest girth, and hind foot length from horn length and circumference. Relationships are approximately linear and correlations (r_p) averaged 0.83.

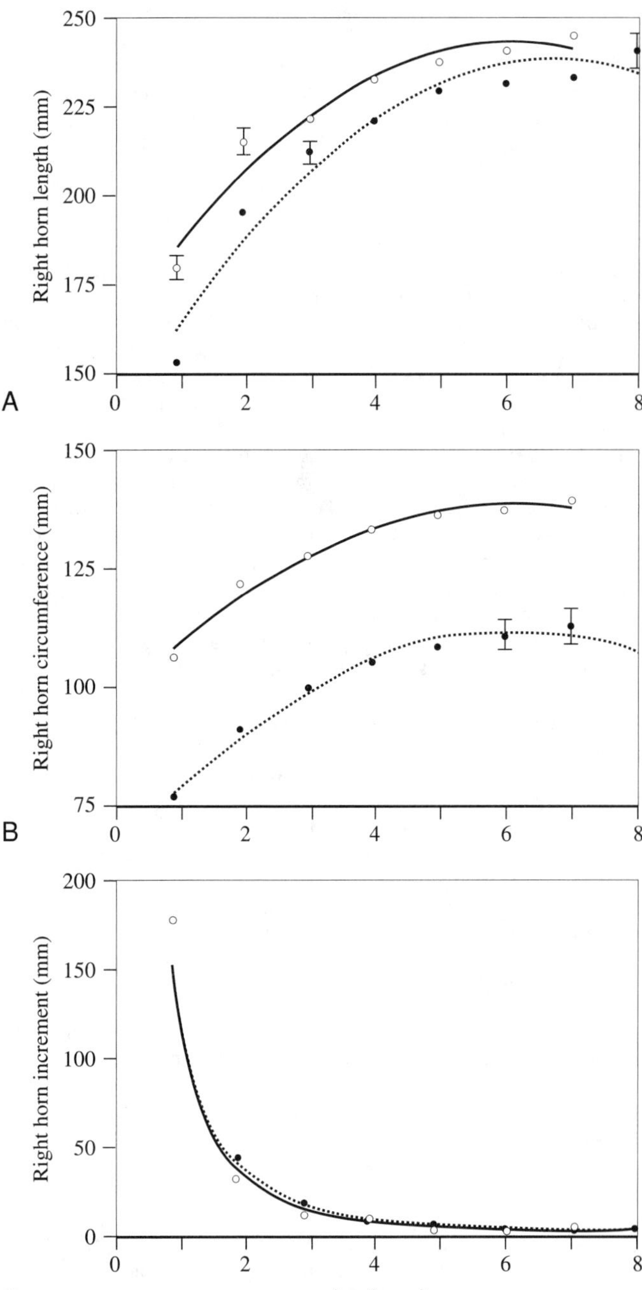

FIGURE 49.7. (A) Average horn length, (B) horn circumference, and (C) annual horn increment of mountain goats at Caw Ridge, Alberta ($n = 259$ individuals). Open circles and broken lines represent males; closed circles and continuous lines represent females. *SE* is shown when large enough for the figure's resolution; fitted lines are second-degree polynomial regressions (A, B) and a power function (C) through the averages and are included to visualize differences between males and females. SOURCE: Data from Côté et al. (1998a).

HABITAT

Foraging Sites. Mountain goats live in some of the most inhospitable terrain in North America, alpine areas close to cliffs or rocky ledges on which they depend to escape predators (Brandborg 1955; McFetridge 1977; Von Elsner-Schack 1986; Haynes 1992; Gross et al. 2002). Weather is typically very harsh, with snow on the ground for 8–9 months of the year and frequent high winds. Goats may seek shelter under rocky ledges and trees during extreme weather. They occur mainly from treeline to the highest alpine meadows (Haynes 1992). In the northern Rocky Mountains, typical elevation ranges from 1500 to 2700 m (Smith 1977), but goats can be seen at >4000 m in Colorado (Hibbs 1967). On the west coast of British Columbia and Alaska, some winter ranges are near sea level (Hebert and Turnbull 1977).

Typical foraging sites for mountain goats are alpine meadows near cliffs (Brandborg 1955; Von Elsner-Schack 1986). Female and juvenile groups rarely wander far from escape terrain or below treeline (McFetridge 1977). Adult males, however, often forage in coniferous forest below treeline (Hebert and Turnbull 1977; Risenhoover and Bailey 1982). Therefore, in some areas, bachelor groups may have access to new-growth forage before adult females in the spring.

Bedding Sites. Selection of bedding sites is important for mountain goats. Areas near or on cliffs with a clear view of the surroundings are normally used (McFetridge 1977). Before lying down, goats often dig a bedding site with their front paws and sometimes throw dirt on themselves (Holroyd 1967). Bedding sites are used repeatedly by different animals. All individuals in the group normally use the same area to bed, and, most of the time, distances between neighbors are only

1–4 m (Côté 1999). Good bedding sites are limited and situated near cliffs, are leveled, and contain 3–10 cm of soft ground (Rideout and Hoffmann 1975). Goats compete for these sites; about 36% ($n = 6775$) of all aggressive interactions occur at resting–ruminating sites, when goats displace each other to occupy the best bedding places (S. D. Côté, unpublished data).

Daily Movements. The extent of daily movements is variable and depends on season and sex. Not surprisingly, movements are shorter in winter when snow is deep (Geist 1964; Hjeljord 1973; Chadwick 1983). Rates of movements while foraging increase with group size (Risenhoover and Bailey 1985). Nursery groups (females, 1- and 2-year-olds of both sexes, and kids) have longer average daily movements than bachelor groups or solitary males. In general, males move <1 km/day, whereas females may move 2–5 km/day or more (S. D. Côté, unpublished data; Singer and Doherty 1985). Foster and Rahs (1985) reported average daily movements of 400 m for canyon-dwelling mountain goats in British Columbia. In contrast, Singer and Doherty (1985) reported hourly movements of 530 and 990 m in males and females, respectively. We hypothesize that females with kids travel more than males as an antipredator strategy, because longer displacements make their location less predictable to predators. Generally, males have small summer home ranges and do not travel much, although some studies have reported extended movements between intensively used areas (Rideout 1974). Conversely, nursery groups generally have larger home ranges than bachelor groups. At Caw Ridge, Alberta, for example, males have summer home ranges of about 5 km^2, whereas females have annual home ranges (100% of locations) of approximately 25 km^2 (S. D. Côté, unpublished data). Rideout (1977) reported annual home range sizes of 21.5 and 24 km^2 for males and females, respectively, in Montana. Although based on small sample sizes, Singer and Doherty (1985) reported annual home ranges of 6.3 km^2 for males and 8.9 km^2 for females in Glacier National Park, Montana. Differences in reported home range sizes could be due to topography. In certain areas, goats may use seasonal home ranges that are a considerable distance from each other. Males could also make extensive movements during the rut in some populations but not in others, depending on the distance between neighboring groups.

Seasonal Movements. Little is known about seasonal movements of mountain goats. Some populations remain in the same area throughout the year, whereas others have distinct summer and winter ranges. In summer, nursery groups normally use all foraging grounds near escape terrain from treeline to the limit of vegetation. Winter ranges, however, are restricted to wind-swept and west/south-facing slopes at treeline and just below treeline near escape terrain (Brandborg 1955; Hjeljord 1973; Smith 1976, 1977; Rideout 1977). Goats may use low-elevation habitats during winter, especially when snow is deep (Hjeljord 1973; Rideout 1974; Smith 1977). Male groups and solitary males use forested areas near treeline throughout the year and may perform important seasonal migrations in some populations (Rideout 1974; Risenhoover and Bailey 1982). Smith (1977) reported that in winter, males were found more often than females in sites with deep snow. On the coast of British Columbia and in Southeast Alaska, goats can winter in coniferous forests at sea level and summer in the mountains (Hebert and Turnbull 1977; Fox 1983). To our knowledge, mountain goats do not have specific rutting ranges.

Feeding Habits. Along the continuum from grazers to browsers, mountain goats are classified as intermediate browsers (Hofmann 1989). They eat a variety of forage (Saunders 1955; Laundré 1994). Diets are similar in summer and winter and are generally dominated by grasses (Saunders 1955; Hibbs 1967; Rideout 1974; Laundré 1994). Laundré (1994) summarized 10 studies on feeding habits of mountain goats and found that summer diet included 52% grass, 30% forb, and 16% browse. Based on averages of the values reported in technical literature, the preferred plant genera in summer were bluegrass (*Poa* sp., 14%), sedges (*Carex* sp., 10%), wheatgrass (*Agropyron* sp., 9%), bluebells (*Mertensia* sp., 6%), fescue (*Festuca* sp., 5%), and hairgrass (*Koeleria* sp., 5%) (Laundré 1994). However, feeding habits at the level of plant species vary tremendously among populations. Goats are generalist herbivores and seem to eat what is available. In the spring, they seek growing, young alpine herbaceous plants (Dailey et al. 1984). In the summer, they also eat large amounts of young leaves of common treeline shrubs such as willows (*Salix* spp., 4%) and dwarf birch (*Betula glandulosa*) (Laundré 1994). Some goats use forested areas in autumn, possibly because of seasonal changes in forage quality (Festa-Bianchet et al. 1994).

In winter, Laundré (1994) reported that the average diet shifted to 60% grass, only 8% forb, and 32% browse. The preferred plant genera were fescue (18%), sedges (8%), wheatgrass (4%), bluegrass (4%), sagebrush (*Artemisia* sp., 3%), hairgrass (1%), and willows (1%). Diet can also vary with snow conditions; Fox and Smith (1988) reported that forbs and ferns decreased in the diet of southeast Alaska goats as snow depth increased to >50 cm. When forage is limited in winter, goats also eat twigs and needles of coniferous trees such as Engelmann spruce (*Picea engelmanii*) and alpine fir (*Abies lasiocarpa*) (Saunders 1955; Geist 1971; Adams and Bailey 1983; Fox and Smith 1988). Harmon (1944) and Fox and Smith (1988) reported heavy use of lichens and mosses in winter by goats in South Dakota and southeast Alaska, respectively.

Salt Licks. Minerals are limited in alpine vegetation and a diet of succulent new-growth vegetation decreases sodium retention by herbivores (Hebert and Cowan 1971a). Mountain goats, therefore, use traditional salt licks where they return regularly during the summer (Singer and Doherty 1985; Hopkins et al. 1992). They normally start visiting the licks in April–May (males) or early June (females) and stop in early autumn (Hebert and Cowan 1971a). Goats sometimes travel long distances to reach mineral licks, sometimes crossing dangerous terrain such as forest or rivers in flood (Holroyd 1967; Hebert and Cowan 1971a; Hopkins et al. 1992). Rates of aggressive encounters are higher at mineral licks than elsewhere, presumably because licks are a limited and defensible resource (Chadwick 1977; Hopkins et al. 1992; Côté 2000b). Over 100 animals can concentrate in 150–200 m^2 and the outcome of social interactions at mineral licks may be different from that seen elsewhere (Côté 2000b).

BEHAVIOR

Activity Patterns. Like all ruminants, mountain goats alternate feeding and resting bouts. Goats typically have six to seven feeding-resting cycles/24 hr. There are peaks of activity in early morning and in the evening (Saunders 1955; Rideout 1974; Romeo and Lovari 1996), but goats also are active during the night (Rideout 1974; Singer and Doherty 1985). Daily activity patterns can be affected by weather conditions (Fox 1977). In the summer, Singer and Doherty (1985) reported increased movements in early morning and late afternoon. The proportion of time that goats spend feeding while active generally increases with group size, mainly because of reduced time spent vigilant (Risenhoover and Bailey 1985; Holmes 1988). During hot midsummer days, activity in the afternoon is often reduced. Long-distance movements (>5 km) can occur at any time of the day.

Social Organization. Male and females are spatially segregated except during the rut (Geist 1964; Chadwick 1977; Stevens 1983; Risenhoover and Bailey 1985). Males are solitary or form bachelor groups of two to six individuals (Risenhoover and Bailey 1982; Haviernick 1996). In late spring, however, bachelor groups can include up to 15 males (Chadwick 1977; S. D. Côté, unpublished data). As noted, females form nursery groups with kids, yearlings, and 2-year-olds of both sexes (Chadwick 1977). Numbers of goats in nursery groups vary substantially depending on population size and on season (Brandborg 1955; Holroyd 1967; Smith 1976, 1977; Chadwick 1977; Singer 1977; Hayden 1984; Masteller and Bailey 1988a).

The mother–kid bond is very strong. Yearlings, and sometimes 2-year-olds, may associate with their mother and occasionally suckle, especially if the mother does not have a kid at heel (Hutchins 1984; Dane 2002). Kids are normally physiologically weaned at about 4 months, and behaviorally weaned at about 1 year of age when the new kid is born (Brandborg 1955; Holroyd 1967; Côté 1999). Kids start eating vegetation when about 1 week old, and by 4–5 weeks of age they feed extensively on vegetation (DeBock 1970; S. D. Côté, pers. obs.). Suckling is greatly reduced by 5–6 weeks of age, although it continues, decreasing in frequency, until behavioral weaning (DeBock 1970; Hutchins 1984). Most males leave nursery groups when they are 2 or 3 years old, sometimes 4 years (Chadwick 1977; Romeo et al. 1997).

Females isolate themselves from other goats to give birth, sometimes tolerating the presence of their previous year's offspring. About 1 week after parturition, in early June, females start to form small groups. By July, nursery groups attain peak size, sometimes including >100 individuals in the largest populations (Hopkins et al. 1992; Côté 1999). In late autumn and winter, nursery groups rarely exceed 30 individuals (Brandborg 1955; Smith 1977; Masteller and Bailey 1988a).

Aggressive Interactions. Compared to other ungulates, mountain goats display a high frequency of intraspecific aggression (Chadwick 1977; Dane 1977; Côté 2000a), even though their dangerous horns can injure or even kill conspecifics (Geist 1964, 1967). Adult females at Caw Ridge, Alberta, had an average of 3.4 aggressive interactions/hr, the highest frequency of intraspecific aggression for any female ungulate for which aggressiveness has been measured in the wild (Fournier and Festa-Bianchet 1995).

Aggressive behaviors include present threat (broadside orientation with apparent size enhanced by arching the back), horn threat (display or aggressive movement of the horns), rush threat (sudden quick movement toward an antagonist) and orientation threat (a lower intensity form of rush threat involving walking) (see Geist 1964 and Chadwick 1977 for complete descriptions). Submissive behaviors include orientation avoidance (avoiding the opponent by walking or staring away from it) and rush avoidance (quickly moving away from the antagonist) (Chadwick 1977). Low-pitched grunts are sometimes associated with aggressive interactions (Rideout and Hoffmann 1975). Escalated fights are rare (Geist 1964). They made up only 1.9% of 4265 aggressive interactions during active bouts at Caw Ridge (S. D. Côté, unpublished data). Fights mostly occur as circle fights, where opponents in present threat circle in an antiparallel position, with the head facing the rump of the antagonist (Chadwick 1977). Body (or horn) contacts and high-intensity, bleatlike vocalizations occur in <0.5% of aggressive interactions (Côté 1999).

The presence of a kid does not affect the aggressive behavior of females (Côté 2000a), which suggests that aggressiveness probably did not evolve for offspring defense in mountain goats as previously suggested by Geist (1974). Females and kids nevertheless remain close to each other most of the time and if separated will bleat loudly to locate each other (Brandborg 1955; Holroyd 1967; Rideout and Hoffmann 1975). Most adult behavior patterns are performed by kids during play.

Dominance Hierarchy. Except during the rut, adult females are dominant over all other age–sex classes (Chadwick 1977; Irby and Fitzgerald 1994) and establish highly linear and stable hierarchies even within large groups (Côté 2000a). Adult females are also more aggressive than other age–sex classes (Risenhoover and Bailey 1985; Masteller and Bailey 1988b; Côté 1999). Female social rank is strongly related to age and does not decrease in the oldest females (Côté 2000a). When age is accounted for, social rank does not vary with body mass, horn length, or body size (Côté 2000a). Dyadic dominance relationships are likely established early in life, when a single year of difference in age results in an important difference in body size, and are thereafter maintained into adulthood even though by then some dominant individuals may be smaller or have smaller horns than their younger subordinates. Females interact more often with individuals of similar ranks than with individuals that are distant in the dominance hierarchy (Côté 2000a).

Rutting. Little information is available on the rutting behavior of mountain goats. Early work by Geist (1964) and Chadwick (1983) suggested that adult males establish a dominance hierarchy during the rut and defend estrous females from other males. Throughout the rut, males dig rutting pits, where they often urinate and paw dirt over their underside and hindquarters (Geist 1964; Holroyd 1967). They also mark grass or twigs by brushing their horns on the vegetation to spread secretions from the supraoccipital glands (Geist 1964; DeBock 1970). At the peak of the rut, males follow receptive females everywhere, often leading to perilous chases on cliff faces. According to Geist (1964), females are very aggressive toward courting males and can only be approached during estrus. Males use the low stretch posture (a slow and submissive approach, crouching and exposing the vulnerable part of their neck) to approach females and can sometimes be hit and injured by the sharp horns of females (Geist 1964, 1967). Males test the receptivity of a female by inhaling the scent of the female's urine in their vomeronasal organ (a behavior known as flehmen or "lip curl") and perform frequent tongue flicking when near females in estrus (Geist 1964; Chadwick 1977). Before mounting, males perform a front leg kick between or along the female's haunches (Geist 1964).

REPRODUCTION

Breeding. Mating season is from late October to early December, normally peaking in mid-November (Brandborg 1955; Geist 1964; Holroyd 1967; DeBock 1970; Chadwick 1983). Timing of the rut may vary according to latitude, but little information on geographic variation is available. Little is known about the mating system of mountain goats. During estrus, which is thought to last about 2 days, males follow females and defend them from other males (Geist 1964). It is not clear whether a male can defend more than one female at a time nor how long a male normally defends a female. Based on the short birth season observed (see below), it is likely that most females in a population attain estrus within a 2-week period. It is not known whether females copulate with more than one male. Large and dominant males likely secure most of the mating opportunities, but yearlings and 2-year-olds, which are fertile, may also mate if the opportunity arises (Henderson and O'Gara 1978). There are no data on male reproductive success, which could only be measured through molecular methods.

Birth Season. Mountain goats give birth from mid-May to early June, and about 80% of kids are normally born within 2 weeks of the first birth (Holroyd 1967; Rideout 1978; Côté and Festa-Bianchet 2001a). At Caw Ridge, the median birthdate ranged from 24 to 27 May and the time from first to last birth varied between 28 and 47 days over 5 years (Côté and Festa-Bianchet 2001a). Although the beginning of the birth season is highly synchronized, there are usually a few late births from mid-June to early July (Fig. 49.8). The gap in parturitions between 6 and 15 June

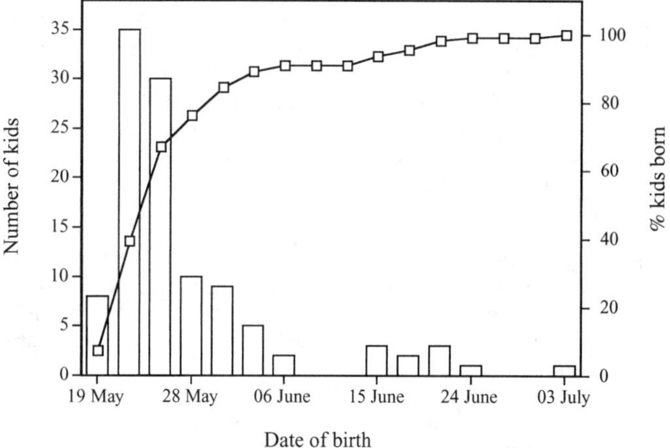

FIGURE 49.8. Birthdates of 109 mountain goat kids during 1993–1997 at Caw Ridge, Alberta. Histograms indicate the number of kids born every 3 days beginning on 15 May, and the solid line shows the cumulative percentage of kids born. SOURCE: Côté and Festa-Bianchet (2001a).

(Fig. 49.8) may suggest that late births result from females that failed to conceive during their first estrus. Based on peak rutting activities and peak birth period, the gestation is approximately 190 days.

Females isolate themselves to give birth (Holroyd 1967). They normally select a site safe from predators in a rocky outcrop or a cliff where they give birth and remain for a few days with the newborn kid (Côté and Festa-Bianchet 2001a). Kids are of the "montane follower" type (Geist 1981) and at about 2–3 days of age they can follow their mother. Some females return to the same birth site every year, but most do not (S. D. Côté, unpublished data). Following birth, females normally eat the placental remains and lick blood drops from the ground and rocks surrounding the birth site (S. D. Côté, pers. obs.). By doing so, they make birth sites more difficult to detect by predators and likely gain a nutritional advantage by recovering proteins.

Primiparity. In the native population of mountain goats at Caw Ridge, Alberta, females produced their first kid at 3–7 years of age (average = 4.6 ± 0.9 years) (Festa-Bianchet et al. 1994; Côté and Festa-Bianchet 2001c). In introduced populations, females can produce their first kid at 2 years of age (Stevens 1983; Bailey 1991), but normally do so at 3 years (Houston and Stevens 1988; Bailey 1991). These results suggest that the physical development of females in native populations is slow (see also Fig. 49.4) (Côté 1999), and that the costs of reproducing at a young age are probably high. There is no evidence that population density or social rank affects age at first reproduction in mountain goats (Côté and Festa-Bianchet 2001c).

MORTALITY

Starvation. Starvation, particularly of kids, is a likely cause of mortality in late winter and spring. However, animals that die from starvation are difficult to find for two reasons. First, starvation is more likely to occur in late winter, when access to mountain goat habitats by researchers is very difficult and limited. Second, most carcasses are eaten by scavengers before they can be found. At Caw Ridge, for example, only three carcasses of non-radio-collared animals were found, but >200 animals disappeared. In the same study, we documented two kids <1 week old that died of starvation (S. D. Côté, unpublished data). Some deaths of starving animals may be attributed to predation because weak animals are more easily taken by predators.

Predation. In native mountain goat herds, predation is likely the most important mortality factor. The most important predators are grizzly bears (*Ursus arctos*), wolves (*Canis lupus*), and cougars (*Puma concolor*) (Table 49.3) (C. A. Smith 1986; Fox and Streveler 1986; Festa-Bianchet et al. 1994; Côté and Beaudoin 1997; Côté et al. 1997). Coyotes (*Canis latrans*) and black bears (*Ursus americanus*) are other potential predators (Brandborg 1955; Smith 1976), and wolverines (*Gulo gulo*) attack goats, especially young kids (Guiguet 1951; Côté et al. 1997). Brandborg (1955) and Smith (1976) reported golden eagles (*Aquila chrysaetos*) preying on kids in Idaho and Montana. Risk of predation appears higher in areas with trees or in krummholz (stunted forest/scrub), which provide cover for ambush predators (Festa-Bianchet et al. 1994; Côté and Beaudoin 1997). When goats cross forested valleys, they normally use traditional and well-marked trails and often run through areas with trees (S. D. Côté, pers. obs.).

Accidental Death. Although accidents are generally not a frequent mortality factor in goats, deaths due to climbing accidents have been repeatedly suggested (Chadwick 1983; Fox and Streveler 1986). Similarly, authors have reported goats being killed by avalanches (Brandborg 1955; Macgregor 1977; Chadwick 1983).

Fatal injuries can also be inflicted during fighting (Geist 1964, 1967), although this phenomenon appears to be rare. Road kills are not a serious concern for most mountain goats because of the scarcity of roads in their natural range.

Parasites and Diseases. Compared to other ungulates, relatively little is known about parasites and diseases in mountain goats. Kerr and Holmes (1966) identified two species of ticks, two species of cestodes, and nine species of nematodes including a species of lungworm (*Protostrongylus* sp.) in mountain goats from Alberta. They examined 10 goats >2 weeks old, and all were infected with parasites. Samuel et al. (1977) examined 53 goats from the same region and found 17 species of helminths (12 nematodes, 5 cestodes) and 1 species of tick. Brandborg (1955) and Boddicker et al. (1971) found a similar number of parasite species in goats from Idaho and Montana, and South Dakota, respectively. Overall, the nematode *Ostertagia circumcincta* had the highest prevalence (Samuel et al. 1977). Detrimental infections of lungworms and of the nematode *Nematodirus maculosus* were reported in introduced goats from South Dakota (Boddicker et al. 1971). Lungworms were common in mountain goats examined by Cooley (1977) in Colorado and Alberta. Protozoan parasites responsible for coccidiosis (*Eimeria* spp.) (Shah and Levine 1964; Todd and O'Gara 1968) and sarcocyst infections (*Sarcocystis* sp.) (Mahrt and Colwell 1980; Foreyt 1989) also occur in mountain goats. Although mountain goats are infected by several species of parasites, there is no evidence that parasitism has a strong effect on individual body condition, survival, or reproduction (Kerr and Holmes 1966). No study, however, has specifically examined the ecological effects of parasites on mountain goats. Indeed, there has been very little research on the parasites of mountain goats since 1980. Most mountain goat populations appear to be free of contagious ecthyma, a viral disease that is more common in bighorn sheep (*Ovis canadensis*) (but see Samuel et al. 1975; Hebert et al. 1977). To our knowledge, there is no clear evidence of disease transmission from domestic livestock to mountain goats, possibly because the opportunities for contact are few. Selenium deficiency myopathy, known as white muscle disease, has been reported in mountain goats from British Columbia (Hebert and Cowan 1971b), and respiratory syncytial virus, which predisposes to respiratory diseases, was isolated from goats in Washington (Dunbar et al. 1986).

Table 49.3. Date of death and predator species for marked mountain goats that were apparently killed by predators at Caw Ridge, Alberta, 1988–2000

Predator	Goat Sex	Goat Age (months)	Date Death Discovered (day/month/year)[a]
Grizzly bear[b]	F	4	03/10/1989
	F	5	10/10/1990
	M	5	23/10/1990 (?)
	M	12	11/06/1991
	F	16	29/09/1991
	F	4	23/09/1992
	M	39	13/09/1996
Wolf[b]	M	4	16/09/1989
	M	6	28/11/1989
	M	3	09/09/1990
	F	8	13/02/1992 (?)
	F	15	30/08/1995
	M	5	19/09/1996 (?)
	F	143	09/05/1999
Cougar	F	7	25/01/1991
	F	14	27/07/1991
	F	3	14/09/1991

Source: Updated from Festa-Bianchet et al. (1994).

[a]Question marks indicate cases when >5 days elapsed between death and recovery of the remains; therefore the cause of death is not certain.

[b]In addition, we observed eight unsuccessful attacks by grizzly bears and three by wolves.

POPULATION DYNAMICS

Productivity. Most females in most populations produce a single kid, but varied frequencies of twinning have been reported. Twins constituted 18% of the kids observed in 1 year in the Stikine River population in British Columbia (Foster and Rahs 1985) and 25–33% over 2 years in the introduced Snake River population in Idaho (Hayden 1984). Triplets were reported during a phase of rapid population growth in the Crazy Mountains of Montana (Lentfer 1955). At Caw Ridge, only 2 cases of twinning were observed out of 300 parturitions. Bailey (1991) reported that only 9% of lactating females had twins in an introduced population in Colorado. In the Olympic Mountains, twinning rate among lactating females increased from 2% to 12% after population density was experimentally lowered (Houston and Stevens 1988). Therefore, litter size may be related to resource availability. The fragmentary information available suggests that twinning is more common in introduced and rapidly growing populations than in either native or established and stable populations.

Kid production increases with age during the first few years following primiparity, then remains stable until about 10 years of age, and finally declines in very old females. At Caw Ridge, kid production increased from 4% for females 3 years old to 50% at 4 years, 74% at 5 years, and 84% at 6 years, remained stable at about 84% until 10 years, then declined to 73% for females older than 10 years (Côté and Festa-Bianchet 2001c; M. Festa-Bianchet and S. D. Côté, unpublished data). Similar age-related patterns in productivity have been reported for other goat populations, although in regions with milder climates the plateau phase in kid production begins at an earlier age than at Caw Ridge. Bailey (1991) grouped adult females into three age classes: 3-year-olds, 4- to 9-year-olds, and 10 years of age and older. He reported respective kid production rates for each age class of 52%, 66%, and 53% for mountain goats introduced to Colorado and 77%, 69%, and 57% for those introduced in the Olympic Mountains. Therefore, age-specific productivity in mountain goats fits the typical, inverse-U shape reported for most ungulates (Gaillard et al. 2000). Kid production increases with social rank of the female, but the positive effect of rank on productivity is much stronger for females 3–5 years of age than for older females (Côté and Festa-Bianchet 2001c). Houston et al. (1989) reported that only females weighing at least 50 kg were lactating in the Olympic Mountains, and most nonlactating females were lighter than this threshold mass.

Partly because of differences in female age structure, kid production can vary considerably from year to year, and variability is presumably greater in populations with a variable frequency of twinning. At Caw Ridge, the proportion of females 3 years and older seen with a kid ranged from 45% to 85% and averaged 63% from 1991 to 2000. If 2-year-old females were included, the proportion with kids ranged from 39% to 71% and averaged 54.5% (Côté et al. 2001).

Other studies reported kid:female ratios in summer of 71:100 (Glacier Park, Montana, Rideout 1974), 77–84:100 (Sapphire Mountains, Montana; Rideout 1974), 55–79:100 (Red Butte Range, Montana; Brandborg 1955), 42–43:100 (Glacier National Park; Singer and Doherty 1985), and 92–100:100 (Flathead National Forest, Montana; Singer and Doherty 1985). Hayden (1984) reported kid:female ratios of 90–114:100 over 2 years in a rapidly growing population in Idaho. Because the sex of mountain goats is not easily distinguishable from a distance, some studies pooled the sexes to calculate age ratios. Swenson (1985) reported ratios of kid:100 adult goats varied from 21 to 60 over 9 years in the introduced Absaroka population in Montana. Unfortunately, age ratios are of extremely limited utility in predicting population changes (Gaillard et al. 2000).

Other than litter size, there is limited evidence of density dependence in mountain goat recruitment, possibly because no study has monitored the same population over a sufficiently wide range of densities. The experimental results obtained by Houston and Stevens (1988) suggest that density dependence is likely a threshold rather than a linear response. However, density dependence was not supported by their statistical analysis, possibly because of the limited number of years of study and imprecision of population counts. Swenson (1985) reported density dependence in population growth over 10 years in an introduced population in southern Montana, but his analysis did not account for harvest rates changing from 5.7% to 23.1%. Bailey (1991) found that kid:older goat ratios decreased with increasing population size in the introduced Sheep Mountain–Gladstone Ridge population in Colorado, but age ratios continued to decline after the population had stabilized. He suggested that there may be long-term effects of mountain goats on their forage base, but also commented that changes in population age structure could affect age ratios. Bailey's (1991) suggestion that mountain goats may have a negative long-term impact on the forage supply of their range is important because it may provide clues to the differences in reproductive performance between colonizing and established populations. At Caw Ridge, where the population in June ranged from 81 to 138 mountain goats, we found no evidence of density dependence in kid production or survival.

Survival. Survival from birth to 1 year of age is generally lower and much more variable than adult survival, as is typical of ungulates (Gaillard et al. 2000). By comparing kid:adult ratios to yearling:adult ratios the following year, Adams and Bailey (1982) estimated that an average of 57% of kids (range = 46–78%) annually survived during a period of 4 years. Smith (1976) estimated that 69% of kids survived to 1 year of age in the Bitterroot Mountains, Montana. At Caw Ridge, kid survival (calculated by comparing the number of kids born to the number of yearlings alive in June the following year) averaged 60%, but varied from 38% to 92% over 11 years (Côté and Festa-Bianchet 2001a, and unpublished data). Survival to 1 year of kids of known sex was 70% for males ($n = 99$) and females ($n = 93$). Kid survival varies according to winter weather. In particular, greater snow depth and longer duration of snow cover have a negative effect on kid survival (Brandborg 1955; Rideout 1974; Smith 1976; Chadwick 1983). The late summer ratio of kids to older goats was negatively affected by snow depth in May in two introduced populations in Colorado (Adams and Bailey 1982; Hopkins et al. 1992), suggesting a negative effect of deep snow in early spring on juvenile survival. Orphaned kids appear to have lower survival than nonorphaned kids, but some orphans survive at least to yearling age (Foster and Rahs 1982; Côté et al. 1998b; Côté and Festa-Bianchet 2001a).

Survival of yearling and adult mountain goats has only been estimated at Caw Ridge and in coastal Alaska (C. A. Smith 1986). Results from both studies (Table 49.4) suggest an age-specific survival pattern similar to that of other ungulates (Gaillard et al. 2000): Survival of yearlings is higher than kid survival but lower than adult survival, male survival is lower than female survival, and survival decreases in older goats of both sexes. Annual survival of adult females (≥2 years of age) on Caw Ridge varied from 89% to 97% (based on an average of 37 females/year), whereas it ranged from 50% to 94% for adult males, but the sample only averaged 16 males/year.

As with bighorn sheep (Ross et al. 1997), the effects of predation on a mountain goat herd may vary substantially according to the presence of individual predators that specialize on this species. Most mountain goat populations are too small to serve as prey base for a population of predators, and a single cougar, bear, or wolf pack that specialized on preying on mountain goats could have a very strong impact on a local herd. Consequently, the effects of predation on mountain goat population dynamics may be density independent.

TABLE 49.4. Annual survival rates of yearling and adult mountain goats in two populations

Population	Sex	Age: Yearling	2–8 Years[a]	9 Years or Older[a]
Alaska	both	0.71 (7)	0.95 (152)	0.68 (25)
Caw Ridge	females	0.82 (57)	0.94 (303)	0.84 (85)
Caw Ridge	males	0.73 (48)	0.78 (161)	0.71 (17)

SOURCE: Data for coastal Alaska are from C. A. Smith (1986), who did not report sex-specific survival rates.

NOTE: Sample sizes (number of goat-years) in parentheses.

[a]To compare with the C. A. Smith (1986) data, we report survival rates for the same age classes for the Caw Ridge, Alberta, population. If calculated for goats ≥10 years of age, survival decreases to 0.76 ($n = 59$) for females and 0.64 ($n = 11$) for males.

Causes of death of mountain goats are very rarely known, except for those with radio-collars, and it is often impossible to distinguish between mortality and dispersal. We documented emigration of four males (2–3 years old) and three females (1–4 years old) from Caw Ridge over 11 years, but all four known immigrants were males, 2 to >5 years old (Côté and Festa-Bianchet 2001d, and unpublished data). Therefore, it is likely that a few of the young goats that disappeared from Caw Ridge, especially the males, may have emigrated, and some may have recruited into other populations. Dispersal of 2-year-old goats also was reported by Stevens (1983).

Sex Ratio. The sex ratio of kids is 1:1, but with age it becomes progressively skewed in favor of females because of the higher mortality rate of males. In the unhunted population of Caw Ridge, 50.2% of 245 kids of known sex were males (Côté and Festa-Bianchet 2001d, and unpublished data), but in most years there were $\leq 1/3$ as many males as females 3 years and older (Fig. 49.9). At Caw Ridge, kid sex ratio is affected by maternal age. Females produce an increasing proportion of sons as they become older, from about 70% daughters for mothers 3–6 years of age to about 80% sons for mothers 11 years of age and older (Côté and Festa-Bianchet 2001d). Because these results have major implications for management of mountain goats, it would be useful to know whether this pattern is repeated in other populations.

Heavily female-biased adult sex ratios have been reported for other populations. Foster and Rahs (1985) observed 30 adult males:100 females in the Stikine River population in British Columbia, Rideout (1974) reported 32 adult males:100 females in the Sapphire Mountains of Montana, and Chadwick (1983) counted 37 adult males:100 females in the Swan Range, Montana. Hayden (1984) reported 32 and 51 adult males:100 females over 2 consecutive years in an introduced population in Idaho. Before the removal experiment, Houston and Stevens (1988) reported 29–51 males:100 females (2 years and older) over 5 years in Olympic National Park.

Longevity. The oldest known individual is often reported as a measure of longevity, but that is a misleading statistic of little practical use because in most cases only a very small proportion of adults will survive to that age. The oldest reported male and female mountain goats were 15.5 (B. L. Smith 1986) and 18 years old (Cowan and McCrory 1970), respectively. Very few goats, however, survive >12 years. At Caw Ridge, age-specific survival data suggest that only 25% of yearling females and 6% of yearling males would survive to age 13 years (assuming that all goats that disappeared died). Because mountain goat females can only produce one (rarely two) young a year, longevity likely has an important effect on their lifetime reproductive success.

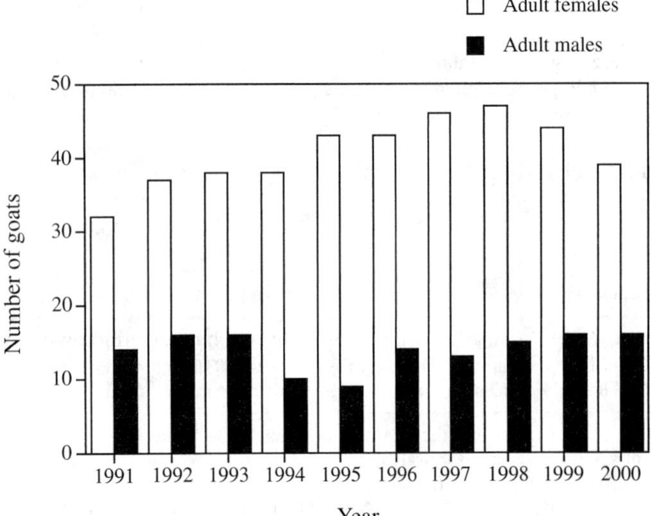

FIGURE 49.9. Adult ($\geq$3 years old) sex ratio in the Caw Ridge mountain goat population, Alberta, in mid-September, 1991–2000.

MANAGEMENT

Capture and Marking. Several capture and marking techniques have been used for mountain goats, including drop nets, foot snares, traps, net guns, and darting with immobilizing drugs either from a helicopter or from the ground. This species, however, is much more sensitive to handling than most other ungulates. Therefore, we recommend that great care should be used in capturing mountain goats, and that the number of captures should be minimized.

A drop net was used to capture goats during 1981–1984 in Washington (Houston et al. 1989) and in 1986–1988 at Caw Ridge (Côté et al. 1998b), without injuring any goats. Drop nets are difficult to operate in strong winds, usually result in only one capture /day, and require several people in place, because as soon as the net is dropped, there should be at least 1 person restraining and blindfolding each goat. During the snow-free season, remotely controlled wooden box traps and self-tripping nylon mesh Clover traps baited with salt have been used successfully (Clover 1956; Côté et al. 1998b). Because goats normally visit salt licks in the early morning, the traps can be left open overnight, but should be checked every 45 min starting at first light (Côté and Festa-Bianchet 2001c). Goats in box traps cannot see outside and remain very calm, but those in Clover traps can become frightened, especially after the rest of the group leaves the area. A total of 465 goats has been captured and handled using these traps at Caw Ridge (Côté et al. 1998b). Goats older than about 14 months should be chemically immobilized before handling. Kids can be handled without drugs, and so can most yearlings until sometime in August when their aggressive behavior and larger body and horn size make them dangerous to handlers. We recommend a long-pole syringe to immobilize trapped adult goats via intramuscular injection of xylazine hydrochloride (5 mg/kg), the effect of which can be reversed by intramuscular injection of 0.7–1.2 mg of idazoxan (Haviernick et al. 1998). A mixture of ketamine hydrochloride and xylazine hydrochloride was used to immobilize 11 goats captured in Clover traps in Glacier National Park, Montana (Singer and Doherty 1985). Immobilized goats should be blindfolded and hobbled, and manipulated for a maximum of 40 min (Côté et al. 1998b; Haviernick et al. 1998; Côté and Festa-Bianchet 2001c). Xylazine is an effective and safe drug for immobilization of mountain goats captured in traps (Haviernick et al. 1998). Goats take 3–9 min to recover after injection of the antagonist idazoxan (Haviernick et al. 1998). Chemical immobilization, however, decreases the probability of kid production the following year by 3- and 4-year-old females, and increases the risk of kid abandonment by females of all ages (Côté et al. 1998b). Not surprisingly, abandonment decreases kid survival (Côté and Beaudoin 1997; Côté et al. 1998b). Although goats in box traps are more calm than those caught in Clover traps, trap type does not affect the efficiency of the drug or the risk of injury (Haviernick et al. 1998). We recommend use of traps to capture mountain goats, but strongly suggest not drugging or manipulating young (3- and 4-year-olds) or lactating females (Côté et al. 1998b).

Free-ranging goats have been caught by darting, but this technique is risky because during the 5–15 min induction, a goat could reach precipitous terrain where handling could be very difficult and, more important, where the drugged goat could fall. Net guns fired from helicopters have also been used to capture mountain goats, notably for transplants (Jorgenson and Quinlan 1996). That technique, however, is costly and involves considerable risk for the animals and the aircraft crew because of the precipitous and rugged habitat of mountain goats. In addition, mountain goats are very sensitive to helicopter disturbance compared to other alpine ungulates (see Censusing) (Foster and Rahs 1983; Côté 1996). We strongly suspect that kid abandonment is likely when lactating females are caught by darting or net-gunning.

Goats can be marked with radio-collars (about 0.8% of body weight) (Côté et al. 1998b), canvas collars, and plastic ear tags. Because uncollared kids may be more likely to survive to 1 year of age than kids with radio-collars (Côté et al. 1998b), radio-collaring kids should be avoided or conducted with caution.

Censusing. Because most mountain goats are in rugged mountainous terrain, most management and research agencies use helicopters or fixed-wing aircraft to survey populations, usually in summer when goats are easy to see (Johnson 1977; Thompson and Baker 1981). Aerial counts of mountain goats, however, have limited precision (Ballard 1977; Hebert and Langin 1982; Houston and Stevens 1988; Williams 1999; Gonzalez-Voyer et al. 2001). In Alberta, the percentage of kids observed during helicopter surveys in July, for example, varied from 41% to 121% of the number in the population, and the total number of goats seen varied from 55% to 83% (average = 70%) of the number present (Gonzalez-Voyer et al. 2001). Cichowski et al. (1994) observed only 19 of 28 marked goats (68%) during aerial surveys in British Columbia. Because they are solitary or form small groups and use forested areas, adult males likely are less observable than other age–sex classes (Risenhoover and Bailey 1982). Adults can be recognized during aerial surveys, but yearlings and kids are difficult to classify; even experienced observers can confuse these two age classes (Gonzalez-Voyer et al. 2001). Summer age ratios, in addition, are very poor predictors of recruitment or of population changes (McCullough 1994), partly because overwinter juvenile survival is highly variable (Gaillard et al. 2000). Nevertheless, Gonzalez-Voyer et al. (2001) concluded that helicopter surveys could detect long-term trends in total population size and number of adults for mountain goats.

Mountain goats appear to be more sensitive to helicopter disturbance than other open-terrain ungulates such as bighorn sheep or caribou (*Rangifer tarandus*) (Miller and Gunn 1979; Foster and Rahs 1983; Stockwell et al. 1991; Côté 1996). Goats showed overt responses to 58% of helicopter flights within 2 km (Côté 1996). When helicopters flew within 500 m, 85% of flights caused the goats to move >100 m or to be alert for >10 min (Côté 1996). Based on these observations, Côté (1996) recommended avoiding helicopter flights within 2 km of mountain goat habitat. Clearly, wildlife managers must take the strong sensitivity of mountain goats to helicopters into account when planning the frequency of aerial censuses.

Ground counts are more precise than aerial surveys, but are time-consuming and difficult to perform in remote areas (Ballard 1977; B. L. Smith 1988). From the ground, mountain goats can be classified as kids, yearlings, 2-year-old females, 2-year-old males, adult females, and adult males (B. L. Smith 1988). Only very experienced observers can reliably distinguish yearling males and females. Where possible, we strongly recommend ground counts over aerial surveys.

Hunting and Harvest. Native Americans hunted mountain goats for their pelts and meat, and used the hide as breast armor and the sharp horn sheaths as weapons (Ballard 1977; Wigal and Coggins 1982). Because of the remoteness and harshness of goat habitat, however, it is unlikely that Native people had a strong impact on most mountain goat populations. Nowadays, mountain goats are primarily hunted in most of their range for their trophy horns. Mountain goat meat is reputed to have an unpleasant and strong taste (Rideout and Hoffmann 1975).

Mountain goats are normally hunted in a limited-entry season, but open seasons exist in some management units in British Columbia. Most hunting seasons are from September to November, but may start as early as August and extend until February in parts of British Columbia. Most sport hunting of mountain goats occurs in British Columbia, with an average of 902 goats harvested each year (range = 706–1029 in 1995–1999), or 1.4–2.5% of the estimated population (Table 49.1) (I. Hatter, Wildlife Branch, British Columbia Environment, pers. commun., 2001). Recent goat harvests in British Columbia are slightly lower than in 1972–1976, when an average of 1079 were harvested annually (Johnson 1977). Elsewhere, hunting intensity can be much higher. In Montana, 211 goats were harvested each year from 1995 to 1999, corresponding to 7 – 9% of the total estimated population (Table 49.1) (J. McCarthy, Montana Fish, Wildlife, and Parks, pers. commun., 2001). Hunter success during 1995–1999 in Montana averaged 77%.

Native populations of mountain goats are very sensitive to overharvest. During the twentieth century, some populations were severely reduced or extirpated through sport hunting, often combined with increased motorized access (Hebert and Turnbull 1977; Hoefs et al. 1977; Johnson 1977; Kuck 1977; K. G. Smith 1988). In Alberta, harvest rates based on minimum population estimates were 4.5–9% during 1973–1987. Hunted populations declined in 1980–1987 despite a >50% reduction in harvest (K. G. Smith 1988). The goat hunting season was closed in Alberta in 1988 (K. G. Smith 1988), but population recovery has been very slow. In 2001, following the recovery of some populations and because of the important interest shown by hunters in this species, wildlife managers reopened the mountain goats hunting season to a few restricted areas. A total of 3.921 applications for 3 tags were received across Alberta.

Management of mountain goats for sport hunting is challenging because different populations appear to have radically different reactions to harvests. Historical population declines likely occurred because mountain goats were managed by applying knowledge obtained from other ungulates, such as bighorn sheep, that are not as susceptible to harvest (K. G. Smith 1988). Introduced mountain goat herds, however, are generally much more productive than native herds and can tolerate much higher harvest levels (Swenson 1985; Houston and Stevens 1988; Williams 1999; but see Côté et al. 2001). Many introduced herds show several years of rapid growth, which can sometimes be sustained through much higher levels of harvest than those possible in native herds. Hayden (1984) reported 20% yearly increases over 12 years in Idaho. In the Absaroka Mountains, Montana, 23 goats introduced in 1956–1958 resulted in a population of at least 86 individuals by 1969, despite 6–9% annual harvests beginning in 1964 (Swenson 1985). Consequently, harvest plans for mountain goats must consider the history of individual populations. Harvest rates <10% were unsustainable in Alberta (K. G. Smith 1988). Recruitment and productivity declined as harvest rates increased in native mountain goat populations in Idaho and British Columbia (Kuck 1977, Hebert 1978).

Native goat populations may be unable to sustain a yearly harvest greater than 2–3%, possibly because kid production is low and age at first reproduction is late (Côté and Festa-Bianchet 2001c). Adams and Bailey (1982) concluded that a yearly 7% harvest was sustainable in an introduced population in Colorado, but their simulation suggested that harvests of 7.5% or more would lead to a population decline. Hebert and Turnbull (1977) suggested that harvest of coastal herds in British Columbia should not exceed 4% of the total population. It is likely that sustainable harvest rates are substantially greater in introduced mountain goat populations with good range conditions and without predators, but the harvest rate of 15–20% recommended by Williams (1999) is unlikely to be sustainable (see also Wigal and Coggins 1982; Swenson 1985; Côté et al. 2001). Although Swenson (1985) and Houston and Stevens (1988) suggested that introduced mountain goat population show evidence of density dependence in reproduction, most studies report no effect of population size on reproductive success or evidence of compensatory reproduction in native populations (Hebert and Turnbull 1977; Kuck 1977; K. G. Smith 1988; Côté and Festa-Bianchet 2001a). Hunting appears to lead to additive mortality in native populations (Hebert and Turnbull 1977; Kuck 1977; C. A. Smith 1986).

Because mountain goats are sensitive to overharvest, we caution wildlife managers to set conservative harvest goals (Côté et al. 2001). We suggest a strategy of tracking harvest (C. A. Smith 1986): Population size should be monitored almost annually and hunting effort adjusted accordingly. Hunting programs that encourage harvest of males may have a lower impact on population dynamics that those that allow either-sex harvest. A male-only hunting program, however, would be difficult because most hunters cannot reliably identify the sex of mountain goats and because adult males are scarce in mountain goat populations. The best management strategy for native populations of mountain goats would combine a 2–3% yearly harvest with a strong encouragement to harvest adult males. Education should be provided to teach hunters how to distinguish males and females. Managers must be prepared to close hunting seasons or drastically reduce the number of permits issued following years when an excessive number of females is harvested.

Hunting programs and quotas must be set on a herd-specific base. The dynamics of neighboring herds can be very different (K. G. Smith 1988) and managers can only control hunter distribution by setting very precise geographic boundaries to hunt areas. For example, in 1986–1987, managers in southwestern Alberta assigned seven permits to a "population" of about 150 goats, but hunters shot seven goats from a readily accessible small herd that could be seen from a major road (J. T. Jorgenson, Alberta Natural Resources Service, pers. commun., 2001).

Artificial Introductions and Reintroductions. Mountain goats have been reintroduced to parts of their historical distribution in Alaska, Idaho, Montana, Washington, and southern Alberta. The success rate of mountain goat transplants has been relatively high, although reintroduced mountain goat populations tend to show a low rate of increase compared to other reintroduced ungulates (Komers and Curman 2000).

Mountain goats have also been introduced in many states outside their original range including Colorado, Oregon, Nevada, South Dakota, Utah, and Wyoming since about 1950 (Johnson 1977). Mountain goats have done well in many introductions because of the absence of predators, good range conditions, and relatively mild climates. In Colorado, for example, goats dispersed and colonized extensive mountain ranges considerable distances from release sites (Hopkins et al. 1992). In some areas, such as in Olympic National Park in Washington, introduced populations of mountain goats expanded and had a negative effect on alpine vegetation, raising important concerns and a debate about the legitimate presence of introduced populations, especially in national parks (Pfitsch and Bliss 1985; Lyman 1994, 1995; Houston 1995; Hutchins 1995). Grazing by mountain goats can alter the fragile alpine plant community. Much of the current debate about introduced mountain goats is due to uncertainties about their historical and fossil distribution (Nagorsen and Keddie 2000). Introduced populations outside the historical range are exotics and incompatible with the mission of national parks to preserve natural biodiversity.

RESEARCH, MANAGEMENT, AND CONSERVATION NEEDS

Despite much progress in the 1990s, the mountain goat remains among the least-understood ungulates in North America. Future research should focus on factors determining mountain goat population dynamics and attempt to develop precise and noninvasive census methods. The causes of changes in population size of mountain goats are poorly understood. In particular, we know little about why native populations are highly susceptible to harvest, whereas some introduced populations are not. Management concerns include late age of primiparity, poor recruitment, and high susceptibility to harvest (Kuck 1977; Adams and Bailey 1982; Swenson 1985; Houston and Stevens 1988; K. G. Smith 1988; Bailey 1991; Festa-Bianchet et al. 1994). Within a population, annual differences in kid survival and reproductive success are substantial but much of that variation is unexplained (Côté 1999; Côté and Festa-Bianchet 2001c). Research is needed to understand why recruitment is often low in mountain goat populations that do not appear to show evidence of density dependence and why goats are unable to sustain lower harvest rates than those commonly applied to other species of similar-sized ungulates. Productivity estimates for mountain goat herds inhabiting areas with contrasting predation risk and climate would provide information on factors determining population dynamics. Information on the lifetime reproductive strategies of females would help us understand the effects of maternal characteristics on recruitment and population growth rates. Because of the harsh environmental conditions of their habitat, mountain goat population dynamics may be more susceptible than other ungulates to density-independent factors such as weather (Rideout 1974; Adams and Bailey 1982; Hopkins et al. 1992). Predation may play an important yet unpredictable role because it appears to be density independent. Local information on movements, migration patterns, and dispersal would be useful for managing mountain goats for hunting. We know very little about dispersal patterns, yet emigration of young goats may have an important effect on population dynamics, particularly in leading to the strongly female-biased adult sex ratio.

Population structure may have a profound effect on the success of different management regimes. In particular, the skewed adult sex ratio that seems typical of unhunted populations (Fig. 49.9) can invalidate harvest regulations based on unrealistic sex ratio assumptions. For example, the Caw Ridge population of about 102 goats (August counts) recruited on average only two 4-year-old males/year, suggesting a sustainable adult male harvest of perhaps 1/year, or 1% of the population.

Mountain goats are sensitive not only to harvesting, but also to disturbance (Pendergast and Bindernagel 1977; Foster and Rahs 1983, 1985; Joslin 1986; Pedevillano and Wright 1987; Côté 1996; Côté and Beaudoin 1997). Industrial and recreational activities in mountain goats range are common, yet information on the effects of these activities on goat behavior and ecology is very scarce. The rapidly increasing frequency of helicopter flights for industrial and recreational activities in mountainous areas necessitates more research on the potential effects of these activities. For example, mining exploration relies on heavy and noisy equipment, flown in by helicopter and used in goat range during the lactation period. The effects of these activities on goats are virtually unknown (Côté 1996). Similarly, the effects of industrial activities that increase accessibility and destroy habitat on or near goat ranges, such as mining, logging, and recreational facilities, are almost unknown and could potentially be important (Pendergast and Bindernagel 1977; Foster and Rahs 1985; Joslin 1986). Critical areas such as winter range, parturition areas, and salt licks must be identified and protected from industrial or recreational activities. Impact of intense wildlife-viewing activities in some areas such as national parks should also be monitored. At the same time, more research is needed about whether mountain goats can habituate to some types of human activities. In some national parks (e.g., Glacier in Montana and Jasper in Alberta), goats are very tolerant of road traffic and of tourists in certain specific locations, but in general goats appear less able than bighorn sheep to habituate to human activities. Penner (1988) reported some habituation to persistent and predictable noise stimuli, but not to stimuli that were unpredictable. Goats appear unable to habituate to helicopter flights; Côté's study (1996) was done on goats that had been exposed to helicopters for at least 25 years.

ACKNOWLEDGMENTS

We thank the wildlife managers who provided us with recent population estimates of mountain goats in their jurisdictions. The Caw Ridge mountain goat project has been funded by the Alberta Conservation Association; the Alberta Natural Resources Service; the Alberta Sports, Recreation, Parks, and Wildlife Foundation; the Alberta Wildlife Enhancement Fund; the Fonds pour la Formation de Chercheurs et Aide à la Recherche (Québec); the Natural Sciences and Engineering Research Council of Canada; the Rocky Mountain Goat Foundation, the Mammal Conservation Trust (U.K.); and the Université de Sherbrooke. We thank C. Beaudoin, F. Fournier, Y. Gendreau, A. Gonzalez-Voyer, S. Hamel, M. Haviernick, K. Smith, and M. Urquhart for their contributions to the Caw Ridge study. This chapter was written while SDC was at the Centre for Ecology and Hydrology in Banchory, Scotland.

LITERATURE CITED

Adams, L. G., and J. A. Bailey. 1982. Population dynamics of mountain goats in the Sawatch Range, Colorado. Journal of Wildlife Management 46:1003–9.

Adams, L. G., and J. A. Bailey. 1983. Winter forages of mountain goats in central Colorado. Journal of Wildlife Management 47:1237–43.

Alaska Department of Fish and Game. 2000. Mountain goat survey-inventory report (Federal Aid in Wildlife Restoration Management Report, Grants W-27-1 and W-27-2). Juneau, AK.

Alberta Fisheries and Wildlife Management Division. 2000. Management plan for mountain goats in Alberta. Unpublished discussion draft, November 2000. Alberta Environment, Edmonton, Canada.

Bailey, J. A. 1991. Reproductive success in female mountain goats. Canadian Journal of Zoology 69:2956–2961.

Ballard, W. 1977. Status and management of the mountain goat in Alaska. Pages 15–23 *in* W. Samuel, and W. G. Macgregor, eds. Proceedings of the first international mountain goat symposium. Kalispell, MT.

Boddiker, M. L., E. J. Hugghins, and A. H. Richardson. 1971. Parasites and pesticide residues of mountain goats in South Dakota. Journal of Wildlife Management 35:94–103.

Brandborg, S. M. 1955. Life history and management of the mountain goat in Idaho. Idaho Wildlife Bulletin 2:1–142.

Byers, J. A. 1997. American pronghorn: Social adaptations and the ghosts of predators past. University of Chicago Press, Chicago.

Carl, G. R., and C. T. Robbins. 1988. The energetic cost of predator avoidance in neonatal ungulates: Hiding versus following. Canadian Journal of Zoology 66:239–46.

Chadwick, D. H. 1977. The influence of mountain goat social relationships on population size and distribution. Pages 74–91 *in* W. Samuel, and W. G. Macgregor, eds. Proceedings of the first international mountain goat symposium. Kalispell, MT.

Chadwick, D. H. 1983. A beast the color of winter. Sierra Club Books, San Francisco.

Cichowski, D. B., D. Haas, and G. Schultze. 1994. A method used for estimating mountain goat numbers in the Babine Mountains Recreation Area, British Columbia. Biennial Symposium of the Northern Wild Sheep and Goat Council 9:56–64.

Clover, M. R. 1956. Single-gate deer trap. California Fish and Game 42:199–201.

Cooley, T. M. 1977. Lungworms in mountain goats. M.Sc. Thesis, Colorado State University, Fort Collins.

Côté, S. D. 1996. Mountain goat responses to helicopter disturbance. Wildlife Society Bulletin 24:681–85.

Côté, S. D. 1999. Dominance sociale et traits d'histoire de vie chez les femelles de la chèvre de montagne (in English). Ph.D. Dissertation, University of Sherbrooke, Sherbrooke, Quebec, Canada.

Côté, S. D. 2000a. Dominance hierarchies in female mountain goats: Stability, aggressiveness and determinants of rank. Behaviour 137:1541–1566.

Côté, S. D. 2000b. Determining social rank in ungulates: A comparison of aggressive interactions recorded at a bait site and under natural conditions. Ethology 106:945–955.

Côté, S. D., and C. Beaudoin. 1997. Grizzly bear (*Ursus arctos*) attacks and nanny–kid separation on mountain goats (*Oreamnos americanus*). Mammalia 61:614–17.

Côté, S. D., and M. Festa-Bianchet. 2001a. Birthdate, mass and survival in mountain goat kids: Effects of maternal characteristics and forage quality. Oecologia 127:230–38.

Côté, S. D., and M. Festa-Bianchet. 2001b. Life-history correlates of horn asymmetry in mountain goats. Journal of Mammalogy 82:389–400.

Côté, S. D., and M. Festa-Bianchet. 2001c. Reproductive success in female mountain goats: The influence of maternal age and social rank. Animal Behaviour 62:173–81.

Côté, S. D., and M. Festa-Bianchet. 2001d. Offspring sex ratio in relation to maternal age and social rank in mountain goats (*Oreamnos americanus*). Behavioral Ecology and Sociobiology 49:260–65.

Côté, S. D., A. Peracino, and G. Simard. 1997. Wolf, *Canis lupus*, predation and maternal defensive behavior in mountain goats, *Oreamnos americanus*. Canadian Field-Naturalist 111:389–92.

Côté, S. D., M. Festa-Bianchet, and K. G. Smith. 1998a. Horn growth in mountain goats (*Oreamnos americanus*). Journal of Mammalogy 79:406–14.

Côté, S. D., M. Festa-Bianchet, and F. Fournier. 1998b. Life-history effects of chemical immobilization and radiocollars on mountain goats. Journal of Wildlife Management 62:745–52.

Côté, S. D., M. Festa-Bianchet, and K. G. Smith. 2001. Compensatory reproduction in harvested mountain goat populations: A word of caution. Wildlife Society Bulletin 29:726–30.

Cowan, I. M., and W. McCrory. 1970. Variation in the mountain goat, *Oreamnos americanus* (Blainville). Journal of Mammalogy 51:60–73.

Cronin, M. A., R. Stuart, B. J. Pierson, and J. C. Patton. 1996. K-casein gene phylogeny of higher ruminants (Pecora, Artiodactyla). Molecular Phylogenetics and Evolution 6:295–311.

Dailey, T. V., N. T. Hobbs, and T. N. Woodard. 1984. Experimental comparisons of diet selection by mountain goats and mountain sheep in Colorado. Journal of Wildlife Management 48:799–806.

Dane, B. 1977. Mountain goat social behavior: Social structure and "play" behavior as affected by dominance. Pages 92–106 *in* W. Samuel, and W. G. Macgregor, eds. Proceedings of the first international mountain goat symposium. Kalispell, MT.

Dane, B. 2002. Retention of offspring in a wild population of ungulates. Behaviour 139:1–21.

Debock, E. A. 1970. On the behavior of mountain goats (*Oreamnos americanus*) in Kootenay National Park. M.Sc. Thesis, University of Alberta, Edmonton, Canada.

Dunbar, M. R., W. J. Foreyt, and J. F. Evermann. 1986. Serologic evidence of respiratory syncytial virus infection in free-ranging mountain goats (*Oreamnos americanus*). Journal of Wildlife Diseases 22:415–16.

Festa-Bianchet, M., M. Urquhart, and K. G. Smith. 1994. Mountain goat recruitment: Kid production and survival to breeding age. Canadian Journal of Zoology 72:22–27.

Foreyt, W. J. 1989. *Sarcocystis* sp. in mountain goat (*Oreamnos americanus*) in Washington: Prevalence and search for the definitive host. Journal of Wildlife Diseases 25:619–22.

Foster, B. R., and E. Y. Rahs. 1982. Implications of maternal separation on overwinter survival of mountain goat kids. Biennial Symposium of the Northern Wild Sheep and Goat Council 3:351–63.

Foster, B. R., and E. Y. Rahs. 1983. Mountain goat response to hydroelectric exploration in northwestern British Columbia. Environmental Management 7:189–97.

Foster, B. R., and E. Y. Rahs. 1985. A study of canyon-dwelling mountain goats in relation to proposed hydroelectric development in northwestern British Columbia, Canada. Biological Conservation 33:209–28.

Fournier, F., and M. Festa-Bianchet. 1995. Social dominance in adult female mountain goats. Animal Behaviour 49:1449–59.

Fox, J. L. 1977. Summer mountain goat activity and habitat preference in coastal Alaska as a basis for the assessment of survey techniques. Pages 190–199 *in* W. Samuel and W. G. Macgregor, eds. Proceedings of the first international mountain goat symposium. Kalispell, MT.

Fox, J. L. 1983. Constraints on winter habitat selection by the mountain goat (*Oreamnos americanus*) in Alaska. Ph.D. Dissertation, University of Washington, Seattle.

Fox, J. L., and C. A. Smith. 1988. Winter mountain goat diets in southeast Alaska. Journal of Wildlife Management 52:362–365.

Fox, J. L., and G. P. Streveler. 1986. Wolf predation on mountain goats in southeastern Alaska. Journal of Mammalogy 67:192–95.

Gaillard, J.-M., M. Festa-Bianchet, N. G. Yoccoz, A. Loison, and C. Toïgo. 2000. Temporal variation in fitness components and population dynamics of large herbivores. Annual Review of Ecology and Systematics 31:367–93.

Gangestad, S. W., and R. Thornhill. 1999. Individual differences in developmental precision and fluctuating asymmetry: A model and its implications. Journal of Evolutionary Biology 12:402–16.

Gatesy, J., G. Amato, E. Vrba, G. Schaller, and R. Desalle. 1997. A cladistic analysis of mitochondrial ribosomal DNA from the Bovidae. Molecular Phylogenetics and Evolution 7:303–19.

Geist, V. 1964. On the rutting behavior of the mountain goat. Journal of Mammalogy 45:551–68.

Geist, V. 1967. On fighting injuries and dermal shields of mountain goats. Journal of Wildlife Management 31:192–94.

Geist, V. 1971. Mountain sheep. University of Chicago Press, Chicago.

Geist, V. 1974. On the relationship of social evolution and ecology in ungulates. American Zoologist 14:205–220.

Geist, V. 1981. On the reproductive strategies in ungulates and some problems of adaptation. Pages 111–132 *in* G. G. E. Scudder and J. L Reveal, eds. Evolution today. Proceedings of the 2nd international congress of systematic and evolutionary biology. Hunts Institute for Botanical Documentation, Carnegie-Mellon University, Pittsburg, PA.

Gonzalez-Voyer, A., M. Festa-Bianchet, and K. G. Smith. 2001. Efficiency of aerial surveys of mountain goats. Wildlife Society Bulletin 29:140–44.

Gross, J. E., M. C. Kneeland, D. F. Reed, and R. M. Reich. 2002. GIS-based habitat models for mountain goats. J. Mamm. 83:218–228.

Groves, P., and G. Shields. 1996. Phylogenetics of the Caprinae based on cytochrome *b* sequence. Molecular Phylogenetics and Evolution 5:467–476.

Guiguet, C. J. 1951. An account of wolverine attacking mountain goat. Canadian Field-Naturalist 65:187.

Harmon, W. H. 1944. Notes on mountain goats in the Black Hills. Journal of Mammalogy 25:149–51.

Hassanin, A., E. Pasquet, and J.-D. Vigne. 1998. Molecular systematics of the subfamily Caprinae (Artiodactyla, Bovidae) as determined from cytochrome *b* sequences. Journal of Mammalian Evolution 5:217–36.

Haviernick, M. 1996. La stratégie alimentaire de la chèvre de montagne (*Oreamnos americanus*): Etude de l'utilisation de l'habitat et du comportement anti-prédateur. M.Sc. Thesis, University of Sherbrooke, Sherbrooke, Quebec, Canada.

Haviernick, M., S. D. Côté, and M. Festa-Bianchet. 1998. Immobilization of mountain goats with xylazine and reversal with idazoxan. Journal of Wildlife Diseases 34:342–47.

Hayden, J. A. 1984. Introduced mountain goats in the Snake River Range, Idaho: Characteristics of vigorous population growth. Biennial Symposium of the Northern Wild Sheep and Goat Council 4:94–119.

Haynes, L. A. 1992. Mountain goat habitat of Wyoming's Beartooth Plateau: Implications for management. Biennial Symposium of the Northern Wild Sheep and Goat Council 8:325–39.

Hebert, D. M. 1978. A systems approach to mountain goat management. Biennial Symposium of the Northern Wild Sheep and Goat Council 2:227–43.

Hebert, D. M., and I. M. Cowan. 1971a. Natural salt licks as a part of the ecology of the mountain goat. Canadian Journal of Zoology 49:605–10.

Hebert, D. M., and I. M. Cowan. 1971b. White muscle disease in the mountain goat. Journal of Wildlife Management 35:752–56.

Hebert, D. M., and H. D. Langin. 1982. Mountain goat inventory and harvest strategies: A reevaluation. Biennial Symposium of the Northern Wild Sheep and Goat Council 3:339–50.

Hebert, D. M., and W. G. Turnbull. 1977. A description of southern interior and coastal mountain goat ecotypes in British Columbia. Pages 126–146 *in* W. Samuel and W. G. Macgregor, eds. Proceedings of the first international mountain goat symposium. Kalispell, MT.

Hebert, D. M., W. M. Samuel, and G. W. Smith. 1977. Contagious ecthyma in mountain goat of coastal British Columbia. Journal of Wildlife Diseases 13:135–36.

Henderson, R. E., and B. W. O'Gara. 1978. Testicular development of the mountain goat. Journal of Wildlife Management 42:921–22.

Hibbs, L. D. 1967. Food habits of the mountain goat in Colorado. Journal of Mammalogy 48:242–48.

Hjeljord, O. 1973. Mountain goat forage and habitat preference in Alaska. Journal of Wildlife Management 37:353–62.

Hoefs, M., G. Lortie, and D. Russell. 1977. Distribution, abundance and management of mountain goats in the Yukon. Pages 47–53 *in* W. Samuel and W. G. Macgregor, eds. Proceedings of the first international mountain goat symposium. Kalispell, MT.

Hofmann, R. R. 1989. Evolutionary steps of ecophysiological adaptation and diversification of ruminants: A comparative view of their digestive system. Oecologia 78:443–57.

Holmes, E. 1988. Foraging behaviors among different age and sex classes of rocky mountain goats. Biennial Symposium of the Northern Wild Sheep and Goat Council 6:13–25.

Holroyd, J. C. 1967. Observations of Rocky Mountain goats on Mount Wardle, Kootenay National Park, British Columbia. Canadian Field-Naturalist 81:1–22.

Hopkins, A., J. P. Fitzgerald, A. Chappell, and G. Byrne. 1992. Population dynamics and behavior of mountain goats using Elliott Ridge, Gore Range, Colorado. Biennial Symposium of the Northern Wild Sheep and Goat Council 8:340–56.

Houston, D. B. 1995. Response to inaccurate data and the Olympic National Park mountain goat controversy. Northwest Science 69:239–40.

Houston, D. B., and V. Stevens. 1988. Resource limitation in mountain goats: A test by experimental cropping. Canadian Journal of Zoology 66:228–38.

Houston, D. B., C. T. Robbins, and V. Stevens. 1989. Growth in wild and captive mountain goats. Journal of Mammalogy 70:412–16.

Hutchins, M. 1984. The mother–offspring relationship in mountain goats (*Oreamnos americanus*). Ph.D. Dissertation, University of Washington, Seattle.

Hutchins, M. 1995. Olympic mountain goat controversy continues. Conservation Biology 9:1324–26.

Irby, M. L., and J. P. Fitzgerald. 1994. Social status and nanny–kid separation in Rocky Mountain goats. Biennial Symposium of the Northern Wild Sheep and Goat Council 9:121–30.

Johnson, R. L. 1977. Distribution, abundance and management status of mountain goats in North America. Pages 1–7 *in* W. Samuel and W. G. Macgregor, eds. Proceedings of the first international mountain goat symposium. Kalispell, MT.

Jorgenson, J. T., and R. Quinlan. 1996. Preliminary results of using transplants to restock historically occupied mountain goat ranges. Biennial Symposium of the Northern Wild Sheep and Goat Council 10:94–108.

Joslin, G. 1986. Mountain goat population changes in relation to energy exploration along Montana's Rocky Mountain front. Biennial Symposium of the Northern Wild Sheep and Goat Council 5:253–69.

Kerr, G. R., and J. C. Holmes. 1966. Parasites of mountain goats in west central Alberta. Journal of Wildlife Management 30:786–90.

Komers, P., and G. P. Curman. 2000. The effect of demographic characteristics on the success of ungulate re-introductions. Biological Conservation 93:187–93.

Kuck, L. 1977. The impact of hunting on Idaho's Pahsimeroi mountain goat herd. Pages 114–25 *in* W. Samuel and W. G. Macgregor, eds. Proceedings of the first international mountain goat symposium. Kalispell, MT.

Laundré, J. W. 1994. Resource overlap between mountain goats and bighorn sheep. Great Basin Naturalist 54:114–21.

Lentfer, J. W. 1955. A two-year study of the Rocky Mountain goat in the Crazy Mountains, Montana. Journal of Wildlife Management 19:417–429.

Lyman, R. L. 1994. The Olympic mountain goat controversy: A different perspective. Conservation Biology 8:898–901.

Lyman, R. L. 1995. Inaccurate data and the Olympic National Park mountain goat controversy. Northwest Science 69:234–238.

Macgregor, W. G. 1977. Status of mountain goats in British Columbia. Pages 24–28 *in* W. Samuel and W. G. Macgregor, eds. Proceedings of the first international mountain goat symposium. Kalispell, MT.

Mahrt, J. L., and D. D. Colwell. 1980. Sarcocystis in wild ungulates in Alberta. Journal of Wildlife Diseases 16:571–76.

Masteller, M. A., and J. A. Bailey. 1988a. Do persisting matrilineal groups partition resources on mountain goat winter ranges? Biennial Symposium of the Northern Wild Sheep and Goat Council 6:26–38.

Masteller, M. A., and J. A. Bailey. 1988b. Agonistic behavior among mountain goats foraging in winter. Canadian Journal of Zoology 66:2585–88.

McCullough, D. R. 1994. What do herd composition counts tell us? Wildlife Society Bulletin 22:295–300.

McFetridge, R. J. 1977. Strategy of resource use by mountain goat nursery groups. Pages 169–73 *in* W. Samuel and W. G. Macgregor, eds. Proceedings of the first international mountain goat symposium. Kalispell, MT.

Mead, J. I., and M. C. Lawler. 1994. Skull, mandible, and metapodials of the extinct Harrington's mountain goat (*Oreamnos harringtoni*). Journal of Vertebrate Paleontology 14:562–76.

Miller, F. L., and A. Gunn. 1979. Responses of Peary caribou and muskoxen to helicopter harassment (Occasional Paper No. 40). Canadian Wildlife Service, Ottawa.

Møller, A. P., and J. P. Swaddle. 1997. Asymmetry, developmental stability and evolution. Oxford University Press, Oxford.

Nagorsen, D. W., and G. Keddie. 2000. Late Pleistocene mountain goats (*Oreamnos americanus*) from Vancouver Island: Biogeographic implications. Journal of Mammalogy 81:666–75.

Pedevillano, C., and R. G. Wright. 1987. The influence of visitors on mountain goat activities in Glacier National Park, Montana. Biological Conservation 39:1–11.

Pendergast, B., and J. Bindernagel 1977. The impact of exploration for coal on mountain goats in northeastern British Columbia. Pages 64–68 *in* W. Samuel and W. G. Macgregor, eds. Proceedings of the first international mountain goat symposium. Kalispell, MT.

Penner, D. F. 1988. Behavioral response and habituation of mountain goats in relation to petroleum exploration at Pinto Creek, Alberta. Biennial Symposium of the Northern Wild Sheep and Goat Council 6:141–58.

Pfitsch, W. A., and L. C. Bliss. 1985. Seasonal forage availability and potential vegetation limitations to a mountain goat population, Olympic National Park. American Midland Naturalist 113:109–21.

Rideout, C. B. 1974. A radio telemetry study of the ecology and behavior of the mountain goat in western Montana. Ph.D Dissertation, University of Kansas, Lawrence.

Rideout, C. B. 1977. Mountain goat home ranges in the Sapphire Mountains of Montana. Pages 201–11 *in* W. Samuel and W. G. Macgregor, eds. Proceedings of the first international mountain goat symposium. Kalispell, MT.

Rideout, C. B. 1978. Mountain goat. Pages 149–59 *in* J. L. D. Schmidt and D. L. Gilbert, eds. Big game of North America: Ecology and management. Stackpole, Harrisburg, PA.

Rideout, C. B., and R. S. Hoffmann. 1975. *Oreamnos americanus*. Mammalian Species 63:1–6.

Risenhoover, K. L., and J. A. Bailey. 1982. Social dynamics of mountain goats in summer: Implications for age ratios. Biennial Symposium of the Northern Wild Sheep and Goat Council 3:364–73.

Risenhoover, K. L., and J. A. Bailey. 1985. Relationships between group size, feeding time, and agonistic behavior of mountain goats. Canadian Journal of Zoology 63:2501–6.

Robbins, C. T. 1993. Wildlife feeding and nutrition. Academic Press, New York.

Romeo, G., and S. Lovari. 1996. Summer activity rhythms of the Rocky Mountain goat *Oreamnos americanus* (de Blainville, 1816). Mammalia 60:496–99.

Romeo, G., S. Lovari, and M. Festa-Bianchet. 1997. Group leaving in mountain goats: Are young males ousted by adult females? Behavioural Processes 40:243–46.

Ross, I., M. Jalkotzy, and M. Festa-Bianchet. 1997. Cougar predation on bighorn sheep in southwestern Alberta during winter. Canadian Journal of Zoology 75:771–75.

Samuel, W. M., G. A. Chalmers, J. G. Stelfox, A. Loewen, and J. J. Thomsen. 1975. Contagious ecthyma in bighorn sheep and mountain goat in western Canada. Journal of Wildlife Diseases 11:26–31.

Samuel, W. M., W. K. Hall, J. G. Stelfox, and W. D. Wishart. 1977. Parasites of mountain goat, *Oreamnos americanus* (Blainville), of west central Alberta with a comparison of the helminths of mountain goat and Rocky Mountain bighorn sheep. Pages 212–25 *in* W. Samuel and W. G. Macgregor, eds. Proceedings of the first international mountain goat symposium. Kalispell, MT.

Saunders, J. K., Jr. 1955. Food habits and range use of the Rocky Mountain goat in the Crazy Mountains, Montana. Journal of Wildlife Management 19:429–37.

Shah, H. L., and N. D. Levine. 1964. *Eimeria oreamni* sp. n. (Protozoa: Eimeriidae) from the Rocky Mountain goat *Oreamnos americanus*. Journal of Parasitology 50:634–35.

Singer, F. J. 1977. Dominance, leadership and group cohesion of mountain goats at a natural lick, Glacier National Park, Montana. Pages 107–13 *in* W. Samuel and W. G. Macgregor, eds. Proceedings of the first international mountain goat symposium. Kalispell, MT.

Singer, F. J., and J. L. Doherty. 1985. Movements and habitat use in an unhunted population of mountain goats, *Oreamnos americanus*. Canadian Field–Naturalist 99:205–17.

Smith, B. L. 1976. Ecology of the Rocky Mountain goat in the Bitterroot Mountains, Montana. M.Sc. Thesis, University of Montana, Missoula.

Smith, B. L. 1977. Influence of snow conditions on winter distribution, habitat use, and group size of mountain goats. Pages 174–89 *in* W. Samuel and W. G. Macgregor, eds. Proceedings of the first international mountain goat symposium. Kalispell, MT.

Smith, B. L. 1986. Longevity of American mountain goats. Biennial Symposium of the Northern Wild Sheep and Goat Council 5:341–46.

Smith, B. L. 1988. Criteria for determining age and sex of American mountain goats in the field. Journal of Mammalogy 69:395–402.

Smith, C. A. 1986. Rates and causes of mortality in mountain goats in southeast Alaska. Journal of Wildlife Management 50:743–46.

Smith, K. G. 1988. Factors affecting the population dynamics of mountain goats in west-central Alberta. Biennial Symposium of the Northern Wild Sheep and Goat Council 6:308–29.

Stevens, V. 1983. The dynamics of dispersal in an introduced mountain goat population. Ph.D. Dissertation, University of Washington, Seattle.

Stevens, V., and D. B. Houston. 1989. Reliability of age determination of mountain goats. Wildlife Society Bulletin 17:72–74.

Stockwell, C. A., G. C. Bateman, and J. Berger. 1991. Conflicts in national parks: A case study of helicopters and bighorn sheep time budgets at the Grand Canyon. Biological Conservation 56:317–28.

Swenson, J. E. 1985. Compensatory reproduction in an introduced mountain goat population in the Absaroka Mountains, Montana. Journal of Wildlife Management 49:837–43.

Thompson, B. C., and B. W. Baker. 1981. Helicopter use by wildlife agencies in North America. Wildlife Society Bulletin 9:319–23.

Todd, K. S., Jr., and B. W. O'Gara. 1968. *Eimeria montanaensis* n. sp. and *E. ernesti* n. sp. (Protozoa, Eimeriidae) from the Rocky Mountain goat *Oreamnos americanus*. Journal of Protozoology 15:808–10.

Von Elsner-Schack, I. 1986. Habitat use by mountain goats, *Oreamnos americanus*, on the eastern slopes region of the Rocky Mountains at Mount Hamell, Alberta. Canadian Field-Naturalist 100:319–24.

Washington Department of Fish and Wildlife. 2000. 2000 game status and trend report. Wildlife Program, Washington Department of Fish and Wildlife, Olympia.

Wigal, R. A., and V. L. Coggins. 1982. Mountain goat. Pages 1008–20 *in* J. A. Chapman and G. A. Feldhamer, eds. Wild mammals of North America: Biology, management, and economics. Johns Hopkins University Press, Baltimore.

Williams, J. S. 1999. Compensatory reproduction and dispersal in an introduced mountain goat population in central Montana. Wildlife Society Bulletin 27:1019–24.

Steve D. Côté, Département de biologie, Université Laval, Sainte-Foy, Quebec, Canada G1K 7P4. Email: steeve.cote@bio.ulaval.ca.

Marco Festa-Bianchet, Département de biologie, Université de Sherbrooke, Sherbrooke, Quebec J1K 2R1, Canada. Email: marco.festa-bianchet@usherbrooke.ca.

50

Muskox

Ovibos moschatus

Anne Gunn
Jan Adamczewski

NOMENCLATURE

Common Name. Muskox
Scientific Name. *Ovibos moschatus*

TAXONOMY

Muskoxen have no close living relatives, and their classification in the subfamily Caprinae reflects affinities among muskoxen, sheep, and goats. Muskoxen were originally grouped with the takin (*Budorcas taxicolor*) in the tribe Ovibovini, based on their relatively similar appearance. However, mitochondrial DNA analyses indicate that the muskox's closest relatives are the gorals (*Naemorhedus* spp.) and that they should both be classified in the Rupicaprini (Groves 1995). Gorals are medium-sized, goatlike ruminants found in rugged wooded mountains of eastern Asia.

Muskox ancestors may have diverged from a sheeplike ungulate found in tropical or subtropical Asian uplands 6–8 million years ago (Harington 1961). Muskoxen may have separated from the takin 9 million years ago and from the gorals 5 million years ago (Groves 1995). As the climate cooled 3–4 million years ago during the Pleistocene, several primitive stem muskox ancestors spread across Asia and Europe and into North America. By the late Pleistocene, those stem muskoxen had given rise to several genera, including *Ovibos,* which was the only one to persist into modern times (Harington 1961).

Between 1 million and 9000 years ago, *Ovibos* muskoxen occurred in Europe, persisting, although never apparently abundant, during the wide climatic swings as glaciers advanced and retreated during the Pleistocene. *Ovibos* reached North America 150,000–250,000 years ago, spreading as part of the Beringian fauna (Lent 1999). Beringia was the swath of grassland steppes stretching as a subcontinent linking Russia and North America (Bliss and Richards 1982). Muskoxen were probably not abundant, as they would have mostly occurred in the smaller patches of wetter habitats with sedges and shrubs (Lent 1999). As the Pleistocene ended about 13,000 years ago, the climate became wetter and sedges and shrubs replaced much of the arid grasslands. Many large mammals disappeared, including the helmeted muskoxen *Boötherium,* but *Ovibos* persisted (Lent 1999).

Despite large environmental changes since the Pleistocene, muskoxen have varied little in appearance or genetics. Tener (1965) concluded that contemporary *Ovibos* is a monotypic species. The differences in appearance and tooth characteristics for muskoxen from the High Arctic and Greenland compared to mainland Canadian muskoxen suggested differentiation, but not enough to warrant subspecies designation. Skull and dental characteristics from contemporary and Pleistocene muskoxen are similar (Harington 1970). Comparisons of DNA sequences from Pleistocene and early Holocene muskox bones also suggest muskoxen changed little, and their variation has not been sufficient to recognize subspeciation (Groves 1995).

A finer resolution of genetic variation through microsatellite DNA analysis revealed that muskoxen from the mainland are distinguishable from muskoxen from the Arctic Islands (van Coeverden de Groot 2000), supporting Tener's (1965) earlier findings. Within the Arctic Islands, muskoxen from Greenland grouped more closely with muskoxen from Ellesmere than with those from the southern Arctic Islands (van Coeverden de Groot 2000). These patterns suggest that the Arctic Islands muskoxen remained isolated in glacial refugia such as Banks Island, which was ice-free for the last 700,000 years of the Pleistocene. Other refugia may have persisted in Greenland and on Ellesmere, Amund Ringness, and Ellef Ringness islands (van Coeverden de Groot 2000). Muskoxen have low genetic variability, although mainland muskoxen are the more variable, with an average heterozygosity $H_e = 0.498$ compared to $H_e = 0.243$ for Arctic Island muskoxen (van Coeverden de Groot 2000). The relatively greater variability is consistent with the likely origin of mainland muskoxen from a larger range south of the continental ice sheet.

DISTRIBUTION

Muskoxen are distributed across the tundra of the circumpolar Arctic. Climate changes following the retreat of the glaciers probably reduced muskox distribution as they disappeared from Russia and Europe thousands of years ago (Lent 1999). Historically and currently, humans have affected muskox distribution through hunting and, more recently, translocations. Hunting had a role in muskox disappearance from Alaska as well as large areas of mainland Canada in the late nineteenth and early twentieth centuries (Barr 1991). In the twentieth century, muskoxen were reintroduced, generally with success, to Alaska, Russia, western Greenland, Norway, and Sweden (Lent 1999). Muskoxen were not apparently native to the Svalbard Islands north of Norway and a small introduction failed (Alendal 1976; Klein and Staaland 1984). In Canada, recolonization of historical ranges has been natural across their large mainland ranges except for the successful introduction to northern Quebec, although muskoxen are not native to that area (Le Hénaff and Crete 1989).

Canada and eastern Greenland have the only endemic muskoxen. In Canada, muskoxen are found on almost all the islands of the Arctic Archipelago (Fig. 50.1), although they periodically disappear from and recolonize the smaller islands (Miller et al. 1977; Gunn and Dragon 2002). Currently, on the large southern Arctic Islands, muskox distribution is essentially island-wide, but in the late 1800s and early 1900s, abundance was low and distribution scattered (Barr 1991). The wide shifts in muskox abundance on the Arctic Islands and in Greenland suggest that environmental conditions there have also varied widely over time.

On the Canadian mainland, muskoxen by the 1990s had recolonized most of their historical ranges extending from the Mackenzie River east almost to the west coast of Hudson Bay (Barr 1991; Fournier and Gunn 1998). However, along the periphery of their ranges, muskoxen are still scarce and their recolonization is slow—perhaps <10 km/year (Fournier and Gunn 1998). From their nadir in the early 1900s, muskoxen have taken nearly a century to reach the northeastern parts of their former range. On Baffin Island, muskoxen are occasional visitors, but have not established populations there or on associated islands in Hudson Bay. Muskoxen dispersing from the

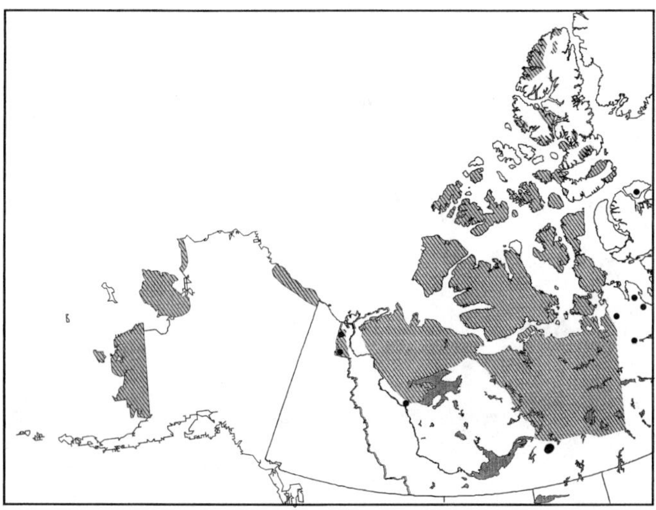

FIGURE 50.1. Distribution of the muskox in North America (*Ovibos moschatus*). Dots show recent range extensions.

reintroduced population in northeastern Alaska have recolonized northwestern Yukon Territory. At least 100 muskoxen have become established along the United States–Canada border and the Babbage River, and animals have been reported as far south as the Old Crow Basin and the Richardson Mountains (P. Reynolds, U.S. Fish and Wildlife Service, pers. commun., 2001).

Muskoxen apparently did not colonize Quebec after the glacial retreat, and so the release of 54 muskoxen by the Quebec government between 1973 and 1984 (Le Hénaff and Crete 1989) was an introduction. The released muskoxen were born at Umingmaqautik in northern Quebec as part of a domestication project started with 13 muskox calves caught on Ellesmere Island in 1967 (Lent 1999). The releases were mostly yearlings or 2-year-olds and were at three sites on the south coast of Baie d'Ungava. By 1986, there were 290 muskoxen (Le Hénaff and Crete 1989). Based on a count of 574 muskoxen in 1994, the number in northern Quebec was likely to total 1000 in 2001 (S. Couturier, Quebec Environment and Wildlife, pers. commun., 2001).

Muskoxen are distributed at four sites along the Alaskan coast (Arctic National Wildlife Refuge in northeastern Alaska, Seward Peninsula, Cape Thompson, and Nelson Island). The muskoxen were reintroduced between 1967 and 1981 (Coady and Hinman 1984) from Nunivak Island, whose original stock came from Greenland in 1935. Muskoxen prehistorically were probably not abundant and by historical times were uncommon and restricted to northern Alaska (Lent 1999). Their distribution contracted until only small herds occurred in mountains. By the late 1800s, the last muskoxen were killed by hunters (Lent 1999). However, the historical distribution is not well known, which adds to uncertainties in predicting the potential for muskox distribution across the state (Danks and Klein 1999).

Elsewhere, introductions and reintroductions have contributed to ecological knowledge of muskoxen. In Russia, muskoxen were reintroduced to the Taimyr Peninsula and Wrangel Island in 1974–1975 with animals from Alaska and Canada (Lent 1999). Muskoxen on Taimyr increased and stabilized at about 1450 in 1998 (Tsarev and Sipko 1999). Muskoxen between 1962 and 1987 were introduced to southwest and western Greenland (Boertmann et al. 1992). Efforts to restore muskoxen to Norway used five different releases of 27 muskox calves from eastern Greenland in the early 1950s. The muskoxen finally established themselves, but deaths were relatively frequent from accidents and poaching. After disturbance by tourists, six muskoxen crossed into Sweden in 1971 (Alendal 1974 in Lent 1999). The 1929 release of muskoxen on an island in the Svalbard Archipelago was initially successful, but the numbers slowly declined from a high of 50 in 1959 and the last muskox had disappeared by the 1970s (Klein and Staaland 1984).

DESCRIPTION

Size. Their stocky build and long, thick hair make muskoxen look more massive than they really are. A slow, ponderous walk and the long, trailing guard hair add to that impression of massiveness. A large adult bull from the Canadian mainland stands about 135 cm at the shoulder and weighs about 400 kg (Tener 1965). Body masses for bulls up to 5 years old from a rapidly expanding population in southwestern Greenland averaged 290 $\pm$32 kg in winter and 321 $\pm$ 11 kg in summer (Olesen et al. 1994). A maximum of just over 400 kg (J. Adamczewski, pers. obs., Saskatoon) seems more common in captive bulls. Maximal masses are not usually reached before 6–8 years of age in captivity, and apparently in the wild as well (Olesen et al. 1994). Five-year-old males held in the University of Saskatchewan farm at Saskatoon reached 347 $\pm$ 6 kg (Chaplin and Stevens 1989).

Cows usually attain 50–70% of the mass of bulls and appear to reach adult mass 1–2 years before bulls. On Victoria Island, cows at least 5 years old were 210–220 kg in September (Adamczewski et al. 1997), which is similar to 220 $\pm$ 31 kg recorded for adult female muskoxen captured in summer in northeastern Alaska (P. Reynolds, U.S. Fish and Wildlife Service, unpublished data, 2001). Full-grown muskox cows in captivity in Alaska varied seasonally from 180 to 230 kg in body mass (White et al. 1989, 1997; Groves 1992). In Saskatoon, 5-year-old cows averaged 260 $\pm$ 7 kg (Chaplin and Stevens 1989). Muskoxen are generally smaller on the Arctic Islands than on the mainland.

Appearance. Muskoxen have a short, broad head held low on a short neck. From the characteristic shoulder hump, the back slopes slightly to the hindquarters. The legs are short and stout with compact rounded hooves, which have hard, sharp outer rims and spongy lateral heels. The ears are short but conspicuous and triangular, and the short tail is buried in the hair. The muzzle is broad and densely covered with short hairs.

Both sexes have horns with a hard keratinous sheath, which grows from the base and is solid from the tip for one third to one half of the length in adults. The horn's core is a bony outgrowth of the skull. The horns of adult bulls meet across the forehead in a massive and heavily ridged "boss" and down past the eyes before curving up (Fig. 50.2). Cows do not have a well-developed boss, and white hair separates the base of the horns, which are more slender and shorter. Horn growth starts when calves are about 1 month old and growth is complete at about 6 years old. The horn tips of cows can be sharp; those of bulls are usually more worn, and older bulls sometimes have horn tips and pieces of the boss lost during fights.

Pelage. Long, blackish brown, coarse guard hairs form a trailing skirt, which almost reaches the ground in Arctic Island muskoxen and halfway down the legs for mainland muskoxen. The guard hairs develop as a beard and mane in adult bulls and contribute to their massive appearance. The short guard hairs on the legs are whitish, and behind the shoulder hump is a white-yellow saddle patch of wooly rather than coarse guard hairs. The face has short, brown, dense hairs; the ears are densely wooly; and the muzzle and udder are also densely covered in short light hair.

Underlying the guard hairs is a remarkable wool layer—qiviut—with fibers as long and fine (13–17.3 μm; Wilkinson 1975) as cashmere and vicuña. Average fiber diameter is less in females and increases with age in both sexes. The qiviut yield is highest from adult females, which, in captivity, annually shed an average of 2.6 kg (White et al. 1991; Rowell et al. 1999). The wool grows from secondary follicles, whereas guard hairs are from the primary skin follicles. The ratio of secondary to primary follicles averages 37:1. Only improved breeds of domestic sheep such as Merino come close to that ratio (Flood et al. 1989b). Similar to domestic sheep, the secondary follicles are tightly packed in the skin (42/mm^2).

The guard hairs grow even during the winter for several years and are shed and replaced continuously. In contrast, the qiviut is annually shed in May and June as wool fibers work their way through the guard hairs appearing as mats and clumps (Flood et al. 1989b). Shedding is

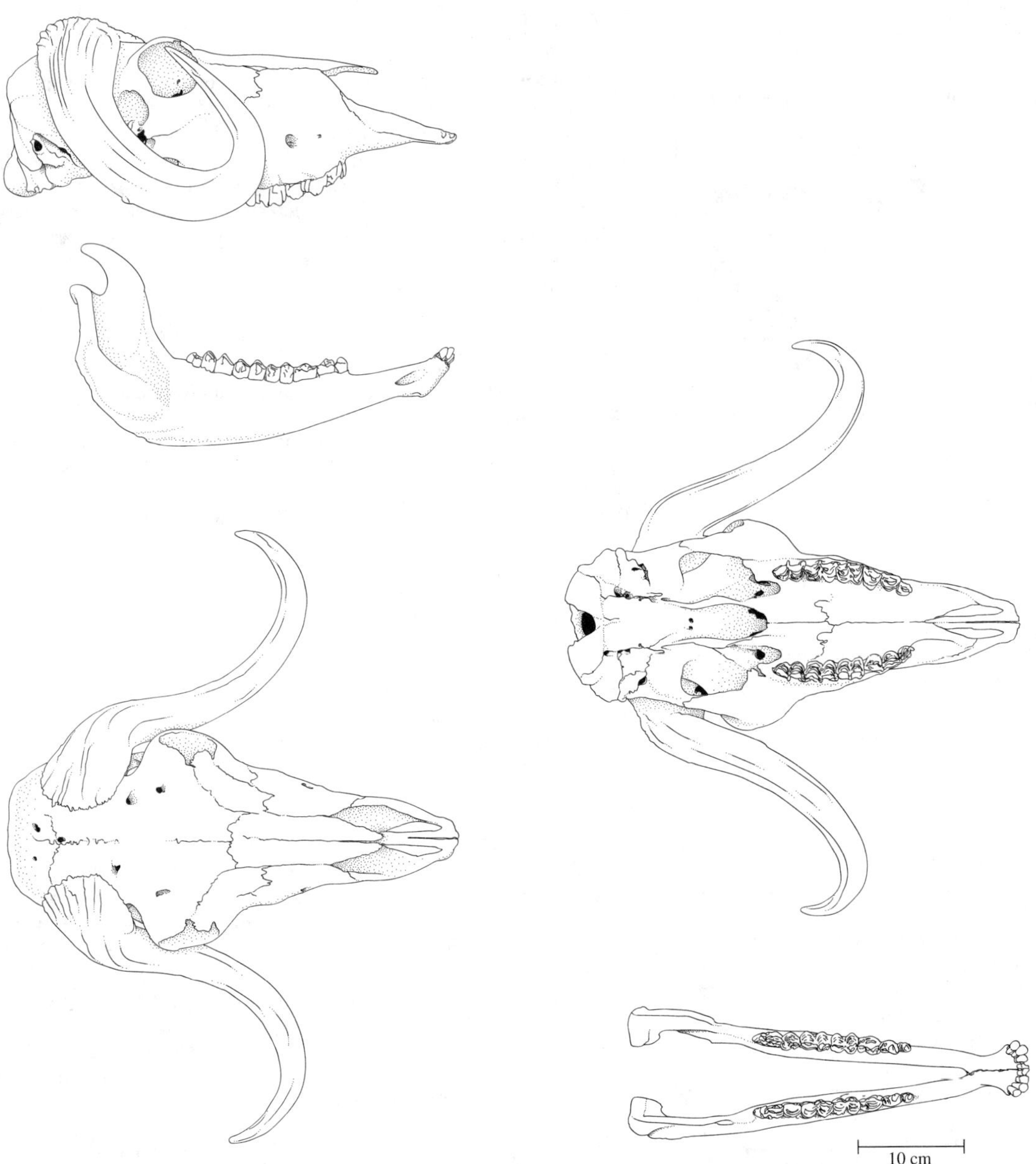

FIGURE 50.2. Skull of the muskox (*Oviobos moaschatus*). Left (from top to bottom): lateral view of cranium, lateral view of mandible, dorsal view of cranium. Right: (top) ventral view of cranium and (bottom) dorsal view of mandible.

delayed in muskoxen in poor nutritional condition, which is why in a herd, muskoxen will be at different stages of losing their qiviut. The hair growth from the secondary follicles starts in April and ceases by November (Flood et al. 1989b). Calves have both pelage layers, but the guard hairs are short and fine. The calf sheds and replaces both layers with a denser wool undercoat and longer guard hairs during the first year. The typical long, coarse guard hairs are not present until the third winter (Tener 1965).

Almost invariably, muskoxen have blackish-brown pelage with white-yellow stockings and saddle and varying amounts of light-colored hair between the horn bosses (Tener 1965). Rare reports include an albino muskox (Tener 1965) and a golden-brown muskox (Gray 1987) in the Canadian High Arctic. Only on the east-central Canadian mainland are there consistent reports of cream-colored muskoxen with darker horns and light-colored legs (Gunn 1985). The sightings were spread 200 km across the Queen Maud Gulf area. Up to 14 blonde bulls and cows were counted among 3751 muskoxen in 1984, including a blonde cow suckling a blonde calf (A. Gunn, unpublished data).

Skin Glands. Despite their name, muskoxen do not have musk glands or sacs. An early description correctly recognized that the musky smell of bulls was from the preputial tube (Hearne 1795 in Lent 1999). Like many ruminants, both muskox sexes have preorbital glands. These glandular sacs are between the lacrimal bone and the skin and open below the eye (Gray et al. 1988). The chemical composition of the glands differs in bulls and cows (Flood et al. 1989a) and their role is not completely clear other than involvement during the rut. Both sexes and all age classes, but especially the bulls, will lower their head to rub the inside of their foreleg against the preorbital gland when agitated by the approach of predators, humans, or unknown muskoxen.

Skull and Dentition. The muskox skull has prominent tubular orbits (Fig. 50.2), which protrude beyond the maxillary plane for up to 5 cm (Harington 1961). Another striking characteristic is the paired horn cores growing from the posterior edge of the frontal bones. As the muskox matures, spongy bone increases around the bases of the bony cores, especially in the bulls. These bony masses support the base of the horns as they flare out across the skull to nearly meet in the middle. This dense bone mass and the exceptionally heavy set cervical vertebrae and the short, heavily muscled neck serve as shock absorbers during head-on clashes. The eyes are large and have the horizontal pupils typical of goats. Muskox peripheral vision is exceptionally wide, as might be expected in an open-country herbivore.

Muskox teeth are typical of those of ruminant grazers, with a relatively broad incisor arcade and high-crowned molariform teeth with crescent-shaped cusps and incisiform canines. The dental formula is I 0/3, C 0/1, P 3/3, M 3/3. The frequency of certain dental anomalies such as missing canines, endostyles on molars, and rotation of maxillary premolars depends on location, suggesting a genetic basis for these anomalies (Tener 1965; Henrichsen 1982).

Less is known about other dental anomalies with environmental or pathological causes and how they interact with genetic traits. Normal tooth wear caused about 65% of the jaw bone or tooth damage found in Canadian and Greenland endemic muskoxen, and 20% of adult muskoxen (>4.5 years) had dental lesions (Henrichsen and Dieterich 1984). Environmental factors may interact with age in causing tooth damage and lesions.

Muskoxen on the mainland have a relatively higher incidence of cracked and broken incisors than those from nearby Victoria Island (A. Gunn, unpublished data), possibly because of browsing on willow twigs of greater diameter. The consequences of tooth damage on forage selection and chewing efficiency are unknown. In moose (*Alces alces*), pathological incisor wear is not yet correlated with either diet or loss of condition (Young and Marty 1986), but Soay sheep (*Ovis aries*), with narrower incisor arcades, survived at a higher rate during a die-off (Illius et al. 1995).

Body Composition and Condition Indices. Muskox body composition is generally similar to that for sheep and cattle and the few differences relate to arctic adaptations (Adamczewski 1995). Those differences include that the pelage accounts for 4.5% of ingesta-free body mass, almost three times the value for a white-tailed deer (*Odocoileus virginianus*) (Adamczewski et al. 1995). Muskoxen can fatten more than any other wild ruminants, except for Svalbard reindeer (*Rangifer tarandus platyrhynchus*). The estimated proportion of fat in the ingesta-free body mass of nonlactating cows in the fall was about 26% (Adamczewski et al. 1997). The energy value of these reserves for late-winter lactation and dietary shortfalls may be proportionately greater than in cervids, including Svalbard reindeer, due to the slow metabolism of muskoxen (Nilssen et al. 1994).

Seasonal amplitude in body mass is large. Muskox cows on Victoria Island had their lowest ingesta-free body mass during early lactation in May (112–116 kg) and the greatest in September (166–176 kg) (Adamczewski et al. 1997). The most rapid rates of fat (242 g/day) and protein (55 g/day) catabolism were during early lactation, and pregnant muskoxen appeared able to retain much of their fat reserves for this energetically demanding period.

Ingesta accounted for 21–28% of body mass in adult cows, with the maximum in November (Adamczewski et al. 1995). Such a large and variable gut fill means that body weight is a weak predictor of fatness. However, estimating relative body condition rarely requires knowing the exact quantities of fat or protein. Practically, mass of the carcass, kidney fat, or an indicator muscle or bone and the depth of subcutaneous back fat are adequate measures of relative condition for particular age and sex classes at different seasons. Combinations of these variables can be used in multiple regressions to estimate body composition (Adamczewski et al. 1995).

Carcass Characteristics. Muskox dressing percentages varied seasonally between 46 ± 0.5% (*SE*) in August and 50 ± 1.2% in November, based on the ratio of carcass mass (body mass less skin, head, hooves, and viscera) to body mass after bleeding out (Gunn et al. 1991). This is similar to dressing percentages of cattle. Muskox carcasses are shoulder heavy and the hind quarters are 36% of carcass weight, which is less than in domestic cattle (Gunn et al. 1991). The intermuscular fat depot was slightly larger than the subcutaneous and abdominal fat depots (Adamczewski et al. 1995), a pattern typical of domestic sheep (Kempster 1980). Like cattle and sheep, some muskox muscles develop considerable intramuscular fat reserves, identifiable as "marbling."

PHYSIOLOGY

Energy Metabolism and Seasonal Adaptations. Muskoxen live in an environment characterized by a short but variable season of plant growth and high diet quality and a long winter of variable availability of relatively low-quality forage. Their adaptations can be summed as both conservatism and flexibility, especially in the cow's allocation of body resources to reproduction (Adamczewski 1995). Suttie and Webster (1998) commented that, of arctic ruminants, muskoxen have the most extreme adaptive strategies and their complex of adaptations (food intake, metabolic and possibly growth cycles) is more than the sum of the individual components. Muskoxen are large-bodied ruminants with a slow metabolism (Nilssen et al. 1994), an associated low requirement for forage, and a capability to digest low-quality forage and accumulate large fat reserves. Maintenance energy requirements in winter are low (White et al. 1994) and their fasting metabolic rate of 0.205 $MJ/kg^{0.75}$ is one of the lowest recorded for ruminants in either summer or winter (Tyler and Blix 1990).

The relatively small size of the liver, heart, and kidneys provide insight into the muskox's slow metabolism. These organs in April (late winter) accounted for 1.7%, 1.2%, and 0.3% of carcass mass, compared to 2.4%, 1.8%, and 0.4% in caribou cows in the eastern Canadian Arctic and 2.9%, 0.7%, and 0.6% in nonlactating, nonpregnant Jersey cows (Adamczewski et al. 1997). These three organs, together with the alimentary tract, can account for 50–60% of whole-body energy metabolism in ruminants (Burrin et al. 1990). The disadvantage of this energy-frugal metabolism may be a weak ability to take full advantage of more favorable conditions for growth. Slow growth and slow regaining of energy reserves are typical of muskoxen (Parker et al. 1990). The relatively small heart may also impose limits on oxygen-demanding activities such as running; muskoxen are clearly not designed for marathons.

Other known or suspected metabolic adaptations include the role of some serum proteins in reducing heat loss (Halikas 1971). Muskoxen legs are relatively uninsulated and are considerably cooler than core body temperatures during cold weather. The serum proteins may reduce blood flow at low temperatures, which would conserve heat. Other adaptations include the muskoxen's apparent tolerance of high serum levels of urea. This suggests a high capacity for recycling urea (Tedesco et al. 1991), which was confirmed in captive muskoxen (P. F. Flood, University of Saskatchewan, unpublished data). Water turnover of muskoxen is among the slowest known in ruminants, somewhat lower than in goats and sheep and much lower than in cervids or cattle (Macfarlane et al. 1971). This trait, combined with a low metabolic rate and high efficiency of urea recycling, confirms that the muskox is adapted to areas with limited availability of free water.

Digestion. Low maintenance energy requirements contribute to a low forage intake in winter (White et al. 1994; Adamczewski et al. 1994a). In comparative feeding trials where mature nonbreeding muskoxen and cattle were fed low-protein grass hay, the muskoxen consumed one third the intake of cattle, even after scaling to metabolic body mass (Adamczewski et al. 1994b). Apparent digestibility of the hay during the feeding trials also was higher for muskoxen, which contributed to the low forage requirement.

Compared to other ruminants, muskoxen have exceptionally long retention times for fibrous forage through the gastrointestinal tract, which helps to explain the high digestibility of low-quality forages. Mean retention time of hay varied from 114 ± 4 (*SE*) hr in March to 95 ± 4 hr in July, using mordant chromium as a marker (Adamczewski et al. 1994b). The long retention time is aided by the large capacity of the rumen and omasum (Staaland et al. 1997). From his evaluation of the muskox alimentary tract, Hofmann (2000) suggested that muskoxen are grazers adapted primarily to a diet of grasses and sedges.

Voluntary forage intake varied seasonally in captive Alaskan muskoxen, with peak dry matter intake in summer (102 ± 7 g/kg$^{0.75}$) and a much lower value in winter (38 ± 3 g/kg$^{0.75}$) on a high-quality diet (White et al. 1994). Seasonal variation was less pronounced on a medium-quality diet and voluntary intake was even lower (32 ± 2 g/kg$^{0.75}$) in winter on a poor-quality diet (Adamczewski et al. 1994a). Seasonal range in forage intake of free-ranging muskoxen likely matches or exceeds the range measured in captive animals, as the forage quality varies widely. Boyd et al. (1996) showed that browse availability increased overall intake in summer in muskoxen.

The endocrine control of altered voluntary food intake and associated changes in fat and protein reserves in ruminants is complex. Changes in day length affect daily melatonin patterns in muskoxen (Tedesco et al. 1992; Tedesco 1996), and these likely affect hormones such as prolactin, growth hormone, and thyroid hormones. Nilssen et al. (1994) measured significantly higher thyroid hormone values in summer in muskoxen than in winter. Thyroid hormones regulate metabolic rates (Nilssen et al. 1994). Serum concentrations of insulin-like growth factor 1 (IGF-1) are strongly seasonal in captive and wild muskoxen, with a fall peak in breeding and nonbreeding animals (Adamczewski et al. 1995). High IGF-1 concentrations correlate well with increased lean body growth (Kerr et al. 1991). The fall surge in IGF-1 may help explain the muskox tendency to gain mass at this time of year.

Blood Chemistry. Most hematological and serum biochemical values from captive muskoxen are similar to those of domestic ruminants, although seasonal variation is high (Tedesco et al. 1991). Values for urea declined as protein intake decreased during winter and muscle catabolism elevated creatinine. Urea values ranged from 7.5 to 22.7 mmol/L and creatinine from 137 to 429 mmol/L over 12 months in captive muskoxen. Muskoxen on Victoria Island had urea and creatinine values of 3.2–3.8 and 253–293 mmol/L, respectively, in May, which is the seasonal low in body condition (Adamczewski 1995). Although these urea:creatinine ratios were low compared to, for example, those of caribou from Banks Island (Larter and Nagy 2000a), body condition of the muskoxen did not indicate malnutrition (Adamczewski 1995; Larter and Nagy 2000a). This suggests that more information is needed to establish baselines in urea:creatinine ratios for muskoxen.

Limited information is published on the physiological consequences of capture and handling of muskoxen. Dieterich and Fowler (1986) mention elevated serum enzymes and glucose. Lund (1992) reported that corralling large numbers of muskoxen during a commercial harvest did not appear to cause additional stress (caused by the herding to reach the corrals) based on serum cortisol levels. However, stress during the herding was not measured (Lund 1992) nor were other measures recorded such as level of creatine kinase, which is elevated during acute muscle exertion or damage (Tedesco et al. 1991). Muskoxen are physiologically ill suited to enforced long-distance movements, and like many wild animals, they show relatively few overt signs of stress or pain, because these are likely to draw the attention of predators. From our experience with captive muskoxen, we suggest that muskoxen under stress may appear calm to the casual observer.

REPRODUCTION

Anatomy. Muskox cows have a uterus similar to that of goats and sheep, except that endometrial pigmentation is found only in parous cows and is associated with lipofiscin granules rather than melanin as in sheep and goats (Rowell et al. 1987). Muskox cows, also only when parous, have a muscular band, which is not found in sheep or goats, running along the uterine horn's border. Its function is unknown, although it may be associated with involution (Rowell et al. 1987).

Breeding Season. Muskoxen on the High Arctic Islands generally breed in August, but those at lower latitudes breed in September (Tedesco 1996; Flood and Tedesco 1997). This pattern is the reverse of the pattern in caribou. The endocrine mechanism behind this pattern in muskoxen is unclear (Flood and Tedesco 1997). The latitudinal pattern centers on the cow's ability to retain ample fat stores for late-winter calving and early lactation, along with their relatively slow growth and regaining of body reserves (Parker et al. 1990). If calving and early lactation can begin before the summer pulse of high-quality nutrition, then both mother and calf (as a functioning ruminant) can take full advantage of it for growth and rebuilding body reserves. Because the summer is shorter at higher latitudes, calving and early lactation must begin earlier or the summer may be too short to permit adequate calf growth and replenishment of fat and protein stores. In caribou, on the other hand, a faster metabolism (Tyler and Blix 1990), smaller fat reserves, and needs for higher nutrient quality may not allow much lactation to proceed before spring plant growth. However, if caribou can take better advantage of ample summer nutrients than muskoxen (Parker et al. 1990), then both mother and calf have sufficient time to grow and replenish body reserves before the onset of winter.

Muskoxen likely have an endogenous circadian pacemaker, which cues melatonin secretion and with it, secretion patterns of prolactin, luteinizing hormone, follicle-stimulating hormone, and steroids (Tedesco 1996). Photoperiodic cues around the spring equinox appear to reset the endogenous clock of muskoxen. Melatonin implants at this time of year shifted the timing of prolactin, gonadotropin, and steroid secretions, wool shedding, and reproductive behavior (Tedesco 1996). Rapidly changing daylength at this time of year occurs at all latitudes, including the High Arctic Islands, which experience long periods of continuous daylight and night (Tedesco 1996). In sheep, a 3-month period of photoperiodic information is sufficient to entrain their circadian rhythm, and muskoxen likely have a similar mechanism to cue their breeding (Tener 1954; Suttie and Webster 1998).

The estrous cycle is about 20 days with short luteal phases; most conceptions are during the first overt estrous cycle and the remainder during the second cycle (Rowell and Flood 1988). The first estrous cycle of the season in captive muskoxen lasted 6.5 days, with one follicular wave. Further cycles, on average, were 22.8 days long, with three to four follicular waves at about 6-day intervals (Hoare et al. 1997). The corpus luteum normally regresses 20–22 weeks into pregnancy, which transiently decreases progesterone regardless of whether the cows were on high- or low-nutrition diets (Rowell et al. 1993, 1997).

Reproductive function is also highly seasonal in males. Mean luteinizing hormone concentrations in captive bulls rose from March to May and declined by September (Tedesco 1996). Testosterone concentrations peaked soon after the summer solstice, followed by enlargement of accessory genital glands, and peak breeding condition was between August and late October (Tedesco 1996). Body mass of breeding males usually peaks just before the breeding season.

Age at Sexual Maturity and Longevity. Under a high plane of nutrition, a few cows breed as yearlings and calve as 2-year-olds in the wild (Jingfors and Klein 1982). Most cows will first be bred in their second year and calve for the first time as 3-year-olds (Adamczewski et al. 1997). In northeastern Alaska, the mean age of first calving was 3.0–3.4 years between 1982 and 1990 (P. Reynolds, U.S. Fish and Wildlife Service, pers. commun., 2000). Four of 17 known-age cows continued to calve when 15–18 years old. Three cows were 17–19 years old when they died (Reynolds 1998). Bulls are physiologically capable of breeding as yearlings in captivity. In the wild, however, they are unlikely to breed until they are 4–8 years old and large enough to secure breeding opportunities from other bulls.

Conception and Pregnancy Rates. Data on pregnancy rates are relatively few except when muskoxen are killed during commercial or scientific harvests or serum is available during livecapture. More often, field observations of calf:cow ratios are used to index productivity, which is then an index that combines pregnancy and early calf survival. Earlier accounts of muskoxen productivity emphasized that pregnancy rates were relatively low, as alternate-year breeding was normal (Tener 1965). Experience with reintroduced muskoxen in Alaska initially showed high ratios (89:100) of calves to cows >2 years old, which suggested calving in successive years (Jingfors and Klein 1982). However, following the initial phase of rapid population increase, productivity declined. Individual variation among radio-collared cows with calves was high, but suggested that initially the cows calved every year (1979–1984) and subsequently calved at 2- or 3-year intervals (Reynolds 1998).

Commercial harvests on Banks Island revealed pregnancy rates as high as 95% in November 1983 (Latour 1987) for 3-year-old and older cows. During the 1990s, however, pregnancy rates annually varied between 35% and 68% (Nagy et al. 2001). Pregnancy rates for cows >3 years old on southern Victoria Island averaged 64% in November 1989–1992 and 58% by April (Adamczewski 1995). On the Canadian mainland, the pregnancy rate was low (36%) in March 1990 based on serum progesterone levels in 36 cows (>3 years old) (Gunn and Fournier 2000).

A possible contribution to high productivity is twinning, but it is extremely rare. On Nelson Island during 1981–1982, in a rapidly increasing population, 4 of 102 pregnant cows had twin fetuses (Dinneford and Anderson 1984). There are scattered observations of liveborn twins, but they are very much the exception (Tener 1965; Pattie 1986).

Productivity is often equated with observed proportion of calves described during aerial surveys at some time during the summer. In populations increasing at or close to the maximum finite rate of increase ($r_m = 1.30$), the proportion of calves can be 23–29% of all muskoxen counted (Le Hénaff and Crete 1989; Reynolds 1998). Calves in July-August on mainland Canada ranged between 10% and 15%, which does not discriminate between either low pregnancy rates or low calf survival (Gunn and Fournier 2000).

In the 1980s and early 1990s, seasonal field collections and observations of muskox research herds in Saskatchewan and Alaska revealed that the differences in the observed reproductive rates could largely be attributed to differences in nutrition. Nutrition, especially during August–October, is the key to whether cows reproduce (White et al. 1989, 1997; Adamczewski et al. 1992; Rowell et al. 1997). Lactating cows lose less of their fat during the winter than between April and May (Adamczewski et al. 1997) by giving priority to lactation in the late-winter period. Metabolizing fat to synthesize milk is more efficient than using body protein (Coppock et al. 1968 in White et al. 1989). Lactating muskox cows compress their compensatory growth from mid-August to early November, which is well after the July peak in forage quality (Adamczewski 1995). Nonlactating cows regained their body reserves during the summer and tended to be fatter than lactating cows in November.

Muskoxen are unlike caribou, which start to gain weight about 1 month after calving during peak plant growth and thus regain condition before the rut (Parker et al. 1990). Muskoxen replenish their body stores later and slower (Parker et al. 1990), although, as in caribou, muskox conception depends on condition. A clear link has been established between the likelihood of fall pregnancy and the cow's condition. A 50% probability of pregnancy was associated with a body mass of 176 kg in captive Alaskan cows (White et al. 1997) and with 183 kg in wild cows (Adamczewski et al. 1998). In both studies, measures of fatness were more strongly linked to the probability of pregnancy than was body mass. Of particular interest is the exceptionally high threshold of fatness apparently needed for fall pregnancy in wild muskoxen, about 22% fat in the ingesta-free body mass, compared to 5–9% in caribou and 9–18% in cattle (Adamczewski et al. 1998). This high threshold is consistent with a conservative approach to breeding. Successful gestation, birth, and initial lactation in winter should not proceed unless the cow has the large fat reserves she will need.

The effects of nutrition on reproduction are not simple in muskoxen. Some cows continue to lactate for 18 months both in the wild and in captivity (Adamczewski et al. 1997; Schulman and White 1997). Although captive calves daily nursed for longer than calves weaned at 9 months, those calves were not smaller (Swingley and White 1999, oral presentation). Prolonged lactation (suckling lasting into the next rut) may influence the likelihood of pregnancy at that time in cows with 5-month-old calves or even in cows with 17-month-old yearlings.

Regression of the corpus luteum of pregnancy in muskoxen, with its transient decrease in progesterone and a presumed switch to placental progesterone (Rowell et al. 1993), may be a further point when the cues of poor nutrition or body condition might end pregnancy. Pregnancy rates declined between November and April on Victoria Island (Adamczewski et al. 1997), which suggests, assuming the decline was not a sampling artifact, that lactating females in poor condition might be susceptible to pregnancy failure. However, evidence for the role of nutritional stress in early abortion or fetus resorption for muskox cows on an experimental low nutritional plane was limited by small sample size (Rowell et al. 1997). This type of postrut regulation of pregnancy rates remains speculative.

We cannot distinguish whether exceptionally low calf numbers (0–5% of cows with calf) in summer seen after weather-related malnutrition (Miller et al. 1977; Gray 1987; Miller 1998; Gunn and Dragon 2002) is a result of failure to conceive, abortion/absorption, or early calf death. Insufficient time for replenishment of female body condition (hence failure to breed) could stem from poor starting condition at the end of winter or from a constrained summer growing season. Both may have occurred in 1967–1968 on Bathurst Island. A severe winter in 1967/68 was followed by an exceptionally cool summer in 1968; Inuit hunters reported muskoxen leaving the island and found carcasses with signs of malnourishment (Gray 1987). Calves were absent that year and the next 2 years, and reproductive behavior in the fall was much reduced (Gray 1987). In 1973, many calves were born on Bathurst Island, only to be lost during their first winter, which was marked by heavy, wet snows and strong winds in the fall (Miller et al. 1977).

Gestation, Calving, and Lactation. Gestation was about 242 days on Banks Island based on observations of calving and estimated peak in conceptions, close to the 235 ± 4 (*SD*) days for captive muskoxen (Rowell et al. 1993). Muskoxen usually calve from mid-April to early June (Tener 1965), although calving dates vary geographically. On Banks Island, 65% of the calves were born by 30 April 1981; 94% by 8 May 1982; and 91% by 5 May 1983 (Latour 1987). Backdating from calf weight suggested that calving had begun about 8 April (Latour 1987). On southern Victoria Island, calving was also mid-April to late May (Adamczewski et al. 1997), but in Alaska, calving occurred in the first 3 weeks of May (Jingfors 1984). However, newborn calves are occasionally seen in June and into July (Tener 1965; A. Gunn, unpublished data). The spread in calving dates reflects the duration of the time that tending behavior by bulls indicates estrous cows. A low percentage of cows in estrus was seen 16 July and 15 August (1–15%) and again 16–30 September. The plateau for 38–40% cows to be in estrus extended from 16 August to 30 September (Forchhammer and Boomsma 1998).

Cows do not leave the herd to calve, and births are relatively rapid. The newborn calves have thick wool insulation and use metabolic heat production from their large deposits of brown adipose tissue to cope with temperatures as low as –30°C and high winds (Blix et al. 1984). They also lie close to and on the downwind side of their mother (Jingfors 1984).

Muskox calves weigh about 10 kg at birth (5.6 ± 0.7% of maternal body mass) in captivity, grow more slowly, and suckle for longer on less concentrated milk than caribou (Parker et al. 1990). In captivity, lactation peaks at 2–3 weeks after calving and declines after 6–8 weeks (White et al. 1989). However, on Victoria Island, Adamczewski (1995) estimated from mammary gland masses that lactation peaked 8 weeks after calving (July) and stayed high during August before declining. This pattern coincided with peak forage quality. The calves are usually weaned between December and February or even later in nonpregnant or poorly fed cows (White et al. 1989; Adamczewski 1995).

Muskox calves are precocial (although more altricial than caribou calves) and, like other precocial mammals, have relatively highly mineralized bones as neonates (Heinrich et al. 1999). Mineralization stiffens the bones, and bone architecture also changes to compensate for mechanical loading as body mass increases eightfold during the first 6 months. Neonate muskoxen have relatively stronger bones with greater resistance to fracture than adults, presumably to compensate for lack of coordination. Neonates also have relatively longer femurs than adults, which would increase stride length (Heinrich et al. 1999).

Muskoxen reach maximum bone growth (based on mandibles) and total body weight at about 5–8 years, but growth rates vary among areas depending on nutrition (Olesen et al. 1994).

ECOLOGY

Habitat. Muskoxen mostly occur on tundra where winter habitats have either shallow snow (20–40 cm) or sufficient relief where wind keeps the snow shallow. Summer habitats include low-lying river valleys or coastal plains with sedge meadows, riparian willows, and gravel bars. Muskoxen use subarctic maritime, continental arctic, and High Arctic habitats, which fall within eight North American ecoregions (Ricketts et al. 1999). Ecoregions are geographically distinct areas that have common environmental conditions, dynamics, and species and they are treated as conservation units on a global scale.

In Canada, muskoxen naturally occur in the Low, Middle, and High Arctic Tundra ecoregions (2,293,372 km^2) and have been introduced to the Eastern Canadian Shield Taiga ecoregion (741,510 km^2; northern Quebec). In northern Alaska, muskox reintroductions have been to the Arctic Foothills and Coastal Tundra ecoregions (227,328 km^2) and in western Alaska, the Beringia Uplands, and Lowlands Tundra (248,410 km^2). Muskoxen occur along and within the treeline transition areas and even further below the treeline north of Great Bear Lake, where they forage in openings where grasses and sedges dominate the vegetation. On the Arctic Islands with their more severe limitations to plant growth, combinations of higher ground influence local weather, which together with glacial till and drainage lead to relatively small pockets of favorable habitat (e.g., Bailey Point, Melville Island) with high muskox densities (Thomas et al. 1981).

Banks Island is an Arctic Island with exceptional muskox habitat. Many areas have rolling hills with well-developed soils from glacial till as glaciers receded earlier from Banks than the other Arctic Islands (M. Raillard, Canadian Wildlife Service, pers. commun., 2001). The island is large enough, and without a highly indented coastline, to have a relatively warmer interior climate. Banks consequently has high plant standing crop for its latitude and high muskox densities. It is uncertain whether forage quality is higher, as the mean values for crude protein recorded elsewhere in the Arctic are within the reported range of annual variation in forage quality (Larter and Nagy 2000b).

Klein (1992) and Nellermann (1997) suggested that muskoxen are relatively unadapted to snow-rich environments. If muskoxen have access to wind-blown slopes, small-scale relief, or vegetation protruding above the snow, however, they are relatively adaptable and select shallow snow to forage. Thresholds of snow depth where muskoxen will crater vary between 20 and 50 cm, depending on snow hardness and density (Thomas and Edmonds 1984). On Nunivak Island, which has a maritime climate with 138 cm of annual snowfall, muskoxen forage on wind-swept dunes and escarpment crests where beach rye grass (*Elymus arenarius*) grows (Lent 1999).

No single meteorological variable entirely delimits muskox range, as there is a trade-off between factors and their interactions. For example, snow depths tend to be less at the higher latitudes, but the plant growth season is shorter. Where deeper snow cover (>25 cm) occurs early in winter (late October), muskoxen have been slow to recolonize or are absent. Examples are the Canadian northeastern mainland and Baffin Island, although the terrain there is frequently rugged. Since muskoxen gain body condition during the fall (Adamczewski 1995; White et al. 1997), heavier snowfall earlier in winter is a likely disadvantage. Based on weather records from 1955 to 1972 (Maxwell 1980), the frequency of snow depth exceeding 25 cm at the end of October is 5% on Banks Island and 0% on Victoria Island, where muskoxen are widespread. By contrast, the frequency of snow depth exceeding 25 cm at the end of October is 50% at Pelly Bay on the northeastern Canadian mainland and 75% at Dewar Lakes on central Baffin Island, where muskoxen have always been scarce or absent. Total annual snowfall for these locations reflects a similar pattern: Banks Island, 84 cm; Victoria Island, 80 cm; Pelly Bay, 125 cm; and Dewar Lakes, 148 cm (Maxwell 1980).

Diet. Arctic plant communities are typically dominated by relatively few species, some of which are nearly ubiquitous. A similar suite of sedges (*Carex* spp.), grasses (*Kobresia* spp.), and deciduous shrubs, especially willows (*Salix* spp.), dominates muskox diets across their circumpolar ranges (Klein 1992; Larter and Nagy 1997). Large rumens, slow passage of digesta, and ability to conserve nitrogen mean that muskoxen are well suited to year-round digestion of sedges and grasses. In graminoids, there is a summer peak in readily digestible cell contents and protein and a late-summer/fall rise in cell wall content, but little indigestible lignin and few secondary compounds that would deter herbivores (Thing et al. 1987; Klein and Bay 1990; Oakes et al. 1989). Willow browse, by contrast, has a short early-summer peak of high cell contents and becomes increasingly lignified at the end of the growing season. Secondary compounds that reduce digestibility, such as tannins, also accumulate in willow, particularly in the twigs. Muskoxen tend to select willow most in early summer (Klein and Bay 1990; Larter and Nagy 1997), although it is eaten in winter.

Graminoids dominate winter diets (Klein 1992; Larter and Nagy 1997). In northern Alaska, muskoxen selected sedges and grasses throughout the winter (Wilson 1992). On Banks Island, sedges increased from about 50% to 80% of the diet during the winter (Nagy et al. 2001). Mosses were neither selected nor avoided, although by late winter they were 25% of the diet based on fecal samples from Alaska (Wilson 1992; Ihl 1999) and 7–22% in the eastern High Arctic (Parker 1978). The actual percentage of moss in diets is likely lower, as these poorly digestible plants are overrepresented in feces (Thing 1984). Willow twigs are also likely overrepresented for the same reason (Klein and Bay 1994).

Snow conditions influence the use of willow in the diet. In late winter in Alaska, where meadows with riparian willows are low lying with deep snow, muskoxen fed along slopes and ridges on prostrate willows, grasses, and moss (Ihl 1999) or along river and creek bluffs (Wilson 1992). Willows were eaten only when they were in shallow snow (Wilson 1992) or as snow conditions changed in late winter. Muskoxen on Banks Island increased their use of willow from <20% to 50% in May (Larter and Nagy 1997). On southern Victoria Island, muskoxen selected the sedge *Carex aquatalis* from shallow-snow areas, but in May, shifted to feeding on prostrate willows on the slopes and ridges where snow was receding (Schaefer 1995). Earlier in winter, those areas that had less vegetation were avoided (Schaefer 1995). In the High Arctic, when winters with deep snow forced muskoxen up onto sparsely vegetated ridges, the percentage of willows in the rumens of muskoxen dead from malnutrition was almost 50% (Parker 1978). Such a diet was likely highly lignified and had low digestibility.

As spring and summer progress, muskoxen selectively feed following plant phenology and hence nutritive value (Robus 1981). In northern Alaska, muskoxen in late May fed on flowering *Eriophorum* heads, then shifted to the new leaves of sedges. By mid-June, they selected young willow leaves and flowering forbs, benefiting from the earlier peak in nitrogen and, later in the summer, from the increase in plant biomass. The muskoxen clipped new-growth willow twigs, but as the twigs became woodier, the muskoxen pulled them through their teeth stripping off the leaves. Throughout August into October, muskoxen continued to feed on willow leaves, dried stems and leaves of forbs, and the green basal parts of legumes (Robus 1981). Further north, on Ellesmere Island, use of willows peaked in June, as the young leaves had lower amounts of secondary compounds (Raillard 1992) and less indigestible lignin (Thing 1984; Klein and Bay 1990). As sedge growth increased and peaked in late July, the muskoxen switched back to sedge meadows.

Summer muskox diets on the Canadian mainland were similar to those on the Alaskan northern coastal mainland (Tener 1965; Gunn and Sutherland 1997), being largely willow with 20–35% graminoids (*Carex* and *Eriophorum*) and forbs, including legumes. On Banks Island, there was greater use of willow in areas with more muskoxen (Larter and Nagy 1997). Further north, on other of the Arctic Islands, muskox ate less willow and more graminoids, which reflects the plant's relative availability (Parker 1978; Nagy et al. 2001).

Although graminoids and willow dominate the diet, muskoxen also select other species. In eastern Greenland, 15 species made up 7–15% of the late-summer diet, half of which were flowering forbs, and 25 species made up 20% of the late winter–spring diet (Thing 1984). Likewise in Alaska, muskoxen select flowering forbs such as *Pedicularis* spp. and several legumes including *Oxytropis* spp. (Robus 1981; Mulder and Harmsen 1995). In summer, herbs generally have higher minerals than deciduous shrubs and grasses (Staaland and Olesen 1999). Forbs are highly digestible (64–70%) and have relatively high peak nitrogen and crude protein values in summer (Robus 1981; Larter and Nagy 1999b). Their dietary representation is often much underestimated in fecal samples due to their digestability.

A comparison of macrominerals in the gastrointestinal tract and forage suggested that they were not limiting in western Greenland (Staaland and Olesen 1999). Sodium was relatively low and potassium high in most forage. Low sodium may influence muskoxen foraging elsewhere, as they use mineral licks in summer on Ellesmere Island and Greenland (Tener 1965; Klein and Thing 1989; Staaland and Olesen 1999). Muskoxen are likely selecting for sodium when they wade in shallow lakes feeding on semisubmergent aquatic plants (Tener 1965; Gunn 1990b). Besides remaining succulent, plants such as *Hippurus vulgaris* are high in sodium, 230 mmol/kg compared to 6–28 mmol/kg in terrestrial forage plants (Staaland and Olesen 1999).

Feeding Behavior. Arctic ruminants face a scarcity of both forage quantity and quality in winter. In summer, forage quantity is not usually limiting, but the pulse of highly digestible nutrients is short. Graminoids are relatively digestible, but store their nutrients below ground for most of the year. Woody dicotyledons have carbohydrate and protein reserves aboveground (cambial layers and buds), but are less digestible, better protected by the plant, and less abundant. The constraints for herbivores are the time and energy required to locate forage and, additionally in winter, to uncover, chew, and then warm it up to core body temperature. Like all ruminants, muskoxen feed selectively on a variety of plants, as small changes in digestibility have proportionally greater effects on nutritional status and thus individual fitness—the multiplier effect (White 1983). Given the Arctic's strongly seasonal plant growth and the tight coupling between muskox nutrition and reproduction, muskoxen can be expected to be strongly selective in their foraging. Our experience with captive muskoxen is that they are selective in their feeding.

Muskox forage selection operates within a hierarchy of scales from the landscape, to the patch, and down to the bite, a pattern common to other ruminants (White 1983; Bailey et al. 1996). In winter, muskoxen minimize energy and time expended on foraging by selecting for greater food abundance (graminoids), especially where the snow is shallow, and they also select for shallower and softer snow across spatial scales (Wilson 1992; Schaefer and Messier 1995a; Larter and Nagy 2000). In western Alaska, muskoxen fed in snow 7–12 cm deep and of 5–10 kg/cm hardness, whereas the feeding site and range had snow depths up to 40 cm and hardness up to 32 kg/cm (Ihl 1999).

Consistency in selection across scales varies when muskoxen face greater trade-offs, such as when snow depths are greater or the snow is harder. On western Alaskan lowlands, where the snow is deep (up to 100 cm), muskoxen selected against graminoids at feeding sites, but selected them in the diet when they were available (Ihl 1999). Muskox selectivity is strongest at the plant level. Although their gastrointestinal anatomy classifies muskoxen as generalist grazers (Hofmann 2000), their mobile lips and tongue allow them to be discriminating in selecting plants and even parts such as graminoid flowers (Adamczewski 1995).

Muskoxen adjusted their feeding behavior to snow conditions (Raillard 1992), but their responses differed across scales (Wilson 1992; Schaefer and Messier 1995a). With thin and soft snow, muskoxen nuzzled it away to reach forage. As snow became deeper, muskoxen pawed the snow away (Schaefer 1995). As depths increased from 5 to 15 cm, muskoxen doubled the rate at which they pawed at the snow, increased the number of pawing bouts and the length of time they fed at a station (without moving their front legs), but decreased the number of stations at each patch (Schaefer 1995). The time spent at each patch (aggregation of feeding stations) declined throughout winter, but increased in spring, when muskoxen did not crater, but fed along snow-free ridges (Schaefer 1995). Muskoxen on Ellesmere Island also fed on the dry slope communities in late winter, but included meadows in mid-April as temperatures exceeded −19°C and snow hardness declined (Raillard 1992). In those meadows, snow was relatively shallow (<20 cm) but wind packed.

Muskoxen displaced smaller individuals from craters on average 1/hr/individual in shallow snow (<10 cm). The frequency of displacements increased with snow depth and herd size (Gray 1987; Schaefer 1995). The displaced muskoxen either moved to another feeding crater (Schaefer 1995) or, half of the time, dug a new crater (Gray 1987). Although this may be a disadvantage of social grouping (Heard 1992), the benefits and costs to the individuals are not yet measured. Smaller bodied individuals have lower energy requirements and may not be at a disadvantage when displaced to another crater. They have narrower mouths and incisor arcades and may be more selective at removing plants than larger mouthed muskoxen.

Generally, muskoxen spend more time feeding in summer than in winter; for example, on Ellesmere, muskoxen spent 54% of their time foraging in late winter (May), compared to 82% in June and 60–70% in July (Raillard 1992). Time allocated to foraging appears to be related not just to forage biomass and forage quality, but probably also to their "internal needs" (Jingfors 1980; Thing 1984; Klein and Bay 1990; Boertmann et al. 1992; Forchhammer and Boomsma 1995; Schaefer 1995). In early and late winter, muskoxen spent long periods lying down (233 and 213 min, respectively; 66% and 58% of their time budgets), which is longer than in most other ruminants (Schaefer and Messier 1995b). Muskoxen needed to spend more time ruminating winter's highly fibrous forage, but as time spent lying increased with snow depths, individuals were also likely minimizing energy expenditures (Forchhammer and Boomsma 1995; Schaefer and Messier 1996).

Less is known about the hierarchy of scale for summer foraging partly because describing diet has relied on identifying plant fragments in fecal samples. This underestimates the relative proportion of dicotyledons, especially forbs, due to their high digestibility (Robus 1981; Forchhammer and Boomsma 1995). Observing time allocated to forage categories and sampling for browsed plants is necessary to document preference for forbs (Forchhammer and Boomsma 1995).

The underlying reasons for forage selection have been simplified to a dichotomy between maximizing energy intake and minimizing energy expenditure (Belovsky 1991). Foraging strategies that are energy maximizing will be expected to have a higher proportion of dicotyledons (forbs and woody browse) with a higher cell content/digestibility ratio. In western Greenland, muskoxen appeared to maximize energy intake from calving to the rut by balancing the time spent foraging on graminoids and shrubs (Forchhammer and Boomsma 1995).

Foraging strategies differ between sex and age classes, as body size, and thus forage intake requirement, differ. In summer, bulls in Greenland spent less foraging time on dicotyledons than did cows, subadults, and yearlings (Forchhammer and Boomsma 1995). On Banks Island, cows consumed more dicotyledons, but their time spent foraging did not differ from that of bulls (Oakes et al. 1989). Calves spent less time foraging than other individuals and foraged on more dicotyledons early in the summer (Oakes et al. 1989).

The trade-offs that muskoxen make between the energetic costs of foraging relative to forage quantity and quality are further complicated by other factors:intra- and interspecific competition, risk of exposure to parasites and predators, and reproductive strategies. Almost nothing is known about how muskoxen balance forage selection and risks of predation and parasites. These trade-offs have been studied in domestic sheep (Hutchings et al. 2000) and research is starting in Greenland (K. Raundrup and S. Clemmensen, University of Copenhagen, pers. commun., 2001).

Responses of plants to muskox foraging have received limited attention despite their likely importance as feedback between muskoxen and their forage. Plants vary in their defenses against herbivory, but typically rely on chemical defenses such as tannins or other secondary compounds, regrowth, or some combination of these (Archer and Tieszen 1980; Van Soest 1982). Arctic graminoids appear to have low silica concentrations (Pinsonneault 1995). On northern Banks Island, 20% of individuals of the legume *Oxytropis viscida* were grazed (Mulder and Harmsen 1995). Removal of growth tips reduced plant growth for the following 2 years, which decreased the probability of regrazing (Mulder and Harmsen 1995).

Graminoids tend to respond to grazing by regrowth (Archer and Tieszen 1980), although experiments with artificial clipping have shown variable responses (Pinsonneault 1995; Smith 1996). Muskoxen increase their intake during summer, but they rarely remove much of the green biomass of graminoids and shrubby browse even when their relative densities are high (Raillard 1992). The density of muskoxen in Sverdrup Pass, northern Ellesmere Island, was relatively low (0.35 muskoxen/km^2), but where they concentrated on the sedge meadows, local density was 6.4 muskoxen/km^2. Even so, they only removed an estimated 4% of the available forage (Raillard 1992).

In the High Arctic, the annual 10 months of muskox grazing when plants were not growing considerably reduced litter accumulation (Raillard 1992); exclosures preventing grazing had almost five times more litter than ungrazed sites. Accumulation of litter stalls the turnover of nutrients for plant growth, and thus litter removal enhances plant productivity. Additionally, muskoxen return nitrogen to the grazed sedge meadows through fecal pellets and urine, although deposition rates are higher on the drier slopes where muskoxen tend to bed (Murray 1991). Simulated grazing (clipping) resulted in compensatory sedge growth even after 3 years, although it did reduce flowering (Raillard 1992). Willows were also less likely to be grazed during the growing season on Ellesmere Island than during the rest of the year. Any effect of browsing on the production of willow leaves was masked by weather, however, as mean summer temperatures annually varied and affected willow growth (Raillard 1992).

Generally, the ability to tolerate the loss of leaves to grazing and the trend in photosynthetic capacity appears highest in perennial graminoids such as *Eriophorum* and *Carex,* intermediate in deciduous shrubs such as *Salix,* and least in evergreen shrubs such as *Ledum* (Archer and Tieszen 1980). Investment in chemicals protecting arctic plants from herbivory shows the opposite pattern:highest in evergreen shrubs, intermediate in deciduous shrubs, and least in the graminoids (Jung et al. 1979). Whereas graminoids responded well to a single clipping with little carryover to a second season, the effects of repeated clipping were still evident a full year after the clipping had ended (Archer and Tieszen 1980). Most tundra vegetation is rarely grazed with this degree of intensity by muskoxen.

The effect of weather on forage growth is among the least known aspects of the relationship between muskoxen and their forage. In the High Arctic, although peak total seasonal standing crops in sedge meadows varied relatively little over 4 years of study, the length of graminoid green shoots was greater in warmer, earlier summers (Muc 1977). Flowering in graminoids and dicotyledons varied considerably with a 1-year lag, as it depended on the previous growing season's temperature (Muc 1977; Svoboda 1977).

Another topic that is only now being investigated in other herbivores is the effect of subclinical parasitism on host foraging dynamics. Subclinical parasitism can reduce forage intake by 15–25% (Sykes 1987) and the consequent reduction in host reproductive fitness has led to the host evolving behavioral strategies to reduce the risk of parasitism. In other herbivores, and at several scales, the herbivore reduces its exposure to parasitism even at the cost of reducing forage intake (Hutchings et al. 1999).

Interspecific Relations. In Alaska, parts of northern Canada, and western Greenland, the relationship between caribou and muskoxen on common range has been a concern since the 1970s (Wilkinson et al. 1976; Coady and Hinman 1984; Gunn and Dragon 1998; Ihl 1999; Nagy et al. 2001). Where muskoxen have increased and caribou declined, competitive exclusion of caribou by muskoxen has been invoked as a factor in the caribou decline. Clear evidence for causality has been elusive, however, because it is difficult to isolate the effects of confounding variables such as weather and predation. Unambiguous evidence of competition would require experimental manipulation, which is rarely feasible. Most studies have drawn inferences from descriptions of habitat use and diets. Hares, geese, and small mammals are also herbivores in tundra ecosystems, but most attention has focused on caribou and muskoxen as potential competitors.

Muskox–caribou relationships operate at different scales of space and time. On the evolutionary time scale, muskoxen and caribou have long coexisted and have evolved different strategies in adapting to arctic ecosystems (Parker et al. 1990; Adamczewski 1995; Suttie and Webster 1998). However, caribou and muskoxen evolved in complex communities that included more species of large herbivores and predators before the last major glaciation. It is by no means certain how current relationships would reflect the "ghost of competition past." As a general rule, the foraging niches of the two species are well separated; by comparison, a greater number of large herbivore species coexist in African grasslands and forests.

Muskoxen have two to three times the body mass of caribou. This difference alone has consequences for the two species's nutritional ecology (Robbins et al. 1995); muskoxen would be expected to be better able to use relatively widespread, low-quality forages, rather than focusing on selection of high-quality forage items. Forage intake in herbivores is scaled to body size and is related to individual fitness, spatial scale for habitat use, and intra- and interspecific relationships (Shipley et al. 1994). Whether the grazer–browser continuum described by Hofmann (1989) explains the ability of ruminants to use particular forages has been questioned (Robbins et al. 1995). However, muskoxen have the digestive anatomy and physiology common to grazers, whereas caribou do not (Staaland et al. 1997; Hofmann 2000). Caribou have relatively high metabolic rates, require food of higher quality, have smaller rumens and higher rates of passage, and, correspondingly, do less well than muskoxen on low-quality forage such as grasses and sedges (Adamczewski 1995). Based on these first principles of body size and digestive physiology and anatomy, muskoxen and caribou are suited to very different foraging strategies.

At a regional scale, trends in caribou and muskox abundance suggest that either their trajectories are independent or relationships are inconsistent. Caribou and muskoxen both increased on the Northwest Territories and Nunavut mainland and Victoria Island during the 1970s (Gunn 1990a; Barr 1991). On the mainland, caribou numbers have continued to increase while some muskox populations have declined in size (Fournier and Gunn 1998). In contrast, the sharp decline in caribou on several of the Arctic Islands coincided with an increase in muskoxen (Nagy et al. 1996; Gunn et al. 2000). Muskox abundance on Banks Island increased from about 3000 in 1972 to 80,000 in 2001 (J. Nagy, Government of Northwest Territories, pers. commun., 2001). From 1972 to 1998, caribou numbers declined by one order of magnitude and then started to increase between 1998 and 2001 (Nagy et al. 2001; J. Nagy, Government of Northwest Territories, pers. commun., 2001). On the western High Arctic Islands, between 1961 and 1997, Peary caribou and muskoxen tended to increase and decrease in synchrony (Gunn and Dragon 2002). These patterns suggest that if competition exists between the two species, its outcome is uncertain and other factors (hunting, weather, predators) play a major role.

Within the regional scale (landscape), a hierarchy of scales describes herbivore foraging behavior that leads to grazing distribution patterns (Bailey et al. 1996). Most studies of muskox–caribou relationships have only dealt with diet or habitat use rather than the mechanisms that lead to diet and habitat selection at different scales. By focusing on diet and habitat use, the studies have not dealt with behavioral exclusion, although Inuit in western Canada and Alaska have identified that caribou avoid muskoxen. The few direct encounters recorded by Ihl (1999) showed little evidence of direct interaction. Inuit hunters have also suggested that the packed and hardened snow typical of the site-intensive foraging of muskoxen would handicap caribou feeding.

On the Canadian Arctic Islands, muskox and Peary caribou habitat use and diet did not significantly overlap (Tener 1965; Parker 1978;

Thomas and Edmonds 1984). On Prince of Wales and Somerset Islands and Boothia Peninsula in the mid-1970s, Russell et al. (1978) documented little overlap between caribou and muskox seasonal ranges, based on distribution of fecal pellets. The exception was Prince of Wales, where caribou and muskoxen fed in summer on willow–moss–lichen patterned ground.

On western Alaska's Seward Peninsula, with 226 mm of average annual precipitation, deep snow and relatively low sedge availability forced muskoxen to crater for forage on the same dryas—lichen heaths or hummocky lichen mats—used by about 2000 domestic reindeer and, in some years, several thousand caribou (Ihl 1999). Muskoxen and reindeer selected feeding sites and craters with the same snow and vegetative characteristics and craters, but had significantly different diets with little overlap. Muskoxen fed mostly on sedges and mosses, whereas almost half the caribou diet was lichen. However, muskoxen fed on 10–15% lichen and if their densities increased, trampling damage and lichen removal might reduce lichen availability. Potential for competition would be greater in this situation if muskox densities or snow depths increased (Ihl 1999).

Environmentally reduced forage availability, when it leads to changes in foraging behavior, can alter individual foraging strategies and relationships among species. Faced with forage depletion, white-tailed deer fed less selectively, increased bite size, and moved less (Kohlmann and Risenhoover 1994). Jenkins and White (1987) described a shift to possible interspecific among between moose, elk (*Cervus elaphus*), and white-tailed deer during winters with deeper than average snowfall. Deeper snow forced muskoxen to move upslope to forage where snow was shallow, which increased the likelihood of diet overlap with caribou (Parker 1978). On Banks Island, during the 1992/93 winter, muskoxen fed more on willow, which is widely used by caribou (Larter and Nagy 1997). Nagy et al. (1996) and Larter and Nagy (1997) suggested that the greatest likelihood of competition between muskoxen and caribou on Banks Island would be during severe winters and with high muskox densities, when the normally divergent feeding niches of the two species might be forced into substantial overlap. In exceptionally severe winters, caribou may have an advantage because of their mobility and ability to select widely scattered patches of better forage. When the snow is less deep, muskoxen may have the advantage, as they are able to make use of relatively abundant low-quality forage in low-lying sedge areas (Ihl 1999).

Willows are a dominant species in the diets of most arctic mammalian or avian herbivores, including northern collared lemmings (*Dicrostonyx groenlandicus*), arctic hares (*Lepus arcticus*) (Smith and Wang 1977), and ptarmigan (*Lagopus* spp.). Larter and Nagy (1997) and Larter (1999) suggested that muskox and arctic hare browsing willow could reduce its availability for Banks Island caribou, which feed on willow in summer. However, given the abundance of willow and its resilience to browsing, this effect is unlikely to be widespread. Less abundant forbs might be a better candidate, particularly given their high nutrient content.

Two key factors affecting muskox and caribou numbers are the effects of wolves (*Canis lupus*) in a two-prey system and hunter harvest rates. On Banks Island, wolf numbers were reduced in the 1950s by poisoning and then increased in the 1990s (Nagy et al. 2001). Muskoxen can be the primary prey of wolves (Marquard-Petersen 1998). A muskox population >50,000 could support many wolves, which may have contributed to the decline of caribou on Banks Island (Nagy et al. 1996). Overall, however, at least the later stages of the decline of caribou on Banks Island were driven by excessive hunting; caribou have always been preferred to muskoxen by hunters from Sachs Harbour (Nagy et al. 1996).

One aspect of the relationship between caribou and muskoxen that has received almost no attention is interspecies transfer of disease and parasites. Korsholm and Olesen (1993) investigated gastrointestinal parasites and reported that reindeer and muskoxen in Greenland carried the same species of abomasal nematodes.

Descriptive studies of habitat use and diet have not really brought us much closer to being able to predict the relationship between caribou and muskox foraging and demography. Overlap in diet and habitat use between caribou and muskoxen is most likely when animal densities are high and in winter, when both species have fewer foraging options. Overlap in diets or habitat use is not in itself evidence for a competitive relationship (Wilkinson et al. 1976). The competitive advantage of one species over another will vary with environmental conditions within and among years, and interactions with other herbivores cannot be discounted, especially if their fluctuations in density coincide. In winter 1993–94, on Banks Island, lemmings, ptarmigan, and hares peaked (based on trapping and harvest records), as did muskox numbers (Nagy et al. 2001). Possibly, weather and climate may entrain the fluctuations in herbivore numbers, as, for example, simultaneous reproductive pauses in hares and muskoxen followed an exceptionally short summer in 1998 on northern Ellesmere Island (Mech 2000). At the very least, we have a large gap in understanding how herbivores interact in the Arctic.

BEHAVIOR

Social Behavior. Muskoxen have a polygynous breeding system with attendant sexual dimorphism in life history traits, including their social behavior. Male and female breeding strategies influence the dynamics of mixed-sex-and-age herds and whether the bulls occur as singles or in bachelor herds. Cows are rarely found alone (<1–5% of sightings), based on radio-collared cows, and those solitary cows either rejoined a herd or died (Reynolds 1993). More cows were seen as singles on the central Arctic mainland, where the lungworm *Umingmakstrongylus pallikuukensis* is prevalent. Possibly, infected cows may not have been able to keep up with a herd all the time (Gunn and Fournier 2000).

During July and August, adult bulls occur either as harem bulls, singles, or in bachelor herds (Reynolds 1993). The larger bodied bulls face a trade-off between foraging and breeding strategies and they may leave the mixed herds to increase forage intake either as solitary individuals or in bachelor groups. Synchrony of feeding bouts constrains the duration of foraging bouts to the requirements of the numerically dominant smaller bodied muskoxen in a mixed herd. Bulls in bachelor herds can spend more time foraging than bulls in mixed-sex herds (Oakes et al. 1989; Cote et al. 1997).

The proportion of single bulls and bachelor herds correlates with the size of the mixed-sex-and-age herds. Where mixed-sex-and-age herds are smaller (10.2 ± 2.11 *SE* individuals) such as on southern Victoria Island, the percentages of single bulls (17.1%) and bachelor herds (4.6%) were significantly lower than from the nearby mainland (Gunn 1992a). On the mainland, muskoxen were in larger herds (19.2 ± 3.09 *SE*), and the proportion of single bulls (35%) and bachelor herds (24%) was higher. The higher number of smaller herds on Victoria Island suggested that a higher proportion of bulls would be breeding, but the dominant bull in mixed-sex-and-age herds on the mainland was in larger herds with more cows.

During the rut, single bulls occur as satellites on the periphery of mixed herds, waiting for breeding opportunities as cows try to evade the dominant bull. Alternatively, the satellite bulls may be testing the herd bull for a chance to replace him as he becomes exhausted during the rut (Smith 1989). Bulls in bachelor herds and harem bulls spent similar proportions of time on foraging and breeding behavior, which suggests that bulls in the bachelor groups are potential breeders and bulls maximize their breeding opportunities by roaming among herds (Forchhammer and Boomsma 1995). Radio-collared harem bulls were 4–9 years old in western Alaska (1982–1986). Younger and older bulls traveled to lower density areas, where acquiring a harem was less competitive, but provided fewer breeding opportunities (Smith 1989).

Herd Size. Herds are generally larger on the Arctic mainland than on the Arctic Islands (Tener 1965). Overall, herds tend to be 5–12 individuals in summer and 12–30 in winter (Reynolds 1993). Infrequently, herds temporally aggregate in response to wolves in their immediate vicinity (Miller et al. 1977). Herds are 1.2–2.3 times larger in winter than summer (Heard 1992). On the Canadian mainland, herd size (excluding single bulls) decreased from April to a minimum (12.3 ± 2.4 *SE*) in July and August and then increased after the rut to 28 ± 1.6 (*SE*) in October and November (Gunn and Fournier 2000). The decrease in bachelor and mixed herd size during July and August is partially due to intolerance between bulls as the frequency of dominance displays increases (Reynolds 1993; Gunn and Fournier 2000).

The tendency for herd sizes to be smaller at higher latitudes correlates with higher wolf density (and caribou density) rather than muskox density or population trend. This correlation led Heard (1992) to argue that muskox herd size was a trade-off between reduced risk of predation in larger herds and greater intraspecific competition for forage. However, the latitudinal trend also suggests that the size of muskox herds is influenced by plant productivity and patch size (Gunn 1992b). On southern Victoria Island, larger groups are found in areas with higher indices of green biomass, which also suggests that the size of forage patches and forage quality may influence herd size (A. Gunn, unpublished data). Snow conditions, especially later in winter, accentuate forage patchiness.

Little is known about whether larger muskox herds have more efficient vigilance or more effective defense against predators. Accounts of predation do not present a strong case for larger herds to be more effective defending against wolves (Gray 1987), although herds with as few as five individuals may be more vulnerable (Reynolds 1993). Larger herds have to contend with a wide range of foraging conditions from the fall to early summer, which lessens the likelihood that larger herds are aggregations imposed by snow conditions. Herds are smaller after severe winters, which argues that intraspecific competition may fragment herds (Miller et al. 1977).

Herds may remain stable for 1–2 months in summer, but they also splinter and merge with neighboring herds (Gray 1987; Reynolds 1993). Only about 11% of consecutive observations of mixed-sex herds did not detect change in herd size during 1987–1990, and most changes only involved one to five individuals (Reynolds 1993). However, consecutive observations of herds and a relatively low proportion of radio-collared muskoxen are unlikely to reveal subgrouping within herds such as matriarchal lines.

Movements and Seasonal Migrations. Muskoxen walk relatively slowly, although they have a faster walk that often breaks into a gallop during charges and chases. They are deceptively agile despite their appearance, and are capable of rapid turns and spins and of climbing steep slopes with apparent ease. Winge (in Harington 1961) described muskoxen as "essentially a mountain dweller, well adapted for climbing on rocks . . . holding on with the feet."

Muskoxen will ford rivers, although reports of swimming are contradictory. Gray (1987) did not observe muskoxen to swim, although sometimes they were swept underwater by a current for a short distance. P. E. Reynolds (U.S. Fish and Wildlife Service, pers. commun., 2001) and Mallory (1995) described muskoxen swimming across rivers. Muskoxen are occasionally marooned on lake islands during summer as the ice breaks up, and they do not swim off despite lack of forage (A. Gunn, unpublished data).

Muskoxen have regular patterns of local movements within habitats between resting and foraging sites and between foraging patches as they track plant growth and relative availability according to snow conditions (Gray 1987; Raillard 1992). Longer distance movements are typical as muskoxen seasonally migrate. On Alaska's North Slope, seasonal ranges were separated by an average 20–30 km (Reynolds 1998). The maximum straight-line distances were up to 114 km between calving and summer range. Monthly rates of travel were highest in summer because summer ranges were larger than winter ranges. Radio-tracking in the central Canadian mainland revealed that seasonal ranges overlapped. Herds concentrated in a large river valley in March and dispersed in April, with most cows moving north to the coast until September, when they returned to the Rae–Richardson River Valley. The maximum straight-line distance between summer and winter locations was 140 km (Gunn and Fournier 2000).

Tener's (1965) comment that muskoxen on the Arctic Islands had shorter seasonal migrations than those on the mainland is supported by data from Banks Island, where radio-tracked muskox cows seasonally migrated over shorter distances (B. McLean, unpublished data). Between October 1985 and August 1987, seasonal ranges were adjacent or overlapping, as 13 of the 20 collared cows remained within a 50-km radius of their capture location and all relocations were within 80 km.

Inuit hunters report that muskoxen migrate to the coast of western Bathurst Inlet in June and return to more elevated rocky areas in late July (Gunn 1990b). The muskoxen were probably following willow phenology, whose leaves unfurl earlier on the coast. Seasonal migrations west of Kugluktuk were more likely a response to snow conditions. The muskoxen winter in a wide valley where the snow is deep but relatively soft and they forage among dense stands of shrubs. However, at the end of March, the snow in the valley becomes denser and crusted and the muskoxen move to higher land toward the coast, where the snow tends to be more shallow and wind packed.

Dispersion is either innate or environmental (Caughley 1977). Adult bulls may be innately likely to disperse as they search for breeding opportunities (Smith 1989). Muskox colonization or recolonization is frequently preceded by sightings of solitary bulls in an area before mixed-sex-and-age herds are seen (Gunn and Case 1984; Smith 1989). Range expansion is a form of environmental dispersal, as it occurs in pulses following periods of increasing densities, which have been most documented in northeastern Alaska (Reynolds 1998). Between 1969 and 1993, the size of the core area increased almost fivefold. Elsewhere, the rate of spread was gauged from distances between sightings on the edge of the distribution over time. Muskoxen were spreading east on Canada's mainland at 13 km/year (Gunn and Case 1984) during the 1970s and 1980s.

How the pulsed emigration of mixed-sex-and-age herds relates to the innate dispersal of adult bulls is uncertain. One male strategy is to roam (Sandell and Liberg 1992), and a bull could subsequently lead a mixed-sex-and-age herd back to areas that they gained familiarity with during their dispersion. Alternatively, the dispersion of solitary bulls does not influence the dispersion of the mixed-sex herds (Smith 1989). Cows apparently do move and return over long distances; a cow born on Nelson Island and collared as a 3-year-old moved at least 320 km before returning to Nelson Island (Patten 1997).

Environmentally caused dispersion also includes unpredictable and irregular movements when deep, wind-packed dense snow and icing force muskoxen to abandon their usual ranges. Dispersal of muskoxen from Bailey Point on Melville Island during the 1996/97 severe winter and the numbers of muskoxen that moved from Bathurst to Cornwallis Island during the severe early snow conditions in 1967, 1973, and 1995 are examples. The total numbers and survival of muskoxen forced to make those movements usually are unknown. F. L. Miller (Canadian Wildlife Service, pers. commun., 2001) found 36 dead muskoxen on the sea-ice 100–800 m from the shoreline of Bathurst Island after the second consecutive severe winter.

POPULATION DYNAMICS

Muskoxen have the demographic characteristics typical of large-bodied herbivores (Gaillard et al. 2000). Large-bodied herbivores typically have high adult survival, low fecundity, late onset of breeding, and a single birth every 1 or 2 years; females have a life span often exceeding 15 years. Juvenile survival contributes more variation in rates of population change and adult survival buffers the variation (Gaillard et al. 2000). However, age classes are not independent, and effects such as weather acting on forage availability can affect individual cohorts and their subsequent offspring. Some cohorts are born during especially favorable years or, as Albon et al. (1987) described it, cohorts can be born with or without a silver spoon in their mouths. Muskoxen appear especially conservative in their life-history strategies when compared to caribou. Cows appear reluctant to gamble on the Arctic's conspicuous environmental stochasticity, as the large, prolonged costs of reproduction must be weighed against the risk of reduced survival.

Mortality. Mortality can be estimated for muskoxen of both sexes up to 2 years of age, as these age classes are recognizable in the field. Calf and yearling survival are calculated from the ratio of calves to yearlings, and yearlings to >2-year-old muskoxen in consecutive years. In northern Alaska, since 1972, mean calf and yearling survival annually varied from 75% to 85% and from 69% to 72%, respectively (Reynolds 1998). Calf and yearling survival were negatively correlated with snow depth, especially after the initial phase of rapid population increase. Rates for grizzly bear (*Ursus arctos*) and wolf predation are unknown.

On Banks Island, calf survival varied annually (mean 55.5% ± 5.9 *SE*) and was reduced during severe winters (Nagy et al. 2001). During the same time period (1986–1999), productivity as indexed by calves:100 cows (>2 years old) annually varied less than calf overwinter survival (41.8% ± 2.13 *SE*). Pregnancy rates measured for 6 of the 13 years with calf productivity and survival data (Nagy et al. 2001) do not show a trend between 1990 and 2000 (2-year-olds, 14% ± 4.46 *SE;* 3-year-olds, 51.7% ± 5.03; ≥4-year-olds, 67.2% ± 4.28). Toward the edge of their range, as in the High Arctic, annual variability in weather increasingly affects calf productivity and survival (Tener 1965; Hubert 1977; Miller et al. 1977).

Adult survival estimated from radio-collared females in northern Alaska was higher (89%) than in subadults (81–83%) as is typical for large-bodied herbivores. However, although subadults and adult bulls are more vulnerable during severe winters with resulting malnutrition (Gunn et al. 1989b), winters can be severe enough to overwhelm age-specific survival differences. Muskoxen of both sexes and all ages die, up to 90% of an island's population. Die-offs have been reported from the High Arctic (Parker et al. 1975; Miller et al. 1977). For example, on Bathurst Island, the estimated 1400 muskoxen in summer 1994 (Miller 1998) collapsed to an estimated 124 ± 45 (*SE*) live muskoxen in 1997 after three consecutive winters with record snowfall. At a minimum, based on carcasses, 650 muskoxen died, an unknown proportion emigrated, and 36 were found dead in small groups on the sea-ice (Miller 1998).

Malnutrition. Muskoxen during severe winters have increasingly depleted body reserves as they expend energy obtaining forage. As the animal's nutritional status declines, rumen microbial numbers decrease with a loss of digestive efficiency. Although forage intake continues, the declining microbial digestive efficiency means digestion is increasingly inadequate (Parker 1978). The severe conditions such as deep snow force the muskoxen to feed on less-digestible forage such as willow rather than more-digestible graminoids. Severe and prolonged malnutrition eventually causes death. Smaller bodied calves with proportionally lower body fat reserves and adult males with depleted reserves after the rut are usually the first to die (Tener 1965; Parker et al. 1975; Gunn et al. 1989b; McLean et al. 1993).

Malnutrition can be a deficiency of micronutrients, but even baseline information is sparse for muskoxen, let alone an understanding of their dynamics. Inevitably, the baseline information is contrasted with values for domestic livestock, which suggests muskoxen in November from Victoria Island have "marginal" liver and serum selenium levels (Blakley et al. 2000). Liver copper but not serum copper levels also were marginal. Increasingly, the realization is growing of the importance of parasitism in the dynamics of micronutrients (van Houtert and Sykes 1996), and further research will likely reveal the interactions.

Predation. Rates and effects of predation on muskoxen at the population scale are unmeasured. Since alternate prey are usually present on muskox ranges, it will be difficult to determine the effects of predation on any species of prey (Holt 1984). Wolf predation on muskoxen is common, with packs or single wolves observed killing adult as well as younger muskoxen, although most attacks are unsuccessful (Gray 1987; Mech and Adams 1999). Muskoxen may be more frequent in the wolf diet where caribou are infrequent or absent. On northern Ellesmere Island and northern and eastern Greenland, muskox occurrence in wolf scats ranges from 65% to 98%, with lemmings and arctic hares as the next most frequent item (Marquard-Petersen 1998). On Banks Island, where caribou are uncommon compared to muskoxen, the frequency of muskoxen in wolf scats and stomachs was 80% (1992–1999), with lemmings being the next most frequent item.

Grizzly bear predation on adult muskoxen is not unusual (Gunn and Miller 1982; Case and Stevenson 1991; Clarkson and Liepins 1993; Gunn and Fournier 2000; P. Reynolds, U.S. Fish and Wildlife Service, pers. commun., 2001). Two instances are from an area where muskoxen are infected with the lungworm *Umingmakstrongylus pallikuukensis,* which compromises breathing and may increase vulnerability when a muskox cannot keep up with a stampeding herd. Newborn calves also cannot keep up with a herd and are easily killed by pursuing bears (Clarkson and Liepins 1993). Deep crusted snow also increased the vulnerability of muskoxen in at least one instance of a grizzly bear killing a bull (Case and Stevenson 1991). However, in Alaska where the lungworm is unrecorded, grizzly bears are increasingly killing muskoxen (P. Reynolds, U.S. Fish and Wildlife Service, pers. commun., 2001). Polar bear (*Ursus maritimus*) predation is apparently rare (Pattie 1986).

Weather-Related Deaths. Most weather-related deaths are indirectly related to malnutrition, but deaths directly caused by weather include lightning strikes (Alendal 1980), muskoxen trapped in their beds by freezing rain (Vibe 1967), and the effect of freeze-up without snowfall. A single case of a bull dead from omasal impaction has been documented, likely from the ingestion of forage at a time when water or snow was unavailable, as a hard freeze-up occurred without snow (Gunn et al. 1991).

Accidents and Injuries. Rut injuries among adult bulls include penetrative horn injuries with subsequent infection and impact injuries including internal ruptures (Wilkinson and Shank 1976; Gunn et al. 1991), but rates are unknown. Other accidents include falling off cliffs and breaking through the ice and drowning. On the western edge of their range, muskoxen on Nunivak Island occasionally leave and become stranded on drifting sea-ice. In the 1993–1994 winter, about 70 muskoxen disappeared on the pack ice (Patten 1997).

Diseases and Parasites. Information about diseases comes sporadically from hunters, from serum samples when muskoxen are immobilized for radio-collaring, and more extensively from commercial harvests. Prevalence of bacterial and viral diseases is mostly low or zero based on serum reactivity (Clausen 1986; Zarnke 1996). In Alaska between 1984 and 1989, muskoxen from Nunivak Island, Seward Peninsula, and northeast Alaska had extremely low prevalence of nine bacterial or viral diseases. Three muskoxen were positive for contagious ecthyma, one was positive for epizootic hemorrhagic virus, and four for parainfluenza 3 virus, all from northeast Alaska. A single muskox was positive for *Brucella suis* IV from Nunivak Island, and clinical signs were confirmed as brucellosis in three bulls at widely separated Canadian locations in the 1980s (Gates et al. 1984; Gunn et al. 1991; B. Elkin, Government of Northwest Territories, pers. commun., 1998).

An exception to the apparently low prevalence of bacterial diseases is *Yersinia pseudotuberculosis* on Banks Island. Exposure to *Yersinia* is high: 76% of 350 muskoxen of all sex and age classes had antibodies in the 1990s when sampled during commercial harvests (Larter and Nagy 1999a). Deaths from acute yersiniosis were first diagnosed in August 1986 (Blake et al. 1991) and rates varied annually between 1987 and 1990, when 84% of the 184 carcasses were adult bulls (McLean et al. 1993). An outbreak in August 1996 involved 66–86 muskoxen of all sex and age classes (Larter and Nagy 1999a).

Muskoxen may have a low resistance to parasites even when in good body condition (Alendal and Helle 1983; Korsholm and Olesen 1993). On Victoria Island, seasonal prevalence of the abomasal nematode *Ostertagia* varied from 52% to 100% (Gunn et al. 1991) and averaged 20% on Banks Island (Nagy et al. 2001). On Bathurst Island when muskoxen were at a low density, the prevalence of *Ostertagia* peaked at 40% in June and July. The prevalence of the protozoan *Eimeria* (100%), the abomasal nematode *Nematodirus* (75%), and the tapeworm *Moniezia* (70%) also peaked in summer, in contrast to *Marshallagia,* which peaked in October–November (20%) (Samuel and Gray 1974). The prevalence of the nematode lungworm *Dictyocaulus* on southeast

Victoria Island and Banks Island was variable annually and highest in yearlings (14–54%) (Gunn et al. 1991; Nagy et al. 2001).

Most knowledge of muskox parasites emphasizes prevalence rather parasite ecology and relationships. An exception is our understanding of the lungworm *Umingmakstrongylus pallikuukensis,* a recently described genus of protostrongylid nematode (Hoberg et al. 1995). The lungworm was serendipitously found during the radio-collaring of muskox cows (Gunn and Wobeser 1992). Kutz (1999) unraveled the life cycle, which is indirect, requiring a gastropod intermediate host for larval development and is sensitive to summer temperatures. The effect on muskox survival and risk of predation is uncertain (Kutz et al. 2001). Ninety-three percent of muskoxen were infected and had a strong local reaction to the lungworm, which caused a nodule of a thick layer of fibrous connective tissue surrounding a mass of coiled adult worms. The nodules may displace and compress the surrounding lung tissue, which could impair breathing, especially during exertion. Numbers of nodules increased with age and were greatest in adult bulls (mean = 106.0, range = 0–258) (Gunn et al. 1991). The geographic distribution of *U. pallikuukensis* appears to be restricted to the western Canadian Arctic mainland extending west of Kugluktuk (Kutz et al. 2001).

Other parasites include protozoa such as *Besnoitia* spp. On the Canadian mainland (near Kugluktuk), three bulls had clinical signs (necrotizing solar lamenitis and necrosis of the scrotal skin), which are the first recorded cases of muskoxen with *Besnoitia* (Gunn et al. 1991). Another protozoan parasite is *Toxoplasma gondii,* and 6.4% of 203 muskoxen in the central Canadian Arctic had antibodies, including 4 of 10 adult cows collected near Kugluktuk (Kutz et al. 2000). However, no positive reactions were found to toxoplasmosis in the serum from 132 muskoxen from northeast Greenland screened for 16 diseases (Clausen 1986).

Arthropod parasites are almost unknown, and the muskox's thick pelage seemingly protects them from mosquitoes (Culicidae) and blackflies (Simulididae). On Victoria Island, 3 of 90 muskoxen collected in late winter had a few warble larvae (*Hypoderma tarandi*). The warbles were about half the size of the larvae seen in caribou at the same time of year, however, suggesting that possibly muskoxen are not a suitable host (Gunn et al. 1991).

Population Regulation. Our experience with muskox population dynamics initially was largely with increasing populations because muskoxen in Alaska are reintroduced and those on the Canadian mainland are recolonizing after unregulated commercial harvesting 100 years ago. Coincidentally, muskoxen on the larger southern Arctic Islands were also increasing after a likely weather-related decline in the late 1800s. On the northern Arctic Islands, muskox numbers fluctuate (Tener 1965; Miller et al. 1977; Gunn and Dragon 2002).

Progress in describing muskox forage intake relative to forage availability at different scales, including seasonal effects, has not yet included describing the relationship between forage intake and rate of population increase. We have a fair understanding of how nutrition relates to cow productivity, and some understanding of the effects of grazing on at least the graminoid component of forage, but functional and numerical responses have not been measured. In a cultivated pasture, muskox forage intake was linear over biomass ranging from 1100 to 1200 kg/ha. This suggests that grazing-down experiments to describe functional response curves need biomass levels to be scaled to accommodate the low rate of forage intake (J. Nishi, Government of Northwest Territories, pers. commun., 2001).

Muskoxen are coupled to forage growth through fidelity to their grazing areas. There are positive feedback loops involving nitrogen replenishment from urinary and fecal nitrogen and compensatory growth by grazed plants. Although those positive feedback loops may dampen environmental variation, high levels of variation overwhelm the feedback and drive fluctuations in population size (Caughley and Gunn 1993). Environmental stochasticity is high. The timing of plant growth and snow-ice-mediated availability are annually highly variable and often correlated. For example in the Canadian Arctic, the least and greatest snow cover depth can vary by 75% on either side of the mean and the earliest and latest dates of snow cover bracket the mean by 2–4 weeks. The coefficient of variation for total 9-month snowfall on Banks Island is 42% and 47% for plant-growing degree-days >5°C (Caughley and Gunn 1993). Muskoxen are cold desert herbivores in a variable environment whose populations are likely more closely coupled to the trajectory of their forage than are caribou.

MANAGEMENT TECHNIQUES

Monitoring Trend in Abundance and Productivity. Stratified linear strip-transect surveys from a fixed-wing aircraft have become standard in Canada to estimate muskox abundance (Graf and Case 1989). These surveys usually cover 15–30% of the survey area and strive for a coefficient of variation for the estimate of 10% or less. Stratification to allocate survey effort proportional to muskox density is based on reconnaissance surveys or previous surveys. Alaskan biologists, who work with smaller survey areas, rely on aerial counts covering all the survey area rather than sample counting and use radio-collared muskoxen for a sightability correction.

Unbounded transect (distance) sampling is yet to be tested to verify that it is more efficient (Buckland et al. 1993), especially in low-density areas, where the proportion of muskoxen seen outside the transect lines may be as high as the proportion seen within the transects. The effective strip width for each survey is calculated from the actual data to minimize bias. Errors in estimating group size in relation to animal distance from the observer can be quantified by calculating the frequency of observed group sizes at various distance classes from the observers. The key to the approach is to determine the appropriate distance classes.

Survey altitudes are usually relatively high (180–300 m above the ground) and strip widths are either 500 m or 1 km on either side of the aircraft. Over areas along and within the treeline, survey width and altitude are reduced. Seasonal timing contributes to effective survey design, but the choice depends on whether accuracy or precision is the objective. Surveying in winter means the muskoxen are more conspicuous against a snow background, but even so, double-counting tests suggest that observers miss 10% of the herds (A. Gunn, unpublished data) and the herds are large, which hinders accurate counting. In midsummer, muskoxen are more dispersed in smaller herds, but less conspicuous against their background.

Calf production, survival, and sex ratios are determined during composition surveys, when muskoxen are classified into eight sex–age classes using horn development and body size. Those classes are calves and yearlings (difficult to reliably determine the sex); 2-year-old males and females; 3-year-old males and females; and 4-year-old and older males and females (Gray 1987; Olesen and Thing 1989). Either a helicopter or snowmachines are used to position observers with a spotting scope or binoculars to classify the muskoxen. Muskox calving is extended over several weeks. The muskoxen are in larger herds at this time of year and newborn calves are easily abandoned during any inadvertent stampedes. However, the later after calving that a productivity survey is undertaken, the more the resulting data will measure productivity moderated by early calf mortality.

Capture and Immobilization. Advantage can be taken of the muskox tendency to group together and face an approaching threat. Dogs or snowmachines (Clausen et al. 1984; Smith et al. 1985; Jingfors and Gunn 1989) are effective in rounding up and holding muskoxen while someone approaches on foot to dart a muskox in the muscular neck or thigh. This approach is less stressful than darting from a helicopter, based on serum biochemical values (A. Gunn, unpublished data). The opioids etorphine (reversed with diprenorphine) or carfentanil (reversed with naltrexone) combined with xylazine (reversed with yohimbine) are appropriate immobilizing drugs (Jingfors and Gunn 1989; Kreeger 1996; Reynolds 1998; Gunn and Fournier 2000).

Age Estimation. Counting cementum annuli is unreliable because annuli split and join (Gronquist and Dinneford 1984; Latour 1987). Counting the incremental lines of horn growth has not been adequately tested and is not convenient, as it requires horn removal and polishing (A. Gunn, unpublished data). More convenient are tooth eruption and wear patterns, which can be used up to the age of 5 years (Tener 1965; Henrichsen and Grue 1980).

RESEARCH, MANAGEMENT, AND CONSERVATION NEEDS

Until the 1980s, management goals in Canada and Alaska generally were to increase muskox numbers. In Alaska, the goals were met through translocations (Coady and Hinman 1984; Lent 1999), and in Canada, through conservative levels of harvesting to allow natural recolonization (Gunn 1984). By the 1990s, Canadian management goals were becoming more diverse. Muskox numbers had generally increased and most former range had been reoccupied. Political developments, particularly land claims and associated boards and councils, also changed how goals were established, with a broadening of input into choice of goals. A recent and still-developing trend is to give hunters an effective voice in wildlife conservation through shared responsibilities with government (co-management).

In Alaska, the federal government now has responsibility for the management of game species on federal lands (i.e., national parks, wildlife refuges, and Bureau of Land Management and Forest Service lands) and the state manages species on state and private lands. The dual state–federal management is complex and not without controversy (Lent 1999). Recommendations from local advisory boards, agencies, and private citizens lead to recommendations for regulations to the Alaska Board of Game and the Federal Subsistence Board. The planning includes village, Native corporation, and reindeer herder association input to establish goals for the different regions.

In Canada, Inuit and Inuvialuit settled land claims with the federal government in 1993 and 1984, respectively, which shifted some power and responsibility to co-management boards. The co-management boards represent both the public interest and aboriginal hunters and make day-to-day management decisions on wildlife. The boards have authority for harvest allocation issues and provide advice on wildlife management issues to the government agencies.

Alaska uses formal plans to develop goals, but in the Northwest Territories and Nunavut, planning is still relatively informal. In the Northwest Territories, co-management plans are developed through intensive consultation within the Inuvialuit land claim area. Once the goals are established for a population, then a series of technical steps is used to implement them (Caughley and Sinclair 1994). The goals include public education about muskoxen in response to concerns about muskoxen and caribou or reindeer and recognition of the potential for nonconsumptive use. In both Alaska and Canada, the goal of sustainable harvesting is implemented from pragmatic flexibility rather than application of theory as outlined in, for example, Caughley (1977) and Milner-Gulland and Mace (1998). Muskoxen are hunted under an annual quota-based system. Annual quotas are typically 3–5% of the most recently estimated number of muskoxen and are allocated as a fixed number of muskoxen of either sex. In Alaska, seasons, bag limits, and sex of muskoxen to be harvested vary by area. In Canada, hunter effort is limited by season and assignment to geographic management units. The management units are arbitrary relative to muskox population structure demographically and genetically and are established on the basis of local hunting areas and muskox distribution. The mean size of the 22 muskox management units in Nunavut is 31,762 ± 21,241 km^2 (*SD*) and the 9 units in the Norhtwest Territories are 40,734 ± 29,356 km^2 is size. The total areas are about one third of each territory's land area.

In Canada, it is the community that decides whether the quota is for subsistence or commercial use. In 1979, commercial use either for sport hunting or meat sales was written into regulations. The scale and the capital investment in commercial harvesting vary from frequent small-scale harvests with individual hunters, to small, portable abattoirs handling up to 100–200 muskoxen, to occasional large-scale harvests taking up to 1800 muskoxen (Gunn et al. 1991; Nagy et al. 2001). Commercial use is at low levels and continues to be irregular largely because markets are still developing for meat and qiviut. The quotas have increased since 1967 to 11,210 tags and 1879 tags available in the Northwest Territories and Nunavut, respectively. In Alaska, the number of tags available is 140 (1997).

Harvest rates are generally conservative, but occasionally quotas are set to reduce a population that is perceived as too large. In practice, it is not simple to judge whether a population is exceeding its food supply. Muskoxen on Banks Island annually increased at an exponential rate ($r = 0.136$), reaching 80,000 in 2001 (J. Nagy, Government of Northwest Territories, pers. commun., 2001). In the mid-1980s, the rate and extent of the increase prompted fears of the value-laden specter of "overpopulation" and the need to "reduce the overpopulation" (Gunn et al. 1989a). The annual quota was raised to 15% of the estimated population with the intent of reducing the rate of increase; however, it was unfeasible to annually harvest 10,000 muskoxen. Subsequently, the population increase has continued (Nagy et al. 2001).

In Alaska, harvest rates are adjusted to maintain muskoxen on Nunivak and Nelson Islands at the stated management goal sizes. Between 1975 and 1980, male-only harvesting failed to reduce muskox numbers on Nunivak Island and led to a skewed sex ratio. Inclusion of females into the harvest quota and removal of individuals for transplants stabilized numbers (Smith 1984).

Harvest rates have also been adjusted downward to avoid accelerating a muskox decline even when the cause of the decline was uncertain. A case in point is the decline of muskoxen west of Kugluktuk, on the Canadian mainland. The 1983 estimate was 1295 ± 279 (*SE*) and in the 1986 the estimate was 1800 ± 290 muskoxen, but by 1994 the number had declined to 974 ± 336 (Gunn and Fournier 2000) and the annual quota was reduced from 50 to 20.

Many uncertainties remain in muskox management. In particular, wildlife managers still struggle with the terms *density dependent* and *density independent,* which constrain our thinking, as they describe effects and not causes. Increasingly, the realization is that a situation is not one or the other, but they coexist in a population and have similar effects (Gaillard et al. 1998), although they lead to dissimilar management prescriptions. Data from northeastern Alaska and Banks Island (Reynolds 1998; Nagy et al. 2001) suggest that as muskox numbers increase and presumably start to affect their forage, resilience (the ability to buffer environmental variation) declines and the annual variability of measures such as calf survival increases.

Research and Management Needs. Understanding interactions between muskoxen and their forage, including how weather affects forage growth and availability, will be key to using harvest rates as a management tool. Currently, harvest rates are conservative, which lessens the risk of inadvertently driving a decline instead of reducing the probability of its occurrence. However, the trend toward increasing commercial use will likely reduce the conservatism, which brings risks, as commercial use of wildlife has been likened to "dancing with the devil" (Hawley 1993). Given uncertainties in our understanding of muskox population dynamics, it will be prudent to integrate precautionary strategies into management planning (e.g., Auster 2001). It also will be prudent to invest more effort in explaining to the public that fluctuating populations are natural, and consequently stable yields are unlikely.

Knowledge of mortality is still mainly restricted to causes rather than rates—such as rates of grizzly bear or wolf predation—and we know little about how parasitic infections interact with either predation or foraging. However, to paraphrase Caughley (1981), we will make less progress understanding ecological systems that include predators preying on herbivores until we understand the dynamics of plant–herbivore relationships. This suggests a need to focus on aspects such as discriminating between environmental variability and the effects of the muskoxen themselves on their forage supply.

Conservation. The Global Conservation Status Rank for muskox is G4: "Apparently Secure—Uncommon but not rare (although it may be rare in parts of its range, particularly on the periphery), and usually widespread" (Natural Heritage Central Databases Association, Association for Biodiversity Information, 2001, www.abi.org). Neither Canadian nor U.S. federal legislation lists muskoxen as a species in trouble. Alaska had about 3600 muskoxen (P. Reynolds, U.S. Fish and Wildlife Service, pers. commun., 2001) and in 2001, Canada had about 166,000 (A. Gunn unpublished data).

The ecoregions where muskoxen occur are at least 95% intact in terms of their habitats (Ricketts et al. 1999), although there are concerns for potential effects of oil and gas development, especially on Alaska's North Slope (Reynolds 1998). Muskoxen are vulnerable to human activities especially during calving, when newborn calves are easily left behind by stampeding herds. Although muskoxen can habituate to humans (Miller and Gunn 1980), too little is known about how to foster such behavioral adaptation to warrant complacency about the effects of human activities, especially how those effects may accumulate. Such activities are not only associated with industrial exploration and development, because hunting and snowmachine traffic can displace muskoxen (Patten 1997). Tourism can also be disturbing, especially through overzealous attempts to photograph muskoxen. The tendency of muskoxen to hold their ground when approached is deceptive as to whether they are stressed or not. There is, perhaps, a natural tendency to get closer and closer to such an enigmatic and enduring symbol of the Arctic.

LITERATURE CITED

Adamczewski, J. Z. 1995. Body composition of muskoxen (*Ovibos moschatus*) and its estimation from condition index and mass measurements. Ph.D. Dissertation, University of Saskatchewan, Saskatoon, Canada.

Adamczewski, J. Z., A. Gunn, B. Laarveld, and P. F. Flood. 1992. Seasonal changes in weight, condition and nutrition of free-ranging and captive muskox females. Rangifer 12:179–83.

Adamczewski, J. Z., R. K. Chaplin, J. A. Schaefer, and P. F. Flood. 1994a. Seasonal variation in intake and digestion of a high-roughage diet by muskoxen. Canadian Journal of Animal Science 74:305–13.

Adamczewski, J. Z., W. M. Kerr, E. F. Lammerding, and P. F. Flood. 1994b. Digestion of low-protein grass hay by muskoxen and cattle. Journal of Wildlife Management 58:679–85.

Adamczewski, J. Z., P. F. Flood, and A. Gunn. 1995. Body composition of muskoxen (*Ovibos moschatus*) and its estimation from condition index and mass measurements. Canadian Journal of Zoology 73:2021–34.

Adamczewski, J. Z., P. F. Flood, and A. Gunn. 1997. Seasonal patterns in body composition and reproduction of female muskoxen (*Ovibos moschatus*). Journal of Zoology (London) 241:245–69.

Adamczewski, J. Z., P. J. Fargey, B. Laarveld, A. Gunn, and P. F. Flood. 1998. The influence of fatness on the likelihood of early-winter pregnancy in muskoxen (*Ovibos moschatus*). Theriogenology 50:605–14.

Albon, S. D., T. H. Clutton-Brock, and F. F. Guinness. 1987. Early development and population dynamics in red deer. II. Density-independent effects and cohort variation. Journal of Animal Ecology 56:69–81.

Alendal, E. 1976. The muskox population (*Ovibos moschatus*) in Svalbard. Norsk Polarinstitutt Årbok 1974:159–74.

Alendal, E. 1980. Twelve muskoxen killed by lightning in the Dovre Mountains, South Norway. Fauna 33:49–51.

Alendal, E., and O. Helle. 1983. Helminth parasites of muskoxen *Ovibos moschatus* in Norway including Spitsbergen and in Norway with a synopsis of parasites reported from this host. Fauna norvegica Series A 4:41–52.

Archer, S., and L. L. Tieszen. 1980. Growth and physiological responses of tundra plants to defoliation. Arctic and Alpine Research 12:531–52.

Auster, P. J. 2001. Defining thresholds for precautionary habitat management actions in a fisheries context. North American Journal of Fisheries Management 21:1–9.

Bailey, D. W., J. E. Gross, E. A. Laca, L. R. Rittenhouse, M. B. Coughenour, D. M. Swift, and P. L. Sims. 1996. Mechanisms that result in large herbivore grazing distribution patterns. Journal of Range Management 49:386–400.

Barr, W. 1991. Back from the brink: The road to muskox conservation in the Northwest Territories. Arctic Institute of North America, University of Calgary, Calgary, Alberta, Canada.

Belovsky, G. E. 1991. Insights for caribou/reindeer management using optimal foraging theory. Rangifer 7:7–23.

Blake, J. E., B. D. McLean, and A. Gunn. 1991. Yersiniosis in free-ranging muskoxen on Banks Island, Northwest Territories, Canada. Journal of Wildlife Diseases 27:527–83.

Blakley, B. R., S. J. Kutz, S. C. Tedesco, and P. F. Flood. 2000. Trace mineral and vitamin concentrations in the liver and serum of wild muskoxen from Victoria Island. Journal of Wildlife Diseases 36:301–7.

Bliss, L. C., and J. H. Richards. 1982. Present-day arctic vegetation and ecosystems as a predictive tool for the arctic-steppe mammoth biome. Pages 241–57 in D. M. Hopkins, Jr., J. V. Matthews, C. E. Schweger, and S. B. Young, eds. Paleoecology of Beringia. Academic Press, New York.

Blix, A. S., H. J. Grav, K. A. Markussen, and R. G. White. 1984. Modes of thermal protection in newborn muskoxen. Page 207 *in* Biological papers, special report No. 4. University of Alaska, Fairbanks.

Boertmann, D., M. Forchhammer, C. R. Olesen, P. Aastrup, and H. Thing. 1992. The Greenland muskox population status since 1989. Rangifer 12:5–12.

Boyd, C. S., W. B. Collins, and P. J. Urness. 1996. Relationship of dietary browse to intake in captive muskoxen. Journal of Range Management 49:2–7.

Buckland, S. T., D. R. Anderson, K. P. Burnham, and J. L. Laake. 1993. Distance sampling: Estimating abundance of biological populations. Chapman and Hall, London.

Burrin, D. G., C. L. Ferrell, R. A. Britton, and M. Bauer. 1990. Level of nutrition and visceral organ size and metabolic activity in sheep. British Journal of Nutrition 64:439–48.

Case, R., and J. Stevenson. 1991. Observations of barren-ground grizzly bear predation, *Ursus arctos* on muskoxen *Ovibos moschatus* in the Northwest Territories. Canadian Field-Naturalist 105:105–6.

Caughley, G. 1977. Analysis of vertebrate populations. Wiley-Interscience, London.

Caughley, G. 1981. What we do not know about the dynamics of large mammals. Chapter 18 *in* C. W. Fowler and T. D. Smith, eds. Dynamics of large mammal populations. Wiley-Interscience, New York.

Caughley, G., and A. Gunn. 1993. Dynamics of large herbivores in deserts: Kangaroos and caribou. Oikos 67:47–55.

Caughley, G., and A. R. E. Sinclair. 1994. Wildlife ecology and management. Blackwell, Oxford.

Chaplin, R. K., and C. S. Stevens. 1989. Growth rates of captive Banks Island muskoxen. Canadian Journal of Zoology 67:A54–55.

Clarkson, P. L., and I. S. Liepins. 1993. Grizzly bear (*Ursus arctos*) predation on muskox (*Ovibos moschatus*) calves near Horton River, Northwest Territories. Canadian Field-Naturalist 107:100–102.

Clausen, B. 1986. Survey for antibodies against various infectious disease agents in muskoxen (*Ovibos moschatus*) from Jamesonland, northeast Greenland. Journal of Wildlife Diseases 22:264–66.

Clausen, B., P. Hjort, H. Strandgaard, and P. L. Sorensen. 1984. Immobilization and tagging of muskoxen (*Ovibos moschatus*) in Jameson Land, Northeast Greenland. Journal of Wildlife Diseases 20:141–45.

Coady, J. W., and R. A. Hinman. 1984. Management of muskoxen in Alaska. Pages 47–51 *in* Biological papers, special report No. 4. University of Alaska, Fairbanks.

Cote, S., J. A. Schaefer, and F. Messier. 1997. Time budgets and synchrony of activities in muskoxen: The influence of sex, age, and season. Canadian Journal of Zoology 75:1628–35.

Danks, F. S., and D. R. Klein. 1999. Development of a muskox habitat map for northern Alaska using GPS. Scientific and social programme, abstracts, 10th arctic ungulate conference (Rangifer Report No. 4). Tromsø, Norway.

Dieterich, R. A., and M. E. Fowler. 1986. Musk-oxen. Pages 996–98 *in* M. E. Fowler, ed. Zoo and wild animal medicine. W. B. Saunders, Philadelphia.

Dinneford, W. B., and D. A. Anderson. 1984. Fetal twinning rates, pregnancy rates, and fetal sex ratios in two Alaskan muskox populations. Pages 64–66 *in* Biological papers, special report No. 4. University of Alaska, Fairbanks.

Flood, P. F., and S. C. Tedesco. 1997. Relationship between conception date and latitude in muskoxen. Rangifer 17:25–30.

Flood, P. F., S. R. Abrams, G. D. Muir, and J. E. Rowell. 1989a. The odour of the muskox: A preliminary investigation. Journal of Chemical Ecology 15:2207–17.

Flood, P. F., M. J. Stalker, and J. E. Rowell. 1989b. The hair follicle density and seasonal shedding cycle of the muskox (*Ovibos moschatus*). Canadian Journal of Zoology 67:1143–47.

Forchhammer, M. C., and J. J. Boomsma. 1995. Foraging strategies and seasonal diet optimization of muskoxen in West Greenland. Oecologia 104:169–80.

Forchhammer, M. C., and J. J. Boomsma. 1998. Optimal mating strategies in nonterritorial ungulates: A general model tested on muskoxen. Behavioral Ecology 9:136–43.

Fournier, B., and A. Gunn. 1998. Muskox numbers and distribution in the Northwest Territories, 1997 (File Report No. 121). Northwest Territories Department of Resources, Wildlife and Economic Development, Yellowknife, Canada.

Gaillard, J.-M., M. Festa-Bianchet, and N. G. Yoccoz. 1998. Population dynamics of large herbivores: Variable recruitment with constant adult survival. Trends in Ecological Evolution 13:58–63.

Gaillard, J.-M., M. Festa-Bianchet, N. G. Yoccoz, A. Loison, and C. Toïgo. 2000. Temporal variation in fitness components and population dynamics of large herbivores. Annual Review of Ecological Systems 31:367–93.

Gates, C. C., G. Wobeser, and L. B. Forbes. 1984. Rangiferine brucellosis in a muskox, *Ovibos moschatus moschatus* (Zimmerman). Journal of Wildlife Diseases 20:234–35.

Graf, R., and R. Case. 1989. Counting muskoxen in the Northwest Territories. Canadian Journal of Zoology 67:1112–15.

Gray, D. R. 1987. The muskoxen of the Polar Bear Pass. National Museum of Natural Sciences. Fitzhenry and Whiteside, Markham, Ontario, Canada.

Gray, D. R., P. F. Flood, and J. E. Rowell. 1988. The structure and function of muskox preorbital glands. Canadian Journal of Zoology 67:1134–42.

Gronquist, R. M., and B. W. Dinneford. 1984. Age determination of muskoxen from dental cementum annuli. Pages 67–68 *in* Biological papers, special report No. 4. University of Alaska, Fairbanks.

Groves, P. 1992. Muskox husbandry; a guide for the care, feeding and breeding of captive muskoxen. Biological papers, special report No. 5. University of Alaska, Fairbanks.

Groves, P. 1995. The takin and muskox: A molecular and ecological evaluation of relationship. Ph.D. Dissertation, University of Alaska, Fairbanks.

Gunn, A. 1984. Aspects of the management of muskoxen in the Northwest Territories. Pages 33–40 *in* Biological papers, special report No. 4. University of Alaska, Fairbanks.

Gunn, A. 1985. Observations of cream-colored muskoxen in the Queen Maud Gulf area of Northwest Territories. Journal of Mammalogy 66:803–4.

Gunn, A. 1990a. The decline and recovery of caribou and muskoxen on Victoria Island. Pages 590–607 *in* C. R. Harington, ed. Canada's missing dimension: Science and history in the Canadian Arctic Islands. Canadian Museum of Nature, Ottawa.

Gunn, A. 1990b. Distribution and abundance of muskoxen between Bathurst Inlet and Contwoyto Lake NWT, 1986. (File Report No.100). Northwest Territories Department of Renewable Resources, Yellowknife, Canada.

Gunn, A. 1992a. Differences in the sex and age composition of two muskox populations and implications for male breeding strategies [Expanded abstract]. Rangifer 12:17–19.

Gunn, A. 1992b. The dynamics of caribou and muskoxen foraging in arctic ecosystems. Rangifer 12:12–15.

Gunn, A., and R. Case. 1984. Numbers and distribution of muskoxen in the Queen Maud Gulf area, July 1982 File Report No. 39. Northwest Territories Department of Renewable Resources, Yellowknife, Canada.

Gunn, A., and J. Dragon. 1998. Abundance and distribution of caribou and muskoxen on Prince of Wales and Somerset islands and Boothia Peninsula, 1995 (File report No. 122). Northwest Territories Department of Resources, Wildlife and Economic Development, Yellowknife, Canada.

Gunn, A., and J. Dragon. 2002. Peary caribou and muskox abundance and distribution on the western Queen Elizabeth Islands, Northwest Territories and Nunavut June–July 1997 (File Report No. 130). Northwest Territories Department of Resources, Wildlife and Economic Development, Yellowknife, Canada.

Gunn, A., and B. Fournier. 2000. Calf survival and seasonal migrations of a mainland muskox population (File Report No. 124). Northwest Territories Department of Resources, Wildlife and Economic Development, Yellowknife, Canada.

Gunn, A., and F. L. Miller. 1982. Muskox bull killed by barren-ground grizzly bear, Thelon Game Sanctuary, N.W.T. Arctic 35:545–46.

Gunn, A., and M. Sutherland. 1997. Muskox diet and sex–age composition in the Central Arctic coastal mainland (Queen Maud Gulf area) 1988–1991 (Manuscript Report No. 95). Northwest Territories Department of Resources, Wildlife and Economic Development, Yellowknife, Canada.

Gunn, A., and G. Wobeser. 1992. Protostrongylid lungworm infection in muskoxen, Coppermine, N.W.T. Rangifer 13:45–48.

Gunn, A., C. C. Shank, and G. Caughley. 1989a. Report of the workshop on management options for rapidly expanding muskox populations using Banks Island as an example. Pages A37–A38 *in* P. F. Flood, ed. Proceedings of the second international muskox symposium. National Research Council of Canada, Ottawa.

Gunn, A., F. L. Miller, and B. McLean. 1989b. Evidence for and possible causes of increased mortality of bull muskoxen during severe winters. Canadian Journal of Zoology 67:1106–11.

Gunn, A., J. Adamczewski, and B. Elkin. 1991. Commercial harvesting of muskoxen in the Northwest Territories. Pages 197–204 *in* L. A. Renecker and R. J. Hudson, eds. Wildlife production: Conservation and sustainable development (Miscellaneous Publication 91–6). University of Alaska, Agricultural and Forestry Experiment Station.

Gunn, A., F. L. Miller, and J. Nishi. 2000. Status of endangered and threatened caribou on Canada's Arctic Islands. Rangifer (Special Issue) 12:39–50.

Halikas, G. C. 1971. Viscous properties of muskox blood. Comparative Biochemistry and Physiology 39A:869–74.

Harington, C. R. 1961. History, distribution and ecology of the muskoxen. M.Sc. Thesis, McGill University, Montreal.

Harington, C. R. 1970. A Pleistocene muskox (*Ovibos moschatus*) from gravels of Illinion age near Nome, Alaska. Journal of Earth Science 7:1326–31.

Hawley, A. W. L. 1993. Commercialization and wildlife management: Dancing with the devil. Krieger, Malabar, FL.

Heard, D. C. 1992. The effect of wolf predation and snow cover on musk-ox group size. American Naturalist 139:190–204.

Heinrich, R. E., C. B. Ruff, and J. Z. Adamczewski. 1999. Ontogenetic changes in mineralization and bone geomety in the femur of muskoxen (*Ovibos moschatus*). Journal of Zoology (London) 247:215–23.

Henrichsen, P. 1982. Population analysis of muskoxen, *Ovibos moschatus* (Zimmermann 1780), based on occurrence of dental anomalies. Säugetierkundliche Mitteilungen 30:260–80.

Henrichsen, P., and R. A. Dieterich. 1984. Dental lesions in muskoxen from natural, introduced and captive populations. Pages 183–85 *in* Biological papers, special report No. 4. University of Alaska, Fairbanks.

Henrichsen, P., and H. Grue. 1980. Age criteria in the muskox (*Ovibos moschatus*) from Greenland. Danish Review of Game Biology 11:1–18.

Hoare, E. K., S. E. Parker, P. F. Flood, and G. P. Adams. 1997. Ultrasonic imaging of reproductive events in muskoxen. Rangifer 17:119–23.

Hoberg, E. P., L. Polley, A. Gunn, and J. S. Nishi. 1995. Umingmakstrongylus pallikuukensis gen. nov. et sp. nov. (Nematoda: Protostrongylidae) from muskoxen, *Ovibos moschatus,* in the central Canadian Arctic, with comments on biology and biogeography. Canadian Journal of Zoology 73:2266–82.

Hofmann, R. R. 1989. Evolutionary steps of ecophysiological adaptation and diversification of ruminants: A comparative view of their digestive system. Oecologia 78:443–57.

Hofmann, R. R. 2000. Functional and comparative digestive system anatomy of arctic ungulates. Rangifer 20:71–82.

Holt, R. D. 1984. Spatial heterogeneity, indirect interactions, and the coexistence of prey species. American Naturalist 124:377–406.

Hubert, B. A. 1977. Estimated productivity of muskox on Truelove Lowland. Pages 467–91 *in* L. C. Bliss, ed. Truelove Lowland, Devon Island, Canada—A High Arctic ecosystem. University of Alberta Press, Edmonton, Canada.

Hutchings, M. R., I. Kyriazakis, I. J. Gordon, and F. Jackson. 1999. Trade-offs between nutrient intake and faecal avoidance in herbivore foraging decisions: The effect of animal parasitic status, level of feeding motivation and sward nitrogen content. Journal of Animal Ecology 68:310–23.

Hutchings, M. R., I. Kyriazakis, T. G. Papachristou, I. J. Gordon, and F. Jackson. 2000. The herbivores' dilemma: Trade-offs between nutrition and parasitism in foraging decisions. Oecologia 124:242–51.

Ihl, C. 1999. Comparative habitat and diet selection of muskoxen and reindeer on the Seward Peninsula, western Alaska. M.Sc. Thesis, University of Alaska, Fairbanks.

Illius, A. W., S. D. Albon, J. M. Pemberton, J. M. Gordan, and T. H. Clutton-Brock. 1995. Selection for foraging efficiency during a population crash in Soay sheep. Journal of Animal Ecology 64:481–92.

Jenkins, K. J., and R. G. White. 1987. Dietary niche relationships among cervids relative to snowpack in northwestern Montana. Canadian Journal of Zoology 65:1397–1401.

Jingfors, K. 1980. Habitat relationships and activity patterns of an introduced muskox population. M.Sc. Thesis, University of Alaska, Fairbanks.

Jingfors, K. 1984. Observations of cow–calf behavior in free-ranging muskoxen. Pages 105–9 *in* Biological papers, special report No. 4. University of Alaska, Fairbanks.

Jingfors, K., and A. Gunn. 1989. The use of snowmachines in the drug immobilization of muskoxen. Canadian Journal of Zoology 67:1120–21.

Jingfors, K. T., and D. R. Klein. 1982. Productivity in recently established muskox populations in Alaska. Journal of Wildlife Management 46:1092–96.

Jung, H. G., G. O. Batzli, and D. S. Siegler. 1979. Patterns in the phytochemistry of Arctic plants. Biochemical Systematics and Ecology 7:203–9.

Kempster, A. J. F. 1980. Fat partition and distribution in the carcasses of cattle, sheep and pigs. Meat Science 5:83–98.

Kerr, D. E., B. Laarveld, M. Fehr, and J. G. Manns. 1991. Profiles of serum IGF-1 calves from birth to 18 months of age and in cows throughout the lactation cycle. Canadian Journal of Animal Science 71:695–706.

Klein, D. R. 1992. Comparative ecological and behavioral adaptations of *Ovibos moschatus* and *Rangifer tarandus*. Rangifer 12:47–55.

Klein, D. R., and C. Bay. 1990. Foraging dynamics of muskoxen in Peary Land, northern Greenland. Holarctic Ecology 13:269–80.

Klein, D. R., and C. Bay. 1994. Resource partitioning by mammalian herbivores in the High Arctic. Oecologia 97:439–50.

Klein, D. R., and H. Staaland. 1984. Extinction of Svalbard muskoxen through competitive exclusion: A hypothesis. Pages 26–31 *in* Biological papers, special report No. 4. University of Alaska, Fairbanks.

Klein, D. R., and H. Thing. 1989. Chemical elements in mineral licks and associated muskoxen feces in Jameson Land, northeast Greenland. Canadian Journal of Zoology 67:1092–95.

Kohlmann, S. G., and K. R. Risenhoover. 1994. Spatial and behavioural response of white-tailed deer to forage depletion. Canadian Journal of Zoology 72:506–13.

Korsholm, H., and C. R. Olesen. 1993. Preliminary investigation on the parasite burden and distribution of endoparasite species of muskox (Ovibos moschatus) and caribou (*Rangifer tarandus groenlandicus*) in West Greenland. Rangifer 13:185–89.

Kreeger, T. J. 1996. Handbook of wildlife chemical immobilization. International Wildlife Veterinary Services, Laramie, WY.

Kutz, S. J. 1999. The biology of *Umingmakstrongylus pallikuukensis,* a lung nematode of muskoxen in the Canadian Arctic: Field and laboratory studies. Ph.D. Dissertation, University of Saskatchewan, Saskatoon, Canada.

Kutz, S. J., B. Elkin, A. Gunn, and J. P. Dubey. 2000. Prevalence of *Toxoplasma gondii* antibodies in muskox (*Ovibos moschatus*) sera from northern Canada. Journal of Parasitology 86:879–82.

Kutz, S. J., E. P. Hoberg, and L. Polley. 2001. A new lungworm in muskoxen: An exploration in arctic parasitology. Trends in Parasitology 17:276–80.

Larter, N. C. 1999. Seasonal changes in arctic hare, *Lepus arcticus,* diet composition and differential digestibility. Canadian Field-Naturalist 113:481–86.

Larter, N. C., and J. A. Nagy. 1997. Peary caribou, muskoxen and Banks Island forage: Assessing seasonal diet similarities. Rangifer 17:9–16.

Larter, N. C., and J. A. Nagy. 1999a. Muskox mortality survey, Banks Island, August 1996 (Manuscript Report No. 117). Northwest Territories Department of Resources, Wildlife and Economic Development, Yellowknife, Canada.

Larter, N. C., and J. A. Nagy. 1999b. Seasonal and annual variability in the quality of forages consumed by Peary caribou and muskoxen on Banks Island. *In* Scientific and social programme, abstracts, 10th arctic ungulate conference (Rangifer Report No. 4). Tromsø, Norway.

Larter, N. C., and J. A. Nagy. 2000a. Annual and seasonal differences in snow depth, density, and resistance in four habitats on southern Banks Island, 1993–1998 (Manuscript Report No. 136). Northwest Territories Department of Resources, Wildlife and Economic Development, Yellowknife, Canada.

Larter, N. C., and J. A. Nagy. 2000b. Overwinter changes in urea nitrogen:creatinine and cortisol:creatinine ratios in urine from Banks Island muskox. Journal of Wildlife Management 65:226–34.

Larter, N. C., and J. A. Nagy. 2001. Seasonal and annual variability in the quality of important forage plants on Banks Island, Canadian High Arctic. Applied Vegetation Science 4:115–28.

Latour, P. B. 1987. Observations on demography, reproduction, and morphology of muskoxen (*Ovibos moschatus*) on Banks Island, Northwest Territories. Canadian Journal of Zoology 65:265–69.

Le Hénaff, D., and M. Crete. 1989. Introduction of muskoxen in northern Quebec: The demographic explosion of a colonizing herbivore. Canadian Journal of Zoology 67:1102–5.

Lent, P. C. 1999. Muskoxen and their hunters: A history. University of Oklahoma Press, Norman.

Lund, D. C. 1992. Humane management of captive muskoxen. Rangifer 12:151–57.

Macfarlane, W. V., B. Howard, H. Haines, P. J. Kennedy, and C. M. Sharpe. 1971. Hierarchy of water and energy turnover of desert mammals. Nature 234:483–84.

Mallory, F. F. 1995. Observations on maternal behaviour in Muskoxen, *Ovibos moschatus,* during a river crossing. Canadian Field-Naturalist 109:264–65.

Marquard-Petersen, U. 1998. Food habits of arctic wolves in Greenland. Journal of Mammalogy 79:236–44.

Maxwell, J. B. 1980. The climate of the Canadian Arctic Islands and adjacent waters, Vol. 1. Atmospheric Environment Service, Environment Canada, Hull, Quebec, Canada.

McLean, B. D., P. Fraser, and J. E. Blake. 1993. Yersiniosis in muskoxen on Banks Island, N.W.T., 1987–1990. Rangifer 13:65–66.

Mech, L. D. 2000. Lack of reproduction in muskoxen and arctic hares caused by early winter? Arctic 53:69–71.

Mech, L. D., and L. G. Adams. 1999. Killing of a muskox, *Ovibos moschatus,* by two wolves, *Canis lupus,* and subsequent caching. Canadian Field-Naturalist 113:673–75.

Miller, F. L. 1998. Status of Peary caribou and muskox populations within the Bathurst Island complex, south-central Queen Elizabeth Islands, Northwest Territories, July 1996 (Technical Report Series No. 317). Canadian Wildlife Service, Ottawa.

Miller, F. L., and A. Gunn. 1980. Behavioural responses of muskox herds to simulation of slinging by helicopter, Northwest Territories. Canadian-Field Naturalist 94:52–60.

Miller, F. L., R. H. Russell, and A. Gunn. 1977. Distributions, movements and numbers of Peary caribou and muskoxen on western Queen Elizabeth Islands (Report Series No. 40). Canadian Wildlife Service, Ottawa.

Milner-Gulland, E. J., and R. E. Mace. 1998. Conservation of biological resources. Blackwell, Cambridge, MA.

Muc, M. 1977. Ecology and primary production of sedge–moss meadow communities, Truelove Lowland. Pages 185–216 *in* L. C. Bliss, ed. Truelove Lowland, Devon Island, Canada—A High Arctic ecosystem. University of Alberta Press, Edmonton, Canada.

Mulder, C. P. H., and R. Harmsen. 1995. The effect of muskox herbivory on growth and reproduction in an arctic legume. Arctic and Alpine Research 27:44–53.

Murray, J. L. 1991. Biomass allocation and nutrient pool in major muskoxen-grazed communities in Sverdrup Pass, 75°N, Ellesmere Island, N.W.T. M.Sc. Thesis, University of Toronto, Toronto.

Nagy, J. A., N. C. Larter, and V. P. Fraser. 1996. Population demography of Peary caribou and muskox on Banks Island, N.W.T., 1982–1992. Rangifer (Special Issue) 9:213–22.

Nagy, J. A., N. Larter, M. Branigan, E. McLean, and J. Hines. 2001. Co-management plan for caribou, muskox, arctic wolves, snow geese, and small herbivores on Banks Island. Recommendation by Sachs Harbour Hunters and Trappers Committee, Inuvialuit Game Council, and Wildlife Management Advisory Council, Northwest Territories, Canada.

Nellermann, C. 1997. Grazing strategies of muskoxen (*Ovibos moschatus*) during winter in Anguajaartorfiup Nunaa in western Greenland. Canadian Journal of Zoology 75:1129–34.

Nilssen, K. J., S. D. Mathiesen, and A. S. Blix. 1994. Metabolic rate and plasma T3 in ad lib fed and starved muskoxen. Rangifer 14:79–81.

Oakes, E. J., R. Harmsen, and C. Eberl. 1989. Sex, age, and seasonal differences in the diets and activity budgets of muskoxen (*Ovibos moschatus*). Canadian Journal of Zoology 70:605–16.

Olesen, C. R., and H. Thing. 1989. Guide to field classification by sex and age of the muskox. Canadian Journal of Zoology 67:1116–19.

Olesen, C. R., H. Thing, and P. Aastrup. 1994. Growth of wild muskoxen under two nutritional regimes in Greenland. Rangifer 14:3–10.

Parker, G. R. 1978. The diets of muskoxen and Peary caribou on some islands in the Canadian High Arctic (Occasional Paper No. 35). Canadian Wildlife Service, Ottawa.

Parker, G. R., D. C. Thomas, E. Broughton, and D. R. Gray. 1975. Crashes of muskox and Peary caribou populations in 1973–74 in the Parry Islands, Arctic Canada (Progress Notes No. 56). Canadian Wildlife Service, Ottawa.

Parker, K., R. G. White, M. P. Gillingham, and D. F. Holleman. 1990. Comparison of energy metabolism in relation to daily activity and milk consumption by caribou and muskox neonates. Canadian Journal of Zoology 68:106–14.

Patten, S. M. 1997. Muskox survey—Inventory management report (Federal Aid in Wildlife Restoration Program Report Project W-24–3 and W-24–4, Study 16). Alaska Department of Fish and Game.

Pattie, D. L. 1986. Muskox density and calf numbers on Devon Island's north coast. Journal of Mammalogy 67:190–91.

Pinsonneault, Y. 1995. Response of arctic sedges to simulated grazing by muskoxen. M.Sc. Thesis, University of Alberta, Edmonton, Canada.

Raillard, M. 1992. Influence of muskox grazing on plant communities of Sverdrup Pass (79°N), Ellesmere Island, N.W.T. Canada. Ph.D. Dissertation, University of Toronto, Toronto.

Reynolds, P. E. 1993. Dynamics of muskox groups in Alaska. Rangifer 13:83–90.

Reynolds, P. E. 1998. Dynamics and range expansion of a reestablished muskox population. Journal of Wildlife Management 62:734–44.

Ricketts, T. H., E. Dinerstein, D. M. Olson, C. Loucks, W. Eichbaum, D. DellaSala, K. Kavanagh, P. Hedao, P. Hurley, K. Carney, R. Abell, and S. Walters. 1999. Terrestrial ecoregions of North America: A conservation assessment. Island Press, Washington, DC.

Robbins, C. T., D. E. Spalinger, and W. van Hoven. 1995. Adaptation of ruminants to browse and grass diets: Are anatomically-based browser–grazer interpretations valid? Oecologia 103:208–13.

Robus, M. A. 1981. Muskox habitat and use patterns in northeastern Alaska. M.Sc. thesis, University of Alaska, Fairbanks.

Rowell, J. E., and P. F. Flood. 1988. Progesterone, oestradiol 17β and LH during the oestrous cycle of muskoxen (*Ovibos moschatus*). Journal of Reproduction and Fertility 84:117–22.

Rowell, J., K. J. Betteridge, G. C. B. Randall, and J. C. Fenwick. 1987. Anatomy of the reproductive tract of the female muskox (*Ovibos moschatus*). Journal of Reproduction and Fertility 80:431–44.

Rowell, J. E., R. A. Pierson, and P. F. Flood. 1993. Endocrine changes and luteal morphology during pregnancy in muskoxen (*Ovibos moschatus*). Journal of Reproduction and Fertility 99:7–13.

Rowell, J. E., R. G. White, and W. E. Hauer. 1997. Progesterone during the breeding season and pregnancy in female muskoxen on different dietary regimens. Rangifer 17:125–29.

Rowell, J. E., C. J. Lupton, M. A. Robertson, J. A. Nagy, and R. G. White. 1999. Objective measures of qiviut fibre from wild muskoxen. *in* Scientific and social programme, abstracts, 10th arctic ungulate conference (Rangifer Report No. 4). Tromsø, Norway.

Russell, R. H., E. J. Edmonds, and J. Roland. 1978. Caribou and muskoxen habitat studies (Environmental–Social Program, Northern Pipelines, ESCOM No. A1-26). Minister of Indian and Northern Affairs and Minister of State, Ottawa.

Samuel, W. M., and D. R. Gray. 1974. Parasitic infection in muskoxen. Journal of Wildlife Management 38:775–82.

Sandell, M., and O. Liberg. 1992. Roamers and stayers: A model on male mating tactics and mating systems. American Naturalist 139:177–89.

Schaefer, J. A. 1995. High Arctic habitat structure and habitat selection by muskoxen (*Ovibos moschatus*): A multiscale approach. Ph.D. Dissertation, University of Saskatchewan, Saskatoon, Canada.

Schaefer, J. A., and F. Messier. 1995a. Habitat selection as a hierarchy: The spatial scales of winter foraging by muskoxen. Ecography 18:333–44.

Schaefer, J. A., and F. Messier. 1995b. Winter foraging by muskoxen: A hierarchial approach to patch residence time and cratering behaviour. Oecologia 104:39–44.

Schaefer, J. A., and F. Messier. 1996. Winter activity of muskoxen in relation to foraging conditions. Ecoscience 3:147–53.

Schulman, A. B., and R. G. White. 1997. Nursing behaviour as a predictor of alternate-year reproduction in muskoxen. Rangifer 17:31–35.

Shipley, L. A., J. E. Gross, D. E. Spalinger, N. T. Hobbs, and B. A. Wunder. 1994. The scaling of intake rate in mammalian herbivores. American Naturalist 143:1055–82.

Smith, D. L. 1996. Muskoxen/sedge meadows interactions, north central Banks Island, Northwest territories, Canada. Ph.D. Dissertation, University of Saskatchewan, Saskatoon, Canada.

Smith, R. F. C., and L. C. H. Wang. 1977. Arctic hares on Truelove Lowland. Pages 461–466 *in* L. C. Bliss, ed. Truelove Lowland, Devon Island, Canada—A High Arctic ecosystem. University of Alberta Press, Edmonton, Canada.

Smith, T. E. 1984. Population status and management of muskoxen on Nunivak Island, Alaska. Pages 52–56 *in* Biological papers, special report No. 4. University of Alaska, Fairbanks.

Smith, T. E. 1989. The role of bulls in pioneering new habitats in an expanding muskox population on the Seward Peninsula, Alaska. Canadian Journal of Zoology 67:1096–1101.

Smith, T. E., C. C. Grauvogel, and D. A. Anderson. 1985. Status and dispersal of an introduced muskox population on the Seward Peninsula (Federal Aid in Wildlife Restoration Project Progress Report, Job 16.1, Projects W-22-4 and W-22-6). Alaska Department of Fish and Game.

Staaland, H., and C. R. Olesen. 1999. Mineral nutrition and alimentary pools in muskoxen and caribou on the Angujaartorfiup Nunaa range in West Greenland. Rangifer 19:33–40.

Staaland, H., J. Z. Adamczewski, and A. Gunn. 1997. A comparison of digestive tract morphology in muskoxen and caribou from Victoria Island, Northwest Territories, Canada. Rangifer 17:17–19.

Suttie, J. M., and J. R. Webster. 1998. Are arctic ungulates physiologically unique? Rangifer 18:99–118.

Svoboda, J. 1977. Ecology and primary production of raised beach communities, Truelove Lowland. Pages 185–216 *in* L. C. Bliss, ed. Truelove Lowland, Devon Island. Canada—A High Arctic ecosystem. University of Alberta Press, Edmonton, Canada.

Swingley, A. B. S., and R. G. White. 1999. Behavioural influence on duration of lactation in muskoxen may not be nutritional. *in* Scientific and social programme, abstracts, 10th arctic ungulate conference (Rangifer Report No. 4). Tromsø, Norway.

Sykes, A. R. 1987. Endoparasites and herbivore nutrition. Pages 211–232 *in* J. B. Hacker and J. H. Ternouth, eds. Nutrition of herbivores. Academic Press, Sydney, Australia.

Tedesco, S. C. 1996. Melatonin and seasonal cycles in muskoxen. Ph.D. Dissertation, University of Saskatchewan, Saskatoon, Canada.

Tedesco, S., S. Buczkowski, J. Adamczewski, J. Archer, and P. F. Flood. 1991. Hematology and serum biochemistry values in muskoxen. Rangifer 11:75–77.

Tedesco, S. C., P. F. Flood, D. J. Morton, and R. J. Reiter. 1992. Seasonal melatonin and luteinizing hormone rhythms in muskoxen at 52°N. Rangifer 12:197–201.

Tener, J. S. 1954. A preliminary study of the musk-oxen of Forsheim Peninsula, Ellesmere Island, N.W.T. (Management Bulletin Series 1, Vol. 9). Canadian Wildlife Service, Ottawa.

Tener, J. S. 1965. Muskoxen (Monograph No. 2). Canadian Wildlife Service, Ottawa.

Thing, H. 1984. Food and habitat selection by muskoxen in Jameson Land, northeast Greenland: A preliminary report. Pages 69–74 *in* Biological papers, special report No. 4. University of Alaska, Fairbanks.

Thing, H., D. R. Klein, K. Jingfors, and S. Holt. 1987. Ecology of muskoxen in Jameson Land, northeast Greenland. Holarctic Ecology 10:95–103.

Thomas, D. C., and E. J. Edmonds. 1984. Competition between caribou and muskoxen, Melville Island, N.W.T., Canada. Pages 93–100 *in* Biological papers, special report No. 4. University of Alaska, Fairbanks.

Thomas, D. C., F. L. Miller, R. H. Russell, and G. R. Parker. 1981. The Bailey Point region and other muskox refugia in the Canadian Arctic: A short review. Arctic 34:34–36.

Tsarev, S. A., and T. P. Sipko. 1999. Dynamics and state of muskox population in Taimyr. Pages XX–XX *in* Scientific and social programme, abstracts, 10th arctic ungulate conference (Rangifer Report No. 4). Tromsø, Norway.

Tyler, N. J. C., and A. S. Blix. 1990. Survival strategies in Arctic ungulates. Rangifer (Special Issue) 3:211–30.

van Coeverden de Groot, P. 2000. Microsatellite variation in the muskox Ovibos moschatus. Ph.D. Dissertation, Queens University, Kingston, Ontario, Canada.

van Houtert, M. F. J., and A. R. Sykes. 1996. Implications of nutrition for the ability of ruminants to withstand gastrointestional nematode infections. International Journal for Parasitology 26:1151–68.

Van Soest, P. J. 1982. Nutritional ecology of the ruminant. Cornell University Press, Ithaca, NY.

Vibe, C. 1967. Arctic animals in relation to climatic fluctuations. Meddelelser om Grønland 170:1–227.

White, R. G. 1983. Foraging patterns and their multiplier effects on productivity of northern ungulates. Oikos 40:377–84.

White, R. W., D. F. Holleman, and B. A. Tiplady. 1989. Seasonal body weight, body condition and lactational trend in muskoxen. Canadian Journal of Zoology 67:1125–33.

White, R. G., B. A. Tiplady, and P. Groves. 1991. Qiviut production from muskoxen. Pages 387–400 *in* R. J. Hudson, K. R. Drew, and L. M. Baskin, eds. Wildlife production systems: Economic utilisation of wild ungulates. Cambridge University Press, Cambridge.

White, R. G., D. F. Holleman, P. Wheat, P. G. Tallas, M. Jourdan, and P. Henrichsen. 1994. Seasonal changes in voluntary intake and digestibility of diets by captive muskoxen. Pages 193–94 *in* Biological papers, special report No. 4. University of Alaska, Fairbanks.

White, R. G., J. E. Rowell, and W. E. Hauer. 1997. The role of nutrition, body condition and lactation on calving success in muskoxen. Journal of Zoology (London) 243:13–20.

Wilkinson, P. F. 1975. The length and diameter of coat fibres of the musk ox. Journal of Zoology (London) 177:363–75.

Wilkinson, P. F., and C. C. Shank. 1976. Rutting-fight mortality among musk oxen on Banks Island, Northwest Territories, Canada. Animal Behaviour 24:756–58.

Wilkinson, P. F., C. C. Shank, and D. F. Penner. 1976. Muskox–caribou summer range relations on Banks Island, N.W.T. Journal of Wildlife Management 40:151–62.

Wilson, K. J. 1992. Spatial scales of muskox resource selection in late winter. M.Sc. Thesis, University of Alaska, Fairbanks.

Young, W. G., and T. M. Marty. 1986. Wear and microwear on teeth of a moose (*Alces alces*) population in Manitoba, Canada. Canadian Journal of Zoology 64:2467–79.

Zarnke, R. L. 1996. Serologic survey of Alaska wildlife for microbial pathogens (Federal Aid in Wildlife Restoration Final Report, Study 18.7, Grants W-23-5, W-24-1 through W-24-4). Alaska, Department of Fish and Game.

Anne Gunn, Wildlife and Fisheries Division, Department of Resources, Wildlife and Economic Development, Government of the Northwest Territories, Yellowknife, Northwest Territories, Canada X1A 3S8. Email: anne_gunn@gov.nt.ca.

Jan Adamczewski, Fish and Wildlife Branch, Department of Environment, Government of Yukon, Watson Lake, Yukon, Canada Y0A 1C0. Email: jan.adamczewski@gov.yk.ca.

51

Mountain Sheep

Ovis canadensis and *O. dalli*

Paul R. Krausman
R. Terry Bowyer

NOMENCLATURE

COMMON NAMES. Mountain sheep, bighorn sheep
SCIENTIFIC NAME. *Ovis canadensis*
SUBSPECIES. *O. c. auduboni,* Audubon's bighorn, Black Hills bighorn, or badland bighorn; *O. c. californiana,* California bighorn; *O. c. canadensis,* Rocky Mountain bighorn; *O. c. nelsoni, O. c. mexicana, O. c. cremnobates,* and *O. c. weemsi,* desert bighorn

It is unlikely all currently recognized subspecies will be maintained when bighorn taxonomy is revised (Wehausen and Ramey 1993; Jessup and Ramey 1995).

COMMON NAMES. Thinhorn sheep, Dall's sheep
SCIENTIFIC NAME. *Ovis dalli*
SUBSPECIES. *O. d. dalli,* Dall's or Alaskan white sheep; *O. d. stonei,* Stone's or black thinhorn sheep (Valdez and Krausman 1999).

Recent summaries of mountain sheep by the authors (Krausman et al. 1999; Valdez and Krausman 1999; Bowyer et al. 2000; Krausman and Shackleton 2000) were used to develop this account.

DISTRIBUTION

Distribution of Rocky Mountain bighorn sheep closely follows the Rocky Mountains, and extends from about 55° N in Alberta and British Columbia, south through Montana, Idaho, Utah, Wyoming, and Colorado, and into northern New Mexico at around 36° N (Stelfox 1971, Clark 1978). California bighorns historically ranged from the eastern slopes of the Coast Mountains in central British Columbia (51° N), south into Washington, Oregon, and Idaho as far as the Sierra Nevada in California (37° N) (Cowan 1940). Nevertheless, from 1900 until 1954, this subspecies was extripated from much of its distribution, especially in the United States. Subsequently, California bighorn sheep have been translocated, mainly from British Columbia, to restock and reestablish populations in California, Oregon, Washington, Idaho, Nevada, and North Dakota (Demarchi and Mitchell 1973). Desert bighorn sheep formerly occupied ranges from Nevada (40° N) to Baja California, Mexico (24° N), and from western Texas, southern New Mexico and Arizona, and western Colorado and Utah to California (Monson 1980) (Fig. 51.1).

Dall's sheep range through rugged and steep mountains in Alaska, Northwest Territories, the Yukon, and British Columbia from 69°40′ to 59°30′ N latitude. Stone's sheep is distributed the farthest southward, with populations in British Columbia (Bowyer et al. 2000) (Fig. 51.1). Unlike bighorn sheep, populations of Dall's sheep have not been reduced markedly and still occur throughout much of their original range.

DESCRIPTION

Mountain sheep are in order Artiodactyla, suborder Ruminantia, and family Bovidae. True sheep of the genus *Ovis* are characterized by the presence of interdigital, inguinal, and preorbital glands and the absence of subcaudal glands and a chin beard (Valdez and Krausman 1999), which separates them from the genus *Capra*.

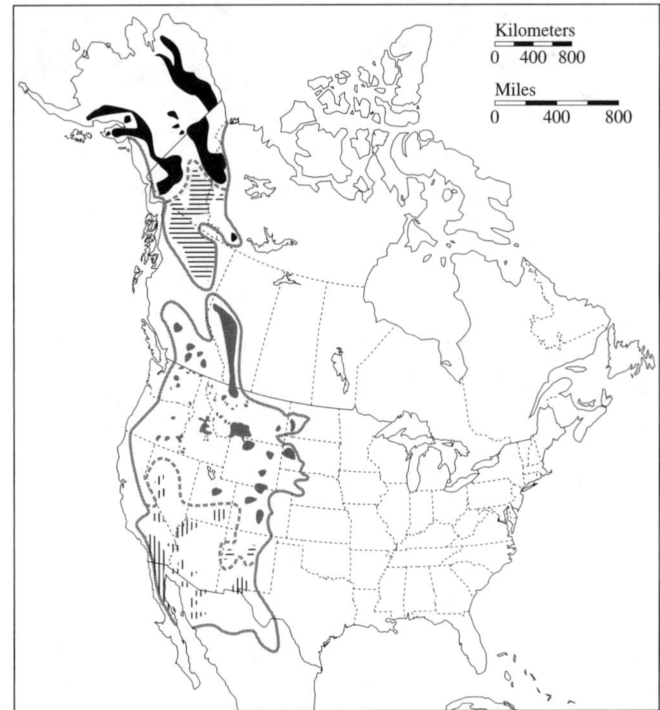

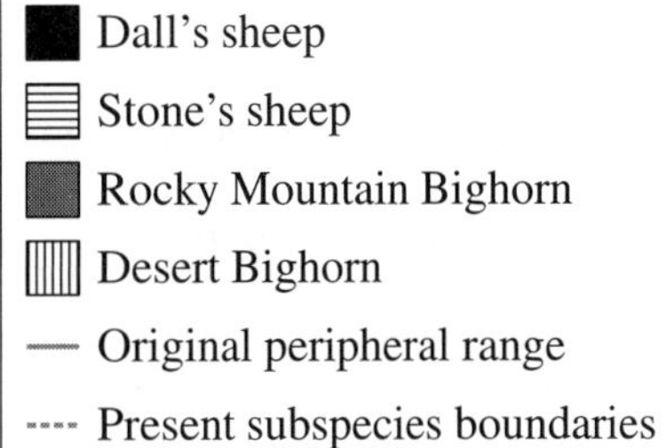

FIGURE 51.1. Distribution of mountain sheep. SOURCE: Data from Valdez and Krausman (1999).

Horns. The most distinguishing feature of mountain sheep is the massive horns of adult males (Fig. 51.2). Bighorns possess more massive, less rugose, and usually more broomed (i.e., broken tips), and less widely expanded horns than thinhorn sheep. Desert bighorn exhibit more diverging horns than northern bighorn sheep. Horns of desert bighorn females are more expanded and curved than those of other subspecies (Valdez and Krausman 1999) (Table 51.1). Horns of Dall's sheep flair more widely than those of bighorn sheep.

FIGURE 51.2. Adult male bighorn sheep. SOURCE: Photo by V. Geist.

Weight. At birth, northern bighorn weigh 2.7–4.5 kg. Adult males average 79 kg (up to 145 kg). Females average 59 kg. At birth, desert bighorn weigh 2.7–4.1 kg. Adult males average 68 kg (up to 104 kg). Females average 52.2 kg. At birth, thinhorn weigh 3.2–4.1 kg. Adult males average 81.7 kg (up to 136.2 kg). Females average 56.8 kg (Valdez and Krausman 1999).

Adult External Measurements. Measurements for bighorn males (mm) are as follows: for males, total length, 1321–1956; tail, 102–152; hind foot, 356–483; shoulder height, 813–1118; for females, total length, 1168–1880; tail, 102–127; hind foot, 279–432; shoulder height, 76–91 (Valdez and Krausman 1999). Measurements for thinhorn males and females, respectively, are as follows (mm): total length, 1300–1780 and 1324–1620 (Cowan 1940; Bunnell and Olsen 1976; Hall 1981); tail, 8–13 and 8–10; hind foot, 38–46 and 28–41 (Valdez and Krausman 1999); and shoulder height, 916–1090 (Cowan 1940; Bunnell and Olsen 1976; Hall 1981) and 787–889 (Valdez and Krausman 1999).

Pelage. The back of the legs, distinct rump patch, and usually part of the muzzle are white. Bighorn are usually brown and have a stripe middorsally of dark body hair across the rump patch to the tip of their tail. Thinhorn are white (i.e., Dall's) or grayish to blackish (Stone's). Markings on Stone's sheep are similar to those of bighorn sheep. Mountain sheep have unbanded awn-type guard hairs, which are round to oval (length ≤32.2 mm and diameter ≤4.8 mm) (Moore et al. 1974). The annual spring molt lasts approximately 1–2 months (Lawson and Johnson 1982). Dall's sheep are white or off-white, but sometimes have black tails. Stone's sheep are gray with white leg trimmings and rump patches similar to bighorn sheep. Fannin's sheep are intermediate in pelage color and markings between Dall's sheep and Stone's sheep, and is most common where the subspecies of thinhorns intergrade in distribution. However, it is not recognized as a subspecies.

Skull and Dentition. The mean skull mass of mature males excluding the lower jaw is 18 kg (Clark 1970). The lambdoidal suture forms a fairly straight line and the upper ends of the premaxillae do not meet the nasals and maxillae (Lawson and Johnson 1982). The infraorbital foramen is small with a well-defined rim, the braincase is pneumatic, and the occipital condyles are enlarged (Figs. 51.3 and 51.4).

Of all ruminants, wild sheep have the largest horns in proportion to body size, about 8–12% of an adult male's body mass (Geist 1966b). Horns begin to be visible at 2 months of age (Hansen 1965). At 5.5 months, the horns are 5–7 cm long (Jones 1959). At 12 months, the basal circumference of male horns is greater than that of female horns and increases at approximately the same growth rate as the length (Lawson and Johnson 1982). The horn is triangular in cross section during the first 2 years. During the third year, the base of the horn swells and the horn loses it flat-sided shape (Lawson and Johnson 1982). Growth is greatest during summer; differential yearly growth produces annual rings. During the first 3–4 years, horn rings are several inches apart and then occur closer and become harder to differentiate (Fig. 51.3) (Taylor 1962; Shackleton et al. 1999).

Female horns are smaller and shorter than those of males. Female horns are relatively thin and gently curved, whereas those of adult males are massive at the base, tapered, and curled in a spiral as they grow. Horn growth each year occurs when a new keratin horn sheath develops over the underlying bone horn-core, which is an extension of the frontal bone. The horn sheath grows beneath the preceding year's horn, so that each year, horn sheaths are grown one inside the other. Except for the first or lamb horn, only a part of the annual horn sheath is exposed (Taylor 1962; Shackleton et al. 1999). When horn growth stops, probably in autumn or early winter, and is then followed by growth of a new sheath the following spring, a distinct break or annual ring is formed.

Horns and horn growth are of value to wildlife biologists for several reasons. First, an individual's age can be determined by counting the horn rings or annuli that develop each year. With long-lived individuals,

TABLE 51.1. Mean, standard deviation (*SD*), and range (in inches; standard unit of measurement) of horn measurements of the 25 longest horned specimens of mountain sheep

	Horn Length		Basal Circumference		Tip-to-Tip Spread	Greatest Spread
	Right	Left	Right	Left		
Dall's sheep						
Mean	44.9	44.5	14.3	14.3	26.9	27.2
SD	2.9	2.6	0.6	0.6	4.0	3.6
Range	38.8–49.5	39.0–47.5	13.0–15.3	13.0–15.3	20.9–34.4	20.9–34.4
Stone's sheep						
Mean	45.0	44.9	14.8	14.8	24.5	26.6
SD	2.2	1.9	0.6	0.6	3.1	2.9
Range	42.1–50.1	41.9–51.6	13.5–16.3	13.5–16.3	19.0–31.4	22.0–31.5
Rocky Mountain and California bighorn						
Mean	43.7	43.7	15.9	15.9	22.0	23.4
SD	2.3	2.2	0.7	0.7	2.8	1.9
Range	39.1–49.5	40.5–49.3	14.8–17.5	14.8–17.4	18.1–28.9	21.5–28.9
Desert bighorn						
Mean	40.6	40.8	15.7	15.7	22.1	22.9
SD	2.1	2.3	0.6	0.6	2.9	2.2
Range	37.0–45.6	36.0–46.3	14.5–16.8	14.6–17.0	16.8–27.4	17.8–27.4

SOURCE: Data from Reneau and Reneau (1993).

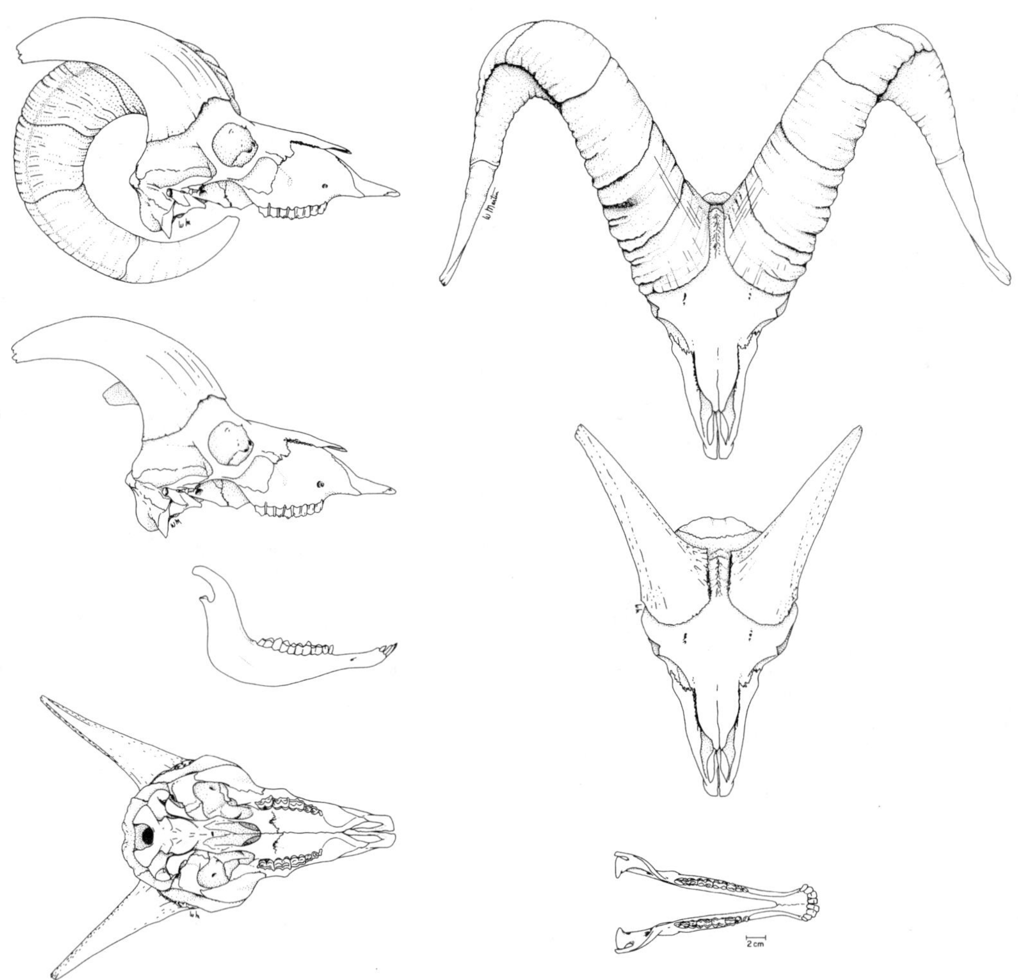

FIGURE 51.3. Skull of the thinhorn sheep (*Ovis dalli*). From top to bottom: (left) lateral view of cranium with one horn removed and (right) dorsal view of cranium with horns; (left) lateral view of cranium with both horns removed to show permanent bony core, and lateral view of mandible, and (right) dorsal view of cranium with horns removed; (left) ventral view of cranium with horns removed and (right) dorsal view of mandible.

however, it is usually easier to estimate the age of males than females. After about 4 or 5 years of age, female horns grow very little, probably because of reproductive costs, so their horn rings become very crowded and hard to distinguish. In males, the first 2 years or more of horn growth can be lost to "brooming" or breakage of the horn tips during fighting, making precise age determination problematic in older animals (Shackleton and Hutton 1971); brooming is less common in Dall's sheep because of the widely flaring tips of the horns. Although only relative measures of annual horn length are possible because all but the first year's horn are partially hidden by preceding ones, measuring annual horn growth can provide insights into an individual's and a population's status.

When the average lengths of the exposed annual horn sheaths are plotted (for males or females), the generalized pattern of horn growth is relatively consistent among bighorn sheep populations. The longest visible horn sheath almost always is grown in the second year of life, after which exposed horn length decreases rapidly. Maximum annual horn growth in the second year and rate of "decline" in annual horn length vary among populations, reflecting primarily environmental rather than genetic differences (Geist 1971; Shackleton 1973, 1976). Relative annual horn growth is initially greater in low-density populations on good range. This is because of rapid population growth, fast individual growth rates, early maturation, early mortality, intense social interaction, and high milk production, though it declines faster than in stable or declining populations at higher density relative to carrying capacity (Geist 1971; Shackleton 1973, 1976; Wishart and Brochu 1982).

Managers may be able to use the average pattern of annual horn sheath growth and body growth to compare populations and evaluate management options (Smith and Wishart 1978; Gilchrist 1992). This can be especially valuable for wildlife managers because size limits for trophy hunting are often described in terms of horn size (e.g., three-quarter curl and full curl) and hence are influenced by horn growth. Horn growth of rams is minimal after ages 7 or 8 years. Horns are fairly distinct between thinhorn and bighorn races of sheep (Table 51.1). Dall's and Stone's sheep exhibit the longest mean horn lengths, followed by Rocky Mountain bighorn and desert bighorn. Rocky Mountain bighorn exhibit the greatest mean basal circumferences, followed by desert bighorn, then Stone's and Dall's sheep. Dall's and Stone's sheep, however, exhibit the longest means of tip-to-tip spread and the greatest spread, followed by Rocky Mountain and desert bighorn (Table 51.1). The longest horns ever recorded in a North American wild sheep (right = 1273.2 mm, left = 1311 mm) are those of the Chadwick ram, a Stone's sheep shot by L. S. Chadwick along the Muskwa River, British Columbia, in 1936 (Valdez and Krausman 1999).

The largest number of record-sized Rocky Mountain bighorn rams recorded since 1975 originated from introduced populations in Montana (Boone 1988; Gilchrist 1992). Twenty-six of the 100 highest scoring Rocky Mountain bighorn rams listed in the 10th edition of the Boone and Crockett Club record book are from Montana (Reneau and Reneau 1993).

North American wild sheep have similar patterns of tooth development (Table 51.2). Deciduous dentition is complete within the first

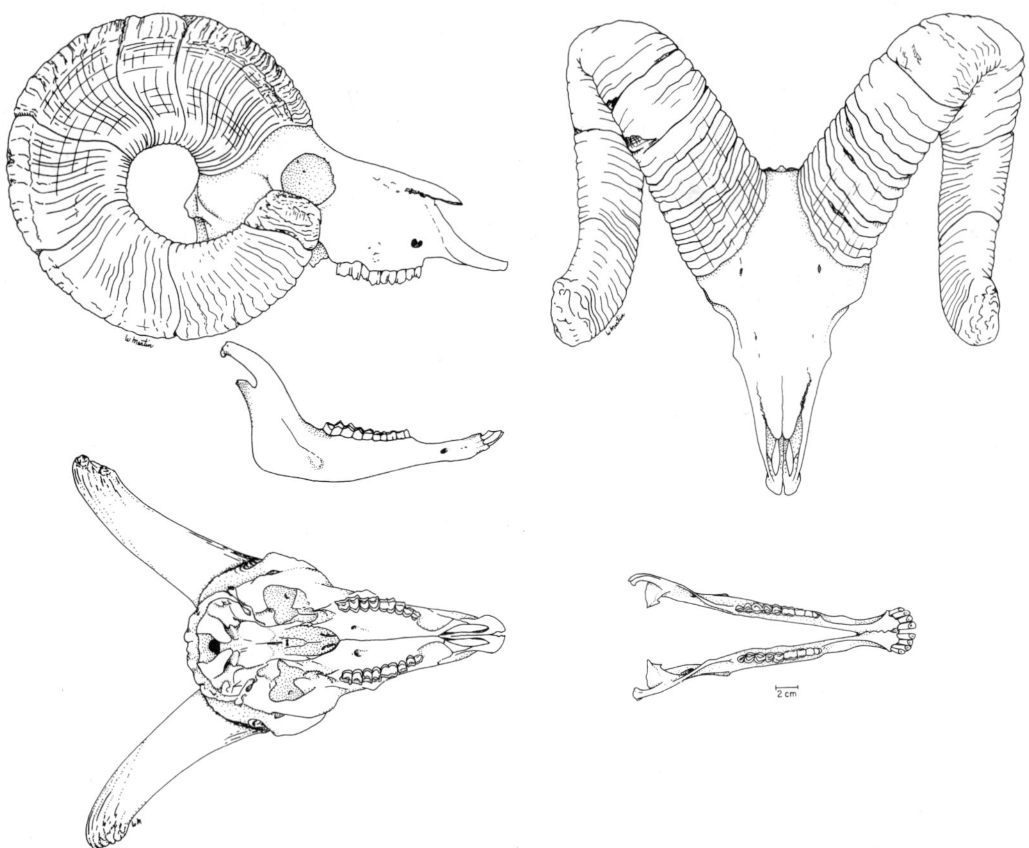

FIGURE 51.4. Skull of the bighorn sheep (*Ovis canadensis*). From top to bottom: (left) lateral view of cranium and lateral view of mandible and (right) dorsal view of cranium; (left) ventral view of cranium with horns removed and (right) dorsal view of mandible. Note the horn annuli, the normal corrugations, and the increment lost to brooming.

TABLE 51.2. Sequence of tooth eruption in the lower jaw of North American wild sheep

Age (months)	Incisors 1	Incisors 2	Incisors 3	Canine 1	Premolars 2	Premolars 3	Premolars 4	Molars 1	Molars 2	Molars 3
6	D	D	D	D	D	D	D	(P)		
12	(P)	D	D	D	D	D	D	P	(P)	
18	P	D	D	D	D	D	D	P	P	
24	P	D	D	D	D	D	P	P	P	
36	P	P	(P)	D	P	P	P	P	P	(P)
42	P	P	(P)	D	P	P	P	P	P	P
44	P	P	P	P	P	P	P	P	P	P

NOTE: D, Deciduous; P, permanent; (P), erupting.

week of life, but permanent dentition is not fully erupted until 4 years of age (Table 51.2). Incisors are spatulate, the lower canine is incisiform, and molars are subhypsodont (i.e., long and broad); they do not contain an open pulp cavity throughout their life, (Geist 1971). Bighorn sheep and thinhorns have 32 teeth. The dental formula is I 0/3, C 0/1, P 3/3, M 3/3. Dental anomalies include deformations, lost teeth, malocclusions, incomplete sets of teeth, and vestigial upper canines (Murie 1944; Deming 1952; Welles and Welles 1961; Lawson and Johnson 1982).

PHYSIOLOGY

Temperature and Blood Chemistry. Lawson and Johnson (1982) reported the temperature and blood chemistry of mountain sheep. The normal range of rectal temperature is 38.3–38.9°C, but it varies according to the ambient temperature, excitability, and diet. The cyclic nature of body temperature in mountain sheep appears to be related to metabolic factors that are physiological adaptations to conserve energy during winter (Franzmann and Hebert 1971).

Chappel and Hudson (1978) reported that the metabolic rate of Rocky Mountain bighorns is lowest at an ambient temperature of –10°C and the thermoneutral zone is –20°C to +10°C. Those authors also observed that moderate wind affected metabolic rate only at ambient temperatures below –20°C.

The blood transferrins and hemoglobins differed among *O. c. canadensis, O. c. mexicana, O. d. dalli,* and *O. d. stonei* (Nadler et al. 1971). Other physiological values, including blood proteins and minerals, white blood cell count, red blood cell count, and packed cell volume, were reported by Franzmann and Thorne (1970), Woolf and Kradel (1970), and Franzmann (1971, 1972).

Hematology and blood chemistry of Dall's sheep are similar to those reported for bighorn (Franzmann 1971; Butcher and Hawkey 1979; Foreyt et al. 1983). Such data, however, may vary with sex, age, physical condition, season, handling, and numerous other factors (Shackleton 1985; Keech et al. 1998). Although some blood variables may correlate with physical condition, direct comparisons among populations using these values should be made with appropriate caution (Keech et al. 1998).

Neural. Mountain sheep depend more on their visual capabilities than on their auditory or olfactory senses. The mean hearing thresholds for desert bighorn sheep were 47, 47, 69, and 89 dB peak equivalent sound pressure level for click, 4000-, 2000-, and 1000-Hz tonebursts, respectively (De Young et al. 1993).

REPRODUCTION

Anatomy. The reproductive anatomy and physiology of wild sheep have not been thoroughly examined. Lawson and Johnson (1982:1038–39), however, summarized accounts from Blom (1968) and Hafez (1968) of domestic sheep anatomy and physiology as applicable to mountain sheep in North America:

> The ovary, oviduct, and uterus are supported by a broad dorsolateral ligament in the region of the ilium. The bipartite uterus resembles a ram's horns with a convexity dorsal. The almond-shaped ovaries are located laterally and in close apposition to the fusion of the uteri in open ovarian bursae. The bursae are pouches derived from the same tissue as the ligament and attach the suspended oviducts to the uteri. The right ovary is most active. Mature corpora lutea are spheroid or oval and the oviduct is pigmented. The endometrium of the uterus is characterized by numerous pigmented caruncles, each generating many cotyledons during pregnancy. Annular folds form the lumen of the cervix, and the hymen is well developed.
>
> There are two inguinal, functional mammaries; supranumerary teats, if present, are located anteriorly to the normal ones. Fine hair covers the teats, and the connective tissue closing the orifice is elastic.
>
> The prostate gland of the ram is disseminate. The scrotum is pendulous and the testes inguinal. Rams have a fibroelastic penis with a filiform appendage, the precesses uretherae, located at the tip, which rotates rapidly during ejaculation to spray the semen in the vagina.

Physiology. Females are monoestrous. The rutting or mating season is that period in which mating activities result in 70% of the lamb production for the following season (Turner and Hansen 1980). Duration of the mating season is longer at lower elevations and southern latitudes, and shorter at higher elevations and more northern latitudes (Bunnell 1982; Thompson and Turner 1982). Rocky Mountain and northern populations of California bighorn mate in late autumn and early winter, with the mating season beginning as early as late October, or, as is typical for most Rocky Mountain populations, in early November (McCann 1956; Buechner 1960; Blood 1963; Geist 1971; Demarchi and Mitchell 1973). Sometimes mating extends into late December and early January, but usually peaks between mid-November and mid-December (Honess and Frost 1942; Smith 1954; Wishart 1958; Buechner 1960; Sugden 1961; Blood 1963; Morgan 1970; Geist 1971; Shackleton 1973). By contrast, desert bighorn exhibit an extended mating period, and births have been documented in the Sonoran Desert in all months except October (Wilson 1968; Leslie and Douglas 1979; Lenarz and Conley 1982; Witham 1983). This variation in mating seasons appears related to environmental conditions at the time of birth. Dall's sheep rut in November and December (Geist 1971; Nichols 1978a), timing reproduction so neonates can be provisioned successfully (Bowyer 1991; Rachlow and Bowyer 1991, 1994; Bowyer et al. 1998).

Before estrus, one to four Graafian follicles develop simultaneously. The estrous cycle in bighorns lasts 28 days, with a receptive period of around 48 hr (Turner and Hansen 1980). Gestation lasts 173–185 days in desert sheep, 173–175 days in California bighorn, 173–176 days in Rocky Mountain sheep (Blunt et al. 1972; Turner and Hansen 1980; Whitehead and McEwen 1980; Sandoval et al. 1984; Shackleton et al. 1984), and 171 days in Dall's sheep (Nichols 1978a). Females in all races usually produce a single lamb each year until old age or death overtakes them. Most males actively mate for only a few years, during which time they may inseminate many females. Successful mating is not entirely a function of sexual maturation, as social hierarchy and behavior play major roles (Geist 1971; Shackleton 1991).

If Dall's sheep are similar to domestic sheep, they have an estrous cycle of about 17 days (Asdell 1964). Photoperiod is likely an important cue in timing of reproduction, and the presence of an adult male and physical condition of the female are proximal stimuli that also affect onset of estrus, which is thought to last for 1 day (Geist 1971; Nichols 1978a).

Breeding. During mating, adult male and female–juvenile groups join for the duration of the rut, which usually takes place within the home range of females (Krausman and Shackleton 2000). An anestrous female may be courted throughout the year by young males, but the males are discouraged, as females avoid them and withdraw from mounting attempts (Welles and Welles 1961; Geist 1968, 1971). During estrus, females are more aggressive. Large-horned, older males are the most dominant and also those most likely to mate (Geist 1971; Hogg 1984, 1987; Shackleton 1991). Females do not show any clear relationships between physical attributes and social status; their dominance hierarchies appear less linear and more subtle than those of males (Eccles and Shackleton 1986; Festa-Bianchet 1991; Hass 1991; Zine and Krausman 2000). Only age and nursing rate have been related to social status in female bighorn, but there is no evidence that a female's status is related either to her reproductive fitness or to her differential investment in male and female offspring (Eccles and Shackleton 1986; Festa-Bianchet 1991; Hass 1991).

Mountain sheep are extremely gregarious and polygamous (Geist 1971). Males search for females in estrus (Smith 1954; McCann 1956; Leslie and Douglas 1979), and on finding a receptive female, dominant males attempt to chase away other males. Males examine females for estrus by sniffing the vulva and tasting urine.

Males often deliver a stiff foreleg kick to females in heat to stimulate a chase, or the ewe may initiate the chase. The chase may or may not be strenuous. If one of the pair tires, the partner usually waits until the other is ready to resume. When the female is sufficiently stimulated, she assumes a position of lordosis. It the male becomes exhausted, nearby subordinates may usurp his position at any time (Blood 1963; Geist 1971).

Dall's sheep possess a tending-bond mating system in which a dominant male guards, tends, and courts a female (Geist 1971; Hirth 1977). After copulation has occurred, the male leaves in search of additional mates.

Ovulation and spermatogenesis usually begin by 18 months of age, but wild bighorn do not become fully sexually active until they are older (Woodgerd 1964; Geist 1971; Blunt et al. 1972). Most females mate first when at least 2.5 years old, and male Rocky Mountain sheep usually do not begin to participate fully in the rut until 7–8 years old, well after puberty (Geist 1971). In expanding populations, however, or in rare instances, female bighorn have given birth to their first lamb at 18 months (Woodgerd 1964; Shackleton 1973; McCutchen 1976; Van Dyke 1978; Sandoval 1981; Morgart and Krausman 1983). Sexual activity can occur much earlier in captivity; captive desert bighorn male lambs began spermatogenesis at 26–28 weeks and exhibited a seasonal spermatogenetic cycle after 21 months of age. In other studies, captive yearling desert bighorn males inseminated all females living in their enclosure (Turner 1976; Blaisdell 1976; McCutchen 1976). Irvine (1969) reported no apparent decrease in spermatogenesis with increasing age and concluded that even the oldest males were capable of breeding.

Average climatic and forage conditions vary relatively predictably and seasonally in most areas inhabited by California and Rocky Mountain bighorn. Variation in these two factors is important for reproduction, especially for the timing of the birth season. Climatic conditions can affect the survival of newborn lambs and the forage quality and quantity that are important for lactation (Geist 1971; Festa-Bianchet 1988a, 1988b). Lambs, however, must also grow large enough to survive their first winter (Festa-Blanchet 1988c). Thus, for bighorns the birth season is a trade-off between young being born early enough for adequate prewinter growth and being born late enough to avoid the thermal stress and poor forage conditions of late winter (Sadler 1987).

Thompson and Turner (1982) assessed temporal variation in parturition seasons for 22 populations of bighorn sheep. In those from northern latitudes, parturition seasons were shorter, later, and cued to brief, relatively predictable periods of vegetation growth (Bunnell 1982).

The birth period of Rocky Mountain and California bighorn begins in early spring (late April or May) usually coinciding with initiation of spring vegetation growth and ameliorating climatic conditions.

Few lambs are born after June (Shackleton et al. 1999). The same factors, however, may operate differently for desert bighorn sheep due to low population density (Lenarz 1979). More importantly, Thompson and Turner (1982) reported poor correlation between the inception and duration of the vegetation growing season and the lambing period in desert bighorns. They, together with Lenarz and Conley (1982), concluded that an extended lambing season was a result of unpredictable precipitation patterns, and consequently plant regrowth, both of which appear essential for maternal and neonatal survival. For desert bighorns, seasonal fluctuations in resources are not as predictable from year to year. Throughout much of the range of desert bighorn, plant productivity is related directly to temporal and spatial precipitation patterns, and these vary considerably and unpredictably. Nonseasonal reproductive behavior may be an adaptive strategy of desert bighorn that ensures lamb survival during periods of varying and unpredictable forage production. An extended lambing period would increase the probability that late gestation and early lactation would coincide with a period of adequate precipitation and forage availability (Leslie and Douglas 1979; Sandoval 1979; Thompson and Turner 1982). Such reproductive responses to unpredictable resources are common in desert vertebrates (Sadler 1987).

There is increasing evidence that adjustment in gestation length may be under proximal control of female Dall's sheep (Rachlow and Bowyer 1991; Berger 1992; Bowyer et al. 1998). For instance, Dall's sheep in interior Alaska delayed onset of lambing 14 days when a spring storm deposited 25 cm of fresh snow during the peak lambing period of the previous year (Fig. 51.5) (Rachlow and Bowyer 1991). There is a trend for date of parturition to be earlier for populations of mountain sheep with increasing north latitude (Bunnell 1980, 1982). Nonetheless, marked interannual differences in the date of birth can occur. Rachlow and Bowyer (1991) reported that median date of birth in a normal year was 18 May, but was 27 May in the year with the late snowstorm. Synchrony of births also differed in these two disparate years. Evidence that birth synchrony in Dall's sheep is related to predation is lacking. Instead, synchronous births result from a limited time in which offspring can develop to a sufficient size in spring and summer to withstand harsh conditions in winter (Rachlow and Bowyer 1991).

Reproduction has not been reported for lambs (i.e., <6 months), but yearlings (>12 and $\geq$24 months) may become pregnant in highly productive populations (Nichols 1978a). Females usually do not begin reproducing, however, until 30 months of age (Geist 1971). Indeed, young:adult female ratios at birth for Dall's sheep in interior Alaska varied from 0.4:1 to 0.6:1, indicating not all adult females reproduced each year, a pattern reported for other arctic ungulates (Murphy and Whitten 1976; Rachlow and Bowyer 1991; Cameron and Ver Hoef 1994). Males can become sexually mature at 18 months, but because of the polygamous mating system, they seldom gain an opportunity to breed until 5–7 years of age (Geist 1971).

Fetal sex ratios, although skewed slightly toward males in free-ranging populations, do not depart significantly from parity (Geist 1971; Nichols 1978). Captive females kept on a high nutritional plane, however, produced proportionally more daughters than sons (Hoefs and Nowlan 1994). Thus, nutritional condition of the female likely affects the rate of reproduction and the sex of her offspring.

Parturition. Most female mountain sheep give birth to only one lamb/year; however, twins do occur (Welles and Welles 1961; Spalding 1966; Geist 1971; Nichols 1978a; Hoefs 1978; Eccles and Shackleton 1979). Birth is relatively fast. Approximately 45 min before parturition, the placental membranes begin to appear, and 10 min before birth, the female begins to pant. Birth occurs in 10–15 min, at which time the female stands to facilitate the final expulsion (Lawson and Johnson 1982).

Females seek steep, rugged terrain where they seclude themselves from other sheep for 1–2 days to give birth (Pitzman 1970; Rachlow and Bowyer 1991, 1994, 1998). In the Little Harquahala Mountains, Arizona, individual females exhibited fidelity to parturition sites and sites did not overlap (Etchberger and Krausman 1999). Neonates are exceptionally precocial and stand within 30 min following parturition; young travel with their mothers within 24 hr of birth (Murie 1944; Pitzman 1970). Weaning generally is completed within 3–5 months (Bunnell and Olsen 1976).

Newborn. After birth, the mother licks the placental fluids from the neonate. Females usually consume the placental membrane (Lawson and Johnson 1982). Bighorn sheep weigh 2.7–4.5 kg at birth (Sugden 1961; Geist 1971; Blunt et al. 1972; McEwan 1975; Jorgensen and Wishart 1984). Dall's sheep weigh 3–4 kg at birth (Bunnell 1980).

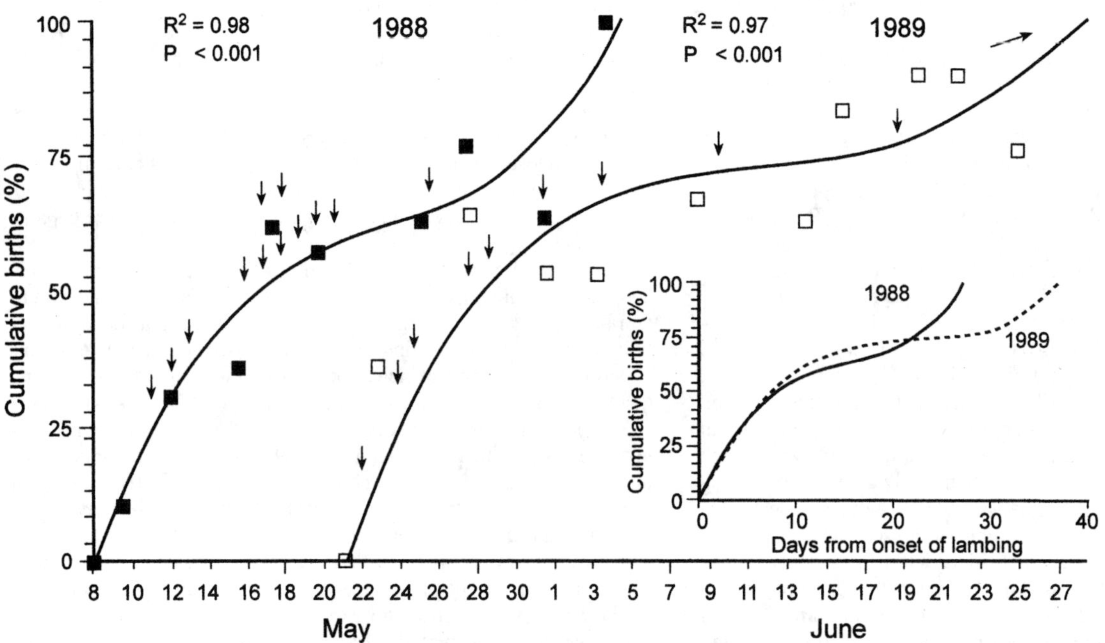

FIGURE 51.5. Cumulative percentage of births determined from young-to-female ratios showing differences in timing and synchrony (inset) of lambing in interior Alaska in 1988 and 1989. Arrows indicate date of birth observed or inferred. SOURCE: Rachlow and Bowyer (1991). Reproduced with permission from Alliance Communications Group.

ECOLOGY

HABITAT: BIGHORN SHEEP

Topography. The general name of "mountain sheep" reflects the bighorn's preference for steep, rugged topography typically found in mountains, though they can meet their requirements in other areas (Geist 1971; Adams et al. 1982; Risenhoover and Bailey 1985; Shackleton 1985). The range of habitat components they can use varies from steep or gentle slopes, broken cliffs, rock outcrops, and canyons and adjacent river benches to mesa tops (Buechner 1960; Sugden 1961; Wilson 1968; Welch 1969; Drewek 1970; Merritt 1974; Stelfox 1975; McQuivey 1978; Holl and Bleich 1983; Etchberger et al. 1989; Wakeling and Miller 1989). Alpine, related slopes, and river benches are mainly used for foraging, whereas cliffs and other precipitous ground supply bighorn with escape terrain (i.e., topography where sheep have a locomotive advantage over predators). These steep habitats provide sheep with their best defense against predators, and bighorns rarely occur far from such security cover (Oldemeyer et al. 1971; Erickson 1972; Pollister 1974; Shank 1979; Hansen 1982; Gionfriddo and Krausman 1986; Krausman and Leopold 1986). Though males may move further away than female–juvenile groups, escape terrain is especially important for females when giving birth (Blood 1961; Drewek 1970; Kornet 1978; Hall 1981; Akeson and Akeson 1992). Although specific characteristics of birth sites for Rocky Mountain and California bighorn have not been studied, parturition microsites for desert bighorn were made up of steep rugged terrain and had less barrel cacti for up to 8 weeks following parturition than sites used by females without lambs. Parturition site fidelity was strong among years with a mean distance between parturition sites for individual females of 450 m (Etchberger and Krausman 1999).

Bighorn occupy an extreme range of elevations. Desert sheep occur from 78 m below sea level in Death Valley, California, to more than 4267 m above sea level in the White Mountains, California. Other subspecies inhabit ranges from 450 to >3300 m above sea level (Welles and Welles 1961; Shackleton et al. 1999). Sheep in most populations make seasonal altitudinal movements, but a few remain at similar elevations year-round (Sugden 1961; Spalding and Bone 1970). Because of these variations, it is difficult to generalize about the elevational preferences of bighorn. A variety of exogenous and endogenous factors may influence elevational use of a particular range, and together reflect the animal's environmental, physiological, and behavioral preferences.

Climate. Areas inhabited by Rocky Mountain and California bighorn are relatively arid. Summers are warm or even hot with highs up to 35°C, and winters are usually cold with temperatures sometimes reaching −40°C (Jones 1959; Smith 1954; Schallenberger 1966). Extreme conditions also prevail in desert habitats, where temperatures as low as minus 29°C have been recorded in winter and >49°C in summer (Hansen 1980a; Sandoval 1980).

Desert bighorn activities are affected by temperature (Chilelli and Krausman 1981). In the Cabeza Prieta National Wildlife Refuge, Arizona, desert bighorn bedded in the shade an average of 7 hr each day when wet bulb temperatures were above 18°C (Simmons 1969). During the hot, dry summer months, heat stress is a serious obstacle to desert bighorn survival; they minimize its effect by limiting their activities, bedding in the shade during the hottest part of the day, and feeding and watering on shaded slopes. Conversely, wind, cold temperatures, and extremes of precipitation levels are limiting to Rocky Mountain bighorn.

Snow accumulation in winter limits habitat use by California and especially Rocky Mountain bighorn. Generally, areas with snow deeper than 30 cm are avoided because deep snow increases the cost of foraging and travel (Stelfox 1975). These activities also can be hampered by snow crusts that develop during freeze–thaw cycles (Sugden 1961; Petocz 1973). Consequently, most Rocky Mountain and California bighorn winter ranges are located in areas with low snow accumulation. Low snow depths can occur as a result of low precipitation, heat gains on south-facing aspects, high winds blowing snow away, or a combination of these factors.

Precipitation in desert ranges is low and unpredictable, ranging from <2.5 cm in the Mojave Desert to >251 cm in the White and San Gabriel Mountains, California (Hansen 1980a). Erratic rainfall patterns alter the significance of any precipitation measurements taken at a single locality. However, rainfall can determine the distribution and production of forage and is an important physical requisite for desert sheep survival in the Southwest (Russo 1956).

Vegetation Cover. Climate, elevation, and latitude vary widely throughout the geographic distribution of bighorn sheep. Such variation is naturally reflected in the structure and floristic composition of vegetation in their habitat (Demarchi 1965; Todd 1972; Goodson 1978; Dale 1987; Risenhoover and Bailey 1985; Krausman et al. 1989; Bleich et al. 1997). Different habitats can meet specific requirements of bighorn activities such as foraging, resting, mating, lambing, thermal regulation, and predator avoidance, so habitat use can vary daily and seasonally as requirements of sheep change (Hansen 1982; Risenhoover and Bailey 1985; Dale 1987).

Generally, bighorn use open habitats such as grasslands and shrub–steppe communities located at various elevations and on different slope gradients. Such open habitats provide bighorn with good visibility, improving their chances of detecting predators, but the sheep are usually not far from cliffs or other precipitous habitat. Open grasslands are used primarily for foraging. They contain grasses such as fescues (*Festuca* spp.), wheatgrasses (*Agropyron* spp.), and ricegrasses (*Oryzopsis* spp.), along with forbs and often shrubs at lower elevations, and sedges at the higher elevations used in summer (Jaeger 1957; Jorgensen and Turner 1975; Kovack 1979; Shackleton et al. 1999).

Throughout much of the desert bighorn's habitat, the vegetation associations are predominantly adapted to dry, rocky, or sandy soils, and plants characteristically have a thickened epidermis and reduced leaf surface (Jaeger 1957). In general, the vegetation is uniformly sparse. Plants are widely spaced by the demands of their root systems in the shallow soils, whereas a rounded canopy results from equal exposure to solar radiation from all sides (Douglas and White 1979; Kelly 1979; Leslie and Douglas 1979; Watts 1979; Krausman et al. 1989). Such plants must withstand severe drought lasting up to several years.

For most bighorn habitat in the Great Basin Desert, sagebrush (*Artemisia* spp.), shadscale (*Atriplex* spp.), blackbrush (*Coleogyne ramosissima*), and cliffrose (*Cowania* spp.) constitute the major browse species. The major grasses include wild rye (*Lolium* spp.), Indian rice (*Oryzopsis hymenoides*), galleta (*Hilaria rigida*), bluegrass (*Poa* spp.), and fescue (Bradley 1964; Hansen 1980a). In the lower elevations of the White, San Gabriel, San Jacinto, and Santa Rosa Mountains, California, their habitat is characterized by lowland browse types. Bighorn habitat extends through piñon (*Pinus* spp.)–juniper (*Juniperus* spp.) and ponderosa pine (*P. ponderosa*) associations. The summer range includes subalpine and alpine biotic communities (McQuivey 1978; DeForge 1980). In the Painted Desert, sagebrush, blackbrush, shadscale, galleta, and piñon–juniper constitute the major vegetation associations used by desert bighorn (Wilson 1968). Characteristic plants of desert bighorn habitat in the Sonoran Desert include paloverde (*Cercidium* spp.), ironwood (*Olynea testota*), saguaro (*Cereus giganteus*), and organpipe (*C. thurberi*) cactus. The major grass species include grama (*Bouteloua* spp.), galleta, and sacaton (*Sporobulus wrightii*) (Mendoza 1976; Seegmiller and Ohmart 1981). Bighorn habitat throughout the Chihuahuan Desert is characterized by few trees, agave (*Agave* spp.), yucca (*Yucca* spp.), small cacti, and numerous spring shrubs (Moore 1958; Sandoval 1979). In general, volcanic soils support relatively homogeneous grasslands, and sedimentary parent material produces creosote (*Larrea tridentata*), mesquite (*Prosopis* spp.), cactus savannas, and agave thickets. Much of the bighorn habitat in Baja California and Sonora in Mexico is characterized by agave, ocotillo (*Fouquieria splendens*), ironwood, cholla (*Opuntia* spp.), acacia, and numerous cacti (Flores et al. 1972; Alvarez 1976).

Bighorn rarely use densely forested areas, probably because forage and visibility are more limited, although trees may be used for shade when bedding or during cold days with high winds (Geist 1971).

Similarly, desert bighorn in the San Gabriel Mountains of California use vegetation for thermal cover (DeForge 1980). In the San Andres Mountains, New Mexico, the piñon–juniper community received little use by desert bighorn, except for a few occasions where they were traveling along established trails close to escape terrain (Sandoval 1979).

Visibility. For bighorn, structure of the vegetation probably is more important than the type of plant species present. Open habitats, with high visibility, were used most by Rocky Mountain bighorn to facilitate detection of predators (Risenhoover and Bailey 1985; Wakelyn 1987). Nonetheless, open habitat may not be used if escape terrain is not readily available (McCann 1956). Visibility is an important habitat feature for bighorn sheep because their predator-evasion strategy involves foraging diurnally in relatively large dispersed groups on open habitat close to escape terrain. Predators are detected visually, and a larger dispersed group of sheep may be more alert to potential predators over a relatively large area. Foraging efficiency was higher when sheep were in large groups and in habitats with greater visibility (Risenhoover and Bailey 1985).

Some areas of the Harquahala Mountains, Arizona, were not used, in part because large boulders obstructed vision (Krausman and Leopold 1986). Desert bighorn in Arizona have abandoned areas because fire suppression allowed vegetation to grow and obstruct visibility (Etchberger et al. 1989, 1990; Krausman et al. 1996). Similarly, DeForge (1980) observed that reduced visibility in maturing chaparral lowered its suitability for bighorn, thus decreasing carrying-capacity potential, and eventually resulting in the total loss of bighorn range.

HABITAT: THINHORN SHEEP

Dall's sheep generally inhabit wind-swept, dry, steep, and rugged mountains characterized by subalpine-grass and low-shrub communities typical of high elevations and high latitudes (Murie 1944; Lord and Luckhurst 1974; Hoefs 1984; Rachlow and Bowyer 1998). Most populations of Dall's sheep are migratory and occupy different ranges in summer and winter, although a few populations are relatively sedentary (Dixon 1938; Geist 1971; Hoefs and Cowan 1979). Typical of other polygynous and sexually dimorphic ruminants, the sexes of adult Dall's sheep spatially segregate around the time of parturition (Bowyer 1984; Bleich et al. 1997; Weckerly 1998; Rachlow and Bowyer 1998). Movements of Dall's sheep between seasonal ranges have been related to plant phenology, temperature, and depth of snow (Hoefs and Cowan 1979). Seasonal movements from 8 to 48 km have been reported. Because summers are short at northern latitudes, Dall's sheep spend most of the year on winter range (males 271–303 days, females 240–263 days; Geist 1971). Wind-swept areas with sufficient forage and suitable escape terrain to elude predators are likely the key elements of winter habitat for Dall's sheep. For instance, Dall's sheep in Kluane National Park, Yukon, spent 70% of their time foraging in areas with <5 cm snow depth and <10% of their time in areas with snow depths >15 cm (Hoefs and Cowan 1979). Primary productivity of plants on winter range (29–120 g/m^2) is an important component of overwinter survival and for production of young (Hoefs and Bayer 1983; Hoefs 1984).

Adult males may occupy a variety of ranges throughout the year including areas inhabited during prerut, early to midwinter, late winter, spring, and summer. They also may move to areas with salt licks (Geist 1971). Ranges were smallest in midwinter (about 0.8 km in diameter) and largest in spring and summer (6 km). Adult females inhabited seasonal ranges in spring, for lambing, during summer, and in winter (Geist 1971). Estimates of home-range size from modern, quantitative methods, however, are unavailable (Kie et al. 1996).

Because of the severity of winters in the Arctic and subarctic, growth and development of young Dall's sheep and replenishment of female body reserves must occur during the short summer (Bunnell 1982; Rachlow and Bowyer 1991, 1994). Moreover, maternal females are likely constrained in their selection of habitat because of the vulnerability of young to predators and to exposure and hypothermia from severe climatic conditions (Frid 1977; Rachlow and Bowyer 1998). Consequently, suitable lambing habitat may be a crucial component affecting the productivity of sheep populations.

Lambing habitat for Dall's sheep in interior Alaska was characterized by steep, rugged terrain intermixed with forage including grasses and dryas (*Dryas* spp.) (Rachlow and Bowyer 1998). Lambing sites typically occurred above 1180 m and were free of snow. A suite of variables is useful in discriminating lambing sites from random sites, including distance to escape terrain, cover of grasses, cover of dryas, slope aspect, slope brokenness, slope steepness, and presence of snow (Fig. 51.6) (Rachlow and Bowyer 1998). Moreover, maternal females altered selection of habitat with the chronology of lambing, with additional variables becoming important at peak lambing that related to climate (windchill and cover from wind provided by browse). Additionally, females selected terrain features more strongly in a year with adequate food, but selected forages in a year with reduced availability of food (Rachlow and Bowyer 1998).

Young Dall's sheep likely acquire home ranges from adults. Females typically have ranges that are similar to those of their mother (or maternal group), whereas males gradually disassociate from their mothers and begin associating with groups of mature males. As with many large mammals, males are the initial dispersers (Geist 1971). Nontheless, Dall's sheep exhibit a high degree of fidelity (males, 88%; females, 90%) to seasonal ranges (Geist 1971).

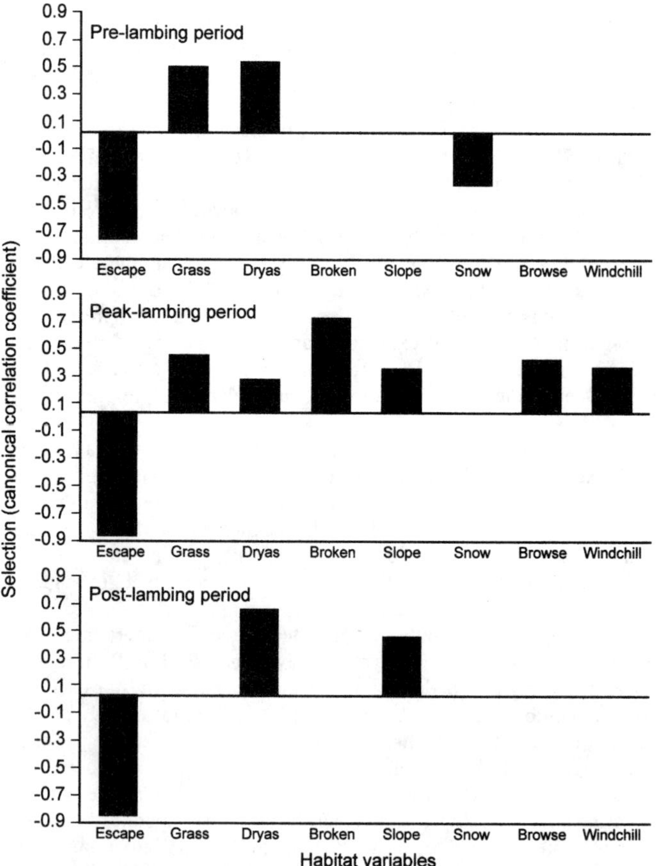

FIGURE 51.6. Selection of habitat by female Dall's sheep during pre-, peak, and postlambing periods in interior Alaska, 1988–1989. Escape, Distance to escape terrain; grass, percentage cover of grasses; dryas, percentage cover of dryas; broken, brokenness of terrain; slope, percentage slope; snow, presence or absence of snow; browse, percentage cover of browse; and windchill, temperature × wind. Canonical correlations indicate the strength and direction of selection. For instance, the negative correlation for distance to escape terrain indicates females used sites closer to such terrain than randomly located sites. The positive correlation for grass indicates there was more of this forage at sites used by females than at random sites. SOURCE: Rachlow and Bowyer (1998).

Activity. The activities of mountain sheep are determined to a large extent by day length, energy demands (Nichols and Bunnell 1999), and availability of forage and water (Krausman et al. 1999; Shackleton et al. 1999). It is difficult to generalize about daily activities for all mountain sheep, but they are basically diurnal. The general pattern of activity involves alternating peaks of foraging and resting or loafing; other activities contribute little to the daily activity budgets (Blood 1963; Augsburger 1970; Geist 1971; Olech 1979; Chilelli and Krausman 1981; Eccles 1983; Seip and Bunnell 1985). Foraging is most frequent around dawn and dusk. In winter, however, the proportion of daylight hours spent feeding increases while resting decreases (Van Dyke 1978) to maintain foraging activity during shorter winter days (Shackleton et al. 1999). This relative increase in foraging activity may be a response to the shorter day length reducing foraging time, winter plants being in poorer nutritional condition and less available, especially for northern populations, or thermal stress in winter, which requires more energy consumption (Van Dyke 1978; Shackleton et al. 1999).

In the Arctic Brooks Range, Dall's sheep were active during all 24 hr of summer daylight, but after darkness began in late summer, activity was confined to the limited daylight hours. Nocturnal activity was reported for desert bighorn as a mechanism for avoiding solar radiation, minimizing water needs, and still obtaining sufficient forage (Simmons 1969). During the hottest and driest part of the year in western Arizona, desert bighorn were active 34% of any hour (Krausman et al. 1985). Diel activity of females relative to the moisture content of forage was not correlated; moisture content of forage was high during each hour of the day for all seasons (Alderman et al. 1989). Others have also documented nocturnal activity of desert bighorns (Monson 1964; Sandoval 1979; Elenowitz 1983), and in the Arctic, where the sun does not rise in midwinter, Dall's sheep have to feed in the dark or very dim twilight (Nichols and Bunnell 1999).

Desert bighorn and northern sheep exhibit similar behavior patterns (Augsburger 1970; Geist 1971). Desert bighorn exhibit similar activity patterns across their range (Wilson 1968; Lopez and Orihuela 1976; Olech 1979; Sandoval 1979; Chilelli and Krausman 1981). Daily activity patterns change seasonally, especially those related to foraging and bedding. In winter, foraging is the primary activity in crepuscular hours, but from 0900 to 1055 hr and from 1200 to 1355 hr, bedding made up more than 50% of winter activity. During summer, more feeding occurred before 0800 hr and after 1600 hr.

Social interactions between desert bighorn sheep were less intense than interactions described by Geist (1971), Olech (1979), and Chilelli and Krausman (1981) for northern sheep. Sexual aggression, contact greeting, dominance display, and resource competition were <1% of the activity of any hour (Chilelli and Krausman 1981). This lower behavioral activity level was likely caused by smaller group sizes, less social facilitation, and reduced play among lambs (Berger 1978).

Beds. Sheep bed in areas with loose surface material, which is cleared away by several swipes of the forefeet (Lawson and Johnson 1982). An oval depression is formed (0.6×0.4 m) from a few centimeters to >0.3 m deep (Jones 1959; Clark 1970), and a new bed is made each time a sheep lies down. The "new" bed may simply be a pawed-out "old" bed (Jones 1959; Welles and Welles 1961). Likely bed sites during the day are along ridges, out in the open, or in caves where visibility is maximized. Toward evening, bed sites are near rugged terrain with protection from cliffs and large rocks (Smith 1954; Jones 1959; Geist 1968; Nichols and Erickson 1969).

Movements. With few exceptions, males and females in most populations migrate between different areas at different seasons of the year, and there are usually at least two seasonal ranges for Rocky Mountain and California bighorn (i.e., winter and summer) and three for desert bighorn (summer, fall–winter, and spring) (Eustis 1962; Geist 1971; Bates et al. 1976; King and Workman 1982; Elenowitz 1983; Krausman et al. 1989). Seasonal movements of bighorn are not attributable to any one specific factor, probably because animal movements are dictated by an individual's response to a variety of stimuli (Leslie 1977). Behavioral, physiological, and environmental factors seem to influence these migrations, including home range knowledge, water and forage availability, lambing and mating activities, season, topography, and age and sex class.

Female desert bighorn have a relatively small home range, especially compared to those occupied by adult male groups (Krausman et al. 1989; Scott et al. 1990). Most desert bighorn populations are restricted to small areas during hot summer months due to suboptimal distribution of resources. As a result, local rainfall patterns significantly affect seasonal range size and movement patterns of desert bighorn. Summer showers that fill natural water catchments allow sheep to use areas not normally available during dry periods.

For most Rocky Mountain and some California bighorn populations, seasonal movements are not just changes in location, but also involve vertical (elevational) migrations. Annual elevational movements between ranges have been examined from a functional and adaptive viewpoint. Hebert (1973) concluded that upward movements in spring and summer allowed bighorn to maintain or prolong a diet composed mainly of new, growing vegetation because the onset of plant phenology is retarded by increased elevations. New-growing vegetation provides high-quality, readily digestible forage; thus, sheep prolong the period of high forage quality by making vertical migrations from low to high elevations in spring and summer. Differences between bighorn populations in the elevation ranges used may affect development of individuals and populations (Klein 1965; Shackleton 1973, 1976).

Factors other than vegetation are also involved in migrations. If suitable lambing grounds are separate from the winter range, the primary stimulus for pregnant females is to move to more secure habitat just before birth to avoid predators. This drive for security from predators can even override selection of forage conditions. Females left areas with good forage and gave birth in areas of lower forage quality but with greater security cover (Festa-Bianchet 1988c). Similar trade-offs have been documented for desert bighorn sheep (Berger 1991; Bleich et al. 1997; Etchberger and Krausman 1999). Traditional use of parturition sites in rugged terrain outweighed the increased need for water by lactating females (Etchberger and Krausman 1999). Climatic factors such as snow accumulation are also believed to stimulate Rocky Mountain and California bighorns to return to lower elevations in late summer and early fall (Sugden 1961; Blood 1963). Avoidance of biting insects may cause movements to more exposed areas (Blood 1963). Availability and distribution of mineral licks may also affect space use by bighorns (Smith 1954).

In general, most Rocky Mountain and many California bighorn populations return to winter ranges in October and November (Woolf et al. 1970; Geist 1971; Becker et al. 1978). Similarly, most sheep leave their winter range in May and June, which for females is most likely a response to lambing and foraging requirements. Perhaps not surprisingly, distances sheep move between seasonal ranges vary (some move up to 70 km), most probably in relation to the availability and distribution of suitable ranges (Blood 1963; McCullough and Schneegas 1966; Berwick 1968; Festa-Bianchet 1986; Hengel et al. 1992). Migratory patterns of desert bighorn also vary considerably, with most populations remaining in isolated areas, whereas others move between mountain ranges, including elevational movements and dispersal from water sources, and movement to seasonally available water sources (McQuivey 1978).

Competition. Exploitation competition arises when two species use common resources that are in short supply or, if the resources are not in short supply, competition arises when the animals seeking those resources nevertheless harm each another in the process (Pianka 1978). Exploitation competition may result in reduced population quality of one or both species through increased mortality, increased dispersal, or decreased reproduction.

Overgrazing by domestic livestock may result in competition for forage, water, and space. Competition is an important factor in the historical decline of populations of desert bighorn in western North America. The effects of diseases, human encroachment, and poaching can be documented more convincingly than the detrimental effects of overgrazing by livestock (Gallizioli 1977).

Overgrazing by livestock results in large-scale changes in plant composition and density, reduction of important plant species used by bighorn, and permanent reduction in the overall carrying capacity for bighorn sheep populations. In addition to reducing bighorn numbers, conversion of primarily grassland communities into predominantly homogeneous shrub communities provides better habitat for increasing deer populations.

Spatial and forage competition between cattle and desert bighorn continues to be a major problem that limits increased sheep population density. Overgrazed ranges also may be the major obstacle to the reintroduction of desert bighorn to historical habitats (Gallizioli 1977; Sandoval 1979). In Nevada, grazed areas supported significantly lower bighorn sheep densities than ungrazed areas. Grazed areas supported an average of 2.3 bighorn/km^2, whereas densities in ungrazed areas averaged 6.6 bighorn/km^2 (McQuivey 1978).

Microscopic analysis of fecal samples from desert bighorn and cattle in the Big Hatchet Mountains, New Mexico, revealed 12 of 18 major forage species were common to both bighorn and cattle. Winter fat (*Ceratoides lanata*) constituted an average of 17.9% and 15.4% of the annual diet of bighorn and cattle, respectively (Bavin 1975). Bavin (1975) concluded that cattle could be serious competitors with bighorn sheep for available forage, especially during years of limited forage production. A 32% dietary overlap (the percentage use in common of shared forage plants) between desert bighorn and cattle in the Peloncillo Mountains, New Mexico, was documented by Elenowitz (1983).

Wilson (1968) reasoned that spatial competition resulted in the displacement of a herd of desert bighorn in Utah. A recognizable group of bighorn occupying an ungrazed study area had been under observation for 5 years. A herd of 30 cattle was introduced into the area for 2 weeks. The bighorn sheep left the area and did not return for 8 months. Similar patterns of livestock and bighorn interactions have been reported by Jones (1980) and Steinkamp (1990). Steinkamp (1990) experimentally introduced cattle into bighorn core areas; the sheep responded by relocating.

In Aravaipa Canyon, Arizona, diet and spatial overlap were low between cattle and desert bighorn, primarily due to cattle preference for level slopes and bighorn use of steep slopes (Dodd and Brady 1986). However, plans were under consideration to redistribute cattle grazing to include areas with moderate slopes, which might have increased overlap and competition (Dodd and Brady 1986). Diets of desert bighorn and cattle in Carrizo Canyon, California, did not overlap because the populations were not sympatric; however, competition could arise if cattle were introduced to bighorn range (Cunningham and Ohmart 1986).

Historically, the most serious ecological competitors with desert bighorn have been domestic sheep and goats. These species have similar feeding habits, forage preferences, and affinities for rough topography, and they harbor parasites and other disease agents detrimental to bighorn. Extirpation of desert bighorn from numerous ranges in California (Weaver 1972), Arizona (Russo 1956; Gallizioli 1977), Nevada (McQuivey 1978; Kelly 1979), New Mexico (Gross 1960; Sandoval 1979), Utah (Wilson 1968; Dean and Spillett 1976), western Texas (Davis and Taylor 1939; Kilpatric 1982), and northwestern Mexico (Mendoza 1976) has been attributed to competition for forage and space and to the transmission of diseases from domestic sheep and goats (Bunch et al. 1999).

Concurrent with introduction of the first domestic sheep in western North America, bighorn sheep died off on a large scale, ostensibly from psoroptic scabies (*Psoroptes ovis*) (Buechner 1960). Mortalities caused by scabies reduced populations of bighorn sheep at the time domestic sheep were introduced at Greybull River, Wyoming (Honess and Frost 1942; Honess and Winter 1956); Rocky Mountain National Park, Colorado (Wright et al. 1933; Packard 1946); Sierra Nevada, California (Jones 1959); and the Owyhee River, Oregon (Bailey 1936). Psoroptic mites on desert bighorn sheep have been reported from the Desert National Wildlife Range, Nevada (Carter 1968; Decker 1970); the San Andres Mountains, New Mexico (Lange et al. 1980; Sandoval 1980); southeastern Utah (Irvine 1969); and western Arizona (deVos et al. 1980; Remington 1981).

Existence of free-ranging, exotic ungulates on occupied or potential bighorn habitat poses a serious threat to desert bighorn sheep survival. The feral burro (*Equus asinus*), aoudad (*Ammotragus lervia*), and Persian wild goat (*Capra aegagrus*) currently occupy historical bighorn habitat or are rapidly radiating into habitat vital to desert bighorn (Sandoval 1979).

The habitat requirements of aoudad (Simpson and Krysl 1981) and Persian wild goats (Bavin 1975) are similar to those of desert bighorn. Seegmiller and Simpson (1979) reported that the initial competition between two ecologically similar species consists of niche overlap and/or behavioral intolerance, shortage of limited resource, and reduced population fitness.

Hardin (1960) proposed the competitive exclusion principle, which states that two species cannot coexist on the same limited resources. Therefore, competition results in reduced population fitness for one or both species through decreased natality, decreased recruitment, increased mortality, or increased dispersal. The survivor is the species able to perpetuate itself under competitive interactions, whereas the other is excluded from the area of sympatry (Seegmiller and Simpson 1979).

Competition between feral burros and desert bighorn has attracted much attention and has been a source of controversy. Many argue that feral burros are disrupting the desert ecosystem and propose strong control and elimination measures. Others consider feral burros to have significant aesthetic value and support their preservation.

Feral burros efficiently use coarse, low-quality forage, but are not primarily browsers by preference. They are opportunistic and prefer grasses when available (Douglas and Norment 1977). McMichael (1964), Walters and Hansen (1978), and Seegmiller and Ohmart (1981) substantiated an overlap of burro and bighorn sheep range and competition for available forage in the Black Mountains, Grand Canyon, and Bill Williams Mountains, Arizona, respectively. McMichael (1964) showed a 50–58% similarity index in burro and desert bighorn diets, Walters and Hansen (1978) indicated a 52% overlap, and Seegmiller and Ohmart (1981) reported a 64% overlap. Based on fecal analysis, Hansen and Martin (1973) found that grasses dominated the diet of feral burros in Grand Canyon, Arizona. Browse species dominated feral burro diets in Death Valley, California (Browning 1960; Moehlman 1974), and the Chemehuevi Mountains, California (Woodward and Ohmart 1976). Douglas and Norment (1977) demonstrated the impact of burros on shrubby species in the Panamint Mountains, Death Valley, California. Of all shrubs, 46% exhibited some evidence of use and the survival of 12% was threatened by severe browsing.

Geist (1985) presented evidence that suggests current species in lower North America survive because of the absence of more capable competitors, and that most Siberian forms have been excluded for thousands of years. Bailey (1980) argued that large mammals in North America may not be robust competitors. As competitors (i.e., burros) return, they can be expected to outcompete bighorn sheep for resources.

Competition between Dall's sheep and other large herbivores with sympatric distributions has not been documented (Bowyer and Leslie 1992). Caribou (*Rangifer tarandus*), moose (*Alces alces*), and mountain goats (*Oreamnos americanus*) use different habitats or select diets different than those of Dall's sheep (Klein 1953; Henshaw 1970; Miquelle et al. 1992).

DIETS

Bighorn sheep diets have been well studied, and research has focused on their nutritional requirements, foraging impacts on vegetation, and competition and dietary overlap with native and exotic herbivores. Various methods were used to study bighorn diets including analysis of rumen contents of collected animals, direct observations of feeding, laboratory studies of captive animals, and detailed analyses of monthly and seasonal diets. Despite all this work, it is difficult to make more than the broadest generalizations about bighorn diets because of the differing methodologies employed (Shackleton et al. 1999).

The general consensus is that bighorn are relatively opportunistic in their diet selection, using whatever palatable species are available to them (Sugden 1961; Todd 1972; Browning and Monson 1980; Shackleton et al. 1999). For example, more than 267 plant species were eaten by California bighorn (Wikeem 1984). Browning and Monson's (1980) review of desert bighorn diets showed they used more than 470 species throughout their distribution. Forbs are most frequently eaten by Rocky Mountain and California bighorn, followed by grasses and lastly browse, whereas browse dominates desert bighorn diets, followed by grasses or forbs, depending on precipitation patterns (Shackleton et al. 1999). Perhaps not surprisingly, the relative amounts of these three forage classes vary significantly in bighorn diets, among populations, within subspecies, and among age–sex classes and individuals (Blood 1967; Sanchez 1976; Brown et al. 1977; Shank 1982).

Besides geographic, taxonomic, and individual variation, diets of bighorn exhibit strong seasonal variation. Most probably this reflects changes in the availability and palatability of forage species and in the nutritional requirements of the animals, although forage nutrient quality does not correlate well with what bighorn eat (Shackleton et al. 1999). In desert regions, seasonal use of different forages has been attributed to precipitation patterns and the effects of soil moisture on vegetation classes (Brown et al. 1977; Krausman et al. 1989). Nonetheless, observed declines in their use of browse could be due to declines in palatability when twigs become coarse and woody, and possibly when they exceed an acceptable diameter (Sandoval 1979).

Studies on the diet of Dall's sheep are not numerous, but there is general agreement that these ungulates feed mostly on graminoids (Luckhurst 1973; Hoefs and Cowan 1979; Seip and Bunnell 1985; Hansen 1996; Bowyer et al. 2000). Nonetheless, Hoefs and Cowan (1979) recorded 110 different plant species in the diet of these herbivores. In the southwest Yukon, Canada, Hoefs and Cowan (1979) noted that reedgrass (*Calamagrostis* spp.) was the most common grass in the diet of sheep, whereas fescue was eaten most often in Alaska (Nichols 1978b, Hansen 1996). Other important grasses eaten by Dall's sheep include bromegrass, bluegrass, and wheatgrass.

Dall's sheep also eat a variety of forbs, especially during late spring and summer (Hoefs and Cowan 1979; Hansen 1996). Browse consumed by Dall's sheep was principally sagebrush in the southern part of their range and willow (*Salix* spp.), dryas, and blueberry (*Vaccinium* spp.) in the north (Hoefs and Cowan 1979; Hansen 1996). Sheep sometimes move to lower elevations in early spring to obtain browse and do so again in early autumn when frost curtails growth of forbs and graminoids at higher elevations.

Hansen (1996) reported strong selection (use greater than availability) for grasses during all seasons, and selection for forbs in spring and summer and for sedges in winter; browse was avoided (use less than availability) throughout the year in the Brooks Range, Alaska. Hansen (1996) cautioned, however, that interannual patterns of diet selection could be affected by 10–15 cm of snow cover. Similarly, Rachlow and Bowyer (1998) noted that a cool summer that limited growth of forage resulted in differences in selection of habitat by Dall's sheep in interior Alaska. Mineral licks also can be important seasonally to Dall's sheep (Geist 1971; Jones and Hansen 1985).

Diet quality, as indexed by fecal crude protein, increased rapidly during early spring, peaked in June, and then began declining by July. Hansen (1996) reported a continued decline in fecal crude protein to prespring levels by October and levels remained low throughout the winter. This same general pattern was evident for digestible nitrogen, digestible energy, and in vitro dry matter digestibility of sheep forages. The growing season in arctic and subarctic environments is very short. The number of days between the last freeze in spring and the first freeze in autumn was 79 days in 1988, but only 33 days in 1989 for Dall's sheep in interior Alaska (Rachlow and Bowyer 1994). The cool, short growing season in 1989 resulted in a marked decrease in grasses and dryas available to Dall's sheep (Rachlow and Bowyer 1998). Thus, Dall's sheep must acquire the resources necessary to meet their need for reproduction and to cope with harsh winter conditions in a narrow window that varies markedly among years (Rachlow and Bowyer 1994). In addition, forage generally was more plentiful as distance from steep, precipitous terrain preferred by Dall's sheep increased (Frid 1997; Rachlow and Bowyer 1994). Hence, diet selection was affected by risk of predation in these mountain ungulates (Rachlow and Bowyer 1998).

BEHAVIOR

Social Organization. Bighorn are highly social animals that spend their life in groups with other sheep, although group integrity remains flexible to some degree throughout the year (Geist 1971; Leslie and Douglas 1979). In general, group composition consists of spatially and sexually segregated units of all-male groups and female–juvenile groups made up of adult females, lambs, and offspring from the previous 1–2 years (Geist 1971). Sandoval (1979) described another group type in desert bighorn: groups of barren adult females, yearlings, and socially mature males.

Group sizes vary by group type, seasonally, and geographically, and range from 2 to >100 individuals (Geist 1971; Ashcroft 1986; Shackleton et al. 1999). Lone bighorns, most often males, are uncommon and probably only transitory (Shackleton 1973; Ashcroft 1986). Groups appear to provide two main advantages over individuals in many species: improved foraging efficiency and predator avoidance (Pulliam and Caraco 1984). In bighorns, predation is probably a major selective force in the formation and maintenance of groups (Jarman 1974; Jarman and Jarman 1979; Berger 1991).

Males and females in most bighorn populations occupy separate, seasonal ranges, although spatial and temporal overlap does occur (Geist and Petocz 1977; Morgantini and Hudson 1981; Krausman et al. 1989; Bleich et al. 1997). Several explanations have been offered for sexual segregation in bighorn (Geist and Petocz 1977; Bleich et al. 1997). The most probable, which also applies to other ungulates, is that these differences in habitat use are most likely due to females selecting secure areas for raising their young and males choosing areas for maximizing body condition (Main and Coblentz 1990; Main et al. 1996). Bleich et al. (1997) supported this and concluded that sexual segregation in mountain sheep most probably results from different reproductive strategies of the sexes. To optimize their fitness, male bighorn grow large bodies and horns to improve their chances in intrasexual competition and hence mating success. As a result, they need high-quality food resources, and because of their larger size can use areas with a higher predation risk. Females are smaller than males and are accompanied by highly vulnerable young, so they cannot afford to use areas with high predation risks. Thus, females may forgo foraging quality for higher security.

All but young males are socially dominant over adult females. Among males, it seems that age, horn size, and probably body size and fighting behavior are positively correlated with dominance status (Geist 1971; Hass and Jenni 1991).

Social Behavior. Social relations begin at birth when a female bighorn and her young must learn to recognize each other. Lactation is costly and females cannot afford to squander scarce resources on strange offspring, whereas the young need to be able to find their mother for food and protection. This mutual recognition, or mother–young bond, seems to develop during the first 1–2 days following birth when the female is mostly isolated from other group members (Shackleton and Haywood 1985; Etchberger and Krausman 1999). Females rely more on their neonate's smell for recognition, invariably sniffing its rear when it begins to suckle. Lambs, by contrast, quickly learn the sound of their mother's voice, and it is interesting to watch when a female gives an alarm bleat. Her young invariably runs straight to her, often briefly suckling, and the female sniffs its rear. The suckle probably reinforces the young's response to the alarm call (Shackleton and Haywood 1985).

Researchers have measured several variables related to nursing in bighorn and reported changes with the lamb's age. General trends are similar, but significant differences occur within and among populations

(Horejsi 1976; Berger 1979a, 1979b; Festa-Bianchet 1988a; Hass 1990). Length of suckling bouts, suckling rates, and total time spent suckling decrease with age as the lamb spends time grazing (Berger 1979a; Festa-Bianchet 1988b). Times spent nursing are also affected by whether they follow bedding periods, and by maternal age, condition, and lungworm loads, but not by sex of offspring.

Adult male bighorn have a diverse repertoire of social behaviors they perform toward females and other males (Geist 1971). Fighting is their most spectacular and well-known behavior. Just before rut, and sometimes in spring, males gather to interact and fight with each other, presumably to determine relative dominance. Fights are usually between pairs of males, but sometimes small groups or "huddles" of three or more males will interact (Geist 1971). When fighting, males rear up on their hind legs and run bipedally toward each other, lunging headfirst at the last moment to clash horns. These echoing clashes can be repeated many times before one of the males is defeated.

The social behavior pattern most often used by bighorns is the "low stretch." An animal, usually a male, approaches or passes another with head low, neck extended, and nose pointed upward. This pattern is used in many social situations, and especially in the rut, when males use it to approach and test females. Females usually respond by squatting and urinating, then the male sniffs the urine and lip curls (flehmen) to test whether the female is coming into estrus. Once a female comes into heat, she is guarded and courted by a dominant male, though the pair is often surrounded by eager, but subordinate, males of various ages (Geist 1971).

Rocky Mountain bighorn use as many as three mating tactics. The most typical is the tending pair (i.e., male defending and copulating with a single female), less frequent is "coursing" (i.e., fighting tending males for temporary access to defended females), and an even rarer tactic is "blocking" (i.e., moving and holding females away from other males beyond the periphery of a traditional tending area) (Geist 1971; Hogg 1987; Shackleton 1991). During actual courtship of an estrous female, the tending male uses a variety of behavior patterns including nosing the female's flank and rear, usually while twisting the top of the head to one side and accompanied by flicking the tongue and vocalizing, gently kicking her with his foreleg, pushing his chest against her rump; and finally attempting to copulate (Geist 1971; Shackleton 1991).

Dall's sheep are an extremely gregarious species and exhibit a high degree of polygamy, with large, dominant males mating most often (Geist 1971). Competition among males for mates has led to the evolution of marked dimorphism in body mass and size of horns between the sexes (Bowyer and Leslie 1992).

Rutting groups of Dall's sheep were composed of as many as 21 individuals, with an overall mean of 3.7 for 166 groups; however, group size declined for males and females by midwinter (Geist 1971). The mean size of 139 maternal bands ranged from about 4 to 10 individuals from prelambing through postlambing in spring and early summer. Group size is related to foraging efficiency. Larger groups spend more of their active time feeding than smaller ones, and spend less time in vigilance or alarm behaviors than smaller groups (Rachlow and Bowyer 1998). Similarly, group size increases with increasing distance to escape terrain, ostensibly an adaptation to lower the risk of predation (Hamilton 1971; Rachlow and Bowyer 1998).

The sexes of Dall's sheep spatially segregate from each other during spring and summer (Rachlow and Bowyer 1998). Causes of sexual segregation in bighorn sheep have been more thoroughly studied than in Dall's sheep, but the explanation is likely similar (Bleich et al. 1997). Females are confined to steep, rugged terrain where they and their neonates can reduce the risk of predation, but where forage is less abundant and of lower quality than on areas inhabited by males. Larger males are less susceptible to predation than females with young, forage in areas with more predators, and obtain the forage required to attain large body size necessary to be effective in combat for mates.

The sexes begin to associate as rut approaches. By that time, males already have formed groups and begun to sort out dominance interactions with relatively low intensity behaviors, which include foreleg kicks, horn displays, and, more rarely, jump-threats and clashes. Males also mount other males in dominance interactions, but the notion that dominant males treat all subordinates as if they were females may not be correct. For instance, dominant males seldom direct courtship behaviors such as tongue flicks toward subordinate males or lick their perianal region, and do not flehmen in response to smaller males or their urine. Dominance mounting is simply a common form of aggression among ungulates (Geist 1971).

Aggressive interactions during rut are vigorous and tend to lack ritualization. Such behaviors include low-stretch postures, vigorous kicking, jump-threats, and forceful clashing of horns. The skulls of males have undergone pneumation to help absorb such forceful horn clashes. Males often display their horns, which serve as indicators of social rank. Aggressive interactions determine which males mate, and such behavior between males over estrous females can be fierce; large males may strike opponents in the side of the body with their horns and even push rivals off cliffs. Vigorous rutting activities by these large males exhaust their energy reserves, and survivorship drops markedly in older age classes of males (Geist 1971).

Geist (1971) provided excellent descriptions of courtship behaviors in mountain sheep. Dominant males approach females in a low-stretch posture while flicking their tongues. Males ascertain the reproductive status of females by licking the perianal region of the female or lapping urine from the ground where a female has urinated. Males then flehmen to determine whether the female is in estrus and receptive (Estes 1973). Courting males may be distracted by females inducing them to investigate an area where a female has urinated, and then moving away while the male flehmens (Geist 1971). Estrous females sometimes elicit courtship from a male by butting and rubbing against him. Males may kick a female with a stiff foreleg during courtship, ostensibly to determine if she is willing to stand for mounting. Copulation cannot occur if a female moves forward, which prevents the male from mating successfully. Males may tend females for 2–3 days before copulation. Some smaller males may continue to direct courtship behaviors toward females, but females typically are not receptive to such advances (Geist 1971). No evidence exists of territoriality in Dall's sheep or bighorn sheep.

Communication. Wild sheep vocalize with more vibrato and in a deeper tone than domestic sheep (Lawson and Johnson 1982). Bleating is a mechanism to maintain contact between sheep, especially females and lambs (Welles and Welles 1961).

Play. Lambs play more with each other or their mothers than adults play (Murie 1944; Jones 1959). Contact play involves similar-aged lambs butting each other, and in locomotor play, all age classes chase each other (Berger 1978). Most play occurs during crepuscular hours (Welles and Welles 1961).

Wariness. Sheep respond to disturbances by assuming an attention posture or an alarm posture, or, if startled at close range, they run. In the attention posture, they stare in the direction of the disturbance (Lawson and Johnson 1982). When alarmed, they may snort, paw the ground, bow their head, or, in the presence of wolves (*Canis lupus*), huddle in a tight circle facing out (Murie 1944; Geist 1971). Females with young lambs are the most wary (Murie 1944). If not hunted, some populations are tolerant of humans and easily become habituated to their presence (Geist 1971).

MORTALITY

Reported mortality for bighorn sheep during their first year of life is as high as 90% (Blood 1961; Morgan 1970; Akeson and Akeson 1992). Most lambs are lost during their first few weeks, often due to predation (Spraker 1974; Stewart 1980; Hoefs et al. 1986; Festa-Bianchet 1988c; Hass 1989). Coyotes (*Canis latrans*) are usually cited as the main predator, but other causes of lamb mortality also occur and seem to be interrelated (Hebert and Harrison 1988). Besides predation, these causes may include pneumonia, weather, inbreeding depression, poor maternal nutrition, poor mothering, and human disturbance (Hass 1989;

Akeson and Akeson 1992). The effect of these mortality agents also may be influenced by underlying factors including birth date, range condition, population density, and quality of security cover (Shackleton et al. 1999).

Predation also may affect the population dynamics of Dall's sheep. These mountain ungulates inhabit steep, rugged terrain, which they readily negotiate to elude, avoid, and outdistance predators. Nonetheless, wolves prey on Dall's sheep, and diets of this canid contained 2–25% of Dall's sheep. Predation by wolves and coyotes may increase during periods of deep snow. Other predators of Dall's sheep include grizzly bears (*Ursus arctos*), black bears (*Ursus americanus*), wolverines (*Gulo gulo*), lynx (*Lynx canadensis*), and golden eagles (*Aquila chrysaetos*). Dall's sheep also may perish from accidental falls and be killed by avalanches (Burles and Hoefs 1984). We concur with Nichols (1978b) that predation does not exert an important influence on populations of Dall's sheep under most circumstances.

Published mortality data are limited for subadult bighorn. For males, Geist (1971) reported no mortality among yearlings and only 3% among 2-year-olds. However, others have found yearling mortality rates of 33% in males and 41% in females, and 16–18% and 41% for 2-year-old males and females, respectively (Stewart 1980; Festa-Bianchet 1989).

With males in most populations, death rates between 3 and 5 years old were similar, ranging from 3% to 14% (Shackleton et al. 1999). For older males, age-specific death rates were higher, but the slope of the increase depended on whether or not the population was hunted. Without hunting, age-specific mortality rates in stable populations increased slowest, reaching more than 60% by age 15 years, and life expectancy was high. In hunted populations living on high-quality ranges and increasing in numbers, male mortality rates increased more rapidly with increasing age, and life expectancy was shorter.

Only limited mortality data are available for females, partly because it is difficult to determine age of females older than 5 years by horn annuli counts. Adult female mortality rates seem to be higher than those of males, and have been estimated to be around 11% (Hansen 1980b; Hengel et al. 1992).

In desert bighorn sheep, Hansen (1980a) reported that the highest adult mortality of 47% occurred in age classes 10–12 years. By contrast, McQuivey (1978) reported that the average age structure of males in Nevada declined gradually through successive years of life, indicating relatively constant mortality rates. Also, he found no single mortality factor affected any age group for males >1 year old. These results closely agree with those of Leslie and Douglas (1979), who noted relatively constant mortality rates for males from the River Mountains, Nevada. Those authors hypothesized that increased mortality of young males could be due to involvement in rutting activities, especially during the prerut period. High levels of prerut activity of socially immature males also were observed by Sandoval (1979).

Survivorship of desert bighorn females did not differ between the Little Harquahala and Harquahala Mountains, Arizona (Krausman et al. 1989). The probability of survival for males, however, was consistently higher in the Harquahala Mountains. Mortality factors were similar between the study areas, except that predation by mountain lion (*Puma concolor*) was higher for female sheep in the Harquahala Mountains.

Disease can be a major mortality factor for bighorn sheep, and large (>50%), sudden population die-offs are common (Buechner 1960; Stelfox 1971; Ryder et al. 1992). Mortality from disease is likely related to stress that interacts with lungworm (*Protostrongylus* spp.) infestation (Ryder et al. 1992). It is unclear to what extent die-offs can be attributed to range conditions or domestic stock (Coggins and Matthews 1992; Dunbar 1992). Festa-Bianchet (1991) claimed lungworm infection is normal for bighorn and higher levels of infection do not predict epizootics or indicate poor population health.

Dall's sheep also are infected or infested with numerous disease organisms and parasites. Lungworm has been reported from Dall's sheep, but there have not been massive die-offs from this parasite and its associated pneumonias or hemorrhagic septicemia, as in bighorn sheep (Gable and Murie 1942; Stelfox 1971; Shackleton 1985). As with other sources of mortality, this disease complex may be more prevalent in populations on a low nutritional plane (Seip and Bunnell 1985). Other nematodes have been reported in Dall's sheep, but caused no obvious pathology (Gibbs and Fuller 1959; Seip and Bunnell 1985). Perhaps low temperatures in winter or a short summer growing season at high latitudes helps limit direct transmission of parasite larvae or does not provide suitable conditions for intermediate hosts of pathogenic parasites that debilitate populations of bighorn sheep. Additionally, domestic sheep and cattle are uncommon throughout the range of Dall's sheep. Parasitic coccidia, however, have been isolated from feces of Dall's sheep (Clark and Colwell 1974). Furthermore, positive titers in blood sera have been reported for several arboviruses, contagious ecthyma, parainfluenza III, epizootic hemorrhagic disease, and Q fever (Zarnke et al. 1983). Serology also indicated that Dall's sheep were exposed to the bacteria *Brucella* spp. and *Campylobacter feti* (Heimer et al. 1982; Foreyt et al. 1983). *Mycoplasma ovipneumoniae* has killed captive Dall's sheep, but was not reported for free-ranging sheep (Zarnke and Rosedal 1989).

Necrosis of horn cores and the mandible are prevalent in Dall's sheep; *Corynebacterium pyogenes, Fusobacterium necropitorum, Proteus* spp., *Micrococci* spp., and *Escherichia* spp. have been associated with infections of lumpy jaw (Glaze et al. 1982; Hoefs et al. 1982; Bunch et al. 1984; Hoefs and Bunch 1992). Causes of horn aberrations in Dall's sheep are uncertain, but may involve *Actinomaayces* spp. (Hoefs and Bunch 1992). Skull asymmetry was caused by osteoporosis in bighorn sheep (Bunch et al. 1990). Dall's sheep may be predisposed to mandibular infections from excessive tooth wear because of wind-blown silt deposited on forage (Hoefs and Bayer 1983). There is no evidence that diseases play a major role in regulating or limiting populations of Dall's sheep.

AGE ESTIMATION

Age of mountain sheep up to 1 year old can be determined based on tooth eruption (Fig. 51.7), size, and external criteria (Hansen 1965). After 1 year, age of wild sheep is determined by horn annuli (Fig. 51.4) up to 7 years. Horn annuli on older males and females are not distinct and often are covered with hair (Geist 1966a; Hemming 1969). Tooth replacement patterns are useful up to 4 years (Table 51.2), but this method is not commonly used (Murie 1944; Hemming 1969).

The sex and age of sheep older than lambs are generally classified by body size and horn length (Fig. 51.7) (Geist 1966a). The horns of males become longer at the base than those of females and develop a curl with increased age. Horns on females retain the shape of lamb horns (Fig. 51.7) and grow a little each year (Lawson and Johnson 1982).

STATUS AND MANAGEMENT

Seton (1929) estimated that in pristine times there were about 2 million mountain sheep in the contiguous United States and another 2 million in Canada and Alaska combined. Seton's (1929) estimate of 4 million sheep is often cited as a reliable approximation of mountain sheep numbers. Valdez (1988), however, doubted wild sheep numbers ever exceeded 500,000 for all North America. Mountain sheep are highly selective in their habitat preferences, and it is probably a misconception that sheep were uniformly distributed throughout the mountains of western North America.

The present distribution ranges of all races of bighorn sheep seem considerably reduced; sheep may be occupying habitats in the most remote portions of their historical distribution (Sugden 1961). Bighorn sheep were quite well adapted to habitats far from the rugged terrain more often considered typical for the species (Buechner 1960; Shackleton 1985; Krausman 1993). Valdez and Krausman (1999) and Krausman (1997a) estimated the number of bighorns in North American in 1991. At that time, Canada supported 12,700 Rocky Mountain bighorn, and 3000 California bighorn sheep. Mexico supported 3500 desert bighorn, and in the contiguous United States there were 25,219 Rocky Mountain bighorn sheep, 4901 California bighorn, and 17,450

FIGURE 51.7. Age classes of bighorn sheep. Note that the classes form a cline in body and horn size. SOURCE: Redrawn from Geist (1968).

desert bighorn. The total number of bighorn sheep in North America has not changed significantly since the 1991 survey (Krausman 1997a).

Dall's sheep far outnumber all others, with over 100,000: there are 9000 in the Yukon, 500 in British Columbia, 14,000 in the Northwest Territories, and > 72,000 in Alaska (Bowyer et al. 2000). There are about 14,500 Stone's sheep (Valdez and Krausman 1999). Dall's and Stone's sheep populations have remained relatively unimpacted by humans and have retained their historical distribution and abundance.

Population Management. Bighorn sheep populations have declined significantly since 1900 with a subsequent reduction in their geographic distribution (Buechner 1960; Bailey 1980; Hansen 1980b). Declines in density and distribution have been attributed to a wide range of human-induced factors, which include overgrazing by livestock and feral burros, diseases transmitted by livestock, road construction, housing developments, canals, fire suppression, and recreational activities (Gallizioli 1977; Krausman et al. 1989; Harris 1992; Krausman 1996). As a result, population management has been directed at reducing human-caused decimating factors and at learning more about other population-limiting factors. Understanding limiting factors is especially important for resource managers charged with managing bighorns because many sheep populations are small and susceptible to extirpation (Berger 1990; Fitzsimmons and Buskirk 1992; Krausman et al. 1993, 1996).

Management efforts have been directed at eliminating domestic stock from bighorn habitat, understanding the effects of diseases on populations, eliminating competition with species such as feral burros, and minimizing human disturbance. One of the most active management programs for bighorn sheep, however, has been the translocation of animals to various locations within their historical ranges. In general, translocations have been successful and many populations have been returned to historical habitats. Human disturbance of translocated and other herds, however, needs to be closely monitored. In general, bighorn are intolerant of human activities, especially outside protected areas, and can have difficulty adjusting to human encroachment. These sheep have traditional range use and movement patterns that are probably passed from adults to young; consequently, bighorn do not adjust well when these patterns are disrupted (Geist 1971). Disturbances, whether specifically directed toward bighorn or not, have adverse effects on sheep populations. They may abandon ranges following increased levels of human activity (Leslie 1977; Hamilton et al. 1982; Krausman et al. 1989).

Currently, there is no reason to attempt to manage populations of Dall's sheep using predator control or by intervening in disease processes; predators or diseases do not regulate their populations often enough to be of consequence. Effective management, then, is restricted largely to regulating sport and subsistence harvest and providing viewing opportunities for aesthetic and other nonconsumptive uses (Bowyer and Leslie 1992).

Dall's sheep are managed by different state, federal, and provincial government agencies within their distribution. In general, Dall's sheep are managed to furnish trophies for hunters; subsistence harvest is thought to be small (Nichols 1978b; Heimer 1985). Although some harvesting of females was permitted in the past, harvests now are restricted mostly to three-fourths- or full-curl males (Elliot 1985; Hoefs and Barichello 1985; Poole and Graf 1985). Total harvest relative to the total size of populations is typically <2% (Bowyer and Leslie 1992). Most large males are harvested by nonresident hunters, who are required to follow a set of regulations that includes hiring guides or outfitters, payment of special fees, and hunting only in specified areas (Bowyer and Leslie 1992).

Nichols (1978a,b) concluded that the harvest of large males did not influence reproduction in populations of Dall's sheep, and this practice was not sufficient to reduce herd numbers. This is not surprising, because large males only use the same ranges as females during rut, and one consequence of sexual segregation is a reduction in competition of large males with females and young for much of the year. Thus, only the harvest of females can affect populations in a density-dependent manner. Density, interacting with climatic conditions, is the primary factor regulating productivity of Dall's sheep populations.

This interaction between density and climate also plays an important role in determining horn size in males and hence the quality of trophies available to harvest. For instance, Bunnell (1978) reported a strong positive relationship between precipitation (an index to primary productivity) and horn growth for male Dall's sheep. Moreover, depressed recruitment was in years in which growth of horns also was diminished (Bunnell 1978). The quality (i.e., growth) of horns of young males is mediated through the physical condition of the females, and this effect can be observed for up to 5 years of age. Annual growth of horns is much greater in males than females, and there is marked interannual variation in growth. Annual growth of horns and body mass were not correlated significantly for males (Bunnell 1978). This analysis, however, was complicated by the horns composing 8–12% of body mass in large males and the fact that some brooming (wear) of horn tips occurred (Geist 1971). Bunnell (1978) interpreted this lack of significance to mean that a male could recover from a period of nutritional deprivation and reattain its body mass, but that this period of hardship was recorded in horns. That the cause of differences in horn size was related to nutrition was demonstrated by greater growth of horns in a male held on a game farm than for individuals in a free-ranging population from which the captive male was obtained (Bunnell 1978). Finally, Bunnell (1978) calculated an index to horn quality, which varied markedly among years. Clearly, there are not populations of high and low quality. Instead, changes in quality within a population vary through time depending on nutrition and environmental conditions.

One manner in which the harvest of large males might affect population demography was proposed by Geist (1971). He argued that young males (i.e., less than a three-quarter curl) would experience high rates of mortality if large males were harvested heavily, because younger individuals would begin participating in rutting activities that lead to high rates of mortality in larger males. Heimer et al. (1982) offered some support for Geist's (1971) hypothesis, but in a reanalysis of those data, Whitten (2001) reached the opposite conclusion. Murphy et al. (1990) reported no relationship between the ratios of old or younger males to females among a number of populations throughout Alaska, and concluded that a reduction in older males via hunting did not affect survivorship of younger males.

One additional way in which the harvest of large males might affect the demographics of a population is via the role of large males in initiation of estrus in females. The presence of a rutting male can hasten onset of estrus in bovids and cervids (Coblentz 1976). Whether young, but sexually mature, males can fulfill this role in Dall's sheep is unknown. Even if estrus is delayed in populations without a sufficient number of large males, whether this would markedly affect timing or synchrony of parturition is uncertain because of the apparent ability of females to adjust the length of gestation (Rachlow and Bowyer 1991; Berger 1992). Moreover, whether there might be a cost (e.g., low mass of a neonate) from adopting such a strategy requires more study. Indeed, this is one of the least known aspects of the biology of Dall's sheep.

Habitat Management. Because bighorn sheep exist in arid environments, habitat-related management generally has been limited to prescribed burns and the provision of water. Prescribed burns are used to control forest and heavy shrub encroachment on bighorn ranges in an attempt to mimic natural fires and maintain open grasslands for foraging and visibility (McWhirter et al. 1992). Water sources have been developed in many areas, but their value to bighorn sheep has been questioned because there is limited empirical evidence demonstrating the benefit of such artificial sources of water (Burkett and Thompson 1994; Broyles 1995; Krausman and Etchberger 1995; Krausman and Czech 1998). However, more research is needed in this area.

The relationship between food supply and population density is recognized as a basic concept of wildlife management. Management of renewable resources is based on the knowledge that there is a limit to the intensity of harvest that each species can tolerate. If this critical level is exceeded, the species will decline and the future annual harvest will diminish. Notwithstanding lower forage production in desert mountain ranges, it is probable that sufficient food supply exists to support more sheep than are currently present. When a certain population has a number of different requirements, the one in shortest supply relative to the demand obviously will be the limiting factor (Leopold 1933). The apparent excess of one life requirement, such as rough topography, cannot make up for the scarcity of another, such as suitable interspersion of food and water.

Until recently, bighorn sheep were often managed on a population-by-population basis, usually within a mountain range. The flatter areas between mountains, however, may act as corridors for sheep to gain access to other ranges for lambing and foraging (Bleich et al. 1990, 1996; Krausman 1997b). Bleich et al. (1990) proposed a model for the conservation of desert bighorn sheep. Schwartz et al. (1986) were among the first to suggest a management strategy based at the landscape level. Overall, they suggested that management should address intermountain travel corridors for bighorn and, where necessary, take steps to minimize potential barriers. Habitat for bighorn sheep still exists in the West, but managers (and the public) have to ensure that sufficient habitat is protected, movement corridors remain open, human disturbance is reduced or kept to a minimum, and transmission of diseases from livestock is eliminated. Only if these are accomplished will efforts to enhance viable populations of bighorn have a chance to be successful.

Because of the remote and rugged terrain occupied by Dall's sheep, manipulation of habitat is rare. Subalpine ranges in British Columbia, Canada, were burned in spring to improve habitat (Elliot 1985; Seip and Bunnell 1985). Elliot (1985) noted that such burning was capable of slowing population declines and enhancing horn size in males. Nutrient quality of burned areas was not superior to that of unburned ranges in spring and summer, but burned areas had a greater quantity of forage than unburned areas in winter (Seip and Bunnell 1985).

Mineral exploration and extraction, road construction, harassment by low-flying aircraft, and other human disturbances of Dall's sheep, especially on lambing grounds, have the potential to affect populations (Nichols 1975; Hoefs and Barichello 1985; Poole and Graf 1985). Nonetheless, most of the range of Dall's sheep remains remote, pristine, and relatively unaffected by human developments or other activities.

Compared with other large mammals in North America, our understanding of the biology of Dall's sheep is incomplete. Much remains to be learned about these unique ungulates, which inhabit mountainous areas of the far north.

RESEARCH NEEDS

In 1957 and 1978, the Desert Bighorn Council and the Northern Wild Sheep and Goat Council were established, respectively. Since then, scientists have regularly published an array of life history data on mountain sheep. These efforts continue, and ideas and concepts are being refined, sharpened, and revised as technology advances and society changes. Radiotelemetry has been a significant tool in ascertaining movement, home range, site fidelity, sexual segregation, and other important information. Satellite telemetry and global positioning systems are the latest technology assisting with these types of data collection and others important to the management of mountain sheep.

Although biologists have acquired a wealth of data, more needs to be obtained. Some of the types of earlier data collection are still needed and will continue, but future and novel research is also necessary toward a better understanding of the ecology of mountain sheep. This should include studies of taxonomy, landscape use, anthropogenic forces, predation, competition, exploitation, translocations, small populations, and natality.

LITERATURE CITED

Adams, L. G., K. L. Risenhoover, and J. A. Bailey. 1982. Ecological relationship of mountain goat and Rocky Mountain bighorn sheep. Proceedings of the Biennial Symposium of the Northern Sheep and Goat Council 3:9–22.

Akeson, J. J., and H. A. Akeson. 1992. Bighorn sheep movements and summer lamb mortality in central Idaho. Proceedings of the Biennial Symposium of the Northern Wild Sheep and Goat Council 8:14–27.

Alderman, J. A., P. R. Krausman, and B. D. Leopold. 1989. Diel activity of female desert bighorn sheep in western Arizona. Journal of Wildlife Management 52:264–71.

Alvarez, T. 1976. Status of bighorns in Baja, California. Desert Bighorn Council Transactions 20:18–21.

Asdell, S. A. 1964. Patterns of mammalian reproduction. Cornell University Press, Ithaca, NY.

Ashcroft, G. E. W. 1986. Sexual segregation and group sizes in California bighorn sheep. M.S. Thesis, University of British Columbia, Vancouver, Canada.

Augsburger, J. G. 1970. Behavior of Mexican bighorn sheep in the San Andres Mountains, New Mexico. M.S. Thesis, New Mexico State University, Las Cruces.

Bailey, J. A. 1980. Desert bighorn, forage competition and zoogeography. Wildlife Society Bulletin 8:208–16.

Bailey, J. A. 1936. The mammals and life zones of Oregon. North American Fauna 55:1–416.

Bates, J. W., Jr., J. C. Pederson, and S. C. Amstrop. 1976. Bighorn sheep range, population trend and movement. Desert Bighorn Council Transactions 20:11–12.

Bavin, R. L. 1975. Ecology and behavior of the Persian ibex in the Florida Mountains, New Mexico. M.S. Thesis, New Mexico State University, Las Cruces.

Becker, K., T. Varcalli, E. T. Thorne, and G. B. Butler. 1978. Seasonal distribution patterns of Whiskey Mountain bighorn sheep. Proceedings of the Biennial Symposium of the Northern Wild Sheep and Goat Council 1:1–16.

Berger, J. 1978. Group size, foraging, and antipredator ploys: An analysis of bighorn sheep decisions. Behavioral Ecology and Sociobiology 4:91–99.

Berger, J. 1979a. Social ontogeny and behavioural diversity: Consequences for bighorn sheep *Ovis canadensis* inhabiting desert and mountain environments. Journal of Zoology (London) 192:251–66.

Berger, J. 1979b. Weaning conflict in desert and mountain sheep (*Ovis canadensis*): An ecological interpretation. Zeitschrift für Tierpsychologie 50:188–200.

Berger, J. 1990. Persistence of different sized populations: An empirical assessment of rapid extinctions in bighorn sheep. Conservation Biology 4:91–98.

Berger, J. 1991. Pregnancy incentives, predation constraints and habitat shifts: Experimental and field evidence for wild bighorn sheep. Animal Behavior 41:61–77.

Berger, J. 1992. Facilitation of reproductive synchrony by gestation adjustment in gregarious mammals: A new hypothesis. Ecology 73:323–29.

Berwick, S. H. 1968. Observations on the decline of the Rock Creek, Montana, population of bighorn sheep. M.S. Thesis, University of Montana, Missoula.

Blaisdell, J. A. 1976. The Lava Beds bighorn: So who worries? Desert Bighorn Council Transactions 20:50.

Bleich, V. C., J. D. Wehausen, and S. A. Holl. 1990. Desert-dwelling mountain sheep: Conservation implications of a naturally fragmented distribution. Conservation Biology 4:383–90.

Bleich, V. C., J. D. Wehausen, R. R. Ramey II, and J. L. Rechel. 1996. Metapopulation theory and mountain sheep: Implications for conservation. Pages 353–73 *in* D. R. McCullough, ed. Metapopulations and Wildlife Conservation. Island Press, Washington, DC.

Bleich, V. C., R. T. Bowyer, and J. D. Wehausen. 1997. Sexual segregation in mountain sheep: Resources or predation? Wildlife Monographs 134:1–50.

Blom, E. 1968. Male reproductive organs. Pages 27–37 *in* E. S. E. Hafez, ed. Reproduction in farm animals. 2nd ed. Lea & Febiger, Philadelphia.

Blood, D. A. 1961. An ecological study of California bighorn sheep (*Ovis canadensis californiana* Douglas) in southern British Columbia. M.S. Thesis, University of British Columbia, Vancouver, Canada.

Blood, D. A. 1963. Some aspects of behavior of a bighorn herd. Canadian Field-Naturalist 77:77–94.

Blood, D. A. 1967. Food habits of the Ashnola bighorn sheep herd. Canadian Field-Naturalist 81:23–29.

Blunt, M. H., H. A. Dawson, and E. T. Thorne. 1972. The birth weights and gestation in captive Rocky Mountain bighorn sheep. Journal of Mammalogy 58:106.

Boone, J. P. 1988. Historical review of exceptional sheep trophies. Pages 121–142 *in* R. Valdez, ed. Wild sheep and wild sheep hunters of the new world. Wild Sheep and Goat International, Mesilla, NM.

Bowyer, R. T. 1984. Sexual segregation in southern mule deer. Journal of Mammalogy 65:410–17.

Bowyer, R. T. 1991. Timing of parturition and lactation in southern mule deer. Journal of Mammalogy 72:138–45.

Bowyer, R. T., and D. M. Leslie, Jr. 1992. *Ovis dalli*. Mammalian Species 393:1–7.

Bowyer, R. T., V. Van Ballenberghe, and J. G. Kie. 1998. Timing and synchrony of parturition in Alaskan moose: Long-term versus proximal effects of climate. Journal of Mammalogy 79:1244–1332.

Bowyer, R. T., D. M. Leslie, Jr., and J. L. Rachlow. 2000. Dall's and Stone's sheep. Pages 491–516 *in* S. Demarais and P. R. Krausman, eds. Ecology and management of large mammals in North America. Prentice-Hall, Upper Saddle Ridge, NJ.

Bradley, W. G. 1964. The vegetation of the Desert Game Range with special reference to the desert bighorn. Desert Bighorn Council Transactions 8:43–67.

Brown, B. W., D. D. Smith, and R. P. McQuivey. 1977. Food habits of desert bighorn sheep in Nevada 1956–1976. Desert Bighorn Council Transactions 21:32–61.

Browning, B. 1960. Preliminary report on the food habits of the wild burro in the Death Valley National Monument. Desert Bighorn Council Transactions 4:88–90.

Browning, B. M., and G. Monson. 1980. Food. Pages 80–99 *in* G. Monson and L. Sumner, eds. The desert bighorn. University of Arizona Press, Tucson.

Broyles, B. 1995. Desert wildlife water developments: Questioning use in the Southwest. Wildlife Society Bulletin 23:663–75.

Buechner, H. K. 1960. The bighorn sheep in the United States, its past, present, and future. Wildlife Monographs 4:1–174.

Bunch, T. D., M. Hoefs, R. L. Glaze, and H. S. Ellsworth. 1984. Further studies in Dall's sheep (*Ovis dalli dalli*) from Yukon Territory, Canada. Journal of Wildlife Diseases 20:125–33.

Bunch, T. D., R. M. Mitchell, and A. Maciulis. 1990. G-banded chromosomes of the Gansu orgali (*Ovis ammon jubata*) and their implications in the evolution of the *Ovis* karyotype. Journal of Heredity 81:227–30.

Bunch, T. D., R. S. Hoffmann, and C. F. Nadler. 1999. Cytogenetics and genetics. Pages 263–76 *in* R. Valdez and P. R. Krausman, eds. Mountain sheep of North America. University of Arizona Press, Tucson.

Bunnell, F. L. 1978. Horn growth and population quality in Dall sheep. Journal of Wildlife Management 42:764–75.

Bunnell, F. L. 1980. Factors controlling lambing period of Dall's sheep. Canadian Journal of Zoology 58:1027–31.

Bunnell, F. L. 1982. The lambing period of mountain sheep: Synthesis, hypothesis, and tests. Canadian Journal of Zoology 60:1–14.

Bunnell, F. L., and N. A. Olsen. 1976. Weights and growth of Dall sheep in Kluane Park Reserve, Yukon Territory, Canada. Canadian Field-Naturalist 90:157–62.

Burkett, D. W., and B. C. Thompson. 1994. Wildlife association with human-altered water sources in semi-arid vegetation communities. Conservation Biology 8:682–90.

Burles, D. W., and M. Hoefs. 1984. Winter mortality of Dall sheep (*Ovis dalli dalli*), in Kluane National Park, Yukon. Canadian Field-Naturalist 98:479–84.

Butcher, P. D., and C. M. Hawkey. 1979. The nature of erythrocyte sickling in sheep. Comparative Biochemistry and Physiology 64A:411–18.

Cameron, R. D., and J. M. Ver Hoef. 1994. Predicating parturition rate of caribou from autumn body mass. Canadian Journal of Zoology 71:480–86.

Carter, B. H. 1968. Scabies in desert bighorn sheep. Desert Bighorn Council Transactions 12:76–77.

Chappel, R. W., and R. J. Hudson. 1978. Winter bioenergetics of Rocky Mountain bighorn sheep. Canadian Journal of Zoology 56:2388–93.

Chilelli, M. E., and P. R. Krausman. 1981. Group organization and activity patterns of desert bighorn sheep. Desert Bighorn Council Transactions 25:17–24.

Clark, G. W., and D. A. Colwell. 1974. *Eimeria dalli*: A new species of protozoan (Eimeriidae) from Dall sheep, *Ovis dalli*. Journal of Protozoology 21:197–99.

Clark, J. L. 1970. The great arc of the wild sheep, 3rd ed. University of Oklahoma Press, Norman.

Clark, J. L. 1978. The great arc of the wild sheep. 4th ed. University of Oklahoma Press, Norman.

Coblentz, B. E. 1976. Functions of scent urination in ungulates with special reference to feral goats (*Capra hircus*). American Naturalist 110:549–57.

Coggins, V. L., and P. E. Matthews. 1992. Lamb survival and herd status of the Lostine bighorn herd following a *Pasteurella* die-off. Proceedings of the Biennial Symposium of the Northern Wild Sheep and Goat Council 8:147–54.

Cowan, I. M. 1940. Distribution and variation in the native sheep of North America. American Midland Naturalist 24:505–80.

Cunningham, S. C., and R. D. Ohmart. 1986. Aspects of the ecology of desert bighorn sheep in Corrizo Canyon, California. Desert Bighorn Council Transactions 30:14–19.

Dale, A. R. 1987. Ecology and behavior of bighorn sheep, Waterton Canyon, Colorado, 1981–1982. M. S. Thesis, Colorado State University, Fort Collins.

Davis, W. B., and W. P. Taylor. 1939. The bighorn sheep of Texas. Journal of Mammalogy 20:440–55.

Dean, H. C., and J. J. Spillett. 1976. Bighorn in Canyonlands National Park. Desert Bighorn Council Transactions 20:15–17.

Decker, J. V. 1970. Scabies in desert bighorn sheep in the Desert National Wildlife Range. Desert Bighorn Council Transactions 14:107–8.

DeForge, J. R. 1980. Population biology of desert bighorn sheep in the San Gabriel Mountains of California. Desert Bighorn Council Transactions 24:29–32.

Demarchi, R. A. 1965. An ecological study of the Ashnola bighorn winter ranges. M.S. Thesis, University of British Columbia, Vancouver, Canada.

Demarchi, R. A., and H. B. Mitchell. 1973. The Chilcotin River bighorn population. Canadian Field Naturalist 87:433–54.

Deming, O. V. 1952. Tooth development of the Nelson bighorn sheep. California Fish and Game Journal 38:523–29.

De Vos, J., R. L. Glaze, and T. D. Bunch. 1980. Scabies (*Psoroptes ovis*) in Nelson desert bighorn of northwestern Arizona. Desert Bighorn Council Transactions 24:44–46.

De Young, D. W., P. R. Krausman, L. E. Weiland, and R. C. Etchberger. 1993. Baseline ABRs in mountain sheep and desert mule deer. International Congress on Noise as a Public Health Problem 6:251–54.

Dixon, J. S. 1938. Birds and mammals of Mount McKinley National Park, Alaska. U.S. National Park Service Fauna Series 3:1–236.

Dodd, N. L., and W. W. Brady. 1986. Cattle grazing influences on vegetation of sympatric desert bighorn range in Arizona. Desert Bighorn Council Transactions 30:8–13.

Douglas, C. L., and C. Norment. 1977. Habitat damage by feral burros in Death Valley. Desert Bighorn Council Transactions 21:23–25.

Douglas, C. L., and L. D. White. 1979. Movements of desert bighorn sheep in the Stubbe Spring Area, Joshua Tree National Monument. Desert Bighorn Council Transactions 23:71–77.

Drewek, J. R. 1970. Population characteristics and behavior of introduced bighorn sheep in Owyhee County, Idaho. M.S. Thesis, University of Idaho, Moscow.

Dunbar, M. R. 1992. Theoretical concepts of disease versus nutrition as primary factors in population regulation of wild sheep. Proceedings of the Biennial Symposium of the Northern Wild Sheep and Goat Council 8:174–92.

Eccles, T. R. 1983. Aspects of social organization and diurnal activity patterns of California bighorn sheep (*Ovis canadensis californiana* Douglas 1829) (Report R-6). British Columbia Ministry of Environment, Fish, and Wildlife, Victoria, Canada.

Eccles, T. R., and D. M. Shackleton. 1979. Recent records of twinning in mountain sheep. Journal of Wildlife Management 43:974–76.

Eccles, T. R., and D. M. Shackleton. 1986. Correlates and consequences of social status in female bighorn sheep. Animal Behavior 34:1391–1401.

Elenowitz, A. S. 1983. Habitat use and population dynamics of transplanted desert bighorn sheep in the Peloncillo Mountains, New Mexico. M.S. Thesis, New Mexico State University, Las Cruces.

Elliot, J. P. 1985. The status of thinhorn sheep (*Ovis dalli*) in British Columbia. Pages 43–47 *in* M. Hoefs, ed. Wild sheep: Distribution, abundance, management and conservation of the sheep of the world and closely related mountain ungulates. Special report. Northern Wild Sheep and Goat Council, Whitehorse, Yukon, Canada.

Erickson, G. L. 1972. The ecology of Rocky Mountain bighorn sheep in the Sun River area of Montana with special reference to summer food habits and range movements (Federal Aid and Wildlife Restoration Project W-120-R-2 and R-3). Montana Fish and Game Department, Helena.

Estes, R. D. 1973. The role of the vomeronasal organ in mammalian reproduction. Mammalia 36:315–41.

Etchberger, R. C., and P. R. Krausman. 1999. Frequency of birth and lambing sites of a small population of mountain sheep. Southwestern Naturalist 44:354–60.

Etchberger, R. C., P. R. Krausman, and R. Mazaika. 1989. Mountain sheep habitat characteristics in the Pusch Ridge Wilderness, Arizona. Journal of Wildlife Management 53:902–7.

Etchberger, R. C., P. R. Krausman, and R. Mazaika. 1990. Effects of fire on desert bighorn sheep habitats. Pages 53–57 *in* P. R. Krausman and N. S. Smith, eds. Managing wildlife in the Southwest. Arizona Chapter, Wildlife Society, Phoenix.

Eustis, G. P. 1962. Winter lamb surveys on the Kofa Game Range. Desert Bighorn Council Transactions 6:83–86.

Festa-Bianchet, M. 1986. Site fidelity and seasonal range use by bighorn rams. Canadian Journal of Zoology 64:2126–32.

Festa-Bianchet, M. 1988a. Seasonal range selection in bighorn sheep conflicts between forage quality, forage quantity, and predator avoidance. Oecologia 75:580–86.

Festa-Bianchet, M. 1988b. Nursing behavior of bighorn sheep: Correlates of ewe age, parasitism, lamb age, birthdate and sex. Animal Behavior 36:1445–54.

Festa-Bianchet, M. 1988c. Birthdate and survival in bighorn lambs (*Ovis canadensis*). Journal of Zoology (London) 214:653–61.

Festa-Bianchet, M. 1989. Survival of male bighorn sheep in southwestern Alberta. Journal of Wildlife Management 53:259–63.

Festa-Bianchet, M. 1991. The social system of bighorn sheep: Grouping patterns, kinship and female dominance rank. *Animal Behavior* 42:71–82.

Fitzsimmons, N. N., and S. W. Buskirk. 1992. Effective population sizes for bighorn sheep. Proceedings of the Biennial Symposium of the Northern Wild Sheep and Goat Council 8:1–7.

Flores, M. G., L. J. Jimenez, S. X. Madrigal, R. F. Moncayo, and R. F. Takaki. 1972. Tipos de vegetacion de la Republica Mexicana. Subsecretaria de Planeacion, Dirección General de Estudios, Dirección Agricola.

Foreyt, W. J., T. C. Smith, J. F. Evermann, and W. E. Heimer. 1983. Hematologic serum chemistry and serologic values of Dall's sheep, *Ovis dalli dalli*, in Alaska, USA. Journal of Wildlife Diseases 19:136–39.

Franzmann, A. W. 1971. Physiologic values of Stone's sheep. Journal of Wildlife Disease 7:139–41.

Franzmann, A. W. 1972. Environmental sources of variation of bighorn sheep physiologic values. Journal of Wildlife Management 36:924–32.

Franzmann, A. W., and D. M. Hebert. 1971. Variation of rectal temperature in bighorn sheep. Journal of Wildlife Management 35:488–94.

Franzmann, A. W., and E. T. Thorne. 1970. Physiologic values in wild bighorn sheep (*Ovis canadensis canadensis*) at capture, after handling and after captivity. American Veterinary Medical Association 57:647–50.

Frid, A. 1997. Vigilance by female Dall's sheep: Interactions between predation risk factors. Animal Behavior 53:799–808.

Gable, F. C., and A. Murie. 1942. A record of lungworms in *Ovis dalli* (Nelson). Journal of Mammalogy 23:220–21.

Gallizioli, S. 1977. Overgrazing on desert bighorn ranges. Desert Bighorn Council Transactions 21:21–23.

Geist, V. 1966a. Validity of horn segment counts in aging bighorn sheep. Journal of Wildlife Management 30:634–35.

Geist, V. 1966b. The evolutionary significance of mountain sheep horns. Evolution 20:558–66.

Geist, V. 1968. On delayed social and physical maturation in mountain sheep. Canadian Journal of Zoology 46:899–904.

Geist, V. 1971. Mountain sheep: A study in behavior and evolution. University of Chicago Press, Chicago.

Geist, V. 1985. On Pleistocene bighorn sheep: Some problems of adaptation and relevance to today's American megafauna. Wildlife Society Bulletin 13:351–59.

Geist, V., and R. G. Petocz. 1977. Bighorn sheep in winter: Do rams maximize reproductive fitness by spatial separation and habitat segregation from ewes? Canadian Journal of Zoology 55:1802–10.

Gibbs, H. C., and W. A. Fuller. 1959. Record of *Wyominia tetoni* Scoot, 1941, from *Ovis dalli* in the Yukon Territory. Canadian Journal of Zoology 37:815.

Gilchrist, D. 1992. Why is Montana the land of the giant rams? Proceedings of the Biennial Symposium of the Northern Wild Sheep and Goat Council 8:8–13.

Gionfriddo, J. P., and P. R. Krausman. 1986. Summer habitat use by mountain sheep. Journal of Wildlife Management 50:331–36.

Glaze, R. L., M. Hoefs, and T. D. Bunch. 1982. Aberrations of the tooth arcade and mandible in Dall's sheep from southwestern Yukon, Canada. Journal of Wildlife Diseases 18:305–10.

Goodson, N. J. 1978. Status of bighorn sheep in Rocky Mountain National Park. M.S. Thesis, Colorado State University, Fort Collins.

Gross, J. E. 1960. History, present, and future status of the desert bighorn sheep (*Ovis Canadensis mexicana*) in the Guadalupe Mountains of southeastern New Mexico and northwestern Texas. Desert Bighorn Council Transactions 4:66–71.

Hafez, E. S. E. 1968. Female reproductive organs. Pages 61–80 *in* E. S. E. Hafez, ed. Reproduction in farm animals, 2nd ed. Lea & Febiger, Philadelphia.

Hall, E. R. 1981. The mammals of North America. 2nd ed. John Wiley, New York.

Hamilton, K. S., S. A. Holl, and C. L. Douglas. 1982. An evaluation of the effects of recreational activity on bighorn sheep in the San Gabriel Mountains, California. Desert Bighorn Council Transactions 26:50–55.

Hamilton, W. D. 1971. Geometry for the selfish herd. Journal of Theoretical Biology 31:295–311.

Hansen, C. G. 1965. Growth and development of desert bighorn sheep. Journal of Wildlife Management 29:387–91.

Hansen, C. G. 1980a. Habitat. Pages 64–79 *in* G. Monson and L. Sumner, eds. The desert bighorn. University of Arizona Press, Tucson.

Hansen, C. G. 1980b. Population dynamics. Pages 217–35 *in* G. Monson and L. Sumner, eds. The desert bighorn. University of Arizona Press, Tucson.

Hansen, M. C. 1982. Status and habitat preferences of California bighorn sheep on Sheldon National Wildlife Refuge, Nevada. M.S. Thesis, Oregon State University, Corvallis.

Hansen, M. C. 1996. Foraging ecology of female Dall's sheep in the Brooks Range, Alaska. Ph.D. Dissertation, University of Alaska, Fairbanks.

Hansen, R. M., and P. S. Martin. 1973. Ungulate diets in the lower Grand Canyon. Journal of Range Management 26:380–81.

Hardin, G. 1960. The competitive exclusion principle. Science 131:1291–97.

Harris, L. K. 1992. Recreation in mountain sheep habitat. Ph.D. Dissertation, University of Arizona, Tucson.

Hass, C. C. 1989. Bighorn lamb mortality: Predation, inbreeding, and population effects. Canadian Journal of Zoology 67:699–705.

Hass, C. C. 1990. Alternative maternal-care patterns in two herds of bighorn sheep. Journal of Mammalogy 71:24–35.

Hass, C. C. 1991. Social status in female bighorn sheep (*Ovis canadensis*): Expression, development and reproductive correlates. Journal of Zoology (London): 225:509–23.

Hass, C. C., and D. A. Jenni. 1991. Structure and ontogeny of dominance relationships among bighorn rams. Canadian Journal of Zoology 69:471–76.

Hebert, D. M. 1973. Altitudinal migration as a factor in the nutrition of bighorn sheep. M.S. Thesis, University of British Columbia, Vancouver, Canada.

Hebert, D. M., and S. Harrison. 1988. The impact of coyote predation on lamb mortality patterns at the Junction Wildlife Management Area. Proceedings of the Biennial Symposium of the Northern Wild Sheep and Goat Council 5:283–91.

Heimer, W. E. 1985. Population status and management of Dall sheep in Alaska. Pages 1–15 *in* M. Hoefs, ed. Wild Sheep: distribution, abundance, management and conservation of the sheep of the world and closely related mountain ungulates. Special report. Northern Wild Sheep and Goat Council, Whitehorse, Yukon, Canada.

Heimer, W. E., R. L. Zarnke, and D. J. Preston. 1982. Disease surveys in Dall sheep in Alaska. Symposium of the Northern Wild Sheep and Goat Council 3:188–97.

Hemming, J. E. 1969. Cemental deposition, tooth succession, and horn development as criteria of age in Dall sheep. Journal of Wildlife Management 33:552–58.

Hengel, D. A., S. H. Anderson, and W. G. Hepworth. 1992. Population dynamics, seasonal distribution and movement patterns of the Laramie Peak bighorn sheep herd. Proceedings of the Biennial Symposium of the Northern Wild Sheep and Goat Council 8:83–96.

Henshaw, J. 1970. Conflict between Dall's sheep and caribou. Canadian Field-Naturalist 84:388–90.

Hirth, D. H. 1977. Social behavior of white-tailed deer in relation to habitat. Wildlife Monographs 53:1–55.

Hoefs, M. 1978. Twinning in Dall sheep. Canadian Field-Naturalist 92:292–93.

Hoefs, M. 1984. Productivity and carrying capacity of a subarctic sheep winter range. Arctic 37:141–47.

Hoefs, M., and N. Barichello. 1985. Distribution, abundance and management of wild sheep in Yukon. Pages 16–34 *in* M. Hoefs, ed. Wild sheep: Distribution, abundance, management and conservation of the sheep of the world and closely related mountain ungulates. Special report. Northern Wild Sheep and Goat Council, Whitehorse, Yukon, Canada.

Hoefs, M., and M. Bayer. 1983. Demographic characteristics of an unhunted Dall sheep, *Ovis dalli dalli*. Canadian Journal of Zoology 61:1346–57.

Hoefs, M., and T. D. Bunch. 1992. Cranial asymmetry in a Dall ram (*Ovis dalli dalli*). Journal of Wildlife Diseases 28:330–32.

Hoefs, M., and Cowan. 1979. Ecological investigation of a population of Dall sheep (*Ovis dalli dalli* Nelson). Syesis 12(Supplement 1):1–81.

Hoefs, M., and U. Nowlan. 1994. Distorted sex ratios in young ungulates: The role of nutrition. Journal of Mammalogy 75:631–36.

Hoefs, M., T. D. Bunch, R. L. Glaze, and H. S. Ellsworth. 1982. Horn aberrations in Dall's sheep (Ovis dalli*)* from Yukon Territory, Canada. Journal of Wildlife Diseases 18:297–304.

Hoefs, M., H. Hoefs, and D. Burles. 1986. Gray wolf, *Canis lupus pambasilens*, in Kluane Lake Area, Yukon. Canadian Field-Naturalist 100:78–84.

Hogg, J. T. 1984. Mating in bighorn sheep: Multiple creative male strategies. Science 225:526–29.

Hogg, J. T. 1987. Intrasexual competition and mate choice in Rocky Mountain bighorn sheep. Ethology 75:119–44.

Holl, S. A., and V. C. Bleich. 1983. San Gabriel mountain sheep: Biological and management considerations (San Bernardino National Forest Administration Report). U.S. Forest Service, San Bernardino, CA.

Honess, R. F., and N. M. Frost. 1942. A Wyoming bighorn sheep study (Bulletin 1). Wyoming Game and Fish Department, Laramie.

Honess, R. F., and K. Winter. 1956. Diseases of wildlife in Wyoming (Bulletin 9). Wyoming Game and Fish Department, Laramie.

Horejsi, B. L. 1976. Suckling and feeding behavior in relation to lamb survival in bighorn sheep (*Ovis canadensis*). Ph.D. Dissertation, University of Calgary, Calgary, Alberta, Canada.

Irvine, C. A. 1969. Factors affecting the desert bighorn in southeastern Utah. Desert Bighorn Council Transactions 13:6–13.

Jaeger, E. C. 1957. The North American deserts. Stanford University Press, Palo Alto, CA.

Jarman, P. J. 1974. The social organization of antelope in relation to their ecology. Behavior 48:215–67.

Jarman, P. J., and M. V. Jarman. 1979. The dynamics of ungulate social organization. Pages 185–220 *in* A. R. E. Sinclair and M. Norton-Griffiths, eds. Serengeti: Dynamics of an ecosystem. University of Chicago Press, Chicago.

Jessup, D. A., and R. R. Ramey II. 1995. Genetic variation of bighorn sheep as measured by blood protein electrophoresis. Desert Bighorn Council Transactions 39:17–25.

Jones, F. L. 1959. A survey of the Sierra Nevada bighorn. Sierra Club Bulletin 35:29–76.

Jones, F. L. 1980. Competition. Pages 197–216 *in* G. Monson and L. Sumner, eds. The desert bighorn. University of Arizona Press, Tucson.

Jones, R. L., and H. C. Hanson. 1985. Mineral licks: Geography and biochemistry of North American ungulates. Iowa State University Press, Ames.

Jorgensen, J. T., and W. D. Wishart. 1984. Growth rates of Rocky Mountain Bighorn sheep on Ram Mountain, Alberta. Northern Wild Sheep and Goat Council Proceedings 4:270–84.

Jorgenson, M. C., and R. E. Turner. 1975. Desert bighorn of the Anza-Borrego Desert State Park. Desert Bighorn Council Transactions 19:51–53.

Keech, M. A., T. R. Stephenson, R. T. Bowyer, V. Van Ballenberghe, and J. Ver Hoef. 1998. Relationships between blood-serum variables and depth of rump fat in Alaskan moose. Alces 34:173–79.

Kelly, W. E. 1979. A comparison of 3 bighorn areas on the Humboldt National Forest. Desert Bighorn Council Transactions 23:37–39.

Kie, J. G., J. A. Baldwin, and C. J. Evans. 1996. CALHOME: A program for estimating animal home ranges. Wildlife Society Bulletin 24:342–44.

Kilpatric, J. 1982. Texas desert bighorn sheep status report, 1982. Desert Bighorn Council Transactions 26:102–4.

King, M. M., and G. W. Workman. 1982. Desert bighorn on BLM lands in southeastern Utah. Desert Bighorn Council Transactions 26:104–6.

Klein, D. R. 1953. A reconnaissance study of the mountain goat in Alaska. M.S. Thesis, University of Alaska, Fairbanks.

Klein, D. R. 1965. Ecology of deer range in Alaska. Ecological Monographs 35:259–84.

Kornet, C. A. 1978. Status and habitat use of California bighorn sheep on Hart Mountain, Oregon. Thesis, Oregon State University, Corvallis.

Kovach, S. D. 1979. An ecological survey of the White Mountain Peak bighorn. Desert Bighorn Council Transactions 23:57–61.

Krausman, P. R. 1993. The exit of the last wild mountain sheep. Pages 242–50 *in* G. P. Nabhan, ed. Counting sheep. University of Arizona Press, Tucson.

Krausman, P. R. 1996. Problems facing bighorn sheep in and near domestic sheep allotments. Pages 59–64 *in* W. D. Edge, ed. Sustaining rangeland ecosystem symposium. Oregon State University, Corvallis.

Krausman, P. R. 1997a. Regional summary. Pages 316–17 *in* D. M. Shackleton, ed. Conservation of wild sheep goats and their relatives: Status survey and conservation action plan for Caprinae. International Union for the Conservation of Nature, Gland, Switzerland.

Krausman, P. R. 1997b. The influence of scale on the management of desert bighorn sheep. Pages 349–67 *in* J. A. Bissonette, ed. Primer in landscape ecology. Springer-Verlag, New York.

Krausman, P. R., and B. Czech. 1998. Water developments and desert ungulates. Pages 138–54 *in* Symposium on environmental, economic and legal issues related to rangeland water developments. Center for the Study of Law, Science, and Technology, Arizona State University, Tempe.

Krausman, P. R., and R. C. Etchberger. 1995. Response of desert ungulates to a water project in Arizona. Journal of Wildlife Management 59:292–300.

Krausman, P. R., and B. D. Leopold. 1986. The importance of small populations of desert bighorn sheep. Transactions of the North American Wildlife and Natural Resource Conference 51:52–61.

Krausman, P. R., and D. M. Shackleton. 2000. Bighorn sheep. Pages 517–44 *in* S. Demarais and P. R. Krausman, eds. Ecology and management of large mammals in North America. Prentice-Hall, Upper Saddle River, NJ.

Krausman, P. R., S. Torres, L. L. Ordway, J. J. Hervent, and M. Brown. 1985. Diel activity of ewes in the little Harquahala Mountains, Arizona. Desert Bighorn Council Transactions 29:24–26.

Krausman, P. R., B. D. Leopold, R. F. Seegmiller, and S. G. Torres. 1989. Relationships between desert bighorn sheep and habitat in western Arizona. Wildlife Monographs 102:1–66.

Krausman, P. R., R. Etchberger, and R. M. Lee. 1993. Persistence of mountain sheep. Conservation Biology 7:219.

Krausman, P. R., G. Long, and L. Tarango. 1996. Desert bighorn sheep and fire, Santa Catalina Mountains, Arizona. Pages 162–68 *in* P. F. Ffolliott, L. F. DeBano, M. B. Baker, Jr., G. J. Gottfried, G. Sols-Garza, C. B. Edminster, D. G. Neary, L. S. Allen, and R. H. Hamre, tech. coords. Effects of fire on the Madrean Province ecosystems (RM-GTR-289). U.S. Forest Service, Fort Collins, CO.

Krausman, P. R., A. V. Sandoval, and R. C. Etchberger. 1999. Natural history of desert bighorn sheep. Pages 139–91 *in* R. Valdez and P. R. Krausman, eds. Mountain sheep of North America. University of Arizona Press, Tucson.

Lange, R. E., A. V. Sandoval, and W. P. Meleney. 1980. Psoroptic scabies in bighorn sheep (*Ovis canadensis mexicana*) in New Mexico. Journal of Wildlife Disease 16:77–82.

Lawson, B., and R. Johnson. 1982. Mountain sheep. Pages 1036–55 *in* J. A. Chapman and G. A. Feldhamer, eds. Wild mammals of North America. Johns Hopkins University Press, Baltimore.

Lenarz, M. S. 1979. Social structure and reproductive strategy in desert bighorn sheep (*Ovis canadensis mexicana*). Journal of Mammalogy 60:671–78.

Lenarz, M. S., and W. Conley. 1982. Reproductive gambling in bighorn sheep (*Ovis*): A simulation. Journal of Theoretical Biology 98:1–7.
Leopold, A. 1933. Game management. Charles Scribners Son's, New York.
Leslie, D. M., Jr. 1977. Home range, group size, and group integrity of the desert bighorn sheep in the River Mountains, Nevada. Desert Bighorn Council Transactions 21:25–28.
Leslie, D. M., and C. L. Douglas. 1979. Desert bighorn of the River Mountains, Nevada. Wildlife Monographs 66:1–56.
Lopez, F. M. C., and V. M. Orihuela G. 1976. Behavior of the desert bighorn (*Ovis canandensis weemsi*) in Baja California. Desert Bighorn Council Transactions 20:24–25.
Lord, T. M., and A. J. Luckhurst. 1974. Alpine soils and plant communities of a Stone sheep habitat in Northeastern British Columbia. Northwest Science 48:38–51.
Luckhurst, A. J. 1973. Stone sheep and their habitat. M.S. Thesis, University of British Columbia, Vancouver, Canada.
Main, M. B., and B. E. Coblentz. 1990. Sexual segregation among ungulates: A critique. Wildlife Society Bulletin 18:204–10.
Main, M. B., F. W. Weckerly, and V. C. Bleich. 1996. Sexual segregation in ungulates: New directions for research. Journal of Mammalogy 77:449–61.
McCann, J. L. 1956. Ecology of mountain sheep. American Midland Naturalist 56:297–324.
McCullough, D. R., and E. R. Schneegas. 1966. Winter observations on the Sierra Nevada bighorn sheep. California Fish and Game Department 52:68–84.
McCutchen, H. E. 1976. Status of Zion National Park desert bighorn restoration project 1975. Desert Bighorn Council Transactions 20:52–54.
McEwan, E. H. 1975. The adaptive significance of the growth patterns in cervids compared with other ungulate species. Zoologicheskii Zhurnal 54:1221–32.
McMichael, T. J. 1964. Relationships between desert bighorn and feral burros in the Black Mountains of Mohave County. Desert Bighorn Council Transactions 8:29–35.
McQuivey, R. P. 1978. The desert bighorn sheep of Nevada (Bulletin 6). Nevada Department of Wildlife Biology, Las Vegas.
McWhirter, D., A. Smith, E. Merrill, and L. Irwin. 1992. Foraging behavior and vegetation responses to prescribed burning on bighorn winter range. Proceedings of the Biennial Symposium of the Northern Wild Sheep and Goat Council 8:264–78.
Mendoza, V. J. 1976. The bighorn sheep of the state of Sonora. Desert Bighorn Council Transactions 20:25–26.
Merritt, M. F. 1974. Measurement of utilization of bighorn sheep habitat in the Santa Rosa Mountains. Desert Bighorn Council Transactions 18:4–17.
Miquelle, D. G., J. M. Peek, and V. Van Ballenberghe. 1992. Sexual segregation in Alaskan moose. Wildlife Monographs 122:1–57.
Moehlman, P. D. 1974. Behavior and ecology of feral asses (*Equus asinus*). Ph.D. Dissertation, University of Wisconsin, Madison.
Monson, G. 1964. Long-distance and nighttime movements of desert bighorn sheep. Desert Bighorn Council Transactions 8:11–17.
Monson, G. 1980. Distribution and abundance. Pages 40–15 *in* G. Monson and L. Sumner, eds. The desert bighorn. University of Arizona Press, Tucson.
Moore, T. D. 1958. Transplanting and observation of transplanted bighorn sheep. Desert Bighorn Council Transactions 2:43–46.
Moore, T. D., L. E. Spence, and C. E. Dugnolle. 1974. Identification of the dorsal guard hairs of some mammals of Wyoming (Bulletin 4). Wyoming Game and Fish Department, Laramie.
Morgan, J. K. 1970. Ecology of the Morgan Creek and East Fork of the Salmon River bighorn sheep herds and management of bighorn sheep in Idaho. M.S. Thesis, Utah State University, Logan.
Morgantini, L. E., and R. J. Hudson. 1981. Sex differential in use of the physical environment by bighorn sheep (*Ovis canadensis*). Canadian Field-Naturalist 95:60–74.
Morgart, J. R., and P. R. Krausman. 1983. Early breeding in bighorn sheep. Southwestern Naturalist 28:460–61.
Murie, A. 1944. The wolves of Mount McKinley. U.S. National Park Service Fauna Series 5:1–238.
Murphy, E. C., and K. R. Whitten. 1976. Dall sheep demography in McKinley Park and a reevaluation of Murie's data. Journal of Wildlife Management 40:597–609.
Murphy, E. C., F. J. Singer, and L. Nichols. 1990. Effects of hunting on survival and productivity of Dall sheep. Journal of Wildlife Management 40:597–609.
Nadler, C. F., A. Wolf, and Harris, K. E. 1971. The transferrins and hemoglobins of bighorn sheep (*Ovis canadensis*), Dall sheep (*Ovis dalli*) and mouflon (*Ovis musimon*). Comparative Biochemical Physiology 40B:567–70.
Nichols, L. 1975. Report and recommendations of the Dall and Stone Sheep Workshop Group. Pages 208–66 *in* J. B. Trefethen, ed. The wild sheep in modern North America. Boone and Crocket Club and Winchester Press, New York.
Nichols, L. 1978a. Dall sheep reproduction. Journal of Wildlife Management 42:570–80.
Nichols, L., Jr. 1978b. Dall's sheep. Pages 173–89 *in* J. L. Schmidt and D. L. Gilbert, eds. Big game of North America: Ecology and management. Wildlife Management Institute and Stackpole Books, Harrisburg, PA.
Nichols, L., and F. L. Bunnell. 1999. Natural history of thinhorn sheep. Pages 23–77 *in* R. Valdez and P. R. Krausman, eds. Mountain sheep in North America. University of Arizona Press, Tucson.
Nichols, L., and Erickson, J. A. 1969. Dall sheep (Federal Aid in Wildlife Restoration Project W-15-R-3 and W-17–1; Work Plan N, Job Nos. 3, 4, 5, 6, 7). Alaska Department of Fish and Game, Fairbanks.
Oldemeyer, J. L., W. J. Barmore, and D. L. Gilbert. 1971. Winter ecology of bighorn sheep in Yellowstone National Park. Journal of Wildlife Management 35:257–69.
Olech, L. A. 1979. Summer activity rhythms of peninsula bighorn sheep in Anza–Barrego Desert State Park, San Diego County, California. Desert Bighorn Council Transactions 23:33–36.
Packard, F. M. 1946. An ecological study of the bighorn sheep in Rocky Mountain National Park, Colorado. Journal of Mammalogy 27:3–28.
Petocz, R. G. 1973. The effect of snow cover on the social behavior of bighorn rams and mountain goats. Canadian Journal of Zoology 51:987–93.
Pianka, E. 1978. Evolutionary ecology. Harper and Row, New York.
Pitzman, M. S. 1970. Birth behavior and lamb survival in mountain sheep in Alaska. M.S. Thesis, University of Alaska, Fairbanks.
Pollister, G. L. 1974. The seasonal distribution and range use of bighorn sheep in the Beartooth Mountains, with special reference to the West Rosebud and Stillwater herds (Federal Aid Wildlife Restoration Project W-120-R-5). Montana Fish and Game Department, Helena.
Poole, K. G., and R. P. Graf. 1985. Status of Dall's sheep in the Northwest Territories, Canada. Pages 35–42 *in* M. Hoefs, ed. Wild sheep: Distribution, abundance, management and conservation of the sheep of the world and closely related mountain ungulates. Special report. Northern Wild Sheep and Goat Council, Whitehorse, Yukon, Canada.
Pulliam, H. R., and T. Caraco. 1984. Living in groups: Is there an optimal group size? Pages 122–47 *in* J. R. Krebs and N. B. Davies, eds. Behavioral ecology: An evolutionary approach, 2nd ed. Blackwell, Oxford.
Rachlow, J. L., and R. T. Bowyer. 1991. Interannual variation in timing and synchrony of parturition in Dall's sheep. Journal of Mammalogy 72:487–92.
Rachlow, J. L., and R. T. Bowyer. 1994. Variability in maternal behavior by Dall's sheep: Environmental tracking or adaptive strategy? Journal of Mammalogy 75:328–37.
Rachlow, J. L., and R. T. Bowyer. 1998. Habitat selection by Dall's sheep (*Ovis dalli*): Maternal trade-offs. Journal of Zoology (London) 245:465–75.
Remington, R. R. 1981. Arizona bighorn sheep status report. Desert Bighorn Council Transactions 25:44–46.
Reneau, J., and S. C. Reneau, eds. 1993. Records of North American big game. Boone and Crockett Club, Missoula, MT.
Risenhoover, K. L., and J. A. Bailey. 1985. Foraging ecology of mountain sheep: Implications for habitat management. Journal of Wildlife Management 49:797–804.
Russo, J. 1956. The desert bighorn in Arizona (Bulletin No. 1). Arizona Game and Fish Department, Phoenix.
Ryder, T. J., E. S. Williams, K. W. Mills, K. H. Bowles, and E. T. Thorne. 1992. Effect of pneumonia on population size and lamb recruitment in Whiskey Mountain bighorn sheep. Proceedings of the Biennial Symposium of the Northern Wild Sheep and Goat council 8:136–46.
Sadler, R. M. 1987. Reproduction in female cervids. Pages 123–44 *in* C. M. Wemmer, ed. Biology and management of the cervidae. Smithsonian Institution Press, Washington, DC.
Sanchez, D. R. 1976. Analysis of stomach contents of bighorn sheep in Baja, California. Desert Bighorn Council Transactions 20:21–22.
Sandoval, A. V. 1979. Preferred habitat of desert bighorn sheep in the San Andres Mountains, New Mexico. M.S. Thesis, Colorado State University, Fort Collins.
Sandoval, A. V. 1980. Management of a psoroptic scabies epizootic in bighorn sheep (*Ovis canadensis mexicana*) in New Mexico. Desert Bighorn Council Transactions 24:21–28.
Sandoval, A. V. 1981. New Mexico bighorn sheep status report. Desert Bighorn Council Transactions 25:66–68.

Sandoval, A. V., R. G. Peterson, J. Haywood, and A. Bottrell. 1984. Gestation period in *Ovis canadensis*. Journal of Mammalogy 65:337–38.

Schallenberger, A. D. 1966. Food habits, range use and interspecific relationships of bighorn sheep in the Sun River area, west-central Montana. M.S. Thesis, Montana State University, Bozeman.

Schwartz, O. A., V. C. Bleich, and S. A. Holl. 1986. Genetics and the conservation of mountain sheep. Biology Conservation 37:179–90.

Scott, J. E., R. R. Remington, and J. C. de Vos, Jr. 1990. Numbers, movements, and disease status of bighorn in southwestern Arizona. Desert Bighorn Council Transactions 34:9–13.

Seegmiller, R. E., and R. D. Ohmart. 1981. Ecological relationships of feral burros and desert bighorn sheep. Wildlife Monographs 78:1–58.

Seegmiller, R. E., and C. D. Simpson. 1979. The Barbary sheep: Some conceptual implications of competition with desert bighorn. Desert Bighorn Council Transactions 23:47–49.

Seip, D. R., and F. L. Bunnell. 1985. Nutrition of Stone's sheep on burned and unburned ranges. Journal of Wildlife Management. 49:397–405.

Seton, E. T. 1929. The bighorn. Pages 519–73 *in* E. T. Seton, ed. Lives of the game animals. Vol. 3, Part 2. Doubleday, Garden City, NY.

Shackleton, D. M. 1973. Population quality and bighorn sheep (*Ovis canadensis canadensis* Shaw). Ph.D. Dissertation, University of Calgary, Alberta, Canada.

Shackleton, D. M. 1976. Variability in physical and social maturation between bighorn sheep populations. Transactions of the Northern Wild Sheep Council 4:1–8.

Shackleton, D. M. 1985. *Ovis canadensis.* Mammalian Species 230:1–9.

Shackleton, D. M. 1991. Social maturation and productivity in bighorn sheep: Are young males incompetent? Applied Animal Behavior Science 29:173–84.

Shackleton, D. M., and J. Haywood. 1985. Early mother–young interactions in California bighorn sheep, *Ovis canadensis californiana.* Canadian Journal of Zoology 63:868–75.

Shackleton, D. M., and D. A. Hutton. 1971. An analysis of the mechanisms of brooming in mountain sheep horns. Zeitschrift für Säugetierkunde 36:342–50.

Shackleton, D. M., R. G. Peterson, J. Haywood, and A. Botrell. 1984. Gestation period in *Ovis canadensis*. Journal of Mammalogy 65:337–38.

Shackleton, D. M., C. C. Shank, and B. M. Wikeen. 1999. Natural history of Rocky Mountain and California bighorn sheep. Pages 78–138 *in* R. Valdez and P. R. Krausman, eds. Mountain sheep of North America. University of Arizona Press, Tucson.

Shank, C. C. 1979. Sexual dimorphism and the ecological niche of wintering Rocky Mountain bighorn sheep. Ph.D. Dissertation, University of Calgary, Alberta.

Shank, C. C. 1982. Age–sex differences in the diets of wintering Rocky Mountain bighorn sheep. Ecology 63:627–33.

Simmons, N. M. 1969. Heat stress and bighorn behavior in the Cabeza Prieta Game Range, Arizona. Desert Bighorn Council Transactions 13:56–63.

Simpson, C. D., and L. J. Krysl. 1981. Status and distribution of Barbary sheep in the Southwest United States. Desert Bighorn Council Transactions 25:9–15.

Smith, D. 1954. The bighorn sheep in Idaho: (Wildlife Bulletin 1). Idaho Department of Fish and Game.

Smith, K. G., and W. D. Wishart. 1978. Further observations of bighorn sheep non-trophy seasons in Alberta and their management implications. Proceedings of the Biennial Symposium of the Northern Wild Sheep and Goat Council 1:52–74.

Spalding, D. J. 1966. Twinning in bighorn sheep. Journal of Wildlife Management 30:207.

Spalding, D. J., and J. N. Bone. 1970. The California bighorn sheep of the south Okanagan Valley, British Columbia (Wildlife Management Publication 3). Fish and Wildlife Branch, Victoria, British Columbia, Canada.

Spraker, T. R. 1974. Lamb mortality. Transactions of the Northern Wild Sheep Council 3:102–3.

Steinkamp, M. J. 1990. The effect of seasonal cattle grazing on California bighorn sheep habitat use. M.S. Thesis, Utah State University, Logan.

Stelfox, J. G. 1971. Bighorn sheep in the Canadian Rockies: A history, 1800–1970. Canadian Field-Naturalist 85:101–22.

Stelfox, J. G. 1975. Range ecology of Rocky Mountain bighorn sheep in Canadian National Parks. Ph.D. Dissertation, University of Montana, Missoula.

Stewart, S. T. 1980. Mortality patterns in a bighorn sheep population. Proceedings of the Biennial Symposium of the Northern Wild Sheep and Goat Council 2:313–30.

Sugden, L. G. 1961. The California bighorn in British Columbia with particular reference to the Churn Creek herd. British Columbia Department of Recreation and Conservation, Victoria, Canada.

Taylor, R. A. 1962. Characteristics of horn growth in bighorn rams. M.S. Thesis, University of Montana, Missoula.

Thompson, R. W., and J. C. Turner. 1982. Temporal geographic variation in the lambing season of bighorn sheep. Canadian Journal of Zoology 60:1781–93.

Todd, J. W. 1972. A literature review of bighorn sheep food habits (Special Report 27). Colorado Department of Game, Fish and Parks and Cooperative Wildlife Research Unit, Fort Collins.

Turner, J. C. 1976. Initial investigations into the reproductive biology of the desert bighorn ram, *Ovis canadensis nelsoni, O. c. cremnobates*. Proceedings of the Biennial Symposium of the Northern Wild Sheep and Goat Council 4:22–25.

Turner, J. C., and C. G. Hansen. 1980. Reproduction. Pages 145–51 *in* G. Monson and L. Sumner, eds. The desert bighorn. University of Arizona Press, Tucson.

Valdez, R. 1988. Wild sheep and wild sheep hunters of the New World. Wild Sheep and Goat International, Mesilla, NM.

Valdez, R., and P. R. Krausman. 1999. Description, distribution and abundance of mountain sheep. Pages 3–22 *in* R. Valdez and P. R. Krausman, eds. Mountain sheep of North America. University of Arizona Press, Tucson.

Van Dyke, W. A. 1978. Population characteristics and habitat utilization of bighorn sheep, Steens Mountain, Oregon. M.S. Thesis, Oregon State University, Corvallis.

Wakeling, B. F., and W. H. Miller. 1989. Bedsite characteristics of desert bighorn sheep in the Superstition Mountains, Arizona. Desert Bighorn Council Transactions 33:6–8.

Wakelyn, L. A. 1987. Changing habitat conditions on bighorn sheep ranges in Colorado. Journal of Wildlife Management 51:904–12.

Walters, J. E., and R. M. Hansen. 1978. Evidence of feral burro competition with desert bighorn sheep in Grand Canyon National Park. Desert Bighorn Council Transactions 22:10–16.

Watts, T. J. 1979. Detrimental movement patterns in a remnant population of bighorn sheep (*Ovis canadensis mexicana*). M.S. Thesis, New Mexico State University, Las Cruces.

Weaver, R. A. 1972. Conclusion of the bighorn investigation in California. Desert Bighorn Council Transactions 16:56–65.

Weckerly, F. L. 1998. Sexual size dimorphism: Influence of mass and mating systems in the most dimorphic mammals. Journal of Mammalogy 79:33–52.

Wehausen, J. D., and R. R. Ramey II. 1993. A morphometric reevaluation of the peninsular bighorn species. Desert Bighorn Council Transactions 37:1–10.

Welch, R. D. 1969. Behavioral patterns of desert bighorn sheep in south-central New Mexico. Desert Bighorn Council Transactions 13:114–29.

Welles, R. E., and F. B. Welles. 1961. The bighorn of Death Valley (Fauna Series 6). U.S. National Park Service.

Whitehead, P. E., and E. H. McEwen. 1980. Progesterone levels in peripheral plasma of Rocky Mountain bighorn ewes (*Ovis canadensis*) during the oestrous cycle and pregnancy. Canadian Journal of Zoology 58:1005–1108.

Whitten, K. R. 2001. Effects of horn-curl regulations on demography of Dall's sheep: A critical review. Alces 483–495.

Wikeem, B. M. 1984. Forage selection by California bighorn sheep and the effects of grazing on an *Artemisia–Agropyron* community in southern British Columbia. Ph.D. Dissertation, University of British Columbia, Vancouver, Canada.

Wilson, L. O. 1968. Distribution and ecology of desert bighorn sheep in southeastern Utah (Publication No. 68-5). Utah Department of Natural Resources and Division of Fish and Game, Salt Lake City.

Wishart, W. D. 1958. The bighorn sheep of the Sheep River Valley. M.S. Thesis, University of Alberta, Edmonton, Canada.

Wishart, W. D., and D. Brochu. 1982. An evaluation of horn and skull characteristics as a measure of population quality in Alberta bighorns. Proceedings of the Biennial Symposium of the Northern Wild Sheep and Goat Council 3:127–42.

Witham, J. H. 1983. Desert bighorn sheep in southwestern Arizona. Ph.D. Dissertation, Colorado State University, Fort Collins.

Woodgerd, W. 1964. Population dynamics of bighorn sheep on Wildhorse Island. Journal of Wildlife Management 28:381–91.

Woodward, S. L., and R. D. Ohmart. 1976. Habitat use and fecal analysis of feral burros (*Equus asinus*), Chemehuevi Mountains, California, 1974. Journal of Range Management 29:482–85.

Woolf, A., and D. C. Kradel. 1970. Hematological values of captive Rocky Mountain bighorn sheep. Journal of Wildlife Disease 6:67–68.

Woolf, A., T. O'Shea, and D. L. Gilbert. 1970. Movements and behavior of bighorn sheep on summer ranges in Yellowstone National Park. Journal of Wildlife Management 34:446–50.

Wright, G. J., J. S. Dixon, and B. H. Thompson. 1933. A preliminary survey of faunal relations in National Parks of the U.S. (Fauna Series 1). U.S. National Park Service.

Zarnke, R. L., and S. Rosendal. 1989. Serologic survey for *Mycoplasma ovipneumoniae* in freeranging Dall sheep (*Ovis dalli*) in Alaska. Journal of Wildlife Diseases 25:612–13.

Zarnke, R. L., C. L. Calisher, and J. Kerschner. 1983. Serologic evidence of arbovirus infections in humans and wild animals in Alaska, USA. Journal of Wildlife Diseases 19:175–79.

Zine, M., and P. R. Krausman. 2000. Behavior of captive mountain sheep in a Mojave Desert environment. Southwestern Naturalist 45:184–95.

Paul R. Krausman, School of Renewable Natural Resources, The University of Arizona, Tucson, Arizona 85721. Email: krausman@ag.arizona.edu.

R. Terry Bowyer, Institute of Arctic Biology, University of Alaska, Fairbanks, Alaska 99775-7000. Email: ffrtb@uaf.edu.

VIII

Introduced Mammals

52

Nutria

Myocastor coypus

Dixie L. Bounds
Mark H. Sherfy
Theodore A. Mollett

NOMENCLATURE

Common Names. Nutria, coypu, swamp beaver, South American beaver, coypus, nutria-rat
Scientific Name. *Myocastor coypus*
Subspecies. *M. c. coypus* and *M. c. melanops* from Chile; *M. c. santacruzae* from Patagonia; *M. c. bonariensis* from Argentina; and *M. c. popelairi* from Bolivia (Osgood 1943; Murua et al. 1981; Willner 1982)

The family Myocastoridae (Rodentia) has one monotypic genus, *Myocastor* (Gosling and Baker 1991; Woods et al. 1992; Whitaker and Hamilton 1998). Some researchers consider myocastorids as a subfamily of the hutias, family Capromyidae (Simpson 1945; Hall 1981; Nowak 1991), whereas others suggest myocastorids and capromyids are subfamilies of American spiny rats, family Echimyidae, based on retention of the deciduous premolar in these taxa (Patterson and Pascual 1968; Patterson and Wood 1982). Several morphological differences, however, prevent grouping myocastorids within either capromyids or echimyids (Woods 1972, 1982; Woods and Howland 1979; Woods et al. 1992). In contrast, Koehler et al. (2000) investigated the genetic variation and systematic relationships among allozymes in 23 species of hystricomorph rodents at 19 presumptive gene loci. They concluded that the nutria does not warrant a distinct family, but should be placed within the guinea pigs, family Caviidae. In 1966, the U.S. National Museum reported it was impossible to separate U.S. nutria into five subspecies (Evans 1970). Evans (1970) suggested the nutria introduced in North America were all of mixed races.

Early Spaniards believed this animal was a form of European otter (*Lutra lutra*), thereupon naming it "nutria," the Spanish name for otter (Kinler et al. 1987). *Myocastor,* the genus name, translates to "mouse beaver," and *coypus,* the species name, means "water-sweeper" (Lowery 1974; Murua et al. 1981).

DISTRIBUTION

Nutria are indigenous to southern Brazil, Bolivia, Paraguay, Uruguay, Argentina, and central and southern Chile (Cabrera 1961; Gosling and Baker 1991). This species has been introduced in North America, Europe, the former Soviet Union, the Middle East, Africa, and Japan (Gosling and Skinner 1984; Kinler et al. 1987; Gosling and Baker 1991; Gebhardt 1996; Guichón and Cassini 2000; Carter and Leonard 2002). European nutria populations occur in the Netherlands, Belgium, West Germany, France and Italy; nutria are most likely extirpated in Great Britain (Gosling and Baker 1991).

In the early twentieth century, the value of nutria pelts was recognized and ranching efforts were initiated in their native range. Historically, the first extensive nutria farms were in South America in the early 1920s, which led to nutria farming in Europe and North America (Evans 1970). Canada had nutria in zoos by 1900 and nutria were found later in three provinces — Quebec, Ontario, and British Columbia (Evans 1970). The first nutria fur farming in North America was established in 1899 at Elizabeth Lake, California. Thereafter, ranches were established in the United States (Fig. 52.1), including California, Washington, Oregon, and Michigan during the early 1930s; New Mexico during the mid-1930s; Louisiana and Ohio in 1937; and Utah in 1939 (Kinler et al. 1987). The federal and state governments released nutria in Alabama, Arkansas, Georgia, Kentucky, Maryland, Mississippi, Oklahoma, Louisiana, and Texas (Evans 1970). Bounds (2000) surveyed 22 states where nutria had been intentionally introduced and 11 adjacent states (e.g., Kennedy and Kennedy 1998) and found nutria were established in 15 states in 1999 (Fig. 52.2). In addition, nutria have been found in New Mexico (R. Beausoleil, New Mexico Game and Fish Department, pers. commun., 2002). In 1999, nutria range included >423,000 ha (1 million acres) of lands managed by the U.S. Fish and Wildlife Service (Bounds 2000).

Figure 52.1. Introduced range of the nutria (*Myocastor coypus*) in North America. Inset: approximate native range in South America. Source: Inset adapted from Packard (1967).

DESCRIPTION

Nutria are hystricomorph rodents with arched bodies, short legs and necks, and stocky builds (Gosling and Skinner 1984). The head is large and almost triangular, with a tapering muzzle, small eyes and ears, and long (up to 130 mm) white vibrissae (Gosling and Baker 1991; Woods et al. 1992). The skull is heavy, with an elongated paraoccipital process, which is anteriorly curved (Woods et al. 1992) (Fig. 52.3). The

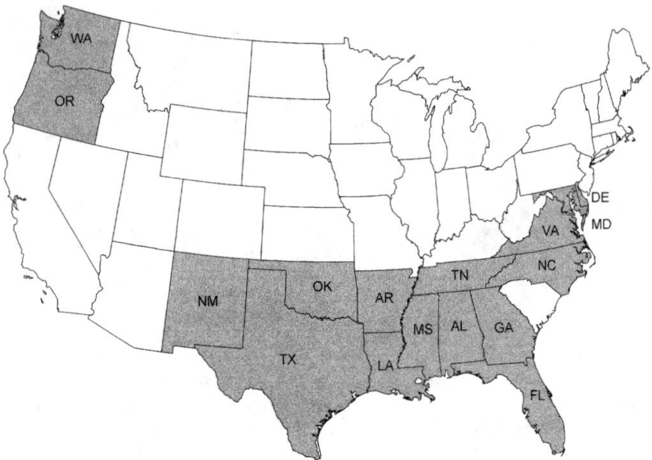

FIGURE 52.2. In 2002, nutria were established in 16 states and were found to no longer occur in some states where they were originally introduced (e.g., California, Idaho, and Michigan). SOURCE: Data from Bounds (2000) and R. Beausoleil, New Mexico Game and Fish Department, pers. commun., 2002.

broad incisors have orange-pigmented anterior surfaces. Body length averages 61 cm, tail length around 33–41 cm; total body length with tail is about 100 cm (Evans 1970; LeBlanc 1994). The average weight of the sexes is similar (about 5.4 kg); however, maximum weight of females is 8.2 kg and that of males is 9.1 kg (Evans 1970; LeBlanc 1994).

Nutria have long, round tails and webbed hind feet. The tail is nearly hairless and may serve an important function in thermoregulation at the high ambient temperatures of subtropical South America (Krattenmacher and Rubsamen 1987). For nutria introduced in colder climates, however, the tail may become a handicap because frost scars may remain purulent for several months and may cause postwinter mortality (Doncaster and Micol 1990).

Nutria have several adaptations for living in aquatic environments. For example, the ears, nostrils, and mouth have valves to keep out water. The eyes, ears, and nostrils are placed high on the dorsal surface of the head, and are exposed as the animal swims (Evans 1970; Gosling and Skinner 1984). Auditory abilities are poorer than those of terrestrial rodents, and vision and olfaction are not well developed (Figurina 1984; Gosling and Baker 1991). Females have four or five pairs of dorsal–lateral thoracic mammae, which may allow the young to suckle while the mother swims, although young are normally suckled on nests rather than in the water (Gosling and Baker 1991; Nowak 1991). The hind feet are much longer than the forefeet and have five digits; four of these are connected by skin and the fifth digit is free and used for grooming (Nowak 1991). The dextrous forepaws have four long, unwebbed digits and a vestigial thumb. Nutria have strong, sharp claws used for digging.

Nutria have 20 teeth with a dental formula I 1/1, C 0/0, P 1/1, M 3/3 (Kinler et al. 1987). The premolars and molars are hypsodont and the incisors are large. The premolars and molars wear to a flat grinding surface with complex infundibular patterns, which change over time. Male incisors (15.1 mm) are slightly wider than female incisors (13.4 mm) (Gosling and Baker 1991).

Fur color of nutria varies from black or dark amber to a light rust or brownish-blond color depending on the animal's environment and ancestry (Evans 1970). Nutria have coarse, stiff guard hairs, which protrude beyond the rest of the fur. Guard hairs are longest and thickest on the back, whereas the fur on the ventral surface is soft and dense, and is used by the fur industry to produce coats (Evans 1970; Gosling and Baker 1991).

The dark brown or green feces or scat are long (11×70 mm for adults) and cylindrical, with fine longitudinal striations (Gosling and Baker 1991). The size and pattern (lines) of nutria feces make them distinguishable from other species, such as muskrat (*Ondatra zibethicus*) or beaver (*Castor canadensis*) (Evans 1970). Nutria deposit feces in water or on land, and fecal deposition does not seem to have any behavioral significance (Gosling and Baker 1991).

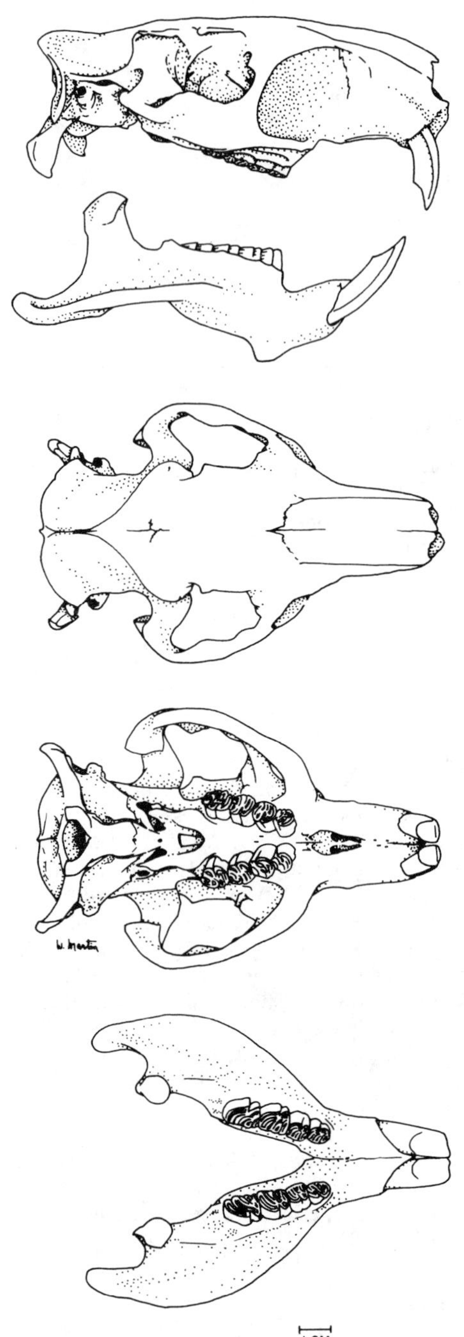

FIGURE 52.3. Skull of the nutria (*Myocastor coypus*). From top to bottom: lateral view of cranium, lateral view of mandible, dorsal view of cranium, ventral view of cranium, dorsal view of mandible.

GENETICS

Chromosomes. Information on the cytogenetics of nutria is limited; however, it is widely accepted that the diploid number of chromosomes is $2n = 42$ (Tsigalidou et al. 1966; George and Weir 1974). The karyotype was initially characterized as containing 7 pairs of metacentric, 4 pairs of telocentric, and 10 pairs of submetacentric chromosomes (Tsigalidou 1967). Subsequently, Kasumova et al. (1976) reported that the karyotype was composed of dibrachial chromosomes, the Y chromosome in males was small and acrocentric, and one pair of chromosomes carried satellites. Kinler et al. (1987) reported that no nucleolar organizer region staining had been conducted with nutria chromosome preparations.

Coat Color Genetics. Most wild and feral nutria pelts have brown or gray underfur with guard hairs in shades of reddish brown to brownish blond (Evans 1970; Maum 1986). Scheelje (1982) described the inheritance of valuable color variants in commercially raised captive nutria. Gosling and Skinner (1984) concluded that the genetics of nutria coat color was complex and achieved through planned matings of color variants appearing spontaneously in captive individuals. For example, there are at least four genetically distinct types of white nutria. Typically, color varieties produced through selective breeding of captive animals include variants of white, black, golden, beige, and others (Gosling and Skinner 1984).

Protein Polymorphisms and Cytogenetics. Pancreatic lipases are categorized as either classical pancreatic lipases (those exhibiting interfacial activation and dependance on colipase in the presence of bile salts) or pancreatic lipase-related proteins (i.e., RP1 or RP2). Nutria produce a classical pancreatic lipase and a novel pancreatic lipase belonging to the RP2 subfamily (Thristrup et al. 1994). Nutria also possess a procolipase, which functions as a specific cofactor for the classical pancreatic lipase, but not for the RP2 lipase (Thristrup et al. 1995). Unlike most mammals, nutria and guinea pigs (*Cavia* spp.) do not produce a pancreatic phospholipase A, and the RP2 lipase functions in its place for phospholipid digestion in hystricomorph rodents (Thristrup et al. 1995). The procolipase and the specific cDNAs encoding the two nutria lipases have been cloned, expressed, and characterized (Thristrup et al. 1995).

Blood Analysis. Several electrophoretic studies have been conducted to evaluate differences in serum and various tissue proteins among age classes and geographic populations. Concentrations of serum globulins and albumin were higher in adults than in fetal nutria, and adult serum contained one additional protein band not found in fetal serum (Brown 1966). Moreover, Brown (1966) noted that there were differences in lipoproteins in adult and fetal sera. Neville et al. (1974) concluded that there is a marked difference in the structure of insulin in nutria and some other hystricomorphic rodents when compared to other mammals and there is considerable polymorphism among the hystricomorphs. Szynkiewicz (1968) identified two immunologically distinct blood groups (CO_1 and CO_2) among animals held in breeding centers in Poland. In addition, he differentiated four phenotypes of beta-globulin subfractions in blood serum collected from nutria at six breeding centers and determined that the phenotypes occurred at significantly different frequencies among the breeding centers (Szynkiewicz 1971). Evaluation of erythrocyte cytoskeleton proteins demonstrated that nutria, unlike other rodents (e.g., Norway rats, *Rattus norvegicus;* and house mice, *Mus musculus*), did not present band 6 and that nutria lack the erythrocyte cytoskeleton protein 4.2, which was detected in the majority of mammals that were screened (Guerra-Shinohara and Barretto 1999).

Morgan et al. (1981) analyzed serum proteins, eye-lens proteins, and serum and liver enzymes to evaluate genetic variation in a nutria population in Maryland. All serum protein systems were monomorphic, and nutria were polymorphic for only three enzymatic systems (lactate dehydrogenase, esterase-1, and esterase-2), with only two alleles observed at each locus. The calculated heterozygosity was $<1.0\%$ (Morgan et al. 1981). In a survey of protein polymorphisms in wild populations of nutria, Ramsey et al. (1985) found that only 3 of the 22 presumptive loci were polymorphic. Average individual heterozygosity was 4.1–5.6%, 0.2%, and 0.0% for populations in Louisiana, England, and Washington, respectively (Ramsey et al. 1985). Typically, mammals have an average heterozygosity of 3.5% (Nevo 1978). The lower level of genetic variation observed in isolated, noncoastal populations and those reduced periodically by adverse environmental conditions may be the result of founder effects and genetic drift (Morgan et al. 1981; Kinler et al. 1987).

TABLE 52.1. Blood chemistry parameters in serum samples taken from nutria (*Myocastor coypus*) collected at four locations in Louisiana during fall, winter, and spring

Parameter	Mean	Range
Electrolyte		
Calcium (mg/dl)	9.6	5.0–15.6
Phosphorus (mg/dl)	7.5	3.7–13.0
Potassium (mEq/L)	6.0	3.8–9.0
Sodium (mEq/L)	147.0	105.0–184.0
Metabolite		
Glucose (mg/dl)	144.0	25.0–350.0
Cholesterol (mg/dl)	71.0	33.0–189.0
Triglycerides (mg/dl)	73.0	17.0–332.0
Urea nitrogen (mg/dl)	18.0	1.0–62.0
Uric acid (mg/dl)	4.0	1.3–9.0
Albumin (g/dl)	3.0	1.8–3.8
Total protein (g/dl)	6.0	3.0–7.6
Enzyme		
Alkaline phosphatase (mU/ml)	262.0	57.0–820.0
Glutamic oxalacetic transaminase (mU/ml)	150.0	7.0–540.0

SOURCE: Adapted from Ramsey et al. (1981).

Ramsey et al. (1981) analyzed factors, such as sex, reproductive status, season, age, habitat, and capture method, that could influence blood chemistry in serum samples taken from nutria collected during the fall, winter, and spring at four locations in Louisiana (Table 52.1). Compared to females, males had higher levels of uric acid and lower levels of cholesterol. Nongravid females tended to have higher cholesterol, total protein, albumin, triglycerides, and blood urea nitrogen, but lower serum enzymatic activities than pregnant animals. Cholesterol and triglycerides varied seasonally in females. The effect of age was significant; sexually immature individuals had lower serum calcium, total protein, albumin, and sodium, but higher phosphorus and alkaline phosphatase than adults. Ramsey et al. (1981) hypothesized that freshwater habitats provided more available calcium and digestible energy than brackish or saline wetlands, and that blood parameters were effective indicators of habitat deterioration.

Jelinik and Glasrova (1982) reported that erythrocytes, hemoglobin, and hematocrits increased uniformly during the neonatal period through approximately 200 days of age. Kinler et al. (1987) compared basic hematological and blood chemistry values observed in nutria collected in Louisiana with Komarek's (1983) values obtained from captive animals in Czechoslovakia. Nutria reared in confinement had higher hematocrits and erythrocyte counts, but lower hemoglobin concentrations and leukocyte counts.

PHYSIOLOGY

Robicheaux (1978) observed that nutria of all age and sex classes in Louisiana experienced a decline in growth rate during the summer. Adult males and females exhibited a greater decrease in total length grown and a greater weight loss compared to immature animals. Dixon et al. (1979) reported that the daily growth rates for males and females were 0.0120 and 0.0116 g/g/day, respectively. Males had a faster initial rate of increase in weight, and reached a greater weight than females. In addition, growth rates were highest during summer and lowest during winter. These investigators concluded that differential seasonal growth rates were attributable to severity of winter weather. They developed the following model to predict weight (W) in nutria in Maryland:

$$W = (W_{\max} - W_0)(1 - e^{-bt}) + W_0$$

where $W_{\max}$ is the maximum body weight (grams), W_0 is the initial weight at birth (227.36 g), e is the base of natural logarithms (2.71828), b is a constant (0.00244 $\pm$ 0.000060/day for males and 0.00247 $\pm$ 0.00012/day for females), and t is time (number of days). Willner et al. (1980) compared body weights predicted by this model against a

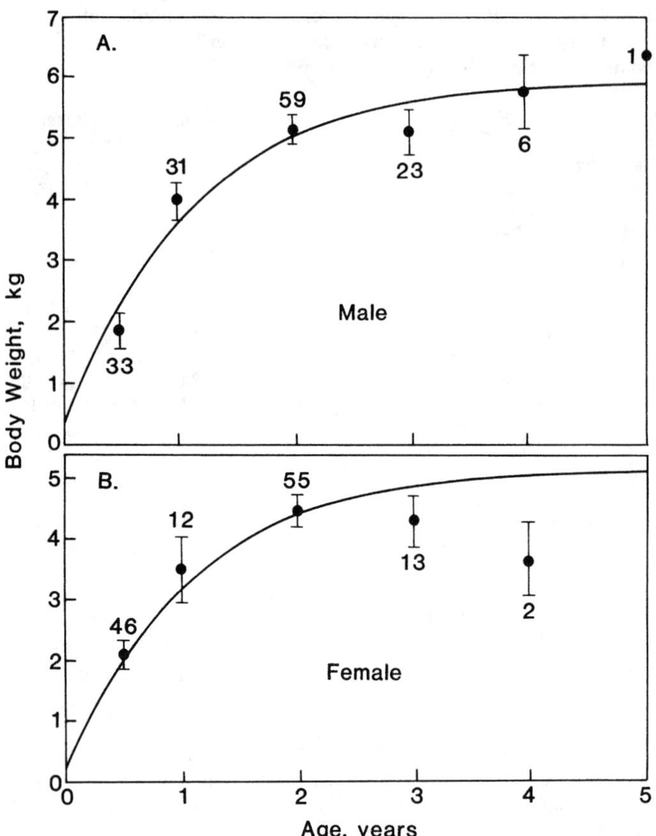

FIGURE 52.4. Relationship between body weight and age in nutria (*Myocastor coypus*). SOURCE: Adapted from Willner et al. (1980).

separate data set generated from the body weights of nutria whose age was estimated by standard methods. They concluded that there were no significant differences between observed and predicted body weights. Moreover, it was demonstrated that the model of Dixon et al. (1979) could reliably predict the weight of males up to 6 years of age; however, the weight of females could be predicted accurately only to 2 years of age (Fig. 52.4).

Body Condition Factors. Techniques to evaluate body condition and physiological responses of nutria include several morphological indicators, such as body fat reserves, adrenal weights, and spleen weights, which are used to calculate various indices to adjust for differences in body size. The left adrenal gland of nutria is kidney bean shaped, and is larger and significantly heavier than the right gland, which is hook shaped (Wilson and Dewees 1962). Mature male nutria have larger adrenal glands than do mature females, with an average adrenal weight to body weight ratio of 0.627 g/kg for males and 0.512 g/kg for females (Wilson and Dewees 1962). Katomski and Ferrante (1974) reported a marked dimorphism in shape and size between the left and right adrenal glands and noted that the ratio of adrenal weight to body weight was greater for adult nutria than for other mammals. The adrenal glands of nutria were composed mostly of cortical tissue. The content of adrenal epinephrine and norepinephrine was relatively low; however, levels of both catecholamines were higher in young nutria than in adults. The predominant hormones were epinephrine in adults and norepinephrine in young (Katomski and Ferrante 1974). Because of this anatomical, age, and sexual dimorphism, it is important that when adrenal weights are used as morphological indicators, the total or mean adrenal weight and the gender of the animal are factored into the analysis.

Gosling (1974) used a fat index (inguinal fat pad weight [g] $\div$ hind foot length [mm]) to evaluate the seasonal body condition of male and female nutria in England. Fat index scores of males were correlated with climatic variation and the availability of food. Males were in the best condition during late summer and exhibited the poorest body condition in late winter. The fat index of pregnant females varied greatly. Nevertheless, Gosling (1974) concluded that there may have been a relationship between the number of females maintaining pregnancies to full term and environmental factors such as climate and food supply. In general, Willner et al. (1979) reported similar findings using a body condition index (body weight [g] $\times 10^5 \div$ the cube of the body length [dm]), an adrenal index (adrenal weight [mg] $\times 10^4 \div$ body weight [g]), and a spleen index (spleen weight [mg] $\times 10^4 \div$ body weight [g]) to assess physiological responses to factors such as season, sex, age, weather conditions, and reproductive status of nutria. Body condition indices showed that nutria in Maryland were in the best condition from May to September and at their poorest body condition in January and February. Willner et al. (1979) suggested that the adrenal index was the best parameter to evaluate body condition, and that the seasonal adrenal responses of males, nonpregnant females, and pregnant females were comparable, with the highest values recorded in January and February. Because age correlated directly with the adrenal index, Willner et al. (1979) hypothesized that nutria may be under more stress as they grow older. There was a direct correlation with freeze-free days, temperature, total precipitation, and the spleen index of pregnant females. Willner (1982) suggested that these results may be due to the fact that Maryland is the northernmost portion of the range of nutria in eastern North America.

Toxicology. Toxicants and insecticides may affect reproduction in nutria. Low doses of chlorinated hydrocarbon insecticides given to caged, pregnant nutria resulted in partial or total abortion of the litters (Evans 1970). Acute lead poisoning of nutria in Poland occurred within 12 hr following the feeding of grass harvested from an airport where petroleum runoff was likely to have occurred (Gorka and Kulczyki 1983). The effects of bromadiolone poisoning in nutria were characterized by erythrocyte damage, saturation of transferrin, and organelle distortion in hepatocytes (Jeantet et al. 1991).

Weather-Related Mortality. Several investigators in Europe, North America, and South America have documented the physical effects of weather on nutria survival and behavior. Ehrlich (1962) proposed that winter mortality of nutria was the result of sheet-ice formation. He demonstrated that established populations in Poland could be maintained if ponds were drained in the fall and a dense growth of vegetation was encouraged to provide shelter. Aliev (1965b) reported that periods of freezing conditions in Russia often resulted in mass mortality. Newson (1966) and Norris (1967a) observed that nutria in England were unable to withstand more than a few days of severe frost and that during extreme winter conditions, mortality was estimated to be 80% of the population. Similar winter-related mortalities have been noted ranging from 71% in west-central France (Doncaster and Micol 1990) to 90% in Maryland (Willner et al. 1979). Several investigators have noted that following periods of freezing temperatures, surviving nutria had frost lesions on portions of the ears, tail, feet, and toes (Axell 1963; Norris 1967b; Gosling 1981b). Incidence of frostbite lesions observed on nutria did not vary between males and females (Doncaster and Micol 1989).

Doncaster and Micol (1990) reported that juvenile nutria were largely absent from late-winter populations and that male-biased mortality was associated with catastrophic events such as cold or flooding during winter. Males may have been more susceptible than females to energy loss during cold weather because their home ranges were larger and they traveled at greater average speeds (Doncaster and Micol 1989). Similarly, Doncaster et al. (1990) observed that nutria under temperate climatic conditions stayed close to water and swam frequently with no marked deviation from these behaviors during winter, despite high mortality among young (male and female) and adult males. Social factors such as a dominance hierarchy may have required the use of waterways even under freezing conditions and resulted in mortalities (Doncaster et al. 1990).

The effect of behavior on maintenance of thermal homeostasis was quantified by indirect calorimetry (Moinard et al. 1992). Huddling behavior was responsible for a 20% decrease in resting metabolic rate at temperatures below thermal neutrality, and females tended to reduce the amount of time they were in cold water, but males did not. These researchers concluded that nutria illustrate a conflict between social and thermoregulatory behaviors, which contributes to juvenile- and male-biased mortality in winter (Doncaster and Micol 1989; Doncaster et al. 1990; Moinard et al. 1992).

McKean (1982) examined cardiovascular adaptations to diving in nutria, observing that heart rate and cardiac output decreased by 78% and 76%, respectively, following submersion. Concurrently, regional blood flow decreased to all organs except the adrenal glands, heart, and lungs; however, blood flow to the brain increased during diving. These circulatory adaptations during diving appeared to reduce the rate of oxygen removal from the blood by 88.9% compared to the predive oxygen consumption rate. Cardiovascular adjustments may be the critical physiological factors that determine duration and activity of diving in nutria.

Data and observations on other aspects of systemic physiology in nutria are limited and often anecdotal. Redding et al. (1968) studied the effect of thyrotrophin-releasing factor (TRF) on plasma thyroid-stimulating hormone (TSH) and concluded that, although nutria exhibited a low sensitivity to TRF, the releasing factor could increase plasma TSH levels. The apparent reduced sensitivity of nutria to TRF may have been partially related to the relatively low concentrations of TSH in the anterior pituitary (Redding et al. 1968). The histology of the medial geniculate body (MGB) of the thalamic auditory center in the brains of different rodents was studied by Figurina (1984) to clarify the relationship between ecology and morphological differences in the auditory system. In terrestrial nocturnal rodents such as the common vole (*Microtus arvalis*) and fat dormouse (*Myoxus* [*Glis*] *glis*), the ventral MGB was the largest, whereas in terrestrial diurnal rodents such as the yellow ground squirrel (*Spermophilus fulvus*) and speckled ground squirrel (*S. suslicus*), the dorsal MGB was the largest. In nutria, the MGB was not as extensively compartmentalized and contained fewer cells than among the other rodents studied. Nutria may not possess the same auditory acuity as terrestrial rodents, and differences in hearing may result from structural differences in the auditory center of the thalamus (Figurina 1984).

REPRODUCTION AND DEVELOPMENT

Sex Determination. Determining the sex of nutria is possible by examining external genitalia. Male genitalia consist of a prepus and penis projected caudally, whereas females have a vulva and a vaginal orifice with a prominent urinary papilla. Sex of embryos (i.e., conceptus in rapid development before formation of all major structures) and fetuses (i.e., conceptus in development after all major structures have formed) may be determined based on the anal–genital distance and the appearance of the external genitalia. When the embryo is positioned dorsally recumbent, the sheath of the penis is located ventral-carnally to, and clearly separated from, the anal opening in males. Willner et al. (1979) reported this distance was approximately 0.5 mm during early embryonic development. The degree of separation increases progressively as the conceptus develops. In females, the urogenital opening appears to be almost contiguous with the anus in early development, but gradually separates as the transition is made from embryo to fetus. The anal–genital distance remains relatively shorter in females than in males throughout fetal and neonatal development.

Reproductive Potential. Male and female nutria exhibit many of the unusual anatomical and physiological characteristics of hystricomorph rodent reproduction (Gosling and Skinner 1984). Nutria become sexually active before reaching adult size and are prolific. For example, Gosling (1974) calculated that a population of nutria in Great Britain that was estimated to be between 8000 and 11,000 animals could increase to 15,000–18,000 individuals in approximately 12 months if not controlled. The number of females in a population could double after about 8 years under continuous mild climatic conditions (Gosling et al. 1983). Brown (1975) reported that in Florida, the total maximum reproductive potential of an adult female would average 15 young/year, whereas Willner et al. (1979) estimated that the adjusted mean number of young produced annually/female was 8.10 in Maryland.

Male Reproductive Anatomy. Detailed descriptions of the male reproductive tract are provided by Hillemann et al. (1958), Stanley and Hillemann (1960), Weir (1974), Willner (1982), and Kinler et al. (1987). The surface of the glans penis is covered with minute scales and contains a cylindrically shaped *os baculum* composed of bone, a transition zone, and a cartilage tip. The baculum is 15–23 mm in total length and has a diameter of 1–4 mm (Willner 1982). The testes of nutria are elliptical and scrotal, with the epididymis attached parallel to the long axis of each testicle. As with many rodents, the inguinal canal remains sufficiently patent to allow retraction of the testes into the abdominal cavity. Brown (1975) reported that the mean testis weight for adult males in Florida was 4.8 ± 0.8 g. Willner et al. (1979) stated that testicular weight increased with age, averaging 2.4 g at 6 months and 14 g at 4 years; after 4 years, testicular weight declined. The ratio of testicular weight to body weight increased with age as well.

Mann and Wilson (1962) analyzed secretions of the male reproductive accessory glands and found the vesicular gland contained high concentrations of fructose, prostate glands were rich in citric acid, and the bulbourethral glands contained relatively higher concentrations of sialic acid and phosphorus. Willner (1982) reported that the secretions from these glands caused the ejaculated seminal fluid to gel and form a copulatory plug. The presence of copulatory plugs can be used to indicate mating activity.

Puberty in Males. The onset of male sexual maturity (i.e., puberty) has been defined as the age at which the penis has become elongated and protrusible, abundant spermatozoa occur in the cauda epididymis and the vas deferens, and sexual activity has begun. In the United States, puberty occurred when individual nutria testicle weight reached about 2–3 g or at approximately 6 months of age (Willner et al. 1979; Kinler 1992). In England, puberty was identified when total combined testicular weight was 5 g (Newson 1966) or at about 5 months of age, when body weight was 1.8–2.2 kg (Laurie 1946). Season of birth modulates the onset of puberty in males. In Louisiana, Evans (1970) observed that males reached puberty at 4–6 months if they were born during summer; however, puberty was delayed until 7–8 months if the males were born during winter. Similarly, in England, Newson (1966) suggested that puberty occurred at 3–4 months for males born during summer, but not until 6–7 months for males born during fall.

Breeding Activity of Males. Based on observation of sexual activity, production of spermatozoa, and pregnancy rates, male nutria remain fertile independent of the season, and will breed throughout the year in the United States (Atwood 1950; Adams 1956; Kays 1956; Peloquin 1969; Evans 1970; Willner et al. 1979; Kinler 1992), England (Newson 1966; Gosling 1974; Gosling and Skinner 1984), and France (Dagault and Saboureau 1990; Doncaster and Micol 1990). Brown (1975) observed large quantities of spermatozoa in the seminiferous tubules and epididymides during each month of the year in adult males, with no detectable seasonal changes in spermatogenic activity. Willner et al. (1979) reported no substantial monthly variation in mean testicular weight, but a significant negative correlation between mean testicular weight and temperature. They concluded that males may reduce reproductive activity during adverse weather conditions. Other studies also have suggested that weather conditions may interrupt or modify nutria reproductive activity (Newson 1966; Evans 1970; Doncaster and Micol 1990).

Dagault and Saboureau (1990) reported that plasma testosterone in feral males varied yearly and ranged from 0.4 to 6.8 ng/ml. Average testosterone levels tended to be lower during December, January, and February than in October and November in wild populations (Dagault and Saboureau 1990). For captive nutria, average testosterone concentrations were greatly elevated compared to levels in wild nutria, often by as much as 20 ng/ml or more, but the same seasonal pattern was evident in wild and captive nutria. In captives, plasma testosterone significantly decreased from 23.3 ng/ml in January to 1.6 ng/ml in March, and these results may reflect reduced food availability. In fact, a 50% reduction in available food resulted in a significant reduction of plasma testosterone, and when ad libitum feeding was resumed, testosterone concentrations returned to control levels (Dagault and Saboureau 1990).

Female Reproductive Anatomy. Anatomy of the female reproductive tract is presented in detail by Hillemann et al. (1958), Stanley and Hillemann (1960), Weir (1974), Willner (1982), and Kinler et al. (1987). The clitoris of nutria is in the vestibule of the vagina and contains a cartilaginous *os clitoris.* Atwood (1950) reported that once the vagina opened at puberty, it remained open throughout life. However, Newson (1966) and Peloquin (1969) noted that during the last month of pregnancy, some constriction was observed. These observations are consistent with observations of Weir (1974) and Roberts and Perry (1974) that no vaginal membrane is formed in nutria. In contrast, Gosling et al. (1981) reported that in England total or near-total closure of the vaginal orifice is normal in wild nutria.

The uterus of the nutria is duplex and supported by the mesometrium, which is attached dorsolaterally to the abdominal wall. There are separate openings from the vagina into the cervix of each distinct uterine horn. Each horn terminates into an oviduct, which is supported by the mesosalpinx. The terminal portion of the oviduct is dilated and forms an infundibulum with fimbria, which do not envelope the ovary. A detailed description of the histological arrangement of the mucosa lining of the female reproductive tract of nutria is provided by Felipe et al. (1998).

Ovaries of nutria are ovoid and not encapsulated. They are supported by the mesovarium, which is attached to the dorsal wall of the abdominal cavity. Kinler et al. (1987) reported that on average the ovaries of mature nutria weigh 186 mg. They average 12 mm in length, and have a mean diameter of 7.5 mm. In nulliparous, sexually mature, captive females, the ovaries were slightly smaller, with an average weight of 110 mg, length of 8.25 mm, and diameter of 4.05 mm (Felipe et al. 1999). The surface of the ovary is relatively smooth and, with the exception of corpora hemorrhagicum, it is rare to observe other structures such as follicles or corpora lutea projecting from the ovarian surface (Weir and Rowlands 1974). Felipe et al. (1999) reported that the ovarian surface of some nutria appeared lobulated.

In-depth descriptions and discussions of the histology of the ovary and its structures are provided by Gluchowski and Maciejowski (1958), Rowlands and Heap (1966), and Felipe et al. (1999, 2000). The ovarian cortex consists mainly of follicles in all stages of development and their derivatives; follicular diameters ranged from 35.8 ± 3.9 μm, observed in primordial follicles, to 1.2 ± 0.01 mm, in preovulatory or Graafian follicles (Felipe et al. 1999). Rowlands and Heap (1966) reported that near the time of ovulation the volume of the mature Graafian follicle was 0.52–1.75 mm^3. Felipe et al. (2000) developed a descriptive series for ovarian follicles based on their qualitative and quantitative characteristics.

The average number of corpora lutea observed in the ovaries ranged from 7.6 (Peloquin 1969) to 8.8 (Rowlands and Heap 1966). Rowlands and Heap (1966) reported that the distribution of corpora lutea was not equal in the majority (67.5%) of the animals. Average number of corpora lutea was 5.0 ± 0.3 on the right and 3.9 ± 0.3 on the left ovaries. The mean volume of corpora lutea (1.0–1.5 mm^3) changed little through the 45-day postmating period (Rowlands and Heap 1966). Between day 45 and day 100 of gestation, luteal growth reached a maximum size of 10–15 mm^3, and at day 110, corpora luteal volume began to decline until parturition. The diameter of well-developed corpora lutea varied greatly in nulliparous, sexually mature females and averaged 0.7 ± 0.1 mm (Felipe et al. 1999). Newson (1966), Rowlands and Heap (1966), and Peloquin (1969) observed that the number of corpora lutea was often greater than the number of embryos in pregnant females. Based on the histological similarity of corpora observed, Rowlands and Heap (1966) concluded that secondary corpora lutea were not formed at any stage of pregnancy in nutria. In contrast, Weir and Rowlands (1974) reported two types of corpora lutea were found in the hystricomorph ovary: (1) true corpora lutea, which developed from the collapsed follicle following ovulation and persisted until the end of pregnancy; and (2) accessory corpora lutea, which developed from follicles that had undergone luteinization during the latter stages of pregnancy. Sometimes only a few accessory corpora lutea were formed in nutria (Weir and Rowlands 1974). More recently, Felipe et al. (1999) reported the presence of accessory or secondary corpora lutea in nulliparous, sexually mature females and noted that accessory corpora lutea could be distinguished from true corpora lutea by their smaller size and the persistent remains of degenerated oocytes and remnants of the zona pellucida.

Puberty in Females. Sexual maturity in female nutria is characterized by opening of the vaginal orifice, onset of estrous cycles, and initiation of sexual activity. Atwood (1950) reported that puberty in females in Louisiana occurred at mean body length of 69.6 ± 4.5 cm and about 5 months of age. He also noted that ovulation may occur before the opening of the vaginal orifice. In farm-reared nutria in Illinois, Wilson and Dewees (1962) found that puberty, as indicated by vaginal opening, occurred at 5 months. Sexual maturity was observed at 4–9 months in females in Oregon (Peloquin 1969); in Maryland, females began breeding at 6 months of age (Willner et al. 1979). Kinler (1992) also concluded that puberty was reached at approximately 6 months. In England, Laurie (1946) reported fecundity (sic) was reached with a body weight of 1.8–2.2 kg at about 5 months. Similarly, Newson (1966) concluded that puberty was attained when the body weight of females reached 1.5–2.5 kg. Gosling (1986) noted that although female nutria were not fully grown until about 15 months of age, they often conceived at about 3 months. As observed in males, the time of year during which females are born may affect the onset of puberty. In Louisiana, puberty occurred at 4–6 months when females were born during summer and at 5–7 months if they were born during winter (Evans 1970). Similarly, in England, females reached puberty at 3–4 months if born in summer and at 6–7 months if born in fall (Newson 1966). Atwood (1950) found that nutritional status may affect the onset of puberty in female nutria. During a time of food shortage, the average weight of adolescent females was reduced by 80% compared to the previous year (Atwood 1950). In this population, puberty did not occur until the females reached the age of 5.4 ± 0.2 months. There was a delay of 1.3 months in puberty resulting from a deficiency of available food. In a series of in vitro experiments, Sirotkin et al. (2000) compared the effects of food restriction on ovarian secretory activity (i.e., secretion of estrogen and progesterone) and the role of insulinlike growth factor-I and cyclic AMP derivatives on the control of ovarian function. Ovarian tissue from underfed female nutria secreted more progesterone and less estrogen than tissue from control animals. In tissue obtained from underfed animals, progesterone secretion was not altered, but secretion of estrogen was stimulated.

Breeding Activity of Females. Female nutria breed throughout the year in Louisiana (Atwood 1950; Adams 1956; Kays 1956; Evans 1970; Kinler 1992), Florida (Brown 1975), Oregon (Peloquin 1969), and Maryland (Willner et al. 1979). Similarly, in England (Newson 1966; Gosling 1974; Gosling and Skinner 1984) and France (Dagault and Saboureau 1990; Doncaster and Micol 1990), female nutria are nonseasonal breeders. During a 2-year study of nutria inhabiting a freshwater marsh in Louisiana, Atwood (1950) found that all adult females were pregnant and the level of reproductive activity was relatively constant. Other investigators working in Louisiana reported pregnancy rates of 80% (Adams 1956) and 90.6% (Kays 1956) in nutria obtained from trappers. Brown (1975) reported that virtually every female collected near Tampa, Florida, was pregnant or lactating and that no obvious seasonal peaks in female breeding activity were evident.

Pregnancy Rate. Numerous other investigations confirm that nutria are nonseasonal breeders and that the rate of pregnancy in most populations of nutria is high. However, considerable variation in pregnancy and parturition rates within and among years has been reported. For example, in Oregon, nutria breed nonseasonally; however, maximal pregnancy rates were observed in March and May as well as a smaller peak in October. Evans (1970) determined that in Louisiana and Texas, 85% of the adult female nutria examined were pregnant, but noted that dates of parturition were affected by extremely hot weather, droughts, hurricanes, and freezing weather. In Maryland, Willner et al. (1979) reported that the pregnancy rate in nutria over a 12-month period was 64.9%. When nutria were examined over three consecutive trapping seasons in Louisiana, variation in pregnancy rates was significant from year to year, with 85.4% of females pregnant in year 1, 76% in year 2, and 78.7% in year 3 (Linscombe et al. 1981). In Louisiana, Kinler et al. (1987) reported that the proportion of pregnant females ranged from 58% in May to 100% in February. Variation in birth rates was related to habitat. Nutria collected from freshwater marshes had parturition peaks during January, June, and November, whereas nutria in brackish marshes had parturition peaks during April, July, October, and November. Kinler (1992) reported that parturition peaks were influenced by climatic factors. In England, Newson (1966) and Gosling (1974) concluded that at any time of the year, most female nutria were pregnant. Gosling et al. (1983) reported that peaks in parturition were determined by the severity of the preceding winter. Peaks in parturition occurred from March to June in years when the winters were mild and in June to August following severe winters. Similarly, in France, Doncaster and Micol (1990) observed that after a moderate winter, peak parturition occurred in April. However, following a severe winter, the first peak in parturition occurred from June to August. Doncaster and Micol (1990) suggested that freezing weather during the winter synchronized parturition the following spring. Newson (1966) hypothesized that one factor that may reinforce synchronized parturition is that the age at which young females first conceive is often roughly equal to the length of gestation. Therefore, female offspring may reach puberty at about the same time that their mother has her first estrus following the birth of a subsequent litter. If this scenario occurs, sexually mature subadult females would be in phase reproductively with their mothers and could become pregnant at approximately the same time as their mothers, thus amplifying any synchrony of parturition resulting from environmental factors.

Estrous Cycle. Nutria are polyestrous, and wild and captive-raised animals display great variation in the length of their estrous cycles. Matthias (1941) and Atwood (1950) reported that nutria undergo a 2- to 4-day estrus approximately 48 hr postpartum. Vaginal smears collected from nutria reared in captivity in Illinois indicated that estrous cycles lasted 5–28 days (Wilson and Dewees 1962). Evans (1970) observed that nutria in Louisiana came into estrus every 24–26 days and remained in heat for 24–48 hr. In Europe, Skowron-Cendrzak (1956) concluded that parity influenced the length of the estrous cycle and the duration of estrus; primiparous nutria had an estrous cycle length of 17 days with a 24-hr period of estrus, whereas multiparous females had estrous cycles of 19 days with durations of estrus averaging 48 hr. In England, Newson (1966) reported that nutria most commonly had estrous cycles ranging from 2 to 4 weeks, with an average of 26 days. British investigators (Newson 1966; Gosling 1980; Gosling and Skinner 1984) also observed that most postpartum females were pregnant and that estrus occurred within 2 days of giving birth. Gosling (1980) calculated the mean postpartum interval (i.e., time from parturition to conception) in nutria to be 2.1 ± 0.8 weeks. With the possible exception of the postpartum estrus, the irregular length of the estrous cycle appears to be the primary basis for suggesting that nutria may be induced ovulators (Matthias 1941; Newson 1966; Brown 1975; Gilbert 1987; Willner 1982; Gosling and Skinner 1984). However, Atwood (1950) noted that ovulation may occur before the opening of the vulva, and Felipe et al. (1999) reported ovulation in nulliparous females.

Gestation. Nutria and other hystricomorphs have extremely long periods of gestation compared to other rodents (Roberts and Perry 1974). Evans (1970) and Kinler (1992) reported that nutria in Louisiana had a 130-day gestation with little variation. In Maryland, the duration of pregnancy in nutria ranged from 130 to 134 days (Willner et al. 1979). Similar periods of gestation have been observed by European investigators. Aliev (1956) monitored 446 pregnant nutria on a fur farm in Azerbaijan and noted that the mean length of gestation was 131 days, but could vary from 128 to 138 days. In England, Newson (1966) studied >400 pregnant females and found gestation was 127–138 days and postpartum conceptions were frequent. Heap and Illingworth (1974) noted that gestation was 130–134 days. Other researchers in Great Britain reported that the average length of gestation was 130 days (Rowlands and Heap 1966), 132 days (Gosling 1974), and 133 days (Gosling and Skinner 1984). These findings are consistent with the gestation period (127–132 days) for nutria in South America (Cabrera and Yepes 1940).

Female Hormones. Rowlands and Heap (1966) reported that during the initial 45 days of pregnancy, plasma progesterone ranged from 3 to 18 ng/ml, but then increased to maximal concentrations of approximately 500 ng/ml between days 65 and 84. Concurrent with the increase in circulating progesterone was a period of rapid luteal growth, which was observed between days 45 and 100 of gestation. By day 110 of pregnancy, luteal regression had commenced and progesterone levels decreased to 15 ng/ml in plasma collected postpartum. Tam (1974) noted that concentrations of progesterone observed in nutria during pregnancy were considerably greater than those reported in a number of other mammalian species. In addition to supplementing progesterone secretion by the primary corpora lutea of pregnancy during gestation (i.e., production of progesterone by secondary corpora lutea and the placenta), a third mechanism involving a plasma-binding globulin may contribute to the high progesterone concentrations quantified during pregnancy in nutria (Tam 1974). Heap and Illingworth (1974) demonstrated that the rate at which progesterone was removed from blood during pregnancy was reduced compared to the removal rate in non-pregnant nutria. This phenomenon appeared to be related to the synthesis of a plasma protein during pregnancy that had a high affinity and moderately high binding capacity for progesterone and its 5-alpha- and 20-alpha-reduced metabolites. The tertiary features of the binding domain of the progesterone-binding protein of nutria displayed unique structures compared to other species (Mais et al. 1995).

Placenta. Complete descriptions of the development and structure of the nutria placenta have been presented by Hillemann and Gaynor (1961), Newson (1966), Hillemann and Ritschard (1967), and Roberts and Perry (1974). Nutria have a chorioallantoic placenta, which consists of four main regions: (1) the trophoblastic area containing the labyrinthine and spongy zones; (2) the subplacenta originating from chorionic ectoderm; (3) the junctional zone, which forms a region of intimate fetal–maternal contact and is composed largely of necrotic tissue derived mainly from the decidua; and (4) the decidua basalis, which is the maternal portion of the placenta (Roberts and Perry 1974). The temporal pattern of placental development in nutria is similar to that of guinea pigs (Hillemann and Gaynor 1961). Roberts and Perry (1974) reported that in nutria the formation of the amniotic cavity, allantois, and chorioallantois occurs at 15–18, 20–25, and 32–39 days after conception, respectively. In the near-term placenta, there is a well-developed maternal decidua basalis or pedicle, which may be up to 5 cm long (Newson 1966). On the maternal side, the placental disk is deeply undercut and attached to the uterine wall by a few large blood vessels running through the pedicle (Newson 1966; Chapman et al. 1980). The complete interstitial implantation and the subplacenta found in nutria are unique to hystricomorphs; no comparable structure has been described in any other mammal (Roberts and Perry 1974). From 12 nutria placentae collected near the time of parturition, Hillemann and Gaynor (1961) reported that placental weights averaged 14.4 g, with a range of 7.5–45.6 g; mean placental length equaled 34.4 mm and ranged from 25 to 46 mm; mean placental width was 27.8 mm, with a range of 22–34 mm; and mean placental thickness was 18.9 mm and varied between 10 and 28 mm. At parturition, all of the pedicles were shed and all that remained were the ruptured stumps of their blood vessels in small groups along the uterine wall. These implantation sites continue to degenerate into pigmented scars, which may remain visible for up to 2 months into the next pregnancy (Newson 1966).

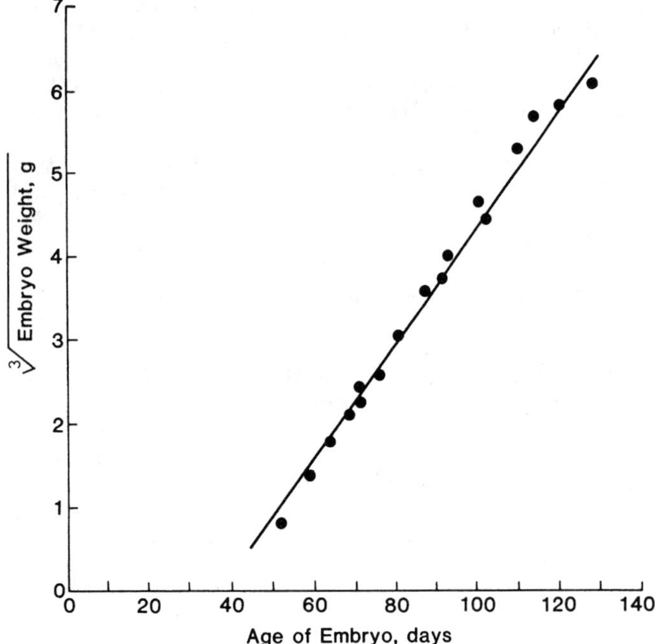

FIGURE 52.5. Relationship between embryo weight and age in nutria (*Myocastor coypus*). SOURCE: Adapted from Chapman et al. (1980).

Embryonic and Fetal Development. Fetal growth rates of hystricomorphs are extremely slow compared to other mammals of similar birth weights, and fetal growth of nutria is intermediate among hystricomorphs (Roberts and Perry 1974). During the first 30 days of pregnancy, Newson (1966) found that nutria embryos developed slowly and remained embedded in the mass of tissue that formed at each implantation site. Using data from 19 litters of known gestational stage, Newson (1966) determined that fetal weight and embryonic age were linearly related (Fig. 52.5). The regression equation describing this relationship can be solved to allow prediction of embryonic age from fetal weight:

$$EA = 43.69 + 14.27 \times 3\sqrt{FW}$$

where EA = embryonic age (days) and FW = fetal weight (grams).

Chapman et al. (1980) and Kinler et al. (1987) concluded that this equation should be applied to embryos older than 50 days. By a modification of Newson's (1966) data, Chapman et al. (1980) established a relationship between embryonic age and length that may be used to estimate the age of embryos collected at or before day 50 of gestation (Table 52.2 or Fig. 52.5). By determining the ages of embryos, researchers have been able to estimate the dates of conception and parturition and apply these data to the dynamics of nutria populations (Newson 1966; Chapman et al. 1980; Gosling et al. 1981; Kinler et al. 1987).

TABLE 52.2. Relationship between age and length in nutria embryos

Length (mm)	Approximate Age (days)
3–5	≤(15 or less
>5–7	>15–20
>7–9	>20–25
>9–10	>25–30
>10–12	>30–35
>12–14	>35–40
>14–20	>40–45
>20–25	>45–55
>25–30	>55–70

SOURCE: Data from Newson (1966) and Chapman et al. (1980).

Descriptions of the embryonic and fetal development of nutria were given by Newson (1966), Roberts and Perry (1974), and Chapman et al. (1980). By the time that the nutria embryo attains a crown-to-rump length of 7 mm, the external features are well developed. These include the points of flexion in the cranial, cervical, dorsal, and lumbosacral regions, with appendage buds and digits evident at the 10-mm length and 20 -to 25-mm stage (Chapman et al. 1980). The pattern of embryological development of nutria corresponds to that of other rodents, except that (1) complete closure of the neural tube occurs much later than in other mammals, (2) the thin roof that forms on the metencephalon of other mammalian embryos is not observed, (3) development of the mesonephrol is delayed, and (4) development of the olfactory nerves may be accelerated.

The distribution of embryos and fetuses is not always uniform between the uterine horns in nutria. Adams (1956) examined 223 pregnant nutria captured in coastal Louisiana and found that the ratio of young developing in the right and left uterine horns was 49:51, respectively. The average number of young in the right horn of individual animals was 2.48, while the left horn contained 2.56 developing kits. A significant difference in fetal distribution between the right and left uterine horns was reported by Willner et al. (1979). They determined that of a total of 211 viable fetuses examined from females trapped in Maryland, 123 (58%) were implanted in the right horns and 88 (42%) in the left horns. Similar results were obtained in England by Rowlands and Heap (1966), who found that in 43 pregnant animals, the mean number of viable and resorbing embryos in the right and left horns was 3.4 ± 0.25 and 2.7 ± 0.20, respectively.

Mortality in Utero. Embryonic and fetal mortality may result in resorption or spontaneous abortions of partial or complete litters. In Louisiana, resorption rates ranged from 1% to 3.5%. Higher rates of resorption resulted from a decrease in available food resources and habitat deterioration (Atwood 1950). Total embryonic losses were estimated to range from 33% to 35% in a nutria population in Oregon, with a mean resorption rate of 24.6% (Peloquin 1969). Evans (1970) estimated that only about 60% of the embryos of nutria in Louisiana and Texas survived to be delivered as viable young. Nutria in Maryland had a mean embryo resorption rate of 9.8%, with the highest rates observed during inclement weather in December and January (Willner et al. 1979). Kinler (1992) reported that nutria experience high rates of embryo mortality and that their overall reproductive success is closely related to habitat conditions. In England, Newson (1966) estimated that during the period between implantation and birth, 50–60% of the embryos and fetuses die, with equal losses due to total resorption or abortion of the litters. Greatest losses occurred between weeks 5 and 10 of pregnancy, and few fetuses died after week 16 (Newson 1966). Similarly, Gosling (1974) reported that total mortality during gestation was 60% if females became pregnant after they were 6 months of age and 80% if conceived when they were <6 months old. The frequency of resorption and abortion increased during winter (Gosling 1974). Resorption or abortion of partial or whole litters occurred if pregnant females were not able to maintain adequate body fat reserves (Gosling 1981a). Subsequently, Gosling (1986) found that young females in better-than-average body condition and due to litter in summer aborted small litters of predominantly female fetuses about week 13 or 14 of gestation. Other females in similar physical condition with predominantly male litters carried them to term. Those females that did abort conceived soon after aborting and the newly conceived litter was significantly larger than the aborted litter (Gosling 1986). Gosling (1986, 1989b) proposed that selective abortion of entire litters by nutria may be a mechanism of adaptive control of offspring production in response to the quality and sex of the fetuses. Although nutria abort readily (Willner 1982), it appears that they can control the timing of parturition to a degree. Gosling et al. (1988b) observed that captive nutria delayed the birthing process for up to 3 days if they were disturbed on the day before the expected date of delivery. They speculated that in the wild, parturient nutria, if detected by a predator, may delay parturition to relocate to a safer location to give birth (Gosling et al. 1988b).

TABLE 52.3. Litter size for native and nonnative nutria populations, in approximate latitudinal order

Litter Size				
Mean	Range	Number of Observations	Data Collection Location	Reference
5.0	—	—	Argentina	Cabrera and Yepes 1940
5.8	3–12	148	Florida	Brown 1975
5.5	1–9	—	Texas	Evans 1970
5.6	2–11	35	Louisiana	Atwood 1950
5.0	1–11	223	Louisiana	Adams 1956
5.0	—	272	Louisiana	Kays 1956
4.8	—	321	Louisiana	Linscombe et al. 1981
4.7	—	216	Louisiana	Linscombe et al. 1981
4.6	—	550	Louisiana	Linscombe etgal. 1981
4.5	1–9	—	Louisiana	Evans 1970
4.5	—	—	Louisiana	Kinler 1992
4.2	1–9	26	Louisiana	Atwood 1950
4.0	1–7	138	Maryland	Willner et al. 1979
4.9	—	60	Oregon	Peloquin 1969
6.0	1–13	—	England	Gosling 1974
5.7	2–12	43	England	Rowlands and Heap 1966
5.5	1–13	—	England	Gosling and Skinner 1984
5.5	—	741	England	Gosling 1986
5.3	1–13	423	England	Newson 1966
5.0	2–11	27	England	Laurie 1946

Litter Size. Litter size of nutria ranges from 1 to 13 kits, with an average of 4 or 5 neonates (Table 52.3). In Louisiana, Atwood (1950) concluded that litter size increased with body size and age of the female. In Maryland, Willner et al. (1979) reported that litter size was significantly related to age of the female. The oldest dams (3 years old) had mean litter sizes of 7 offspring, whereas younger females had litter sizes averaging 4 young. The smallest mean litter sizes occurred concurrently with the lowest winter temperatures (Willner et al. 1979). Average litter size decreased from 5.6 to 4.2 young when food availability was reduced due to habitat deterioration. Linscombe et al. (1981) demonstrated that destruction of habitat significantly reduced nutria populations and resulted in smaller litter sizes. In England, mean litter size in primiparous females also was smaller than the mean litter sizes of multiparous nutria. Largest litters were produced in the summer and smallest litters in the winter (Newson 1966; Gosling 1974, 1986).

Neonates. Nutria kits at birth weigh 170–332 g (Atwood 1950; Adams 1956; Newson 1966; Evans 1970; Gosling and Skinner 1984) and have a total length of 31 cm (Atwood 1950). Gosling et al. (1984) reported that among newborns escaping perinatal mortality, males had higher weights (224 g) than females (208 g). The young are highly precocial, are born fully furred with their eyes open, can swim shortly after birth, and average about 225 g in body weight (Newson 1966; Gosling and Skinner 1984; Webster et al. 1985; Gosling and Baker 1991). Neonates have functional incisors and premolars and are active within 1 hr after birth (Gosling and Baker 1991). Newborns can survive if artificially weaned at 5 days (Newson 1966), but usually nurse for about 7–8 weeks (Gosling 1980). Peloquin (1969) characterized the pelt of the neonate as soft and downy hair, with the tail hair remaining silky through the first 4 weeks postpartum. Neonates have been observed swimming (Atwood 1950; Evans 1970) and eating plant material within 24 hr of birth (Atwood 1950). Gosling (1974) reported that kits began eating solid food within a few days of birth, and by 2–3 weeks postpartum were eating relatively large quantities of solid food. There is little care provided by the mother to neonates or young (Warkentin 1968). Young are generally nursed in a den, but when the young are outside the den, the mother does not remain near them. While nutria nurse their young, little attempt is made to protect them from predators. Warkentin (1968) found that other members of a group would watch over young. Young nutria tend to stay in groups outside of the den. Mothers do not groom their young (Warkentin 1968). Peloquin (1969) found that during the first 5 months of life, nutria growth was rapid. The color transition of the incisor teeth associated with puberty has been described by Ehrlich (1966).

Lactation. Nutria have four to five mammary glands arranged in two parallel rows located dorsolaterally on each side of the body. The teats protrude above the surface of the skin, but are usually buried in the fur. Inactive mammary glands are small, flat, and whitish, but appear pinkish as they develop during pregnancy or while undergoing involution. Lactating glands are markedly larger, thicker, appear to be lobulated, and have a creamy color. A complete description of the mammary gland, including physical measurements, may be found in Gosling (1980). Evans (1970) reported that female nutria begin producing milk as early as 1–2 weeks before parturition and that lactation continues until 6–8 weeks postpartum. In nutria in Louisiana, most young are weaned at about 5 weeks; however, some young nurse a few weeks longer (Evans 1970). In England, Gosling (1974) estimated that juvenile nutria suckle for about 6–7 weeks. Subsequently, Gosling (1980) and Gosling et al. (1984) found that the duration of lactation in nutria was 7.7 ± 1.0 weeks. Male offspring spent relatively more time suckling the highest yielding teats, whereas females suckled more from the lowest yielding glands (Gosling et al. 1984). Ehrlich (1958) found that nutria milk contained 41.5% dry matter and 58.5% water. The percentage composition of whole nutria milk for fat, protein, ash, and sugar determined by individual assays was 27.9%, 13.7%, 1.3%, and 0.5%, respectively.

ECOLOGY

Habitat. Nutria inhabit coastal aquatic and wetland habitats, and are generally associated with either riverine habitats and the adjoining fringe marshes or with extensive marshes lacking a major river channel. North American populations occur in freshwater and brackish habitats (Bounds 2000), although freshwater habitats appear to be preferred where both are equally available (Dozier 1985; Nowak 1991). Although several studies have described the plant communities of nutria habitats, information is lacking on selection among available habitats (Coriel et al. 1988). Most studies have described the most frequently used plant communities within a given habitat where high density populations occur, as in coastal brackish marsh. Patterns of selection among habitats also have been inferred from harvest records, but these are influenced by density, trapping effort, and capture success, which may be poor indicators of habitat preference (Van Horne 1983).

Coriel et al. (1988) found that the dominant plants in nutria habitats also were most often represented in nutria diets. For example, nutria in Louisiana are found primarily in coastal marshes dominated by saw grass (*Cladium jamaicense*), giant cutgrass (*Zizaniopsis miliacea*), southern bulrush (*Scirpus californicus*), alligator grass (*Alternanthera philoxeroides*), cattail (*Typha domingensis* and *T. latifolia*), and maidencane (*Panicum hemitomon*) (Atwood 1950; Dozier 1985). Louisiana nutria generally occur in the highest densities in freshwater marshes and river and swamp margins to brackish or salt marshes (Dozier 1985). Coreil et al. (1988) examined habitat selection of radio-collared nutria within the intermediate marsh zone in Louisiana and found that the most preferred plant communities were southern bulrush and cattail/wiregrass (*Spartina patens*).

At Blackwater National Wildlife Refuge (NWR), Maryland, the highest nutria densities occurred in open brackish marsh consisting primarily of three-square bulrush (*Scirpus olneyi*), cattail, and saltmarsh bulrush (*S. robustus*) (Willner et al. 1979). Although wooded habitats are not used extensively by nutria, they may be important sources of cover in the northern portions of the range. Brackish marshes at Blackwater NWR are bordered by moist loblolly pine (*Pinus taeda*) woods; nutria use of these woods increases during severe winter weather (D. L. Bounds et al., University of Maryland–Eastern Shore, unpublished data, 2002). There also is some indication that bark and roots of woody shrubs may be an important source of forage during winter. Based on behavioral observations of nutria pairs before and after they entered wooded areas, Warkentin (1968) suggested that nutria may copulate in forests. Forested habitat is generally restricted to the periphery of marshes used by nutria, and is not considered a habitat of primary importance. However, the abundant forested wetlands on the Delmarva Peninsula are recognized as a potentially significant source of cover for nutria occupying marsh fringes.

Dense herbaceous vegetation also is an important source of winter cover, and colonies of male and female nutria are more likely to use this type of cover than lone young or single-sex groups (Ehrlich 1962). In Florida, Brown (1975) reported that nutria population densities were four times higher in nutrient-rich dairy farm runoff ponds than in unpolluted ponds, due in part to a high production of water hyacinth (*Eichornia crassipes*) in enriched ponds. Canals, tidal creeks, and other drainage channels are also important habitat features. Micol and Doncaster (1987) found that nutria use of small canals declined in winter as the canals filled with vegetative debris and became prone to freezing.

Habitats used by European nutria populations are similar to those of North American nutria. Before their eradication in Great Britain, nutria were found in reed beds, marshes, river banks, and fens (Laurie 1946; Gosling and Baker 1991). Dominant vegetation in these communities included reeds (*Phragmites communis*), reed meadow-grass (*Glyceria maxima*), cattails, and sedges (*Carex* spp.). Transitional habitats between open water and extensive reed beds were used more for nest building than were sedge transition or marsh interior habitats (Ryszkowski 1966). Gosling and Baker (1991) described nutria densities as moderate in brackish marshes, found lower densities in inland ponds and streams, and suggested that most lowland river valleys in England could probably support nutria.

Nutria habitat in South America includes swamps and marshes in and along rivers and lakes (Larrison 1943). In the Argentinean Pampas, Guichón and Cassini (2000) found the primary habitat requirement for nutria was availability of grasslands used for cattle production. These lands, however, were grazed because they were too wet for timber or crop production, and probably would have been used by nutria in the absence of cattle. Guichón and Cassini (2000) also found that nutria distribution was negatively related to indicators of human presence including roads, docks, and houses.

Nutria subspecies may have different tolerance levels for salinity and temperature. For example, *Myocastor coypus bonariensis* is native to northern Argentina and was introduced to the British Isles, where frost lesions and winter mortality were common (Gosling and Skinner 1984). The subspecies native to Patagonia, *M. c. santacruzae,* is apparently more tolerant of cold and icy conditions. Larrison (1943) suggested that nutria in Chile and Tierra del Fuego were found in estuaries of glacier-fed streams, with nutria colonies often observed swimming near ice blocks and glaciers. South American populations occur primarily in freshwater habitats, although two subspecies (*M. c. melanops* of southern Chile and *M. c. bonariensis*) are tolerant of saline water (Gosling and Skinner 1984; Nowak 1991).

Burrows and Nests. Nutria construct burrow systems in river banks and impoundment dikes. Burrows may consist of a simple tunnel or a complex system of passages and chambers extending up to 15 m from the bank entrance (Schitoskey et al. 1972; Nowak 1991). Chambers within burrow systems are usually spherical and may contain crude nests with large amounts of vegetation (Gosling and Skinner 1984). Warkentin (1968) described simple burrows with one or two openings, or complex systems with three to five entrances above water and two below water. Complex systems were occupied by family groups, and developed from simple burrows initially constructed by pairs. In Europe, burrows occurred among roots of ash (*Fraxinus excelsior*) and poplar (*Populus* spp.) trees lining canal banks (Doncaster and Micol 1989; Doncaster et al. 1990).

Nutria increase their use of burrows during severe winter weather. Atwood (1950) found that nutria constructed burrows when mean temperatures were <4°C. Swank and Petrides (1954) found that nutria in Texas seldom used burrows, but more often used beds beneath dense overhanging or collapsed vegetation for resting, feeding, and giving birth. Dozier (1985) reported that nutria in Louisiana burrowed and constructed dens, but did so less frequently than muskrats. Conversely, Larrison (1943) stated that nutria burrowing activity exceeded that of muskrats in Oregon, where temperatures are cooler. In a survey of beaver lodge occupancy at Caddo Lake, Texas, King et al. (1998) observed 57 nutria at 18 beaver lodges over a 2-year period. During severe winter weather, nutria are opportunistic in seeking nontraditional sources of cover, such as straw stacks, farm buildings, rabbit burrows, and uplands (Norris 1967a).

ACTIVITY

Home Range Estimates. Nutria home range estimates vary according to season, sex, and habitat. Home range estimates tend to be smaller for females and juveniles, populations at high densities, and nutria in habitats with adequate resources (Doncaster and Micol 1989; Gosling and Baker 1989a, 1991). Gosling and Baker (1989a, 1991) found that nutria in marginal habitats such as edges of drainage ditches had larger home ranges (46.3 and 93.9 ha for females and males, respectively) than nutria in preferred habitats such as marsh and fen (4 and 7 ha for females and males, respectively). Doncaster and Micol (1989) found a mean home range of 5.7 ha for males and 2.5 ha for females in France. Males used longer canal lengths (1913 m) than females (670 m). Lohmeier (1981) reported the mean home range for four nutria in Mississippi was 2.31 ha from June to October and was almost exclusively open water. The mean minimum convex polygon home range estimate for Maryland nutria ($n = 40$) was 0.098 km^2 (Ras 1999). Coreil et al. (1988) recorded the two largest home range estimates during winter ($\bar{x} = 138$ ha), whereas the two smallest home range estimates occurred during summer ($\bar{x} = 7$ ha). Two females tracked by Doncaster and Micol (1989) in consecutive springs had larger home range estimates when population density was lower, but maintained the same center of use. Adjacent home ranges generally overlap, with only the nest site being defended (Ryszkowki 1966; Lohmeier 1981; Coreil et al. 1988).

Activity Patterns. Nutria are active throughout the 24-hr day, but activity is generally greatest from dusk to dawn. Chabreck (1962) found that the number of trail crossings was highest during sunrise and nighttime. Activity was higher 1 hr before than 1 hr after sunrise. The number of crossings was also higher for the postsunset than the presunset hour. Similarly, Palomares et al. (1994) reported that winter activity peaks occurred about 1 hr after sunset and 2 hrs before sunrise. Ras (1999) found that the net linear distance between locations of radio-collared nutria in Maryland was greater for nocturnal periods than diurnal periods.

Observations by Norris (1967b) also showed that activity was greatest during crepuscular periods, especially for juveniles. Gosling (1979) found that the interval length between peak activity and sunrise was inversely related to temperature, with huddling being common during cold predawn hours.

Time budget data collected by Doncaster et al. (1990) showed that swimming bouts were longer and less frequent during spring and shorter and more frequent during summer. However, total time spent swimming did not differ between spring and summer. Palomares et al. (1994) found that of active animals, 79% were active at night and 23% during the day.

Nocturnal activity may be greatly reduced during severe winter weather. In an enclosure with nutria, Gosling (1981b) often noted a complete absence of tracks in snow that had fallen the previous day. Baker and Clarke (1988) also noted that nutria tracks were frequently absent after heavy snows, indicating that nutria were sedentary during both day and night. Gosling et al. (1980a) believed that diurnal activity increased during cold weather, based on increased harvest by shooting during these periods. The number of animals visible in Louisiana ponds increased throughout the day from sunrise to sunset from January to July. Animals generally spent daytime hours sunning or sleeping and nighttime hours eating (Warkentin 1968). Moinard et al. (1992) suggested that males spent a greater proportion of time actively defending territories and swimming than females.

Nutria generally have discrete centers of activity with the exception of occasional long-distance movements by adult males. For example, Warkentin (1968) reported that most marked nutria were observed within 300 m of their initial capture location. The farthest distance a nutria moved was 600 m. Kays (1956) recaptured 87 animals, of which only 8 had moved >1200 m from the initial capture site. In a 6-month period, recapture distances from the release site were less than 100 m for 50% of males and less than 50 m for 50% of females. The mean distance between mark and recapture locations of Maryland nutria was 0.15 km and the range of distances was 0.03–1.5 km (Ras 1999).

Movements and activity are often centered on ditch, canal, or river banks. Doncaster et al. (1990) found that nocturnal telemetry locations were almost exclusively within 10 m of a canal, regardless of whether the animal was active (85%) or inactive (89%). Nutria were more likely to be found close to canals during dry conditions. Use of canals also seemed to be related to social status, with locations of dominant males in the canal more frequent (56%) than for subdominant males (6%).

Doncaster and Micol (1989) calculated a mean movement speed of 107 m/hr, although males (121 m/hr) tended to move faster than females (96 m/hr). During extremely cold weather, the limited activity that does occur is usually centered around burrows. Doncaster and Micol (1990) observed that grazing on vegetation during winter was heavier near burrow entrances than in adjacent areas. Similarly, movements during warm weather are more likely to be associated with presence of cover. Warkentin (1968) observed that diurnal movements often were in areas with shade or high duckweed (*Lemna minor, Wolfia* sp., and *Wolfiella floridana*) cover.

Intensive trapping activity does not appear to increase movement distances by nutria. Linscombe et al. (1981) marked animals from June through October on a 324-ha study area in Louisiana. Of 310 recaptures during the commercial trapping season, 279 were within the study area. Most of the nutria that moved were within 400 m of the study area, with the greatest distance moved 3.2 km. Similarly, Simpson and Swank (1979) trapped continuously from April to September, but four radio-collared animals did not leave the study area while trapping was ongoing.

Dispersal and Immigration. Although the implications of understanding dispersal for control efforts are recognized (Norris 1967b; Gosling et al. 1988a), little information is available about dispersal ecology other than documentation of long-distance movements. Doncaster and Micol (1989) defined transients as marked animals for which no fate is documented; these accounted for 63% of the population in France. However, dispersal and trap avoidance cannot be separated using this criterion, and trap avoidance is a problem in describing population dynamics for nutria using mark–recapture data (Simpson and Swank 1979).

Long-distance movements have been documented for both sexes, although they are more common among males (Norris 1967b). Gosling and Baker (1991) found that 25% of males and 9% of females dispersed >4 km. The longest distances were 9 km for males and 7 km for females. Kays (1956) documented one animal that moved 29.6 km in 187 days and another that moved 24.1 km in 67 days. Animals that moved outside the study area were all males weighing 3.2–5.0 kg, and had moved 0.6–9.0 km from their initial capture site (Norris 1967b). Doncaster and Micol (1989) found that the range of long-distance movements was 0.6–3.5 km, with the majority occurring during March and April when male survival was lowest. Wolfe and Bradshaw (1986) moved 10 adult nutria 2.7 or 3.3 km from their home pond; 5 of these animals (4 females, 1 male) returned to the home pond within 15 days, 4 (1 female, 3 males) remained at the release site, and 1 (male) moved 1.8 km to a new area.

Seasonal reproductive activity and weather patterns may also influence long-distance movements. Norris (1967b) observed slight increases in the number of immigrant females just before peak littering periods, suggesting that females may move before giving birth. They concluded that when severe weather leads to reproductive synchrony, timing of the first estrus may coincide with littering because time to sexual maturity approximates the gestation period. Thus, parturition periods also may reflect peak periods of movement.

Reggiani et al. (1995) found that immigration was highest from July to November, particularly after a severe winter when population density was depressed. Habitat heterogeneity also may determine the level of immigration that occurs, especially where optimal and suboptimal habitats are interspersed. Aliev (1965a) believed that drought was a driving factor behind a 60-km nutria range expansion along a network of lakes in Armenia. Similarly, Kuhn and Peloquin (1974) believed periodic flooding during winter facilitated the rapid spread of Oregon's nutria population. Long-distance movements of males may occur in the absence of deteriorating habitat conditions, particularly along rivers, streams, and channels. Ryszkowski (1966) believed that dispersal was a mechanism to alleviate high-density populations and their pressure on food resources. This influence may occur in one of two ways: (1) individuals disperse in response to aggressive behavior before habitat damage occurs, or (2) high population densities are tolerated, but habitat deterioration eventually forces nutria to emigrate to new areas.

FEEDING HABITS

Nutria are almost exclusively herbivorous, relying extensively on herbaceous marsh plants as their primary source of forage (Ashbrook 1948; Atwood 1950; Shirley et al. 1981). Occasional feeding on freshwater mussels has been noted (Gosling 1974), but animal matter is not usually a component of the diet. Use of individual plant species and plant parts varies seasonally and is somewhat predictable based on phenology. For example, nutria feed on belowground plant material such as roots and rhizomes during winter, whereas they feed on aboveground vegetation during summer.

Feeding Platforms. Nutria often consume foods at fixed feeding stations, which may consist of a stationary object, a river bank, or a raised platform constructed of discarded plant material (Atwood 1950; Warkentin 1968). The typical platform is about 51–76 cm across and extends 15–23 cm above the water surface (Atwood 1950). Nutria often carry and eat larger plant material such as cattail at feeding stations, whereas finer species, including submerged aquatics, are consumed on encounter (Warkentin 1968). Nutria frequently cut cattail stalks, consume the basal and belowground portions, and discard the tops or use them to construct a platform (Swank and Petrides 1954). Atwood (1950) observed that nutria used feeding platforms more often during cold weather, and suggested the platforms allow the animals to have free use of their front feet for manipulating food. Warkentin (1968) did not observe nutria using or constructing nests or platforms, but did observe them sunning and feeding on floating objects or roots of willow trees. Platforms also may be used by groups of nutria that huddle together during winter to conserve body heat (Atwood 1950).

Foraging Activity. The front feet are used extensively in foraging to uproot tubers, manipulate stems, reach high plants such as seed heads of grain crops, and hold food items during consumption (Hailman 1961; Gosling and Baker 1991). The dominant foraging activity during winter, particularly in the northern portion of the range, is excavation of belowground tubers and rhizomes of marsh plants. Nutria break tall upright stems, clip plants about 2.5–7.6 cm above the water surface, and turn the plant vertically to consume the succulent basal portions (Milne and Quay 1966). Grazing on submerged aquatic plants and the newly emerging stems of marsh plants also is common. Nutria float on the water surface and either reach down with their front feet or submerge their head to consume submerged aquatic plants (Milne and Quay 1966). In the Argentinean Pampas, D'Adamo et al. (2000) found that 87.2% of nutria foraging activity occurred within 10 m of water.

Diet and Food Selection. Diet composition is a thoroughly studied component of nutria ecology. There is substantial variation in diet among seasons, habitats, and regions, with the dominant herbaceous species generally being common in nutria diets (Table 52.4). The ability of nutria to exploit a variety of wetland plants for food has enabled them to survive in many areas where they have been introduced (Abbas 1991). The trophic plasticity of nutria is demonstrated by the temporal and spatial balance among three feeding tactics: grazing, feeding on aquatic plants, and feeding on roots (Abbas 1991).

Nutria stomach contents are less diverse than available plants in the habitats in which they forage. For example, three to four plant species accounted for 50% of nutria diets in Argentina, despite the presence of 53 plant taxa in the study area (Borgnia et al. 2000). Wilsey et al. (1991) found that six plant taxa accounted for 88% of nutria diets in Louisiana freshwater wetlands, including two duckweeds (*Lemna minor* and *Spirodela polyrrhiza*), Pontederiacea, *Alternanthera philoxeroides,* Poaceae, and Cyperaceae. Other aquatic plants, such as water hyacinths, may be significant components of the diet, particularly in the southern portions of the North American range where seasonal availability is higher (Brown 1975). Shirley et al. (1981) found 17 plant taxa in stomachs of nutria from freshwater marshes in Louisiana, with 6 taxa dominating: *Eleocharis palustris, Hydrocotyle* spp., *Alternanthera philoxeroides, Bidens laevis, Sacciolepis striata,* and *Sagittaria falcata.* A dominant plant on the study area, *E. palustris,* was the most important food, occurring in 98% of stomachs and accounting for 29% of stomach contents on an annual basis.

Aboveground and belowground portions of plants accounted for 87.5% and 12.5% of the diet, respectively, in Louisiana brackish marsh (Chabreck et al. 1981; Chabreck 1992). Studies in Great Britain (Gosling 1974; Gosling and Baker 1991), South America (Murua et al. 1981), and Maryland (Willner et al. 1979) have documented similar patterns. Roots dominated nutria diets at Blackwater NWR, except in April, when 90% of the diet was stems (Willner et al. 1979). Leaves were eaten from May to October, but were always <20% of the diet. *Scirpus olneyi* constituted 80% of the annual diet (range 46% in August to 97% in February). Conversely, aboveground portions of *Scirpus olneyi* were consumed mainly in summer by Louisiana nutria, with little use during fall and winter (Chabreck et al. 1981). Other significant plants consumed were *Phragmites communis* and panic grasses (*Panicum* spp.); *Phragmites* was most abundant in May and October diets, and panic grass was most abundant in November and April diets (Willner et al. 1979). Louisiana nutria consume *Eleocharis palustris* to a much greater degree than do Maryland nutria, reflecting the greater availability of this species in Louisiana freshwater marshes (Willner et al. 1979; Shirley et al. 1981).

Nutria diets may be augmented by fruits, such as seed pods of yellow water lily (*Nuphar lutea*) during autumn (Gosling and Baker 1991), but fruits generally represent a low portion of the diet. In Argentina, Borgnia et al. (2000) found that hydrophilic monocots (40–60%) were the main foods consumed on an annual basis, followed by terrestrial monocots (30–35%), hydrophilic dicots (5%), and terrestrial dicots (0–2%). High invertebrate association with floating aquatics may lead to incidental consumption, which could contribute to the protein and fat content of the diet (Shirley et al. 1981; Wilsey et al. 1991). Algae are consumed (Willner et al. 1979), but few studies have quantified their use. Agricultural crops, such as corn, sugar beets, and cabbage, also may be substantial components of the diet (Swank and Petrides 1954; Kuhn and Peloquin 1974; Willner et al. 1979).

Woody plants are common in the diet, but they generally are consumed at low rates relative to herbaceous species. High tide bush (*Iva frutescens*) was the only woody species documented in Maryland nutria diets (Willner et al. 1979). Similarly, Wilsey et al. (1991) found that woody tissue constituted <1% of the diet in Louisiana forested wetlands. Warkentin (1968) documented nutria feeding on black willow (*Salix nigra*) bark and leaves of red maple (*Acer rubrum*). Nutria may feed extensively on bark and seedlings of baldcypress (*Taxodium distichum*) (Conner and Toliver 1987). During a prolonged flood in France, bark of poplar saplings and fallen branches was stripped at water level (Doncaster and Micol 1990).

Nutria have been described as opportunistic foragers, based on the seasonal variation in selection of plant species and plant parts (Gosling 1974; Willner 1982). Wilsey et al. (1991) conducted a preference analysis of nutria foods in Louisiana freshwater forested wetlands and concluded that duckweeds were preferred during spring and summer, but were consumed in proportion to availability during autumn and winter. Other preferred foods included grasses during spring and sedges during summer and autumn. In Chile, Murua et al. (1981) found that aquatic plants dominated the diet where they were available, potentially offsetting the level of feeding on roots of marsh plants. On ponds dominated by cattail (*Typha latifolia*) in Texas, nutria used few other species as long as cattail remained available (Swank and Petrides 1954). Giant cutgrass (*Zizaniopsis milacea*) was also a favored food, but water shield (*Brasenia schreberi*), arrow arum (*Peltandra virginica*), American lotus (*Nelumbo lutea*), and smartweed (*Polygonum* sp.) were not used extensively. Roots and shoots of pickerel weed (*Pontederia cordata*), white water lily (*Nymphaea elegans*), and *Panicum* spp. were used. In Maryland, Willner et al. (1979) believed three-square rush was preferred, cattail was consumed in proportion to availability, and saltgrass (*Distichlis spicata*) was avoided, although they did not conduct a formal preference analysis.

Some plant species seem to be actively avoided by nutria. Llewellyn and Shaffer (1993) concluded that *Justicia lanceolata, Polygonum punctatum,* and *Typha domingensis* were avoided despite their abundance in areas of intensive foraging. Similarly, Wilsey et al. (1991) concluded that Pontederiaceae were avoided during spring, as were *Polygonum* sp., *Bidens laevis, Hygrophila lacustris,* and *Saururus cernuus* throughout the year. Chabreck et al. (1981) found that *Juncus roemerianus* was almost completely avoided in Louisiana brackish marsh, and two other dominant species, *Spartina patens* and *Distichlis spicata,* often were not eaten. In Louisiana, *Alternanthera philoxeroides* was a highly preferred food in freshwater marshes (Shirley et al. 1981), but was avoided in freshwater forested wetlands (Wilsey et al. 1991). Ryszkowski (1966) referred to plants that are poisonous to nutria (but did not name them), and stated that species richness of poisonous plants increased from one to six as intensity of foraging on favored species increased.

Little information is available regarding nutritional requirements of nutria, but feeding patterns are strongly linked to seasonal variation in food quality (Gosling and Baker 1991). Proximate analysis revealed protein content was about twice as high in freshwater as in brackish diets, which may be a cause of lower population densities in brackish marshes (Wilsey and Chabreck 1991). Feeding on leaves of common reed is highest from midsummer to autumn, and is correlated with carbohydrate content (Gosling and Baker 1991). Coprophagy was documented by Gosling (1979), who observed that plant material eaten during the night passed through the gut, was reingested the next day, and was eliminated the following night. Total daily food consumption

TABLE 52.4. Herbaceous wetland plants consumed by nutria (*Myocastor coypus*) in North America, by habitat and location

Latin Name	Common Name	Location[a]	Habitat[b]	Source[c]
Alternanthera philoxeroides	Alligator weed	AL, LA	BM, FM, FO	10, 11, 12, 13, 15
Bacopa monnieri	Water hyssop	LA, NC	BM, FM	1, 5, 9
Bidens cernua	Bur marigold	OR	FM	7
Bidens laevis	Smooth beggartick	LA	FM	10, 13, 15
Brasenia schreberi	Water shield	LA	FM	1
Castalia odorata	White waterlily	LA	FM	1
Cladium jamaicense	Saw grass	LA	FM	1
Cyperacea	Sedges	LA	FO	14
Distichlis spicata	Spike grass	LA, NC	BM	5, 13
Eichornia crassipes	Water-hyacinth	LA	FM, FO	1, 13
Eleocharis equisetoides	Jointed spikerush	LA	FM	1
Eleocharis palustris	Common spikerush	LA, MD	BM, FM	8, 9, 10, 13, 15
Eleocharis quadrangulata	Squarestem spikerush	LA	FM	1
Hydrocotyle spp.	Water pennywort	LA	FM	10, 13, 15
Juncus roemerianus	Black needlerush	NC	FM	5
Lemna minor	Duckweed	LA	FO, FP	6, 13, 14
Nymphaea advena	Spatterdock	LA	FM	1
Nymphaea odorata	Pond lily	NC	FM	5
Nymphaea elegans	White waterlily	TX	FP	2
Panicum hemitomon	Maidencane	LA	FM	1, 11
Panicum spp.	Panic grasses	MD, TX	BM, FO, FP	2, 8, 13
Phragmites communis	Common reed	AL, LA, MD	BM	4, 8, 12
Poaceae	Grasses	LA	FO	13
Pontederia cordata	Pickerelweed	TX	FP	2
Pontederia lanceolata	Pickerelweed	LA	FO, FM	1, 11, 13
Rhynchospora sp.	Beaked rush	LA	FM	1
Sacciolepis striata	American cupscale	LA	FM	10, 13, 15
Sagittaria falcata	Bull's tongue	LA, NC	BM, FM	1, 5, 10, 11, 15
Sagittaria graminea	Arrowhead	LA	FM	1, 11
Sagittaria latifolia	Arrowhead	LA, NC	FM	5, 16
Sagittaria lancifolia	Arrowhead	LA	FM	13
Sagittaria platyphylla	Arrowhead	LA	FM	16
Scirpus americanus	Three-square rush	AL, NC	BM, FM	3, 5, 12
Scirpus californicus	Southern bulrush	LA	BM	1
Scirpus olneyi	Three-square rush	LA, MD, NC	BM	3, 4, 8, 9, 13
Scirpus robustus	Salt marsh bulrush	LA	BM	9, 13
Scirpus validus	Soft-stemmed bulrush	NC	FM	5
Solidago sempervirens	Seaside goldenrod	LA	BM	9, 13
Sparganium simplex	Bur-reed	OR	FM	7
Spartina alterniflora	Saltmarsh cordgrass	NC	BM	5
Spartina cynosuroides	Big cordgrass	LA, MD	BM	4, 8, 13
Spartina patens	Saltmeadow cordgrass	LA, NC	BM	4, 5, 13
Spirodela polyrrhiza	Duckweed	LA	FO	13, 14
Typha spp.	Cattail	LA, MD, NC, TX	BM, FM, FP	1, 2, 5, 8
Wolfia spp.	Duckweed	LA	FP	6
Wolfiella floridana	Duckweed	LA	FP	6
Zizaniopsis mileacea	Giant cut-grass	AL, LA, TX	FM, BM, FP	1, 2, 12

[a]AL, Alabama; LA, Louisiana; MD, Maryland; NC, North Carolina; OR, Oregon; TX, Texas.
[b]BM, Brackish marsh; FM, freshwater marsh; FO, forested wetland; FP, freshwater pond.
[c]1. Atwood 1950; 2. Swank and Petrides 1954; 3. Hailman 1961; 4. Harris and Webert 1962; 5. Milne and Quay 1966; 6. Warkentin 1968; 7. Wentz 1971; 8. Willner et al. 1979; 9. Chabreck et al. 1981; 10. Shirley et al. 1981; 11. Dozier 1985; 12. Lueth 1985; 13. Wilsey and Chabreck 1991; 14. Wilsey et al. 1991; 15. Chabreck 1992; 16. Llewellyn and Shaffer 1993.

was estimated as about 25% of body weight (Gosling 1974), although consumption by lactating females was 63% higher than that by nonlactating females (Gosling et al. 1984). D'Adamo et al. (2000) found that nutria in Argentina spent most of their active time foraging (80.5%).

POPULATION ECOLOGY

Density Estimates. A wide range of values for population density has been reported (Table 52.5). Density usually is highest in autumn and winter and lowest in spring, after trapping season and winter mortality have occurred. For example, Doncaster and Micol (1989) reported density of 4.27/ha in winter in France and a density of 0.9/ha in spring. Peak densities in winter occurred when smaller canals either froze or accumulated leaves and detritus, concentrating nutria on larger canals. Similarly, density was 3.65/ha in autumn and 0.72/ha during summer in Italy (Reggiani et al. 1995). A portion of the Oregon population achieved its highest density (138/ha) during summer and its lowest density during winter (0.6/ha) (Wentz 1971). However, this was a consequence of a flood cycle that concentrated animals during the summer drought and enabled dispersal during winter. Other than this population, the highest density reported in North America was 24/ha for the Blackwater NWR population in Maryland (Willner et al. 1979) and 25/ha for a polluted cattle sewage lagoon in Florida (Brown 1975). An unpolluted Florida pond had a density of about 6/ha, suggesting that nutrient enrichment may be a significant predictor of high-density nutria populations (Brown 1975).

Sex and Age Ratios. In most populations of nutria, there is a slight preponderance of males, although several studies have shown that sex ratios become skewed toward females in response to intensive trapping

TABLE 52.5. Nutria population density estimates by location, season, and habitat, in approximate latitudinal order

Density (nutria/ha)	Location	Season	Habitat	Source
138	Oregon	Summer (low water)	Freshwater marsh	Wentz 1971
43.7	Louisiana	Fall	Freshwater marsh	Kinler et al. 1987
24.7	Florida	Annual estimate	Cattle sewage lagoon	Brown 1975
24	Mississippi	Summer–fall	Freshwater pond	Lohmeier 1981
24	Louisiana	Annual estimate	Brackish marsh	Linscombe et al. 1981
2.7–21.4	Maryland	Annual estimate	Brackish marsh	Willner et al. 1979
0.5–16.0	Maryland	Annual estimate	Brackish marsh	Willner et al. 1979
5.9	Florida	Annual estimate	Unpolluted pond	Brown 1975
0.1–1.29	Louisiana	Annual estimate	Brackish marsh	Valentine et al. 1972
0.6	Oregon	Winter (high water)	Freshwater marsh	Wentz 1971
6	France	Winter	Freshwater canals	Micol and Doncaster 1987
4.28	France	Fall	Freshwater canals	Doncaster and Micol 1990
4.27	France	Fall	Freshwater canals	Doncaster and Micol 1989
3.65	Italy	Fall	Freshwater marsh	Reggiani et al. 1995
1	France	Early winter	Freshwater canals	Micol and Doncaster 1987
0.91	France	Spring	Freshwater canals	Doncaster and Micol 1989
0.72	Italy	Summer	Freshwater marsh	Reggiani et al. 1995
0.7	France	Spring	Freshwater canals	Micol and Doncaster 1987
0.54	France	Spring	Freshwater canals	Doncaster and Micol 1990

(Table 52.6). This effect is due partly to the higher susceptibility of males to trapping, which is a consequence of the larger home range and movement rates of males (Gosling and Baker 1989a, 1991). Consequently, sex ratios obtained from harvest data typically show a higher proportion of males than exists in the population as a whole. For example, Linscombe et al. (1981) found that a sex ratio of 1.16M:1F dropped to its lowest point (0.75M:1F) after 3 years of trapping. Norris (1967b) reported that sex ratio of trapped nutria varied greatly, usually favoring males during short trapping periods. However, when the population was evaluated over a longer time frame ($n > 8000$ animals), a balanced sex ratio was evident. Adams (1956) also reported that the sex ratio was influenced by the method of capture. Livecapture methods generated a ratio of 1M:1F, whereas commercial trapping resulted in a ratio of 1.3M:1F. Willner et al. (1979) found that the fetal sex ratio of Maryland nutria favored females (0.8M:1F), but the adult population favored males (1.1M:1F). The discrepancy between these studies may reflect a longer history and greater intensity of trapping in the Louisiana population (Gosling and Baker 1989a). Fetal and embryonic

TABLE 52.6. Ratio of males to females in utero, at birth, and in the general population of nutria

Time of Life Cycle	Male-to-Female Ratio	Reference
In utero	1.0:1.0	Adams 1956
	0.8:1.0	Willner et al. 1979
	1.1:1.0	Gosling 1986
At birth	1.0:1.0	Skrzydlewski 1966
	1.1:1.0	Doncaster and Micol 1990[a]
	1.2:1.0	Gosling and Baker 1991
General population	1.0:1.0	Adams 1956[b]
	1.3:1.0	Adams 1956[c]
	1.0:1.0	Norris 1967b
	1.2:1.0	Brown 1975
	1.2:1.0	Willner et al. 1979
	1.2:1.0	Linscombe et al. 1981[d]
	0.8:1.0	Linscombe et al. 1981[e]
	1.0:1.0	Doncaster and Micol 1990[f]
	0.3:1.0	Doncaster and Micol 1990[g]
	0.8:1.0	Gosling and Baker 1991[b]

[a]Estimate based on farm-raised nutria.
[b]Estimate based on livetrapping.
[c]Estimate based on commercial trapping.
[d]Estimate made before intense trapping.
[e]Estimate made after 3 years of trapping.
[f]Estimate made before catastrophic cold and flooding.
[g]Estimate made after catastrophic cold and flooding.

sex ratios near unity have been reported by Adams (1956) in Louisiana and Skrzydlewski (1966) and Gosling (1986) in Europe. Newson (1969) determined that males outnumber females at birth, with sex ratio rapidly approaching unity due to higher postpartum survival of females. Doncaster and Micol (1990) reported that adult sex ratio was biased toward females in spring (0.3M:1F) and in winter following a prolonged flood (0.43M:1F). Sex ratio was biased toward males in November and December each year (2M:1F or 3M:1F). However, among juveniles, sex ratio was biased toward males in summer and females in autumn (Doncaster and Micol 1989). In a livetrapped population in Great Britain, Gosling et al. (1981) found that the sex ratio was skewed toward females throughout the year, and Gosling and Baker (1991) reported a sex ratio of 1.2M:1F at birth and 0.77M:1F for adults. Among the trapped animals, 6-month-olds tended to be dominated by males, but by 15 months of age, females were caught more frequently due to depletion of males by trapping (Gosling and Baker 1989a).

Doncaster and Micol (1990) believed that the polygynous mating system of nutria created a male bias in surviving young, male bias in mortality, and an overall female bias in the population. This theory is supported by Gosling et al. (1984), who found no indication of a difference in birth weight between males and females, although males tended to be heavier among the young that survived to start nursing. Gosling (1983) also studied the adaptive control of offspring sex ratios by female nutria.

Age ratios are similarly affected by intensity of harvest, as well as seasonal reproductive activity and winter severity. Doncaster and Micol (1990) found that few juveniles were present in February after a severe winter (4% of the population <4 months old), whereas there was a high proportion of juveniles entering the winter following a synchronizing event (35% of the population <4 months old). Similarly, Doncaster and Micol (1989) reported juveniles were 44% of the June captures after a mild winter, but after a severe winter, juveniles were absent in May and 26% of the population in July.

Survival and Population Dynamics. Severity of winter weather and intensity of trapping were the two major factors affecting dynamics of the nutria population in the United Kingdom. Low recruitment and reduced population size typically are observed during the spring after a severe winter, particularly when trapping activity also is high (Gosling et al. 1981). Gosling (1981a) and Gosling et al. (1983) described an index of winter severity that incorporates the cumulative effects of extended cold weather on nutria as

$$CRS = \sum_{i=1}^{n} x_i^2$$

where x is the length of a run of freezing days (defined as daily minimum temperature <0°C and a daily maximum temperature <5°C), n is the number of runs each winter, and *CRS* stands for "cumulative runs squared." Negative correlations have been shown between this index and littering rate of females during the subsequent spring, as well as inguinal fat mass of adult males and females. During the eradication campaign in Great Britain, recruitment was negatively related to trapping intensity and *CRS* (Gosling and Baker 1987). Reggiani et al. (1995) reported that nutria population density remained constant between November and March when winter weather was mild, but decreased by 44–64% after severe winters.

Few efforts have been made to quantify seasonal or annual survival of nutria. Reggiani et al. (1995) calculated Jolly–Seber survival rates for seven trapping sessions. The highest survival rates occurred from July to November (0.87–1.00) and the lowest survival occurred from November to May (0.64–0.74). Doncaster and Micol (1989) calculated Jolly–Seber survival rates for two summer and one autumn seasons. Survival for males (0.78–0.99) was substantially higher than for females (0.54–0.68) during summer, but female survival (0.97) was higher during autumn than male survival (0.67). This population decreased by 71% during a cold winter and 69% during a mild winter with prolonged flooding, with population peaks occurring in autumn (Doncaster and Micol 1990).

Gosling (1974) believed the principal factors contributing to increasing nutria populations in the early 1970s were improved reproductive success and juvenile survival, both of which were closely related to climatic variation and effects of climate on food availability. Juvenile survival may be higher during mild winters because young can more easily excavate roots and rhizomes. Juveniles are weaned at 2–3 weeks, and thus ability to forage is important to their survival.

Newly introduced populations (Gosling 1974) show a pattern of stable to slowly increasing numbers for several years, followed by rapid population expansion. This same pattern occurred in the British population when intensive trapping was stopped and winter weather became milder. However, range expansion did not occur concurrently with dramatic increases in abundance during the early 1970s. The species occupied the same range during the intital stages of population growth, and expanded geographically in later years.

Survival to 3 years of age in a fenced population studied by Gosling and Baker (1991) was 8.2% for males and 14.7% for females, whereas in an intensively trapped wild population survival to 3 years was 0.6% for males and 1.7% for females. They also reported that survival to adulthood of young born before an extremely cold winter may be as low as 5%, whereas nearly 90% of young born in April may survive to adulthood.

Population Modeling. Several simulation models have been developed to describe nutria population dynamics. The first was by Gosling et al. (1983), who generated a simulation model for nutria in Great Britain using empirically derived estimates of fecundity, recruitment, and mortality. The model was based on necropsies of >20,000 animals killed in a control trapping operation during a 12-year period, and was designed to illustrate the relative influence of winter weather and trapping effort on nutria populations. Demographic and reproductive parameters were estimated as follows: (1) monthly littering frequencies were estimated based on projections of parturition month from known litter age and the proportion of females lactating; (2) litter size for month of parturition was projected forward, and the number of female embryos/litter was pooled by month of parturition to obtain averages for cold and mild winters (based on *CRS*); (3) recruitment curves were based on Gosling et al. (1981), and derived from monthly estimates of age ratio, birth rate, and survival to adulthood; (4) adult mortality was number of adults killed each month by trappers; and (5) population sizes were estimated based on retrospective census (Gosling et al. 1981). The model showed that littering frequency and litter size were influenced by severe winter weather, but when expressed as proportion of surviving animals that recruit, recruitment was not influenced by winter severity. However, when expressed as proportion of the number born, recruitment rate was up to two times higher for young born during mild winters than for young born during severe winters (Gosling et al. 1983). Although the simulation model could not provide an unbiased test of variation in mortality with winter severity, Gosling et al. (1983) reported no independent indication that such variation existed. The simulation model was used to predict population responses to various levels of trapping under assumed climatic conditions. This was used as a decision tool in implementing nutria control in Great Britain.

Reeves and Usher (1989) also used modeling to describe the spread of the nutria population in Great Britain. Their model required speculation about various parameters, such as carrying capacity, diffusivity, population growth rate, diffusion threshold, and threshold for breeding/detection. They concluded that accurate estimates were most important for diffusivity, population growth rate, and diffusion threshold, but that Gosling's (1985) data were adequate to model dynamics. Another limitation was the inability to account for trapping mortality, as the model may not accurately represent situations where management is reducing carrying capacity or population growth rate.

Carter et al. (1999) generated a three-phase model to predict changes in nutria population size and rate of marsh loss in Louisiana based on nutria demographic parameters. The model addressed interactions among nutria demographics, reproduction, food consumption, and marsh loss and predicted a stable population size of 4.6 nutria/ha. Survivorship parameters had to be adjusted significantly upward to make the model achieve the rapid population growth demonstrated in the field, suggesting that field studies have either underestimated true survival or have overestimated true population density. Carter et al. (1999) concluded that the best-known parameters are those from captive populations, and data are needed from free-ranging animals to improve accuracy of models.

BEHAVIOR

Although there is no comprehensive study of nutria behavior in the wild (Kinler et al. 1987; D'Adamo et al. 2000), existing studies point to their adaptability. Nutria have several generalist characteristics including tolerance of high densities of conspecifics, ability to exploit a variety of habitats and food resources, and an ability to manipulate their surroundings to promote survival. For example, Atwood (1950) described four methods nutria use to manipulate habitats: (1) burrowing during cold weather in dikes, levees, or ditchbanks; (2) building platforms of compacted vegetation where animals huddle together; (3) creating surface nests of matted vegetation; and (4) establishing feeding stations on floating objects. Ehrlich (1966) and Gosling (1979) described nutria as curious, gregarious, and easily handled when in captivity.

Nutria are mainly herbivorous (Ashbrook 1948; Atwood 1950; Shirley et al. 1981), but will eat large quantities of freshwater mussels when they are available (Gosling 1974). Abbas (1991) stated the generalist feeding strategy of nutria contributed to its spread in French wetlands. Hailman (1961) and Gosling (1979) described feeding behaviors and forepaw use as stereotyped. Often the feeding behavior of nutria may be wasteful of aquatic vegetation in that only the roots of plants are consumed and the stems discarded (Harris and Webert 1962).

On land, nutria amble in a slow, waddling gait unless they are frightened; they move quickly by bounding when disturbed. Nutria can climb steep banks, some trees, and over wire fences. They are excellent swimmers, using their large, webbed hind feet to move through the water with powerful alternate thrusts. Nutria most often swim with their head and most of their back above water, with their tails floating on the surface. When disturbed, they will swim rapidly underwater for cover, may became completely immobile and hide in submerged aquatic vegetation, and may stay submerged for >1 min.

Courtship and Mating. Courtship includes calls produced by both sexes as they chase each other on land or in the water. Playful behavior also is common before the female goes into heat. Males may sometimes squirt urine or seminal fluid on the female during courtship (Evans 1970). Nutria have a polygynous mating system (Gosling 1977; Doncaster and Micol 1989; Gosling and Baker 1989a, 1991). Females live in kin groups when populations are at moderate or high densities, whereas at low population densities, females may be solitary. Males tend to disperse as young adults, and as adults they compete for exclusive access to groups of females. Males may defend groups of related females at high population densities or defend multiple isolated females at lower population densities (Gosling and Baker 1989a). Consistent with a polygynous mating system, fighting is more common among males than females, males are more likely to disperse from their natal area, and males are more vulnerable to mortality factors (Doncaster and Micol 1990).

Social Organization. Warkentin (1968) suggested that wild nutria live in social groups with an alpha female dominating an alpha male, all other nutria in the group being equally subordinate. The alpha female showed aggressive behavior to the alpha male and all subordinate females, whereas the alpha male exhibited aggressive behavior to all subordinate males and females. When the alpha female was in estrus, she became submissive to the alpha male (Warkentin 1968). Nutria have a social organization with clans of related females (Gosling 1981b); entire colonies may feed or sun together. Ehrlich (1966) suggested that nutria exhibited group territoriality and social dominance, and that colonies comprised both sexes with some free-ranging males.

Anal Glands. Males have a large anal gland for scent marking (Willner 1982). The gland is positioned ventral to the anus and is surrounded by muscle. Gosling (1977) estimated the mean weight of adult male anal glands was 12.2 g. Later, Gosling and Wright (1994) described the development and function of the anal gland and reported that in males >1 year old, the mean gland weight was 10.2 ± 4.2 g. Females also have an anal gland for marking sites within their home range; however, in females, the gland is much smaller, with a mean weight of 4.1 g (Gosling 1977). Gosling and Wright (1994) reported that despite year-round breeding activity, the anal gland of nutria in England showed regular seasonal variation, with marked enlargement in October through December. No relationship was observed between gland size and mate availability. Seasonal changes in anal gland activity are linked to intrasexual competition, and an increase in scent marking may keep the biological costs of resource defense within physiological economic limits (Gosling and Wright 1994).

Adaptability. Nutria do not hibernate and have been introduced in a variety of geographic ranges with markedly different macroclimates. Freezing weather results in frost lesions and large-scale mortality (Ehrlich 1967; Gosling 1981b; Gosling and Baker 1989a; Doncaster and Micol 1990). Many rodent species use huddling behavior to survive low temperatures (Hart 1971; Martin et al. 1980; Vickery and Millar 1984). Nutria are gregarious, and huddling behavior in the wild during cold weather is likely to conserve energy resources and effectively increase the chances of survival (Moinard et al. 1992).

Vocalizations and Communication. Warkentin (1968) reported that alpha animals chased subordinate nutria and vocalized with a "mooing" sound, whereupon the subordinate would flee by diving and swimming under water. Nutria mothers communicate with their offspring in a soft "maaw" contact call, and adults may make loud "maawk" calls (Evans 1970; Gosling and Baker 1991).

Scent marking may be a primary form of communication. Warkentin (1968) observed nutria marking areas by dragging the genitalia (penis extruded) over logs, rocks, or other objects. Males also raised their tails, backed toward the object to be marked, and urinated backward in spurts up to 9 m. Both sexes have a well-developed anal gland, which is extruded and wiped over the ground whenever an animal gets in or out of the water (Gosling and Baker 1991).

Interactions with Other Species. Nutria populations in North America often occupy habitats that are similar to those inhabited by muskrats. State and federal natural resource managers have expressed concerns about competition between native muskrat populations and introduced nutria populations (Bounds 2000). Nutria may compete with muskrats for food resources and damage native habitats that affect a variety of wildlife and fish species (Bounds 2000). However, relatively little quantitative research is available regarding nutria effects on other species. Woods et al. (1992) suggested the primary basis for competition is similarity in food habits in some areas where nutria and muskrat ranges overlap. However, Atwood (1950) and Dozier (1985) concluded that there was little competition between muskrats and nutria in Louisiana due to habitat segregation, with muskrats primarily occupying brackish marshes and nutria favoring freshwater marshes. Dozier (1985) described muskrat diets in Louisiana as consisting primarily of finer marsh grasses such as three-cornered grasses, whereas nutria favor coarser grasses, such as pickerel weed (*Pontederia lanceolata*), bull's tongue, arrowhead (*S. graminea*), and maidencane. Nutria in Louisiana (Dozier 1985) and North Carolina (Milne and Quay 1966) also may rely more on aboveground plant parts than do muskrats, particularly during summer. Indirect evidence suggests that nutria may negatively affect muskrats in some areas. For example, Valentine et al. (1972) reported opposite trends in nutria and muskrat populations on Sabine National Wildlife Refuge, Louisiana. Muskrats were abundant in the mid-1940s and scarce in the early 1960s, whereas nutria were rare in 1946 and abundant in 1961. Similarly, Blackwater NWR has observed a decrease in muskrat populations with an increase in nutria (Bounds and Carowan 2000).

Although there is little empirical evidence for direct interactions between nutria and other species, substantial habitat alterations induced by nutria could affect a variety of marsh wildlife. This potential interaction was demonstrated by Lueth (1985), who introduced two nutria to a 100 ft^2 enclosure on the Mobile Delta, Alabama. After nutria had almost completely removed vegetation, blue-winged teal (*Anas discors*) and shorebirds began using the enclosure. However, waterfowl or shorebird use was not observed in adjacent areas that were not affected by nutria.

MORTALITY

Life Span. Estimates of the life span of nutria are variable and, in some cases, controversial. Atwood (1950) conducted an extensive study of the life history of nutria in coastal Louisiana and estimated that <10% of the animals lived to 2.5 years of age. Evans (1970) suggested that captive nutria may survive for 15–20 years, but estimated that most wild nutria in Louisiana died within approximately 2 years after their birth. Based on Newson (1969), Gosling et al. (1981) reported that <2.0% of nutria in England survived beyond 3 years in populations that were subjected to intensive trapping pressure. In such populations, only 6.9% of deaths were not the result of trapping. Gosling and Baker (1981) concluded that the best available estimate of potential longevity was 6.3 ± 0.4 years. In a nutria population subjected to cage trapping, Gosling and Baker (1981) observed that none of the animals survived >5.0 years. Females had a slightly higher survival rate (1–2%) in captive and wild populations. No animals >6 years old were observed during the course of a 3-year study in Maryland (Willner et al. 1983). Considering the body of published literature, it is likely few nutria in captivity or in the wild live more than 6 years. In the wild, the majority of nutria die during their first 2 years of life, and those surviving beyond growth maturity constitute no more than 15–20% of the total population.

Environmental Factors. In the northern latitudes of the United States (Evans 1970; Willner et al. 1979; Willner 1982; Kinler et al. 1987), England (Newson 1966; Norris 1967a; Gosling 1974), France (Doncaster and Micol 1989, 1990), Poland (Ehrlich 1962), and Russia (Aliev 1965b), the major environmental factors responsible for nutria mortality are temperatures below or near freezing and associated starvation. Gosling et al. (1983) concluded that such climatic conditions were the main environmental constraint on nutria populations in eastern England. Under climatic conditions in Louisiana, where more diverse habitats are found, environmental factors affecting nutria survival were more complex (Kinler et al. 1987). Extreme summer or winter temperatures resulted in mortalities. Moreover, additional factors such as habitat quality, vegetation composition, water level fluctuations due to tides and rainfall, or catastrophic events such as droughts or hurricanes contributed to nutria mortalities independently or in concert with the effects of temperature.

Human Factors. In locations where nutria are actively harvested, such as the United States (Evans 1970; Willner et al. 1979; Kinler et al. 1987) and England (Newson 1966; Gosling 1974), humans are the greatest source of mortality through regulated harvest for fur or by coordinated efforts to control or eradicate nutria. A good example of the impact of humans on nutria populations is offered by Barlow (1969), who reported that even in their native habitats in South America, intense poaching of nutria in some areas has placed them in danger of extirpation, whereas populations flourished in other native areas with lower harvest activity. The influence of nonlethal trap types on differential survival rate was investigated by Chapman et al. (1978). They determined that nutria captured in cage traps experienced 53% mortality during the posttrapping period compared to 74% mortality of animals captured by foot-hold traps. They concluded that, when using nonlethal trapping techniques, the type of trap has a significant impact on later survival rates.

Predation. Predation on nutria varies with geographic location and often is influenced by the population density of nutria. Gosling et al. (1981) observed that adult nutria were too large to be preyed on by most British predators. Young nutria in England were preyed on by red fox (*Vulpes vulpes*), stoat (*Mustela erminea*), gray heron (*Ardea cinerea*), marsh harrier (*Circus aeruginosus*), and owls (Ellis 1965). Predators in the former USSR were documented by Aliev (1965b) to include the golden jackal (*Canis aureus*), feral dog, wolf (*Canis lupus*), jungle cat (*Felis chaus*), harrier, tawny owl (*Strix aluco*), black-billed magpie (*Pica pica*), and hooded crow (*Corvus cornix*). Willner (1982) reported that within the natural range of nutria in South America, caymans (*Caiman latirostris, C. sclerops, C. niger*) were the major predators; however, jaguar (*Panthera onca*), puma (*Puma concolor*), ocelot (*Leopardus pardalis*), and the little spotted cat (*L. tigrinus*) also prey on nutria. In coastal Louisiana, Kinler et al. (1987) reported that the alligator (*Alligator mississippiensis*) was the single most significant nonhuman predator of nutria, with predation occurring in fresh, intermediate, and brackish marshes (McNease and Joanen 1977). Other researchers have reported nutria have been consumed by alligators in the United States (Wolfe et al. 1987; Chabreck 1996). The extent to which alligators use nutria as a food source appears to be proportional to the size of the nutria population. During periods when nutria populations were estimated at 74,000 on the Sabine National Wildlife Refuge in Louisiana, nutria remains were identified in 56% of the stomachs of alligators examined (Valentine et al. 1972). In contrast, when nutria populations were estimated to be <10,000 animals on the refuge, <7.0% of alligator stomachs contained nutria. Additional predators common to the marshes of the Gulf Coast states in the United States include garfish (*Lepisosteus* sp.), cottonmouths (*Agkistrodon piscivorous*), and red shouldered hawks (*Buteo lineatus*) (Warkentin 1968). Evans (1970) reported feral dogs and cats as well as turtles, large snakes, large fish, and some birds of prey may on occasion prey on young nutria or older, sick, or injured individuals. Nutria remains have been found in the nests of bald eagles (*Haliaeetus leucocephalus*) in Maryland (Willner 1982) and Louisiana (Dugoni 1980). Examination of barn owl (*Tyto alba*) pellets in an area of Louisiana with abundant nutria failed to provide evidence that these owls prey on nutria (Jemison and Chabreck 1962).

Diseases. Clinical signs of infectious diseases are rarely observed in nutria with adequate food supplies and healthy teeth. Brown (1975) observed that nutria populations inhabiting the sewage lagoons of dairy farms in Florida were thriving, healthy, and apparently not affected by the microorganisms, parasites, and pollutants common to their habitat. Willner (1982) observed that many of the female nutria in Maryland exhibited a yellow-white vaginal discharge and concluded that the condition was caused by salmonelloses. In contrast, Howerth et al. (1994) reported that *Salmonella* spp. were not isolated from the feces of nutria in Louisiana. In the same group of nutria, 14% of the animals were positive for *Chlamydia psittaci* antibodies. None of the animals in this study were seropositive for *Francisella tularensis*, the causative agent of tularemia.

In south-central Louisiana, Roth et al. (1962) isolated *Leptospira* from 8 of 26 nutria. Antibodies against the *bataviae* serovar of *Leptospira paidjan* were present in 38% of these animals. Evans (1970) reported that leptospirosis, hemorrhagic septicemia, and paratyphoid occur in populations of nutria in Louisiana, and that these diseases may be responsible for spontaneous abortions and reproductive failure. In Louisiana, Howerth et al. (1994) observed antibodies against the *L. interrogans* serovar *canicola* in 7% of the nutria captured by trapping. Although male and female animals of all age classes were collected from East Baton Rouge, Iberville, Tangipahoa, and St. Helena Parishes, the seropositive cases were limited to juvenile females trapped in East Baton Rouge Parish.

During serological screening of nutria from Norfolk and Suffolk in England, Waitkins et al. (1985) determined that 24% of the animals exhibited positive evidence of leptospiral infection, with the majority of the seropositive animals reacting to the *L. interrogans* serovar *icterohaemorrhagiae.* This serovar has been isolated from nutria as long as 1 week postmortem (Wanyangu et al. 1986); similarly infected carcases may be a potential source of infection for susceptible scavengers or humans. Subsequently, Wanyangu et al. (1987) concluded that slide agglutination methods alone were not sufficient for screening and routine diagnosis of leptospirosis and that seropositive cases should be confirmed by organism isolation techniques. Although the clinical consequences of leptospirosis, hemorrhagic septicemia, and paratyphoid infection have been well documented in a number of domestic, feral, and wildlife species, the impact of these diseases on reproduction and health of nutria has not been established.

Serum samples from nine wildlife species were screened for parapoxvirus between 1984 and 1995 in Japan (Inoshima et al. 1999); none of the nutria were seropositive. Howerth et al. (1994) screened nutria in Louisiana for encephalomyocarditis virus antibodies, but did not observe any seropositive animals. Although the incidence of encephalitis in nutria in the United States has not been documented, Evans (1970) hypothesized that nutria populations may be susceptible to the same etiological agents responsible for encephalitis afflicting other North American wildlife species. Considering the endemic distribution of equine encephalomyelitis in the nutria populations of South America (Page et al. 1957), this hypothesis may be supported by future research.

Renal neoplasia was described in 15 of 9400 (0.16%) wild and captive nutria in East Anglia (Keymer et al. 1999). The tumors were bilateral in approximately 50% of the cases, with no significant relationship among occurrence, sex, or age of the animals. However, tumors were observed most frequently in captive nutria. All of the tumors were epithelial and consistent with adenomata or adenocarcinomata neoplasia. The incidence of malignancy was low, with the majority of the tumors classified as benign; no unequivocal evidence of metastasis was established.

In Louisiana, 2 of 32 animals (adult females) were seropositive for *Toxoplasma gondii* (Howerth et al. 1994). Howerth et al. (1994) concluded that nutria meat should be thoroughly cooked for human consumption and that caution should be exercised when feeding uncooked meat to animals susceptible to toxoplasmosis. Holmes et al. (1976) reported that 27% of the nutria collected in the Norfolk and Suffolk regions of Great Britain had antibody titers against *T. gondii.* Howerth et al. (1994) suggested that their failure to detect *Giardia* spp. in nutria in Louisiana may have resulted from the use of feces for examination rather than mucosal scrapings. The infection rate of nutria in the British Isles by coccidia (*Eimeria* spp.) was investigated by Ball and Lewis (1984) and determined to be 6%, based on sporulated oocyst morphology in fecal samples. Additional information on coccidia in nutria is presented by Prasad (1960).

Antibody titers against *Coxiella burnetii* were detected at a relatively low incidence (13%) in the serum of nutria in Japan (Ejercito et al. 1993). Such seropositive animals may be a potential source of the rickettsial infection Q fever in susceptible individuals in domestic animal and human populations (Ejercito et al. 1993).

Parasites. Babero and Lee (1961) studied the helminths of nutria and documented that there were 11 trematode (fluke), 21 cestode (tapeworm), and 31 nematode (roundworm) species commonly infesting populations in Louisiana. In southern Louisiana, *Strongyloides myopotami* was found in 80–90% of the nutria population (Evans 1970). The skin rash occasionally contracted by trappers or individuals handling nutria pelts in the Gulf Coast area known as "nutria itch" is caused by *S. myopotami* and a similar rash known as "swimmers itch" is caused by *Schistosoma mansoni* (Little 1965; Evans 1970). Evans (1970) reported that the nematodes were enzootic in nutria and were ubiquitous in waters of the Gulf Coastal marshes that nutria inhabit. Evans (1970) also noted that *Strongyloides* spp. may adversely affect nutria reproduction and can cause mass mortality in young and old age classes. Willner (1982) reported the cestode *Taenia taeniformes* was identified in the livers of nutria inhabiting marshes on the Eastern Shore of Maryland.

Nagahana et al. (1977) investigated the Chinese liver fluke (*Clonorchis sinensis*) and concluded that nutria were not a host reservoir for this cestode. Nutria were a definitive host for *Fasicola hepatica* in France. Based on screening for antibodies against this trematode (54.9% seropositive), the incidence of flukes in the liver (40.6%), and the number of fecal samples containing eggs (38.6%), Menard et al. (2000) suggested that nutria may play a role in the maintenance and dissemination of *F. hepatica* in the environment. Helminths identified in nutria in Chile included nematodes (*Graphidioides myocastoris, Trichuris myocastoris,* and *Dipetalonema travassosi*), a trematode (*Hippocrepis myocastoris*), and cestodes (*Rodontolepis* spp.) (Babero et al. 1979).

Various lice, ticks, and fleas also infest nutria. Miller (1956) identified biting lice (*Pitrufquenia coypus*) with a predilection for the nasal area of nutria in Louisiana. Similar infestations of *P. coypus* were observed in nutria in England (Newson and Holmes 1968). Season appeared to modulate the rate of infestation, with the highest incidence of infestation occurring in late winter.

Harman et al. (1984) studied infestation by the American dog tick (*Dermacentor variabilis*) on nutria in marsh habitat at Blackwater NWR, Maryland. The overall prevalence of infestation was low (12.3%), with ticks found attached to ears of nutria during spring and summer; peak infestations occurred in May. Intensity of infestation was not correlated with body size or sex and ranged from 1 to 40 ticks/animal; most infested individuals harbored 1–3 ticks. In England, ticks (*Ixodes ricinus, I. arvicolae, I. hexagonus,* and *I. trianguliceps*) were localized primarily on the ears, but also were found on the lips and eyelids of nutria (Newson and Holmes 1968). The flea *Ceratophyllus gallinae* has been identified on nutria in England (George 1964).

Chabreck et al. (1977) reported chronic dermatitis in nutria in Louisiana that was initiated by the achenes (small, dry, one-seeded fruit) of the smooth beggartick plant. Achenes became tangled in nutria fur during mid-to-late November; awns of the achenes then embeded in the skin, and sores and lesions later developed. Skin located on the thoracic region was most commonly affected (Chabrek et al. 1977). Nutria skin is extremely sensitive to foreign objects that are either on it or penetrate it (Kinler et al. 1981). Bacterial and fungal infections develop secondarily and result in loss of the epidermis. Depression and loss of appetite often accompanied the dermatitis. By the traditional trapping season (December–February), dermatitis often was observed in 90% of the nutria inhabiting areas with high densities of beggarticks Pelt value from animals with the condition was significantly reduced (Kinler and Chabreck 1978; Kinler et al. 1981, 1987).

Diseases and Parasites of Captive-Reared Nutria. The majority of literature on infectious diseases of captive nutria comes from countries where nutria are raised commercially. Most of the diseases are similar to those reported for wild nutria and are associated with predisposing factors such as overcrowding, poor sanitation, and suboptimal nutrition.

Nutria in captivity frequently suffer from bacterial pneumonia (Pridham et al. 1966), and this disease is one of the major bacterial diseases in nutria (Martino and Stanchi 1994). Martino and Stanchi (1994) concluded that *Streptococcus zooepidemicus* was the most common etiological agent, although other bacteria also have been isolated. *S. durans* has been identified in ranched nutria (Pridham and Thackeray 1959). Wibawan et al. (1993) investigated the distribution of the polysaccharide and protein antigens of streptococcal isolates from farm-raised nutria. They concluded that the group B streptococci from nutria differed from bovine and human group B streptococci, and neither was involved in cross-infection among animals nor in that between nutria and humans (Wibawan et al. 1993).

Steffin (1955) isolated *Salmonella typhimurium* from captive nutria, who were listless, emaciated, and dying. Internal signs included hemorrhagic mucous membranes, hypertrophy of the spleen and lymph nodes, degeneration of the kidneys and liver, and edema of the intestinal mucosa. When nutrition and sanitation were improved, the incidence and spread of the disease was reduced (Steffin 1955). Other potential pathogenic organisms identified in captive nutria included the protozoan *Cryptosporidium parvum* isolated from nutria on Czechoslovakian fur farms (Pavlasek and Kozakiweicz 1991) and the nematodes *Strongyloides* spp. found in ranch nutria (Pridham and Thackeray 1959) and *Baylisascaris* spp. identified as the etiological agent in an epidemic of cerebral nematodiasis that afflicted nutria <1 year old in a zoological exhibit in the United States (Dade et al. 1977).

Pridham et al. (1966) reported that nutria kept in pens commonly suffered from hepatitis, nephritis, and neoplasms, and concluded that the cause of these diseases was a lack of sanitary conditions. In Hungary, viral hepatitis associated with abortion was diagnosed on fur farms. Morbidity rate was 50% and observed only in pregnant animals. Electron microscopy verified that the nuclear inclusions contained viral particles, the cause of death was viral hepatitis, and the etiological agent was an adenovirus (Dobos-Kovacs and Skulteti 1983).

AGE ESTIMATION

Atwood (1950) used total length of 74.9 cm as a criterion for adult nutria, as did Warkentin (1968). Evans (1970) stated that nutria do not reach full size until the age of 1.5 years and some animals not until they are 2.5 years old.

Nutria become sexually mature before reaching full size (Kinler et al. 1987). Reggiani et al. (1995) found that males and females became sexually mature between 2 and 3.2 kg body weight. No individuals weighing more than 3.2 kg were ever found to be nonreproductive, nor were any individuals weighing <2 kg found in the sexually mature age class. The age of sexual maturity for males and females varies seasonally (Newson 1966; Evans 1970) and with habitat quality (Atwood 1950; Peloquin 1969).

Willner et al. (1983) used body length and hind foot length to develop a model for estimating age. This model, however, was based on captive-raised nutria and has not been validated for wild populations. Adams (1956) suggested age of nutria could be estimated based on hind foot length, with that of immatures (birth to 3 months) ≤10.9 cm, that of subadults (>3–5 months) 11.2–12.4 cm, and that of adults (>5 months) >12.7 cm. However, Ras (1999) found the age estimation criteria recommended by Adams (1956) for nutria in Louisiana were not accurate for nutria in Maryland. Ras (1999) found that females with frost-damaged tails or ear-tags from the previous year had hind foot lengths ranging from 10.5 to 14.0 cm, with a mean of 12.0 cm, and the length for males ranged from 10.0 to 14.0 cm, with a mean of 12.5 cm. Ras (1999) concluded that hind foot length alone was not a reliable technique for estimating age of nutria in Maryland.

Gosling and Baker (1991) stated that the second molar erupts at 3 months of age and the third molar erupts at 6 months and is fully in place by 1 year. Aliev (1965b) studied captive nutria and developed an age estimation technique based on tooth eruption and wear. Willner et al. (1979) estimated age of >600 nutria in Maryland using tooth eruption, tooth wear, and body weights. However, there may be significant differences in tooth wear depending on the habitat and vegetation. For example, Kinler et al. (1987) described significantly greater tooth wear in nutria found in brackish marshes than in freshwater marshes in

Louisiana, and suggested a less-fibrous diet in freshwater areas resulted in reduced molar wear.

Eye lens weight also has been used to estimate age of nutria (Gosling et al. 1980b; Gorostiague and Regidor 1993). Using 199 known-age animals, Gosling et al. (1980b) developed a technique for estimating age based on the following relationship ($r = .99$) between dry eye lens weight and age in months: $\log(\text{age} + K) = 0.511 + 0.013X$, where $X =$ dry eye lens weight in milligrams, and $K = 4.34$ months or the gestation period. This linear equation may be rearranged to solve for age as follows:

$$\text{Age} = 10^{0.511+0.013(\text{lens weight})} - 4.34 \text{ months.}$$

Estimating age of wild nutria without sacrificing the animal to collect an eye lens or to extract teeth is difficult for several reasons. Body size is of limited value for estimating age because an asymptote is reached fairly early in the growth of nutria. In addition, body weight varies throughout the life of a nutria based on diet, and body weight tends to decline in old age (L. M. Gosling, University of Newcastle, pers. commun., 2002). Some researchers have attempted to develop models for predicting age-specific body weights without having to estimate the age of nutria (Willner et al. 1980). Development of additional techniques to effectively determine age of live, wild nutria is needed.

ECONOMIC STATUS

Historically, nutria were introduced in North America for fur farming and were originally considered a desirable species as a furbearer (Evans 1970; Nichols and Chabreck 1974). Currently, nutria have a dual role as a valued furbearer in some areas and a problem species in others (Evans 1970; Webster et al. 1985; Bó et al. 1992; Colantoni 1993; Drost and Fellers 1995; Bounds 2000; Guichón and Cassini 2000). As with many invasive species, once established, nutria become problematic in areas where they are nonnative, whereas they are valued assets in their indigenous regions (D'Adamo et al. 2000). Within their native range in South America, nutria are a desirable wildlife species and provide a significant revenue source for local people (Bó et al. 1992). For example, in Argentina, nutria are a highly valued animal in the furbearer trade, with an average of 2.5 million furs exported each year from 1975 to 1985 (Marchetti and Morello 1991). Many of the countries where nutria were introduced now consider the animals to be a pest species with adverse ecological and economic consequences for native natural resources (Gosling 1981a; Grace 1992; Drost and Fellers 1995; Gebhardt 1996; Jouventin et al. 1997; Bounds 2000).

In countries where nutria were introduced, the animals have destroyed agricultural crops including sugar cane, rice, and corn. They have impaired natural habitats and plant communities, competed with native wildlife for natural resources, and damaged water control structures such as impoundments, dikes, and levees (Ensminger 1956; Ehrlich and Jedynak 1962; Cotton 1963; Ellis 1963; Norris 1967a; Evans 1970; Gosling 1974; Kuhn and Peloquin 1974; Boorman and Fuller 1981; Lavenceau 1981; Linscombe et al. 1981; Willner 1982; Webster et al. 1985; Conner et al. 1986; Kinler et al. 1987; Abbas 1988; Baker and Clarke 1988; Gosling and Baker 1991; Grace 1992; Drost and Fellers 1995; Bounds 2000; Bounds and Carowan 2000). During the late 1950s, nutria caused extensive damage to agricultural crops in Texas and Louisiana, and these states asked the U.S. government to assist them in controlling populations (Evans 1970). However, researchers studying nutria in their native range reported that nutria do not damage agricultural crops and are not considered a pest species for agrosystems (D'Adamo et al. 2000; Guichón and Cassini 2000). Similarly, Borgnia et al. (2000) found that agricultural crops were a minimal component of their diet and that nutria were not a pest in agricultural areas in the Pampas Plain of Argentina.

During the 1940s, nutria were promoted as "weed cutters" and were sold throughout the southeastern United States as a way to control undesirable vegetation. However, nutria are often responsible for maintaining large areas of wetlands as mudflats in an undesirable and unvegetated condition (Chabreck et al. 1959; Harris and Webert 1962; Linscombe et al. 1981; Fuller et al. 1985; Hess et al. 1997). In Louisiana, wetland restoration projects allocated almost 50% of their budgets to nutria-exclusion devices (Llewellyn and Shaffer 1993). Nutria inhibited baldcypress regeneration in Louisiana (Blair and Langlinais 1960; Conner et al. 1986; Conner and Toliver 1990; Myers et al. 1995). In many areas, nutria threaten the viability of coastal marshes and wetlands (Chabreck 1988; Grace 1992; Llewellyn and Shaffer 1993; Hess et al. 1997; Ford and Grace 1998). As a result, several countries have attempted to control or eradicate nutria, including Great Britain, Italy, France, Holland, Germany, and the United States (Gosling 1989a; Abbas 1991; Reggiani et al. 1993; Bounds and Carowan 2000). Linscombe (1992) suggested that decreases in nutria harvest rates correspond directly with substantial increases in wetland loss rates. For example, historically, when nutria pelts have yielded \$8/pelt or more, the number of nutria harvested has exceeded 1.8 million animals/year in Louisiana (Fig. 52.6).

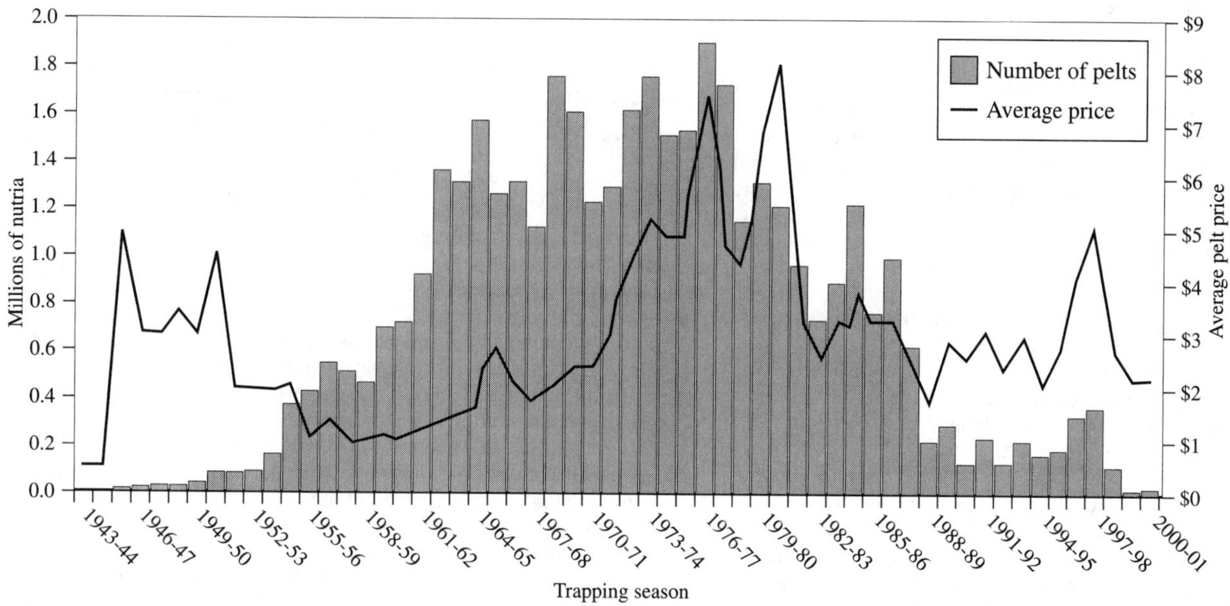

FIGURE 52.6. Number of nutria (*Myocastor coypus*). harvested and average pelt prices in Louisiana from 1943 to 2001. SOURCE: Data from Comparative Takes of Fur Animals, Louisiana Department of Wildlife and Fisheries Annual Reports, 1944–2001.

However, during 1990, pelt prices decreased to <$2/pelt and the number of nutria harvested also dramatically declined. Herbivory by animals such as nutria may increase marsh subsidence by reducing belowground plant production; lowering root and rhizome production; removing aboveground plant tissues, which increases light penetration and soil temperatures; and increasing decomposition rates, which has dramatic effects on wetland soils because of their high organic content (Godshalk and Wetzel 1978; Wetzel 1983; Llewellyn and Shaffer 1993; Ford and Grace 1998).

DAMAGE

Agricultural Damage. Diversity of feeding habits has allowed nutria to exploit a variety of agricultural crops, resulting in significant damage (Gosling 1974). Damage is most prevalent in areas where cropland is adjacent to marshes or other aquatic habitats supporting nutria, and particularly where nutria are abundant. Crops depredated by nutria include cereal grains, corn, and root crops. In Europe, crops damaged by nutria include cabbage, sugar beets, kale, mangolds, rutabagas, potatoes, brussels sprouts, and cereal grains (Laurie 1946; Norris 1967a, 1967b). Kuhn and Peloquin (1974) stated that nutria damage most crops grown in Oregon, including alfalfa, grass seed, wheat, barley, oats, corn, carrots, beets, cauliflower, cucumbers, and melons. Little crop damage was attributed to a low-density nutria population in California (Schitoskey et al. 1972). Crop damage is a serious problem in Texas and Louisiana, where rice, corn, cabbage, and sugar cane are extensively used by nutria (Swank and Petrides 1954; Schitoskey et al. 1972; Kinler et al. 1987).

Crop damage may occur predominantly with introduced nutria populations, but there is little evidence for agricultural impacts in South American nutria populations (Guichón and Cassini 2000; D'Adamo et al. 2000). Larrison (1943) provided an anecdotal report of nutria being killed by Argentinean farmers because they are a "great agricultural pest." However, more recent studies have provided little evidence for significant agricultural impacts. Guichón and Cassini (2000) concluded that Argentinean nutria were not distributed as expected for an agricultural pest, because cultivated lands were not associated with abundant nutria sign. They believed nutria do not constitute a pest in their natural range because there is greater availability of natural food near water and severe winters are absent. However, there has been little research on nutria population limitations in their natural range. Guichón and Cassini (2000) found no evidence of nutria impact on crops, low consumption of crops, and use of pastures in higher proportions than croplands. Similarly, Borgnia et al. (2000) found that six cultivated crops accounted for <2% of the diet. A cultivated terrestrial monocot, *Lolium multiflorum,* was consumed significantly more when 1 m from water than when 5 m from water, whereas *Eleocharis bonariensis,* a native hydrophilic monocot, was consumed equally regardless of distance from water. Total consumption of the agricultural crop was substantially higher than the native plant even at 5 m. Foraging near shelter (waterways) may be a factor in nutria food selection. When availability of food near water decreases, nutria tend to feed farther from the water's edge, and crop damage increases (Borgnia et al. 2000).

Marsh Damage. Due to their destructive foraging habits, nutria have been implicated in marsh loss throughout their introduced range. Exclosure studies have documented several negative effects of nutria on marsh integrity, including reduced plant biomass (Fuller et al. 1985; Foote and Johnson 1992), reduced species richness and evenness (Shaffer et al. 1992), and increased erosion rates (Visser 1994; G. M. Haramis, U.S.Geological Survey, pers. commun., 1998). Favored nutria forage plants usually dominate inside exclosures, whereas tough, fibrous plants not extensively used by nutria dominate outside exclosures (Fuller et al. 1985; Shaffer et al. 1992). Linscombe et al. (1981) also documented increased abundance of *Scirpus olneyi* and decreased coverage of unvegetated areas as a Louisiana nutria population declined. Reduced plant species richness and diversity have been correlated with density of nutria sign (Nyman et al. 1993). Ellis (1963) noted that flowering rush (*Butomus umbellatus*) and cowbane (*Cicuta virosa*) became extremely rare in the presence of nutria herbivory. In many North American marsh ecosystems, nutria foraging compromises the integrity of the root mat, leading to erosion and subsidence. Even where the root mat is not damaged, extensive loss of aboveground vegetation may reduce the ability of marshes to trap sediment, reduce the amount of light reaching the soil surface, reduce litter accumulation, and may ultimately impede soil genesis (Ford and Grace 1998; Llewellyn and Shaffer 1993). Marsh vegetation responds rapidly to changes in nutria abundance, particularly where marsh loss has not been complete (Linscombe et al. 1981; Foote and Johnson 1992; G. M. Haramis, U.S.Geological Survey, pers. commun., 1998). However, marsh loss in some ecosystems, such as Maryland's Lower Eastern Shore, is permanent due to the narrow range of marsh elevations that support vegetation. Grace (1992) noted that herbivory is complex and, at some levels, can enhance productivity. However, interaction of herbivory with other factors enhances the rate of marsh loss, particularly in the presence of abundant or increasing nutria populations. Modeling by Carter et al. (1999) suggested that biomass demands of nutria can exceed sustainable levels during biomass lows in winter. Because nutria populations in Louisiana do not change rapidly during winter, they may be sustained near levels that cause collapse of a marsh ecosystem.

Other Damage. Nutria burrows have significantly affected the integrity of marsh management structures, such as dikes and levees. Extensive burrowing may make these structures susceptible to collapse due to other stressors, such as flooding or vehicular traffic (Gebhardt 1996). Burrowing activity may also destabilize natural embankments, leading to erosion. This type of damage is difficult to quantify because it is not often readily visible (Gebhardt 1996). Nutria also gnaw on seismic cables left overnight on the ground. Gunn and Schmidly (1984) investigated several deterrents to gnawing, and found that a physical deterrent (stainless steel and vinyl-plastic jackets) eliminated damage. Chemical deterrents were ineffective because the valvular mouthparts close during gnawing, preventing chemoreception.

Restoration of baldcypress stands has been compromised by nutria foraging on seedlings. Conner and Toliver (1987) found that nutria ate 85–100% of spring-planted seedlings, but fall plantings were not damaged. Seedlings were typically uprooted, and the bark and tap root were consumed (Blair and Langlinais 1960; Conner and Toliver 1987; King et al. 1998). Tree shelters and castor oil-based materials are effective deterrents, although their effectiveness is limited by the height to which they are applied (Myers et al. 1995). Nutria damage native plant species, compete with and displace muskrats, reduce cover and food for migratory birds, reduce quality of wetland habitat as nurseries for fin fish and shellfish, damage trees, and enhance spread of purple loosestrife (*Lythrum salicaria*), an invasive herbaceous wetland plant (Bounds 2000). Accounts of populations early in colonization (e.g., Laurie 1946) generally do not describe extensive damage, although prevalence of damage to crops, dikes, levees, and native plant communities usually precedes frequent sightings as the first indication that a nutria problem exists (Gosling 1974).

MANAGEMENT

Radiotelemetry Techniques. Using radiotelemetry to track nutria movements is challenging for several reasons. First, proper fit of radio-collars is important yet difficult because nutria tend to gain weight during summer and lose weight during winter; these fluctuations may be up to 2 kg (S. Daugherty, University of Maryland–Eastern Shore, unpublished data, 2002). Radio-collars that are too tight may become ingrown, and collars that are too loose may easily slip over the head or allow the nutria to put a foot or limb through the collar, which generally results in mortality. Second, most radio-collar materials (hard plastic, nylon webbing, latex tubing, and nylon harnesses) may cause dermatitis, abrasions, or lesions around the neck, potentially causing animals to discard the collars (Coreil and Perry 1977). Evans et al. (1971) radio-collared 50 animals, and only 2 did not show adverse reactions to the radio-collars. Coreil and Perry (1977) found the preferred radio-collar was nylon-covered rubber tubing with a 60-g transmitter, fitted to 14% greater than neck circumference, which allowed nutria to feed and groom, but prevented their feet from getting caught under the collar. Other collars at this tightness, but lacking the nylon-covered tubing, allowed feet to get caught, and a tighter fit resulted in abrasions. Lohmeier (1981) used nylon-covered tubing over collars and documented one case (out of four animals) in which tooth marks suggested an adult female had torn a transmitter loose from the collar. Ras (1999) used rubber tubing around radio-collars in Maryland, but found that nutria shredded the tubing by scratching. The problem was addressed by covering the collars with brown plastic beads (Ras 1999).

The combination of adverse reactions to radio-collars and potentially high mortality rates has led to difficulty in monitoring the same individuals throughout the year. For example, year-long data were only obtained for one of seven animals radio-collared by Coreil et al. (1988) and only 1 of 73 animals radio-collared by Ras (1999).

In addition, difficulty obtaining locations at night may compromise accuracy of home range estimates for this predominantly nocturnal species. In Great Britain, the study area of Gosling and Baker (1989a) contained adequate roads to allow telemetry locations to be obtained from vehicles. However, other portions of the nutria range, such as Maryland and Louisiana, are characterized by extensive roadless marsh, which provides little opportunity for vehicle-based telemetry.

Anesthesia. Van Forest (1980) reported that the combination of 10.1–19.6 mg/kg of ketamine hydrochloride (ketamine) and 2.0 mg of xylazine hydrochloride (xylazine) produced sufficient muscle relaxation and analgesia for surgical amputation of a nutria's tail. Induction of anesthesia occurred within 5 min after intramuscular injection. Duration of anesthesia varied from 15 to 40 min, with recovery in 1.5–8 hr. At reduced dosages, this combination of drugs has been used to immobilize nutria for fitting animals with radio-collars (Bó et al. 1994). The average doses of ketamine and xylazine were 4.07 ± 0.52 mg/kg and 0.5 ± 0.08 mg/kg, respectively. Induction required 7.3 ± 2.0 min after intramuscular injection, and arousal from sedation occurred 23.5 ± 0.3 min after administration. Recovery was observed 46.0 ± 2.5 min after injection of the drugs, and no adverse responses were observed in any of the animals. If required, an additional injection of 0.75 mg/kg of ketamine and 0.15 mg/kg of xylazine may be given to maintain anesthesia (Bó et al. 1994).

Signs of Presence. Various indirect indicators have been used to determine whether nutria are present in an area, assess relative habitat use, and index population size. Gosling and Baker (1991) described the following indicators of nutria presence: (1) burrows; (2) resting structures; (3) bare "runs" about 15 cm wide, leading from climb-outs at the bank edge to a complex network away from the bank; (4) feeding remains scattered near open water, with distinctive paired crescentic incisor marks on food items; (5) excavations for roots (especially in winter); and (6) unique feces. In responding to a U.S. nationwide survey, 70 state or federal biologists reported having nutria problems on lands under their management. Of these, 99% reported visual sightings, 77% reported vegetation damage, 69% reported scat and tracks, and 44% reported hearing vocalizations (Bounds 2000). Guichón and Cassini (2000) assessed population status in Argentina through abundance of burrows, nests, food remnants, and trails. Milne and Quay (1966) separated muskrat and nutria use areas based on several factors. Because there is little overlap in incisor widths between species (3.8–7.0 mm for muskrat and 5.0–15.0 for adult nutria), their foraging areas usually could be distinguished using tooth marks on wide-stemmed vegetation. On narrow-stemmed vegetation, nutria tend to bite stems off cleanly and muskrats tend to tear stems. Other signs that differed between species included droppings (soft and unconsolidated for muskrat; firm, cylindrical, and striated for nutria), footprints (small, unwebbed hind feet for muskrat; larger, webbed hind feet for nutria), and tail drags (showing distinct dorsoventral flattening for muskrat; larger and rounded for nutria).

Marking and Population Estimation. Several attempts have been made to generate estimates of nutria population sizes using mark–recapture methods, but most of these have encountered significant methodological problems. Much of the information on historical population trends comes from agency records of harvest rates by trappers (Valentine et al. 1972; Kuhn and Peloquin 1974) and from qualitative assessments based on abundance of sign. A variety of capture techniques has been tested to determine the catch per unit of effort in different seasons (Table 52.7).

Hair bleach can be used to mark animals for visual "recapture." Johnson (1992) tested this technique on 38 nutria, using time–area counts on 50-m wide transects from airboats. However, a significant drawback was obstruction of visual observations by dense cover. Only 5% of the marked group was observed, making population estimation impossible. Lazarns and Rowe (1975) attempted freeze-marking with dichlorodifluoromethane, but did not recommend further application of this technique with nutria. Other common marking techniques include use of Monel ear tags (Simpson and Swank 1979; Ras 1999), aluminum tags, and web-clip codes (Reggiani et al. 1995). Evans et al. (1971) reviewed numerous methods and concluded marking nutria is problematic. Gross physiological rejection was observed for almost all materials applied to skin, even with antibiotic treatment. Most markers caused sloughing or infection. A #3 self-piercing Monel tag inserted through the hind foot webbing was the only reliable, long-lasting marker. Visibility is improved when tags are placed in the ears, but retention is lower because of frostbite damage. Toe-clip and web-clip methods worked, but these marks are not always recognized by untrained trappers and could be confused with trap-induced injuries. Most external markers

TABLE 52.7. Catch per unit effort (CPUE) for nutria, by type of trap and season

Type of Trap	CPUE	Season	Source
Unbaited cage	0.17	Fall–winter	Linscombe et al. 1981
Baited cage	0.02–0.08	Fall–spring	Doncaster and Micol 1988
Baited cage	0.15–0.20	Late summer	Doncaster and Micol 1988
Baited cage	12.9	Annual average(5 yr)	Baker and Clarke 1988
Baited cage	0.47	Annual average(1 yr)	Reggiani et al. 1995
Baited cage	0.047	Spring–summer	Simpson and Swank 1979
Baited cage on platform	0.28	Summer–fall	Lohmeier 1981
Baited cage on platform	24.2	Annual average(5 yr)	Baker and Clarke 1988
Unknown live trap	0.08	Late summer	Johnson 1992
Victor #1.5 Leghold	0.062	Winter (2 yr)	Palmisano and Dupuie 1975
Victor #2 Leghold	0.048	Winter (2 yr)	Palmisano and Dupuie 1975
Victor #2 Leghold (brackish marsh)	0.13	Winter (2 yr)	Linscombe 1976
Victor #2 Leghold (fresh marsh)	0.27	Winter (2 yr)	Linscombe 1976
Conibear 220	0.024	Winter (2 yr)	Palmisano and Dupuie 1975
Conibear 220 (brackish marsh)	0.097	Winter (2 yr)	Linscombe 1976
Conibear 220 (fresh marsh)	0.16	Winter (2 yr)	Linscombe 1976

intended for observation at a distance, including flags, buttons, and disks, were rapidly sloughed. Powdered aluminum pigment was reliable as a fecal marker when mixed with baits, but does not provide the identification of individuals necessary for mark–recapture studies.

Marking the calcified tissues of captive nutria with vital fluorescent markers was investigated by Pascal et al. (1989). These researchers fed nutria a diet consisting solely of pieces of carrots coated with the following compounds at the noted concentrations: alizarin (2.0 g/kg), xylenol orange (1.8 g/kg), tetracycline (0.5 and 1.0 g/kg), and fluorescein (0.125, 0.25, and 0.5 g/kg). After 3 days of treatment, calcified tissues were examined *in toto* under ultraviolet light, and histological sections of these tissues were observed by fluorescent optical microscopy. The mandibles, skull, and incisors were the most clearly marked tissues, with the best *in toto* results obtained from teeth and microscopic results from mandibular sections. The rank of the most successful treatments for tissue labeling was as follows: 0.5 g/kg fluorescein, 1.8 g/kg xylenol orange, and 2.0 g/kg alizarin. Tissues of young nutria were more consistently and clearly marked than those of older animals. However, longevity of the labels was 20–29 weeks in mature nutria and 12–14 weeks in younger individuals. Fluorescent marking of nutria in France was done by placing carrots coated with either fluorescein (0.5 g/kg bait) or xylenol orange (2.0 g/kg bait) along a 1.5-km section on the bank of a tributary of the Loire River (Fichet and Pascal 1989a). During the 20-consecutive-day trapping period after vital marking, 100% of the nutria captured ($n = 98$) exhibited labeling of the mandibles and incisors on examination by fluorescent optical microscopy. The investigators observed that the phalanges were rarely marked and concluded that phalanx removal and examination was not a reliable method for observing labeling in live animals. Nevertheless, they proposed that nonlethal determination of marking could be accomplished by the removal and examination of a fragment of the incisor for the presence of the label. They also demonstrated that xyenol orange, but not fluorescein, can cross the placenta and mark the dentin of the lower incisors of fetal nutria (Fichet and Pascal 1989b).

Reggiani et al. (1995) used closed population models (program CAPTURE; Otis et al. 1978) to estimate nutria density. They deployed 40–42 baited box traps on a grid in a 37.5-ha study area, and trapped 10- to 12-day sessions with 110–122 days between sessions. Recapture probabilities were sufficient (0.12–0.42) to generate population estimates with reasonable precision using the jackknife estimator in program CAPTURE. Problems with trap avoidance or heterogeneity of capture rates have hindered most other efforts to estimate nutria population sizes. For example, Doncaster and Micol (1988) found that males were strongly trap prone and females slightly trap shy, which, coupled with small sample sizes, led to difficulty using the Lincoln–Peterson estimator. They proposed several alternative approaches using regression models, but these were largely unreliable despite marking 80–90% of the population.

Because of the difficulties in using mark–recapture methods for nutria, a retrospective census approach was used to estimate nutria abundance in Great Britain (Gosling 1981c; Gosling et al. 1981). This technique assumes that all deaths are recorded and that age at death is known. From these data, each animal's life can be extrapolated back in time, ultimately allowing the population's composition to be described. During a control trapping program, Gosling (1981c) conducted such a survey, assigning each animal to a month of birth using eye lens weight (Gosling et al. 1980b). Because winters were mild during the trapping period, the assumption of all deaths being recorded was believed to be reasonable. Shortcomings of the technique included the necessity to assume some demographic unknowns and the ability to only estimate historical population trends.

Trapping Methods. Most nutria populations in North America and Europe are trapped, either by fur trappers or as part of population control programs. Consequently, much descriptive data on capture rates of various trap types are available, although comparatively little effort has gone into rigorous testing of trapping methods and strategies. Many investigators have concluded that the initial capture of a given animal is not particularly difficult, although recapture of marked animals has proven problematic. Trapping generally relies on interception, with cage, foothold, or body-gripping traps placed on active runs or climb-outs where nutria enter a water body (Robicheaux and Linscombe 1978; Gosling et al. 1988a). Because nutria tend to be wary of traps set on land (Warkentin 1968), cage traps often are baited with corn, bread, or carrots (e.g., Laurie 1946; Norris 1967b; Evans et al. 1971; Schitoskey et al. 1972). Baiting land-set traps has shown marginal improvement in capture success, although it may be most effective for traps set off runs and during winter (Norris 1967b; Gosling et al. 1988a). Floating rafts holding six to eight baited cage traps are an effective alternative to land sets, particularly where few topographical features are available to guide animals to traps. Baker and Clarke (1988) found that traps on rafts were 50% more likely to catch nutria than baited cage traps set on land. Evans et al. (1971) found that baited traps on rafts produced 80% of the recaptures. However, use of rafts may be limited in areas of strong water current (Schitoskey et al. 1972). Baker and Clarke (1988) and Norris (1967b) reported that capture rates were highest at the beginning of a trapping campaign, but decreased dramatically after the first few months. However, these observations may be artifacts of population size rather than behavior. Because of their greater movement rates, males tend to be more trap prone than females (Doncaster and Micol 1989).

Several studies have reported problems with trap-induced injuries. Dixon et al. (1979) found that 31% of leg-trapped nutria had broken legs when recaptured, and of these, 80% had either lost a leg or were otherwise crippled when recaptured. They also reported a significant effect of trapping method (foothold versus live trap) on rate of weight gain, but the effect was more pronounced in a mild than in a severe winter. Similarly, Chapman et al. (1978) found that survival of cage-trapped animals was nearly twice as high as that of foothold-trapped animals. Baker and Clarke (1988) reported a mortality rate of 0.3% for cage traps set on land, due to hypothermia or drowning after an unexpected rise in water level. However, no mortality occurred in raft traps. Evans et al. (1971) reported a 15% mortality rate for nutria in land-set cage traps (8% drowning during flash floods, 2% feral dog attacks, 3% heat exposure, 2% cold exposure).

Control Methods. The most comprehensive effort to control a nutria population occurred in Great Britain during the 1960s–1980s (Norris 1967a, 1967b; Gosling and Baker 1987, 1989b; Gosling et al. 1988a, Gosling 1989a). This effort initially consisted of employing 6–12 trappers at an effort of over 600,000 trap-nights on about 6848 km^2. The trappers harvested a total of 40,461 animals at a cost of about \$5/animal (Norris 1967a). This effort, however, was unsuccessful in eradicating the population, as a series of mild winters, combined with an untimely reduction in the trapper force, resulted in a resurgence of the population in the 1970s. Consequently, an organization called Coypu Control was formed in 1971, employed 5–20 trappers until 1979, and tested whether nutria could be eradicated in discrete, representative areas (Gosling et al. 1988a). By 1975, necropsies had been completed on over 30,000 animals, allowing the team to build demographic models that incorporated reproductive parameters (Gosling and Baker 1987; Gosling 1989a). Using these models, an eradication campaign was designed, consisting of 24 trappers averaging 200,000 trap-nights/year. The eradication campaign started in 1980, and the last breeding group was found and trapped in 1987. Eradication was declared a success after 21 subsequent months of trapping with no captures (Gosling 1989a). Lessons learned from the British experience include the following: (1) a flaw in the initial strategy was that trappers spent too much time clearing low-density areas rather than attempting to maximize capture rates, (2) there was initially little understanding of how trapping intensity and severity of winter weather interacted, (3) immigration from adjacent source populations must be considered in designing eradication efforts, (4) boredom and declining motivation at the prospect of losing employment were potential issues among trappers toward the end of the campaign, and (5) effective eradication efforts depend on knowledge of reproductive and demographic parameters, their density dependence, and how they respond to intensive harvest and weather conditions (Gosling et al. 1988a; Gosling and Baker 1989b).

The British experience illustrates the need for strategic planning, sound scientific data, and sustainable sources of funding to eradicate problematic nutria populations. Linscombe et al. (1981) pointed out the importance of implementing population controls while desirable marsh components (e.g., *Scirpus olneyi*) are increasing in abundance, as efforts are less likely to succeed once substantial nutria impacts are evident. Although no successful eradication campaign has been conducted in North America, efforts under way in Maryland (Bounds and Carowan 2000) seek to develop similar eradication strategies applicable to North America.

Poison. A poison used to kill nutria is zinc phosphide. Evans (1970) reported that baits consisting of a 5.1-cm piece of carrot containing zinc phosphide at the concentration of 0.75% by weight would kill even the largest nutria. Zinc phosphide toxicity results from liberation of phosphine gas by the acid pH in the stomach. The gas produces direct gastrointestinal irritation and cardiovascular collapse. At the toxic dose (approximately 40 mg/kg) in animals with a full stomach, onset of intoxication is rapid, with death occurring from respiratory arrest (Aiello 1998).

Habitat Management. A variety of techniques has been employed to manage marshes that contain nutria, and these generally mirror marsh management to benefit muskrats (Kinler et al. 1987). These management strategies include periodic burning, water level control, and ditching. These methods focus on improving production of desirable plant species such as *Scirpus olneyi* and enhancing trapper access to the marsh. Water level management may encourage revegetation of marshes denuded by nutria (Willner 1982), but such techniques are limited to marshes with containment dikes and water control structures. Owing to the rapid disappearance of marsh habitat in areas with abundant nutria populations, marsh management often takes the form of managing nutria to mitigate these losses. Management to benefit muskrats may have ancillary benefits for nutria, but declines in fur prices and trapper effort have lessened the impetus to manage for sustainable nutria populations. Many state and federal initiatives are focusing on control or eradication of exotic invasive species for the purpose of restoring native fish, wildlife, and plant communities. Consequently, management targeted at nutria is increasingly focused on minimizing their undesirable effects on marshes and marsh-dependent wildlife.

RESEARCH AND MANAGEMENT NEEDS

Much of the published literature on nutria in North America is descriptive in nature, providing data on feeding habits, habitats, reproduction, and other aspects of the species's ecology. Negative effects of nutria on agricultural and ecological resources also have been described. Although few studies have quantified nutria damage, there is substantial evidence that it is pervasive and of great magnitude. Given this evidence and the recent emphasis on effectively managing invasive and exotic species (Everett and Sherfy 2001), further investigation of nutria impacts should be of relatively low priority. Instead, the most significant research needs in North America center on developing and implementing effective and efficient methods for controlling or eradicating nutria populations. Specifically, the following areas warrant further investigation:

1. Habitat selection. Many studies have investigated nutria habitat use, but few have included analysis of habitat preferences or how habitat selection patterns change with respect to season and population density.
2. Movements and home range. Much anecdotal information exists on movement distances of nutria, but few investigators have been able to monitor animals for sufficient time periods to develop estimates of home range. Of practical importance to control efforts is the question of how movement patterns change in response to intensive harvest.
3. Reproductive physiology. Estimates of reproductive parameters (e.g., birth rate, pregnancy rate) are available. Other parameters, such as the endocrine regulation of reproduction, the mechanism of ovulation, or factors modulating synchony of reproductive activity, have not been fully elucidated. The density dependence of these parameters has not been adequately addressed.
4. Trapping strategies. Successful eradication in Great Britain demonstrated the importance of developing a trapping strategy that is tailored to local patterns of immigration and reproduction. What is unknown is the degree to which these results are applicable to North American populations.
5. Telemetry techniques. Studies of nutria movement patterns are complicated by difficulties with achieving precise fit of neck-mounted radio-collars and by the frequency with which animals are either underground or in the water. Telemetry methods that account for these factors are needed.
6. Age-estimation techniques. Reliable techniques for determining age of live nutria in the field are needed. Use of eye lens weight is an accurate estimation technique, but requires sacrifice of the animal. Estimating age of nutria by tooth wear is also not feasible with live nutria in the field.

In areas where nutria have been introduced, most managers would now prefer to eradicate the species because of the damage to native animal and plant communities. However, the only country to successfully eradicate nutria is England (Gosling 1989a). Within the United States, two different approaches are being used to control or eradicate nutria populations. In Maryland, the approach is similar to that used in Great Britain; funds are used to hire federal trappers to harvest nutria throughout the year with the goal of eradication. In contrast, in Louisiana, funds are being used to pay private trappers an incentive to trap nutria with the goal of controlling nutria numbers and limiting the damage to coastal wetlands.

The Delmarva (Delaware–Maryland–Virginia) Peninsula includes Delaware and those portions of Maryland and Virginia east of Chesapeake Bay. Within this region, the Lower Eastern Shore of Maryland has experienced substantial marsh damage due to an abundant population of nutria. Additionally, corn, soybean, and small-grain farming are dominant land uses in the area. Because much tillable land is in close proximity to tidal marshes, there is potential for nutria-induced crop damage to increase. Marshes of the Lower Eastern Shore are being rapidly converted to open water, and this habitat change is due in part to extensive herbivory by nutria. Consequently, the Maryland Nutria Partnership was formed in 1997 to develop management strategies for nutria. The Partnership consists of federal, state, and private organizations with the common goal of determining the most effective way to reduce the nutria population to a point where sustainable reproduction does not occur. In a 1998 Pilot Program Proposal, the Partnership outlined a 3-year plan consisting of research on nutria life history and control strategies, experiments with wetland restoration methods, and public outreach (Bounds and Carowan 2000). The proposal was subsequently approved by Congress and President Clinton through Public Law 105–322, which authorized the Secretary of the Interior Department to spend $2.9 million on nutria control in Maryland (Bounds and Mollett 2000).

The Pilot Program was implemented in August 2000 by hiring 12 trappers, one trapper supervisor, and two graduate students. The Partnership initially selected three study areas: Blackwater National Wildlife Refuge (federal), Fishing Bay Wildlife Management Area (state), and Tudor Farms (private). Within each area, three study sites were selected, each approximately 250 ha. Among the three sites, one was designated an experimental control in which no harvest occurred. The other two were designated as treatment sites. During the project's first year (2001), demographic parameters were measured on all nine sites and reproductive parameters were measured on the treatment sites ($n = 6$). Demographic research consisted of using mark–recapture methods to estimate population size, survival rates, and trapping efficiencies and radiotelemetry methods to estimate movement patterns. Reproductive studies included collecting a sample of animals each month for necropsies to determine pregnancy rates, litter size, body condition, and presence of disease or parasites. In May 2002, the

Partnership moved from the research phase of the project to the management phase and implemented intensive nutria harvest. Ultimately, the Partnership plans to apply its strategies in an eradication effort throughout the Delmarva range of nutria and offer management suggestions to other regions of the United States where nutria populations are causing marsh and agricultural damage.

In contrast to the approach being used in Maryland, the state of Louisiana is relying on private, licensed trappers to control nutria populations. In Louisiana, nutria are concentrated in the fresh, intermediate, and brackish marshes of the coastal zone. A fur market developed in the 1960s, and from that time until the late 1980s, 1–1.5 million nutria were harvested annually in Louisiana (Hess et al. 1997). Whenever the annual harvest of nutria falls below 500,000 nutria, which it has every year since 1988, the Louisiana Department of Wildlife and Fisheries has documented habitat damage in the coastal wetlands. In 1997, funding for the Nutria Harvest and Wetland Demonstration Project was approved under the Coastal Wetlands Planning, Protection, and Restoration Act (CWPPRA), also known as the Breaux Bill. The project consisted of three components: (1) develop nutria meat marketing activities, (2) conduct a coast-wide nutria herbivory survey, and (3) provide incentive payments to trappers and nutria meat processors. In April 2002, another CWPPRA project, entitled the Coastwide Nutria Control Program, was approved for implementation. Beginning with the trapping season in 2002, private trappers were paid \$4.00/nutria tail. Each trapper was required to purchase a trapping license and have an approved application with a landowner/manager signature and a detailed legal description of the property. As trappers turn in nutria tails, the data is recorded in a database. In addition, trappers may receive \$1/nutria by providing the carcass in the round to a nutria processing dealer, thereby potentially yielding \$5/nutria.

Managers in Maryland and Louisiana are directing their efforts to controlling or eradicating nutria. Different strategies are being employed and the results of these approaches will provide useful information to other managers in the United States who want to control or eradicate this invasive species.

ACKNOWLEDGMENTS

We thank Dr. L. Morris Gosling for reviewing the draft manuscript and Greg Linscombe of the Louisiana Department of Wildlife and Fisheries for providing the figure on nutria harvest and pelt prices. We also thank Ann Rasberry for assistance with graphics and Daphne Chatham, Shawn O'Brien, and Alma Ramirez for administrative assistance with the manuscript.

LITERATURE CITED

Abbas, A. 1988. Impact du ragondin (*Myocastor coypus Molina*) sur une culture de mais (*Zea mays* L.) dans le marais Poitevin. Acta Oecologica/Oecologica Applicata 9:173–89.

Abbas, A. 1991. Feeding strategy of coypu (*Myocastor coypus*) in central western France. Journal of Zoology (London) 224:385–401.

Adams, W. H. Jr. 1956. The nutria in coastal Louisiana. Proceedings of the Louisiana Academy of Science 19:28–41.

Aiello, S. E., ed. 1998. The Merck veterinary manual, 8th ed. Merck, Whitehouse Station, NJ.

Aliev, F. F. 1956. Theoretical and practical foundations of coypu (*Myocastor coypus* Molina) raising in Azerbaijan. Trudy Instituta Zoologii (Azerbaijan) 19:5–96.

Aliev, F. F. 1965a. Dispersal of nutria in the USSR. Journal of Mammalogy 46:101–2.

Aliev, F. F. 1965b. Extent and causes of nutria mortality in the water bodies of southern U.S.S.R. Mammalia 29:435–37.

Ashbrook, F. G. 1948. Nutrias grow in United States. Journal of Wildlife Management 12:87–95.

Atwood, E. L. 1950. Life history studies of nutria, or coypu, in coastal Louisiana. Journal of Wildlife Management 14:249–65.

Axell, H. E. 1963. Coypu (*Myocastor coypus*) at Minsmere during the frost of January and February, 1963. Transactions of the Suffolk Naturalists' Society 12:257–59.

Babero, B. B., and J. W. Lee. 1961. Studies on the helminths of nutria, *Myocastor coypus* (Molina), in Louisiana with check-list of other worm parasites from this host. Journal of Parasitology 47:378–90.

Babero, B. B., C. Cabello, and J. Kinoed. 1979. Helmintofauna de Chile. Part 5: Nuevos parasitos del coipo *Myocastor coypus* (Molina, 1782). Bulletin of Chile Parasitology 34:26–31 (In Spanish with English summary).

Baker, S. J., and C. N. Clarke. 1988. Cage trapping coypus (*Myocastor coypus*) on baited rafts. Journal of Applied Ecology 25:41–48.

Ball, S. J., and D. C. Lewis. 1984. *Eimeria* (Protozoa: Coccidia) in wild populations of some British rodents. Journal of Zoology 202:373–81.

Barlow, J. C. 1969. Observations on the biology of rodents in Uruguay. Life Sciences Contributions, Royal Ontario Museum 75:1–59

Blair, R. M., and M. J. Langlinais. 1960. Nutria and swamp rabbits damage baldcypress plantings. Journal of Forestry 58:388–89.

Bó, R., R. Quintana, J. Merler, P. Minotti, A. Malvarez, and G. De Villafañe. 1992. Problems in the conservation of mammals in the lower delta region of the Paraná River: Evaluation of the current situation using a combined methodology. Noragric Occasional Papers Series C, Development and Environment 11:142–52.

Bó, R. F., F. Palomares, J. F. Beltrán, G. de Villafañe, and S. Moreno. 1994. Immobilization of coypus (*Myocastor coypus*) with ketamine hydrochloride and xylazine hydrochloride. Journal of Wildlife Diseases 30:596–98.

Boorman, L. A., and R. M. Fuller. 1981. The changing status of reedswamp in the Norfolk Broads. Journal of Applied Ecology 18:241–69.

Borgnia, M., M. L. Galante, and M. H. Cassini. 2000. Diet of the coypu (nutria, *Myocastor coypus*) in agro-systems of Argentinean Pampas. Journal of Wildlife Management 64:354–61.

Bounds, D. L. 2000. Nutria: An invasive species of national concern. Wetland Journal 12:9–16.

Bounds, D. L., and G. A. Carowan, Jr. 2000. Nutria: A nonnative nemesis. Transactions of the North American Wildlife and Natural Resources Conference 65:405–13.

Bounds, D. L., and T. A. Mollett. 2000. Can nutria be eradicated in Maryland? Proceedings of the Vertebrate Pest Conference 19:121–26.

Brown, L. E. 1966. An electrophoretic comparison of the serum proteins of fetal and adult nutria (*Myocastor coypus*). Comparative Biochemistry and Physiology 19:479–81.

Brown, L. N. 1975. Ecological relationships and breeding biology of the nutria (*Myocastor coypus*) in the Tampa, Florida, area. Journal of Mammalogy 56:928–30.

Cabrera, A. 1961. Catálogo de los mamiferos de Ámerica del Sur. Revisto del Mususeo Argentino de Ciencias Nataturales "Bernardo Rivadavia" 4:1–732.

Cabrera, A., and J. Yepes. 1940. Mamiferos Sud-Americanos (oida, costumbres y description). Compania Argentina de Edes, Buenos Aires, Argentina.

Carter, J., and B. P. Leonard. 2002. A review of the literature on the worldwide distribution, spread of, and efforts to eradicate the coypu (*Myocastor coypus*). Wildlife Society Bulletin 30:162–75.

Carter, J. A., L. Foote, and L. A. Johnson-Randall. 1999. Modeling the effects of nutria (*Myocastor coypus*) on wetland loss. Wetlands 19:209–219.

Chabreck, R. H. 1962. Daily activity of nutria in Louisiana. Journal of Mammalogy 43:337–44.

Chabreck, R. H. 1988. Coastal marshes. University of Minnesota Press, Minneapolis.

Chabreck, R. H. 1992. Food habits of nutria. Pages 76–78 *in* M. C. Landin, ed. Proceedings of the 13th annual conference of the Society of Wetland Scientists. New Orleans, LA.

Chabreck, R. H. 1996. Regurgitation by the American alligator. Herpetological Review 27:185–86.

Chabreck, R. H., C. M. Hoffpauir, and F. J. Webert. 1959. A study of nutria exclosures in southwest Louisiana. Louisiana Wildlife Fisheries Commission, New Orleans.

Chabreck, R. H., R. B. Thompson, and A. B. Ensminger. 1977. Chronic dermatitis in nutria in Louisiana. Journal of Wildlife Diseases 13:333–34.

Chabreck, R. H., J. R. Love, and G. Linscombe. 1981. Foods and feeding habits of nutria in brackish marsh in Louisiana. Pages 531–43 *in* J. A. Chapman and D. Pursley, eds. Worldwide furbearer conference proceedings. Frostburg, MD.

Chapman, J. A., G. R. Willner, K. R. Dixon, and D. Pursley. 1978. Differential survival rates among leg-trapped and live-trapped nutria. Journal of Wildlife Management 42:926–28.

Chapman, J. A., J. C. Lanning, G. R. Willner, and D. Pursley. 1980. Embryonic development and resorption in feral nutria (*Myocastor coypus*) from Maryland. Mammalia 44:371–79.

Colantoni, L. O. 1993. Ecología poblacional de la nutria (*Myocastor coypus*) en la provincia de Buenos Aires. Flora y Fauna Silvestres 1:1–25 (In Spanish).

Conner, W. H., and J. R. Toliver. 1987. The problem of planting Louisiana swamplands when nutria (*Myocastor coypus*) are present. Pages 42–49 *in* N. R. Holler, ed. Proceedings of the third eastern wildlife damage control conference. Gulf Shores, AL.

Conner, W. H., and J. R. Toliver. 1990. Long-term trends in the bald-cypress (*Taxodium distichum*) resource in Louisiana (U.S.A.). Forest Ecology and Management 33/ 34:543–57.

Conner, W. H., J. R. Toliver, and F. H. Sklar. 1986. Natural regeneration of baldcypress (*Taxodium distichum*) in a Louisiana swamp. Forest Ecology and Management 24: 305–17.

Coreil, P. D., and H. R. Perry, Jr. 1977. A collar for attaching radio transmitters to nutria. Proceedings of the Annual Conference of the Southeastern Association of Fish and Wildlife Agencies 31:254–58.

Coreil, P. D., P. J. Zwank, and H. R. Perry, Jr. 1988. Female nutria habitat use in the intermediate marsh zone of coastal Louisiana. Proceedings of the Louisiana Academy of Science 51:21–30.

Cotton, K. E. 1963. The coypu. River Board Association Yearbook 11:31–39.

D'Adamo, P. M., L. Guichón, R. F. Bó, and M. H. Cassini. 2000. Habitat use by coypu *Myocastor coypus* in agro-systems of the Argentinean Pampas. Acta Theriologica 45:25–33.

Dade, A. W., J. F. Williams, A. L. Trapp, and W. H. Ball. 1977. Cerebral nematodiasis in captive nutria. Journal of the American Veterinary Medical Association 171:885–86.

Dagault, N., and M. Saboureau. 1990. Caracteristiques de la reproduction du myocastor (*Myocastor coypus* M.) mâle dans la region du marais Poitevin. Canadian Journal of Zoology 68:1584–89.

Dixon, K. R., G. R. Willner, J. A. Chapman, W. C. Lane, and D. Pursley. 1979. Effects of trapping and weather on body weights of feral nutria in Maryland. Journal of Applied Ecology 16:69–76.

Dobos-Kovacs, M., and J. Skulteti. 1983. Viral hepatitis associated with abortion in coypu. Maky. Allator. Lapja. 38:176–79.

Doncaster, C. P., and T. Micol. 1988. Comparison of three absolute estimates of coypu abundance from cage trapping. Acta Oecologica/Oecologia Generalis 9:89–99.

Doncaster, C. P., and T. Micol. 1989. Annual cycle of a coypu (*Myocastor coypus*) population: Male and female strategies. Journal of Zoology (London) 217:227–40.

Doncaster, C. P., and T. Micol. 1990. Response by coypus to catastrophic events of cold and flooding. Holarctic Ecology 13:98–104.

Doncaster, C. P., E. Dumonteil, H. Barre, and P. Jouventin. 1990. Temperature regulation of young coypus (*Myocastor coypus*) in air and water. American Journal of Physiology 259:R1220–27.

Dozier, H. L. 1985. The present status and future of nutria in the southeast states. Proceedings of the Annual Conference of the Southeastern Association of Game and Fish Commissioners 5:368–73.

Drost, C. A., and G. M. Fellers. 1995. Non-native animals on public lands. Pages 440–42 *in* E. T. LaRoe, G. S. Farris, C. E. Puckett, P. D. Doran, and M. J. Mac, eds. Our living resources: A report to the nation on the distribution, abundance, and health of U.S. plants, animals, and ecosystems. U.S. Department of the Interior, Washington, DC.

Dugoni, J. A. 1980. Habitat utilization, food habits and productivity of nesting southern bald eagles in Louisiana. M.S. Thesis, Louisiana State University, Baton Rouge.

Ehrlich, S. 1958. The biology of the nutria. Bamidgeh 10:36–43, 60–70.

Ehrlich, S. 1962. Experiment on the adaptation of nutria to winter conditions. Journal of Mammalogy 43:418.

Ehrlich, S. 1966. Ecological aspects of reproduction in nutria *Myocastor coypus* Mol. Mammalia 30:142–52.

Ehrlich, S. 1967. Field studies in the adaptation of nutria to seasonal variations. Mammalia 31:347–60.

Ehrlich, S., and K. Jedynak. 1962. Nutria influence on bog lake in northern Pomorze, Poland. Hydrobiologia, 19:273–97.

Ejercito, C. L. A., L. Cai, K. K. Htwe, M. Taki, T. Inoshima, T. Kondo, C. Kano, S. Abe, K. Shirota, T. Sugioto, T. Yamaguchi, H. Fukushi, N. Minamoto, T. Kinjo, E. Isogai, and K. Hirai. 1993. Serological evidence of *Coxiella burnetti* infection in wild animals in Japan. Journal of Wildlife Diseases 29:481–84.

Ellis, E. A. 1963. Some effects of selective feeding by the coypu (*Myocastor coypus* Molina) on the vegetation of Broadland. Transactions of the Norfolk and Norwich Naturalists' Society 20:32–35.

Ellis, E. A. 1965. The Broads. Collins, London.

Ensminger, A. B. 1956. The economic status of nutria in Louisiana. Proceedings of the Southeastern Association of Game and Fish Commissioners 9:185–88.

Evans, J. 1970. About nutria and their control (Resource Publication 86). Bureau of Sport Fisheries and Wildlife, Denver, CO.

Evans, J., J. O. Ellis, R. D. Nass, and A. L. Ward. 1971. Techniques for capturing, handling, and marking nutria. Proceedings of the Annual Conference of the Southeastern Association of Game and Fish Commissioners 25:295–315.

Everett, R. A., and M. H. Sherfy. 2001. The Chesapeake Bay: A model for regional approaches to the prevention and control of aquatic non-indigenous species. Transactions of the North American Wildlife and Natural Resources Conference 66:611–24.

Felipe, A., S. Callejas, and J. Cabodevila. 1998. Anatomicohistological characteristics of female genital tubular organs of the South American nutria (*Myocastor coypus*). Zentralblatt für Vetinärmedizin. C: Anatomie, Histologie, Embryologie 27:245–50.

Felipe, A., J. Cabodevila, and S. Callejas. 1999. Anatomicohistological characteristics of the ovary of the coypu (*Myocastor coypus*). Zentralblatt für Vetinärmedizin. C: Anatomie, Histologie, Embryologie 28:89–95.

Felipe, A., M. T. Teruel, S. Callejas, and J. Cabodevila. 2000. Typological series for ovarian follicles of sexually mature *Myocastor coypus* (coypus). Biocell 24:97–106.

Fichet, E., and M. Pascal. 1989a. Marquage collectif de rongeurs sauvages au moyen de fluoromarqueurs vitaux des tissus calcifies. Canadian Journal of Zoology 67:847–54 (In French with English summary).

Fichet, E., and M. Pascal. 1989b. Vital marking in utero of calcified tissues in *Myocastor coypus*. Mammalia 53:451–56.

Figurina, I. I. 1984. Cytoarchitecture of the medial geniculate body in different rodents. Vestnik Leningradskogo Universiteta Seriya 3 Biologiya 9:66–71 (In Russian with English summary).

Foote, A. L., and L. A. Johnson. 1992. Plant stand development in Louisiana coastal wetlands: Nutria grazing effects on plant biomass. Pages 265–71 *in* M. C. Landin, ed. Proceedings of the 13th annual conference of the Society of Wetland Scientists. New Orleans, LA.

Ford, M. A., and J. B. Grace. 1998. Effects of vertebrate herbivores on soil processes, plant biomass, litter accumulation and soil elevational changes in a coastal marsh. Journal of Ecology 86:974–82.

Fuller, D. A., C. E. Sasser, W. B. Johnson, and J. G. Gosselink. 1985. The effects of herbivory on vegetation on islands in Atchafalaya Bay, Louisiana. Wetlands 4:105–14.

Gebhardt, H. 1996. Ecological and economic consequences of introductions of exotic wildlife (birds and mammals) in Germany. Wildlife Biology 2:205–11.

George, R. S. 1964. *Ceratophyllus g. gallinae* (Schrank) (Siphonaptera: Cetatophyllidae) from a British coypu (*Myocaster coypus* Molina) (Rodentia: Capromyidae). Entomology Gazette 15:40–41.

George, W., and B. J. Weir. 1974. Hystricomorph chromosomes. Symposia of the Zoological Society of London 34:79–108.

Gilbert, F. F. 1987. Methods for assessing reproductive characteristics of furbearers. Pages 180–89 *in* M. Novak, J. A. Baker, M. E. Obbard, and B. Malloch, eds. Wild furbearer management and conservation in North America. Ontario Ministry of Natural Resources, Toronto.

Gluchowski, W., and J. Maciejowski. 1958. Investigations on factors controlling fertility in the coypu. Part 2: Attempts at determining the potential fertility, based on histological studies of the ovary. Annales Universitatis Mariae Curie-Sklodowska, Sectio E Agricultura 13:345–61.

Godshalk, G. L., and R. G. Wetzel. 1978. Decomposition of aquatic angiosperms. III. *Zostera marina* L. and a conceptual model of decomposition. Aquatic Botany 5:329–54.

Gorka, T., and J. Kulczyki. 1983. A case of acute lead poisoning in the coypu by Pb compounds. Medycyna Weterynaryjna 39:535–36 (In Polish with English summary).

Gorostiague, M., and H. A. Regidor. 1993. La captura comercial del coypo *Myocastor coypus* (Mammalia: Myocastoridae) en Laguna Adela, Argentina. Studies on Neotropical Fauna and Environment 28:57–63.

Gosling, L. M. 1974. The coypu in East Anglia. Transactions of the Norfolk and Norwich Naturalist's Society 23:49–59.

Gosling, L. M. 1977. Coypu, *Myocastor coypus*. Pages 256–65 *in* G. B. Corbet and H. N. Southern, eds. The handbook of British mammals. Blackwell, Oxford.

Gosling, L. M. 1979. The twenty-four hour activity cycle of captive coypus (*Myocastor coypus*) Journal of Zoology (London) 187:341–67.

Gosling, L. M. 1980. The duration of lactation in feral coypus (*Myocastor coypus*). Journal of Zoology (London) 191:461–74.

Gosling, L. M. 1981a. Climatic determinants of spring littering by feral coypus, *Myocastor coypus*. Journal of Zoology (London) 195:281–88.

Gosling, L. M. 1981b. The effect of cold weather on success in trapping feral coypus (*Myocastor coypus*). Journal of Applied Ecology 18:467–70.

Gosling, L. M. 1981c. The dynamics and control of a feral nutria population. Pages 1806–25 *in* J. A. Chapman and D. Pursley, eds. Worldwide furbearer conference proceedings. Frostburg, MD.

Gosling, L. M. 1983. The adaptive control of offspring sex ratio by female coypus (*Myocastor coypus*). American Zoologist 23:934.

Gosling, L. M.. 1985. Coypus in East Anglia (1970–1984). Transactions of the Norfolk and Norwich Naturalist's Society 27:151–53.

Gosling, L. M. 1986. Selective abortion of entire litters in the coypu: Adaptive control of offspring production in relation to quality and sex. American Naturalist 127:772–95.

Gosling, L. M. 1989a. Extinction to order. New Scientist 121:44–49.

Gosling, L. M. 1989b. The reproductive success of female coypus in a feral population: Variation in relation to age, fat reserves, and the availability of mates. Zoological Journal of the Linnean Society 95:149.

Gosling, L. M., and S. J. Baker. 1981. Coypu (*Myocastor coypus*) potential longevity. Journal of Zoology (London) 197: 285–89.

Gosling, L. M., and S. J. Baker. 1987. Planning and monitoring an attempt to eradicate coypus from Britain. Symposia of the Zoological Society of London 58:99–113.

Gosling, L. M., and S. J. Baker. 1989a. Demographic consequences of differences in the ranging behaviour of male and female coypus. Pages 155–67 *in* R. J. Putman, ed. Mammals as pests. Chapman and Hall, London.

Gosling, L. M., and S. J. Baker. 1989b. The eradication of muskrats and coypus from Britain. Biological Journal of the Linnean Society 38:39–51.

Gosling, L. M., and S. J. Baker. 1991. Coypu. Pages 267–75 *in* G. B. Corbet and S. Harris, eds. The handbook of British mammals. Blackwell, Oxford.

Gosling, L. M., and J. R. Skinner. 1984. Coypu. Pages 246–51 *in* I. L. Mason, ed. Evolution of domesticated animals. Longman, London.

Gosling, L. M., and K. M. H. Wright. 1994. Scent marking and resource defence by male coypus (*Myocastor coypus*). Journal of Zoology (London) 234:423–36.

Gosling, L. M., G. E. Guyon, and K. M. H. Wright. 1980a. Diurnal activity of feral coypus (*Myocastor coypus*) during the cold winter of 1978–9. Journal of Zoology (London) 192:143–46.

Gosling, L. M., L. W. Huson, and G. C. Addison. 1980b. Age estimation of coypus (*Myocastor coypus*) from eye lens weight. Journal of Applied Ecology 17:641–47.

Gosling, L. M., A. D. Watt, and S. J. Baker. 1981. Continuous retrospective census of the East Anglian coypu population between 1970 and 1979. Journal of Animal Ecology 50:885–901.

Gosling, L. M., S. J. Baker, and J. R. Skinner. 1983. A simulation approach to investigating the response of a coypu population to climatic variation. EPPO Bulletin 13:183–92.

Gosling, L. M., S. J. Baker, and K. M. H. Wright. 1984. Differential investment by female coypus (*Myocastor coypus*) during lactation. Symposia of the Zoological Society of London 51:273–300.

Gosling, L. M., S. J. Baker, and C. N. Clarke. 1988a. An attempt to remove coypus (*Myocastor coypus*) from a wetland habitat in East Anglia. Journal of Applied Ecology 25:49–62.

Gosling, L. M., K. M. H. Wright, and G. D. Few. 1988b. Facultative variation in the timing of parturition by female coypus (*Myocastor coypus*), and the cost of delay. Journal of Zoology (London) 214:407–15.

Grace, J. B. 1992. The impact of nutria (*Myocastor coypus*) on Gulf Coastal wetlands: Symposium introduction. Pages 70–74 *in* M. C. Landin, ed. Proceedings of the 13th annual conference of the Society of Wetland Scientists. New Orleans, LA.

Guerra-Shinohara, E. M., and O. C. Barretto. 1999. The erythrocyte cytoskeleton protein 4.2 is not demonstrable in several mammalian species. Brazilian Journal of Medical Research 32:683–87.

Guichón, M. L., and M. H. Cassini. 2000. Local determinants of coypu distribution along the Luján River, eastcentral Argentina. Journal of Wildlife Management 63:895–900.

Gunn, S. J., and D. J. Schmidly. 1984. Preventing nutria damage to seismic cables with chemical and physical deterrents. Texas Journal of Science 36:205–13.

Hailman, J. P. 1961. Stereotyped feeding behavior of a North Carolina nutria. Journal of Mammalogy 42:269.

Hall, E. R. 1981. The mammals of North America, 2nd ed., Vol. 2. John Wiley, New York.

Harman, D. M., G. R. Willner, and J. A. Chapman. 1984. Frequency and distribution of the American dog tick on the nutria in Maryland. American Midland Naturalist 3:81–85.

Harris, V. T., and F. Webert. 1962. Nutria feeding activity and its effect on marsh vegetation in southwestern Louisiana (Special Scientific Report-Wildlife No. 64). U.S. Fish and Wildlife Service.

Hart, J. S. 1971. Rodents. Pages 2–130 *in* G. C. Whittow, ed. Comparative physiology of thermoregulation, Vol: 2 Mammals. Academic Press, New York.

Heap, R. B., and D. V. Illingworth. 1974. The maintenance of gestation in the guinea-pig and other hystricomorph rodents: Changes in the dynamics of progesterone metabolism and the occurrence of progesterone-binding globulin (PBG). Symposia of the Zoological Society of London 34:385–415.

Hess, I. D., W. Conner, and J. Visser. 1997. Nutria—Another threat to Louisiana's vanishing coastal wetlands. Aquatic Nuisance Species Digest 2(1):4–5.

Hillemann, H. H., and A. I. Gaynor. 1961. The definitive architecture of the placentae of nutria, *Myocastor coypus* (Molina). American Journal of Anatomy 109:299–318.

Hillemann, H. H., and R. L. Ritschard. 1967. Comparative fibroarchitecture in the mammalian placenta and adnexa. Transactions of the American Microscopic Society 86:184–94.

Hillemann, H. H., A. I. Gaynor, and H. P. Stanley. 1958. The genital system of nutria (*Myocastor coypus*). Anatomical Record 130:515–28.

Holmes, R. G., O. Illman, and J. K. A. Beverley. 1976. Toxoplasmosis in coypu. Veterinary Record 101:74–75.

Howerth, E. W., A. J. Reeves, M. R. McElveen, and F. W. Austin. 1994. Survey for selected diseases in nutria (*Myocastor coypus*) from Louisiana. Journal of Wildlife Diseases 30:450–53.

Inoshima, Y., S. Shimizu, N. Minamoto, K. Hirai, and H. Sentsui. 1999. Use of protein AG in an enzyme-linked immunosorbent assay for screening for antibodies against parapoxvirus in wild animals in Japan. Clinical and Diagnostic Laboratory Immunology 6:388–91.

Jeantet, A. Y., M. Truchet, G. Naulleau, and R. Martoja. 1991. Cytological effects of bromadiolone on some organs or tissues (liver, kidney, spleen, blood) of coypu (*Myocastor coypus*). Comptes Rendus de l'Academie des Sciences. Serie III: Sciences de la Vie (Paris) 312:149–56.

Jelinik, P., and M. Glasrova. 1982. The red blood picture in male coypu in the post-natal period. Veterinarni Medicina 27:337–48 (In Czech with English summary).

Jemison, E. S., and R. H. Chabreck. 1962. Winter barn owl (*Tyto alba*) foods in a Louisiana coastal marsh. Wilson Bulletin 74:95–96.

Johnson, L. A. 1992. Use of a mark–visual recapture technique to estimate the relative abundance of nutria. Pages 857–60 *in* M. C. Landin, ed. Proceedings of the 13th annual conference of the Society of Wetland Scientists, New Orleans, LA.

Jouventin, P., T. Micol, C. Verheyden, and G. Guédon. 1997. Le Ragondin. Biologie et méthodes de limitation des populations. Association de Coordination Technique Agricole (ACTA), Paris.

Kasumova, N. I., S. I. Radzhabli, and G. K. Kuliev. 1976. Cytogenic study of nutria. Part 1: Somatic and meiotic cells of standard and white nutria. Genetika 12:174–76 (In Russian with English summary).

Katomski, P. A., and F. L. Ferrante. 1974. Catecholamine content and histology of the adrenal glands of nutria (*Myocastor coypus*). Comparative Biochemistry and Physiology 48A:539–46.

Kays, C. E. 1956. An ecological study with emphasis on nutria (*Myocastor coypus*) in the vicinity of Price Lake, Rockefeller Refuge, Cameron Parish, Louisiana. M.S. Thesis, Louisiana State University, Baton Rouge.

Kennedy, M. L., and P. K. Kennedy. 1998. First record of nutria, *Myocastor coypus* (Mammalia: Rodentia), in Tennessee. Brimleyana 25:156–57.

Keymer, I. F., G. A. Wells, and H. L. Ainsworth. 1999. Renal neoplasia in coypus (*Myocastor coypus*). Veterinary Journal 158:144–51.

King, S. L., B. D. Keeland, and J. L. Moore. 1998. Beaver lodge distributions and damage assessments in a forested wetland ecosystem in the southern United States. Forest Ecology and Management 108:1–7.

Kinler, N. 1992. Biology and ecology of nutria. Page 75 *in* M. C. Landin, ed. Proceedings of the 13th annual conference of the Society of Wetland Scientists. New Orleans, LA.

Kinler, N., and R. H. Chabreck. 1978. Nutria pelt damage from *Bidens laevis*. Proceedings of the Annual Conference of the Southeastern Association of Fish and Wildlife Agencies 32:369–77.

Kinler, N. W., G. Linscombe, and R. H. Chabreck. 1981. Smooth beggartick, its distribution, control and impact on nutria in coastal Louisiana. Pages 142–54 *in* J. A. Chapman and D. Pursley, eds. Worldwide furbearer conference proceedings. Frostburg, MD.

Kinler, N. W., G. Linscombe, and P. R. Ramsey. 1987. Nutria. Pages 326–43 *in* M. Novak, J. A. Baker, M. E. Obbard, and B. Malloch, eds. Wild furbearer management and conservation in North America. Ontario Trappers Association, Toronto.

Koehler, N., M. H. Gallardo, L. C. Contreras, and J. C. Torres-Mura. 2000. Allozymic variation and systematic relationships in Octodontidae and allied taxa (Mammalia, Rodentia). Journal of Zoology 252:243–50.

Komarek, J. 1983. Some biochemical and haematological values for the blood of nutria (*Myocastor coypus*). Veterinari Medicina 28:351–55 (In Czech with English summary).

Krattenmacher, R., and K. Rubsamen. 1987. Thermoregulatory significance of non-evaporative heat loss from the tail of the coypu (*Myocastor coypus*) and the tammer-wallaby (*Macropus eugenii*). Journal of Thermoregulatory Biology 12:15–18.

Kuhn, L. W., and E. P. Peloquin. 1974. Oregon's nutria problem. Proceedings of the Vertebrate Pest Conference 6:101–5.

Larrison, E. J. 1943. Feral coypus in the Pacific Northwest. Murrelet 24:3–9.

Laurie, E. M. O. 1946. The coypu (*Myocastor coypus*) in Great Britain. Journal of Animal Ecology 15:22–34.

Lavenceau, P. 1981. Ragondins: Bilan des luttes organisées en Gironde 1979/1980. La Defense des Vegetaux 207:33–58.

Lazarns, A. B., and F. P. Rowe. 1975. Freeze-marking rodents with a pressurized refrigerant. Mammal Review 5:31–34.

LeBlanc, D. 1994. Nutria (B 71–80). Animal and Plant Health Inspection Service, Animal Damage Control, Port Allen, LA.

Linscombe, G. 1976. An evaluation of the No. 2 Victor and 220 Conibear traps in coastal Louisiana. Proceedings of the Southeastern Association of Game and Fish Commissioners 30:560–66.

Linscombe, G. 1992. Current status of nutria populations and potential solutions. Pages 272–74 *in* M. C. Landin, ed. Proceedings of the 13th annual conference of the Society of Wetland Scientists. New Orleans, LA.

Linscombe, G., N. Kinler, and V. Wright. 1981. Nutria population density and vegetative changes in brackish marsh in coastal Louisiana. Pages 129–41 *in* J. A. Chapman and D. Pursley, eds. Worldwide furbearer conference proceedings. Frostburg, MD.

Little, M. D. 1965. Dermatitis in a human volunteer infected with *Strongyloides* of nutria and raccoon. American Journal of Tropical Medicine and Hygiene 14:1007–9.

Llewellyn, D. W., and G. P. Shaffer. 1993. Marsh restoration in the presence of intense herbivory: The role of *Justicia lanceolata* Chapm. Small. Wetlands 13:176–84.

Lohmeier, L. 1981. Home range, movements, and population density of nutria on a Mississippi pond. Journal of the Mississippi Academy of Sciences 26:50–54.

Lowery, G. H., Jr. 1974. The mammals of Louisiana and its adjacent waters. Louisiana State University Press, Baton Rouge.

Lueth, F. X. 1985. The first year of nutria investigations on the Mobile Delta. Proceedings of the Annual Conference of the Southeastern Association of Game and Fish Commissioners 3:98–104.

Mais, D. E., J. S. Hayes, R. B. Heap, and M. W. Wang. 1995. Specific interactions of progestin and anti-progestin with progesterone antibodies, plasma binding proteins and the human recombinant receptor. Journal of Steroid Biochemistry and Molecular Biology 54:63–69.

Mann, T., and E. D. Wilson. 1962. Biochemical observations on the male accessory organs of nutria, *Myocastor coypus* (Molina). Journal of Endocrinology 25:407–8.

Marchetti, B., and J. Morello. 1991. Sustainable development in Argentina. Centro de Estudios avanzados, Universidad de Buenos Aires, Buenos Aires, Argentina.

Martin, R. A., M. Fiorentini, and F. Conners. 1980. Social facilitation of reduced oxygen consumption in *Mus musculus* and *Meriones unquiculatus*. Comparative Biochemistry and Physiology 65:519–22.

Martino, P., and N. Stanchi. 1994. Epizootic pneumonia in nutria. Zentralblatt für Veterinärmedizin. B 41:561–66.

Matthias, K. E. 1941. Nutria, profitable fur discovery. American Fur Breeder 14:18–20.

Maum, D. G. 1986. Geographic variation in pelt quality of nutria (*Myocastor coypus*) from coastal Louisiana. M.S. Thesis, Louisiana Technical University, Ruston.

McKean, T. 1982. Cardiovascular adjustments to laboratory diving in beavers and nutria. American Journal of Physiology 242:R434–40.

McNease, L., and T. Joanen. 1977. Alligator diets in relation to marsh salinity. Proceedings of the Annual Conference of the Southeastern Association of Fish and Wildlife Agencies 31:36–40.

Menard, A., M. L'Hostis, G. Leray, S. Marchandeau, M. Pascal, N. Roudot, V. Michel, and A. Chauvin. 2000. Inventory of wild rodents and lagomorphs as natural hosts of *Fasciola hepatica* on a farm located in a humid area in Loire Atlantique (France). Parasite (Paris) 7:77–82.

Micol, T., and C. P. Doncaster. 1987. Seasonal changes in a wild coypu population. Mammalia 51:475.

Miller, A. 1956. The biting louse, *Pitrufquenia coypus* Marelli, on nutria in Louisiana. Journal of Parasitology 42:583.

Milne, R. C., and T. L. Quay. 1966. The foods and feeding habits of the nutria on Hatteras Island, North Carolina. Proceedings of the Annual Conference of the Southeastern Association of Game and Fish Commissioners 20:112–23.

Moinard, C., C. P. Doncaster, and H. Barré. 1992. Indirect calorimetry measurements of behavioral thermoregulation in a semiaquatic social rodent, *Myocastor coypus*. Canadian Journal of Zoology 70:907–11.

Morgan, R. P., II., G. R. Willner, and J. A. Chapman. 1981. Genetic variation in Maryland nutria, *Myocastor coypus*. Pages 30–37 *in* J. A. Chapman and D. Pursley, eds. worldwide furbearer conference proceedings. Frostburg, MD.

Murua, R., O. Neumann, and I. Dropelmann. 1981. Food habits of *Myocastor coypus* in Chile. Pages 544–58 *in* J. A. Chapman and D. Pursley, eds. worldwide furbearer conference proceedings. Frostburg, MD.

Myers, R. S., G. P. Shaffer, and D. W. Llewellyn. 1995. Baldcypress (*Taxodium distichum* (L. Rich.) restoration in southeast Louisiana: The relative effects of herbivory, flooding, competition, and macronutrients. Wetlands 15:141–48.

Nagahana, M., R. Hatsushika, M. Shimizu, and S. Kawakami. 1977. Does the nutria, *Myocastor coypus,* serve as the reservoir of the liver fluke, *Clonorchis sinensis?* Japanese Journal of Parasitology 26:41–45 (In Japanese with English summary).

Neville, R. W. J., B. J. Weir, and N. R. Lazarus., 1974. Hystricomorph insulin. Symposia of the Zoological Society of London 34:417–35.

Nevo, E. 1978. Genetic variation in natural populations: Patterns and theory. Theoretical Population Biology 13:121–77.

Newson, R. M. 1966. Reproduction in the feral coypu (*Myocastor coypus*). Symposia of the Zoological Society of London 15:323–34.

Newson, R. M. 1969. Population dynamics of the coypu, *Myocastor coypus* (Molina), in eastern England. Pages 203–4 *in* K. Petrusewics and L. Ryszkowky, eds. Energy flows through small mammal populations. Polish Scientific Publishers, Warsaw.

Newson, R. M., and R. G. Holmes. 1968. Some ectoparasites of the coypu (*Myocastor coypus*) in eastern England. Journal of Animal Ecology 37:471–81.

Nichols, J. D., and R. H. Chabreck. 1974. A survey of fur resources of the Atchafalaya River flood plain in Louisiana. Proceedings of the Southeastern Association of Game and Fisheries Commisioners 27:157–63.

Norris, J. D. 1967a. A campaign against feral coypus (*Myocastor coypus* Molina) in Great Britain. Journal of Applied Ecology 4:191–99.

Norris, J. D. 1967b. The control of coypus (*Myocastor coypus* Molina) by cage trapping. Journal of Applied Ecology 4:167–89.

Nowak, R. M. 1991. Walker's mammals of the world, 5th ed. Johns Hopkins University Press, Baltimore.

Nyman, J. A., R. H. Chabreck, and N. W. Kinler. 1993. Some effects of herbivory and 30 years of weir management on emergent vegetation in brackish marsh. Wetlands 13:165–75.

Osgood, Wilfred H. 1943. The mammals of Chile. Field Museum of Natural History Zoological Series 30:1–268.

Otis, D. L., K. P. Burnham.G. C. White, and D. R. Anderson. 1978. Statistical inference from capture data on closed animal populations. Wildlife Monographs 62.

Packard, R. L. 1967. Octodontoid, Bathyergoid, and Ctenodactlyoid rodents. Pages 273–290 *in* S. Anderson, and J. K. Jones Jr., eds. Recent mammals of the world. The Ronald Press Co., New York, NY.

Page, C. A., V. T. Harris, and J. Durand. 1957. A survey of virus in nutria. Southwestern Louisiana Journal 1:207–10.

Palmisano, A. W., and H. H. Dupuie. 1975. An evaluation of steel traps for taking fur animals in coastal Louisiana. Proceedings of the Southeastern Association of Game and Fish Commissioners 29:342–47.

Palomares, F., R. Bó, J. F. Beltrán, G. Villafañe, and S. Moreno. 1994. Winter circadian activity pattern of free-ranging coypus in the Paraná River delta, eastern Argentina. Acta Theriologica 39:83–88.

Pascal, M., E. Fichet, H. Burin des Roziers, and P. Douville. 1989. Methode de marquage des tissues calcifies du ragondin au moyen d'appats additionnes de fluoromarqueurs vitaux. Revue d'Ecologiela Terre et la Vie 44:191–200 (In French with English summary).

Patterson, B., and R. Pascual. 1968. New echimyid rodents from the Oligocene of Patagonia, and a synopsis of the family. Breviora 301:1–14.

Patterson, B., and A. E. Wood. 1982. Rodents from the Deseadan Oligocene of Bolivia and relationships of the Caviomorph. Bulletin of the Museum of Comparative Zoology (Cambridge; MA) 149:371–543.

Pavlasek, I., and B. Kozakiewicz. 1991. Coypus (*Myocaster coypus*) as a new host of *Cryptosporidium parvum* (Apicomplexa: Cryptsporidiidae). Folia Parasitology 38:90.

Peloquin, E. P. 1969. Growth and reproduction of the feral nutria *Myocastor coypus* (Molina) near Corvallis, Oregon. M.S. Thesis, Oregon State University, Corvallis.

Prasad, H. 1960. Two new species of coccidia of the coypu. Journal of Protozoology 7:207–10.

Pridham, T. J., and E. L. Thackeray. 1959. The isolation of a streptococcus, Lancefield's group D, from nutria. Canadian Journal of Comparative Medicine and Veterinary Science 23:81–83.

Pridham, T. J., J. Bud, and L. H. Karstad. 1966. Common diseases of fur bearing animals. Part 2: Diseases of chinchillas, nutria and rabbits. Canadian Journal of Veterinary Medicine 7:84–87.

Ramsey, P. R., R. M. Edmunds, G. Linscombe, and N. W. Kinler. 1981. Factors influencing blood chemistry in nutria. Pages 325–42 *in* J. A. Chapman and D. Pursley, eds. Worldwide furbearer conference proceedings. Frostburg, MD.

Ramsey, P. R., R. M. Edmunds, G. Linscombe, and N. W. Kinler. 1985. Protein polymorphisms in feral populations of nutria (*Myocastor coypus*). Acta Zoologica Fennica 170:35–36.

Ras, L. B. 1999. Population estimates and movements of nutria (*Myocastor coypus*) at Tudor Farms, Dorchester County, Maryland. M.S. Thesis, University of Maryland–Eastern Shore, Princess Anne.

Redding, T. W., Z. Itoh, and A. V. Schally. 1968. Effect of thyrotropin (TSH)-releasing factor (TRF) on plasma TSH levels in nutria (*Myocastor coypus*). General and Comparative Endocrinology 12:391–94.

Reeves, S. A., and M. B. Usher. 1989. Application of a diffusion model to the spread of an invasive species: The coypu in Great Britain. Ecological Modelling 47:217–32.

Reggiani, G., L. Boitani, S. D'Antoni, and R. De Stefano. 1993. Biology and control of the coypu in the Mediterranean area. Supplementi Richerche di Biologia della Selvaggina 21:67–100.

Reggiani, G., L. Boitani, and R. De Stefano. 1995. Population dynamics and regulation in the coypu *Myocastor coypus* in central Italy. Ecography 18:138–46.

Roberts, C. M., and J. S. Perry. 1974. Hystricomorph embryology. Symposia of the Zoological Society of London 34:333–60.

Robicheaux, B. L. 1978. Ecological implication of variably spaced ditches on nutria in a brackish marsh. Rockefeller Refuge, Louisiana. M.S. Thesis, Louisiana Technical University, Ruston.

Robicheaux, B., and G. Linscombe. 1978. Effectiveness of live-traps for capturing furbearers in a Louisiana coastal marsh. Proceedings of the Annual Conference of the Southeastern Association of Fish and Wildlife Agencies 32:208–12.

Roth, E. E., W. V. Adams, G. E. Sanford, B. Greer, and R. Mayeux. 1962. *Leptospira paidjan* (Bataviae serogroup) isolated from nutria in Louisiana. Public Health Reports 77:583–87.

Rowlands, I. W., and R. B. Heap. 1966. Histological observations on the ovary and progesterone levels in the coypu, *Myocastor coypus*. Symposia of the Zoological Society of London 15:335–52.

Ryszkowski, L. 1966. The space organization of nutria (*Myocastor coypus*) populations. Symposia of the Zoological Society of London 18:259–65.

Scheelje, R. 1982. Sumpfbiber: Zucht und Haltung. Annimal Verlag, Burgdorf, Germany.

Schitoskey, F., Jr., J. Evans, and G. K. La Voie. 1972. Status and control of nutria in California. Pages 15–17 *in* R. E. Marsh, ed. Proceedings of the fifth vertebrate pest conference. Fresno, CA.

Shaffer, G. P., C. E. Sasser, J. G. Gosselink, and M. Rejmánek. 1992. Vegetation dynamics in the emerging Atchafalaya Delta, Louisiana, USA. Journal of Ecology 80:677–87.

Shirley, M. G., R. H. Chabreck, and G. Linscombe. 1981. Foods of nutria in fresh marshes of southeastern Louisiana. Pages 517–30 *in* J. A. Chapman and D. Pursley, eds. Worldwide furbearer conference proceedings. Frostburg, MD.

Simpson, G. G. 1945. The principles of classification and a classification of mammals. Bulletin of the American Museum of Natural History 85:1–350.

Simpson, T. R., and W.G. Swank. 1979. Trap avoidance by marked nutria: A problem in population estimation. Proceedings of the Annual Conference of the Southeastern Association of Fish and Wildlife Agencies 33:11–14.

Sirotkin, A. V., D. Mertin, K. Suvegova, A. V. Mararevich, H. G. Genieser, M. R. Luck, and L. V. Osadchuk. 2000. Effect of restricted food intake on production, catabolism, and effects of IGF-I and cyclic nucleotides in cultured ovarian tissue of domestic nutria (*Myocastor coypus*). General and Comparative Endocrinology 117:207–127.

Skowron-Cendrzak, A. 1956. Sexual maturation and reproduction in *Myocastor coypus*, Part 1: The oestrus cycle. Folia Biologica (Prague) 4:119–38.

Skrzydlewski, A. 1966. Postnatal development of nutria. Roczniki Wyzszej Szkoly Rolniczej no Poznaniu 32:317–36.

Stanley, H. P., and H. H. Hillemann. 1960. Histology of the reproductive organs of nutria, (*Myocastor coypus*, Molina). Journal of Morphology 106:277–99.

Steffin, J. 1955. Observations on diseases occurring in mass breeding of coypu. Medycyna Weterynaryjna 11:270–75.

Swank, W. G., and G. A. Petrides. 1954. Establishment and food habits of the nutria in Texas. Ecology 35:172–76.

Szynkiewicz, E. 1968. Studies on antigenic differentiation of blood in the coypu (*Myocastor coypus* Molina 1792). Proceedings of the European Conference on Animal Blood Groups and Biochemical Polymorphism 11:567–70.

Szynkiewicz, E. 1971. Investigations on differentiation of beta globulin subfractions in blood serum of nutria (*Myocastor coypus* Molina 1792). Genetica Polonica 12:465.

Tam, W. H. 1974. The synthesis of progesterone in some hystricomorph rodents. Symposia of the Zoological Society of London 34:363–84.

Thristrup, K., R. Verger, and F. Carriere. 1994. Evidence for a pancreatic lipase subfamily with new kinetic properties. Biochemistry 33:2748–56.

Thristrup, K., F. Carriere, S. A. Hjorth, P. B. Rasmussen, P. F. Neilsen, C. Ladefoged, L. Thim, and E. Bel. 1995. Cloning and expression in insect cells of two pancreatic lipases and a procolipase from *Myocastor coypus*. European Journal of Biochemistry 227:186–93.

Tsigalidou, V. 1967. The chromosomes of the nutria. Rivista Zootecnica (Special Issue: International Symposium on Zootechnology 1967:421–22.

Tsigalidou, V., A. G. Simotas, and A. Fasoulas. 1966. Chromosomes of the coypus (*Myocastor coypus* Molina). Nature 211:994–95.

Valentine, J. M., Jr., J. R. Walther, K. M. McCartney, and L. M. Ivy. 1972. Alligator diets on the Sabine National Wildlife Refuge. Journal of Wildlife Management 36:809–15.

Van Forest, A. 1980. Use of ketamine/xylazine combination for tail amputation in nutria (*Myocastor coypus*). Journal of Zoo Animal Medicine 11:19–20.

Van Horne, B. 1983. Density as a misleading indicator of habitat quality. Journal of Wildlife Management 47:893–901.

Vickery, W. L., and J. S. Millar. 1984. The energetics of huddling by endotherms. Oikos 43:88–93.

Visser, J. M. 1994. Erosion increased by nutria grazing. Louisiana Environmentalist 2:18–21.

Waitkins, S. A., S. Wanyangu, and M. Palmer. 1985. The coypu as a rodent reservoir of leptospira infection in Great Britain. Journal of Hygiene 95:409–17.

Wanyangu, S. W., S. A. Waitkins, and M. F. Palmer. 1986. Isolation of leptospires from a one week dead coypu (*Myocastor coypus* Molina). International Journal of Zoonoses 13:236–40.

Wanyangu, S. W., M. F. Palmer, W. J. Zochowski, and S. A. Waitkins. 1987. Comparison of the DIFCO and Patoc 1 slide antigens in the screening of leptospirosis. Comparative Immunology, Microbiology, and Infectious Diseases 10:155–61.

Warkentin, M. J. 1968. Observations on the behavior and ecology of the nutria in Louisiana. Tulane Studies in Zoology and Botany 15:10–17.

Webster, W. D., J. F. Parnell, and W. C. Briggs, Jr. 1985. Mammals of the Carolinas, Virginia, and Maryland. University of North Carolina Press, Chapel Hill.

Weir, B. J. 1974. Reproductive characteristics of hystricomorph rodents. Symposia of the Zoological Society of London 34:265–301.

Weir, B. J., and I. W. Rowlands. 1974. Functional anatomy of the hystricomorph ovary. Symposia of the Zoological Society of London 34:303–32.

Wentz, W. A. 1971. The impact of nutria (*Myocastor coypus*) on marsh vegetation in the Willamette Valley, Oregon. M.S. Thesis, Oregon State University, Corvallis.

Wetzel, R. G. 1983. Limnology, 2nd ed. Saunders, New York.

Whitaker, J. O., Jr., and W. J. Hamilton, Jr. 1998. Mammals of the Eastern United States, 3rd ed. Cornell University Press, Ithaca, NY.

Wibawan, I. W. T., C. Lammler, and J. Smola. 1993. Properties and type antigen patterns of group B streptococcal isolates from pigs and nutrias. Journal of Clinical Microbiology 31:762–64.

Willner, G. R. 1982. Nutria. Pages 1059–76 *in* J. A. Chapman and G. A. Feldhamer, eds. Wild mammals of North America. Johns Hopkins University Press, Baltimore.

Willner, G. R., J. A. Chapman, and D. Pursley. 1979. Reproduction, physiological responses, food habits, and abundance of nutria on Maryland marshes. Wildlife Monographs 65:1–43.

Willner, G. R., K. R. Dixon, J. A. Chapman, and J. R. Stauffer, Jr. 1980. A model for predicting age-specific body weights of nutria without age determination. Journal of Applied Ecology 7:343–47.

Willner, G. R., K. R. Dixon, and J. A. Chapman. 1983. Age determination and mortality of the nutria (*Myocastor coypus*) in Maryland, U.S.A. Zeitschrift für Säugetierkunde 48:19–34.

Wilsey, B. J., and R. H. Chabreck. 1991. Nutritional quality of nutria diets in three Louisiana wetland habitats. Northeast Gulf Science 12:67–72.

Wilsey, B. J., R. H. Chabreck, and R. G. Linscombe. 1991. Variation in nutria diets in selected freshwater forested wetlands of Louisiana. Wetlands 11:263–78.

Wilson, E. D., and A. A. Dewees. 1962. Body weights, adrenal weights and oestrous cycles of nutria. Journal of Mammalogy 43:362–64.

Wolfe, J. L., and D. K. Bradshaw. 1986. Homing behavior in nutria. Journal of the Mississippi Academy of Sciences 31:1–4.

Wolfe, J. L., D. K. Bradshaw, and R. H. Chabreck. 1987. Alligator feeding habits: new data and a review. Northeast Gulf Science 9:1–8.

Woods, C. A. 1972. Comparative myology of jaw, hyoid, and pectoral appendicular regions of New and Old World hystricomorph rodents. Bulletin of the American Museum of Natural History 147:115–98.

Woods, C. A. 1982. The history and classification of South American hystricognath rodents: reflections on the far away and long ago. Pages 377–92 *in* M. A. Mares and H. H. Genoways, eds. Mammalian biology in South America, (Special publications series). Pymatuning Laboratory of Ecology, University of Pittsburgh, Pittsburgh, PA.

Woods, C. A., and E. B. Howland. 1979. Adaptive radiation of capromyid rodents: Anatomy of the masticatory apparatus. Journal of Mammalogy 60:95–116.

Woods, C. A., L. Contreras, G. Willner-Chapman, and H. P. Widden. 1992. *Myocastor coypus*. Mammalian Species 398:1–8.

Dixie L. Bounds, U.S. Geological Survey, Maryland Cooperative Fish and Wildlife Research Unit, University of Maryland–Eastern Shore, Princess Anne, Maryland 21853. Email: dlbounds@mail.umes.edu.

Mark H. Sherfy, U.S. Geological Survey, Northern Prairie Wildlife Research Center, Jamestown, North Dakota 58401. Email: msherfy@usgs.gov.

Theodore A. Mollett, Department of Agriculture, University of Maryland–Eastern Shore, Princess Anne, Maryland 21853. Email: tamollett@mail.umes.edu.

53

Wild Horse

Equus caballus and Allies

Stephen H. Jenkins
Michael C. Ashley

NOMENCLATURE

Common Names. Feral horse, wild horse, mustang
Scientific Name. *Equus caballus*

Common Names. Feral burro, feral donkey
Scientific Name. *Equus asinus* (syn *E. africanus* for populations of the wild ass)

Free-roaming horses (Fig. 53.1) and burros in western North America and elsewhere are properly called feral horses and burros because they are descended from domesticated animals (Berger 1986). However, these animals are commonly called wild horses and burros by wildlife managers as well as members of the general public. The last subspecies of *Equus caballus* to go extinct in the wild was the takhi (Przewalski's horse), but these horses have been maintained and bred in captivity and are being reintroduced in Russia, Mongolia, and China (Moehlman 2000).

HISTORY AND ORIGIN

The study of horse evolution has blossomed in recent years because of improvements in technology that have allowed for reinterpretation of equid evolution based on older specimens and recent finds. Additional fossil discoveries and technological advances in chemistry and physics have improved methods of dating strata (Flynn et al. 1984; MacFadden and Dobie 1998) and inferring ancient climates and ecologies (MacFadden and Cerling 1994; Wang et al. 1994; Bocherens et al. 1996; Bryant et al. 1996; MacFadden and Cerling 1996; Latorre et al. 1997; Sharp and Cerling 1998; MacFadden 2000).

The perception of equid evolution has changed considerably since the nineteenth century when workers such as O. C. Marsh depicted evolution of horses as a ladder-like, orthogenetic process in which one species smoothly graded into the next, more advanced species (Gould 1987). The modern view is that equid evolution has been bushy (cladogenetic) with many evolutionary dead-ends (Fig. 53.2) (Simpson 1951; MacFadden 1985). Many species and even genera coexisted during the Miocene and the Pliocene. The distribution of more advanced evolutionary traits was not confined to any single lineage, with specializations of morphology and diet appearing in a number of species in different genera (Simpson 1951; Woodburne and MacFadden 1982; MacFadden 1986b, 1992a). For example, the dietary shift from browsing to grazing was not complete, as grazing and browsing species coexisted during the Miocene (Fig. 53.2) (MacFadden 1985, 1988a). Interestingly, it had been assumed that the evolution to hypsodonty (increased height of tooth crowns) and elongation of teeth marked a change to obligate grazing in fossil equids (Rensberger et al. 1984; Janis and Ehrhardt 1988; MacFadden 1988b; Janis 1990). However, recent work by MacFadden et al. (1999), who studied carbon isotope composition and microwear patterns of 5-million-year-old fossil teeth, revealed that two of six sympatric hypsodont equid species were browsers and a third included a large percentage of forbs and browse in its diet. MacFadden et al. (1999) inferred that this indicated resource partitioning among species morphologically adapted for grazing. Changes in limb morphology were also not tied to changes in diet, as three-toed browsers (anchitheres) and grazers (merychippines) coexisted during the Miocene, as did one- and three-toed grazers from the mid-Pliocene to the Pleistocene (Fig. 53.2) (Simpson 1951; MacFadden 1985).

Horse phylogeny has undergone many revisions with new fossil discoveries and the advent of new technologies (Woodburne and MacFadden 1982; MacFadden 1985, 1988a, 1997; Hooker 1994; Prado and Alberdi 1996), but there is still much to be resolved about the evolution of fossil horse species and the phylogenetic relationships of fossil and present-day equids (MacFadden 1992a). Modern molecular techniques have been applied to improve our understanding of equid and perissodactyl evolution, but have yet to fully resolve the relationships of extant equids (George and Ryder 1986; Wijers et al. 1993; Breen et al. 1994; Babini et al. 1995; Ishida et al. 1995; Jordana et al. 1995; Oakenfull and Clegg 1998; Holmes and Ellis 1999; Norman and Ashley 2000).

Origins. The earliest fossil members of the family Equidae were of the genus *Hyracotherium,* originally identified as *Eohippus* in North America until later comparisons with Old World fossil species of *Hyracotherium* placed them in the same genus (Simpson 1951). Species of these small browsing mammals, 25–50 cm tall, were Holarctic in distribution during the early Eocene, 57 to 52 million years ago (mya) (MacFadden 1988a), but the precise center of their origin has not been determined (MacFadden 1992a). These animals had four toes on the front feet and three on the rear (Simpson 1951), and were relatively plantigrade (Radinsky 1966). Based on dental morphology, they were browsers, likely subsisting on a diet of leaves, forbs, and fruit (Janis and Ehrhardt 1988; Janis 1990). By the middle Eocene (ca. 50 mya), continental drift had closed the DeGeer Bridge, isolating North American hyracotheres from the remaining Old World species (McKenna 1975). The Paleotheriidae, the Eurasian group of horselike perissodactyls that evolved from the remaining hyracotheres, went extinct in the late Oligocene (MacFadden 1988a). The result was that the family Equidae evolved in isolation in North America for about 26 million years (Simpson 1951; MacFadden 1985, 1988a,1992a). Almost all of the equid genera evolved in North America until the Miocene, Pliocene, and Pleistocene, when new dispersal allowed for radiation in both the Old and New Worlds (Fig. 53.2).

Ancient Dispersal. Following closing of the DeGeer Bridge, which linked North America to northern Europe, there were three major periods of dispersal from North America to the Old World across the Bering Land Bridge into eastern Siberia during times of low sea level and two dispersals to South America across the Isthmus of Panama. At the beginning of the Miocene (ca. 24 mya), *Anchitherium,* a three-toed browsing species, dispersed across the Bering Land Bridge to the Old World (Simpson 1951; MacFadden 1988a). Old-world *Anchitherium* species dispersed throughout Eurasia and into Africa, persisting until the late Miocene (MacFadden 1992a).

The next major dispersal of equids to the Old World came during the middle Miocene (ca. 10 mya), when periods of low sea level again

FIGURE 53.1. A band of feral horses in west-central Nevada. SOURCE: Photo by M. C. Ashley.

opened up the Bering Land Bridge, allowing the three-toed hipparions, both browsing and grazing species, to disperse to the Old World, where they gave rise to many new species in Eurasia (Woodburne and Burnor 1980; MacFadden 1992a). At the same time, this period saw the diversification of the tribe Equini, monodactyl grazing equids, in North America. As a result, the Miocene marked the acme of equid diversity between about 15 and 8 mya, but the Old World hipparion horses went extinct by the end of the Pliocene (MacFadden 1985).

Dispersal of the monodactyl equids from North America spanned the end of the Pliocene and the early Pleistocene (8–5 mya; MacFadden 1992a). The *Onohippidion, Hippidion,* and *Parahippidion* genera of equids dispersed to South America about 3–2 mya with the formation of the Isthmus of Panama, only to go extinct by the end of the Pleistocene. The genus *Equus* first appeared in the fossil record at about 3.7 mya, dispersed across Beringia, and spread throughout Eurasia and Africa about 2.5 mya (Lindsay et al. 1980). *Equus* had also dispersed across the Panamanian Land Bridge to South America, making it the most widespread equid genus. By the end of the Pleistocene, however, *Equus* was no longer present in its original home of North America (Simpson 1951). The causes of its extinction in North America are debatable, but it is likely that a combination of changes in climate and vegetation (Webb 1977; Hulbert 1993; MacFadden and Cerling 1996; Sharp and Cerling 1998; MacFadden 1992b) and the arrival of humans eliminated horses in North America along with many other megafaunal species (Martin and Klein 1984; Azzaroli 1991; MacFadden 1992a).

Morphology. The most notable evolutionary changes in equids were in their feet, teeth, and body size. The number of functional toes decreased from 4 in the manus and 3 in the pes (4/3) during the Eocene to 3/3 in *Mesohippus* in the Oligocene, and eventually to 1/1 beginning in the Pliocene (Simpson 1951). Such reductions in the number of functional toes neither occurred all at once nor were universal within genera (Simpson 1951; MacFadden 1992a). With these changes came reductions and fusions of metatarsal and metacarpal bones, thickening of many of the remaining structures, and general lengthening of the legs. These changes greatly improved the cursorial ability of many horse species as locomotion progressed from plantigrade to digitigrade and eventually unguligrade (MacFadden 1992a). The three-toed *Merychippus* was the first of the equids to be truly unguligrade (Thomason 1986). It must be noted, however, that changes in the number of toes and the progression to unguligrade locomotion were not paired events (MacFadden 1992a).

Hyracotheres had low-crowned, lobed molars and premolars suitable for a diet of fruits, leaves, and forbs (Rensberger et al. 1984; Janis and Ehrhardt 1988; MacFadden 1988b; Janis 1990). As species evolved toward grazing, equid teeth increased in crown height (hypsodonty) and changed in the occlusal surface from enamel lobes to ridges of enamel and cementum, more suitable for grinding. Tooth length also increased, accommodating for extra wear from silica found in the tissues of many grasses (Rensberger et al. 1984; Janis and Ehrhardt 1988; MacFadden 1988b; Janis 1990). Modern horses are obligate grazers, with grasses making up the majority of their diet (see below).

Average body size of equids increased from the Eocene through the Pleistocene, but there were sometimes considerable differences among contemporary species (Fig. 53.3) (MacFadden 1986a; Alberdi et al. 1995). Although equid body size increased in general over time, markedly so in the late Miocene, members of some genera became smaller in body size (MacFadden 1986a; Alberdi et al. 1995).

DISTRIBUTION

Feral horses and burros are widely distributed on public rangelands in 10 western U.S. states (Fig. 53.4). Most of these animals occur in locations identified and administered as herd management areas by the Bureau of Land Management (BLM). BLM personnel estimated that there were about 42,000 feral horses and 5000 feral burros on public

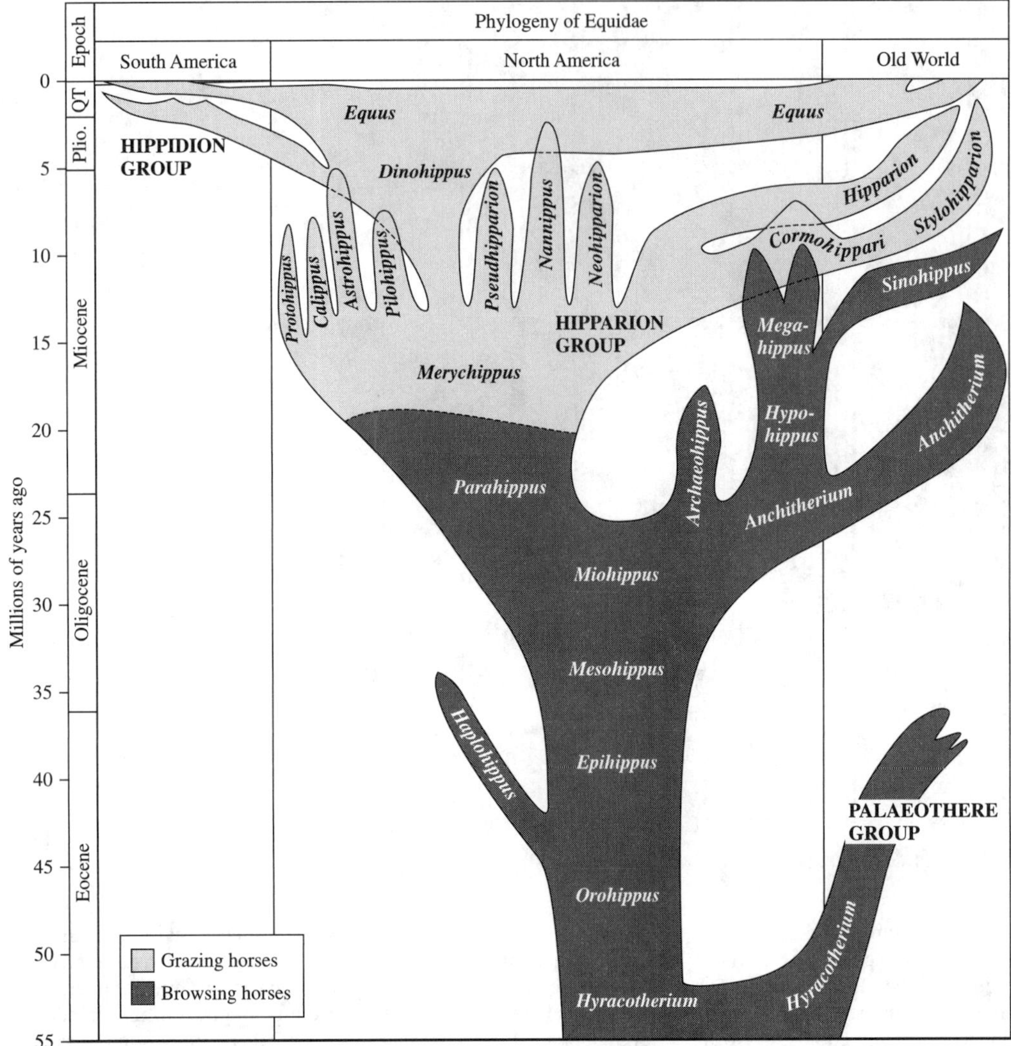

FIGURE 53.2. Fifty-seven million years of evolutionary history of the Equidae. SOURCE: From MacFadden (1985), with permission of the Paleontological Society.

lands under their jurisdiction in 1999, compared to 53,000 horses and 6800 burros in 1976 (Fig. 53.5) (Slade and Godfrey 1982). As in 1976, about half of the horses occurred in Nevada in 1999 and most of the burros were in Arizona and California. In 1996, there were about 1800 feral horses and 200 feral burros on public lands administered by the U.S. Forest Service in seven western states (U.S. Department of the Interior and U.S. Department of Agriculture, 1997).

Feral horses also occur on barrier islands such as Assateague Island, Maryland (Rubenstein 1986), and Cumberland Island, Georgia (Goodloe et al. 2000). Feral donkeys occur on Ossabaw Island, Georgia (McCort 1980), and St. John Island in the Caribbean (Rudman 1998). Outside North America, feral equids are widespread, occurring in Australia (Choquenot 1991), China (Gao and Gu 1989), France (Feh 1999), New Zealand (Cameron et al. 2000), and Russia (Paklina and Klimov 1990), for example. Research in diverse habitats provides a wealth of information for comparative studies of behavior, population dynamics, and social organization (e.g., Linklater 2000).

DESCRIPTION

Adult feral horses (Fig. 53.6) have the dental formula I 3/3, C 0–1/0–1, P 3/3, M 3/3 (American Association of Equine Practitioners 1981). There is a pronounced diastema (gap) between the incisors and cheek teeth. Canines, if present, occur at the anterior end of the diastema at varied distances from the third incisors. They are generally smaller in females. The cheek teeth are hypsodont with large occlusal surfaces and ridges of cementum, enamel, and dentin, suitable for processing a diet of grasses containing variable amounts of silica (Rensberger et al. 1984; Janis and Ehrhardt 1988; Janis 1990).

As stated above, the leg morphology of horses has undergone considerable evolutionary change: reduction of the number of toes, reduction and fusion of the ulna to the radius and of the fibula to the tibia, and thickening and extension of the metapodials (equivalents of carpal and metacarpal bones in humans; MacFadden 1992a). Modifications of the digital ligaments in modern horses gave rise to "springing" locomotion in which a large proportion of the energy generated when a horse's foot impacts the ground is stored in the elastic ligaments. Energy is returned when the foot leaves the ground and the hoof automatically springs back. Thus, elongation of the leg and modification of ligaments worked together to improve the efficiency of locomotion of modern horses compared to their ancestors (MacFadden 1992a).

Size varies within and among populations. Differences are likely attributable to different environmental conditions, such as amount and quality of food and water available, and to the varieties of individuals introduced to different sites over the years. Males ($\overline{X} = 444 \pm 40$ kg) weighed slightly but not significantly more than females ($\overline{X} = 413 \pm 54$ kg) in a population in northwest Nevada (Berger 1986). The African

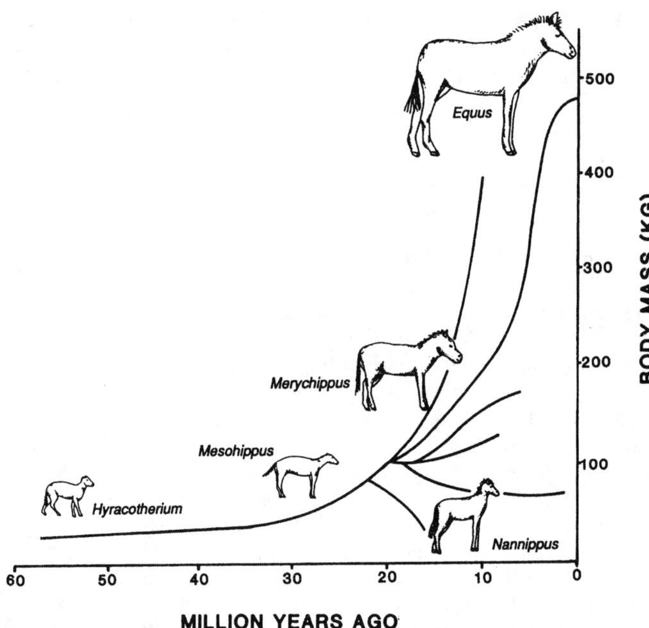

FIGURE 53.3. General trends in the evolution of body size of five genera of fossil horses, based on 40 North American species from most of the principal clades. There was very little change in body size during the first half of horse evolution until the late Cenozoic, when major diversification of body size occurred. SOURCE: From MacFadden (1988b), with permission of Plenum Press.

wild ass (*Equus africanus*), progenitor of the domestic burro, has an approximate body weight of 250 kg and a shoulder height of 125 cm (Nowak 1991). Domestic breeds range from 80 to 150 cm in shoulder height (Nowak 1991). Mass of feral burros in the Virgin Islands ranged from 90 to 115 kg (Turner et al. 1996).

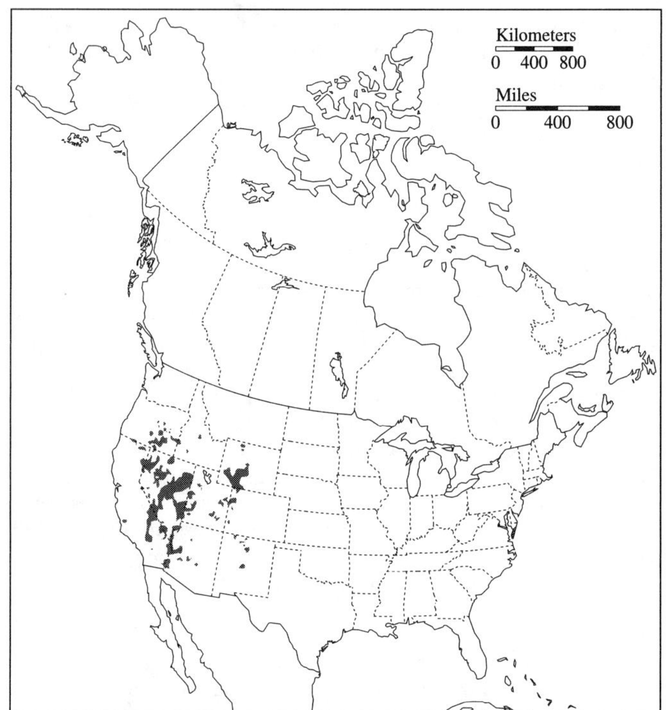

FIGURE 53.4. Distribution of feral horses and burros. Small populations on barrier islands in the western Atlantic are not shown. SOURCE: Data from the National Wild Horse and Burro Program Office, Bureau of Land Management.

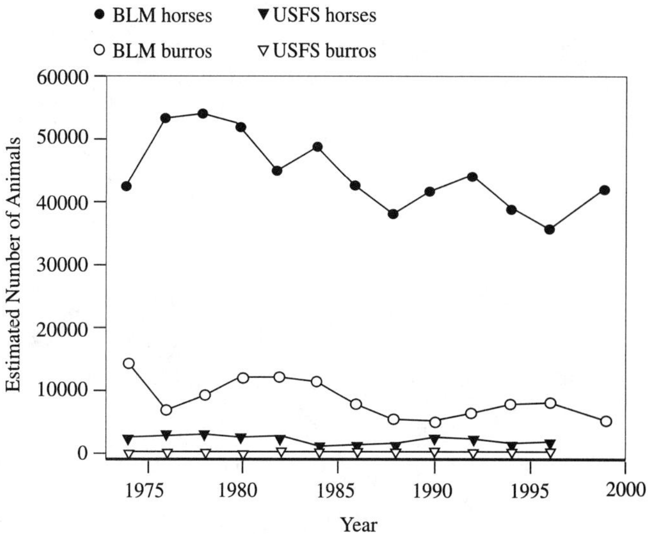

FIGURE 53.5. Estimated numbers of feral horses and burros on lands administered by the Bureau of Land Management (BLM) and U.S. Forest Service (USFS) in the United States, 1974–1999. SOURCE: Data from the National Wild Horse and Burro Program Office, Bureau of Land Management.

Feral horses exhibit the full range of colors and patterns found in domestic breeds, likely due to numerous introductions during the last 150–200 years or lack of natural selection based on these traits (Berger 1986). Individual populations often differ in the variety of color patterns present, but can include appaloosas, bays, blacks, browns, buckskins, chestnuts, duns, grays, palominos, and pintos. Some populations retain vestiges of the original Spanish introductions: striped legs, a dorsal stripe, and slightly convex nasal bone structure.

GENETICS

Genetic and biochemical investigations of the horse have been done almost exclusively with domestic animals. From these studies have come hundreds of allozyme characterizations as well as microsatellite and gene sequences suitable for use as molecular markers (Benson et al. 1999), which have been useful in studies of feral horses. Goodloe et al. (1991) used allozyme data to assess the conservation value (Moritz 1994) of feral horse populations on four barrier islands in the eastern United States. These populations were genetically similar to various breeds of domestic horses. Based on this and their calculations of minimum effective population size, Goodloe et al. (1991) argued that the populations could be reduced in size to minimize deleterious impacts on native vegetation without detrimental loss of genetic diversity.

Early studies of feral horse parentage used blood-group and biochemical markers to rule out putative sires. Bowling and Touchberry (1990) found that about 30% of foals were sired by males other than the dominant band stallion, whereas Duncan et al. (1984) found far less reproductive success of subordinate and bachelor males. Since then, species-specific microsatellite markers (Ellegren et al. 1992; Breen et al. 1994; Guerin et al. 1994; Marklund et al. 1994; Van Haeringen et al. 1994; Bowling et al. 1997) have become an effective new tool in parentage determination for horses. Ashley (2000) used these markers to show that subordinate and bachelor males were responsible for 31% and 21% of reproduction, respectively, in a population of approximately 225 feral horses in the Great Basin.

As in parentage determination, blood-group and biochemical markers have been used to study population genetics of feral horses (Plotka et al. 1988; Morris 1993). Bowling (1994) studied feral horses from seven locations in Oregon and Nevada and found that genetic variation and levels of heterozygosity of feral horses were not significantly different from those of 16 domestic horse breeds. Furthermore,

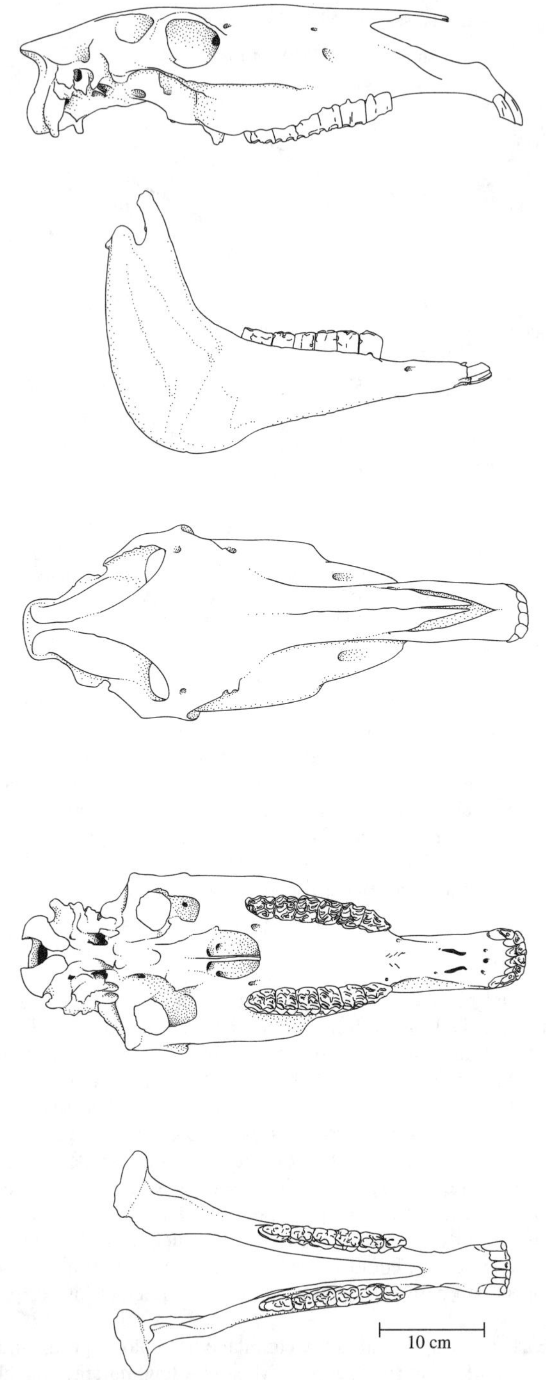

FIGURE 53.6. Skull of a feral horse. From top to bottom: lateral view of cranium, lateral view of mandible, dorsal view of cranium, ventral view of cranium, dorsal view of mandible.

Bowling (1994) constructed a phylogeny from the molecular data that conformed reasonably well to historical information about origins of the domestic breeds and the feral populations. More recently, Cothran and Singer (2000) used microsatellites to demonstrate high levels of genetic diversity and heterozygosity in a closely managed population. They inferred that there was no evidence of inbreeding in this population. Microsatellite markers have also been used to document population genetic structure due to use of specific water holes after spring snowmelt by two subpopulations of a herd managed as one unit in west-central Nevada (Ashley 2000). Although the two subpopulations had broadly overlapping ranges in winter, they were genetically distinct because of their separation during the breeding season. Management of feral horses for genetic health draws on the results of such studies.

REPRODUCTION

Feral horses may be born at any time of the year (Berger 1986), although most populations exhibit "well defined breeding and foaling season[s]" (Kirkpatrick and Turner 1986:224). At Sable Island in the North Atlantic, 45% of 220 births over 4 years occurred in May, and 95% of all births occurred between April and July (Lucas et al. 1991). At Garfield Flat in east-central Nevada, 60% of 217 births over 5 years occurred in April or May, and 97% of all births occurred between March and July (Ashley 2000). These examples illustrate the consistency in seasonal patterns of reproduction reported by Kirkpatrick and Turner (1986) for three additional populations. Feral burros exhibit similar seasonality of reproduction, at least in Australia (Choquenot 1991). Seasonality of foaling is less pronounced in domestic horses (Kirkpatrick and Turner 1986).

As the mean gestation period is about 336 days (Lush 1945 in Kiltie 1982), feral horses typically experience postpartum estrus (Berger 1986). The estrous cycle lasts 20–22 days, with behavioral estrus during the last 5–8 days of the cycle. Ovulation occurs 1 or 2 days before the end of estrus (Turner and Kirkpatrick 1986; Bennett and Hoffmann 1999). If fertilization does not occur, females may continue to cycle through the end of August (in the Northern Hemisphere), although Tyler (1972) and Berger (1986) reported that almost all mares conceived during their first or second estrus. Nelson (1980) and Berger (1983b) reported year-round mating behavior of feral horses in New Mexico and Nevada, respectively, but Kirkpatrick and Turner (1983) found that horses at Pryor Mountain, Montana, exhibited behavioral estrus without ovulation during fall.

Infanticide and induced abortion occur in many species of mammals. They have been interpreted as tactics that enhance the reproductive success of individual males that gain control of pregnant or lactating females, by causing these females to enter estrus sooner than in the absence of infanticide or induced abortion (Bruce 1969; Hiraiwa-Hasegawa 1988; Swenson et al. 1997). Berger (1983b) presented evidence of induced abortion in a population of feral horses in northern Nevada. He studied 11 cases in which males took over bands in which females were presumed to be pregnant as a result of mating with the prior stallion associated with that band. In most cases, there were forced copulations by the new males and circumstantial evidence of abortion of the original fetus and impregnation by the new male. This was the first report of induced abortion in a wild ungulate, although Duncan (1982) observed infanticide in feral horses in the Camargue region of France. However, Kirkpatrick and Turner (1991a) found no evidence of induced abortion in feral horses at Assateague Island National Seashore. They analyzed steroid metabolites in urine and feces as more definitive evidence of pregnancy than the absence of behavioral estrus, which was the basis for Berger's (1983b) work. There were also differences in population structure, social organization, and environmental conditions that might have accounted for the different results of Berger and of Kirkpatrick and Turner.

Lucas et al. (1991) provided the most extensive data on rates of fetal loss for feral horses. They sampled estrogen in feces of mares for 4 years at Sable Island, Nova Scotia. In mares that gave birth, estrogen levels in feces were consistently elevated above baseline after day 120 of gestation. Based on repeated samples from known individuals, they estimated that 26% of pregnant mares lost their fetuses after day 120. The total rate of fetal loss likely was higher, because most loss occurs earlier than this in domestic horses (Woods et al. 1987). There was substantial year-to-year variation in rates of fetal loss at Sable Island. Lucas et al. (1991) did not ascribe this to a particular environmental factor, although they suggested that severe weather in winter and early spring may have induced nutritional stress, hence abortion, in pregnant mares. They also found that rates of fetal loss were 70% for yearlings versus 5.6% for 4-year-olds, with intermediate values for intermediate age classes. Noninvasive methods for assessing the pregnancy status of feral horses by sampling feces and urine for steroid metabolites have become well standardized (Kirkpatrick et al. 1990, 1993; Kirkpatrick and Turner 1991a), which should facilitate additional detailed studies of reproduction in wild populations.

Early work on free-ranging domestic (Speelman et al. 1944) and feral horses (Boyd 1979; National Research Council 1980; Seal and Plotka 1983) suggested that mares typically produce their first foals at age 3 years. Potential rates of population growth implied by this age of sexual maturation seemed inconsistent with annual increases of 20–30% estimated for some populations based on aerial surveys, leading some to question the accuracy of the surveys (Wolfe 1980). Subsequent work, however, showed that 2-year-olds produce foals in many populations (e.g., Berger 1986; Wolfe et al. 1989; Garrott et al. 1991a; Ashley 2000). In 31 of 38 herds in Nevada, Garrott et al. (1991a) found evidence of foaling by 10% to >50% of 2-year-olds. Feral burros are also capable of first foaling at age 2 years (Johnson et al. 1987). Two important exceptions to foaling by 2-year-olds occur in populations of feral horses that have been intensively studied at Pryor Mountain in Montana (Garrott and Taylor 1990) and Assateague Island off the coast of Maryland (Keiper and Houpt 1984). Surprisingly, Bennett and Hoffmann (1999) perpetuated the misconception that feral horses rarely reproduce successfully at age 2 years.

Several authors have presented data on age-specific fecundity of feral horses (Seal and Plotka 1983; Keiper and Houpt 1984; Berger 1986; Siniff et al. 1986; Wolfe et al. 1989; Garrott and Taylor 1990; Garrott et al. 1991a; Lucas et al. 1991; Kirkpatrick 1995; Ashley 2000) and burros (Johnson et al. 1987). Various methods were used to assess successful reproduction of free-ranging horses in these studies. Garrott et al. (1991a) used the largest sample size in their analyses: >15,000 horses captured in culling operations in Nevada and Oregon between 1979 and 1987. They compared three sources of data: lactation status of mares captured at any time of year, lactation status of mares captured in June, and serum progesterone levels of mares captured in fall and winter. The estimated relationship between proportion of females foaling and age based on lactation status of mares captured throughout the year underestimated foaling rates for all ages. Garrott et al. (1991a) concluded that foaling rates of females in their prime, from 6 to 15 years of age, were between 80% and 90% and that foaling rates declined gradually with age for older females. By contrast, foaling rates for females at Pryor Mountain, Montana, were 41% for 3- to 5-year-olds, 55% for 5- to 10-year-olds, and 47% for horses >10 years old (Garrott and Taylor 1990).

Foaling rates as high as 80–90%/year imply that gestation and lactation may occur simultaneously. This may entail significant physiological stress on females, especially during harsh winters or at high population densities. Keiper and Houpt (1984) tested this idea by comparing foaling rates of horses in the Assateague Island National Seashore (AINS) and the Chincoteague National Wildlife Refuge (CNWR). Both of these locations are on Assateague Island, a barrier island off the coast of Virginia and Maryland. Most foals are removed from CNWR shortly after birth, but foals are typically not removed from the AINS population. Average foaling rate was 57.1%/year for AINS, but 74.4%/year for CNWR, suggesting that lactation by AINS mares may have inhibited their ability to produce foals the following year (Keiper and Houpt 1984). Kirkpatrick and Turner (1991b) used noninvasive sampling of steroid metabolites to study the mechanism for this difference between the two populations. They found no evidence for a higher rate of abortion for mares at AINS, and suggested that lactational anestrus accounted for the lower foaling rate at AINS. By contrast, Wolfe et al. (1989) found no difference in pregnancy rates of lactating and nonlactating mares in populations in three western states in the United States and Garrott and Taylor (1990) found that mares at Pryor Mountain had the same probability of foaling regardless of whether they produced a foal during the previous year. In the latter case, however, primiparous females had a lower probability of producing a foal the year after their first offspring than did multiparous females after successful foaling (Garrott and Taylor 1990). The Pryor Mountain data are interesting because they are inconsistent with a hypothesis of alternate-year foaling even though overall foaling rates were relatively low.

The most detailed study of maternal investment and reproductive success of feral horses was done by Cameron et al. (2000) at Kaimanawa, New Zealand. As females aged, they were more likely to give birth in successive years. Older mares also had higher average foaling rates and their offspring had lower mortality rates in the first year of life. Behavioral differences between young and old mares were subtle, leading Cameron et al. (2000) to conclude that their results were inconsistent with the hypothesis that parental investment increases with age due to a decrease in residual reproductive value. Instead, they argued that older females were more successful at raising offspring because of their greater experience.

Although Garrott (1991b) found little evidence of skewed sex ratios at birth in an analysis of data for >10,000 foals captured and removed from 74 herd management areas in Nevada, Wyoming, and Oregon, more recent detailed analyses provide convincing evidence for adjustments of sex ratios consistent with the Trivers–Willard hypothesis (Trivers and Willard 1973). This hypothesis predicts that, in polygynous species, mothers in good condition should produce relatively more sons than mothers in poor condition if "(1)... the condition of the young at the end of parental investment will tend to be correlated with the condition of the mother during parental investment; (2)... these differences... tend to endure into adulthood; and (3)... adult males will be differentially helped in reproductive success by slight advantages in condition" (Cameron et al. 1999:472). Cameron et al. (1999) argued that feral horses are especially suitable for testing the Trivers–Willard hypothesis. Several lines of evidence for feral horses and other equids support this hypothesis. Saltz and Rubenstein (1995) reported that young and old female Asiatic wild asses (*Equus hemionus*) produced relatively more females, whereas intermediate-aged females produced relatively more males in a population introduced into the Negev Desert in Israel. Young and old females should have been in poorer condition in this harsh desert environment. Monard et al. (1997) studied a population of feral horses in the Camargue in France that increased rapidly over a period of 8 years before management by culling was initiated. Body condition of the horses was used to classify years into two categories, good and poor. Mares produced fewer males in poor years than in good years; this effect was especially pronounced if their previous offspring was a male. The study of Monard et al. (1997) is not completely satisfying, however, because they did not relate the condition of individual mares to their likelihood of producing male or female foals. By contrast, Cameron et al. (1999) showed for their New Zealand population that individual mares in poor body condition at the time of conception were much more likely to produce female foals than were mares in fair to good condition. Their analyses implied that sex ratio adjustment occurred at conception rather than by differential abortion during gestation. Cameron and Linklater (2000) showed further that mares in good condition gave more maternal care to sons than to daughters. Mares in poor condition showed the opposite pattern, consistent with another prediction of the Trivers–Willard hypothesis. This relationship only existed for comparisons at the individual level (comparison of investment by the same mares that had both sons and daughters), not for those at the population level, attesting to the importance of detailed, long-term studies at appropriate scales for testing important hypotheses in behavioral ecology.

ECOLOGY

Population Growth. Early estimates of foaling rates and survival probabilities of feral horses in the western United States suggested that maximum potential rates of population growth should be no greater than about 17%/year (Conley 1979; National Research Council 1980; Wolfe 1980). However, aerial surveys of two populations in Oregon implied growth rates of about 20%/year (Eberhardt et al. 1982). Garrott et al. (1991b) analyzed aerial survey data for 22 populations. They used the index-removal method, which is based on counts before and after a known number of animals are removed from a population, to show that sighting probabilities were typically >85% for these horse populations. There was no evidence of a temporal trend in sighting probabilities, which would have biased estimates of growth rate. For 12 populations with ≥5 years of data, annual growth rates ranged from 15% to 27% ($\overline{X} = 21\%$). The inconsistency between potential rates of growth based on demographic data and higher rates estimated from aerial surveys evidently occurred because investigators such as Wolfe (1980) underestimated foaling rates, especially of 2-year-olds, and age-specific survival probabilities (Garrott et al. 1991b).

Choquenot (1990) found that two populations of feral burros in Australia grew at 23%/year and 28%/year following reductions by culling of at least 40%. Perryman and Muchlinski (1987) reported annual growth rates up to 29% for feral burros in the southwestern United States. They provided useful data on reproduction and age structure for a population in southern California, but their demographic analysis of these data was seriously flawed (Jenkins 1989). In contrast to these high rates of population growth, Saltz and Rubenstein (1995) reported that there were only 16 adult females in a population of Asiatic wild ass 10 years after a reintroduction of 14 adult females and 14 adult males in the Negev Desert in Israel.

Demography. A full understanding of the population dynamics of feral horses and burros requires long-term data on survival and reproduction of known individuals, so that age-specific survival probabilities and fecundities can be estimated. To the best of our knowledge, only three such data sets are available: for the Granite Range in northwestern Nevada (Berger 1986), for Pryor Mountain in Montana (Garrott and Taylor 1990), and for Garfield Flat in west-central Nevada (Ashley 2000). Similar data may be available for Assateague Island National Seashore (Keiper and Houpt 1984; Kirkpatrick 1995), but these data have not been published. Garrott and Taylor (1990) summarized 11 years of data through 1986 for Pryor Mountain. BLM personnel and other researchers have continued intensive study of this population, which will provide extremely valuable long-term demographic information once the data are published (L. Coates-Markle, Billings Field Office of BLM, pers. commun., 2001).

Garrott and Taylor (1990) used the survivorship model of Siler (1979) and Eberhardt (1985) to fit their data on age-specific survival for Pryor Mountain. This nonlinear model has four parameters representing survival probability from birth to age 1 year, annual survival probability of prime-age adults, modal age of senescence, and the standard deviation in the age of senescence. After excluding 1 year in which overwinter mortality was unusually high, the model fit the Pryor data reasonably well, with estimates of juvenile survival probability of 0.94, adult survival probability of 0.98, modal age of senescence of 21.4 years, and standard deviation of this age of 2.37. For comparison, Ashley (2000) estimated these values to be 0.93, 0.97, 20.0, and 4.90 for feral horses at Garfield Flat, Nevada. Though slight, these differences produced significantly lower survivorship for horses at Garfield Flat through age 21 years (Ashley 2000). The lower survivorship at Garfield Flat was combined with first foaling at age 2 years and higher foaling rates for older age classes, resulting in a higher potential rate of increase than for horses at Pryor Mountain.

Garrott and Taylor (1990) found significantly higher annual survival probabilities of females than of males at Pryor Mountain in most years, but not in a year of unusually high overwinter mortality (see below). On a much larger spatial scale, Garrott (1991b) analyzed age-specific sex ratios of about 60,000 horses removed in management operations in Nevada, Wyoming, and Oregon between 1976 and 1987. Among foals, there were approximately equal numbers of males and females. Proportions of males decreased from age 0 to age 5 years, and then increased with older ages until males exceeded females among horses >10 years old. These results suggested that mortality rates were greater for males than for females through age 5 years but greater for females after age 5 years.

Estimates of age-specific survival probabilities and fecundities can be used to develop a Leslie matrix model of population growth (Caswell 2001). This yields estimates of population growth rate and stable age distribution assuming the demographic parameters remain constant. More importantly, this modeling approach leads to estimates of relative sensitivities (called elasticities) of population growth rate to small changes in each of the demographic parameters. For feral horses, population growth rate is much more sensitive to changes in age-specific survival probabilities, especially of younger age classes, than to changes in foaling rates. For example, at Garfield Flat, total elasticity with respect to the full set of survival probabilities across all age classes is 83%, whereas total elasticity with respect to foaling rates of all age classes is 17%. This implies that manipulation of survival probabilities (by culling) will likely be more effective in controlling population growth than manipulation of fecundities (by fertility control).

Population Limitation and Regulation. Population modeling can be used to evaluate the efficacy of alternative management strategies if it incorporates information on variability as well as mean values of demographic parameters. At Pryor Mountain, annual survival probability averaged across all age classes varied between 0.87 and 0.99 for 10 years, but was only 0.49 between 1977 and 1978 (Garrott and Taylor 1990). This unusually high mortality rate was due to "alternating periods of heavy snow accumulations followed by abnormally warm temperatures creating several layers of ice within a deep layer of snow… [which] made foraging exceedingly difficult" (Garrott and Taylor 1990:606). Similarly, Berger (1983a) reported that the majority of the horses that died during his study in the Granite Range of Nevada apparently succumbed to severe winter weather at high elevations. He also described a group of foals that died after becoming mired in mud at a water hole during a drought. Across a broader region, approximately 15% of feral horses in northern Nevada died during the severe winter of 1992/93 (T. Pogacnik, National Wild Horse and Burro Program of BLM, pers. commun., 2001).

Clearly, density-independent factors influence the population dynamics of feral horses. Do density-dependent factors also influence them? Choquenot (1991) experimentally tested for density dependence in the demography of feral donkeys in Australia. He compared a population at a density of about 3.3 animals/km^2 to a population in similar habitat reduced to about 1.5 animals/km^2 by culling (this was an unreplicated experiment). The low-density population increased by 20% between 1986 and 1987, whereas the high-density population decreased by 3% during the same period. The primary demographic difference between the populations was that juvenile mortality was much higher in the high-density population (62% mortality rate for the first half-year of life vs. 21% for the low-density population). By contrast, there was no difference in fecundity between the two populations. Choquenot (1991) attributed density-dependent juvenile mortality to nutritional stress on lactating females. Freeland and Choquenot (1990) documented that the mechanism for this nutritional stress was a less diverse diet for the high-density population, which was dominated by a fibrous species of grass with low concentrations of nitrogen and other nutrients.

The feral donkeys studied by Choquenot (1991) and Freeland and Choquenot (1990) experienced no predation, so it was fairly straightforward to relate density-dependent juvenile mortality to intraspecific competition for food. In other cases, predation may be an important factor regulating feral equid populations. Turner et al. (1992) studied a population of feral horses on the border of California and Nevada that remained relatively constant from 1986 to 1991. Average mortality rate of foals was 73%, much higher than values reported for other populations of feral horses. Much of the foal mortality was due to predation by mountain lions (*Puma concolor*), which switched to mule deer (*Odocoileus hemionus*) during winter. These results suggest that mule deer supported a relatively large density of mountain lions, enabling mountain lions to regulate the feral horse population. A similar example was described by Greger and Romney (1999), who studied a population at the Nevada Test Site that declined from 65 horses in 1992 to 36 in 1998. Average annual foal mortality was about 88%, and indirect evidence suggested that mountain lions were responsible for much of this mortality.

Jenkins (2000) examined the possibility of density dependence in population growth rates of feral horses using time-series data for seven populations. Despite the large amount of effort devoted to aerial counts of feral horses by BLM personnel, relatively few data sets met minimal criteria for statistical analyses of density dependence. Nevertheless, based on a meta-analysis of these limited data, Jenkins (2000) reported a negative relationship between growth rate and population size, suggesting a general pattern of density dependence for feral horses. However, further unpublished analyses called into question the pattern of density dependence because of imprecision in the population estimates. In summary, there is limited experimental evidence of population regulation through intraspecific competition for food in feral donkeys (Choquenot 1991), indirect evidence for population regulation through predation by mountain lions for two populations of feral horses (Turner et al. 1992; Greger and Romney 1999), but little or no basis for generalizing these results.

Feeding Ecology, Resource Partitioning, and Competitive Relationships. Feral horses and burros receive a great deal of attention because of their potential interactions with other species on rangelands in the western United States. Interactions with wild and domestic ungulates have been of primary concern, although Beever (1999) assessed a wide range of community characteristics at sites with and without feral horses in the Great Basin. Despite the work by Beever, the questions of most concern have been: (1) Do feral horses have negative impacts on populations of wildlife or free-ranging domestic animals? (2) If so, what are the implications of these impacts for management? Surprisingly, the first question has not been definitively answered despite many studies on feeding habits of feral horses and other ungulates. Therefore, answers to the second question have been based largely on conjectures about biology combined with prejudices about policy.

Horses are large herbivores and use hindgut (postgastric) fermentation in an enlarged colon and cecum to acquire energy from the fiber in their diets (Janis 1976). Feral horses typically coexist with ruminant artiodactyls, which use foregut (pregastric) fermentation for dealing with high-fiber diets. Ruminants are more efficient at extracting energy and nutrients from vegetation, but horses have faster passage rates than similar-sized ruminants (e.g., cattle), especially for low-quality (high-fiber) foods (Hanley 1982; Demment and Van Soest 1985; Duncan et al. 1990; Illius and Gordon 1992). The implications of this trade-off between efficiency and passage rate are that horses require larger quantities of food, but can use lower quality food, than similar-sized and smaller ruminants. Because plants such as forbs with low to intermediate fiber content are typically scarcer and more sparsely distributed than low-quality plants like grasses, the fundamental differences in digestive physiology between ruminants and equids may facilitate coexistence, with ruminants specializing on higher quality, more sparsely distributed plant species and equids specializing on poorer quality, more abundant species (Bell 1970; Duncan et al. 1990; Illius and Gordon 1992). However, these differences in digestive physiology also suggest that equids may have an advantage over ruminants when food is abundant, but be at a disadvantage when food is scarce. Furthermore, the lower digestive efficiency of equids implies that they need to feed longer each day to meet their energy requirements. Duncan (1985) reported that horses fed 15 hr/day vs. 8–10 hr/day for cattle (Arnold and Dudzinski 1978). This may expose native equids to greater risk of predation; indeed, predation appears to have a greater impact on zebras (*Equus burchelli*) than on wildebeest (*Connochaetes taurinus*) in the Serengeti of East Africa (Schaller 1972; Sinclair 1985).

Numerous researchers have used microhistological analyses of feces to compare diets of horses, cattle, and various native ungulates in western North America. Most of these studies show that horses and free-ranging cattle feed primarily on grasses, whereas smaller, coexisting native ruminants include higher percentages of forbs and shrubs in their diets (Hubbard and Hansen 1976; Salter and Hudson 1979, 1980; Hanley and Hanley 1982; Wagner 1983; Krysl et al. 1984; McInnis and Vavra 1987; Smith et al. 1998). Graminoids often constitute >80% of horse diets, regardless of season, although in New Mexico, Hansen (1976) and Stephenson et al. (1985) reported substantially greater use of forbs and shrubs by horses than did other studies. However, the results of Stephenson et al. (1985) were based on small sample sizes in winter only.

As discussed above, differences among ungulates in digestive physiology (pregastric vs. postgastric fermentation) and body size lead to the prediction that large ungulates should be less selective and should tolerate lower quality items in their diets than smaller ungulates, especially in ruminants. The general pattern that horses and free-ranging cattle include higher percentages of grass in their diets than smaller, coexisting ruminants is consistent with this prediction. This pattern also implies greater dietary overlap between horses and cattle than between horses and smaller ungulates such as pronghorn (*Antilocapra americana*) and mule deer (Hubbard and Hansen 1976; Salter and Hudson 1980; Stephenson et al. 1985; McInnis and Vavra 1987). Dietary overlap between horses and elk (*Cervus elaphus*) is typically somewhat less than that between horses and cattle, but much greater than that between horses and mule deer or pronghorn (Hubbard and Hansen 1976; Hansen and Clark 1977; Hansen et al. 1977; Salter and Hudson 1980; Stephenson et al. 1985).

It may be tempting to infer from these kinds of data that the potential for interspecific competition between feral horses and native ungulates is relatively small, but that the potential for competition between feral horses and cattle is greater. However, this ignores two additional specific aspects of resource partitioning and one important general point. The specific issues are:

1. Sympatric ungulates may have similar diets, but forage in different habitats within a common range. If so, then dietary overlap overestimates the potential for exploitative competition. For example, Salter and Hudson (1980) found that the spatial distributions of horses and cattle at their study site in Alberta overlapped very little, although diets were quite similar, at least in summer. Cattle used areas in summer that had been used by horses earlier in the year, but in most habitat types, there was little evidence of a negative impact of grazing by horses on food availability for cattle. Salter and Hudson (1980:266) concluded that "potential for competition appeared highest between horses and cattle but grazing relationships were complex." Similarly, Miller (1983) found that horses and cattle in the Red Desert of Wyoming occurred in the same areas and had similar diets in fall, but not other seasons, suggesting that the potential for competition between these two species was higher in fall than in other seasons.
2. In addition to exploitative competition for food, horses may compete with cattle and native ungulates for access to limited water supplies. Berger (1985) reported that feral horses were socially dominant to native ungulates (mule deer, pronghorn, and bighorn sheep [*Ovis canadensis*]) in 19 of 20 observed interactions. Miller (1983) found that horses and cattle deterred pronghorn from approaching water sources, but there was no consistent tendency for horses to dominate cattle at water sources or vice versa. By contrast, Krysl et al. (1983) reported that aggression by horses toward cattle kept cattle away from artificial water sources in an enclosure study in Wyoming. However, neither Miller (1983) nor Krysl et al. (1983) presented quantitative data to support their conclusions. Thus, the potential clearly exists for interference competition between feral horses and other ungulates, especially when water is limited, but there is little concrete documentation of such competition.

More generally, none of these studies of resource partitioning demonstrate competition between feral horses and sympatric native or domestic species. Interspecific competition occurs when two species have negative effects on each other's population sizes. Common mechanisms of competition are exploitation of resources in short supply and interference with access to such resources, which cause reduced birth rates or increased mortality rates of the affected species. In the case of domestic species such as cattle, for which birth rates and mortality rates are largely controlled by managers even though animals are free-ranging, competitive effects could be expressed in reduced weight gain in the presence of feral horses (Wagner 1983). In his summary of an extensive review of feral horse management by the National Research Council (1980, 1982), Wagner (1983) concluded that there was little definitive evidence of competition between feral horses and native or domestic ungulates, despite the common assumption that rangelands are severely degraded by horse grazing. There is still little concrete evidence of such competition.

Feral burros have broader diets than do feral horses, with browse species often constituting >50% of burro diets (Browning 1960; Woodward and Ohmart 1976; Seegmiller and Ohmart 1981; but see Hansen and Martin [1973] for a report of burros in the Grand Canyon using primarily grass). Unlike feral horses, burros typically occupy habitat that is marginal for cattle, so possible competitive interactions between burros and cattle have been of less concern to managers than interactions between horses and cattle (Wagner 1983). However, burros coexist with bighorn sheep in some areas, so wildlife biologists have studied potential impacts of burros on sheep. The most detailed study of resource partitioning between feral burros and bighorn sheep was done by Seegmiller and Ohmart (1981) in western Arizona. During the single year of their study, the burro population was about four times as large as the bighorn sheep population. There was a high degree of spatial overlap throughout the year, and moderate dietary overlap. Seegmiller and Ohmart (1981) saw no evidence of aggressive interactions between the species while foraging or drinking. Although they speculated that burros might negatively affect bighorn sheep populations through overuse of shared food plants, their evidence was indirect and circumstantial (Wagner 1983).

Impacts of Horses on Vegetation. It is generally easier to examine the impacts of large mammalian herbivores on vegetation using exclosures than to study competition between wide-ranging herbivores experimentally. Beever and Brussard (2000) compared vegetation at four springs in the Clan Alpine Mountains in the Great Basin of Nevada; two of these springs had been protected from horse grazing for 8 years by exclosures and all of them had been protected from cattle grazing for 15 years. Cover and diversity of plants were much greater at the ungrazed sites than at the grazed sites. Beever and Brussard (2000) reported similar patterns for an exclosure in a high-elevation meadow in the Seven Troughs Range in Nevada. By contrast, Fahnestock and Detling (1999) compared vegetation in grazed and ungrazed sites in the Pryor Mountain Wild Horse Range in Montana, and found that plant cover differed more in the same plots between a wet and a dry year than between grazing treatments in the same year. The effect of grazing was greater in dry, lowland habitats than in more mesic, upland habitats. In particular, the most abundant grass in the lowlands, *Pseudoroegneria spicata,* had 12% less cover in long-term grazed sites than in exclosures that prevented grazing for 1–2 years before vegetation measurements. Furthermore, Fahnestock and Detling (2000) showed that this grass as well as two common upland plant species were shorter in areas subjected to long-term grazing than in areas protected from grazing, whereas there were no consistent physiological differences between individual plants in grazed and ungrazed areas. They suggested that these morphological responses to herbivory probably represented phenotypic plasticity rather than genetic changes in plant populations protected from grazing.

Through selective foraging, herbivores may alter competitive relationships between plant species, which may in turn influence other populations of herbivores. Reiner and Urness (1982) grazed horses on pastures in northern Utah to see whether consumption of grass by horses might reduce competition between grass and bitterbrush (*Purshia tridentata*), thus enhancing growth of bitterbrush. They found significantly greater twig growth of bitterbrush in pastures grazed by horses than in control pastures. In a similar study, Austin et al. (1994) found that grazing by horses caused increased production of big sagebrush (*Artemisia tridentata*) compared to plots used by horses and mule deer and by deer alone and to plots with no ungulate herbivory. They suggested that horses, because of their strong preference for grasses, might be used as a management tool to enhance production of browse for wildlife species. Of course, horse numbers were controlled at fixed levels in their experiments, which would not happen under natural conditions.

Turner (1987) and Furbish and Albano (1994) studied effects of grazing by feral horses on distribution and abundance of salt marsh grasses on two barrier islands. At Cumberland Island, Georgia, feeding and trampling by horses reduced the abundance of *Spartina alterniflora* growing at low tidal levels. However, a grass more characteristic of higher levels, *Distichlis spicata,* did not expand its range to lower levels when *Spartina alterniflora* was removed (Turner 1987). By contrast, at Assateague Island, *Distichlis spicata* increased substantially in abundance at low tidal levels over a period of 10–20 years coincident with a marked increase in the feral horse population on the island (Furbish and Albano 1994). Horses preferentially graze on *Spartina alterniflora.* Furbish and Albano argued that this reversed the normal competitive dominance of *Spartina alterniflora* over *Distichlis spicata* at low tide levels, accounting for the spread of the latter species with increased impact of horses. At Cumberland Island, by contrast, the habitat occupied by *Spartina alterniflora* was simply not suitable for *Distichlis spicata,* even after the former was reduced in abundance by grazing (Furbish and Albano 1994).

BEHAVIOR

Horses are gregarious and form harem bands, aggregations of bachelors, or occasionally mixed-sex, nonreproductive groups. In most cases, the basic reproductive unit is the band, consisting of one or more sexually mature females, their juvenile offspring, and one to three mature males (Berger 1986; Rubenstein 1986; Linklater 2000). Horses form within-band dominance hierarchies, with mature females tending to be the most dominant animals, followed by mature males, then offspring (Rubenstein 1986). Spatially, mature females and their offspring form the core of the band, with mature males occupying the periphery, often several meters away. In multiple-male bands, the dominant male still occupies a position apart from the mares and offspring, but closer to the females than subordinate males. Except for birth, juvenile dispersal, and mortality, bands are relatively stable from year to year in most populations, retaining the same adult males and females. Mares or stallions occasionally change bands (Berger 1986; Ashley 2000) and sometimes a bachelor will successfully depose a band stallion (Berger 1986). Bachelor males also form groups that are less stable and can change in composition daily. Solitary males may also be encountered and are usually older animals. Bands are generally nonterritorial, with overlapping home ranges (Berger 1986; Linklater 2000), although exceptions exist in some populations on eastern barrier islands (Rubenstein 1986). On a larger scale, bands may aggregate to form temporary herds at watering and feeding sites (Miller 1983).

Within bands, horses of all ages and sexes interact (Fraser 1992). Allogrooming can occur between members of any sex or age class (Fraser 1992), except for band stallions in multiple-male bands (M. Ashley, unpublished data). Allogrooming is done with the teeth, usually on the neck and withers, whereas self-grooming is accomplished by rolling in a dust pit or wallow or by rubbing against a fixed object (Fraser 1992). Male foals and juveniles romp and spar with each other, as do young males in bachelor bands, likely a preparation for combat as adults (Berger 1986).

Mating Behavior. Courtship behavior between a stallion and mare in a band often begins with nose-to-nose contact, frequently progressing to allogrooming at the withers. The male will then sniff the genitals of the mare. If the mare is receptive, she will remain close to the male and eventually lift her tail and allow the male to mount. Copulation is brief, 20–30 sec. The male may exhibit flehmen, head raising accompanied by curling of the upper lip, one or more times during the encounter. The entire sequence may last 2–3 min. Conversely, copulation between a mare and an extraband male, "sneaking," can occur with reduced or no precopulatory courtship behavior, reducing the total time of the encounter to <1 min (M. Ashley, pers. obs.).

A receptive mare may copulate several times with a band stallion during the mating season (Berger 1986). A mare that is not receptive can usually rebuff the copulatory attempts of a band stallion or invading male by moving away, neighing, and even biting or kicking. In some cases, one or more band mares will aid the unreceptive mare in rebuffing a copulation attempt. In some extreme cases, a male will persist in his advances and forcibly overcome the unreceptive female to copulate (Berger 1986). Berger (1983b) inferred that stallions harassed or forced copulation with newly acquired mares to induce abortion, thereby eliminating a fetus sired by another male and allowing the female to come into estrus. However, Kirkpatrick and Turner (1991a) found no evidence of either forced copulation or stress-induced abortion in the population of feral horses on Assateague Island.

Dispersal. Both sexes disperse from their natal bands, generally at 1–3 years of age (Berger 1986; Rutberg and Keiper 1993). Usually, dispersing females are quickly integrated into existing bands or are captured by a bachelor or bachelors to form a new band (Berger 1986; Rubenstein 1986). In some cases, a dispersing female may go unaccompanied for several days before being discovered by a bachelor or band stallion (Berger 1986). Dispersing males usually join a bachelor band until they form a new band by acquiring a dispersing female, stealing a female from an established band, or taking over an existing band by force. A bachelor may also integrate into an existing band as a subordinate stallion (Berger 1986; Rubenstein 1986; Ashley 2000). In rare instances, dispersing males and females form temporary, nonbreeding bands (Rubenstein 1986). Females tend to disperse in a manner that reduces the potential for inbreeding (Monard and Duncan 1996). This is accomplished by selecting bands without familiar males, although dispersing females often join bands with sisters or half-sisters, which presumably makes it easier for them to integrate into the new bands. Monard and Duncan (1996) also found through parentage analysis that females that conceived before dispersal had mated with unfamiliar stallions rather than male members of their natal band.

Communication and Agonistic Behavior. Feral horses produce the same range of vocalizations as their domestic counterparts, including neighs, grunts, and squeals (Fraser 1992; Rubenstein and Hack 1992). The length and intensity of these sounds vary by sex and situation. Neighs serve as location calls between pair-bonded animals that have been separated or to locate conspecifics at a distance. Grunts accompany feeding in both sexes, but may also be given by males before sexual encounters (Fraser 1992). Snorts are made by members of both sexes in response to threat and between males during agonistic encounters (Rubenstein and Hack 1992). Squeals are most often made by males during agonistic encounters, but can also be made by unreceptive females when resisting males attempting copulation. Olfactory cues serve to identify males and can immediately resolve dominance conflicts if males retain knowledge of prior conflicts and their outcomes, whereas vocalizations by males serve as honest signals of fighting ability (Rubenstein and Hack 1992). Honest vocal signaling limits the percentage of agonistic encounters between males that escalate to extreme physical contact (19%), thereby reducing the incidence of injury while still resolving dominance contests (Rubenstein and Hack 1992).

The majority of aggression in feral horses is between males (Berger 1986). Band stallions aggressively defend their mares from bachelors or other band stallions (Berger 1986; Rubenstein 1986; Ashley 2000). Aggressive male–male encounters are usually limited to threatening postures, with necks arched and ears back (Berger 1986), or squeals and screams (Rubenstein and Hack 1992). In multiple-male bands, periodic bouts of snorting and stomping erupt between the band stallions, likely as an affirmation of the male hierarchy. Such bouts seldom result in a change of status or the ouster of a male (M. Ashley, unpublished data). Aggressive encounters that escalate to fighting can lead to injury or death. In the population he studied, Berger (1986) noted that 95% of adult males bore combat scars and that as many as 3% of males died from combat-related injuries. Combat behaviors include pushing, kicking with forelegs or hind legs, and biting (Berger 1986; Rubenstein and Hack 1992). Biting can inflict the most serious injuries, especially when directed at the opponent's forelegs. Wounds to the legs can become infected, impair mobility, and even lead to death. Combat terminates when one horse retreats (Berger 1986; Rubenstein and Hack 1992).

Aggression between females is usually limited to threatening postures, but can escalate to kicks and bites to the neck or body (Berger 1986; Ashley 2000). Mares often tolerate interactions between their foal and one or more foals from other bands. In contrast, band stallions often chase away foals or juveniles from other bands (M. Ashley, pers. obs.). Nonsexual aggression between males and females can be mild, in the form of snaking, a head-down, neck-elongated herding behavior, to moderate, in the form of minor bites to the body by a band stallion (Berger 1986). Sexual aggression by males ranges from moderate mating harassment (Berger 1986; Rubenstein 1986; Linklater et al. 1999) to forced copulation (Berger 1983b), although Kirkpatrick and Turner (1991a) questioned the significance of the latter. Linklater et al. (1999) reported that greater harassment of mares by stallions in multiple-male bands was associated with poorer body condition, higher parasite levels in feces, and lower reproductive success than for mares in single-male bands or unaffiliated mares.

Social Organization. In most cases, feral horses fit the resource-defense model of polygyny (Emlen and Oring 1977) in which one or more males defend one or more females to obtain the majority of copulations and consequently the greatest amount of reproductive success (Berger 1986; Rubenstein 1986; Asa 1999; Feh 1999). The majority of reproductive success in multiple-male bands usually accrues to the dominant stallion (Bowling and Touchberry 1990; Ashley 2000). Subordinate stallions, in turn, gain far more reproductive success than bachelor stallions. Band stallions, whether in single- or multiple-male bands, defend against intrusion, takeover, or theft of females by outside males (Berger 1986; Rubenstein 1986; Feh 1999). During the mating season, a band stallion may traverse hundreds of meters to encounter bachelor males or stallions from other bands. In multiple-male bands, the subordinate stallion is often the first to defend the band from outside males while the dominant stallion herds the mares and young away. The dominant stallion often joins in defense of the band when encounters persist or escalate (Berger 1986).

Feh (1999, 2001) and Linklater and Cameron (2000a) disagreed about the causes and consequences of associations among stallions in multiple-male bands. Feh (1999) suggested that these associations were mutualistic, with the dominant stallion sacrificing some reproductive success in exchange for help in contests with extraband males. Linklater and Cameron (2000a) argued that the evidence for cooperation between stallions in multiple-male bands was not convincing. They proposed instead that multiple-male bands were simply a consequence of females forming consort relationships with multiple males at the time of band establishment. Their perspective is similar to that of Jamieson (1989), who disputed adaptationist interpretations of helping behavior in cooperatively breeding birds (Emlen et al. 1991). Linklater and Cameron (2000b) concluded that more data are needed to distinguish between the mutualism hypothesis of Feh (1999), a mate parasitism hypothesis (Berger 1986), and their consort hypothesis for the existence of multiple-male bands in feral horses. Feh (2001) reiterated her case for cooperative alliances between stallions in the Camargue in France and suggested that there might be more flexibility in social behavior of feral horses than argued by Linklater (2000). Future researchers who hope to resolve this controversy may benefit from keeping in mind that hypotheses about proximate causes (e.g., mechanisms) and about ultimate causes (e.g., adaptive significance) are complementary, not mutually exclusive (Holekamp and Sherman 1989).

AGE ESTIMATION

Age of horses is typically estimated by tooth eruption and wear (American Association of Equine Practitioners 1981). The deciduous incisors erupt during the first 10–12 months of life. At age 2–2½ years, the permanent central incisors (I1) have erupted. The intermediate incisors (I2) erupt at 3–3½ years of age. The lateral incisors (I3) emerge at about 4½ years. Canine teeth (C1) usually appear in males, but only occasionally in females. These teeth appear as permanent teeth only and usually erupt after the animal is 4 years old (American Association of Equine Practitioners 1981).

Age then can be approximated by tooth wear and the presence/absence of age-specific tooth characteristics. At about age 7 years, a distinctive point or hook appears on the posterior edge of the upper lateral incisor (I^3) from uneven wear because of a temporarily smaller occlusal surface ("table") of the lower lateral incisor (American Association of Equine Practitioners 1981). This condition disappears by about age 8 years, but reappears at about age 11 years, again usually disappearing the following year. As horses age, the tables of the incisors progress from an oval shape in foals and young animals, with greater width than breadth, to a circular shape by about age 12 years, to a triangular shape from ages 14 to 17 years, and back to an oval, with greater breadth than width, past age 20 years (Fig. 53.7). Another distinct age-specific characteristic is Galvayne's groove, which appears at the gum line of I^3 at age 10 years, has progressed about half-way down the tooth by age 15 years, and runs the length of the tooth by age 20 years (Fig. 53.8). Accompanying the increased length of Galvayne's groove is an increasing anterior divergence from perpendicularity of the occlusal angle of the incisors (Fig. 53.8) (American Association of Equine Practitioners 1981). It is important to note that using dental characteristics to assess the age of a horse can have its drawbacks. There is variation in the timing of tooth eruption, which, in combination with the time of year the horse is examined, can lead to errors in age determination, often resulting in assignment of an animal to an older age class (Garrott 1991a). Beyond age 4 years, differences in the amount of annual tooth wear can also lead to similar errors (Garrott 1991a).

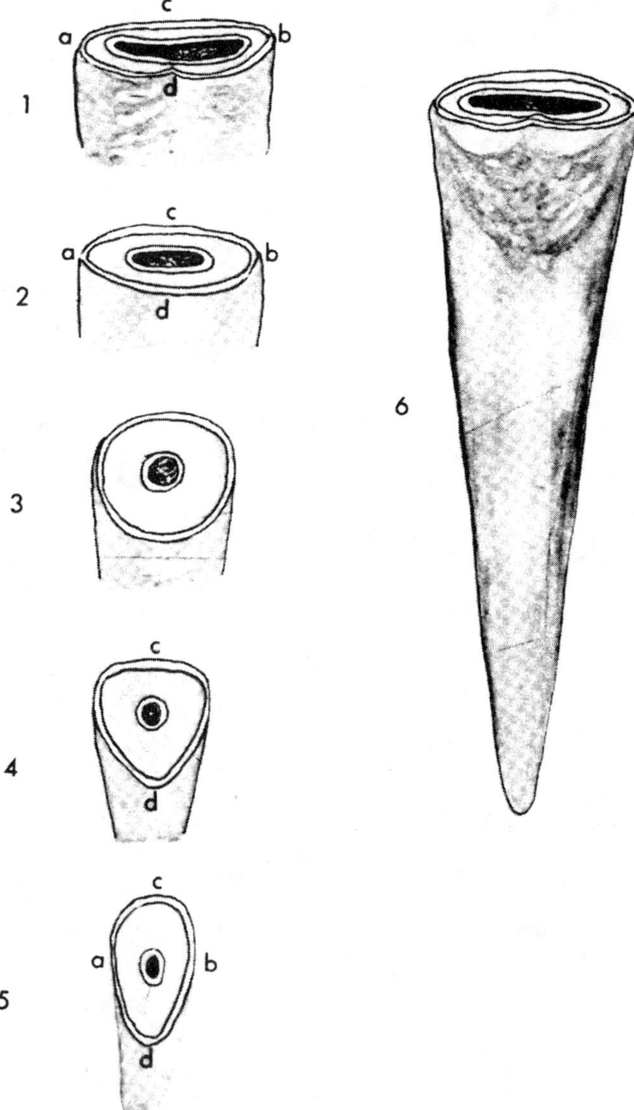

FIGURE 53.7. Schematic drawing of the central incisor of the horse. (1) After eruption, at age 2–2 ½ years, the tooth is wider in the transverse (a–b) than distal–proximal (c–d) direction. (2) The occlusal table at age 6 years, somewhat narrower from a to b and wider from c to d than at eruption. (3) The round appearance of the tooth at age 10–12 years. (4) The triangular appearance of the tooth at age 14–17 years. (5) The appearance of the tooth at ages >20 years, now narrower from a to b than from c to d. (6) Profile of the exposed incisor. SOURCE: Adapted from American Association of Equine Practitioners (1981) with permission.

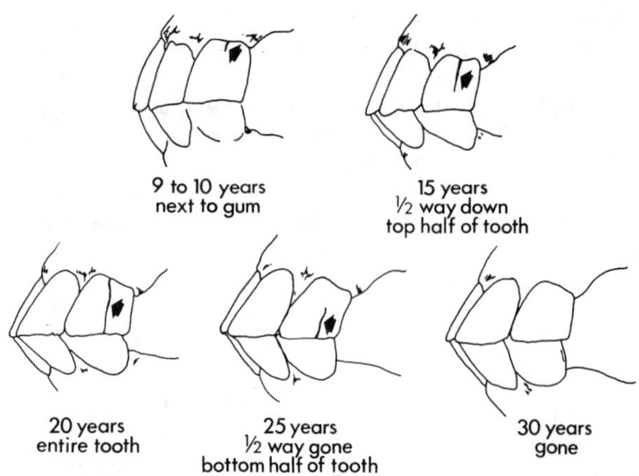

FIGURE 53.8. The progression of Galvayne's groove down I^3 and the anterior progress of the incisor angle of occlusion with advancing age in the horse. SOURCE: Adapted from American Association of Equine Practitioners (1981) with permission.

ECONOMIC STATUS AND MANAGEMENT

The Wild and Free-Roaming Horse and Burro Act of 1971 (PL 92-195) elevated feral horses and burros to the status of a protected species in the United States, prohibiting their harassment or destruction. The BLM became responsible for the active management of most of the protected feral horse populations with the passage of the Public Rangelands Improvement Act (PL 95-514) in 1978. Because of this, feral horses and burros became a major component of multiple-use plans for public lands administered by the BLM. Managers ostensibly consider the goal of "arriving at a natural thriving ecological balance" (U.S. Department of the Interior, Bureau of Land Management 1992:4) in setting appropriate management levels for specific herd management areas. However, the National Research Council (1982) concluded that sociopolitical as well as ecological factors influenced decisions about appropriate management levels for feral horses and burros (Wagner 1983; Boyles 1986), and this still appears to be true.

Because many populations of feral horses have no major predators and are well below ecological carrying capacity, they may increase by >20%/year (Eberhardt et al. 1982; Garrott et al. 1991b). Without control, population sizes can rapidly exceed appropriate management levels in particular herd management areas. This may lead to habitat alteration or degradation (Boyles 1986; Turner 1987; Rogers 1991; Beever and Brussard 2000) because overgrazing by large herbivores has been linked to changes in plant communities (Bisigato and Bertiller 1997; Manzano and Navár 2000) and degradation of riparian and stream ecosystems (Armour et al. 1994). The major method of population control is capture and removal of excess animals from the range (Garrott et al. 1991b). Captured horses that are not suitable for removal, usually older animals, are returned to the range except when an animal that is severely injured or in extremely poor physical condition is euthanized.

In 1973, the BLM started the Adopt-a-Horse-or-Burro Program in which animals removed from public rangeland are offered for adoption to the general public. This program has become quite popular and successful. In 2001, the average adoption fee was about $185 for horses and $135 for burros. Adopters may receive title to their animals after 1 year. From fiscal year (FY)1996 to 2000, >40,000 horses and burros were removed from managed populations in the western United States and offered to the public for adoption. Of these animals, 93% were adopted, with 41% of adoptions occurring east of the Mississippi River. Animals that are not adopted remain in the care of the BLM for the rest of their lives. The entire process of bringing a feral horse from the range through successful adoption cost an average of $1400/horse in FY2000.

The annual budget of the Wild Horse and Burro Program was $21 million in FY 2000 (T. Pogacnik, National Wild Horse and Burro Program of BLM, pers. commun., 2001). Between 1980 and 1987, spending by the U.S. government averaged $58 million/year for grazing management, $10 million/year for feral horses and burros, and $27 million/year for threatened and endangered species (Berger 1991). This means that federal spending to benefit livestock and feral horses and burros was 2.5 times as great as spending for all threatened and endangered species. By 1992–1995, the situation had changed dramatically, with annual expenses for rangelands of $92 million, for feral horses and burros of $17 million, and for threatened and endangered species of $265 million (Berger and Snow 2001). However, >40% of the expenditures for threatened and endangered species were devoted to chinook (*Oncorhynchus tshawytscha*) and sockeye salmon (*O. nerka*), and annual expenditures for most other native species individually were far less than those for livestock and feral equids.

Fertility control is seen as an alternative or additional means of reducing the rate of population growth, which in turn would reduce the frequency of removals and the cost of managing populations of feral horses (Garrott et al. 1992; Kirkpatrick 1995). Bowling and Touchberry (1990), Eagle et al. (1992, 1993), and Plotka et al. (1992) did field studies in the late 1980s involving hormonal implants in mares and sterilization of dominant stallions, but current research focuses on immunocontraception. A safe and effective 1-year female contraceptive has been developed, which uses a protein vaccine from pig zona pellucida to stimulate an autoimmune response in mares against their own ova, greatly reducing reproduction without long-term impacts on ovarian function (Kirkpatrick et al. 1992, 1997; Turner et al. 1996, 1997). Currently, the treatment is acceptably effective for only 1 year, but research is underway to develop a single-shot, multiyear treatment. There is still debate about the efficacy of fertility reduction as the sole means of population control because high levels of treatment are necessary to bring about the desired results (Caughley et al. 1992; Garrott et al. 1992; Hobbs et al. 2000). Other concerns include (1) females that are spared the costs of reproduction may have better body condition and higher survival probabilities, (2) recruitment may be enhanced through increased juvenile survival (Twigg and Williams 1999), (3) changes of reproductive behavior in reproductively suppressed females may affect male behavior and survival (Muller et al. 1997), and (4) selective contraception of animals with strong immune responses by immunocontraceptive treatment may lead to an increase in immunologically inferior animals in the remaining population (Nettles 1997). Immunocontraception is appealing to managers and the public because it is presumably less stressful to treated individuals than sterilization or use of hormonal implants because the immunocontraceptive can be delivered by injection, perhaps even remotely (Turner et al. 1996). However, the short duration of effectiveness of immunocontraceptives with present technology limits their effectiveness at controlling population growth. Also, the elasticity analysis outlined in the Demography section suggested that growth rates of feral horse populations are relatively insensitive to alterations in fecundity.

Feral horses, often called "wild horses" by the public, will likely remain part of the faunal landscape of North America for the foreseeable future. The basis of their support lies in their status as beloved pets and as icons of the romanticized "Old West." In a Web search in early 2001, we found >130 horse-rescue organizations operating in the United States. Although most cater to domestic horses and some to specific breeds, many also champion feral horses. There are also several organizations devoted specifically to preservation and protection of feral horses and burros. These include three national organizations and at least five devoted to individual populations. Feral horses also fall under the aegis of many of the more general wildlife-protection, humane society, and animal-rights organizations. Despite the growing concern of many for the preservation of habitats, there are few organizations directly opposed to widespread maintenance of feral horse populations. One, the Public Lands Foundation, advocates the establishment of a small number of feral horse reserves and the elimination of the remaining feral horses from public lands. In addition to emotional, philosophical, and environmental issues related to feral horses and burros, the adoption program has significant economic impacts through revenue generated by the care, feeding, housing, training, transportation, and showing of adopted animals.

RESEARCH NEEDS

Feral horses and burros are intensively and expensively managed by federal agencies in the United States (Berger 1991). Much of this management, however, rests on assumptions about impacts of equids on plant communities and on other herbivores that have not been tested experimentally. Therefore, we believe that the highest priority for future research on feral horses and burros should be placed on rigorous experimental studies of interactions of feral horses and burros with other species. These interactions include impacts of horses and burros on plant community structure and composition, which may be complex (e.g., Furbish and Albano 1994); competitive and facilitative effects on other species of herbivores; and predator–prey relationships involving horses and burros. Such experiments will not be easy to do because they may require large-scale removals of horses from some areas and because large spatial and temporal variation in environmental conditions make it difficult to achieve necessary replication to test hypotheses adequately. However, Hobbs et al. (1996a, 1996b) provided a model of how these experiments might be done in their studies of competition between elk and cattle, as did Fahnestock and Detling (1999, 2000) in their work with exclosures and Choquenot (1991) in his comparison of a culled and an unculled population of feral burros in Australia (although the latter experiment was unreplicated). Furthermore, intensive management of feral horses and burros by the BLM and the U.S. Forest Service provides opportunities for testing hypotheses using principles of scientific adaptive management (Walters and Holling 1990). These opportunities have not yet been exploited to any significant extent.

There are also opportunities for better understanding foraging ecology and behavior of feral horses and burros by using optimal foraging theory as a framework for testing hypotheses. Virtually all of the research on feeding habits of feral horses and burros has been purely descriptive, and much of it has relied on microhistological examination of feces rather than direct observations of behavior. By contrast, Belovsky and colleagues developed a powerful theoretical approach to understanding diet selection by generalist herbivores (Belovsky 1984, 1986a, 1986b), but have not yet applied this approach to an ungulate that uses postgastric fermentation, like feral horses and burros. Testing Belovsky's linear programming model with data for horses or burros might provide new insight about the strengths and limitations of the model because of their digestive systems, and might also contribute to a better understanding of potential competition between feral equids and other herbivores (Belovsky 1997).

At the population level, we believe that one of the major issues that deserves attention in future research is the magnitude and characteristics of dispersal of horses or burros between populations. This is important because of the recent interest in managing populations of feral horses and burros for genetic diversity (Coates-Markle 2000), which has been stimulated in part by emphasis of groups interested in horse protection on genetic viability of individual populations as a justification for minimal management of these populations. Information on genetic diversity of feral horses is rapidly accumulating, but the implications of patterns of genetic diversity among populations cannot be fully understood without more and better data on dispersal.

Recent research has shown that horses and burros may be model species for answering fundamental questions about behavioral and evolutionary ecology (e.g., Berger 1986; Cameron et al. 1999; MacFadden et al. 1999). Studies of the population and community ecology of feral horses and burros have not been as successful at improving understanding of important general issues, but more careful testing of hypotheses about population dynamics and interactions with other species may not only improve the scientific basis for managing feral horses and burros, but also contribute to basic knowledge about population and community ecology in general.

ACKNOWLEDGMENTS

We thank J. Berger, T. Pogacnik, and B. Wade for comments on this chapter.

LITERATURE CITED

Alberdi, M. T., J. L. Prado, and E. Ortiz-Juaregizar. 1995. Patterns of body size changes in fossil and living Equini (Perissodactyla). Biological Journal of the Linnean Society 54:349–70.

American Association of Equine Practitioners, ed. 1981. Official guide for determining the age of the horse, 4th ed. Fort Dodge Laboratories, Fort Dodge, IA.

Armour, C., D. Duff, and W. Elmore. 1994. The effects of livestock grazing on western riparian and stream ecosystem. Fisheries 19:9–12.

Arnold, G. W., and M. L. Dudzinski. 1978. Ethology of free-ranging domestic animals. Elsevier, New York.

Asa, C. S. 1999. Male reproductive success in free-ranging feral horses. Behavioral Ecology and Sociobiology 47:89–93.

Ashley, M. C. 2000. Feral horses in the desert: Population genetics, demography, mating, and management. Ph.D. Dissertation, University of Nevada, Reno.

Austin, D. D., P. J. Urness, and S. L. Durham. 1994. Impacts of mule deer and horse grazing on transplanted shrubs for revegetation. Journal of Range Management 47:8–11.

Azzaroli, A. 1991. Ascent and decline of monodactyl equids: A case for prehistoric overkill. Annales Zoologici Fennici 28:151–63.

Babini, S., L. C. Sheppard, and H. H. Hobart. 1995. A tale of two chromosomes: Przewalski's horse and karyotypic evolution in Equidae. Cytogenetics and Cell Genetics 71:398–99.

Beever, E. A. 1999. Species- and community-level responses to disturbance imposed by feral horse grazing and other management practices. Ph.D. Dissertation, University of Nevada, Reno.

Beever, E. A., and P. F. Brussard. 2000. Examining ecological consequences of feral horse grazing using exclosures. Western North American Naturalist 60:236–54.

Bell, R. H. V. 1970. The use of the herb layer by grazing ungulates in the Serengeti. British Ecological Society Symposium 10:111–23.

Belovsky, G. E. 1984. Herbivore optimal foraging: A comparative test of three models. American Naturalist 124:97–115.

Belovsky, G. E. 1986a. Generalist herbivore foraging and its role in competitive interactions. American Zoologist 26:51–69.

Belovsky, G. E. 1986b. Optimal foraging and community structure: Implications for a guild of generalist grassland herbivores. Oecologia 70:35–52.

Belovsky, G. E. 1997. Optimal foraging and community structure: The allometry of herbivore food selection and competition. Evolutionary Ecology 11:641–72.

Bennett, D., and R. S. Hoffmann. 1999. *Equus caballus*. Mammalian Species 628:1–14.

Benson, D. A., M. S. Boguski, D. J. Lipman, J. Ostell, B. F. F. Ouellette, B. A. Rapp, and D. L. Wheeler. 1999. GenBank. Nucleic Acids Research 27:12–17.

Berger, J. 1983a. Ecology and catastrophic mortality in wild horses: Implications for interpreting fossil assemblages. Science 220:1403–4.

Berger, J. 1983b. Induced abortion and social factors in wild horses. Nature 303:59–61.

Berger, J. 1985. Interspecific interactions and dominance among wild Great Basin ungulates. Journal of Mammalogy 66:571–73.

Berger, J. 1986. Wild horses of the Great Basin: Social competition and population size. University of Chicago Press, Chicago.

Berger, J. 1991. Funding asymmetries for endangered species, feral animals, and livestock. Bioscience 41:105–6.

Berger, J., and K. Snow. 2001. Endangered species and the decline of America's western legacy: What do changes in funding reflect? Bioscience 51:591–93.

Bisigato, A. J., and M. B. Bertiller. 1997. Grazing effects on patchy dryland vegetation in northern Patagonia. Journal of Arid Environments 36:639–53.

Bocherens, H., G. Pacaud, P. A. Lazarev, and A. Mariotti. 1996. Stable isotope abundances (13C, 15N) in collagen and soft tissues from Pleistocene mammals from Yakutia: Implications for the paleobiology of the Mammoth Steppe. Paleogeography, Paleoclimatology, and Paleoecology 126:31–44.

Bowling, A. T. 1994. Population genetics of Great Basin feral horses. Animal Genetics 25(Supplement 1):67–74.

Bowling, A. T., and R. W. Touchberry. 1990. Parentage of Great Basin feral horses. Journal of Wildlife Management 54:424–29.

Bowling, A. T., M. L. Eggleston-Stott, G. Byrns, R. S. Clark, S. Dileanis, and E. Wictum. 1997. Validation of microsatellite markers for routine horse parentage testing. Animal Genetics 28:247–52.

Boyd, L. 1979. The mare–foal demography of feral horses in Wyoming's Red Desert. Pages 185–204 *in* R. H. Denniston, ed. Symposium on the ecology and behavior of wild and feral equids. University of Wyoming, Laramie.

Boyles, J. S. 1986. Managing America's wild horses and burros. Journal of Equine Veterinary Science 6:261–65.

Breen, M., P. Downs, Z. Irwin, and K. Bell. 1994. Intrageneric amplification of horse microsatellite markers with emphasis on Przewalski's horse (*Equus przewalskii*). Animal Genetics 25:401–5.

Browning, B. 1960. Preliminary report of the food habits of the wild burro in the Death Valley National Monument. Transactions of the Desert Bighorn Council 4:88–90.

Bruce, H. M. 1969. A block to pregnancy in the mouse caused by proximity of strange males. Journal of Reproduction and Fertility 1:96–103.

Bryant, J. D., P. N. Froelich, W. J. Showers, and B. J. Genna. 1996. Biologic and climatic signals in the oxygen isotopic composition of Eocene–Oligocene equid enamel phosphate. Paleogeography, Paleoclimatology, and Paleoecology 126:75–89.

Cameron, E. Z., and W. L. Linklater. 2000. Individual mares bias investment in sons and daughters in relation to their condition. Animal Behaviour 60:359–67.

Cameron, E. Z., W. L. Linklater, K. J. Stafford, and C. J. Veltman. 1999. Birth sex ratios relate to mare condition at conception in Kaimanawa horses. Behavioral Ecology 10:472–75.

Cameron, E. Z., W. L. Linklater, K. J. Stafford, and E. O. Minot. 2000. Aging and improving reproductive success in horses: Declining residual reproductive value or just older and wiser? Behavioral Ecology and Sociobiology 47:243–49.

Caswell, H. 2001. Matrix population models, 2nd ed. Sinauer, Sunderland, MA.

Caughley, G., R. Pech, and D. Grice. 1992. Effect of fertility control on a population's productivity. Wildlife Research 19:623–27.

Choquenot, D. 1990. Rate of increase for populations of feral donkeys in northern Australia. Journal of Mammalogy 71:151–55.

Choquenot, D. 1991. Density-dependent growth, body condition, and demography in feral donkeys: Testing the food hypothesis. Ecology 72:805–13.

Coates-Markle, L. 2000. Summary recommendations—BLM wild horse and burro population viability forum, April 21, 1999 (Resource Notes 35). National Science and Technology Center, Bureau of Land Management.

Conley, W. 1979. The potential for increase in horse and ass populations: A theoretical analysis. Pages 221–34 *in* R. H. Denniston, ed. Symposium on the ecology and behavior of wild and feral equids. University of Wyoming, Laramie.

Cothran, E. G., and F. J. Singer. 2000. Analysis of genetic variation in the Pryor Mountain wild horse herd. Pages 91–104 *in* F. J. Singer and K. A. Schoenecker, eds. Manager's summary—Ecological studies of the Pryor Mountain Wild Horse Range, 1992–1997. U.S. Geological Survey, Midcontinent Ecological Science Center, Fort Collins, CO.

Demment, M. W., and P. J. Van Soest. 1985. A nutritional explanation for body-size patterns of ruminant and nonruminant herbivores. American Naturalist 125:641–72.

Duncan, P. 1982. Foal killing by stallions. Applied Animal Ethology 8:567–70.

Duncan, P. 1985. Time-budgets of Camargue horses. III. Environmental influences. Behaviour 92:188–208.

Duncan, P., C. Feh, J. C. Gleize, P. Malkas, and A. M. Scott. 1984. Reduction of inbreeding in a natural herd of horses. Animal Behaviour 32:520–27.

Duncan, P., T. J. Foose, I. J. Gordon, C. G. Gakahu, and M. Lloyd. 1990. Comparative nutrient extraction from forages by grazing bovids and equids: A test of the nutritional model of equid/bovid competition and coexistence. Oecologia 84:411–18.

Eagle, T. C., E. D. Plotka, R. A. Garrott, D. B. Siniff, and J. R. Tester. 1992. Efficacy of chemical contraception in feral mares. Wildlife Society Bulletin 20:211–16.

Eagle, T. C., C. S. Asa, R. A. Garrott, E. D. Plotka, D. B. Siniff, and J. R. Tester. 1993. Efficacy of dominant male sterilization to reduce reproduction in feral horses. Wildlife Society Bulletin 21:116–21.

Eberhardt, L. L. 1985. Assessing the dynamics of wild populations. Journal of Wildlife Management 49:997–1012.

Eberhardt, L. L., A. K. Majorowicz, and J. A. Wilcox. 1982. Apparent rates of increase for two feral horse herds. Journal of Wildlife Management 46:367–74.

Ellegren, H., M. Johansson, K. Sandberg, and L. Andersson. 1992. Cloning of highly polymorphic microsatellites in the horse. Animal Genetics 23:133–42.

Emlen, S. T., and L. W. Oring. 1977. Ecology, sexual selection, and the evolution of mating systems. Science 197:215–23.

Emlen, S. T., H. K. Reeve, P. W. Sherman, P. H. Wrege, F. L. W. Ratnieks, and J. Shellman-Reeve. 1991. Adaptive versus nonadaptive explanations of behavior: The case of alloparental helping. American Naturalist 138:259–70.

Fahnestock, J. T., and J. K. Detling. 1999. The influence of herbivory on plant cover and species composition in the Pryor Mountain Wild Horse Range, USA. Plant Ecology 144:145–57.

Fahnestock, J. T., and J. K. Detling. 2000. Morphological and physiological responses of perennial grasses to long-term grazing in the Pryor Mountains, Montana. American Midland Naturalist 143:312–20.

Feh, C. 1999. Alliances and reproductive success in Camargue stallions. Animal Behaviour 57:705–13.

Feh, C. 2001. Alliances between stallions are more than just multimale groups: Reply to Linklater & Cameron (2000). Animal Behaviour 61:F27–30.

Flynn, J. J., B. J. MacFadden, and M. C. McKenna. 1984. Land-mammal ages, faunal heterochrony, and temporal resolution in Cenozoic terrestrial sequences. Journal of Geology 92:687–705.

Fraser, A. F. 1992. The behaviour of the horse. C.A.B. International, Wallingford, UK.

Freeland, W. J., and D. Choquenot. 1990. Determinants of herbivore carrying capacity: Plants, nutrients, and *Equus asinus* in northern Australia. Ecology 71:589–97.

Furbish, C. E., and M. Albano. 1994. Selective herbivory and plant community structure in a mid-Atlantic salt marsh. Ecology 75:1015–22.

Gao, X., and J. Gu. 1989. The distribution and status of the Equidae in China. Acta Theriologica Sinica 9:269–74.

Garrott, R. A. 1991a. Bias in aging feral horses. Journal of Range Management 44:611–13.

Garrott, R. A. 1991b. Sex ratios and differential survival of feral horses. Journal of Animal Ecology 60:929–36.

Garrott, R. A., and L. Taylor. 1990. Dynamics of a feral horse population in Montana. Journal of Wildlife Management 54:603–12.

Garrott, R. A., T. C. Eagle, and E. D. Plotka. 1991a. Age-specific reproduction in feral horses. Canadian Journal of Zoology 69:738–43.

Garrott, R. A., D. B. Siniff, and L. L. Eberhardt. 1991b. Growth rates of feral horse populations. Journal of Wildlife Management 55:641–48.

Garrott, R. A., D. B. Siniff, J. R. Tester, T. C. Eagle, and E. D. Plotka. 1992. A comparison of contraceptive technologies for feral horse management. Wildlife Society Bulletin 20:318–26.

George, M., Jr., and O. A. Ryder. 1986. Mitochondrial DNA evolution in the genus *Equus*. Molecular Biology and Evolution 3:535–46.

Goodloe, R. B., R. J. Warren, E. G. Cothran, S. P. Bratton, and K. A. Trembicki. 1991. Genetic variation and its management applications in eastern U.S. feral horses. Journal of Wildlife Management 55:412–21.

Goodloe, R. B., R. J. Warren, D. A. Osborn, and C. Hall. 2000. Population characteristics of feral horses on Cumberland Island, Georgia and their management implications. Journal of Wildlife Management 64:114–21.

Gould, S. J. 1987. Life's little joke. Natural History 96(4):16–25.

Greger, P. D., and E. M. Romney. 1999. High foal mortality limits growth of a desert feral horse population in Nevada. Great Basin Naturalist 59:374–79.

Guerin, G., M. Bertaund, and Y. Amiques. 1994. Characterization of seven new horse microsatellites: HMS1, HMS2, HMS3, HMS5, HMS6, HMS7, and HMS8. Animal Genetics 25:62.

Hanley, T. A. 1982. The nutritional basis for food selection by ungulates. Journal of Range Management 35:146–51.

Hanley, T. A., and K. A. Hanley. 1982. Food resource partitioning by sympatric ungulates on Great Basin rangeland. Journal of Range Management 35:152–58.

Hansen, R. M. 1976. Foods of free-roaming horses in southern New Mexico. Journal of Range Management 29:347.

Hansen, R. M., and R. C. Clark. 1977. Foods of elk and other ungulates at low elevations in northwestern Colorado. Journal of Wildlife Management 41:76–80.

Hansen, R. M., and P. S. Martin. 1973. Ungulate diets in the lower Grand Canyon. Journal of Range Management 26:380–81.

Hansen, R. M., R. C. Clark, and W. Lawhorn. 1977. Foods of wild horses, deer and cattle in the Douglas Mountain area, Colorado. Journal of Range Management 30:116–18.

Hiraiwa-Hasegawa, M. 1988. Adaptive significance of infanticide in primates. Trends in Ecology and Evolution 3:102–5.

Hobbs, N. T., D. L. Baker, G. D. Bear, and D. C. Bowden. 1996a. Ungulate grazing in sagebrush grassland: Effects of resource competition on secondary production. Ecological Applications 6:218–27.

Hobbs, N. T., D. L. Baker, G. D. Bear, and D. C. Bowden. 1996b. Ungulate grazing in sagebrush grassland: Mechanisms of resource competition. Ecological Applications 6:200–217.

Hobbs, N. T., D. C. Bowden, and D. L. Baker. 2000. Effects of fertility control on populations of ungulates: General, stage-structured models. Journal of Wildlife Management 64:473–91.

Holekamp, K. E., and P. W. Sherman. 1989. Why male ground squirrels disperse. American Scientist 77:232–39.

Holmes, E. C., and S. A. Ellis. 1999. Evolutionary history of MHC class I genes in the mammalian order Perissodactyla. Journal of Molecular Evolution 49:316–24.

Hooker, J. J. 1994. The beginnings of the equoid radiation. Zoological Journal of the Linnean Society 112:29–63.

Hubbard, R. E., and R. M. Hansen. 1976. Diets of wild horses, cattle, and mule deer in the Piceance Basin, Colorado. Journal of Range Management 29:389–92.

Hulbert, R. C., Jr. 1993. Taxonomic evolution in North American Neogene horses (subfamily Equinae): The rise and fall of an adaptive radiation. Paleobiology 19:216–34.

Illius, A. W., and I. J. Gordon. 1992. Modelling the nutritional ecology of ungulate herbivores: Evolution of body size and competitive interactions. Oecologia 89:428–34.

Ishida, N., T. Oyunsuren, S. Mashima, H. Mukoyama, and N. Saitou. 1995. Mitochondrial DNA sequences of various species of the genus *Equus* with special reference to the phylogenetic relationship between Przewalskii's [*sic*] wild horse and domestic horse. Journal of Molecular Evolution 41:180–88.

Jamieson, I. 1989. Behavioral heterochrony and the evolution of birds' helping at the nest: An unselected consequence of communal breeding? American Naturalist 133:394–406.

Janis, C. 1976. The evolutionary strategy of the Equidae and the origins of rumen and cecal digestion. Evolution 30:757–74.

Janis, C. 1990. The correlation between diet and dental wear in herbivorous mammals, and its relationship to the determination of diets in extinct species. Pages 241–59 *in* A. J. Boucot, ed. Evolutionary paleobiology of behavior and coevolution. Elsevier, Amsterdam.

Janis, C. M., and D. Ehrhardt. 1988. Correlation of relative muzzle width and relative incisor width with dietary preferences in ungulates. Zoological Journal of the Linnean Society 95:267–84.

Jenkins, S. H. 1989. Comments on an inappropriate population model for feral burros. Journal of Mammalogy 70:667–70.

Jenkins, S. H. 2000. Density dependence in population dynamics of feral horses (Resource Notes 26). National Science and Technology Center, Bureau of Land Management.

Johnson, R. A., S. W. Carothers, and T. J. McGill. 1987. Demography of feral burros in the Mohave Desert. Journal of Wildlife Management 51:916–20.

Jordana, J., P. M. Pares, and A. Sanchez. 1995. Analysis of genetic relationships in horse breeds. Journal of Equine Veterinary Science 15:320–28.

Keiper, R., and K. Houpt. 1984. Reproduction in feral horses: An eight-year study. American Journal of Veterinary Research 45:991–95.

Kiltie, R. A. 1982. Intraspecific variation in the mammalian gestation period. Journal of Mammalogy 63:646–52.

Kirkpatrick, J. F. 1995. Management of wild horses by fertility control: The Assateague experience (Scientific Monograph Series 25). U.S. Department of the Interior, National Park Service.

Kirkpatrick, J. F., and J. W. Turner Jr. 1983. Seasonal ovarian function in feral mares. Journal of Equine Veterinary Science 3:113–18.

Kirkpatrick, J. F., and J. W. Turner Jr. 1986. Comparative reproductive biology of North American feral horses. Journal of Equine Veterinary Science 6:224–30.

Kirkpatrick, J. F., and J. W. Turner Jr. 1991a. Changes in herd stallions among feral horse bands and the absence of forced copulation and induced abortion. Behavioral Ecology and Sociobiology 29:217–20.

Kirkpatrick, J. F., and J. W. Turner, Jr. 1991b. Compensatory reproduction in feral horses. Journal of Wildlife Management 55:649–52.

Kirkpatrick, J. F., S. E. Shideler, and J. W. Turner, Jr. 1990. Pregnancy determination in uncaptured feral horses based on steroid metabolites in urine-soaked snow and free steroids in feces. Canadian Journal of Zoology 68:2576–79.

Kirkpatrick, J. F., I. M. K. Liu, J. W. Turner, Jr., R. Naugle, and R. Keiper. 1992. Long-term effects of porcine zonae pellucidae immunocontraception on ovarian function in feral horses (*Equus caballus*). Journal of Reproduction and Fertility 94:437–44.

Kirkpatrick, J. F., B. L. Lasley, S. E. Shideler, J. F. Roser, and J. W. Turner, Jr. 1993. Non-instrumented immunoassay field tests for pregnancy detection in free-roaming feral horses. Journal of Wildlife Management 57:168–73.

Kirkpatrick, J. F., J. W. Turner Jr., I. K. M. Liu, R. Fayrer-Hosken, and A. T. Rutberg. 1997. Case studies in wildlife immunocontraception: Wild and feral equids and white-tailed deer. Reproduction, Fertility and Development 9:105–10.

Krysl, L. J., G. E. Plumb, M. E. Hubbert, B. F. Sowell, T. K. Jewett, M. A. Smith, and J. W. Waggoner Jr. 1983. Foraging behavior and water use of horses and cattle in the Wyoming Red Desert. Prairie Naturalist 15:29–34.

Krysl, L. J., M. E. Hubbert, B. F. Sowell, G. E. Plumb, T. K. Jewett, M. A. Smith, and J. W. Waggoner. 1984. Horses and cattle grazing in the Wyoming Red Desert, I. Food habits and dietary overlap. Journal of Range Management 37:72–76.

Latorre, C., J. Quade, and W. C. McIntosh. 1997. The expansion of C4 grasses and global change in the late Miocene: Stable isotope evidence from the Americas. Earth and Planetary Science Letters 146:83–96.

Lindsay, E. H., N. D. Opdyke, and N. D. Johnson. 1980. Pliocene dispersal of the horse *Equus* and late Cenozoic mammalian dispersal events. Nature 287:135–38.

Linklater, W. L. 2000. Adaptive explanation in socio-ecology: Lessons from the Equidae. Biological Reviews 75:1–20.

Linklater, W. L., and E. Z. Cameron. 2000a. Distinguishing cooperation from cohabitation: The feral horse case study. Animal Behaviour 59:F17–21.

Linklater, W. L., and E. Z. Cameron. 2000b. Tests for cooperative behaviour between stallions. Animal Behaviour 60:731–43.

Linklater, W. L., E. Z. Cameron, E. O. Minot, and K. J. Stafford. 1999. Stallion harassment and the mating system of horses. Animal Behaviour 58:295–306.

Lucas, Z., J. I. Raeside, and K. J. Betteridge. 1991. Non-invasive assessment of the incidences of pregnancy and pregnancy loss in the feral horses of Sable Island. Journal of Reproduction and Fertility, Supplement 44:479–88.

Lush, J. L. 1945. Animal breeding plans. Iowa State College Press, Ames.

MacFadden, B. J. 1985. Patterns of phylogeny and rates of evolution in fossil horses: Hipparions from the Miocene and Pliocene of North America. Paleobiology 11:245–57.

MacFadden, B. J. 1986a. Fossil horses from "Eohippus" (Hyracotherium) to *Equus*: Scaling, Cope's law, and the evolution of body size. Paleobiology 12:355–69.

MacFadden, B. J. 1986b. Late Hemphillian monodactyl horses (Mammalia, Equidae) from the Bone Valley Formation of central Florida. Journal of Paleontology 60:466–75.

MacFadden, B. J. 1988a. Fossil horses from "Eohippus" (Hyracotherium) to *Equus,* 2: Rates of dental evolution revisited. Biological Journal of the Linnean Society 35:37–48.

MacFadden, B. J. 1988b. Horses, the fossil record, and evolution: A current perspective. Pages 131–58 *in* M. K. Hecht, B. Wallace, and G. T. Prance, eds. Evolutionary biology, Vol. 22. Plenum Press, New York.

MacFadden, B. J., ed. 1992a. Fossil horses: Systematics, paleobiology, and evolution of the family Equidae. Cambridge University Press, Cambridge.

MacFadden, B. J. 1992b. Interpreting extinctions from the fossil record: Methods, assumptions, and case examples using horses (family Equidae). Pages 17–45 *in* M. J. Novacek and Q. D. Wheeler, eds. Extinction and phylogeny. Columbia University Press, New York.

MacFadden, B. J. 1997. Pleistocene horses from Tarija, Bolivia, and validity of the genus †Onohippidium (Mammalia: Equidae). Journal of Vertebrate Paleontology 17:199–218.

MacFadden, B. J. 2000. Middle Pleistocene climate change recorded in fossil mammal teeth from Tarija, Bolivia, and upper limit of the Ensendan land-mammal age. Quaternary Research 54:121–31.

MacFadden, B. J., and T. E. Cerling. 1994. Fossil horses, carbon isotopes and global change. Trends in Ecology and Evolution 9:481–86.

MacFadden, B. J., and T. E. Cerling. 1996. Mammalian herbivore communities, ancient feeding ecology, and carbon isotopes: A ten million-year sequence from the Neogene of Florida. Journal of Vertebrate Paleontology 16:103–15.

MacFadden, B. J., and J. L. Dobie. 1998. Late Miocene three-toed horses *Protohippus* (Mammalia: Equidae) from southern Alabama. Journal of Paleontology 72:149–52.

MacFadden, B. J., N. Solounias, and T. E. Cerling. 1999. Ancient diets, ecology, and extinction of 5-million-year-old horses from Florida. Science 283:824–27.

Manzano, M. G., and J. Navár. 2000. Processes of desertification by goats overgrazing in the Tamaulipan thornscrub (*matorral*) in north-eastern Mexico. Journal of Arid Environments 44:1–17.

Marklund, S., H. Ellegren, S. Eriksson, K. Sandberg, and L. Andersson. 1994. Parentage testing and linkage analysis in the horse using a set of highly polymorphic microsatellites. Animal Genetics 25:19–23.

Martin, P. S., and R. G. Klein, eds. 1984. Quaternary extinctions: A prehistoric revolution. University of Arizona Press, Tucson.

McCort, W. D. 1980. The behavior and social organization of feral asses (*Equus asinus*) on Ossabaw Island, Georgia. Ph.D. Dissertation, Pennsylvania State University, University Park.

McInnis, M. L., and M. Vavra. 1987. Dietary relationships among feral horses, cattle, and pronghorn in southeastern Oregon. Journal of Range Management 40:60–66.

McKenna, M. C. 1975. Fossil mammals and early Eocene North Atlantic land continuity. Annals of the Missouri Botanical Garden 62:335–53.

Miller, R. 1983. Habitat use of feral horses and cattle in Wyoming's Red Desert. Journal of Range Management 36:195–99.

Moehlman, P. D. 2000. Conservation issues for wild zebra, asses, and horses in Africa and Asia (Resource Notes 24). National Science and Technology Center, Bureau of Land Management.

Monard, A. M., and P. Duncan. 1996. Consequences of natal dispersal in female horses. Animal Behaviour 52:565–79.

Monard, A. M., P. Duncan, H. Fritz, and C. Feh. 1997. Variations in the birth sex ratio and neonatal mortality in a natural herd of horses. Behavioral Ecology and Sociobiology 41:243–49.

Moritz, C. 1994. Defining "evolutionary significant units" for conservation. Trends in Ecology and Evolution 9:373–75.

Morris, B. 1993. A survey of blood types in Nevada wild horses. A report commissioned by the Bureau of Land Management, Winnemucca District Office.

Muller, L. I., R. J. Warren, and D. L. Evans. 1997. Theory and practice of immunocontraception in animals. Wildlife Society Bulletin 25:504–14.

National Research Council. 1980. Wild and free-roaming horses and burros: Current knowledge and recommended research. National Academy of Sciences Press, Washington, DC.

National Research Council. 1982. Wild and free-roaming horses and burros/Final report. National Academy Press, Washington, DC.

Nelson, K. J. 1980. Sterilization of dominant males will not limit feral horse populations (USDA Forest Service Research Paper RM-226). Rocky Mountain Forest and Range Experiment Station, Fort Collins, CO.

Nettles, V. F. 1997. Potential consequences and problems with wildlife contraceptives. Reproduction, Fertility and Development 9:137–43.

Norman, J. E., and M. V. Ashley. 2000. Phylogenetics of Perissodactyla and tests of the molecular clock. Journal of Molecular Evolution 50:11–21.

Nowak, R. M. 1991. Walker's mammals of the world, 5th ed., Vol. II. Johns Hopkins University Press, Baltimore.

Oakenfull, E. A., and J. B. Clegg. 1998. Phylogenetic relationships within the genus *Equus* and the evolution of "alpha" and "theta" globin genes. Journal of Molecular Evolution 47:772–83.

Paklina, N. V., and V. V. Klimov. 1990. Social organization of a population of feral horses *Equus caballus* on the Yuzhny Island (The Manych-Gudilo Lake). Zoologicheskii Zhurnal 69:107–16.

Perryman, P., and A. Muchlinski. 1987. Population dynamics of feral burros at the Naval Weapons Center, China Lake, California. Journal of Mammalogy 68:435–38.

Plotka, E. D., T. C. Eagle, S. J. Gaulke, J. R. Tester, and D. B. Siniff. 1988. Hematologic and blood chemical characteristics of feral horses from three management areas. Journal of Wildlife Diseases 24:231–39.

Plotka, E. D., D. N. Vevea, T. C. Eagle, J. R. Tester, and D. B. Siniff. 1992. Hormonal contraception of feral mares with silastic rods. Journal of Wildlife Diseases 28:255–62.

Prado, J. L., and M. T. Alberdi. 1996. A cladistic analysis of the tribe Equini. Paleontology 39:663–80.
Radinsky, L. R. 1966. The adaptive radiation of the phenacodontid condylarths and the origin of the Perissodactyla. Evolution 20:408–17.
Reiner, R. J., and P. J. Urness. 1982. Effect of grazing horses managed as manipulators of big game winter range. Journal of Range Management 35:567–71.
Rensberger, J. M., A. Forsten, and M. Fortelius. 1984. Functional evolution of the cheek tooth pattern and chewing direction in Tertiary horses. Paleobiology 10:439–52.
Rogers, G. M. 1991. Kaimanawa feral horses and their environmental impacts. New Zealand Journal of Ecology 15:49–64.
Rubenstein, D. I. 1986. Ecology and sociality in horses and zebras. Pages 282–302 *in* D. I. Rubenstein and R. W. Wrangham, eds. Ecological aspects of social evolution. Princeton University Press, Princeton, NJ.
Rubenstein, D. I., and M. A. Hack. 1992. Horse signals: The sounds and scents of fury. Evolutionary Ecology 6:254–60.
Rudman, R. 1998. The social organisation of feral donkeys (*Equus asinus*) on a small Caribbean island (St. John, US Virgin Islands). Applied Animal Behaviour Science 60:211–28.
Rutberg, A. T., and R. R. Keiper. 1993. Proximate causes of natal dispersal in feral ponies: Some sex differences. Animal Behaviour 46:969–75.
Salter, R. E., and R. J. Hudson. 1979. Feeding ecology of feral horses in western Alberta. Journal of Range Management 32:221–25.
Salter, R. E., and R. J. Hudson. 1980. Range relationships of feral horses with wild ungulates and cattle in western Alberta. Journal of Range Management 33:266–71.
Saltz, D., and D. I. Rubenstein. 1995. Population dynamics of a reintroduced Asiatic wild ass (*Equus hemionus*) herd. Ecological Applications 5:327–35.
Schaller, G. B. 1972. The Serengeti lion: A study of predator–prey relations. University of Chicago Press, Chicago.
Seal, U. S., and E. D. Plotka. 1983. Age-specific pregnancy rates in feral horses. Journal of Wildlife Management 47:422–29.
Seegmiller, R. F., and R. D. Ohmart. 1981. Ecological relationships of feral burros and desert bighorn sheep. Wildlife Monographs 78:1–58.
Sharp, Z. D., and T. E. Cerling. 1998. Fossil isotope records of seasonal climate and ecology: Straight from the horse's mouth. Geology 26:219–22.
Siler, W. 1979. A competing-risk model for animal mortality. Ecology 60:750–57.
Simpson, G. G. 1951. Horses: The story of the horse family in the modern world and through sixty million years of history. Oxford University Press, Oxford.
Sinclair, A. R. E. 1985. Does interspecific competition or predation shape the African ungulate community? Journal of Animal Ecology 54:899–918.
Siniff, D. B., J. R. Tester, and G. L. McMahon. 1986. Foaling rate and survival of feral horses in western Nevada. Journal of Range Management 39:296–97.
Slade, L. M., and E. B. Godfrey. 1982. Wild horses. Pages 1089–98 *in* J. A. Chapman and G. A. Feldhamer, eds. Wild mammals of North America: Biology, management, and economics. Johns Hopkins University Press, Baltimore.
Smith, C., R. Valdez, J. L. Holechek, P. J. Zwank, and M. Cardenas. 1998. Diets of native and non-native ungulates in southcentral New Mexico. Southwestern Naturalist 43:163–69.
Speelman, S. R., W. M. Dawson, and W. R. Phillips. 1944. Some aspects of fertility in horses raised under western range conditions. Journal of Animal Science 3:233–41.
Stephenson, T. E., J. L. Holechek, and C. B. Kuykendall. 1985. Diets of four wild ungulates on winter range in northcentral New Mexico. Southwestern Naturalist 30:437–42.
Swenson, J. E., F. Sandegren, A. Söderberg, A. Bjärvall, R. Franzén, and P. Wabakken. 1997. Infanticide caused by hunting of male bears. Nature 386:450–51.
Thomason, J. J. 1986. The functional morphology of the manus in the tridactyl equids *Merychippus* and *Mesohippus*: Paleontological inferences from neontological models. Journal of Vertebrate Paleontology 6:143–61.
Trivers, R. L., and D. E. Willard. 1973. Natural selection of parental ability to vary the sex ratio of offspring. Science 179:90–92.
Turner, J. W., Jr., and J. F. Kirkpatrick. 1986. Hormones and reproduction in feral horses. Journal of Equine Veterinary Science 6:250–58.
Turner, J. W., Jr., M. L. Wolfe, and J. F. Kirkpatrick. 1992. Seasonal mountain lion predation on a feral horse population. Canadian Journal of Zoology 70:929–34.
Turner, J. W., Jr., I. K. M. Liu, and J. F. Kirkpatrick. 1996. Remotely delivered immunocontraception in free-roaming feral burros (*Equus asinus*). Journal of Reproduction and Fertility 107:31–35.
Turner, J. W., Jr., I. K. M. Liu, A. T. Rutberg, and J. F. Kirkpatrick. 1997. Immunocontraception limits foal production in free-roaming feral horses in Nevada. Journal of Wildlife Management 61:873–80.
Turner, M. G. 1987. Effects of grazing by feral horses, clipping, trampling, and burning on a Georgia salt marsh. Estuaries 10:54–60.
Twigg, L. E., and C. K. Williams. 1999. Fertility control of overabundant species; can it work for feral rabbits? Ecology Letters 2:281–85.
Tyler, S. J. 1972. The behaviour and social organisation of the New Forest ponies. Animal Behaviour Monographs 5:85–196.
U.S. Department of the Interior and U.S. Department of Agriculture. 1997. The 10th and 11th report to Congress on the administration of the Wild Free-Roaming Horses and Burros Act for fiscal years 1992–1995 (1997-573-153/40527). U.S. Government Printing Office, Washington, DC.
U.S. Department of the Interior, Bureau of Land Management. 1992. Strategic plan for management of wild horses and burros on public lands. U.S. Government Printing Office, Washington, DC.
Van Haeringen, H., A. T. Bowling, M. L. Scott, J. A. Lenstra, and K. A. Zwaagstra. 1994. A highly polymorphic horse microsatellite locus—VHL20. Animal Genetics 25:207.
Wagner, F. H. 1983. Status of wild horse and burro management on public rangelands. Transactions of the North American Wildlife and Natural Resources Conference 48:116–33.
Walters, C. J., and C. S. Holling. 1990. Large-scale management experiments and learning by doing. Ecology 71:2060–68.
Wang, Y., T. E. Cerling, and B. J. MacFadden. 1994. Fossil horses and carbon isotopes: New evidence for Cenozoic dietary, habitat, and ecosystem changes in North America. Paleogeography, Paleoclimatology, and Paleoecology 107:269–79.
Webb, S. D. 1977. A history of savannah vertebrates in the New World. Part I: North America. Annual Review of Ecology and Systematics 8:355–80.
Wijers, E. R., C. Zijlstra, and J. A. Lenstra. 1993. Rapid evolution of horse satellite DNA. Genomics 18:113–17.
Wolfe, M. L., Jr. 1980. Feral horse demography: A preliminary report. Journal of Range Management 33:354–60.
Wolfe, M. L., L. C. Ellis, and R. MacMullen. 1989. Reproductive rates of feral horses and burros. Journal of Wildlife Management 53:916–24.
Woodburne, M. O., and R. L. Burnor. 1980. On superspecific groups of some Old World hipparionine horses. Journal of Paleontology 54:1319–48.
Woodburne, M. O., and B. J. MacFadden. 1982. A reappraisal of the systematics, biogeography, and evolution of fossil horses. Paleobiology 8:315–27.
Woods, G. L., C. B. Baker, J. L. Baldwin, B. A. Ball, J. Bilinski, W. L. Cooper, W. B. Ley, E. C. Mank, and H. N. Erb. 1987. Early pregnancy loss in brood mares. Journal of Reproduction and Fertility (Supplement) 35:455–59.
Woodward, S. L., and R. D. Ohmart. 1976. Habitat use and fecal analysis of feral burros (*Equus asinus*), Chemehuevi Mountains, California, 1974. Journal of Range Management 29:482–85.

Stephen H. Jenkins, Department of Biology, University of Nevada, Reno, Nevada 89557. Email: jenkins@unr.edu.

Michael C. Ashley, U.S. Department of Agriculture, Agriculture Research Service, Reno, Nevada 89512. Email: dr_mc_ashley@hotmail.com.

54

Feral Hog

Sus scrofa

John R. Sweeney
James M. Sweeney
Sheron W. Sweeney

NOMENCLATURE

COMMON NAMES. Feral hog, feral swine, feral pig, piney woods rooters, razorback, wild boar, wild hog, and woods hogs
SCIENTIFIC NAME. *Sus scrofa*
SUBSPECIES. Mayer and Brisbin (1991) described the 23 normally recognized geographic subspecies of Eurasian wild boar, and noted that domestic swine are most commonly given the trinomial designation of *Sus scrofa domesticus,* with *S. s. domestica* also often cited. However, there is continued debate over whether domestic species should be given scientific names, let alone subspecific designations, because of the artificial nature of their genetics. Mayer and Brisbin (1991) pointed out that the feral hog takes this controversy one step further, in that it is once again subjected to natural selection. In addition, feral hog populations are subjected to the regular introduction of genetic material from escaped domestic stock and from animals of unknown lineage that have been transplanted to an area. Therefore, the best option appears to be to avoid the use of subspecific nomenclature for feral swine and simply use *Sus scrofa.*

The family Suidae first appeared in the early Oligocene in or around India (Pilgrim 1941). During the following epochs, suids spread throughout Europe, central Asia, and Africa. The genus *Sus* first appears in fossil records of the middle Miocene, and occurred in Europe and Africa by the upper Pliocene (Osborn 1910). In addition to this natural migration, humans, in domesticating swine, introduced pigs throughout the remainder of the Old World and most of the New World, which resulted in the nearly worldwide distribution of suids today.

This chapter will be limited primarily to the "feral hog" of domestic origin but living in a wild state, as opposed to the "European wild boar" (original wild stock). Although morphologically distinct, both the feral hog and European wild boar are recognized as *Sus scrofa.* Hybrids of feral hog × European wild boar will also be considered as feral hogs. All are often referred to as simply wild swine. The only other species of wild swine in North America is the collared peccary or javelina (*Tayassu tajacu*).

DISTRIBUTION

Feral hogs in North America are believed to have originated from domestic hogs introduced by early settlers from Europe. The first introductions probably were made by Columbus in 1493 in the West Indies and DeSoto in 1593 in Florida (Towne and Wentworth 1950). However, there is evidence that the genus *Sus* was present in the United States long before the introduction of domestic hogs by Columbus. An incisor, two canines, three cheek teeth, and a fragment of the occiput were recovered from the 10-Mile Rock archeological dig in northwestern Arkansas (Quinn 1970). Carbon-14 dating of these remains (possibly *Sus indicus*) indicates they were present before the 1490s. Domestic hogs reportedly were also released in California by Spanish explorers in 1769 (Hutchinson 1946) and in the Hawaiian Islands by Captain Cook in 1778 (Kramer 1971). For a detailed state-by-state history of free-ranging hog introductions in the United States, see Mayer and Brisbin (1991).

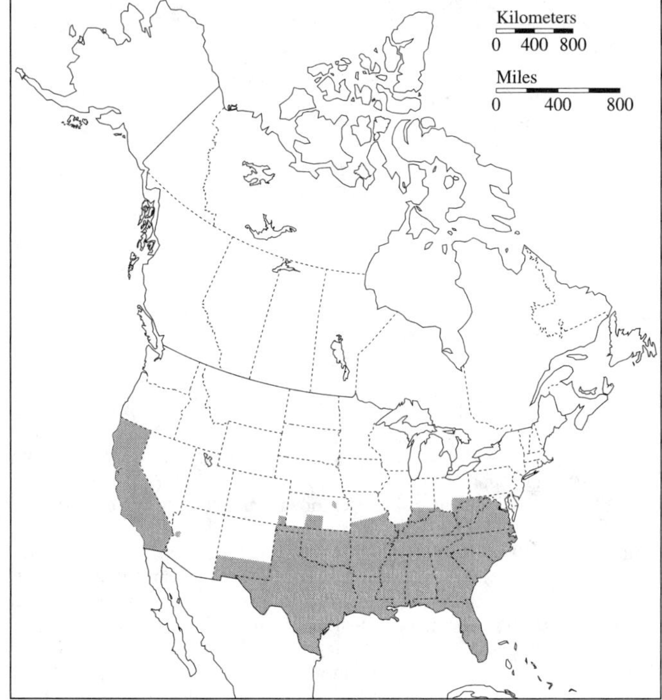

FIGURE 54.1. Distribution of the feral hog (*Sus scrofa*).

Surveys conducted in 1981 and 1988 by Mayer and Brisbin (1991) revealed that feral hogs occurred in 19 states, primarily the southeastern states, and California. By 1993, feral hogs were reported in 23 states and populations were increasing rapidly (Miller 1993; Taylor 1993; Gipson et al. 1995; Wagner 1995; Stevens 1996). According to Gipson et al. (1998), who conducted the most recent survey of the status of feral hogs in North America, they currently occur in 23 contiguous states. They are found in eastern Colorado; two isolated areas in Kansas; across southern Missouri, Illinois, and Indiana; in southeastern Ohio; and in isolated populations throughout Kentucky and West Virginia (Fig. 54.1). Texas has the highest densities of feral hogs, with current population estimates of >1 million animals (Mapston 1997). Florida's population densities are second only to those of Texas, with hogs found in all counties (Belden 1997). In most states, feral hog populations are distributed unevenly, with densities varying from abundant to sparse, depending on the quality of the habitat and the history of the local population. Wild boar or feral hogs also are being held in enclosed hunting preserves or private collections in Alaska, Oregon, Idaho, North and South Dakota, Nebraska, Minnesota, Iowa, Michigan, Pennsylvania, New York, Vermont, and New Hampshire (Nettles 1997). This species is the most abundant free-ranging introduced ungulate in the United States (Mayer and Brisbin 1991). Current estimates place the U.S. population at up to 4 million animals (Nettles 1997; Pimentel et al. 2000).

FIGURE 54.2. The pelage of feral hogs (*Sus scrofa*) is usually coarser and denser than that of domestic hogs, providing greater resistance to the elements. The general body shape of the feral hog lies somewhere between that of the round, "fattened" domestic hog and the streamlined, sloping shape of the European wild hog.

FIGURE 54.3. The prominent pelage color of feral hogs (*Sus scrofa*) is black, with mottled black and brown or black and white being the next most common.

The often-practiced open-range policy of hog management has resulted in the continued "release" of domestic hogs into many feral hog populations. The European wild boar has also been released in several locations within the continental United States, particularly in the Appalachian Mountains (Lewis et al. 1965; Howe and Bratton 1976). Although no longer a practice, trapping and relocation of feral hogs by state wildlife agencies also aided in the early dispersal of feral hogs. Gipson et al. (1998) felt there has been a concentrated effort by hog hunters to release wild hogs in new areas of the United States during the last 10 years. They cited this, along with escapes from hunting preserves, as the two key factors in the rapid increase of feral hog populations in the central United States. Therefore, feral hog populations today are a mixture of European wild boars, recent domestic hogs, and feral hogs.

DESCRIPTION

The physical appearance of a feral hog is intermediate between a domestic hog and the European wild hog (Fig. 54.2). Early studies (Hanson and Karstad 1959; Golley 1962) described feral hogs as thinner, appearing more streamlined in body shape, with longer and sharper tusks and coarser coats than domestic hogs.

Mayer and Brisbin (1991) conducted an exhaustive statistical analysis of the morphology of domestic swine, feral hogs, wild boar × feral hog hybrids, and Eurasian wild boar. They found that the four morphotypes can be reliably distinguished only by a combination of morphological characteristics, including skull morphology, external body dimensions and proportions, coat coloration, and hair morphology. Any of these four morphological characteristics used alone is often inadequate to determine the type of *Sus scrofa*.

Size and Weight. Body size and weight of feral hogs are highly variable, depending on the genetic history of the population, the local environment in which the population exists, and the length of time the population has been feral (Brisbin et al. 1977a; Mayer and Brisbin 1991). Short-term feral hog populations (<100 years since domestication) were heavier and had greater total body length than feral hogs that had been free-ranging for >100 years.

Mayer and Brisbin (1991) found that although feral hogs tended to have the lowest averages for total length and head–body length measurements of the four morphotypes studied, they also had the highest maxima and the greatest variation in these two dimensions. Feral hog males and females averaged 137.3 and 128.7 cm, respectively, in head–body length (Mayer and Brisbin 1993). Sweeney (1970) reported ranges of mean total lengths for adult boars and sows from a recently feral population as 177–192 and 135–185 cm, respectively. Wild boar and hybrid morphotypes tend to have proportionally greater shoulder heights than feral hogs as a result of their relatively shorter body lengths (Mayer and Brisbin 1991). Feral hog males and females averaged 76.2 and 68.3 cm, respectively, in shoulder height (Mayer and Brisbin 1993).

Feral hogs also had more-variable body weights, including the smallest averages for males and females—about 60 and 50 kg, respectively—and also the largest overall weight of any of the three types of wild pigs (Mayer and Brisbin 1991). Studies have reported mean weights of adult males from 36 to 114 kg and adult females from 34 to 92 kg (Sweeney 1970; Wood and Brenneman 1977; Mayer and Brisbin 1991; Ilse and Hellgren 1995b). The largest feral hog on record was a male from the Santee River Delta, South Carolina, which weighed 406 kg (Rutledge 1965). The smallest feral hogs, including miniature swine, inhabited an extremely overpopulated Georgia coastal island, and averaged only 23 kg in weight and 96 cm in total length (Brisbin et al. 1977a).

Pelage. Feral hog coat coloration patterns resemble the variety seen in domestic swine; however, the dominant pelage color is black (Mayer and Brisbin 1991). Spotted black (Fig. 54.3) and red/brown or black and white are the next most common patterns. Brown or roan is common, and occasionally feral hogs are all white. Only rarely has the Hampshire color phenotype (generally gray to black with a white shoulder band) been reported (Golley 1962; Sweeney 1970; Brisbin et al. 1977a; Mayer and Brisbin 1991). No obvious sexual dimorphism is apparent in coat coloration (Mayer and Brisbin 1991). The striped pattern in juvenile coats, typical of wild boars, has been reported in at least eight feral hog populations (Mayer and Brisbin 1991). The frequency of striped juveniles in these populations varied from infrequent (only one or two out of hundreds) to about 50%. The pelage of feral hogs is usually coarser and denser than that of domestic hogs. It provides a greater resistance to chilling, probably a selective advantage in the feral state (Foley et al. 1971; Hansen et al. 1972). Curly, wool-like underfur occurs in populations of all three wild morphotypes (Mayer and Brisbin 1991). However, only 30.4% of feral hogs examined exhibited this trait; all were from nine long-term feral populations in California, Florida, Georgia, and Hawaii. Hair is longer in the winter than in the summer (Mayer and Brisbin 1991). Feral hogs and domestic swine typically have solid bristle or guard hair coloration, whereas the bristles of wild boar and hybrids are usually black with a white to dark tan distal tip. Because splitting can cause the appearance of a slightly lighter tip, feral hog bristles often appear to be black with a dark brown tip (Mayer and Brisbin 1991).

Skull Morphology. Depth of the dorsal profile often has been used to compare wild and domestic swine. Mayer and Brisbin (1991) found that domestic swine had a significantly deeper dorsal profile than free-ranging swine in the three oldest age classes. This difference increased with age. There were no significant differences among the three wild morphotypes for any age class. Mayer and Brisbin (1991) showed that feral hogs, however, did have a higher average dorsal profile (about 14 mm) than either wild boar (10 mm) or hybrids (8 mm). The angle of the occipital wall in relation to the plane of the palate tended to be vertical or posteriorly inclined for wild boar, whereas the other morphotypes had occipital walls that were anteriorly inclined. Mayer and Brisbin (1991) also reported that the angle of inclination for feral hogs averaged between 65° and 75°, which was less than those of hybrids (77–80°) and greater than those of domestic swine (60–65°). Feral hogs generally have the smallest skulls of the four morphotypes. The average length of the rostrum of a feral hog is between that of a domestic animal and a wild boar or hybrid. The skull of a domestic pig generally is higher and broader than any of the free-ranging forms. Mayer and Brisbin (1991) found that the most successful method of skull morphotype classification was discriminant function analysis of cranial measurements. However, cranial morphology appears to be strongly influenced by nutrition (Mayer and Brisbin 1991). They found significant differences in populations of feral swine based on the length of time they had been in a feral state. Adult and subadult hogs in short-term feral populations (<100 years since domestication) exhibited larger averages in almost all cranial measurements than feral hogs that had been essentially free from the influences of domestication for >100 years.

Dentition. The dental formula of the pig is I 3/3, C 1/1, P 4/4, M 3/3 = 44 (Fig. 54.4). Replacement of the deciduous teeth is usually complete by 20–22 months of age. The occlusal surfaces of cheek teeth are bunodont. The upper and lower canines of the boar are curved backward and outward, with the lower canines aligned in front of the upper canines. Friction between the upper and lower canines produces a sharp edge on the lower teeth. Tusks of adult boars may grow to 8–9 cm long (lower canine from gum line to tip along outer curve) (Pullar 1953; Barrett 1978). Canines of adult sows are usually smaller and differ from those of males (see Sex Determination). The canines are open rooted and continually grow (Sisson and Grossman 1938).

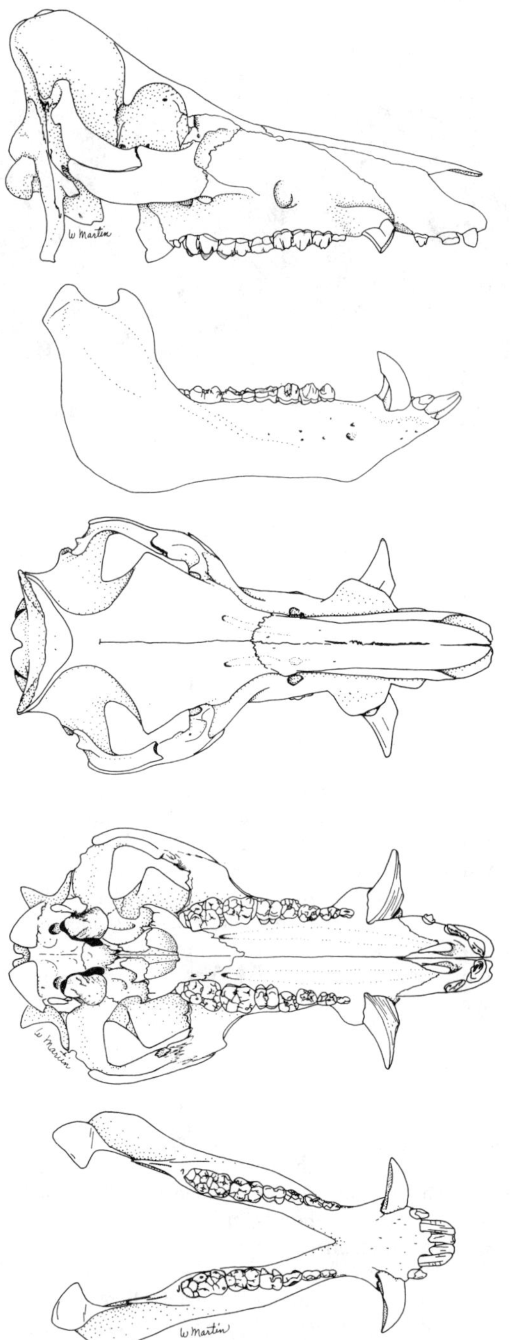

FIGURE 54.4. Skull of the feral hog (*Sus scrofa*). From top to bottom: lateral view of cranium, lateral view of mandible, dorsal view of cranium, ventral view of cranium, dorsal view of mandible.

GENETICS

Little definitive work has been completed on the genetics of feral hogs. However, comparisons of the karyotypes of domestic hogs, European wild hogs, and domestic × European hybrids may be instructive. McFee et al. (1966) reported that the normal complement of chromosomes for European wild hogs was $2n = 36$. Similar results were found in a study on the cytogenetics of hogs on the Tellico Wildlife Management Area in Tennessee (Rary et al. 1968). Pure-strain domestic hogs consistently had $2n = 38$ chromosomes, pure-strain European wild hogs had 36, and their hybrid cross produced fertile offspring with a diploid chromosome number of 37. Many populations of feral hogs have a history of crossbreeding to varying extents with introduced European wild hogs, and these hybrids have 36, 37, or 38 chromosomes (Mayer and Brisbin 1991).

Rary et al. (1968) hypothesized that the difference in the normal diploid number of chromosomes could be accounted for by centric fusion of two pairs of telocentric chromosomes, resulting in an evolutionary reduction from $2n = 38$ for domestic hogs to $2n = 36$ for European wild hogs. Alternatively, centric splitting of chromosomes 35 and 36 in the European wild hog could have increased the total number of chromosomes in the domestic hog. Hsu and Mead (1969) supported the latter explanation, saying that $2n = 38$ is most likely the result of a centric split of a pair of metacentric chromosomes producing two pairs of telocentrics.

Genetic polymorphism was examined in two populations of feral hogs in South Carolina (Brisbin et al. 1977b). Horizontal starch gel electrophoresis was used to examine 20 proteins in feral hogs collected from the Ossabaw Island and Savannah River populations (Selander et al. 1971; Manlove et al. 1976). Allele frequencies differed between the two populations at two variable loci: phosphoglucose isomerase and 6-phosphogluconate dehydrogenase. Five percent of the loci were polymorphic in the Ossabaw Island population, whereas 20% were polymorphic in the Savannah River population. Average heterozygosity across all loci was $H = 0.005$ for the Ossabaw Island population and $H = 0.035$ for the Savannah River population.

PHYSIOLOGY

The amount of definitive research on the physiology of feral hogs continues to grow. However, it is still useful to draw inferences from the vast amount of information available on the physiological processes of domestic swine (Frandson 1965; Irving 1966; Mount 1968; Pond and Maner 1974; Sorenson 1979) to provide insight into feral hogs. Caution should be exercised, however, because notable differences also occur.

In a study of a feral hog population that has been extant in a wild state for several hundred years, Graves and Graves (1977) found that they are more efficient than domestic hogs in storing and mobilizing energy. This is an advantage for wild hogs in coping with cold periods, and especially in meeting the selective pressures of extreme seasonal variation in food availability (Martin and Herbein 1976). Also, as noted earlier, the pelage of feral hogs is usually denser than that of domestic hogs, which aids in resistance to chilling.

Domestic hogs generally have low tolerance of food with high salt content. However, Graves and Graves (1977) and Roark (1977) suggested feral hogs on Ossabaw Island and along coastal South Carolina were more capable of handling salt in their diets, because much of their food was obtained in salt marshes during the summer. Subsequent research by Zervanos et al. (1983) and Zervanos and Naveh (1988) supported these ideas. Zervanos et al. (1983) reported that Ossabaw Island hogs compensated for the large quantities of dietary salt by drinking limited quantities of fresh water and reducing salt uptake through the intestine. Furthermore, Zervanos and Naveh (1988) indicated feral hogs inhabiting coastal South Carolina and Ossabaw Island have acclimated to water stress (drought and high-salt diet) through development of a more efficient kidney structure compared to captive feral hogs raised under ad libitum water conditions.

Feral hogs have been assessed as a possible indicator species for monitoring environmental contaminants (Stribling 1986). Because of their omnivorous feeding habits and nomadic movements, feral hogs "sample" their environment more completely than animals with restricted movements and diets. Highly significant correlations between levels of cesium-137 in muscle and feces of a population of swine inhabiting a river swamp system that was receiving coolant waters from nuclear reactors suggested that collections of feral hog feces, especially in the summer, could be used to monitor cesium-137 level in these animals on this site.

The ability to accumulate, store, and use body fat is important to free-living mammals that may have to face periods of uncertain food availability or environmental stress (Young 1976). Various measures of total body fat have been used to monitor a number of ungulate species, and backfat thickness has long been used in domestic breeds of swine to establish carcass composition (Skelley and Handlin 1967). However, no such index is available for feral hogs. Stribling et al. (1984) examined feral swine, and found that no marrow index, combination of external measurements, or body weight adequately predicted body fat reserves. Brisket fat thickness was the only measure that showed even limited correlation as a predictor of total body fat reserves. This relationship probably varies considerably between populations (Stribling et al. 1984).

REPRODUCTION

Puberty. There have been no detailed studies to determine the estrous cycle and length of gestation in feral hogs. However, female feral hogs are polyestrus, and are likely similar to domestic swine, which have an estrous cycle of approximately 21 days and a gestation period of about 112–114 days. Spermatozoa are first present in the testes of male domestic hogs from 4.6 to 5.2 months of age, depending on variations in nutrition and breed (Phillips and Zeller 1943). In inbred male domestic hogs, spermatozoa are first present at 6 months of age (Hauser et al. 1952). Male European wild hogs, however, reach maturity at the age of 7 months (Henry 1966).

In general, feral hogs reach puberty at 5–10 months of age. Feral boars attained puberty at 5–7 months old in South Carolina (Sweeney et al. 1970), around 6 months old in California (Barrett 1978), and 20–22 weeks old in Florida (Belden and Frankenberger 1989). However, feral boars seldom played an important role in breeding until they were 12–18 months old because of the dominance of older males (Barrett 1978). Females in these populations attained puberty at 10 months (Sweeney et al. 1970), 3–13 months (Barrett 1978), and 20–23 weeks (Belden and Frankenberger 1989).

Litter Size. In seven studies in the Southeast (South Carolina, Florida, Texas) and California, litter size ranged from 1 to 12, with mean litter size ranging from 3.0 to 8.4 (Springer 1977; Barrett 1978; Sweeney et al. 1979; Belden and Frankenberger 1989; Hellgren 1993; Taylor et al. 1998; Gabor et al. 1999). Older females (≥21 months) are often more productive (Belden and Frankenberger 1989; Hellgren 1993; Taylor et al. 1998). In addition to age, female body size and weight have a positive relationship with litter size. Population density is negatively related with litter size in other parts of this species's world distribution (Fernandez-llario and Mateos-Quesada 1998).

Average fetal counts should be adjusted to account for postnatal loss from stillborn hogs. The ratio of the number of fetuses to the number of corpora lutea averaged 66% in California (Barrett 1978); 69% (Sweeney et al. 1979) and 83% (Sweeney 1979) in South Carolina and Georgia; 78% in Texas (Taylor et al. 1998); and 89% in Florida (Belden and Frankenberger 1989). Asdell (1964) stated that stillborn loss in domestic hogs averaged 6%. A similar, perhaps larger, loss may occur in feral hogs.

Average litter size for feral hogs is generally larger than that reported for European wild hogs, but within the lower limits of the range for domestic hogs. Asdell (1964) noted the average for European wild hogs was 4–5 young/sow. Henry (1966) found that European wild hogs in Tennessee had an average litter size of 4.8. Pine and Gerdes (1973) reported that European wild hogs in California had an average litter size of 4.2. Nalbandov (1976) noted that average litter size of domestic hogs ranged from 6.6 to 21.2, depending on breed. Barrett's (1978) review indicated that modern breeds of domestic hog can produce as many as 24 young/litter. In studies of reproduction in penned feral and domestic swine by Hagen and Kephart (1980) and Hagen et al. (1980), feral hogs had smaller litter sizes because of a lower ovulatory rate rather than higher embryonic mortality or limited uterine capacity.

Farrowing. The feral hog is physiologically capable of breeding year-round, and farrowing has been observed throughout the year (Sweeney et al. 1979). However, there are often two peaks in farrowing activity, with one often more pronounced than the other. There is a prominent peak in farrowing during winter (November–March) and a secondary peak in July (Barrett 1978; Sweeney et al. 1979; Taylor et al. 1998). Belden and Frankenberger (1989) reported the same primary peak in winter, and a secondary peak in April. Barrett (1978) also reported a population with two equal, small rises in farrowing activity in July and November. This pattern is similar to the farrowing activity of European wild hogs in Tennessee, where peaks occur in early summer (May–June) and midwinter (January–February) (Henry 1966). Taylor et al. (1998) suggested that, at least in Texas, heat plays a role in the seasonality of breeding. Peak breeding coincided with peaks in bimodal precipitation and moderate temperatures in the region. Belden and Frankenberger (1989) attributed the bimodal reproductive pattern to nutrition, with the fall mast crop and "spring green-up" strongly influencing timing of peaks.

Although feral hogs are capable of producing two litters/year, individuals in most populations usually produce only one litter/year (Taylor et al. 1998). Two litters do occur, especially if nutritious foods have been available year-round, or in some cases where the entire first litter was lost soon after parturition (Barrett 1978; Taylor et al. 1998). Successful hog reproduction is directly related to acorn production (Matschke 1964; Baber and Coblentz 1987). However, because acorns are low in protein and lactating sows and growing piglets need protein, it can be limiting (Belden and Frankenberger 1989). Warren and Ford (1997) documented a mean ovulation rate twice that normally observed coincident with significantly higher kidney fat and a unique preponderance of grape (*Vitis* sp.) leaves in the diet. They suggested that reproduction in feral hogs may be highly responsive to short-term dietary changes.

ECOLOGY

Range and Habitat. Until recently, it was believed the range of feral hogs in North America was limited by two major factors: land use, particularly as it affects cover, and climate (Hanson and Karstad 1959). Cover is essential for feral hogs, and habitat use is directly proportional to the density of cover (Barrett 1978). Besides providing food and protection from disturbance, cover is important in reducing heat loss by decreasing air flow and reradiating thermal energy (Tregear 1965; Moen 1968). Climate may also play an important role. Increased frost penetration may increase difficulty in rooting (Hanson and Karstad 1959). However, during this time of year, hard mast—acorns of oaks (*Quercus* spp.) and nuts of hickories (*Carya* spp.)—becomes available and rooting activity often is greatly reduced in preference for these nutritious and palatable foods. Recent establishment of free-ranging feral hog populations in Canada suggests that they are not as limited by cold as previously suspected. However, the primary distribution of feral hogs is in the southern half of the United States.

Within their ranges, feral hogs have adapted to a variety of habitat types. By far the most commonly used habitats are mesic, such as bottomland hardwood forests with permanent water. Feral hogs in the upper coastal plain of South Carolina used the river swamp and associated bottomlands almost exclusively, except in late winter and early summer, when a shift in movement occurred into adjacent upland pine plantations (Sweeney 1970; Kurz and Marchinton 1972; Crouch 1983; Hughes 1985). Feral hogs in the lower coastal plain of South Carolina often used fresh- and brackish-water marshes, but rarely used salt marshes (Wood and Brenneman 1980). Upland pine habitats were used in proportion to their availability, probably mainly for bedding. Similarly, Golley (1962) noted that feral hogs in Georgia resided primarily in the swamps and marshes of the coastal plain, ranging out to the uplands and pinewoods less frequently.

Feral hogs on the Welder Wildlife Refuge in southern Texas used most habitats in proportion to availability (Ilse and Hellgren 1995a). Use of wetter habitat types, such as halophyte communities, lakes and ponds, and riparian communities, increased during summer, but in proportion to availability (Ilse and Hellgren 1995a). Bedding areas were frequently located in the more open areas along the periphery of dense patches of woody vegetation. Increased canopy cover at heights ≥1 m suggested that feral hogs selected bedding areas for shade and thermal regulation rather than for cover. The tall grasses and forbs constituting much of the bedding areas frequently were manipulated into nestlike structures (Ilse and Hellgren 1995a).

Seasonal changes in habitat use are apparently linked with changes in food availability (Sweeney 1970; Kurz and Marchinton 1972; Graves and Graves 1977; Barrett 1978; Crouch 1983; Hughes 1985). Feral hogs on Ossabaw Island, off the coast of Georgia, moved out of their preferred oak/hickory habitat into surrounding salt marshes only in late winter when food became scarce in the forested areas (Graves and Graves 1977). In maritime coastal South Carolina, upland hardwoods were used only when oak mast was available (Wood and Brenneman 1980). In another area, abundant acorn mast in the fall concentrated feral hog activity in the bottomlands along the Savannah River; a distinct shift in movement into the upland pine plantations in June corresponded with the ripening of plums (Kurz and Marchinton 1972). Irrigated pastures and boulder washes were used as alternate habitats by California feral hogs for nocturnal feeding during dry periods (Barrett 1971).

Herd Composition. Female feral hogs characteristically travel in groups of up to three related generations (Barrett 1978; Gabor et al. 1999). Individuals in these groups have highly overlapping home ranges (>50%) and are frequently found in close proximity to each other (Gabor et al. 1999). These groups, sometimes referred to as "sounders," appear to be territorial, with little overlap between adjacent groups (Gabor et al. 1999). Group sizes typically are eight or fewer individuals, usually with no more than three adults. Individual sows have been reported with one to nine piglets (Kurz and Marchinton 1972; Graves 1984). Graves and Graves (1977) reported the number of piglets with two sows ranged from 3 to 19 (mean = 7). The second sow in a group often had experienced high piglet mortality. Ilse and Hellgren (1995b) reported many sightings of individual hogs, and Gabor et al. (1999) indicated that 41% of their sightings were of lone animals. The majority of these lone animals were adult males. Adult males typically are territorial, with home range overlaps of only 10–18% (Gabor et al. 1999), except when associated with breeding groups (Kurz and Marchinton 1972; Graves and Graves 1977; Barrett 1978; Gabor et al. 1999). The social pattern described here does not preclude occasionally large sounders or congregations of large numbers of feral hogs on a particularly choice and localized food source. Gabor et al. (1999) reported a sounder that included 28 individuals: 7 adult sows, 4 subadults, and 17 juveniles. Barrett (1978) noted herds of as many as 97 feral hogs feeding in an irrigated pasture.

Feral piglets in South Carolina remained together as a group until they weighed 25–35 kg (Kurz and Marchinton 1972). Weaning occurred at about 3 months of age. Dispersal was a gradual process consisting of forays of increasing duration away from the family group, with final separation occurring at 5–10 months (Crouch 1983; Hughes 1985). Barrett (1978) reported similar dispersal behavior with final separation at 9–12 months. In southern Texas, young males were commonly captured and observed with their maternal group until they reached 40–50 kg, which usually was about 16 months of age (Gabor et al. 1999).

The sex ratio of feral hogs within different size or sex groups is often the expected 1:1. Reported exceptions include a preponderance (3:1) of large (>64 kg) males on Ossabaw Island, South Carolina (Graves and Graves 1977); a greater number of males (2:1) in the 7- to 12-month age group along the Savannah River (Sweeney 1970); a significantly male-biased fetal sex ratio (1.27:1.00) in a Texas population (Taylor et al. 1998); a preponderance of males in the juvenile and subadult age classes (1.50:1.00) and a preponderance of females in the adult age classes (0.89:1.00) in a hunted population in Florida (Belden and Frankenberger 1989); and a slightly higher number of adult females (50.4%) on the Dye Creek Preserve in California.

Movements. Feral hogs generally drift or are almost nomadic, but within a given home range area. As a result, annual home ranges are larger than seasonal ranges, which may or may not overlap. Movement among areas within the home range apparently is linked primarily with seasonal changes in food availability, and secondarily with thermoregulation. Daily activity patterns, however, are influenced by ambient temperature, with diurnal or crepuscular activity common throughout most of the year and nocturnal activity often predominant in summer (Kurz and Marchinton 1972; Crouch 1983; Hellgren 1993; Ilse and Hellgren 1995a). Nocturnal activity may be evident, even in winter, in populations of feral hogs subjected to regular hunting (Hanson and Karstad 1959).

The lack of sweat glands causes hogs to rely on behavioral thermoregulation (Mount 1968). Hughes (1985) noted an increased selection of hardwood habitats and hardwood ecotones in spring and summer, and suggested that availability of water and cool areas for thermoregulation were important environmental factors in the selection of these habitats.

Reported annual home ranges for feral hogs monitored for ≥1 year vary dramatically, from 123 to more than 5000 ha. Individual home ranges of feral hogs in South Carolina have ranged from 123 to 2443 ha, with means for various studies ranging from 203 to 1305 ha (Kurz and Marchinton 1972; Wood and Brennemen 1980; Hughes 1985). Reported mean home ranges in Texas have varied from 280 to 590 ha, with at least two individuals exhibiting home ranges of 2600 ha (Ellisor 1973; Mansouri and DeYoung 1987; Ilse and Hellgren 1995b; Gabor et al. 1999). Observations of feral hogs in California indicated home ranges of at least 1000 to >5000 ha (Barrett 1978).

Home range size in feral hogs is believed to be influenced by a number of factors. These include the following:

1. Habitat. Home ranges are larger in semiarid habitats (Barrett 1978; Gabor et al. 1999), where the extent of bottomland hardwoods has declined (Crouch 1983), or in less spatially complex habitats (Wood and Brennemen 1980).

2. Food. Availability (Graves 1984; Mansouri and DeYoung 1987) and distribution (Hellgren 1993) of food are important.
3. Season. Increased movements occur in the fall (Barrett 1978; Hughes 1985), in the summer (Ilse and Hellgren 1995b), and on moonlit nights (Kurz and Marchinton 1972).
4. Sex. Males are more mobile than females (Barrett 1971, 1978; Wood and Brennemen 1980; Hughes 1985; Gabor et al. 1999), particularly when females are farrowing (Kurz and Marchinton 1972, Barrett 1978).
5. Body mass. If body mass is taken into account, there are no sexual differences in movements (Saunders and McLeod 1999).
6. Population density. Home range size is inversely related to population density (Saunders and McLeod 1999).

FEEDING HABITS

Feral hogs are omnivorous and their feeding habits are opportunistic, with grazing and rooting predominating. There is no major variation in feeding habits between sexes or age classes. The composition of their diet changes seasonally and with food availability. They exhibit no particular preferences for a given plant or animal species, except that vegetation is preferred over animal matter, and mast, both soft and hard, is preferred over other plant food items. Roots and tubers are used throughout the year and are particularly important in the summer. Animal matter (both vertebrate and invertebrate) and mushrooms are consumed all year, but are only a minor part of the diet. Snakes are the only vertebrate consumed in all seasons, and earthworms are the most common invertebrate. Most feeding habits studies have not determined whether animal matter in the diet of hogs was the result of predation or scavenging.

In general, the spring diet is dominated by green herbaceous materials (such as grasses, new stems, and leaves) and some forbs, roots, or tubers. Summer diet is dominated by soft and hard mast, roots, and tubers. In the fall, mast is important, as are roots, tubers, and corn. Winter diets are again dominated by grasses and forbs, depending on moisture and ambient temperature; otherwise hogs continue to use acorns, roots and tubers, and agricultural grains. For additional information on feeding habits of feral hogs consult Hanson and Karstad (1959), Roark (1977), Springer (1977), Thompson (1977), Barrett (1978), Everitt and Alaniz (1980), Graves (1984), Hellgren (1993), and Taylor and Hellgren (1997).

Many studies of feeding habits have suggested the potential for competition with domestic livestock or wildlife species such as white-tailed deer (*Odocoileus virginianus*), wild turkeys (*Meleagris gallopavo*), javelinas, black bears (*Ursus americanus*), squirrels (*Sciurus* spp.), chipmunks (*Tamias striatus*), skunks, raccoons (*Procyon lotor*), opossums (*Didelphis virginiana*), foxes, bobcats (*Lynx rufus*), waterfowl, and sandhill cranes (*Grus canadensis*) (Thompson 1977; Wood and Lynn 1977; Everitt and Alaniz 1980; Yarrow 1987; Beach 1993; Synatzske 1993). Most competition with other wildlife centers on the use of mast. The degree of competition depends on the relative abundance of the mast supply and alternate sources of nutritionally adequate foods during periods of poor mast production. However, studies actually documenting competitive interactions have not been conducted.

Rooting by feral hogs has positive and negative effects on growth and survival of some trees, soils and soil processes, and the distribution of native and exotic grasses. Negative effects of rooting are amplified when population densities are high. The most widespread and costly type of forest damage done by feral hogs is depredation of commercial pine or hardwood seedlings to the extent it causes serious problems with forest regeneration (Wahlenberg 1946; Wakely 1954; Lipscomb 1989; Mayer et al. 2000). Regeneration of hardwoods can also be restricted through consumption of acorns (Lucas 1977). In some cases, depredation on agricultural crops can be a problem (Pine and Gerdes 1973; Springer 1977; Thompson 1977).

Equally important is the potential for significant ecological impacts. Feral hog rooting may temporarily reduce plant cover, but it also increases microhabitat diversity and species richness in returning vegetation communities (Kotanen 1995; Arrington et al. 1999). Rooting also increases American beech (*Fagus grandifolia*) shoot elongation because of enhanced litter decomposition and nutrient mobilization in soils (Lacki and Lancia 1983, 1986). In contrast, rooting may reduce litter-layer mass and decrease food sources for macroinvertebrates (Wood and Barrett 1979). Rooting can be a serious problem in destabilizing surface soils along stream channels and road banks (Lucas 1977) and sand dunes in coastal areas (Wood and Barrett 1979). However, Baron (1982) considered extensive rooting on a sand barrier island off the coast of Mississippi to have an insignificant impact on the vegetation when compared to other environmental factors such as overwash from hurricanes, fires, drought, and shifting sand, which keep the island in early stages of succession.

Feeding habits of feral hogs may have a serious impact on endangered species in a community. Rooting can destroy rare plants, particularly where they occur in localized populations. In addition, although relatively small amounts of animal matter are consumed, the impact on endangered species could be significant either by direct predation or destabilization of critically important habitats (Singer et al. 1984; Taylor and Hellgren 1997). Nest predation has also been cited as a concern by Thompson (1977), Weinwandt (1977), and Tolleson et al. (1993).

BEHAVIOR

Farrowing. Farrowing nests of feral hogs are usually shallow depressions in the ground with or without bedding material, located in shaded areas on high, dry ground (Hanson and Karstad 1959). Bedding materials include pine straw (dead, fallen pine needles) (Sweeney 1970; Kurz and Marchinton 1972; Wood and Brenneman 1980); grasses, leaves, and other vegetation (Barrett 1978); and broom sedge (*Andropogon virginicus*) (Sweeney 1970).

Feral sows may leave the nest during daylight hours in the first 2 weeks after parturition, during which time the newborn piglets remain closely huddled in the nest or explore the nearby area (Kurz and Marchinton 1972; Crouch 1983). At about 3 weeks of age, piglets begin following the sow in her daily movements (Kurz and Marchinton 1972; Barrett 1978; Crouch 1983).

On one occasion, a feral sow was observed to nurse another sow's litter (Kurz and Marchinton 1972). Observations of the pseudoparent 1 week earlier indicated that she was very close to parturition. Although there was no evidence of her litter, her physical condition while nursing indicated that parturition had occurred. The parent feral sow did not display aggression toward this sow (Kurz and Marchinton 1972).

Breeding Groups. Barrett (1978) observed four different feral sows come into estrus. As many as 10 boars gathered around each sow for about 2 days. Once the sows were receptive, copulation took place as often as every 10 min. Unless the weather was very hot or very cold, breeding continued through day and night. Kurz and Marchinton (1972) observed copulation in two groups of feral hogs, each consisting of four boars and four sows. Although the boars fought among themselves, the sows appeared to be rooting for food. This feeding response was believed to be displacement behavior, as the sows moved much more rapidly than usual through the area.

There are two types of fighting between feral boars in breeding groups: the establishment of dominance, and the display or maintenance of dominance (Kurz and Marchinton 1972). In the establishment of dominance, feral boars charge head-on, pushing with great force. Two competing boars place their heads along the neck and shoulder region of the opponent. Constant circling, upward slashing with the tusks, and variable amounts of vocalization are all part of the pattern. Fighting continues until one animal squeals and retreats. Displays for the maintenance of dominance, however, last only a matter of seconds. Boars display by facing each other with erect manes and open mouths. There is often much loud roaring and coughing. These displays are usually terminated with one or two lunges by the dominant boar (Kurz and Marchinton 1972; Barrett 1978).

Breeding activity is generally a combination of courtship, male fighting, displacement behavior, and copulation. After copulation, dominant boars usually lie down nearby, permitting subordinate boars to breed the receptive sow. Homosexual activity often occurs, especially in young males (Barrett 1978).

Senses. The feral hog's senses of smell and hearing are well developed (Graves 1984), and it depends more on these than its vision to keep in contact with its environment (Wesley and Klopfer 1962). Populations of feral hogs in California (Barrett 1978) and South Carolina were easily approached from downwind, even in open country. A common vocalization by feral hogs is an alarm grunt given by the individual that first scents an intruder. This alarm causes almost immediate flight response in the rest of the herd (Crouch 1983; Hughes 1985). Other vocalizations include those common to domestic hogs.

Scent Marking. John Mayer (Wildlife Ecologist, Environmental Protection Department, Westinghouse Savannah River Company, Aiken, SC, pers. commun.) described three glands that are used by feral hogs in scent marking: (1) Tusk glands are found in the upper lip above the "tusks" or canine teeth in boars. Boars mark with these glands by rubbing their lips against an object or by champing their jaws to produce foamed saliva, which contains the scent. Adult boars exhibit this behavior when in social situations with other boars or sows. When a boar "tusks" a tree, he is rubbing the tusk gland against the bark of a tree to leave its scent. The sharp canine teeth located in the same area of the snout being rubbed against the bark inadvertently damage the tree. Boars typically tend to tusk trees when in solitary situations and produce foamed saliva during social interactions. (2) The preputial gland is a diverticulum or blind sac, which is connected to the urethra in the preputial area of the penis. Both urine and cells shed from the walls of the urethra collect in this blind sac, undergo decomposition, and produce a foul-smelling fluid, which can be detected as the "boar odor." This is released whenever a mature male feral hog either intentionally or inadvertently marks with this gland during urination. Such scent marking is also observed in conjunction with metacarpal gland marking, usually after the boar has marked the spot with its metacarpal glands. (3) The metacarpal glands are on the posterior part of the front feet. A boar feral hog marks with its metacarpal glands by leaning forward and pawing the ground with the bottoms of its front feet. Typically, boars mark with their metacarpal glands when they are participating in breeding groups or when they detect the presence of scent marking by another boar.

Scent posts are commonly used, often in conjunction with body scratching (Barrett 1978). Mayer and Brisbin's (1986) observations of captive feral boars suggested that scent-marking behavior varies according to dominance rank and the immediate social environment of the animal. They found the dominant male scent marked more than the subordinate male when both were present with a sow. When alone, the dominant male did not scent mark, whereas the subordinate male scent marked at about the same intensity as when paired with the dominant male (Mayer and Brisbin 1986).

MORTALITY

The most important factor responsible for mortality in feral hogs is hunting. Natural predation usually plays only a minor role. Intermediate is loss to diseases and parasites, particularly through an interaction with age.

Hunting. Hunting of feral hogs may take place to control nuisance animals, for sport, or through poaching. Populations in good habitat apparently can withstand annual harvests of 50–70% (Barrett and Pine 1980), and many populations will not be controlled by normal sport hunting. However, heavy hunting has caused hog populations to decline below sustainable levels in Florida (Belden and Frankenberger 1989; Belden 1997). For control purposes, hunting at night (Brown 1985) or with dogs (Sanders 1997) can be very effective in reducing feral hog populations. Because most states do not consider feral hogs as game animals, the impact of unrecorded hunting on feral hog populations has not been estimated throughout most of the range. However, this is considered to be insignificant to nonexistent in most states.

Predation. In general, predation is a minor mortality factor, particularly in the removal of adult hogs. Alligators (*Alligator mississippiensis*), black bears, and mountain lions (*Puma* [*Felis*] *concolor*) have all been reported as predators of an occasional adult feral hog (Hamilton 1941; Hanson and Karstad 1959; Wood and Brenneman 1977). Bobcats and coyotes (*Canis latrans*) have been reported as potentially significant predators in some cases on feral hogs (Hanson and Karstad 1959; Wood and Brenneman 1977; Barrett 1978), but much of this may be consumption of carrion rather than predation.

Parasites and Diseases. The combined mortality because of parasites and diseases, including trauma (other than hunting or predation) and starvation, impinges more heavily on feral hog populations than predation. Disease seems to interact with age, increasing the impact of this mortality factor in animals older than 2 years or younger than 6 months (Sweeney 1970; Wood and Brenneman 1977; Nettles 1997).

Pseudorabies, or Aujeszky disease, and swine brucellosis are the diseases of feral swine that have elicited the greatest concern because of their potential impacts on commercial swine production. Pseudorabies is a herpesvirus occurring naturally in several species of domestic and wild mammals (Tozzini et al. 1982; Davidson and Nettles 1997). The disease causes high mortality among young (<4 weeks) domestic hogs, abortion or macerated fetuses in pregnant sows, and productivity losses in growing pigs. Domestic swine have been considered the principal reservoir of the disease (Tozzini et al. 1982). However, as the eradication program in domestic swine nears completion, the implications of a reservoir of this disease in feral hogs becomes more important (Nettles and Erickson 1984). Pseudorabies is not a zoonotic disease, so humans are not affected. Seropositive feral hogs have been detected among free-living populations in 11 states: Alabama, Arkansas, California, Florida, Georgia, Hawaii, Louisiana, Mississippi, Oklahoma, South Carolina, and Texas (Nettles and Erickson 1984; Pirtle et al. 1989; van der Leek et al. 1993a). The primary mode of transmission of this disease is through sexual contact (Romero et al. 1997); however, it can also be transmitted intranasally and by cannibalism (Hahn et al. 1997; Nettles 1997).

Brucellosis is an infectious, zoonotic disease caused by five known species of *Brucella* (Drew et al. 1992), with *Brucella suis* infecting swine. States that have infected populations include Arkansas, California, Florida, Georgia, Hawaii, Louisiana, South Carolina, and Texas (Nettles 1989). Infections in domestic swine are usually chronic, and are characterized by abortion and infertility in sows, orchitis in boars, piglet mortality, and lameness (Davidson and Nettles 1997). *Brucella suis* infections in humans cause flulike symptoms characterized by fever chills, headaches, and general weakness (Davidson and Nettles 1997). The mortality rate is very low, but the disease often is prolonged and debilitating. Increasing hog populations and increasing harvest rates will lead to greater exposure of this disease to hunters (Drew et al. 1992; van der Leek et al. 1993b). Future management of these two diseases in wild populations may include administration of oral vaccines, but the effectiveness of such a program remains questionable (Nettles 1997). Awareness of disease risks posed by wild swine has increased significantly (15.5%) among swine producers and somewhat among wildlife management professionals. The general population of hunters, however, remains unaware of the disease implications of feral hogs (Nettles 1997). Regulation of translocation of feral hogs has increased dramatically. Essentially all movement of these animals is now under jurisdiction of the state–federal eradication programs for swine brucellosis and pseudorabies (Nettles 1997).

Evidence of vesicular stomatitus has been found in feral hog populations in Arkansas, Florida, Georgia, and Louisiana (Hanson and Karstad 1956; Jonkers 1967; Stallknecht et al. 1986; Comer et al. 1990, 1992; Corn et al. 1990). Infections typically have been very localized and have not spread to adjacent populations in spite of its persistence in the areas in which it has been recorded (Jonkers 1967; Stallknecht et al. 1986). Apparently, a precise set of ecological conditions must be met for this virus to exist in a given area, or potential reservoirs must be very short ranged. More recent studies have shown arthropod transmission to be important (Stallknecht et al. 1987, 1993; Corn et al. 1990; Comer et al. 1990, 1992). Stallknecht et al. (1999) determined that sufficient virus is shed from infected swine for contact or mechanical transmission to occur. The lesions from this virus are indistinguishable from those of other important vesicular diseases of swine (Davidson and Nettles 1997). Therefore, discovery of fluid-filled blisters in feral swine should be reported promptly to the state veterinarian.

Other diseases reported in feral hogs in North America include bovine tuberculosis in California (Clark et al. 1983; Davidson and Nettles 1997); eastern encephalomyelitis in Georgia and Florida (Elvinger et al. 1996); enterovirus in Kansas (Gipson et al. 1999); hog cholera, eradicated from the United States in 1978, with no known current infestations (Nettles et al. 1989); leptospirosis in Oklahoma (New et al. 1994); porcine parvovirus in Oklahoma and Kansas (New et al. 1994; Gipson et al. 1999); porcine reproductive and respiratory syndrome in Oklahoma (Saliki et al. 1998); and swine influenza in Oklahoma and Kansas (New et al. 1994; Gipson et al. 1999).

Levels and types of parasites and diseases differ between feral hogs and domestic swine most likely because of differences in behavior, habitats, nutrition levels, and genetic background. Free-ranging feral hogs do not have as much opportunity to transmit parasites or diseases as do domestic hogs confined in pastures.

At least 25 species of internal parasites have been reported from wild swine, including 2 protozoa, 4 trematodes, 2 cestodes, 1 acanthocephalan, and 16 nematodes (Nettles 1989). Parasites, such as ascarid worms, that have no intermediate host, but require the accidental ingestion of an embryonic ovum passed to the ground by a previous host, are uncommon in feral hogs. They become abundant only when feral hog populations are high. However, parasites whose life cycles include passage through an intermediate host commonly consumed as food by feral hogs, such as *Metastrongylus* lungworms via earthworms, are much more prevalent. These parasites are able to maintain higher levels of infection rates as feral hog populations decline. Numerous studies of parasitism in feral hogs have been conducted, and high abundances of some parasites have been recorded. Typically, however, parasitism does not appear to cause any significant levels of morbidity or mortality in feral swine. Parasitism most likely causes mortality only as an additional stressor in conjunction with poor nutrition, severe environmental conditions, or some other debilitating factor.

The most commonly reported parasites in feral hogs are lungworms, Ixod ticks, roundworms (*Ascaris suum*), and kidney worms (*Stephanurus dentatus*). A more complete listing of parasites found in feral swine is presented in Table 54.1.

Three parasites of potential public health significance (*Sarcoptes scabiei, Spirometra* sp., and *Toxoplasma gondii*) have been reported in feral hogs, although infection rates were very low. *Trichinella spiralis* was either not found in swine surveyed (Coombs and Springer 1974; Smith et al. 1982; Corn et al. 1986; Gajadhar et al.1997) or was present at a very low (1.4%) muscle tissue infection rate (Nettles 1989). Gray et al. (1999) recovered spargana (the second-stage larvae of the pseudophyllidean tapeworm *Spirometra* spp.) from 5% of feral hogs inhabiting 20 Florida counties and from seven feral hogs in Texas. Although sparganosis in humans is rare in the United States, the potential for migration through a variety of tissues makes sparganosis a potentially life-threatening infection.

Populations of feral hogs along the coastal mountains of California were studied to determine the prevalence of, and the associated demographic and environmental risk factors for, the shedding of *Cryptosporidium parvum* oocysts and *Giardia* sp. cysts (Atwill et al. 1997). Given their propensity to focus their activity in riparian areas, feral hogs may serve as a source of protozoan contamination for surface water (Atwill et al. 1997).

Mortality due to trauma was considered a major factor in loss of piglets in California, whereas starvation was considered the most important cause of death in older animals (Barrett 1978). Accidents resulting in the loss of piglets included crushing by the sow while in the nest, tusking or trampling by boars, blindness caused by dry grass florets impacting in the eyes, and separation from the sow and littermates resulting in exposure and starvation. Starvation of piglets can result when sows are on a low plane of nutrition, leading to poorer quality and/or quantity of milk. In older feral hogs, tooth deterioration is a significant mortality factor. All hogs ≥4 years old had one or more periodontal abscesses, leading to bone decay and generalized septicemia (Barrett 1978).

TABLE 54.1. Parasites of feral hogs

Parasite	Source[a]
Ectoparasites	
Anoplura	
Haematopinus suis	1, 6, 10, 13
Acarina	
Ixodid ticks	1, 4, 12
Amblyomma americanum	9, 10, 13
A. cajennense	4
A. maculatum	4, 9, 10, 13
Demodex phylloides	13
Dermacentor variabilis	4, 9, 10, 13
Ixodes scapularis	4, 9, 13
Sarcoptes scabiei	13
Endoparasites	
Protozoa	
Coccidia	4, 7, 13
Sarcocystis sp.	2, 4, 10, 13
Toxoplasma gondii	3, 13
Trematoda	
Brachylaima virginianum	13
Fasciola hepatica	13
Paragonimus kellicotti	13
Cestoda	
Spirometra sp.	7, 13
Acanthocephala	
Macracanthorhynchus hirundinaceus	13
Nematoda	
Ascaris suum	4, 6, 11, 12, 13
Ascarops strongylina	11, 13
Capillaria putorii	13
Globocephalus urosubulatus	4, 11, 12, 13
Gongylonema pulchrum	4, 11, 13
Haemonchus contortus	13
Hyostrongylus rubidus	13
Metastrongylus spp.	1, 6, 10, 12, 13
M. apri	4, 5, 11, 13
M. pudendotectus	4, 5, 11, 13
M. salmi	4, 5
Oesophagostomum brevicaudum	13
O. dentatum	13
O. quadrispinulatum	11, 13
Physaloptera sp.	13
Physocephalus sexalatus	4, 11, 13
Stephanurus dentatus	4, 10, 11, 12, 13
Strongyloides ransomi	4, 13
Trichuris suis	6, 13

[a] 1. Barrett 1978; 2. Barrows et al. 1981; 3. Clark et al. 1983; 4. Coombs and Springer 1974; 5. Forrester et al. 1982; 6. Gipson et al. 1999; 7. Gray et al. 1999; 8. Greiner et al. 1982; 9. Greiner et al. 1984; 10. Hanson and Karstad 1959; 11. Pence et al. 1988; 12. Ruddle 1975; 13. Smith et al. 1982.

AGE ESTIMATION

Several techniques have been used for determining ages of feral hogs. The pattern of tooth eruption and wear appears to be the most reliable and easiest to apply.

Tooth Eruption and Wear. A detailed schedule has not been published on the tooth eruption pattern in feral hogs, based on known-age specimens. The most commonly used schedules have been those published for domestic hogs (Sisson and Grossman 1938) and European wild hogs (Matschke 1967). Whereas tooth eruption and replacement varies

TABLE 54.2. Average periods of tooth eruption and replacement in domestic swine

Tooth	Eruption	Replacement[a]
I_1	2–4 weeks	12 months
I_2 upper	2–3 months	16–20 months
I_2 lower	1½–2 months	16–20 months
I_3	Before birth	8–10 months
C	Before birth	9–10 months
P_1	5 months	
P_2	5–7 weeks	12–15 months
P_3 upper	4–8 days	12–15 months
P_3 lower	2–4 weeks	12–15 months
P_4 upper	4–8 days	12–15 months
P_4 lower	2–4 weeks	12–15 months
M_1	4–6 months	
M_2	8–12 months	
M_3	18–20 months	

SOURCE: Sisson and Grossman (1938:488). Reproduced with permission from Elsevier.

[a]P_1 and all molars have no deciduous precursors, and therefore are not replaced.

TABLE 54.3. Average periods of tooth eruption and replacement in European wild hogs

Age[a]	Deciduous Teeth[b]	Permanent Teeth[b]
0–6 days	$i\frac{3}{3}, c\frac{1}{1}$	
7–22 days	$i\frac{3}{3}, c\frac{1}{1}, p\frac{3}{4}$	
23–40 days	$i\frac{1\,3}{1\,3}, c\frac{1}{1}, p\frac{3}{3\;4}$	
6–7 weeks	$i\frac{1\,3}{1\,3}, c\frac{1}{1}, p\frac{3\;4}{3\;4}$	
7–19 weeks	$i\frac{123}{123}, c\frac{1}{1}, p\frac{234}{234}$	
20–33 weeks	$i\frac{123}{123}, c\frac{1}{1}, p\frac{234}{234}$	$P\frac{1}{1}, M\frac{1}{1}$
30–51 weeks	$i\frac{1\,2}{1\,2}, p\frac{234}{234}$	$I\frac{3}{3}, C\frac{1}{1}, P\frac{1}{1}, M\frac{1}{1}$
12–15 months	$i\frac{2}{2}, p\frac{234}{234}$	$I\frac{1\,3}{1\,3}, C\frac{1}{1}, P\frac{1}{1}, M\frac{1\,2}{1\,2}$
14–18 months	$i\frac{2}{2}$	$I\frac{1\,3}{1\,3}, C\frac{1}{1}, P\frac{1234}{1234}, M\frac{1\,2}{1\,2}$
18–22 months	$i\frac{2}{}$	$I\frac{1\;\;3}{123}, C\frac{1}{1}, P\frac{1234}{1234}, M\frac{1\,2}{1\,2}$
21–26 months		$I\frac{123}{123}, C\frac{1}{1}, P\frac{1234}{1234}, M\frac{1\,2}{123}$
>26 months		$I\frac{123}{123}, C\frac{1}{1}, P\frac{1234}{1234}, M\frac{123}{123}$

SOURCE: Matschke (1967:112).

[a]Age intervals overlap due to wide variations in timing of eruption of some individual teeth. For example, $I\frac{2}{}$ replaces its deciduous counterpart as early as 21 months or as late as 26 months.

[b]Lowercase letters indicate deciduous teeth; uppercase letters indicate permanent teeth.

considerably among animals (Matschke 1967), in general, for each respective tooth, eruption often occurs at an earlier age in domestic hogs (Table 54.2) than in European wild hogs (Table 54.3), particularly $I\frac{2}{2}$ and $M\frac{3}{3}$. However, with the exception of $p\frac{4}{4}$, the sequence of eruptions is the same. In the domestic hog, $p\frac{4}{}$ erupts before $p\frac{}{4}$, which is the reverse of the pattern seen in European wild hogs.

In feral hogs, $I\frac{3}{}$ and $P\frac{}{1}$ frequently do not erupt. Barrett (1978) reported that 45 of 100 skulls and lower jaws examined were missing one to three of these four teeth. Normally, these eruption failures were bilateral. In feral hogs along the Savannah River in South Carolina, $P\frac{}{1}$ failed to erupt >50% of the time.

Pine and Gerdes (1973) found that the tooth eruption pattern for the European wild hog gave satisfactory estimates when used to determine ages for feral hogs in Monterey County, California. Similarly, the true ages of tagged feral piglets on the Dye Creek Preserve in California corresponded best with the tooth eruption pattern of European wild hogs. Use of this schedule slightly overestimated age of feral piglets (Barrett 1978). However, the tooth eruption pattern for domestic hogs provided the most accurate schedule for estimating the age of feral hogs along the Savannah River, South Carolina. Limited data suggested that in this instance the domestic schedule slightly underaged feral hogs (Sweeney 1970). Sisson and Grossman's (1938) schedule provides estimated ages for domestic hogs ≤20 months, and Matschke's (1967) schedule provides age estimates for European wild hogs ≤26 months of age. Barrett (1978) used the degree of eruption and wear of $M\frac{3}{3}$ to group animals roughly into yearly age classes from 3 to 6 years (Table 54.4).

More recently, Mayer and Brisbin (1991) assigned wild hogs to age classes that represented postnatal development and growth, rather than absolute ages, because of the differences in rates of eruption and growth of domestic swine, European wild boar, and wild boar × feral hog hybrids (Table 54.5). This may be the best method for determining relative age of individuals in populations where lineage is uncertain. Mayer and Brisbin (1991) also compared their relative age classes and the known absolute age ranges described by Sisson and Grossman (1938) and Matschke (1967) (Table 54.6).

Fetal Development. Henry (1968) described fetal development in European wild hogs and found that crown–rump measurements of known-age fetuses were very similar to those of domestic swine (Table 54.7). The most accurate method of estimating ages of fetuses is to compare crown–rump measurements to those listed in Table 54.7 and interpolate for ages not listed. Error by this method averaged 1.3 days or 2.2%. Henry (1968) also described changes in morphology of fetuses between 30 and 110 days of gestation, which he used to estimate fetal age within 10 days.

Other Criteria. Counts of cementum annuli and use of morphometric relationships have been examined as potential criteria for determining ages of feral hogs. However, cementum annuli counts probably are reliable only in subalpine areas (Choquenot and Saunders 1993) and most likely are unreliable in feral hog populations inhabiting most North American environments. Morphometric criteria are affected significantly by environmental variation in nutrition and most likely require local calibration (Choquenot and Saunders 1993). Eye lens weight, or possibly protein content of the lens, offers another criterion for age determination of feral hogs. In a controlled experiment with pen-raised European wild hogs fed three different diets, Matschke (1963) determined that eye lens weight was correlated with body weight and, as such, was not a reliable indicator of age. However, in a study conducted on feral hogs in South Carolina, a significant correlation ($r = .95, p < .001$) was found between eye lens weight and age of feral hogs, even when body weight failed as an indicator of age (Sweeney et al. 1970). More recently, researchers working with other mammals

TABLE 54.4. Estimated ages of feral hogs over 26 months of age by examination of the third molars

Third Molars	Age (Months)
Present, but <75% erupted	26–30
75–95% erupted	30–36
100% erupted	36+
Cusps 25–50% worn	48+
Cusps 60–90% worn	60+
Cusps completely worn (first molars often lost)	72+

SOURCE: Barrett (1978:289).

TABLE 54.5. Relative age classes in *Sus scrofa* as defined by Mayer and Brisbin (1991)

Age Class		
Number	Name	Erupted Teeth[a]
1	Neonate/infant	$i\frac{3}{3}\ c\frac{1}{1}$ to $i\frac{123}{123}\ c\frac{1}{1}\ p\frac{234}{234}$
2	Juvenile	$i\frac{123}{123}\ c\frac{1}{1}\ p\frac{234}{234}\ P\frac{1}{1}\ M\frac{1}{1}$ to $i\frac{12}{12}\ I\frac{3}{3}\ C\frac{1}{1}\ p\frac{234}{234}\ P\frac{1}{1}\ M\frac{1}{1}$
3	Yearling	$i\frac{123}{123}\ C\frac{1}{1}\ p\frac{234}{234}\ P\frac{1}{1}\ M\frac{12}{12}$ to $I\frac{123}{123}\ C\frac{1}{1}\ P\frac{1234}{1234}\ M\frac{12}{12}$
4	Subadult	$I\frac{123}{123}\ C\frac{1}{1}\ P\frac{1234}{1234}\ M\frac{12}{123^{b}}$ to $I\frac{123}{123}\ C\frac{1}{1}\ P\frac{1234}{1234}\ M\frac{123^{c}}{123}$
5	Adult	$I\frac{123}{123}\ C\frac{1}{1}\ P\frac{1234}{1234}\ M\frac{123}{123}$

[a]Lowercase letters indicate deciduous teeth; uppercase letters indicate permanent teeth.
[b]Lower third molar starting to erupt.
[c]Upper third molar starting to erupt.

TABLE 54.6. Comparisons of the relative age classes (Table 54.5) defined by Mayer and Brisbin (1991:82) and known absolute age ranges of domestic and hybrid wild swine

	Age (months)	
Age Class	Domestic Swine[a]	European Wild Hog[b]
1	0–8	0–8
2	8–12	8–12
3	12–18	12–22
4	18–20	21–30
5	20+	36+

[a]Sisson and Grossman (1938).
[b]Matschke (1967). Mayer and Brisbin (1991) consider these as wild boar × hog hybrids.

have had better success in predicting age by measuring the tyrosine content in the lens, rather than using lens weight (Dimmick and Pelton 1994). Although this also may be a potential improvement in age predictions in feral hogs, use of the eye lens requires special handling and detailed lab work, which often will be impractical for management purposes. As a result, examining the tooth eruption pattern remains the best field technique.

SEX DETERMINATION

It is often important to identify the sex of skeletal remains, and in feral hogs this can be done by examination of the canines and skull. Mayer and Brisbin (1988) found that mean size of male canines was larger than those of females. The degree in overlap, however, caused size differences alone to be inadequate for conclusive identification of sex. However, the enamel on the canines extended along the entire length of the crowns and roots in males. In female canines, the enamel did not extend along the roots, but created a distinct enamocementum junction line. In addition, the upper canines of males and the associated socket (alveolus) had a distinct trapezoidal cross section, whereas those of females had a more triangular cross section. The enamocementum line was present in females by 14–18 months of age, whereas the distinct

TABLE 54.7. Crown–rump measurements of fetal European hogs and domestic hogs

	Domestic Hog[a]			European Wild Hog			
		Length (mm)				Length (mm)	
Age (days)	No. of Fetuses	Average	Range[b]	No. of Sows	No. of Fetuses	Average	Range
20	38	9.4	—	—	—	—	—
21	13	15.2	—	—	—	—	—
22	13	11.5	—	—	—	—	—
28	10	23.3	—	—	—	—	—
30	4	25.3	23.0–26.5	—	6	20.2	19–21
32	20	28.1	—	—	—	—	—
33	—	—	—	1	4	24.0	23–25
40	37	48.5	43.3–49.7	2	14	47.1	43–50
50	22	82.0	78.0–85.7	2	9	82.7	78–86
60	22	119.2	116.7–120.9	2	10	111.9	103–121
64	—	—	—	1	7	131.6	129–137
68	—	—	—	1	7	148.7	143–154
70	34	158	153–171	1	4	159.5	156–162
80	32	176	137–189	2	9	185.1	180–190
90	10	194	173–208	2	13	209.0	193–217
100	47	226	212–239	2	4	226.5	210–244
101	—	—	—	1	1	207.0	—
110	28	239	212–261	2	5	240.8	219–255

SOURCE: Henry (1968:967).
[a]From Warwick (1928:69).
[b]Range is for averages of groups of fetuses.

trapezoidal shape of male upper canines was evident at 30–51 weeks old. Mayer and Brisbin (1988) also noted a distinct bony ridge over the alveolus of the upper canine in males, but not in females. However, this dimorphism is less evident in younger age classes because males have not had time to develop this ridge (Mayer and Brisbin 1988).

ECONOMIC STATUS

Reflecting its variable status as a game animal or a pest, the feral hog has positive and negative economic impacts. Its greatest negative economic impact is probably the cost related to the detrimental effects of its omnivorous diet and feeding behavior (see Feeding Habits). Based on a conservative estimate of annual environmental and crop damages of approximately $200/animal, Pimentel et al. (2000) calculated the yearly damage from feral hogs in the United States to be approximately $800 million. This excludes the significant impact of the environmental damages and diseases that cannot be translated easily into market-based values (Pimentel et al. 2000). Some national wildlife refuges have had to resort to extensive fencing to protect crops planted for waterfowl management (Thompson 1977). Feral hogs represent increased maintenance costs on management areas where rooting has resulted in destruction of dikes, roads, trails, and recreation areas (Pine and Gerdes 1973; Belden and Frankenberger 1977; Thompson 1977). These programs represent a constant economic drain on both labor and support budgets.

Another area of economic concern is the potential of feral hog populations to serve as a reservoir for certain endemic diseases and parasites transmissible to humans and domestic livestock. Feral hogs may represent a greater hazard in this respect than other species of wildlife (Hanson and Karstad 1959) and may pose a significant challenge in current attempts to eradicate endemic diseases from domestic swine herds (van der Leek 1989). Feral hogs also may spread exotic diseases introduced into the United States. For example, a primary threat is the introduction of foot-and-mouth disease and/or hog cholera from uncooked animal products (e.g., intestinal casings) discarded by people involved in the illegal transportation of narcotics from Central and South America (Mebus 1989).

Although in most situations, feral hogs are not considered a financial asset, they have the potential of providing increased commercial gain because of their productivity and adaptability to different habitats. Increasing numbers of game ranchers in Texas recognize the feral hog as a marketable product (Springer 1977). In Texas, prices for hog hunts ranged from $25 to $1000 and averaged $169 in 1991 (Sanders 1997). The Dye Creek Preserve in California uses commercial guided hunts as an economical approach to controlling feral hogs (Barrett 1978). Feral hog hunting and its related activities brought in an estimated $10.1 million in Florida in 1988 (Degner 1989), and sale of feral hogs as commercial livestock brought in about $21,000 (Degner 1989). Exportation to Europe brought up to $1.30/kg at wild game processing plants in Texas in 1991 (Sanders 1997). "Hog claims," which came into use in the 1800s (Lucas 1977) as a method of recognizing local ownerships, were much more prevalent in the past than they are today. In the few areas that still recognize hogs raised under open range, claim owners may charge for hunting privileges on their claims, or livetrap and sell adult boars to hunting preserves (Belden and Frankenberger 1977).

Degner et al. (1983) and Degner (1989) pointed out that the damage caused by feral hogs may outweigh any commercial return for landowners. The extent of economic gain is quite variable depending on the recreational/market demand and the cost of management.

MANAGEMENT

Management of feral hogs varies substantially throughout their range. In most locations, management was nonexistent until local populations reached densities high enough to cause noticeable environmental damage by rooting or significant competition for mast with other, more favored wildlife species. Then, management usually took the form of control or attempted eradication. With the absence of large predators, feral hogs have the capability of quickly reaching high population densities in favorable habitat. Therefore, herd increase is rarely a management problem.

Federal. The U.S. Forest Service recognizes all feral hogs as estray domestic livestock and, as such, considers them trespassers on national forest lands (Lucas 1977; Belden and Frankenberger 1989; Belden 1993). The U.S. Fish and Wildlife Service recognizes them as feral animals that can, and often do, cause damage to wildlife habitat and refuge infrastructure such as roads and foods plots (Lucas 1977; Thompson 1977; Dale Hall, Deputy Director, U.S. Fish and Wildlife Service, Atlanta, pers. commun.). Both agencies consider the species a nuisance and pursue programs of control to reduce populations to a minimum. The U.S. Fish and Wildlife Service encourages control, primarily through incidental harvest during standard legal hunting seasons. Control efforts on national forests include a process of trapping, notification of potential owners, and ultimate removal (Ron Archuleta, National Assistant Wildlife Program Leader, U.S. Forest Service, Atlanta, pers. commun.).

State. Management of feral hogs is inconsistent at the state level of government as well. Although most states do not consider the feral hog a game species, some do. California, Florida, Hawaii, North Carolina, Tennessee, and West Virginia consider the feral hog a game animal, at least in some areas. Seasons and bag limits in these states usually are liberal. In most other states, the feral hog is unprotected, and is considered the property of the land owner. Hunting on private lands usually is year-round, with no limit and under the control of the landowner. Hunting on public lands may be only slightly more restricted to certain seasons, but usually with no limit. Fourteen states have a ban on the importation or introduction of feral hogs (Mackey 1992). In some states where they are considered domestic livestock, regulatory authority rests with the state department of agriculture, and in others, regulatory authority is unclear.

When managing the feral hog as a game species, it may be difficult to maintain hog populations under heavy hunting (Belden and Frankenberger 1989). Until recently, to solve this problem, Florida annually relocated feral hogs to high-use public hunting areas. They obtained these hogs primarily from state parks, where they had become a nuisance. However, this has become unfeasible (Belden and Frankenberger 1977, 1989). Other management strategies employed in Florida to prevent depletion of hog populations have included shortening the season, limiting or excluding the use of dogs, and setting a limit on the total number of hogs to be harvested, after which the season is closed. These regulations generally have been successful only in those areas that contained an abundance of escape cover (Belden and Frankenberger 1989). The best approach to allow for sustainable harvest may be to develop regulations that permit the harvest of 50–80% of the younger age classes in late summer and early fall (Belden 1993). Harvest at this time is advantageous because the subadult population is highest in relation to the adult breeding population, and harvest has the least impact on suckling sows. Also, this would permit the removal of animals immediately before acorn drop, thereby increasing available supplies of this important food to the surviving population (Belden 1993).

Private. Management of feral hogs by private landowners may be no more than periodic releases on land leased by private hunt clubs, use of "hog claims," or detailed management of existing populations on commercial hunting areas.

Releases of wild hogs by members of private hunt clubs have been to initiate new populations or, theoretically, to improve the local population's genetic stock. The animals released are usually purported to be European wild hogs, but often are of domestic origin. Such releases ordinarily involve only a few animals, but over time have been a major factor in range expansion. Release of feral hogs is illegal in many states and strongly discouraged in all others due to their significant potential for detrimental impacts.

As noted, "hog claims" are still recognized in some states with feral hogs. These claims provide income for rural families through the sale of feral boars for relocation onto hunt clubs or through the leasing of hunting privileges on the area covered by the claim (Belden and Frankenberger 1977, 1989).

Some feral hog populations are managed much more intensively to provide commercial hunting. A prime example of this is the Dye Creek Preserve in California. All feral hog hunts are guided and closely monitored. The goal of management in this case is to create a population sex and age structure that will ensure the continued production of adult trophy boars while maintaining the population at or below the estimated carrying capacity of 6 feral hogs/km^2 (Barrett 1978).

Management Guidelines. There are as many management options for the feral hog as there are managers confronted with the problem. When developing management guidelines, whether on federal, state, or private lands, one must first compare the feral hog's potential detrimental impact to its potential benefits. This should include careful consideration of such negative factors as (1) adverse impacts of feeding habits on local flora and fauna, especially endangered species; (2) contribution to soil erosion and stream siltation; (3) potential spread of disease to other animal species, especially domestic livestock; and (4) possible confrontation with other outdoor recreational activities. These factors should be weighed against a critical review of positive factors, including (1) the potential of the feral hog as another game species, (2) hunter interest, and (3) economic gain to the surrounding area. Also, local traditions must not be overlooked and costs of implementing and maintaining each management option must be detailed.

An effective management program must have (1) a clear population density goal based on landowner objectives and acceptable levels of impacts; (2) a well-designed inventory scheme, which includes regular monitoring of population density and welfare, disease levels, and environmental impacts; and (3) plans for population control developed ahead of time. Automatic cameras with infrared motion detectors may be used at bait stations to successfully monitor feral hog population densities using a version of the capture–mark–recapture estimate (Sweitzer et al. 1997b).

Landowners have few options available to control feral hog populations on their land (Zivin et al. 2000). They can trap and kill animals themselves, at some cost; or, if the government allows, and there is a market, they can sell hunting rights. Relocations of feral hogs should be discouraged unless the animals have been thoroughly tested for diseases and parasites.

Livetrapping with cage or pen traps very often has limited success (Brown 1985), and the animals must be dealt with once trapped. Traps used are typically box traps or panel traps with multiple entry doors (Sweitzer et al. 1997c). Chemical restraint, used primarily for research, has typically been most successful with xylazine hydrochloride in combination with either ketamine hydrochloride (Baber and Coblentz 1982) or, more recently, a 1:1 mixture of tileamine hydrochloride and zolazepam hydrochloride (Sweitzer et al. 1997a). In areas in Texas that produce sheep and goats, the most effective capture technique for hogs is use of neck snares (Littauer 1993). In areas where there is a market for live hogs, as in Texas, some of the cost can be recovered (Sanders 1997).

Hunting, especially hunting with dogs, can be one of the most effective means for controlling a problem hog population (Sanders 1997). However, evidence suggests that feral hog populations can readily withstand annual harvests of 50–70% on a sustained basis (Barrett and Pine 1980).

In areas where hunting is not a viable alternative, night shooting over bait has proven successful for eradication of small feral hog populations from a localized area (Brown 1985). Aerial hunting is another option that can quickly reduce a hog population, but this only works in areas where a significant portion of the population is visible from the air, and is very costly (about $250/hr in 1997) (Sanders 1997). Hog-proof fencing can sometimes keep feral hogs out of an area, but fencing is costly, and must be maintained regularly (Sanders 1997). Littauer (1993) provided an excellent review of the various control techniques for feral hogs.

RESEARCH, MANAGEMENT, AND CONSERVATION NEEDS

Feral hogs are the most abundant introduced ungulate in North America, and their populations continue to expand. Their reproductive potential may be four times greater than that of native ungulates (Taylor et al. 1998). Although feral hogs have positive attributes, they also have significant potential for negative impacts on the environment, native wildlife, and domestic livestock, especially as hog populations increase in density. Emphasis for future research, therefore, should be directed toward developing effective methods of population monitoring and control. Specific information needs suggested by recent and current researchers include information in the following areas.

Population ecology:

- Survival rates of cohorts of feral swine and environmental factors influencing them (Taylor et al. 1998)
- Better techniques for age estimation and population censuses (Higginbotham 1993)
- Detailed reproductive and demographic data to develop growth models (Hellgren 1993)
- Measurement of competitive relationships between feral hogs and native ungulates (Hellgren 1993; Taylor et al. 1998)
- Environmental factors influencing age at puberty and age-specific reproductive potential (Peine and Farmer 1990)
- Impacts of nutrition and other environmental factors on timing of estrus and litter size (Fernandez-Llario and Mateos-Quesada 1998)

Biological controls:

- Development and efficacy of species-specific biological controls that will not affect domestic swine (Peine and Farmer 1990)
- Development of oral contraceptives (Nettles 1997)
- Development of vaccines (Wardley 1989)
- Development of an immunosterilization or use of transgenic males (Bazer 1989)
- Development of oral baits for delivery of desired chemicals (Fletcher 1989; Fletcher et al. 1990)

Diseases:

- Continued research into the epidemiology of feral hog diseases with potential impact on domestic swine (Stallknecht et al. 1993; van der Leek et al. 1993a; Nettles 1997)
- Development of effective disease monitoring and risk assessment methodologies
- Better description of dispersal and the environmental factors influencing this behavior (Gabor et al. 1999)

Impacts:

- Impacts of feral hog feeding behavior on endangered plants and animals (Sweitzer 1998)
- The significance of feral hogs as prey for native predators (Sweitzer 1998)
- Aspects of human dimensions, including translocation, impacts of changing regulations, education programs, and alternative markets (Higginbotham 1993; Nettles 1997)

LITERATURE CITED

Arrington, D. A., L. A. Toth, and J. W. Koebel, Jr. 1999. Effects of rooting by feral hogs *Sus scrofa* L. on the structure of a floodplain vegetation assemblage. Wetlands 19:535–44.

Asdell, S. A. 1964. Patterns of mammalian reproduction, 2nd ed. Cornell University Press, Ithaca, NY.

Atwill, E. R., R. A. Sweitzer, M. D. G. C. Pereira, I. A. Gardner, D. V. Vuren, and W. M. Boyce. 1997. Prevalence of and associated risk factors for shedding *Cryptosporidium parvum* oocysts and *Giardia* cysts within feral pig populations in California. Applied and Environmental Microbiology 63:3946–49.

Baber, D. W., and B. E. Coblentz. 1982. Immobilization of feral pigs with a combination of ketamine and xylazine. Journal of Wildlife Management 46:557–59.

Baber, D. W., and B. E. Coblentz. 1987. Diet nutrition and conception in feral pigs on Santa Catalina Island. Journal of Wildlife Management 51:306–7.

Baron, J. 1982. Effects of feral hogs (*Sus scrofa*) on the vegetation of Horn Island, Mississippi. American Midland Naturalist 107:202–5.

Barrett, R. H. 1971. Ecology of the feral hog in Tehama County, California. Ph.D. Dissertation. University of California, Berkeley.

Barrett, R. H. 1978. The feral hog on the Dye Creek Ranch, California. Hilgardia 46:283–355.

Barrett, R. H., and D. S. Pine. 1980. History and status of wild pigs, *Sus scrofa*, in San Benito County, California. California Fish and Game 67:105–17.

Barrows, P. L., H. M. Smith, A. K. Prestwood, and J. Brown. 1981. Prevalence and distribution of *Sarcocystis* sp. among wild swine of southeastern United States. Journal of the American Veterinary Medical Association 179:1117–18.

Bazer, F. W. 1989. Biological control of feral pigs. Pages 54–56 *in* Proceedings of the feral pig symposium. Livestock Conservation Institute, Madison, WI.

Beach, R. 1993. Depredation problems involving feral hogs. Pages 67–75 *in* C. W. Hanselka and J. F. Cadenhead, eds. Feral swine: A compendium for resource managers: Proceedings of a conference. Texas Agricultural Extension Service, College Station.

Belden, R. C. 1993. Feral hogs: The Florida experience. Pages 101–6 *in* C. W. Hanselka and J. F. Cadenhead, eds. Feral swine: A compendium for resource managers: Proceedings of a conference. Texas Agricultural Extension Service, College Station.

Belden, R. C. 1997. Wild hog management in Florida. Pages 2.11–2.15 *in* K. L. Schmitz, ed. Proceedings: National feral swine symposium. U.S. Department of Agriculture Animal and Plant Health Inspection Service.

Belden, R. C., and W. B. Frankenberger. 1977. Management of feral hogs in Florida: Past, present and future. Pages 5–10 *in* G. W. Wood, ed. Research and management of wild hog populations: Proceedings of a symposium. Belle W. Baruch Forest Science Institute, Clemson University, Clemson, SC.

Belden, R. C., and W. B. Frankenberger. 1989. History and biology of feral swine. Pages 3–10 *in* Proceedings of the feral pig symposium. Livestock Conservation Institute, Madison, WI.

Brisbin, I. L., Jr., R. A. Geiger, H. B. Graves, J. E. Pinder, III., J. M. Sweeney, and J. R. Sweeney. 1977a. Morphological characterizations of two populations of feral swine. Acta Theriologica 22:75–85.

Brisbin, I. L., M. W. Smith, and M. H. Smith. 1977b. Feral swine studies at the Savannah River Ecology Laboratory: An overview of program goals and design. Pages 71–90 *in* G. W. Wood, ed. Research and management of wild hog populations: Proceedings of a symposium. Belle W. Baruch Forest Science Institute, Clemson University, Clemson, SC.

Brown, L. N. 1985. Elimination of a small feral swine population in an urbanizing section of central Florida. Florida Scientist 48:120–23.

Cabon, K. 1959. Problem der Altersbestimmung beim Wildschwein *Sus scrofa* nach der Methode von Dub. Acta Theriologica 4:188–20.

Choquenot, D., and G. Sanders. 1993. A comparison of three ageing techniques for feral pigs from subalpine and semi-arid habitats. Wildlife Research 20:163–71.

Clark, R. K., D. A. Jessup, D. W. Hird, R. Ruppanner, and M. E. Meyer. 1983. Serologic survey of California wild hogs for antibodies against selected zoonotic disease agents. Journal of the American Veterinary Medical Association 183:1248–51.

Comer, J. A., R. B. Tesh, G. B. Modi, J. L. Corn, and V. F. Nettles. 1990. Vesicular stomatitus virus, New Jersey serotype: Replication in and transmission by *Lutzomyia shannoni* (Diptera: Psychodidae). American Journal of Tropical Medicine and Hygiene 42:483–90.

Comer, J. A., J. L. Corn, D. E. Stallknecht, J. G. Landgraf, and V. F. Nettles. 1992. Titers of vesicular stomatitus virus, New Jersey serotype, in naturally infected male and female *Lutzomyia shannoni* (Diptera: Psychodidae) in Georgia. Journal of Medical Entomology 29:368–70.

Coombs, D. W., and M. D. Springer. 1974. Parasites of feral pig × European wild boar hybrids in southern Texas. Journal of Wildlife Diseases 10:436–41.

Corn, J. L., P. K. Swiderek, B. O. Blackburn, G. A. Erickson, A. B. Thiermann, and V. F. Nettles. 1986. Survey of selected diseases in wild swine in Texas. Journal of the American Veterinary Medical Association 189:1029–32.

Corn, J. L., J. A. Comer, G. A. Erickson, and V. F. Nettles. 1990. Isolation of vesicular stomatitus virus New Jersey serotype from phlebotomine sand flies in Georgia. American Journal of Tropical Medicine and Hygiene 42:476–82.

Crouch, L. C., Jr. 1983. Movements of and habitat utilization by feral hogs at the Savannah River Plant, South Carolina. M.S. Thesis, Clemson University, Clemson, SC.

Davidson, W. R., and V. F. Nettles. 1997. Field manual of wildlife diseases in the southeastern United States, 2nd ed. Southeastern Cooperative Wildlife Disease Study, University of Georgia, Athens.

Degner, R. L. 1989. Economic importance of feral swine in Florida. Pages 39–41 *in* Proceedings of the feral pig symposium. Livestock Conservation Institute, Madison, WI.

Degner, R. L., L. W. Rodan, W. K. Mathis, and E. P. J. Gibbs. 1983. The recreational and commercial importance of feral swine in Florida: Relevance to the possible introduction of African swine fever into the U.S.A. Preventive Veterinary Medicine 1:371–81.

Dimmick, R. W., and M. R. Pelton. 1994. Criteria of sex and age. Pages 169–214 *in* T. A. Bookhout, ed. Research and management techniques for wildlife and habitats, 5th ed. The Wildlife Society, Bethesda, MD.

Drew, M. L., D. A. Jessup, A. A. Burr, and C. E. Franti. 1992. Serologic survey of brucellosis in feral swine, wild ruminants, and black bear of California, 1977 to 1989. Journal of Wildlife Diseases 28:355–63.

Ellisor, J. E. 1973. Feral hog studies. Final report (Federal Aid in Wildlife Restoration Act, Project No. W-101-R-4). [As cited in Hellgren 1993.]

Elvinger, F., C. A. Baldwin, A. D. Liggett, K. N. Tang, and D. E. Stallknecht. 1996. Prevalence of exposure to eastern encephalomyelitis virus in domestic and feral swine in Georgia. Journal of Veterinary Diagnostic Investigation 8:481–84.

Everitt, J. H., and M. A. Alaniz. 1980. Fall and winter diets of feral pigs in south Texas. Journal of Range Management 33:126–29.

Fernandez-Llario, P., and P. Mateos-Quesada. 1998. Body size and reproductive parameters in the wild boar *Sus scrofa*. Acta Theriologica 43:439–44.

Fletcher, W. O. 1989. Oral baits capable of delivering vaccines to wild swine. Page 57 *in* Proceedings of the feral pig symposium. Livestock Conservation Institute, Madison, WI.

Fletcher, W. O., T. E. Creekmore, M. S. Smith, and V. F. Nettles. 1990. A field trial to determine the feasibility of delivering oral vaccines to wild swine. Journal of Wildlife Diseases 26:502–10.

Foley, C. W., R. W. Seerley, W. J. Hansen, and S. E. Curtis. 1971. Thermoregulatory responses to cold environment by neonatal wild and domestic piglets. Journal of Animal Science 32:926–29.

Forrester, D. J., J. H. Porter, R. C. Belden, and W. B. Frankenberger. 1982. Lungworms of feral swine in Florida. Journal of the American Veterinary Medical Association 181:1278–80.

Frandson, R. D. 1965. Anatomy and physiology of farm animals. Lea & Febiger, Philadelphia.

Gabor, T. M., E. C. Hellgren, R. A. Van Den Bussche, and N. J. Silvy. 1999. Demography, sociospatial behaviour and genetics of feral pigs (*Sus scrofa*) in a semi-arid environment. Journal of Zoology (London) 247:311–22.

Gajadhar, A. A., J. R. Bisaillon, and G. D. Appleyard. 1997. Status of *Trichinella spiralis* in domestic swine and wild boar in Canada. Canadian Journal of Veterinary Research 61:256–59.

Gipson, P. S., R. Matlack, D. P. Jones, H. J. Able, and A. E. Hynek. 1995. Feral pigs, *Sus scrofa*, in Kansas. Proceedings of the North American Prairie Conference: Prairie Diversity 14:93–96.

Gipson, P. S., B. Hlavachick, and T. Berger. 1998. Range expansion by wild hogs across the central United States. Wildlife Society Bulletin 26:279–86.

Gipson, P. S., J. K. Veatch, R. S. Matlack, and D. P. Jones. 1999. Health status of a recently discovered population of feral swine in Kansas. Journal of Wildlife Diseases 35:624–27.

Golley, F. B. 1962. Mammals of Georgia. University of Georgia Press, Athens.

Graves, H. B. 1984. Behavior and ecology of wild and feral swine (*Sus scrofa*). Journal of Animal Science 58:482–92.

Graves, H. B., and K. L Graves. 1977. Some observations of biobehavioral adaptations of swine. Pages 103–10 *in* G. W. Wood, ed. Research and management of wild hog populations: Proceedings of a symposium. Belle W. Baruch Forest Science Institute, Clemson University, Clemson, SC.

Gray, M. L., F. Rogers, S. Little, M. Puette, D. Ambrose, and E. P. Hoberg. 1999. Sparganosis in feral hogs (*Sus scrofa*) from Florida. Journal of the American Veterinary Medical Association 215:204–8.

Greiner, E. C., C. Taylor III, W. B. Frankenberger, and R. C. Belden. 1982. Coccidia of feral swine from Florida. Journal of the American Veterinary Medical Association 181:1275–77.

Greiner, E. C., P. P. Humphrey, R. C. Belden, W. B. Frankenberger, D. H. Austin, and P. J. Gibbs. 1984. Ioxid ticks on feral swine in Florida. Journal of Wildlife Diseases 20:114–19.

Hagen, D. R., and K. B. Kephart. 1980. Reproduction in domestic and feral swine. I. Comparison of ovulatory rate and litter size. Biology of Reproduction 22:550–52.

Hagen, D. R., K. B. Kephart, and P. J. Wangsness. 1980. Reproduction in domestic and feral swine. II. Interrelationships between fetal size and spacing and litter size. Biology of Reproduction 23:929–34.

Hahn, E. C., G. R. Page, P. S. Hahn, K. D. Gillis, C. Romero, J. A. Annelli, and E. P. J. Gibbs. 1997. Mechanisms of transmission of Aujeszky's disease virus originating from feral swine in the USA. Veterinary Microbiology 55:123–30.

Hamilton, W. J., Jr. 1941. Notes on some mammals of Lee County, Florida. American Midland Naturalist 25:686–91.
Hansen, W. J., C. W. Foley, R. W. Seerley, and S. E. Curtis. 1972. Pelage traits in neonatal wild, domestic and crossbred piglets. Journal of Animal Science 34:100–103.
Hanson, R. P., and L. Karstad. 1956. Enzootic vesicular stomatitis. Proceedings of the Annual Meeting of the U.S. Livestock Sanitation Association 60:288–92.
Hanson, R. P., and L. Karstad. 1959. Feral swine in the southeastern United States. Journal of Wildlife Management 23:64–74.
Hauser, E. R., G. E. Dickerson, and D. T. Mayer. 1952. Reproductive development and performance of inbred and crossbred boars (Research Bulletin 503). Missouri Agricultural Experiment Station.
Hellgren, E. C. 1993. Biology of feral hogs (*Sus scrofa*) in Texas. Pages 50–58 *in* C. W. Hanselka and J. F. Cadenhead, eds. Feral swine: A compendium for resource managers: Proceedings of a conference. Texas Agricultural Extension Service, College Station.
Henry, V. G. 1966. European wild hog hunting season recommendations based on reproductive data. Proceedings of the Southeastern Association of Fish and Game Commissioners 20:139–45.
Henry, V. G. 1968. Fetal development in European wild hogs. Journal of Wildlife Management 32:966–70.
Higginbotham, B. 1993. Feral swine: Research needs and summary. Pages 158–60 *in* C. W. Hanselka and J. F. Cadenhead, eds. Feral swine: A compendium for resource managers: Proceedings of a conference. Texas Agricultural Extension Service, College Station.
Howe, T. D., and S. P. Bratton. 1976. Winter rooting activity of the European wild boar in the Great Smoky Mountains National Park. Castanea 41:256–64.
Hsu, T. C., and R. A. Mead. 1969. Mechanisms of chromosomal changes in mammalian speciation. Pages 8–17 *in* K. Benirschke, ed. Comparative mammalian cytogenetics. Springer-Verlag, Berlin.
Hughes, T. W. 1985. Home range, habitat utilization, and pig survival of feral swine on the Savannah River Plant. M.S. Thesis, Clemson University, Clemson, SC.
Hutchinson, C. B., ed. 1946. California agriculture. University of California Press, Berkeley.
Ilse, L. M., and E. C. Hellgren. 1995a. Resource partitioning in sympatric populations of collared peccaries and feral hogs in southern Texas. Journal of Mammalogy 76:784–99.
Ilse, L. M., and E. C. Hellgren. 1995b. Spatial use and group dynamics of sympatric collared peccaries and feral hogs in southern Texas. Journal of Mammalogy 76:993–1002.
Irving, L. 1956. Physiological insulation in bare-skinned swine. Applied Physiology 9:414–20.
Irving, L. 1966. Adaptations to cold. Scientific American 214:94–101.
Jonkers, A. H. 1967. The epizootiology of the vesicular stomatitus virus: A reappraisal. American Journal of Epidemiology 86:286–91.
Kotanen, P. M. 1995. Responses of vegetation to a changing regime of disturbance: Effects of feral pigs in a California coastal prairie. Ecogeography 18:190–99.
Kramer, R. J. 1971. Hawaiian land mammals. Charles E. Tuttle, Rutland, VT.
Kurz, L. C., and R. L. Marchinton. 1972. Radiotelemetry studies of feral hogs in South Carolina. Journal of Wildlife Management 36:1240–48.
Lacki, M. J., and R. A. Lancia. 1983. Changes in soil properties of forests rooted by wild boar. Proceedings of the Annual Conference of the Southeastern Association of Fish and Wildlife Agencies 37:228–36.
Lacki, M. J., and R. A. Lancia. 1986. Effects of wild pigs on beech growth in Great Smoky Mountains National Park. Journal of Wildlife Management 50:655–59.
Lewis, J. C., G. Matschke, and R. Murry. 1965. Hog Subcommittee report to the chairman of the Forest Game Committee, Southeastern Section, The Wildlife Society. Unpublished report.
Lipscomb, D. J. 1989. Impacts of feral hogs on longleaf pine regeneration. Southern Journal of Applied Forestry 13:177–81.
Littauer, G. A. 1993. Control techniques for feral hogs. Pages 139–48 *in* C. W. Hanselka and J. F. Cadenhead, eds. Feral swine: A compendium for resource managers: Proceedings of a conference. Texas Agricultural Extension Service, College Station.
Lucas, F. G. 1977. Feral hogs: Problems and control on national forest lands. Pages 17–22 *in* G. W. Wood, ed. Research and management of wild hog populations: Proceedings of a symposium. Belle W. Baruch Forest Science Institute, Clemson University, Clemson, SC.
Mackey, W. 1992. A survey on wild hogs of the United States. Minnesota Board of Animal Health. [As cited in Miller 1993.]
Manlove, M. N., J. C. Avise, H. O. Hillestad, P. R. Ramsey, M. H. Smith, and D. O. Straney. 1976. Starch gel electrophoresis for the study of population genetics in white-tailed deer. Proceedings of the Southeastern Association of Fish and Game Commissioners 29:392–402.
Mansouri, A., and C. A. DeYoung. 1987. Feral hog fidelity to home range after exposure to supplemental feed. Texas Journal of Agriculture and Natural Resources 1:46–49.
Mapston, M. E. 1997. Feral hog control in Texas. Pages 2.1–2.3 *in* K. L. Schmitz, ed. Proceedings: National feral swine symposium. USDA, Animal and Plant Health Inspection Service.
Martin, R. L., and J. G. Herbein. 1976. A comparison of the enzyme levels and in vitro utilization of various substrates for lipogenesis in pair-fed lean and obese pigs. Proceedings of the Society for Experimental Biology and Medicine 151:231–35.
Matschke, G. H. 1963. An eye lens–nutrition study of penned European wild hogs. Proceedings of the Southeastern Association of Fish and Game Commissioners 17:20–27.
Matschke, G. H. 1964. The influence of oak mast on European wild hog reproduction. Proceedings of the Annual Conference of the Southeastern Association of Game and Fish Commissioners 18:35–39.
Matschke, G. H. 1967. Aging European wild hogs by dentition. Journal of Wildlife Management 31:109–13.
Mayer, J. J., and I. L. Brisbin, Jr. 1986. A note on scent-marking behavior of 2 captive-reared feral boars. Applied Animal Behaviour Science 16:85–90.
Mayer, J. J., and I. L. Brisbin, Jr. 1988. Sex identification of *Sus scrofa* based on canine morphology. Journal of Mammalogy 69:408–12.
Mayer, J. J., and I. L. Brisbin, Jr. 1991. Wild pigs of the United States: Their history, comparative morphology, and current status. University of Georgia Press, Athens.
Mayer, J. J., and I. L. Brisbin, Jr. 1993. Distinguishing feral hogs from introduced wild boar and their hybrids: A review of past and present efforts. Pages 28–49 *in* C. W. Hanselka and J. F. Cadenhead, eds. Feral swine: A compendium for resource managers: Proceedings of a conference. Texas Agricultural Extension Service, College Station.
Mayer, J. J., E. A. Nelson, and L. D. Wike. 2000. Selective depredation of planted hardwood seedlings by wild pigs in a wetland restoration area. Ecological Engineering 15:S79–85.
McFee, A. F., M. W. Banner, and J. M. Rary. 1966. Variation in chromosome number among European wild pigs. Cytogenetics 5:75–81.
Mebus, C. A. 1989. Potential role of feral pigs in the spread of foreign animal disease. Pages 34–36 *in* Proceedings of the feral pig symposium. Livestock Conservation Institute, Madison, WI.
Miller, J. E. 1993. A national perspective on feral swine. Pages 9–16 *in* C. W. Hanselka and J. F. Cadenhead, eds. Feral swine: A compendium for resource managers: Proceedings of a conference. Texas Agricultural Extension Service, College Station.
Moen, A. N. 1968. Surface temperatures and radiant heat loss from white-tailed deer. Journal of Wildlife Management 32:338–44.
Mount, L. E. 1968. Adaptation of swine. Pages 277–91 *in* E. S. E. Hafez, ed. Adaptation of domestic animals. Lea & Febiger, Philadelphia.
Nalbandov, V. 1976. Reproductive physiology of mammals and birds. W. H. Freeman, San Francisco.
Nettles, V. F. 1989. Diseases of wild swine. Pages 16–18 *in* Proceedings of the feral pig symposium. Livestock Conservation Institute, Madison, WI.
Nettles, V. F. 1997. Feral swine: Where we've been, where we're going. Pages 1.1–1.9 *in* K. L. Schmitz, ed. Proceedings: National feral swine symposium. USDA, Animal and Plant Health Inspection Service.
Nettles, V. F., and G. A. Erickson. 1984. Pseudorabies in wild swine. Proceedings of the Annual Meeting of the U.S. Animal Health Association 88:505–6.
Nettles, V. F., J. L. Corn, G. A. Erickson, and D. A. Jessup. 1989. A survey of wild swine in the United States for evidence of hog cholera. Journal of Wildlife Diseases 25:61–65.
New, J. C., Jr., K. Delozier, C. E. Barton, P. J. Morris, and L. N. D. Potgieter. 1994. A serologic survey of selected viral and bacterial diseases of European wild hogs, Great Smoky Mountains National Park, USA. Journal of Wildlife Diseases 30:103–6.
Osborn, H. F. 1910. The age of mammals in Europe, Asia, and North America. Macmillan, New York.
Peine, J. D., and J. A. Farmer. 1990. Wild hog management program at Great Smoky Mountains National Park. Proceedings of the Vertebrate Pest Conference 14:221–27.
Pence, D. B., R. J. Warren, and C. R. Ford. 1988. Visceral helminth communities of an insular population of feral swine. Journal of Wildlife Diseases 24:105–12.
Phillips, R. W., and J. H. Zeller. 1943. Sexual development in small and large types of swine. Anatomical Record 85:387–400.
Pilgrim, G. F. 1941. The dispersal of the Artiodactyla. Biological Reviews 16:134–63.

Pimentel, D., L. Lach, R. Zuniga, and D. Morrison. 2000. Environmental and economic costs of nonindigenous species in the United States. Bioscience 50:53–65.

Pine, D. S., and G. L. Gerdes. 1973. Wild pigs in Monterey County, California. California Fish Game 59:126–37.

Pirtle, E. C., J. M. Sacks, V. F. Nettles, and E. A. Rollor, III. 1989. Prevalence and transmission of pseudorabies virus in an isolated population of feral swine. Journal of Wildlife Diseases 25:605–7.

Pond, W. G., and J. H. Maner. 1974. Swine production in temperate and tropical environments. W. H. Freeman, San Francisco.

Pullar, E. M.. 1953. The wild (feral) pigs of Australia: Their origin, distribution and economic importance. Memoirs of the National Museum of Victoria 18:7–23.

Quinn, J. H. 1970. Special note. Society of Vertebrate Paleontology News Bulletin 8:33.

Rary, I. M., V. G. Henry, G. M. Matschke, and R. L. Murphee. 1968. The cytogenetics of swine in the Tellico Wildlife Management Areas, Tennessee. Journal of Heredity 59:201–4.

Roark, D. N. 1977. Stomach analyses of feral hogs at Hobcaw Barony, Georgetown, South Carolina. M.S. Thesis, Clemson University, Clemson, SC.

Romero, C. H., P. Meade, J. Santagata, K. Gillis, G. Lollis, E. C. Hahn, and E. P. J. Gibbs. 1997. Genital infection and transmission of pseudo rabies virus in feral swine in Florida, USA. Veterinary Microbiology 55:131–39.

Ruddle, W. D. 1975. Helminth parasites of feral swine from the Aransas National Wildlife Refuge. M.S. Thesis, Texas A&M University, College Station.

Rutledge, A. 1965. Demons of the delta. Sports Afield 153:68–69, 167–70.

Saliki, J. T., S. J. Rodgers, and G. Eskew. 1998. Serosurvey of selected viral and bacterial diseases in wild swine from Oklahoma. Journal of Wildlife Diseases 34:834–38.

Sanders, A. 1997. Are feral hogs driving you hog wild? Cattleman 84:40–48.

Saunders, G., and S. McLeod. 1999. Predicting home range size from the body mass or population densities of feral pigs, *Sus scrofa* (Artiodactyla: Suidae). Australian Journal of Ecology 24:538–43.

Selander, R. K., M. H. Smith, Y. Yang, W. E. Johnson, and J. B. Gentry. 1971. IV Biochemical polymorphism and systematics of the genus *Peromyscus*. I, Variation in the old field mouse (*Peromyscus polionotus*). Pages 49–90 *in* M. R. Wheeler, ed. Studies in Genetics VI. University of Texas, Austin.

Singer, F. J., W. T. Swank, and E. E. C. Clebsch. 1984. Effects of wild pig rooting in a deciduous forest. Journal of Wildlife Management 48:464–73.

Sisson, S., and J. D. Grossman. 1938. The anatomy of the domestic animals, 3rd ed. W. B. Saunders, Philadelphia.

Skelley, G. C., and D. L. Handlin. 1967. Evaluating pork carcasses (Bulletin 535). South Carolina Agricultural Experiment Station.

Smith, H. M., Jr., W. R. Davidson, V. F. Nettles, and R. R. Gerrish. 1982. Parasitisms among wild swine in southeastern United States. Journal of the American Veterinary Medical Association 181:1281–84

Sorenson, A. M., Jr. 1979. Animal reproduction principles and practices. McGraw-Hill, New York.

Springer, M. D. 1977. Ecologic and economic aspects of wild hogs in Texas. Pages 37–46 *in* G. W. Wood, ed. Research and management of wild hog populations: Proceedings of a symposium. Belle W. Baruch Forest Science Institute, Clemson University, Clemson, SC.

Stallknecht, D. E., V. F. Nettles, G. A. Erickson, and D. A. Jessup. 1986. Antibodies to vesicular stomatitus virus in populations of feral swine in the United States. Journal of Wildlife Diseases 22:320–25.

Stallknecht, D. E., W. O. Fletcher, G. A. Erickson, and V. F. Nettles. 1987. Antibodies to vesicular stomatitus virus New Jersey type in feral and domestic sentinel swine. American Journal of Epidemiology 125:1058–65.

Stallknecht, D. E., D. M. Kavanaugh, J. L. Corn, K. A. Eernisse, J. A. Comer, and V. F. Nettles. 1993. Feral swine as a potential amplifying host for vesicular stomatitus virus New Jersey serotype on Ossabaw Island, Georgia. Journal of Wildlife Diseases 29:377–83.

Stallknecht, D. E., E. W. Howerth, C. L. Reeves, and B. S. Seal. 1999. Potential for contact and mechanical vector transmission of vesicular stomatitus virus New Jersey in pigs. American Journal of Veterinary Research 60:43–48.

Stevens, R. L. 1996. The feral hog in Oklahoma. Samuel Roberts Noble Foundation, Ardmore, OK.

Stribling, H. L., I. L. Brisbin, Jr., J. R. Sweeney., and L. A. Stribling. 1984. Body fat reserves and their prediction in two populations of feral swine. Journal of Wildlife Management 48:635–39.

Stribling, H. L., I. L. Brisbin, Jr., and J. R. Sweeney. 1986. Radiocesium concentrations in two populations of feral hogs. Health Physics 50:852–54.

Sweeney, J. M. 1970. Preliminary investigations of a feral hog (*Sus scrofa*) population on the Savannah River Plant, South Carolina. M.S. Thesis, University of Georgia, Athens.

Sweeney, J. M., E. E. Provost, and J. R. Sweeney. 1970. A comparison of eye lens weight and tooth irruption pattern in age determination of feral hogs (*Sus scrofa*). Proceedings of the Southeastern Association of Fish and Game Commissioners 24:285–91.

Sweeney, J. M., J. R. Sweeney, and E. E. Provost. 1979. Reproductive biology of a feral hog population. Journal of Wildlife Management 43:555–59.

Sweeney, J. R. 1979. Ovarian activity in feral swine. Bulletin of the South Carolina Academy of Science 41:74.

Sweitzer, R. A. 1998. Conservation implications of feral pigs in island and mainland ecosystems, and a case study of feral pig expansion in California. Proceedings of the Vertebrate Pest Conference 18:26–34.

Sweitzer, R. A., G. S. Ghneim, I. A. Gardner, D. VanVuren, B. J. Gonzales, and W. M. Boyce. 1997a. Immobilization and physiological parameters associated with chemical restraint of wild pigs with Telazol® and xylazine hydrochloride. Journal of Wildlife Diseases 33:198–205.

Sweitzer, R. A., B. J. Gonzales, I. A. Gardner, D. Van Vuren, and W. M. Boyce. 1997b. Population densities and disease surveys of wild pigs in the coast ranges of central and northern California. Pages 3.26–3.46 *in* K. L. Schmitz, ed. Proceedings: National feral swine symposium. U.S. Department of Agriculture, Animal and Plant Health Inspection Service.

Sweitzer, R. A., B. J. Gonzales, I. A. Gardner, D. VanVuren, J. D. Waithman, and W. M. Boyce. 1997c. A modified panel trap and immobilization technique for capturing multiple wild pig. Wildlife Society Bulletin 25:699–705.

Synatzske, D. R. 1993. The ecological impacts of feral swine. Pages 59–66 *in* C. W. Hanselka and J. F. Cadenhead, eds. Feral swine: A compendium for resource managers: Proceedings of a conference. Texas Agricultural Extension Service, College Station.

Taylor, R. B. 1993. History and distribution of feral hogs in Texas. Pages 17–27 *in* C. W. Hanselka and J. F. Cadenhead, eds. Feral swine: A compendium for resource managers: Proceedings of a conference. Texas Agricultural Extension Service, College Station.

Taylor, R. B., and E. C. Hellgren. 1997. Diet of feral hogs in the western South Texas plains. Southwestern Naturalist 42:33–39.

Taylor, R. B., E. C. Hellgren, T. M. Gabor, and L. M. Ilse. 1998. Reproduction of feral pigs in southern Texas. Journal of Mammalogy 79:1325–31.

Thompson, R. L. 1977. Feral hogs on national wildlife refuges. Pages 11–16 *in* G. W. Wood, ed. Research and management of wild hog populations: Proceedings of a symposium. Belle W. Baruch Forest Science Institute, Clemson University, Clemson, SC.

Tolleson, D., D. Rollins, W. Pinchak, M. Ivy, and A. Hierman. 1993. Impact of feral hogs on ground-nesting game birds. Pages 76–83 *in* C. W. Hanselka and J. F. Cadenhead, eds. Feral swine: A compendium for resource managers: Proceedings of a conference. Texas Agricultural Extension Service, College Station.

Towne, C. W., and E. N. Wentworth. 1950. Pigs from cave to cornbelt. University of Oklahoma Press, Norman, OK.

Tozzini, F., A. Poli, and G. D. Croce. 1982. Experimental infection of European wild swine (*Sus scrofa* L.) with pseudorabies virus. Journal of Wildlife Diseases 18:425–28.

Tregear, R. T. 1965. Hair density, wind speed, and heat loss in mammals. Journal of Applied Physiology 20:796–801.

van der Leek, M. L. 1989. Wild swine: Resource or risk? Pages 11–15 *in* Proceedings of the feral pig symposium. Livestock Conservation Institute, Madison, WI.

van der Leek, M. L., H. N. Becker, E. C. Pirtle, P. Humphrey, C. L. Adams, B. P. All, G. A. Erickson, R. C. Belden, W. B. Frankenberger, and E. P. J. Gibbs. 1993a. Prevalence of pseudorabies (Aujeszky's disease) virus antibodies in feral swine in Florida. Journal of Wildlife Diseases 29:403–9.

van der Leek, M. L., H. N. Becker, P. Humphrey, C. L. Adams, R. C. Belden, W. B. Frankenberger, and P. L. Nicoletti. 1993b. Prevalence of *Brucella* sp. antibodies in feral swine in Florida. Journal of Wildlife Diseases 29:410–15.

Wagner, S. 1995. Hogs: Feral swine are beginning to wreak havoc on Oklahoma's outdoors. Outdoor Oklahoma 51:24–28.

Wahlenberg, W. G. 1946. Longleaf pine. Charles Lathrop Pack Forestry Foundation, Washington, DC.

Wakely, P. C. 1954. Planting southern pines (Agricultural Monograph 18). U.S. Forest Service.

Wardley, R. C. 1989. Vaccination of feral populations. Pages 58–62 *in* Proceedings of the feral pig symposium. Livestock Conservation Institute, Madison, WI.

Warren, R. J., and C. R. Ford. 1997. Diets, nutrition, and reproduction of feral hogs on Cumberland Island, Georgia. Proceedings of the Annual Conference of the Southeastern Association of Fish and Wildlife Agencies 51:285–96.

Warwick, B. L. 1928. Prenatal growth of swine. Journal of Morphology and Physiology 46:59–84.

Weinwandt, T. A. 1977. Pigs. Pages 182–212 *in* Unit plan for management of Mono Island Forest Reserve. Forest Task Force, Puerto Rico Department of Natural Resources, San Juan, Puerto Rico.

Wesley, F., and F. D. Klopfer. 1962. Visual discrimination learning in swine. Zeitschrift für Tierpsychologie 19:93–104.

Wood, G. W., and R. H. Barrett. 1979. Status of wild pigs in the United States. Wildlife Society Bulletin 7:237–46.

Wood, G. W., and R. E. Brenneman. 1977. Research and management of feral hogs on Hobcaw Barony. Pages 23–35 *in* G. W. Wood, ed. Research and management of wild hog populations: Proceedings of a symposium. Belle W. Baruch Forest Science Institute, Clemson University, Clemson, SC.

Wood, G. W., and R. E. Brenneman. 1980. Feral hog movements and habitat use in coastal South Carolina. Journal of Wildlife Management 44:420–27.

Wood, G. W., and T. E. Lynn, Jr. 1977. Wild hogs in southern forests. Southern Journal of Applied Forestry 1:12–17.

Yarrow, G. K. 1987. The potential for interspecific resource competition between white-tailed deer and feral hogs in the post oak savannah region of Texas. D. F. Dissertation, Stephen F. Austin State University, Nacogdoches, TX.

Young, R. A. 1976. Fat, energy, and mammalian survival. American Zoologist 16:699–710.

Zervanos, S. M., and S. Naveh. 1988. Renal structural flexibility in response to environmental water stress in feral hogs. Journal of Experimental Zoology 247:285–88.

Zervanos, S. M., W. D. McCort, and H. B. Graves. 1983. Salt and water balance of feral versus domestic hampshire hogs. Physiological Zoology 56:167–77.

Zivin, J., B. M. Hueth, and D. Zilberman. 2000. Managing a multiple-use resource: The case of feral pig management in California rangeland. Journal of Environmental Economics and Management 39:189–204.

John R. Sweeney, Department of Aquaculture, Fisheries, and Wildlife, Clemson University, Clemson, South Carolina 29634-0362. Email: jrswny@clemson.edu.

James M. Sweeney, Warnell School of Forest Resources, University of Georgia, Athens, Georgia 30602-2152. Email: jsweeney@smokey.forestry.uga.edu.

Sheron W. Sweeney, National Council for Air and Stream Improvement, Inc., Clemson University, Clemson, South Carolina 29634-0362. Email: sswny@clemson.edu.

55

Nonnative Large Mammals in North America

James G. Teer

Introductions and releases of nonnative wildlife ("exotics") are considered by ecologists and conservationists to be among the most serious problems affecting native species and ecosystem integrity, with negative impacts on biodiversity. David McDowell and David Brackett, both with the International Union for the Conservation of Nature (the World Conservation Union), summed up the problems with nonnative species as follows: "Alien species constitute a major threat to ecosystem integrity around the world. Species that have been either intentionally or accidentally introduced outside their natural ranges have caused species extinctions and/or significant alteration of ecosystems. Predation, introduction of diseases, competition for food and other resources, hybridization, and habitat degradation are among their devastating effects" (Rubic and Lee 1997).

There have been noteworthy publications issued since the 1970s on nonnative species and their interactions with native species and their habitats. These include publications on chital or axis deer (*Axis axis*) (Ables 1977), blackbuck antelope (*Antelope cervicapra*) (Mungall 1978), Barbary or aoudad sheep (*Ammotragus lervia*) (Ogren 1965; Simpson 1980), nilgai antelope (*Boselaphus elaphus*) (Sheffield et al. 1983; Teer 1995), fallow deer (*Dama dama*) (Chapman and Chapman 1975, 1980), sika deer (*Cervus nippon*) (Feldhamer et al. 1978; Feldhamer and Armstrong 1993), and feral swine (*Sus scrofa*) (Mayer and Brisbin 1991; Taylor 1991). Research and management papers increased after about 1980 because of development of commercial or fee-hunting and game ranching for meat, hides, medicines, and cosmetics. Much information on nonnative ungulates is summarized in *Exotics on the Range* (Mungall and Sheffield 1994).

I address history and general problems of introductions of nonnative ungulates and describe population dynamics and impacts on native wildlife by nilgai antelope on a 2145-ha cattle ranch in south Texas.

HISTORY AND REASONS FOR RELEASES OF NONNATIVE WILDLIFE

Translocations of nonnative wildlife have been made in the New World since European settlement. As Europeans colonized new lands, domestic livestock and plants they had known or used were brought with them or sent for after settlement. Animals and plants brought into the New World and used for food at the time of settlement of North America were native to Europe. Ungulates from Africa and Asia came later.

Enormous and costly efforts were made to transplant wild birds, fish, and mammals. These efforts have persisted to date. As sources and capture, holding, and transport facilities improved, introductions of large mammals (big game) increased. Because wild ungulates were important as food and as animals of the chase, many species were tried. Most failed; a few succeeded.

Statistical data on introductions are rare, incomplete, or seriously out of date. Only after nonnative species have established themselves and spread are they reported, and these are usually reported because of interactions with native species and habitats. At least 93 species of wildlife were translocated between 1973 and 1986 in Australia, Canada, Hawaii, New Zealand, and the United States (Griffith et al. 1989). Ninety percent were game species, but threatened and endangered species were also moved. An average of 700 translocations were made worldwide annually. Of these, 98% were made in the United States and Canada. These data do not include the movements of exotic ungulates in the United States and Canada by the private sector for recreational hunting and for production of meat, hides, and traditional Asian medicinal materials. Numbers would no doubt double if these were included (Teer 1991).

Introductions between continents and translocations of large mammals within continents have increased enormously for use in game farms and ranches in Africa, Asia, Europe, Australia, New Zealand, and the Oceanic Islands. More than 500,000 cervids are husbanded on game farms in New Zealand to supply venison to Europeans and antler velvet to Asians. Benson (1991) estimated more than 8200 game ranches are operated in South Africa to supply recreational hunting, brood stock, and game viewing to international and local participants. By the end of 1999, there were around 9000 game ranches in South Africa. Ranchers are changing production systems from domestic livestock to combinations of wildlife and livestock because wildlife now offers more profit than livestock alone (J. duP. Bothma, pers. commun., 2001). Another 17 million ha had cattle and enough game in combination on 15,000 ranches to be operated as commercial wildlife units (Standard Bank 2000). Many of these ranches are introducing nonnative species onto their properties. The increase of game ranches portends the spread of nonnative species throughout the world.

Most alien animals were deliberately brought to the New World. However, many vertebrates, insects, and plant materials entered the New World as stowaways on ships, airplanes, and their cargo. The brown tree snake (*Boiga irrugularis*) in the Mariana Islands and the imported fire ant (*Solenopsis invicta*) in the southern United States are current examples of stowaways among a host of unwanted aliens now causing enormous mortality of native wildlife including neonates of large mammals.

Introductions of wildlife into the New World were made for various reasons. Songbirds and fishes were imported to satisfy nostalgia for animals of immigrants' gardens and, in some cases, to provide familiar table food. The English sparrow (*Passer domesticus*) and the European starling (*Sturnis vulgaris*) were introduced by early English settlers who wished to have familiar ties with their former homes. A few species, some desirable, many undesirable, resulted from a desire to "improve on nature." Many ungulates from distant lands were introduced by hunters who had become acquainted with them on safaris. They were introduced and released to serve recreational hunting interests (Teer 1979a; Mungall and Sheffield 1994). Commercial hunting of exotics has been a powerful stimulus for the great numbers of species that were introduced into North America.

Game parks and "exotic ranches" were established in the United States, Canada, and Mexico after World War II. Large mammals are stocked and displayed often without regard to ecological integrity or habitat needs of the species. They are fed and husbanded much the same as domestic livestock. The public, fascinated by large mammals, enjoys drivethrough game parks and recreational hunting. The provision of animals to stock such parks and the development of commercial hunting have resulted in the enormous business and spread of exotic animals (Teer 1993).

Several species of the deer family (Cervidae) are tractable, and, under semiconfinement, produce valuable products—meat, hides, medicines, and totems—for a ready market. Several species of deer are traded around the world much the same as livestock. Animals are bred and managed with attention to genetic lines, nutrition, diseases, and productivity. Modern methods of managing domestic livestock, such as uses of shepherds and guard dogs, artificial insemination, vaccines, and prophylactic drugs, are now used in husbandry and production of large mammals on game farms and ranches.

Martin (1984) suggested that loss of the great assemblage of large mammals at the end of the last Ice Age, some 20,000 years ago, was caused by humans. Empty niches presumably provided opportunities for introduced species. Whether losses of the great mammalian fauna were human induced, as Martin suggests, or whether habitats decreased with encroachment of the ice sheet is now a moot point because many habitats are stocked with alien animals.

With presumed unoccupied niches as the reason, a few states sought large mammals to release in arid and mountainous habitats of the western United States. New Mexico, for example, searched for suitable large mammals for release in the state (Wood et al. 1970). Free-ranging populations of gemsbok (*Oryx gazella*), ibex (*Ibex* spp.), and aoudad or Barbary sheep resulted. The latter two species are no longer desired because of competition with desert bighorn sheep (*Ovis canadensis*) and mule deer (*Odocoileus hemionus*) for forage and because of the infections of parasites and diseases introduced with the exotics. Several hundred gemsbok are taken by hunters each year in desert habitats, including those on White Sands Missile Range in New Mexico, one of the original release sites.

Because federal quarantine laws prohibit direct release of introduced brood stock taken abroad, trades with zoos and private game parks are sometimes made by private landowners for exotic ungulates (Teer 1984). Zoos use the introduced animals for display while they meet lifetime quarantine requirements. Progeny of the original stock can be released. Zoos are now reluctant to enter into such trades with the private sector. They are much more conservation minded than they were previously. Furthermore, brood stocks of many species are now readily available from the private sector.

Few exotic species remain captives. Individuals ultimately escape confinement and a few become established in wild stocks. Others are deliberately released on private farms and ranches, and others escape and spread from private game parks. Transplants of several nonnative large mammal species are now established in about every nation or region of the world in untended and free-ranging populations. Texas, Florida, and California are among the states conducive to successful introductions. Large numbers of foreign ungulates in these states attest to their suitability for exotic species, perhaps largely due to benign, tropical climates.

Introductions are seldom made after prior studies have been conducted to determine a species' requirements and potential impacts on indigenous fauna and flora. The consequences of a species' release on native species and their habitats are seldom known, nor are its habitat requirements known. Perhaps with the exception of some cervids and bovids, especially sheep and goats, few ungulates have sufficient tolerance and ecological plasticity for success in diverse environments.

A vigorous commercial trade in nonnative ungulates exists in North America and many other regions of the globe. Just about any species of bovid or cervid is available from private sources. Fee hunting, game ranching and farming, and the loss of appreciation for ecological purity are responsible for the dilemma surrounding nonnative species that conservationists now find in the natural world.

In summary, wildlife has followed the spread of humans wherever they have gone. Exotic ungulates are serious threats to biological diversity. In moving wildlife about, consideration is seldom given to the purity of nature, functional ecosystems, or possibilities of harm to native species and their habitats. Curiously, most cultures do not perceive exotics as threats, although most societies favor conservation of wild, living resources and environmental quality and work diligently to preserve them.

SPECIES AND NUMBERS OF NONNATIVE UNGULATES IN NORTH AMERICA

Nonnative plants and animals have been introduced in nearly every region of the world. Exotics have produced a Noah's Ark of animals, which in many cases are harmful and unwanted. Perhaps the country with the most nonnative free-ranging ungulates, certainly the best known, is New Zealand (Howard 1965; Caughley 1970, 1983). Now, however, the United States and Canada vie with New Zealand in numbers of species and individuals of nonnative ungulates. Texas is undoubtedly the center of exotic ungulates in the United States (Traweek 1995) because of the vast acreages of cattle ranches with extensive habitats similar to those of ungulates in Asia and Africa.

Teer (1991) surveyed the status and uses of exotics in the United States and Canada. Fallow deer and sika deer were the most numerous nonnative species. North American elk or wapiti (*Cervus elaphus*) were the most numerous native species on game farms. Respondents reported exotics were used mostly for producing meat and associated by-products; however, most species were used in recreational hunting, where they occurred in free-ranging stocks.

Inventories of nonnative species have been conducted in Texas at about 5-year intervals since 1963 (Jackson 1964). Traweek (1995) reported a total of 124 species or varieties of nonnative ungulates on 637 ranches in 155 of the 254 counties in Texas. Cervids and bovids were most common. However, representatives of most families of ungulates were identified in the survey. Ninety-one percent of the exotics occurred on ranches in the Edwards Plateau and Rio Grande Plains of south Texas.

Nonnative ungulates in Texas are hunted in fee-hunting systems or produced for sale of brood stock. Exotic ungulates have a special attraction to ranchers and others with large landholdings. Fees for hunting are expensive. Axis deer, blackbuck antelope, fallow deer, and sika deer bring as much as $3500 per animal. Black (*Biceros bicornis*) and white rhinoceroses (*Ceratotherium simum*); giraffes (*Giraffa camelopardalis*); horses, asses, and zebras (*Equus* spp.); camels (*Camelus* spp.) and their allies (*Lama* spp.); and scarce and endangered species such as bongo (*Tragelaphus euryceros*), gaur (*Bos gaurus*), yak (*Bos grunniens*), Ankole cattle (*Bos taurus*), sitatunga (*Tragelaphus spekei*), and kudu (*Tragelaphus strepsiceros*) are kept primarily for sale of brood stock and viewing. Some owners of private property serve conservation interests by featuring threatened or endangered species in their collections. Of the 195,423 animals counted by Traweek (1995) in his survey, 118,265 were confined and 77,218 were estimated to be free ranging. Nilgai antelope, axis deer or chital, aoudad or Barbary sheep, sika deer, fallow deer, and blackbuck antelope (Fig. 55.1) made up more than 95% of the free-ranging nonnative species in Texas (Table 55.1).

Species of ungulates from the Indian subcontinent constitute most of the exotics that have escaped and now exist in free-ranging populations in Texas. Most releases of African species in North America have failed to become established in free-ranging stocks. Reasons for the success or failure of releases relate to habitats of the species, climatic factors, and the ability of the transplanted animals to withstand predation, diseases, exploitation by humans, and other sources of mortality that they have not encountered in their native ranges.

Using climagraphs and comparisons of vegetation, Ables and Ramsey (1974) examined similarities in habitats and climate for several species of Indian ungulates and found a strong relationship between temperature and precipitation regimes in the nonnatives' old and new ranges. Thirty degrees north latitude passes through the southern United States and India, and runs through the hot, arid regions of North Africa. Quite a few genera of woody vegetation, *Acacia* among the most numerous and widespread, are present in the savannas and semiarid grasslands of the three continents. The reverse introductions of North American mammals into Asia or Africa have not been made. Argentina, on the other hand, is home to a healthy population of red deer and wapiti.

Surveys of exotics by the Texas Parks and Wildlife Department were conducted through the mail and, in some cases, by visits to individual ranches. Undoubtedly, errors occurred in the surveys through misidentification of species, races, and crosses and in numbers of

FIGURE 55.1. Common nonnative species released in North America: (A) Blackbuck antelope; (B) axis deer; (C) red deer; (D) aoudad or Barbary sheep; (E) fallow deer (white phase); (F) sika deer. Red, fallow, and sika deer are the nonnative species most commonly stocked on game farms and ranches because of their tractability and use in recreational hunting and production of meat and antler velvet.

TABLE 55.1. Major free-ranging nonnative ungulates in Texas

Species	Number	Percentage
Nilgai antelope	28,468	37
Axis deer	22,331	29
Aoudad sheep	9,826	13
Sika deer	5,697	7
Fallow deer	5,251	7
Blackbuck antelope	4,757	6
Other	740	1

SOURCE: Adapted from Traweek (1995).

animals in confinement and free ranging. Many ranchers do not know the sources of their stock, and censuses of free-ranging stocks are always subject to errors. However, errors are largely those of omission, and numbers reported by the respondents must be considered as minimal estimate.

PROBLEMS ASSOCIATED WITH NONNATIVE SPECIES

Releases of nonnative large mammals often exhibit one or more of the following results (Teer 1979b, 1997): (l) rapid increases in numbers of the exotic beyond the habitat's ability to support them and native species, (2) reduction of food resources leading to poor pasturage and mortality of both the native and nonnative species, (3) introductions of diseases and parasites that cause harm to native species, (4) interbreeding of closely related taxa (species and subspecies and occasionally genera), (5) changes in functional ecoystems and compromises in biodiversity, and (6) an accommodation with native species with success of nonnatives in free-ranging populations.

Competition for Food and Other Resources. Sheffield et al. (1983), Baccus et al. (1985), Feldhamer and Armstrong (1993), and many others have reported competitive relationships between exotic and native species. Although food resources are usually involved, competition may invoke aggressive behavior when exotics and natives vie for territories, favored habitats, and water. Studies of diets of native and nonnative ungulates have shown that production and survival of white-tailed deer are influenced by competition (Armstrong and Harmel 1981; Sheffield et al. 1983; Harmel 1989, 1997). Sika deer, axis deer, and aoudad sheep displaced white-tailed deer in 39-ha enclosures on the Kerr Wildlife Management Area in the Edwards Plateau of Texas (Feldhamer and Armstrong 1993). Sika deer, introduced on the eastern shore of Maryland in 1916, outnumber white-tailed deer and are displacing them in areas of sympatry (Flyger 1960; Feldhamer et al. 1978; Keiper 1985).

Introductions of Diseases and Parasites with Their Hosts. Introductions of diseases and parasites with nonnative ungulates and their transfer to native wildlife and domestic livestock are major problems. Brucellosis (Davis 1990), chronic wasting disease (Miller 1999), bovine tuberculosis (Hunter 1996; Schmitt et al. 1997), and meningeal worms (*Parelaphostrongylus tenuis* and *Elaphostrongylus cervi*) (Oates et al. 1999) are among the pathogens and parasites discovered in both penned and free-ranging native and introduced ungulates. Discoveries of bovine tuberculosis in seven herds of red deer in Ontario and in two herds of fallow deer in British Columbia resulted in condemnation and destruction of the infected herds (Geist 1995). Free-ranging herds of white-tailed deer infested with CWD are being destroyed in several states.

Interbreeding of Exotics. Cervids are species of choice on game farms and are the source of crosses between closely related species. Red deer, wapiti, sika, sambar, and maral, all contained in the complex of *Cervus elaphus,* readily interbreed when confined. Biochemical tests showed that 30% of elk and red deer in New Zealand game farms had genetic markers of both. Wild stocks of red deer and sika deer have hybridized in Scotland, where no genetically pure individuals of either species remain (Wheaton et al. 1993). Intergeneric crosses of wild sheep are also known to occur, and even crosses of domestic and wild species of these bovids occur.

Destruction of Habitat of Sympatric Species. As apex or keystone animals, ungulates (herbivores) fashion habitats for sympatric species as well as for themselves (Teer 1997). Exotic species released in habitats with full complements of native large-mammal populations can have devastating effects on biodiversity. Howard (1965) and Caughley (1970) reported on destruction of habitat and losses of biodiversity in Australia and New Zealand because of excessive numbers of various species of introduced deer and Himalayan thar (*Hemitragus jemlahicus*). Overuse of forage, soil erosion, and destruction of commercially important forest products and food plants resulted. Overgrazing and browsing by wild and domestic large mammals are identified as among, if not the major, causes of degradation of rangeland habitats throughout the world. Habitats of forest species are also affected by ungulates. Explosive numbers of white-tailed deer in North America resulted in forest habitat modifications and reduction of nesting sites of breeding birds that used it (Balda 1975; Casey and Hein 1983; McShea and Rappole 1992, 1997).

LAWS REGULATING EXOTICS IN NORTH AMERICA

Federal regulations governing imports and interstate commerce in exotic animals are not adequate to police or control imports of nonnative species in North America (Teer 1991). Operating quarantine stations at ports of entry, the U.S. Department of Agriculture screens introduced animals for unwanted pathogens and parasites (Parker 1968). It also has jurisdiction over interstate movements and animal health.

Functional ecosystems and purity of nature do not seem to be a matter of concern. Most states have statutes and regulations for controlling unwanted species that are or could be harmful as predators, hosts of diseases and parasites, and dangers to human health and agriculture. Interstate commerce in the exotic mammal industry is poorly if at all controlled by most states.

Most states, Texas among them, consider exotic wildlife the same as they do domestic livestock. They are considered private property and can be used for any purpose without restrictions on seasons, bag limits, or methods of taking. Regulations governing game ranching and farming, where they exist at all, have many similarities to those of animal agriculture. Most states are now reviewing their legislation and regulations with the view to strengthening them (Teer 1991), mainly because of the threats of diseases and parasites to native species.

Several organizations protect the interests of the exotic animal industry in the United States. The Exotic Wildlife Association, a national organization of breeders and traders in exotic ungulates, is attentive to any legislation or regulation that restricts trade or uses of exotic animals. Canada has a similar organization allied to agriculture. Game farming and ranching has gained momentum in North America despite some notable transfers of diseases and parasites from nonnative to native species and domestic livestock.

NILGAI ANTELOPE: A CASE STUDY OF A NONNATIVE UNGULATE

Origin, Numbers, and Distribution of Nilgai Antelope in Texas. About 40 nilgai antelope or blue bull (Fig. 55.2), a large bovid of the Indian Subcontinent, were released on the Norias Division of King Ranch in two transplants in the mid-1920s and early 1930s (Sheffield et al. 1983). Records of the releases and their subsequent increase are sketchy. King Ranch family members state that Caesar Kleberg, then a U.S. congressman, stocked nilgai to increase hunting opportunity on the ranch. Stock was obtained from the Brookfield Zoo, Chicago, and from sources in Mexico.

Nilgai were seldom seen on the ranch until about the end of World War II, when they reached the accelerated stage of population growth. They became common in brushland and semiopen, savannah-like habitats on and off King Ranch. They spread widely from the release sites on the Norias Division and now occupy about 35,000 km^2 of rangeland habitats, primarily in Kleberg, Kenedy, and Willacy Counties (Fig. 55.3). They are most numerous on several large cattle ranches

FIGURE 55.2. The nilgai antelope or blue bull is perhaps the most abundant, free-ranging, nonnative ungulate in the United States. About 30,000 occur in south Texas on large cattle ranches, where they have spread from the release site on the King Ranch. Only the axis and sika deer can rival it in numbers in free-ranging herds.

FIGURE 55.4. Nilgai prefer a matrix of open and brush-covered habitat quite similar to tropical savannas. Using mechanical and chemical methods, savanna-like habitat has been created in south Texas in alternate strips or checkerboard patterns of brush and open areas.

in the three counties: King, Kenedy, El Saus, Yturria, Armstrong, and Robert East Ranches. They occur in sparing and scattered numbers in the northern reaches of their distribution about as far north as northern Kleberg County.

Herds of young bulls with up to 80 members, and herds of cows and calves almost as large, are commonly seen in late summer and cool seasons of the year. However, older bulls (aged by almost black or grayish pelage) are solitary or occur in much smaller herds throughout the warmer seasons. Both sexes defecate at the very same sites for many months. Feces build up in mounds to 20 cm deep and 1–2 m in diameter. Presumably, nilgai mark their ranges or territories; however, I have not seen aggression at these sites and have seen different bulls and cows use them. Bulls sometimes kill other bulls in fights, as evidenced by scars and wounds usually about the necks and shoulders of dead animals.

Curiously, very few occur west of U.S. Highway 77 in south Texas on the above-mentioned ranches, where habitat appears to be similar. Traffic on the highway may be a hindrance to them, and regular 142-cm-high stock fences may also deter them. Nilgai can jump fences when pressed, but prefer to negotiate them by using scuttle holes under them.

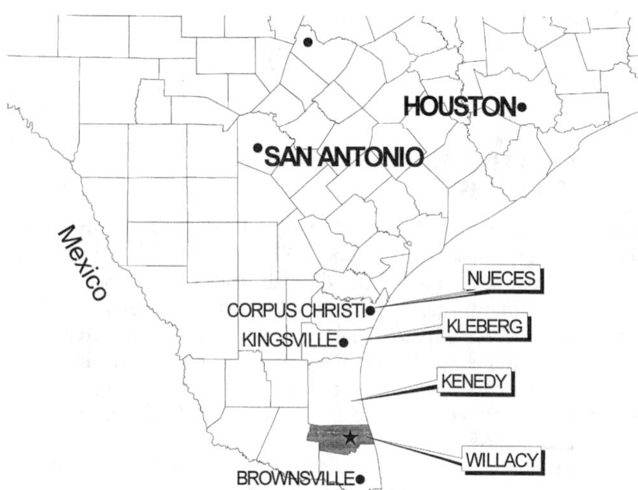

FIGURE 55.3. Having spread from the King Ranch where they were released, the principal range of the nilgai antelope is now widespread in Willacy, Kenedy, and Kleberg Counties in south Texas. It continues to spread naturally, and brood stock is being sold by breeders to ranches primarily in Texas, but also elsewhere in the United States and Mexico.

According to aerial censuses of the herd conducted by the Texas Parks and Wildlife Department, there were 36,000 in 1988 and 28,000 in 1994 in south Texas (Traweek 1995). The difference in numbers is probably due to vagaries in weather conditions and errors in visibility of animals in aerial censuses. Whatever the number, it is very large and all are established in free-ranging herds. All came from the approximately 40 nilgai released in the original two transplants. A few have been moved from Norias to other ranches in south Texas and the Edwards Plateau, but these transplants have not resulted in large, free-ranging herds. Some were transplanted to other states in the United States and to Mexico and Canada. They have not thrived and remain in low numbers or have died out.

I studied nilgai herd dynamics on a 2145-ha study ranch (hereafter called SR) from 1978 through 2000. SR joins the Norias Division of King Ranch on its southern boundary, about 30 km northeast of Raymondville in Willacy County, Texas.

Woody vegetation consists largely of scrub brush habitats dominated by mesquite (*Prosopis glandulosa*), liveoak (*Quercus virginianus*), at least five species of *Acacia,* and several other species of brush. Ranchers manage brush to increase forage for cattle and to produce habitat for game birds and mammals. Chemicals, cattle (grazing), and mechanical treatments are principal tools in habitat management. Alternate strips of open areas and brush simulate savannah habitats that many species of large mammals, especially plains game, prefer (Fig. 55.4). Similarly, blocks of open and brush-covered range cleared in a checkerboard pattern provide much the same effect. White-tailed deer (*Odocoileus virginianus*) and nilgai antelope prefer this habitat above all others in south Texas.

SR is operated for recreational hunting of white-tailed deer, bobwhite quail (*Colinus virginianus*), Rio Grande turkeys (*Meleagris gallapavo*), javelina (*Tayassu tajacu*), and waterfowl and to demonstrate conservation practices in land management. These species and mourning doves (*Zenaida macroura*) are abundant or occur seasonally on SR. The endangered ocelot (*Felis pardalis*) and many species of tropical birds reach their northern limits on SR. Cattle are stocked primarily to manage vegetation for wildlife habitat. Harvest of game on the ranch is low. Fewer than 10 white-tailed deer are taken annually. SR is not easily reached, and is excellent habitat for native species.

I conducted annual aerial transect censuses from helicopters of nilgai antelope and white-tailed deer on SR in October and November of each year for 20 years (Table 55.2). The same transects were permanently marked and used each year from 1981 through 2000. Density of nilgai on SR doubled from 11/km^2 to 21/km^2 between 1981 and 2000. The ratio of white-tailed deer to nilgai was 1.4 to 1 in the first 12 years of

TABLE 55.2. Densities (animals/km2) of white-tailed deer and nilgai antelope on a 2145-ha cattle ranch in Willacy County, Texas, 1981–2000

Years	White-Tailed Deer	Nilgai Antelope
1981–83	14	8
1984–86	12	7
1987–89	12	10
1990–92	8	8
1993–95	11	8
1996–98	12	18
1999–2000	11	21
Average	11	11

NOTE: Aerial (helicopter) censuses were conducted in early fall on permanent transects. Census results are presented in 3-year averages to dampen annual fluctuations.

the study. After 1992, nilgai outnumbered white-tailed deer by a ratio of 1.4 to 1.

The change in numbers of the two species is even more striking when biomass is compared. Mean weights of adult male and female white-tailed deer and nilgai antelope are 57 and 205 kg, respectively (Sheffield et al. 1983). Compared on the basis of biomass, nilgai are now 3.3 times more abundant than white-tailed deer.

Competition Among Nilgai, White-Tailed Deer, and Cattle. An influx of nilgai from adjoining ranches occurs and the resident population of nilgai on SR increases during October through early March. Reasons for the movements are not understood. As many as 800 nilgai may occur on SR where a few weeks before less than half that number may have been present. Quantity and quality of range forage on SR may be superior to that of surrounding areas. Disturbance by hunters on neighboring ranches may also be a factor.

Diets of nilgai, cattle, and white-tailed deer overlap to some degree (Table 55.3). During those years when nilgai build to large numbers in the cool seasons, wildlife habitat is affected. Adjustments in stocking rates of cattle are made to protect habitat of bobwhite quail and other seed-eating, ground-nesting birds. Forbs are particularly vulnerable to overgrazing, and as they are favored items in the diets of seed-eating birds and white-tailed deer, Overuse of them can be critical.

To determine the grazing pressure being put on the range, animal unit equivalent values of cattle, deer, and nilgai antelope were calculated on the basis of weights of the three species. Equivalencies were then compared to the stocking rate for the area recommended by the Texas Natural Resources Conservation Commission.

Eight deer or 2.2 nilgai antelope are equivalent in weight to a 454-kg cow with calf at side. Obviously, these values are not related precisely to diets of the three species. Animal unit equivalencies are instructive when used to relate numbers of animals to carrying capacity of the region, but must be used for what they are, comparisons of biomass of herbivores with somewhat similar diets.

The combined stocking rate of the three ungulates is about one animal unit/7.3 ha on SR. Carrying capacity of rangeland in the region is about one animal unit/10 ha (Texas Natural Resources Conservation Commission, pers. commun., 1985). SR is overstocked accordingly by about 50% of the recommended rate and nilgai antelope are largely responsible.

TABLE 55.3. Diets of white-tailed deer, nilgai antelope, and cattle

Animal	Grasses	Forbs	Browse	Mesquite Beans
White-tailed deer	23	60	13	4
Nilgai antelope	60	25	6	9
Steer	95	3	1	1

SOURCE: Adapted from Sheffield et al. (1983).
NOTE: Of the three species, deer and nilgai have quite similar preferences for forage, and thus are the most competitive.

FIGURE 55.5. Nilgai antelope are being cropped by shooting from helicopters on several large ranches to reduce competition with white-tailed deer and domestic livestock. Carcasses are collected by truck and taken to a portable abattoir, where they are inspected by a licensed veterinarian, butchered, chilled, and transported to markets. Upscale restaurants in the western United States buy most of the meat.

Carrying capacity of the range for wildlife may vary from year to year depending on rainfall and stocking rates of domestic livestock. Adjustments in livestock numbers are the most efficient method of managing grazing pressure. SR owners elected to change from a cow–calf operation to using stockers (yearlings pastured for weight gain). Stockers provide flexibility in stocking rates as animals may be put on or taken off the range in response to range forage conditions. Furthermore, a cow–calf system is expensive to maintain if the base herd must be sold as a result of drought, markets, or other factors.

The owners of SR have also elected to remove nilgai through commercial cropping. Nilgai are not prized as a trophy. Only the males have horns. Their horns are relatively short, up to about 40 cm, and lack the twists, whorls, and curves of desired trophy species. Many ranches offer nilgai for recreational hunting for about $1000 for a packaged, 3-day, guided hunt. Recreational hunting has not reduced nilgai numbers, and it is now doubtful whether hunting can control their number over the region. The nilgai is a fecund animal, commonly producing twins, and has few natural predators. Freezing temperatures accompanied by ice on vegetation is about the only extensive mortality factor affecting nilgai numbers in south Texas.

Some large ranches are now taking several hundred nilgai each year by shooting from helicopters for their meat and hides. They are collected by truck (Fig. 55.5), taken to a portable abattoir, butchered, chilled, and transported to markets. Ranchers are paid $0.30–$0.56/kg, field dressed. The state requires a licensed veterinarian to inspect the carcasses for parasites and diseases. Handling and transport facilities are also regulated. The meat is sold for up to $8.00/kg for the loin and other prime cuts to upscale restaurants in the United States.

The SR has cropped 895 nilgai for meat from 1994 through 2000. Yet the population remains higher than the white-tailed deer population from which fewer than 10 bucks are taken each year. Nilgai antelope have added another layer to management of wildlife on one of the most productive game ranges in south Texas.

MANAGEMENT AND RESEARCH NEEDS

The nilgai antelope is a case study of influences of a large, nonnative ungulate on a native ungulate, the white-tailed deer, and its habitat. From two releases on a large cattle ranch about 70 years ago, nilgai have increased to about 25,000–30,000 and are continuing to increase in density and distribution despite recreational hunting and their being cropped for meat. They now occupy an area about the size of a state in the midwestern United States, and where they occur, are now about twice as abundant as white-tailed deer. Stocking rates of cattle and nilgai must be adjusted to protect wildlife habitat and native species.

Movements and releases of nonnative ungulates are major problems in conservation of native wildlife and its habitat. Negative interactions with native fauna include competition for forage and other resources, interbreeding of closely related forms, transfer of pathogens and parasites to free-ranging native stock, and degradation of ranges and their ability to sustain native species.

When numbers of a nonnative species are low, they may be tolerated and sometimes accepted for the economic value they offer. However, when a species becomes as numerous as the nilgai has on SR and on many other ranches of the region, competition for forage becomes a major factor in the ranches' economic viability. Ranchers are then forced to reduce nilgai numbers through reduction harvests and by reduction of domestic livestock.

Societal interests in exotic animals include nostalgic interest, curiosity in strange animals, and economic gain. Society and especially landowners should be educated about the risks that introduced fauna and flora portend to native species. Federal and state import and quarantine regulations are not adequate to prevent introductions of unwanted species and especially to prevent introductions of diseases and parasites with their nonnative hosts. Introductions of exotics are expensive and most fail. Careful studies of a species, its native range, and its proposed release sites should be conducted. Agencies of state and federal governments should be discouraged from promoting and permitting introductions of nonnative species. Statutes and regulations should be reviewed and improved to prevent harm to native wild and domestic animals and habitats.

ACKNOWLEDGMENTS

Max Traweek of the Texas Parks and Wildlife Department and William J. Sheffield, Elizabeth C. Mungall, and Eugene Fuchs, all former students at Texas A&M University, made significant contributions to the understanding of the status and management of exotic ungulates in Texas. They and other students at Texas A&M University were part of a focused effort on exotics in Texas, which were, and are, a major conservation problem in the United States and elsewhere in the world. Eugene Fuchs provided several photographs. Misty Smith, secretary in the Department of Wildlife and Fisheries Sciences at Texas A&M University, was helpful in preparation of the manuscript. I am truly grateful for their contributions.

LITERATURE CITED

Ables, E. D. 1977. The axis deer in Texas (Kleberg Studies in Natural Resources). Texas A&M University, College Station.

Ables, E. D., and C. W. Ramsey. 1974. Indian mammals on Texas rangelands. Journal of the Bombay Natural History Society 71:18–25.

Armstrong, W. E., and D. E. Harmel. 1981. Exotic mammals: Competing with the natives (LF C2000-103). Texas Parks and Wildlife Department, Austin.

Baccus, J. T., D. E. Harmel, and W. E. Armstrong. 1985. Management of exotic deer in conjunction with white-tailed deer. Pages 213–26 *in* W. L. Beasom and S. F. Roberson, eds. Game harvest management. Caesar Kleberg Wildlife Research Institute, Kingsville, TX.

Balda, R. P. 1975. Vegetation structure and breeding bird diversity. Pages 59–80 *in* D. R. Smith, ed. Proceedings of the symposium on management of forest and range habitats for nongame birds (General Technical Report WO-1). U.S. Department of Agriculture, Washington, DC.

Benson, D. E. 1991. Values and management of wildlife and recreation on private land in South Africa. Wildlife Society Bulletin 19:497–510.

Casey, D., and D. Hein. 1983. Effects of heavy browsing on a bird community in deciduous forest. Journal of Wildlife Management 47:829–36.

Caughley, G. 1970. Eruptions of ungulate populations with emphasis on Himalayan thar in New Zealand. Ecology 51:53–72.

Caughley, G. 1983. The deer wars: The story of deer in New Zealand. Heinemann, Auckland, New Zealand.

Chapman, D., and N. Chapman. 1975. Fallow deer: Their history, distribution, and biology. Terrance Dalton, Laverham, Suffolk, UK.

Chapman, N. G., and D. I. Chapman. 1980. The distribution of fallow deer: A worldwide review. Mammal Review 10:61–138.

Davis, D. S. 1990. Brucellosis in wildlife. Pages. 321–34 *in* K. Nielsen and J. R. Duncan, eds. Animal diseases. CRC Press, Boca Raton, FL.

Feldhamer, G. A., and W. E. Armstrong. 1993. Interspecific competition between four exotic species and native artiodactyls in the United States. Transactions of the North American Wildlife and Natural Resources Conference 58:468–78.

Feldhamer, G. A., J. A. Chapman, and R. A. Miller. 1978. Sika deer and white-tailed deer on Maryland's eastern shore. Wildlife Society Bulletin 6:155–57.

Flyger, V. 1960. Sika deer on islands in Maryland and Virginia. Journal of Mammalogy 41:140.

Geist, V. 1995. North American policies of wildlife conservation. Pages 77–129 *in* V. Geist and I. McTaggart-Cowan, eds. Wildlife conservation policy. Detselig Enterprises, Calgary, Alberta, Canada.

Griffith, B., J. M. Scott, J. W. Carpenter, and C. Reed. 1989. Translocations as a species conservation tool: Status and strategy. Science 245:477–80.

Harmel, D. E. 1989. The influence of exotic artiodactyls on white-tailed deer production and survival (Federal Aid Report, P.R. Project No. W-109-R-l2, Final Report). Texas Parks and Wildlife Department, Austin.

Harmel, D. E. 1997. The influence of fallow deer and aoudad sheep on white-tailed deer production and survival (Federal Aid Report, Federal Aid Grant No. W-127-R-5, Big Game Research and Surveys). Texas Parks and Wildlife Department, Austin.

Howard, W. E. 1965. Control of introduced mammals in New Zealand (No. 45). New Zealand Department of Scientific and Industrial Research Information Service.

Hunter, D. L. 1996. Tuberculosis in free-ranging and captive cervids. Revue Scientific et Technique de l'Office International des Epizootics 15:171–81.

Jackson, A. 1964. Texotics. Texas Game and Fish Commission Magazine 22:7–11.

Kieper, R. R. 1985. Are sika deer responsible for the decline of white-tailed deer on Assateague Island, Maryland? Wildlife Society Bulletin 13:144–46.

Martin, P. 1984. Prehistoric overkill: The global model. Pages 354–453 *in* P. Martin, ed. Quaternary extinctions: A prehistoric revolution. University of Arizona Press, Tucson.

Mayer, J. J., and L. L. Brisbin, Jr. 1991. Wild pigs in the United States. University of Georgia Press, Athens.

McShea, W. J., and J. H. Rappole. 1992. White-tailed deer as keystone species within forest habitats of Virginia. Virginia Journal of Science 43:177–86.

McShea, W. J., and J. H. Rappole. 1997. Herbivores and the ecology of forest understory birds. Pages 298–309 *in* W. J. McShea, H. B. Underwood, and J. H. Rappole, eds. The science of overabundance: Deer ecology and population management. Smithsonian Institution Press, Washington, DC.

Miller, M. W. 1999. Chronic wasting disease surveillance and epidemiology: Western United States. Pages 608–10 *in* M. W. Miller, Chairman. Report of the Committee on Wildlife Diseases. U.S. Animal Health Association, Richmond, VA.

Mungall, E. C. 1978. The Indian blackbuck: A Texas view (Kleberg Studies in Natural Resources). Texas A&M University, College Station.

Mungall, E. C., and W. J. Sheffield. 1994. Exotics on the range. The Texas example. Texas A&M University Press, College Station.

Oates, D. W., M. C. Sterner, and D. J. Steffen. 1999. Meningeal worm in free-ranging deer in Nebraska. Journal of Wildlife Diseases 35:101–4.

Ogren, H. A. 1965. Barbary sheep (Bulletin No. 13). New Mexico Department of Game and Fish, Santa Fe.

Parker, R. L. 1968. Quarantine and health problems associated with introductions of exotic animals. Pages 21–22 *in* Introduction of exotic animals: Ecologic and socioeconomic considerations. Caesar Kleberg Research Program in Wildlife Ecology, Texas A&M University, College Station.

Rubic, C. D. A., and G. O. Lee, eds. 1997. Conserving vitality and biodiversity. Proceedings, the World Conservation Congress workshop on alien invasive species. IUCN Species Survival Commission and North American Wetlands Conservation Council (Canada), Canadian Wildlife Service, Environment Canada, Ottawa.

Schmitt, S. M., S. D. Fitzgerald, T. M. Cooley, C. S. Bruning-Fann, L. Sullivan, D. Berry, T. Carlson, R. B. Minnis, J. B. Payeur, and J. Sikarskie. 1997. Bovine tuberculosis in free-ranging white-tailed deer in Michigan. Journal of Wildlife Diseases 33:749–58.

Sheffield, W. J., B. A. Fall, and B. A. Brown. 1983. The nilgai antelope in Texas. Kleberg Studies in Natural Resources, Texas A&M University, College Station.

Simpson, C. D., ed. 1980. Proceedings. Symposium on the ecology and management of Barbary sheep. Department of Range and Wildlife Management, Texas Tech University, Lubbock.

Standard Bank. 2000. The game industry: Delicately poised. SA Game and Hunt 6:21, 23.

Taylor, R. 1991. The feral hog in Texas (Federal Aid Report Series No. 28). Texas Parks and Wildlife Department, Austin.

Teer, J. G. 1979a. Commercial uses of game animals on rangelands of Texas. Journal of Animal Science 40:1000–1007.

Teer, J. G. 1979b. Introduction of exotic animals. Pages 172–77 *in* R. D. Teague and E. Decker, eds. Wildlife conservation: Principles and practices. Wildlife Society, Washington, DC.

Teer, J. G. 1984. Exotic animals: Conservation implications. Pages 31–37 *in* S. Kinwood, ed. Proceedings of the 44th annual conference, International Union of Directors of Zoological Gardens. San Antonio, TX.

Teer, J. G. 1991. Non-native large ungulates in North America. Pages 55–66 *in* L. A. Renecker and R. J. Hudson, eds. Wildlife Production: Conservation and sustainable development. (AFES Miscellaneous Publication 91-6). University of Alaska, Fairbanks.

Teer, J. G. 1993. Commercial utilization of wildlife: Has its time come? Pages 73–83 *in* A. W. L. Hawley, ed. Commercialization and wildlife management: Dancing with the devil. Krieger, Malabar, FL.

Teer, J. G. 1995. Exotic animals: Conservation implications. Pages 235–46 *in* V. Geist and I. McTaggart Cowan, eds. Wildlife conservation policy. Detselig, Calgary, Alberta, Canada.

Teer, J. G. 1997. Management of ungulates and the conservation of biodiversity. Pages 424–64 *in* C. H. Freese, ed. Harvesting wild species. Implications for biodiversity conservation. Johns Hopkins University Press, Baltimore.

Traweek, M. A. 1995. Statewide census of exotic big game animals (Job No. 21, Federal Aid Project No. W-127-R-3). Texas Parks and Wildlife Department, Austin.

Wheaton C., M. Pybus, and K. Blakely. 1993. Agency perspectives on private ownership of wildlife in the United States and Canada. Transactions of the North American Wildlife and Natural Resources Conference 58:487–94.

Wood, J. E., R. J. White, and J. L. Durham. 1970. Investigations preliminary to the release of exotic ungulates in New Mexico (Bulletin No. 13). New Mexico Department of Game and Fish, Santa Fe.

James G. Teer, Department of Wildlife and Fisheries Sciences, Texas A&M University, College Station, Texas 77845. Email: jteer@wfscgate.tamu.edu.

Appendixes, Glossary, Index

Appendix 1

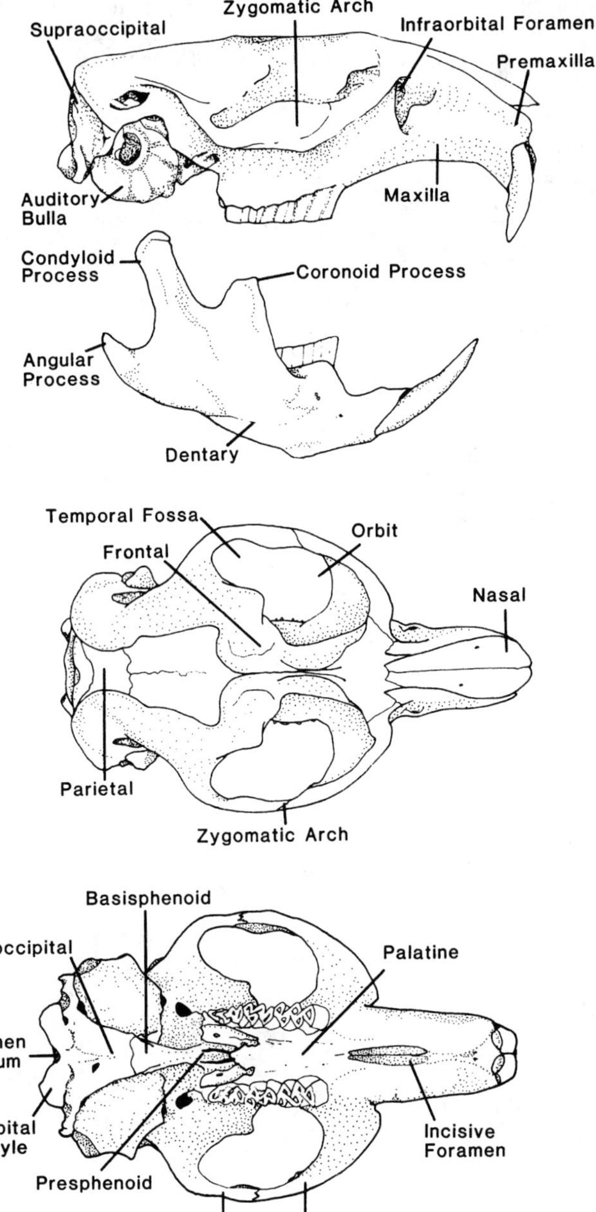

Cranium and mandible of a muskrat (*Ondatra zibethicus*), showing the location of bones commonly cited in the text. Top: lateral view of cranium and mandible. Middle: dorsal view of cranium. Bottom: ventral view of cranium.

Appendix 2

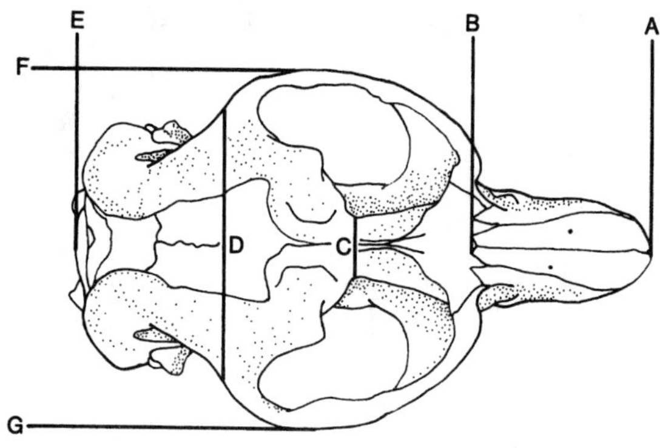

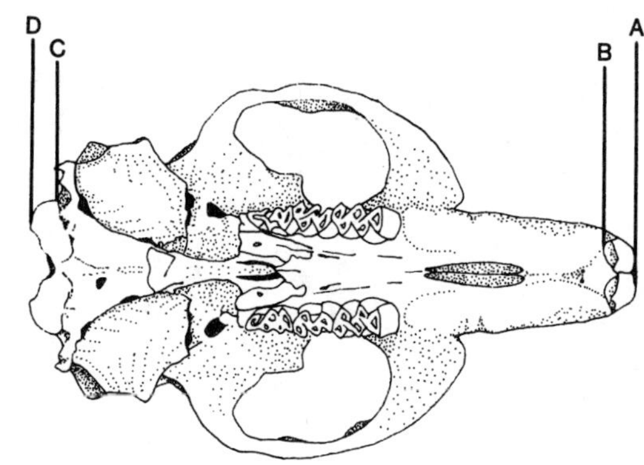

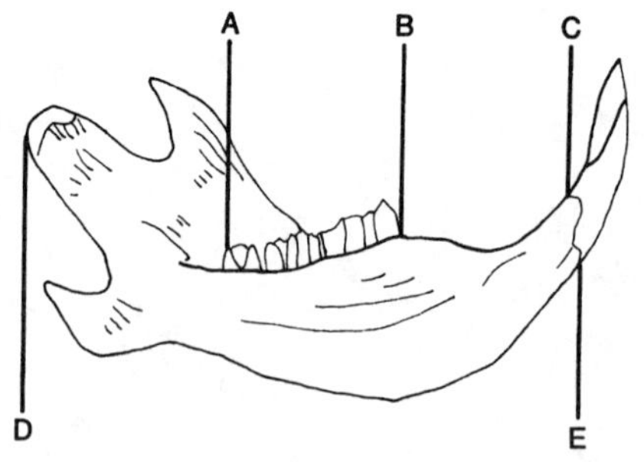

Cranium and mandible of a muskrat (*Ondatra zibethicus*), showing the common measurements taken from a mammalian skull. Top: dorsal view of cranium. Middle: ventral view of cranium. Bottom: lateral view of mandible.

top
greatest length of skull = A → E
nasal length = A → B
least interorbital breadth = C
breadth of braincase = D
zygomatic breadth = F → G

middle
condylobasal length = A → D
basal length = A → C
basilar length = B → C

bottom
mandibular diastema = B → C
mandible length = D → E
mandibular tooth row = A → B

Glossary

ABOMASUM. The fourth, or true, stomach of a ruminant.

ACETABULUM. The socket in the pelvic girdle at the point where the ilium, ischium, and pubis meet, and into which the head of the femur articulates.

ADULT PELAGE. The hair covering characteristic of an adult.

AGENESIS. Lack of, faulty, or incomplete development.

ALPHA FEMALE. Dominant female.

ALPHA MALE. Dominant male.

ALTRICIAL. Pertaining to newborn mammals that initially are completely dependent upon parental care for survival; usually born blind and naked.

ANESTRUS. The period of sexual quiescence between two periods of sexual activity in cyclically breeding mammals.

ANKYLOSED. United or jointed together, as bones or hind parts joined to form a single bone or part.

ANOREXIA. Loss of appetite.

ANTHROPOGENIC. Relating to natural processes of animal behavior interpreted in human terms.

ASPHYXIA. Deficiency of oxygen, excess of carbon dioxide in the blood.

ATAXIA. Lack of normal coordination of voluntary muscular movements.

BACULUM. A bone in the penis of certain mammals, principally carnivores, rodents, and primates.

BALEEN. Whalebone; the cornified epithelial plates suspended from the upperjaws of mysticete whales, used to strain food from the ocean waters.

BEDDING SITES. Sites at which an animal sleeps (usually refers to ungulates).

BIFURCATED. Divided into two branches, forked.

BLASTOCYST. An early stage of an embryo.

BOLE. The trunk or stem of a tree.

BOSS. A knoblike process on the horns of a mammal.

BRADYCARDIA. An abnormally slow heart beat.

BREACHING. The leaping of a whale out of the water.

BUREAU OF BIOLOGICAL SURVEY. The precursor of the United States Fish and Wildlife Service.

BURSA. A sac or saclike cavity.

CALCAR. A spurlike projection on the ankle of chiropterans.

CALLOSITIES. Barnacles and other attachments on a whale, usually in the facial region.

CARNASSIAL TEETH. The last upper premolar and first lower molar of many carnivorous mammals with secondont dentition; the largest pair of bladelike teeth that occlude with a scissorslike action.

CATABOLISM. Destructive or aerobic metabolism.

CEMENTUM. The layer of bonelike material covering the root of a tooth, sometimes termed *cement*. It is deposited throughout life and can be used to determine age in many species.

CERVICAL CANAL. The passage to the uterus.

CERVIX. The necklike, constricted portion of the uterus.

CHEEK TEETH. Collectively, the premolars and molars, or any teeth posterior to the canines (generally used when premolars and molars cannot be readily differentiated).

CINERADIOGRAPH. A photograph made by a process similar to an X-ray process.

CIRCADIAN. Pertaining to behavioral or physiological rhythms associated with a 24-hour cycle.

CIRQUE. Geologically, a steep, hollow excavation on a mountainside due to glacial erosion.

CLAVICLES. Bones that make up part of the pectoral girdle.

COIL-SPRING TRAP. A type of steel leg trap that utilizes coil springs to maintain pressure on the trap jaws.

COLLOID. A solution with small, insoluble particles that remain in suspension.

COLOSTRUM. A specialized secretion of the mammary glands that is produced during the first few days after parturition. It contains. a high concentration of protein and antibodies.

COMMERCIAL. Pertaining to certain mammals harvested for profit; those wild mammals of monetary value.

CONDYLOBASAL. Pertaining to the length of the skull from the anterior edge of premaxilla to the posterior edge of the occipital condyles (see appendix 1).

CONIBEAR TRAP. A steel kill-trap used primarily to collect squirrels, muskrats, and the smaller carnivores.

CONSPECIFICS. Individuals of the same species.

COPROPHAGY. Reingestion of fecal material (also referred to as refection).

COPULATION. Coitus; the union of male and female reproductive organs.

CORACOID. One of the elements of the pectoral girdle in lower vertebrates and monotremes, rudimentary and fused to the scapula in most marsupial and placental mammals.

CORPUS ALBICANS. The degenerated corpus luteum, formed after birth of the fetus or after the egg fails to implant in the uterus.

CORPUS LUTEUM. A mass of yellowish, glandular tissue formed from the Graafian follicle after ovulation.

CORPUS UTERI. The main part or body of the uterus.

CORTEX. The outer or superficial part of an organ or structure.

CREPUSCULAR. Pertaining to the twilight periods of dusk and dawn.

CURSORIAL. Pertaining to running locomotion.

CUTICLE. The thin outer layer of a hair.

DBH. Diameter at breast height (refers to measurement of trees).

DEME. An isolated population.

DEW CLAWS. Vestigial digits on the foot of certain mammals; the second and fifth toes.

DEWLAP. The pendulous fold of skin under the neck of some mammals, notably the moose.

DICHROMISM. The ability of a structure to present two different colors when viewed from different directions.

DIGITIGRADE. Pertaining to walking on the digits, with the wrist and heel bones held off the ground.

DILAMBDODONT TOOTH. A tooth characterized by a W-shaped ectoloph on the occlusal surface.

DIURNAL. Active during daylight hours.

DOMAIN. The area in which a mammal lives.

DRESSED WEIGHT. The weight of a mammal after removal of all visceral organs.

ECOTONE. A zone of transition between habitat types.
EMBRYO. An unborn animal.
ENDOGENOUS. Growing from or on the inside, as catabolic products excreted by a normal organism.
ENVIRONMENTAL IMPACT STATEMENT (EIS). A written report required by governmental agencies stating the impact of various types of proposed alteration, such as a power plant or highway, on the environment.
EPIPUBIC BONES. Paired bones that project anteriorly from the pelvic girdle into the abdominal body wall of most marsupials and monotremes.
ESTIVATE. To pass the summer in a dormant state.
ESTRUS. The receptive period of female mammals; the rut; the whole estrous cycle.
EXCRESCENCE. Abnormal growth or increase.
EXOSTOSIS. A spur or growth from a bone or tooth root.
EXTINCTION. Complete destruction of a mammalian species.
EXTIRPATION. Destruction of a mammalian population in a specifically defined area.
FARROWING. The birth of a swine.
FATHOM. A measure of length containing six feet.
FEEDING CRATERS. Formed by ungulates as they dig through snow for food.
FETUS. Unborn mammal (embryo).
FLUKE. Tailfin of a whale.
FODDER. Coarse food for cattle, horses, and sheep.
FORAGE. The plant food of certain mammals, such as deer.
FOSSORIAL. Pertaining to or the adaptation for existence under ground.
FUSIFORM. Cigar or torpedo shaped.
GAMBOL. To bound or spring, to frisk or play.
GENETIC RELATEDNESS. The degree to which different populations of the same species are related to one another.
GEOLOGIC TIME TABLE. A time table based on geology which dates from the Recent Epoch (less than 600,000 years ago) to the Precambrian Period (4.5 billion years ago).
GESTATION PERIOD. The period of embryonic development during which the developing zygote is in the uterus; the period between fertilization and parturition.
GROUNDFAST ICE. Ice that forms on the bottom of an arctic ocean, lake, or stream.
GUARD HAIR. Coarse hair that extends beyond the underfur of mammals. It is often clipped when the pelt is prepared as a garment.
HABITUATION. The tendency to utilize the same area repeatedly.
HAUL-OUT SITE. An area where pinnipeds leave the water.
HEMATOCRIT. The relative amounts of plasma and corpuscles in the blood.
``HERDING.'' The forced grouping of a large number of ungulates.
HETEROTHERMIC. A homeotherm in which body temperature may fluctuate; adaptive hypothermia.
HIBERNACULA. Places where a bat may hibernate or spend time in a torpid state.
HIBERNATE. To pass the winter in a dormant state.
HOG-DRESSED WEIGHT. The weight of an animal with the heart and lungs remaining (often used synonymously with field dressed).
HOLARCTIC REGION. Collectively, the Nearctic and Palearctic faunal regions.
HOMEOTHERM. A warm-blooded animal.
HOMEOTHERMIC. Able to maintain constant body termperature within the range of tolerance of ambient temperatures.
HYDROPHYTIC. Descriptive of a plant that grows in water or saturated soil.
HYPERCAPNIA. Tolerance of excess carbon dioxide in the tissues of the body.
HYPOPHYSIS. The body of the pituitary gland.
HYPOXIA. Deficiency of oxygen in the tissues of the body.
HYPSODONT. Pertaining to a high-crowned tooth (opposite of brachydont).
HYSTRICOMORPHS. Old and New World rodents in which the infraorbital foramen is greatly enlarged. Typically, it includes species such as the porcupine and nutria.
ILIUM. The most dorsal of the three bones in each half of the pelvic girdle; the pelvic bone that articulates with the sacral vertebrae.
INCISIFORM. Referring to chisel-shaped teeth, generally canines with the same structure as incisors.
INDUCED OVULATION. Ovulation that requires the act of copulation to occur, as in the Felidae.
INFUNDIBULA. The hollow, conical process of gray matter to which the pituitary body is attached (also refers to inner crests of molariform dentition).
INGUINAL. Of or near the groin.
INGUINAL CANAL. In male mammals, a small opening in the musculature of the abdominal wall on either side at the base of the scrotum through which the testes move out of the abdominal cavity into the scrotum.
INSULAR. Refers to a population separated from other populations of the same species, usually by a barrier.
INTROMISSION. The act of passing sperm into the female during copulation.
INTUMESCENT. Swollen or inflamed; a tumor.
ISCHIUM. The most posterior and ventral of the three bones in each half of the pelvic girdle.
JUVENILE. A generalized age category between immature and adult; may or may not be sexually mature.
KERATINOUS. Impregnated with keratin, a tough, fibrous protein especially abundant in the epidermis and epidermal derivatives.
KITCHEN MIDDENS. A mound of shells, animal bones, and other refuse such as often marks the location of a prehistoric settlement.
KNOT. A unit of speed equal to one nautical mile per hour.
LANUGO. Fine, soft hair on the embryonic fetus; a type of villus.
LAPAROTOMY. A cut through the abdominal wall, as in some types of surgery.
LINEAL. Directly descended.
LONG-SPRING TRAP. A type of steel leg-trap in which the spring is a long metal strap projecting away from the jaws; often used for fox and coyote.
LOPHODONT. Pertaining to a tooth that has an occlusal surface pattern consisting of ridges formed by the elongation and fusion of cusps.
MACULA UTRICULI. A portion of the inner ear, important in orientation.
MAMMAE. A gland for secreting milk, present in female mammals but rudimentary in males.
MANUBRIUM. The uppermost portion of the sternum which articulates with the clavicles.
MARSUPIUM. The external pouch formed by a fold of skin and supported by epipubic bones in the abdominal wall, found in most marsupials and in some monotremes (echidnas). It encloses the mammary glands and serves as an incubation chamber.
MEDULLA. The central portion of a structure composed of distinct concentric layers or regions, e.g., the medulla of a hair, ovary, kidney, or adrenal gland.
MELON. A fatty deposit on the facial area responsible for the prominently bulging forehead of many delphinids.
MESIC. Pertaining to environmental conditions with medium moisture supply, as opposed to hydric (wet) or xeric (dry).
METACONID. In the lower cheek teeth, a cusp on the posterior, lingual side of the trigonid area of the crown.
METALOPH. A cusp-formed elongate ridge on upper molariform tooth.
MYSTACIAL. Pertaining to a stripe or fringe of hairs suggestive of a mustache, as in certain mustelids.
NASOPHARYNX. The upper portion of the pharynx, continuous with the nasal passages.
NAUTICAL MILE (NM). A unit of distance for sea and air navigation, equal to 1,852 meters.
NEGRI BODY. An inclusion found in the nerve cells in rabies.

NESTLING. Neonate or young animal.
NULLIPAROUS. Never having given birth.
NURSERY GROUNDS. Traditional areas used for parturition and initial rearing of young (often used in reference to pinnipeds).
OLFACTORY. Relating to the sense of smell.
OMASUM. The division between the reticulum and the abomasum in the stomach of a ruminant; the third stomach.
OS CLITORIDIS. A small, sesamoid bone present in the clitoris of the females of some mammal species, homologous to the baculum in males.
OVIDUCT. The Fallopian tube; the duct that carries the egg from the ovary to the uterus.
OVULATION. The process by which an egg is released from the ovary into the oviduct.
PACK. A group of wild carnivores living and hunting together.
PAINTS. Patches or spots of three colors.
PAIRED SPERM. Two sperm that are joined.
PANMICTIC. Exhibiting random or nonselective mating within a breeding population.
PAROUS. Having produced young or given birth.
PARTURITION. The process by which the embryo of therian mammals separates from the mother's uterine wall and is born.
PELAGE. Collectively, all the hairs on a mammal.
PERINEUM. The area between the anus and the vulva in a female, or the anus and the scrotum in a male.
PHARYNGEAL POUCHES. Outgrowths of ectoderm on both sides of the pharynx which meet the corresponding visceral furrows and give rise to the visceral clefts in vertebrate embryos.
PIEBALD. Having patches or spots of two colors (usually used in relation to cervids).
PINNA. The external ear.
PISCIVOROUS. Fish eating.
PLANTAR. Relating to the bottom (sole) of the foot.
PODS. Groups of marine mammals (seals, whales, etc.).
POLYMORPHONUCLEAR LEUKOCYTE. A neutrophil or other leukocyte with a distinctly lobed nucleus.
POSTPARTUM. Following birth.
PREBAITING. Placing bait at a trap or station prior to trapping.
PREPUTIAL. Pertaining to modified sebaceous glands.
PRIMARY FOLLICLES. Structures in the ovary that contain eggs awaiting development.
PRIME FUR. A pelt of the highest quality.
PRIMORDIAL. Earliest formed, most primitive.
PURKINJE SHIFT. A shift of the region of apparent maximal spectral luminosity from yellow with the light-adapted eye toward violet with the dark-adapted eye, associated with predominance of cone vision in lighter and rod vision in darker illumination.
PUSH-UP. A mass of frozen vegetation over a hole in the ice used by muskrats for access to the water.
RAPHE. A seamlike joining of two lateral halves of an organ.
``RENDEZVOUS SITE.'' Aggregation areas of wolf pups or other immature canids.
RETE. A network (generally of blood vessels or nerve fibers).
RETE MIRABILE. ("Wonderful net") a dense network of blood vessels important in oxygen and heat exchange.
RETICULUM. The second division in the stomach of a ruminant, or "cud-chewing" mammal.
RIPARIAN. Adjacent to a body of water.
ROLLING SITES. Areas about two square meters in size matted down by frolicking otter.
RUGOSE. Ridged; full of wrinkles.
RUMEN. The first division in the stomach of a ruminant, or "cud-chewing" mammal.
SAGITTAL. Pertaining to the medial dorsal ridge of the cranium.
SCALLOPED. Having segments or projections forming an edge.
SCATS. Feces or droppings.
SCRAPES. Areas scraped bare (usually by otters), distinguished from haul-outs by the absence of food remains or scats.
SECTORIAL. Cutting or shearing (same as carnassial).
SELENODONT. Having a crown pattern of molariform teeth characterized by longitudinally oriented, crescent-shaped ridges.
SEROTINOUS. Late or delayed in development.
SERRATED. Notched along the edge.
SOMATIC TISSUE. Tissue of the body, as opposed to germ tissue.
SPERMATOGENESIS. Production of sperm.
SUBADULT. A general age category between immature and adult, juvenile.
SUBCUTANEOUS. Beneath the skin.
SUBNIVEAN. Beneath the snow.
SUCCESSION. The orderly process of replacement of one community with another.
SUPRAOCCIPITAL. Pertaining to the dorsal portion of the occipital bone.
SUSTAINED YIELD. Yield from a renewable resource that is produced by a rate of harvest that is less than or equal to the net rate of productivity of the resource.
TEMPORAL RIDGES. A pair of ridges on the top of the cranium of many mammal species.
THERMAL NEUTRAL ZONE. The range of temperatures in which an endotherm expends little or no energy in regulating body temperature.
THERMOGENIC. Related to the production of heat.
THERMOREGULATION. The maintenance of a fairly constant body temperature through heat production, heat transfer, and other physiological processes.
TORPID. Dormant or inactive, usually accompanied by decreased body temperature.
TUBOUTERINE JUNCTION. The junction of the uterus and Fallopian tubes in female mammals.
UNGULATES. Hoofed mammals of the orders Perissodactyla and Artiodactyla.
VAGINA. The portion of the female reproductive tract that receives the male's penis during copulation; the canal between the vulva and uterus.
VASOCONSTRICTION. Reduction in the diameter of blood vessels; reduced flow of blood.
VESTIBULE. A cavity or space serving as an entrance to another cavity or space.
VIXEN. A female fox.
VULVA. The external genital organs of the female.
WHELP. To give birth to; the young of any carnivore.
WISENT. The European bison, *Bison bonasus*.
XEROPHYTIC. Adapted for environmental conditions of limited water supply.
YEARLING. A general age class between immature and adult (usually used in reference to a cervid that is $1\frac{1}{2}$ years of age).
ZONA PELLUCIDA. The layer surrounding the blastula during early cleavage stages.

Index